I	Ib	IIb	IIIa	IVa	Va	VIa	VIIa	0
								2 `0` **He** 4.002602
			5 `+3` **B** 10.811 2-3	**6** `+2 +4` **C** 12.011 2-4	**7** `±1 ±2 ±3 +4 +5` **N** 14.00674 2-5	**8** `-2` **O** 15.9994 2-6	**9** `-1` **F** 18.9984032 2-7	**10** `0` **Ne** 20.1797 2-8
			13 `+3` **Al** 26.981539 2-8-3	**14** `+2 +4` **Si** 28.0855 2-8-4	**15** `±3 +5` **P** 30.973762 2-8-5	**16** `+4 +6 -2` **S** 32.066 2-8-6	**17** `±1 +5 +7` **Cl** 35.4527 2-8-7	**18** `0` **Ar** 39.948 2-8-8
27 `+2 +3`	**28** `+2 +3` **Ni** 58.69 -8-16-2	**29** `+1 +2` **Cu** 63.546 -8-18-1	**30** `+2` **Zn** 65.39 -8-18-2	**31** `+3` **Ga** 69.723 -8-18-3	**32** `+2 +4` **Ge** 72.61 -8-18-4	**33** `±3 +5` **As** 74.92159 -8-18-5	**34** `+4 +6 -2` **Se** 78.96 -8-18-6	**35** `±1 +5` **Br** 79.904 -8-18-7
								36 `0` **Kr** 83.80 -8-18-8
45 `+2 +4`	**46** `+2 +4` **Pd** 106.42 -18-18-0	**47** `+1` **Ag** 107.8682 -18-18-1	**48** `+2` **Cd** 112.411 -18-18-2	**49** `+3` **In** 114.82 -18-18-3	**50** `+2 +4` **Sn** 118.710 -18-18-4	**51** `±3 +5` **Sb** 121.75 -18-18-5	**52** `+4 +6 -2` **Te** 127.60 -18-18-6	**53** `±1 +5 +7` **I** 126.90447 -18-18-7
								54 `0` **Xe** 131.29 -18-18-8
77 `+2 +4`	**78** `+2 +4` **Pt** 195.08 -32-16-2	**79** `+1 +3` **Au** 196.96654 -32-18-1	**80** `+1 +2` **Hg** 200.59 -32-18-2	**81** `+1 +3` **Tl** 204.3833 -32-18-3	**82** `+2 +4` **Pb** 207.2 -32-18-4	**83** `+3 +5` **Bi** 208.98037 -32-18-5	**84** `+2 +4` **Po** (208.9824) -32-18-6	**85** `±1 +5 +7` **At** (209.9871) -32-18-7
								86 `0` **Rn** (222.0176) -32-18-8

Noble
Gases

64	**65** `+3` **Tb** 158.92534 -27-8-2	**66** `+3` **Dy** 162.50 -28-8-2	**67** `+3` **Ho** 164.93032 -29-8-2	**68** `+3` **Er** 167.26 -30-8-2	**69** `+3` **Tm** 168.93421 -31-8-2	**70** `+2 +3` **Yb** 173.04 -32-8-2	**71** `+3` **Lu** 174.967 -32-9-2
96	**97** `+3 +4` **Bk** (247.0703) -27-8-2	**98** `+3` **Cf** (251.0796) -28-8-2	**99** `+3` **Es** (252.083) -29-8-2	**100** `+3` **Fm** (257.0951) -30-8-2	**101** `+2 +3` **Md** (258.10) -31-8-2	**102** `+2 +3`	**103** `+3`

THE MERCK INDEX

ELEVENTH EDITION

THE
MERCK INDEX

AN ENCYCLOPEDIA OF
CHEMICALS, DRUGS, AND BIOLOGICALS

ELEVENTH EDITION

Susan Budavari, *Editor*
Maryadele J. O'Neil, *Associate Editor*
Ann Smith, *Assistant Editor*
Patricia E. Heckelman, *Editorial Assistant*

Published by
MERCK & CO., INC.
RAHWAY, N.J., U.S.A.

1989

The Merck Index is published on a non-profit basis as a
service to the scientific community.

Merck & Co., Inc.
Rahway, New Jersey, U.S.A.

MERCK SHARP & DOHME
West Point, Pa.

MERCK SHARP & DOHME INTERNATIONAL
Rahway, N.J.

MERCK SHARP & DOHME RESEARCH LABORATORIES
Rahway, N.J./West Point, Pa.

MSD AGVET DIVISION
Woodbridge, N.J.

HUBBARD FARMS, INC.
Walpole, N.H.

MERCK CHEMICAL MANUFACTURING DIVISION
Woodbridge N.J.

MERCK PHARMACEUTICAL MANUFACTURING DIVISION
Rahway, N.J.

CALGON CORPORATION
Water Management Division
Pittsburgh, Pa.
Calgon Vestal Laboratories
St. Louis, Mo.

KELCO DIVISON
San Diego, Ca.

1st Edition—1889
2nd Edition—1896
3rd Edition—1907
4th Edition—1930
5th Edition—1940
6th Edition—1952
7th Edition—1960
8th Edition—1968
9th Edition—1976
10th Edition—1983
11th Edition—1989

Library of Congress Catalog
Card Number 89-60001
ISBN Number 911910-28-X

Printed in the U.S.A
First Printing—November 1989
Second Printing—February 1990

EDITOR'S PREFACE

In the 100 years since The Merck Index was first published in 1889, it has evolved from a 170-page alphabetic catalog of chemicals and drugs sold by Merck, into an authoritative and widely quoted 2300-page multidisciplinary encyclopedia. Changes in size, scope and emphasis have been made with each edition, both to keep pace with the progress in science and to accommodate the diverse interests of an ever broadening and sophisticated readership.

The Eleventh Edition is intended to continue the tradition of serving chemists, pharmacists, physicians and others in allied professions as a handy one-volume compendium of information on the most important chemicals, drugs and biological substances, to which they can turn for quick and reliable answers to questions arising in the course of their work.

Every effort has been made to critically select the entries of sustained interest from previous editions and to add the most significant compounds that have emerged since publication of the Tenth Edition in 1983. New entries include substances used against AIDS and cancer, drugs with novel mechanisms of action, products of recombinant technology, unique naturally-occurring compounds, and chemicals of environmental concern.

Among the features introduced for this edition is a Therapeutic Category and Biological Activity Index, which permits the reader an additional entry point to drug monographs. There are also expanded Substance—Chemical Abstracts Registry Number compilations. Several of the tables have been extensively revised and reformatted for easier usage, a number of new tables have been added, whereas the Organic Name Reactions section has not been included.

While recent editions of The Merck Index have been published every seven to eight years, the Editorial Staff has sought the means to bring new information to the public between printed editions. In 1984, one goal was realized with the introduction of THE MERCK INDEX ONLINE. This enhanced and fully searchable electronic version of the monograph section is updated semiannually and is available internationally through four major online database vendors. Methods are also being explored for making the structures fully searchable and for distributing the database in a number of additional electronic forms.

This volume, like its predecessors, as well as its online counterpart, has again been published by Merck & Co., Inc., as a service to the scientific community and it is hoped that this edition will prove to be a valuable resource to all who consult it.

Susan Budavari, *Editor*
Merck Sharp & Dohme Research Laboratories
Rahway, New Jersey 07065

ACKNOWLEDGMENTS

The publication of this edition of The Merck Index would not have been achieved without the hard work and close collaboration of a great number of talented and dedicated people with diverse expertise. While it is not possible to mention every individual who participated, the Editorial Staff wishes to name and acknowledge the major contributors.

We are especially indebted to our coworkers: *Jo Ann Gallipeau* for assistance in the researching and writing of the manuscript and for diligently inputting the chemical structures and the major portion of the text; *Elizabeth V. Gannon*, who shared in the data entry and who performed most of the documentation work needed for efficient operation; *Edward Hendrzak*, who helped update and revise the tabular material; and *Sandra Spitzer McKelvey*, who effectively facilitated and coordinated all essential activities involving the Editorial Staff and outside vendors.

Our appreciation is extended to *Martha Windholz*, Editor of the Ninth and Tenth Editions, who provided the benefit of her vast experience, who read and commented on a large number of the new monographs and who made many helpful suggestions during the preparation of this edition.

We are also grateful to members of Corporate Computer Resources, and in particular to *Nancy Jarossy*, for coordination of the necessary computer operations; to *Linda Lojko* and *Joel Flamholz* for custom programming and laboratory assistance; to *Benjamin Hickey, James Polashock, John Flanagan* and *Bill Conte* for computer hardware support; and also to *John Reminger* of the Research Photolab for providing photographs of structures; and to *Gary Zelko*, of the Professional Handbooks Department, for his continual support, cooperation, and management of all business aspects of this edition.

Special tribute is due to all previous editorial staff members who painstakingly compiled the first ten editions of The Merck Index, and to the numerous Merck colleagues and other individuals who generously lent their technical expertise, who reviewed critical monographs, suggested new entries, consulted on the design of various tables and indices, and who recommended useful revisions to the manuscript.

Finally, we wish to express our gratitude to *Dr. Myra N. Williams* for her advice, enthusiasm, and encouragement throughout every phase of this endeavor.

TABLE OF CONTENTS

EXPLANATORY NOTES

More than 10,000 concise descriptions of significant chemicals, drugs, and biological substances are included in the monograph section of the Eleventh Edition of The Merck Index. The entries have been selected on the basis of present or past importance and interest, and are not a list of Merck & Co., Inc. products. Since the publication of the Tenth Edition in 1983, hundreds of new monographs have been written, almost 5000 monographs have been revised and updated, and more than 500 have either been deleted or have been combined with other monographs. (*Note*: A list of the monographs that appeared in *either* the Ninth Edition, 1976, *or* the Tenth Edition, but which are not included in the Eleventh can be found on page CI-350.) While entries are generally limited to single substances, there are a number of monographs devoted to families of natural substances such as attacins, cecropins, periplanones, etc. However, multi-component drugs are, for the most part, excluded. Although the information contained in the monographs is extracted from the published literature, the length of the monograph or the number of references included is not necessarily related to the importance of a compound, but may simply be an indication of the amount of relevant information currently available.

The general monograph format that was used in previous editions has been retained. A typical drug monograph is illustrated and the components are identified. While all the possible categories of information in a monograph are described below, it must be emphasized that monographs differ markedly in length and in distribution of text and not all categories are present in every monograph.

Monograph Number. Sequential accession numbers are assigned to monographs. Items in the indices are referenced to these numbers rather than to monograph titles or to page numbers. (*Note*: Monograph numbers in the Eleventh Edition do not correspond to Tenth Edition numbers.)

Monograph Title. Titles, arranged in alphabetical order, are usually generic (USAN, INN, or BAN), trivial, or simple chemical names. Trademarks (designated by ®) are used for a small number of entry titles, primarily when generic terms are not available.

Chemical Abstracts Name(s). The first synonym in **boldface italic** is the uninverted form of the name used by Chemical Abstracts Service (CAS) during the ninth and/or subsequent Collective Index Periods (CIPs). If present, the second synonym in **boldface italic** is the uninverted form of an earlier CIP name. In this edition of The Merck Index, stereochemical descriptors have been included, when possible, to indicate the particular isomer being discussed. The Chemical Abstracts Names (and Registry Numbers starting on page REG-1) are provided as an aid in locating the compound of interest in both the printed and electronic versions of Chemical Abstracts and can thus serve as an entry point to further literature searching.

5460. Lovastatin. *[1S-[1α(R*),3α,7β,8β(2S*,4S*),8aβ]]-2-Methylbutanoic acid 1,2,3,7,8,8a-hexahydro-3,7-dimethyl-8-[2-(tetrahydro-4-hydroxy-6-oxo-2H-pyran-2-yl)ethyl]-1-naphthalenyl ester;* (1S,3R,7S,8S,8aR)-1,2,3,7,8,8a-hexahydro-3,7-dimethyl-8-[2-[(2R,4R)-tetrahydro-4-hydroxy-6-oxo-2H-pyran-2-yl]ethyl]-1-naphthalenyl (S)-2-methylbutyrate; 1,2,6,7,8,8a-hexahydro-β,δ-dihydroxy-2,6-dimethyl-8-(2-methyl-1-oxobutoxy)-1-naphthaleneheptanoic acid δ-lactone; 2β,6α-dimethyl-8α-(2-methyl-1-oxobutoxy)-mevinic acid lactone; mevinolin; 6α-methylcompactin; monacolin K; MK 803; Mevacor; Mevinacor; Mevlor. $C_{24}H_{36}O_5$; mol wt 404.55. C 71.25%, H 8.97%, O 19.78%. Fungal metabolite, potent inhibitor of HMG-CoA reductase, the rate controlling enzyme in cholesterol biosynthesis. Isoln from *Monascus ruber:* A. Endo, *J. Antibiot.* **32,** 852 (1979); from *Aspergillus terreus:* R. L. Monaghan *et al., U.S. pat.* **4,231,938** (1980 to Merck & Co.). Structure and biochemical properties: A. W. Alberts *et al., Proc. Nat. Acad. Sci. USA* **77,** 3957 (1980). Synthesis: M. Hirama, M. Iwashita, *Tetrahedron Letters* **24,** 1811 (1983); D. L. J. Clive *et al., J. Am. Chem. Soc.* **110,** 6914 (1988). Biosynthesis: M. D. Greenspan, J.B. Yudkovitz, *J. Bacteriol.* **162,** 704 (1985); R. N. Moore *et al., J. Am. Chem. Soc.* **107,** 3694 (1985). HPLC determn in plasma and bile: R. J. Stubbs *et al., J. Chromatog.* **383,** 438 (1986). Stimulation of receptor-medicated clearance of low density lipoproteins: D. W. Bilheimer *et al., Proc. Nat. Acad. Sci. USA* **80,** 4124 (1983). Effects on lipoprotein metabolism: S. M. Grundy, G. L. Vega, *J. Lipid Res.* **26,** 1464 (1985). Multicenter clinical comparison with gemfibrozil, *q.v.:* M. J. Tikkanen *et al., Am. J. Cardiol.* **62,** 35J (1988). Review of syntheses: T. Rosen, C. H. Heathcock, *Tetrahedron* **42,** 4909-4951 (1986). Review of clinical experience: J. A. Tobert, *Circulation* **76,** 534-538 (1987); *idem, Am. J. Cardiol.* **62,** 28J-34J (1988).

White crystals, mp (under N_2): 174.5°. $[α]_D^{25}$ +323° (c = 0.5 g in 100 ml acetonitrile). uv max: 231, 238, 247 nm ($A^{1\%}$ 532, 621, 418). Freely sol in chloroform; sol in acetone; sparingly sol in acetonitrile, methanol, ethanol. Insol in water. LD_{50} orally in mice: > 1000 mg/kg (Endo).

THERAP CAT: Antihypercholesterolemic.

Alternate Name(s). Other chemical names identifying the entry, trivial names, experimental drug codes, and current (and former) trademarks are in lightface roman. Trademarks are indicated by first letter capitalization; absence of capitalization, however, does not exclude the possibility that a name may now be a proprietary name or may once have been the subject of proprietary rights. If known, the company associated with a particular trademark is listed alongside the trademark in the Cross Index of Names. Names appearing elsewhere in the monograph in boldface italic also appear in the Cross Index of Names.

Molecular Formula, Molecular Weight, % Composition. Elements in the molecular formula are listed according to the Hill convention (C,H, then other elements in alphabetical order). This information and molecular weight are provided for all compounds having a *specific* known structure.

Literature References. This section contains a concise reference history of the compound. Frequently, a brief description or capsule statement is provided, although in some monographs, particularly those on significant biological substances, a lengthier description is given. References to isolation, preparation/synthesis, patent information, and structural studies are cited. Patent numbers are provided merely as a source of preparative information, however, an attempt has been made, whenever possible, to cite the United States product patent in monographs which are new for the Eleventh Edition. As with earlier editions, references to pharmacological, clinical, toxicological, and toxicity studies have been included. Review articles, where available, are usually cited at the end of the references. When a review pertains to a group of closely related compounds or to a family of natural substances, it is generally given only in the monograph for the parent element or compound. Literature references are cited in the conventional manner; journal abbreviations (with notable exceptions listed in the Table of Abbreviations and Selected Definitions, p. xiii) generally correspond to those in Chemical Abstracts Service Source Index (CASSI).

Structure. Structural depictions, including stereochemistry if relevant, are included in almost 6000 monographs. In instances where the structure shown does not correspond precisely to the monograph title, the structure has been so labeled. Standard conventions of heavy wedges and dotted lines to show bonds directed above or below the plane of the paper are used where appropriate. In polypeptide representations, amino acid residues are assumed to be L unless specified otherwise. In addition, more than 2000 monographs contain line formulas showing molecular arrangements.

Physical Data. Data are cited as found in the literature. When conflicting data have been published, the alternate values and sources are given. Whenever possible, the color of a substance is stated, but the absence of color (white, colorless) is often omitted. Temperatures are given in degrees Celsius (centigrade) unless otherwise noted. When solubilities are determined at room temperature (about 25° C), the temperature is generally omitted. When optical rotations are measured in water, the solvent is usually not specified. For ultraviolet absorption measurements, the solvent is provided within parentheses.

As in the previous edition, an effort has been made to provide toxicity data (e.g. LD_{50}, LC_{50}, etc.) and to identify the source of this information. *Caution* and/or *Human Toxicity* statements are also provided for a number of substances. Specific caution statements are given for drugs and compounds on the U.S. Government's Schedules of Controlled Substances, for additives controlled by the Food and Drug Administration, and for compounds listed as suspected or confirmed carcinogens in the *Fourth Annual Report on Carcinogens* issued in 1985 by the U.S. Department of Health and Human Services. *Note*: Absence of toxicity data or cautions does not imply that toxic effects do not exist.

Derivatives. When derivatives (isomers, salts, etc) of the title compound are described in a monograph, the information appears in the paragraph(s) directly following the physical data. Data for derivatives are presented in the same format as for the parent compound. A derivative molecular formula is listed in the Formula Index only if there is a chemical name, generic name, or trademark associated with it.

Use. Descriptions of specific uses which are not medical or veterinary therapeutic applications are summarized under this heading.

Therapeutic Category and Therapeutic Category (Veterinary). Wherever possible, the editors have adhered to the categories of activity proposed by the USAN Council in describing therapeutic indications of drugs. However, in instances where the USAN Council has listed a mechanism of action, rather than a therapeutic category, a specific therapeutic category which most closely describes the indication claimed by the manufacturer or reported in the clinical literature has been assigned. In general, mechanism information, when known, has been included in the body of the monograph.

Indices. More than 60,000 synonyms, including titles, CAS names, alternate names, trademarks, and derivatives are contained in the Cross Index of Names, and over 10,000 entries appear in the Formula Index. Each entry directs the reader to the number of the monograph in which the substance of interest is described. The effort to match trademarks with company ownership, begun in the Ninth Edition, has been greatly expanded for this edition. In the Cross Index of Names an abbreviated form of the company name appears in brackets following the trademark. (*Note*: Due to reorganizations or mergers, some company names may have changed since the initial matching process was completed *and* some trademarks may have been reassigned.) A list of company addresses appears in an updated and expanded Company Register that begins on page MISC-9.

A Therapeutic Category and Biological Activity Index of human drugs appears for the first time in this edition. The categories are arranged alphabetically; cross references are provided. Many drugs are posted both under mechanism of action and under the corresponding therapeutic category. Whenever appropriate, subcategories have been developed by grouping compounds according to structural similarities. A master list of all headings can be found on pages THER-1 to THER-4.

Two CAS Registry Number indices are also provided; one arranged alphabetically by substance name (*see* pages REG-1 to REG-47); the second by ascending Registry Number order (*see* pages REG-48 to REG-103). Registry numbers are provided for title substance and for selected derivatives.

Although The Merck Index has a strong medical character, it is not intended as an official therapeutic guide. Inclusion of a drug in this book is not an endorsement, but merely a statement of the fact that such an entity exists. THERAPEUTIC CATEGORY and THERAPEUTIC CATEGORY (VETERINARY) paragraphs are intended only as summary statements of major pharmacological properties or indications for the individual drugs. For additional information on uses, dosage, side effects and adverse reactions, readers should consult pertinent scientific and professional publications and product circulars published by the respective manufacturers.

Great care has been taken to assure the accuracy of the information contained in The Merck Index. However, the Editorial Staff and the Publisher cannot be responsible for errors in publication or for any consequences arising from use of the information published in The Merck Index. Accordingly, reference to original sources is encouraged as is reporting of errors and omissions in order to assure that appropriate changes may be made in the next edition.

Abbreviations and Selected Definitions

A	adenine, adenosine or absorbance (extinction)
Å	Angstrom units
AAT	α_1-antitrypsin
Ab	antibody
abs	absolute; absorption
abs config	absolute configuration
abstr	abstract
Abu	α-aminobutyric acid
Ac	acetyl CH_3CO-
acc	according
ACE	angiotensin converting enzyme
Acetyl CoA	acetyl coenzyme A
Ac_2O	acetic anhydride
AcOEt	ethyl acetate
AcOH	acetic acid
ACP	acyl carrier protein
A.C.S.	American Chemical Society
ACTH	adrenocorticotropic hormone
AcylSCoA	acyl coenzyme A
ADCC	antibody-dependent cellular cytotoxicity
add(n)	adding, addition
Ade	adenine
ADH	alcohol dehydrogenase
ADP	adenosine diphosphate
AEC	(United States) Atomic Energy Commission
Ag	antigen
AI	anaphylotoxin inhibitor
AIDS	acquired immunodeficiency syndrome
Ala	alanine
alc, alcoh	alcohol(ic); ethanol; ethyl alcohol
ALG	anti-lymphocyte globulin
alk	alkali(ne)
$[\alpha]_D^{25}$	specific optical rotation at 25°C for D (sodium) line; absence of brackets indicates optical rotation of a liquid in a 1 decimeter cell-neat
ALS	anti-lymphocyte serum, amyotrophic lateral sclerosis
a_M	molar absorptivity (concentration in g-moles/l)
amorph	amorphous
AMP	adenosine 5'-monophosphate (adenylic acid)
cAMP	cyclic AMP
amps	ampules
amt(s)	amount(s)
ANF	anti-nuclear factor; atrial natriuretic factor
anhydr	anhydrous
Ann.	*Justus Liebig's Annalen der Chemie*
anti-	stereochemical descriptor restricted to bicyclo[X,Y,Z]anes in which the Z bridge reference group is oriented away from the X bridge
APhA	American Pharmaceutical Association
approx	approximate(ly)
APT	alum precipitated toxoid
aq	aqueous
Ar	aryl
A.R.	analytical reagent
ARC	AIDS related complex
Archiv Exp. Pathol. Pharmakol.	*Naunyn Schmiedebergs Archiv für Experimentelle Pathologie und Pharmakologie*
ArCO-	aromatic acyl radical
Arg	arginine
Asa	β-carboxyaspartic acid
Asn	asparagine

Asp	aspartic acid
assoc	association
assocd	associated
A.S.T.M.	American Society for Testing Materials
as-, asym-	asymmetrical, unsymmetrical
at.	atomic
ATCC	American Type Culture Collection
atm, atmos	atmosphere(s), atmospheric
at.no.	atomic number
ATP	adenosine triphosphate
ATPase	adenosine triphosphatase
at.wt.	atomic weight
B	base. Example: if the formula of an alkaloid is $C_{21}H_{23}NO_5$ the abbreviated formula for the hydrochloride may be written $B.HCl$ instead of $C_{21}H_{23}NO_5.HCl$
BAN	British Approved Name
B cell	bone marrow or bursa of Fabricius derived cell
BCG	Bacillus Calmette-Guerin
Bé	Baumé (a specific gravity scale)
Beilstein	*Beilsteins Handbuch der Organischen Chemie*, a comprehensive German encyclopedia of organic chemistry (Springer-Verlag)
Belg. pat.	Belgian patent
Ber	*Chemische Berichte* (Berichte der Deutschen Chemischen Gesellschaft)
BGG	bovine gamma globulin
biol	biological
B.I.O.S.	British Intelligence Objectives Subcommittee
BOC	*tert*-butoxycarbonyl
B.O.D.	biochemical oxygen demand
boil.	boiling
bp	basepair; boiling point; boils; boils at; boiling at (always followed by a figure denoting temperature; the pressure, if different from one atm, is indicated by a subscript. Example: bp_{70} 48° means *boils at 48 °C if the pressure is 70 mm Hg*)
B.P.	*British Pharmacopeia*
B.P.C.	*British Pharmaceutical Codex*
Brit. pat.	British patent
BSA	bovine serum albumin
Btu	British thermal units
Bu	butyl (*normal*-butyl)
Bz	benzoyl C_6H_5CO-; BzH benzaldehyde; BzOH benzoic acid
BzH	benzaldehyde
BzOH	benzoic acid
c	concentration by volume (after optical rotations only). Example: $[\alpha]_D^{25} + 14°$ (c = 2.5 in abs alcohol), meaning 2.5 g of the substance dissolved in 100 ml abs alcohol; when no solvent is given, the solvent is water
C	cytosine or cytidine or complement
°C	Centigrade degrees, Celsius degrees
C_p	heat capacity (constant pressure)
ca.	(*circa*) about
C.A.	*Chemical Abstracts*
cal	calorie(s)
calc(d)	calculate, calculated
Can. pat.	Canadian patent
Cbz	carbobenzoxy

cbc	complete blood count
cc	cubic centimeter(s) (milliliter)
CCK	cholecystokinin
CD	circular dichroism
CDP	cytidine diphosphate
CEA	carcinoembryonic antigen
cf.	*(confer)* compare
CFA	complete Freund's adjuvant (same as FCA)
CFT	complement fixation test
chem	chemical
Chem. Commun.	*J. Chem. Soc., Chem. Commun.*
Ci	curie
C.I.	*Colour Index* (British)
cis-	configuration in which the two priority groups are on the same side of the reference plane
cm	centimeter(s)
CM-cellulose	*O*-(carboxymethyl) cellulose
CMI	cell-mediated immunity
CML	cell-mediated lymphocytotoxicity
CMP	cytidine 5'-monophosphate (cytidylic acid)
CNS	central nervous system
CoA or CoASH	coenzyme A
coll. vol.	collective volume
compd	compound
compn	composition
Con A	concanavalin A
conc(d)	concentrated
concn	concentration
config	configuration
constit	constituent
contd	continued
contg	containing
CoQ	coenzyme Q (ubiquinone)
cor(r)	corrected
corresp	corresponding, corresponds
cp	centipoise
C.P.	chemically pure
cpd	compound
crit press	critical pressure
crit temp	critical temperature
cryst	crystalline or crystals
crystn	crystallization
CSF	colony stimulating factor
CTFA	The Cosmetic, Toiletry and Fragrance Assoc.
CTP	cytidine triphosphate
Cys	cysteine
Cyt	cytosine
d	density; specific gravity (d^{19} specific gravity at 19° referred to water at 4°). Also 2'-deoxyribo-
d-	*dextro* (rotatory), refers to optical rotation, indicating that a soln of the substance is capable of turning the plane of polarized light to the right
D-	*dextro* (in configurational sense only). Used before carbohydrates and amino acids to show that the groups at the significant asymmetric carbon atom are placed at the right. In carbohydrate nomenclature the configuration of the highest numbered asymmetric carbon atom determines the prefix that is used. Carbohydrate nomenclature is based upon the glyceric aldehydes, the dextrorotatory isomer being by convention designated D-glyceric aldehyde. In the amino acid field, it is the

	configuration of the lowest numbered asymmetric carbon atom, i.e., the α-carbon atom, that determines the prefix, as in D-alanine
Da	daltons
DEAE cellulose	*O*-(diethylaminoethyl)cellulose
dec, decomp	decompose(s)
decompn	decomposition
deg	degree
deliquesc	deliquescent
Δ (delta)	indicates the locant of the double bond
deriv	derivative
determn	determination
DFP	diisopropyl fluorophosphate or diisopropyl
(DIFP,DIPF)	phosphofluoridate
Dha	dihydroalanine
Dhb	dehydrobutyrine, β-methyl-dehydroalanine
diff	difference
dil(d), (n)	dilute, diluted, dilution
distln	distillation
dl-	racemic; optically inactive by external
DL-	compensation as contrasted with *meso-*
dm	decimeter(s)
DMARD	disease modifying antirheumatic drug
DMF	dimethylformamide
DMSO	dimethylsulfoxide
DNA	deoxyribonucleic acid
cDNA	complementary DNA
mtDNA	mitochondrial DNA
DNAase	deoxyribonuclease
DNFB	2,4-dinitro-1-fluorobenzene
DNP	2,4-dinitrophenyl or 2,4-dinitrophenol
Dopa	dihydroxyphenylalanine
dp(DP)	degree of polymerization (number of monomeric units in the polymer)
D.R.P.	*(Deutsches Reichs-Patent)* German patent
ds	double stranded
DTT	dithiothreitol
dyn	dynes
(*E*)-	*entgegen* (German for opposite). Geometric stereodescriptor used for substances having achiral elements resulting from double bonds where the groups of highest priority are on the opposite sides of the vertical reference plane; equivalent to *trans* in simple cases, opposite of (*Z*)
$E_{1cm}^{1\%}$	the absorbance of a solution containing one gram per 100 ml contained in a cell having an absorption path of one centimeter
E_M	molar extinction coefficient (conc. in g-moles/l)
EAA	excitatory amino acid
EAC	erythrocyte coated by antibody and complement
EAE	experimental allergic encephalomyelitis
EC	electron capture
ECF-A	eosinophil chemotactic factor of anaphylaxis
ECG	electrocardiogram
E.C.No.	Enzyme Commission Number
ed.	edition

ED	effective dose	Fortschr.	Fortschritte der Chemie
Ed(s)	editor(s)	Chem. Org.	Organischer Naturstoffe
EDTA	ethylenediaminetetraacetic acid	Naturst.	(Progress in the Chemistry of
EEG	electroencephalogram		Organic Natural Products,
e.g.	(exempli gratia) for example		Springer-Verlag)
EGF	epidermal growth factor	fp	freezing point
eidem	the same (authors), plural of idem	FP	flavoprotein
EKG	electrocardiogram	Fr. pat.	French patent
ELISA	enzyme-linked immunosorbent assay	Frdl.	P. Friedlander Fortschritte der Teerfarbenfabrikation, a
emf	electromotive force		collection of patents (Springer-
en	ethylenediamine (in formulas)		Verlag)
endo-	stereochemical descriptor restricted	Fru	fructose
	to bicyclo[X,Y,Z]anes in which	FSH	follicle-stimulating hormone
	the reference group is on the side	FT	Fourier transform
	of the reference plane opposite to	g	gram(s)
	the Z bridge	G	guanine or guanosine
EPA	Environmental Protection Agency	G-1-P	glucose 1-phosphate
EPO	erythropoietin, European Patent	G-3-P	gyceraldehyde 3-phosphate
	Office	G-6-P	glucose 6-phosphate
ϵ (epsilon)	molar extinction coefficient (conc.	GABA	γ-aminobutyric acid
	in g-moles/l); dielectric constant	gal	gallon(s)
eq	equation	Gal	galactose
equilib	equilibrium	GalNAc	N-acetyl-D-galactosamine
equiv	equivalent	γ (gamma)	microgram(s)
esp	especially	GC	gas chromatography
esu	electrostatic units of electrical	GDH	glutamate dehydrogenase
	charge; the amount of electrical	GDP	guanosine diphosphate
	charge which in a vacuum will	gem-	geminal, indicates that the two
	repel a like charge at a distance		substituents in a disubstituted
	of one centimeter with a force of		compound are on the same atom)
	one dyne	geol	geological
Et	ethyl C_2H_5—	Ger. pat.	German patent
η (eta)	viscosity	GH	growth hormone
et al.	(et alii) and others	G.I.	gastrointestinal
etc.	et cetera	g/l	grams per liter
Et_2O	ether	Gla	γ-carboxyglutamic acid
EtOH	ethyl alcohol	Glc	glucose
Eur. pat.		GLC	gas-liquid chromatography
Appl.	European patent application	GlcA	gluconic acid
ev	electron volt	GlcN	glucosamine
evac	evacuated	GlcNAc	N-acetyl-D-glucosamine
evapn	evaporation	GlcUA	glucuronic acid
exo-	stereochemical descriptor restricted	Gln	glutamine
	to bicyclo[X,Y,Z]anes in which	Glc	glucose
	the reference group is on the	Glu	glutamic acid
	same side of the reference plane	GluA	glucuronic acid
	as the Z bridge	Gly	glycine
		Glycerol-3-P	glycerol 3-phosphate
exptl(ly)	experimental(ly)	Gmelin's	Gmelin's Handbuch der
ext(d)	extract, extracted		Anorganischen Chemie, a
extern	externally		comprehensive German
°F	Fahrenheit degrees; also Fourneau		encyclopedia of inorganic
F-1-P	fructose 1-phosphate		chemistry (Verlag Chemie)
F-6-P	fructose 6-phosphate	GMP	guanosine 5'-monophosphate
FA	fatty acid		(guanylic acid)
FAD (FADH2)	flavin adenine dinucleotide (reduced form)	cGMP	cyclic GMP
FCA	Freund's complete adjuvant (same	GM-CSF	granulocyte macrophage colony
	as CFA)		stimulating factor
Fd	ferredoxin	gov't	government
F.D.A.	Food and Drug Administration	GRF	Growth hormone releasing factor
	(U.S.A.)	GSH	glutathione (reduced)
FD & C	Food, Drug and Cosmetic (Act)–	GSSG	glutathione (oxidized)
	U.S.A.	GTP	guanosine triphosphate
FDNB	1-fluoro-2,4-dinitrobenzene	G.U.	genitourinary
FDP	fructose 1,6-diphosphate	Gua	guanine
ff	following	GVH	graft-versus-host
FFA	free fatty acid	habit.	habitat
FFC	free from chlorine	Har	homoarginine
FIA	Freud's incomplete adjuvant (same	Hb	hemoglobin
	as ICFA)	HbCO	carbon monoxide hemoglobin
FIAT	Field Information Agency,	HbO_2	oxyhemoglobin
	Technical (U.S. reports)	Hcy	homocysteine
FMN (FMNH2)	flavin mononucleotide (reduced	HDL	high density lipoproteins
	form); same as riboflavin	HeLa	cells of the first continuously
	phosphate		cultured (human cervical)
Fmoc	9-fluorenylmethoxycarbonyl		carcinoma strain

HGG	human gamma globulin		I.U.P.A.C.	International Union of Pure and Applied Chemistry
HGH	human growth hormone		i.v.	intravenous
His	histidine		**Japan. Kokai**	Japanese patent (unexamined)
HIV	human immunodeficiency virus		**Japan.** pat.	Japanese patent
HLA	human leukocyte antigen		K	dissociation constant, equilibrium constant
HMG-CoA	3-hydroxy-3-methylglutaryl coenzyme A		°K	degrees Kelvin
Houben	a German collection of medicinal patents		Kb	kilobase
			kcal	kilocalorie(s)
Houben Weyl	*Houben-Weyl Methoden der Organischen Chemie*, a German collection of preparative methods in organic chemistry (Thieme)		K cell	killer cell
			kg	kilogram(s)
			αKG	α-ketoglutarate
HPLC	high performance (pressure, power) liquid chromatography		KLH	keyhole limpet hemocyanin
			KPA	prourokinase, kidney plasminogen activator
hr	hour		l	liter
HSA	human serum albumin		*l-*	*levo*(rotatory), the opposite of *d*, *q.v.*
Hse	homoserine			
HSV	herpes simplex virus		L-	*levo* (in configurational sense only), the opposite of D, *q.v.*
HT	hydroxytryptamine (serotonin)			
HTLV	human T lymphotropic virus see also HIV		Lac	lactose
			LAD	lymphocyte-activating determinant
Hyl	hydroxylysine		λ (lambda)	wavelength
Hyp	hydroxyproline		LATS	long acting thyroid stimulator
i-	optically inactive by internal compensation as *i*-inositol; archaic for *meso-*		lb	pound(s)
			LC	Lethal Concentration; LC$_{50}$, a concentration which is lethal to 50% of the animals tested; liquid chromatography
I	inosine			
Ia	I-region antigen			
IACR	International Association of Cancer Registries		LCM	lymphocyte choriomeningitis
			LD	Lethal Dose; LD$_{50}$, a dose which is lethal to 50% of the animals tested
IARC	International Agency for Research on Cancer			
IARC Monographs	*IARC Monographs on the Evaluation of Carcinogenic Risk of Chemicals to Man*		LDH	lactate dehydrogenase
			LDL	low density lipoproteins
ibid	(*ibidem*) at the same place		Leu	leucine
I.C.C.	Interstate Commerce Commission		LH	luteinizing hormone (same as ICSH)
ICFA	incomplete Freund's adjuvant (same as FIA)			
			ln	natural logarithm
ICSH	interstitial cell-stimulating hormone (same as LH)		LNPF	lymph node permeability factor
			loc. cit.	(*loco citato*) in the place cited
idem	the same (author); plural: *eidem*, the same (authors)		log	logarithm (common)
			l.o.i.	limit of impurities
IDP	inosine diphosphate		LPS	lipopolysaccharide
i.e.	(*id est*) that is		Lys	lysine
IEF	Isoelectric focusing		m	meter; given after mass number signifies metastable isomer
IF, IFN	interferon			
i.g.	intragastric		*m-*	*meta*
Ig	immunoglobulin		*M*	molar (concentration)
I.G. Farben	*Interessengemeinschaft der Farbenindustrie, Aktiengesellschaft*- the German dye trust		Mab, mAb	monoclonal antibody
			MAC	maximum allowable concentration
			MAF	macrophage activating factor
			MAO	monoamine oxidase
IGF-1	insulin like growth factor 1		MAOI	monoamine oxidase inhibitor
IL	interleukin		mass spec	mass spectrometry
Ile	isoleucine		max	maximum, maxima
i.m.	intramuscular		Mb	myoglobin
IMP	inosine 5'-monophosphate (inosinic acid)		MbO$_2$	oxymyoglobin
			M.C.A.	Manufacturing Chemists Association (U.S.A.)
incl	including			
incompat	incompatibility		mcg	microgram
INN	International Nonproprietary Name		mCi	millicurie
inorg	inorganic			
insol	insoluble		M$_D$	molecular rotation $\dfrac{[\alpha]_D \times \text{mol wt}}{100}$
intern	internal			
Intl	International		MDH	malate dehydrogenase
i.p.	intraperitoneal		Me	methyl CH$_3$—
IR	infrared		Me$_2$CO	acetone
Ir genes	immune response genes		MeOH	methyl alcohol
ISO	Internal Organization for Standardization		*Mellor's*	*Mellor's Comprehensive Treatise on Inorganic and Theoretical Chemistry* (Longmans)
isoln	isolation			
ITP	inosine triphosphate or idiopathic thrombocytopenic purpura		mEq	milli-equivalent (0.001 of an equivalent)
I.U.	international unit		Met	methionine
I.U.C.	International Union of Chemistry		fMet	*N*-formylmethionine

MetHb	methemoglobin	OA	ovalbumin
meV	millielectron volts	OAA	oxaloacetate
mfg, manuf	manufacturing	OD	optical density
mfr	manufacture	*op. cit.*	(*opere citato*) in the work cited
mg	milligram	org	organic
MHC	major histocompatibility complex	OSHA	Occupational Safety and Health Act
μCi	microcurie		
μg	microgram	OsM	osmolar
microcryst	microcrystalline	oz	ounce(s)
MIF	migration inhibition factor	P *or* p	concentration by weight (after optical rotations only)
min	minimum; also minute(s)		
MIS	Mullerian inhibiting substance	p, pp	page(s)
misc	miscible	*p-*	*para*
mixt	mixture	P_1	inorganic phosphate
ml	milliliter (cubic centimeter)	Pa	pascal
MLD	minimum lethal dose	PABA	*p*-aminobenzoic acid
MLR	mixed lymphocyte reaction	PAF	platelet-activating factor
mm	millimeter	PAGE	polyacrylamide gel electrophoresis
mμ	millimicron(s), nanometer	*passim*	here and there, scattered
mol wt	molecular weight	pat.	patent
Monatsh.	*Monatshefte für Chemie*	PB report	Publication Board Report (United States Department of Commerce, Scientific and Industrial Reports)
mp	melting point; melts, melting at, when followed by a figure denoting temperature		
		PCA	passive cutaneous anaphylaxis
M_r	relative molecular mass	PCT	Patent Co-operation Treaty
ms-	*meso-* (internally compensated)	PEP	phosphoenolpyruvate
MS	mass spectrometry	petr, petrol	petroleum
MSH	melanocyte-stimulating hormone (melanotropin)	PFC	plaque-forming cell
		3PG	3-phosphoglycerate
n	index of refraction (n_D^{20} for 20° and sodium light); normal, as *n*-propyl	PG	prostaglandin
		PGA	pteroylglutamic acid (folic acid)
		PGP	3-phosphoglyceroyl phosphate
N	normal (equivalents per liter, as applied to concentration); nitrogen (as in *N*-methyl-pyridine)	pH	acid-base scale; log of reciprocal of hydrogen ion concentration
		PHA	phytohemagglutin
		Phe	phenylalanine
NAcneu	*N*-acetylneuraminic acid	physiol	physiological
NAD⁺(NADH)	nicotinamide adenine dinucleotide (reduced form)	pK	log of the reciprocal of the dissociation constant, 1/log K
		PKU	phenylketonuria
NADP⁺(NADPH)	nicotinamide adenine dinucleotide phosphate (reduced form)	PMN	polymorphonuclear leukocyte
		PP_1	inorganic pyrophosphate
NANSAIDS	nonaspirin nonsteroidal anti-inflammatory drugs	ppm	parts per million
		ppt(g)	precipitate, precipitating
NBS	National Bureau of Standards	pptd	precipitated
NCTC	*National Collection of Type Cultures*	PQ	plastoquinone
		Pr	propyl (normal)
NDP	nucleoside 5'-phosphate	prepd	prepared
Neth. pat.		prepn	preparation
Appl.	Netherlands patent application	press.	pressure
N.F.	*National Formulary*	Pro	proline
ng	nanogram (10⁻⁹ grams)	PRPP	5-phosphoribosyl 1-pyrophosphate
NGF	nerve growth factor	ψ (psi)	pseudo
NIOSH	National Institute for Occupational Safety and Health	pt	point
		PTH	parathyroid hormone
nm	nanometers	Q	coenzyme Q (ubiquinone)
NMN⁺(NMNH)	nicotinamide mononucleotide (reduced form)	*q.q.v.*	(*quae vide*) which see, plural
		q.v.	(*quod vide*) which see
NMP	nucleoside 5'-monophosphate	r	"roentgen" unit of radiation. That quantity of x or gamma radiation which produces one esu of charge in one cubic centimeter of air under standard conditions, i.e., the associated corpuscular emission per 0.001293 g of air (1 cc at 0° and 760 mm) produces, in air, ions carrying one esu
NMR	nuclear magnetic resonance		
N.N.D.	*New and Nonofficial Drugs* (Lippincott, 1959-1964)		
N.N.R.	*New and Nonofficial Remedies* (Lippincott, 1933-1958)		
no.	number		
nor-	(*Nitrogen ohne Radikal*) a prefix indicating a parent compound (no longer limited to nitrogenous compounds)		
		r-	racemic
		R	alkyl, univalent hydrocarbon radical (or hydrogen)
NRDC	National Research Development Corporation	(*R*)-	*rectus* (right). Absolute term describing the spatial arrangement about an asymmetric carbon when the observed order of decreasing priority of the groups is clockwise
NSAID	nonsteroidal anti-inflammatory drugs		
NSC	National Service Center		
NTP	nucleoside 5'-triphosphate		
o-	*ortho*		
O	denoting attachment to oxygen, as in *O*- acetylhydroxylamine		

RAST	radioallergosorbent test
rbc	red blood cell, red blood count
RCO-	aliphatic acyl radical
recryst(n)	recrystallize, recrystallization
ref	reference
rep [REP]	"roentgen equivalent physical" means a dose of ionizing radiation capable of producing energy absorption of 93 ergs per gram of tissue
resp	respectively
R_f or R_F	(in paper chromatography) ratio of movement of the band to the front of the solvent
RIA	radioimmunoassay
Rib	D-ribose
RIST	radioimmunosorbent test
RNA	ribonucleic acid
hnRNA	heterogeneous nuclear RNA
mRNA	messenger RNA
nRNA	nuclear RNA
rRNA	ribosomal RNA
snRNA	small nuclear RNA
tRNA	transfer (soluble) RNA
RNase	ribonuclease
RTECS	*Registry of Toxic Effects of Chemical Substances*
s-	symmetrical
S	denoting attachment to sulfur as *S*-methylcysteine; *Streptomyces*, used only in genus and species names
(*S*)-	*sinister* (left) (opposite of (*R*))
S.A.E.	Society of Automotive Engineers
SAM	*S*-adenosylmethionine
sapon(if)	saponification
satd	saturated
s.c.	subcutaneous
S.D.	Sprague Dawley
SDS	sodium dodecyl sulphate
sec	second(s)
sec-	secondary
sepn	separation
Ser	serine
SI	International System of Units
SLE	systemic lupus erythematosus
sol; soly	soluble; solubility
solidif	solidifies, solidification
soln	solution
sp.	species; specific
spec	spectroscopy
sp gr	specific gravity
spp.	species (plural)
sq	square
sqq	(*sequentia*) and following
SRBC	sheep red blood cells
SRS-A	slow reacting substance of anaphylaxis
ss	single stranded
SSPE	subacute sclerosing panencephalitis
S.T.P.	standard temperature and pressure
subl	sublimes
suppl	supplement
sym-	symmetrical
syn-	stereochemical descriptor restricted to bicyclo[X,Y,Z]anes in which the Z bridge reference group is oriented toward the X bridge
T	thymine or thymidine
$T_{1/2}$	half-life
tabl	tablet(s)
TB, tb	tuberculosis
TCA	trichloroacetic acid, tricarboxylic acid
T cell	thymus derived cell
TDL	thoracic duct lymphocyte
TDP	thymidine diphosphate
TEAE cellulose	*O*-(triethylaminoethyl)cellulose
tech	technical
temp	temperature
tert-	tertiary
TH	thyrotropic hormone
THF	tetrahydrofuran
THFA	tetrahydrofolic acid
Thr	threonine
Thy	thymine
TLC	thin-layer chromatography
TMP	thymidine 5'-monophosphate
TMV	tobacco mosaic virus
TNF	tumor necrosis factor
TNP	trinitrophenyl
TPA	tissue plasminogen activator
TPP or ThPP	thiamine pyrophosphate
trans-	configuration in which the two priority groups are on the opposite sides of the reference plane
Tris	tris(hydroxymethyl)aminomethane or 2-amino-2-hydroxymethyl-propane-1,3-diol
Trp	tryptophan
TSH	thyroid-stimulating hormone
TSTA	tumor-specific transplantation antigens
TTP or dTTP	thymidine triphosphate
Tyr	tyrosine
U	uracil or uridine
UDP	uridine diphosphate
UDP-gal	uridine diphosphogalactose
UDP-glc	uridine diphosphoglucose
U.K.	United Kingdom
UMP	uridine 5'-monophosphate (uridylic acid)
uncor(r)	uncorrected
uns-, *unsym-*	unsymmetrical, asymmetrical
Ura	uracil
U.S.A.E.C.	United States Atomic Energy Commission
USAN	United States Adopted Names
U.S.D.	*United States Dispensatory*
U.S.D.A.	*United States Department of Agriculture*
U.S.P.	*United States Pharmacopeia*
U.S. pat.	United States patent
UTP	uridine triphosphate
uv	ultraviolet
v	volt(s)
v-	vicinal (adjacent)
Val	valine
var	variety
viz.	(*videlicet*) that is to say; namely
VLDL	very low density lipoproteins
vol	volume
vs	versus
v/v	% "volume in volume" expresses the number of milliliters of an active constituent in 100 milliliters of solution
wbc	white blood cell, white blood count
WHO	World Health Organization
wks	weeks
wt	weight
w/v	percent "weight in volume" expresses the number of grams of an active constituent in 100 milliliters of solution, and is used regardless of whether water or another liquid is the solvent
w/w	percent "weight in weight" expresses the number of grams of an active constituent in 100 grams of solution or mixture
XMP	xanthosine monophosphate

yr(s)	year(s)	~	approximately
Z	carbobenzoxy	≅	approximately equal
(Z)-	*zusammen* (German for together). Opposite of (*E*)-. Equivalent to *cis-* in simpler cases	>	greater than
		<	less than
Z. Physiol. Chem	*Hoppe-Seyler's Zeitschrift für Physiologische Chemie*		

MONOGRAPHS

THE MERCK INDEX

OF CHEMICALS, DRUGS, AND BIOLOGICALS

A

1. Abamectin. *Avermectin B₁; 5-O-demethylaver-* mectin A₁ₐ and 5-O-demethyl-25-de(1-methylpropyl)-25-(1-methylethyl)avermectin A₁ₐ (4:1); MK-936; Affirm; Agrimek; Avomec; Vertimec. Mixture of avermectins, q.v., containing at least 80% of avermectin B₁ₐ ($C_{48}H_{72}O_{14}$) and not more than 20% of avermectin B₁ᵦ ($C_{47}H_{70}O_{14}$). Isoln from *Streptomyces avermitilis:* G. Albers-Schönberg *et al.,* **Ger. pat. 2,717,040;** *eidem,* **U.S. pat. 4,310,519** (1977, 1982 both to Merck & Co.). Separation of components: T. W. Miller *et al., Antimicrob. Ag. Chemother.* **15,** 368 (1979); by semi-preparative HPLC: C. C. Ku *et al., J. Liq. Chromatog.* **7,** 2905 (1984). Structure determn: G. Albers-Schönberg *et al., J. Am. Chem. Soc.* **103,** 4216 (1981). Absolute configuration: J. P. Springer *et al., ibid.* 4221. Partial synthesis of B₁ₐ: K. C. Nicolaou *et al., ibid.* **106,** 4189 (1984). Total synthesis: S. Hanessian *et al., ibid.* **108,** 2776 (1986). Anthelmintic activity in animals: L. S. Blair, W. C. Campbell, *J. Parasitol.* **64,** 1032 (1978); J. R. Egerton *et al., Antimicrob. Ag. Chemother.* **15,** 372 (1979); K. S. Todd *et al., Am. J. Vet. Res.* **45,** 976 (1984). Pesticidal activity: I. Putter *et al., Experientia* **37,** 963 (1981); R. A. Dybas, A. St. J. Green, *Brit. Crop. Prot. Conf. - Pests Dis.* **1984,** 947. Control of red imported fire ants: J. A. Greenblatt *et al., J. Agr. Entomol.* **3,** 233 (1986). Interaction with GABA receptors: S. S. Pong *et al., J. Neurochem.* **34,** 351 (1980); T. N. Mellin *et al., Neuropharmacol.* **22,** 89 (1983); receptor binding studies: S. S. Pong, C. C. Wang, *ibid.* **19,** 311 (1980); G. Drexler, W. Sieghart, *Eur. J. Pharmacol.* **99,** 269 (1984). Fate in soil and plants: D. L. Bull *et al., J. Agr. Food Chem.* **32,** 94 (1984); H. A. Moye *et al., ibid.* **35,** 859 (1987); M. S. Maynard *et al., ibid.* **37,** 178, 184 (1989). *Review:* J. R. Babu, "Avermectins: Biological and Pesticidal Activities" in *Biologically Active Natural Products,* ACS Symp. Series 380, H. G. Cutler, Ed. (American Chemical Society, Washington, D.C., 1988) pp 91-108. Reviews of modes of action: C. C. Wang, S. S.

Pong, *Prog. Clin. Biol. Res.* **97,** 373-395 (1982); D. J. Wright, *Biochem. Soc. Trans.* **15,** 65-67 (1987). Review of insecticidal activities and use: L. Strong, T. A. Brown, *Bull. Ent. Res.* **77,** 357-389 (1987).

Odorless, off-white to yellow crystals from methanol, mp 150-155° (dec). [α]ᴅ +55.7° ±2° (c = 0.87 in CHCl₃). uv max (methanol): 237, 245, 253 nm (log ε 4.48, 4.53, 4.34). Vapor pressure: 1.5×10^{-9} torr. Soly at 21° (μg/l): water 10; (mg/ml): acetone 100; n-butanol 10; chloroform 25; cyclohexane 6; ethanol 20; isopropanol 70; kerosene 0.5; methanol 19.5; toluene 350. Hydrolysis does not occur in aq soln at pH 3, 5, 7. LD₅₀ (technical grade) orally in sesame oil in mouse, rat: 13.5, 10.0 mg/kg; dermally in rabbit: >2000 mg/kg. LD₅₀ in mallard duck, bobwhite quail: 84.6, >2000 mg/kg. LC₅₀ (96 hr) in rainbow trout, bluegill: 3.6, 9.6 μg/l; LC₅₀ (48 hr) in *Daphnia magna* 0.34 μg/l (Merck Technical Data Sheet).

USE: Acaricide; insecticide.

THERAP CAT (VET): Anthelmintic.

2. Abietic Acid. *[1R-(1α,4aβ,4bα,10aα)]-1,2,3,4,4a,-4b,5,6,10,10a-Decahydro-1,4a-dimethyl-7-(1-methylethyl)-1-phenanthrenecarboxylic acid; 13-isopropylpodocarpa-7,13-dien-15-oic acid;* sylvic acid. $C_{20}H_{30}O_2$; mol wt 302.44. C 79.42%, H 9.99%, O 10.58%. A widely available organic acid, prepared by isomerization of rosin: Harris, Sanderson, *Org. Syn.,* **coll. vol. IV,** 1 (1963); Fieser, Fieser, *The Chemistry of Natural Products Related to Phenanthrene* (New York, 3rd ed., 1949). Synthesis from dehydroabietic acid: A. W. Burgstahler, L. W. Worden, *J. Am. Chem. Soc.* **83,** 2587 (1961); E. Wenkert *et al., ibid.* **86,** 2038 (1964). Chromatographic study: A. G. Douglas, T. G. Powell, *J. Chromatog.* **43,** 241 (1969). Metabolism in rabbits: Y. Asakawa *et al., Xenobiotica* **16,** 753 (1986).

Monoclinic plates from alcohol + water, mp 172-175°. [α]ᴅ²⁴ −106° (c = 1 in abs alc). uv max: 235, 241.5, 250 nm (ε 19500, 22000, 14300). Insol in water. Sol in alc, benzene, chloroform, ether, acetone, carbon disulfide, dil NaOH soln. Commercial abietic acid made by heating rosin alone or with acids may be glassy or partly crystalline, usually of yellow color and melting as low as 85°.

Methyl ester *see* methyl abietate.

USE: Manufacture of esters (ester gums), e.g., methyl, vinyl and glyceryl esters for use in lacquers and varnishes. Manufacture of "metal resinates", soaps, plastics and paper sizes. Assists growth of lactic and butyric acid bacteria.

3. Abikoviromycin. *7-Ethylidene-1a,2,3,7-tetrahydrocyclopent[b]oxireno[c]pyridine; 4,4a-epoxy-5-ethylidene-2,3,4,4a-tetrahydro-5H-1-pyridine; abicoviromycin; latum-*

component B₁ₐ, R=C₂H₅   component B₁ᵦ, R=CH₃

Consult the cross index before using this section.

cidin. $C_{10}H_{11}NO$; mol wt 161.20. C 74.51%, H 6.88%, N 8.69%, O 9.93%. Antiviral antibiotic produced by *Streptomyces abikoensis* and *Streptomyces rubescens*. Chromatographic isoln from broth cultures: Umezawa *et al., Japan. Med. J.* **4**, 331 (1951); *C.A.* **46**, 7167 (1952); Umezawa, **Japan.** pat. 6200('54) (to Nippon). Identity with latumcidin: Sakagami *et al., J. Antibiot.* **11A**, 231 (1958). Structure: Gurevich *et al., Tetrahedron Letters* **1968**, 2209. Stereochemistry: Kono *et al., J. Antibiot.* **23**, 572 (1970); Gurevich *et al., Khim. Prir. Soedin.* **7**, 104 (1971), *C.A.* **75**, 5752e (1971). Crystal and molecular structure of the selenate: Y. Kono *et al., Acta Crystallog.* Sect. B, **27**, 2341 (1971). *In vitro* antiviral activity: V. M. Roikhel, N. A. Zeitlenok, *Antibiotiki (Moscow)* **14**, 969 (1969), *C.A.* **72**, 19394q (1969).

Highly unstable and polymerizes promptly on isolation even at $-50°$; however, it can be handled in dilute solutions and in the form of its salts. uv max (neutral ethanol or $0.1N$ KOH): 218, 244, 289 nm (log ε 3.83, 3.99, 3.94); (0.1N HCl) 236, 341 nm (log ε 3.99, 4.05).

4. Abrin. Agglutinin; toxalbumin. A toxic lectin and hemagglutinin obtained from seeds of *jequirity, Abrus precatorius* L., *Leguminosae*, a common vine of tropical countries, also found in central and southern Florida. Isoln and purification: J. Y. Lin *et al., J. Formosan Med. Assoc.* **68**, 518 (1969), *C.A.* **72**, 98695 (1970); *eidem, Toxicon* **9**, 97 (1971). The high toxicity of abrin was originally believed to result from its hemagglutinating activity, but subsequent studies have shown that separate proteins are responsible for the toxicity and agglutination: S. Olsnes, A. Pihl, *Eur. J. Biochem.* **35**, 179 (1973). Five glycoproteins have been purified from the seeds of *A. precatorius*: *Abrus agglutinin* and the toxic principles, abrins a-d. *Abrus agglutinin* is a tetramer of 134,900 Da; is non-toxic to animal cells and a potent hemagglutinator. Abrins a through d (mol wt 63,000-67,000 Da) are composed of two disulfide-linked polypeptide chains. The smaller A-chain inhibits protein synthesis and causes cell death; the larger B-chain binds to the cell plasma membrane. Purification of major components: C. H. Wei *et al., J. Biol. Chem.* **249**, 3061 (1974). Crystallographic study: C. H. Wei, J. R. Einstein, ibid. 2985. Improved purification, properties, crystallography of *Abrus agglutinin*: C. H. Wei *et al., ibid.* **250**, 4790 (1975). Physical properties of the toxic principles: M. S. Herrmann, W. D. Behnke, *Biochim. Biophys. Acta* **621**, 43 (1980); *eidem, ibid.* **667**, 397 (1981). Isoln and purification of all five proteins: J. Y. Lin *et al., Toxicon* **19**, 41 (1981). Amino acid sequence of the A-chain of abrin-a and comparison with ricin: G. Funatsu *et al., Agr. Biol. Chem.* **52**, 1095 (1988). Antitumor effects in animals: V. V. S. Reddy, M. Sirsi, *Cancer Res.* **29**, 1447 (1969); J. Y. Lin *et al., Nature* **227**, 292 (1970); O. Fodstad *et al., Cancer Res.* **37**, 4559 (1977). Immunoelectron microscopy studies of abrin toxic action on tumor cells: C. T. Lin *et al., J. Ultrastruc. Res.* **73**, 310 (1980). Studies on toxicity and binding kinetics: M. Witten *et al., Exp. Cell Biol.* **49**, 306 (1981); C. E. Bennett *et al., ibid.* 319. Use of A-chain in cell-type-specific cytotoxic agents known as "immunotoxins": A. J. Cumber *et al., Methods in Enzymology* **112**, 207 (1985). *See also* Ricin, Lectins.

Yellowish-white powder. Sol in solns of sodium chloride, usually with turbidity. The toxic portion is heat-stable to incubation at 60° for 30 min; at 80°, most of the toxicity is lost in 30 min. LD_{50} i.p. in mice: 0.020 mg/kg, J. Y. Lin *et al., J. Formosan Med. Assoc.* **68**, 322 (1969), *C.A.* **71**, 121926 (1969).

Caution: Seeds of *A. precatorius* are extremely toxic; one seed, if thoroughly masticated, can cause fatal poisoning, *cf.* J. M. Kingsbury, *Poisonous Plants of the United States and Canada* (Prentice-Hall, New Jersey, 1964) p 303; K. Genest *et al., Arzneimittel-Forsch.* **21**, 888 (1971).

Note: Do not confuse with abrine, q.v.

USE: Exptly in cancer research.

5. Abrine. *N-Methyl-L-tryptophan;* α-methylamino-β-(3-indole)propionic acid. $C_{12}H_{14}N_2O_2$; mol wt 218.25. C 66.03%, H 6.47%, N 12.84%, O 14.66%. Not to be confused with the albuminous substance abrin, q.v. Obtained from the seeds of *Abrus precatorius* L., *Leguminosae* (jequirity): Hoshino, *Ann.* **520**, 31 (1935). Synthesis: Miller, Robson, *J. Chem. Soc.* **1938**, 1910. Configuration: Cahill, Jackson, *J. Biol. Chem.* **126**, 29 (1938).

Prisms from water, dec 295°. $[\alpha]_D^{21}$ +44° (0.28 g in 10 ml 0.5N HCl). One gram dissolves in about 100 ml methanol, slightly sol in water, insol in ether. Sol in dil acids, alkalies.

Hydrochloride, $C_{12}H_{14}N_2O_2 \cdot HCl$, needles, mp 222°, soluble in water.

Nitrate, $C_{12}H_{14}N_2O_2 \cdot HNO_3$, needles, dec 143°.

Acetyl derivative, $C_{14}H_{16}N_2O_3$. mp 176°. $[\alpha]_D^{25}$ $-148°$ (43 mg in 5 ml 0.1N NaOH).

6. Abscisic Acid. *[S-(Z,E)]-5-(1-Hydroxy-2,6,6-trimethyl-4-oxo-2-cyclohexen-1-yl)-3-methyl-2, 4-pentadienoic acid;* abscisin II; dormin; ABA. $C_{15}H_{20}O_4$; mol wt 264.31. C 68.16%, H 7.63%, O 24.21%. An abscission-accelerating plant hormone. Presence of the naturally occurring (+)-*cis,trans*-form (also designated as (S)-abscisic acid) in sycamore, birch, rose leaves, cabbage, potato, lemon, avocado: Cornforth *et al., Nature* **210**, 627; **211**, 742 (1966). Identity with dormin: Cornforth *et al., Nature* **205**, 1269 (1965); *idem, Tetrahedron, Suppl.* No. 8, Part II, 603 (1967). Isoln from young cotton fruit: Ohkuma *et al., Science* **142**, 1592 (1963). Identification of (+)-abscisic acid in rat, pig brain: M.-Th. Le Page-Degivry *et al., Proc. Nat. Acad. Sci. USA* **83**, 1155 (1986). Synthesis of ($\pm$)-*cis,trans*-forms: Cornforth *et al., Nature* **206**, 715 (1965); of (+)-*cis,trans*- and ($-$)-*cis,trans*-forms: F. Kienzle *et al., Helv. Chim. Acta* **61**, 2616 (1978); of ($\pm$)-*trans,cis*-form: Ohkuma, *Agr. Biol. Chem.* **29**, 962 (1965); **30**, 434 (1966); Findlay, MacKay, *Can. J. Chem.* **49**, 2369 (1971). Structure: Cornforth *et al., Nature* **206**, 715 (1965); Ohkuma *et al., Tetrahedron Letters* **1965**, 2529. Absolute configuration of (+)-*cis,trans*-form: Cornforth *et al., Chem. Commun.* **1967**, 114; revised stereochemistry: Ryback, *ibid.* **1972**, 1190; Koreeda *et al., J. Am. Chem. Soc.* **95**, 239 (1973). Crystal and molecular structure: H. W. Schmalle *et al., Acta Crystallogr.* **B33**, 2218 (1977). Effect on seed set in wheat: J. M. Morgan, *Nature* **285**, 655 (1980). *Review:* Addicott, Lyon, *Ann. Rev. Plant Physiol.* **20**, 139 (1969). Review of role in root growth regulation: P. E. Pilet, P. W. Barlow, *Plant Growth Regul.* **6**, 217-265 (1987); of metabolism: J. A. D. Zeevaart, R. A. Creelman, *Ann. Rev. Plant Physiol. Plant Mol. Biol.* **39**, 439-473 (1988). Book: *Abscisic Acid,* F. T. Addicott, Ed. (Praeger, New York, 1983) 608 pp.

(+)-*cis,trans*-Form, crystals from ethyl acetate + hexane, mp 161-163°. Sublimes at 120°. $[\alpha]_D^{20}$ +411.1° (c = 1 in ethanol). $[\alpha]_D^{20}$ +426.5° (c = 1 in 0.005N methanolic H_2SO_4). Sol in aq NaHCO$_3$, chloroform, acetone, ethyl acetate, ether; slightly sol in benzene, water; sparingly sol in petr ether. uv max (methanol): 252 nm (ε 25200).

($-$)-*cis,trans*-Form, mp 162-163°. $[\alpha]_D^{20}$ $-426.2°$ (c = 1 in 0.005N H$_2$SO$_4$).

($\pm$)-*cis,trans*-Form, crystals, mp 188-190°.

Note: **Abscisin I,** an abscission-accelerating substance. Nomenclature: Ohkuma *et al., Science* **142**, 1592 (1963). Isoln from mature fruit walls of cotton: Liu, Carns, *Science* **134**, 384 (1961). Crystals, mp 197-198°. uv max (methanol): 250 nm. Acidic reaction. Sol in chloroform, dil

NaOH; slightly sol in dimethyl ether. Practically insol in dil HCl.

7. Absinthin. [3S-(3α,3aα,6β,6aα,6bβ,7α,7aβ,8α,-10aβ,11β,13aα,13bα,13cβ,14bβ)]-3,3a,4,5,6,6a,6b,7,7a,8,9,-10,10a,13a,13c,14b-Hexadecahydro-6,8-dihydroxy-3,6,8,11,-14,15-hexamethyl-2H-7,13b-ethenopentaleno[1'',2'':6,7;5'',4'':6',7']dicyclohepta[1,2-b:1',2'-b']difuran-2,12-(11H)-dione; absinthiin; absynthin. $C_{30}H_{40}O_6$; mol wt 496.62. C 72.55%, H 8.12%, O 19.33%. Chief bitter principle of wormwood, *Artemisia absinthium L., Compositae.* Isoln by chromatography: Herout *et al., Coll. Czech. Chem. Commun.* **21**, 1485 (1956); see also *Chem. & Ind. (London)* **1955**, 569. Structural studies: Novotny *et al., ibid.* **1958**, 465; *Coll. Czech. Chem. Commun.* **25**, 1492 (1960); Vokac *et al., Tetrahedron Letters* **1968**, 3855. Structure: J. Beauhaire *et al., ibid.* **21**, 3191 (1980).

Very bitter orange needles from abs ether, mp 179-180° (dec).

Solvated crystals from benzene, decomp 165°. $[\alpha]_D^{20}$ +180.0° (c = 1.9 in $CHCl_3$). Bitterness threshold 1:70,000.

8. Absinthium. Wormwood; Absinthe; Armoise. Dried leaves and flowering tops of *Artemisia absinthium L., Compositae. Habit.* Grows as weed in Europe, U.S., Canada, North and West Asia, Africa. Has been used as stomachic tonic and anthelmintic. *Constit.* Absinthin, anabsinthin, dark green or brown volatile oil (chiefly thujone). E. Guenther, *The Essential Oils,* V, 487 (Van Nostrand, New York 1952). Isolation of various constituents: Cekan, Herout, *Coll. Czech. Chem. Commun.* **21**, 79 (1956); Herout *et al., ibid.* 1485. Characterization of oil: T. Sacco, F. Chialva, *Planta Med.* **54**, 93 (1988).

Very strong odor, acrid taste.

USE: As flavoring in alcoholic beverages, *e.g. vermouth,* which is a blend of white wines, contg traces of absinthium and other flavors. *Caution:* Ingestion of the volatile oil or of the liqueur, absinthe, may cause G.I. symptoms, nervousness, stupor, convulsions, death.

9. Acacetin. 5,7-Dihydroxy-2-(4-methoxyphenyl)-4H-1-benzopyran-4-one; 5,7-dihydroxy-4'-methoxyflavone; apigenin-4'-methyl ether. $C_{16}H_{12}O_5$; mol wt 284.26. C 67.60%, H 4.26%, O 28.14%. The aglycon of linarin, *q.v.,* and of acaciin. Isoln from linarin: Zemplén, Bognar, *Ber.* **74B**, 1818 (1941). From acaciin: Hattori, *Acta Phytochim.* **2**, 105 (1925). Isoln from *Robinia pseudoacacia L., Leguminosae:* Nakazawa, Matsuura, *J. Pharm. Soc. Japan* **73**, 481 (1953). Structure: Baker *et al., J. Chem. Soc.* **1951**, 691. Synthesis: Robinson, Venkataraman, *ibid.* **1926**, 2348; Zemplén, Bognar, *Ber.* **76B**, 452 (1943); Narasimhachari, Seshadri, *Proc. Indian Acad. Sci.* **30A**, 151 (1949); Simpson, *Sci. Proc. Roy. Dublin Soc.* **27**, 111 (1956), *C.A.* **51**, 8082a (1957).

Yellow needles from 95% alcohol, mp 263°. Sol in hot alc, practically insol in ether. Sol in alkalies with yellow color. Diacetate, $C_{20}H_{16}O_7$, lustrous needles from alc, mp 203°.

7-Rhamnoglucoside, $C_{28}H_{32}O_{14}$, *acaciin.* From *Robinia pseudacacia L., Leguminosae:* Freudenberg, Hartmann, *Ann.* **587**, 207 (1954). Structure: Zemplén, Mester, *Magyar Kém.*

Folyoirat **56**, 2 (1950), *C.A.* **45**, 7977d (1951). Needles from pyridine + water, mp 263°. $[\alpha]$ −85.3° (pyridine); −99.5° (glacial acetic acid). Sparingly soluble in cold, more sol in boiling water; slightly sol in organic solvents.

10. Acacia. Gum Arabic. Estimations of mol wt range from about 240,000: Oakley, *Trans. Faraday Soc.* **31**, 136 (1935), to 580,000: Anderson *et al., Carbohyd. Res.* **3**, 308 (1967). According to the U.S.P., acacia is the dried gummy exudation from the stems and branches of *Acacia senegal* (L.) Willd., *Leguminosae,* or other African species of Acacia. According to C. L. Mantell, *The Water-Soluble Gums* (New York, 1947), Kordofan gum (hashab geneina), the gum from *Acacia verek* Guill. & Perr. from plantations in the Kordofan province (Sudan) is considered the best commercial variety. Grades of Kordofan gum which are clear, white (sun bleached) and tasteless are preferred for food prepns and pharmaceuticals. (There is a close relationship between color and flavor due to the presence of tannins.) Acacia was originally thought to be composed only of (−)-arabinose, (+)-galactose, (−)-rhamnose, (+)-glycuronic acid. Revised composition and structural studies: Anderson *et al., J. Chem. Soc. (C)* **1966**, 1959. *See also* Swenson *et al., J. Polym. Sci. Part A-2* **6**, 1593 (1968). General review: Anderson, Dea, *J. Soc. Cosmet. Chem.* **22**, 61-76 (1971). Review of use as food additive: D. M. W. Anderson, *Food Addit. Contam.* **3**, 225-230 (1986).

Occurs in spheroidal tears up to 32 mm in diameter. Also flakes and powder. Solns of gum from *Acacia verek* are levorotatory; other acacia species are dextrorotatory: Hamy, *Bull. Sci. Pharmacol.* **35**, 421 (1928). Specific gravity: 1.35-1.49 (samples dried at 100° are heavier). Moisture content usually varies from 13-15%. U.S.P. limit 15%. Material containing less than 12% chips easily and produces dust during transportation. Insol in alcohol. Almost completely sol in twice its weight of water. 100 grams of a satd soln contains 37 g at 25°; 38 g at 50°; 40 g at 90°: Taft, Malm, *Trans. Kans. Acad. Sci.* **32**, 49 (1929). Aq soln acid to litmus. Also sol in glycerol and in propylene glycol, but prolonged heating (several days) may be necessary for complete solution (about 5%).

Incompat. Precipitates or jellies result upon addition of solns of ferric salts, borax, basic lead acetate (lead subacetate, but not neutral lead acetate), alcohol, sodium silicate, gelatin, ammoniated tincture of guaiac.

USE: As mucilage, excipient for tablets, size, emulsifier, thickener, also in candy, other foods; as colloidal stabilizer. In the manufacture of spray-dried "fixed" flavors—stable, powdered flavors used in packaged dry-mix products (puddings, desserts, cake mixes) where flavor stability and long shelf life are important.

11. Acacic Acid. (3β,16β,21β)-Trihydroxyolean-12-en-28-oic acid. $C_{30}H_{48}O_5$; mol wt 488.68. C 73.73%, H 9.90%, O 16.37%. Isoln from pods of *Acacia concinna* D.C., *Leguminosae:* Varshney, Shamsuddin, *Tetrahedron Letters* **1964**, 2055; from bark of *A. concinna* D.C.: R. Banerji, S. K. Nigam, *J. Ind. Chem. Soc.* **57**, 1043 (1980). Structure and stereochemistry: Varshney *et al., ibid.* **1965**, 1187. Revised structure: A. K. Barua *et al., Trans. Bose Res. Inst., Calcutta* **39**, 61 (1976), *C.A.* **87**, 53460c (1977).

Needles from methanol, mp 280-281°.

Methyl ester, $C_{31}H_{50}O_5$, needles from methanol, mp 223-224°.

Diacetyl lactone, $C_{34}H_{50}O_6$, crystals, mp 235-236°.

12. Acarbose. *O-4,6-Dideoxy-4-[[[1S-(1α,4α,5β,6α)]-4,5,6-trihydroxy-3-(hydroxymethyl)-2-cyclohexen-1-yl]amino]-α-D-glucopyranosyl-(1→4)-O-α-D-glucopyranosyl-(1→4)-D-glucose;* Bay-g-5421; Glucobay. $C_{25}H_{43}NO_{18}$; mol wt 645.63. C 46.51%, H 6.71%, N 2.17%, O 44.61%. Pseudotetrasaccharide containing an unsaturated cyclitol moiety. An α-glucosidase inhibitor that reduces sugar absorption in the gastrointestinal tract. Isoln from strains of *Actinoplanes:* W. Frommer *et al.,* **Ger. pat. 2,347,782;** *eidem,* **U.S. pat. 4,062,950** (1975, 1977 both to Bayer). Total synthesis: S. Ogawa, Y. Shibata, *Chem. Commun.* **1988**, 605. Biosynthetic studies: U. Degwert *et al., J. Antibiot.* **40**, 855 (1987). Glucosidase inhibition studies: D. D. Schmidt *et al., Naturwiss.* **64**, 535 (1977); W. Puls *et al., ibid.* 536. Use in treatment of diabetic adults: D. Sailor, G. Roder, *Arzneimittel-Forsch.* **30**, 2182 (1980); H. Laube *et al., ibid.* 1154. Long-term study in sulfonylurea-treated diabetics: H. Vierhapper *et al., Diabetologia* **20**, 586 (1981). Potential use in prophylaxis of dental caries: N. E. Fiehn, D. Moe, *Scand. J. Dent. Res.* **90**, 124 (1982). Review of pharmacodynamics, pharmacokinetics and therapeutic potential: S. P. Clissold, C. Edwards, *Drugs* **35**, 214-243 (1988).

$[\alpha]_D^{18}$ +165° (water).
THERAP CAT: Antidiabetic.

13. Acebutolol. *N-[3-Acetyl-4-[2-hydroxy-3-[(1-methylethyl)amino]propoxy]phenyl]butanamide; 3'-acetyl-4'-[2-hydroxy-3-(isopropylamino)propoxy]butyranilide;* 1-(2-acetyl-4-n-butyramidophenoxy)-2-hydroxy-3-isopropylaminopropane; 5'-butyramido-2'-(2-hydroxy-3-isopropylaminopropoxy)acetophenone; Prent. $C_{18}H_{28}N_2O_4$; mol wt 336.43. C 64.26%, H 8.39%, N 8.33%, O 19.02%. Cardioselective β-adrenergic blocker. Prepn: Wooldridge, Basil, **S. Afr. pat. 68 08,345** corresp to **U.S. pat. 3,857,952** (1969, 1974 both to May & Baker). Pharmacology: Cuthbert, Owusu-Ankomah, *Brit. J. Pharmacol.* **43**, 639 (1971); Basil *et al., ibid.* **48**, 198 (1973); Lewis *et al., Brit. Heart J.* **35**, 743 (1973). Crystal structure: A. Carpy *et al., Acta Crystallogr.* **B35**, 185 (1979). Review of pharmacology, therapeutic efficacy in hypertension, angina pectoris, arrhythmia: B. N. Singh *et al., Drugs* **29**, 531-569 (1985); of pharmacokinetics, clinical pharmacology and clinical efficacy: G. DeBono *et al., Am. Heart J.* **109**, 1211-1223 (1985).

Crystals, mp 119-123°.
Hydrochloride, $C_{18}H_{29}ClN_2O_4$, **IL-17803A, M & B 17803A, Neptall, Sectral.** Crystals from anhydr methanol-anhydr diethyl ether, mp 141-143°.
THERAP CAT: Antihypertensive, antianginal, antiarrhythmic.

14. Acecainide. *4-(Acetylamino)-N-[2-(diethylamino)ethyl]benzamide;* 4'-[[2-(diethylamino)ethyl]carbamoyl]acetanilide; N-acetylprocainamide. $C_{15}H_{23}N_3O_2$; mol wt 277.37. C 64.95%, H 8.36%, N 15.15%, O 11.54%. Metabolite of procainamide, *q.v.* Prepn: E. C. Schreiber, **Ger. pat. 2,062,978** (1971 to Squibb), *C.A.* **75**, 76427 (1972). Pharmacology studies: R. D. Reynolds, B. L. Kamath, *Eur. J. Pharmacol.* **59**, 115 (1979); R. D. Reynolds, R. J. Gorczynski, *J. Pharmacol. Exp. Ther.* **212**, 579 (1980). Pharmacokinetics: M. Wierzchowiecki *et al., Int. J. Clin. Pharmacol.*

Ther. Toxicol. **18**, 272 (1980). Clinical pharmacology and anti-arrhythmic efficacy: J. Kluger *et al., Am. J. Cardiol.* **45**, 1250 (1980); R. A. Winkle *et al., ibid.* **47**, 123 (1981). HPLC determn in plasma: D. Raphanaud *et al., Ther. Drug Monit.* **8**, 365 (1986).

Hydrochloride, $C_{15}H_{24}ClN_3O_2$, **ASL-601, NAPA.** Cryst, mp 190-193°.
THERAP CAT: Cardiac depressant (anti-arrhythmic).

15. Acecarbromal. *N-[(Acetylamino)carbonyl]-2-bromo-2-ethylbutanamide;* N-acetyl-N-bromodiethylacetyl-urea; acetylbromodiethylacetylcarbamide; N-acetyl-N'-α-bromo-α-ethylbutyrylcarbamide; acetylcarbromal; Abasin; Carbased; Sedamyl; Acetyl Adalin. $C_9H_{15}BrN_2O_3$; mol wt 279.14. C 38.72%, H 5.42%, Br 28.63%, N 10.04%, O 17.19%. $(C_2H_5)_2CBrCONHCONHCOCH_3$. Prepd by acetylating carbromal with acetic anhydride in the presence of $ZnCl_2$: **Ger. pat. 327,129;** *see also* H. P. Kaufmann, *Arzneimittelsynthese* (Springer, 1953).

Crystals, slightly bitter taste, mp 109°. Slightly sol in water; freely sol in alcohol, ethyl acetate.
Caution: This substance may be habit forming and is listed in the U.S. Code of Federal Regulations, Title 21 Part 329.1 (1987).
THERAP CAT: Sedative, hypnotic.

16. Acedapsone. *N,N'-(Sulfonyldi-4,1-phenylene)bisacetamide; 4',4'''-sulfonylbis[acetanilide];* bis(4-acetamidophenyl)sulfone; 4,4'-diacetyldiaminodiphenyl sulfone; 4,4'-diacetylaminodiphenyl sulfone; N,N'-diacetyl-4,4'-diaminodiphenyl sulfone; DADDS; diacetyldapsone; sulfadiamine; 1399F; CI 556; Hansolar; Rodilone. $C_{16}H_{16}N_2O_4S$; mol wt 332.37. C 57.82%, H 4.85%, N 8.43%, O 19.25%, S 9.65%. Prepn: Fromm, Wittmann, *Ber.* **41**, 2270 (1908); Raiziss *et al., J. Am. Chem. Soc.* **61**, 2763 (1939); Elslager *et al., J. Med. Chem.* **12**, 357 (1969). Properties: Elslager, Worth, *Nature* **206**, 630 (1965).

Crystalline solid, mp 289-292°. uv max (methanol): 256, 284 nm (ε 25500, 36200). Soly in water: 0.003 mg/ml; in 40% benzyl benzoate-60% castor oil: 0.026 mg/ml.
THERAP CAT: Antimalarial; antibacterial (leprostatic).

17. Acediasulfone. *N-[4-[(4-Aminophenyl)sulfonyl]phenyl]glycine;* N-p-sulfanilylphenylglycine; p-amino-p'-(carboxymethylamino)diphenyl sulfone; 4-carboxymethylamino-4'-aminodiphenylsulfone; diaminodiphenylsulfone-N-acetic acid. $C_{14}H_{14}N_2O_4S$; mol wt 306.35. C 54.89%, H 4.60%, N 9.15%, O 20.89%, S 10.47%. Prepn: Jackson, *J. Am. Chem. Soc.* **70**, 680 (1948); **Swiss pats. 254,803** and **278,482** (1949, 1952, to Cilag Ltd.); Rawlins, **U.S. pat. 2,589,211** (1952 to Parke, Davis).

Crystals, mp 194°. Sol in methanol, dil sodium hydroxide, acetone.
Sodium salt, $C_{14}H_{13}N_2NaO_4S$, **Sulfon-Cilag.** Ingredient of *Ciloprin.*
Morpholine salt, $C_{18}H_{23}N_3O_5S$, **Bentrofene.** Glittering crystals, mp 133-135° (dec). Prepn: Martin, Habicht, **U.S. pat. 2,751,382** (1956 to Cilag Ltd.).
THERAP CAT: Antibacterial.

18. Acefylline. *1,2,3,6-Tetrahydro-1,3-dimethyl-2,6-dioxopurine-7-acetic acid;* carboxymethyltheophylline; 7-theophyllineacetic acid. $C_9H_{10}N_4O_4$; mol wt 238.20. C 45.38%, H 4.23%, N 23.52%, O 26.87%. Prepd from theophylline and chloroacetic acid in hot dil NaOH: **Ger. pat. 352,980** (1922 to E. Merck); *Frdl.* **14**, 1320. Prepn of free acid and sodium salt: Milletti, Virgili, *Chimica (Milan)* **6**, 394 (1951), *C.A.* **46**, 8615h (1952).

Crystals from water, mp 271°.

Sodium salt, $C_9H_9N_4NaO_4$, *Aminodal.* Silky needles, mp > 300°.

THERAP CAT: Diuretic; cardiotonic; bronchodilator.

19. Acefylline Piperazine. *1,2,3,6-Tetrahydro-1,3-dimethyl-2,6-dioxo-7H-purine-7-acetic acid compd with piperazine;* acepifylline; piperazine theophylline-7-acetate; 7-theophyllineacetic acid piperazine salt; piperazine theophylline ethanoate; Dynaphylline; Etaphydel; Etaphylline; Etafillina; Etophylate. $C_{13}H_{20}N_6O_4$; mol wt 324.34. C 48.14%, H 6.22%, N 25.91%, O 19.73%. Undefined mixture of the 1:1 and 2:1 salts which contains 75-78% theophylline acetic acid and 22-25% anhydrous piperazine. Active metabolite is acefylline, *q.v.* Prepn: Baisse, *Bull. Soc. Chim. France* **1949**, 769. Pharmacokinetics: J. Zuidema, F. W. H. Merkus, *Pharm. Weekbl.* **113**, 605 (1978); S. Sved *et al., Biopharm. Drug Dispos.* **2**, 177 (1981). HPLC determn in serum: J. Zuidema, F. W. H. Merkus, *J. Chromatog.* **145**, 489 (1978). GC determn in urine: J. Zuidema, H. Hilbers, *ibid.* **182**, 445 (1980).

n = 1 or 2

THERAP CAT: Bronchodilator.

20. Aceglatone. D-*Glucaric acid di-γ-lactone diacetate;* 2,5-di-*O*-acetyl-D-glucaro-1,4:6,3-dilactone; 2,5-di-*O*-acetyl-D-glucosaccharo-1,4:6,3-dilactone; Aceglaton; Glucaron. $C_{10}H_{10}O_8$; mol wt 258.19. C 46.52%, H 3.90%, O 49.58%. Prepn and structure: Hirasaka, Umemoto, *Chem. Pharm. Bull.* **13**, 325 (1965), *C.A.* **63**, 3024h (1965); Ishidate *et al.,* **Japan.** pat. **14,956**('67) (1967 to Tokyo Biochem. Res. Com.), *C.A.* **68**, 78558m (1968). Pharmacological studies: Iida *et al., Japan. J. Pharmacol.* **15**, 88 (1965), *C.A.* **63**, 5961g (1965). Review and toxicity data: *Japan. Med. Gaz.* **8**(8), 15 (1971).

White, odorless and tasteless crystalline powder, mp 185-186° (Hirasaka); from 2:1 ethanol-ethyl acetate 192° (dec) *(Japan. Med. Gaz.).* Sol in dimethylformamide, sparingly sol in acetone, slightly sol in dioxane, methanol and ethanol. Practically insol in water. LD_{50} in mice, rats (g/kg): > 20,

> 10 orally; > 20, > 10 s.c.; 5.80-6.35, 6.10-6.15 i.p. *(Japan. Med. Gaz.).*

THERAP CAT: Antineoplastic.

21. Aceglutamide. N^2-*Acetyl*-L-*glutamine;* α-*N*-acetyl-L-glutamine; Acutil-S. $C_7H_{12}N_2O_4$; mol wt 188.18. C 44.68%, H 6.43%, N 14.88%, O 34.01%. Prepn: P. Karrer *et al., Helv. Chim. Acta* **9**, 301 (1926); **Brit.** pat. **792,576** (1958 to Merck & Co.), *C.A.* **53**, 2109a (1959); I. J. Maschler, N. Lichtenstein, *Biochim. Biophys. Acta* **57**, 252 (1962). Stability study: G. Sekules, G. Guadagnini, *Farmaco Ed. Prat.* **21**, 22 (1966). NMR study: W. Voelter *et al., Z. Naturforsch. B* **26**, 213 (1971). Fermentation study: T. Nakanishi, *J. Ferment. Technol.* **56**, 573 (1978). Prepn of the aluminum complex: T. Kagawa *et al.,* **Ger.** pat. **2,127,176** corresp to **U.S.** pat. **3,787,466** (1971, 1974 both to Kyowa). Physico-chemical properties: E. Hayakawa *et al., Yakugaku Zasshi* **97**, 731 (1977), *C.A.* **87**, 141198 (1977). Effect on exptl chronic gastric ulcer: H. Tanaka *et al., Oyo Yakuri* **7**, 1035 (1973), *C.A.* **81**, 33283v (1974). Cytoprotective effect: H. Tanaka, *Arzneimittel-Forsch.* **36**, 1485 (1986).

Crystals from ethanol, mp 197°. $[\alpha]_D^{20}$ −12.5° (c = 2.9 in water).

Aluminum complex, $C_{35}H_{59}Al_3N_{10}O_{24}$, *pentakis (N^2-acetyl-*L-*glutaminato)tetrahydroxytrialuminum, aceglutamide aluminum, KW-110, Glumal.* White powder, mp 221° (dec). Sol in water. Practically insol in methanol, ethanol, acetone. LD_{50} in male mice, rats (g/kg): 14.3, > 14.5 orally; 5.0, 4.2 i.p.; 0.46, 0.40 i.v. (Kagawa).

THERAP CAT: Free acid as nootropic; aluminum complex as anti-ulcerative.

22. Acemetacin. *1-(4-Chlorobenzoyl)-5-methoxy-2-methyl-1H-indole-3-acetic acid carboxymethyl ester;* [[1-(4-chlorobenzoyl)-5-methoxy-2-methylindol-3-yl]acetoxy]-acetic acid; TV 1322; Acemix; Rantudil; Rheumibis. $C_{21}H_{18}ClNO_6$; mol wt 425.91. C 59.22%, H 6.63%, Cl 8.32%, N 3.29%, O 22.54%. Deriv of indomethacin, *q.v.* Prepn: K. H. Boltze *et al.,* **Ger.** pat. **2,234,651** corresp to **U.S.** pat. **3,910,-952** (1972, 1975 both to Troponwerke). Series of articles on chemistry, analysis, pharmacodynamics, toxicology and clinical trials: *Arzneimittel-Forsch.* **30**, 1313-1468 (1980). Toxicity data: H. Jacobi, H.-D. Dell, *ibid.* 1398.

Very fine pale yellow crystals from petr ether, mp 150-153°. LD_{50} in male, female mice, male, female rats (mg/kg): 55.5, 18.42, 24.2, 30.1 orally; 34.1, 51.1, 38.1, 28.3 i.v. (Jacobi, Dell).

THERAP CAT: Anti-inflammatory.

23. Acenaphthene. *1,2-Dihydroacenaphthylene; peri-ethylenenaphthalene;* 1,8-ethylenenaphthalene. $C_{12}H_{10}$; mol wt 154.21. C 93.46%, H 6.54%. Occurs in coal tar. Isoln: Ges. f. Teerverwertung, **Ger.** pat. **277,110**; *Chem. Zentr.* **1914,** II, 597. In petroleum residues: Orloff *et al., C.A.* **31**, 2800⁹ (1937). By passing ethylene and benzene or naphthalene though a red hot tube: Berthelot, *Bull. Soc. Chim.* [2] **7**, 274; **8**, 226, 245 (1867). By heating tetrahydroacenaphthene with sulfur to 180°: Braun *et al., Ber.* **55**, 1694 (1922). From acenaphthenone or acenaphthenequinone by high pressure hydrogenation in decalin with nickel at 180-240°: Braun, Bayer, *Ber.* **59**, 921, 923 (1926). From acenaphthenone oxime: Morgan, Stanley, *J. Soc. Chem. Ind. (London)* **44**, 494T (1925).

Consult the cross index before using this section.

Orthorhombic bipyramidal needles from alcohol. d 1.189. mp 95°. bp 279°. uv spectrum: Seshan, *Proc. Indian Acad. Sci.* **A3**, 148 (1936). Factors influencing the uv spectrum: Jones, *J. Am. Chem. Soc.* **67**, 2127 (1945). Insol in water. One gram dissolves in 31 ml alcohol, 56 ml methanol, 25 ml propanol, 2.5 ml chloroform, 5 ml benzene or toluene. 3.2 g are sol in 100 ml glacial acetic acid. Forms water-sol, cryst complexes with desoxycholic acid, containing two molecules of the bile acid as a rule. The complexes crystallize when concentrated solns of the proper amount of the components in alcohol or dioxane are allowed to cool slowly.

USE: Dye intermediate; manuf plastics; insecticide; fungicide.

24. Acenocoumarol. *4-Hydroxy-3-[1-(4-nitrophenyl)-3-oxobutyl]-2H-1-benzopyran-2-one; 3-(α-acetonyl-p-nitrobenzyl)-4-hydroxycoumarin; 3-(α-p-nitrophenyl-β-acetylethyl)-4-hydroxycoumarin; acenocoumarin; nicoumalone; G-23350; Sinthrome; Sintrom.* $C_{19}H_{15}NO_6$; mol wt 353.32. C 64.58%, H 4.28%, N 3.96%, O 27.17%. Prepn: Stoll, Litvan, **U.S.** pat. **2,648,682** (1953 to Geigy). Pharmacology: Jindal, Shah, *Arzneimittel-Forsch.* **16**, 878 (1966).

Crystals, mp 196-199°. Almost insol in water or organic solvents. Forms water-sol salt with alkalies. LD_{50} in mice: 114.7 mg/kg i.p.

THERAP CAT: Anticoagulant.

25. Aceperone. *N-[[1-[4-(4-Fluorophenyl)-4-oxobutyl]-4-phenyl-4-piperidinyl]methyl]acetamide; N-[[1-[3-(p-fluorobenzoyl)propyl]-4-phenyl-4-piperidyl]methyl]acetamide;* 1-(p-fluorophenyl)-4-(4-phenyl-4-acetamidomethylpiperidino)-1-butanone; 1-[γ-(4-fluorobenzoyl)propyl]-4-acetamidomethyl-4-phenylpiperidine; N-[1-[3-(p-fluorobenzoyl)propyl]-4-phenylpiperidin-4-ylmethyl]acetamide; 4'-fluoro-4-(4-acetamidomethyl-4-phenylpiperido)butyrophenone; acetabutone; acetobuton; R 3248. $C_{24}H_{29}FN_2O_2$; mol wt 396.51. C 72.70%, H 7.37%, F 4.79%, N 7.07%, O 8.08%. Prepd by reacting γ-chloro-4-fluorobutyrophenone with 4-acetamidomethyl-4-phenylpiperidine: P. A. J. Janssen, **Belg.** pat. **606,849** (1961), corresp to **U.S.** pat. **3,083,205** (1961, 1963 both to Janssen). Crystal structure: N. Van Opdenbosch *et al., Acta Crystallogr.* **33B**, 171 (1977).

Crystals, mp 97-100°.

26. Acephate. *Acetylphosphoramidothioic acid O,S-dimethyl ester;* Ortho 12420; Orthene. $C_4H_{10}NO_3PS$; mol wt 183.16. C 26.23%, H 5.50%, N 7.65%, O 26.21%, P 16.91%, S 17.50%. Prepn and activity: Magee, **U.S.** pats. **3,716,600** and **3,845,172** (1973, 1974, both to Chevron).

White solid, mp 64-68° (impure). Very sol in water; moderately sol in acetone, alc. Low soly in aromatic solvents. LD_{50} orally in rats: 700 mg/kg, Magee, *loc. cit.* (1973).

USE: Contact and systemic insecticide.

27. Acepromazine. *1-[10-[3-(Dimethylamino)propyl]-10H-phenothiazin-2-yl]ethanone; 10-[3-(dimethylamino)-propyl]phenothiazin-2-yl methyl ketone;* 2-acetyl-10-(3-dimethylaminopropyl)phenothiazine; 3-acetyl-10-(3-dimethylaminopropyl)phenothiazine; 10-(3-dimethylaminopropyl)-phenothiazine-3-ethylone; acetazine; acetopromazine; acetylpromazine; 1522CB; Vetranquil. $C_{19}H_{22}N_2OS$; mol wt 326.47. C 69.90%, H 6.79%, N 8.58%, O 4.90%, S 9.82%. Prepn: Schmitt *et al., Bull. Soc. Chim. France* **1957**, 938.

Orange-colored oil, $bp_{0.5}$ 220-240°.

Maleate, $C_{23}H_{26}N_2O_5S$, *Atravet, Plegicil, Notensil, Soprontin.* Yellow crystals from ethyl acetate, mp 135-136°. Sol in water. pH of 0.1% aq soln 5.2. LD_{50} in rats (calcd as free base): 130 mg/kg orally; 70 mg/kg i.v.

THERAP CAT (VET): Tranquilizer.

28. Acerin. An extract from the dried fruit of the Norway maple *Acer plantanoides* L., *Aceraceae.* Freshly prepd solutions of acerin destroy E. *coli* and vaccinia virus in 5 min but are less effective on the Staphylococcus Twort virus. Phagicidal action also exists even in the presence of the host cells. Extraction and properties: G. Fischer, *Acta Pathol. Microbiol. Scand.* **31**, 433 (1952); **34**, 482 (1954); Fischer *et al., Experientia* **10**, 329 (1954).

Practically insol in methanol, ethanol, ether, acetone. Does not dialyze through a cellophane membrane at room temp but passes through a Seitz filter. Sodium chloride precipitates the active principle in acerin preparations as does also normal serum with 1:800 solns of acerin. Strong acerin preparations acquire reddish-brown color after treatment with alkali and oxidation in air. In general, the substance behaves similarly to vegetable tannins. It is practically nontoxic to rabbits and mice when administered s.c.

29. Acerola. The ripe fruit of *Malpighia punicifolia* L., *Malpighiaceae,* also called West Indian Cherry. *Habit.* Central America, Puerto Rico. Probably the richest natural source of ascorbic acid. Analysis of pitted fruit: Water 92.28%; ascorbic acid 1690 mg/100 g; vitamin A 11.0 mg/100 g; niacin 407 γ/100 g; vitamin B_6 8.7 γ/100 g; thiamine·HCl 30.0 γ/100 g; fluorine 10 γ/100 g; carbohydrates 6.4%. pH of juice 3.3. The pits comprise 19.25% of the fruit. *Ref:* Derse, Elvehjem, *J. Am. Med. Assoc.* **156**, 1501 (1954). Commercial aspects: *J. Agr. Food Chem.* **2**, 1155 (1954). *Review:* Moscoso, *Econ. Bot.* **10**, 280 (1956). Method of preparing acerola juice concentrates: Morse, **U.S.** pats. **3,012,942-3** (1961 to Nutrilite Products).

THERAP CAT: Nutrient.

30. Acesulfame. *6-Methyl-1,2,3-oxathiazin-4(3H)-one 2,2-dioxide;* 6-methyl-3,4-dihydro-1,2,3-oxathiazin-4-one 2,2-dioxide; acetosulfam. $C_4H_5NO_4S$; mol wt 163.15. C 29.44%, H 3.09%, N 8.59%, O 39.23%, S 19.65%. Non-nutritive artificial sweetener. Prepn: K. Clauss, H. Jensen, **Ger.** pat. **2,001,017**; *eidem,* **U.S.** pat. **3,689,486** (1971, 1972 to Hoechst); K. Clauss *et al., Z. Lebensm.-Unters. Forsch.* **162**, 37 (1976). Crystal structure: E. F. Paulus, *Acta Crystallogr.* **B31**, 1191 (1975). HPLC analysis: H. Grosspietsch, H. Hachenberg, *Z. Lebensm.-Unters. Forsch.* **171**, 41 (1980).

Needles from benzene or chloroform, mp 123-123.5°.

Potassium salt, $C_4H_4KNO_4S$, *acesulfame-K, Sunette.* mp 250°. Very sol in water, DMF, DMSO. Sol in alc, glycerin-water.

USE: Potassium salt as sweetener for foods, cosmetics.

31. Acetal. *1,1-Diethoxyethane;* diethylacetal; acetaldehyde diethyl acetal; ethylidene diethyl ether. $C_6H_{14}O_2$; mol wt 118.17. C 60.98%, H 11.94%, O 27.08%. $CH_3CH-(OC_2H_5)_2$. Made from acetaldehyde and alcohol in the presence of anhydrous calcium chloride or of small quantities of mineral acid: H. Adkins, B. H. Nissen, *J. Am. Chem. Soc.* **44**, 2750 (1922); *eidem, Org. Syn.* **coll. vol. I**, 1 (2nd ed., 1941).

Volatile liquid. d_4^{20} 0.8254. bp_{760} 102.7°; bp_{200} 66.3°; bp_{60} 39.8°; bp_{40} 31.9°; bp_{20} 19.6°; bp_{10} +8.0°; bp_5 −2.3°; $bp_{1.0}$ −23°. n_D^{20} 1.38193. Flash pt, closed cup: 97°F (36°C). uv spectrum: Purvis, *J. Chem. Soc.* **127**, 9 (1925). 100 g water dissolve 5 g acetal. Misc with alcohol, 60% alcohol, ether. Sol in heptane, methylcyclohexane, propyl-, isopropyl-, butyl-, isobutyl alcohol, and ethyl acetate. Tends to polymerize on standing. Stable to alkalies. LD_{50} orally in rats: 4.57 g/kg, H. F. Smyth *et al., J. Ind. Hyg. Toxicol.* **31**, 60 (1949).

USE: Solvent; in synthetic perfumes such as jasmine; in organic syntheses.

THERAP CAT: Hypnotic.

32. Acetaldehyde. Ethanal; "aldehyde"; acetic aldehyde; ethylaldehyde. C_2H_4O; mol wt 44.05. C 54.53%, H 9.15%, O 36.32%. CH_3CHO. Produced by oxidation of alcohol with $Na_2Cr_2O_7$ and H_2SO_4; usually from acetylene, dil H_2SO_4 and mercuric oxide as catalyst; also by passing alcohol vapor over a heated metallic catalyst. Lab procedure from ethanol: Wertheim, *J. Am. Chem. Soc.* **44**, 2658 (1922); Fricke, Havestadt, *Angew. Chem.* **36**, 546 (1923); Gattermann-Wieland, *Praxis des organischen Chemikers* (de Gruyter, Berlin, 40th ed., 1961) p 180; from acetylene: Gattermann-Wieland, *op. cit.* 183; from paraldehyde: A. I. Vogel, *Practical Organic Chemistry* (Longmans, London, 3rd ed., 1959) p 324; by catalytic oxidation of ethylene in aq soln: J. Smidt *et al., Angew Chem.* **71**, 176 (1959); by oxidation of ethylene in fuel cells in the gas phase: K. Otsuka *et al., Chem. Commun.* **1988**, 1272. Manuf: Faith, Keyes & Clark's *Industrial Chemicals,* F. A. Lowenheim, M. K. Moran, Eds. (Wiley-Interscience, New York, 4th ed., 1975) pp 1-7. Toxicity data: Smyth, *Arch. Ind. Hyg. Occup. Med.* **4**, 119 (1951). *Review:* H. J. Hagemeyer in Kirk-Othmer *Encyclopedia of Chemical Technology* **vol. 1** (Wiley-Interscience, New York, 3rd ed., 1978) pp 97-112.

Flammable liquid; characteristic, pungent odor. d_4^{16} 0.788. mp −123.5°. bp 21°. n_D^{20} 1.3316. Flash pt, closed cup: −36°F (−38°C). Miscible with water, alcohol. *Keep cold. Chill thoroughly before opening.* LD_{50} orally in rats: 1930 mg/kg (Smyth).

Human Toxicity: General narcotic action. Large doses may cause death by respiratory paralysis. Symptoms of chronic intoxication resemble those of chronic alcoholism, *cf. Clinical Toxicology of Commercial Products,* R. E. Gosselin *et al.,* Eds. (Williams & Wilkins, Baltimore, 4th ed., 1976) Section II, p 124.

USE: Manuf paraldehyde, acetic acid, butanol, perfumes, flavors, aniline dyes, plastics, synthetic rubber; silvering mirrors, hardening gelatin fibers. *Caution:* Irritating to mucous membranes.

33. Acetaldehyde Ammonia. *1-Aminoethanol;* α-aminoethyl alcohol; aldehyde ammonia. C_2H_7NO; mol wt 61.08. C 39.32%, H 11.55%, N 22.93%, O 26.19%. $CH_3CH-(OH)NH_2$. Prepd from acetaldehyde and ammonia: Aschan, *Ber.* **48**, 874 (1915).

Crystals; gradually turns yellow to brown in air. mp 97°. bp 110°, partly decomposing. Freely sol in water, slightly in ether. *Protect from light and air.*

USE: For preparing pure acetaldehyde; in organic syntheses. *Caution:* Irritates eyes, mucous membranes.

34. Acetaldehyde Sodium Bisulfite. *1-Hydroxyethanesulfonic acid sodium salt.* $C_2H_5NaO_4S$; mol wt 148.11. C 16.22%, H 3.40%, Na 15.52%, O 43.21%, S 21.65%. $CH_3-CH(OH)SO_3Na$.

Hemihydrate, crystals. Decomposed by acids. Freely sol in water; insol in alcohol.

USE: Making pure acetaldehyde; in organic synthesis. *Caution:* Irritates skin, mucous membranes.

35. Acetaldoxime. *Acetaldehyde oxime;* aldoxime; ethylidenehydroxylamine. C_2H_5NO; mol wt 59.07. C 40.66%, H 8.53%, N 23.72%, O 27.09%. $CH_3CH=NOH$. Prepn: Dunstan, Dymond, *J. Chem. Soc.* **61**, 470 (1892). Manuf: Donaruma, U.S. pat. **2,763,686** (1956 to du Pont).

Two crystalline modifications, mp 12° (β-form), mp 46.5° (α-form). d 0.966. bp 114.5°. n_D^{20} 1.415. Dec by aq HCl into acetaldehyde and hydroxylamine. Very sol in water, alcohol, ether.

36. Acetamide. Acetic acid amide. C_2H_5NO; mol wt 59.07. C 40.66%, H 8.53%, N 23.72%, O 27.09%. CH_3-CONH_2. Prepd by fractional distillation of ammonium acetate: Coleman, Alvarado, *Org. Syn.;* **coll. vol. I**, 3 (2nd ed., 1941); Gattermann-Wieland, *Praxis des organischen Chemikers* (40th ed., 1961) p 118; Vogel, *Practical Organic Chemistry* (3rd ed., 1959) p 401. Prepn from methyl acetate, W. P. Munro *et al.,* U.S. pat. **2,106,697** (1936 to Calco Chem.); from ethyl acetate, Vogel, *op. cit.,* p 403. Studies of acetamide as an ionizing solvent: Jauder, Winkler, *J. Inorg. Nucl. Chem.* **9**, 24, 32, 39 (1959). Toxicological study: Weisburger *et al., Toxicol. Appl. Pharmacol.* **14**, 163 (1969).

Deliquescent hexagonal crystals. Odorless when pure, but frequently has a mousy odor. d_4^{20} 1.159. mp 81°. bp_{760} 222°; bp_{100} 158°; bp_{40} 136°; bp_{20} 120°; bp_{10} 105°; bp_5 92°. n_D^{78} 1.4274. Neutral reaction. Kb at 25° = 3.1 × 10^{-15}. One gram dissolves in 0.5 ml water, 2 ml alcohol, 6 ml pyridine. Sol in chloroform, glycerol, hot benzene.

USE: Solvent; molten acetamide is an excellent solvent for many organic and inorganic compounds. Solubilizer; renders sparingly soluble substances more soluble in water by mere addition or by fusion. Plasticizer; stabilizer. Manuf methylamine, denaturing alcohol. In organic syntheses. *Caution:* Mild irritant.

37. Acetamidine Hydrochloride. *Ethanimidamide hydrochloride;* ethanamidine hydrochloride; α-amino-α-iminoethane hydrochloride; ethenylamidine hydrochloride; acediamine hydrochloride; SN 4455. $C_2H_7ClN_2$; mol wt 94.55. C 25.41%, H 7.46%, Cl 37.50%, N 29.63%. Prepd by passing HCl into a soln of acetonitrile in abs alcohol, then passing NH_3 into the reaction mixture: Pinner, *Ber.* **16**, 1654 (1883); **17**, 178 (1884); Dox, *Org. Syn.,* **coll. vol. I**, 5 (New York, 2nd ed., 1941). *See also* Fargher, *J. Chem. Soc.* **117**, 674 (1920). *Review:* Shriner, Neumann, *Chem. Rev.* **35**, 351-425 (1944).

$$\begin{array}{c} NH \\ \parallel \\ CH_3CNH_2 \cdot HCl \end{array}$$

Long prisms from alcohol, somewhat deliquescent, mp 174° (Fargher); mp 164-166° (Dox). Very sol in water. Sol in alcohols. Practically insol in acetone, ether. Should be stored in a closed container and in a cool place. If alkali is added to an aq soln, the free base is liberated.

Free base, $C_2H_6N_2$. uv max: 224 nm (ε 4000). pK₁ (25°): 12.1. Has a strong alkaline reaction and on slight warming dissociates into ammonia and acetic acid.

USE: In the synthesis of imidazoles, pyrimidines, triazines. *Caution:* Irritates mucous membranes, skin. Avoid tasting, swallowing, inhalation of the dust, skin contact.

38. ε-Acetamidocaproic Acid. *6-(Acetylamino)hexanoic acid; 6-acetamidohexanoic acid;* acetaminocaproic acid; acexamic acid; *N-acetyl-6-aminohexanoic acid.* $C_8H_{15}NO_3$; mol wt 173.21. C 55.47%, H 8.73%, N 8.09%, O 27.71%. $CH_3CONH(CH_2)_5COOH$. Prepn: Offe, *Z. Naturforsch.* **2b**, 182 (1947); **Fr. pat. M2332** (1964 to Rowa), *C.A.* **61**, 577b (1964). Anti-inflammatory activity of salts: O. Guillard *et al., Pharmacology* **34**, 296 (1987). Clinical comparison with ranitidine in gastroduodenal ulcer: M. J. Varas Lorenzo, *Curr. Ther. Res.* **39**, 19 (1986).

Crystals from acetone, mp 104-105.5° (**Fr. pat. M2332**), mp 112° (Offe).

Sodium salt, $C_8H_{14}NNaO_3$, *Plastenan.*

Zinc salt, $C_{16}H_{28}NO_3Zn$, *zinc acexamate, Copinal.*
THERAP CAT: Anti-inflammatory; zinc salt as antiulcerative.

39. Acetamidoeugenol. *N,N-Diethyl-2-[2-methoxy-4-(2-propenyl)phenoxy]acetamide; 2-(4-allyl-2-methoxyphenoxy)-N,N-diethylacetamide; N,N-diethyl-2-methoxy-4-allylphenoxyacetamide; 2-methoxy-4-allylphenoxyacetic acid N,N-diethylamide;* 2-M-4-A; G 29505; Detrovel; Estil. $C_{16}H_{23}NO_3$; mol wt 277.35. C 69.28%, H 8.36%, N 5.05%, O 17.31%. Prepn: **Brit.** pats. **792,490; 837,995** (1958, 1960 to Geigy).

$$OCH_2CON(C_2H_5)_2$$
$$OCH_3$$
$$CH_2CH=CH_2$$

Oil. $bp_{0.001}$ 143-146°; n_D^{25} 1.5300. Soluble in water, pH of 5% aq soln 7.3.
THERAP CAT: Anesthetic (intravenous).

40. Acetaminophen. *N-(4-Hydroxyphenyl)acetamide; 4'-hydroxyacetanilide;* p-hydroxyacetanilide; *p-acetamidophenol;* p-acetaminophenol; *p-acetylaminophenol; N-acetyl-p-aminophenol;* paracetamol; Abensanil; Acamol; Acetalgin; Alpiny; Amadil; Anaflon; Anhiba; Apamide; APAP; Ben-u-ron; Bickie-mol; Calpol; Captin; Cetadol; Dafalgan; Datril; Dial-a-gesic; Dirox; Disprol; Doliprane; Dolprone; Dymadon; Enelfa; Eneril; Eu-Med; Exdol; Febrilix; Finimal; Gelocatil; Hedex; Homoolan; Korum; Lyteca; Momentum; Naprinol; Nobedon; Ortensan; Pacemo; Paldesic; Panadol; Panaleve; Panasorb; Panets; Panex; Panofen; Parelan; Paraspen; Parmol; Pasolind; Pasolind N; Salzone; Tabalgin; Tapar; Temlo; Tempra; Tralgon; Tylenol; Valadol. $C_8H_9NO_2$; mol wt 151.16. C 63.56%, H 6.00%, N 9.27%, O 21.17%. Prepn from *p*-nitrophenol: Morse, *Ber.* **11,** 232 (1878); Tingle, Williams, *Am. Chem. J.* **37,** 63 (1907); from *p*-aminophenol: Lumière et al., *Bull. Soc. Chim. France* [3] **33,** 785 (1905); Fierz-David, Kuster, *Helv. Chim. Acta* **22,** 94 (1939); Wilbert, De Angelis, **U.S.** pat. **2,998,450** (1961 to Warner-Lambert); Bergmann, **Ger.** pat. **453,577;** *Chem. Zentr.* **1928, I,** 2663; *Frdl.* **16,** 238; from *p*-hydroxyacetophenone hydrazone: Pearson *et al., J. Am. Chem. Soc.* **75,** 5907 (1953). Toxicity data: G. A. Starmer *et al., Toxicol. Appl. Pharmacol.* **19,** 20 (1971); D. C. Dahlin, S. D. Nelson, *J. Med. Chem.* **25,** 885 (1982). Evaluation of renal effects: D. P. Sandler *et al., N. Engl. J. Med.* **320,** 1238 (1989). Comprehensive description: J. E. Fairbrother in *Analytical Profiles of Drug Substances* vol. **3,** K. Florey, Ed. (Academic Press, New York, 1974) pp 1-109. Review of pharmacology: B. Ameer, D. J. Greenblatt, *Ann. Int. Med.* **87,** 202-209 (1977). Review of acetaminophen-induced hepatotoxicity: J. A. Hinson, *Rev. Biochem. Toxicol.* **2,** 103-129 (1980); *idem, Life Sci.* **29,** 107-116 (1981). For proposed toxic metabolite *see* Acetimidoquinone.

$$CH_3CONH \quad OH$$

Large monoclinic prisms from water, mp 169-170.5°. d_4^{21} 1.293. uv max (ethanol): 250 nm (ϵ 13,800). Very slightly sol in cold water, considerably more sol in hot water. Sol in methanol, ethanol, dimethylformamide, ethylene dichloride, acetone, ethyl acetate. Slightly sol in ether. Practically insol in petr ether, pentane, benzene. LD_{50} in mice (mg/kg): 338 orally (Starmer), 500 i.p. (Dahlin, Nelson).
USE: Manuf azo dyes, photographic chemicals.
THERAP CAT: Analgesic. Antipyretic.

41. Acetaminosalol. *2-Hydroxybenzoic acid 4-(acetylamino)phenyl ester; p-acetamidophenyl salicylate; acetyl-aminophenyl salicylate; acetyl-p-aminosalol; p-acetylaminophenol salicylic acid ester; phenetsal;* Salophen; Phenosal. $C_{15}H_{13}NO_4$; mol wt 271.26. C 66.41%, H 4.83%, N 5.16%, O

23.59%. Prepn: Brewster, *J. Am. Chem. Soc.* **40,** 1136 (1918).

$$COO \quad NHCOCH_3$$
$$OH$$

Crystals from hot ethanol, mp 187°. Practically insol in petr ether, cold water, more sol in warm water. Sol in alcohol, ether, benzene. Incompatible with alkalies and alkaline solns which dissolve it with decompn. The alkaline soln gradually becomes blue when boiled, the blue color being discharged upon continued boiling and again produced upon cooling and exposure to air.
THERAP CAT: Analgesic; antipyretic; anti-inflammatory.
THERAP CAT (VET): Analgesic, antipyretic.

42. Acetanilide. *N-Phenylacetamide;* antifebrin; acetylaniline; acetylaminobenzene. C_8H_9NO; mol wt 135.16. C 71.09%, H 6.71%, N 10.36%, O 11.84%. Usually prepd from aniline and acetic acids: A. I. Vogel, *Practical Organic Chemistry* (London, 3rd ed., 1959) p 577. From aniline and acetyl chloride: Gattermann-Wieland, *Praxis des Organischen Chemikers* (Berlin, 40th ed., 1961) p 114. From aniline, acetone and ketene: Hurd, *Org. Syn.,* **coll. vol. I** (New York, 2nd ed., 1941) p 332. Monograph: Gross, *Acetanilid* (Hillhouse Press, New Haven, 1946).

$$NHCOCH_3$$

Orthorhombic plates, scales from water. mp 113-115°; bp 304-305°. Slightly burning taste. Appreciably volatile at 95°. d_4^{15} 1.219. Kb at 28°: 1×10^{-13}. One gram dissolves in 185 ml water, 20 ml boiling water, 3.4 ml alcohol, 3 ml methanol, 0.6 ml boiling alcohol, 3.7 ml chloroform, 4 ml acetone, 5 ml glycerol, 8 ml dioxane, 18 ml ether, 47 ml benzene. Very sparingly sol in petr ether. Chloral hydrate increases the soly of acetanilide in water. LD_{50} orally in rats: 800 mg/kg.
Pharmaceut. Incompat. Ethyl nitrate, amyl nitrite, acid solns of nitriles, alkalies (aniline liberated), alkali bromide and iodides in aq soln (insol compounds formed); chloral hydrate, phenol, resorcinol, thymol (these produce a liquid or a soft mass on trituration); ferric salts (red color or precipitate).
USE: Manufacture of other medicinals and of dyes; stabilizer for H_2O_2 soln; as addition to cellulose ester varnishes.
THERAP CAT: Antipyretic, analgesic.
THERAP CAT (VET): Antipyretic, analgesic.

43. p-Acetanisidine. *N-(4-Methoxyphenyl)acetamide; p-acetanisidide;* methacetin; *p-methoxyacetanilide.* $C_9H_{11}NO_2$; mol wt 165.19. C 65.44%, H 6.71%, N 8.48%, O 19.37%. $CH_3OC_6H_4NHCOCH_3$. Prepn by acetylation of *p*-anisidine: Reverdin, Bucky, *Ber.* **39,** 2569 (1906); from *p*-acetylanisole and hydroxylamine-*O*-sulfonic acid: Sanford *et al., J. Am. Chem. Soc.* **67,** 1941 (1945).
Cryst powder, feeble, bitter taste, mp 130-132°. Slightly sol in water; sol in alcohol, acetone, chloroform, dil acids, alkalies.

44. Acetarsone. *[3-(Acetylamino)-4-hydroxyphenyl]arsonic acid; N-acetyl-4-hydroxy-m-arsanilic acid;* 3-acetamido-4-hydroxyphenylarsonic acid; 3-acetamido-4-hydroxybenzenearsonic acid; acetarsol; acetphenarsine; Ehrlich 594; Fourneau 190; 190 F; F 190; Stovarsol; Amarsan; Arsaphen; Dynarsan; Goyl; Kharophen; Limarsol; Malagride; Gynoplix; Oralcid; Devegan; Orarsan; Osarsal; Osarsol; Osvarsan; Paroxyl; Sanogyl; Spirocid; S.V.C.; Monargan; Ginarsol; Stovarsolan. $C_8H_{10}AsNO_5$; mol wt 275.08. C 34.93%, H 3.66%, As 27.23%, N 5.09%, O 29.08%. The commercial product contains from 27.1 to 27.4% As. Various synthetic routes: Raiziss, Fisher, *J. Am. Chem. Soc.* **48,** 1323 (1926); Hewitt, King, *J. Chem. Soc.* **1926,** 823. Toxicity data: H. H. Anderson, C. D. Leake, *Proc. Soc. Exp. Biol. Med.* **27**, 267 (1930).

Stout prisms from water, dec 240-250°; slight acid taste. Slightly sol in water; freely sol in solns of alkalies and alkali carbonates. Stable at ordinary temps. MLD in rabbits, cats (mg/kg): 125-150, 150-175 orally (Anderson, Leake).

Calcium salt (with hexosediphosphoric acid), *Realphene.* Bismuth salt, *Bistovol.*

Diethylamine salt, $C_{12}H_{21}AsN_2O_5$, *Acetarsin, Acetilarsano, Acetylarsan, Arsaphenan, Golarsyl, Syntharsol.* Prepn: **Brit. pat. 224,764** (1954 to Rhone-Poulenc).

Arecoline salt, *see* Drocarbil.

THERAP CAT: Antiprotozoal (Trichomonas). Diethylamine salt formerly as antisyphilitic.

THERAP CAT (VET): Antiprotozoal (Histomonas) in turkeys. Antibacterial (spirocheticide) in poultry. Has been used as a tonic.

45. Acetazolamide. *N-[5-(Aminosulfonyl)-1,3,4-thiadiazol-2-yl]acetamide; 5-acetamido-1,3,4-thiadiazole-2-sulfonamide;* 2-acetylamino-1,3,4-thiadiazole-5-sulfonamide; acetazoleamide; #6063; carbonic anhydrase inhibitor no. 6063; Acetamox; Atenezol; Cidamex; Défiltran; Diacarb; Diamox; Didoc; Diluran; Diureticum-Holzinger; Diuriwas; Diutazol; Donmox; Edemox; Fonurit; Glaupax; Glupax; Natrionex; Nephramid. $C_4H_6N_4O_3S_2$; mol wt 222.25. C 21.61%, H 2.72%, N 25.21%, O 21.60%, S 28.86%. Prepn: Roblin, Clapp, *J. Am. Chem. Soc.* **72**, 4890 (1950); Clapp, Roblin, **U.S. pat. 2,554,816** (1951 to Am. Cyanamid); **U.S. pat. 2,980,679** (1961 to Omikron-Gagliardi Soc. Fatto).

Crystals from water, mp 258-259° (effervescence). Weak acid. pKa 7.2. Sparingly sol in cold water.

Sodium salt, *Vetamox.*

THERAP CAT: Carbonic anhydrase inhibitor, diuretic; treatment of glaucoma.

THERAP CAT (VET): Diuretic, carbonic anhydrase inhibitor.

46. Acetiamine. *Ethanethioic acid S-[1-[2-(acetyloxy)ethyl]-2-[[(4-amino-2-methyl-5-pyrimidinyl)methyl]formylamino]-1-propenyl] ester; thioacetic acid S-ester with N-[(4-amino-2-methyl-5-pyrimidinyl)methyl]-N-(4-hydroxy-2-mercapto-1-methyl-1-butenyl)formamide O,S-diacetate;* N-[(4-amino-2-methyl-5-pyrimidinyl)methyl]-*N*-(4-hydroxy-2-mercapto-1-methyl-1-butenyl)formamide *O,S*-diacetate; 3-acetylthio-4-[(4-amino-2-methyl-5-pyrimidinyl)methyl-*N*-formylamino]-3-pentenyl acetate; 5-acetoxy-3-acetyl-thio-2-[(4-amino-2-methyl-5-pyrimidinyl)methyl-*N*-formylamino]-2-pentene; diacethiamine; *O,S*-diacetylthiamine; vitamin B₁ *O,S*-diacetate; D.A.T.; Thianeuron. $C_{16}H_{22}N_4O_4S$; mol wt 366.45. C 52.44%, H 6.05%, N 15.29%, O 17.46%, S 8.75%. A fat-soluble deriv. of vitamin B₁. Prepn: Matsukawa, Kawasaki, *J. Pharm. Soc. Japan* **23**, 705, 709 (1953), *C.A.* **48**, 7017e (1954); Takamizawa *et al.*, *Bull. Chem. Soc. Japan* **36**, 1214 (1963). Synthesis and pharmacology: Gauthier *et al.*, *Ann. Pharm. Franc.* **21**, 655 (1963). Clinical studies: Wagner, Wagner, *Arzneimittel-Forsch.* **16**, 1643 (1966); Blum, Thomas, *Pharmacol. Clin.* **2**, 177 (1970).

Colorless prisms from benzene-petr ether, mp 122-123° (dec); from water, mp 123-124°. Soluble in water, methanol, ethanol.

Hydrochloride, $C_{16}H_{23}ClN_4O_4S$, *Nevriton Comprimés.*

THERAP CAT: Vitamin (enzyme cofactor).

47. Acetic Acid Glacial. Aci-Jel. $C_2H_4O_2$; mol wt 60.05. C 40.00%, H 6.71%, O 53.29%. CH_3COOH. Obtained in the destructive distillation of wood; from acetylene and water, via acetaldehyde by oxidation with air. Manuf processes: Bhattacharyya, Sourirajan, *J. Appl. Chem. (London)* **6**, 442 (1956); *eidem, ibid.* **9**, 126 (1959); Elce *et al.*, **U.S. pat. 2,800,504** (1957 to Distillers Co.); Wirth, **U.S. pat. 2,818,428** (1957 to British Petroleum); McKusick and Hoover, **U.S. pats. 2,940,913** and **2,940,914** (both 1960 to Du Pont); Faith, Keyes & Clark's *Industrial Chemicals,* F. A. Lowenheim, M. K. Moran, Eds. (Wiley-Interscience, New York, 4th ed., 1975) pp 8-15. Toxicity data: H. F. Smyth *et al.*, *Arch. Ind. Hyg. Occup. Med.* **4**, 119 (1951). *Review:* F. S. Wagner in Kirk-Othmer *Encyclopedia of Chemical Technology* vol. **1** (Wiley-Interscience, New York, 3rd ed., 1978) pp 124-147.

Liquid; pungent odor. *Produces burns of the skin!* $d^{16.67}$ (liq) 1.053; $d^{16.60}$ (solid) 1.266; d_{25}^{25} 1.049. bp 118°. mp 16.7°. n_D^{20} 1.3718. Flash pt, closed cup: 103°F (39°C). *Caution: Flammable!* Contracts slightly on freezing. It is an excellent solvent for many organic compounds; also dissolves phosphorus, sulfur and halogen acids. Miscible with water, alcohol, glycerol, ether, carbon tetrachloride. Practically insol in carbon disulfide. Weakly ionized in aq solns: pKa 4.74. pH of aq solns $1.0M = 2.4$; $0.1M = 2.9$; $0.01M = 3.4$. LD_{50} in rats (g/kg): 3.53 orally (Smyth).

Incompat. Carbonates, hydroxides, many oxides, and phosphates, etc.

Caution: Ingestion may cause severe corrosion of mouth and G.I. tract, with vomiting, hematemesis, diarrhea, circulatory collapse, uremia, death. Chronic exposure may cause erosion of dental enamel, bronchitis, eye irritation, *cf.* Patty's *Industrial Hygiene and Toxicology* vol. **2C**, G. D. Clayton, F. E. Clayton, Eds. (Wiley-Interscience, New York, 3rd ed., 1982) p 4909-4911.

USE: Manuf various acetates, acetyl compounds, cellulose acetate, acetate rayon, plastics and rubber in tanning; as laundry sour; printing calico and dyeing silk; as acidulant and preservative in foods; solvent for gums, resins, volatile oils and many other substances. Widely used in commercial organic syntheses. Pharmaceutic aid (acidifier).

THERAP CAT (VET): Vesicant, caustic, destruction of warts.

48. Acetic Anhydride. Acetic oxide; acetyl oxide. $C_4H_6O_3$; mol wt 102.09. C 47.06%, H 5.92%, O 47.02%. $(CH_3CO)_2O$. Equivalent to 117.64% acetic acid. Made formerly from sod. acetate and acetyl or sulfuryl chloride; now usually obtained from acetaldehyde or acetic acid: Faith, Keyes & Clark's *Industrial Chemicals,* F. A. Lowenheim, M. K. Moran, Eds. (Wiley-Interscience, New York, 4th ed., 1975) pp 16-20. Of industrial importance is also the ketene process, starting with the thermal decomposition of acetone: Schmidlin, Bergmann, *Ber.* **43**, 2821 (1910).

Very refractive liquid; strong acetic odor. *Readily combustible. Fire hazard.* Flash pt 130°F. d_4^{15} 1.080. mp $-73°$. bp 139°. n_D^{20} 1.3904. Slowly soluble in water, forming acetic acid; with alcohol forms ethyl acetate; sol in chloroform, ether. LD_{50} orally in rats 1.78 g/kg, H. F. Smyth *et al.*, *Arch. Ind. Hyg. Occup. Med.* **4**, 119 (1951).

USE: Manuf acetyl compounds, cellulose acetates. As acetulizer and solvent in examining wool fat, glycerol, fatty and volatile oils, resins; detection of rosin. Widely used in organic syntheses, e.g., as dehydrating agent in nitrations, sulfonations and other reactions where removal of water is necessary. *Caution:* Produces irritation and necrosis of tissues in liquid or in vapor state. Avoid contact with skin, eyes.

49. Acetimidoquinone. *N-(4-oxo-2,5-cyclohexadien-1-ylidene)acetamide;* N-acetyl-*p*-benzoquinonimine; N-acetylimidoquinone; NAPQI. $C_8H_7NO_2$; mol wt 149.14. C 64.42%, H 4.73%, N 9.39%, O 21.46%. Proposed "ultimate" toxic metabolite of acetaminophen, *q.v.* Initially prepared by oxidation of acetaminophen with lead tetraacetate and characterized as a Diels-Alder adduct: I. C. Calder *et al.*, *J. Med. Chem.* **16**, 499 (1973). Electrochemical generation in buffer: D. J. Miner, P. T. Kissinger, *Biochem. Pharmacol.* **28**, 3285 (1979). Synthesis in stable benzene soln and reactivity: I. A. Blair *et al., Tetrahedron Letters* **21**, 4947 (1980). Prepn in cryst form, decomposition kinetics, preliminary toxicological studies: D. C. Dahlin, S. D. Nelson, *J. Med. Chem.* **25**, 885 (1982). Microsomal reactivity study: G. B. Corcoran, *Adv. Exp. Med. Biol.* **136B**, 1085 (1982). Review of acetaminophen-induced hepatotoxicity: J. A. Hinson *et al., Life Sci.* **29**, 107-116 (1981).

Yellow cubic cryst, mp 74-75°. Sublimes at 45-50° (0.07 mm Hg). uv max (*n*-hexane): 263, 376 nm (ϵ 3.3 × 10^4, 1.6 × 10^2 M^{-1} cm^{-1}). LD$_{50}$ in male BALB/c mice: 20 mg/kg i.p., D. C. Dahlin, S. D. Nelson, *loc. cit.*

50. Acetoacetanilide. *3-Oxo-N-phenylbutanamide;* α-acetylacetanilide; acetoacetic anilide; β-ketobutyranilide. C$_{10}$H$_{11}$NO$_2$; mol wt 177.20. C 67.78%, H 6.26%, N 7.91%, O 18.06%. C$_6$H$_5$NHCOCH$_2$COCH$_3$. Prepd by the reaction of ketene dimer with aniline: Chick, Wilsmore, *J. Chem. Soc.* **1908**, 946; Boese, *Ind. Eng. Chem.* **32**, 16 (1940); Williams, Krynitsky, *Org. Syn.* **coll. vol. III**, 10 (1955). By the reaction of aniline with ethyl acetoacetate: Knorr, *Ann.* **26**, 69 (1886); Roos, *Ber.* **21**, 624 (1883); Knorr, Reuter, *Ber.* **27**, 1169 (1894). Enol form (unstable) prepd by pouring an alkaline soln of the keto form in cold, dilute H$_2$SO$_4$.
Keto form: Leaflets from dilute alcohol, mp 85°. Slightly soluble in water; sol in alcohol, chloroform, ether, hot benzene, hot petr ether, acids, or alkali hydroxide solns. Gives a violet color with ferric chloride.
USE: Manuf of yellow dyes, such as Hansa and benzidine yellows. In rubber compounding. In organic syntheses.

51. Acetoacetic Acid. *3-Oxobutanoic acid;* diacetic acid; acetylacetic acid; acetonecarboxylic acid; 3-ketobutyric acid; 2-ketobutyric acid. C$_4$H$_6$O$_3$; mol wt 102.09. C 47.06%, H 5.92%, O 47.02%. CH$_3$COCH$_2$COOH. Prepd by hydrolysis of ethyl acetoacetate: Krueger, *J. Am. Chem. Soc.* **74**, 5536 (1952).
Crystals from ether, mp 36-37°. A strong, but unstable acid. At 100° violently decomposes into acetone and CO$_2$. Miscible with water, alc.
USE: In organic syntheses. *Caution:* Severely irritating to skin, mucous membranes.

52. Acetobromglucose. α-D-*Glucopyranosyl bromide 2,3,4,6-tetraacetate;* α-acetobromoglucose; 2,3,4,6-tetraacetyl-α-D-glucopyranosyl bromide. C$_{14}$H$_{19}$BrO$_9$; mol wt 411.21. C 40.89%, H 4.66%, Br 19.43%, O 35.02%. Prepd by the action of hydrogen bromide upon anhydrous glucose in acetic anhydride: Redemann, Niemann, *Org. Syn.* **coll. vol. III**, 11 (1955). *See also* Fischer, *Ber.* **49**, 584 (1916); Freudenberg, *Ber.* **60**, 241 (1927); Gattermann-Wieland, *Praxis des organischen Chemikers* (de Gruyter, Berlin, 40th ed., 1961) p 340.

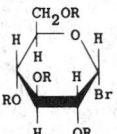

R = COCH$_3$

Crystals from isopropyl ether, mp 88-89°. $[\alpha]_D^{19}$ +199.3° (c = 3 in chloroform); $[\alpha]_D^{15}$ +230.3° (c = 9 in benzene). Best stored in a vacuum desiccator. Dec on contact with water. One gram dissolves in 20 ml absolute ethanol, more soluble in methanol. Freely sol in ether, chloroform, acetone, ethyl acetate, benzene. Slightly sol in petr ether.

53. Acetohexamide. *4-Acetyl-N-[(cyclohexylamino)-carbonyl]benzenesulfonamide; 1-[(p-acetylphenyl)sulfonyl]-3-cyclohexylurea;* 3-cyclohexyl-1-(p-acetylphenylsulfonyl)-

urea; *N-(p-acetylbenzenesulfonyl)-N'-cyclohexylurea;* cyclamide; tsiklamid; Dimelor; Dymelor; Dimelin; Ordimel. C$_{15}$H$_{20}$N$_2$O$_4$S; mol wt 324.42. C 55.54%, H 6.21%, N 8.64%, O 19.73%, S 9.89%. Prepn: **Brit.** pat. **912,789** (1962 to Lilly); Marshall *et al., J. Med. Chem.* **6**, 60 (1963). Comprehensive description: C. E. Shafer in *Analytical Profiles of Drug Substances* vol. 1, K. Florey, Ed. (Academic Press, New York, 1972) pp 1-14.

CH$_3$CO—⟨ ⟩—SO$_2$NHCONH—⟨ ⟩

Crystals from 90% aq ethanol, mp 188-190° (**Brit.** pat. **912,789**); from dil ethanol, mp 175-177° (Marshall *et al.*). Sol in pyridine; slightly sol in alcohol, chloroform. Insol in water, ether.
THERAP CAT: Antidiabetic.

54. Acetohydroxamic Acid. *N-Hydroxyacetamide; N-acetylhydroxylamine; acetic acid oxime; AHA; Lithostat.* C$_2$H$_5$NO$_2$; mol wt 75.07. C 32.00%, H 6.71%, N 18.66%, O 42.63%. CH$_3$CONHOH. Urease inhibitor. Prepn: A. Miolati, *Ber.* **25**, 699 (1892); W. M. Wise, W. W. Brandt, *J. Am. Chem. Soc.* **77**, 1058 (1955); H. A. Staab *et al., Ber.* **95**, 1275 (1962); G. Sosnovsky, J. A. Krogh, *Synthesis* **1980**, 654. Inhibition of urease activity: K. Kobashi *et al., Biochim. Biophys. Acta* **65**, 380 (1962); W. N. Fishbein *et al., Nature* **208**, 46 (1965); W. N. Fishbein, P. P. Carbone, *J. Biol. Chem.* **240**, 2407 (1965); D. P. Griffith *et al., Invest. Urol.* **11**, 234 (1973). Metabolism: E. Wolpert *et al., Proc. Soc. Exp. Biol. Med.* **136**, 592 (1971); W. N. Fishbein *et al., J. Pharmacol. Exp. Ther.* **186**, 173 (1973). Pharmacokinetics: S. Feldman *et al., Invest. Urol.* **15**, 498 (1978). Clinical studies in treatment of kidney stones: D. B. Griffith *et al., J. Urol.* **119**, 9 (1978); A. Martelli *et al., Urology* **17**, 320 (1981).
mp 89-92°. pKa 8.70. pH (aq soln): 9.39.
THERAP CAT: Antiurolithic. Antibacterial adjunct (urinary tract infection).

55. Acetoin. *3-Hydroxy-2-butanone;* 2,3-butanolone; acetyl methyl carbinol; dimethylketol; γ-hydroxy-β-oxobutane. C$_4$H$_8$O$_2$; mol wt 88.10. C 54.53%, H 9.15%, O 36.32%. CH$_3$CH(OH)COCH$_3$. A product of fermentation, also in cream ripened for churning. Obtained by the action of sorbose bacterium or *Mycoderma aceti* on 2,3-butanediol or by the action of fungi, such as *Aspergillus, Penicillium, Mycoderma* on sugar cane juice: Browne, *J. Am. Chem. Soc.* **28**, 467 (1906). By action of yeast on diacetyl: Nagelschmidt, *Biochem. Z.* **186**, 317 (1927). From diacetyl by partial reduction with Zn and acid: Diels, Stephan, *Ber.* **40**, 4338 (1907). Fermentation process: Vergnaud, U.S. pat. **2,529,-061** (1950 to Usines de Melle).
Liquid. Pleasant odor. mp 15°. d$_4^7$ 0.9972. bp$_{760}$ 148°. $n_D^{17.3}$ 1.4190. Miscible with water, alcohol. Sparingly sol in ether, petr ether. Reduces Fehling's soln forming acetic acid. Forms a solid dimer C$_8$H$_{16}$O$_4$ on standing or on treatment with granulated zinc. The dimer is easily converted back to the monomer by melting, distilling or dissolving.

56. Acetol®. The name ACETOL appears in the literature covering a variety of substances.
According to earlier indices to *Chemical Abstracts* it is synonymous with 1-hydroxy-2-propanone or hydroxyacetone.
According to *Gehes Codex* (5th ed, 1929) ACETOL is registered in Canada as a name for acetylsalicylic acid by the Ontario College of Pharmacy upon stipulation by the Canadian Pharmaceutical Association.
In W. Gardner's *Chemical Synonyms and Trade Names* (5th ed 1948) ACETOL is also found as a proprietary name to designate cellulose acetate material in the form of flake, and as a synonym for acetal or diethyl aldehyde used as a hypnotic, and last as a trademark for a liquid fuel for internal combustion engines containing 80% alcohol and 20% other materials including acetylene.

57. Acetomeroctol. *(Acetato-O)[2-hydroxy-5-(1,1,3,3-tetramethylbutyl)phenyl]mercury; 2-(acetoxymercuri)-4-(1,1,-3,3-tetramethylbutyl)phenol;* Merbak. C$_{16}$H$_{24}$HgO$_3$; mol wt

464.96. C 41.33%, H 5.20%, Hg 43.15%, O 10.32%. Prepn: Niederl, Shukis, *J. Am. Chem. Soc.* **66**, 844 (1944).

Crystals, mp 158°. Practically insol in water. Sol in alc, ether, chloroform. Sparingly sol in benzene. Marketed as a 1:1000 soln contg 50% (v/v) alc and 10% (v/v) acetone.
THERAP CAT: Topical anti-infective.

58. Acetone. *2-Propanone;* dimethyl ketone; β-keto-propane; pyroacetic ether. C_3H_6O; mol wt 58.08. C 62.04%, H 10.41%, O 27.55%. CH_3COCH_3. Obtained by fermentation (by-product of butyl alcohol manufacture) or by chemical synthesis from isopropanol (as chief product); from cumene (by-product in phenol manufacture); from propane (by-product of oxidation-cracking); Faith, Keyes & Clark's *Industrial Chemicals*, F. A. Lowenheim, M. K. Moran, Eds. (Wiley-Interscience, New York, 4th ed., 1975) pp 21-25 . Toxicity: Smyth *et al.*, *Ind. Hyg. J.* **23**, 95 (1962). *Review:* Weiss, *Chem. Eng. News* **36**, 79 (1958); D. L. Nelson, B. P. Webb in Kirk-Othmer *Encyclopedia of Chemical Technology* vol. 1 (Wiley-Interscience, New York, 3rd ed., 1978) pp 179-191.
Volatile, *highly flammable* liquid; characteristic odor; pungent, sweetish taste. d_{25}^{25} 0.788. bp 56.5°. mp $-94°$. n_D^{20} 1.3591. Flash pt, closed cup: 0°F ($-18°C$). Miscible with water, alcohol, dimethylformamide, chloroform, ether, most oils. *Keep away from fire!* Keep away from plastic eyeglass frames, jewelry, pens and pencils, rayon stockings and other rayon garments. LD_{50} in rats: 10.7 ml/kg orally (Smyth).
Caution: Prolonged or repeated topical use may cause erythema, dryness. Inhalation may produce headache, fatigue, excitement, bronchial irritation, and, in large amounts, narcosis. Serious poisoning rare.
USE: Solvent for fats, oils, waxes, resins, rubber, plastics, lacquers, varnishes, rubber cements. Manuf methyl isobutyl ketone, mesityl oxide, acetic acid (ketene process), diacetone alcohol, chloroform, iodoform, bromoform, explosives, aeroplane dopes, rayon, photographic films, isoprene; storing acetylene gas (takes up about 24 times its vol of the gas); extraction of various principles from animal and plant substances; in paint and varnish removers; purifying paraffin; hardening and dehydrating tissues. Pharmaceutic aid (solvent).

59. Acetone Cyanohydrin. *2-Hydroxy-2-methylpropanenitrile;* *2-methyllactonitrile;* α-hydroxyisobutyronitrile. C_4H_7NO; mol wt 85.10. C 56.45%, H 8.28%, N 16.46%, O 18.80%. $(CH_3)_2C(OH)CN$. Prepd by adding acetone to sodium or potassium cyanide in water and treating with H_2SO_4 at below 20°: Welch, Clemo, *J. Chem. Soc.* **1928**, 2629; Cox, Stormont, *Org. Syn.* **coll. vol. II**, 7 (1943). Continuous production from HCN and aq acetone: G. Barsky, U.S. pat. 2,731,490 (1956).
Liquid; d_{25}^{25} 0.9267; d_4^{19} 0.932. mp $-19°$. bp_{15} 81°; bp_{20} 88-90°; bp_{23} 82°; bp_{760} 95°. n_d^{19} 1.40002; n_D^{15} 1.3980. Freely sol in water and in the usual organic solvents but not in petr ether and carbon disulfide. LD_{50} orally in rats: 0.17 g/kg, H. F. Smyth *et al.*, *Am. Ind. Hyg. Assoc. J.* **23**, 95 (1962).
USE: In preparative organic chemistry for transcyanohydrination, such as preparing the 17-monocyanohydrin of a 3,17-diketo steroid by hydrogen cyanide exchange with the reagent, *e.g.* Ercoli, de Ruggieri, *J. Am. Chem. Soc.* **75**, 650 (1953). *Caution:* May be slightly irritating to skin, mucous membranes. Decomposes readily to form hydrogen cyanide, *q.v.*, which is highly toxic.

60. Acetonedicarboxylic Acid. *3-Oxopentanedioic acid;* *3-oxoglutaric acid;* β-ketoglutaric acid. $C_5H_6O_5$; mol wt 146.10. C 41.10%, H 4.14%, O 54.76%. CO_2HCH_2-$COCH_2CO_2H$. Prepd by the action of fuming sulfuric acid on citric acid: Willstätter, Pfannenstiel, *Ann.* **422**, 5 (1921);

Ingold, Nickolls, *J. Chem. Soc.* **121**, 1642 (1922); R. Adams *et al.*, *Org. Syn.* **coll. vol. I**, 10 (2nd ed., 1941). By the action of *Aspergillus niger* on ammonium citrate mixed with some citric acid: Walker *et al.*, *J. Chem. Soc.* **1927**, 3050; also by *B. pyocyaneus:* Butterworth, Walker, *Biochem. J.* **23**, 931 (1929). Decomposition studies: Wiig, *J. Phys. Chem.* **32**, 961 (1928). Intermediate in synthesis of tropinone: R. Robinson, *J. Chem. Soc.* **111**, 762 (1917); C. Schöpf, G. Lehmann, *Ann.* **518**, 1 (1921).
Needles from ethyl acetate, mp 138° (dec). The pure compound may be stored over P_2O_5 for several months. Crude material dec after a few hours. Decomposed by hot water, acids or alkalies to CO_2 and acetone. K (25°): 7.9 × 10^{-4}. Very sol in water and alcohol; slightly sol in ethyl acetate, ether. Insol in chloroform, ligroin, benzene.
USE: In organic synthesis.

61. Acetone Sodium Bisulfite. *2-Hydroxy-2-propane-sulfonic acid sodium salt;* acetone sulfite. $C_3H_7NaO_4S$; mol wt 162.15. C 22.22%, H 4.35%, Na 14.18%, O 39.47%, S 19.78%. $(CH_3)_2C(OH)SO_3Na$.
Crystals, fatty feel, slight SO_2 odor. Freely sol in water, sparingly in alcohol. Decomposed by acids.
USE: In photography; manuf of pure acetone; dyeing and printing texiles; in organic syntheses.

62. Acetonitrile. Methyl cyanide; cyanomethane; ethanenitrile. C_2H_3N; mol wt 41.05. C 58.51%, H 7.37%, N 34.12%. CH_3CN. Small amounts occur in coal tar. Obtained commercially as a byproduct in manuf of acrylonitrile, *q.v.*: R. A. Smiley in Kirk-Othmer *Encyclopedia of Chemical Technology*, **vol. 15** (Wiley-Interscience, New York, 3rd ed., 1978) p 896. Prepn by dehydration of acetamide: Adkins, Nissen, *J. Am. Chem. Soc.* **46**, 143 (1924); A. I. Vogel, *Practical Organic Chemistry* (Longmans, London, 3rd ed., 1959) p 407; Gattermann-Wieland, *Praxis des organischen Chemikers* (de Gruyter, Berlin, 40th ed., 1961) p 125; or from acetylene and ammonia: **Ger.** pat. 365,432; *Chem. Zentr.* **1924 I**, 2398; **1925 II**, 1563. Review of purification methods: *Techniques of Chemistry, vol. II, Organic Solvents*, J. A. Riddick, W. B. Bunger, Eds. (Wiley, 3rd ed., 1970) pp 798-805. Toxicity: H. F. Smyth, C. P. Carpenter, *J. Ind. Hyg. Toxicol.* **30**, 63 (1968).
Liquid. Ether-like odor. *Poisonous!* Burns with a luminous flame. Flash pt 12.8°C (55°F). d_D^{15} 0.78745; d_4^{30} 0.7138. mp $-45°$. bp_{760} 81.6°. n_D^{15} 1.34604, n_D^{30} 1.33934. Dielectric constant at 20° = 38.8. Surface tension at 20° = 29.04 dynes/cm. Misc with water, methanol, methyl acetate, ethyl acetate, acetone, ether, acetamide solutions, chloroform, carbon tetrachloride, ethylene chloride and many unsaturated hydrocarbons. Immiscible with many saturated hydrocarbons (petroleum fractions). Dissolves some in organic salts, *e.g.*, silver nitrate, lithium nitrate, magnesium bromide. Constant boiling mixture with water contains 16% H_2O and bp 76°. LD_{50} orally in rats: 3800 mg/kg (Smyth).
USE: In organic synthesis as starting material for acetophenone, α-naphthaleneacetic acid, thiamine, acetamidine. To remove tars, phenols, and coloring matter from petroleum hydrocarbons which are not soluble in acetonitrile. To extract fatty acids from fish liver oils and other animal and vegetable oils. Can be used to recrystallize steroids. As an indifferent medium in physicochemical investigations. Wherever a polar solvent having a rather high dielectric constant is required. As medium for promoting reactions involving ionization. As a solvent in non-aqueous titrations. As a non-aqueous solvent for inorganic salts. *Caution:* Avoid breathing vapors. May cause skin irritation.

63. Acetonylacetone. *2,5-Hexanedione;* α,β-diacetyl-ethane. $C_6H_{10}O_2$; mol wt 114.14. C 63.13%, H 8.83%, O 28.03%. $CH_3COCH_2CH_2COCH_3$. Prepn by decarboxylation of diacetosuccinic ester (from sodium acetoacetic ester and iodine): Paal, *Ber.* **18**, 58 (1885); Knorr, *Ber.* **22**, 2100 (1889); by hydrolysis of 2,5-dimethylfuran: Perkins, Toussaint, U.S. pat. 2,052,652 (1936 to Union Carbide). Toxicity data: H. F. Smyth, C. P. Carpenter, *J. Ind. Hyg. Toxicol.* **26**, 269 (1944).
Liquid, gradually turns yellow. d_4^{20} 0.970. mp $-9°$. bp 188°, n_D^{20} 1.449. Misc with water, alcohol, ether. LD_{50} orally in rats: 2.7 g/kg (Smyth, Carpenter).

Caution: A mild local irritant. High concns cause narcosis.

64. Acetophenazine. *1-[10-[3-[4-(2-Hydroxyethyl)-1-piperazinyl]propyl]-10H-phenothiazin-2-yl]ethanone; 10-[3-[4-(2-hydroxyethyl)-1-piperazinyl]propyl]phenothiazin-2-yl methyl ketone;* 2-acetyl-10-[3-[4-(β-hydroxyethyl)piperazinyl]propyl]phenothiazine; 1-(2-hydroxyethyl)-4-[3-(2-acetyl-10-phenothiazyl)propyl]piperazine. $C_{23}H_{29}N_3O_2S$; mol wt 411.57. C 67.12%, H 7.10%, N 10.21%, O 7.78%, S 7.79%. Prepd from 1-(2-hydroxyethyl)piperazine and 10-(3-chloropropyl)-2-acetylphenothiazine: Sherlock, Sperber, **U.S. pat. 2,985,654** (1961 to Schering).

Dimaleate, $C_{31}H_{37}N_3O_{10}S$, *Tindal.* Crystals from ethanol, mp 167-168.5°. LD_{50} orally in rats: 433 mg/kg, E. W. Schafer, *Toxicol. Appl. Pharmacol.* **21,** 315 (1972).

THERAP CAT: Antipsychotic.

65. Acetophenone. *1-Phenylethanone;* phenyl methyl ketone; acetylbenzene; hypnone. C_8H_8O; mol wt 120.15. C 79.97%, H 6.71%, O 13.32%. $C_6H_5COCH_3$. Made from benzene and acetylchloride in presence of aluminum chloride; catalytically from acetic and benzoic acids. Prepn from benzene and acetic anhydride: Adams, *J. Am. Chem. Soc.* **46,** 1889 (1924); A. I. Vogel, *Practical Organic Chemistry* (Longmans, London, 3rd ed., 1959) p 730; Gattermann-Wieland, *Die Praxis des Organischen Chemikers* (de Gruyter, Berlin, 40th ed., 1961) p 297.

Liquid. Forms laminar crystals at low temp. mp 20.5°. d_{15}^{15} 1.033. bp 202°. n_D^{20} 1.5339. Flash pt, closed cup: 221°F (105°C). Slightly sol in water; freely in alcohol, chloroform, ether, fatty oils, glycerol. Sol in concd H_2SO_4 with orange color. LD_{50} orally in rats: 0.90 g/kg, H. F. Smyth, C. P. Carpenter, *J. Ind. Hyg. Toxicol.* **30,** 63 (1948).

USE: In perfumery to impart an orange-blossom-like odor; catalyst for the polymerization of olefins in organic syntheses, esp. as photosensitizer.

THERAP CAT: Hypnotic.

66. Acetosulfone Sodium. *N-[[5-Amino-2-[(4-aminophenyl)sulfonyl]phenyl]sulfonyl]acetamide monosodium salt; N-6-sulfanilylmetanilylacetamide N-sodium salt; N¹-acetyl-6-(sulfanilylmetanilamido)sodium;* sodium 2-N-acetylsulfamyl-4,4'-diaminodiphenylsulfone; sodium 4,4'-diaminodiphenylsulfone-2-N-acetylsulfonamide; sulfadiasulfone sodium; acetosulphone; Internal Antiseptic no. 307; I.A. 307; Promacetin. $C_{14}H_{14}N_3NaO_5S_2$; mol wt 391.41. C 42.96%, H 3.61%, N 10.74%, Na 5.88%, O 20.44%, S 16.39%. Prepn: Tullar, **U.S. pat. 2,358,365** (1944 to Parke, Davis).

Crystals. Soly in water about 3%. mp ~285° (free sulfonamide).

THERAP CAT: Antibacterial (leprostatic).

67. Acetotoluide. $C_9H_{11}NO$; mol wt 149.20. C 72.45%, H 7.43%, N 9.39%, O 10.72%. Prepn of *m-, o-,* and *p*-acetotoluidides, from *m-, o-,* or *p*-toluidine and acetic acid: Gasopoulos, *Ber.* **59B,** 2187 (1926).

m-Acetotoluide, N-(3-methylphenyl)acetamide, m-acetotoluidide, m-tolylacetamide, aceto-m-aminotoluene, m-methylacetanilide, N-acetyl-m-toluidine. Needles, mp 65.5°, bp 303°. Slightly sol in water; freely sol in alc, ether.

o-Acetotoluide, crystals, mp 110°, bp 296°. Slightly sol in water; sol in alc, chloroform, ether. Determn of soly at various temps: Hall *et al., J. Phys. Chem.* **37,** 1087 (1933).

p-Acetotoluide, crystals, mp 153°, bp 307°. Very slightly sol in water; sol in alc, ether, ethyl acetate, glacial acetic acid.

68. Acetoxime. *2-Propanone oxime;* acetone oxime; β-isonitrosopropane. C_3H_7NO; mol wt 73.09. C 49.30%, H 9.65%, N 19.17%, O 21.89%. $(CH_3)_2C{=}NOH$. Prepd by shaking an aqueous soln of hydroxylamine with acetone and extracting with ether: W. L. Semon, *Org. Syn.* **coll. vol. I,** 318 (2nd ed., 1941).

Columnar prisms. Volatilizes quickly in air. Odor resembling that of chloral hydrate. d_4^{62} 0.9113. mp 60°. bp_{728} 134.8°. Neutral reaction. Ka at 25° = 0.596 × 10⁻¹²; Kb at 25° = 0.648 × 10⁻¹². uv spectrum: Hartley, Dobbie, *J. Chem. Soc.* **77,** 326 (1900). Freely soluble in water, alcohol, ether, petr ether. Can be extracted with ether from neutral water solns, but not from alkaline or acidic solns.

69. Acetoxolone. *3-(Acetyloxy)-11-oxoolean-12-en-29-oic acid; 3β-hydroxy-11-oxoolean-12-en-30-oic acid acetate;* 3-O-acetyl-18β-glycyrrhetic acid; acetylglycyrrhetinic acid; glycyrrhetic acid acetate; $C_{32}H_{48}O_5$; mol wt 512.73. C 74.96%, H 9.44%, O 15.60%. Deriv of enoxolone, *q.v.;* homolog of carbenoxolone, *q.v.* Prepn: J. M. Beaton, F. S. Spring, *J. Chem. Soc.* **1955,** 3126. Prepn of the aluminum salt: **S. Afr. pat. 70 06901** corresp to A. Bonati, **U.S. pat. 3,764,618** (1971, 1973 both to Dott. Inverni & Della Beffa). NMR study of the acetate: M. Mousseron-Canet *et al., Bull. Soc. Chim. France* **12,** 4668 (1967). Pharmacology of the aluminum salt: C. Capra, *Filoterapia* **40,** 8 (1969), *C.A.* **73,** 64754 (1970). Absorption in rats: B. D. Cameron *et al., Arzneimittel-Forsch.* **26,** 1680 (1976). Analysis of gastric protective effect: P. Altmayer *et al., ibid.* **31,** 853 (1981).

Cryst, mp 322-325°. $[\alpha]_D^{20}$ +141°.

Aluminium salt, $C_{96}H_{141}AlO_{15}$, *Oriens.* White powder, mp 286-290°. $[\alpha]_D^{20}$ +126°±2° (c = 1 in chloroform). Insol in water, sol in most organic solvents. LD_{50} in male rats: >3300 mg/kg orally and i.p., A. Bonati, *loc. cit.*

THERAP CAT: Anti-ulcerative.

70. 21-Acetoxypregnenolone. *21-(Acetyloxy)-3β-hydroxypregn-5-en-20-one; 3β,21-dihydroxypregn-5-en-20-one 21-acetate;* 5-pregnene-3β,21-diol-20-one 21-acetate; 21-acetoxy-5-pregnen-3-ol-20-one; 3-hydroxy-21-acetoxy-5-pregnen-20-one; prebediolone acetate; A.O.P.; Acetoxanon; Artisone acetate. $C_{23}H_{34}O_4$; mol wt 374.50. C 73.76%, H 9.15%, O 17.09%. Intermediate product in the synthesis of desoxycorticosterone acetate: Steiger, Reichstein, *Helv. Chim. Acta* **20,** 1164 (1937). Prepd by reacting 5,6-pregnen-3-ol-20-one with bromine in chloroform and treating the obtained oily bromide with potassium acetate in alcohol: Herloff, Inhoffen, **U.S. pat. 2,409,043** (1946 to Schering).

Needles from acetone. Becomes opaque at about 80° or on prolonged standing. mp 184-185°. Very slightly sol in ether, in pentane; sol in chloroform, toluene. The esterified keto group is relatively stable to the action of chromic acid in cold and to such reducing agents as potassium iodide or zinc. On heating with aluminum isopropoxide in isopropyl alcohol it is reduced to 5-pregnene-3,20,21-triol.

THERAP CAT: Anti-inflammatory.

71. Acetozone. *Acetyl benzoyl peroxide;* benzoyl acetyl peroxide; benzozone. $C_9H_8O_4$; mol wt 180.15. C 60.00%, H 4.48%, O 35.52%. Prepd from benzaldehyde and acetic anhydride in the presence of O-contg gas and dibenzoyl peroxide: Carruthers, U.S. pat. **1,985,886** (1935 to Carbide & Carbon Chemicals); Appell, U.S. pat. **3,397,245** (1968 to Koppers).

Crystals, mp 36-37°. bp_{19} 130°. Decomposes on contact with moisture or in a warm place. Soluble in carbon tetrachloride, chloroform, ether, oils. Marketed as a mixture with an equal part of an absorbent powder. *Incompat.* Water, alkalies, heat.

USE: Germicide; in bleaching of flour and food oils. *Caution:* Strong oxidizing agent. Can cause severe skin burns.

72. Acetrizoate Sodium. *3-(Acetylamino)-2,4,6-triiodobenzoic acid sodium salt; 3-acetamido-2,4,6-triiodobenzoic acid sodium salt;* acetrizoic acid sodium salt; Acétiodone; Bronchoselectan; Cystokon; Diaginol; Iodopaque; Pyelokon-R; Salpix; Thixokon; Tri-Abrodil; Triopac; Triurol; Urokon Sodium; Vesamin; Visotrast. $C_9H_5I_3NNaO_3$; mol wt 578.90. C 18.67%, H 0.87%, I 65.77%, N 2.42%, Na 3.97%, O 8.29%. Prepn: Wallingford *et al., J. Am. Chem. Soc.* **74**, 4365 (1952); Wallingford, *J. Am. Pharm. Assoc., Sci. Ed.* **42**, 721 (1953). Toxicity data: J. O. Hoppe *et al., J. Pharmacol. Exp. Ther.* **116**, 394 (1956).

Not isolated from soln. Marketed as a sterile 30% aq soln, d about 1,200, pH between 4.8 and 5.2. The free acid is a white powder, decompn 278-283°. Sol in alcohol. Solubility in water at 25°: 94.2 g/100 ml. Very slightly sol in chloroform. Practically insol in benzene. LD_{50} in rabbits, cats, dogs (mg/kg): 5200, 5600, 6300 i.v. (Hoppe).

Opacoron is a mixture of acetrizoate sodium and meglumine acetrizoate, *q.v.*

THERAP CAT: Diagnostic aid (radiopaque medium).

73. Acetulan®. Acetylated lanolin alcohols. Prepn: Conrad, Motiuk, *J. Soc. Cosmet. Chem.* **6**, 344 (1955); *eidem,* U.S. pat. **2,725,334** (1955 to American Cholesterol Prod.). Pale yellow, practically odorless liquid. Sp gr at 25°: 0.867. Neutral to litmus. Acid no. 0.35. Hydroxyl no. 2.0. Saponification no. 190.0. Hydrophobic, practically insol in water with no emulsification. Miscible with mineral oil, castor oil, vegetable oils, isopropanol, 95% ethanol, isopropyl myristate, isopropyl palmitate, butyl stearate.

USE: In cosmetic formulations and shampoo.

74. Aceturic Acid. *N-Acetylglycine;* acetamidoacetic acid; acetylaminoacetic acid; acetylglycocoll; ethanoylaminoethanoic acid. $C_4H_7NO_3$; mol wt 117.10. C 41.02%, H 6.03%, N 11.96%, O 40.99%. $CH_3CONHCH_2COOH$. Prepd by warming glycine with a slight excess of acetic anhydride in benzene: Radenhausen, *J. Prakt. Chem.* [2] **52**, 437 (1895); by warming glycine in glacial acetic acid with the stoichiometric amount of acetic anhydride: Dakin, *J. Biol. Chem.* **82**, 443 (1929).

Long needles from water, mp 206-208°. K at 25° = 2.3 × 10^{-4}. Sol in water at 15° = 2.7%. Moderately soluble in alcohol. Slightly sol in acetone, chloroform, glacial acetic acid. Practically insol in ether, benzene. Forms stable salts with organic bases.

75. Acetylacetone. *2,4-Pentanedione;* diacetylmethane. $C_5H_8O_2$; mol wt 100.11. C 59.98%, H 8.06%, O 31.96%. $CH_3COCH_2COCH_3$. Made by the action of sodium on ethyl acetate and acetone or by the action of anhydrous aluminum chloride on acetyl chloride in the presence of an inert solvent: Combes, *Compt. Rend.* **103**, 814 (1886); *Ann. Chim. (Paris)* **12**, 199 (1887); Claisen, *Ber.* **38**, 695 (1905); Hunt, U.S. pat. **2,737,528** (1956 to Shawinigan Chem.); Georgieff, U.S. pat. **2,834,811** (1958); Gattermann-Wieland, *Praxis des Organischen Chemikers* (de Gruyter, Berlin, 40th ed, 1961) p 219.

Colorless or slightly yellow, flammable liquid; pleasant odor. d 0.976. bp 140.5°. mp −23°. n_D^{20} 1.4512. One part dissolves in about 8 parts of water. Miscible with alcohol, benzene, chloroform, ether, acetone, glacial acetic acid. LC_{50} (4 hrs) in rats: 1000 ppm: Carpenter, *J. Ind. Hyg. Toxicol.* **31**, 343 (1949).

USE: Forms organometallic complexes which are used as gasoline additives, lubricant additives, driers for varnishes and printer's inks, fungicides, insecticides, colors. *Caution:* Mild irritant to skin, mucous membranes.

76. Acetyl Bromide. C_2H_3BrO; mol wt 122.96. C 19.54%, H 2.46%, Br 64.99%, O 13.01%. CH_3COBr. Prepd from phosphorus tribromide and glacial acetic acid or acetic anhydride: Burton, Degering, *J. Am. Chem. Soc.* **62**, 227 (1940).

Fuming liquid; *irritating to the eyes.* bp 76°. d^9 1.52. mp −96°. Violently decomposed by water, or by alcohol; miscible with ether, chloroform, benzene. *Protect from water.*

77. α-Acetylbutyrolactone. *3-Acetyldihydro-2(3H)-furanone;* α-(2-hydroxyethyl)acetoacetic acid γ-lactone; α-acetyl-γ-hydroxybutyric acid γ-lactone; α-acetobutyrolactone. $C_6H_8O_3$; mol wt 128.12. C 56.24%, H 6.29%, O 37.46%. Prepn from sodium acetoacetate and ethylene oxide in abs alcohol: Knunyantz *et al., Compt. Rend. Acad. Sci. (U.R.S.S.)* [N.S.] **1**, 312 (1934), *C.A.* **28**, 4383 (1934); Feofilaktov, Onishchenko, *J. Gen. Chem. USSR* **9**, 304 (1939), *C.A.* **34**, 378 (1940); Forman; Johnson, U.S. pats. **2,397,134** and **2,443,827** (1946, 1948, both to U.S. Industrial Chemicals); *see also* Matukawa *et al.,* **Japan.** pat. **134,284**; *C.A.* **35**, 7421 (1941).

Liquid; ester-like odor. Solubility in water 20% v/v; soly of water in lactone 12% v/v. d_4^{20} 1.1846; d_{20}^{20} 1.185-1.189; bp_{30} 142-143°; bp_{18} 130-132°; bp_5 107-108°. n_D^{20} 1.4562 (Knunyantz *et al.*); n_D^{20} 1.4590 (Feofilaktov *et al.*). Acquires a blue to blue-purple color when in contact with iron.

USE: In synthesis of 3,4-disubstituted pyridines; of 5-(β-hydroxethyl)-4-methylthiazole. *Caution:* Irritating to skin, mucous membranes.

78. Acetylcarnitine. *2-(Acetyloxy)-3-carboxy-N,N,N-trimethyl-1-propanaminium hydroxide inner salt; (3-carboxy-2-hydroxypropyl)trimethylammonium hydroxide inner salt acetate;* carnitine acetyl ester; vitamin B_T acetate. $C_9H_{17}NO_4$; mol wt 203.24. C 53.19%, H 8.43%, N 6.89%, O 31.49%. $(CH_3)_3N^+CH_2CH(OCOCH_3)CH_2COO^-$. O-Acetyl deriv of carnitine, *q.v.* Prepn of racemate: R. Engeland, *Ber.* **42**, 2457 (1909); of L-form: R. Krimberg, W. Wittandt,

Biochem. Z. **251,** 229 (1932); E. Strack *et al., Z. Physiol. Chem.* **238,** 183 (1936). Resolution of isomers: K. Brendel, R. Bressler, *Biochim. Biophys. Acta* **137,** 98 (1967). Extraction from calf brain tissue: E. A. Hosein, A. Orzeck, *Int. J. Neuropharmacol.* **3,** 71 (1964). Vitamin B_T activity: P. K. Bhattacharyya *et al., Arch. Biochem. Biophys.* **54,** 424 (1955). Neuropharmacology of isomers: E. Tempesta *et al., Neuropharmacology* **24,** 43 (1985). Determn in urine by carboxylic acid analyzer and reversed-phase column chromatography: K. Kidouchi *et al., J. Chromatog.* **423,** 297 (1987). Clinical evaluation of L-form in Alzheimer's disease: G. Acierno, *Clin. Ter.* **105,** 135 (1983); in senile brain: E. Bonavita, *Int. J. Clin. Pharmacol. Ther. Toxicol.* **24,** 511 (1986); in depression: E. Tempesta *et al., Drugs Exp. Clin. Res.* **13,** 417 (1987). Review of pharmacology in the CNS: L. Janiri, E. Tempesta, *Int. J. Clin. Pharmacol. Res.* **3,** 295-306 (1983).

DL-Form hydrochloride, $C_9H_{18}ClNO_4$, crystals, mp 187-188° (dec).

L-Form, $C_9H_{17}NO_4$, hygroscopic crystals, mp 145° (dec). $[\alpha]_D^{20} -19.52°$ (c = 6). Very sol in water, alcohol. Practically insol in ether.

L-Form hydrochloride, $C_9H_{18}ClNO_4$, *acetyl-L-carnitine chloride, levacecarnine hydrochloride, ST-200, Alcar, Branigen, Nicetile.* Stable crystals, mp 187° (dec). Also reported as small rod-shaped crystals from butanol, mp 181° (Strack). $[\alpha]_D^{20} -26.9°$ (c = 9). $[\alpha]_D^{25} -28°$ (c = 2 in water). Very sol in water, sol in alcohol. Practically insol in ether.

THERAP CAT: Nootropic.

79. Acetyl Chloride. C_2H_3ClO; mol wt 78.50. C 30.60%, H 3.85%, Cl 45.17%, O 20.38%. CH_3COCl. Prepd from acetic acid and chlorine in the presence of phosphorus; from acetic acid and salts of chlorosulfonic acid; from sodium acetate and sulfuryl chloride. Details of prepn: Gattermann-Wieland, *Praxis des Organischen Chemikers* (de Gruyter, Berlin, 40th ed., 1961) p 111; A. I. Vogel, *Practical Organic Chemistry* (Longmans, London, 3rd ed., 1959) p 367. Lab prepn from acetic anhydride and calcium chloride: Gmünder, *Helv. Chim. Acta* **36,** 2021 (1953).

Flammable liquid; pungent odor. *Extremely irritating to the eyes.* Imparts a green tinge to a colorless flame. d 1.104. mp −112°. bp 52°. n_D^{20} 1.3898. Decomposed violently by water or alc. Miscible with benzene, chloroform, ether, glacial acetic acid, petr ether. *Protect from water.*

USE: Acetylating agent; in testing for cholesterol, determination of H_2O in organic liquids. *Caution:* Irritant, corrosive. Causes severe burns. Avoid contact with skin, eyes, mucous membranes.

80. Acetylcholine Bromide. *2-(Acetyloxy)-N,N,N-trimethylethanaminium bromide;* Pragmoline; Tonocholin B. $C_7H_{16}BrNO_2$; mol wt 226.14. C 37.18%, H 7.13%, Br 35.34%, N 6.20%, O 14.15%. $(CH_3)_3N(Br)CH_2CH_2OOC-CH_3$. From trimethylamine and β-bromoethyl acetate: Fourneau, Page, *Bull. Soc. Chim. France* [4] **15,** 544 (1914).

Deliquesc crystals. Very soluble in cold water; decomposed by hot water or alkalies; soluble in alcohol; practically insol in ether. *Keep tightly closed.*

THERAP CAT: Cholinergic.

81. Acetylcholine Chloride. *2-(Acetyloxy)-N,N,N-trimethylethanaminium chloride;* Acecoline; Arterocoline; Miochol; Ovisot. $C_7H_{16}ClNO_2$; mol wt 181.68. C 46.28%, H 8.88%, Cl 19.52%, N 7.71%, O 17.61%. $(CH_3)_3N(Cl)CH_2-CH_2OOCCH_3$. Prepd from trimethylamine and β-chloroethyl acetate: Fourneau, Page, *Bull. Soc. Chim. France* [4] **15,** 544 (1914).

Crystalline powder, very deliquescent, mp 149-152°. Very sol in cold water, alcohol; decomposed by hot water or alkalies. Practically insol in ether. *Keep tightly closed.* LD_{50} s.c. in rats: 250 mg/kg.

THERAP CAT: Cholinergic.

82. Acetylcysteine. *N-Acetyl-L-cysteine;* L-α-acetamido-β-mercaptopropionic acid; *N*-acetyl-3-mercaptoalanine; Airbron; Broncholysin; Brunac; Fabrol; Fluatox; Fluimucil; Fluimucetin; Fluprowit; Inspir; Mucocedyl; Mucolator; Mucolyticum; Mucomyst; Muco Sanigen; Mucosolvin; Mucret; NAC; Neo-Fluimucil; Parvolex; Respaire (obsolete); Tixair. $C_5H_9NO_3S$; mol wt 163.20. C 36.79%, H 5.56%, N 8.58%, O 29.41%, S 19.65%. Prepn: Smith, Gorin,

J. Org. Chem. **26,** 820 (1961). Prepn and use in treatment of respiratory diseases: Martin, Waller, U.S. pat. **3,184,505** (1965 to Mead Johnson). Effect in corneal healing: F. Menna *et al., Bull. Mem. Soc. Fr. Opthalmol.* **94,** 425 (1982); G. Petroutsos *et al., Ophthal. Res.* **14,** 241 (1982). Toxicity: E. I. Goldenthal, *Toxicol. Appl. Pharmacol.* **18,** 185 (1971). Multicenter clinical trial in acetaminophen overdose: M. J. Smilkstein *et al., N. Engl. J. Med.* **319,** 1557 (1988). *Review:* G. R. McKinney, G. M. Sisson, in *Pharmacological and Biochemical Properties of Drug Substances* vol. 2, M. E. Goldberg, Ed. (Am. Pharm. Assoc., Washington, DC, 1979) pp 479-488.

$$HSCH_2CHCOOH$$
$$|$$
$$NHCOCH_3$$

Crystals from water, mp 109-110°. LD_{50} orally in rats: 5050 mg/kg (Goldenthal).

THERAP CAT: Mucolytic. Corneal vulnerary. Antidote to acetaminophen poisoning.

83. Acetyldigitoxins. $C_{43}H_{66}O_{14}$; mol wt 807.00. C 64.00%, H 8.24%, O 27.76%. Obtained by enzymatic hydrolysis of lanatoside A (*q.v.*). Composed of the aglycone digitoxigenin and 3 mols digitoxose, to one of which an acetyl group is attached. Acetyldigitoxin-α differs from acetyldigitoxin-β in the position of the acetyl group. The β-form is obtained either by splitting off the glucose residue from lanatoside A by means of enzymes, or by extraction from the leaves of *Digitalis ferruginea* L., Scrophulariaceae. The α-form can be obtained from acetyldigitoxin-β by heating in an anhydrous or aq organic solvent at pH 3.5 to 8: Stoll, Kreis, U.S. pat. **2,776,963** (1957 to Sandoz).

α-Form, *Acylanid.* Platelets from methanol, mp 217-221°. $[\alpha]_D^{20} +5.0°$ (c = 0.7 in pyridine). Slightly sol in chloroform.

β-Form, stout, solvated prisms from methanol losing their solvent of crystn in desiccator. When dry, dec 225°. $[\alpha]_D^{20} +16.7°$ (pyridine). One gram dissolves in 7 to 9 ml chloroform, in 150 ml methanol, in 220 ml amyl alc; almost insol in ether, water (one gram dissolves in 200 liters H_2O at 20°).

THERAP CAT: Cardiotonic.

84. Acetylene. *Ethyne;* ethine. C_2H_2; mol wt 26.02. C 92.26%, H 7.74%. HC≡CH. Manuf from calcium carbide and water: Eastman, U.S. pat. **3,017,259** (1962 to Texaco); from methane: Anderson, U.S. pat. **3,051,639** (1962 to Union Carbide). Review of manuf processes: Faith, Keyes & Clark's *Industrial Chemicals,* F. A. Lowenheim, M. K. Moran, Eds. (Wiley-Interscience, New York, 4th ed., 1975) pp 26-35. *Review:* Nieuwland, Vogt, *The Chemistry of Acetylene* (Reinhold, New York, 1945) pp 1-219. Comprehensive monograph in 2 vols: S. A. Miller, *Acetylene* (Academic Press, New York, 1965); several authors in Kirk-Othmer *Encyclopedia of Chemical Technology* vol. 1 (Wiley-Interscience, New York, 3rd ed., 1978) pp 192-243.

Gas; odor not unpleasant when pure, but disagreeable when impure (due to phosphine). *Toxic when inhaled.* mp −81° (subl). At 0° liquifies at 21.5 atm; below 37° (crit temp) liquifies at 68 atm. One liter at 0° and 760 mm weighs 1.165 g; d gas (air = 1) 0.90. Burns brilliantly in air with very sooty flame. Heat of combustion 313 cal. Not explosive at ordinary atmospheric pressure, but at 2 atms or more it is explosive by spark or decomposition. Mixture with air containing more than 3% or less than 65% gas is explosive, max being 1 vol gas and 12.5 vol air. Forms insoluble explosive compounds with copper and silver; hence copper or brass containers must be avoided. One vol dissolves in 1 vol water, in 6 vols glacial acetic acid or alcohol; soluble in ether, benzene. Acetone dissolves 25 vols acetylene at 15° and 760 mm; but 300 vols at 12 atm. LC in rats: 900,000 ppm, Riggs, *Proc. Soc. Exp. Biol. Med* **22,** 269 (1925).

Caution: A simple asphyxiant. High concentrations cause narcosis. 20% concn may cause dyspnea, headache; 40% or more concn may cause collapse.

USE: Illuminant, oxyacetylene welding, cutting, and soldering metals, signalling; pptg metals, particularly Cu; manuf acetaldehyde, acetic acid; fuel for motor boats.

85. Acetylene Dibromide. *1,2-Dibromoethene; 1,2-*

dibromoethylene; sym-*dibromoethylene.* $C_2H_2Br_2$; mol wt 185.87. C 12.92%, H 1.09%, Br 85.99%. BrCH=CHBr. Prepd by reaction of tetrabromoethylene with Zn and alc followed by sepn of *cis-* and *trans-*forms by fractional distillation: Noyes *et al., J. Am. Chem. Soc.* **72**, 33 (1950).

Liquid, gradually decomposed by air, moisture, or light. d_4^{17} 2.21. n_D^{20} 1.5428. Practically insol in water; sol in many organic solvents. LD_{50} in rats: 117 mg/kg orally.

*cis-*Form, n_D^{25} 1.5370.

*trans-*Form, n_D^{25} 1.5440.

Caution: Narcotic in high concns.

86. Acetylene Dichloride. *1,2-Dichloroethene; 1,2-dichloroethylene;* sym-*dichloroethylene; Dioform.* $C_2H_2Cl_2$; mol wt 96.95. C 24.78%, H 2.08%, Cl 73.14%. ClCH=CHCl. Prepn: Bordner, U.S. pat. **2,504,919** (1950 to du Pont). Prepn of *trans-*form: Adler, U.S. pat. **2,440,997** (1948 to Stockholms Superfosfot Fabriks Aktiebolag). Sepn of *cis-* and *trans-*forms by fractional distillation: Wood, Dickinson, *J. Am. Chem. Soc.* **61**, 3259 (1939); Johnsen, Fitzpatrick, *Rec. Trav. Chim.* **70**, 823 (1951); Truce, Barney, *J. Org. Chem.* **27**, 128 (1962). Toxicity: D. Gradiski *et al., Eur. J. Toxicol.* **7**, 247 (1974); K. J. Freundt *et al., Toxicology* **7**, 141 (1977). *Review:* V. L. Stevens in Kirk-Othmer *Encyclopedia of Chemical Technology,* Vol. **5** (Wiley-Interscience, New York, 3rd ed., 1979) pp 742-745. Toxicity: D. Gradiski *et al., Eur. J. Toxicol.* **7**, 247 (1974); K.J. Freundt *et al., Toxicology* **7**, 242 (1977). *Review:* V. L. Stevens in Kirk-Othmer *Encyclopedia of Chemical Technology* vol. **5** (Wiley-Interscience, New York, 3rd ed., 1979) pp 742-745.

Liquid; ethereal, slightly acrid odor; gradually decomposed by air, light and moisture, forming HCl. d about 1.28. bp about 55°. Insol in water. Sol in alc, ether and most other organic solvents. LD_{50} i.p. in mice: ~2150 mg/kg (Gradiski).

*cis-*Form, mp −81.5°. bp_{745} 59.6°, bp_{760} 60°. n_D^{25} 1.4435.

*trans-*Form, mp −49.4°. bp_{745} 47.2°. Noticeably subject to air oxidation. LD_{50} in rats (ml/kg): 1.0 orally; 60 i.p.; in mice (ml/kg): 3.2 i.p. (Freundt).

Caution: May cause respiratory irritation, narcosis: *Patty's Industrial Hygiene & Toxicology* Vol. **2B**, G. D. Clayton, F. E. Clayton, Eds. (Wiley-Interscience, New York, 3rd ed., 1981) pp 3550-3553.

USE: Solvent for fats, phenol, camphor, etc.

87. Acetyleneurea. *Tetrahydroimidazo[4,5-d]imidazole-2,5-(1H,3H)-dione; glycoluril;* acetylenediureine; glyoxaldiureine; acetylene carbamide; $C_4H_6N_4O_2$; mol wt 142.12. C 33.80%, H 4.26%, N 39.42%, O 22.52%. Prepd by the sodium amalgam reduction of allantoin: Biltz, Schiemann, *J. Prakt. Chem.* **113**, 77 (1926); from glyoxal and urea: Reibnitz, U.S. pat. **2,731,472** (1956 to BASF).

Crystals, dec at about 300°. Slightly sol in cold, more sol in hot water; sol in warm mineral acids, warm ammonia.

88. Acetyl Iodide. C_2H_3IO; mol wt 169.96. C 14.13%, H 1.78%, I 74.68%, O 9.41%. CH_3COI. Prepd from acetyl chloride and HI: Gustus, Stevens, *J. Am Chem. Soc.* **55**, 374 (1933).

Liquid; suffocating odor; fumes and turns brown in air. bp 108°, bp_{735} 104-106°; d_4^{20} 2.0674; n_D^{20} 1.5491. Dec by water or alcohol; sol in benzene, ether. *Keep tightly closed and protected from light.*

Caution: A strong irritant. Avoid contact with skin. Vapors can cause pulmonary edema.

89. Acetylleucine Monoethanolamine. *N-Acetyl-*DL-*leucine compound with 2-aminoethanol (1:1);* monoethanolamine DL-acetylleucinate; DL-acetylleucine monoethanolamine salt; monoethanolamine salt of α-acetamidoisocaproic acid; RP 7452; Tanganil. $C_{10}H_{22}N_2O_4$; mol wt 234.29. C 51.26%, H 9.46%, N 11.96%, O 27.32%. Prepn: Gailliot *et al.*, U.S. pat. **2,941,924** (1960 to Rhône-Poulenc).

$$(CH_3)_2CHCH_2CHCOOH \cdot H_2NCH_2CH_2OH$$
$$|$$
$$NHCOCH_3$$

Crystals, mp about 150°. Soly in water: > 20%; slight soly in alcohol: ~1%. pH of 10% aq soln: about 6.

THERAP CAT: Antivertigo agent.

90. N-Acetylmethionine. $C_7H_{13}NO_3S$; mol wt 191.26. C 43.96%, H 6.85%, N 7.33%, O 25.10%, S 16.77%. CH_3S-$CH_2CH_2CH(NHCOCH_3)COOH$. Prepn of racemic mixture: Wheeler, Ingersoll, *J. Am. Chem. Soc.* **73**, 4604 (1951); Callanan, Patton, U.S. pat. **2,745,873** (1956 to Distillers Products). Resolution: Wheeler, Ingersoll, *loc. cit.;* Gillingham, U.S. pat. **3,028,395** (1962 to Parke, Davis); Tullar, U.S. pat. **3,056,799** (1962 to Sterling Drug); Brit. pat. **1,072,876** (1967 to Tanabe Seiyaku), *C.A.* **67**, 114015x (1967).

DL-Form, *Methionamine.* Crystals from water, mp 114-115°.

D(+)-Form, crystals from water or ethyl acetate, mp 104-105°. $[\alpha]_D^{25}$ +20.3° (c = 4 in water).

L(−)-Form, crystals, mp 104°. $[\alpha]_D^{25}$ −20.3°.

THERAP CAT: DL-Form as lipotropic.

91. 5-Acetyl-2-methoxybenzaldehyde. 3-Acetyl-6-methoxybenzaldehyde. $C_{10}H_{10}O_3$; mol wt 178.18. C 67.40%, H 5.66%, O 26.94%. A natural plant growth inhibitor found in the leaves of *Encelia farinosa* A. Gray, *Compositae.* Isoln and synthesis from 3-acetyl-6-methoxybenzonitrile: Gray, Bonner, *J. Am. Chem. Soc.* **70**, 1249 (1948).

Needles from alcohol or ether, mp 144°. Sublimes without decomp. Emits fragrant odor on prolonged heating. Sol in hot water, warm ether, alcohol, benzene, chloroform; practically insol in cold water, 5% HCl, 5% NaOH, petr ether, carbon tetrachloride; sol with orange color in concd mineral acids.

92. Acetyl Nitrate. *Acetic acid anhydride with nitric acid.* $C_2H_3NO_4$; mol wt 105.05. C 22.87%, H 2.88%, N 13.33%, O 60.92%. CH_3COONO_2. Prepn from acetic anhydride and N_2O_5: Boh, *Annales de Chimie* [11] **20**, 437 (1945).

Fuming, mobile, hygroscopic liquid. Should be colorless. d_4^{15} 1.24. bp_{70} 22°. Although it may be stored in the dark over P_2O_5 for the weekend, it should be used in *statu nascendi* to avoid explosions. Always explodes when heated suddenly over 60° or when in contact with HgO. Explosions have occurred upon contact with ground glass surfaces: König, *Angew. Chem.* **67**, 157 (1955).

USE: In nitrations, especially to introduce a single nitro group in an ortho position on an aromatic ring. *Caution:* Irritant, corrosive.

93. N-Acetylpenicillamine. *N-Acetyl-3-mercapto-*D-*valine.* $C_7H_{13}NO_3S$; mol wt 191.25. C 43.96%, H 6.85%, N 7.33%, O 25.10%, S 16.76%. Prepn: Crooks in *The Chemistry of Penicillin* (Princeton Univ. Press, 1949) p 470; Sheehan *et al.*, U.S. pats. **2,477,148** and **2,496,416** (1949, 1950, both to Merck & Co.).

*dl-*Form, crystals from hot water, mp 183°.

*d-*Form, crystals from water, mp 189-190°; $[\alpha]_D^{25}$ +18° (50% ethanol).

94. Acetylpheneturide. *N-[(Acetylamino)carbonyl]-α-ethylbenzeneacetamide; 1-acetyl-3-(2-phenylbutyryl)urea; N-(α-ethylphenylacetyl)-N'-acetylurea;* Crampol. $C_{13}H_{16}$-N_2O_3; mol wt 248.27. C 62.89%, H 6.50%, N 11.28%, O

19.33%. Synthesis: Umemoto *et al.*, *Yakugaku Zasshi* **83**, 753 (1963), *C.A.* **59**, 13849d (1963); Takamatsu *et al.*, U.S. pat. **3,110,728** (1963 to Dainippon). Pharmacology and toxicology: Nakamura *et al.*, *Arzneimittel-Forsch.* **18**, 524 (1968).

Crystals from dil alcohol, mp 100-101°. LD_{50} orally in mice: 4.73 mMole/kg (1.17 g/kg), Nakamura *et al.*, *loc. cit.*
THERAP CAT: Anticonvulsant.

95. Acetylsalicylsalicylic Acid. *2-(Acetyloxy)benzoic acid 2-carboxyphenyl ester; 2-hydroxybenzoic acid acetate 2-carboxyphenyl ester;* salicylacetylsalicylic acid; Diplosal Acetate. $C_{16}H_{12}O_6$; mol wt 300.26. C 64.00%, H 4.03%, O 31.97%. Prepd by treating salicylsalicylic acid *q.v.*, with acetic anhydride or by fractionating an acetylsalicylic acid melt: Ger. pat. **236,196** (1908 to Boehringer).

Leaflets from dilute acetic acid, mp 159° [mp 152°, Rudoi, *Soviet Med.* 1941, no. 12, 26-27]. Practically insol in water, benzene. One part dissolves in 3 parts boiling alcohol, in 70 parts boiling ether, in 45 parts boiling benzene. Sol in alkaline solns, but is saponified by heating in alk soln. On acidification of the soln salicylic acid is precipitated.
THERAP CAT: Analgesic.

96. Acetyl Sulfamethoxypyrazine. *N-[(4-Aminophenyl)sulfonyl]-N-(3-methoxypyrazinyl)acetamide; N^1-acetyl-N^1-(3-methoxypyrazinyl)sulfanilamide;* 2-(N^1-acetylsulfanilamido)-3-methoxypyrazine; Acetylazide. $C_{13}H_{14}N_4O_4S$; mol wt 322.35. C 48.44%, H 4.38%, N 17.38%, O 19.85%, S 9.95%. Prepn: Camerino, Palamidessi, *Gazz. Chim. Ital.* **90**, 1815 (1960); *eidem*, U.S. pat. **3,098,069** (1963 to Soc. Farmaceutici Italia).

Crystals, mp 199°.
THERAP CAT: Antibacterial.

97. N^4-Acetylsulfanilamide. *N-[4-(Aminosulfonyl)phenyl]acetamide; p-*sulfamylacetanilide. $C_8H_{10}N_2O_3S$; mol wt 214.24. C 44.85%, H 4.71%, N 13.07%, O 22.40%, S 14.97%. Formed in the animal body where as much as 60% of ingested sulfanilamide is acetylated by the liver; some extrahepatic acetylation occurs also. Prepd in the laboratory by condensing acetanilide with chlorosulfonic acid, and treating the resulting N^4-acetylsulfanilyl chloride with ammonia. The formation of N^4-acetylsulfanilamide is a step in the manufacture of sulfanilamide: Miller *et al.*, *J. Am. Chem. Soc.* **61**, 1198 (1939); *cf.* Smiles, Stewart, *Org. Syn.* coll. vol. I, 8 (1932).

Needles from alcohol + water, mp 216°. Slightly sol in water, alcohol. Max soly is reached in a 65% alcoholic soln.

98. N-Acetylsulfanilic Acid. *4-(Acetylamino)benzenesulfonic acid; N-p-*acetylanilinosulfonic acid; acetanilidsulfonic acid; 4-acetamidobenzenesulfonic acid. $C_8H_9NO_4S$;

mol wt 215.23. C 44.64%, H 4.21%, N 6.51%, O 29.74%, S 14.90%. Prepd by the action of oleum or sulfuric acid on acetanilide in acetic anhydride: Junghahn, Neumann, *Ber.* **33**, 1366 (1900); Söll, Stutzer, *Ber.* **42**, 4539 (1909).

Dihydrate, crystals. Freely sol in water, sol in alcohol, sparingly sol in glacial acetic acid, insol in ether. Slowly hydrolyzes in aq soln.
Sodium salt, $C_8H_8NNaO_4S$, *Cosaprin*. Prisms, soluble in water, moderately sol in alcohol, insol in ether.
Ethyl ester, $C_{10}H_{13}NO_4S$, needles from benzene, mp 115°.

99. N-Acetylsulfanilyl Chloride. *4-(Acetylamino)benzenesulfonyl chloride; p-*acetamidobenzenesulfonyl chloride; *p-*acetaminobenzenesulfonyl chloride; acetanilide-*p-*sulfonyl chloride; ASC; Dagenan chloride. $C_8H_8ClNO_3S$; mol wt 233.68. C 41.12%, H 3.45%, Cl 15.17%, N 5.99%, O 20.54%, S 13.72%. Prepd by sulfonation of acetanilide: Stewart, *J. Chem. Soc.* **121**, 2558 (1922); Smiles, Stewart, *Org. Syn.* coll. vol. I, 8 (2nd ed., 1941).

Thick, light tan prisms from benzene, mp 149°. Has slight odor of acetic acid. Benzene and ether are poor solvents for purification by recrystn. Much better results are obtained using chloroform or ethylene dichloride. *Cf.* Northey, *Sulfonamides* (New York, 1948) p 12.
USE: Intermediate in the prepn of sulfanilamide and its derivs. *Caution:* Irritant to skin, eyes, and mucous membranes of nose and throat.

100. Acetyl Sulfisoxazole. *N-[(4-Aminophenyl)sulfonyl]-N-(3,4-dimethyl-5-isoxazolyl)acetamide; N^1-acetyl-N^1-(3,4-dimethyl-5-isoxazolyl)sulfanilamide;* N^1-monoacetyl sulfisoxazole; Gantrisin Acetyl. $C_{13}H_{15}N_3O_4S$; mol wt 309.35. C 50.47%, H 4.89%, N 13.58%, O 20.69%, S 10.37%. Prepd by the action of acetic anhydride on sulfisoxazole suspended in acetone and pyridine: Hoffer, U.S. pat. **2,721,200** (1955 to Hoffmann-La Roche).

Tasteless crystals, mp 192-195°. Solubilities at 25° (mg/ml): 0.07 in water; 4.93 in methanol; 5.7 in 95% ethanol; 0.94 in ether; 29.0 in chloroform.
THERAP CAT: Antibacterial.
THERAP CAT (VET): Antibacterial, orally or parenterally, chiefly for urinary or respiratory tract infections. Also used for foot rot, mastitis (locally).

101. Acetyltannic Acid. Diacetyltannic acid; tannyl acetate; Acetannin; Tannigen. Acetylation of tannin: Ciusa, Sollazzo, *Ann. Chim. Appl.* **33**, 72 (1942); *C.A.* **38**, 5794⁴ (1944).

Yellowish- or grayish-white odorless powder; darkens on exposure to light. Softens under water at about 70°. Slightly soluble in water or alcohol; sol in ethyl acetate, solns of borax or sodium phosphate, or with gradual decompn in solns of alkali hydroxides and carbonates. *Incompat.* Alkalies, iron salts.
THERAP CAT: Astringent (intestinal).
THERAP CAT (VET): Astringent (intestinal) for foals, calves, dogs.

102. Achillea. Milfoil; yarrow; thousand-leaf. Flowering herb of *Achillea millefolium* L., *Compositae. Habit.* Europe, Asia, naturalized in U.S. Extensively used as herbal remedy by many cultures. Constituents of plant and of the volatile oil may vary depending on origin of plant. Identification of constituents: Pailer, Kump, *Monatsh.* **90**, 395 (1959); *eidem, Arch. Pharm.* **293**, 646 (1960); A. J. Falk *et al., J. Nat. Prod.* **37**, 598 (1974); R. F. Chandler *et al., J. Pharm. Sci.* **71**, 690 (1982). Isoln of anti-inflammatory constituents: A. S. Goldberg, *J. Pharm. Sci.* **58**, 938 (1969). Phytopharmacology: J. P. Tewari *et al., Ind. J. Med. Sci.* **28**, 331 (1974). Review of medicinal uses and composition: R. F. Chandler *et al., Econ. Bot.* **36**, 203-223 (1982).

103. Acid Fuchsin. *2-Amino-5-[(4-amino-3-sulfophenyl)(4-imino-3-sulfo-2,5-cyclohexadien-1-ylidene)methyl]-3-methylbenzenesulfonic acid disodium salt; C.I. Acid Violet 19;* C.I. 42685; 2-amino-α^5-(4-amino-3-sulfophenyl)-α^5-(4-imino-3-sulfo-2,5-cyclohexadien-1-ylidene)-3,5-xylenesulfonic acid disodium salt; acid magenta; acid rubin; fuchsin(e) acid; acid roseine. $C_{20}H_{17}N_3Na_2O_9S_3$; mol wt 585.54. C 41.02%, H 2.93%, N 7.18%, Na 7.85%, O 24.59%, S 16.43%. The trisulfonic acid disodium salt of magenta I, *q.v.* Prepn: *Colour Index* vol. 4 (3rd ed., 1971) p 4399. Brief review: H. J. Conn's *Biological Stains,* R. D. Lillie, Ed. (Williams & Wilkins, Baltimore, 9th ed., 1977) pp 285-286.

Olive to dark olive-green coarse powder. Absorption max: 540-545 nm (10 mg/l of 0.01% HCl). One gram dissolves in 7 ml water. Slightly sol in alc. Dil aq solns are purplish red. Very dil aq solns (1:10,000) are decolorized by drops of concd aq NaOH solns, but not returned by concd HCl. The decolorized soln neutralized with NaOH is called the *Andrade indicator.* Changes from red to colorless at pH 12-14.

USE: As pH indicator; biological stain.

104. Acid Violet 7B. *N-[4-[Bis[4-[ethyl(3-sulfophenyl)-amino]phenyl]methylene]-2,5-cyclohexadien-1-ylidene]-N-methylmethanaminium hydroxide inner salt monosodium salt;* C.I. Acid Violet 25; C.I. 42745. $C_{37}H_{36}N_3NaO_6S_2$; mol wt 705.81. C 62.96%, H 5.14%, N 5.95%, Na 3.26%, O 13.60%, S 9.09%. Prepd by sulfonation of the condensation product from *p*-dimethylaminobenzoyl chloride with *N*-ethyldiphenylamine: Müller, **Ger.** pat. **34,463**; *Frdl.* **1**, 88; **U.S.** pat. **353,266**; Turnbull, Evans, **U.S.** pat. **1,402,195.** *See also: Colour Index* vol. 4 (3rd ed., 1971) p 4401.

Violet powder. Sol in water, alcohol with deep blue color and violet tinge. Insol in ether. Absorption max (water): 607.5 nm.

USE: As dye; in inks and stains.

105. Acifluorfen. *5-[2-Chloro-4-(trifluoromethyl)-phenoxy]-2-nitrobenzoic acid.* $C_{14}H_7ClF_3NO_5$; mol wt 361.66. C 46.49%, H 1.94%, Cl 9.82%, F 15.76%, N 3.87%, O 22.12%. Selective pre- and post-emergence herbicide. Prepn: H. O. Bayer *et al.,* **Ger.** pat. **2,311,638** corresp to H. O. Bayer, R. Y. Yih, **U.S.** pats. **3,928,416, 4,063,929** (1973,

1975, 1977, all to Rohm & Haas). Synthesis and activity: W. O. Johnson *et al., J. Agr. Food Chem.* **26**, 285 (1978).

Off-white solid, mp 151.5-157°.
Sodium salt, $C_{14}H_6ClF_3NNaO_5$, *RH 6201, scifluorfen, Blazer.* White powder, mp 124-125°. Soly in water > 25%. LD_{50} orally in rats: 1300 mg/kg, W. O. Johnson *et al., loc. cit.*

USE: Herbicide. *Caution:* Skin and eye irritant. "Do not incorporate in soil. Do not mix with oils, surfactants, liquid fertilizers or other pesticides".

106. Acipimox. *5-Methylpyrazinecarboxylic acid 4-oxide;* 2-carboxy-5-methylpyrazine 4-oxide; K 9321; Olbemox; Olbetam. $C_6H_6N_2O_3$; mol wt 154.13. C 46.76%, H 3.93%, N 18.17%, O 31.14%. Prepn: V. Ambrogi *et al.,* **Ger.** pat. **2,319,834**; *eidem,* **U.S.** pat. **4,002,750** (1973, 1977 both to Carlo Erba). Prepn and toxicology: *eidem, Eur. J. Med. Chem.* **15**, 157 (1980). Pharmacological profile: P. P. Lovisolo *et al., Pharmacol. Res. Commun.* **13**, 151, 163 (1981). Pharmacokinetics: L. M. Fuccella *et al., Clin. Pharmacol. Ther.* **28**, 790 (1980); L. Musatti *et al., J. Int. Med. Res.* **9**, 381 (1981). Mechanism of action: K. Aktories *et al., Arzneimittel-Forsch.* **33**, 1525 (1983).

Crystals from water, mp 177-180°. LD_{50} orally in mice: 3500 mg/kg (Ambrogi).

THERAP CAT: Antihyperlipoproteinemic.

107. Acitretin. *(all-E)-9-(4-Methoxy-2,3,6-trimethylphenyl)-3,7-dimethyl-2,4,6,8-nonatetraenoic acid;* etretin; Ro 10-1670; Neotigason; Soriatane. $C_{21}H_{26}O_3$; mol wt 326.44. C 77.27%, H 8.03%, O 14.70%. Synthetic retinoid; free acid form and major metabolite of etretinate, *q.v.* Prepn: W. Bollag *et al.,* **Ger.** pat. **2,414,619**; *eidem,* **U.S.** pat. **4,105,681** (1974, 1978 both to Hoffmann-La Roche). Teratogenicity study: A. Kistler, H. Hummler, *Arch. Toxicol.* **58**, 50 (1985). HPLC determn in plasma: N. R. Al-Mallah *et al., Anal. Letters* **21**, 1603 (1988). Pharmacokinetics in humans: F. G. Larsen *et al., Pharmacol. Toxicol.* **62**, 159 (1988). Clinical evaluation in cutaneous lupus erythematosus: T. Ruzicka *et al., Arch. Dermatol.* **124**, 897 (1988). Review of clinical pharmacology: A. Vahlquist, O. Rollman, *Dermatologica* **175**, Suppl. 1, 20-27 (1987). Review of clinical studies in psoriatic and nonpsoriatic dermatoses: J.-M. Geiger, B. M. Czarnetzki, *ibid.* **176**, 182-190 (1988).

Crystals from hexane, mp 228-230°. LD_{50} i.p. in mice (mg/kg): > 4000 (1 day), 700 (10 days), 700 (20 days) (Bollag, 1978).

THERAP CAT: Antipsoriatic.

108. Aclacinomycins. Antitumor antibiotic complex of the anthracycline group, produced by *Streptomyces galilaeus.* Thirteen yellow and seven red-colored components have been identified. Isoln of the major components, aclacinomycins A and B: H. Umezawa *et al.,* **Ger.** pat. **3,532,568** corresp to **U.S.** pat. **3,988,315** (1974, 1976 both to Microbiochem. Res. Found.). *See also:* T. Oki *et al., J. Antibiot.* **28**, 830 (1975). Structure, taxonomy, production, properties:

eidem, ibid. **32,** 791, 801 (1979). The aglycone portion of aclacinomycin A is known as *aklavinone.* Isoln of aklavinone: J. J. Gordon *et al., Tetrahedron Letters* **8,** 28 (1960). Synthesis of racemic aklavinone: B. A. Pearlman *et al., J. Am. Chem. Soc.* **103,** 4248 (1981); P. A. Confalone, G. Pizzolato, *ibid.* 4251; R. K. Boeckman, F. W. Sum, *ibid.* **104,** 4604 (1982). Synthesis of optically active aklavinone: A. S. Kende, J. P. Rizzi, *ibid.* **103,** 4247 (1981); J. M. McNamara, Y. Kishi, *ibid.* **104,** 7371 (1982). *In vitro* metabolism of aclacinomycins: T. Komiyama *et al., Gann* **70,** 395 (1979), *C.A.* **92,** 15143x (1980). HPLC determn of aclacinomycin A and its metabolites: T. Ogasawara *et al., J. Antibiot.* **34,** 47, 52 (1981). Pharmacokinetics: M. J. Egorin *et al., Cancer Chemother. Pharmacol.* **8,** 41 (1982). Series of articles on absorption, excretion, distribution and general pharmacology: *Japan. J. Antibiot.* **33,** 169-213 (1980), *C.A.* **93,** 19316, 125326, 125649, 142986 (1980). Immunological study: M. Ishizuka *et al., J. Antibiot.* **34,** 331 (1981). Clinical studies: R. Maral, *Drugs Exp. Clin. Res.* **9,** 375 (1983); R. P. Warrell, Jr., S. J. Kempin, *Am. J. Clin. Oncol.* **6,** 81 (1983); A. Y. Bedikian *et al., ibid.* 187 (1983). Review of pharmacology of aclacinomycin A: T. Oki, *Anthracyclines [Proc. Workshop],* S. T. Crooke, S. D. Reich, Eds. (Academic Press, New York, 1980) pp 323-342, *C.A.* **93,** 160778h (1980).

aclacinomycin A

Aclacinomycin A, $C_{42}H_{53}NO_{15}$, *2-ethyl-1,2,3,4,6,11-hexahydro-2,5,7-trihydroxy-6,11-dioxo-4-[[2,3,6-trideoxy-4-O-[2,6-dideoxy-4-O-[(2R-trans)-tetrahydro-6-methyl-5-oxo-2H-pyran-2-yl]-α-L-lyxo-hexopyranosyl]-3-(dimethylamino)-α-L-lyxo-hexopyranosyl]oxy]-1-naphthacenecarboxylic acid methyl ester, antibiotic MA 144A1, NSC-208734, aclarubicin, Jaclacin.* Yellow microcryst powder from chloroform/hexane, mp 151-153° (dec). $[\alpha]_D^{24}$ −11.5° (c = 1 in methylene chloride). uv max (methanol): 229.5, 259, 289.5, 431 nm ($E_{1cm}^{1\%}$ 550, 326, 135, 161); (0.1N HCl) 229.5, 258.5, 290, 431 nm ($E_{1cm}^{1\%}$ 571, 338, 130, 161); (0.1N NaOH) 239, 287, 523 nm ($E_{1cm}^{1\%}$ 450, 113, 127). Sol in CHCl₃, ethyl acetate, insol in ethyl ether, n-hexane, petr ether. Addn of alkali to aq solns gives an intense reddish-purple color; in conc HCl the soln is yellow. LD_{50} in mice (mg/kg): 22.6 i.p., 33.7 i.v. (Oki).

Aclacinomycin A hydrochloride, $C_{42}H_{54}ClNO_{15}$, *aclarubicin hydrochloride, Aclacinon, Aclaplastin.*

Aclacinomycin B, $C_{42}H_{51}NO_{15}$, *4-[[[2'''',3''-anhydro]-O-3,6-dideoxy-α-L-erythro-hexopyranos-4-ulos-1-yl-(1 → 4)-O-2,6-dideoxy-α-L-lyxo-hexopyranosyl-(1 → 4)-2,3,6-trideoxy-3-(dimethylamino)-α-L-lyxo-hexopyranosyl]oxy]-2-ethyl-1,2,3,4,6,11-hexahydro-2,5,7-trihydroxy-6,11-dioxo-1-naphthacenecarboxylic acid methyl ester, antibiotic MA 144B1.* Yellow microcryst powder from chloroform/hexane, mp 163-167° (dec). $[\alpha]_D^{24}$ +3° (c = 1 in methylene chloride). Other physical properties similar to aclacinomycin A. LD_{50} in mice (mg/kg): 13.7 i.p., 16.4 i.v. (Oki).

THERAP CAT: Antineoplastic.

109. Aclatonium Napadisilate. *2-[2-(Acetyloxy)-1-oxopropoxy]-N,N,N-trimethylethanaminium 1,5-naphthalenedisulfonate (2:1); 1,5-naphthalenedisulfonic acid bis [2-[2-(acetyloxy)-1-oxopropoxy]-N,N,N-trimethylethanaminium]-ion (2⁻);* acetyllactoylcholine 1,5-naphthalenedisulfonate; choline 1,5-naphthalenedisulfonate (2:1), dilactate, diacetate; TM 723; Abovis. $C_{30}H_{46}N_2O_{14}S_2$; mol wt 722.83. C 49.85%, H 6.42%, N 3.87%, O 30.99%, S 8.87%. Spasmolytic agent related structurally to acetylcholine chloride, *q.v.* Prepn: K. Miura *et al.,* Ger. pat. **2,425,983** corresp to U.S. pat. **3,903,-137** (1973, 1975 both to Toyama). Synthesis and smooth muscle activity: *eidem, Yakugaku Zasshi* **99,** 1245 (1979), *C.A.* **92,** 146197r (1980). Series of articles on metabolism, pharmacology, toxicity studies: *Oyo Yakuri* **18,** 695-942 (1979), *C.A.* **92,** 191015r, 191281z, 191282a, 208995k-7n (1980).

Crystals, mp 189-191°. LD_{50} in mice: 15 g/kg orally, 826 mg/kg s.c., K. Miura *et al., loc. cit.;* in dogs: >10 g/kg orally, A. Takai *et al., Oyo Yakuri* **19,** 93 (1980), *C.A.* **93,** 613367 (1980).

THERAP CAT: Cholinergic.

110. Aconine. *(1α,3α,6α,14α,15α,16β)-20-Ethyl-1,6,-16-trimethoxy-4-(methoxymethyl)aconitane-3,8,13,14,15-pentol;* Aconitysat. $C_{25}H_{41}NO_9$; mol wt 499.59. C 60.10%, H 8.27%, N 2.80%, O 28.82%. Obtained by hydrolysis of aconitine: Schulze, *Arch. Pharm.* **244,** 165 (1906); Schneider, *Ber.* **89,** 768 (1956). Structure: Schneider, Tausend, *Ann.* **628,** 114 (1959); McCaldin, Marion, *Can. J. Chem.* **37,** 1071 (1959).

Amorphous powder, mp 132°. Bitter taste. $[\alpha]_D$ +23°. pK 9.52. Absorption spectrum: Nath, *J. Indian Chem. Soc.* **32,** 75 (1955). Very sol in water, alcohol; moderately sol in chloroform; slightly sol in benzene. Practically insol in ether, petr ether.

Hydrochloride dihydrate, $C_{25}H_{42}ClNO_9.2H_2O$, crystals, mp 175-176°. $[\alpha]_D$ −8°.

Hydrobromide sesquihydrate, $C_{25}H_{42}BrNO_9.1\frac{1}{2}H_2O$, crystals from water, mp 225°.

THERAP CAT: Antipyretic.

111. Aconite. Monkshood; wolf's-bane; friar's cowl; mouse-bane. Dried tuberous root of *Aconitum napellus* L., *Ranunculaceae. Habit.* Mountainous regions of Europe, Asia, and North America. *Constit.* 0.4-0.8% aconitine; aconine; napelline (isoaconitine, pseudoaconitine); picraconitine; aconitic acid; itaconic acid; succinic acid; malonic acid; fat; levulose. *Quite toxic. Ref:* Freudenberg, *Ber.* **69,** 1962 (1936); Rogers, Freudenberg, *ibid.* **70,** 349 (1937); Lascombes, *Ann. Pharm. Franc.* **16,** 429 (1958). *See also* "Aconite," A. Ph. A. Monograph No. 1 (Am. Pharm. Assoc., Washington, D.C., 1938).

THERAP CAT: Antipyretic.

THERAP CAT (VET): Formerly used as an antihypertensive.

112. Aconitic Acid. *1-Propene-1,2,3-tricarboxylic acid;* equisetic acid; citridic acid; achilleic acid. $C_6H_6O_6$; mol wt 174.11. C 41.39%, H 3.47%, O 55.14%. Found in leaves and tubers of *Aconitum napellus* L., *Ranunculaceae,* in various species of *Achillea (Compositae)* and *Equisetum (Equisetaceae),* in beet root, and in sugar cane. Can be prepd commercially from calcium magnesium aconitate recovered from sugar cane juice: McCalip, Seibert, *Ind. Eng. Chem.* **33,** 637 (1941); from molasses: Regna, Bruins, *ibid.* **48,** 1268 (1956). Most of the commercial aconitic acid is, however, manufactured by sulfuric acid dehydration of citric acid: Bruce, *Org. Syn. coll. vol.* **II,** p 12 (1943); by using methanesulfonic acid instead of sulfuric: Cranston, U.S. pat. **2,727,066** (1955 to Daniel F. Kelly). Aconitic acid prepd by any of the above methods has the *trans*-configuration which is the form described here.

Leaflets, plates from water. Decompn 198-199° (capillary inserted in oil bath at 190°); decompn 204-205° (capillary in oil bath at 195°); decompn 209° (electrically heated bar). K_1 at 25° = 1.58 × 10⁻³; K_2 = 3.5 × 10⁻⁵. One gram dissolves in 5.5 ml water at 13°, in 2 ml water at 25°. Soluble in 2 parts of 88% alcohol at 12°. Slightly sol in ether.

Triethyl ester, bp_5 155°.
Tributyl ester, bp_3 190°.

USE: Manuf itaconic acid. As plasticizer for buna rubber and plastics. Used in form of triethyl or tributyl ester.

113. Aconitine. *(1α,3α,6α,14α,15α,16β)-20-Ethyl-1,6,-16-trimethoxy-4-(methoxymethyl)aconitane-3,8,13,14,15-pentol 8-acetate 14-benzoate.* $C_{34}H_{47}NO_{11}$; mol wt 645.72. C 63.24%, H 7.34%, N 2.17%, O 27.26%. Several isomers from *Aconitum napellus* L., *Ranunculaceae* and other aconites. Majima *et al., Ber.* **57,** 1456 (1924); *Proc. Imp. Acad. Tokyo* **5,** 415 (1929); Freudenberg, *Ber.* **69,** 1964 (1936); Swanson *et al., Aconite,* A. Ph. A. Monograph no. 1 (1938); *Methods of Analysis, A.O.A.C.,* 8th ed., 598, 651 (1955). Structure: Wiesner *et al., Coll. Czech. Chem. Commun.* **28,** 2462 (1963); Wiesner *et al., Can. J. Chem.* **47,** 2734 (1969). Stereochemistry: Bachelor *et al., Tetrahedron Letters* **1960,** no. 10, 1; Gilman, Marion, *ibid.* **1961,** 923; Tsuda, Marion, *Can. J. Chem.* **41,** 1634 (1963); Birnbaum *et al., Tetrahedron Letters* **1971,** 867. Toxicity: Dybing *et al., Acta Pharmacol. Toxicol.* **7,** 337 (1951).

Hexagonal plates, mp 204°. $[\alpha]_D$ +17.3° (chloroform). Aq soln is alkaline to litmus. pK 5.88. K 1.3 × 10⁻⁶. One gram dissolves in 2 ml chloroform, 7 ml benzene, 28 ml abs alcohol, 50 ml ether, 3300 ml water. Slightly sol in petr ether. *Poisonous!* LD_{50} in mice (mg/kg): 0.166 i.v.; 0.328 i.p.; approx 1 orally (Dybing).

Hydrobromide hemipentahydrate, $C_{34}H_{48}BrNO_{11}$·2½H_2O, hexagonal tablets from water, mp 200-207° (sinters at 160°). Amorphous form mp 115-120°. Also crystallizes from ethanol + ether with ½H_2O, mp 206-207°; $[\alpha]_D$ −30.8°. *Violent poison! Ref:* Paech, Tracey, *Modern Methods of Plant Analysis* **vol. IV** (Springer-Verlag, Berlin, 1955) p 375.

Hydrochloride hemipentahydrate, $C_{34}H_{48}ClNO_{11}$·2½H_2O, crystals, mp 149-153°. mp 194-5° (dry). $[\alpha]_D$ −30.9°. *Violent poison! Ref:* Paech, Tracey, *loc. cit.*

Nitrate, $C_{34}H_{48}N_2O_{14}$, crystals, mp about 200° (dec). $[\alpha]_D^{20}$ −35° (c = 2 in water). *Violent poison!* One gram dissolves in 10 ml boiling water. Less sol in cold water. Sol in alcohol.

USE: Used in producing heart arrhythmia in experimental animals: Boyadzhiev, *C.A.* **73,** 86256e (1970).
THERAP CAT: Antipyretic.

114. Aconitine, Amorphous. Mild aconitine. Mixture of amorphous alkaloids from *Aconitum napellus* L., *Ranunculaceae.* Contains aconitine, mesaconitine, hypaconitine, neopelline, *l*-ephedrine, sparteine, neoline and napelline. *Ref:* Rogers, Freudenberg, *Ber.* **70,** 349 (1937); *eidem, J. Am. Chem. Soc.* **59,** 2572 (1937).

Yellowish-white, amorphous powder. *Violent poison!* Insol in water; sol in alcohol, chloroform, ether, dil acids. MLD s.c. in rats: 0.175 mg/kg, *Handbook of Toxicology* **Vol. 1,** W. S. Spector, Ed. (Saunders, Philadelphia, 1956) pp 12-13.

Caution: A highly toxic alkaloid. Can be absorbed through skin.

115. Aconitum Ferox. Indian aconite; bish; visha; bishma; bikhroot. Tuber of *Aconitum ferox* Wall., *Ranunculaceae. Habit.* Nepal, Himalaya Mountains, India. *Constit.* Pseudoaconitine, *q.v.* Most powerful of all the aconites. Used by natives as arrow poison in big game hunting; in neuralgia and rheumatism. *Ref: Aconite,* A. Ph. A. Monograph no. 1, p 78 and *passim* (Am. Pharm. Assoc., Washington, D.C., 1938); N. B. Dutt, *Indian J. Pharm.* **1,** 81-84 (1939); *C.A.* **34,** 6768 (1940).

Caution: Extremely toxic. A very small dose can cause fatal cardiac depression.

116. Acranil. *1-[(6-Chloro-2-methoxy-9-acridinyl)-amino]-3-(diethylamino)-2-propanol dihydrochloride;* 5-[(γ-diethylamino-β-hydroxypropyl)amino]-3-methoxy-8-chloroacridine dihydrochloride; SKF 16214-A2; SN 186. $C_{21}H_{28}Cl_3N_3O_2$; mol wt 460.82. C 54.74%, H 6.12%, Cl 23.08%, N 9.12%, O 6.94%. Prepn: Mietzsch, Mauss, **Ger. pat. 553,072,** *Frdl.* **19,** 1167; **Brit. pat. 363,392; U.S. pat. 2,113,-357** (1930, 1930, 1938 all to I. G. Farben.); Maghidson, Grigorovski, *Ber.* **69,** 396 (1936); *J. Gen. Chem. (USSR)* **6,** 806 (1936). Antiviral activity: T. E. Gláz *et al., Antimicrob. Ag. Chemother.* **3,** 537 (1973). Interferon induction and radioprotective activity: T. E. Glaz, M. Talas, *Arch. Virol.* **48,** 375 (1975); M. Talas, E. Szolgay, *ibid.* **56,** 309 (1978).

Bitter, yellow crystals, mp 237-239° (dec). Sol in water. Free base, $C_{21}H_{26}ClN_3O_2$, yellow crystals, mp 105-107°. Sparingly sol in ether.
THERAP CAT: Antiprotozoal (Giardia).

117. Acridine. Dibenzo[b,e]pyridine; 10-azaanthracene. $C_{13}H_9N$; mol wt 179.21. C 87.12%, H 5.06%, N 7.82%. Occurs in coal tar. Isoln: Graebe, Caro, *Ann.* **158,** 265 (1871); from high boiling tar oils: Wirth, **Ger. pat. 440,771** (1926). Prepn from *N*-phenylanthranilic acid: Perkin, Clemo, **Brit. pat. 214,756** (1923); from Ca-anthranilate: Koller, Krakauer, *Monatsh.* **50,** 51 (1928); by passing benzylaniline vapor over red-hot platinum wire: Meyer, Hofmann, *ibid.* **37,** 698 (1916); *cf.* Ullmann, *Ber.* **40,** 2521 (1907). Absorption spectrum: Pinnow, *J. Prakt. Chem.* [2] **66,** 276 (1902). Toxicity: S. D. Rubbo, *Brit. J. Exp. Pathol.* **28,** 1 (1947). *Reviews:* A. Albert, *The Acridines* (St. Martin's Press, New York, 2nd ed., 1966); R. M. Acheson, *Acridines* (Interscience, New York, 2nd ed., 1973).

Orthorhombic plates, needles from diluted alcohol. There

are five crystalline forms of acridine: mp 106°, 110° (stable forms); 109°, 109.5°, > 110° (unstable forms). Begins to sublime at 100°. bp_{760} 346° (also reported as above 360°). Volatile with steam. Slightly sol in boiling water; freely sol in alcohol, ether, hydrocarbons, carbon disulfide. Dilute solns of acridine and its salts have a violet and green fluorescence, respectively. Weak base, colors litmus paper blue, forms yellow crystalline salts with mineral acids. Forms colored quaternary ammonium compounds (acridinium compounds), by the action of alkyl and aryl halides and sulfates. LD_{50} s.c. in mice: 0.40 g/kg (Rubbo).

Caution: Strongly irritating to skin, mucous membranes: *Clinical Toxicology of Commercial Products,* R. E. Gosselin *et al.,* Eds. (Williams & Wilkins, Baltimore, 5th ed., 1984) Section II, p 384.

USE: Manuf dyes and intermediates; some dyes derived from it are used as antiseptics, e.g. 9-aminoacridine, acriflavine and proflavine. The hydrochloride has been used as reagent for cobalt, iron and zinc.

118. Acriflavine. *3,6-Diamino-10-methylacridinium chloride mixt with 3,6-acridinediamine;* neutral acriflavine; euflavine; trypaflavine; neutroflavine; gonacrine. Contains 30-40% unmethylated compd. Prepn: L. Benda, *Ber.* **45,** 1787 (1912); M. Gaillot, *Quart. J. Pharm. Pharmacol.* **7,** 63 (1934); J. Marshall, *ibid.* 514. Activity studies: C. H. Browning, W. Gilmour, *J. Path. Bact.* **18,** 144 (1913); C. H. Browning *et al., Brit. Med. J.* **1,** 73 (1917); H. Berry, *Quart. J. Pharm. Pharmacol.* **14,** 149 (1941); H. McIlwain, *Biochem. J.* **35,** 1311 (1941). Toxicology: J. Ungar *et al., J. Pharmacol. Exp. Ther.* **80,** 217 (1944). *Reviews:* A. Adrien, *The Acridines* (St. Martin's Press, New York, 2nd ed., 1966); R. M. Acheson, *Acridines* (Interscience, New York, 2nd ed., 1973).

Deep orange, granular powder. One gram dissolves in about 3 ml water. Incompletely sol in alcohol; nearly insol in ether, chloroform, fixed oils. Aqueous solns are reddish orange and fluoresce on dilution. pH (1% soln) ~3.5.

Dihydrochloride, *acid acriflavine, acid trypaflavine, acriflavine hydrochloride, Panflavin.* Deep reddish-brown, crystalline powder. pH (1% soln) ~1.5.

THERAP CAT: Anti-infective.

THERAP CAT (VET): Antiseptic; locally in wounds in trichomoniasis of bulls.

119. Acrilan®. A brand of acrylic fiber, a copolymer of acrylonitrile and of a minor constituent with mildly basic character. Prepn of such copolymers: Mowry, Craig, U.S. pat. **2,744,086** and Craig, U.S. pat. **2,749,325** (both 1956 to Chemstrand). *Review:* R. W. Moncrieff, *Man-Made Fibres* (John Wiley & Sons, New York, 4th ed., 1963) pp 471-482.

Solid, d^{25} 1.17. Fiber decomposes before it melts. Sticks at 245°. Prolonged exposure in air at elevated temps causes some yellowing. Fabrics are not readily ignited and their rate of burning is less than that of cotton, viscose rayon, or acetate rayon. Fiber is practically insoluble in, and unaffected by common solvents. Has fairly good resistance to weak alkalies, very good resistance to mineral acids, sunlight, weathering. Has good dyeing properties. Its wear resistance is better than that of wool, but generally greatly inferior to that of nylon or Terylene. It is unaffected by mildew, molds, moth and carpet beetle larvae.

USE: Fabrics, sweaters, blankets, carpets.

120. Acrisorcin. *4-Hexyl-1,3-benzenediol compound with 9-acridinamine (1:1); 9-aminoacridine compound with*

4-hexylresorcinol; 9-aminoacridinium 4-hexylresorcinolate; Sch-7056; Akrinol. $C_{25}H_{28}N_2O_2$; mol wt 388.49. C 77.29%, H 7.27%, N 7.21%, O 8.24%. Prepd from 9-aminoacridine and 4-hexylresorcinol: Seneca, *Antibiot. & Chemother.* **11,** 587 (1961).

Yellow crystals.
THERAP CAT: Antifungal.

121. Acrivastine. *(E,E)-3-[6-[1-(4-Methylphenyl)-3-(1-pyrrolidinyl)-1-propenyl]-2-pyridinyl]-2-propenoic acid; (E)-6-[(E)-3-(1-pyrrolidinyl)-1-p-tolylpropenyl]-2-pyridineacrylic acid;* BW 270C; BW 825C; BW A825C; Semprex. $C_{22}H_{24}N_2O_2$; mol wt 348.44. C 75.83%, H 6.94%, N 8.04%, O 9.18%. Nonsedating type histamine H_1-receptor antagonist; analog of triprolidine, *q.v.* Prepn: G. G. Coker, J. W. A. Findlay, Eur. pat. **Appl. 85,959** (1983 to Wellcome); J. W. A. Findlay, G. G. Coker, U.S. pat. **4,501,893** (1985). Pharmacodynamics and pharmacokinetics in humans: A. F. Cohen *et al., Eur. J. Clin. Pharmacol.* **28,** 197 (1985). Evaluation of CNS effects: A. F. Cohen *et al., Clin. Pharmacol. Ther.* **38,** 381 (1985). Clinical trials in idiopathic urticaria: J. G. Gibson *et al., Dermatologica* **169,** 179 (1984); H. Neittaanmaki *et al., ibid.* **177,** 98 (1988); in allergic rhinitis: T. G. Gibbs *et al., J. Int. Med. Res.* **16,** 413 (1988).

Crystals from isopropanol, mp 222° (dec).
Combination with pseudoephedrine, *Duact.*
THERAP CAT: Antihistaminic.

122. Acrolein. *2-Propenal;* acrylic aldehyde; acrylaldehyde; acraldehyde; Aqualin. C_3H_4O; mol wt 56.06. C 64.27%, H 7.19%, O 28.54%. CH_2=CHCHO. Prepd industrially by passing glycerol vapors over magnesium sulfate heated to 330-340°. Lab prepn by heating a mixture of anhydr glycerol, acid potassium sulfate and potassium sulfate in the presence of a small amount of hydroquinone and distilling in the dark: H. Adkins, W. H. Hartung, *Org. Syn.* **coll. vol. I,** 15 (1941). Formation from glycerol by the action of *B. amaracrylus:* Voisenet, *Compt. Rend.* **188,** 941, 1271 (1929); by *B. welchii:* Humphreys, *J. Infect. Dis.* **35,** 282; *Chem. Zentr.* **1925,** II, 309. *Review:* L. G. Hess *et al.,* in Kirk-Othmer *Encyclopedia of Chemical Technology* **vol. 1** (Wiley-Interscience, New York, 3rd ed., 1978) pp 277-297.

Flammable liquid. Pungent odor. Irritates eyes and mucosa. Unstable, polymerizes (especially under light or in the presence of alkali or strong acid) forming disacryl, a plastic solid. d^0 0.8621; d^{20} 0.8389; d^{50} 0.8075. mp −88°. bp_{760} 52.5°; bp_{200} 17.5°; bp_{100} +2.5°; bp_{60} −7.5°; $bp_{1.0}$ −64.5°. n_D^{19} 1.4022. Flash pt, open cup: <0°F (−18°C). Vapor press. at 20°: 210 mm Hg. Absorption spectrum: Lüthy, *Z. Physik. Chem.* **107,** 291, 298 (1923). Sol in 2 to 3 parts water; in alcohol, ether. LD_{50} orally in rats: 0.046 g/kg, H. F. Smyth *et al., Arch. Ind. Hyg. Occup. Med.* **4,** 119 (1951).

Caution: Irritates skin, mucous membranes. Vapors cause lacrimation. A weak sensitizer; inhalation may cause asthmatic reaction. Inhalation of high concns causes pulmonary edema, cf. *Clinical Toxicology of Commercial Products,* R. E. Gosselin *et al.,* Eds. (Williams & Wilkins, Baltimore, 4th ed., 1976) Section II, p 125.

USE: Manuf colloidal forms of metals; making plastics, perfumes; warning agent in methyl chloride refrigerant. Has been used in military poison gas mixtures. Used in organic syntheses. Aquatic herbicide.

123. Acrylamide. *2-Propenamide.* C_3H_5NO; mol wt 71.08. C 50.69%, H 7.09%, N 19.71%, O 22.51%. $CH_2=CHCONH_2$. Prepd from acrylonitrile by treatment with H_2SO_4 or HCl: Bayer, *Angew. Chem.* **61**, 240 (1949); Weisgerber, U.S. pat. **2,535,245** (1950 to Hercules). *Reviews:* Carpenter, Davis, *J. Appl. Chem. (London)* **7**, 671 (1957); D. C. MacWilliams in Kirk-Othmer, *Encyclopedia of Chemical Technology* **vol. 1** (Wiley-Interscience, New York, 3rd ed., 1978) pp 298-311. Toxicity: R. E. Peterson, N. K. Sheth, *Toxicol. Appl. Pharmacol.* **33**, 142 (1975).

Monomer, flake-like crystals from benzene. d_4^{30} 1.122. mp 84.5°. bp_2 87°; bp_5 103°; bp_{25} 125°. Solubilities in g/100 ml solvent at 30°: water 215.5; methanol 155; ethanol 86.2; acetone 63.1; ethyl acetate 12.6; chloroform 2.66; benzene 0.346; heptane 0.0068. The solid may be stored in a cool, dark place. Readily polymerizes at the mp or under uv light. Commercial solns of the monomer may be stabilized with hydroquinone, *tert*-butylpyrocatechol, *N*-phenyl-2-naphthylamine or other antioxidants. LD_{50} i.p. in mice: 170 mg/kg (Peterson, Sheth).

Polymer, various forms, sol and insol in water, are obtained by heating with various polymerization catalysts: C. E. Schildknecht, *Vinyl and Related Polymers* (Wiley, New York, 1952) pp 314-322.

Caution: Highly toxic and irritant. Causes CNS paralysis. Can be absorbed through unbroken skin.

124. Acrylic Acid. *2-Propenoic acid;* vinylformic acid. $C_3H_4O_2$; mol wt 72.06. C 50.00%, H 5.60%, O 44.40%. $CH_2=CHCO_2H$. Prepd by hydrolysis of acrylonitrile: Kaszuba, *J. Am. Chem. Soc.* **67**, 1227 (1945) or by oxidation of acrolein: U.S. pats. **1,911,219** (1933 to Rohm & Haas); **2,288,566** (1942 to Acrolein Corp.); **2,341,339** (1944 to Distillers). Various other syntheses, *see Org. Syn.* coll. vol. III, 30-34 (1955). *Review:* J. W. Nemec, W. Bauer in Kirk-Othmer *Encyclopedia of Chemical Technology* **vol. 1** (Wiley-Interscience, New York, 3rd ed., 1978) pp 330-354.

Corrosive liquid; acrid odor and fumes. d_4^{16} 1.0621. mp 14°. bp 141.0°; bp_{400} 122.0°; bp_{200} 103.3°; bp_{100} 86.1°; bp_{40} 66.2°; bp_{10} 39.0°; bp_5 27.3°. n_D^{20} 1.4224. Flash pt, open cup: 155°F (68°C). K at 25° = 5.6 × 10^{-5}. Miscible with water, alc, ether. Polymerizes readily in the presence of oxygen. LD_{50} orally in rats: 2.59 g/kg, H. F. Smyth *et al.*, *Am. Ind. Hyg. Assoc. J.* **23**, 95 (1962).

USE: In the manuf of plastics. *Caution:* Strong irritant.

125. Acrylonitrile. *2-Propenenitrile;* vinyl cyanide; cyanoethylene; Acritet; Fumigrain; Ventox. C_3H_3N; mol wt 53.06. C 67.90%, H 5.70%, N 26.40%. $CH_2=CHCN$. Prepn by dehydration of ethylene cyanohydrin or acrylamide: Moureu, *Ann. Chem. Phys.* [7] **2**, 186 (1893). Manuf by ammoxidation of propylene: Faith, Keyes & Clark's *Industrial Chemicals*, F. A. Lowenheim, M. K. Moran, Eds. (Wiley-Interscience, New York, 4th ed., 1975) pp 46-49. Toxicity: H. F. Smyth, C. P. Carpenter, *J. Ind. Hyg. Toxicol.* **30**, 63 (1948). Causes acute and chronic adrenocortical insufficiency: S. Szabo *et al.*, *Lab. Invest.* **42**, 533 (1980); eidem, *J. Appl. Toxicol.* **4**, 131 (1984). Review of carcinogenicity studies: *IARC Monographs* **19**, 73-133 (1979). Comprehensive review and bibliography: *The Chemistry of Acrylonitrile* (Am. Cyanamid, New York, 2nd ed., 1959) 272 pp; L. T. Groet in Kirk-Othmer *Encyclopedia of Chemical Technology* **vol. 1** (Wiley-Interscience, New York, 3rd ed., 1978) pp 414-426.

Explosive, flammable and toxic liquid. Should be stored and used in closed systems whenever possible. Work areas should be adequately ventilated, and should be free from open lights, flames, and equipment that is not explosion-proof. *Handle in hood.* May polymerize spontaneously, particularly in the absence of oxygen or on exposure to visible light. Polymerizes violently in the presence of concentrated alkali. On standing may slowly develop a yellow color particularly after excessive exposure to light. bp_{760} 77.3°; bp_{500} 64.7°; bp_{250} 45.5°; bp_{100} 23.6°; bp_{50} 8.7°. mp −83.55°. d_4^{20} 0.8060; d_4^{25} 0.8004. n_D^{25} 1.3888. Flash pt, open cup: 32°F (0°C). Explosive mixtures in air at 25°: 3.05% low limit; 17.0% upper limit. At 20° 7.35 parts dissolve in 100 parts water and 3.1 parts water dissolve in 100 parts acrylonitrile. Miscible with most organic solvents. LD_{50} orally in rats: 0.093 g/kg (Smyth, Carpenter).

Caution: Highly toxic through cyanide effect; irritating to eyes and skin: *Clinical Toxicology of Commercial Products*, R. E. Gosselin *et al.*, Eds. (Williams & Wilkins, Baltimore, 5th ed., 1984) Section II, p 215. This substance may reasonably be anticipated to be a carcinogen: *Fourth Annual Report on Carcinogens* (NTP 85-002, 1985) p 15.

USE: Manufacture of acrylic fibers. In the plastics, surface coatings, and adhesives industries. As a chemical intermediate in the synthesis of antioxidants, pharmaceuticals, dyes, surface-active agents, etc. In organic synthesis to introduce a cyanoethyl group. As a modifier for natural polymers. As a pesticide fumigant for stored grain. Experimentally to induce adrenal hemorrhagic necrosis in rats.

126. Actaplanins. A-4696; Kamoran. Complex of glycopeptide antibiotics produced by *Actinoplanes missouriensis.* Six actaplanins (A, B_1, B_2, B_3, C_1, G) have been isolated and characterized as having a central peptide core with the amino sugar *ristosamine* and up to four neutral sugars attached. Isoln: R. L. Hamill *et al.*, *Ger.* pat. **2,209,018** (1972 to Lilly), *C.A.* **77**, 138338n (1972); A. P. Raun, U.S. pat. **3,816,618**; R. L. Hamill *et al.*, U.S. pat. **4,115,552** (1974, 1978 to Lilly). Chemical characterization: M. Debono *et al.*, *J. Antibiot.* **37**, 85 (1984). ^{1}H NMR studies and structures: A. H. Hunt *et al.*, *J. Org. Chem.* **49**, 635 (1984); eidem, ibid. 641. Growth promotant activity: C. Tsaltos *et al.*, *Bull. Hellenic Vet. Med. Soc.* **33**, 139 (1982). Use to increase milk production in ruminants: C. C. Scheifinger, **Eur.** pat. **Appl. 63,491** corresp to U.S. pat. **4,430,328** (1982, 1984 to Eli Lilly). Determn in milk: K. H. Hahne *et al.*, *Milchwissenschaft* **39**, 473 (1984).

Actaplanin		R_1	R_2
	A :	mannosylglucose	mannose
	B_1 :	rhamnosylglucose	mannose
	B_2 :	glucose	mannose
	B_3 :	mannosylglucose	H
	C_1 :	rhamnosylglucose	H
	G :	glucose	H

Hydrochloride, white cryst solid, mp > 220°. Approx mol wt 1158. $[\alpha]_D^{25}$ −42.3° (c = 1 in water). uv max (acidic and neutral solns): 276 nm ($E_{1cm}^{1\%}$ 65). Sol in water. Insol in most organic solvents. Stable over pH 1.0 to 10.0 up to 27°.

THERAP CAT (VET): Growth stimulant.

127. ACTH. *Corticotropin;* adrenocorticotrop(h)in; corticotrophin; adrenocorticotrop(h)ic hormone of the pituitary gland; Acethropan; Acortan; Acorto; Acthar; Acton; Actonar; Adrenomone; Alfatrofin; Cibacthen; Corstiline; Cortiphyson; Cortrophin; Isactid; Reacthin; Solacthyl; Tubex. Pituitary hormone which stimulates the secretion of adrenal cortical steroids and induces growth of the adrenal cortex. Occurs also in female human urine and in serum of pregnant mares. Isoln procedure from swine pituitaries: Sayers *et al.*, *J Biol. Chem.* **149**, 425 (1943); *Proc. Soc. Exp. Biol. Med.* **52**, 199 (1943); from sheep pituitaries: Li *et al.*, *J. Biol. Chem.* **149**, 413 (1943); Li, *J. Am. Chem. Soc.* **74**, 2124 (1952); from human pituitaries: Pickering *et al.*, *Biochim. Biophys. Acta* **74**, 763 (1963). Purification: Johnson, U.S.

pat. **3,124,509** (1964 to Upjohn). Corticotropin is a single chain polypeptide containing 39 amino acids. The first 24 residues are identical in all species. *In vivo* studies show that this portion of the peptide chain is responsible for the biological activity and that the remaining residues, while not necessary for the hormonal action, are essential in the species immunological specificity of ACTH. The term α-ACTH is used to distinguish it from the pepsin- or acid-degraded product, β-ACTH. Prepn of active cleavage products by hydrolysis and simultaneous dialysis: Wettstein, Benz, U.S. pat. **2,734,015** (1956 to Ciba). Amino acid sequence of bovine ACTH: Li *et al.*, *J. Am. Chem. Soc.* **80,** 2587 (1958); proposed structure of human ACTH: Lee *et al.*, *J. Biol. Chem.* **236,** 2970 (1961). Revised amino acid sequences of porcine and human ACTH: Riniker *et al.*, *Nature New Biol.* **235,** 114 (1972). Revised sequences of ovine and bovine ACTH: Li, *Biochem. Biophys. Res. Commun.* **49,** 835 (1972). Synthesis of the revised human corticotropin: Sieber *et al.*, *Helv. Chim. Acta* **55,** 1243 (1972); solid phase synthesis: Yamashiro, Li, *J. Am. Chem. Soc.* **95,** 1310 (1973).

White powder. Freely sol in water. Partly precipitated at the isoelectric point (pH 4.65-4.80). Appreciably soluble in 60 to 70% alcohol or acetone. Almost completely precipitated in 2.5% trichloroacetic acid soln. Also precipitated from dilute soln by 20% sulfosalicylic acid and by 5% lead acetate soln. Solns are stable to heat. An aq soln buffered to pH 7.5 may be put in a boiling water bath for at least 120 minutes. In 0.10 molar HCl an 0.2% soln retains its biological potency when kept at 100° for 60 min, but in 0.10 molar NaOH the activity is lost within 30 min. At 60° at 0.2% soln at pH 10.8 maintains its potency for 60 min. In general the substance is more stable in acid soln. Pure ACTH has 150 to 200 potency units per mg. The biological activity is not destroyed by digestion with pepsin, and hydrolysis products have the biological activity of ACTH. One U.S.P. unit or one international unit or one Armour unit or one potency unit denotes the same activity.

Corticotrophin zinc hydroxide suspension, Cortrophin-Z. ACTH absorbed on zinc hydroxide. Used as an aqueous suspension.

THERAP CAT: Hormone.

THERAP CAT (VET): Stimulates glucocorticoid production.

128. Actin. One of the major proteins of muscle and an important component of all eukaryotic cells. Formerly believed to be a single, highly conserved protein in all cell types, but multiple forms have been shown to exist. Muscle actin, or α-*actin*, is found in differentiated muscle cells; β-*actin* and γ-*actin* are present in all non-muscle cell types. All three forms contain equimolar amounts of N-methylhistidine, although α-actin differs from β- and γ-actins by several peptides; the two "non-muscle" actins are nearly identical. Several "minor actins" are also known. The various forms, although similar in activity, mol wt and amino acid composition, show immunological differences and are considered to be synthesized under the control of different genes. In addition to its role in muscle relaxation and contraction (together with myosin, *q.v.*), actin is involved in a variety of cellular events: cell movement, cytokinesis, phagocytosis, exocytosis and chromosome movement. Identification in muscle: F. B. Straub, *Stud. Szeged* **2,** 3 (1942). Isoln: Feuer *et al.*, *Hung. Acta Physiol.* **1,** 150 (1948). Separation of actin from muscle material: A. Szent-Gyorgyi, U.S. pat. **2,742,456** (1956 to Armour). In soln, actin and myosin combine to give *actomyosin:* Tonomura *et al.*, *J. Biol. Chem.* **237,** 1074 (1962). Actin, in the absence of salts, exists in a globular form, designated as *G-actin;* in the presence of ATP and potassium, sodium, or magnesium chlorides it polymerizes to a fibrous form, *F-actin.* Removal of bound ATP results in complete loss of polymerizability: Straub, Feuer, *Biochim. Biophys. Acta* **4,** 455 (1950). Structure of G-actin: Nagy, Jencks, *Biochemistry* **1,** 987 (1962); of F-actin: Hanson, Lowy, *J. Mol. Biol.* **6,** 46 (1963). Complete amino acid sequence of rabbit skeletal muscle actin: Elzinga *et al.*, *Proc. Nat. Acad. Sci. USA* **70,** 2687 (1973). Identification and characterization of multiple forms of actin: J. I. Garrels, W. Gibson, *Cell* **9,** 793 (1976). Partial amino acid sequence of calf brain actin: R. C. Lu, M. Elzinga, *Biochemistry* **16,** 5801 (1977). Comparison of actins

from calf thymus, bovine brain, mouse cells and rabbit skeletal muscle: J. Vandeker, K. Weber, *Eur. J. Biochem.* **90,** 451 (1978). Isoln, characterization of porcine brain actin: J. P. Weir, D. W. Frederik, *Arch. Biochem.* **203,** 1 (1980). *In vivo* and *in vitro* synthesis of multiple forms of rat brain actin: E. Palmer, J. L. Saborio, *J. Biol. Chem.* **253,** 7482 (1978). Evidence for control of synthesis of human heart and platelet actins by different genes: M. Elzinga *et al.*, *Science* **191,** 94 (1976). Immunological differences between cardiac muscle, skeletal muscle and brain actins: J. L. Morgan *et al.*, *Proc. Nat. Acad. Sci. USA* **77,** 2069 (1980). History of discovery: Finck, *Science* **160,** 332 (1968). *Reviews:* Several authors, *Biochemistry of Muscle Contraction,* J. Gergely, Ed. (Little, Brown, Boston, 1964); Laki, "Actin" in *Contractile Proteins and Muscle,* K. Laki, Ed. (Marcel Dekker, New York, 1971) pp 97-133; E. D. Korn, *Proc. Nat. Acad. Sci. USA* **75,** 588-599 (1978); R. Kassab *et al.*, *Biochimie* **63,** 273-289 (1981); S. Highsmith, *Biochem. Biophys. Acta* **639,** 31-39 (1981); E. D. Korn, *Physiol. Rev.* **62,** 672-737 (1982).

129. Actinium. Ac; at. wt 227.0278; at. no. 89. Principal isotope 227. A trivalent element homologous with lanthanum. Discovered by Debierne in 1899. $T_{1/2}$ 21.6 years; emits soft β-rays and about 1.2% α-rays. Originally extracted from uranium ores, now prepd by neutron bombardment of radium. A second isotope isolated from natural sources is ^{228}Ac or *mesothorium II.* (MsTh^2), $T_{1/2}$ 6.13 hrs; decays by β-emission to ^{228}Th; produced by β-decay of ^{228}Ra (mesothorium I). Known isotopes (mass numbers): 221-231. Review of actinium and the actinides: *Comprehensive Inorganic Chemistry* vol. 5, J. C. Bailar, Jr. *et al.*, Eds. (Pergamon Press, Oxford, 1973) *passim.*

Cubic crystal, d 10.07. mp 1050°. (bp *ca.* 3300°). Actinium is more basic than lanthanum and is a strong electropositive element. Forms insol salts, e.g. carbonate, oxalate, etc. just as the lanthanides.

Caution: Radiation hazard.

130. Actinobolin. *[3R-[3α,4α(S*),4aβ,5β,6α]]-2-Amino-N-(3,4,4a,5,6,7-hexahydro-5,6,8-trihydroxy-3-methyl-1-oxo-1H-2-benzopyran-4-yl)propanamide; 4-(2-amino-propionamido)-3,4,4a,5,6,7-hexahydro-5,6,8-trihydroxy-3-methylisocoumarin.* $C_{13}H_{20}N_2O_6$; mol wt 300.31. C 51.99%, H 6.71%, N 9.33%, O 31.97%. Antibiotic produced by *Streptomyces griseoviridus* var *atrofaciens;* (+)-form is naturally occurring. Isoln and characterization: T. H. Haskell, Q. R. Bartz, *Antibiot. Ann.* **1958-1959,** 505; T. H. Haskell *et al.*, U.S. pat. **3,043,830** (1962 to Parke, Davis). Structural studies: Struck *et al.*, *Tetrahedron Letters* **1967,** 1589; Munk *et al.*, *J. Am. Chem. Soc.* **89,** 4158 (1967). Structure: *eidem, ibid.* **90,** 1087 (1968). Absolute configuration and chemistry: Antosz *et al.*, *ibid.* **92,** 4933 (1970). Crystal structure: J. B. Wetherington, J. W. Moncrief, *Acta Crystallogr.* **B31,** 501 (1975). Total synthesis: M. Yoshioka *et al.*, *J. Am. Chem. Soc.* **106,** 1133 (1984); *eidem, Heterocycles* **21,** 151 (1984); R. S. Garigipati *et al.*, *J. Am. Chem. Soc.* **107,** 7790 (1985). Antibacterial activity and toxicity: R. F. Pittillo *et al.*, *Antibiot. Ann.* **1958-1959,** 497. Series of articles on antineoplastic activity: *ibid.* 515-532. Inhibition of protein synthesis: D. Smithers *et al.*, *Mol. Pharmacol.* **5,** 433 (1969). Experimental use as cariostat: D. E. Hunt *et al.*, *J. Dent. Res.* **50,** 371 (1971). Reduction of periodontal syndrome in mice, rats: J. H. Shaw, J. K. Ivimey, *Arch. Oral Biol.* **18,** 357 (1973).

Amorphous, fluffy, very hygroscopic powder. $[\alpha]_D^{28}$ +59° (c = 0.5 in pH 7 phosphate buffer). pKa: 7.5, 8.8. Freely sol in water; mod sol in methanol, ethanol. Unstable in aqueous solns pH >7. uv max: 262 nm (0.1N HCl); 288 nm (0.1N NaOH). LD$_{50}$ in mice, rats (mg/kg): 800 ±27, 1550 ±26 i.v. (Pittillo).

Acetate, C$_{15}$H$_{24}$N$_2$O$_8$, needles from ethanol; softens at 130°, resolidifies at 145°, melts at 263-266°. [α]$_D^{26}$ +58°. pKa: 4.6, 7.5, 8.8. Very sol in water; sol in warm methanol, ethanol, acetone; sparingly sol in ethyl acetate. uv max: 264 nm (pH 7 phosphate buffer).

Hydrochloride, C$_{13}$H$_{21}$ClN$_2$O$_6$, hygroscopic. [α]$_D^{22}$ +59° (c = 0.41) (natural); [α]$_D^{22}$ +55° (c = 0.47) (synthetic).

131. Actinodaphnine. *6,7,7a,8-Tetrahydro-11-methoxy-5H-benzo[g]-1,3-benzodioxolo[6,5,4-de]quinolin-10-ol; 10-methoxy-1,2-(methylenedioxy)-6aα-noraporphin-9-ol; 1,2-methylenedioxy-9-hydroxy-10-methoxynoraporphine.* C$_{18}$H$_{17}$NO$_4$; mol wt 311.32. C 69.44%, H 5.50%, N 4.50%, O 20.56%. In bark of *Actinodaphne hookeri* Meissn., *Lauraceae.* Isoln: Krishna, Ghose, *J. Indian Chem. Soc.* **9**, 429 (1932). Structure: Ghose *et al.*, *Helv. Chim. Acta* **17**, 919 (1934). Photolytic synthesis: M. S. Premila *et al.*, *Indian J. Chem.* **13**, 945 (1975).

Needles from alcohol, mp 211°. [α]$_D^{20}$ +33° (ethanol). Sol in acetone, alcohol, benzene, chloroform. Moderately sol in ether. Practically insol in water.

Hydrochloride, C$_{18}$H$_{17}$NO$_4$.HCl, needles from alcohol-ether, mp 281° (dec). [α]$_D^{20}$ +9°.

Methyl ether, C$_{19}$H$_{19}$NO$_4$, needles from ether + alc, mp 114-115°. Synthesis: Hey, Lobo, *J. Chem. Soc.* **1954**, 2246.

132. Actinomycetin. Bacteriolytic cell-free fluid of culture filtrates of actinomycetes. Produced by most species of *Streptomyces;* studied mostly in *Streptomyces albus.* Protein-like in nature. Stated to consist of a bactericidal fatty acid fraction and an enzyme: Welsch, *J. Bacteriol.* **42**, 801 (1941); **44**, 571 (1942); **53**, 101 (1947). Dissolves dead gram-negative, and, with more difficulty, dead gram-positive organisms. Also dissolves living organisms in aq suspensions, such as *Staph. aureus.* In answer to the question whether actinomycetin deserves the name "antibiotic", *see* the polemic between Hoogerheide and Welsch, *Bot. Rev.* **10**, 599 (1944) and *J. Bacteriol.* **53**, 101 (1947). Purification and prepn from *Streptomyces albus:* Ghuysen, *Belg. pats.* **517,-191** and **521,114** (both 1953 to Soc. Belge de l'Azote), *C.A.* **53**, 5600g, 12593e (1959). *Review:* Caltrider in *Antibiotics* **vol. 1**, D. Gottlieb, P. D. Shaw, Eds. (Springer Verlag, New York, 1967) pp 681-683.

Sol in water. Pptd by alcohol, acetone, ammonium sulfate and other protein precipitants. Destroyed by strong acid. More stable at pH 10 than at pH 4. Rapidly inactivated by heat, 60-70° destroys activity.

133. Actinomycin F$_1$. Actinomycin KS4; KS4. An antibiotic produced by cultures of actinomycin C-elaborating strains of Streptomyces, such as Streptomyces BOP 476 (NRRL 2580) and *Streptomyces chrysomallus* (NRRL 2250). Considered to be a derivative of actinomycin C$_2$; amino acid analysis indicates that the prolines present in actinomycin C$_2$ are substituted by sarcosine in actinomycin F$_1$. Production by controlled biosynthesis: Schmidt-Kastner, *Naturwiss.* **43**, 131 (1956); *idem, Ann. N.Y. Acad. Sci.* **89**, 299 (1960); Schmidt-Kastner, Hackmann, U.S. pat. **3,219,544** (1965 to Bayer).

THERAP CAT: Antineoplastic.

134. Actinoquinol. *8-Ethoxy-5-quinolinesulfonic acid.* C$_{11}$H$_{11}$NO$_4$S; mol wt 253.29. C 52.16%, H 4.38%, N 5.53%, O 25.28%, S 12.66%. Prepn: Orlov, Bogdanov, *J. Appl. Chem. USSR* **5**, 803 (1932), *C.A.* **27**, 724^8 (1933); Ghosh, Roy, *J. Indian Chem. Soc.* **22**, 39 (1945); Dorier *et al., Trav.*

Soc. Pharm. Montpellier **14**(3), 152 (1954), *C.A.* **50**, 1011a (1956).

Brown needles from water, mp 286-288° (dec). Sol in dil NaHCO$_3$.

Sodium salt, C$_{11}$H$_{10}$NNaO$_4$S, *sodium etoquinol, etoquinol sodium, Corodenin, Uviban.*

THERAP CAT: Sodium salt as ultraviolet screen.

135. Actinorhodine. C$_{32}$H$_{26}$O$_{14}$; mol wt 634.56. C 60.57%, H 4.13%, O 35.30%. Antibiotic pigment produced by *Streptomyces coelicolor*, found in woods near Göttingen, Germany: Brockmann, Pini, *Naturwiss.* **34**, 190 (1947); Brockmann *et al., Ber.* **83**, 161 (1950). Tentative structure: Brockmann, Hieronymus, *Ber.* **88**, 1379 (1955); Brockmann *et al., Naturwiss.* **49**, 131 (1962). Structure: Brockmann, *Angew. Chem.* **76**, 863 (1964); Brockmann *et al., Ann.* **698**, 209 (1966). Stereochemistry: Zeeck, Christiansen, *Ann.* **724**, 172 (1969). Biosynthesis and determn of point of dimerization: C. P. Gorst-Allman *et al., J. Org. Chem.* **46**, 455 (1981).

Fine red needles from dioxane, dec 270°. Absorption max (dioxane): 560, 523 nm. Sol in pyridine, piperidine, tetrahydrofuran, dioxane, phenol; slightly sol in alcohol, acetic acid, acetone. Red soln with acetone, blue soln with pyridine. Practically insol in aq acid; sol in aq alkali with bright blue color.

136. Actiphenol. *4-[2-(2-Hydroxy-3,5-dimethylphenyl)-2-oxoethyl]-2,6-piperidinedione; 3-(3-hydroxy-3,5-dimethylphenacyl)glutarimide; β-(3,5-dimethyl-2-hydroxybenzoylmethyl)glutarimide; C-73.* C$_{15}$H$_{17}$NO$_4$; mol wt 275.29. C 65.44%, H 6.22%, N 5.09%, O 23.25%. Metabolic product found in culture filtrates of cycloheximide-producing actinomycetes *(Streptomyces albulus):* Highet, Prelog, *Helv. Chim. Acta* **42**, 1523 (1959); Rao, *J. Org. Chem.* **25**, 661 (1960). Synthesis: Johnson, *ibid.* **27**, 3658 (1962). May be prepd from cycloheximide: Highet, Prelog, *loc. cit.* Structure-activity studies: Ennis, *Biochem. Pharmacol.* **17**, 1197 (1968).

Fluffy needles from methylene chloride + methanol. mp 200-202°. Pure crystals are white or colorless. uv max: 262, 345 nm (ϵ 10,870, 4550). Exhibits bright yellow fluorescence under u.v. Soluble in aqueous NaOH (forming bright yellow solns), in anhydr alcohol, tetrahydrofuran. Insol in aq solns of sodium bicarbonate.

Acetate, C$_{17}$H$_{19}$NO$_5$, crystals from methylene chloride + ether, mp 155.5°.

137. Activins. Polypeptide hormone, identified in ovarian follicular fluid, which selectively stimulates secretion of FSH, *q.v.* Dimer composed of β subunits of inhibin, *q.v.* (either β_A or β_B), mol wt 24,000 daltons. Isolation and characterization of the heterodimer, *activin AB*: N. Ling *et al.*, *Nature* **321**, 779 (1986); of the homodimer, *activin A*: W. Vale *et al.*, *ibid.* 776; N. Ling *et al.*, *Biochem. Biophys. Res. Commun.* **138**, 1129 (1986). Comparison of FSH stimulation with gonadotropin releasing hormone (LH-RH, *q.v.*): R. H. Schwall *et al.*, *ibid.* **151**, 1099 (1988). Approach to nonradiometric assays: R. H. Schwall *et al.*, *Prog. Clin. Biol. Res.* **285**, 205 (1988). Brief review of chemical properties and biological activity: S.-Y. Ying, *Proc. Soc. Exp. Biol. Med.* **186**, 253-264 (1987).

138. ACV. (*S*)-*N*-[*N*-(*5-Amino-5-carboxy-1-oxopentyl*)-L-*cysteinyl*]-D-*valine*; *N*-[*N*-(L-*5-amino-5-carboxyvaleryl*)-L-*cysteinyl*]-L-*cysteinyl*-D-valine. $C_{14}H_{25}N_3O_6S$; mol wt 363.43. C 46.27%, H 6.93%, N 11.56%, O 26.41%, S 8.82%. Biosynthetic precursor of penicillins and cephalosporins. Isoln from mycelial extracts of *Penicillium chrysogenum*: H. R. V. Arnstein *et al.*, *Biochem. J.* **76**, 353 (1960); structure determn of tripeptide: H. R. V. Arnstein, D. Morris, *ibid.* 357. Isoln from *Cephalosporium* sp., identification and optical configuration of constituent amino acids: P. B. Loder, E. P. Abraham, *ibid.* **123**, 471 (1971). Configuration of peptide from *P. chrysogenum*, identity with peptide from *Cephalosporium*: P. Adriaens *et al.*, *Antimicrob. Ag. Chemother.* **8**, 638 (1975). Synthesis of labelled tripeptide and role in penicillin biosynthesis: P. A. Fawcett *et al.*, *Biochem. J.* **157**, 651 (1976). Synthesis of ACV as the disulfide: S. Wolfe, M. G. Jokinen, *Can. J. Chem.* **57**, 1388 (1979). Cellfree conversion of ACV directly into isopenicillin N: J. O'Sullivan *et al.*, *Biochem. J.* **184**, 421 (1979); T. Konomi *et al.*, *ibid.* 427. Confirmation by ^{13}C NMR: J. E. Baldwin *et al.*, *Chem. Commun.* **1980**, 1271. *Reviews*: D. J. Aberhart, *Tetrahedron* **33**, 1545-1559 (1977); E. P. Abraham, *Japan. J. Antibiot.* **30**, Suppl., S1-S26 (1977); S. Wolfe *et al.*, *Science* **226**, 1386-1392 (1984).

139. Acyclovir. *2-Amino-1,9-dihydro-9-[(2-hydroxyethoxy)methyl]-6H-purin-6-one*; *acycloguanosine*; *9-[(2-hydroxyethoxy)methyl]guanine*; BW 248U; Wellcome-248U; Aclovir; Vipral; Virorax; Zovirax. $C_8H_{11}N_5O_3$; mol wt 225.21. C 42.66%, H 4.92%, N 31.10%, O 21.31%. Orally active acyclic nucleoside with inhibitory activity towards several herpes viruses. Prepn: H. J. Schaeffer, *Ger. pat.* **2,539,963**; *idem*, U.S. pat. **4,199,574** (1974, 1980 to Wellcome). Convenient synthesis from guanine: H. Matsumoto *et al.*, *Chem. Pharm. Bull.* **36**, 1153 (1988). Selectivity of action: G. B. Elion *et al.*, *Proc. Nat. Acad. Sci. USA* **74**, 5716 (1977). Chemistry, antiviral activity, metabolism: H. J. Schaeffer *et al.*, *Nature* **272**, 583 (1978). *In vitro* activity: P. Collins, D. J. Bauer, *J. Antimicrob. Chemother.* **5**, 431 (1979). Effect on herpes simplex infections in mice: H. J. Field *et al.*, *Antimicrob. Ag. Chemother.* **15**, 554 (1979); on herpes zoster in immunocompromised patients: H. H. Balfour *et al.*, *N. Engl. J. Med.* **308**, 1448 (1983). Treatment of primary episodes of genital herpes simplex infection: Y. J. Bryson *et al.*, *ibid.* 916; of recurrent genital herpes: S. E. Straus *et al.*, *ibid.* **310**, 1545 (1984); J. M. Douglas *et al.*, *ibid.* 1551. Symposia on pharmacology and clinical studies: *Am. J. Med.* **73**, Suppl. 1A, 1-392 (1982); *J. Antimicrob. Chemother.* **12**, Suppl. B, 1-202 (1983); *Scand. J. Infect. Dis.* **Suppl. 47**, 1-176 (1985).

Crystals from methanol, mp 256.5-257°. LD_{50} in mice (mg/kg): > 10,000 orally; 1000 i.p. (Schaeffer).

THERAP CAT: Antiviral.

140. Adamantane. *Tricyclo[3.3.1.1³,⁷]decane*; *diamantane* (obsolete). $C_{10}H_{16}$; mol wt 136.23. C 88.16%, H 11.84%. Isoln from petroleum from Moravia province in Czechoslovakia: Landa, Machacek, *Coll. Czech. Chem. Commun.* **3**, 1 (1933); from American petroleum: Mair *et al.*, *Anal. Chem.* **31**, 2082 (1959). Synthesis: Prelog, Seiwerth, *Ber.* **74**, 1769 (1941); Stetter *et al.*, *Ber.* **89**, 1922 (1956); by aluminum chloride-catalyzed isomerization of tetrahydrodicyclopentadiene: Yan, Shreh, *Bull. Inst. Chem. Acad. Sinica*, **Dec. 1965**, no. 11, pp 79-81. Two step synthesis starting with dicyclopentadiene: Schleyer, *J. Am. Chem. Soc.* **79**, 3292 (1957); Schleyer, Donaldson, *ibid.* **82**, 4645 (1960); Ludwig U.S. pat. **2,937,211** (1960 to du Pont). *Reviews*: Stetter, *Angew. Chem.* **66**, 217 (1954); Fort, Schleyer, *Chem. Revs.* **64**, 277 (1964).

Crystallizes at −30°; can be purified by recrystallization from acetone or by sublimation. mp 269.6-270.8°.
Note: The name diamantane has been abandoned as a synonym for adamantane and proposed as the name for the second member of the adamantane series, congressane: Vogl, Anderson, *Tetrahedron Letters* **1966**, 415.
USE: The diamine as curing agent for epoxy resins [U.S. pat. **3,053,907** (1962 to du Pont)].

141. Adenine. *1H-Purin-6-amine*; 6-aminopurine; 6-amino-1*H*-purine; 6-amino-3*H*-purine; 6-amino-9*H*-purine; 1,6-dihydro-6-iminopurine; 3,6-dihydro-6-iminopurine; Leuco-4. $C_5H_5N_5$; mol wt 135.14. C 44.45%, H 3.73%, N 51.82%. Also referred to as *vitamin B₄*: Lecoq, *Int. Z. Vitaminforsch.* **27**, 291 (1957). Widespread throughout animal and plant tissues combined with niacinamide, D-ribose, and phosphoric acids; a constituent of nucleic acids and coenzymes, such as codehydrase I and II, adenylic acid, coalaninedehydrase. Isoln from bovine pancreas: Kossel, *Ber.* **18**, 79, 1928 (1885). Syntheses: Fischer, *ibid.* **30**, 2226 (1897); Traube, *Ann.* **331**, 64 (1904); Hoffer, *Jubilee Vol. Emil Barell* **1946**, 428-434; Taylor *et al.*, *Ciba Foundation Symposium, Chem. and Biol. Purines* **1957**, 20, *C.A.* **53**, 6238b (1959); Bredereck *et al.*, *Angew. Chem.* **71**, 524 (1959); Morita *et al.*, *Chem. & Ind. (London)* **1968**, 1117; Sekiya, Suzuki, *Chem. Pharm. Bull.* **20**, 209 (1972); N. J. Kos *et al.*, *J. Org. Chem.* **44**, 3140 (1979). *Review*: Ts'o, "Bases, Nucleosides and Nucleotides" in *Basic Principles in Nucleic Acid Chemistry* vol. 1, P. O. P. Ts'o, Ed. (Academic Press, New York, 1974) pp 453-584. *See also* Nucleic Acids.

Trihydrate, orthorhombic needles. Anhydr at 110°, dec 360-365°, subl 220°. uv max (pH 7.0): 207, 260.5 nm ($\epsilon \times 10^{-3}$ 23.2, 13.4). One gram of anhydr compd dissolves in 2000 ml water, 40 ml boiling water; slightly sol in alc; practically insol in ether, $CHCl_3$. Aq solns are neutral. Combines with acids and bases. LD_{50} orally in rats: 745 mg/kg, Philips *et al.*, *J. Pharmacol. Exp. Ther.* **104**, 20 (1952).
Hydrochloride hemihydrate, monoclinic prisms. One gram dissolves in 42 ml water.
Sulfate dihydrate, crystals. One gram dissolves in 150 ml water; slightly sol in alc.
USE: In microbial determination of niacin; in research on heredity, virus diseases, and cancer.

142. A-Denopterin. *N-[4-[[1-(2,4-Diamino-6-pteridinyl)ethyl]methylamino]benzoyl]-L-glutamic acid*; *4-amino-*

N^{10},9-dimethylpteroylglutamic acid. $C_{21}H_{24}N_8O_5$; mol wt 468.48. C 53.84%, H 5.16%, N 23.92%, O 17.08%. Prepn: Hultquist *et al., J. Am. Chem. Soc.* **71**, 619 (1949); Hultquist, Smith, **Brit. pat.** 667,098 (1952 to Am. Cyanamid).

Dihydrate, yellow-orange microcrystals. uv max (0.1N NaOH): 255, 306, 369 nm (ε 25,000, 25,000, 9000); (0.1N HCl): 244, 316 nm (ε 22,000, 14,000).

143. Adenosine. *9-β-D-Ribofuranosyl-9H-purin-6-amine;* 6-amino-9-β-D-ribofuranosyl-9H-purine; 9-β-D-ribofuranosidoadenine; adenine riboside; Adenocard. C_{10}-$H_{13}N_5O_4$; mol wt 267.24. C 44.94%, H 4.90%, N 26.21%, O 23.95%. Nucleoside; widely distributed in nature. From yeast nucleic acid: Levene and Bass, *Nucleic Acids* (New York, 1931) p 163. Structure: Levene, Tipson, *J. Biol. Chem.* **94**, 809 (1932); Bredereck, *Ber.* **66**, 198 (1933); *Z. Physiol. Chem.* **223**, 61 (1934); Gulland, Holiday, *J. Chem. Soc.* **1936**, 765. *Cf.* Szent-Györgyi, *J. Physiol.* **68**, 213 (1930); Lythgoe *et al., J. Chem. Soc.* **1947**, 355; **1948**, 965. Synthesis: Davoll *et al., ibid.* **1948**, 967; H. Vorbrueggen, K. Krolikiewicz, *Angew. Chem. Intl. Ed.* **14**, 421 (1975). Crystal structure: T. F. Lai, R. E. Marsh, *Acta Crystallogr.* **B28**, 1982 (1972). Conformational properties: D. B. Davies, A. Rabczenko, *J. Chem. Soc. Perkin Trans. 2* **1975**, 1703. Symposium on cardiac electrophysiology, pharmacology and clinical efficacy in supraventricular tachycardia: *Prog. Clin. Biol. Res.* **230**, 1-395 (1987). *Reviews: see* Adenine; Nucleic Acids.

Crystals from water, mp 234-235°. $[\alpha]_D^{11}$ −61.7° (c = 0.706 in water); $[\alpha]_D^9$ −58.2° (c = 0.658 in water). uv max: 260 nm (ε 15100). Practically insol in alcohol.

THERAP CAT: Anti-arrhythmic.

144. Adenosine Diphosphate. *Adenosine 5'-(trihydrogen diphosphate);* ADP; adenosine 5'-pyrophosphoric acid; 5'-adenylphosphoric acid; adenosinediphosphoric acid. $C_{10}H_{15}N_5O_{10}P_2$; mol wt 427.22. C 28.11%, H 3.54%, N 16.39%, O 37.45%, P 14.50%. Formed from ATP in the muscle by the enzyme adenosinetriphosphatase upon stimulation of the muscle unless hydrolysis is prevented by injection of magnesium sulfate. Prepd by hydrolysis of ATP by means of adenosinetriphosphatase from lobster or rabbit muscle: LePage, *Biochem. Prepns.* **1**, 1 (1949). Synthesis: Chambers *et al., J. Am. Chem. Soc.* **82**, 970 (1960). *See also* the ref and data *under* Adenosine Triphosphate.

Barium salt, $Ba_3(C_{10}H_{12}N_5O_{10}P_2)_2$. The purity of the prepns can be checked by analyses for nitrogen, ribose (orcinol reaction), total phosphorus, easily hydrolyzable phosphorus, and inorganic phosphorus. ADP should give a ratio of 2:1 for total organic phosphorus to easily hydrolyzable phosphorus, *see* LePage, *loc. cit.* a_M (molar absorbancy of pure ADP): 15.4 × 10³ at 259 nm and pH 7.0.

145. Adenosine Triphosphate. *Adenosine 5'-(tetrahydrogen triphosphate);* ATP; adenosine 5'-triphosphoric acid; Adephos; Adetol; Atipi; Atriphos; Striadyne; Triadenyl; Triphosaden. $C_{10}H_{16}N_5O_{13}P_3$; mol wt 507.21. C 23.68%, H 3.18%, N 13.81%, O 41.01%, P 18.32%. Coenzyme valuable in the transfer of phosphate bond energy. Mammalian skeletal muscle at rest contains 350-400 mg ATP per 100 g. Upon stimulation of the muscle the ATP is hydrolyzed to ADP by the myosin-actin complex unless hydrolysis is prevented by injection of magnesium sulfate. Isoln from rabbit muscle: LePage, *Biochem. Prepn.* **1**, 5 (1949); *cf.* Fiske, Subbarow, *Science* **70**, 381 (1929); Lohmann, *Biochem. Z.* **223**, 460 (1931); **254**, 381 (1932); Barrenscheen, Filz, *ibid* **250**, 281 (1932); Kerr, *J. Biol. Chem.* **139**, 121 (1941); Needham, *Biochem. J.* **36**, 113 (1942). Synthetic routes: Tanaka, Honjo, **U.S. pat.** 3,079,379 (1963 to Takeda). Reviews of biosynthesis: Racker, *Advan. Enzymol.* **23**, 323-399 (1961); Deamer, *J. Chem. Ed.* **46**, 198-206 (1971). Reviews of nucleotide coenzymes: Bock in *The Enzymes* **vol. 2A**, P. D. Boyer *et al.*, Eds. (Academic Press, New York, 2nd ed., 1960) pp 3-38; A. M. Michelson, *The Chemistry of Nucleosides and Nucleotides* (Academic Press, New York, 1963) pp 153-250; D. W. Hutchinson, *Nucleotides and Coenzymes* (John Wiley, New York, 1964) pp 36-82.

Free ATP, isolated as a glass (by treatment of the Ba salt with H_2SO_4 and treating the concd aq soln with acetone). $[\alpha]_D^{22}$ −26.7° (c = 3.095). a_M(molar absorbancy): 15.4 × 10³ at 259 nm and pH 7.0. Freely sol in water. A 1% aq soln has a pH of about 2 and is stable at 0° for several hrs. ATP is a tetrabasic acid and after hydrolytic cleavage it is hexabasic. It is usually pptd as the dibarium salt with 4 or 6 mols of water of crystn which can be removed by prolonged drying at 100° over P_2O_5 and in a vacuum. The anhydr salt is stable, but the hydrated salt slowly decomp forming 5'-adenylic acid and barium pyrophosphate. For use the barium salt is converted to the sodium or potassium salt by treatment with sodium or potassium sulfate in HCl soln. At pH 6.8-7.4 the Na salt is stable in aq soln provided the product is pure. Ba^{2+} catalyzes breakdown. Injectable solns of the disodium salt are marketed as *Trinosin, Adetphos, Circulen "Kyorin", Atenen.*

USE: In biochemical research. To inhibit enzymatic

browning of raw edible plant materials, such as sliced apples, potatoes, etc.

THERAP CAT: Coenzyme.

146. S-Adenosylmethionine. *5'-[(3-Amino-3-carboxypropyl)methylsulfonio]-5'-deoxyadenosine hydroxide, inner salt;* S-(5'-desoxyadenosin-5'-yl)-L-methionine; "active methionine"; ademetionine; AdoMet; SAMe. $C_{15}H_{22}N_6O_5S$; mol wt 398.44. C 45.22%, H 5.56%, N 21.09%, O 20.08%, S 8.05%. Physiological methyl donor involved in enzymatic transmethylation reactions; present in all living organisms. Enzymatic prepn: G. L. Cantoni, *J. Am. Chem. Soc.* **74,** 2942 (1952). Isoln from cultures of *Torulopsis utilis:* F. Schlenk, R. E. DePalma, *J. Biol. Chem.* **229,** 1037 (1957). Synthesis of S-adenosyl-D,L-methionine: J. Baddiley, G. A. Jamieson, *J. Chem. Soc.* **1954,** 4280. Synthesis of optically active chloride: C. H. Shunk, J. W. Richter, U.S. pat. **2,969,353** (1961 to Merck & Co.). Fermentative production by *Aspergillus tamarii* IAM 2137: H. Kusakabe *et al.,* **Japan.** pat. **72 37,037** (1972 to Yamaza Shoyu), *C.A.* **78,** 2745f (1973); production by yeasts: T. Tsuchida *et al.,* **Japan. Kokai 75 82,288;** *eidem,* U.S. pat. **3,962,034** (1975, 1976 both to Ajinomoto). Pharmacology: G. Stramentinoli *et al., Minerva Med.* **66,** 1541 (1975). Clinical efficacy in hypertriglyceridemia: F. Consolo *et al., Curr. Ther. Res.* **22,** 75 (1977); in hepatic steatosis: R. Mazzanti *et al., ibid.* **25,** 25 (1979); G. Mascio *et al., Gazz. Med. Ital.* **140,** 37 (1981). Clinical antidepressant effects: A. Agnoli *et al., J. Psychiat. Res.* **13,** 43 (1976); J. F. Lipinski *et al., Am. J. Psychiat.* **141,** 448 (1984). Anti-inflammatory activity in rats: M. Gualano *et al., Int. J. Tissue React.* **7,** 41 (1985). Clinical comparison with ibuprofen in osteoarthritis: S. Glorioso *et al., Int. J. Clin. Pharm. Res.* **5,** 39 (1985). Review of preparative methods and physiological activity: S. Shimizu, H. Yamada, *Trends Biotechnol.* **2,** 137 (1984). Books: *The Biochemistry of Adenosylmethionine,* F. Salvatore *et al.,* Eds. (Columbia Univ. Press, New York, 1977) 588 pp; *Biochemistry of S-Adenosylmethionine and Related Compounds,* E. Usdin *et al.,* Eds. (Macmillan Press, London, 1982) 760 pp.

Chloride, $C_{15}H_{23}ClN_6O_5S$, uv max (water): 260 nm. $[\alpha]_D^{25}$ +32° (c = 3.3 in water).

Disulfate ditosylate, $C_{29}H_{42}N_6O_{19}S_5$, *Gumbaral, Samyr.* LD_{50} in mice (mg/kg): 560 i.v., 2500 i.p., >6000 orally (Stramentinoli).

THERAP CAT: Anti-inflammatory. Treatment of chronic liver disease.

147. 3'-Adenylic Acid. Adenosine 3'-monophosphate; adenosine-3'-phosphoric acid; adenosine-3'-monophosphoric acid; adenylic acid b; yeast adenylic acid; synadenylic acid; h-adenylic acid. $C_{10}H_{14}N_5O_7P$; mol wt 347.23. C 34.59%, H 4.06%, N 20.17%, O 32.26%, P 8.92%. Early prepns from yeast nucleic acid: Levene, Bass, *Nucleic Acids,* (Chemical Catalogue Co., New York, 1931). Early work probably done on mixtures of 2'- and 3'-adenylic acids; both compds isomerize readily to form an equilibrium mixture under acid conditions: Carter, Cohn, *Fed. Proc.* **8,** 190 (1949); Baddily, in *The Nucleic Acids* vol. 1, E. Chargaff, J. N. Davidson, Eds. (Academic Press, New York, 1955) pp 165-168; A. M. Michelson, *The Chemistry of Nucleosides and Nucleotides* (Academic Press, New York, 1963) pp 100-106. Synthesis: Brown, Todd, *J. Chem. Soc.* **1952,** 44. Structure of dihydrate: Brown *et al., Nature* **172,** 1184

(1953); Sundaralingam, *Acta Cryst.* **21,** 495 (1966). *Reviews:* see Adenosine; Nucleic Acids.

Long, fine, colorless, needles. mp 197° (dec). Dihydrate, d (calc) 1.698. Absorption spectrum: Voet *et al., Biopolymers* **1,** 193 (1963). Difficultly sol in boiling water. Gives quantitative yield of furfural when distilled with 20% HCl for 3 hrs (different from 5'-adenylic acid which yields only traces). *cf.* Hoffman, *J. Biol. Chem.* **73,** 15 (1927); Embden, Schmidt, *Z. Physiol. Chem.* **181,** 130 (1929).

148. 5'-Adenylic Acid. Muscle adenylic acid; ergadenylic acid; t-adenylic acid; adenosine 5'-monophosphate; adenosine phosphate; adenosine-5'-phosphoric acid; adenosine-5'-monophosphoric acid; A5MP; AMP; NSC-20264; Adényl; Cardiomone (Na salt); Lycedan; My-B-Den; Myoston; Phosaden. $C_{10}H_{14}N_5O_7P$; mol wt 347.23. C 34.59%, H 4.06%, N 20.17%, O 32.25%, P 8.92%. Nucleotide; widely distributed in nature. Prepn from tissues: Embden, Zimmerman, *Z. Physiol. Chem.* **167,** 137 (1927); Embden, Schmidt, *ibid.* **181,** 130 (1929); *cf.* Kalckar, *J. Biol. Chem.* **167,** 445 (1947). Prepn by hydrolysis of ATP with barium hydroxide: Kerr, *J. Biol. Chem.* **139,** 131 (1941). Synthesis: Baddiley, Todd, *J. Chem. Soc.* **1947,** 648. Commercial prepn by enzymatic phosphorylation of adenosine. Monograph on synthesis of nucleotides: G. R. Pettit, *Synthetic Nucleotides* **vol. 1** (Van Nostrand-Reinhold, New York, 1972) 252 pp. Crystal structure: Kraut, Jensen, *Acta Cryst* **16,** 79 (1963). *Reviews:* see Adenosine; Nucleic Acids.

Crystals from water + acetone, mp 196-200°. $[\alpha]_D^{20}$ −47.5° (c = 2, 2% NaOH); $[\alpha]_D^{20}$ −26.0° (c = 2, 10% HCl). pK_1 = 3.8; pK_2 = 6.2. a_M (molar absorbancy): 15.4×10^3 at 259 nm (pH 7.0). Readily sol in boiling water. The compound is readily deaminated by nitrous acid to form inosinic acid; less rapidly hydrolyzed than 3'-adenylic acid by sulfuric acid. Furfural is formed only in traces on distillation with 20% HCl, *cf.* Levene, Bass, *Nucleic Acids* (New York, 1931) pp 230-232. Absorption spectrum: Kalckar, *loc. cit.*

THERAP CAT: Nutrient.

149. Adhatoda. Malabar nut; arusa; adulsa; vasaca; adhatodai; bakash. Leaves of *Adhatoda vasica* Nees, *Acanthacae. Habit.* East India. *Constit.* Adhatodic acid, vasicine. *Ref:* Krishnaswami, David, *Indian J. Pharm.* **2,** 141 (1940).

THERAP CAT: Expectorant.

150. Adinazolam. *8-Chloro-N,N-dimethyl-6-phenyl-*

4H-[1,2,4]triazolo[4,3-a][1,4]benzodiazepine-1-methan-amine; 8-chloro-1-[(dimethylamino)methyl]-6-phenyl-*4H-s*-triazolo[4,3-*a*][1,4]benzodiazepine; U-41123. $C_{19}H_{18}ClN_5$; mol wt 351.84. C 64.86%, H 5.16%, Cl 10.08%, N 19.90%. Triazolobenzodiazepine with antidepressant and anxiolytic properties. Dimethylamino derivative of alprazolam, *q.v.* Prepn: H. Allgeier, A. Gagneux, *Ger. pat.* **2,201,210** corresp to **Brit.** pat. **1,393,256** (1972, 1975 both to Ciba-Geigy); J. B. Hester, Jr., **U.S.** pat. **4,250,094** (1984 to Upjohn); *idem, J. Heterocyclic Chem.* **17**, 575 (1980). Chromatographic determn in human plasma: G. W. Peng, *J. Pharm. Sci.* **73**, 1173 (1984). Pharmacological profile: R. A. Lahti *et al., Neuropharmacology* **22**, 1277 (1983). Biological activity and metabolism: V. H. Sethy *et al., J. Pharm. Pharmacol.* **36**, 546 (1984). Use as antipsychotic: R. Lahti, R. E. Pyke, **Ger.** pat. **3,401,290** corresp to **U.S.** pat. **4,472,397** (both 1984 to Upjohn). Preliminary clinical study in severe depression: R. E. Pyke *et al., Psychopharmacol. Bull.* **19**, 96 (1983).

Crystals from ethyl acetate, mp 171-172.5° (Hester, U.S. pat.). Also reported as mp 165-166° (Allgeier, Gagneaux). Monomethanesulfonate, $C_{20}H_{22}ClN_5O_3S$, *adinazolam mesylate, Deracyn.* Crystals from methanol-ether, mp 230-244°.

THERAP CAT: Antidepressant.

151. Adiphenine Hydrochloride. *α-Phenylbenzeneacetic acid 2-(diethylamino)ethyl ester hydrochloride;* 2-diethylaminoethyl diphenylacetate hydrochloride; diphenylacetyldiethylaminoethanol hydrochloride; Trasentine hydrochloride; Diphacil hydrochloride; Spasmolytin; Difacil hydrochloride; Patrovina. $C_{20}H_{26}ClNO_2$; mol wt 347.87. C 69.05%, H 7.53%, Cl 10.19%, N 4.03%, O 9.20%. Antimuscarinic antispasmodic. Prepd by the action of diphenylacetyl chloride or diphenyl ketene on diethylaminoethanol: **Swiss** pat. **190,541** (1937 to Ciba); **Ger.** pats. **626,539; 653,778** (1937). Crystal and molecular structure: Guy, Hamor, *J. Chem. Soc. Perkin Trans. II* **1973**, 942.

Crystals, mp 113-114°. Freely sol in water; very sparingly sol in alcohol, ether. A 5% aq soln is neutral to litmus. Methyl bromide, $C_{21}H_{28}BrNO_2$, *Lunal.* LD_{50} i.v. in rabbits: 22.5-27.5 mg/kg.

THERAP CAT: Anticholinergic.

152. Adipic Acid. *Hexanedioic acid;* 1,4-butanedicarboxylic acid. $C_6H_{10}O_4$; mol wt 146.14. C 49.31%, H 6.90%, O 43.79%. $HOOC(CH_2)_4COOH$. Found in beet juice. Prepn from cyclohexanol: Bouveault, Locquin, *Bull. Soc. Chim.* [4] **3**, 438 (1908); Ellis, *Org. Syn.* coll. vol. **I**, 18 (2nd ed., 1941); Feagen, Copenhaver, *J. Am. Chem. Soc.* **62**, 869 (1940); **U.S.** pats. **2,191,786** (1940); **2,196,357** (1940); Zilberman *et al., J. Appl. Chem. (USSR)* **29**, 621 (1956). Convenient lab prepn from cyclohexanone: L. F. Fieser, *Organic Experiments* (Heath, Boston, 1964) pp 106-108. Prepn by one-step oxidation of cyclohexane: Onopchenko, Schulz, *J. Org. Chem.* **38**, 3729 (1973); Tanaka, 167th Am. Chem. Soc. Meet. (Los Angeles, March-April, 1974) *Abstracts of Papers*, p 29. Manuf: Faith, Keyes & Clark's *Industrial Chemicals*, F. A. Lowenheim, M. K. Moran, Eds. (Wiley-Interscience, New York, 4th ed., 1975) pp 50-54. *Review:* D. E. Danly, C. R. Campbell in Kirk-Othmer *Encyclopedia of Chemical Tech-*

nology vol. **1** (Wiley-Interscience, New York, 3rd ed., 1978) pp 510-531.

Monoclinic prisms from ethyl acetate, from water or from acetone + petr ether. d_4^{25} 1.360. mp 152°. bp_{760} 337.5°; bp_{100} 265°; bp_{40} 240.5°; bp_{20} 222°; bp_{10} 205.5°; bp_5 191°; $bp_{1.0}$ 159.5°. K_1 (25°) = 3.90 × 10^{-5}; K_2 = 5.29 × 10^{-6}. Absorption spectrum: Ramart-Lucas, Salmon-Legagneur, *Compt. Rend.* **189**, 916 (1929). Freely soluble in methanol, ethanol; sol in acetone. 100 ml of a satd aq soln contains 1.44 g; 100 ml of boiling water dissolves 160 g; 100 parts of ether dissolve 0.633 parts (w/w) at 19°. Slightly sol in cyclohexane. Practically insol in benzene, petr ether. pH of satd aq soln at 25° = 2.7; pH of 0.1% soln = 3.2.

Dimethyl ester, $C_8H_{14}O_4$, liquid, solidifies at 0°, bp_{10} 112°. Diethyl ester, $C_{10}H_{18}O_4$, *ethyl adipate.* Liquid, bp 240-245°. d^{20} 1.009. Insol in water; sol in alcohol and many other organic solvents.

USE: Manuf artificial resins, plastics (nylon), urethan foams. Used as acidulant in baking powders instead of tartaric acid, cream of tartar, and phosphates because adipic acid is not hygroscopic. As an intermediate in lubricating oil additives.

153. Adipsin. Serine protease homolog synthesized by mammalian adipocytes and secreted into the bloodstream; also produced by sciatic nerve. Developmentally regulated glycoprotein; mol wt ranges from 37-44 kilodaltons depending on the degree of glycosylation. Circulating levels of adipsin protein are reduced in certain genetic and experimental models of obesity. Originally identified as a 28 kDa protein encoded by a cDNA clone corresponding to messenger RNA specifically induced during adipocyte differentiation: B. M. Spiegelman *et al., J. Biol. Chem.* **258**, 10083 (1983). Nucleotide analysis and predicted amino acid sequence: K. S. Cook *et al., Proc. Nat. Acad. Sci. USA* **82**, 6480 (1985). Structure of adipsin gene: H. Y. Min, B. M. Spiegelman, *Nucleic Acids Res.* **14**, 8879 (1986). Isoln from adipose tissue, reportedly the primary site of synthesis, and from sciatic nerve: K. S. Cook *et al., Science* **237**, 402 (1987). Studies in murine obesity models and potential biological role: J. S. Flier *et al., ibid.* 405.

154. Adlumidine. *[S-(R*,R*)]-6-(5,6,7,8-Tetrahydro-6-methyl-1,3-dioxolo[4,5-g]isoquinolin-5-yl)furo[3,4-e]-1,3-benzodioxol-8(6H)-one.* $C_{20}H_{17}NO_6$; mol wt 367.34. C 65.39%, H 4.66%, N 3.81%, O 26.13%. An alkaloid present in the *d*-form in the entire plant of *Adlumia fungosa* (Ait.) Greene (*A. cirrhosa* Raf.), *Corydalis thalictrifolia* Franch. and *C. incisa* (Thumb.) Pers., *Fumariaceae.* Schlotterbeck, Watkins, *J. Am. Chem. Soc.* **25**, 596 (1903); R. H. F. Manske, *Can. J. Res.* **21B**, 111 (1943). *l*-Adlumidine was isolated from *Corydalis sempervirens* (L.) Pers., *C. scouleri* Hook., and *C. crystallina* Engelm., *Fumariaceae:* R. H. F. Manske, *ibid.* **8**, 407 (1932); **14B**, 347 (1936); **17B**, 57 (1939). Structure: *idem, J. Am. Chem. Soc.* **72**, 3207 (1950). Stereochemistry: Blaha *et al., Coll. Czech. Chem. Commun.* **29**, 2328 (1964); Snatzke *et al., Tetrahedron* **25**, 5059 (1969). *Review:* Stanek, Manske in *The Alkaloids* vol. **IV**, R. H. F. Manske, H. L. Holmes, Eds. (Academic Press, New York, 1954) pp 167-198. Stereoisomer of bicuculline, *q.v.*

d-adlumidine

Rhombic plates from chloroform + methanol, mp 236-237°. $[\alpha]_D^{25}$ +116.2° (c = 22 in chloroform). pKa 4.27. Practically insol in water. Very sparingly sol in alcohol, ether, hexane.

l-Adlumidine, *capnoidine.* Stout prisms from chloroform

+ methanol, mp 238°. $[\alpha]_D^{22}$ −113.2° (in chloroform). pKa 4.24. Practically insol in water. Sparingly sol in methanol; sol in chloroform.

155. Adlumine. *6-(1,2,3,4-Tetrahydro-6,7-dimethoxy-2-methyl-1-isoquinolinyl)furo[3,4-e]-1,3-benzodioxol-8(6H)-one.* $C_{21}H_{21}NO_6$; mol wt 383.39. C 65.78%, H 5.52%, N 3.65%, O 25.04%. An alkaloid present in the entire plant of some *Fumariaceae* species. *d*-Adlumine was isolated from *Adlumina fungosa* (Ait.) Greene (*A. cirrhosa* Raf.); *l*-adlumine was isolated from *Corydalis scouleri* Hook. and *C. sempervirens* (L.) Pers., *C. ophiocarpa* Hook. f. et Thoms *Fumariaceae:* Manske, *Can. J. Res.* **8**, 210 (1933); **14B**, 347 (1936); **16B**, 81 (1938); **17B**, 51 (1939). Preliminary stereochemical studies of *d*-form: Safe, Moir, *Can. J. Chem.* **42**, 160 (1964). Revised stereochemistry: Blaha *et al.*, *Coll. Czech. Chem. Commun.* **29**, 2328 (1964); Snatzke *et al.*, *Tetrahedron* **25**, 5059 (1969). *Review:* Stanek, Manske in *The Alkaloids* Vol. **IV**, R. H. F. Manske, H. L. Holmes, Eds. (Academic Press, New York, 1954) pp 167-198.

d-adlumine

l-Form, hexagonal plates from chloroform + methanol, mp 180°. $[\alpha]_D^{22}$ −42.5° (c = 0.8 in chloroform). *dl*-Form, stout short prisms, mp 190°.

156. Adonis vernalis. False hellebore; vernal pheasant's eye; bird's eye. Herb of *Adonis vernalis* L., *Ranunculaceae*. *Habit.* Lime-rich soil of Eastern and Southern Europe, especially Hungary and Serbia. Also found in Labrador, Asian U.S.S.R. *Constit.* Adonidoside, adonitoxin *q.v.;* acetyladonitoxin *see* adonitoxin; adonivernoside; adonitol *q.v.;* adonivernith; vernadin; strophanthidin *q.v.;* 16-hydroxystrophanthidin *see* strophanthidin; vernadigin *q.v.;* aconitic acid (?) *q.v.;* 2,6-dimethoxyquinone *q.v.;* phytosterols *q.v.;* choline *q.v.;* resin. Physiologically active extracts are known as *Adovern; Adonilen; Adonisid; Adonigen; Adonival.* Monograph: Jean Sassard, *Essai sur les propriétés et les modes d'application de l'Adonis vernalis* (Lyon 1930); Heyl *et al.*, *J. Am. Chem. Soc.* **40**, 436 (1918); Karrer, *Festschrift E. C. Barell* **1936**, 240; Meyer, *Pharmazie* **4**, 432 (1949); Reinhard, Hufschmidt, *Deut. Apoth. Ztg.* **98**, 1197 (1958).

THERAP CAT: Cardiotonic.

157. Adonitol. *Ribitol;* adonite. $C_5H_{12}O_5$; mol wt 152.15. C 39.47%, H 7.95%, O 52.58%. A pentitol from *Adonis vernalis* L., *Ranunculaceae;* also from *A. amurensis* L., *Ranunculaceae:* Santavy, Reichstein, *Pharm. Acta Helv.* **23**, 153 (1948).

Large, optically inactive crystals. mp 102°. Does not reduce Fehling's soln. Freely soluble in water, hot alcohol; insol in ether.

158. Adonitoxin. *3-[(6-Deoxy-α-L-mannopyranosyl)-oxy]-14,16-dihydroxy-19-oxocard-20(22)-enolide.* $C_{29}H_{42}O_{10}$; mol wt 550.63. C 63.25%, H 7.69%, O 29.06%. From *Adonis vernalis* L., *Ranunculaceae:* Katz, Reichstein, *Pharm.*

Acta Helv. **22**, 437 (1947); *C.A.* **43**, 1790i (1949); Pitra, Cekan, *Coll. Czech. Chem. Commun.* **26**, 1551 (1961). Structure: Poláková, Cekan, *Chem. & Ind. (London)* **1963**, 1766.

Crystals from chloroform + methanol + ether, decomp 251-258°.

3-*O*-Acetyladonitoxin, $C_{31}H_{44}O_{11}$, needles from dil methanol, prisms from methanol+ ether, decomp 213-219°. $[\alpha]_D^{22}$ −19.9° (c = 0.33 in methanol).

159. Adrafinil. *2-[(Diphenylmethyl)sulfinyl]-N-hydroxyacetamide;* 2-(benzhydrylsulfinyl)acetohydroxamic acid; CRL 40028; Olmifon. $C_{15}H_{15}NO_3S$; mol wt 289.35. C 62.27%, H 5.22%, N 4.84%, O 16.59%, S 11.08%. α-Adrenergic agonist. Prepn, pharmacology: L. Lafon, **Belg.** pat. **846,880;** U.S. pat. **4,066,686** (1977, 1978 both to Lafon). Mode of action: J. Duteil *et al.*, *Eur. J. Pharmacol.* **59**, 121 (1979); C. Rozé *et al.*, *Arch. Int. Pharmacodyn. Ther.* **265**, 119 (1983). Psychopharmacology in mice: F. A. Rambert *et al.*, *J. Pharmacol.* **17**, 37 (1986).

Crystals from ethyl acetate-isopropyl alcohol, mp 159-160°. Soly in water <1 g/l. LD_{50} in mice (mg/kg): <2048 i.p.; 1950 gastric admin (Lafon).

THERAP CAT: Treatment of depression.

160. Adrenal Cortical Extract. ACE; Cortical; Corticopan; Cortidin; Cortidyn; Cortine Naturelle; Cortisural; Escatin; Eschatin; Eucortone; Maxicortex; Mencortex; Pancorton; Supracort. Isoln and purification: Swingle, Pfiffner, U.S. pats. **2,074,492** and **2,074,493** (both 1937 to Parke, Davis).

THERAP CAT: Corticosteroid.

THERAP CAT (VET): Has been used for bovine ketosis.

161. Adrenalone. *1-(3,4-Dihydroxyphenyl)-2-(methylamino)ethanone; 3',4'-dihydroxy-2-(methylamino)acetophenone;* 3,4-dihydroxy-α-methylaminoacetophenone; adrenone; 4-methylaminocetopyrocatechol; Kephrine; Stryphnone; Stypnone. $C_9H_{11}NO_3$; mol wt 181.19. C 59.66%, H 6.12%, N 7.73%, O 26.49%. Intermediate in some epinephrine manuf processes. Prepd from 4-chloroacetylpyrocatechol and methylamine: Ger. pat. **152,814** (Hoechst); Stolz, *Ber.* **37**, 4152 (1904); by heating α-(*p*-toluene sulfonylmethylamino)-3,4-dimethoxyacetophenone with HCl under pressure: **Ger.** pat. **277,540** (Bayer); *Frdl.* **12**, 764; *Chem. Zentr.* **1914, II**, 740.

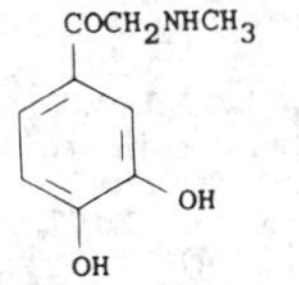

Needles, dec 235-236°. Sparingly sol in water, alc, ether. Hydrochloride, $C_9H_{12}ClNO_3$, *Stryphnonasal*. Crystals, mp 243°. Freely sol in water; sol in alcohol. Insol in ether. Aq solns are neutral to litmus.

THERAP CAT: Hemostatic.

162. Adrenochrome. *2,3-Dihydro-3-hydroxy-1-methyl-1H-indole-5,6-dione; 3-hydroxy-1-methyl-5,6-indolinedione.* $C_9H_9NO_3$; mol wt 179.17. C 60.33%, H 5.06%, N 7.82%, O 26.79%. Pigment obtained by the oxidation of epinephrine. Prepn using Ag_2O as oxidizing agent: MacCarthy, *Chim. Ind. Paris* **55**, 435 (1946). Structure: Green, Richter, *Biochem. J.* **31**, 596 (1937); Harley-Mason, *Experientia* **4**, 307 (1948). Probably has the zwitterionic quinonimine structure shown here. *Review:* Heacock, *Chem. Rev.* **59**, 181-237 (1959). Review, as a psychotomimetic agent: Heacock, *Chim. Ther.* **6**, 300 (1971).

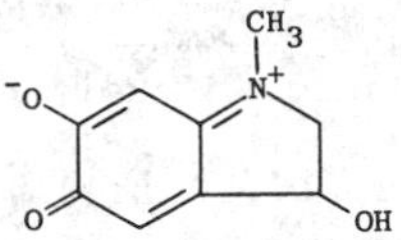

Hemihydrate, brilliant red crystals from methanol-formic acid. Dec 115-120°. Absorption max: 220, 302, 485 nm ($E_{1cm}^{5\%}$ 5.53, 2.45, 1). Well-formed and well-dried crystals can be kept in a vacuum desiccator for several weeks. Easily oxidized to melanin. Freely sol in water; fairly sol in alcohol. Almost insol in benzene and ether. Solns are unstable. Optimum pH of water soln 4.0.

Oxime, $C_9H_{10}N_2O_3$, sesquihydrate, orange needles from water, mp 278°. Much more stable than adrenochrome.

Monosemicarbazone, $C_{10}H_{12}N_4O_3$, *carbazochrome, Adrenoxyl, Cromadrenal, Cromosil.* Orange-red crystals from alcohol, mp 203° (dec). Prepn: Dechamps *et al.*, U.S. pat. **2,506,294** (1950 to Société Belge de l'Azote). *See also* Carbazochrome Salicylate.

Thiosemicarbazone, mp 215-220°. Description: Fleischhacker, Barsel, U.S. pat. **2,712,024** (1955 to International Hormones).

THERAP CAT: Monosemicarbazone as hemostatic.

163. Adrenoglomerulotropin. *2,3,4,9-Tetrahydro-6-methoxy-1-methyl-1H-pyrido[3,4-b]indole;* aldosterone-stimulating hormone; ASH; 1-methyl-6-methoxy-1,2,3,4-tetrahydro-2-carboline. $C_{13}H_{16}N_2O$; mol wt 216.27. C 72.19%, H 7.46%, N 12.95%, O 7.40%. Found in extracts of pineal gland tissue: Farrell, *Circulation* **21**, 1009 (1960); Farrell, McIsaac, *Arch. Biochem. Biophys.* **94**, 543 (1961). Synthesis from 5-methoxytryptamine and acetaldehyde: McIsaac, *Biochim. Biophys. Acta* **52**, 607 (1961); Meek *et al.*, *Chem. & Ind. (London)* **1964**, 622.

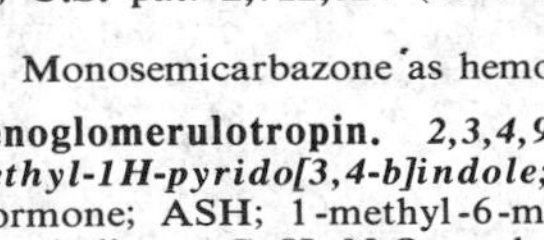

Crystals, mp 150-151°. uv max (ethanol): 225, 280 nm (log ε 4.34, 3.86); (0.1N HCl): 220, 273 nm (log ε 4.40, 3.86).

164. Adrenolutin. *1-Methyl-1H-indole-3,5,6-triol;* N-methyl-5,6-dihydroxyindoxyl; 3,5,6-trihydroxy-N-methylindole. $C_9H_9NO_3$; mol wt 179.17. C 60.33%, H 5.06%, N 7.82%, O 26.79%. Alkaline rearrangement product of adrenochrome. Isoln and identification: Lund, *Acta Pharmacol. Toxicol.* **5**, 75 (1949). Structure and synthesis: Balsinger *et*

al., *Helv. Chim. Acta* **36**, 708 (1953); Heacock, Mahon, *Can. J. Chem.* **36**, 1550 (1958). NMR studies and its existence in the ketonic form in solution: Powell, Heacock, *Chim. Ther.* **7**, 133 (1972).

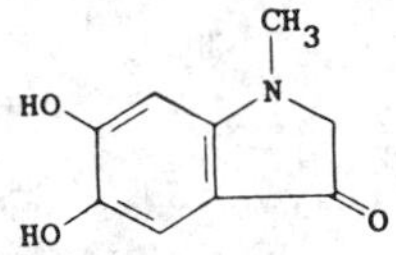

Monohydrate, bright yellow prisms from water, mp 236° (decompn); mp 195° (Balsinger *loc. cit.*).

Anhydrous, bright yellow prisms, mp 245°.

165. Adrenosterone. *Androst-4-ene-3,11,17-trione;* Reichstein's substance G. $C_{19}H_{24}O_3$; mol wt 300.38. C 75.97%, H 8.05%, O 15.98%. Prepn: Reichstein, *Helv. Chim. Acta* **19**, 29, 1107 (1936); Kendall *et al.*, *J. Biol. Chem.* **116**, 267 (1936); Reichstein, *Helv. Chim. Acta* **20**, 817, 953 (1937). Total synthesis: Velluz *et al.*, *Compt. Rend.* **250**, 1293 (1960). Crystal structure: Ohrt *et al.*, *Acta Cryst.* **19**, 479 (1965). Metabolism: Bradlow *et al.*, *Steroids* **10**, 233 (1967).

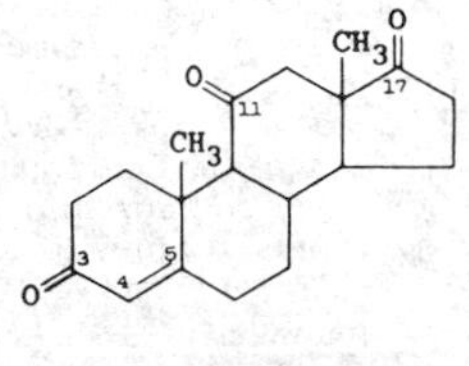

Needles from alcohol, mp 220-224°. Sublimes in high vacuum. Soly in water: 9.85 (23°), 15.2 (37°) mg/100 ml. Sol in alcohol, acetone, ether. $[\alpha]_D^{20}$ +262° (abs alcohol); $[\alpha]_{5461}^{25}$ +364° ± 5° (c = 0.18 in abs alc). uv max: 235 nm.

166. AET. *Carbamimidothioic acid 2-aminoethyl ester dihydrobromide; 2-(2-aminoethyl)-2-thiopseudourea dihydrobromide;* S-(2-aminoethyl)isothiuronium bromide hydrobromide; β-aminoethylisothiuronium bromide hydrobromide; Antirad; Surrectan; Antiradon. $C_3H_{11}Br_2N_3S$; mol wt 281.04. C 12.82%, H 3.95%, Br 56.87%, N 14.95%, S 11.41%. Prepd by refluxing thiourea with 2-bromoethylamine HBr in isopropanol: Clinton *et al.*, *J. Am. Chem. Soc.* **70**, 950 (1948); Funahashi, Miyano, *J. Agr. Chem. Soc. Japan* **27**, 775 (1953), *C.A.* **49**, 15737b (1955); Doherty *et al.*, *J. Am. Chem. Soc.* **79**, 5667 (1957).

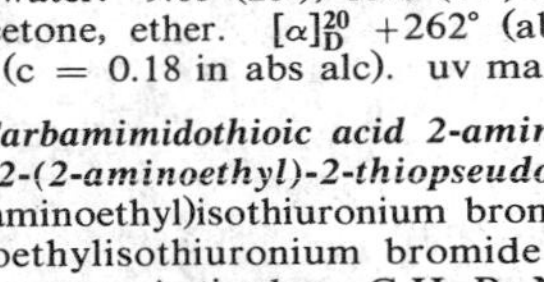

Crystals from abs ethanol + ethyl acetate, mp 194-195°. Hygroscopic, cakes together, and is converted in significant amounts to 2-aminothiazoline, a transformation that can be detected by a drop in melting point by as much as 30°. The animal organism appears to convert it to 2-mercaptoethylguanidine hydrobromide. LD_{50} in mice i.v., s.c., i.p., orally: 100, 280, 480, 1600 mg/kg, *C.A.* **67**, 72261s (1967).

THERAP CAT: Radioprotective agent.

167. Affinin. *N-(2-Methylpropyl)-2,6,8-decatrienamide; N-isobutyl-2,6,8-decatrienamide; N-isobutyldeca-trans-2-cis-6-trans-8-trienamide;* spilanthol. $C_{14}H_{23}NO$; mol wt 221.33. C 75.97%, H 10.47%, N 6.33%, O 7.23%. Insecticidal lipid amide isolated from *Heliopsis longipes* (A. Gray) Blake, *Compositae.* Isoln and structure: Acree *et al.*, *J. Org. Chem.* **10**, 236, 449 (1945). Identity with spilanthol: Jacobson, *Chem. & Ind. (London)* **1957**, 50. Stereochemistry and synthesis: Crombie, Krasinski, *ibid.* **1962**, 983; Crombie *et al.*, *J. Chem. Soc.* **1963**, 4970.

Pale yellow viscous oil, $bp_{0.2}$ 141°; $bp_{0.3-0.5}$ 160-165°. mp 23°. n_D^{25} 1.5134. Soluble in organic solvents; practically insoluble in aq alkali and acid. uv max (95% ethanol): 228.5 nm (ϵ 33,700).

USE: Insecticide synergist.

168. Aflatoxins B. Aflatoxins are a closely related group of secondary fungal metabolites shown to be mycotoxins. They are produced by *Aspergillus flavus* Link ex Fries, the causative principle of turkey "X" disease; and by *Aspergillus parasiticus:* Sargeant *et al., Nature* **192**, 1096 (1961); Hesseltine *et al., Proc. 1st U.S. Japan Conf. Toxic Microorganisms, Honolulu, Hawaii 1968,* p 202. Aflatoxins have been reported to naturally occur in peanuts, peanut meal, cottonseed meal, corn, dried chili peppers etc. However, the growth of the mold itself does not always indicate the presence of toxin since the yield of aflatoxin depends on growth conditions such as moisture, temperature, substrates, and aeration as well as genetic requirements. These heterocycles are now characterized as aflatoxins B_1, B_2, G_1, G_2, M_1 and M_2 (milk toxins) and B_{2a}, G_{2a}: Büchi, Rae in *Aflatoxin,* L. Goldblatt, Ed. (Academic Press, New York, 1969) pp 55-75. Toxic material is separated chromatographically into 4 distinct compounds based on fluorescent color (blue = B, green = G with subscripts relating to relative mobility): Nesbitt *et al., Nature* **195**, 1062 (1962); Sargeant *et al., Chem. & Ind. (London)* **1963**, 53. B_1 is one of the most potent environmental mutagens and carcinogens known. B_2 and G_2 are the less toxic dihydro derivs of B_1 and G_1: Asao *et al., J. Am. Chem. Soc.* **85**, 1706 (1963); **87**, 882 (1965), and B_{2a} and G_{2a} are the 2-hydroxy derivs of B_2 and G_2: Dutton, Heathcote, *Biochem. J.* **101**, 21P (1967); *Chem. & Ind. (London)* **1968**, 418. Isoln of B_3 (*parasiticol*), a possible metabolite of G_1, from *A. flavus:* J. G. Heathcote, M. F. Dutton, *Tetrahedron* **25**, 1497 (1969). Aflatoxins R_0 (*aflatoxicol*), P_1, Q_1, RB_1, RB_2, and D_1 are also known: P. F. Schuda, *Topics in Current Chemistry* **91**, 77-106 (1980). Total syntheses of aflatoxins B_1 and G_1: Buchi *et al., J. Am. Chem. Soc.* **88**, 4534 (1966); Knight *et al., Chem. Commun.* **1966**, 706; Büchi, Weinreb, *J. Am. Chem. Soc.* **93**, 746 (1971). The extreme toxicity and carcinogenicity of aflatoxins may be due to their inhibition of nucleic acid synthesis with their direct interaction with enzymes involved or by a toxin-DNA template: Clifford, Rees, *Nature* **209**, 312 (1966); Sporn *et al., Science* **151**, 1539 (1966). *See also* Wogan *et al., Food Cosmet. Toxicol.* **12**, 681 (1974). Inhibition of salt-induced conversion of B-DNA to Z-DNA by aflatoxin B_1: A. Nordheim *et al., Science* **219**, 1434 (1983). Prepn of the exo-8,9-epoxide of B_1, the metabolic deriv thought to be responsible for B_1's carcinogenicity: S. W. Baertschi *et al., J. Am. Chem. Soc.* **110**, 7929 (1988). uv spectrum of B_2: Hartley *et al., Nature* **198**, 1056 (1963). Toxicity data: Carnaghan *et al., ibid.* **200**, 1101 (1963); G. Büchi *et al., Life Sci.* **13**, 1143 (1973). Review and evaluation of studies of carcinogenic action in laboratory animals and humans: *IARC Monographs* **10**, 51-72 (1976). Comprehensive reviews: Goldblatt, *Econ. Bot.* **22**, 51-62 (1968); Detroy *et al.,* "Aflatoxin and Related Compounds" in *Microbial Toxins,* Vol. VI, A. Ciegler *et al.,* Eds. (Academic Press, New York, 1971) pp 3-178; W. F. Busby, Jr., G. N. Wogan, "Aflatoxins" in *Chemical Carcinogens* 2, 2nd ed., C. E. Searle, Ed.,

A.C.S. Monograph Series, no. **182** (American Chemical Society, Washington DC, 1984) pp 945-1136. Review of chemistry and syntheses: P. F. Schuda, *loc. cit.*

Aflatoxin B_1, $C_{17}H_{12}O_6$, *2,3,6aα,9aα-tetrahydro-4-methoxycyclopenta[c]furo[3',2':4,5]furo[2,3-h][1]benzopyran-1,11-dione.* Crystals, mp 268-269°. Exhibits blue fluorescence. $[\alpha]_D$ −558° (c = 0.1 in $CHCl_3$). $[\alpha]_D$ −480° (c = 0.1 in DMF). uv max (ethanol): 223, 265, 362 nm (ϵ 25600, 13800, 21800). LD_{50} orally in day old duckling: 18.2 μg/50 gm body wt (Carnaghan); i.p. in newborn mice: 9.50 mg/kg body wt (Büchi).

Aflatoxin B_2, $C_{17}H_{14}O_6$, *2,3,6aα,8,9,9aα-hexahydro-4-methoxycyclopenta[c]furo[3',2':4,5]furo[2,3-h][1]benzopyran-1,11-dione.* The 8,9-dihydro deriv of aflatoxin B_1. Crystals, mp 286-289°. Exhibits blue fluorescence. $[\alpha]_D$ −492° (c = 0.1 in $CHCl_3$). uv max (ethanol): 265, 363 nm (ϵ 11700, 23400). LD_{50} orally in day old duckling: 84.8 μg/50 gm body wt (Carnaghan).

Note: The aflatoxins may reasonably be anticipated to be carcinogens: *Fourth Annual Report on Carcinogens* (NTP 85-002, 1985) p 18.

169. Aflatoxins G. Toxic metabolites of fungi *Aspergillus flavus* Link ex Fries and *Aspergillus parasiticus.* Total synthesis of G_1: Büchi, Weinreb, *J. Am. Chem. Soc.* **93**, 746 (1971). Physical and chemical data: Hartley *et al., Nature* **198**, 1056 (1963); Asao, *J. Am. Chem. Soc.* **87**, 882 (1965). Toxicity data: R. B. A. Carnaghan *et al., Nature* **200**, 1101 (1963). For general refs *see* Aflatoxins B.

Aflatoxin G_1

Aflatoxin G_1, $C_{17}H_{12}O_7$. *3,4,7aα,10aα-Tetrahydro-5-methoxy-1H,12H-furo[3',2':4,5]furo[2,3-h]pyrano[3,4-c]-[1]benzopyran-1,12-dione.* Crystals, mp 244-246°. Exhibits green fluorescence. $[\alpha]_D$ −556° (chloroform). uv max (ethanol): 243, 257, 264, 362 nm (ϵ 11500, 9900, 10000, 16100). LD_{50} orally in day old duckling: 39.2 μg/50 gm body wt (Carnaghan).

Aflatoxin G_2. $C_{17}H_{14}O_7$. *3,4,7aα,9,10,10aα-Hexahydro-5-methoxy-1H,12H-furo[3',2':4,5]furo[2,3-h]pyrano[3,4-c]-[1]benzopyran-1,12-dione.* The 9,10-dihydro deriv of aflatoxin G_1. Crystals, mp 237-240°. Exhibits green-blue fluorescence. $[\alpha]_D$ −473° (c = 0.084 in chloroform). uv max (ethanol): 265, 363 nm (ϵ 9700, 21000). LD_{50} orally in day old duckling: 172.5 μg/50 gm body wt (Carnaghan).

Note: The aflatoxins may reasonably be anticipated to be carcinogens: *Fourth Annual Report on Carcinogens* (NTP 85-002, 1985) p 18.

170. Aflatoxins M. Highly toxic 4-hydroxylated aflatoxin B derivs found in the milk of cows fed toxic meal. Initial reports: Allcroft, Carnaghan, *Vet. Rec.* **75**, 259 (1963). Isoln: De Iongh *et al., Nature* **202**, 466 (1964); Allcroft *et al., ibid.* **209**, 154 (1966). Isoln and structure of M_1 and M_2: C. W. Holzapfel, P. S. Steyn, *Tetrahedron Letters* **1966**, 2799. Total synthesis of M_1: Büchi, Weinreb, *J. Am. Chem. Soc.* **91**, 5408 (1969); **93**, 746 (1971); G. Büchi *et al., ibid.* **103**, 3497 (1981). Carcinogenicity studies: Wogan, Paglialunga, *Food Cosmet. Toxicol.* **12**, 381 (1974). For general refs *see* Aflatoxins B.

Aflatoxin B_1

Aflatoxin M_1

Aflatoxin M_1. $C_{17}H_{12}O_7$. *2,3,6a,9a-Tetrahydro-9a-hydroxy-4-methoxycyclopenta[c]furo[3',2':4,5]furo[2,3-h][1]benzopyran-1,11-dione; 4-hydroxyaflatoxin B_1.* Crystals from methanol, mp 299° (dec). Exhibits blue-violet fluorescence. $[\alpha]_D$ −280° (c = 0.1 in DMF). uv max (ethanol): 226, 265, 357 nm (ε 23100, 11600, 19000). LD_{50} orally in day old Pekin ducklings: 16.6 μg/duckling (Holzapfel, Steyn).

Aflatoxin M_2. $C_{17}H_{14}O_7$. *2,3,6a,8,9,9a-Hexahydro-9a-hydroxy-4-methoxycyclopenta[c]furo[3',2':4,5]furo[2,3-h][1]benzopyran-1,11-dione; 4-hydroxyaflatoxin B_2.* The 8,9-dihydro deriv of aflatoxin M_1. Exhibits violet fluorescence. Crystals from methanol-chloroform, mp 293 (dec). uv max (ethanol): 221, 264, 357 nm (ε 20000, 10900, 21000). LD_{50} orally in day old Pekin ducklings: 62 μg/duckling (Holzapfel, Steyn).

Note: The aflatoxins may reasonably be anticipated to be carcinogens: *Fourth Annual Report on Carcinogens* (NTP 85-002, 1985) p 18.

171. Afloqualone. *6-Amino-2-(fluoromethyl)-3-(2-methylphenyl)-4(3H)-quinazolinone;* 6-amino-2-fluoromethyl-3-(o-tolyl)-4(3H)-quinazolinone; HQ 495; Aroft; Arofuto. $C_{16}H_{14}FN_3O$; mol wt 283.30. C 67.83%, H 4.98%, F 6.71%, N 14.83%, O 5.65%. A centrally acting muscle relaxant. Prepn: I. Inoue *et al.*, Ger. pat. 2,449,113; *eidem*, U.S. pat. 3,966,731 (1975, 1976 to Tanabe); J. Tani *et al.*, *J. Med. Chem.* **22**, 95 (1979). Pharmacology: T. Ochiai, R. Ishida, *Japan. J. Pharmacol.* **31**, 491 (1981); **32**, 427 (1982). Metabolism: N. Otsuka *et al.*, *J. Pharmacobio-Dyn.* **5**, S-59 (1982); S. Furuuchi *et al.*, *Drug Metab. Dispos.* **11**, 371 (1983).

Pale yellow prisms from 2-propanol, mp 195-196°. LD_{50} in mice (mg/kg): 315.1 i.p. (Tani).

THERAP CAT: Muscle relaxant (skeletal).

172. Agar. Agar-agar; gelose; Japan agar; Bengal isinglass; Ceylon isinglass; Chinese isinglass; Japan isinglass; Layor Carang. A polysaccharide complex extracted from the agarocytes of algae of the *Rhodophyceae.* Predominant agar-producing genera are *Gelidium, Gracilaria, Acanthopeltis, Ceramium, Pterocladia* found in the Pacific and Indian Oceans and Japan Sea. Can be separated into a neutral gelling fraction, *agarose,* and a sulfated non-gelling fraction, *agaropectin*: Araki, *J. Chem. Soc. Japan* **58**, 1338 (1937). Structure believed to be a complex range of polysaccaride chains having alternating α-(1 → 3) and β-(1 → 4) linkages and varying in total charge content; three extremes of structure noted, namely neutral agarose, pyruvated agarose having little sulfation, and a sulfated galactan: Duckworth *et al.*, *Carbohyd. Res.* **16**, 189, 435, 446 (1971). Reviews: V. J. Chapman, *Seaweeds and Their Uses* (Pitman Publ., New York, 1952) pp 89-123; Humm, *Econ. Bot.* **1**, 317 (1947); Mori, *Advan. Carbohyd. Chem.* **8**, 317 (1953); Selby, Wynne, in *Industrial Gums,* R. L. Whistler, Ed. (Academic Press, N.Y., 2nd ed., 1973) pp 29-48.

Transparent, odorless, tasteless strips or coarse or fine powder. Insol in cold water, alc; slowly sol in hot water to a viscid soln. A 1% soln forms a stiff jelly on cooling.

USE: Substitute for gelatin, isinglass, etc. in making emulsions including photographic, gels in cosmetics, and as thickening agent in foods esp. confectionary and dairy products; in meat canning; in production of medicinal encapsulations and ointments; as dental impression mold base; as corrosion inhibitor; sizing for silks and paper; in the dyeing and printing of fabrics and textiles; in adhesives. In nutrient media for bacterial cultures.

THERAP CAT: Cathartic.

THERAP CAT (VET): Laxative in dogs, cats. Demulcent.

173. Agaric. Larch agaric; touch wood; white agaric; purging agaric; amadou; German fungus. The dried fruit body of *Fomes laricis* (Jacq.) Murrill *(Polyporus officinalis* Fries), *Polyporaceae,* deprived of its outer rind. *Habit.* European and Asiatic Russia. Grows upon various species of *Pinus, Larix,* and *Picea. Constit.* 14-16% Agaricic acid; agaricoresin, agaricol, phytosterin, ricinoleic acid, cetyl alcohol, glucose, malic acid, carbohydrates. Isoln and separation of some constituents: Valentin, Knütter, *Pharm. Zentralhalle* **96**, 478 (1957). Impregnated with potassium nitrate and dried, it constitutes punk or tinder.

Light, fibrous, grayish-white to pale brown, spongy, friable pieces of irregular shape; feeble odor and bitter, acrid, yet somewhat sweetish taste.

USE: Anhidrotic.

174. Agaricic Acid. *2-Hydroxy-1,2,3-nonadecanetricarboxylic acid;* agaric acid; agaricin; agaricinic acid; laricic acid; *n*-hexadecylcitric acid; α-cetylcitric acid. $C_{22}H_{40}O_7$; mol wt 416.56. C 63.43%, H 9.68%, O 26.89%. Active principle of *Fomes laricis* (Jacq.) Murrill *(Polyporus officinalis* Fries), *Polyporaceae:* Thoms, Vogelsang, *Ann.* **357**, 145 (1907). Attempted syntheses: Evans, *J. Chem. Soc.* **1959**, 1313; Graf, Liu, *Arch. Pharm.* **306**, 366 (1973). Activity studies: Bacchi *et al.*, *J. Bacteriol.* **98**, 23 (1969).

Sesquihydrate, odorless, almost tasteless, crystalline powder. Anhydrous, mp 142° (dec). $[\alpha]_D^{19}$ −9° (NaOH). Slightly sol in cold water, chloroform or ether; freely sol in boiling water, alkalies, hot glacial acetic acid. One gram dissolves in 180 ml cold, 10 ml boiling alcohol.

Note: The name agaric acid has also been used for a lanostane-like mixture [*C.A.* **68**, 29894j (1968)].

USE: Has been used as antiperspirant.

175. Agaritine. L-*Glutamic acid 5-[2-[4-(hydroxymethyl)phenyl]hydrazide]; β-N-[γ-L(+)-glutamyl]-4-hydroxymethylphenylhydrazine.* $C_{12}H_{17}N_3O_4$; mol wt 267.28. C 53.92%, H 6.41%, N 15.72%, O 23.94%. Constituent of the commercial, edible mushroom *Agaricus bisporus* (Lange) Sing. *[Psalliota hortensis* Cooke var. *bispora], Agaricaceae (Agaricales).* Isoln and structure: Levenberg, *Fed. Proc.* **19**, 6 (1960); *J. Am. Chem. Soc.* **83**, 503 (1961); Daniels *et al.*, *ibid.* 3333; *see also* Levenberg in *Methods Enzymol.* **17** (Part A), 877 (1970). Synthesis: Kelly *et al.*, *J. Org. Chem.* **27**, 3229 (1962); Hinman, Kelly, U.S. pats. 3,274,232; 3,288,848 (both 1966 to Upjohn).

Glistening crystals from dil alcohol, dec 205-209°. $[\alpha]_D^{25}$ +7° (c = 0.8). uv max (water): 237.5, 280 nm (ε 12,000, 1,400). Very freely sol in water, practically insol in the usual anhydr organic solvents. pKa in water: 3.4 and 8.86.

176. Agmatine. *4-(Aminobutyl)guanidine;* 1-amino-4-guanidobutane. $C_5H_{14}N_4$; mol wt 130.19. C 46.12%, H 10.84%, N 43.04%. Decarboxylated arginine. Found in pollen of *Ambrosia artemisifolia* L., *Compositae,* in ergot, in sponges, in herring sperm, in octopus muscle: Heyl, *J. Am. Chem. Soc.* **41**, 681 (1919); Kossel, *Z. Physiol. Chem.* **66**, 257 (1910); Irvin, Wilson, *J. Biol. Chem.* **127**, 565 (1939). Prepn of salts: **Ger. pat.** 463,576 (1928 to Schering-Kahlbaum AG). Synthesis: Kossel, *Z. Physiol. Chem.* **68**, 170 (1910); Odo, *J. Chem. Soc. Japan* **67**, 132 (1946); Dose, *Ber.* **90**, 1251 (1957).

$C_5H_{14}N_4 \cdot H_2SO_4$, needles from dil methanol, mp 231°. Fairly sol in water, nearly insol in alcohol.

$C_5H_{14}N_4 \cdot 2HCl \cdot 2AuCl_3$, yellow needles from water, dec 223°.

177. Agroclavine. *8,9-Didehydro-6,8-dimethylergoline.* $C_{16}H_{18}N_2$; mol wt 238.32. C 80.63%, H 7.61%, N 11.76%. A non-peptide ergot alkaloid obtained from cultures of fungi parasitic on *Elymus mollis* Trin.: Abe *et al.*, **Japan.** pat. **178,336('49)** (to Takeda), *C.A.* **45**, 6352c (1951); *Ann. Rep. Takeda Res. Lab.* **10**, 145, 167, 171 (1951); **Japan.** pat. **7498**- **('54);** *C.A.* **50**, 6000^b (1956); **U.S.** pat. **2,835,675** (1958). Found in fungi parasitic on *Pennisetum typhoideum* Rich.: Stoll *et al.*, *Helv. Chim. Acta* **37**, 1815 (1954). Structure and stereochemistry: Schreier, *ibid.* **41**, 1984 (1958). Biosynthesis: Floss *et al.*, *J. Am. Chem. Soc.* **90**, 6500 (1968). Synthesis: Plieninger *et al.*, *Ann.* **743**, 95 (1971). Metabolism: Ramstad, *Lloydia* **31**, 327 (1968).

Rods from ether, dec 198-203°; needles from acetone, dec 205-206°. $[\alpha]_D^{20}$ −155° (c = 0.9 in chloroform); $[\alpha]_D^{20}$ −182° (c = 0.5 in pyridine). uv max: 225, 284, 293 nm (ε 4.47, 3.88, 3.81). Freely sol in alc, chloroform, pyridine; sol in benzene, ether; very slightly sol in water.

178. Agrocybin. *8-Hydroxy-2,4,6-octatriynamide.* C_8-H_5NO_2; mol wt 147.13. C 65.30%, H 3.43%, N 9.52%, O 21.75%. $CH_2OHC \equiv C - C \equiv C - C \equiv C - CONH_2$. Antibiotic substance produced by the basidiomycete *Agrocybe dura*: Kavanagh *et al.*, *Proc. Nat. Acad. Sci. USA* **36**, 102 (1950). Structure and synthesis: Bu'Lock *et al.*, *Chem. & Ind. (London)* **1954**, 990; Ashworth *et al.*, *J. Chem. Soc.* **1958**, 950. Origin of the carbon skeleton: E. R. H. Jones *et al.*, *J. Chem. Res. (M)* **1977**, 744.

Crystals from 20% alc or ether, mp 140° (conflagrates). Stable in air for about one day after which the crystals turn black and become insol in ether. Sol in alc, acetone, ether, chloroform, methyl isobutyl ketone; slightly sol in water; practically insol in hexane. Absorption spectrum: Ashworth *et al.*, *loc. cit.* Has antifungal activity. Causes dermatitis. Because of high toxicity and inactivation by blood no therapeutic tests have been attempted.

179. Ahistan. *10-(Dimethylamino)acetyl-10H-phenothiazine; 10-(N,N-dimethylglycyl)phenothiazine;* Histantin (Richter). $C_{16}H_{16}N_2OS$; mol wt 284.40. C 67.57%, H 5.67%, N 9.85%, O 5.63%, S 11.28%. Prepd from phenothiazine and chloroacetyl chloride, followed by treatment with dimethylamine: Cusic, **U.S.** pat. **2,694,705** (1954 to Searle).

Crystals, mp 144-145°.
Hydrochloride, $C_{16}H_{16}N_2OS \cdot HCl$. Crystals, mp 230-231°.
THERAP CAT: Antihistaminic.

180. AICAR. *5-Amino-1-(5-O-phosphono-β-D-ribofuranosyl)-1H-imidazole-4-carboxamide; 5-amino-1-ribofuranosylimidazole-4-carboxamide 5'-phosphate;* 5-amino-4-imidazolecarboxamide ribonucleotide; 5-amino-4-imidazolecarboxamide ribotide; 5-amino-1-(5'-phosphofuranoribosyl)-4-imidazolecarboxamide. $C_9H_{15}N_4O_8P$; mol wt 338.22. C 31.96%, H 4.47%, N 16.57%, O 37.85%, P 9.16%. Synthesis from AIR: Lukens, Buchanan, *J. Am. Chem. Soc.* **79**, 1511 (1957); *J. Biol. Chem.* **234**, 1791 (1959).

Barium salt, $C_9H_{13}BaN_4O_8P$. uv max (pH 7.0): 269 nm (ε 12600); *see* Flaks *et al.*, *J. Biol. Chem.* **228**, 201 (1957). In the presence of Bratton-Marshall reagents [*ibid.* **128**, 537 (1939)] AICAR is converted to a purple dye; absorption max: 540 nm (ε 26400).

181. AIR. *1-(5-O-Phosphono-β-D-ribofuranosyl)-1H-imidazol-5-amine; 5-amino-1-ribofuranosylimidazole 5'-phosphate;* 5-aminoimidazole ribonucleotide; 5-aminoimidazole ribotide; 5-amino-1-(5'-phosphofuranoribosyl)imidazole. $C_8H_{14}N_3O_7P$; mol wt 295.19. C 32.55%, H 4.78%, N 14.23%, O 37.94%, P 10.49%. An intermediate in the biosynthesis of purines. Synthesis from formylglycinamide ribonucleotide: Levenberg, Buchanan, *J. Biol. Chem.* **224**, 1005 (1957); Lukens, Buchanan, *ibid.* **234**, 1791 (1959).

Isolated in soln only or as the barium salt. The ultraviolet absorption spectrum at pH 7 shows no selective absorption between 215 and 300 nm, although a moderately strong general absorption is evident below 250 nm. In the presence of Bratton-Marshall reagents [*J. Biol. Chem.* **128**, 537 (1939)] AIR is converted into an orange chromophor (absorption max 500 nm).

182. Ajacine. *4-[[[2-(Acetylamino)benzoyl]oxy]methyl]-20-ethyl-1,6,14,16-tetramethoxyaconitane-7,8-diol; N-acetylanthranilic acid ester of lycoctonine.* $C_{34}H_{48}N_2O_9$; mol wt 628.74. C 64.95%, H 7.70%, N 4.46%, O 22.90%. From seeds of larkspur *(Delphinium ajacis* L., *Ranunculaceae)*: Keller, Völker, *Arch. Pharm.* **251**, 207 (1913); Hunter, *Quart. J. Pharm.* **17**, 302 (1944); Goodson, *J. Chem. Soc.* **1944**, 108; **1945**, 245; from *D. barbeyi*: Cook, Beath, *J. Am. Chem. Soc.* **74**, 1411 (1952). Structure: Kuzovkov, Platonova, *Zh. Obshch. Khim.* **29**, 2782 (1959). For structure *see* Lycoctonine.

Needles from 70% alcohol, mp 154°. $[\alpha]_D^{22}$ +49.5° (c = 2 in abs alcohol); $[\alpha]_D^{16}$ +53° (c = 0.66 in chloroform). uv max: 223, 252, 310 nm (ε 28,400; 16,600; 5400). Slightly sol in water; sol in ether, alcohol with blue fluorescence.

183. Ajaconine. $C_{27}H_{33}NO_3$; mol wt 359.49. C 73.50%, H 9.25%, N 3.90%, O 13.35%. From seeds of larkspur, *Delphinium ajacis* L., *Ranunculaceae:* Keller, Volker, *Arch. Pharm.* **251**, 207 (1913); Goodson, *J. Chem. Soc.* **1945**, 245. Structure: Dvornik, Edwards, *Tetrahedron* **14**, 54 (1961); Nabors *et al.*, *Tetrahedron Letters* **1969**, 2445. Stereochemistry: Solo, Pelletier, *Chem. & Ind. (London)* **1960**, 1108; Whalley, *Tetrahedron* **18**, 43 (1962). Synthetic studies: Nabors *et al.*, *ibid.* **27**, 2385 (1971). Rearrangement: S. W. Pelletier, N. V. Mody, *J. Am. Chem. Soc.* **101**, 492 (1979).

Prisms from dil alcohol, mp 172°. $[\alpha]_D^{18}$ −119° (c = 2 in anhydr alcohol).

Sulfate heptahydrate, hexagonal plates from dil acetone, mp 113°. $[\alpha]_D^{20}$ +5.5° (c = 2).

184. Ajmaline. *Ajmalan-17,21-diol;* rauwolfine; Gilu-rytmal; Cardiorythmine; Ritmos; Tachmalin. $C_{20}H_{26}N_2O_2$; mol wt 326.42. C 73.59%, H 8.03%, N 8.58%, O 9.80%. From roots of *Rauwolfia serpentina* (L.) Benth. *(Ophioxylon serpentinum* L.), *Apocynaceae.* Isolation: S. Siddiqui, R. H. Siddiqui, *J. Indian Chem. Soc.* **8**, 667 (1931); **9**, 539 (1932); **12**, 37 (1935); L. van Itallie, A. J. Steenhauer, *Arch. Pharm.* **270**, 313 (1932). Structure: A. Chatterjee, S. Bose, *J. Indian Chem. Soc.* **31**, 17 (1954); F. A. L. Anet *et al., J. Chem. Soc.* **1954**, 1242. Stereochemistry: M. F. Bartlett *et al., J. Am. Chem. Soc.* **84**, 622 (1962). Synthesis: S. Masamune *et al., ibid.* **89**, 2506 (1967); E. E. Van Tamelen, L. K. Oliver, *ibid.* **92**, 2136 (1970); K. Mashimo, Y. Sato, *Chem. Pharm. Bull.* **18**, 353 (1970). Physico-chemical properties: A. Petter, *Arzneimittel-Forsch.* **24**, 874 (1974). Antiarrhythmic activity: A. Petter, K. Engelmann, *ibid.* 876. *Reviews:* R. Robinson in *Festschrift Arthur Stoll* (Birkhäuser-Verlag, Basel, 1957) pp 457-467; A. Koskinen, M. Lounasmaa in *Progress in the Chemistry of Natural Products*, **vol. 43**, W. Herz *et al.,* Eds. (Springer-Verlag, New York, 1983) pp 268-346.

Pale amber, solvated, tetragonal prisms from methanol, $C_{20}H_{26}N_2O_2\cdot CH_3OH$, mp 158-160°. $[\alpha]_D^{18}$ +131° (c = 0.4 in chloroform). Anhydr mp 205-207°. $[\alpha]_D^{20}$ +144° (c = 0.8 in chloroform). uv max (ethanol): 247, 295 nm (log ϵ 3.94, 3.49). Sol in methanol, ethanol, ether, chloroform; slightly sol in water.

Hydrochloride dihydrate, $C_{20}H_{26}N_2O_2\cdot 2HCl\cdot 2H_2O$, hex-agonal bipyramidal crystals from water, mp 140°. $[\alpha]_D^{18}$ +96.6°. One gram dissolves in 40 ml water.

THERAP CAT: Antihypertensive. Antiarrhythmic.

185. Ajoene. *2-Propenyl 3-(2-propenylsulfinyl)-1-pro-penyl disulfide;* 4,5,9-trithiadodeca-1,6,11-triene 9-oxide. $C_9H_{14}OS_3$; mol wt 234.39. C 46.12%, H 6.02%, O 6.83%, S 41.03%. Antithrombotic principle found in garlic (*Allium sativum* L., *Liliaceae*). Formed from allicin, *q.v.,* in the approximate ratio of 4:1 (*E:Z*). The (Z)-isomer appears to be more bioactive. Isoln and effect on human platelet aggregation *in vitro:* R. Apitz-Castro *et al., Thromb. Res.* **32**, 155 (1983). [*Note:* Structure initially assigned as *allyl 1,5-hexa-dienyltrisulfide*.] Structural study, isoln of isomers and synthesis of (*E,Z*)-form: E. Block *et al., J. Am. Chem. Soc.* **106**, 8295 (1984). Structure determn and structure-activity study: *eidem, ibid.* **108**, 7045 (1986). Synthesis, not claimed: E. Block, S. Ahmad, U.S. pat. 4,643,994 (1987 to Res. Found. SUNY). Mechanism of antiplatelet activity: R. Apitz-Castro *et al., Biochem. Biophys. Res. Commun.* **141**, 145 (1986). Synergism with other antiplatelet compounds: *eidem, Thromb. Res.* **42**, 303 (1986). *In vitro* antifungal activity: S. Yoshida *et al., Appl. Environ. Microbiol.* **53**, 615 (1987).

(Z)-Ajoene

(E)-Ajoene

(*E*)-Form, colorless, odorless oil. uv max: 240 nm.

186. Ajowan Oil. Ptychotis oil. Obtained by distilla-tion of seeds of *Carum copticum* (DC.) Benth. & Hook. (*Ptychotis ajowan* DC.), *Umbelliferae.* Contains thymol, α-pinene, *p*-cymene, dipentene, γ-terpinene; fixed oil content includes petroselenic acid, oleic acid, linoleic acid, resin acids, and palmitic acid. *Ref:* Farook *et al. J. Sci. Food Agr.* **4**, 132 (1953); E. Guenther, *The Essential Oils* IV, 551-2 (Van Nostrand, New York, 1950).

Almost colorless or brownish liquid; thyme odor; sharp burning taste. d 0.910-0.930. Rotation 0° to +5°. n_D^{20} 1.498-1.504. Sol in 1-2.5 vol and more of 80% alcohol. The oil is toxic to earthworms, while alcoholic extracts are toxic to staphylococci and *Escherichia coli.* Dil extracts are useful as expectorants: Umanskii, Krutik, *Farmatsiya* **8**, no. 6, 19 (1945), *C.A.* **41**, 2209 (1947).

187. Ajugarins. Diterpenes isolated from the leaves of *Ajuga remota* (*Labiatae*). Five different ajugarins have been isolated and identified; of these ajugarins I-III possess anti-feedant activity against the African army worm, ajugarin-IV has insecticidal activity, ajugarin V is inactive. Isoln and structure of ajugarins I-III: I. Kubo *et al., Chem. Commun.* **1976**, 949; of ajugarin IV: *eidem, ibid.* **1982**, 618; of ajugarin V: *eidem, Chem. Letters* **1983**, 223. *See also:* **Japan. Kokai** 82 48,979 (1982 to Otsuka). X-ray crystal structure of 12-bromoajugarin-I and absolute configuration: I. Kubo *et al., Chem. Commun.* **1980**, 897; ^{13}C-NMR shift data: J. M. Luteijn *et al., Org. Magn. Reson.* **19**, 95 (1982). Synthetic approaches: D. J. Goldsmith *et al., J. Org. Chem.* **43**, 3182 (1978); J. M. Luteijn, Ae. de Groot, *Tetrahedron Letters* **1981**, 789. Total synthesis of ajugarin-I: S. V. Ley *et al., Chem. Commun.* **1983**, 503.

Ajugarin I

R = CH₂COOCH₃

Ajugarin-I, $C_{24}H_{34}O_7$, *[1R-(1α,4aβ,5β,6α,8α,8aα)]-4-[2-[8-(acetyloxy)-8a-[(acetyloxy)methyl]octahydro-5,6-dimeth-ylspiro(naphthalene-1(2H),2'-oxiran)-5-yl]ethyl]-2(5H)-furanone.* mp 155-157°. uv max (methanol): 212 nm (ϵ 10000).

Ajugarin-II, $C_{22}H_{32}O_6$, 8-hydroxyanalog of ajugarin I. mp 188-189°.

Ajugarin-III, $C_{24}H_{36}O_8$, 1,1-dihydroxy analog of ajugarin I. mp 243-245°.

Ajugarin-IV, $C_{23}H_{34}O_6$, mp 119-120.5°. $[\alpha]_D$ −57.5° (c = 0.06 in CHCl₃). uv max (ethanol): 215 nm (ϵ 17000).

Ajugarin-V, $C_{22}H_{32}O_5$, mp 217-218°. $[\alpha]_D$ −13.5° (c = 0.18 in CHCl₃). uv max (ethanol): 210 nm (ϵ 11000).

188. Aklomide. *2-Chloro-4-nitrobenzamide;* Aklomix. $C_7H_5ClN_2O_3$; mol wt 200.60. C 41.91%, H 2.51%, Cl

17.68%, N 13.97%, O 23.93%. Prepn from 2-chloro-4-nitrobenzoylchloride and ammonium carbonate: Grohmann, *Ber.* **24**, 3813 (1891); Cosulich *et al., J. Am. Chem. Soc.* **75**, 4675 (1953). Used vs. infections caused by protozoa and bacteria: **Belg. pat. 608,763** (1962 to Salsbury's Labs).

Gray scales from alcohol, mp 172°.
Ingredient of *Novastat.*
THERAP CAT (VET): Coccidiostat.

189. Akuammicine. *2,16,19-20-Tetradehydrocuran-17-oic acid methyl ester.* $C_{20}H_{22}N_2O_2$; mol wt 322.39. C 74.51%, H 6.88%, N 8.69%, O 9.93%. From the seeds of *Picralima klaineana,* Pierre, *Apocynaceae* found in the Gold Coast: Henry, Sharp, *J. Chem. Soc.* **1927**, 1950; Henry, *ibid.* **1932**, 2759. Structure: Aghoramurthy, Robinson, *Tetrahedron* **1**, 172 (1957); Bernauer *et al., Helv. Chim. Acta* **43**, 717 (1960); Edwards, Smith, *J. Chem. Soc.* **1961** 152. Derivs: Robinson, Thomas, *ibid.* **1955**, 2049. Total synthesis: Kutney, Fuller, *Heterocycles* **3**, 197 (1975).

Plates from ethanol+ water, mp 182°. $[\alpha]_D^{16}$ −745° (c = 0.994 in ethanol). pKa 7.45. uv max (ethanol): 227, 300, 330 nm (log ϵ 4.09, 4.07, 4.24).
Hydrochloride dihydrate, $C_{20}H_{22}N_2O_2 \cdot HCl \cdot 2H_2O$, leaflets from alcohol or water, mp 171°. $[\alpha]_D^{21}$ −610° (c = 1.430 in ethanol).
Perchlorate monohydrate, $C_{20}H_{22}N_2O_2 \cdot HClO_4 \cdot H_2O$, needles from ethanol + water, mp 134-136°.
Hydriodide monohydrate, $C_{20}H_{22}N_2O_2 \cdot HI \cdot H_2O$, square plates from water, mp 128°.
Methiodide, crystals from water, mp 252°.
Nitrate, needles from hot water, mp 182.5°.

190. Akuammine. Vincamajoridine. $C_{22}H_{26}N_2O_4$; mol wt 382.44. C 69.09%, H 6.85%, N 7.33%, O 16.73%. From seeds of *Picralima nitida* (Stapf.) Th. & H. Dur. (*P. klaineana* Pierre), *Apocynaceae:* Henry, Sharp, *J. Chem. Soc.* **131**, 1950 (1927); Henry, *ibid.* **1932**, 2759. Identity with vincamajoridine from *Vinca major* L., *Apocynaceae:* Janot *et al., Experientia* **11**, 343 (1955). Structure and chemistry: Millson *et al., ibid.* **9**, 89 (1953); Saxton, *Quart. Rev.* **10**, 141 (1956); Joule, Smith, *J. Chem. Soc.* **1962**, 312. Revised structure and absolute configuration: Olivier *et al., Bull. Soc. Chim. France* **1965**, 868.

Crystals, decomp 255°. $[\alpha]_D^{16}$ −105.3° (c = 1.31 in pyridine). pKa 7.5. uv max (ethanol): 245, 312 nm (log ϵ 3.96, 3.66). Slightly sol in water, cold alcohol; sol in boiling alcohol, chloroform, acetone.
Hydrochloride, $C_{22}H_{26}N_2O_4 \cdot HCl \cdot C_2H_5OH$, crystals, mp 232-233°.

Methyl ether, $C_{23}H_{28}N_2O_4$, needles from acetone, mp 242-243°. uv max (ethanol): 244, 308 nm (ϵ 11,500, 4200).

191. AL 721. Active lipid; Altrigen. Membrane fluidizer composed of neutral glycerides, phosphatidylcholine, and phosphatidylethanolamine in a ratio of 7:2:1. Prepn from hen egg yolk: M. Shinitzky *et al.,* **Eur.** pat. **Appl. 74,-251;** *eidem,* **U.S. pat. 4,474,773** (1983, 1984 both to Yeda Res. Dev.). Effect on drug withdrawal symptoms in morphine-addicted mice: D. S. Heron *et al., Eur. J. Pharmacol.* **83**, 253 (1982). Characterization and mode of action study: M. Lyte, M. Shinitzky, *Biochim. Biophys. Acta* **812**, 133 (1985). Effect on HIV-1 (HTLV-III/LAV) infectivity *in vitro*: P. S. Sarin *et al., N. Engl. J. Med.* **313**, 1289 (1985). Effect on lymphocyte responsiveness in the elderly: H. Rabinowich *et al., Mech. Ageing Devel.* **40**, 131 (1987).
USE: Nutritional supplement

192. Alacepril. *(S)-N-[1-[3-(Acetylthio)-2-methyl-1-oxopropyl]-*L-*prolyl]-*L-*phenylalanine;* 1-(D-3-acetylthio-2-methylpropanoyl)-L-prolyl-L-phenylalanine; *N*-[1-[(S)-3-mercapto-2-methylpropionyl]-L-prolyl]-3-phenyl-L-alanine acetate (ester); DU-1219; Cetapril. $C_{20}H_{26}N_2O_5S$; mol wt 406.50. C 59.09%, H 6.45%, N 6.89%, O 19.68%, S 7.89%. Angiotensin-converting enzyme inhibitor. Prepn: T. Sawayama *et al.,* **Japan. Kokai 80 9058;** *eidem,* **U.S. pat. 4,248,-883** (1980, 1981 both to Dainippon Pharm.). Pharmacology in animals: K. Takeyama *et al., Arzneimittel-Forsch.* **35**, 1502 (1985); *eidem, ibid.* 1507. Series of articles on pharmacology, metabolism, enzyme inhibiting activity: *ibid.* **36**, 47-83 (1986). Metabolism to captopril, *q.v.*: K. Matsumoto *et al., ibid.* 40. HPLC determn of metabolites in plasma and urine: K. Hayashi *et al., J. Chromatog.* **338**, 161 (1985). Pharmacokinetics in humans: K. Onoyama *et al., Clin. Pharmacol.* **38**, 462 (1985). Preclinical evaluation in essential hypertension: K. Mizuno *et al., Res. Commun. Chem. Pathol. Pharmacol.* **49**, 175 (1985); H. Shionoiri *et al., Curr. Ther. Res.* **38**, 537 (1985). Series of articles on toxicology: M. Iida *et al., Yakuri to Chiryo* **13**, 7033-7121 (1985), *C.A.* **104**, 21888h-21890c; **105**, 396v (1986).

Crystals from ethanol/*n*-hexane, mp 155-156°. $[\alpha]_D^{25}$ −81.3° (c = 1.02 in ethanol). LD_{50} in rats, mice (mg/kg): > 5000, > 5000 orally; > 3000, > 3000 s.c.; ~2000, ~3000 i.p. (Iida, pp 7033-40).
THERAP CAT: Antihypertensive.

193. Alachlor. *2-Chloro-N-(2,6-diethylphenyl)-N-(methoxymethyl)acetamide; 2-chloro-2',6'-diethyl-N-(methoxymethyl)acetanilide;* metachlor; CP 50144; Lasso; Alanex. $C_{14}H_{20}ClNO_2$; mol wt 269.77. C 62.33%, H 7.47%, Cl 13.14%, N 5.19%, O 11.86%. Pre-emergence herbicide. Prepn: **Neth. pat. Appl. 6,602,564** (1967 to Monsanto), *C.A.* **67**, 99832r (1967). Activity: D. M. Evans, *Chem. & Ind. (London)* **1969**, 615. Soil degradn: R. S. Hargrove, M. G. Merkle, *Weed Sci.* **19**, 652 (1971). Mechanism of action: L. M. Deal, F. D. Hess, *ibid.* **28**, 168 (1980).

Cryst solid, mp 40-41°. $d_{15.6}^{25}$ 1.133. Soly in water at 23°: 140 mg/l. Sol in ether, acetone, benzene, ethanol, ethyl acetate. Hydrolyzed under strong acid or alkaline conditions. LD_{50} orally in rats: 1200 mg/kg (Evans).

Note: The EPA has determined that alachlor is oncogenic in rats and mice: *Fed. Reg.* **50,** 1115 (1985).
USE: Herbicide.

194. Alafosfalin. *[R-(R*,S*)]-[1-[(2-Amino-1-oxopropyl)amino]ethyl]phosphonic acid;* 1R-1-(L-alanylamino)-ethylphosphonic acid; alaphosphin; Ro 03-7008. $C_5H_{13}N_2$-O_4P; mol wt 196.14. C 30.62%, H 6.68%, N 14.28%, O 32.63%, P 15.79%. Synthetic phosphonodipeptide with antibacterial activity. Prepn: F. R. Atherton *et al.,* **Ger. pat. 2,602,193;** *eidem,* **U.S. pat. 4,016,148** (1976, 1977 both to Hoffmann-La Roche); J. G. Allen *et al., Nature* **272,** 56 (1978); F. R. Atherton *et al., Antimicrob. Ag. Chemother.* **15,** 677 (1979). Improved process: E. K. Baylis, **Eur. pat. Appl. 10,872;** *eidem,* **U.S. pat. 4,331,591** (1980, 1982 both to Ciba-Geigy). Separation of diastereoisomers: J. Szewczyk *et al., Experientia* **38,** 983 (1982). Antibacterial spectrum: F. R. Atherton *et al., Antimicrob. Ag. Chemother.* **15,** 684 (1979); W. H. Traub, *Chemotherapy* **26,** 103 (1980). Synergism with β-lactams: H. B. Maruyama *et al., Antimicrob. Ag. Chemother.* **16,** 444 (1979); F. R. Atherton *et al., ibid.* **20,** 470 (1981); M. Arisawa *et al., ibid.* **21,** 706 (1982). Pharmacokinetics: J. D. Allen, L. J. Lees, *ibid.* **17,** 973 (1980). Comprehensive review: C. H. Hassall in *Antibiotics VI,* F. E. Hahn, Ed. (Springer-Verlag, New York, 1983) pp 1-11.

$$\begin{array}{c} CH_3 \\ | \\ NH_2CHCONHCH \\ | \\ HO-P=O \\ | \\ OH \end{array}$$

Crystals from ethanol-water, mp 295-296° (dec). $[\alpha]_D^{20}$ −44.0° (c = 1 in H_2O).

195. L-Alanine. L-Ala (IUPAC Abbrev.); L-α-alanine; L-α-aminopropionic acid; *(S)*-2-aminopropanoic acid. C_3-H_7NO_2; mol wt 89.09. C 40.44%, H 7.92%, N 15.72%, O 35.92%. $CH_3CH(NH_2)COOH$. An amino acid classified as non-essential for the maintenance of growth in rats. Obtained from proteins. Prepn of DL-form: Zelinsky, Stadnikoff, *Ber.* **41,** 2061 (1908); Benedict, *J. Am. Chem. Soc.* **51,** 2277 (1929); Cocker, Lapwroth, *J. Chem. Soc.* **1931,** 1399; Kendall, McKenzie, *Org. Syn. Coll. Vol. I* (2nd Ed., 1941) p 21. Laboratory procedure making use of acid hydrolysis of silk fibroin (degummed white silk): Stein *et al., J. Biol. Chem.* **143,** 121 (1942); Stein *et al., ibid.* **154,** 191 (1944); Stein, Moore, *Biochem. Prepn.* **1,** 9 (1949).

Orthorhombic crystals from water, decomp 297°. d 1.401. $[\alpha]_D^{25}$ +2.8° (c = 6 in H_2O); +9.55° (HCl). Soly (g/l) in water at 0°: 127.3; at 25°: 166.5; at 50°: 217.9; at 75°: 285.1; at 100°: 373.0. Soly in cold 80% ethanol: 0.2%. Insol in ether.

Hydrochloride, $C_3H_7NO_2$·HCl, prisms, decomp 204°. $[\alpha]_D^{26}$ +8.5° (c = 9.3).

DL-Form, orthorhombic bipyramidal needles or rods from water. d 1.424. Sublimes above 200°. Decomp 264-296° depending on rate of heating. pK_1 2.35; pK_2 9.87. Soly in water at 0°: 121 g/l, at 25°: 167 g/l, at 50°: 231 g/l, at 75°: 319 g/l, at 100°: 440 g/l; in ethanol 0.0087 g/100 g at 25°. Insol in ether.

196. β-Alanine. Beta-alanine; β-aminopropionic acid; 3-aminopropanoic acid; 3-aminopropionic acid; Abufène. $C_3H_7NO_2$; mol wt 89.09. C 40.44%, H 7.92%, N 15.72%, O 35.92%. $NH_2CH_2CH_2CO_2H$. Prepd by the action of KOBr and KOH upon succinimide: Clarke, Behr, *Org. Syn.* **16,** 1 (1936). By the action of liq ammonia upon methyl acrylate: Morsch, *Monatsh.* **63,** 220 (1933), cf. *C.A.* **41,** 4104 (1947); by the addition of NH_4OH to acrylonitrile: Ford *et al., J. Am. Chem. Soc.* **69,** 844 (1947). By electrolytic oxidation of 3-amino-1-propanol in H_2SO_4 using Pb electrodes without diaphragm: *Jubilee Vol. Émil Barell* **1946,** 85-91. For industrial methods of prepn *see* several pats. by T. L. Gresham to B. F. Goodrich. Prepn from ethylene cyanohydrin (β-hydroxypropionitrile): Boatright, **U.S. pat. 2,734,081** (1956 to Am. Cyanamid); from β-aminopropionitrile: Ford, *Org. Syn. coll. vol. III,* 34 (1955). Improved process: Beutel, Klemchuk, **U.S. pat. 2,956,080** (1960 to Merck & Co.).

Orthorhombic bipyramidal crystals from water, decomp 207° (very rapid heating). Decomp 197-198° (Ford, *Org. Syn. loc. cit.*). Slightly sweet taste. pK_1 3.60; pK_2 10.19. pH of 5% aq soln: 6.0 to 7.3. Freely sol in water, slightly in alcohol. Practically insol in ether, acetone.

Hydrochloride, $C_3H_7NO_2$·HCl, plates, leaflets, mp 122.5°. Freely sol in water, less sol in alcohol, insol in ether.

Platinichloride, $2C_3H_7NO_2$·2HCl·$PtCl_4$, yellow leaflets from alcohol + HCl, decomp 210°. Freely sol in water, sparingly in abs alcohol.

USE: In the synthesis of pantothenic acid and derivatives; as buffer in electroplating.

197. L-Alanosine. *3-(Hydroxynitrosoamino)-L-alanine;* L-2-amino-3-(hydroxynitrosoamino)propanoic acid; L-2-amino-3-[(N-nitroso)hydroxylamino]propionic acid. C_3H_7-N_3O_4; mol wt 149.11. C 24.16%, H 4.73%, N 28.18%, O 42.92%. Antibiotic substance from the fermentation of *Streptomyces alanosinicus* n. sp. Prepn: **Neth. pat. Appl. 6,509,543** corresp to J. Thiemann, Y. S. K. Murthy, **U.S. pat. 3,676,490** (1966, 1972 to Lepetit). The first natural product found to have a *N*-nitrosohydroxylamino group on an aliphatic chain. Isoln and structure: Coronelli *et al., Farmaco Ed. Sci.* **21,** 269 (1966). Characterization: Thiemann, Beretta, *J. Antibiot.* **19A,** 155 (1966). Structure and synthesis of L-, D- and DL-forms: Lancini *et al., Tetrahedron Letters* **1966,** 1769; *eidem, Farmaco Ed. Sci.* **24,** 169 (1969). Synthesis of L-form: Isowa *et al., Bull. Chem. Soc. Japan* **46,** 1847 (1973). Improved synthesis of DL-form: Eaton *et al., J. Med. Chem.* **16,** 289 (1973). Pharmacology: Murthy *et al., Nature* **211,** 1198 (1966). Mode of action studies: Gale *et al., Biochem. Pharmacol.* **17,** 363 (1968).

$$\begin{array}{c} NH_2 \\ | \\ HONCH_2 \cdots C \cdots COOH \\ | \quad\quad | \\ NO \quad\quad H \end{array}$$

Crystalline powder from slightly acidic water, dec 190°. $[\alpha]_D$ +8°, −46°, −37.8° (in 1N HCl, 0.1N NaOH, water). uv max (0.1N HCl): 228 nm ($E_{1cm}^{1\%}$ 505); in 0.1N NaOH: 250 nm ($E_{1cm}^{1\%}$ 630). pKa 4.8. Slightly sol in water; practically insol in the common organic solvents. Sol in alkaline and acidic solns, from which it ppts by adjusting the pH between 4 and 6. LD_{50} in mice (mg/kg): 600 i.p.; 300 i.v. (Thiemann, Murthy).

D-Form, mp 183°. $[\alpha]_D$ +45° (c = 0.5N in 1N HCl).
USE: Experimental insect reproduction inhibitor, Kenaga, *J. Econ. Entomol.* **62,** 1006 (1969).

198. Alantolactone. *[3aR-(3aα,5β,8aβ,9aα)]-3a,5,6,7,-8,8a,9,9a-Octahydro-5,8a-dimethyl-3-methylenenaphtho-[2,3-b]furan-2(3H)-one; 8β-hydroxy-4αH-eudesm-5-en-12-oic acid γ-lactone;* helenin; alant camphor; elecampane camphor; inula camphor; Eupatal. $C_{15}H_{20}O_2$; mol wt 232.31. C 77.55%, H 8.68%, O 13.77%. A terpene from roots of *Inula helenium* L., *Compositae:* Kallen, *Ber.* **6,** 1506 (1873); **9,** 154 (1876); Ruzicka *et al., Helv. Chim. Acta* **14,** 397, 1090 (1931); **16,** 268 (1933). Structure: Marshall, Cohen, *J. Org. Chem.* **29,** 3727 (1964). Stereoselective synthesis: Marshall *et al., J. Am. Chem. Soc.* **88,** 3408 (1966).

Crystals from alcohol, mp 78-79°. bp 275°. $[\alpha]_D$ +175° (chloroform). uv max (ethanol): 212 nm (ε 9500). Volatile with steam. Freely sol in alcohol, chloroform, benzene, ether, oils. Practically insol in water.

THERAP CAT: Anthelmintic (Nematodes).

199. Alazopeptin. L-Alanyl-(6-diazo-5-oxo)-L-norleucyl-(6-diazo-5-oxo)-L-norleucine. $C_{15}H_{20}N_6O_5$; mol wt 364.37. C 49.45%, H 5.53%, N 23.06%, O 21.96%. Tumor-inhibiting antibiotic produced by *Streptomyces griseoplanus* from soil near Williamsburg, Iowa: De Voe *et al., Antibiot.*

Ann. **1956-7**, p 730. Peptide consisting of one mole α-alanine and two moles of a C_6 diazo keto amino acid oxidizable to glutamic acid: E. L. Patterson *et al.*, *Antimicrob. Ag. Chemother.* **1965**, 115-118. Anti-tumor activity: T. Hata *et al.*, *J. Antibiot.* **26**, 181 (1973). Toxicity: J. B. Thiersch, *Proc. Soc. Exp. Biol. Med.* **97**, 888 (1958).

$$N_2 = CHCOCH_2CH_2CHNHCOCHCH_2CH_2COCH = N_2$$

with COOH and NHCH_2CH = CH_2 substituents

Monohydrate, crystals from dilute acetone. Poor stability. Has no definite melting point. $[\alpha]_D^{25}$ +9.5° (c = 4.7 in H_2O). uv max (pH 7.0 phosphate buffer): 242, 274 nm ($E_{1cm}^{1\%}$ 321, 549). Freely sol in water. Somewhat sol in acetic acid, formamide, DMSO, aq solns of methanol, ethanol, acetone. Practically insol in anhyd alcohols, acetone, ethyl acetate, ether. LD_{50} in rats: 150 mg/kg (Thiersch).

200. Albaspidin. *2,2'-Methylenebis[6-butyryl-3,5-dihydroxy-4,4-dimethyl-2,5-cyclohexadien-1-one];* polystichalbin; methylenebis(butyrylfilicinic acid); albaspidin-BB. $C_{25}H_{32}O_8$; mol wt 460.51. C 65.20%, H 7.00%, O 27.80%. Found in the rhizomes of the male fern, *Aspidium filix mas* (L.) Schott., *Dryopteris filix mas* (L.) Schott., *A. spinulosum, Polypodiaceae* and other ferns. Isoln from *Aspidium* extract: Boehm, *Ann.* **318**, 305 (1901); McGookin *et al.*, *J. Chem. Soc.* **1953**, 1828; *see also* Tryon *et al.*, *Phytochemistry* **12**, 683 (1973). Isoln as one component of a homologous mixture: Penttila, Sundman, *Acta Chem. Scand.* **18**, 344 (1964). Synthesis: Riedl, Mitteldorf, *Ber.* **89**, 2595 (1956); Inagaki *et al.*, *J. Pharm. Soc. Japan* **76**, 1258 (1956). Biosynthetic studies: Penttila, Fales, *J. Am. Chem. Soc.* **88**, 2327 (1966).

Crystals from ethanol or methanol, mp 149°. Freely sol in chloroform; moderately sol in ether, benzene; sparingly sol in alc, acetone, glacial acetic acid, sol in KOH solns; very sparingly sol in Na_2CO_3 solns. Practically insol in methanol.

201. Albendazole. *[5-(Propylthio)-1H-benzimidazol-2-yl]carbamic acid methyl ester;* methyl 5-(propylthio)-2-benzimidazolecarbamate; 5-(propylthio)-2-carbomethoxyaminobenzimidazole; SKF 62979; Valbazen; Zental. $C_{12}H_{15}N_3O_2S$; mol wt 265.33. C 54.32%, H 5.70%, N 15.84%, O 12.06%, S 12.08%. Prepn: R. J. Gyurik, V. J. Theodorides, U.S. pat. **3,915,986** (1975 to SmithKline). Anthelmintic activity: V. J. Theodorides, *Experientia* **32**, 702 (1976); D. A. Denham *et al.*, *J. Helminthol.* **54**, 199 (1980). Comparative effect vs cestodes: W. S. Evans *et al.*, *J. Parasitol.* **66**, 935 (1980). Efficacy studies: J. C. Williams *et al.*, *Am. J. Vet. Res.* **42**, 318 (1981); P. Stevenson *et al.*, *Vet. Rec.* **109**, 82 (1981). Clinical trial in humans: P. Pene *et al.*, *Am. J. Trop. Med. Hyg.* **31**, 263 (1982).

Colorless cryst, mp 208-210°.
THERAP CAT (VET): Anthelmintic.

202. Albizziin. *3-[(Aminocarbonyl)amino]-L-alanine; 2-amino-3-ureidopropionic acid.* $C_4H_9N_3O_3$; mol wt 147.14. C 32.65%, H 6.17%, N 28.56%, O 32.62%. $NH_2CONHCH_2$-$CH(NH_2)COOH.$ Constituent of several plants belonging to the family *Mimosaceae.* First obtained from the seeds of *Albizzia julibrissin* Durazz., *Mimosaceae;* Gmelin *et al.*, *Z. Naturforsch.* **13b**, 252 (1958); *Z. Physiol. Chem.* **314**, 28 (1959). Structure: Kjaer *et al.*, *Experientia* **15**, 253 (1959).

Synthesis: Kjaer, Larsen, *Acta Chem. Scand.* **13**, 1565 (1959); Rudinger *et al.*, *Coll. Czech. Chem. Commun.* **25**, 2022 (1960).

Needles from dil alcohol, decomp 218-220°. $[\alpha]_D^{25}$ −66.2° (c = 4 in H_2O); $[\alpha]_D^{24}$ −22.2° (c = 4.24 in 1.0N HCl); $[\alpha]_D^{24}$ +3.2° (c = 4.7 in 1.0N NaOH). ORD data: Kjaer *et al.*, *Acta Chem. Scand.* **18**, 2412 (1964).

203. Albofungin. *13-Amino-3,4,8a,13-tetrahydro-1,-15,16-trihydroxy-4-methoxy-12-methyl-1H-xantheno[4',-3',2':4,5][1,3]benzodioxino[7,6-g]isoquinoline-14,17(2H,-9H)-dione.* $C_{27}H_{24}N_2O_9$; mol wt 520.50. C 62.30%, H 4.65%, N 5.38%, O 27.67%. Antifungal antibiotic produced by *Streptomyces albus* var *fungistaticus (fungatus)* Solovyeva et Rudaya. Isoln procedure: Khokhlov, Liberman, *Proc. Symp. Antibiotics Prague* (May 1959) p 81. Structure: Gurevich *et al.*, *Tetrahedron Letters* **1972**, 1751. Stereochemistry: Gurevich *et al.*, *ibid.* **1974**, 2801. Thought to be identical with *BA-180265* described by Liu *et al.*, in *Antimicrob. Ag. Chemother.* **1962**, 767. Chemical and biological properties: Gurevich *et al.*, *Antibiotiki* **17**, 771 (1972).

Crystals from nitromethane, mp 304-307°. Also reported as dec 190°. $[\alpha]_D^{20}$ −670° (DMF). uv max (ethanol): 228, 254, 303, 376 nm (log ε 4.58, 4.58, 4.19, 4.42). Practically insol in water, petr ether. Sparingly sol in alcohol; quite sol in chloroform, dichloroethane, acetone, chlorobenzene, formamide, dimethylformamide, glacial acetic acid.

204. Albomycin. Iron-containing antibiotic produced by *Actinomyces subtropicus:* Gauze, Braznikova, *Novosti Med.* **23**, 3 (1951). Used as the sulfate. Consists of six components, α, β, γ, δ_1, δ_2, ε, of which δ_1 and δ_2 are the main components. δ_2 is the unstable, highly active albomycin A_1; the other components are its degradation products: Braznikova *et al.*, *Biokhimiya* **22**, 111 (1959); Turková *et al.*, *Antibiotiki* **7**, 878 (1962). Early studies proposed cyclic hexapeptide structures containing three serines and three ornithine derivatives for δ_1-, δ_2- and ε-albomycin (*cf.* Ferrichromes). Structural studies: *eidem.*, *Coll. Czech. Chem. Commun.* **30**, 118 (1965); Poddubnaya, el'Naggar, *Zh. Obshch. Khim.* **38**, 450 (1968). Conflicting structural analysis shows a 3:1 ornithine : serine ratio: Maehr, Pitcher, *J. Antibiot.* **24**, 830 (1971). Similarity to or identity with grisein, *q.v.*: Stapley, Ormond, *Science* **125**, 587 (1957); Turková *et al.*, *Coll. Czech. Chem. Commun.* **31**, 2444 (1966). Review and antibacterial spectrum of albomycin and other hydroxamic acids: Gauze, *Brit. Med. J.* **2**, 1177 (1955); Bhuyan in *Antibiotics* **1**, D. Gottlieb, P. D. Shaw, Eds. (Springer-Verlag, New York, 1967) pp 153-155; Maehr, *Pure Appl. Chem.* **28**, 603-636 (1971); Emery, *Advan. Enzymol. Relat. Areas Mol. Biol.* **35**, 135-185 (1971).

Sulfate, amorphous red powder. uv max: 283 nm ($E_{1cm}^{1\%}$ 880). Freely sol in water. Slightly sol in methanol. Practically insol in most other organic solvents. Aq solns have a bright orange color. Effective against penicillin-resistant pneumococci and staphylococci. The toxicity is comparable to that of penicillin.

Note: Not to be confused with albamycin.

205. Alborixin. Antibiotic S 14750A. $C_{48}H_{84}O_{14}$; mol wt 885.20. C 65.13%, H 9.57%, O 25.30%. Polycyclic polyether antibiotic ionophore, isolated from a strain of *Streptomyces albus.* Initial description and x-ray structure: M. Alléaume *et al.*, *Chem. Commun.* **1975**, 411. Production from *Streptomyces hygroscopicus* and use: M. Kuhn, H. D. King, **Ger.** pat. **2,608,337** (1976 to Sandoz), *C.A.* **86**, 28491k (1977). Isoln from *S. albus*, structure, properties: P. Gachon *et al., J. Antibiot.* **29**, 603 (1976); C. Delhomme *et al., ibid.* 692. NMR spectrum and solution conformation: N. A. Rodios, M. J. O. Anteunis, *Bull. Soc. Chim. Belg.* **88**, 279 (1979). Revised structure: H. Seto *et al., J. Antibiot.* **32**, 970 (1979). *In vitro* study: M. Chapel *et al., ibid.* 740. Effects on cardiovascular function and plasma cation concentration in the dog: N. Moins *et al., J. Cardiovasc. Pharmacol.* **1**, 659 (1979).

White amorphous powder, mp 100-105°. $[\alpha]^{20}$ $-7°$ (c = 4 in acetone). pKa 10.02 (25° in methanol). LD_{50} in mice: 15 mg/kg s.c. (Delhomme). LD_{50} originally reported as 150 mg/kg, corrected in *J. Antibiot.* **30**, Suppl. (1977), facing p 77-12.

206. Albumen. Egg white; dried egg white. By convention the pure protein is spelled "albumin" and commercial egg white is called "albumen". The word "albumen" goes back to Roman times and is classical Latin for egg white: *Ref:* Plinius (Major), *Historia Naturalis* **28**, 6, 18 paragraph 66. Constitutes about six parts by weight of a hen's egg (wet basis). Average weight of raw egg white 32.9 g. Albumen contains 75% ovalbumin *(q.v.)*, ovoconalbumin, ovomucoid, ovomucin, ovoglobulin, lysozyme *(q.v.)* and avidin *(q.v.)*. Monograph: A. L. Romanoff, A. J. Romanoff, *The Avian Egg* (Wiley, New York, 1949) 918 pp.

Raw egg albumen, clear, colloidal, flowing, limpid mass. White rubbery solid when denatured. d_{25}^{25} 1.035. n_D^{25} 1.356. Coagulating temp 61°. mp -0.42. pH 7.6.

Dried albumen, yellow, transparent, amorphous lumps or scales, or yellow powder. Decomposes in moist air. In water it swells at first, then dissolves gradually. The soln coagulates at 61° (because of denaturation of the proteins). Albumen solns are also denatured on contact with the following chemicals: Salts of copper, iron, mercury and silver; H_2O_2, phenol, picric acid, alum, tannic acid, formaldehyde, ether, alcohol.

USE: For clarifying and refining wines and vinegars. As textile dye mordant, in lithography as vehicle for substances that sensitize plates, in gilding leather (Venetian decorations), stamping with gold and bronze powder (especially in book binding). In adhesives and veneers. Ingredient of compositions used in sizing and in making papers. As activating agent for certain enzymes. Formerly in sugar refining. Ingredient of bakery products, confectionery, food preparations. In pharmaceutical compounding, i.e. to make various albuminates. As analytical reagent in testing for ionic mercury, foreign oil in olive oil, candy or wine colorings.

THERAP CAT: Antidote to mercury poisoning.

207. Albumin(s). A group of proteins characterized by heat coagulability and solubility in dilute salt soln. Found in nearly all living body tissues. The most notable albumins are ovalbumin *(q.v.)*, serum albumin *(q.v.)*, lactalbumin, grain and soybean albumins.

Sol in pure water; coagulated (denatured) by heat. Precipitate from solution only at high ammonium sulfate concentrations (70-100% of saturation).

208. Albumin Tannate. Tannin albuminate; Albutannin; Tannalbin. Contains about 50% tannin.

Yellowish-white, odorless powder. Insol in water, alcohol, chloroform, ether; decomposed by alkali hydroxides and carbonates.

THERAP CAT: Astringent; antidiarrheal.

THERAP CAT (VET): Astringent, antidiarrheal.

209. Albuterol. α^1-[[(1,1-Dimethylethyl)amino]methyl]-4-hydroxy-1,3-benzenedimethanol; α^1-[tert-butylamino)methyl]-4-hydroxy-m-xylene-α,α'-diol; 2-(tert-butylamino)-1-(4-hydroxy-3-hydroxymethylphenyl)ethanol; 4-hydroxy-3-hydroxymethyl-α-[(tert-butylamino)methyl]-benzyl alcohol; salbutamol; AH 3365. $C_{13}H_{21}NO_3$; mol wt

239.31. C 65.24%, H 8.85%, N 5.85%, O 20.06%. Prepn: L. H. C. Lunts *et al., S. Afr.* pat. **67** 05,591; *eidem, U.S.* pat. **3,644,353** (1968, 1972, to Allen and Hanburys). Synthesis and structure-activity studies: Collin *et al., J. Med. Chem.* **13**, 674 (1970). Absolute configuration of optical isomers: Hartley, Middlemiss, *ibid.* **14**, 895 (1971). Pharmacology: Brittain *et al., Nature* **219**, 862 (1968); Callum *et al., Brit. J. Pharmacol.* **35**, 141 (1969). Metabolism: Martin *et al., Eur. J. Pharmacol.* **14**, 183 (1971). Clinical studies in asthma: S. Chodosh, *Arch. Intern. Med.* **138**, 1394 (1978); R. W. Light *et al., ibid.* **139**, 639 (1979). Use in prevention of premature labor: G. J. Addis, *Lancet* **1**, 42 (1981). *Review: Drugs* **1**, 274-302 (1971); R. T. Brittain, D. M. Harris, in *Pharmacological and Biochemical Properties of Drug Substances* vol. 1, M. E. Goldberg, Ed. (Am. Pharm. Assoc., Washington, DC, 1977) pp 257-276. Comprehensive description: H. Y. Aboul-Enein *et al.,* in *Analytical Profiles of Drug Substances* vol. 10, K. Florey, Ed. (Academic Press, New York, 1981) pp 665-689.

Crystalline powder from ethanol-ethyl acetate or ethyl acetate-cyclohexane, mp 151° (Lunts); 157-158° (Collin). Sol in most organic solvents.

Sulfate, $C_{26}H_{44}N_2O_{10}S$, *Aerolin, Asmaven, Broncovaleas, Cetsim, Cobutolin, Ecovent, Proventil, Salbulin, Salbumol, Salbutine, Salbuvent, Sultanol, Venetlin, Ventodisks, Ventolin, Ventolin Obstetric Injection, Volmax.*

THERAP CAT: Bronchodilator; tocolytic.

210. Albutoin. *5-(2-Methylpropyl)-3-(2-propenyl)-2-thioxo-4-imidazolidinone; 3-allyl-5-isobutyl-2-thiohydantoin;* 3-allyl-5-*sec*-butyl-2-thiohydantoin; CO-ORD; Euprax. $C_{10}H_{16}N_2OS$; mol wt 212.33. C 56.57%, H 7.60%, N 13.20%, O 7.53%, S 15.10%. Prepn: Oba *et al., C.A.* **46**, 3885f (1952).

Crystals, mp 210-211°.

USE: Fog inhibitor in photography.

THERAP CAT: Anticonvulsant.

211. Alcian Blue. *C.I. Ingrain Blue 1;* C.I. 74240; alcian blue 8GX; michrome no. 24. When R = toluyl, the mol wt is about 1300. Discovered by Haddock and Wood in 1944: Haddock, *Research* **1**, 685 (1948); **Brit.** pats. **586,340** and

X = an onium group, e.g.

(R = alkyl or aryl)

587,636 (both to ICI). During dyeing the X groups split off. Prepn: *Colour Index* **vol. 4** (3rd ed.) p 4620.

Greenish-black crystals with metallic sheen. When R = toluyl the solubility in water at 20° is about 9.5% w/w giving bright greenish-blue solutions. Soly in absolute ethanol about 6.0, Cellosolve 6.0, ethylene glycol 3.25. Practically insoluble in xylene.

USE: Gelling agent for lubricating fluids. As bacterial stain. To dye histiocytes and fibroblasts *in vivo*: McManus, Bailie, *Fed. Proc.* **22**, no. 2, part I, p 190 (March-April 1963).

212. Alclofenac. *3-Chloro-4-(2-propenyloxy)benzene-acetic acid; [4-(allyloxy)-3-chlorophenyl]acetic acid;* W 7320; Allopydin; Argun; Epinal; Medifenac; Mervan; Neoston; Prinalgin; Reufenac; Zumaril. $C_{11}H_{11}ClO_3$; mol wt 226.66. C 58.29%, H 4.89%, Cl 15.64%, O 21.17%. Prepn: Buu-Hoi *et al.*, **Belg. pat. 704,368** (1968 to Madan). Pharmacological studies: *eidem, Naturwiss.* **56**, 330 (1969); Lambotte, *Arzneimittel-Forsch.* **20**, 569 (1970); Lambelin *et al., ibid.* 610: Metabolism: Roncucci *et al., ibid.* 631. Toxicity data: Lambelin *et al., ibid.* 618. Clinical studies: Van Hoek, *Curr. Ther. Res.* **12**, 551 (1970).

$$CH_2=CHCH_2O-\bigcirc-CH_2COOH$$

Prisms from cyclohexane, mp 92-93°. LD_{50} in mice, rats (mg/kg): 1100, 1050 orally; 600, 630 s.c.; 550, 530 i.p. Monoethanolamine complex, $C_{13}H_{18}ClNO_4$, *Mirvan A.*

THERAP CAT: Analgesic, antipyretic, anti-inflammatory.

213. Alclometasone. *7-Chloro-11,17,21-trihydroxy-16-methylpregna-1,4-diene-3,20-dione;* 7α-chloro-16α-methyl-prednisolone. $C_{22}H_{29}ClO_5$; mol wt 408.92. C 64.62%, H 7.15%, Cl 8.67%, O 19.56%. Non-fluorinated corticosteroid with low systemic effects. Prepn: M. J. Green *et al.*, **U.S. pat. 4,076,708;** M. J. Green, H. J. Shue, **U.S. pat. 4,124,707** (both 1978 to Schering); M. J. Green *et al., J. Steroid Biochem.* **11**, 61 (1979); H. J. Shue *et al., J. Med. Chem.* **23**, 430 (1980). Topical anti-inflammatory activity of the 17,21-dipropionate: B. Lutsky *et al., Arzneimittel-Forsch.* **29**, 992 (1979); M. J. Green *et al., ibid.* **30**, 1618 (1980).

Crystals from acetone/hexane, mp 176-179°. $[\alpha]_D^{26}$ +47.5° (c = 0.3 in DMF). uv max (methanol): 242 nm (ε 15500).

17,21-Dipropionate, $C_{28}H_{37}ClO_7$, *Sch 22219, Aclosone, Aclovate, Almeta, Delonal, Ecoderm, Legederm, Miloderme, Modrasone, Perderm, Vaderm.* Crystals from acetone/methanol/isopropyl ether, mp 212-216°. $[\alpha]_D^{26}$ +42.6° (c = 0.3 in DMF). uv max (methanol): 242 nm (ε 15600).

THERAP CAT: Topical anti-inflammatory.

214. Alcuronium. *4,4'-Didemethyl-4,4'-di-2-propenyltoxiferine I; N,N'-diallylnortoxiferinium;* diallylbis(nortoxiferine); diallylnortoxiferine; diallyltoxiferine. $[C_{44}H_{50}N_4\text{-}O_2]^{2+}$. Prepn of dichloride and diiodide: Boller *et al.*, **U.S. pat. 3,080,373** (1963 to Hoffmann-La Roche).

Dichloride, $C_{44}H_{50}Cl_2N_4O_2$, *N,N'-diallylnortoxiferinium dichloride, Ro 4-3816, Alloferin, Toxiferene.* Crystals from methanol or ethanol. Compd contains 5 moles of water of crystn after equilibration in air; $[\alpha]_D^{22}$ −348° (methanol); uv max (methanol): 292 nm (ε 43,000).

Diiodide, $C_{44}H_{50}I_2N_4O_2$, solid. uv max (methanol): 291 nm (ε 39,900).

THERAP CAT: Dichloride as skeletal muscle relaxant.

215. C_{14}-Aldehyde. *2-Methyl-4-(2,6,6-trimethyl-1-cyclohexen-1-yl)-3-butenal;* allo-β-C_{14}-aldehyde; α,2,6,6-tetramethyl-1-cyclohexene-3-butenal. $C_{14}H_{22}O$; mol wt 206.32. C 81.50%, H 10.75%, O 7.76%. Usually prepared from β-ionone and methyl or ethyl chloroacetate: Milas *et al., J. Am. Chem. Soc.* **70**, 1584 (1948); Stieg, Gillis, **U.S. pat. 2,987,550** (1961 to Pfizer); Oediger *et al., Ber.* **97**, 549 (1964). May also exist in its isomeric form α,2,6,6-*tetramethyl-1-cyclohexene-2-butenal,* named also α,2,6,6-*tetramethyl-1-cyclohexene-1-crotonaldehyde.*

Crystals from cooled pentane, mp about 0° ± 1.0°. $bp_{1.0}$ 103-110°. d_{25}^{25} 0.956. n_D^{20} 1.5112. uv max: 232 nm ($E_{1cm}^{1\%}$ 967).

USE: In the synthesis of vitamin A.

216. Aldicarb. *2-Methyl-2-(methylthio)propanal O-[(methylamino)carbonyl]oxime; 2-methyl-2-(methylthio)propionaldehyde O-(methylcarbamoyl)oxime;* UC 21149; Temik. $C_7H_{14}N_2O_2S$; mol wt 190.25. C 44.19%, H 7.41%, N 14.72%, O 16.82%, S 16.85%. $CH_3SC(CH_3)_2CH=NOCONHCH_3$. Synthesis: Payne *et al., J. Agr. Food Chem.* **14**, 356 (1966). C^{14}-labeled synthesis: Bartley *et al., ibid.* 604. Metabolism: Bartley *et al., ibid.* **18**, 446 (1970). Mass spectrum: Benson, Damico, *J. Assoc. Offic. Anal. Chem.* **1968**, 347. Crystal and molecular structure: F. Takusagawa, R. A. Jacobson, *J. Agr. Food Chem.* **25**, 333 (1977). *Review:* Romine, *Anal. Methods Pestic. Plant Growth Regul.* **7**, 147-162 (1973).

Crystals from isopropyl ether, mp 99-100°. Soly at 25° (w/w) in water: 0.6%; acetone: 35%; benzene: 15%; xylene: 5%; methylene chloride: 30%. LD_{50} orally in female rats: ~1 mg/kg, Weiden *et al., J. Econ. Entomol.* **58**, 154 (1965).

USE: Insecticide, acaricide, nematocide.

217. Aldol. *3-Hydroxybutanal; 3-hydroxybutyraldehyde;* acetaldol. $C_4H_8O_2$; mol wt 88.10. C 54.53%, H 9.15%, O 36.32%. $CH_3CH(OH)CH_2CHO$. Manuf by condensation of acetaldehyde in aq NaOH: Alhéritière, Gobron, **U.S. pat. 2,713,598** (1955 to Usines de Melle).

Colorless, thick liquid. d^{16} 1.109. bp_{20} 83°; dec about 85°. Miscible with water, alcohol, ether. LD_{50} orally in rats: 2.18 g/kg, H. F. Smyth *et al., J. Ind. Hyg. Toxicol.* **31**, 60 (1949).

USE: Manuf rubber vulcanizers, accelerators and age resisters; in perfumes; ore flotation.

THERAP CAT: Hypnotic, sedative.

218. Aldosterone. *(11β)-11,21-Dihydroxy-3,20-dioxopregn-4-en-18-al;* 3,20-diketo-11β,18-oxido-4-pregnene-

18,21-diol; Aldocorten; Aldocortin; Electrocortin. $C_{21}H_{28}$-O_5; mol wt 360.44. C 69.97%, H 7.83%, O 22.20%. Adrenocortical steroid which exerts regulatory influence on metabolism of electrolytes and water. Isoln: Simpson et al., Experientia **9**, 333 (1953); Helv. Chim. Acta **37**, 1163 (1954); Mattox et al., J. Am. Chem. Soc. **75**, 4869 (1953); Harman et al., ibid. **76**, 5035 (1954). Solutions contain an equilibrium mixture of the aldehyde and the hemiacetal, the equilibrium favoring the latter. Structure: Tait et al., Experientia **10**, 132 (1954); Helv. Chim. Acta **37**, 1200 (1954). Crystal structure and molecular conformation: Duax, Hauptmann, J. Am. Chem. Soc. **94**, 5467 (1972). ^{13}C-NMR spectrum: P. Gerard, Org. Magn. Resonance **11**, 478 (1978). Total synthesis: Schmidlin et al., Helv. Chim. Acta **40**, 1438 (1957); Johnson et al., J. Am. Chem. Soc. **80**, 2585 (1958); **85**, 1409 (1963). Three-step synthesis from corticosterone: Barton, Beaton, ibid. **82**, 2640 (1960); **83**, 4083 (1961). Alternate synthesis: D. H. R. Barton et al., J. Chem. Soc., Perkin Trans. I **1975**, 2243; M. Miyano, J. Org. Chem. **46**, 1846 (1981). Biosynthesized in the zona glomerulosa and transported chiefly by albumin. In man, 400 μg secreted normally in one day. Secretion influenced by ACTH, growth hormone, plasma sodium and potassium, and the renin-angiotensin system. Causes reabsorption of Na^+, Cl^-, and HCO_3^- and diuresis of K^+. Review: L. F. Fieser, M. Fieser, Steroids (Reinhold, New York, 1959) pp 701-720.

Hydrated crystals from dilute acetone, mp 108-112° (when anhydr mp 164°). $[\alpha]_D^{23}$ +152.2° (anhydr; c = 2 in acetone). $[\alpha]_D^{25}$ +161° (c = 0.1 in chloroform). uv max: 240 nm (log ϵ 4.20 for the monohydrate; ϵ_{mol} 15,000 for the anhydr).

21-Acetate, $C_{23}H_{30}O_6$, flat needles from acetone + ether, mp 198-199°, $[\alpha]_D^{24}$ +121.7° (c = 0.71 in chloroform). Synthesis: Wettstein et al.; Jeger, U.S. pats. **3,002,972** and **3,014,029** (both 1958 to Ciba).

Minimum observable activity of the free alcohol in the urinary sodium retention assay occurs at between 0.05 and 0.01 γ per rat, while at least 16 γ of desoxycorticosterone acetate is required for the same level of activity.

THERAP CAT: Mineralcorticoid.

THERAP CAT (VET): Mineralcorticoid.

219. Aldrin. *1,2,3,4,10,10-Hexachloro-1,4,4a,5,8,8a-hexahydro-1,4:5,8-dimethanonaphthalene;* HHDN; compd 118; Octalene. $C_{12}H_8Cl_6$; mol wt 364.93. C 39.50%, H 2.21%, Cl 58.30%. Activity: C. W. Kearns et al., J. Econ. Entomol. **42**, 127 (1949). Prepn of aldrin and *endo,endo-*isomer: Lidov, U.S. pat. **2,635,977** (1953 to Shell). Alternate syntheses: Schmerling, U.S. pat. **2,911,447** (1959 to Universal Oil Prod.); Korte, Rechmeier, Ann. **656**, 131 (1962).

Crystals, mp 104°. Vapor press at 20°: 7.5 $\times$ 10^{-5} mm Hg. Very sol in most organic solvents and insol in water. Stable in presence of organic and inorganic alkalies; stable to the action of hydrated metal chlorides. LD_{50} orally in male, female rats: 39, 60 mg/kg, T. B. Gaines, Toxicol. Appl. Pharmacol. **14**, 515 (1969).

*endo,endo-*Isomer, *isodrin, compd 711.* Crystals, mp 240-242°. LD_{50} orally in male, female rats: 15, 7.0 mg/kg, T. B. Gaines, loc. cit.

Caution: Poisoning may occur by ingestion, inhalation, skin absorption. Severe symptoms may result from ingestion or percutaneous absorption of 1 to 3 g, especially in presence of liver disease. *Acute toxicity:* Renal damage, tremors, ataxia, convulsions followed by CNS depression, respiratory failure, death. *Chronic toxicity:* Prolonged exposure may cause hepatic damage, cf. Patty's Industrial Hygiene and Toxicology vol. 2B, G. D. Clayton, F. E. Clayton, Eds. (Wiley-Interscience, New York, 3rd ed., 1981) pp 3702-3707.

USE: Formerly as insecticide; manuf and use has been discontinued in the U.S.

220. Aletris. Star grass; starwort; true unicorn root; blazing star; colic root. Rhizome of Aletris farinosa L., Liliaceae. Habit. Eastern U.S., Ontario. Constit. Starch, diosgenin. Isoln of sapogenin: Marker et al., J. Am. Chem. Soc. **62**, 2620 (1940). Pharmacological studies: Butler, Costello, J. Am. Pharm. Assoc., Sci. Ed. **33**, 177 (1944).

THERAP CAT: Antiflatulent.

221. Aleuritic Acid. DL-*erythro-9,10,16-Trihydroxyhexadecanoic acid;* 9,10,16-trihydroxypalmitic acid; 8,9,15-trihydroxypentadecane-1-carboxylic acid. $C_{16}H_{32}O_5$; mol wt 304.42. C 63.12%, H 10.60%, O 26.28%. One of the constituent acids of shellac. Obtained in 43% yield from dewaxed shellac: Schaeffer, Gardner, Ind. Eng. Chem. **30**, 333 (1938); Gidvani, J. Chem. Soc. **1944**, 306; Sengupta, Bose, J. Sci. Ind. Res. (India) **11B**, 458 (1952). The acid obtained from shellac is optically inactive, although it contains two asymmetric carbon atoms. It has been shown to be the DL-*erythro* or dl-*cis* form, and is the only form described here. Synthesis of diastereoisomers: Mitter et al., Sci. Cult. (Calcutta) **8**, 273 (1942); Hunsdiecker, Ber. **76**, 142 (1943); **77**, 185 (1944); Baudart, Compt. Rend. **221**, 205 (1945).

Crystals from dilute ethanol, mp 100-101°. Sol in methanol. Forms a crystalline sodium salt.

Methyl ester, $C_{17}H_{34}O_5$, fine feathery needles, mp 72-73°; $bp_{0.1}$ 235°. Sol in methanol, ethanol, chloroform, acetone. Less sol in benzene. Insol in petr ether.

Ethyl ester, $C_{18}H_{36}O_5$, needles from dil ethanol, mp 59°. Hydrazide, $C_{16}H_{34}N_2O_4$, crystals from abs ethanol, mp 139-140°.

222. Alexidine. *N,N''-Bis(2-ethylhexyl)-3,12-diimino-2,4,11,13-tetraazatetradecanediimidamide; 1,1'-hexamethylenebis[5-(2-ethylhexyl)biguanide];* Win 21904; Sterwin 904; Bisguadine. $C_{26}H_{56}N_{10}$; mol wt 508.81. C 61.37%, H 11.09%, N 27.53%. Prepn from 1,1'-hexamethylenebis(3-cyanoguanide) and 2-ethylhexylamine hydrochloride: Fr. pat. **1,463,818** (1965 to Sterling Drug). Evaluation as antimicrobial agent: McNamara et al., J. Soc. Cosmet. Chem. **16**, 499 (1965).

Dihydrochloride, crystals from methanol + ether, mp 220.6-223.4°.

THERAP CAT: Antibacterial.

223. Alexitol Sodium. Sodium polyhydroxyaluminum monocarbonate hexitol complex; aluminum sodium carbonate hexitol complex; Actal. Probable structure and properties: Gwilt et al., J. Pharm. Pharmacol. **10**, 770 (1958). Stabilization of an antacid prepn contg gelatinous aluminum hydroxide with a hexitol (sorbitol or mannitol): Alford, U.S. pat. **2,999,790** (1962 to Sterling Drug).

Hexitol

Tasteless, odorless powder. Dec without melting when strongly heated. Practically insol in water. Readily sol in dil acids. At normal temps can be stored indefinitely without apparent change in its physical or chemical properties.
THERAP CAT: Antacid.

224. Alfadolone Acetate. *21-(Acetyloxy)-3-hydroxypregnane-11,20-dione; 3α,21-dihydroxy-5α-pregnane-11,20-dione 21-acetate;* 21-acetoxy-3α-hydroxy-5α-pregnane-11,20-dione; alphadolone acetate; GR 2/1574. $C_{23}H_{34}O_5$; mol wt 390.52. C 70.74%, H 8.77%, O 20.49%. Prepn: Brown, Kirk, *J. Chem. Soc. (C)* **1969**, 1653; Davis *et al.*, **Ger. pat. 2,030,402** and **S. Afr. pat. 70 03,861** (both 1971 to Glaxo), *C.A.* **75,** 20793n, 64114w (1971).

Crystals from acetone-hexane, mp 175-177°. $[\alpha]_D^{26}$ +97° (c = 1.02 in chloroform).
Mixture with alfaxalone, *see* Alfaxalone for trademarks and additional refs.
THERAP CAT: Combination with alfaxalone as anesthetic (intravenous).
THERAP CAT (VET): Combination with alfaxalone as anesthetic (intravenous).

225. Alfaprostol. *[1R-[1α(Z),2β(S*),3α,5α]]-7-[2-(5-Cyclohexyl-3-hydroxy-1-pentynyl)-3,5-dihydroxycyclopentyl]-5-heptenoic acid methyl ester;* 18,19,20-trinor-17-cyclohexyl-13,14-didehydro-PGF$_{2α}$ methyl ester; Ro 22-9000; K 11941; Alfavet; Alphacept. $C_{24}H_{38}O_5$; mol wt 406.56. C 70.90%, H 9.42%, O 19.68%. Synthetic analog of prostaglandin F$_{2α}$, *q.v.* Synthesis: C. Gandolfi *et al.*, **Ger. pat. 2,539,116**; *eidem*, **U.S. pat. 4,035,413** (1976, 1977 both to Carlo Erba). Efficacy in bovine estrus synchronization: E. Schilling *et al.*, *Theriogenology* **18,** 413 (1982). Treatment of infertility in cows caused by persistent corpus luteum: G. Maffeo *et al.*, *Prostaglandins* **25,** 541 (1983).

THERAP CAT (VET): Estrus control.

226. Alfaxalone. *3-Hydroxypregnane-11,20-dione;* alphaxalone; GR 2/234. $C_{21}H_{32}O_3$; mol wt 332.49. C 75.86%, H 9.70%, O 14.44%. Prepn: Nagata *et al.*, *Helv. Chim. Acta* **42,** 1399 (1959); Browne, Kirk, *J. Chem. Soc. (C)* **1969,** 1653; Davis *et al.*, **Ger. pat. 2,030,402** and **S. Afr. pat. 70 03,861** (both 1971 to Glaxo), *C.A.* **75,** 20793n, 64114w (1971). Mass spectral data: Ende, Spiteller, *Monatsh.* **102,** 929 (1971). Review of pharmacology and clinical efficacy of mixture with alfadolone acetate, *q.v.*: *Postgrad. Med. J.* **48,** Suppl. 2, 1-139 (1972). *See also* Child *et al.*; Campbell *et al.*; *Brit. J. Anaesth.* **43,** 2-24 (1971); *Lancet* **1,** 888 (1972).

Colorless prisms from ether, mp 172-174°. $[\alpha]_D^{26}$ +113.4° (c = 1.2 in chloroform).
Mixture with alfadolone acetate (3:1), *alphadione, alfadione, CT-1341, Alfathesin, Althesin, Aurantex, Saffan.* (Alfaxalone is the more active component.) LD$_{50}$ i.v. in mice: 54.7 mg/kg.
THERAP CAT: Combination with alfadolone acetate as anesthetic (intravenous).
THERAP CAT (VET): Combination with alfadolone acetate as anesthetic (intravenous).

227. Alfentanil. *N-[1-[2-(4-Ethyl-4,5-dihydro-5-oxo-1H-tetrazol-1-yl)ethyl]-4-(methoxymethyl)-4-piperidinyl]-N-phenylpropanamide;* N-[1-[2-(4-ethyl-5-oxo-2-tetrazolin-1-yl)ethyl]-4-(methoxymethyl)-4-piperidyl]propionanilide. $C_{21}H_{32}N_6O_3$; mol wt 416.52. C 60.56%, N 7.74%, N 20.18%, O 11.52%. Tetrazole derivative of fentanyl, *q.v.* Prepn: F. Janssens, **Ger. pat. 2,819,873**; *idem*, **U.S. pat. 4,167,574** (1978, 1979 both to Janssen); F. Janssens *et al.*, *J. Med. Chem.* **29,** 2290 (1986). Determn in human plasma by GC: T. J. Gillespie *et al.*, *J. Anal. Toxicol.* **5,** 133 (1981); by radioimmunoassay: M. Michaels *et al.*, *J. Pharm. Pharmacol.* **35,** 86 (1983). Pharmacology in animals: C. J. E. Niemegeers, P. A. J. Janssen, *Drug. Dev. Res.* **1,** 83 (1981); in humans: B. Kay, B. Pleuvry, *Anaesthesia* **35,** 952 (1980). Cardiovascular effects during surgery: P. S. Sebel *et al.*, *Brit. J. Anaesth.* **54,** 1185 (1982). Clinical evaluation as analgesic component of anesthesia: B. Kay *et al.*, *Ann. Roy. Coll. Surg. Engl.* **65,** 316 (1983); as anesthetic induction agent: J. Nauta *et al.*, *Anesth. Analg. (Cleveland)* **61,** 267 (1982); M. E. Sinclair, G. M. Cooper, *Anaesthesia* **38,** 435 (1983). Review of clinical pharmacokinetics: L. E. Mather, *Clin. Pharmacokinet.* **8,** 422 (1983); of clinical trials of alfentanil and sufentanil: S. De Lange, *Mt. Sinai J. Med.* **50,** 312 (1983).

Hydrochloride monohydrate, $C_{21}H_{33}ClN_6O_3 \cdot H_2O$, *alfentanil hydrochloride, R 39209, Rapifen, Alfenta.* Crystals from 2-propanone, mp 140.8° (Janssens, 1979); also reported as crystals from acetone, mp 138.4° (Janssens, 1986). Sol in water. LD$_{50}$ i.v. in rats, dogs (mg/kg): 47.5, 20 (Niemegeers, Janssen).
Note: This is a controlled substance (opiate) listed in the U.S. Code of Federal Regulations, Title 21 Part 1308.12 (1987).
THERAP CAT: Narcotic analgesic.
THERAP CAT (VET): Narcotic analgesic.

228. Alfuzosin. *N-[3-[(4-Amino-6,7-dimethoxy-2-quinazolinyl)methylamino]propyl]tetrahydro-2-furancarboxamide;* N_1-(4-amino-6,7-dimethoxyquinazol-2-yl)-N_1-methyl-N_2-(tetrahydrofuroyl-2)-propylenediamine; SL-77.499. $C_{19}H_{27}N_5O_4$; mol wt 389.45. C 58.60%, H 6.99%, N 17.98%, O 16.43%. α$_1$-Adrenoceptor antagonist structurally similar to prazosin, *q.v.* Prepn: P. M. J. Manoury, **Ger. pat. 2,904,445**; *idem*, **U.S. pat. 4,315,007** (1979, 1982 both to Synthelabo); and antihypertensive activity in rats: P. M. Manoury *et al.*, *J. Med. Chem.* **29,** 19 (1986). Pharmacology: A. G. Ramage, *Eur. J. Pharmacol.* **129,** 307 (1986). HPLC determn in biological fluids: P. Guinebault *et al.*, *J. Chromatog.* **353,** 361 (1986). Pharmacology in humans: A. H. Deering, *Br. J. Clin. Pharmacol.* **25,** 417 (1988). Clinical evaluation in essential hypertension: S. Leto Di Priolo *et al.*, *Eur. J. Clin. Pharmacol.* **35**, 25 (1988); A. K. Ghosh, S. Ghosh, *Ger. Cardiovasc. Med.* **1**, 81 (1988).

Hydrochloride, $C_{19}H_{28}ClN_5O_4$, *SL-77.499-10*, *Alfoten*, *Xatral*. Crystals from ethanol + ether, mp 225° (Manoury, 1986), also reported earlier as mp 235° (dec) (Manoury, 1982). pKa 8.13.

THERAP CAT: Antihypertensive.

229. Algestone. *16α,17-Dihydroxypregn-4-ene-3,20-dione;* 16α,17-dihydroxyprogesterone; 4-pregnene-16α,17α-diol-3,2-dione; alphasone. $C_{21}H_{30}O_4$; mol wt 346.45. C 72.80%, H 8.73%, O 18.47%. Prepn from 16-dehydropro-gesterone: Inhoffen *et al.*, *Ber.* **87**, 593 (1954); Cooley *et al.*, *J. Chem. Soc.* **1955**, 4373; Allen, Bernstein, *J. Am. Chem. Soc.* **78**, 1909 (1956); Hydorn *et al.*, *Steroids* **3**, 493 (1964). Manuf from 16-dehydroprogesterone: Colton, U.S. pat. **2,727,909** (1955 to Searle); Hydorn *et al.*, U.S. pat. **3,165,-541** (1965 to Olin Mathieson); from its 16α-acetate: Diassi, U.S. pat. **3,027,384** (1962 to Olin Mathieson).

Needles from ethanol + dichloromethane, mp 225°. $[\alpha]_D^{22}$ +95° (c = 0.81 in $CHCl_3$). uv max: 240 nm (ϵ 16,600).

Cyclic acetal with acetone, $C_{24}H_{34}O_4$, *16α,17-[(1-methyl-ethylidene)bis(oxy)]pregn-4-ene-3,20-dione, algestone aceton-ide, alphasone acetonide, 16α,17α-isopropylidenedioxyproges-terone.* Needles from aq ethanol, mp 210°. $[\alpha]_D^{20}$ +137° (c = 0.7 in $CHCl_3$). Prepn: Cooley *et al.*, *loc. cit;* Fried *et al.*, *Chem. & Ind. (London)* **1961**, 465.

16α-Methyl ether, $C_{22}H_{32}O_4$, *17-hydroxy-16α-methoxy-pregn-4-ene-3,20-dione.* Crystals from 95% ethanol, mp 142-143°. $[\alpha]_D^{23}$ +60° (c = 0.15 in $CHCl_3$). uv max: 234 nm (ϵ 15,400). Prepd from the free diol via *16α,17-dihydroxy-pregn-4-ene-3,20-dione cyclic borate:* Fried, U.S. pat. **3,006,930** (1961 to Olin Mathieson).

THERAP CAT: Acetonide as a topical anti-inflammatory.

230. Algestone Acetophenide. *[16α(R)]-16,17-[(1-Phenylethylidene)bis(oxy)]pregn-4-ene-3,20-dione; 16α,17-dihydroxyprogesterone cyclic acetal with acetophe-none; 16α,17α-dihydroxyprogesterone acetophenide; alpha-sone acetophenide; P-DHP; SQ 15101; Deladroxone* (obso-lete); *Droxone* (obsolete); *Neolutin Depositum.* $C_{29}H_{36}O_4$; mol wt 448.58. C 77.64%, H 8.09%, O 14.27%. Progesto-gen. Prepn: J. Fried, U.S. pat. **2,941,997** (1960 to Olin Mathieson); J. Fried *et al.*, *Chem. & Ind. (London)* **1961**, 465. Improved prepn: Irish pat. **6,800,457**; S. J. Brancato *et al.*, U.S. pat. **3,488,347** (1968, 1970 both to Smith Kline & French). In estrus synchronization: J. N. Wiltbank *et al.*, *J. Anim. Sci.* **26**, 764 (1967). Clinical evaluations of combina-tion with estradiol enanthate as injectable contraceptive: R. Plesner, *Acta Endocrinol.* **61**, 494 (1969); *idem, ibid.* **65**, 683 (1970); R. Recio *et al.*, *Contraception* **33**, 579 (1986).

Crystals from 95% ethanol, mp 150-151°. $[\alpha]_D^{23}$ +51° ($CHCl_3$). Stable to boiling mineral acids; readily cleaved by warming with formic acid, with subsequent deformylation.

THERAP CAT: In treatment of acne.

231. Algin. *Alginic acid sodium salt;* sodium alginate; sodium polymannuronate; Alto; Alman; Alloid; Allose; Kelgin; Minus; Protanal. A gelling polysaccharide extracted from giant brown seaweed (giant kelp, *Macrocystis pyrifera* (L.) Ag., *Lessoniaceae)* or from horsetail kelp (*Laminaria digitata* (L.) Lamour, *Laminariaceae)* or from sugar kelp (*Laminaria saccharina* (L.) Lamour). Process of manuf: Tseng, *Chem. Met. Eng.* **52**, 97 (1945); Mantell, *The Water-Soluble Gums* (New York, 1947); Green, U.S. pat. **2,036,934** (1936 to Kelco); Gloahec, Herter, U.S. pat. **2,128,551** (1938 to Algin Corp. of America). Wound healing properties and use in hemostatic dressings: J. H. M. Miller, **Brit.** pat. **1,328,088** (1973 to Wallace, Cameron & Co.), *C.A.* **80**, 6974u (1974). Series of articles on hemostatic effects: *Yakugaku Zasshi* **101**, 452-469 (1981), *C.A.* **95**, 35654e; 35655f; 35405z (1981). Immunoadjuvant effect: G. H. Scherr, A. S. Mar-kowitz, U.S. pat. **3,075,883** (1963 to Consolidated Labs.). Clinical comparison with alum immunoadjuvant: G. Bruno *et al.*, *Ann. Allergy* **56**, 384 (1986). Review of structural studies: D. A. Rees, E. J. Welsh, *Angew. Chem. Int. Ed.* **16**, 214 (1977). Review of production, properties and use in the food industry: A. Askar, *Alimenta* **21**, 165-169 (1982). *Reviews:* McNeely, Pettitt, in *Industrial Gums,* R. L. Whist-ler, Ed. (Academic Press, New York, 2nd ed., 1973) pp 49-81; I. W. Cottrell, J. K. Baird, "Gums" in Kirk-Other *Encyclopedia of Chemical Technology* vol. 12 (Wiley-Inter-science, New York, 3rd ed., 1980) pp 48-51.

Cream-colored powder. Sol in water, forming a viscous, colloidal soln. Insol in alcohol and in hydro-alcoholic solns in which the alcohol content is > 30% w/w. Insol in chloro-form, ether, in aq acid solns when the pH is below 3.

USE: In the manufacture of ice cream where it serves as a stabilizing colloid, insuring creamy texture and preventing the growth of ice crystals. In drilling muds; in coatings; in the flocculation of solids in water treatment; as sizing agent; thickener; emulsion stabilizer; suspending agent in soft drinks; in dental impression preparations. Pharmaceutic aid (suspending agent).

THERAP CAT: Hemostatic agent.

232. Alginic Acid. Norgine; polymannuronic acid; Sa-zio. Mol wt about 240,000. A hydrophilic, colloidal poly-saccharide obtained from seaweeds which, in the form of mixed salts of calcium, magnesium, and other bases, occurs as a structural component of the cell wall. Isoln from fronds of *Laminaria digitata* (L.) Edmonson, *Laminariaceae:* Stan-ford, *J. Chem. Soc.* **44**, 943 (1883); Krefting, *ibid.* **25**, 403 (1931); from *Macrocystis pyrifera* (L.) C. Ag., *Lesso-niaceae:* Nelson, Cretcher, *J. Am. Chem. Soc.* **51**, 1914 (1929). Polysaccharides similar to alginic acid are also se-creted by the bacteria *Pseudomonas aeruginosa* and *Azoto-bacter vinelandii:* G. H. Cohen, D. B. Johnstone, *J. Bacter-iol.* **88**, 329 (1964); L. R. Evans, A. Linker, *ibid.* **116**, 915 (1973). Alginic acid is a linear polymer of β-(1→4)-D-mannosyluronic acid and α-(1→4)-L-gulosyluronic acid residues, the relative proportions of which vary with the botanical source and state of maturation of the plant. Clini-cal use of combination with antacid in gastric reflux eso-phagitis: D. Y. Graham *et al.*, *Curr. Ther. Res.* **22**, 653 (1977). Clinical evaluation of calcium salt as hemostatic dressing: T. Gilchrist, A. M. Martin, *Biomaterials* **4**, 317 (1983); A. R. Groves, J. C. Lawrence, *Ann. Roy. Coll. Surg. Engl.* **68**, 27 (1986). Review of structure studies and of secondary and tertiary structure in solutions and gels: D. A. Rees, E. J. Welsh, *Angew. Chem. Int. Ed.* **16**, 214 (1977).

Review of production, properties and use in the food indus-try: A. Askar, *Alimenta* **21**, 165-169 (1982).

Very slightly sol in water. Tasteless. Capable of absorb-ing 200-300 times its weight of water and salts to the extent of 60%. Resists hydrolysis. Sol in alkaline solns. pH of a 3 in 100 suspension in water is between 2.0 and 3.4.

Calcium salt, *Sorbsan*. Forms gelatinous precipitate in water.

Potassium salt, *Stercofuge*.

Sodium salt, *See* Algin.

USE: Sizing paper and textiles; as binder for briquettes; manuf artificial horn, ivory, celluloid; emulsionizing mineral oils; mucilage. Additional uses are described under Algin.

THERAP CAT: Calcium salt as hemostatic.

233. Alibendol. *2-Hydroxy-N-(2-hydroxyethyl)-3-methoxy-5-(2-propenyl)benzamide; 5-allyl-2-hydroxy-N-(2-hydroxyethyl)-m-anisamide; 2-hydroxy-3-methoxy-5-allyl-N-(β-hydroxyethyl)benzamide;* EB 1856; FC 54; H 3774; Cebera. $C_{13}H_{17}NO_4$; mol wt 251.29. C 62.13%, H 6.82%, N 5.58%, O 25.47%. Amide analog of eugenol, *q.v.* Prepn: **Fr. pat. 1,584,715** corresp to F. Clémence, O. Le Martret, **U.S. pat. 3,668,238** (1967, 1972 both to Roussel). Synthesis and pharmacologic activity: F. Clémence *et al., Chim. Ther.* **5**, 188 (1970).

Cryst from benzene, mp 95°. uv max (ethanol): 316, 218 nm. LD_{50} in Swiss male mice (mg/kg): > 3000 orally; > 2000 s.c.; 209 i.p.; 217 i.v., F. Clémence *et al., loc. cit.*

THERAP CAT: Choleretic; antispasmodic.

234. Alinidine. *N-(2,6-Dichlorophenyl)-4,5-dihydro-N-2-propenyl-1H-imidazol-2-amine; 2-(N-allyl-2,6-dichloroanilino)-2-imidazoline; 2-[N-allyl-N-(2,6-dichlorophenyl)amino]-2-imidazoline;* ST 567. $C_{12}H_{13}Cl_2N_3$; mol wt 270.16. C 53.35%, H 4.85%, Cl 26.25%, N 15.55%. Analog of clonidine, *q.v.*, with specific bradycardic activity. Prepn: H. Stähle *et al.,* **Ger. pat. 1,958,201;** *eidem,* **U.S. pat. 3,708,-485** (1971, 1973 both to Boehringer, Ing.); *eidem, J. Med. Chem.* **23**, 1217 (1980). Heart rate reduction without β-receptor blockade in animals: W. Kobinger *et al., Arch. Pharmacol.* **306**, 255 (1979); in humans: D. W. G. Harron *et al., J. Cardiovasc. Pharmacol.* **4**, 213 (1982). Mode of action study: J. S. Millar, E. M. Vaughan Williams, *Lancet* **1**, 1291 (1981). HPLC determn in human plasma: U.-W. Wiegand *et al., J. Chromatog.* **223**, 238 (1981). Pharmacokinetics in humans: *eidem, J. Cardiovasc. Pharmacol.* **4**, 59 (1982). Metabolism: D. Arndts, H. J. Forster, *Eur. J. Drug Metab. Pharmacokinet.* **6**, 313 (1981). Comparison with propranolol in angina: J. Schurmans *et al., Eur. J. Clin. Pharmacol.* **23**, 389 (1982). Hemodynamic effects in angina or infarction: M. L. Simoons, P. G. Hugenholtz, *Eur. Heart J.* **5**, 227 (1984). Review of pharmacology and potential therapeutic uses: D. W. G. Harron, R. G. Shanks, *ibid.* **6**, 722-729 (1985).

Crystals, mp 130-131° (Stähle, **U.S. pat.**); also reported as mp 127-129° (Stähle, *J. Med. Chem.*). pka 10.42.

Hydrobromide, $C_{12}H_{14}BrCl_2N_3$, crystals from methanol + water, mp 193-194°.

235. Alitame. *L-α-Aspartyl-N-(2,2,4,4-tetramethyl-3-*

thietanyl)-D-alaninamide; L-aspartyl-D-alanine-*N*-(2,2,4,4-tetramethylthietan-3-yl)amide; 3-(L-aspartyl-D-alaninami-do)-2,2,4,4-tetramethylthietane; CP-54802. $C_{14}H_{25}N_3O_4S$; mol wt 331.43. C 50.74%, H 7.60%, N 12.68%, O 19.31%, S 9.67%. Dipeptide amide reported to be approx 2000 times sweeter than sucrose. Prepn: T. M. Brennan, M. E. Hen-drick, **Eur. pat. Appl. 34,876;** *eidem,* **U.S. pat. 4,411,925** (1981, 1983 both to Pfizer). Prepn of aromatic sulfonic acid salts: C. Sklavounos, **U.S. pat. 4,375,430** (1983 to Pfizer).

USE: Non-nutritive sweetener.

236. Alizapride. *6-Methoxy-N-[[1-(2-propenyl)-2-pyr-rolidinyl]methyl]-1H-benzotriazole-5-carboxamide; N-[(1-allyl-2-pyrrolidinyl)methyl]-6-methoxy-1H-benzotriazole-5-carboxamide.* $C_{16}H_{21}N_5O_2$; mol wt 315.39. C 60.93%, H 6.71%, N 22.21%, O 10.15%. Prepn: **Belg. pat. 825,605** corresp to G. Bulteau *et al.,* **U.S. pat. 4,039,672** (1975, 1977 both to Soc. Etudes Sci. Ind. de Ile de France). Series of articles on pharmacokinetics, bioavailability, pharmaco-dynamics, clinical studies: *Sem. Hop. Paris* **58**, 323-374 (1982).

Cryst from acetone, mp 139°. LD_{50} i.v. in mice (5 days): 92.7 mg/kg (Bulteau).

Hydrochloride, $C_{16}H_{22}ClN_5O_2$, *Nausilen, Plitican, Ver-gentan.* Cryst from methanol/methyl ethyl ketone, mp 206-208°.

THERAP CAT: Neuroleptic; anti-emetic.

237. Alizarin. *1,2-Dihydroxy-9,10-anthracenedione; 1,2-dihydroxyanthraquinone;* C.I. Mordant Red 11; C.I. Pigment Red 83; C.I. 58000. $C_{14}H_8O_4$; mol wt 240.20. C 70.00%, H 3.36%, O 26.64%. Occurs in the root of the madder plant (*Rubia tinctorum* L., *Rubiaceae;* Krappwurzel) in combination with 2 mols glucose, called ruberythric acid. Was known and used in ancient Egypt, Persia, and India. Synthesized from 2-anthraquinonesulfonic acid sodium salt : Caro *et al., Ber.* **3**, 359 (1870); Perkin, *Ber.* **9**, 281 (1876). Historical *review:* Fieser, *J. Chem. Ed.* **7**, 2609 (1930). Lab-oratory prepn: Gattermann-Wieland, *Laboratory Methods of Organic Chemistry* (New York, 1937). Modern methods of manufacture: Pohl, *Ullmann's Enzyklopädie der technischen Chemie* vol. I, p 200; Fierz-David and Blangey, *Grundle-gende Operationen der Farbenchemie* (Vienna, 5th ed., 1943). *See also Colour Index* vol. 4, (3rd ed., 1971) p 4513.

Orthorhombic, orange needles by sublimation or from abs alc. Solvated scales from dil alc or by evaporation from ether. Sublimes at 110° (2 mm Hg). mp 290°. bp 430°. Absorption spectrum: Moir, *J. Chem. Soc.* **1927**, 1810. Soly in water at 18°: 2.1 × 10⁻⁶ mols/l; at 25°: 2.5 × 10⁻⁶ mols/l. Sol in 300 parts boiling water; moderately sol in alcohol, freely in hot methanol and in ether at 25°. Also sol in benzene, toluene, xylene, pyridine, carbon disulfide, gla-cial acetic acid. Sol in water solns of alkalies with blue color, but without fluorescence. Fluorescent solns indicate unchanged 2-anthraquinone sodium sulfonate.

1-Methyl ether, $C_{15}H_{10}O_4$, orange needles with $1H_2O$ from dil methanol. When dried at 100° mp 179°.

2-Methyl ether, $C_{15}H_{10}O_4$, orange needles from alcohol, mp 231°.

Dimethyl ether, $C_{16}H_{12}O_4$, golden-yellow needles from alcohol, mp 215°.

USE: In the manufacture of acid and chrome dyes for wool; acid-base indicator (in 0.5% alcoholic soln; pH: yellow 5.5, red 6.8); in spot tests as reagent for aluminum, indium, mercury, zinc, and zirconium; biological stain.

238. Alizarin Cyanine Green F. *2,2-[(9,10-Dihydro-9,10-dioxo-1,4-anthracenediyl)diimino]bis(5-methylbenzenesulfonic acid) disodium salt; 6,6'-(1,4-anthraquinonylenediimino)di-m-toluenesulfonic acid disodium salt;* D & C Green No. 5; C.I. Acid Green 25; C.I. 61570. $C_{28}H_{20}N_2Na_2O_8S_2$; mol wt 622.58. C 54.02%, H 3.24%, N 4.49%, Na 7.39%, O 20.56%, S 10.30%. Discovered by R. E. Schmidt in 1894: *Colour Index* vol. 4 (3rd ed., 1971) p 4541.

Green powder. Slightly sol in acetone, alc, pyridine. Insol in chloroform, toluene. Dull blue soln in conc H_2SO_4, turning turquoise on dilution. Absorption spectra: C. F. H. Allen *et al.*, *J. Org. Chem.* **7**, 63, 169 (1942).

USE: For nylon sutures: *Fed. Regist.* **42**, 52395 (1977). Permitted for use in drugs and cosmetics, excluding use in eye area: *ibid.* **47**, 49628 (1982).

239. Alizarine Blue. *5,6-Dihydroxynaphtho[2,3-f]quinoline-7,12-dione;* 7,8-dihydroxy-5,6-phthalylquinoline; Alizarin Blue R; C.I. 67410. $C_{17}H_9NO_4$; mol wt 291.25. C 70.10%, H 3.12%, N 4.81%, O 21.97%. Prepn from 3-nitroalizarin, glycerol, and concd sulfuric acid: Auerbach, *J. Chem. Soc.* **35**, 799 (1879); *Colour Index* vol. 4 (3rd ed., 1971) p 4567.

Lustrous brownish violet needles from benzene, mp 268-270°. Practically insol in water; sparingly sol in alc, ether; slightly sol in cold benzene; sol in amyl alcohol, glacial acetic acid, hot benzene.

USE: As indicator in saturated alcoholic soln. pH: pink 0.0 to yellow 1.6; yellow 6.0 to green 7.6.

240. Alizarine Orange. *1,2-Dihydroxy-3-nitro-9,10-anthracenedione; 1,2-dihydroxy-3-nitroanthraquinone;* 3-nitroalizarin; C.I. Mordant Orange 14; C.I. 58015. $C_{14}H_7NO_6$; mol wt 285.20. C 58.96%, H 2.47%, N 4.91%, O 33.66%. From 3-bromoalizarin in acetic acid with nitric acid: Barnett, Cook, *J. Chem. Soc.* **121**, 1376 (1922). Additional prepns: *Colour Index* vol. 4 (3rd ed., 1971) p 4514.

Orange needles or plates from acetic acid, dec 244°. Sublimes with partial decompn. Yellow in organic solvents, purple-red in dil aq alkali, orange in H_2SO_4.

USE: Dyes cloth mordanted with Al orange, with Fe red to violet. As indicator in satd alc soln. pH 2.0-4.0, color change from golden orange to flat yellow (in water). pH 5.0-6.5 from yellow to purplish red.

241. Alizarine Yellow R. *2-Hydroxy-5-[(4-nitrophenyl)azo]benzoic acid; C.I. Mordant Orange I;* 5-(p-nitrophenylazo)salicylic acid; *p*-nitrobenzeneazosalicylic acid; Alizarine Yellow RW; C.I. 14030. $C_{13}H_9N_3O_5$; mol wt 287.23. C 54.36%, H 3.16%, N 14.63%, O 27.85%. Prepd by coupling diazotized *p*-nitroaniline with salicylic acid: Armento, U.S. pat. **2,746,955** (1956 to General Aniline and Film); *Colour Index* vol. 4 (3rd ed., 1971) p 4058.

Orange-brown needles from dil glacial acetic acid, dec 253-254°. Sol in water, alc; slightly sol in acetone, Cellosolve; practically insol in other org solvents.

Sodium salt, $C_{13}H_8N_3NaO_5$, *Mordant Yellow 3R.* Brownish yellow powder. Sol in water.

USE: Sodium salt used as indicator in 0.1% aq soln pH: yellow to red, range 10.2 to 12.0.

242. Alkanet. Alkanna; orcanette; dyer's alkanet; anchusa; orkanet. The root of *Alkanna tinctoria* Tausch, *Boraginaceae*. *Habit.* Asia Minor, Hungary, Greece, Mediterranean region. *Constit.* Alkannin (the coloring principle) and tannin. Extraction of roots: Betrabet, Chakravarti, *J. Indian Inst. Sci.* **16A**, 51 (1933); *C.A.* **28**, 1038 (1934); Majamdar, Chakravarti, *J. Indian Chem. Soc.* **17**, 272 (1940); *C.A.* **34**, 6262 (1940).

Alkannin paper, anchusin paper, Boettger's paper. White paper impregnated with a 1% alc tincture of alkanet root and dried. A blue paper can be made from the red paper by treating the latter with a 1% Na_2CO_3 soln. This paper acts nearly like litmus paper. Alkalies = green or blue; acids = red. pH: 8.0 red; 10.0 blue.

USE: For coloring wines, cosmetics, confectionery. Alkannin paper as indicator.

THERAP CAT: Astringent.

243. Alkannin. *(S)-5,8-Dihydroxy-2-(1-hydroxy-4-methyl-3-pentenyl)-1,4-naphthalenedione; (−)-5,8-dihydroxy-2-(1-hydroxy-4-methyl-3-pentenyl)-1,4-naphthoquinone;* anchusa red; anchusin; alkanna red; alkanet extract; (1-hydroxy-3-isohexenyl)naphthazarine; 2-(1-hydroxy-4-methyl-3-pentenyl)-5,8-dihydroxy-1,4-naphthoquinone; C.I. Natural Red 20; C.I. 75530. $C_{16}H_{16}O_5$; mol wt 288.29. C 66.66%, H 5.59%, O 27.75%. Isoln from the root of *Alkanna tinctoria* Tausch, *Boraginaceae*: Brockmann, *Ann.* **521**, 1 (1936); Toribara, Underwood, *Anal. Chem.* **21**, 1352 (1949). Abs config: Arakawa, Nakagaki, *Chem. & Ind. (London)* **1961**, 947.

Brownish-red prisms with a metallic sheen from benzene, mp 149°. Can be sublimed in high vac at 140-150°. $[\alpha]_{Cd}^{20}$

—165° (benzene); —226° (chloroform). Also reported as —254° ± 7° (chloroform) (Toribara). Sol in organic solvents, sparingly sol in water. Buffered aq solns are red at pH 6.1; purple at pH 8.8; blue at pH 10.0. LD_{50} in male mice: 3.0 ± 1.0 g/kg, *C.A.* **76**, 122513j.

(+)-Form, *shikonin*.
(±)-Form, *shikalkin*.

USE: Red dye for cosmetics and food; spectrophotometric microdetermination of beryllium.

THERAP CAT: Astringent.

244. Alkofanone. *3-[(4-Aminophenyl)sulfonyl]-1,3-diphenyl-1-propanone; 3-phenyl-3-sulfanilylpropiophenone;* p-aminophenyl(2-benzoyl-1-phenylethyl)sulfone; Nu 404; Ro 2-0404; Alfone. $C_{21}H_{19}NO_3S$; mol wt 365.46. C 69.02%, H 5.24%, N 3.83%, O 13.13%, S 8.78%. Prepn: Goldberg *et al.*, *Jubilee Volume Emil Barell* (Basle, 1946) p 341; Goldberg, U.S. pat. **2,421,836** (1947 to Hoffmann-La Roche).

Decomp 221-223°. Slightly soluble in acetone, dioxane. Practically insol in most of the common organic solvents.

THERAP CAT: Antidiarrheal.

245. Alkyd Resins. Reaction product of a dibasic acid and a polyol, to which sufficient monofunctional acid or alcohol has been added to prevent gelation during processing. *Review:* Kraft, *J. Am. Oil Chem. Soc.* **39**, 501 (1962).

USE: Plastics, paint vehicles, decorative coatings.

246. Allantoin. *(2,5-Dioxo-4-imidazolidinyl)urea; 5-ureidohydantoin;* glyoxyldiureide; cordianine; Psoralon; Septalan. $C_4H_6N_4O_3$; mol wt 158.12. C 30.38%, H 3.82%, N 35.44%, O 30.36%. Product of purine metabolism. Prepd synthetically by the oxidation of uric acid with alkaline potassium permanganate: *Org. Syn. coll. vol. II*, 23 (1943). By heating urea with dichloroacetic acid: C. N. Zellner, J. R. Stevens, U.S. pat. **2,158,098** (1939 to Merck & Co.) Acetyl derivs: Biltz, Loewe, *J. Prakt. Chem.* **141**, 291 (1934). Optically active forms have been obtained by extraction procedures.

Racemic form, monoclinic plates or prisms from water, mp 238°. One gram dissolves in 190 ml water, 500 ml alcohol; more sol in hot water and hot alcohol. Almost insol in ether. pH of satd water soln: 5.5.

THERAP CAT: Topical vulnerary; skin ulcer therapy.

THERAP CAT (VET): Has been used topically to stimulate healing of suppurating wounds, resistant ulcers.

247. Allenolic Acid. *6-Hydroxy-2-naphthalenepropanoic acid;* 2-hydroxy-6-naphthalenepropionic acid; amphi-hydroxynaphthyl-β-propionic acid. $C_{13}H_{12}O_3$; mol wt 216.23. C 72.21%, H 5.59%, O 22.20%. Prepn from the corresponding methoxy compd: Jacques, Horeau, *Bull. Soc. Chim. France* **1948**, 714.

Crystals from aq methanol, mp 180-181°. Sol in alcohol, methanol, pyridine.

USE: In biochemical research, in the prepn of estrogenic compds.

248. Allethrins. Allyl cinerins. Synthetic analogs of the naturally occurring insecticides cinerin, jasmolin, and pyre-

thrin (q.v.). *Review:* Barthel, *World Rev. Pest Contr.* **6**, 59 (1967).

Allethrin I, $C_{19}H_{26}O_3$, *2,2-dimethyl-3-(2-methyl-1-propenyl)cyclopropanecarboxylic acid 2-methyl-4-oxo-3-(2-propenyl)-2-cyclopenten-1-yl ester, allethrolone ester of chrysanthemummonocarboxylic acid, Pynamin, Pyresyn.* R = CH_3. Synthesis: Schechter, *J. Am. Chem. Soc.* **71**, 1517, 3165 (1949); Stansbury, Guest, U.S. pat. **2,768,965** (1956 to UCC). Commercial process: Sanders, Taff, *Ind. Eng. Chem.* **46**, 414 (1954). The commercial product, a viscous liquid, is a mixture of 8 optically active isomers: d_{20}^{20} 1.010; n_D^{20} 1.5040; n_D^{30} 1.5023. Practically insol in water. Sol in alcohol, petr ether, kerosene, carbon tetrachloride, ethylene dichloride, nitromethane. Incompatible with alkalies. LD_{50} orally in mice: 480 mg/kg.

Allethrin II, $C_{20}H_{26}O_5$, *3-(3-methoxy-2-methyl-3-oxo-1-propenyl)-2,2-dimethylcyclopropanecarboxylic acid 2-methyl-4-oxo-3-(2-propenyl)-2-cyclopenten-1-yl ester, allethrolone ester of chrysanthemumdicarboxylic acid monomethyl ester.* R = $COOCH_3$. Synthesis: Matsui, Yamada, *Agr. Biol. Chem.* **27**, 373 (1963); Matsui, Meguro, *ibid.* **28**, 27 (1964); Elliott, Janes, *Chem. & Ind. (London)* **1969**, 270; Kobayashi *et al.*, *Agr. Biol. Chem.* **35**, 1961 (1971); Sugiyama *et al.*, *ibid.* **36**, 565 (1972). Oily pale yellow liq, the less toxic of the two forms. $n_D^{20.6}$ 1.5156. uv max (ethanol): 232 nm (ε 23,000).

USE: Insecticide. *Caution:* Toxic symptoms similar to those of the pyrethrins, q.v.

249. Allicin. *2-Chloro-1-sulfinothioic acid S-2-propenyl ester; thio-2-propene-1-sulfinic acid S-allyl ester.* $C_6H_{10}OS_2$; mol wt 162.27. C 44.41%, H 6.21%, O 9.86%, S 39.52%. An antibacterial principle of garlic (*Allium sativum* L., *Liliaceae*). Isoln and antibacterial activity: C. J. Cavallito, J. H. Bailey, *J. Am. Chem. Soc.* **66**, 1950 (1944). Structure: Cavallito *et al.*, *ibid.* 1952. Synthesis: Stoll, Seebeck, *Experientia* **6**, 330 (1950); Cavallito, Small, U.S. pat. **2,508,745** (1950 to Sterling Drug). Antifungal activity: Y. Yamada, K. Azuma, *Antimicrob. Ag. Chemother.* **11**, 743 (1977).

Yellow liquid. True odor of garlic. Decomp on distilling. d_4^{20} 1.112. n_D^{20} 1.561. Soly in water at 10° about 2.5% w/w. pH about 6.5. Upon standing an oily precipitate forms from aq solns. Miscible with alcohol, ether, benzene; fairly insol in the Skellysolves; unstable to hot alkali; stable to acids. LD_{50} in mice (mg/kg): 60 i.v.; 120 s.c. (Cavallito, Bailey).

250. Allidochlor. *2-Chloro-N,N-di-2-propenylacetamide;* α- chloro-N,N-diallylacetamide; N,N-diallyl-2-chloroacetamide; CDAA; CP 6343; Randox. $C_8H_{12}ClNO$; mol wt 173.65. C 55.33%, H 6.97%, Cl 20.42%, N 8.07%, O 9.21%. Selective pre-planting or pre-emergence herbicide. Prepd from chloroacetyl chloride and diallylamine: Speziale, Hamm, *J. Am. Chem. Soc.* **78**, 2556 (1956); eidem, *J. Agr. Food Chem.* **4**, 518 (1956); Bruce, Hanslick, U.S. pat. **2,844,629** (1958 to Am. Home Prods.); Hamm, Speziale, U.S. pat. **2,864,683** (1958 to Monsanto). Toxicity: G. W. Bailey, J. L. White, *Residue Rev.* **10**, 97 (1965).

Liquid. $bp_{0.3}$ 74°; bp_1 115-117°; bp_2 92°. n_D^{25} 1.4932. Slightly sol in water (2%), sol in alcohol, hexane, xylene. LD_{50} orally in rats: 700 mg/kg (Bailey, White).

Caution: A strong irritant. Can be absorbed through skin.

USE: Herbicide.

251. Alliin. *3-(2-Propenylsulfinyl)-L-alanine; 3-((S)-allylsulfinyl)-L-alanine;* S-allyl-L-cysteine sulfoxide. $C_6H_{11}NO_3S$; mol wt 177.22. C 40.66%, H 6.26%, N 7.91%, O 27.08%, S 18.09%. Constituent of garlic, *Allium sativum* L., *Liliaceae;* also found in other *Allium* spp. Isoln: Stoll, Seebeck, *Helv. Chim. Acta* **31**, 189 (1948). Synthesis: *eidem, Experientia* **6**, 330 (1950); *Helv. Chim. Acta* **34**, 481 (1951).

$$CH_2=CHCH_2SCH_2CHCOOH$$

Hemihydrate, odorless, bunched needles from dil acetone, mp 164-166° (effervescence). $[\alpha]_D^{20}$ +63.5° (c = 2). Freely sol in water. Practically insol in abs ethanol, chloroform, acetone, ether, benzene. Upon cleavage by the specific enzyme alliinase, an odor of garlic develops, and the fission products show antibacterial action similar to allicin, *q.v.*

252. Allobarbital. *5,5-Di-2-propenyl-2,4,6(1H,3H,5H)-pyrimidinetrione; 5,5-diallylbarbituric acid;* allobarbitone; Malilum; Diadol; Dial. $C_{10}H_{12}N_2O_3$; mol wt 208.21. C 57.68%, H 5.81%, N 13.46%, O 23.05%. Prepn: U.S. pat. **1,042,265** (1912). Acute toxicity: F. Sandberg, *Acta Physiol. Scand.* **24**, 7 (1951).

Crystals, leaflets. Slightly bitter taste, mp 171-173°. One part dissolves in about 300 parts water, 50 parts boiling water, 20 parts cold alcohol, 20 parts ether; very sol in hot alcohol and in acetone; sol in ethyl acetate. Insol in aliphatic hydrocarbons. A satd aq soln is acid to litmus. LD_{50} i.p. in rats: 127.3 mg/kg (Sandberg).

Caution: May be habit forming. This is a controlled substance (depressant) listed in the U.S. Code of Federal Regulations, Title 21 Parts 329.1 and 1308.13 (1987).

THERAP CAT: Sedative, hypnotic.

253. Allocholesterol. *Cholest-4-en-3β-ol;* coprostenol; 4:5-coprosten-3-ol. $C_{27}H_{46}O$; mol wt 386.64. C 83.87%, H 11.99%, O 4.14%. Prepn: Windaus, *Ann.* **453**, 101 (1927); Schoenheimer, Evans, *J. Biol. Chem.* **114**, 567 (1936). Sepn from cholesterol: Stoll, *Z. Physiol. Chem.* **246**, 10 (1937).

Needles from ether-methanol, mp 132°. $[\alpha]_D^{23}$ +43.7° (c = 1 in benzene). Freely sol in benzene, acetone, ether, chloroform, dioxane, pyridine; less sol in methanol, alcohol. Pptd by digitonin. Deep-red color with 90% trichloroacetic acid.

Acetate, $C_{29}H_{48}O_2$, long needles from dil methanol, mp 85°.

254. Alloclamide. *4-Chloro-N-[2-(diethylamino)ethyl]-2-(2-propenyloxy)benzamide; 2-(allyloxy)-4-chloro-N-[2-(diethylamino)ethyl]benzamide.* $C_{16}H_{23}ClN_2O_2$; mol wt 310.84. C 61.83%, H 7.46%, Cl 11.41%, N 9.01%, O 10.29%. Prepn: Mauvernay, U.S. pat. **3,160,557** (1964 to C.E.R.M.).

Hydrochloride, an ingredient of *Dégryp.* Crystals from abs alcohol + ether, mp 125-127°. Sol in abs alcohol. LD_{50} orally in mice: 740 mg/kg.

THERAP CAT: Antitussive, antihistaminic.

255. Allocryptopine. *5,7,8,15-Tetrahydro-3,4-dimethoxy-6-methylbenzo[e][1,3]dioxolo[4,5-k][3]benzazecin-14(6H)-one.* $C_{21}H_{23}NO_5$; mol wt 369.40. C 68.28%, H 6.28%, N 3.79%, O 21.66%. Isomeric with cryptopine, *q.v.* Obtained from *Chelidonium majus* L., *Bocconia cordata* Willd., *Sanguinaria canadensis* L. and allied *Papaveraceae:* Manske in Manske, Holmes, *The Alkaloids* vol. **IV** (Academic Press, New York, 1954) p 159. Structure: Gadamer, *Arch. Pharm.* **257**, 298 (1919). Exists in two isomeric modifications designated as α- and β-forms. Identity of α-form with α-fagarine: Redemann *et al., J. Am. Chem. Soc.* **71**, 1030 (1949). Synthesis: Haworth, Perkin, *J. Chem. Soc.* **1926**, 445; Bentley, Murray, *ibid.* **1963**, 2497.

α-Allocryptopine, α-fagarine. Crystals from ethanol, mp 160-161°. Soluble in alcohol, chloroform, ether, ethyl acetate and dil acids.

β-Allocryptopine, crystallizes with ½ mol of alcohol or ethyl acetate. Melts after expelling the solvent, at 169-171°.

256. Allocupreide Sodium. *3-[[(2-Propenylamino)thioxomethyl]amino]benzoic acid monocopper(1+) monosodium salt; m-[(N-allyl-1-mercaptoformimidoyl)amino]benzoic acid copper derivative sodium salt; 3-[1-(3-allyl-S-cupropseudothioureido)]benzoic acid sodium salt; m-[[(allylimino)mercaptomethyl]amino]benzoic acid S-copper derivative sodium salt;* sodium 3-(3-allyl-S-cuproisothioureido)benzoate; sodium 3-(3-allyl-S-cupropseudothioureido)benzoate; cuprothiosinamine m-benzoate sodium; cupralylnatrium; cupralylsodium; Cuprelon; Cupralene; Cuprion; Ebesal. $C_{11}H_{10}CuN_2NaO_2S$; mol wt 320.81. C 41.18%, H 3.14%, Cu 19.81%, N 8.73%, Na 7.17%, O 9.97%, S 10.00%. Prepd by heating m-aminobenzoic acid with allyl isothiocyanate and treating the resulting m,ω-allylthioureidobenzoic acid with cuprous chloride: Bockmühl, Fritzsche, Ger. pat. **551,421** (1932 to I. G. Farben).

Dark brown powder. Freely sol in water. Insol in alcohol, ether.

THERAP CAT: Antirheumatic.

257. Allopregnane. *5α-Pregnane.* $C_{21}H_{36}$; mol wt 288.50. C 87.42%, H 12.58%. Prepn from corticosterone: Steiger, Reichstein, *Helv. Chim. Acta* **21**, 161 (1938); from allopregnan-3-one: Ruzicka *et al., ibid.* **22**, 1294 (1939); from 21-hydroxyallopregnane: Plattner *et al., ibid.* **27**, 1177 (1944); Casanova, Reichstein, *ibid.* **32**, 647 (1949); by degradation of conessine: Haworth *et al., J. Chem. Soc.* **1949**, 831;

by Hofmann decomposition of 3β- and 3α-dimethylamino-pregnanes: Haworth et al., ibid. **1953**, 1110.

Crystals from acetone + methanol, mp 84-85°. $[\alpha]_D^{19}$ +18.4° (c = 1.69 in chloroform).

258. Allopregnane-3α,20α-diol. 5α-*Pregnane-3α,20α-diol;* 3α,20α-dihydroxy-5α-pregnane. $C_{21}H_{36}O_2$; mol wt 320.50. C 78.69%, H 11.32%, O 9.98%. Progesterone metabolite. Isoln from human pregnancy urine: Beall, *Biochem. J.* **31**, 35 (1937); from bulls' urine: Marker et al., *J. Am. Chem. Soc.* **60**, 2931 (1938). Prepn by reduction of 5α-pregn-1-ene-3,20-dione with lithium aluminum hydride: Schütt, Tamm, *Helv. Chim. Acta* **41**, 1751 (1958).
Crystals from methanol, mp 243-245°. $[\alpha]_D^{20}$ +17° (c = 0.148 in ethanol).
Diacetate, $C_{25}H_{40}O_4$, crystals, mp 139.5-140.5°. $[\alpha]_D^{20}$ +18° (c = 0.408 in benzene).

259. Allopregnane-3α,20β-diol. 5-*Pregnane-3,20-diol;* 3α,20β-dihydroxy-5α-pregnane. $C_{21}H_{36}O_2$; mol wt 320.50. C 78.69%, H 11.32%, O 9.98%. Progesterone metabolite. Prepn by hydrogenation of progesterone: Marker, Lawson, *J. Am. Chem. Soc.* **61**, 588 (1939); by catalytic hydrogenation of epiallopregnane-3-ol-20-one: Marker, U.S. pat. **2,231,019** (1941 to Parke, Davis); by reduction of 5α-pregn-1-ene-3,20-dione with lithium aluminum hydride: Schütt, Tamm, *Helv. Chim. Acta* **41**, 1751 (1958).
Needles from acetone, mp 207-209°. $[\alpha]_D^{26}$ +12° (c = 1.132 in chloroform).

260. Allopregnane-3β,20α-diol. 5α-*Pregnane-3β,20α-diol;* 3β,20α-dihydroxy-5α-pregnane. $C_{21}H_{36}O_2$; mol wt 320.50. C 78.69%, H 11.32%, O 9.98%. Progesterone metabolite. Isoln from pregnant mares' urine: Brooks et al., *Biochem. J.* **51**, 694 (1952). Prepn by heating 3α,20α-dihydroxy-5α-pregnane with sodium, or by hydrogenation of 5α-pregnane-20-ol-3-one in acid: Marker, U.S. pats. **2,196,220** and **2,250,962** (1940, 1941, both to Parke, Davis); by reduction of allopregnan-3β-ol-20-one acetate with sodium: Klyne, Barton, *J. Am. Chem. Soc.* **71**, 1500 (1949); by reduction of 5,16-pregnadien-3β-ol-20-one with Zn + acetic acid: Ercoli, De Ruggieri, *Farm. Sci. Tec. Pavia* **7**, 11 (1952), *C.A.* **46**, 10186b (1952).
Crystals from acetone, mp 218-219°. $[\alpha]_D^{22}$ +23° (c = 0.93 in chloroform).
Diacetate, $C_{25}H_{40}O_4$, crystals from petr ether, mp 164-165°. $[\alpha]_D^{19}$ +0.8° (c = 1.2 in chloroform).
Dibenzoate, $C_{35}H_{44}O_4$, crystals, mp 170.5-172°. $[\alpha]_D^{20}$ +27.7° (c = 0.84 in chloroform).

261. Allopregnane-3β,20β-diol. 5-*Pregnane-3,20-diol;* 3β,20β-dihydroxy-5α-pregnane. $C_{21}H_{36}O_2$; mol wt 320.50. C 78.69%, H 11.32%, O 9.98%. Progesterone metabolite. Isoln from pregnant mares' urine: Brooks et al., *Biochem. J.* **51**, 694 (1952). Prepn by catalytic reduction of allopregnane-3,20-dione: Marker et al., **Brit. pat. 512,940** (1939 to Parke, Davis); by hydrogenation of pregn-5-ene-3β-ol-20-one: Klyne, Barton, *J. Am. Chem. Soc.* **71**, 1500 (1949); by hydrogenation of pregn-5-ene-3β,20β-diol: Klyne, Miller, *J. Chem. Soc.* **1950**, 1972.
Leaflets from ethyl acetate + petr ether, mp 194.5-195.5°. $[\alpha]_D$ +4.4° (c = 1.04 in chloroform).
Diacetate, $C_{25}H_{40}O_4$, needles from ethyl acetate + petr ether, mp 141-142.5°. $[\alpha]_D^{20}$ +21° (c = 1 in chloroform).
Dibenzoate, $C_{35}H_{44}O_4$, crystals, mp 237.5-239°. $[\alpha]_D$ −10.1° (c = 2.08 in chloroform).

262. Allopregnane-3β,21-diol-11,20-dione. 3β,21-*Dihydroxy-5α-pregnane-11,20-dione;* 3β,21-dihydroxy-11,20-dioxo-5α-pregnane; Kendall's compound H; Reichstein's

substance N. $C_{21}H_{32}O_4$; mol wt 348.47. C 72.38%, H 9.26%, O 18.37%. Isoln from adrenal glands: Steiger, Reichstein, *Helv. Chim. Acta* **21**, 546 (1938). Prepn by reduction of allopregnan-21-ol-3,11,20-trione acetate: Mancera et al., *J. Am. Chem. Soc.* **77**, 5669 (1955).
Hydrated crystals from dil acetone; anhyd shiny spears from benzene or acetone, mp 189-191°. $[\alpha]_D^{19}$ +93.8° (c = 1.4 in abs ethanol).
3,21-Diacetate, $C_{25}H_{36}O_6$, clusters of needles from ether-pentane, mp 148-149.5°. $[\alpha]_D^{18}$ +77.5° ± 2.5° (acetone); $[\alpha]_D^{19}$ +85.6 ± 2° (dioxane).

263. Allopregnane-3β,17α-diol-20-one. 3β,17-*Dihydroxy-5α-pregnan-20-one;* 3β,17-dihydroxy-20-oxo-5α-pregnane; 17-(1-ketoethyl)androstane-3,17-diol; Reichstein's substance L; Wintersteiner's compound G. $C_{21}H_{34}O_3$; mol wt 334.48. C 75.40%, H 10.25%, O 14.35%. Isoln from adrenal cortex: Wintersteiner, Pfiffner, *J. Biol. Chem.* **116**, 291 (1936). Structure: Reichstein, Gätzi, *Helv. Chim. Acta* **21**, 1497 (1938). Prepn from allopregnan-3β-ol-20-one 3-acetate: Rosenkranz et al., *J. Am. Chem. Soc.* **72**, 4081 (1950); from pregn-5-ene-3β-ol-20-one: Ramirez, Stafie, ibid. **77**, 134 (1955).
Crystals from abs alcohol, mp 264-266°. $[\alpha]_D^{21}$ +30.6° (c = 0.54 in abs alcohol).
3-Acetate, $C_{23}H_{36}O_4$, crystals from acetone, mp 187-189°. $[\alpha]_D^{20}$ +18° (acetone).

264. 3,20-Allopregnanedione. 5α-*Pregnane-3,20-dione;* 3,20-dioxo-5α-pregnane. $C_{21}H_{32}O_2$; mol wt 316.47. C 79.69%, H 10.19%, O 10.11%. From pregnancy urine: Hartmann, Locher, *Helv. Chim. Acta* **18**, 160 (1935); Lieberman et al., *J. Biol. Chem.* **172**, 263 (1948). Prepn from pregna-4,11-diene-3,20-dione: Shoppee, Reichstein, *Helv. Chim. Acta* **24**, 356 (1941); by oxidation of allopregnan-3β-ol-20-one with chromium trioxide: Billeter, Miescher, ibid. **30**, 1409 (1947); by hydrogenation of pregnenolone: Pappas, Nace, *J. Am. Chem. Soc.* **81**, 4556 (1959); from funtumine: Janot et al., *Bull. Soc. Chim. France* **1960**, 1669.
Crystals from methylene chloride + hexane, mp 200°. $[\alpha]_D$ +125° (c = 1.2 in chloroform).
USE: In the synthesis of progesterone, **Fr. pat. 845,034**, *C.A.* **34**, 8184² (1940).

265. Allopregnane-3β,11β,17α,20β,21-pentol. 5α-*Pregnane-3β,11β,17,20β,21-pentol;* 3β,11β,17,20β,21-pentahydroxy-5α-pregnane; 17-(1,2-dihydroxyethyl)androstane-3,11,17-triol; Kendall's compound D; Reichstein's substance A; Wintersteiner's compound A. $C_{21}H_{36}O_5$; mol wt 368.50. C 68.44%, H 9.85%, O 21.71%. Occurs in adrenal cortex: Mason et al., *J. Biol. Chem.* **114**, 613 (1936). Prepn from 21-acetoxy-3β,17α-dihydroxy-5α-pregnane-11,20-dione + lithium aluminum hydride: Klyne, Ridley, *J. Chem. Soc.* **1956**, 4825; by hydrogenation of cortisol with platinum oxide: Caspi, *J. Org. Chem.* **24**, 669 (1959).
Crystals from methanol, mp 159-163°, resolidifies and mp 215-216°. $[\alpha]_D^{22}$ +9.8° (c = 0.56 in methanol).

266. Allopregnane-3β,17α,20β,21-tetrol. *Pregnane-3,17,20,21-tetrol;* 3β,17,20β,21-tetrahydroxy-5α-pregnane; 17-(1,2-dihydroxyethyl)androstane-3,17-diol; Reichstein's substance K. $C_{21}H_{36}O_4$; mol wt 352.52. C 71.55%, H 10.29%, O 18.15%. Isoln from adrenal cortex: Steiger, Reichstein, *Helv. Chim. Acta* **21**, 546 (1938). Prepn by reduction of allopregn-17-ene-3,17-diol 3,21-diacetate with osmium tetroxide: Serini et al., *Ber.* **72**, 391 (1939).
Leaflets from dilute methanol, mp 198-200°. $[\alpha]_D^{21}$ −1.0° (c = 1 in abs ethanol).
3,20,21-Triacetate, $C_{27}H_{42}O_7$, rods from ether + pentane, mp 179°. $[\alpha]_D^{19}$ +53.2° (c = 1.9 in acetone).

267. Allopregnane-3α,11β,17α,21-tetrol-20-one. 3α,-*11β,17,21-Tetrahydroxy-5α-pregnan-20-one;* 3α,11β,17,21-tetrahydroxy-20-oxo-5α-pregnane; 17-(1-keto-2-hydroxyethyl)androstane-3,11,17-triol; 3α-allotetrahydrocortisol; Kendall's compound C; Reichstein's substance C; Wintersteiner's compound D. $C_{21}H_{34}O_5$; mol wt 366.48. C 68.82%, H 9.35%, O 21.83%. Isoln from adrenal cortex: Mason et al., *J. Biol. Chem.* **124**, 459 (1938); Kuizenga, Cartland, *Endocrinology* **24**, 526 (1939); v. Euw, Reichstein, *Helv. Chim. Acta* **25**, 988 (1942); v. Euw et al., ibid. **41**, 1516 (1958). Prepn by hydrogenation of cortisol with rhodium:

Caspi, *J. Org. Chem.* **24**, 669 (1959); from bismethylenedi-oxyhydrocortisone: Fukushima, Daum, *ibid.* **26**, 520 (1961).

Crystals from methanol, mp 244-245°. $[\alpha]_D^{21}$ +59.7° (c = 0.34 in methanol).

3,21-Diacetate, $C_{25}H_{38}O_7$, crystals, decomp 204-205°. $[\alpha]_D^{20}$ +73.8°; $[\alpha]_{542}^{20}$ +90.5° (dioxane).

268. Allopregnane-3β,11β,17α,21-tetrol-20-one.
3β,-11β,17,21-Tetrahydroxy-5α-pregnan-20-one; 3β,11β,17,21-tetrahydroxy-20-oxo-5α-pregnane; Reichstein's substance V. $C_{21}H_{34}O_5$; mol wt 366.48. C 68.82%, H 9.35%, O 21.83%. Isoln from adrenal cortex: von Euw, Reichstein, *Helv. Chim. Acta* **25**, 988 (1942). Prepn from 5α-pregnan-3β-ol-11,20-dione: Chamberlin, Chemerda, *J. Am. Chem. Soc.* **77**, 1221 (1955); by hydrogenation of cortisol with rhodium: Caspi, *J. Org. Chem.* **24**, 669 (1959); from bismethyl-enedioxyhydrocortisone: Fukushima, Daum, *ibid.* **26**, 520 (1961).

Needles from dilute methanol, decomp 220-225°. $[\alpha]_D^{13}$ +50.7° (c = 0.8672 in dioxane).

3,21-Diacetate, $C_{25}H_{38}O_7$, flat prisms from acetone-ether, decomp 225-227°. $[\alpha]_D^{18}$ +62.6° (c = 1.101 in dioxane).

269. Allopregnane-3β,17α,20α-triol.
5α-Pregnane-3β,17,20α-triol; 3β,17,20α-trihydroxy-5α-pregnane; Reichstein's substance O. $C_{21}H_{36}O_3$; mol wt 336.50. C 74.95%, H 10.78%, O 14.26%. Isoln from adrenal glands: Steiger, Reichstein, *Helv. Chim. Acta* **21**, 546 (1938). Prepn from allopregn-16-en-3β-ol-20-one: Plattner *et al., ibid.* **31**, 2210 (1948); Julian *et al.,* U.S. pat. 2,662,904 (1953 to Glidden).

Leaflets from dil methanol, mp 222-223°. $[\alpha]_D^{20}$ −12.55° (c = 1.195 in methanol).

3,20-Diacetate, $C_{25}H_{40}O_5$, crystals by sublimation, mp 252°. $[\alpha]_D^{21}$ −30.1° (c = 1.097 in acetone).

270. Allopregnane-3β,17α,20β-triol.
5α-Pregnane-3β,-17,20β-triol; 3β,17,20β-trihydroxy-5α-pregnane; 17-(1-hy-droxyethyl)androstane-3,17-diol; Reichstein's substance J. $C_{21}H_{36}O_3$; mol wt 336.50. C 74.95%, H 10.78%, O 14.26%. Isoln from adrenal cortex: Reichstein, *Helv. Chim. Acta* **19**, 1126 (1936); Steiger, Reichstein, *ibid.* **21**, 546 (1938). Prepn from 3β-acetoxy-16α,17α-oxido-20-oxo-5-allopregnane: Plattner *et al., ibid.* **31**, 2210 (1948); Brit. pat. 665,254 (1952 to Ciba); from 3β-acetoxy-17α-hydroxyallopregnan-20-one: Fukushima, Meyer, *J. Org. Chem.* **23**, 174 (1958).

Rhomboid needles from dil acetone, mp 216-217°. Sublimes in high vacuum. $[\alpha]_D^{20}$ −7.9° (c = 1.77 in abs alc).

3,20-Diacetate, $C_{25}H_{40}O_5$, crystals from pentane, mp 161-162°. $[\alpha]_D^{20}$ +24.6° (c = 2.11 in acetone).

271. Allopregnane-3β,17α,21-triol-11,20-dione.
3β,-17,21-Trihydroxy-5α-pregnane-11,20-dione; 3β,17,21-trihy-droxy-11,20-dioxo-5α-pregnane; 17-(1-keto-2-hydroxyeth-yl)androstane-3,17-diol-11-one; Kendall's compound G; Reichstein's substance D; Wintersteiner's compound B. $C_{21}H_{32}O_5$; mol wt 364.47. C 69.20%, H 8.85%, O 21.95%. Occurs in adrenal cortex: Kuizenga, Cartland, *Endocrinology* **24**, 526 (1939); von Euw, Reichstein, *Helv. Chim. Acta* **25**, 988 (1942); **41**, 1516 (1958). Prepn by reduction of allopregnane-17α,21-diol-3,11,20-trione: Kaufmann, Pataki, *Experientia* **7**, 260 (1951); from 5α-pregnan-3β-ol-11,20-dione: Chamberlin, Chemerda, *J. Am. Chem. Soc.* **77**, 1221 (1955).

Crystals from abs alc, mp 238-242°. $[\alpha]_D^{20}$ +61.8° (c = 1.07 in dioxane).

3,21-Diacetate, $C_{25}H_{36}O_7$, crystals, mp 223-224°. $[\alpha]_D^{20}$ +72.3° (dioxane).

272. Allopregnane-3β,11β,21-triol-20-one.
3β,11β,21-Trihydroxy-5α-pregnan-20-one; 3β,11β,21-trihydroxy-20-oxo-5α-pregnane; Reichstein's substance R. $C_{21}H_{34}O_4$; mol wt 350.48. C 71.96%, H 9.78%, O 18.26%. Isolation from adrenal glands: Reichstein, von Euw, *Helv. Chim. Acta* **21**, 1197 (1938); Reichstein, *ibid.* 1490. Partial synthesis by hydrogenation of corticosterone acetate: Pataki *et al., J. Biol. Chem.* **195**, 751 (1952). Prepn from 3β,11β-dihydroxy-alloetiocholanic acid: Lardon, Reichstein, *Helv. Chim. Acta* **37**, 443 (1954); from hydrocortisone: Mancera *et al, J. Am. Chem. Soc.* **77**, 5669 (1955); from 3β,21-diacetoxy-17(20)-allopregnene + osmium tetroxide and triethylamine oxide peroxide: Schneider, Hanze, U.S. pat. 2,769,823 (1956 to Upjohn).

Needles from alc, mp 202-204°. $[\alpha]_D$ +110° (ethanol).

3,21-Diacetate, $C_{25}H_{38}O_6$, crystals from acetone + ether, mp 170-172°. $[\alpha]_D^{22}$ +82.5° (c = 1.38 in dioxane); $[\alpha]_D^{20}$ +101° (acetone).

273. Allopregnane-3β,17α,21-triol-20-one.
3β,17,21-Trihydroxy-5α-pregnan-20-one; 3α,17,21-trihydroxy-20-oxo-5α-pregnane; Reichstein's substance P. $C_{21}H_{34}O_4$; mol wt 350.48. C 71.96%, H 9.78%, O 18.26%. Isoln from adrenal glands: Steiger, Reichstein, *Helv. Chim. Acta* **21**, 546 (1938); Reichstein, Gatzi, *ibid.* 1185. Partial synthesis from cholesterol: Reichstein, von Euw, *ibid.* **24**, 401 (1941). Prepn from allopregnan-3β-ol-20-one: Rosenkranz *et al., J. Am. Chem. Soc.* **72**, 4081 (1950); from 3β-acetoxy-21-bromo-17α-hydroxy-2-oxoallopregnane: Kaufmann *et al.,* U.S. pat. 2,596,562 (1952 to Syntex).

Pointed needles from abs ethanol, decomp 230-239°. $[\alpha]_D^{20}$ +48.0° (c = 0.938 in abs ethanol). Freely sol in alcohol, acetone. Sparingly sol in ether, water.

3,21-Diacetate, $C_{25}H_{38}O_6$, crystals from benzene, mp 208-210°. $[\alpha]_D^{20}$ +41.5° (chloroform).

274. Allopregnane-3α-ol-20-one.
3α-Hydroxy-5α-preg-nan-20-one; 3α-hydroxy-20-oxo-5α-pregnane; epiallopreg-nan-3-ol-20-one. $C_{21}H_{34}O_2$; mol wt 318.48. C 79.19%, H 10.76%, O 10.05%. Isoln from pregnancy urine of women: Marker *et al., J. Am. Chem. Soc.* **59**, 616 (1937); Lieberman *et al., J. Biol. Chem.* **172**, 263 (1948); Davis, Plotz, *Acta Endocrinol.* **21**, 245 (1956). Prepn from pregnenolone: Fleischer *et al., J. Am. Chem. Soc.* **60**, 79 (1938); Schwenk *et al.,* U.S. pat. 2,180,614 (1940 to Schering); by reduction of allopregnane-3,20-dione: Soloway *et al., J. Am. Chem. Soc.* **75**, 2356 (1953).

Crystals from abs alc, mp 176-178°. $[\alpha]_D$ +87.7° (abs alc). Acetate, $C_{23}H_{36}O_3$, crystals from aq ethanol, mp 141-142°. $[\alpha]_D^{22}$ +94.5° (abs ethanol).

275. Allopregnan-3β-ol-20-one.
3β-Hydroxy-5α-preg-nan-20-one; 3β-hydroxy-20-oxo-5α-pregnane. $C_{21}H_{34}O_2$; mol wt 318.48. C 79.19%, H 10.76%, O 10.05%. Isoln from adrenal cortex: von Euw, Reichstein, *Helv. Chim. Acta* **24**, 885 (1941); from corpus luteum: Butenandt, Mamoli, *Ber.* **68**, 1847 (1935); Prelog, Meister, *Helv. Chim. Acta* **32**, 2435 (1949); from human pregnancy urine: Lieberman *et al., J. Biol. Chem.* **172**, 263 (1948); from human placenta: Pearlman, Cerceo, *ibid.* **194**, 807 (1952). Prepn from 20-methyl-Δ20-allopregnen-3β-ol: Koechlin, Reichstein, *Helv. Chim. Acta* **27**, 549 (1944); from pregnenolone: Mancera *et al., J. Org. Chem.* **16**, 192 (1951); Pappas, Nace, *J. Am. Chem. Soc.* **81**, 4556 (1959); by redn of allopregnane-3,20-dione with sodium borohydride: Mancera *et al., ibid.* **75**, 1286 (1953); from progesterone 20-cycloethylene ketal: Sondheimer, Klibansky, *Tetrahedron* **5**, 15 (1959).

Plates from dil methanol, mp 194-195°. $[\alpha]_D^{27}$ +91.2° (c = 0.4 in ethanol).

Acetate, $C_{23}H_{36}O_3$, plates from methanol, mp 144-146°. $[\alpha]_D^{25}$ +69° (chloroform).

276. Allopregnane-20α-ol-3-one.
20α-Hydroxy-5α-pregnan-3-one; 20α-hydroxy-3-oxo-5α-pregnane. $C_{21}H_{34}O_2$; mol wt 318.48. C 79.19%, H 10.76%, O 10.05%. Prepn from funtumidine: Janot *et al., Bull. Soc. Chim. France* **1960**, 1669.

Crystals, mp 179°. $[\alpha]_D$ +36.6° (c = 0.8 in chloroform).

Acetate, $C_{23}H_{36}O_3$, crystals from ethanol, mp 155°. $[\alpha]_D$ +24° (c = 1.5 in chloroform).

277. Allopregnan-20β-ol-3-one.
20β-Hydroxy-5α-pregnan-3-one; 20β-hydroxy-3-oxo-5α-pregnane. $C_{21}H_{34}O_2$; mol wt 318.48. C 79.19%, H 10.76%, O 10.05%. Prepn from allopregnane-3β,20β-diol-3-acetate: Rubin *et al., J. Am. Chem. Soc.* **73**, 2338 (1951); from isofuntumidine: Janot *et al., Bull. Soc. Chim. France* **1960**, 1669.

Crystals from heptane, mp 185°. $[\alpha]_D$ +20° (c = 1.2 in chloroform).

Acetate, $C_{23}H_{36}O_3$, crystals from ethanol, mp 148°. $[\alpha]_D$ +57° (c = 19 in chloroform).

278. Allopurinol.
1,5-Dihydro-4H-pyrazolo[3,4-d]pyri-midin-4-one; 1H-pyrazolo[3,4-d]pyrimidin-4-ol; 4-hydroxy-pyrazolo[3,4-d]pyrimidine; HPP; BW 56158; Al-100; Adenock; Aloral; Alosilot; Allo-Puren; Allozym; Allural; Alu-

line; Anoprolin; Anzief; Apulonga; Apurol; Apurin; Bleminol; Bloxanth; Caplenal; Cellidrin; Cosuric; Dabroson; dura
AL; Embarin; Epidropal; Epuric; Foligan; Geapur; Gichtex;
Hamarin; Hexanurat; Ketanrift; Ketobun-A; Ledopur; Lopurin; Lysuron; Miniplanor; Monarch; Nektrohan; Remid;
Riball; Suspendol; Takanarumin; Urbol; Uricemil; Uripuri-
nol; Urobenyl; Urosin; Urtias 100; Xanturat; Zyloprim;
Zyloric. $C_5H_4N_4O$; mol wt 136.11. C 44.13%, H 2.96%, N
41.17%, O 11.76%. Xanthine oxidase inhibitor which de-
creases uric acid production. Prepn: Robins, *J. Am. Chem.
Soc.* **78**, 784 (1956); Schmidt, Druey, *Helv. Chim. Acta* **39**,
986 (1956); Druey, Schmidt, **U.S.** pat. **2,868,803** (1959 to
Ciba); **Brit.** pat. **798,646** (1958 to Wellcome Found.);
Hitchings, Falco, **U.S.** pat. **3,474,098** (1969 to Burroughs
Wellcome). Physiological and biochemical studies: Hitch-
ings, in *Biochem. Aspects Antimetab. Drug Hydroxylation*, D.
Shugar, Ed. (Academic Press, London, 1969) pp 11-22, *C.A.*
75, 3531h (1971). Clinical trial in treatment of renal calculi:
M. J. V. Smith, *J. Urol.* **117**, 690 (1977); B. Ettinger *et al.*, *N.
Engl. J. Med.* **315**, 1386 (1986). Use in hyperuricemia and
gout: G. R. Boss, J. E. Seegmiller, *ibid.* **300**, 1459 (1977).
Effect on renal function in treatment of gout: T. Gibson,
Ann. Rheum. Dis. **41**, 59 (1982). Comprehensive description:
S. A. Benezra, T. R. Bennett in *Analytical Profiles of Drug
Substances* vol. **7**, K. Florey, Ed. (Academic Press, New
York, 1978) pp 1-17.

Crystals, mp above 350°. uv max (0.1*N* NaOH): 257 nm
(ε 7200); (0.1*N* HCl): 250 nm (ε 7600); (methanol): 252 nm
(ε 7600). Soly in mg/ml at 25°: water 0.48; *n*-octanol
<0.01; chloroform 0.60; ethanol 0.30; DMSO 4.6. pKa
10.2.

THERAP CAT: Treatment of hyperuricemia and chronic
gout. Antiurolithic.

279. D-**Allose.** β-D-Allopyranose. $C_6H_{12}O_6$; mol wt
180.16. C 40.00%, H 6.71%, O 53.29%. An aldohexose
sugar. Obtained from leaves of *Protea rubropilosa:* Beylis,
Perold, *Chem. Commun.* **1971**, 597. Prepd by reduction of
D-allonic lactone with sodium amalgam: Levene, Jacobs,
Ber. **43**, 3147 (1910). Synthesis: Hughes, Speakman, *J.
Chem. Soc.* **1965**, 2236; Baker *et al.*, *Carbohyd. Res.* **24**, 192
(1972).

Crystals from dilute methanol, mp 128-128.5°. Shows
mutarotation: $[\alpha]_D^{20}$ +0.58° (4 min) → +3.26° (10 min) →
+14.41° (final value, 20 hr; c = 5). Soluble in water; practically insol in alcohol.
Phenylosazone, $C_{18}H_{22}N_4O_4$, mp 178°.

280. Allotetrahydrocortisone. *3α,17,21-Trihydroxy-5α-
pregnane-11,20-dione;* Reichstein's substance "Dehydro-C";
11-dehydro C; allopregnane-3α,17α,21-triol-11,20-dione.
$C_{21}H_{32}O_5$; mol wt 364.47. C 69.20%, H 8.85%, O 21.95%.
Isoln from beef adrenals: v. Euw *et al.*, *Helv. Chim. Acta* **41**,
1516 (1958). Synthesis of the 3α,21-diacetate: Nagata *et al.*,
ibid. **42**, 1399 (1959).

Rhombohedric leaflets from acetone + water, mp 212-
214°.
3α,21-Diacetate, $C_{25}H_{36}O_7$, rhombohedric platelets, mp
225-227°; $[\alpha]_D^{25}$ +94.2° ±1.5° (c = 1.45 in dioxane).

281. Alloxan. *2,4,5,6(1H,3H)-pyrimidinetetrone;* 2,4,-
5,6-tetraoxohexahydropyrimidine; mesoxalylurea; mesoxalylcarbamide. $C_4H_2N_2O_4$; mol wt 142.07. C 33.81%, H
1.42%, N 19.72%, O 45.05%. Found by Liebig in mucus
excreted during dysentery. Prepn by direct oxidation of uric
acid: G. Brugnatelli, *Ann. Chim. Phys. [2]* **8**, 201 (1818); J.
Liebig, F. Wöhler, *Ann.* **26**, 241 (1838); H. Biltz, M. Hehn,
ibid. **413**, 60 (1917). Prepd from alloxantin: J. Liebig, *ibid.*
147, 366 (1868); W. W. Hartman, O. E. Sheppard, *Org. Syn.
coll. vol. III*, 37 (1955). *See also* A. V. Holmgren, W. W.
Wenner, *ibid.* **coll. vol. IV**, 23 (1963). Produces diabetes in
animals by selective necrosis of pancreatic islet β-cells: J. S.
Dunn, N. G. B. McLetchie, *Lancet* **2**, 384 (1943); W. B.
Kennedy, F. D. W. Lukens, *Proc. Soc. Exp. Biol. Med.* **57**,
143 (1944). Mechanism of action study: L. Boquist, *Acta
Pathol. Microbiol. Scand. Sect. A* **88**, 201 (1980). *In vitro*
antineoplastic activity: P. Grobon, *Compt. Rend. Ser. D.*
280, 2413 (1975). Antibacterial and antifungal activity: J.
D. Douros, A. F. Kerst, **Japan. Kokai 72 4900**; *eidem*, **U.S.**
pat. **3,728,454** (1972, 1973 both to Gates Rubber Co.).
Toxicological study in mice: B. A. Waisbren, *Proc. Soc. Exp.
Biol. Med.* **67**, 154 (1948).

Anhydrous, orthorhombic crystals from anhydr acetone
or glacial acetic acid or by sublimation in vacuo. Turns pink
at 230° and decomp at 256°. Acid to litmus. K at 25° =
2.32 × 10⁻⁷. Absorption spectrum: Hartley, *J. Chem. Soc.*
87, 1802, 1808 (1905). Freely sol in water. Hot aqueous
solns are yellow and become colorless on cooling. Aqueous
solns, after being in contact with the human skin for some
time, give it a red color and a disagreeable odor. Sol in
acetone, alcohol, methanol, glacial acetic acid; slightly sol in
chloroform, petr ether, toluene, ethyl acetate and acetic anhydride. Insol in ether.
Tetrahydrate, large triclinic prisms or oblique monoclinic
rhombs from water.
Monohydrate, by heating the tetrahydrate at 100° or by
exposing it to the air. Forms triclinic pinacoidal crystals.
Compound with urea, $CH_4N_2O \cdot C_4H_2N_2O_4 \cdot H_2O$, minute,
yellow needles; red at 170°, decomp at 185-186°.
USE: In production of diabetes in experimental animals; in
nutrition experiments; in organic syntheses.

282. Alloxantin. *5,5'-Dihydroxy-[5,5'-bipyrimidine]-
2,2',4,4',6,6'(1H,1'H,3H,3'H,5H,5'H)-hexone; 5,5'-dihy-
droxy-5,5'-bibarbituric acid;* uroxin. $C_8H_6N_4O_8$; mol wt
286.16. C 33.58%, H 2.11%, N 19.58%, O 44.73%. Prepd
from uric acid: Nightingale, *Org. Syn.* **coll. vol. III**, 42
(1955); from alloxan monohydrate: Tipson, *ibid.*, **coll. vol.
IV**, 25 (1963). Structure: Singh, *Acta Cryst.* **19**, 767 (1965).

Dihydrate, cryst powder. On exposure to air becomes red. Becomes yellow at 225°; dec at 253-255°. Sparingly sol in cold water, alcohol or ether. Its aq soln is acid; reduces Ag salts and gives a blue ppt with $Ba(OH)_2$ soln. *Keep tightly closed.* Emits toxic fumes upon decompn.

Caution: Ingestion causes disturbed carbohydrate metabolism leading to diabetes, *cf. Dangerous Properties of Industrial Materials*, N. I. Sax, Ed. (Van Nostrand Reinhold, New York, 4th ed., 1975) p 366.

283. Allura® Red AC. *6-Hydroxy-5-[(2-methoxy-5-methyl-4-sulfophenyl)azo]-2-naphthalenesulfonic acid disodium salt; 6-hydroxy-5-[(6-methoxy-4-sulfo-m-tolyl)azo]-2-naphthalenesulfonic acid disodium salt; 1-[(6-methoxy-4-sulfo-m-tolyl)azo]-2-naphthol-6-sulfonic acid disodium salt;* FD &C Red No. 40; C.I. 16035; C.I. Food Red 17. $C_{18}H_{14}N_2Na_2O_8S_2$; mol wt 496.42. C 43.55%, H 2.84%, N 5.64%, Na 9.26%, O 25.79%, S 12.92%. Prepn: **Brit. pat. 1,164,249,** *C.A.* **72,** 68195j (1970) corresp to Rast, Steiner, **U.S. pat. 3,519,617** (1969, 1970, both to Allied Chem.).

Dark red powder. Soly at 25°: water, 22.5%; 50% alcohol, 1.3%.

USE: Color additive. Approved by FDA for use in foods, drugs, cosmetics.

284. Allyl Alcohol. *2-Propen-1-ol;* 1-propenol-3; vinyl carbinol. C_3H_6O; mol wt 58.08. C 62.04%, H 10.41%, O 27.55%. $CH_2=CHCH_2OH$. Prepd by heating glycerol with formic acid: O. Kamm, C. S. Marvel, *Org. Syn.* coll. vol. I, 42 (2nd ed., 1941); *cf.* Delaby, Dubois, *Compt. Rend.* **188,** 710 (1929).

Colorless liquid; pungent, mustard-like odor, irritating to the eyes. d_4^{20} 0.8540. bp 96-97° (mp −50°). n_D^{20} 1.41345. Absorption spectrum: Lüthy, *Z. Phys. Chem.* **107,** 289 (1923). Flash pt 70°F (open cup); 75°F (closed cup); autoignition temp 713°F. 72.3% of the alcohol with 27.7% water forms a constant boiling mixture, bp at 87.5°. Miscible with water, alcohol, chloroform, ether, petr ether. *Keep tightly closed.* Upon storage for several years allyl alcohol polymerizes and a thick syrup is formed (insol in water, sol in chloroform) which on treatment with ether yields a brittle resinoid mass: Blicke, *J. Am. Chem. Soc.* **45,** 1563 (1923). LD_{50} orally in rats: 64 mg/kg, Smyth, Carpenter, *J. Ind. Hyg. Toxicol.* **30,** 63 (1948).

USE: Manuf allyl compds, war gas, resins, plasticizers.

Caution: Causes severe irritation of mucous membranes, eyes: E. Browning, *Toxicity and Metabolism of Industrial Solvents* (Elsevier, New York, 1965) pp 377-381, 401-411.

285. Allylamine. *2-Propen-1-amine;* 3-aminopropylene. C_3H_7N; mol wt 57.09. C 63.11%, H 12.36%, N 24.53%. $CH_2=CHCH_2NH_2$. Manuf from allyl chloride and ammonia: Ploetz, **U.S. pat. 2,915,385** (1959 to Feldmühle Papier- und Zellstoffwerke).

Liquid; burning taste; strong ammonia odor causing sneezing and tears. d_{20}^{20} 0.760. bp 55-58°. n_D^{20} 1.4186. Flash pt, closed cup: 10°F (−12°C). Misc with water, alc, chloroform, ether. *Keep tightly closed.* LD_{50} i.p. in mice: 49 mg/kg, C. H. Hine *et al., Arch. Environ. Health* **1,** 34 (1960).

Caution: A strong eye and respiratory tract irritant; intolerable at 14 ppm, *cf.* Patty's *Industrial Hygiene and Toxicol-*

ogy, **vol. 2B,** G. D. Clayton, F. E. Clayton, Eds. (Wiley-Interscience, New York, 3rd ed., 1981) pp 3157-3158.

USE: In the manuf of mercurial diuretics.

286. Allyl Bromide. *3-Bromo-1-propene;* 3-bromopropylene; bromallylene. C_3H_5Br; mol wt 120.99. C 29.78%, H 4.17%, Br 66.05%. $CH_2=CHCH_2Br$. Prepd from hydrobromic acid and allyl alcohol: O. Kamm, C. S. Marvel, *Org. Syn.* coll. vol. I, 27 (2nd ed., 1941); from triphenylphosphite, allyl alcohol and benzyl bromide: Landauer, Rydon, *J. Chem. Soc.* **1953,** 2224.

Colorless liquid; unpleasant, pungent odor. d_4^{20} 1.398. bp_{760} 71.3° (mp −119°). n_D^{20} 1.46545. Slightly sol in water; miscible with alcohol, chloroform, ether, carbon disulfide, carbon tetrachloride. *Keep tightly closed.*

USE: Manuf synthetic perfumes, other allyl compounds.

Caution: Produces irritation of eyes, respiratory passages.

287. Allyl Chloride. *3-Chloro-1-propene;* 3-chloropropylene; Chlorallylene. C_3H_5Cl; mol wt 76.53. C 47.08%, H 6.59%, Cl 46.33%. $CH_2=CHCH_2Cl$. Manuf by chlorination of propylene: Samples, Hilbert, **U.S. pat. 3,054,831** (1962 to Union Carbide). Prepd from diphenylphosphite, allyl alcohol and benzyl chloride: Landauer, Rydon, *J. Chem. Soc.* **1953,** 2224. Toxicity data: H. F. Smyth, C. P. Carpenter, *J. Ind. Hyg. Toxicol.* **30,** 63 (1948).

Liquid; unpleasant, pungent odor. d_4^{20} 0.938. bp 44-45°. mp −134.5°. n_D^{20} 1.4154. Flash pt, closed cup: −25°F (−31°C). Slightly sol in water; misc with alcohol, chloroform, ether, petrol ether. *Keep tightly closed.* LD_{50} orally in rats: 0.7 g/kg (Smyth, Carpenter).

USE: In the synthesis of allyl compds. *Caution:* Produces irritation of eyes, respiratory passages. Readily absorbed through skin. *Acute* exposure can cause unconsciousness; *chronic* exposure, injury to liver, kidneys.

288. Allyl Cyanide. *3-Butenenitrile;* vinylacetonitrile; β-butenenitrile. C_4H_5N; mol wt 67.09. C 71.61%, H 7.51%, N 20.88%. $CH_2=CHCH_2CN$. Found in some mustard oils. Prepd by treating dry CuCN with allyl bromide: Bruylants, *Bull. Soc. Chim. Belg.* **31,** 175 (1922); Supniewski, Salzberg, *Org. Syn.* coll. vol. I, 46 (2nd ed., 1941).

Liquid. Agreeable onion-like odor. Stable to heat. d_4^{20} 0.8341. mp −87°. bp_{760} 119°; bp_{400} 98°; bp_{200} 78°; bp_{100} 60.2°; bp_{60} 48.8°; bp_{40} 40.0°; bp_{20} 26.6°; bp_{10} 14.1°; bp_5 +2.9°; $bp_{1.0}$ −19.6°. n_D^{20} 1.4060. Absorption spectrum in hexane, water and dil alkali: Bruylants, Castille, *op. cit.* **34,** 265; *Chem. Zentr.* **1926,** I, 1962. LD_{50} orally in rats: 0.115 g/kg, H. F. Smyth *et al., Am. Ind. Hyg. Assoc. J.* **30,** 470 (1969).

289. Allylestrenol. *17-(2-Propenyl)estr-4-en-17-ol; 17-allylestr-4-en-17β-ol;* 17α-allyl-17-hydroxy-19-nor-4-androstene; 17-hydroxy-17α-allyl-4-estrene; Gestanin; Gestanol; Orageston; Gestanyn; Gestanon; Turinal. $C_{21}H_{32}O$; mol wt 300.47. C 83.94%, H 10.73%, O 5.32%. Prepn: **Brit. pat. 841,411** (1960 to Organon).

Crystals, mp 79.5-80°. Practically insol in water. Sol in alcohol, ether, acetone, chloroform. Sensitive to oxidizing agents.

THERAP CAT: Progestogen.

290. Allyl Ether. *3,3′-Oxybis-1-propene;* diallyl ether. $C_6H_{10}O$; mol wt 98.14. C 73.43%, H 10.27%, O 16.30%. $(CH_2=CHCH_2)_2O$. Prepn from allyl alcohol in the presence of $CuCl \cdot H_2SO_4$: Stephenson, **Brit. pat. 913,919** (1962 to Monsanto); from allyl alcohol with K_2CO_3 and allyl bromide: Riemschneider, Kötzsch, *Monatsh.* **90,** 787 (1959). Toxicity: H. F. Smyth *et al., J. Ind. Hyg. Toxicol.* **31,** 60 (1949).

Liquid; radish-like odor. d_0^{18} 0.805. bp 94°. n_D^{20} 1.4240.

Practically insol in water; miscible with alc, ether. LD_{50} orally in rats: 0.32 g/kg (Smyth).

Caution: An irritant. Can be absorbed through skin.

291. Allyl Ethyl Ether. *3-Ethoxy-1-propene;* ethyl allyl ether. $C_5H_{10}O$; mol wt 86.13. C 69.72%, H 11.70%, O 18.58%. $CH_2=CHCH_2OCH_2CH_3$. Prepd from allyl bromide and sodium alcoholate: Brühl, *Ann.* **200,** 139 (1880).

Liquid. d 0.765. $bp_{742.9}$ 66-67°. n_D^{20} 1.3881. Practically insol in water; miscible with alc, ether.

Caution: Irritating to eyes, skin, mucous membranes.

292. Allyl Iodide. *3-Iodo-1-propene;* 3-iodopropylene. C_3H_5I; mol wt 167.99. C 21.45%, H 3.00%, I 75.55%. $CH_2=CHCH_2I$. Prepd from allyl alcohol, methyl iodide, and triphenyl phosphite: Landauer, Rydon, *J. Chem. Soc.* **1953,** 2224; Rydon, Landauer, *Brit.* pat. **695,468** (1953 to Nat. Res. Dev. Corp.).

Yellowish liquid; darkens on exposure to light and air, liberating iodine. Unpleasant, pungent odor. d^{12} 1.848. bp 103°. n_D^{21} 1.5540. Practically insol in water; miscible with alc, chloroform, ether. *Keep tightly closed and protected from light.*

Caution: See Allyl Chloride.

293. Allyl Isothiocyanate. *3-Isothiocyanato-1-propene; isothiocyanic acid allyl ester;* allyl isosulfocyanate; volatile oil of mustard; Redskin. C_4H_5NS; mol wt 99.15. C 48.45%, H 5.08%, N 14.13%, S 32.34%. $CH_2=CHCH_2NCS$. Isolated from *Brassica nigra* (L.) Koch, *Cruciferae* (black mustard seed), or prepd from allyl iodide and potassium thiocyanate: Dulière, *J. Pharm. Belg.* **2,** 981 (1920), *C.A.* **15,** 571[4] (1921).

Colorless or pale yellow, very refractive liquid; very pungent, irritating odor; acrid taste. d 1.013-1.020. bp 148-154°. Optically inactive. n_D^{20} 1.5268-1.5280. Slightly sol in water; misc with alcohol and most organic solvents. One ml dissolves in 10 ml 70% alcohol. LD_{50} orally in rats: 339 mg/kg, P. M. Jenner *et al., Food Cosmet. Toxicol.* **2,** 327 (1964).

USE: Manuf flavors; war gas.

THERAP CAT: Counterirritant.

THERAP CAT (VET): Counterirritant.

294. Allylprodine. *1-Methyl-4-phenyl-3-(2-propenyl)-4-piperidinol propanoate;* α-3-allyl-1-methyl-4-phenyl-4-propionoxypiperidine; α-3-allyl-1-methyl-4-phenyl-4-piperidinol propionate; NIH 7440; Ro 2-7113; Alperidine. $C_{18}H_{25}NO_2$; mol wt 287.39. C 75.22%, H 8.77%, N 4.87%, O 11.13%. Prepn: Ziering *et al., J. Org. Chem.* **22,** 1521 (1957); Lee, Ziering, U.S. pat. **2,798,073** (1957 to Hoffmann-La Roche).

Hydrochloride, $C_{18}H_{26}ClNO_2$, crystals from acetone + methanol, mp 186-187°.

Caution: May be habit forming. This is a controlled substance (opiate) listed in the U.S. Code of Federal Regulations, Title 21 Part 1308.11 (1987).

THERAP CAT: Narcotic analgesic.

295. Allyl Sulfide. *3,3'-Thiobis-1-propene;* diallyl sulfide; thioallyl ether; "oil garlic". $C_6H_{10}S$; mol wt 114.20. C 63.10%, H 8.83%, S 28.08%. $(CH_2=CHCH_2)_2S$. Prepared from allyl bromide and sulfur in soln of Na in liquid NH_3: Brandsma, Wijers, *Rec. Trav. Chim.* **82,** 68 (1963).

Liquid, garlic odor. bp 139°. d_4^{27} 0.888. n_D^{27} 1.4877. Practically insoluble in water; miscible with alc, chloroform, ether, carbon tetrachloride.

USE: Manuf flavors.

296. Allylurea. *N-2-Propenylurea;* allylcarbamide. $C_4H_8N_2O$; mol wt 100.12. C 47.98%, H 8.06%, N 27.98%, O 15.98%. $CH_2=CHCH_2NHCONH_2$. Prepn: Neville, Mc-Gee, *Can. J. Chem.* **41,** 2123 (1963).

Crystals, mp 85°. Freely sol in water, alc. Practically insol in chloroform, ether, toluene, carbon disulfide.

USE: Manuf allylthiourea; corrosion inhibitors.

297. Almagate. *[Carbonato(2)]heptahydroxy(aluminum)trimagnesium dihydrate; carbonic acid, aluminum-magnesium complex;* Almax. $C_2H_{14}Al_2Mg_6O_{20}.4H_2O$; mol wt 629.98. C 3.81%, H 3.52%, Al 8.57%, Mg 23.15%, O 60.95%. $Al_2Mg_6(OH)_{14}(CO_3)_2.4H_2O$. A crystalline hydrated aluminum-magnesium hydroxycarbonate. Prepn: R. G. W. Spickett *et al.,* Belg. pat. **873,843;** *eidem,* U.S. pat. **4,447,417** (1979, 1984 both to Anphar). Crystal structure: J. Moragues *et al., Arzneimittel-Forsch.* **34,** 1346 (1984). pH curves, acid consuming capacity: J. E. Beneyto *et al., ibid.* 1350. Effect of proteolytic enzymes, polypeptides: J. E. Beneyto, J. L. Fábregas, *ibid.* 1357. Serum levels of magnesium, aluminum not increased by high doses of almagate: J. Jauregui, J. Segura, *ibid.* 1364. Pharmacology: P. R. Beckett *et al., ibid.* 1367. Clinical trials: A. Suau *et al., ibid.* 1380.

Dehydrates at 510°K. USP acid neutralization capacity 28.3 mEq HCl/g.

THERAP CAT: Antacid.

298. Alminoprofen. α-*Methyl-4-[(2-methyl-2-propenyl)amino]benzeneacetic acid;* 2-(p-methylallylaminophenyl)-propionic acid. EB 382; Minalfene. $C_{13}H_{17}NO_2$; mol wt 219.29. C 71.20%, H 7.82%, N 6.39%, O 14.59%. Prepn: **Fr.** pat. **2,137,211** corresp to E. Bouchera, U.S. pat. **3,957,-850** (1971, 1976 both to Bouchera). Synthesis and pharmacology: B. Dumaitre *et al., Eur. J. Med. Chem.* **14,** 207 (1979). Pharmacokinetic study: A. Premel-Cabic *et al., Eur. J. Clin. Pharmacol.* **18,** 419 (1980).

Cryst from cyclohexane, mp 107°. LD_{50} in mice: 2400 mg/kg orally, B. Dumaitre *et al., loc. cit.*

THERAP CAT: Anti-inflammatory; analgesic.

299. Almitrine. *6-[4-[Bis(4-fluorophenyl)methyl]-1-piperazinyl]-N,N'-di-2-propenyl-1,3,5-triazine-2,4-diamine; 2,4-bis[allylamino]-6-[4-[bis(p-fluorophenyl)methyl]-1-piperazinyl]-s-triazine;* S 2620. $C_{26}H_{29}F_2N_7$; mol wt 477.57. C 65.39%, H 6.12%, F 7.96%, N 20.53%. Prepn: G. Regnier, R. Canevari, Ger. pat. **1,947,332;** *eidem,* U.S. pat. **3,647,-794** (1970, 1972 both to Sci. Union et Cie.-Soc. Franç. Recher. Méd.). Physical properties and toxicity: G. Regnier *et al., Experientia* **29,** 814 (1972). Pharmacology: M. Laubie, F. Diot, *J. Pharmacol. (Paris)* **3,** 363 (1972); M. Laubie, H. Schmitt, *Eur. J. Pharmacol.* **61,** 125 (1980). Evaluation of combination with raubasine, *q.v.,* in cerebral ischemia in rats: M. G. Borzeix, J. Cahn, *Arzneimittel-Forsch.* **37,** 491 (1987). Clinical effect on hypoxemia in chronic obstructive pulmonary disease: R. C. Bell *et al., Ann. Int. Med.* **105,** 342 (1986); B. Gothe *et al., Am. J. Med.* **84,** 436 (1988).

Crystals, mp 181°.

Dimethanesulfonate, $C_{28}H_{37}F_2N_7O_6S_2$, *almitrine dimesylate, Vectarion.* mp 243° (dec). uv max (ethanol): 227, 246 nm (log ε 4.52, 4.53). LD_{50} in mice: 210 mg/kg i.v., 390 mg/kg i.p., >2 g/kg orally (Regnier).

THERAP CAT: Respiratory stimulant.

300. Almond, Bitter. Ripe seed of *Prunus amygdalus*

Stokes var *amara* Focke (*P. communis* Arcang. var *amara* (Focke) Schneid.), *Rosaceae.* *Habit.* Italy, Spain, and Southern France. *Constit.* 35-50% fixed oil, about 3% amygdalin, proteins, emulsin (synaptase), sugar. *Ref:* E. W. Eckey, *Vegetable Fats and Oils* (Reinhold, New York, 1954) p 455.

USE: Preparing amygdalin, essential and expressed oils, almond and bitter-almond water; in perfumery and in manuf of liqueurs. *Caution:* Cyanide poisoning from ingestion of burnt bitter almonds has been reported: *C.A.* **50**, 6666a (1956).

301. Almond, Sweet. Jordan almond. Ripe seed of *Prunus amygdalus* Stokes var *dulcis* (D.C.) Baill. (*P. communis* Arcang. var *dulcis* (Focke) Schneid.); *P. amygdalus* var *sativa* (Focke), *Rosaceae.* *Habit.* Italy, Spain and Southern France. *Constit.* About 50% fixed oil, proteins, emulsin, sugar. *Ref:* E. W. Eckey, *Vegetable Fats and Oils* (Reinhold, New York, 1954) p 455; Subrahmanyam, Achaya, *J. Sci. Food Agr.* **8**, 657 (1957).

USE: In perfumery and confectionery; preparing expressed oil of almond, almond milk, almond meal, etc.

302. Aloe. Dried latex of leaves of Curaçao Aloe (*Aloe barbadensis* Miller, *Aloe vera* Linné) or Cape Aloe (*Aloe ferox* Miller and hybrids), family *Liliaceae.* *Habit.* Curaçao, in Dutch West Indies; Cape, in Southern Africa. *Constit.* Curaçao 18-25% aloin (curaçaloin), Cape 4.5-9% aloin (capaloin); all the aloes, moreover, contain resin, emodin and volatile oil. Aloe yields not less than 50% of water-sol extractive. Structure of *aloeresin A*, a major constituent of Cape Aloe: P. Gramatica *et al., Tetrahedron Letters* **23**, 2423 (1982). *Review: Chem. Week* **78**(1), 44 (Jan. 7, 1956).

303. Aloe-Emodin. *1,8-Dihydroxy-3-(hydroxymethyl)-9,10-anthracenedione; 1,8-dihydroxy-3-(hydroxymethyl)-anthraquinone;* 3-hydroxymethylchrysazin; rhabarberone. $C_{15}H_{10}O_5$; mol wt 270.23. C 66.67%, H 3.73%, O 29.60%. Occurs in the free state and as a glycoside in *Rheum* (rhubarb), in senna leaves and in various species of *Aloe* (*Liliaceae*). Isolation: Condo-Vissicchio, *Arch. Pharm.* **247**, 81 (1909); Mary *et al., J. Am. Pharm. Assoc.* **45**, 229 (1956). Prepn: Cahn, Simonsen, *J. Chem. Soc.* **1932**, 2573; Hay, Haynes, *ibid.* **1956**, 3141; Bapat *et al., Tetrahedron Letters* no. 5, 15 (1960).

Orange needles from toluene, mp 223-224°. Sublimes in CO_2 stream. Absorption spectrum: Stone, Furman, *J. Am. Chem. Soc.* **68**, 2742 (1946). Freely sol in hot alcohol, in ether, in benzene with yellow color, in ammonia water and in sulfuric acid with crimson color.

Trimethyl ether, $C_{18}H_{16}O_5$, orange needles from acetic acid, mp 163°.

Triacetate, $C_{21}H_{16}O_8$, yellow needles from benzene, mp 175-177°.

Note: See also Emodin.

THERAP CAT: Cathartic.

304. Aloin. *10-Glucopyranosyl-1,8-dihydroxy-3-(hydroxymethyl)-9(10H)-anthracenone;* 1,8-dihydroxy-3-hydroxymethyl-10-(6-hydroxymethyl-3,4,5-trihydroxy-2-pyranyl)anthrone; 10-(1',5'-anhydroglucosyl)-aloe-emodin-9-anthrone; barbaloin. $C_{21}H_{22}O_9$; mol wt 418.39. C 60.28%, H 5.30%, O 34.42%. Isoln from various species of aloe (Cape, Uganda and Socotrine): Léger, *Ann. Chim.* **6**, 318 (1916); **8**, 265 (1918); Cahn, Simonsen, *J. Chem. Soc.* **1932**, 2573; Rosenthaler, *Arch. Pharm.* **270**, 214 (1932); Mühlemann, *Pharm. Acta Helv.* **27**, 17 (1952); Böhme, Bertram, *Arch. Pharm.* **288**, 510 (1955). The molecule is built from aloe-emodin, *q.v.,* and from glucose. Structure: Hay, Haynes, *J. Chem. Soc.* **1956**, 3141.

Lemon-yellow crystals, mp 148-149°. Quickly forms a monohydrate, mp 70-80°. Slight odor of aloe; bitter taste. Solubility at 18°: 57% in pyridine, 7.3% in glacial acetic acid, 5.4% in methanol, 3.2% in acetone, 2.8% in methyl acetate, 1.9% in ethanol, 1.8% in water, 1.6% in propanol, 0.78% in ethyl acetate, 0.27% in isopropanol. Very slightly soluble in isobutanol, chloroform, carbon disulfide, ether. Incompatible with alkali hydroxides, tannin, ferric chloride.

THERAP CAT: Cathartic.

THERAP CAT (VET): Purgative.

305. Aloxidone. *5-Methyl-3-(2-propenyl)-2,4-oxazolidinedione; 3-allyl-5-methyl-2,4-oxazolidinedione;* allomethadione; Malidone; Malazol. $C_7H_9NO_3$; mol wt 155.15. C 54.19%, H 5.85%, N 9.03%, O 30.94%. Prepn: Davies, Hook, **Brit.** pats. 627,971; 632,423 (1949 to British Schering).

Oily liquid. $bp_{0.5}$ 86-87°; $bp_{1.8}$ 88.90°. n_D^{20} 1.4710. Slightly sol in water; sol in alcohol. Not miscible with vegetable oils. Forms a sodium salt.

THERAP CAT: Anticonvulsant.

306. Aloxiprin. *2-(Acetyloxy)benzoic acid polymer with aluminum oxide;* polyoxyaluminum acetylsalicylate; Alaprin; Lyman; Palaprin; Rumatral; Superpyrin. A polymeric condensation product of aluminum oxide and aspirin. Approximate formula, $Al_3O_2[C_6H_4(OOCCH_3)COO]_5$. Prepd from aluminum isopropoxide and aspirin: Cummings *et al., J. Pharm. Pharmacol.* **15**, 56 (1963).

Powder with greater bulk density and relatively low free salicylic acid content as compared with other aluminum salicylates. Rates of soln of aspirin from aloxiprin in buffer solns of pH 2-8 are lower than those of aspirin, particularly in more acid buffers.

THERAP CAT: Analgesic.

307. Alphaprodine. *1,3-Dimethyl-4-phenyl-4-piperidinol propanoate;* α-1,3-dimethyl-4-phenyl-4-piperidinyl propionate; α-1,3-dimethyl-4-phenyl-4-propionoxypiperidine; α-prodine. $C_{16}H_{23}NO_2$; mol wt 261.35. C 73.53%, H 8.87%, N 5.36%, O 12.24%. Prepn of alphaprodine and its stereoisomer **betaprodine** as hydrochlorides: Lee, Ziering, U.S. pat. **2,498,433** (1950 to Hoffmann-La Roche); Beckett *et al., J. Pharm. Pharmacol.* **9**, 939 (1957); Ziering *et al., J. Org. Chem.* **22**, 1521 (1957). Abs config of alphaprodine: F. R. Ahmed *et al., Chem. & Ind. (London)* **1959**, 485; of betaprodine: *eidem, ibid.* **1962**, 97. Metabolism of both isomers: M. M. Abdel-Monem *et al., J. Med. Chem.* **15**, 494 (1972). Pharmacology and toxicology of alphaprodine: G. M. Gruber *et al., J. Pharmacol. Exp. Ther.* **99**, 312 (1950).

dl-Form hydrochloride, $C_{16}H_{24}ClNO_2$, *Nu-1196, Nisentil, Nisintel, Prisilidene.* Crystals from acetone, mp 220-221°. Slightly saline taste. Freely sol in water, alc, chloroform. Practically insol in ether. pH of 1% aq soln 4.5-5.2. LD_{50} in mice: 54 mg/kg i.v., 73 mg/kg i.p.; in rats: 22 mg/kg i.p. (Gruber).

Betaprodine *dl*-form hydrochloride, $C_{16}H_{24}ClNO_2$, β-*prodine, NU-1779.* Crystals from methyl ethyl ketone, mp 195-196°; from acetone + methanol, mp 199-200°.

Caution: May be habit forming. This is a controlled substance (opiate) listed in the U.S. Code of Federal Regulations, Title 21 Parts 1308.11 and 1308.12 (1985).

THERAP CAT: Narcotic analgesic.

308. Alpidem. *6-Chloro-2-(4-chlorophenyl)-N,N-dipropylimidazo[1,2-a]pyridine-3-acetamide;* SL 80.0342; Ananxyl. $C_{21}H_{23}Cl_2N_3O$; mol wt 404.34. C 62.38%, H 5.73%, Cl 17.54%, N 10.39%, O 3.96%. Non-benzodiazepine tranquilizer. Prepn: J. P. Kaplan, P. George, **Eur.** pat. **Appl. 50,563;** *eidem,* **U.S.** pat. **4,382,938** (1982, 1983 both to Synthelabo). Pharmacology and pharmacokinetics in humans: B. Saletu *et al., Int. Clin. Psychopharmacol.* **1,** 145 (1986). HPLC determn in human plasma: V. Ascalone *et al., J. Chromatog.* **414,** 101 (1987). Clinical evaluation of tranquilizing vs hypnotic effects: B. Saletu *et al., Curr. Ther. Res.* **40,** 769 (1986). Pilot study in anxiety: S. F. Langer *et al., Psychopharmacol. Bull.* **24,** 161 (1988).

mp 140-141°.
THERAP CAT: Anxiolytic.

309. Alpiropride. *4-Amino-2-methoxy-5-[(methylamino)sulfonyl]-N-[[1-(2-propenyl)-2-pyrrolidinyl]methyl]benzamide;* (±)-*N-[(1-allyl-2-pyrrolidinyl)methyl]-4-amino-5-(methylsulfamoyl)-o-anisamide; N-(1-allyl-2-pyrrolidinylmethyl)-2-methoxy-4-amino-5-methylsulfamoylbenzamide;* RIV-2093; Revistel. $C_{17}H_{26}N_4O_4S$; mol wt 382.48. C 53.39%, H 6.85%, N 14.65%, O 16.73%, S 8.38%. Dopamine D_2-receptor antagonist structurally similar to sulpiride, *q.v.* Prepn: J. Perrot, M. Thominet, **Eur.** pat. **Appl. 47,207;** *eidem,* **U.S.** pat. **4,550,179** (1982, 1985 both to Soc. d'Etudes Sci. Ind. l'Ile de France). Dopamine receptor binding study: P. Sokoloff *et al., Arch. Pharmacol.* **327,** 221 (1984). HPLC determn in plasma and urine: F. Bressolle *et al., J. Chromatog.* **343,** 443 (1985). Pharmacokinetics and metabolism in humans: F. Bressolle *et al., J. Pharm. Clin.* **4,** 261 (1985). Preliminary study in experimentally induced migraine: A. Bès *et al., Int. J. Clin. Pharm. Res.* **6,** 189 (1986).

Crystals from absolute ethanol, mp 168.5-169°. LD_{50} in male mice (mg/kg): 44 i.v.; 184 i.p.; 204 s.c.; 3600 orally (Perrot, Thominet, 1985).
THERAP CAT: Antimigraine.

310. Alprazolam. *8-Chloro-1-methyl-6-phenyl-4H-[1,2,4]triazolo[4,3-a][1,4]benzodiazepine; 8-chloro-1-methyl-6-phenyl-4H-s-triazolo[4,3-a][1,4]benzodiazepine;* D 65MT; U-31889; Alplax; Tafil; Trankimazin; Xanax; Xanor. $C_{17}H_{13}ClN_4$; mol wt 308.77. C 66.13%, H 4.24%, Cl 11.48%, N 18.15%. Prepn: J. B. Hester, **Ger.** pat. **2,012,190;** *idem,* **U.S.** pat. **3,987,052** (1970, 1976 both to Upjohn); J. B. Hes-

ter *et al., Tetrahedron Letters* **1971,** 1609; A. Walser, G. Zenchoff, *J. Med. Chem.* **20,** 1694 (1977). Central depressant activity: R. Nakajima *et al., Japan. J. Pharmacol.* **21,** 497 (1971). Pharmacology: V. H. Sethy, *Arch. Pharmacol.* **301,** 157 (1978). Clinical studies: L. F. Fabre, *Curr. Ther. Res.* **19,** 661 (1976); J. B. Cohn, *J. Clin. Psychiat.* **42,** 347 (1981). Pharmacokinetics: D. R. Abernethy *et al., ibid.* **44,** 45 (1983). Review of pharmacokinetics, clinical efficacy, and mechanism of action: J. A. Fawcett, H. M. Kravitz, *Pharmacotherapy* **2,** 242-254 (1982); of pharmacology and efficacy in anxiety and depression: *Drugs* **27,** 132 (1984).

Crystals from ethyl acetate, mp 228-228.5°. Sol in alc. Insol in water. uv max (ethanol): 222 nm (ε 40250). LD_{50} in mice, rats (mg/kg): 1020, > 2000 orally; 540, 610 i.p. (Nakajima).

Caution: May be habit forming. This is a controlled substance (depressant) listed in the U.S. Code of Federal Regulations, Title 21 part 1308.14 (1985).

THERAP CAT: Anxiolytic.

311. Alprenolol. *1-[(1-Methylethyl)amino]-3-[2-(2-propenyl)phenoxy]-2-propanol; 1-(o-allylphenoxy)-3-(isopropylamino)-2-propanol;* H 56/28. $C_{15}H_{23}NO_2$; mol wt 249.34. C 72.25%, H 9.30%, N 5.62%, O 12.83%. β-Adrenergic blocker. Prepn: Brandstrom *et al., Acta Pharm. Suecica* **3,** 303 (1966); **Neth.** pat. **Appl. 6,605,692** (1966 to AB Hassle), *C.A.* **66,** 46214p (1967); **Neth.** pat. **Appl. 6,612,958** (1967 to ICI), *C.A.* **67,** 99851w (1967). Pharmacology: Marmo, *Clin. Ter.* **56,** 121-176 (1971). Series of articles on clinical effect in myocardial infarction: *Acta Med. Scand.* **1984,** Suppl. 680, 1-64.

Hydrochloride, $C_{15}H_{24}ClNO_2$, *Apllobal, Aprobal, Aptine, Aptol Duriles, Gubernal, Regletin, Yobir.* Crystals from ethyl acetate, mp 107-109°. LD_{50} in mice, rats, rabbits (mg/kg): 278.0, 597.0, 337.3 orally (Marmo).

THERAP CAT: Antihypertensive; antianginal; antiarrhythmic.

312. Alsactide. *1-β-Alanine-17-[N-(4-aminobutyl)-L-lysinamide]-α[1-17]-corticotropin; 1-β-alanine-17-[L-2,6-diamino-N-(4-aminobutyl)hexanamide]-α[1-17]-corticotropin; [β-ala¹]corticotropin-(1-17)-heptadecapeptide-4-amino-N-butylamide; [β-ala¹,lys¹⁷]ACTH[1-17]-4-amino-N-butylamide; alisactide;* HOE-433; Synchrodyn. $C_{99}H_{155}N_{29}O_{21}S$; mol wt 2119.55. C 56.10%, H 7.37%, N 19.16%, O 15.85%, S 1.51%. Short-chain ACTH analog. Synthesis: R. Geiger, H. G. Schroeder, **Ger.** pat. **1,954,794;** *eidem,* **U.S.** pat. **3,-749,704** (1971, 1973, both to Hoechst); R. Geiger, *Ann.* **750,** 165 (1971). Pharmacology: J. Sandow *et al., Arch. Pharmacol.* **297,** Suppl. 2, R41 (1977). Dose dependent effects on cortisol and aldosterone plasma levels in man: A. Angeli *et al., Horm. Metab. Res.* **13,** 24 (1981).

β–Ala-Tyr-Ser-Met-Glu-His-Phe-Arg-Trp-Gly-

Lys-Pro-Val-Gly-Lys-Lys-Lys-NH-$(CH_2)_4$-NH$_2$

Acetate, $C_{99}H_{155}N_{29}O_{21}S.C_2H_3O_2$, $[\alpha]_D^{20}$ −68.6° (c = 0.5 in 1% acetic acid).

THERAP CAT: Diagnostic aid (adrenal function).

313. Alstonidine. *3β-(Hydroxymethyl)-2α-methyl-4β-[(9-methyl-9H-pyrido[3,4-b]indol-1-yl)methyl]-2H-pyran-*

5-carboxylic acid methyl ester. $C_{22}H_{24}N_2O_4$; mol wt 380.43. C 69.45%, H 6.36%, N 7.36%, O 16.82%. From the bark or root of *Alstonia constricta* F. Muell., *Apocynaceae:* Hesse, *Ann.* **205**, 360 (1880); Svoboda, *J. Am. Pharm. Assoc.* **46**, 508 (1957). Structure: Boaz *et al., ibid.* 510. Stereochemistry: Crow *et al., Aust. J. Chem.* **23**, 2489 (1970).

Fine crystals from ether, mp 188-190°. Practically insol in water; sol in alcohol, chloroform, ether, acetone, 5% HCl. uv max (methanol): 238, 291, 360 nm (log ε 4.66, 4.25, 3.74). The acid solutions are fluorescent.

Acetyl deriv trihydrate, $C_{24}H_{26}N_2O_5\cdot3H_2O$, fine needles from dil acetone, mp 92-96°.

314. Alstonine. *3,4,5,6,16,17-Hexadehydro-16-(methoxycarbonyl)-19α-methyl-20α-oxayohimbanium.* $C_{21}H_{20}N_2O_3$; mol wt 348.39. C 72.39%, H 5.79%, N 8.04%, O 13.78%. From *Alstonia constricta* F. Mull., *Rauwolfia vomitoria* Afzel., *Rauwolfia obscura* K. Schum., *Vinca rosea* L., and other *Apocynaceae:* Sharp, *J. Chem. Soc.* **1934**, 287; Schlittler *et al., Helv. Chim. Acta* **35**, 271 (1952); Pillay, Kumari, *J. Sci. Ind. Res.* **20B**, 458 (1961); *C.A.* **56**, 7425b (1962). Structure: Bader, *Helv. Chim. Acta* **36**, 215 (1953). Stereoisomer of serpentine (alkaloid) *q.v.* Stereochemistry: Wenkert, Roychaudhuri, *J. Am. Chem. Soc.* **79**, 1519 (1957); **80**, 1613 (1958); Shamma, Richey, *ibid.* **85**, 2507 (1963). Biosynthesis: Woodward, *Angew. Chem.* **68**, 13 (1956).

Fine, yellow needles from acetone, dec above 300° (brown at 200°, black at 260°). uv max (methanol): 252, 289, 309, 336, 369 nm (log ε 4.54, 4.08, 4.36, 3.39, 3.60). Quickly decomposes on standing in organic solvents, yielding red solns with blue fluorescence.

Hydrochloride, $C_{21}H_{20}N_2O_3\cdot HCl$, yellow plates from abs alcohol, dec 286°. $[\alpha]_D$ +132° (c = 1.06). Sol in water.

315. Althea. Marshmallow. Dried root of *Althaea officinalis* L., *Malvaceae* (marshmallow), deprived of brown, corky layer. *Habit.* Europe, Western and Northern Asia, naturalized in Eastern U.S. *Constit.* Asparagine, 25-35% mucilage, sugar, pectin.

USE: Demulcent.

316. Althiazide. *6-Chloro-3,4-dihydro-3-[(2-propenylthio)methyl]-2H-1,2,4-benzothiadiazine-7-sulfonamide 1,1-dioxide; 3-[(allylthio)methyl]-6-chloro-3,4-dihydro-2H-1,2,4-benzothiadiazine-7-sulfonamide 1,1-dioxide;* 3-allylthiomethyl-6-chloro-7-sulfamoyl-3,4-dihydrobenzothiadiazine 1,1-dioxide; 6-chloro-3,4-dihydro-7-sulfamoyl-3-(2-thiapent-4-enyl)-2H-1,2,4-benzothiadiazine 1,1-dioxide; altizide; P 1779. $C_{11}H_{14}ClN_3O_4S_3$; mol wt 383.91. C 34.41%, H 3.68%, Cl 9.24%, N 10.95%, O 16.67%, S 25.06%. Prepn: **Brit.** pat. **902,658** (1962 to Pfizer).

Solid, mp 206-207°.
Combination with spironolactone, **Aldactazine.**
THERAP CAT: Diuretic, antihypertensive.

317. Altrenogest. *17-Hydroxy-17-(2-propenyl)estra-4,9,11-trien-3-one; 17α-allyl-17-hydroxyestra-4,9,11-trien-3-one;* 13β-methyl-17α-allyl-$\Delta^{4,9,11}$-gonatriene-17β-ol-3-one; allytrenbolone; A-35957; RU-2267; Regumate. $C_{21}H_{26}O_2$; mol wt 310.44. C 81.25%, H 8.44%, O 10.31%. Synthetic oral progestogen. Prepn: **Neth.** pat. **Appl.** 6,401,555 corresp to G. Nomine *et al.,* U.S. pat. 3,257,278 (1964, 1966 both to Roussel); D. Bertin *et al.,* U.S. pat. 3,478,067 (1967 to Roussel). Effect on estrous cycle and fertility of mares: E. L. Squires *et al., J. Anim. Sci.* **49**, 729 (1979). Synchronization of estrus in swine: R. R. Kraeling *et al., ibid.* **52**, 831 (1981); in gilts: V. G. Pursel *et al., ibid.* 130.

Cryst, mp 120°. $[\alpha]_D^{20}$ −72° (c = 0.5 in ethanol).
THERAP CAT (VET): Progestogen.

318. Altretamine. *N,N,N′,N′,N″,N″-Hexamethyl-1,3,5-triazine-2,4,6-triamine;* 2,4,6-tris(dimethylamino)-s-triazine; hemel; hexamethylmelamine; HMM; ENT 50852; NSC-13875; Hexalen; Hexastat. $C_9H_{18}N_6$; mol wt 210.27. C 51.40%, H 8.63%, N 39.97%. An antitumor agent which also acts as a chemosterilant for male houseflies and other insects. Synthesis: A. W. Hofmann, *Ber.* **18**, 2755 (1885); D. W. Kaiser *et al., J. Am. Chem. Soc.* **73**, 2984 (1951); Y. Bessiere-Chretien, H. Serne, *Bull. Soc. Chim. France* **6**, 2039 (1973). Production process: H. von Brachel, H. Kindler, **Ger.** pat. **1,240,870** corresp to U.S. pat. 3,424,752 (1967, 1969 both to Casella Farbwerke). Chemosterilant effect: S. C. Chang *et al., Science* **144**, 57 (1964); A. De Milo *et al., J. Econ. Entomol.* **65**, 1548 (1972). Chromatographic detn in plasma or serum: A. Hulshoff *et al., J. Chromatog.* **181**, 363 (1980). Disposition and metabolism in rabbits and humans: M. M. Ames *et al., Cancer Res.* **39**, 5016 (1979). Evaluation in breast cancer: C. J. Fabian *et al., Cancer Treat. Rep.* **63**, 1359 (1979); in ovarian cancer: J. T. Wharton *et al., Am. J. Obstet. Gynecol.* **133**, 833 (1979). Mammalian toxicity study: R. L. Jasper *et al., Fed. Proc.* **24**, 641 (1965). Toxicological studies in dogs and mice: D. C. Thake *et al., Gov. Rep. Announce. Index (USA)* **79**, 92 (1979), *C.A.* **91**, 168372a (1979). *In vitro* cytotoxicity study: M. D'Incalci *et al., Brit. J. Cancer* **41**, 630 (1980).

Needles from abs ethanol, mp 172-174°. uv max (ethanol): 226 nm (ε 49,400), H. J. Anderson *et al., Can. J. Chem.* **49**, 2315 (1971). LD_{50} in rats, guinea pigs (mg/kg): 350, 255 orally (Jasper).
USE: As exptl insect chemosterilant.
THERAP CAT: Antineoplastic.

319. D-Altrose. D-Altropyranose. $C_6H_{12}O_6$; mol wt 180.16. C 40.00%, H 6.71%, O 53.29%. Prepd by the reduction of D-altronic lactone with sodium amalgam: Levene, Jacobs, *Ber.* **43**, 3143 (1910).

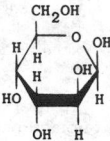

Prisms from dilute alcohol, mp 103-105°. $[\alpha]_D^{20}$ +32.6° (c = 7.6). Soluble in water; practically insol in alcohol. Phenylosazone, $C_{18}H_{22}N_4O_4$, mp 178°.

320. Aluminon. *5-[(3-Carboxy-4-hydroxyphenyl)(3-carboxy-4-oxo-2,5-cylohexadien-1-ylidene)methyl]-2-hydroxybenzoic acid triammonium salt; 3-[bis(3-carboxy-4-hydroxyphenyl)methylene]-6-oxo-1,4-cyclohexadiene-1-carboxylic acid triammonium salt;* ammonium salt of aurintricarboxylic acid; Lysofon. $C_{22}H_{23}N_3O_9$; mol wt 473.43. C 55.81%, H 4.89%, N 8.88%, O 30.42%. Prepd by reacting sodium nitrite with salicylic acid and adding formaldehyde, then treating with ammonia: G. B. Heisig, W. M. Lauer, *Org. Syn.* **coll. vol. I,** 54 (2nd ed., 1941).

Yellowish-brown, glassy powder. Freely sol in water.
USE: Forms brilliantly colored lakes with aluminum, chromium, iron, beryllium. Generally used for the detection and colorimetric estimation of aluminum in water, foods, tissues. *Cf.* Scherrer, Mogerman, *J. Res. Nat. Bur. Stand.* **21,** 105 (1938).
THERAP CAT: Pharyngeal aerosol spray.

321. Aluminum. Aluminium. Al; at. wt 26.98154; at. no. 13; valence 3. One naturally occurring isotope: [27]Al. In addition, six radioactive isotopes and one isomer are known; the most important, [26]Al (found in meteors), decays with emission of β^+ and γ-radiation, $T_{1/2}$ 7.4 × 10^5 years. One of the most abundant metals in earth's crust: 8.8 % by wt; occurs in nature primarily in combination with silica, also as oxide (*see* Aluminum Silicate; Aluminum Oxide). First obtained in impure form by Oersted in 1825; prepd as metal powder by Wöhler in 1827. *Reviews* of aluminum, its alloys and compds: Brandt, "Aluminum and Aluminum Alloys" in *Proc. Met. Soc. Conf.* **Vol. 40,** E. D. Verink, Ed. (Gordon & Breach, New York, 1966); *Aluminum,* 3 Vols. K. R. Van Horn, Ed. (American Society for Metals, Metal Park, Ohio, 1967); Wade, Bannister, "Aluminum, Gallium, Indium and Thallium" in *Comprehensive Inorganic Chemistry,* **Vol. 1,** J. C. Bailar, Jr. *et al.,* Eds. (Pergamon Press, Oxford, 1973) pp 993-1064.
Tin-white, malleable, ductile metal, with somewhat bluish tint; capable of taking brilliant polish which is retained in dry air. In moist air, oxide film forms which protects metal from corrosion. Available in bars, leaf, powder, sheets, or wire. d 2.70. mp 660°. bp 2327°. Does not vaporize even at high temps, but finely divided aluminum dust is easily ignited, and may cause explosions. Reacts with dil HCl, H_2SO_4, KOH and NaOH with evolution or hydrogen. Reduces the cations of many heavy metals to the metallic state E°(aq) Al^{3+}/Al −1.66 V. Solns of Al^{3+} in dil HCl or neutral or slightly acid solns of most aluminum salts, yield with Na_2S a white ppt soluble in excess of Na_2S. Dil neutral soln of aluminum salts yields white gelatinous ppt on boiling with sodium acetate.

USE: As the pure metal or as alloys (magnalium, aluminum bronze, etc.) for aircraft, utensils, apparatus, electrical conductors; instead of copper in dental alloys. The coarse powder is used in aluminothermics (thermite process); the fine powder as flashlight in photography; in explosives, fireworks and in aluminum paints; for absorbing occluded gases in manuf of steel. In testing for Au, As, Hg; coagulating colloidal solns of As or Sb; pptg Cu; reducer for determining nitrates and nitrites; instead of Zn for generating hydrogen in testing for As. Forms complex hydrides with lithium and boron, such as $LiAlH_4$, which are very useful in preparative organic chemistry.
Human Toxicity: Evidence that dust may cause "aluminosis" or aluminum pneumoconiosis not definitive: E. Browning, *Toxicity of Industrial Metals* (Appleton-Century-Crofts, New York, 2nd ed., 1969) pp 3-22.

322. Aluminum Acetate Solution. *Burow's Solution; Domeboro.* Contains about 5% neutral aluminum acetate. $Al(C_2H_3O_2)_3$ = 0.66% Al, 4.4% acetic acid (1.25% Al_2O_3). Prepn from a dry mix consisting of an alkali metal acetate, acetic acid and a dibasic aluminum acetate stabilized with boric acid: Gibbons *et al.,* U.S. pat. **2,824,042** (1958).
Colorless liquid; slight odor of acetic acid. d about 1.002. pH of 1:20 aq soln 4.2.
THERAP CAT: Astringent, antiseptic.
THERAP CAT (VET): External astringent, antiseptic.

323. Aluminum Acetotartrate. *Essitol.* Consists of approx 70% basic aluminum acetate and 30% tartaric acid. Prepn: *Gmelin's Aluminum* (8th ed.) **35B,** 304 (1934).
Crystals or powder; slight acetic odor; astringent, acidulous taste. On long keeping loses acetic acid and becomes incompletely sol. Slowly sol in cold water; practically insol in alcohol. Aq solns: *liquor aluminii aceticotartarici; Alactor; Alsol.*
THERAP CAT: Astringent, antiseptic.

324. Aluminum Alkyls. Highly reactive compounds of the form R_3Al, *trialkylaluminum;* $[R_nAlX_{3-n}]_2$, n = 1, 2, *alkylaluminum halide;* or $R_3Al_2X_3$, *alkylaluminum sesquihalide.* (The sesquihalides are actually equilibrium mixtures of $R_3Al_2X_3$, $[R_2AlX]_2$ and $[RAlX_2]_2$.) First prepn of aluminum alkyls: ethylaluminum sesquiiodide $[(C_2H_5)_3Al_2I_3]$ prepd by Hallwachs, Schaferik, *Ann.* **109,** 207 (1859); trimethylaluminum, $[(CH_3)_3Al]_2$, and triethylaluminum, $[(C_2H_5)_3Al]_2$, prepd by Buckton, Odling, *Ann.* suppl. **4,** 109 (1865). The alkylaluminum halides are prepd from alkyl halides and aluminum, or by halogenation of trialkylaluminums. The trialkylaluminums are prepd from aluminum, hydrogen and olefins; this improved "direct synthesis" was developed by Ziegler and co-workers. Comprehensive reviews of prepn, properties and chemistry: Schultz, *Advan. Chem. Ser.* **23,** 163-171 (1959); Ziegler, "Organo-Aluminum Compounds" in *Organometallic Chemistry,* H. Zeiss, Ed., A.C.S. Monograph Series no. **147** (Reinhold, New York, 1960) p 194-269; Köster, Binger, *Advan. Inorg. Chem. Radiochem.* **7,** 263-348 (1965); T. Mole, E. A. Jeffery, *Organoaluminum Compounds* (Elsevier, New York, 1972) 465 pp; Wade, Banister in *Comprehensive Inorganic Chemistry* **vol 1,** J. C. Bailar, Jr. *et al.,* Eds. (Pergamon Press, Oxford, 1973) pp 1058-1064.
Trialkylaluminum compounds are colorless liquids at room temperature. Must be stored in an inert atm; sensitive to oxidation and hydrolysis in air; the lighter trialkylaluminums ignite spontaneously in air. The low mol wt, linear-chain alkyl compounds exist as dimers; the branched-chain alkyl compounds exist primarily as monomers. Among the industrially important trialkyl aluminums are *triethylaluminum* (dimeric liq, d^{25} 0.832, bp_{760} 194°, bp_{13} 100°) and *triisobutylaluminum* (primarily monomeric liq, d^{25} 0.781; bp_{10} 86°, mp 6°).
Alkylaluminum halides are colorless, volatile liquids or low melting solids. Strongly associated forming dimers; held together by bridging halogen bonds. Less sensitive than trialkylaluminums to oxidation on exposure to air. Halogen aluminum bonds cleaved by water, alcohol. Industrially important halides include *chlorodiethylaluminum,* $[(C_2H_5)_2AlCl]_2$, (liq, d^{25} 0.961, bp_{50} 127°, bp_{17} 100°, $bp_{1.9}$ 60°) and *dichloroethylaluminum* $[C_2H_5AlCl_2]_2$, (solid, d^{50} 1.207, mp 32°, bp_{50} 113°; bp_{30} 100°, bp_5 60°).

USE: Catalyst; with compds of early transition metals as Ziegler-Natta polymerization catalysts; intermediates in organic syntheses.

325. Aluminum Ammonium Sulfate.
Burnt ammonium alum; exsiccated ammonium alum. $AlH_4NO_8S_2$; mol wt 237.14. Al 11.38%, H 1.70%, N 5.91%, O 53.98%, S 27.04%. $AlNH_4(SO_4)_2$. Prepn: *Gmelin's, Aluminum* (8th ed) **35B**, pp 508-515 (1934). Crystal structure of dodecahydrate: Larson, Cromer, *Acta Cryst.* **22**, 793 (1967).

About 97-98% pure, the balance is chiefly excess Al_2O_3. White powder. One gram dissolves in about 20 ml cold, 1.5 ml boiling water, usually incompletely; practically insol in alcohol. *Keep well closed.*

Dodecahydrate, *ammonium alum.* It is about 99.5% pure. Colorless crystals, white granules or powder; styptic taste. d 1.65. mp 94.5°. At about 250° becomes anhydr; decomposes above 280°. One gram dissolves in 7 ml water, 0.5 ml boil. water; freely soluble in glycerol; practically insol in alc. The aq soln is acid to litmus; pH of 0.05 molar soln 4.6.

USE: Purifying drinking water; in baking powders; dyeing and printing fabrics; manuf pigments, lakes, artificial gems, paper, vegetable glue, marble and porcelain cements; fireproofing; tanning; electrolytic copperplating. *See also* Aluminum Potassium Sulfate.

THERAP CAT: Astringent, styptic.

THERAP CAT (VET): Astringent, styptic.

326. Aluminum Antimonide.
AlSb; mol wt 148.74. Al 18.14%, Sb 81.86%. Prepn by fusion of the component elements: Kover, *Compt. Rend.* **243**, 648 (1956). Prepn of high purity crystals, resistance measurements, activation energies: Justi, Lautz, *Abhandl. Braunschweig. Wiss. Ges.* **5**, 36 (1953); *C.A.* **48**, 1756i (1954). Zone melting to remove impurities and to improve electr properties: Scheel, *Z. Metallkunde* **46**, 58 (1955), *C.A.* **49**, 4486f (1955). Oxidation by water: Rudorff, Kohlmeyer, *Z. Metallkunde* **45**, 608 (1954), *C.A.* **49**, 7432c (1955).

Lattice constant: 6.1361×10^{-8} cm. mp 1050°. Electrical properties: Kover, *loc. cit.* Absorption const, *n*, width of forbidden zone (opt & electr) = 0.78; 3.4; 16; 1.65 e.v.: Oswald, Schade, *Z. Naturforsch.* **9a**, 611 (1954); *C.A.* **49**, 52a (1955). Width of forbidden band = 1.6 e.v. at 0°K: Blunt et al., *Phys. Rev.* **96**, 578 (1954); *C.A.* **49**, 2861b (1955). Other electrical properties: Willardson et al., *J. Electrochem. Soc.* **101**, 354 (1954).

USE: In semiconductor research.

327. Aluminum Benzoate.
$C_{21}H_{15}AlO_6$; mol wt 390.30. C 64.62%, H 3.87%, Al 6.91%, O 24.59%. $Al(C_6H_5COO)_3$. Prepn: Kränzlein et al., *Ger.* pat. **569,946** (1933 to I. G. Farbenind.); *Frdl.* **19**, 597.

Cryst powder. Very slightly sol in water.

328. Aluminum Bis(acetylsalicylate).
Bis[2-(acetyloxy)-benzoato-O']hydroxyaluminum; hydroxybis(acetylsalicylato)aluminum; aluminum diacetylsalicylate; aluminum acetylsalicylate N.F.; acetylsalicylate aluminum; aluminum diaspirin; aluminum aspirin; aspirin aluminum; monohydroxyaluminum diacetylsalicylate; monohydroxyaluminum bis-(acetylsalicylate); Rumasal. $C_{18}H_{15}AlO_9$; mol wt 402.30. C 53.74%, H 3.76%, Al 6.71%, O 35.79%. Prepn: Wolf, *U.S.* pat. **1,967,649** (1934 to Chinoin); *Brit.* pat. **826,299** (1959 to Lewis-Howe).

Amorphous white powder or granules. Practically insol in water, alcohol, ether. Decomposes in dil acids or alkalies and alkali carbonates.

See also Dihydroxyaluminum Acetylsalicylate.

THERAP CAT: Analgesic, antipyretic.

329. Aluminum Borate.
Occurs in nature as the mineral *eremeyevite* or *jeremejevite*, $Al_2O_3 \cdot B_2O_3$. Prepd by heating Al_2O_3 with B_2O_3; when heated to 1000° $2Al_2O_3 \cdot B_2O_3$ is formed; when heated to 1100° $9Al_2O_3 \cdot 2B_2O_3$ is formed: Scholze, *Z. Anorg. Allgem. Chem.* **284**, 272 (1956).

$2Al_2O_3 \cdot B_2O_3$, mol wt 273.56. Al 39.45%, B 7.91%, O 52.64%. Needles, mp approx 1050°. Practically insol in water.

$9Al_2O_3 \cdot 2B_2O_3$, mol wt 1056.92. Al 45.95%, B 4.09%, O 49.96%. Needles, mp approx 1440°. Practically insol in water.

USE: Catalyst for polymerizations; in glass manuf.

330. Aluminum Boroformate.
Aluminum boro-formicicum. Contains about 33% Al_2O_3, 20% H_3BO_3, 15% formic acid, 32% H_2O. Prepn: *Hager's Handb. Pharm. Praxis* **1**, (Aluminum) p 368 (1930).

Lustrous scales. Freely sol in water, alc. The aq soln is acid, has a sweetish astringent taste and is not precipitated by alkalies.

331. Aluminum Borohydride.
Aluminum tetrahydroborate. AlB_3H_{12}; mol wt 71.53. Al 37.71%, B 45.38%, H 16.91%. $Al(BH_4)_3$. Prepn from alkali metal hydride and aluminum halide: Schlesinger et al., *J. Am. Chem. Soc.* **75**, 209 (1953); Hinkamp, Hnizda, *Ind. Eng. Chem.* **47**, 1560 (1955); Kollonitsch, Fuchs, *Nature* **176**, 1081 (1955); Hinkamp, *U.S.* pat. **2,854,312** (1958 to Ethyl Corp.); Schechter, *U.S.* pat. **2,913,306** (1959 to Callery Chem.); from trimethylaluminum and diborane: Schlesinger et al., *J. Am. Chem. Soc.* **61**, 536 (1939).

Liquid. mp −64.5°; bp 44.5°; bp_{119} 0°. Reacts vigorously with water and hydrogen chloride to liberate hydrogen; ignites in air; decomposes slowly even at room temp evolving hydrogen. Forms addn products with dimethyl ether, trimethylamine and ammonia.

USE: Reducing agent; prepn of borohydrides of heavy metals; fuel for jet engines and rockets.

332. Aluminum Bromide.
$AlBr_3$; mol wt 266.72. Al 10.11%, Br 89.89%. Prepd from aluminum and bromine: Nicholson et al., *Inorg. Syn.* **3**, 30 (1950).

White to yellowish-red very hygroscopic lumps, mp 97°; bp reported within the range 250-270°; d_4^{18} 3.205. Fumes strongly in air; combines with water *with violence. Keep tightly closed and protect from moisture.* Sol in many organic solvents such as benzene, nitrobenzene, toluene, xylene, simple hydrocarbons.

Hexahydrate, colorless to slightly yellow deliquesc crystals. mp 93°; d 2.5. Sol in water, alcohol, ether, carbon disulfide. *Keep well closed.*

USE: *Anhydrous* form as acid catalyst in organic syntheses. It is similar to anhydr $AlCl_3$ but is more reactive and more sol in organic media.

333. Aluminum *tert*-Butoxide.
2-Methyl-2-propanol aluminum salt. $C_{12}H_{27}AlO_3$; mol wt 246.32. C 58.51%, H 11.05%, Al 10.95%, O 19.49%. $Al[OC(CH_3)_3]_3$. Prepd from aluminum *tert*-butyl alcohol and mercuric chloride: Wayne, Adkins, *Org. Syn. coll. vol. III*, 48 (1955).

Powder; can be recryst from benzene. Sublimes at 180°. Pure compd does not melt or decomp upon heating up to 300° in a sealed tube, but traces of moisture or *tert*-butyl alc cause it to melt at 160-200°. Very sol in organic solvents; approx 9 g dissolves in 5 g ethyl propionate at 120°.

USE: Reagent for oxidation of alcohols to ketones; in dealcoholation of orthoesters.

334. Aluminum Calcium Hydride.
Calcium tetrahydroaluminate; calcium aluminum hydride. Al_2CaH_8; mol wt 102.10. Al 52.85%, Ca 39.26%, H 7.89%. $Ca(AlH_4)_2$. Prepd by the interaction of aluminum chloride and calcium hydride in tetrahydrofuran: Schwab, Wintersberger, *Z. Naturforsch.* **8b**, 690 (1953); Conn, Taylor, *U.S.* pat. **2,999,005** (1961 to Merck & Co.).

Slate-gray mass. The dry pulverized material can ignite spontaneously in moist air and is best handled under dry nitrogen. Reacts violently with water, the ensuing conflagration resembles a display of fireworks. Slightly less violent reaction with alcohols. Sol in dry tetrahydrofuran; practically insol in dry ether, dioxane, benzene.

USE: Reducing agent for aldehydes, ketones, acid chlorides. Also in the reduction of esters to alcohols, nitriles to amines, aromatic nitro compounds to azo compounds.

335. Aluminum Carbide. C_3Al_4; mol wt 143.91. Al 74.96%, C 25.04%. Al_4C_3. Prepd by heating aluminum powder with carbon: Becher in *Handbook of Preparative Inorganic Chemistry* vol. 1, G. Brauer, Ed. (Academic Press, New York, 2nd ed., 1963) p 832.

Yellow hexagonal crystals or powder. mp 2100°; decomposes above 2200°. d 2.36. Decomposed by water with evolution of methane (fire hazard).

USE: In generating methane; reducing metal oxides; in manuf of aluminum nitride.

336. Aluminum Cesium Sulfate. $AlCsO_8S_2$; mol wt 352.01. Al 7.66%, Cs 37.76%, O 36.36%, S 18.22%. $CsAl(SO_4)_2$. Prepn: *Gmelin's, Aluminum* (8th ed.) **35B**, pp 529-531 (1934).

Dodecahydrate, *cesium alum.* Crystals; mp reported from 105-122°. Slightly sol in cold, freely in hot water; practically insol in alcohol.

337. Aluminum Chlorate. $AlCl_3O_9$; mol wt 277.35. Al 9.73%, Cl 38.35%, O 51.92%. $Al(ClO_3)_3$. Occurs as hexahydrate and nonahydrate. Prepn: *Gmelin's Aluminum* (8th ed.) **35B**, 216-217 (1934).

Nonahydrate, *mallebrin.* Deliquesc crystals. Freely sol in water; sol in alc. *Keep well closed.*

USE: Disinfectant; ClO_2 manuf; prevention of yellowing of acrylic fibers.

THERAP CAT: Antiseptic, astringent.

338. Aluminum Chloride. $AlCl_3$; mol wt 133.34. Al 20.23%, Cl 79.77%. Prepd from aluminum metal in a heated stream of HCl gas: Gattermann-Wieland, *Praxis des Organischen Chemikers* (de Gruyter, Berlin, 40th ed., 1961) p 295; H. J. Becher in *Handbook of Preparative Inorganic Chemistry,* **vol. 1**, G. Brauer, Ed. (Academic Press, New York, 2nd ed., 1963) p 812. Manuf: Faith, Keyes & Clark's *Industrial Chemicals,* F. A. Lowenheim, M . K. Moran, Eds. (Wiley-Interscience, New York, 4th ed., 1975) pp 72-75. Monograph: C. A. Thomas, *Anhydrous Aluminum Chloride in Organic Chemistry,* A.C.S. Monograph Series no. 87 (Reinhold, New York, 1941).

White when pure; ordinarily gray or yellow to greenish. Fumes in air; strong odor of HCl; when heated in small quantities volatilizes without melting. Combines with water *with explosive violence* and liberation of much heat. Freely sol in many organic solvents, such as benzophenone, benzene nitrobenzene, carbon tetrachloride, chloroform. *Keep tightly closed and protected from moisture.* For physical properties *see* C. A. Thomas, *loc. cit.*

Hexahydrate, *Aluwets, Anhydrol, Driclor.* Colorless crystals, or white or slightly yellow deliquesc, cryst powder; odorless or slight HCl odor. One gram dissolves in 0.9 ml water, 4 ml alc; sol in ether, glycerol, propylene glycol. *Keep well closed.*

USE: The *anhydrous* form suitable as an acid catalyst, esp in Friedel-Crafts type reactions; in cracking of petroleum; in manuf rubbers, lubricants. The *hexahydrate* form used in preserving wood; disinfecting stables, slaughterhouses, etc.; in deodorants and antiperspirant preparations; refining crude oil; dyeing fabrics; manuf parchment paper. *Caution: Anhydrous* form is a strong irritant.

THERAP CAT: The *hexahydrate* as a topical astringent.

339. Aluminum Diacetate. *Bis(aceto-O)hydroxyaluminum; hydroxybis(acetato)aluminum;* basic aluminum acetate; aluminum subacetate; aluminum hydroxyacetate; Lenicet; Casil. $C_4H_7AlO_5$; mol wt 162.08. C 29.64%, H 4.36%, Al 16.64%, O 49.36%. $Al(OH)(CH_3CO_2)_2$. Prepd from aluminum hydroxide and acetic acid or from sodium acetate and aluminum chloride hexahydrate: Hood, Ihde, *J. Am. Chem. Soc.* **72**, 2094 (1950). Other methods of prepn: *Gmelin's Aluminum* (8th ed.) **35B**, p 296 (1934). Also prepd in aq solution, *see* Aluminum Subacetate Solution.

White curdy precipitate or white amorphous powder. Material that has been oven-dried at 110° is practically insoluble in water. Freshly prepared material forms numerous hydrates and is quite sol in water. Greatest soly is obtained by formation in solution. The pharmacist's stock soln (*see* Aluminum Subacetate Solution) contains about 8% Al(OH)-($CH_3CO_2)_2$, while commercial solns used for waterproofing, contain 22 to 25% $Al(OH)(CH_3CO_2)_2$. When heated, the salt loses acetic acid, and aluminum to oxygen bonding occurs, resulting in a very insol compd of disputed nature, *see Gmelin's, loc. cit.* and Thomas, *Paper Trade J.* **100**, 36 (1935). Aq solns are generally acid to litmus; gradually become turbid and gelatinous. Sometimes a more basic salt precipitates out and settles to the bottom of the container. Increasing the pH to a marked degree will clear up an old soln due to formation of sodium acetate and sodium aluminate. Urea and thiourea have been suggested as stabilizers for aq solns.

USE: Manuf color lakes; mordant in dyeing; in waterproofing and fireproofing fabrics (solns for these purposes are known as *red liquor* or *mordant rouge* because they were originally used for preparing red color lakes); in antiperspirant formulations; as disinfectant by embalmers. Dusting powder.

340. Aluminum Ethoxide. *Ethanol aluminum salt;* aluminum ethylate. $C_6H_{15}AlO_3$; mol wt 162.15. C 44.44%, H 9.33%, Al 16.63%, O 29.60%. $Al(OC_2H_5)_3$. Prepd by reacting aluminum powder with absolute ethanol in xylene using small amounts of mercuric chloride and iodine as catalysts: Meerwein, Schmidt, *Ann.* **444**, 232 (1925); *Newer Methods of Preparative Organic Chemistry* (Interscience, New York, 1948) p 132; *see also* Farbwerke vorm. Meister, Lucius und Brüning, *Ger. pat.* 286,596; *J. Soc. Chem. Ind. (London)* **34**, 1168 (1915); Adkins, *J. Am. Chem. Soc.* **44**, 2178 (1922). Laboratory procedure: Gattermann-Wieland, *Praxis des Organischen Chemikers* (de Gruyter, Berlin, 40th ed., 1961) p 333.

Liquid. $bp_{6.8}$ 200°; bp_3 175-180°. Slowly solidifies to a whole white solid, mp 140°. May crystallize with alcohol of crystallization. Decomposed by water. Slightly sol in hot xylene, chlorobenzene, other high boiling solvents.

USE: In the reduction of aldehydes and ketones; as catalyst for polymerizations.

341. Aluminum Fluoride. Aluminum trifluoride. AlF_3; mol wt 83.98. Al 32.13%, F 67.87%. Prepd by heating $(NH_4)_3AlF_6$ to red heat in a stream of nitrogen: Witt, Barrow, *Trans. Faraday Soc.* **55**, 730 (1959); Kwasnik in *Handbook of Preparative Inorganic Chemistry,* Vol. 1, G. Brauer, Ed. (Academic Press, New York, 2nd ed., 1963) p 225. Review: Kemmitt, Sharp, *Advan. Fluorine Chem.* **4**, 154-155 (1965).

Hexagonal crystals. Sublimes (760 mm) 1272°. Soly in water (25°) 0.559 g/100 ml. Sparingly sol in acids and alkalies, even hot concd H_2SO_4 has little effect. Hydrolyzed by superheated steam at 300-400°.

Monohydrate, *fluellite,* orthorhombic crystals, d 2.17.

Trihydrate, usually $AlF_3.3.1H_2O$. Prepn: Ehret, Frere, *J. Am. Chem. Soc.* **67**, 64 (1945). Loses water at 100°, more at 200°. It does not seem possible to obtain the anhydrous compd free from oxides by dehydration of the hydrates.

Human Toxicity: Less toxic on ingestion than other fluorides because of slight solubility.

USE: In ceramics, as flux in metallurgy, in aluminum manufacture, as inhibitor of fermentation, as catalyst in organic reactions.

342. Aluminum Hexafluorosilicate. Aluminum fluosilicate; aluminum silicofluoride. $Al_2F_{18}Si_3$; mol wt 480.23. Al 11.24%, F 71.22%, Si 17.55%. $Al_2(SiF_6)_3$. Occurs in nature as *topaz,* $Al_2SiO_4(OH,F)_2$. Prepn: Sanfourche, Krapivine, *Compt. Rend.* **208**, 2080 (1939).

Nonahydrate, hexagonal prisms. Easily sol in water; aq soln decomposes on heating or neutralization. Solid loses water on heating to temperatures below 500°, leaving a hexahydrate form; decomposes completely on heating to 1000°.

USE: Protection and preservation of construction materials; manuf of glass.

343. Aluminum Hexaurea Sulfate Triiodide. *Hexakis-(urea-O)aluminum sulfate (triiodide)(1:1:1);* aluminum hexacarbamide sulfate triiodide; Hexadine-S; Alure. $C_6H_{24}-AlI_3N_{12}O_{10}S$; mol wt 864.14. C 8.34%, H 2.80%, Al 3.12%, I 44.06%, N 19.45%, O 18.52%, S 3.71%. $Al[CO(NH_2)_2]_6$-SO_4I_3. Prepn: Barbieri, *Atti. Reale. Acad. Lincei.* **24**, 919 (1915); Morris *et al., Ind. Eng. Chem.* **45**, 1013 (1953).

Crystals. Soly in water at 25°: 590 g/l.

USE: Decontamination of drinking water in emergencies. Used in amounts sufficient to yield 8 ppm of active iodine.

A tablet contg 30 mg plus 82 mg $Na_2H_2PO_7$ plus 4 mg talc will decontaminate one quart of water. Such tablets after 7 days of storage at 60° retained 97% of their original active iodine. More stable than tetraglycine hydroperiodide.

344. Aluminum Hydride. AlH_3; mol wt 29.99. Al 89.92%, H 10.08%. Prepd by treating lithium hydride with an ether solution of aluminum chloride: Finholt *et al., J. Am. Chem. Soc.* **69**, 1199-1203 (1947).

Colorless solid, nonvolatile, probably highly polymerized and containing residual ether which cannot be completely removed.

USE: As catalyst for polymerizations; reducing agent. Lithium aluminum hydride is a more useful reagent because of its greater solubility.

345. Aluminum Hydroxide. Aluminum hydrate; aluminum trihydrate; hydrated alumina. AlH_3O_3; mol wt 77.99. Al 34.58%, H 3.88%, O 61.54%. $Al(OH)_3$. Prepn and properties: Dominé-Berges, *Ann. Chim.* [12], **5**, 106 (1950); Hennig, *Chem. Tech.* **1**, 66 (1949), *C.A.* **44**, 965f (1950); *Gmelin's Aluminum* (8th ed.) **35B**, pp 98-132 (1934); Becher in *Handbook of Preparative Inorganic Chemistry*, vol. 1, G. Brauer, Ed. (Academic Press, New York, 2nd ed., 1963) pp 820-821; Wagner, *ibid.*, vol. 2 (1965) pp 1652-1654. Clinical comparison with calcium carbonate as phosphate binder in chronic renal failure: R. H. K. Mak *et al., Brit. Med. J.* **291**, 623 (1985).

Usually obtained as a white, bulky, amorphous powder. Practically insol in water, but sol in alkaline aq solns or in HCl, H_2SO_4 and other strong acids in the presence of some water. Forms gels on prolonged contact with water. Absorbs acids, CO_2.

USE: Adsorbent; emulsifier; ion-exchanger; in chromatography; mordant in dyeing; filtering medium; manuf glass, fire clay, paper, pottery, printing inks, lubricating compositions, detergents; waterproofing fabrics; in antiperspirants, dentifrices; used in pharmacy as the gel or dried gel, *algedrate*. Some commercial names for such prepns are: *Antidiar; Creamalin; Cremorin; Chefarox; Alternagel; Al-U-Creme; Alucol; Aludrox; Amphojel; Alkagel; Collumol; Aldrox; Aludyal; Hycolal; Hydrolum; Merlum; Pepsamar; Fluagel; Gelumina; Uracid; Vanogel*.

THERAP CAT: Antacid. Antihyperphosphatemic.

THERAP CAT (VET): Vaccine adjuvant. Gastric antacid, and gastrointestinal protective in dogs.

346. Aluminum Hydroxychloride. *Basic aluminum chloride; aluminum chlorohydroxide; aluminum chlorohydrate; Astringen; Chlorhydrol; Hyperdrol; Locron; Phosphonorm.* The generally accepted empirical formula is $Al_2(OH)_5Cl.2H_2O$. Prepd by electrolyzing solns of suitable Al salts: **Fr. pat. 837,862** and **Brit. pat. 509,815** (both 1939 to I. G. Farben.); H. Huehn, W. Haufe, **U.S. pat. 2,392,531** (1946); Andersen, **U.S. pat. 2,492,085** (1949 to Elizabeth Arden). Structural studies and physicochemical properties: D. L. Teagarden *et al., J. Pharm. Sci.* **70**, 758, 762 (1981). Phosphate binding activity and use in hyperphosphatemia: **Ger. pat. 3,147,869** (1983 to Nefro-Pharma), *C.A.* **99**, 110747a (1983).

Glassy solid. Dissolves in water, forming slightly turbid colloidal solns (up to 55% w/w). pH of 15% aq soln ~4.3.

USE: In antiperspirants.

THERAP CAT: Astringent. Treatment of hyperphosphatemia.

347. Aluminum Hypophosphite. $AlH_6O_6P_3$; mol wt 221.94. Al 12.16%, H 2.73%, O 43.25%, P 41.86%. $Al(PO_2H_2)_3$. Prepd by heating $Al(OH)_3$ or a solution of an aluminum salt with hypophosphorous acid or sodium hypophosphite: Everest, *J. Chem. Soc.* **1952**, 2945.

Cryst powder; decomposes without melting at approx 220° with evolution of phosphine. Practically insol in water; sol in warm NaOH soln, dilute sulfuric and dilute or concd hydrochloric acid.

USE: In acrylonitrile polymer fiber finishes.

348. Aluminum Iodide. AlI_3; mol wt 407.73. Al 6.61%, I 93.39%. Prepn from aluminum and iodine: Watt, Hall, *Inorg. Syn.* **4**, 117 (1953); H. J. Becher in *Handbook of Preparative Inorganic Chemistry* vol. 1, G. Brauer, Ed. (Aca-

demic Press, New York, 2nd ed., 1963) p 814; Wilson, Worrall, *J. Chem. Soc. (A)* **1968**, 316.

White leaflets if pure; commercial grade yellowish- to blackish-brown lumps. mp 191°; bp 382°; d^{17} 3.948. Fumes in moist air; strong exothermic reaction with water. *Keep tightly closed and protected from light.* Sol in carbon disulfide, alcohol, ether, liquid ammonia.

Hexahydrate, yellowish, deliquesc cryst powder. Sol in water, alcohol, ether. *Keep tightly closed.*

USE: Catalyst in organic reactions.

349. Aluminum Isopropoxide. *2-Propanol aluminum salt;* aluminum isopropylate. $C_9H_{21}AlO_3$; mol wt 204.23. C 52.93%, H 10.36%, Al 13.21%, O 23.50%. $Al[OCH(CH_3)_2]_3$. Prepd from aluminum and isopropyl alcohol in the presence of mercuric chloride: Young *et al., J. Am. Chem. Soc.* **58**, 100 (1936); by adding excess isopropyl alcohol to a benzene soln of $AlCl_3$ at 6°: Teichner, *Compt. Rend.* **237**, 810 (1953). Forms trimers and tetramers: Shiner *et al., J. Am. Chem. Soc.* **85**, 2318 (1963); Oliver *et al., J. Inorg. Nucl. Chem.* **31**, 1609 (1969); Worrall, *J. Chem. Ed.* **46**, 510 (1969). Toxicity: Smyth *et al., Am. Ind. Hyg. Assoc. J.* **30**, 470 (1969). *Review:* Whitaker, *Advan. Chem. Series* **23**, 184-189 (1959).

Hygroscopic white solid, mp 119°. Solidifies rather slowly after distillation. bp_{10} 135°; $bp_{7.5}$ 131°; $bp_{5.5}$ 125.5°; $bp_{2.5}$ 113°; $bp_{1.5}$ 106°; $bp_{0.5}$ 94°. Sol in ethanol, isopropanol, benzene, toluene, chloroform, carbon tetrachloride, petroleum hydrocarbons. Decomposed by water. LD_{50} orally in rats: 11.3 g/kg (Smyth).

USE: Meerwein-Ponndorf reactions; alcoholysis and ester exchange; synthesis of higher alkoxides, chelates, and acylates; formation of aluminum soaps, formulation of paints; waterproofing finishes for textiles.

350. Aluminum Lactate. Aluctyl. $C_9H_{15}AlO_9$; mol wt 294.18. C 36.74%, H 5.14%, Al 9.17%, O 48.95%. Prepn from lactic acid and aluminum isopropoxide or aluminum chloride: Rai *et al., J. Prakt. Chem.* **20**, 105 (1963); from lactic acid and aluminum foil: Jones, Cluskey, *Cereal Chem.* **40**, 589 (1963).

Powder. Freely sol in water.

USE: In foam fire extinguishers; in dental-impression materials.

351. Aluminum Lithium Hydride. *Lithium tetrahydroaluminate;* lithium aluminum hydride; lithium aluminohydride; lithium alanate. AlH_4Li; mol wt 37.94. Al 71.11%, H 10.62%, Li 18.29%. $LiAlH_4$. Prepd by treating lithium hydride with an ether soln of $AlCl_3$: Finholt, *et al., J. Am. Chem. Soc.* **69**, 1199 (1947). Crystal structure: Sklar, Post, *Inorg. Chem.* **6**, 669 (1967).

Microcrystalline white powder when pure; gray when aluminum impurity present. Monoclinic crystals. d 0.92. Stable in dry air at room temperature, decomp above 125°, slowly loses hydrogen at 120°, decomp in moist air, may ignite on grinding in air. Soly (parts/100 parts solvent): 30 (ether); 13 (tetrahydrofuran); 10 (dimethylcellosolve); 2 (dibutyl ether); 0.1 (dioxane). Reacts rapidly with water and alcohols; reduces aldehydes, ketones, acid chlorides and esters to alcohols; nitriles to amines; aromatic nitro compounds to azo compounds. Does not attack olefinic double bonds unless they are conjugated with a phenyl group and a carbonyl or nitrile group.

USE: Reducing agent; in preparation of other hydrides. *Review:* J. S. Pizey, *Synthetic Reagents,* **Vol. 1** (John Wiley, New York, 1974) pp 101-294.

352. Aluminum Magnesium Silicate. Magnesium aluminum silicate. $Al_2MgO_8Si_2$; mol wt 262.45. Al 20.56%, Mg 9.26%, O 48.77%, Si 21.41%. $MgAl_2(SiO_4)_2$. Occurs in nature in the minerals: *colerainite, leuchtenbergite, pyrope,* *saponite, sapphirine, sheridanite, zebedassite.* Prepn: **Brit. pat. 834,517** (1960 to Fuji Chem.)

Hydrates, *Ervasil, Gelusil, Ultin.*

USE: As suspending agent, thickening agent.

THERAP CAT: Antacid.

353. Aluminum β-Naphtholdisulfonate. *2-Hydroxy-naphthalenedisulfonic acid aluminum salt;* Alumnol. $C_{30}H_{18}Al_2O_{21}S_6$; mol wt 960.83. C 37.50%, H 1.89%, Al 5.62%, O 34.97%, S 20.02%. $Al_2[C_{10}H_5(OH)(SO_3)_2]_3$. Prepd from barium β-naphtholdisulfonate and aluminum sulfate: *Hagers' Handb. Pharm. Praxis* **Band 1**, 204 (Berlin, 1930).

Fine, almost white, powder. Sol in 1.5 parts water; sol in glycerol, slightly in alcohol; practically insol in ether.

THERAP CAT: Astringent, antiseptic.

THERAP CAT (VET): Has been used as an internal astringent, antiseptic.

354. Aluminum Nicotinate. *3-Pyridinecarboxylic acid aluminum salt;* tris(nicotinato)aluminum; nicotinic acid aluminum salt; Alunitine; Nicalex. $C_{18}H_{12}AlN_3O_6$; mol wt 393.30. C 54.97%, H 3.08%, Al 6.86%, N 10.68%, O 24.41%. Each tablet of 625 mg is a complex consisting of aluminum nicotinate, nicotinic acid and aluminum hydroxide equivalent in activity to 500 mg nicotinic acid. Manuf of suitable compositions: Miale, **U.S. pat. 2,970,082** (1961 to Walker Labs.).

THERAP CAT: Peripheral vasodilator, antilipemic.

355. Aluminum Nitrate. AlN_3O_9; mol wt 213.00. Al 12.67%, N 19.73%, O 67.60%. $Al(NO_3)_3$. Occurs in several states of hydration of which the nonahydrate is the most stable. Prepn: *Gmelin's, Aluminum* (8th ed.) **35B**, p 149-152 (1934).

Nonahydrate, deliquesc crystals; mp 73°; dec at 135°. Very sol in water, alc; very slightly sol in acetone; almost insol in ethyl acetate and pyridine. The aq soln is acid. *Keep well closed.* LD_{50} orally in rats: 4.28 g/kg, Smyth *et al., Am. Ind. Hyg. Assoc. J.* **30**, 470 (1969).

USE: Tanning leather; antiperspirant; corrosion inhibitor; extraction of uranium; nitrating agent.

356. Aluminum Nitride. AlN; mol wt 40.99. Al 65.82%, N 34.18%. Prepd commercially by heating bauxite and coal in a stream of nitrogen. Laboratory prepn from powdered aluminum metal: Becher in *Handbook of Preparative Inorganic Chemistry* **Vol. 1**, G. Brauer, Ed. (Academic Press, New York, 2nd ed., 1963) p 827.

Orthorhombic or hexagonal, bluish-white crystals. A yellow modification (?) has been reported. In moist air, odor of ammonia. d_4^{25} 3.05. Hardness no. 9 to 10 on Mohs' scale. mp 2150-2200° at 4.3 atm. Spec heat at 0°: 0.180 cal/g/°C; at 100°: 0.207 cal/g/°C; at 500°: 0.313 cal/g/°C. Heat of formation: −74 kcal/mol. Decomposed by water into $Al(OH)_3$ and NH_3.

USE: In semiconductor electronics; in steel manuf.

357. Aluminum Oleate. *9-Octadecenoic acid aluminum salt; oleic acid aluminum salt.* $C_{54}H_{99}AlO_6$; mol wt 871.36. C 74.43%, H 11.45%, Al 3.10%, O 11.02%. $[CH_3(CH_2)_7-CH=CH(CH_2)_7COO]_3Al$. Prepd from freshly pptd $Al_2(OH)_6$ and oleic acid: Stich, *Pharm. Zentralhalle* **63**, 261 (1922), *C.A.* **16**, 2755 (1922).

Yellowish, viscid mass. Practically insol in water; sol in alcohol, benzene, ether, oil turpentine.

USE: In oil or turpentine soln as lacquer for metals, as size, waterproofing agent, drier, for paints, high-pressure and high-temp greases for thickening lubricating oils.

358. Aluminum Oxalate. $C_6Al_2O_{12}$; mol wt 318.03. C 22.66%, Al 16.97%, O 60.37%. $Al_2(C_2O_4)_3$. Prepn: **Brit. pats. 348,789** and **348,790** (both 1930 to I.G. Farben).

Hydrate, powder. Practically insol in water, alc; sol in mineral acids.

USE: Mordant in printing textiles, dyeing cotton.

359. Aluminum Oxide. Alumina. Al_2O_3; mol wt 101.94. Al 52.91%, O 47.08%. Occurs in nature as the minerals: *bauxite, bayerite, boehmite, corundum, diaspore, gibbsite.* Prepn and properties: *Mellor's* **vol. V**, 263-273 (1929); *Gmelin's, Aluminum* (8th ed.) **35B**, pp 7-98 (1934); Becher in *Handbook of Preparative Inorganic Chemistry* **vol. 1**, G. Brauer, Ed. (Academic Press, New York, 2nd ed., 1963) pp 822-823; Wagner, *ibid.* **vol. 2** (1965) pp 1660-1663.

Approximate characteristics of native aluminum oxide: White cryst powder. d_4^{20} 4.0. mp about 2000°. Very hard, about 8.8 on Moh's scale. An electrical insulator; electrical resistivity at 300° about 1.2×10^{13} ohms-cm. Practically insol in water. Slowly sol in aq alkaline solns with the formation of hydroxides. Practically insol in non-polar organic solvents.

USE: As adsorbent, desiccant, abrasive; as filler for paints and varnishes; in manuf of alloys, ceramic materials, electrical insulators and resistors, dental cements, glass, steel, artificial gems; in coatings for metals, etc.; as catalyst for organic reactions. The minerals *corundum* (hardness = 9) and *Alundum* (obtained by fusing bauxite in an electric furnace) are used as abrasives and polishes; in manuf of refractories. Aluminum oxide is also used in chromatography, see Aluminum Oxide (Brockmann).

360. Aluminum Oxide (Brockmann). Activated aluminum oxide. Aluminum oxide suitable for chromatographic adsorption. Prepd from carefully screened aluminum hydroxide having a particle size which will pack easily in a column without obstructing the seepage of liquids (typical sieve analysis: All through 150 mesh, 40% on 200 mesh, 40% on 325 mesh; Zechmeister and Cholnoky, *Die Chromatographische Adsorptionsmethode,* 2nd ed., p 45, state that the material should have an average particle size of 7 microns). The aluminum hydroxide is dehydrated and calcined at about 900° in a CO_2 stream which tends to coat the individual Al_2O_3 particles with a thin layer of aluminum oxycarbonate approximating the formula $[Al_2(OH)_6]_2CO_3.H_2O$. Water content and alkalinity are then adjusted by washing with dilute acids. To test the suitability of the material a mixture of 2.5 mg lycopene and 2.5 mg carotene is dissolved in 50 ml of a 1:3 benzene-petr ether mixture. 10 ml of this soln is poured onto a column of the aluminum oxide, 10 cm high and 1 cm diameter, and washed down with a 1:1 mixture of benzene and petr ether. The column must always be covered with fluid. For the washing, 40 ml of the benzene-petr ether mixture must suffice to remove all of the yellow carotene from the column, while the lycopene should remain attached to the aluminum oxide as a luminous, red zone. For a similar test using crude phytol, which also contains chlorophyll, see Rosin, *Reagent Chemicals and Standards,* (New York, 2nd ed., 1946) p 25; as aluminum oxide for chromatography in later editions of Rosin's book; see also Brockmann, *Z. Physiol. Chem.* **241**, 104 (1936). Monograph: *Activated Alumina. Its Properties and Uses.* Aluminum Co. of America (Pittsburgh, 1949).

361. Aluminum Palmitate. *Hexadecanoic acid aluminum salt;* palmitic acid aluminum salt. $C_{48}H_{93}AlO_6$; mol wt 793.25. C 72.68%, H 11.82%, Al 3.40%, O 12.10%. $[CH_3(CH_2)_{14}COO]_3Al$. Prepd from $AlCl_3$ and palmitic acid: Mehrotra, Rai, *J. Indian Chem. Soc.* **39**, 1 (1962).

White to yellow mass or powder. Practically insol in water or alcohol; when fresh, dissolves in petrol ether or oil turpentine.

USE: Thickening petroleum and lubricants; water-proofing fabrics; sizing and glazing paper and leather.

362. Aluminum Phosphate. Aluminum orthophosphate. AlO_4P; mol wt 121.95. Al 22.12%, O 52.48%, P 25.40%. $AlPO_4$. Occurs in nature as the minerals *angelite; coeruleolactite; evansite; lucinite; metavariscite; sterretite; variscite; vashegyite; wavellite; zepharovicht.* Prepn from $NaAlO_2$ and H_3PO_4: Becher in *Handbook of Preparative Inorganic Chemistry* **vol. 1**, G. Brauer, Ed. (Academic Press, New York, 2nd ed., 1963) p 831.

White, infusible powder. mp above 1460°; d^{23} 2.56. Practically insol in water or acetic acid. Very slightly sol in concentrated hydrochloric acid and nitric acid. Isomorphous with quartz.

USE: As cement in admixture with calcium sulfate and sodium silicate; as flux for ceramics; dental cements; for special glasses. Used in pharmacy as the gel or dried gel, such as *Aluphos, Fosfalugel, Ulcocid, Phosphaljel, Phosphalugel, Phosphalutab*.

THERAP CAT: Antacid.

363. Aluminum Phosphide. Celphos; Detia; Phostoxin. AlP; mol wt 57.96. Al 46.55%, P 53.44%. Prepd from red phosphorus and aluminum powder: White, Bushey, *J. Am. Chem. Soc.* **66**, 1666 (1944); *Inorg. Syn.* **4**, 23 (1953); Montignie), *Bull. Soc. Chim. France* **1946**, 276; from Al and Zn_2P_3: Wang *et al.*, *J. Inorg. Nucl. Chem.* **25**, 326 (1963). Use as insecticidal fumigant: W. Freyberg, W. Haupt, U.S. pat. **2,117,158** (1938 to Freyberg).

Dark gray or dark yellow crystals. Cubic zinc blende structure. Must be protected from moist air since it reacts readily to produce phosphine which is highly toxic. d_4^{15} 2.85 (Montignie); d 2.40 (Wang *et al.*). Does not melt or decompose thermally at temps up to 1000°. Treatment with water and acid produces phosphine in quantitative yields.

USE: Source of phosphine; in semiconductor research; as fumigant.

364. Aluminum Potassium Sulfate. $AlKO_8S_2$; mol wt 258.20. Al 10.45%, K 15.14%, O 49.57%, S 24.83%. KAl-$(SO_4)_2$. Prepn: *Gmelin's, Aluminum* (8th ed.) **35B**, 453-477 (1934). Crystal structure: Manoli *et al.*, *Bull. Soc. Chim. France* **1970**, 98; of dodecahydrate: Larson, Cromer, *Acta Cryst.* **22**, 793 (1967).

Anhydrous, *burnt alum* or *exsiccated alum*. Usually about 97-98% pure. White powder; styptic taste; attracts moisture from the air. One gram dissolves in about 20 ml cold or 1 ml boiling water, usually incompletely; practically insol in alcohol. *Keep well closed.*

Dodecahydrate, *alum, potassium alum, kalinite*. The technical product is also known as *alum flour, alum meal, cube alum*. It is about 99.5% pure. Colorless, odorless, hard, large, transparent crystals or cryst fragments or white, cryst powder; sweetish, astringent taste. Stable at ordinary temp; when kept for a long time at 60-65° (or over H_2SO_4) loses 9 H_2O which is reabsorbed on exposure to air. Becomes anhyd at about 200°; at higher temp loses SO_3 and becomes basic and incompletely sol in water. d 1.725. mp 92.5°. One gram dissolves in 7.2 ml water, in 0.3 ml boiling water; freely sol in glycerol; insol in alcohol. The aq soln is acid; pH of 0.2 molar aq soln 3.3.

USE: In dyeing; printing fabrics; manuf dyes, lakes, paper, vegetable glue, marble cement, porcelain cement, explosives; in tanning, hardening gelatin, baking powders, purifying water, clarifying sugar, hardening plaster casts; electrolytic copperplating; as catalyst in synthesis of ammonia; as mordant in staining with carmine, eosine, hematoxylin; clarifier (as alumina cream); identifying coloring matters; hardening agent in microscopy.

THERAP CAT: Astringent.
THERAP CAT (VET): Astringent.

365. Aluminum Rubidium Sulfate. AlO_8RbS_2; mol wt 304.58. Al 8.86%, O 42.03%, Rb 28.06%, S 21.05%. RbAl-$(SO_4)_2$. Prepn: *Gmelin's Aluminum* (8th ed) **35B**, p 525-527 (1934). Crystal structure of dodecahydrate: Larson, Cromer, *Acta Cryst.* **22**, 793 (1967).

Dodecahydrate, *rubidium alum*. Crystals, mp reported from 99° to 109°. Sol in water; practically insol in alc.

366. Aluminum Salicylates, Basic. Basic aluminum salicylates. $(C_7H_5O_3)_nAl(OH)_{3-n}.xH_2O$.

Monosalicylate, $C_7H_7AlO_5$, *Alunozal*. Prepn: **Ger.** pat. **354,698** and **Brit.** pat. **182,446** (both 1922 to Soc. Chim. des Usines du Rhône), *Frdl.* **14**, 1234 and *C.A.* **16**, 4301[4] (1922). Fine, colorless powder, occasionally pink.

Disalicylate, $C_{14}H_{11}AlO_7.H_2O$. $[(C_7H_5O_3)_2Al(OH)(H_2O)]$, *Baluvet*. Prepn: Burrows, Wark, *J. Chem. Soc.* **1928**, 222. Pink needles. Sparingly sol in water. Readily loses 1 mol H_2O on drying at 100°, a second mol is driven off with difficulty.

Note: The neutral salicylate, $(C_7H_5O_3)_6Al_2.3H_2O$, is called *Saluminium insolubile* or *insoluble Salumin*.

THERAP CAT: Antidiarrheal.
THERAP CAT (VET): Antidiarrheal.

367. Aluminum Selenide. Al_2Se_3; mol wt 290.82. Al 18.55%, Se 81.45%. Prepn from aluminum and selenium: Waitkins, Shutt, *Inorg. Syn.* **2**, 184 (1946); Becher in *Handbook of Preparative Inorganic Chemistry* vol. 1, G. Brauer, Ed. (Academic Press, New York, 2nd ed., 1963) p 825.

Yellowish to light-brown powder. d_4^{15} 3.437. Unstable in air. Decomp in water and acid.

USE: In prepn of hydrogen selenide; in semiconductor research.

368. Aluminum Silicate. Al_2O_5Si; mol wt 162.05. Al 33.30%, O 49.37%, Si 17.33%. Al_2SiO_5. Usually contains some water. Polymorphous; the three naturally occurring forms are *andalusite, cyanite, sillimanite*. Other aluminum silicate minerals are *anauxite, dickite, kaolinite, kochite, mullite, newtonite, pyrophyllite, takizolite, termierite, ton*. For prepn and properties see *Gmelin's Aluminum* (8th ed.) **35B**, p 313-317 (1934).

USE: In dental cements, glass industry; manuf of semiprecious stones, enamels, ceramics, and colored lakes; paint filler; in washing compounds.

369. Aluminum Sodium Sulfate. $AlNaO_8S_2$; mol wt 242.10. Al 11.14%, Na 9.50%, O 52.87%, S 26.49%. NaAl-$(SO_4)_2$. Prepn: *Gmelin's Aluminum* (8th ed) **35B**, p 378-384 (1934).

Dodecahydrate, *sodium alum, soda alum*. Colorless crystals or white granules or powder. d 1.61. mp about 60°. Sol in 1 part water; practically insol in alc.

USE: Industrially, like aluminum potassium sulfate.

THERAP CAT: Astringent.
THERAP CAT (VET): Astringent.

370. Aluminum Stearate. *Octadecanoic acid aluminum salt; stearic acid aluminum salt;* aluminum tristearate. C_{54}-$H_{105}AlO_6$; mol wt 877.35. C 73.92%, H 12.06%, O 10.94%, Al 3.08%. $[CH_3(CH_2)_{16}COO]_3Al$. Prepn: Gilmour *et al.*, *J. Chem. Soc.* **1956**, 1972.

Hard material, passes into plastic form on heating. mp 117-120°. Practically insol in water. When freshly made, sol in alcohol, benzene, oil turpentine, mineral oils. Forms cryst pyridine complex.

USE: Waterproofing fabrics, ropes; in paint and varnish driers; thickening lubricating oils; in cements; in light-sensitive photographic compositions.

371. Aluminum Subacetate Solution. Essigsäure Tonerde. Contains about 8% aluminum diacetate, $Al(OH)(CH_3-CO_2)_2$. Prepd from aluminum sulfate, acetic acid, calcium carbonate and water. Details of prepn: *USP XVIII*, pp 29-30. Several other routes of prepn, *see* Aluminum Diacetate.

Clear, colorless liquid. Odor of acetic acid. d 1.045. Acid to litmus. Gradually becomes turbid and colloidal. Sometimes a more basic salt precipitates out and settles to the bottom of the container. Heat promotes the decomposition. *Keep in a cool place and discard when cloudy.*

USE: As mordant in cotton dyeing, in printing fabrics, in water- and fireproofing fabrics, in waterproofing tissues, in formulations for antiperspirants and embalming fluids. *See also* Aluminum Diacetate. *Caution:* Ingestion of large doses may cause severe nausea, vomiting, diarrhea, melena, hematemesis.

THERAP CAT: Astringent, antiseptic.
THERAP CAT (VET): Astringent.

372. Aluminum Sulfate. $Al_2O_{12}S_3$; mol wt 342.14. Al 15.77%, O 56.12%, S 28.11%. $Al_2(SO_4)_3$. Occurs in nature as the mineral *alunogenite*. Prepn: *Gmelin's Aluminum* (8th ed.) **35B**, pp 267-269 (1934).

White, lustrous crystals, pieces, granules, or powder. d 1.61. Melts when gradually heated. At 250° loses its water; decomposes at red heat. Sol in 1 part water; practically insol in alcohol. On long boiling of the aq soln, insol basic salt ppts. The aq soln is acid.

Octadecahydrate. The commercial product is also known as *cake alum* or *patent alum*. It is about 99.5% pure. The article of commerce usually contains 5 to 10% less water than theory.

USE: Tanning leather, sizing paper, mordant in dyeing;

purifying water; manuf lakes, aluminum resinate; fireproofing and waterproofing cloth; clarifying oils and fats; treating
sewage; waterproofing concrete; deodorizing and decolorizing petroleum; antiperspirants; agricultural pesticides; manuf
aluminum salts.
THERAP CAT: Anti-infective.

373. Aluminum Sulfide. Al_2S_3; mol wt 150.14. Al
35.93%, S 64.07%. Prepd by heating aluminum with sulfur:
Flahaut, *Compt. Rend.* **232**, 334 (1951); Becher in *Handbook
of Preparative Inorganic Chemistry* vol. 1, G. Brauer, Ed.
(Academic Press, New York, 2nd ed., 1963) p 823.
Yellowish-gray, compact lumps; H_2S odor; decomposes in
moist air to a gray powder. d 2.02. mp 1100°. Hydrolyzed
by water to $Al(OH)_3$ and H_2S. *Keep tightly closed.*

374. Aluminum Tartrate. *2,3-Dihydroxybutanedioic
acid aluminum salt.* $C_{12}H_{12}Al_2O_{18}$; mol wt 498.19. C
28.93%, H 2.43%, Al 10.83%, O 57.81%. $Al_2(C_4H_4O_6)_3$.
Usually contains some water. Prepn: Goldman, *Biochem.
Z.* **133**, 459 (1922); **Brit.** pat. **348,790** (1930 to I. G. Farben.).
Odorless granules. Slowly sol in cold, readily in hot water; sol in ammonia.
USE: In textile dyeing.

375. Aluminum Thiocyanate. Aluminum sulfocyanate.
$C_3AlN_3S_3$; mol wt 201.22. C 17.91%, Al 13.40%, N 20.88%,
S 47.81%. $Al(CNS)_3$. Prepn: *Gmelin's, Aluminum* (8th ed.)
35B, p 306 (1934).
USE: Aq solution used as mordant in dye industry.

376. Aluminum Zinc Sulfate. $Al_2O_{16}S_4Zn$; mol wt
503.58. Al 10.72%, O 50.84%, S 25.47%, Zn 12.98%. Al_2-
$Zn(SO_4)_4$. Prepn: *Gmelin's Alumimum* (8th ed.) **35B**, p 610
(1934).
Tetracosahydrate, *zinc alum.* Crystals; sol in water.

377. Alverine. *N-Ethyl-N-(3-phenylpropyl)benzenepro-
panamine; N-ethyl-3,3'-diphenyldipropylamine; bis(γ-phenylpropyl)ethylamine; di(phenylpropyl)ethylamine; Sestron
base.* $C_{20}H_{27}N$; mol wt 281.43. C 85.35%, H 9.67%, N
4.98%. Prepd from phenylpropyl chloride and ethylamine in
alkaline medium: Külz *et al., Ber.* **72**, 2165 (1939); alternate
route: Stühner, Elbrächter, *Arch. Pharm.* **287**, 139 (1954).

Liquid, $bp_{0.3}$ 165-168°; bp_{13} 210-215°.
Hydrochloride, sirup, sol in water.
Citrate, $C_{26}H_{35}NO_7$, *Antispasmin, Calmabel, Gamatran,
Profenil, Profenine, Prophelan, Proverine, Sestron, Spacolin,
Spasmaverine.* Crystals, freely sol in water. A 10% aq soln
is neutral to litmus.
THERAP CAT: Anticholinergic.

378. Amanitin. Toxic group from the poisonous mushroom *Amanita phalloides* (Fr.) Secr., Agaricaceae. Comprised of α-, β-, γ-amanitin and *amanin.* Isoln of α- and
β-amanitin: Wieland, *Ann.* **564**, 152 (1949). Prepn of
α-amanitin from β-amanitin: Wieland, Boehringer, *ibid.*
635, 178 (1960). Structure: Wieland, *Pure Appl. Chem.* **9**,
145 (1964); Wieland, Gebert, *Ann.* **700**, 157 (1967). Inhibits
protein synthesis of mammalian cells. In contrast to phalloidin, *q.v.*, amanitin has a delayed action; even at high doses
the lethal interval is >15 hrs. α-Amanitin is the major
poisonous constituent of *A. phalloides;* it is 10-20 times more
toxic than phalloidin. Review of the chemistry and toxicology of the toxins of *Amanita phalloides:* Wieland, Wieland,
Pharmacol. Rev. **11**, 87-107 (1959); *see also* T. Wieland,
Fortschr. Chem. Org. Naturst. **25**, 214-250 (1967); T. Wieland, H. Faulstich, *Crit. Rev. Biochem.* **5**, 185-260 (1978).
Book: H. Faulstich *et al., Amanita Toxins and Poisoning:
International Amanita Symposium* (Lubrecht & Cramer,
Heidelberg, 1980) 246 pp.

α-Amanitin, $C_{39}H_{54}N_{10}O_{13}S$. R = NH_2. Needles from
methanol, mp 254-255°. $[α]_D^{20}$ +191°. uv max: 302 nm.
LD_{50} i.p. in albino mice: 0.1 mg/kg, Wieland, Wieland, *loc.
cit.*
β-Amanitin, $C_{39}H_{53}N_9O_{14}S$. R = OH. Needles from
methanol, mp 300°. uv max: 302 nm. Sol in water, methanol, ethanol, aqueous butanol. LD_{50} i.p. in albino mice: 0.4
mg/kg, Wieland, Wieland, *loc. cit.*
Human Toxicity: Highly toxic. Causes salivation, vomiting, bloody stools, cyanosis, muscular twitching, convulsions. Can be fatal.
USE: As a tool in molecular biology.

379. Amanozine. *N-Phenyl-1,3,5-triazine-2,4-diamine;
2-amino-4-anilino-s-triazine;* 4-amino-2-anilino-1,3,5-tri-
azine; 2-amino-6-anilino-s-triazine; *N-phenylformoguan-
amine; W-1191-2; Urofort.* $C_9H_9N_5$; mol wt 187.20. C
57.74%, H 4.85%, N 37.41%. Prepd by the action of formic
acid on phenylbiguanide: Papini, Folena, *Gazz. Chim. Ital.*
80, 837 (1950); **Austrian** pat. **168,063** (1951 to Gedeon
Richter), *C.A.* **47**, 8097h (1953); from phenylbiguanide and
ethyl formate in methanol in 67% yield: Overberger, Shapiro, *J. Am. Chem. Soc.* **76**, 93 (1954).

Crystals from dioxane or 50% ethanol, mp 235-236°.
Precipitated by alkalies.
Hydrochloride, $C_9H_9N_5$·HCl, long needles, dec 258-260°.
THERAP CAT: Diuretic.

380. Amantadine. *Tricyclo[3.3.1.1^{3,7}]decan-1-amine;
1-adamantanamine;* 1-aminoadamantane; 1-aminodiaman-
tane (obsolete); 1-aminotricyclo[3.3.1.1^{3,7}]decane. $C_{10}H_{17}N$;
mol wt 151.26. C 79.40%, H 11.34%, N 9.26%. Synthesis
(two different routes): Stetter *et al., Ber.* **93**, 226 (1960); *cf.
Angew. Chem.* **71**, 429-430 (1959); from adamantyl chloride:
Gerzon *et al., J. Med. Chem.* **6**, 760 (1963). Prepn: **Brit.**
pat. **925,728** corresp to **U.S.** pat. **3,152,180** (1963, 1964 both
to Studiengesellschaft Kohle); **Belg.** pat. **646,581** (1964 to
Du Pont), *C.A.* **63**, 14726h (1965). Pharmacology and toxicology: Vernier *et al., Toxicol. Appl. Pharmacol.* **15**, 642
(1969). Metabolism: Bleidner *et al., J. Pharmacol. Exp.
Ther.* **150**, 484 (1965). Clinical trial in prophylaxis of influenza A infection: R. Dolin *et al., N. Engl. J. Med.* **307**, 580
(1982). *See also* Adamantane, Rimantadine. Comprehensive description: J. Kirschbaum in *Analytical Profiles of
Drug Substances* vol. 12, K. Florey, Ed. (Academic Press,
New York, 1983) pp 1-36.

Hexakistetrahedral crystals by sublimation, mp 160-190°
(closed tube). Also reported as mp 180-192°. Sparingly sol
in water.
Hydrochloride, $C_{10}H_{18}ClN$, *EXP-105-1, NSC-83653,*

Amazolon, Mantadix, Mantadan, Mantadine, Midantan, Mydantane, Symmetrel, Virofral. Crystals from abs ethanol + anhydr ether, dec 360°. Sol in water (at least 1:20), alc. Practically insol in ether. LD$_{50}$ orally in mice, rats: 700, 1275 mg/kg (Vernier).

Sulfate, *Contenton, PK-Merz, Trivaline.*

THERAP CAT: Antiviral (Influenza A); antiparkinsonian; treatment of drug-induced extrapyramidal reactions.

381. Amantanium Bromide. *N,N-Dimethyl-N-[2-[(tricyclo[3.3.1.1^{3,7}]dec-1-ylcarbonyl)oxy]ethyl]-1-decanaminium bromide;* 2-(1'-adamantanecarbonyloxy)ethyldimethyldecylammonium bromide; decyl(2-hydroxyethyl)dimethylammonium bromide 1-adamantanecarboxylate; Amantol. C$_{25}$-H$_{46}$BrNO$_2$; mol wt 456.55. C 65.77%, H 10.16%, Br 17.50%, N 3.07%, O 3.50%. Prepn: R. A. Bauman, U.S. pat. **3,928,-411** (1975 to Colgate-Palmolive); and activity: L. Rovati *et al.,* U.S. pat. **4,288,609** (1981 to Rotta).

Crystals, mp 182-184°. LD$_{50}$ orally in mice: 910 mg/kg (Rovati).

THERAP CAT: Antiseptic.

382. Amaranth (Dye). *3-Hydroxy-4-[(4-sulfo-1-naphthalenyl)azo]-2,7-naphthalenedisulfonic acid trisodium salt; C.I. Acid Red 27;* C.I. 16185; FD & C Red No. 2; Red no. 2; trisodium salt of 1-(4-sulfo-1-naphthylazo)-2-naphthol-3,6-disulfonic acid. C$_{20}$H$_{11}$N$_2$Na$_3$O$_{10}$S$_3$; mol wt 604.49. C 39.74%, H 1.83%, N 4.63%, Na 11.41%, O 26.47%, S 15.91%. Prepd by coupling diazotized α-naphthylamine-4-sulfonic acid with β-naphthol-3,6-disulfonic acid: Knecht, *J. Soc. Dyers Colour.* **2**, 24 (1886); Farbw. Hoechst, Ger. pat. **3229;** Brit. pat. **1715;** BASF, Ger. pat. **5411;** U.S. pat. **204,799.** Metabolism: J. L. Radomski, T. J. Mellinger, *J. Pharmacol. Exp. Ther.* **136**, 259 (1962). *See also Colour Index* **vol. 4** (3rd ed., 1971) p 4093.

Dark, reddish-brown powder. Absorption max (water): 522.5 nm. One gram dissolves in about 15 ml water. Also reported as 7.20 g/100 ml H$_2$O at 26°. Very slightly soluble in alcohol. The aq soln is vivid red (1 cm layer). Discharge white by hydrosulfite on wool and silk. HCl does not change the color intensity of the soln, NaOH increases it. The aq soln is stable toward light.

USE: Dyeing wool and silk bright bluish-red from an acid bath. As indicator in hydrazine titrations. In color photography. *Caution:* Banned by the FDA in 1976 for use in foods, drugs and cosmetics.

383. Amaranth (Plant). Genus of the *Amaranthaceae* L. family which contains approx 60 species having worldwide distribution. Many species are considered weeds but *Amaranthus caudatus* L. (*love-lies-bleeding*), *A. hybridus* variety *hypochondriacus* L. (*prince's feather*), *A. tricolor* have been cultivated as ornamentals. *A. retroflexus* L. and some of the other weedy species are known as *pigweed, redroot* and *water hemp.* *A. spinosus* L. has been used in the treatment of gonorrhea: W. H. Brown, *Useful Plants of the Philippines* **1** (Philippines Dept. Agr. and Natl. Resources, Manila, 1951) pp 510-515; as a poultice in the treatment of inflammation, bruises and eczema: T. H. P. de Tavera, *The Medicinal*

Plants of the Philippines (P. Blakiston's Son, Philadelphia, 1901) pp 200-202. Most species are hardy, herbaceous and fast-growing cereal-like plants. Leaves and grain are used for food in parts of South America, Africa and Asia. Plants are high in protein; the amino acid composition is complementary to that of wheat. The grain was a basic food in pre-Columbian South and Central America and was important in Aztec ritual. Grain amaranths (*A. hypochondriacus, A. cruentas, A. caudatus*) produce large seedheads containing many edible seeds. The seed can be made into low gluten flour, cooked into gruel or popped like corn. Compositional study of grain: R. Becker *et al., J. Food Sci.* **46,** 1175 (1981); K. C. Pant, *Nutr. Rep. Int.* **28,** 1445 (1983). Properties of seed starches: P. V. S. Rao, J. K. Goering, *Cereal Chem.* **47,** 655 (1970); Y. Tomita *et al., J. Nutr. Sci. Vitaminol.* **27,** 471 (1981); saccharides: K. Lorenz, M. Gross, *Nutr. Rep. Int.* **29,** 721 (1984); lipids: F. I. Opute, *J. Exp. Botany* **30,** 601 (1979). Baking potential of flour: K. Lorenz, *Starch* **33,** 149 (1981). Most amaranth species have edible leaves with mild spinach-like flavor. *A. cruentas, A. dubius, A. hybridus, A. lividus* and *A. tricolor* are some of the species grown as vegetables. Magnesium and copper content of leaf: N. M. Guttiker *et al., J. Nutr. Diet.* **3,** 4 (1966); vitamin A content: C. N. Rao, *ibid.* **4,** 10 (1967); amino acids: I. G. Vasi, V. P. Kalintha, *J. Inst. Chemists (India)* **52,** 13 (1980). Nutritive value of leaf protein concentrate: P. R. Cheeke *et al., Can. J. Animal Sci.* **61,** 199 (1981). Toxicology: R. M. Hill, P. D. Rawate: *J. Agr. Food Chem.* **30,** 465 (1982). Book: J. N. Cole, *Amaranth, From the Past for the Future* (Rodale Press, Emmaus, Pennsylvania, 1979) 311 pp; Nat'l. Academy of Sciences Report: *Amaranth, Modern Prospects for an Ancient Crop* (Nat'l. Academy Press, Washington, D.C., 1984) 80 pp.

384. Amarogentin. *[4aS-(4aα,5β,6α)]-3,3',5-Trihydroxy[1,1'-biphenyl]-2-carboxylic acid 2-ester with 5-ethenyl-6-(β-D-glucopyranosyloxy)-4,4a,5,6-tetrahydro-1H,3H-pyrano[3,4-c]pyran-1-one;* sweroside-2'-(3'',5'',3'''-trihydroxydiphenyl)-2''-carboxylic acid ester. C$_{29}$H$_{30}$O$_{13}$; mol wt 586.56. C 59.38%, H 5.16%, O 35.46%. Strongly bitter glucoside. Isoln from *Swertia chirata* Buch-Ham., *Gentianaceae:* Korte, *Ber.* **88,** 704 (1955); **89,** 2404 (1956); from *Swertia japonica* Makino: Inouye *et al., Chem. Pharm. Bull.* **18,** 1856 (1970). Occurrence in gentianaceous plants: Inouye, Nakamura, *J. Pharm. Soc. Japan* **91,** 755 (1971), *C.A.* **75,** 95431b (1971). Structure: *eidem, Tetrahedron Letters* **1968,** 4919; *eidem, Tetrahedron* **27,** 1951 (1971).

Colorless needles, mp 229-230° (monohydrate). [α]$_D^{20}$ −116.6° (methanol). uv max: 230, 266, 306 nm (log ε 4.46, 4.07, 3.68). Slightly sol in benzene; water; freely sol in acetone, anhydr dioxane, tetrahydrofuran, methanol, ethanol. Practically insol in petr ether, ether, cyclohexane, CHCl$_3$.

385. Amarolide. *2α,11α-Dihydroxypicrasane-1,12,16-trione.* C$_{20}$H$_{28}$O$_6$; mol wt 364.44. C 65.91%, H 7.74%, O 26.34%. Isoln from *Ailanthus altissima* (Mill.) Swingle (*A. glandulosa* Desf.), *Simaroubaceae,* and proposed structure: Casinovi *et al., Tetrahedron Letters* **1965,** 2273. Isoln from *Castela nicholsoni* Hook, *Simaroubaceae,* and revised structure: Stöcklin *et al., ibid.* **1970,** 2399.

mp 253-255°. $R^1 = R^2 = H$. Practically insol in weak alkaline soln, slightly sol in sodium carbonate, easily dissolved by sodium hydroxide.

Monoacetate, $C_{22}H_{30}O_7$, mp 264-265° for $R^1 = H$, $R^2 = CH_3CO$. When $R^1 = CH_3CO$, $R^2 = H$, mp 225°.

Diacetate, $C_{24}H_{32}O_8$. $R^1 = R^2 = CH_3CO$, mp 269-270° (from alcohol). Also reported as mp 260-262°.

386. Ambazone. *2-[4-[(Aminoiminomethyl)hydrazo-no]-2,5-cyclohexadien-1-ylidene]hydrazinecarbothioamide; [(4-oxo-2,5-cyclohexadien-1-ylidene)amino]guanidine thio-semicarbazone;* 1-amidinohydrazono-4-thiosemicarbazono-2,5-cyclohexadiene; *p*-benzoquinone amidinohydrazone thiosemicarbazone; benzoquinone guanylhydrazone thio-semicarbazone; Iversal; Primal; Promassol. $C_8H_{11}N_7S$; mol wt 237.30. C 40.49%, H 4.67%, N 41.32%, S 13.51%. Prepn from *p*-benzoquinone and aminoguanidine followed by treatment of the reaction product with thiosemicarbazide in acid soln: Petersen, Domagk, *Naturwiss.* **41**, 10 (1954); *eidem,* **Ger.** pat. **965,723** and **Brit.** pat. **774,794** (both 1957 to Bayer).

Monohydrate, brown or copper-colored crystals from water, dec 195°. Sparingly soluble in water. Moderately sol in alcohol, acetone. Freely sol in DMF, dilute acids.
THERAP CAT: Antibacterial.

387. Ambenonium Chloride. *N,N'-[(1,2-Dioxo-1,2-ethanediyl)bis(imino-2,1-ethanediyl)]bis[2-chloro-N,N-di-ethylbenzenemethanaminium] dichloride;* [oxalylbis(imino-ethylene)]bis[(o-chlorobenzyl)diethylammonium chloride]; *N,N'*-bis[2-diethylaminoethyl]oxamide bis[2-chlorobenzyl chloride]; Win 8077; Mytelase; Mysuran. $C_{28}H_{42}Cl_4N_4O_2$; mol wt 608.51. C 55.27%, H 6.96%, Cl 23.31%, N 9.21%, O 5.26%. Prepn: F. K. Kirchner, **Ger.** pat. **1,024,517** (1958 to Sterling Drug), *C.A.* **54**, 18366g (1960). Determn in serum: C. Tharasse-Bloch *et al., J. Chromatog.* **421**, 407 (1987). Pharmacokinetics in dogs: *eidem, Arzneimittel-Forsch.* **39**, 257 (1989).

Crystals, mp 196-199°. Freely soluble in water.
THERAP CAT: Cholinesterase inhibitor.

388. Amber. Succinite; Baltic amber; Bernstein. A fossil resin from the extinct pine tree *Pinites succinifera* (Goepp.) Conway, Pinaceae. Found along the Baltic coast, also mined in Samland (East Prussia). Baltic amber contains: C 79%, H 10.5%, O 10.5%; succinic acid 3-8%; α-amyrin 20-30%. *Refs:* Plonait, *Angew. Chem.* **48**, 184 (1935); Schmid and co-workers: *Ann.* **503**, 269 (1933); *Monatsh.* **63**, 210 (1933); **65**, 348 (1935); **72**, 290, 311 (1939). *Review:* Berthelot, *Chim. & Ind.* **50**, 78-9 (1943); Frondel, *Econ. Bot.* **22**, 371 (1968). Infrared spectroscopy of different varieties of powdered amber: C. W. Beck *et al., Nature* **201**, 256 (1964). Chemical constitution: J. B. Lambert, J. S. Frye, *Science* **217**, 55 (1982).

Pale-yellow to reddish-brown resin. Transparent or cloudy (due to enclosed air bubbles and free succinic acid). Brittle; conchoidal fracture. d 1.05 to 1.10. Harder than most other resins. n_D 1.539-1.545. Softens at 150°, mp 350-375° giving off a choking, aromatic odor. When rubbed it is a good generator of static electricity.
USE: The best quality is machined into beads and other personal ornaments. For teething strings. Also used for making mouthpieces of tobacco pipes and cigarette holders. Small pieces are pressed into "ambroid" and then used for the same purpose. Impure material goes into the manufacture of "amber" varnishes. *See also* Oil of Amber.

389. Ambergris. Concretion from intestinal tract of the sperm whale, *Physeter catodon* L., Physeteridae. Found in tropical seas or seashores. Perfumers have used ambergris for centuries for its desirable odoriferous and fixative properties. Three major components isolated are the triterpene alcohol ambrein, epicoprostanol and coprostanone: Ruzicka, Lardon, *Helv. Chim. Acta* **29**, 912 (1946); Lederer *et al., ibid.,* 1354; Hardwick, Laws, *The Analyst* **76**, 662 (1951). Ambergris falls under the Marine Mammal Protection Act of 1972 and is illegal to import in the U.S.A. Analytical method for the detection and identification of ambergris: T. F. Governo *et al., J. Assoc. Offic. Anal. Chem.* **60**, 160 (1977). Hypothesis on the biological origin of ambergris: P. A. Dubois, *Parfums, Cosmet., Aromes* **19**, 35 (1978).

Gray to black, waxy mass; characteristic odor. d 0.8-0.92. mp about 60°; flammable; almost completely volatile by heat. Insoluble in water or in alkali hydroxides; sol in hot alcohol, chloroform, ether, fats, volatile oils.
USE: Chiefly in perfumery as tincture and essence for fixing delicate odors.

390. Amberlite®. Synthetic, high capacity cation and anion exchange resins and ion exchange-impregnated papers. *Reviews:* Nachod, Schubert, *Ion Exchange Technology* (Academic Press, New York, 1956); Kunin, *Ion Exchange Resins* (John Wiley, New York, 1958); Kunin, *Elements of Ion Exchange* (Reinhold, New York, 1960); Rieman, Walton, *Ion Exchange in Analytical Chemistry* (Pergamon Press, 1970); Dorfner, *Ion Exchangers Properties and Applications* (Ann Arbor Science, 1972).

391. Ambrosin. *3,3a,4,5,6,6a,9a,9b-Octahydro-6,9a-di-methyl-3-methyleneazuleno[4,5-b]furan-2,9-dione; 6β-hydr-oxy-4-oxo-10αH-ambrosa-2,11(13)-dien-12-oic acid* γ-*lac-tone.* $C_{15}H_{18}O_3$; mol wt 246.31. C 73.15%, H 7.36%, O 19.49%. From herb of *Ambrosia maritima* L., Compositae: Abu-Shady, Soine, *J. Am. Pharm. Assoc.* **42**, 387 (1953); **43**, 365 (1954). Structure: Bernardi, Büchi, *Experientia* **13**, 466 (1957); Sorm *et al., Coll. Czech. Chem. Commun.* **24**, 1548 (1959); Nerz *et al., J. Am. Chem. Soc.* **84**, 2601 (1962). Absolute configuration: Emerson *et al., Tetrahedron Letters* **1966**, 6151. Stereospecific total synthesis of (±)-form: P. A. Grieco *et al., J. Am. Chem. Soc.* **99**, 7393 (1977).

Crystals from alc, mp 146°. $[\alpha]_D^{22}$ −154.50° (c = 2 in alcohol). uv max (95% ethanol): 217, 324 nm (ϵ 13,465; 36). Very sol in chloroform; sol in alc, benzene, sparingly sol in ether, petr ether. Practically insol in water, cold dil NaOH, dil acids.
(±)-Form, mp 188-190°.

392. Ambroxol. *4-[[(2-Amino-3,5-dibromophenyl)-methyl]amino]cyclohexanol; N-(trans-p-hydroxycyclohexyl)-(2-amino-3,5-dibromobenzyl)amine;* NA 872. $C_{13}H_{18}Br_2$-N_2O; mol wt 378.11. C 41.29%, H 4.80%, Br 42.27%, N 7.41%, O 4.23%. Metabolite of bromhexine, *q.v.* Structure: E. Schraven *et al., Eur. J. Pharmacol.* **7**, 445 (1967). Synthesis: J. Keck, *Ann.* **707**, 107 (1967); **Fr.** pat. **1,522,709**; J.

Keck *et al.,* U.S. pat. **3,536,713** (1968, 1970 both to Thomae). Toxicity: S. Püschmann, R. Engelhorn, *Arzneimittel-Forsch.* **28**, 889 (1978). Series of articles on pharmacology, metabolism, and clinical studies: *ibid.* 889-935; on pharmacology and clinical efficacy of combination with amoxicillin, *q.v.,* in bronchopulmonary disease: *ibid.* **37**, 965-971 (1987). Symposium on pharmacology and efficacy in multicenter studies: *Respiration* **51**, Suppl. **1**, 1-68 (1987).

Hydrochloride, $C_{13}H_{19}Br_2ClN_2O$, *Bronchopront, Duramucal, Fluibron, Fluixol, Frenopect, Lindoxyl, Muco-Burg, Mucosolvan, Mucoclear, Mucovent, Pect, Stas-Hustenloser, Surbronc, Surfactal.* Crystals from ethanol, mp 233-234.5° (dec). LD_{50} in mice, rats (mg/kg): 268, 380 i.p.; 2720, 13400 orally (Püschmann, Engelhorn).

THERAP CAT: Expectorant.

393. Ambucaine. *4-Amino-2-butoxybenzoic acid 2-diethylaminoethyl ester;* 2-butoxy-4-aminobenzoic acid β-diethylaminoethyl ester; β-diethylaminoethyl 2-butoxy-4-aminobenzoate; 2-diethylaminoethyl 4-amino-2-*n*-butoxybenzoate; ambutoxate. $C_{17}H_{28}N_2O_3$; mol wt 308.41. C 66.20%, H 9.15%, N 9.08%, O 15.56%. Prepn: Büchi *et al., Helv. Chim. Acta* **34**, 1002 (1951); Clinton *et al., J. Am. Chem. Soc.* **74**, 592 (1952); Clinton, Laskowski, U.S. pat. **2,689,248** (1954 to Sterling Drug); Grimme, Schmitz, *Ber.* **87**, 179 (1954).

Hydrochloride, $C_{17}H_{29}ClN_2O_3$, *Win 3706, Sympocaine.* Crystals, mp 127°.
Dihydrochloride, crystals, decomp 156.8-159°.
THERAP CAT: Local anesthetic.

394. Ambucetamide. *α-(Dibutylamino)-4-methoxybenzeneacetamide;* α-*dibutylamino-α-(p-methoxyphenyl)acetamide;* α-*p*-methoxyphenyl-α-di-*n*-butylaminoacetamide; A 16; Dibutamide; Bersen; Meritin. $C_{17}H_{28}N_2O_2$; mol wt 292.41. C 69.82%, H 9.65%, N 9.58%, O 10.94%. Prepn starting with dibutylamine, anisaldehyde and potassium cyanide: Janssen, *J. Am. Chem. Soc.* **76**, 6192 (1954).

Rods from ethanol + 10% ether. mp 134° (first reported as mp 125-127°). Practically insol in water. Sol in ethanol, isopropanol, glacial acetic acid.
THERAP CAT: Antispasmodic.

395. Ambuphylline. *3,7-Dihydro-1,3-dimethyl-1H-purine-2,6-dione compd with 2-amino-2-methyl-1-propanol (1:1);* theophylline aminoisobutanol; theophylline compd with 2-amino-2-methyl-1-propanol; bufylline (rescinded USAN); buthoid; Butaphyllamine. $C_{11}H_{19}N_5O_3$; mol wt 269.30. C 49.06%, H 7.11%, N 26.01%, O 17.82%. Prepn: Shelton, U.S. pat. **2,404,319** (1946 to Wm. S. Merrell). Toxicity studies: Thompson, Warren, *J. Clin. Med.* **31**, 1337 (1946).

Crystals, mp 254-256°. Soly in water about 55%. LD_{50} i.v. in rabbits: 163 mg/kg; orally in mice: 600 mg/kg, (Thompson, Warren).
THERAP CAT: Bronchodilator.

396. Ambuside. *4-Chloro-6-[(3-hydroxy-2-butenylidene)amino]-N[1]-2-propenyl-1,3-benzenedisulfonamide; N'-allyl-4-chloro-6-[(3-hydroxy-2-butenylidene)amino]-m-benzenedisulfonamide;* 2-(allylsulfamoyl)-5-chloro-4-sulfamoyl-N-(3-hydroxy-2-butenylidene)aniline; 2-allylsulfamyl-4-sulfamyl-5-chloro-N-(3-hydroxy-2-butenylidene)-aniline; 5-allylsulfamoyl-2-chloro-4-(3-hydroxybut-2-enylideneamino)benzenesulfonamide; EX 4810; Hydrion; Novohydrin. $C_{13}H_{16}ClN_3O_5S_2$; mol wt 393.86. C 39.64%, H 4.10%, Cl 9.00%, N 10.67%, O 20.31%, S 16.28%. Prepn: Robertson, Fr. pat. **1,331,680** (1963 to Lakeside), corresp to U.S. pat. **3,188,329** (1965 to Colgate-Palmolive); Robertson *et al., J. Med. Chem.* **8**, 90 (1965).

Crystals, mp 205-207°. uv max: 343 nm (ε 32900).
THERAP CAT: Diuretic; antihypertensive.

397. Ambutonium Bromide. *γ-(Aminocarbonyl)-N-ethyl-N,N-dimethyl-γ-phenylbenzenepropanaminium bromide;* 3-(carbamoyl-3,3-diphenylpropyl)ethyldimethylammonium bromide; 4-dimethylamino-2,2-diphenylbutyramide ethyl bromide; R 100. $C_{20}H_{27}BrN_2O$; mol wt 391.35. C 61.38%, H 6.96%, Br 20.42%, N 7.16%, O 4.09%. Prepn: Janssen *et al., Arch. Int. Pharmacodyn.* **103**, 82 (1955).

Crystals, mp 228-229° (dec).
Combination with oxazepam, *q.v., Praxiten SP.*
THERAP CAT: Anticholinergic.

398. Amcinonide. *21-(Acetyloxy)-16,17-[cyclopentylidenebis(oxy)]-9-fluoro-11-hydroxypregna-1,4-diene-3,20-dione;* 9-fluoro-11β,16α,17,21-tetrahydroxypregna-1,4-diene-3,20-dione cyclic 16,17-acetal with cyclopentanone, 21-acetate; CL-34699; Amciderm; Cyclocort; Penticort. $C_{28}H_{35}FO_7$; mol wt 502.59. C 66.92%, H 7.02%, F 3.78%, O 22.28%. Prepn: W. Shultz *et al.,* Ger. pat. **2,437,847** (1975 to Am. Cyanamid), *C.A.* **83**, 10608g (1975). Bioavailability study: R. Woodford, J. M. Haigh, *Curr. Ther. Res.* **26**, 301 (1979). Therapeutic use in eczematoid conditions: G. L. Rocha *et al., ibid.* **19**, 538 (1976); E. W. Rosenberg, *Cutis* **24**, 642 (1979).

THERAP CAT: Glucocorticoid.

399. Amdinocillin. *6-[[(Hexahydro-1H-azepin-1-yl)-methylene]amino]-3,3-dimethyl-7-oxo-4-thia-1-azabicyclo-[3.2.0]heptane-2-carboxylic acid;* 6-[(hexahydro-1H-azepin-1-yl)methyleneamino]penicillanic acid; mecillinam; FL 1060; Ro 10-9070; Selexidin. $C_{15}H_{23}N_3O_3S$; mol wt 325.43. C 55.36%, H 7.12%, N 12.91%, O 14.75%, S 9.85%. Semisynthetic antibiotic related to penicillin. Prepn: F. J. Lund, **Ger.** pat. **2,055,531** corresp to **U.S.** pat. **3,957,764** (1971, 1976 both to Leo Pharm.). Synthesis and chemical properties: H. B. König *et al., Arzneimittel-Forsch.* **33**, 88 (1983). X-ray structural study: J. W. Krajewski *et al., J. Antibiot.* **34**, 282 (1981). *In vitro* study: B. Chattopadhyay, I. Hall, *J. Antimicrob. Chemother.* **5**, 549 (1979). Activity against gram-negative bacteria: D. S. Reeves, *J. Antimicrob. Chemother.* **3**, Suppl. B, 5 (1977). Studies on mechanism of action: B. G. Spratt, *ibid.* 13. Metabolism: A. P. Ball *et al., ibid.* **4**, 241 (1978). Use in urinary tract infections and septicemia: N. Frimodt-Miller, T. J. Ravn, *Infection* **7**, 35 (1979). Determn in plasma and urine by HPLC: T. L. Lee, M. A. Brooks, *J. Chromatog.* **227**, 137 (1982). Pharmacokinetics in man: B. R. Meyers *et al., Antimicrob. Ag. Chemother.* **23**, 827 (1983). Symposium on pharmacology, pharmacokinetics, clinical studies: *Am. J. Med.* **75**, no. 2A, 1-138 (1983). Review of pharmacology and clinical efficacy: H. C. Neu, *Pharmacother.* **5**, 1-10 (1985).

Crystals from methanol-acetone, mp 156° (dec). $[\alpha]_D^{20}$ +285° (c = 1 in 0.1N HCl). Sol in water.
Pivaloyloxymethyl ester, *see* Amdinocillin Pivoxil.
THERAP CAT: Antibacterial.

400. Amdinocillin Pivoxil. *6-[[(Hexahydro-1H-azepin-1-yl)methylene]amino]-3,3-dimethyl-7-oxo-4-thia-1-azabicyclo[3.2.0]heptane-2-carboxylic acid (2,2-dimethyl-1-oxopropoxy)methyl ester;* pivaloyloxymethyl 6-[(hexahydro-1H-azepin-1-yl)methyleneamino]penicillanate; pivamdinocillin; pivmecillinam; FL 1039; Selexid (susp.). $C_{21}H_{33}N_3O_5S$; mol wt 439.57. C57.38%, H 7.57%, N 9.56%, O 18.20%, S 7.29%. Semi-synthetic antibiotic related to penicillin. Pivaloyloxymethyl ester of amdinocillin, *q.v.* Prepn: F. J. Lund, **Ger.** pat. **2,055,531** corresp to **U.S.** pat. **3,957,764** (1971, 1976 both to Leo Pharm.). HPLC stability analysis: R. B. Hagel, E. H. Waysek, *J. Chromatog.* **178**, 97 (1979). Bacteriological and pharmacokinetic study: T. Damsgaard *et al., J. Antimicrob. Chemother.* **5**, 267 (1979). Metabolism: J. D. Anderson, M. A. Adams, *Chemotherapy* **25**, 1 (1979). Clinical study: B. T. Andersen *et al., Infection* **8**, 27 (1980). Toxicity study: S. Sato *et al., Takeda Kenkysho Ho* **35**, 179 (1976), *C.A.* **86**, 115154w (1977).

Crystals from cyclohexane, mp 118.5-119.5°. $[\alpha]_D^{20}$ +231° (c = 1 in 96% ethanol). LD_{50} in mice, rats (mg/kg): 475-480, 465 i.v.; 1736-1930, 1935-2100 s.c.; 3020, 9500-10,000 orally (Sato).
Hydrochloride, $C_{21}H_{34}ClN_3O_5S$, **Melysin, Negaxid, Selexid** (tabl.). Crystals from methanol-diisopropyl ether, mp 172-173°. $[\alpha]_D^{20}$ +219° (c = 1 in 0.1N HCl).
THERAP CAT: Antibacterial.

401. Americium. Am; at. wt (most stable isotope) 243; at. no. 95; valence 3, 4, 5, 6. Completely man-made element. Isotopes (mass numbers): 237-246; all are radioactive. First isotope prepared: ^{241}Am (T$_{\frac{1}{2}}$ 458 years; α-emitter); prepd in 1944 by Seaborg, James, Morgan, by bombardment of ^{234}Pu with α particles, *see* G.T. Seaborg, J. J.

Katz, Eds., *The Transuranium Elements* (McGraw-Hill, New York, 1949) p 1525. Also formed spontaneously by β-decay of ^{241}Pu. Isoln: Armstrong *et al., A. I. Ch. E. J.* **3**, 286 (1957); Coleman, *J. Inorg. Nucl. Chem.* **3**, 327 (1957). Prepn of americium metal by reduction of americium trifluoride with barium: Westrum, Eyring, *J. Am. Chem. Soc.* **73**, 3396 (1951); Cunningham, Lohr, *ibid.* 2026. Americium is the most isolated superconductor in the periodic table; its superconductivity is allowed by its special nonmagnetic ground state: J. L. Smith, R. G. Haire, *Science* **200**, 535 (1978). Clinical application in bone mineral determn: E. G. De Puey *et al., J. Nucl. Med.* **16**, 891 (1975); in cancer radiotherapy: R. Nath *et al., Int. J. Rad. Oncol. Biol. Phys.* **14**, 969 (1988). *Reviews:* J. J. Katz, G. T. Seaborg, *The Chemistry of the Actinide Elements* (John Wiley, New York, 1957) pp 331-372; Keenan, *J. Chem. Ed.* **36**, 27-31 (1959); C. Keller, *The Chemistry of the Transuranium Elements* (Verlag Chemie, Weinheim, English Ed., 1971) pp 485-527; *Comprehensive Inorganic Chemistry* vol. **5**, J. C. Bailar, Jr. *et al.,* Eds. (Pergamon Press, Oxford, 1973) *passim; Handb. Exp. Pharmakol.* **36**, 689-940. *See also* the metabolism study from an unusual case of internal contamination of ^{241}Am in man: N. Cohen *et al., Science* **206**, 64 (1979).

Silvery metal. Two allotropic forms: α-form, double close packed hexagonal structure, d 13.67, transition pt 1074°; β-form, face-centered cubic structure. mp 1175°.
Trivalent americium is the most common in aq soln. Color light pink changing to yellow with increasing concn. Sharp absorption peak at 5027 Å. Americium dioxide, AmO_2, is obtained by ignition of most trivalent Am compounds.
Tetravalent americium is known only in the solid state. When AmO_2 or AmF_3 is treated with fluoride, solid AmF_4 results.
Pentavalent and hexavalent americium compds obtained in soln are doubly oxygenated and have the general formula AmO_2^{+n}, where n = +1 for Am(V) and n = +2 for Am(VI). Hexavalent americium is yellow or light brown in dilute perchloric or nitric acid, green in fluoride solns and dark brown in sulfuric acid. A deep red ion complex is formed in bicarbonate-carbonate solns.
THERAP CAT: ^{241}Am as diagnostic aid (bone mineral analyzer); as antineoplastic (radiation source).

402. Ametryn. *N-Ethyl-N'-(1-methylethyl)-6-(methylthio)-1,3,5-triazine-2,4-diamine;* 2-(ethylamino)-4-(isopropylamino)-6-(methylthio)-s-triazine; 2-ethylamino-4-isopropylamino-6-methylmercapto-s-triazine; 2-methylthio-4-ethylamino-6-isopropylamino-s-triazine; ametryne; G-34162; Ametrex; Evik; Gesapax. $C_9H_{17}N_5S$; mol wt 227.35. C 47.55%, H 7.54%, N 30.81%, S 14.11%. Prepn: Gysin, Knuesli, **Swiss** pat. **337,019**, *C.A.* **57**, 14226c (1962); Rufener *et al.,* **U.S.** pat. **3,558,622** (1959, 1971 both to Geigy). Acute toxicity: T. B. Gaines, R. E. Linder, *Fundam. Appl. Toxicol.* **7**, 299 (1986).

Crystals, mp 88-89°. Aqueous soly data: Ward, Weber, *J. Agr. Food Chem.* **16**, 959 (1968). LD_{50} in adult male, female rats (mg/kg): 508, 590 orally (Gaines, Linder).
USE: Herbicide.

403. Amezinium Methyl Sulfate. *4-Amino-6-methoxy-1-phenylpyridazinium methyl sulfate;* LU 1631; Regulton; Risumic; Supratonin. $C_{12}H_{15}N_3O_5S$; mol wt 313.33. C 46.00%, H 4.83%, N 13.41%, O 25.53%, S 10.23%. Sympathomimetic agent with vascular and cardiac activity. Prepn: F. Richenender, R. Kropp, **Ger.** pat. **1,912,941**; *eidem,* **U.S.** pat. **3,631,038** (1970, 1971 both to BASF). Series of articles on synthesis, pharmacology, mechanism of action, metabolism, pharmacokinetics, bioavailability, clinical trials: *Arzneimittel-Forsch.* **31**, 1527-1671 (1981). Acute toxicity data: H. J. Teschendorf, *ibid.* 1580. HPLC determn in human plasma: D. Hotz, E. Brode, *J. Chromatog.* **277**, 217 (1983).

Disposition and identification of major metabolites in rats: K. Nambu *et al.*, *Arzneimittel-Forsch.* **38**, 909 (1988).

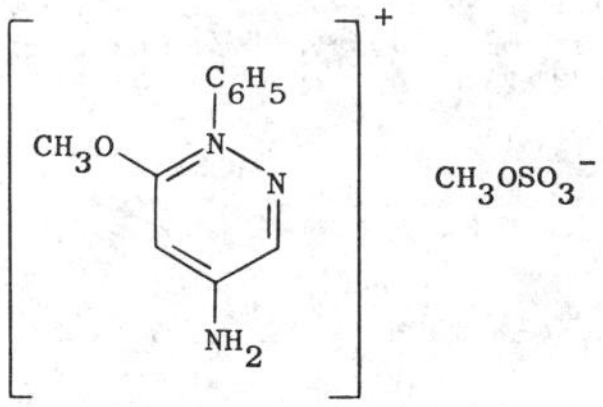

Crystals from water, mp 176° (dec). LD_{50} in mice, rats (mg/kg): 1630, 1410 orally; 40.4, 45.5 i.v. (Teschendorf).

THERAP CAT: Antihypotensive.

404. Amfenac. *2-Amino-3-benzoylbenzeneacetic acid;* 2-amino-3-benzoylphenylacetic acid. $C_{15}H_{13}NO_3$; mol wt 255.28. C 70.58%, H 5.13%, N 5.49%, O 18.80%. Prepn: W. J. Welstead, H. W. Moran, Ger. pat. **2,324,768**, *C.A.* **80**, 59708s (1974) and U.S. pat. **4,045,576** (1973, 1977 both to A. H. Robbins). Synthesis and anti-inflammatory activity: W. J. Welstead *et al.*, *J. Med. Chem.* **22**, 1074 (1979). Anti-inflammatory, analgesic, antipyretic activities: H. Fujimura *et al.*, *Oyo Yakuri* **22**, 381 (1981), *C.A.* **97**, 49427m (1982). Platelet aggregation inhibition: *eidem*, *ibid.* 399, *C.A.* **97**, 49502g (1982). Effect on polymorphonuclear leukocytes: T. Matsumoto *et al.*, *Pharmacol. Res. Commun.* **14**, 523 (1982). Toxicity data: L. F. Sancilio *et al.*, 170th Am. Chem. Soc. Meet. (Chicago, Aug. 1975), Abstracts of Papers, MEDI 17.

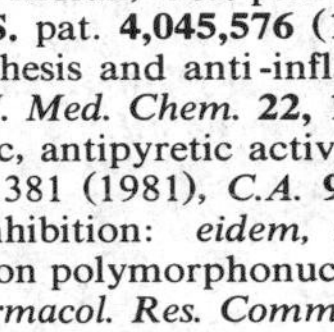

mp 121-123° (dec). LD_{50} in mice, rats (mg/kg): 615, 311 orally (Sancilio).

Sodium salt monohydrate, $C_{15}H_{12}NNaO_3 \cdot H_2O$, *AHR-5850D*, *Fenamate*, *Fenazox.* Yellow solid from ethanol/-isopropyl ether, mp 254-255.5°.

THERAP CAT: Anti-inflammatory.

405. Amicarbalide. *3,3'-(Carbonyldiimino)bisbenzene-carboximidamide; N,N'-bis[3-(aminoiminomethyl)phenyl]-benzenecarboximidamide;* 3,3'-diamidinocarbanilide; *N,N'*-bis(*m*-amidinophenyl)urea; *N,N'*-di(*m*-amidinophenyl)urea. $C_{15}H_{16}N_6O$; mol wt 296.34. C 60.80%, H 5.44%, N 28.36%, O 5.40%.

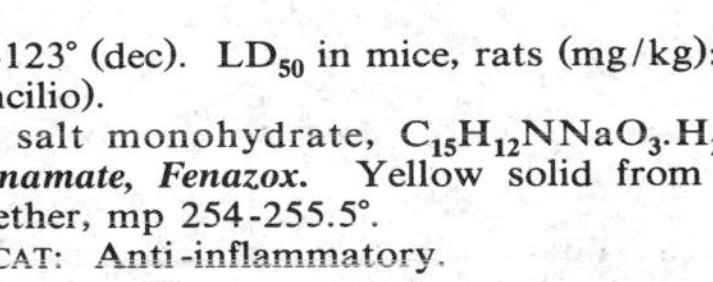

Diisethionate, $C_{19}H_{28}N_6O_9S_2$, *M & B 5062A*, *Diampron.* Prepn: Ashley *et al.*, *Nature* **185**, 461 (1960); Berg, **Belg.** pat. **585,595** (1960 to May & Baker); *idem*, *J. Chem. Soc.* **1961**, 5097. Crystals, mp 209° (decomp 256°). Soly in water: ~100 g/100 ml. LD_{50} s.c. in mice: 120 mg/kg.

THERAP CAT (VET): Babesiacide in cattle.

406. Amicetin. *[2R-[2α(S*),5β,6α]]-4-[(2-Amino-3-hydroxy-2-methyl-1-oxopropyl)amino]-N-[1-[5-[[4,6-di-deoxy-4-(dimethylamino)-α-D-glucopyranosyl]oxy]tetrahy-dro-6-methyl-2H-pyran-2-yl]-1,2-dihydro-2-oxo-4-pyrimi-dinyl]benzamide.* $C_{29}H_{42}N_6O_9$; mol wt 618.71. C 56.30%, H 6.84%, N 13.58%, O 23.27%. Antibiotic substance produced by *Streptomyces vinaceus-drappus* isolated from soil near Kalamazoo, Mich.: De Boer *et al.*, *J. Am. Chem. Soc.* **75**, 499, 5864 (1953); McKormich, Hoehn, *Antibiot. & Chemother.* **3**, 718 (1953); De Boer, Hinman, U.S. pats. **2,909,463**, **2,909,464** and **2,909,517** (all 1959 to Upjohn). Structure:

Haskell, *J. Am. Chem. Soc.* **80**, 747 (1958); Stevens *et al.*, *J. Org. Chem.* **27**, 2991 (1962). Structure of *amosamine*: Stevens *et al.*, *J. Am. Chem. Soc.* **85**, 1552 (1963). Configuration: Hanessian, Haskell, *Tetrahedron Letters* **1964**, 2451. Synthesis of amosamine: Stevens *et al.*, *J. Org. Chem.* **31**, 2822 (1966).

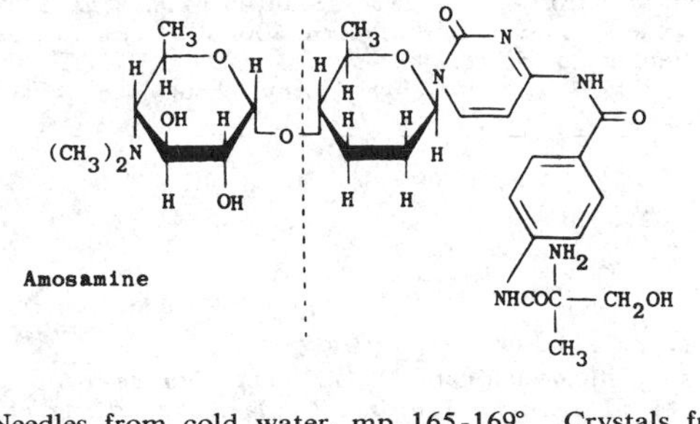

Amosamine

Needles from cold water, mp 165-169°. Crystals from warm water, dec 243-244°. pKa = 10.4, 7.0. $[\alpha]_D^{25}$ +116.5° (c = 0.5 in 0.1N HCl). uv max in water: 305 nm ($E_{1cm}^{1\%}$ 465); in 0.1N HCl: 316 nm ($E_{1cm}^{1\%}$ 433); in 0.1N NaOH: 322 nm ($E_{1cm}^{1\%}$ 470). Soly in water at 25° = 2 mg/ml. Slightly sol in common organic solvents.

Dihydrochloride, $C_{29}H_{44}Cl_2N_6O_9$, crystals from methanol + acetone, mp 190-192°. $[\alpha]_D^{24.5}$ +117° (c = 0.4 in alc).

407. Amicibone. *1-[2-(Hexahydro-1H-azepin-1-yl)eth-yl]-2-oxocyclohexanecarboxylic acid phenylmethyl ester;* 2-(β-hexamethyleniminoethyl)cyclohexanone-2-carboxylic acid benzyl ester; *Biotussal; Pectipront.* $C_{22}H_{31}NO_3$; mol wt 357.48. C 73.91%, H 8.74%, N 3.92%, O 13.43%. Prepn: **Belg.** pats. **639,472**; **639,473**; Frank *et al.*, **Austrian** pat. **237,593** (all 1964 to Biochemie GmbH), *C.A.* **62**, 5207e (1965).

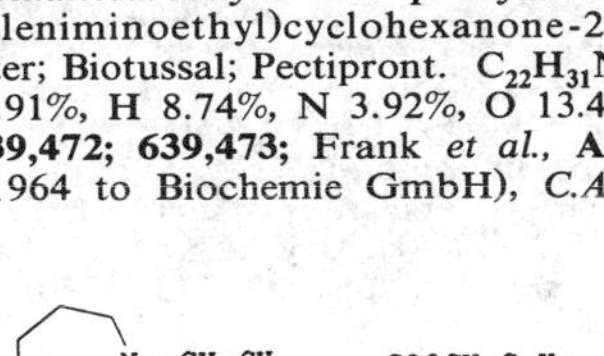

Hydrochloride, $C_{22}H_{31}NO_3 \cdot HCl$. Crystalline powder, mp 136-137°. Sol in water, ethanol and other common organic solvents.

THERAP CAT: Antitussive.

408. Amicoumacin A. *3-Amino-2,3-dideoxy-N^6-[1-(3,4-dihydro-8-hydroxy-1-oxo-1H-2-benzopyran-3-yl)-3-methylbutyl]hexaramide.* $C_{20}H_{29}N_3O_7$; mol wt 423.47. C 56.73%, H 6.90% N 9.92%, O 26.45%. Major component of a complex of antibiotics produced by *Bacillus pumilus* BN-103. Also exhibits anti-inflammatory activity *in vivo.* Isoln and biol activity: **Japan. Kokai 83 18,379** (1983 to Meiji Seika), *C.A.* **99**, 20906x (1983); J. Itoh *et al.*, *J. Antibiot.* **34**, 611 (1981); *eidem*, *Agr. Biol. Chem.* **46**, 1255 (1982). Structure: *eidem*, *ibid.* 2659. Use as acaricide: **Japan. Kokai 83 216,107** (1983 to Meiji Seika), *C.A.* **100**, 116494b (1984).

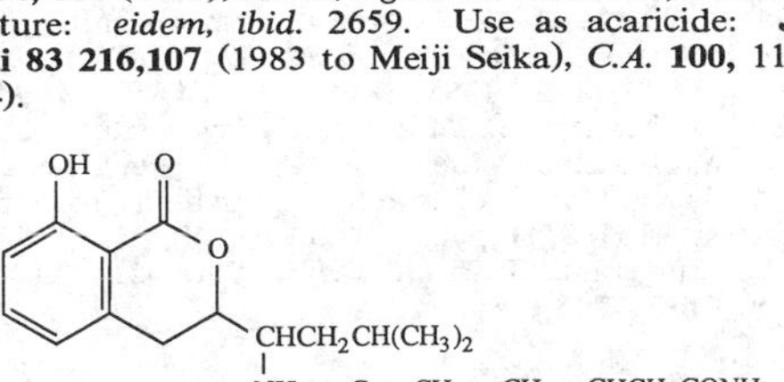

Hydrochloride, colorless powder, mp 132-135° (dec). uv max (methanol): 208, 247, 315 nm (ε 27300, 6400, 4380). $[\alpha]_D^{23}$ −97.2° (c = 1.0 in methanol). LD_{50} orally in mice: 132 mg/kg (Itoh).

409. Amidephrine. *N-[3-[1-Hydroxy-2-(methylamino)-*

ethyl]phenyl]methanesulfonamide; 3'-[1-hydroxy-2-(methyl-amino)ethyl]methanesulfonanilide; MJ 1996. $C_{10}H_{16}N_2O_3S$; mol wt 244.31. C 49.16%, H 6.60%, N 11.47%, O 19.65%, S 13.12%. Prepn: Larsen, Uloth, **Fr.** pat. **M3027** (1965 to Mead Johnson), *C.A.* **62**, 13091a (1965); Uloth *et al., J. Med. Chem.* **9**, 88 (1966). Stimulates adrenergic α-receptors. Pharmacology studies of the racemate: Dungan *et al., Int. J. Neuropharmacol.* **4**, 219 (1965); Stanton *et al., ibid.* 235; of the isomers: Larsen, Lish, *Nature* **203**, 1283 (1964); Buchthal, Jenkinson, *Eur. J. Pharmacol.* **10**, 293 (1970). Toxicology: Weikel, Harper, *Toxicol. Appl. Pharmacol.* **23**, 589 (1972).

Crystals, mp 159-161°. pKa 9.1.

Monomethanesulfonate, $C_{11}H_{20}N_2O_6S_2$, *amidephrine mesylate, MJ 5190, Dricol, Fentrinol, Nalde.* Crystals from ethanol, mp 207-209°. LD_{50} in female rats: 13-36 mg/kg orally; 5-25 mg/kg i.p.

THERAP CAT: Vasoconstrictor; nasal decongestant.

410. Amidinomycin. *3-Amino-N-(3-amino-3-imino-propyl)cyclopentanecarboxamide; N-(2'-amidinoethyl)-3-aminocyclopentanecarboxamide;* myxoviromycin. $C_9H_{18}N_4O$; mol wt 198.27. C 54.52%, H 9.15%, N 28.26%, O 8.07%. Antibiotic substance produced by *Streptomyces flavochromogenes* isolated from Japanese soil (Shiuoka Prefecture). Isoln and structure: S. Nakamura *et al., J. Antibiot.* **14A**, 103 (1961); S. Nakamura, *Chem. Pharm. Bull.* **9**, 641 (1961). Identity with myxoviromycin: S. Nakamura *et al., J. Antibiot.* **14A**, 163 (1961). Prepn: Katsube, Saito, **Japan.** pat. **21,418('68)** (to Sumitomo), *C.A.* **70**, 87135q (1969). Synthesis of amidinomycin and *trans* isomer: H. Paul *et al., Arch. Pharm.* **301**, 512 (1968). Crystal and molecular structure: M. Kaneda *et al., J. Antibiot.* **33**, 778 (1980).

Sulfate, $C_9H_{18}N_4O.H_2SO_4$, plates or needles from water + methanol, dec 285-288°. $[\alpha]_D^{21}$ −3.9° (c = 3). Absorption spectra: S. Nakamura, *loc. cit.* Soluble in water. Practically insol in ether, benzene, ethyl acetate, methanol, ethanol, butanol, acetone.

THERAP CAT: Antiviral.

411. Amidochlor. *N-[(Acetylamino)methyl]-2-chloro-N-(2,6-diethylphenyl)acetamide; N-acetamidomethyl-2-chloro-2',6'-diethylacetanilide;* MON-4621; Limit. $C_{15}H_{21}ClN_2O_2$; mol wt 296.80. C 60.70%, H 7.13%, Cl 11.95%, N 9.44%, O 10.78%. Plant growth regulator for cool season grasses. Prepn: **Neth.** pat. **Appl. 7,207,261;** K. W. Ratts, **U.S.** pat. **3,830,841** (1972, 1974 both to Monsanto); K. W. Ratts, J. P. Chupp, *J. Org. Chem.* **39**, 3745 (1974). Use as plant growth regulator: K. W. Ratts *et al., U.S.* pat. **3,829,-306** (1974 to Monsanto). Effect on growth and seedhead suppression of annual bluegrass: A. M. Petrovic *et al., Agron. J.* **77**, 670 (1985); of wild and cultivated proso millet: J. L. Carpenter, H. J. Hopen, *HortScience* **20**, 942 (1985); on established turfgrass: P. C. Bhowmik, *Proc. 5th Int. Turf-grass Res. Conf.* 735 (1985).

Crystals from methanol, mp 148-149°.

USE: Turf growth regulator.

412. Amido-G-Acid. *7-Amino-1,3-naphthalenedisul-fonic acid;* 2-naphthylamine-6,8-disulfonic acid; amino-G-acid. $C_{10}H_9NO_6S_2$; mol wt 303.32. C 39.60%, H 2.99%, N 4.62%, O 31.65%, S 21.14%. Prepd by sulfonation of β-naphthylamine: Fierz-David, Braunschweig, *Helv. Chim. Acta* **6**, 1146 (1923).

Tetrahydrate, fine monoclinic needles. Sol in water, less sol in alc. Soly in water at 20°: 9.24 g in 100 g of satd soln.

USE: Manufacture of dyes.

413. Amidomycin. $C_{40}H_{68}N_4O_{12}$; mol wt 796.98. C 60.28%, H 8.60%, N 7.03%, O 24.09%. Antibiotic substance produced by an unidentified *Streptomyces* culture (PRL 1642). Composed of 4 moles each of D-(−)-valine and D-(−)-α-hydroxyisovaleric acid, linked alternately by ester and amide bonds to form a 24-membered ring: Vining, Taber, *Can. J. Chem.* **35**, 1109 (1957). Structure studies: Shemyakin *et al., Tetrahedron Letters* **1963**, 351, *Tetrahedron* **19**, 995 (1963).

Needles from dilute ethanol or petr ether, mp 192°. $[\alpha]_D^{26}$ +19.2° (c = 1.2 in ethanol). Neutral reaction. Practically insol in water. Readily sol in most organic solvents. Primarily active against yeasts.

414. Amido-R-Acid. *3-Amino-2,7-naphthalenedisul-fonic acid;* 2-naphthylamine-3,6-disulfonic acid. $C_{10}H_9-NO_6S_2$; mol wt 303.32. C 39.60%, H 2.99%, N 4.62%, O 31.65%, S 21.14%. Prepd by treating 2-hydroxy-3,6-naphthalenedisulfonic acid with ammonium sulfite and ammonium hydroxide: Petitcolas, Josué, *Bull. Soc. Chim. France* **1952**, 89.

Crystals or powder. Soluble in water. Solutions show a violet-blue fluorescence.

USE: Manufacture of dyes.

415. Amifloxacin. *6-Fluoro-1,4-dihydro-1-(methyl-amino)-7-(4-methyl-1-piperazinyl)-4-oxo-3-quinolinecarb-oxylic acid;* Win 49375. $C_{16}H_{19}FN_4O_3$; mol wt 334.35. C 57.48%, H 5.73%, F 5.68%, N 16.76%, O 14.35%. Fluorinated quinolone antibacterial. Prepn: M. P. Wentland, D. M. Bailey, **Eur.** pat. **Appl. 90,424;** *eidem,* **U.S.** pat. **4,499,091** (1983, 1985 both to Sterling); *eidem, J. Med. Chem.* **27**, 1103 (1984); D. T. W. Chu, *J. Heterocycl. Chem.* **22**, 1033 (1985). *In vitro* activity against *Pseudomonas aeruginosa:* K. D. Thompson *et al., Antimicrob. Ag. Chemother.* **26**, 275 (1984). *In vitro* and *in vivo* antimicrobial spectrum: J. B. Cornett *et*

al., ibid. **27**, 4 (1985). Metabolism and disposition in animals: J. A. Johnson, D. P. Benziger, *ibid.* 774. Comparison
of activity with other nalidixic acid analogs: C. M. Bassey *et
al., J. Antimicrob. Chemother.* **17**, 623 (1986).

Crystals from DMF, mp 299-301° (dec).
Monomethanesufonate, $C_{17}H_{23}FN_4O_6S$, *amifloxacin mesylate, WIN 49375-3.* Crystals from aqueous acetonitrile, mp
287-289° (dec).
THERAP CAT: Antibacterial.

416. Amikacin. *O-3-Amino-3-deoxy-α-D-glucopy-
ranosyl-(1→6)-O-[6-amino-6-deoxy-α-D-glucopyranosyl-
(1→4)]-N¹-(4-amino-2-hydroxy-1-oxobutyl)-2-deoxy-D-
streptamine;* 1-N-[L(−)-4-amino-2-hydroxybutyryl]kana-
mycin A; Lukadin. $C_{22}H_{43}N_5O_{13}$; mol wt 585.62. C 45.12%,
H 7.40%, N 11.96%, O 35.52%. Semisynthetic aminoglycoside antibiotic derived from kanamycin A. Prepn: Kawaguchi *et al., J. Antibiot.* **25**, 695 (1972); Kawaguchi, Naito, **Ger.
pat. 2,234,315** corresp to U.S. pat. **3,781,268** (both 1973 to
Bristol-Myers). Biological formation from kanamycin A:
L. M. Cappelletti, R. Spagnoli, *J. Antibiot.* **36**, 328 (1983).
Microbiological evaluation: Price *et al., ibid.* **25**, 709 (1972).
Pharmacokinetics: Cabana, Taggart, *Antimicrob. Ag. Chemother.* **3**, 478 (1973). *In vitro* studies: Yu, Washington,
ibid. **4**, 133 (1973); Bodey, Stewart, *ibid.* 186. Pharmacology
in humans: Bodey *et al., ibid.* **5**, 508 (1974). Toxicity stud-
ies: Fujisawa *et al., J. Antibiot.* **27**, 677 (1974). *Review:* K.
A. Kerridge in *Pharmacological and Biochemical Properties of
Drug Substances* vol. 1, M. E. Goldberg, Ed. (Am. Pharm.
Assoc., Washington, DC, 1977) pp 125-153. Comprehensive description: P. M. Monteleone *et al.* in *Analytical Profiles of Drug Substances* vol. **12**, K. Florey, Ed. (Academic
Press, New York, 1983) pp 37-71.

White crystalline powder from methanol-isopropanol, mp
203-204° (sesquihydrate). $[\alpha]_D^{23}$ +99° (c = 1.0 in water).
LD_{50} in mice of solns pH 6.6, pH 7.4 (mg/kg): 340, 560 i.v.
(Kawaguchi).
Sulfate, $C_{22}H_{47}N_5O_{21}S_2$, *Amikin, Amiklin, BB-K8, Biklin,
Fabianol, Kaminax, Mikavir, Novamin, Pierami.* Amor-
phous form, dec 220-230°. $[\alpha]_D^{22}$ +74.75° (water).
THERAP CAT: Antibacterial.

417. Amiloride. *3,5-Diamino-N-(aminoiminomethyl)-
6-chloropyrazinecarboxamide; N-amidino-3,5-diamino-6-
chloropyrazinecarboxamide; N-amidino-3,5-diamino-6-
chloropyrazinamide;* 1-(3,5-diamino-6-chloropyrazinecarb-
oxyl)guanidine; 1-(3,5-diamino-6-chloropyrazinoyl)guani-
dine; guanamprazine; amipramidin; amipramizide. C_6H_8-
ClN_7O; mol wt 229.65. C 31.38%, H 3.51%, Cl 15.44%, N
42.70%, O 6.97%. Prepn: Cragoe, **Belg. pat. 639,386** (1964
to Merck & Co.), *C.A.* **62**, 14698f (1965). NMR study on
tautomerism and conformation: R. L. Smith *et al., J. Am.
Chem. Soc.* **101**, 191 (1979). Pharmacology: Baer *et al., J.
Pharmacol. Exp. Ther.* **157**, 472 (1967); Baba *et al., Clin.
Pharmacol. Ther.* **9**, 318 (1968); Lant *et al., ibid.* **10**, 50
(1969). Metabolism: Weiss *et al., ibid.* 401. *Reviews:* H. L.
Macfie *et al., Drug Intell. Clin. Pharmacol.* **15**, 94-98 (1981);
D. E. Hyams, *Int. Congr. Symp. Ser.-R. Soc. Med.* **44**, 65-73

(1981). Comprehensive description of hydrochloride: D. J.
Mazzo in *Analytical Profiles of Drug Substances,* vol. **15**, K.
Florey, Ed. (Academic Press, New York, 1986) pp 1-34.

Solid, mp 240.5-241.5°.
Hydrochloride dihydrate, $C_6H_9Cl_2N_7O.2H_2O$, *MK-870,
Amikal, Arumil, Colectril, Midamor, Modamide.* Crystalline
solid, dec 285-288°. Anhydr, mp 293.5°. uv max (water):
212, 285, 362 nm ($E_{1cm}^{1\%}$ 642, 555, 617). pKa 8.7. Freely sol
in DMSO, slightly sol in water, isopropanol, ethanol. Practically insol in acetone, chloroform, diethyl ether, ethyl
acetate.
Mixture with hydrochlorothiazide, *co-amilozide, Amilco,
Aquaretic, Ecodurex, Grodurex, Hexarese, Hydrocomp, Moduretic, Moduretik, Normetic, Normorix.*
THERAP CAT: Potassium-sparing diuretic.

418. Aminacrine. *9-Acridinamine;* 5-aminoacridine;
9-aminoacridine. $C_{13}H_{10}N_2$; mol wt 194.23. C 80.39%, H
5.19%, N 14.42%. Prepd from 9-chloroacridine which is
obtained by heating *N*-phenylanthranilic acid with PCl_5: A.
Albert, B. Ritchie, *Org. Syn.* **coll. vol. III,** 53 (1955).

Sulfur-yellow needles from alcohol or acetone, mp 241°.
Moderately strong base. K (25°): 3×10^{-5}. Freely sol in
alcohol; slightly sol in chloroform, toluene, pyridine; sol in
acetone.
Hydrochloride, $C_{13}H_{11}ClN_2$, *Monacrin, Acramine Yellow.*
Pale yellow crystals. Neutral reaction. One of the most
highly fluorescent substances. One gram dissolves in 300 ml
water giving a faintly yellow soln showing bluish-violet fluo-
rescence. LD_{50} orally in mice: 78 mg/kg, D. C. Brodie, E.
Lowenhaupt, *J. Am. Pharm. Assoc.* **38**, 498 (1949).
THERAP CAT: Anti-infective.
THERAP CAT (VET): Local antiseptic.

419. Amine 220®. *2-(8-Heptadecenyl)-4,5-dihydro-
1H-imidazole-1-ethanol; 2-(8-heptadecenyl)-2-imidazoline-
1-ethanol;* 1-(2-hydroxyethyl)-2-(8-heptadecenyl)-2-imid-
azoline. $C_{22}H_{42}N_2O$; mol wt 350.57. C 75.37%, H 12.08%,
N 7.99%, O 4.56%. Prepared from oleic acid and 2-[(2-ami-
noethyl)amino]ethanol: Stromberg, Hughes, and Hughes,
Stromberg, U.S. pats. **2,987,515** and **3,020,276** (1961 and
1962, both to Petrolite Corp.).

USE: Fungicide, emulsifier, soil stabilizer.

420. Amineptine. *7-[(10,11-Dihydro-5H-dibenzo[a,d]-
cyclohepten-5-yl)amino]heptanoic acid.* $C_{22}H_{27}NO_2$; mol wt
337.47. C 78.30%, H 8.07%, N 4.15%, O 9.48%. Prepn: C.
Malen *et al.,* **Ger. pat. 2,011,806** corresp to U.S. pats. **3,-
758,528** and **3,821,249** (1970, 1973 and 1974, all to Sci.
Union et Cie Soc. Franc. Recher. Med.). Kinetics: C.
Sbarra *et al., J. Chromatog.* **162**, 31 (1970). Biochemical and
pharmacological study: R. Samanin *et al., J. Pharm. Pharmacol.* **29**, 555 (1977). Pharmacodynamics: J. Dankova *et
al., Eur. J. Pharmacol.* **42**, 113 (1977); J. C. Poignant, A.
Avril, *Arzneimittel-Forsch.* **28**, 267 (1978). Clinical study:
P. Van Amerongden, *Encephale* **4**, 131 (1978).

Hydrochloride, $C_{22}H_{28}ClNO_2$, S-1694, Maneon, Survector. Crystals from distilled water, mp 226-230°.

THERAP CAT: CNS stimulant.

421. Aminitrozole. N-(5-Nitro-2-thiazolyl)acetamide; 2-acetamido-5-nitrothiazole; 2-acetylamino-5-nitrothiazole; acinitrazole; Tritheon; Trichorad; Trichoral; Gynofon; Enheptin-A; Pleocide. $C_5H_5N_3O_3S$; mol wt 187.19. C 32.08%, H 2.69%, N 22.45%, S 17.13%, O 25.64%. Prepd by nitration of 2-acetamidothiazole: Ganapathi, Venkataraman, Proc. Indian Acad. Sci. **22A**, 343 (1945); Bellavita, Ann. Chim. Applicata **38**, 449 (1948); Hurd, Wehrmeister, J. Am. Chem. Soc. **71**, 4007 (1949); Yamamoto, J. Pharm. Soc. Japan **72**, 1017 (1952). As an antihistomonad: Waletzky, Marson, U.S. pat. **2,531,756** (1950 to Am. Cyanamid).

Needles from alc, elongated plates from acetic acid. mp 264-265°. The commercial product may be yellow. Sol in aq solns of NaOH and NH₃ with deep orange color.

THERAP CAT: Antiprotozoal (Trichomonas).

THERAP CAT (VET): Antihistomonad for turkeys.

422. p-Aminoacetanilide. N-(4-Aminophenyl)acetamide; 4'-aminoacetanilide; acetyl-p-phenylenediamine. $C_8H_{10}N_2O$; mol wt 150.18. C 63.98%, H 6.71%, N 18.66%, O 10.65%. Prepd by catalytic hydrogenation of 4'-nitroacetanilide: Atkinson et al., J. Chem. Soc. **1954**, 2023. Crystal and molecular structure: M. Haisa et al., Acta Crystallogr. **33B**, 2449 (1977).

White or slightly reddish crystals, mp 163.5-166.0°; darkens in air. Slightly soluble in cold water; freely soluble in hot water, alc, ether.

USE: Intermediate in the manufacture of azo dyes, pharmaceuticals.

423. Aminoacetonitrile. Glycinonitrile; cyanomethylamine; glycine nitrile. $C_2H_4N_2$; mol wt 56.07. C 42.84%, H 7.19%, N 49.97%. H_2NCH_2CN. Prepd by the action of alcoholic HCl or H_2SO_4 on dimolecular or trimolecular methyleneaminoacetonitrile: Anslow, King, J. Chem. Soc. **1929**, 2465.

Oily liquid. bp_{15} 58° (partial decompn).

Hydrochloride, $C_2H_4N_2$.HCl, hygroscopic crystals from alcohol, dec 165°.

Sulfates: $C_2H_4N_2$.H_2SO_4, mp 101° and $(C_2H_4N_2)_2$.H_2SO_4, dec 165°. Leaflets from dil alcohol, very sol in water, slightly sol in alcohol, insol in ether.

424. Aminoacetophenone. C_8H_9NO; mol wt 135.16. C 71.09%, H 6.71%, N 10.36%, O 11.84%. Prepn of m-, o-, and p-isomers: Grammaticakis, Compt. Rend. **235**, 516 (1952); Braude et al., J. Chem. Soc. **1954**, 3586. Prepn of m-isomer: Tinsley, U.S. pat. **2,797,244** (1957 to Union Carbide); of p-isomer: Norman et al., Can. J. Chem. **40**, 1547 (1962).

m-Aminoacetophenone, 1-(3-aminophenyl)ethanone,

3'-aminoacetophenone, m-aminoacetylbenzene. Yellow leaflets, mp 98-99°. Partly volatile in steam. LD_{50} orally in rats: 1.87 g/kg, H. F. Smyth et al., Arch. Ind. Hyg. Occup. Med. **10**, 61 (1954).

o-Aminoacetophenone, yellow oily liquid. bp_{760} 250-252° (some dec); bp_{17} 135°. Volatile with steam. Practically insol in water; sol in alc.

p-Aminoacetophenone, yellow needles, pleasant, characteristic odor. mp 106°. bp 293-295°. Sparingly sol in cold, freely in hot water; sol in alc, ether, HCl; sparingly sol in benzene. LD_{50} i.p. in rats: 260 mg/kg, J. M. Vandenbelt et al., J. Pharmacol. Exp. Ther. **80**, 31 (1944).

425. D-Amino Acid Oxidase. DAAO. Flavoprotein that catalyzes the oxidative deamination of D-amino acids to the corresponding α-keto acids. Found in the kidney and liver of nearly all mammals studied. Isoln from pig kidney: Krebs, Enzymologia **7**, 53 (1939); Straub, Nature **141**, 603 (1938); Negelein, Brömel, Biochem. Z. **300**, 225 (1939); Massey et al., Biochim. Biophys. Acta **48**, 1 (1961); from sheep kidney: Burton, Methods Enzymol. **2**, 199 (1955). Physical properties: Yagi et al., J. Biochem. **61**, 580 (1967). Partial structure: Kotaki et al., ibid. 598. Studies on specificity and inhibition: Dixon, Kleppe, Biochim. Biophys. Acta **96**, 368 (1965). Reaction intermediates: K. Yagi, Front. Physiochem. Biol., Proc. Int. Symp. **1977**, B. Pullman, Ed. (Academic Press, New York, 1978) pp 299-308. Review: Meister, Wellner in The Enzymes **7**, P. Boyer et al., Eds. (Academic Press, New York, 1963) pp 634-648.

426. L-Amino Acid Oxidase. LAAO. Flavoprotein that catalyzes the oxidative deamination of L-amino acids to the corresponding α-keto acids. Found in microorganisms and in animal tissue, esp in kidney and liver. Occurs also in many snake venoms. Isoln from rat kidney: Blanchard et al., J. Biol. Chem. **161**, 583 (1945). Isoln of crystalline enzyme from the venom of the eastern diamondback rattlesnake (Crotalus adamanteus): Wellner, Meister, J. Biol. Chem. **235**, 2013 (1960). Structural studies: DeKok, Rawitch, Biochem. **8**, 1405 (1969). Studies on inhibitors: DeKok, Veeger, Biochim. Biophys. Acta **167**, 35 (1968). Mechanism of reversible activation-deactivation: C. J. Coles et al., Flavins Flavoproteins, Proc. Int. Symp. 6th **1978**, K. Yagi, T. Yamano, Eds. (Japan Sci. Soc. Press, Tokyo, 1980) pp 101-105. Multiple conformational states: D. Wellner, L. A. Lichtenberg, Dev. Biochem. **21**, 78 (1982). Review: Meister, Wellner in The Enzymes **7**, P. Boyer et al., Eds. (Academic Press, New York, 1963) pp 609-634.

427. α-Aminoadipic Acid. 2-Aminohexanedioic acid. $C_6H_{11}NO_4$; mol wt 161.16. C 44.71%, H 6.88%, N 8.69%, O 39.71%. $HO_2CCH(NH_2)CH_2CH_2CH_2CO_2H$. An amino acid isolated from Cholera vibrio: Blass, Macheboeuf, Helv. Chim. Acta **29**, 1315 (1946). Occurrence in proteins or protein containing material: Windsor, J. Biol. Chem. **192**, 595 (1951). Synthesis: Dieckmann, Ber. **38**, 1656 (1905); T. P. Waalkes et al., J. Am. Chem. Soc. **72**, 5760 (1952); A. I. Scott, T. J. Wilkinson, Synth. Commun. **10**, 127 (1980).

Forms a monohydrate if crystallized from water below 20°. Anhydrous platelets from water above 20°. mp 206° (effervescence). One gram dissolves in 450 ml water. Sparingly sol in alcohol, ether.

Diethyl ester, $C_{10}H_{19}NO_4$, bp_{13} 155-156°.

428. 1-Aminoanthraquinone. 1-Amino-9,10-anthracenedione. $C_{14}H_9NO_2$; mol wt 223.22. C 75.32%, H 4.06%, N 6.28%, O 14.33%. Prepd by reduction of 1-nitroanthraquinone: Graham, Hort, U.S. pat. **2,874,168** (1959 to General Aniline & Film).

Ruby-red crystals. mp about 250°; also stated as 243°.

Practically insoluble in water; freely soluble in alcohol, benzene, chloroform, ether, glacial acetic acid, HCl.

USE: In the mfg of dyes and pharmaceuticals.

429. 1-Aminoanthraquinone-2-carboxylic Acid. *1-Amino-9,10-dihydro-9,10-dioxo-2-anthracenecarboxylic acid.* $C_{15}H_9NO_4$; mol wt 267.23. C 67.41%, H 3.39%, N 5.24%, O 23.97%. Prepd by reduction of 1-nitroanthraquinone-2-carboxylic acid with sodium sulfide: Terres, *Ber.* **46**, 1639 (1913); by boiling 1-nitro-2-methylanthraquinone with alcoholic potassium hydroxide: Scholl, *Monatsh.* **34**, 1011 (1913); by heating 1-nitro-2-bromomethylanthraquinone with sodium acetate and *o*-dichlorobenzene: Locher, Fierz, *Helv. Chim. Acta* **10**, 667 (1927).

Red needles from nitrobenzene, mp 295-296°. Soluble in aniline, boiling nitrobenzene, in sodium hydroxide and aq pyridine with formation of a deep red soln; in concd sulfuric acid with formation of a yellowish-brown soln; slightly sol in ether, benzene, alcohol. Insol in water, ligroin.

Methyl ester, $C_{16}H_{11}NO_4$, red needles from glacial acetic acid, mp 228°.

Ethyl ester, $C_{17}H_{13}NO_4$, mp 198°.

Phenyl ester, $C_{21}H_{13}NO_4$, reddish-golden crystals from glacial acetic acid, mp 198°.

USE: For the detection of traces of aluminum, magnesium, or zinc.

430. *p*-Aminoazobenzene. *4-(Phenylazo)benzenamine; C.I. Solvent Yellow 1; p*-(phenylazo)aniline; *p*-aminodiphenylimide; aniline yellow; C.I. 11000. $C_{12}H_{11}N_3$; mol wt 197.23. C 73.07%, H 5.62%, N 21.31%. Prepd from aniline, $NaNO_2$, and HCl: A. I. Vogel, *A Text-Book of Practical Organic Chemistry* (Longmans, Green & Co., New York, 3rd ed., 1956) p 627; W. R. Hydro, T. L. Willard, U.S. pat. **2,894,942** (1959 to Goodrich); *Colour Index* vol. **4** (3rd ed., 1971) p 4014.

Brownish-yellow needles with a bluish cast. mp 128°. bp above 360°. Slightly sol in water; freely sol in alc, benzene, chloroform, ether. It reduces alcoholic ammoniacal $AgNO_3$.

USE: In form of its salts in dyeing; intermediate in manuf of Acid Yellow, diazo dyes and indulines.

431. *o*-Aminoazotoluene. *2-Methyl-4-[(2-methylphenyl)azo]benzenamine; C.I. Solvent Yellow 3;* 4-(*o*-tolylazo)-*o*-toluidine; 4'-amino-2,3'-dimethylazobenzene; 5-(*o*-tolylazo)-2-aminotoluene; toluazotoluidine; C.I. 11160. $C_{14}H_{15}N_3$; mol wt 225.28. C 74.64%, H 6.71%, N 18.65%. Prepd from *o*-toluidine, $NaNO_2$, and HCl: Shulman, U.S. pat. **2,538,431** (1951 to Pfister Chem. Works); *Colour Index* vol. **4** (3rd ed., 1971) p 4017. Crystal and molecular structure: S. Kurosaki *et al.*, *Acta Crystallogr.* **32B**, 3160 (1976). Carcinogenic activity: Maini, Stich, *J. Nat. Cancer Inst.* **26**, 1413 (1961).

Golden crystals, mp 101-102°. Practically insoluble in water; sol in alc, ether, chloroform. Absorption max (50% alcoholic 1N HCl): 326, 490 nm (ϵ 19,000, 2,500), Sawicki, *J. Org. Chem.* **21**, 605 (1956).

432. *m*-Aminobenzoic Acid. *3-Aminobenzoic acid.* $C_7H_7NO_2$; mol wt 137.13. C 61.31%, H 5.15%, N 10.21%, O

23.33%. Prepn: Toland, Heaton, U.S. pat. **2,878,281** (1959 to California Res. Corp.); Neilson *et al.*, *J. Chem. Soc.* **1962**, 371.

Cryst, mp 174°. d 1.51. Aq soln turns brown on standing in air. Forms soluble salts with mineral acids; slightly sol in cold, freely in boiling water or alcohol; sol in ether.

433. *o*-Aminobenzoic Acid. *2-Aminobenzoic acid; anthranilic acid.* Empirical formula: *See m*-isomer above. Found to be the same as *vitamin* L_1: Nakahara *et al.*, *Sci. Papers Inst. Phys. Chem. Res.* (Tokyo) **42**, 39 (1945), *C.A.* **41**, 6317d (1947). *See also* Vitamins L. Prepd by reduction of *o*-nitrobenzoic acid: Neilson *et al.*, *J. Chem. Soc.* **1962**, 371. Purification: Sugihara, Newman, *J. Org. Chem.* **21**, 1445 (1956).

White to pale yellow, cryst powder; sweetish taste. mp 144-146°. Sparingly sol in cold, freely in hot water, alcohol, ether. The solns in alcohol or ether and particularly in glycerol exhibit an amethyst fluorescence.

THERAP CAT (VET): The cadmium salt has been used as an ascaricide in swine.

434. *p*-Aminobenzoic Acid. *4-Aminobenzoic acid;* vitamin B_x; bacterial vitamin H^1; chromotrichia factor; antichromotrichia factor; trichochromogenic factor; anticanitic vitamin; PABA; Amben; Paraminol; Sunbrella. $C_7H_7NO_2$; mol wt 137.13. C 61.31%, H 5.15%, N 10.21%, O 23.33%. Widely distributed in nature as a B complex factor. Baker's yeast contains 5 to 6 ppm, brewer's yeast from 10 to 100 ppm. Occurs free and in ester form. Prepn: Toland, Heaton, U.S. pat. **2,878,281** (1959 to Calif. Research Corp.); Spiegler, U.S. pat. **2,947,781** (1960 to du Pont); Nielson *et al.*, *J. Chem. Soc.* **1962**, 371. Purification: Lyding, U.S. pat. **2,735,865** (1956 to Heyden Chem.). Antirickettsial activity: M. L. Robbins *et al.*, *J. Immunol.* **64**, 431 (1950). Toxicity studies: C. C. Scott, E. B. Robbins, *Proc. Soc. Exp. Biol. Med.* **49**, 184 (1942); R. K. Richards, *Fed. Proc.* **1**, 71 (1942); G. Cronheim, *ibid.* **10**, 289 (1952).

Monoclinic prisms from dil alcohol. May turn slightly yellow on prolonged exposure to light and air. mp 187.0°-187.5°. pKa: 4.65, 4.80; pH (0.5% soln): 3.5. uv max (water): 266 nm ($E_{1cm}^{1\%}$ 1070); (isopropanol): 288 nm ($E_{1cm}^{1\%}$ 137). One gram dissolves in 170 ml water at 25°, in 90 ml boiling water; in 8 ml alcohol, in 50 ml ether. Sol in ethyl acetate, glacial acetic acid; slightly sol in benzene; practically insol in petr ether. Incompatible with ferric salts and oxidizing agents. LD_{50} in mice, rats (g/kg): 2.85, >6.0 orally (Scott, Robbins). LD_{50} in rabbits (g/kg): 2.0 i.v. (Richards); 1.83 orally (Cronheim).

Ingredient in *Pabanol, PreSun.*

Diethylamine salt, $C_{11}H_{18}N_2O_2$, *Navanide*. Crystals from acetone, mp 170-173°. Very soluble in water.

Potassium salt, *see* Potassium *p*-Aminobenzoate

Note: *p*-Aminobenzoic acid is also known as *anti-gray-hair factor* (in rats only).

USE: Manuf various esters (local anesthetics), folic acid, and azo dyes; in sunburn preventives. Used in laboratories as sulfonamide antagonist.

THERAP CAT: Ultraviolet screen. Formerly as antirickettsial.

THERAP CAT (VET): Has been used in *eczema nasi* ("collie nose") in dogs.

435. 2-Aminobenzothiazole. *2-Benzothiazolamine.* $C_7H_6N_2S$; mol wt 150.20. C 55.97%, H 4.03%, N 18.65%, S 21.35%. Prepd from 2-chlorobenzothiazole by treatment with alcoholic ammonia at 150-160°: Hofmann, *Ber.* **12**, 1129 (1880); **13**, 11 (1881). From benzothiazole by boiling with hydroxylamine in water or in 2N NaOH: Skraup, *Ann.* **419**, 65 (1919).

Leaflets from water, mp 132°. Distills without decompn. Very sparingly sol in water; freely in alcohol, ether, chloroform; sol in concd acids. LD_{50} i.v. in mice: 126 mg/kg, E. F. Domino *et al.*, *J. Pharmacol. Exp. Ther.* **105**, 486 (1952).
USE: In the prepn of azo dyes.

436. 6-Aminobenzothiazole. *6-Benzothiazolamine.* $C_7H_6N_2S$; mol wt 150.20. C 55.97%, H 4.03%, N 18.65%, S 21.35%. Prepd from 2,5-diaminothiophenol by boiling with concd formic acid or from 6-nitrobenzothiazole by reduction with tin and HCl: Mylius, *Thesis* (Berlin, 1883).

Prisms from water, mp 87°. Sol in alcohol. Practically insol in water and ether.
USE: In the prepn of azo dyes.

437. N-(p-Aminobenzoyl)glutamic Acid. *N-(4-Aminobenzoyl)-L-glutamic acid;* $C_{12}H_{14}N_2O_5$; mol wt 266.25. C 54.13%, H 5.30%, N 10.52%, O 30.05%. Prepd by catalytic hydrogenation (Pd charcoal in AcOH) of the corresponding p-nitrobenzylglutamic acid: Auhagen, *Z. Physiol. Chem.* **277**, 203 (1943). *See also* Landsteiner, Scheer, *J. Exp. Med.* **55**, 781 (1932); *J. Immunol.* **29**, 371 (1935); Carter, Stevens, *J. Biol. Chem.* **138**, 627 (1941). Inhibition of dihydrofolate reductase: G. C. K. Roberts *et al.*, *Biochemistry* **13**, 5351 (1974).

L-Form, crystals from water, mp 173°. $[\alpha]_D^{20}$ −16° (c = 2 in 0.1N HCl).
D-Form, crystals from water, mp 172°. $[\alpha]_D^{20}$ +16.7° (c = 2 in 0.1N HCl).
DL-Form, crystals from water, mp 197°.

438. 2-Amino-1-butanol. 2-Amino-n-butyl alcohol. $C_4H_{11}NO$; mol wt 89.14. C 53.89%, H 12.44%, N 15.72%, O 17.95%. $CH_3CH_2CH(NH_2)CH_2OH$. Prepd from 2-nitro-1-butanol by reduction or catalytic hydrogenation: Stiénon, *Chem. Zentr.* **1902, I**, 717; Vanderbilt, Haas, U.S. pat. 2,174,242 (1940); Johnson, Degering, *J. Org. Chem.* **8**, 7 (1943).
Liquid. d_{20}^{20} 0.944. mp −2°. bp_{760} 178°; bp_{10} 79-80°. n_D^{20} 1.453. Miscible with water. Sol in alcohols. pH of 0.1 molar aq soln 11.1.
Hydrochloride, $C_4H_{11}NO.HCl$, deliquesc needles.
Platinichloride monohydrate, $(C_4H_{11}NO)_2.2HCl.PtCl_4.H_2O$, yellow leaflets, dec 189-190°. Moderately sol in water, freely sol in alcohol.
Oxalate, $(C_4H_{11}NO)_2.C_2H_2O_4$, mp 176°.
USE: In the synthesis of surface-active agents, vulcanization accelerators, pharmaceuticals. As emulsifying agent for cosmetic creams and lotions; mineral oil and paraffin wax emulsions, leather dressings, textile specialties, polishes, cleaning compounds, so-called soluble oils.

439. α-Aminobutyric Acid. *2-Aminobutanoic acid;* α-amino-n-butyric acid. $C_4H_9NO_2$; mol wt 103.12. C 46.59%, H 8.80%, N 13.58%, O 31.03%. $CH_3CH_2CH(NH_2)$-

COOH. Prepn of the DL-form from α-bromobutyric acid and ammonia: Fisher, Mounegrat, *Ber.* **33**, 2388 (1900); from potassium cyanide, ammonium chloride and propionaldehyde: Zelinsky, Stadnikow, *Ber.* **41**, 2062 (1908); by reduction of α-oxobutyric acid: Knoop, Oesterlin, *Z. Physiol. Chem.* **148**, 305 (1925); from α-ethylacetoacetic ester and hydrazoic acid: Schmidt, *Ber.* **57**, 706 (1924). The L(+)-form has been isolated from proteins: Oikawa, *Chem. Zentr.* **1926**, 1, 148. Configuration: Clough, *J. Chem. Soc.* **113**, 544, 551; Levene, *Chem. Rev.* **2**, 203 (1926); Vogler, *Helv. Chim. Acta* **30**, 1766 (1947).
DL-Form, crystals, mp 304° (begins to sublime when heated above 300°). Sol in water. One liter of water will dissolve 210.5 g at 25°. Sparingly sol in alcohol. One liter of boiling ethanol dissolves about 1.8 g. Insol in ether.
DL-Form ethyl ester, viscous liquid. bp_{11} 61°. Sol in water and in organic solvents.
L-Form, leaflets from dil alc. Sweet taste. mp 270-280° (depending on speed of heating, *see* Vogler, *loc. cit.*). $[M]_D$ +21.2° (5N HCl); $[M]_D$ +43.3° (glacial acetic acid). $[\alpha]_D^{16}$ +8.40° (c = 4); $[\alpha]_D^{16}$ +18.65° (c = 4.8 in 6N HCl).
L-Form hydrochloride, needles; $[\alpha]_D^{19}$ +12.90° (c = 3.64). Readily sol in water.

440. β-Aminobutyric Acid. *3-Aminobutanoic acid;* β-amino-n-butyric acid. $C_4H_9NO_2$; mol wt 103.12. C 46.59%, H 8.80%, N 13.58%, O 31.03%. $CH_3CH(NH_2)CH_2$-COOH. Prepd from pyrotartaric acid diamide: Weidel, Roithmer, *Monatsh.* **17**, 185 (1896); from β-chlorobutyric acid ethyl ester and alcoholic NH_3: Balbiano, *Ber.* **13**, 312 (1880); from crotonic acid and concd NH_3: Engel, *Bull. Soc. Chim.* [2] **50**, 102 (1888); Curtius, Gumlich, *J. Prakt. Chem.* [2] **70**, 204 (1904); Stadnikow, *Chem. Zentr.* **1909, II**, 1988; *see* Fischer, Roeder, *Ber.* **34**, 3755 (1901) footnote; Stoermer, Robert, *Ber.* **55**, 1038 (1922). Prepn of HCl salt from β-aminobutyronitrile: Bruylants, *Bull. Soc. Chim. Belg.* **32**, 259 (1923); *Chem. Zentr.* **1924, I**, 1668. By reducing acetoacetic ester phenylhydrazone and saponifying the resulting ester: Fischer, Groh, *Ann.* **383**, 338 (1911). The D(−)-form has been obtained by hydrolysis of its ester: Fischer, Scheibler, *ibid.* 346.
DL-Form, crystals from alcohol, mp 193-094°. Practically tasteless. Sol in water. One liter of water dissolves 1250 g. Insol in cold absolute alcohol and ether.
Methyl ester, $C_5H_{11}NO_2$, odoriferous liquid; d^{20} 0.993; bp_{13} 54-55°. Sol in water, alcohol, ether, petr ether.
D-Form, prisms from methanol. Dec near 220° without melting. $[\alpha]_D^{20}$ −35.20° (p = 10).

441. γ-Aminobutyric Acid. *4-Aminobutanoic acid;* γ-amino-n-butyric acid; piperidic acid; GABA; Gamarex; Gammalon. $C_4H_9NO_2$; mol wt 103.12. C 46.59%, H 8.80%, N 13.58%, O 31.03%. $H_2NCH_2CH_2CH_2COOH$. Nonprotein amino acid that functions as a neurotransmitter. Prepn from succinimide: Tafel, Stern, *Ber.* **33**, 2224 (1900); from piperylurethan and fuming nitric acid: Schotten, *Ber.* **16**, 643 (1883); Abderhalden, *Chem. Zentr.* **1926, II**, 779; from N-(β-bromoethyl)phthalimide and sodiomalonic ester: Aschan, *Ber.* **24**, 2450 (1891); from γ-chlorobutyronitrile and potassium phthalimide: Gabriel, *Ber.* **22**, 3335 (1889); **23**, 1771 (1890); DeWitt, *Org. Syn.* **coll. vol. II**, 25 (1943). Review of biochemical pharmacology: R. Tapia in *Handbook of Psychopharmacology*, L. L. Iversen *et al.*, Eds. (Plenum, New York, 1975) pp 1-58; L. L. Iversen in *Psychopharmacology: A Generation of Progress*, M. A. Lipton *et al.*, Eds. (Raven, New York, 1978) pp 25-38; C. C. Mao, E. Costa, *ibid.* pp 307-318.
Leaflets from methanol + ether, needles from water + alcohol, mp 202° (dec on rapid heatng). Ka 3.7×10^{-11}; Kb 1.7×10^{-10} at 25°. Freely sol in water; insol or poorly sol in other solvents. On melting it dec forming pyrrolidone and water.
Hydrochloride, $C_4H_9NO_2.HCl$, crystals, mp 135-136°.
Ethyl ester, $C_6H_{13}NO_2$, liquid, bp_{12} 76°.
THERAP CAT: Antihypertensive.

442. ε-Aminocaproic Acid. *6-Aminohexanoic acid;* epsilon-aminocaproic acid; epsilcapramin; CY 116; Ipsilon; Hemocaprol; Capralense; Caprocid; Capramol; Amicar; EACA Kabi; EACS; Afibrin; Epsikapron; Epsamon; Hepin. $C_6H_{13}NO_2$; mol wt 131.17. C 54.94%, H 9.99%, N 10.68%,

O 24.40%. H$_2$N(CH$_2$)$_5$COOH. Prepn from ε-benzoylaminocapronitrile: Ruzicka, Hugoson, *Helv. Chim. Acta* **4**, 472 (1921); from ε-caprolactam: Galat, Mallin, *J. Am. Chem. Soc.* **68**, 2729 (1946); Koch, U.S. pat. **2,453,234** (1948 to Am. Enka); J. C. Eck, *Org. Syn.* coll. **vol. II**, 28 (1943); C. Y. Meyers, L. E. Miller, *ibid.* coll. **vol. IV**, 39 (1963); Garmaise *et al.*, *Can. J. Chem.* **34**, 743 (1956); Sulzbacher, *Mfg. Chemist* **33**, 463 (1963); from 1-hydroxycyclohexylhydroperoxide: Minisci, Portolani, *Gazz. Chim. Ital.* **89**, 1941 (1959). Metabolic studies in man: McNicol *et al.*, *J. Lab. Clin. Med.* **59**, 15 (1962).

Crystals from alcohol, mp 204-206°. pK$_1$ 4.43; pK$_2$ 10.75. Freely sol in water; sparingly in methanol. Practically insol in ethanol. LD$_{50}$ in rats (g/kg): 7.0 i.p.; ~3.3 i.v., D. W. Hallesy *et al.*, *Pharmacologist* **3**, 62 (1961).

Hydrobromide, C$_6$H$_{13}$NO$_2$·HBr, crystalline powder from alcohol + ether, mp 105°.

Hydrochloride, C$_6$H$_{13}$NO$_2$·HCl, mp 128-129°.

THERAP CAT: Hemostatic.

443. Aminocarb. *4-(Dimethylamino)-3-methylphenol methylcarbamate (ester); methylcarbamic acid 4-(dimethylamino)-m-tolyl ester;* A 363; Bay 44646; ENT 25784; Matacil. C$_{11}$H$_{16}$N$_2$O$_2$; mol wt 208.26. C 63.44%, H 7.74%, N 13.45%, O 15.37%. Prepn: **Brit. pat. 913,439**, *C.A.* **58**, 10122e (1963) and R. Heiss *et al.*, **Ger. pat. 1,145,162**, *C.A.* **59**, 9885b (1963) (1962 and 1963 to Bayer). Metabolism: A. Strother, *Toxicol. Appl. Pharmacol.* **21**, 112 (1972). Photochemistry: J. B. Addison *et al.*, *Bull. Environ. Contam. Toxicol.* **11**, 250 (1974); J. B. Addison, *ibid.* **27**, 250 (1981).

Crystals, mp 93-94°. uv max (ethanol): 248.5 nm (ε 6.67 × 10^4). Slightly sol in water; moderately sol in aromatic solvents; sol in most polar organic solvents. LD$_{50}$ orally in male, female rats: 40, 38 mg/kg, T. B. Gaines, *Toxicol. Appl. Pharmacol.* **14**, 515 (1969).

USE: Insecticide.

444. 7-Aminocephalosporanic Acid. *3-[(Acetyloxy)-methyl]-7-amino-8-oxo-5-thia-1-azabicyclo[4.2.0]oct-2-ene-2-carboxylic acid; 7-amino-3-(hydroxymethyl)-8-oxo-5-thia-1-azabicyclo[4.2.0]oct-2-ene-2-carboxylic acid acetate ester;* 3-acetoxymethyl-7-aminoceph-3-em-4-oic acid; 7-ACA. C$_{10}$H$_{12}$N$_2$O$_5$S; mol wt 272.30. C 44.11%, H 4.44%, N 10.29%, O 29.38%, S 11.78%. Starting material for semisynthetic cephalosporins. Obtained by mild acid hydrolysis of cephalosporin C (*q.v.*): Loder *et al.*, *Biochem. J.* **79**, 408 (1961); Morin *et al.*, *J. Am. Chem. Soc.* **84**, 3400 (1962); Morin *et al.*, **Belg. pat. 615,955** (1962 to Lilly), *C.A.* **58**, 11373c (1963); by enzymatic hydrolysis of cephalosporin C: Walton, U.S. pat. **3,239,394** (1966 to Merck & Co.). Improved prepn: Fechtig *et al.*, *Helv. Chim. Acta* **51**, 1108 (1968). Review of preparative methods: Huber *et al.*, in *Cephalosporins and Penicillins*, E. H. Flynn, Ed. (Academic Press, New York, 1972) pp 27-73.

Crystals. Isoelectric pt pH 3.5. R$_f$ 0.14 in 1-butanol-ethanol-water (4:1:5 by vol).

445. 4-Amino-4'-chlorodiphenyl. *4'-Chloro[1,1'-biphenyl]-4-amine;* 4'-chloroxenylamine; 4'-chloro-4-aminodiphenyl; *p'-chloro-p-phenylaniline; p-amino-p'-chlorobiphenyl.* C$_{12}$H$_{10}$ClN; mol wt 203.67. C 70.76%, H 4.95%, Cl 17.41%, N 6.88%. Prepd from 4'-amino-4-biphenyldiazonium chloride by Sandmeyer's reaction: Gelmo, *Ber.* **39**, 4176 (1906); by chlorination of 4-nitrodiphenyl and reduc-

tion of 4-chloro-4'-nitrodiphenyl with iron and HCl in ethanol: Belcher *et al.*, *J. Chem. Soc.* **1953**, 1334.

Crystals from light petroleum, mp 134° (Gelmo); mp 128° (Belcher). uv max (0.1N HCl): 254 nm (ε 22,090). Sol in warm alcohol, ether, benzene, acetone, glacial acetic acid. Practically insol in water and alkalies.

USE: As a reagent in the determination of sulfur, in coal, rubber, etc.

446. Aminochlorthenoxazin. *6-Amino-2-(2-chloroethyl)-2,3-dihydro-4H-1,3-benzoxazin-4-one;* 2-(β-chloroethyl)-2,3-dihydro-4-oxo-6-amino-1,3-benzoxazine. C$_{10}$H$_{11}$ClN$_2$O$_2$; mol wt 226.68. C 52.99%, H 4.89%, Cl 15.64%, N 12.36%, O 14.12%. Prepn: Baroli *et al.*, *Arzneimittel-Forsch.* **13**, 884 (1963).

Crystalline powder, mp 164°. Practically insol in water. LD$_{50}$ orally in mice: 10,000 mg/kg.

Hydrochloride, C$_{10}$H$_{12}$Cl$_2$N$_2$O$_2$, A350, ICI 350. Crystals, mp 209-210°. LD$_{50}$ orally in mice: 2250 mg/kg. Principal ingredient of *Dereuma*.

THERAP CAT: Antipyretic, analgesic.

447. Aminochromes. Family of highly colored 2,3-dihydroindole-5,6-quinones obtained by oxidative cyclization of catecholamines: Sobotka, Austin, *J. Am. Chem. Soc.* **73**, 3077 (1951). Best represented by a zwitterionic structure with the substituents determined by the catecholamine used, e.g. for epinephrine as starting material, R = CH$_3$, R' = H, and R'' = OH. Reviews: Sobotka *et al.*, *Fortschr. Chem. Org. Naturst.* **14**, 217 (1957); Heacock, *Advan. Heterocycl. Chem.* **5**, 205 (1965); Heacock, Powell, *Progr. Med. Chem.* **9**, 275 (1972).

Numerous physiological activities, such as hallucinogenic, hemostatic, radioprotective, have been ascribed to aminochromes.

See also adrenochrome, adrenolutin, carbazochrome salicylate.

THERAP CAT: Hemostatic.

448. 2-Amino-4,6-dichlorophenol. 2,4-Dichloro-6-aminophenol; 4,6-dichloro-o-aminophenol. C$_6$H$_5$Cl$_2$NO; mol wt 178.02. C 40.48%, H 2.83%, Cl 39.83%, N 7.87%, O 8.99%. Prepd by reduction of the corresponding nitrophenol: F. Fischer, *Z. Chem.* [N.F.] **4**, 386 (1868); *Ann.*, suppl vol. **7**, 189 (1870); Katz, Cohen, *J. Org. Chem.* **19**, 758 (1954). Purification: J. Meyer, *Helv. Chim. Acta* **41**, 1890 (1958).

Long needles from carbon disulfide, warts from benzene, mp 95-96°. Sublimes (0.06 torr) 70-80° (bath temp). Freely sol in benzene, somewhat less in carbon disulfide, much less in petr ether. The stability of the free base (snow-white when pure) seems to be impaired by impurities.

Hydrochloride, $C_6H_5Cl_2NO.HCl$, crystals, dec 280-285°. Very stable when pure. The commercial product may be dark brown. Sol in water, alcohol. Precipitated from aq soln by the addition of concd HCl.

USE: Important azo-dye intermediate.

449. 2-Amino-1,2-diphenylethanol. *β-Amino-α-phenylbenzeneethanol;* 1,2-diphenyl-2-hydroxyethylamine; *sym*-diphenylethanolamine; *β*-amino-*α*-hydroxybibenzyl; *α*-hydroxy-*β*-aminodibenzyl; 2-hydroxy-1,2-diphenylethylamine; 1-hydroxy-2-aminodiphenylethane. $C_{14}H_{15}NO$; mol wt 213.27. C 78.84%, H 7.09%, N 6.57%, O 7.50%. C_6H_5-CH(NH$_2$)CH(OH)C$_6$H$_5$. The racemic *erythro*-form, the higher melting isomer, has been found effective in relieving pain due to pressure on nerves. *erythro*-Form prepd from benzylamine and benzaldehyde in aq alc NaOH: Söderbaum, *Ber.* **28**, 2522 (1895); Erlenmeyer, Jr., *Ber.* **30**, 1525 (1897); **32**, 2378 (1899); *Ann.* **307**, 114, 131 (1899); **337**, 212 (1904); by reduction of α-benzoin oxime with zinc dust in NaOH at 100°: Irvine, Fyfe, *J. Chem. Soc.* **105**, 1649 (1914); by catalytic reduction of β-benzyl monoxime: Rabe, *Ber.* **45**, 2166 (1912). *See also* Read, Steele, *J. Chem. Soc.* **1927**, 910; Ingersoll, *J. Am. Chem. Soc.* **50**, 2264 (1928); Read *et al., J. Chem. Soc.* **1929**, 2305. Absolute configuration of *erythro*- and *threo*-forms: Weijlard *et al., J. Am. Chem. Soc.* **73**, 1216 (1951); Lyle, Lacroix, *J. Org. Chem.* **28**, 900 (1963).

dl-erythro-Form, monoclinic prismatic needles from alcohol, mp 165°. Insol in water; sol in hot alcohol; slightly sol in ether.

dl-erythro-Form hydrochloride, $C_{14}H_{15}NO.HCl$, monoclinic leaflets or rods, softens at 210°, dec 234°. Freely sol in water.

dl-threo-Form, white crystals, mp 126-128°. Hydrochloride, mp 200-201°.

450. 1-[(2-Aminoethyl)amino]-2-propanol. *N-(2-Hydroxypropyl)ethylenediamine; Monolene.* $C_5H_{14}N_2O$; mol wt 118.18. C 50.81%, H 11.94%, N 23.71%, O 13.54%. CH$_3$CH(OH)CH$_2$NHCH$_2$CH$_2$NH$_2$. Prepn: Kitchen, Pollard, *J. Org. Chem.* **8**, 342 (1943).

Viscous liquid; mild ammoniacal odor. bp$_{3.0}$ 94°; bp$_{10.0}$ 112°. d$_4^{25}$ 0.9837. n$_D^{25}$ 1.4738.

Dihydrochloride, $C_5H_{16}Cl_2N_2O$, mp 184.7-185.0°.

USE: Rapid curing agent in the manuf of epoxy resins.

451. 2-Amino-2-ethyl-1,3-propanediol. $C_5H_{13}NO_2$; mol wt 119.16. C 50.39%, H 11.00%, N 11.76%, O 26.85%. An amino glycol prepd by reduction of catalytic hydrogenation of the corresp nitro compound: Vanderbilt, Hass, U.S. pat. **2,174,242** (1940); Johnson, Degering, *J. Org. Chem.* **8**, 7 (1943). Manuf: McMillan, U.S. pat. **2,485,982** (1949 to Comm. Solvents).

$$CH_2OHCCH_2OH$$

with NH$_2$ above and CH$_2$CH$_3$ below

Crystalline mass, mp 37.5-38.5°. (The commercial product may be a viscous liquid.) d$_{20}^{20}$ 1.099. n$_D^{20}$ 1.490. bp$_{10}$ 152-153°. Miscible with water. Sol in alcohols. pH of 0.1 molar aq soln 10.8.

USE: In the synthesis of surface-active agents, vulcanization accelerators, pharmaceuticals. As emulsifying agent for cosmetic creams and lotions, mineral oil and paraffin wax emulsions, leather dressings, textile specialties, polishes, cleaning compounds, so-called soluble oils. For absorbing CO_2 and H_2S from industrial gases.

452. Aminoglutethimide. *3-(4-Aminophenyl)-3-ethyl-2,6-piperidinedione; 2-(p-aminophenyl)-2-ethylglutarimide; 3-ethyl-3-(p-aminophenyl)-2,6-dioxopiperidine; Cytadren; Elipten; Orimeten.* $C_{13}H_{16}N_2O_2$; mol wt 232.27. C 67.22%, H 6.94%, N 12.06%, O 13.78%. Adrenocortical suppressant that also inhibits conversion of androgens to estrogens by the aromatase enzyme system. Prepn: Hoffmann, Urech, U.S. pat. **2,848,455** (1958 to Ciba). Metabolism: Douglas, Nicholls, *J. Pharm. Pharmacol. Suppl.* **17**, 115S (1965).

Mass spectrum: Ruecker, Bohn, *Arch. Pharm. (Weinheim)* **302**, 204 (1969). Resolution and abs config of antipodes: Finch *et al., Experientia* **31**, 1002 (1975). Review of its role as an inhibitor of adrenocortical steroidogenesis: Touitou *et al., Biomedicine* **18**, 185-191, 272-278 (1973). Clinical studies: Küchel, *Pharmacol. Clin.* **2**, 138 (1970). Use in treatment of Cushing's syndrome: R. I. Misbin *et al., J. Clin. Pharmacol.* **16**, 645 (1976); of metastatic breast cancer: R. F. Asbury *et al., Cancer* **47**, 1954 (1981). Comparative clinical trial with tamoxifen: I. E. Smith *et al., Brit. Med. J.* **283**, 1432 (1981). Hematologic toxicity study: A. A. Messeih *et al., Cancer Treat. Rep.* **69**, 1003 (1985). Clinical trial in advanced prostatic carcinoma: R. Murray, P. Pitt, *Eur. J. Cancer Clin. Oncol.* **21**, 453 (1985). Comprehensive guide to therapeutic use: *Pharmanual (Basel)* vol. 2, R. J. Santen, I. C. Henderson, Eds. (Karger, Basel, 1981) 160 pp. Reviews of mechanisms of action, endocrinological effects and clinical experience in breast cancer: R. C. Stuart-Harris, I. E. Smith, *Cancer Treat. Rev.* **11**, pp 189-204 (1984); A. L. Harris, *Exp. Cell. Biol.* **53**, pp 1-8 (1985). Comprehensive description: H. Y. Aboul-Enein in *Analytical Profiles of Drug Substances* vol. 15, K. Florey, Ed. (Academic Press, New York, 1986) pp 35-69.

Crystals from methanol or ethyl acetate, mp 149-150°. Freely sol in most organic solvents; poorly sol in ethyl acetate, 0.1N HCl and absolute ethanol; readily sol in acetone and 100% acetic acid. Practically insol in water.

Hydrochloride, $C_{13}H_{17}ClN_2O_2$, mp 223-225°. Freely sol in water.

THERAP CAT: In Cushing's syndrome and other adrenal hormone disorders. Palliative treatment of breast and prostatic cancer. Formerly as anticonvulsant.

453. Aminoguanidine. *Hydrazinecarboximidamide; guanylhydrazine.* CH_6N_4; mol wt 74.09. C 16.21%, H 8.16%, N 75.63%. Prepn: J. Thiele, *Ann.* **270**, 1 (1892); G. B. L. Smith, E. Anzelmi, *J. Am. Chem. Soc.* **57**, 2730 (1935). Reviews of chemistry of aminoguanidine and related compounds: E. Lieber, G. B. L. Smith, *Chem. Rev.* **25**, 213-271 (1939); of preparative methods: F. Kurzer, L. E. A. Godfrey, *Chem. & Ind. (London)* **1962**, 1584-1595. Prevention of glucose-derived aortic collagen cross-linking in diabetic rats: M. Brownlee *et al., Science* **232**, 1629 (1986).

$$H_2N-C-NHNH_2$$

with NH below

Crystals. Sol in water, alc. Practically insol in ether. Aq soln is strongly alkaline and reddens on standng in air; ammonia is evolved on heating.

Hydrochloride, CH_7ClN_4, large prisms from dil alc, mp 163°. Very sol in water; sol in alc. Practically insol in ether.

Dihydrochloride, $CH_8Cl_2N_4$, crystals, mp 183-183.5°.

Hemisulfate monohydrate, $C_2H_{14}N_8O_4S.H_2O$, needles, dec 207-208°. Becomes anhyd at 100° in a vacuum, and melts at 206°. Very sol in water.

Sulfate, $CH_8N_4O_4S$, large platelets, mp 161°.

454. p-Aminohippuric Acid. *N-(4-Aminobenzoyl)glycine; N-(p-aminobenzoyl)aminoacetic acid; PAH.* C_9H_{10}-N_2O_3; mol wt 194.19. C 55.66%, H 5.19%, N 14.43%, O 24.72%. Prepd by reduction of p-nitrohippuric acid: Shimizu, Okano, *J. Pharm. Soc. Japan* **73**, 523 (1953).

Needles from hot water, mp 198-199°. Soluble in alc,

chloroform, benzene, acetone. Practically insol in water, ether, carbon tetrachloride.

Sodium salt, $C_9H_8N_2NaO_3$, *Nephrotest.* Soluble in water. Aq solns are alkaline and must be buffered with citric acid to pH 7.0 for i.v. injection.

THERAP CAT: Diagnostic aid (renal function).

455. 3-Amino-4-hydroxybutyric Acid. *3-Amino-4-hydroxybutanoic acid; γ-hydroxy-β-aminobutyric acid;* GOBAB. $C_4H_9NO_3$; mol wt 119.12. C 40.33%, H 7.61%, N 11.76%, O 40.29%. HOCH$_2$CH(NH$_2$)CH$_2$COOH. Prepn: Jollès, Fromageot, *Bull. Soc. Chim. France* **1951**, 862; Chibnall *et al., Biochem. J.* **68**, 122 (1958); Piskov, *Zh. Obshch. Khim.* **32**, 3407 (1962); Nagai *et al., Arzneimittel-Forsch.* **17**, 1575 (1967); Kondo, Tanaka, **Japan.** pat. **12127**('**68**) (1968 to Kaken Kagaku), *C.A.* **70**, 77328r (1969).

Prismatic crystals, mp 216°. Also reported as mp 228° (Nagai); 232-233.5° (Chibnall).

Hydrochloride, $C_4H_{10}ClNO_3$, mp 156°.

THERAP CAT: Anti-inflammatory, antifungal.

456. 4-Amino-3-hydroxybutyric Acid. *4-Amino-3-hydroxybutanoic acid; γ-amino-β-hydroxybutyric acid; buksamin;* GABOB; Gabomade; Gamibetal. $C_4H_9NO_3$; mol wt 119.12. C 40.33%, H 7.62%, N 11.76%, O 40.30%. H$_2$N-CH$_2$CH(OH)CH$_2$COOH. Prepn of DL-form: Tomita, Z. *Physiol. Chem.* **124**, 255 (1923); Balenovic *et al., J. Org. Chem.* **19**, 1589 (1964); Hayashi *et al.*, **Japan.** pat. **58 772,** *C.A.* **53**, 1172d (1959); Sakai *et al.*, **Japan.** pat. **62 12,264** (to Kaken), *C.A.* **59**, 9805e (1963); **Spanish** pat. **278,780** (1963 to Antonio Gallardo, S.A.), *C.A.* **60**, 2779a (1964); Hayashi, **French** pat. **1,348,105** (1964 to Kaken), *C.A.* **60**, 11956b (1964); D'Alo, Masserini, *Farmaco Ed. Sci.* **19**, 30 (1964); M. Pinza, G. Pifferi, *J. Pharm. Sci.* **67**, 120 (1978). Prepn of L-form: Tomita, Sendju, *Z. Physiol. Chem.* **169**, 270 (1927); Kaneko, Yoshida, *Bull. Chem. Soc. Japan* **35**, 1153 (1962); M. E. Jung, T. J. Shaw, *J. Am. Chem. Soc.* **102**, 6304 (1980); of D-form: Tomita, Sendju, *loc. cit.* Improved synthesis of L-form: S. Takano *et al., Tetrahedron Letters* **28**, 1783 (1987). Purification: Yamagiwa, Tanaka, **Japan.** pat. **63 24,365** (to Kaken), *C.A.* **60**, 10515a (1964).

DL-Form, crystals from dil alc, dec 218°. Fairly readily sol in water; very sparingly sol in methanol, alc, ether, chloroform, ethyl acetate.

D(+)-Form, crystals from water, dec 214°. [α]$_D^{20}$ +18.3° (p = 2 in H$_2$O).

L(−)-Form, crystals from water or water + ethanol, dec 212°, also reported as dec 216-217°. [α]$_D^{20}$ −21.06° (p = 2 in H$_2$O); −20.7° (c = 1.83 in H$_2$O).

THERAP CAT: Anticonvulsant.

457. α-Aminoisobutyric Acid. *2-Methylalanine;* 2-aminoisobutyric acid; 2-amino-2-methylpropanoic acid. C_4H_9-NO$_2$; mol wt 103.12. C 46.59%, H 8.80%, N 13.58%, O 31.03%. (CH$_3$)$_2$C(NH$_2$)COOH. Prepd by the treatment of acetone with hydrocyanic acid and then with alcoholic ammonia (Strecker synthesis): Tiemann, Friedländer, *Ber.* **14**, 1970 (1881), *see also* p 1965; Marckwald *et al., Ber.* **24**, 3283 (1891); Bailey, Randolph, *Ber.* **41**, 2507 (1908); Clarke, Bean, *Org. Syn. coll. vol.* **II**, 29 (1943); or directly with ammonium cyanide: Gulewitsch, *Ber.* **33**, 1900 (1900) or with a mixture of KCN and NH$_4$Cl: Zelinsky, Stadnikow, *Ber.* **39**, 1726 (1906); *cf.* Hellsing, *Ber.* **37**, 1921 (1904); and subsequent hydrolysis of the nitrile formed. By heating dimethylhydantoin (obtained from acetone, hydrocyanic and cyanic acids) with concd HCl: Urech, *Ann.* **164**, 268 (1872), *cf.* Heilpern, *Monatsh.* **17**, 241 (1896).

Monoclinic prisms, tables, mp 335° (sealed capillary). Begins to sublime at 280°. Sweetish taste. Absorption spectrum: Ley, Arends, *Ber.* **61**, 219 (1928); Abderhalden, Rossner, *Z. Physiol. Chem.* **176**, 253 (1928). Freely sol in water. Difficultly sol in alcohol; insol in ether.

Hydrochloride, $C_4H_9NO_2$.HCl, platelets from water, dec 236-237°. Readily sol in water, methanol, alcohol.

458. δ-Aminolevulinic Acid. *5-Amino-4-oxopentanoic acid.* $C_5H_9NO_3$; mol wt 131.13. C 45.79%, H 6.92%, N 10.68%, O 36.60%. H$_2$NCH$_2$COCH$_2$CH$_2$COOH. Tetrahydropyrrole precursor; intermediate in heme, chlorophyll biosynthesis. Prepn as hydrochloride by the hydrogenation of oxaloacetone: R. W. Wynn, A. H. Corwin, *J. Org. Chem.*

15, 203 (1950); from furfurylamine: A. A. Marei, R. H. Raphael, *J. Chem. Soc.* **1958**, 2624; by a malonic ester synthesis: D. P. Tschudy, A. Collins, *J. Org. Chem.* **24**, 556 (1959); from 5-hydroxy-2-pyridone: C. Herdeis, A. Dimmerlung, *Arch. Pharm.* **317**, 304 (1984). Effect on dark-grown barley seedlings: E. C. Sisler, W. H. Klein, *Physiol. Plant.* **16**, 315 (1963); A. K. Stobart, I. Ameen-Bukhari, *Biochem. J.* **222**, 419 (1984). Accumulation in light-treated barley seedlings: S. Gough, *Carlsberg Res. Commun.* **43**, 497 (1978). Biosynthesis and metabolism: S. I. Beale, *Plant Physiol.* **48**, 316 (1971). Use of labelled acid in porphyrin biosynthesis studies: C. A. Rebeiz *et al., ibid.* **46**, 543 (1970). Enhancement of chlorophyll formation: S. Ochiai, E. Hase, *Plant Cell Physiol.* **11**, 663 (1970). Proposed use as photodynamic herbicide: C. A. Rebeiz *et al., Enzyme Microb. Technol.* **6**, 390 (1984); *Chem. & Eng. News* **62**, 8 (Sept. 24, 1984).

Hydrochloride, $C_5H_{10}ClNO_3$, needles from methanol-ether, mp 144-147° (decomp) (Marei, Raphael). uv max (water): 266.5 nm (ε 23.0); after addition of 2 equivalents of NaOH, uv max: 276 nm (ε 2000).

USE: Experimental herbicide.

459. β-Amino-α-methylphenethyl Alcohol. *β-Amino-α-methylbenzeneethanol;* 1-amino-1-phenyl-2-propanol; 1-phenyl-1-amino-2-propanol. $C_9H_{13}NO$; mol wt 151.20. C 71.49%, H 8.67%, N 9.26%, O 10.58%. Prepn from α-isonitroso-α-phenylacetone: Sichner, Pankova, *Coll. Czech. Chem. Commun.* **20**, 1419 (1955); from an oxazole: Viscontini, *Helv. Chim. Acta* **44**, 636 (1961).

DL-*erythro*-Form, so-called *dl-norisoephedrine.* Crystals, mp 85°.

DL-*erythro*-Hydrochloride, $C_9H_{13}NO.HCl$, crystals from ethanol + ether, mp 170-171°.

460. 2-Amino-2-methyl-1,3-propanediol. $C_4H_{11}NO_2$; mol wt 105.14. C 45.69%, H 10.55%, N 13.32%, O 30.44%. An amino glycol prepd by reduction of the corresponding nitro compd: Vanderbilt, Hass, U.S. pat. **2,174,242** (1940); Johnson, Degering, *J. Org. Chem.* **8**, 7 (1943).

Crystalline mass, mp 109-111°. bp$_{10}$ 151-152°. 250 grams dissolve in 100 ml water at 20°. Sol in alcohols. pH of 0.1 molar aq soln 10.8.

USE: In the synthesis of surface-active agents, vulcanization accelerators, pharmaceuticals. As emulsifying agent for cosmetic creams and lotions, mineral oil and paraffin wax emulsions, leather dressings, textile specialties, polishes, cleaning compounds, so-called sol oils. Absorbent for acidic gases.

461. 2-Amino-2-methyl-1-propanol. $C_4H_{11}NO$; mol wt 89.14. C 53.89%, H 12.44%, N 15.72%, O 17.95%. An amino alcohol prepd by reduction of the corresp nitro compound: Vanderbilt, Hass, U.S. pat. **2,174,242** (1940); Johnson, Degering, *J. Org. Chem.* **8**, 7 (1943).

Crystalline mass, mp 30-31°. (The commercial product may be a viscous liquid.) d$_{20}^{20}$ 0.934. n$_D^{20}$ 1.449. bp$_{760}$ 165°; bp$_{10}$ 67.4°. Miscible with water. Sol in alcohols. pH of 0.1 molar aq soln 11.3.

USE: In the synthesis of surface-active agents, vulcanization accelerators, pharmaceuticals. As emulsifying agent for cosmetic creams and lotions, mineral oil and paraffin wax emulsions, leather dressings, textile specialties, polishes, cleaning compounds, so-called soluble oils. Absorbent for acidic gases. For use as a drug *see* Pamabrom.

462. 2-Amino-4-methylthiazole. *4-Methyl-2-thiazol-amine;* 4-methyl-2-thiazolylamine; Normotiroide. C_4H_6-N_2S; mol wt 114.18. C 42.07%, H 5.30%, N 24.54%, S 28.09%. Prepn: Sprague *et al., J. Am. Chem. Soc.* **68**, 2155 (1946); Rossi, *Gazz. Chim. Ital.* **85**, 898 (1955).

Crystals, mp 45-46°. bp_{20} 124-126°; $bp_{0.4}$ 70°. Very sol in water, alcohol, ether.

THERAP CAT: Antihyperthyroid.

463. Aminometradine. *6-Amino-3-ethyl-1-(2-propen-yl)-2,4(1H,3H)-pyrimidinedione; 1-allyl-6-amino-3-ethyl-uracil;* 1-allyl-6-amino-3-ethyl-2,4(1H,3H)-pyrimidinedi-one; 1-allyl-3-ethyl-6-amino-1,2,3,4-tetrahydro-2,4-pyri-midinedione; aminometramide; Mictine; Katapyrin; Min-card; Catapyrin. $C_9H_{13}N_3O_2$; mol wt 195.22. C 55.37%, H 6.71%, N 21.53%, O 16.39%. Prepn: Papesch, Schroeder, U.S. pat. **2,650,922** (1953 to Searle).

Monohydrate, crystals from water, mp 75-115°. When anhydrous, mp 143-144°.

THERAP CAT: Diuretic.

464. 3-Amino-2-naphthoic Acid. *3-Amino-2-naphtha-lenecarboxylic acid;* 3-aminoisonaphthoic acid. $C_{11}H_9NO_2$; mol wt 187.19. C 70.58%, H 4.85%, N 7.48%, O 17.09%. Prepd from 3-hydroxy-2-naphthoic acid: C. F. H. Allen, A. Bell, *Org. Syn. coll. vol. III*, 78 (1955).

Yellow scales from dil alcohol, mp 214°. Sol in alcohol, ether. Solns are yellow with greenish fluorescence.

Sodium salt, $C_{11}H_8NNaO_2$, leaflets, very sparingly sol in water, alcohol.

Ethyl ester, $C_{13}H_{13}NO_2$, yellow needles from dil alcohol. Sol in the usual organic solvents, mp 115-115.5°.

USE: In the determination of copper, nickel, cobalt.

465. 4-Amino-1-naphthol. *4-Amino-1-naphthalenol;* 4-hydroxy-α-naphthylamine. $C_{10}H_9NO$; mol wt 159.18. C 75.45%, H 5.70%, N 8.80%, O 10.05%. Prepd by treating α-naphthol with benzenediazonium chloride and reducing the benzeneazo-α-naphthol with sodium hydrosulfite: Co-nant *et al., Org. Syn.* **3**, 7 (1923); *cf.* Fieser, Fieser, *J. Am. Chem. Soc.* **57**, 493 (1935). From α-naphthylhydroxylamine by rearrangement in acetone: Neunhoffer, Liebich, *Ber.* **71B**, 2247 (1938).

Needles. Unless kept absolutely dry, it acquires a violet discoloration on storage and oxidizes to 1,4-naphthoqui-none. Usually isolated as the hydrochloride, $C_{10}H_9NO.HCl$, needles, very sol in water.

N-Acetyl deriv, $C_{12}H_{11}NO_2$, *4-acetamido-1-naphthol, naphthacetol,* needles from alcohol, mp 188°, sparingly sol in water, sol in alcohol.

USE: Polymerization inhibitor; prepn of 2-allyl-4-amino-1-naphthol hydrochloride (an antihemorrhagic cpd); prepn of 1,4-naphthoquinone.

466. 1-Amino-2-naphthol-4-sulfonic Acid. *4-Amino-3-hydroxy-1-naphthalenesulfonic acid;* 1,2,4-acid. $C_{10}H_9$-NO_4S; mol wt 239.25. C 50.20%, H 3.79%, N 5.86%, O 26.75%, S 13.40%. Prepd by treatment of nitroso-β-naph-thol with sodium bisulfite and sulfuric acid: L. F. Fieser, *Org. Syn. coll. vol. II*, 42 (1943); *cf. idem, J. Am. Chem. Soc.* **57**, 494 (1935).

White or gray needles, usually contg 0.5 H_2O. May turn pink on exposure to light, especially when moist. Insol in water, alcohol, ether, benzene; sol in hot sodium bisulfite soln; sol in alkaline soln but such solns oxidize quickly on exposure to air yielding a brown substance which is sol in hot water giving a green soln. Controlled oxidation with HNO_3 yields ammonium 1,2-naphthoquinone-4-sulfonate, *see* Fieser, *loc. cit.*

Sodium salt, $C_{10}H_8NNaO_4S$, needles, sol in hot water with blue fluorescence.

USE: Manuf of azo dyes, 1,2-naphthoquinone-4-sulfonic acid.

467. 1-Amino-2-naphthol-6-sulfonic Acid. *5-Amino-6-hydroxy-2-naphthalenesulfonic acid.* $C_{10}H_9NO_4S$; mol wt 239.24. C 50.21%, H 3.79%, N 5.86%, O 26.75%, S 13.40%. Prepd by reduction of sodium 1-nitroso-2-naphthol-6-sul-fonate with zinc and acetic acid: Fierz-David *et al., Helv. Chim. Acta* **29**, 1765 (1946).

Needles or prisms. Slightly sol in boiling water, less sol in alc. Practically insol in ether.

Sodium salt hemipentahydrate, $C_{10}H_8NNaO_4S.2\frac{1}{2}H_2O$, *Eikonogen.* Powder. Sol in water. Practically insol in alc, ether. Strong reducing action on silver salts.

USE: Free acid formerly used in manuf of azo dyes. Sodi-um salt used as photographic developer and for detection of potassium.

468. 6-Aminonicotinic Acid. *6-Amino-3-pyridinecar-boxylic acid;* 6-amino-3-carboxypyridine. $C_6H_6N_2O_2$; mol wt 138.12. C 52.17%, H 4.38%, N 20.28%, O 23.17%. Prepd by heating 6-chloronicotinic acid with ammonia: Marck-wald, *Ber.* **26**, 2188 (1893); **27**, 1319 (1894); Räth, Prauge, *Ann.* **467**, 4 (1928); Johnson *et al., J. Biol. Chem.* **153**, 37 (1944).

Dihydrate, crystals from dil acetic acid. Dec above 300° yielding 2-aminopyridine and CO_2. Sparingly sol in most solvents.

Potassium salt, $C_6H_5KN_2O_2$, crystals, freely sol in water, insol in alcohol.

Hydrochloride, $C_6H_6N_2O_2 \cdot HCl$, needles, freely sol in water, slightly sol in alcohol.

469. 2-Amino-5-nitrothiazole. *5-Nitro-2-thiazolamine;* Enheptin; Entramin. $C_3H_3N_3O_2S$; mol wt 145.15. C 24.83%, H 2.08%, N 28.95%, S 22.09%, O 22.05%. Prepd by deacetylation of 2-acetamido-5-nitrothiazole: Hubbard; Steahly, U.S. pats. **2,573,641; 2,573,656; 2,573,657** (all 1951 to Monsanto).

Greenish-yellow to orange-yellow fluffy powder, dec 202°. Slightly bitter taste. uv max (0.0005% in water): 386 nm (ϵ 0.540); min: 295 nm. Very sparingly soluble in water. (To make a water soln, a 15% stock soln in propylene glycol should first be prepd.) One gram dissolves in 150 g of 95% alcohol; in 250 g ether. Almost insol in chloroform. Sol in dilute mineral acids.

THERAP CAT (VET): Antihistomonad in turkeys, chickens. For trichomoniasis in pigeons.

470. 6-Aminopenicillanic Acid. *6-Amino-3,3-dimethyl-7-oxo-4-thia-1-azabicyclo[3.2.0]heptane-2-carboxylic acid;* 6-APA; penicin; penin. $C_8H_{12}N_2O_3S$; mol wt 216.28. C 44.43%, H 5.59%, N 12.96%, O 22.19%, S 14.83%. Intermediate in the manuf of synthetic penicillins. Obtained from cultures of *Penicillium chrysogenum* in the absence of side chain precursors: Batchelor *et al., Nature* **183,** 257 (1959); Doyle *et al.,* **Belgian** pat. **569,728** (1959); U.S. pat. **2,941,995** (1960 to Beecham); U.S. pat. **2,951,839** (1960); U.S. pat. **3,071,575** (1963); Johnson, Hardcastle; Nettleton *et al.,* U.S. pats. **3,008,955/6** (1961 to Beecham); Robinson *et al.,* U.S. pats. **3,014,845/6** (1961). From cultures of *Pleurotus ostreatus:* Brandl, Kleiber, U.S. pat. **3,109,779** (1963 to Biochemie GmbH). Synthesis: Sheehan, Henery-Logan, *J Am. Chem. Soc.* **81,** 5838 (1959); **84,** 2983 (1962); Sheehan, U.S. pat. **3,028,379** (1962). Synthesis of derivs: Sheehan, U.S. pat. **2,934,540** (1960 to Res. Corp.). From phenoxymethylpenicillin by treatment with kidney enzymes: Weitnauer, U.S. pat. **3,070,511** (1962 to Lepetit). From penicillin G benzyl ester: Hoover, U.S. pat. **3,107,250** (1963 to SKF). Improved prepn: Weissenburger, van der Hoeven, U.S. pat. **3,499,909** (1970 to Koninklijke Nederlandische Gisten Spiritusfabriek).

Non-hygroscopic crystals from water + HCl, dec 209-210°. $[\alpha]_D^{31}$ +273° (c = 1.2 in 0.1N HCl).

USE: In manuf of synthetic penicillins. Basic patent: Sheehan, U.S. pat. **3,159,617** (1964 to Arthur D. Little).

471. Aminopentamide. α-*[2-(Dimethylamino)propyl]-α-phenylbenzeneacetamide; 4-dimethylamino-2,2-diphenylvaleramide;* α,α-diphenyl-γ-dimethylaminovaleramide; BL 139; Centrine; Valeramide-OM. $C_{19}H_{24}N_2O$; mol wt 296.40. C 76.99%, H 8.16%, N 9.45%, O 5.40%. Prepn similar to that of methadone, starts by condensing diphenylmethyl cyanide with chlorodimethylaminopropane: Walton, *et al., J. Chem. Soc.* **1949,** 648; Cheney *et al., J. Org. Chem.* **17,** 770 (1952); Specter, U.S. pat. **2,647,926** (1953 to Bristol). Prepn of optical isomers: Wheatley *et al., J. Org. Chem.* **19,** 794 (1954). *See also* Moffett, Aspergren, *J. Am. Chem. Soc.* **79,** 4451 (1957).

dl-Form, long prisms from dilute alcohol, mp 183-184°. Practically insol in water.

d-Form, crystals from petr ether, mp 136.5-137.5° (sintering). $[\alpha]_D^{23}$ +98.9° (methanol).

l-Form, crystals from petr ether, mp 136.5-137.5° (sintering). $[\alpha]_D^{23}$ −101.9° (methanol).

dl-Form hydrochloride, $C_{19}H_{24}N_2O \cdot HCl$, slightly deliquescent leaflets from alcohol + ether, dec 190-191°. Bitter taste. Soluble in water, alcohol. A 1% aq soln has a pH of 6.8. LD_{50} in mice: 34.7 mg/kg i.v.; 396 mg/kg orally, Cazort, *J. Pharmacol. Exp. Ther.* **100,** 325 (1950).

dl-Form acid sulfate; (bisulfate), $C_{19}H_{24}N_2O \cdot H_2SO_4$, deliquescent crystals from isopropanol + ethyl acetate, mp 185-187°. The commercial medicinal grade, mp 178-181°. Bitter taste. uv max (1% H_2SO_4): 258.5 nm ($A_{1cm}^{1\%}$ 10.3); min: 249 nm. Freely sol in water, alcohol. Very slightly sol in chloroform. Practically insol in ether. pH of 2.5% aq soln 1.3-2.2.

THERAP CAT: Anticholinergic.

THERAP CAT (VET): Anticholinergic; anti-emetic; anticonvulsant.

472. *m*-Aminophenol. 3-Amino-1-hydroxybenzene; 3-hydroxyaniline. C_6H_7NO; mol wt 109.12. C 66.03%, H 6.47%, N 12.84%, O 14.66%. Manuf by reduction of *m*-nitrophenol: Freifelder, Robinson, U.S. pat. **3,079,435** (1963 to Abbott).

Crystals, mp 122-123°. Sol in 40 parts cold water, freely in hot water, alcohol, ether, amyl alcohol; slightly in benzene, very slightly in petr ether. LD_{50} i.p. in mice: 4.5 mg/20 g, Koelzer, Giesen, *Z. Naturforsch.* **6b,** 183 (1951).

USE: Dye intermediate, manuf *p*-aminosalicylic acid.

473. *o*-Aminophenol. *2-Aminophenol;* 2-amino-1-hydroxybenzene; 2-hydroxyaniline. Empirical formula: See *m*-isomer. Manuf by reduction of *o*-nitrophenol: Freifelder, Robinson, U.S. pat. **3,079,435** (1963 to Abbott).

Crystals, rapidly becoming brown, mp 170-174°; sublimes. One gram dissolves in 50 ml cold water, 23 ml alcohol; freely soluble in ether, very slightly in benzene. *Keep tightly closed and protected from light.*

Hydrochloride, $C_6H_7NO \cdot HCl$, crystals readily becoming gray on exposure to light. Freely sol in water or alcohol.

USE: Manuf azo and sulfur dyes; dyeing furs and hair. Hydrochloride used in dyeing fur, hair, leather, etc. *Protect from light.*

474. *p*-Aminophenol. *p*-Hydroxyaniline; 4-amino-1-hydroxybenzene; Activol; Azol; Certinal; Citol; Paranol; Rodinal; Unal; Ursol P. Empirical formula: *see m*-isomer above. Usually prepd by the reduction of *p*-nitrophenol: *BIOS Final Report 986;* Freifelder, Robinson, U.S. pat. **3,079,435** (1963 to Abbott).

Orthorhombic plates from water. *Deteriorates under the influence of air and light.* mp 189.6-190.2°. The commercial product is usually pink, mp 186°. Can be sublimed at 0.3 mm and 110° without decompn. bp_{760} 284° (decompn); $bp_{8.0}$ 167°; $bp_{3.0}$ 150°; $bp_{0.3}$ 130.2°. Kb at 15° = 6.6 × 10^{-9}. Forms salts with acids and bases. Soly in water: 0.39% at 13°; 0.65% at 24°; 0.80% at 30°; 1.5% at 50°; 4.7% at 80°; 8.5% at 96°. Soly in ethyl methyl ketone: 9.3% at 58.5°; in abs ethanol: 4.5% at 0°. Practically insol in benzene, chloroform.

Hydrochloride, $C_6H_7NO \cdot HCl$, cryst powder; gradually becomes darker. Decomp about 306°. Very sol in water; sol in alc.

USE: Photographic developer; intermediate in the manufacture of sulfur and azo dyes; in dyeing furs and feathers. *Caution:* May cause skin sensitization, dermatitis. Inhalation can cause asthma, methemoglobin formation.

475. *p*-Aminophenylacetic Acid. *4-Aminobenzeneacetic*

acid; p-amino-α-toluic acid. $C_8H_9NO_2$; mol wt 151.16. C 63.56%, H 6.00%, N 9.27%, O 21.17%. Prepd by the reduction of p-nitrophenylacetic acid with hydrogen sulfide in the presence of ammonia: G. R. Robertson, *Org. Syn.* coll. vol. I, 52 (2nd ed., 1941).

Plates, leaflets from water, mp 199-200° (dec). Moderately sol in hot water; sol in alcohol, in alkalies.

Hydrochloride, $C_8H_9NO_2.HCl$, rods from HCl, freely sol in water, sol in alcohol (about 3% w/w).

Ethyl ester, $H_2NC_6H_4CH_2CO_2C_2H_5$, platelets from water, mp 51°.

N-Benzoyl deriv, needles from alcohol, mp 205-206°.

476. 4-Amino-3-phenylbutyric Acid. *β-(Aminomethyl)benzenepropanoic acid; β-(aminomethyl)hydrocinnamic acid; 4-amino-3-phenylbutanoic acid; β-phenyl-γ-aminobutyric acid; phenygam; Phenigam; Fenigam; Phenigama; Fenigama; PhGABA.* $C_{10}H_{13}NO_2$; mol wt 179.21. C 67.02%, H 7.31%, N 7.82%, O 17.85%. Prepn starting with α,γ-diamino-β-phenylpropane: Jackson, Kenner, *J. Chem. Soc.* **1928**, 1657. Alternate prepn: Cologne, Pouchol, *Bull. Soc. Chim. France* **1962**, 598. Structure-activity studies: Khaunina, *Farmakol. Toksikol.* **31**, 202 (1968). Activity studies of the isomers: *eidem, Byull. Eksp. Biol. Med.* **72**, 49 (1971), *C.A.* **76**, 81208t (1972).

mp 250-253° (dec).

Hydrobromide, $C_{10}H_{13}NO_2.HBr$, irregular platelets from benzene, mp 114°. LD_{50} i.p. in mice: 900 mg/kg.

THERAP CAT: Mood elevator, tranquilizer.

477. Aminophylline. *3,7-Dihydro-1,3-dimethyl-1H-purine-2,6-dione compd with 1,2-ethanediamine (2:1); theophylline compd with ethylenediamine;* theophylline ethylenediamine; theophyllamine; Carena; Inophylline; Metaphyllin; Theophyldine; Aminocardol; Aminodur; Ammophyllin; Cardiofilina; Cardophylin; Phylcardin; Tefamin; Cardiomin; Grifomin; Minaphil; Pecram; Peterphyllin; Phyllocontin; Somophyllin; Stenovasan; Theodrox; Cardophyllin; Diophyllin; Etilen-Xantisan Tabl.; Genophyllin; Phyllindon; Theolamine; Euphyllin; Theomin; TH 100; Variaphylline LA. $C_{16}H_{24}N_{10}O_4$; mol wt 420.44. C 45.71%, H 5.75%, N 33.32%, O 15.22%. $(C_7H_8N_4O_2)_2.C_2H_4(NH_2)_2.2H_2O$. Prepd from theophylline and aq ethylenediamine: Grüter, U.S. pat. **919,161** (1909 to Byk). Toxicity data: C. R. Thompson, M. R. Warren, *J. Lab. Clin. Med.* **31**, 1337 (1946). Comprehensive description: K. D. Thakker, L. T. Grady, in *Analytical Profiles of Drug Substances* **vol. 11**, K. Florey, Ed. (Academic Press, New York, 1982) pp 1-44.

Dihydrate, white or slightly yellowish granules or powder. Slight ammoniacal odor. Bitter taste. Gradually absorbs carbon dioxide from the air and becomes incompletely sol due to liberation of theophylline. One gram dissolves in about 5 ml water, but the soln may become turbid on standing. Insol in alcohol, ether. *Keep tightly closed.* LD_{50} orally in mice: 540 mg/kg (Thompson, Warren).

Note: Euphyllin was originally described as a compd consisting of 3 mols theophylline and 2 mols of ethylenediamine, *see* Grüter, *Therap. Monatsh.* **24**, 613 (1910).

Human Toxicity: Acute poisoning: Causes restlessness, anorexia, nausea, fever, vomiting, dehydration. Eventual bloody, syrupy or "coffee-ground" vomitus, tremors, delirium and coma can occur. May result in cardiovascular and respiratory collapse, shock, cyanosis and death, *Clinical Toxicology of Commercial Products,* R. E. Gosselin *et al.,*

Eds. (Williams & Wilkins, Baltimore, 4th ed., 1976) Section III, pp 16-20.

THERAP CAT: Bronchodilator.

THERAP CAT (VET): Smooth muscle relaxant. For heaves in horses. Diuretic in dogs with congestive heart failure.

478. 2-Amino-4-picoline. *4-Methyl-2-pyridinamine;* 2-amino-4-methylpyridine; α-amino-γ-picoline; 4-methyl-2-aminopyridine; W 45 Raschig. $C_6H_8N_2$; mol wt 108.14. C 66.64%, H 7.46%, N 25.91%. Prepd by heating 4-picoline with sodamide in xylene: Seide, *Ber.* **57**, 791 (1924); **58**, 1733 (1925); Räth, *ibid.* 347; Kakimoto, Nishie, *C.A.* **50**, 14744g (1956). Formulations: von Haxthausen *et al.*, U.S. pat. **2,937,118** (1960 to Raschig). Pharmacology: von Haxthausen, *Arch. Exp. Pathol. Pharmakol.* **226**, 163; **227**, 234 (1955); Marchetti *et al.*, *Arch. Int. Pharmacodyn. Ther.* **143**, 385 (1963).

Leaflets from petr ether, mp 100-100.5°. bp_{11} 115-117°. Sublimes on slow heating. Freely sol in water, lower alcohols, dimethylformamide, coal tar bases. Slightly sol in petr ether, aliphatic hydrocarbons. Component of *Ascensil, Askensil.*

Hydrochloride, $C_6H_8N_2.HCl$, prisms from alcohol, mp 176-177°. Freely sol in water, alcohol.

Camphorsulfonate, $C_{16}H_{24}N_2O_4S$, *Piricardio, Varunax.*

THERAP CAT: Analgesic; cardiac stimulant.

479. Aminopromazine. *N,N,N',N'-Tetramethyl-3-(10H-phenothiazin-10-yl)-1,2-propanediamine; 10-[2,3-bis-(dimethylamino)propyl]phenothiazine;* proquamezine; RP 3828; Tetrameprozine. $C_{19}H_{25}N_3S$; mol wt 327.50. C 69.68%, H 7.69%, N 12.83%, S 9.79%. Prepn: Jacob *et al., Compt. Rend.* **243**, 1637 (1956); Horclois, **Brit.** pat. **800,635** (1958 to Rhône-Poulenc).

Fumarate, $C_{42}H_{54}N_6O_4S_2$, *Lispamol, Lorusil, Spamol.* Crystals, sensitive to light, dec 166-170°. Soly at 20° (g/100 ml): water 9.0; methanol 5.0; ethanol 0.5. Very slightly sol in isopropanol, acetone. Practically insol in benzene, ether. pH of a 2% aq soln 5.0 to 7.0.

THERAP CAT: Antispasmodic.

THERAP CAT (VET): Antispasmodic.

480. 2-Aminopropanol. *2-Amino-1-propanol;* 2-aminopropyl alcohol; β-propanolamine; 2-hydroxyisopropylamine. C_3H_9NO; mol wt 75.11. C 47.97%, H 12.08%, N 18.65%, O 21.30%. $CH_3CH(NH_2)CH_2OH$. Obtained together with isopropylamine on reduction of acetylcarbinol-oxime with sodium amalgam in dilute acetic acid: Gabriel, *Ber.* **49**, 2121 (1916); from DL-alanylglycine by reduction with sodium and abs alcohol: Abderhalden, Schwab, *Z. Physiol. Chem.* **143**, 292 (1925); by boiling the ethyl ester of DL-acetylalanine with sodium in abs alcohol: Karrer, *Helv. Chim. Acta* **4**, 98 (1921); as hydrolytic cleavage product of ergonovine and ergometrinine: Stoll, Hofmann, *ibid.* **26**, 956 (1943); by catalytic hydrogenation of the ethyl ester of alanine: Adkins, Pavlic, *J. Am. Chem. Soc.* **69**, 3039 (1947).

dl-Form, liquid, fishy odor. bp 173-176°. Freely sol in water, alcohol, ether.

Hydrochloride, $C_3H_9NO.HCl$, leaflets from abs alcohol + acetone, dec 86-87.5°.

481. 3-Aminopropionitrile. *3-Aminopropanenitrile;* β-aminopropionitrile. $C_3H_6N_2$; mol wt 70.09. C 51.40%, H 8.63%, N 39.97%. $H_2NCH_2CH_2CN$. Prepd by the reaction

of acrylonitrile with ammonia: Buc *et al., J. Am. Chem. Soc.*
67, 93 (1945); *Org. Syn.* **coll. vol. III,** 93 (1955); Weijlard,
Sullivan, **U.S. pat. 2,742,491** (1956 to Merck & Co.).

Liquid. Amine odor. bp_{760} 185°; bp_{20} 87-89°; bp_5 66-69°;
bp_3 50-55°. n_D^{20} 1.4396. The pure, anhydrous material may
be stored in tightly stoppered bottles for several months.
Storing under refrigeration is recommended to avoid development of pressure in the bottles. Polymerization takes
place slowly during storage in an open container, or very
rapidly in the presence of acid or acidic compds.

USE: Intermediate in the manufacture of β-alanine and
pantothenic acid.

482. p-Aminopropiophenone. *1-(4-Aminophenyl)-1-
propanone;* ethyl *p*-aminophenyl ketone; PAPP. $C_9H_{11}NO$;
mol wt 149.19. C 72.45%, H 7.43%, N 9.39%, O 10.72%.
Prepd by the action of propionyl chloride on aniline in carbon bisulfide in the presence of aluminum chloride: Kun-
ckell, *Ber.* **33,** 2641 (1900); Derrick, Bornemann, *J. Am.
Chem. Soc.* **35,** 1283 (1913); Hartung, Foster, *J. Am. Pharm.
Assoc.* **35,** 15 (1946).

Yellow needles from water, mp 140°. Sol in water, alc.
Hydrochloride, $C_9H_{11}NO.HCl$, needles, yellowish cast, mp
198-199°. Freely sol in water. LD_{50} i.v. in dogs: 7 mg/kg.
Sulfate, $(C_9H_{11}NO)_2.H_2SO_4$, yellow plates from alcohol,
dec 25°.

THERAP CAT: Antidote (cyanide).

483. α-(α-Aminopropyl)benzyl Alcohol. *α-(1-Amino-
propyl)benzenemethanol;* 2-amino-1-phenyl-1-butanol;
β-amino-α-phenylbutyl alcohol; 1-phenyl-1-hydroxy-2-
amino-*n*-butane. $C_{10}H_{15}NO$; mol wt 165.23. C 72.69%, H
9.15%, N 8.48%, O 9.68%. Prepn of the *dl-threo*-form:
Abrams, Kipping, *J. Chem. Soc.* **1936,** 1480; Rebstock *et al.,
J. Am. Chem. Soc.* **73,** 3666 (1951). Prepn of the *dl-erythro*-
form: Hartung *et al., ibid.* **52,** 3317 (1930); Rebstock *et al.,
loc. cit. See also* Beilstein **13,** suppl. 3, 1791-1792 (1973).

dl-threo-Form, thick, shiny plates from benzene + petr
ether, mp 79-80° (Abrams, Kipping); 78-79° (Rebstock *et
al.*). Freely sol in alcohol; moderately sol in chloroform,
benzene; sparingly sol in petr ether.

dl-threo-Form hydrochloride, $C_{10}H_{15}NO.HCl$, lustrous
prisms, mp 195-196° (Abrams, Kipping); 204-205° (Rebstock *et al.*). Freely sol in water; sol in alcohol; slightly sol
in chloroform.

dl-erythro-Form, mp 80.5-81°; hydrochloride, mp 242°.

484. Aminopropylon. *N-(2,3-Dihydro-1,5-dimethyl-3-
oxo-2-phenyl-1H-pyrazol-4-yl)-2-(dimethylamino)propan-
amide; N-(antipyrinyl)-2-(dimethylamino)propionamide; 4-
[2-(dimethylamino)propionamido]antipyrine; 4-[α-(dimethylamino)propionamido]-2,3-dimethyl-1-phenyl-3-pyrazo-
lin-5-one;* Amipylo. $C_{16}H_{22}N_4O_2$; mol wt 302.39. C 63.55%,
H 7.33%, N 18.53%, O 10.58%. Prepn: Takahashi *et al., J.
Pharm. Soc. Japan* **75,** 1431 (1955), *C.A.* **50,** 10086d (1956);
Ogiu, Takahashi, **Japan.** pat. **8770('58)** (to Res. Found. for
Practical Life), *C.A.* **54,** 4621f (1960).

Prisms from benzene, mp 181°. Very soluble in water.
THERAP CAT: Analgesic.

485. Aminopterin. *N-[4-[[(2,4-Diamino-6-pteridinyl)-
methyl]amino]benzoyl]-L-glutamic acid;* 4-aminofolic acid;
4-aminopteroylglutamic acid; 4-amino-PGA. $C_{19}H_{20}N_8O_5$;
mol wt 440.43. C 51.82%, H 4.58%, N 25.44%, O 18.16%.
Prepd from 2,4,5,6-tetraminopyrimidine sulfate, 2,3-di-
bromopropionaldehyde and *p*-aminobenzoylglutamic acid:
Seeger *et al., J. Am. Chem. Soc.* **69,** 2567 (1947); from 6-
(bromomethyl)-2,4-diaminopteridine HBr: Piper, Montgomery, *J. Heterocycl. Chem.* **11,** 279 (1974). Purification:
Loo, *J. Med. Chem.* **8,** 139 (1965). Inhibition of dihydrofolate reductase: J. S. Erickson *et al., J. Biol. Chem.* **247,**
5661 (1972). Use as rodenticide: **U.S. pat. 2,575,168** (1951
to Am. Cyanamid).

Dihydrate, clusters of yellow needles. uv max of the 0.75
hydrate (0.1N NaOH): 261, 282, 373 (log ϵ 4.41, 4.39, 3.91).
USE: Rodenticide.

486. α-Aminopyridine. *2-Pyridinamine;* 2-aminopyri-
dine. $C_5H_6N_2$; mol wt 94.11. C 63.80%, H 6.43%, N
29.77%. Prepd from pyridine and sodamide: Tschitschibabin, *Chem. Zentr.* **1915, I,** 1065; Wibaut, *Rec. Trav. Chim.*
42, 240 (1923); A. I. Vogel, *Practical Organic Chemistry,*
(Longmans, London, 3rd ed., 1959) p 1007; Gattermann-
Wieland, *Praxis des Organischen Chemikers,* (de Gruyter,
Berlin, 40th ed., 1961) p 316.

Leaflets, or large crystals. mp 58.1°. bp 210.6°. Soluble in
water, alcohol, benzene, ether, hot petrol ether.
N-Acetyl deriv, mp 71°.
USE: Manuf pharmaceuticals, esp antihistaminic drugs.

487. β-Aminopyridine. *3-Pyridinamine;* 3-aminopyri-
dine. $C_5H_6N_2$; mol wt 94.11. C 63.80%, H 6.43%, N
29.77%. Prepn from nicotinamide: Allen, Wolf, *Org. Syn.,*
coll. vol. IV, 45 (1963).
Crystals, mp 64°. bp 250-252°. Sol in water, alcohol, benzene, ether; insol in petr ether.
N-Acetyl deriv, mp 133°.
USE: In the manufacture of drugs and dyes.

488. Aminopyrine. *4-(Dimethylamino)-1,2-dihydro-
1,5-dimethyl-2-phenyl-3H-pyrazol-3-one; 4-(dimethylamino)antipyrine;* dimethylaminophenyldimethylpyrazolone;
4-dimethylamino-2,3-dimethyl-1-phenyl-3-pyrazolin-5-
one; aminophenazone; amidazophen; dipyrine; dimethylaminophenazone; amidopyrine; Dipirin; Brufaneuxol; Mamallet-
A; Pyramidon; Amidopyrazoline; Polinalin; Amidofebrin;
Anafebrina; Netsusarin; Novamidon; Pyradone; Piridol;
Itamidone; Febrinina; Dimapyrin; Dimethylamino-analgesine. $C_{13}H_{17}N_3O$; mol wt 231.29. C 67.50%, H 7.41%, N
18.17%, O 6.92%. Prepn: *Beilstein* **25,** 452 (1936); T. Takahashi *et al., J. Pharm. Soc. Japan* **76,** 1180 (1956); T. Takahashi, K. Kenematsu, *Chem. Pharm. Bull.* **6,** 98 (1958).
Metabolism: G. F. Lockwood, J. B. Houston, *J. Pharm.*

Pharmacol. **31**, 787 (1979). Hepatotoxicity: E. Bien *et al.*, *Pharmazie* **36**, 492 (1981).

Leaflets from ligroin, mp 107-109°. The aq soln is slightly alkaline to litmus. Soly in water is increased by the addition of sodium benzoate. One gram dissolves in 1.5 ml alcohol, 12 ml benzene, 1 ml chloroform, 13 ml ether, 18 ml water. Stable in air, but affected by light. Readily attacked by mild oxidizing agents in the presence of water. LD_{50} orally in rats: 1.7g/kg, Hart, *J. Pharmacol. Exp. Ther.* **89**, 205 (1947); for additional toxicity data *see* Tubaro *et al.*, *Arzneimittel-Forsch.* **20**, 1024 (1970).
Hydrochloride, $C_{13}H_{17}N_3O$.HCl, deliquescent prisms, mp 143-144°. Freely sol in water with acid reaction.
Bicamphorate, $C_{23}H_{33}N_3O_5$, *Pyramidon bicamphorate.* Cryst powder, mp 94°. Sol in water with gradual decompn; sol in alcohol.
Salicylate, $C_{20}H_{23}N_3O_4$, *Pyramidon salicylate.* Cryst powder, mp 70°. One gram dissolves in 16 ml water, about 6 ml alcohol.
Human Toxicity: Agranulocytosis may occur: E. Urbach, H. L. Goldburgh, *J. Am. Med. Assoc.* **131**, 893 (1946); G. Discombe, *Brit. Med. J.* **1952**, 1270.
THERAP CAT: Antipyretic, analgesic.
THERAP CAT (VET): Analgesic, antipyretic.

489. Aminoquinuride. *N,N'-Bis(4-amino-2-methyl-6-quinolinyl)urea;* bis(2-methyl-4-amino-6-quinolyl)urea; di(4-aminoquinald-6-yl)urea; aminoquincarbamide; aminochinuride; aminokinuride; Surfen. $C_{21}H_{20}N_6O$; mol wt 372.42. C 67.72%, H 5.41%, N 22.57%, O 4.30%. Prepn: Jensch, *Ger.* pat. **591,480** (1934 to I. G. Farbenind). Oncogenic and heparin-neutralizing properties: D. T. Hunter, J. M. Hill, *Nature* **191**, 1378 (1961).

Crystals from butanol, dec 255° (effervescence).
THERAP CAT: Antiseptic.

490. Aminorex. *4,5-Dihydro-5-phenyl-2-oxazolamine; 2-amino-5-phenyl-2-oxazoline;* aminoxafen; aminoxaphen. McN 742. $C_9H_{10}N_2O$; mol wt 162.19. C 66.65%, H 6.22%, N 17.27%, O 9.87%. Prepn: Poos *et al.*, *J. Med. Chem.* **6**, 266 (1963); Poos, **U.S.** pat. **3,161,650** (1964 to McNeil).

Crystals from benzene, mp 136-138°.
Fumarate, *Menocil, Apiquel.*
THERAP CAT: Anorexic.

491. p-Aminosalicylic Acid. *4-Amino-2-hydroxybenzoic acid;* 4-aminosalicylic acid; Deapasil; Apas; Apacil; Hellipidyl; **PAS**; PAS-C; Pamacyl; Parasal; Pascorbic; Pasolac; Parasalicil; Parasalindon; Pasnodia; Propasa; Rezipas; Sanipirol-4; Pamisyl; Para-Pas. $C_7H_7NO_3$; mol wt 153.13. C 54.90%, H 4.61%, N 9.15%, O 31.34%. Prepn: **Ger.** pat. **50,835** (1889); **U.S.** pat. **427,564** (1890); Sheehan, *J. Am. Chem. Soc.* **70**, 1665 (1948); Erlenmeyer *et al.*, *Helv. Chim. Acta* **31**, 988 (1948); Wenis, Gardner, *J. Am. Pharm. Assoc.* **38**, 9 (1949); Centolella, **U.S.** pat. **2,844,625** (1958 to Miles). Comprehensive description: M. M. A. Hassan *et al.*, *Analytical Profiles of Drug Substances* vol. **10**, K. Florey, Ed. (Academic Press, New York, 1981) pp 1-27.

Minute crystals from alcohol, mp 150-151° with effervescence. uv max (0.1N HCl): 265, 300 nm. pKa 3.25. pH of 0.1% aq soln: 3.5. One gram dissolves in about 500 ml water, in 21 ml alcohol. Slightly sol in ether, practically insol in benzene. Sol in dilute nitric acid, dil sodium hydroxide. At temps above 40°, aq solns of PAS and its hydrochloride are readily decarboxylated to give brown solns consisting mainly of *m*-aminophenol. LD_{50} orally in mice: 4 g/kg, E. M. Bavin *et al.*, *J. Pharm. Pharmacol.* **2**, 764 (1950).
Hydrochloride, $C_7H_6NNaO_3$.HCl, crystals, dec 224°.
Sodium salt dihydrate, $C_7H_6NNaO_3.2H_2O$, *Aminox* (also as Ca salt), *Gabbropas, Pasalon, Pasalon-Rakeet, Passodico, Salvis, Pasid, Tubersan, Sanipirol, Pamisyl Sodium, Pasmed Sodium, Aminopar, Paramycin, Nippas, Lepasen, Paramisan, Bactylan, Osacyl, Entepas, Enteropas, PAS sodium dihydrate.* pH of 1% soln ∼ 7. One gram dissolves in 2 ml water. Very sparingly sol in acetone. Practically insol in ether, chloroform, benzene. Solns of the sodium salt are more resistant to heat than PAS, but sterilization by bacterial filtration is recommended.
Calcium salt, $C_{14}H_{12}CaN_2O_6$, *P.A.C., Paracipan, Pasura Calcium, Pasmicina, Aminacyl.* Bittersweet crystals. One gram dissolves in about 7 ml water, slightly sol in alc. Aq solns dec slowly and darken in color. Also marketed as the hemihydrate and the trihydrate.
Potassium salt, $C_7H_6NO_3K$, *Paskalium, Paskate.* Crystals, freely sol in water. pH of 1% soln about 7. Reported to cause less gastric irritation than the free acid or the sodium salt.
Ethyl ester, $C_9H_{11}NO_3$, needles from alc, mp 115°. Prepd by reduction of the ethyl ester of 4-nitrosalicylic acid.
Phenyl ester, *see* Phenyl *p*-Aminosalicylic Acid.
THERAP CAT: Antibacterial (tuberculostatic).

492. p-Aminosalicylic Acid Hydrazide. *4-Amino-2-hydroxybenzoic acid hydrazide;* Apacizin; Apacizina. $C_7H_9N_3O_2$; mol wt 167.17. C 50.29%, H 5.43%, N 25.14%, O 19.14%. Prepn: Drain *et al.*, *J. Chem. Soc.* **1949**, 1498; Magrane, **Span.** pat. **206,645** (1952), *C.A.* **49**, 5529e (1955); **Japan.** pat. **7472('54)**, *C.A.* **50**, 9947c (1956).

Needles from alcohol, mp 190-200°. Slightly soluble in water, somewhat more in ethanol.
THERAP CAT: Antibacterial (tuberculostatic).

493. 4-Amino-2-sulfobenzoic Acid. *p*-Amino-*o*-sulfobenzoic acid. $C_7H_7NO_5S$; mol wt 217.20. C 38.71%, H 3.25%, N 6.45%, O 36.83%, S 14.76%. Prepd by treating the potassium salt of 4-nitro-2-sulfobenzoic acid with ammonium sulfide: Hart, *Am. Chem. J.* **1**, 351 (1879/80); Kastle, *ibid.* **44**, 490 (1910).

Needles from water. Soly in water: 3 g/l at 25°. Almost insol in alcohol, ether.
USE: Proposed as alkalimetric standard.

494. 2-Aminothiazole. *2-Thiazolamine;* Abadol; Basedol. $C_3H_4N_2S$; mol wt 100.14. C 35.98%, H 4.03%, N 27.97%, S 32.02%. Prepn from vinyl acetate: Christiansen, U.S. pat. **2,242,237** (1941 to Squibb); Kyrides, U.S. pat. **2,330,223** (1943 to Monsanto); cf. Skrimshire, **Brit.** pat. **540,032** (1941 to B.D.H.). Prepd also by condensing tribromoparaldehyde with thiourea: Leitch, Brickman, U.S. pats. **2,230,962** (1941); **2,339,083** (1944 to Mallinckrodt). Prepn from paraldehyde and thiourea: Erlenmeyer *et al., Helv. Chim. Acta* **38**, 1293 (1955). Toxicology: W. B. Deichmann *et al., J. Ind. Hyg. Toxicol.* **30**, 71 (1948).

Crystals from benzene + petr ether, mp 93°. Distills at 3 mm without decompn. Sol in hot water. Slightly sol in cold water, alc, ether. Freely sol in dil HCl and in 20% H_2SO_4. LD_{50} orally in rats: 0.48 g/kg, W. B. Deichmann *et al., loc. cit.*

Hydrochloride monohydrate, $C_3H_4N_2S.HCl.H_2O$, needles, freely sol in water. Also used as the acid tartrate. Crystals, sol in water.

THERAP CAT: Thyroid inhibitor.

495. 2-Amino-1,1,3-tricyanopropene. *2-Amino-1-propene-1,1,3-tricarbonitrile;* 2-amino-1,3,3-tricyano-2-propene; malononitrile dimer. $C_6H_4N_4$; mol wt 132.12. C 54.54%, H 3.05%, N 42.41%. $(CN)_2C=C(NH_2)CH_2CN$. Prepn: Carboni *et al., J. Am. Chem. Soc.* **80**, 2838 (1958); Carboni, U.S. pat. **2,719,861** (1955 to du Pont); Coenen, **Ger.** pat. **922,531** (1955 to Bayer).

Rod-like crystals from water, mp 170-173°.

496. Amioca. *Amylopectin.* A non-linear polymer of glucose, obtained from waxy corn: Caldwell, *Converter* **17**, No. 11, 12, 14 (1943); *Mfg. Confectioner* **33**, No. 12, 15, 17 (1943), *C.A.* **38**, 1137[1] (1944); Schopmeyer, *Food Ind.* **17**, 1476 (1945), *C.A.* **40**, 2331[3] (1946). *Review:* Powell in *Industrial Gums,* R. L. Whistler, Ed. (Academic Press, New York, 2nd ed., 1973) pp 567-576.

497. Amiodarone. *(2-Butyl-3-benzofuranyl)[4-[2-(diethylamino)ethoxy]-3,5-diiodophenyl]methanone; 2-butyl-3-benzofuranyl 4-[2-(diethylamino)ethoxy]-3,5-diiodophenyl ketone;* 2-butyl-3-[3,5-diiodo-4-(β-diethylaminoethoxy)-benzoyl]benzofuran. $C_{25}H_{29}I_2NO_3$; mol wt 645.32. C 46.53%, H 4.53%, I 39.33%, N 2.17%, O 7.44%. Benzofuran derivative capable of blocking both α- and β-adrenoceptors. Prepn: **Fr.** pat. **1,339,389;** R. Tondeur, F. Binon, U.S. pat. **3,248,401** (1963, 1966 to Soc. Belge l'Azote Prod. Chim. Marly). Pharmacology: R. Charlier *et al., Arch. Int. Pharmacodyn.* **139**, 234 (1962); A. Baudine *et al., Arch. Int. Pharmacodyn. Ther.* **169**, 469 (1967); R. Charlier *et al., Arzneimittel-Forsch.* **18**, 1408 (1968). Metabolism: J. Broekhuysen *et al., Arch. Int. Pharmacodyn. Ther.* **177**, 340 (1969). Comparative antidysrhythmic profile: C. Labrid *et al., ibid.* **249**, 87 (1981). Symposia on pharmacology, pharmacokinetics, clinical studies: *Am. Heart J.* **106**, 787-964 (1983); *Drugs* **29**, Suppl. 3, 1-56 (1985). Review of pharmacology, clinical efficacy and safety: J. W. Mason, *N. Engl. J. Med.* **316**, 455-466 (1987).

Hydrochloride, $C_{25}H_{30}ClI_2NO_3$, *L-3428, Amiodar, Ancoron, Angiodarona, Atlansil, Cordarex, Cordarone, Cordarone X, Miocard, Miodaron, Ortacrone, Ritmocardyl, Rythmarone, Trangorex.* Crystalline powder, mp 156°.

THERAP CAT: Anti-arrhythmic; anti-anginal.

498. Amiphenazole. *5-Phenyl-2,4-thiazolediamine; 2,4-diamino-5-phenylthiazole;* DAPT; phenamizole; Dizol;

Daptazole; Daptazile; Fenamizol. $C_9H_9N_3S$; mol wt 191.26. C 56.52%, H 4.75%, N 21.98%, S 16.77%. Prepd by the interaction of thiourea and α-bromobenzyl cyanide in alcohol: Davies *et al., J. Chem. Soc.* **1950**, 3491; from thiourea and α-cyanobenzyl benzenesulfonate in acetone: Dodson, Turner, *J. Am. Chem. Soc.* **73**, 4517 (1951).

Flakes from water or dilute alcohol, dec 163-164°. Turns brown on exposure to light and air.

Hydrobromide, $C_9H_9N_3S.HBr$, prisms from alc, dec >250°. Freely sol in hot water, moderately in cold water.

Benzenesulfonate, $C_{15}H_{15}N_3O_3S_2$, crystals from alcohol + ether, dec 261-262°. Practically insol in water, most organic solvents. Slightly sol in alcohol.

THERAP CAT: Narcotic antagonist.

THERAP CAT (VET): Barbiturate and morphine antagonist.

499. Amiprilose. *3-O-[3-(Dimethylamino)propyl]-1,2-O-(1-methylethylidene)-α-D-glucofuranose;* 1,2-O-isopropylidene-3-O-[3'-(N,N-dimethylamino)propyl]-α-D-glucofuranose. $C_{14}H_{27}NO_6$; mol wt 305.37. C 55.06%, H 8.91%, N 4.59%, O 31.44%. Synthetic substituted monosaccharide with immunomodulatory activity. Prepn: **Belg.** pat. **823,313;** P. Gordon, U.S. pat. **3,939,146** (1975, 1976 both to Strategic Med. Res.). Comparative *in vitro* effects on lymphocyte and macrophages: J. W. Hadden *et al., Int. J. Immunopharmacol.* **1**, 17 (1979). Prepn, properties and pharmacokinetics in dogs: E. R. Garrett *et al., J. Pharm. Sci.* **71**, 387 (1982). Human pharmacokinetics: *eidem, J. Pharmacokinet. Biopharm.* **10**, 247 (1982). *In vitro* effect on phagocytes: C. J. Morrison *et al., Antimicrob. Ag. Chemother.* **26**, 74 (1984).

Colorless, viscous oil. n^{25} 1.4687.

Hydrochloride, $C_{14}H_{28}ClNO_6$, *SM-1213, Therafectin.* Crystals from methanol, mp 181-183°. Sol in water, methanol, hot ethanol.

THERAP CAT: Immunomodulator.

500. Amisometradine. *6-Amino-3-methyl-1-(2-methyl-2-propenyl)-2,4(1H,3H)-pyrimidinedione; 6-amino-3-methyl-1-(2-methylallyl)uracil;* 1-methallyl-3-methyl-6-aminotetrahydro-2,4-pyrimidinedione; 6-amino-1-methallyl-3-methylpyrimidine-2,4-dione; aminoisometradine; Rolicton. $C_9H_{13}N_3O_2$; mol wt 195.22. C 55.37%, H 6.71%, N 21.53%, O 16.39%. Prepn: Papesch, Schroeder, U.S. pat. **2,729,669** (1956 to Searle).

Crystals, mp 175°. Soly in water (25°): 2.0 g/100 ml. Freely sol in alcohol, acetone; insol in ether. LD_{50} in mice: 610 mg/kg orally; 415 mg/kg i.p.

THERAP CAT: Diuretic.

501. Amisulpride. *4-Amino-N-[(1-ethyl-2-pyrrolidinyl)methyl]-5-(ethylsulfonyl)-2-methoxybenzamide;* 4-amino-N-[(1-ethyl-2-pyrrolidinyl)methyl]-5-(ethylsulfonyl)-o-anisamide; aminosultopride; DAN 2163; Socian; Solian. $C_{17}H_{27}N_3O_4S$; mol wt 369.48. C 55.26%, H 7.37%, N 11.37%, O 17.32%, S 8.68%. Neuroleptic agent, analog of sulpiride, sultopride, *q.q.v.* Prepn and pharmacology: M. Thominet *et al.,* **Belg. pat.** 872,585; *eidem,* **U.S. pat.** 4,401,-822 (1979, 1983 both to Soc. Etudes Sci. Ind. l'Ile-de-France). Dopamine antagonist activity in rats: P. Protais *et al., Neuropharmacol.* **24,** 861 (1985); in mice: M. Vasse *et al., Arch. Pharmacol.* **329,** 108 (1985). Clinical study in schizophrenics: K. Mann *et al., Pharmacopsychiat.* **17,** 111 (1984).

CONHCH$_2$ — N — OCH$_3$ — C$_2$H$_5$

C$_2$H$_5$O$_2$S — NH$_2$

Crystals from acetone, mp 126-127°. LD$_{50}$ in male mice (mg/kg): 56-60 i.v.; 175-180 i.p.; 224-250 s.c.; 1024-1054 orally (Thominet).

THERAP CAT: Antipsychotic.

502. Amiton. *S-[2-(Diethylamino)ethyl]phosphorothioic acid O,O-diethyl ester; O,O-diethyl S-(β-diethylamino)ethyl phosphorothiolate;* Inferno; Metramac; Tetram. $C_{10}H_{24}$-NO_3PS; mol wt 269.35. C 44.59%, H 8.98%, N 5.20%, O 17.82%, P 11.50%, S 11.91%. Prepn: Ghosh, Newman, *Chem. & Ind. (London)* **1955,** 118; Fukuto, Stafford, *J. Am. Chem. Soc.* **79,** 6083 (1957); Lorenz, Schrader, **U.S. pat.** 3,082,240 (1963 to Bayer).

C$_2$H$_5$O — P — O — SCH$_2$CH$_2$N(C$_2$H$_5$)$_2$ — C$_2$H$_5$O

Liquid. bp$_{0.01}$ 76°; bp$_{0.2}$ 110°. n_D^{27} 1.4655. LD$_{50}$ orally in rats: 5.4 mg/kg, Frawley *et al., Toxicol. Appl. Pharmacol.* **5,** 605 (1963).

Acid oxalate (hydrogen oxalate), $C_{10}H_{24}NO_3PS \cdot C_2H_2O_4$, crystals from isopropanol + ether, mp 98-99°. LD$_{50}$ orally in male albino rats: 9 mg/kg, Shaffer, West, *Toxicol. Appl. Pharmacol.* **2,** 1 (1960).

USE: Contact insecticide, miticide. *Caution:* Cholinesterase inhibitor.

503. Amitraz. *N'-(2,4-Dimethylphenyl)-N-[[(2,4-dimethylphenyl)imino]methyl]-N-methylmethanimidamide; N-methyl-N'-2,4-xylyl-N-(N-2,4-xylylformimidoyl)formamidine; N,N-di-(2,4-xylyliminomethyl)methylamine;* 2-methyl-1,3-di-(2,4-xylylimino)-2-azapropane; 1,5-di-(2,4-dimethylphenyl)-3-methyl-1,3,5-triazapenta-1,4-diene; N-methylbis(2,4-xylyliminomethyl)amine; BTS 27419; U 36059; ENT 27967; BAAM; Mitaban; Mitac; Taktic. C_{19}-$H_{23}N_3$; mol wt 293.41. C 77.78%, H 7.90%, N 14.32%. Prepn: I. R. Harrison *et al.,* **Ger. pat.** 2,061,132 corresp to **U.S. pats.** 3,781,355, 3,864,497 (1971, 1973, 1975 all to Boots). Activity: *eidem, Pestic. Sci.* **3,** 679 (1972); **4,** 901 (1973). Bacterial degradation: P. B. Baker, D. R. Woods, *J. Appl. Bacteriol.* **42,** 187 (1977); E. R. Allcock, D. R. Woods, *ibid.* **44,** 383 (1978). Metabolism: C. O. Knowles, H. J. Benezet, *J. Environ. Sci. Health* **16,** 547 (1981). Use in treatment of demodecosis in dogs: H. Farmer, A. A. Seawright, *Aust. Vet. J.* **56,** 537 (1980).

CH$_3$ — CH$_3$ — N=CH—N—CH=N — H$_3$C — CH$_3$ — CH$_3$

White monoclinic needles, mp 86-87°. Soly in water: 1

ppm. Sol in most org solvents. Unstable to acidic pH. LD$_{50}$ orally in mice, rats: 1600, 400 mg/kg (Harrison, second U.S. patent).

USE: Acaricide, insecticide.

THERAP CAT: Scabicide.

THERAP CAT (VET): In treatment of demodectic mange in dogs.

504. Amitriptyline. *3-(10,11-Dihydro-5H-dibenzo-[a,d]cyclohepten-5-ylidene)-N,N-dimethyl-1-propanamine; 10,11-dihydro-N,N-dimethyl-5H-dibenzo[a,d]cycloheptene-Δ$^{5,\gamma}$-propylamine;* 5-(γ-dimethylaminopropylidene)-5H-dibenzo[a,d]-10,11-dihydrocycloheptene; 10,11-dihydro-5-(γ-dimethylaminopropylidene)-5H-dibenzo[a,d]cycloheptene; 5-(3-dimethylaminopropylidene)dibenzo[a,d][1,4]-cycloheptadiene; Ro 4-1575; Triptisol; Tryptanol; Sarotex; Seroten; Laroxyl; Redomex; Adepril. $C_{20}H_{23}N$; mol wt 277.39. C 86.59%, H 8.36%, N 5.05%. Prepn: Hoffsommer *et al., J. Org. Chem.* **27,** 4134 (1962); **28,** 1751 (1963); *J. Med. Chem.* **8,** 555 (1965); Engelhardt *et al.,* **Belg. pat.** 584,-061 (1960 to Merck & Co.); **Brit. pats.** 858,187-8 (both 1961 to Hoffmann-La Roche); Tristram, Tull, **U.S. pat.** 3,205,264 (1965 to Merck & Co.). Toxicity data: A. Tobe *et al., Arzneimittel-Forsch.* **31,** 1278 (1981). Comprehensive description: K. W. Blessel *et al.* in *Analytical Profiles of Drug Substances* Vol. 3, K. Florey, Ed. (Academic Press, New York, 1974) pp 127-148.

CHCH$_2$CH$_2$N(CH$_3$)$_2$

Hydrochloride, $C_{20}H_{24}ClN$, *Amitid, Amitril, Deprex, Domical, Elavil, Endep, Euplit, Lentizol, Miketorin, Saroten, Sylvemid, Tryptizol.* Minute crystals, mp 196-197°. Freely sol in water, chloroform, alc. uv max (methanol): 240 nm (ε 13800). pKa 9.4. LD$_{50}$ in mice, rats (mg/kg): 350, 380 orally; 65, 75 i.p. (Tobe).

THERAP CAT: Antidepressant.

505. Amitriptylinoxide. *3-(10,11-Dihydro-5H-dibenzo[a,d]cyclohepten-5-ylidene)-N,N-dimethyl-1-propanamine N-oxide; 10,11-dihydro-N,N-dimethyl-5H-dibenzo[a,d]cycloheptene-Δ$^{5,\gamma}$-propylamine N-oxide;* amitriptyline N-oxide; Ambivalon; Equilibrin. $C_{20}H_{23}NO$; mol wt 293.41. C 81.87%, H 7.90%, N 4.77%, O 5.45%. Centrally acting metabolite of amitriptyline, *q.v.* Prepn: J. B. Pedersen, **Brit. pat.** 991,651 corresp to **U.S. pat.** 3,299,139 (1965, 1967 both to Dumex). Series of articles on pharmacology, pharmacokinetics, metabolism, clinical studies, toxicity studies, teratological studies: *Arzneimittel-Forsch.* **28,** 1873-1926 (1978). HPLC determn: K. M. Jensen, *J. Chromatog.* **183,** 321 (1980). Neuropharmacology: J. Hyttel *et al., Acta Pharmacol. Toxicol.* **47,** 53 (1980).

CHCH$_2$CH$_2$N(CH$_3$)$_2$ ↓ O

Cryst, mp 228-230°. (Dihydrate, mp 102-103°.) LD$_{50}$ in mice, guinea pigs, rabbits, dogs (mg/kg): between 330-460 orally; in mice, rats: 87, 25 i.v.; 320, 110 i.p., H. Friehe, R. Fontaine, *Arzneimittel-Forsch.* **28,** 1898 (1978).

THERAP CAT: Antidepressant.

506. Amitrole. *1H-1,2,4-Triazol-3-amine;* 3-amino-1H-1,2,4-triazole; aminotriazole; ATA; ENT 25445; Amizol; Cytrol; Weedazol. $C_2H_4N_4$; mol wt 84.08. C 28.57%, H 4.80%, N 66.64%. Non-selective post-emergence, translocated herbicide. Prepn: G. Sjostedt, L. Gringas, *Org. Syn.* **coll. vol. III,** 95 (1955). Use as herbicide: Allen, **U.S. pat.** 2,670,282 (1954 to Am. Chemical Paint Co.). Antithyroid activity: T. H. Jukes, C. B. Shaffer, *Science* **132,** 296 (1960).

Consult the cross index before using this section.

Review: E. Kröller, Residue Rev. **12**, 163-192 (1966). Review of carcinogenicity studies: IARC Monographs **7**, 31-43 (1974).

Crystals from abs ethanol, mp 159°. Soluble in water, methanol, ethanol, chloroform. Sparingly sol in ethyl acetate. Insol in ether, acetone. Aq solns are neutral. LD_{50} in mice, rats (g/kg): 14.7, 25.0 orally (Kröller).

Hydrochloride, $C_2H_4N_4$.HCl, crystals from alcohol, mp 153°.

Note: This substance may reasonably be anticipated to be a carcinogen: Fourth Annual Report on Carcinogens (NTP 85-002, 1985) p 23.

USE: Herbicide.

507. Amixetrine. 1-[2-(3-Methylbutoxy)-2-phenylethyl]pyrrolidine; 1-[β-(isopentyloxy)phenethyl]pyrrolidine; N-(2-phenyl-2-isoamyloxy)ethylpyrrolidine. $C_{17}H_{27}NO$; mol wt 261.42. C 78.11%, H 10.41%, N 5.36%, O 6.12%. Prepn by reacting styrene and isoamyl alcohol with t-butyl hypobromite and then condensing the product with pyrrolidine: Mauvernay, Busch, Ger. pat. **1,811,767** (1969 to C.E.R.M.), C.A. **72**, 21607e (1970). Identification of urinary metabolites in humans and dogs: M. Constantin et al., Arzneimittel-Forsch. **26**, 80 (1976). Pharmcokinetics: M. Constantin, J. F. Pognat, ibid. **28**, 646 (1978).

bp_2 121°. $n_D^{21.6}$ 1.4978.

Hydrochloride, $C_{17}H_{28}ClNO$, Somagest. Crystals, mp 150°.

THERAP CAT: Anti-inflammatory; anticholinergic.

508. Amlexanox. 2-Amino-7-(1-methylethyl)-5-oxo-5H-[1]benzopyrano(2,3-b)pyridine-3-carboxylic acid; 2-amino-7-isopropyl-5-oxo-5H-[1]benzopyrano[2,3-b]pyridine-3-carboxylic acid; 2-amino-7-isopropyl-1-azaxanthone-3-carboxylic acid; amoxanox; AA-673; CHX-3673; Elics; Solfa. $C_{16}H_{14}N_2O_4$; mol wt 298.30. C 64.42%, H 4.73%, N 9.39%, O 21.45%. Orally active lipoxygenase inhibitor. Prepn: A. Nohara et al., Belg. pat. **864,647**; U.S. pat. **4,143,042** (1978, 1979 both to Takeda); eidem, J. Med. Chem. **28**, 559 (1985). Series of articles on toxicity, teratogenicity, mutagenicity, reproductive and developmental effects: Yakuri to Chiryo **13**, 4873-4931 (1985), C.A. **104**, 61684p-61691p (1986). Mode of action: T. Saijo et al., Int. Arch. Allergy Appl. Immunol. **79**, 231 (1986).

Crystals from dimethylformamide, mp > 300°.

THERAP CAT: Anti-allergic; anti-asthmatic.

509. Amlodipine. 2-[(2-Aminoethoxy)methyl]-4-(2-chlorophenyl)-1,4-dihydro-6-methyl-3,5-pyridinedicarboxylic acid 3-ethyl 5-methyl ester; (±)-2-[(2-aminoethoxy)methyl]-4-(2-chlorophenyl)-3-ethoxycarbonyl-5-methoxycarbonyl-6-methyl-1,4-dihydropyridine; UK-48340. $C_{20}H_{25}ClN_2O_5$; mol wt 408.88. C 58.75%, H 6.16%, Cl 8.67%, N 6.85%, O 19.56%. Dihydropyridine calcium channel blocker with activity residing mainly in the (−)-isomer. Prepn: S. F. Campbell et al., Eur. pat. Appl. **89,167**; eidem, U.S. pat. **4,572,909** (1983, 1986 both to Pfizer). Synthesis of racemate and enantiomers; preliminary pharmacology: J. E. Arrowsmith et al., J. Med. Chem. **29**, 1696 (1986). Antag-

onist activity: R. A. Burges et al., J. Cardiovasc. Pharmacol. **9**, 110 (1987). Pharmacokinetics: J. K. Faulkner et al., Brit. J. Clin. Pharmacol. **22**, 21 (1986). Metabolism in animals: A. P. Beresford et al., Xenobiotica **18**, 169 (1988); in man: A. P. Beresford et al., ibid. 245. Effects on renal system in hypertensive men: G. P. Reams et al., Am. J. Kidney Dis. **10**, 446 (1987). GC determn in plasma: A. P. Beresford et al., J. Chromatog. **420**, 178 (1987). Clinical evaluation in hypertension: J. Webster et al., Brit. J. Clin. Pharmacol. **24**, 713 (1987).

(±)-Form maleate, $C_{24}H_{29}ClN_2O_9$, UK-48340-11. White crystals from ethyl acetate, mp 178-179°.

THERAP CAT: Antianginal; antihypertensive.

510. Ammonia. H_3N; mol wt 17.03. N 82.25%, H 17.75%. NH_3. Manufactured from water gas (obtained by blowing steam through incandescent coke) as source of hydrogen, and from producer gas (obtained from steam and air through incandescent coke), as source of nitrogen by the Haber-Bosch process. Manuf from natural gas: Faith, Keyes & Clark's Industrial Chemicals, F. A. Lowenheim, M. K. Moran, Eds. (Wiley-Interscience, New York, 4th ed., 1975) pp 83-92. Historical monograph: A. Mittasch, Geschichte der Ammoniaksynthese (Verlag Chemie, 1951). Reviews of prepn, properties and chemistry: Several authors in Mellor's Vol. VIII, supplement I, Nitrogen part 1 (1964) pp 240-369; Jones in Comprehensive Inorganic Chemistry Vol. 2, J. C. Bailar, Jr. et al., Eds. (Pergamon Press, Oxford, 1973) pp 199-227; J. R. LeBlanc et al., in Kirk-Othmer Encyclopedia of Chemical Technology vol. 2 (Wiley-Interscience, New York, 3rd ed., 1978) pp 470-516.

Colorless gas; very pungent odor (characteristic of drying urine). Lower limit of human perception: 0.04 g/cubic meter or 53 ppm. One liter of the gas weighs 0.7714 g. d 0.5967 (air = 1). mp −77.7°. bp_{760} −33.35°. Densities of liq NH_3 (temp; press.): 0.6818 (−33.35°; 1 atm); 0.6585 (−15°; 2.332 atm); 0.6386 (0°; 4.238 atm); 0.6175 (15°; 7.188 atm); 0.5875 (35°; 13.321 atm). Critical temp 132.4°; critical press. 111.5 atm. Heat capacity (25°) 8.38 cal/mole/deg. Mixtures of ammonia and air will explode when ignited under favorable conditions: Angew. Chem. **43**, 302 (1930), but ammonia is generally regarded as nonflammable. Corrosive, alkaline gas. pH of 1.0N aq soln 11.6; of 0.1N aq soln 11.1; of 0.01N aq soln 10.6. Water at 0° holds 47%, at 15° 38%, at 20° 34%, at 25° 31%, at 30° 28%, at 50° 18%. d_4^{20} (aq solns): 0.9939 (1%); 0.9811 (4%); 0.9651 (8%); 0.9362 (16%); 0.9229 (20%); 0.9101 (24%); 0.8980 (28%). fp (aq solns): −2.9° (4%); −8.1° (8%); −23.1° (16%); −34.9° (20%); −44.5° (24%); −69.2° (28%). Solution of NH_3 in water is exothermic. 95% alcohol at 20° holds 15%, at 30° 11%. Abs ethanol at 0° 20%, at 25° 10%. Methanol at 25° 16%. It is also sol in chloroform and ether. Liquid ammonia produces low temps by its own evaporation. Heat of vaporization: 5.581 kcal/mole. It is a good solvent for many elements and compds. Usually marketed in liquefied form in steel cylinders or as ammonia water (aqua ammonia, ammonium hydroxide) in drums and bottles.

Caution: Inhalation of concd vapor causes edema of respiratory tract, spasm of the glottis, asphyxia. Treatment must be prompt to prevent death, cf. Patty's Industrial Hygiene and Toxicology vol. 2B, G. D. Clayton, F. E. Clayton, Eds. (Wiley-Interscience, New York, 3rd ed., 1981) pp 3045-3052.

USE: Manuf nitric acid, explosives, synthetic fibers, fertilizers. In refrigeration. In the chemical industry.

511. Ammoniacum. Gum ammoniac. A gum-resin exuded from the flowering and fruiting stem of Dorema

ammoniacum, D. Don, and probably other *Umbelliferae.*
Habit. Persia, Northern India, Southern Siberia, Africa.
Constit. 1.3-6.7% volatile oil; 50-70% resin; 18-26% gum;
ash content about 2%, may be as high as 10%; salicylic acid;
aminoresinol. *Ref:* W. Sandermann, *Naturharze, Terpen-
tinöl, Tallöl* (Springer, 1960) pp 82-83.

Irregular rounded tears, yellowish or brownish outside
and whitish within; brittle when cold, but soft when warm;
also masses, darker in color and less homogeneous; peculiar
odor; slightly sweetish, bitter, somewhat acrid taste. mp
45-55°. Acid no. 60-80. Sapon no. 97-114. d 1.207. Partly
soluble in water, alcohol, ether, vinegar or alkali soln; forms
emulsions with water.

USE: Ingredient of porcelain cements.

THERAP CAT: Diaphoretic, emmenagogue.

512. Ammonia Water—10%. Colorless liquid; very
pungent odor. d_{25}^{25} 0.957.

Caution: Irritating to eyes, mucous membranes.

THERAP CAT: Reflex respiratory stimulant.

THERAP CAT (VET): Externally on bites and stings. As a
rubefacient on bruises, sprains. Inhalant as a stimulant.
Internally as an antacid and carminative.

513. Ammonia Water, Stronger. Ammonium hydroxide; aqua ammonia; "Spirit of Hartshorn". A soln of 28-
29% NH_3 in water.

Colorless liquid; intense, pungent, suffocating odor; acrid
taste; strong alkaline reaction. d_{25}^{25} about 0.90, 26°Bé. It
dissolves copper, zinc. Fumes are formed when ammonia
water is brought near volatile acids. *Keep cool in strong
glass, plastic or rubber-stoppered bottles not completely filled.*
Caution: Reaction with H_2SO_4 or other strong mineral acids
is exothermic; mixture becomes boiling hot.

USE: Detergent, removing stains, bleaching, calico printing, extracting plant colors (cochineal, archil, etc.) and alkaloids; manuf ammonium salts, aniline dyes, and a wide variety of other uses.

514. Ammonium Acetate. Acetic acid ammonium salt.
$C_2H_7NO_2$; mol wt 77.08. C 31.16%, H 9.15%, N 18.17%, O
41.51%. CH_3COONH_4. Commercial product contains
95-97% salt with acetic acid and some water. Prepd from
acetic acid and NH_3: Zuffanti, *J. Am. Chem. Soc.* **63**, 3123
(1941). Toxicity data: Welch *et al., J. Lab. Clin. Med.* **29**,
809 (1944).

Deliquesc crystals or crystalline masses. Slight acetous
odor. d 1.07. mp 114° (Zuffanti, *loc. cit.).* Tends to lose
NH_3. Sol in less than 1 part water; freely sol in alc; slightly
sol in acetone. Very concd aq soln is slightly acid; a 0.5
molar aq soln has pH 7.0. *Keep cool and tightly closed.* LD
i.v. in mice: 1.8 mg (NH_4^+)/20g (Welch).

Ammonium acetate solution, Spirit of Mindererus. Colorless, clear liquid; acid reaction to litmus. Prepd from 1 g
ammonium carbonate, 20 ml dil acetic acid (6%). Contains
6.5-7.5% CH_3COONH_4.

USE: Preserving meats, dyeing, stripping; as a reagent in
anal. chemistry, *e.g.,* for determining Pb, Fe; separating
$PbSO_4$ from other sulfates.

THERAP CAT: Diuretic.

THERAP CAT (VET): Formerly as diuretic, antipyretic.

515. Ammonium Benzoate. $C_7H_9NO_2$; mol wt 139.15.
C 60.42%, H 6.52%, N 10.07%, O 23.00%. $C_6H_5COONH_4$.
It is about 99% pure. Manuf from benzoic acid and NH_3:
Spina, **U.S.** pat. **1,704,636** (1929 to Hooker Electrochem.).

Lamellar crystals or crystalline powder; odorless or faint
benzoic acid odor; gradually loses NH_3 on exposure to air.
d 1.26. mp 198°. One gram dissolves in 4.7 ml water, 1.2 ml
boil. water, 36 ml alcohol, 8 ml boil. alcohol, 8 ml glycerol.
The aq soln is slightly acid. *Keep well closed. Incompat.*
Ferric salts, acids, alkali hydroxides or carbonates.

USE: To preserve glue and latex.

THERAP CAT: Urinary anti-infective.

516. Ammonium Bicarbonate. Acid ammonium carbonate; ammonium hydrogen carbonate. CH_5NO_3; mol wt
79.06. C 15.19%, H 6.38%, N 17.72%, O 60.71%. NH_4-
HCO_3. Occurs in the urine of alligators: Coulson, Hernan-
dez, *Proc. Soc. Exp. Biol. Med.* **88**, 682 (1955). Usually
prepd by passing an excess of carbon dioxide through concd
ammonia water. Manuf: **Brit.** pat. **304,872** (1927 to I.G.

Farben.); Brooks, **Brit.** pat. **742,386** (1955 to I.C.I.). *See
also* Ammonium Carbonate.

Shiny, hard, colorless or white prisms or crystalline mass.
Faint odor of ammonia. Comparatively stable at room
temp. Volatile with decompn at about 60°. The white fumes
given off consist of NH_3 21.5%, CO_2 55.7%, H_2O vapor
22.8%. Rate of decompn increases as temp rises. mp 107.5°
(very rapid heating). Soly in water: 14% (10°); 17.4% (20°);
21.3% (30°). Decomposed by hot water. One gram dissolves
in 10 ml glycerol (pharmaceutical grade). pH of 0.1N soln
in water at 25° = 7.8. Insol in alcohol and acetone. Negative heat of solution.

Pharmaceutical Incompat: Acids, caustic alkalies.

USE: In baking powder formulations; in cooling baths (one
kg dissolved in 5 liters H_2O at 17° lowers it to 7°); in fire extinguishers; manuf porous plastics, ceramics; manuf dyes,
pigments; in compost heaps to accelerate decompn; as fertilizer; for defatting textiles; in cold wave solns; in chrome
leather tanning; to remove gypsum from heat exchanges and
other processing equipment.

THERAP CAT: Expectorant.

THERAP CAT (VET): Expectorant. Used in bloat, colic.

517. Ammonium Bifluoride. Acid ammonium fluoride;
ammonium hydrogen fluoride. H_5F_2N; mol wt 57.05. H
8.83%, F 66.61%, N 24.55%. NH_4HF_2. Prepn from hydrofluoric acid and NH_3: Hassel, Luzanski, *Z. Kristallogr.* **A83**,
449 (1932); *Gmelin's, Ammonium* (8th ed.) **23**, 148 (1936).

Orthorhombic crystals which readily etch glass. d 1.5.
mp 124.6°. Freely sol in water. *Keep in plastic, rubber, wool
or paraffined containers and well closed. See also* Ammonium
Fluoride.

USE: In manuf of Mg and Mg alloys; in brightening of Al;
for purifying and cleansing various parts of beer-dispensing
apparatus, tubes, etc., sterilizing dairy and other food equipment; in glass and porcelain industries; as mordant for aluminum; as a "sour" in laundering cloth. In lab production
of HF.

518. Ammonium Bimalate. *Hydroxybutanedioic acid
monoammonium salt; l-malic acid monoammonium salt;*
ammonium acid malate. $C_4H_9NO_5$; mol wt 151.12. C
31.79%, H 6.00%, N 9.27%, O 52.94%. NH_4OOCCH_2-
CH(OH)COOH or HOOCCH$_2$CH(OH)COONH$_4$. Prepn:
Kendrick, *Ber.* **30**, 1749 (1897); Holmberg, *ibid.* **61**, 1885
(1928).

Orthorhombic disphenoidal crystals. mp 160-161°. d
1.15. Sol in 3 parts water; slightly sol in alcohol. $[\alpha]_D^{18}$
−6.4° (c = 6).

519. Ammonium Binoxalate. Ammonium acid oxalate;
ammonium hydrogen oxalate. $C_2H_5NO_4$; mol wt 107.07. C
22.43%, H 4.71%, N 13.08%, O 59.77%. $NH_4OOCCOOH$.
Prepn of monohydrate: Dehn, Heuse, *J. Am. Chem. Soc.* **29**,
1137 (1907). *Review: Gmelin's, Ammonium* (8th ed.) **23**,
pp 405-406 (1936).

Monohydrate, rhombic crystals. *Poisonous.* d 1.56. Sol in
25 parts water; slightly sol in alcohol.

USE: To remove ink stains.

520. Ammonium Bisulfate. Acid ammonium sulfate;
ammonium hydrogen sulfate. H_5NO_4S; mol wt 115.11. H
4.38%, N 12.17%, O 55.60%, S 27.86%. NH_4HSO_4. Prepn:
Gmelin's, Ammonium (8th ed.) **23**, 293-298 (1936).

Deliquesc crystals. mp about 147°. d 1.787. Freely sol in
water; practically insol in alcohol, acetone, pyridine. *Keep
well closed.*

USE: In hair-waving preparations; as catalyst for organic
reactions.

521. Ammonium Bisulfide. Ammonium hydrogen sulfide; ammonium hydrosulfide; ammonium sulfhydrate. H_5-
NS; mol wt 51.11. H 9.86%, N 27.41%, S 62.73%. NH_4HS.
Prepd by mixing stoichiometric amts of NH_3 and H_2S gases
at 0°: Thomas, Riding, *J. Chem. Soc.* **123**, 1181 (1923).

White, tetragonal or orthorhombic crystals. d 1.17. Sublimes *in vacuo.* Decomposes into H_2S and NH_3 rather easily
at room temp when in crystal form. The commercial product is furnished in porcelain-like lumps which are more
stable and can be stored in a closed bottle at a cool place.
Dissociation press. at room temp about 350 mm Hg. Freely
sol in water or alcohol giving colorless solns which turn

yellow rapidly. Decomposed by boiling water. Soly in water (0°): 128.1 g/100 g H_2O. Slightly sol in acetone; almost insol in ether, benzene.

USE: In lubricants. *Caution:* Very irritating to skin; penetrates more rapidly than hydrogen sulfide and may be fatal.

522. Ammonium Bisulfite. H_5NO_3S; mol wt 99.11. H 5.09%, N 14.13%, O 48.43%, S 32.35%. NH_4HSO_3. Prepn: *Gmelin's, Ammonium* (8th ed.) **23,** 259-260 (1936).

Crystals. Soly in water (g/100 ml H_2O): 267 (10°); 620 (60°). *Keep well closed.* Sold as solution only.

USE: Preservative.

523. Ammonium Bitartrate. *(R)-2,3-Dihydroxybutane-dioic acid monoammonium salt;* L-tartaric acid monoammonium salt; ammonium acid tartrate; ammonium hydrogen tartrate. $C_4H_9NO_6$; mol wt 167.12. C 28.75%, H 5.43%, N 8.38%, O 57.44%. $NH_4OOCCH(OH)CH(OH)COOH$. Prepd from ammonium tartrate and tartaric acid: Dulk, *Ann.* **2,** 39 (1832).

Odorless crystals. d 1.68. Sol in 45.6 parts water at 15°; freely sol in hot water, alkalies, alkali carbonates; practically insol in alc. $[\alpha]_D^{20}$ +26.0° (c = 1.5 in water): Long, *J. Am. Chem. Soc.* **23,** 813 (1901).

524. Ammonium Borate. *Ammonium tetraborate;* ammonium biborate; ammonium metaborate. $B_4H_8N_2O_7$; mol wt 191.36. B 22.61%, H 4.21%, N 14.64%, O 58.53%. $(NH_4)_2B_4O_7$. Prepn: *Gmelin's, Ammonium* (8th ed.) **23,** 323 (1936).

Tetrahydrate, tetragonal crystals. Sol in water; practically insol in alcohol.

USE: Fireproofing wood and textiles; in electrolytic condensers.

525. Ammonium Bromide. BrH_4N; mol wt 97.96. Br 81.58%, H 4.12%, N 14.30%. NH_4Br. Contains 99-99.5% NH_4Br. Prepn and properties: *Gmelin's, Ammonium* (8th ed.) **23,** 203-218 (1936); Richards, "Ammonium Bromide" in *Mellor's,* **Vol. VIII,** supplement I, *Nitrogen* (part 1) 433-447 (1964).

White, odorless, slightly hygroscopic crystals or granules; pungent, saline taste; slowly becomes yellowish in air; sublimes at high temp without melting. d^{25} 2.429. Freely sol in water, methanol, ethanol, acetone; slightly sol in ether; practically insol in ethyl acetate. *Keep well closed.*

Incompat. Acids, acid salts, spirit nitrous ether, alkaloids; salts of lead, mercury, silver.

USE: Manuf of photographic films, plates, and papers; in process engraving and lithography; fireproofing of wood; in corrosion inhibitors.

THERAP CAT: Sedative.

THERAP CAT (VET): Sedative.

526. Ammonium Caprylate. *Octanoic acid ammonium salt;* caprylic acid ammonium salt. $C_8H_{19}NO_2$; mol wt 161.24. C 59.59%, H 11.88%, N 8.69%, O 19.85%. C_7H_{15}-COONH₄. Prepn from the acid and ammonia: McMaster, Magill, *J. Am. Chem. Soc.* **38,** 1793 (1916); Stumpf, *Am. Paint J.* **38** (45), 60 (1954).

Monoclinic crystals from ether + alcohol. Somewhat hygroscopic. Decomposes on standing and develops odor of caprylic acid. mp 70-85°. Easily hydrolyzed by water. Freely sol in glacial acetic acid and ethanol; less sol in methanol; slightly sol in acetone and ethyl acetate; practically insol in chloroform, benzene. *Keep well closed.*

USE: In photographic emulsions; as insecticide and nematocide; in manuf of zinc caprylate.

527. Ammonium Carbamate. "Anhydride" of ammonium carbonate; ammonium aminoformate. $CH_6N_2O_2$; mol wt 78.06. C 15.39%, H 7.73%, N 35.89%, O 40.99%. NH_2-COONH₄. Prepd from dry ice and liq ammonia: Brooks, Audrieth, *Inorg. Syn.* **2,** 85 (1946).

Cryst powder; ammonia odor; gradually loses ammonia in the air changing to ammonium carbonate. Volatilizes at about 60°. Freely sol in water; sol in alcohol.

USE: Ammoniating agent, less vigorous than NH_3.

528. Ammonium Carbonate. Hartshorn. A mixture of ammonium bicarbonate and ammonium carbamate, obtained by subliming a mixture of ammonium sulfate and

calcium carbonate. Contains 30-34% NH_3, about 45% CO_2. See *USP* **XVII,** p 941. *Review:* Allen, "Ammonium Carbonate" in *Mellor's* **Vol. VIII,** supplement 1, *Nitrogen* (part 1) 459-468 (1964).

Colorless, hard, translucent, crystalline masses, white cubes or powder; strong odor of ammonia; sharp taste and alkaline reactions. Decomposes on exposure to air with loss of NH_3 and CO_2, becoming white and powdery and converting into ammonium bicarbonate. Volatilizes at about 60°. Slowly sol in 4 parts water; decomposed by hot water; the carbamate portion dissolves in alcohol. *Keep tightly closed in a cool place. Incompat.* Acids and acid salts; salts of iron, zinc; alkaloids, alum, calomel, tartar emetic.

USE: In baking powders, for washing and defatting woolens, tanning; as mordant in dyeing; manuf rubber articles, casein glue, casein colors; in fire extinguishers, smelling salts; for separating cacao constituents; as a reagent in analytical chemistry. Pharmaceutic aid (source of ammonia).

THERAP CAT: Expectorant.

THERAP CAT (VET): Expectorant. Has been used as a carminative and stomachic.

529. Ammonium Ceric Nitrate. *Ammonium hexanitra-tocerate(IV);* ammonium nitratocerate(IV); ceric ammonium nitrate. $CeH_8N_8O_{18}$; mol wt 548.26. Ce 25.56%, H 1.47%, N 20.44%, O 52.53%. $(NH_4)_2Ce(NO_3)_6$. Prepd by dissolving $CeO.H_2O$ in hot concd HNO_3: Smith *et al., Ind. Eng. Chem., Anal. Ed.* **8,** 449 (1936); Smith, Fly, *Anal. Chem.* **21,** 1233 (1949); Smith, *Talanta* **10,** 709 (1963).

Small, orange-red, monoclinic crystals. Very sol in water.

USE: As a standard in oxidimetry; catalyst for polymerization of olefins.

530. Ammonium Cerous Sulfate. Ammonium disulfa-tocerate(III); cerous ammonium sulfate. $CeH_4NO_8S_2$; mol wt 350.29. Ce 40.00%, H 1.15%, N 4.00%, O 36.54%, S 18.30%. $NH_4Ce(SO_4)_2$. Prepd by slow evaporation of a soln contg equimolar amounts of $(NH_4)_2SO_4$ and $CeSO_4$: Blandin, Rerat, *Compt. Rend.* **242,** 1740 (1956).

Tetrahydrate, monoclinic crystals. Sol in water.

531. Ammonium Chloride. Ammonium muriate; sal ammoniac; salmiac; Amchlor; Darammon. ClH_4N; mol wt 53.50. Cl 66.28%, H 7.54%, N 26.18%. NH_4Cl. Contains 99.5-99.8% NH_4Cl; principal impurity is NaCl. Prepn and properties: *Gmelin's, Ammonium* (8th ed.) **23,** pp 150-184 (1936); Kane, "Ammonium Chloride" in *Mellor's* **Vol. VIII,** supplement 1, *Nitrogen* (part 1), 378-432 (1964). Manuf: Faith, Keyes & Clark's *Industrial Chemicals,* F. A. Lowenheim, M. K. Moran, Eds. (Wiley-Interscience, New York, 4th ed., 1975) pp 93-96.

Colorless, odorless crystals or cryst masses; or white, granular powder; cooling, saline taste; somewhat hygroscopic. Tendency to cake. d^{25} 1.5274. Sublimes without melting. Soly in water (w/w): 22.9% (0°); 26.0% (15°); 28.3% (25°); 39.6% (80°). Strongly endothermic. HCl and NaCl decrease soly in water. pH of aq solns (25°): 1% 5.5; 3% 5.1; 10% 5.0. Sol in methanol, ethanol; almost insol in acetone, ether, ethyl acetate. LD_{50} i.m. in rats: 30 mg/kg, *RTECS* **Vol. I,** R. J. Lewis, R. L. Tatken, Eds. (1979) pp 99.

Incompat. Alkalies and their carbonates; lead and silver salts.

USE: As a flux for coating sheet iron with zinc; tinning; in dry and Leclanché batteries; dyeing, freezing mixtures, electroplating, to clean soldering irons, safety explosives, lustering cotton, tanning; in washing powders; manuf dyes; in cement for iron pipes; for snow treatment (slows melting on ski slopes).

THERAP CAT: Systemic acidifier.

THERAP CAT (VET): Expectorant; diaphoretic; acidifying diuretic.

532. Ammonium Chromate(VI). Neutral ammonium chromate. $CrH_8N_2O_4$; mol wt 152.09. Cr 34.20%, H 5.30%, N 18.42%, O 42.08%. $(NH_4)_2CrO_4$. Prepn: *Gmelin's, Chromium* (8th ed.) **52,** part B, pp 707-712 (1962).

Yellow acicular crystals; loses some NH_3 in air; decomposes at 185°. d 1.8. Soly in water: 19.78% (0°); 41.20% (75°). Sparingly sol in liquid ammonia, acetone. Slightly sol in methanol. Practically insol in ethanol. The aq soln is alkaline. *Keep well closed.*

USE: Sensitizing gelatin in photography, in textile printing pastes, in fixing chromate dyes on wool, and as a reagent in analytical chemistry.

533. Ammonium Chromic Sulfate. Ammonium disulfatochromate(III); chromic ammonium sulfate. $CrH_4NO_8S_2$; mol wt 262.17. Cr 19.84%, H 1.54%, N 5.34%, O 48.82%, S 24.46%. $NH_4Cr(SO_4)_2$. Prepd by crystallization from a soln contg equimolar amounts of $Cr_2(SO_4)_3$ and $(NH_4)_2SO_4$: Howarth in *Chromium*, vol. I, M. J. Udy, Ed., A.C.S. Monograph Series, no. **132** (Reinhold, New York, 1956) p 287; electrolytic prepn: Nishihara *et al.*, *Japan. pat.* **2164**-('60); *C.A.* **55**, 5200e (1961).

Dodecahydrate, *chrome alum ammonium.* Small dark-violet or violet-blue, octahedral, cubic crystals; ruby-red by transmitted light. mp 94°; loses $9H_2O$ on melting and the remaining H_2O by 300°. d^{25} 1.72. Readily sol in water; slightly sol in alcohol. Aq soln is violet when cold, green when hot.

USE: Mordant in textile industry; in manuf of electrolytic Cr metal.

534. Ammonium Citrate, Dibasic. Diammonium citrate; citric acid diammonium salt. $C_6H_{14}N_2O_7$; mol wt 226.19. C 31.86%, H 6.24%, N 12.39%, O 49.52%. Prepn: Heldt, *Ann.* **47**, 157 (1843).

$$\begin{array}{l} CH_2COONH_4 \\ | \\ HOCCOOH \\ | \\ CH_2COONH_4 \end{array}$$

Granules or crystals; acid reaction. d 1.48. Sol in about 1 part water, slightly in alc. pH of $0.1M$ soln in H_2O = 4.3.

USE: For the determination of phosphate, esp in fertilizers.

535. Ammonium Cobaltous Phosphate. Cobaltous ammonium phosphate. CoH_4NO_4P; mol wt 171.96. Co 34.28%, N 8.15%, H 2.34%, O 37.22%, P 18.01%. NH_4CoPO_4. Prepd by reaction of a cobaltous salt with $(NH_4)_3PO_4$, $(NH_4)_2HPO_4$, or H_3PO_4 and NH_3: Grat-Cabanac, *Bull. Microscop. Appl.* **8**, 97 (1958); Salutsky *et al.*; McCullogh, Salutsky, U.S. pats. **3,126,254**; **3,141,732** (both 1964 to W. R. Grace).

Hydrate, red to violet powder, or monoclinic rectangular lamellae. Practically insol in water; sol in acids.

USE: As pigment in ceramic glazes and vitreous enamels; as temp indicator in textile industry; in fertilizers for plant nutrition; in Co analysis.

536. Ammonium Cobaltous Sulfate. Ammonium disulfatocobaltate(II); cobaltous ammonium sulfate. $CoH_8N_2O_8S_2$; mol wt 287.14. Co 20.53%, H 2.81%, N 9.76%, O 44.58%, S 22.33%. $(NH_4)_2Co(SO_4)_2$. Prepd by mixing equimolar amounts of solns of $(NH_4)_2SO_4$ and $CoSO_4$: Malard, *Bull. Soc. Chim. France* **1961**, 2296; *Gmelin's, Cobalt* (8th ed.) **58**, part A, p 435 (1932); suppl, pp 811-815 (1961).

Hexahydrate, red monoclinic prismatic crystals. d 1.90. Sol in water; almost insol in alcohol.

537. Ammonium Cupric Chloride. *Ammonium tetrachlorocuprate(II);* ammonium chlorocuprate; cupric ammonium chloride. $Cl_4CuH_8N_2$; mol wt 241.45. Cl 58.75%, Cu 26.32%, H 3.34%, N 11.60%. $(NH_4)_2CuCl_4$. Prepd by evaporation of a 2:1 soln of NH_4Cl and $CuCl_2$: Chrobak, *Bull. Int. Acad. Polonaise* **1929A**, 361; *C.A.* **24**, 3688[7] (1930); Willet, *J. Chem. Phys.* **41**, 2243 (1964).

Yellow, hygroscopic, orthorhombic crystals. Sol in water. *Keep well closed.*

Dihydrate, *ammonium tetrachlorodiaquocuprate(II).* Blue to bluish-green tetragonal, rhombododecahedral crystals. d 2.0. Becomes anhydr at 110-120°; decomposes on stronger heating. Sol in water, alcohol, liquid NH_3; the aq soln is acid to litmus.

USE: Analytical reagent; formerly for determining carbon in iron and steel.

538. Ammonium Dichromate(VI). Ammonium bichromate. $Cr_2H_8N_2O_7$; mol wt 252.10. Cr 41.26%, H 3.20%, N 11.11%, O 44.42%. $(NH_4)_2Cr_2O_7$. Prepd from ammonium sulfate and sodium dichromate or by the interaction of ammonia gas and chromic acid in soln: Hartford, *Ind. Eng. Chem.* **41**, 1993 (1949); *Chromium Chemicals*, Publication **52** (Mutual Chem. Div., Allied Chem.) pp 34-39. *Review:* M. J. Udy, *Chromium*, Vol. **1**, ACS Monograph Series no. **132** (Reinhold, New York, 1956) passim.

Bright orange-red crystals. *Flammable.* Odorless and non-hygroscopic. Crystal system: monoclinic. Crystal habit: prismatic. d^{25} 2.155. Bulk density: 82 lbs/cu ft. Dec at about 180°. Decomposition becomes self-sustaining at about 225° with spectacular swelling and evolution of heat and nitrogen, leaving Cr_2O_3. Heat of soln -23.0 cal/g. Very sol in water. Soly in water (w/w): 15.16% (0°); 26.67% (20°); 36.99% (40°); 46.14% (60°); 54.20% (80°); 60.89% (100°). Acid reaction. A 1% soln has a pH of 3.95 and a 10% soln has a pH of 3.45.

USE: Source of pure nitrogen (esp in the laboratory); in pyrotechnics (Vesuvius fire); in lithography and photo engraving; in special mordant, catalysts, and porcelain finishes; intermediate in the manufacture of pigments; of magnetic recording materials. *Caution:* Causes skin irritation, ulceration, "chrome sores", perforation of nasal septum, pulmonary irritation.

539. Ammonium Dithiocarbamate. *Dithiocarbamic acid monoammonium salt;* ammonium sulfocarbamate. $CH_6N_2S_2$; mol wt 110.19. C 10.90%, H 5.48%, N 25.42%, S 58.20%. NH_2CSSNH_4. Prepd from CS_2 and NH_3: Mathes, *Inorg. Syn.* **3**, 48 (1950); Redemann *et al.*, *Org. Syn. coll. vol. III*, 763 (1955); Gatlow, Hahnkamm, *Z. Anorg. Allgem. Chem.* **364**, 161 (1969). Crystal structure: Cappuchi *et al.*, *Chem. Commun.* **1966**, 441.

Yellow, lustrous almost odorless, orthorhombic crystals when fresh. Undergoes a reversible, exothermic transition at 63°; mp 99° (dec). d_4^{20} 1.451. Sol in water. Decomposes in air, and is then no longer clearly sol; acquires an odor of H_2S; the decomposition products contain ammonium thiocyanate, ammonium sulfide, etc. *Keep in tightly closed bottles.*

USE: Instead of H_2S or $(NH_4)_2S$ for pptg metals in chemical analysis; synthesis of heterocyclic compounds.

540. Ammonium Ferric Chromate. Ferric ammonium chromate. $Cr_2FeH_4NO_8$; mol wt 305.91. Cr 34.00%, Fe 18.26%, H 1.32%, N 4.58%, O 41.84%. $NH_4Fe(CrO_4)_2$. Prepd by addn of NH_3 to an aq soln of $Fe(NO_3)_3.6H_2O$ and CrO_3: Weinland, Mergenthaler, *Ber.* **57B**, 776 (1924).

Carmine-red, microcryst powder. Practically insol in water.

541. Ammonium Ferric Citrate. Ferric ammonium citrate. Compounds of NH_3, iron, and citric acid of undetermined structure. Prepd by addn of $Fe(OH)_3$ to an aq soln of citric acid and NH_3: Kruse, Mounce, U.S. pat. **2,644,828** (1953 to Mallinckrodt); prepn of isotonic solns: Hammarlund *et al.*, *Pharm. Acta Helv.* **35**, 593 (1960).

Ammonium ferric citrate, brown, *soluble ferric citrate.* Contains about 9% NH_3, 16.5-18.5% Fe, and about 65% hydrated citric acid. Reddish-brown granules, garnet-red transparent scales, or brownish-yellow powder. Odorless or slight NH_3 odor; saline, ferruginous taste. Very deliquesc. Reduced to ferrous salt by light. Extremely sol in water; practically insol in alcohol. *Keep well closed and protected from light.*

Ammonium ferric citrate, green. Contains about 7.5% NH_3, 14.5-16% Fe, and about 75% hydrated citric acid. Green transparent, deliquesc scales, pearls, granules, or powder. Odorless; mild ferruginous taste. More readily reduced to the ferrous salt by light than the brown form. Very sol in water; practically insol in alc. *Keep well closed and protected from light.*

USE: For blueprints; in photography.

THERAP CAT: Hematinic.

THERAP CAT (VET): In iron deficiency anemia.

542. Ammonium Ferric Oxalate. *Triammonium tris-[ethanedioato(2—)-O,O']ferrate(3—);* ammonium trioxalatoferrate(III); ferric ammonium oxalate. $C_6H_{12}FeN_3O_{12}$; mol wt 374.04. C 19.27%, H 3.23%, Fe 14.93%, N 11.24%, O 51.33%. $(NH_4)_3Fe(C_2O_4)_3$. Prepn: *Gmelin's, Iron* (8th ed.) **59**, part B, pp 1020-1021 (1932).

Hydrate, bright-green, monoclinic, prismatic crystals.

$d^{17.5}$ 1.78. Affected by light. Loses 3 H_2O by 100°, decomp at 160-170°. Very sol in water; practically insol in alcohol. *Protect from light.*

USE: In photography, blueprints; in coloring of Al and Al alloys.

543. Ammonium Ferric Sulfate. Ferric ammonium sulfate. $FeH_4NO_8S_2$; mol wt 266.01. Fe 21.00%, H 1.52%, N 5.27%, O 48.12%, S 24.10%. $NH_4Fe(SO_4)_2$. Prepn: *Gmelin's, Iron* (8th ed.) 59, part B, pp 1010-1018 (1932).

Dodecahydrate, *ferric alum, iron alum.* Colorless to pale-violet, transparent, efflorescent, octahedral crystals. Odorless; acid styptic taste. mp about 37°; d 1.71. Very sol in water; practically insol in alcohol; pH of $0.1M$ aq soln 2.5.

USE: As analytical reagent; mordant in dyeing and printing textiles.

THERAP CAT: Astringent, styptic.

544. Ammonium Ferricyanide. *Triammonium hexakis-(cyano-C)ferrate(3—); ammonium hexacyanoferrate(III).* $C_6H_{12}FeN_9$; mol wt 266.08. C 27.08%, H 4.55%, Fe 20.99%, N 47.38%. $(NH_4)_3Fe(CN)_6$. Prepn: *Gmelin's, Iron* (8th ed.) 59, part B, p 1027 (1932).

Trihydrate, red crystals. Freely sol in water; practically insol in alcohol. *Protect from light.*

545. Ammonium Ferrocyanide. *Triammonium hexakis-(cyano-C)ferrate(4—); ammonium hexacyanoferrate (II).* $C_6H_{16}FeN_{10}$; mol wt 284.12. C 25.36%, H 5.68%, Fe 19.66%, N 49.30%. $(NH_4)_4Fe(CN)_6$. Prepn: *Gmelin's, Iron* (8th ed.) 59, part B, p 1024 (1932); Lux in *Handbook of Preparative Inorganic Chemistry,* Vol. 2, G. Brauer, Ed. (Academic Press, New York, 2nd ed., 1965) pp 1509-1510.

Trihydrate, yellow cryst powder. Loses NH_3 on exposure to air and light. Decomposes on heating. Freely sol in water; practically insol in alcohol. *Protect from light.*

546. Ammonium Ferrous Sulfate. Ferrous ammonium sulfate; Mohr's salt. $FeH_8N_2O_8S_2$; mol wt 284.05. Fe 19.66%, H 2.84%, N 9.86%, O 45.06%, S 22.57%. $(NH_4)_2Fe-(SO_4)_2$. Manuf from Fe, H_2SO_4 and NH_3: Demmerle *et al., Ind. Eng. Chem.* 42, 9 (1950); from pickling waste: Brundin, U.S. pat. 2,694,657 (1954 to Ekstrand and Tholand).

Hexahydrate, pale blue-green crystals or cryst powder. Slowly oxidizes and effloresces in air. d_4^{20} 1.86. Sol in water; practically insol in alcohol. *Keep well closed and protected from light.* LD_{50} orally in rats: 3.25 g/kg, H. F. Smyth *et al., Am. Ind. Hyg. Assoc. J.* 30, 470 (1969).

USE: In photography; as analytical standard; as polymerization catalyst; in dosimeters.

547. Ammonium Fluoride. Neutral ammonium fluoride. FH_4N; mol wt 37.04. F 51.30%, H 10.88%, N 37.82%. NH_4F. Prepd by passing ammonia gas into ice-cooled 40% hydrofluoric acid or by heating 1 part NH_4Cl with 2.25 parts NaF and separatng the ammonium fluoride by sublimation: Kwasnik in *Handbook of Preparative Inorganic Chemistry* vol. 1, G. Brauer, Ed. (Academic Press, New York, 2nd ed., 1963) p 183. *Review:* Steele, "Ammonium Fluoride" in *Mellor's* vol. VIII, supplement I, *Nitrogen* (part 1), 370-377 (1964).

Deliquescent leaflets or needles. Hexagonal prisms by sublimation. Occurs commercially as a granular powder. d 1.015. On heating dec into NH_3 and HF. *Corrodes glass.* Soly in water (0°): 100 g/100 ml. Decomposed by hot water into NH_3 and ammonium bifluoride. $(NH_4F.HF)$. Cannot be obtained by evapn of its aq soln. Slightly sol in alcohol. The aq soln is acid. May be stored in iron vessels. *Incompat.* Quinine salts; soluble calcium salts.

USE: Etching and frosting glass; as antiseptic in brewing beer; preserving wood; in printing and dyeing textiles; as mothproofing agent. *Caution:* Ingestion produces nausea, salivation, vomiting, abdominal pain, diarrhea, hemorrhagic gastroenteritis, muscular weakness, tremors, convulsions, vascular collapse. Increased respiration is followed by depression, death. *Chronic toxicity:* mottling of enamel, generalized osteosclerosis, calcification in tendons and ligaments; synostoses.

548. Ammonium Formate. *Formic acid ammonium salt.* CH_5NO_2; mol wt 63.06. C 19.05%, H 7.99%, N 22.21%, O 50.75%. $HCOONH_4$. Prepd from formic acid and NH_3:

Zuffanti, *J. Am. Chem. Soc.* 63, 3123 (1941); from methyl-formate and NH_3: Kelly, Cuthbert, U.S. pat. 3,122,584 (1964 to Allied Chem.).

Deliquesc crystals or granules. d 1.27. mp 116°. Sol in less than its own wt of water; sol in alc. *Keep tightly closed.*

USE: In chemical analysis, especially to ppt base metals from salts of the "noble" metals.

549. Ammonium Hexafluoroaluminate. *Triammonium hexafluoroaluminate(3—);* ammonium cryolite; ammonium aluminum fluoride; ammonium fluoaluminate. $AlF_6H_{12}N_3$; mol wt 195.10. Al 13.83%, F 58.43%, H 6.20%, N 21.54%. $(NH_4)_3AlF_6$. Prepn from ammonium fluoride and aluminum hydroxide: v. Helmolt, *Z. Anorg. Allgem. Chem.* 3, 127 (1893); Petersen, *J. Prakt. Chem.* [2] 40, 55 (1889); Kwasnik in *Handbook of Preparative Inorganic Chemistry,* vol. 1, G. Brauer, Ed. (Academic Press, New York, 2nd ed. 1963) p 236.

Cubic crystals, d 1.78. Thermally stable to above 100°. Freely sol in water. Does not attack glass.

USE: Prepn of pure aluminum fluoride.

550. Ammonium Hexafluorogallate. *Triammonium hexafluorogallate(3—);* ammonium hexafluogallate. $F_6Ga-H_{12}N_3$; mol wt 237.84. F 47.93%, Ga 29.31%, H 5.09%, N 17.67%. $(NH_4)_3GaF_6$. Prepd from $Ga(OH)_3$, HF and NH_4F: Hannebohn, Klemm, *Z. Anorg. Allgem. Chem.* 229, 341 (1936); Kwasnik in *Handbook of Preparative Inorganic Chemistry,* vol. 1, G. Brauer, Ed., (Academic Press, New York, 2nd ed., 1963) p 228.

Octahedra. On heating in air changes to Ga_2O_3; on heating *in vacuo* at 200° for several hours forms GaN.

USE: In the prepn of GaF_3.

551. Ammonium Hexafluorophosphate. Ammonium hexafluophosphate; ammonium phosphorus hexafluoride. F_6H_4NP; mol wt 163.02. F 69.93%, H 2.47%, N 8.59%, P 19.00%. NH_4PF_6. Prepn: Lange, Müller, *Ber.* 63, 1063 (1930); v. Krueger, *Ber.* 65, 1265 (1932); Woyski, *Inorg. Syn.* 3, 111 (1950); Kwasnik in *Handbook of Preparative Inorganic Chemistry,* Vol. 1, G. Brauer, Ed. (Academic Press, New York, 2nd ed., 1963) p 195.

Square leaflets or tables, seldom rectangular plates; cubic system. d_4^{18} 2.180. Decomposes on heating to a relatively high temp without prior melting. Soly in water (20°): 74.8 g/100 ml. Sol in acetone, methanol, ethanol, methyl acetate. Does not etch glass at room temp. Slowly hydrolyzed by boiling with strong acids.

552. Ammonium Hexafluorosilicate. *Diammonium hexafluorosilicate(2—);* ammonium fluosilicate; ammonium silicofluoride. $F_6H_8N_2Si$; mol wt 178.16. F 63.99%, H 4.52%, N 15.72%, Si 15.77%. $(NH_4)_2SiF_6$. Occurs in nature as the mineral *cryptohalite.* Prepn: *Gmelin's, Ammonium* (8th ed.) 23, pp 414-415 (1936). Crystal structure: Schlemper *et al., J. Chem. Phys.* 44, 2499 (1966); 45, 408 (1966).

Odorless cryst powder. Two modifications at room temp: stable, cubic phase; metastable, trigonal phase. Freely sol in water; almost insol in alcohol, acetone. LD orally in guinea pigs: 150 mg/kg, Simonin, Pierron, *C. R. Soc. Biol.* 124, 133 (1937).

USE: In pesticides; in soldering flux; etching glass. *Caution:* Toxic symptoms similar to sodium fluoride.

553. Ammonium Hypophosphite. H_6NO_2P; mol wt 83.04. H 7.28%, N 16.87%, O 38.54%, P 37.31%. $NH_4H_2-PO_2$. Prepn: *Gmelin's, Ammonium* (8th ed.) 23, 416 (1936).

Hygroscopic and deliquesc crystals or white granules. Decomposes when heated with evolution of phosphine which ignites spontaneously. One gram dissolves in about 1 ml water, 0.2 ml boiling water, 20 ml alcohol; freely sol in boiling alcohol, practically insol in acetone. The aq soln is practically neutral. *Keep well closed.*

USE: As catalyst in polyamide manuf.

554. Ammonium Iodide. H_4IN; mol wt 144.96. H 2.78%, I 87.55%, N 9.66%. NH_4I. Contains about 98% NH_4I; about 1% ammonium hypophosphite is usually added as a preservative. Prepd from ammonia, iodine and hydrogen peroxide; from ammonia and hydrogen iodide; or from ammonium carbonate and hydrogen iodide: Wulff, Cameron, *Z. Physik. Chem.* B10, 350 (1930); Schmeisser in *Hand-*

Consult the cross index before using this section.

book of Preparative Inorganic Chemistry, vol. 1, G. Brauer, Ed. (Academic Press, New York, 2nd ed., 1963) p 289. *Review:* Richards "Ammonium Iodide" in *Mellor's* vol. VIII, Supplement I, *Nitrogen* (part 1) 448-458 (1964).

White, odorless, very hygroscopic, tetragonal crystals or granular powder, sharp saline taste. Becomes yellow to brown on exposure to air and light because of liberation of iodine. When heated it partly decomposes and partly sublimes. d^{25} 2.5142. One gram dissolves in 0.6 ml water, 0.5 ml boiling water, 3.7 ml alc, 1.5 ml glycerol, 2.5 ml methanol. Except in the presence of a stabilizer, such as ammonium hypophosphite, the aq soln will quickly become yellow. The aq soln is nearly neutral to litmus. pH of $0.1M$ soln about 4.6. *Keep tightly closed and protected from light.*

USE: In photographic chemicals.

555. Ammonium Lactate. DL-Lactic acid ammonium salt. $C_3H_9NO_3$; mol wt 107.11. C 33.64%, H 8.47%, N 13.08%, O 44.81%. $CH_3CH(OH)COONH_4$. Prepd by neutralizing DL-lactic acid with NH_4OH: Costello, Filachione, *J. Am. Chem. Soc.* **75**, 1242 (1953).

Crystals from propanol, mp 91-94°. Sol in water, glycerol, 95% alc; slightly sol in methanol; practically insol in ethyl, n-propyl, isopropyl, and n-butyl alcohols, ether, acetone, ethyl acetate. For a 78.8% by wt soln: n_D^{20} 1.4543, n_D^{25} 1.4536, n_D^{40} 1.4503; d_4^{20} 1.2006, d_4^{25} 1.1984, d_4^{40} 1.1904.

THERAP CAT (VET): Has been used for bovine ketosis.

556. Ammonium Magnesium Chloride. Magnesium ammonium chloride. Cl_3H_4MgN; mol wt 148.73. Cl 71.52%, H 2.71%, Mg 16.35%, N 9.42%. $MgNH_4Cl_3$.

Hexahydrate, deliquesc crystals. Sol in 6 parts water. *Keep well closed.*

USE: Prepn of magnesia mixture and anhydrous magnesium chloride.

557. Ammonium Mandelate. *α-Hydroxybenzeneacetic acid monoammonium salt;* mandelic acid ammonium salt. $C_8H_{11}NO_3$; mol wt 169.18. C 56.79%, H 6.56%, N 8.28%, O 28.37%. $C_6H_5CH(OH)COONH_4$. Prepd from mandelic acid and NH_3 or excess of strong NH_3-water: Tabern *et al.,* U.S. pat. 2,220,692 (1941 to Abbott); Baker, U.S. pat. 2,209,314 (1941 to Squibb).

Very deliquescent, cryst powder; odorless or with slight odor; discolors in light. Very sol in water, sparingly in alc. The aq soln is slightly acid to litmus. *Keep tightly closed and protected from light.*

THERAP CAT: Urinary anti-infective.

THERAP CAT (VET): Has been used as a urinary antiseptic.

558. Ammonium Mercuric Chloride. Ammonium tetrachloromercurate(II); mercuric ammonium chloride. $Cl_4H_8HgN_2$; mol wt 378.52. Cl 37.47%, H 2.13%, Hg 53.00%, N 7.40%. $(NH_4)_2HgCl_4$.

Dihydrate, powder. *Poison!* Sol in water; partly sol in alcohol.

THERAP CAT (VET): As ointment in chronic eczema. Antifungal.

559. Ammonium Molybdate(VI). Ammonium paramolybdate. $H_{24}Mo_7N_6O_{24}$; mol wt 1163.89. H 2.08%, Mo 57.71%, N 7.22%, O 32.99%. $(NH_4)_6Mo_7O_{24}$. Prepn and structure: Guiter, *Compt. Rend.* **220**, 146 (1945); Lindqvist, *Acta Chem. Scand.* **2**, 88 (1948).

Tetrahydrate. Colorless or slightly greenish or yellowish crystals. Sol in 2.3 parts water; practically insol in alcohol. pH of 5% aq soln 5.0 to 5.5.

USE: In photography and for decorating ceramics; for detecting and determining phosphates, arsenates, lead; also as reagent for alkaloids and many other substances.

560. Ammonium Nickel Sulfate. Ammonium disulfatonickelate(II); nickel ammonium sulfate. $H_8N_2NiO_8S_2$; mol wt 286.89. H 2.81%, N 9.77%, Ni 20.46%, O 44.62%, S 22.35%. $Ni(NH_4)_2(SO_4)_2$.

Hexahydrate, bluish-green crystals; somewhat efflorescent. d 1.923. Sol in 10.3 parts water, 5 parts water at 60°; practically insol in alcohol. pH of $0.1M$ aq soln 4.6.

USE: Electroplating metals.

561. Ammonium Nitrate. $H_4N_2O_3$; mol wt 80.05. H 5.04%, N 35.00%, O 59.96%. NH_4NO_3. Manuf: Faith,

Keyes & Clark's *Industrial Chemicals,* F. A. Lowenheim, M. K. Moran, Eds. (Wiley-Interscience, New York, 4th ed., 1975) pp 97-102. Comprehensive description of manuf processes and physical data: B. T. Federoff *et al., Encyclopedia of Explosives and Related Items,* vol. I (Picatinny Arsenal, Dover, N.J., 1960) pp A311-A379. *Review:* Thatcher, "Ammonium Nitrate" in *Mellor's* vol. VIII, supplement I, *Nitrogen* (part 1) 506-562 (1964).

Odorless, transparent, hygroscopic, deliquesc crystals or white granules. Five solid phases exist at normal pressure. Orthorhombic at room temp. d. 1.72. Dec at about 210°, mostly into H_2O and N_2O. One gram dissolves in 0.5 ml water, 0.1 ml boil. water, about 20 ml alc, about 8 ml methanol. pH of $0.1M$ soln in water: 5.43. *Keep well closed.*

Note: Disastrous explosions ascribed to ammonium nitrate occurred in 1947 at Texas City and at Brest.

USE: Making nitrous oxide (laughing gas); in freezing mixtures, safety explosives; matches; pyrotechnics; in fertilizers.

THERAP CAT (VET): Has been used as an expectorant, urinary acidifier.

562. Ammonium Nitroferricyanide. *Diammonium pentakis(cyano-C)nitrosylferrate(2−);* ammonium pentacyanonitrosylferrate(III); ammonium nitroprusside. $C_5H_8FeN_8O$; mol wt 252.02. C 23.83%, H 3.20%, Fe 22.16%, N 44.46%, O 6.35%. $(NH_4)_2Fe(CN)_5NO$. Prepn and structure: Burrows, Turner, *J. Chem. Soc.* **115**, 1429 (1919).

Red to brownish-red crystals. Freely sol in water, alc.

563. Ammonium Oleate. *(Z)-9-Octadecenoic acid ammonium salt; oleic acid ammonium salt;* ammonia soap. $C_{18}H_{37}NO_2$; mol wt 299.48. C 72.19%, H 12.45%, N 4.68%, O 10.68%. $CH_3(CH_2)_7CH=CH(CH_2)_7COONH_4$. Prepd from oleic acid and excess 28-30% NH_3 soln: Stumpf, *Am. Paint J.* **38**, no. 45, 60, 64, 68 (1954), *C.A.* **48**, 12428b (1954).

Yellowish-brown paste, softens at 50-55°F, mp 70-72°F. At 80°F: sol in water; slightly sol in acetone, ethanol, methanol, benzene, CCl_4, xylene, naphtha. Sol in water at 212°F, acetone at 134°F, ethanol at 172°F, methanol at 148°F, benzene at 176°F, CCl_4 at 170°F, xylene at 180°F, naphtha at 160°F.

USE: Detergent, solidifying alcohol, demonstrating liquid crystals according to Lehmann.

564. Ammonium Osmium Chloride. Ammonium hexachloroosmate(IV); ammonium chloroosmate; osmium ammonium chloride. $Cl_6H_8N_2Os$, mol wt 439.02. Cl 48.46%, H 1.84%, N 6.38%, Os 43.32%. $(NH_4)_2OsCl_6$. Prepn: Dwyer, Hogarth, *Inorg. Syn.* **5**, 206 (1957).

Red powder, or dark red, octahedral crystals. Soluble in water or alcohol.

565. Ammonium Oxalate. *Ethanedioic acid diammonium salt.* $C_2H_8N_2O_4$; mol wt 124.10. C 19.36%, H 6.49%, N 22.58%, O 51.57%. $NH_4OOCCOONH_4$. Commercial product is 98-99% pure. Prepn of monohydrate from aq oxalic acid and NH_3 or ammonium carbonate: Bérard, *Ann. Chim.* **73**, 277 (1810). *Review: Gmelin's, Ammonium* (8th ed.) **23**, pp 400-407 (1936). Crystal structure of monohydrate: Robertson, *Acta Cryst.* **18**, 410, 417 (1965).

Monohydrate, orthorhombic, odorless crystals or granules. *Poisonous!* d 1.50. One gram dissolves in 20 ml water, 2.6 ml boil water; slightly sol in alcohol. The aq soln is practically neutral. pH of 0.1 M soln 6.4.

USE: Manuf explosives; electrolytic detinning of iron; in dyeing, metal polishes; for detection and determination of Ca, Pb, and rare earth metals.

566. Ammonium Palmitate. *Hexadecanoic acid ammonium salt; palmitic acid ammonium salt.* $C_{16}H_{35}NO_2$; mol wt 273.45. C 70.27%, H 12.90%, N 5.12%, O 11.70%. $CH_3(CH_2)_{14}COONH_4$. Prepn from palmitic acid and excess 28-30% NH_3 soln: Stumpf, *Am. Paint J.* **38**, No. 45, 60, 64, 68 (1954), *C.A.* **48**, 12428b (1954); from palmitic acid and ammonium carbonate: Reiling, U.S. pat. 3,053,867 (1962 to Boston Chem. Prods.).

Yellow-white powder, softens at 38-40°F, mp 70-73°F. At 80°F: sol in water; slightly sol in benzene, xylene; practically insol in acetone, ethanol, methanol, CCl_4, naphtha. Sol in water at 212°F, acetone at 134°F, ethanol at 172°F, methanol at 148°F, benzene at 176°F, CCl$_4$ at 170°F, xylene at 180°F, naphtha at 160°F.

USE: Waterproofing fabrics, thickening lubricants.

567. Ammonium Pentachlorozincate. Zinc ammonium chloride. Cl$_5$H$_{12}$N$_3$Zn; mol wt 296.79. Cl 59.74%, H 4.08%, N 14.16%, Zn 22.03%. ZnCl$_2$.3NH$_4$Cl. Prepn: Klug, Alexander, *J. Am. Chem. Soc.* **66**, 1056 (1944).

Hygroscopic, orthorhombic, bipyramidal crystals. d 1.81. Sublimes at 340° without melting if absolutely dry. Very sol in water.

USE: In manuf of dry cells; as flux for welding, soldering, galvanizing.

568. Ammonium Perchlorate. ClH$_4$NO$_4$; mol wt 117.49. Cl 30.17%, H 3.43%, N 11.92%, O 54.47%. NH$_4$-ClO$_4$. Prepn: *Gmelin's, Ammonium* (8th ed.) **23**, pp 196-200 (1936). Review of decompn and combustion: Jacobs, Whitehead, *Chem. Rev.* **69**, 551-590 (1969).

Orthorhombic crystals. Dec on heating. d 1.95. Freely sol in water; sol in methanol; slightly sol in ethanol, acetone; almost insol in ethyl acetate, ether.

USE: In explosives, pyrotechnic compositions, jet and rocket propellants.

569. Ammonium Peroxydisulfate. Ammonium persulfate. H$_8$N$_2$O$_8$S$_2$; mol wt 228.20. H 3.53%, N 12.28%, O 56.09%, S 28.10%. (NH$_4$)$_2$S$_2$O$_8$. Available oxygen 7.01%. Contains, when recently made, 95 to 98% (NH$_4$)$_2$S$_2$O$_8$. Prepd by anodic oxidation of a satd (NH$_4$)$_2$SO$_4$ soln: Feher in *Handbook of Preparative Inorganic Chemistry*, Vol. 1, G. Brauer, Ed. (Academic Press, New York, 2nd ed., 1963) p 390.

Odorless platelike or prismatic (monoclinic) crystals, or white granular powder. Stable for months when pure and dry; dec in presence of moisture, gradually evolving ozone-contg oxygen; dec on heating, evolving O$_2$ and forming (NH$_4$)$_2$S$_2$O$_7$. d 1.98. Strong oxidizing agent. Freely sol in water; aq soln is acid and dec slowly at room temp and rapidly at higher temp evolving O$_2$ and forming NH$_4$HSO$_4$. *Keep dry, in a cool place and protected from organic matter.* LD$_{50}$ orally in rats: 820 mg/kg, Smyth *et al., Am. Ind. Hyg. Assoc. J.* **30**, 470 (1969).

USE: As oxidizer and bleacher; to remove hypo; reducer and retarder in photography; in dyeing, manuf aniline dyes; oxidizer for copper; etching zinc; decolorizing and deodorizing oils; electroplating; washing infected yeast; removing pyrogallol stains; making soluble starch; depolarizer in electric batteries; in anal. chemistry chiefly for detection and determination of manganese.

570. Ammonium Phosphate, Dibasic. Secondary ammonium phosphate; diammonium hydrogen phosphate; Fyrex. H$_9$N$_2$O$_4$P; mol wt 132.07. H 6.87%, N 21.21%, O 48.46%, P 23.48%. (NH$_4$)$_2$HPO$_4$. The grade used medicinally is 98-99% pure. Prepn: *Gmelin's, Ammonium* (8th ed.) **23**, pp 422-426 (1936).

Odorless crystals or cryst powder; saline, cooling taste; gradually loses about 8% NH$_3$ on exposure to air. One gram dissolves in 1.7 ml water, 0.5 ml boil water; practically insol in alcohol, acetone. pH about 8. *Keep well closed.*

USE: Fireproofing textiles, paper, wood, and vegetable fibers; impregnating lamp wicks; preventing afterglow in matches; flux for soldering tin, copper, brass, and zinc; purifying sugar; in yeast cultures; in dentifrices; in corrosion inhibitors; in fertilizers.

571. Ammonium Phosphate, Monobasic. Ammonium biphosphate; ammonium dihydrogen phosphate; primary ammonium phosphate. H$_6$NO$_4$P; mol wt 115.03. H 5.26%, N 12.18%, O 55.63%, P 26.93%. (NH$_4$)H$_2$PO$_4$. Prepn: *Gmelin's, Ammonium* (8th ed.) **23**, pp 426-429 (1936).

Odorless crystals or white, cryst powder; stable in air. d 1.80. One gram dissolves in about 2.5 ml water; slightly sol in alcohol; practically insol in acetone. pH of 0.2 molar aq soln 4.2.

USE: As baking powder with sodium bicarbonate; in fermentations (yeast cultures, etc.); fireproofing of paper, wood, fiberboard, etc.

572. Ammonium Phosphite. H$_9$N$_2$O$_3$P; mol wt 116.05. H 7.81%, N 24.14%, O 41.36%, P 26.69%. (NH$_4$)$_2$HPO$_3$.

Prepn: *Gmelin's, Ammonium* (8th ed.) **23**, pp 416-417 (1936).

Monohydrate, deliquesc crystals. Sol in water. *Keep tightly closed.*

USE: As reducing agent; corrosion inhibitor for lubricating grease.

573. Ammonium Phosphomolybdate. *Ammonium molybdophosphate.* H$_{12}$Mo$_{12}$N$_3$O$_{40}$P; mol wt 1876.49. H 0.64%, Mo 61.36%, N 2.24%, O 34.11%, P 1.65%. (NH$_4$)$_3$PO$_4$.-12MoO$_3$. Prepn and constitution: Illingworth, Keggin, *J. Chem. Soc.* **1935**, 575; Thistlewaite, *Analyst* **72**, 531 (1947); Healy, *Radiochim. Acta* **3**, 100 (1964).

Yellow, heavy, cryst powder. Solubility in water (20°): 0.2 ± 0.1 g/l. Practically insol in nitric acid; sol in fixed alkali hydroxides.

USE: In phosphorus analysis; as cation-exchanger.

574. Ammonium Phosphotungstate. *Ammonium tungstophosphate;* ammonium phosphowolframate. H$_{12}$N$_3$O$_{40}$-PW$_{12}$; mol wt 2931.29. H 0.41%, N 1.43%, O 21.83%, P 1.06%, W 75.26%. (NH$_4$)$_3$PO$_4$.12WO$_3$. Prepn: Wu, *J. Biol. Chem.* **43**, 189 (1920); Healy, *Radiochim. Acta* **2**, 146 (1964).

Dihydrate, microcryst powder. Soly in water (20°): 0.15 g/l. Freely sol in fixed alkali hydroxide solns.

USE: As ion-exchanger.

575. Ammonium Picrate. *2,4,6,-Trinitrophenol ammonium salt; picric acid ammonium salt;* ammonium picronitrate; ammonium carbazoate. C$_6$H$_6$N$_4$O$_7$; mol wt 246.14. C 29.28%, H 2.46%, N 22.76%, O 45.50%. Prepn: Berl, Berl, U.S. pat. **2,350,322** (1944). Crystal structure: Maartmann-Moe, *Acta. Cryst.* **25B**, 1452 (1969).

Bright yellow, bitter scales or orthorhombic crystals. *Explodes easily from heat or shock.* d 1.72. Soly in water at 20°: about 1 g/100 ml. Slightly sol in alc. "Red modification" is not a distinct polymorph, but a slightly contaminated form of the yellow salt: Mitchell, Bryant, *J. Am. Chem. Soc.* **65**, 128 (1943). Physiologic effects and protective measures in cases of exposure to ammonium picrate: Foulger, *U.S. Armed Forces Med. J.* **4**, 1425 (1953).

USE: In explosives, fireworks, rocket propellants.

576. Ammonium Platinic Chloride. Ammonium hexachloroplatinate(IV); ammonium chloroplatinate; platinic ammonium chloride. Cl$_6$H$_8$N$_2$Pt; mol wt 443.91. Cl 47.90%, H 1.82%, N 6.31%, Pt 43.97%. (NH$_4$)$_2$PtCl$_6$.

Orange-red crystals or yellow powder. d 3.06. Slightly sol in water; practically insol in alcohol.

USE: Platinum plating; manuf spongy platinum.

577. Ammonium Platinous Chloride. Ammonium tetrachloroplatinate(II); ammonium chloroplatinite; ammonium platinochloride; platinous ammonium chloride. Cl$_4$H$_8$-N$_2$Pt; mol wt 373.00. Cl 38.01%, H 2.16%, N 7.51%, Pt 52.32%. (NH$_4$)$_2$PtCl$_4$.

Dark ruby-red crystals. Sol in water.

USE: In photography.

578. Ammonium Salicylate. *2-Hydroxybenzoic acid monoammonium salt; salicylic acid monoammonium salt;* Salicyl-Vasogen. C$_7$H$_9$NO$_3$; mol wt 155.15. C 54.19%, H 5.85%, N 9.03%, O 30.94%. Prepd from salicylic acid and aq NH$_3$: Cahours, *Ann.* **52**, 336 (1844).

Odorless, lustrous crystals or white, cryst powder. Discolors on exposure to light; loses some NH_3 on long exposure to air. Readily discolored by iron compounds. One gram dissolves in 1 ml water, 3 ml alcohol. The aq soln is slightly acid. *Keep in the dark and protected from contamination by iron.*

THERAP CAT: Analgesic; topically to loosen psoriatic scales.

579. Ammonium Selenate. $H_8N_2O_4Se$; mol wt 179.04. H 4.50%, N 15.65%, O 35.75%, Se 44.10%. $(NH_4)_2SeO_4$. Prepd by treating a soln of selenic acid with ammonia: Retgers, *Z. Phys. Chem.* **8**, 36 (1891); King, *J. Phys. Chem.* **41**, 797 (1937). Crystal structure: Gatlow, *Acta Cryst.* **15**, 419 (1962).

Monoclinic crystals. d_4^{20} 2.194. Decomposed by heat. 117 parts dissolve in 100 parts water at 7°, 197 parts in 100 parts at 160°; sol in glacial acetic acid; insol in alcohol, acetone, ammonia.

580. Ammonium Selenite. $H_8N_2O_3Se$; mol wt 163.04. H 4.95%, N 17.18%, O 29.44%, Se 48.43%. $(NH_4)_2SeO_3$. Prepd by dissolving selenious acid in a slight excess of concd aqueous ammonia and evaporating: Berzelius, *Acad. Handl. Stockholm* **39**, 13 (1818); Nilson, *Bull. Soc. Chim.* [2] **21**, 253 (1874); *ibid.* [2] **23**, 262 (1875).

White or slightly reddish crystals. Sol in water. Dec by heat. Deliquescent.

USE: In manuf red glass; reagent for alkaloids.

581. Ammonium Sodium Phosphate. Sodium ammonium phosphate; microcosmic salt; salt of phosphorus; phosphorsalz. H_5NNaO_4P; mol wt 137.02. H 3.68%, N 10.22%, Na 16.78%, O 46.71%, P 22.61%. $NaNH_4HPO_4$.

Tetrahydrate, odorless, monoclinic crystals or granules. Efflorescent in air with loss of some NH_3. d 1.544. mp about 80° when rapidly heated. Prolonged heating produces $NaPO_3$. Sol in about 5 parts cold, 1 part boiling water; practically insol in alcohol. pH of 5% aq soln 7.8 to 8.2.

USE: As reagent for determination of Mg, Zn, and in blowpipe analysis. For standardizing uranium solns. In the prepn of ammonium phosphate and ammonium molybdate stock solns.

582. Ammonium Stearate. *Octadecanoic acid ammonium salt; stearic acid ammonium salt.* $C_{18}H_{39}NO_2$; mol wt 301.50. C 71.70%, H 13.04%, N 4.65%, O 10.61%. CH_3-$(CH_2)_{16}COONH_4$. Prepn from stearic acid and excess 28-30% NH_3 soln: Stumpf, *Am. Paint J.* **38**, No. 45, 60, 64, 68 (1954); *C.A.* **48**, 12428b (1954); from stearic acid and ammonium carbonate: Reiling, *U.S. pat.* **3,053,867** (1962 to Boston Chem. Prod.).

Yellow-white powder, softens at 35-40°F., mp 70-75°F. At 80°F: sol in methanol, ethanol; slightly sol in water, benzene, xylene, naphtha; practically insol in acetone, CCl_4. Sol in water at 212°F, acetone at 134°F, ethanol at 172°F, methanol at 148°F, benzene at 176°F, CCl_4 at 170°F, xylene at 180°F, naphtha at 160°F.

USE: In vanishing creams, in waterproofing cements.

583. Ammonium Sulfamate. *Sulfamic acid monoammonium salt; AMS; Amcide; Ammate.* $H_6N_2O_3S$; mol wt 114.13. H 5.30%, N 24.55%, O 42.06%, S 28.10%. NH_4SO_3-NH_2. Weed-killing prepn: Cupery, Tanberg, *U.S. pat.* **2,277,744** (1941 to du Pont). Prepn from ammonia and sulfamic acid: Sisler, Audrieth, *Inorg. Syn.* **2**, 180 (1946).

Hygroscopic crystals (large plates). mp 131°, dec 160°. Extremely soluble in water, liquid NH_3; slightly soluble in ethanol; moderately soluble in glycerol, glycol, formamide. pH of 0.2 molar soln in H_2O = 4.9. Aq solns are stable to boiling. "Ammate" brand weed killer is a brownish-gray crystalline, hygroscopic material contg a minimum of 80% ammonium sulfamate. LD_{50} orally in rats: 3.0 g/kg, Ambrose, *J. Ind. Hyg. Toxicol.* **25**, 26 (1943).

Toxicity: Limited animal expts indicate a low order of toxicity. Ingestion may cause G.I. disturbances, N. I. Sax, *Dangerous Properties of Industrial Materials* (Reinhold, New York, 1968) p 412.

USE: In the manuf of fire-retardant compositions, for flameproofing textiles and paper products; in the manuf of weed killing compositions; in electroplating solns; for the generation of nitrous oxide gas.

584. Ammonium Sulfate. *Sulfuric acid diammonium salt;* mascagnite. $H_8N_2O_4S$; mol wt 132.14. H 6.10%, N 21.20%, O 48.43%, S 24.27%. $(NH_4)_2SO_4$. The "pure" grade of commerce contains at least 99% $(NH_4)_2SO_4$. Prepn: *Gmelin's, Ammonium* (8th ed.) **23**, 261-280 (1936). Manuf: Faith, Keyes & Clark's *Industrial Chemicals*, F. A. Lowenheim, M. K. Moran, Eds. (Wiley-Interscience, New York, 4th ed., 1975) pp 103-108. *Review:* Call, "Ammonium Sulfate" in *Mellor's* Vol. VIII, Supplement I, *Nitrogen* (part 1) 473-505 (1964). Solubility studies: A. C. D. Rivett, *J. Chem. Soc.* **121**, 379 (1922); R. M. Caven, T. C. Mitchell, *J. Chem. Soc.* **125**, 1428 (1924).

Odorless, orthorhombic crystals or white granules. d 1.77. Dec above 280°. Soly in water (g $(NH_4)_2SO_4$ per 100 g of satd soln): 41.22 (0°); 43.47 (25°); 50.42 (100°) (Rivett). Soly in water (g/100 g H_2O): 70.6 (0°); 76.7 (25°); 103.8 (100°) (Caven, Mitchell). Insol in alcohol, acetone. pH of 0.1 molar aq soln 5.5.

USE: Manuf ammonia alum; in the manuf of H_2SO_4 to free it from nitrogen oxides; analytical uses; freezing mixtures; flameproofing fabrics and paper; manuf viscose silk; tanning, galvanizing iron; in fractionation of proteins. The commercial grade is used as fertilizer.

585. Ammonium Sulfide. True ammonium sulfide. H_8N_2S; mol wt 68.15. H 11.83%, N 41.11%, S 47.05%. $(NH_4)_2S$. Prepn: *Gmelin's, Ammonium* (8th ed.) **23**, p 246 (1936). Prepn of high purity soln: Johnson *et al.*, *Chemist-Analyst* **53**, 46 (1964).

Forms crystals below −18°; at high temps it dec into NH_4HS, NH_3, polysulfides, etc.

USE: To apply patina to bronze; in photographic developers, in textile manufacture, in trace metal analysis.

586. Ammonium Sulfide Solution, Red. Ammonium polysulfide soln. Contains about 8% NH_3 and 22% sulfur, corresponding to about 30% $(NH_4)_2S_3$.

Clear, red liquid; odor of NH_3 and H_2S. d about 1.10. Miscible with water. Strong alkaline reaction. *Keep tightly closed and in a cool place.*

587. Ammonium Sulfide Solution, Yellow. Ammonium bisulfide soln; ammonium sulfhydrate soln; ammonium hydrosulfide soln. Contains 8-9% NH_3 and about the same percentage of sulfur, corresponding to 16-20% as $(NH_4)_2S$.

When freshly made it is an almost colorless liquid, but soon acquires a yellow color which intensifies with age, due to formation of some polysulfide. Odor of NH_3 and H_2S; strong alkaline reaction. *Keep tightly closed.*

588. Ammonium Sulfite. $H_8N_2O_3S$; mol wt 116.14. H 6.94%, N 24.12%, O 41.33%, S 27.60%. $(NH_4)_2SO_3$. Prepn: *Gmelin's, Ammonium* (8th ed.) **23**, pp 256-258 (1936).

Monohydrate, efflorescent crystals. Under the influence of air and heat loses all its water of crystallization and is gradually oxidized to $(NH_4)_2SO_4$. Sol in water; almost insol in alcohol, acetone. Its aq soln is alkaline to litmus. *Keep tightly closed and in a cool place.*

USE: In photography; as reducing agent; in bricks for blast-furnace linings; in lubricants for metal cold-working.

589. Ammonium Tetrachloroaluminate. Aluminum ammonium chloride; ammonium chloroaluminate. $AlCl_4$-H_4N; mol wt 186.84. Al 14.44%, Cl 75.91%, H 2.16%, N 7.50%. NH_4AlCl_4. Prepn from $AlCl_3$ and NH_4Cl: Friedman, Taube, *J. Am. Chem. Soc.* **72**, 2236 (1950).

mp 304°. Sol in water and ether.

USE: In the processing of furs.

590. Ammonium Tetrachlorozincate. $Cl_4H_8N_2Zn$; mol wt 243.28. Cl 58.30%, H 3.31%, N 11.52%, Zn 26.87%. $(NH_4)_2ZnCl_4$. Prepd by dissolving 70 g $ZnCl_2$ and 30 g NH_4Cl in 29 ml hot water and crystallizing: Meerburg, *Z. Anorg. Allgem. Chem.* **37**, 199 (1903); Wagenknecht, Juza in *Handbook of Preparative Inorganic Chemistry*, Vol. 1, G. Brauer, Ed. (Academic Press, New York, 2nd ed., 1963) p 1072.

White, thin, shiny platelets. Orthorhombic bipyramidal. Hygroscopic. d 1.879. mp about 150° (dec). Sublimes at 341° without melting, if absolutely dry. Very sol in water with absorption of heat.

591. Ammonium Thiocyanate. *Thiocyanic acid ammonium salt;* ammonium rhodanide; ammonium sulfocyanate; ammonium sulfocyanide. CH_4N_2S; mol wt 76.12. C 15.78%, H 5.30%, N 36.80%, S 42.12%. NH_4SCN. The usual grade is 98-99% pure. Prepn: *Gmelin's, Ammonium* (8th ed.) **23**, pp 372-383 (1936).

Deliquesc crystals. mp about 149°. Freely sol in water, ethanol; sol in methanol, acetone; practially insol in $CHCl_3$, ethyl acetate. *Keep tightly closed.*

USE: In matches; double-dyeing fabrics; photography; improving and increasing strength of silks weighted with tin salts; producing grayish-black coating on Zn; manuf transparent artificial resins, thiourea; in pesticides. Detection and determn of small quantities of Fe; determn of Ag, Hg, etc.

592. Ammonium Thiosulfate. Ammonium hyposulfite. $H_8N_2O_3S_2$; mol wt 148.21. H 5.44%, N 18.90%, O 32.39%, S 43.27%. $(NH_4)_2S_2O_3$. Prepn: *Gmelin's, Ammonium* (8th ed.) **23**, pp 304-306 (1936). Crystal structure: S. T. Teng *et al., Acta Crystallogr.* **B35**, 1682 (1979).

Crystals, dec at 150°. Sol in water; insol in alc, ether. USE: To clean "white" metal; in photography; in lubricants for metal cold-working.

593. Ammonium Titanium Oxalate. Ammonium oxodioxalatotitanate(IV); titanium ammonium oxalate; titanyl ammonium oxalate. $C_4H_8N_2O_9Ti$; mol wt 276.02. C 17.40%, H 2.92%, N 10.15%, O 52.17%, Ti 17.35%. $(NH_4)_2$-$TiO(C_2O_4)_2$.

Monohydrate, crystals or crystalline powder. Very sol in water.

USE: As mordant in dyeing cotton and leather.

594. Ammonium Tungstate(VI). Ammonium paratungstate. $H_{40}N_{10}O_{41}W_{12}$; mol wt 3043.44. H 1.32%, N 4.60%, O 21.55%, W 72.52%. $(NH_4)_{10}W_{12}O_{41}$. Prepn: Pilloton, U.S. pat. **3,077,379** (1963 to Union Carbide).

Pentahydrate, plates. Freely sol in water; practically insol in alcohol.

USE: Manuf of tungsten alloys.

595. Ammonium Uranate(VI). Ammonium diuranate. $H_8N_2O_7U_2$; mol wt 624.22. H 1.29%, N 4.49%, O 17.94%, U 76.28%. $(NH_4)_2U_2O_7$. Incorrectly called "*Uranium Yellow*"; the true uranium yellow is sodium uranate. Prepn: Miller, Armstrong, U.S. pat. **2,466,118** (1949 to USAEC); Tridot, *Ann. Chim.* [12] **5**, 358 (1950).

Reddish-yellow, amorphous powder. Practically insol in water, alkalies; sol in acids or in ammonium carbonate soln. USE: Painting black on porcelain; as reagent for alkaloids; prepn of UO_2 and UF_4.

596. Ammonium Uranium Carbonate. Ammonium dioxotricarbonatouranate(VI); uranium ammonium carbonate; uranyl ammonium carbonate. $C_3H_{16}N_4O_{11}U$; mol wt 522.26. C 6.90%, H 3.09%, N 10.73%, O 33.70%, U 45.59%. May contain $2H_2O$.

Yellow crystals; decomposed on exposure to air. Sol in water.

USE: In uranium yellow glazes.

597. Ammonium Uranium Fluoride. *Triammonium pentafluorodioxouranate;* ammonium dioxopentafluorouranate (VI); uranium ammonium fluoride; uranyl ammonium fluoride. $F_5H_{12}N_3O_2U$; mol wt 419.19. F 22.66%, H 2.89%, N 10.03%, O 7.63%, U 56.79%. $UO_2(NH_4)_3F_5$. Usually contains some water of crystallization. Prepn and crystal structure: Nguyen-Quy-Dao, *Bull. Soc. Chim. France* **1968**, 3543; *idem, Acta. Cryst.* **25B**, 67 (1969).

Greenish-yellow, monoclinic, cryst powder. Freely sol in water; practically insol in alcohol.

USE: Has been used in x-ray work because of its fluorescence under these rays.

598. Ammonium Valerate. *Pentanoic acid ammonium salt; valeric acid ammonium salt;* ammonium valerianate. $C_5H_{13}NO_2$; mol wt 119.16. C 50.39%, H 11.00%, N 11.76%, O 26.86%. $CH_3(CH_2)_3COONH_4$. Prepn: *Hager's Handb. Pharm. Praxis* **Vol. II** (Springer-Verlag, Berlin, 1969) p 1060; Zuffanti, *J. Am. Chem. Soc.* **63**, 3123 (1941).

Very hygroscopic crystals, mp 108°. Very readily sol in water, alc; sol in ether. *Keep tightly closed.*

THERAP CAT: Formerly as sedative.

599. Ammonium Vanadate(V). Ammonium metavanadate. H_4NO_3V; mol wt 116.99. H 3.45%, N 11.97%, O 41.03%, V 43.55%. NH_4VO_3. Prepn: Baker *et al., Inorg. Syn.* **3**, 117 (1950).

White or slightly yellow, cryst powder. Sol in 165 parts water; more sol in hot water, in dil ammonia. Loses water and ammonia on heating. LD_{50} orally in rats: 0.16 g/kg, H. F. Smyth *et al., Am. Ind. Hyg. Assoc. J.* **30**, 470 (1969).

USE: In dyeing and printing on woolens; staining wood black; manuf vanadium black and "indelible ink"; producing vanadium luster on pottery; as photographic developer; in hematoxylin staining in microscopy; as reagent in analytical chemistry.

600. Ammonium Zirconyl Carbonate. Ammonium tricarbonatozirconate. $C_3H_{13}N_3O_{10}Zr$; mol wt 342.37. C 10.52%, H 3.83%, N 12.27%, O 46.73%, Zr 26.64%. $(NH_4)_3$-$ZrOH(CO_3)_3$. Prepd from $ZrO_2CO_2.8H_2O$, NH_4HCO_3 and $(NH_4)_2CO_3$: Blumenthal, *J. Chem. Ed.* **39**, 604 (1962).

Dihydrate, large prisms from water. Unstable in air, gradually evolving carbon dioxide and ammonia. Sol in water. Marketed as aq soln: d_{24}^{24} 1.238. The aq soln is moderately stable at room temp, although, after standing a month or so, it is likely to deposit hydrous zirconia. At temps above 60° aq solns dec rapidly.

USE: Water-repellent. Leaves a zirconium residue after the evaporation of ammonia and carbon dioxide, without leaving other cations or anions.

601. Amobarbital. *5-Ethyl-5-(3-methylbutyl)-2,4,6-(1H,3H,5H)-pyrimidinetrione; 5-ethyl-5-isopentylbarbituric acid;* 5-ethyl-5-isoamylbarbituric acid; barbamil; amylobarbitone; 5-isoamyl-5-ethylbarbituric acid; pentymal; Somnal; Dormytal; Isomytal; Eunoctal; Amal; Mylodorm; Sednotic; Amasust; Stadadorm; Amytal. $C_{11}H_{18}N_2O_3$; mol wt 226.27. C 58.39%, H 8.02%, N 12.38%, O 21.21%. Prepn: U.S. pat. **1,514,573** (1924). Metabolism: Frey, Magnussen, *Arzneimittel-Forsch.* **16**, 612 (1966). Toxicity data: K. Irrgang, *ibid.* **15**, 688 (1965).

Slightly bitter crystals, mp 156-158°. One gram dissolves in 1300 ml water, in 5 ml alc, in 17 ml chloroform, in 6 ml ether. Freely sol in benzene; sol in alkaline solns. Insol in petr ether, aliphatic hydrocarbons. A satd aq soln is acid to litmus paper. LD_{50} in mice (mg/kg): 212 s.c. (Irrgang).

Sodium salt, $C_{11}H_{17}N_2NaO_3$, *Amytal Sodium, Amsebarb, Barbamyl, Dorminal, Inmetal.* Hygroscopic, friable granules of powder. Slightly bitter taste. Very soluble in water; sol in alcohol (1:1). Practically insol in ether.

Caution: May be habit forming. This is a controlled substance (depressant) listed in the U.S. Code of Federal Regulations, Title 21 Parts 329.1, 1308.12 and 1308.13 (1987).

THERAP CAT: Sedative, hypnotic.

THERAP CAT (VET): Sedative, hypnotic.

602. Amodiaquin. *4-[(7-Chloro-4-quinolinyl)amino]-2-[(diethylamino)methyl]phenol; 4-[(7-chloro-4-quinolyl)-amino]-α-(diethylamino)-o-cresol;* 7-chloro-4-(3-diethylaminomethyl-4-hydroxyanilino)quinoline; 7-chloro-4-(3-diethylaminomethyl-4-hydroxyphenylamino)quinoline; 4-(3'-diethylaminomethyl-4'-hydroxyanilino)-7-chloroquinoline; SN 10751. $C_{20}H_{22}ClN_3O$; mol wt 355.86. C 67.50%, H 6.23%, Cl 9.96%, N 11.81%, O 4.50%. Prepd from 4,7-dichloroquinoline and 4-acetamido-α-diethylamino-o-cresol: Burckhalter *et al., J. Am. Chem. Soc.* **70**, 1363 (1948); U.S. pats. **2,474,819; 2,474,821** (1949 to Parke, Davis). Alternate synthesis from 2-aminomethyl-*p*-aminophenol and 4,7-dichloroquinoline: Natarajan, Lan, *Arzneimittel-Forsch.* **22**, 1230 (1972).

Crystals from Cellosolve, mp 208°.

Dihydrochloride dihydrate, $C_{20}H_{24}Cl_3N_3O.2H_2O$, *CAM-AQ1, Camoquin, Flavoquine, Miaquin.* Yellow, bitter crystals, dec 150-160°. uv max: 224 and 342 nm ($E_{1cm}^{1\%}$ 394-410). Soluble in water. Sparingly sol in alc; very slightly sol in benzene, chloroform, ether. pH of 1% aq soln 4.0 to 4.8. Dihydrochloride hemihydrate, yellow crystals from methanol, mp 243°. Slightly sol in water, alc.

THERAP CAT: Antimalarial.

603. Amolanone. *3-[2-(Diethylamino)ethyl]-3-phenyl-2(3H)-benzofuranone;* 3-(2-diethylaminoethyl)-2-oxo-3-phenyl-2,3-dihydrobenzofuran; γ-diethylamino-α-(o-hydroxyphenyl)-α-phenylbutyric acid lactone; amocaine; AP-43; Amethone. $C_{20}H_{23}NO_2$; mol wt 309.39. C 77.64%, H 7.49%, N 4.53%, O 10.34%. Prepn: Weston, Brownell, *J. Am. Chem. Soc.* **74**, 653 (1952).

Crystals from petr ether, mp 43-44°. bp$_{2.0}$ 192-194°. $n_D^{24.5}$ 1.5614.

Hydrochloride, $C_{20}H_{23}NO_2$.HCl, crystals from isopropanol, mp 152-153°. Sol in water.

THERAP CAT: Local anesthetic.

604. Amoproxan. *3,4,5-Trimethoxybenzoic acid 1-[(3-methylbutoxy)methyl]-2-(4-morpholinyl)ethyl ester; 3,4,5-trimethoxybenzoic acid 1-[(isopentyloxy)methyl]-2-morpholinoethyl ester;* α-[(isopentyloxy)methyl]-4-morpholineethanol 3,4,5-trimethoxybenzoate; 4-[3-(isoamyloxy)-2-(3,4,5-trimethoxybenzoyloxy)propyl]tetrahydro-1,4-oxazine; 3,4,5-trimethoxybenzoic acid 2-(isopentyloxy)-1-(morpholinomethyl)ethyl ester. $C_{22}H_{35}NO_7$; mol wt 425.53. C 62.10%, H 8.29%, N 3.29%, O 26.32%. Prepn: Mauvernay *et al.,* **Brit.** pat. **1,211,307** corresp to U.S. pat. **3,781,432** (1970, 1973 to C.E.R.M.) and U.S. pat. **3,790,569** (1974). Pharmacology: Duchene-Marullaz *et al., Therapie* **24**, 665 (1969); Pernod *et al., ibid.* 675; Roulet *et al., ibid.* **27**, 93 (1972).

Hydrochloride, $C_{22}H_{36}ClNO_7$, *730-CERM, Mederel.* Crystals from anhydr isopropyl alcohol, mp 145°. Sol in water, alcohol; very slightly sol in ethyl acetate.

THERAP CAT: Antiarrhythmic.

605. Amorolfine. *cis-4-[3-[4-(1,1-Dimethylpropyl)-phenyl]-2-methylpropyl]-2,6-dimethylmorpholine;* 4-[3-(4-tert-amylphenyl)-2-methylpropyl]-2,6-dimethylmorpholine; (±)-cis-2,6-dimethyl-4-[2-methyl-3-(p-tert-pentylphenyl)-propyl]morpholine. $C_{21}H_{35}NO$; mol wt 317.51. C 79.44%, H 11.11%, N 4.41%, O 5.04%. Broad spectrum topical antimycotic, morpholine deriv. Prepn: A. Pfiffner, K. Bohnen, **Ger.** pat. **2,752,096**; A. Pfiffner, **U.S.** pat. **4,202,894** (1978, 1980 both to Hoffmann-La Roche). *In vitro* comparative antifungal spectrum: S. Shadomy *et al., Sabouraudia* **22**, 7

(1984). Mechanism of action: A. Polak-Wyss *et al., ibid.* **23**, 433 (1984). Inhibitory effect on germ tube formation in *Candida albicans:* M. Schaude *et al., Mykosen* **30**, 28 (1987); on *Candida* adherence: V. Vuddhakul *et al., J. Antimicrob. Chemother.* **21**, 755 (1988).

bp$_{0.036}$ 134°.

Hydrochloride, $C_{21}H_{36}ClNO$, *Ro 14-4767/002.*

THERAP CAT: Topical antifungal.

606. Amoscanate. *4-Isothiocyanato-N-(4-nitrophenyl)-benzeneamine; isothiocyanic acid p-(p-nitroanilino)phenyl ester;* 4-isothiocyanato-4'-nitrodiphenylamine; nithiocyamine; C-9333-Go; CGP-4540. $C_{13}H_9N_3O_2S$; mol wt 271.29. C 57.55%, H 3.34%, N 15.49%, O 11.80%, S 11.82%. Analog of nitroscanate, *q.v.* Prepn: K. Antos *et al.,* **Ger.** pat. **1,932,690** (1970 to Cesk. Akad. Ved), *C.A.* **72**, 100265 (1970); S. Rajappa *et al., J. Chem. Soc. Perkins Trans I* **1979**, 2001; N. Viswanathan, R. C. Desai, *Indian J. Chem.* **20B**, 308 (1981). Anthelmintic activity: H. P. Striebel, *Experientia* **32**, 457 (1976); K. R. Middleton *et al., ibid.* **35**, 243 (1979); H. G. Sen, B. N. Deb, *Am. J. Trop. Med. Hyg.* **30**, 992 (1981). HPLC determn in human plasma: W. M. Kofi-Tsekpo, C. W. Karekezi, *Drugs Exp. Clin. Res.* **14**, 31 (1988). Clinical pharmacology: A. B. Vaidya *et al., Brit. J. Clin. Pharmacol.* **4**, 463 (1977). Mutagenicity study: B. S. Reddy *et al., Antimicrob. Ag. Chemother.* **22**, 707 (1982). Brief review: J. I. Bruce, *Int. J. Parisitol.* **17**, 131-140 (1987).

Crystals from acetone, mp 196-198°.

THERAP CAT: Anthelmintic (Schistosoma).

607. Amosulalol. (±)-5-[1-Hydroxy-2-[[2-(2-methoxyphenoxy)ethyl]amino]ethyl]-2-methylbenzenesulfonamide. $C_{18}H_{24}N_2O_5S$; mol wt 380.46. C 56.82%, H 6.36%, N 7.36%, O 21.03%, S 8.43%. Sulfonamide-substituted phenylethanolamine with α- and β-adrenergic blocking activity. Prepn: K. Imai *et al.,* **Ger.** pat. **2,843,016**; *eidem,* **U.S.** pat. **4,217,-305** (1979, 1980 both to Yamanouchi). Cardiovascular pharmacology and receptor blocking activity in dogs: T. Takenaka *et al., Eur. J. Pharmacol.* **85**, 35 (1982). GC determn in urine: H. Kamimura *et al., J. Chromatog.* **275**, 81 (1983). Pharmacokinetics in humans: M. Nakashima *et al., Clin. Pharmacol. Ther.* **36**, 436 (1984). Metabolism in humans: H. Kamimura *et al., Xenobiotica* **15**, 413 (1985). Comparison of receptor blocking activity of isomers: K. Honda *et al., J. Pharmacol. Exp. Ther.* **236**, 776 (1986).

Monohydrochloride, $C_{18}H_{25}ClN_2O_5S$, *YM-09538, Lowgan.* Colorless crystals, mp 158-160°. pK$_1$ 7.4; pK$_2$ 10.2.

R-(−)-Form hydrochloride, mp 158°. $[\alpha]_D^{20}$ −30.4° (c = 1 in methanol).

S-(+)-Form hydrochloride, mp 158°. $[\alpha]_D^{20}$ +30.7° (c = 1 in methanol).

THERAP CAT: Antihypertensive.

608. Amotriphene. *β-[Bis(4-methoxyphenyl)methylene]-4-methoxy-N,N-dimethylbenzeneethanamine; 2,3,3-tris(p-methoxyphenyl)-N,N-dimethylallylamine;* 3-dimethylamino-1,1,2-tris(4-methoxyphenyl)-1-propene; aminoxytriphene; Win 5494; Myordil. $C_{26}H_{29}NO_3$; mol wt 403.50. C 77.39%,

H 7.24%, N 3.47%, O 11.90%. Prepn: Elpern, **U.S.** pat. **3,010,965** (1961 to Sterling Drug).

Hydrochloride, $C_{26}H_{29}NO_3$·HCl, crystals from absolute ethanol, mp 182-184°. LD_{50} orally in mice: 385 mg/kg; i.v. 30 mg/kg.

THERAP CAT: Coronary vasodilator.

609. Amoxapine. *2-Chloro-11-(1-piperazinyl)dibenz-[b,f][1,4]oxazepine;* CL 67772; Asendin; Demolox; Moxadil. $C_{17}H_{16}ClN_3O$; mol wt 313.79. C 65.07%, H 5.14%, Cl 11.30%, N 13.39%, O 5.10%. Prepn: J. Schmutz *et al., Helv. Chim. Acta* **50**, 245 (1967); *eidem, Chim. Ther.* **2**, 424 (1967); C. F. Howell *et al.,* **Fr.** pat. **1,508,536,** *C.A.* **70,** 57923c (1969) and **U.S.** pat. **3,663,696** (1968, 1972 both to Am. Cyanamid). Analysis of amoxapine and its metabolites by GLC: T. B. Cooper, R. G. Kelly, *J. Pharm. Sci.* **68,** 216 (1979). X-ray crystallography: D. B. Cosulich, F. M. Lovell, *Acta Crystallogr.* **B33,** 1147 (1977). Pharmacology: E. N. Greenblatt *et al., Arch. Int. Pharmacodyn. Ther.* **233,** 107 (1978). Clinical study: R. Takahashi *et al., J. Int. Med. Res.* **7,** 7 (1979). *Review:* T. A. Ban, *Psychopharmacol. Bull.* **15,** 22-25 (1979); E. N. Greenblatt *et al.,* in *Pharmacological and Biochemical Properties of Drug Substances* **vol. 2,** M. E. Goldberg, Ed. (Am. Pharm. Assoc., Washington, DC, 1979) pp 1-18. Review of pharmacology and therapeutic efficacy: S. G. Jue *et al., Drugs* **24,** 1-23 (1982).

Cryst from benzene/petr ether, mp 175-176°. LD_{50} in mice (mg/kg): 122 i.p.; 112 orally (Howell, U.S. patent).

THERAP CAT: Antidepressant.

610. Amoxicillin. *[2S-[2α,5α,6β(S*)]]-6-[[Amino(4-hydroxyphenyl)acetyl]amino]-3,3-dimethyl-7-oxo-4-thia-1-azabicyclo[3.2.0]heptane-2-carboxylic acid; (−)-6-[2-amino-2-(p-hydroxyphenyl)acetamido]-3,3-dimethyl-7-oxo-4-thia-1-azabicyclo[3.2.0]heptane-2-carboxylic acid;* 6-[D(−)-α-amino-p-hydroxyphenylacetamido]penicillanic acid; α-amino-p-hydroxybenzylpenicillin; 6-(p-hydroxy-α-aminophenylacetamido)penicillanic acid; p-hydroxyampicillin; amoxycillin; AMPC; Alfamox; Almodan; Amocilline; Amolin; Amopenixin; Amoxi; Amoxipen; Anemolin; Aspenil; Betamox; Bristamox; Cabermox; Cuxacillin; Delacillin; Efpenix; Grinsil; Ibiamox; Ospamox; Optium; Piramox; Simoxil; Sumox. $C_{16}H_{19}N_3O_5S$; mol wt 365.41. C 52.59%, H 5.24%, N 11.50%, O 21.89%, S 8.77%. Semi-synthetic antibiotic related to penicillin. Prepn: Nayler, Smith, **Brit.** pat. **978,178** (1964 to Beecham) corresp to **U.S.** pat. **3,192,198** (1965); Long, Nayler, **Ger.** pat. **1,942,693** and **Brit.** pat. **1,241,844** (1970 and 1971 to Beecham), *C.A.* **72,** 90447q (1970). Resolution of isomers: Long *et al., J. Chem. Soc. (C)* **1971,** 1920. Series of articles on activity, pharmacology, absorption and excretion: *Antimicrob. Ag. Chemother.* **1970,** 407-430. Comprehensive description: P. K. Bhattacharyya, W. M. Cort in *Analytical Profiles of Drug Substances* **vol. 7,** K. Florey, Ed. (Academic Press, New York, 1978) pp 19-41.

Review of antibacterial activity, pharmacokinetics and therapeutic use: R. N. Brogden *et al., Drugs* **18,** 169-184 (1979).

Trihydrate, *BRL 2333, Agram, Amodex, Amoxibiotic, Amoxidal, Amoxidin, Amoxil, Amoxillat, Amoxi-Wolff, Amoxypen, Ardine, AX 250, Clamoxyl, Dura AX, Hiconcil, Infectomycin, Larocin* (obsolete), *Larotid, Moxal, Moxaline, Neamoxyl, Pamocil, Pasetocin, Penamox, Polymox, Raylina, Robamox, Sawacillin, Sigamopen, Silamox, Trimox, Uro-Clamoxyl, Utimox, Widecillin, Wymox, Zamocillin.* Off-white crystalline powder. $[α]_D^{20}$ +246° (c = 0.1). uv max (ethanol): 230, 274 nm (ε 10850, 1400); (0.1N HCl): 229, 272 nm (ε 9500, 1080); (0.1N KOH): 248, 291 (ε 2200, 3000). Soly (mg/ml): water 4.0; methanol 7.5; abs ethanol 3.4. Insol in hexane, benzene, ethyl acetate, acetonitrile. Hydrochloride trihydrate, browns at 90°, then dehydrates, dec 216-218°. β-Naphthalenesulfonate trihydrate, mp 194° (dec).

THERAP CAT: Antibacterial.

THERAP CAT (VET): Antibacterial.

611. Ampelopsin. *(2R-trans)-2,3-Dihydro-3,5,7-tri-hydroxy-2-(3,4,5-trihydroxyphenyl)-4H-1-benzopyran-4-one; 3,3',4',5,5',7-hexahydroxyflavanone;* ampeloptin; dihydromyricetin. $C_{15}H_{12}O_8$; mol wt 320.25. C 56.25%, H 3.78%, O 39.97%. From leaves of *Ampelopsis meliaefolia* Kudo, *Vitaceae:* Kotake, Kubota, *Ann.* **544,** 253 (1940); from bark of *Pinus contoria* Dougl., *Pinaceae:* Hergert, *J. Org. Chem.* **21,** 534 (1956); **U.S.** pat. **2,870,165** (1959 to Rayonier); from *Erythrophleum africanum* (Welw.) Harms, *Caesalpiniaceae:* Hansel, Klaffenbach, *Arch. Pharm.* **294,** 158 (1961). Synthesis from myricetin, *q.v.:* Kotake, Kubota, *J. Inst. Polytech., Osaka City Univ.* **1,** no. 2, 47 (1950), *C.A.* **46,** 2052e (1952).

Hemipentahydrate, needles from water, mp 245-246°. Hexaacetate, $C_{27}H_{24}O_{14}$, needles from alc, mp 174-175°. Hexabenzoate, $C_{57}H_{42}O_{14}$, needles from alc, mp 174°.

612. Amperozide. *4-[4,4-Bis(4-fluorophenyl)butyl]-N-ethyl-1-piperazinecarboxamide;* FG-5606; Hogpax. $C_{23}H_{29}F_2N_3O$; mol wt 401.50. C 68.80%, H 7.28%, F 9.46%, N 10.47%, O 3.99%. Piperazine deriv with effects on stress-induced disorders. Prepn: A. K. K. Björk *et al.,* **Ger.** pat. **2,941,880;** **U.S.** pat. **4,308,387** (1980, 1981 both to Ferrosan). Effects on psychosomatic disorders in man: E. Christensson *et al., 3rd World Congr. Biol. Psychiatr.,* Abstr. F 78 (Stockholm, 1981). Clinical effects in pig prodn: A. Björk *et al., Proc. Int. Pig Vet. Soc. Congr.,* pp 315, 316, 317 (Mexico, 1982).

Hydrochloride, $C_{23}H_{30}ClF_2N_3O$, mp 177-178°.

THERAP CAT (VET): Anti-aggressive.

613. Amphecloral. α-*Methyl-N-(2,2,2-trichloroethyli-dene)benzeneethanamine;* α-methyl-*N*-(2,2,2-trichloroethyli-dene)phenethylamine; *N*-[2-(1-phenylpropyl)]-2,2,2-trichlo-roethylidenimine; amfecloral; Acutran. $C_{11}H_{12}Cl_3N$; mol wt 264.60. C 49.93%, H 4.57%, Cl 40.20%, N 5.29%. Prepn: Cavallito, U.S. pat. **2,923,661** (1960 to Irwin, Neisler).

dl-Form, bp$_{0.5}$ 95°. n_D^{25} 1.530.
d-Form, $[\alpha]_D$ +49.9°±0.3° (c = 5 in dioxane).
THERAP CAT: Anorexic.

614. Amphenidone. *1-(3-Aminophenyl)-2(1H)-pyridin-one; 1-(m-aminophenyl)-2(1H)-pyridone;* Dornwal. $C_{11}H_{10}$-N_2O; mol wt 186.21. C 70.95%, H 5.41%, N 15.05%, O 8.59%. Prepn: Scudi *et al.*, U.S. pat. **2,947,754** (1960 to Wallace & Tiernan).

Crystals, mp 182.5-184.5°. LD$_{50}$ orally in mice, rats: 1300, 3200 mg/kg, Plekss *et al.*, *Fed. Proc.* **19**, 390 (1960).
THERAP CAT: Sedative; hypnotic.

615. Amphenone B. *3,3-Bis[4-aminophenyl]-2-butan-one;* amphenone; 2-oxo-3,3-bis[*p*-aminophenyl]butane. C_{16}-$H_{18}N_2O$; mol wt 254.32. C 75.56%, H 7.13%, N 11.02%, O 6.29%. Prepn: Allen, Corwin, *J. Am. Chem. Soc.* **72**, 117 (1950); U.S. pat. **2,539,388** (1951). Structure: Bencze, Allen, *J. Org. Chem.* **22**, 352 (1957). Shows antiestrogenic activity in the chick oviduct test: Hertz *et al.*, *Recent Progr. Horm. Res.* **11**, 119-147 (1955). Decreases adrenal action. Review: *Subsidia Medica* **10**, 99-102 (1958).

Crystals, mp 137.5-138°.
Dihydrochloride, $C_{16}H_{20}Cl_2N_2O$, crystals from ethanol, dec 272-275°. Soluble in water.
Note: Formerly a pinacolone structure was assigned to amphenone B: *1,2-Bis[p-aminophenyl]-2-methyl-1-pro-panone.*
USE: In biological research.

616. Amphetamine. (±)-α-*Methylbenzeneethanamine; dl-α-methylphenethylamine;* 1-phenyl-2-aminopropane; (phenylisopropyl)amine; β-aminopropylbenzene; racemic desoxy-nor-ephedrine; Actedron; Allodene; Adipan; Sym-patedrine; Psychedrine; Isomyn; Isoamyne; Mecodrin; Norephedrane; Novydrine; Elastonon; Ortédrine; Phene-drine; Profamina; Propisamine; Sympamine; Simpatedrin. $C_9H_{13}N$; mol wt 135.20. C 79.95%, H 9.69%, N 10.36%. Prepn: U.S. pats. **1,879,003** (1932); **1,921,424** (1933); **2,015,408** (1935); Hartung, Munch, *J. Am. Chem. Soc.* **53**, 1875 (1931). Demonstration of stereospecific binding sites for (+)-^{3}H-amphetamine in hypothalamic membranes and correlations with anorexic potency of phenylethylamines: S. M. Paul *et al.*, *Science* **218**, 487 (1982). Toxicity data: M. R. Warren, H. W. Werner, *J. Pharmacol. Exp. Ther.* **85**, 119 (1945); W. A. Behrendt, R. Deininger, *Arzneimittel-Forsch.* **13**, 711 (1963). Series of articles on the biochemical and behavioral effects of amphetamines in man and animals: *Handb. Exp. Pharmakol.* **45**, 3-304 (1977); *Handb. Psycho-*

pharmacol. **11**, 1-98 (1978). Review of use and abuse: J. P. Morgan, *Substance Abuse: Clinical Problems and Perspec-tives,* J. H. Lowinson, P. Ruiz, Eds. (Williams & Wilkins, Baltimore, 1981) pp 167-184. Books: C. D. Leake, *The Amphetamines: Their Actions and Uses* (Thomas, Springfield, 1958) 167 pp; O. J. Kalant, *The Amphetamines: Toxicity and Addiction* (Thomas, Springfield, 1966) 151 pp.

Mobile liquid. Amine odor. Acrid, burning taste. Vola-tilizes slowly at room temp. d$_4^{25}$ 0.913. bp$_{760}$ 200-203°; bp$_{13}$ 82-85°. Slightly soluble in water; sol in alc, ether; readily sol in acids. Aq solns are alkaline to litmus. LD$_{50}$ in rats (mg/kg): 180 s.c. (Warren, Werner).
Sulfate, $C_{18}H_{28}N_2O_4S$, *Alentol, Benzedrine, Psychoton, Simpamina.* Crystals. Slightly bitter taste followed by a sensation of numbness. mp above 300° (dec). One part dissolves in 8.8 parts water, 515 parts 95% alc. A soln of 1 g/10 ml water has a pH 5-6. LD$_{50}$ in mice, rats (mg/kg): 24.2, 55 orally (Behrendt, Deininger).
Phosphate, $C_9H_{16}NO_4P$, *Actemin, Aktedron, Monophos, Profetamine Phosphate, Racephen, Raphetamine Phosphate.* Crystals, bitter taste. Sinters at about 150°. Dec around 300°. More sol in water than amphetamine sulfate. Slightly sol in alcohol. Practically insol in benzene, chloroform, eth-er. The pH of a 10% soln is about 4.6. Prepn: Goggin, U.S. pat. **2,507,468** (1950 to Clark & Clark).
d-Form tannate, *tanphetamin, Synatan.* Prepn: Cavallito, U.S. pat. **2,950,309** (1960 to Irwin, Neisler and Co.).
d-Form sulfate, see Dextroamphetamine Sulfate.
l-Form, *levamphetamine, levamfetamine.*
l-Form succinate, *Cydril.*
Note: This is a controlled substance (stimulant) listed in the U.S. Code of Federal Regulations, Title 21 Part 1308.12 (1987).
THERAP CAT: CNS stimulant; anorexic.
THERAP CAT (VET): CNS stimulant, in narcotic poisoning, anesthetic collapse, in depression from encephalitis.

617. Amphetaminil. α-*[(1-Methyl-2-phenylethyl)ami-no]benzeneacetonitrile;* N-(α-methylphenethyl)-2-phenylgly-cinonitrile; α-phenyl-α-(β-phenylisopropylamino)acetoni-trile; α-phenyl-α-(1-methyl-2-phenyl)ethylaminoacetoni-trile; α-phenyl-α-N-(1-phenylisopropyl)aminoacetonitrile; AN 1; Aponeuron. $C_{17}H_{18}N_2$; mol wt 250.33. C 81.56%, H 7.25%, N 11.19%. Prepd by reaction of DL-β-phenyliso-propylamine with sodium cyanide and benzaldehyde or with α-phenyl-α-bromoacetonitrile: Klosa, **Ger.** pat. **1,112,987** (1959), *C.A.* **56**, 3409d (1962); idem, *J. Prakt. Chem.* **20**, 275 (1963). Pharmacology: Dominok, Oelssner, *Acta Biol. Med. Ger.* **20**, 625 (1968); Beyer *et al.*, *Deut. Apoth-Ztg.* **111**, 677, 680 (1971). Metabolic studies: Remberg *et al.*, *Arch. Toxi-col.* **29**, 153 (1972). Chemistry: Beyrich *et al.*, *Pharmazie* **27**, 28 (1972); Gloeckl, Beyrich, *ibid.* 95.

Crystals from ethanol-water, mp 85-87°.
Hydrochloride, $C_{17}H_{19}ClN_2$, sinters at 100-104°, mp 134-136°.
THERAP CAT: Psychotropic.

618. Amphomycin. Amfomycin; glumamycin. $C_{58}H_{91}$-$N_{13}O_{20}$; mol wt 1290.46. C 53.98%, H 7.11%, N 14.11%, O 24.80%. Polypeptide antibiotic active against gram positive bacteria. Produced by *Streptomyces canus* from soil collec-ted near Syracuse, N.Y.: B. Heinemann *et al.*, *Antibiot. & Chemother.* **3**, 1239 (1953). Production: *eidem*, U.S. pat. **3,126,317** (1964 to Bristol-Myers). Structure and identity with glumamycin: M. Bodanszky *et al.*, *J. Am. Chem. Soc.* **95**, 2352 (1973). Pharmacology and toxicity: D. E. Tisch *et al.*, *Antibiot. Ann.* **1954-1955**, 1011. Mechanism of action: H. Tanaka *et al.*, *Biochem. Biophys. Res. Commun.* **86**, 902

(1979). Use to improve feed efficiency in ruminant animals: M. Gordon, G. J. Christie, **Ger.** pat. **3,027,370** corresp to U.S. pat. **4,414,206** (1981, 1983 to Bristol-Myers).

$(CH_3CH_2CH(CH_2)_5CH=CHCH_2CO\text{---}Asp\text{-}MeAsp\text{-}Asp\text{-}Gly\text{-}Asp\text{-}Gly\text{-}Dab^e\text{-}Val\text{-}Pro$
$|$
CH_3
$\lfloor Pip\text{-}Dab^t \rfloor$

Dab^e = D-erythro-α,β-diaminobutyric acid
Dab^t = L-threo-α,β-diaminobutyric acid
Pip = D-pipecolic acid

Acidic, surface-active polypeptide. $[\alpha]_D^{25}$ +7.5° (c = 1 at pH 6). Isoelectric point 3.5-3.6. Soluble in water and the lower alcohols; insol in nonpolar solvents. Aq solns at neutral pH are stable at room temp for at least one month. Forms sodium and calcium salts. LD_{50} for the calcium salt: 120.2 mg/kg i.v. Induces hemolysis. Active against gram-positive bacteria. Suggested as a topical agent for animal and plant infections.

Sodium salt, amorphous solid, sol in water. LD_{50} in mice (mg/kg): 177.8 i.v. (Tisch).

Calcium salt, crystalline solid, sol in water, methyl alcohol. Used in combination with other antibiotics as antibacterial ingredient in topical anti-inflammatory preparations. LD_{50} in mice (mg/kg): 120.2 i.v. (Tisch).

THERAP CAT: Antibacterial.

619. Amphotalide. *2-[5-(4-Aminophenoxy)pentyl]-1H-isoindole-1,3(2H)-dione; N-[5-(p-aminophenoxy)pentyl]-phthalimide;* 1-p-aminophenoxy-5-phthalimidopentane; amphothalide; RP 6171; Schistomide. $C_{19}H_{20}N_2O_3$; mol wt 324.37. C 70.35%, H 6.22%, N 8.64%, O 14.80%. Prepn: Ashley *et al., J. Chem. Soc.* **1959**, 3880; Barber *et al.,* **Brit.** pat. **769,706** (1957 to May & Baker).

[chemical structure: N-CH2CH2CH2CH2CH2O—phenyl—NH2]

Crystals from ethanol, mp 113-114°.
THERAP CAT: Anthelmintic (Schistosoma).

620. Amphotericin B. Amphozone; Fungizone; Fungilin; Ampho-Moronal. $C_{47}H_{73}NO_{17}$; mol wt 924.11. C 61.09%, H 7.96%, N 1.51%, O 29.43%. Polyene antibiotic produced by *Streptomyces nodosus* M4575 obtained from soil of the Orinoco river region of Venezuela: Gold *et al., Antibiot. Ann.* **1955-1956**, 579; Vandeputte *et al., ibid.* 587; Dutcher *et al., ibid.* **1956-1957**, 866; Walters *et al., J. Am. Chem. Soc.* **79**, 5076 (1957); Dutcher *et al.,* U.S. pat. **2,908,-611** (1959 to Olin Mathieson). Structure studies: Borowski *et al., Tetrahedron Letters* **1965**, 473. Carbon skeleton, ring size, and partial structure: Cope *et al., J. Am. Chem. Soc.* **88**, 4228 (1966). Complete structure: Mechlinski *et al., Tetrahedron Letters* **1970**, 3873; Borowski *et al., ibid.* 3909; R. C. Pandey, K. L. Rinehart, *J. Antibiot.* **29**, 1035 (1976). Synthetic approaches: K. C. Nicolaou *et al., Chem. Commun.* **1986**, 413. Total synthesis: K. C. Nicolaou *et al., J. Am. Chem. Soc.* **109**, 2821 (1987). Mechanism of action: R. W. Holz in *Antibiotics* vol. 5(pt. 2), F. E. Hahn, Ed. (Springer-Verlag, New York, 1979) pp 313-340. Toxicity: G. R. Keim *et al., Science* **179**, 584 (1973). Comprehensive description: I. M. Asher *et al.,* in *Analytical Profiles of Drug Substances* vol. 6, K. Florey, Ed. (Academic Press, New York, 1977) pp 1-42.

[chemical structure of Amphotericin B]

Deep yellow prisms or needles from DMF. Dec gradually above 170°. uv max (methanol): 406, 382, 363, 345 nm. $[\alpha]_D^{24}$ +333° (acidic DMF); −33.6° (0.1N methanolic HCl). Insol in water at pH 6 to 7. Soly at pH 2 or pH 11 in water: about 0.1 mg/ml. Water soly increased by sodium desoxycholate. Soly in DMF 2 to 4 mg/ml; in DMF + HCl: 60 to 80 mg/ml; in DMSO: 30 to 40 mg/ml. Solids and solns appear stable for long periods between pH 4 and 10 when stored at moderate temps out of contact with light and air. LD_{50} in mice (mg/kg): 88 i.p., 4 i.v. (Keim).

THERAP CAT: Antifungal.
THERAP CAT (VET): Antifungal, used for systemic mycoses.

621. Ampicillin. *6-[(Aminophenylacetyl)amino]-3,3-dimethyl-7-oxo-4-thia-1-azabicyclo[3.2.0]heptane-2-carboxylic acid;* 6-[D(−)-α-aminophenylacetamido]penicillanic acid; D(−)-α-aminobenzylpenicillin; ampicillin A; Ay 6108; BRL 1341; P 50; Adobacillin; Alpen; Amblosin; Amfipen; Amipenix S; Ampi-Bol; Ampicin; Ampicina; Ampilar; Ampimed; Ampipenin; Ampi-Tablinen; Amplisom; Amplital; Ampy-Penyl; Austrapen; Binotal; Bonapicillin; Britacil; Copharcilin; Doktacillin; Grampenil; Guicitrina; Marisilan; Nuvapen; Pen-Bristol; Penbritin; Penbrock; Pénicline; Penstabil; Pentrex; Pentrexyl; Polycillin; Ponecil; QI Damp; Rosampline; Synpenin; Tokiocillin; Totacillin; Totalciclina; Totapen; Ultrabion; Viccillin. $C_{16}H_{19}N_3O_4S$; mol wt 349.42. C 55.00%, H 5.48%, N 12.02%, O 18.32%, S 9.18%. Orally active, broad spectrum antibiotic; semi-synthetic deriv of penicillin. Prepn: Doyle *et al.,* U.S. pat. **2,985,648** (1961); *eidem,* **Brit.** pat. **902,703** (1962 to Beecham); Doyle *et al., J. Chem. Soc.* **1962**, 1440; Kaufmann, Bauer, U.S. pat. **3,079,-307** (1963 to Bayer); Johnson, Wolfe, and Johnson, Hardcastle, U.S. pats. **3,140,282** and **3,157,640** (both 1964 to Bristol-Myers); Grant, Alburn, U.S. pat. **3,144,445** (1964 to Am. Home Prods.); Dane, Dockner, *Ber.* **98**, 789 (1965); F. Kajfez *et al., J. Heterocycl. Chem.* **13**, 561 (1976). Prepn of dl-form: **Brit.** pat. **958,824** (1964 to Pfizer). Comprehensive description: E. Ivashkiv in *Analytical Profiles of Drug Substances* vol. 2, K. Florey, Ed. (Academic Press, New York, 1973) pp 1-61.

[chemical structure of Ampicillin]

Monohydrate, *Redicilin.* Crystals from water, dec 202°. $[\alpha]_D^{21}$ +281° (water). Sparingly sol in water at room temp.

Potassium salt, $C_{16}H_{18}KN_3O_4S$, *Suractin.*

Sesquihydrate, dec 199-202°. $[\alpha]_D^{20}$ +283.1° (water).

Trihydrate, *Acillin, Amcap, Amcill, Amperil, Ampichel, Ampikel, Ampinova, Amplin, Cetampin, Cymbi, Divercillin, Lifeampil, Morepen, Pen A, Pensyn, Princillin, Principen, Ro-Ampen, Trafarbiot, Ukopen, Vidopen.*

Anhydrous form, *Ampicillin B, Omnipen, Orbicilina.* Crystals, dec 199-202°. $[\alpha]_D^{23}$ +287.9° (water). Compared to hydrated form, anhydr compd is more stable on storage, less sol in water, dimethyl sulfoxide, and has a different crystal structure. Can be converted to ampicillin A by hydration.

Sodium salt, $C_{16}H_{18}NaN_3O_4S$, *sodium ampicillin, Alpen-N, Amcill-S, Ampilag, Cilleral, Domicillin, Omnipen-N, Pen A/N, Penbritin-S, Penialmen, Polycillin-N, Principen/N.*

L(+)-Form, crystals, dec at about 205°. $[\alpha]_D^{20}$ +209° (c = 0.2 in water). Less active as an antibiotic than D(−)-form.

THERAP CAT: Antibacterial.
THERAP CAT (VET): Antibacterial.

622. Ampligen. Poly I . poly (C$_{12}$,U); poly(rI) . poly-

$[r(C_{12}U)_n]$; $rI_n \cdot r(C_{12}U)_n$. Synthetic, mismatched double-stranded RNA capable of inducing interferons, *q.v.* Consists of a strand of 5'-inosinic acid (I) units hydrogen bonded to a second strand of 5'-cytidylic (C)- 5'uridylic (U) acid units in a 12C:U ratio. Prepn and pharmacology: P. O. P. Ts'o *et al., Mol. Pharmacol.* **12**, 299 (1976). Antiviral activity: W. A. Carter *et al., ibid.* 440; against HIV: D. C. Montefiori, W. M. Mitchell, *Proc. Nat. Acad. Sci. USA* **84**, 2985 (1987); D. C. Montefiori *et al., Antiviral Res.* **9**, 47 (1988). Induction of interferon: D. A. Stringfellow, S. D. Weed, *Antimicrob. Ag. Chemother.* **17**, 988 (1980). Mechanism of action study: W. G. Hearl, M. I. Johnston, *Biochem. Biophys. Res. Commun.* **138**, 40 (1986). Antitumor effects in comparison with interferon: H. R. Hubbell *et al., Cancer Res.* **44**, 3252 (1984); and immunomodulatory effects: H. R. Hubbell *et al., ibid.* **45**, 2481 (1985); W. A. Carter *et al., J. Biol. Response Modif.* **4**, 613 (1985). HPLC determn in plasma: M. G. Rosenblum, L. Cheung, *J. Liq. Chromatog.* **9**, 2869 (1986). Clinical evaluation in metastatic cancers: I. Brodsky *et al., J. Biol. Response Modif.* **4**, 669 (1985); in AIDS or AIDS related complex: W. A. Carter *et al., Lancet* **1**, 1286 (1987).

623. Amprolium. *1-[(4-Amino-2-propyl-5-pyrimidinyl)methyl]-2-methylpyridinium chloride; 1-[(4-amino-2-propyl-5-pyrimidinyl)methyl]-2-picolinium chloride;* Corid. $C_{14}H_{19}ClN_4$; mol wt 278.78. C 60.32%, H 6.87%, Cl 12.71%, N 20.10%. Prepn: Rogers *et al., J. Am. Chem. Soc.* **82**, 2974 (1960); Rogers, Sarett, U.S. pat. **3,020,277** (1962 to Merck & Co.), *see also* U.S. pat. **3,020,200**. Formulations as poultry feed: Rogers, Sarett, U.S. pat. **3,065,132** (1962 to Merck & Co.).

Hydrochloride, $C_{14}H_{20}Cl_2N_4$, *Amprol.* Crystals from methanol + ethanol, dec 248-249°. Freely sol in water, methanol, 95% ethanol, dimethylformamide. Sparingly sol in abs ethanol. Practically insol in isopropanol, butanol, dioxane, acetone, ethyl acetate, acetonitrile, isooctane. pH of 10% aq soln 2.5-3.0.

THERAP CAT (VET): Coccidiostat.

624. Amprotropine Phosphate. *α-(Hydroxymethyl)-benzeneacetic acid 3-(diethylamino)-2,2-dimethylpropyl ester phosphate;* 3-diethylamino-2,2-dimethylpropyl tropate phosphate; phosphate of the *dl*-tropic acid ester of 3-diethylamino-2,2-dimethyl-1-propanol; Syntropan. $C_{18}H_{32}N-O_7P$; mol wt 405.42. C 53.32%, H 7.96%, N 3.45%, O 27.62%, P 7.64%. Prepn: U.S. pats. **1,932,341** (1933); **1,987,546** (1935).

Bitter crystals, mp 142-145°. Freely sol in water; slightly sol in abs alcohol; insol in chloroform, ether. The aq soln is acid to litmus.

THERAP CAT: Anticholinergic.

625. Ampyrone. *4-Amino-1,2-dihydro-1,5-dimethyl-2-phenyl-3H-pyrazol-3-one; 4-aminoantipyrine;* 4-amino-2,3-dimethyl-1-phenyl-3-pyrazolin-5-one; 1,5-dimethyl-2-phenyl-4-aminopyrazolone. $C_{11}H_{13}N_3O$; mol wt 203.24. C 65.00%, H 6.45%, N 20.68%, O 7.87%. Prepn: Waser, *Helv. Chim. Acta* **8**, 117 (1925); Freedman, Sherndal, U.S. pat. **1,877,166** (1933 to H. A. Metz Labs.); Turpechen, Kallro,

C.A. **36**, 426 (1942); Takahashi, Ogyu, **Japan.** pat. **3663('54)** (to Nippon New Drug), *C.A.* **50**, 2687c (1956). Metabolite of aminopyrine, *q.v.:* E. S. Vesell *et al., Clin. Pharmacol. Ther.* **20**, 661 (1976); G. F. Lockwood, J. B. Houston, *J. Pharm. Pharmacol.* **31**, 787 (1979).

Pale yellow crystals from benzene, mp 109°. Sol in water, ethanol, benzene; sparingly sol in ether.

626. Amrinone. *5-Amino-(3,4'-bipyridin)-6(1H)-one; 3-amino-5-(4-pyridinyl)-2(1H)-pyridinone;* Win 40680; Inocor; Wincoram. $C_{10}H_9N_3O$; mol wt 187.20. C 64.16%, H 4.84%, N 22.45%, O 8.55%. Non-glycosidic cardioactive agent with selective inotropic properties and vasodilatory activity. Prepn: G. Y. Lesher, C. J. Opalka, U.S. pat. **4,004,012** (1977 to Sterling). Cardiotonic activity: A. A. Alousi *et al., Fed. Proc.* **37**, 3692 (1978). Hemodynamic assessment: J. R. Benotti *et al., N. Engl. J. Med.* **21**, 1373 (1978). Clinical pharmacology: D. Davolos *et al., Circulation* **58**, 183 (1978). Effect on intractable myocardial failure in man: T. H. Le Jemtel *et al., ibid.* **59**, 1098 (1979). Review of pharmacology: A. E. Farah, A. A. Alousi, *Life Sci.* **22**, 1139-1148 (1978). Review of pharmacology and clinical efficacy: M. B. Bottorff *et al., Pharmacotherapy* **5**, 227-237 (1985). Reviews: A. A. Alousi, J. Edelson in *Pharmacological and Biochemical Properties of Drug Substances* **vol. 3**, M. E. Goldberg, Ed. (Am. Pharm. Assoc., Washington, DC, 1981) pp 120-147; A. Ward *et al., Drugs* **26**, 468 (1983).

Cryst from DMF, mp 294-297° (dec).
THERAP CAT: Cardiotonic.

627. Amsacrine. *N-[4-(9-Acridinylamino)-3-methoxyphenyl]methanesulfonamide;* 4'-(9-acridinylamino)methane-sulfon-*m*-anisidide; *m*-AMSA; CI-880; NSC-249992; SN-11841; Amekrin; Amsidine; Amsidyl; Lamasine. $C_{21}-H_{19}N_3O_3S$; mol wt 393.46. C 64.10%, H 4.87%, N 10.68%, O 12.20%, S 8.15%. Cytostatic agent with antiviral and immunosuppressive properties. Prepn: B. F. Cain *et al., J. Med. Chem.* **18**, 1110 (1975). Synthesis and biological properties of spin-labeled amsacrine: B. K. Sinha *et al., ibid.* **19**, 994 (1976). Mechanism of action: W. R. Wilson, *Chem. New Zeal.* **37**, 148 (1973). Pharmacologic disposition: R. L. Cysyk *et al., Drug Metab. Disp.* **5**, 579 (1977). Antiviral activity: D. M. Byrd, *Ann. N.Y. Acad. Sci.* **284**, 463 (1977). Immunosuppressive properties: B. C. Baguley *et al., Eur. J. Cancer* **10**, 169 (1974). Exptl antitumor properties: B. F. Cain, G. J. Atwell, *ibid.* 539. Efficacy in adult acute leukemia: S. S. Legha *et al., Ann. Int. Med.* **93**, 17 (1980). Toxicologic studies: K. L. Pavkov *et al., U.S. NTIS Rep.* PB-298106 (1979) 284 pp, *C.A.* **92**, 34335c (1980). Review of pharmacology and clinical efficacy: J. Hornedo, D. A. Van Echo, *Pharmacother.* **5**, 78-90 (1985).

LD$_{50}$ in male, female CDF$_1$ mice: 810 mg/m^2; 729 mg/m^2 orally (Pavkov).

Hydrochloride, C$_{21}$H$_{20}$ClN$_3$O$_3$S, *NSC-141549*. Crystals, mp 197-199°. LD$_{50}$ i.p. in mice: approx 60 mg/kg (Byrd).

Methanesulfonate, C$_{22}$H$_{23}$N$_3$O$_6$S$_2$, *NSC-156303*. Crystals, mp 292-293°. LD$_{50}$ i.p. in mice: approx 24 mg/kg (Byrd).

THERAP CAT: Antineoplastic.

628. Amsonic Acid. *2,2'-(1,2-Ethenediyl)bis[5-aminobenzenesulfonic acid]; 4,4'-diamino-2,2'-stilbenedisulfonic acid.* C$_{14}$H$_{14}$N$_2$O$_6$S$_2$; mol wt 370.41. C 45.40%, H 3.81%, N 7.56%, O 25.92%, S 17.31%. Prepn: Bender, Schultz, *Ber.* **19**, 3234 (1886); H. E. Fierz-David, L. Blangey, *Grundlegende Operationen der Farbenchemie* (Springer-Verlag, Vienna, 7th ed., 1947) p 161; Spiegler, U.S. pat. **2,784,220** (1957 to du Pont).

Yellow needles; very slightly sol in water. Forms sparingly water-sol, crystalline salts with many bisquaternary ammonium bases.

USE: In manuf of dyes, bleaching agents.

629. Amygdalin. *[(6-O-β-D-Glucopyranosyl-β-D-glucopyranosyl)oxy]benzeneacetonitrile; amygdaloside; mandelonitrile-β-gentiobioside; D-mandelonitrile-β-D-glucoside-6-β-D-glucoside; NSC-15780.* C$_{20}$H$_{27}$NO$_{11}$; mol wt 457.42. C 52.51%, H 5.95%, N 3.06%, O 38.47%. The name amygdalin is currently used interchangeably with *laetrile.* Cyanogenic glycoside which occurs in seeds of *Rosaceae;* principally in bitter almonds; also in peaches and apricots. Most common constituent of *Laetrile®* preparations. Structure and synthesis: W. N. Haworth, B. Wylam, *J. Chem. Soc.* **123**, 3120 (1923); Kuhn, *Ber.* **56**, 857 (1923); R. Campbell, W. N. Haworth, *J. Chem. Soc.* **125**, 1337 (1924); Hudson, *J. Am. Chem. Soc.* **46**, 483 (1924); Zemplén, Kunz, *Ber.* **57**, 1357 (1924); Kuhn, Sobotka, *ibid.* 1767; Baumann, Pigman, *The Carbohydrates,* W. Pigman, Ed. (Academic Press, New York, 1957) p 550. Enzymic hydrolysis studies: Haisman, Knight, *Biochem. J.* **103**, 528 (1967). The term Laetrile® has also been applied to *mandelonitrile β-glucuronide.* Purported prepn: E. T. Krebs, E. T. Krebs, Jr., **Brit.** pat. **788,855** (1958) and U.S. pat. **2,985,664** (1961). Synthesis, characterization and comparison of mandelonitrile β-glucuronide with amygdalin: C. Fenselau *et al., Science* **198**, 625 (1977). Pharmacology and cyanide toxicity studies of amygdalin (laetrile): C. G. Moertel *et al., J. Am. Med. Assoc.* **245**, 591 (1981); M. M. Ames *et al., Cancer Chemother. Pharmacol.* **6**, 51 (1981). Pharmacokinetics: A. G. Rauws *et al., Arch. Toxicol.* **49**, 311 (1982). Amygdalin (laetrile) is a toxic drug that is not effective as a cancer treatment: C. G. Moertel *et al., N. Engl. J. Med.* **306**, 201 (1982). Review of the controversial use of amygdalin (laetrile): V. Herbert, *Am. J. Clin. Nutr.* **32**, 1121-1158 (1979).

Trihydrate, orthorhombic columns from water, mp 200°; mp about 220° when anhydr. The once melted and solidif substance remelts at 125-130°. [α]$_D^{20}$ −42° (anhydr basis). One gram dissolves in 12 ml water, in 900 ml alcohol, in 11 ml boiling alcohol. Very sol in boiling water; almost insol in ether. pH of satd aq soln ～7.

Note: The misleading term *vitamin B$_{17}$,* has sometimes been applied to amygdalin.

630. n-Amylamine. *1-Pentanamine; pentylamine;* 1-aminopentane. C$_5$H$_{13}$N; mol wt 87.16. C 68.90%, H 15.03%, N 16.07%. CH$_3$(CH$_2$)$_3$CH$_2$NH$_2$. Prepd by reduction of valeronitrile with LiAlH$_4$ and with LiAlH$_4$-AlCl$_3$: Nystrom, *J. Am. Chem. Soc.* **77**, 2544 (1955).

Liquid. d^{19} 0.766. bp 104°. mp −55°. Very sol in water; sol in alcohol; miscible with ether.

Caution: A strong irritant.

631. Amylase. Enzymes catalyzing the hydrolysis of α-1→4 glucosidic linkages of polysaccharides such as starch, glycogen, or their degradation products. *Endoamylases* attack the α-1→4 linkage at random. A single type of endoamylase is known, *i.e.,* α-amylases (*α-1,4-glucan 4-glucanohydrolases*), so named, because the reducing hemiacetal group liberated by the hydrolysis has α optical configuration and mutarotates downward. The more common α-amylases include those isolated from human saliva, human, hog and rat pancreas, *Bacillus subtilis, B. coagulans, Aspergillus oryzae (Taka-amylase), A. candidus, Pseudomonas saccharophila,* and barley malt. *Exoamylases* attack the α-1→4 linkages only from the non-reducing outer polysaccharide chain ends. Those breaking every glucosidic bond to produce solely α-glucose are known as *glucoamylases* (*γ-amylases*). Those breaking every alternate bond to produce maltose are known as β-amylases (*α-1,4-glucan maltohydrolases*). Exoamylases are exclusively of vegetable or microbial origin. *Reviews:* Fischer, Stein, "α-Amylases" and French, "β-Amylases", in *The Enzymes,* Vol. 4, P. D. Boyer *et al.,* Eds., (Academic Press, New York, 2nd ed., 1960) pp 313-343, 345-368; J. A. Thoma *et al., ibid.* Vol. V (3rd ed., 1971) pp 115-189; W. M. Fogarty, C. T. Kelly, *Microbial Enzymes and Bioconversions,* A. H. Rose, Ed. (Academic Press, New York, 1980) pp 115-170.

USE: In starch processing, brewing, distilling, baking, animal feed, sewage treatment.

632. α-Amylase (Bacterial). Agrozyme; Amylo-Liquifase; Biolase; Diasmen; Grozyme; Rapidase; Superclastase. Usually prepd using *Bacillus subtilis, e.g.,* by submerged fermentation of wheat bran and peanut meal medium by a strain of *B. subtilis:* Babbar *et al., Biochim. Biophys. Acta* **65**, 347 (1962). Purification and crystallization of α-amylase: from *B. subtilis,* Stein, Fischer, *Helv. Chim. Acta* **40**, 529 (1957), Babbar *et al., loc. cit.;* from *B. subtilis* or *B. mesentericus,* Meyer *et al., Experientia* **3**, 411 (1947). Mol wt about 15,500 (from *B. subtilis*) about 48,700 (from *B. stearothermophilus*). *Reviews:* Fischer, Stein, "α-Amylases" in *The Enzymes* vol. 4, P. D. Boyer *et al.,* Eds. (Academic Press, New York, 2nd ed., 1960) pp 313-343; M. Dixon, E. Webb, *Enzymes* (Academic Press, New York, 2nd ed., 1964). *See also* Amylase.

Needle-shaped crystals (from *B. subtilis*).

633. α-Amylase (Swine Pancreas). Amylopsin; Buclamase; Fortizyme; Maxilase. Mol wt about 45,000: Danielsson, *Nature* **160**, 899 (1947). Prepn from hog pancreas and crystallization: Caldwell *et al., J. Am. Chem. Soc.* **74**, 4033 (1952). *Review:* Fischer, Stein, "α-Amylases" in *The Enzymes,* vol. 4, P. D. Boyer *et al.,* Eds. (Academic Press, New York, 2nd ed., 1960) pp 313-343; M. Dixon, E. Webb, *Enzymes* (Academic Press, New York, 2nd ed., 1964). *See also* Amylase.

Crystals. Ultraviolet absorption spectra and other properties: Caldwell *et al., loc. cit.*

THERAP CAT: Enzyme (digestive aid, anti-inflammatory).

634. β-Amylase (Sweet Potato). Mol wt about 152,000. Prepn of crystalline material from sweet potatoes: Balls *et al., J. Biol. Chem.* **173**, 9 (1948). Review: French, "β-Amylases" in *The Enzymes* vol 4, P. D. Boyer *et al.,* Eds. (Academic Press, New York, 2nd ed., 1960) pp 345-368; M. Dixon, E. Webb, *Enzymes* (Academic Press, New York, 2nd ed., 1964).

Crystallized from ammonium sulfate. Rapidly grown crystals are usually twelve-sided figures, but when grown slowly in the cold, they often appear as tetragonal prisms capped by pyramids of the same order. *See also* Amylase.

635. Amylbenzene. *Pentylbenzene; n-*amylbenzene; 1-phenylpentane. $C_{11}H_{16}$; mol wt 148.24. C 89.12%, H 10.88%. $C_6H_5(CH_2)_4CH_3$. Prepd by the action of benzylmagnesium chloride on *n-*butyl *p-*toluenesulfonate: Rossander, Marvel, *J. Am. Chem. Soc.* **50,** 1491 (1928); Gilman, Heck, *ibid.* **50,** 2223 (1928); Gilman, Robinson, *Org. Syn.* **coll. vol. II,** 47, (1943).

Liquid. mp −78.25°. bp_{760} 202.2°; bp_{10} 81°. d_4^{20} 0.8594. n_D^{20} 1.48849: Vogel, *J. Chem. Soc.* **1948,** 607. Insol in water; sol in alcohol; miscible with ether, benzene.

636. d-Amyl Bromide. *1-Bromo-2-methylbutane; d-pri-act-*amyl bromide; L-1-bromo-2-methylbutane; optically active amyl bromide. $C_5H_{11}Br$; mol wt 151.06. C 39.75%, H 7.34%, Br 52.91%. $CH_3CH_2CH(CH_3)CH_2Br$. Prepn: Marckwald, *Ber.* **37,** 1038 (1904); Jones, *J. Chem. Soc.* **87,** 135 (1905); Frohardt *et al., J. Am. Chem. Soc.* **81,** 5500 (1959).

Liquid. d_4^{20} 1.221. bp 120-121°. n_D^{25} 1.4425; $[\alpha]_D^{28}$ +4.69° (neat): Frohardt *et al., loc. cit.* Practically insol in water; sol in alcohol, ether.

637. n-Amyl Bromide. *1-Bromopentane.* $C_5H_{11}Br$; mol wt 151.06. C 39.75%, H 7.34%, Br 52.91%. $CH_3(CH_2)_3$-CH_2Br. Prepn: Fournier, *Bull. Soc. Chim. France* **35,** 623 (1906); Lindstone, Morris, *Chem. & Ind. (London)* **1958,** 560.

Liquid. d_4^{15} 1.2237. bp_{740} 129.7° (mp −95°). n_D^{20} 1.4444. Practically insol in water; sol in alcohol; miscible with ether.

638. tert-Amyl Bromide. *2-Bromo-2-methylbutane;* 2-bromoisopentane. $C_5H_{11}Br$; mol wt 151.06. C 39.75%, H 7.34%, Br 52.91%. $(CH_3)_2CBrCH_2CH_3$. Prepd from trimethylethylene and dry HBr at −78°: Michael, Weiner, *J. Org. Chem.* **4,** 531 (1939).

Liquid, bp 107.4°. d_0^0 1.2439. n_D^{20} 1.4430.

639. n-Amyl Butyrate. *Butanoic acid pentyl ester.* $C_9H_{18}O_2$; mol wt 158.23. C 68.31%, H 11.47%, O 20.22%. $CH_3CH_2CH_2COOCH_2(CH_2)_3CH_3$. Prepn: Gartenmeister, *Ann.* **233,** 269 (1886).

Liquid. Apricot-like odor. mp −73.2°. d_0^0 0.8832; d_4^{15} 0.8713. bp_{760} 185°. n_D^{20} 1.4110. Soly in water (50°): 0.54 g/l; very sol in alc, ether. LD_{50} orally in rats: 12,210 mg/kg, P. M. Jenner *et al., Food Cosmet. Toxicol.* **2,** 327 (1964).

USE: Has been used in such flavors as apricot, pineapple, pear, plum, and sparingly in some perfume compositions.

640. n-Amyl Caproate. *Hexanoic acid pentyl ester;* pentyl hexanoate; *n-*caproic acid *n-*amyl ester. $C_{11}H_{22}O_2$; mol wt 186.29. C 70.92%, H 11.90%, O 17.18%. $CH_3(CH_2)_4$-$COOCH_2(CH_2)_3CH_3$. Prepn: Simonini, *Monatsh.* **13,** 320 (1892).

Liquid, bp 222-227°.

641. Amyl Carbamate, Tertiary. *2-Methyl-2-butanol carbamate;* carbamic acid 1,1-dimethylpropyl ester; 1,1-dimethylpropyl carbamate; Aponal. $C_6H_{13}NO_2$; mol wt 131.17. C 54.94%, H 9.99%, N 10.68%, O 24.39%. H_2N-$COOC(CH_3)_2C_2H_5$. Prepn: Saunders *et al., J. Am. Chem. Soc.* **73,** 3796 (1951).

Crystals, mp 94-95°. Slightly sol in water; sol in alc. THERAP CAT: Hypnotic.

642. Amyl Chloride. *1-Chloropentane; n-*amyl chloride; *n-*butylcarbonyl chloride. $C_5H_{11}Cl$; mol wt 106.60. C 56.34%, H 10.40%, Cl 33.26%. $CH_3(CH_2)_3CH_2Cl$. Prepd from 1-pentanol and concd HCl in sealed tube at 120°: Conant, Kirner, *J. Am. Chem. Soc.* **46,** 245 (1924); with HCl and zinc chloride: Clark, Streight, *Trans. Roy. Soc. Can.* [3] **23, III** 77 (1929); Vogel, *J. Chem. Soc.* **1943,** 638, 640; for prepn without admixed 2- and 3-chloropentanes *see* Whitmore's procedure from 1-pentanol with $SOCl_2$ and pyridine: Whitmore *et al., J. Am. Chem. Soc.* **60,** 2540 (1938); Mixer, Young, *ibid.* **78,** 3382 (1956).

Liquid. d_4^{20} 0.8828. mp −99°. bp_{760} 107.8°. n_D^{20} 1.41280. Flash pt, closed cup: 55°F (13°C). Miscible with alc, ether; insol in water. Forms a constant boiling mixture with water, bp 82°, with ethanol bp 72.5°.

643. 6-n-Amyl-m-cresol. *5-Methyl-2-pentylphenol;* 6-pentyl-m-cresol. $C_{12}H_{18}O$; mol wt 178.26. C 80.85%, H 10.18%, O 8.97%. Prepn: Coulthard, *J. Chem. Soc.* **1930,** 280.

Solid or liquid, mp 24°, bp_{15} 137-139°. Practically insol in water; sol in ethanol, acetone, ether, alkali: Baker, *Drug & Cosmet. Ind.* **34,** 464 (1934).

USE: Antiseptic, germicide and mold preventive.

644. Amylene. *2-Methyl-2-butene;* β-isoamylene; trimethylethylene. C_5H_{10}; mol wt 70.13. C 85.63%, H 14.37%. $(CH_3)_2C=CHCH_3$. Prepd by dehydration of *tert-*amyl alcohol in the presence of *p-*toluenesulfonic acid: Applequist, Babad, *J. Org. Chem.* **27,** 288 (1962); by disproportionation of isobutene with propylene or 2-butene: Banks, Regier, *Ind. Eng. Chem., Prod. Res. Dev.* **10,** 46 (1971).

Liquid, bp 37.5-38.5°. Highly flammable, flash pt 0°F. d_4^{15} 0.66. Disagreeable odor. Polymerizes on long standing. Practically insol in water; miscible with alc, ether. *Caution:* A simple asphyxiant.

645. Amylene Dichloride. *2,3-Dichloro-2-methylbutane; tert-*amylene dichloride; trimethylethylene dichloride. $C_5H_{10}Cl_2$; mol wt 141.04. C 42.57%, H 7.15%, Cl 50.28%. Prepd from trimethylethylene (2-methyl-2-butene) by treatment with Cl_2 or SO_2Cl_2: Ger. pat. 251,100 (1911); **258,555** (1912); from neopentyl iodide and chlorine: Beringer, Schultz, *J. Am. Chem. Soc.* **77,** 5533 (1955).

*dl-*Form, liquid. d_4^{15} 1.0696. bp_{760} 138°; bp_{20} 37°. n_D^{18} 1.4450. Insol in water; sol in alcohol, ether.

646. n-Amyl Ether. *1,1'-Oxybispentane; pentyl ether;* amyl oxide; diamyl ether. $C_{10}H_{22}O$; mol wt 158.28. C 75.88%, H 14.01%, O 10.11%. $[CH_3(CH_2)_3CH_2]_2O$. Prepn from amyl alcohol with concd H_2SO_4: Hinton, Nieuwland, *Proc. Indiana Acad. Sci.* **42,** 109 (1933), *C.A.* **27,** 5716 (1933).

Liquid, bp 186.75°. mp −69.43°. d_4^{20} 0.78326, d_4^{25} 0.77924. Flash pt, closed cup: 57°C. n_D^{20} 1.41195, n_D^{25} 1.40985: Dreisbach, Martin, *Ind. Eng. Chem.* **41,** 2875 (1949). Practically insol in water; miscible with alcohol, ether.

USE: Industrial solvent. *Caution:* Vapors narcotic in high concns.

647. tert-Amyl Isovalerate. *3-Methylbutanoic acid 1,1-dimethylpropyl ester;* isovaleric acid *tert-*pentyl ester; *tert-*pentyl isovalerate. $C_{10}H_{20}O_2$; mol wt 172.26. C 69.72%, H 11.70%, O 18.58%. $(CH_3)_2CH$-$CH_2COOC(CH_3)_2C_2H_5$. Prepd from trimethylethylene, isovaleric acid and zinc chloride: *Beilstein* **vol. 2,** 312.

Liquid, valerian odor and taste, bp 173-174°. d_0^0 0.8729, d_0^{14} 0.8608. Slightly sol in water; freely sol in alc; miscible with oils.

648. n-Amyl Mercaptan. *1-Pentanethiol;* amyl thioalcohol. $C_5H_{12}S$; mol wt 104.21. C 57.62%, H 11.61%, S 30.77%. $CH_3(CH_2)_3CH_2SH$. Prepn: Cossar *et al., J. Org. Chem.* **27,** 93 (1962).

Liquid, bp 123-124°, n_D^{25} 1.4439. Penetrating, unpleasant odor. d_4^{20} 0.857. Practically insol in water; sol in alc.

USE: In organic syntheses. *Caution:* Mild irritant to skin, mucous membranes. May cause skin sensitization.

649. Amyl Nitrite. Mixture of isomers containing not less than 97.0% and not more than 100.0% of $C_5H_{11}NO_2$. Consists chiefly of isoamyl nitrite $[(CH_3)_2CHCH_2CH_2ONO]$, but other isomers are also present. The N.F. grade has d_{25}^{25} 0.870-0.876. *See* Isoamyl Nitrite.

650. Amylocaine Hydrochloride. *1-(Dimethylamino)-*

2-methyl-2-butanol benzoate hydrochloride; 1-(dimethylaminomethyl)-1-methylpropyl benzoate hydrochloride; methylethyldimethylaminomethylcarbinol benzoyl ester hydrochloride; amyleine hydrochloride; Stovaine. $C_{14}H_{22}ClNO_2$; mol wt 271.80. C 61.87%, H 8.16%, Cl 13.05%, N 5.15%, O 11.77%. Prepd by benzoylation of 1-(dimethylamino)-2-methyl-2-butanol: Fourneau, Ribas, *Bull. Sci. Pharmacol.* **35**, 273 (1928), *C.A.* **22**, 2919[1] (1928).

Crystals, bitter taste, followed by temporary numbness of the tongue. Dec 177-179°. One gram dissolves in 2 ml water, in 3.3 ml abs ethanol. Practically insol in ether. A 5% aq soln is faintly acid to litmus and neutral to Congo red.

THERAP CAT: Local anesthetic.

651. Amyloid. Amyloid substance; AS. Tissue deposits occurring in animals and humans, primarily composed of unique proteinaceous fibrils. The term "amyloid" was coined by Virchow in 1853 to reflect the starch-like darkening of the deposits after iodine staining. Deposition of the relatively inert fibrils by different pathogenic mechanisms leads to excessive accumulation, with pressure atrophy and death resulting from interference with physiological processes of affected organs. This disease complex is referred to as amyloidosis or the β-fibrilloses; it occurs in association with inflammatry conditions such as rheumatoid arthritis, with immunocyte-derived dyscrasias, and with tumors of various types. Aging is thought to occur in association with amyloid deposition in the brain, heart and pancreas; systemic amyloidosis has been implicated in "senile" cardiomyopathy as well as in presenile dementia (Alzheimer's syndrome) and senile dementia, *cf.* T. I. Mandybur, *Neurology* **25**, 120 (1975). Human amyloid fibrils are characterized by green polarization birefringence after staining with Congo red, by constant tryptophan content and by an x-ray diffraction crystallographic pattern indicating a β-pleated sheet conformation: E. D. Eanes, G. G. Glenner, *J. Histochem. Cytochem.* **16**, 673 (1968). Of the two chemically distinguishable classes of amyloid fibrils, one has as its major component a fragment of immunoglobulin (*q.v.*) light chain and is designated *AL-protein:* G. G. Glenner *et al., Science* **172**, 1150 (1971); **174**, 712 (1971); and the other a non-immunoglobulin protein designated *AA-protein:* E. P. Benditt, N. Eriksen, *Am. J. Pathol.* **65**, 231 (1971); M. Levin *et al., J. Clin. Invest.* **51**, 2773 (1972); D. Ein *et al., J. Biol. Chem.* **247**, 5653 (1972); M. Levin *et al., J. Exp. Med.* **138**, 373 (1973). In addition to the fibrils, a constant constituent of amyloid deposits has been found to be a normal serum globular glycoprotein called *P-component* or *SAP;* the staining of amyloid deposits with periodic acid-Schiff has been attributed to this substance, *cf.* H. Haupt *et al., Z. Physiol. Chem.* **353**, 1841 (1972). For a discussion of isoln and structure of P-component and its relationship to C-reactive protein, *q.v., see* several authors in *Ann. N.Y. Acad. Sci.* **389** (1982) 482 pp. Comprehensive review of amyloid deposits and amyloidosis: G. G. Glenner, *N. Engl. J. Med.* **302**, 1283-1292, 1333-1343 (1980). Review of amyloidosis: P. Lovisetto *et al., Minerva Med.* **71**, 1793-1813 (1980). *See also* Bence-Jones Proteins.

652. Amylpenicillin Sodium. *3,3-Dimethyl-7-oxo-6-[(1-oxohexyl)amino]-4-thia-1-azabicyclo[3.2.0]heptane-2-carboxylic acid monosodium salt;* sodium *n*-amylpenicillinate; penicillin dihydro F sodium; flavacidin; flavicin. C_{14}-$H_{21}N_2NaO_4S$; mol wt 336.39. C 49.98%, H 6.29%, N 8.33%, Na 6.84%, O 19.03%, S 9.53%. Antibiotic produced by the mold *Aspergillus flavus* from a Czapek-Dox medium supplemented with corn steep liquor or by *Penicillium chrysogenum* Q176 or by *P. notatum* strains: McKee, MacPhillamy, *Proc. Soc. Exp. Biol. Med.* **53**, 247 (1943); Bush, Goth, *J. Pharmacol. Exp. Ther.* **78**, 164 (1943); McKee *et al., J. Bacteriol.* **47**, 187 (1944); Bush *et al., J. Pharmacol. Exp. Ther.* **84**, 264 (1945); Fried *et al., J. Biol. Chem.* **163**, 341 (1946); *see also* Wintersteiner under "Flavacidin" in *Chemistry of Penicillin*

(Princeton, 1949). Prepn by hydrogenation of 2-pentenylpenicillin: Catch *et al.,* **Brit.** pat. **584,852**; Cook, Heilbron in *Chemistry of Penicillin* (Princeton, 1949). Characterization and antibacterial activity: Leigh, *Nature* **163**, 95 (1949).

Monohydrate, flat, blunt-ended needles from moist acetone or moist ethyl acetate. When anhydrous, mp 188° (dec). $[\alpha]_D^{23}$ +319°. Very soluble in water.

653. α-Amyrin. *Urs-12-en-3β-ol;* α-amyrenol; viminalol. $C_{30}H_{50}O$; mol wt 426.70. C 84.44%, H 11.81%, O 3.75%. Occurs mostly as acetate in latex of rubber trees, in latex from *Ficus variegata* Blume, *Moraceae,* also in *Balanophora elongata* Blume, *Balanophoraceae,* and in *Erythroxylum coca* Lam. var. *novogranatense* Morris, and var. *spruceanum* Burck, *Erythroxylaceae.* Isoln from *Manila elemi:* Vesterberg, Westerlind, *Ann.* **428**, 247 (1922). Structural studies: Spring, Vickerstaff, *J. Chem. Soc.* **1937**, 249; Beynon *et al., ibid.* **1938**, 1233; Meisels *et al., Helv. Chim. Acta* **32**, 1075 (1949), **38**, 1298 (1955); Melera *et al., ibid.* **39**, 441 (1956). Identity with viminalol: Soldin, Marais, *J. Pharm. Soc.* **55**, 452 (1966). Formation from ursolic acid: Goodson, *J. Chem. Soc.* **1938**, 999; from boswellic acid: Ruzicka, Wirz, *Helv. Chim. Acta* **22**, 948 (1939). Partial synthesis from glycyrrhetic acid and stereochemistry: Corey, Cantrall, *J. Am. Chem. Soc.* **81**, 1745 (1959). *Reviews:* J. Simonsen, W. C. J. Ross, *The Terpenes,* vol. **IV** (University Press, Cambridge, 1957) pp 116-148.

Needles from alcohol, mp 186°. bp $_{0.7}$ 243°. $[\alpha]_D^{17}$ +91.6° (c = 1.3 in benzene). Sol in 22 parts 98% alc. Sol in ether, benzene, chloroform, glacial acetic acid. Slightly sol in petr ether.

Acetate, $C_{32}H_{52}O_2$, leaflets from petr ether, mp 227°. $[\alpha]_D^{20}$ +76.35° (c = 0.572 in $CHCl_3$).

Benzoate, $C_{37}H_{54}O_2$, prisms from benzene + acetone, mp 195-196°. $[\alpha]_D^{10}$ +94.6° (c = 1.9 in $CHCl_3$).

654. β-Amyrin. *Olean-12-en-3β-ol;* β-amyrenol. C_{30}-$H_{50}O$; mol wt 426.70. C 84.44%, H 11.81%, O 3.75%. Occurs together with α-amyrin. Isoln and structural studies: *see* α-amyrin. *See also* Vesterberg, *Bull. Soc. Chim.* **37**, 742 (1925); Horrmann, Firzlaff, *Arch. Pharm.* **268**, 64 (1930); Ruzicka, Marxer, *Helv. Chim. Acta* **22**, 195 (1939); Jeger, Ruzicka, *ibid.* **28**, 209 (1945); Prelog *et al., ibid.* **29**, 360 (1946). Conversion of δ-amyrene to β-amyrin: Barton *et al., J. Chem. Soc.* **1938**, 1031. Biogenetic-type total synthesis: van Tamelen *et al., J. Am. Chem. Soc.* **94**, 8229 (1972). Biosynthesis from squalene: Suga *et al., Chem. Letters* **1972**, 129, 313.

Needles from petr ether or alc, mp 197-197.5°. bp$_{0.8}$ 260°. $[\alpha]_D^{19}$ +99.8° (c = 1.3 in benzene). Somewhat less soluble than the α-form. Soluble in 37 parts of 98% alc.

Acetate, $C_{32}H_{52}O_2$, prisms from petr ether, mp 241°. $[\alpha]_D^{17}$ +79° (c = 0.9 in benzene).

Palmitate, $C_{46}H_{80}O_2$, *balanophorin*. mp 77°. $[\alpha]_D^{15}$ +54.5° (c = 1.1 in benzene). Occurs in *Balanophora elongata* Blume, *Balanophoraceae*, in *Erythroxylum coca* Lam. var *novogranatense* Morris, and var *spruceanum* Burck, *Erythroxylaceae*, in latex from *Ficus variegata* Blume, *Moraceae*.

Di-β-amyrin ether, $C_{60}H_{98}O$, mp 135-136°: Rollett, *Monatsh.* **47**, 437 (1926).

655. Anabasine. *3-(2-Piperidinyl)pyridine; 2-(3-pyridyl)piperidine;* neonicotine. $C_{10}H_{14}N_2$; mol wt 162.24. C 74.03%, H 8.70%, N 17.27%. In *Anabasis aphylla* L., *Chenopodiaceae:* Orechoff, Menschikoff, *Ber.* **64**, 266 (1931); in *Nicotiana glauca* Graham, *Solanaceae:* Smith, *J. Am. Chem. Soc.* **57**, 959 (1935); Pyriki, Oehler, *Pharmazie* **9**, 685 (1954). Synthesis: Späth, *Ber.* **70B**, 70 (1937). Extracted on a large scale in Russia. Industrial extraction processes: Sadykov, Timbekov, *J. Appl. Chem. USSR* **29**, 148 (1956). Abs config: Lukes *et al.*, *Coll. Czech. Chem. Commun.* **27**, 751 (1962).

Liquid, bp 270-272°; bp$_{14}$ 145-147°; bp$_2$ 105°. Freezes at 9°. d$_4^{20}$ 1.0455. n$_D^{20}$ 1.5430. $[\alpha]_D^{20}$ −83.1°. Sol in water and in most organic solvents.

Hydrochloride, $[\alpha]_D$ +16.5° (c = 10 in water).

USE: Insecticide. *Caution: Acute and subacute toxicity:* increased salivation, vertigo, confusion, disturbed vision and hearing, photophobia, cold extremities, nausea, vomiting, diarrhea, syncope, clonic spasms.

656. Anabsinthin. *3,3a,4,5,6,6a,6b,7,7a,8,9,10,10a,13a,-13c,14b-Hexadecahydro-6-hydroxy-3,6,8,11,14,15-hexamethyl-2H-8,15-epoxy-7,13b-ethanopentaleno(1'',2'':6,7;5'',-4'':6',7')dicyclohepta(1,2-b:1',2'-b')difuran-2,12(11H)-dione;* anabsynthin. $C_{30}H_{40}O_6$; mol wt 496.62. C 72.55%, H 8.12%, O 19.33%. A bitter principle isolated from *Artemisia absinthium* L., *Compositae* (wormwood). Isoln: F. Sorm *et al.*, *Chem. & Ind. (London)* **1955**, 569; V. Herout *et al.*, *Coll. Czech. Chem. Commun.* **21**, 1485 (1956). Structure: L. Novotny *et al.*, *Chem. & Ind. (London)* **1958**, 465; V. Herout *et al.*, *Coll. Czech. Chem. Commun.* **25**, 1492 (1960); J. Beauhaire *et al.*, *Tetrahedron Letters* **21**, 3191 (1980).

Anhydrous form, crystals from benzene or isopropanol + diisopropyl ether, or by drying over phosphorus pentoxide in vacuo, mp 267°. $[\alpha]_D^{20}$ + 113° (c = 1.85 in chloroform). Monohydrate, crystals from methanol, mp 210°.

657. Anacardic Acid. Principal constituent of cashew nut-shell liquid, *Anacardium occidentalis* L., *Anacardiaceae*, member of the family of non-isoprenoid long-chain phenols. (*See also* Urushiol.) Anacardic acid is a mixture of *2-hydroxy-6-alkylbenzoic acids* in which the alkyl chain (C_{11} or higher) is fully saturated (I) or is a monoene (II), a diene (III) or a triene (IV). The name anacardic acid is also used in the literature to designate one component of the mixture: *6-pentadecyl-2-hydroxybenzoic acid* (I) (*6-pentadecylsalicylic acid*). Another component of the mixture *6-(8-pentadecenyl)-2-hydroxybenzoic acid* (II), is also known as *ginkgoic acid.* Isoln and structure: Städeler, *Ann.* **63**, 137 (1847); Ruhemann, Skinner, *Ber.* **20**, 1861 (1887); Haagen, Smit, *K. Akad. Wetensch.* **34**, 165 (1931); Backer, Haack, *Rec. Trav. Chim.* **660**, 61 (1941); Kremers, U.S. pat. **2,431,127** (1947). Structure of side chain: Sletzinger, Dawson, *J. Org. Chem.* **14**, 670, 849 (1949). Structure of unsaturated components: J. H. P. Tyman, N. Jacobs, *J. Chromatog.* **54**, 83 (1971). GLC analysis of components: J. H. P. Tyman, *ibid.* **111**, 285 (1975). Synthesis and prostaglandin synthetase inhibiting activity of I and II: Y. Yamagiwa *et al.*, *Tetrahedron* **43**, 3387 (1987); of I: I. Kubo *et al.*, *Chem. Letters* **1987**, 1101. Antitumor activity of I: H. Itokawa *et al.*, *Chem. Pharm. Bull.* **35**, 3016 (1987). *Review:* J. H. P. Tyman, *Chem. Soc. Rev.* **8**, 499-537 (1979).

I   R = $(CH_2)_{14}CH_3$
II   R = $(CH_2)_7CH=CH(CH_2)_5CH_3$
III   R = $(CH_2)_7CH=CHCH_2CH=CH(CH_2)_2CH_3$
IV   R = $(CH_2)_7CH=CHCH_2CH=CHCH_2CH=CH_2$

Mixture, crystals from acetone, mp 34-37°. Sparingly sol in water; freely sol in alc, ether, petr ether. Forms a water-soluble sodium salt.

6-Pentadecyl component (I), needles from hexane, mp 90.2-91.5°.

6-(8-Pentadecenyl) component (II), needles, mp 45.3-48° (subl).

658. Anagestone. *17-Hydroxy-6-methylpregn-4-en-20-one;* 6α-methyl-4-pregnen-17α-ol-20-one. $C_{22}H_{34}O_2$; mol wt 330.49. C 79.95%, H 10.37%, O 9.68%. Prepn of free alc, acetate, propionate, hexanoate, and cyclopentylpropionate: Belg. pat. **624,370** (1963 to Ortho), *C.A.* **60**, 10764g (1964).

Crystals from methanol + dichloromethane, mp 190-193°. $[\alpha]_D^{20}$ +51° (in chloroform).

Acetate, $C_{24}H_{36}O_3$, *Anatropin.* Crystals from methanol + dichloromethane, mp 173-175°. $[\alpha]_D^{20}$ +24° (in chloroform).

Propionate, $C_{25}H_{38}O_3$, crystals from methanol + dichloromethane, mp 143-144°. $[\alpha]_D^{20}$ +35° (in chloroform).

Hexanoate, $C_{28}H_{44}O_3$, crystals from methanol + dichloromethane, mp 78-79°. $[\alpha]_D^{20}$ +20°.

Cyclopentylpropionate, $C_{30}H_{46}O_3$, crystals, mp 93-95°.

THERAP CAT: The acetate as a progestogen.

659. Anagrelide. *6,7-Dichloro-1,5-dihydroimidazo[2,1-b]quinazolin-2(3H)-one;* 6,7-dichloro-1,2,3,5-tetrahydroimidazo[2,1-b]quinazolin-2-one. $C_{10}H_7Cl_2N_3O$; mol wt 256.09. C 46.90%, H 2.75%, Cl 27.69%, N 16.41%, O 6.25%.

Phosphodiesterase inhibitor with antiplatelet activity. Prepn: W. N. Beverung, A. Partyka, U.S. pats. 3,932,407; RE 31,617; T. A. Jenks *et al.*, U.S. pat. 4,146,718; R. R. Crenshaw, T. A. Montzka, U.S. pat. 4,208,521 (1976, 1984, 1979, 1980 all to Bristol-Myers); H. Yamaguchi, F. Ishikawa, *J. Heterocycl. Chem.* **18**, 67 (1981). Antithrombotic and platelet aggregation inhibiting properties *in vitro* and *in vivo:* J. S. Fleming, J. P. Buyniski, *Thromb. Res.* **15**, 373 (1979). Inhibition of platelet cAMP phosphodiesterase *in vitro:* S. S. Tang, M. M. Frojmovic, *J. Lab. Clin. Med.* **95**, 241 (1980). Synergism with thromboxane synthetase inhibitors *in vitro:* J. B. Smith, *Thromb. Res.* **28**, 477 (1982). Clinical inhibition of platelet production: W. A. Andes *et al.*, *Thromb. Haemostas.* **52**, 325 (1984). GC-MS determn in human plasma: E. H. Kerns *et al.*, *J. Chromatog.* **416**, 357 (1987). Clinical evaluation in thrombocytosis associated with chronic myeloproliferative diseases: M. N. Silverstein *et al.*, *N. Engl. J. Med.* **318**, 1292 (1988).

Monohydrochloride hemihydrate, $C_{10}H_8Cl_3N_3O.\frac{1}{2}H_2O$, *BL-4162A*, *BMY 26538-01.* Crystals from ethanolic HCl, mp >280°.

THERAP CAT: Antithrombotic.

660. Anagyrine. *[7R-(7α,7aβ,14α)]-7,7a,8,9,10,11,13,-14-Octahydro-7,14-methano-4H,6H-dipyrido[1,2-a:1',2'-e]-[1,5]diazocin-4-one;* monolupine; rhombinin. $C_{15}H_{20}N_2O$; mol wt 244.33. C 73.73%, H 8.25%, N 11.47%, O 6.55%. Found in seeds of *Anagyris foetida* L., *Leguminosae* and in gorse *(Ulex europaeus* L., *Leguminosae).* Isoln: Ing, *J. Chem. Soc.* **1933**, 504; Orekhov *et al.*, *Ber.* **67**, 1394 (1934); Couch, *J. Am. Chem. Soc.* **61**, 3327 (1939); Briggs, Russell, *J. Chem. Soc.* **1942**, 507; Galinsky, Stern, *Ber.* **77**, 132 (1944); Faugeras, *Ann. Pharm. Fr.* **29**, 241 (1971). Absolute configuration: Okuda *et al.*, *Chem. & Ind. (London)* **1961**, 1116; Okuda *et al.*, *Chem. Pharm. Bull.* **13**, 491 (1965). Synthesis: van Tamelen, Baran, *J. Am. Chem. Soc.* **80**, 4659 (1958); Goldberg, Lipkin, *J. Org. Chem.* **37**, 1823 (1972).

Pale yellow glass. bp_4 210-215°; bp_{12} 260-270°. $[\alpha]_D^{25}$ −168° (c = 4.8 in ethanol). Sol in water, alcohol, chloroform; slightly sol in ether, benzene.

Hydrochloride trihydrate, $C_{15}H_{21}ClN_2O.3H_2O$, crystals, mp 235-236° (mp 296° when dry). $[\alpha]_D^{25}$ −142.5° (c = 5). Freely sol in water.

661. Anatabine. *1,2,3,6-Tetrahydro-2,3'-bipyridine;* 2-(3-pyridyl)-1,2,3,6-tetrahydropyridine; 1,2,3,6-tetrahydro-2-(3-pyridyl)pyridine. $C_{10}H_{12}N_2$; mol wt 160.21. C 74.96%, H 7.55%, N 17.49%. The most abundant of the minor alkaloids of tobacco: Späth, Kesztler, *Ber.* **70**, 239, 704, 2450 (1937). Fresh *Nicotiana tabacum*, the species most commonly used for the production of cigarette tobacco, contains 3.9% anatabine. Configuration: Lukes *et al.*, *Coll. Czech. Chem. Commun.* **27**, 751 (1962). Total synthesis of *dl*-form: Quan *et al.*, *J. Org. Chem.* **30**, 2769 (1965). Biosynthesis: E. Leete, S. Slattery, *J. Am. Chem. Soc.* **98**, 6326 (1976). Biomimetic synthesis: E. Leete, M. E. Mueller, *ibid.* **104**, 6440 (1982).

l-Form, liquid. d_4^{19} 1.091. bp_{10} 145-146°. n_D^{20} 1.5676. $[\alpha]_D^{17}$ −177.8°. Misc with water. Sol in alc, ether, benzene. *dl*-Form, liquid. $bp_{6.5}$ 136°.

662. Anazolene Sodium. *4-Hydroxy-5-[[4-(phenylamino)-5-sulfo-1-naphthalenyl]azo]-2,7-naphthalenedisulfonic acid trisodium salt;* 4'-anilino-8-hydroxy-1,1'-azonaphthalene-3,5',6-trisulfonic acid trisodium salt; trisodium 4'-anilino-8-hydroxy-1,1'-azonaphthalene-3,6,5'-trisulfonate; 1-naphthol-3,6-disulfonic acid-8-azo-4'-[N-phenyl-1'-naphthylamine]-8'-sulfonic acid trisodium salt; C.I. Acid Blue 92; C.I. 13390; Coomassie Blue; Coomassie Blue Medicinal; Coomassie Blue RL; Sulfon Acid Blue R; Wool Blue RL; Pontacyl Fast Blue R. $C_{26}H_{16}N_3Na_3O_{10}S_3$; mol wt 695.61. C 44.89%, H 2.32%, N 6.04%, Na 9.92%, O 23.00%, S 13.83%. Prepn: Ulrich, U.S. pat. 611,664 (1897 to Bayer); Ger. pat. 108,546 (1899 to Farbwerke Mülheim); *Frdl.* **5**, 497; *Beilstein* **16**, EII, 226; *Colour Index* vol. **4** (3rd ed., 1971) p 4053. Properties and biological behavior: S. H. Taylor, J. M. Thorp, *Brit. Heart J.* **21**, 492 (1959). Clinical applications: S. H. Taylor, J. P. Shillingford, *ibid.* 497; I. S. Menzies, *J. Clin. Pathol.* **19**, 179 (1966).

Reddish-black powder. Soluble in water, acetone, Cellosolve, giving a reddish-blue soln. Slightly sol in alc. Absorption max (water): 565-570 nm; in acetone 585 nm; in human plasma 580-590 nm ($E_{1cm}^{1\%}$ about 600). Aq solns are stable and are not affected by light. Solns up to 10% do not stain the skin appreciably. LD_{50} i.v. in mice: 450 mg/kg (Taylor, Thorp).

THERAP CAT: Diagnostic aid (cardiac output, blood volume determination).

663. Ancitabine. *2,3,3a,9a-Tetrahydro-3-hydroxy-6-imino-6H-furo[2',3':4,5]oxazolo[3,2-a]pyrimidine-2-methanol;* 2,2'-anhydro-(1β-D-arabinofuranosyl)cytosine; 2,2'-O-cyclocytidine; $O^{2,2'}$-cyclocytidine; ancytabine; anhydroara C. $C_9H_{11}N_3O_4$, mol wt 225.20. C 48.00%, H 4.92%, N 18.66%, O 28.42%. A cytostatic agent and intermediate in the synthesis of cytarabine, *q.v.* Prepn of the hydrochloride: E. R. Walwick *et al.*, *Proc. Chem. Soc.* **1959**, 84; T. Y. Shen, W. V. Ruyle, U.S. pat. 3,463,850 (1969 to Merck & Co.); L. B. Orgel, R. A. Sanchez, Ger. pat. 2,027,305 corresp to U.S. pat. 3,658,788 (1970, 1972 both to Salk Institute); I. L. Doerr, J. J. Fox, *J. Org. Chem.* **32**, 1462 (1972); E. K. Hamamura *et al.*, *J. Med. Chem.* **19**, 654 (1976). General pharmacological properties: H. Hirayama *et al.*, *Oyo Yakuri* **6**, 1259 (1972), *C.A.* **79**, 49175f (1973). Metabolism: D. H. W. Ho, *Drug Metab. Dispos.* **1**, 752 (1973). Biochemical study: *idem*, *Biochem. Pharmacol.* **23**, 1235 (1974). Pharmacokinetic study: H. S. Chen, J. F. Gross, *Cancer Chemother. Pharmacol.* **2**, 85 (1979). Clinical studies: J. Z. Finklestein *et al.*, *Cancer Treat. Rep.* **63**, 1331 (1979); T. Miale *et al.*, *ibid.* 1913. Toxicity studies: K. Sugihara *et al.*, *Oyo*

Yakuri **8**, 1469 (1974); H. Hirayama *et al., ibid.* 1693, *C.A.* **83**, 71766, 37747 (1975).

Hydrochloride, $C_{19}H_{12}ClN_3O_4$, NSC-145668, *Cyclo-C.* Cryst, mp 248-250° (dec). $[\alpha]_D^{23}$ —21.8° (c = 2.0 in water). uv max (pH 1-7): 262, 231 nm (ε 10600, 9400). Treatment of a soln with an equiv of NaOH rapidly converts the nucleoside to arabinosylcytosine, *cf.* I. L. Doerr, J. J. Fox, *loc. cit.*

THERAP CAT: Antineoplastic.

664. Ancrod. *Agkistrodon serine proteinase; agkistrodon rhodostoma venom protease;* A 38414 [enzyme]; Abbott 38414; Arvin; Arwin; Venacil (formerly). Defibrinating enzyme composed of approx 64% protein and 36% carbohydrate, with mol wt of about 35,400, isolated from the venom of the Malaysian pit-viper, *Agkistrodon rhodostoma* (Boie): Neth. pat. Appl. **6,502,120** corresp to H. A. Reid *et al.,* U.S. pat. **3,657,416** (1965, 1972 to Nat. Res. Dev. Corp.); K. E. Chan *et al.,* *Brit. J. Haematol.* **11**, 646 (1965). Initial purification: M. P. Esnouf, G. W. Tunnah, *Brit. J. Haematol.* **13**, 581 (1967). Improved purification, chemical composition: C. Nolan *et al., Methods Enzymol.* **45**, 205 (1976). Effects on prothrombin and fibrinogen metabolism in man: W. R. Bell *et al., J. Lab. Clin. Med.* **91**, 592 (1978). Series of articles on hematology, defibrinogenation effects on peripheral circulation and on exptl glomerulonephritis: *J. Med. Enzymol.* **3**, 510-538 (1980), *C.A.* **93**, 215548-50, 231022 (1980). Clinical studies: G. Trübestein, *Angiology* **32**, 699 (1981); J. A. Dormandy, *ibid.* 710; G. D. Lowe *et al., ibid.* **33**, 46 (1982). *Review:* W. R. Bell, *CRC Handbook Ser. Clin. Lab. Sci. Sect. I* **3**, 301-324 (1980).

Colorless substance when pure, having a light powdery texture when in the freeze-dried state. Soluble in physiological saline. Absorbable on weakly basic anion exchange materials.

THERAP CAT: Anticoagulant.

665. Ancymidol. *α-Cyclopropyl-α-(4-methoxyphenyl)-5-pyrimidinemethanol; α*-cyclopropyl-4-methoxy-*α*-(pyrimidin-5-yl)benzyl alcohol; EL-531; A-Rest; Reducymol. $C_{15}H_{16}N_2O_2$; mol wt 256.31. C 70.29%, H 6.29%, N 10.93%, O 12.48%. Prepn: J. D. Davenport *et al., Fr. pat.* **1,569,940** corresp to H. M. Taylor *et al.,* U.S. pat. **3,818,009** (1969, 1974 to Lilly). GLC determn: R. Frank, E. W. Day, *Anal. Methods Pestic. Plant Growth Regul.* **8**, 475 (1976); S. D. West, E. W. Day, *J. Assoc. Off. Anal. Chem.* **60**, 904 (1977). Studies on the specificity and site of action: R. C. Coolbaugh *et al., Plant Physiol.* **62**, 571 (1978).

Cryst solid, mp 110-111°. Vapor press. at 50°: <1 × 10^{-6} mm Hg. Soly in water at 25°: ~650 mg/l. Freely sol in acetone, methanol, ethyl acetate, chloroform, acetonitrile. Moderately sol in aromatic hydrocarbons; slightly sol in saturated hydrocarbons. LD_{50} orally in rats: 4500 mg/kg, *RTECS Vol. II*, R. J. Lewis, R. L. Tatken, Eds. (1980) p 532.

USE: Plant growth regulator.

666. Andrographolide. *3-[2-[Decahydro-6-hydroxy-5-(hydroxymethyl)-5,8a-dimethyl-2-methylene-1-naphthalenyl]ethylidene]dihydro-4-hydroxy-2(3H)-furanone; 3α,14,15,-18-tetrahydroxy-5β,9βH,10α-labda-8(20),12-dien-16-oic acid γ-lactone.* $C_{20}H_{30}O_5$; mol wt 350.44. C 68.54%, H 8.63%, O 22.83%. Bicyclic diterpenoid lactone found in leaves of *Andrographis paniculata* Nees, *Acanthaceae:* Chakravarti, *J. Chem. Soc.* **1952**, 1697; Kondi *et al., Ann. Rep. Itsuu Lab.* **1**, 25 (1950), *C.A.* **47**, 3280i (1953). Structure: Schwyzer *et al., Helv. Chim. Acta* **34**, 652 (1951); Cava, Weinstein, *Chem. & Ind. (London)* **1959**, 851; Chan *et al., ibid.* **1960**, 22; Cava *et al., Tetrahedron* **18**, 397 (1962); Cava *et al., Chem. & Ind. (London)* **1963**, 167; Cava *et al., Tetrahedron* **21**, 2617 (1965).

Rhombic prisms or plates from ethanol or methanol, mp 230-231°. $[\alpha]_D^{17}$ —126.6° ± 2° (in glacial acetic acid). d_4^{21} 1.2317. uv max: 223 nm (log ε 4.09). Sparingly sol in water; sol in acetone, methanol, chloroform, ether.

Triacetyl derivative, $C_{26}H_{36}O_8$, fine needles from alcohol + ether, mp 126-126.5°.

667. Androisoxazole. *17-Methylandrostano[3,2-c]isoxazol-17-ol; 2,3,3a,3b,4,5,5a,6,10,10a,10b,11,12,12a-tetradecahydro-1,10a,12a-trimethyl-1H-cyclopenta[7,8]phenanthro-[2,3-c]isoxazol-1-ol; 17β-*hydroxy-17α-methylandrostano-[3,2-c]isoxazole; 17α-methylandrostan[3,2-c]isoxazol-17β-ol; Neo-Ponden. $C_{21}H_{31}NO_2$; mol wt 329.47. C 76.55%, H 9.48%, N 4.25%, O 9.71%. Prepn: Zderic *et al., Chem. & Ind. (London)* **1960**, 1625.

Crystals, mp 169-170°. $[\alpha]_D$ +19°. uv max (ethanol): 226 nm (log ε 3.71).

THERAP CAT: Anabolic.

668. Androstane. *5α-Androstane;* etioallocholane. $C_{19}H_{32}$; mol wt 260.45. C 87.62%, H 12.37%. From androstane-3,17-dione: Butenandt, Tscherning, *Z. Physiol. Chem.* **229**, 185 (1934). From androstane-3,17-diol: Steiger, Reichstein, *Helv. Chim. Acta* **20**, 817 (1937). From Δ^{16}-androstene: Prelog *et al., ibid.* **27**, 66 (1944).

Leaflets from acetone-methanol, mp 50-50.5°. Sublimes at 60° and 0.003 mm Hg. $[\alpha]_D^{16}$ +2° (c = 1.2 in chloroform). Sol in acetone, alc, methanol, ether, petr ether, chloroform.

17-Amino-HCl, dec 345°: Marker, *J. Am. Chem. Soc.* **58**, 480 (1936).

669. Androstane-3β,11β-diol-17-one. *3,11-Dihydroxyandrostan-17-one.* $C_{19}H_{30}O_3$; mol wt 306.43. C 74.47%, H 9.87%, O 15.66%. First obtained by degradation of allopregnane-3β,11β,17α,20β,21-pentol (Reichstein's Substance A): Reichstein, *Helv. Chim. Acta* **19**, 402 (1936), but later isolated in small quantities directly from extracts of the adrenal cortex: Reichstein, von Euw, *ibid.* **21**, 1197 (1938); *ibid.* **24**, 879 (1941). It is uncertain whether this substance occurs in the fresh adrenal gland; it may possibly occur by oxidation or decompn during the isolation procedure: Reichstein, Shoppee, *Vitamins & Hormones* **I**, 368 (1943). Structure-activity study: S. Sassa *et al., J. Biol. Chem.* **254**, 10011 (1979). Chromatographic studies: A. Kerebel *et al., J. Chromatog.* **140**, 229 (1977); J. T. Lin, E. Heftmann, *ibid.* **237**, 215 (1982).

Needles from acetone + ether, mp 235-238°. $[\alpha]_D^{20}$ +84.5° (ethanol); $[\alpha]_D^{19}$ +81.3° (dioxane); $[\alpha]_{545}^{19}$ +105° (dioxane). Precipitated by digitonin.

3-Acetate, $C_{21}H_{32}O_4$, needles from actone + ether, mp 230-231°. $[\alpha]_D^{19}$ +70.5° (dioxane); $[\alpha]_{546}^{19}$ +87.1° (dioxane). Diacetate, $C_{23}H_{34}O_5$, crystals, mp 154-156°.

670. Androstenediol. *Androst-5-ene-3β,17β-diol;* Δ^5-*androstene-3β,17β-diol.* $C_{19}H_{30}O_2$; mol wt 290.43. C 78.57%, H 10.41%, O 11.02%. Obtained from dehydroandrosterone: Butenandt, Hanisch, *Ber.* **68**, 1859 (1935); Z. *Physiol. Chem.* **237**, 89 (1935); Ruzicka, Wettstein, *Helv. Chim. Acta* **18**, 1264 (1935); *FIAT Final Report* 996, 45 (1947); Levy, Kapp, U.S. pat. **2,521,586** (1950 to Nopco Chem.).

Leaflets from acetone + petr ether, or from methanol or ethyl acetate. Sublimes in high vacuum. mp 184°. $[\alpha]_D^{18}$ −55.5° (c = 0.4 in isopropanol). Insol in water.

3-Acetate, $C_{21}H_{32}O_3$, crystals from hexane, mp 147-148°. 17-Acetate, $C_{21}H_{32}O_3$, crystals from hexane, mp 146.5-148.5°, $[\alpha]_D^{18}$ −62.4° (alc). Diacetate, $C_{23}H_{34}O_4$, leaflets from hexane, mp 165-166°, $[\alpha]_D^{18}$ −56.5° (alc). 17-Benzoate, $C_{26}H_{34}O_3$, crystals from methanol, mp 220-222°.

3-Acetate-17-benzoate, $C_{28}H_{36}O_4$, crystals, mp 180-182°. Dipropionate, *Bisexovis, Stenandiol.*

THERAP CAT: Anabolic.

671. 4-Androstene-3,17-dione. Androtex. $C_{19}H_{26}O_2$; mol wt 286.40. C 79.68%, H 9.15%, O 11.17%. Prepd from 5-dehydroandrosterone: Butenandt, Kudszus, Z. *Physiol. Chem.* **237**, 75 (1935); Ruzicka, Wettstein, *Helv. Chim. Acta* **18**, 986 (1935). *See also* U.S. pats. **2,384,335; 2,175,220; 2,194,235**, and U. Westphal, H. Hellmann, *Ber.* **70**, 2136 (1937). From 5-androsten-3-ol-17-one: F. Galinovsky, *Ber.* **74**, 1624 (1941). From 4-androsten-17-ol: R. E. Marker *et al., J. Am. Chem. Soc.* **62**, 223 (1940); U.S. pat. **2,397,424**. From 5,6-dibromoandrostane-3,17-dione: P. L. Julian *et al., J. Am. Chem. Soc.* **67**, 1728 (1945); U.S. pat. **2,374,683**. From 17-hydroxyprogesterone: D. A. Prins, T. Reichstein, *Helv. Chim. Acta* **24**, 951 (1941). Isolated in small amounts from adrenal cortex: von Euw, T. Reichstein, *ibid.* 879; it is possible that it does not occur in the adrenal gland, but originates by oxidation of Reichstein's Substance S during work-up.

Dimorphous. Needles from acetone, mp 142-144°; crystals from hexane, mp 173-174°. $[\alpha]_D^{30}$ +191° (alc). uv max 235 nm.

Dioxime, $C_{19}H_{28}N_2O_2$, crystals, mp 143°.

3-Semicarbazone, $C_{20}H_{29}N_3O_2$, crystals, dec 245°.

672. Androst-16-en-3-ol. 3α-Hydroxy-5α-androst-16-ene; Δ^{16}-androsten-3-ol. $C_{19}H_{30}O$; mol wt 274.45. C 83.15%, H 11.02%, O 5.83%. A major constituent of boar pheromone, having a pronounced musk-like odor. Isoln from swine testes: V. Prelog, L. Ruzicka, *Helv. Chim. Acta* **27**, 61 (1944). Prepn: V. Prelog *et al., ibid.* 66; J. Fishman *et al., J. Org. Chem.* **28**, 1443 (1963). Physiological role as a sex attractant for pigs: D. B. Gower, *J. Steroid Biochem.* **3**, 45 (1972). Use in pig artificial insemination: D. R. Melrose *et al.*, **Ger.** pat. **1,937,264** corresp to U.S. pat. **3,681,490** (1970, 1972 both to Nat. Res. Dev. Corp.). *In vivo* metabolism in boar testes: Y. A. Saat *et al., Biochem. J.* **144**, 347 (1974). It has also been detected in human male axillary sweat, but has no androgenic activity: B. W. L. Brooksbank *et al., Experientia* **30**, 864 (1974). Radioimmunoassay: D. C. Bickell, D. B. Gower, *J. Steroid Biochem.* **7**, 451 (1976). Receptor studies: J. N. Gennings *et al., Biochim. Biophys. Acta* **496**, 547 (1977). Biosynthetic studies: E. L. Hurden *et al., J. Endocrinol.* **81**, 161P (1979); G. M. Cook, D. B. Gower, *ibid.* **88**, 409 (1981). Discovery of the presence of androst-16-en-3-ol in truffles *(Tuber melanosporum)* has been offered as an explanation for the ability of pigs to detect truffles growing as deep as 1 meter underground: R. Claus *et al., Experientia* **37**, 1178 (1981).

Cryst, mp 142.5-143°. Purified by sublimation in high vacuum and recryst from acetone. $[\alpha]_D^{20}$ +13.1° (c = 0.957 in chloroform). Gives a blue color in the Kägi-Miescher test, *cf. Helv. Chim. Acta* **22**, 683 (1939).

USE: As an aid to estrus determn in pig artificial insemination.

673. Androsterone. *3α-Hydroxy-5α-androstan-17-one; cis*-androsterone; 3α-hydroxy-17-androstanone; androstan-3(α)-ol-17-one; 3(α)-hydroxyetioallocholan-17-one; 3-epihydroxyetioallocholan-17-one. $C_{19}H_{30}O_2$; mol wt 290.43. C 78.57%, H 10.41%, O 11.02%. Isolation from male urine after removal of the phenolic estrogen fraction: Butenandt, Tscherning, Z. *Physiol. Chem.* **229**, 167 (1934); v. Euw, Reichstein, *Helv. Chim. Acta* **25**, 988 (1942). Prepn from cholesterol: Ruzicka, *ibid.* **17**, 1389 (1934); Marker, *J. Am. Chem. Soc.* **57**, 1755 (1935); Schoeller *et al.*, U.S. pat. **2,232,-735** (1941 to Schering).

Crystals from acetone-ether, mp 185-185.5°. Sublimes in high vacuum. $[\alpha]_D^{20}$ +94.6° (c = 0.7 in abs alc). $[\alpha]_D^{15}$ +87.8° (c = 1.5 in dioxane). Not precipitated by digitonin. Barely soluble in water. Sol in most organic solvents.

Acetate, $C_{21}H_{32}O_3$, crystals from ether, sublimes in high vac, mp 165°, $[\alpha]_D^{14}$ +76.7° (c = 2.04 in dioxane); $[\alpha]_D^{25}$ +86° (c = 2 in ethanol).

Propionate, $C_{22}H_{34}O_3$, mp 151-152°.

Benzoate, $C_{26}H_{34}O_3$, mp 178°.

674. Anemonin. *1,7-Dioxadispiro[4.0.4.2]dodeca-3,9-diene-2,8-dione;* 1,2-dihydroxy-1,2-cyclobutanediacrylic acid di-γ-lactone; Anemone camphor; Pulsatilla camphor. $C_{10}H_8O_4$; mol wt 192.16. C 62.50%, H 4.20%, O 33.30%.

Found in *Anemone pulsatilla* L. and other *Ranunculaceae*. Its precursor in plants is protoanemonin. Isoln from *Ranunculus acer:* Zecher, Wohlmuth, *Sci. Pharm.* **22**, 95 (1954); *C.A.* **48**, 13169b (1954). Structure: Moriarty *et al., J. Am. Chem. Soc.* **87**, 3251 (1965); Romain, *Diss. Abstr. B* **27**, 3867 (1967). Synthesis: Sugiyama *et al., C.A.* **67**, 116-604n (1967).

Crystals from petr ether, mp 157-158°. Volatile with steam. Slightly sol in cold, more in hot water; sol in hot alcohol, chloroform, alkalies with yellow color; practically insol in ether. LD_{50} i.p. in mice: 150 mg/kg, R. Brodersen, A. Kjaer, *Acta Pharmacol.* **2**, 109 (1946).

Note: Not to be confused with anemonine which is 5-(carboxymethyl)-1,1-dimethylimidazolium hydroxide inner salt.

675. Anethole. *1-Methoxy-4-(1-propenyl)benzene; p-propenylanisole;* anise camphor; Monasirup. $C_{10}H_{12}O$; mol wt 148.20. C 81.04%, H 8.16%, O 10.80%. Chief constituent of anise, star anise and fennel oils: Monad, de Dortan, *Ind. Parfum.* **5**, 401 (1950); Naves, Tucakov, *Compt. Rend.* **248**, 843 (1959). Separation of *cis* and *trans* isomers: Naves *et al., ibid.* **246**, 1734 (1958); *Bull. Soc. Chim. France* **1958**, 566; Naves, *Helv. Chim. Acta* **43**, 230 (1960); Ferroni *et al., Gazz. Chim. Ital.* **92**, 1198 (1962). Synthesis: Mueller, Röscheisen, *Ber.* **90**, 543 (1957); R. J. DePasquale, *Synth. Commun.* **10**, 225 (1980). Toxicity: J.-R. Boissier *et al., Therapie* **22**, 309 (1967). *Review:* Wagner, *Mfg. Chemist* **23**, 56 (1952).

trans-Isomer, crystalline mass at 20-21°; mp 21.4°. Liquid above 23°. d_4^{20} 0.9883. $bp_{2.3}$ 81-81.5°. n_D^{20} 1.56145. uv max (ethanol): 259 nm (ϵ 22,300). Practically insol in water. Misc with ether, chloroform; sol in benzene, ethyl acetate, acetone, carbon disulfide, petr ether; 1 ml dissolves in 2 ml alc. LD_{50} i.p. in rats: 900 mg/kg (Boissier).

cis-Isomer: d_4^{20} 0.9878. $bp_{2.3}$ 79-79.5°. n_D^{20} 1.55455. uv max (ethanol): 253.5 nm (ϵ 18500). LD_{50} i.p. in rats: 93 mg/kg (Boissier).

USE: Manuf anisaldehyde; flavoring agent; in perfumery, particularly for soap and dentifrices; sensitizer in bleaching colors in color photography; as an imbedding material in microscopy. Pharmaceutic aid (flavor).

THERAP CAT (VET): Has been used as a carminative.

676. Anethole Trithione. *5-(p-Methoxyphenyl)-3H-1,2-dithiole-3-thione;* 3-(p-anisyl)trithione; 3-(p-methoxyphenyl)-4,5-dithiacyclopent-2-ene-1-thione; 3-(p-anisyl)-4,5-dithiacyclopent-2-ene-1-thione; (p-methoxyphenyl)trithiopropene; trithio-p-methoxyphenylpropene; 5-(p-methoxyphenyl)-1,2-dithiacyclopent-4-ene-3-thione; 3-(p-methoxyphenyl)trithione; trithioanethole; Heporal; Mucinol; Trithio; Sufralem; Tiotrifar; Felviten; Sulfogal; Sulfarlem. $C_{10}H_8OS_3$; mol wt 240.36. C 49.98%, H 3.37%, O 6.66%, S 39.98%. Prepn: Böttcher, Lüttringhaus, *Ann.* **557**, 89 (1947); Gaudin, Lozac'h, *Compt. Rend.* **224**, 557 (1947); Lüttringhaus *et al., Ann.* **560**, 201 (1948); Gaudin, U.S. pats. **2,556,963, 2,688,620** (1951, 1954); Böttcher, Ger. pats. **855,865, 869,-799** and **874,447** (1952 and 1953); Thuiller, Vialle, *Bull. Soc. Chim. France* **1959**, 1398.

Orange-colored prisms from butyl acetate. Very bitter taste. mp 111°. Practically insol in water. Sol in pyridine, chloroform, benzene, dioxane, carbon disulfide. Slightly sol in ether, acetone, ethyl acetate, acetic acid, alc, cyclohexane, petr ether.

Oxime, $C_{10}H_9NO_2S_2$, yellow needles, mp 170°. Soluble in dioxane.

Methiodide, yellow crystals, mp 189°.

THERAP CAT: Choleretic.

677. Angelica. Fruit or root of *Angelica archangelica* L. (*A. officinalis* Moench.), *Umbelliferae. Habit.* Europe, Asia. *Constit.* Root: Volatile oil (0.3-1%), angelic acid, 6% resin, angelicol, angelicin, xanthotoxol, starch, osthole, osthenol, archangelicin, archangin, sitosterol, and acids such as aconitic, malic, quinic, chlorogenic, caffeic, fumaric, citric, angelic, and oxalic. Fruit: About 1% volatile oil, bitter substance, coumarins, resin. *Refs:* Späth, Pesta, *Ber.* **67**, 853 (1934); Späth, Vierhapper, *ibid.* **70**, 248 (1937); Svendsen, *C.A.* **52**, 2173g (1958); Sroka, *Apoth.-Ztg.* **61**, 37 (1949).

THERAP CAT: Carminative, diaphoretic, diuretic.

678. Angelic Acid. *(Z)-2-Methyl-2-butenoic acid; cis*-2-dimethylcrotonic acid; 2-methylisocrotonic acid; *cis*-2,3-dimethylacrylic acid. $C_5H_8O_2$; mol wt 100.11. C 59.98%, H 8.06%, O 31.96%. Stereoisomer of tiglic acid. Found in ester form in sumbul root, *Angelica archangelica* L., *Umbelliferae* and together with tiglic acid esters in the oil of the Roman camomile, *Anthemis nobilis* L., *Compositae.* Isoln from seeds of *Schoenocaulon officinale* (Lindl.) A. Gray, *Liliaceae* (cevadilla seeds) by alkaline hydrolysis of cevadine: Stoll, Seebeck, *Helv. Chim. Acta* **35**, 1275 (1952). Synthesis by *trans* addition of bromine to tiglic acid: Buckles, Mock, *J. Org. Chem.* **15**, 680 (1950). Review and bibliography: Buckles *et al., Chem. Rev.* **55**, 659 (1955).

Monoclinic rods, needles, plates; mp 45°. Spicy odor. *Vesicant.* d_4^{47} 0.983. bp_{760} 185°; bp_{12} 86°. Sublimes. Volatile with steam. n_D^{47} 1.4434. K at 25° = 5.0 × 10⁻⁵. uv max (H_2O): 217 nm (ϵ 5.15 × 10³). Molar heat of combustion 626.6 kcal. Sparingly soluble in cold water, freely sol in hot water. Sol in alcohol, ether. Prolonged boiling of aq soln causes isomerization to tiglic acid; the process is speeded up by traces of bromine and sunlight, also by strong mineral acids or alks. Dry crystals of angelic acid have been stored in bottles for years without evidence of isomerization.

Calcium salt dihydrate, $Ca(C_5H_7O_2)_2.2H_2O$, leaflets. Much more soluble in water than calcium tiglate: 100 parts of aq soln satd at 17.5° contains 23 parts of anhydr calcium angelate.

Amide, C_5H_9NO, crystals, mp 127-128°.

Methyl ester, $C_6H_{10}O_2$, liquid; d_4^{20} 0.9413; bp_{764} 128°; n_D^{20} 1.4321.

Ethyl ester, $C_7H_{12}O_2$, liquid; $d_D^{19.5}$ 0.9178; bp_{760} 141.5°, bp_{11} 49°. n_D^{20} 1.4304. Heat of formn at constant vol 963.1 kcal, at constant press. 964.2 kcal.

679. Angelica Lactone. *5-Methyl-2-furanone.* $C_5H_6O_2$; mol wt 98.10. C 61.21%, H 6.17%, O 32.62%. Exists in three forms. Prepn of α and β-forms: Wolff, *Ann.* **229**, 250 (1885); Thiele, *Ann.* **319**, 184 (1901); v. Auwers, *Ber.* **56**, 1672 (1923); J. H. Helberger *et al., Ann.* **561**, 215 (1949). Prepn of α'-form: J. P. Wineburg *et al., J. Heterocycl. Chem.* **12**, 749 (1975); V. Jäger, H. J. Günther, *Tetrahedron Letters* **1977**, 2543; R.A. Amos, J. A. Katzenellenbogen, *J. Org. Chem.* **43**, 560 (1978). Toxicity data for α-form: E. J. Moran *et al., Drug Chem. Toxicol.* **3**, 249 (1980).

α-form   β-form   α'-form

α-Form, *5-methyl-2(3H)-furanone*, Δ^2-*angelica lactone*, γ-*methyl-β,γ-crotonolactone*, *4-hydroxy-3-pentenoic acid* γ-*lactone*. Volatile needles, mp 18°. d_4^{20} 1.084. bp_{12} 56°. n_{He}^{20} 1.4476. One gram dissolves in 20 ml water at 15°. Heating with triethylamine soln converts it to the β-form. LD_{50} orally in mice: 2800 mg/kg (Moran).

β-Form, *5-methyl-2(5H)-furanone*, Δ^1-*angelica lactone*, γ-*methyl-α,β-crotonolactone*, *4-hydroxy-2-pentenoic acid* γ-*lactone*. Liquid. Not solidified at −17°. d_4^{20} 1.076. bp_{751} 208-209°. bp_{10} 87°. n_{He}^{20} 1.4603. Sol in water. Forms a dimer. More stable than α-form.

α'-Form, *dihydro-5-methylene-2(3H)-furanone*, γ-*methylene-γ-butyrolactone*. bp_{17} 80°.

680. Angiotensin. Formerly *hypertensin* and *angiotonin*. Pressor substance formed by the action of renin, *q.v.*, on a plasma substrate, *angiotensinogen* (*renin substrate*, *hypertensinogen*). The substance so formed is a decapeptide called *angiotensin I* which is converted to the active pressor agent, *angiotensin II*, by the splitting off of the C-terminal His-Leu residues by the "converting enzyme" or *angiotensinase*. The octapeptide angiotensin II differs among species only in the amino acid residue in position 5 being either Val or Ile. *See* reviews for refs to isoln and synthesis. Angiotensin II acts directly on the adrenal gland to stimulate the release of aldosterone, *q.v.* Rapid liquid phase synthesis of a protected angiotensin II: S. Nozaki, I. Muramatsu, *Bull. Chem. Soc. Japan* **55**, 2165 (1982). Extraction and characterization of angiotensins I and II from rat brain: D. Ganten *et al., Science* **221**, 869 (1983). *Reviews:* M. Bodanszky, M. A. Ondetti, *Peptide Synthesis* (John Wiley, New York, 1966) pp 215-223; E. Schröder, K. Lübe, *The Peptides* vol. **II** (English ed., New York, 1966) pp 4-62; Bumpus, Smeby, "Angiotensin" in *Renal Hypertension*, I. Page, J. McCubbin, Eds. (Year Book Medical Publishers, Chicago, 1968) pp 62-98; Lee, "Angiotensin" in *Renin and Hypertension* (Williams & Wilkins, Baltimore, 1969) pp 32-94; G. M. Molinatti, P. Limone, *Minerva Med.* **72**, 715-732 (1981); J. A. Oliver *et al., Ann. N.Y. Acad. Sci.* **394**, 275-277 (1982); M. Marin-Grez, *Biochem. Pharmacol.* **31**, 3941-3947 (1982). *See also* Tonin.

Asp-Arg-Val-Tyr-Ile-His-Pro-Phe

angiotensin II (horse)

Angiotensins are very stable. Hydrolyzed by strong acids and bases and above pH 9.5. Sol in organic solvents, in aq solns pH 5-8. In high dilution, lost by absorption on walls of glass vessels.

Amide, $C_{49}H_{70}N_{14}O_{11}$, *5-valine-angiotensin II amide*, *angiotensin II aspartic-β-amide 5-valine*, *1-asparagine-5-valine-angiotensin II*, *Val⁵-angiotensin II-asp¹-β-amide*, *Hypertensin*, *Ipertensina*.

THERAP CAT: Amide as vasoconstrictor.

681. Angostura Bark. Cusparia bark; Carony bark; *Cortex Angosturae; Cortex Cuspariae*. The bark of the tree *Galipea officinalis* Hancock, habitat Venezuela, or the bark of the tree *Cusparia febrifuga* Humb., or *C. trifoliata* Engl., *Rutaceae*, habitat Brazil. *Ref:* Meyer, *Pharm. Ztg.* **80**, 120 (1935).

Unpleasant, musty odor. Bitter, slightly aromatic taste with pungent after-taste. The bitterness is due mainly to the bitter principle angosturin, $C_9H_{12}O_5$. The bark contains also the quinoline alkaloids cusparine, cuspareine; galipine, galipoidine, and galipoline.

THERAP CAT: Antipyretic.

682. Anhalamine. *1,2,3,4-Tetrahydro-6,7-dimethoxy-8-isoquinolinol*; 6,7-dimethoxy-8-hydroxy-1,2,3,4-tetrahydroisoquinoline. $C_{11}H_{15}NO_3$; mol wt 209.24. C 63.14%, H 7.23%, N 6.69%, O 22.94%. From *Lophophora williamsii* (Lemaire) Coutl. (*Anhalonium lewinii* Henn.), *Cactaceae*: Kauder, *Arch. Pharm.* **237**, 190 (1899); Späth, Becke, *Mo-*

natsh. **66**, 327 (1935). Structure: *eidem, Ber.* **67**, 2100 (1934). Synthesis: Späth, Roder, *Monatsh.* **43**, 93 (1922); Brossi *et al., Helv. Chim. Acta* **47**, 2089 (1964); **49**, 403 (1966). Biosynthetic studies: Kapadia *et al., J. Am. Chem. Soc.* **92**, 6943 (1970). *Review:* Manske in R. H. F. Manske, H. L. Holmes, *The Alkaloids* vol. **IV** (Academic Press, New York, 1954) pp 8-14.

Crystals, mp 189-191°. uv max (ethanol): 274 nm (log ϵ 2.90). Almost insol in cold water, cold alcohol, ether. Sol in hot water, alcohol, acetone, dil acids.

Hydrochloride dihydrate, $C_{11}H_{15}NO_3.HCl.2H_2O$, crystals from water, mp 258°.

683. Anhalonidine. *1,2,3,4-Tetrahydro-6,7-dimethoxy-1-methyl-8-isoquinolinol*; 6,7-dimethoxy-8-hydroxy-1-methyl-1,2,3,4-tetrahydroisoquinoline. $C_{12}H_{17}NO_3$; mol wt 223.24. C 64.55%, H 7.68%, N 6.27%, O 21.50%. From mescal buttons, the buds of *Lophophora williamsii* (Lemaire) Coult. (*Anhalonium lewinii* Henn.) *Cactaceae:* Heffter, *Ber.* **27**, 2975 (1894); **29**, 221 (1896); Kauder, *Arch. Pharm.* **237**, 190 (1899). Structure and synthesis from 3-acetoxy-4,5-dimethoxy-N-acetylphenethylamine: Späth, Passl, *Ber.* **65**, 1778 (1932); from mescaline: Brossi *et al., Helv. Chim. Acta* **47**, 2089 (1964). Biosynthesis: Kapadia *et al., J. Am. Chem. Soc.* **92**, 6943 (1970).

Small octahedra from benzene, mp 160-161°. uv max (ethanol): 270 nm (log ϵ 2.81). Strong base. Freely sol in water, alcohol, chlorofom, hot benzene. Sparingly sol in ether. Insol in petr ether. Solns of anhalonidine acquire a reddish color on standing.

684. Anhalonine. *6,7,8,9-Tetrahydro-4-methoxy-9-methyl-1,3-dioxolo[4,5-h]isoquinoline*. $C_{12}H_{15}NO_3$; mol wt 221.25. C 65.14%, H 6.83%, N 6.33%, O 21.69%. From mescal buttons [*Lophophora williamsii* (Lemaire) Coult. (*Anhalonium lewinii* Henn.), *Cactaceae*] also in *Ariocarpus*, in *Gymnocalycium gibbosum*. Synthesis of *dl*-form and resolution: Späth, Kesztler, *Ber.* **68**, 1663 (1935); Brossi *et al., J. Am. Chem. Soc.* **93**, 6248 (1971). Configuration: Battersby, Edwards, *J. Chem. Soc.* **1960**, 1214.

Rhombic needles from petr ether, mp 86°, $bp_{0.02}$ 140°. $[\alpha]_D^{25}$ −63.8° (methanol); −56.3° (chloroform). Very sol in alcohol, ether, chloroform, benzene, petr ether.

Hydrochloride, $C_{12}H_{15}NO_3.HCl$, orthorhombic prisms, dec 255°; freely sol in hot water. Aq soln is neutral.

685. Anilazine. *4,6-Dichloro-N-(2-chlorophenyl)-1,3,5-triazin-2-amine*; *2,4-dichloro-6-(o-chloroanilino)-s-triazine*; (o-chloroanilino)dichlorotriazine; Dyrene. $C_9H_5-Cl_3N_4$; mol wt 275.51. C 39.23%, H 1.83%, Cl 38.60%, N 20.34%. Prepn: C. N. Wolf, U.S. pat. **2,720,480** (1955 to Ethyl Corp.); E. G. Hill, E. Clinton, U.S. pat. **2,820,032** (1958); K. H. Rattenburg *et al.*, U.S. pat. **3,074,946** (1963 to Chemagro). Toxicity study: S. D. Cohen, S. D. Murphy, *J. Agr. Food Chem.* **21**, 140 (1973).

White to tan crystals, mp 159-160°. Insol in water. Soly at 30° (g/100 ml): toluene, 5; xylene, 4; acetone, 10. Subject to hydrolysis; not compatible with oils and alkaline materials. LD_{50} orally in rats: >5000 mg/kg, Mobay Technical Information Sheet, Jan. 1979.

USE: Fungicide.

686. Anileridine. *1-[2-(4-Aminophenyl)ethyl]-4-phenyl-4-piperidinecarboxylic acid ethyl ester; 1-(p-aminophenethyl)-4-phenylisonipecotic acid ethyl ester;* ethyl 1-(4-aminophenethyl)-4-phenylisonipecotate; *N-[β-(p-aminophenyl)ethyl]-4-phenyl-4-carbethoxypiperidine; N-β-(p-aminophenyl)ethylnormeperidine; Leritine; Nipecotan; Alidine; Apodol.* $C_{22}H_{28}N_2O_2$; mol wt 352.46. C 74.96%, H 8.01%, N 7.95%, O 9.08%. Synthesis: Weijlard *et al., J. Am. Chem. Soc.* **78**, 2342 (1956); U.S. pat. **2,966,490** (1960 to Merck & Co.).

Dihydrochloride, $C_{22}H_{28}N_2O_2 \cdot 2HCl$, crystals from methanol + ether, dec 280-287° (free base, mp 83°). Freely soluble in water, methanol. Solubility in ethanol: 8 mg/g. pH of aq solns 2.0 to 2.5. Solns are stable at pH 3.5 and below. At pH 4 and higher the insol free base is precipitated. uv max (pH 7 in 90% methanol contg phosphate buffer): 235, 289 nm ($A_{1cm}^{1\%}$ 293, 34.5). Distribution coefficient between water at pH 3.6 and *n*-butanol = 0.9.

Caution: May be habit forming. This is a controlled substance (opiate) listed in the U.S. Code of Federal Regulations, Title 21 Part 1308.12 (1985).

THERAP CAT: Narcotic analgesic.

687. Aniline. *Benzenamine;* aniline oil; phenylamine; aminobenzene; aminophen; kyanol. C_6H_7N; mol wt 93.12. C 77.38%, H 7.58%, N 15.04%. First obtained in 1826 by Unverdorben from dry distillation of indigo. Runge found it in coal tar in 1834. Fritzsche, in 1841, prepared it from indigo and potash and gave it the name aniline. Manuf from nitrobenzene or chlorobenzene: Faith, Keyes & Clark's *Industrial Chemicals*, F. A. Lowenheim, M. K. Moran, Eds. (Wiley-Interscience, New York, 4th ed., 1975) pp 109-116. Procedures: A. I. Vogel, *Practical Organic Chemistry* (Longmans, London, 3rd ed., 1959) p 564; Gattermann-Wieland, *Praxis des organischen Chemikers* (de Gruyter, Berlin, 40th ed., 1961) p 148. Brochure *"Aniline"* by Allied Chemical's National Aniline Division (New York, 1964) 109 pp, gives reactions and uses of aniline (877 references).

Oily liquid; colorless when freshly distilled, darkens on exposure to air and light. *Poisonous!* Characteristic odor and burning taste; combustible; volatile with steam. d_{20}^{20} 1.022. bp 184-186°. Solidif −6°. Flash pt, closed cup: 169°F (76°C). n_D^{20} 1.5863. pKb 9.30. pH of 0.2 molar aq soln 8.1. One gram dissolves in 28.6 ml water, 15.7 ml boil. water; misc with alcohol, benzene, chloroform, and most other organic solvents. Combines with acids to form salts. It dissolves alkali or alkaline earth metals with evolution of hydrogen and formation of anilides, e.g., C_6H_5NHNa. *Keep well closed and protected from light. Incompat.* Oxidizers, albumin, solns of Fe, Zn, Al, acids, and alkalies. LD_{50} orally in rats: 0.44 g/kg, K. H. Jacobson, *Toxicol. Appl. Pharmacol.* **22**, 153 (1972).

Hydrobromide, $C_6H_7N \cdot HBr$, white to slightly reddish, cryst powder, mp 286°. Darkens in air and light. Sol in water, alc. *Protect from light.*

Hydrochloride, $C_6H_7N \cdot HCl$, crystals, mp 198°. d 1.222. Darkens in air and light. Sol in about 1 part water; freely sol in alc. *Protect from light.*

Hydrofluoride, $C_6H_7N \cdot HF$, cryst powder. Turns gray on standing. Freely sol in water; slightly sol in cold, freely in hot alc.

Nitrate, $C_6H_7N \cdot HNO_3$, crystals, dec about 190°. d 1.36. Discolors in air and light. Sol in water, alc. *Protect from light.*

Hemisulfate, $C_5H_7N \cdot \frac{1}{2}H_2SO_4$, cryst powder. d 1.38. Darkens on exposure to air and light. One gram dissolves in about 15 ml water; slightly sol in alc; practically insol in ether. *Protect from light.*

Acetate, $C_6H_5NH_2 \cdot HOOCCH_3$. Prepd from aniline and acetic acid: Vignon, Evieux, *Bull. Soc. Chim. France* [4] **3**, 1012 (1908). Colorless liquid. d 1.070-1.072. Darkens with age; gradually converted to acetanilide on standing. Misc with water, alc.

Oxalate, $C_6H_5NH_2 \cdot HOOCCOOH \cdot H_2NC_6H_5$. Prepd from aniline and oxalic acid in alc soln: Hofmann, *Ann.* **47**, 37 (1843). Triclinic rods from water, mp 174-175°. Readily sol in water; sparingly sol in abs alc; practically insol in ether.

Human Toxicity: Intoxication may occur from inhalation, ingestion, or cutaneous absorption. *Acute:* cyanosis, methemoglobinemia, vertigo, headache, mental confusion. *Chronic:* anemia, anorexia, wt loss, cutaneous lesions, *Clinical Toxicology of Commercial Products*, R. E. Gosselin *et al.*, Eds. (Williams & Wilkins, Baltimore, 4th ed., 1976) Section III, pp 29-35.

USE: Manuf dyes, medicinals, resins, varnishes, perfumes, shoe blacks; vulcanizing rubber; as solvent. Hydrochloride used in manuf of intermediates, aniline black and other dyes, in dyeing fabrics or wood black.

688. Aniline Mustard. *N,N-Bis(2-chloroethyl)benzenamine; N,N-bis(2-chloroethyl)aniline;* phenylbis[2-chloroethylamine]; *β,β'-dichlorodiethylaniline;* Lymphochin; Lymphocin; Lymphoquin. $C_{10}H_{13}Cl_2N$; mol wt 218.12. C 55.06%, H 6.01%, Cl 32.51%, N 6.42%. Prepd by the action of phosphorus pentachloride on *N,N-bis-[2-hydroxyethyl]-aniline (phenyldiethanolamine)*: Robinson, Watt, *J. Chem. Soc.* **1934**, 1538; Korshak, Strepikheev, *J. Gen. Chem. USSR* **14**, 312 (1944).

Stout prisms from methanol, mp 45°. bp_{14} 164°; $bp_{0.5}$ 110°. Sol in hot methanol, ethanol. Very slightly sol in ether.

Hydrochloride, $C_{10}H_{14}Cl_3N$, crystals. *Vesicant.* Freely sol in water. Sol in alcohol.

USE: In cancer research.

689. Anilinephthalein. *3,3-Bis(4-aminophenyl)-1(3H)-isobenzofuranone; 3,3-bis(p-aminophenyl)phthalide.* $C_{20}H_{16}N_2O_2$; mol wt 316.34. C 75.93%, H 5.10%, N 8.86%, O 10.11%. Prepn: Hubacher, *J. Am. Chem. Soc.* **73**, 5885 (1951).

Crystals from methanol, mp 203°. Sol in acetone. Moderately sol in methanol, ethyl acetate. Insol in water, ether, benzene.

Consult the cross index before using this section.

690. 1-Anilino-8-naphthalenesulfonate. *8-(Phenyl-amino)-1-naphthalenesulfonic acid; 8-anilino-1-naphtha-lenesulfonic acid;* ANS; phenylperi acid. $C_{16}H_{13}NO_3S$; mol wt 299.35. C 64.20%, H 4.38%, N 4.68%, O 16.03%, S 10.71%. Hydrophobic fluorescent probe for protein studies; originally used in prepn of aniline dyes. Prepd from 8-aminonaphthalenesulfonic acid, aniline and anilinium hydrochloride. Fluorescence spectrum: L. Stryer, *Science* **162**, 526 (1968). Binding to bovine serum albumin: G. Weber, L. B. Young, *J. Biol. Chem.* **239**, 1415 (1964); to lipoproteins: R. A. Muesing, T. Nishida, *Biochemistry* **10**, 2952 (1971); to chymotrypsin: J. D. Johnson *et al., ibid.* **18**, 1292 (1979); L. D. Weber *et al., ibid.* 1297. Interaction with rod outer segment membrane: U. P. Andley, B. Chakrabarti, *ibid.* **20**, 1687 (1981).

Magnesium salt, $C_{32}H_{24}MgN_2O_6S_2$, green crystals from water. uv max: 350 nm (ϵ 4.95 × 10^3).
Ammonium salt, $C_{16}H_{16}N_2O_3S$, mp 242-244°.
USE: Fluorescent probe. In protein conformation studies. Magnesium salt as visualization reagent for proteins.

691. A-Ninopterin. *N-[p-[[1-(2,4-Diamino-6-pteridin-yl)ethyl]amino]benzoyl]glutamic acid;* 4-amino-9-methyl-pteroylglutamic acid. $C_{20}H_{22}N_8O_5$; mol wt 454.46. C 52.86%, H 4.88%, N 24.66%, O 17.60%. Prepn: Hultquist, Smith, **Brit.** pat. **667,098** (1952 to Am. Cyanamid).

692. Aniracetam. *1-(4-Methoxybenzoyl)-2-pyrrolidin-one;* 1-*p*-anisoyl-2-pyrrolidinone; Ro 13-5057; Draganon; Sarpul. $C_{12}H_{13}NO_3$; mol wt 219.24. C 65.74%, H 5.98%, N 6.39%, O 21.89%. Cognition enhancer related to piracetam, *q.v.* Prepn: **Japan Kokai** 79 117,468; E. Kyburz, W. Asch-wanden, **U.S.** pat. **4,369,139** (1979, 1983 both to Hoffmann-La Roche). Effects on learning and memory in rats: R. Cumin *et al., Psychopharmacology* (Berlin) **78**, 104 (1982); K. Yamada *et al., Pharmacol. Biochem. Behav.* **22**, 645 (1985); in monkeys and pigeons: M. J. Pontecorvo, H. L. Evans, *ibid.* 745. Clinical evaluation in geriatric patients: P. Foltyn *et al., Arzneimittel-Forsch.* **33**, 865 (1983); in Alzheimer disease: L. B. Sourander *et al., Psychopharmacology* (Berlin) **91**, 90 (1987).

Crystals from ethanol, mp 121-122°. LD_{50} in rats, mice (mg/kg): ~4500, > 5000 orally (Cumin).
THERAP CAT: Nootropic.

693. *p*-Anisaldehyde. *4-Methoxybenzaldehyde;* anisic aldehyde. $C_8H_8O_2$; mol wt 136.14. C 70.57%, H 5.92%, O 23.50%. Metabolic product of the odoriferous fungus *Lentinus lepidus* Fr.: Birkinshaw *et al., Biochem. J.* **38**, 131 (1944); of wood-rotting fungus *Polyporus benzoinus* (Wahl.) Fr.: Birkinshaw *et al., ibid.* **50**, 509 (1952); of *Daldalea juniperina* Murr.: Birkinshaw, Chaplen, *ibid.* **60**, 255 (1955). Prepn: Niedzielski, Nord, *J. Am. Chem. Soc.* **63**, 1462 (1941); Sisti *et al., J. Org. Chem.* **27**, 279 (1962).

Oily liquid, bp 248°, $bp_{1.5}$ 89-90°. mp 0°. d_4^{15} 1.119. n_D^{13} 1.5764. Volatile in steam. Very slightly sol in water; misc with alc, ether. LD_{50} orally in rats: 1510 mg/kg, P. M. Jenner *et al., Food Cosmet. Toxicol.* **2**, 327 (1964).
USE: Perfumery and toilet soaps; odor resembles that of coumarin, but the aldehyde must be mixed with other odorous substances to yield an agreeable odor. Also used in organic syntheses.

694. Anise. Anise seed. Dried ripe fruit of *Pimpinella anisum* L., *Umbelliferae. Habit.* Western Asia, Egypt; cultivated in Southern Europe, India and U.S. *Constit.* 1.5-3.5% volatile oil, starch, protein, fixed oil.
USE: Manuf oil of anise; as condiment and flavor in foods or beverages.
THERAP CAT: Carminative.
THERAP CAT (VET): Carminative, flavoring agent.

695. Anise Alcohol. *4-Methoxybenzenemethanol; p-methoxybenzyl alcohol;* anisyl alcohol. $C_8H_{10}O_2$; mol wt 138.16. C 69.54%, H 7.30%, O 23.16%. Prepd by reduction of anisaldehyde with triisobutyl aluminum and with amine boranes: Ziegler *et al., Ann.* **623**, 9 (1959); Chamberlain, Schechter, **U. S.** pat. **2,898,379** (1959 to Callery Chemical).

Liquid, bp 259°. mp 24-25°, solidif 17°. d_{15}^{15} 1.113. Practically insoluble in water; freely soluble in alc, ether. LD_{50} orally in rats: 1.2 ml/kg, Woodart *et al., J. Pharmacol. Exp. Ther.* **93**, 26 (1948).

696. *p*-Anisic Acid. *4-Methoxybenzoic acid.* $C_8H_8O_3$; mol wt 152.14. C 63.15%, H 5.30%, O 31.55%. Prepd from methoxybenzene: Gross *et al., Ber.* **96**, 1382 (1963). Manuf by catalytic oxidation of *p*-methoxytoluene: **Brit.** pat. **798,-619** (1958 to General Electric); Cotterill *et al., **Brit.** pat. **842,998** (1960 to I.C.I.).

Needles, mp 184° (subl). bp 275-280°. d 1.385. Soluble in 2500 parts water; more sol in boil. water; freely sol in alc, chloroform, ether, ethyl acetate.

697. Anisidine. C_7H_9NO; mol wt 123.15. C 68.27%, H 7.37%, N 11.37%, O 12.99%. Prepn of *p*-isomer: Cahours, *Ann.* **74**, 298 (1850); of *o*-isomer: Mühlhäuser, *Ann.* **207**, 235 (1881); of *m*-isomer: Kadaba, Massie, *J. Org. Chem.* **22**, 333 (1957).

m-Anisidine, 3-methoxybenzenamine, 3-methoxyaniline, 3-aminoanisole. Pale yellow, oily liquid. Remains fluid even at −10°. bp 251°, bp₂ 81-86°. Sparingly sol in water; sol in alc, acids.

o-Anisidine. 2-methoxybenzenamine. Yellowish liquid; becomes brownish on exposure to air. Volatile with steam. bp 225°. mp +5°. d¹⁵₁₅ 1.098. Practically insol in water. Miscible with alc, ether. *Keep well closed and protected from light.*

p-Anisidine. 4-methoxybenzenamine. Crystals, mp 57°. bp 246°. Sparingly sol in water; freely sol in methanol, ethanol.

Note: o-Anisidine and its hydrochloride may reasonably be anticipated to be carcinogens: *Fourth Annual Report on Carcinogens* (NTP 85-002, 1985) p 24.

USE: In the manuf of azo dyes.

698. Anisindione. *2-(4-Methoxyphenyl)-1H-indene-1,3(2H)-dione;* 2-p-anisyl-1,3-indandione; 2-(p-methoxyphenyl)-1,3-indandione; SPE 2792; Miradon; Unidone. $C_{16}H_{12}O_3$; mol wt 252.26. C 76.18%, H 4.80%, O 19.03%. Prepn: Koelsch, *J. Am. Chem. Soc.* **58**, 1331 (1936); Horeau, Jacques, *Bull. Soc. Chim. France* **1948**, 53; Sperber, U.S. pat. 2,899,358 (1959 to Schering).

Pale yellow crystals from acetic acid or ethanol, mp 156-157°.

THERAP CAT: Anticoagulant.

699. Anisole. *Methoxybenzene.* C_7H_8O; mol wt 108.13. C 77.75%, H 7.46%, O 14.80%. $C_6H_5OCH_3$. Prepn from phenol and dimethyl sulfate: Ullmann, *Ann.* **327**, 114 (1903); Graebe, *Ann.* **340**, 204 (1905); G. S. Hiers, F. D. Hager, *Org. Syn.* **coll. vol. I**, 58 (2nd ed., 1941); from bromobenzene: Agfa, **Ger.** pat. **411,052;** *Chem. Zentr.* **1925, I,** 2411; **Frdl. 15,** 193; by passing methyl chloride into a suspension of sodium phenolate in liquid ammonia: White *et al., J. Am. Chem. Soc.* **46**, 965 (1924); from phenol, methyl iodide and potassium carbonate in dimethylformamide: Brieger *et al., J. Chem. Eng. Data* **13**, 581 (1968). Forms oils or resins by condensation with formaldehyde: **Ger.** pats. **403,264; 406,152;** *Chem. Zentr.* **1925, I,** 307, 1816; **Frdl. 14,** 626, 627. Absorption spectrum: Scheibe, *Ber.* **59**, 2625 (1926). Soly in glycerol, see McEwen, *J. Chem. Soc.* **123**, 2285 (1923). Toxicity studies: J. M. Taylor *et al., Toxicol. Appl. Pharmacol.* **6**, 378 (1964).

Liquid. Agreeable aromatic odor. d¹⁸₄ 0.9956; d⁴⁵₄ 0.9701. mp −37.3°. bp₇₆₀ 155.5°; bp₁₀₀ 93.0°; bp₄₀ 70.7°; bp₂₀ 55.8°; bp₁₀ 42.2°; bp₅ 30.0°; bp₁.₀ 5.4°. n²⁰_D 1.51791. Sol in alcohol and ether; insol in water. LD₅₀ orally in rats: 3700 mg/kg (Taylor).

USE: In perfumery, in organic syntheses.

700. Anisomycin. *1,4,5-Trideoxy-1,4-imino-5-(4-methoxyphenyl)-D-xylo-pentitol 3-acetate; [2R-(2α,3α,4β)]-2-[(4-methoxyphenyl)methyl]-3,4-pyrrolidinediol 3-acetate;* 2-p-methoxyphenylmethyl-3-acetoxy-4-hydroxypyrrolidine; Flagecidin. $C_{14}H_{19}NO_4$; mol wt 265.30. C 63.38%, H 7.22%, N 5.28%, O 24.12%. Protein synthesis inhibiting antibiotic isolated from *Streptomyces griseolus* and *S. roseochromogenes:* Sobin, Tanner, Jr., *J. Am. Chem. Soc.* **76**, 4053 (1954); Tanner *et al.*, **U.S.** pat. **2,691,618** (1954 to Pfizer). Activity: J. E. Lynch *et al., Antibiot. & Chemother.* **4**, 844, 899 (1954). Structure and stereochemistry: Beereboom *et al., J. Org. Chem.* **30**, 2334 (1965); Schaefer, Wheatley, *ibid.* **33**, 166 (1968); Butler, *ibid.* 2136. Biosynthesis: Butler, *ibid.* **31**, 317 (1966). Total synthesis: Oida, Ohki, *Chem. Pharm. Bull.* **16**, 2086 (1968); *ibid.* **17**, 1405 (1969); Felner, Schenker, *Helv. Chim. Acta* **53**, 754 (1970). Chiral synthesis: J. P. H. Verheyden *et al., Pure Appl. Chem.* **50**, 1363 (1978). Stereospecific total synthesis: D. P. Schumacher, S. S. Hall, *J. Am. Chem. Soc.* **104**, 6076 (1982). Mechanism of action: A. Jiménez, D. Vázquez in *Antibiotics* vol. 5(pt. 2), F. E. Hahn,

Ed. (Springer-Verlag, New York, 1979) pp 1-19. Solubility and stability data: *Antibiot. Ann.* **1954-55,** pp 809-810. Prepn of deacetylanisomycin from anisomycin: Nickell *et al.*, **U.S.** pat. **2,935,444** (1960 to Pfizer).

Long needles from ethyl acetate or water, mp 140-141°. [α]²³_D −30° (methanol). uv max: 224, 277, 283 nm (ε 10800, 1800, 1600). Base is moderately sol in water; sol in lower alcohols, esters, ketones, chloroform; slightly sol in benzene, toluene and hexane. Aq solns are stable over a wide pH range at room temp.

Hydrochloride, $C_{14}H_{20}ClNO_4$, crystals from ethyl acetate + ethanol, mp 187-188°. Very sol in water.

Deacetylanisomycin, $C_{12}H_{17}NO_3$, mp 176-179°. [α]²⁵_D −20.0° (methanol), pK 9.2.

USE: Anisomycin and deacetylanisomycin in the eradication of bean mildew; to inhibit other pathogenic fungi in plants.

THERAP CAT: Antiprotozoal (Trichomonas).

701. Anisotropine Methylbromide. *endo-8,8-Dimethyl-3-[(1-oxo-2-propylpentyl)oxy]-8-azoniabicyclo[3.2.1]octane bromide; 3α-hydroxy-8-methyl-1αH,5αH-tropanium bromide 2-propylvalerate;* 8-methyltropinium bromide 2-propylvalerate; 8-methyl-3-(2-propylpentanoyloxy)tropinium bromide; octatropine methylbromide; Lytispasm; Valpin. $C_{17}H_{32}BrNO_2$; mol wt 362.37. C 56.35%, H 8.90%, Br 22.05%, N 3.87%, O 8.83%. Prepn: Weiner, Gordon, **U.S.** pat. **2,962,499** (1960 to Endo Labs.). Metabolism: Shindo *et al., Chem. Pharm. Bull.* **19**, 513 (1971).

Crystals from acetone, mp 329°.
Methyl chloride, $C_{17}H_{32}ClNO_2$, crystals from acetone, mp 289°.

THERAP CAT: Anticholinergic.

702. o-(p-Anisoyl)benzoic Acid. *2-(4-Methoxybenzoyl)benzoic acid;* S 23/46. $C_{15}H_{12}O_4$; mol wt 256.25. C 70.30%, H 4.72%, O 24.98%. Prepn from phthalic anhydride and anisole: Meyer, Turnau, *Monatsh.* **30**, 486 (1909). Alternate route: Arcus, Marks, *J. Chem. Soc.* **1956**, 1627.

Leaflets from water. Stout crystals from alcohol or toluene, mp 146°. Very sparingly sol in water. Freely sol in alc, ether, toluene, chloroform, glacial acetic acid.

Sodium salt, $C_{15}H_{11}NaO_4$, needles. Freely sol in water. Sol in alcohol.

USE: The sodium salt has been proposed as a sweetening agent. Rated approximately 150 times as sweet as cane sugar. Bitter taste if used in concns exceeding 0.2 g/liter. *Review:* Möhler, *Z. Lebensm.-Untersuchung u. Forschung* **90**, 431 (1950), *C.A.* **44**, 8558d (1950).

703. p-Anisoyl Chloride. *4-Methoxybenzoyl chloride.* $C_8H_7ClO_2$; mol wt 170.59. C 56.32%, H 4.14%, Cl 20.78%, O 18.76%. $CH_3OC_6H_4COCl.$ Prepd from p-anisic acid and thionyl chloide: Vanderhaeghe *et al.*, *J. Pharm. Pharmacol.* **6**, 119 (1954).

Crystals or liquid. mp 22°. bp about 262-263° with slight decompn. Dec by water or alcohol; sol in acetone, benzene.

Caution: Vapors can cause serious eye burns. Sealed containers may explode, because slow decompn at room temp may build up pressure. *Ref:* Carroll, *Chem. & Eng. News* **38**, 40 (Aug. 22, 1960).

704. Annatto. Arnotta; annotta. Coloring matter from seeds of *Bixa orellana* L., *Bixaceae.* Extraction from seed: Barnett, Espoy, U.S. pat. **2,815,287** (1957); Kocher, U.S. pat. **2,831,775** (1958). Contains bixin and several yellow to orange-red pigments which give carotene reactions: Diemair *et al.*, *Naturwiss.* **39**, 211 (1933).

Sol in alcohol, ether, oils.

USE: Dyeing silk orange in hot soap bath; coloring butter, margarine, cheese and oils; manuf wood stains, varnishes.

705. Annotinine. $C_{16}H_{21}NO_3$; mol wt 275.34. C 69.79%, H 7.69%, N 5.09%, O 17.43%. Isoln from *Lycopodium annotinum* L., *Lycopodiaceae:* Manske, Marion, *Can. J. Res.* **21B**, 92 (1943). Structure: Przybylska, Marion, *Can. J. Chem.* **35**, 1075 (1957); Wiesner *et al.*, *Tetrahedron* **4**, 87 (1958); Przybylska, Ahmed, *Acta Cryst.* **11**, 718 (1958). Stereochemistry: Wiesner *et al.*, *Tetrahedron Letters* **1961**, 187; Ho, *ibid.* **1969**, 1307. Synthesis: Wiesner *et al.*, *Can. J. Chem.* **47**, 433 (1969). Biogenesis: Leete, *Tetrahedron* **3**, 313 (1958).

Brilliant prisms from chloroform+ methanol, mp 232°. Sol in chloroform, dil HCl. Sparingly sol in methanol.

706. *p*-Anol. *4-(1-Propenyl)phenol; p-propenylphenol;* 4-hydroxy-1-propenylbenzene. $C_9H_{10}O$; mol wt 134.17. C 80.56%, H 7.51%, O 11.93%. Prepn from anethole: Stoermer, Kahlert, *Ber.* **34**, 1812 (1901); from 4-propenylphenylmagnesium bromide: Quelet, *Bull. Chim. Soc.* [4] **45**, 268 (1929).

Leaflets from boiling water. Weak odor of cloves, resembling that of eugenol. Spicy, pungent taste. mp 93-94°. bp_{760} 250° (dec); bp_{14} 138-140°. Slightly sol in hot water. Freely sol in dimethylformamide, other organic solvents. Sol in aq solns of KOH and NaOH.

USE: Intermediate in the synthesis of estrogens.

707. Anot. *3-Amino-2-methyl-5-nitrobenzamide; 3-amino-5-nitro-o-toluamide.* $C_8H_9N_3O_3$; mol wt 195.18. C 49.23%, H 4.65%, N 21.53%, O 24.59%. Major metabolite of chickens on a zoalene diet. Isoln procedure: Thiegs *et al.*, *J. Agr. Food Chem.* **9**, 201 (1961).

Golden-yellow crystals, mp 198.5-199.5°. Slightly sol in water (less than 1%).

708. Anserine. *N-β-Alanyl-3-methyl-L-histidine; α-(β-aminopropionylamino)-β-(1-methyl-5-imidazolyl)propionic acid; 3-methyl-N,α-(β-alanyl)-L-histidine.* $C_{10}H_{16}N_4O_3$; mol wt 240.26. C 49.99%, H 6.71%, N 23.32%, O 19.98%. First found in muscles of geese, later in many more animals,

but not in man, beef, horse, or dog. Isoln: Ackermann, *Z. Physiol. Chem.* **183**, 1 (1929); Wolff, Wilson, *J. Biol. Chem.* **109**, 565 (1935). Synthesis: Behrens, du Vigneaud, *J. Biol. Chem.* **120**, 517 (1937); Rinderknecht *et al.*, *J. Org. Chem.* **29**, 1968 (1964). *Review:* M. Guggenheim, *Die biogenen Amine* (Karger, Basel—New York, 4th ed., 1951) pp 439-446, 502.

L-Form, needles from dil alc, dec 240-242°. $[\alpha]_D^{30}$ +12.3° (c = 5). pK_1 2.64; pK_2 7.04; pK_3 9.49. The dried material is very hygroscopic. Freely soluble in water. Slightly sol in methanol, even less in ethanol.

Nitrate, $C_{10}H_{16}N_4O_3 \cdot HNO_3$, needles from dil methanol, mp 226-228°.

709. Antazoline. *4,5-Dihydro-N-phenyl-N-(phenylmethyl)-1H-imidazole-2-methanamine; 2-(N-benzylanilinomethyl)-2-imidazoline;* phenazoline; *2-(N-phenyl-N-benzylaminomethyl)imidazoline;* imidamine; 5512-M; Antistine; Antistin; Histostab; Antastan; Antasten; Antihistal; Azalone; Ben-a-hist. $C_{17}H_{19}N_3$; mol wt 265.35. C 76.94%, H 7.22%, N 15.84%. Prepd by the condensation of benzylaniline with 2-(chloromethyl)imidazoline: Miescher, Klarer, U.S. pat. **2,449,241** (1948 to Ciba).

Crystals, mp 120-122°.

Hydrochloride, $C_{17}H_{20}ClN_3$, *Fenazolina, Histazine.* Bitter crystals producing temporary numbness of the tongue. mp 237-241°. uv max: 242 nm ($E_{1cm}^{1\%}$ 495 to 515); min 222 nm. One gram dissolves in 40 ml water, in 25 ml alc. Practically insol in ether, benzene, chloroform. pH (1% aq soln): 6.3.

Phosphate, $C_{17}H_{19}N_3 \cdot H_3PO_4$, crystals. Bitter taste. mp 194-198°. Sol in water. Sparingly sol in methanol. Practically insol in benzene, ether. pH (2% aq soln): 4.5.

THERAP CAT: Antihistaminic.

THERAP CAT (VET): Antihistaminic.

710. Antheridiol. *3,22,23-Trihydroxy-7-oxostigmasta-5,24(28)-dien-29-oic acid γ-lactone; 3β,22,23-trihydroxy-24-(carboxymethylene)cholest-5-en-7-one 23,24-lactone;* Hormone A. $C_{29}H_{42}O_5$; mol wt 470.66. C 74.01%, H 8.99%, O 17.00%. The first specific functioning sex hormone to be identified in the plant kingdom. A diffusible substance secreted by the female mycelium of the filamentous water molds *Achlya bisexualis* and *A. ambisexualis,* which induces the growth of antheridial hyphae in the male plant, thereby initiating sexual reproduction in the species. Isoln and preliminary chemical data: Raper, Haagen-Smit, *J. Biol. Chem.* **143**, 311 (1942). Isoln of crystalline material: McMorris, Barksdale, *Nature* **215**, 320 (1967). Structure: Arsenault *et al.*, *J. Am. Chem. Soc.* **90**, 5635 (1968). Synthesis: Edwards *et al.*, *ibid.* **91**, 1248 (1969); Fried, Edwards, U.S. pat. **3,547,911** (1970 to Syntex). Stereochemical proposal: Green *et al.*, *Tetrahedron* **27**, 1199 (1971). Absolute stereochemistry and alternate syntheses: Edwards *et al.*, *Tetrahedron Letters* **1972**, 791; T. C. McMorris *et al.*, *ibid.* **2673**; *eidem, J. Org. Chem.* **39**, 669 (1974). *Reviews:* Barksdale, *Ann. N.Y. Acad. Sci.* **144**, 313-319 (1967); *idem, Science* **166**, 831-837 (1969); T.C. McMorris, *Lipids* **13**, 716-722 (1978).

Colorless crystals from methanol, mp 250-255°. uv max (ethanol): 220 nm (ε 17,000). Sol in hot methanol. Slightly sol in chloroform; very slightly sol in water. Has extremely high specific biological activity even at conc of 2×10^{-8} mg/ml (McMorris, Barksdale, *loc. cit.*).

USE: Control and regulation of plant fertility: Fried, Edwards, *loc. cit.*

711. Anthiolimine. *Mercaptobutanedioic acid antimony(3+) lithium salt (3:1:6); 2,2',2''-[stibilidynetris(thio)]-trisbutanedioic acid hexalithium salt; mercaptosuccinic acid antimonate(III) hexalithium salt;* mercaptosuccinic acid S-antimony derivative lithium salt; lithium antimoniothiomalate; lithium antimony thiomalate; Anthiomaline. $C_{12}H_9$-$Li_6O_{12}S_3Sb$; mol wt 604.78. C 23.83%, H 1.50%, Li 6.88%, O 31.75%, S 15.91%, Sb 20.13%. Prepn: Delepine, Gaillot, U.S. pat. **2,060,181** (1937 to Rhône-Poulenc).

$$\left[Sb \left[S - \underset{\underset{CH_2COOLi}{|}}{CH COOLi} \right]_3 \right]$$

Enneahydrate, hygroscopic powder. Very soluble in water; very slightly sol in alcohol, ether.

THERAP CAT: Anthelmintic (Trematodes).

712. Anthracene. $C_{14}H_{10}$; mol wt 178.22. C 94.34%, H 5.66%. Obtained from coal tar: Dumas, Laurent, *Ann.* **5**, 10 (1833); Laurent, *Ann.* **34**, 287 (1840); Anderson, *Ann.* **122**, 294 (1862); *J. Chem. Soc.* **15**, 44 (1862); Auerbach, *Das Anthracen und seine Derivate* (Braunschweig, 1880); Perkin, *J. Soc. Arts* **27**, 572 (1879); Lunge, *Coal Tar and Ammonia* (1916); Barnett, *Anthracene and Anthraquinone* (London, 1921); Nanson, *Textile Colorist* **48**, 605, 678, 751 (1926); **49**, 19, 246, 557, 593 (1927); Houben, Fischer, *Das Anthracen und die Anthrachinone* (Leipzig, 1929); Borrmann, *Der Teer* (Leipzig, 1940); Schumann, *Kokereiteer* (Stuttgart, 1940). Extensive patent literature on purification. Prepn of very pure anthracene from synthetic anthraquinone: Clar, *Ber.* **72**, 1645 (1939). *Review:* E. Clar, *Polycyclic Hydrocarbons* 2 vols. (Academic Press, New York, 1964).

Monoclinic plates from alc. Sublimes. When pure, colorless with violet fluoresence; when impure (due to tetracene, naphthacene), yellow with green fluorescence. Strongly triboluminescent and triboelectric. d_4^{27} 1.25. mp 218°. bp_{760} 342°. Absorption spectrum: Clar, *Ber.* **65**, 506 (1932). Less soluble than the isomeric phenanthrene. Insol in water; one gram dissolves in 67 ml abs alcohol, 70 ml methanol, 62 ml benzene, 85 ml chloroform, 200 ml ether, 31 ml carbon disulfide, 86 ml carbon tetrachloride, 125 ml toluene. Anthracene darkens in sunlight. According to Downs, U.S. pat. **1,303,639** (1919), when solns of crude anthracene in coal tar naphtha are exposed to ultraviolet irradiation, the anthracene is precipitated as dianthracene (*para*-anthracene) which is reconverted to anthracene by sublimation. Forms molecular addn products with nitro compounds. Picric acid complex, mp 139°; *sym*-trinitrobenzene complex, mp 164°; trinitrotoluene complex, mp 162°.

USE: Important source of dyestuffs (manuf anthraquinone, alizarin dyes).

713. Anthragallol. *1,2,3-Trihydroxy-9,10-anthracene-*

dione; 1,2,3-trihydroxyanthraquinone; anthragallic acid; anthracene brown. $C_{14}H_8O_5$; mol wt 256.20. C 65.63%, H 3.15%, O 31.22%. From gallic acid and benzoic acid with sulfuric acid at 125° or from phthalic anhydride and pyrogallol with sulfuric acid at 160°: Seuberlich, *Ber.* **10**, 39 (1877). Other methods: Kubota, Perkin, *J. Chem. Soc.* **127**, 1889 (1925); Perkin, Story, *ibid.* **1929**, 1399; Cross, Perkin, *ibid.* **1930**, 292.

Brown crystals, mp 312-313°. Sublimes 290°. Absorption spectrum: Meyer, Fischer, *Ber.* **46**, 85 (1913). Slightly sol in water, chloroform; sol in alcohol, ether, glacial acetic acid. Its soln in concd H_2SO_4 is reddish-brown. Greenish-brown soln in ammonia water changes to blue on heating. Forms salts with Na, K, Ba, Tl, Pb, La, Ce, Nd, Co.

Trimethyl ether, $C_{17}H_{14}O_5$, yellow needles from benzene + petr ether, mp 168°. Insol in water solns of alkalies.

714. Anthralin. *1,8-Dihydroxy-9(10H)-anthracenone; 1,8-dihydroxyanthrone;* dithranol; Anthra-Derm; Antraderm; Batidrol; Cignolin; Cigthranol; Dithrocream; Psoradrate; Psoriacide. $C_{14}H_{10}O_3$; mol wt 226.22. C 74.33%, H 4.45%, O 21.22%. Prepd from chrysazin by reduction with hydrogen and nickel catalyst at high pressure: Zahn, Koch, *Ber.* **71B**, 172 (1938). Revised structure: H. M. Avdovich, G. A. Neville, *Can. J. Spectrosc.* **25**, 110 (1980); F. R. Ahmed, *Acta Crystallogr.* **B36**, 3184 (1980).

Lemon yellow leaflets or needles from ligroin, mp 176-181°. Practically insol in water. Freely sol in chloroform. Sol in acetone, benzene, pyridine, oils. Slightly sol in alcohol, ether, glacial acetic acid. Sol in dil NaOH with yellow color and green fluorescence. Under the influence of air, alkaline solns turn red and lose their fluorescence.

Triacetate, $C_{20}H_{16}O_6$, *1,8,9-anthracenetriol triacetate, 1,8,9-triacetoxyanthracene, Exolan.*

THERAP CAT: Antipsoriatic.
THERAP CAT (VET): Antifungal.

715. Anthramycin. *3-(5,10,11,11a-Tetrahydro-9,11-dihydroxy-8-methyl-5-oxo-1H-pyrrolo[2,1-c][1,4]benzodiazepin-2-yl)-2-propenamide; 5,10,11,11a-tetrahydro-9,11-dihydroxy-8-methyl-5-oxo-1H-pyrrolo[2,1-c][1,4]benzodiazepin-2-acrylamide.* $C_{16}H_{17}N_3O_4$; mol wt 315.32. C 60.94%, H 5.43%, N 13.33%, O 20.30%. From *Streptomyces refuineus* var *thermotolerans,* NRRL 3143: Leimgruber *et al., J. Am. Chem. Soc.* **87**, 5791 (1965); Berger *et al.,* U.S. pat. **3,361,742** (1968 to Hoffmann-La Roche). Also produced by *Streptomyces spadicogriseus:* N. Komatsu *et al., J. Antibiot.* **33**, 54 (1980). Structure: Leimgruber *et al., J. Am. Chem. Soc.* **87**, 5793 (1965). Synthesis: *eidem, ibid.* **90**, 5641 (1968); Batcho, Leimgruber, U.S. pat. **3,524,849** (1970 to Hoffmann-La Roche). Activity studies: Horwitz, Grollman, *Antimicrob. Ag. Chemother.* **1968**, 21. Biosynthesis: L. H. Hurley *et al., J. Am. Chem. Soc.* **97**, 4372 (1975). *Review:* Kohn in *Antibiotics* vol. 3, J. W. Corcoran, F. E. Hahn, Eds. (Springer-Verlag, New York, 1975) pp 3-11.

Yellow prisms from acetone + water, dec 188-194°. uv max (acetonitrile): 235, 333 nm (ε 18,200, 31,800). $[\alpha]_D^{25}$ +930° (DMF).

Methyl ether hydrate, $C_{17}H_{19}N_3O_4 \cdot H_2O$, pale yellow needles from methanol + water, dec above 120°.

THERAP CAT: Antineoplastic.

716. Anthranol. 9-*Anthracenol;* 9-*anthrol;* 9-hydroxy-anthracene. $C_{14}H_{10}O$; mol wt 194.22. C 86.57%, H 5.19%, O 8.24%. Prepd by dissolving anthrone in 5 to 10% boiling NaOH soln, cooling to −5° and pouring in cooled 5% H_2SO_4: Meyer, *Ann.* **379,** 56 (1911). Also prepared from anthraquinone.

Needles from glacial acetic acid, plates from dil alcohol. Orthorhombic, mp 120° (if bath is heated to 110°). Sinters at 120°, mp 152° (if bath is cold at start). Absorption spectrum (of acetate): Barnett *et al., J. Chem. Soc.* **1928,** 885. Sol in most solvents with blue fluorescence. The solid is fairly stable when kept dry. In solution it changes quickly to anthrone. Recrystallizations of anthranol may lead to a less pure compound.

Acetate, $C_{16}H_{12}O_2$, crystals from petr ether, mp 134°.

Benzoate, $C_{21}H_{14}O_2$, crystals from pyridine + alcohol, mp 170-172°.

Methyl ether, $C_{15}H_{12}O$, crystals from alcohol, mp 97-98°.

Ethyl ether, $C_{16}H_{14}O$, crystals from methanol, mp 73°.

USE: In the manuf of dyes.

717. Anthraquinone. 9,10-*Anthracenedione;* 9,10-an-thraquinone; 9,10-dioxoanthracene; Morkit. $C_{14}H_8O_2$; mol wt 208.20. C 80.76%, H 3.87%, O 15.37%. Produced industrially from phthalic anydride and benzene in the presence of aluminum chloride by a Friedel-Crafts reaction: Klipstein, *Ind. Eng. Chem.* **18,** 1327 (1926). From anthracene with vanadium pentoxide, sodium chlorate, glacial acetic and sulfuric acids: *Org. Syn.* **coll. vol. II,** 554 (1943). Convenient lab procedure: L. F. Fieser, *Organic Experiments* (Heath & Co., Boston, 1964) pp 195-200. *Reviews:* de Barry, Barnett, *Anthracene and Anthraquinone* (London, 1921); Phillips, *Chem. Rev.* **6,** 157 (1929); Houben, Fischer, *Das Anthracen und die Anthrachinone* (Leipzig, 1929); R. H. Chung in Kirk-Othmer *Encyclopedia of Chemical Technology,* **Vol. 2** (Wiley-Interscience, New York, 3rd ed., 1978) pp 700-707.

Light yellow, slender monoclinic prisms by sublimation *in vacuo*. Almost colorless, orthorhombic, bipyramidal crystals from $H_2SO_4 + H_2O$. d_4^{20} 1.42-1.44. mp 286°. bp_{760} 377°. Absorption spectrum: Flexser *et al., J. Am. Chem. Soc.* **57,** 2103 (1935). Insol in water. Solubility (g/100 g) in alc at 18° 0.05; at 25° 0.44; in boiling alc 2.25; in ether at 25° 0.11; in chloroform at 20° 0.61; at 40° 1.00; at 60° 1.60; in benzene at 20° 0.26; at 40° 0.50; at 60° 1.00; at 80° 1.80; in toluene at 25° 0.30.

USE: Important starting material for the manufacture of vat dyes. Usually marketed as 1-anthraquinonesulfonic acid

sodium salt, or as 2-anthraquinonesulfonic acid sodium salt (called "silver salt"), or as 1,5-dihydroxyanthraquinone (called "anthrarufin"). Anthraquinone itself is used to make seeds distasteful to birds: *Chem. & Eng. News* **33,** 4998 (1955). *Caution:* Low systemic toxicity, but may cause skin irritation, sensitization.

718. Anthrarobin. 1,2,10-*Anthracenetriol;* 1,2,10-anthratriol; 3,4-dihydroxyanthranol; desoxyalizarin; leucoalizarin. $C_{14}H_{10}O_3$; mol wt 226.22. C 74.33%, H 4.45%, O 21.22%. Prepd from alizarin, ammonia, and zinc dust: *Hagers Handb. Pharm. Praxis* **Vol. 1,** 466 (1930).

Yellow-brown to dark brown crystals; easily oxidized to alizarin. mp 208°; not sublimable. Slightly sol in water or benzene; freely sol in alcohol, ether, acetone, glacial acetic acid. Dissolves in alkalies with greenish-yellow color; in concd H_2SO_4 with golden-yellow color.

THERAP CAT: Parasiticide.

719. Anthrarufin. 1,5-*Dihydroxy-9,10-anthracenedione;* 1,5-dihydroxyanthraquinone. $C_{14}H_8O_4$; mol wt 240.20. C 70.00%, H 3.36%, O 26.64%. Prepd from 1,5-anthraquinone potassium disulfonate: Fierz-David, Blangey, *Farbenchemie,* (Vienna, 5th ed, 1943), pp 224-225.

Green to yellow crystals from acetic acid, mp 280°. Sublimes at 120°. Absorption spectrum in H_2SO_4: Meyer, Fischer, *Ber.* **46,** 85 (1913). Sol in concd H_2SO_4, in aq KOH soln (reddish color); insol in aq Na_2CO_3, NH_3, $Ba(OH)_2$. Moderately sol in alcohol, slightly in water.

USE: Important intermediate in the manuf of alizarin and indanthrene dyestuffs. Forms insol Ba and Ca lakes; has been proposed as analytical reagent for the detection of Ca.

720. Anthrimide. 1,1'-*Iminobis-9,10-anthracenedione;* 1,1'-iminodianthraquinone; dianthraquinonylamine; dianthrimide. $C_{28}H_{15}NO_4$; mol wt 429.41. C 78.31%, H 3.52%, N 3.26%, O 14.90%. Prepd by heating 1-aminoanthraquinone with 1-chloroanthraquinone in the presence of copper: Bayer & Co., **Ger. pat.** 162,823; *Chem. Zentr.* **1905,** II, 1206; Eckert, Steiner, *Monatsh.* **35,** 1129 (1914); *FIAT Final Rept.* no. 1313, **vol. II** (1948), p 137.

Deep red needles from chlorobenzene, rhombs from nitrobenzene. Also described as coppery red needles with metallic sheen. Practically insol in low boiling organic solvents. Sparingly sol in aniline, nitrobenzene, chlorobenzene, quinoline. Sol in concd sulfuric acid first giving a scarlet-red color which turns olive-green on standing.

USE: Determination of boron. Used as a 0.02% soln in concd H_2SO_4. The olive-green soln turns blue in the presence of boron. The blue soln has an absorption max at 620 nm. Also used in the determination of silicon.

721. Anthrone. 9(10H)-*Anthracenone;* 9,10-dihydro-9-

oxoanthracene; Carbothrone. $C_{14}H_{10}O$; mol wt 194.22. C 86.57%, H 5.19%, O 8.24%. Prepn from anthraquinone by reduction with tin, glacial acetic and hydrochloric acid: *Org. Syn. coll. vol. I*, 60 (1941). By cyclization of *o*-benzylbenzoic acid with liquid HF: Fieser, Hershberg, *J. Am. Chem. Soc.* **61**, 1278 (1939).

Orthorhombic needles from benzene + petr ether, mp 155°. Absorption spectrum: Martin, *Ann. Combustibles Liq.* **12**, 967 (1937). Sol in most organic solvents without fluorescence. Any fluorescence present is due to anthranol. Tendency to change to anthraquinone. Equilibrium in abs alc: 89% anthrone; 11% anthranol.

Addition compound with 4 mols desoxycholic acid, $C_{110}H_{170}O_{17}$; mp 179°.

USE: In organic syntheses; in the colorimetric determination of sugar and animal starch in body fluids.

722. α-Antiarin. 3-[(6-*Deoxy*-β-D-*gulopyranosyl*)*oxy*]-5,12,14-*trihydroxy*-19-*oxocard*-20(22)-*enolide*. $C_{29}H_{42}O_{11}$; mol wt 566.63. C 61.47%, H 7.47%, O 31.06%. From latex (arrow poison) of the upas tree, *Antiaris toxicaria* Lesch., *Moraceae*, found in Indonesia: Taylor, *Brit. J. Pharmacol.* **8**, 237 (1953). Older literature and isoln: Doebel *et al., Helv. Chim. Acta* **31**, 688 (1948); Dolder *et al., ibid.* **38**, 1364 (1955). Structure of aglycone: Martin, Tamm, *ibid.* **42**, 696 (1959). The sugar portion of α-antiarin is antiarose (D-gulomethylose), while that of the isomeric β-antiarin is L-rhamnose. Structure: Juslén *et al., ibid.* **45**, 2285 (1962). Toxicity: Chen, Henderson, *J. Pharmacol. Exp. Ther.* **150**, 53 (1965).

Tetrahydrate, crystals (six-sided leaflets) from water. Anhydrous after 3 hrs at 110° in high vac, dec 238-240°. $[\alpha]_D^{17}$ −3.9° (c = 0.905 in methanol). Sol in water, alcohol, slightly in ether. uv max: 305, 217 nm (log ε 1.8, 4.08). Mean LD (anhydr compd) in cats (mg/kg): 0.116 i.v. (Chen, Henderson).

Tribenzoate, $C_{50}H_{54}O_{14}$, needles from methanol, mp 240°. $[\alpha]_D^{21}$ +3.0° (acetone). Practically insol in water, very sparingly sol in methanol, ethanol, ether. Freely sol in acetone, chloroform.

723. Antigens. Substances which have the ability of inducing the formation of antibodies and of reacting with antibodies in animals and man. In order to induce antibody formation an antigen must (a) be foreign to the animal's own circulation or contain some foreign group, (b) have a molecular weight of at least 10,000-15,000 and (c) be metabolized or at least partially susceptible to the hydrolytic enzymes of the test animal: Campbell, Bulman, *Fortschr. Chem. Org. Naturst.* **9**, 443-484 (1952). *Review:* several authors in *Immunogenicity*, F. Borek, Ed. (Elsevier, New York, 1972).

724. Antimony. Stibium; regulus of antimony. Sb; at. wt 121.75; at. no. 51; valence 3, 5. Two naturally occurring

isotopes: 121 (57.25%); 123 (42.75%); artificial radioactive isotopes: 112-120; 122; 124-135; isotopes 122 and 124 are useful radioactive tracers. First accurate description of antimony by Thölde (Basil Valentine) in 1604. Antimony ore is mined in China, Mexico, and Bolivia. The antimony of commerce is about 99% pure. Prepn in the laboratory by reduction of Sb_2O_5 with KCN: Schenk in *Handbook of Preparative Inorganic Chemistry* vol. 1, G. Brauer, Ed. (Academic Press, New York, 2nd ed., 1963) p 606. *Review:* Smith, "Arsenic, Antimony and Bismuth" in *Comprehensive Inorganic Chemistry* vol. 2, J. C. Bailar, Jr. *et al.,* Eds. (Pergamon Press, Oxford, 1973) pp 547-683; S. C. Carapella in Kirk-Othmer *Encyclopedia of Chemical Technology* vol. **3** (Wiley-Interscience, New York, 3rd ed., 1978) pp 96-105.

Silver-white, lustrous, hard, brittle metal; scale-like cryst structure; or dark gray, lustrous powder. Is not tarnished in dry air and only slowly in moist air. d 6.68. mp 630°. bp 1635°; also reported to be 1440°: *Gmelin's, Antimony* (8th ed.) **18B**, p 69 (1949). sp heat 0.049; electrical resistivity 39 μ-ohm-cm at 0°. Not affected by cold dil acids; attacked by hot concd H_2SO_4; readily by aqua regia. Nitric acid, depending on the concn, converts it to antimonous or antimonic oxide. When finely divided it reacts with hot concd HCl. Qualitative analysis for antimony: react with slight excess of HCl with aid of HNO_3; pour soln in large vol of water; white ppt forms which becomes orange-red on addn of H_2S and is sol in ammonium sulfide. LD_{50} i.p. in rats: 100 mg/kg, Bradley, Fredrick, *Ind. Med.* **10**, *Ind. Hyg. Sect.* **2**, 15 (1941).

Human Toxicity: Antimony and its compds have been reported to cause dermatitis, keratitis, conjunctivitis and nasal septal ulceration by contact, fumes or dust. *Caution:* Avoid conditions in which nascent hydrogen will react with antimony to form stibine (SbH_3), which is extremely toxic (nausea, vomiting, headache, hemolysis, hematuria, abdominal pain, death). Stibine can be liberated from storage batteries when nascent hydrogen reacts, in an acid medium, with antimony present in the battery plates: E. Browning, *Toxicity of Industrial Metals* (Appleton-Century Crofts, New York, 2nd ed., 1969) pp 23-38.

USE: In manufacture of alloys, such as Britannia or Babbitt metal, hard lead, white metal, type, bullets and bearing metal; in fireworks; for thermoelectric piles, blackening iron, coating metals, etc.

725. Antimony Chloride Oxide. Antimony oxychloride; basic antimony chloride; powder of Algaroth; Mercurius vitae. ClOSb; mol wt 173.22. Cl 20.47%, O 9.24%, Sb 70.29%. SbOCl. Prepd from $SbCl_3$ and water: Schenk in *Handbook of Preparative Inorganic Chemistry,* vol 1, G. Brauer, Ed. (Academic Press, N.Y., 2nd ed., 1963) p 611.

Monoclinic crystals or crystalline powder. Hydrolyzed by water to Sb_2O_3. Sol in HCl, tartaric acid, CS_2; practically insol in alcohol, ether. Heating to 250° results in formation of $Sb_2O_5Cl_2$; above 320° Sb_2O_3 is formed.

726. Antimony Dichlorotrifluoride. *Antimony chloride fluoride;* antimony dichlorofluoride; antimony trifluorodichloride. Cl_3F_3Sb; mol wt 249.67. Cl 28.40%, F 22.83%, Sb 48.77%. $SbCl_2F_3$. Prepd according to the exothermic reaction $SbF_3 + Cl_2 \rightarrow SbCl_2F_3$: Henne, *Org. Reactions* **2**, 61 (1944); Kwasnik in *Handbook of Preparative Inorganic Chemistry* Vol. 1, G. Brauer, Ed. (Academic Press, New York, 2nd ed., 1963) pp 200-201. Prepd from $SbCl_5$ and ClF_3: Dehnicke, Weidlein, *Z. Anorg. Allgem. Chem.* **323**, 267 (1963).

Viscous liquid. Can be stored in iron vessels.

USE: As catalyst in the manufacture of organic fluorine compds: Slesser, Schram, *Preparation, Properties, and Technology of Fluorine and Organic Fluoro Compounds* (New York, 1951) *passim. Caution:* Highly toxic.

727. Antimony Pentachloride. Cl_5Sb; mol wt 299.05. Cl 59.28%, Sb 40.72%. $SbCl_5$. Usually prepd by passing chlorine into molten antimony trichloride. Laboratory procedure: Schenk in *Handbook of Preparative Inorganic Chemistry* Vol. 1, G. Brauer, Ed. (Academic Press, New York, 2nd ed., 1963) p 610.

Colorless to yellow, oily liquid. Fumes in air. Cannot be distilled at atmospheric pressures without complete decompn. d_4^{16} 2.358; d_4^{36} 2.319; d_4^{52} 2.289; d_4^{78} 2.231. mp 3.5°. bp_{14} 68°; bp_{22} 79°; bp_{30} 92°; bp_{55} 85°; bp_{68} 102.5°. Dipole moment: 1.14 at 16.3° in CCl_4. Mono- and tetrahydrates are formed in the presence of small amounts of water. Large amounts of water cause hydrolysis to Sb_2O_5. Sol in hydrochloric acid, chloroform, carbon tetrachloride.

USE: As catalyst when replacing a fluorine substituent with chlorine in organic compounds.

728. Antimony Pentafluoride. F_5Sb; mol wt 216.76. F 43.83%, Sb 56.17%. SbF_5. Lewis acid. Prepd industrially (in aluminum apparatus) according to the equation $SbCl_5 + 5HF \rightarrow 5HCl + SbF_5$: Ruff, Plato, *Ber.* **37**, 673 (1904); Perkins, Irwin, U.S. pat. **2,410,358** (1946). Laboratory procedure using SbF_3 and F_2: Woolf, Greenwood, *J. Chem. Soc.* **1950**, 2200; Kwasnik in *Handbook of Preparative Inorganic Chemistry*, **Vol. 1**, G. Brauer, Ed. (Academic Press, New York, 2nd ed., 1963) p 200. *Reviews:* Burg in *Fluorine Chemistry*, **Vol. 1**, J. Simons, Ed. (Academic Press, New York, 1950) pp 104-106; Kemmitt, Sharp, *Advan. Fluorine Chem.* **4**, 210-211 (1965).

Hygroscopic, corrosive, moderately viscous liquid. *Poisonous, attacks skin.* mp 8.3°; bp 141°. $d^{25.8}$ 3.097; density data: Hoffman, Jolly, *J. Phys. Chem.* **61**, 1574 (1957). Reacts violently with water. Also forms a solid dihydrate, which reacts violently with more water to form a clear soln. Slowly hydrolyzed in NaOH solns forming $Sb(OH)_6^-$. Forms solids with sulfur chloride, carbon disulfide, benzene, toluene, petr ether (resin formation), ether, alc, acetone, ethyl acetate. Glacial acetic acid gives a clear soln. Slowly corrodes glass, copper, lead. May be stored in aluminum vessels.

USE: In the fluorination of organic compds, *see* the monograph *Preparation, Properties and Technology of Fluorine and Organic Fluoro Compounds*, C. Slesser, S. R. Schram, Eds. (McGraw-Hill, New York, 1951) 868 pp.

729. Antimony Pentasulfide. Golden antimony sulfide; antimonic sulfide; antimonial saffron; antimony red. S_5Sb_2; mol wt 403.82. S 39.70%, Sb 60.30%. Sb_2S_5. Prepn: *Gmelin's, Antimony* (8th ed.) **18B**, pp 534-539 (1949). Existence of Sb^{5+} sulfide doubted: Birchall, Della Valle, *Chem. Commun.* **1970**, 675.

Orange-yellow, odorless powder. Insol in water; sol in concd HCl with evolution of H_2S; sol in solns of alkali hydroxides or sulfides; forming sulfantimoniates. *Incompat.* Acids, metal salts. LD_{50} i.p. in rats: 1.5 g/kg, Bradley, Fredrick, *Ind. Med.* **10**, *Ind. Hyg. Sect.* **2**, 15 (1941).

USE: As pigment; vulcanizing and coloring rubber; manuf matches and fireworks.

730. Antimony Pentoxide. Antimonic oxide; "stibic" anhydride; antimonic "acid". O_5Sb_2; mol wt 323.52. O 24.73%, Sb 75.27%. Sb_2O_5. Prepn: Schenk in *Handbook of Preparative Inorganic Chemistry*, **vol. 1**, G. Brauer, Ed. (Academic Press, New York, 2nd ed., 1963) p 616.

Yellowish powder. Cubic. d 3.78. Loses oxygen at 300° or higher. Studies have shown that the compound does not correspond fully to Sb_2O_5, but that it is always somewhat hydrated. Slightly soluble in water; practically insol in HNO_3. Slowly dissolves in warm HCl or in warm KOH soln. LD_{50} in rats: 4 g/kg, Bradley, Frederick, *Ind. Med.* **10**, *Ind. Hyg. Sect.* **2**, 15 (1941).

USE: Fire retardant in clothing.

731. Antimony Potassium Oxalate. *Ethanedioic acid anhydride with antimonic acid (3:1) tripotassium salt; tripotassium tris(oxalato)antimonate(3−)*; potassium oxalatoantimonate(III); "antimony salt". $C_6K_3O_{12}Sb$; mol wt 503.12. C 14.32%, K 23.31%, O 38.16%, Sb 24.20%. $K_3[Sb(OOC-COO)_3]$. Prepn of trihydrate: Graddon, *J. Inorg. Nuclear Chem.* **3**, 308 (1956-1957).

Trihydrate, cryst powder. Sol in water. *Poisonous!*

USE: A mordant in dyeing and printing fabrics instead of tartar emetic.

732. Antimony Potassium Tartrate. *Bis[μ-[2,3-dihydroxybutanedioato(4−)-01,02:03,04]]-diantimonate dipotassium trihydrate (stereoisomer)*; tartar emetic; tartrated antimony; tartarized antimony; potassium antimonyltartrate. $C_8H_4K_2$-$O_{12}Sb_2.3H_2O$; mol wt 667.86. C 14.39% H 1.51%, K 11.71%, O 35.93%, Sb 36.46%. Manuf from potassium bitartrate and metallic antimony in the presence of HNO_3 or solid antimony oxide: Davies, U.S. pats. **2,335,585** (1943 to Am. Cream

Tartar), **2,391,297** (1945 to Stauffer Chem.). Structural studies: P. Pfeiffer, E. Schmitz, *Pharmazie* **4**, 451 (1949); E. Chinoporos, N. Papathanasopoulos, *J. Phys. Chem.* **65**, 1643 (1961); D. Grdenic, B. Kamenar, *Acta Crystallogr.* **19**, 197 (1965); R. Iyer *et al., J. Inorg. Nucl. Chem.* **34**, 3351 (1972). Toxicity: N. Ercoli, *Proc. Soc. Exp. Biol. Med.* **129**, 284 (1968). Anthelmintic activity: Z. Farid *et al., Trans. Roy. Soc. Trop. Med. Hyg.* **66**, 119 (1972); R. R. C. New *et al., Nature* **272**, 56 (1978). Oxidimetric determn: Y. A. Gawargious *et al., Pharmazie* **41**, 59 (1980).

Transparent crystals (effloresce on exposure to air) or powder. Sweetish, metallic taste. *Poisonous!* d 2.6. $[\alpha]_D^{20}$ +140.69° (c = 2 in water), +139.25° (c = 2 in glycerol). One gram dissolves in 12 ml water, 3 ml boiling water, 15 ml glycerol. Insol in alcohol. The aq soln is slightly acid. LD_{50} in mice (mg/kg): 55 s.c.; 65 i.v. (Ercoli).

Incompat. Mineral acids, tannic acid, gallic acids, alkali hydroxides and carbonates, lead and silver salts, mercury bichloride, lime water, albumin, soap.

USE: As mordant in the textile and leather industry.

THERAP CAT: Anthelmintic (Schistosoma).

THERAP CAT (VET): Has been used as a parasiticide (gastrointestinal and blood parasites); as an expectorant, and as a ruminatoric.

733. Antimony Sodium Gluconate. *Gluconic acid antimony sodium derivative*; antimony gluconate sodium; antimony gluconate complex sodium salt; sodium antimony gluconate; sodium stibogluconate. Compounds contg both tri- and pentavalent antimony are known.

Trivalent antimony compound, $C_6H_8NaO_7Sb$, *trivalent sodium antimonyl gluconate, Triostam, T.S.A.G.* Mol wt 336.88. C 21.39%, H 2.39%, Na 6.82%, O 33.25%, Sb 36.14%. Prepn: Das Gupta, *Indian J. Pharm.* **15**, 84 (1953); Chyan *et al., C.A.* **52**, 9963d (1958); Axon *et al.,* U.S. pat. **3,306,921** (1967 to Burroughs Wellcome). Amorphous powder. Sol in water. pH of 2% aq soln 9-10; the soln at this pH is unstable and should be adjusted with gluconic acid to pH 6-7 for stability. Solns should be prepd immediately before use.

Pentavalent antimony compound, *pentavalent sodium antimonyl gluconate, Myostibin, Pentostam, Solustibosan, Solyusurmin, Stibanate, Stibanose, Stibatin, Stibinol.* Contains 28-29.5% Sb. Prepd from Sb_2O_5, gluconic acid and NaOH: Bose, Ghosh, *Indian J. Pharm.* **11**, 155 (1949); Datta, Ghosh, *Sci. Cult.* **11**, 699 (1945-6). Crystals. Freely sol in water. pH of 10% aq soln 5.4-5.6. Ampuled solutions should be stored at room temp.

THERAP CAT: Trivalent compd as an anthelmintic (Schistosoma); pentavalent compd as an antiprotozoal (Leishmania).

734. Antimony Sodium Tartrate. Antimony sodium oxide L(+)-tartrate; sodium antimonyl tartrate; Emeto-Na; Stibunal. $C_4H_4NaO_7Sb$; mol wt 308.83. C 15.56%, H 1.31%, Na 7.45%, O 36.27%, Sb 39.43%. (SbO)NaC$_4H_4O_6$. Prepn: Fargher, Gray, *J. Pharmacol. Exp. Ther.* **18**, 341 (1921).

Hygroscopic, transparent or whitish scales or powder. Sweetish taste. Sol in 1.5 parts water. Practically insol in 90% alc. Aq solns are slightly acid to litmus. LD i.v. in mice: 25 mg/kg (Fargher, Gray).

THERAP CAT: Anthelmintic (Schistosoma).

THERAP CAT (VET): Leishmanicide, schistosomicide.

735. Antimony Sodium Thioglycollate. *[(5-Oxo-1,3,2-oxathiostibolan-2-yl)thio]acetic acid sodium salt*; mercaptoacetic acid antimony derivative sodium salt; antimony sodium thioacetate. $C_4H_4NaO_4S_2Sb$; mol wt 324.96. C 14.78%, H 1.24%, Na 7.08%, O 19.70%, S 19.73%, Sb 37.47%. Prepn of the free acid: Klason, Carlson, *Ber.* **39,** 732 (1906); of the salt: Myers, *J. Lab. Clin. Med.* **6,** 359 (1921).

Hygroscopic prismatic crystals turning pink under the influence of light. May develop slight odor of mercaptan on standing. Freely sol in water. Practically insol in alc. Decomposed by excess alkali; incompatible with alkalies.

THERAP CAT: Anthelmintic (Schistosoma).

736. Antimony Sulfate. Antimonous sulfate; antimony trisulfate. $O_{12}S_3Sb_2$; mol wt 531.72. O 36.11%, S 18.09%, Sb 45.80%. $Sb_2(SO_4)_3$. Prepn: Schenk in *Handbook of Preparative Inorganic Chemistry* vol. **1,** G. Brauer, Ed. (Academic Press, New York, 2nd ed., 1963) pp 618-619.

Crystalline powder or lumps; deliquesc in air. *Poisonous!* d 3.62. With a little water forms a solid mass which dissolves in more water, but excess of water changes it into an insol basic salt; sol in dil acids. *Keep well closed.*

737. Antimony Thioglycollamide. *Thioantimonic acid tris(2-amino-2-oxoethyl) ester;* mercaptoacetamide antimony derivative; antimony thioglycollic acid triamide. $C_6H_{12}N_3$-O_3S_3Sb; mol wt 392.09. C 18.38%, H 3.07%, N 10.72%, O 12.24%, S 24.54%, Sb 31.05%. $Sb(SCH_2CONH_2)_3$. Prepd by the action of ammonia on ethyl antimony thioglycollate in absolute alc. The ethyl antimony glycollate is prepd from ethyl thioglycollate and antimony trioxide: Rowntree, Abel, *J. Pharmacol.* **2,** 109 (1910). *cf.* Christiansen, *Organic Derivatives of Antimony,* A.C.S. Monograph Series No. 24, (New York, 1925).

Crystals, mp 140°. One gram dissolves in about 200 ml water. Slightly sol in alc. Insol in ether. Incompatible with alkalies.

THERAP CAT: Anthelmintic (Schistosoma).

738. Antimony Tribromide. Br_3Sb; mol wt 361.51. Br 66.32%, Sb 33.68%. $SbBr_3$. Best prepd from the elements: Schenk in *Handbook of Preparative Inorganic Chemistry* vol. **1,** G. Brauer, Ed. (Academic Press, New York, 2nd ed., 1963) p 613.

Orthorhombic bipyramidal needles. Less hygroscopic than $SbCl_3$. d_{23}^{23} 4.148. mp 96°. bp_{749} 288°. Cryoscopic constant: 26.7. Specific heat: 0.0709 at 33°. Critical temp: 904.5°; crit press. 56 atm. Dipole moment: 2.47. Dec by light, water, alc. Sol in dil HCl, HBr, carbon disulfide, acetone, benzene, chloroform.

739. Antimony Trichloride. *Trichlorostibine;* Butter of antimony. Cl_3Sb; mol wt 228.13. Cl 46.63%, Sb 53.37%. $SbCl_3$. Prepd from the elements: Hensgen, *Rec. Trav. Chim.* **9,** 301 (1890); Kendall *et al., J. Am. Chem. Soc.* **45,** 967 (1923); Schenk in *Handbook of Preparative Inorganic Chemistry* vol. **1,** G. Brauer, Ed. (Academic Press, New York, 2nd ed., 1963) p 608. Purification: Werner, *Z. Anorg. Allgem. Chem.* **181,** 154 (1929); Schaafsma, U.S. pat. **2,324,240** (1943 to Shell).

Orthorhombic, deliquesc needles from carbon disulfide or by sublimation at 100°. Fumes in air. *Poisonous!* d_4^{20} 3.14. mp 73°. bp 223.5°; bp_{70} 143.5°; bp_{11} 102°. Cryoscopic constant: 18.4. Heat of fusion at 73.2°: 13.29 cal/g. Specific heat at 33°: 0.110 cal/g/°C. Dipole moment: 3.12 at 25° in CS_2. One gram dissolves in 10.1 ml H_2O at 25°. Gradual hydrolysis to SbOCl. Considerably more sol in dil HCl. Sol in alcohol, benzene, carbon disulfide, dioxane, chloroform (about 22%), ether, acetone, carbon tetrachloride. A satd soln in carbon tetrachloride is about 1.1 molar. Insol in pyridine, quinoline, other organic bases.

USE: As of antimony trichloride soln; also as reagent for chloral, aromatic hydrocarbons, and vitamin A. For mol wt determinations. In chemical microscopy for the identification of drugs (forms adducts and addition compds). Commercially as a mordant, in making other antimony salts, in organic syntheses, as catalyst. *Caution:* Irritant, corrosive to skin.

THERAP CAT (VET): Escharotic; dehorning agent (calves, goats).

740. Antimony Trichloride Solution. Antimony chloride solution; liquid butter of antimony. Contains 36-38% $SbCl_3$ and 10-12% free HCl.

Yellow to reddish-brown, clear, somewhat oily, strongly caustic liquid. d 1.5.

USE: Bronzing iron, especially gun barrels; mordant for patent leather and in dyeing; coloring zinc black; manuf lakes, particularly from dye woods; furniture polishes.

741. Antimony Trifluoride. Antimony fluoride; antimonous fluoride. F_3Sb; mol wt 178.76. F 31.89%, Sb 68.11%. SbF_3. Prepd by dissolving Sb_2O_3 in aq HF and evaporating the water: Berzelius, *Pogg. Ann.* **1,** 34 (1824); Söll, *FIAT-Review* **23,** 276 (1947); Andersen *et al., Acta Chem. Scand.* **7,** 236 (1953); Kwasnik in *Handbook of Preparative Inorganic Chemistry,* Vol. 1, G. Brauer, Ed. (Academic Press, New York, 2nd ed., 1963) p 199. Industrial prepn from $SbCl_3$ and HF: Midgeley *et al.,* U.S. pat. **2,024,-654** (1935). *Reviews:* Burg in *Fluorine Chemistry,* Vol. 1, J. H. Simons, Ed. (Academic Press, New York, 1950) p 106; Kemmitt, Sharp, *Advan. Fluorine Chem.* **4,** 211 (1965).

Orthorhombic, deliquesc crystals. *Poisonous! Irritates the skin.* d_{20}^{20} 4.379. mp 292°. bp 376°. Also reported as bp 319° (Andersen). Soly in water (g/100 ml): 443 (20°); 562 (30°). Dissolves in water with limited hydrolysis, readily forms complexes such as $[SbF_4]^-$ and many sol salts. For the purpose of halogen exchange fluorination, the compd must be dry: Henne, U.S. pat. **2,082,161** (1937). May be stored in glass vessels or steel drums. *Keep well closed.*

USE: To catalyze fluorinations by HF, manuf chlorofluorides, in dyeing, usually in form of double salts, e.g., antimony sodium fluoride or antimony fluoride and ammonium sulfate double salt, manuf pottery and porcelains.

742. Antimony Triiodide. I_3Sb; mol wt 502.52. I 75.77%, Sb 24.23%. SbI_3. Prepd by the interaction of antimony and iodine in boiling benzene or tetrachloroethane: Bailar, Cundy, *Inorg. Syn.* **1,** 104 (1939); Scattergood, *J. Chem. Ed.* **20,** 40 (1943).

Ruby-red, trigonal crystals (yellowish-green modifications have been observed). d_4^{17} 4.921. The tendency to sublime becomes noticeable at 100°. mp 168°. bp 420°. Critical temp 1101°; crit press. 55 atm. Dipole moment: 1.58. Dec by water and air to SbOI (antimony oxyiodide). Sol in alcohol, acetone, carbon disulfide, HCl, soln of KI. Insol in carbon tetrachloride.

743. Antimony Trioxide. Diantimony trioxide; flowers of antimony; Senarmontite; Valentinite; Exitelite; Weisspiessglanz. O_3Sb_2; mol wt 291.52. O 16.47%, Sb 83.53%. Sb_2O_3. Laboratory prepn from $SbCl_3$ and water: Schenk in *Handbook of Preparative Inorganic Chemistry,* vol. **1,** G. Brauer, Ed. (Academic Press, New York, 2nd ed., 1963) p 615. Obtained from antimony ore minerals by a volatilization (roasting) process: L. D. Freedman in Kirk-Othmer *Encyclopedia of Chemical Technology,* vol. **3** (Wiley-Interscience, New York, 3rd ed., 1978) pp 107-108.

Crystals, polymorphic. mp 655°. bp 1425°; bp_{210} 870°. Sublimes in high vacuum at 400°. Exists in the vapor phase as Sb_4O_6. Heat capacity at 21° (294.4°K): 24.11 cal/g-atom/°C, Anderson, *J. Am. Chem. Soc.* **52,** 2712 (1930). Heat of vaporization: 17.82 kcal/mol. Slightly sol in water, dilute H_2SO_4, or dilute HNO_3. Soly in dil HCl (0.1 moles HCl/kg H_2O): ~1 × 10⁻⁴ g-atoms Sb/kg H_2O, Gayer, Garrett, *ibid.* **74,** 2353 (1952); soly increases with increasing HCl concn: Lea, Wood, *J. Chem. Soc.* **125,** 137 (1924). Sol in solns of alkali hydroxides or sulfides, in warm soln of tartaric acid, or of bitartrates. LD_{50} orally in rats: > 20 g/kg, Smyth, *et al., J. Ind. Hyg. Toxicol.* **30,** 63 (1948).

USE: Manuf tartar emetic; as paint pigment; in enamels and glasses; as mordant; in flame-proofing canvas.

744. Antimony Triselenide. Sb_2Se_3; mol wt 480.40. Sb 50.69%, Se 49.31%. Prepd by the action of hydrogen selenide on a soln of potassium antimonyl tartrate: Moser, Atynsky, *Monatsh.* **45,** 235 (1924); Berzelius cited in *Mellor's* Vol. **10,** 793 (1930); by direct union of the elements: Chretien, *Compt. Rend.* **142,** 1339, 1412 (1906); Parravano, *Gazz. Chim. Ital.* **43,** I, 210 (1913); Berzelius, *loc. cit.*

Gray powder. mp 605° (Berzelius); mp 611° (Chretien); mp 617° (Parravano); mp 572° [Chikashige, Fujita, *Mem.*

Coll. Science Kyoto **2**, 233 (1917)]. Very slightly sol in water. Forms a brown soln with hot potash lye.

745. Antimony Trisulfide. Antimonous sulfide; antimony sulfide; needle antimony; antimony glance. S_3Sb_2; mol wt 339.72. S 28.32%, Sb 71.68%. Sb_2S_3. Occurs in nature as the mineral *stibnite*. Prepn: Donges, Fricke, *Z. Anorg. Chem.* **253**, 2 (1945); Gagliardi, Pilz, *Z. Anal. Chem.* **136**, 344 (1952); *Gmelin's, Antimony* (8th ed.) **18B**, pp 503-524 (1949).

Gray, lustrous, cryst masses or grayish-black powder. Also exists in a red modification. mp 550°. Practically insol in water; sol in concd HCl with evolution of H_2S; sol in solutions of the fixed alkali hydroxides. LD i.p. in rats, 1.0 g/kg: Bradley, Fredrick, *Ind. Med.* **10**, *Ind. Hyg. Sect.* **2**, 15 (1941).

USE: In pyrotechnics, Bengal fires; manuf ruby glass, matches, explosives; as a pigment in paints.

746. Antimycin A_1. *3-Methylbutanoic acid 3-[[3-(formylamino)-2-hydroxybenzoyl]amino]-8-hexyl-2,6-dimethyl-4,9-dioxo-1,5-dioxonan-7-yl ester; isovaleric acid, 8-ester with 3-formamido-N-(7-hexyl-8-hydroxy-4,9-dimethyl-2,6-dioxo-1,5-dioxonan-3-yl)salicylamide.* $C_{28}H_{40}N_2O_9$; mol wt 548.62. C 61.30%, H 7.35%, N 5.11%, O 26.25%. Antibiotic substance produced by *Streptomyces* spp. Isoln: Dunshee *et al., J. Am. Chem. Soc.* **71**, 2436 (1949); Lockwood *et al., Phytopathology* **44**, 438 (1954); U.S. pat. **2,657,170** (1953 to Wisconsin Alumni Res. Found.). Structure: van Tamelen *et al., J. Am. Chem. Soc.* **83**, 1639 (1961); Birch *et al., J. Chem. Soc.* **1961**, 889. Absolute config.: Kinoshita *et al., J. Antibiot.* **25**, 373 (1972).

Crystals from ethyl acetate + Skellysolve B, mp 149-150°. $[\alpha]_D^{26}$ +76° (c = 1 in chloroform). uv max (alc): 226, 320 nm (log ε 4.54, 3.68). Freely sol in alc, ether, acetone, chloroform. Very slightly sol in petr ether, benzene, carbon tetrachloride. Practically insol in water and in 5% aq solns of hydrochloric acid, sodium carbonate and sodium bicarbonate. In aq sodium hydroxide a milky suspension is formed. This clears on warming, but all potency is lost.

USE: Experimentally as fungicide, insecticide, miticide.

747. Antimycin A_3. *3-Methylbutanoic acid 8-butyl-3-[[3-(formylamino)-2-hydroxybenzoyl]amino]-2,6-dimethyl-4,9-dioxo-1,5-dioxonan-7-yl ester; isovaleric acid, 8-ester with N-(7-butyl-8-hydroxy-4,9-dimethyl-2,6-dioxo-1,5-dioxonan-3-yl)-3-formamidosalicylamide;* blastmycin. $C_{26}H_{36}N_2O_9$; mol wt 520.56. C 59.99%, H 6.97%, N 5.38%, O 27.66%. Antifungal antibiotic substance produced by *Streptomyces blastmyceticus* from Japanese soil: Watanabe *et al., J. Antibiot.* **10A**, 39 (1957). Sepn from complex: Harada *et al., ibid.* **11**, 32 (1958). Identity of blastmycin with antimycin A_3: Liu, Strong, *J. Am. Chem. Soc.* **81**, 4387 (1959). Structural studies: Yonehara, Takeuchi, *J. Antibiot.* **11**, 254 (1958). Structure: van Tamelen *et al., J. Am. Chem. Soc.* **83**, 1639 (1961). Abs config.: Kinoshita *et al., J. Antibiot.* **25**, 373 (1972). Total synthesis of diastereomeric mixture: eidem, *ibid.* **22**, 580 (1969); of natural form: eidem, *ibid.* **24**, 724 (1971); eidem, *Bull. Chem. Soc. Japan* **46**, 1279 (1973). Improved synthesis: S. Aburaki, M. Kinoshita, *ibid.* **52**, 198 (1979). Chemical studies: Endo, Yonehara, *J. Antibiot.* **23**, 91 (1970).

Needles from benzene + petr ether, mp 170.5-171.5° (Liu); 174-174.5° (Kinoshita). $[\alpha]_D^{26}$ +64.3° (chloroform) (Liu); $[\alpha]_D^{24}$ +80° (Kinoshita). uv max (methanol): 225, 320 nm (log ε 4.52, 3.86). Moderately sol in methanol, ethanol, ether; freely sol in acetone, ethyl acetate, benzene, chloroform, carbon tetrachloride. Slightly sol in hexane, cyclohexane; very sparingly sol in petr ether. Practically insol in water. LD_{50} in mice: 1.8 mg/kg i.p.; 1.6 mg/kg s.c.

748. Antipyrine. *1,2-Dihydro-1,5-dimethyl-2-phenyl-3H-pyrazol-3-one; 2,3-dimethyl-1-phenyl-3-pyrazolin-5-one;* phenazone; 1,5-dimethyl-2-phenyl-3-pyrazolone; phenyldimethylpyrazolon(e); dimethyloxychinizin; dimethyloxyquinazine; Analgesine; Anodynine; Dolo-Med-Much; Oxydimethylquinizine; Parodyne; Phenylone; Sedatine. $C_{11}H_{12}N_2O$; mol wt 188.22. C 70.19%, H 6.43%, N 14.88%, O 8.50%. Prepn: Müller *et al., Monatsh.* **89**, 23 (1958); *Hagers Handb. Pharm. Praxis*, Vol. 6 (Springer Verlag, Berlin, 1977) p 571. Toxicity: Hart, *J. Pharmacol. Exp. Ther.* **89**, 205 (1947). Metabolism: H. Uchino *et al., Xenobiotica* **13**, 155 (1983). Clinical comparison with paracetamol: H. Quiding *et al., Int. J Oral Surg.* **11**, 304 (1982). Use as indicator of hepatic drug metabolism: E. S. Vessel, *Clin. Pharm. Ther.* **26**, 275 (1979); G. C. Farrell, L. Zaluzny, *Brit. J. Clin. Pharmacol.* **18**, 559 (1984).

Tabular crystals or white powder; slightly bitter taste. mp 111-113°. One gram dissolves in less than 1 ml water, 1.3 ml alcohol, 1 ml chloroform, 43 ml ether. The aq soln is neutral to litmus. LD_{50} orally in rats: 1.8 g/kg (Hart).

Incompat. Acids, alkalies, alum, ammonia water, amyl nitrite, benzoates, betanaphthol, phenol, calomel, chloral hydrate, copper sulfate, ferric chloride, ferrous sulfate, chromium trioxide (chromic acid), cinchona alkaloids, hydrocyanic acid, iodides, iodine, lead subacetate, mercuric chloride, orthoform, potassium permanganate, resorcinol, sod. bicarbonate, sod. salicylate (in powder), soln arsenic and mercury iodide, spirit nitrous ether (unless prescribed with sod. bicarbonate), syrup ferrous iodide, tartar emetic, tannic acid, thymol, urethane, infusions of catechu, cinchona, rose leaves and uva ursi; tinctures of catechu, ferric chloride, cinchona, hamamelis, iodine, kino, and rhubarb.

Acetylsalicylate, $C_{20}H_{20}N_2O_5$, *Acetopyrine, Acopyrine, Acetasol.* Crystals, mp 63-65°. One part dissolves in 400 parts of water, in 20 parts of 2% aq sod. bicarbonate soln. Freely sol in hot water, cold alc, chloroform; sparingly sol in ether. *Note:* Not to be confused with antipyrine salicylacetate.

Mandelate, $C_{19}H_{20}N_2O_4$, *antipyrine amygdalate, Tussol.* Cryst powder, mp 52-55°. One gram dissolves in 15 ml water, 4 ml alc, 26 ml ether.

Methylethylglycolate, $C_{16}H_{22}N_2O_4$, *antipyrine 2-hydroxy-2-methylbutyrate, Astrolin.* Cryst powder, mp 64-65.5°. Soluble in water, alc.

Salicylacetate, $C_{20}H_{20}N_2O_6$, *α-carboxy-o-anisic acid compound with antipyrine, Pyrosal.* Crystals, mp 149-150°. Bitter, acid taste. Sparingly soluble in water; soluble in alc. *Note:* Not to be confused with antipyrine acetylsalicylate.

THERAP CAT: Analgesic.

THERAP CAT (VET): Has been used as an antipyretic, analgesic and in laminitis of horses.

749. Antipyrine Salicylate. *2-Hydroxybenzoic acid compd with 1,2-dihydro-1,5-dimethyl-2-phenyl-3H-pyrazol-3-one (1:1);* Phenazone salicylate; Saliphenazon; Salazolon; Salipyrazolon; Salipyrine. $C_{18}H_{18}N_2O_4$; mol wt 326.34. C 66.24%, H 5.56%, N 8.59%, O 19.61%. $C_{11}H_{12}N_2O \cdot C_6H_4$-(OH)COOH. Prepd by fusing antipyrine and salicylic acid: *Hagers Handb. Pharm. Praxis* Vol. 6, (Springer Verlag, Berlin, 1977) p 574.

Slightly sweet, cryst powder. mp 91-92°. One gram dissolves in 200 ml water, 40 ml boil. water; freely sol in alc, chloroform, sparingly in ether.

THERAP CAT: Analgesic.

THERAP CAT (VET): Has been used as an antipyretic, analgesic in dogs.

750. Antireticular Cytotoxic Serum. ACS; cytoxin-anticytotoxin serum; Bogomolets' serum; Bogoserum; Sarvinal. Biologic synthesis: Loiseleur, *Ann. Inst. Pasteur* **91**, 445 (1956). Purification: *C.A.* **52**, 6574e (1958). Stimulating effect on reticuloendothelial system: I. P. Miagkaia, *Zh. Mikrobiol. Epidemiol. Immunobiol.* **1978**, 82.

751. α_1-Antitrypsin. *α_1-Trypsin inhibitor; α_1-protein-ase inhibitor; α_1-protease inhibitor; alpha$_1$-antitrypsin;* AAT; A1AT; A1PI; Prolastin. Serum glycoprotein synthesized by the liver. Major serine protease inhibitor (serpin) in mammalian plasma; primarily inhibits neutrophil elastase. Consists of a single polypeptide chain of 394 amino acid residues and 3 carbohydrate side chains linked to asparagine residues. Mol wt ~52,000 daltons. Highly pleomorphic protein with a number of genetic variants. Characterization of trypsin inhibitor in alpha$_1$-fraction of human serum: K. Jacobsson, *Scand. J. Clin. Invest.* **7**, Suppl. 14, 57-102 (1955). Isoln from human serum: F. C. Moll *et al., J. Biol. Chem.* **233**, 121 (1958); H. F. Bundy, J. W. Mehl, *ibid.* **234**, 1124 (1959). Purification: I. P. Crawford, *Arch. Biochem. Biophys.* **156**, 215 (1973); M. H. Coan, W. J. Brockway, U.S. pat. **4,379,087** (1983 to Cutter). Structural study of active site: D. Johnson, J. Travis, *J. Biol. Chem.* **253**, 7142 (1978). Complete amino acid sequence and review of genetic variants: R. W. Carrell *et al., Nature* **298**, 329 (1982). Crystal structure: H. Loebermann *et al., J. Mol. Biol.* **177**, 531 (1984). Cloning and expression of human AAT gene: M. Courtney *et al., Proc. Nat. Acad. Sci. USA* **81**, 669 (1984). AAT deficiency has been associated with degenerative lung disease: C.-B. Laurell, S. Eriksson, *Scand. J. Clin. Lab. Invest.* **15**, 132 (1963); with hepatic cirrhosis: H. L. Sharp, *Gastroenterology* **70**, 611 (1976). Inhibition of leukocyte elastase: K. Ohlsson, *Scand. J. Clin. Lab. Invest.* **28**, 251 (1971); K. Beatty *et al., J. Biol. Chem.* **255**, 3931 (1980). Review of human phenotypes and relationship to disease: J. O. Morse, *N. Engl. J. Med.* **299**, 1045-1048; 1099-1105 (1978). Clinical evaluation in AAT deficiency: M. D. Werwers *et al., ibid.* **316**, 1055 (1987); N. Konietzko *et al., Deutsch. Med. Wochenschr.* **113**, 369 (1988). Symposium on structure, function and clinical applications: *Am. J. Med.* **84**, Suppl. 6A, 1-90 (1988). Review of role in emphysema: P. J. Stone, *Clin. Chest Med.* **4**, 405-412 (1983); J. A. Pierce, *J. Am. Med. Assoc.* **259**, 2890-2895 (1988).

THERAP CAT: Treatment of emphysema associated with inherited α_1-antitrypsin deficiency.

752. Antivenin (*Crotalidae*) **Polyvalent.** North and South American antisnakebite serum. An antitoxin obtained from serum of horses that have been immunized with venom of four crotaline snakes (*Crotalus atrox, C. adamanteus, C. durissus terrificus, Bothrops atrox*). *Ref:* Gingrich, Hohenadel, in *Venoms,* E. E. Buckley, N. Porges, Eds. (Publ. 44 of Am. Assoc. Adv. of Sci., Washington, 1956) p 381; Keegan, *ibid.,* p 413; *Behringwerk-Mitteilungen* (Sonderband: *Die Gift-Schlangen der Erde,* Marburg, 1963) 464 pp.

THERAP CAT: Antidote (vs snake venom).

THERAP CAT (VET): To neutralize snake venom.

753. Antivirin. AV. A virus inhibitory factor produced by various cells including HeLa-S$_3$, Hep. no. 2, KB, FL, L, G$_2$V, Chang's liver cells, human embryo lung, mouse embryo fibroblast, chick embryo fibroblast: Y. Seto, S. Toyoshima, *Progr. Antimicrob. Anticancer Chemother., Proc. Int. Congr. Chemother., 6th,* **vol. I** (Univ. Park Press, Baltimore, 1970) pp 108-113. The antiviral spectrum of this inhibitor is wide, and unlike interferon, its activity is stable after proteolytic enzyme treatment and heating. AV is produced in the absence of interferon inducers and does not possess such strict host specificity as interferon. It exerts its effects at any early stage of virus replication. The active part of AV is of smaller molecular size than interferon. Nature and mechanism of action: Toyoshima *et al., Interferon, Proc. Symp.* **1969,** Y. Nagano, Ed. (Igaku Shoin, Ltd., Tokyo, 1970) pp 185-193.

754. Antrafenine. *2-[[7-(Trifluoromethyl)-4-quinolin-yl]amino]benzoic acid 2-[4-[3-(trifluoromethyl)phenyl]-1-piperazinyl]ethyl ester;* 2-[4-(α,α,α-trifluoro-*m*-tolyl)-1-piperazinyl]ethyl-*N*-(7-trifluoromethyl-4-quinolyl)anthranilate; SL 73-033; Stakane. C$_{30}$H$_{26}$F$_6$N$_4$O$_2$; mol wt 588.57. C 61.22%, H 4.45%, F 19.37%, N 9.52%, O 5.44%. Analgesic with minimal anti-inflammatory and antipyretic activity, related structurally to floctafenine, *q.v.* Prepn: P. R. L. Giudicelli *et al.,* **Ger. pat. 2,415,982** corresp to U.S. pat. **3,935,229** (1974, 1976 both to Synthelabo); P. M. Manoury *et al., J. Med. Chem.* **22**, 554 (1979). Pharmacokinetic study: L. G. Dring *et al., Brit. J. Pharmacol.* **63**, 368P (1978). Metabolism: V. Rovei *et al., Ann. Chim.* **67**, 733 (1977). Pharmacologic study: R. D. Sofia *et al., Pharmacol. Res. Commun.* **11**, 179 (1979).

Crystals from isopropyl alc, mp 88°. LD$_{50}$ in mice: 4000 mg/kg orally, P. M. Manoury *et al., loc. cit.*

THERAP CAT: Analgesic.

755. ANTU. *1-Naphthalenylthiourea;* 1-(1-naphthyl)-2-thiourea; α-naphthylthiourea; *N*-1-naphthylthiourea; α-naphthylthiocarbamide; krysid; chemical 109; Anturat; Bantu; Rattrack. C$_{11}$H$_{10}$N$_2$S; mol wt 202.27. C 65.31%, H 4.98%, N 13.85%, S 15.85%. Prepd from α-naphthylamine and ammonium or potassium or sodium thiocyanate: de Clermont, Wehrlin, *Bull. Soc. Chim.* [2] **26**, 125 (1876); Alvarez, *C.A.* **42**, 4560 (1948) from α-naphthylisothiocyanate by treatment with alcoholic ammonia: Dyson, Hunter, *Chem. News* **134**, 4 (1927). Rodenticide activity discovered by Richter, *J. Am. Med. Assoc.* **129**, 927 (1945).

Prisms from alc. *Toxic!* Bitter taste. mp 198°. Solubility in water at 25°: 0.06 g/100 ml; in acetone: 2.43 g/100 ml; in triethylene glycol: 8.6 g/100 ml. Fairly sol in hot alc.

USE: Rodenticide. Specific control for the adult Norway rat. *Toxicity:* Less toxic to other rat species. Safe for domestic animals; induces vomiting in dogs. Produces massive pulmonary edema and pleural effusion in exptl animals, Fairhall, *Industrial Toxicology* (Hafner Publishing Co., New York, 2nd ed., 1969) p 300.

756. Apalcillin. *6-[[[[(4-Hydroxy-1,5-naphthyridin-3-yl)carbonyl]amino]phenylacetyl]amino]-3,3-dimethyl-7-oxo-4-thia-1-azabicyclo[3.2.0]heptane-2-carboxylic acid.* C$_{25}$H$_{23}$N$_5$O$_6$S; mol wt 521.56. C 57.57%, H 4.44%, N 13.43%, O 18.41%, S 6.15%. Semi-synthetic antibiotic related to penicillin. Prepn of the potassium salt: H. Tobiki *et al.,* **S. Afr. pat. 72 05,865** corresp to U.S. pat. **3,864,329** (1973, 1975 both to Sumitomo); of the sodium salt: Y. Hirotada *et al.,* U.S. pat. **4,005,075** (1977 to Sumitomo); H. Tobiki *et al., Yakugaku Zasshi* **100**, 49 (1980), *C.A.* **93**, 95179x (1980). Microbiological evaluation: H. Noguchi *et al., Antimicrob. Ag. Chemother.* **9**, 262 (1976). Analytical studies: I. Umeda *et al., Yakugaku Zasshi* **99**, 717 (1979), *C.A.* **92**, 11287z (1980). Series of articles on pharmacology, exptl and clinical effects: *Chemotherapy (Tokyo)* **26**, Suppl. 2, 111-223 (1978). Pharmacokinetics in man: U. Busch *et al., Arzneimittel-Forsch* **32**, 1131 (1982).

Sodium salt, C$_{25}$H$_{22}$N$_5$NaO$_6$S, *PC 904, Lumota, Palcin.* White cryst, sol in water.

THERAP CAT: Antibacterial.

757. Apamin. $C_{79}H_{131}N_{31}O_{24}S_4$; mol wt 2027.38. C 46.80%, H 6.51%, N 21.42%, O 18.94%, S 6.33%. The smallest neurotoxic polypeptide known and the only one whose interaction with the spinal cord is well established. Comprises about 2% by wt of the dried venom of *Apis mellifica (mellifera)*, the honey bee. Isoln: E. Habermann, K. G. Reiz, *Naturwiss.* **51**, 61 (1964); *eidem, Biochem. Z.* **341**, 451 (1965). Structure: P. Haux *et al., Z. Physiol. Chem.* **348**, 737 (1967); R. Shipolini *et al., Chem. Commun.* **1967**, 679. Conformation: R. C. Hider, U. Ragnarsson, *FEBS Letters* **111**, 189 (1980); B. Busetta, *ibid.* **112**, 138 (1980). Solution structure: J. H. B. Pease, D. E. Wemmer, *Biochemistry* **27**, 8491 (1988). Solid-phase synthesis: J. van Rietschoten *et al., Eur. J. Biochem.* **56**, 35 (1975); B. E. B. Sandberg, U. Ragnarsson, *Int. J. Pept. Protein Res.* **11**, 238 (1978). Pharmacology: Wellheoner, *Arch. Pharmakol. Exp. Pathol.* **262**, 29 (1969). Biochemistry: E. Habermann, K. G. Reiz, *Biochem. Z.* **343**, 192 (1965). Action on CNS: E. Habermann, D. Cheng-Raude, *Toxicon* **13**, 465 (1975). *Review:* E. Habermann, *Science* **177**, 314-322 (1972); C. Granier, J. van Rietschoten, in *Natural Toxins*, D. Eaker, T. Wadström, Eds. (Pergamon, New York, 1980) pp 481-486; E. Habermann, *Pharmacol. Ther.* **25**, 255-270 (1984).

```
Cys-Asn-Cys-Lys-Ala-Pro-Glu-Thr-Ala-Leu-Cys-Ala-Arg-Arg-Cys-Gln
|                                                            Gln
|                                                            HisNH₂
```

Highly basic compd. Pharmacologic activity destroyed by oxidn with performic acid. LD_{50} i.v. in mice: 4 mg/kg (Habermann, Reiz).

758. Apazone. *5-(Dimethylamino)-9-methyl-2-propyl-1H-pyrazolo[1,2-a][1,2,4]benzotriazine-1,3(2H)-dione;* 3-dimethylamino-7-methyl-1,2-(n-propylmalonyl)-1,2-dihydro-1,2,4-benzotriazine; azapropazone; AHR 3018; Mi 85; Azapren; Rheumox; Sinnamin; Cinnamin. $C_{16}H_{20}N_4O_2$; mol wt 300.37. C 63.98%, H 6.71%, N 18.65%, O 10.65%. Prepn: Fr. pat. **1,440,629** corresp to I. Molnar *et al.*, U.S. pats. **3,349,088** and **3,482,024** (1966, 1967, 1969, all to Siegfried); Mixich, *Helv. Chim. Acta* **51**, 532 (1968). Pharmacological and toxicological studies: Jahn, Adrian, *Arzneimittel-Forsch.* **19**, 36 (1969). Metabolism: Mixich, *Helv. Chim. Acta* **55**, 1031 (1972). HPLC determn in plasma and urine: B. J. Kline *et al., Arzneimittel-Forsch.* **33**, 504 (1983).

mp 228°.
Dihydrate, *Mi 85Di, Prolixan, Tolyprin.* Almost colorless crystals, mp 247-248°.
THERAP CAT: Anti-inflammatory; analgesic.

759. Aphidicolin. *Tetradecahydro-3,9-dihydroxy-4,11b-dimethyl-8,11a-methano-11aH-cyclohepta[a]naphthalene-4,9-dimethanol;* ICI 69653; NSC-234714. $C_{20}H_{34}O_4$; mol wt 338.49. C 70.97%, H 10.12%, O 18.91%. Novel tetracyclic diterpene antibiotic with antiviral and antimitotic properties, isolated from *Cephalosporium aphidicola* Petch. It is a specific inhibitor of DNA α-polymerase. Description and x-ray crystallographic determn of structure: K. M. Brundret *et al., Chem. Commun.* **1972**, 1027. Prepn and antiviral activity: A. Barrow *et al.*, Ger. pat. **2,216,205** corresp to U.S. pat. **3,761,512** (both 1973 to ICI). Structure and abs config: W. Dalziel *et al., J. Chem. Soc. Perkin Trans. I* **1973**, 2841. Total synthesis of (±)-aphidicolin: B. M. Trost *et al., J. Am. Chem. Soc.* **101**, 1328 (1979); J. E. McMurry *et al., ibid.* 1331; E. J. Corey *et al., ibid.* **102**, 1743 (1980); J. E. McMurry *et al., Tetrahedron* **37**, Suppl. 1, 319 (1981). Antiviral effects *in vitro* and *in vivo*: R. A. Bucknall *et al., Antimicrob. Ag. Chemother.* **4**, 294 (1973). Antimitotic activity: S. Ikegami *et al., Nature* **275**, 458 (1978). Effects on DNA

synthesis in mouse sarcoma: S. Seki *et al., Biochim. Biophys. Acta* **610**, 413 (1980). Enzymatic determn method: G. Pedrali-Noy *et al., J. Biochem. Biophys. Methods* **4**, 113 (1981). Induction of differentiation of human myeloid leukemia cells: J. Griffin *et al., Exp. Hematol.* **10**, 774 (1982). Effects on cell proliferation: G. Iliakis *et al., Int. J. Radiat. Biol.* **42**, 417 (1982); on γ-irradiated human fibroblasts: P. J. Smith, M. C. Paterson, *Biochim. Biophys. Acta* **739**, 17 (1983).

Cryst from ethyl acetate, mp 227-232°. $[\alpha]_D^{27}$ +12° (c = 1 in methanol).
(±)-Form, white cryst, mp 218-220°.
USE: As a biological tool in studies of cell proliferation and differentiation.

760. Apholate. *2,2,4,4,6,6-Hexakis(1-aziridinyl)-2,2,-4,4,6,6-hexahydro-1,3,5,2,4,6-triazatriphosphorine;* hexakis-(1-aziridinyl)phosphonitrile; 1-aziridinylphosphonitrile trimer. $C_{12}H_{24}N_9P_3$; mol wt 387.32. C 37.21%, H 6.25%, N 32.55%, P 23.99%. Prepn: Rätz, Grundmann, U.S. pat. **2,858,306** (1958 to Olin Mathieson).

Crystals from heptane, mp 147.5°. LD_{50} orally in male, female rats: 98, 113 mg/kg, T. B. Gaines, *Toxicol. Appl. Pharmacol.* **14**, 515 (1969).
USE: Insect chemosterilant.

761. Aphylline. *[7R-(7α,7aβ,14α,14aα)]-Dodecahydro-7,14-methano-2H,6H-dipyrido[1,2-a:1',2'-e][1,5]diazocin-6-one.* $C_{15}H_{24}N_2O$; mol wt 248.36. C 72.54%, H 9.74%, N 11.28%, O 6.44%. From *Anabasis aphylla* L., *Chenopodiaceae.* Extraction procedure and structure: Orechoff, Menschikoff, *Ber.* **64**, 266 (1931); **65**, 234 (1932). Structure: Galinovsky, Jarish, *Monatsh.* **84**, 199 (1953). Stereochemistry: Edwards *et al., Can. J. Chem.* **32**, 235 (1954); Galinovsky *et al., Monatsh.* **86**, 1014 (1955). Synthesis: Bohlmann *et al., Ber.* **90**, 653 (1957).

Crystals, mp 52-57°. bp_4 200°. $[\alpha]_D^{20}$ +10.3° (c = 20 in methanol). Soluble in the usual organic solvents.
Hydrochloride, $C_{15}H_{24}N_2O.HCl$, rhombohedra from alc, mp 209°. $[\alpha]_D^{20}$ +14° (c = 25). Freely soluble in water.

762. Apicycline. *α-[4-(Dimethylamino)-1,4,4a,5,5a,6,-11,12a-octahydro-3,6,10,12,12a-pentahydroxy-6-methyl-1,11-dioxo-2-naphthacenecarboxamido]-4-(2-hydroxyethyl)-1-piperazineacetic acid;* N-[(2-hydroxyethyl)piperazinocarboxymethyl]tetracycline; RIT 1140; Traserit. $C_{30}H_{38}N_4O_{11}$; mol wt 630.66. C 57.14%, H 6.07%, N 8.88%, O 27.91%. Semi-synthetic antibiotic related to tetracycline. Prepn: Rondelet, Neth. pat. Appl. **6,515,688**, *C.A.* **66**, 46260a (1967) and Belg. pat. **673,130**, *C.A.* **68**, 104840u (1968) (both 1966 to Recherche & Ind. Ther.); Valcavi *et al., Farmaco Ed. Sci.*

21, 775 (1966); Rondelet, Froment, *Ann. Pharm. Franc.* **26,** 63 (1968).

Yellow, amorphous powder with slight amine odor, mp 144.5° (dec). Freely sol in water. $[\alpha]_D$ −123° (c = 0.5 in methanol); $[\alpha]_D$ −133° (c = 0.5 in water).

THERAP CAT: Antibacterial.

763. Apigenin. *5,7-Dihydroxy-2-(4-hydroxyphenyl)-4H-1-benzopyran-4-one; 4′,5,7-trihydroxyflavone;* 2-(p-hydroxyphenyl)-5,7-dihydroxychromone; pelargidenon 1449; Versulin. $C_{15}H_{10}O_5$; mol wt 270.23. C 66.67%, H 3.73%, O 29.60%. The aglucon of apiin and of apigenin-7-glucoside. From apiin by boiling with acids, from apigenin-7-glucoside by enzymatic hydrolysis with emulsin or by boiling with 15% H_2SO_4. Isoln and structure: Czajkowski *et al., Ber.* **33,** 1992 (1900); Schmid, Waschkau, *Monatsh.* **49,** 83 (1928); Baker *et al., J. Chem. Soc.* **1963,** 1477. Synthesis: Hutchins, Wheeler, *ibid.* **1939,** 91; Farooq *et al., Arch. Pharm.* **292,** 792 (1959).

Yellow needles from aqueous pyridine, mp 345-350°. uv max (ethanol): 269, 340 nm (ε 18,800; 20,900). Practically insoluble in water; moderately sol in hot alcohol. Sol in dil KOH with intense yellow color.

USE: Has been used to dye Cr mordanted wool yellow. The color is fast to soap.

764. Apigetrin. *7-(β-D-Glucopyranosyloxy)-5-hydroxy-2-(4-hydroxyphenyl)-4H-1-benzopyran-4-one;* apigenin-7-D-glucoside; 7-D-glycosylapigenin; cossmetin. $C_{21}H_{20}O_{10}$; mol wt 432.39. C 58.33%, H 4.66%, O 37.00%. Isoln from flowers of *Anthemis nobilis* L., *Compositae:* Power, Browning, *J. Chem. Soc.* **105,** 1833 (1914); from parsley: Nordström, Swain, *Chem. & Ind. (London)* **1953,** 85; *J. Chem. Soc.* **1953,** 2764.

Pale yellow, cryst powder, dec 178-180°. Astringent taste. uv max (ethanol): 335, 268 nm. Sol in water, dil alcohol.

765. Apiin. *7-[(2-O-D-Apio-β-D-furanosyl-β-D-glucopyranosyl)oxy]-5-hydroxy-2-(4-hydroxyphenyl)-4H-1-benzopyran-4-one; 4′,5,7-trihydroxyflavone-7-apiosylglucoside;* apigenin-7-apiosylglucoside; apioside. $C_{26}H_{28}O_{14}$; mol wt 564.48. C 55.32%, H 5.00%, O 39.68%. Isoln from parsley and from celery: Vongerichten, *Ber.* **33,** 2334, 2904 (1900); Gupta, Seshadri, *Proc. Indian Acad. Sci.* **35A,** 242 (1952), *C.A.* **47,** 3306c (1953); Rahman, *Z. Naturforsch.* **13b,** 201 (1958). Structure: Vongerichten, *Ann.* **318,** 121 (1901); Marchlewski, Skarzynski, *Biochem. Z.* **297,** 56 (1938); Hemming, Ollis, *Chem. & Ind. (London)* **1953,** 85.

Crystals from ethanol, mp 230-232°. uv max (96% ethanol): 267.5, 342.5 nm (log ε 4.18, 4.30). Sol in hot water, hot alcohol; practically insol in ether. Sol in Na_2CO_3 or NH_3 solns with intense yellow color, in NaOH with pale yellow color.

766. Apiole (Dill). *4,5-Dimethoxy-6-(2-propenyl)-1,3-benzodioxole; 1-allyl-2,3-dimethoxy-4,5-(methylenedioxy)-benzene;* dill apiole. $C_{12}H_{14}O_4$; mol wt 222.24. C 64.85%, H 6.35%, O 28.80%. Occurs in dill oil, *Anethum graveolus* L., *Umbelliferae.* Isoln: G. Ciamician, P. Silber, *Ber.* **29,** 1799 (1896); D. B. Spoelstra, *Rec. Trav. Chim.* **48,** 372 (1929). Structure: H. Thoms, *Arch. Pharm.* **242,** 344 (1904). Synthesis: W. Baker *et al., J. Chem. Soc.* **1934,** 1681; F. Dallacker, *Ber.* **102,** 2663 (1969); J. R. Cannon *et al., J. Sci. Soc. Thailand* **6,** 59 (1980). Synergistic activity with insecticides: E. P. Lichtenstein *et al., J. Agr. Food Chem.* **22,** 658 (1974); S.S. Tomar *et al., Agr. Biol. Chem.* **43,** 1479 (1979).

Oil, mp 29.5°. bp 285°. n_D^{17} 1.5305; d_4^{15} 1.1598.

767. Apiole (Parsley). *4,7-Dimethoxy-5-(2-propenyl)-1,3-benzodioxole; 1-allyl-2,5-dimethoxy-3,4-methylenedioxy-benzene;* parsley apiole; apiol; apioline; parsley camphor. $C_{12}H_{14}O_4$; mol wt 222.23. C 64.85%, H 6.35%, O 28.80%. Occurs in parsley oil, *Petroselinum sativum:* Blanchet, Sell, *Ann.* **6,** 259 (1833); Vongerichten, *Ber.* **9,** 1477 (1876); Kolesnikov *et al., Aptechnoe Delo* 7(4), 27 (1958), *C.A.* **54,** 12491e (1960). Structure: Ciamician, Silber, *Ber.* **23,** 2283 (1890); Thoms, *ibid.* **36,** 1714 (1903). Synthesis: Baker, Savage, *J. Chem. Soc.* **1938,** 1602; F. Dallacker, *Ber.* **102,** 2663 (1969). Synergistic activity with insecticides: E. P. Lichtenstein *et al., J. Agr. Food Chem.* **22,** 658 (1974).

Crystals; faint parsley odor. mp 29.5°. bp 294°. n_D^{20} 1.536-1.538. Insol in water; sol in alcohol, benzene, chloroform, ether, acetone, oils. *Keep in a cool place.*

768. Apiose. D-*Apiose;* tetrahydroxyisovaleraldehyde; 3-C-(hydroxymethyl)-D-glyceroaldotetrose. $C_5H_{10}O_5$; mol wt 150.13. C 40.00%, H 6.71%, O 53.29%. First found in parsley in which it occurs as the flavinoid glycoside apiin, *q.v.* Isoln from apiin: Vongerichten, *Ann.* **318,** 126 (1901); **321,** 74 (1902); Hemming, Ollis, *Chem. & Ind. (London)* **1953,** 85. From the rubber plant, *Hevea brasiliensis, Euphorbiaceae:* Patrick, *Nature* **178,** 216 (1956). Discussion of structure and isoln from the Australian marine plant *Posidonia australis* Kon., *Potamogetonaceae:* Bell, *Methods in Carbohydrate Chemistry* vol. I (Academic Press, New York, 1962) pp 260-263. Synthesis: Gorin, Perlin, *Can. J. Chem.* **36,** 480 (1958); Khalique, *J. Chem. Soc.* **1962,** 2515; Ezekiel *et al., Tetrahedron Letters* **1969,** 1635. Synthesis of L-form: Weygand, Schmiechen, *Ber.* **92,** 535 (1959); of DL-form: Kinoshita, Miwa, *Carbohyd. Res.* **28,** 175 (1973); Y. Araki *et al., ibid.* **58,** C4 (1977); of D- and L-forms: P. Ho, *Can. J. Chem.* **57,** 381 (1979). Chemistry, configuration and synthesis studies: Williams, Jones, *ibid.* **42**, 69 (1964); Hulyalker *et al., ibid.* **43**, 2085 (1965). *Review:* Watson, Orenstein, *Advan. Carbohyd. Chem. Biochem.* **31**, 135-184 (1975).

Syrup. $[\alpha]_D^{15}$ +5.6°; $[\alpha]_D^{19}$ +9.1°. Soluble in water.

D-Apiose di-*O*-isopropylidene, $C_{11}H_{18}O_5$, plates from water containing a trace of NH_3, mp 81-83°. $[\alpha]_D^{20}$ +55.5° (c = 1.1 in ethanol).

769. Aplasmomycin. ICI 122,378. $C_{40}H_{60}BNaO_{14}$; mol wt 798.72. C 60.15%, H 7.57%, B 1.35%, Na 2.88%, O 28.04%. Antibiotic produced by *Streptomyces griseus* strain SS-20, obtained from shallow sea mud: Y. Okami *et al., J. Antibiot.* **29**, 1019 (1976); *eidem,* **Japan. Kokai 77 108901** (1977 to Microbiochem. Res. Found.), *C.A.* **88**, 35843 (1978). It inhibits the growth of gram-positive bacteria *in vitro* and is active vs *Plasmodium berghei in vivo.* Aplasmomycin is a symmetrical dimer related to boromycin, *q.v.,* the only other known natural product that contains boron. Structure, x-ray crystallography: H. Nakamura *et al., J. Antibiot.* **30**, 714 (1977). Total synthesis of (+)-form: E. J. Corey *et al., J. Am. Chem. Soc.* **104**, 6816, 6818 (1982); T. Nakata *et al., Tetrahedron Letters* **27**, 6341, 6345 (1986); J. D. White *et al., J. Am. Chem. Soc.* **108**, 8105 (1986). Use as growth promotant in ruminants: D. H. Davies *et al., Eur. pat. Appl. 2893; eidem,* **U.S. pat. 4,225,593** (1979, 1980 both to ICI). Two minor components, aplasmomycins B and C, have also been isolated from the fermentation; both aplasmomycin C and aplasmomycin B show ionophoric properties, mediating net K^+ transport across a bulk phase: K. Sato *et al., J. Antibiot.* **31**, 632 (1978). NMR analysis of aplasmomycin and deboroaplasmomycin: T. S. S. Chen *et al., ibid.* **33**, 1316 (1980). Comparative anti-anaerobic activity of aplasmomycin: K. Watanabe *et al., Antimicrob. Ag. Chemother.* **19**, 519 (1981). Review of biosynthetic studies: H. G. Floss, C. Chang in *Antibiotics IV*, J. W. Corcoran, Ed. (Springer-Verlag, New York, 1981) pp 203-210.

(+)-aplasmomycin

Colorless needles, mp 283-285° (dec). $[\alpha]_D^{22}$ +225° (c = 1.24 in chloroform). Very lipophilic. Practically insol in water. LD_{50} i.p. in mice: 125 mg/kg (Okami).

Silver salt, $C_{40}H_{60}AgBO_{14}$, colorless needles, mp 218-220°. $[\alpha]_D^{23}$ +194° (c = 0.34 in chloroform).

USE: Growth promotant in ruminants.

770. Apoatropine. *endo-α-Methylenebenzeneacetic acid 8-methyl-8-azabicyclo[3.2.1]oct-3-yl ester; 1αH,5αH-tropan-3α-ol atropate;* atropamine; atropyltropeine. $C_{17}H_{21}NO_2$; mol wt 271.35. C 75.24%, H 7.80%, N 5.16%, O 11.79%. Occurs in root of *Atropa belladonna* L., Solanaceae. Also obtained from atropine by splitting off water or by total synthesis: Ladenburg, *Ann.* **217**, 102 (1883); Merck, *Arch.*

Pharm. **230**, 134 (1892); **231**, 110 (1893); Hesse, *Ann.* **261**, 87 (1891); **271**, 124 (1892); **277**, 290 (1893). Isoln by chromatography: Steinegger, Phokas, *Pharm. Acta Helv.* **31**, 284 (1956).

Prisms from chloroform, mp 62°. Absorption spectrum: Gompel, Henri, *Compt. Rend.* **156**, 1543 (1913). Freely sol in alcohol, ether, chloroform, benzene, carbon disulfide; slightly in petr ether, isoamyl alcohol. Almost insol in water. LD_{50} in mice: 160 mg/kg orally, 14.1 mg/kg i.p.

Hydrochloride, $C_{17}H_{21}NO_2 \cdot HCl$, scales, mp 239°. Sol in hot water; sparingly sol in alc, acetone; nearly insol in ether.

Sulfate pentahydrate, $(C_{17}H_{21}NO_2)_2 \cdot H_2SO_4 \cdot 5H_2O$, crystals, sparingly sol in water.

Toxicity: Death by respiratory arrest occurs in animals with relatively small doses. It is more active by mouth than by injection.

THERAP CAT: Antispasmodic.

771. Apocodeine. *(R)-5,6,6a,7-Tetrahydro-10-methoxy-6-methyl-4H-dibenzo[de,g]quinolin-11-ol; 10-methoxy-6aβ-aporphin-11-ol.* $C_{18}H_{19}NO_2$; mol wt 281.34. C 76.84%, H 6.81%, N 4.98%, O 11.37%. A monomethyl ether of apomorphine (*q.v.*). Prepd by heating codeine with oxalic acid: Folkers, *J. Am. Chem. Soc.* **58**, 1814 (1936); by heating codeine with phosphoric acid: Small *et al., J. Org. Chem.* **5**, 344 (1940). Configuration: Corrodi, Hardegger, *Helv. Chim. Acta* **38**, 2038 (1955). Total synthesis and pharmacology: Neumeyer *et al., J. Med. Chem.* **16**, 1223 (1973).

Small prisms from methanol (lose solvent at 80°/2 mm), mp 124° (dry). $[\alpha]_D^{24}$ −97° (c = 0.45). Slightly soluble in water; sol in alcohol, ether, dilute acids.

Hydrochloride, $C_{18}H_{20}ClNO_2$, crystals from alcohol-ether, softening 140°, dec 260-263°. $[\alpha]_D^{22}$ −43° (c = 0.51). Very sol in water; sol in alcohol.

THERAP CAT: Emetic.

772. Apocynin. *1-(4-Hydroxy-3-methoxyphenyl)ethanone;* acetovanillon; 4-hydroxy-3-methoxyacetophenone. $C_9H_{10}O_3$; mol wt 166.17. C 65.05%, H 6.07%, O 28.88%. From rhizome of Canadian hemp, *Apocynum cannabinum* L., Apocynaceae and from *A. androsaemifolium*: Finnemore, *J. Chem. Soc.* **93**, 1513, 1520 (1908). Also from the essential oil (butter) of the rhizomes of *Iris* spp., Iridaceae: Naves, *Helv. Chim. Acta* **32**, 1351 (1949).

Fine needles from water, mp 115°. bp 295-300°. Faint vanilla odor. Slightly sol in cold, freely in hot water, alc, benzene, chloroform, ether. Practically insol in petr ether.

773. Apocynum androsaemifolium. Dogbane; spreading dogbane; bitter root; milk ipecac; wild ipecac; rheumatism weed. Root of *Apocynum androsaemifolium* L., Apocynaceae. *Habit.* North America. *Constit.* Apocynin, apocynein, apocynamarin, volatile oil.

THERAP CAT: Cardiotonic.

774. Apocynum cannabinum. Canadian hemp; American Indian hemp; black Indian hemp; Indian physic; Indian dogbane. Dried rhizome and roots of *Apocynum cannabinum* L., *Apocynaceae.* *Habit.* U.S. *Constit.* Cynotoxin, apocyncein, apocynin, cymarin, resin, tannin, bitter extractive, starch.

Note: Not to be confused with *Apocynum androsaemifolium* which has few of its properties.

THERAP CAT: Cardiotonic.

775. Apo-β-erythroidine. *4,5,7,8,9,12-Hexahydro-11H-pyrano[3,4-d]pyrrolo[3,2,1-jk][1]benzazepin-11-one.* $C_{15}H_{15}NO_2$; mol wt 241.28. C 74.66%, H 6.27%, N 5.81%, O 13.26%. A degradation product of β-erythroidine. Prepd by heating β-erythroidine to 120° with phosphoric or sulfuric acid: Sauvage *et al., Science* **109**, 627 (1947); by reacting β-erythroidine with concd hydrobromic acid at 100°: Koniuszy, Folkers, *J. Am. Chem. Soc.* **73**, 333 (1951). Synthesis and structure: Blake *et al., J. Am. Chem. Soc.* **87**, 1397 (1965).

Crystals, mp 128-129°, also reported as mp 132-132.5°. uv max (ethanol): 345, 240 nm (ε 3500, 24,500).

776. Apomorphine. *5,6,6a,7-Tetrahydro-6-methyl-4H-dibenzo[de,g]quinoline-10,11-diol;* 6aβ-aporphine-10,11-diol. $C_{17}H_{17}NO_2$; mol wt 267.31. C 76.38%, H 6.41%, N 5.24%, O 11.97%. Synthetic opiate obtained by treating morphine with concd HCl: Small *et al., J. Org. Chem.* **5**, 344 (1940); by heating morphine with $ZnCl_2$: Mayer, *Ber.* **4**, 121 (1871). Prepn of methylbromide: Pschorr, *Ger. pat.* **158,-620** (1905 to Riedel), *Frdl.* **7**, 699. Structure: R. Pschorr, *Ber.* **40**, 1984 (1907). Configuration: H. Corrodi, E. Hardegger, *Helv. Chim. Acta* **38**, 2038 (1955). Total synthesis of (±)-form: J. L. Neumeyer *et al., J. Pharm. Sci.* **59**, 1850 (1970); J. L. Neumeyer, U.S. pat. **3,717,639** (1973 to A. D. Little); J. L. Neumeyer *et al., J. Med. Chem.* **16**, 1223 (1973). Synthesis of (+)- and (−)-forms: V. J. Ram, J. L. Neumeyer, *J. Org. Chem.* **46**, 2830 (1981). Toxicity data: J. G. Cannon *et al., J. Med. Chem.* **15**, 348 (1972); J. Z. Ginos *et al., ibid.* **18**, 1194 (1975). Review of pharmacology and neurochemistry: G. DiChiara, G. L. Gessa, *Adv. Pharmacol. Chemother.* **15**, 87-160 (1978). *Reviews:* F. C. Colpaert *et al., Int. Rev. Neurobiol.* **19**, 225-268 (1976); J. L. Neumeyer *et al.,* in *Apomorphine and Other Dopaminomimetics* vol. 1, G. L. Gessa, G. U. Corsini, Eds. (Raven, New York, 1981) p 1-17.

Hexagonal plates from chloroform and petr ether, dec 195°; subl in high vacuum. Oxidizes rapidly in air and becomes green. Sol in alcohol, acetone, chloroform. Slightly sol in water, benzene, ether, petr ether. Solns darken rapidly. Kb 1 × 10^{-7}; Ka 1.2 × 10^{-9}. uv max (98% alc): 336, 399 nm. LD_{50} i.p. in mice: 600 μmoles/kg (Cannon).

Hydrochloride, $C_{17}H_{18}ClNO_2$, small crystals (usually hemihydrate). Dec and turn green on exposure to light and air. $[\alpha]_D^{25} -48°$ (c = 1.2). uv spectrum: Csokan, *Z. Anal. Chem.* **124**, 344 (1942). pH of aq soln (1 in 300) = 4.8. One gram dissolves in 50 ml water, 17 ml water at 80°, 50 ml

alcohol. Very slightly sol in chloroform and ether. LD_{50} i.p. in mice: 145 μg/g (Ginos).

Diacetylapomorphine, $C_{21}H_{21}NO_4$, mp 127-128°, $[\alpha]_D^{24} -88°$ (c = 1.12 in 0.1N HCl).

Methylbromide, $C_{18}H_{20}BrNO_2$, *N-methylapomorphinium bromide, Euporphin, Bromophin.* Crystals, mp 180°. Freely sol in water, alc. Practically insol in ether.

THERAP CAT: Emetic.

THERAP CAT (VET): Emetic, expectorant in dogs.

777. Apoquinine. *3,10-Didehydro-10,11-dihydrocinchonan-6',9-diol;* apocupreine. $C_{19}H_{22}N_2O_2$; mol wt 310.38. C 73.52%, H 7.14%, N 9.03%, O 10.31%. By demethylation of quinine: Henry, Solomon, *J. Chem. Soc.* **1934**, 1923; Butler, Cretcher, *J. Am. Chem. Soc.* **57**, 1083 (1935).

α-Form, prisms from ether, dec 184°. $[\alpha]_D^{20} -215°$ (alc). Readily sol in alc, sparingly sol in ether or acetone.

Hydrochloride, $C_{19}H_{22}N_2O_2$·HCl, clusters of needles from aq alc, dec 273°. $[\alpha]_D^{15} -164°$.

778. Aporeine. *(S)-6,7,7a,8-Tetrahydro-7-methyl-5H-benzo[g]-1,3-benzodioxolo[6,5,4-de]quinoline;* aporheine; 1,2-methylenedioxyaporphine; (+)-roemerine. $C_{18}H_{17}NO_2$; mol wt 279.32. C 77.39%, H 6.13%, N 5.01%, O 11.46%. From *Papaver dubium* L., *Papaveraceae:* Pavesi, *Gazz. Chim. Ital.* **37 I**, 629 (1907); **44 I**, 398 (1914). Structure and identity with (+)-roemerine: Slavik, *Coll. Czech. Chem. Commun.* **28**, 1738 (1963). Synthesis of *dl*-roemerine: Marion, Grassie, *J. Am. Chem. Soc.* **66**, 1290 (1944).

Needles from ether + petroleum ether, mp 102°. $[\alpha]_D^{22} +80°$ (c = 0.50 in ethanol). uv max: 262, 315 nm (log ε 4.3, 3.7). pK 6.1. Sol in ether, methanol, ethanol, chloroform; slightly sol in petr ether. Practically insol in water, alkali.

Hydrochloride, $C_{18}H_{17}NO_2$·HCl, leaflets from ethanol or water, mp 266-267°. Slightly sol in ethanol, water.

Methiodide, $C_{18}H_{17}NO_2$·CH_3I, crystals from boiling methanol, mp 232-233°.

779. Apraclonidine. *2,6-Dichloro-N¹-(4,5-dihydro-1H-imidazol-2-yl)-1,4-benzenediamine;* 2,6-dichloro-N'-2-imidazolidinylidene-1,4-benzenediamine; 2-[(4-amino-2,6-dichlorophenyl)imino]imidazolidine; *p*-aminoclonidine; aplonidine; NC 14. $C_9H_{10}Cl_2N_4$; mol wt 245.11. C 44.10%, H 4.11%, Cl 28.93%, N 22.86%. α_2-Adrenergic agonist; structural analog of clonidine, *q.v.* Synthesis: B. Rouot, G. LeClerc, *Bull. Soc. Chim. France* Pt. 2, 520 (1979). Prepn (not claimed) and use in treatment of intraocular pressure: B. M. York, Jr., U.S. pat. **4,517,199** (1985 to Alcon). Pharmacology: B. Rouot *et al., C. R. Acad. Sci.* **286**, 909 (1978). Receptor binding studies: D. C. U'Prichard, *Prog. Clin. Biol. Res.* **71**, 53 (1981). D. C. Stump, D. E. MacFarlane, *J. Lab. Clin. Med.* **102**, 779 (1983). Clinical pharmacology: D. A. Abrams *et al., Arch. Ophthalmol.* **105**, 1205 (1987). Clinical evaluation in treatment of intraocular pressure: A. L. Robin *et al., ibid.* 1208.

Solid, mp > 230°.
Hydrochloride, $C_9H_{11}Cl_3N_4$, *ALO 2145, Iopidine.*
Dihydrochloride, $C_9H_{12}Cl_4N_4$, uv max (ethanol): 254, 304 nm (ϵ 1800, 2500).
THERAP CAT: Treatment of post-surgical elevated intraocular pressure.

780. Apramycin. *O-4-Amino-4-deoxy-α-D-glucopyranosyl-(1→8)-O-(8R)-2-amino-2,3,7-trideoxy-7-(methylamino)-D-glycero-α-D-allo-octodialdo-1,5:8,4-dipyranosyl-(1→4)-2-deoxy-D-streptamine;* 4-O-[3α-amino-6α-[(4-amino-4-deoxy-α-D-glucopyranosyl)oxy]-2,3,4,4aβ,6,7,8aα-octahydro-8β-hydroxy-7β-(methylamino)pyranopyrano[3,2-*b*]pyran-2α-yl]-2-deoxy-D-streptamine; nebramycin factor 2; EL-857/820; EL-857; 47657; Ambylan; Apralan. $C_{21}H_{41}N_5O_{11}$; mol wt 539.60. C 46.74%, H 7.66%, N 12.98%, O 32.62%. Broad spectrum aminocyclitol antibiotic and component of the nebramycin complex, produced by a strain of *Streptomyces tenebrarius:* R. Q. Thompson, E. A. Presti, *Antimicrob. Ag. Chemother.* **1967,** 332; W. M. Stark, U.S. pat. **3,691,279** (1972 to Lilly). Structure, abs config and properties: S. O'Connor *et al., J. Org. Chem.* **41,** 2087 (1976). ^{13}C-NMR: E. Wenkert, E. W. Hagaman, *ibid.* 701. *In vitro* activity: R. Ryden *et al., J. Antimicrob. Chemother.* **3,** 609 (1977). Synthetic studies: H. C. Jarrell, W. A. Szarek, *Carbohyd. Res.* **67,** 43 (1978); *eidem, Can. J. Chem.* **56,** 144 (1978); **57,** 924 (1979). *See also* tobramycin.

Monohydrate from aqueous ethanol, mp 245-247°. p*K*a (H_2O): 8.5, 7.8, 7.2, 6.2, 5.4. Very soluble in water; slightly sol in lower alcohols.
THERAP CAT: Antibacterial.
THERAP CAT (VET): Antibacterial antibiotic.

781. Aprindine. *N-(2,3-Dihydro-1H-inden-2-yl)-N',N'-diethyl-N-phenyl-1,3-propanediamine; N,N-diethyl-N'-2-indanyl-N'-phenyl-1,3-propanediamine; N-[3-(diethylamino)propyl]-N-phenyl-2-indanamine;* AC 1802; Lilly 99170; Compound 99170. $C_{22}H_{30}N_2$; mol wt 322.49. C 81.94%, H 9.38%, N 8.68%. Long-acting membrane-stabilizing anti-arrhythmic agent. Prepn: P. M. Vanhoof, P. M. Clarebout, *Belg.* pat. **760,048** corresp to U.S. pat. **3,923,814** (1971, 1975 both to Christiaens). Toxicological study: A. Georges *et al., Arzneimittel-Forsch.* **23,** 519 (1973). Pharmacokinetics: L. Didion *et al., Thérapie* **29,** 221 (1974); K. E. Wirth *et al., Herz* **8,** 302 (1983); T. Kazuki *et al., Eur. J. Clin. Pharmacol.* **26,** 129 (1984). HPLC determn in plasma: T. Kobari *et al., J. Chromatog.* **278,** 220 (1983). Clinical studies: D. P. Zipes *et al., Am. J. Cardiol.* **40,** 586 (1977); P. J. Troup, D. P. Zipes, *Am. Heart J.* **97,** 322 (1979). Reviews: P. Danilo, *ibid.* 119; I. Stoel, S. F. Hagemeijer, *Eur. Heart J.* **1,** 147 (1980). General review of anti-arrhythmic agents: L. H. Opie, *Lancet* **1,** 861 (1980).

Hydrochloride, $C_{22}H_{31}ClN_2$, *Compound 83846, Amidonal,*

Aspenon, Fibocil, Fiboran, Ritmusin. Crystals from benzene, mp 120-121°.
THERAP CAT: Cardiac depressant (anti-arrhythmic).

782. Aprobarbital. *5-(1-Methylethyl)-5-(2-propenyl)-2,4,6(1H,3H,5H)-pyrimidinetrione; 5-allyl-5-isopropylbarbituric acid;* Alurate; Numal; Allypropymal; Isonal (Swedish); Aprozal. $C_{10}H_{14}N_2O_3$; mol wt 210.23. C 57.13%, H 6.71%, N 13.33%, O 22.83%. Prepn: U.S. pat. **1,444,802** (1923). Toxicity data: H. H. Frey, *Arzneimittel-Forsch.* **12,** 389 (1962).

Slightly bitter crystals, mp 140-141.5°. Almost insoluble in water, petr ether, aliphatic hydrocarbons. Sol in alcohol, chloroform, ether, acetone, benzene, glacial acetic acid, also in solns of fixed alkali hydroxides. A satd aq soln is acid to litmus. LD_{50} i.p. in mice: 200 mg/kg (Frey).
Sodium salt, $C_{10}H_{13}N_2NaO_3$, *aprobarbital sodium, sodium 5-allyl-5-isopropylbarbiturate, Alurate Sodium, Somnipron.* Hygroscopic powder. Slightly bitter taste. Very soluble in water; slightly sol in alc. Practically insol in ether. Aq solns are alkaline to litmus.
Caution: May be habit forming. This is a controlled substance (depressant) listed in the U.S. Code of Federal Regulations, Title 21 Parts 329.1 and 1308.13 (1987).
THERAP CAT: Sedative, hypnotic.

783. Apronalide. *N-(Aminocarbonyl)-2-(1-methylethyl)-4-pentenamide; (2-isopropyl-4-pentenoyl)urea;* allylisopropylacetylurea; Apronal; Isodormid; Sedormid. $C_9H_{16}N_2O_2$; mol wt 184.23. C 58.67%, H 8.75%, N 15.21%, O 17.37%. Methods of prepn: Slotta, *Grundriss der modernen Arzneistoff-Synthese* (Stuttgart, 1931) p 35.

$$CH_2=CHCH_2CHCONHCONH_2$$
$$|$$
$$CH(CH_3)_2$$

Crystals. Practically tasteless. mp 194°. One gram dissolves in 3000 ml cold water, in 210 ml boiling water. One part dissolves in 10 parts alc, 75 parts ether. Decolorizes $KMnO_4$ solns.
Caution: This substance may be habit forming and is listed in the U.S. Code of Federal Regulations, Title 21 Part 329.1 (1987).
THERAP CAT: Sedative, hypnotic.

784. Aprotinin. *Pancreatic basic trypsin inhibitor;* pancreatic trypsin inhibitor (Kunitz); Bayer A-128; Riker 52G; RP 9921; Antagosan; Antikrein; Fosten; Iniprol; Kir Richter; Onquinin; Repulson; Trasylol; Trazinin; Zymofren. $C_{284}H_{432}N_{84}O_{79}S_7$; mol wt 6511.83. C 52.38%, H 6.69%, N 18.07%, O 19.41%, S 3.45%. A kallikrein inhibitor which also inhibits plasmin, trypsin, chymotrypsin and various intracellular proteases. Polypeptide found in tissues and blood, but in highest concentration in bovine parotid gland, pancreas and lung. Initial description: H. Kraut *et al., Z. Physiol. Chem.* **192,** 1 (1930). Isoln from bovine pancreas: M. Kunitz, J. H. Northrup, *J. Gen. Physiol.* **19,** 991 (1936). Isoln from bovine parotid glands: H. Kraut, R. Körbel, U.S. pat. **2,890,986** (1959 to Bayer). Aprotinin is a single chain polypeptide containing 58 amino acids of known sequence: B. Kassell *et al., Biochem. Biophys. Res. Commun.* **18,** 255 (1965); F. A. Anderer, S. Hörnle, *Z. Naturforsch.* **20b,** 457, 462 (1965); B. Kassell, M. Laskowski, *Biochem. Biophys. Res. Commun.* **20,** 463 (1966). Two-dimensional ^{1}H-NMR studies: K. Nagayama, K. Wüthrich, *Eur. J. Biochem.* **114,** 365 (1981); G. Wagner *et al., ibid.* 375. General review: I. Trautschold *et al., Biochem. Pharmacol.* **16,** 59-72 (1957). Review of therapeutic action: G. Haberland, R. McConn, *Fed. Proc.* **38,** 2760-2767 (1979). Review of use as a proteolytic inhibitor in radioimmunoassays of polypeptide hormones: E. S. Zyzner, *Life Sci.* **28,** 1861-1866 (1981).

Review of biochemistry and applications: H. Fritz, G. Wunderer, *Arzneimittel-Forsch.* **33**, 479-494 (1983).

```
         ┌──66──            ┌──36──
Arg-Pro-Asp-Phe-Cys-Leu-Glu-Pro-Pro-Tyr-Thr-Gly-Pro-Cys-Lys-Ala-
 1         5                      14      16

                                ┌──51──
Arg-Ile-Ile-Arg-Tyr-Phe-Tyr-Asn-Ala-Lys-Ala-Gly-Leu-Cys-Gln-Thr
 17             23                      30      32

       14──┐
Phe-Val-Tyr-Gly-Gly-Cys-Arg-Ala-Lys-Arg-Asn-Asn-Phe-Lys-Ser-Ala-
 33             36                              48

   30──┐        5──┐
Glu-Asp-Cys-Met-Arg-Thr-Cys-Gly-Gly-Ala
 49      51          55          58
```

uv max (pH 5.9): 280 nm. Isoelec pt pH 10.5. Stable in neutral or acid media at high temp. Irreversible changes in molecular structure occur in strongly alkaline media (pH > 12). Partially and reversibly denatured on treatment with 8*M* urea. May be kept at room temp in physiological saline soln for > 1 yr without detrimental effects. LD_{50} i.v. in mice: 2.5 million kallikrein inhibitor units/kg (Trautschold).

USE: Proteolytic inhibitor in radioimmunoassays of polypeptide hormones.

THERAP CAT: Enzyme inhibitor (protease).

785. APSAC. Anisoylated plasminogen streptokinase activator complex; anistreplase; BRL-26921; Eminase. Binary complex of streptokinase and human plasminogen in which the catalytic site is reversibly blocked by a *p*-methoxybenzoyl group without affecting the fibrin-binding site. Prepn and fibrinolytic activity: R. A. G. Smith, J. B. Winchester, **Eur. pat. Appl. 28,489** (1981 to Beecham). Deacylation of the fibrin-bound complex initiates thrombolysis: R. A. G. Smith *et al., Nature* **290**, 505 (1981); D. Matsuo *et al., Thromb. Res.* **24**, 347 (1981). Effect on hemostasis in humans: D. H. Staniforth *et al., Eur. J. Clin. Pharmacol.* **24**, 751 (1983). Pharmacokinetics: M. Been *et al., Int. J. Cardiol.* **11**, 53 (1986). Evaluation of thrombolytic efficacy in myocardial infarction: W. Kasper *et al., Am. J. Cardiol.* **58**, 418 (1986); in major pulmonary embolism: J. H. Brett *et al., Aust. N.Z. J. Med.* **17**, 77 (1987). Controlled clinical trial in acute myocardial infarction: S. Ikram *et al., Brit. Med. J.* **293**, 786 (1986). Review of mechanism of action, pharmacology and therapeutic use: J. P. Monk, R. C. Heel, *Drugs* **34**, 25-49 (1987).

THERAP CAT: Thrombolytic.

786. Apyrase. An enzyme found in plants, especially potatoes. Cofactor: Ca^{2+}. Hydrolyzes acid anhydride links. Substrate: Adenosine triphosphate, adenosine diphosphate (giving adenylate and phosphate). *Ref:* Krishnan, *Arch. Biochem.* **20**, 272 (1949). Partial purification and properties of potato apyrase: Liebecq *et al., Bull. Soc. Chim. Biol.* **45**, 573 (1963).

787. Aquocobalamin. *Cobinamide dihydroxide, monohydrate, dihydrogen phosphate ester, mono(inner salt), 3'-ester with 5,6-dimethyl-1-α-D-ribofuranosyl-1H-benzimidazole;* vitamin B_{12b}; aquocobamide; α-(5,6-dimethylbenzimidazolyl)aquocobamide.

Hydroxide form *see* Hydroxocobalamin.

788. Arabinose. L-Arabinose; pectin sugar. $C_5H_{10}O_5$; mol wt 150.13. C 40.00%, H 6.71%, O 53.29%. Widely distributed in plants, usually in the form of complex polysaccharides. Also in mycobacteria. Isoln from mesquite gum: Anderson, Sands, *J. Am. Chem. Soc.* **48**, 3172 (1926); *Org. Syn. coll. vol. I,* 16 (2nd ed., 1941); White, *J. Am. Chem. Soc.* **69**, 715 (1947); from western red cedar *(Thuja plicata* Don., *Cupressaceae):* Anderson, Erdtman, *ibid.* **71**, 2927 (1949); from sapote gum: White, *ibid.* **75**, 257 (1953); from heartwood of port orford cedar *(Chamaecyparis lawsoniana* (Murr.) Parl, *Cupressaceae):* Kritchevsky, Anderson, *ibid.* **77**, 3391 (1955). Structure: Wolfrom, Christman, *ibid.* **58**, 39 (1936). Synthesis: Hough, Jones, *J. Chem. Soc.* **1951**, 1122.

β-L-arabinose

Orthorhombic bisphenoidal crystals, mp 157-160°. Shows mutarotation. $[\alpha]_D^{12}$ +173° (6 min) → $[\alpha]_D^{20}$ +105.1° (22½ hrs c = 3). One gram dissolves in about 1 ml water, about 250 ml 90% alc. Ka at 17°, -3.7×10^{-13}. Reduces Fehling's soln. Forms furfurol on heating to 200° in closed tube also contg water.

USE: As culture medium for certain bacteria.

789. Arabitol. *Arabinitol;* 1,2,3,4,5-pentanepentol; arabite. $C_5H_{12}O_5$; mol wt 152.15. C 39.47%, H 7.95%, O 52.58%. D-Form obtained by reduction of D-arabinose or D-lyxose with sodium amalgam: Ruff, *Ber.* **32**, 555 (1899); Ruff, Ollendorff, *Ber.* **33**, 1802 (1900); Bertrand, *Bull. Soc. Chim.* [3] **15**, 593 (1896). Production by fermentation from molasses using *Saccharomyces rouxii* and *Saccharomyces mellis:* Lavin, Holloway, **U.S. pat. 2,934,474** (1960 to Comm. Solvents). L-Form obtained by reduction of L-arabinose with sodium amalgam: Kiliani, *Ber.* **20**, 1234, 1571 (1887). DL-Form obtained from equal parts of D- and L-arabitol: Ruff, *loc. cit.* Synthesis: Lespian, *Compt. Rend.* **206**, 1773 (1938).

```
        CH₂OH              CH₂OH
    HO─C─H             H─C─OH
    H─C─OH             HO─C─H
    H─C─OH             HO─C─H
        CH₂OH              CH₂OH

    D-Arabitol          L-Arabitol
    ═══════             ═══════
```

D-Form, big prismatic crystals. Sweet taste. mp 103°. $[\alpha]_D^{20}$ +7.7° (c = 9.26 in satd borax soln). Freely sol in water. One part dissolves in 48 parts of 90% alc at 12°.

Pentanitrate, $C_5H_7N_5O_{15}$, sirup, sol in alcohol, ether, acetone. Strongly reduces Fehling's soln.

L-Form, wart-like crystals, sweet taste. mp 102°. Weakly levorotatory in satd borax soln. Freely sol in water and in boiling 90% alc. One part dissolves in 46 parts of 90% alc at 12°. Does not reduce Fehling's soln.

DL-Form, prisms from 90% alc. mp 105-106°. One part dissolves in 66 parts of 90% alc at 12°.

790. D-Araboflavin. *1-Deoxy-1-(3,4-dihydro-7,8-dimethyl-2,4-dioxobenzo[g]pteridin-10(2H)-yl)-D-arabinitol; 7,8-dimethyl-10-(arabino-2,3,4,5-tetrahydroxypentyl)benzo-[g]pteridine-2,4(3H,10H)-dione;* 6,7-dimethyl-9-D-arabofla-vin. $C_{17}H_{20}N_4O_6$; mol wt 376.36. C 54.25%, H 5.36%, N 14.89%, O 25.51%. An antagonist of riboflavine. Prepd from D-(2-amino-4,5-dimethylphenyl)arabinamine and alloxan: Kuhn, Weygand, *Ber.* **68**, 1286 (1935); Sahashi *et al., Bull. Inst. Phys. Chem. Res.* (Tokyo) **24**, 72 (1948), *C.A.* **42**, 5458 (1948).

Orange-yellow needles from dil acetic acid. Bitter taste, mp 302-303°. $[\alpha]_D^{20}$ +78.6° (c = 0.509 in 0.1N NaOH); $[\alpha]_D^{20}$ −441° (c = 0.253 in 0.2N NaOH satd with borax).

Tetraacetate, $C_{25}H_{28}N_4O_{10}$, flat yellow prisms from ethyl acetate, mp 221-222°.

791. Arachidic Acid. *Eicosanoic acid;* arachic acid. $C_{20}H_{40}O_2$; mol wt 312.52. C 76.86%, H 12.90%, O 10.24%. $CH_3(CH_2)_{18}COOH$. Fatty acid found in peanut oil, vegetable and fish oils, etc. Also obtained by hydrogenation of arachidonic acid: Ege *et al.*, *J. Am. Chem. Soc.* **83**, 3080 (1961). Synthesis: Linstead *et al.*, *J. Chem. Soc.* **1955**, 1097. *Reviews:* A .W. Ralston, *Fatty Acids and Their Derivatives* (Wiley & Sons, New York, 1948) pp 43-46; K. S. Markley, *Fatty Acids*, Part I (Interscience, New York, 2nd ed., 1960) pp 164-167, 398-400.

Crystals from alc, mp 75.5°. bp_{760} about 328° with some decompn; $bp_{1.0}$ 205°. d_4^{100} 0.8240. n_D^{100} 1.4250. Practically insoluble in water; sparingly sol in cold water; freely in hot abs alcohol, benzene, chloroform, ether, petr ether.

Methyl ester, $C_{21}H_{42}O_2$, mp 47°; $bp_{1.0}$ 180°.
Ethyl ester, $C_{22}H_{44}O_2$, mp 41°; $bp_{0.3}$ 177°.

792. Arachidonic Acid. *5,8,11,14-Eicosatetraenoic acid.* $C_{20}H_{32}O_2$; mol wt 304.46. C 78.89%, H 10.60%, O 10.51%. An essential fatty acid, *q.v.*, and a precursor in the biosynthesis of prostaglandins, thromboxanes, and leukotrienes, *q.q.v.* Structure: Mowry *et al.*, *J. Biol. Chem.* **142**, 679 (1942); Arcus, Smedley-Maclean, *Biochem. J.* **37**, 1 (1943). Occurs in liver, brain, glandular organs, and depot fats of animals, in small amounts in human depot fats, and is a constituent of animal phosphatides. Isolation from liver lipids: Brown, *J. Biol. Chem.* **80**, 455 (1928); from beef suprarenal phosphatides: Ault, Brown, *ibid.* **107**, 615 (1934); Shinowara, Brown, *ibid.* **134**, 331 (1940). *See also* Dolby *et al.*, *Biochem. J.* **34**, 1422 (1940). Synthesis from 2-propargyloxytetrahydropyran and 1-heptyne: Goldberg, Rachlin, U.S. pat. **2,934,570** (1960 to Hoffmann-La Roche); Rachlin *et al.*, *J. Org. Chem.* **26**, 2688 (1961). Alternate route: Osbond, Wickens, *Chem. & Ind. (London)* **1959**, 1288; Ege *et al.*, *J. Am. Chem. Soc.* **83**, 3080 (1961). *Reviews:* K. S. Markley, *Fatty Acids*, Part I (Interscience, New York, 2nd ed., 1960) pp 164-167, 398-400; T. K. Schaaf, *Ann. Rep. Med. Chem.* **12**, 182-190 (1977); B. B. Weksler, *N. Engl. Soc. Allergy Proc.* **2**, 56-61 (1981); N. A. Nelson *et al.*, *Chem. & Eng. News* **60**, 30-44 (Aug. 16, 1982). Series of articles on metabolism, role in inflammation and therapeutic implications: *Drugs* **33**, Suppl. 1, 2-66 (1987).

Liquid. mp −49.5°. Neutralization value 184.20. Iodine value 333.50. n_D^{20} 1.4824.

Methyl ester, $C_{21}H_{34}O_2$, bp_2 200-205°, $bp_{0.001}$ 127-128°. n_D^{20} 1.4797 to 1.4810.

THERAP CAT: Nutrient (essential fatty acid).

THERAP CAT (VET): With linoleic and linolenic acids, in eczema and dermatitis in dogs and swine.

793. Aralia. Spikenard; American spikenard; spignet; petty-morrel; spice berry. Dried rhizome and roots of *Aralia racemosa* L., Araliaceae. *Habit.* Northeastern U.S. *Constit.* Starch, pectin, sugar, resin. The active constituent appears to be a volatile oil of unknown constitution.

Ingredient of Compound White Pine Syrup: *N. F. XI*.

794. Aramite®. *Sulfurous acid 2-chloroethyl 2-[4-(1,1-dimethylethyl)phenoxy]-1-methylethyl ester; sulfurous acid 2-(p-tert-butylphenoxy)-1-methylethyl 2-chloroethyl ester;* 2-chloroethyl 1-methyl-2-(p-tert-butylphenoxy)ethyl sulfite; 2-(p-tert-butylphenoxy)isopropyl 2-chloroethyl sulfite; β-chloroethyl β-(p-tert-butylphenoxy)-α-methylethyl sulfite; compound 88R; ENT 16519; Aratron; Niagaramite; Ortho-Mite. $C_{15}H_{23}ClO_4S$; mol wt 334.87. C 53.80%, H 6.92%, Cl 10.59%, O 19.11%, S 9.58%. Prepn: Harris *et al.*, U.S. pat. **2,529,494** (1950 to U.S. Rubber). Toxicology: B. L. Oser, M. Oser, *Toxicol. Appl. Pharmacol.* **2**, 441 (1960).

Carcinogenicity studies: *eidem, ibid.* **4**, 70 (1962); Truhaut *et al.*, *C.R. Acad. Sci., Ser. D* **281**, 599 (1975).

Liquid. d 1.1450 to 1.1620. mp −31.7°. $bp_{0.1}$ 175°; $bp_{7.0}$ 200-210°. Vapor tension at 25°: < 10 mm. n_D^{20} 1.5100 to 1.5118; n_D^{27} 1.5075. Practically insol in water. Misc with many organic solvents. Soly in petroleum oils decreases rapidly with decreasing temperatures. LD_{50} orally in rats: 3.90±0.28 g/kg (Oser, Oser).

Caution: In undiluted form may cause skin irritation. Large doses may cause CNS depression. This substance may reasonably be anticipated to be a carcinogen: *Fourth Annual Report on Carcinogens* (NTP 85-002, 1985) p 26.

USE: Miticide.

795. Araroba. Goa powder; crude chrysarobin; Brazil powder; ringworm powder; "Arariba". Found in cavities in the trunk of *Andira araroba*, Aguiar (*Vouacapoua araroba* [Aguiar] Lyons), *Leguminosae* and freed as much as possible from fragments of wood. *Habit.* In damp forests of Brazil and Bahia. *Constit.* About 50% chrysarobin; chrysophanic acid, gum, about 2% resin, ararobinol—said to be the methyl ether of emodin.

Brown-yellow, slightly cryst, rough, odorless powder; becomes brownish on exposure to air.

THERAP CAT: Dermatologic.

796. Arbaprostil. *(5Z,11α,13E,15R)-11,15-Dihydroxy-15-methyl-9-oxoprosta-5,13-dien-1-oic acid; 15(R)-15-methylprostaglandin E₂;* 15(R)-15-methyl-PGE_2; (E,Z)-(1R,2R,3R)-7-[3-hydroxy-2-[(3R)-(3-hydroxy-3-methyl-1-octenyl)]-5-oxocyclopentyl]-5-heptenoic acid; U42842; Arbacet. $C_{21}H_{34}O_5$; mol wt 366.50. C 68.82%, H 9.35%, O 21.83%. Antisecretory, cytoprotective derivative of prostaglandin E_2, *q.v.* The introduction of a methyl group at the C-15 position of PGE_2 interferes with inactivation by 15-hydroxyprostaglandin dehydrogenase. Synthetic studies: G. Bundy *et al.*, *Ann. N.Y. Acad. Sci.* **180**, 76 (1971); E. W. Yankee, G. Bundy, *J. Am. Chem. Soc.* **94**, 3651 (1972). Total synthesis of arbaprostil and its S-epimer; epimerization: E. W. Yankee *et al.*, *ibid.* **96**, 5865 (1974). Prepd, not claimed: G. L. Bundy, U.S. pats. **3,728,382; 3,904,679** (1973, 1975 both to Upjohn). Arbaprostil is a prodrug for its S-epimer; epimerization is promoted by acid. HPLC separation of epimers: R. K. Lustgarten, *J. Pharm. Sci.* **65**, 1533 (1976); G. E. Peng, V. K. Sood, *J. Liquid Chromatog.* **6**, 1499 (1983). Kinetics of epimerization: M. V. Merritt, G. E. Bronson, *ibid.* **100**, 1891 (1978); *eidem, ibid.* **102**, 346 (1980). Pharmacology: J. R. Weeks *et al.*, *J. Pharmacol. Exp. Ther.* **186**, 67 (1973). Clinical study in duodenal ulcers: G. Vantrappen *et al.*, *Gastroenterology* **83**, 357 (1982); in prevention of aspirin-induced gastric mucosal injury: D. A. Gilbert *et al.*, *ibid.* **86**, 339 (1984).

(15S)-Methyl ester, $C_{22}H_{36}O_5$, colorless oil, $[\alpha]_D$ −79° (c = 1.3 in chloroform). uv max (basic ethanol): 278 nm (ε = 25250).

(15R)-Methyl ester, $C_{22}H_{36}O_5$, colorless oil, $[\alpha]_D$ −74° (c = 1.0 in chloroform). uv max (basic ethanol): 278, 343sh nm (ε = 25200, 699).

THERAP CAT: Anti-ulcerative.

797. Arbekacin. *(S)-O-3-Amino-3-deoxy-α-D-glucopyranosyl-(1 → 6)-O-[2,6-diamino-2,3,4,6-tetradeoxy-α-D-erythro-hexopyranosyl-(1 → 4)]-N¹-(4-amino-2-hydroxy-1-oxobutyl)-2-deoxy-D-streptamine;* 1-N-[(S)-4-amino-2-hydroxybutyryl]dibekacin; 1-N-[(S)-4-amino-2-hydroxybutyryl]-3',4'-dideoxykanamycin B; habekacin; AHB-DKB; HABA-DKB; HBK; 1665-RB; $C_{22}H_{44}N_6O_{10}$; mol wt 552.63. C 47.82%, H 8.02%, N 15.21%, O 28.95%. Derivative of kanamycin B, *q.v.*, closely related to dibekacin, *q.v.* Prepn and antibacterial activity: S. Kondo *et al.*, *J. Antibiot.* **26**, 412 (1973); H. Umezawa *et al.*, *Ger. pat.* **2,350,169**; *eidem*, U.S. pat. **4,107,424** (1974, 1978 both to Microbiochem. Res. Found.). Mechanism of action: N. Tanaka *et al.*, *Antimicrob. Ag. Chemother.* **24**, 797 (1983); K. Matsunaga *et al.*, *J. Antibiot.* **37**, 596 (1984). Toxicology: J. Stewens *et al.*, *Arzneimittel-Forsch.* **35**, 1440 (1985); M. Kurebe *et al.*, *ibid.* **36**, 1511 (1986). HPLC determn in biological fluids: I. Komiya *et al.*, *Chemotherapy* **34**, Suppl. 1, 82 (1986). Series of articles on antibacterial activity, pharmacodynamics, pharmacokinetics, and clinical evaluations: *ibid.*, pp 1-670.

Dicarbonate, $C_{24}H_{48}N_6O_{16}$, colorless, crystalline powder, mp 178° (dec). $[\alpha]_D^{24}$ +86.8° (c = 0.77 in water). LD_{50} i.v. in mice: >150 mg/kg (Umezawa, 1978).

THERAP CAT: Antibacterial.

798. Arborescin. *5,6,6a,7,9a,9b-Hexahydro-1,4a,7-trimethyl-3H-oxireno[8,8a]azuleno[4,5-b]furan-8(4aH)-one; 1,10-epoxy-6β-hydroxy-1β,5β,7α-guiai-3-en-12-oic acid γ-lactone.* $C_{15}H_{20}O_3$; mol wt 248.31. C 72.55% H 8.12%, O 19.33%. Ancient Greek contraceptive. Isoln from *Artemisia arborescens* L. and *Matricaria globifera* (Thunb.) Druce, *Compositae*: Meisels, Weizmann, *J. Am. Chem. Soc.* **75**, 3865 (1953); Cekan *et al.*, *Coll. Czech. Chem. Commun.* **25**, 2553 (1960). Structure: Bates *et al.*, *Tetrahedron Letters* **1963**, 1127; M. Suchy *et al.*, *Coll. Czech. Chem. Commun.* **29**, 1829 (1964). Synthesis and revised structure: M. Ando *et al.*, *Chem. Letters* **1978**, 727.

Crystals from ethanol, mp 140-142°. $[\alpha]_D^{20}$ +64° (c = 0.72 in chloroform).

799. Arbutin. *4-Hydroxyphenyl-β-D-glucopyranoside;* hydroquinone-*β*-D-glucopyranoside; hydroquinone glucose; arbutoside; ursin; Uvasol. $C_{12}H_{16}O_7$; mol wt 272.25. C 52.94%, H 5.92%, O 41.14%. From dried leaves of *Bergenia crassifolia* (L.) Fritsch, *Saxifragaceae*: Tschitschibabin *et al.*, *Ann.* **479**, 303 (1930); from leaves of blueberry, cranberry, and pear trees (*Pyrus communis* L., *Rosaceae*): Urban, Rogowski, *Arch. Exp. Pathol. Pharmakol.* **211**, 194 (1950); Friedrich, *Pharmazie* **15**, 650 (1960); from leaves of cowberry (*Vaccinium vitis-idaea* L., *Ericaceae*) and bearberry (*Arctostaphylos uva-ursi* Spreng., *Ericaceae*): Friedrich, *Naturwiss.* **48**, 304 (1961); Kraus, *Pharmazie* **19**, 41 (1964). Synthesis from acetobromglucose and hydroquinone: Mannich, *Arch. Pharm.* **250**, 547 (1912); from β-D-glucose pentaacetate + hydroquinone monobenzyl ether in the presence of $POCl_3$: Jarrett, U.S. pat. **3,201,385** (1965 to Polaroid). Frequently occurs together with methylarbutin in plants, particularly those of the family *Ericaceae*.

Forms an unstable form which appears to be converted to the stable form by melting. Needles from hot ethyl acetate, unstable form mp 165°; stable form mp 199.5-200°: Lindpainter, *Arch. Pharm.* **277**, 398 (1940). $[\alpha]_D^{25}$ −64° (c = 3). Sol in water and in alc. Very hygroscopic. Easily hydrolyzed by dil acids, or by emulsin yielding 1 mol D-glucose and 1 mol hydroquinone. Technique of fermentative degradation: Helferich, Reischel, *Ann.* **533**, 278 (1938). Gallotannin prevents enzymes such as β-glucosidase from splitting arbutin which explains why crude plant extracts are more effective medicinally than pure arbutin: Friedrich, *Arch. Pharm.* **288**, 583 (1955). Forms a complex with hexamethylenetetramine which may be used to separate it from methylarbutin.

USE: Stabilizer for color photographic images.

THERAP CAT: Diuretic, urinary anti-infective.

800. Areca. Betel nuts; pinang. Nuts (seeds) of *Areca catechu* L., *Palmaceae*. *Habit.* East Indies. *Constit.* Arecoline, arecaidine, guvacine, guvacoline, arecolidine, choline, about 15% red tannin, about 14% fat. *Reviews*: L. Marion "The Alkaloids of Areca Nut" in Manske-Holmes, *The Alkaloids* Vol. I (Academic Press, New York, 1950) pp 171-175; Raghavan, Baruah, *Econ. Bot.* **12**, 315 (1958).

Hard and heavy; round-conical and depressed at base. Extern. brown, mottled with fawn color; intern. brownish-red with whitish veins; astringent taste; the fresh nuts have a faint, cheese-like odor.

THERAP CAT: Anthelmintic.

THERAP CAT (VET): Has been used for tapeworms, ascarids.

801. Arecaidine. *1,2,5,6-Tetrahydro-1-methyl-3-pyridinecarboxylic acid; 1,2,5,6-tetrahydro-1-methylnicotinic acid;* arecaine; methylguvacine. $C_7H_{11}NO_2$; mol wt 141.17. C 52.81%, H 8.23%, N 8.80%, O 30.15%. From seeds of *Areca catechu* L., *Palmaceae* (betel nuts). Synthesis: Wohl, Johnson, *Ber.* **40**, 4712 (1907); Hess, Leibbrandt, *ibid.* **51**, 806 (1918); Freudenberg, *ibid.* **976**; Maurit, Preobrazhenskii, *Zh. Obshch. Khim.* **28**, 968 (1958), *C.A.* **52**, 17263 (1958); Merck, Ger. pat. **485,139**, *C.A.* **24**, 919 (1930).

Plates from dil alc, dec 232° (after drying at 102°). pH of 0.1 molar soln 5.6. Freely soluble in water, and dilute alc; almost insol in abs alc, chloroform, ether, benzene.

Hydrochloride, $C_7H_{11}NO_2.HCl$, mp 251°; needles, dec 263° (rapid heating). Freely sol in water.

Hydrobromide, $C_7H_{11}NO_2.HBr$, crystals from methanol, dec 249°.

802. Arecoline. *1,2,5,6-Tetrahydro-1-methyl-3-pyridinecarboxylic acid methyl ester; methyl 1,2,5,6-tetrahydro-1-methylnicotinate; methyl 1-methyl-Δ³,⁴-tetrahydro-3-pyridinecarboxylate; methyl N-methyltetrahydronicotinate;* arecaline; arecholine; methylarecaidin. $C_8H_{13}NO_2$; mol wt 155.19. C 61.91%, H 8.44%, N 9.03%, O 20.62%. Alkaloid from seeds of the betel nut palm *Areca catechu* L., *Palmaceae* (catechu): E. Johns, *Arch. Pharm.* **229**, 673 (1891). By synthesis: F. Chemnitius, *J. Prakt. Chem.* **117**, 147 (1926); Ger. pat. **485,139**; Mannich, *Ber.* **75B**, 1480 (1942); Maurit, Preobrazhenskii, *Zh. Obshch. Khim.* **28**, 968 (1958), *C.A.* **52**, 17263 (1958); Knox, U.S. pat. **2,506,458** (1950 to Nopco). Improved synthesis from nicotinic acid: I. A. Kozello *et al.*, *Khim.-Farm. Zh.* **10**, 90 (1976), *C.A.* **86**, 171205a (1977). Effect on human serial learning: N. Sitaram *et al.*, *Science* **201**, 274 (1978). *Review*: R. B. Burrows, *Progress in Drug*

Research vol. 17, E. Jucker, Ed. (Birkhaüser Verlag, Basel, 1973) pp 108-210.

Oily liquid, bp 209°, bp$_7$ 92-93°, bp$_{12}$ 105°. Volatile with steam. Strong base. pK 6.84. n_D^{20} 1.4302. d^{20} 1.0495. Miscible with water, alc, ether. Soluble in chloroform. LD$_{50}$ in mice, dogs (mg/kg): 100, 5 s.c. (Burrows).

Hydrochloride, $C_8H_{14}ClNO_2$, needles, mp 158°. Sol in water and alc.

Hydrobromide, $C_8H_{14}BrNO_2$, bitter, optically inactive crystals, mp 169-171°. One gram dissolves in about 1 ml water, 10 ml alc, 2 ml boiling alc; slightly soluble in chloroform, ether. The aq soln is practically neutral.

THERAP CAT: Has been used as anthelmintic (Cestodes).

THERAP CAT (VET): Teniacide for dogs, poultry. Cathartic in horses. Ruminatoric in cattle.

803. Arecoline *p*-Stibonobenzoic Acid. *p*-Stibonobenzoic acid methyl 1,2,5,6-tetrahydro-1-methylnicotinate; 4-carboxyphenylstibonic acid *N*-methyltetrahydropyridinecarboxylic acid methyl ester; *N*-methyltetrahydromethylnicotinate *p*-carboxyphenylstibonic acid; Anthelin. $C_{15}H_{20}$-NO$_7$Sb; mol wt 448.64. C 40.28%, H 4.49%, N 3.12%, O 24.96%, Sb 27.14%. Prepn: Kartsonis, Austin, U.S. pat. 2,557,353 (1951 to Jensen-Salsbery Labs.).

Also ingredient of *Anthol*.
THERAP CAT (VET): Anthelmintic.

804. Argatroban. *1-[5-[(Aminoiminomethyl)amino]-1-oxo-2-[[(1,2,3,4-tetrahydro-3-methyl-8-quinolinyl)sulfonyl]amino]pentyl]-4-methyl-2-piperidinecarboxylic acid; (2R,-4R)-4-methyl-1-[N^2-[(1,2,3,4-tetrahydro-8-quinolinesulfonyl)-L-arginyl]-2-piperidinecarboxylic acid; (2R,4R)-4-methyl-1-[(S)-N^2-[[(R,S)-1,2,3,4-tetrahydro-3-methyl-8-quinolinyl]sulfonyl]arginyl]pipecolic acid; argipidine; MQPA.* $C_{23}H_{36}N_6O_5S$; mol wt 508.64. C 54.31%, H 7.13%, N 16.52%, O 15.73%, S 6.30%. Synthetic thrombin inhibitor. Prepn of argatroban and stereoisomers: R. Kikumoto *et al.*, **Eur. pat. Appl. 8,746**; S. Okamoto *et al.*, **U.S. pat. 4,258,192** (1980, 1981 both to Mitsubishi Chem. Ind.); S. Okamoto *et al.*, *Biochem. Biophys. Res. Commun.* **101**, 440 (1981). Comparison with heparin, *q.v.*, of antithrombotic effect in animals: T. Kumada, Y. Abiko, *Thromb. Res.* **24**, 285 (1981). Stereoselective inhibition of thrombin: R. Kikumoto *et al.*, *Biochemistry* **23**, 85 (1984). Clinical evaluation in hemodialysis: K. Ota *et al.*, *Proc. Eur. Dial. Transplant. Assoc.* **20**, 144 (1983); in disseminated intravascular coagulation: K. Kumon *et al.*, *Crit. Care Med.* **12**, 1039 (1984).

Crystals from ethanol, mp 188-191°.
Monohydrate, *DK-7419, MCI-9038, MD-805, OM-805,*

Novastan, Slonnon. Crystals from aq ethanol, mp 176-180°. $[\alpha]_D^{27}$ +76.1° (c = 1 in 0.2N HCl).
THERAP CAT: Antithrombotic.

805. Arginine. Arg (IUPAC abbrev.); 2-amino-5-guanidinovaleric acid. $C_6H_{14}N_4O_2$; mol wt 174.20. C 41.36%, H 8.10%, O 18.37%, N 32.16%. An amino acid classified as essential with respect to its growth effect in rats. The physiologically active L(+)-form is readily obtained by hydrolysis of proteins. Isoln from gelatine: Brand, Sandberg, *Org. Syn. coll. vol. II*, 49 (1943). In industrial practice it is precipitated from gelatine hydrolyzate as the flavianate (2,4-dinitro-1-naphthol-7-sulfonate); *cf.* Cox, *J. Biol. Chem.* **78**, 475 (1928); Vickery, *ibid.* **132**, 325 (1940). L-Arginine may be obtained from L-ornithine and cyanamide in aq soln in the presence of some Ba(OH)$_2$: Schulze, Winterstein, *Ber.* **32**, 3191 (1899); *Z. Physiol. Chem.* **34**, 134 (1902).

L-arginine

L-Arginine (the natural product), prisms containing 2 mol H$_2$O from water; anhydrous monoclinic plates from 66% alcohol. The dihydrate becomes anhyd at 105°, browns at 230°. Dec 244°. $[\alpha]_D^{20}$ +26.9° (c = 1.65 in 6.0N HCl); $[\alpha]_D^{20}$ +12.5° (c = 3.5 in water); $[\alpha]_D^{20}$ 11.8° (c = 0.87 in 0.5N NaOH). pK$_1$ 2.18; pK$_2$ 9.09; pK$_3$ 13.2. Absorption spectrum: Castille, Ruppol, *Bull. Soc. Chim. Biol.* **10**, 643 (1928); *Chem. Zentr.* **1928, II**, 622. The satd aq soln contains 15% (w/w) L-arginine at 21°. Sparingly sol in alc. Insol in ether. Strongly alkaline, its aq solns absorb CO$_2$ from the air.

Hydrochloride, $C_6H_{15}ClN_4O_2$, *Argivene, R-gene*. Plates, prisms from alc. Sinters at 218°, solidifies again at 225°. Dec 235°. $[\alpha]_D^{20}$ +12.0° (c = 4). $[\alpha]_D^{21}$ +21.9° (c = 12 in dil HCl). Sol in water, slightly sol in hot alc.
THERAP CAT: Ammonia detoxicant (hepatic failure); diagnostic aid (pituitary function).

806. Arginine Glutamate. *Glutamic acid compd with L-arginine*; Ginamate; Glutargin; Modumate. $C_{11}H_{23}N_5O_6$; mol wt 321.33. C 41.11%, H 7.22%, N 21.80%, O 29.88%. [H$_2$NC(NH)]HNCH$_2$CH$_2$CH$_2$CH(NH$_2$)COOH.HOOCCH$_2$-CH$_2$CH(NH$_2$)COOH. Prepd from L-arginine and L-glutamic acid: Barker, Chang, U.S. pat. 2,851,482 (1958 to General Mills). Activity studies: Weisburger *et al.*, *Toxicol. Appl. Pharmacol.* **14**, 163 (1969); Yamamoto, Weisburger, *Life Sci.* **9(II)**, 285 (1970). Crystal and molecular structure: T. N. Bhat, M. Vijayan, *Acta Crystallogr.* **B33**, 1754 (1977).

Crystals, dec 193-194.5°. Supplied as a 25% (w/v) soln in water for injection; each 100 ml represents 13.5 g of arginine and 11.5 g of glutamic acid.
THERAP CAT: Ammonia detoxicant (hepatic failure).

807. Argol. Argilla vini; argil; arcilla; Weinstein; wine lees; crude cream of tartar; crude potassium bitartrate. Formed in the secondary fermentation of grapes for wine. Contains over 40% tartaric acid, potassium bitartrate, calcium. *Ref*: R. Pasternack in Kirk-Othmer *Encyclopedia of Chemical Technology* (Interscience, New York, 1954) vol. 13, p 649.

Gray or reddish-brown crystalline crusts.
USE: Mordant; manufacturing tartaric acid; vinegar from malt; in fertilizers.

808. Argon. Ar; at. wt 39.948; at. no. 18. Three stable isotopes: 36 (0.337%); 38 (0.063%); 40 (99.600%); artificial, radioactive isotopes: 33; 35; 37; 39; 41; 42. Abundance in earth's crust: 4 × 10^{-4}%; concentration in the atmosphere: 0.93% by vol; cosmic abundance: ~1.5 × 10^5 atom/10^6 atoms of Si. Elemental, monoatomic, gaseous constituent of air, discovered by Rayleigh and Ramsay in 1894. Although molecular ions, hydrates and clathrates of argon have been observed, it should be considered a "noble", chemically inert gas, due to its electronic structure. The outer *p* subshell is entirely filled: $1s^2 2s^2 2p^6 3s^2 3p^6$. Obtained commercially during liquid-air manufacture. A radioactive isotope ^{41}Ar (110 min, β^-) is found in the air surrounding atomic reactors, when n,γ reaction is possible. Monograph: *Argon, Helium and the Rare Gases,* **Vols. 1, 2,** G. A. Cook, Ed. (Interscience, New York, 1961) 818 pp. *Review:* Cockett, Smith, "The Monatomic Gases" in *Comprehensive Inorganic Chemistry,* **vol. 1,** J. C. Bailar, Jr. *et al.,* Eds. (Pergamon Press, Oxford, 1973) pp 139-211.

Colorless, odorless, inert gas. d^0 (gas) 1.784 g/l. bp $-185.86°$ ($87.29°$K); crit temp $-122.3°$; critical press. 48.3 atm. Crystallizes in a face-centered cubic lattice; triple pt $-189.37°$ ($83.78°$K). d (solid at triple pt) 1.623. Atomic radius 1.92×10^{-8} cm. Soly of gas in water at $20°$: 33.6 cc/kg water. Also sol in organic liquids.

USE: In fluorescent tubes analogous to neon lights, but produces a bluish-purplish light; in rectifier tubes; in thermometers above mercury; in lasers; wherever an inert atmosphere is desired and the much cheaper nitrogen cannot be used; in ionization chambers and particle counters. The isotope ^{40}Ar is always found in minerals contg potassium, since it is a product of ^{40}K decay. Measuring the amount of ^{40}Ar and ^{40}K can be used for determining the geologic age of minerals and meteors. *Human Toxicity:* Simple asphyxiant.

809. Aricine. *16,17-Didehydro-10-methoxy-19α-methyloxayohimban-16-carboxylic acid;* quinovatine; cinchovatine; heterophylline. $C_{22}H_{26}N_2O_4$; mol wt 382.44. C 69.09%, H 6.85%, N 7.33%, O 16.73%. Originally found in Cusco bark, the bark of *Cinchona pelletierana* Wedd., *Rubiaceae:* Pelletier, Coriol, *J. Pharm. Chim.* [2] **15,** 565 (1829). Also from *Rauwolfia canescens* L., *R. heterophylla* Roem. & Schult., *R. sellowii* Muell. Argov. and *Aspidosperma marcgravianum* Woodson., *Apocynaceae:* Stoll *et al., Helv. Chim. Acta* **38,** 270 (1955); Hochstein *et al., J. Am. Chem. Soc.* **77,** 3551 (1955); Hochstein, *ibid.* **77,** 5744 (1955); Gilbert *et al., J. Org. Chem.* **27,** 4702 (1962). Structure: Stoll *et al., loc. cit.* Stereochemistry: Neuss, Boaz, *J. Org. Chem.* **22,** 1001 (1957); Shamma, Richey, *J. Am. Chem. Soc.* **85,** 2507 (1963).

Orthorhombic, elongated prisms from methanol, dec 188°. Sublimes at 0.01 mm and 180°. $[\alpha]_D^{20}$ $-91°$ (c = 1.4 in chloroform); $[\alpha]_D^{20}$ $-63°$ (c = 1.5 in pyridine); $[\alpha]_D^{20}$ $-57°$ (ethanol). uv max (ethanol): 229, 281 nm (log ϵ 4.54, 3.97). pKa in 80% methyl cellosolve 5.80; in 1:1 DMF-water 6.8. Practically insol in water. Sol in 100 parts of 90% alcohol, 33 parts ether. Very sol in chloroform.

Hydrochloride, $C_{22}H_{26}N_2O_4$·HCl, square plates from methanol + acetone, dec 241-254°. $[\alpha]_D^{20}$ $-5°$ (c = 0.9 in 50% ethanol).

Hydrobromide, $C_{22}H_{26}N_2O_4$·HBr, needles from methanol, dec 262-263°.

Note: Aricine does not seem to share the hypotensive and sedative properties of reserpine: Hochstein *et al., J. Am. Chem. Soc.* **77,** 3551 (1955).

810. Aristolochic Acid. *8-Methoxy-6-nitrophenanthro-[3,4-d]-1,3-dioxole-5-carboxylic acid;* 3,4-methylenedioxy-8-methoxy-10-nitro-1-phenanthrenecarboxylic acid; aristolochic acid-I; aristolochine. $C_{17}H_{11}NO_7$; mol wt 341.29. C 59.83%, H 3.25%, N 4.10%, O 32.82%. One of a group of fourteen known, substituted 1-phenanthrenecarboxylic acids, aristolochic acids, that occur in *Aristolochiaceae* and in butterflies feeding on these plants. They are often accompanied by *aristololactams,* twelve of which have been characterized. Isoln: Gänshirt, *Pharmazie* **8,** 584 (1953); Pailer *et al., Monatsh.* **86,** 676 (1955); Coutts *et al., J. Pharm. Pharmacol.* **11,** 607 (1959); Kupchan, Doskotch, *J. Med. Pharm. Chem.* **5,** 657 (1962). Structure: Pailer *et al., Monatsh.* **87,** 249 (1956). Synthesis: Kupchan, Wormser, *J. Org. Chem.* **30,** 3792 (1965). Biosynthesis: Spenser, Tiwari, *Chem.*

Commun. **1966,** 55. Acute toxicity: U. Mengs, *Arch. Toxicol.* **59,** 328 (1987). *Review:* D. B. Mix *et al., J. Nat. Prod.* **45,** 657-666 (1982).

Shiny brown leaflets from DMF + hot water, dec 281-286°. uv max (ethanol): 390, 318, 250 nm (ϵ 6500; 12000; 27000). Slightly sol in water; sol in alcohol, chloroform, ether, acetone, acetic acid, aniline, alkalies. Practically insol in benzene, carbon disulfide. LD_{50} in male, female mice, male, female rats (mg/kg): 38.4, 70.1, 82.5, 74.0 i.v.; 55.9, 106.1, 203.4, 183.9 orally (Mengs).

Methyl ester, $C_{18}H_{13}NO_7$, orange-yellow rods from hot methanol, mp 285-286°.

811. Armepavine. *(R)-4-[(1,2,3,4-Tetrahydro-6,7-dimethoxy-2-methyl-1-isoquinolinyl)methyl]phenol; 6,7-dimethoxy-1-p-hydroxybenzyl-2-methyl-1,2,3,4-tetrahydroisoquinoline.* $C_{19}H_{23}NO_3$; mol wt 313.38. C 72.82%, H 7.40%, N 4.47%, O 15.32%. Isoln from *Papaver armeniacum* (L.) DC., *Papaveraceae:* Konowalowa *et al., Ber.* **68,** 2161 (1935); from *Euonymus europaea* L., *Celastraceae:* Bishay *et al., J. Pharm. Pharmacol.* **23** (Suppl), 233S (1971); from *Rhamnus frangula* L., *Rhamnaceae:* Pailer, Haslinger, *Monatsh.* **103,** 1399 (1972). Structure: *eidem, J. Gen. Chem. USSR* **10,** 641 (1940). Synthesis: Marion *et al., J. Org. Chem.* **15,** 216 (1950); Gibson *et al., J. Hetercyclic Chem.* **3,** 99 (1966). Resolution: Knabe, Horn, *Arch. Pharm.* **300,** 547 (1967); Farber, Giacomazi, *Chem. & Ind. (London)* **1968** (2), 57.

dl-Form hydrate, needles from acetone + ether; loses its water around 100°, then melts at 140°. Freely sol in alc, acetone, chloroform; slightly sol in ether. Almost insol in water.

R(−)-Form, mp 149-150°. $[\alpha]_D^{20}$ $-117.6°$ ($CHCl_3$) (Knabe); $[\alpha]_D^{27}$ $-103.2°$ (Farber).

R(−)-Form 10-camphorsulfonate hydrate, $C_{29}H_{39}NO_7S$·H$_2$O, mp 130°. $[\alpha]_D^{20}$ $-65°$ (MeOH).

dl-Form hydrochloride, $C_{19}H_{23}NO_3$·HCl, needles from alcohol + ether, mp 152°. Freely soluble in water and alc.

812. Armstrong's Acid. *1,5-Naphthalenedisulfonic acid.* $C_{10}H_8O_6S_2$; mol wt 288.30. C 41.66%, H 2.80%, S 22.24%, O 33.30%. Prepd by sulfonation of naphthalene: Lynch, Scanlan: *Ind. Eng. Chem.* **19,** 1010 (1927).

Crystals. Soluble in water, alc; practically insol in ether.

813. Arnica. Arnica flowers; leopard's bane; wolf's bane; mountain tobacco. Dried flowerheads of *Arnica montana* L., *Compositae.* *Habit.* Northern Europe. *Constit.*

0.5-1% volatile oil; arnicin, arnisterol (arnidiol), anthoxanthine, tannin, resin. *Review:* Faber, *Pharmazie* **8**, 179, 286, 340 (1953).

THERAP CAT: Topical counterirritant.

THERAP CAT (VET): Counterirritant.

814. Arogenic Acid. α-*Amino-1-carboxy-4-hydroxy-2,5-cyclohexadiene-1-propanoic acid;* L-(8S)-β-(1-carboxy-4-hydroxy-2,5-cyclohexadien-1-yl)alanine; arogenate; pretyrosine. $C_{10}H_{13}NO_5$; mol wt 227.22. C 52.86%, H 5.76%, N 6.17%, O 35.21%. Precursor in the biosynthesis of L-phenylalanine and L-tyrosine, distributed widely in nature. Pseudomonad bacteria and plants use both the prephenate and arogenate pathways for L-tyrosine biosynthesis. In cyanobacteria, coryniform bacteria and at least one yeast organism the arogenate pathway has been identified as the sole route of L-tyrosine biosynthesis. Identification of arogenate (pretyrosine) in blue-green algae biosynthesis of L-tyrosine: S. L. Stenmark *et al., Nature* **247**, 290 (1974). Dual enzymic routes to L-tyrosine and L-phenylalanine via arogenate (pretyrosine) in *P. aeruginosa:* N. Patel *et al., J. Biol. Chem.* **252**, 5839 (1977). Isoln and prepn: R. A. Jensen *et al., J. Bacteriol.* **132**, 896 (1977). Confirmation that arogenate is an obligatory intermediate of L-tyrosine biosynthesis: A. M. Fazel *et al., Proc. Nat. Acad. Sci. USA* **77**, 1270 (1980). Structure and config: L. O. Zamir *et al., J. Am. Chem. Soc.* **102**, 4499 (1980). Synthesis via immobilized microbial proteins: *eidem, Bioorg. Chem.* **11**, 32 (1982). Arogenate pathway of tyrosine and phenylalanine biosynthesis in *P. aureofaciens:* B. Keller *et al., J. Gen. Microbiol.* **128**, 1199 (1982).

Unstable. Quantitatively converted to phenylalanine at acidic pH. Forms a disodium salt that is stable in the lyophilized state at basic pH (7.5) and room temperature, but is thermally unstable.

815. Arotinolol. ($\pm$)-5-[2-[[3-[(1,1-Dimethylethyl)-amino]-2-hydroxypropyl]thio]-4-thiazolyl]-2-thiophenecarboxamide. 2-(3'-*tert*-butylamino-2'-hydroxypropylthio)-4-(5'-carbamoyl-2'-thienyl)thiazole. $C_{15}H_{21}N_3O_2S_3$; mol wt 371.53. C 48.49%, H 5.70%, N 11.31%, O 8.61%, S 25.89%. Propanolamine deriv with α- and β-adrenergic blocking activity. Prepn: T. Hibino *et al.,* **Ger. pat. 2,341,-753;** *eidem,* U.S. pat. **3,932,400** (1974, 1976 both to Sumitomo); Y. Hara *et al., J. Pharm. Sci.* **67**, 1334 (1978). Characterization of adrenoceptor blocking effects *in vivo:* A. Miyagishi *et al., Arch. Int. Pharmacodyn. Ther.* **271**, 249 (1984). Possible anti-anginal effects: M. Sakanashi *et al., Oyo Yakuri* **28**, 709 (1984); *C.A.* **102**, 39669z (1985). Long-term antihypertensive effect in rats: K. Kishi *et al., J. Pharmacobiodyn.* **8**, 50 (1985). Comparison with timolol, *q.v.,* of effect on intraocular pressure and hemodynamics: M. Nakashima *et al., Eur. J. Clin. Pharmacol.* **28**, 391 (1985).

Crystals from chloroform/petroleum ether, mp 148-149°. Hydrochloride, $C_{15}H_{22}ClN_3O_2S_3$, *S 596, ARL, Almarl.* Crystals from methanol/water, mp 234-235.5° (dec). LD_{50} in mice (mg/kg): 86 i.v., >360 i.p., >5000 orally (Hara).

THERAP CAT: Antihypertensive; anti-anginal; anti-arrhythmic.

816. Arprinocid. 9-[(2-Chloro-6-fluorophenyl)meth-

yl]-9H-purin-6-amine; 9-(2-chloro-6-fluorobenzyl)adenine; MK-302; Arpocox. $C_{12}H_9ClFN_5$; mol wt 277.69. C 51.90%, H 3.27%, Cl 12.77%, F 6.84%, N 25.22%. Prepn: E. P. Lira *et al.,* **Fr. pat. 2,128,600** corresp to U.S. pats. **3,846,426** and **3,953,597** (1972, 1974, 1976, all to Int. Minerals and Chem.); G. D. Hartman *et al., J. Org. Chem.* **43**, 960 (1978). Determination in feed: D. W. Fink *et al., J. Assoc. Off. Anal. Chem.* **61**, 1078 (1978); **62**, 1 (1979). Evaluation of anticoccidial efficacy: P. Schindler *et al., Poult. Sci.* **58**, 23 (1979); L. R. McDougald, J. K. Johnson, *ibid.* 72.

Cryst from methanol/water, mp 245-246°.

THERAP CAT (VET): Coccidiostat.

817. Arsacetin. [4-(Acetylamino)phenyl]arsonic acid; p-acetamidobenzenearsonic acid; N-acetyl-p-aminobenzene-arsonic acid; N-acetylarsanilic acid. $C_8H_{10}AsNO_4$; mol wt 259.09. C 37.09%, H 3.89%, As 28.91%, N 5.41%, O 24.70%. Prepn of free acid and sodium salt: Bart, *Ann.* **429**, 55 (1922). Toxicity: A. Gros, *Biochem. Z.* **184**, 360 (1927).

Sodium salt tetrahydrate, $C_8H_9AsNNaO_4.4H_2O$, cryst powder. *Poisonous!* One gram dissolves in 10 ml cold water, 3 ml hot water. LD_{50} i.v. in rabbits: 550 mg/kg (Gros). *Note:* Sodium salt was formerly referred to as arsacetin.

THERAP CAT: Sodium salt formerly as antisyphilitic.

818. Arsanilic Acid. (4-Aminophenyl)arsonic acid; p-aminobenzenearsonic acid; atoxylic acid; AS 101. C_6H_8-$AsNO_3$; mol wt 217.04. C 33.20%, H 3.72%, N 6.45%, As 34.51%, O 22.11%. Prepd by heating aniline and arsenic acid: Lewis, Cheetham, *Org. Syn. coll. vol. I* (2nd ed., 1941) p 70; Hoffmann, Green, U.S. pat. **3,763,201** (1973); from p-nitroaniline: Bart, *Ann.* **429**, 96 (1922); Gattermann-Wieland, *Praxis des Organischen Chemikers* (de Gruyter, Berlin, 40th ed., 1961) p 254. Toxicology in swine: Ledet *et al., Clin. Toxicol.* **6**, 439 (1973).

Needles from water or alcohol. Slightly sol in cold water, alcohol or acetic acid; sol in hot water, amyl alcohol, in solns of alkali carbonates; moderately sol in concd mineral acids; insol in acetone, benzene, chloroform, ether, or in moderately dil mineral acids. LD_{50} orally in male rats: >1000 mg/kg, E. I. Goldenthal, *Toxicol. Appl. Pharmacol.* **18**, 185 (1971).

USE: Manuf medicinal arsenicals.

THERAP CAT (VET): Growth promotant; to improve feed efficiency. To control swine dysentery.

819. Arsenamide. 2,2'-[[[4-(Aminocarbonyl)phenyl]-arsinidene]bis(thio)]bisacetic acid; [[(p-carbamoylphenyl)-arsylene]dithio]diacetic acid; bis[carboxymethylmercapto](p-carbamylphenyl)-arsine; p-[bis(carboxymethylmercapto)-

arsino]benzamide; dithioglycolyl *p*-arsenobenzamide; 4-carbamylphenyl bis[carboxymethylthio]arsenite; thioarsenite; thiacetarsamide; Caparsolate; Caparside. $C_{11}H_{12}AsNO_5S_2$; mol wt 377.26. C 35.02%, H 3.21%, As 19.86%, N 3.71%, O 21.21%, S 17.00%. Prepd by the condensation of *p*-arsenosobenzamide and thioglycolic acid: Gough, King *J. Chem. Soc.* **1930,** 669; Maren, *J. Am. Chem. Soc.* **68,** 1864 (1946).

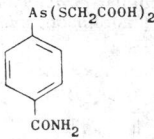

White crystalline powder. Sparingly sol in cold water; appreciably sol in water above 90°. Sparingly sol in cold methanol, ethanol; very sol in these solvents when they are warm. Insol in warm isopropyl ether. pKa = 4. The disodium salt is stoichiometrically formed in aq soln at pH 7-8.

THERAP CAT (VET): Anthelmintic for heartworm infection in dogs, especially the adult worm.

820. Arsenic. Grey arsenic; metallic arsenic; arsen (German). As; at. wt 74.9216; at. no. 33; valence 3, 5. One natural isotope: 75; artificial, radioactive isotopes: 68-74; 76-81. Arsenic compds were described and used in antiquity, their reduction to the element was known to medieval alchemists, and the first precise directions for the prepn of As are found in Paracelsus' writings (*ca.* 1520 A.D.). Arsenic occurs probably throughout the universe. Meteorites contain from 0.0005 to 0.1% As. Occurrence in the earth's crust: 0.0005%. Found to a small extent as the element, mostly as an arsenide of true metals. Usually produced as the trioxide when smelting ores for such metals. Reduction with carbon (sugar charcoal) and sublimation in N_2 current yields very pure As: Krepelka, *Coll. Czech. Chem. Commun.* **2,** 255 (1930); E. H. Archibald, *The Preparation of Pure Inorganic Substances* (Wiley, New York, 1932) p 269. Other methods: Schenk in *Handbook of Preparative Inorganic Chemistry* vol. 1, G. Brauer, Ed. (Academic Press, New York, 2nd ed., 1963) pp 591-593. *Reviews: Gmelin's, Arsenic* (8th ed.) **17,** 475 (1952); Smith, "Arsenic, Antimony and Bismuth" in *Comprehensive Inorganic Chemistry* vol. 2, J. C. Bailar, Jr. *et al.*, Eds. (Pergamon Press, Oxford, 1973) pp 547-683. Review of carcinogenicity studies of arsenic and arsenic compds: *IARC Monographs* **23,** 39-141 (1980).

Gray, shiny, brittle, metallic-looking rhombohedra. Can be heated to burn in air with bluish flame, giving off an odor of garlic and dense white fumes of As_2O_3. Loses its luster on exposure to air, forming a black modification + As_2O_3. Brinell hardness: 147; Mohs' scale: 3.5. d_4^{25} 5.727. Sublimes$_{760}$ 615° without melting. Vaporization becomes apparent at 100° and is already rapid at 450°. mp 818° at 36 atm. Heat of vaporization 11.2 kcal/g-atom. Heat of sublimation 30.5 kcal/g-atom. Heat of fusion: 22.4 kcal/g-atom: *Gmelin's, loc. cit.* pp 135-136. Also reported: heat of fusion: 6.620 kcal/g-atom; heat of sublimation 7.63 kcal/g-atom: D. R. Stull, G. C. Sinke, *Thermodynamic Properties of the Elements*, Advances in Chemistry Series **18** (A.C.S., Washington, 1956) pp 11, 44. Specific heat: 0.0822 for 0° to 100°. Dielectric constant = 10.23 at 20° and 60 cycles. Electrical and magnetic properties of crystalline As: Taylor *et al.*, *J. Phys. Chem. Solids* **26,** 69 (1965). Insol in water; not attacked by cold H_2SO_4 or HCl; converted by HNO_3 or hot H_2SO_4 into arsenous or arsenic acid.

A yellow modification which has no metallic properties is obtained by sudden cooling of As-vapor. This yellow arsenic is converted back to the gray modification upon very short exposure to ultraviolet light.

Note: In German and other languages *Arsenik* means arsenic trioxide.

Caution: Most forms of arsenic are toxic. *Acute* symptoms following ingestion relate to irritation of the G.I. tract: nausea, vomiting, diarrhea which can progress to shock and death. *Chronic* poisoning can result in exfoliation and pigmentation of skin, herpes, polyneuritis, altered hematopoiesis, degeneration of liver and kidneys: Vallee *et al.*, *Arch.*

Ind. Health **21,** 132 (1960); E. Browning, *Toxicity of Industrial Metals* (Appleton-Century-Crofts, New York, 2nd ed., 1969) pp 36-60. This substance and certain arsenic compounds have been listed as known carcinogens: *Fourth Annual Report on Carcinogens* (NTP 85-002, 1985) p 26.

USE: In metallurgy for hardening copper, lead, alloys. In the manufacture of certain types of glass. The artificial isotope ^{76}As as radioactive tracer in toxicology.

821. Arsenic Acid. Orthoarsenic acid. AsH_3O_4; mol wt 141.93. As 52.78%, H 2.13%, O 45.09%. H_3AsO_4. Exists only as the hemihydrate. Excessive drying produces As_2O_5.-5/3H_2O. Conveniently prepd from As_2O_3 and HNO_3: Simon, Thaler, *Z. Anorg. Allgem. Chem.* **161,** 143 (1927); **246,** 19 (1941).

Hemihydrate, hygroscopic crystals. *Poisonous!* Converted to As_2O_5 by heating above 300°. Freely sol in water, alcohol, glycerol. LD_{50} i.v. in rabbits: 6 mg/kg, Joachimoglu, *Biochem. Z.* **70,** 144 (1915).

USE: In the manuf of arsenates.

822. Arsenic Disulfide. *Arsenic sulfide;* red arsenic sulfide; realgar; red orpiment; ruby arsenic; red arsenic glass; C.I. Pigment Yellow 39; C.I. 77085. As_4S_4; mol wt 427.92. As 70.03%, S 29.97%. Prepn: *Gmelin's, Arsenic* (8th ed.) **17,** pp 417-422 (1952); Schenk in *Handbook of Preparative Inorganic Chemistry* vol 1, G. Brauer, Ed. (Academic Press, New York, 2nd ed., 1963) p 603.

Deep red, lustrous monoclinic crystals. mp 320°; bp 565°; d 3.5. Ignites at high temp. Practically insol in water; sol in alkali hydroxides; decomposed by HNO_3; very slightly sol in hot CS_2 and benzene.

USE: As pigment in painting; in fireworks as blue fire and to give an intense white flame; manuf shot; calico printing and dyeing; tanning and depilating hides.

823. Arsenic Hemiselenide. As_2Se; mol wt 228.78. As 65.49%, Se 34.51%. Prepd by melting arsenic and selenium in the correct proportions in a sealed tube filled with nitrogen: Szarvasy, *Ber.* **28,** 2654 (1895); **30,** 1244 (1897); *Gmelin's, Arsenic* (8th ed.) **17,** p 462 (1952).

Black crystals with a metallic luster. Insol in the usual organic and inorganic solvents. Dec in boil. alkali hydroxide solns; slowly decompd by concd hydrochloric and sulfuric acids.

USE: Manuf of glass.

824. Arsenic Pentafluoride. AsF_5; mol wt 169.91. As 44.09%, F 55.91%. Prepd from As + F_2: Ruff *et al.*, *Z. Anorg. Allgem. Chem.* **206,** 59 (1932); Seel, Detmer, *ibid.* **301,** 113 (1959); Kwasnik in *Handbook of Preparative Inorganic Chemistry* vol. 1, G. Brauer, Ed. (Academic Press, New York, 2nd ed., 1963) p 198. *Reviews:* Burg in *Fluorine Chemistry* vol. I, J. Simons, Ed. (Academic Press, New York, 1950) p 102; Kemmitt, Sharp, *Advan. Fluorine Chem.* **4,** 208 (1965).

Colorless gas. Forms white clouds in moist air. $d_{liq}^{25.8}$ 2.33. mp −79.8°. bp −53.2°. Instantly hydrolyzed by water. Sol in alcohol, ether, benzene. Dry AsF_5 does not attack glass, but a minute trace of moisture or HF catalyzes the etching reaction to the point of total destruction.

825. Arsenic Pentaselenide. As_2Se_5; mol wt 544.62. As 27.51%, Se 72.49%. Prepd by melting the correct proportions of arsenic and selenium in a sealed tube containing nitrogen: Szarvasy, *Ber.* **28,** 2654 (1895); **30,** 1244 (1897); by the action of an arsenic salt on a solution of hydrogen selenide: Moser, Atynsky, *Monatsh.* **45,** 235 (1925).

Black, brittle solid with a metallic luster. *Poisonous!* Dec when heated in air. Sol in alkali hydroxides and sulfides; insol in water, dilute acids, concd hydrochloric acid, alcohol, ether, carbon disulfide; dec in nitric acid.

826. Arsenic Pentasulfide. Diarsenic pentasulfide. As_2S_5; mol wt 310.12. As 48.31%, S 51.69%. Prepd from H_3AsO_4 and H_2S: Schenk in *Handbook of Preparative Inorganic Chemistry* vol 1, G. Brauer, Ed. (Academic Press, New York, 2nd ed., 1963) p 603.

Brownish-yellow, glassy, amorphous, highly refractive mass. Insol in water; sol in alkalies and alkali sulfides. Dec into As_2O_3, S, and As_2S_3 when boiled with water.

USE: In thin sheets as a light filter; in pigments.

827. Arsenic Pentoxide. Arsenic acid anhydride. As_2O_5; mol wt 229.82. As 65.20%, O 34.81%. Prepn: *Gmelin's, Arsenic* (8th ed.) **17**, pp 273-277 (1952).

Amorphous lumps or powder. *Poisonous!* Gradually deliquesces on exposure to air. Freely sol in water or alc. Combines very slowly with H_2O to form H_3AsO_4. *Keep well closed.*

USE: Manuf of colored glass; in adhesives for metals; in wood preservatives; in weed control; as fungicide.

828. Arsenic Tribromide. $AsBr_3$; mol wt 314.66. As 23.81%, Br 76.19%. Prepn: Oddo, Giachery, *Gazz. Chim. Ital.* **53**, 56 (1923); Schenk in *Handbook of Preparative Inorganic Chemistry* vol. **1**, G. Brauer, Ed. (Academic Press, New York, 2nd ed., 1963) p 597.

Deliquescent, orthorhombic prisms. *Intensely poisonous!* d_4^{25} 3.397; d_4^{50} (liq) 3.3282; d_4^{75} 3.2623; d_4^{100} 3.1995. mp 31.1°. bp_{760} 221°; bp_{11} 89°. Dipole moment 1.66. Heat of fusion 8.93 cal/g. Fumes in moist air. Dec by water with the formation of As_2O_3 and HBr. Miscible with ether, benzene. Sol in hydrocarbons, chlorinated hydrocarbons, carbon disulfide, oils and fats.

829. Arsenic Trichloride. Butter of arsenic; fuming liquid arsenic. $AsCl_3$; mol wt 181.28. As 41.32%, Cl 58.68%. Prepn: Smith, *Ind. Eng. Chem.* **11**, 109 (1919); Reisener in *Ullman's Encyklopädie der Technischen Chemie* vol. **3**, (Urban & Schwarzenberg, Munich, 1953) p 850; Schenk in *Handbook of Preparative Inorganic Chemistry* vol. **1**, G. Brauer, Ed. (Academic Press, New York, 2nd ed., 1963) p 596.

Oily liquid. *Intensely poisonous!* Fumes in air. d_4^{25} 2.1497. mp −16°. bp 130.21°; bp_{11} 25°. Dipole moment: 2.17. Heat of vaporization 8.9 kcal/mol. Specific heat: 3.19 cal/mol/°C. n_D^{20} 1.6006. Dec by water to form $As(OH)_3$ and HCl. One mol $AsCl_3$ can be dissolved in 9 mols H_2O, this soln (d 1.53) may be diluted again with another 9 mols H_2O giving a soln with d 1.346. Further dilution results in the precipitation of As_2O_3. Also dec by ultraviolet light. Miscible with, or solvent for, chloroform, carbon tetrachloride, ether, iodine, phosphorus, sulfur, alkali iodides, oils and fats.

USE: In the ceramic industry; in syntheses of chlorine-contg arsenicals, e.g., chloro derivs of arsine.

830. Arsenic Trifluoride. AsF_3; mol wt 131.91. As 56.79%, F 43.21%. Prepn: Russell *et al., J. Am. Chem. Soc.* **63**, 2825 (1941); Woolf, Greenwood, *J. Chem. Soc.* **1950**, 2200; Hoffman, *Inorg. Syn.* **4**, 151 (1953); Kwasnik in *Handbook of Preparative Inorganic Chemistry* vol. **1**, G. Brauer, Ed. (Academic Press, New York, 2nd ed., 1963) pp 197-198. *Review:* Kemmitt, Sharp, *Advan. Fluorine Chem.* **4**, 208-209 (1965).

Mobile liquid. *Very poisonous!* Fumes in air. Etches glass. d_{15}^{15} 2.73. mp −5.95°. bp 57.8°. Also reported as mp −8.5°; bp 63°. Kwasnik, *loc. cit.* Hydrolyzed by water. Sol in alcohol, ether, benzene. May be stored in iron vessels.

Human Toxicity: Extremely toxic.

831. Arsenic Triiodide. AsI_3; mol wt 455.67. As 16.44%, I 83.56%. Prepd from the elements or from $AsCl_3$ and KI: Bailar, *Inorg. Syn.* **I**, 103 (1939).

Orange-red, trigonal rhombohedra from acetone. Reacts slowly with O_2 from air, liberating iodine. d_4^{25} 4.688. Some tendency to sublime below 100°. mp 140.9° forming a red liquid. bp_{760} about 400°. One gram dissolves in 12 ml water forming a yellow soln. Does not hydrolyze rapidly and may be recovered from the water soln unchanged within 5 hrs. Aqueous solns are strongly acid (pH of $0.1N$ soln about 1.1) and ultimately form HI and As_2O_3, although an equilibrium $AsI_3 + 3H_2O = H_3AsO_3 + 3HI$ has been observed. Freely sol in carbon disulfide, chloroform, benzene, toluene, xylene. Less sol in alc, ether.

THERAP CAT: Formerly in dermatitides.

832. Arsenic Trioxide. Arsenous acid; arsenous acid anhydride; arsenous oxide; arsenic sesquioxide; white arsenic. As_2O_3; mol wt 197.82. As 75.74%, O 24.26%. Prepn: Schenk in *Handbook of Preparative Inorganic Chemistry* vol. **1**, G. Brauer, Ed. (Academic Press, New York, 2nd ed., 1963) p 600.

White or transparent, glassy, amorphous lumps or cryst powder. Two cryst modifications, *claudetite* (monoclinic, mp 313°) and *arsenolite* (cubic, mp 275°). bp 465°. *Intensely poisonous!* When slowly heated sublimes unchanged; when rapidly heated the cryst sublimes without fusion, while the amorphous first fuses, then sublimes. Sparingly and extremely slowly sol in cold water; sol in 15 parts boiling water, in dil HCl, in alkali hydroxide or carbonate solns; practically insol in alc, chloroform, ether. LD_{50} orally in mice, rats: 39.4, 15.1 mg/kg, J. Harrison *et al., Arch. Ind. Health* **17**, 118 (1958).

Incompat. Tannic acid, infusion cinchona and other vegetable astringent infusions and decoctions; iron in soln.

USE: It is the primary material for all arsenic compounds. Used in manuf of glass, Paris green, enamels, weed killers, metallic arsenic; for preserving hides; killing rodents, insects; in sheep dips and weed killers; textile mordant.

THERAP CAT (VET): Formerly as parasiticide, also for non-parasitic skin and blood diseases; in rheumatism, asthma and heaves, and as an alterative.

833. Arsenic Triselenide. Arsenious selenide; arsenous selenide. As_2Se_3; mol wt 386.70. As 38.74%, Se 61.26%. Prepd by melting arsenic and selenium in the correct proportions: Uelsmann, *Ueber Selenverbindungen*, Göttingen (1860); *Ann.* **116**, 122 (1860); by the action of an arsenic salt on a soln of hydrogen selenide: Moser, Atynsky, *Monatsh.* **45**, 235 (1925).

Dark brown solid. d 4.75. mp 260°. Sol in nitric acid, alkali-lye, alkali sulfide solutions; insol in water.

834. Arsenic Trisulfide. Yellow arsenic sulfide; orpiment; auripigment; arsenic yellow; king's yellow; king's gold. As_2S_3; mol wt 246.00. As 60.90%, S 39.10%. The article of commerce contains much less sulfur than theory. Prepn: *Gmelin's, Arsenic* (8th ed.) **17**, pp 422-433 (1952).

Yellow or orange powder. mp reported from 300° to 325°. d 3.46. Practically insol in water; sol in alkalies, alkali sulfides or carbonates; slowly sol in hot HCl; dec by HNO_3.

USE: Manuf of glass, particularly infrared-transmitting glass; manuf of oil cloth, linoleum; in electrical semiconductors, photoconductors; as pigment; for depilating hides; in pyrotechnics.

835. Arsenious Acid Solution. Arsenic chloride solution. *Poisonous!* Prepared with 1 g As_2O_3, 5 ml dil HCl and water to 100 ml.

THERAP CAT (VET): Has been used in skin and blood disorders.

836. Arsenoacetic Acid. Arsenodiacetic acid; diarsenoacetic acid. $C_4H_6As_2O_4$; mol wt 267.93. C 17.93%, H 2.26%, As 55.92%, O 23.89%. $HOOCCH_2As=AsCH_2-COOH$. Prepd by dissolving sodium arsonoacetate and sodium hypophosphite in 15% H_2SO_4: Palmer, *J. Am. Chem. Soc.* **45**, 3023 (1923); *Org. Syn.* **coll. vol. I,** 73 (1941).

Minute yellow needles, begin to dec at about 205°, but do not melt below 260°. Practically insol in water and common organic solvents; readily sol in pyridine, dil sodium hydroxide and dil sodium carbonate solns.

Disodium salt, $C_4H_4As_2Na_2O_4$, *disodium arsenoacetate, Tonarsan.* Yellow powder, freely sol in water. Has been used as injectable tonic for horses. *Compare* Arsonoacetic Acid (Disodium Salt).

837. Arsine. Arsenic trihydride; hydrogen arsenide. AsH_3; mol wt 77.93. As 96.12%, H 3.88%. Prepn: Wendt, Landauer, *J. Am. Chem. Soc.* **42**, 930 (1920); Jolly, Drake, *Inorg. Syn.* **7**, 41 (1963); Schenk in *Handbook of Preparative Inorganic Chemistry* vol **1**, G. Brauer, Ed. (Academic Press, New York, 2nd ed., 1963) pp 593-595.

Colorless, neutral gas. Disagreeable garlic odor. *Very poisonous!* mp −117°. bp −62.5°. Dissociation pressure at 0° = 0.806 atm. Decomposes when heated at 300°, depositing arsenic which volatilizes at 400°. On exposure to light, moist arsine decomposes quickly depositing shiny black arsenic. Slightly sol in water. Aq solns are neutral. Traces are best removed by absorption in potassium permanganate soln or in bromine water.

Caution: Inhalation of as little as 0.5 ppm may be dangerous. Headache, vomiting, anorexia, paresthesia, abdominal pains, chills, hematemesis, hemoglobinuria and anuria may occur within a few hours after exposure. Death results from renal failure, pulmonary edema: Doig, *Lancet* **2**, 88 (1958); Vallee *et al.*, *Arch. Ind. Health* **21**, 132 (1960); E. Browning, *Toxicity of Industrial Metals* (Appleton-Century-Crofts, New York, 2nd ed., 1969) pp 53-60.

838. Arsonoacetic Acid. $C_2H_5AsO_5$; mol wt 183.97. C 13.06%, H 2.74%, As 40.72%, O 43.48%. $H_2O_3AsCH_2$-COOH. Prepn from sodium arsenite and sodium chloroacetate: Palmer, *J. Am. Chem. Soc.* **45**, 3023 (1923); *Org. Syn. coll. vol.* **I**, 73 (1941).

Shiny plates, mp 152°. Very sol in water, alc; sparingly sol in hot glacial acetic acid. Practically insol in petr ether, benzene, acetone, chloroform, ethyl acetate.

Barium salt, $C_4H_4As_2Ba_3O_{10}$, feathery needles. Slightly sol in boiling water.

Sodium salt monohydrate, $C_2H_3AsNa_2O_5.H_2O$, *disodium arsonoacetate, disodium acetarsenate, acetarsonic acid disodium salt, sodium acetoarsinate, Aricyl.* Crystals. Freely sol in water.

THERAP CAT (VET): Disodium salt has been used to treat anaplasmosis, and as a stimulant in nervous diseases.

839. Arsphenamine. *4,4'-(1,2-Diarsenediyl)bis[2-aminophenol] dihydrochloride; 4,4'-arsenobis(2-aminophenol) dihydrochloride;* arsenphenolamine hydrochloride; 3,3'-diamino-4,4'-dihydroxyarsenobenzene dihydrochloride; 606; Arsaminol; Ehrlich 606; Kharsivan; Salvarsan; Sanluol. $C_{12}H_{14}As_2Cl_2N_2O_2$; mol wt 438.99. C 32.83%, H 3.22%, As 34.13%, Cl 16.15%, N 6.38%, O 7.29%. Prepd by Ehrlich in 1909. Prepn: Ehrlich, Bertheim, U.S. pat. **986,148** (1911 to Hoechst); Christiansen, *J. Am. Chem. Soc.* **42**, 2402 (1920). Toxicity data: J. F. Schamberg *et al.*, *Am. J. Syph. Neurol.* **18**, 37 (1934).

Light yellow, somewhat hygroscopic powder. It is odorless, or has a slight odor of the precipitant used. Oxidizes on exposure to air, becoming darker and more toxic. *It is poisonous!* Soluble in water, alc or glycerol, very slightly in chloroform or ether. Its aq soln is acid; pH about 3. For medicinal use, it must first be completely "neutralized" with 8.4 ml $1N$ NaOH/gm. LD_{100} i.v. in rats: 140 mg/kg (Schamberg).

THERAP CAT: Formerly as antisyphilitic.
THERAP CAT (VET): Formerly in contagious bovine pleuropneumonia.

840. Arsthinol. *N-[2-Hydroxy-5-[4-(hydroxymethyl)-1,3,2-dithiarsolan-2-yl]phenyl]acetamide; 3-acetamido-4-hydroxydithiobenzenearsonous acid, cyclic (hydroxymethyl)-ethylene ester;* 2-(3'-acetamido-4'-hydroxyphenyl)-1,3-dithia-2-arsacyclopentane-4-methanol; 3-acetamido-4-hydroxydithiobenzenearsonous acid cyclic 3-hydroxypropylene ester; 2-acetylamino-4-(methylolcycloethylene-dimercaptoarsine)phenol; Mercaptoarsenol; Balarsen. $C_{11}H_{14}AsNO_3S_2$; mol wt 347.27. C 38.04%, H 4.06%, As 21.57%, N 4.03%, O 13.82%, S 18.47%. Prepn: Friedheim, U.S. pats. **2,593,434** (1952) and **2,772,303** (1956).

Minute crystals, mp 163-166°. Soly in 95% ethanol about 2.7% w/v. Very slightly sol in water, ether.

THERAP CAT: Antiamebic.

841. Arteether. *[3R-(3α,5aβ,6β,8aβ,9α,10α,12β,-12aR*)]-10-Ethoxydecahydro-3,6,9-trimethyl-3,12-epoxy-12H-pyrano[4,3-j]-1,2-benzodioxepin;* dihydroartemisinin ethyl ether; dihydroqinghaosu ethyl ether; SM227. $C_{17}H_{28}O_5$; mol wt 312.41. C 65.36%, H 9.03%, O 25.61%.

Derivative of artemisinin, *q.v.* Prepn: Y. Li *et al.*, *Acta Pharm. Sinica* **16**, 429 (1981). Synthesis and antimalarial activity: China Cooperative Research Group on Qinghaosou, *J. Tradit. Chinese Med.* **2**, 9 (1982); A. Brossi *et al.*, *J. Med. Chem.* **31**, 645 (1988).

White crystalline solid, mp 80-82°. $[α]_D^{21}$ 154.5° (c = 1.0 in $CHCl_3$).

THERAP CAT: Antimalarial.

842. Artemether. *[3R-(3α,5aβ,6β,8aβ,9α,10α,12β,-12aR*)]-Decahydro-10-methoxy-3,6,9-trimethyl-3,12-epoxy-12H-pyrano[4,3-j]-1,2-benzodioxepin;* dihydroartemisinin methyl ether; dihydroqinghaosu methyl ether; o-methyldihydroartemisinin; SM 224. $C_{16}H_{26}O_5$; mol wt 298.38. C 64.41%, H 8.78%, O 26.81%. Derivative of artemisinin, *q.v.* Prepn: Y. Li *et al.*, *K'o Hseuh T'ung Pao* **24**, 667 (1979), *C.A.* **91**, 211376u (1979); *eidem*, *Acta Pharm. Sinica* **16**, 429 (1981). Absolute configuration: X.-D. Luo *et al.*, *Helv. Chim Acta* **67**, 1515 (1984). NMR spectral study: F. S. El-Feraly *et al.*, *Spectrosc. Lett.* **18**, 843 (1985). Inhibition of protein synthesis: H. M. Gu *et al.*, *Biochem. Pharmacol.* **32**, 2463 (1983). Antimalarial activity: S. Thaithong, G. H. Beale, *Bull. WHO* **63**, 617 (1985); and pharmacology: W. Peters *et al.*, *Ann. Trop. Med. Parasitol.* **80**, 483 (1986). Clinical trials in falciparum malaria: W. Tongyin, X. Ruchang, *J. Tradit. Chinese Med.* **5**, 240 (1985). Toxicity data: *China Cooperative Research Group on Qinghaosou, ibid.* **2**, 31 (1982). Series of articles on chemistry, pharmacology and antimalarial efficacy: *ibid.* 3-50.

Crystals, mp 86-88°. $[α]_D^{19.5}$ 171° (c = 2.59 in $CHCl_3$). LD_{50} i.m. in mice: 263 mg/kg (China Cooperative Research Group on Qinghaosou).

THERAP CAT: Antimalarial.

843. Artemisin. *3a,5,5a,9b-Tetrahydro-4-hydroxy-3,5a,9-trimethylnaphtho[1,2-b]furan-2,8(3H,4H)-dione; 6α,8α-dihydroxy-3-oxoeudesma-1,4-dien-12-oic acid 12,6-lactone;* 8-hydroxysantonin. $C_{15}H_{18}O_4$; mol wt 262.29. C 68.68%, H 6.92%, O 24.40%. From the closed, unexpanded flower heads of several *Artemisia* spp, especially *Artemisia maritima* L., and *A. cina* Berg., *Compositae* ("wormseed"). Found in the mother liquors from the extraction of santonin. Isoln: E. Merck, *Merck's Jahresber.* **1894**, 3; *Chem. Zentr.* **1895**, I, 436. Structure: Sumi, *J. Am. Chem. Soc.* **80**, 4869 (1958); Cocker, McMurry, *Tetrahedron* **8**, 181 (1960). Stereochemistry: Bolt *et al.*, *J. Chem. Soc.* **1963**, 5235. Synthesis of (+)-artemisin: Nakazaki, Naemura, *Tetrahedron Letters* **1966**, 2615.

Crystals from abs ethanol or ethyl acetate. Bitter taste. Turns yellow on exposure to light. mp 203°. $bp_{0.1}$ 260°. Sublimes$_{760}$ 170-175°. $[\alpha]_D^{23}$ $-84.9°$ (c = 3 in 95% ethanol). One gram dissolves in 60 ml boiling water, in 3 ml boiling alc. Sol in ethyl acetate. Practically insol in petr ether. Somewhat sol in chloroform.

844. Artemisinin. *[3R-(3α,5aβ,6β,8aβ,9α,12β,12aR*)]-Octahydro-3,6,9-trimethyl-3,12-epoxy-12H-pyrano[4,3-j]-1,2-benzodioxepin-10(3H)-one;* artemisine; arteannuin; huanghuahaosu; QHS; qinghaosu; qing hau sau. $C_{15}H_{22}O_5$; mol wt 282.35. C 63.81%, H 7.86%, O 28.33%. Active antimalarial constituent of the traditional Chinese medicinal herb *Artemisia annua* L., *Compositae*, which has been known for almost 2000 years as *Qinghao*. Isolated in 1972 and shown to be a sesquiterpene lactone with a peroxide moiety: *K'o Hsueh T'ung Pao* **22**, 142 (1977), *C.A.* **87**, 98788g (1977); L. Jing-Ming et al., *Acta Chim. Sinica* **37**, 129 (1979), *C.A.* **92**, 94594w (1980). Total synthesis and absolute configuration: G. Schmid, W. Hofheinz, *J. Am. Chem. Soc.* **105**, 624 (1983). NMR spectral study: F. S. El-Feraly et al., *Spectrosc. Letters* **18**, 843 (1985). Antimalarial activity: Qinghaosu Antimalaria Coordinating Research Group, *Chinese Med. J.* **92**, 811 (1979); L. J. Bruce-Chwatt, *Brit. Med. J.* **284**, 767 (1982). J.-B. Jiang et al., *Lancet* **2**, 285 (1982). Preliminary mode of action study: D. S. Ellis et al., *Ann. Trop. Med. Parasitol.* **79**, 367 (1985). Clinical trial in comparison with artemether, q.v.: W. Tongyin, X. Ruchang, *J. Tradit. Chinese Med.* **5**, 240 (1985). Toxicity data: China Cooperative Research Group on Qinghaosu, *ibid.* **2**, 31 (1982). Series of articles on chemistry, pharmacology, and antimalarial efficacy: *ibid.* 3-50. Brief reviews: H. Koch, *Pharm. Int.* **2**, 184-185 (1981); D. L. Klayman, *Science* **228**, 1049-1054 (1985). Review of chemistry, pharmacology and clinical applications: X. D. Luo, C. C. Shen, *Med. Res. Rev.* **7**, 29-52 (1987).

Needles, mp 156-157°. $[\alpha]_D^{17}$ +66.3°. (c = 1.64 in CHCl$_3$). Sol in most aprotic solvents. Slightly sol in oil. LD$_{50}$ in mice (mg/kg): 5105 orally; 2800 i.m.; 1558 i.p. (Koch). LD$_{50}$ in mice, rats (mg/kg): 4228, 5576 orally; 3840, 2571 i.m. (China Cooperative Research Group on Qinghaosu).

Dihydro derivative, $C_{15}H_{24}O_5$, *Dihydroartemisinin, dihydroqinghaosu.* Main metabolite of artemisinin, arteether, artemether, artesunate, qqv. HPLC determn: X.-D. Luo et al., *Chromatographia* **23**, 112 (1987).

THERAP CAT: Antimalarial.

845. Artesunate. *[3R-(3α,5aβ,6β,8aβ,9α,10β,12β,-12aR*)]-Butanedioic acid mono(decahydro-3,6,9-trimethyl-3,12-epoxy-12H-pyrano[4,3-j]-1,2-benzodioxepin-10-yl) ester;* artesunic acid; dihydroqinghaosu hemisuccinate; C_{19}-$H_{28}O_8$; mol wt 384.43. C 59.36%, H 7.34%, O 33.30%. Derivative of artemisinin, q.v. Prepn: China Cooperative Research Group on Qinghaosu, *J. Tradit. Chinese Med.* **2**, 9 (1982). Absolute configuration: X.-D. Luo et al., *Helv. Chim. Acta* **67**, 1515 (1984). GC/mass spec. determn.: A. D. Theoharides et al., *Anal. Chem.* **60**, 115 (1988). Pharmacology: Y. Zhao, *J. Trop. Med. Hyg.* **88**, 391 (1985). Antimalarial activity: W. Peters et al., *Ann. Trop. Med. Parasitol.* **80**, 483 (1986); A. J. Lin et al., *J. Med. Chem.* **30**, 2147

(1987). Inhibition of cytochrome oxidase: Y. Zhao et al., *J. Nat. Prod.* **49**, 139 (1986). Toxicology: China Cooperative Research Group on Qinghaosu, *J. Tradit. Chinese Med.* **2**, 31 (1982). Series of articles on chemistry, pharmacology, and antimalarial efficacy: *ibid.* 3-50.

Fine white crystalline powder.

Sodium salt, $C_{19}H_{27}NaO_8$, *SM804.* Poor stability in aqueous solutions. LD$_{50}$ in mice (mg/kg): 520 i.v.; 475 i.m. (China Cooperative Research Group); also reported as 699 ± 58.5 i.v. (Zhao, 1985).

THERAP CAT: Antimalarial.

846. Asafetida. Devil's dung; food of the gods; asafoetida; asant. Gum-resin obtained as an exudation of the decapitated rhizome and roots of *Ferula assafoetida* L., *F. foetida* Regel and some other species of *Ferula, Umbelliferae*. *Habit.* Iran, Turkestan, Afghanistan. *Constit.* 6-17% ethereal oil; 40-60% resin, consisting of ester of asaresinotannol and ferulic acid; pinene, vanillin, about 25% gum. Three sulfur-containing compounds isolated from asafetida resin are: *1-methylpropyl 1-propenyl disulfide, 1-(methylthio)-propyl 1-propenyl disulfide* and *1-methylpropyl 3-(methylthio)-2-propenyl disulfide;* the latter two have pesticidal properties. Studies of various constituents: Bézanger-Beauquesne, Chosson, *Ann. Pharm. Franc.* **16**, 665 (1958); Caglioti et al., *Helv. Chim. Acta* **41**, 2278 (1958); **42**, 2557 (1959). Isoln and structure elucidation of three sulfur-containing components: H. Naimie et al., *Coll. Czech. Chem. Commun.* **37**, 1166 (1972). Synthesis and configuration of sulfur-containing constituents: J. Meijer, P. Vermeer, *Rec. Trav. Chim.* **93**, 242 (1974); A. Kjac et al., *Acta Chem. Scand. B* **30**, 137 (1976). *Review:* Howes, *Econ. Bot.* **4**, 313 (1950); Subrahmanyam et al., *J. Sci. Ind. Res.* **13A**, 382-386 (1954).

Soft mass or irregular lumps or "tears"; garlic-like odor; bitter acrid taste. When triturated with water, it makes a milky emulsion. The N. F. requires not less than 50% alcohol-soluble extractives. *See also* Abiotyl.

USE: In India, Iran, etc., as condiment and flavoring for foods; an ingredient in Worcestershire sauce. A 2% suspension as repellent against dogs, cats, rabbits, deer.

THERAP CAT: Carminative; antispasmodic; expectorant.

THERAP CAT (VET): Has been used as a carminative, and externally to prevent bandage chewing by dogs.

847. Asaprol. *2-Hydroxy-1-naphthalenesulfonic acid calcium salt;* calcium 2-hydroxy-1-naphthalenesulfonate; calcium β-naphthol-α-monosulfonate; calcium 2-naphthol-1-sulfonate; 2-naphthol-1-sulfonic acid calcium salt; Abrastol; Calcinaphthol. $C_{20}H_{14}CaO_8S_2$; mol wt 486.54. C 49.37%, H 2.90%, Ca 8.24%, O 26.31%, S 13.18%. Prepn: *Hagers Handb. Pharm. Praxis* **Band 2**, 204 (Berlin, 1930).

Trihydrate, reddish-white, odorless powder. Dec at about 50°. One gram dissolves in 1.5 ml water, 3 ml alc.

USE: Has been used instead of gypsum to plaster wines.

848. Asarinin. *[1R-(1α,3aα,4β,6aα)]-5,5'-(Tetrahydro-1H,3H-furo[3,4-c]furan-1,4-diyl)bis-1,3-benzodioxole; l-asarinin;* (−)-episesamin; xanthoxylin S. $C_{20}H_{18}O_6$; mol wt 354.34. C 67.79%, H 5.12%, O 27.09%. Naturally occurring l-form isolated from *Xanthoxylum clava-herculis* L. (*X. carolinianum* Lam.), *Rutaceae; Asarum sieboldi* Miguel var.

seulensis Nakai, *A. blumei* Duch., *Aristolochiaceae:* Colton, *Am. J. Pharm.* **52**, 191 (1880); Eberhardt, *ibid.* **62**, 231 (1890); Gordin, *J. Am. Chem. Soc.* **28**, 1649 (1906); H. Dieterle *et al.*, *Arch. Pharm.* **269**, 384 (1931); Huang-Minlon, *Ber.* **70**, 951 (1937); T. Kaku *et al.*, *Keijo J. Med.* **9**, 1 (1934); *C.A.* **32**, 9090[1] (1938). Structure: H. Dieterle, K. Schwenger, *Arch. Pharm.* **277**, 33 (1939). Synthesis of *dl*-form: M. Beroza, M. S. Schechter, *J. Am. Chem. Soc.* **78**, 1242 (1956); K. Freudenberg, E. Fischer, *Ber.* **89**, 1230 (1956); D. R. Stevens, D. A. Whiting, *Tetrahedron Letters* **27**, 4629 (1986). Diastereoisomeric with sesamin, *q.v.* Stereochemistry of the *d*-form: K. Freudenberg, G. S. Sidhu, *ibid.* **94**, 851 (1961). Antitubercular activity: Ramaswamy, *Naturwiss.* **44**, 380 (1957).

Crystals from alc, mp 121°. $[\alpha]_D^{20}$ −118.6°; $[\alpha]_D^{23}$ −122° (chloroform). Practically insol in water. Freely sol in boiling methanol, alcohol, chloroform, acetone, benzene.

d-Form, *episesamin*, (+)-*episesamin*. Crystals from ethanol, mp 121.5°. $[\alpha]_D^{25}$ +124° (chloroform).

dl-Form, crystals, mp 134-135°.

849. Asarones. *1,2,4-Trimethoxy-5-(1-propenyl)benzene;* 2,4,5-trimethoxy-1-propenylbenzene; asarin; asarum camphor; asarabacca camphor. $C_{12}H_{16}O_3$; mol wt 208.25. C 69.21%, H 7.74%, O 23.05%. From root of *Asarum europaeum* L., *Aristolochiaceae* by distillation with water. Also found in the ethereal oils of *A. europaeum* and *A. arifolium* L., *Aristolochiaceae* and in *Acorus calamus* L., *Araceae*. Occurs in nature as a mixture of two isomeric forms, α-*asarone* being the (*E*)- or *trans*-isomer, β-*asarone*, the (*Z*)- or *cis*-isomer. The unqualified term asarone is often used synonymously with α-asarone. Isoln: Gattermann, Eggers, *Ber.* **32**, 289 (1899). Early syntheses: Seshadri, Thiruvengadam, *Proc. Indian Acad. Sci.* **32A**, 110 (1950); Sharma, Dandiya, *Indian J. Appl. Chem.* **32**, 236 (1969). Stereochemistry of isomers: Baxter *et al.*, *Can. J. Chem.* **40**, 154 (1962). Insect chemosterilant activity of β-asarone: B. P. Saxena *et al.*, *Nature* **270**, 512 (1977); *see also* G. Motolesy *et al.*, *Z. Naturforsch.* **35B**, 1449 (1980). Stereospecific synthesis of β-asarone: M. T. S. Hsia *et al.*, *177th Am. Chem. Soc. Meet.* (Honolulu, April 1979), *Abstracts of Papers*, PEST 98. Synthetic and HPLC study of α- and β-asarone: L. Gracza, *Arch. Pharm.* **314**, 972 (1981).

α−asarone   β−asarone

α-Asarone, needles from light petroleum, mp 62-63°. bp 296°. n_D^{11} 1.5719. Practically insol in water; sol in alcohol, ether, glacial acetic acid, carbon tetrachloride, chloroform, petr ether.

850. Asarum. Wild ginger; Canada snakeroot; Indian ginger. Dried rhizome and roots of *Asarum canadense* L., *Aristolochiaceae*. *Habit.* Canada to N. Carolina and Kansas. *Constit.* Acrid resin, arom. volatile oil, methyl eugenol.

851. Asbestos. Amianthus. Fibrous mineral silicates. Divided into two groups: *serpentine* and *amphibole*. Most common form is *chrysotile* [$Mg_6(Si_4O_{10})(OH)_8$], the fibrous form of serpentine (*see also* magnesium silicates). Subdivi-

sions of amphibole are *anthophyllite* [$(Mg,Fe)_7(Si_8O_{22})(OH)_2$] (low iron content); *amosite* [$Fe_5Mg_2(Si_8O_{22})(OH)_2$]; *actinolite* [$Ca_2(Mg,Fe)_5(Si_8O_{22})(OH)_2$]; *tremolite* [$Ca_2Mg_5(Si_8O_{22})$-$(OH)_2$]; *crocidolite* or *blue asbestos* [$Na_2Fe_3^{2+}Fe_2^{3+}(Si_8O_{22})$-$(OH)_2$]. Reviews of carcinogenicity and toxicology: T. J. Haley, *J. Pharm. Sci.* **64**, 1435-1449 (1975); *IARC Monographs* **14**, 1-106 (1977); *Arch. Pathol. Lab. Med.* **106**, 541-596 (1982). Review of properties and industrial applications: W. C. Streib in Kirk-Othmer *Encyclopedia of Chemical Technology* vol. 3 (Wiley-Interscience, New York, 3rd ed., 1978) pp 267-283.

Fire resistant fibers. Chrysotile attacked by acid; amphiboles, acid resistant.

Caution: Occupational exposure to the dust can result in mesothelioma, squamous cell carcinoma and adenocarcinoma of the lung after a long latent period. This substance has been listed as a known carcinogen: *Fourth Annual Report on Carcinogens* (NTP 85-002, 1985) p 29.

USE: Heat-resistant insulators, cements, furnace and hot pipe coverings, inert filler medium (laboratory & commercial), fireproof gloves, clothing, brake linings. NaOH treated asbestos, *Ascarite*, has been used to absorb CO_2 in combustion analysis.

852. Ascaridole. *1-Methyl-4-(1-methylethyl)-2,3-dioxabicyclo[2.2.2]oct-5-ene;* 1,4-peroxido-*p*-menthene-2; Ascarisin. $C_{10}H_{16}O_2$; mol wt 168.23. C 71.39%, H 9.59%, O 19.02%. An organic peroxide which constitutes 60-80% of oil of chenopodium. Synthesis from α-terpinene by treatment with oxygen, chlorophyll, and light: Schenck, Ziegler, *Naturwiss.* **1944**, 157. Purification: Beckett *et al.*, *J. Pharm. Pharmacol.* **7**, 55 (1955).

Liquid; unstable; prone to explode when heated or when treated with organic acids. d_4^{20} 1.0103; d_{20}^{20} 1.0113. mp +3.3°. $bp_{0.2}$ 39-40°. $[\alpha]_D^{20}$ ±0.00. Sol in hexane, pentane, ethanol, toluene, benzene, castor oil.

THERAP CAT: Has been used as anthelmintic (Nematodes).
THERAP CAT (VET): Anthelmintic.

853. Asclepias. Pleurisy root; butterfly weed. Dried root of *Asclepias tuberosa* L., *Asclepiadaceae*. *Habit.* Ontario to Minnesota. *Constit.* Asclepiadin, resins, volatile oil.

854. Asclepias syriaca. Milkweed; silkweed; wild cotton. Root of *Asclepias syriaca* L. (*A. cornuti* Decaisne), *Asclepiadaceae*. *Habit.* Canada to North Carolina and Kansas. *Constit.* Asclepiadin, asclepion—a bitter principle; tannin, volatile oil.

855. Ascorbic Acid. L-*Ascorbic acid;* vitamin C; L-xyloascorbic acid; 3-oxo-L-gulofuranolactone (enol form); L-3-ketothreohexuronic acid lactone; antiscorbutic vitamin; cevitamic acid; Cebid; Cebion; Cantaxin; Celaskon; Cevalin; Cevatine; Cevimin; Cevitex; Cewin; Cipca; Cebicure; C-Vimin; Cevitamin; Testascorbic; Allercorb; Cecon; Cetebe; Ce-Vi-Sol; Ascorin; Ascorteal; Cegiolan; Adenex; Ascorvit; Cevex; Lemascorb; Ciamin; Hybrin; Vitacee; Cantan; Catavin C; Celin; Cenetone; Cescorbat; Cereon; Cergona; Cetemican; Cetamid; Planavit C; Colascor; Concemin; Duoscorb; Scorbacid; Davitamon C; Proscorbin; Redoxon; Scorbu-C; Ribena; Vicelat; Vitacin; Vitacimin; Vitascorbol; Xitix; Cevitan; Laroscorbine. $C_6H_8O_6$; mol wt 176.12. C 40.91%, H 4.58%, O 54.51%. Widely distributed in the plant and animal kingdom. Good sources are citrus fruits, hip berries, acerola, fresh tea leaves. Isolated from the adrenal cortex of ox and later from lemons and paprika (originally called hexuronic acid): Szent-Györgyi, *Biochem. J.* **22**, 1387 (1928); Haworth, Szent-Györgyi, *Nature* **131**, 24 (1933). Structure studies: Herbert *et al.*, *J. Chem. Soc.* **1933**, 1270. Crystal structure: Hvoself, *Acta Chem. Scand.* **18**, 841 (1964). Synthesis: Ault *et al.*, *J. Chem. Soc.* **1933**, 1419;

Reichstein *et al.*, *Helv. Chim. Acta* **16**, 561, 1019 (1933); **17**, 311, 510 (1934); Bakke, Theander, *Chem. Commun.* **1971**, 175; R. J. Ferrier, R. H. Furneaux, *ibid.* **1977**, 332; T. C. Crawford, R. Breitenbach, *ibid.* **1979**, 388; G. Andrews *et al.*, *ibid.* 740. Review of syntheses: T. C. Crawford, S. A. Crawford, *Advan. Carbohyd. Chem.* **37**, 79-155 (1980). *Reviews:* E. L. Hirst, *Fortschr. Chem. Org. Naturst.* **2**, 132-159 (1939); H. R. Rosenberg, *Chemistry and Physiology of the Vitamins* (Interscience, New York, 1945); *Ann. N.Y. Acad. Sci.* **92**, entitled "Vitamin C", J. J. Burns, Ed. (1961) pp 1-332; A. K. Sim, *Chem. & Ind. (London)* **1972**, 160-165; *Ann. N.Y. Acad. Sci.* **258**, entitled "Second Conference on Vitamin C", C. G. King, J. J. Burns, Eds. (1975) pp 1-552. Discussion of the use of ascorbic acid in the treatment of the common cold: Karlowski *et al.*, *J. Am. Med. Assoc.* **231**, 1038 (1975); Dykes, Meier, *ibid.* 1073; Pauling, *Med. Trib.* **17**, 1, 18 (March 24, 1976); 37 (April 7, 1976). Clinical applications in immunology, lipid metabolism and cancer: *Int. J. Vit. Nutr. Res.* **1982**, Suppl. 23, 294 pp. Comprehensive description: I. A. Al-Meshal, M. M. A. Hassan, in *Analytical Profiles of Drug Substances* vol. **11**, K. Florey, Ed. (Academic Press, New York, 1982) pp 45-78. Review of biochemistry, physiology and clinical uses: M. Levine, *N. Engl. J. Med.* **314**, 892-902 (1986).

Crystals (usually plates, sometimes needles, monoclinic system). mp 190-192° (some dec). Pleasant, sharp acidic taste. Stable to air when dry. In impure prepns and in many natural products the vitamin oxidizes on exposure to air and light. d 1.65. $[\alpha]_D^{25}$ +20.5° to +21.5° (c = 1); $[\alpha]_D^{23}$ +48° (c = 1 in methanol). pH = 3 (5 mg/ml); pH = 2 (50 mg/ml); pK_1 = 4.17; pK_2 = 11.57. uv max: 245 nm (acid soln); 265 nm (neutral soln). Redox potential of first stage at pH 5.0 is $\epsilon_0{}'$ = +0.127 v. One gram dissolves in about 3 ml water, 30 ml alc, 50 ml abs alc, 100 ml glycerol U.S.P., 20 ml propylene glycol. Soly in water: 80.0% at 100°; 40.0% at 45°. Insol in ether, chloroform, benzene, petr ether, oils, fats, fat solvents. Possesses relatively strong reducing power, decolorizes many dyes. Aq solns are rapidly oxidized by air. The reaction is accelerated by alkalies, iron, copper. Forms stable metal salts, *see* Sodium Ascorbate. *Pharmaceutical Incompat.* Vitamin C should not be formulated with sodium salicylate, sodium nitrite, theobromine sodium salicylate, methenamine: R. K. Aliev, M. A. Etinger, *Aptechnoe Delo* **2**, 7 (1952), *C.A.* **46**, 7707b (1952).

Calcium hypophosphite, $C_6H_9CaO_8P$, *asphocalcium, Calscorbat, Calscorbate.*

Note: One unit (U.S.P. or international) is the vitamin C activity of 0.05 mg of the U.S.P. ascorbic acid reference standard.

USE: As antimicrobial and antioxidant in foodstuffs.

THERAP CAT: Treatment of vitamin C deficiency.

THERAP CAT (VET): Treatment of vitamin C deficiency in primates, guinea pigs, fish.

856. Ascorbigen. *2-C-(1H-Indol-3-ylmethyl)-β-L-lyxo-3-hexulofuranosonic acid γ-lactone mixt with 2-C-(1H-indol-3-ylmethyl)-β-L-xylo-3-hexulofuranosonic acid γ-lactone.* $C_{15}H_{15}NO_6$; mol wt 305.28. C 59.01%, H 4.95%, N 4.59%, O 31.45%. Isoln from savoy cabbage juice: Procházka, Sanda, *Coll. Czech. Chem. Commun.* **25**, 270 (1960). Prepn from L-ascorbic acid + glucobrassicin: Gmelin, Virtanan, *Suomen Kemistilehti* **34B**, 15 (1961), *C.A.* **55**, 17774f (1961); from L-ascorbic acid + gramine methiodide: Procházka, *Coll. Czech. Chem. Commun.* **28**, 544 (1963). Prepn from L-ascorbic acid + 3-hydroxymethylindole; forms two diastereoisomers: *ascorbigen A* as the main product and *ascorbigen B* as a minor product: Kiss, Neukom, *Helv. Chim. Acta* **49**, 989 (1966); Kiss, *Angew. Chem.* **78**, 1066 (1966).

ascorbigen A   ascorbigen B

Ascorbigen A, amorphous powder, sinters about 65°. $[\alpha]_D^{25}$ +11.0° (c = 2 in ethanol). uv max (ethanol): 220, 273-274, 280, 290 nm.

Ascorbigen B, yellow amorphous powder, sinters about 70°. $[\alpha]_D^{25}$ +12.5° (methanol).

857. Asiaticoside. *2,3,23-Trihydroxyurs-12-en-28-oic acid O-6-deoxy-α-L-mannopyranosyl-(1→4)-O-β-D-glucopyranosyl-(1→6)-O-β-D-glucopyranosyl ester;* Centelase Dermatologico; Madecassol. $C_{48}H_{78}O_{19}$; mol wt 959.15. C 60.11%, H 8.20%, O 31.69%. Active principle of *Centella asiatica* (L.) Urban, *Umbelliferae.* Trisaccharide moiety linked to the aglycone: *asiatic acid.* Isoln: J.-E. Bontems, *Bull. Sci. Pharmacol.* **49**, 186 (1941), *C.A.* **38**, 4094 (1944); idem, *Gaz. Med. Madagascar* **5**, 29 (1942). Structure of asiatic acid: J. Polonsky, *Bull. Soc. Chim. France* **1953**, 173. Structure of asiaticoside: P. Boiteau *et al.*, *Nature* **163**, 258 (1949); J. Polonsky *et al.*, *Bull. Soc. Chim. France* **1959**, 880. Metabolism: L. F. Chasseaud *et al.*, *Arzneimittel-Forsch.* **21**, 1379 (1971). Wound healing properties: H. Rosen *et al.*, *Proc. Soc. Exp. Biol. Med.* **125**, 279 (1967). Clinical study: J.-P. Bosse *et al.*, *Ann. Plast. Surg.* **3**, 13 (1979). Review on asiatic acid: J. L. Simonsen, W. C. J. Ross, *The Terpenes* vol. **5** (University Press, Cambridge, 1957) pp 58-67.

Minute needles from 60% methanol, mp 230-233°. Insol in water. Sol in alcohol, pyridine. $[\alpha]_D^{20}$ −14° (alc).

THERAP CAT: Vulnerary.

858. L-Asparaginase. L-Asparagine amidohydrolase; colaspase; L-asnase; E.C. 3.5.1.1; MK-965; NSC-109229; Re 82-TAD-15; Crasnitin; Elspar; Kidrolase; Leucogen; Leunase. An enzyme which catalyzes the hydrolysis of L-asparagine to L-aspartate and ammonia. Monomer consists of four subunits of mol wt about 33,000 each, for a unit mol wt of 133,000 ± 5000. It is widely distributed, occurring in fungi, *Penicillium camemberti* and *Aspergillus niger;* in yeasts; in bacteria, *Escherichia coli, Mycobacteria phlei, Pseudomonas fluorescens* (prepn also contains glutaminase activity), *Bacillus coagulans, Brucella abortus, Serratia marcescens;* in plants, barley rootlets; in animal tissues; and in sera of guinea pigs and rodents known as *Cavioidea.* Pure enzyme has been obtained from guinea pig serum, yeast, *B. coagulans* and *E. coli* with the latter yielding two different asparaginases. Purif, chemical properties of L-asparaginase from *E. coli:* Ho *et al., J. Biol. Chem.* **245**, 3708 (1970). Primary structure of asparaginase from *E. coli:* T. Maita, G. Matsuda, *Z. Physiol. Chem.* **361**, 105 (1980). Experimental and clinical effects: *Recent Results Cancer Res.* **vol. 33**, E. Grundmann, H. F. Oettgen, Eds., (Springer-Verlag, New York, 1970). *Reviews:* Adamson, Fabro, *Cancer Chemother. Rep.* **52**, part I, 617 (1968); Capizzi *et al., Ann. Rev. Med.* **21**, 433 (1970); Wriston, Yellin, *Advan. Enzymol. Relat. Areas Mol. Biol.* **39**, 185 (1973); R. L. Capizzi, Y. C. Cheng, *Enzymes as Drugs,* J. S. Holcenberg, J. Roberts, Eds. (Wiley, New York, 1981) pp 1-24.

White crystalline powder. Freely sol in water where it appears to be global in shape. $[\alpha]_D^{20}$ −30° to −32°. uv max (0.03M sodium phosphate, pH 7.3): 278 nm (A$_{1cm}^{1\%}$ 7.1 ± 0.15). Active at pH 5-9. Practically insol in methanol, acetone, chloroform.

THERAP CAT: Antineoplastic (acute leukemia).

859. Asparagine. Asn (IUPAC abbrev.); L-β-asparagine; D-β-asparagine; α-aminosuccinamic acid; aspartic acid β-amide; altheine; asparamide; agedoite. $C_4H_8N_2O_3$; mol wt 132.12. C 36.36%, H 6.10%, N 21.20%, O 36.33%. NH_2-$COCH_2CH(NH_2)COOH$. One of the non-essential amino acids. L-Asparagine is much more common than D-asparagine and occurs as the monohydrate. Isolation from sprouting vetch (*Vicia sativa* L., *Leguminosae)*: Piria, *Ann.* **68**, 343 (1848); Piutti, *Ber.* **19**, 1691 (1886); *Gazz. Chim. Ital.* **18**, 463 (1888), *C.A.* **19**, 3285 (1925); from white lupine and soybean seedlings: Vickery *et al., J. Biol. Chem.* **145**, 45 (1942). Synthesis of DL-asparagine starting with ethyl oxaloacetate: Cocker, *J. Chem. Soc.* **1940**, 1489. Configuration: Timmermans *et al., Bull. Soc. Chim. Belg.* **48**, 33 (1939). Structure: Steward, Thompson, *Nature* **169**, 739 (1952).

L-Asparagine monohydrate, the common asparagine. Orthorhombic bisphenoidal crystals, mp 234-235° (bath preheated to 226°). d$_4^{15}$ 1.543. Acid to litmus. pK$_1$ 2.02; pK$_2$ 8.80. $[\alpha]_D^{20}$ −5.42° (c = 1.3); $[\alpha]_D^{20}$ +20.0° (c = one mol in 1M HCl); $[\alpha]_D^{20}$ −9.3° (c = one mol in 1M NaOH). A satd aq soln contains (monohydrate) 0.95% w/w at 0°; 1.79% at 10.5°; 3.53% at 28°; 5.73% at 40°; 8.97% at 50°; 27.93% at 78°; 52.75% at 100°. Practically insol in methanol, ethanol, ether, benzene. Sol in acids, alkalies.

D-Asparagine monohydrate, crystals, mp 215°. $[\alpha]_D^{20}$ +5.41° (c = 1.3).

860. Asparagus. Shoot of *Asparagus officinalis* L., *Liliaceae. Habit.* Europe; cultivated everywhere. *Constit.* Asparagine, tyrosine, succinic acid, arginine, α-aminodimethyl-γ-butyrothetin (a methylsulfonium deriv of methionine), fat, sugar. In some humans, ingestion of asparagus is followed by excretion of a substance that produces a characteristic strong odor in the urine: M. Nencki, *Arch. Exp. Pathol. Pharmacol.* **28**, 206 (1891); C. Gautier, *C.R. Soc. Biol.* **89**, 239 (1923); A. C. Allison, K. G. McWhirter, *Nature* **178**, 748 (1956). The odor-causing substance was originally thought to be methyl mercaptan, but subsequent investigation using GC-mass spectrometry has suggested that S-methylthioacrylate and S-methyl 3-(methylthio)thiopropionate are the malodorous agents, *cf* R. H. White, *Science* **189**, 810 (1975).

861. Aspartame. N-L-α-Aspartyl-L-phenylalanine 1-methyl ester; 3-amino-N-(α-carboxyphenethyl)succinamic acid N-methyl ester; APM; SC 18862; Canderel; Equal; NutraSweet; Sanecta; Tri-Sweet. $C_{14}H_{18}N_2O_5$; mol wt 294.30. C 57.14%, H 6.16%, N 9.52%, O 27.18%. Dipeptide ester about 160 times sweeter than sucrose in aqueous solution. Prepn: Davey *et al., J. Chem. Soc. (C)* **1966**, 555; Schlatter, **S. Afr.** pat. **67 02,190** corresp to **U.S.** pat. **3,492,-131** (1968, 1970 to Searle); H. Pietsch, *Tetrahedron Letters* **1976**, 4053; K. J. Vinick, S. Jung, *ibid.* **23**, 1315 (1982); utilizing immobilized enzyme technology: C. Fuganti, P. Grasselli, *ibid.* **27**, 3191 (1986). Structure-taste relationship: Mazur *et al., J. Am. Chem. Soc.* **91**, 2684 (1969). Potential as a low-calorie sweetener: Cloninger, Baldwin, *Science* **170**,

81 (1970). Metabolism: Oppermann *et al., J. Nutr.* **103**, 1454, 1460 (1973).

$$H_2N \!-\!\!\!-\! CHCONHCHCH_2 \quad \text{---} \quad \text{(ring)}$$

Colorless needles from water, mp 246-247°. $[\alpha]_D^{22}$ − 2.3° (1N HCl).

USE: Sweetener.

862. Aspartic Acid. Asp (IUPAC abbrev.); aminosuccinic acid; asparagic acid; asparaginic acid; 2-aminobutanedioic acid; Asparaginsäure (German). $C_4H_7NO_4$; mol wt 133.10. C 36.09%, H 5.30%, N 10.52%, O 48.08%. HOOC-$CH_2CH(NH_2)COOH$. Commonly occurs in the L-form. One of the nonessential amino acids. Occurs in animals and plants, esp in young sugar cane and in sugar beet molasses. Usually obtained by hydrolysis of asparagine: Cocker, *J. Chem. Soc.* **1940**, 1489. Prepd from ammonia and fumaric acid: Enkvist, *Ber.* **72B**, 1927 (1939). Synthesis of DL-form: Dunn, Smart, *Org. Syn.,* **coll. vol. IV**, 55 (1963). Resolution of DL-form: Harada, *Bull. Chem. Soc. Japan* **37**, 1383 (1964). Synthesis of optically active forms: Harada, Matsumoto, *J. Org. Chem.* **31**, 2985 (1966); **32**, 1790 (1967); Vigneron *et al., Bull. Soc. Chim. France* **1972**, 3836. Crystal structure: B. Dawson, *Acta Crystallogr.* **33B**, 882 (1977). L-Form as excitatory neurotransmitter: H. McLennan, H. V. Wheal, *Can. J. Physiol. Pharmacol.* **54**, 70 (1976). Brief review of role of EAAs: P. J. Roberts, S. W. Davies, *Biochem. Soc. Trans.* **15**, 218-219 (1986). *Review:* Greenstein, Winitz, *Chemistry of the Amino Acids* **3** (John Wiley, New York, 1961) pp 1856-1878.

L-Form, orthorhombic bisphenoidal leaflets or rods, mp 270-271° (sealed capillary, preheated bath). d$^{12.5}$ 1.661. $[\alpha]_D^{20}$ +25.0° (c = 1.97 in 6N HCl). pK$_1'$ 1.88; pK$_2'$ 3.65; pK$_3'$ 9.60. Soly in water at 20°, 30°: 1 g/222.2 ml, 1 g/149.9 ml. Forms supersatd solns easily. More soluble in salt solns; sol in acids, alkalies; insol in alcohol.

D-Form, $[\alpha]_D^{27}$ −23.0° (c = 2.30 in 6N HCl).

Potassium salt (L-form), *Aspara K.*

Mixture of the potassium salt and magnesium salt, *Spartase, Trophicard, Aspara, Aspartat.*

Ferrous salt tetrahydrate, $C_8H_{12}FeN_2O_8.4H_2O$, *ferrous aspartate, Sideryl, Spartocine.* Prepn: Gaudin, **Brit.** pat. **910,321** (1962).

Calcium salt, *calcium aspartate, asparaginate calcium, Calciretard.*

Compound with L-arginine, $C_{10}H_{21}N_5O_6$, *arginine L-aspartate, Dynamisan, Sargenor.*

Compound with L-ornithine, L-*ornithine* L-*aspartate, Ormeta, Orparan.*

THERAP CAT: Roborant.

863. Aspergillic Acid. 1-Hydroxy-6-(1-methylpropyl)-3-(2-methylpropyl)-2(1H)-pyrazinone; 6-sec-butyl-1-hydroxy-3-isobutyl-2(1H)-pyrazinone; 6-sec-butyl-3-isobutylpyrazinol 1-oxide; 2-hydroxy-3-isobutyl-6-(1-methylpropyl)pyrazine 1-oxide; 3-isobutyl-6-sec-butyl-2-hydroxypyrazine 1-oxide. $C_{12}H_{20}N_2O_2$; mol wt 224.30. C 64.25%, H 8.99%, N 12.49%, O 14.27%. Antibiotic substance produced by *Aspergillus flavus:* White, Hill, *J. Bacteriol.* **45**, 433 (1943); Dunn *et al., J. Chem. Soc.* **1949** Suppl. S126. Extraction and purifn: Dutcher, *J. Biol. Chem.* **171**, 321 (1947). Structure: *idem, ibid.* **232**, 785 (1958). Biosynthesis: MacDonald, *ibid.* **236**, 512 (1961); **237**, 1977 (1962). Synthesis of racemate: Chigira *et al., Bull. Chem. Soc. Japan* **39**, 632 (1966); Masaki *et al., J. Org. Chem.* **31**, 4143 (1966); Ohta, Fujii, *Chem. Pharm. Bull.* **17**, 851 (1969). Industrial prepn: Omata, Ueno, **Japan.** pat. **13,794('65)** (to Dainippon), *C.A.* **63**, 11589b (1965). Proposed as a hypotensive: Jones, Martin, **U.S.** pat. **3,720,768** (1973 to Abbott). *Reviews:* MacDonald, "Aspergillic Acid and Related Compounds" in *Antibiotics,* D. Gottlieb, P. Shaw, Eds. (Springer-Verlag, New York, 1967) **vol. II**, pp 43-51; Wilson, "Miscellaneous Aspergillus Toxins", in *Microbial Toxins,* A. Ciegler *et al.,* Eds. (Academic Press, New York, 1971) **vol. VI**, pp 207-235.

Pale yellow needles having odor similar to black walnuts, mp 97-99° (methanol). pK'a 5.5. $[\alpha]_D^{18}$ +13.3° (c = 3.9 in ethanol). uv max (water pH 8): 328, 235 nm (ϵ 8,500, 10,500). Slightly sol in cold water; sol in dil acids and alkalies, alcohol, ether, acetone, benzene, chloroform, pyridine.

864. Aspergillin. The inappropriate designation of a number of different antibacterial agents produced by *Aspergilli* as "aspergillin" has resulted in confusion. The legitimate contender for this name by prior use is an allomelanin produced by mature spores of *A. niger:* Hugouneng, Florence, *Bull. Soc. Chim. Biol.* **2**, 133 (1920). This pigment is a mixture of chemically similar macromolecules of which the fundamental monomer is probably the perylene unit substituted with several oxygenated functions. *Review:* R. A. Nicolaus, *Melanins* (Hermann, Paris, 1968) pp 130-142.

Dried pigment, as shiny black blocks, does not melt but decomposes above 200°. Slightly sol in polar solvents. uv max (2% aqueous soln): 295, 450 nm.

865. Asperlicin. *[2S-[2α,9β,9(R*),9aβ]]-6,7-Dihydro-7-[[2,3,9,9a-tetrahydro-9-hydroxy-2-(2-methylpropyl)-3-oxo-1H-imidazo[1,2-a]indol-9-yl]methyl]quinazolino[3,2-a][1,4]benzodiazepine-5,13-dione.* $C_{31}H_{29}N_5O_4$; mol wt 535.60. C 69.52%, H 5.46%, N 13.07%, O 11.95%. Naturally occurring, nonpeptide cholecystokinin (CCK) antagonist. Produced by several strains of *Aspergillus alliaceus* along with minor related compounds known as asperlicin B, C, D and E. Isoln from fermentation cultures of *A. alliaceus* Thom and Church, ATCC 20655 and ATCC 20656: R. L. Monaghan *et al.*, **Eur. pat. Appl.** 116,150; *eidem*, **U.S. pat.** **4,530,790** (1984, 1985 both to Merck & Co.). Fermentation, isoln and bioactivity: M. A. Goetz *et al.*, *J. Antibiot.* **38**, 1633 (1985); of asperlicins B, C, D and E: *eidem, ibid.* **41**, 875 (1988). Structure elucidation: J. M. Liesch *et al., ibid.* **38**, 1638 (1985); of asperlicins B, C, D and E: *eidem, ibid.* **41**, 878 (1988). Biosynthetic study: D. R. Houck *et al., ibid.* 882. Total synthesis of asperlicins C and E: M. G. Bock *et al., J. Org. Chem.* **52**, 1644 (1987). Pharmacology of asperlicin: R. S. L. Chang *et al., Science* **230**, 177 (1985). Effect on pancreatic enzyme secretion *in vitro:* K. A. Zucker *et al., Surgery* **102**, 163 (1987). Effect in exptl pancreatitis in rats: J. R. Wisner, Jr., I. G. Renner, *Pancreas* **3**, 174 (1988).

White crystals, mp 211-213°. $[\alpha]_D^{26.5}$ -185.3° (c = 1.10 in methanol). uv max (methanol): 310.5 nm (ϵ 4075). Sol in methylene chloride, acetone and lower alcohols. Insol in water.

Asperlicin B, $C_{31}H_{29}N_5O_5$, colorless powder. uv max (methanol): 310 nm (ϵ 4000).

Asperlicin C, $C_{25}H_{18}N_4O_2$, off-white powder. uv max (methanol): 222, 268, 278, 310 nm (ϵ 56300, 15100, 14000, 4650).

Asperlicin D, $C_{25}H_{18}N_4O_2$, colorless powder. uv max (methanol): 222, 290, 310 nm (ϵ 61060, 14600, 5920).

Asperlicin E, $C_{25}H_{18}N_4O_3$, off-white powder. uv max (methanol): 227, 268, 324 nm (ϵ 38450, 8680, 3200).

866. Asperuloside. *[2aS-(2aα,4aα,5α,7bα)]-4-[(Acetyloxy)methyl]-5-(β-D-glucopyranosyloxy)-2a,4a,5,7b-tetrahydro-1H-2,6-dioxacyclopent[cd]inden-1-one;* rubichloric acid. $C_{18}H_{22}O_{11}$; mol wt 414.38. C 52.17%, H 5.35%, O 42.48%. From herb of *Asperula odorata* L., *Galium aparine* L., *Rubiaceae.* Occurs also in *Coprosma* spp. Isoln: Hérissey, *Compt. Rend.* **180**, 1695 (1925); Briggs, Nicholls, *J. Chem. Soc.* **1954**, 3940. Structure: Grimshaw, *Chem. & Ind. (London)* **1961**, 403; Briggs *et al., J. Chem. Soc.* **1965**, 2595.

Needles from alcohol or acetone, mp 131-132°. $[\alpha]_D^{25}$ -198.6° (c = 1.44 in water). Absorption spectra: Briggs, Cain, *J. Chem. Soc.* **1954**, 4182. Sol in water, methanol, ethanol, acetone, ethyl acetate, dioxane, pyridine, acetic acid. Practically insol in ether, benzene, chloroform, ligroin.

867. Asphalt. Asphaltum; mineral pitch; Judean pitch; bitumen. Bituminous substance resulting from petroleum by evaporation of lighter hydrocarbons and partial oxidation of the residue. Occurs in West Indies (chiefly Trinidad), Venezuela, Dead Sea, Switzerland, etc.

The "Syriac" asphalt (from the Dead Sea) forms deep black, shining, brittle masses of conchoidal fracture; faint, pitch-like odor and luster. Burns with a bright flame. d 1.00-1.18. Insol in water, alc, acids, alkalies; sol in oil turpentine, petroleum, CS_2, chloroform, ether, acetone.

USE: Making roads, roofs; making tanks watertight.

868. Aspidin. *2-[[2,6-Dihydroxy-4-methoxy-3-methyl-5-(1-oxobutyl)phenyl]methyl]-3,5-dihydroxy-4,4-dimethyl-6-(1-oxobutyl)-2,5-cyclohexadien-1-one; 3'-[(5-butyryl-2,4-dihydroxy-3,3-dimethyl-6-oxo-1,4-cyclohexadien-1-yl)methyl]-2',4'-dihydroxy-6'-methoxy-5'-methylbutyrophenone;* polystichin. $C_{25}H_{32}O_8$; mol wt 460.51. C 65.20%, H 7.00%, O 27.80%. Active principle of fern root: Boehm, *Ann.* **302**, 171 (1898); **329**, 321 (1903); from *Dryopteris austriaca* (Jacq.) Woynar, *Polypodiaceae:* Aebi *et al., Helv. Chim. Acta* **40**, 266 (1957). Structure: *eidem, ibid.* 569. Synthesis: Riedl, Mitteldorf, *Ber.* **89**, 2595 (1956).

Crystals from ethanol, mp 124-125°. uv max (cyclohexane): 230, 290 nm (ϵ 25500, 21300). Soluble in ether, benzene, chloroform; less sol in petr ether; sparingly sol in methanol, ethanol, acetone.

THERAP CAT: Anthelmintic (Cestodes).

869. Aspidinol. *1-(2,6-Dihydroxy-4-methoxy-3-methylphenyl)-1-butanone; 2',6'-dihydroxy-4'-methoxy-3'-methyl-1-butyrophenone;* 4-butyryl-2-methylphloroglucinol 1-methyl ether; 4-butyryl-3,5-dihydroxy-1-methoxy-2-methylbenzene. $C_{12}H_{16}O_4$; mol wt 224.25. C 64.27%, H 7.19%, O 28.54%. Occurs in extracts of male fern: Boehm, *Ann.* **318**, 247 (1901); Hausmann, *Arch. Pharm.* **237**, 559 (1899). Isoln from *Dryopteris austriaca* (Jacq.) Woynar, *Polypodiaceae:* Aebi *et al., Helv. Chim. Acta* **40**, 266 (1957). Synthesis from 2-methylphloroglucinol 1-methyl ether: Karrer, Widmer, *Helv. Chim. Acta* **3**, 392 (1920); Riedl, Mitteldorf, *Ber.* **89**, 2589 (1956).

Needles or prisms from benzene. mp 156-161°. Freely sol in alc, ether, chloroform, acetone; sparingly sol in water, benzene. Less soluble in ligroin than pseudoaspidinol, mp 116.5°. Sol in NaOH solns. Practically insol in Na_2CO_3 solns.

THERAP CAT: Anthelmintic (Cestodes).

870. Aspidium. Male fern; male shield-fern; filix mas (B.P.). Rhizome and stipes of *Dryopteris filix-mas* (L.) Schott., *Polypodiaceae*. *Habit.* North America, Northern Asia, Europe, Northern Africa. *Constit.* Filicic and flavaspidic acids, volatile oil, asbaspidin, filicin, filmaron, filix red, resin. It yields not less than 6.5% oleoresin (U.S.P.).

THERAP CAT: Anthelmintic.

THERAP CAT (VET): See Oleoresin of Aspidium.

871. Aspidosperma. Quebracho. Dried bark of *Aspidosperma quebracho-blanco* Schlecht., *Apocynaceae*. *Habit.* Argentina. *Constit.* 0.3-1.4% alkaloids—aspidospermine, aspidospermatine, aspidosamine, quebrachine, quebrachamine, hypoquebrachine, quebrachol, quebrachit; tannin.

USE: Chlorinated quebracho can be used to control nematodes and other parasitic worms in soils: Santmyer, U.S. pat. **2,799,612** (1957 to Monsanto).

THERAP CAT: Respiratory stimulant.

872. Aspidospermine. *1-Acetyl-17-methoxyaspidospermidine*. $C_{22}H_{30}N_2O_2$; mol wt 354.48. C 74.54%, H 8.53%, N 7.90%, O 9.03%. In *Aspidosperma quebracho-blanco* Schlect., *Vallesia dichotoma* Ruiz & Pav., and *Vallesia glabra* (Cav.) Link, *Apocynaceae*: G. Fraude, *Ber.* **11**, 2189 (1878); Hesse, *Ann.* **211**, 249 (1882); Ewins, *J. Chem. Soc.* **105**, 2738 (1914); Deulofeu *et al., ibid.* **1940**, 1051; Holker *et al., J. Org. Chem.* **24**, 314 (1959). Structure: Conroy *et al., Tetrahedron Letters* no. 11, 4 (1959). Crystal structure: Mills, Nyburg, *J. Chem. Soc.* **1960**, 1458. Stereochemistry: Smith, Wrobel, *ibid.* 1463; Craven, Zacharias, *Experientia* **24**, 770 (1968). Biogenesis: Robinson, *Tetrahedron Letters* no. 18, 1F4 (1959). Synthesis of *dl*-form: Stork, Dolfini, *J. Am. Chem. Soc.* **85**, 2872 (1963); Ban *et al., Tetrahedron Letters* **1965**, 2261; Stevens *et al., Chem. Commun.* **1971**, 857; S. F. Martin *et al., J. Am. Chem. Soc.* **102**, 3294 (1980). Aspidospermine has been reported as having a wide variety of pharmacological properties, including diuretic and respiratory stimulant activity. Biological and phytochemical evaluation: R. L. Lyon *et al., J. Pharm. Sci.* **62**, 218 (1973). Microbial transformation: S. K. Lin *et al., ibid.* **64**, 2021 (1975).

Needles or prisms from alc, needles from petr ether. mp 208°. Sublimes 180°. bp$_2$ 220°. $[\alpha]_D^{15}$ —100.2° (alc); $[\alpha]_D$ —93° (chloroform). uv max (methanol): 218, 255, 280-290 nm (log ϵ 4.52, 4.04, 3.53-3.40). One gram dissolves in 60 ml water, 50 ml alc, 100 ml ether. Also sol in benzene, chloroform, petr ether. LD$_{50}$ in mice: 40 mg/kg i.p., *RTECS* Vol. 1, R. J. Lewis, R. L. Tatken, Eds. (1979) p 156.

N-Formyl-*N*-deacetylaspidospermine, $C_{21}H_{28}N_2O_2$, *vallesine*. Structure: Taylor *et al., Helv. Chim. Acta* **42**, 2750 (1959). Long, fine needles from acetone, mp 154-156°. $[\alpha]_D^{24}$ —91 ± 2° (c = 1.814 in abs alc). uv max: 211, 250 nm (log ϵ 4.47, 3.94).

873. Aspirin. *2-(Acetyloxy)benzoic acid; salicylic acid acetate*; 2-acetoxybenzoic acid; acidum acetylsalicylicum;

acetylsalicylic acid; Acetilum Acidulatum; Acenterine; Aceticyl; Acetophen; Acetosal; Acetosalic Acid; Acetosalin; Acetylin; Acetyl-SAL; Acimetten; Acylpyrin; A.S.A.; Asatard; Aspro; Asteric; Caprin; Claradin; Colfarit; Contrheuma retard; Cosprin; Delgesic; Duramax; ECM; Ecotrin; Empirin; Encaprin; Endydol; Entrophen; Enterosarine; Helicon; Levius; Longasa; Measurin; Neuronika; Platet; Rhodine; Salacetin; Salcetogen; Saletin; Solprin; Solpyron; Xaxa. $C_9H_8O_4$; mol wt 180.15. C 60.00%, H 4.48%, O 35.53%. Prepn: C. Gerhardt, *Ann.* **87**, 149 (1853). Manuf from salicylic acid and acetic anhydride: Faith, Keyes & Clark's *Industrial Chemicals*, F. A. Lowenheim, M. K. Moran, Eds. (Wiley-Interscience, New York, 4th ed., 1975) pp 117-120. Crystallization from acetone: Hamer, Phillips, U.S. pat. **2,890,240** (1959 to Monsanto). Novel process involving distillation: Edmunds, U.S. pat. **3,235,583** (1966 to Norwich Pharm.). Crystal structure: P. J. Wheatley, *J. Chem. Soc. (Suppl.)* **1964**, 6036. Toxicity data: E. R. Hart, *J. Pharmacol. Exp. Ther.* **89**, 205 (1947). Evaluation as a risk factor in Reye's syndrome: P. J. Waldman *et al., J. Am. Med. Assoc.* **247**, 3089 (1982). Review of clinical trials in prevention of myocardial infarction and stroke: P. C. Elwood, *Drugs* **28**, 1-5 (1984). Symposium on aspirin therapy: *Am. J. Med.* **74**, no. 6A, 1-109 (1983). Comprehensive description: K. Florey, Ed. in *Analytical Profiles of Drug Substances*, vol. 8 (Academic Press, New York, 1979) pp 1-46. Monograph: M. J. H. Smith, P. K. Smith, *The Salicylates* (Interscience, New York, 1966) 313 pp. Book: *Acetylsalicylic Acid*, H. J. M. Barnett *et al.*, Eds. (Raven, New York, 1982) 278 pp.

Monoclinic tablets or needle-like crystals. d 1.40. mp 135° (rapid heating); the melt solidifies at 118°. uv max (0.1*N* H_2SO_4): 229 nm (E$_{1cm}^{1\%}$ 484); (CHCl$_3$): 277 nm (E$_{1cm}^{1\%}$ 68). Is odorless, but in moist air it is gradually hydrolyzed into salicylic and acetic acids and acquires the odor of acetic acid. Stable in dry air. pK (25°) 3.49. One gram dissolves in 300 ml water at 25°, in 100 ml water at 37°, in 5 ml alcohol, 17 ml chloroform, 10-15 ml ether. Less soluble in anhydr ether. Decomp by boiling water or when dissolved in solns of alkali hydroxides and carbonates. LD$_{50}$ orally in mice, rats (g/kg): 1.1, 1.5 (Hart).

Guaiacol ester, $C_{16}H_{14}O_5$, *guacetisal, Broncaspin, Guaiaspir*.

Methyl ester, see Methyl Acetylsalicylate.

Phenyl ester, see Phenyl Acetylsalicylate.

Inorganic salts of acetylsalicylic acid are soluble in water (esp the Ca salt, *q.v.*), but are decomposed quickly.

Pharmaceutical Incompat. (from *Remington's Pharmaceutical Sciences*): Aspirin forms a damp to pasty mass when triturated with acetanilide, phenacetin, antipyrine, aminopyrine, methenamine, phenol or phenyl salicylate. Powders containing aspirin with an alkali salt such as sodium bicarbonate become gummy on contact with atmospheric moisture. Hydrolysis occurs in admixture with salts contg water of crystallization. Solns of the alkaline acetates and citrates, as well as alkalies themselves, dissolve aspirin but the resulting solns hydrolyze rapidly to form salts of acetic and salicylic acids. Sugar and glycerol have been shown to hinder this decomp. Aspirin very slowly liberates hydriodic acid from potassium or sodium iodide. Subsequent oxidation by air produces free iodine.

THERAP CAT: Analgesic; antipyretic; anti-inflammatory.

THERAP CAT (VET): Analgesic; antipyretic; antirheumatic; anticoagulant.

874. Aspoxicillin. *[2S-(2α,5α,6β)]-N-Methyl-D-asparaginyl-N-(2-carboxy-3,3-dimethyl-7-oxo-4-thia-1-azabicyclo[3.2.0]hept-6-yl)-D-2-(4-hydroxyphenyl)glycinamide*; 6-[D-2-(D-2-amino-3-*N*-methylcarbamoylpropionamido)-2-*p*-hydroxyphenylacetamido]penicillanic acid; (2S,5R,6R)-6-[(2R)-2-[(2R)-2-amino-3-(methylcarbamoyl)propionamido]-2-(*p*-hydroxyphenyl)acetamido]-3,3-dimethyl-7-oxo-4-thia-1-azabicyclo[3.2.0]heptane-2-carboxylic acid; N^4-methyl-D-asparaginylamoxicillin; ASPC; TA-058; Doyle. $C_{21}H_{27}N_5O_7S$; mol wt 493.53. C 51.11%, H 5.51%, N 14.19%, O 22.69%, S 6.50%. Semisynthetic penicillin. Prepn: M. Kawazu et al., Ger. pat. 2,638,067; eidem, U.S. pat. 4,053,609 (both 1977 to Tanabe); prepn and antibacterial activity: M. Wagatsuma et al., J. Antibiot. 36, 147 (1985). Mechanism of action study: T. Nishino et al., Chemotherapy (Tokyo) 33, 132 (1985), C.A. 103, 34792v (1985). Toxicological study: M. Takeshita et al., Oyo Yakuri 30, 687 (1985), C.A. 104, 101990u (1986). HPLC determn in serum: J. Knöller et al., Zentralblatt Bakteriol. Mikrobiol. Hyg. 265, 176 (1987). Clinical evaluation in ocular infections: M. Ooishi et al., Acta Med. Biol. 34, 1 (1986). Series of articles on antibacterial activity, pharmacology and clinical efficacy: Chemotherapy (Tokyo) 32, Suppl. 2, 1-791 (1984).

Colorless crystalline powder, mp 195-198° (dec).
THERAP CAT: Antibacterial.

875. Astacin. β,β-Carotene-3,3',4,4'-tetrone; 3,4,3',4'-tetraketo-β-carotene; 3,3'-dihydroxy-2,3,2',3'-tetradehydro-β,β-carotene-4,4'-dione; astacene. $C_{40}H_{48}O_4$; mol wt 592.78. C 81.04%, H 8.16%, O 10.80%. Red carotenoid pigment isolated from biological material originating from crustacea, algae, sponges, protozoa, fish and reptiles. Small amounts were isolated from the fat of mammals (whales, Balaenoptera musculus). Occurs together with astaxanthin from which it is formed by autoxidation. Appears to be an artifact rather than a natural product. Isoln from lobster shells: Kuhn, Lederer, Ber. 66, 488 (1933). Structure: Karrer et al., Helv. Chim. Acta 17, 412, 745 (1934); 18, 96 (1935); 19, 479 (1936). Total synthesis: J. B. Davis, B. C. L. Weedon, Proc. Chem. Soc. 1960, 182; E. Widmer et al., Helv. Chim. Acta 65, 671 (1982). Prepn by autoxidation of canthaxanthin: R. D. G. Cooper et al., J. Chem. Soc. Perkin Trans. I 1975, 2195.

Purple needles or leaflets with metallic luster, sometimes bent into sickle shape, esp when crystallized from pyridine + water, mp 240-243° (slow heating in evac tube, Karrer), mp 228° (Kuhn). Absorption max (pyridine): 500 nm. Practically insol in water; freely sol in chloroform, pyridine, dioxane, carbon disulfide and dil aq alkali; slightly sol in benzene, ethyl acetate, glacial acetic acid; nearly insol in ether, petr ether, methanol.
Diacetate, $C_{44}H_{52}O_6$, black to violet needles from pyridine + water, dec 235°.
Dipalmitate (astacein), $C_{72}H_{108}O_6$, almost square red leaflets from petr ether, mp 121°.

876. Astatine. At; at. no. 85. Radioactive halogen; one of the rarest elements in nature. Radioisotopes range in mass number from 200 to 219; naturally occurring isotopes found in uranium ores: 215, 217, 218, 219 (longest-lived natural isotope, $T_{1/2}$ 0.9 min); most stable artificial isotopes: 209 ($T_{1/2}$ 5.5 hrs); 210 ($T_{1/2}$ 8.3 hrs); 211 ($T_{1/2}$ 7.21 hrs). First convincingly identified by Corson et al., Phys. Rev. 58, 672 (1940). [209]At, [210]At, [211]At are obtained artificially by alpha bombardment of bismuth: Kelley, Segre, Phys. Rev. 75, 999 (1949); Johnson et al., J. Chem. Phys. 17, 1, (1949); Neumann, J. Inorg. & Nucl. Chem. 4, 349 (1957). Reviews: Hyde, J. Chem. Ed. 36, 15 (1959); Haissinsky, Adloff, Ra-

diochemical Survey of the Elements (Elsevier, 1965) pp 11-12; Appelman, "Astatine" in MTP Int. Rev. Sci.: Inorg. Chem., Ser. One, vol. 3, V. Gutmann, Ed. (Butterworths, London, 1972) 181-198; Downs, Adams, "Chlorine, Bromine, Iodine and Astatine" in Comprehensive Inorganic Chemistry vol. 2, J. C. Bailar, Jr. et al., Eds. (Pergamon Press, Oxford, 1973) pp 1107-1594.
More metallic than iodine. Soluble in organic solvents. Due to short half-life of isotopes, only a few physical properties can be measured. Chemical studies are carried out on trace amounts ($< 10^{-8}M$ solns). HAt, CH_3At, AtCl, AtBr, AtI have been identified by time-of-flight mass spectrometry: Appelman et al., Inorg. Chem. 5, 766 (1966). Oxidation states: -1, 0, $+5$; existence of $+1$, $+3$, $+7$ oxidation states uncertain; $E^0(aq)$ At/At⁻ 0.3 V: Appleman, J. Am. Chem. Soc. 83, 805 (1961). When fed to guinea pigs, is found in the thyroid gland. Mammary and pituitary tumors can be induced with a single injection.

877. Astaxanthin. 3,3'-Dihydroxy-β,β-carotene-4,4'-dione; 3,3'-dihydroxy-4,4'-diketo-β-carotene; ovoester. $C_{40}H_{52}O_4$; mol wt 596.82. C 80.49%, H 8.78%, O 10.72%. Carotenoid pigment found mostly in animal organisms, but also occurring in plants; thought to be the precursor of astacin, q.v. Structure and isoln from lobster eggs: Kuhn, Sörensen, Angew. Chem. 51, 465 (1938); Ber. 71, 1879 (1938). Isoln from other animal organisms: Kuhn et al., Ber. 72, 1688 (1939). Occurrence in plants: Tischer, Z. Physiol. Chem. 267, 281 (1941); in the flower petals of Adonis annua L., Ranunculaceae: Seybold, Goodwin, Nature 184, 1714 (1959). Occurs also in the red feathers of birds of the Laniarius spp: Z. Physiol. Chem. 288, 20 (1951). Stereochemistry and spectra: Grangaud, Compt. Rend. 242, 1767 (1956). Abs config: Andrewes et al., Acta Chem. Scand. B 28, 730 (1974). See also T. W. Goodwin, Carotenoids (Chemical Publ. Co., New York, 1954) pp 167-172. Synthesis: R. D. G. Cooper et al., J. Chem. Soc. Perkin Trans. I 1975, 2195; F. Kienzle, H. Mayer, Helv. Chim. Acta 61, 2609 (1978); E. Widmer et al., ibid. 64, 2405 (1981); H. Mayer et al., ibid. 2419.

Needles from acetone/light petroleum, mp 182-183°. uv max: (CS_2) 503 nm; (methanol) 472 nm; (hexane) 466-467 nm; (chloroform) 485 nm (Cooper et al.). Also reported as shiny purple platelets with gold luster from pyridine, mp 216° (some decompn). Readily sol in pyridine, from which it can be cryst by the addn of water (Kuhn, Sörensen).
Diacetate, $C_{44}H_{56}O_6$, stout, blue-black needles, mp 203-205° (vac).

878. Astemizole. 1-[(4-Fluorophenyl)methyl]-N-[1-[2-(4-methoxyphenyl)ethyl]-4-piperidinyl]-1H-benzimidazol-2-amine; 1-(p-fluorobenzyl)-2-[[1-(p-methoxyphenethyl)-4-piperidyl]amino]benzimidazole; R 43512; Astemisan; Hismanal; Histamen; Histaminos; Histazol; Kelp; Laridal; Metodik; Novo-Nastizol A; Paralergin; Retolen; Waruzol. $C_{28}H_{31}FN_4O$; mol wt 458.59. C 73.34%, H 6.81%, F 4.14%, N 12.22%, O 3.49%. Nonsedating-type histamine H_1-receptor antagonist. Prepn: F. Janssens et al., Eur. pat. Appl. 5318; eidem, U.S. pat. 4,219,559 (1979, 1980 both to Janssen). Pharmacology: J. Van Wauwe et al., Arch. Int. Pharmacodyn. Ther. 251, 39 (1981); A. Wauquier, C. J. E. Niemegeers, Eur. J. Pharmacol. 72, 245 (1981). In vitro and in vivo binding characteristics: P. M. Laduron et al., Mol. Pharmacol. 21, 294 (1982). Effect on human psychomotor performance: T. Seppala, K. Savolainen, Curr. Ther. Res. 31, 638 (1982). Clinical study in treatment of hay fever: J. Callier et al., ibid. 29, 24 (1981). Mutagenicity study: P. Vanparys et al., Arch. Toxicol. 50, 167 (1982). Review of pharmacology and clinical trials: D. M. Richards et al., Drugs 28, 38-61 (1984).

Crystals, mp 149.1°.

THERAP CAT: Anti-allergic; antihistaminic.

879. Atenolol. *4-[2-Hydroxy-3-[(1-methylethyl)amino]propoxy]benzeneacetamide; 2-[p-[2-hydroxy-3-(isopropylamino)propoxy]phenyl]acetamide;* 1-*p*-carbamoylmethylphenoxy-3-isopropylamino-2-propanol; ICI 66082; AteHexal; Atenol; Ibinolo; Myocord; Prenormine; Seles Beta; Selobloc; Tenoblock; Tenormin; Unibloc; Uniloc. $C_{14}H_{22}N_2O_3$; mol wt 266.34. C 63.13%, H 8.33%, N 10.52%, O 18.02%. Cardioselective β-adrenergic blocker. Prepn: Barrett *et al.*, **Ger.** pat. 2,007,751; *eidem*, **U.S.** pats. 3,663,607 and 3,836,671 (1970, 1972, 1974, all to I.C.I.). Pharmacology and clinical studies: Giudicelli *et al.*, *Compt. Rend. Soc. Biol.* **167**, 232 (1973); Hansson *et al.*, *Acta Med. Scand.* **194**, 549 (1973); Amery *et al.*, *N. Engl. J. Med.* **290**, 284 (1974). Clinical trial in treatment of alcohol withdrawal syndrome: M. L. Kraus *et al.*, *ibid.* **313**, 905 (1985). *Review:* J. D. Fitzgerald in *Pharmacological and Biochemical Properties of Drug Substances* vol. 2, M. E. Goldberg, Ed. (Am. Pharm. Assoc., Washington, DC, 1979) pp 98-147; E. Marmo, *Drugs Exp. Clin. Res.* **6**, 639-663 (1980). Symposium on clinical studies: *Drugs* **25**, Suppl. 2, 1-346 (1983).

Crystals from ethyl acetate, mp 146-148°. LD_{50} in mice, rats (mg/kg): 2000, 3000 orally; 98.7, 59.24 i.v. (Fitzgerald).

THERAP CAT: Antihypertensive, antianginal, antiarrhythmic.

880. Athamantin. *3-Methylbutanoic acid 8,9-dihydro-8-[1-methyl-1-(3-methyl-1-oxobutoxy)ethyl]-2-oxo-2H-furo[2,3-h]-1-benzopyran-9-yl ester; isovaleric acid diester with 8,9-dihydro-9-hydroxy-8-(1-hydroxy-1-methylethyl)-2H-furo[2,3-h]-1-benzopyran-2-one;* 2,3-dihydro-3,4-dihydroxy-2-(1-hydroxy-1-methylethyl)-5-benzofuranacrylic acid δ-lactone diisovalerate. $C_{24}H_{30}O_7$; mol wt 430.48. C 66.96%, H 7.02%, O 26.02%. From *Peucedanum oreoselinum* (L.) Moench (*Athamanta oreoselinum* L.), *Ammi visnaga* Lam., *Umbelliferae*. Isoln: Schnedermann, Winckler, *Ann.* **51**, 315 (1844). Structure: Halpern *et al.*, *Helv. Chim. Acta* **40**, 758 (1957). Absolute configuration: Nakazaki *et al.*, *Tetrahedron Letters* **1966**, 4735.

Fine needles from petr ether, mp 58-60°; sublimes in high vacuum at 180-200°. $[\alpha]_D^{20}$ +88° (c = 1.145 in glacial acetic acid); $[\alpha]_{546}^{22}$ +129° (c = 0.5575 in methanol); $[\alpha]_{546}^{21}$ +73.9° (c = 1.024 in chloroform). $[M]_D$ +440° (methanol); +258° (chloroform). uv max (96% ethanol): 217, 322 nm (log ϵ 4.18, 4.17). Practically insol in water; sol in alc, ether, chloroform.

881. Atisine. Anthorine. $C_{22}H_{33}NO_2$; mol wt 343.49. C 76.92%, H 9.68%, N 4.08%, O 9.32%. From roots of the "atis" plant, *Aconitum heterophyllum* Wall., and from *A. anthora* L., *Ranunculaceae:* Broughton, *Blue Book of East India Cinchona Cultivation* **1877**, 133; Goris, Metin, *Compt.*

Rend. **180**, 968 (1925); Lawson, Topps, *J. Chem. Soc.* **1937**, 1640. Structure: Wiesner *et al.*, *Chem. & Ind. (London)* **1954**, 132; Pelletier, Jacobs, *J. Am. Chem. Soc.* **76**, 4496 (1954). Structure and stereochemistry: Dvornik, Edwards, *Can. J. Chem.* **42**, 137 (1964). Partial synthesis: Pelletier, Jacobs, *J. Am. Chem. Soc.* **78**, 4144 (1956); Pelletier, Parthasarathy, *Tetrahedron Letters* **1963**, 205. Racemic synthesis and resolution: Nagata *et al.*, *J. Am. Chem. Soc.* **85**, 2342 (1963); **89**, 1499 (1967); Masamune, *ibid.* **86**, 291 (1964); Guthrie *et al.*, *Tetrahedron Letters* **1966**, 4645. ^{13}C-NMR study of epimers: N. V. Mody, S. W. Pelletier, *Tetrahedron* **34**, 2421 (1978).

Solid, mp 57-60°. Distills in high vacuum at a bath temp of 140°. Strong base: pK 12.2.

Hydrochloride, $C_{22}H_{33}NO_2 \cdot HCl$, flat needles from dil alc, dec 311-312°. $[\alpha]_D^{25}$ +28°.

882. Atmosphere. Composition of Earth's atmosphere at surface for midlatitudes (in ppm): N_2 780,840, O_2 209,460, Ar 9340, CO_2 325, Ne 18.18, He 5.24, CH_4 1.4, Kr 1.14, H_2 0.5, N_2O 0.25, Xe 0.087, O_3 0.025, H_2O variable. Monograph on chemistry of atmosphere: J. Heicklen, *Atmospheric Chemistry* (Academic Press, New York, 1976) 406 pp. Other planets of our solar system have different atmospheres. The atmosphere of Jupiter consists largely of ammonia and methane, and those of Saturn, Uranus, and Neptune are practically all methane. Exact data in *Landolt-Börnstein*, 6th ed., **vol. III** (Springer, 1952) p 59; G. P. Kuiper, *The Atmospheres of the Earth and the Planets* (University of Chicago Press, 1949).

883. Atractyloside. *(2β,4α,15α)-15-Hydroxy-2-[[2-O-(3-methyl-1-oxobutyl)-3,4-di-O-sulfo-β-D-glucopyranosyl]oxy]-19-norkaur-16-en-18-oic acid dipotassium salt;* potassium atractylate; atractylin (C_{30} glucoside). $C_{30}H_{44}K_2O_{16}S_2$; mol wt 803.00. C 44.87%, H 5.52%, K 9.74%, O 31.88%, S 7.99%. Toxic principle isolated from the thistle *Atractylis gummifera* L., *Compositae:* M. Lefranc, *Compt. Rend.* **67**, 954 (1868). Structure and stereochemistry of the aglycone, *atractyligenin:* F. Piozzi *et al.*, *Tetrahedron* **Suppl. 8 II**, 515 (1966); total synthesis of ($\pm$)-form: A. K. Singh *et al.*, *J. Am. Chem. Soc.* **109**, 6187 (1987). Structure and stereochemistry of atractyloside: *eidem*, *Gazz. Chim. Ital.* **97**, 935 (1967). Toxicity studies: G. Cascio *et al.*, *Boll. Soc. Ital. Biol. Sper.* **44**, 253 (1968). *Review: Atractyloside: Chemistry, Biochemistry, and Toxicology,* R. Santi, S. Luciani, Eds. (Piccin Medical Books, Padova, Italy, 1978) 136 pp.

Crystals, dec 174°. $[\alpha]_D^{20}$ −53° (water). *Highly toxic!* Has strychnine-like action; produces convulsion of a hypoglycemic nature: R. Santi, G. Cascio, *C.A.* **50**, 7320i (1956); R. Santi, *C.A.* **52**, 15733c (1958). LD_{50} i.m. in rats: 431 mg/kg (Cascio).

884. Atracurium Besylate. *2,2'-[1,5-Pentanediylbis[oxy(3-oxo-3,1-propanediyl)]]bis[1-[(3,4-dimethoxyphenyl)methyl]-1,2,3,4-tetrahydro-6,7-dimethoxy-2-methylisoquinolinium] dibenzenesulfonate;* 2-(2-carboxyethyl)-1,2,3,4-tetrahydro-6,7-dimethoxy-2-methyl-1-veratrylisoquinolinium benzenesulfonate pentamethylene ester; *N,N'*-dimethyl-*N,N'*-(4,10-dioxa-3,11-dioxotridecylene)-1,13-bis-tetrahydropapaverinium dibenzenesulfonate; BW 33A; Wellcome 33-A-74; Tracrium. $C_{65}H_{82}N_2O_{18}S_2$; mol wt 1243.51. C 62.78%, H 6.65%, N 2.25%, O 23.16%, S 5.16%. Non-depolarizing neuromuscular blocking agent. Prepn: J. B. Stenlake *et al., Ger. pat.* 2,655,883; *eidem, U.S. pat.* 4,179,507 (1977, 1979 both to Burroughs Wellcome). Pharmacology: R. Hughes, D. J. Chapple, *Brit. J. Anaesth.* 53, 31 (1981). Clinical pharmacology: S. J. Basta *et al., Anesth. Analg.* 61, 723 (1982). Neuromuscular effects in man: R. L. Katz *et al., ibid.* 730. Metabolic studies: E. A. Neill, D. J. Chapple, *Xenobiotica* 12, 203 (1982). Pharmacokinetics: S. Ward *et al., Brit. J. Anaesth.* 55, 113 (1983). Use during halothane anesthesia in humans: J. A. Stirt *et al., Anesth. Analg.* 62, 207 (1983). Symposium on pharmacology, metabolism, clinical studies: *Brit. J. Anaesth.* 55, Suppl. 1, 1S-139S (1983); on pharmacokinetics and clinical experience: *ibid.* 58, Suppl. 1, 1S-113S (1986).

Off-white powder, mp 85-90°. Softens at 60°.
THERAP CAT: Muscle relaxant (skeletal).

885. Atranorin. *3-Formyl-2,4-dihydroxy-6-methylbenzoic acid 3-hydroxy-4-(methoxycarbonyl)-2,5-dimethylphenyl ester;* atranoric acid. $C_{19}H_{18}O_8$; mol wt 374.33. C 60.96%, H 4.85%, O 34.19%. From a number of lichens. Belongs to the group of lichen acids. Isoln: Paterno, Ogliarori, *Gazz. Chim. Ital.* 1877, 189; Hesse, *Ber.* 30, 357, 1983 (1897); St. Pfau, *Helv. Chim. Acta* 9, 650 (1926). Synthesis: Neelakantan *et al., Tetrahedron Letters* 1962, 287. TLC analysis: R. Klee, L. Steubing, *J. Chromatogr.* 129, 478 (1976); J. L. Ramaut *et al., ibid.* 155, 450 (1978).

Bitter crystals or cryst powder from chloroform, mp 195°. Practically insol in water; slightly sol in alc; sol in boiling benzene or in chloroform; also sol in alkalies giving a yellow soln.

886. Atrazine. *6-Chloro-N-ethyl-N'-(1-methylethyl)-1,3,5-triazine-2,4-diamine; 2-chloro-4-ethylamino-6-isopropylamino-s-triazine;* G 30027; AAtrex; Atranex; Gesaprim; Primatol A. $C_8H_{14}ClN_5$; mol wt 215.68. C 44.55%, H 6.54%, Cl 16.44%, N 32.47%. Prepn: Gysin, Knüsli, **Swiss** pats. 342,784-5 (1960 to Geigy), *C.A.* 55, 5552d (1961); Mel'nikov *et al., Khim. Prom.* 1961, 703, *C.A.* 58, 526c (1963); Andriska *et al., Hung. pat.* 149,189 (1962 to Nehézvegyipari Kutato Intézet), *C.A.* 58, 13972c (1963); Mildner, **Fr. pat.** 1,317,812 (1963 to Radonja Kemijska Ind.), *C.A.* 59, 8765h (1963).

Crystals, mp 171-174°. Soly at 25° in water 70 ppm; ether 12,000 ppm; chloroform 52,000 ppm; methanol 18,000 ppm. Stable in slightly acidic or basic media; hydrolyzed to inac-

tive hydroxy deriv by alkali or mineral acids. LD_{50} orally in mice: 1750 mg/kg, S. Dalgaard-Mikkelsen, E. Poulsen, *Pharmacol. Rev.* 14, 225 (1962).
USE: Selective herbicide. *Toxicity:* Inhalation hazard is low in humans. Acutely poisoned sheep and cattle show muscular spasms, fasciculations, stiff gait, increased respiratory rates. Adrenal degeneration, congestion of lungs, liver, kidneys observed. No apparent skin irritation or other toxic manifestations in humans: *Clinical Toxicology of Commercial Products,* R. E. Gosselin *et al.,* Eds. (Williams & Wilkins, Baltimore, 4th ed., 1976) Section II, p 207.

887. Atrial Natriuretic Factor. *Atriopeptin;* ANF; ANP; auriculin; cardionatrin. Potent natriuretic and vasodilatory peptide or mixture of homologous peptides derived from a common precursor and isolated from the atrium of the mammalian heart. Involved in hormonal regulation of extracellular fluid volume and blood pressure homeostasis. Stored in specific granules of atrial cardiocytes, ANF produces rapid diuresis when injected into laboratory animals: A. J. de Bold *et al., Life Sci.* 28, 89 (1981); R. Garcia *et al., Experientia* 38, 1071 (1982); A. J. de Bold, *Can. J. Physiol. Pharmacol.* 60, 324 (1982). Isoln of rat ANF (rANF): *idem, Proc. Soc. Exp. Biol. Med.* 170, 133 (1982); G. Thibault *et al., Hypertension* 5, Suppl. I, I-75 (1983); of human ANF (hANF): N. C. Trippodo *et al., ibid.,* I-81. Preliminary characterization of high and low molecular weight components: *eidem, Proc. Soc. Exp. Biol. Med.* 170, 502 (1982). Amino acid comp of low mol wt component: A. J. de Bold, T. G. Flynn, *Life Sci.* 33, 297 (1983); G. Thibault *et al., FEBS Letters* 164, 286 (1983). A number of biologically active ANF peptides containing 21-33 amino acids have been identified (mol wt ~4000). Amino acid sequence of rANF peptides: T. G. Flynn *et al., Biochem. Biophys. Res. Commun.* 117, 859 (1983); D. M. Geller *et al., ibid.* 120, 333 (1984); M. G. Currie *et al., Science* 223, 67 (1984). Synthesis of bioactive rANF: N. G. Seidah *et al., Proc. Nat. Acad. Sci. USA* 81, 2640 (1984); S. A. Atlas *et al., Nature* 309, 727 (1984). hANF differs from rANF by one amino acid substitution. Identification of 3 peptide components of hANF and sequence of 28 amino acid α-hANP: K. Kangawa, H. Matsuo, *Biochem. Biophys. Res. Commun.* 118, 131 (1984). The high mol wt component *(atriopeptigen* or *preproANF)* has been identified as the precursor for the active, low mol wt forms. Specific proteolytic conversion is required for biological activation: D. M. Geller *et al., Biochem. Biophys. Res. Commun.* 121, 802 (1984); T. Oshima *et al., Circ. Res.* 54, 612 (1984). Structure of rat preproANF: M. Maki *et al., Nature* 309, 722 (1984). Cloning and nucleotide sequence of cDNA for rat preproANF: M. Yamanaka *et al., ibid.* 719; for human preproANF: S. Oikawa *et al., ibid.* 724. ANF is stored in the rat atrium as a 126 amino acid peptide (proANF or γ-rANP): K. Kangawa *et al., Nature* 312, 1523 (1984). Biosynthesis, processing and secretion of rat proANF: K. D. Bloch *et al., Science* 230, 1168 (1985). Identification of 28 amino acid α-rANP as major plasma component and comparison with human: A. Miyata *et al., Biochem. Biophys. Res. Commun.* 129, 248 (1985). Comparison of diuretic effect of ANF with furosemide: H. Sonnenberg *et al., Can. J. Physiol. Pharmacol.* 59, 1278 (1981). Studies on the mechanism of natriuresis: *eidem, ibid.* 60, 1149 (1982); R. Keeler, *ibid.* 1078; M. J. F. Camargo *et al., Am. J. Physiol.* 246, F447 (1984). Inhibition of aldosterone production: K. Atarashi *et al., Science* 224, 992 (1984). Vasorelaxant activity: M. G. Currie *et al., ibid.* 221, 71 (1983); R. T. Grammer *et al., Biochem. Biophys. Res. Commun.* 116, 696 (1983); H. D. Kleinert *et al., Hypertension* 6, Suppl. I, I-143 (1984). Vasodilator profile of synthetic peptide: R. Garcia *et al., Biochem. Biophys. Res. Commun.* 119, 685 (1984); R. J. Winquist *et al., Eur. J. Pharmacol.* 102, 169 (1984). Direct renal effect of synthetic ANF: A. A. Seymour *et al., Life Sci.* 36, 33 (1985). Determn of ANF in rat atria by radioimmunoassay: J. Gutkowska *et al., Proc. Soc. Exp. Biol. Med.* 176, 105 (1984). Evidence of cardiac secretion of ANF: E. A. Espiner *et al., Lancet* 2, 398 (1985). Clinical pharmacology of synthetic peptide: A. M. Richards *et al., ibid.* 1, 545 (1985). Symposium on biochemistry, activity and clinical pharmacology: *J. Hypertension* 4, Suppl. 2, S1-S157 (1986). *Reviews:* M. A. Napier, E. H. Blaine in *Ann. Rep. Med. Chem.* 19, D. M.

Bailey, Ed. (Academic Press, New York, 1984) pp 256-262; P. Needleman *et al.*, *Trends Pharmacol. Sci.* **5**, 506-509 (1984); P. Needleman, J. E. Greenwald, *N. Engl. J. Med.* **314**, 828-834 (1986); U. Ackermann, *Clin. Chem.* **32**, 241-247 (1986); J. D. Baxter *et al.*, *Biotechnology* **6**, 529-546 (1988).

Ser-Leu-Arg-Arg-Ser-Ser-Cys-Phe-Gly-Gly-Arg-Met-Asp-Arg-Ile-Gly

└── S ── S ──┐   Ala

Try-Arg-Phe-Ser-Asn-Cys-Gly-Leu-Gly-Ser-Gln

α-hANP

888. Atrolactamide. *α-Hydroxy-α-methylbenzeneacetamide; α-hydroxy-α-phenylpropionamide; 2-hydroxy-2-phenylpropionamide; 2-phenyl-2-hydroxypropionamide;* M-144; Themisone. $C_9H_{11}NO_2$; mol wt 165.19. C 65.44%, H 6.71%, N 8.48%, O 19.37%. Prepd from atrolactic acid by treatment with ammonia water: McKenzie, Smith, *J. Chem. Soc.* **121**, 1356 (1922); from the nitrile by treatment with fuming HCl: Staudinger, Ruzicka, *Ann.* **380**, 291 (1911).

$$CH_3 - \underset{\underset{OH}{|}}{\overset{\overset{C_6H_5}{|}}{C}} - CONH_2$$

DL-Form, leaflets from dichloroethylene, mp 102°. Freely sol in water.
L-Form, leaflets from benzene, mp 62.5-63.5°. $[\alpha]_D^{15}$ +12.8° (c = 2.2 in acetone). Freely sol in water; sol in alcohol, acetone.
THERAP CAT: Anticonvulsant.

889. Atrolactic Acid. *α-Hydroxy-α-methylbenzeneacetic acid; α-methylmandelic acid;* 2-phenyllactic acid; *α-hydroxy-α-phenylpropionic acid;* 2-hydroxy-2-phenylpropionic acid; 2-phenyl-2-hydroxypropionic acid. $C_9H_{10}O_3$; mol wt 166.17. C 65.05%, H 6.07%, O 28.88%. Prepd from acetophenone cyanohydrin: McKenzie, Clough, *J. Chem. Soc.* **101**, 393 (1912); Freudenberg *et al.*, *Ann.* **501**, 213 (1932); Eliel, Freeman, *Org. Syn.* **coll. vol. IV**, 58 (1963).

$$CH_3 - \underset{\underset{OH}{|}}{\overset{\overset{C_6H_5}{|}}{C}} - COOH$$

DL-Form, hemihydrate, orthorhombic crystals from water, mp 88-90° (softening at 75°). Anhydr, mp 94.5° (obtained by drying the hemihydrate at 55° and 0.5 mm Hg). K at 25° = 3.41 × 10⁻⁴. Soly of the anhydr acid in one liter of water at 18° = 17.04 g, at 25° = 21.17 g, at 30° = 25.65 g. Much more sol in boiling water. Slightly sol in petr ether.

890. Atropic Acid. *α-Methylenebenzeneacetic acid; α-phenylacrylic acid.* $C_9H_8O_2$; mol wt 148.15. C 72.96%, H 5.44%, O 21.60%. Prepn: Normant, Maitte, *Bull. Soc. Chim. France* **1956**, 1439.

$$C_6H_5 - \underset{}{\overset{\overset{COOH}{|}}{C}} = CH_2$$

Tabular or acicular crystals; volatilizes in steam. mp 106-107°. bp about 267° with partial decompn. Sol in 790 parts water; sol in alc, benzene, chloroform, ether, CS_2.
Caution: Irritating to skin, mucous membranes.

891. Atropine. *endo-(±)-α-(Hydroxymethyl)benzeneacetic acid 8-methyl-8-azabicyclo[3.2.1]oct-3-yl ester; 1αH,5αH-tropan-3α-ol (±)-tropate;* dl-hyoscyamine; tropic acid ester with tropine; dl-tropyl tropate; tropine tropate. $C_{17}H_{23}NO_3$; mol wt 289.38. C 70.56%, H 8.01%, N 4.84%, O 16.59%. Parasympatholytic alkaloid isolated from *Atropa belladonna* L., *Datura stramonium* L., and other *Solanaceae.*

Extraction procedure: Chemnitius, *J. Prakt. Chem.* **116**, 276 (1927). During extraction, partial racemization of the *l*-hyoscyamine takes place which is completed by treatment with dil alkali on heating in chloroform soln: Schneider, *Arch. Pharm.* **284**, 306 (1951). Structure and synthesis: Ladenburg, *Ann.* **217**, 75 (1883); Willstätter, *Ber.* **31**, 1537 (1898); *idem.*, *Ann.* **326**, 23 (1903); Schwenker *et al.*, *Ber.* **99**, 2407 (1966). Prepn of the sulfate: **Ger. pat.** 247,455 (1912 to Hoffmann-La Roche), *Frdl.* **11**, 1022. Clinical use of the mucate: J. S. Heron, *Can. Med. Assoc. J.* **72**, 302 (1955). Use as antidote to cholinesterase inhibitors: R. V. Brown, *Brit. J. Pharmacol.* **15**, 170 (1960). Effect on cardiac arrhythmias: P. Schweitzer, H. Mark, *Am. Heart J.* **100**, 119, 255 (1980). Pharmacokinetics and pharmacodynamics: P. H. Hinderling *et al.*, *J. Pharm. Sci.* **74**, 703, 711 (1985). Toxicity: R. L. Cahen, K. Tvede, *J. Pharmacol. Exp. Ther.* **105**, 166 (1952); Goldenthal, *Toxicol. Appl. Pharmacol.* **18**, 185 (1971). Review of clinical use in anesthesia: L. E. Shutt, J. B. Bowes, *Anaesthesia* **34**, 476-490 (1979). Comprehensive description: A. A. Al-Badr, F. J. Muhtadi in *Analytical Profiles of Drug Substances* vol. **14**, K. Florey, Ed. (Academic Press, New York, 1985) pp 325-389.

Long, orthorhombic prisms from acetone, mp 114-116°. Sublimes in high vacuum at 93-110°. pK 4.35; pH of 0.0015 molar soln 10.0. Absorption spectra: Dobbie, Fox, *J. Chem. Soc.* **103**, 1194 (1913); Fischer, *Arch. Exp. Pathol. Pharmakol.* **170**, 623 (1933). One gram dissolves in 455 ml water, 90 ml water at 80°, 2 ml alc, 1.2 ml alc at 60°, 27 ml glycerol, 25 ml ether, 1 ml chloroform; also sol in benzene, dil acids. LD_{50} orally in rats: 750 mg/kg (Cahen, Tvede).
Hydrochloride, $C_{17}H_{24}ClNO_3$, granular crystals, mp 165°. Sol in water, alc. pH of 0.05 molar soln 5.8.
Methylbromide, $C_{18}H_{26}BrNO_3$, *Tropin.* Crystals, mp 222-223°. Sol in 1 part water; slightly sol in alc. Almost insol in chloroform, ether.
Methylnitrate, $C_{18}H_{26}N_2O_6$, *methylatropine nitrate, Atro-Dote, Eumydrin, Metropine, Harvatrate, Metanite, Ekomine.* Crystals, mp 163°. Freely sol in water or alc, very slightly in chloroform, ether.
Mucate, *atropine hyperduric.* Mucic acid salt of atropine.
Sulfate monohydrate, $C_{34}H_{48}N_2O_{10}.S.H_2O$, *Atrophate, Atropisol, Atrosed.* Granules or powder, mp 190-194°. Almost inactive optically. Very bitter. pH ~5.4. One gram dissolves in 0.4 ml water; 5 ml cold, 2.5 ml boil. alc; in 2.5 ml glycerol, 420 ml chloroform, 3000 ml ether. Bitterness threshold 1:10,000. Incompatible with alkalies, tannin, salts of mercury or gold, vegetable decoctions or infusions, borax, bromides, iodides, benzoates. LD_{50} orally in rats: 622 mg/kg (Goldenthal).
Human Toxicity: Causes blurred vision, suppressed salivation, vasodilation, hyperpyrexia, excitement, agitation, and delirium, *Clinical Toxicology of Commercial Products,* R. E. Gosselin *et al.*, Eds. (Williams & Wilkins, Baltimore, 5th Ed., 1984) Section III, pp 47-50.
THERAP CAT: Anticholinergic. Mydriatic. In preanesthetic medication.
THERAP CAT (VET): Anticholinergic. Mydriatic. Antispasmodic. Antidote to organophosphous insecticides.

892. Atropine N-Oxide. *endo-(±)-α-(Hydroxymethyl)benzeneacetic acid 8-methyl-8-azabicyclo[3.2.1]oct-3-yl ester 8-oxide; 1αH,4αH-tropan-3α-ol (±)-tropate 8-oxide;* atropine aminoxide; genatropine; aminoxytropine tropate. $C_{17}H_{23}NO_4$; mol wt 305.36. C 66.86%, H 7.59%, N 4.59%, O 20.96%. Prepd by hydrogen peroxide oxidation of atropine or racemization of hyoscyamine N-oxide: Polonovski, Polonovski, *Bull. Soc. Chim. France* **39**, 1147 (1926).

Crystalline powder, mp 127-128°, dec 135°. Very hygroscopic. Soluble in alc, chloroform; practically insol in ether. Hydrochloride, $C_{17}H_{24}ClNO_4$, *Tropinox*, *Xtro*. Prisms from abs alc, mp 192-193°.

THERAP CAT: Anticholinergic.

893. Attacins. Immune protein P5. A major class of non-specific antibacterial proteins induced in the hemolymph of the pupae of silk moths, particularly *Hyalophora cecropia*, as part of the immune response. (*See also* Cecropins.) Originally designated as immune *protein P5*, attacins are a heterogeneous family consisting of four basic (A,B,C,-D) and two acidic (E,F) forms with similar amino acid sequences and mol wts of 20,000-23,000 daltons. Initial identification and isolation of immune response proteins: I. Faye *et al., Infect. Immun.* **12,** 1426 (1975). Purification and preliminary characterization: A. E. Pye, H. G. Boman, *ibid.* **17,** 408 (1977). Amino acid sequences and antibacterial activity: D. Hultmark *et al., EMBO J.* **2,** 571 (1983). Structural study on attacin F: Å. Engström *et al., ibid.* **3,** 2065 (1984). Mode of action: P. Engström *et al., ibid.* 3347.

894. Aucubin. *1,4a,5,7a-Tetrahydro-5-hydroxy-7-(hydroxymethyl)cyclopenta[c]pyran-1-yl-β-D-glucopyranoside;* rhinanthin; aucuboside. $C_{15}H_{22}O_9$; mol wt 346.33. C 52.02%, H 6.40%, O 41.58%. From leaves, roots, stalks and seeds of *Aucuba japonica* Thunb., *Cornaceae;* also occurs in 75 different plants: Paris, Chaslot, *Ann. Pharm. Franc.* **13,** 648 (1955). Isoln: Trim, Hill, *Biochem. J.* **50,** 310 (1952); Rombouts, Links, *Experientia* **12,** 78 (1956). Structure: Haegele *et al., Tetrahedron Letters* **1961,** 110; Birch *et al., J. Chem. Soc.* **1961,** 5194.

Crystals from ethanol + ether, mp 181°. $[\alpha]_D^{21}$ −163.1° (c = 1.6). Soluble in water, alc, methanol; practically insol in chloroform, ether, petr ether. Absorption spectrum: Trim, Hill, *loc. cit.*

895. Auranofin. *(1-Thio-β-D-glucopyranase-2,3,4,6-tetraacetato-S)(triethylphosphine)gold; (2,3,4,6-tetra-O-acetyl-1-thio-β-D-glucopyranosato-S)(triethylphosphine)gold; (1-thio-β-D-glucopyranosato)(triethylphosphine)gold 2,3,4,6-tetracetate;* SKF 39162; Aktil; Crisinar; Crisofin; Ridaura; Ridauran. $C_{20}H_{34}AuO_9PS$; mol wt 678.49. C 35.41%, H 5.05%, Au 29.03%, O 21.22%, P 4.56%, S 4.73%. Orally active gold coordination complex. Prepn: E. R. McGusty, B. M. Sutton, **Ger.** pat. **2,051,495;** *eidem,* **U.S.** pat. **3,635,-945** (1971, 1972 both to SKF). Series of articles on metabolism, distribution, pharmacokinetics, pharmacology, and clinical studies: *J. Rheumatol.* **6,** Suppl. 5, 1-164 (1979). Use in rheumatoid arthritis: A. E. Finkelstein *et al., ibid.* **7,** 160 (1980). Inhibitory effect on HeLa cells: T. M. Simon *et al., Cancer* **44,** 1965 (1979). Screening trial with mouse lymphocyte leukemia: *eidem, Cancer Res.* **41,** 94 (1981). Toxicity study: B. H. Payne, D. T. Walz, *Vet. Pathol.* **15,** Suppl. 5, 1 (1978). *Review:* D. T. Walz in *Pharmacological and Biochemical Properties of Drug Substances* vol. 2, M. E. Goldberg, Ed. (Am. Pharm. Assoc., Washington, DC, 1979) pp 400-406. Symposium: *J. Rheumatol.* **9,** Suppl. 8 (1982) pp 1-209. Review of pharmacology and clinical studies: M. Chaffman *et al., Drugs* **27,** 378 (1984).

R = $OCOCH_3$

Colorless crystals, mp 110-111°. LD_{50} in rats, mice (mg/kg): 265, 310 orally (Payne, Walz).

THERAP CAT: Antirheumatic.

896. Aurantiogliocladin. *2,3-Dimethoxy-5,6-dimethyl-2,5-cyclohexadiene-1,4-dione;* 2,3-dimethoxy-5,6-dimethyl-*p*-benzoquinone. $C_{10}H_{12}O_4$; mol wt 196.20. C 61.21%, H 6.17%, O 32.62%. Antibiotic substance produced by a *Gliocladium* sp similar to *Gliocladium roseum* (Link.) Bainier: Brian *et al., Experientia* **7,** 266 (1951). Structure: Vischer, *J. Chem. Soc.* **1953,** 815. Synthesis: Baker *et al., ibid.* 820; Seshadri, Venkatasubramanian, *ibid.* **1959,** 1660. Biosynthesis: Pettersson, *Acta Chem. Scand.* **19,** 1827 (1965).

Bright orange plates from petr ether, mp 62.5°. Absorption max: 407, 275 nm (ε 1,200; 38,400). Sol in chloroform, ethanol, pyridine. Gives a violet color with concd H_2SO_4.

Dihydro deriv, $C_{10}H_{14}O_4$, *gliorosein*, *(+)-trans-2,3-dimethoxy-5,6-dimethylcyclohex-2-ene-1,4-dione.* Structure: Grave, *J. Chem. Soc. (C)* **1966,** 985. Needles, mp 48°. $[\alpha]_D^{22}$ +125° (c = 0.2 in CHCl₃). uv max: 289 nm (ε 24,320).

Quinhydrone deriv, *rubrogliocladin.* Dark red needles from petr ether, mp 74°. Absorption max (alcohol): 407, 275 nm (ε 650, 22,000). Sol in most organic solvents, water.

897. Aureothin. *2-Methoxy-3,5-dimethyl-6-[tetrahydro-4-[2-methyl-3-(4-nitrophenyl)-2-propenylidene]-2-furanyl]-4H-pyran-4-one;* mycolutein. $C_{22}H_{23}NO_6$; mol wt 397.41. C 66.49%, H 5.83%, N 3.52%, O 24.16%. Antibiotic by-product of aureothricin, *q.v.,* isolated from the culture of *Streptomyces thioluteus.* Isoln: K. Maeda, *J. Antibiot.* **6A,** 137 (1953); (as mycolutein): H. Schmitz, R. Woodside, *Antibiot. & Chemother.* **5,** 652 (1955). Structure: Y. Hirata *et al., Tetrahedron* **14,** 252 (1961). Identity with mycolutein: J. L. Schwartz *et al., J. Antibiot.* **29,** 236 (1976). Biosynthesis: R. Cardillo *et al., Tetrahedron* **30,** ·459 (1974); M. Yamazaki *et al., Chem. Pharm. Bull.* **23,** 569 (1975). Synthesis: Y. Shizuri *et al., Chem. Letters* **1987,** 1381.

Yellow prisms, mp 158°. $[\alpha]_D^{18}$ +51° (chloroform). Sol in methanol, ethanol, acetone, chloroform, tetrahydrofuran. Practically insol in water, nonpolar solvents. uv max (ethanol): 257, 346 nm (log ε 4.39, 4.27).

898. Aureothricin. *N-(4,5-Dihydro-4-methyl-5-oxo-1,2-dithiolo[4,3-b]pyrrol-6-yl)propanamide;* 4-methyl-6-propionamido-1,2-dithiolo[4,3-*b*]pyrrol-5(4*H*)-one; 5-methyl-3-propionamidopyrrolin-4-one[4,3-*d*]-1,2-dithiole; propionopyrrothine. $C_9H_{10}N_2O_2S_2$; mol wt 242.36. C 44.61%, H 4.16%, N 11.56%, O 13.21%, S 26.46%. Antibiotic substance produced by *Streptomyces sp* 26A (Mitaka, Tokyo, June 1947) which resembles *Nocardia farcinicus* or *Streptomyces lipmanii:* Umezawa *et al., Japan. Med. J.* **1,** 512 (1948); Umezawa *et al., J. Antibiot.* **2,** Suppl. A, 105 (1949); Maeda, *ibid.* **2,** 793 (1949); Maeda, *Japan. Med. J.* **2,** 85 (1949); Celmer *et al., J. Am. Chem. Soc.* **74,** 6304 (1952). Structure: Celmer, Solomons, *ibid.* **77,** 2861 (1955). Prepn: Celmer, **U.S.** pat. **2,752,359** (1956 to Pfizer). Total synthesis: Schmidt, Geiger, *Angew. Chem.* **74,** 328 (1962).

Crystals, usually solvated. Stable in air. When anhydr dec 260-270° (sublimes near 200°). uv max: 248, 312, 388 nm (ϵ 6,100, 3,900, 11,000). Practically insol in water; slightly sol in ethyl acetate, butyl acetate, acetone, benzene, ether, alc.

899. Aurin. *4-[Bis(4-hydroxyphenyl)methylene]-2,5-cyclohexadien-1-one;* 4-(*p,p'*-dihydroxybenzhydrylidene)-2,5-cyclohexadien-1-one; *p*-rosolic acid; corallin; C.I. 43800. $C_{19}H_{14}O_3$; mol wt 290.30. C 78.60%, H 4.86%, O 16.53%. Prepd by heating a mixture of phenol, oxalic and sulfuric acids: Zulkowsky, *Ann.* **194,** 109, 122 (1878); **202,** 179 (1880); *Monatsh.* **16,** 358 (1895); *Colour Index* **vol. 4** (3rd ed., 1971) pp 4407. By heating phenol with formic acid and stannous chloride: Nencki, Schmid, *J. Prakt. Chem.* [2] **23,** 549 (1881). By heating phenol and carbon tetrachloride to 140-160° in presence of $AlCl_3$ or $ZnCl_2$ and treating the reaction product with water: Heumann, **Ger.** pat. **68,976** (*Frdl.* **3,** 103).

Deep red (garnet-like) crystals with metallic luster. Orthorhombic bipyramidal. Technical product may occur as yellowish-brown pieces with dark green metallic fracture. Dec 308-310°. Not volatile in vacuo at 180°. Absorption max (KOH): 534.6, 479.5 nm. Practically insol in water (0.12%) and in benzene. Freely sol in alc, giving a golden yellow solution, in aq or alc NaOH and KOH giving a carmine-red solution, in concd or 70% H_2SO_4, in HCl, in $HClO_4$ giving a yellow to orange soln. Moderately soluble in glacial acetic acid; slightly sol in ether, chloroform.

Sodium salt (yellow corallin), yellowish pieces with greenish, metallic luster; sol in water, giving a carmine-red soln.

USE: Dye intermediate.

900. Aurodox. *1-Methylmocimycin;* antibiotic X-5108; goldinodox; goldinomycin; X-5108. $C_{44}H_{62}N_2O_{12}$; mol wt 810.99. C 65.17%, H 7.71%, N 3.45%, O 23.67%. Antibiotic produced by *Streptomyces goldiniensis* var. *goldiniensis:* J. Berger, **Ger.** pat. **2,140,322** corresp to **U.S.** pat. **3,708,577** (both 1972 to Hoffmann-La Roche). Isoln, production and properties: J. Berger *et al., J. Antibiot.* **26,** 15 (1973). Improved production: J. Unowsky, D. C. Hoppe, *ibid.* **31,** 662 (1978). Structure: H. Maehr *et al., J. Am. Chem. Soc.* **95,** 8449 (1973). Absolute stereochemistry: *eidem, ibid.* **96,** 4034 (1974). Biosynthesis: C-M. Liu *et al., J. Antibiot.* **30,** 416 (1977); **32,** 414 (1979). Stereospecific total synthesis: R. E. Dolle, K. C. Nicolaou, *J. Am. Chem. Soc.* **107,** 1691, 1695 (1985). Growth promoting activity: W. L. Marusich *et al.,*

Poultry Sci. **53,** 636 (1974). Review of chemistry: H. Maehr *et al., Can. J. Chem.* **58,** 501 (1980).

Yellow amorphous solid. Weakly acidic. Sol in methanol, ethyl acetate, $CHCl_3$, acetone, methylene chloride. Insol in water. pKa = 6.1. Unstable in acidic or basic soln, but can be kept in aq soln at pH 7-9 for 4 hrs at 25° without significant loss of activity. Solns kept at 25° or heated to 56° are 2-10 times more stable in methanol than in water or pH 9 borate buffer. LD_{50} in mice (mg/kg): > 1000 s.c., > 4000 orally (Berger *et al.*).

Sodium salt, $C_{44}H_{61}N_2NaO_{12}$, yellow solid. $[\alpha]_D^{25}$ −82.8° (c = 0.52 in ethanol). uv max (0.1*N* HCl): 334, 233, 206 nm ($E_{1cm}^{1\%}$ 403, 610, 500); (0.1*N* KOH): 327, 231 nm ($E_{1cm}^{1\%}$ 416, 647). Sol in water, methanol, ethanol, isopropanol, butanol, DMF. Slightly sol in amyl alcohol, THF, dioxane. Insol in benzene, chloroform, ethyl ether, petr ether.

THERAP CAT (VET): Poultry growth promotant.

901. Aurothioglucose. *(1-Thio-*D-*glucopyranosato)-gold;* (1-D-glucosylthio)gold; gold thioglucose; Aureotan; Solganal; Solganal B; Aurumine; Oronol. $C_6H_{11}AuO_5S$; mol wt 392.22. C 18.36%, H 2.83%, Au 50.25%, O 20.40%, S 8.17%. Prepd by adding a soln of gold bromide to an aq soln of thioglucose containing sulfur dioxide; after refluxing, the gold thioglucose is pptd with alcohol. The product is purified by dissolving in water and reprecipitating with alc: Lebeau-Janot, *Traité de Pharmacie Chimique* II, 661 (1956). Clinical comparison with auranofin, *q.v.,* in rheumatoid arthritis: P. L. C. M. Van Riel *et al., Clin. Rheumatol.* **5,** 359 (1986).

Yellow crystals. Slight mercaptan-like odor. Sol in water with decompn; slightly sol in propylene glycol. Insol in alc and in most other organic solvents and in vegetable oils. Suspension in vegetable oil, **Solganal B Oleosum.**

USE: Exptly to produce obesity in animals: *Chem. & Eng. News* **32,** 22 (1954).

THERAP CAT: Antirheumatic.

902. Aurothioglycanide. *(2-Mercapto-N-phenylacetamidato-O,S)gold; [[(phenylcarbamoyl)methyl]thio]gold;* α-auromercaptoacetanilide; 2-mercaptoacetanilide *S*-gold derivative; aurothioglycolic acid anilide; aurothioglycolanilide; Lauron. C_8H_8AuNOS; mol wt 363.23. C 26.45%, H 2.22%, Au 54.24%, N 3.86%, O 4.40%, S 8.83%. Prepd by the action of aurous bromide on thioglycolic acid anilide: Lewenstein, **U.S.** pat. **2,451,841** (1948).

Grayish-yellow powder, mp 238-241°. Insol in water, benzene, ether, chloroform, acids, bases. Marketed as a suspension in sesame oil.

THERAP CAT: Antirheumatic.

903. Avermectins. AVM; C-076. A group of broad-spectrum antiparasitic antibiotics which are derivs of pentacyclic 16-membered lactones related to the milbemycins, *q.v.* Isoln from a novel actinomycete, *Streptomyces avermitilis,* and separation of major components A_{1a}, A_{2a}, B_{1a}, B_{2a} and minor components A_{1b}, A_{2b}, B_{1b}, B_{2b}: G. Albers-Schönberg et al., **Ger.** pat. **2,717,040;** eidem, **U.S.** pat. **4,310,519** (1977, 1982 both to Merck & Co.); R. W. Burg et al., Antimicrob. Ag. Chemother. **15,** 361 (1979); T. W. Miller et al., ibid. 368. Antiparasitic activity: S. R. Egerton et al., ibid. 372. Mechanism of action: L. C. Fritz et al., Proc. Nat. Acad. Sci. USA **76,** 2062 (1979). Pesticidal activity: I. Putter et al., Experientia **37,** 963 (1981). Structure determn: G. Albers-Schönberg et al., J. Am. Chem. Soc. **103,** 4216 (1981). Absolute configuration of avermectin B_{1a} and aglycon B_{2a}: J. P. Springer et al., ibid. 4221. Prepn of aglycons: H. Mrozik et al., J. Org. Chem. **47,** 489 (1982). Approaches to synthesis: eidem, Tetrahedron Letters **23,** 2377 (1982); S. Hanessian et al., ibid. **27,** 5071 (1986); R. Baker et al., J. Chem. Soc. Perkin Trans. I, 85 (1988). Total synthesis of A_{1a} ($C_{49}H_{74}O_{14}$): S. J. Danishefsky et al., J. Am. Chem. Soc. **109,** 8119 (1987). Biosynthetic studies: D. E. Cane et al., ibid. **105,** 4110 (1983); M. D. Schulman e al., J. Antibiot. **39,** 541 (1986); T. S. Chen et al., Arch. Biochem. Biophys. **269,** 544 (1989). Review: W. C. Campbell et al., "The Discovery of Ivermectin and Other Avermectins", in Pesticide Synthesis Through Rational Approaches, ACS Symp. Series **255,** P. S. Magee et al., Eds. (American Chemical Society, Washington, D.C., 1984) pp 1-20; M. H. Fisher, H. Mrozik, "The Avermectin Family of Macrolide Antibiotics" in Macrolide Antibiotics, S. Omura, Ed. (Academic Press, New York, 1984) pp 553-606; Southwest Entomol. **1985,** Suppl. 7, pp 1-51; J. R. Babu, "Avermectins: Biological and Pesticidal Activities", in Biologically Active Natural Products, ACS Symp. Series **380,** H. G. Cutler, Ed. (American Chemical Society, Washington, D.C., 1988) pp 91-108. For related structures see abamectin, ivermectin.

Avermectin A_{1a}

Avermectin $A_{1a/b}$, $[\alpha]_D^{27}$ +68.5 ± 2° (c = 0.77 in chloroform). uv max (methanol): 237, 243, 252 nm (ϵ 28700, 31275, 20290).
Avermectin $A_{2a/b}$, $[\alpha]_D^{27}$ +48.8 ± 2° (c = 1.64 in chloroform). uv max (methanol): 237, 243, 245 nm (ϵ 28800, 31740, 20425).
Avermectin $B_{1a/b}$, see abamectin.
Avermectin $B_{2a/b}$, $[\alpha]_D^{27}$ +38.3 ± 2° (c = 0.87 in chloroform). uv max (methanol): 237, 243, 252 nm (ϵ 27580, 30590, 20060).

904. Avidin. A factor isolated from raw egg white, capable of producing biotin deficiency in rats and chicks. Occurs in the white portion of eggs of birds and amphibia, and probably in the genital tract of all animals. Destroyed by cooking or irradiation. Isoln: Eakin et al., J. Biol. Chem. **136,** 801 (1940); Pennin᷃ ᷃n et al., J. Am. Chem. Soc. **64,** 469 (1942); Fraenkel-Con aι eι al., Arch. Biochem. Biophys. **39,** 80, 97 (1952). Improved purification and crystallization: Green et al., Biochem. J. **118,** 67, 71 (1970). Structure is a glycoprotein containing four essentially identical subunits: Green, ibid. **92,** 16c (1964). The combined mol wt of the subunits is about 66,000. Each subunit is a single polypeptide chain containing 128 amino acid residues with alanine at the N-terminal, glutamic acid at the C-terminal, and a carbohydrate moiety attached at the asparaginyl residue, position 17. Complete amino acid sequence of the protein sub-

unit: DeLange, Huang, J. Biol. Chem. **246,** 698 (1971). Studies on biotin inactivation by avidin: Becker, Wilchek, Biochim. Biophys. Acta **264,** 165 (1972). Review: Green, Advan. Protein Chem. **29,** 85-133 (1975).

905. Avilamycin. Polyether antibiotic complex produced by Streptomyces viridochromogenes, ETH 23575 (NRRL 2860); belongs to the orthosomycin family of antibiotics, members of which contain one or more ortho ester linkages associated with carbohydrate residues. Isoln: E. Gaeumann et al., Ger. pat. **1,116,864;** eidem, U.S. pat. **3,-131,126** (1961, 1964 both to Ciba-Geigy); and characterization: F. Buzzetti et al., Experientia **24,** 320 (1968). Mechanism of action studies: H. Wolf, FEBS Letters **36,** 181 (1973). Avilamycin A is the main component; avilamycins B through N are also known and are derivatives of A differing at the C-45 linkage and/or the C-56 ketone adduct. Isoln of avilamycin C: W. Heilman et al., Helv. Chim. Acta **62,** 1 (1979). Structural study of A and C: eidem, ibid.; W. Keller-Schierlein et al., ibid. 7; E. Kupfer et al., ibid. **65,** 3 (1982). Isoln and structure of avilamycins F through N: J. L. Mertz et al., J. Antibiot. **39,** 877 (1986). GC determn of avilamycin residues in swine tissue and fluids: G. Formica, C. Giannone, J. Assoc. Off. Anal. Chem. **69,** 763 (1986). Synthesis of the A-B fragment of avilamycins A and C: P. Jütten et al., Tetrahedron **43,** 4133 (1987); of the disaccharide C-D fragment: J.-M. Beau et al., Tetrahedron Letters **28,** 1105 (1987). Use as a feed additive: F. Knusel et al., U.S. pat. **4,185,091** (1980 to Ciba-Geigy); in prevention of swine dysentery: E. E. Ose, U.S. pat. **4,436,734** (1984 to Lilly). Effect on feed conversion efficiency in swine: D. J. Jones et al., J. Anim. Sci. **65,** 881 (1987).

Avilamycin A

Colorless, needle-shaped crystals from acetone/ether, mp 188-189.5°. $[\alpha]_D^{20}$ +0.8° (c = 1.165 in abs ethanol) and −7.7° (c = 1.083 in chloroform).
Avilamycin A, $C_{61}H_{88}Cl_2O_{32}$, LY 048 740. Colorless needles from chloroform/petr ether, mp 181-182° (1-2.H_2O). uv max (methanol): 227, 286 br nm (log ϵ 4.15, 3.33).
Avilamycin C, $C_{61}H_{90}Cl_2O_{32}$. Dihydrate, colorless fine plates from acetone/ether, mp 188-189°. $[\alpha]_D^{20}$ −4.8° (c = 1.44 in chloroform). uv max (methanol): 228, 284 br nm (log ϵ 4.12, 3.33).

THERAP CAT (VET): Growth promotant.

906. Avoparcin. AV-290; C-254; CL 81,588; LL-AV290; Avotan. Animal feed glycopeptide antibiotic complex produced by Streptomyces candidus. Prodn: M. P. Kunstmann, J. N. Porter, U.S. pat. **3,338,786** (1967 to Am. Cyanamid). Fermentation, isoln, characterization: M. P. Kunstmann et al., Antimicrob. Ag. Chemother. **24** (1968). Isoln, purif of major components: W. J. McGahren et al., J.Antibiotic. **36,** 1671 (1983). HPLC separation of α-, β-avoparcin: F. Sztaricskai et al., ibid. 1691. Structural studies: J. J. Hlavka *et al., Tetrahedron Letters* **1974**, 175; W. J. McGahren *et al., J. Am. Chem. Soc.* **101**, 2237 (1979). Structure: *eidem, ibid.* **102**, 1671 (1980). Stereochemistry and epimerization: G. A. Ellestad *et al., J. Antibiot .* **36**, 1683 (1983). Relationship between structure, antibacterial activity: S.W. Fesik *et al., Mol. Pharmacol.* **25**, 275 (1984). Effect on resistance patterns in chickens: J. R. Walton, *Zentralbl. Veterinaermed. (B)* **25**, 290 (1978), *C.A.* **89**, 173507 (1978). Prevention of exptly induced necrotic enteritis in chickens: J. F. Prescott, *Avian Dis.* **23**, 1072 (1979). Effect on feedlot performance: R. J. Johnson *et al., J. Anim. Sci.* **48**, 1338 (1979).

α-avoparcin R = H
β-avoparcin R = Cl

White, hygroscopic, amorphous solid; no definitive mp. uv max: 280 nm in neutral or acidic solns; 300 nm in basic solns. Sol in water, DMF, DMSO. Max stability of aq solns is at pH 4-8. Moderately sol in methanol. LD_{50} in mice, rats, and chickens: > 10,000 mg/kg orally. (Am. Cyanamid, company literature)

α-Avoparcin, $C_{89}H_{101}ClN_9O_{36}$, $[\alpha]_D^{25}$ −96° ±2° (c = 0.62 in 0.1N HCl). uv max (0.1N HCl): 280 nm ($E_{1cm}^{1\%}$ 42.0).
β-Avoparcin, $C_{89}H_{100}Cl_2N_9O_{36}$, $[\alpha]_D^{25}$ −102° ±2° (c = 0.65 in 0.1N HCl). uv max (0.1N HCl): 280 nm ($E_{1cm}^{1\%}$ 44.0).
THERAP CAT (VET): Antibacterial. Growth promotant.

907. Azacitidine. *4-Amino-1-β-D-ribofuranosyl-1,3,5-triazine-2(1H)-one;* 5-azacytidine; ladakamycin; U-18,496; NSC-102816; Mylosar. $C_8H_{12}N_4O_5$; mol wt 244.21. C 39.34%, H 4.96%, N 22.94%, O 32.76%. Glycosyl deriv of *5-azocytosine* with antibiotic and antitumor activity. Chemical synthesis: A. Piskala, F. Sorm, *Coll. Czech. Chem. Commun.* **29**, 2060 (1964); **Neth. pat. Appl. 6,403,587** corresp to F. Sorm, A. Piskala, **U.S. pat. 3,350,388** (1965, 1967 both to Cesk. Acad. Ved.). Production by *Streptoverticillium ladakanus* and biological activity: L. J. Hanka *et al., Antimicrob.Ag. Chemother.* **1966**, 619. Isoln and structure: M. E. Bergy, R. R. Herr, *ibid.* 625. Pharmacokinetics: C. J. Kelly *et al., Cancer Treat. Rep.* **62**, 1025 (1978). Metabolism: P. G. Plagemann *et al., Cancer Res.* **38**, 2458 (1978). Use in acute leukemia: J. H. Saiki *et al., Cancer* **42**, 2111 (1979); P. L. Lomen *et al., Neoplasma* **27**, 101 (1980). Azacitidine has been shown to selectively increase γ-globulin synthesis in a patient with β-thalassemia: T. J. Ley, *N. Engl. J. Med.* **307**, 1469 (1982). Toxicology study: P. E. Palm, C. J. Kensler, *U.S. Clearinghouse Fed. Sci. Tech. Inform.,* **PB-194791** (1970) 191 pp., *C.A.* **75**, 33704j (1971). Carcinogenicity study: G. D. Stoner *et al., Cancer Res.* **33**, 3069 (1973). *Reviews:* C. C. Cheng, *Ann. Rep. Med. Chem.* **7**, 129-144 (1972); A. Cibak, *Oncology* **30**, 405-422 (1974).

Crystals from methanol, mp 228-230°. $[\alpha]_D^{25}$ +39° (c = 1 in water). uv max (water): 241 nm (ε 8,767); (0.01N HCl): 249 nm (ε 3,077); (0.01N KOH): 223 nm (ε 24,200). LD_{50} in mice: 115.9 mg/kg i.p.; 572.3 mg/kg orally, P. E. Palm, C. J. Kensler, *loc. cit.*
THERAP CAT: Antineoplastic.

908. Azacosterol. *17β-[[3-(Dimethylamino)propyl]-methylamino]androst-5-en-3β-ol;* N-methyl-N-[3-(dimethylamino)propyl]-17β-aminoandrost-5-en-3β-ol; 20,25-diazacholesterol; diazasterol. $C_{25}H_{44}N_2O$; mol wt 388.62. C 77.26%, H 11.41%, N 7.21%, O 4.12%. Prepn: Counsell *et al., J. Med. Pharm. Chem.* **5**, 1224 (1962); Counsell, Klimstra, **U.S. pat. 3,084,156** (1963 to Searle).

Crystals from acetone + methanol, mp 146-148°. $[\alpha]_D$ −54.5°.
Dihydrochloride, $C_{25}H_{46}Cl_2N_2O$, *SC 12937, Azasterol, Ornitrol.* $[\alpha]_D$ − 32°.
THERAP CAT: Hypocholesteremic.
THERAP CAT (VET): Avian chemosterilant.

909. Azacyclonol. α,α-*Diphenyl-4-piperidinemethanol;* α-(4-piperidyl)benzhydrol; diphenyl (γ-pyridyl)carbinol; MER-17; gamma-pipradrol; Psychosan; Frenoton; Ataractan; Calmeran. $C_{18}H_{21}NO$; mol wt 267.36. C 80.86%, H 7.92%, N 5.24%, O 5.98%. Prepd by hydrogenation of α,α-diphenyl-4-pyridinemethanol: Schumann *et al.,* **U.S. pat. 2,804,422** (1957 to Merrell).

Crystals, mp 160-161°.
Hydrochloride monohydrate, $C_{18}H_{22}ClNO.H_2O$, *Frenquel.* Crystals from butanone, mp 283-285°. Slightly bitter taste. Moderately sol in water.
THERAP CAT: Hydrochloride as tranquilizer.

910. Azadirachtin. $C_{35}H_{44}O_{16}$; mol wt 720.73. C 58.33%, H 6.15%, O 35.52%. A tetranortriterpinoid isolated from the seeds of the neem tree, *Azadirachta indica* A. Juss. *(Melia azadirachta* L.), Meliaceae, and the chinaberry tree, *M. azedarach* L. Highly active insect feeding deterrent and growth regulator. Isoln from *A. indica* and identification as feeding inhibitor in locusts: J. H. Butterworth, E. D. Morgan, *Chem. Commun.* **1968**, 23; from *M. azedarach:* E. D. Morgan, M. D. Thornton, *Phytochemistry* **12**, 391 (1973). Partial synthesis: D. Pflieger *et al., Tetrahedron Letters* **28**, 1519 (1987). Structural studies: J. H. Butterworth *et al., J. Chem. Soc. Perkin Trans. II* **1972**, 2445. ^{1}H- and ^{13}C-NMR data and structure: P. R. Zanno *et al., J. Am. Chem. Soc.* **97**, 1975 (1975); K. Nakanishi in *Recent Advances in Phytochemistry,* vol. 9, V. C. Runeckles, Ed. (Plenum Press, New York, 1975) pp 283-298. Revised structure: W. Kraus *et* al., *Tetrahedron Letters* **26**, 6435 (1985); H. B. Broughton *et al.*, *Chem. Commun.* **1986**, 46. Isoln by HPLC: E. C. Uebel *et al.*, *J. Liq. Chromatog.* **2**, 875 (1979); J. D. Warthen, Jr. *et al.*, *ibid.* **7**, 591 (1984). Antifeedant activity in locusts: J. S. Gill, C. T. Lewis, *Nature* **232**, 402 (1971); in fall army worms, cotton bollworms: J. A. Klocke, I. Kubo, *Entomol. Exp. Appl.* **32**, 299 (1982). Insect ecdysis and growth inhibition: H. Rembold, K. P. Sieber, *Z. Naturforsch.* **36C**, 466 (1981); I. Kubo, J. A. Klocke, *Agr. Biol. Chem.* **46**, 1951 (1982); K. P. Sieber, H. Rembold, *J. Insect. Physiol.* **29**, 523 (1983). Series of articles on chemistry and activity: *Natural Pesticides from the Neem Tree*, Proc. 1st Int. Neem Conf., 1980, H. Schmutterer *et al.*, Eds. (German Agency for Technical Cooperation, Eschborn, 1981) 291 pp.

Microcrystalline powder from carbon tetrachloride, mp 154-158°. $[\alpha]_D -53°$ (c = 0.5 in $CHCl_3$). uv max (methanol): 217 nm (ϵ 9100).
USE: Experimentally as insect control agent.

911. Azafrin. *(5R,6R)-5,6-Dihydro-5,6-dihydroxy-10'-apo-β,ψ-carotenoic acid*; escobedin. $C_{27}H_{38}O_4$; mol wt 426.57. C 76.02%, H 8.98%, O 15.00%. Carotenoid-carboxylic acid from roots of the South American plant "Aza-franillo," *Escobedia scabrifolia* Ruiz & Pav., and *Escobedia laevis* Cham. & Schlecht., *Scrophulariaceae*. Isoln: R. Kuhn *et al.*, *Ber.* **64**, 333 (1931); *ibid.* **65**, 1873 (1932). Structure: R. Kuhn, A. Deutsch, *ibid.* **66**, 883 (1933); R. Kuhn, H. Brockmann, *ibid.* **67**, 885 (1934); *Ann.* **516**, 104 (1935). Absolute configuration (5R:6R): W. Eschenmoser, C. H. Eugster, *Helv. Chim. Acta* **58**, 1722 (1975).

Orange-colored prisms from toluene, mp 213°. $[\alpha]_{6438}^{20}$ −75° (c = 0.28 in alcohol). Absorption max (chloroform): 458, 428 nm. Practically insol in water. Sol in dil NaOH or Na_2CO_3 solns, in chloroform, alc, acetic acid and benzene; sparingly sol in ether.
Methyl ester, $C_{28}H_{40}O_4$, reddish-yellow leaflets from ether, mp 191°. $[\alpha]_{6438}^{22} -32°$ (chloroform).

912. 8-Azaguanine. *5-Amino-1,4-dihydro-7H-1,2,3-triazolo[4,5-d]pyrimidin-7-one; 5-amino-1,6-dihydro-7H-v-triazolo[4,5-d]pyrimidine-7-one; 5-amino-1H-v-triazolo[d]-pyrimidin-7-ol; 5-amino-7-hydroxy-1H-v-triazolo[d]pyrimidine*; pathocidin; guanazolo. $C_4H_4N_6O$; mol wt 152.12. C 31.58%, H 2.65%, N 55.25%, O 10.52%. Triazolo analog of guanine. Prepd from 2,4,5-triamino-6-hydroxypyrimidine: Roblin, Jr. *et al.*, *J. Am. Chem. Soc.* **67**, 290 (1945); from 2-amino-5-nitro-4-oxopyrimidine: H. U. Blank *et al.*, *J. Org. Chem.* **35**, 1131 (1970). Formation from guanine by *Streptomyces albus* and identity with pathocidin: K. Hira-sawa, K. Isono, *J. Antibiot.* **31A**, 628 (1978). The first purine analog to show carcinostatic effects in murine malignancies; is readily incorporated into ribonucleic acids. Review: R. E. Parks, Jr., K. C. Agarwal, *Handb. Exp. Pharmacol.* **38**, pt. 2, 458 (1975); D. Grunberger, G. Grunberger in

Antibiotics vol. **5**(pt. 2), F. E. Hahn, Ed. (Springer-Verlag, New York, 1979) pp 110-123.

Crystals from dil aq NaOH. Dec above 300° without melting. Absorption spectrum: L. F. Cavalieri *et al. J. Am. Chem. Soc.* **70**, 3875 (1948). Soluble in dil caustic, in dil acids. Practically insol in water, alcohol, ether.
USE: Purine antimetabolite.

913. Azamethonium Bromide. *2,2'-(Methylimino)bis-[N-ethyl-N,N-dimethylethanaminium]dibromide; [(methyl-imino)diethylene]bis[ethyldimethylammonium bromide]; 3-methyl-3-azapentane-1,5-bis(ethyldimethylammonium) dibromide; N,N,N',N'-3-pentamethyl-N,N'-diethyl-3-aza-pentylene-1,5-diammonium dibromide*; pentamethazene dibromide; Präparat 9295; Ciba 9295; Pendiomid; Pentamin; Azameton; Azamethone; Ganlion; Pentaméthazène. $C_{13}H_{33}$-Br_2N_3; mol wt 391.25. C 39.91%, H 8.50%, Br 40.85%, N 10.74%. Prepd from pentamethyl diethylenetriamine and ethyl bromide: Marxer, Miescher, *Helv. Chim. Acta* **34**, 924 (1951); Swiss pat. **284,212**, C.A. **48**, 7625 (1954) and U.S. pat. **2,654,785** (1952, 1953 both to Ciba). Ganglion blocking agent. Pharmacological studies: H. J. Bein, R. Meier, *Experientia* **6**, 351 (1950); *eidem, Schweiz. Med. Wochenschr.* **81**, 446 (1951); I. V. Uranov, *Farmakol. Toksikol.* **21**, 110 (1958).

Prisms or rhombohedra from alcohol + ethyl acetate, mp 212-215°. Freely sol in water. pH of satd aq soln about 6.5. LD_{50} in mice (mg/kg): 2500 orally, 60 i.v., I. V. Uranov, *loc. cit.* LD_{50} in rabbits (mg/kg): 3000 orally; 160 s.c.; 75 i.v., H. J. Bein, R. Meier, *loc. cit.* (1951).
THERAP CAT: Antihypertensive.

914. Azanidazole. *4-[2-(1-Methyl-5-nitro-1H-imid-azol-2-yl)ethenyl]-2-pyrimidinamine; (E)-2-amino-4-[2-(1-methyl-5-nitroimidazol-2-yl)vinyl]pyrimidine*; nitromidine; F-4; Triclose. $C_{10}H_{10}N_6O_2$; mol wt 246.23. C 48.78%, H 4.09%, N 34.13%, O 13.00%. Prepn and use: A. Garzia, Ger. pat. **2,358,483**; *eidem*, U.S. pats. **3,882,105** and **3,969,-520** (1974, 1975 and 1976, all to Chemoterapico). Toxicological and teratological studies: R. Tammiso *et al.*, *Arznei-mittel-Forsch.* **28**, 2251 (1978).

Bright yellow odorless powder, mp 232-235°. Sol in DMF, DMSO, mineral oils and acids. Slightly sol in dioxane and acetone. LD_{50} in mice, rats (mg/kg): 5100, 7600 orally; 590, 860 i.p. (Tammiso).
THERAP CAT: Antiprotozoal (Trichomonas).

915. Azaperone. *1-(4-Fluorophenyl)-4-[4-(2-pyridin-yl)-1-piperazinyl]-1-butanone; 4'-fluoro-4-[4-(2-pyridyl)-1-piperazinyl]butyrophenone; 1-[3-[4-(4-fluorobenzoyl)propyl]-4-(2-pyridyl)piperazine*; R-1929; Stresnil; Suicalm. $C_{19}H_{22}$-FN_3O; mol wt 327.41. C 69.70%, H 6.77%, F 5.80%, N 12.83%, O 4.89%. Prepn: Janssen, U.S. pat. **2,979,508** (1961). Synthesis of labeled compound: Soudijn, van Wijn-gaarden, *J. Lab. Comp.* **4**, 159 (1968). Distribution and metabolism studies in rat and pig: Heykants *et al.*, *Arznei-*

mittel-Forsch. **21**, 982, 1263, 1357 (1971). Veterinary clinical studies: Symoens, van den Brande, *Vet. Rec.* **85**, 64 (1969). Crystal structure: M. H. J. Koch *et al., Acta Crystallogr.* **33B**, 1975 (1977).

Crystals, mp 73-75°.
THERAP CAT (VET): Sedative; tranquilizer.

916. Azaserine. L-*Serine diazoacetate(ester); O*-diazo-acetyl-L-serine; Cl 337; CN 15757; P 165. $C_5H_7N_3O_4$; mol wt 173.13. C 34.68%, H 4.08%, N 24.27%, O 36.97%. N_2-CHCOOCH$_2$CH(NH$_2$)COOH. Antibiotic substance produced by a *Streptomyces* sp. or by synthesis: Bartz *et al., Nature* **173**, 72 (1954); Fusari *et al., J. Am. Chem. Soc.* **76**, 2878, 2881 (1954); Moore *et al., ibid.*, 2884; Nicolaides *et al., ibid.*, 2887; T. J. Curphey, D. S. Daniel, *J. Org. Chem.* **43**, 4666 (1978); Ehrlich *et al.*, Moore *et al.*, U.S. pats. **2,996,435** and **3,030,388** (1961, 1962, both to Parke, Davis). Retards growth of transplantable neoplasms in animals: Stock *et al., Nature* **173**, 71 (1954). Pathology and pharmacology studies: S. S. Sternberg, F. S. Philips, *Cancer* **10**, 889 (1957). Crystal and molecular structure: A. Fitzgerald, L. H. Jensen, *Acta Crystallogr.* **B34**, 828 (1978). Review of carcinogenicity studies: *IARC Monographs* **10**, 73-77 (1976). *Review:* Pettillo, Hunt, *Antibiotics*, **vol. 1**, D. Gottlieb, P. Shaw, Eds. (Springer-Verlag, New York, 1967) pp 481-493.

Orthorhombic, pale yellow to green crystals from 90% ethanol, dec 146-162°. $[\alpha]_D^{27.5} -0.5°$ (c = 8.46 in H_2O at pH 5.18). uv max (pH 7): 250.5 nm (E$^{1\%}_{1cm}$ 1140); in 0.1*N* NaOH: 252 nm (E$^{1\%}_{1cm}$ 1230). Very sol in water; slightly sol in methanol, abs ethanol acetone, but sol in warm aq solns of these solvents. pKa 8.55. Aq solns are most stable at pH 8. Incompatible with acids. LD$_{50}$ orally in mice, rats: 150, 170 mg/kg/day, S. S. Sternberg, F. S. Philips, *loc. cit.*
THERAP CAT: Antifungal; antineoplastic.

917. Azatadine. *6,11-Dihydro-11-(1-methyl-4-piperidinylidene)-5H-benzo[5,6]cyclohepta[1,2-b]pyridine;* 4-aza-5-(*N*-methyl-4-piperidinylidene)-10,11-dihydro-5*H*-dibenzo[*a,d*]cycloheptene. $C_{20}H_{22}N_2$; mol wt 290.41. C 82.72%, H 7.64%, N 9.64%. Prepn: F. J. Villani, **Belg.** pat. **647,043** (1964 to Scherico) corresp to U.S. pats. **3,326,924** and **3,419,565** (1967, 1968, both to Schering); *see also* U.S. pat. **3,357,986** (1967 to Schering); F. J. Villani *et al., J. Med. Chem.* **15**, 750 (1972). Pharmacology: S. Tozzi *et al., Agents Actions* **4**, 264 (1974); T. Yanagita *et al., Jitchuken, Zenrinsho Kenkyuho* **1**, 79 (1976), *C.A.* **84**, 38611x (1976). Toxicity studies: H. Tanaka *et al., ibid.* 173 (1975), *C.A.* **84**, 25951h (1976).

Crystals from isopropyl ether, mp 124-126°.
Dimaleate, $C_{28}H_{30}N_2O_8$, *Sch 10649, Bonamid, Idulian, Optimine, Zadine.* Crystals from ethyl acetate-methanol, mp 152-154°.
THERAP CAT: Antihistaminic.

918. Azathioprine. *6-[(1-Methyl-4-nitro-1H-imidazol-5-yl)thio]-1H-purine;* 6-(1-methyl-4-nitro-5-imidazolyl)mercaptopurine; azothioprine; BW 57-322; NSC-39084; Azamune; Azanin; Azoran; Imuran; Imurel. $C_9H_7N_7O_2S$; mol wt 277.29. C 38.98%, H 2.55%, N 35.36%, O 11.54%, S 11.57%. Prepn: Hitchings, Elion, U.S. pat. **3,056,785** (1962 to Burroughs Wellcome). Pharmacokinetics: T. L. Ding, L. Z. Benet, *Drug. Metab. Dispos.* **7**, 373 (1979). Use in com-

bination with cyclophosphamide and hydroxychloroquine, *q.q.v.* for the treatment of rheumatoid arthritis: D. J. McCarty, G. F. Carrera, *J. Am. Med. Assoc.* **248**, 1718 (1982). Carcinogenicity studies: P. S. Mitrou *et al., Arzneimittel-Forsch.* **29**, 483, 662 (1979). Comprehensive description: W. P. Wilson, S. A. Benezra in *Analytical Profiles of Drug Substances* **vol. 10**, K. Florey, Ed. (Academic Press, New York, 1981) pp 29-53. Review of pharmacology and clinical efficacy in renal transplantation: G. L. C. Chan *et al., Pharmacotherapy* **7**, 165-177 (1987).

Pale yellow crystals from 50% aq acetone, dec 243-244°. uv max (methanol): 276 nm (ϵ 1.82 $\times$ 10^4); (0.1*N* HCl): 280 nm (ϵ 1.73 $\times$ 10^4); (0.1*N* NaOH): 285 nm (ϵ 1.55 $\times$ 10^4). pKa$_2$ 8.2. Slightly sol in water, chloroform, ethanol. Sodium salt, *Imurek.*
Note: This substance has been listed as a known carcinogen: *Fourth Annual Report on Carcinogens* (NTP 85-002, 1985) p 31.
THERAP CAT: Immunosuppressive; antirheumatic.

919. 6-Azathymine. *6-Methyl-1,2,4-triazine-3,5(2H,-4H)-dione;* 5-methyl-6-azauracil; 3,5-dihydroxy-6-methyl-1,2,4-triazine. $C_4H_5N_3O_2$; mol wt 127.10. C 37.80%, H 3.97%, N 33.06%, O 25.18%. Prepn: Thiele, Bailey, *Ann.* **303**, 82 (1898); Bailey, *Am. Chem. J.* **28**, 386 (1902); Bougault, Daniel, *Compt. Rend.* **186**, 1216 (1928); Chang, *J. Org. Chem.* **23**, 1951 (1958).

Crystals from water. mp 210-212°. pKa 7.6. uv max (0.1*N* HCl): 261 nm (ϵ 5,200); in 0.1*N* NaOH: 246 nm (ϵ 4,770). Its deoxyriboside, azathymidine, markedly inhibits the biosynthesis of DNA by neoplastic and normal cells *in vitro.*

920. 6-Azauridine. *2-β-D-Ribofuranosyl-1,2,4-triazine-3,5(2H,4H)-dione;* 6-azauracil riboside; 3,5-dioxo-2,3,4,5-tetrahydro-1,2,4-triazine riboside; AzUR; Ribo-Azauracil. $C_8H_{11}N_3O_6$; mol wt 245.19. C 39.19%, H 4.52%, N 17.14%, O 39.15%. Biosynthesis by *E. coli* in the presence of 6-azauracil: Skoda *et al., Experientia* **13**, 150 (1957). Chemical synthesis: Prystas *et al., Chem. & Ind. (London)* **1961**, 947; Sorm *et al., Czech.* pat. **92,697** (1959), *C.A.* **54**, 15411 (1960); **Brit.** pat. **827,441** (1960 to Spofa); Cristescu, *Rev. Roum. Chem.* **13**, 365 (1968). Crystal structure and conformation: Schwalbe *et al., Biochem. Biophys. Res. Commun.* **44**, 57 (1971). Pharmacology: Smahel *et al., Neoplasma* **18**, 435 (1971), *C.A.* **76**, 20997c (1972).

Crystals from ethanol, ether, mp 160-161°. $[\alpha]_D^{24}$ −132° (pyridine). uv max (water): 262 nm (ϵ 6100). pK 6.70.

2',3',5'-Triacetate, $C_{14}H_{17}N_3O_9$, *azaribine*, *2-(2,3,5-tri-O-acetyl-β-D-ribofuranosyl)-as-triazine-3,5(2H,4H)-dione*, *CB 304*, *Triazure*. LD_{50} in mice, rats: 7.8, 12.0 g/kg orally, Plevova *et al.*, *Toxicol. Appl. Pharmacol.* **17**, 511 (1970).

THERAP CAT: Base as antineoplastic; triacetate as an antipsoriatic.

921. Azelaic Acid. *Nonanedioic acid;* 1,7-heptanedicarboxylic acid; lepargylic acid; anchoic acid; Skinoren. $C_9H_{16}O_4$; mol wt 188.22. C 57.43%, H 8.57%, O 34.00%. $HOOC(CH_2)_7COOH$. Prepd by disruptive oxidation of ricinoleic acid: Hill, McEwen, *Org. Syn.* **coll. vol. II**, 53 (1943). Occurs in rancid oleic acid. Clinical comparison with tetracycline in acne: P. T. Bladon *et al.*, *Brit. J. Dermatol.* **114**, 493 (1986). Review of clinical studies in hyperpigmentary disorders: A. S. Breathnach, M. Nazzaro-Porro, *ibid.* **111**, 115-120 (1984); of pharmacology and clinical uses: M. Nazzaro-Porro, *J. Am. Acad. Dermatol.* **17**, 1033-1041 (1987).

Monoclinic prismatic needles, mp 106.5°. Distills above 360° with partial anhydride decompn. bp_{100} 286.5°; bp_{50} 2-65°; bp_{15} 237°; bp_{10} 225°. $d_4^{110.6}$ 1.0291. One liter of water dissolves 1.0 g at 1.0°; 2.4 g at 20°; 8.2 g at 50°; 22 g at 65°. Freely sol in boiling water, in alcohol. 1000 g of ether dissolves 18.8 g at 11° and 26.8 g at 15°. pK_1 (25°) 4.53; pK_2 5.33.

Dimethyl ester, $C_{11}H_{20}O_4$, liquid, d_4^{20} 1.0026. mp −3.9°. bp_8 140°.

THERAP CAT: Antiacne.

922. Azelastine. *4-[(4-Chlorophenyl)methyl]-2-(hexahydro-1-methyl-1H-azepin-4-yl)-1(2H)-phthalazinone;* 4-(p-chlorobenzyl)-2-(hexahydro-1-methyl-1H-azepin-4-yl)-1(2H)-phthalazinone; 4-(p-chlorobenzyl)-2-(N-methylperhydroazepin-4-yl)-1(2H)-phthalazinone. $C_{22}H_{24}ClN_3O$; mol wt 381.91. C 69.19%, H 6.33%, Cl 9.28%, N 11.00%, O 4.19%. Orally active H_1-histamine receptor antagonist. Prepn: **Belg.** pat. **778,269**; D. Vogelsang *et al.*, **U.S.** pat. **3,813,384** (1972, 1974 both to Asta-Werke AG). Synthesis and x-ray structure determn: G. Scheffler *et al.*, *Arch. Pharm.* **321**, 205 (1988). Pharmacology: K. Tasaka, M. Akagi, *Arzneimittel-Forsch.* **29**, 488 (1979). Pharmacology and toxicology: H. J. Zechel *et al.*, *ibid.* **31**, 1184 (1981). Series of articles on pharmacokinetics, pharmacology and toxicology: *ibid.* 1184-1238. Mechanism of action studies: N. Chand *et al.*, *Int. J. Immunopharmacol.* **7**, 833 (1985); *eidem*, *Br. J. Pharmacol.* **87**, 443 (1986). HPLC determn in plasma: J. Pivonka *et al.*, *J. Chromatog.* **420**, 89 (1987). Clinical evaluations in asthma: H. Magnussen, *Chest* **91**, 855 (1987); M. K. Albazzaz, K. R. Patel, *Thorax* **43**, 306 (1988).

Oil. Sol in methylene chloride.

Monohydrate, $C_{22}H_{24}ClN_3O.H_2O$, crystals from alcohol/-water. Two distinct crystal forms have been identified (Scheffler).

Monohydrochloride, $C_{22}H_{25}Cl_2N_3O$, *A-5610*, *W-2979M*, *E-0659*, *Azeptin*. Crystals from alcohol, mp 225-229°. LD_{50} in male, female mice, male, female rats (mg/kg): 36.5, 35.5, 26.9, 30.3 i.v.; 56.4, 42.8, 43.2, 46.6 i.p.; 63.0, 54.2, 66.5, 59.6 s.c.; 124, 139, 310, 417 orally (Zechel).

THERAP CAT: Anti-allergic; anti-asthmatic; antihistaminic.

923. 2-Azetidinecarboxylic Acid. $C_4H_7NO_2$; mol wt 101.10. C 47.52%, H 6.98%, N 13.86%, O 31.65%. From *Convallaria majalis* L. (lily-of-the-valley) and *Polygonatum officinale* Moensh., *Liliaceae*: Fowden, *Biochem. J.* **64**, 323 (1956); Fowden, Bryant, *ibid.* **70**, 626 (1958). Structure: Virtanen, *Angew. Chem.* **67**, 619 (1955); Fowden, *loc. cit.*

Biosynthesis: Linko, *Acta Chem. Scand.* **12**, 101 (1958); Fowden, Bryant, *Biochem. J.* **71**, 210 (1959); Fowden, *ibid.* **71**, 643 (1959); Leete, *J. Am. Chem. Soc.* **86**, 3162 (1964). Shows growth-inhibitory activity on cultures of *Escherichia coli* and on germinating seeds of different species in which it does not normally occur; Fowden, Richmond, *Biochim. Biophys. Acta* **71**, 459 (1963). Acts as a proline analog where a stoichiometric replacement of proline occurred with production of abnormal proteins having impaired biological activity: Peterson, Fowden, *Nature* **200**, 148 (1963).

Crystals from 95% hot methanol, discolors at 200° and darkens until 310° when heating stopped. $[\alpha]_D^{20}$ −108° (c = 3.6). Unstable to mineral acids. Soluble in cold and hot water. Practically insol in abs ethanol.

924. Azidamfenicol. *2-Azido-N-[2-hydroxy-1-(hydroxymethyl)-2-(4-nitrophenyl)ethyl]acetamide;* D-(−)-*threo*-1-p-nitrophenyl-2-azidoacetylaminopropane-1,3-diol; D-(−)-*threo*-2-azidoacetamido-1-p-nitrophenyl-1,3-propanediol; azidoamphenicol; Leukomycin N. $C_{11}H_{13}N_5O_5$; mol wt 295.25. C 44.74%, H 4.44%, N 23.72%, O 27.10%. Semi-synthetic antibiotic related to chloramphenicol, *q.v.* Prepn: Meiser, Domagk, **U.S.** pat. **2,882,275** (1959 to Bayer).

Crystals from ethylene chloride, mp 107°. $[\alpha]_D^{20}$ −20° (c = 1.6 in ethyl acetate). pH 7.1. Sol in water up to 2%.

THERAP CAT: Antibacterial.

925. Azidocillin. *6-[(Azidophenylacetyl)amino]-3,3-dimethyl-7-oxo-4-thia-1-azabicyclo[3.2.0]heptane-2-carboxylic acid;* α-azidobenzylpenicillin; 6-[D-α-azidophenylacetamido]penicillanic acid; SPC 97D; BRL 2351. $C_{16}H_{17}N_5O_4S$; mol wt 375.42. C 51.19%, H 4.56%, N 18.66%, O 17.05%, S 8.54%. Semi-synthetic antibiotic related to penicillin. Prepn: Sjöberg, Ekström *et al.*, **Brit.** pat. **940,488** corresp. to **U.S.** pat. **3,293,242** (1963 and 1966 to Beecham); Ekström *et al.*, *Acta Chem. Scand.* **19**, 281 (1965). Pharmacology: Hanssen *et al.*, *Antimicrob. Ag. Chemother.* **1967**, 568; Tunevall, Frisk, *ibid.* 573. Chemistry: Sjöberg *et al.*, *ibid.* 560. Metabolic studies: Ramsey *et al.*, *Arzneimittel-Forsch.* **22**, 1962 (1972).

Sodium salt, $C_{16}H_{16}N_5NaO_4S$, *Globacillin*, *Longatren*. Potassium salt, $C_{16}H_{16}KN_5O_4S$, *Nalpen*. mp 194° (dec).

THERAP CAT: Antibacterial.

THERAP CAT (VET): Antimicrobial. (In mastitis, as feed supplement).

926. Azinphos-methyl. *Phosphorodithioic acid O,O-dimethyl S-[(4-oxo-1,2,3-benzotriazin-3(4H)-yl)methyl] ester; phosphorodithioic acid O,O-dimethyl ester, S-ester with 3-mercaptomethyl-1,2,3-benzotriazin-4(3H)-one;* Bayer 17147; ENT 23233; R 1582; Cotnion-methyl; Gusathion M; Guthion. $C_{10}H_{12}N_3O_3PS_2$; mol wt 317.34. C 37.85%, H 3.81%, N 13.24%, O 15.13%, P 9.76%, S 20.21%. Prepn: Lorenz,

U.S. pat. 2,758,115 (1956 to Bayer). Activity: E. E. Ivy *et al., J. Econ. Entomol.* **48**, 293 (1955).

Crystals from methanol, mp 73-74°. d_4^{20} 1.44; n_D^{76} 1.6115. Soly in water at 25°: 33 mg/l. Sol in methanol, ethanol, propylene glycol, xylene, other organic solvents. Solns in ethanol and propylene glycol are stable for at least 3 weeks. Unstable at temperatures > 200°. Hydrolyzes in acid or cold alkali. LD_{50} in female rats: 11 mg/kg orally; 220 mg/kg dermally, T. B. Gaines, *Toxicol. Appl. Pharmacol.* **14**, 515 (1969).

O,O-Diethyl analog, $C_{12}H_{16}N_3O_3PS_2$, *azinphos-ethyl, Bayer 16259, ENT 22014, R 1513, Gusathion A, Ethyl Guthion.* Colorless needles, mp 53°, $bp_{0.001}$ 111°. d_4^{20} 1.284; n_D^{53} 1.5928. Vapor press at 20°: 2.2×10^{-7} mm Hg.

USE: Insecticide; acaricide. *Caution:* A cholinesterase inhibitor.

927. Azintamide. *2-[(6-Chloro-3-pyridazinyl)thio]-N,N-diethylacetamide; N,N*-diethyl-2-[6-(3-chloropyridazin-yl)thio]acetamide; *N,N*-diethyl-2-[6-(3-chloropyridazinyl)-mercapto]acetamide; (3-chloro-3-pyridazinylthio)acetic acid diethylamide; ST 9067; Oragallin. $C_{10}H_{14}ClN_3OS$; mol wt 259.77. C 46.24%, H 5.43%, Cl 13.65%, N 16.17%, O 6.16%, S 12.35%. Prepn: Kloimstein *et al., Arzneimittel-Forsch.* **14**, 261 (1964); Kloimstein *et al,* **Ger.** pat. **1,188,604** (1965 to Lentia).

Bitter tasting microcrystals from acetone, mp 97-98°. Freely sol in benzene, chloroform, ethyl acetate, acetone. Solubility in water 0.5%. LD_{50} orally in mice, rats: 2.34, 1.55 g/kg.

THERAP CAT: Choleretic.

928. Azithromycin. *[2R-(2R*,3S*,4R*,5R*,8R*,10R*,-11R*,12S*,13S*,14R*)]-13-[(2,6-Dideoxy-3-C-methyl-3-O-methyl-α-L-ribo-hexopyranosyl)oxy]-2-ethyl-3,4,10-trihy-droxy-3,5,6,8,10,12,14-heptamethyl-11-[[3,4,6-trideoxy-3-(dimethylamino)-β-D-xylo-hexopyranosyl]oxy]-1-oxa-6-aza-cyclopentadecan-15-one;* N-methyl-11-aza-10-deoxo-10-di-hydroerythromycin A; 9-deoxo-9a-methyl-9a-aza-9a-homoerythromycin A; CP 62993; XZ-450; Sumamed. $C_{38}H_{72}N_2O_{12}$; mol wt 748.99. C 60.94%, H 9.69%, N 3.74%, O 25.63%. Semi-synthetic macrolide antibiotic; related to erythromycin A, *q.v.* Prepn: **Belg.** pat. **892,357**; G. Kobrehel, S. Djokic, **U.S.** pat. **4,517,359** (1982, 1985 both to Sour Pliva). Antibacterial spectrum: S. C. Aronoff *et al., J. Antimicrob. Chemother.* **19**, 275 (1987); and mode of action: J. Retsema *et al., Antimicrob. Ag. Chemother.* **31**, 1939 (1987). Pharmacokinetics in animals: A. E. Girard *et al., ibid.* 1948.

Crystals, mp 113-115°. $[\alpha]_D^{20}$ −37° (c = 1 in $CHCl_3$). THERAP CAT: Antibacterial.

929. Azlocillin. *3,3-Dimethyl-7-oxo-6-[[[[(2-oxo-1-imidazolidinyl)carbonyl]amino]phenylacetyl]amino]-4-thia-1-azabicyclo[3.2.0]heptane-2-carboxylic acid;* D-α-[(imidazo-lidin-2-on-1-yl)carbonylamino]benzylpenicillin; Bay-e 6905. $C_{20}H_{23}N_5O_6S$; mol wt 461.50. C 52.05%, H 5.02%, N 15.18%, O 20.80%, S 6.95%. Semi-synthetic, broad-spectrum acylureido penicillin. Prepn: H. Disselnkotter, K. G. Metzger, **Fr.** pat. **2,100,682** corresp to **U.S.** pat. **3,933,795** (1971, 1976, both to Bayer); H. B. Konig *et al., Eur. J. Med.-Chim. Ther.* **17**, 59 (1982). *In vitro* studies: D. Stewart *et al., Antimicrob. Ag. Chemother.* **11**, 865 (1977). *In vitro* and *in vivo* activity: G. K. Daikos *et al., Curr. Chemother., Proc. 10th Int. Congr. Chemother.* , 1977 (Amer. Soc. Microbiol., Washington, D.C., 1978) **1**, pp 626-8. Pharmacokinetics: P. Fiegel, K. Becker, *Antimicrob. Ag. Chemother.* **14**, 288 (1978). Comparison with other penicillins: J. M. Andrews, K. A. Bedford, *ibid.* 559. Clinical studies: H. Lode *et al., Infection* **5**, 163 (1977); E. B. Helm *et al., Deutsch. Med. Wochenschr.* **102**, 1211 (1977). Series of articles on antibacterial activity, pharmacology, and clinical trials: *Arzneimittel-Forsch.* **29**, 1915-2032 (1979); *Infection* **10**, Suppl. 3, S121-S266 (1982); *J. Antimicrob. Chemother.* **11**, Suppl. B, 1-239 (1983).

Sodium salt, $C_{20}H_{22}N_5NaO_6S$, *Azlin, Securopen.* Pale yellow cryst, sol in water, methanol, DMF. Slightly sol in ethanol, isopropanol.

THERAP CAT: Antibacterial.

930. Azobenzene. *Diphenyldiazene;* diphenyl diimide; azobenzide; azobenzol; benzeneazobenzene. $C_{12}H_{10}N_2$; mol wt 182.22. C 79.09%, H 5.53%, N 15.38%. Made by reducing nitrobenzene with Fe in NaOH soln or by electrolytic reduction of nitrobenzene in dil alkali. Lab procedure: A. I. Vogel, *Practical Organic Chemistry* (Longmans, London, 3rd ed., 1959) p 631. Review of carcinogenicity studies: *IARC Monographs* **8**, 75 (1975).

Orange-red leaflets. d 1.20. mp 68°. bp 293°. Insol in water; sol in alcohol, ether, glacial acetic acid. LD_{50} orally in rats: 1000 mg/kg, *RTECS Vol. I*, R. J. Lewis, R. L. Tatken, Eds. (1980) p 209.

USE: Acaricide.

931. 2,2′-Azobisisobutyronitrile. *2,2′-Azobis[2-methylpropanenitrile];* AIBN; α,α′-azodiisobutyronitrile; 2,2′-dicyano-2,2′-azopropane; Porofor-57. $C_8H_{12}N_4$; mol wt 164.21. C 58.51%, H 7.37%, N 34.12%. Prepn by oxidation of α,α′-hydrazobutyric acid dinitrile: Overberger *et al., J. Am. Chem. Soc.* **71**, 2661 (1949); Horner, Schwenk, *Ann.* **566**, 69 (1950); from 2-aminoisobutyronitrile + NaOCl: **Brit.** pat. **672,106** (1952 to Rohm & Haas); Anderson, **U.S.** pat. **2,711,405** (1955 to du Pont); from 2,2′-dichloro-2,2′-azopropane: Benzing, *Ann.* **631**, 1 (1960); **Brit.** pat. **929,182** (1963 to Monsanto).

Crystals from ethanol + water, dec 107°. uv max (ethanol): 345 nm. Soly in methanol at 0°, 20°, 40°: 1.8, 4.96, 16.06 g/100 ml. Soly in ethanol at 0°, 20°, 40°: 0.58, 2.04,

7.15 g/100 ml. LD_{50} orally in mice: 0.7 g/kg. In the organism, forms HCN which is found in blood, liver and brain: Rusin, *C.A.* **56**, 2682f (1962). Can explode when dissolved in acetone: Carlisle, *Chem. & Eng. News* **27**, 150 (1949).

USE: Blowing agent for elastomers and plastics. Initiator for free radical reactions: Griesbaum *et al.*, *J. Org. Chem.* **30**, 261 (1965).

932. Azodicarbonamide. *Diazenedicarboxamide; 1,1'-azobisformamide;* azodicarboxamide; azobiscarbonamide; azobiscarboxamide; 1,1'-azobiscarbamide. $C_2H_4N_4O_2$; mol wt 116.08. C 20.69%, H 3.47%, N 48.27%, O 27.57%. NH_2-CON=NCONH_2. Prepn: J. Thiele, *Ann.* **270**, 1 (1892); **271**, 127 (1892); T. Curtis, K. Heindenreich, *J. Prakt. Chem.* [2] **52**, 454 (1895); J. P. Picard, J. L. Boivin, *Can. J. Chem.* **29**, 223 (1951). Crystal structure: D. T. Cromer, A. C. Larson, *J. Chem. Phys.* **60**, 176 (1974). Safety assessment as a flour-maturing agent: B. L. Oser *et al.*, *Toxicol. Appl. Pharmacol.* **7**, 445 (1965).

Orange-red crystals, mp 225° (dec). *Flammable!* Sol in hot water; insol in cold water, alc.

USE: As blowing and foaming agent for plastics; as maturing and bleaching agent in cereal flour.

933. Azolitmin. Purified coloring matter from litmus: I. M. Kolthoff, C. Rosenblum, *Acid-Base Indicators* (Macmillan Co., New York, 1937) pp 160-162, 174, 208, 355, 361, 365-366, 368-369, 373, 377, 387.

Dark violet scales or dark red powder. Sparingly soluble in water; insol in alc; freely sol in dil alkali hydroxides or carbonates. The indicator solution is prepared by dissolving 0.5 g in 80 ml of warm H_2O, then adding 20 ml alcohol.

USE: Indicator instead of litmus. pH: 4.5 red, 8.3 blue. Usable with most mineral, some organic acids (not hydroxy acids) and some alkaloids; also used for preparing litmus media for bacteriologic purposes.

934. Azomycin. *2-Nitro-1H-imidazole.* $C_3H_3N_3O_2$; mol wt 113.08. C 31.86%, H 2.67%, N 37.16%, O 28.30%. Antibiotic substance produced by an unidentified *Streptomyces* sp. (resembling *Nocardia mesenterica* in some aspects) from soil collected at Shiba Shirokane Daimachi (Japan): Maeda *et al.*, *J. Antibiot.* **6A**, 182 (1953); Okami *et al., ibid.* **7A**, 53 (1954); Nakamura, Umezawa, *ibid.* **8A**, 66 (1955). Structure: Nakamura, *Pharm. Bull.* **3**, 379 (1955). Synthesis: Beaman *et al.*, *J. Am. Chem. Soc.* **87**, 389 (1965); Lancini, Lazzari, *Experientia* **21**, 83 (1965).

Crystals from methanol, dec 283°. uv max (ethanol): 313 nm ($E_{1cm}^{1\%}$ 915); in 0.1N NaOH: 374 nm (ϵ 12,750). Soluble in methanol, ethanol, acetone, ethyl acetate, butyl acetate, alkaline water. Practically insol in ether, petr ether, chloroform, acidic water. LD_{50} i.v. in mice: 80 mg/kg, Maeda *et al., loc. cit.*

Human Toxicity: Irritating to skin.

935. Azosemide. *2-Chloro-5-(1H-tetrazol-5-yl)-4-[(2-thienylmethyl)amino]benzenesulfonamide; 2-chloro-5-(2H-tetrazol-5-yl)-N⁴-2-thenylsulfanilamide;* 5-(4'-chloro-2'-thenylamino-5'-sulfamoylphenyl)tetrazole; Ple-1053; Diart; Diurapid. $C_{12}H_{11}ClN_6O_2S_2$; mol wt 370.83. C 38.87%, H 2.99%, Cl 9.56%, N 22.66%, O 8.63%, S 17.29%. Prepn: A. Popelak *et al.*, *Ger. pat.* **1,815,922** corresp to U.S. pat. **3,665,002** (1968, 1972 both to Boehringer, Mann.). Separation and analysis in blood and urine by HPLC: R. Seiwell, C. Brater, *J. Chromatog.* **182**, 257 (1980). Sites of action: C. Brater, *Clin. Pharmacol. Ther.* **25**, 428 (1979). Clinical and pharmacological studies: F. Krueck *et al.*, *Eur. J. Clin. Pharmacol.* **14**, 153 (1978). Diuretic effect in animals: J. Greven, O. Heidenreich, *Arzneimittel-Forsch.* **31**, 346, 350 (1981).

Crystals, mp 218-221°.

THERAP CAT: Diuretic.

936. Azosulfamide. *6-(Acetylamino)-3-[[4-(aminosulfonyl)phenyl]azo]-4-hydroxy-2,7-naphthalenedisulfonic acid disodium salt;* disodium 2-(4'-sulfamylphenylazo)-7-acetamido-1-hydroxynaphthalene-3,6-disulfonate; Prontosil Soluble; Prontosil S; Neoprontosil; Drometil. $C_{18}H_{14}N_4Na_2$-$O_{10}S_3$; mol wt 588.52. C 36.73%, H 2.40%, N 9.52%, Na 7.82%, O 27.19%, S 16.35%. Prepd by diazotizing sulfanilamide and coupling with the proper aminonaphtholdisulfonic acid: Mietzsch, Klarer, **Ger. pat.** **638,701** (1936); **U.S.** pats. **2,123,634** (1938); **2,148,910** (1939).

Reddish-brown powder. Soly in water (g/100 ml): 6.6 at 15°; 11.5 at 20°; 18.1 at 40°. Practically insol in abs alcohol, ether, acetone, chloroform, benzene.

THERAP CAT: Antibacterial.

THERAP CAT (VET): Formerly as an antimicrobial.

937. Azoxybenzene. *Diphenyldiazene 1-oxide;* azoxybenzide. $C_{12}H_{10}N_2O$; mol wt 198.22. C 72.71%, H 5.09%, N 14.13%, O 8.07%. Prepd by reduction of nitrobenzene with sodium arsenite: Bigelow, Palmer, *Org. Syn.* **coll. vol. II,** 57 (1943); also by treatment of nitrobenzene with glucose in alkaline medium: Opolonick, *Ind. Eng. Chem.* **27**, 1045 (1935). Industrial prepn by heating nitrobenzene with molasses and NaOH in high flash naphtha: **Ger. pat.** **228,722.** Also prepd by catalytic reduction of nitrobenzene: Busch, Schulz, *Ber.* **62**, 1458 (1929). By peracetic acid oxidation of azobenzene: D'Ans, Kneip, *Ber.* **48**, 1145 (1915); *cf.* Greenspan, *Ind. Eng. Chem.* **39**, 847 (1947). Prepn of *cis-* and *trans*-forms by perbenzoic acid oxidation of azobenzene: G. M. Badger *et al.*, *J. Chem. Soc.* **1953**, 2143.

trans-Azoxybenzene

trans-Form, pale yellow orthorhombic needles. d_4^{26} 1.1590. d_4^{50} 1.1373. mp 36°. Slightly volatile with steam; easily volatile in superheated steam at 140-150°. Absorption spectrum: Müller, *Ann.* **493**, 166 (1932). Insol in water; sol in alcohol, ether. 100 parts of abs alc satd at 16° contain 17.5 parts (w/w) azoxybenzene. 100 g ligroin dissolves 43.5 g azoxybenzene at 15°.

Note: The stereoisomer, the so-called *cis*-azoxybenzene, melts at 87°.

USE: In organic syntheses.

938. Aztreonam. *[2S-[2α,3β(Z)]]-2-[[[1-(2-Amino-4-thiazolyl)-2-[(2-methyl-4-oxo-1-sulfo-3-azetidinyl)amino]-2-oxoethylidene]amino]oxy]-2-methylpropanoic acid;* azthreonam; SQ 26776; Azactam; Azonam; Aztreon; Nebactam; Primbactam. $C_{13}H_{17}N_5O_8S_2$; mol wt 435.44. C 35.86%, H 3.94%, N 16.08%, O 29.40%, S 14.72%. The first totally synthetic monocyclic β-lactam (*monobactam*) antibiotic. It has a high degree of resistance to β-lactamases and shows specific activity vs aerobic gram-negative rods. Prepn: R. B. Sykes *et al.*, **Neth. pat. Appl. 8,100,571** (1981 to Squibb), *C.A.* **96**, 181062x (1982). Fast-atom-bombardment mass spectra: A. I. Cohen *et al.*, *J. Pharm. Sci.* **71**, 1065 (1982). Activity vs gram-negative bacteria: R. B. Sykes *et al.*, *Antimicrob. Ag. Chemother.* **21**, 85 (1982). Series of articles on structure-activity, *in vitro* and *in vivo* properties, pharmacokinetics: *J. Antimicrob. Chemother.* **8**, Suppl. E, 1-148 (1981). Toxicology: G. R. Keim *et al.*, *ibid.* 141. Mechanism of action study: A. D. Russell, J. R. Furr, *ibid.* **9**, 329 (1982). Comparative stability to renal dipeptidase: H. Mikami *et al.*, *Antimicrob. Ag. Chemother.* **22**, 693 (1982). Human pharmacokinetics: E. A. Swabb *et al.*, *ibid.* **21**, 944 (1982). Clinical evaluation in urinary tract infection: C. Donadio *et al.*, *Drugs Exp. Clin. Res.* **13**, 167 (1987). Comprehensive description: K. Florey in *Analytical Profiles of Drug Substances* **vol. 17**, K. Florey, Ed. (Academic Press, New York, 1988) pp 1-39.

White crystalline, odorless powder, dec 227°. Very slightly sol in ethanol, slightly sol in methanol, sol in DMF, DMSO. Practically insol in toluene, chloroform, ethyl acetate.
Disodium salt, $C_{13}H_{15}N_5Na_2O_8S_2$. LD_{50} (mg/kg): 3300 i.v. in mice; 6600 i.p. in rats (Keim).
THERAP CAT: Antibacterial.

939. Azulene. *Cyclopentacycloheptene;* bicyclo-[5.3.0]-deca-2,4,6,8,10-pentaene; bicyclo-[0.3.5]-deca-1,3,5,7,9-pentaene. $C_{10}H_8$; mol wt 128.16. C 93.71%, H 6.29%. Prepn from octahydronaphthalene: Plattner, Pfau, *Helv. Chim. Acta* **20**, 224 (1937); from indan: Plattner, Magyar, *ibid.* **25**, 581 (1942).

Intensely blue leaflets or monoclinic plates from alc. Odor of naphthalene. mp 98.5-99°. Absorption spectrum: Susz *et al.*, *Helv. Chim. Acta* **20**, 469 (1937); Plattner, *ibid.* **24**, 283 (1941). Insol in water; sol in the usual organic solvents, in concd mineral acids with decompn.
Sodium sulfonate, *Azusalen.*
THERAP CAT: Sodium sulfonate as antacid.

940. Azure A. *3-Amino-7-(dimethylamino)phenothiazin-5-ium chloride; 7-(dimethylamino)-3-imino-3H-phenothiazine hydrochloride;* 3-amino-7-dimethylaminophenazathionium chloride; *asym*-dimethyl-3,7-diaminophenazathionium chloride; *asym*-dimethylthionine chloride; methylene azure A; C.I. 52005. $C_{14}H_{14}ClN_3S$; mol wt 291.79. C 57.62%, H 4.84%, Cl 12.15%, N 14.40%, S 10.99%. One of the main constituents of methylene azure. Prepn: Bernth-

sen, *Ann.* **230**, 169 (1885); Kehrmann, *Ber.* **39**, 1804 (1906); MacNeal, Killian, *J. Am. Chem. Soc.* **48**, 740 (1926). Review: H. J. Conn's *Biological Stains*, R. D. Lillie, Ed. (Williams & Wilkins, Baltimore, 9th ed., 1977) pp 420-421, 603. See also: *Colour Index* **vol. 4** (3rd ed., 1971) p 4469.

Green glistening crystals or dark green powder. Soluble in water (blue soln); sparingly sol in alc. Absorption max: 620-634 nm (3 ml of a soln of 50 mg in 250 ml H_2O diluted to 200 ml; read in a Beckman spectrophotometer 1 cm cell).
USE: Biological stain.

941. Azure B. *3-(Dimethylamino)-7-(methylamino)-phenothiazin-5-ium chloride; 7-(dimethylamino)-3-(methylimino)-3H-phenothiazine hydrochloride;* 3-methylamino-7-dimethylaminophenazathonium chloride; trimethyldiaminophenazathonium chloride; trimethylthionine chloride; methylene azure B; C.I. 52010. $C_{15}H_{16}ClN_3S$; mol wt 305.83. C 58.91%, H 5.27%, Cl 11.59%, N 13.74%, S 10.48%. One of the main constituents of methylene azure. Prepn: Bernthsen, *Ann.* **230**, 169 (1885); Kehrmann, *Ber.* **39**, 1804 (1906); Kehrmann, Duttenhöfer, *ibid.* 925, 1403; Kehrmann *et al.*, *ibid.* **46**, 2137 (1913); MacNeal, Killian, *J. Am. Chem. Soc.* **48**, 740 (1926). *Review:* H. J. Conn's *Biological Stains*, R. D. Lillie, Ed. (Williams & Wilkins, Baltimore, 9th ed., 1977) pp 421-423, 603-604. *See also: Colour Index* **vol. 4** (3rd ed., 1971) p 4470.

Green glistening crystals or dark green powder. Sol in water (blue soln); sparingly sol in alc. Absorption max: 648-655 nm (3 ml of a soln of 50 mg in 250 ml H_2O dil to 200 ml; read in a Beckman spectrophotometer 1 cm cell).
USE: Biological stain.

942. Azure C. *3-Amino-7-(methylamino)phenothiazin-5-ium chloride;* 3-imino-7-(methylamino)-3H-phenothiazine hydrochloride; 3-amino-7-methylaminophenazathonium chloride; monomethyldiaminodiphenazothionium chloride; monomethylthionine chloride; C.I. 52002. $C_{13}H_{12}ClN_3S$; mol wt 277.79. C 56.21%, H 4.35%, Cl 12.76%, N 15.13%, S 11.54%. Prepn: Holmes, French, *Stain Techn.* **1**, 17 (1926). *Review:* H. J. Conn's *Biological Stains*, R. D. Lillie, Ed. (Williams & Wilkins, Baltimore, 9th ed., 1977) pp 419-420, 604. *See also: Colour Index* **vol. 4** (3rd ed., 1971) p 4469.

Green glistening crystals or dark green powder. Sol in water (blue soln); sparingly sol in alc. Absorption max: 608-622 nm (3 ml of a soln of 50 mg in 250 ml H_2O dil to 200 ml; read in a Beckman spectrophotometer 1 cm cell).
USE: Biological stain.

B

943. Babbitt Metal. Originally an alloy of 69% Zn, 19% Sn, 4% Cu, 3% Sb, and 5% Pb. This name now covers a variety of formulations, such as lead base babbitt, silver base babbitt, tin base babbitt, cadmium base babbitt, arsenic base babbitt. Generally and better described as "white metal bearing alloys".
USE: For machinery bearings.

944. Bacampicillin. *6-[(Aminophenylacetyl)amino]-3,3-dimethyl-7-oxo-4-thia-1-azabicyclo[3.2.0]heptane-2-carboxylic acid 1-[(ethoxycarbonyl)oxy]ethyl ester;* (2S,5R,-6R)-6-[(R)-(2-amino-2-phenylacetamido)]-3,3-dimethyl-7-oxo-4-thia-1-azabicyclo[3.2.0]heptane-2-carboxylic acid ester with ethyl 1-hydroxyethyl carbonate; 6-[(R)-2-amino-2-phenylacetamido]penicillanic acid [1-(ethoxycarbonyl-oxy)ethyl] ester; 1'-ethoxycarbonyloxyethyl 6-(D-α-amino-phenylacetamido)penicillanate. $C_{21}H_{27}N_3O_7S$; mol wt 465.53. C 54.18%, H 5.85%, N 9.02%, O 24.06%, S 6.89%. Semi-synthetic antibiotic related to penicillin. Prepn: B. A. Ekstrom, B. O. H. Sjoberg, **Ger. pat. 2,144,457** corresp to **U.S. pats. 3,873,521** and **3,939,270** (1972, 1975 and 1976, all to Astra). *In vitro* and *in vivo* study: N. O. Bodin *et al., Antimicrob. Ag. Chemother.* **8,** 518 (1975). Pharmacokinetics: M. Rozencweig *et al., Clin. Pharmacol. Ther.* **19,** 592 (1976); T. Bergan, *Antimicrob. Ag. Chemother.* **13,** 971 (1978). Animal studies: C. Carbon *et al., Scand. J. Infect. Dis. (Suppl.)* **14,** 127 (1978); T. Bergan, I. Versland, *ibid.* 135; U. Forsgren *et al., ibid.* 207. Clinical studies: P. V. Maesen *et al., J. Antimicrob. Chemother.* **2,** 279 (1976); J. Sjövall, *J. Int. Med. Res.* **5,** 313 (1977); C. Ekedahl *et al., Scand. J. Infect. Dis. (Suppl.)* **14,** 279 (1978). Toxicology: M. Edanaga *et al., Chemotherapy (Tokyo)* **27,** Suppl. 4, 17 (1979).

Hydrochloride, $C_{21}H_{28}ClN_3O_7S$, *Ambacamp, Ambaxin, Bacacil, Bacampicine, Penglobe, Spectrobid.* White cryst from acetone-petr ether, mp 171-176° (dec). $[\alpha]_D^{20}$ +161.5°, also reported as +173° (Bodin *et al., loc. cit.*). Sol in water. LD_{50} in mice (mg/kg): 8529 orally; 176 i.p.; 9475 s.c.; 184 i.v. (Edanaga).
THERAP CAT: Antibacterial.

945. Bacillus thuringiensis. Agritol; Bactospeine; Bakthane; Biotrol; BT; Dipel; Larvatrol; Skeetal; Sporeine; Teknar; Thuricide; Tribactur. A gram-positive spore-forming bacterium which causes the death of lepidopterous larvae. First isolated by Ishiwata in 1902 from dying silkworm larvae. Later isolated as *B. thuringiensis,* Berliner, from the larvae of the flour moth, *Ephestia kuehniella* Zell: Berliner, *Z. Angew. Entomol.* **2,** 29 (1915). The insecticidal action is ascribed to protein crystals produced by the bacillus: Hannay, *Nature* **172,** 1004 (1953). Structure and function of the protein crystal: K. W. Nickerson, *Biotechnol. Bioeng.* **22,** 1305 (1950). In addition to the crystalline parasporal body, called the endotoxin, there is another toxic entity, called the exotoxin, which is released in the culture medium during bacterial growth. Isoln and stucture of the exotoxin: Sebesta *et al., Coll. Czech. Chem. Commun.* **34,** 891 (1969); Farkas *et al., ibid.* 1118. Partial synthesis: Sorm, *Pure Appl. Chem.* **25,** 253 (1971). *Reviews:* Heimpel, *Ann. Rev. Entomol.* **12,** 287 (1967); Rogoff, Yousten, *Ann. Rev. Microbiol.* **23,** 357 (1969); Lecadet in *Microbial Toxins* **vol. 3,** T. C. Montie *et al.,* Eds. (Academic Press, New York, 1970) pp 437-471; L. A. Bulla *et al., Crit. Rev. Microbiol.* **8,** 147-204 (1980).
The commercial product is composed of live spores of *B.*

thuringiensis, Berliner. Non-toxic to humans and animals: Fisher *et al., J. Agr. Food Chem.* **7,** 687 (1959).
USE: Microbial insecticide.

946. Bacilysin. *N-L-Alanyl-3-(5-oxo-7-oxabicyclo[4.-1.0]hept-2-yl)-L-alanine; α-[(2-amino-1-oxopropyl)amino]-5-oxo-7-oxabicyclo[4.1.0]heptane-2-propanoic acid; α-(2-aminopropionamido)-5-oxo-7-oxabicyclo[4.1.0]heptane-2-propionic acid;* bacillin; tetaine. $C_{12}H_{18}N_2O_5$; mol wt 270.28. C 53.33%, H 6.71%, N 10.36%, O 29.60%. Antibiotic produced by the soil bacillus NCTC 7197: Gilliver *et al.,* in *Antibiotics* **vol. I,** Florey *et al.,* Eds. (Oxford, 1949) p 458. Production by *Bacillus subtilis* and purification: Rogers *et al., Biochem. J.* **97,** 573 (1965). Identity with tetaine: K. Kaminski, T. Sololowska, *J. Antibiot.* **26,** 184 (1973). Identity with bacillin: K. Atsumi *et al., ibid.* **28,** 77 (1975). Improved isoln: Walker, Abraham, *ibid.* **118,** 557 (1970). Structural study: Rogers *et al., ibid.* **97,** 579 (1965). Final structure: Walker, Abraham, *ibid.* **118,** 563 (1970).

White amorphous powder. Freely sol in water; sol in 80% alc; sparingly sol in abs alc. Stable in aq soln at 100° for 5 min at pH 7; becomes inactive at pH 2 or pH 9.

947. Bacimethrin. *4-Amino-2-methoxy-5-pyrimidine-methanol;* 4-amino-5-hydroxymethyl-2-methoxypyrimidine. $C_6H_9N_3O_2$; mol wt 155.16. C 46.44%, H 5.85%, N 27.08%, O 20.62%. Antibiotic substance produced by *Bacillus megatherium* from Japanese soil. Isoln and structure: Tanaka *et al., J. Antibiot.* **14A,** 161 (1961); **15A,** 191, 197 (1962). Prepn: **Brit. pat. 884,772** (1961 to Merck & Co.); Koppel *et al., J. Org. Chem.* **27,** 1492, 3614 (1962). Biological studies: Nishimura, Tanaka, *J. Antibiot.* **16A,** 179 (1963).

Needles from methanol, or ethanol + ether, mp 174°. uv max (water): 227, 271 nm (ε 7600, 7300); (0.1N HCl): 229, 261 nm (ε 8400, 9500). Sol in water, methanol, acetic acid. Mod sol in ethanol, pyridine. Practically insol in other organic solvents. LD_{50} in mice: 300 mg/kg i.v. and i.p.

948. Bacitracin. Altracin; Ayfivin; Fortracin; Penitracin; Topitracin; Zutracin. Antibiotic polypeptide complex produced by *Bacillus subtilis* and *licheniformis:* Johnson *et al., Science* **102,** 376 (1945); Anker *et al., J. Bacteriol.* **55,** 249 (1948). Commercial bacitracin is a mixture of at least nine bacitracins. Purification of bacitracin by carrier displacement method: Porath, *Acta Chem. Scand.* **6,** 1237 (1952). Purification with ion exchange resin: Chaiet, Cochrane, **U.S. pat. 2,915,432** (1959 to Merck & Co.). Production: Johnson, Meleney, **U.S. pat. 2,498,165** (1950 to the U.S. Secy. of War); Freaney, Allen, **U.S. pat. 2,828,246** (1958 to Commercial Solvents). Solubilities: Weiss *et al., Antibiot. Chemother.* **7,** 374 (1957). Preliminary structure studies: Hausmann *et al., J. Am. Chem. Soc.* **77,** 723 (1955); Lockhart *et al., Biochem. J.* **61,** 534 (1955); Stoffel, Craig, *J. Am. Chem. Soc.* **83,** 145 (1961). Structure of bacitracin A: Ressler, Kashelikar, *ibid.* **88,** 2025 (1966); Galardy *et al., Biochemistry* **10,** 2429 (1971). Synthetic studies: Munekata *et al., Bull. Chem. Soc. Japan* **46,** 3187, 3835 (1973). Mechanism of action: Storm, *Ann. N.Y. Acad. Sci.* **235,** 387 (1974). Comprehensive description: G. A. Brewer in *Analytical Profiles of Drug Substances* **vol. 9,** K. Florey, Ed. (Academic Press, New York, 1980) pp 1-69. *Reviews:* Craig *et al.,* "Bacitracin" in G. E. W. Wolstenholme, C. M. O'Connor,

Ciba Foundation Symposium on Amino Acids and Peptides with Antimetabolic Activity (Little, Brown, Boston, 1958) pp 226-246; E. D. Weinberg in *Antibiotics,* D. Gottlieb, P. D. Shaw, Eds. (Springer-Verlag, New York, 1967) **I,** pp 90-99; **II,** pp 240-245; D. R. Storm, W. A. Toscano, Jr. in *Antibiotics* vol. **5,** pt. 1, F. E. Hahn, Ed. (Springer-Verlag, New York, 1979) pp 1-17.

bacitracin A

Grayish-white powder. Very bitter taste. Sol in water, alcohol. Practically insol in ether, chloroform, acetone. Stable in acid soln; unstable in alkaline solns. Potency loss probably due to transformation of bacitracin A to bacitracin F, latter having little antimicrobial activity.

Bacitracin A, $C_{66}H_{103}N_{17}O_{16}S$, major component of commercial bacitracin.

THERAP CAT: Antibacterial.

THERAP CAT (VET): Antibacterial, usually used locally. Growth promotant. Enteric infections.

949. Bacitracin Methylenedisalicylic Acid. *Bacitracin methylenebis[2-hydroxybenzoate];* bacitracin methylenedisalicylate. Prepd by the reaction of two moles methylenedisalicylic acid and one mole bacitracin: Baron, U.S. pat. **2,774,-712** (1956 to S. B. Penick & Co.). Contains about 18 bacitracin units/mg (minimum 14 units).

White to grayish-brown powder. Slightly disagreeable odor. Somewhat less bitter than bacitracin. uv max (dil acetic acid): 318 nm, min: 280 nm. Solubility in water about 50 mg/ml. pH of satd aq soln 3.5 to 5.0. Sol in pyridine, ethanol, less sol to insol in acetone, ether, chloroform, pentane, benzene. Sol in dil aq alkali at pH 6 and higher, increasingly insol at pH 6 to 3.

Sodium salt, creamy-white powder. Potency about 10.8 bacitracin units/mg. Sol in water. pH 2% aq soln 9.5.

THERAP CAT (VET): Feed additive.

950. Baclofen. *β-(Aminomethyl)-4-chlorobenzenepropanoic acid; β-(aminomethyl)-p-chlorohydrocinnamic acid; γ-amino-β-(p-chlorophenyl)butyric acid; β-(4-chlorophenyl)GABA;* Ba 34647; Baclon; Lioresal; Myospan. $C_{10}H_{12}ClNO_2$; mol wt 213.67. C 56.21%, H 5.66%, Cl 16.59%, N 6.56%, O 14.97%. Prepn: Keberle *et al.,* **Swiss** pat. **449,046** (1968 to Ciba), *C.A.* **69,** 106273f (1968); Uchimaru *et al.,* **Japan** pat. **16,692** ('70) (to Daiichi), *C.A.* **73,** 77617w (1970). Clinical studies: Hudgson, Weightman, *Brit. Med. J.* **IV,** 15 (1971). Metabolism: H. Yamamoto *et al., Oyo Yakuri* **14,** 97, 109, 115 (1977). Toxicity study: T. Tadokoro *et al., Osaka Daigaku Igaku Zasshi* **28,** 265 (1976), *C.A.* **88,** 183016u (1978). Comprehensive description: S. Ahuja in *Analytical Profiles of Drug Substances* vol. **14,** K. Florey, Ed. (Academic Press, New York, 1985) pp 527-548. Book: *Baclofen: A Broader Spectrum of Activity,* I. T. Boyle *et al.,* Eds. (Longman, Harlow, England, 1980) 52 pp.

Crystals from water, mp 206-208° (Keberle); 189-191°, (Uchimaru). LD_{50} in male mice, rats (mg/kg): 45, 78 i.v.; 103, 115 s.c.; 200, 145 orally (Tadokoro).

Hydrochloride, $C_{10}H_{13}Cl_2NO_2$, mp 179-181°.

THERAP CAT: Muscle relaxant (skeletal).

951. Bacteriorhodopsin. BR. The only protein of the purple membrane of *Halobacterium halobium* and other halophilic bacteria. It is a rhodopsin-like pigment containing retinal linked to lysine through a Schiff's base and has a mol wt of approx 26,000. Bacteriorhodopsin functions as an energy transducer or "proton pump"; unlike animal rhodopsins, it uses light energy to generate an electrochemical gradient and this stored energy is used by the cell for ATP synthesis and other important energy-requiring functions. Discovery in the purple membrane of *Halobacterium halobium:* D. Oesterlen, W. Stoekenius, *Nature New Biol.* **233,** 149 (1971); A. E. Blaurock, W. Stoekenius, *ibid.* 152. Isoln and identification in *H. cutirubrum:* S. C. Kushwaha, M. Kates, *Biochim. Biophys. Acta* **316,** 235 (1973). Description of functions: D. Oesterhelt, W. Stoekenius, *Proc. Nat. Acad. Sci. USA* **70,** 2853 (1973). Proposed mechanism of the "proton pump": K. Schulten, P. Tavan, *Nature* **272,** 85 (1978). Structural elucidation: Y. A. Ovchinnikov *et al., FEBS Letters* **100,** 219 (1979); H. G. Khorana *et al., Proc. Nat. Acad. Sci. USA* **76,** 5046 (1979). Three-dimensional crystallographic study: H. Michel, D. Oesterhelt, *ibid.* **77,** 1283 (1980). Series of articles on structure, biosynthesis, and energy transduction: *Photochem. Photobiol.* **33,** 417-608 (1981). Review of energy transduction: H. V. Westerhoff, Z. Dancshazy, *Trends Biochem. Sci.* **9,** 112 (1984). Comprehensive reviews: W. Stoekenius *et al., Biochim. Biophys. Acta* **505,** 215-278 (1979); W. Stoekenius, R. A. Bogomolni, *Ann. Rev. Biochem.* **51,** 587-616 (1982).

Bacteriorhodopsin is not bleached by light in the same fashion as animal rhodopsins; only small spectral changes are caused by transfer from light to dark. When exposed to light, BR has a broad light absorption maximum at 568 nm that slowly shifts back to 558 nm in the dark. The 558 and 568 nm states are known as the dark-adapted and light-adapted states, respectively. Light-adapted BR, upon absorption of a light quantum, undergoes a series of reversible changes, having intermediate forms with absorption maxima of 590, 550, and 412 nm. uv max (2M HCl): 565 nm. Slow bleaching occurs when membrane suspensions are illuminated in the presence of hydroxylamine or sodium borohydride. When hydroxylamine is used, the uv max is shifted to 360 nm. The bleached membrane is called the "apomembrane"; the protein after loss of the chromophore is termed *bacterioopsin.*

USE: As a tool in biological energy transduction research.

952. Badische Acid. *7-Amino-1-naphthalenesulfonic acid;* 2-naphthylamine-8-sulfonic acid. $C_{10}H_9NO_3S$; mol wt 223.25. C 53.80%, H 4.06%, N 6.28%, O 21.50%, S 14.36%. Prepd by sulfonation of 2-naphthylamine: Green, Vakil, *J. Chem. Soc.* **113,** 35 (1918); Hennion, Schmidle, *J. Am. Chem. Soc.* **65,** 2468 (1943).

Needles or prisms. Sol in 1680 parts water, in alkalies; slightly sol in alc.

953. Bagasse. Fibrous residue from cane sugar mfg operations, which contains 50% cellulose, 25% pentosans, and 25% lignin: Wiggins, *Sugar J.* **16,** No. 8, pp 18, 22, 24 (1954), *C.A.* **48,** 9729h (1954). Review of properties and uses: Vo Tong Xuan, Samaniego, *Sugar News* **46,** 276, 284 (1970); M. A. Clarke in Kirk-Othmer *Encyclopedia of Chemical Technology,* vol. **3** (Wiley-Interscience, New York, 3rd ed., 1978) pp 434-438.

USE: Raw material for pulp, paper and board. *Caution:* Inhalation of bagasse dust may cause pneumonitis, asthma.

954. Baicalein. *5,6,7-Trihydroxy-2-phenyl-4H-1-benzopyran-4-one; 5,6,7-trihydroxyflavone;* noroxylin. $C_{15}H_{10}O_5$; mol wt 270.23. C 66.67%, H 3.73%, O 29.60%. From roots of *Scutellaria baicalensis.* Isoln and structure: Bargellini, *Gazz. Chim. Ital.* **49,** II, 47 (1919); Shibata *et al., Acta Phytochim.* **1,** 109 (1923). Synthesis: Sastri, Seshadri, *Proc. Indian Acad. Sci.* **23A,** 262 (1946), *C.A.* **41,** 449 (1947);

Schönberg et al., J. Am. Chem. Soc. 77, 5390 (1955); Jouanne, Mentzer, Compt. Rend. 254, 727 (1962); Agasimundin, Siddappa, J. Chem. Soc. Perkin Trans. I 1973, 503. Pharmacology: Koda et al., C.A. 75, 47200d (1971).

Yellow prisms from alc, dec 264-265°. uv max (ethanol): 324, 276 nm (log ε 4.18, 4.42). Sol in alcohol, methanol, ether, acetone, ethyl acetate, hot glacial acetic acid. Sparingly sol in chloroform, nitrobenzene. Practically insol in water. Sol in dil NaOH with greenish-brown color. Concd H_2SO_4 gives yellow color, green fluorescence.
THERAP CAT: Astringent.

955. Bakankosin. *4-Ethenyl-3-(β-D-glucopyranosyloxy)-3,4,4a,5,6,7-hexahydro-8H-pyrano[3,4-c]pyridin-8-one;* bacancosin; Bakankoside. $C_{16}H_{23}NO_8$; mol wt 357.37. C 53.77%, H 6.49%, N 3.92%, O 35.82%. Nitrogenous glucoside from seed of *Strychnos vacacoua* Baill., *Loganiaceae.* Ref: E. Bourquelot, H. Hérissey, Compt. Rend. 144, 575 (1907); 147, 750 (1908); Arch. Pharm. 247, 56 (1909); K. Balenovic et al., Helv. Chim. Acta 35, 2519 (1952). Proposed structure: G. Büchi, R. E. Manning, Tetrahedron Letters no. 26, 5 (1960); H. C. Beyerman et al., Bull. Soc. Chim. France 1961, 1812; H. Inouye et al., Chem. Pharm. Bull. 24, 1406 (1976). Synthesis and structure: L. F. Tietze, Tetrahedron Letters 29, 2535 (1976).

Hydrate, bitter crystals, mp 157°; when anhydr mp 200°, also reported as mp 211-212°. $[\alpha]_D$ −197°. uv max (alcohol): 235 nm (log ε 4.2). Freely sol in water, alc; slightly sol in ethyl acetate.

956. Balsam Canada. Canada turpentine; balsam of fir. Improperly *"Balm of Gilead"*. Liquid oleoresin from *Abies balsamea* (L.), Mill., *Pinaceae.* Habit. Canada and Northern U.S. to Va., west to Minnesota. Constit. 27.5% Volatiles (pinene, nopinene, β-phellandrene), 44.5% resin acid (13% abietic, 8% neoabietic), 27% neutral resinous compounds. Ref: Lombard et al., Peintures, Pigments, Vernis 34, 106 (1958), C.A. 52, 12420 (1958).
Yellowish to greenish, viscid, transparent, slightly fluorescent liquid; agreeable, aromatic pine-like odor; bitter taste; on exposure to air gradually solidifies to a solid, noncryst mass. d 0.987-0.994. n_4^{20} 1.52-1.54. $[\alpha]_D^{20}$ +1° to +4°. Acid no. 84-87. Sapon no. 89.4-95.7 (2 g in xylene for 1 hr). Insoluble in water; miscible with benzene, chloroform, xylene, ethyl acetate, oil of cedar; completely sol or almost sol in ether, oil turpentine; about 90% dissolves in alcohol or petr ether.
USE: Cement for lenses; manuf fine lacquers; for mounting in microscopy.

957. Balsam Gurjun. Wood oil; "East Indian Copaiba". Oleoresin from various species of *Dipterocarpus, Dipterocarpaceae.* Habit. Eastern India, Burma. Constit. About 75% volatile oil, boiling at about 255°; gurjunic acid, resin, bitter substance sol in water. Isoln of an optically inactive sesquiterpene δ-elemene, from gurjun balsam: Gough, Powell, Tetrahedron Letters 1961, 763.
Clear liquid, light brown by transmitted light with green-

ish fluorescence; somewhat bitter, but not very acrid taste. Acid no. 5-15. Sapon no. 10-20 (2 g in 20 g xylene for 1 hr). d 0.95-0.97; $[\alpha]_D$ −23° to −70°; n_D^{28} 1.510-1.516. Insol in water; completely sol in benzene, chloroform; incompletely in alcohol, ether, carbon disulfide, petr ether.
USE: As of copaiba; adapted for varnishes and lacquers, particularly for articles to be exposed to a temp of 80° or so.

958. Balsam Mecca. Balm of Gilead; balsam of Gilead; balsán-Katél; Duhnul-balsan. Balsam obtained from twigs of *Commiphora opobalsamum* (Kunth.) Engl. (*Balsamodendrum gilaedense* Kunth.), *Burseraceae.*
Light-colored, mobile to viscid, turbid, brownish-red liquid; aromatic odor. Insoluble in water; sol in alcohol, benzene, chloroform, acetone, glacial acetic acid, carbon disulfide, oil turpentine, ether.
USE: In perfumery.

959. Balsam Peru. Peruvian balsam; Indian balsam; China oil; Black balsam; Honduras balsam; Surinam balsam. From *Toluifera pereirae* (Klotzsch) Baill. (*Myroxylon pereirae* Klotzsch), *Leguminosae.* Habit. Central America (San Salvador) in forests near Pacific coast. Constit. 50-60% cinnamein—esters of cinnamic and benzoic acid; about 28% resin, styracine, vanillin. Ref: Bergemann, Pharmazie 5, 494 (1950); Cortesi, Bull. Soc. Pharm. Bordeaux 89, 141 (1951), C.A. 46, 1718b (1952).
Dark brown, viscid liquid; pleasant aromatic odor; warm, bitter taste and persistent aftertaste. d 1.150-1.170. Insol in water, olive oil; sol in alcohol, chloroform and glacial acetic acid, usually with a slight opalescence. Partly sol in ether, petr ether.
USE: In perfumery and some chocolate flavorings; also in masking of odors.
THERAP CAT: Scabicide; skin ulcer therapy.
THERAP CAT (VET): Miticide. To aid in healing of indolent wounds.

960. Balsam Tolu. Thomas balsam; opobalsam; resin Tolu. From *Toluifera balsamum* L. (*Myroxylon toluiferum* H. B. K.), *Leguminosae.* Habit. South America (Venezuela, Colombia, Peru) on elevated plains and mountains. Constit. 12-15% free cinnamic and benzoic acids; about 40% benzyl, etc., esters of these acids (5.2-13.4% cinnamein); 1.5-3% volatile oil. Ref: Rosenthaler, Pharm. Ztg.-Nachr. 88, 716 (1952), C.A. 47, 5076 (1953).
Yellowish-brown or brown, semifluid or nearly solid, resinous mass; aromatic odor and taste; brittle when cold. Insol in water; almost insol in petr ether. Sol in alcohol, benzene, chloroform, ether, glacial acetic acid, partially in carbon disulfide or NaOH.
USE: In perfumery, confectionery and chewing gums; in pharmacy, as ingredient and vehicle for expectorants.
THERAP CAT: Expectorant.
THERAP CAT (VET): Has been used as an expectorant.

961. Balsam Traumatic. Friar's balsam; Turlington's balsam. Composed of 100 parts benzoin, 35 storax, 35 balsam Tolu, 16 balsam Peru, 8 aloe, 8 myrrh, 4 angelica, and alcohol to make 1000.
THERAP CAT: Skin protective; expectorant.
THERAP CAT (VET): Has been used for chronic bronchitis. Wound antiseptic and styptic.

962. Bambermycins. Moenomycin; flavophospholipol; Flavomycin. Antibiotic complex comprised of at least 4 active components, moenomycins A, B_1, B_2 and C, with *moenomycin A* being the major component. Obtained from cultures of *Streptomyces bambergiensis, S. ghanaensis, S. ederensis, S. geysiriensis* and related strains: Wallhäuser et al., Antimicrob. Ag. Chemother. 1965, 734 sqq.; Neth. pat. Appl. 6,602,132, C.A. 66, 74946x (1967) and Lindner, Wallhäuser, U.S. pat. 3,674,866 (1966, 1972 both to Hoechst). Structural studies: Tschesche et al., Ann. 720, 58 (1968); Tetrahedron Letters 1968, 2905; 1969, 141; Huber, J. Antibiot. 25, 1226 (1972); P. Welzel et al., Tetrahedron Letters 1973, 227; F. J. Witteler et al., ibid. 1979, 3493; P. Welzel et al., Tetrahedron 39, 1583 (1983). Structure of moenomycin A: P. Welzel et al., Angew. Chem. Int. Ed. 20, 121 (1981). Synthesis of *moenocinol*, the lipid moiety: Tschesche, Reden, Ann. 1974, 853; P. J. Kocienski, J. Org. Chem. 45, 2037 (1980); R. M. Coates, M. W. Johnson, ibid. 2685.

Review: G. Huber in *Antibiotics* **vol. 5**, pt. 1, F. E. Hahn, Ed. (Springer-Verlag, New York, 1979) pp 135-153.

moenomycin A

Complex is a colorless, amorphous solid, no definite mp, decompn starting at 200°C. uv max (water pH 7): 258 nm (E$_{1cm}^{1\%}$ 60). Sol in water, methanol, DMF; less sol in ethanol, propanol; slightly sol in ether, ethyl acetate. Insol in benzene, chloroform. Stable in neutral aq and methanolic solns; slowly decomp in acid and alkaline solns. LD$_{50}$ in mice (mg/kg): > 2000 orally, s.c. and i.p.; 1400 i.v. (Lindner, Wallhäuser).

THERAP CAT: Antibacterial.

THERAP CAT (VET): Antibacterial; feed additive (poultry, swine and calves).

963. Bambuterol. *Dimethylcarbamic acid 5-[2-[(1,1-dimethylethyl)amino]-1-hydroxyethyl]-1,3-phenylene ester;* 1-[bis(3',5'-N,N-dimethylcarbamoyloxy)phenyl]-2-N-*tert*-butylaminoethanol; (±)-5-[2-(*tert*-butylamino)-1-hydroxyethyl]-*m*-phenylene bis(dimethylcarbamate); terbutaline bisdimethylcarbamate. C$_{18}$H$_{29}$N$_3$O$_5$; mol wt 367.45. C 58.84%, H 7.95%, N 11.44%, O 21.77%. Ester prodrug of the β_2-adrenergic agonist terbutaline, *q.v.* Prepn: O. A. T. Olsson *et al.,* **Eur.** pat. **Appl. 43,807**; *eidem,* **U.S.** pat. **4,451,-663** (1982, 1984 both to AB Draco). Pharmacology: O. A. T. Olsson, L.-A. Svensson, *Pharm. Res.* **1984,** 19. Review of pharmacology and metabolism: L.-A. Svensson, *Acta Pharm. Suec.* **24,** 333 (1987). LC determn: O. Wannerberg, B. Persson, *J. Chromatog.* **435,** 199 (1988). Clinical evaluation in asthma: T. Sandström *et al., Respiration* **53,** 31 (1988). Clinical comparison with terbutaline: G. Persson *et al., Eur. Respir. J.* **1,** 223 (1988).

Hydrochloride, C$_{18}$H$_{30}$ClN$_3$O$_5$, *KWD-2183, Bambec.* THERAP CAT: Bronchodilator.

964. Bamethan. α-*[(Butylamino)methyl]-4-hydroxy-benzenemethanol;* α-*[(butylamino)methyl]-p-hydroxybenzyl alcohol;* 1-(p-hydroxyphenyl)-2-butylaminoethanol; 1-(4-hydroxyphenyl)-1-hydroxy-2-butylaminoethane; 2-butyl-amino-1-*p*-hydroxyphenylethanol; Butyl-Nor-Sympatol. C$_{12}$H$_{19}$NO$_2$; mol wt 209.28. C 68.86%, H 9.15%, N 6.69%, O 15.29%. Prepn: Corrigan *et al., J. Am. Chem. Soc.* **67,** 1894 (1945); Kovács, *Pharm. Zentralhalle* **92,** 193 (1953), *C.A.* **49,**

936 (1955). Metabolism of sulfate: Hajos, Szporny, *Arznei-mittel-Forsch.* **18,** 1212 (1968).

Crystals, mp 123.5-125°.
Hydrochloride, C$_{12}$H$_{19}$NO$_2$.HCl, mp 109-110°.
Sulfate, C$_{24}$H$_{40}$N$_2$O$_8$S, *Vasculat, Vasculit, Vascunicol, Rotesar, Butedrin, Vaskulat, Bupatol, Garmian.*
THERAP CAT: Vasodilator.

965. Bamifylline. *7-[2-[Ethyl(2-hydroxyethyl)amino]-ethyl]-3,7-dihydro-1,3-dimethyl-8-(phenylmethyl)-1H-purine-2,6-dione; 8-benzyl-7-[2-[ethyl(2-hydroxyethyl)amino]-ethyl]theophylline;* 8-benzyl-7-[N-ethyl-N-(β-hydroxyethyl)aminoethyl]theophylline; benzetamophylline; 8102 CB; Bamiphylline. C$_{20}$H$_{27}$N$_5$O$_3$; mol wt 385.47. C 62.32%, H 7.06%, N 18.17%, O 12.45%. Prepn: *Belg.* pat. **602,888** (1961 to Christiaens), *C.A.* **56,** 5981c (1962). Pharmacology: Georges *et al., Therapie* **17,** 211 (1962). Metabolism: Dodion *et al., Arzneimittel-Forsch.* **19,** 785 (1969). Toxicological studies: Georges *et al., ibid.* **18,** 460 (1968).

Crystals, mp 80-80.5°.
Hydrochloride, C$_{20}$H$_{28}$ClN$_5$O$_3$, *AC-3810, BAX 2739Z, Briofil, Trentadil.* Crystals, mp 185-186°. LD$_{50}$ in mice, rats (mg/kg): 246, 1139 orally; 89, 131 i.p.; 67, 65 i.v. (Georges).
THERAP CAT: Bronchodilator.

966. Bamipine. *1-Methyl-N-phenyl-N-(phenylmethyl)-4-piperidinamine; 4-N-benzylanilino-1-methylpiperidine;* N-phenyl-N-benzyl-4-amino-1-methylpiperidine; Soventol. C$_{19}$H$_{24}$N$_2$; mol wt 280.40. C 81.38%, H 8.63%, N 9.99%. Prepn: Kallischnigg, **U.S.** pat. **2,683,714** (1954 to Knoll). Toxicity studies: Hanna, *Toxicol. Appl. Pharmacol.* **3,** 393 (1961).

Crystals from butanone or methanol, mp 115°.
Dihydrochloride, C$_{19}$H$_{26}$Cl$_2$N$_2$, *Taumidrine.* Crystals from alc, mp 189° (dec).
THERAP CAT: Antihistaminic.

967. Baptigenin. *7-Hydroxy-3-(3,4,5-trihydroxyphenyl)-4H-1-benzopyran-4-one; 7,3',4',5'-tetrahydroxyisoflavone.* C$_{15}$H$_{10}$O$_6$; mol wt 286.23. C 62.94%, H 3.52%, O 33.54%. The aglucon of baptisin. From baptisin by acid hydrolysis or vacuum sublimation. From *radix Baptisiae:* Fischer, Ehrlich, *C.A.* **31,** 4449[3] (1937). Structure: Böhm, *Arzneimittel-Forsch.* **10,** 472 (1960). Synthesis: Farkas *et al., Ber.* **96,** 1865 (1963). *See also* Pseudobaptigenin.

Needles from dil ethanol, mp 284-285°. Begins to sublime at 240°. Sublimes in oil pump vacuum at 180-200°. uv max (ethanol): 270.2, 247 nm. Practically insol in water, ammonia water; slightly sol in dil alc, hot glacial acetic acid; sol in acetone, in NaOH solns.

Tetraacetylbaptigenin, $C_{23}H_{18}O_{10}$, needles from methanol, mp 214°.

Tetrabenzoylbaptigenin, $C_{43}H_{26}O_{10}$, prisms from methanol, mp 191-192°.

Tetramethylbaptigenin, $C_{19}H_{18}O_6$, crystals from methanol, mp 144-145°.

968. Baptisia. Wild indigo; indigo weed; false indigo; yellow indigo. From root of *Baptisia tinctoria* R. Br., *Leguminosae. Habit.* North America. *Constit.* Baptin—a purgative glucoside, baptisin—a bitter glucoside, baptitoxine (identical with cytisine), an alkaloid.

THERAP CAT: Anti-infective.

969. Barban. *(3-Chlorophenyl)carbamic acid 4-chloro-2-butynyl ester; m-chlorocarbanilic acid 4-chloro-2-butynyl ester;* chloro-2-butynyl *m*-chlorocarbanilate; 4-chloro-2-butynyl N-(3-chlorophenyl)carbamate; barbamate; barbane; chlorinat; CS-847; Carbyne. $C_{11}H_9Cl_2NO_2$; mol wt 258.11. C 51.19%, H 3.51%, Cl 27.47%, N 5.43%, O 12.40%. Prepn: Hopkins *et al., J. Org. Chem.* **24,** 2040 (1959); U.S. pat. 2,906,614 (1959 to Spencer Chem.); Baskakov *et al., Zh. Obshch. Khim.* **33,** 46 (1963). Activity as an herbicide: Hoffmann *et al., Weeds* **8,** 198 (1960). Metabolism: Grunow *et al., Food Cosmet. Toxicol.* **8,** 277 (1970). *Review:* Abel, *Rep. Progr. Appl. Chem.* **47,** 552 (1962).

NHCOOCH$_2$C≡CCH$_2$Cl

Cl

Crystals from *n*-hexane + benzene, mp 75-76°. Practically insol in water (soly at 25° = 11 ppm); slightly sol in hexane; readily sol in benzene, ethylene dichloride. Hydrolyzed by alkali with liberation of the terminal chlorine. Hydrolysis under acidic conditions gives 3-chloroacrylic acid. LD_{50} orally in rats: 600 mg/kg, Bailey, White, *Residue Rev.* **10,** 97 (1965).

USE: Selective herbicide for wild oats. *Caution:* May cause skin irritation.

970. Barbasco. Name applied in the Spanish-speaking countries of the New World to many unrelated plants used to poison or stun fish. In Mexico it usually means roots of *Dioscorea composita* Hemsl., or of *Dioscorea tepinapensis* Uline, *Dioscoreaceae* which yield up to 5% of their dry weight in diosgenin: *Chem. Week* **79,** no. 2, p 20 (July 14, 1956). Isoln procedure: Julian, U.S. pat. 3,019,220 (1962 to Julian Labs.). Book: D. G. Coursey, *Yams* (Longmans, London, 1967) 230 pp, an account of the nature, origins, cultivation and utilization of the useful members of the *Dioscoreaceae.* Compare Yam, Mexican.

971. Barberry Bark. Berberis bark; jaundice berry; woodsour; sowberry; pepperidge bush; sour-spine. Root bark of *Berberis vulgaris* L., *Berberidaceae. Habit.* Europe and Western Asia; also U.S. (New England States, Pennsylvania and Virginia). *Constit.* Berberine, berbamine, oxyacanthine, tannin, wax, fat, resin. *Ref:* Neugebauer, Brunner, *Pharm. Zent.* **80,** 113 (1939).

972. Barbital. *5,5-Diethyl-2,4,6(1H,3H,5H)-pyrimidinetrione; 5,5-diethylbarbituric acid;* barbitone; diethylmalonylurea; Veronal; Malonal; Veroletten; Sédeval; Dormonal; Hypnogène; Deba; Vespéral; Uronal. $C_8H_{12}N_2O_3$; mol wt 184.19. C 52.16%, H 6.57%, N 15.21%, O 26.08%. Prepd by the condensation of the diethyl ester of diethylmalonic acid with urea in sodium ethoxide soln: Fischer, Dilthey, *Ann.* **335,** 334 (1904); Ger. pat. 146,496 (1903), *Frdl.* **7,** 651, *Chem. Zentr.* **1903,** II, 1483; Fischer, v. Mering, *Therapie der Gegenwart,* March 1903; *Therapeutische Monatsh.* **17,** 208 (1903), *Chem. Zentr.* **1903,** I, 1155.

Faintly bitter needles (trigonal in the stable phase) from water, mp 188-192°. Can be sublimed *in vacuo.* Acid to litmus. K at 25° = 3.7 × 10⁻⁸. One gram dissolves in about 130 ml water, 13 ml boiling water, 14 ml alcohol, 75 ml chloroform, 35 ml ether. Sol in acetone, ethyl acetate, alkalies, petr ether, acetic acid, amyl alcohol, pyridine, aniline, nitrobenzene. LD orally in mice: 600 mg/kg.

Sodium salt, $C_8H_{11}N_2NaO_3$, *barbital sodium, sodium 5,5-diethylbarbiturate, barbitone sodium, soluble barbital, sodium diethylmalonylurea, Veronal sodium, Medinal, Embinal.* Bitter crystals or powder. One gram dissolves in 5 ml water, 2.5 ml boiling water, 400 ml alc. Aq soln is alkaline to litmus and phenolphthalein. pH of 0.1 molar aq soln, 9.4.

Caution: May be habit forming. This is a controlled substance (depressant) listed in the U.S. Code of Federal Regulations, Title 21 Parts 329.1 and 1308.14 (1987).

THERAP CAT: Sedative, hypnotic.

THERAP CAT (VET): Sedative, hypnotic.

973. Barbituric Acid. *2,4,6(1H,3H,5H)-pyrimidinetrione;* malonylurea; 2,4,6-trioxohexahydropyrimidine. $C_4H_4N_2O_3$; mol wt 128.09. C 37.50%, H 3.15%, N 21.87%, O 37.47%. Prepn from hydurilic acid + nitric acid: Baeyer, *Ann.* **127,** 199 (1863); *ibid.* **130,** 129 (1864). Structure: Mulder, *Ber.* **6,** 1233 (1873). Prepd from ethyl malonate and urea using sodium ethoxide as a condensing agent: Dickey, Gray, *Org. Syn.* coll. vol. II, 60 (1943). Crystal structure: Bolton, *Nature* **201,** 987 (1964). Unsubstituted barbituric acid has no hypnotic properties. *Review:* Carter, *J. Chem. Ed.* **28,** 524 (1951).

Dihydrate, rhombs from water. mp about 248° when anhydrous, with some decompn. Strong acid. K at 25° = 9.9 × 10⁻⁵. uv spectrum: Hartley, *J. Chem. Soc.* **87,** 1808 (1905). Difficultly sol in cold water; freely sol in hot water, in dil acids. Forms salts with metals. LD_{50} orally in male rats: > 5000 mg/kg, E. I. Goldenthal, *Toxicol. Appl. Pharmacol.* **18,** 185 (1971).

USE: Manuf plastics, pharmaceuticals.

974. Barium. Ba; at. wt 137.33; at. no. 56; valence 2. An alkaline earth metal. Abundance in earth's crust 0.05% by wt. Isotopes: 138 (71.66%); 137 (11.32%); 136 (7.81%); 135 (6.59%); 134 (2.42%); 132 (0.097%); 130 (0.101%). Occurs in barite and witherite. First prepared as a mercury amalgam by Davy in 1808. Toxicity studies of barium compds: Syed, Hosain, *Toxicol. Appl. Pharmacol.* **22,** 150 (1972). *Reviews: Gmelin's Handb. Anorg. Chem., Barium* (8th ed.) **30,** (1960); Goodenough, Stenger, "Magnesium, Calcium, Strontium, Barium and Radium" in *Comprehensive Inorganic Chemistry* Vol. **1,** J. C. Bailar, Jr. *et al.,* Eds. (Pergamon Press, Oxford, 1973) pp 591-664; C. J. Kunesh in Kirk-Othmer *Encyclopedia of Chemical Technology* vol. **3** (Wiley-Interscience, New York, 3rd ed., 1978) pp 457-463.

Yellowish-white, slightly lustrous lumps; body-centered cubic structure; somewhat malleable; very easily oxidizable; must be kept under petroleum or other oxygen-free liquid to exclude air. d 3.6. mp approx 710°. bp approx 1600°. E⁰ (aq) Ba²⁺/Ba −2.91 V. Description of reactions which are characteristic of alkaline earth metals *see* Calcium. Solns of sol barium salts give a white ppt with H_2SO_4 or sol sulfates; they also color nonluminous flame green.

USE: Carrier for radium. The β- and γ-radiation emitted by ¹⁴⁰Ba + ¹⁴⁰La makes a large contribution to the activity of the fission products of uranium rods during the first few weeks after their withdrawal from the reactor. Alloys of Ba with Al or Mg are used as getters in electronic tubes. The emissions from ^{133}Ba and ^{137m}Ba are used as standards in γ-spectrometry: Haissinsky, Adloff, *Radiochemical Survey of the Elements* (Elsevier, 1965) pp 12-14. *Caution:* All water or acid soluble barium compounds are *poisonous!*

975. Barium Acetate. $C_4H_6BaO_4$; mol wt 255.45. C 18.81%, H 2.37%, Ba 53.77%, O 25.05%. $Ba(C_2H_3O_2)_2$. Prepn: *Gmelin's, Barium* (8th ed.) **30**, 315 (1932) and supplement, 478 (1960).

Monohydrate. *Poisonous!* d 2.19. Loses its H_2O of hydration at 110°. One gram dissolves in 1.5 ml cold or boiling water, in 700 ml alc. The aq soln is neutral or slightly acid to litmus. LD_{50} in ICR mice: 23.31 mg Ba^{2+}/kg i.v., Syed, Hosain, *Toxicol. Appl. Pharmacol.* **22**, 150 (1972).

USE: Mordant for printing fabrics; in lubricating oil and grease; as catalyst for organic reactions.

976. Barium Benzenesulfonate. *Benzenesulfonic acid barium salt.* $C_{12}H_{10}BaO_6S_2$; mol wt 451.70. C 31.91%, H 2.23%, Ba 30.41%, O 21.25%, S 14.20%. Prepn: Freund, *Ann.* **120**, 76 (1861).

Monohydrate, white, nacreous leaflets. *Poisonous!* Freely sol in water; slightly sol in alc.

USE: Lubricating oil additives.

977. Barium Bromate. $BaBr_2O_6$; mol wt 393.19. Ba 34.93%, Br 40.65%, O 24.42%. $Ba(BrO_3)_2$. Prepd from potassium bromate and barium chloride: Pearce, Russell, *Inorg. Syn.* **2**, 20 (1946).

Monohydrate, monoclinic crystals from hot water. *Poisonous!* May develop slight odor of bromine on long standing. d 3.99. Dec. at 260°. Sol in water (g/100 ml): 0.44 (10°); 0.96 (30°); 5.39 (100°). Sol in acetone. Practically insol in alc, most other organic solvents.

USE: In the prepn of rare earth bromates; as corrosion inhibitor for low-C steel.

978. Barium Bromide. $BaBr_2$; mol wt 297.19. Ba 46.22%, Br 53.78%. Prepn: *Gmelin's, Barium* (8th ed.) **30**, 223 (1932) and supplement 380-381 (1960).

Dihydrate, crystals or granules; loses $1H_2O$ at 75° and all the H_2O at 120°. *Poisonous!* mp about 850° when anhydr. Very sol in water; sol in methanol; almost insol in ethanol, ethyl acetate, acetone, dioxane.

USE: In the manuf of other bromides; in the prepn of phosphors.

979. Barium Carbonate. $CBaO_3$; mol wt 197.37. C 6.09%, Ba 69.58%, O 24.32%. $BaCO_3$. Occurs in nature as the mineral *witherite*. Prepn: *Gmelin's, Barium* (8th ed) **30**, 301-303 (1932) and supplement, 186-188, 461-466 (1960). The barium carbonate of commerce is made by precipitation and is 98-99% pure. Manuf: Faith, Keyes & Clark's *Industrial Chemicals*, F. A. Lowenheim, M. K. Moran, Eds. (Wiley-Interscience, New York, 4th ed., 1975) pp 121-125.

White, heavy powder. *Poisonous!* d (witherite) 4.2865. At about 1300° dec into BaO and CO_2. Almost insol in water, 0.024 g in a liter; slightly sol (1:1000) in CO_2-water; sol in dil HCl, HNO_3 or acetic acid; also sol in soln NH_4Cl or NH_4NO_3. LD_{50} orally in rats: 800 mg/kg.

Human Toxicity: Acute: excessive salivation, vomiting, colic, violent diarrhea, convulsive tremors, increased blood pressure, hemorrhages in G.I. tract and kidneys, muscular paralysis.

USE: Rat poison; in ceramics, paints, enamels, marble substitutes, rubber; manuf of paper, barium salts, electrodes, optical glasses; as an analytical reagent.

980. Barium Chlorate. $BaCl_2O_6$; mol wt 304.27. Ba 45.14%, Cl 23.31%, O 31.55%. $Ba(ClO_3)_2$. Prepn: Vanino, *Handb. Präp. Chem., Anorgan. Teil* (2.Aufl., Stuttgart, 1925) p 297; Schmeisser in *Handbook of Preparative Inorganic Chemistry*, Vol. 1, G. Brauer, Ed. (Academic Press, New York, 2nd ed., 1963) p 314. Large-scale process: Munroe, *Chem. Met. Eng.* **23**, 188 (1920). Also prepd by electrolysis of barium chloride.

Monohydrate, monoclinic prismatic crystals. *Poisonous!* d 3.179. Loses its water of hydration at 120°, begins to give off oxygen at 250°, mp 414°. Freely sol in water; sol in hydrochloric acid; moderately sol in ethylamine; very sparingly sol in alc, somewhat more in acetone. Practically insol in

ethyl acetate, pyridine. Fire hazard when in contact with combustible material.

USE: In pyrotechnics (green fire); manuf of explosives and matches; mordant in dyeing.

981. Barium Chloride. $BaCl_2$; mol wt 208.27. Ba 65.95%, Cl 34.05%. Prepn: *Gmelin's, Barium* (8th ed) **30**, 171-175 (1932) and supplement, 179-181, 324-325 (1960). Toxicity studies: I. B. Syed, F. Hosain, *Toxicol. Appl. Pharmacol.* **22**, 150 (1972). Induction of arrhythmia in exptl animals: F. W. Eichbaum, *Basic Res. Cardiol.* **68**, 73 (1973).

Dihydrate, crystals or granules or powder; bitter salty taste. d 3.86; mp 963. *Poisonous!* Very sol in water; sol in methanol. Practically insol in ethanol, acetone, ethyl acetate. LD_{50} in ICR mice (mg Ba^{2+}/kg): 19.2 i.v. (Syed, Hosain).

USE: Manuf pigments, color lakes, glass, mordant for acid dyes; weighting and dyeing textile fabrics; in Al refining; as pesticide; boiler compds for softening water; tanning and finishing leather.

THERAP CAT (VET): Formerly used as purgative in horses, ruminatoric in cattle.

982. Barium Chromate(VI). C.I. 77103; C.I. Pigment Yellow 31; Baryta yellow; lemon yellow; permanent yellow; Steinbühl yellow; ultramarine yellow. $BaCrO_4$; mol wt 253.37. Ba 54.21%, Cr 20.53%, O 25.26%. Prepn: Beyer, Rieman, *J. Am. Chem. Soc.* **65**, 971 (1943); *Colour Index* vol. **4** (3rd ed., 1971) p 4656.

Yellow, heavy, monoclinic, orthorhombic crystals. *Poisonous!* d 4.50. Practically insol in water, dil acetic or chromic acids; dissolved or dec by mineral acids.

USE: As a pigment almost entirely in anticorrosion jointing pastes to prevent electro-chemical corrosion at junctions of dissimilar metals; some use in artists' colors and in coloring glass, ceramics, porcelain. Also used in metal primers, pyrotechnic compositions.

983. Barium Cyanide. C_2BaN_2; mol wt 189.40. C 12.68%, Ba 72.52%, N 14.79%. $Ba(CN)_2$. Prepn: **Brit. pat.** 602,393 (1948 to I.C.I.); *Gmelin's, Barium* (8th ed) **30**, 327 (1932) and supplement 483 (1960).

Crystals; slowly dec in air. *Very poisonous!* Very sol in water; sol in alcohol.

USE: In electroplating processes; in metallurgy.

984. Barium Dithionate. Barium "hyposulfate". BaO_6S_2; mol wt 297.48. Ba 46.17%, O 32.27%, S 21.55%. BaS_2O_6. Prepn: Pfanstiel, *Inorg. Syn.* **2**, 170 (1946).

Dihydrate, crystals. *Poisonous!* d 4.54. Loses SO_2 on heating above 150° forming $BaSO_4$. Sol in 4 parts water; more sol in hot water; slightly sol in alcohol.

985. Barium Ferrocyanide. *Barium hexacyanoferrate-(II).* $C_6Ba_2FeN_6$; mol wt 486.68. C 14.81%, Ba 56.45%, Fe 11.48%, N 17.27%. $Ba_2Fe(CN)_6$. Prepn: Grat-Cabanac, *Bull. Soc. Chim. France* **1956**, 1743.

Hexahydrate, yellowish rectangular monoclinic crystals; loses most of water at 40° becoming colorless. Dec at 80°, evolving HCN. Almost insol in water, alcohol.

986. Barium Fluoride. BaF_2; mol wt 175.36. Ba 78.33%, F 21.67%. Prepd by dissolving $BaCO_3$ in excess HF, evaporating to dryness, and heating to red heat: W. Olbrich, *Thesis* (Technische Hochschule, Breslau, 1929), p 2; Kwasnik in *Handbook of Preparative Inorganic Chemistry* vol. **1**, G. Brauer, Ed., (Academic Press, New York, 2nd ed., 1963) p 234.

Transparent cubic crystals (fluorite lattice). *Poisonous!* d 4.83. mp 1353. bp 2260. Soly in water (g/l): 1.586 (10°); 1.607 (20°); 1.620 (30°). Also sol in hydrochloric, nitric, hydrofluoric acids and in aq solns of ammonium chloride. May be stored in glass bottles.

USE: As a flux and opacifier in vitreous enamels; in the manuf of carbon brushes for D.C. motors and generators; in heat-treating metals; in embalming; in glass manuf.

987. Barium Formate. $C_2H_2BaO_4$; mol wt 227.40. C 10.56%, H 0.89%, Ba 60.40%, O 28.14%. $Ba(HCOO)_2$. Prepn: *Gmelin's, Barium* (8th ed) **30**, 311 (1932), and supplement, 477 (1960).

Crystals. *Poisonous!* d 3.21. Soluble in 4 parts cold, 3 parts boiling water; practically insol in alcohol.

988. Barium Hexafluorosilicate. Barium fluosilicate; barium silicofluoride. BaF_6Si; mol wt 279.45. Ba 49.15%, F 40.79%, Si 10.05%. $BaSiF_6$. Prepd from $BaCl_2$ and H_2SiF_6: Truchot, *Compt. Rend.* **98**, 821 (1884); Hoffmann, Gutowsky, *Inorg. Syn.* **4**, 145 (1953).

Orthorhombic needles. d_4^{21} 4.29. Dec at 300°. Heat of formation (solid) -677.42 kcal/mol. Soly in water (g/100 ml H_2O): 0.015 (0°); 0.0235 (25°); 0.091 (100°). Prolonged contact with water produces hydrolysis which is much accelerated by the presence of alkali. Slightly sol in dil acids; sol in ammonium chloride soln; practically insol in alc.

USE: Prepn of silicon tetrafluoride; as pesticide. *Caution:* Highly toxic, especially when brought into soln by alkali.

989. Barium Hydroxide. Barium hydrate; caustic baryta. BaH_2O_2; mol wt 171.38. Ba 80.15%, H 1.18%, O 18.67%. $Ba(OH)_2$. Prepn: *Gmelin's, Barium* (8th ed) **30**, 106-111 and supplement 175-177, 289 (1960).

Monohydrate, *dried barium hydroxide.* Usually contains 92-95% $Ba(OH)_2.H_2O$. White powder. *Poisonous!* d 3.743. Slightly sol in water; sol in dil acids.

Octahydrate, transparent crystals or white masses. Very alkaline; rapidly absorbs CO_2 from air, becoming incompletely sol in water. *Poisonous!* mp 78°. Freely sol in water, methanol; slightly sol in ethanol; practically insol in acetone. *Keep tightly closed.*

USE: In manuf of alkali, glass; in synthetic rubber vulcanization, in corrosion inhibitors, drilling fluids, lubricants, pesticides, sugar industry; boiler scale remedy; refining animal and vegetable oils; softening water; fresco painting.

990. Barium Hypophosphite. $BaH_4O_4P_2$; mol wt 267.34. Ba 51.38%, H 1.51%, O 23.94%, P 23.17%. $Ba(H_2PO_2)_2$. Prepd by treating white phosphorus with barium hydroxide: $8P + 3Ba(OH)_2.8H_2O + H_2O \rightarrow 3Ba(H_2PO_2)_2 + H_2O + 2PH_3$: Rose, *Pogg. Ann.* **9**, 370 (1827); Klement in *Handbook of Preparative Inorganic Chemistry,* Vol. 1, G. Brauer, Ed. (Academic Press, New York, 2nd ed., 1963) p 557.

Monohydrate, monoclinic platelets with nacreous sheen from hot water. *Poisonous!* d_4^{17} 2.90. Soly (g/100 ml H_2O): 28.6 (17°); 33.3 (100°). Practically insol in alcohol.

USE: In nickel plating.

991. Barium Iodate. BaI_2O_6; mol wt 487.18. Ba 28.19%, I 52.10%, O 19.71%. $Ba(IO_3)_2$. Prepn: Lambert, Yasada, *Inorg. Syn.* **7**, 13 (1963).

Monohydrate, crystals; becomes anhyd at 130°. d 5.00. *Poisonous!* Sol in 3350 parts water at 25°, 625 parts boiling water; sol in HCl or HNO_3; practically insol in alcohol.

992. Barium Iodide. BaI_2; mol wt 391.18. Ba 35.11%, I 64.89%. Prepn: *Gmelin's, Barium* (8th ed) **30**, 238-239 (1932) and supplement, 394 (1960).

Dihydrate, colorless, odorless, transparent crystals or white granules; rapidly becomes reddish in the air due to liberation of iodine. *Poisonous!* d 5.15. Freely sol in water; the aq soln is neutral or slightly alkaline. Sol in alcohol, acetone. *Keep well closed and protected from light.*

USE: In the manuf of other iodides.

993. Barium Manganate(VI). Manganese green; Cassel's green; Rosenstiehl's green. $BaMnO_4$; mol wt 256.29. Ba 53.60%, Mn 21.43%, O 24.97%. Prepn: Schlesinger, Stems, *J. Am. Chem. Soc.* **46**, 1965 (1924); Jellinek, *J. Inorg. Nucl. Chem.* **13**, 329 (1960); R. S. Nyholm, P. R. Woolliams, *Inorg. Syn.* **11**, 56 (1968); Bayan, Aymonino, *Ber.* **101**, 3337 (1968); H. Firouzabadi, Z. Mostafavipoor, *Bull. Chem. Soc. Japan* **56**, 914 (1983).

Dark blue-green crystals. d 4.85. Disproportionates in water or dil acid to form $Ba(MnO_4)_2$ and MnO_2.

USE: As pigment in fresco painting instead of Scheele's green because not so poisonous as latter. As mild oxidizing agent in organic synthesis.

994. Barium Mercuric Iodide. Barium tetraiodomercurate(II); mercuric barium iodide. $BaHgI_4$; mol wt 845.65. Ba 16.24%, Hg 23.72%, I 60.03%.

Yellow or reddish, deliquesc crystals; said to become red even when kept in sealed tubes. *Poison!* Very sol in water or alc. *Keep well closed!*

USE: As an aq soln known as *Rohrbach's Soln,* for separating minerals of different densities the concd soln has a d of 3.5; also for microchemical detection of alkaloids.

995. Barium Nitrate. BaN_2O_6; mol wt 261.38. Ba 52.55%, N 10.72%, O 36.73%. $Ba(NO_3)_2$. Prepn: *Gmelin's, Barium* (8th ed) **30**, 149-151 (1932) and supplement, 178-179, 305 (1960).

Crystals or cryst powder. *Poisonous!* d 3.24. mp about 590°; dec at higher temp. Freely sol in water; very slightly sol in alcohol, acetone. LD_{50} i.v. in ICR mice: 20.10 mg Ba^{2+}/kg, Syed, Hosain, *Toxicol. Appl. Pharmacol.* **22**, 150 (1972).

USE: Manuf BaO_2; pyrotechnics for green fire; green signal lights; in the vacuum-tube industry.

996. Barium Nitrite. BaN_2O_4; mol wt 229.38. Ba 59.88%, N 12.21%, O 27.90%. $Ba(NO_2)_2$. Prepn: *Gmelin's, Barium* (8th ed) **30**, 144-147 (1932) and suppl, 303 (1960).

Monohydrate, crystals. *Poisonous!* d 3.187. Sol in water; practically insol in alc.

USE: In diazotization reactions; prevention of corrosion of steel bars; in explosives.

997. Barium Oxalate. *Ethanedioic acid barium salt.* C_2BaO_4; mol wt 225.38. C 10.66%, Ba 60.95%, O 28.40%. BaC_2O_4. Prepn: *Gmelin's, Barium* (8th ed) **30**, 320-323 (1932) and supplement, 480-482 (1960).

Monohydrate, cryst powder. *Poisonous!* d 2.66. Sol in 10,000 parts cold, 5000 parts boiling water; sol in dil HNO_3 or HCl.

998. Barium Oxide. Barium monoxide; barium protoxide; calcined baryta. BaO; mol wt 153.36. Ba 89.57%, O 10.43%. Prepn: Ehrlich in *Handbook of Preparative Inorganic Chemistry,* vol. 1, G. Brauer, Ed. (Academic Press, New York, 2nd ed., 1963) p 933.

White to yellowish-white powder or lumps. Very alkaline; absorbs moisture and CO_2 on exposure to air. *Poisonous!* On contact with water it forms $Ba(OH)_2$ with evolution of much heat. At 450° combines with oxygen to form BaO_2, which is reduced to BaO above 600°. d 5.7. mp about 1920°. Sol in water, dil acids. Slowly, but considerably sol in methanol, ethanol forming barium alcoholate. *Keep tightly closed.*

USE: Porous grades are marketed especially for drying gases and solvents (particularly alcohols, aldehydes and petroleum solvents). Swells, but does not become sticky upon absorption of moisture. Used in manuf of lubricating oil detergents. Also used for making barium methoxide.

999. Barium Perchlorate. $BaCl_2O_8$; mol wt 336.27. Ba 40.85%, Cl 21.09%, O 38.06%. $Ba(ClO_4)_2$. Prepn: *Gmelin's, Barium* (8th ed) **30**, 218 (1932) and supplement, 373 (1960).

Trihydrate, crystals. *Poisonous!* Sol in water, methanol; slightly sol in ethanol, ethyl acetate, acetone; practically insol in ether.

USE: In the determination of ribonuclease; as absorbent for water in C and H analysis.

1000. Barium Permanganate. Barium manganate(VII). $BaMn_2O_8$; mol wt 375.22. Ba 36.61%, Mn 29.28%, O 34.12%. $Ba(MnO_4)_2$. Prepn: Lux in *Handbook of Preparative Inorganic Chemistry,* vol. 1, G. Brauer, Ed. (Academic Press, New York, 2nd ed., 1963) p 1462.

Brownish-violet to black crystals. *Poisonous!* d 3.77. Sparingly sol in water; dec by alcohol.

USE: As dry cell depolarizer.

1001. Barium Peroxide. Barium dioxide; barium superoxide. BaO_2; mol wt 169.36. Ba 81.11%, O 18.89%. Prepn: *Gmelin's, Barium* (8th ed) **30**, 92-98 (1932) and supplement, 177, 296 (1960). The article of commerce contains about 85% BaO_2; the remainder is chiefly BaO.

White or grayish-white, heavy powder. *Poisonous!* Dec slowly in air. Insol in water, but slowly dec by contact with it; dec by dil acid or CO_2 in presence of H_2O forming H_2O_2. Combines with water to form octahydrate. *Keep well closed.*

USE: Bleaching animal substances, vegetable fibers and straw; glass decolorizer; manuf H_2O_2 and oxygen; dyeing and printing textiles; with powdered aluminum in welding; in cathodes; in igniter compositions. Oxidizing agent in organic synthesis.

1002. Barium Phosphate, Dibasic. Secondary barium phosphate. $BaHO_4P$; mol wt 233.35. Ba 58.86%, H 0.43%, O 27.43%, P 13.28%. $BaHPO_4$. Prepn: *Gmelin's, Barium* (8th ed) **30**, 340 (1932) and supplement, 492 (1960).
Crystals. *Poisonous!* d 4.16. Practically insol in water; sol in dil HCl or HNO_3.
USE: In fireproofing compositions; in prepn of phosphors.

1003. Barium Platinous Cyanide. Barium tetracyano-platinate(II); barium cyanoplatinate(II); barium platinocyanide; platinous barium cyanide. C_4BaN_4Pt; mol wt 436.66. C 11.00%, Ba 31.46%, N 12.83%, Pt 44.71%. $BaPt(CN)_4$.
Tetrahydrate, large dichroic crystals; yellowish-green by transmitted, bluish-violet by reflected light. *Poison!* d 3.05. Sol in about 35 parts water; more sol in hot water.
USE: An aq soln mixed with some adhesive and painted on paper or wood exhibits luminescence when exposed to the invisible ultraviolet rays of the spectrum or to Roentgen, radium, or cathode rays; hence used in radiography for making x-ray screens.

1004. Barium Selenide. BaSe; mol wt 216.32. Ba 63.50%, Se 36.50%. Prepn: Henglein, *Z. Anorg. Chem.* **120**, 77 (1921); *Z. Electrochem.* **30**, 11 (1924); *Gmelin's, Barium* (8th ed) **30**, 290 (1932) and supplement, 453 (1960).
Cubic microcrystalline powder. d 5.02. Turns red on exposure to air. Dec in water.
USE: In photocells, semiconductors.

1005. Barium Silicide. $BaSi_2$; mol wt 193.48. Ba 70.99%, Si 29.01%. Prepn: *Gmelin's, Barium* (8th ed) **30**, 330 (1932) and supplement, 486 (1960).
Metal-like, gray lumps; quite permanent in dry air, but dec by moisture with evolution of H_2. Melts at white heat.
USE: Deoxidizing and desulfurizing steel and for other metallurgical purposes.

1006. Barium Sulfate. Blanc fixe; Actybaryte; Bakontal; Baridol; Baritop; Barosperse; Citobaryum; Esophotrast; E-Z-HD; E-Z-Paque; Intestibar; Microbar; Micropaque; Microtrast; Neobar; Oratrast; Polybar; Prontobario; Radiopaque; Raybar; Telebar; Unibaryt. BaO_4S; mol wt 233.43. Ba 58.84%, O 27.42%, S 13.74%. $BaSO_4$. Occurs in nature as the mineral *barite*; also as *barytes, heavy spar.* Prepn: *Gmelin's, Barium* (8th ed) **30**, 262-267 (1932) and supplement, 182-186, 412-414 (1960). Review of clinical use: R. F. Theoni, A. R. Margulis, *Radiology* **167**, 7 (1988).
Fine, heavy, odorless powder or polymorphous crystals, d 4.25-4.5. Dec above 1600°. Practically insol in water (one gram dissolves in 400,000 parts), dil acids, alc. Sol in hot concd H_2SO_4.
USE: Manuf photographic papers, artificial ivory, cellophane; filler for rubber, linoleum, oil cloth, polymeric fibers and resins, paper, lithographic inks; as a water-color pigment for colored paper, in wallpaper; as a size for modifying the colors of other pigments; in heavy concrete for radiation shield.
THERAP CAT: Diagnostic aid (radiopaque medium).
THERAP CAT (VET): X-ray contrast medium.

1007. Barium Sulfide. BaS; mol wt 169.42. Ba 81.08%, S 18.92%. The commercial article contains 80-90% BaS. Prepn: Ehrlich in *Handbook of Preparative Inorganic Chemistry,* **vol. 1,** G. Brauer, Ed. (Academic Press, New York, 2nd ed., 1963) p 938.
Heavy, grayish-white or pale yellow powder. d 4.36. mp > 2000°. Oxidizes in dry air; slowly decomp in damp air into carbonate, etc., with evolution of H_2S. *Poisonous!* Slightly sol in cold, more in hot water; sol in NH_4Cl soln; decomposed by acids, even CO_2, forming H_2S. *Keep in well-closed containers.*
USE: As depilatory; in luminous paints; manuf lithopone; vulcanizing rubber, generating H_2S.
THERAP CAT (VET): Depilatory.

1008. Barium Sulfide, Black. Black ash. This is the crude barium sulfide obtained by strong heating of a mixture of barium sulfate mineral (barite) and charcoal. Contains 60-70% BaS; the remainder is chiefly barium sulfate and carbon.
USE: Prepn of barium salts.

1009. Barium Sulfite. BaO_3S; mol wt 217.42. Ba 63.18%, O 22.08%, S 14.75%. $BaSO_3$. Prepn: *Gmelin's, Barium* (8th ed.) **30**, 261 (1932) and supplement, 411 (1960).
Odorless crystals or powder; gradually oxidizes in air to $BaSO_4$. *Poisonous!* Slightly sol in water; practically insol in alcohol; dec by acids with liberation of SO_2.
USE: In paper manufacture.

1010. Barium Thiocyanate. Barium sulfocyanate; barium sulfocyanide; barium rhodanide. $C_2BaN_2S_2$; mol wt 253.52. C 9.47%, Ba 54.18%, N 11.05%, S 25.30%. Ba(SCN)$_2$. Prepn: Herstein, *Inorg. Syn.* **3**, 24 (1950).
Deliquesc crystals. *Poisonous!* Very sol in water, sol in acetone, methanol, ethanol. *Keep well closed.*
Trihydrate, needle-shaped crystals (from water).
USE: In dyeing; in photography; as dispersing agent for cellulose; in prepn of thiocyanates of other metals.

1011. Barium Thiosulfate. Barium hyposulfite. BaO_3S_2; mol wt 249.48. Ba 55.06%, O 19.24%, S 25.70%. BaS_2O_3. Prepn: *Gmelin's, Barium* (8th ed.) **30**, 283 (1932) and supplement, 449 (1960).
Monohydrate, cryst powder. *Poisonous!* Very slightly sol in water; practically insol in alc, ether, acetone, CCl_4, CS_2.
USE: Manuf explosives, matches; as an iodometry standard; in photographic diffusion-transfer process.

1012. Barium Titanate(IV). Barium metatitanate. BaO_3Ti; mol wt 233.26. Ba 58.89%, O 20.58%, Ti 20.53%. $BaTiO_3$. Usually prepd by calcining (at about 1300°) an intimate mixture of titanium dioxide and barium carbonate. Prepn from Ti oxalate and a barium compd: Lynd, Merker, U.S. pat. 2,758,911 (1956 to National Lead). Wet process by the addn of a titanium ester, such as tetrapropyl titanate, to an aq soln of barium hydroxide: Flaschen, *J. Am. Chem. Soc.* **77**, 6194 (1955). Prepn of high purity material by ignition of barium titanyloxalate: Clabaugh *et al., J. Res. NBS* **56**, 289 (1956). Prepn by ignition of Ba and Ti alcoholates in an organic solvent: DiVita, Fischer, U.S. pat. 2,985,504 (1961 to U.S. Dept. of the Army).
Exists in five cryst modifications. The tetragonal form (obtained by the wet process) appears to have the most desirable electric properties and is described here: d 6.08. mp 1625°. Curie point 120°. Has ferroelectric and piezoelectric properties. Becomes permanently polarized when exposed to high voltage direct current, provided the temperature is never allowed to rise above Curie pt. Has high dielectric properties which can be influenced by temp, voltage, and frequency.
USE: In electronic devices, e.g., as voltage-sensitive dielectric in so-called dielectric amplifiers, in computer elements, magnetic amplifiers, memory devices.

1013. Barium Uranium Oxide. Barium uranate(VI); barium diuranate; uranium barium oxide. BaO_7U_2; mol wt 725.50. Ba 18.93%, O 15.44%, U 65.63%. BaU_2O_7. Prepn: Allpress, *J. Inorg. Nucl. Chem.* **26**, 1847 (1964); Klima *et al., ibid.* **28**, 1861 (1966).
Orange or yellow powder. Practically insol in water; sol in acids.
USE: In painting on porcelain.

1014. Barthrin. *2,2-Dimethyl-3-(2-methyl-1-propenyl)-cyclopropanecarboxylic acid (6-chloro-1,3-benzodioxol-5-yl) methyl ester; 2,2-dimethyl-3-(2-methylpropenyl)cyclopropanecarboxylic acid 6-chloropiperonyl ester;* 6-chloropiperonyl 2,2-dimethyl-3-(2-methylpropenyl)cyclopropanecarboxylate; 6-chloropiperonyl chrysanthemumate; chrysanthemummonocarboxylic acid 6-chloropiperonyl ester. $C_{18}H_{21}ClO_4$; mol wt 336.83. C 64.19%, H 6.29%, Cl 10.53%, O 19.00%. Prepn: Barthel, Alexander, *J. Org. Chem.* **23**, 1012 (1958); Barthel *et al.,* U.S. pat. 2,886,485 (1959). Shows low mammalian toxicity.

Oily liquid, bp$_{0.2}$ 155-171°, bp$_{0.7}$ 184-206°. n_D^{25} 1.5383. Sol in kerosine.

USE: Insecticide.

1015. Basic Aluminum Carbonate Gel. Basaljel. An aluminum hydroxide—aluminum carbonate gel.

THERAP CAT: Phosphorus-binding agent in prevention of recurrent renal phosphatic calculi; antacid.

1016. Basic Lead Carbonate. *C.I. Pigment White 1;* C.I. 77597; lead subcarbonate; white lead; flake lead; ceruse; cerussa; bleiweiss (German). Approx (PbCO$_3$)$_2$.Pb(OH)$_2$. Occurs in nature as the mineral *hydrocerussite. See: Mellor's* **vol. VII,** 837 (1927); *Colour Index* **vol. 4** (3rd ed., 1971) p 4676.

White, heavy powder. *Poisonous!* Dec at 400°, leaving a residue of PbO. Insol in water or alcohol; sol in acetic acid or dil HNO$_3$ with effervescence.

USE: Pigment in oil paints and water colors; in cements; for making putty and lead carbonate paper; in the processing of parchment.

1017. Basswood. Linden tree. Also called *Tilia americana* L., *Tiliaceae.* Grows in mountainous woods from Canada to Georgia and west to Texas.

Various decoctions of flowers, trees, bark and wood are used in American folk medicine for disorders of the bile and liver. Preparation of active extracts from the obscure *"Tilia alburnum"*: Lafon, U.S. pat. **3,030,271** (1962 to S. A. Orsymonde).

1018. Batrachotoxin. *Batrachotoxinin A 20-(2,4-dimethyl-1H-pyrrole-3-carboxylate);* 3α,9α-epoxy-14β,18β-(epoxyethano-N-methylimino)-5β-pregna-7,16-diene-3β,-11α,20α-triol, 20α-ester with 2,4-dimethylpyrrole-3-carboxylic acid. C$_{31}$H$_{42}$N$_2$O$_6$; mol wt 538.69. C 69.12%, H 7.86%, N 5.20%, O 17.82%. The most toxic of the steroidal alkaloids extracted from the skin of five species of neotropical poison-dart frogs, genus *Phyllobates.* Its name is derived from the Greek word "batrachos" meaning frog. Isoln and tentative structure: Marki, Witkop, *Experientia* **19,** 329 (1963); Daly *et al., J. Am. Chem. Soc.* **87,** 124 (1965). Contains unprecedented structural features for a naturally occurring pregnane deriv: 9α-OH; 3α,9α-oxide; 3β-hemiketal; seven-membered 14β,18β-heterocyclic ring; Δ^{16}-unsaturation; comparable pyrrole ester attached to a 20α-OH. Structure, partial synthesis from batrachotoxinin A, *q.v.:* T. Tokuyama *et al., ibid.* **91,** 3931 (1969). Synthetic studies: J. F. W. Keana, R. R. Schumaker, *J. Org. Chem.* **41,** 3840 (1976); P. Magnus *et al., Chem. Commun.* **1985,** 1185. Mode of action: Albuquerque, *Fed. Proc.* **31,** 1133 (1972); *see also* J. W. Daly *et al., Science* **208,** 1383 (1980). Toxicology: J. Daly, B. Witkop, *Clin. Toxicol.* **4,** 331 (1971). Review of chemistry and pharmacology: Albuquerque *et al., ibid.* **172,** 995 (1971); J. W. Daly, *Fortschr. Chem. Org. Naturst.* **41,** 206-227 (1982). Review of effects on sodium channel: G. B. Brown, *Int. Rev. Neurobiol.* **29,** 77-116 (1988).

[α]$_{584}^{24}$ −5° to −10°; [α]$_{300}^{24}$ −260° (c = 0.23 in methanol). uv max (methanolic HCl): 267, 234 nm (ε 5100, 9200). Venom has no effect on intact skin but it causes a long last-

ing pungent pain not unlike a bee's sting when in contact with even the smallest scratch. Furthermore, consumption of material exposed to the poison is dangerous only when one has an oral scratch or digestive tract ulcer. Toxic effect is caused by selective and irreversible increase in permeability of membranes to Na ions. LD$_{50}$ s.c. in mice: 2 μg/kg (Daly, Witkop).

1019. Batrachotoxinin A. *1,2,3,4,7a,10,11,11a,12,13-Decahydro-14-(1-hydroxyethyl)-2,11a-dimethyl-7H-9,11b-epoxy-13a,5a-propenophenanthro[2,1-f][1,4]oxazepine-9,-12(8H)-diol;* 3α,9α-epoxy-14β,18β-(epoxyethano-N-methylimino)-5β-pregna-7,16-diene-3β,11α,20α-triol. C$_{24}$H$_{35}$NO$_5$; mol wt 417.55. C 69.04%, H 8.45%, N 3.35%, O 19.16%. Least active of the four apparent major toxic steroidal alkaloids, batrachotoxin, *q.v., isobatrachotoxin, pseudobatrachotoxin* and batrachotoxinin A, isolated fom skin extracts of Colombian arrow poison frogs, genus *Phyllobates.* Isoln: Marki, Witkop, *Experientia* **19,** 329 (1963); Daly *et al., J. Am. Chem. Soc.* **87,** 124 (1965). Structure: T. Tokuyama *et al., ibid.* **90,** 1917 (1968); Karle, Karle, *Acta Crystallogr.* **25B,** 428 (1969). Mass spec. and NMR data: T. Tokuyama *et al., J. Am. Chem. Soc.* **91,** 3931 (1969). Absolute configuration: R. D. Gilardi, *Acta Crystallogr.* **B26,** 440 (1970). Partial synthesis: R. Imhof *et al., Helv. Chim. Acta* **55,** 1151 (1972). Toxicology: J. Daly, B. Witkop, *Clin. Toxicol.* **4,** 331 (1971).

LD$_{50}$ s.c. in mice: 1000 μg/kg (Daly, Witkop).

1020. Batroxobin. *Bothrops atrox serine proteinase; Bothrops venom proteinase;* reptilase R; Botropase; Defibrase; Reptilase. Thrombin-like enzyme from the venom of *Bothrops atrox* (Linn.), a pit viper found in several varieties in Southern and Central America. Prepn of conc extract from *B. jararaca* venom and coagulative properties: D. von Klobusitzky, *Arch. Exp. Pathol. Pharmakol.* **179,** 204 (1935). The enzyme from *B. atrox* venom is hemostatic at low doses and acts as a blood anti-coagulant at higher doses. Composition and manuf: E. E. Percs *et al.,* **Ger.** pat. **2,201,993** corresp to **U.S.** pat. **3,849,252** (1972, 1974 both to Pentapharm). Studies on the subunit structure of fibrin produced by batroxobin (reptilase): P. Mattock, M. P. Esnouf, *Nature New Biol.* **233,** 277 (1971). *In vitro* blood-clotting activity: K. O. Wik *et al., Brit. J. Haematol.* **23,** 37 (1972). Electron microscopic study: S. Ishimaru *et al., Thromb. Haemost.* **45,** 276 (1981). Clinicopharmacological study in normal humans: K. Fukutake *et al., Nippon Ketsueki Gakkai Zasshi* **44,** 1178 (1981), *C.A.* **96,** 62775x (1982). Review of characterization, properties: K. G. Stocker, G. H. Barlow, *Aktuel. Probl. Angiol.* **26,** 45-62 (1975).

Rectangular crystals. Sol in physiological saline. Practically insol in distilled water. Forms complexes with phenol and phenol derivs that are practically insol in water.

THERAP CAT: Hemostatic (in peripheral arterial circulatory disorders).

1021. Batyl Alcohol. *3-(Octadecyloxy)-1,2-propanediol;* monooctadecyl ether of glycerol. C$_{21}$H$_{44}$O$_3$; mol wt 344.56. C 73.20%, H 12.87%, O 13.93%. CH$_2$OHCHOHCH$_2$O-(CH$_2$)$_{17}$CH$_3$. Isoln from shark liver oils: Heilbron, Owens, *J. Chem. Soc.* **1928,** 942; Davies *et al., ibid.* **1933,** 165; Nakamiya, *C.A.* **33,** 8175 (1939); from yellow bone marrow: Holmes *et al., J. Am. Chem. Soc.* **63,** 2607 (1941); from coral-reef-building animals: Kind, Bergmann, *J. Org. Chem.* **7,** 424 (1942). Synthesis from allyl octadecyl ether: Davies *et al., J. Chem. Soc.* **1930,** 2542; Kornblum, Holmes, *J. Am. Chem. Soc.* **64,** 3045 (1942). Synthesis of optically active batyl alc: Baer, Fischer, *J. Biol. Chem.* **140,** 397 (1941).

Glistening plates from dilute acetone, mp 70.5-71°. $[\alpha]_D^{20}$ +1.14° (c = 6.6 in CHCl₃). Sol in the usual fat solvents.

Bis(*p*-nitrobenzoate), $C_{35}H_{50}N_2O_9$, pale yellow needles from methanol, soft at 63°, mp 65-66°.

THERAP CAT (VET): Has been recommended for bracken fern poisoning in cattle.

1022. Bayberry Bark. Candleberry bark; myrica; myrtle wax; berry wax; tallow shrub. Dried root bark of *Myrica cerifera* L., *Myricaceae*. *Habit.* Maryland to Florida, west to Texas and Arkansas. *Constit.* Acrid and astringent resin, myricic acid, tannin, red coloring matter, gum, starch. Bayberry wax consists of palmitic, myristic and lauric acid esters.

THERAP CAT: Astringent; emetic.

1023. BCG. An abbreviation for *Bacillus Calmette-Guérin*, a strain of *Mycobacterium tuberculosis.* Prepn of BCG labeled with C^{14}: Pasquier, Kurylowicz, *Rev. Immunol.* **20**, 245 (1956). Has the potential to act as a non-specific immunopotentiating agent that stimulates the whole range of immune responses. Role as an immunotherapeutic agent: Bartlett and Zbar *et al., J. Nat. Cancer Inst.* **48**, 245, 1441, 1709 (1972); **49**, 119 (1972). Comparison of potentiation of specific tumor immunity in mice: M. T. Scott, R. Bomford, *ibid.* **57**, 555 (1976). Review of use in Hodgkin's disease: H. Zywicka-Lopaciuk *et al., Arch. Immunol. Ther. Exp.* **29**, 739-755 (1981); in operable lung cancer: B. H. Stack *et al., Thorax* **37**, 588-593 (1982). Review of use as anti-tumor agent and of the interaction of BCG with cells of the mammalian immune system: M. Davies, *Biochim. Biophys. Acta* **651**, 143-174 (1982). Review of pathogenesis of tuberculosis and the effectiveness of BCG vaccination: H. G. ten Dam, A. Pio, *Tubercle* **63**, 225-233 (1982).

THERAP CAT: BCG vaccine as active immunizing agent.

THERAP CAT (VET): Vaccination vs. tuberculosis (cattle). Use prohibited where bovine tuberculosis eradication programs are in progress.

1024. Bebeerine. $1'\alpha$-*6,6'-Dimethoxy-2,2'-dimethyltubocuraran-7',12'-diol; d*-bebeerine; chondodendrine; pelosine. $C_{36}H_{38}N_2O_6$; mol wt 594.68. C 72.70%, H 6.44%, N 4.71%, O 16.14%. From bark of *Nectandra rodioei* Hook., *Lauraceae:* Maclagan, *Ann.* **48**, 106 (1843). From root of *Chondodendron microphyllum* (Eichl.) Moldenke and stem of *Ch. candicans* (Rick ex. D.C.) Sandwith, *Menispermaceae* and *radix pareirae bravae:* King, *J. Chem. Soc.* **1940**, 737. Structure: Faltis *et al., Ber.* **69**, 1269 (1936); King, *J. Chem. Soc.* **1939**, 1157. Stereochemistry: King, *ibid.* **1948**, 265.

Needles from methanol, mp 215°. $[\alpha]_D^{20}$ +345.7° (c = 0.4 in 1*N* HCl). Sol in benzene, chloroform, pyridine.

Hydrochloride, $C_{36}H_{38}N_2O_6 \cdot HCl$. mp 260°. $[\alpha]_D^{20}$ +294° (c = 0.7). Sol in water and alcohol.

THERAP CAT: Antimalarial.

1025. Bebeeru Bark. Greenheart; bibiru bark; sipiri bark. From *Nectandra rodioei* Hook., *Lauraceae. Habit.* British Guiana. *Constit.* Bebeerine, sipirine, bebeeric acid, tannin, resin.

1026. Becanthone. *1-[[2-[Ethyl(2-hydroxy-2-methyl-propyl)amino]ethyl]amino]-4-methyl-9H-thioxanthen-9-one;* becantone. $C_{22}H_{28}N_2O_2S$; mol wt 384.55. C 68.71%, H 7.34%, N 7.29%, O 8.32%, S 8.34%. Prepn: Blanz, French, *J. Med. Chem.* **6**, 185 (1963).

Hydrochloride, $C_{22}H_{29}ClN_2O_2S$, *Win 13820, Loranil.* mp 157.6-160.4°.

THERAP CAT: Anthelmintic (Schistosoma).

1027. Beclamide. *3-Chloro-N-(phenylmethyl)propanamide; N-benzyl-3-chloropropionamide; N-benzyl-β-chloropropanamide; benzchlorpropamide; chloroethylphenamide;* Chloracon; Hibicon; Neuracen; Nidrane; Nydrane; Posedrine; Seclar. $C_{10}H_{12}ClNO$; mol wt 197.66. C 60.76%, H 6.12%, N 7.09%, Cl 17.94%, O 8.09%. $ClCH_2CH_2CONH-CH_2C_6H_5$. Prepd by the action of β-chloropropionyl chloride on benzylamine in cooled water at pH 8: Cassell, Kushner, U.S. pat. **2,569,288** (1951 to Am. Cyanamid); Kushner *et al., J. Org. Chem.* **16**, 1283 (1951).

Large crystals from methanol, mp 94°. Slightly sol in water (0.005 to 0.01%). Moderately sol in the lower alcs. Dec in hot aq acid soln and in hot aq alkaline soln.

THERAP CAT: Anticonvulsant.

1028. Beclobrate. ($\pm$)-*2-[4-[(4-Chlorophenyl)methyl]-phenoxy]-2-methylbutanoic acid ethyl ester;* ($\pm$)-*2-methyl-2-[p-(p'-chlorobenzyl)phenoxy]butyric acid ethyl ester; ethyl* ($\pm$)-*2-[[α-(p-chlorophenyl)-p-tolyl]oxy]-2-methylbutyrate;* SGD 24774; Beclipur; Beclosclerin; Turec. $C_{20}H_{23}-ClO_3$; mol wt 346.85. C 69.26%, H 6.68%, Cl 10.22%, O 13.84%. Diphenylmethane derivative. Prepn: K. Thiele *et al.,* Ger. pat **2,461,069**; *eidem,* U.S. pat. **4,483,999** (1975, 1984 both to Siegfried); *eidem, Arzneimittel-Forsch.* **29**, 711 (1979). Preliminary clinical trial in hyperlipidemia: C. Najemnik *et al., ibid.* **31**, 2168 (1981). Series of articles on pharmacology: *ibid.* **33**, 1464-1472 (1983). Metabolism: W. Roth *et al., ibid.* **35**, 244 (1985). Pharmacokinetics: I. Gikalov, U. Ifflaender, *ibid.* **37**, 1065 (1987).

Oil, $bp_{0.01-0.1}$ 200-204°. LD_{50} orally in mice: 8000 mg/kg (Thiele, 1979).

THERAP CAT: Antihyperlipoproteinemic.

1029. Beclomethasone. *9-Chloro-11,17,21-trihydroxy-16-methylpregna-1,4-diene-3,20-dione; 9α-chloro-16β-methyl-1,4-pregnadiene-11β,17α,21-triol-3,20-dione; 9α-chloro-16β-methylprednisolone.* $C_{22}H_{29}ClO_5$; mol wt 408.93. C 64.62%, H 7.15%, Cl 8.67%, O 19.56%. Glucocorticoid. Prepn of free alcohol and 21-acetate: **Brit.** pat. **912,378** (1962 to Merck & Co.); of 21-acetate: **Brit.** pat. **901,093** (1962 to Schevico). Symposium on clinical studies: *Brit. J. Clin. Pharmacol.* **4**, Suppl. 3, 249S-312S (1977). Use in chronic asthma in children: M. Rao *et al., J. Asthma* **19**, 21 (1982); in treatment of asthma in steroid-independent adults: V. A. Malfitan, *Clin. Ther.* **4**, 472 (1982). Review of use in rhinitis: P. Small *et al., Ann. Allergy* **49**, 127 (1982); of pharmacology, side effects and use in asthma and allergic rhinitis: R. N. Brogden *et al., Drugs* **28**, 99-126 (1984).

Dipropionate, $C_{28}H_{37}ClO_7$, *Sch 18020W, Aldecin, Anceron, Andion, Beclacin, Becloforte, Beclomet, Beclorhinol, Becloval, Beclovent, Becodisks, Beconase, Beconasol, Becotide, Clenil-A, Entyderma, Inalone O, Inalone R, Korbutone, Propaderm, Rino-Clenil, Vancenase, Vanceril, Viarex, Viarox, Sanasthmax, Sanasthmyl.*

THERAP CAT: Antiallergic, antiasthmatic (inhalant). Topical anti-inflammatory.

1030. Beclotiamine. *3-[(4-Amino-2-methyl-5-pyrimidinyl)methyl]-5-(2-chloroethyl)-4-methylthiazolium chloride;* 5-chloroethylthiamine; chlorothiamine; clotiamine; CT; Cocciden. $C_{12}H_{16}Cl_2N_4S$; mol wt 319.25. C 45.15%, H 5.05%, Cl 22.21%, N 17.55%, S 10.04%. Antithiamine compound derived from thiamine; developed as an anticoccidial for chickens: Inone *et al., J. Japan. Vet. Med. Assoc.* **20,** 293 (1967); **Brit. pat. 1,141,055** and Matsui *et al.,* **Japan. pat. 14,276('71)** (1969 and 1971 to Sankyo), *C.A.* **71,** 111875z (1969); **75,** 36098a (1971); Matsuzawa, *Japan. J. Vet. Sci.* **34,** 157 (1972), *C.A.* **78,** 23930c (1973). Metabolism and mechanism of action studies: Shindo, Komai, *J. Vitaminol.* **18,** 41-62, 102, 172, 218 (1972). Safety studies: Matsuzawa *et al., Sankyo Kenkyusho Nempo* **23,** 192 (1971), *C.A.* **77,** 28983m (1972).

Hydrochloride, $C_{12}H_{17}Cl_3N_4S$. LD_{50} i.v. in mice, rats: 33.6, 60.5 mg/kg.

Naphthalene-1,5-disulfonate salt, $C_{22}H_{24}Cl_2N_4O_6S_3$, mp 276° (dec). Soly in water: 0.06 g/100 ml. LD_{50} i.p. in mice, rats: 294, 585 mg/kg.

THERAP CAT (VET): Coccidiostat.

1031. Beeswax. Yellow beeswax. A substance obtained from bee honeycombs. Consists of esters of straight-chain monohydric alcohols with even-numbered carbon chains from C_{24} to C_{36} esterified with straight-chain acids also having even numbers of C atoms up to C_{36} (some C_{18} hydroxy acids). Examples of such esters are triacontanol hexadecanoate and hexacosanol hexacosanoate. These esters are mixed with about 20% (w/w) of hydrocarbons having odd-numbered straight carbon chains from C_{21} to C_{33}. Propolis, pigments and unidentified substances amount to about 6%. Composition: D. T. Downing *et al., Aust. J. Chem.* **14,** 253 (1961); Callow, *Bee World* **44,** 95 (1963). Brief review: C. S. Letcher in Kirk-Othmer *Encyclopedia of Chemical Technology* Vol. **24** (Wiley-Interscience, New York, 3rd ed., 1984) pp 466-467.

Yellowish to brownish-yellow, soft to brittle; honey-like odor; slight balsamic taste. d 0.95-0.960. mp 62-65°. Saponification number 84. Acid number 20. Practically insol in water. Slightly sol in cold alc; sol in hot alc, chloroform, benzene, ether, carbon disulfide.

White beeswax, white wax, bleached yellow wax, bleached beeswax. Prepd by oxidizing yellow beeswax cakes with peroxide or in sunlight. Yellowish-white. Properties similar to those of yellow beeswax, except for a slightly different taste. Preferred to yellow beeswax in cosmetics.

USE: Manuf of wax paper, candles, cosmetics; modeling artificial fruits and flowers; in process engraving; shoe polish. Pharmaceutic aid (in ointments, plasters).

1032. Befunolol. *1-[7-[2-Hydroxy-3-[(1-methylethyl)-amino]propoxy]-2-benzofuranyl]ethanone;* 7-[2-hydroxy-3-(isopropylamino)propoxy]-2-benzofuranyl methyl ketone; 2-acetyl-7-(2-hydroxy-3-isopropylaminopropoxy)benzofuran. $C_{16}H_{21}NO_4$; mol wt 291.36. C 65.96%, H 7.26%, N 4.81%, O 21.97%. β-Adrenergic blocker. Prepn: K. Ito *et al.,* **Ger. pat. 2,223,184;** *eidem,* **U.S. pat. 3,853,923** (1972, 1974 both to Kakenyaku Kako). Prepn and activity of optical isomers: J. Nakano *et al., Chem. Pharm. Bull.* **36,** 1399 (1988). Determ of befunolol and metabolites in human plasma: K. Kawahara, T. Ofuji, *J. Chromatog.* **168,** 266

(1979). Absorption, distribution, excretion in rats: H. Kitagawa *et al., Oyo Yakuri* **17,** 383, 393 (1979), *C.A.* **91,** 13374w-5x (1979). General pharmacology: S. Masumoto *et al., Iyakuhin Kenkyu* **10,** 741 (1979), *C.A.* **92,** 88015s (1980). Pharmacodynamics: S. Harada *et al., Arch. Int. Pharmacodyn. Ther.* **252,** 262 (1981).

Crystals from cyclohexane/acetone, mp 115°. LD_{50} in mice: 100-105 mg/kg i.v. (Ito).

Hydrochloride, $C_{16}H_{22}ClNO_4$, *BFE 60, Benfuran, Bentos, Bentox, Glauconex.* Crystals from ethyl acetate, mp 163°.

(*S*)-(−)-Form hydrochloride, pale yellow prisms from isopropyl alcohol, mp 151-152°. $[\alpha]_D$ −15.5° (c = 1 in methanol).

(*R*)-(+)-Form hydrochloride, pale yellow prisms from isopropyl alcohol, mp 151°. $[\alpha]_D$ +15.3° (c = 1 in methanol).

THERAP CAT: Antiglaucoma agent.

1033. Behenic Acid. *Docosanoic acid.* $C_{22}H_{44}O_2$; mol wt 340.57. C 77.58%, H 13.02%, O 9.40%. $CH_3(CH_2)_{20}CO_2H$. Minor constituent of most seed fats, animal milk fats and marine animal oils. Large amts (~50%) are found in (hydrogenated?) jamba oil, mustard seed oil and rape oil: Sudborough *et al., J. Indian Inst. Sci.* **9A,** 25 (1926). Prepn from erucic acid by catalytic reduction: Morgan, Holmes, *J. Soc. Chem. Ind. (London) Trans. Commun.* **44,** 491 (1925). Brief review: E. S. Lower, *Mfg. Chem.* **56**(7), 44-46 (1985).

Waxy solid, mp 79.95°. Neutralization value 164.73. bp_{60} 306°. d_4^{100} 0.8221; n_D^{100} 1.4270. One hundred grams of 90% ethanol dissolve 0.102 g of behenic acid at 17°; at 25° 0.218 g of the acid dissolves in 100 ml of 91.5% ethanol, 0.116 g in 100 ml of 86.2% ethanol, 0.011 g in 100 ml of 63.07% ethanol. One hundred grams of ether dissolves 0.1922 g of behenic acid at 16°.

Methyl ester, $C_{23}H_{46}O_2$, mp 54°.

Ethyl ester, $C_{24}H_{48}O_2$, mp 50°. $bp_{0.2}$ 185°.

Amide, $C_{22}H_{45}NO$, mp 111-112°.

USE: In lubricating oils, as solvent evaporation retarder in paint removers. Amide as anti-foam in the manuf of detergents, in floor polishes, in dripless candles.

1034. Belladonna. Deadly nightshade; banewort; death's herb; dwale; poison black cherry. Dried leaves and root of *Atropa belladonna* L., *Solanaceae. Habit.* Southern and Central Europe, Asia Minor, Algeria; cultivated in North America. On extraction it yields atropine, hyoscyamine, scopolamine, asparagine, choline, chrysatropic acid, atroscine, leucatropic acid, phytosterol. Leaves usually contain 0.3-0.5% total alkaloids; dried root contains 0.4-0.7% total alkaloids. *See:* Romeike, *Pharmazie* **8,** 729 (1953); Phokas, Steinegger, ibid. **11,** 652 (1956); *Pharm. Acta Helv.* **31,** 284 (1956).

Human Toxicity: Toxicity based on content of atropine, *q.v.,* and related alkaloids.

THERAP CAT: Anticholinergic.

THERAP CAT (VET): Anticholinergic; antispasmodic.

1035. Belladonnine. *1,2,3,4-Tetrahydro-1-phenyl-1,4-naphthalenedicarboxylic acid bis(8-methyl-8-azabicyclo-[3.2.1]oct-3-yl) ester;* isatropylditropeine; tropyl isatropate; ditropyl isatropate. $C_{34}H_{42}N_2O_4$; mol wt 542.69. C 75.24%, H 7.80%, N 5.16%, O 11.79%. From *Atropa belladonna* L. and allied *Solanaceae.* Probably formed during the process of extraction. Upon hydrolysis the substance extracted from plants yields β-isatropic acid and tropine: Küssner, *Arch. Pharm.* **276,** 617 (1938); Hotovy *et al.,* **U.S. pat. 2,734,062** (1956 to E. Merck).

(tropine ester)
(tropine ester)

Crystals from ethyl acetate, mp 129°. Sparingly sol in water, petr ether. Sol in alcohol, benzene, chloroform, ethyl acetate.

Bisulfate monohydrate, $C_{34}H_{44}N_2O_8S.H_2O$, **Bellacristin.** Crystals, sol in water.

Bis[ethobromide] tetrahydrate, $C_{38}H_{52}Br_2N_2O_4.4H_2O$, **belladonnine bis[bromoethylate] tetrahydrate, C100.** A quaternary tetrahydrate salt, crystals from water + acetone, mp 98-101°. uv max: 258, 261 nm. $E_{1cm}^{1\%}$ 4.8. Sol in water, methanol, ethanol. Practically insol in ether, petr ether, benzene, ethyl acetate.

1036. Bemegride. *4-Ethyl-4-methyl-2,6-piperidinedione; 3-ethyl-3-methylglutarimide;* 4-ethyl-4-methyl-2,6-dioxopiperidine; 2,6-dioxo-4-methyl-4-ethylpiperidine; methetharimide; *β,β-methylethylglutarimide; NP 13;* Mikedimide; Eukraton; Malysol; Megimide. $C_8H_{13}NO_2$; mol wt 155.19. C 61.91%, H 8.44%, N 9.03%, O 20.62%. Prepn: Thole, Thorpe, *J. Chem. Soc.* **99,** 439 (1911); Sircar, *ibid.* **1937,** 602, 604; Benica, Wilson, *J. Am. Pharm. Assoc. [Sci. Ed.]* **39,** 451 (1950); Lukes, Ferles, *Chem. Listy* **49,** 510 (1955), *C.A.* **49,** 10290 (1955). Pharmacology: Delay *et al., Presse Med.* **64,** 1525 (1956); Oberdorf, Meyer, *Arch. Exp. Pathol. Pharmakol.* **238,** 128 (1960); Kretzschmar *et al., Arch. Int. Pharmacodyn. Ther.* **174,** 318 (1968). Metabolism: Nicholls, *Nature* **185,** 927 (1960). Reduction of barbiturate anesthesia in mice: V. K. Patel *et al., Psychopharmacol.* **71,** 21 (1980). Postsynaptic effects: P. W. Gage, P. Sah, *Brit. J. Pharmacol.* **75,** 493 (1982).

Platelets from water or from acetone + ether, mp 127°. Sublimes at 100° and 2 mm press. Sol in water, acetone. LD_{50} in mice, rats: 18.8, 17.0 mg/kg i.v.

THERAP CAT: Stimulant (central). Analeptic (in barbiturate poisoning).

THERAP CAT (VET): Analeptic and CNS stimulant. To counteract barbiturate poisoning.

1037. Benactyzine. *α-Hydroxy-α-phenylbenzeneacetic acid 2-(diethylamino)ethyl ester; benzilic acid β-diethylaminoethyl ester;* β-diethylaminoethyl benzilate; 2-diethylaminoethyl diphenylglycolate. $C_{20}H_{25}NO_3$; mol wt 327.41. C 73.36%, H 7.70%, N 4.28%, O 14.66%. An antagonist of acetylcholine in the central and peripheral nervous systems. Prepn: Horenstein, Pahlicke, *Ber.* **71,** 1654 (1938); Blicke, Maxwell, *J. Am. Chem. Soc.* **64,** 428 (1942); Hill, Holmes, U.S. pat. **2,394,770** (1946 to Am. Cyanamid). Toxicity and pharmacodynamics: Fournier, Petit, *Therapie* **17,** 1245 (1962). Metabolism: Eldeson *et al., Arch. Int. Pharmacodyn. Ther.* **187,** 139 (1970). Crystal and molecular structure determn by x-ray diffraction: T. J. Petcher, *J. Chem. Soc., Perkin Trans. II* **1974,** 1151.

Crystals, mp 51°.

Hydrochloride, $C_{20}H_{26}ClNO_3$, *AY-5406-1, Actozine, Amizil, Arcadine, Cafron, Cedad, Cevanol, Fobex, Ibiotyzil, Lucidil, Nervacton, Neuroleptone, Nutinal, Parasan, Parpon, Phobex, Suavitil, Tranquillin.* Crystals from acetone, mp 177-

178°. Soly in water (25°): 14.9/100 ml. Practically insol in ether.

Methobromide, $C_{21}H_{28}BrNO_3$, *Finalin, Spatomac.* Crystals from alcohol + ether, mp 169-170°.

THERAP CAT: Antidepressant; anticholinergic.

1038. Benalaxyl. *N-(2,6-Dimethylphenyl)-N-(phenylacetyl)-DL-alanine methyl ester;* methyl N-phenylacetyl-N-2,6-xylyl-DL-alaninate; N-(2,6-dimethylphenyl)-N-(1-carbomethoxyethyl)phenylacetamide; M 9834; Galben. $C_{20}H_{23}NO_3$; mol wt 325.41. C 73.82%, H 7.12%, N 4.30%, O 14.75%. Acylalanine fungicide systemically active against phytopathogens of the order *Peronosporales.* Prepn: E. Boson *et al.,* Ger. pat. **2,903,612;** *eidem,* U.S. pat. **4,425,357** (1978, 1984 both to Montedison). Activity and toxicology: P. Bergamaschi *et al., Proc. Brit. Crop Prot. Conf.-Pests Dis.* **1981,** 19. Residue accumulation in edible crops: P. Cabras *et al., J. Agr. Food Chem.* **33,** 86 (1985). GC-MS study: B. D. Ripley, *ibid.* 560.

Crystals from ligroin, mp 78-80°. LD_{50} in rats (mg/kg): 4200 orally; 1100 i.p. LC_{50} (96 hr) in rainbow trout: 3.75 mg/l (Bergamaschi).

USE: Agricultural fungicide.

1039. Benapryzine. *α-Hydroxy-α-phenylbenzeneacetic acid 2-(ethylpropylamino)ethyl ester; benzilic acid 2-(ethylpropylamino)ethyl ester;* 2-(ethyl-n-propylamino)ethyl α,α-diphenylglycolate. $C_{21}H_{27}NO_3$; mol wt 341.46. C 73.87%, H 7.97%, N 4.10%, O 14.06%. Prepn: **Neth. pat. Appl. 6,409,- 696** corresp to Mehta, Bainbridge, U.S. pat. **3,746,743** (1963, 1973 to Beecham). Pharmacology: Brown *et al., Nature* **223,** 416 (1969); Barthelemy *et al., Therapie* **27,** 369 (1972); Brown *et al., Brit. J. Pharmacol.* **47,** 476 (1973). Metabolism: Jeffery *et al., Xenobiotica* **1,** 169 (1971).

Hydrochloride, $C_{21}H_{28}ClNO_3$, *AP 1288, BRL 1288.* Rectangular plates from butanone, mp 164-166°. LD_{50} in mice: 400 mg/kg s.c.; 500 mg/kg orally.

THERAP CAT: Anticholinergic.

1040. Bence-Jones Proteins. Low molecular weight, heat-sensitive proteins isolated from urine of multiple myeloma patients: Caputo, *Aminoacidosi* **7,** 163-77 (1957), *C.A.* **54,** 6841 (1960). Used as a model in structural studies of immunoglobulins, they are structurally homogeneous within each patient and antigenically related. Crystallographic structural study: E. E. Aeola *et al., Biochemistry* **19,** 432 (1980). They constitute the light-chain component of myeloma globulin and have been implicated in the pathogenesis of amyloidosis, *cf.* G. G. Glenner, *N. Engl. J. Med.* **302,** 1283 (1980); G. G. Glenner *et al., Amyloid and Amyloidosis* (Excerpta Medica, Amsterdam, 1980) p 351. For structural studies and reviews *see* Immunoglobulins.

Coagulates on heating at 45-55°. Redissolves partially or wholly on boiling. Amino acid composition in g-%: Arginine 5.6, histidine 3.3, lysine 9.5, tyrosine 6.4, tryptophan 3.4, phenylalanine 4.2, cystine 2.6, methionine 1.9, serine 8.4, threonine 8.1, leucine 7.3, isoleucine 2.2, valine 9.8, glycine 8.6, alanine 3.1, proline 6.9, glutamic acid 7.5, aspartic acid 9.2. The molecule also shows a carbohydrate content of 9.4% (mostly glucose, mannose, and glucosamine).

1041. Bencyclane. *N,N-Dimethyl-3-[[1-(phenylmethyl)-cycloheptyl]oxy]-1-propanamine; 3-[(1-benzylcycloheptyl)-oxy]-N,N-dimethylpropylamine;* 1-benzyl-1-(3-dimethylaminopropoxy)cycloheptane; N-[3-(1-benzylcycloheptyloxy)-

propyl]-*N*,*N*-dimethylamine; benzcyclan. $C_{19}H_{31}NO$; mol wt 289.45. C 78.84%, H 10.80%, N 5.53%, O 4.84%. Prepn: Pallos *et al.*, **Hung.** pat. **151,865** (1965 to EGYT), *C.A.* **62,** 16125b (1965). Series of articles on pharmacology: *Arzneimittel-Forsch.* **20,** 1337-1460 (1970). Toxicity: E. Komlos, L. E. Petöcz, *ibid.* 1338.

bp₃ 146-156°.
Fumarate, $C_{23}H_{35}NO_5$, *EGYT 201, Angiociclan, Dantrium, Dilangio, Fludilat, Fluxema, Halidor, Vasorelax.* Crystals from 25% ethanol, mp 131-133°. Soly in water at 25°: 1 g/100 ml; in hot water: 2 g/100 ml. Poorly sol in acetone, easily in alc. uv spectrum: Simonyi *et al.*, *Acta Pharm. Hung.* **36,** 257 (1966). At pH 3.4-6.6, uv max (ethanol): 207 nm. LD_{50} in mice, rats (mg/kg): 445.6, 414 orally; 49.9, 41.3 i.v.; 132, 86.3 i.p.; 203, 257 s.c. (Komlos, Petöcz).
THERAP CAT: Vasodilator (peripheral, cerebral).

1042. Bendazac. *[[1-(Phenylmethyl)-1H-indazol-3-yl]oxy]acetic acid; [(1-benzyl-1H-indazol-3-yl)oxy]acetic acid;* bendazolic acid; bindazac; AF 983; Versus; Zildasac. $C_{16}H_{14}N_2O_3$; mol wt 282.30. C 68.08%, H 5.00%, N 9.92%, O 17.00%. Prepn from 1-benzylindazol-3-ol sodium and chloroacetonitrile and subsequent hydrolysis: Palazzo, Silvestrini, **S. Afr.** pat. **67 03,206**; Palazzo, U.S. pat. **3,470,194** (1967, 1969, both to Francesco Angelini). Pharmacology: Silvestrini *et al.*, *Inflammation Biochem. Drug Interaction, Proc. Int. Symp.* **1968,** A. Bertelli, Ed. (Excerpta Med., Amsterdam, 1970) p 283. Mechanism of action: Silvestrini *et al.*, *Arzneimittel-Forsch.* **20,** 250 (1970). Photoprotective capacity: Fuga *et al.*, *Ann. Ital. Dermatol. Sper.* **24,** 205 (1970). Use in treatment of cataracts: M. Testa *et al.*, *Lancet* **1,** 849 (1982). Toxicity: *Rx Bulletin* **3,** 147 (1972).

Crystals from ethanol, mp 160°. Practically insol in water; sol in chloroform, acetone. uv max: 306 nm ($E_{1cm}^{1\%}$ 191). LD_{50} in mice, rats (mg/kg): 380, 304 i.v.; 355, 388 i.p.; 440, 910 s.c.; 1105, ~1200 orally (*Rx Bulletin*).
THERAP CAT: Anti-inflammatory.

1043. Bendazol. *2-(Phenylmethyl)-1H-benzimidazole; 2-benzylbenzimidazole;* 2-benzylbenziminazole; bendazole; Tromasedan. $C_{14}H_{12}N_2$; mol wt 208.25. C 80.74%, H 5.81%, N 13.45%. Prepn: Walther, v. Pulawski, *J. Prakt. Chem.* [2] **59,** 253 (1899); Feitelson, Rothstein, *J. Chem. Soc.* **1958,** 2426; Partridge, Turner, *ibid.* **1958,** 2086.

Needles from benzene, mp 187°. Practically insol in water. Freely sol in glacial acetic acid. Sol in alcohol, hot benzene, propylene glycol.
Hydrochloride, $C_{14}H_{13}ClN_2$, *Dibasol, Dibasole.* Needles from dil HCl, mp 175° (Feitelson, Rothstein, *loc. cit.*).
THERAP CAT: Vasodilator.

1044. Bendiocarb. *2,2-Dimethyl-1,3-benzodioxol-4-ol methylcarbamate; methylcarbamic acid 2,3-(isopropylidenedioxy)phenyl ester;* NC 6897; Ficam. $C_{11}H_{13}NO_4$; mol wt 223.23. C 59.18%, H 5.87%, N 6.28%, O 28.67%. Prepn: Gates, Gillon, **S. Afr.** pat. **68 00736,** *C.A.* **71,** 38941m (1969), corresp to U.S. pat. **3,736,338** (1968, 1973, both to Fisons). Insecticidal activity: Story, *Int. Pest Contr.* **14,** 6 (1972).

White solid, mp 129-130°. Soly in water: 40 ppm; in hexane: 350 ppm. LD_{50} orally in mammals: 35-100 mg/kg.
USE: Contact insecticide.

1045. Bendroflumethiazide. *3,4-Dihydro-3-(phenylmethyl)-6-(trifluoromethyl)-2H-1,2,4-benzothiadiazine-7-sulfonamide 1,1-dioxide;* 3-benzyl-6-trifluoromethyl-3,4-dihydro-7-sulfamoyl-2H-1,2,4-benzothiadiazine 1,1-dioxide; 3-benzyl-3,4-dihydro-7-sulfamyl-6-trifluoromethyl-1,2,4-benzothiadiazine 1,1-dioxide; bendrofluazide; benzydroflumethiazide; benzylhydroflumethiazide; Aprinox; Benuron; Benzyl-Rodiuran; Berkozide; Bristuric; Bristuron; Centyl; Flumesil; Naturetin; Naturine; Neo-Naclex; Niagaril; Nikion; Orsile; Pluryle; Plusuril; Poliuron; Relan Beta; Salural; Salures; Sinesalin; Sodiuretic; Urlea. $C_{15}H_{14}F_3N_3O_4S_2$; mol wt 421.41. C 42.75%, H 3.35%, F 13.53%, N 9.97%, O 15.19%, S 15.22%. Prepn: Holdrege *et al.*, *J. Am. Chem. Soc.* **81,** 4807 (1959); Goldberg, U.S. pat. **3,265,573** (1966 to Squibb). Comprehensive description: K. Florey, F. M. Russo-Alesi, in *Analytical Profiles of Drug Substances* vol. **5,** K. Florey, Ed.(Academic Press, New York, 1976) pp 1-19.

Crystals, mp 221-223°. Also reported as mp 226-227°. uv max (methanol): 208, 273, 326 nm ($E_{1cm}^{1\%}$ 745, 565, 96). Insol in water, chloroform, benzene, ether. Sol in acetone, alcohol.
THERAP CAT: Diuretic, antihypertensive.

1046. Benexate Hydrochloride. *trans-2-[[[4-[[(Aminoiminomethyl)amino]methyl]cyclohexyl]carbonyl]oxy]benzoic acid phenylmethyl ester monohydrochloride;* benzyl salicylate *trans-4-(guanidinomethyl)cyclohexanecarboxylate* hydrochloride; (2'-benzyloxycarbonyl)phenyl *trans-4-(guanidinomethyl)cyclohexanecarboxylate* hydrochloride. $C_{23}H_{28}ClN_3O_4$; mol wt 445.95. C 61.95%, H 6.33%, Cl 7.95%, N 9.42%, O 14.35%. Synthetic protease inhibitor with antiulcer activity. Prepn: **Belg.** pat. **885,263**; M. Muramatsu *et al.*, U.S. pat. **4,348,410** (1981, 1982 to Nippon Chemiphar; Teikoku Chem. Ind.). Prepn and protease inhibition: T. Satoh *et al.*, *Chem. Pharm. Bull.* **33,** 647 (1985). Prepn of the clathrate compound with β-cyclodextrin: **Japan. Kokai 83 38,250**; M. Shinoda *et al.*, U.S. pat. **4,478,995** (1983, 1984 both to Teikoku Chem. Ind.). Effect on gastric secretion and exptl ulcers in rats: S. Okabe *et al.*, *Oyo Yakuri* **27,** 829 (1984), *C.A.* **101,** 143881c (1984); I. Tanaka, H. Tagami, *Nippon Yakurigaku Zasshi* **85,** 167 (1985), *C.A.* **103,** 64660t (1985); F. Hirose *et al.*, *Yakuri to Chiryo* **15,** 4749 (1987), *C.A.* **108,** 137769a (1988).

Crystals from methanol + ether, mp 83°.
Compd with β-cyclodextrin (1:1), $C_{65}H_{98}ClN_3O_{39}$, *TA-903, benexate-CD, Lonmiel, Ulgut.*
THERAP CAT: Antiulcerative.

1047. Benfluorex. *2-[[1-Methyl-2-[3-(trifluoromethyl)-phenyl]ethyl]amino]ethanol benzoate (ester); 2-[[α-methyl-m-(trifluoromethyl)phenethyl]amino]ethanol benzoate (ester);*

1-(m-trifluoromethylphenyl)-2-(β-benzoyloxyethyl)amino-propane; N-(2-benzoyloxyethyl)norfenfluramine; benflura-mate; S 780; Se 780; Minolip. $C_{19}H_{20}F_3NO_2$; mol wt 351.38. C 64.94%, H 5.74%, F 16.22%, N 3.99%, O 9.11%. Prepn from 1-(m-trifluoromethyl)-2-(β-hydroxyethyl)aminopro-pane and benzoyl chloride: L. Beregi et al., Fr. pat. **1,517,-587** corresp to U.S. pat. **3,607,909** (1968, 1971, both to Sci. Union et Cie, Soc. Franc. Rech. Med.). Metabolism studies: A. H. Beckett et al., J. Pharm. Pharmacol. **23**, 950 (1971); **24**, 281 (1972). Pharmacology: D. N. Brindley et al., ibid. **28**, 670 (1976); P. Pritchard et al., ibid. **29**, 343 (1977); ei-dem, Biochem. J. **166**, 639 (1977).

Colorless oil.
Hydrochloride, $C_{19}H_{21}ClF_3NO_2$, S 992, JP 992, Mediator, Mediaxal. Crystals from ethyl acetate, mp 161-162°.
THERAP CAT: Antihyperlipoproteinemic.

1048. Benfluralin. *N-Butyl-N-ethyl-2,6-dinitro-4-(tri-fluoromethyl)benzenamine; N-butyl-N-ethyl-α,α,α-trifluoro-2,6-dinitro-p-toluidine; N-butyl-N-ethyl-2,6-dinitro-4-tri-fluoromethylaniline; benefin; bethrodine; EL-110; Balan; Balfin; Benefex; Quilan.* $C_{13}H_{16}F_3N_3O_4$; mol wt 335.29. C 46.57%, H 4.81%, F 17.00%, N 12.53%, O 19.09%. Selective pre-emergence herbicide. Prepn: Q. F. Soper, U.S. pat. **3,257,190** (1966 to Lilly). Activity: E. F. Alder et al., Proc. Northeast Weed Control Conf. **15**, 298 (1961). Environ-mental fate: T. Golab et al., J. Agr. Food Chem. **18**, 838 (1970). Soil degradn: J. H. Miller et al., Weed Sci. **23**, 211 (1975); R. L. Zimdahl, S. M. Gwynn, ibid. **25**, 247 (1977).

Yellow-orange cryst solid, mp 65-66.5°. Vapor press at 30°: 3.89×10^{-4} mm Hg. Soly in water at 25°: < 1 mg/l. Sol in most organic solvents; less sol in ethanol. Decomp in uv light. LD_{50} orally in female rats: > 10,000 mg/kg, E. I. Goldenthal, Toxicol. Appl. Pharmacol. **18**, 185 (1971).
USE: Herbicide.

1049. Benfotiamine. *Benzenecarbothioic acid S-[2-[[(4-amino-2-methyl-5-pyrimidinyl)methyl]formylamino]-1-[2-(phosphonooxy)ethyl]-1-propenyl] ester; thiobenzoic acid S-ester with N-[(4-amino-2-methyl-5-pyrimidinyl)methyl]-N-(4-hydroxy-2-mercapto-1-methyl-1-butenyl)formamide O-phosphate; S-benzoylthiamine monophosphate; 6-(2-methyl-6-amino-5-pyrimidinyl)-5-aza-5-formyl-4-methyl-3-benzoylthio-3-hexenyl phosphate; Biotamin; Neurostop; Vitanevril.* $C_{19}H_{23}N_4O_6PS$; mol wt 466.47. C 48.92%, H 4.97%, N 12.01%, O 20.58%, P 6.64%, S 6.88%. Prepn: Ito et al., Ger. pat. **1,130,811** (1962), C.A. **57**, 13764h (1962); Sunagawa et al., Japan. pats. **13,481-4('62)** (all to Sankyo), C.A. **59**, 11519d-h (1963). Changes in crystal system (α-, γ- and δ-forms): Hamanaka et al., Japan. pat. **16,042('62)** (to Sankyo), C.A. **59**, 94b (1963); Ito et al., C.A. **59**, 3920a (1963). Conversion of γ-form to δ-form: Hamanaka et al., Japan. pat. **11,040('62)** (to Sankyo), C.A. **59**, 10077g (1963).

Crystals, dec 165°, 160-162°, 155-160°, 195° (δ-form).
THERAP CAT: Vitamin B_1 source.

1050. Benfuracarb. *2-Methyl-4-(1-methylethyl)-7-oxo-8-oxa-3-thia-2,4-diazadecanoic acid 2,3-dihydro-2,2-di-methyl-7-benzofuranyl ester; 2,3-dihydro-2,2-dimethylben-zofuran-7-yl N-(N-isopropyl-N-ethoxycarbonylethylami-nosulfenyl)-N-methylcarbamate; 2,3-dihydro-2,2-dimethyl-7-benzofuranyl N-[N-[2-(ethoxycarbonyl)ethyl]-N-iso-propylsulfenamoyl]-N-methylcarbamate; ethyl N-[2,3-di-hydro-2,2-dimethylbenzofuran-7-yloxycarbonyl(methyl)-aminothio]-N-isopropyl-β-alaninate; OK 174; Oncol.* $C_{20}H_{30}N_2O_5S$; mol wt 410.53. C 58.51%, H 7.37%, N 6.82%, O 19.49%, S 7.81%. Broad spectrum soil and foliar insecticide; cholinesterase inhibitor. Prepn and insecticidal properties: A. Tanaka et al., Belg. pat. **890,162**; T. Goto et al., U.S. pat. **4,413,005** (1982, 1983 both to Otsuka). Biological activity and field trials: T. Goto et al., Proc. 10th Conf. Int. Congr. Plant Prot. **1**, 360 (1983). Absorption, metabolism in plants: A. Tanaka et al., J. Agr. Food Chem. **33**, 1049 (1985). Brief review, field trials: P. Cagnieul, Def. Veg. **38**, 3-10 (1984).

Viscous brown-red liquid. d^{20} 1.17. Almost insol in wa-ter (8 mg/l at 20°). Sol in organic solvents. LD_{50} in male rats, mice, dogs (mg/kg): 138, 175, 300 orally (Goto).
USE: Insecticide.

1051. Benfurodil Hemisuccinate. *Butanedioic acid mono[1-[5-(2,5-dihydro-5-oxo-3-furanyl)-3-methyl-2-ben-zofuranyl]ethyl] ester; succinic acid monoester with 4-[2-(1-hydroxyethyl)-3-methyl-5-benzofuranyl]-2(5H)-furanone; 2-[1-(succinoyloxy)ethyl]-3-methyl-5-(2-oxo-2,5-dihydro-4-furyl)benzo[b]furan; 2-(1-hydroxyethyl)-β-(hydroxy-methyl)-3-methyl-5-benzofuranacrylic acid γ-lactone hy-drogen succinate; benzofurodil; Eucilat; Eudilat.* $C_{19}H_{18}O_7$; mol wt 358.33. C 63.68%, H 5.06%, O 31.25%. Prepn: J. Schmitt, **Fr.** pat. **1,408,721** corresp to U.S. pat. **3,355,463** (1965, 1967 to Clin-Byla). Prepn and toxicity: J. Schmitt et al., Bull. Soc. Chim. France **1967**, 74. Synthesis and phar-macology: eidem, Chim. Ther. **1**, 305 (1966).

White to pale yellow crystals from ethyl acetate, mp 144°. Sol in alkaline solns. LD_{50} in mice: 550 mg/kg orally (Schmitt).
THERAP CAT: Cardiotonic, vasodilator.

1052. Benmoxine. *Benzoic acid 2-(1-phenylethyl)hydra-zide; 1-(benzoyl)-2-(α-methylbenzyl)hydrazine; 1-(α-meth-ylbenzyl)-2-benzoylhydrazine; Neuralex.* $C_{15}H_{16}N_2O$; mol wt 240.29. C 74.97%, H 6.71%, N 11.66%, O 6.66%. Prepn: Bettinetti, Farmaco Ed. Sci. **16**, 823 (1961); **Fr.** pat. **1,314,-362** (1963 to I.C.I.), C.A. **59**, 2716a (1963), corresp to **Brit.** pat. **919,491**.

Crystals from isopropyl ether or petr ether + benzene, mp 93-94°.
THERAP CAT: Antidepressant.

1053. Benomyl. *[1-[(Butylamino)carbonyl]-1H-benz-imidazol-2-yl]carbamic acid methyl ester; 1-(butylcarbamo-yl)-2-benzimidazolecarbamic acid methyl ester; methyl 1-*

(butylcarbamoyl)-2-benzimidazolecarbamate; F-1991; Benlate. $C_{14}H_{18}N_4O_3$; mol wt 290.32. C 57.92%, H 6.25%, N 19.30%, O 16.53%. Prepn: C. D. Adams, R. Schlatter, **Ger. pat. 1,956,157,** corresp to U.S. pat. **3,738,995** (1970, 1973, both to du Pont). The degradation product, methyl 2-benzimidazolecarbamate, is thought to be the active component; *see* carbendazim. Antifungal activity and mode of action studies: Kilgore, White, *Bull. Environ. Contam. Toxicol.* **5,** 67 (1970); Maxwell, Brody, *Appl. Microbiol.* **21,** 944 (1971); Bartels-Schooley, MacNeill, *Phytopathology* **61,** 816 (1971). Activity as catalyst in biological oxidation of sewage and fertilizers: Kouba, U.S. pat. **3,649,530** (1972 to du Pont). Toxicity: E. W. Schafer, *Toxicol. Appl. Pharmacol.* **21,** 315 (1972). Metabolism: J. A. Gardiner *et al., J. Agr. Food Chem.* **22,** 419 (1974). Mutagenicity study: J. P. Seiler, *Mutat. Res.* **32,** 151 (1975); H. Sherman *et al., Toxicol. Appl. Pharmacol.* **32,** 305 (1975). *Review:* W. E. Bleidner *et al., Anal. Methods Pestic. Plant Growth Regul.* **10,** 157-171 (1978).

CONH(CH₂)₃CH₃ [structure]

White cryst solid. Sol in chloroform; insol in water or oil. LD_{50} orally in rats: >9590 mg/kg (Schafer).

USE: Fungicide; ascaricide.

THERAP CAT (VET): Anthelmintic.

1054. Benorylate. *2-(Acetyloxy)benzoic acid 4-(acetylamino)phenyl ester; salicylic acid acetate, ester with 4-hydroxyacetanilide;* 2-acetoxy-4'-(acetamino)phenylbenzoate; *p*-acetamidophenyl acetylsalicylate; 4'-(acetamido)phenyl-2-acetoxybenzoate; fenasprate; Win 11450; Benoral; Benortan; Quinexin; Salipran. $C_{17}H_{15}NO_5$; mol wt 313.32. C 65.17%, H 4.83%, N 4.47%, O 25.53%. Prepn: **Neth. pat. Appl. 6,504,517** (1965 to Sterwin), *C.A.* **64,** 8097c (1966), corresp to **Brit. pat. 1,101,747** (1968); Robertson, U.S. pat. **3,431,293** and Miller, **Brit. pat. 1,168,289** (both 1969 to Sterling Drug). Pharmacological investigations: Rosner *et al., Therapie* **23,** 525 (1968); Raab, *Arzneimittel-Forsch.* **21,** 1662 (1971). Pharmacokinetic studies in animals: Liss, Palme, *ibid.* **19,** 1177 (1969). Clinical evaluation: Bain, Burt, *Clin. Trials J.* **7,** 307 (1970); Cardoe, *ibid.* 313.

[structure] OCOCH₃ / COO / NHCOCH₃

Crystals from methanol or ethanol, mp 175-176°. LD_{50} in mice, rats (mg/ml): 2000, ~10,000 orally; 1255, 1830 i.p. (Robertson).

THERAP CAT: Analgesic, anti-inflammatory, antipyretic.

1055. Benoxaprofen. *2-(4-Chlorophenyl)-α-methyl-5-benzoxazoleacetic acid;* Compound 90459; Coxigon; Opren; Oraflex; Uniprofen. $C_{16}H_{12}ClNO_3$; mol wt 301.74. C 63.69%, H 4.01%, Cl 11.75%, N 4.64%, O 15.91%. Prepn: D. Evans *et al.,* **Ger. pat. 2,324,443** corresp to U.S. pat. **3,912,748** (1973, 1975 both to Lilly); D. W. Dunwell *et al., J. Med. Chem.* **18,** 53 (1975). Analysis by HPLC: S. W. McKay *et al., J. Chromatog.* **170,** 482 (1979). Pharmacokinetics: D. H. Chatfield *et al., Brit. J. Clin. Pharmacol.* **4,** 579 (1977). Metabolism: D. H. Chatfield, J. N. Green, *Xenobiotica* **8,** 133 (1978). Effect on lysosomal enzyme secretion: R. J. Smith, *J. Pharmacol. Exp. Ther.* **207,** 618 (1978). Dissolution and absorption in humans: A. S. Ridolfo *et al., J. Pharm. Sci.* **68,** 850 (1979). Series of clinical studies: *Rheumatol. Rehabil.* **17,** 254-264 (1978).

[structure with Cl, O, N, CHCOOH, CH₃]

Cream solid from ethanol, mp 189-190°. LD_{50} in mice: 800 mg/kg orally, D. Evans *et al., loc. cit.*

THERAP CAT: Anti-inflammatory; analgesic.

1056. Benoxinate. *4-Amino-3-butoxybenzoic acid 2-(diethylamino)ethyl ester;* 3-butoxy-4-aminobenzoic acid 2-(diethylamino)ethyl ester; 2-(diethylamino)ethyl 4-amino-3-*n*-butoxybenzoate; oxibuprokain; oxybuprocaine. $C_{17}H_{28}N_2O_3$; mol wt 308.41. C 66.20%, H 9.15%, N 9.08%, O 15.56%. Prepn: **Brit. pat. 654,484** (1951 to Wander).

[structure] COOCH₂CH₂N(C₂H₅)₂ / OCH₂CH₂CH₂CH₃ / NH₂

bp₂ 215-218°.

Hydrochloride, $C_{17}H_{29}ClN_2O_3$, *Conjuncain, Cebesine, Dorsacaine, Novesine, Benoxil, Lacrimin.* Crystals, mp about 155° (*U.S.P.* **XVI,** 1960), also reported as 157-160° (*N.N.R.* 1955). Very sol in water, chloroform; sol in alc; practically insol in ether. pH of aq solns, 4.5-5.2.

THERAP CAT: Anesthetic (topical).

1057. Benperidol. *1-[1-[4-(4-Fluorophenyl)-4-oxobutyl]-4-piperidinyl]-1,3-dihydro-2H-benzimidazol-2-one; 1-[1-[3-(p-fluorobenzoyl)propyl]-4-piperidyl]-2-benzimidazolinone;* 1-[1-[4-(p-fluorophenyl)-4-oxobutyl]piperidin-4-yl]-2-benzimidazolinone; benzperidol; McN-JR 4584; R 4584; Anquil; Frénactil; Frenactyl; Glianimon. $C_{22}H_{24}FN_3O_2$; mol wt 381.45. C 69.27%, H 6.34%, F 4.98%, N 11.02%, O 8.39%. Prepn: **Belg. pat. 626,307** (1963 to Janssen), *C.A.* **60,** 10690c (1964), corresp. to **Brit. pat. 989,755.**

[structure] CH₂CH₂CH₂-C(=O)-F / N / N / O / NH

Solid, mp 170-171.8°.

Hydrochloride hydrate, solid, mp 134-142°.

THERAP CAT: Antipsychotic.

1058. Benproperine. *1-[1-Methyl-2-[2-(phenylmethyl)-phenoxy]ethyl]piperidine; 1-[1-methyl-2-[(α-phenyl-o-tolyl)-oxy]ethyl]piperidine;* 1-[2-(2-benzylphenoxy)-1-methylethyl]piperidine. $C_{21}H_{27}NO$; mol wt 309.43. C 81.51%, H 8.80%, N 4.53%, O 5.17%. Prepn: **Brit. pat. 914,008** corresp to Rubinstein, U.S. pat. **3,117,059** (1962, 1964 to Aktieselskabet Pharmacia). Pharmacology: Yamatsu *et al., Japan. J. Pharmacol.* **17,** 538 (1967); Tellini, De Fina, *Boll. Chim. Farm.* **109,** 476 (1970).

[structure] CH₃ / N-CHCH₂O / CH₂C₆H₅

Liquid. bp₀.₂ 159-161°.

Trihydrogen phosphate, $C_{21}H_{30}NPO_5$, *Blascorid, Pirexyl.*

Crystals from abs ethanol, mp 150-152°. LD$_{50}$ in albino mice: 192 mg/kg i.p.; 1365 mg/kg orally.

Pamoate, C$_{65}$H$_{70}$N$_2$O$_8$, *benzproperine ambonate, Tussafug.*

THERAP CAT: Antitussive.

1059. Benserazide. DL-*Serine 2-[(2,3,4-trihydroxyphenyl)methyl] hydrazide; N-(DL-seryl)-N'-(2,3,4-trihydroxybenzyl) hydrazine.* C$_{10}$H$_{15}$N$_3$O$_5$; mol wt 257.25. C 46.69%, H 5.88%, N 16.33%, O 31.10%. Peripheral decarboxylase inhibitor. Prepn: **Belg. pat. 619,015;** Hegedüs, Zeller, **U.S. pat. 3,178,476** (1962, 1965 both to Hoffmann-La Roche). Inhibition of dopa decarboxylase *in vitro* and *in vivo:* W. P. Burkard *et al., Experientia* **18,** 411 (1962); *eidem, Arch. Biochem. Biophys.* **107,** 187 (1964). Synergy with dopa *in vivo:* G. Bartholini *et al., Nature* **215,** 852 (1967); G. Bartholini, A. Pletscher, *J. Pharmacol. Exp. Ther.* **161,** 14 (1968). Clinical comparison with carbidopa, *q.v.:* I. Kuruma *et al., J. Pharm. Pharmacol.* **24,** 289 (1972). Clinical and pharmacokinetic profile: M.-H. Marion *et al., Adv. Neurol.* **45,** 493 (1987). Review of pharmacology and clinical efficacy of decarboxylase inhibitors: R. M. Pinder *et al., Drugs* **11,** 329-377 (1976).

HOCH$_2$CHCONHNHCH$_2$

Hydrochloride, C$_{10}$H$_{16}$ClN$_3$O$_5$, *Ro 4-4602.* White crystalline powder, mp 146-148°. Sol in water.

Combination with levodopa, *Madopar.*

THERAP CAT: In combination with levodopa as antiparkinsonian.

1060. Bentazon. *3-(1-Methylethyl)-1H-2,1,3-benzothiadiazin-4(3H)-one 2,2-dioxide; 3-isopropyl-1H-2,1,3-benzothiadiazin-4(3H)-one 2,2-dioxide;* bentazone; bendioxide; Basagran. C$_{10}$H$_{12}$N$_2$O$_3$S; mol wt 240.28. C 49.99%, H 5.03%, N 11.66%, O 19.98%, S 13.34%. Selective post-emergence herbicide. Prepn: A. Zeidler *et al., S. Afr. pat.* **67 05,164** corresp to U.S. pat. **3,708,277** (1968, 1973, both to BASF); G. Hamprecht *et al., U.S. pat.* **3,822,257** and **Ger. pat. 2,357,063,** *C.A.* **83,** 131644z (1974, 1975, both to BASF). Fate, metabolism and toxicity in a model ecosystem: G. M. Booth *et al., J. Environ. Qual.* **2,** 408 (1973).

White crystalline powder, mp 137-139°. Soly (w/w) at 20°: water 0.05%; acetone 150.7%; benzene 3.3%; chloroform 18%; ethanol 86.1%. LD$_{50}$ orally in rats: 1100 mg/kg, *RTECS vol.* II, E. J. Fairchild, Ed. (1977) p 206.

USE: Herbicide.

1061. Bentiromide. *(S)-4-[[2-(Benzoylamino)-3-(4-hydroxyphenyl)-1-oxopropyl]amino]benzoic acid; (S)-p-(α-benzamido-p-hydroxyhydrocinnamamido)benzoic acid; N-benzoyl-L-tyrosyl-p-aminobenzoic acid;* BTPABA; E-2663; Chymex; PFT Roche. C$_{23}$H$_{20}$N$_2$O$_5$; mol wt 404.42. C 68.31%, H 4.98%, N 6.93%, O 19.78%. Synthetic chymotrypsin-labile peptide, used in diagnosis of exocrine pancreatic disease. Prepn: P. L. De Benneville, N. J. Greenberger, **Ger. pat. 2,156,835;** *eidem,* **U.S. pat. 3,745,212** (1972, 1973 both to Rohm & Haas). Synthesis, *in vitro* and *in vivo* data: P. L. De Benneville *et al., J. Med. Chem.* **15,** 1098 (1972). Early clinical studies: K. Gyr *et al., Schweiz. Med. Wochenschr.* **105,** 1717 (1975); W. Bornschein *et al., Clin. Chim. Acta* **67,** 21 (1976). Pediatric study: G. Dockter *et al., Eur. J. Pediatrics* **135,** 277 (1981). Human toxicity study, effects of renal insufficiency on pancreatic function test: C. Lang *et al., J. Clin. Chem. Clin. Biochem.* **18,** 551 (1980).

Cryst from methanol/water, mp 240-242°. [α]$_D^{25}$ +72.3° (c = 1 in DMF) (U.S. pat.); also reported as [α]$_D^{25}$ +87° (c = 1 in DMF) (De Benneville).

THERAP CAT: Diagnostic aid (pancreatic function).

1062. Bentonite. Wilkinite. A colloidal native hydrated aluminum silicate (clay) found in the midwest of the U.S.A. and in Canada. Consists principally of montmorillonite, Al$_2$O$_3$.4SiO$_2$.H$_2$O. Usually contains some magnesium, iron, and calcium carbonate. *Review:* J. Alexander, *Ind. Eng. Chem.* **16,** 1140 (1924).

The color in the massive condition varies from yellowish-white to almost black. The powder is cream colored to pale brown. It has the property of forming highly viscous suspensions or gels with not less than ten times its weight of water. The property of forming gels is very much increased by the addition of small amounts of alkaline substances such as magnesium oxide.

USE: As of Fuller's earth; as emulsifier for oils; as a base for plasters. Pharmaceutic aid (suspending agent).

1063. β-Benzalbutyramide. *3-Methyl-4-phenyl-3-butenamide; 3-methyl-4-phenyl-3-butenoic acid amide;* β-benzylidenebutyramide; Kata-Lipid. C$_{11}$H$_{13}$NO; mol wt 175.22. C 75.40%, H 7.48%, N 7.99%, O 9.13%. Prepn of the acid: Canonica *et al., Farmaco Ed. Sci.* **14,** 112 (1959); Datta, Bagchi, *J. Org. Chem.* **25,** 932 (1960). Clinical studies in atherosclerotic patients: F. Del Regno, G. Capobianco, *Minerva Med.* **54,** 2917 (1963); A. Iacobelli *et al., G. Arterioscler.* **4,** 277 (1966); M. G. Piccardo *et al., ibid.* 294. Metabolism: L. Canonica *et al., J. Biol. Chem.* **243,** 1645 (1968). Inhibition of cholesterol biosynthesis: D. Giorgini, G. Porcellati, *Farmaco Ed. Sci.* **24,** 392 (1969). Clinical evaluation in hyperlipidemia: U. Marini *et al., Farmaco Ed. Prat.* **36,** 532 (1981).

Crystals, mp 132-134°. uv max: 249 nm (log ε 4.1).

THERAP CAT: Antihyperlipoproteinemic.

1064. Benzal Chloride. *(Dichloromethyl)benzene; benzylidene chloride;* benzyl dichloride; α,α-dichlorotoluene; benzylene chloride. C$_7$H$_6$Cl$_2$; mol wt 161.03. C 52.21%, H 3.76%, Cl 44.04%. C$_6$H$_5$CHCl$_2$. Obtained by chlorination of toluene. Lab procedure: A. I. Vogel, *Practical Organic Chemistry* (Longmans, London, 3rd ed., 1959) p 539; Gattermann-Wieland, *Praxis des organischen Chemikers* (de Gruyter, Berlin, 40th ed., 1961) p 184.

Very refractive liquid; fumes in air; vapors irritate the eyes; pungent odor. d 1.26. bp 205°. mp −17°. Insol in water; freely sol in alcohol, ether.

USE: Manuf benzaldehyde, cinnamic acid.

1065. Benzaldehyde. Benzoic aldehyde; artificial essential oil of almond. C$_7$H$_6$O; mol wt 106.12. C 79.22%, H 5.70%, O 15.08%. Occurs in kernels of bitter almonds; made synthetically from benzal chloride and lime or by oxidation of toluene. Laboratory prepn from benzal chloride: A. I. Vogel, *Practical Organic Chemistry* (Longmans, London, 3rd ed., 1959) p 693; Gattermann-Wieland, *Praxis des organischen Chemikers* (de Gruyter, Berlin, 40th ed., 1961) p 184. Toxicity data: P. M. Jenner *et al., Food Cosmet. Toxicol.* **2,** 327 (1964). *Review:* A. E. Williams in Kirk-Othmer *Encyclopedia of Chemical Technology* **vol. 3** (Wiley-Interscience, New York, 3rd ed., 1978) pp 736-743.

Strongly refractive liquid, becoming yellowish on keeping; characteristic odor of volatile oil of almond; burning aromatic taste. Oxidizes in air to benzoic acid; volatile with steam. d $_4^{15}$ 1.050; 1.043 at 25°. bp 179°. mp −56.5°. Flash pt 62°. n_D^{20} 1.5456. Sol in 350 parts water; miscible with alcohol, ether, oils. It reduces ammoniacal $AgNO_3$, but not Fehling's soln. *Keep tightly closed and protected from light.* LD$_{50}$ in rats, guinea pigs (mg/kg): 1300, 1000 orally (Jenner).

Note: Benzaldehyde FFC designates a grade of benzaldehyde free from chlorine.

USE: Manufacture of dyes, perfumery, cinnamic and mandelic acids, as solvent; in flavors. *Caution:* Narcotic in high concns. May cause contact dermatitis.

1066. Benzalkonium Chloride. Benirol; BTC; Capitol; Cequartyl; Drapolene; Drapolex; Enuclen; Germinol; Germitol; Osvan; Paralkan; Roccal; Rodalon; Zephiran Chloride; Zephirol. A mixture of alkyldimethylbenzylammonium chlorides of the general formula in which R represents a mixture of the alkyls from C_8H_{17} to $C_{18}H_{37}$. Acute toxicity data: L. M. Cummins, E. T. Kimura, *Toxicol. Appl. Pharmacol.* **20,** 89 (1971). Review of antimicrobial activity and use: W. Gump, "Disinfectants and Antiseptics" in Kirk-Othmer *Encyclopedia of Chemical Technology* vol. 7 (Wiley-Interscience, New York, 3rd ed., 1979) pp 815-818.

White or yellowish-white, amorphous powder or gelatinous pieces. Aromatic odor, very bitter taste. Very sol in water, alcohol, acetone; slightly sol in benzene. Almost insol in ether. The aq soln is slightly alkaline to litmus and foams strongly when shaken. LD$_{50}$ orally in rats: 400 mg/kg (Cummins, Kimura).

Incompatible with anionic detergents, such as soap, and with nitrates. A white precipitate is formed in a 1:3000 aq soln of benzalkonium chloride when nitrates are present in concns greater than the equiv. of 0.5% ammonium nitrate.

Dibromide, *Callusolve.* Adduct of benzalkonium chloride with bromine which is insol in water and non-ionized.

USE: Cationic surface active agent and germicide. Pharmaceutic (preservative).

THERAP CAT: Topical antiseptic.

THERAP CAT (VET): Antiseptic for skin preoperatively or for wounds, burns, etc. Udder wash.

1067. Benzamide. Benzoylamide. C_7H_7NO; mol wt 121.13. C 69.41%, H 5.82%, N 11.56%, O 13.21%. C_6H_5-$CONH_2$. Prepd from benzoyl chloride and ammonium carbonate. Lab prepn: Gattermann-Wieland, *Praxis des organischen Chemikers* (de Gruyter, Berlin, 40th ed., 1961) p 119. Alternate procedure using concd ammonia soln: A. I. Vogel, *Practical Organic Chemistry* (Longmans, London, 3rd ed., 1959) p 797.

Crystals. d^4 1.341. mp 130°. bp 288°. One gram dissolves in 74 ml water, more sol in boiling water, in 6 ml alc, 3.3 ml pyridine. Sol in hot benzene, slightly in ether; sol in ammonia with formation of a small quantity of benzonitrile.

N-chloro deriv, mp 116°. Sol in water; insol in alcohol or benzene.

1068. Benzanilide. *N*-Phenylbenzamide; *N*-benzoylaniline. $C_{13}H_{11}NO$; mol wt 197.23. C 79.16%, H 5.62%, N 7.10%, O 8.11%. $C_6H_5CONHC_6H_5$. Prepd by the treatment of aniline with benzoic acid: Hübner, *Ann.* **208,** 291 (1881); Nägeli, *Bull. Soc. Chim.* [3] **11,** 892 (1894); Webb, *Org. Syn.* **7,** 6 (1927); **coll. vol. I** (2nd ed., 1941) p 82.

Leaflets from alc. d 1.315. mp 163°. Sublimes. Distills

without decompn. bp$_{10}$ 117-119°. Absorption spectrum: Crymble, *J. Chem. Soc.* **99,** 459 (1911). Insoluble in water. One gram dissolves in 60 ml alc, 7 ml boiling alc. Slightly sol in ether.

USE: Manuf dyes and perfumes.

1069. 1,2-Benzanthracene. *Benz[a]anthracene;* 2,3-benzphenanthrene; tetraphene; benzanthrene; naphthanthracene. $C_{18}H_{12}$; mol wt 228.28. C 94.70%, H 5.30%. Occurs in coal tar: Cook *et al., J. Chem. Soc.* **1933,** 395. Synthesis from naphthalene and phthalic anhydride: Elbs, *Ber.* **19,** 2209 (1886). From *o*-toluylnaphthalene: Fieser, Dietz, *Ber.* **62,** 1827 (1929). From phenanthrene and succinic anhydride: Haworth, Mavin, *J. Chem. Soc.* **1933,** 1012. Absorption spectrum: Capper, Marsh, *J. Chem. Soc.* **1926,** 726; Clar, *Ber.* **65,** 507 (1932); Mayneord, Roe, *Proc. Roy. Soc. London* A152, 299 (1935). *Review:* E. Clar, *Polycyclic Hydrocarbons* (Academic Press, New York, 1964) 2 vols.

Plates from glacial acetic acid or alc. Greenish-yellow fluorescence. Sublimes. mp 155-157° (Fieser, Dietz); mp 160° [Fieser, Hershberg, *J. Am. Chem. Soc.* **59,** 2502 (1937)]; mp 167° (I. G. Farbenind., **Ger.** pats. **481,819; 486,766).** Difficultly sol in boiling alc; sol in most other organic solvents. Insol in water.

Note: This substance may reasonably be anticipated to be a carcinogen: *Fourth Annual Report on Carcinogens* (NTP 85-002, 1985) p 33.

1070. Benzanthrone. *7H-Benz[de]anthracen-7-one.* C_{17}-$H_{10}O$; mol wt 230.25. C 88.67%, H 4.38%, O 6.95%. Prepd by heating a reduction product of anthraquinone with sulfuric acid and glycerol: Macleod, Allen, *Org. Syn.* **14,** 4 (1934); *cf.* U.S. pat. **1,626,392** (1927).

Pale yellow needles from alcohol or xylene. mp 170°. Absorption spectrum: Clar, *Ber.* **65,** 846 (1932). Solubility at 20° = 0.52 g in 100 g glacial acetic acid; 1.61 g in 100 g benzene; 2.05 g in 100 g chlorobenzene. Solution in H_2SO_4 is orange with green fluorescence.

1071. Benzarone. *(2-Ethyl-3-benzofuranyl)(4-hydroxyphenyl)methanone; 2-ethyl-3-benzofuranyl p-hydroxyphenyl ketone;* 2-ethyl-3-(4-hydroxybenzoyl)benzofuran; 2-ethyl-3-(*p*-hydroxybenzoyl)coumarone; L 2197; Fragivix; Fragivil; Vasoc; Venagil. $C_{17}H_{14}O_3$; mol wt 266.28. C 76.67%, H 5.30%, O 18.03%. Prepn: Buu-Hoi, Beaudet, U.S. pat. **3,012,042** (1961 to Soc. Belge l'Azote Prod. Chim. Marly). Pharmacological studies: Barchewitz *et al., Arzneimittel-Forsch.* **22,** 553 (1972). Site of action studies: I. Filipovic, E. Buddecke, *ibid.* **29,** 1578 (1979); L. O. Zwillenberg *et al., ibid.* **32,** 1114 (1982). HPLC determination in serum: H. Vergin, G. Bishop, *J. Chromatog.* **183,** 383 (1980).

Crystals, mp 124.3°.

THERAP CAT: Capillary protectant.

1072. Benzathine. *N,N'-Dibenzylethylenediamine; N,N'-bis(phenylmethyl)-1,2-ethanediamine;* 1,2-bis(benzylamino)ethane; DBED. $C_{16}H_{20}N_2$; mol wt 240.34. C 79.95%, H 8.39%, N 11.66%. Prepd from N,N'-dibenzenesulfonyl-N,N'-dibenzylethylenediamine by heating with concd HCl in bomb tube at 170-180°: Bleier, *Ber.* **32**, 1829 (1899); from ethylene chloride and benzylamine: Frost *et al., J. Am. Chem. Soc.* **71**, 3842 (1949); by reduction of N,N'-dibenzylideneethylenediamine: **Ger. pat. 98,031** (1898 to Schering); *Chem. Zentr.* **1898**, II, 743; Van Alphen, *Rec. Trav. Chim.* **54**, 93 (1935); Lob, *ibid.* **55**, 859 (1936); Szabo *et al., Antibiot. & Chemother.* **1**, 499 (1951); Rebenstorf, **U.S. pat. 2,773,098** (1956 to Upjohn).

$$CH_2NH-CH_2C_6H_5$$
$$|$$
$$CH_2NH-CH_2C_6H_5$$

Oily liquid, mp 26°. d_4^{20} 1.024. bp_{12} 212-213°; bp_4 195°. n_D^{20} 1.5624. Insol in water. Freely sol in the usual organic solvents except CS_2, with which it gives a solid addition product.
Diacetate, $C_{16}H_{20}N_2 \cdot 2CH_3COOH$, needles, mp 111°. Soly in water: 253 mg/ml.
Dihydrobromide, $C_{16}H_{20}N_2 \cdot 2HBr$, shiny leaflets, dec 300°. Soly in water: 30 mg/ml.
Dihydrochloride, $C_{16}H_{20}N_2 \cdot 2HCl$, shiny leaflets, dec 298°. Soly in water: 23.9 mg/ml.
Sulfate, $C_{16}H_{20}N_2 \cdot H_2SO_4$, crystals, dec 247-250°. Soly in water: 15.8 mg/ml.
Disalicylate, $C_{16}H_{20}N_2 \cdot 2C_6H_5(OH)COOH$, crystals, mp 85°. Soly in water: 2.43 mg/ml.
Penicillin salt, benzathine penicillin, *see* separate entries.
USE: Manuf of a repository form of penicillin.

1073. Benzbromarone. *(3,5-Dibromo-4-hydroxyphenyl)-(2-ethyl-3-benzofuranyl)methanone; 3,5-dibromo-4-hydroxyphenyl 2-ethyl-3-benzofuranyl ketone;* 3-(3,5-dibromo-4-hydroxybenzoyl)-2-ethylbenzofuran; 2-ethyl-3-benzofuranyl 4-hydroxy-3,5-dibromophenyl ketone; 2-ethyl-3-(3,5-dibromo-4-hydroxybenzoyl)benzofuran; 2-ethyl-3-(3,5-dibromo-4-hydroxybenzoyl)oxaindene; 2-ethyl-3-benzofuryl 3,5-dibromo-4-hydroxyphenyl ketone; L 2214; MJ 10061; Azubromaron; Besuric; Desuric; Max-Uric; Minuric; Narcaricin; Normurat; Uricovac; Urinorm. $C_{17}H_{12}Br_2O_3$; mol wt 424.11. C 48.15%, H 2.85%, Br 37.68%, O 11.32%. Prepn: **Belg. pat. 553,621** (1957 to Labaz); Buu-Hoi *et al., J. Chem. Soc.* **1957**, 625; Buu-Hoi, Beaudet, **U.S. pat. 3,012,042** (1961 to Soc. Belge l'Azote Prod. Chim. Marly). Structure-activity study: Delbarre *et al., Chim. Ther.* **3**, 470 (1968). Pharmacology of hypouricemic effect: D. S. Sinclair, I. H. Fox, *J. Rheumatol.* **2**, 437 (1975). Pharmacokinetic and clinical studies: T. F. Yu, *ibid.* **3**, 305 (1976). Pharmacokinetics and biotransformation in man: H. Ferber *et al., Eur. J. Clin. Pharmacol.* **19**, 431 (1981). HPLC determn in serum: H. Vergin *et al., J. Chromatog.* **183**, 383 (1980).

Yellowish prisms, mp 151°.
THERAP CAT: Uricosuric.

1074. Benzene. Benzol; cyclohexatriene. C_6H_6; mol wt 78.11. C 92.25%, H 7.75%. Discovered by Faraday in compressed oil gas in 1825. Obtained in the coking of coal and in the production of illuminating gas from coal. Purification by washing with water: **Brit. pat. 863,711** (1961 to Schloven-Chemie and H. Koppers GmbH), *C.A.* **55**, 16971f (1961). Lab prepn from aniline: Gattermann-Wieland, *Praxis des organischen Chemikers* (de Gruyter, Berlin, 40th ed., 1961) p 247. Production of pure benzene: French, *Ind. Chemist* **39**, 9-12 (1963). Manuf: Faith, Keyes & Clark's *Industrial Chemicals,* F. A. Lowenheim, M. K. Moran, Eds. (Wiley-Interscience, New York, 4th ed., 1975) pp 126-137. Physical properties: Thorne *et al., Ind. Eng. Chem. Anal. Ed.* **17**, 481 (1945). Solubility studies: F. P. Schwarz, *Anal.*

Chem. **52**, 10 (1980). Toxicity data: Kimura *et al., Toxicol. Appl. Pharmacol.* **19**, 699 (1971). Review of toxicology: E. Browning, *Toxicity and Metabolism of Industrial Solvents* (Elsevier, New York, 1965) pp 3-65; R. Snyder *et al., Rev. Biochem. Toxicol.* **3**, 123-154 (1981). *Review:* W. P. Purcell in Kirk-Othmer *Encyclopedia of Chemical Technology* **vol. 3** (Wiley-Interscience, New York, 3rd ed., 1978) pp 744-771.

Clear, colorless, highly flammable liquid; characteristic odor. d_4^{15} 0.8787. bp 80.1°. mp +5.5°. n_D^{20} 1.50108. Flash pt, closed cup: 12°F (−11°C). Soly in water at 23.5°C (w/w): 0.188%. Miscible with alcohol, chloroform, ether, carbon disulfide, carbon tetrachloride, glacial acetic acid, acetone, oils. *Keep in well-closed containers in a cool place and away from fire.* LD_{50} orally in young adult rats: 3.8 ml/kg (Kimura).
Sodium deriv, C_6H_5Na, *phenyl sodium.* Prepn: Schlosser, *Angew. Chem.* **76**, 267 (1964). Solid mass, dec by water, acids, alkalies. Sol in liquid ammonia, tetrahydrofuran.
Human Toxicity: Acute (from ingestion or inhalation): Iirritation of mucous membranes, restlessness, convulsions, excitement, depression. Death may follow from respiratory failure. *Chronic:* Bone marrow depression and aplasia; rarely, leukemia. Harmful amts may be absorbed through skin. Benzene has been listed as a known carcinogen: *Fourth Annual Report on Carcinogens* (NTP 85-002, 1985) p 34.
USE: Manuf of medicinal chemicals, dyes and many other organic compounds, artificial leather, linoleum, oil cloth, airplane dopes, varnishes, lacquers; as solvent for waxes, resins, oils, etc.
THERAP CAT (VET): Destroys screwworm larvae in wounds.

1075. Benzenearsonic Acid. Phenylarsonic acid. $C_6H_7AsO_3$; mol wt 202.03. C 35.67%, H 3.49%, As 37.08%, O 23.76%. $C_6H_5AsO(OH)_2$.
Cryst powder. mp 158-162° with dec. Sol in 40 parts water, 50 parts alcohol; insol in chloroform.
USE: Reagent for tin.

1076. Benzeneboronic Acid. *Phenylboronic acid;* phenylboric acid; phenylboron dihydroxide. $C_6H_7BO_2$; mol wt 121.94. C 59.10%, H 5.79%, B 8.87%, O 26.24%. Prepd by the reaction of phenylmagnesium bromide with methyl borate: Washburn *et al., Advances in Chemistry Series* **23**, 102-128 (1959). Crystal and molecular structure: S. J. Rettig, J. Trotter, *Can. J. Chem.* **55**, 3071 (1977).

Crystals from water. Spontaneous conversion to C_6H_5BO, *benzene boronic anhydride* or *phenylboroxide* on standing in dry air, although the best method is by azeotropic dehydration with toluene. mp 215-216° (anhydride). Ka 13.7. Dipole moment 1.72 (dioxane). Soly in water at 25°: 2.5%, benzene: 1.75%, xylene: 1.2%, ether: 30.2%, methanol: 178%.

1077. Benzenestibonic Acid. *Dihydroxyphenylstibine oxide;* phenylstibonic acid; phenylstibinic acid. $C_6H_7O_3Sb$; mol wt 248.87. C 28.95%, H 2.84%, O 19.29%, Sb 48.92%. $C_6H_5SbO(OH)_2$. Arylstibonic acids exist in the solid state as polymers of high mol wt, associated by H-bonds. The ions exist probably in the form $[ArSb(OH)_5]^-$. Benzenestibonic acid is prepd from benzenediazonium chloride and antimony trioxide in the presence of alkali: Schmidt, *Ann.* **421**, 176, 188 (1920); **Ger. pat. 254,421;** *Chem. Zentr.* **1913**, I, 345; *Frdl.* **11**, 1084. From the double compd of benzenediazonium chloride and antimony trichloride by the action of alkali: May, *J. Chem. Soc.* **101**, 1033 (1912); **U.S. pat. 1,260,707.**
Minute crystals from acetic acid, larger diamond-shaped, glistening crystals from alcohol-chloroform mixture on the addition of water. Not melted at 250°, dec at higher temps. Insol in water; slightly sol in alc; more sol in acetone, in

alcohol + benzene, alcohol + ether, alcohol + chloroform; sol in warm chloroform or amyl acetate; freely sol in aq soln of sodium hydroxide, sodium carbonate, and ammonia.

1078. Benzenesulfonic Acid. $C_6H_6O_3S$; mol wt 158.17. C 45.56%, H 3.82%, O 30.35%, S 20.27%. $C_6H_5SO_3H$. Made by treating benzene with fuming H_2SO_4: Gattermann-Wieland, *Praxis des organischen Chemikers* (de Gruyter, Berlin, 40th ed., 1961) p 168.

Sesquihydrate, deliquescent plates, tablets, mp 43-44°. When anhydrous mp 50-51°, also reported as 65-66°. K at 25° = 2 × 10⁻¹. Freely sol in water, alc; slightly sol in benzene; insol in ether, carbon disulfide. *Keep well closed.*

Ethyl ester, $C_8H_{10}O_3S$, *ethyl benzenesulfonate.* Colorless to slightly yellow, almost odorless liq. bp₁₅ 156°. d₄¹⁷ 1.219. Slightly sol in water; misc with alc, benzene, chloroform, ether.

Sodium salt, $C_6H_5NaO_3S$, *sodium benzenesulfonate, sodium benzosulfonate.* Crystals. Sol in water.

USE: Manuf phenol by fusion with NaOH. *Caution:* Highly irritant to skin, eyes, mucous membranes.

1079. Benzenesulfonic Anhydride. $C_{12}H_{10}O_5S_2$; mol wt 298.33. C 48.31%, H 3.38%, O 26.82%, S 21.50%. $(C_6H_5-SO_2)_2O$. Prepd by heating benzenesulfonic acid with excess phosphorus pentoxide mixed with inert support: Field, *J. Am. Chem. Soc.* **74,** 394 (1952).

Light tan solid, mp 60-85° (softens about 55°); may be recrystallized from ether, mp 88-91° (softens about 75°). Liquefies upon exposure to air for 2.5 hrs. Explosion occurs when mixed with 90-95% H_2O_2.

USE: In Friedel-Crafts sulfone synthesis; in other sulfonylation reactions.

1080. Benzenesulfonyl Chloride. Benzene sulfonechloride; benzenesulfonic (acid) chloride. $C_6H_5ClO_2S$; mol wt 176.62. C 40.80%, H 2.85%, Cl 20.08%, O 18.12%, S 18.15%. Prepd from benzene and chlorosulfonic acid or from the sodium salt of benzenesulfonic acid and PCl₅ or POCl₃: R. Adams, C. S. Marvel, *Org. Syn. coll. vol. I,* (2nd ed., 1964) p 84; H. T. Clarke *et al., ibid.* p 85.

Colorless, oily liquid. d₁₅¹⁵ 1.3842. Solidifies 0°, mp 14.5°. bp₇₆₀ 251-252° (decompn); bp₁₀₀ 177°; bp₁₀ 120°. Insol in and stable toward cold water; sol in ether, alc.

1081. 1,2,4-Benzenetriol. Hydroxyhydroquinone; hydroxyquinol. $C_6H_6O_3$; mol wt 126.11. C 57.14%, H 4.80%, O 38.06%. Prepd by hydrolysis of the triacetate with H_2SO_4 in methanol, the triacetate being formed by the action of acetic anhydride on quinone: Vliet, *Org. Syn.* **coll. vol. I** (2nd ed, 1941) p 317.

Monoclinic prismatic leaflets from ether, mp 141°. Freely sol in water, alcohol, ether, ethyl acetate; almost insol in chloroform, carbon disulfide, ligroin, benzene.

Trimethyl ether, $C_9H_{12}O_3$, bp 247°.

Triacetate, $C_{12}H_{12}O_6$; needles from abs alc, mp 96.5-97.0°. Readily hydrolyzed by acids or alkalies.

USE: In gas analysis; 1,2,4-benzenetriol in alkaline soln is just as good an absorbent for oxygen as is pyrogallol.

1082. Benzestrol. *4,4'-(1,2-Diethyl-3-methyl-1,3-propanediyl)bisphenol;* 3-ethyl-2,4-bis(p-hydroxyphenyl)hexane; Octofollin; Ocestrol; Chemestrogen. $C_{20}H_{26}O_2$; mol wt 298.41. C 80.49%, H 8.78%, O 10.72%. Eight isomers are possible. The compound accepted for medical use and des-

ignated benzestrol is the racemic mixture "B-2" of Stuart. Synthesis: Stuart *et al., J. Am. Chem. Soc.* **68,** 729 (1946). Improved method: Kiprianov, Kutsenko, *Zh. Prikl. Khim. (Leningrad)* **31,** 665 (1958).

Crystals, mp 162-166°. Freely sol in acetone, ether, ethanol, methanol and dil sodium hydroxide soln; sol in vegetable oils; moderately sol in glacial acetic acid; slightly sol in benzene, chloroform, petr ether, and dil ethanol; practically insol in water and in dil mineral acids. Neutral to litmus in 75% alc soln.

Dibenzoate, crystals from ethanol, mp 118-120°.

Dimethyl ether, crystals from ethanol, mp 56°.

THERAP CAT: Estrogen.

1083. Benzethonium Chloride. *N,N-Dimethyl-N-[2-[2-[4-(1,1,3,3-tetramethylbutyl)phenoxy]ethoxy]ethyl]benzenemethanaminium chloride; benzyldimethyl[2-[2-(p-1,1,3,3-tetramethylbutylphenoxy)ethoxy]ethyl]ammonium chloride;* diisobutylphenoxyethoxyethyl dimethyl benzyl ammonium chloride; Hyamine 1622; Phemerol chloride; Phemeride; Phemithyn; Quatrachlor; Solamin. $C_{27}H_{42}ClNO_2$; mol wt 448.10. C 72.37%, H 9.45%, Cl 7.91%, N 3.13%, O 7.14%. Prepn: U.S. pats. **2,115,250** (1938); **2,170,111** (1939); **2,229,024** (1941).

Monohydrate, thin, hexagonal plates from chloroform + ether. Sinters slightly at 120°; mp 164-166° (hot stage). Very sol in water giving a foamy, soapy soln. Sol in alcohol, acetone, chloroform. The pH of a 1% aq soln is between 4.8 and 5.5. Mineral acids and many salt solns precipitate benzethonium chloride from solns more concd than 2% as an oil which crystallizes on drying. Incompatible with soap, anionic detergents. LD₅₀ in rats: 420 mg/kg.

Human Toxicity: Ingestion may cause vomiting, collapse, convulsions, coma.

THERAP CAT: Topical anti-infective.

THERAP CAT (VET): Topical antiseptic.

1084. Benzetimide. *3-Phenyl-3-[1-(phenylmethyl)-4-piperidinyl]-2,6-piperidinedione; 2-(1-benzyl-4-piperidyl)-2-phenylglutarimide;* 1-benzyl-4-(2,6-dioxo-3-phenyl-3-piperidyl)piperidine. $C_{23}H_{26}N_2O_2$; mol wt 362.45. C 76.21%, H 7.23%, N 7.73%, O 8.83%. Prepn: Janssen, U.S. pat. **3,125,578** (1964 to Janssen); Hermans *et al., J. Med. Chem.* **11,** 797 (1968). Synthesis of labelled compound: van Wijngaarden, Soudijn, *J. Label. Compounds* **1,** 207 (1965). Resolution of isomers and pharmacology: van Wijngaarden, *Life Sci.* **8,** 517 (1969); *ibid.* **9,** part 1, 489 (1970); Janssen *et al., Arzneimittel-Forsch.* **21,** 1365 (1971). Metabolism: van Wijngaarden, Soudijn, *Life Sci.* **7,** 225 (1968).

Crystals, mp 156-159°.

Hydrochloride, $C_{23}H_{27}ClN_2O_2$, *R4929, Dioxatrine, Spasmentral.* Crystals, mp 299-301.5°. LD₅₀ i.v. in rats, mice: 37.6, 46.0 mg/kg.

l-Form, mp 180.5-182° (from toluene). [α]D²⁰ −124° (chloroform). LD₅₀ i.v. in mice: 38.5 mg/kg.

d-Form, *see* Dexetimide.
THERAP CAT: Hydrochloride as an anticholinergic.
THERAP CAT (VET): Antidiarrheal.

1085. Benzhydrylamine. α-*Phenylbenzenemethanamine;*
1,1-diphenylmethylamine; α-aminodiphenylmethane. C_{13}-$H_{13}N$; mol wt 183.24. C 85.20%, H 7.15%, N 7.64%. Prepd by boiling benzophenone oxime with zinc dust and ammonia: Scholl, *Ber.* **60**, 1247 (1927).

$$C_6H_5-CH-C_6H_5$$
$$\overset{NH_2}{|}$$

Hexagonal plates from water, mp 34°. d_4^{22} 1.0635 (supercooled liq). bp_{763} 304.1°; bp_{12} 166°. n_D^{99} 1.59631 (supercooled liq). Strong base. Absorbs CO_2 from the air forming a substance that melts at 91°. Slightly sol in water.
Hydrochloride, $C_{13}H_{13}N.HCl$, needles, mp 293°. Sparingly sol in cold water.

1086. Benzidine. *[1,1'-Biphenyl]-4,4'-diamine; p-*diaminodiphenyl. $C_{12}H_{12}N_2$; mol wt 184.23. C 78.23%, H 6.56%, N 15.21%. Discovered by Zinin in 1845. Produced by reduction of nitrobenzene with Zn and NaOH; the resultant hydrazobenzene is heated with acid. Lab proc: A. I. Vogel, *Practical Organic Chemistry* (Longmans, London, 3rd ed., 1959) p 633; Gattermann-Wieland, *Praxis des organischen Chemikers* (de Gruyter, Berlin, 40th ed., 1961) p 165. Review of carcinogenicity studies: *IARC Monographs* **1**, 80-86 (1972).

$$H_2N-\langle\rangle-\langle\rangle-NH_2$$

White or slightly-reddish, cryst powder; darkens on exposure to air and light. *Poisonous!* mp 115-120° when slowly heated; 128° when anhydr and rapidly heated. bp about 400°. Soly: one gram dissolves in 2500 ml cold, 107 ml boiling water, 5 ml boiling alcohol, 50 ml ether. *Keep well closed and protected from light.*
Dihydrochloride, $C_{12}H_{12}N_2.2HCl$, crystals. Sol in water, alcohol.
Toxicity: Solid and vapor rapidly absorbed through skin; on ingestion may produce nausea, vomiting, liver and kidney damage. This substance has been listed as a known carcinogen: *Fourth Annual Report on Carcinogens* (NTP 85-002, 1985) p 37.
USE: Manuf dyes; as a reagent for H_2O_2 in milk and for detection of blood. Dihydrochloride used for quantitative determn of sulfates, and as reagent for metals.

1087. Benzil. *Diphenylethanedione; bibenzoyl;* dibenzoyl; diphenylglyoxal; diphenyl-α,β-diketone. $C_{14}H_{10}O_2$; mol wt 210.22. C 79.98%, H 4.80%, O 15.22%. $C_6H_5COCOC_6H_5$. Prepd by the oxidation of benzoin, $C_6H_5CH(OH)COC_6H_5$, with HNO_3 or with a copper sulfate-pyridine mixture: Adams, Marvel, *Org. Syn.* **vol. I**, p 25 (1921); Clarke, Dreger, *ibid.* **coll. vol. I**, 80 (87, 2nd ed); Hatt *et al., J. Chem. Soc.* **1936**, 93; L. F. Fieser, *Experiments in Organic Chemistry* (Boston, 3rd ed, 1955) p 173; *Organic Experiments* (Boston, 1964) p 214.
Yellow prisms (trigonal trapezohedral) from alcohol. d_4^{15} 1.23. mp 95°. bp_{760} 346-348°; bp_{12} 188°. Absorption spectrum: Hantzsch, Schwiete, *Ber.* **49**, 216 (1916). uv max (ethanol): 260 nm (ϵ 22,000). Infrared in chloroform: 5.93; 6.22; 6.85 μ. Insol in water; sol in alcohol, ether, chloroform, ethyl acetate, benzene, toluene, nitrobenzene.
α-Benzilmonoxime, $C_{14}H_{11}NO_2$, leaflets from 30% alc, mp 137-138°.
β-Benzilmonoxime, $C_{14}H_{11}NO_2$, solvated needles from benzene, when dry mp 113-114°.
Disemicarbazone, $C_{16}H_{16}N_6O_2$, leaflets from alc, dec 243-244°.
USE: In organic syntheses.

1088. Benzil Dioxime. *Diphenylethanedione dioxime; diphenylglyoxime.* $C_{14}H_{12}N_2O_2$; mol wt 240.25. C 69.99%, H 5.03%, N 11.66%, O 13.32%. Three isomers occur: α or *anti*, β or *syn*, γ or *amphi*. Prepn of α-form from benzil and

hydroxylamine hydrochloride: Brady, Perry, *J. Chem. Soc.* **127**, 2874 (1925); F. J. Welcher, *Organic Analytical Reagents* vol. III (Van Nostrand, New York, 1947) pp 224-227; Boyer *et al., J. Am. Chem. Soc.* **77**, 5688 (1955). Prepn of β-form: Brady, Perry, *loc. cit.;* Boyer *et al., loc. cit.* Prepn of γ-form: Welcher, *loc. cit.,* pp 227-228; *see also* Boyer *et al., loc. cit.*

$$C_6H_5-C-C-C_6H_5 \qquad C_6H_5-C-C-C_6H_5$$
$$\underset{HON\quad NOH}{} \qquad \underset{NOH\quad HON}{}$$
$$\text{α-form} \qquad\qquad \text{β-form}$$

α-Form, crystals from methanol, mp 238-240°, Boyer *et al., loc. cit.,* also reported as mp 243-244°, Meisenheimer, Lamparter, *Ber.* **57**, 276 (1924). Practically insol in water, ether, glacial acetic acid; slightly sol in alc; readily sol in NaOH solns.
β-Form, crystals, mp 212-214°, Meisenheimer, Lamparter, *loc. cit.*

1089. Benzilic Acid. α-*Hydroxy-*α*-phenylbenzeneacetic acid; diphenylglycolic acid.* $C_{14}H_{12}O_3$; mol wt 228.24. C 73.67%, H 5.30%, O 21.03%. Prepd from benzil by the action of concd aq or alc KOH: Adams, Marvel, *Org. Syn.* **vol. 1**, p 29 (1921); Ballard, Dehn, *ibid.* **coll. vol. I**, 82; Kao, Ma, *J. Chem. Soc.* **1931**, 443.

$$\langle\rangle-\overset{OH}{\underset{COOH}{C}}-\langle\rangle$$

Monoclinic needles from water. Bitter taste. mp 150°. Melt is deep red at higher temp. K at 25° = 9.2×10^{-4}. Slightly sol in cold water; freely sol in hot water, alc, ether.
Potassium salt, $C_{14}H_{11}O_3K$, crystals, very sol in water, alc.
Lead salt, $(C_{14}H_{11}O_3)_2Pb$, amorphous precipitate. Upon heating it becomes a red liquid.
Methyl ester, $C_{15}H_{14}O_3$, mp 74-75°. bp_{13} 187°.

1090. Benzilonium Bromide. *1,1-Diethyl-3-[(hydroxydiphenylacetyl)oxy]pyrrolidinium bromide;* 1-ethyl-3-pyrrolidinyl benzilate ethyl bromide; 3-benzoyloxy-1,1-diethylpyrrolidinium bromide; benzilic acid 1-ethyl-3-pyrrolidinyl ester ethyl bromide; CI-379; PU-239; Minelsin; Ortyn retard; Minelcin; Portyn; Ulcoban. $C_{22}H_{28}BrNO_3$; mol wt 434.39. C 60.83%, H 6.50%, Br 18.40%, N 3.23%, O 11.05%. Prepd from 1-ethyl-3-hydroxypyrrolidine and ethyl benzilate followed by reaction with ethyl bromide: Bowman *et al.,* **Brit.** pat. 821,436 (1956 to Parke, Davis); Ryan, Ainsworth, *J. Org. Chem.* **27**, 2901 (1962).

$$\left[\begin{array}{c} \overset{C_2H_5\quad C_2H_5}{\underset{+}{N}} \\ \text{(pyrrolidinium ring)} \\ OH \\ (C_6H_5)_2C-COO \end{array}\right] Br^-$$

Crystals, mp 203-204°. LD_{50} orally in rats: 1.86 g/kg, Reichertz, Schliva, *Arzneimittel-Forsch.* **12**, 414 (1962).
THERAP CAT: Anticholinergic.

1091. Benzimidazole. Benziminazole; 1,3-benzodiazole; azindole; benzoglyoxaline; *N,N'*-methenyl-*o*-phenylenediamine. $C_7H_6N_2$; mol wt 118.13. C 71.16%, H 5.12%, N 23.72%. Prepd by the reaction of *o*-phenylenediamine with formic acid: Wundt, *Ber.* **11**, 826 (1878); Wagner, Millett, *Org. Syn.* **coll. vol. II**, 65 (1943). *Reviews:* J. B. Wright, *Chem. Rev.* **48**, 397-541 (1951); K. Hofmann, *Imidazole and its Derivatives* (Interscience, New York, 1953) p 247 sqq.

Consult the cross index before using this section.

Tabular crystals. Orthorhombic and monoclinic modifications. mp 170.5°. bp$_{760}$ above 360°. Weak base. pKa = 5.48 at 25°. Absorption spectrum: Steck *et al.*, *J. Am. Chem. Soc.* **70**, 3406 (1948). Dipole moment: 3.93 (dioxane). Sparingly sol in cold water, more sol in hot water. Freely sol in alcohol, sparingly sol in ether. Practically insol in benzene, petr ether. One gram dissolves in 2 g of boiling xylene. Sol in aq solns of acids and strong alkalies. High degree of chemical stability.

1092. 2-Benzimidazolethiol. *1,3-Dihydro-2H-benzimidazole-2-thione;* 2-mercaptobenzimidazole. $C_7H_6N_2S$; mol wt 150.22. C 55.97%, H 4.03%, N 18.65%, S 21.35%. Prepn from o-phenylenediamine and potassium ethyl xanthate: Van Allan, Deacon, *Org. Syn.* **coll. vol. IV,** 569 (1963).

Glistening platelets from 95% ethanol, mp 303-304°. Slightly sol in water; sol in methanol, ethanol.

1093. Benziodarone. *2-Ethyl-3-benzofuranyl 4-hydroxy-3,5-diiodophenyl methanone;* 2-ethyl-3-(3',5'-diiodo-4'-hydroxybenzoyl)benzofuran; 2-ethyl-3-(3',5'-diiodo-4'-hydroxybenzoyl)oxaindene; 2-ethyl-3-benzofuryl 3',5'-diiodo-4'-hydroxyphenyl ketone; 2-ethyl-3-(3,5-diiodo-4-hydroxyphenyl)coumarone; 2-ethyl-3-(4-hydroxy-3,5-diiodobenzoyl)benzofuran; 2329 Labaz; L 2329; Dilafurane; Dila-Vasal; Algocor; Amplivix; Cardivix; Retrangor. $C_{17}H_{12}I_2O_3$; mol wt 518.11. C 39.41%, H 2.34%, I 48.99%, O 9.26%. Description: Charlier, *Acta Cardiol.*, Suppl. 7, 1-60 (1959). Synthesis: Beaudet, Henaux, **Belg.** pat. **553,621; Brit.** pat. **836,272** (1957, 1960, both to Labaz); Buu-Hoi, Beaudet, **U.S.** pat. **3,012,042** (1961 to Soc. Belge Azote Prod. Chim. Marly).

Yellowish powder, mp 167°. Soly in water at 25° about 0.2%, at 45° about 1.0%. Sol in chloroform, acetone.
THERAP CAT: Coronary vasodilator.

1094. Benznidazole. *2-Nitro-N-(phenylmethyl)-1H-imidazole-1-acetamide;* N-benzyl-2-nitroimidazole-1-acetamide; Ro 7-1051; Radanil. $C_{12}H_{12}N_4O_3$; mol wt 260.26. C 55.38%, H 4.65%, N 21.53%, O 18.44%. Prepn: **Brit.** pat. **1,138,529** (1966 to Hoffmann-La Roche), *C.A.* **71**, 3383d (1969). *In vitro* effect on *T. cruzi*: S. Yoneda *et al.*, *Experientia* **33**, 1201 (1977). Pharmacokinetics: J. Raaflaub, W. H. Ziegler, *Arzneimittel-Forsch.* **29**, 1611 (1979). Antibacterial activity and electron affinity: A. V. Reynolds, *J. Pharm. Pharmacol.* **31**, Suppl., 29P (1979). Mutagenicity study: C. E. Voogd *et al.*, *Mutat. Res.* **66**, 207 (1979). *In vitro* toxicity study: G. E. Adams *et al.*, *J. Nat. Cancer Inst.* **64**, 555 (1980).

Crystals from ethanol, mp 188.5-190°. uv max (ethanol): 313 nm (ε 7,600). Soly in water at 37°: 40 mg/100 ml, J. Raaflaub, W. H. Ziegler, *loc. cit.*

THERAP CAT: Antiprotozoal (Trypanosoma).

1095. Benzo Azurine G. *3,3'-[(3,3'-Dimethoxy[1,1'-biphenyl]-4,4'-diyl)bis(azo)]bis[4-hydroxy-1-naphthalenesulfonic acid] disodium salt;* disodium o-dianisidinediazobis-(1-naphthol-4-sulfonate); C.I. Direct Blue 8; C.I. 24140. $C_{34}H_{24}N_4Na_2O_{10}S_2$; mol wt 758.69. C 53.82%, H 3.19%, N 7.39%, Na 6.06%, O 21.09%, S 8.45%. Prepared from diazotized o-dianisidine and sodium 4-hydroxy-1-naphthalenesulfonate: F. J. Welcher, *Organic Analytical Reagents* vol. 4 (D. Van Nostrand Co., New York, 1948) pp 337-338. *See also Colour Index* vol. 4 (3rd ed., 1971) p 4202.

Bluish-black powder. Sol in water, aq sodium hydroxide (red soln), sulfuric acid (blue soln), Cellosolve; very slightly sol in ethanol; practically insol in other organic solvents.
USE: As a dye: *Colour Index* vol. 2 (3rd ed., 1971) p 2223. As analytical reagent in detection of Mg: Welcher, *loc. cit.*

1096. Benzoctamine. *N-Methyl-9,10-ethanoanthracene-9(10H)-methanamine;* 1-(methylaminomethyl)dibenzo[b,e]-bicyclo[2.2.2]octadiene; 9-(methylaminomethyl)-9,10-dihydro-9,10-ethanoanthracene. $C_{18}H_{19}N$; mol wt 249.34. C 86.70%, H 7.68%, N 5.62%. Prepn: **Belg.** pat. **610,863** corresp to P. Schmidt *et al.*, **U.S.** pat. **3,399,201** (1962, 1968 to Ciba); Wilhelm, Schmidt, *Helv. Chim. Acta* **52**, 1385 (1969). Pharmacology: Keberle *et al.*, *Present Status Psychotropic Drugs* (Excerpta Medica, New York, 1969) p 123; Baltzer, Bein, *Arch. Int. Pharmacodyn Ther.* **201**, 25 (1973). Toxicity: E. I. Goldenthal, *Toxicol. Appl. Pharmacol.* **18**, 185 (1971).

Hydrochloride, $C_{18}H_{20}ClN$, *BA 30803, Tacitin.* Crystals, mp 320-322°. pKa: 7.6. LD$_{50}$ orally in rats: 700 ±170 mg/kg (Goldenthal).
THERAP CAT: Anxiolytic, muscle relaxant (skeletal).

1097. Benzodepa. *[Bis(1-aziridinyl)phosphinyl]carbamic acid phenylmethyl ester;* benzyl [bis(1-aziridinyl)phosphinyl]carbamate; benzyl bis(ethylenimido)phosphorourethan; benzcarbimine; AB-103; NSC-37096; Dualar. $C_{12}H_{16}N_3O_3P$; mol wt 281.26. C 51.24%, H 5.73%, N 14.94%, O 17.07%, P 11.01%. Prepn: Bardos *et al.*, *Nature* **183**, 399 (1959); Papanastassiou, Bardos, *J. Med. Pharm. Chem.* **5**, 1000 (1962).

Crystals from benzene + cyclohexane, mp 134-135°. Practically insol in water. Soluble in fat solvents, peanut oil, other oils.
THERAP CAT: Antineoplastic.

1098. Benzofuran. Coumarone; cumarone. C_8H_6O; mol wt 118.13. C 81.33%, H 5.12%, O 13.54%. Constituent of coal tar. Isoln from coal tar oils: Kraemer, Spilker, *Ber.* **23**, 78 (1890); **23**, 2261 (1900); Breston, Gauger, *Am. Gas Assoc. Proc.* **28**, 492 (1946). Synthesis by heating phenoxyacetaldehyde with zinc chloride and acetic acid: Stoermer, *Ber.* **30**, 1703 (1897); *Ann.* **312**, 261 (1900). The benzofuran used in the manuf of coumarone-indene resins is derived from the crude heavy solvent naphtha fraction of coal tar light oil, obtained as a byproduct in the coking of bituminous coal.

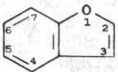

Oil. Aromatic odor. Not solid at −18°. bp$_{760}$ 173-175°; bp$_{15}$ 62-63°. Volatile with steam. d$_4^{22.7}$ 1.0913. n$_D^{16.3}$ 1.56897; n$_D^{22.7}$ 1.565. Insol in water, aq alkaline solns; miscible with benzene, petr ether, abs alcohol, ether. Slowly polymerizes on standing. For additional physical data *see* Breston, Gauger, *loc. cit.*

USE: Manuf of coumarone-indene resins.

1099. Benzoguanamine. *6-Phenyl-1,3,5-triazine-2,4-diamine;* 2,4-diamino-6-phenyl-s-triazine; 4,6-diamino-2-phenyl-s-triazine. C$_9$H$_9$N$_5$; mol wt 187.20. C 57.74%, H 4.85%, N 37.41%. Prepd from benzonitrile and dicyandiamide in the presence of Na and liquid ammonia: Jones, U.S. pat. **2,735,850** (1956 to British Oxygen Co.); from benzonitrile and dicyandiamide in the presence of KOH and methyl Cellosolve: Simons, Saxton, *Org. Syn.* **33**, 13 (1953).

Crystals. d$_4^{25}$ 1.40. mp 227-228°. uv max (ethanol): 249 nm (ε 25,000). Sol in alcohol, ether, dil HCl; partially sol in dimethylformamide; practically insol in acetone, chloroform, ethyl acetate. Soly in water at 22° 0.06%; at 100° 0.6%.

USE: In the manuf of thermosetting resins, pesticides, pharmaceuticals and dyestuffs.

1100. Benzohydrol. α-*Phenylbenzenemethanol;* benzhydrol; diphenylcarbinol. C$_{13}$H$_{12}$O; mol wt 184.23. C 84.75%, H 6.57%, O 8.68%. (C$_6$H$_5$)$_2$CHOH. Prepd by reducing benzophenone with zinc dust in strongly alkaline soln: Wiselogle, Sonneborn, *Org. Syn.* coll. vol. I (2nd ed, 1941) p 90. By reducing benzophenone with magnesium in methanol: Zechmeister, Rom, *Ann.* **468**, 123 (1929).

Needles from ligroin, mp 69°. bp$_{748}$ 298°; bp$_{20}$ 180°; bp$_{13}$ 176°. Absorption spectrum: Orndorff *et al., J. Am. Chem. Soc.* **49**, 1542 (1927). One gram dissolves in 2 liters of water at 20°. Freely sol in alcohol, ether, chloroform and carbon bisulfide; almost insol in cold ligroin.

Diphenylmethyl ether, (C$_6$H$_5$)$_2$CHOCH(C$_6$H$_5$)$_2$, monoclinic crystals from benzene, mp 110°; bp$_{15}$ 267°.

USE: In organic syntheses.

1101. Benzoic Acid. Benzenecarboxylic acid; phenylformic acid; dracylic acid. C$_7$H$_6$O$_2$; mol wt 122.12. C 68.84%, H 4.95%, O 26.20%. Occurs in nature in free and combined forms. Gum benzoin may contain as much as 20%. Most berries contain appreciable amounts (around 0.05%). Excreted mainly as hippuric acid by almost all vertebrates, except fowl. Mfg processes include the air oxidation of toluene, the hydrolysis of benzotrichloride, and the decarboxylation of phthalic anhydride: Faith, Keyes & Clark's *Industrial Chemicals*, F. A. Lowenheim, M. K. Moran, Eds. (Wiley-Interscience, New York, 4th ed., 1975) pp 138-144. Lab prepn from benzyl chloride: A. I. Vogel, *Practical Organic Chemistry* (Longmans, London, 3rd ed, 1959) p 755; from benzaldehyde: Gattermann-Wieland, *Praxis des organischen Chemikers* (de Gruyter, Berlin, 40th ed, 1961) p 193. Prepn of ultra-pure benzoic acid for use as titrimetric and calorimetric standard: Schwab, Wicher, *J. Res. Nat. Bur. Standards* **25**, 747 (1940). *Review:* A. E. Williams in Kirk-Othmer *Encyclopedia of Chemical Technology* vol. 3 (Wiley-Interscience, New York, 3rd ed., 1978) pp 778-792.

Monoclinic tablets, plates, leaflets. d 1.321 (also reported as 1.266). mp 122.4°. Begins to sublime at ∼100°. bp$_{760}$ 249.2°; bp$_{400}$ 227°; bp$_{200}$ 205.8°; bp$_{100}$ 186.2°; bp$_{60}$ 172.8°; bp$_{40}$ 162.6°; bp$_{20}$ 146.7°; bp$_{10}$ 132.1°. Volatile with steam. Flash pt 121-131°. K at 25°: 6.40 × 10^{-5}; pH of satd soln at 25°: 2.8. Soly in water (g/l) at 0° = 1.7; at 10° = 2.1; at 20° = 2.9; at 25° = 3.4; at 30° = 4.2; at 40° = 6.0; at 50° = 9.5; at 60° = 12.0; at 70° = 17.7; at 80° = 27.5; at 90° = 45.5; at 95° = 68.0. Mixtures of excess benzoic acid and water form two liquid phases beginning at 89.7°. The two liquid phases unite at the critical soln temp of 117.2°. Composition of critical mixture: 32.34% benzoic acid, 67.66% water: see Ward, Cooper, *J. Phys. Chem.* **34**, 1484 (1930). One gram dissolves in 2.3 ml cold alc, 1.5 ml boiling alc, 4.5 ml chloroform, 3 ml ether, 3 ml acetone, 30 ml carbon tetrachloride, 10 ml benzene, 30 ml carbon disulfide, 23 ml oil of turpentine; also sol in volatile and fixed oils, slightly in petr ether. The soly in water is increased by alkaline substances, such as borax or trisodium phosphate, *see also* Sodium Benzoate.

Barium salt dihydrate, C$_{14}$H$_{10}$BaO$_4$.2H$_2$O, *barium benzoate.* Nacreous leaflets. *Poisonous!* Soluble in about 20 parts water; slightly sol in alc.

Calcium salt trihydrate, C$_{14}$H$_{10}$CaO$_4$.3H$_2$O, *calcium benzoate.* Orthorhombic crystals or powder. d 1.44. Soluble in 25 parts water; very sol in boiling water.

Cerium salt trihydrate, C$_{21}$H$_{15}$CeO$_6$.3H$_2$O, *cerous benzoate.* White to reddish-white powder. Sol in hot water or hot alc.

Copper salt dihydrate, C$_{14}$H$_{10}$CuO$_4$.2H$_2$O, *cupric benzoate.* Light blue, cryst powder. Slightly soluble in cold water, more in hot water; sol in alc or in dil acids with separation of benzoic acid.

Lead salt dihydrate, C$_{14}$H$_{10}$O$_4$Pb.2H$_2$O, *lead benzoate.* Cryst powder. *Poisonous!* Slightly sol in water.

Manganese salt tetrahydrate, C$_{14}$H$_{10}$MnO$_4$.4H$_2$O, *manganese benzoate.* Pale-red powder. Sol in water, alc. Also occurs with 3H$_2$O.

Nickel salt trihydrate, C$_{14}$H$_{10}$NiO$_4$.3H$_2$O, *nickel benzoate.* Light-green odorless powder. Slightly sol in water; sol in ammonia; dec by acids.

Potassium salt trihydrate, C$_7$H$_5$KO$_2$.3H$_2$O, *potassium benzoate.* Crystalline powder. Sol in water, alc.

Silver salt, C$_7$H$_5$AgO$_2$, *silver benzoate.* Light-sensitive powder. Sol in 385 parts cold water, more sol in hot water; very slightly sol in alc.

Uranium salt, C$_{14}$H$_{10}$O$_6$U, *uranium benzoate, uranyl benzoate.* Yellow powder. Slightly sol in water, alc.

Caution: Mild irritant to skin, eyes, mucous membranes.

USE: Preserving foods, fats, fruit juices, alkaloidal solns, etc; manuf benzoates and benzoyl compds, dyes; as a mordant in calico printing; for curing tobacco. As standard in volumetric and calorimetric analysis. Pharmaceutic aid (antifungal).

THERAP CAT (VET): Has been used with salicylic acid as a topical antifungal.

1102. Benzoic Anhydride. Benzoic acid anhydride. C$_{14}$H$_{10}$O$_3$; mol wt 226.22. C 74.33%, H 4.46%, O 21.22%. (C$_6$H$_5$CO)$_2$O. Prepd by heating benzoic acid with a trace of phosphoric acid: Clarke, Rahrs, *Org. Syn.* **3**, 2 (1923); coll. vol. I (Wiley, New York, 2nd ed., 1941) p 91.

Orthorhombic bipyramidal prisms from benzene + petr ether. d$_4^{15}$ 1.1989. mp 42°. bp$_{760}$ 360°; bp$_{200}$ 299.1°; bp$_{60}$ 252.7°; bp$_{40}$ 239.8°; bp$_{20}$ 218°; bp$_{10}$ 198°; bp$_5$ 180°; bp$_{1.0}$ 143.8°. n$_D^{15}$ 1.57665. Almost insol in water (0.01 g/l); sol in alcohol, chloroform, acetone, ethyl acetate, benzene, toluene, xylene, ether, glacial acetic acid, acetic anhydride; moderately sol in petr ether; stable in water and in cold alkaline solns.

USE: In organic syntheses, as benzoylating agent in the manufacture of pharmaceuticals, dyes, and intermediates.

1103. Benzoin. *2-Hydroxy-1,2-diphenylethanone;* benzoylphenylcarbinol; α-hydroxy-α-phenylacetophenone; bitter-almond-oil camphor. C$_{14}$H$_{12}$O$_2$; mol wt 212.22. C 79.22%,

H 5.70%, O 15.08%. Prepd by treating an alcoholic soln of benzaldehyde with an alkali cyanide: Adams, Marvel, *Org. Syn.* vol. **1**, p 33 (1921); **coll. vol. I**, 88; Arnold, Fuson, *J. Am. Chem. Soc.* **58**, 1295 (1936); L. F. Fieser, *Organic Experiments* (D. C. Heath & Co., Boston, 1964) pp 211-214.

dl-Form, six-sided monoclinic prisms from alcohol, mp 137°. bp$_{768}$ 344°. bp$_{12}$ 194°. Reduces Fehling's soln. uv max (ethanol): 247 nm (ϵ 14,500); infrared in chloroform: 2.88; 5.93; 6.21; 6.28, 6.85 μ. Soluble in 3335 parts water, more in hot water, in 5 parts pyridine; sol in acetone, in boiling alc; slightly in ether.

Methyl ether, $C_{15}H_{14}O_2$, needles, mp 49°.
Ethyl ether, $C_{16}H_{16}O_2$, needles, mp 62°.
l-Form, needles, mp 132°. $[\alpha]_D^{12}$ −118° (c = 1.2 in acetone).
d-Form, needles, mp 132°. $[\alpha]_D^{12}$ +120.5° (c = 1.2 in acetone).

Note: Not to be confused with Gum Benzoin.
USE: In organic syntheses.

1104. Benzoin Oxime. *2-Hydroxy-1,2-diphenylethanone oxime;* 2-hydroxy-2-phenylacetophenone oxime. $C_{14}H_{13}$-NO_2; mol wt 227.25. C 73.99%, H 5.76%, N 6.16%, O 14.08%. Two isomers occur: α or *anti* and β or *syn*. Both have been prepd from benzoin and hydroxylamine hydrochloride. Prepn of α-form: Werner, Detscheff, *Ber.* **38**, 69 (1905); F. J. Welcher, *Organic Analytical Reagents* vol. III (Van Nostrand, New York, 1947) pp 239-251. Prepn of β-form: Werner, Detscheff, *loc. cit.* Configuration of α- and β-forms: Meisenheimer, Meis, *Ber.* **57**, 289 (1924).

α-Form, **Cupron.** Prisms from benzene, mp 151-152°. Darkens on exposure to light. Slightly sol in water; sol in alc, aq ammonium hydroxide soln. *Protect from light.*
β-Form, prismatic crystals from ether. Ether of crystn is lost on standing in air. Ether-free compd, mp 99°.
USE: The α-form is used in the detection and determination of Cu, Mo, and W: Welcher, *loc. cit.*

1105. 6,7-Benzomorphan. *1,2,3,4,5,6-Hexahydro-2,6-methano-2,6-benzazocine;* 6,7-benzmorphan. $C_{12}H_{15}N$; mol wt 173.26. C 83.19%, H 8.73%, N 8.08%. Parent compound of a series of potent analgetics. Synthesis: Kanematsu *et al.*, *J. Am. Chem. Soc.* **90**, 1064 (1968); Kanematsu *et al.*, *J. Med. Chem.* **12**, 405 (1969). Review of derivative syntheses and pharmacology: May, Sargent, "Morphine and its Modifications", in *Analgetics*, G. de Stevens, Ed. (Academic Press, New York, 1965) pp 145-171.

C.A. numbering

alternate numbering

Hydrochloride, $C_{12}H_{15}N \cdot HCl$, needles from methanol-acetone, mp 261-262°.

1106. Benzonatate. *4-(Butylamino)benzoic acid 3,6,9,-12,15,18,21,24,27-nonaoxaoctacos-1-yl ester;* nonaethyleneglycol monomethyl ether *p-n*-butylaminobenzoate; *p*-butylaminobenzoic acid ω-O-methylnonaethyleneglycol ester; benzononatine; Exangit; Tessalon; Tessalon-Ciba; Ventussin. $C_{30}H_{53}NO_{11}$; mol wt 603.73. C 59.68%, H 8.85%, N

2.32%, O 29.15%. Prepn: Matter, U.S. pat. **2,714,608** (1955 to Ciba).

Colorless to faintly yellow oil. Soluble in most organic solvents except aliphatic hydrocarbons.
THERAP CAT: Antitussive.

1107. Benzonitrile. Phenyl cyanide; cyanobenzene. C_7H_5N; mol wt 103.12. C 81.53%, H 4.89%, N 13.59%. C_6H_5CN. Prepd by heating Na benzenesulfonate with NaCN or by adding benzenediazonium chloride soln to a hot aq NaCN soln contg $CuSO_4$ and distilling. Lab prepn: A. I. Vogel, *Practical Organic Chemistry* (Longmans, London, 3rd ed, 1959) p 608.

Liquid, odor of volatile oil of almond. bp$_{760}$ 190.7°, bp$_{100}$ 123.5°, bp$_{10}$ 69.2°, bp$_1$ 28.2°. mp −12.75°. d$_{15}^{15}$ 1.010. n$_D^{20}$ 1.5289. Fire pt 167°F. Slightly sol in cold water; sol to the extent of 1% in water at 100°; miscible with common organic solvents.
USE: Solvent.

1108. Benzophenone. *Diphenylmethanone;* diphenyl ketone; benzoylbenzene. $C_{13}H_{10}O$; mol wt 182.21. C 85.69%, H 5.53%, O 8.78%. $C_6H_5COC_6H_5$. Prepd by the Friedel-Crafts ketone synthesis from benzene and benzoyl chloride in the presence of $AlCl_3$: Marvel, Sperry, *Org. Syn.* **coll. vol. I** (Wiley, New York, 2nd ed., 1941) p 95. By decarboxylation of *o*-benzoylbenzoic acid in the presence of copper catalyst: L. F. Fieser, *Organic Experiments* (D. C. Heath & Co., Boston, 1964) pp 201-203.

Stable form (there are two other labile forms), orthorhombic bisphenoidal prisms from alcohol or ether. Geranium-like odor. d$_4^{18}$ 1.1108. d$_4^{50}$ 1.0869. n$_D^{45.2}$ 1.5975. Absorption spectrum: Purvis, McCleland, *J. Chem. Soc.* **101**, 1516 (1912). mp 48.5°. bp$_{760}$ 305.4°; bp$_{400}$ 276.8°; bp$_{200}$ 249.8°; bp$_{100}$ 224.4°; bp$_{60}$ 208.2°; bp$_{40}$ 195.7°; bp$_{20}$ 175.8°; bp$_{10}$ 157.6°; bp$_5$ 141.7°; bp$_{1.0}$ 108.2°. Insoluble in water. One gram dissolves in 7.5 ml alcohol, 6 ml ether; sol in chloroform.

Oxime, $C_{13}H_{11}NO$, for prepn see *Org. Syn.* **10**, 10. Crystals from ligroin, mp 143-144°. Freely sol in ether, acetone.
USE: Fixative for heavy perfumes, such as geranium, new-mown hay, especially when used in soaps. In the manuf of antihistamines, hypnotics, insecticides.

1109. Benzophenone-6. *Bis(2-hydroxy-4-methoxyphenyl)methanone;* 2,2'-dihydroxy-4,4'-dimethoxybenzophenone; Uvinul D49. $C_{15}H_{14}O_5$; mol wt 274.26. C 65.69%, H 5.15%, O 29.16%. Preparation: Grover *et al.*, *J. Chem. Soc.* **1955**, 3982; Hardy, Forster, U.S. pat. **2,773,903** (1956 to Am. Cyanamid); Dayan, Roberts; Hahn, Stanley, U.S. pats. **2,853,522** and **2,853,523** (both 1958 to Gen. Aniline); Hosler, Storfer, U.S. pat. **2,928,878** (1960 to Am. Cyanamid).

Crystals, mp 139-140°. uv max: 284, 340 nm (log ϵ 4.12, 4.12).
USE: As ultraviolet light absorber, esp in paints, plastics.

1110. Benzopinacol. *1,1,2,2-Tetraphenyl-1,2-ethanediol;* benzpinacone; tetraphenylethylene glycol. $C_{26}H_{22}O_2$; mol wt 366.44. C 85.22%, H 6.05%, O 8.73%. $(C_6H_5)_2C(OH)$-$C(OH)(C_6H_5)_2$. Prepd by photochemical reduction of benzophenone: Bachmann, *Org. Syn.* **coll. vol. II**, 71 (1943); L. F. Fieser, *Organic Experiments* (D. C. Heath & Co., Boston, 1964) p 203. By the action of phenylmagnesium bromide on benzil: Acree, *Ber.* **37**, 2761 (1904).

Monoclinic prisms, may contain 1 mol C_6H_6 when crystallized from benzene. mp 197° (open capillary, rapid heating); mp 222° (copper block). On heating to mp it dec to benzophenone and benzohydrol. Soluble in 11.5 parts boiling glacial acetic acid; in 26 parts boiling benzene; in 39 parts boiling 95% alcohol. Freely sol in ether, carbon bisulfide, chloroform.

1111. Benzopurpurine 4B. *3,3'-[(3,3'-Dimethyl[1,1'-biphenyl]-4,4'-diyl)bis(azo)]bis[4-amino-1-naphthalenesulfonic acid] disodium salt; C.I. Direct Red 2;* disodium *o*-tolidinediazobis(1-naphthylamine-4-sulfonate); azamin 4B; C.I. 23500; eclipse red; fast scarlet; Paper Red 4B; Sultan Red 4B; Cotton Red 4B. $C_{34}H_{26}N_6Na_2O_6S_2$; mol wt 724.74. C 56.35%, H 3.62%, N 11.60%, Na 6.34%, O 13.25%, S 8.85%. Prepn from diazotized *o*-tolidine and sodium 4-amino-1-naphthalenesulfonate: **Ger. pat. 35,615** (1885 to A. G. für Anilinfabrikation Berlin), *Frdl.* **1,** 473; F. J. Welcher, *Organic Analytical Reagents* vol. 4 (Van Nostrand Co., New York, 1948) pp 338-339; *Colour Index* vol. 4 (3rd ed., 1971) p 4189. *See also:* H. J. Conn's *Biological Stains,* R. D. Lillie, Ed. (Williams & Wilkins, Baltimore, 9th ed., 1977) p 154.

Brown powder. Soluble in water, sodium hydroxide, sulfuric acid, ethanol, acetone, Cellosolve; practically insol in other organic solvents.

USE: For dyeing primarily cotton and viscose rayon: *Colour Index* vol. **2** (3rd ed., 1971) p 2101. As an analytical reagent in detection of Al, Mg, Hg, Ag, U; as a biological stain; as pH indicator, violet 1.2 to red 4.0.

1112. 1,2-Benzopyran. *2H-1-Benzopyran;* α-*5:6-benzo-pyran;* 1,2-chromene; 3-chromene. C_9H_8O; mol wt 132.15. C 81.79%, H 6.10%, O 12.11%. Prepn from *cis-o*-hydroxycinnamyl alcohol: Chatterjea, *J. Indian Chem. Soc.* **36,** 76 (1959), *C.A.* **54,** 519b (1960); from phenyl propargyl ether in diethylphenylamine: Iwai, Ide, *Chem. Pharm. Bull.* **11,** 1042 (1963); Iwai, Iwade, **Japan.** pat. **22,587('63)** (to Sankyo), *C.A.* **60,** 2901e (1964). Earlier prepns: Maitte, *Ann. Chim. (Paris)* **9,** 431 (1954); Normant, Maitte, *Compt. Rend.* **234,** 1787 (1952).

Liquid; bp_{102} 132°, bp_{15} 92-92.5°, bp_{13} 91°, bp_9 77°. d^{16} 1.0993. n_D^{24} 1.5869; n_D^{16} 1.5923 (Maitte, *loc. cit.*); also reported as 1.5837 (**Japan.** pat. *see above*).

1113. Benzo[a]pyrene. 3,4-Benzpyrene. $C_{20}H_{12}$; mol wt 252.30. C 95.21%, H 4.79%. Formerly called *1,2-benzpyrene.* Occurs in coal tar. Isoln by fractionation: Cook *et al., J. Chem. Soc.* **1933,** 395; by adsorption and fluorimetric determn: Hieger, *Am. J. Cancer* **29,** 705 (1937); Winterstein *et al., Z. Physiol. Chem.* **230,** 158, 169 (1934); *Naturwiss.* **22,** 237 (1934). Synthesis from pyrene and succinic anhydride: Cook, Hewett, *J. Chem. Soc.* **1933,** 398; Fieser, Fieser, *J. Am. Chem. Soc.* **57,** 782 (1935); Winterstein *et al., Ber.* **68,** 1079 (1935); Fieser *et al., J. Am. Chem. Soc.* **57,** 1509 (1935). Absorption spectrum: Mayneord, Roe, *Proc. Roy. Soc. London* **A152,** 299 (1935). *Review:* Clar, *Polycyclic Hydrocarbons* (Academic Press, New York, 1964), 2 vols. Review of carcinogenicity studies: *IARC Monographs* **3,** 91-136 (1973). Inhibition of the mutagenicity of the ultimate carcinogenic metabolite of benzo[a]pyrene by ellagic acid: A. W. Wood *et al., Proc. Nat. Acad. Sci. USA* **79,** 5513 (1982). Study of the reaction between this metabolite (*benzo[a]pyrene-7,8-diol 9,10-epoxide*) and ellagic acid: J. M. Sayer *et al., J. Am. Chem. Soc.* **104,** 5562 (1982). *Review:* D. H. Phillips, *Nature* **303,** 468-472 (1983).

Yellowish plates, needles from benzene + methanol, mp 179-179.3°. Crystals may be monoclinic or orthorhombic. bp_{10} 310-312°. Violet fluorescence of dil benzene solns: Hieger, *loc. cit.* Soluble in benzene, toluene, xylene; sparingly sol in alc, methanol. Practically insol in water.

Note: This substance may reasonably be anticipated to be a carcinogen: *Fourth Annual Report on Carcinogens* (NTP 85-002, 1985) p 39.

1114. Benzo[e]pyrene. 1,2-Benzpyrene; 4,5-benzpyrene. $C_{20}H_{12}$; mol wt 252.30. C 95.21%, H 4.79%. Constituent of coal tar. Isoln: Cook *et al., J. Chem. Soc.* **1933,** 396. Synthesis: Cook, Hewett, *ibid.* 398; Clar, *Ber.* **76,** 609 (1943); Buchta, Kröger, *Ann.* **705,** 190 (1967). NMR spectra: Cobb, Memory, *J. Chem. Phys.* **47,** 2020 (1967). *Review:* Clar, *Polycyclic Hydrocarbons* vol. 2, (Academic Press, New York, 1964) p 127.

Prisms or plates from benzene, mp 178-179°. Can be separated from the isomeric 3,4-benzpyrene by its more stable and less soluble picrate.

Picrate, $C_{20}H_{12}\cdot C_6H_3N_3O_7$, red needles, mp 229-230°.

1115. Benzo[f]quinoline. β-Naphthoquinoline; 5,6-benzoquinoline; naphthopyridine. $C_{13}H_9N$; mol wt 179.21. C 87.12%, H 5.06%, N 7.82%. Prepn from from β-naphthylamine by the Skraup reaction: Knueppel, *Ber.* **29,** 703 (1896); Clem, Hamilton, *J. Am. Chem. Soc.* **62,** 2349 (1940); Uhle, Jacobs, *J. Org. Chem.* **10,** 76 (1945).

Crystals from alcohol + water, mp 93°. bp_{721} 349-350°. uv max (ethanol): 347, 331, 316, 266, nm (log ε 3.54, 3.41, 3.18, 4.06). Practically insol in water; sol in dil acids; very sol in alcohol, ether, benzene.

USE: As a reagent for the determination of cadmium which is pptd as $(C_{13}H_9N)_2H_2(CdI_4)$ from dil nitric or sulfuric acid soln in the presence of potassium iodide.

1116. Benzoquinonium Chloride. *N,N'-[(3,6-Dioxo-1,4-cyclohexadiene-1,4-diyl)bis(imino-3,1-propanediyl)]bis[N,N-diethylbenzenemethanaminium]dichloride; [2,5-p-benzoquinonylenebis(iminotrimethylene)]bis[benzyldiethylammonium chloride];* 2,5-bis(3-diethylaminopropylamino)benzoquinone bis(benzyl chloride); WIN 2747; Mytolon; Amilyt. $C_{34}H_{50}Cl_2N_4O_2$; mol wt 617.72. C 66.11%, H 8.16%, Cl 11.48%, N 9.07%, O 5.18%. Prepn: Cavallito *et al., J. Am. Chem. Soc.* **72,** 2661 (1950).

Red crystals from ethanol, mp 191-195°. Freely sol in water. Solns can be sterilized by autoclaving and may be stored at room temp for prolonged periods without loss of potency.

THERAP CAT: Skeletal muscle relaxant.

1117. Benzoresorcinol. *(2,4-Dihydroxyphenyl)phenyl-methanone; 2,4-dihydroxybenzophenone; 4-benzoylresorcinol; resbenzophenone; benzophenone-1; Uvinul 400.* $C_{13}H_{10}$-O_3; mol wt 214.21. C 72.89%, H 4.71%, O 22.41%. Prepd from benzoyl chloride and resorcinol: Dobner, *Ann.* **210**, 246 (1881); by Fries rearrangement: Amin, Shah, *J. Indian Chem. Soc.* **29**, 351 (1952).

Needles from hot water, mp 144-145°. Practically insol in cold water; easily sol in alcohol, ether, glacial acetic acid; scarcely sol in cold benzene.

USE: Ultraviolet light absorber, esp in paints and plastics.

1118. Benzothiazole. C_7H_5NS; mol wt 135.18. C 62.19%, H 3.73%, N 10.36%, S 23.72%. Prepd from *N,N*-di-methylaniline and sulfur: Knowles, Watt, *J. Org. Chem.* **7**, 56 (1942). Chemistry of benzothiazoles, their use as carbonyl equivalents and in carbon-carbon bond formation: E. J. Corey, D. L. Boger, *Tetrahedron Letters* **1978**, 5, 9, 13.

Liquid. Odor similar to that of quinoline. Volatile with steam. d_4^{20} 1.246. bp_{765} 227-228°; bp_{34} 131°. n_D^{20} 1.6379. Slightly sol in water. Freely sol in alcohol, carbon disulfide. LD$_{50}$ i.v. in mice: 95±3 mg/kg, E. F. Domino *et al., J. Pharmacol. Exp. Ther.* **105**, 486 (1952).

USE: In organic synthesis.

1119. 1*H*-Benzotriazole. *1,2,3-Benzotriazole; benztriazole; azimidobenzene; benzisotriazole.* $C_6H_5N_3$; mol wt 119.12. C 60.49%, H 4.23%, N 35.28%. Prepd by the action of nitrous acid on *o*-phenylenediamine: Ladenburg, *Ber.* **9**, 219 (1876); Damschroder, Peterson, *Org. Syn.* **coll. vol. III**, 106 (1955).

Needles from benzene, mp 98.5°. bp_{15} 204°; $bp_{2.0}$ 159°. *Caution:* May explode during vacuum distillation, *see Chem. & Eng. News* **34**, 2450 (1956). Sparingly sol in water. Sol in alc, benzene, toluene, chloroform, DMF.

1120. Benzotrichloride. *(Trichloromethyl)benzene; α,α,α-trichlorotoluene; phenylchloroform; ω,ω,ω-trichlorotoluene; benzenyl trichloride; toluene trichloride.* $C_7H_5Cl_3$; mol wt 195.48. C 43.01%, H 2.58%, Cl 54.41%. Produced by chlorination of boiling toluene in the presence of light and of 2%

phosphorus trichloride: Swarts, *Bull. Soc. Chim. Belg.* **31**, 375 (1922); Conklin, U.S. pats. **1,828,858-9** (1931 to Solvay Process Co.). Commercial grades may contain hydrochloric acid, benzylidene chloride, and benzyl chloride. Purification procedures: Holleman, deMooy, *Rec. Trav. Chim.* **33**, 25, 33 (1914); Britton, U.S. pat. **1,804,458** (1931 to Dow); *Chem. Zentr.* **1931**, II, 497. Toxicity data: H. F. Smyth *et al., Arch. Ind. Hyg. Occup. Med.* **4**, 119 (1951).

Liquid. Fumes in air. mp −5.0°. d_4^{20} 1.3756. bp_{760} 220.8°. bp_{60} 129°; bp_{25} 105°; bp_{10} 89°. n_D^{20} 1.55789. Unstable. Hydrolyzes in the presence of moisture, forming benzoic and hydrochloric acids. Insol in water. Sol in alcohol, benzene, ether, many other organic solvents. LD$_{50}$ orally in rats: 6.0 g/kg (Smyth).

Caution: Vapors highly irritant to skin, mucous membranes. Large doses have caused CNS depression in exptl animals. This substance may reasonably be anticipated to be a carcinogen: *Fourth Annual Report on Carcinogens* (NTP 85-002, 1985) p 40.

USE: In dye chemistry. In organic syntheses (source of benzenyl group).

1121. Benzotrifluoride. *(Trifluoromethyl)benzene; α,α,α-trifluorotoluene; phenylfluoroform.* $C_7H_5F_3$; mol wt 146.11. C 57.54%, H 3.45%, F 39.01%. Prepd by the action of hydrogen fluoride on benzotrichloride: Simons, Lewis, *J. Am. Chem. Soc.* **60**, 492 (1938); Simons, *Ind. Eng. Chem.* **32**, 178 (1940); by the action of antimony trifluoride on benzotrichloride: Swarts, *Bull. Sci. Acad. Roy. Belg.*, [4] **35**, 375 (1898); [5] **6**, 389 (1920); [5] **13**, 175 (1927); Holt *et al.*, U.S. pat. **2,058,453** (1937 to Kinetic Chemicals).

Liquid. mp −29.05°. d^{20} 1.1886. bp 103.46°. $n_D^{13.3}$ 1.41486. Soluble in many organic solvents.

USE: In dye chemistry; in the manuf of substituted benzotrifluorides contg an ethylenic group, used in high polymer chemistry; in dielectric fluids, such as transformer oils.

1122. Benzoxiquine. *8-Quinolinol benzoate (ester); 8-benzoyloxyquinoline; 8-hydroxyquinoline benzoate (ester); benzoxyline; Dioxyline.* $C_{16}H_{11}NO_2$; mol wt 249.26. C 77.09%, H 4.45%, N 5.62%, O 12.84%. Prepn from 8-quinolinol and benzoyl chloride: Bedall, Fischer, *Ber.* **14**, 1366 (1881).

Crystals from abs alc, mp 118-120°. Practically insol in water; sol in alcohol, ether.

THERAP CAT: Disinfectant.

1123. Benzoxonium Chloride. *N-Dodecyl-N,N-bis(2-hydroxyethyl)benzenemethanaminium chloride; benzyldodec-ylbis(2-hydroxyethyl)ammonium chloride; dodecyldi(β-hydroxyethyl)benzylammonium chloride; D-301; ZY 15021; Absonal V; Bialcol; Bradophen; Orofar.* $C_{23}H_{42}ClNO_2$; mol wt 400.04. C 69.06%, H 10.58%, Cl 8.86%, N 3.50%, O 8.00%. Quaternary ammonium salt with antimicrobial activity. Prepn: Swiss pat. **306,648** (1955 to Ciba), *C.A.* **51**, 2023g (1957). Antimicrobial activity and toxicology: H. Goeth *et al., Arzneimittel-Forsch.* **9**, 622 (1959). Potentiometric determn: S. Pinzauti, E. La Porta, *Analyst* **102**, 938

(1977). Clinical evaluation in dental plaque control: U. P. Saxer *et al., J. Clin. Periodontol.* **9**, 162 (1982). *In vitro* antibacterial spectrum: M. Cortat, P. Fels, *Arzneimittel-Forsch.* **37**, 463 (1987). Clinical evaluation in pharyngeal infections: M. A. Weibel *et al., ibid.* 467.

Colorless powder from ethyl ether, mp 107-109°. Sol in water, alcohol, benzene, toluene, chlorobenzene. LD_{50} orally in rats: 750 mg/kg (Goeth).

THERAP CAT: Antiseptic.

THERAP CAT (VET): Antiseptic.

1124. Benzoyl Chloride. Benzenecarbonyl chloride. C_7H_5ClO; mol wt 140.57. C 59.81%, H 3.59%, Cl 25.22%, O 11.38%. Prepd by partial hydrolysis of benzotrichloride: Davies, Dick, *J. Chem. Soc.* **1932**, 2808; by chlorination of benzaldehyde: Wöhler, *Ann.* **3**, 262 (1832); from benzoic acid and PCl_5 or from benzoic acid and phosgene: Uwarow, Stepanow, **U.S.S.R.** pat. **56,693** (1936). Lab prepn from benzoic acid and thionyl chloride: A. I. Vogel, *Practical Organic Chemistry* (Longmans, London, 3rd ed., 1959) p 792; Gattermann-Wieland, *Praxis des organischen Chemikers*, (de Gruyter, Berlin, 40th ed., 1961) p 112.

Liquid. Penetrating odor. d_4^{25} 1.2070. mp −1.0°. bp_{760} 197.2°; bp_{35} 100°; bp_{15} 82.3°; bp_9 71°; bp_3 49°. n_D^{20} 1.55369. Dipole moment 3.28. Flash pt 88° (190.4°F). Parachor 289.8. Dec by water and alc. Miscible with ether, benzene, carbon disulfide, oils.

USE: For acylation, i.e., introduction of the benzoyl group into alcohols, phenols, and amines (Schotten-Baumann reaction); in the manuf of benzoyl peroxide and of dye intermediates. In organic analysis for making benzoyl derivatives for identification purposes. *Caution:* Lacrimator. Irritating to skin, eyes, mucous membranes.

1125. Benzoylecgonine. *3-(Benzoyloxy)-8-methyl-8-azabicyclo[3.2.1]octane-2-carboxylic acid; 3β-hydroxy-1αH,5αH-tropane-2β-carboxylic acid benzoate;* ecgonine benzoate. $C_{16}H_{19}NO_4$; mol wt 289.34. C 66.42%, H 6.62%, N 4.84%, O 22.12%. Cocaine metabolite. Isoln from coca leaves: De Jong, *Rec. Trav. Chim.* **42**, 980 (1923). Prepn from *l*-ecgonine and benzoic anhydride: *idem, ibid.* **66**, 544 (1947).

Tetrahydrate, orthorhombic prisms or needles from water, mp 86-92° (dec 195°, dry). $[α]_D^{15}$ −45° (c = 3 in abs alc). Soluble in alc, hot H_2O.

1126. Benzoyl Isothiocyanate. Benzoylthiocarbimide. C_8H_5NOS; mol wt 163.20. C 58.87%, H 3.09%, N 8.58%, O 9.80%, S 19.65%. C_6H_5CONCS. Prepd by refluxing a mixture of KSCN and benzoyl chloride in benzene at 110-120°: Ambelang, Johnson, *J. Am. Chem. Soc.* **61**, 632 (1939). From NH_4SCN, acetone and benzoyl chloride: Frank, Smith, *Org. Syn.* **28**, 89 (1948).

Liquid; bp_{18} 133-137°; bp_{10} 119°. $d_4^{18.3}$ 1.2142; d_4^{16} 1.197. $n_D^{18.3}$ 1.6382. Reacts with a primary or secondary amine to form a benzoylthiourea, which is readily hydrolyzed by alkalies to the free thiourea.

1127. Benzoylpas. *4-(Benzoylamino)-2-hydroxybenzoic*

acid; N-benzoyl-*p*-aminosalicylic acid; 4-benzamidosalicylic acid. $C_{14}H_{11}NO_4$; mol wt 257.25. C 65.36%, H 4.31%, N 5.45%, O 24.88%. Prepn: Drain *et al., J. Chem. Soc.* **1949**, 1498; **Brit.** pat. **676,363** (1952 to Wander). Prepn of sodium and calcium salts: Suddaby, Sumpter, **Brit.** pat. **711,163** (1954 to Herst).

Crystals, mp 260-261°.

Sodium salt, $C_{14}H_{10}NNaO_4$, *BPAS.* White powder.

Calcium salt pentahydrate, $C_{28}H_{20}CaN_2O_8$·$5H_2O$, *benzoylpas calcium, Benzapas, Benzacyl, B-Paracipan, Therapas.* White powder or crystals. *Iso-Benzacyl* is a mixture of the calcium salt and isoniazid, *q.v.*

THERAP CAT: Antibacterial (tuberculostatic).

1128. Benzoyl Peroxide. *Dibenzoyl peroxide;* benzoyl superoxide; Acetoxyl; Acnegel; Benoxyl; Benzagel 10; Benzaknen; Debroxide; Desanden; Lucidol; Nericur; Oxy-5; Oxy-L; PanOxyl; Peroxydex; Persadox; Persa-gel; Sanoxit; Theraderm; Xerac BP 5; Xerac BP 10. $C_{14}H_{10}O_4$; mol wt 242.22. C 69.42%, H 4.16%, O 26.42%. $(C_6H_5CO)_2O_2$. Prepd by interaction of benzoyl chloride and a cooled soln of sodium peroxide. Laboratory procedure: A. I. Vogel, *Practical Organic Chemistry* (Longmans, London, 3rd ed., 1954) p 807; Gattermann-Wieland, *Praxis des organischen Chemikers* (de Gruyter, Berlin, 40th ed., 1961) p 115.

Crystals. mp 103-106°. *May explode when heated.* Sparingly sol in water or alcohol; sol in benzene, chloroform, ether. One gram dissolves in 40 ml carbon disulfide, in about 50 ml olive oil.

USE: Oxidizing agent in bleaching oils, flour, etc.; catalyst in the plastics industry; initiator in polymerization.

THERAP CAT: Keratolytic.

1129. 3,4-Benzphenanthrene. *Benzo[c]phenanthrene.* $C_{18}H_{12}$; mol wt 228.29. C 94.70%, H 5.30%. Synthesis by a Pschorr reaction from diazotized α-(2-naphthyl)-2-aminocinnamic acid: Cook, *J. Chem. Soc.* **1931**, 2524. From diphenylmethylsuccinic anhydride: Hewett, *J. Chem. Soc.* **1936**, 599. By double ring closure of β-benzohydrylglutaric acid: Newman, Joshel, *J. Am. Chem. Soc.* **60**, 487 (1938). From 4-keto-1,2,3,4-tetrahydrophenanthrene: Bachmann, Edgerton, *ibid.* **62**, 2970 (1940).

Needles from alcohol, mp 68° (needles or leaflets from petr ether, fine needles from alcohol + acetone). Absorption maxima: Mayneord, Roe, *Proc. Roy. Soc. London* **A158**, 63 (1937).

Picrate, $C_{24}H_{15}N_3O_7$, red needles, mp 128°.

1130. Benzphetamine. *N,α-Dimethyl-N-(phenylmethyl)-benzeneethanamine;* N-benzyl-*N,α*-dimethylphenethylamine; *d*-N-methyl-N-benzyl-β-phenylisopropylamine. $C_{17}H_{21}N$; mol wt 239.35. C 85.30%, H 8.84%, N 5.85%. Prepd from *d*-desoxyephedrine: Heinzelman, Aspergren, **U.S.** pat. **2,789,138** (1957 to Upjohn).

Liquid. $bp_{0.02}$ 127°. n_D^{19} 1.5515. Practically insol in water. Sol in methanol, ethanol, ether, chloroform, acetone, benzene.

Hydrochloride, $C_{17}H_{22}ClN$, **Didrex, Inapetyl.** Crystals from ethyl acetate, mp 129-130°. Dextrorotatory. Sol in water, 95% ethanol.

Caution: Excessive use may lead to tolerance and physical dependence. This is a controlled substance (stimulant) listed in the U.S. Code of Federal Regulations, Title 21 Part 1308.13 (1985).

THERAP CAT: Anorexic.

1131. Benzpiperylon. *1,2-Dihydro-2-(1-methyl-4-piperidinyl)-5-phenyl-4-(phenylmethyl)-3H-pyrazol-3-one; 4-benzyl-1-(1-methyl-4-piperidyl)-3-phenyl-3-pyrazolin-5-one;* 1-(N-methyl-4-piperidyl)-3-phenyl-4-benzyl-5-pyrazolone; benzpiperilone; KB-95; Benzometan; Humedil; Reublonil; Telon. $C_{22}H_{25}N_3O$; mol wt 347.44. C 76.05%, H 7.25%, N 12.10%, O 4.61%. Prepd from (N-methylpiperidyl)hydrazine and ethyl 2-benzyl-3-oxo-3-phenylpropionate: A. Ebnöther *et al., Helv. Chim. Acta* **42,** 1201 (1959); E. Jucker *et al.,* **Swiss.** pats. 346,885, 346,886 (1960 both to Sandoz), *C.A.* **58,** 5696a,b (1963). *See also: eidem,* **U.S.** pat. **2,903,460** (1959 to Sandoz). Pharmacological properties, toxicity: A. Cerletti *et al., Boll. Chim. Farm.* **102,** 602 (1963).

Crystals from ethanol, dec 181-183°. pK_1 6.73, pK_2 9.13. LD_{50} in mice, rats, rabbits (mg/kg): 160, 160, 83 i.v.; 1880, 2700, 1700 orally (Cerletti).

THERAP CAT: Analgesic. Anti-inflammatory.

1132. Benzpyrinium Bromide. *3-[[(Dimethylamino)carbonyl]oxy]-1-(phenylmethyl)pyridinium bromide; 1-benzyl-3-hydroxypyridinium bromide dimethylcarbamate;* 1-benzyl-3-(dimethylcarbamyloxy)pyridinium bromide; benzstigminum bromidum; Stigmenene bromide; Stigmonene bromide. $C_{15}H_{17}BrN_2O_2$; mol wt 337.22. C 53.42%, H 5.08%, Br 23.70%, N 8.31%, O 9.49%. Prepd from 3-hydroxypyridine dimethylcarbamate and benzyl bromide: Wuest, **U.S.** pat. **2,489,247** (1949 to Wm. R. Warner).

Crystals, mp 114-115°. Freely sol in water, alc. Practically insol in ether. A 1% aq soln has a pH of 4.5 to 5.5. uv max: 269 nm ($E_{1cm}^{1\%}$ 136).

THERAP CAT: Cholinergic.

1133. Benzquinamide. *2-(Acetyloxy)-N,N-diethyl-1,3,4,-6,7,11b-hexahydro-9,10-dimethoxy-2H-benzo[a]quinolizine-3-carboxamide; N,N-diethyl-1,3,4,6,7,11b-hexahydro-2-hydroxy-9,10-dimethoxy-2H-benzo[a]quinolizine-3-carboxamide acetate;* 2-acetoxy-3-(N,N-diethylcarboxamido)-9,10-dimethoxy-1,2,3,4,6,7-hexahydro-11bH-benzopyridocoline; 2-hydroxy-3-diethylcarbamyl-9,10-dimethoxy-1,2,3,4,6,7-hexahydro-11bH-benzoquinolizine acetate; BZQ; NSC 64375; P-2647; Emete-Con; Emeticon; Promecon; Quantril. $C_{22}H_{32}N_2O_5$; mol wt 404.49. C 65.32%, H 7.97%, N 6.93%, O 19.78%. Prepn: Trettner, U.S. pat. **3,053,845** (1962 to Pfizer). Pharmacology: Scriabine *et al., J. Am. Med. Assoc.* **184,** 276 (1963); Kadzielawa, Gumulka, *Arch. Int. Pharmacodyn. Ther.* **163,** 139 (1966). Metabolic studies: Koe, Pinson, *J. Med. Chem.* **7,** 635 (1964); Wiseman *et al., Biochem. Pharmacol.* **13,** 1421 (1964). Toxicity: E. I. Goldenthal, *Toxicol. Appl. Pharmacol.* **18,** 185 (1971).

Crystals from diisopropyl ether, mp 130-131.5°. LD_{50} orally in rats: 990 mg/kg; i.p. in mice: 376 mg/kg (Golden-thal).

THERAP CAT: Antipsychotic, anti-emetic.

1134. Benzthiazide. *6-Chloro-3-[[(phenylmethyl)thio]-methyl]-2H-1,2,4-benzothiadiazine-7-sulfonamide 1,1-dioxide; 3-[(benzylthio)methyl]-6-chloro-2H-1,2,4-benzothiadiazine-7-sulfonamide 1,1-dioxide;* 3-[(benzylthio)methyl]-6-chloro-7-sulfamoyl-2H-benzo-1,2,4-thiadiazine 1,1-dioxide; 6-chloro-7-sulfamoyl-3-benzylthiomethyl-2H-1,2,4-benzothiadiazine 1,1-dioxide; benzothiazide; Aquatag; Dihydrex; Diucen; Edemex; ExNa; Exosalt; Fovane; Freeuril; Hy-Drine; Lemazide; Proaqua; Urese. $C_{15}H_{14}ClN_3O_4S_3$; mol wt 431.96. C 41.71%, H 3.27%, Cl 8.21%, N 9.73%, O 14.82%, S 22.27%. Prepn: J. M. McManus *et al.,* **136th Am. Chem. Soc. Meet.** (Atlantic City, Sept. 1959) *Abstr. of Papers,* pp 13-O.

Crystals, bitter taste, mp 238-239°. Practically insol in water, sol in alkaline solns. LD_{50} in mice, rats: > 5,000, > 10,000 mg/kg orally; 410, 422 mg/kg i.v., P'an *et al., J. Pharmacol. Exp. Ther.* **128,** 122 (1960).

THERAP CAT: Diuretic, antihypertensive.

1135. Benztropine Mesylate. *3-(Diphenylmethoxy)-8-methyl-8-azabicyclo[3.2.1]octane methanesulfonate; 3α-(diphenylmethoxy)-1αH,5αH-tropane methanesulfonate;* benztropine methanesulfonate; tropine benzohydryl ether methanesulfonate; Cogentin; Cogentinol; Cobrentin methanesulfonate. $C_{22}H_{29}NO_4S$; mol wt 403.53. C 65.48%, H 7.24%, N 3.47%, O 15.86%, S 7.95%. Prepd by the action of diphenyldiazomethane on tropine: Phillips, U.S. pat. **2,595,405** (1952 to Merck & Co.). Crystal structure: P. G. Jones *et al., Acta Crystallogr.* **B34,** 3125 (1978).

Crystals from acetone + ether, mp 143°. uv max: 259 nm (E_M = 437). Soluble in water. pH about 6.

THERAP CAT: Anticholinergic.

1136. Benzydamine. *N,N-Dimethyl-3-[[1-(phenylmethyl)-1H-indazol-3-yl]oxy]-1-propanamine; 1-benzyl-3-[3-(dimethylamino)propoxy]-1H-indazole;* 1-benzyl-1H-indazol-3-yl 3-(dimethylamino)propyl ether; benzindamine. $C_{19}H_{23}N_3O$; mol wt 309.40. C 73.75%, H 7.49%, N 13.58%, O 5.17%. Prepn: **Fr.** pat. **1,382,855;** Palazzo, U.S. pat. **3,318,905** (1964, 1967 both to Angelini Francesco); Palazzo *et al., J. Med. Chem.* **9,** 38 (1966). Pharmacology: Lisciani *et al., Eur. J. Pharmacol.* **3,** 157 (1968). Metabolism: Catanese *et al., Arzneimittel-Forsch.* **16,** 1354 (1966); Kataoka *et al., Chem. Pharm. Bull.* **19,** 1511 (1971). Toxicology: B. Silvestrini *et al., Toxicol. Appl. Pharmacol.* **10,** 148 (1967). Series of articles on pharmacology: *Arzneimittel-Forsch.* **37,** 587-646 (1987).

bp$_{0.05}$ 160°.

Hydrochloride, $C_{19}H_{24}ClN_3O$, *Afloben, Andolex, Benalgin, Benzyrin, Difflam, Dorinamin, Enzamin, Epirotin, Imotryl, Indolin, Ririlim, Riripen, Salyzoron, Tamas, Tantum, Verax.* Crystals, mp 160°. uv max: 306 nm (E$^{1\%}_{1cm}$ 160). Very sol in water; rather sol in ethanol, chloroform, *n*-butanol. LD$_{50}$ in mice, rats (mg/kg): 110, 100 i.p.; 515, 1050 orally (Silvestrini).

THERAP CAT: Analgesic; anti-inflammatory; antipyretic.

1137. Benzyl Acetate. *Acetic acid phenylmethyl ester; acetic acid benzyl ester.* $C_9H_{10}O_2$; mol wt 150.17. C 71.98%, H 6.71%, O 21.31%. $C_6H_5CH_2OOCCH_3$. Occurs in a number of plants, particularly jasmine: S. Arctander, *Perfume and Flavor Materials of Natural Origin* (Elizabeth, N.J., 1960) pp 313-314. Prepd from benzyl chloride, acetic acid or sodium acetate and triethylamine: Merker, Scott, *J. Org. Chem.* **26**, 5180 (1961); Hennis *et al., Ind. Eng. Chem., Prod. Res. Develop.* **6**, 193 (1967).

Liquid; pear-like odor. bp 213°, bp$_{102}$ 134°. mp −51°. d$_4^{25}$ 1.050. n_D^{20} 1.5232, n_D^{25} 1.4998. Flash pt, closed cup: 216° F (102° C). Practically insol in water; misc with alcohol, ether. LD$_{50}$ orally in rats: 2490 mg/kg, P. M. Jenner *et al., Food Cosmet. Toxicol.* **2**, 327 (1964).

USE: In perfumery, solvent for cellulose acetate and nitrate. *Caution:* If ingested can cause G.I. irritation with vomiting and diarrhea. Also irritating to skin, eyes, respiratory tract.

1138. Benzyl Alcohol. *Benzenemethanol;* phenylcarbinol; phenylmethanol; α-hydroxytoluene. C_7H_8O; mol wt 108.13. C 77.75%, H 7.46%, O 14.80%. $C_6H_5CH_2OH$. Constituent of jasmine, hyacinth, ylang-ylang oils, Peru and Tolu balsams, storax, where it occurs in ester form also. Originally prepd by the Cannizzaro reaction from benzaldehyde + KOH: Cannizzaro, *Ann.* **88**, 129 (1853); *cf.* Hickinbottom, *Reactions of Organic Compds.* (Longmans, London, 3rd ed., 1957) p 251; A. I. Vogel, *Practical Organic Chemistry* (Longmans, London, 3rd ed., 1959) p 711; Gattermann-Wieland, *Praxis des organischen Chemikers* (de Gruyter, Berlin, 40th ed., 1961) p 193. Produced on a large scale by the action of sodium or potassium carbonate on benzyl chloride: Ger. pat. 484,662; *Chem. Zentr.* **1930**, I, 1052; *Frdl.* **16**, 426; *cf.* Kirk-Othmer *Encyclopedia of Chemical Technology* vol. 3 (Interscience, New York, 1964) pp 442-449. Toxicity: Smyth *et al., Arch. Ind. Hyg. Occup. Med.* **4**, 119 (1951).

Liquid. Faint aromatic odor. Sharp burning taste. d$_4^{20}$ 1.04535; d$_4^{25}$ 1.04156. mp −15.19°. bp$_{760}$ 204.7°; bp$_{400}$ 183.0°; bp$_{200}$ 160.0°; bp$_{100}$ 141.7°; bp$_{60}$ 129.3°; bp$_{40}$ 119.8°; bp$_{20}$ 105.8°; bp$_{10}$ 92.6°; bp$_5$ 80.8°; bp$_{1.0}$ 58.0°. n_D^{20} 1.54035; n_D^{25} 1.53837: Dreisbach, Martin, *Ind. Eng. Chem.* **41**, 2875 (1941). Absorption spectrum: Brode, *J. Phys. Chem.* **30**, 61 (1926). Vapor density 3.72 (air = 1.00). Flash pt, closed cup 213°F, open cup 220°F. Autoignition temp 817°F. One gram dissolves in about 25 ml water. One volume dissolves in 1.5 vols of 50% ethyl alcohol. Misc with abs and 94% alcohol, ether, chloroform. LD$_{50}$ orally in rats: 3.1 g/kg (Smyth).

USE: Manuf other benzyl compds. Pharmaceutic aid (antimicrobial). Solvent for gelatin, casein (when hot), solvent for cellulose acetate, shellac. Used in perfumery and in flavoring (mostly in form of its aliphatic esters). In microscopy as embedding material.

THERAP CAT (VET): Has been used for relief from pruritis.

1139. Benzylamine. *Benzenemethanamine;* aminotoluene; phenylmethylamine; moringine. C_7H_9N; mol wt 107.15. C 78.46%, H 8.46%, N 13.08%. Prepn from benzylchloride and ammonia: Mason, *J. Chem. Soc.* **63**, 1311 (1893); by redn of benzonitrile: Carothers, Jones, *J. Am. Chem. Soc.* **47**, 3051 (1925); from benzyl bromide + acetamide: Erikson, *Ber.* **59**, 2665 (1926); from *N*-benzylphthalimide + hydrazine hydrate: Ing, Manske, *J. Chem. Soc.* **129**, 2348

(1926). Identity with moringine: Chakravarti, *Bull. Calcutta School Trop. Med.* **3**, 162 (1955); *C.A.* **50**, 16891e (1956).

Liquid; strongly alkaline reaction. bp 185°; bp$_{12}$ 90°. d$_4^{19}$ 0.983. n_D^{20} 1.5401. Miscible with water, alcohol, ether. Hydrochloride, C_7H_9N.HCl, crystals, mp 253°. Hydroiodide, C_7H_9N.HI, leaflets, mp 162°. Picrate, $C_7H_9N.C_6H_3N_3O_7$, needles from alc, dec 198°.

USE: In organic synthesis. *Caution:* Highly irritating to skin, mucous membranes.

1140. Benzylaniline. *N-Phenylbenzenemethanamine; N-phenylbenzylamine;* benzylphenylamine. $C_{13}H_{13}N$; mol wt 183.24. C 85.20%, H 7.15%, N 7.65%. $C_6H_5CH_2NHC_6H_5$. Prepn from benzyl alc and aniline in the presence of KOH: Sprinzak, *J. Am. Chem. Soc.* **78**, 3207 (1956); from benzaldehyde and aniline in the presence of NaBH$_4$: Schellenberg, *J. Org. Chem.* **28**, 3259 (1963).

Prisms, mp 37-38°. bp 306-307°. Practically insol in water; sol in alcohol, chloroform, ether.

1141. Benzyl Benzoate. *Benzoic acid phenylmethyl ester;* benzoic acid benzyl ester; benzylbenzenecarboxylate; Ascabin; Venzonate; Vanzoate; Ascabiol; Benylate; Scabanca. $C_{14}H_{12}O_2$; mol wt 212.24. C 79.22%, H 5.70%, O 15.08%. $C_6H_5COOCH_2C_6H_5$. Contained in Peru and Tolu balsams. Prepd by the action of sodium benzylate on benzaldehyde: Kamm, Kamm, *Org. Syn.* **coll. vol. I**, 104 (2nd ed., 1941); by the dry esterification of sodium benzoate and benzyl chloride in the presence of triethylamine: Thorp, Nottorf, *Ind. Eng. Chem.* **39**, 1300 (1947). Toxicity studies: Graham, Kuizenga, *J. Pharmacol. Exp. Ther.* **84**, 358 (1945). Comprehensive description: M. M. A. Hassan, J. S. Mossa, in *Analytical Profiles of Drug Substances*, vol. 10, K. Florey, Ed. (Academic Press, New York, 1981) pp 55-74.

Leaflets or oily liq; faint, pleasant, aromatic odor; sharp burning taste. mp 21°. d$_4^{25}$ 1.118. bp 323-324°. bp$_{16}$ 189-191°. bp$_{4.5}$ 156°. Sparingly volatile with steam. n_D^{21} 1.5681. Insol in water or glycerol; miscible with alc, chloroform, ether, oils. LD$_{50}$ orally in rats, mice, rabbits, guinea pigs (g/kg): 1.7, 1.4, 1.8, 1.0, Draize *et al., J. Pharmacol. Exp. Ther.* **93**, 26 (1948).

Toxicity: In exptl animals, ingestion causes progressive incoordination, excitation, convulsions, death. May cause skin irritation in humans. *Caution:* Avoid contact with eyes, *Clinical Toxicology of Commercial Products*, R. E. Gosselin *et al.*, Eds. (Williams & Wilkins, Baltimore, 4th ed., 1976) Section II, p 137.

USE: As solvent of cellulose acetate, nitrocellulose and artificial musk; substitute for camphor in celluloid and plastic pyroxylin compds; perfume fixative; in confectionery and chewing gum flavors.

THERAP CAT: Scabicide, pediculicide.

THERAP CAT (VET): Acaricide, pediculicide. *Contraindicated* in cats.

1142. Benzyl Bromide. *(Bromomethyl)benzene;* α-bromotoluene; ω-bromotoluene. C_7H_7Br; mol wt 171.04. C 49.15%, H 4.13%, Br 46.72%. $C_6H_5CH_2Br$. Prepd by the action of bromine on toluene in ultraviolet light: v. Konek, Loczka, *Ber.* **57**, 679 (1924); Zelinsky, Ger. pat. 478,084; *Chem. Zentr.* **1929 II**, 1216; *Frdl.* **16**, 335; by the action of bromine on dibenzyl ether: Lachman, *J. Am. Chem. Soc.* **45**, 2359 (1923).

Lacrimatory liquid. mp −3.9°. bp 198-199°. bp$_{80}$ 127°. d$_0^{22}$ 1.4380; d^{17} 1.443; d$_4^{64}$ 1.3886. Slowly decomp by water. *Caution:* Intensely irritating to skin, eyes, mucous membranes. Large doses cause CNS depression.

1143. Benzyl Chloride. *(Chloromethyl)benzene;* α-chlorotoluene. C_7H_7Cl; mol wt 126.58. C 66.42%, H 5.57%, Cl 28.01%. $C_6H_5CH_2Cl$. Made by cautious chlorination of toluene: A. I. Vogel, *Practical Organic Chemistry* (Longmans, London, 3rd ed., 1959) p 538; Gattermann-Wieland, *Praxis des organischen Chemikers* (de Gruyter, Berlin, 40th ed., 1961) p 92. Manuf: Faith, Keyes & Clark's *Industrial*

Chemicals, F. A. Lowenheim, M. K. Moran, Eds. (Wiley-Interscience, New York, 4th ed., 1975) pp 145-148.

Very refractive liquid; rather unpleasant, irritating odor. d_{20}^{20} 1.100. bp 179°. mp -48° to -43°. n_D^{15} 1.5415. Insol in water. Miscible with alcohol, chloroform, ether. Rapidly dec when heated in the presence of iron.

Caution: Intensely irritating to skin, eyes, mucous membranes. Large doses cause CNS depression.

USE: Manuf benzyl compds, perfumes, pharmaceutical products, dyes, synthetic tannins, artificial resins.

1144. Benzyl Cinnamate. *3-Phenyl-2-propenoic acid phenylmethyl ester; trans-cinnamic acid benzyl ester;* cinnamein. $C_{16}H_{14}O_2$; mol wt 238.27. C 80.65%, H 5.92%, O 13.43%. $C_6H_5CH=CHCOOCH_2C_6H_5$. Constituent of storax, Peru and Tolu balsams: Tschirch, Trog, *Arch. Pharm.* **232,** 70 (1894); Tschirch, Oberländer, *ibid.* 559. Prepn: Volwiler, Vliet, *J. Am. Chem. Soc.* **43,** 1672 (1921); Eliel, Anderson, *ibid.* **74,** 547 (1952); Bender, Zerner, *ibid.* **84,** 2550 (1962).

Crystals from 95% ethanol; sweet odor of balsam. mp 39°, also reported as mp 33-34°, 34°, 34.5° (Volwiler and Vliet, Eliel and Anderson, Bender and Zerner, *see above*). Dec on distillation at ordinary pressure; $bp_{0.5}$ 154-157°, bp_5 195-200°, bp_{22} 228-230°. Practically insol in water; propylene glycol and glycerin; sol in alc, ether, oils. LD_{50} orally in rats: 5530 mg/kg, P. M. Jenner *et al., Food Cosmet. Toxicol.* **2,** 327 (1964).

USE: In artificial flavors, in perfumes, mainly as a fixative.

1145. Benzyl Cyanide. *Benzeneacetonitrile;* phenylacetonitrile; α-tolunitrile; ω-cyanotoluene. C_8H_7N; mol wt 117.14. C 82.02%, H 6.02%, N 11.96%. $C_6H_5CH_2CN$. Occurs in garden cress and other plants; made from benzyl chloride, and NaCN: Adams, Thal, *Org. Syn.* **vol. 2,** 9 (1922); **coll. vol. I,** 101 (107 in 2nd ed.).

Oily liquid, aromatic odor. d_{15}^{15} 1.0214. mp -23.8°. bp_{760} 233.5°; bp_{100} 161.8°; bp_{20} 119.4°; $bp_{1.0}$ 60°. n_D^{25} 1.52105. Insoluble in water, miscible with alc, ether.

1146. Benzyl Ether. *1,1'-[Oxybis(methylene)]bis[benzene];* dibenzyl ether. $C_{14}H_{14}O$; mol wt 198.25. C 84.81%, H 7.12%, O 8.07%. $(C_6H_5CH_2)_2O$. Prepd by: Lachman, *J. Am. Chem. Soc.* **45,** 2356 (1923); Staab, Wendel, *Ber.* **93,** 2902 (1960); Lichtenberger, Tritsch, *Bull. Soc. Chim. France* **1961,** 363. Manuf by reduction of benzaldehyde in the presence of $[Co(CO)_4]_2$: Wender, Orchin, U.S. pat. **2,614,107** (1952 to U.S.A. as represented by the Secy. of Agr.). Physical properties: Svirbely *et al., J. Am. Chem. Soc.* **71,** 507 (1949); Dreisbach, Martin, *Ind. Eng. Chem.* **41,** 2875 (1949). Miscibility: Jackson, Drury, *ibid.* **51,** 1491 (1959).

Unstable liquid, bp 295-298° (with dec), bp_{21} 173-174°; bp_2 125.5-126.5°. Appears to dec slowly at ordinary temps. d_4^{35} 1.0341; d_4^{25} 0.99735; d_4^{20} 1.00142; d_4^{15} 1.0482. n_D^{25} 1.5601 (Svirbely *et al.*), 1.53851 (Dreisbach, Martin); n_D^{20} 1.54057 (Dreisbach, Martin), 1.566 (Lichtenberger, Tritsch). Practically insol in water; miscible with ethanol, ether, chloroform, acetone.

USE: Plasticizer for nitrocellulose; solvent in perfumery.

1147. Benzyl Ethyl Ether. *(Ethoxymethyl)benzene.* $C_9H_{12}O$; mol wt 136.19. C 79.37%, H 8.88%, O 11.75%. $C_6H_5CH_2OC_2H_5$. Preparation from sodium ethoxide and benzyl bromide: Letsinger, Pollart, *J. Am. Chem. Soc.* **78,** 6079 (1956); by reduction of benzaldehyde diethyl acetal with $LiAlH_4-AlCl_3$: Eliel, Rerick, *J. Org. Chem.* **23,** 1088 (1958).

Oily liquid, aromatic odor. bp 186°; bp_{10} 65°. d 0.949, n_D^{20} 1.4955. Volatile with steam. Practically insol in water; miscible with alcohol, ether.

1148. Benzyl Formate. *Formic acid phenylmethyl ester; formic acid benzyl ester.* $C_8H_8O_2$; mol wt 136.14. C 70.57%, H 5.92%, O 23.50%. $HCOOCH_2C_6H_5$. Prepn from formic acid and benzyl alcohol: Mailhe, *Chem. Ztg.* **35,** 508 (1911).

Liquid; pleasant fruity odor. d 1.081. bp 203°. Practically insol in water; sol in alcohol.

USE: Solvent for cellulose esters; in perfumery.

1149. Benzyl Fumarate. *(E)-2-Butenedioic acid bis(phenylmethyl) ester; fumaric acid dibenzyl ester;* dibenzyl fumarate. $C_{18}H_{16}O_4$; mol wt 296.31. C 72.96%, H 5.44%, O

21.60%. $C_6H_5CH_2OOCCH=CHCOOCH_2C_6H_5$. Prepd from fumaric acid and benzyl alcohol: Volwiler, Vliet, *J. Am. Chem. Soc.* **43,** 1672 (1921).

Cryst powder, mp 58.5-59.5°. bp_5 210-211°. Practically insol in water; sol in alcohol, chloroform, ether, oils.

USE: In room spray deodorant: Kulka, U.S. pat. **3,077,-457** (1963 to Fritzsche Bros.).

1150. Benzylhydrochlorothiazide. *6-Chloro-3,4-dihydro-3-(phenylmethyl)-2H-1,2,4-benzothiadiazine-7-sulfonamide 1,1-dioxide; 3-benzyl-6-chloro-3,4-dihydro-2H-1,2,4-benzothiadiazine-7-sulfonamide 1,1-dioxide;* 6-chloro-7-sulfamoyl-3-benzyl-3,4-dihydro-1,2,4-benzothiadiazine 1,1-dioxide; 3-benzyl-6-chloro-3,4-dihydro-7-sulfamoyl-1,2,4-benzothiadiazine 1,1-dioxide; Behyd. $C_{14}H_{14}ClN_3O_4S_2$; mol wt 387.87. C 43.35%, H 3.64%, Cl 9.14%, N 10.83%, O 16.50%, S 16.53%. Prepn: Werner *et al., J. Am. Chem. Soc.* **82,** 1161 (1960); Novello *et al., J. Org. Chem.* **25,** 970 (1960); Ugi, U.S. pat. **3,108,097** (1963).

Crystals from acetic acid + water, mp 260-262°. Also reported as crystals from water, mp 269°.

THERAP CAT: Antihypertensive; diuretic.

1151. Benzylideneacetone. *4-Phenyl-3-buten-2-one;* benzalacetone; methyl styryl ketone; cinnamyl methyl ketone; acetocinnamone. $C_{10}H_{10}O$; mol wt 146.18. C 82.15%, H 6.90%, O 10.94%. $C_6H_5CH=CHCOCH_3$. Prepd by condensing acetone and benzaldehyde by means of aq alkali: Drake, Allen, *Org. Syn.* **3,** 17 (1923).

Lustrous plates on vacuum distillation. Coumarin type odor. Purification by steam distillation: Fromm, Haas, *Ann.* **394,** 291 (1912). mp 41.5°. d_{15}^{15} 1.0377; $d_4^{45.2}$ 1.0097. bp_{760} 261°; bp_{200} 211°; bp_{100} 187.8°; bp_{40} 161.3°; bp_{20} 143.8°; bp_{10} 127.4°; bp_5 112.2°; $bp_{1.0}$ 81.7°. $n_D^{45.9}$ 1.5836. Absorption spectrum: Baker, *J. Chem. Soc.* **91,** 1492 (1907); Baly, Schafer, *ibid.* **93,** 1813 (1908). Freely sol in alcohol, benzene, chloroform, ether; sparingly sol in water, petr ether.

USE: In perfumery, organic syntheses.

1152. Benzylideneaniline. *N-(Phenylmethylene)benzenamine;* benzalaniline. $C_{13}H_{11}N$; mol wt 181.23. C 86.15%, H 6.12%, N 7.73%. $C_6H_5N=CHC_6H_5$. Prepd by the action of aniline on benzaldehyde: Bigelow, Eatough, *Org. Syn.* **coll. vol. I** (2nd ed., 1941) p 80. Molecular structure: M. Traetteberg *et al., J. Mol. Struct.* **48,** 395 (1978).

Plates from 85% alc. Has been obtained as the yellow modification (needles from CS_2). The colorless form melts at 48°, after solidifying melts again at 56°. d_4^{50} 1.045. bp_{760} 300°. Absorption spectrum: Baly *et al., J. Chem. Soc.* **97,** 590 (1910). Sol in alcohol, chloroform, acetic anhydride, carbon disulfide.

1153. Benzylimidobis(p-methoxyphenyl)methane. *N-[Bis(4-methoxyphenyl)methylene]benzenemethanamine.* $C_{22}H_{21}NO_2$; mol wt 331.40. C 79.73%, H 6.39%, N 4.23%, O 9.66%. Prepd by heating *p,p'*-dimethoxybenzophenone, thionyl chloride and benzylamine: Schönberg, Urban, *Ber.* **67,** 1999 (1934).

Pale-yellow crystals, mp 89-91°. Soluble in ether, chloroform. Slightly sol in petr ether.

USE: Detection of elementary sulfur.

1154. Benzyl Methyl Ether. *(Methoxymethyl)benzene;* methyl benzyl ether. $C_8H_{10}O$; mol wt 122.16. C 78.65%, H 8.25%, O 13.10%. $C_6H_5CH_2OCH_3$. Prepd from benzyl chloride, methanol, and NaOH: Olson *et al., J. Am. Chem. Soc.* **69**, 2451 (1947).

Liquid. d 0.987. bp 174°. Practically insol in water; sol in alcohol, ether.

1155. Benzylmorphine. *7,8-Didehydro-4,5-epoxy-17-methyl-3-(phenylmethoxy)morphinan-6-ol.* $C_{24}H_{25}NO_3$; mol wt 375.45. C 76.77%, H 6.71%, N 3.73%, O 12.78%. Prepn from morphine and benzyl chloride in the presence of sodium ethylate: **Ger. pat.** **91,813** (1897 to E. Merck), Frdl. **4**, 1245; from morphine and benzyldimethylphenylammonium hydroxide: Rodionow, *Bull. Soc. Chim. France* [4] **45**, 109 (1929). Properties of hydrochloride: Dott, *Pharm. J.* **115**, 757 (1925); *see also* **Ger. pat.** above.

Needles from water, prisms or platelets from ether, chloroform. mp 132°. Soluble in 2500 parts cold water, more sol in boiling water; 1.6 parts dissolves in 100 parts ether; readily sol in 50% alcohol, benzene.

Hydrochloride, $C_{24}H_{26}ClNO_3$, *Peronine.* Monohydrate, needles from water. Sol at 15° in 133 parts water, 218 parts 95% alcohol, 390 parts chloroform, 100 parts methanol; slightly sol in abs alcohol; very slightly sol in ether, acetone, isoamyl alcohol.

Caution: May be habit forming. This is a controlled substance (opium derivative) listed in the U.S. Code of Federal Regulations, Title 21 Part 1308.11 (1985).

THERAP CAT: Narcotic analgesic.

1156. Benzylpenicillinic Acid. *3,3-Dimethyl-7-oxo-6-[(phenylacetyl)amino]-4-thia-1-azabicyclo[3.2.0]heptane-2-carboxylic acid;* free benzylpenicillin; free penicillin G; free penicillin II. $C_{16}H_{18}N_2O_4S$; mol wt 334.38. C 57.47%, H 5.43%, N 8.38%, O 19.14%, S 9.59%. Obtained by extraction at ice temp of the acidified (pH 2) aq soln of sodium benzylpenicillin with ether or chloroform. Shows correct analytical composition only if moisture is excluded completely during its isoln from the ether extract (by lyophilizing from benzene). When kept dry the acid retains its antibiotic potency for limited periods, but it is rapidly inactivated by small amounts of water. The product formed on slow inactivation by moisture seems to be benzylpenicilloic acid, *cf.* Clarke *et al., The Chemistry of Penicillin* (Princeton, 1949). Prepn of methyl ester: **Dan. pat.** **85,976** (1958 to Leo Pharm.). Pharmacokinetics in humans: M. Barza, L. Weinstein, *Clin. Pharmacokinet.* **1**, 297 (1976).

Amorphous white powder. $[\alpha]_D^{20}$ +269° (methanol soln 50 ml, prepd from 350 mg benzylpenicillin sodium). pK (water) at 5°: 2.74; (water) at 25°: 2.76; (80% EtOH) at 5°: 4.84. Sparingly sol in water. Sol in methanol, ethanol, ether, ethyl acetate, benzene, chloroform, acetone. Insol in petr ether.

Note: Trenner, Buhs, *J. Am. Chem. Soc.* **70**, 2897 (1948), obtained cryst benzylpenicillinic acid containing one mol of diisopropyl ether of crystn. The crystals were stable to air. mp 80-87°. $[\alpha]_D^{23}$ +241° (phosphate buffer at pH 7).

Methyl ester, crystals from carbon tetrachloride, mp 97°. $[\alpha]_D^{20}$ +328° (methanol); $[\alpha]_D^{20}$ +286° (chloroform). uv max: 252.5, 258, 264.5, 268.5 nm (E_M 290, 250, 180, 120). Practically insol in water. Freely sol in methanol, benzene, chloroform, amyl acetate; sol in peanut oil. Hydrolyzes in aqueous bicarbonate soln to an active principle, presumably benzylpenicillin sodium.

THERAP CAT: Antibacterial.

1157. Benzylpenicillin Sodium. *[2S-(2α,5α,6β)]-3,3-Dimethyl-7-oxo-6-[(phenylacetyl)amino]-4-thia-1-azabicyclo[3.2.0]heptane-2-carboxylic acid monosodium salt;* sodium penicillin G; penicillin G sodium; sodium penicillin II; sodium benzylpenicillinate; benzylpenicillinic acid sodium salt; penicillin; American penicillin; Nalpen G; Novocillin; Penilaryn; Pen-A-Brasive; Veticillin. $C_{16}H_{17}N_2NaO_4S$; mol wt 356.38. C 53.92%, H 4.81%, N 7.86%, Na 6.45%, O 17.96%, S 9.00%. Industrial production of antibiotic from *Penicillium chrysogenum:* Fleming, *Brit. J. Exp. Pathol.* **10**, 226 (1929); Clutterbuck *et al., Biochem. J.* **26**, 1907 (1932); Chain *et al., Lancet* **2**, 226 (1940); Abraham *et al., ibid.* **2**, 177 (1941); Clarke *et al., The Chemistry of Penicillin* (Princeton, 1949). Biosynthesis from amino acids: Arnstein, Grant, *Biochem. J.* **57**, 353, 360 (1954). Production process: Demain, Somerson, **U.S. pat.** **3,024,169** (1962 to Merck & Co.). Synthesis: Sheehan, **U.S. pat.** **3,159,617** (1964 to Arthur D. Little); M. D. Bachi, R. Breiman, *J. Chem. Soc. Perkin Trans I* **1980**, 11. Solubilities: Weiss *et al., Antibiot. & Chemother.* **7**, 374 (1957). HPLC determn in body fluids: C. van Gulpen *et al., J. Chromatog.* **381**, 365 (1986). Review of early literature: E. Chain, *Ann. Rev. Biochem.* **17**, 657 (1948).

Crystals from methanol + ethyl acetate. d 1.41. $[\alpha]_D^{24.8}$ +301° (c = 2.0). uv max (water soln): 252, 258.6, 264.4 nm (E_M about 300, 240, 180). pK at 25° = 2.76; pH 5.5-6.5. One I.U. or U.S.P. penicillin unit is equiv to 0.6 μg (0.0005988 mg) of benzylpenicillin sodium. One mg of benzylpenicillin sodium is equiv to 1670 units. Very sol in water, isotonic NaCl soln, and glucose solns. Also sol in alcohol, glycerol, and other primary alcohols. Practically insol in acetone, ether, $CHCl_3$, ethyl and amyl acetate. Insol in fixed oils and in liq petrolatum. Pure cryst benzylpenicillin sodium is stable. At pH 5.5 to 6.0 refrigerated solns may be kept for a few days. Benzylpenicillin is also precipitated by many metal ions. *See also* Procaine Penicillin G.

Pharmaceutical Incompatibilities: Must not be formulated with acids, salts of heavy metals, glycerol, naphthalene oils, resorcinol, ZnO, vitamin B_1, procaine, ephedrine, iodine and iodides, alcohol and oxidizing agents. *Ref:* R. K. Aliev, M. A. Etinger, *Aptechnoe Delo* **2**, 7 (1952), *C.A.* **46**, 7707 (1952).

THERAP CAT: Antibacterial.

THERAP CAT (VET): Antibacterial.

1158. o-Benzylphenol. *2-(Phenylmethyl)phenol; α-phenyl-o-cresol;* (2-hydroxydiphenyl)methane. $C_{13}H_{12}O$; mol wt 184.23. C 84.75%, H 6.57%, O 8.68%. Prepn from benzyl bromide and sodium phenoxide: Kornblum, Lurie, *J. Am. Chem. Soc.* **81**, 2711 (1959).

Crystals or liquid. mp 20.2-20.9°. bp_{10} 154-156°; $bp_{1.0}$ 121-123°. n_D^{19} 1.59945. Practically insol in water; sol in organic solvents, in fixed alkali hydroxide solns.

1159. p-Benzylphenol. *4-(Phenylmethyl)phenol; α-phenyl-p-cresol;* (4-hydroxydiphenyl)methane. Empirical formula: *See* preceding o-isomer. Prepd from phenol and benzyl chloride in the presence of zinc chloride: Ziegenbein *et al., Ber.* **88**, 1906 (1955).

Crystals, mp 84°. bp 322°; bp_4 154-157°. Slightly sol in cold water; moderately sol in hot water; sol in organic solvents, glacial acetic acid, alkali hydroxide solns.

USE: Germicide, antiseptic, preservative; also in organic syntheses.

1160. Benzyl Salicylate. *2-Hydroxybenzoic acid phenylmethyl ester; salicylic acid benzyl ester.* $C_{14}H_{12}O_3$; mol wt 228.24. C 73.67%, H 5.30%, O 21.03%. Prepd from sodium salicylate and benzyl chloride: Volwiler, Vliet, *J. Am. Chem. Soc.* **43,** 1672 (1921).

Thick liquid, slight, pleasant odor. d^{20} 1.175. bp_{25} 208°. Slightly sol in water; miscible with alcohol or ether.

USE: As fixer in perfumery; in sunscreen preparations.

1161. Benzylsulfamide. *4-[(Phenylmethyl)amino]benzenesulfonamide;* N^4-benzylsulfanilamide; RP 46; M & B 125; Septazine; Setazine; Chemodyn; Proseptazine. $C_{13}H_{14}N_2O_2S$; mol wt 262.32. C 59.52%, H 5.38%, N 10.68%, O 12.20%, S 12.22%. Prepd by reduction of the anil (Schiff base) resulting from the reaction of benzaldehyde with sulfanilamide: Goissedet *et al., Compt. Rend. Soc. Biol.* **121,** 1082 (1936).

Crystals from dioxane + water, mp 175°. Very slightly sol in water. Somewhat more sol in alc. Sol in acetone and dioxane. Neutral reaction. Breaks down in the body with the liberation of free sulfanilamide.

THERAP CAT: Antibacterial.

1162. Benzyl Sulfide. *1,1'-[Thiobis(methylene)]bisbenzene;* dibenzylsulfide. $C_{14}H_{14}S$; mol wt 214.32. C 78.45%, H 6.59%, S 14.96%. ($C_6H_5CH_2)_2S$. Prepn: Runge *et al., J. Prakt. Chem.* **11,** 284 (1960). Manuf from benzyl chloride and Na_2S: Stucker, Brennan, U.S. pat. **2,755,305** (1956 to Pure Oil Co.).

Plates, mp 49°. Practically insol in water; sol in alc, ether.

1163. p-(Benzylsulfonamido)benzoic Acid. *4-[[(Phenylmethyl)sulfonyl]amino]benzoic acid; p-α-toluenesulfonamidobenzoic acid; 4'-carboxyphenylmethanesulfonanilide;* Carinamide; Caronamide; Staticin. $C_{14}H_{13}NO_4S$; mol wt 291.32. C 57.72%, H 4.49%, N 4.81%, O 21.97%, S 11.01%. Prepn: Brit. pat. **666,546** (1952 to Sharp & Dohme).

Crystals, mp 229-230°. uv max (0.1N NaOH): 280-281 nm. Moderately sol in alcohol, acetone, propylene glycol. Very slightly sol in water, carbon tetrachloride, chloroform, benzene. Both the mono- and disodium salts are very sol in water. Aq solns may be prepared by dissolving the acid in an equivalent amount of sodium carbonate or sodium hydroxide soln. It is relatively stable to hydrolysis with acid or base. LD_{50} orally in mice: 2.45 g/kg; i.v. in mice, dogs, rabbits: 1.405, 1.575, 1.32 g/kg, Beyer *et al., J. Pharmacol. Exp. Ther.* **91,** 263 (1947).

THERAP CAT: Adjunct in penicillin therapy.

1164. S-Benzylthiuronium Chloride. *Carbamimidothioic acid phenylmethyl ester monohydrochloride.* $C_8H_{11}ClN_2S$; mol wt 202.72. C 47.40%, H 5.47%, Cl 17.49%, S 15.82%. Prepd by refluxing a mixture of 126 g benzyl chloride, 76 g thiourea, and 200 ml alcohol for 0.5 hr: Donleavy, *J. Am. Chem. Soc.* **58,** 1004 (1936).

Crystals from alc or dil HCl, mp 172-174°. Metastable form, mp 146-148°. Can be converted to the higher melting form by dissolving in alcohol and seeding with crystals, mp 172-174°.

USE: Reagent for cobalt and nickel; identification and separation of carboxylic, sulfinic, and sulfonic acid.

1165. Benzylurea. *(Phenylmethyl)urea;* benzylcarbamide. $C_8H_{10}N_2O$; mol wt 150.18. C 63.98%, H 6.71%, N 18.66%, O 10.65%. $C_6H_5CH_2NHCONH_2$. Prepn: Neville, McGee, *Can. J. Chem.* **41,** 2123 (1963).

Crystals, mp 147-148°; dec at 200°. One gram dissolves in 60 ml warm water, 33 ml acetone; slightly sol in benzene, ether.

1166. Bephenium. *N,N-Dimethyl-N-(2-phenoxyethyl)-benzenemethanaminium.* Prepn of salts: Copp, U.S. pat. **2,918,401** (1959 to Burroughs Wellcome); Copp, Ger. pat. **1,117,600** (1961 to Wellcome Foundation), *C.A.* **56,** 14165i (1962).

Hydroxynaphthoate, $C_{28}H_{29}NO_4$, *benzyldimethyl(2-phenoxyethyl)ammonium 3-hydroxy-2-naphthoate,* Alcopar(a), Befeniol, Lecibis, Nemex. Crystals, mp 170-171°.

Pamoate, $C_{57}H_{60}N_2O_8$, *benzyldimethyl(2-phenoxyethyl)-ammonium 4,4'-methylenebis[3-hydroxy-2-naphthoate],* bephenium embonate, Frantin. Dihydrate, pale yellow solid from water, mp 144-146°.

Chloride, $C_{17}H_{22}ClNO$, crystals from acetone, mp 135-136°.

Bromide, $C_{17}H_{22}BrNO$, crystals from isopropanol + ethyl acetate, mp 144.5-146°.

Iodide, $C_{17}H_{22}INO$, crystals from methanol + ether, mp 146-147°.

THERAP CAT: Anthelmintic (Nematodes).

THERAP CAT (VET): Anthelmintic.

1167. Bepridil. *β-[(2-Methylpropoxy)methyl]-N-phenyl-N-(phenylmethyl)-1-pyrrolidineethanamine;* 1-[2-(N-benzylanilino)-1-(isobutoxymethyl)ethyl]pyrrolidine; 1-isobutoxy-2-pyrrolidino-3-N-benzylanilinopropane; 3-isobutoxy-2-pyrrolidino-N-phenyl-N-benzylpropylamine. $C_{24}H_{34}N_2O$; mol wt 366.54. C 78.64%, H 9.35%, N 7.64%, O 4.37%. Calcium channel blocker with antianginal and antiarrhythmic properties. Prepn: R. Y. Mauvernay *et al.,* Ger. pat. **2,310,918;** eidem, U.S. pat. **3,962,238** (1972, 1976 both to CERM). Structure corrected in U.S. pat. re-issued as **Re. 30,577** (1981). Comparative effects on cardiac activity in dogs: M. T. Michelin *et al., Therapie* **32,** 485 (1977). Mechanism of action studies: S. Vogel *et al., J. Pharmacol. Exp. Ther.* **210,** 378 (1979); C. Labrid *et al., ibid.* **211,** 546 (1979). Efficacy estimation: J. C. Canicave *et al., Therapie* **35,** 607 (1980). Comparative antidysrhythmic profile: C. Labrid *et al., Arch. Int. Pharmacodyn. Ther.* **249,** 87 (1981). Comparison of hemodynamic and coronary changes: J. P. Merillon *et al., Therapie* **36,** 123 (1981).

Viscous liquid, $bp_{0.1}$ 184°, $bp_{0.5}$ 192°. n_D^{20} 1.5538.

Hydrochloride monohydrate, $C_{24}H_{35}ClN_2O.H_2O$, *CERM 1978, Angopril, Bepadin, Cordium, Vascor.* Crystals, mp

91° ±2°. LD$_{50}$ in mice (mg/kg): 1955 orally, 23.5 i.v., **Ger. pat. 2,802,864** (1977 to CERM).

THERAP CAT: Antianginal.

1168. Berbamine. *6,6',7-Trimethoxy-2,2'-dimethyl-berbaman-12-ol.* C$_{37}$H$_{40}$N$_2$O$_6$; mol wt 608.71. C 73.00%, H 6.62%, N 4.60%, O 15.77%. From *Berberis vulgaris* L. and *Berberis aquifolium* Pursh, *Berberidaceae:* Rüdel, *Arch. Pharm.* **229**, 631 (1891); Pommerehne, *ibid.* **233**, 127 (1895). From *Atherosperma moschatum* Labill., *Monimiaceae:* Bick *et al., Aust. J. Chem.* **9**, 111 (1956). Isomer of oxyacanthine: Santos, *C.A.* **24**, 1647 (1930). Structure: v. Bruchhausen *et al., Ann.* **507**, 144 (1933); Inubushi, *J. Pharm. Soc. Japan.* **72**, 220 (1952), *C.A.* **47**, 6429e (1953).

Crystals from petr ether, mp 197-210°. [α]$_D^{20}$ +114.6° (chloroform); pKa 7.33 at 20° in methanol. Slightly sol in water; sol in alcohol, ether, chloroform, petr ether.

Hydrochloride, C$_{37}$H$_{40}$N$_2$O$_6$·HCl.4H$_2$O. [α]$_D^{20}$ +63.2°.
Dihydrochloride, C$_{37}$H$_{40}$N$_2$O$_6$·2HCl, dec 270°.
Hydrobromide, C$_{37}$H$_{40}$N$_2$O$_6$·HBr, dec 283°.
Methyl ether, C$_{38}$H$_{42}$N$_2$O$_6$, mp 182°. [α]$_D$ +132°. Occurs naturally as the alkaloid *isotetrandrine.*

1169. Berberine. *5,6-Dihydro-9,10-dimethoxybenzo[g]-1,3-benzodioxolo[5,6-a]quinolizinium; 7,8,13,13a-tetradehydro-9,10-dimethoxy-2,3-(methylenedioxy)berbinium;* umbellatine. [C$_{20}$H$_{18}$NO$_4$]$^+$; mol wt 336.37. Alkaloid isolated from *Hydrastis canadensis* L., *Berberidaceae;* found in many other plants: Wasicky, Joachimowitz, *Arch. Pharm.* **255**, 497 (1917); Stermitz, *J. Pharm. Sci.* **56**, 760 (1967); Kawashima, *J. Pharm. Soc. Japan* **89**, 1386 (1969), *C.A.* **72**, 15706v (1970). Structure: Perkin, Robinson, *J. Chem. Soc.* **97**, 305 (1910). Identity with umbellatine: Govindachari *et al., Proc. Indian Acad. Sci.* **47A**, 41 (1958), *C.A.* **52**, 14630f (1958). Biosynthesis: Gear, Spenser, *Can. J. Chem.* **41**, 783 (1963). Pharmacology: Fukuda *et al., Chem. Pharm. Bull.* **18**, 1299 (1970); Shanbhag *et al., Japan. J. Pharmacol.* **20**, 482 (1970), *C.A.* **74**, 97717c (1971). Synthetic studies: Sainsbury *et al., Tetrahedron* **25**, 1881 (1969). Total synthesis of berberine iodide: Kametani *et al., J. Chem. Soc. (C)* **1969**, 2036. *Review:* Hahn, Ciak, in *Antibiotics* vol. 3, J. W. Corcoran, F. E. Hahn, Eds. (Springer-Verlag, New York, 1975) pp 577-584.

Yellow needles from ether, mp 145°: Gadamer, *Arch. Pharm.* **243**, 33 (1905). Dissolves slowly in water with alkaline reaction; behaves as a quaternary base, forming salts by replacement of the OH group: Perkin, Jr., *J. Chem. Soc.* **113**, 503 (1918). uv max: 265, 343 nm. pK 2.47. K 3.35 × 10^{-3}.

A tautomeric pseudobase and an imino aldehyde form are indicated: Pictet, Gams, *Compt. Rend.* **152**, 102 (1911); **153**, 386 (1911); Perkin *et al., J. Chem. Soc.* **127**, 740 (1925); Skinner, *ibid.* **1950**, 823.

Hydroxide hemihendecahydrate, C$_{20}$H$_{19}$NO$_5$.5½H$_2$O, long, silky, yellow needles. Can be dried to the hemipentahydrate at 100°, begins to dec at 110°. One gram dissolves in 20 ml water, 100 ml alcohol; much more sol in hot liquids; slightly sol in acetone, ether, benzene, chloroform.

Acid sulfate, C$_{20}$H$_{19}$NO$_8$S, *berberine bisulfate.* Yellow needles. Sol in about 100 parts water; slightly sol in alc.

Chloride dihydrate, C$_{20}$H$_{18}$NO$_4$.Cl.2H$_2$O, yellow crystals. Slightly sol in cold, freely in boiling water. Practically insol in cold alcohol, chloroform, ether.

Sulfate trihydrate, (C$_{20}$H$_{18}$NO$_4$)$_2$.SO$_4$.3H$_2$O, yellow needles. Sol in about 30 parts water; sol in alcohol. LD$_{50}$ in mice: 24.3 mg/kg i.p. Activity studies: Amin *et al., Can. J. Microbiol.* **15**, 1067 (1969). Pharmacology: Sabir, Bhide, *Indian J. Physiol. Pharmacol.* **15**, 111 (1971).

THERAP CAT: Bitter stomachic. Antibacterial. Antimalarial, antipyretic.

1170. Berberis. Holly-leaved barberry; Oregon grape root; mountain grape. Dried rhizome and roots of species of section of *Mahonia* DC. of the genus *Berberis* L., *Berberidaceae.* *Habit.* U.S. and British Columbia. *Constit.* Berberine, berbamine, oxyacanthine, phytosterol, sugar.

THERAP CAT: Bitter. Antipyretic.

1171. Berberis Aristata. Indian barberry; ruswut; rusat. Dried stem of *Berberis aristata* DC., *Berberidaceae.* *Habit.* India, Ceylon. *Constit.* Berberine and other alkaloidal substances, tannin, resin, starch.

THERAP CAT: Bitter. Antipyretic.

1172. Berbine. *5,8,13,13a-Tetrahydro-6H-dibenzo[a,g]-quinolizine;* tetrahydroprotoberberine. C$_{17}$H$_{17}$N; mol wt 235.31. C 86.77%, H 7.28%, N 5.95%. Proposed as parent substance for naming berberine and similar alkaloids: Awe, *Arch. Pharm.* **270**, 156 (1932). Synthesis: Chakravarti *et al., J. Chem. Soc.* **1927**, 2278.

Clusters of needles from ether, mp 85°. Strong base, dissolves readily in alcohol, ether. Insol in petr ether.

Picrate, mp 151°, sol in hot alcohol.

1173. Bergapten(e). *4-Methoxy-7H-furo[3,2-g][1]benzopyran-7-one;* 5-methoxpsoralen; bergaptan; heraclin; majudin; 5-MOP; Psoraderm. C$_{12}$H$_8$O$_4$; mol wt 216.19. C 66.67%, H 3.73%, O 29.60%. Naturally occurring analog of psoralen and isomer of methoxsalen, q.q.v., found in a wide variety of plants. It was first isolated from oil of bergamot from *Citrus bergamia* Risso, *Aurantiodiae:* Pomeranz, *Monatsh.* **12**, 379 (1891), **14**, 28 (1893). Isoln from *Fagara xanthoxyloides* Lam., *Rutaceae:* H. Thoms, E. Baetcke, *Ber.* **44**, 3326 (1911); **45**, 3705 (1912). Synthesis: E. Späth *et al., Ber.* **70**, 478 (1937); W. N. Howell, R. Robertson, *J. Chem. Soc.* **1937**, 293; G. Caporale, *Farmaco Ed. Sci.* **13**, 784 (1958); V. K. Ahluwalia *et al., Indian J. Chem.* **7**, 831 (1969). Use in photochemotherapy of psoriasis: H. Hönigsmann *et al., Brit. J. Dermatol.* **101**, 369 (1979). Mutagenicity studies: B. R. Scott *et al., Mutat. Res.* **39**, 29 (1976); M. J. Ashwood-Smith *et al., Nature* **285**, 407 (1980). Phototoxicity study: A. Kornhauser *et al., Science* **217**, 733 (1982).

Needles from alcohol, mp 188° (sublimes). Practically insol in boiling water; slightly sol in glacial acetic acid, chloroform, benzene, warm phenol. Sol in abs alcohol: 1 part in 60. Its soln in sulfuric acid is yellow-gold.

USE: Has been used to promote tanning in suntan preparations.

THERAP CAT: Antipsoriatic.

1174. Bergenin. *3,4,4a,10b-Tetrahydro-3,4,8,10-tetrahydroxy-2-(hydroxymethyl)-9-methoxypyrano[3,2-c][2]benzopyran-6(2H)-one; 4-methoxy-2-[tetrahydro-3,4,5-trihydroxy-6-(hydroxymethyl)pyran-2-yl]-α-resorcylic acid δ-lactone;*

bergenit; vakerin; ardisic acid B; cuscutin; peltophorin. C_{14}-$H_{16}O_9$; mol wt 328.27. C 51.22%, H 4.91%, O 43.86%. From root of *Saxifraga (Bergenia) crassifolia* L., and from rhizome of *S. sibirica* L. *Saxifragaceae:* Morelle, *Compt. Rend.* **93**, 646 (1881); *Ber.* **14**, 2694 (1881); Ssadikow, Guthner, *Biochem. Z.* **190**, 340 (1927); Tschitschibabin *et al., Ann.* **469**, 93 (1929). Identity with vakerin: Carruthers *et al., Chem. & Ind. (London)* **1957**, 76. Identity with ardisic acid B: Hung, Chu, *C.A.* **52**, 15827h (1958). Identity with cuscutin: Jain, Mishra, *Indian J. Chem.* **1**, 499 (1963). Identity with peltophorin: Joshi, Kamat, *Naturwiss.* **56**, 89 (1969). Structure: Posternak, Dürr, *Helv. Chim. Acta* **41**, 1159 (1958); Fujise *et al., Bull. Chem. Soc. Japan* **32**, 97 (1959). Synthesis and structure: Hay, Haynes, *J. Chem. Soc.* **1958**, 2231. Biosynthesis: Wenkert, *Chem. & Ind. (London)* **1959**, 906.

Crystals from methanol, mp 238°. uv max: 275, 220 nm (log ε 3.92, 4.42). $[\alpha]_D^{18}$ −37.7° (c = 1.96 in ethanol); $[\alpha]_D^{24}$ −45.3° (c = 0.51 for anhydr in water). Freely sol in water; sol in alcohol.

Monohydrate, crystals from water, mp 140°. Slightly sol in water, freely in alcohol.

1175. Berkelium. Bk; at. wt (most stable known isotope) 247; at. no. 97; valence 3, also 4. A man-made radioactive element; second element in the curide series. First produced in 1950 by helium ion bombardment of ^{241}Am yielding ^{243}Bk ($T_{1/2}$ 4.6 hrs): Thompson *et al., Phys. Rev.* **77**, 838 (1950). Twelve isotopes have been produced, mass numbers 240-251. ^{249}Bk ($T_{1/2}$ 314 days, β-emitter) produced in subgram amounts from ^{239}Pu by multiple neutron capture in high-flux nuclear reactors. Prepn of metal and determn of crystal structure: Fahey, *U.S. At. Energy Comm.* **TID-25741** (1971) 119 pp, *C.A.* **76**, 9882r (1972); Peterson *et al., J. Inorg. Nucl. Chem.* **33**, 3345 (1971). *Reviews:* Cunningham, *J. Chem. Ed.* **36**, 32-37 (1959); M. Haissinsky, J.-P. Adloff, *Radiochemical Survey of the Elements* (Elsevier, New York, 1965) pp 14-15; C. Keller, *The Chemistry of the Transuranium Elements* (Verlag Chemie, Weinheim, English Ed., 1971) pp 553-566; Silva, "Trans-Curium Elements" in *MTP Int. Rev. Sci.: Inorg. Chem., Ser. One* vol. **8**, A. G. Maddock, Ed. (University Park Press, Baltimore, 1972) pp 71-105; *Comprehensive Inorganic Chemistry* vol. **5**, J. C. Bailar, Jr. *et al.,* Eds. (Pergamon Press, Oxford, 1973) *passim;* several authors, *Handb. Exp. Pharmakol.* **36**, 689-928 (1973).

Metal. Two allotropic forms: α-form, double hexagonal close packed structure, d (calc) 14.78 g/cc; β-form, face centered cubic structure, d (calc) 13.25 g/cc. mp 986 ± 25° (Fahey, *loc. cit.*). Changes from the trivalent to the tetravalent state under the influence of oxidizing agents. In the trivalent state, its chemical properties are very close to those of curium. Can be separated from other transuranium elements by ion-exchange or by extraction of Bk(IV) with dioctylphosphoric acid in heptane: Haissinsky, Adloff, *loc. cit.*

1176. Berninamycin. *Berninamycin A.* $C_{51}H_{50}N_{14}O_{16}S$; mol wt 1147.10. C 53.40%, H 4.39%, N 17.09%, O 22.32%, S 2.79%. Major component of cyclic peptide antibiotic complex also containing minor component *Berninamycin B* ($C_{59}H_{74}N_{14}O_{22}S$). Isoln from *Streptomyces bernesis* Dietz: M. E. Bergy *et al.,* **U.S. pat. 3,689,639** (1972). Inhibition of protein synthesis: F. Reusser, *Biochemistry* **8**, 3303 (1969); J. Thompson *et al., J. Gen. Microbiol.* **128**, 875 (1982). Structural studies and characterization of degradation product berninamycinic acid, a novel sulfur containing moiety: J. M. Liesch *et al., J. Am. Chem. Soc.* **98**, 299 (1976); J. M. Liesch *et al., ibid.* 8237; J. M. Liesch, K. L. Rinehart, Jr., *ibid.* **99**, 1645 (1977). Revised structure: H. Abe *et al., Tetrahedron Letters* **29**, 1401 (1988). Biosynthetic study: C.

J. Pearce, K. L. Rinehart, Jr., *J. Am. Chem. Soc.* **101**, 5069 (1979). Use as growth permittant: R. M. Pellegrino, **Eur. pat. Appl. 112,233** (1984 to Merck & Co.). Synthesis of berninamycinic acid: T. R. Kelly *et al., Tetrahedron Letters* **25**, 2127 (1984).

White crystals, mp >290°. Sol in DMF, methanol, ethanol, propanol, butanol. Relatively insol in water, ether, cyclohexane, benzene, acetone, ethyl acetate. uv max (methanol): 208, 236 nm (A = 62.4, 64.4).

Berninamycinic acid, $C_{12}H_6N_2O_5S$, golden needles, mp 210° (dec). pKa 5.8. Insol in water. uv max (0.01N HCl): 228, 272 nm (ε 13500, 13500); uv max (0.01N NaOH): 232, 294 nm (ε 9500, 13000).

Note: The characteristics of berninamycin are similar to those previously reported in the literature for **theiomycetin:** M. Shibata, *Takeda Kenkyusho Nempo* **18**, 44 (1959), *C.A.* **54**, 19840c (1960).

1177. Beryllium. Glucinium. Be; at. wt 9.01218; at. no. 4; valence 2. Group 2a. Estimates of abundance in earth's crust vary from 2 to 10 ppm. Natural isotopes: 9 (100%); radioactive isotopes (mass numbers): 6-8; 10-12. Oxide discovered by Vauquelin in 1797; free metal isolated by Wöhler and Bussy in 1828. Produced industrially from *beryl* ($3BeO.Al_2O_3.6SiO_2$); also found in *phenacite* (Be_2SiO_4), *chrysoberyl* ($BeO.Al_2O_3$). Precious forms of beryl: *emerald, aquamarine.* Reviews of beryllium and its compounds: Kjellgren, "Beryllium" in *Rare Metals Handbook,* C. A. Hampel, Ed. (Reinhold, New York, 1954) pp 31-55; D. A. Everest, *Chemistry of Beryllium* (Elsevier, New York, 1964) 151 pp. *Review:* Pinto, Greenspan, "Beryllium" in *Modern Materials,* vol. **6**, B. W. Gonser, Ed. (Academic Press, New York, 1968) pp 319-372; D. A. Everest, "Beryllium" in *Comprehensive Inorganic Chemistry,* J. C. Bailar, Jr. *et al.,* Eds. (Pergamon Press, Oxford, 1973) pp 531-590; J. Ballance *et al.,* in Kirk-Othmer *Encyclopedia of Chemical Technology* vol. **3** (Wiley-Interscience, New York, 3rd ed., 1978) pp 803-823. Review of carcinogenicity studies of beryllium and beryllium compds: *IARC Monographs* **1**, 17-28 (1972); *ibid.* **23**, 143-204 (1980). Review of health effects of beryllium and its compds: *Beryllium: Its Industrial Hygiene Aspects,* H. E. Stokinger, Ed. (Academic Press, New York, 1966) 318 pp.

Gray metal; close-packed hexagonal structure; anisotropic; high permeability to X-rays. mp 1287°. bp 2500° (extrapolated). d 1.8477. Heat capacity at constant pressure (30°) 0.437 cal/g/°C: Walker *et al., J. Chem. Eng. Data* **7**, 595 (1962). Latent heat of fusion: 3.5 kcal/mole. Brinell hardness: 60-125. Chemical properties similar to aluminum; metal resistant to attack by acid due to the formation of a thin oxide film. E° (aq) Be/Be²⁺ 1.85 V (calc.). Finely divided or amalgamated metal reacts with HCl, dil H_2SO_4 and dil HNO_3; attacked by strong bases with evolution of H_2.

Caution: Death may result from short exposure to very low concns of the element and its salts. Contact dermatitis, chemical conjunctivitis, corneal burns, non-healing ulceration at site of injury, subcutaneous nodules may occur following exposure. *Acute:* Pneumonitis may result from single exposure to beryllium and occasionally is fatal. *Chronic:* Pulmonary granulomatous disease may appear in 3 months to 15 years, often after short exposure to low concn. Uncertainty as to complete recovery. Death rate about 25%. *See:* J. Schubert, "Beryllium and Berylliosis" in *Sci. Am.* **199**, no. 2, pp 27-33 (1958). This substance and certain beryllium compounds may reasonably be anticipated to be carcinogens: *Fourth Annual Report on Carcinogens* (NTP 85-002, 1985) p 42.

USE: Source of neutrons when bombarded with alpha particles according to the equation $^9_4Be + ^4_2He \rightarrow ^{12}_6C + ^1_0n$. This yields about 30 neutrons per million alpha particles. Also as neutron reflector and neutron moderator in nuclear reactors. In beryllium copper and beryllium aluminum alloys (by direct reduction of beryllium oxide with carbon in the presence of Cu or Al). In radio tube parts. In aerospace structures. In inertial guidance systems.

1178. Beryllium Acetate. *Acetic acid beryllium salt.* C_4-H_6BeO_4; mol wt 127.10. C 37.80%, H 4.76%, Be 7.09%, O 50.35%. Prepn: Besson, Hardt, *Compt. Rend.* **237**, 1525 (1953).

Crystals, dec at 60-100° when heated slowly, 150-180° when heated rapidly. Dissolves slowly with hydrolysis in boiling water; practically insol in abs alc and other common organic solvents.

1179. Beryllium Acetate, Basic. *Hexakis(acetato)oxotetraberyllium;* beryllium oxide acetate. $C_{12}H_{18}Be_4O_{13}$; mol wt 406.32. C 35.47%, H 4.47%, Be 8.87%, O 51.19%. Be_4O-$(OCOCH_3)_6$. Prepn: Urbain, Lacombe, *Compt. Rend.* **133**, 874 (1901); **134**, 772 (1902); Haber, van Oordt, *Z. Anorg. Chem.* **40**, 465 (1904); Moeller *et al., Inorg. Syn.* **3**, 9 (1950).

Tetrahedra from chloroform, d 1.25. mp 285-286°. bp 330-331°. Practically insol in water. Hydrolyzed by hot water and by dil acids. Sol in hot glacial acetic acid; in the usual organic solvents except alcohol, ether.

1180. Beryllium Acetylacetonate. *Bis(2,4-pentanediona-to)beryllium.* $C_{10}H_{14}BeO_4$; mol wt 207.23. C 57.95%, H 6.81%, Be 4.35%, O 30.88%. Prepd by the action of 2,4-pentanedione on an soln of beryllium hydroxide in dil acetic acid or on a soln of beryllium chloride in presence of ammonia: Biltz, *Ann.* **331**, 336 (1904); Parsons, *J. Am. Chem. Soc.* **26**, 732 (1904); Arch, Young, *Inorg. Syn.* **2**, 17 (1946); from 2,4-pentanedione and beryllium sulfate in NaOH soln: Jones, *J. Am. Chem. Soc.* **81**, 3188 (1959).

Monoclinic crystals. mp 108°. bp 270°. d_4^{20} 1.168. Practically insol in water; hydrolyzed by boiling water. Freely sol in alc, acetone, ether, benzene, CS_2, other organic solvents.

1181. Beryllium Borohydride. *Beryllium tetrahydrobo-rate(1−).* B_2BeH_8; mol wt 38.72. B 55.88%, Be 23.29%, H 20.82%. $Be(BH_4)_2$. Prepn from diborane and dimethylberyllium: Burg, Schlesinger, *J. Am. Chem. Soc.* **62**, 3425 (1940); from $LiBH_4$ and $BeCl_2$: Schlesinger *et al., ibid.* **75**, 212 (1953).

Similar to aluminum borohydride, but less volatile. Spontaneously flammable. Sublimes at 91.3°. Dec above 123°

without melting. Reacts vigorously with water, HCl to liberate hydrogen.

1182. Beryllium Bromide. $BeBr_2$; mol wt 168.85. Be 5.34%, Br 94.66%. Prepn: Ehrlich in *Handbook of Preparative Inorganic Chemistry,* vol. **1**, G. Brauer, Ed. (Academic Press, New York, 2nd ed., 1963) p 891. Review of beryllium halides: Bell, *Advan. Inorg. Chem. Radiochem.* **14**, 255-332 (1972).

Orthorhombic crystals, d 3.465. mp 506-509°; also reported as 488°: Bell, *loc. cit.* Sublimes at 473°. bp 520°. Very hygroscopic. Freely sol in water. By saturating the concd viscous soln with HBr, the tetrahydrate is formed. Sol in ethanol, in pyridine (185.6 g/l), in ethyl bromide (1.0 g/l). Forms addition compounds with amines, alcohols. *Keep tightly closed.*

1183. Beryllium Carbide. CBe_2; mol wt 30.04. C 39.98%, Be 60.01%. Be_2C. Prepn: Coobs, Koshuba, *J. Electrochem. Soc.* **99**, 115 (1952); Mallett *et al., ibid.* **101**, 298 (1954); Ehrlich in *Handbook of Preparative Inorganic Chemistry,* vol. **1**, G. Brauer, Ed. (Academic Press, New York, 2nd ed., 1963) p 899.

Brick-red or yellow-red octahedra, d 1.90, dec above 2100°. Very slowly dec by water, somewhat faster by mineral acids and quickly by alkalies with the evolution of methane.

USE: Nuclear reactor core material: Schwartz, U.S. pat. 3,170,812 (1965 to USAEC).

1184. Beryllium Chloride. $BeCl_2$; mol wt 79.93. Be 11.28%, Cl 88.72%. Prepn from the elements: Tannenbaum, *Inorg. Syn.* **5**, 22 (1957); from BeO, Cl_2 and C: Ehrlich in *Handbook of Preparative Inorganic Chemistry,* vol. **1**, G. Brauer, Ed. (Academic Press, New York, 2nd ed., 1963) p 889. Toxicity data: K. W. Cochran *et al., Fed. Proc.* **9**, 264 (1950). Review of beryllium halides: Bell, *Advan. Inorg. Chem. Radiochem.* **14**, 255-332 (1972).

White to faintly yellow, very deliquesc, orthorhombic crystals or cryst mass. Reported mp ranges from 399.2° to 440°. 399.2° is considered to be the most reliable (Bell). bp 482.3°. Sublimes *in vacuo* at 300°. d 1.90. Very sol in water with evolution of heat; the aq soln is strongly acid. Sol in alcohol, ether, pyridine, CS_2. Insol in benzene, toluene. *Keep tightly closed.*

Tetrahydrate, monoclinic deliquesc platelets. Has been reported to have $4\frac{1}{2}H_2O$: Semenenko, Turova, *Russ. J. Inorg. Chem.* **10**, 42 (1965). LD_{50} in guinea pigs, rats (mg Be/kg): 63, 0.6 i.p. (Cochran).

USE: Manuf of beryllium. *Anhydrous* form used as acid catalyst in organic reactions, similar to $AlCl_3$.

1185. Beryllium Fluoride. BeF_2; mol wt 47.01. Be 19.17%, F 80.83%. Prepd by heating ammonium fluoroberyllate $(NH_4)_2BeF_4$: Lebeau, *Compt. Rend.* **126**, 1418 (1898); Kwasnik in *Handbook of Preparative Inorganic Chemistry,* vol. **1**, G. Brauer, Ed. (Academic Press, New York, 2nd ed., 1963) p 231. Review of prepn and properties of beryllium halides: Bell, *Advan. Inorg. Chem. Radiochem.* **14**, 255-332 (1972).

Glassy hygroscopic mass (tetragonal system). True mp 555°; becomes free-flowing about 800°. Sublimes at 1036° under 1 mm press. in the presence of beryllium. d_4^{25} 1.986. Very freely sol in water; sparingly sol in alc; more sol in a mixture of alc and ether; insol in anhydr HF.

USE: Manuf of Be and Be alloys; manuf of glass; in nuclear reactors.

1186. Beryllium Formate. *Formic acid beryllium salt.* $C_2H_2BeO_4$; mol wt 99.05. C 24.25%, H 2.04%, Be 9.10%, O 64.61%. $Be(OOCH)_2$. Prepn: Besson, Hardt, *Compt. Rend.* **238**, 355 (1954).

Powder, decomp above 250° to the basic formate, Be_4O-$(HCOO)_6$, which sublimes without melting at about 320°. Very slowly hydrolyzed by water. Practically insol in the usual organic solvents. Sol in hot pyridine, but on cooling a pyridine complex crystallizes from the soln.

1187. Beryllium Hydride. BeH_2; mol wt 11.03. Be 81.69%, H 18.31%. Lower purity material prepd by treating dimethylberyllium with $LiAlH_4$ in ether: Barbaras *et al., J. Am. Chem. Soc.* **73**, 4585 (1951); higher purity by pyrolysis

Consult the cross index before using this section.

of di-*tert*-butylberyllium: Coates, Glocking, *J. Chem. Soc.* **1954**, 2526; Head *et al.*, *J. Am. Chem. Soc.* **79**, 3687 (1957); from triphenyl phosphine and beryllium borohydride: Banford, Coates, *J. Chem. Soc.* **1964**, 5591.

White solid. Higher purity material is inert to laboratory air. Loss of hydrogen at 190-200° negligible, rapid at 220°. Reacts slowly with water, rapidly with dil acids. Insol in ether, toluene, isopentane. Reacts with diborane to form beryllium borohydride.

1188. Beryllium Hydroxide. BeH_2O_2; mol wt 43.03. Be 20.95%, H 4.69%, O 74.37%. $Be(OH)_2$. Prepn: Ehrlich in *Handbook of Preparative Inorganic Chemistry*, vol. **1**, G. Brauer, Ed. (Academic Press, New York, 2nd ed., 1963) p 894.

Amorphous powder or crystals. d 1.92. Amphoteric. Very slightly sol in water and dil alkali. Sol in hot concd NaOH soln and acids.

USE: Manuf of beryllium and beryllium oxide.

1189. Beryllium Iodide. BeI_2; mol wt 262.85. Be 3.43%, I 96.57%. Prepn: Messerknecht, Biltz, *Z. Anorg. Chem.* **148**, 152 (1925); Ehrlich in *Handbook of Preparative Inorganic Chemistry*, vol. **1**, G. Brauer, Ed. (Academic Press, New York, 2nd ed., 1963) p 892. Review of beryllium halides: Bell, *Advan. Inorg. Chem. Radiochem.* **14**, 255-332 (1972).

Needles, mp 480°, bp 488°. Very hygroscopic. Sublimes *in vacuo*. Reacts violently with water, giving off HI. Absorbs ammonia. Dissolves in alcohols, amines, with the formation of addition compds. *Keep tightly closed.*

1190. Beryllium Nitrate. BeN_2O_6; mol wt 133.03. Be 6.78%, N 21.06%, O 72.16%. $Be(NO_3)_2$. Prepn: *Gmelin's, Beryllium* (8th ed.) **26**, 102-104 (1930).

Trihydrate, white to slightly yellow, deliquesc cryst mass. mp about 60°. Very sol in water, alcohol. *Keep well closed in a cool place.* LD_{50} i.p. in guinea pigs: 50 mg/kg, *Handbook of Toxicology* vol. **1**, W. S. Spector, Ed. (Saunders, Philadelphia, 1956) pp 46-47.

USE: Stiffening mantles in gas and acetylene lamps.

1191. Beryllium Nitride. Be_3N_2; mol wt 55.06. Be 49.11%, N 50.88%. Prepn: Ehrlich in *Handbook of Preparative Inorganic Chemistry*, vol. **1**, G. Brauer, Ed. (Academic Press, New York, 2nd ed., 1963) p 898; Langsdorf, Jr., U.S. pat. **2,567,518** (1951 to USAEC).

White crystals to grayish white powder; mp 2200 ± 40°. Volatile at bp, on further heating it dissociates into Be and N_2. Oxidized in air at 600°. Dec slowly by water, quickly by acids and alkalies with the evolution of ammonia.

1192. Beryllium Oxide. Beryllia. BeO; mol wt 25.01. Be 36.03%, O 63.97%. Prepn: *Gmelin's, Beryllium* (8th ed.) **26**, 82-91 (1930); Ehrlich in *Handbook of Preparative Inorganic Chemistry*, vol. **1**, G. Brauer, Ed. (Academic Press, New York, 2nd ed., 1963) p 893. Review: Lillie, *USAEC* **UCRL 6457**, 23 pp (1961).

Light, amorphous powder. mp 2530°. Very sparingly sol in water; slowly sol in concd acids or solns of fixed alkali hydroxides. After ignition it is almost insol in these solvents. Pure (100%) BeO insulates electrically like a ceramic, but conducts heat like a metal. Electrical resistivity in ohm-cm: $> 10^{16}$. Dielectric const at 8.5 gigacycles: 6.57.

USE: Manuf of beryllium oxide ceramics, glass; in nuclear reactor fuels and moderators; catalyst for organic reactions.

1193. Beryllium Perchlorate. $BeCl_2O_8$; mol wt 207.93. Be 4.33%, Cl 34.10%, O 61.56%. $Be(ClO_4)_2$. Prepn: *Gmelin's, Beryllium* (8th ed.) **26**, 121 (1930).

Tetrahydrate, very hygroscopic crystals. Holds its water of crystn tenaciously. Soly in water: 148.6 g/100 ml.

1194. Beryllium Potassium Fluoride. *Potassium tetrafluoroberyllate.* BeF_4K_2; mol wt 163.21. Be 5.52%, F 46.57%, K 47.91%. K_2BeF_4. Prepn: *Gmelin's, Beryllium* (8th ed.) **26**, 172 (1930). Review: See Beryllium Fluoride.

Hard masses. Sol in water, practically insol in alc.

1195. Beryllium Potassium Sulfate. $BeK_2O_8S_2$; mol wt 279.34. Be 3.23%, K 27.99%, O 45.83%, S 22.96%. $BeSO_4 \cdot K_2SO_4$. Prepn: *Gmelin's, Beryllium* (8th ed.) **26**, 174 (1930).

Dihydrate, brilliant crystals. Sol in water, concd K_2SO_4 solns; practically insol in alc.

USE: In chromium- and silver-plating.

1196. Beryllium Selenate. BeO_4Se; mol wt 151.97. Be 5.93%, O 42.11%, Se 51.96%. $BeSeO_4$. Prepn: *Gmelin's, Beryllium* (8th ed.) **26**, 144 (1930).

Tetrahydrate, orthorhombic crystals, d 2.03. Changes to the dihydrate at 100° and becomes anhydr at 300°. Freely sol in water; aq solns of beryllium selenate are good solvents for beryllium oxide.

1197. Beryllium Sodium Fluoride. *Sodium tetrafluoroberyllate.* BeF_4Na_2; mol wt 131.01. Be 6.88%, F 58.01%, Na 35.11%. Na_2BeF_4. Prepn: *Gmelin's, Beryllium* (8th ed.) **26**, 169 (1930). Review: See Beryllium Fluoride.

Orthorhombic or monoclinic crystals. mp about 350°. Sol in water.

1198. Beryllium Sulfate. BeO_4S; mol wt 105.07. Be 8.58%, O 60.91%, S 30.51%. $BeSO_4$. Prepn: *Gmelin's, Beryllium* (8th ed.) **26**, 130-141 (1930).

Tetrahydrate, crystals. d 1.71. At about 100° loses $2H_2O$. Very sol in water; practically insol in alc. LD_{50} i.v. in mice: 0.5 mg Be/kg, White *et al.*, *J. Pharmacol. Exp. Ther.* **102**, 88 (1951).

1199. Bestrabucil. *(17β)-Estra-1,3,5(10)-triene-3,17-diol 3-benzoate 17-[[4-[4-[bis(2-chloroethyl)amino]phenyl]-1-oxobutoxy]acetate];* 3-benzoyloxy-1,3,5(10)-estratriene-17β-[4-[p-[bis(2-chloroethyl)amino]phenyl]butyryloxy]-acetate; KM 2210; $C_{41}H_{47}Cl_2NO_6$; mol wt 720.73. C 68.33%, H 6.57%, Cl 9.84%, N 1.94%, O 13.32%. Benzoyl esters of estradiol-chlorambucil conjugate. Prepn: K. Asano *et al.*, **Ger.** pat. **2,932,607**; *eidem*, U.S. pat. **4,332,797** (1980, 1982 both to Kureha). Physicochemical properties: H. Wada *et al.*, *Iyakuhin Kenkyu* **17**, 703 (1984), *C.A.* **105**, 214009v (1986). Antitumor activity and pharmacokinetics: T. Kubota *et al.*, *Japan. J. Clin. Oncol.* **16**, 357 (1986). Series of articles on pharmacology: Oyo Yakuri **27**, 87-115 (1984); *C.A.* **100**, 168530e, 168531f; **101**, 973y, 33399u (1984).

White crystals, mp 110-111°. Soly at 24° (g/ml): chloroform 1; benzene 2; 1,4-dioxane 2; tetrahydrofuran 2; carbon disulfide 10; ethyl acetate 14; acetone 16; DMSO 23; carbon tetrachloride 50; ether 2500; sesame oil, cyclohexane, ethanol and water: all > 10000. LD_{50} orally in mice: > 10 g/kg (Kubota).

THERAP CAT: Antineoplastic.

1200. Betahistine. *N-Methyl-2-pyridineethanamine; 2-[2-(methylamino)ethyl]pyridine;* [2-(2-pyridyl)ethyl]methylamine. $C_8H_{12}N_2$; mol wt 136.19. C 70.55%, H 8.88%, N 20.57%. Prepn: Löffler, *Ber.* **37**, 161 (1904); Walter *et al.*, *J. Am. Chem. Soc.* **63**, 2771 (1941).

Liquid. bp_{30} 113-114°. Soluble in water, alcohol, ether, chloroform.

Dihydrochloride, $C_8H_{14}Cl_2N_2$, *Betaserc, Serc, Vasomotal.* Crystals from alc, mp 148-149°.

Maleate, $C_{12}H_{16}N_2O_4$, *Suzutolon.*

Mesylate, $C_9H_{16}N_2O_3S$, *Medan, Menitazine, Merislon, Remark, Ribrain, Tenyl.*

Dimesylate, $C_{10}H_{20}N_2O_6S_2$, *Aequamen, Melopat.*

THERAP CAT: Vasodilator.

1201. Betaine. *1-Carboxy-N,N,N-trimethylmethanaminium hydroxide inner salt;* (carboxymethyl)trimethylammonium hydroxide inner salt; glycine betaine; glycocoll betaine; lycine; oxyneurine; trimethylglycine hydroxide inner salt; trimethylglycocoll anhydride. $C_5H_{11}NO_2$; mol wt 117.15. C 51.26%, H 9.46%, N 11.96%, O 27.32%. Widely distributed in plants and animals: M. Guggenheim, *Die biogenen Amine* (S. Karger, Basel, 4th ed., 1951) pp 240-242. Prepn via or from hydrochloride: Stoltzenberg, *Z. Physiol. Chem.* **92**, 445 (1914); Edsall, *J. Am. Chem. Soc.* **65**, 1767 (1943). Prepn of the hydrochloride: Kuhn, Ruelius, *Ber.* **83**, 420 (1950); of the hydrate: Vassel, U.S. pat. **2,800,502** (1957 to International Minerals and Chemical Corp.). Structure of hydrate: Leifer, Lippincott, *J. Am. Chem. Soc.* **79**, 5098 (1957). Prepn of the sodium aspartate: Thuiller, Fr. pat. **M2462**, *C.A.* **61**, 8405F (1964); Belg. pat. **638,361** (both 1964 to Labs. Rolland); Cote, Fr. pat. **1,356,945** (1964), *C.A.* **61**, 7098f (1964).

$$(CH_3)_3\overset{+}{N}CH_2COO^-$$

Deliquescent scales or prisms dec around 310°. (Isomerizes at the mp to methyl ester of dimethylaminoacetic acid.) Sweet taste. Solubility (g/100 g solvent): water, 160; methanol, 55; ethanol, 8.7. Sparingly sol in ether. Betaine yields trimethylamine with concd KOH.

Monohydrate, *(carboxymethyl)trimethylammonium hydroxide, trimethylglycine hydroxide.* Formed by crystn of betaine from aq solvents. pH of satd soln about 8.0. Loses water at 100° forming inner salt again. Mixture with glycocyamine, *Betasyamine, Betacyamine.*

Hydrochloride, $C_5H_{12}ClNO_2$, *1-carboxy-N,N,N-trimethyl-methanaminium chloride, (carboxymethyl)trimethylammonium chloride, acidol, pluchine.* Monoclinic crystals from alc, dec 227-228° (Stoltzenberg); also reported as dec 232° (Kuhn, Ruelius). Soly in water at 25° = 64.7 g/100 ml, in ethanol 5.0 g/100 ml. pH of 5% aq soln 1.0. Practically insol in chloroform, ether.

Sodium aspartate, $C_9H_{17}N_2NaO_6$, *Somatyl.* Crystals, pptd from aq reaction mixture with acetone, hygroscopic, contg about 15% water (Thuiller, Belg. pat.). Also obtained as the trihydrate (Cote). Dec 160-170°. Soluble in water. Practically insol in acetone.

USE: In soldering, resin curing fluxes, organic synthesis.
THERAP CAT: Hepatoprotectant.

1202. Betamethasone. *9-Fluoro-11,17,21-trihydroxy-16-methylpregna-1,4-diene-3,20-dione;* 9α-fluoro-16β-methylprednisolone; 16β-methyl-9α-fluoro-Δ¹-hydrocortisone; 16β-methyl-9α-fluoroprednisolone; betadexamethasone; flubenisolone; β-methasone; Sch 4831; NSC-39470; beta-Corlan; Becort; Betasolon; Betnelan; Betnesol tablets; Celestan; Celestene; Celestone; Dermabet; Diprolene; Visubeta. $C_{22}H_{29}FO_5$; mol wt 392.45. C 67.32%, H 7.45%, F 4.84%, O 20.38%. Prepn: Taub *et al., J. Am. Chem. Soc.* **80**, 4435 (1958); Oliveto *et al., ibid.* 6688; Taub *et al., ibid.* **82**, 4012 (1960). U.S. pat. **3,053,865** (1962 to Merck & Co.); Amiard *et al.,* U.S. pat. **3,104,246** (1963 to Roussel-UCLAF). Also prepared from hecogenin. Comprehensive description of the dipropionate ester: M. G. Ferrante, B. C. Rudy in *Analytical Profiles of Drug Substances* vol. **6**, K. Florey, Ed. (Academic Press, New York, 1977) pp 43-60.

Crystals from ethyl acetate, mp 231-234° (dec). $[\alpha]_D$ +108° (acetone). uv max (methanol): 238 nm (ε 15200).

21-Acetate, $C_{24}H_{31}FO_6$, *Betafluorene.* Hexagonal prisms from acetone + ether, mp 205-208° (Taub); also reported as mp 196-201° (Oliveto). $[\alpha]_D$ +140° (chloroform). uv max (methanol): 238 nm (ε 14800).

21-Adamantoate, $C_{33}H_{43}FO_6$, *Betsovet.* Prepn: Philips,

Page 184   *Consult the cross index before using this section.*

English, Ger. pat. **2,232,827** (1973 to Glaxo), *C.A.* **78**, 97893q (1973).

17-Benzoate, $C_{29}H_{33}FO_6$, *W 5975, Bebate, Beben, Benisone, Euvaderm, Flurobate, Parbetan, Uticort.* Crystals from acetone-ether, mp 225-228°. $[\alpha]_D^{24}$ +63.5° (dioxane). Synthesis and activity: Ercoli *et al., J. Med. Chem.* **15**, 783 (1972). See also Cullen, *Curr. Ther. Res.* **15**, 243 (1973).

17,21-Dipropionate, $C_{28}H_{37}FO_7$, *Sch 11460, Diproderm, Diprophos, Diprosis, Diprosone, Maxivate, Rinderon-DP.* Powder, mp 170-179° (dec). $[\alpha]_D^{26}$ +65.7° (dioxane). uv max (methanol): 238 nm (ε 15700).

17-Valerate, $C_{27}H_{37}FO_6$, *Bedermin, Betnesol-V, Betneval, Betnovate, Bextasol, Celestan-V, Celestoderm-V, Dermosol, Dermovaleas, Ecoval 70, Hormezon, Tokuderm, Valisone.* Needles from acetone + petr ether, mp 183-184°. $[\alpha]_D$ +77° (dioxane). uv max: 239 nm (ε 15920). Neth. pat. Appl. **6,406,615** (1964 to Glaxo).

17,21-Divalerate, $C_{32}H_{45}FO_7$, *Betadival.*

21-Phosphate disodium salt, $C_{22}H_{28}FNa_2O_8P$, *betamethasone 21-(dihydrogen phosphate) disodium salt, Bentelan, Betnesol Injectable, Durabetason, Vista-Methasone.*

THERAP CAT: Glucocorticoid.

THERAP CAT (VET): Glucocorticoid.

1203. Betasine. *β-Amino-4-hydroxy-3,5-diiodobenzenepropanoic acid;* β-amino-4-hydroxy-3,5-diiodohydrocinnamic acid; β-(4-hydroxy-3,5-diiodophenyl)-β-alanine; β-(4-hydroxy-3,5-diiodophenyl)-β-aminopropionic acid; β-diiodotyrosine; betasinum; betazine. $C_9H_9I_2NO_3$; mol wt 433.92. C 24.96%, H 2.10%, I 58.63%, N 3.24%, O 11.09%. Prepn: Rodionov *et al., Zh. Obshch. Khim.* **27**, 2234 (1957); Suvorov *et al.,* U.S.S.R. pat. **104,779** (1957), *C.A.* **51**, 8794b (1957).

Crystals, mp 178-179° (dec). Sparingly soluble in water; sol in aqueous solutions of hydrochloric acid, caustic alkalies and ammonia. Practically insol in organic solvents.

Ammonium salt, crystals, mp 151-152° (dec). Difficultly soluble in cold water.

THERAP CAT: Iodine source.

1204. Betaxolol. *1-[4-[2-(Cyclopropylmethoxy)ethyl]-phenoxy]-3-[(1-methylethyl)amino]-2-propanol;* (±)-1-(isopropylamino)-3-[p-(cyclopropylmethoxyethyl)phenoxy]-2-propanol. $C_{18}H_{29}NO_3$; mol wt 307.44. C 70.32%, H 9.51%, N 4.56%, O 15.61%. Cardioselective β_1-adrenergic blocker. Prepn: P. M. J. Manoury *et al.,* Ger. pat. **2,649,605** corresp to U.S. pat. **4,252,984** (1977, 1981 both to Synthelabo). Blood concn and pharmacodynamic effects: S. J. Warrington *et al., Brit. J. Clin. Pharmacol.* **10**, 449 (1980). Pharmacokinetics: G. Bianchetti *et al., Arzneimittel-Forsch.* **30**, 1912 (1980). Cardiovascular effects in normal volunteers: P. J. Cadigan *et al., Brit. J. Clin. Pharmacol.* **9**, 569 (1980). Efficacy and pharmacokinetics: K. Balnave *et al., ibid.* **11**, 171 (1981). Use in treatment of glaucoma: A. R. Berrospi, H. M. Leibowitz, *Arch. Ophthalmol.* **100**, 943 (1982). Antihypertensive effect: M. Pathe *et al., Therapie* **37**, 75 (1982). *Book: Betaxolol and Other β₁-Adrenoceptor Antagonists,* P. L. Morselli *et al.,* Eds. (Raven Press, New York, 1983) 385 pp.

Crystals from petr ether, mp 70-72°.
Hydrochloride, $C_{18}H_{30}ClNO_3$, *SLD-212, SL-75212, Betoptic, Betoptima, Kerlone.* Crystals from acetone, mp 116°. LD_{50} in mice (mg/kg): 944 orally; 37 i.v. (Manoury).

THERAP CAT: Antihypertensive; antiglaucoma agent.

1205. Betazole. *1H-Pyrazole-3-ethanamine; 3-(2-aminoethyl)pyrazole;* 3-(β-aminoethyl)pyrazole; ametazole; gastramine. $C_5H_9N_3$; mol wt 111.15. C 54.03%, H 8.16%, N 37.81%. Prepd by the catalytic reduction of 3-pyrazoleacet-aldehyde hydrazone: Jones, Mann, *J. Am. Chem. Soc.* **75**, 4048 (1953); Jones, U.S. pat. **2,785,177** (1957 to Lilly).

H
N—N

CH_2CH_2NH_2

Viscous liquid. bp$_{0.5}$ 118-123°.
Dihydrochloride, $C_5H_{11}Cl_2N_3$, *Histimin, Histalog.* Crystals from ethanol, mp 224-226°. Sol in water. Practically insol in chloroform. Aq solns are acid to litmus.
THERAP CAT: Diagnostic aid (gastric secretion stimulant).

1206. Betel. Dried leaves of *Piper betle* L., *Piperaceae. Habit.* India, Ceylon, Malay Archipelago. *Constit.* 0.2-1% volatile oil, chavibetol, chavicol, cadinene, allylpyrocatechol. *Ref:* Ueda, Sasaki, *J. Pharm. Soc. Japan* **71**, 559 (1951), *C.A.* **45**, 9137g (1951).
THERAP CAT: Counterirritant.

1207. Bethanechol Chloride. *2-[(Aminocarbonyl)oxy]-N,N,N-trimethyl-1-propanaminium chloride; carbamate of (2-hydroxypropyl)trimethylammonium chloride;* (2-hydroxy-propyl)trimethylammonium chloride carbamate; 2-carbam-oyloxypropyltrimethylammonium chloride; carbamylmeth-ylcholine chloride; urethan of β-methylcholine chloride; Duvoid; Urecholine chloride; Mechothane; Myocholine; Mictone; Myotonine chloride; Uro-Carb. $C_7H_{17}ClN_2O_2$; mol wt 196.68. C 42.74%, H 8.71%, Cl 18.03%, N 14.25%, O 16.27%. Prepn: Dalmer, Diehl, U.S. pat. **1,894,162** (1933); Major, Bonnett, U.S. pat. **2,322,375** (1943 to Merck & Co.).

$$\left[\begin{array}{c} CH_3CH—CH_2N^+(CH_3)_3 \\ | \\ O—CO—NH_2 \end{array}\right] Cl^-$$

Hygroscopic crystals, slight amine odor, dec 218-219°. One gram dissolves in 0.6 ml of water, in 12.5 ml of 95% alc. pH of an 0.5% aq soln 5.5-6.0. Aq solns may be sterilized by autoclaving at 120° for 20 minutes.
THERAP CAT: Cholinergic.
THERAP CAT (VET): Cholinergic. Has been used in urolithiasis of cats and in atonic conditions of the gut.

1208. Bethanidine. *N,N'-Dimethyl-N''-(phenylmeth-yl)guanidine; 1-benzyl-2,3-dimethylguanidine; N-benzyl-N',N''-dimethylguanidine.* $C_{10}H_{15}N_3$; mol wt 177.24. C 67.76%, H 8.53%, N 23.71%. Adrenergic neuron blocking agent. Prepn: Walton, Ruffell, U.S. pat. **3,168,562** (1965 to Wellcome Found.); Brit. pats. **1,084,461** and **1,111,564** (1967 and 1968 to Wellcome Found.). Pharmacology: A. L. A. Boura, A. F. Green, *Brit. J. Pharmacol.* **20**, 36 (1963); J. A. Oates *et al., Ann. N.Y. Acad. Sci.* **179**, 302 (1971). Pharmacokinetics: C. N. Corder, *J. Clin. Pharmacol.* **19**, 428 (1979). HPLC determn in plasma: J. R. Shipe *et al., Clin. Chem.* **29**, 1793 (1983). Anti-arrhythmic activity: M. B. Bacaner, D. G. Benditt, *Am. J. Cardiol.* **50**, 728 (1982); J. C. Somberg *et al., ibid.* **54**, 343 (1984). *Review:* A. F. Green, *Brit. J. Clin. Pharmacol.* **13**, 25-34 (1982).

CH_2NHC=NCH_3
NHCH_3

Crystals from methanol + ether, mp 195-197°.
Hydriodide, $C_{10}H_{15}N_3 \cdot HI$, crystals from ethanol + ether, mp 141-146°.
Sulfate, $C_{20}H_{32}N_6O_4S$, *benzaidin, BW 467C60, Bendogen, Benzoxine, Betaling, Betanidol, Esbatal, Eusmanid, Hypersin, Tenathan.* LD$_{50}$ in mice (mg/kg): 12 i.v.; 150 i.p.; 260 s.c.; 520 by stomach intubation (Boura, Green).
THERAP CAT: Antihypertensive.

1209. Betonicine. *trans-2-Carboxy-4-hydroxy-1,1-di-methylpyrrolidinium hydroxide, inner salt; l-N,N-dimethyl-4-hydroxypyrrolidine-2-carboxylic acid betaine; l-1-meth-yl-4-hydroxypyrrolidine-2-carboxylic acid methylbetaine; l-4-hydroxystachydrine; l-4-hydroxyproline betaine.* $C_7H_{13}NO_3$; mol wt 159.18. C 52.81%, H 8.23%, N 8.80%, O 30.15%. Occurs in *Croton gubouga* S. Moore; in *Stachys offi-cinalis* (L.) Trev. (*Betonica officinalis* L.), *Labiatae;* in *Achil-lea moschata* Jacq. and *A. millefolium* L., *Compositae:* Goodson, Clewer, *J. Chem. Soc.* **115**, 923 (1919); Guggen-heim, *Die biogenen Amine* (S. Karger, New York, 1951) p 246; Miller, Chow, *J. Am. Chem. Soc.* **76**, 1353 (1954); Pail-er, Kump, *Monatsh.* **90**, 396 (1959); *Arch. Pharm.* **293**, 646 (1960). Stereoisomeric with turicine (*cis*-form). Synthesis: Patchett, Witkop, *J. Am. Chem. Soc.* **79**, 185 (1957).

H_3C   CH_3
N^+   H
OH
H   COO^-

Blunt prisms from ethanol, dec 254-256°. Sweet taste. $[\alpha]_D^{20}$ −34.2° (c = 1.0). Readily sol in water or hot alcohol, slightly in cold alcohol; practically insol in benzene, ether, chloroform, carbon tetrachloride.
Hydrochloride, $C_7H_{13}NO_3 \cdot HCl$, crystals from ethanol, dec 216-217° (after drying). $[\alpha]_D^{20}$ −24.2° (c = 0.0892 in water).
Aurichloride, $C_7H_{13}NO_3 \cdot HAuCl_4$, scaly clusters from water, mp 242°.

1210. Betoxycaine. *3-Amino-4-butoxybenzoic acid 2-[2-(diethylamino)ethoxy]ethyl ester;* 2-diethylaminoeth-oxyethyl 3-amino-4-butoxybenzoate. $C_{19}H_{32}N_2O_4$; mol wt 352.47. C 64.74%, H 9.15%, N 7.95%, O 18.16%. Prepn and properties: E. Cuingnet, U.S. pat. **3,209,022** (1965 to Corbi-ere). Ionization potential: A. Cier *et al., Thérapie* **26**, 941 (1971). Investigations of possible antifibrillatory properties: P. Amaud *et al., Compt. Rend. Soc. Biol.* **159**, 2427 (1965); G. Faucon *et al., Thérapie* **21**, 1253 (1966).

COOCH_2CH_2OCH_2CH_2N(C_2H_5)_2

NH_2
OCH_2CH_2CH_2CH_3

Monohydrochloride, $C_{19}H_{33}ClN_2O_4$, *Millicaine.* White crystals from acetone, mp 117°. Soluble in methanol.
THERAP CAT: Local anesthetic.

1211. Betula. European white birch. Bark and leaves of *Betula alba* L., *Betulaceae. Habit.* Europe and Northern Asia, also America, north of Pennsylvania. *Constit.* 10-15% Betulin (betula camphor), betuloresinic acid, essential oil, saponins, betulol (sesquiterpene alcohol), apigenin dimethyl ether, betuloside, gaultherin, methyl salicylate, ascorbic acid. *Ref:* Kreitmair, *Pharmazie* **8**, 534 (1953).
USE: Pharmaceutic aid (flavor).

1212. Betulin. *Lup-20(29)-ene-3,28-diol;* lup-20(30)-ene-3β,28-diol; trochol; betulinol; betulol. $C_{30}H_{50}O_2$; mol wt 442.70. C 81.39%, H 11.38%, O 7.23%. In the outer portion of the bark of white birch (up to 24%), in other barks, and in lignite. Botanical distribution: Steiner, *Mol-isch Festschrift* (1936). Isoln: Lowitz, *Crell's Chem. Ann.* **1**, 312 (1788); Steiner, *loc. cit;* Ruzicka, Isler, *Helv. Chim. Acta* **19**, 506 (1936); from *Lemaireocereus griseus* Britton et Rose, *Cactaceae:* Djerassi *et al., J. Am. Chem. Soc.* **78**, 2312 (1956). Structure: Ames *et al., J. Chem. Soc.* **1951**, 450; Davy *et al., ibid.* **1951**, 2696, 2702. Stereochemistry: Guider *et al., ibid.* **1953**, 3024; Das, *Chem. & Ind. (London)* **1971**, 1331. *Review:* J. Simonsen, W. C. J. Ross, *The Terpenes* vol. **IV** (Cambridge Univ. Press, 1957) pp 187-328.

Crystals from methanol-chloroform, mp 248-251°; sublimes at 240° at 0.01 mm. Solvated needles from alc contg one mol EtOH. After drying sublimes at 170-180° (bath temp) at 0.08 mm. uv max (H_2SO_4): 316 nm. $[\alpha]_D^{15}$ +20° (c = 2 in pyridine). Sparingly sol in cold water, petr ether, carbon disulfide. One part is sol in 149 parts alc, 251 ether, 113 chloroform, 417 benzene. Sol in acetic acid.

Diacetate, $C_{34}H_{54}O_4$, mp 223-224°. $[\alpha]_D^{20}$ +22° (c = 1.2 in $CHCl_3$). $d_4^{28.5}$ 0.9635; $n_D^{28.5}$ 1.4661.

1213. Bevantolol. *1-[[2-(3,4-Dimethoxyphenyl)ethyl]-amino]-3-(3-methylphenoxy)-2-propanol; 1-[(3,4-dimethoxyphenethyl)amino]-3-(m-tolyloxy)-2-propanol.* $C_{20}H_{27}$-NO_4; mol wt 345.44. C 69.54%, H 7.88%, N 4.05%, O 18.53%. Cardioselective β_1-adrenergic blocker. Prepn: **Belg. pat. 790,165;** A. Holmes, R. F. Meyer, **U.S. pat. 3,857,891** (1973, 1974 both to Parke, Davis & Co.); M. L. Hoefle *et al., J. Med. Chem.* **18,** 148 (1975). Pharmacology in animals: S. G. Hastings *et al., Arch. Int. Pharmacodyn. Ther.* **226,** 81 (1977). Cardiovascular effects in animals: I. D. Dukes, E. M. Vaughan Williams, *Brit. J. Pharmacol.* **84,** 365 (1985). Cardioselectivity in asthmatic patients: C.-G. Löfdahl *et al., Pharmacotherapy* **4,** 205 (1984). GC determn in plasma: E. J. Randinitis *et al., J. Chromatog.* **308,** 345 (1984). Pharmacokinetics in humans: P. Vermeij, P. van Brummelen, *Eur. J. Clin. Pharmacol.* **30,** 375 (1986). Comparative clinical trial with hydrochlorothiazide in hypertension: C. P. Lucas *et al., Clin. Ther.* **8,** 49 (1985). Symposium on pharmacology and clinical efficacy: *Am. J. Cardiol.* **58,** 1E-44E (1986).

Hydrochloride, $C_{20}H_{28}ClNO_4$, *CI-775, Ranestol, Sentiloc, Vantol.* Crystals from acetonitrile, mp 137-138°.

THERAP CAT: Antianginal; antihypertensive; antiarrhythmic.

1214. Bevonium Methyl Sulfate. *2-[[(Hydroxydiphenylacetyl)oxy]methyl]-1,1-dimethylpiperidinium methyl sulfate (salt); 2-(hydroxymethyl)-1,1-dimethylpiperidinium methyl sulfate benzilate;* piribenzil methyl sulfate; benzilic acid ester with 2-(hydroxymethyl)-1,1-dimethylpiperidinium methyl sulfate; CG 201; Acabel. $C_{23}H_{31}NO_7S$; mol wt 465.58. C 59.33%, H 6.71%, N 3.01%, O 24.06%, S 6.89%. Prepn: **Belg. pat. 616,951** (1962 to Gruenenthal), *C.A.* **58,** 7914d (1963); Beckmann, *Arzneimittel-Forsch.* **16,** 910 (1966). Series of publications on pharmacology, toxicology, clinical trials, *see ibid.* 901-988.

Crystals from petr ether, mp 134-135°.

THERAP CAT: Anticholinergic, antispasmodic, bronchodilator.

1215. Bezafibrate. *2-[4-[2-[(4-Chlorobenzoyl)amino]-ethyl]phenoxy]-2-methylpropanoic acid; 2-[p-[2-(p-chloro-*

benzamido)ethyl]phenoxy]-2-methylpropionic acid; α-[4-(4-chlorobenzoylaminoethyl)phenoxy]isobutyric acid; BM 15,075; Befizal; Bezalip; Bezatol; Cedur; Difaterol. $C_{19}H_{20}$-$ClNO_4$; mol wt 361.83. C 63.07%, H 5.57%, Cl 9.80%, N 3.87%, O 17.69%. Prepn: E. Witte *et al.,* **Ger. pat. 2,149,-070;** *eidem,* **U.S. pat. 3,781,328** (both 1973 to Boehringer, Mann.). Pharmacology: R. Zimmerman *et al., Atherosclerosis* **29,** 477 (1978). Clinical studies: A. G. Olsson, P. D. Lang, *ibid.* **31,** 421, 429 (1978); P. Wahl *et al., Deut. Med. Wochenschr.* **103,** 1233 (1978). Review of pharmacodynamics and therapeutic use: J. P. Monk, P. A. Todd, *Drugs* **33,** 539-576 (1987).

Crystals from acetone, mp 186°.
THERAP CAT: Antihyperlipoproteinemic.

1216. Bezitramide. *1-[1-(3-Cyano-3,3-diphenylpropyl)-4-piperidinyl]-1,3-dihydro-3-(1-oxopropyl)-2H-benzimidazol-2-one; 1-[1-(3-cyano-3,3-diphenylpropyl)-4-piperidyl]-3-propionyl-2-benzimidazolinone;* 1-(3-cyano-3,3-diphenyl-propyl)-4-(2-oxo-3-propionyl-1-benzimidazolinyl)piperidine; benzitramide; R-4845; Burgodin. $C_{31}H_{32}N_4O_2$; mol wt 492.63. C 75.58%, H 6.55%, N 11.37%, O 6.50%. Prepn: **Belg. pat. 633,495** corresp to P. A. J. Janssen, **U.S. pat. 3,196,157** (1963, 1965 both to Janssen). Pharmacology and clinical data: P. A. J. Janssen *et al., Arzneimittel-Forsch.* **21,** 862 (1971); W. K. P. Amery *et al., ibid.* 868; H. Knape, *Anaesthesist* **21,** 251 (1972).

White crystalline powder, mp 145-149°. Also reported as pale yellow amorphous powder, mp 124.5-126°. Solubility of > 1 g/100 ml in ethyl acetate, acetone, benzene, chloroform. Almost insol in water, dilute acids. LD_{50} orally in mice, rats: 2101, 141 mg/kg (Janssen).

Caution: May be habit forming. This is a controlled substance (opiate) listed in the U.S. Code of Federal Regulations, Title 21 Part 1308.12 (1987).

THERAP CAT: Narcotic analgesic.

1217. Bialamicol. *3,3'-Bis[(diethylamino)methyl]-5,5'-di-2-propenyl-[1,1'-biphenyl]-4,4'-diol; 5,5'-diallyl-α,α'-bis(diethylamino)-m,m'-bitolyl-4,4'-diol; 6,6'-diallyl-α,α'-bis(diethylamino)-4,4'-bi-o-cresol;* biallylamicol; SN 6771; PAA-701; Camoform. $C_{28}H_{40}N_2O_2$; mol wt 436.62. C 77.02%, H 9.23%, N 6.42%, O 7.33%. Prepd by treating 2,2'-diallyl-*p,p'*-biphenol with diethylamine and formaldehyde: Burckhalter *et al., J. Am. Chem. Soc.* **68,** 1894 (1946); Rawlins *et al.,* **U.S. pat. 2,459,338** (1949 to Parke, Davis).

Dihydrochloride, $C_{28}H_{40}N_2O_2\cdot2HCl$, minute crystals, mp 209-210°. Sol in water.
THERAP CAT: Anti-amebic.

1218. Bibenzonium Bromide. *2-(1,2-Diphenylethoxy)-N,N,N-trimethylethanaminium bromide; [2-(1,2-diphenylethoxy)ethyl]trimethylammonium bromide;* trimethyl-(1,2-diphenylethoxy)ethylammonium bromide; 2-trimethylamino-1,2-diphenyldiethyl ether bromide; 2-(dimethylamino)-1,2-diphenyldiethyl ether methyl bromide; ES 132; Sedobex;

Lysobex; Lysibex; Thoragol; Lysbex; Medipectol. $C_{19}H_{26}$-BrNO; mol wt 364.34. C 62.64%, H 7.19%, Br 21.93%, N 3.85%, O 4.39%. Prepn: Suter, Kündig, U.S. pat. 2,912,429 (1959 to Eprova).

$$[C_6H_5CH_2CHOCH_2CH_2N(CH_3)_3]^+ \ Br^-$$
$$C_6H_5$$

Crystals from ethyl acetate, mp 144-147°. Sol in cold water, methanol, and ethanol; practically insol in hot ethyl acetate, benzene, ether and ligroine.
THERAP CAT: Antitussive.

1219. Bibenzyl. *1,1'-(1,2-Ethanediyl)bisbenzene;* dibenzyl; *sym*-diphenylethane; 1,2-diphenylethane. $C_{14}H_{14}$; mol wt 182.25. C 92.26%, H 7.74%. Prepn by the reduction of benzil or benzoin: Clemmensen, *Ber.* **47**, 688 (1914); Gattermann-Wieland, *Praxis des organischen Chemikers* (de Gruyter, Berlin, 40th ed., 1961) p 333; from stilbene, Kleiderer, Kornfeld, *J. Org. Chem.* **13**, 455 (1948).

Monoclinic prisms from methanol, mp 52.0-52.5°. bp$_{760}$ 284°. d$_4^0$ 1.104; d$_4^{25}$ 0.9782; d$_4^{58}$ 0.958. Heat of combustion (25°): 9909.9 cal (15°)/g. Moderately sol in alcohol; freely sol in carbon disulfide, ether, chloroform, amyl acetate; sol in liquid sulfur dioxide; practically insol in water, liquid ammonia.

1220. Bibrocathol. *4,5,6,7-Tetrabromo-2-hydroxy-1,3,2-benzodioxabismole;* tetrabromopyrocatechol bismuth derivative; bismuth tetrabromopyrocatechol; Cabis bromatum; Bibrocathin; Tetraform; Bismucatebrol; Noviform; Novoform. $C_6HBiBr_4O_3$; mol wt 649.74. C 11.09%, H 0.16%, Bi 32.17%, Br 49.20%, O 7.39%. Prepn: Hundrup, *Arch. Pharm. Chemi* **54**, 537 (1947), *C.A.* **42**, 2727a (1948). Antibacterial activity: Frank, Stark, *Pharm. Acta Helv.* **29**, 283 (1954). Identity tests: Hakkesteegt, *Pharm. Weekbl.* **99**, 922 (1964).

Yellow, odorless, tasteless powder. Practically insol in water; slightly sol in alcohol, ether. Dec in acid, alkalies.
THERAP CAT: Topical antiseptic.

1221. Bicine. *N,N-Bis(2-hydroxyethyl)glycine;* di(hydroxyethyl)glycine; *N,N*-bis(hydroxyethyl)aminoacetic acid; *N,N*-di(hydroxyethyl)aminoacetic acid; diethylolglycine; 2-HxG. $C_6H_{13}NO_4$; mol wt 163.18. C 44.16%, H 8.03%, N 8.58%, O 39.22%. (HOCH$_2$CH$_2$)$_2$NCH$_2$COOH. One of the zwitterionic amino acids known as "Good" buffers, active in the pH range 6-8.5. Prepn by hydrolysis of its lactone (obtained from glycine and ethylene oxide): Pascal, *Compt. Rend.* **245**, 1318 (1957); from diethanolamine and bromoacetic acid: N. E. Good *et al., Biochemistry* **5**, 467 (1966). Crystal structure: V. Cody *et al., Acta Crystallogr.* **33B**, 905 (1977). Use as sequestering agent: A. Grawitz, *Rev. Tech. Ind. Cuir* **65**, 187, 190 (1973), *C.A.* **80**, 49281h (1974). Temperature effects on pKa: M. L. Soni, R. C. Kapoor, *Int. J. Quant. Chem.* **20**, 385 (1981). Use as a buffer: A. L. Remisov, *Biokhimia* **25**, 323 (1960); S. Ito *et al., Histochem. J.* **16**, 489 (1984).
Crystals from dil ethanol, mp 193-195° (slight decompn). pKa (20°): 8.35. pKa$_2$ (0.1M): 0°, 8.7; 20°, 8.35; 37°, 8.2. ΔpKa/°C −0.018. Saturated aq soln is 1.1M at 0°. Slightly sol in water. Forms a water-soluble sodium salt.

USE: Biological buffer and chelating agent.

1222. Bicozamycin. *Bicyclomycin;* aizumycin; 8,10-diaza-6-hydroxy-5-methylene-1-(2-methyl-1,2,3-trihydroxy-propyl)-2-oxabicyclo[4.2.2]decan-7,9-dione; antibiotic 5879; WS-4545 antibiotic; Bacteron; Bacfeed. $C_{12}H_{18}N_2O_7$; mol wt 302.28. C 47.68%, H 6.00%, N 9.27%, O 37.05%. Cyclic peptide antibiotic substance. Prepn by fermentation of *Streptomyces sapporensis:* T. Miyoshi *et al.,* **Ger.** pat. 2,150,593; H. Imanaka *et al.,* **U.S.** pat. 3,923,790 (1972, 1975 both to Fujisawa). Isoln and characterization: T. Miyoshi *et al., J. Antibiot.* **25**, 569 (1972). Isoln from *S. aizunensis* and identity with antibiotic 5879: S. Miyamura *et al., ibid.* **26**, 479 (1973). Bicyclomycin has a unique chemical structure which bears no relation to any group of known antibiotics. Structural elucidation: T. Kamiya *et al., ibid.* **25**, 576 (1972). Crystal and molecular structure: Y. Tokuma *et al., Bull. Chem. Soc. Japan* **47**, 18 (1974). *In vitro* and *in vivo* activity studies: M. Nishida *et al., J. Antibiot.* **25**, 582 (1972). Metabolism: *eidem, ibid.* 594. Mechanism of action: N. Tanaka *et al., ibid.* **29**, 155 (1976); N. Tanaka in *Antibiotics* vol. 5(pt. 1), E. Hahn, Ed. (Springer, New York, 1979) p 18. Synthetic approaches: L. V. Dunkerton, R. M. Ahmed, *Tetrahedron Letters* **21**, 1803 (1980); R. M. Williams, *ibid.* **22**, 2341 (1981); S. Nakatsuka *et al., ibid.* 4973; J. H. Hoare, P. Yates, *Chem. Commun.* **1981**, 1126. Total synthesis of (±)-bicyclomycin: S. Nakatsuka *et al., Tetrahedron Letters* **24**, 5627 (1983); of (+)-bicyclomycin: R. M. Williams *et al., J. Am. Chem. Soc.* **106**, 5749 (1984); *eidem, ibid.* **107**, 3253 (1985). Review of synthetic, mechanistic and biological studies: R. M. Williams, C. A. Durham, *Chem. Rev.* **88**, 511-540 (1988).

Monoclinic crystals from ethanol, mp 188-191° (dec); rhombic crystals from methanol + acetone, mp 187-189° (dec) (Imanaka), also reported as mp 166-170° (Nakatsuka *et al.*). [α]$_D^{23}$ +63.5° (methanol). Weakly basic substance. Soly in water: 192 mg/ml. Sol in methanol; sparingly sol in ethanol. Slightly sol in acetone. Practically insol in chloroform, ethyl acetate, benzene, *n*-hexane. Unstable in alkaline soln. LD$_{50}$ in mice: >4 g/kg (Williams, Durham).
THERAP CAT (VET): Antibacterial; feed additive (livestock).

1223. Bicuculline. *[R-(R*,S*)]-6-(5,6,7,8-Tetrahydro-6-methyl-1,3-dioxolo[4,5-g]isoquinolin-5-yl)furo[3,4-e]-1,3-benzodioxol-8(6H)-one.* $C_{20}H_{17}NO_6$; mol wt 367.34. C 65.39%, H 4.67%, N 3.81%, O 26.13%. Alkaloid naturally occurring in the *d*-form; found in *Dicentra cucullaria* (L.) Bernh., *Adlumia fungosa* (Ait.) Greene, *Fumariaceae,* and several *Corydalis* species: Manske, *Can. J. Res.* **7**, 265 (1932); **8**, 210, 407 (1933); **9**, 436 (1933); Edwards, Handa, *Can. J. Chem.* **39**, 1801 (1961). Synthesis of *dl*-form: Groenewoud, Robinson, *J. Chem. Soc.* **1936**, 199. Resolution of isomers: Haworth *et al., Nature* **165**, 529 (1950). Stereoisomer of adlumidine, *q.v.,* and of capnoidine: Manske, *J. Am. Chem. Soc.* **72**, 3207 (1950). Preliminary stereochemical studies: Safe, Moir, *Can. J. Chem.* **42**, 160 (1964). Revised stereochemistry: Blaha *et al., Coll. Czech. Chem. Commun.* **29**, 2328 (1964); Snatzke *et al., Tetrahedron*

25, 5059 (1969). Crystal and molecular structure: Gorinsky, Moss, *J. Cryst. Mol. Struct.* **3**, 299 (1973). Shows GABA *(q.v.)* antagonist activity: Curtis *et al.*, *Nature* **226**, 1222 (1970).

Elongated plates from chloroform-methanol, mp 215°; mp also reported as 177°, solidifies and remelts 193-195°: Manske, *Can. J. Res.* **21B**, 13 (1943). $[\alpha]_D^{25}$ +130.5° (CHCl$_3$). uv max (acidified ethanol): 225, 296, 324 nm (ϵ 36700, 6390, 5870). pKa 4.84. Sol in benzene, chloroform, ethyl acetate. Sparingly sol in alc and ether.

1224. Bietamiverine. *α-Phenyl-1-piperidineacetic acid 2-(diethylamino)ethyl ester; β-diethylaminoethyl phenylpiperidinoacetate; β-diethylaminoethyl α-(1-piperidyl)phenylacetate.* C$_{19}$H$_{30}$N$_2$O$_2$; mol wt 318.45. C 71.66%, H 9.50%, N 8.80%, O 10.05%. Prepd by reacting piperidine with phenylchloroacetic acid ethyl ester in chloroform: Reetz, **Ger.** pat. **859,892** (1952 to Nordmark-Werke); from β-diethylaminoethyl phenylbromoacetate and piperidine in chloroform: Blicke *et al.*, *J. Am. Chem. Soc.* **76**, 3161 (1954); from ethyl α-(1-piperidyl)phenylacetate and β-diethylaminoethyl chloride: Moffett, Hart, *J. Am. Pharm. Assoc. Sci. Ed.* **42**, 717 (1953). Clinical tests: Ciravegna *et al.*, *Clin. Ter.* **57**, 227 (1971).

$$C_6H_5CHCOOCH_2CH_2N(C_2H_5)_2$$

Liquid, d$_4^{25}$ 1.0184. bp$_1$ 65°. n$_D^{25}$ 1.5070.
Hydrochloride, C$_{19}$H$_{30}$N$_2$O$_2$·HCl, crystals from methyl isobutyl ketone, mp 187-189°.
Dihydrochloride, C$_{19}$H$_{32}$Cl$_2$N$_2$O$_2$, *Novosparol, Spasmaparid, Spasmisolvina, Spasmo-Paparid.* Crystals from ethanol + ether, mp 194-195°.
THERAP CAT: Antispasmodic.

1225. Bietanautine. *1,2,3,6-Tetrahydro-1,3-dimethyl-2,6-dioxo-7H-purine-7-acetic acid compd with 2-(diphenylmethoxy)-N,N-dimethylethanamine (2:1); 2-(benzhydryloxy)-N,N-dimethylethylamine bis(theophylline 7-acetate); O-benzhydryldimethylaminoethanol bis(theophylline 7-acetate); etanautine; Nautamine.* C$_{35}$H$_{41}$N$_9$O$_9$; mol wt 731.79. C 57.45%, H 5.65%, N 17.23%, O 19.68%. C$_{18}$H$_{20}$N$_8$O$_8$·C$_{17}$H$_{21}$NO. Prepn: Mizier, **U.S.** pat. **2,942,000** (1960 to Delagrange).

$$(C_6H_5)_2CHOCH_2CH_2N(CH_3)_2$$

Crystals, mp 168-170°. Sol in alc; sparingly sol in water.
THERAP CAT: Antihistaminic. Anti-emetic. Antiparkinsonian.

1226. Bietaserpine. *1-[2-(Diethylamino)ethyl]-11,17-dimethoxy-18-[(3,4,5-trimethoxybenzoyl)oxy]yohimban-16-carboxylic acid methyl ester; 1-[2-(diethylamino)ethyl]-18β-hydroxy-11,17α-dimethoxy-3β,20α-yohimban-16β-carboxylic acid methyl ester, 3,4,5-trimethoxybenzoate ester; 1-[2-(diethylamino)ethyl]reserpine.* C$_{39}$H$_{53}$N$_3$O$_9$; mol wt 707.84. C 66.17%, H 7.55%, N 5.94%, O 20.34%. Prepn: Buzas, Régnier, *C. R. Acad. Sci.* **250**, 1340 (1960); Buzas *et al.*, **Fr.** pat. **1,256,524** (1961 to Dautreville & Lebas and Andre Buzas), *C.A.* **57**, 2273e (1962); Buzas *et al.*, **Fr.** pat. **M102**, *C.A.* **58**, 9159a (1963), and **Brit.** pat. **894,866** (1961 and 1962, both to Soc. Nogentaise de Prods. Chim. and Andre Buzas). Pharmacology: Garattini *et al.*, *J. Pharm. Pharmacol.* **13**, 548 (1961); Quevauviller *et al.*, *Therapie* **18**, 1429 (1963); Guerrin *et al.*, *C. R. Soc. Biol.* **158**, 1096 (1964); P. Berthaux, M. Neuman, *Arzneimittel-Forsch.* **14**, 1040 (1964).

Free base, $[\alpha]_D^{17}$ −121° (c = 2 in CHCl$_3$). Sol in dil acids and the usual organic solvents except cyclohexane and petr ether.
Bitartrate, *DL 152, Tensibar.* Dec 145-150°. LD$_{50}$ in mice: 620 mg/kg orally; 430 mg/kg i.p.; 215 mg/kg i.v., P. Berthaux, M. Neuman, *loc. cit.*
THERAP CAT: Antihypertensive.

1227. Bifemelane. *N-Methyl-4-[2-(phenylmethyl)phenoxy]-1-butanamine; 4-(o-benzylphenoxy)-N-methylbutylamine; 2-(4-methylaminobutoxy)diphenylmethane; 2-benzyl-1-[4-(methylamino)butoxy]benzene.* C$_{18}$H$_{23}$NO; mol wt 269.39. C 80.25%, H 8.61%, N 5.20%, O 5.94%. Monoamine oxidase inhibitor. Prepn: R. Kikumoto *et al.*, **Ger.** pat. **2,627,227;** eidem, **U.S.** pat. **4,091,114** (1976, 1978 both to Mitsubishi); and antidepressant activity: R. Kikumoto *et al.*, *J. Med. Chem.* **24**, 145 (1981). Pharmacology: A. Tobe *et al.*, *Arzneimittel-Forsch.* **31**, 1278 (1981). Effects on experimental amnesia in rats: A. Tobe *et al.*, *Japan. J. Pharmacol.* **39**, 153 (1985). Effects on neuronal activity in cats: M. Egawa *et al.*, *Neuropharmacol.* **26**, 379 (1987). Inhibition of MAO: M. Naoi *et al.*, *J. Neurochem.* **50**, 243 (1988).

Hydrochloride, C$_{18}$H$_{24}$ClNO, *E-0687, MCI-2016, Alnert, Celeport.* Crystals from acetone, mp 117-121°. LD$_{50}$ in mice, rats (mg/kg): 1000, 1080 orally; 173, 130 i.p. (Tobe, 1981).
THERAP CAT: Nootropic.

1228. Bifenox. *5-(2,4-Dichlorophenoxy)-2-nitrobenzoic acid methyl ester; methyl 5-(2,4-dichlorophenoxy)-2-nitrobenzoate; 2,4-dichlorophenyl 3-(methoxycarbonyl)-4-nitrophenyl ether; MC 4379; Modown.* C$_{14}$H$_9$Cl$_2$NO$_5$; mol wt 342.14. C 49.15%, H 2.65%, Cl 20.72%, N 4.09%, O 23.38%. Pre-emergence herbicide. Prepn and herbicidal activity: **Belg.** pat. **749,444;** R. J. Thiessen, **U.S.** pat. **3,652,645** (1970, 1972 both to Mobil). Comparison with other herbicides in corn: W. M. Dest *et al.*, *Proc. Northeast. Weed Sci. Soc.* **27**, 31 (1973). Metabolism in soil, plants: G. R. Leather, C. L. Foy, *Pestic. Biochem. Physiol.* **7**, 437 (1977). Brief description: R. H. Dreger, *Weeds Today* **8**(2), 18 (1977). *Review:* P. J. Kruger *et al.*, *Proc. 12th Brit. Weed Control Conf.* **2**, 839-845 (1974).

Yellow tan crystals, mp 84-86°. Practically insol in water (0.35 ppm at 25°). Soly in xylene at 25°: 30%. Vapor pressure at 30°: 2.4 × 10^{-6} mm Hg. LD$_{50}$ orally in rats, mice: >6400, 4556 mg/kg; LC$_{50}$ in pheasants, wild ducks: >5000 ppm (Kruger).
USE: Herbicide.

1229. Bifenthrin. *[1α,3α(Z)]-(±)-3-(2-Chloro-3,3,3-trifluoro-1-propenyl)-2,2-dimethylcyclopropanecarboxylic acid (2-methyl[1,1'-biphenyl]-3-yl)methyl ester; 2-methylbi-*

phenyl-3-ylmethyl-(Z)-(1RS)-cis-3-(2-chloro-3,3,3-trifluoroprop-1-enyl)-2,2-dimethylcyclopropanecarboxylate; biphenate; biphenthrin; biphentrin; FMC-54800; Brigade; Talstar; Capture. $C_{23}H_{22}ClF_3O_2$; mol wt 434.89. C 66.28%, H 5.10%, Cl 8.15%, F 13.11%, O 7.36%. Third generation synthetic pyrethroid. Prepn: J. F. Engel, **Eur. pat. Appl.** **3,336;** idem, **U.S. pat. 4,238,505** (1979, 1980 both to FMC); E. L. Plummer et al., Pestic. Sci. **14,** 560 (1983). Physical properties, toxicology and review of field studies: H. J. H. Doel et al., Med. Fac. Landouww. Rijksuniv. Gent **49,** 929 (1984). Insecticidal activity: M. S. Mulla, H. A. Darwazeh, Bull. Soc. Vector Ecol. **10,** 1 (1985); M. S. Hamed, C. O. Knowles, J. Econ. Entomol. **81,** 1295 (1988). Field studies: Y. Antignus et al., Ann. Appl. Biol. **110,** 557 (1987); J. T. Trumble et al., J. Econ. Entomol. **81,** 608 (1988).

Light brown viscous oil, mp 51-66°. d^{25} 1.212 g/ml. Vapor pressure at 25°: 1.81×10^{-7} Torr. Sol in methylene chloride, chloroform, acetone, ether, toluene. Slightly sol in heptane, methanol. Soly in water: < 0.1 ppb. LD_{50} orally in rats: 54.5 mg/kg; LD_{50} dermally in rabbits: > 2000 mg/kg (Doel).

USE: Insecticide, acaricide.

1230. Bifidus Factor. Lactobacillus bifidus factor; Lactobacillus bifidus growth factor. A factor found in human milk and causing a predominant occurrence of L. bifidus in the intestinal tract of breast-fed infants: Petuely, Kristen, Oesterr. Z. Kinderheilk. u. Kinderfürsorge **6,** 173 (1951); Petuely, Naturwiss. **40,** 349 (1953). Essential growth factor for L. bifidus var Penn: György in Ciba Found. Symp., Chemistry and Biology of Mucopolysaccharides (Little, Brown, Boston, 1958) pp 140-156. Isoln from human milk: György et al., Arch. Biochem. Biophys. **48,** 193, 202, 209, 214 (1954); György et al., U.S. pat. 2,786,051 (1957 to Am. Home Prod.); from L. bifidus cultured together with Escherichia coli: Kludas, U.S. pat. 2,962,424 (1960 to J. Carl Pflüger). Isoln from carrots and identification of the major and minor bifidus factor: Samejima et al., Chem. Pharm. Bull. **19,** 166, 178, 186 (1971); Z. Tamura et al., Proc. Japan. Acad. **48,** 138, 144 (1972). Prepn from porcine gastric mucosa and use as dietetic adjuvant in infant food: **Fr. pat. 2,101,032,** C.A. **78,** 28221g (1970) and P. C. Wirth, **Ger. pat. 2,040,268,** C.A. **78,** 43859r (1970) (both 1972 to Sogeras).

USE: As adjuvant in powdered milk formulas for infants.

1231. Bifluranol. 4,4'-(1-Ethyl-2-methyl-1,2-ethanediyl)bis[2-fluorophenol]; erythro-3,3'-difluoro-4,4'-dihydroxy-α-ethyl-α'-methyldibenzyl; BX-341; Prostarex. C_{17}-$H_{18}F_2O_2$; mol wt 292.33. C 69.85%, H 6.20%, F 13.00%, O 10.95%. Orally active antiprostatic agent. Prepn: J. C. Turner, R. P. Chan, **Ger. pat. 2,110,428** corresp to **U.S. pat.** **4,051,263** (1971, 1977 both to Biorex). Antiprostatic activity: J. B. Dekanski, Brit. J. Pharmacol. **71,** 11 (1980). Chemistry, disposition in animals: D. J. Pope et al., J. Pharm. Pharmacol. **33,** 297 (1981). Metabolism: eidem, ibid. 302.

Cryst from toluene, mp 158-159°.
THERAP CAT: Anti-androgen.

1232. Bifonazole. 1-([1,1'-Biphenyl]-4-ylphenylmethyl)-1H-imidazole; (±)-1-(p,α-diphenylbenzyl)imidazole; Bay h 4502; Amycor; Azolmen; Bedriol; Mycospor; Mycosporan. $C_{22}H_{18}N_2$; mol wt 310.39. C 85.13%, H 5.84%, N 9.02%. Antimycotic deriv of imidazole. Prepn: E. Regel et

al., **Ger. pat. 2,461,406;** eidem, **U.S. pat. 4,118,487** (1976, 1978 both to Bayer). Series of articles on in vitro and in vivo antimycotic efficacy, microscopic studies, pharmacokinetics, efficacy in dermatomycoses and comparison with clotrimazole and miconazole, q.q.v.: Arzneimittel-Forsch. **33,** 517-551, 745-754 (1983). Toxicology: G. Schlüter, ibid. 739.

Crystals from acetonitrile, mp 142°. Very lipophilic. Sol in alcohols, DMF, DMSO. Soly in water at pH 6: < 0.1 mg/100 ml. Stable in aq soln at pH 1-12. LD_{50} in male mice, rats (mg/kg): 2629, 2854 orally (Schlüter).
THERAP CAT: Antifungal.

1233. Biguanide. Imidodicarbonimidic diamide; guanylguanidine; amidinoguanidine; diguanide. $C_2H_7N_5$; mol wt 101.12. C 23.75%, H 6.98%, N 69.27%. Preparation from dicyanodiamide: Rackmann, Ann. **376,** 169 (1910); Karipides, Fernelius, Inorg. Syn. **7,** 56, 58 (1963). Crystal structure: S. R. Ernst, F. W. Cagle, Acta Crystallogr. **33B,** 235 (1977); S. R. Ernst, ibid. 237.

Crystals from alc, mp 130°; dec rapidly at about 142°. Sol in water, alcohol. Insol in ether, benzene, chloroform. The aq soln dec on standing or heating. Aq solns are alkaline. Biguanide is a stronger base (pK_1 = 11.52; pK_2 = 2.93) than ammonia (pK = 9.61).

Hydrochloride, $C_2H_7N_5 \cdot HCl$, needles, mp 235°, very sol in water, slightly sol in alc.

Neutral sulfate, $(C_2H_7N_5)_2 \cdot H_2SO_4 \cdot 2H_2O$, large crystals, sol in water.

Acid sulfate, $C_2H_7N_5 \cdot H_2SO_4 \cdot H_2O$, rhombic prisms, sol in water, the soln giving an acid reaction.

USE: Sulfates in the determination of copper and nickel.

1234. Bikhaconitine. (1α,6α,14α,16β)-20-Ethyl-1,6,16-trimethoxy-4-(methoxymethyl)aconitane-8,13,14-triol 8-acetate 14-(3,4-dimethoxybenzoate); acetylveratroylbikhaconine. $C_{36}H_{51}NO_{11}$; mol wt 673.82. C 64.17%, H 7.63%, N 2.08%, O 26.12%. From roots of Aconitum spicatum Stapf., Ranunculaceae: Dunstan, Andrews, J. Chem. Soc. **87,** 1636 (1905). Structure: Tsuda, Marion, Can. J. Chem. **41,** 3055 (1963).

Prisms from n-hexane, mp 163.5-164°.
Monohydrate, needles from ether or dil methanol, mp 105-110°. $[\alpha]_D$ + 16° (c = 1.6 in ethanol). Sol in alcohol, chloroform, ether, dil acids. Practically insol in water, petr ether.

1235. Bilirubin. 2,17-Diethenyl-1,10,19,22,23,24-hexahydro-3,7,13,18-tetramethyl-1,19-dioxo-21H-biline-8,12-dipropanoic acid; 1,10,19,22,23,24-hexahydro-2,7,13,17-tetramethyl-1,19-dioxo-3,18-divinylbiline-8,12-dipropionic acid; 1,3,6,7-tetramethyl-4,5-dicarboxyethyl-2,8-divinyl-(b-13)-dihydrobilenone; bilirubin IXα. $C_{33}H_{36}N_4O_6$; mol wt 584.65. C 67.79%, H 6.21%, N 9.58%, O 16.42%. Principal pigment of bile and constituent of many biliary calculi. Major endproduct of the biological breakdown of heme, q.v. Bilirubin is the chromophore responsible for coloration in various forms of jaundice. Also found in blood serum, where it exists in four major forms: unconjugated bilirubin, the monoglucuronide, the diglucuronide, and albumin-bound bilirubin. Most easily obtained from ox gallstones which are largely calcium bilirubinate: Städeler, *Ann.* **132**, 323 (1864); Küster, *Z. Physiol. Chem.* **94**, 136 (1915); **99**, 86 (1917); **121**, 80 (1922); Küster, Haas, *ibid.* **141**, 279 (1924); Fischer, *ibid.* **3**, 204 (1911); Fischer, Hess, *ibid.* **194**, 193 (1931). Isoln from pig bile: Gibson, Lowe, *J. Biol. Chem.* **123**, XLI (1938); Gray *et al., J. Chem. Soc.* **1961**, 2264, 2268; from ox bile: Libowitzky, *Z. Physiol. Chem.* **263**, 267 (1940). Industrial isoln from ox bile using chlorobenzene as extractant: U.S. pats. **2,166,073; 2,331,574; 2,363,471; 2,386,716.** Structure and synthesis: Fischer, Plieninger, *Naturwiss.* **30**, 382 (1942); *Z. Physiol. Chem.* **274**, 231 (1942); *cf.* Fischer-Orth, *Die Chemie des Pyrrols* II, 1, 621 (Leipzig, 1937); Gray *et al., Nature* **181**, 183 (1958); *eidem, J. Chem. Soc.* 2276. Structure: Fog, Jellum, *Nature* **198**, 88 (1963). Configuration: Kuenzle *et al., Biochem. J.* **133**, 364 (1973). X-ray analysis and structure: Bonnett *et al., J. Chem. Soc. Perkin Trans. II,* **1972**, 902, 1335; *Nature* **262**, 326 (1976). NMR conformation studies: D. Kaplan, G. Navon, *J. Chem. Soc. Perkin Trans. II* **1981**, 1374. Separation of bilirubin species in serum and bile by reversed-phase HPLC: J. J. Lauff *et al., J. Chromatog.* **226**, 391 (1981); *eidem, Clin. Chem.* **29**, 800 (1983). Clinical importance of albumin-bound bilirubin: J. S. Weiss *et al., N. Engl. J. Med.* **309**, 148 (1983). Comprehensive reviews: Lemberg, Legge, *Hematin Compounds and Bile Pigments* (New York, 1949); With, *Bile Pigments* (Academic Press, New York, 1968).

Light orange to deep reddish-brown monoclinic rhomboid, prisms, plates from chloroform. Gradually blackens on heating and does not melt. The greenish solns show a red fluorescence in ultraviolet light. A 0.001% soln in chloroform shows selective abs from 490 to 400 nm with a max at 453 nm; ϵ mM 60.7 ± 0.8. Practically insol in water. Sol in benzene, chloroform, chlorobenzene, carbon disulfide, acids, alkalies; slightly sol in alcohol, ether. Spreads on water. Penetrates into cholesterol, octadecylamine, and protein monolayers: Stenhagen, Rideal, *Biochem. J.* **33**, 1591 (1939).

1236. Biliverdine. *3,18-Diethenyl-1,19,22,24-tetrahydro-2,7,13,17-tetramethyl-1,19-dioxo-21H-biline-8,12-dipropanoic acid; 1,19,22,24-tetrahydro-2,7,13,17-tetramethyl-1,19-dioxo-3,18-divinylbiline-8,12-dipropionic acid;* 1,3,6,7-tetramethyl-4,5-dicarboxyethyl-2,8-divinylbilenone; 4,5-di(2-carboxyethyl)-1,3,6,7-tetramethyl-2,8-divinylbilatriene; dehydrobilirubin; uteroverdine; oöcyan. $C_{33}H_{34}N_4O_6$; mol wt 582.63. C 68.02%, H 5.88%, N 9.62%, O 16.48%. Precursor of bilirubin. Formed in the body from hemoglobin. The bile of amphibia and of birds contains biliverdine only. Does not occur in normal human bile or normal human serum, but regularly accompanies bilirubin in the serum of patients with carcinomatous obstruction of the bile duct, and frequently in that of patients with liver cirrhosis, catarrhal jaundice, and bile duct occlusion by gallstones. Can be obtained by autooxidation of bilirubin in alkaline soln but the yield is poor: Lemberg, *Biochem. J.* **28**, 978 (1934); better yields by oxidation of bilirubin with ferric chloride in glacial acetic acid: Lemberg, *Ann.* **499**, 25 (1932); by coupled oxidation of hemoglobin and ascorbic acid: Lemberg *et al., Biochem. J.* **35**, 363 (1941); by oxidation of bilirubin with ferric chloride in methanol: Gray *et al., J. Chem. Soc.* **1961**, 2264. Crystal and molecular structure: W. S. Sheldrick, *J. Chem. Soc., Perkin Trans. 2* **1976**, 1457. Comprehensive reviews: Lemberg, Legge, *Hematin Compounds and Blue Pigments* (New York, 1949); With, *Bile Pigments* (Academic Press, New York, 1968).

Dark green plates or prisms with violet surface color from methanol. Does not melt, blackens and dec above 300°. Absorption spectrum: Gray *et al., loc. cit.* Soluble in methanol, ether, chloroform, carbon disulfide, benzene, soln of alkali hydroxides. Gives the later color changes of the Gmelin test starting with green.

Dimethyl ester, $C_{35}H_{38}N_4O_6$, green crystals from chloroform + petr ether. When produced from bilirubin, mp 215-223°; synthetic 206-209°; from hemin 208°; from hemoglobin 216°.

Dimethyl ester ferrichloride, $C_{35}H_{39}Cl_4FeN_4O_6$ (green hemin ester), pleochroitic elongated platelets, no definite mp. Ferric chloride and HCl can be removed by washing the chloroform soln with water.

1237. Binapacryl. *3-Methyl-2-butenoic acid 2-(1-methylpropyl)-4,6-dinitrophenyl ester; 3-methylcrotonic acid 2-sec-butyl-4,6-dinitrophenyl ester;* 2-sec-butyl-4,6-dinitrophenyl 3,3-dimethylacrylate; 2-sec-butyl-4,6-dinitrophenyl 3-methyl-2-butenoate; 2-sec-butyl-4,6-dinitrophenyl 3-methylcrotonate; 2-sec-butyl-4,6-dinitrophenyl senecioate; 3,3-dimethylacrylic acid 2-sec-butyl-4,6-dinitrophenyl ester; 4,6-dinitro-2-sec-butylphenyl β,β-dimethylacrylate; 3-methyl-2-butenoic acid 2-sec-butyl-4,6-dinitrophenyl ester; dinoseb methacrylate; senecioic acid 2-sec-butyl-4,6-dinitrophenyl ester; ENT 25793; HOE 2784; Acricid; Ambox; Endosan; Morocide; Niagara 9044. $C_{15}H_{18}N_2O_6$; mol wt 322.31. C 55.89%, H 5.63%, N 8.69%, O 29.78%. Prepn: Belg. pat. **630,947** corresp to Scherer *et al.,* U.S. pat. **3,370,-085** (1963, 1968, both to Hoechst).

Prisms, mp 70°. d_4^{20} 1.25-1.28. Vapor press. at 60°: 1 × 10^4 mm Hg. Very sol in acetone, xylene; sol in ethanol, kerosene. Practically insol in water. LD_{50} orally in female, male rats: 58, 63 mg/kg, T. B. Gaines, *Toxicol. Appl. Pharmacol.* **14**, 515 (1969).

USE: Fungicide, miticide.

1238. Binedaline. *N,N,N'-Trimethyl-N'-(3-phenyl-1H-indol-1-yl)-1,2-ethanediamine;* 1-[[2-(dimethylamino)ethyl]methylamino]-3-phenylindole; binodaline. $C_{19}H_{23}N_3$; mol wt 293.41. C 77.78%, H 7.90%, N 14.32%. Substituted hydrazine with anti-depressant activity. Prepn: F. Schatz *et al.,* Ger. pat. **2,512,702** corresp to U.S. pat. **4,204,998** (1975, 1980 both to Siegfried); F. Schatz *et al., Arzneimittel-Forsch.* **30**, 919 (1980). Inhibits monoamine uptake in rat brain: D. P. Benfield, D. K. Luscombe, *J. Pharm. Pharmacol.* **33**, Suppl., 42P (1981); *eidem, Arzneimittel-Forsch.* **33**, 847 (1983). Pharmacology, toxicity: U. Jahn *et al., ibid.* 726. Biochemical studies: J. Maj *et al., ibid.* 841. Binding in blood, serum: D. Morin *et al., J. Pharm. Sci.* **74**, 727 (1985). Comparison with amitriptyline in volunteers: P. H. Joubert *et al., Eur. J. Clin. Pharmacol.* **27**, 667 (1985); J. Longmore *et al., Brit. J. Clin. Pharmacol.* **19**, 295 (1985).

H₃C — N — CH₂CH₂N(CH₃)₂

Crystals from petroleum, mp 52-53°.
Hydrochloride, C₁₉H₂₄ClN₃, *RU-39780, Sgd-Scha 1059, Ixprim.* Colorless crystals from isopropanol, mp 195-196°. uv max (0.1N HCl): 222, 263 nm (ε 28000, 14500). LD₅₀ in male, female mice, male, female rats (mg/kg): 760, 770, 1380, 1160 orally; 54.0, 54.0, 27.2, 26.0 i.v. (Jahn *et al.*).
    THERAP CAT: Antidepressant.

1239. Binifibrate. *3-Pyridinecarboxylic acid 2-[2-(4-chlorophenoxy)-2-methyl-1-oxopropoxy]-1,3-propanediyl ester;* trihydroxypropane 2-p-chlorophenoxyisobutyrate-1,3-dinicotinate; 2-(p-chlorophenoxy)-2-methylpropionic acid ester with 1,3-dinicotinoyloxy-2-propanol; glyceryl 2-p-chlorophenoxyisobutyrate-1,3-dinicotinate; WAC 104; Biniwas. C₂₅H₂₃ClN₂O₇; mol wt 498.92. C 60.19%, H 4.65%, Cl 7.10%, N 5.61%, O 22.45%. Structurally related to clofibrate, *q.v.*. Prepn: R. Andreoli, X. Cicera, **Span.** pat. 463,-218 (1978 to Soc. Espan. Espec. Farm.-Ter.), *C.A.* **90,** 72067h (1979). Prepn and pharmacology: *eidem,* **Span.** pat. **488,665** corresp to **Belg.** pat. **884,722** (both 1980 to Soc. Espan. Espec. Farm.-Ter.). Microhemorheological properties: L. Bruseghini *et al., Arzneimittel-Forsch.* **33,** 854 (1983). Preliminary clinical study: M. Dalmau *et al., ibid.* 858.

Yellowish-white crystals from alcohol-isopropyl ether, mp 100°. Saponification number 166. LD₅₀ orally in mice and rats: >4000 mg/kg (Andreoli).
    THERAP CAT: Antihyperlipoproteinemic.

1240. Biocytin. *N⁶-[5-(Hexahydro-2-oxo-1H-thieno[3,4-d]imidazol-4-yl)-1-oxopentyl]-L-lysine; ε-N-biotinyl-L-lysine;* biotin complex of yeast. C₁₆H₂₈N₄O₄S; mol wt 372.48. C 51.59%, H 7.58%, N 15.04%, S 8.61%, O 17.18%. A naturally occurring complex of biotin. Contains 65.6% biotin. Isoln: Wright *et al., J. Am. Chem. Soc.* **72,** 1048 (1950); **74,** 1996 (1952). Structure: Peck *et al., ibid.* **72,** 1048 (1950); **74,** 1999 (1952). Synthesis: Wolf *et al., ibid.* **74,** 2002 (1952); Weijlard *et al., ibid.* **76,** 2505 (1954); Wolf, Folkers, **U.S.** pat. **2,710,298** (1955 to Merck & Co.). Purification: McCormick, Föry, *Methods Enzymol.* **18** (pt A), 413 (1970). *Review:* A. F. Wagner, K. Folkers, *Vitamins and Coenzymes* (Wiley, New York, 1964) pp 138-159.

Crystals, mp 241-243°. Upon rapid crystn from dil methanol or dil acetone, mp 228-230° (dec); upon slow crystn sinters at 227°. mp 245-252° (dec, microblock). Crystals from water, mp 228.5°. [α]²⁵_D +53° (c = 1.05 in 0.1N NaOH). Infrared absorption spectrum: *J. Am. Chem. Soc.* **74,** 2001 (1952). Freely sol in water, glacial acetic acid. Less sol in alc. Practically insol in acetone and most other organic solvents. When subjected to strong acid hydrolysis (at least 3N at 120° for one hour) biocytin yields biotin and L-lysine. Forms a crystalline hydrochloride. Biocytin is

characterized microbiologically by its availability as a source of biotin to *Lactobacillus casei, L. delbrückii* LD 5, *L. acidophilus, Streptococcus fecalis* R, *Neurospora crassa,* and *Saccharomyces carlsbergensis* and by its unavailability as a source of biotin to *Lactobacillus arabinosus, L. pentosus,* and *Leuconostoc mesenteroides* P-60.

1241. Bioflavonoids. Vitamin P complex; citrus flavonoid compounds; Arliflav; C.V.P.; Pecitrol Veinogène. A group of compounds which contribute to the maintenance of normal blood vessel conditions by decreasing capillary permeability and fragility. Widely distributed among plants: J. B. Harborne, *Comparative Biochemistry of the Flavonoids* (Academic Press, New York, 1967). Biosynthesis: Grisebach, Barz, *Naturwiss.* **56,** 538 (1969). High concentrates can be obtained from all citrus fruits, rose hips, and black currants. Commercial methods extract the rinds of oranges, tangerines, lemons, limes, kumquats, and grapefruits. Solvents used in the extraction processes are aqueous alkalies, hot water, or water-miscible organic solvents, such as isopropanol: Freedman *et al.,* **U.S.** pat. **2,888,381** (1959 to U.S. Vitamin). Early interest in the compounds developed because of their synergistic effect with ascorbic acid. Other pharmacologic effects such as inhibition of adrenaline autoxidation and of enzyme action are also under study: several authors in *The Pharmacology of Plant Phenolics,* J. W. Fairbairn, Ed. (Academic Press, New York, 1959); several authors in *Angiologica* **9,** 133-446 (1972). Metabolism: DeEds, "Flavonoid Metabolism" in *Comprehensive Biology,* **Vol. 20,** M. Florkin, E. H. Stotz, Eds. (Elsevier, New York, 1968). *Reviews:* Scarborough, Bacharach, *Vitam. Horm. (New York)* **7,** 1-55 (1949); Baier *et al., Ann. N.Y. Acad. Sci.* **61,** (Art. 3), 637-736 (1955); T. Robinson, *The Organic Constituents of Higher Plants* (Burgess, Minneapolis, 1967) pp 178-209; H. Geiger, C. Quinn in *Flavonoids,* J. B. Harborne *et al.,* Eds. (Academic Press, New York, 1975) pp 692-742.
    THERAP CAT: Capillary protectant.

1242. Biopterin. *[S-(R*,S*)]-2-Amino-6-(1,2-dihydroxypropyl)-4(1H)-pteridinone; 1-(2-amino-4-hydroxy-6-pteridinyl)-1,2-propanediol;* 2-amino-4-hydroxy-6-(1,2-dihydroxypropyl)pteridine; pterin HB₂. C₉H₁₁N₅O₃; mol wt 237.22. C 45.57%, H 4.67%, N 29.53%, O 20.23%. A pteridine widely distributed in nature; naturally occurring as the L-erythro-form. Considered as a growth factor for some insects; *see also* Neopterin. Isoln from human urine: Patterson *et al., J. Am. Chem. Soc.* **77,** 3167 (1955); **78,** 5871 (1956); **Brit.** pat. **814,462** (1959 to Am. Cyanamid); from drosophila: H. S. Forrest, H. K. Mitchell, *J. Am. Chem. Soc.* **77,** 4865 (1955); from queen bee jelly: Butenandt, Rembold, *Z. Physiol. Chem.* **311,** 79 (1958). Absolute configuration: E. L. Patterson *et al., J. Am. Chem. Soc.* **78,** 5871 (1956). Synthesis: E. L. Patterson *et al., ibid.* 5868; Tschesche *et al., Ann.* **658,** 193 (1962); Rembold, Metzger, *Ber.* **96,** 1395 (1963); Viscontini *et al., Helv. Chim. Acta* **55,** 570, 574 (1972); B. Schircks *et al., ibid.* **60,** 211 (1977); T. Sugimoto *et al., Bull. Chem. Soc. Japan* **53,** 2344 (1980). Synthesis from neopterin, *q.v.:* A. Kaiser, H. P. Wessel, *Helv. Chim. Acta* **70,** 766 (1987). Biosynthesis: G. Kapatos *et al., Science* **213,** 1129 (1981).

Minute, yellow, spherical crystals from water. Chars without melting at 250-280°. [α]²⁴_D −50° (c = 0.4 in 0.1N HCl); [α]²⁴_D −26° (c = 0.92 in 0.1N NaOH). uv max (0.08N HCl): 247 nm (ε 11,000). Soly in water: 0.7 mg/ml (20°); 4 mg/ml (90°). Soly in alc, ether, acetone, benzene: <0.1 mg/ml; in 1N NaOH, 1N HCl: >25 mg/ml. Fluoresces with a blue color in alkaline soln.

1243. Bioresmethrin. *(1R-trans)-2,2-Dimethyl-3-(2-methyl-1-propenyl)cyclopropanecarboxylic acid [5-(phenylmethyl)-3-furanyl]methyl ester; trans-(+)-2,2-dimethyl-3-(2-methylpropenyl)cyclopropanecarboxylic acid (5-benzyl-3-*

furyl)methyl ester; 5-benzyl-3-furylmethyl-(+)-*trans*-chrysanthemate; *d-trans*-[(5-benzyl-3-furyl)methyl]chrysanthemumate; (+)-*trans*-resmethrin; NRDC 107; NIA-18739; SBP-1390; Resbuthrin; Biobenzyfuroline. $C_{22}H_{26}O_3$; mol wt 338.45. C 78.07%, H 7.74%, O 14.18%. Potent synthetic pyrethroid insecticide: M. Elliott *et al., Nature* **213**, 493 (1967); M. Elliott *et al., Pestic. Sci.* **2**, 243 (1971). Prepn: M. Elliott, N. F. Janes, *Fr.* pat. **1,503,260**; *eidem,* U.S. pats. **3,465,007** and **3,542,928** (1967, 1969, 1970, all to Nat. Res. Dev. Corp.). Mammalian metabolism: C. O. Abernathy, J. E. Casida, *Science* **179**, 1235 (1973). Control of cockroaches: P. R. Chadwick *et al., J. Med. Entomol.* **13**, 625 (1977). Acute toxicity: T. B. Gaines, R. E. Linder, *Fundam. Appl. Toxicol.* **7**, 299 (1986). Brief review: D. S. Gunew, *Anal. Methods Pestic. Plant Growth Regul.* **10**, 19-29 (1978).

$bp_{0.0008}$ 174°. n_D^{20} 1.5346. $[\alpha]_D^{20}$ −7.8° (c = 5 in acetone). Sol in most organic solvents. Insol in water. LD_{50} in adult male, female rats (mg/kg): 1244, 1721 orally (Gaines, Linder).

USE: Insecticide.

1244. Biotin. *Hexahydro-2-oxo-1H-thieno[3,4-d]imidazole-4-pentanoic acid; cis*-tetrahydro-2-oxothieno[3,4-*d*]-imidazoline-4-valeric acid; *cis*-hexahydro-2-oxo-1*H*-thieno[3,4]imidazole-4-valeric acid; vitamin H; Coenzyme R; Bios II; Bioepiderm. $C_{10}H_{16}N_2O_3S$; mol wt 244.31. C 49.16%, H 6.60%, N 11.47%, O 19.65%, S 13.12%. Growth factor present in minute amounts in every living cell. Plays an indispensable role in numerous naturally occurring carboxylation reactions. Occurs mainly bound to proteins or polypeptides. The richest sources are liver, kidney, pancreas, yeast, and milk. The biotin content of cancerous tumors is higher than that of normal tissue. Biotin combines with the proteinaceous substance, avidin, in raw egg-white and becomes inactive. When on diets contg large amounts of raw egg-white the rat or chick develops characteristic skin lesions and growth is retarded; these symptoms can be prevented by feeding additional biotin. Isoln from egg yolk: Kögl, Tönnis, *Z. Physiol. Chem.* **242**, 43 (1936); from liver: György *et al., J. Biol. Chem.* **131**, 745 (1939); du Vigneaud *et al., ibid.* **140**, 643 (1941). Identity of biotin with vitamin H: du Vigneaud *et al., Science* **92**, 62 (1940). Structure: du Vigneaud *et al., J. Biol. Chem.* **146**, 475 (1942). First synthesis: Harris *et al., J. Am. Chem. Soc.* **67**, 2096 (1945). Configuration: Traub, *Nature* **178**, 649 (1956). Stereospecific total synthesis of the naturally occurring *d*-form: P. N. Confalone *et al., J. Am. Chem. Soc.* **97**, 5936 (1975); *eidem, J. Org. Chem.* **42**, 1630 (1977); T. Ogawa *et al., Carbohyd. Res.* **57**, C31 (1977); F. G. M. Vogel *et al., Ann.* **1980**, 1972; R. R. Schmidt, M. Maier, *Synthesis* **1982**, 747. Synthesis of the *dl*-form: P. N. Confalone *et al., Helv. Chim. Acta* **59**, 1005 (1976); M. Marx *et al., J. Am. Chem. Soc.* **99**, 6794 (1977); A. Fliri, K. Hohenlohe-Oehringen, *Ber.* **113**, 607 (1980); Ph. Rossy *et al., Tetrahedron Letters* **22**, 3493 (1981). Biosynthetic studies: R. J. Parry, M. G. Kunitani, *J. Am. Chem. Soc.* **98**, 4024 (1976); R. J. Parry, M. V. Naidu, *Tetrahedron Letters* **1980**, 4783. Coordination properties: H. Siegel, *Experientia* **37**, 789 (1981). Mechanism of action: M. J. Cravey, H. Kohn, *J. Am. Chem. Soc.* **102**, 3928 (1980); D. L. Vesely, *Science* **216**, 1329 (1982). Reviews: A. F. Wagner, K. Folkers, *Vitamins and Coenzymes* (Wiley, New York, 1964) pp 138-159; Harris in *The Vitamins*, vol. II, W. H. Sebrell, R. S. Harris, Eds. (Academic Press, New York, 2nd ed., 1968) pp 261-359; D. B. McCormick, *Nutr. Rev.* **33**, 97-102 (1975). Review of assay methods: P. György, *The Vitamins* vol. VII, P. Gyorgy, W. N. Pearson, Eds. (1967) pp 303-313.

Characteristics of biotin isolated from liver or milk (Kögl's β-biotin): fine long needles, mp 232-233°. $[\alpha]_D^{21}$ +91° (c = 1 in 0.1*N* NaOH). Isoelec pt pH 3.5. pH of 0.01% aq soln 4.5. Soly at 25° (mg/100 ml): water ~22; 95% alc ~ 80. More sol in hot water and in dil alkali. Insol in other common organic solvents. The pure compd is stable to air and temp. Moderately acid and neutral solns are stable several months; alkaline solns are less stable, but appear reasonably stable up to a pH of about 9. Aq solns are very susceptible to mold growth. Acidic solns can be heat sterilized. Incompatible with nitrous acid, oxidizing agents, formaldehyde, chloramine T, strong acid or alkali.

1245. Biotin *l*-Sulfoxide. *[3aS-(3aα,4β,5β,6aα)]-Hexahydro-2-oxo-1H-thieno[3,4-d]imidazole-4-pentanoic acid 5-oxide;* AN factor. $C_{10}H_{16}N_2O_4S$; mol wt 260.31. C 46.14%, H 6.20%, N 10.76%, O 24.59%, S 12.32%. Isolated from *Aspergillus niger* culture filtrate where growth had taken place in the presence of pimelic acid; also from milk residue concentrates or by synthesis: Melville, *J. Biol. Chem.* **208**, 495 (1954); Wright *et al., J. Am. Chem. Soc.* **76**, 4163 (1954).

Polymorphic plates from water, mp 238° (some dec). $[\alpha]_D^{20}$ −39.5° (c = 1.01 in 0.1*N* NaOH). Ineffective in curing the biotin deficiency syndrome when administered to rats at a level 100 times the effective dose of biotin.

1246. Biperiden. *α-Bicyclo[2.2.1]hept-5-en-2-yl-α-phenyl-1-piperidinepropanol; α-5-norbornen-2-yl-α-phenyl-1-piperidinepropanol;* 3-piperidino-1-phenyl-1-bicyclohepten-yl-1-propanol; 1-bicycloheptenyl-1-phenyl-3-piperidino-propanol; 3-piperidino-1-phenyl-(Δ^5-bicyclo[2.2.1]hepten-2-yl)-1-propanol; KL 373. $C_{21}H_{29}NO$; mol wt 311.45. C 80.98%, H 9.39%, N 4.50%, O 5.14%. Prepn: Klavehn, U.S. pat. **2,789,110** (1957 to Knoll).

Crystals, mp 112-116°; also reported as 101° (Klavehn). Sparingly sol in water; slightly sol in ethanol; readily sol in methanol.

Hydrochloride, $C_{21}H_{30}ClNO$, Akineton, Akinophyl. Crystals, dec ~275°; also reported as mp 238° (Klavehn). LD_{50} in mice: 545 mg/kg orally, 56 mg/kg i.v.

THERAP CAT: Anticholinergic, antiparkinsonian.

1247. Biphenamine. β-Diethylaminoethyl 2-hydroxy-3-phenylbenzoate; 2-diethylaminoethyl 3-phenylsalicylate; 2-diethylaminoethyl 2-hydroxy-3-biphenylcarboxylate; 3-phenylsalicylic acid 2-diethylaminoethyl ester; xenysalate. $C_{19}H_{23}NO_3$; mol wt 313.38. C 72.82%, H 7.40%, N 4.47%, O 15.32%. Prepn: Sahyun, U.S. pat. **2,594,350** (1952).

Oily liquid. Sol in water.

Hydrochloride, $C_{19}H_{24}ClNO_3$, *Sebaclen, Sebaklen*.
Ingredient of *Alvinine*.

THERAP CAT: Topical anesthetic; antibacterial; antifungal.

1248. p-Biphenylamine. *[1,1'-Biphenyl]-4-amine;* *p*-aminobiphenyl; *p*-aminodiphenyl; anilinobenzene; xenyl-amine. $C_{12}H_{11}N$; mol wt 169.22. C 85.17%, H 6.55%, N 8.28%. Prepn from diazoaminobenzene: Heusler, *Ann.* **260**, 232 (1890). Review of carcinogenicity studies: *IARC Monographs* **1**, 74-79 (1972).

Leaflets from alc or water. mp 53°. Volatile with steam. Slightly sol in cold water, readily sol in hot water, alcohol, chloroform.

Note: This substance has been listed as a known carcinogen: *Fourth Annual Report on Carcinogens* (NTP 85-002, 1985) p 20.

USE: In the detection of sulfates. Formerly as rubber anti-oxidant. As carcinogen in cancer research.

1249. 2,4'-Biphenyldiamine. Diphenyline; 2,4'-diphen-yldiamine; *o,p'*-dianiline; 2,4'-diaminodiphenyl. $C_{12}H_{12}N_2$; mol wt 184.23. C 78.23%, H 6.56%, N 15.21%. Preparation by reducing azobenzene with tin and hydrochloric acid: Fischer, *Monatsh.* **6**, 547 (1885).

Needles from dil alc, mp 45°. bp 363°. Very slightly sol in alc or ether.

USE: Detection of tungsten; manuf azo dyes.

1250. Bipiperidyl Mustard. *1,1'-Bis(2-chloroethyl)-4,4'-bipiperidine;* BPM. $C_{14}H_{26}Cl_2N_2$; mol wt 293.29. C 57.33%, H 8.94%, Cl 24.18%, N 9.55%. Obesifying agent causing rapid lipid deposition in mice: Rutman *et al., Science* **153**, 1000 (1966). Activity shown only by the active form, prepd by neutralizing the conjugate acid form and generating a bicyclic ammonium deriv at pH 10. Metabolic studies: *eidem, Trans. N.Y. Acad. Sci.* **30**, 244 (1967).

LD_{50} i.p. in mice: 30-35 mg/kg.

1251. Birch Tar Oil, Empyreumatic. Oil white birch; oleum rusci. Obtained by destructive distillation of the bark and wood of *Betula alba* L., *Betulaceae* (white birch). *Constit.* Oil turpentine, other isomeric hydrocarbons, various empyreumatic resins, guaiacol, cresol, pyrocatechol, betulin.

Black, viscid liquid; characteristic odor. d 0.926-0.955. Partly sol in alcohol; sol in chloroform, fats, oils.

USE: For preserving leather and wood.

1252. Birch Tar Oil, Rectified. "Essential oil of birch wood". Obtained from the empyreumatic oil by steam distillation. *Constit.* Phenol, cresol, xylenol, guaiacol, creosol, pyrocatechol.

Dark brown liquid. d 0.886-0.950. One ml dissolves in about 3 ml abs alcohol; sol in benzene, chloroform, ether, glacial acetic acid, carbon disulfide, oil turpentine.

THERAP CAT: Dermatologic.

THERAP CAT (VET): Has been used extern. in skin diseases.

1253. Bisacodyl. *4,4'-(2-Pyridylmethylene)bisphenol di-acetate;* bis(*p*-acetoxyphenyl)-2-pyridylmethane; (4,4'-di-acetoxydiphenyl)(2-pyridyl)methane; 2-(4,4'-diacetoxydi-phenylmethyl)pyridine; Bicol; Broxalax; Contalax; DAMP; Deficol; Dulcolan; Dulcolax; Durolax; Endokolat; Eulaxan; Godalax; Laco; Laxadin; Laxagetten; Laxanin N; Laxorex;

Nigalax; Perilax; Pyrilax; Stadalax; Telemin; Theralax; Ulcolax; VDH. $C_{22}H_{19}NO_4$; mol wt 361.38. C 73.11%, H 5.30%, N 3.88%, O 17.71%. Prepn: **Brit. pat. 730,243** (1955 to Thomae). Pharmacology and toxicology: Schmidt, *Arzneimittel-Forsch.* **3**, 19 (1953).

Tasteless crystals, mp 138°. Practically insol in water and alkaline solns. Sol in acids, alc, acetone, propylene glycol, and other organic solvents. LD_{50} orally in rats: > 3 g/kg (Schmidt).

Complex with tannin, *bisacodyl tannex, Clysodrast*.

THERAP CAT: Cathartic.

1254. Bis(4-amino-1-anthraquinonyl)amine. *1,1'-Iminobis[4-amino-9,10-anthracenedione];* 4,4'-diamino-1,1'-dianthraquinonylamine; 4,4'-diamino-1,1'-anthrimide; 4,4'-diamino-1,1'-iminobisanthraquinone. $C_{28}H_{17}N_3O_4$; mol wt 459.44. C 73.19%, H 3.73%, N 9.15%, O 13.93%. Prepn: **Ger. pat. 255,822** (1913); *Frdl.* **11**, 615; Tinker, Stallmann, **U.S. pat. 2,420,022** (1947), *C.A.* **41**, 5730 (1947).

On heating with fuming sulfuric acid (30% SO_3) at 100° yields a dark powder, sol in water or alc with bluish-black coloration, used as dyestuff for wool. On heating with molten potassium hydroxide at 190° and oxidizing the reaction product with sodium hypochlorite, an olive-green vat dye is obtained.

USE: In the textile industry; in the determination of boron: Beckett, Webster, *Analyst* **68**, 306 (1943), *C.A.* **38**, 38 (1944).

1255. Bisantrene. *9,10-Anthracenedicarboxaldehyde bis-[(4,5-dihydro-1H-imidazol-2-yl)hydrazone];* 9,10-anthracenedicarboxaldehyde bis(2-imidazolin-2-ylhydrazone). $C_{22}H_{22}N_8$; mol wt 398.47. C 66.31%, H 5.56%, N 28.12%. Prepn: K. C. Murdock *et al.,* **Ger. pat. 2,850,822**; K. C. Murdock, F. E. Durr, **U.S. pat. 4,258,181** (1979, 1981 both to Am. Cyanamid); and antitumor activity: K. C. Murdock *et al., J. Med. Chem.* **25**, 505 (1982). Crystallographic characterization of hydrochloride: R. B. Bates *et al., Acta Crystallogr.* **C42**, 186 (1986); C. G. Pierpont, S. A. Lang, Jr., *ibid.* 1085. HPLC determn in plasma: Y.-M. Peng *et al., J. Chromatog.* **233**, 235 (1982). Antitumor activity *in vitro*: J. D. Cowan *et al., Invest. New Drugs* **1**, 139 (1983); *in vivo*: R. V. Citarella *et al., Cancer Res.* **42**, 440 (1982). Clinical pharmacokinetics: K. Lu *et al., Cancer Chemother. Pharmacol.* **16**, 156 (1986). Clinical trials in breast cancer: H.-Y. Yap *et al., Cancer Res.* **43**, 1402 (1983); C. K. Osborne *et al., Cancer Treat Rep.* **68**, 357 (1984).

Dihydrochloride, $C_{22}H_{24}Cl_2N_8$, *NSC-337766, CL 216942, ADAH, ADCA, Orange Crush, Zantrène*. Crystalline orange solid from ethanol. Hemihydrate, mp 288-289° (dec). uv max (H_2O) 260, 415 nm (ε 72700, 16300).

THERAP CAT: Antineoplastic.

1256. 2,5-Bis(1-aziridinyl)-3,6-bis(2-methoxyethoxy)-1,4-benzoquinone. *2,5-Bis(1-aziridinyl)-3,6-bis(2-methoxyethoxy)-2,5-cyclohexadiene-1,4-dione;* Bayer E 39 soluble. $C_{16}H_{22}N_2O_6$; mol wt 338.35. C 56.79%, H 6.55%, N 8.28%, O 28.37%. Prepn: Gauss, Petersen, *Angew. Chem.* **69**, 252 (1957); **Brit. pat.** *793,796* (1958 to Bayer). Antineoplastic activity *in vitro*: M. Akhtar *et al.*, *Can. J. Chem.* **53**, 2891 (1975); J. S. Driscoll *et al.*, *J. Pharm. Sci.* **68**, 185 (1979).

Gray needles from petr ether, mp 79-80.5°: Gauss, Petersen, *Med. Chem.* (Bayer) **7**, 649 (1963), *C.A.* **60**, 2875i (1964).

1257. Bisbentiamine. *N,N'-[Dithiobis[2-[2-(benzoyloxy)ethyl]-1-methyl-2,1-ethenediyl]]bis[N-[(4-amino-2-methyl-5-pyrimidinyl)methyl]formamide]; N,N'-[dithiobis-[2-(2-hydroxyethyl)-1-methylvinylene]]bis[N-[(4-amino-2-methyl-5-pyrimidinyl)methyl]formamide] dibenzoate;* O-benzoylthiamine disulfide; Béprocin; Beston. $C_{38}H_{42}N_8O_6S_2$; mol wt 770.95. C 59.20%, H 5.49%, N 14.54%, O 12.45%, S 8.32%. Prepn: **Brit. pat.** *922,444* (1963 to Tanabe Seiyaku) corresp to **U.S. pat.** *3,109,000* (1963); Yurugi *et al.*, *Ann. Rep. Takeda Res. Lab.* **27**, 25 (1968), *C.A.* **70**, 96741m (1968); Imai, Miyaji, **Ger. pat.** *1,954,519* (1971 to Hitachi), *C.A.* **75**, 63821n (1971).

Prisms from alc, mp 146-147°.
THERAP CAT: Vitamin B_1 source.

1258. Bis(p-chlorophenoxy)methane. *1,1'-[Methylenebis(oxy)]bis[4-chloro]benzene;* di-(p-chlorophenoxy)methane; DCPM; K 1875; Neotran. $C_{13}H_{10}Cl_2O_2$; mol wt 269.12. C 58.02%, H 3.75%, Cl 26.35%, O 11.89%. Prepd from p-chlorophenol and dichloromethane: Moyl, **U.S. pat.** *2,503,207* (1950 to Dow); from sodium phenolate and dichloromethane: Miron, Lowry, *J. Am. Chem. Soc.* **73**, 1872 (1951).

Crystals from petr ether. mp 69.7-70.2°. bp$_6$ 189-194°. Solubilities at 25° in g/100 ml: acetone 189; benzene 40; carbon tetrachloride 28; methanol 0.5; ether 87. Practically insol in water, petr oils. LD$_{50}$ orally in rats: 5.8 g/kg; lethal concn for rats in air: 18,000 ppm.

USE: Miticide. *Toxicity:* No record of human poisoning. Animal expts show low toxicity, but large doses have caused liver injury. Effects may be similar to, but less severe than, DDT, *q.v.*, Spencer *et al.*, *Arch. Ind. Hyg. Occup. Med.* **1**, 341 (1950).

1259. Bisdequalinium Chloride. *6,7,8,9,10,11,12,13,14,-15,16,17,24,25,26,27,28,29,30,31,32,33-Docosahydro-35,37-dimethyl-5,34:18,23-diethenodibenzo[b,r][1,5,16,20]tetraazacyclotriacontine-23,34-diium dichloride; N^1,N^1-decamethylene-N^4,N^4-decamethylenebis[4-aminoquinaldinium chlo-*

ride]; 1,1'-decamethylene-4,4'-(1,10-decamethylenediimino)bis[quinaldinium chloride]; Salvizol. $C_{40}H_{58}Cl_2N_4$; mol wt 665.84. C 72.16%, H 8.78%, Cl 10.65%, N 8.41%. Prepn: Stark, **Brit. pat.** *895,090* (1960).

THERAP CAT: Antiseptic, disinfectant.

1260. Bis(p-dimethylaminobenzylidene)benzidine. *N,N'-Bis[[4-(dimethylamino)phenyl]methylene][1,1'-biphenyl]-4,4'-diamine;* 4,4'-di(4-dimethylaminobenzylideneamino)biphenyl; bis(4-dimethylaminobenzylidene)-p,p'-diaminodiphenyl; *N,N'*-bis(p-dimethylaminobenzal)benzidine. $C_{30}H_{30}N_4$; mol wt 446.57. C 80.68%, H 6.77%, N 12.55%. Prepn from hydrazobenzene and p-dimethylaminobenzaldehyde: Sachs, Whittaker, *Ber.* **35**, 1435 (1902).

Yellow crystals from nitrobenzene; mp 318°. Reacts with solutions of the alkali tungstates to give a cinnabar-red precipitate.

USE: Determination of tungsten: Hovorka, *Coll. Czech. Chem. Commun.* **10**, 518 (1938), *C.A.* **33**, 1624 (1939).

1261. Bis(1,2-dimethylpropyl)borane. Disiamylborane; di-*sec*-isoamylborine; bis(3-methyl-2-butyl)borane. $C_{10}H_{23}B$; mol wt 154.10. C 77.94%, H 15.04%, B 7.01%. Prepn: Brown, Zweifel, *J. Am. Chem. Soc.* **83**, 1241 (1961). Use in selective reductions: G. W. Kabalka, C. F. Lane, *Chem. Tech.* **6**, 324 (1976).

Crystals, mp 35-40°. Unstable to air.
USE: Selective reagent for steric control of hydroboration of olefins.

1262. Bis(2-ethylhexyl) Phthalate. *1,2-Benzenedicarboxylic acid bis(2-ethylhexyl) ester;* di(2-ethylhexyl) phthalate; dioctyl phthalate; Octoil. $C_{24}H_{38}O_4$; mol wt 390.54. C 73.80%, H 9.81%, O 16.39%. Prepn: Garner, Watson, **U.S. pat.** *2,508,911* (1950 to Shell); **Brit. pat.** *747,260* (1956 to Chemische Werke Hüls).

Note: This substance may reasonably be anticipated to be a carcinogen: *Fourth Annual Report on Carcinogens* (NTP 85-002, 1985) p 83.
USE: In vacuum pumps.

1263. Bis(2-ethylhexyl) Sebacate. *Decanedioic acid bis-(2-ethylhexyl) ester;* di(2-ethylhexyl) sebacate; Octoil S; Plexol 201. $C_{26}H_{50}O_4$; mol wt 426.66. C 73.19%, H 11.81%, O 15.00%. $CH_3(CH_2)_3CH(C_2H_5)CH_2OOC(CH_2)_8COOCH_2$-$CH(C_2H_5)(CH_2)_3CH_3$. Prepn: Bruno, **U.S. pat.** *2,628,249* (1953 to Pittsburgh Coke & Chemical); **Brit. pat.** *747,260* (1956 to Chemische Werke Hüls).

Liquid, d$_{25}^{25}$ 0.9119, n$_D^{25}$ 1.4496.

USE: In vacuum pumps.

1264. Bismark Brown R. *4,4'-[(4-Methyl-1,3-phenyl-ene)bis(azo)]bis[6-methyl-1,3-benzenediamine] dihydrochloride; C.I. Basic Brown 4; 5,5'-[(4-methyl-m-phenylene)bis-(azo)]bis[toluene-2,4-diamine] dihydrochloride; C.I. 21010;* Bismark Brown 53; Vesuvine. $C_{21}H_{26}Cl_2N_8$; mol wt 461.41. C 54.67%, H 5.68%, Cl 15.37%, N 24.29%. Prepd by reaction of toluene-2,4-diamine HCl with nitrous acid: *Colour Index* vol. 4 (3rd ed., 1971) p 4154.

Dark brown solid. Very sol in water (yellowish-brown soln); sol in ethanol, Cellosolve; slightly sol in acetone; practically insol in benzene. In concd H_2SO_4 gives brown soln which on dilution turns reddish-brown; in concd HNO_3 gives violet soln which turns brown.

Free base, $C_{21}H_{24}N_8$, *C.I. Solvent Brown 12, C.I. 21010:1.* mp 130-135°. Very slightly sol in water; sol in ethanol, acetone.

USE: As textile dye, leather dye, biological stain.

1265. Bismark Brown Y. *4,4'-[1,3-Phenylenebis(azo)]-bis[1,3-benzenediamine] dihydrochloride; C.I. Basic Brown 1;* 4,4'-[m-phenylenebis(azo)]bis[m-phenylenediamine] dihydrochloride; C.I. 21000; phenylene brown. $C_{18}H_{20}Cl_2N_8$; mol wt 419.33. C 51.56%, H 4.81%, Cl 16.91%, N 26.72%. Prepd by reaction of m-phenylenediamine HCl and nitrous acid: F. J. Welcher, *Organic Analytical Reagents* (D. Van Nostrand Co., New York, 1948) p 339; *Colour Index* vol. 4 (3rd ed., 1971) p 4154.

Blackish-brown powder. Very sol in water; slightly sol in ethanol, Cellosolve; practically insol in acetone, benzene carbon tetrachloride. In concd H_2SO_4 gives brown soln; in concd nitric acid gives orange soln which turns yellow.

USE: As a textile dye, biological stain.

1266. Bis(1-methylamyl) Sodium Sulfosuccinate. *Sulfo-butanedioic acid 1,4-bis(1-methylpentyl) ester sodium salt;* dihexyl sodium sulfosuccinate; Aerosol MA; Alphasol MA. $C_{16}H_{29}NaO_7S$; mol wt 388.46. C 49.47%, H 7.53%, Na 5.92%, O 28.83%, S 8.25%. The bis(1-methylamyl) ester of sulfosuccinic acid monosodium salt, perhaps in admixture with the dihexyl ester. Prepd by the action of the appropriate alcohols on maleic anhydride followed by addition of sodium bisulfite: Jaeger, U.S. pats. **2,028,091; 2,176,423;** **Brit.** pat. **446,568;** Fr. pat. **776,495** (to Am. Cyanamid).

$$CH_3$$
$$|$$
$$H_2CCOOCH(CH_2)_3CH_3$$
$$|$$
$$NaO_3SCHCOOCH(CH_2)_3CH_3$$
$$|$$
$$CH_3$$

Available as white, slightly hygroscopic, wax-like pellets. Must be soaked to dissolve in cold water. Dissolves rapidly in hot water. Solubility in water at 25° = 343 g/l; at 70° = 447 g/l. Maximum concn of electrolyte soln in which 1% of the wetting agent is sol: 2% NaCl; 2% NH_4Cl; 14% $(NH_4)_2$-HPO_4; 3% $NaNO_3$ (slightly turbid); 3% Na_2SO_4. Also sol in pine oil, oleic acid, acetone, kerosene, carbon tetrachloride, 2B ethanol, benzene, hot olive oil, glycerol. Insol in liquid

petrolatum. Stable in acid and neutral solns, hydrolyzes in alkaline solns.

USE: Wetting agent.

1267. Bis[methylthio]methane. Bis[methylmercapto]-methane; methylenebis[methyl sulfide]. $C_3H_8S_2$; mol wt 108.21. C 33.30%, H 7.45%, S 59.25%. $(CH_3S)_2CH_2$. Odorous principle of the white truffle *Tuber magnatum* (Pico) Vitt., *Tuberaceae.* Isoln: Fiecchi *et al., Tetrahedron Letters* **1967,** 1681. Synthesis from methyl mercaptan: Böhme, Marx, *Ber.* **74,** 1672 (1941).

Oily liquid. Odor in high dilutions reminiscent of white truffles. The odor of the neat liquid resembles that of freshly prepd mustard without its acrid and irritating qualities. bp 148-149°. Mass spectra: Fiecchi, *loc. cit.*

1268. Bismuth. Bi; at. wt 208.9804; at. no. 83; valence 3, 5. One naturally occurring isotope: 209; artificial radioactive isotopes: 199-208; 210-215. Confused with tin until 1450. First isolated by Hillot in 1737. It was, however, Geoffrey the Younger who clearly proved its individuality in 1753. Pott and Bergmann are named as the scientific discoverers. Occurrence in the earth's crust: approx 0.2 ppm. Obtained as a byproduct from the processing of lead, copper, and tin ores. *Reviews: Nouveau Traité de Chimie Minérale,* tome **11,** P. Pascal, Ed. (Masson, Paris, 1958); *Gmelin's, Bismuth* (8th ed.) **19,** pp 1-104 (1927); supplement, pp 1-621 (1964); Smith, "Arsenic, Antimony and Bismuth" in *Comprehensive Inorganic Chemistry* vol. 2, J. C. Bailar, Jr. *et al.,* Eds. (Pergamon Press, Oxford, 1973) pp 547-683; S. C. Carapella, H. E. Howe in Kirk-Othmer *Encyclopedia of Chemical Technology* vol. 3 (Wiley-Interscience, New York, 3rd ed., 1978) pp 912-921.

Grayish-white with reddish tinge and bright metallic luster; soft and brittle; superficially oxidized by air, frequently becoming iridescent. mp 271°; contracts when melted. bp 1420°; 1490°: *Gmelin's, loc. cit.* p 43 (1927). d_4^{20} 9.78; d_4^{271} 10.07. Poor conductor of electricity. Has the greatest Hall effect of any metal, *i.e.,* its resistance increases when placed in a magnetic field. Attacked by dil HNO_3, hot H_2SO_4, concd HCl. Cold solns of Bi give a white ppt with NaOH, turning yellow on boiling; with HCl a white ppt sol in excess of acid. The solns in HCl or HNO_3 yield with much water a white ppt blackened by H_2S (different from Sb).

Precipitated bismuth, prepd by treating a soln of bismuth chloride in HCl with hypophosphorous acid, washing and drying. Contains not less than 98.5% metallic bismuth. Dull gray powder. The particles are of no greater diameter than 15 microns (0.015 mm). Easily dispersed in water.

D'Arcet metal-fusible, an alloy of 49.2% Bi, 32.2% Pb and 18.4% Sn. Whitish-gray metal, mp 96-97°.

USE: Manuf Bi salts, fusible alloys, stereotype metal, fusible boiler plugs, electric fuses, low-melting solders; tempering baths for steel; "silvering" mirrors; in dental technique.

THERAP CAT (VET): Has been used externally in dusting powders for indolent, moist or suppurating lesions; internally as a protectant of the gastrointestinal lining, and as an x-ray contrast medium. Has also been recommended to treat buccal warts in dogs.

1269. Bismuth Aluminate. *Aluminum bismuth oxide;* almuth; Bisminat. $Al_6Bi_2O_{12}$; mol wt 771.88. Al 20.97%, Bi 54.15%, O 24.87%. $Bi_2(Al_2O_4)_3$. Prepn: Roques, U.S. pat. **2,901,316** (1959 to Etablissements Roques).

Decahydrate, extremely light powder. Practically insol in water.

THERAP CAT: Antacid.

1270. Bismuth Bromide. Bismuth tribromide. $BiBr_3$; mol wt 448.75. Bi 46.57%, Br 53.43%. Prepn from the elements: Schenk in *Handbook of Preparative Inorganic Chemistry* vol. 1, G. Brauer, Ed. (Academic Press, New York, 2nd ed., 1963) p 623.

Yellowish crystals; odor of HBr. d 5.7. mp 218°. bp 441°. Dec by water, forming BiOBr; sol in solns of KI, KBr, KCl, and in dil HCl. Practically insol in alc. *Keep tightly closed.*

1271. Bismuth Bromide Oxide. Basic bismuth bromide; bismuth oxybromide; bismuthyl bromide; bismuth "subbromide". BiBrO; mol wt 304.92. Bi 68.54%, Br 26.21%, O 5.25%. BiOBr. Prepn: Schenk in *Handbook of Preparative*

Inorganic Chemistry **vol. 1,** G. Brauer, Ed. (Academic Press, New York, 2nd ed., 1963) p 624.

Crystals or amorphous powder. Very stable, melts at red heat. Practically insol in water, alc; sol in HCl, HBr, HNO$_3$.

USE: Manuf of dry cell cathodes.

1272. Bismuth Butylthiolaurate. *2-(Butylthio)dodecanoic acid bismuth basic salt;* Bispecia; Neocardyl. C$_{16}$H$_{33}$-BiO$_4$S; mol wt 530.49. C 36.23%, H 6.27%, Bi 39.39%, O 12.06%, S 6.05%. CH$_3$(CH$_2$)$_9$CH(SC$_4$H$_9$)COOBi(OH)$_2$. Prepn: P. Lebeau, M. M. Janot, *Traité Pharm. Chim.* **Vol. II** (Masson, Paris, 4th ed., 1955-1956) p 435.

Yellowish powder. Practically insol in water; sol in oils. Dec in acids, liberating hydrogen sulfide.

THERAP CAT: Formerly as antisyphilitic.

1273. Bismuth Chloride. Bismuth trichloride. BiCl$_3$; mol wt 315.37. Bi 66.27%, Cl 33.73%. Prepn from the elements: Schenk in *Handbook of Preparative Inorganic Chemistry* **vol. 1,** G. Brauer, Ed. (Academic Press, New York, 2nd ed., 1963) p 621.

White to yellowish, deliquesc crystals; HCl odor. d 4.75. mp about 230°. Sublimes at about 430°. bp 447°. Dec by water or aq alc into BiOCl; sol in HCl, HNO$_3$, abs alcohol, acetone, ethyl acetate. *Keep tightly closed.*

USE: In the manuf of other bismuth salts; as catalyst for organic reactions.

1274. Bismuth Chloride Oxide. Basic bismuth chloride; bismuth oxychloride; bismuthyl chloride; bismuth "subchloride"; pearl white; blanc d'Espagne; blanc de perle; Chlorbismol. BiClO; mol wt 260.48. Bi 80.24%, Cl 13.61%, O 6.14%. BiOCl. Prepn: Schenk in *Handbook of Preparative Inorganic Chemistry* **vol. 1,** G. Brauer, Ed. (Academic Press, New York, 2nd ed., 1963) p 622.

Fine powder or tetragonal crystals. Melts at low red heat; d 7.72. Practically insol in water, alc; sol in HCl, HNO$_3$.

USE: In face powders; as pigment; manuf artificial pearls, dry-cell cathodes.

THERAP CAT: Formerly as antisyphilitic.

1275. Bismuth Ethyl Camphorate. *d-Camphoric acid ethyl ester bismuth salt;* bismuth(III) salt of *d*-camphoric acid monoethyl ester. C$_{36}$H$_{57}$BiO$_{12}$; mol wt 890.82. C 48.54%, H 6.45%, Bi 23.46%, O 21.55%. Prepd from sodium ethyl camphorate and bismuth nitrate: W. M. Lauter, H. A. Braun, *J. Am. Pharm. Assoc.* **25,** 394 (1936).

Amorphous solid, faint aromatic odor. Softens at 55°. mp 61-67° (clear melt). Practically insol in water. Sol in chloroform, ether, ethylene dichloride, oils. The sol$_y$ in vegetable oils is increased by the addition of camphor. MLD i.m. in rats: 250 mg Bi/kg (Lauter, Braun).

THERAP CAT: Formerly as antisyphilitic.

1276. Bismuth Fluoride. Bismuth trifluoride. BiF$_3$; mol wt 266.00. Bi 78.57%, F 21.43%. Prepd by dissolving bismuth oxide or hydroxide in aqueous HF. May also be prepd by reduction of BiF$_5$ with CO$_2$ in very dil H$_2$ at 80-150° in a Pt tube: Muir *et al., J. Chem. Soc.* **39,** 33 (1881); v. Wartenberg, *Z. Anorg. Allgem. Chem.* **244,** 344 (1940); Aurivillius, *Acta Chem. Scand.* **9,** 1206 (1955). *Review:* Kemmitt, Sharp, *Advan. Fluorine Chem.* **4,** 213 (1965).

White to gray dimorphic crystals. d 8.3. mp 725-730°. Volatilizes at higher temps slowly and without decompn. Practically insol in water. Sol in concd HF with the formation of complexes. *See also* Bismuth Pentafluoride.

USE: In the prepn of bismuth pentafluoride.

1277. Bismuth Hydroxide. Bismuth hydrate. BiH$_3$O$_3$; mol wt 260.02. Bi 80.38%, H 1.16%, O 18.46%. Bi(OH)$_3$. Prepn: *Gmelin's, Bismuth* (8th ed.) **19,** pp 119-122 (1927).

White to yellowish-white, amorphous powder. d^{15} 4.962.

Practically insol in water. When freshly pptd it is soluble in glycerol in presence of NaOH; sol in acids. Readily loses one mol H$_2$O forming the metahydroxide which is yellow.

USE: As absorbent for rutin and quercetin; in hydrolysis of ribonucleic acid; in separation of plutonium from irradiated uranium.

1278. Bismuthine. Bismuth trihydride; Bismutan. BiH$_3$; mol wt 212.00. Bi 98.58%, H 1.42%. Prepn: Wiberg, Möditzer, *Z. Naturforsch.* **12b,** 123 (1957); Amberger, *Ber.* **94,** 1447 (1961).

Thermally unstable liquid, bp 16.8° (est).

USE: In manuf of Ge or Si semiconductors, Scott *et al.,* U.S. pat. **2,910,394** (1959 to Int'l Standard Electric).

1279. Bismuth Iodide. Bismuth triiodide. BiI$_3$; mol wt 589.76. Bi 35.44%, I 64.56%. Best prepd from the elements: Watt *et al., Inorg. Syn.* **4,** 114 (1953).

Black, minute, hexagonal crystals with metallic sheen. d$_4^{17}$ 5.778, also reported as d 5.64. Sublimes at 439° (760 mm), dec at 500°. Practically insol in cold water, dec slowly in hot water. Sol in liquid ammonia, abs ethanol (about 3.5% at 20°), in aq solns of KI, HI, and HCl. The solid is slowly converted to Bi(IO$_3$)$_3$ upon prolonged exposure to air.

1280. Bismuth Iodide Oxide. Basic bismuth iodide; bismuth oxyiodide; bismuthyl iodide; bismuth "subiodide". BiIO; mol wt 351.92. Bi 59.39%, I 36.07%, O 4.55%. BiOI. Prepn: *Gmelin's, Bismuth* (8th ed) **19,** p 159 (1927); Schenk in *Handbook of Preparative Inorganic Chemistry* **vol. 1,** G. Brauer, Ed. (Academic Press, New York, 2nd ed., 1963) p 625.

Brick-red, heavy, odorless powder or copper-colored crystals. d 7.92. Fuses at red heat with partial decompn. Practically insol in water, alcohol, CHCl$_3$; sol in HCl; dec by HNO$_3$ or alkali.

USE: Manuf of dry cell cathodes.

THERAP CAT: Anti-infective.

1281. Bismuth Iodosubgallate. *(Gallato)hydroxyiodobismuth;* (galloyloxy)hydroxyiodobismuthine; bismuth oxyiodogallate; Airoform; Airogen. C$_7$H$_6$BiIO$_6$; mol wt 522.04. C 16.11%, H 1.16%, Bi 40.04%, I 24.31%, O 18.39%. C$_6$H$_2$-(OH)$_3$COOBi(OH)I. Prepn: *Hagers Handb. Pharm. Praxis* **Band I,** 1565 (Berlin, 1930).

Dark gray, odorless, tasteless, bulky powder. Dec by water, acids. Practically insol in alc, ether, chloroform; sol in solns of alkali hydroxides. *Protect from light.*

THERAP CAT: Anti-infective.

THERAP CAT (VET): *See* Bismuth. For external use only.

1282. Bismuth Nitrate. Bismuth trinitrate. BiN$_3$O$_9$; mol wt 395.02. Bi 52.91%, N 10.64%, O 36.45%. Bi(NO$_3$)$_3$. Prepn: *Gmelin's, Bismuth* (8th ed.) **19,** pp 126-129 (1927).

Pentahydrate, lustrous, hygroscopic crystals; acid reaction; odor of nitric acid. d 2.83. Sol in water containing nitric acid; dec by water alone into subnitrate; sol in glycerol, dil acids including acetic acid, acetone. Practically insol in alcohol, ethyl acetate.

USE: Prepn of other bismuth salts, luminous paints; pptn of alkaloids.

1283. Bismuth Oleate. *Oleic acid bismuth salt;* bismuth trioleate. C$_{54}$H$_{99}$BiO$_6$; mol wt 1053.36. C 61.57%, H 9.47%, Bi 19.84%, O 9.11%. [CH$_3$(CH$_2$)$_7$CH=CH(CH$_2$)$_7$COO]$_3$Bi; Prepd from Bi$_2$O$_3$, acetic anhydride, and oleic acid: Considine, *Brit.* pat. **947,749** (1964 to Metal & Thermit).

Solid. Sol in about 1500 parts benzene.

USE: In catalysts for manufacture of aldehydes and alcohols by the oxo process: **Brit.** pat. **907,027** (1962 to Esso); Aldridge, Cull, **Ger.** pat. **1,153,006** (1963 to Esso), *C.A.* **60,** 2763 (1964).

1284. Bismuth Oxalate. Oxalic acid bismuth salt. C$_6$-Bi$_2$O$_{12}$; mol wt 682.06. C 10.57%, Bi 61.28%, O 28.15%. Bi$_2$(C$_2$O$_4$)$_3$. Prepn: Vanino, Zumbusch, *Ber.* **41,** 3994 (1908); Soerbye, Kruse, *Acta Chem. Scand.* **16,** 1662 (1962).

Powder. *Poisonous!* Practically insol in water or alcohol; sol in moderately dil HCl or HNO$_3$.

1285. Bismuth Oxide. Bismuth trioxide; bismuthous oxide; bismuth yellow. Bi$_2$O$_3$; mol wt 466.00. Bi 89.70%, O

10.30%. Occurs in nature as the mineral *bismite*. Prepn: *Gmelin's, Bismuth* (8th ed.) **19**, p 109-113 (1927).

Yellow, heavy, odorless powder or monoclinic crystals; stable in air. Practically insol in water; sol in HCl or HNO_3.

USE: In disinfectants, magnets, glass, rubber vulcanization; in fireproofing of papers and polymers; in catalysts.

THERAP CAT: Astringent.

1286. Bismuth Pentafluoride. BiF_5; mol wt 304.00. Bi 68.75%, F 31.25%. Prepd according to the equation $BiF_3 + F_2 \rightarrow BiF_5$: v. Wartenberg, *Z. Anorg. Allgem. Chem.* **244**, 344 (1940); Kwasnik in *Handbook of Preparative Inorganic Chemistry*, **vol. 1**, G. Brauer Ed. (Academic Press, New York, 2nd ed., 1963) pp 202-203. Prepn from Bi and F_2: Fischer, Rodzitis, *J. Am. Chem. Soc.* **81**, 6375 (1959); Kemmitt, Sharp, *Advan. Fluorine Chem.* **4**, 212 (1965); Hebecker, *Z. Anorg. Allgem. Chem.* **384**, 111 (1971).

Body-centered tetragonal crystals, sublimes at 120°. d 5.55. Very sensitive to moisture, discolors quickly in moist air. Violent reaction with water forming BiF_3 and ozone. Reacts with liquid petrolatum above 50°.

USE: Fluorinating agent. *Caution:* Highly toxic and irritating to mucous membranes, skin, eyes, respiratory tract.

1287. Bismuth Phosphate. Bismugel. BiO_4P; mol wt 303.98. Bi 68.75%, O 21.05%, P 10.19%. $BiPO_4$. Prepn: Schenk in *Handbook of Preparative Inorganic Chemistry*, **vol. 1**, G. Brauer, Ed. (Academic Press, New York, 2nd ed., 1963) p 626.

Odorless powder or monoclinic crystals. Does not melt on heating. d^{15} 6.323. Slightly sol in water and dil acids; not hydrolyzed by boiling water. Practically insol in alcohol, acetic acid; sol in concd HNO_3 and HCl.

USE: In the separation of plutonium from fission products; manuf of optical flint glass.

THERAP CAT: Antacid; protectant.

1288. Bismuth Potassium Iodide. *Potassium heptaiodobismuthate(4—).* BiI_7K_4; mol wt 1253.82. Bi 16.67%, I 70.86%, K 12.47%. K_4BiI_7. Prepn: *Gmelin's, Potassium* (8th ed.) **22**, p 1068 (1936).

Red crystals. Partly dec by water; completely sol in alkali iodide solns.

USE: Pptn of vitamins, particularly thiamine HCl, and antibiotics from soln.

1289. Bismuth Potassium Tartrate. Potassium bismuthotartrate; potassium bismuth tartrate; potassium bismuthyl tartrate; tartaric acid bismuth complex potassium salt. The bismuth content of the complex salts of bismuth and alkali tartrate may be made to range within wide limits (35-75% Bi). The medicinal grade contains 60-64% bismuth. Prepn: Kober, U.S. pat. **1,663,201** (1928 to Searle); *Hagers Handb. Pharm. Praxis* 1st Suppl., 316 (Berlin, 1949).

Odorless powder; sweetish taste; darkens on exposure to light. Sol in 2 parts water; practically insol in alc or other organic solvents. The aq soln is slightly alkaline. Dec by mineral acids.

THERAP CAT: Formerly as antisyphilitic.

1290. Bismuth Selenide. Bismuth triselenide. Bi_2Se_3; mol wt 654.88. Bi 63.83%, Se 36.17%. Prepn by heating a stoichiometric mixture at 475° in evacuated tube: Dönges, *Z. Anorg. Allgem. Chem.* **265**, 56 (1951); Konorov, *Zh. Tekh. Fiz.* **26**, 1394 (1956); from BiO, Se and sodium oxalate at high press.: Cambi, Elli, *Chim. Ind. (Milan)* **50**, 94 (1968).

Black crystals. Rhombohedral and hexagonal crystal structure. d_4^{20} 7.70 (also reported as 6.82). mp 710°. Heat of formation: —13.9 kcal/mol. Heat of combustion: 296.4 kcal/mol. Dec when heated in air. Dec by concd nitric acid and aqua regia. Insol in water.

USE: In semiconductor research.

1291. Bismuth Sodium Iodide. *Disodium pentaiodobismuthate(2—)*; sodium pentaiodobismuthate(III); sodium bismuth iodide; sodium iodobismuthite; Aniobi; Bismjol. BiI_5Na_2; mol wt 889.53. Bi 23.50%, I 71.34%, Na 5.17%. Na_2BiI_5. Prepn: Gurchot *et al.*, *J. Pharmacol.* **45**, 427 (1932).

Tetrahydrate, odorless, red crystals; astringent taste. mp 93° (dec). Soluble in water, alcohol, acetone, ethyl acetate; slightly sol in ether; hydrolyzes in soln.

THERAP CAT: Formerly as antisyphilitic.

1292. Bismuth Sodium Tartrate. Sodium bismuth tartrate; sodium bismuthyl tartrate; tartaric acid bismuth complex sodium salt; Natrol; Tartrol. The bismuth content of the complex salts of bismuth and alkali tartrate may be made to range within wide limits (35-75% Bi). The article generally used contains 70-74% bismuth. Prepn: Kober, U.S. pat. **1,663,201** (1928 to Searle); *Hagers Handb. Pharm. Praxis* 1st Suppl., 316 (Berlin, 1949).

Odorless, tasteless powder; discolors in light. Sol in about 3 parts water; insol in alc or other organic solvents. The aq soln is slightly alkaline.

THERAP CAT: Formerly as antisyphilitic.

THERAP CAT (VET): *See* Bismuth.

1293. Bismuth Sodium Triglycollamate. *Nitrilotriacetic acid bismuth complex sodium salt;* Bistrimate. $C_{24}H_{28}BiN_4Na_7O_{25}$; mol wt 1142.28. C 25.23%, H 2.47%, Bi 18.29%, N 4.91%, Na 14.09%, O 35.01%. A double salt of sodium bismuthyl triglycollamate and disodium triglycollamate. Prepn: Lehman, Sproull, U.S. pat. **2,348,984** (1944).

$$\overset{+}{HN}\begin{cases} CH_2COOBiO \\ CH_2COONa \\ CH_2COO^- \end{cases} \quad 3\overset{+}{HN}\begin{cases} CH_2COONa \\ CH_2COONa \\ CH_2COO^- \end{cases}$$

Crystals. Slightly salty taste. Very sol in water. Insol in acetone, ether, benzene. The pH of a 2% aq soln is between 7 and 8.

THERAP CAT: Lupus erythematosus suppressant.

1294. Bismuth Subacetate. *Acetic acid basic bismuth salt;* bismuth acetate basic; bismuth acetate oxide; bismuth oxide acetate; bismuthyl acetate. $C_2H_3BiO_3$; mol wt 284.04. C 8.46%, H 1.06%, Bi 73.58%, O 16.90%. $CH_3COOBiO$. Prepn: Hofmann, *Ann.* **223**, 110 (1884); Aurivillius, *Acta Chem. Scand.* **9**, 1213 (1955). Crystal structure: Aurivillius, *loc. cit.*

Thin crystal plates. Slight acetic odor. Practically insol in water; sol in glacial acetic acid, dil HCl or HNO_3.

1295. Bismuth Subcarbonate. *Bismuth carbonate, basic;* bismuth oxycarbonate. CBi_2O_5; mol wt 510.01. C 2.35%, Bi 81.96%, O 15.69%. $(BiO)_2CO_3$. Prepn: *Gmelin's, Bismuth* (8th ed.) **19**, p 178 (1927).

Hemihydrate, odorless, tasteless powder. Practically insol in water or alc; sol in mineral acids, in concd acetic acid.

USE: In admixture with other substances in glazes on ceramics; for artificial horn products; for pearly surfaces for plastics.

THERAP CAT: Topical protectant.

THERAP CAT (VET): *See* Bismuth.

1296. Bismuth Subcitrate Sol (Dried). Colloidal bismuth subcitrate; CBS; Barrier; De-Nol; De-Noltab; Duosol; Telen; Ulcerone. Cytoprotective, polynuclear colloidal complex of the potassium ammonium salt of oxohydroxocitratobismuthate (III). Approx mol form $K_3(NH_4)_2[Bi_6O_3(OH)_5(C_6H_5O_7)_4]$. General prepn of colloidal bismuth solution: L. Vanino, cited in *Mellor's* **vol. IX**, 598 (1929). Manufacturing processes: C. J. McLoughlin, R. B. Himstedt, Ger. pat. **2,501,787**; P. J. H. Bos *et al.*, Eur. pat. Appl. **75,992**; *eidem*, U.S. pat. **4,801,608** (1975, 1985, 1989 all to Gist-Brocades). Pharmacology: T. R. Wilson, *Postgrad. Med. J.* **51**, Suppl. 5, 18, 22 (1975); J. Wieriks *et al.*, *Scand. J. Gastroenterol.* **17**, Suppl. 80, 11 (1982). Mechanism of action studies: D. R. Williams, *J. Inorg. Nucl. Chem.* **39**, 711 (1977); S. P. Lee, *Scand. J. Gastroenterol.* **17**, Suppl. 80, 17 (1982). Clinical comparisons with cimetidine in duodenal ulcer: G. Vantrappen *et al.*, *Gut* **21**, 329 (1980); I. Hamilton *et al.*, *ibid.* **27**, 106 (1986). Clinical efficacy in *Campylobacter pylori* associated gastritis and ulcer: T. Rokkas *et al.*, *ibid.* **29**, 1386 (1988); B. J. Marshall *et al.*, *Lancet* **2**, 1437 (1988). Symposia on pharmacology and clinical efficacy: *Scand. J. Gastroenterol.* **21**, Suppl. 122, 1-54 (1986); *Digestion* **37**, Suppl. 2, 1-64 (1987).

White, amorphous powder. Sol in water, dil alkali, ammonia. Aq soln is neutral to litmus. Dec by mineral acids.

Note: This substance is widely referred to in the literature by the incorrect name, *tripotassium dicitrato bismuthate* or *TDB.*

THERAP CAT: Antiulcerative.

1297. Bismuth Subgallate. *Gallic acid bismuth basic salt;* bismuth gallate, basic; B.S.G.; Dermatol. $C_7H_5BiO_6$; mol wt 394.09. C 21.33%, H 1.28%, Bi 53.03%, O 24.36%. Prepd from bismuth nitrate and gallic acid in an acetic acid medium. The acetic acid may be replaced with a glycol or mannitol: Pfeiffer, Schmitz, *Pharmazie* **5**, 517 (1950).

Hydrate, bright yellow, odorless, tasteless powder. Practically insol in water, alc, chloroform, ether; sol in dil alkali hydroxide solns, in hot mineral acids with decompn. Forms a water-sol sodium salt which has an alkaline reaction.
Note: Major active ingredient of *Bongast.*
THERAP CAT: Astringent; antacid; protective.
THERAP CAT (VET): *See* Bismuth.

1298. Bismuth Subnitrate. *Bismuth hydroxide nitrate oxide; bismuth nitrate, basic;* bismuth oxynitrate; bismuth subnitricum; bismuthyl nitrate; bismuth white; magistery of bismuth; novismuth; paint white; Spanish white. A basic salt, the compn of which varies with the conditions of preparation. Contains 70 to 74% Bi or 79 to 82% Bi_2O_3. Prepd by partial hydrolysis of $Bi(NO_3)_3$: *Gmelin's, Bismuth* (8th ed.) **19**, pp 132-135 (1927); *Traité Pharm. Chim.* vol. 1, P. Lebeau, M. M. Janot, Eds. (Masson, Paris, 1956) p 371; *Handbuch der Pharmazie,* vol. **4**(1), H. Thoms, Ed. (Urban & Schwarzenberg, Berlin, 1927) p 325.
Odorless, tasteless, heavy, slightly hygroscopic, microcrystalline powder. Dec to Bi_2O_3 and nitrogen oxides when heated to red heat. Practically insol in water, alc; sol in dil HCl and HNO_3. *Keep well closed and protect from light. Incompat.* Alkaline bicarbonates, soluble iodides, gallic acid, calomel, salicylic acid, tannin, sulfur.
USE: Manuf bismuth fluxes for enamels; in cosmetics.
THERAP CAT: Antacid.
THERAP CAT (VET): *See* Bismuth.

1299. Bismuth Subsalicylate. *(2-Hydroxybenzoato-O¹)- oxobismuth; 2-hydroxybenzoic acid bismuth (3+) salt, basic;* bismuth salicylate, basic; basic bismuth salicylate; oxo(salicylato)bismuth; salicylic acid basic bismuth salt; Bismogenol "Tosse" Inj.; Stabisol. $C_7H_5BiO_4$; mol wt 362.11. C 23.22%, H 1.39%, Bi 57.72%, O 17.67%. HOC$_6$H$_4$COOBiO. Prepn: Fischer, Grützner, *Arch. Pharm.* **231**, 680 (1893).
Microscopic prisms. Almost insol in water or alc; dec by boiling water, alkalies into a more basic salt.
USE: To impart pearly surface to cellulose-base, polystyrene, and phenol-formaldehyde resins, Deutsch *et al.,* U.S. pat. **2,816,044** (1957 to Brit. Resin Prods.). In heat-sensitive coating for copying paper, Van Dam, U.S. pat. **2,897,090** (1959 to Anken Chem. & Film). To stabilize tolylenediamine mixtures against deterioration and discoloration, Powers, U.S. pat. **3,138,641** (1964 to Mobay).
THERAP CAT: Lupus erythematosus suppressant.
THERAP CAT (VET): *See* Bismuth.

1300. Bismuth Sulfate. $Bi_2O_{12}S_3$; mol wt 706.18. Bi 59.19%, O 27.19%, S 13.62%. $Bi_2(SO_4)_3$. Prepn: *Gmelin's, Bismuth* (8th ed.) **19**, p 170 (1927).
White crystals. Dec by water or alc into a basic salt; sol in acids.
USE: In analysis of other metallic sulfates.

1301. Bismuth Sulfide. Bi_2S_3; mol wt 514.20. Bi 81.29%, S 18.71%. Occurs in nature as the mineral *bismuthinite.* Prepn: *Gmelin's, Bismuth* (8th ed.) **19**, pp 162-167 (1927).
Blackish-brown, orthorhombic, bipyramidal crystals. Practically insol in water, ethyl acetate; sol in HNO_3, HCl.

1302. Bismuth Tannate. Tanbismuth; tannic acid bismuth derivative. Contains about 36% bismuth equivalent to 40% Bi_2O_3. Prepd from freshly pptd bismuth hydroxide and tannin: *Hagers Handb. Pharm. Praxis* **Band 1**, 685 (Berlin, 1930).
Yellow or brownish-yellow powder. Practically insol in water, alcohol, ether.
THERAP CAT: Astringent; protective.

1303. Bismuth Telluride. Tellurobismuthite. Bi_2Te_3; mol wt 800.83. Bi 52.20%, Te 47.80%. Prepd by heating stoichiometric amounts of the elements to 475° for several days in an evacuated glass or quartz tube: Dönges, *Z. Anorg. Allgem. Chem.* **265**, 56 (1951). Prepn of single crystals: Ainsworth, *Proc. Phys. Soc. (London)* **B69**, 606 (1956); in zone-melting apparatus: Harmon *et al., J. Phys. Chem. Solids* **2**, 181 (1957). Review of different methods: Minden, *Sylvania Technologist* **11** (no. 1), 13-25 (1958).
Gray hexagonal platelets. d 7.642. mp 585°. Single crystals have been grown by the Czochralski technique in which a hydrogen atmosphere was used to minimize the evaporation of tellurium. Since the crystals cleave readily along the (0001) basal hexagonal plane, it is mechanically easier to orient the seed so that the growth direction is in this plane rather than normal to it. The resulting crystals grow more readily along the basal plane, so that they have an oval cross section, often with a characteristic notch. All crystals so pulled are the *P* type. Heat of formation: −8 kcal/mol. Resistivity: 0.00033 ohm-cm. Thermal conductivities at room temp: $\lambda_0 = 0.015$ watt/cm-deg; $\lambda_e = 1.4 \times 10^{-3}$ watt/cm-deg. Energy gap: 0.15 ev. Electron mobility: 800 cm²/volt-sec. Hole mobility: 400 cm²/volt-sec.
USE: In electronics as semiconductor.

1304. Bismuth Tetroxide. "Bismuth peroxide". Bi_2O_4; mol wt 482.00. Bi 86.72%, O 13.28%. It may contain 1 or 2 mols of H_2O. Prepn: *Gmelin's, Bismuth* (8th ed.) **19**, 115 (1927); Schenk in *Handbook of Preparative Inorganic Chemistry* vol. 1, G. Brauer, Ed. (Academic Press, New York, 2nd ed., 1963) p 629.
Orange-red to yellowish-brown heavy powder. Slowly dec by water without dissolving. Dec and dissolved by hot mineral acids.
USE: In lubricants for metal extrusion.

1305. Bismuth Tribromophenate. *Tris(2,4,6-tribromophenoxy)bismuthine;* bismuth tribromophenol; tribromophenolbismuth; Sigmaform; Xeroform. $C_{18}H_6BiBr_9O_3$; mol wt 1198.47. C 18.04%, H 0.50%, Bi 17.44%, Br 60.01%, O 4.01%. $(Br_3C_6H_2O)_3Bi$. Prepd from sodium tribromophenolate and bismuth nitrate: Kollo, *Pharm. Post.* **45**, 1013-1014, *C.A.* **7**, 2831¹ (1913); *Hagers Handb. Pharm. Praxis* **Band 1**, 686 (Berlin, 1930); *N.N.R.* **1951**, pp 66, 505. [In early publications bismuth tribromophenate was described as a basic bismuth salt $(Br_3C_6H_2O)_2BiOH.Bi_2O_3$: *Hagers Handb., loc. cit.*]
Amorphous yellow powder. Stable below 120°. Neutral to moistened litmus paper. Slightly sol in water, alcohol, chloroform, vegetable oils. Dec by alkalies and strong acid.
Ref: N.N.R. **1951**, p 505.
THERAP CAT: Anti-infective.
THERAP CAT (VET): *See* Bismuth.

1306. Bismuth Valerate, Basic. *Valeric acid bismuth basic salt;* basic bismuth valerate; bismuth oxide *n*-valerate; bismuth valerate oxide; bismuth valerianate; bismuthyl valerate. $C_5H_9BiO_3$; mol wt 326.10. C 18.41%, H 2.78%, Bi 64.08%, O 14.72%. $CH_3(CH_2)_3COOBiO$. Prepn: *Hagers Handb. Pharm. Praxis* **Band I**, 687 (Berlin, 1930); Aurivillius, *Acta Chem. Scand.* **9**, 1213 (1955). Crystal structure: Aurivillius, *loc. cit.*
Monohydrate, powder. Valeric acid odor. Practically insol in water, alc; sol in dil HCl or HNO_3. *Keep well closed.*

1307. Bis(1-naphthylmethyl)amine. *N-(1-Naphthalenylmethyl)-1-naphthalenemethanamine;* di(α-naphthylmethyl)-amine; α-dinaphthomethylamine. $C_{22}H_{19}N$; mol wt 297.38. C 88.85%, H 6.44%, N 4.71%. Prepd by catalytic hydrogenation of α-naphthonitrile: Rupe, Becherer, *Helv. Chim. Acta* **6**, 880 (1923).

Pale yellow crystals from petr ether, mp 62°.
USE: In determination of nitrates in fertilizers: Konek, *Z. Anal. Chem.* **97**, 416 (1934), *C.A.* **28**, 5779 (1934). The acetate has been proposed as a sensitive reagent for nitric acid.

1308. Bisobrin. *1,1'-(1,4-Butanediyl)bis[1,2,3,4-tetrahydro-6,7-dimethoxyisoquinoline]; meso-1,1'-tetramethylene-bis[1,2,3,4-tetrahydro-6,7-dimethoxyisoquinoline].* $C_{26}H_{36}$-N_2O_4; mol wt 440.59. C 70.87%, H 8.26%, N 6.35%, O 14.52%. Prepn: Craig, Nabenhauer, U.S. pat. **2,659,728** (1953 to SKF); Schor, Weiner, S. Afr. pat. **67 07,730** (1968 to Endo), *C.A.* **70**, 115024y (1969). Pharmacology: Schor *et al., Chem. Contr. Fibrinolysis-Thrombolysis* **1970**, 113-134; Ambrus *et al., Curr. Ther. Res.* **12**, 451 (1970).

Dilactate, $C_{32}H_{48}N_2O_{10}$, *EN-1661L.* Crystals, mp 217-218°.
Dihydrochloride, $C_{26}H_{28}Cl_2N_2O_4$, mp 260-261° (dec).
THERAP CAT: Fibrinolytic.

1309. Bisoprolol. *1-[4-[[2-(1-Methylethoxy)ethoxy]methyl]phenoxy]-3-[(1-methylethyl)amino]-2-propanol;* (±)-1-[[α-(2-isopropoxyethoxy)-p-tolyl]oxy]-3-(isopropylamino)-2-propanol; (±)-1-[p-(2-isopropoxyethoxymethyl)-phenoxy]-3-(isopropylamino)-2-propanol; EMD 33 512. $C_{18}H_{31}NO_4$; mol wt 325.45. C 66.43%, H 9.60%, N 4.30%, O 19.66%. Cardioselective β_1-adrenergic blocker. Prepn: **Belg.** pat. **859,425**; R. Jonas *et al.*, U.S. pats. **4,171,370, 4,258,062** (1978, 1979, 1981 all to E. Merck). β_1-Selectivity in animals: H.-J. Schliep, J. Harting, *J. Cardiovasc. Pharmacol.* **6**, 1156 (1984); in humans: A. E. Tattersfield *et al., Brit. J. Clin. Pharmacol.* **18**, 343 (1984). β_1-Adrenoceptor binding affinity of (−)-form: A. J. Kaumann, H. Lemoine, *Arch. Pharmacol.* **331**, 27 (1985); X. L. Wang *et al., Eur. J. Pharmacol.* **114**, 157 (1985). HPLC determn in plasma and urine: K. U. Bühring, A. Garbe, *J. Chromatog.* **382**, 215 (1986). HPLC resolution of enantiomers: O. Weller *et al., ibid.* **403**, 263 (1987). Cardiovascular effects in patients with heart disease and bronchitis: P. Dorow, U. Tönnesmann, *Eur. J. Clin. Pharmacol.* **27**, 135 (1984). CNS effects in comparison with pindolol, *q.v.:* R. Görtelmeyer, I. Klingmann, *Arzneimittel-Forsch.* **35**, 1707 (1985). Comparison with atenolol, *q.v.,* of anti-anginal effects: R. S. Kohli *et al., Eur. Heart J.* **6**, 845 (1985); of antihypertensive effects: F. R. Bühler *et al., J. Cardiovasc. Pharmacol.* **8**, Suppl. 11, S122 (1986). Use in treatment of glaucoma: J. Harting, A. Fuchs, U.S. pat. **4,522,829** (1985 to E. Merck).

Hemifumarate, $C_{36}H_{64}N_2O_{12}$, *Concor, Detensiel, Emconcor, Emcor, Euradal, Isoten, Monocor, Soprol.* Crystals, mp 100°. Sol in ethanol.
THERAP CAT: Antihypertensive.

1310. Bisoxatin Acetate. *2,2-Bis[4-(acetyloxy)phenyl]-*

2H-1,4-benzoxazin-3(4H)-one; 2,2-bis(4-acetoxyphenyl)-3-oxo-2,3-dihydrobenz-1,4-oxazine; Wy 8138; Laxonalin; Maratan; Talsis; Tasis. $C_{24}H_{19}NO_6$; mol wt 417.40. C 69.06%, H 4.59%, N 3.36%, O 23.00%. Prepd from 2,2-dichloro-3-oxo-2,3-dihydrobenz-1,4-oxazine and phenylacetate or by acetylation of the diphenol: Seeger, U.S. pat. **3,006,917** (1961 to Thomae). Toxicity: E. I. Goldenthal, *Toxicol. Appl. Pharmacol.* **18**, 185 (1971).

Crystals from ethanol, mp 190°. LD_{50} orally in rats, mice: 8000, > 10,000 mg/kg (Goldenthal).
THERAP CAT: Cathartic.

1311. Bisphenol A. *4,4'-(1-Methylethylidene)bisphenol; 4,4'-isopropylidenediphenol;* 2,2-bis(4-hydroxyphenyl)propane. $C_{15}H_{16}O_2$; mol wt 228.28. C 78.92%, H 7.07%, O 14.02%. Manuf from phenol and acetone: Jansen, U.S. pat. **2,468,982** (1949); Faith, Keyes & Clark's *Industrial Chemicals,* F. A. Lowenheim, M. K. Moran, Eds. (Wiley-Interscience, New York, 4th ed., 1975) pp 149-152. *Review: Chem. & Eng. News* **41**, 35 (June 3, 1963); *ibid.* **51**, 5 (July 16, 1973).

Crystals or flakes. Mild phenolic odor. mp 150-155° (solidification range). bp_4 220°. Dec above 8 mm pressure when heated above 220°. Practically insol in water. Sol in aq alkaline solns, alcohol, acetone. Slightly sol in carbon tetrachloride.
USE: In the manuf of epoxy resins and polycarbonates. As fungicide.

1312. Bisphenol B. *4,4'-(1-Methylpropylidene)bisphenol; p,p'-sec-butylidenediphenol;* 2,2-bis(4-hydroxyphenyl)butane. $C_{16}H_{18}O_2$; mol wt 242.30. C 79.30%, H 7.49%, O 13.21%. May be prepd from phenol and ethyl methyl ketone: Jansen, U.S. pat. **2,468,982** (1949 to Goodrich).

Crystals or tan granules, mp 118.9-121.7°. Approximate soly per 100 g: acetone 266 g; benzene 2.3 g; carbon tetrachloride < 0.1 g; ether 133 g; methanol 166 g; V.M.P. naphtha < 0.1 g; water < 0.1 g.
USE: In the manufacture of phenolic resins.

1313. 1,4-Bis(trichloromethyl)benzene. *α,α,α,α',α',α'-Hexachloro-p-xylene;* p-bis(perchloromethyl)benzene; Bitriben; Hetol. $C_8H_4Cl_6$; mol wt 312.86. C 30.71%, H 1.29%, Cl 68.00%. Usually obtained by photochlorination of p-xylene: McBee *et al., Ind. Eng. Chem.* **39**, 298 (1947); *eidem* in Slesser, Schram, *Preparation, Properties, and Technology of Fluorine and Organic Fluoro Compounds* (McGraw-Hill, New York, 1951) pp 207-221; Ross *et al., J. Am. Chem. Soc.* **75**, 4697 (1953); Rabjohn, *ibid.* **76**, 5479 (1954); Harvey *et al., J. Appl. Chem.* **4**, 319 (1954); Ligett, U.S. pat. **2,654,789** (1953 to Ethyl Corp.).

Crystals from hexane or ether, mp 108-110°.

Note: The name Hetol was previously used to designate sodium cinnamate.

USE: Insecticide.

THERAP CAT (VET): Fasciolicide.

1314. Bis(triphenylphosphine)dicarbonylnickel. *Dicarbonylbis(triphenylphosphine)nickel;* dicarbonyldi(triphenylphosphino)nickel. $C_{38}H_{30}NiO_2P_2$; mol wt 639.28. C 71.39%, H 4.73%, Ni 9.18%, O 5.01%, P 9.69%. $[(C_6H_5)_3P]_2Ni(CO)_2$. Obtained by reacting one mole of nickel carbonyl with two moles of triphenylphosphine: Reppe, Schweckendiek, *Ann.* **560**, 104 (1948); Rose, Statham, *J. Chem. Soc.* **1950**, 69. Also by reacting CO with a methanol soln of triphenylphosphine and reduced Ni at 170° and 60 atm: Yamamoto, Oku, *Bull. Chem. Soc. Japan* **27**, 382 (1954), *C.A.* **49**, 6856 (1955).

Crystals from benzene, mp 210-215° (Rose); mp 206-209° (Reppe). Catalytic activity decreases with storage.

USE: Catalyst in polymerization of acetylene to benzene and styrene, trimerization of ethynyl compds, cyclization of butadiene.

1315. Bitertanol. *β-[(1,1'-Biphenyl)-4-yloxy]-α-(1,1-dimethylethyl)-1H-1,2,4-triazol-1-ethanol;* 1-(biphenyl-4-yloxy)-3,3-dimethyl-1-(1H-1,2,4-triazol-1-yl)butan-2-ol; biloxazol; BAY KWG 0599; Baycor; Sibutol. $C_{20}H_{23}N_3O_2$; mol wt 337.42. C 71.19%, H 6.87%, N 12.45%, O 9.48%. Prepn: **Belg. pat. 814,831**; W. Kramer *et al.*, **U.S. pat. 3,952,002** (1974, 1976 both to Bayer AG). Physical properties and fungicidal activity: W. Brandes *et al.*, *Pflanzenschutz-Nachr.* **32**, 1 (1979). Mechanism of action: P. Kraus, *ibid.* **17**; S. V. Overton *et al.*, *J. Horticult. Sci.* **63**, 183 (1988). GC determn in plants, soil and water: R. Brennecke, *Pflanzenschutz-Nachr.* **38**, 33 (1985). Resolution of diastereomers: R. S. Burden *et al.*, *J. Chromatog.* **391**, 273 (1987). *In vitro* activity: T. B. Sutton *et al.*, *Plant Dis.* **69**, 700 (1985). Field use in control of apple diseases: K. S. Yoder, *ibid.* **66**, 580 (1982); W. F. S. Schwabe, *Pflanzenschutz-Nachr.* **35**, 125 (1982); in wheat seed pretreatment: J. A. Hoffmann, D. V. Sisson, *Plant Dis.* **71**, 839 (1987).

The marketed product is a mixture of diastereomers. Colorless crystals, mp 125-129°. Vapor pressure at 20°: 10^{-5}mbar. Soly at 20° (g/100 g solvent): water 0.0005, ligroin (80-110°) 0-1, propan-2-ol 1-5, toluene 1-5, cyclohexane 5-10, methylene chloride 10-20. Stable in aqueous acid and alkaline solns. LD_{50} in rats, male and female mice (mg/kg): > 5000, 4488, 4202 orally (Brandes).

USE: Agricultural fungicide.

1316. Bithionol. *2,2'-Thiobis[4,6-dichlorophenol];* TBP; bis(2-hydroxy-3,5-dichlorophenyl)sulfide; XL-7; Actamer; Bithin; Lorothidol. $C_{12}H_6Cl_4O_2S$; mol wt 356.07. C 40.48%, H 1.70%, Cl 39.83%, O 8.99%, S 9.01%. Prepn from 2,4-halo substituted phenol with S halides: Muth, **Ger. pat. 583,055** (1933 to I. G. Farbenind.), *C.A.* **28**, 179 (1934); Copper, Godfrey, **U.S. pat. 2,849,494** (1958 to Monsanto). Comprehensive review and bibliography: Shumard *et al.*, *Soap, Sanit. Chemicals* **29**, no. 1, 34-37, 90 (1953). Anthelmintic activity of the sulfoxide: J. Guilhon, M. Graber, *Bull. Acad. Vet. Fr.* **52**, 225 (1979). Metabolism of the sulfoxide in humans: M. Sakamoto *et al.*, *J. Toxicol. Sci.* **6**, 307 (1981).

Crystals, mp 188°. d_4^{25} 1.73. pK_1 4.82; pK_2 10.50. Vapor press. at 37°: 1.1×10^{-9} mm Hg. Practically insol in water (0.0004% at 25°). Sol in dil caustic solns. A 4% NaOH soln will dissolve 16.2% bithionol. Soly (g/100 ml): acetone 15.0; polysorbate 80 19.0; dimethylacetamide 72.5; lanolin at 42° 5.0; pine oil 4.0; corn oil 1.0; propylene glycol 0.5; 70% ethanol 0.3.

Sodium salt, $C_{12}H_4Cl_4Na_2O_2S$, *bithionolate sodium, Vancide BN,* Harvey *et al.,* **U.S. pat. 3,024,163** (1962 to Vanderbilt Co.).

Sulfoxide, $C_{12}H_6Cl_4O_3S$, *BTS, Bitin-S, Disto-5.*

USE: Surfactant-formulated antimicrobial against bacteria, molds and yeast. Banned by FDA from use in cosmetics. Proposed as agricultural fungicide.

THERAP CAT: Topical anti-infective.

THERAP CAT (VET): Anthelmintic; antiseptic.

1317. Bitolterol. *4-Methylbenzoic acid 4-[2-[(1,1-dimethylethyl)amino]-1-hydroxyethyl]-1,2-phenylene ester; p-toluic acid 4-[2-(t-butylamino)-1-hydroxethyl]-o-phenylene ester; α-[(tert-butylamino)methyl]-3,4-dihydroxybenzyl alcohol 3,4-di-p-toluate.* $C_{28}H_{31}NO_5$; mol wt 461.57. C 72.86%, H 6.77%, N 3.04%, O 17.33%. $β_2$-Adrenergic agonist; a diester of *N-tert*-butylnorepinephrine. Prepn: H. Minatoya *et al.*, **Fr. pat. 2,042,295** (1971 to Sterling), *C.A.* **76**, 14129e (1972); B. F. Tullar *et al.*, *J. Med. Chem.* **19**, 834 (1976). Pharmacology: H. Minatoya, *J. Pharmacol. Exp. Ther.* **206**, 515 (1978). Metabolism and excretion: T. Aimoto *et al.*, *Xenobiotica* **9**, 173 (1979). Comparative clinical studies with isoproterenol, *q.v.*, in asthma: J. L. Pinnas *et al.*, *J. Allergy Clin. Immunol.* **79**, 768 (1987); R. A. Nathan *et al.*, *ibid.* **822**. Review: S. B. Walker *et al.*, *Pharmacotherapy* **5**, 127-137 (1985).

Methanesulfonate, $C_{29}H_{35}NO_8S$, *bitolterol mesylate, Win 32784, Effectin, Tornalate.* Crystalline solid, mp 170-172°.

THERAP CAT: Bronchodilator.

1318. Bitoscanate. *1,4-Diisothiocyanatobenzene;* phenylene-1,4-diisothiocyanate; isothiocyanic acid p-phenylene ester; Jonit. $C_8H_4N_2S_2$; mol wt 192.24. C 49.98%, H 2.09%, N 14.57%, S 33.35%. Prepn: Lieber, Slutkin, *J. Org. Chem.* **27**, 2214 (1962). Prepn and use as anthelmintic: **Fr. pat. M1652** and **Brit. pat. 1,001,314** (1963, 1965 to Hoechst). Purification: Soeder, Laemmer, **Ger. pat. 1,172,664** (1964 to Hoechst). Series of articles on clinical studies: *Progress in Drug Research* **vol. 19**, E. Jucker, Ed. (Birkhäuser Verlag, Basel and Stuttgart, 1975) pp 2-107.

Tasteless, odorless, colorless needles, from acetic acid or acetone, mp 132°.

THERAP CAT: Anthelmintic (Nematodes).

1319. Biuret. *Imidodicarbonic diamide;* carbamylurea. $C_2H_5N_3O_2$; mol wt 103.09. C 23.30%, H 4.89%, N 40.77%, O 31.04%. $NH_2CONHCONH_2$. Prepd on a large scale by the action of heat on urea. Other syntheses are based on the treatment of urea with inorganic halides, such as thionyl chloride, on the interaction of urea and cyanic acid, and on the ammonolysis of allophanic esters and related compds. Has weak bacteriostatic and diuretic properties in rats, also causes a fall in blood pressure, but produces strong irritation of the urinary tract. Increases the activity of pepsin. Comprehensive review and bibliography: Kurzer, *Chem Rev.* **56**, 95-197 (1956).

Hygroscopic, elongated plates from ethanol. d_4^{-5} 1.467. When crystallized from water, $5C_2H_5N_3O_2.4H_2O$ is formed, which becomes anhydr at 110°, then dec at about 193°. Pyrolysis at higher temps yields melamine. Soly (g/100 g soln) in water at 25° = 2.01; at 50° = 7; at 75° = 20; at 105.5° = 53.5. Freely sol in alc, very slightly sol in ether. Aq solns treated with cupric sulfate and NaOH give a reddish-violet color (biuret reaction, details: Kurzer, *loc. cit.*, p 181).

1320. Bixin. *6,6'-Diapo-ψ,ψ-carotenedioic acid monomethyl ester.* $C_{25}H_{30}O_4$; mol wt 394.49. C 76.11%, H 7.66%, O 16.22%. Carotenoid carboxylic acid isolated from seeds of *Bixa orellana* L., *Bixaceae:* Kuhn, Ehmann, *Helv. Chim. Acta* **12**, 904 (1929). The *cis* form occurring in the plant is unstable and changes to the stable *trans* form (complete conversion on standing in chloroform soln in the presence of iodine). Isoln of *trans* form from *Aristolochia cymbifera* Mart., *Aristolochiaceae:* Green *et al.*, *Helv. Chim. Acta* **37**, 1717 (1954). Structure: Kuhn *et al.*, *ibid.* **11**, 427 (1928); **12**, 64, 904 (1929); *Ber.* **64**, 1732 (1931); *Ber.* **65**, 646, 1873 (1932). Stereochemistry: Barber *et al.*, *Proc. Chem. Soc.* **1960**, 23. Synthesis of methyl *cis*-bixin: Pattenden *et al.*, *J. Chem. Soc. (C)* **1970**, 235. Reviews: Karrer, Jucker, *Carotenoids* (Elsevier, New York, 1950) pp 256-271; Zechmeister, *Fortsch. Chem. Org. Naturst.* **18**, 320-333 (1960). The following data are for the stable *trans* form; also called *isobixin*.

Orange to purple plates from acetone, dec 217°. Absorption max (chloroform): 509.5; 475; 443 nm. Considerably less sol in organic solvents than the labile form.

Methyl ester, $C_{26}H_{32}O_4$, *methyl bixin.* Broad, blue to purple needles from benzene, mp 203°. Absorption max (benzene): 475 nm (ε 125,000). Synthesis: Isler *et al.*, *Helv. Chim. Acta* **40**, 1242 (1957); Buchta, Andree, *Ber.* **92**, 3111 (1959).

Ethyl ester, $C_{27}H_{34}O_4$, *ethyl bixin.* Red rhombic crystals from ethanol, mp 138°. Freely sol in chloroform, ethyl acetate, acetone.

USE: The ethyl ester as a suspension in vegetable oil for coloring foods and drugs. Imparts a golden yellow color.

1321. Blackstrap Molasses. The final, unpurified mother liquor in the cane sugar industry from which no more sugar can be crystallized by factory methods. Rich in inorganic constituents, may analyze as much as 10% ash upon ignition. Sucrose content about 30%.

1322. Blancophor® R. *2,2'-(1,2-Ethenediyl)bis[5-[[(phenylamino)carbonyl]amino]benzenesulfonic acid] disodium salt;* 4,4'-bis(3-phenylureido)-2,2'-stilbenedisulfonic acid disodium salt; [stilbene-(4,4')]bis[ω-phenylurea]-2,2'-disulfonic acid disodium salt; C.I. Fluorescent Brightener 30; C.I. 40600; Blancol C; Blankophor R; Leucophor R; Lumisol RV; Phorwite RN; Photine R; Pontamine White BR; Tintophen X. $C_{28}H_{22}N_4Na_2O_8S_2$; mol wt 652.62. C 51.53%, H 3.40%, N 8.59%, Na 7.05%, O 19.61%, S 9.82%. Prepd by treating 4,4'-diamino-2,2'-stilbenedisulfonic acid with phenyl isocyanate in aq soln at 40°: *Ger. pat.* **746,569** and *Fr. pat.* **878,155** (to I. G. Farbenind.); *Brit. pat.* **683,895** (1952 to Bayer); *FIAT Final Rept.* No. 1302 (Sept. 15, 1947); *Colour Index* Vol. 4 (3rd ed., 1971) p 4371.

USE: Fluorescent dye for cellulose, protein fibers, nylon, wool or paper.

1323. Blasticidin S. *4-[[3-Amino-5-[(aminoiminomethyl)methylamino]-1-oxopentyl]amino]-1-(4-amino-2-oxo-1(2H)-pyrimidinyl)-1,2,3,4-tetradeoxy-β-D-erythro-hex-2-enopyranuronic acid;* 4-[3-amino-5-(1-methylguanidino)-valeramido]-1-(4-amino-2-oxo-1(2H)-pyrimidinyl)-1,2,3,4-tetradeoxy-β-D-erythro-hex-2-enopyranuronic acid; 1-(1'-cytosinyl)-4-[L-3'-amino-5'-(1''-N-methylguanidino)-valerylamino]-1,2,3,4-tetradeoxy-β-D-erythro-hex-2-enuronic acid. $C_{17}H_{26}N_8O_5$; mol wt 422.46. C 48.33%, H 6.20%, N 26.53%, O 18.94%. Nucleoside antibiotic produced by *Streptomyces griseochromogenes:* Isoln and antimicrobial activity: Takeuchi *et al.*, *J. Antibiot.* **11A**, 1 (1958); Sumiki, Umezawa, **Japan.** pat. **16,449**('60) (1961 to Japan Antibiot. Res. Assoc.), *C.A.* **55**, 21474e (1961). Structure: Otake *et al.*, *Agr. Biol. Chem.* **30**, 132 (1966); Fox, Watanabe, *Tetrahedron Letters* **1966**, 897. Abs config: Yonehara *et al.*, *ibid.* **1966**, 3785. Synthesis of the unsaturated carbohydrate: Goody *et al.*, *ibid.* **1970**, 293; of the cytosinine moiety: Fox, Watanabe, *Pure Appl. Chem.* **28**, 475 (1971); Kondo *et al.*, *Tetrahedron Letters* **1972**, 1881; *eidem, Tetrahedron* **29**, 1801 (1973). Crystal and molecular structure: V. Swaminathan *et al.*, *Biochim. Biophys. Acta* **655**, 335 (1981).

blastidic acid   cytosinine

Needles from water, dec 235-236°. $[\alpha]_D^{11}$ +108.4° (water). uv max (0.1N HCl): 275 nm ($E_{1cm}^{1\%}$ 349); (0.1N NaOH): 266-270 nm ($E_{1cm}^{1\%}$ 266). Sol in water, acetic acid; practically insol in methanol, ethanol, acetone, benzene, ether, ethyl acetate, chloroform, carbon tetrachloride, cyclohexane, xylene, pyridine, dioxane. LD_{50} i.v. in mice: 2.82 mg/kg (Takeuchi).

Hydrochloride, crystals, dec 224-225°.

USE: Antifungal against rice blast disease in Japan.

1324. Bleomycins. NSC-125066; Bleo. A group of related glycopeptide antibiotics isolated from *Streptomyces verticillus:* Umezawa, *Antimicrob. Ag. Chemother.* **1965**, 1079. Purification and separation into bleomycins A and B and their components: Umezawa *et al.*, *J. Antibiot.* **19**, 200, 210 (1966); T. Takita *et al.*, *ibid.* **21**, 79 (1968); **22**, 237 (1969). Bleomycin A_2 is the main component of the bleomycin employed clinically. Total structure elucidation: T. Takita *et al.*, *ibid.* **25**, 755 (1972). Revised structure: *eidem, ibid.* **31**, 801 (1978). Bleomycins differ from one another in the terminal amine and show varying biological activity. Structures of the terminal amines: Fujii *et al.*, *ibid.* **26**, 398 (1973). Synthesis of new bleomycins: T. Takita *et al.*, *ibid.* 254. Total synthesis of bleomycin A_2: *eidem, Tetrahedron Letters* **23**, 521 (1982); Y. Aoyagi *et al.*, *J. Am. Chem. Soc.* **104**, 5537 (1982). Improved total synthesis: S. Saito *et al.*, *J. Antibiot.* **36**, 92 (1983). Biosynthesis: Fujii *et al.*, *ibid.* **27**, 73 (1974). Bleomycins are believed to react with DNA and cause strand scission; they have also been shown to have a type of oxygen transferase activity. Mechanism of action studies: R. M. Burger *et al.*, *Life Sci.* **28**, 715 (1981); N. Marugesan *et al.*, *J. Biol. Chem.* **257**, 8600 (1982). Coordination chemistry: J. C. Dabrowiak, *J. Inorg. Biochem.* **13**, 317 (1980). Clinical pharmacology: S. T. Crooke, *Cancer Chemother.* **3**, 343 (1981). Characterization of analogs: N. J. Oppenheimer *et al.*, *J. Biol. Chem.* **257**, 1606 (1982). Total synthesis of *deglycobleomycin*, a biologically active deriv lacking sugars: Y. Aoyagi *et al.*, *J. Am. Chem. Soc.* **104**, 5237 (1982). *Reviews:* H. Umezawa, *Pure Appl. Chem.* **28**, 665-680 (1971); C. W. Haidle, R. S. Lloyd, *Antibiotics vol.* 5(pt. 2), F. E. Hahn, Ed. (Springer-Verlag, New York, 1979) pp 124-154; H. Umezawa, *Anticancer Agents Based on Na-*

tural Product Models, J. M. Cassady, J. D. Douros, Eds. (Academic Press, New York, 1980) pp 147-166.

Colorless or yellowish powder which becomes bluish depending on Cu content. Very sol in water, methanol; slightly sol in ethanol. Practically insol in acetone, ethyl acetate, butyl acetate, ether. uv max: 244-248, 289-294 nm ($E_{1cm}^{1\%}$ 121-148, 102-121.5).

Sulfate, *Blenoxane*.

Bleomycin A₂, $C_{55}H_{84}N_{17}O_{21}S_3$, N^I-[3-(dimethylsulfonio)-propyl]bleomycinamide. R = $(CH_3)_2S^+CH_2CH_2CH_2NH$—.
THERAP CAT: Antineoplastic.

1325. Bolandiol. *Estr-4-ene-3β,17β-diol*; 3β,17β-dihydroxyestr-4-ene. $C_{18}H_{28}O_2$; mol wt 276.40. C 78.21%, H 10.21%, O 11.58%. Prepn: Colton, U.S. pat. **2,843,608** (1958 to Searle).

Crystals from dil acetone and from ethyl acetate + petr ether, mp 169-172°.

Dipropionate, $C_{24}H_{36}O_4$, *norpropandrolate*, 3β,17β-dipropionyloxy-4-estrene, *SC-7525, Anabiol, Storinal*.
THERAP CAT: Anabolic.

1326. Bolasterone. *17-Hydroxy-7,17-dimethylandrost-4-en-3-one*; 7α,17-dimethyltestosterone; *Myagen.* $C_{21}H_{32}O_2$; mol wt 316.47. C 79.70%, H 10.19%, O 10.11%. Epimeric with calusterone, *q.v.* Prepn: Belg. pat. **610,385** corresp to Babcock, Campbell, U.S. pat. **3,341,557** (1962, 1967 both to Upjohn); Campbell, Babcock, *Hormonal Steroids*, Proc. 1st Int. Congr., **2**, L. Martini, A. Pecile, Eds. (Academic Press, New York, 1965) pp 59-67. Activity studies: Stucki *et al.*, *ibid.* 119-132.

Crystals, mp 163-165°.
THERAP CAT: Anabolic.

1327. Boldenone. *17-Hydroxyandrosta-1,4-dien-3-one;* 1,4-androstadien-17β-ol-3-one; 3-oxo-17β-hydroxy-1,4-androstadiene; dehydrotestosterone. $C_{19}H_{26}O_2$; mol wt 286.40. C 79.68%, H 9.15%, O 11.17%. Anabolic steroid.

Prepn: Meystre *et al., Helv. Chim. Acta* **39,** 734 (1956); Florey, U.S. pat. **2,875,196** (1949 to Olin Mathieson); Nobile, U.S. pat. **2,837,464** (1958 to Schering). Metabolism: Galletti, Gardi, *Steroids* **18,** 39 (1971).

Crystals, mp 164-166°. $[\alpha]_D^{25}$ +25° (in chloroform).

Acetate, $C_{21}H_{28}O_3$, crystals, mp 151-153°. Prepn: Florey, *loc. cit.*

10-Undecenoate, $C_{30}H_{44}O_3$, *boldenone undecylenate, Ba-29038, Parenabol.*
THERAP CAT: Androgen.
THERAP CAT (VET): Anabolic.

1328. Boldine. *5,6,6a,7-Tetrahydro-1,10-dimethoxy-6-methyl-4H-dibenzo[de,g]quinoline-2,9-diol; 1,10-dimethoxy-6aα-aporphine-2,9-diol;* 1,10-dimethoxy-2,9-dihydroxy-aporphine; 2,6-dihydroxy-3,5-dimethoxyaporphine. $C_{19}H_{21}NO_4$; mol wt 327.37. C 69.70%, H 6.47%, N 4.28%, O 19.55%. Isoln from boldo (*Peumus boldus* Molina, *Monimiaceae*): Bourgoin, Verne, *J. Pharm. Chim.* **16,** 191 (1872); from *Laurelia novaezelandiae* A. Cunn.: Bernauer, *Helv. Chim. Acta* **50,** 1583 (1967). Structure: Warnat, *Ber.* **58,** 2768; **59,** 85 (1926); Späth, Tharrer, *ibid.* **66,** 904 (1933); Schlittler, *ibid.* 988. Synthesis of *dl*-form: S. M. Kupchan *et al., Chem. Commun.* **1976,** 91. Biosynthetic study: D. S. Bhakuni *et al., J. Chem. Soc. Perkin Trans. I* **1977,** 706. Detailed description of the boldo brush, its ingredients and pharmacological properties: H. Schindler, *Arzneimittel-Forsch.* **7,** 747 (1957).

d-Form, crystals from ether. mp 162-164°. $[\alpha]_D^{25}$ +127° (c = 0.1 in alcohol). Very slightly sol in water or ether; sol in alcohol, chloroform, dil acids.

d-Form hydrochloride, crystals from methanol-ether and methanol, mp 212-220°.

dl-Form, mp 159-162°.

Dimethyl ether, *see* Glaucine.

Note: Boldine is used as an ingredient in choleretics and laxatives. The total alkaloids from boldo are also marketed as a grayish-white to yellow-green, bitter powder; almost insol in water; sol in alc, chloroform, slightly in ether.

1329. Boldo. Boldu; boldea; boldus; boldoa. Leaves of *Peumus boldus* Molina (*Boldu boldus* [Molina] Lyons, *Boldea fragrans* Gay), *Monimiaceae. Habit.* Peru, Chile. *Constit.* About 2% volatile oil, about 0.1% boldine. *Ref:* Schindler, *Arzneimittel-Forsch.* **7,** 747 (1957).
THERAP CAT: In hepatic dysfunction, cholelithiasis.

1330. Bole, Armenian. Bolus armena; bolus rubra; red bole. A red variety of clay (aluminum silicate) contg naturally occurring ferric oxide (hematite). Originally found in Armenia.

Reddish, soft, unctuous pieces; adheres to the tongue; easily reduced to powder. Insol in usual solvents. d 1.9-2.0.
USE: For coloring powders, cements, and as a pigment.
THERAP CAT: Adsorbant; protectant.

1331. Boleko Oil. Isano oil. Oil extracted from the nuts of the tree *Ongokea gore* (Hua) Pierre, *Olacaceae*, growing in equatorial Africa. Contains fatty acids (as glycerides): isanic acid 46%, isanolic acid 44%. *Refs:* Scher, *Arch. Pharm.* **287**, 548 (1954); Dupont *et al., Bull. Soc. Chim. France* 1957, 1495; De Vries, *Oléagineux* **12**, 427 (1957); Kneeland *et al., J. Am. Oil Chem. Soc.* **35**, 361 (1958). Extraction procedure: Lambert, U.S. pat. **2,800,492** (1957 to UCB). Structure of acids: Gunstone, Sealy, *J. Chem. Soc.* **1963**, 5772; Morris, *ibid.* 5779.

Typical characteristics: Acid no. 1-10; ester no. 183.5; sapon no. 185-200; iodine no. 254-259; d_4^{20} 0.973-0.983; n_D^{20} 1.505-1.509; viscosity at 25° = 7-10 poise. Soluble in acetone, benzene, ethyl ether, carbon tetrachloride, chloroform. Slightly sol in petr ether, ethanol, hexane. The viscosity increases upon heating and the oil is transformed into a rubber-like mass. Does not polymerize when heated to a moderate temp; when heated rapidly to temp above 200° polymerization becomes so rapid as to become explosive. Polymerization is explained by the acetylenic structure of isanic acid.

USE: In fire retardant paints.

1332. Bombesin. *2-L-Glutamine-6-L-asparaginealytesin.* $C_{71}H_{110}N_{24}O_{18}S$; mol wt 1619.82. C 52.64%, H 6.85%, N 20.75%, O 17.78%, S 1.98%. Pharmacologically active tetradecapeptide found in skins of European amphibians of the family Discoglossidae, principally *Bombina bombina* and *Bombina variegata variegata*. Isoln and structure: A. Anastasi *et al., Experientia* **27**, 166 (1971). Pharmacological activity: V. Erspamer *et al., J. Pharm. Pharmacol.* **22**, 875 (1970). Synthesis: L. Bernardi *et al., Experientia* **27**, 873 (1971). It is a potent stimulant of gastric and pancreatic secretions in mammals; a bombesin-like immunoreactive peptide is found in both brain and gut, *cf.* J. M. Polak *et al., Lancet* **1**, 1109 (1976). Other actions include hypertensive reactions, antidiuresis, and hyperglycemic activity. Bombesin has been shown to have a strong effect on core temperature lowering in rats: M. Brown *et al., Science* **196**, 996 (1977). High levels of intracellular bombesin have also been found in human small-cell lung carcinoma: T. W. Moody *et al., ibid.* **214**, 1246 (1981). Brief review of chemistry, isoln, purification: V. Mutt, *Biochem. Soc. Trans.* **8**, 11 (1980). Effect on suppression of food intake: J. Gibbs *et al., Nature* **282**, 208 (1979). Pharmacological reviews: G. Bertaccini, *Pharmacol. Rev.* **28**, 127 (1976); S. R. Bloom, J. M. Polak, *Advan. Clin. Chem.* **21**, 177 (1980).

```
5-oxo-Pro-Gln-Arg-Leu-Gly-Asn-Gln-Trp
                                      |
   H2N-Met-Leu-His-Gly-Val-Ala
```

Hydrochloride, $C_{71}H_{112}Cl_2N_{24}O_{18}S$, cryst from 99% ethanol, mp 185° (dec). $[\alpha]_D^{24}$ −20.6° (c = 0.65 in DMF-HMPT 8:2).

1333. Bomyl®. *3-[(Dimethoxyphosphinyl)oxy]-2-pentenedioic acid dimethyl ester; 3-hydroxy-2-pentenedioic acid dimethyl ester dimethyl phosphate; 3-hydroxyglutaconic acid dimethyl ester dimethyl phosphate;* dimethyl 1,3-bis(carbomethoxy)-1-propen-2-yl phosphate; dimethyl 3-hydroxyglutaconate dimethyl phosphate; ENT 24833; GC 3707; Swat. $C_9H_{15}O_8P$; mol wt 282.20. C 38.31%, H 5.36%, O 45.36%, P 10.98%. Prepn: Gilbert, U.S. pat. **2,891,887** (1959 to Allied). Activity: P. E. Newallis *et al., J. Agr. Food Chem.* **15**, 940 (1967). Toxicity: E. W. Schafer, *Toxicol. Appl. Pharmacol.* **21**, 315 (1972).

Liquid; bp$_{17}$ 155-164°; bp$_2$ 155-165° (tech grade). Miscible with methanol, ethanol, acetone, xylene. Practically insol in water, petr ether, kerosene. LD$_{50}$ orally in rats: 32 mg/kg (Schafer).

USE: Insecticide.

1334. Bongkrekic Acid. *[R-[R*,S*-(E,Z,Z,E,E,Z,E)]]-20-(Carboxymethyl)-6-methoxy-2,5,17-trimethyl-2,4,8,10,-14,18,20-docosaheptaenedioic acid;* 3-carboxymethyl-17-methoxy-6,18,21-trimethyldocosa-2,4,8,12,14,18,20-hep-

taenedioic acid; BA. $C_{28}H_{38}O_7$; mol wt 486.61. C 69.11%, H 7.87%, O 23.02%. One of the two toxic antibiotic principles produced by *Pseudomonas cocovenenans* on partially defatted coconut; the other being toxoflavin, *q.v.* Name derived from "bongkrek", a molded coconut product from Indonesia which becomes highly poisonous when *P. cocovenenans* outgrows the mold. Isoln: van Veen, Mertens, *Rec. Trav. Chim.* **53**, 257 (1934); **54**, 373 (1935); Nugteren, Berends, *ibid.* **76**, 13 (1957). Purification and properties: Lijmbach *et al., Tetrahedron* **26**, 5993 (1970). Structural studies: *eidem, ibid.* **27**, 1839 (1971). Revised structure: De Bruijn *et al., ibid.* **29**, 1541 (1973). Absolute configuration: Zylber *et al., Experientia* **29**, 387 (1973). Influence on carbohydrate metabolism: van Veen, Mertens, *Arch. Neer. Physiol.* **21**, 73 (1936), *C.A.* **30**, 3880[9] (1936); inhibition of adenine nucleotide translocation: Henderson, Lardy, *J. Biol. Chem.* **245**, 1319 (1970); Klingenberg *et al., Biochem. Biophys. Res. Commun.* **39**, 363 (1970).

White, amorphous solid, mp 50-60°. uv max (methanol): 237, 267 nm (ε 32000, 36700). $[\alpha]_D^{25}$ +162.5°. LD$_{50}$ i.v. in mice: 1.41 mg/kg (Lijmbach).

1335. Bopindolol. *(±)1-[[(1,1-Dimethylethyl)amino]-3-[(2-methyl-1H-indol-4-yl)oxy]-2-propanol benzoate ester;* *(±)-1-(tert-butylamino)-3-[(2-methylindol-4-yl)oxy]-2-propanol benzoate (ester).* $C_{23}H_{28}N_2O_3$; mol wt 380.49. C 72.61%, H 7.42%, N 7.36%, O 12.61%. Nonselective β-adrenergic blocker. Prepn: F. Troxler, F. Seemann, **Ger. pat. 2,635,209**; *eidem*, U.S. pat. **4,340,541** (1977, 1982 both to Sandoz). Clinical pharmacology: P. van Brummelen *et al., Eur. J. Clin. Pharmacol.* **22**, 491 (1982). HPLC determn in plasma: C. J. Oddie *et al., J. Chromatog.* **273**, 469 (1983). Effect on plasma lipid fractions: P. van Brummelen *et al., Brit. J. Clin. Pharmacol.* **17**, 86 (1984). Pharmacokinetics: R. Platzer *et al., Clin. Pharmacol. Ther.* **36**, 5 (1984). Clinical trials in hypertension: U. L. Hulthen *et al., J. Cardiovascular Pharmacol.* **5**, 426 (1983); W. J. Schiess *et al., Eur. J. Clin. Pharmacol.* **27**, 529 (1984).

Sol in ether, methylene chloride. LD$_{50}$ i.v. in mice: 17 mg/kg (Troxler, Seemann, 1977).

Hydrogen maleate, $C_{27}H_{32}N_2O_7$, LT 31-200, Sandonorm. THERAP CAT: Antihypertensive.

1336. Boric Acid. Boracic acid; orthoboric acid; Borofax. BH_3O_3; mol wt 61.84. B 17.50%, H 4.88%, O 77.62%. H_3BO_3. Occurs in nature as the mineral *sassolite*. Manuf: Faith, Keyes & Clark's *Industrial Chemicals*, F. A. Lowenheim, M. K. Moran, Eds. (Wiley-Interscience, New York, 4th ed., 1975) pp 153-158.

Colorless, odorless, transparent crystals, or white granules or powder; slightly unctuous to the touch. mp about 171°. Phase diagram for the $B_2O_3 \cdot H_2O$ system: Kracek *et al., Am. J. Sci.* **35A**, 143 (1938). Volatile with steam. pH: 5.1 (0.1 molar). One gram dissolves in 18 ml cold, 4 ml boiling water, in 18 ml cold, 6 ml boiling alcohol, in 4 ml glycerol; soly in water is increased by HCl, citric or tartaric acids. Soly of boric acid in glycerol solns of various concns: Sciarra, Elliott, *J. Am. Pharm. Assoc. (Sci. Ed.)* **49**, 116 (1960). LD_{50} orally in rats: 5.14 g/kg, Smyth *et al.*, *Am. Ind. Hyg. Assoc. J.* **30**, 470 (1969).

Incompat. Alkali carbonates and hydroxides.

Human Toxicity: Ingestion or absorption may cause nausea, vomiting, diarrhea, abdominal cramps, erythematous lesions on skin and mucous membranes, circulatory collapse, tachycardia, cyanosis, delirium, convulsions, coma. Death has occurred from < 5 g in infants and from 5 to 20 g in adults. Chronic use may cause borism (dry skin, eruptions, gastric disturbances): E. Browning, *Toxicity of Industrial Metals* (Appleton-Century-Crofts, New York, 2nd ed., 1969) pp 90-97.

USE: For weatherproofing wood and fireproofing fabrics; as a preservative; manuf cements, crockery, porcelain, enamels, glass, borates, leather, carpets, hats, soaps, artificial gems; in nickeling baths; cosmetics; printing and dyeing; painting; photography; for impregnating wicks; electric condensers; hardening steel. Also used as insecticide for cockroaches and black carpet beetles.

THERAP CAT: Astringent, antiseptic.

THERAP CAT (VET): Antibacterial and antifungal. Used chiefly in aqueous solution or powders for external use.

1337. Boric Anhydride. *Boron oxide; boron trioxide; boric oxide; boron sesquioxide.* B_2O_3; mol wt 69.64. B 31.07%, O 68.93%. Improperly called *anhydrous boric acid* or *fused boric acid.* Prepn of crystalline form: McCulloch, *J. Am. Chem. Soc.* **59**, 2650 (1937).

Colorless, brittle, vitreous, semitransparent, hygroscopic lumps or hard, white crystals. d (amorph) 1.8; d (cryst) 2.46. mp (cryst) 450°. Slowly sol in 30 parts cold, or 5 parts boiling water; sol in alcohol, glycerol. *Keep dry.*

USE: In metallurgy; in analysis of silicates to determine SiO_2 and alkalies; in blowpipe analysis.

1338. Borneol. *endo-1,7,7-Trimethylbicyclo[2.2.1]heptan-2-ol; endo-2-bornanol; endo-2-camphanol; endo-2-hydroxycamphane; bornyl alcohol; Baros camphor; Sumatra camphor; Borneo camphor; Dryobalanops camphor; Bhimsaim camphor; Malayan camphor; camphol.* $C_{10}H_{18}O$; mol wt 154.24. C 77.86%, H 11.76%, O 10.37%. The dextrorotatory form which predominates occurs in the oil from *Dryobalanops aromatica* Gaertn., *Dipterocarpaceae*, and in many other plants; the levorotatory form comes from *Blumea balsamifera* (L.) DC., *Compositae:* E. Gildemeister, F. Hoffman, *Die Atherischen Ole* (Schimmel, Leipzig, 3rd ed., 1928) pp 475-481. Racemic borneol is prepd synthetically by reduction of camphor: Truett, Moulton, *J. Am. Chem. Soc.* **73**, 5913 (1951); Ziegler *et al.*, *Ann.* **623**, 9 (1959); Ziegler, *Brit.* pat. **803,178** (1958); from pinene: Schwyzer, *Pharm. Ztg.* **75**, 1275 (1930). Configuration (isoborneol = exo-form; borneol = endo-form): Toivonen *et al.*, *Acta Chem. Scand.* **3**, 991 (1949). *Review:* J. L. Simonsen, *The Terpenes* vol. II (University Press, Cambridge, 2nd ed., 1949) pp 349-365.

d-borneol

d-Form, hexagonal plates from petr ether, mp 208°. Peculiar peppery odor and burning taste somewhat resembling that of mint. Sublimes, but is less volatile than camphor. d_4^{20} 1.011. bp 212°. $[\alpha]_D^{20}$ +37.7° (c = 5 in alc); $[\alpha]_{546}^{22}$ +44.4° (c = 0.5 in toluene). Almost insol in water. Sol in alc (176 parts dissolve in 100 parts w/w of abs alc), ether, petr ether (about 1:6), benzene (about 1:5), toluene, acetone, decalin, tetralin. LD orally in rabbits: 2 g/kg.

l-Form, hexagonal plates, mp 204°. bp_{779} 210°. $[\alpha]_D^{20}$ −37.7° (c = 5 in alc); $[\alpha]_{546}^{22}$ −44.4° (c = 0.5 in toluene).

dl-Form, mp 206-207°.

USE: Primarily in the manuf of its esters. Some free borneol and isoborneol is used in perfumery and in incense making. *Caution:* May cause nausea, vomiting, mental confusion, dizziness, convulsions.

1339. Bornyl Acetate. *1,7,7-Trimethylbicyclo[2.2.1]heptan-2-ol acetate; borneol acetate.* $C_{12}H_{20}O_2$; mol wt 196.28. C 73.43%, H 10.27%, O 16.30%. $CH_3COOC_{10}H_{17}$. Prepn of the *d*-form by acylation of *d*-borneol: Shishido *et al.*, *J. Am. Chem. Soc.* **82**, 125 (1960).

d-Form, crystals, mp 29°, also reported as mp 26.5°, Considine, *J. Org. Chem.* **25**, 671 (1960). bp 225-226°, bp_8 92-93°. d 0.99. n_D^{22} 1.4623. $[\alpha]_D$ +44.38°, Haller, *Compt. Rend.* **109**, 29 (1889); $[\alpha]_D$ +41.2°, Considine, *loc. cit.*; α_D^{14} +44.72° (neat), Shishido, *loc. cit.* Very slightly sol in water; sol in alcohol, ether.

l-Form, crystals, mp 27°. bp_{14} 103°. $[\alpha]_D$ −44.45°, Haller, *loc. cit;* $[\alpha]_D$ −42.0°, Considine, *loc. cit.*

dl-Form, crystals, mp 7.0°, Considine, *loc. cit.*

1340. d-Bornyl α-Bromoisovalerate. *2-Bromo-3-methylbutanoic acid 1,7,7-trimethylbicyclo[2.2.1]hept-2-yl ester; 2-bromo-3-methylbutyric acid ester with d-borneol; borneol α-bromoisovalerate; d-bornyl 2-bromo-3-methylbutyrate; Brovalol; Eubornyl; Valisan.* $C_{15}H_{25}BrO_2$; mol wt 317.27. C 56.78%, H 7.94%, Br 25.19%, O 10.09%. $(CH_3)_2CHCHBr-COOC_{10}H_{17}$. Prepd from *d*-borneol and α-bromoisovaleryl chloride or α-bromovaleric acid with mineral acid: *Beilstein* vol. 6, p 79.

Oily liquid, bp 163°. d 1.18. Practically insol in water; sol in alc, chloroform, ether.

THERAP CAT: Sedative, hypnotic.

1341. Bornyl Chloride. *endo-2-Chloro-1,7,7-trimethylbicyclo[2.2.1]heptane;* pinene hydrochloride; 2-chlorobornane; 2-chlorocamphane; "terpene" hydrochloride; "turpentine camphor". $C_{10}H_{17}Cl$; mol wt 172.19. C 69.55%, H 9.92%, Cl 20.53%. Prepd from α-pinene: Zeiss, Zwanzig, *J. Am. Chem. Soc.* **79**, 1733 (1957); Hückel, Gelchsheimer, *Ann.* **625**, 12 (1959); by chlorination of camphane: Gandini, *Gazz. Chim. Ital.* **66**, 357 (1936). Configuration: Kwart, *J. Am. Chem. Soc.* **75**, 5942 (1953); Kwart, Null, *ibid.* **78**, 5943 (1956). *Review:* J. L. Simonsen, *The Terpenes* vol. II (University Press, Cambridge, 2nd ed., 1949) pp 340-349.

Crystals from alc, mp 132°. Odor resembling camphor. bp 207-208°. When prepd from optically active sources it has the same sign as the hydrocarbon from which it is prepd: Thurber, Thielke, *J. Am. Chem. Soc.* **53**, 1032 (1931). Practically insol in water; sol in alc, ether.

THERAP CAT: Antiseptic.

1342. d-Bornyl Isovalerate. *(1R-endo)-3-Methylbutanoic acid 1,7,7-trimethylbicyclo[2.2.1]hept-2-yl ester; isovaleric acid d-bornyl ester; borneol isovalerate; Bornyval.* $C_{15}H_{26}O_2$; mol wt 238.36. C 75.58%, H 11.00%, O 13.42%. $(CH_3)_2CH-CH_2COOC_{10}H_{17}$. Prepd from *d*-borneol and isovaleric acid: Siedler, *Pharm. Ztg.* **48**, 772 (1903); Vavon, Peignier, *Bull. Soc. Chim. France* **39**, 924 (1926).

Liquid; odor and taste of valerian and camphor. d 0.955. bp 255-260°. $[\alpha]_D^{20}$ +27-28°. Practically insol in water; sol in alcohol or ether.

THERAP CAT: Sedative.

1343. Bornyl Salicylate. *2-Hydroxybenzoic acid 1,7,7-trimethylbicyclo[2.2.1]hept-2-yl ester; salicylic acid bornyl ester; borneol salicylate; Salit.* $C_{17}H_{22}O_3$; mol wt 274.35. C 74.41%, H 8.08%, O 17.50%. $HOC_6H_4COOC_{10}H_{17}$. Prepd from borneol and salicylic acid: *Hager's Handb. Pharm. Praxis* **Band 1**, 210 (Berlin, 1930).

Brown, oily liquid. Slight odor and taste. Practically insol in water; miscible with alcohol, ether, chloroform, oils; slightly sol in glycerol; dec by alkalies and the intestinal fluids into its constituents.

THERAP CAT: Counterirritant.

1344. Boromycin. $C_{45}H_{74}BNO_{15}$; mol wt 879.91. C 61.43%, H 8.48%, B 1.23%, N 1.59%, O 27.27%. Antibiotic produced by *Streptomyces antibioticus* ETH 28829: Hütter *et al.*, *Helv. Chim. Acta* **50**, 1533 (1967). The first known natural product in which boron has been found. It is a complex of boric acid with a tetradentate organic complexing agent and yields D-valine, boric acid and a polyhydroxy macrolide-type compound upon hydrolysis. Structure: Dunitz *et al., ibid.* **54**, 1709 (1971). Mechanism of action studies: Pache, Zähner, *Arch. Mikrobiol.* **67**, 156 (1969). Activity as a coccidiostat: Miller, Burg, U.S. pat. **3,864,479** (1975 to Merck & Co.). Biosynthesis: T. S. S. Chen *et al., J. Org. Chem.* **46**, 2661 (1981). Partial synthesis: M. A. Avery *et al., Tetrahedron Letters* **22**, 3123 (1981). Synthesis of C-3′ to C-17′: S. Hanessian *et al., J. Am. Chem. Soc.* **103**, 6243 (1981); of C-3 to C-17: *eidem, Can. J. Chem.* **61**, 634 (1983); of C-1 to C-17: J. D. White *et al., J. Am. Chem. Soc.* **105**, 6517 (1983). Total synthesis: *eidem, ibid.* **111**, 790 (1989). *Review:* Pache in *Antibiotics* vol. 3, J. W. Corcoran, F. E. Hahn, Eds. (Springer-Verlag, New York, 1975) pp 585-587.

Colorless crystals from methanol, mp 223-228° (dec). $[\alpha]_D$ 63.5° (c = 0.55 in CHCl$_3$). No uv absorption between 210 and 400 nm. LD$_{50}$ orally in mice: 180 mg/kg (Hütter).

1345. Boron. B; at. no. 5; at. wt 10.81; valence 3. Two naturally-occurring isotopes: 10; 11; three short-lived, artificial isotopes: 8, 12, 13. Occurrence in the earth's crust about 0.001% in the form of its compounds, never as the element. First obtained by Moissan in 1895 by reduction of boric anhydride (B$_2$O$_3$) with magnesium in a thermite-type reaction: Moissan, *Ann. Chim. Phys.* [7] **6**, 296 (1895), still a good method for large quantities of relatively impure boron. Prepn of high purity crystalline boron by vapor phase reduction of boron trichloride with hydrogen on electrically heated filaments in a flow system: Stern, Lynds, *J. Electrochem. Soc.* **105**, 676 (1958). Reviews of prepn and properties of boron and its compds: *Boron, Metallo-Boron Compounds and Boranes,* R. M. Adams, Ed. (Interscience, New York, 1964) 765 pp; *The Chemistry of Boron and Its Compounds,* E. L. Muetterties, Ed. (John Wiley, New York, 1967) 699 pp; Greenwood, "Boron" in *Comprehensive Inorganic Chemistry* vol. 1, J. C. Bailar, Jr. *et al.*, Eds. (Pergamon Press, Oxford, 1973) pp 655-991; J. G. Bower in Kirk-Othmer *Encyclopedia of Chemical Technology* vol. 4 (Wiley-Interscience, New York, 3rd ed., 1978) pp 62-66. Review of synthesis and applications of vinylic organoboranes: H. C. Brown, J. B. Campbell, *Aldrichim. Acta* **14**, 3-11 (1981).

Polymorphic; α-rhombohedral form, clear red crystals, 12 atoms/unit cell, d 2.46; β-rhombohedral form, black, 105 atoms/unit cell, d 2.35; α-tetragonal form, black, opaque crystals with metallic luster, 50 atoms/unit cell, d 2.31; other crystal forms known but not entirely characterized. Amorphous form, black or dark brown powder, d 2.350. Crystals are almost as hard as diamond. mp ~2200°. Vapor pressure at 2413°K (2140°): 1.56 × 10^{-5} atm, Searcy, Myers, *J. Phys. Chem.* **61**, 957 (1957). Heat capacity at 25° of amorphous form: 2.858 cal/g-atom/°C; of β-rhombohedral form: 2.650 cal/g-atom/°C. Feeble conductor of electricity at room temp, good conductor at high temps. Admixture of traces of carbon improves conductivity. Self-limiting reaction with oxygen due to formation of B$_2$O$_3$ film; oxide coating evaporates above 1000°. Reacts with fluorine

at room temp. Insol in water. Unaffected by aq hydrochloric and hydrofluoric acids. When finely divided, it is sol in boiling nitric and sulfuric acids and in most molten metals, such as copper, iron, magnesium, aluminum, calcium. Reacts vigorously with fused sodium peroxide, or with a fusion mixture of sodium carbonate and potassium nitrate.

USE: In nuclear chemistry as neutron absorber, in Ignitron rectifiers, in alloys, usually to harden other metals.

1346. Boron Carbide. Norbide. CB$_4$; mol wt 55.29. C 21.72%, B 78.28%. B$_4$C. Usually prepd in an electric furnace at 2500° according to the equation 2B$_2$O$_3$ + 7C → B$_4$C + 6CO: Ridgway, *Trans. Electrochem. Soc.* **66**, 117-133 (1934); also formed by reducing boric anhydride with magnesium in the presence of carbon: Dawihl, **Ger.** pat. **752,324** (1942 to Krupp); *BIOS* rept. no. 925, p 22 (1947). Lab prepn by the reduction of boron trichloride with hydrogen in the presence of carbon or hydrocarbons: *Bell Labs. Record* **28**, 477 (1950). Comprehensive monograph: P. W. Gilles, *High Temperature Chemistry of the Binary Compounds of Boron,* Advances in Chemistry Series **no. 32**, (ACS, Washington, D.C., 1961).

Black shiny rhombohedra or octahedra. d$_4^{25}$ 2.508-2.512. mp 2350° (no decompn); bp > 3500°. Its hardness is less than that of industrial diamonds, but higher than the hardness of silicon carbide: ca 5,000 kg/mm^2, on Mohs' hardness scale = 9.3. Less brittle than most ceramics. Remarkably resistant to chemical action. Not attacked by hot HF, HNO$_3$ or HCrO$_4$. Decomposed by molten alkalis at red heat. Does not burn in oxygen flame.

USE: Abrasive. In the manuf of hard and chemicals-resistant ceramics or wear-resistant tools. Finely pulverized B$_4$C can be molded under (considerable) pressure and heat.

1347. Boron Monoxide. (BO)$_x$; B 40.34%, O 59.66%. Prepd in quantitative yield by dehydration of tetrahydroxyboron at 250° at reduced pressure: Wartik, Apple, *J. Am. Chem. Soc.* **77**, 6400 (1955); prepn and proposed structure: McCloskey *et al., ibid.* **83**, 4750 (1961).

Fluffy white solid. Hygroscopic, converts back to tetrahydroxyboron on reaction with water. One gram dissolves in 100 ml methanol, in 100 ml warm ethanol or 100 ml warm isopropyl alcohol. Practically insol in dimethylamine and methyl borate. Vaporizes at 1300-1500° into gaseous B$_2$O$_2$.

1348. Boron Nitride. BN; mol wt 24.83. B 43.58%, N 56.42%. Prepd by igniting compds of boron with compds of nitrogen: Taylor, U.S. pat. **2,855,316** (1958 to Carborundum Co.). *Reviews:* Giardini, *U.S. Bur. Mines, Inform. Circ.* No. **7664**, 13 pp (1953); K. Niedenzu, J. W. Dawson, *Boron-Nitrogen Compounds* (Academic Press, New York, 1965) pp 147-153.

Crystals with hexagonal, graphite lattice is most common form. *Borazon,* cubic crystalline modification, is probably the hardest substance known. There exists also an amorphous modification. mp 3000°. Begins to sublime at a temp slightly below 3000°. Begins to dissociate *in vacuo* at about 2700°. The chemical behavior of BN is dependent on the method of prepn. Not attacked by mineral acids, water; in general resistant to chemical attack. Hot concd alkali cleaves boron-nitrogen bond. Oxidation in air begins above 1200°. *See* Niedenzu, Dawson, *loc. cit.*

USE: Manuf of alloys; in semiconductors, nuclear reactors, lubricants.

1349. Boron Tribromide. BBr$_3$; mol wt 250.57. B 4.32%, Br 95.68%. Prepn: Gamble, *Inorg. Syn.* **3**, 27 (1950);

Becher in *Handbook of Preparative Inorganic Chemistry*, **Vol. 1**, G. Brauer, Ed. (Academic Press, New York, 2nd ed., 1963) p 781. Review of boron halides: Massey, *Advan. Inorg. Chem. Radiochem.* **10**, 1-152 (1967).

Colorless, fuming liquid; dec by water or alcohol. mp −46.0°; bp 90°. d^0 2.698. Vapor pressure data: Barber *et al.*, *J. Chem. Eng. Data* **9**, 137 (1964).

USE: Manuf of diborane; ultra high purity boron.

1350. Boron Trichloride. BCl_3; mol wt 117.19. B 9.23%, Cl 90.77%. Prepn: Gamble, *Inorg. Syn.* **3**, 27 (1950). *Reviews:* Gerrard, Lappert, *Chem. Rev.* **58**, 1081-1111 (1958); Massey, *Advan. Inorg. Chem. Radiochem.* **10**, 1-152 (1967).

Colorless, fuming liquid at low temp; dec by water or alcohol. bp 12.5°. mp −107°. d_4^{12} 1.35. d^0 1.3728: Ward, *J. Chem. Eng. Data* **14**, 167 (1969).

USE: Manuf and purification of boron; as catalyst for organic reactions; in semiconductors; in bonding of iron, steels; in purification of metal alloys to remove oxides, nitrides and carbides.

1351. Boron Trifluoride. BF_3; mol wt 67.82. B 15.95%, F 84.05%. A strong Lewis acid. Prepn: Swinehart, U.S. pats. **2,148,514, 2,196,907** (1939, 1940 to Harshaw Chemical); Booth, Wilson, *Inorg. Syn.* **1**, 21 (1939); Kwasnik in *Handbook of Preparative Inorganic Chemistry*, **Vol. 1**, G. Brauer, Ed. (Academic Press, New York, 2nd ed., 1963) pp 219-222; Wiesboeck, U.S. pat. **3,690,821** (1972 to U.S. Steel). *Dihydrate:* McGrath *et al.*, *J. Am. Chem. Soc.* **66**, 1263 (1944). *Reviews:* Booth, Martin, *Boron Trifluoride and Its Derivatives* (John Wiley & Sons, 1949), 296 pp; Booth in *Fluorine Chemistry*, **Vol. 1**, J. Simons, Ed. (Academic Press, New York, 1950) pp 201-224; Topchiev *et al.*, *Boron Fluoride and Its Compounds as Catalysts in Organic Chemistry* (Pergamon Press, 1959) 326 pp; Martin in Kirk-Othmer, *Encyclopedia of Chemical Technology*, **Vol. 9** (Interscience, New York, 2nd ed., 1966) pp 554-562; Massey, *Advan. Inorg. Chem. Radiochem.* **10**, 1-152 (1967).

Colorless gas. Pungent, suffocating odor. *Corrosive to skin. Avoid inhalation!* Forms dense white fumes in moist air. bp −127.1°. bp −100.4°. d_4 (−100.4°; liq) 1.57. d (gas at STP) 3.07666 g/l. Soly in water (0°): 332 g/100 g; some hydrolysis occurs to form fluoboric and boric acids. Soly in anhydrous H_2SO_4: 1.94 g/100 g acid. Forms solid complex with nitric acid $(HNO_3.2BF_3)$. Sol in most saturated and halogenated hydrocarbons and in aromatic compds. Polymerizes unsaturated molecules. Easily forms coordination complexes with molecules having at least one unshared pair of electrons. Reacts with incandescence when heated with alkali metals or alkaline earth metals except magnesium.

USE: To protect molten magnesium and its alloys from oxidation; as a flux for soldering magnesium; as a fumigant; in ionization chambers for the detection of weak neutrons. By far the largest application of boron trifluoride is in catalysis with and without promoting agents. *Caution:* May be irritating to eyes, mucous membranes.

1352. Boron Trifluoride Etherate. *Boron fluoride ethyl ether;* boron fluoride etherate; ethyl ether-boron trifluoride complex. $C_4H_{10}BF_3O$; mol wt 141.94. C 33.85%, H 7.10%, B 7.62%, F 40.16%, O 11.27%. $(CH_3CH_2)_2O.BF_3$. Prepd by vapor-phase reaction of anhydr ether with BF_3: Laubengayer, Finlay, *J. Am. Chem. Soc.* **65**, 884 (1943).

Fuming liquid, immediately hydrolyzed by moisture in air. d_4^{25} 1.125. bp 125.7°. mp −60.4°. n_D^{20} 1.348. Heat of formation: 12.5 kcal. Heat of soln at 0° in ether: 2.7 kcal.

USE: Catalyst in acetylation, alkylation, polymerization, dehydration, and condensation reactions. *Caution:* On decomposition forms highly toxic fumes of fluorides.

1353. Bostrycoidin. *6,9-Dihydroxy-7-methoxy-3-methylbenz[g]isoquinoline-5,10-dione;* 5,8-dihydroxy-6-methoxy-3-methyl-2-aza-9,10-anthraquinone. $C_{15}H_{11}NO_5$; mol wt 285.25. C 63.16%, H 3.89%, N 4.91%, O 28.04%. Antibiotic substance produced by *Fusarium bostrycoides:* Hamilton *et al.*, *Antibiot. & Chemother.* **3**, 853 (1953); Cajori *et al.*, *J. Biol. Chem.* **208**, 107 (1954). Structure: Arsenault, *Tetrahedron Letters* **1965**, 4033. Synthesis: D. W. Cameron *et al.*, ibid. **21**, 5089 (1980).

Dark red crystals; mp 243-244°, changes to purple in alkaline medium, yellow in acid medium. Stable at room temp and withstands autoclaving. Insol in water; sol in abs ethanol and in 60% ethanol; moderately sol in dioxane, benzene, acetone, chloroform, carbon tetrachloride. Soluble in aq sodium carbonate solns. Partly sol in corn oil. Active *in vitro* against *Mycobacterium tuberculosis*. Serum seems to interfere.

1354. β-Boswellic Acid. *3-Hydroxyurs-12-en-23-oic acid.* $C_{30}H_{48}O_3$; mol wt 456.68. C 78.89%, H 10.59%, O 10.51%. Occurs as the acetate in frankincense *(olibanum)* from *Boswellia carterii, Burseraceae*. The β-form is predominant and is accompanied by small amounts of α- and γ-boswellic acid. Isoln from olibanum tears: Winterstein, Stein, *Z. Physiol. Chem.* **208**, 9 (1932); Beton *et al.*, *J. Chem. Soc.* **1956**, 2904. Early structural studies: Simpson, Williams, *ibid.* **1938**, 686, 1712; Ruzicka, Wirz, *Helv. Chim. Acta* **22**, 948 (1939); **23**, 132 (1940); Ruzicka *et al.*, *ibid.* **27**, 1859 (1944). Revised structure and stereochemistry: Beton *et al.*, *loc. cit.*; Allan, *Chimia* **17**, 382 (1963); *idem, Phytochemistry* **7**, 963 (1968). Review: J. Simonsen, W. C. J. Ross, *The Terpenes*, **vol. 5** (University Press, Cambridge, 1957) pp 68-74.

Long prisms from methanol, mp 228-232° with preliminary sintering. $[\alpha]_D$ +107° (c = 0.75 in $CHCl_3$), Beton, *loc. cit.* 100 ml of boiling methanol will dissolve 8 grams of β-boswellic acid. Sol in chloroform, ether, acetone, alc.

Acetate, $C_{32}H_{50}O_4$, prisms, mp 275-278°, $[\alpha]_D$ +63° (c = 1.88 in $CHCl_3$).

Methyl ester, $C_{31}H_{50}O_3$, mp 195-196°, $[\alpha]_D$ +111° (c = 1.6 in $CHCl_3$).

1355. Bottromycin. B-mycin. A complex of five antibiotics of which the main active component is bottromycin A_2. Produced by *Streptomyces bottropensis* and *S. canadensis*: Waisvisz *et al.*, *J. Am. Chem. Soc.* **79**, 4520, 4522, 4524 (1957); Miller *et al.*, *Antimicrob. Ag. Chemother.* **1967**, 407. Production: **Brit. pat. 762,736** (1956 to Koninklijke Nederlandsche Gist en Spiritus-Fabriek); Umesawa *et al.*, **Japan. pat. 10,998('68)** (1968 to Microbiochem. Res. Found.), *C.A.* **69**, 85439x (1968); Hata *et al.*, **U.S. pat. 3,650,904** (1972). Partial structure: Waisvisz, Hoeven, *J. Am. Chem. Soc.* **80**, 38 (1958). Sepn and structures of bottromycins A_1, A_2, and B: Nakamura *et al.*, *J. Antibiot.* **18A**, 47, 60 (1965); **19A**, 10 (1966). Isoln and characterization of bottromycins A_2, B_2, C_2: Nakamura *et al.*, *ibid.* **20A**, 1 (1967). Revised structure of bottromycin A_2: Takahashi *et al.*, *ibid.* **29A**, 1120 (1976); D. Schipper, *ibid.* **36A**, 1076 (1983). Mode of action of bottromycin A_2: T. Otaka, A. Kaji, *J. Biol. Chem.* **251**, 2299 (1976); *idem, FEBS Letters* **123**, 173 (1981).

bottromycin A$_2$

Bottromycin A$_2$, C$_{42}$H$_{62}$N$_8$O$_7$S, *bottromycic A$_2$ acid methyl ester.*

1356. Bradykinin. Kallidin I; kallidin-9; callidin I; BRS 640. C$_{50}$H$_{73}$N$_{15}$O$_{11}$; mol wt 1060.25. C 56.64%, H 6.94%, N 19.82%, O 16.60%. A tissue hormone belonging to a group of hypotensive peptides known as plasma kinins. First obtained by incubation with the venom of *Bothrops jararaca* or with crystalline trypsin: Rocha e Silva *et al., Cien Cult. (Sao Paulo)* **1**, 32 (1949); *eidem, Am. J. Physiol.* **156**, 261 (1949); Prado *et al., Arch. Biochem.* **27**, 410 (1950); Werle *et al., Biochem. Z.* **320**, 372 (1950). Large scale prepn from whole bovine plasma: Hamberg, Deutsch, *Arch. Biochem. Biophys.* **76**, 262 (1958). Formed by proteolysis of a precursor in the globulin fraction of plasma referred to as kininogen by the action of enzymes such as trypsin, plasmin, and plasma kallikrein, *q.q.v.* Acts on smooth muscle, dilates peripheral vessels, increases capillary permeability. Also is a potent pain-producing agent. Structure: Elliott *et al., Biochem. Biophys. Res. Commun.* **3**, 87 (1960); Werle *et al., Z. Physiol. Chem.* **326**, 174 (1961). Synthesis: Boissonnas *et al., Helv. Chim. Acta* **43**, 1349 (1960); **45**, 170 (1962); Merrifield, *Biochemistry* **3**, 1385 (1964); Young *et al., J. Chem. Soc. (C)* **1971**, 46; Bajusz *et al., Hung. pat.* 3840 (1972 to Gyogyszerkutato Intézet), *C.A.* **77**, 20027g (1972); Sipos, *Ger. pat.* 2,212,787 corresp to U.S. pat 3,714,140 (1972, 1973 both to Squibb); Corley *et al., Biochem. Biophys. Res. Commun.* **47**, 1353 (1972); N. S. S. Kumari *et al., Indian J. Chem.* **17B**, 152 (1979). *Review:* Schröder, Hempel, *Experientia* **20**, 529 (1964).

Arg-Pro-Pro-Gly-Phe-Ser-Pro-Phe-Arg

Amorphous precipitate. $[\alpha]_D^{25}$ −76.5° (c = 1.37 in 1N acetic acid). Sol in glacial acetic acid, in a 10% soln of trichloroacetic acid, 70% ethanol, in hot methanol; less sol in 90% ethanol or cold methanol. Almost insol in acetone, chloroform, ethyl ether, ethyl methyl ketone, petr ether, butanol, amyl alcohol, ethyl acetate.
THERAP CAT: Vasodilator.

1357. Brallobarbital. 5-(2-*Bromo-2-propenyl*)-5-(2-*propenyl*)-2,4,6(1H,3H,5H)-*pyrimidinetrione*; 5-*allyl*-5-(2-*bromoallyl*)*barbituric acid;* Vesperone. C$_{10}$H$_{11}$BrN$_2$O$_3$; mol wt 287.14. C 41.83%, H 3.86%, Br 27.83%, N 9.76%, O 16.72%. Prepn: Morren, *Belg. pat.* 497,501 (1950), *C.A.* **49**, 1100e (1955). Metabolism: Keding, Schmidt, *Arzneimittel-Forsch.* **19**, 342 (1969).

Solid, mp 168-169°.
Note: This is a controlled substance (depressant) listed in the U.S. Code of Federal Regulations, Title 21 Part 1308.13 (1987).
THERAP CAT: Sedative, hypnotic.

1358. Brandy. A potable alcoholic liquid distilled from wine or from the fermented juices of peaches, cherries, apples, or other fruit. To meet the specifications of the National Formulary, brandy must have been obtained by distillation of fermented juice from sound ripe grapes and contain between 48 and 54% ethanol (v/v), d$_4^{15}$ 0.921-0.933. It must have been stored in wooden containers for a period of not less than 2 years. Unlike whisky, brandy is never made from cereal mash or from potatoes (as some brands of vodka).
THERAP CAT: Sedative, peripheral vasodilator.

1359. Brassidic Acid. *trans-13-Docosenoic acid.* C$_{22}$H$_{42}$O$_2$; mol wt 338.58. C 78.08%, H 12.46%, O 9.45%. CH$_3$-(CH$_2$)$_7$CH=CH(CH$_2$)$_{11}$COOH. Prepn by isomerization of its *cis*-isomer (erucic acid): Reimer, Will, *Ber.* **19**, 3320 (1886); Skellon, *Mfg. Chemist* **33**, 405 (1962). Synthesis and separation from erucic acid by crystn: Bowman, *Nature* **163**, 95 (1949).
Thin, cryst plates from alc, mp 61-62°. bp$_{30}$ 282°. n$_D^{57}$ 1.448. Practically insol in water; sparingly sol in cold alcohol; sol in ether. Solubility at 10°, 0°, −10°, −20°, and −30° in methanol, ethyl acetate, ether, acetone, toluene, *n*-heptane: Kobl, *Diss. Abstr.* **20**, 82 (1959).

1360. Brassinolide. (2α,3α,5α,22R,23R,24S)-2,3,22,23-*Tetrahydroxy-B-homo-7-oxaergostan-6-one;* 2α,3α,22,23-tetrahydroxy-24-methyl-B-homo-7-oxa-5α-cholestan-6-one. C$_{28}$H$_{48}$O$_6$; mol wt 480.69. C 69.96%, H 10.06%, O 19.97%. Plant hormone; natural steroid containing a seven-membered B-ring lactone, that promotes both cell elongation and cell division. Over ten brassinosteroids have been isolated and characterized from sources such as pollen, seedling, leaf. Isoln, structure and activity of brassinolide from rape pollen, *Brassica napus* L.: M. D. Grove *et al., Nature* **281**, 216 (1979). Stereoselective synthesis: S. Fung, J. B. Siddall, *J. Am. Chem. Soc.* **102**, 6580 (1980); M. Ishiguro *et al., Chem. Commun.* **1980**, 962; J. R. Donaubauer *et al., J. Org. Chem.* **49**, 2833 (1984); S. Takatsuto *et al., J. Chem. Soc. Perkin Trans. I* **1984**, 139. Synthesis of two stereoisomers: M. J. Thompson *et al., J. Org. Chem.* **44**, 5002 (1979). Structure-activity relationship of brassinosteroids: S. Takatsuto *et al., Phytochem.* **22**, 2437 (1983); interaction with cytokinin: C. Schlagnhaufer *et al., Physiol. Plant.* **60**, 347 (1984); bioassay: K. Wada *et al., Agr. Biol. Chem.* **48**, 719 (1984).

Crystals from methanol, mp 274-275°. $[\alpha]_D^{27}$ +16°.
USE: Plant growth regulator.

1361. Brayera. Kousso; kosso; cusso; koso; cousso; kouso; kusso. Dried panicles of pistillate flowers of *Hagenia abyssinica* J. F. Gmel. (*Brayera anthelmintica* Kunth), Rosaceae. Habit. Abyssinia. *Constit.* Kosin, kossein, kosidin, protokosin, kosotoxin, volatile oil, tannin. *Ref:* Leisenring, *Arch. Pharm.* **232**, 50 (1894); Kondakow, *ibid.* **237**, 481 (1899); Hess, Todd, *J. Chem. Soc.* **1937**, 562; Birch, Todd, *ibid.* **1952**, 3102.
THERAP CAT: Anthelmintic (Cestodes).
THERAP CAT (VET): Has been used as a teniacide.

1362. Brazilin. 7,11b-*Dihydrobenz[b]indeno[1,2-d]pyran-3,6a,9,10(6H)-tetrol;* brasilin; C.I. Natural Red 24; C.I. 75280. C$_{16}$H$_{14}$O$_5$; mol wt 286.27. C 67.13%, H 4.93%, O 27.94%. (May crystallize as the mono- or hemihydrate). From *Caesalpinia echinata* Lam. (Brazil-wood), or *C. sappan* L. (sappan-wood), Leguminosae. Isoln and structure: Perkin *et al., J. Chem. Soc.* **1928**, 1504; Pfeiffer *et al., Ber.* **63**, 1301 (1930). Synthesis of (±)-form: Dann, Hofmann, *Ann.* **667**, 116 (1963); Kirkiacharian, Billet, *Bull. Soc. Chim.*

France **1972**, 3292. Synthesis, resolution: Morsingh, Robinson, *Tetrahedron* **26**, 281 (1970). Stereochemistry: Craig *et al., J. Org. Chem.* **30**, 1573 (1965); *Colour Index* vol. 4 (3rd ed., 1971) p 4628. *Review:* Robinson, *Bull. Soc. Chim. France* **1958**, 125-134.

Amber-yellow crystals; turn orange in air and light. Dec above 130°. Sol in water, freely in alcohol, ether, also in alkali hydroxide solns with carmine-red color. *Protect from air and light.*

USE: Chiefly as a dye. Has also been recommended as indicator in acid-base titrations; acids = yellow, alkalies = carmine-red.

1363. Brefeldin A. *1,6,7,8,9,11a,12,13,14,14a-Decahydro-1,13-dihydroxy-6-methyl-4H-cyclopent[f]oxacyclotridecin-4-one; γ,4-dihydroxy-2-(6-hydroxy-1-heptenyl)-4-cyclopentanecrotonic acid λ-lactone; ascotoxin; cyanein;* decumbin. $C_{16}H_{24}O_4$; mol wt 280.37. C 68.54%, H 8.63%, O 22.83%. A fungal metabolite which is a macrocyclic lactone exhibiting a wide range of antibiotic activity. Produced by *Penicillium brefeldianum* Dodge: E. Haerri *et al., Helv. Chim. Acta* **46**, 1235 (1963). Also produced by *P. decumbens:* V. L. Singleton *et al., Nature* **181**, 1072 (1958); *P. cyaneum:* V. Betina *et al., Folia Microbiol.* **7**, 353 (1962). Structure: H. P. Sigg, *Helv. Chim. Acta* **47**, 1401 (1964). Abs configuration: H. P. Weber *et al., ibid.* **54**, 2763 (1971). Synthesis of (±)-form: E. J. Corey, R. H. Wollenberg, *Tetrahedron Letters* **1976**, 4705; E. J. Corey *et al., ibid.* **1977**, 2243; R. Baudouy *et al., ibid.* 2973; P. A. Bartlett, F. R. Green, *J. Am. Chem. Soc.* **100**, 4548 (1978); A. E. Greene *et al., ibid.* **102**, 7583 (1980); M. Honda *et al., Tetrahedron Letters* **1981**, 2679. Total synthesis of (+)-form: T. Kitahara *et al., ibid.* **1979**, 3021. Biosynthesis: B. E. Cross, P. Hendley, *Chem. Commun.* **1975**, 124; C. R. Hutchinson *et al., J. Am. Chem. Soc.* **103**, 2474, 2477 (1981); M. Sunagawa *et al., J. Antibiot.* **36**, 25 (1983). Antifungal activity: V. Betina *et al., ibid.* **17A**, 93 (1964); anti-HeLa cell effect: *eidem, Naturwiss.* **49**, 241 (1962). *See also* W. Keller-Schierlein, "Chemistry of Macrolide Antibiotics" in *Fortschr. Chem. Org. Naturst.* **30**, 313-445 (1973).

Colorless prisms from methanol/ether, mp 204-205°. $[\alpha]_D^{22}$ +96±2° (c = 1.08 in methanol). uv max (ethanol): 215 nm (log ε 4.05). LD_{50} in mice: >200 mg/kg i.p. (Haerri).

1364. Bretylium Tosylate. *2-Bromo-N-ethyl-N,N-dimethylbenzenemethanaminium 4-methylbenzenesulfonate; (o-bromobenzyl)ethyldimethylammonium p-toluenesulfonate; N-ethyl-N-o-bromobenzyl-N,N-dimethylammonium tosylate;* Bretylan; Bretylate; Bretylol; Darenthin; Ornid. $C_{18}H_{24}$-$BrNO_3S$; mol wt 414.37. C 52.18%, H 5.84%, Br 19.28%, N 3.38%, O 11.58%, S 7.74%. Prepn: Copp, Stephenson, U.S. pat. **3,038,004** (1962 to Burroughs Wellcome). Metabolism: R. Kuntzman *et al., Clin. Pharmacol. Ther.* **11**, 829 (1970). Review of pharmacology: R. H. Heissenbuttel, J. T. Bigger, *Ann. Intern. Med.* **91**, 229-238 (1979); R. J. Lee *et al.,* in *Pharmacological and Biochemical Properties of Drug Substances* vol. 2, M. E. Goldberg, Ed. (Am. Pharm. Assoc., Washington, DC, 1979) pp 148-164. Comprehensive description: J. E. Carter *et al.,* in *Analytical Profiles of Drug Substances* vol. 9, K. Florey, Ed. (Academic Press, New York, 1980) pp 71-86.

Crystalline powder, mp 97-99°. Extremely bitter taste. uv max: 278, 271, 264 nm. Freely sol in water, methanol, ethanol; practically insol in ether, ethyl acetate, hexane. LD_{50} orally in mice: 400 mg/kg; i.m. in rats: 250 mg/kg, Goldenthal, *Toxicol. Appl. Pharmacol.* **18**, 185 (1971).

THERAP CAT: Anti-adrenergic; cardiac depressant (anti-arrhythmic).

1365. Brevetoxins. BTX. Structurally unique neurotoxins produced by the "red-tide" dinoflagellate *Ptychodiscus brevis* Davis (*Gymnodinium breve* Davis). Dense growths of these algae have been responsible for massive fish kills, mollusk poisoning and human food poisoning in the Gulf of Mexico and along the Florida coast, *cf. Marine Natural Products,* P. J. Scheuer, Ed. (Academic Press, New York, 1978). Unlike previously isolated dinoflagellate toxins, such as saxitoxin, *q.v.,* which are water-soluble sodium channel blockers, the brevetoxins are lipid-soluble sodium channel activators. Isoln of brevetoxins A, B, and C and structure of B, the major component: Y. Y. Lin *et al., J. Am. Chem. Soc.* **103**, 6773 (1981). Structure of C: J. Golik *et al., Tetrahedron Letters* **23**, 2535 (1982). Structure of A, the most potent toxin: Y. Shimizu *et al., J. Am. Chem. Soc.* **108**, 514 (1986). Synthetic approaches to brevetoxin B: K. C. Nicolau *et al., Chem. Commun.* **1985**, 1359. Absolute configuration: Y. Shimizu *et al., Chem. Commun.* **22**, 1656 (1987). Biosynthetic study: H. N. Chou, Y. Shimizu, *J. Am. Chem. Soc.* **109**, 2184 (1987). Series of articles on pharmacology of brevetoxins: *Toxicon* **23**, 469-524 (1985). Review of chemistry: K. Nakanishi, *ibid.* 473. For additional information on red tide algae, *see Toxic Dinoflagellate Blooms,* D. L. Taylor, H. H. Seliger, Eds. (Elsevier, New York, 1979) pp 327-354.

brevetoxin B : R = —CH_2—$\overset{CH_2}{\underset{}{C}}$CHO

brevetoxin C : R = —CH_2—$\overset{CH_2}{\underset{}{C}}$$CH_2Cl$

Brevetoxin A, $C_{49}H_{70}O_{13}$, *GB-1.* Fine prisms from acetonitrile, mp 197-199°; 218-220° (double melting point). LC_{100} in guppies: 4 ng/ml (Shimizu).

Brevetoxin B, $C_{50}H_{70}O_{14}$, *BTX-B, GB-2, T 34, T 47.* Needles from acetonitrile, mp 270° (dec). uv max (methanol): 208 nm (ε 16000, enal). LC_{50} (1 hr) in fresh water "zebra" fish, *Brachydanio rerio:* 16 ng/ml (Lin).

Brevetoxin C, $C_{49}H_{69}ClO_{14}$, *BTX-C.* uv max (methanol): 208 nm (ε 11300, ene-lactone). LC_{50} (1 hr) in fresh water "zebra" fish: 30 ng/ml (Lin).

USE: As tools in neurochemical research.

1366. Brilliant Blue FCF. *N-Ethyl-N-[4-[[4-[ethyl[(3-sulfophenyl)methyl]amino]phenyl](2-sulfophenyl)methylene]-2,5-cyclohexadien-1-ylidene]-3-sulfobenzenemethanaminium hydroxide inner salt, disodium salt;* FD & C Blue No. 1; C.I. Acid Blue 9; C.I. Food Blue 2; C.I. 42090. $C_{37}H_{34}N_2Na_2O_9$-S_3; mol wt 792.85. C 56.05%, H 4.32%, N 3.53%, Na 5.80%, O 18.16%, S 12.13%. Discovered by Sandmeyer in 1896: *Colour Index* vol. 4 (3rd ed., 1971) p 4385. Also prepared as the diammonium salt. Metabolism: S. M. Hess, O. G. Fitzhugh, *J. Pharmacol. Exp. Ther.* **114**, 38 (1955); J. P. Brown *et al., Food Cosmet. Toxicol.* **18**, 1 (1980). Toxicology: W. A. Mannell, H. C. Grice, *J. Pharm. Pharmacol.* **16**, 56

(1964); W. H. Hansen *et al., Toxicol. Appl. Pharmacol.* **8**, 29 (1966). Review of carcinogenicity studies: *IARC Monographs* **16**, 171-186 (1978).

Reddish-violet powder or granules with a metallic lustre. Absorption max: 630 nm. Sol in water, ethanol; insol in vegetable oils. Pale amber soln in conc H_2SO_4, changing to yellow then greenish blue on dilution. LD_{50} s.c. in mice: 4.6 g/kg, E. Gross, *Z. Krebsforsch.* **64**, 287 (1961).

USE: Approved by FDA for use in food, drugs and cosmetics excluding use in eye area: *Fed. Regist.* **47**, 42563 (1982). Biological stain; textile dye; wood stain; indicator.

1367. Brilliant Green. *N-[4-[[4-(Diethylamino)phenyl]-phenylmethylene]-2,5-cyclohexadien-1-ylidene]-N-ethylethanaminium sulfate (1:1);* C.I. Basic Green 1; C.I. 42040; Malachite Green G; Ethyl Green; Emerald Green; Diamond Green G; Fast Green J; Solid Green. $C_{27}H_{34}N_2O_4S$; mol wt 482.64. C 67.19%, H 7.10%, N 5.80%, O 13.26%, S 6.64%. Prepn: Doebner, *Ber.* **13**, 2222 (1880); Fischer, *ibid.* **14**, 2521 (1881); *Colour Index,* **vol. 4** (3rd ed., 1971) p 4382. Toxicity: Anderson, *Proc. Soc. Exp. Biol. Med.* **31**, 825 (1934). Antibacterial activity: O. H. Paetzold, *Arch. Klin. Exp. Dermatol.* **224**, 90 (1966); *idem, Arch. Dermatol. Forsch.* **243**, 1 (1972). *Review:* H. J. Conn's *Biological Stains,* R. D. Lillie, Ed. (Williams & Wilkins, Baltimore, 9th ed., 1977) pp 251-252, 580.

Minute, glistening, golden crystals. Soluble in water or alcohol with green color. Absorption max: 623 nm. Soln changes color from yellow to green at pH 0.0 to 2.6. LD_{100} i.v. in mice: 3 mg/kg (Anderson).

USE: Dyeing silk, wool, leather, jute and cotton yellowish-green; manuf green ink; biological stain; indicator.

THERAP CAT: Antiseptic.

THERAP CAT (VET): Antiseptic for external and internal (oral) use. In wounds and scours.

1368. Brodifacoum. *3-[3-(4'-Bromo[1,1'-biphenyl]-4-yl)-1,2,3,4-tetrahydro-1-naphthalenyl]-4-hydroxy-2H-1-benzopyran-2-one;* PP-581; WBA 8119; Talon; Ratak+. $C_{31}H_{23}BrO_3$; mol wt 523.44. C 71.13%, H 4.43%, Br 15.27%, O 9.17%. Prepn: M. R. Hadler, R. S. Shadbolt, **Ger.** pat. **2,424,806** corresp to **U.S.** pat. **3,957,824** (1975, 1976 to Ward, Blenkinsop & Co.); R. S. Shadbolt *et al., J. Chem. Soc. Perkin Trans. I* **1976**, 1190. Anticoagulant activity: M. R. Hadler, R. S. Shadbolt, *Nature* **253**, 275 (1975). Field trials: B. D. Rennison, A. C. Dubock, *J. Hyg.* **80**, 77 (1978); F. P. Rowe *et al., ibid.* **81**, 197 (1978).

Off-white powder, mp 228-230°. Insol in water. Slightly sol in alc, benzene; sol in acetone, chloroform. LD_{50} orally in rats: 270 μg/kg, *RTECS* **Vol. I**, R. J. Lewis, R. L. Tatken, Eds. (1980) p 541.

USE: Rodenticide.

1369. Brodimoprim. *5-[(4-Bromo-3,5-dimethoxyphenyl)methyl]-2,4-pyrimidinediamine;* 2,4-diamino-5-(4-bromo-3,5-dimethoxybenzyl)pyrimidine; Ro 10-5970. $C_{13}H_{15}$-BrN_4O_2; mol wt 339.19. C 46.03%, H 4.46%, Br 23.56%, N 16.52%, O 9.43%. Dihydrofolate reductase inhibitor; structural analog of trimethoprim, *q.v.* Prepn: M. Hoffer, I. Kompis, **Ger.** pat. **2,452,889** (1975 to Hoffmann-La Roche), *C.A.* **83**, 97361 (1975); I. Kompis, **U.S.** pat. **4,024,145** (1977 to Hoffmann-La Roche); I. Kompis, A. Wick, *Helv. Chim. Acta* **60**, 3025 (1977). Antibacterial activity: G. Giammanco *et al., Drugs. Exptl. Clin. Res.* **9**, 721 (1983). Comparison with trimethoprim as inhibitor of dihydrofolate reductase: R. L. Then *et al., Rev. Infect. Dis.* **4**, 372 (1982); R. L. Then, F. Hermann, *Chemotherapy (Basel)* **30**, 18 (1984). Antimycobacterial activity *in vitro:* J. K. Seydel *et al., ibid.* **29**, 249 (1983). Penetration into canine bone tissue: P. Iverson, P. O. Madsen, *Acta Pharmacol. Toxicol.* **51**, 446 (1982). Pharmacokinetics in human serum, skin blister fluid: T. Kalager *et al., Chemotherapy (Basel)* **31**, 405 (1985). Clinical evaluation in respiratory infections: H. A. Salmi *et al., Drugs Exptl. Clin. Res.* **12**, 349 (1986).

Crystals from methanol, mp 225-228°. pKa 7.15.

THERAP CAT: Antibacterial.

1370. Bromacil. *5-Bromo-6-methyl-3-(1-methylpropyl)-2,4(1H,3H)-pyrimidinedione; 5-bromo-3-sec-butyl-6-methyluracil;* 5-bromo-6-methyl-3-(1-methylpropyl)uracil; du Pont herbicide 976; Hyvar; Uragon; Urox B. $C_9H_{13}BrN_2O_2$; mol wt 261.11. C 41.40%, H 5.02%, Br 30.60%, N 10.73%, O 12.25%. Prepn: Loux, **U.S.** pat **3,235,357** (1966 to du Pont). *Review:* Pease, Deye, *Anal. Methods Pestic., Plant Growth Regul., Food Additives* **5**, 335 (1967). Toxicology: H. Sherman, A. M. Kaplan, *Toxicol. Appl. Pharmacol.* **34**, 189 (1975).

White crystalline solid, mp 157.5-160°. Vapor press. at 100°: 8 × 10⁻⁴ mm Hg. Soly in water at 20°: 815 mg/l. Moderately sol in strong aq bases, acetone, acetonitrile, ethanol. LD_{50} orally in rats: 5200 mg/kg, H. Sherman, A. M. Kaplan, *loc. cit.*

USE: Herbicide.

1371. Bromadiolone. *3-[3-(4'-Bromo[1,1'-biphenyl]-4-*

yl)-3-hydroxy-1-phenylpropyl]-4-hydroxy-2H-1-benzopy-ran-2-one; 3-[α-[p-(p-bromophenyl)-β-hydroxyphenethyl]-benzyl]-4-hydroxycoumarin; LM-637; Maki; Bromone; Super-Caid; Super-Rozol. $C_{30}H_{23}BrO_4$; mol wt 527.42. C 68.32%, H 4.40%, Br 15.15%, O 12.13%. Anticoagulant rodenticide. Prepn: E. Boschetti *et al.*, **Ger. pat. 1,959,317;** *eidem,* **U.S. pat. 3,764,693** (1970, 1973 both to Lipha). Activity studies: *eidem, Chim. Ther.* **7,** 20 (1972); M. Grand, *Phytiat.-Phytopharm.* **25,** 69 (1976); R. E. Marsh, *Bull. OEPP* **7,** 495 (1977); R. Redfern, J. E. Gill, *J. Hyg.* **84,** 263 (1980). Pharmacokinetics: K. Nahas, *Pharmacol. Res. Commun.* **19,** 767 (1987). HPLC determn of diastereoisomers in animal tissues: K. Hunter *et al., J. Chromatog.* **435,** 83 (1988).

White to offwhite powder, mp 200-210°. UV max (ethanol): 260 nm ($E_{1cm}^{1\%}$ 538-582). pKa (21°) 4.04. Soly at 20-25° (g/l): dimethylformamide 730.0; ethyl acetate 25.0; acetone 22.3; chloroform 10.1; ethanol 8.2; methanol 5.6; ethyl ether 3.7; hexane 0.2; water 0.019. LD_{50} in rats, mice (mg/kg): 1.125, 1.75 orally (Grand).
USE: Rodenticide.

1372. Bromal. *Tribromoacetaldehyde.* C_2HBr_3O; mol wt 280.78. C 8.55%, H 0.36%, Br 85.39%, O 5.70%. Br_3CCHO. Preparation from ethanol and bromine: Löwig, *Ann.* **3,** 288 (1832); from chloral and a bromide: Müller, **U.S. pat. 2,053,964** (1936 to Winthrop); from paraldehyde and bromine: Long, Howard, *Org. Syn.* **17,** 18 (1937).

Yellowish, oily liquid; forms with water bromal hydrate which is solid at temps below 50°. d 2.66. bp about 174° with decompn. Sol in water, alcohol or ether.

1373. Bromal Hydrate. *2,2,2-Tribromo-1,1-ethanediol; tribromoacetaldehyde hydrate.* $C_2H_3Br_3O_2$; mol wt 298.79. C 8.04%, H 1.01%, Br 80.24%, O 10.71%. $Br_3CCH(OH)_2$. Prepd from bromal and water: Löwig, *Ann.* **3,** 288 (1832). Structure: Jain, Soundararajan, *Tetrahedron* **20,** 1589 (1964).

Deliquescent crystals, mp 53.5°. Odor of chloral and pungent taste. Dipole moment in benzene, 2.56D. Sol in water, alcohol, chloroform, ether, glycerol. *Keep tightly closed in a cool place.*

Caution: This substance may be habit forming and is listed in the U.S. Code of Federal Regulations, Title 21 Part 329.1 (1987).

1374. Bromazepam. *7-Bromo-1,3-dihydro-5-(2-pyridinyl)-2H-1,4-benzodiazepin-2-one; 7-bromo-5-(2-pyridyl)-3H-1,4-benzodiazepin-2(1H)-one;* Ro 5-3350; Compendium; Creosedin; Durazanil; Lexotam; Lexomil; Lexotan; Lexotanil; Normoc. $C_{14}H_{10}BrN_3O$; mol wt 316.16. C 53.18%, H 3.19%, Br 25.28%, N 13.29%, O 5.06%. Prepn: Berger *et al.,* **Belg. pat. 619,101;** *eidem,* Fryer *et al.,* **U.S. pat. 3,100,770** (1962, 1963 to Hoffmann-La Roche); *idem, J. Pharm. Sci.* **53,** 264 (1964); **U.S. pats. 3,182,065; 3,182,067** (both 1965 to Hoffmann-La Roche). Pharmacology: Korol, Brown, *Pharmacology* **1,** 115 (1968). Metabolism: Schwartz *et al., J. Pharm. Sci.* **62,** 1776 (1973); *Drug Metab. Dispos.* **2,** 31 (1974). Evaluation as pre-anesthesia medication: P. Chalmers, J. N. Horton, *Anesthesia* **39,** 370 (1984). Multicenter clinical comparison with lorazepam, *q.v.:* G. J. Cordingley *et al., Curr. Med. Res. Opin.* **9,** 505 (1985). Evaluation of adverse effects and withdrawal reactions: R. Fontaine *et al., Psychopharmacol. Bull.* **21,** 91 (1985). Toxicity: E. I. Goldenthal, *Toxicol. Appl. Pharmacol.* **18,** 185 (1971). Comprehensive description: M. M. Hassan, M. A. Abounassif in *Analytical Profiles of Drug Substances* vol. **16,** K. Florey, Ed. (Academic Press, New York, 1987) pp 1-51.

Colorless prisms from acetone, mp 237-238.5° (dec). LD_{50} orally in rats: 3050 ±405 mg/kg (Goldenthal).

Note: This is a controlled substance (depressant) listed in the U.S. Code of Federal Regulations, Title 21 Part 1308.14 (1985).

THERAP CAT: Anxiolytic.

1375. Bromcresol Green. *4,4'-(3H-2,1-Benzoxathiol-3-ylidene)bis[2,6-dibromo-3-methylphenol] S,S-dioxide; α,α-bis(3,5-dibromo-4-hydroxy-o-tolyl)-α-hydroxytoluenesulfonic acid, γ-sultone;* 3,3',5,5'-tetrabromo-m-cresolsulfonphthalein. $C_{21}H_{14}Br_4O_5S$; mol wt 698.05. C 36.13%, H 2.02%, Br 45.79%, O 11.46%, S 4.59%. Prepd by adding bromine to a suspension of *m*-cresolsulfonphthalein in glacial acetic acid: Clark, Lubs, *J. Wash. Acad. Sci.* **5,** 610 (1915); **6,** 481 (1916); *J. Bact.* **2,** 110 (1917); Cohen, *Biochem. J.* **16,** 31 (1922); **17,** 535 (1923); Cohen, *Public Health Repts.* **38,** 814 (1923); **41,** 3051 (1926); *Proc. Soc. Exp. Biol. Med.* **20,** 124 (1922); Orndorff, Purdy, *J. Am. Chem. Soc.* **48,** 2216 (1926).

Minute, slightly yellow crystals from acetic acid, mp 218-219°. Sparingly sol in water. Readily sol in alcohol, ether, ethyl acetate. Fairly sol in benzene. Very sensitive to alkalies, tap water being sufficiently alkaline to give the characteristic blue-green color. pK = 4.7. To prepare a soln for use as pH indicator, dissolve 0.10 g in 7.15 ml $N/50$ NaOH and dil with water to 250 ml. To prepare a soln for use as indicator in volumetric work, dissolve 0.1 g in 250 ml alc.
USE: As indicator, pH 3.8 yellow; pH 5.4 blue-green.

1376. Bromcresol Purple. *4,4'-(3H-2,1-Benzoxathiol-3-ylidene)bis[2-bromo-6-methylphenol] S,S-dioxide; α-(5-bromo-4-hydroxy-m-tolyl)-α-(3-bromo-5-methyl-4-oxo-2,5-cyclohexadien-1-ylidene)-o-toluenesulfonic acid; 5,5'-dibromo-o-cresolsulfonphthalein.* $C_{21}H_{16}Br_2O_5S$; mol wt 540.24. C 46.68%, H 2.99%, Br 29.59%, O 14.81%, S 5.94%. Prepd by treating o-cresol red with bromine in glacial acetic acid: *See refs under* Bromcresol Green.

Minute, slightly yellow crystals, mp 241-242°. Practically insol in water. Sol in alcohol, dil alkalies. pK = 6.3. To prepare a soln for use as pH indicator, dissolve 0.10 g in 9.25 ml $N/50$ NaOH and dil with water to 250 ml. To prepare a soln for use as indicator in volumetric work, dissolve 0.05 g in 250 ml alc.

USE: As indicator, pH 5.2 yellow; pH 6.8 purple.

1377. Bromelain. Bromelin; Ananase; Extranase; Inflamen; Traumanase. Protein-digesting and milk-clotting enzymes found in pineapple fruit juice and stem tissue. Enzymes from the two sources are distinguished as fruit bromelain and stem bromelain. First isolns of fruit bromelain: Marcano, *Bull. Pharm.* **5**, 77 (1891); Chittenden, *Trans. Conn. Acad. Sci.* **8**, 281 (1892). From pineapple juice by precipitation with acetone and also with ammonium sulfide: Heinicke, U.S. pat. 3,002,891 (1961 to Pineapple Res. Inst.). Discovery in stem tissue: *idem, Science* **118**, 753 (1953). Stem bromelain has mol wt of about 33,000 and is probably the first proteolytic enzyme of plant origin to be established as a glycoprotein: Murachi *et al., Biochemistry* **3**, 48 (1964); Ota *et al., ibid.* 180. Purification of crude prepns: Gibian, Bratfisch, U.S. pat. 2,950,227 (1960 to Schering AG). Reviews: Balls *et al., Ind. Eng. Chem.* **33**, 950 (1941); of stem bromelain: Murachi, "Structure and Function of Stem Bromelain," in *Proteins, Structure and Function,* vol. 2, M. Funatsu *et al.,* Eds. (Kodansha, Tokyo, Wiley, New York, 1972) pp 47-101.

Unlike papain, fruit bromelain does not disappear as the fruit ripens. Fruit and stem bromelains are acidic and basic proteins, resp. uv max (stem): 280 nm (A$_{1cm}^{1\%}$ 20.1).

USE: Tenderizing meat, chill-proofing beer, production of protein hydrolyzates.

THERAP CAT: Anti-inflammatory.

1378. Bromethalin. *N-Methyl-2,4-dinitro-N-(2,4,6-tribromophenyl)-6-(trifluoromethyl)benzenamine;* EL 614; Vengeance. C$_{14}$H$_7$Br$_3$F$_3$N$_3$O$_4$; mol wt 578.23. C 29.08%, H 1.22%, Br 41.50%, F 9.86%, N 7.27%, O 11.07%. Rodenticide which acts by uncoupling oxidative phosphorylation. Single-feed toxicant, effective against warfarin-resistant strains of mice, rats. Prepn and identification as rodenticide: B. A. Dreikorn, U.S. pat. 4,187,318 (1980 to Eli Lilly). Toxicology: B. A. Dreikorn *et al., Proc. Brit. Crop Prot. Conf. - Pests Dis.* **1979**, 491. Field trials: S. Kesyakova, I. Nikiforov, *Vet.-Med. Nauki* **20**, 72 (1983), *C.A.* **100**, 152450t (1984). Review: B. A. Dreikorn, G. O. P. O'Doherty, "The Discovery and Development of Bromethalin, an Acute Rodenticide with a Unique Mode of Action", in *Pesticide Synthesis Through Rational Approaches,* ACS Symp Ser. **255**, P. S. Magee *et al.,* Eds. (Am. Chem. Soc., Washington, D.C., 1984) pp 45-63.

Pale yellow odorless crystals from ethanol, mp 150-151°. Sol in chloroform, acetone; moderately sol in aromatic hydrocarbons. Insol in water. LD$_{50}$ in mice, rats, cats, dogs (mg/kg): 2, 5, 2, 5 orally (Dreikorn).

USE: Rodenticide.

1379. Bromhexine. *2-Amino-3,5-dibromo-N-cyclohexyl-N-methylbenzenemethanamine; 3,5-dibromo-N$^\alpha$-cyclohexyl-N$^\alpha$-methyltoluene-α,2-diamine;* N-cyclohexyl-N-methyl-2-(2-amino-3,5-dibromo)benzylammonium; N-(2-amino-3,5-dibromobenzyl)-N-methylcyclohexylammonium; Auxit; Tossimex. C$_{14}$H$_{20}$Br$_2$N$_2$; mol wt 376.14. C 44.70%, H 5.36%, Br 42.49%, %, N 7.45%. Prepn: Keck, *Ann.* **662**, 171 (1963); K. Thomae, Belg. pat. 625,022 (1963), *C.A.* **61**, 5564 (1964). Pharmacology: R. Engelhorn, S. Püschmann, *Arzneimittel-Forsch.* **13**, 474 (1963). Metabolism: R. Jauch *et al., ibid.* **25**, 1954 (1975). Toxicity: Boyd, Sheppard, *Arch. Int. Pharmacodyn. Ther.* **163**, 284 (1966).

Crystals from abs ethanol, dec 237.5-238°. One gram dissolves in 250 ml water or 250 ml 10% ethanol. Approx LD$_{50}$ orally in rabbits: > 10 g/kg (Boyd, Sheppard).

Hydrochloride, C$_{14}$H$_{21}$Br$_2$ClN$_2$, *NA 274, Bisolvon, Ophtosol.*

THERAP CAT: Expectorant; mucolytic.

1380. Bromic Acid. BrHO$_3$; mol wt 128.92. Br 61.99%, H 0.78%, O 37.23%. HBrO$_3$. Prepd from Ba(BrO$_3$)$_2$ and H$_2$SO$_4$: Burchard, *Z. Physik. Chem.* **2**, 814 (1888); Schmeisser in *Handbook of Preparative Inorganic Chemistry,* vol. 1, G. Brauer, Ed. (Academic Press, 2nd ed, 1963) p 315.

Known in aq soln only. *Keep refrigerated!* Colorless soln, turns yellow on standing at room temp, even in the dark. Careful evapn *in vacuo* at −12° yields about a 50% soln. Dec on heating to 100°. Can be dil with cold water.

USE: Oxidizing agent. *Caution:* Highly irritating to skin, eyes, mucous membranes.

1381. Bromindione. *2-(4-Bromophenyl)-1H-indene-1,3-(2H)-dione;* 2-(4-bromophenyl)-1,3-dioxohydrindene; HL 255; MG 2555; Fluidane; Halinone. C$_{15}$H$_9$BrO$_2$; mol wt 301.16. C 59.83%, H 3.01%, Br 26.54%, O 10.63%. Prepn: Cavallini *et al., Farmaco Ed. Sci.* **10**, 710 (1955); Freedman *et al.,* U.S. pat. 2,847,474 (1958 to U.S. Vitamin).

Crystals from ligroin, mp 137-139°.

THERAP CAT: Anticoagulant.

1382. Bromine. Br; at. wt 79.904; at. no. 35; valences 1 to 7; elemental state: Br$_2$. A halogen. Abundance in igneous rock: 1.6 × 10⁻⁴% by wt; in seawater: 0.0065% by weight. Natural isotopes: 79 (50.54%); 81 (49.46%); known isotopes range in mass no. from 74-90; radioactive tracer elements: 77, 80, 80m (metastable), 82. Discovered by Balard in 1826. Isolated from natural brines (salt lakes) and ocean water: Seaton, *Chem. & Met. Eng.* **38**(11), 638 (1931); Robertson, *Ind. Eng. Chem.* **34**, 133 (1942); *Chem. & Met. Eng.* **52**(10), 134 (1945); Gale, Pearson, U.S. pat. 2,251,353 (1941 to Am. Potash and Chem.); Faith, Keyes & Clark's *Industrial Chemicals,* F. A. Lowenheim, M. K. Moran, Eds. (Wiley-Interscience, New York, 4th ed., 1975) pp 159-163. Prepn of ultra-pure bromine for research purposes: Baxter *et al., J. Am. Chem. Soc.* **34**, 260 (1912); Noyes, *ibid.* **45**, 1194 (1923); Hönigschmid, Zintl, *Ann.* **433**, 216 (1923). Book: Z. E. Jolles, Ed., *Bromine and its Compounds* (E. Benn, London, 1966) 940 pp. Reviews: *MTP Int. Rev. Sci.: Inorg. Chem., Ser. One,* vol. 3, V. Gutmann, Ed. (Butterworths, London, 1972); Downs, Adams, "Chlorine, Bromine, Iodine and Astatine" in *Comprehensive Inorganic Chemistry,* vol. 2, J. C. Bailar, Jr. *et al.,* Eds. (Pergamon Press, Oxford, 1973) pp 1107-1594; C. E. Reineke in Kirk-Othmer *Encyclopedia of Chemical Technology* vol. 4 (Wiley-Interscience, New York, 3rd ed., 1978) pp 226-243.

Dark reddish-brown, volatile, diatomic liquid; suffocating odor; vaporizes rapidly at room temperature. mp −7.25° (265.90°K); bp 59.47° *(JANAF Thermochemical Tables);* 58.78° (Mellor's Suppl. II, Part I, "The Halogens"); d$_4^{25}$ 3.1023; crit temp: 315°; crit pressure: 102 atm. Heat capacity at constant pressure (liq, 25°) 18.089 cal/mole deg: Hildenbrand *et al., J. Am. Chem. Soc.* **80**, 4129 (1958). Vapor pressure data: A. N. Nesmeyanov, *Vapor Pressure of the Chemical Elements,* R. Gary, Ed. (Elsevier, New York, 1963) pp 354-58. Total soly in water (25°): 0.2141 moles/l with formation of 0.00115 moles/l of HOBr; freely sol in alc, ether, $CHCl_3$, CCl_4, CS_2, concd HCl, aq solns of bromides. Less reactive than chlorine; E^0 (aq) $\frac{1}{2}Br_2/Br^-$ 1.065 V; dissociation energy (25°): 46.072 kcal. For reactions, complexes and formation of monatomic bromine *see* Chlorine. Incompatible with alkali hydroxides; arsenites; ferrous, mercurous salts; hypophosphites and other oxidizable substances. *Burns and blisters the skin. Keep sealed or glass-stoppered. When handling bromine always keep ammonia water within reach.*

Human Toxicity: Ingestion of soln may cause severe gastroenteritis and death. Dermal contact with concd solns may cause corrosion. Serious irritation of respiratory tract mucosa may follow inhalation of vapor. *Antidote:* For ingestion, ammonia or thiosulfates. *See:* Patty's *Industrial Hygiene and Toxicology* vol. **2B**, G. D. Clayton, F. E. Clayton, Eds. (Wiley-Interscience, New York, 3rd ed., 1981) pp 2965-2972.

USE: In water disinfection; bleaching fibers and silk; manuf medicinal bromine compds, dyestuffs.

1383. Bromine Pentafluoride. BrF_5; mol wt 174.92. Br 45.69%, F 54.31%. Prepd by fluorination of bromine at 200° in iron or Monel metal apparatus: Ruff, Menzel, *Z. Anorg. Allgem. Chem.* **202**, 49 (1931); Kwasnik in *Handbook of Preparative Inorganic Chemistry*, Vol. **1**, G. Brauer, Ed. (Academic Press, New York, 2nd ed., 1963) pp 158-159; from F_2 and KBr: Hyde, Boudakian, *Inorg. Chem.* **7**, 2648 (1968). *Reviews:* Kemmitt, Sharp, *Advan. Fluorine Chem.* **4**, 243-244 (1965); Stein, "Physical and Chemical Properties of Halogen Fluorides" in *Halogen Chemistry*, Vol. **1**, V. Gutmann, Ed. (Academic Press, New York, 1967) pp 133-224; Meinert, *Z. Chem.* **7**, 41-57 (1967).

Liquid. Fumes in air. mp $-60.5°$. bp 40.76°. d^{25} 2.4604. Trouton constant 23.7. Thermostable up to 460°. Does not attack quartz when dry. *Produces an explosion on contact with water.* Very reactive, usually with conflagration.

Caution: Corrosive and irritating to eyes, skin, mucous membranes.

1384. Bromine Trifluoride. BrF_3; mol wt 136.92. Br 58.37%, F 41.63%. Prepd by fluorination of bromine at +80°: Lebeau, *Compt. Rend.* **141**, 1018 (1905); Prideaux, *J. Chem. Soc.* **89**, 316 (1906); Ruff, Braida, *Z. Anorg. Allgem. Chem.* **206**, 59 (1932); **214**, 91 (1933); Simons, *Inorg. Syn.* **3**, 184 (1950); Kwasnik in *Handbook of Preparative Inorganic Chemistry*, Vol. **1**, G. Brauer, Ed. (Academic Press, New York, 2nd ed., 1963) pp 156-157. *Reviews:* Kemmitt, Sharp, *Advan. Fluorine Chem.* **4**, 244-245 (1965); Stein, "Physical and Chemical Properties of Halogen Fluorides" in *Halogen Chemistry*, Vol. **1**, V. Gutmann, Ed. (Academic Press, New York, 1967) pp 133-224; Meinert, *Z. Chem.* **7**, 41-57 (1967).

Colorless liquid; also reported to be pale yellow. Long prisms when solid. mp 8.77°. bp 125.75°. d^{25} 2.8030. *Smokes in air. Attacks skin. Very reactive.*

USE: Solvent for fluorides. *Caution:* Corrosive and irritating to skin, eyes, mucous membranes, respiratory tract.

1385. Bromisovalum. *N-(Aminocarbonyl)-2-bromo-3-methylbutanamide;* (α-*bromoisovaleryl)urea;* 2-monobromo-isovalerylurea; α-bromo-β-dimethylpropanoylurea; bromvaletone; B.V.U.; Bromural; Bromisoval; Uvaleral; Bromuvan; Somnurol; Brovalurea; Dormigene; Isobromyl; Alluval; Pivadorm. $C_6H_{11}BrN_2O_2$; mol wt 223.08. C 32.30%, H 4.97%, Br 35.82%, N 12.56%, O 14.34%. $(CH_3)_2CHCHBr$-$CONHCONH_2$. Obtained by the interaction of urea with α-bromoisovaleryl bromide: U.S. pat. **914,518** (1909). Toxicity: R. I. Mrongovius *et al., Clin. Exp. Pharmacol. Physiol.* **3**, 443 (1976).

Practically tasteless needles, mp 147-149°. Sublimes. Slightly sol in cold water, but freely sol in hot water. Also readily sol in alcohol, ether and in alkaline solns. LD_{50} in male mice (mmoles/kg): 3.25 i.p. (Mrongovius).

Caution: This substance may be habit forming and is listed in the U.S. Code of Federal Regulations, Title 21 Part 329.1 (1987).

THERAP CAT: Sedative, hypnotic.

1386. N-Bromoacetamide. Acetobromamide. C_2H_4Br-NO; mol wt 137.98. C 17.41%, H 2.92%, Br 57.92%, N 10.15%, O 11.60%. $CH_3CONHBr$. Prepd by treating a cooled (0-5°) soln of acetamide dissolved in bromine with ice

cold aq 50% KOH, allowing to stand for 2-3 hr at 0-5°, treating with NaCl, and extracting with chloroform: Oliveto, Gerold, *Org Syn.* **31**, 17 (1951).

Needles from chloroform + hexane, mp 102-105°. Sol in warm water, freely sol in cold ether. Unstable to light and heat.

Monohydrate, rectangular plates from hydr ether, mp 70-80°. Freely sol in cold water, alcohol, ether, less freely in chloroform.

USE: Brominating agent; in oxidation of primary and secondary alcohols.

1387. p-Bromoacetanilide. *N-(4-Bromophenyl)acetamide; 4'-bromoacetanilide;* monobromoacetanilide; bromoanilide; Bromoantifebrin; Asepsin; Antisepsin. C_8H_8BrNO; mol wt 214.07. C 44.88%, H 3.77%, Br 37.33%, N 6.54%, O 7.47%. Prepd by bromination of acetanilide: Remmers, *Ber.* **7**, 346 (1874); Merker, Vona, *J. Chem. Ed.* **26**, 613 (1949); Knowles, Alt, **U.S. pat. 3,012,035** (1959 to Monsanto).

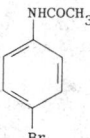

Crystals, from 95% alc, mp 168° (with previous softening). d 1.72. Practically insol in cold water; sparingly sol in hot water; sol in benzene, chloroform, ethyl acetate; moderately sol in alcohol.

THERAP CAT: Analgesic, antipyretic.

1388. Bromoacetic Acid. $C_2H_3BrO_2$; mol wt 138.96. C 17.28%, H 2.18%, Br 57.51%, O 23.03%. $BrCH_2COOH$. Prepn by bromination of acetic acid: Perkin, Duppa, *Ann.* **108**, 106 (1858); from chloroacetic acid and HBr: Lake, Asadorian; Asadorian, Burk, **U.S. pats. 2,553,518; 3,130,222** (1951, 1964, both to Dow); from glycolic acid and HBr: Johnston, **U.S. pat. 2,876,255** (1959 to Ethyl Corp.).

Hygroscopic crystals, mp 50°. bp 208°. d 1.93. Very sol in water, alcohol. *Protect from air and moisture.*

Caution: Irritant and corrosive to skin, mucous membranes.

1389. Bromoacetone. *1-Bromo-2-propanone.* C_3H_5BrO; mol wt 136.99. C 26.30%, H 3.68%, Br 58.34%, O 11.68%. CH_3COCH_2Br. Prepn by bromination of acetone: Emmerling, Wagner, *Ann.* **204**, 29 (1880); Catch *et al., J. Chem. Soc.* **1948**, 272; Ross **U.S. pat. 2,452,154** (1948 to Colgate-Palmolive-Peet); by bromination acetone enol acetate: Magerlein, **U.S. pat. 2,752,341** (1956 to Upjohn).

Liquid, mp $-36.5°$. bp 137°. bp_{50} 63.5-64°. d^{23} 1.634. n_D^{15} 1.4697. Turns violet rapidly even in absence of air. Sparingly sol in water; sol in alcohol, acetone. *Keep tightly closed and protected from light.*

USE: Chemical war gas. *Caution:* Violent lacrimator.

1390. p-Bromoacetophenone. *1-(4-Bromophenyl)ethanone;* methyl *p*-bromophenyl ketone. C_8H_7BrO; mol wt 199.05. C 48.27%, H 3.54%, Br 40.15%, O 8.04%. BrC_6H_4-$COCH_3$. Prepd from bromobenzene and acetic anhydride in CS_2 in the presence of anhyd $AlCl_3$: Adams, Noller, *Org. Syn.* **5**, 17 (1925).

Leaflets from alc. mp 54°. bp_{736} 255.5°; bp_{15} 130°; bp_7 117°. Easily volatile with steam. Sol in alcohol, ether, glacial acetic acid, benzene, petr ether, carbon disulfide.

Oxime, C_8H_8BrNO, needles from dil alc, mp 128.5°.

1391. ω-Bromoacetophenone. *2-Bromo-1-phenylethanone;* phenacyl bromide. C_8H_7BrO; mol wt 199.05. C 48.27%, H 3.54%, Br 40.15%, O 8.04%. $C_6H_5COCH_2Br$. Prepn from acetophenone and bromine: Rother, Reid, *J. Am. Chem. Soc.* **41**, 77 (1919); Cowper, Davidson, *Org. Syn.* **19**, 24 (1939); Shevchuk, Dombrovskii, *Zh. Obshch. Khim.* **33**(4), 1135 (1963).

Crystals, mp 50°. d 1.65. bp_{20} 133-135°. Irritating vapors. *Lacrimator!* Practically insol in water; freely sol in alcohol, benzene, chloroform, ether.

Caution: Highly irritating to skin, eyes, mucous membranes.

1392. p-Bromoaniline. *4-Bromobenzeneamine;* 4-bromoaniline. C_6H_6BrN; mol wt 172.04. C 41.89%, H 3.52%, Br 46.45%, N 8.14%. Prepd by steam distilling sodium hydroxide and p-bromoacetanilide: Scott, *J. Chem. Soc.* **123**, 3199 (1923); by direct bromination of aniline: Kosolapoff, *J. Am. Chem. Soc.* **75**, 3596 (1953).

Rhombic crystals from dil alc; mp 66-66.5°. $d_4^{99.6}$ 1.4970 (liq). Very sol in alcohol and ether, insol in cold water. Kb 5.2×10^{-11} at 25° (Kb 1.04×10^{-10} at 25°).
USE: In prepn of azo dyes; condensed with formaldehyde in prepn of dihydroquinazolines.

1393. 5-Bromoanthranilic Acid. *2-Amino-5-bromobenzoic acid;* 5-bromo-2-aminobenzoic acid. $C_7H_6BrNO_2$; mol wt 216.04. C 38.91%, H 2.80%, Br 36.99%, N 6.48%, O 14.81%. Prepd by bromination of anthranilic acid: Wheeler, *J. Am. Chem. Soc.* **31**, 565 (1909); Wheeler, Oates, *ibid.* **32**, 771 (1910).

Crystals, mp 218-219°. Very slightly sol in water, moderately sol in alcohol, ether, chloroform, benzene, acetic acid, freely sol in acetone.
USE: Determination of cobalt, copper, nickel, and zinc.

1394. Bromobenzene. Monobromobenzene; phenyl bromide. C_6H_5Br; mol wt 157.02. C 45.89%, H 3.21%, Br 50.90%. Prepd industrially by the action of bromide on benzene in the presence of iron powder: Gattermann-Wieland, *Praxis des organischen Chemikers* (de Gruyter, Berlin, 40th ed., 1961) p 95; alternate procedure using pyridine as halogen carrier: A. I. Vogel, *Practical Organic Chemistry* (Longmans, London, 3rd ed., 1959) p 535.

Mobile liquid. Aromatic odor. d_4^0 1.5220; d_4^{10} 1.5083; d_4^{15} 1.5017; d^{20} 1.4952; d_4^{30} 1.4815; d_4^{71} 1.426. mp −30.6°. bp_{760} 156.2°; bp_{400} 132.3°; bp_{200} 110.1°; bp_{40} 68.6°; bp_{20} 53.8°; bp_{10} 40.0°; bp_5 27.8°; $bp_{1.0}$ +2.9°. n_D^{15} 1.5625; n_D^{20} 1.5602. Flash pt 51°. Fire pt 155°. Critical temp 397°; crit press. 33,912 mm (44.6 atm). Viscosity at 20° = 1.124 cp. Vapor density (air = 1): 5.41. Specific heat at 26.84° = 0.2368. Heat of melting 16.186 cal/g at 15°. Practically insol in water (0.045 g/100 g at 30°). Miscible with chloroform, benzene, petr hydrocarbons. Sol in alc (10.4 g/100 g at 25°), in ether (71.3 g/100 g at 25°).
USE: In organic synthesis, especially to make phenyl magnesium bromide; as solvent, especially for crystns on a large scale and where a heavy liquid is desirable; as additive to motor oils. *Caution:* Irritating to skin.

1395. p-Bromobenzenesulfonyl Chloride. $C_6H_4BrClO_2S$; mol wt 255.54. C 28.20%, H 1.58%, Br 31.27%, Cl 13.88%, O 12.52%, S 12.55%. Prepd from sodium p-bromobenzenesulfonate and phosphorus pentachloride: Marvel, Smith, *J. Am. Chem. Soc.* **45**, 2696 (1923).

Needles from ligroin, mp 74.5°.
USE: Identification of amines.

1396. p-Bromobenzoic Acid. $C_7H_5BrO_2$; mol wt 201.03. C 41.82%, H 2.51%, Br 39.75%, O 15.92%. Prepd by oxidation of p-bromotoluene with potassium permanganate: Hale, Thorp, *J. Am. Chem. Soc.* **35**, 269 (1913).

Needles from ether, leaflets from water or 90% alcohol. mp 251-253°. Slightly sol in hot water. Sol in alc, ether.
Methyl ester, $C_6H_4BrCO_2CH_3$, needles from ether or leaflets from dil alcohol, mp 81°. d 1.689. Sol in alc, ether.
Ethyl ester, $C_6H_4BrCO_2C_2H_5$, liquid; $bp_{737.4}$ 262°.
Phenyl ester, $C_{13}H_9BrO_2$, scales, mp 117°. Sol in alc, ether, chloroform, CS_2, benzene; less sol in petr ether.
USE: For the detection of strontium; in org syntheses.

1397. p-Bromobenzyl Bromide. *1-Bromo-4-(bromomethyl)benzene;* p,α-dibromotoluene. $C_7H_6Br_2$; mol wt 249.95. C 33.63%, H 2.42%, Br 63.94%. Prepd by photobromination of p-bromotoluene: Weizmann, Patai, *J. Am. Chem. Soc.* **68**, 150 (1946). Alternate procedure: Goerner, Nametz, *ibid.* **73**, 2940 (1951).

Crystals from alc. Agreeable aromatic odor. mp 61°. bp_{12} 115-124°. Sol in water, cold alc, more sol in hot alc, ether, carbon disulfide, benzene, glacial acetic acid.
USE: Identification of aromatic carboxylic acids. *Caution:* Irritating to eyes, mucous membranes of nose and throat. Action similar to, but less intense than, benzylbromide, *q.v.*

1398. p-Bromobenzyl Chloride. *1-Bromo-4-(chloromethyl)benzene;* p-bromo-α-chlorotoluene; α-chloro-4-bromotoluene. C_7H_6BrCl; mol wt 205.50. C 40.91%, H 2.94%, Br 38.89%, Cl 17.26%. Prepd from p-bromobenzene, paraformaldehyde and $SnCl_4$: Quelet, *Bull. Soc. Chim.* **41**, 329 (1927); by benzoyl peroxide catalyzed chlorination of p-bromotoluene with sulfhydryl chloride: Goerner, Nametz, *J. Am. Chem. Soc.* **73**, 2940 (1951).

Needles from alc, mp 40-41°. bp_{12} 105-115°; bp_{27} 136-139°. Freely sol in hot alc.

1399. p-Bromobenzyl Chloroformate. *Carbonochloridic acid (4-bromophenyl)methyl ester;* chloroformic acid p-bromobenzyl ester. $C_8H_6BrClO_2$; mol wt 249.51. C 38.51%, H 2.42%, Br 32.03%, Cl 14.21%, O 12.83%. A stable, low-melting cryst product obtained from p-bromobenzyl alcohol.

USE: In the carbobenzyloxy method of prepn of amino acids and peptides; gives higher melting and better cryst products than the corresp carbobenzyloxy compds. *Ref:* Channing *et al.*, *Nature* **167**, 487 (1951).

1400. α-Bromobenzyl Cyanide. *α-Bromobenzeneacetonitrile;* α-bromophenylacetonitrile; α-bromo-α-tolunitrile; B.B.C.; C.A.; Camite. C_8H_6BrN; mol wt 196.05. C 49.01%, H 3.08%, Br 40.76%, N 7.15%. $C_6H_5CHBrCN$. The practical industrial prepn consists of three steps: (1) Chlorination of toluene to form benzyl chloride, (2) conversion of benzyl chloride to benzyl cyanide by the action of sodium cyanide in alcoholic soln, (3) bromination of the benzyl cyanide with bromine vapor in the presence of sunlight: Steinkopf *et al.*, *Ber.* **53**, 1146 (1920); Nekrassov, *J. Prakt. Chem.* **119**, 108 (1928).

Crystalline mass, mp 29°. Odor of soured fruit. bp_{760} 242° (dec); bp_{12} 132-134°. d_4^{29} 1.539. Vapor density 6.8 (air = 1). Vapor pressure at 20° = 0.012 mm; at 30° = 0.028 mm. Slightly sol in water. Freely sol in alcohol, ether, chloroform, acetone, and other common organic solvents; also sol in phosgene, chloropicrin, benzyl cyanide. Lethal concn:

0.90 mg/l (30 min.), A. M. Prentiss, *Chemicals in War* (McGraw-Hill, New York, 1937) p 141.
USE: War gas. *Caution:* Strong lacrimator.

1401. α-Bromobutyric Acid. *2-Bromobutanoic acid; dl-2-bromobutyric acid.* $C_4H_7BrO_2$; mol wt 167.01. C 28.76%, H 4.22%, Br 47.85%, O 19.16%. $CH_3CH_2CHBrCOOH$. Prepd by bromination of butyric acid: Naumann, *Ann.* **119**, 115 (1861); Stevens, Holland, *J. Org. Chem.* **18**, 1112 (1953); Smissman, *J. Am. Chem. Soc.* **76**, 5805 (1954).
Oily liquid, mp −4°. bp$_{250}$ 181-182°, bp$_{25}$ 127-128°. d_4^4 1.5855, d_{15}^{15} 1.5735, d_{20}^{20} 1.5669, d_{25}^{25} 1.5620: Perkin, *J. Chem. Soc.* **65**, 402 (1894). Boils with decompn at ordinary pressure. Sol in 15 parts water; sol in alcohol, ether. *Corrosive!* LD$_{50}$ orally in mice: 310 mg/kg, J. L. Morrison, *J. Pharmacol. Exp. Ther.* **86**, 336 (1946).
Ethyl ester, $C_6H_{11}BrO_2$, *ethyl α-bromobutyrate.* Liquid, bp 177-178° with slight dec. d_{20}^{20} 1.329. Insol in water; misc with alc, ether. *Protect from light. Lachrimator!*

1402. 3-Bromo-d-camphor. *3-Bromo-1,7,7-trimethylbicyclo[2.2.1]heptan-2-one; 3-bromo-d-2-bornanone.* $C_{10}H_{15}$-BrO; mol wt 231.14. C 51.96%, H 6.54%, Br 34.58%, O 6.92%. Of the two configurations found, the *endo*-form is more stable than the *exo*-form: Lowry *et al., J. Chem. Soc.* **121**, 633 (1922); Cookson, *ibid.* **1954**, 282. Prepn of the *endo*-form by bromination of *d*-camphor: Kipping, Pope, *ibid.* **63**, 548 (1893); *cf.* Woods, Roberts, *J. Org. Chem.* **22**, 1124 (1957). Prepn of *exo*-form by isomerization of *endo*-form: Lowry *et al., loc. cit.* Configuration: Wiebenga, Krom, *Rec. Trav. Chim.* **65**, 663 (1946); Cookson, *loc. cit. Review:* J. L. Simonsen, Ed., *The Terpenes* vol. II (University Press, Cambridge, 2nd ed., 1949), pp 401-404.

endo-form   *exo*-form

endo-Form, α-bromo-d-camphor, 3α-bromo-d-camphor, *bromated camphor, camphor monobromated.* Crystals from benzene, mp 76°. Camphor-like odor and taste. Discolors on prolonged exposure to light. d 1.449. [α]$_D^{20}$ +122.7° (14.5 g/100 g benzene soln), Cutter *et al., J. Chem. Soc.* **127**, 1260 (1925). uv max (cyclohexane): 307.5 nm (log ε 1.98), Cookson, *loc. cit.* Sublimes, bp 274°. Almost insol in water: 1 g dissolves in 6.5 ml alcohol, 0.5 ml chloroform, 1.6 ml ether; sol in olive oil, slightly in glycerol.
When phenol, chloral hydrate, salol, menthol, or thymol is triturated with bromocamphor the mixture melts; these compds, however, are not incompatible.
exo-Form, α'-bromo-d-camphor, 3β-bromo-d-camphor. Needles from methanol or ethanol, mp 78.5°. d 1.484. [α]$_D^{20}$ −42.1° (14.5 g/100 g benzene soln), Cutter *et al., loc. cit.* uv max (cyclohexane): 312 nm (log ε 1.95), Cookson, *loc. cit.*
THERAP CAT: Topical counterirritant.

1403. α-Bromo-n-caproic Acid. *2-Bromohexanoic acid.* $C_6H_{11}BrO_2$; mol wt 195.06. C 36.94%, H 5.68%, Br 40.97%, O 16.40%. $CH_3(CH_2)_3CHBrCOOH$. Prepd by heating bromine with *n*-caproic acid in the presence of phosphorus trichloride as catalyst: Clarke, Taylor, *Org. Syn.* **4**, 9 (1924). Resoln into optically active forms by means of the strychnine salt: Levene *et al., J. Biol. Chem.* **75**, 352 (1927); Levene, Mardashew, *ibid.* **117**, 707 (1937).
Liquid. bp$_{760}$ 240°; bp$_{30}$ 148-153°; bp$_{17}$ 128-136°; bp$_8$ 116-125°. Soluble in alcohol, ether.
Ethyl ester, $C_8H_{15}BrO_2$, liquid, odor of anise oil. bp$_{760}$ 205-210°; bp$_{32}$ 99-102°; bp$_{11}$ 103°; bp$_9$ 95-96°.

1404. Bromocriptine. *(5'α)-2-Bromo-12'-hydroxy-2'-(1-methylethyl)-5'-(2-methylpropyl)ergotaman-3',6',18-trione; 2-bromoergocryptine; 2-bromo-α-ergokryptin; CB-154.* $C_{32}H_{40}BrN_5O_5$; mol wt 654.62. C 58.71%, H 6.16%, Br 12.21%, N 10.70%, O 12.22%. Dopamine receptor agonist; derivative of the ergotoxin group of ergot alkaloids. Prepn: Flückiger *et al.,* Ger. pat. **1,926,045** corresp to U.S. pats. **3,752,814** and **3,752,888** (1969, 1973, 1973, all to Sandoz).

Activity studies: Flückiger, Wagner, *Experientia* **24**, 1130 (1968); Yanai, Nagasawa, *J. Nat. Cancer Inst.* **45**, 1105 (1970); Stähelin *et al., Experientia* **27**, 915 (1971). Relationship of stereochemistry and biological activity: H. P. Weber, *Adv. Biochem. Psychopharmacol.* **23**, 25 (1980); N. Camerman, A. Camerman, *Mol. Pharmacol.* **19**, 517 (1981). Endocrine profile: Del Pozo *et al., Schweiz. Med. Wochenschr.* **103**, 847 (1973). As an immunomodulator: E. Nagy *et al., Immunopharmacology* **6**, 231 (1983). Use in obstetrics and gynecology: R. A. Kinch, *Fertil. Steril.* **33**, 463 (1980); in parkinsonism: R. C. Duvoisin *et al., Adv. Biochem. Psychopharmacol.* **23**, 271 (1980); M. M. Hoen, *J. Am. Geriatric Soc.* **29**, 251 (1981); in treatment of pituitary tumors: R. F. Spark *et al., J. Am. Med. Assoc.* **247**, 311 (1982). Clinical studies in Parkinson's disease: M. Gawel *et al., Advan. Neurol.* **45**, 535 (1986); E. N. H. Jansen, J. D. Meerwaldt, *ibid.* 539. Review of pharmacology: Flückiger, *Triangle (Engl. Ed.)* **14**, 153-157 (1975); D. Parkes, *Advan. Drug Res.* **12**, 247-344 (1977); M. O. Thorner *et al., Bromocriptine: A Clinical and Pharmacological Review* (Raven Press, New York, 1980) 181 pp. Comprehensive description: D. A. Giron-Forest, W. D. Schönleber in *Analytical Profiles of Drug Substances* Vol. **8**, K. Florey, Ed. (Academic Press, New York, 1979) pp 47-81. Review of therapeutic applications in endocrine and neurological diseases: K. Y. Ho, M. O. Thorner, *Drugs* **36**, 67-82 (1988).

Crystals from methyl ethyl ketone-isopropyl ether, mp 215-218° (dec). [α]$_D^{20}$ −195° (c = 1 in methylene chloride). LD$_{50}$ in rabbits (mg/kg): >1000 orally, 12.0 i.v. (Flückiger, first U.S. patent).
Methanesulfonate, $C_{33}H_{44}BrN_5O_8S$, *CB-154 mesylate, Parlodel, Pravidel, Serono-Bagren.* Crystals from methyl ethyl ketone, mp 192-196° (dec). [α]$_D^{20}$ +95° (c = 1 in methanol-methylene chloride). Soly in mg/ml at 25°: methanol 910; ethanol 23.0; water 0.8; chloroform 0.45; benzene <0.1; hexane <0.1. pKa 4.90.
THERAP CAT: Enzyme inhibitor (prolactin). Antiparkinsonian.

1405. Bromodiphenhydramine. *2-[(4-Bromophenyl)-phenylmethoxy]-N,N-dimethylethanamine; 2-(p-bromo-α-phenylbenzyloxy)-N,N-dimethylethylamine; β-(p-bromobenzhydryloxy)ethyldimethylamine; β-dimethylaminoethyl p-bromobenzhydryl ether; bromdiphenhydramine; bromazine; bromanautine; Bromo-Benadryl; Deserol; Histabromamine.* $C_{17}H_{20}BrNO$; mol wt 334.28. C 61.08%, H 6.03%, Br 23.91%, N 4.19%, O 4.79%. Prepn: Rieveschl, U.S. pat. **2,527,963** (1950 to Parke, Davis).

Hydrochloride, $C_{17}H_{21}BrClNO$, *Ambodryl.* Crystals from isopropanol, mp 144-145°.
THERAP CAT: Antihistaminic.

1406. Bromofenofos. *3,3',5,5'-Tetrabromo-(1,1'-biphenyl)-2,2'-diol mono(dihydrogen phosphate); 3,3',5,5'-tetrabromo-2,2'-biphenyldiol mono(dihydrogen phosphate);*

4,4′,6,6′-tetrabromobiphenyl-2,2′-diol mono(dihydrogen phosphate); bromophenophos; bromphenphos; PH 1882; Acedist. $C_{12}H_7Br_4O_5P$; mol wt 581.79. C 24.77%, H 1.21%, Br 54.94%, O 13.75%, P 5.32%. Prepn: **Neth. pat. Appl. 6,505,635** corresp to S. Van der Meer *et al.*, U.S. pat. **3,482,-016** and **3,662,035** (1966, 1969, 1972 to Chemiefarma); S. Van der Meer, H. Pouwels, *J. Med. Chem.* **12,** 534 (1969). Activity: J. Guilhon *et al.*, *Bull. Acad. Vet. Fr.* **43,** 67 (1970). Toxicology: J. M. Poul, M. Dagorn, *Recl. Med. Vet.* **158,** 363 (1982).

Cryst, no mp, dec > 350°.
THERAP CAT (VET): Flukicide.

1407. Bromoform. *Tribromomethane.* $CHBr_3$; mol wt 252.77. C 4.75%, H 0.40%, Br 94.85%. Prepd from acetone and sodium hypobromite: Günther, *Jahresber. Fortschr. Chem.* **1887,** 741 (Beilstein, **vol. 1,** 68); Kergomard, *Bull. Soc. Chim. France* **1961,** 2360. Toxicity data: Kutob, Plaa, *Toxicol. Appl. Pharmacol.* **4,** 354 (1962).

Heavy liquid; chloroform odor; sweetish taste. bp 149-150°. mp +7.5°. d_4^{15} 2.9035. n_D^{15} 1.6005. Sol in about 800 parts water; miscible with alc, benzene, chloroform, ether, petr ether, acetone, oils. Gradually dec, acquiring a yellow color; air and light accelerate the decompn. Commercial prepn is generally preserved by the addition of 3-4% alc. d 2.6-2.7. *Keep in well-closed containers, protected from light. Incompat.* Caustic alkalies. LD_{50} s.c. in mice: 7.2 mmol/kg (Kutob, Plaa).

Caution: This substance may be habit forming and is listed in the U.S. Code of Federal Regulations, Title 21 Part 329.1 (1987).

USE: In separating mixtures of minerals.

THERAP CAT: Has been used as sedative, hypnotic; antitussive.

1408. α-Bromoisobutyric Acid. *2-Bromo-2-methylpropanoic acid.* $C_4H_7BrO_2$; mol wt 167.01. C 28.76%, H 4.22%, Br 47.85%, O 19.16%. $(CH_3)_2CBrCOOH$. Prepd by bromination of isobutyric acid: Markownikow, *Ann.* **153,** 228 (1870); Smissman, *J. Am. Chem. Soc.* **76,** 5805 (1954).

Crystals, mp 48-49°. bp 198-200°, bp_{20} 110-116°. d 1.52. Sparingly sol in cold water; sol in alcohol, ether. Dec by hot water into the hydroxy acid.

1409. α-Bromoisovaleric Acid. *2-Bromo-3-methylbutanoic acid.* $C_5H_9BrO_2$; mol wt 181.04. C 33.17%, H 5.01%, Br 44.14%, O 17.68%. $(CH_3)_2CHCHBrCOOH$. Prepn of *dl*-form: Ley, Popow, *Ann.* **174,** 61 (1874); Smissman, *J. Am. Chem. Soc.* **76,** 5805 (1954); Marvel, *Org. Syn.* **coll. vol. III,** 848 (1955). Prepn of *d*- and *l*-forms: Berlingozzi, Furia, *Gazz. Chim. Ital.* **56,** 82 (1926); Berlingozzi, Lenoci, *ibid.* **68,** 721 (1938). Configuration of *d*-form: Brewster *et al.*, *Nature* **166,** 179 (1950).

dl-Form, prisms from ether or chloroform, mp 44°. bp about 230° with some decompn. Sparingly sol in water; sol in alcohol, ether.

d-Form, crystals from petroleum ether, mp 43-44°. $[\alpha]_D^{20}$ +22.6° (c = 4 in benzene).

l-Form, crystals from petroleum ether, mp 43-44°. $[\alpha]_D^{20}$ −22.4° (c = 4 in benzene).

1410. β-Bromoisovaleric Acid. *3-Bromo-3-methylbutanoic acid.* $C_5H_9BrO_2$; mol wt 181.04. C 33.17%, H 5.01%, Br 44.14%, O 17.68%. $(CH_3)_2CBrCH_2COOH$. Prepn from β,β-dimethylacrylic acid and HBr: Auwers, *Ber.* **28,** 1130 (1895); from β-isovalerolactone and HBr: Gresham *et al.*, *J. Am. Chem. Soc.* **76,** 486 (1954).

Needles from ligroin, mp 73-74°. Slightly sol in water; sol in alcohol, ether, benzene.

1411. Bromolysergide. *(8β)-2-Bromo-9,10-didehydro-N,N-diethyl-6-methylergoline-8-carboxamide;* D-2-bromo-

lysergic acid diethylamide; 2-bromo-*N,N*-diethyl-D-lysergamide; bromo-LSD; BOL-148. $C_{20}H_{24}BrN_3O$; mol wt 402.35. C 59.70%, H 6.01%, Br 19.86%, N 10.44%, O 3.98%. Prepd by bromination of lysergide, *q.v.*: Troxler, Hofmann, *Helv. Chim. Acta* **40,** 2160 (1957). Serotonin antagonist without the hallucinogenic activity of LSD: J. P. Bennett, S. H. Snyder, *Brain Res.* **94,** 523 (1975).

Needles from ether, mp 120-127°. $[\alpha]_D^{20}$ +15° (c = 0.5 in pyridine); $[\alpha]_D^{20}$ +53° (c = 0.5 in chloroform). uv max: 240, 301 nm (log ε 4.28, 3.95).

1412. p-Bromomandelic Acid. *4-Bromo-α-hydroxybenzeneacetic acid; p-bromophenylglycolic acid.* $C_8H_7BrO_3$; mol wt 231.06. C 41.59%, H 3.06%, Br 34.58%, O 20.77%. Prepd by condensing bromobenzene with ethyl oxomalonate in presence of boron trifluoride: Riebsomer *et al.*, *J. Am. Chem. Soc.* **60,** 2974 (1938).

Crystals, mp 117-118°. Slightly sol in water. Zirconium salt. Soly in water at 25°: 0.0446 g/l. USE: Analytical reagent for zirconium.

1413. 1-Bromonaphthalene. α-Bromonaphthalene. $C_{10}H_7Br$; mol wt 207.07. C 58.00%, H 3.41%, Br 38.59%. Prepd by dropping bromine into a mixture of naphthalene and carbon tetrachloride, distilling the CCl_4, heating the residue with NaOH, and fractionating the liquid under reduced pressure: Clarke, Schram, *Org. Syn.* **1,** 35 (1921); Clarke, Brethen, *ibid.* **coll. vol. I,** 121 (2nd ed., 1941). Also prepd by heating naphthalene (liquid or gaseous) with bromine; 2-bromonaphthalene also being produced, esp with increasing temps: Wibaut, *Chem. Weekbl.* **39,** 326, 328 (1942); Suyver, Wibaut, *Rec. Trav. Chim.* **64,** 65 (1945).

Oily liquid at room temp. More pungent odor than naphthalene. d_4^{20} 1.4834; d_4^{25} 1.4785; d_4^{30} 1.4732. When solid it exists in two forms: mp 0.2-0.7° and mp 6.2°. Darkens on standing when distilled at 760 mm, but remains colorless when distilled at 16 mm. Volatile with steam. bp_{760} 281.1°; bp_{400} 252.0°; bp_{200} 224.2°; bp_{100} 198.8°; bp_{60} 183.5°; bp_{40} 170.2°; bp_{20} 150.2°; bp_{10} 133.6°; bp_5 117.5°; $bp_{1.0}$ 84.2°. $n_D^{16.5}$ 1.66011. Slightly sol in water; miscible with alcohol, ether, benzene, chloroform. Absorption spectrum: DeLaszlo, *Proc. Roy. Soc. London [A]* **111,** 356 (1926).

Compd with 1,3,5-trinitrobenzene, $C_{10}H_7Br.C_6H_3N_3O_6$, lemon-yellow needles, mp 137°.

Picrate, $C_{10}H_7Br.C_6H_3N_3O_7$, yellow needles, mp 135°. USE: Immersion fluid in the determination of the refractive index of crystals. For the determination of water in alc by the cloud point method. For refractometric fat determinations. Mixed with polymerized castor oil as a general immersion oil in microscopy.

1414. 2-Bromonaphthalene. β-Bromonaphthalene. $C_{10}H_7Br$; mol wt 207.09. C 58.00%, H 3.41%, Br 38.59%.

Prepn: Liebermann, Palm, *Ann.* **183,** 267 (1876); Vingiello *et al., J. Chem. Ed.* **40,** 544 (1963).

Crystals from alcohol, mp 59°; also reported as mp 54-56° (Vingiello *et al., loc. cit.*). bp 281-282°, bp$_{4.5}$ 122-127°. d 1.60. Slightly sol in water; sol in 8 parts alcohol; very sol in ether, chloroform, benzene.

1415. *p*-Bromophenacyl Bromide. *2-Bromo-1-(4-bromophenyl)ethanone;* 2,4'-dibromoacetophenone; *p,α-dibromoacetophenone.* $C_8H_6Br_2O$; mol wt 277.96. C 34.57%, H 2.18%, Br 57.50%, O 5.76%. Prepd by adding bromine to *p*-bromoacetophenone in glacial acetic acid at below 20°: Langley, *Org. Syn. coll. vol.* **I,** 127 (2nd ed., 1941).

Crystals from ethanol, mp 109-110°. Very sol in warm alcohol.

USE: Identification of carboxylic acids.

1416. Bromophenol. C_6H_5BrO; mol wt 173.02. C 41.65%, H 2.91%, Br 46.19%, O 9.25%. Prepn of *m*-isomer: Wurster, Nölting, *Ber.* **7,** 904 (1874); Carpenter *et al., J. Org. Chem.* **16,** 586 (1951); of *o*- and *p*-isomers: Medola, Streatfield, *J. Chem. Soc.* **73,** 681 (1898); Kaeding, Lindstrom, U.S. pat. **2,805,263** (1957 to Dow).

m-Bromophenol, *3-bromophenol.* Crystals, mp 33°, also reported as mp 31°. bp 235-236°, bp$_3$ 88-89°. Sol in alcohol, ether, alkalies.

o-Bromophenol, *2-bromophenol.* Yellow to red oily liquid; unpleasant odor. d ~1.5. bp 194°. mp 6°. Sol in water; miscible in chloroform, ether. Fusion with NaOH gives resorcinol.

p-Bromophenol, *4-bromophenol.* Tetragonal bipyramidal crystals from chloroform or ether. mp 64°. bp 238°. d^{15} 1.840; d^{80} 1.5875. Small amounts of water depress the mp considerably and may prevent crystallization. Absorption spectrum: Ley, *Z. Physik. Chem.* **94,** 412 (1920). Soluble in about 7 parts water; freely sol in alc, chloroform, ether, glacial acetic acid.

p-Form methyl ether, C_7H_7BrO, *p-bromoanisole.* Crystals, mp 9-10°. bp 223°.

p-Form ethyl ether, C_8H_9BrO, *p-bromophenetole.* Crystals, mp 4°. bp 233°.

USE: *p*-Form as disinfectant.

1417. *p*-Bromophenylhydrazine. $C_6H_7BrN_2$; mol wt 187.06. C 38.53%, H 3.77%, Br 42.72%, N 14.98%. Prepn from 4-bromophenyldiazonium chloride by modified Fischer phenylhydrazine synthesis: Chattaway, Humphrey, *J. Chem. Soc.* **1927,** 1323.

Needles from water, mp 108-109°. Sol in alcohol, ether, chloroform, benzene; moderately sol in ligroin; slightly sol in water.

USE: In prepn of indoleacetic acid derivatives; in the study of transosazonation of sugar phenylosazones.

1418. *p*-Bromophenyl Isocyanate. *1-Bromo-4-isocyanatobenzene;* 4-bromophenylcarbimide; *p*-bromophenylcarbonimide; *p*-bromocarbanil. C_7H_4BrNO; mol wt 198.03. C 42.45%, H 2.04%, Br 40.36%, N 7.07%, O 8.08%. Prepd by heating phenyl isocyanate dibromide: Curtius, *J. Prakt. Chem.* **87,** 517 (1913); by distilling *p*-bromophenylurethan and P_2O_5: Dennstedt, *Ber.* **13,** 228 (1880).

Needles, pungent odor, mp 42°. bp$_{14}$ 158°. Very sol in ether.

USE: Prepn of bromophenylurea and urethan derivatives.

1419. Bromophos. *Phosphorothioic acid O-(4-bromo-2,5-dichlorophenyl) O,O-dimethyl ester;* O-(4-bromo-2,5-dichlorophenyl)-O,O-dimethylphosphorothioate; bromophos-methyl; ENT 27162; OMS 658; S 1942; Nexion. $C_8H_8BrCl_2O_3PS$; mol wt 365.98. C 26.25%, H 2.20%, Br 21.83%, Cl 19.37%, O 13.12%, P 8.46%, S 8.76%. Prepn: **Belg.** pat. **625,198,** *C.A.* **60,** 13187a (1964) and R. Sehring, K. Zeile, U.S. pats. **3,227,610, 3,275,718** (1963, 1966; 1966 all to Boehringer, Ing.). Metabolism in rats: M. Stiasni *et al., J. Agr. Food Chem.* **15,** 474 (1967). *Review:* D. Eichler, *Residue Rev.* **41,** 65-112 (1972).

Yellow crystals, mp 53-54°. bp$_{0.01}$ 140-142°. Vapor press at 20°C: 1.3×10^{-4} mm Hg. Soly in water at 25°: 40 mg/l. Sol in CCl_4, ether, toluene. Stable in soln up to pH 9. Non-corrosive. LD$_{50}$ orally in male, female rats: 1600, 1730 mg/kg, T. B. Gaines, *Toxicol. Appl. Pharmacol.* **14,** 515 (1969).

O,O-Diethyl analog, $C_{10}H_{12}BrCl_2O_3PS$, *bromophos-ethyl,* *ENT 27258, OMS 659, S 2225, Nexagan.* Colorless to pale yellow oil, bp$_{0.001}$ 122-133°. Vapor press at 30°: 4.6×10^{-5} mm Hg. Soly in water at 25°: 2 mg/l.

USE: Insecticide, acaricide.

1420. Bromopride. *4-Amino-5-bromo-N-[2-(diethylamino)ethyl]-2-methoxybenzamide;* 4-amino-5-bromo-N-[2-(diethylamino)ethyl]-*o*-anisamide; Emepride; Emoril. $C_{14}H_{22}BrN_3O_2$; mol wt 344.26. C 48.84%, H 6.44%, Br 23.21%, N 12.21%, O 9.30%. Prepn: H. Mori, K. Shibata, **Ger.** pat. **2,119,724** (1971 to Teikoku Hormone Manuf. Co.), *C.A.* **76,** 99375e (1972). Action on guinea-pig ileum: J. Fontaine, J. J. Reuse, *Arch. Int. Pharmacodyn. Ther.* **213,** 322 (1975). Effects on intestinal peristalsis in dogs: P. Sava *et al., Arzneimittel-Forsch.* **29,** 799 (1979). Clinical study: M. Fischer *et al., Fortschr. Med.* **97,** 883 (1979). Pharmacokinetics: P. W. Lücker *et al., Arzneimittel-Forsch.* **33,** 453 (1983).

Hydrochloride, $C_{14}H_{23}BrClN_3O_2$, *Cascapride, Plesium, Praiden, Valopride, Viaben.*

Dihydrochloride monohydrate, $C_{14}H_{24}BrCl_2N_3O_2 \cdot H_2O$, *Opridan.*

THERAP CAT: Anti-emetic.

1421. β**-Bromopropionic Acid.** *3-Bromopropanoic acid.* $C_3H_5BrO_2$; mol wt 152.99. C 23.55%, H 3.29%, Br 52.24%, O 20.92%. $BrCH_2CH_2COOH$. Prepd by the action of hydrobromic acid on acrylic acid, on hydracrylic acid, and on ethylene cyanohydrin: Kendall, McKenzie, *Org. Syn.* **3,** 25 (1923). Prepn of the ethyl ester: *eidem, ibid.* 51; of the methyl ester: Mozingo, Patterson, *ibid.* **20,** 64 (1940).

Plates from CCl_4. mp 62.5°. K at 25°: 9.8×10^{-5}. Sol in water, alcohol, ether, chloroform, benzene. Aq alkalies hydrolyze β-bromopropionic acid to hydracrylic acid.

Ethyl ester, $C_5H_9BrO_2$, liquid. Pungent odor. Becomes yellow on exposure to light. d_4^{18} 1.4123. bp_{44} 112°; bp_{12} 70°. n_D^{18} 1.4569. Insol in water; miscible with alchol, ether. *Protect from light.*

Methyl ester, $C_4H_7BrO_2$, liquid. d^{15} 1.4897. bp_{27} 80°. bp_{18} 65°.

1422. Bromopropylate. *4-Bromo-α-(4-bromophenyl)-α-hydroxybenzeneacetic acid 1-methyl ethyl ester; 4,4'-dibromobenzilic acid isopropyl ester;* isopropyl 4,4'-dibromobenzilate; phenisobromolate; ENT 27552; GS 19851; Acarol; Folbex VA; Neoron. $C_{17}H_{16}Br_2O_3$; mol wt 428.14. C 47.69%, H 3.77%, Br 37.33%, O 11.21%. Contact acaricide with residual activity. Prepn: *Fr. pat.* **1,504,969** corresp to K. Gubler, *U.S. pat.* **3,639,446** (1967, 1972 to Geigy). Persistence in soil: W. B. Wheeler *et al., J. Environ. Qual.* **2,** 115 (1973).

Cryst solid, mp 77°. Vapor press at 20°: 5.1×10^{-8} mm Hg. Soly in water at 20°: < 5 mg/l. Readily sol in most organic solvents. LD_{50} orally in rats: 5000 mg/kg, *RTECS* **vol. I,** R. J.. Lewis, R. L. Tatken, Eds. (1980) p 276.

USE: Acaricide.

1423. Bromosalicylchloranilide. *5-Bromo-N-(4-chlorophenyl)-2-hydroxybenzamide; 5-bromo-4'-chlorosalicylanilide; N-5*-bromosalicyloyl-*p*-chloroaniline; Multifungin. $C_{13}H_9BrClNO_2$; mol wt 326.60. C 47.81%, H 2.78%, Br 24.47%, Cl 10.86%, N 4.29%, O 9.80%. Prepn: Schuler, *U.S. pat.* **2,802,029** (1957 to Knoll); Lemaire *et al., J. Pharm. Soc.* **50,** 831 (1961).

Crystals, mp 238-243°.

THERAP CAT: Antifungal.

1424. 5-Bromosalicylhydroxamic Acid. *5-Bromo-N,2-dihydroxybenzamide;* Brosalamid; Bromocyl. $C_7H_6BrNO_3$; mol wt 232.05. C 36.23%, H 2.61%, Br 34.44%, N 6.04%, O 20.69%. Prepd by direct bromination of salicylhydroxamic acid in acetic acid: Urbanski *et al., Nature* **170,** 753 (1952); *Bull. Acad. Polon. Sci.,* cl. III, **1,** 319 (1953); Hornung, Krakowska, *Gruzlica* **20,** 469 (1952).

Crystals from alcohol, dec 232°. Very sparingly sol in water. Forms a water-soluble sodium salt.

THERAP CAT: Antibacterial (tuberculostatic).

1425. 5-Bromosalicylic Acid Acetate. *2-(Acetyloxy)-5-bromobenzoic acid;* acetyl-5-bromosalicylic acid; bromoaspirin. $C_9H_7BrO_4$; mol wt 259.06. C 41.72%, H 2.72%, Br 30.85%, O 24.70%. Prepn: Robertson, *J. Chem. Soc.* **81,** 1482 (1902).

Minute crystals from alcohol, mp 168-169°. Tastes slightly sour with sweet aftertaste. One gram dissolves in about 1500 ml water, 9 ml alcohol, 21 ml ether.

THERAP CAT: Analgesic.

1426. Bromosaligenin. *5-Bromo-2-hydroxybenzenemethanol; 5-bromo-2-hydroxybenzyl alcohol;* 5-bromosaligenin; Bromsalizol. $C_7H_7BrO_2$; mol wt 203.04. C 41.40%, H 3.48%, Br 39.36%, O 15.76%. Prepd by splitting bromosalicin with emulsin: Visser, *Arch. Pharm.* **235,** 551 (1897); by the action of bromine on saligenin in water: Auwers, Büttner, *Ann.* **302,** 138 (1898); Adams *et al., J. Am. Chem. Soc.* **45,** 2419 (1923); Dunning *et al., ibid.* **58,** 1567 (1936); by the electrolytic reduction of 5-bromo-2-hydroxybenzoic acid: Mettler, *Ber.* **39,** 2939 (1906).

Shiny, pearly leaflets, mp 109°. Solubility in water at 25°: 0.70% w/w; considerably more soluble in hot water. Soly in olive oil: 4.8% w/w; freely sol in alc, ether, ethyl acetate; moderately sol in chloroform, benzene.

THERAP CAT: Anti-inflammatory.

1427. Bromosuccinic Acid. *Bromobutanedioic acid;* monobromosuccinic acid. $C_4H_5BrO_4$; mol wt 197.00. C 24.39%, H 2.56%, Br 40.57%, O 32.49%. $HOOCCH_2CHBr$-COOH. Prepn of *dl*-form by bromination of succinic acid: Kekulé, *Ann.* **117,** 120 (1861); from fumaric acid and HBr: Fittig, *Ann.* **188,** 42 (1877); by bromination of succinyl bromide or succinic anhydride: Hughes, Watson, *J. Chem. Soc.* **1930,** 1733. Prepn of *l*-form from *l*-aspartic acid: Walden, *Ber.* **29,** 133 (1896); Karrer *et al., Helv. Chim. Acta* **30,** 271 (1947).

dl-Form, crystals, mp 161°. d 2.07. Soluble in 5.5 parts water; sol in alc.

d-Form, crystals, mp 172°. $[\alpha]_D^{15}$ +41.9° (c = 5); $[\alpha]_D^{20}$ +67.9° (ether). *Ref:* Levene, Mikeska, *J. Biol. Chem.* **55,** 795 (1923); Clough, *J. Chem. Soc.* **129,** 1674 (1926).

l-Form, crystals, dec 177-178°. $[\alpha]_D$ −43.8° (c = 6), −65.0° (c = 6 in abs alc), −73.5° (c = 6 in acetone).

1428. N-Bromosuccinimide. *1-Bromo-2,5-pyrrolidinedione;* succinbromimide; NBS. $C_4H_4BrNO_2$; mol wt 178.00. C 26.99%, H 2.26%, Br 44.90%, N 7.87%, O 17.97%. Prepn: K. Ziegler *et al., Ann.* **551,** 109 (1942). In bromination of olefins: *eidem, ibid.* 80. The method has been extended to other classes of compds by the use of catalysts: H. Schmid, P. Karrer, *Helv. Chim. Acta* **29,** 573 (1946); H. Schmid, *ibid.* 1144. In oxidation of aldehydes: Y. F. Cheung, *Tetrahedron Letters* **1979,** 3809. Review of uses: C. Djerassi, *Chem. Rev.* **43,** 271-317 (1948); R. Filler, *ibid.* **63,** 21-43 (1963); J. S. Pizey, *Synthetic Reagents,* **vol. 2** (John Wiley, New York, 1974) pp 1-63.

Orthorhombic bisphenoidal crystals, mp 173-175° (slight

decompn). Faint odor of bromine. d 2.098. Soly (g/100 g of solvent at 25°): water 1.47; *tert*-butanol 0.73; acetone 14.40; carbon tetrachloride 0.02; hexane 0.006; glacial acetic acid 3.10.

Caution: Highly irritating to eyes, skin, mucous membranes.

USE: In bromination of olefins; in oxidation of alcohols to aldehydes and ketones and of aldehydes to acid bromides.

1429. Bromotoluene. C_7H_7Br; mol wt 171.04. C 49.15%, H 4.13%, Br 46.72%. Prepn of *o*-isomer: Bourgeois, *Ber.* **28**, 2312 (1895). Prepn of *m*-isomer: Acree, *Ber.* **37**, 994 (1904). Kohn, Bum, *Monatsh.* **33**, 924 (1912); Feitler, Z. *Physik. Chem.* **4**, 77 (1889). Prepn of *m*-, *o*- and *p*-isomers: Bigelow *et al.*, *Org. Syn. coll. vol.* I (2nd ed., 1941) pp 133, 135, 136 resp. Isomerization of *p*-isomer yields mixture of monobromotoluenes containing 48.2% of *o*-isomer: Crump, U.S. pat. **3,077,503** (1963 to Dow). Separation of *o*- and *p*-isomers by clathration: Coscia, U.S. pat. **3,114,784** (1963 to Am. Cyanamid).

m-Bromotoluene, 1-bromo-3-methylbenzene, 3-bromotoluene, m-tolylbromide. Liquid. d_4^{20} 1.4099; d_4^{58} 1.309; d_4^{184} 1.201. mp −39.8°. bp$_{760}$ 183.7°; bp$_{400}$ 160°; bp$_{200}$ 138°; bp$_{100}$ 117.8°; bp$_{60}$ 104.1°; bp$_{40}$ 93.9°; bp$_{20}$ 78.1°; bp$_{10}$ 64.0°; bp$_5$ 50.8°; bp$_{1.0}$ 14.8°. n_D^{20} 1.551. Absorption spectrum: Purvis, *J. Chem. Soc.* **99**, 1706, 1710 (1911). Sol in alc, ether, benzene.

o-Bromotoluene, colorless liq. d_{15}^{15} 1.431. bp 181°. mp −26°. n_D^{20} 1.555. Practically insol in water; misc with alc, benzene.

p-Bromotoluene, crystals from abs alc. d_{35}^{35} 1.3959; d_{50}^{50} 1.3856; d_{100}^{100} 1.3637; d^{184} 1.1931. mp 28.5°. bp$_{760}$ 184.5°; bp$_{100}$ d 116.4°; bp$_{60}$ 102.3°; bp$_{40}$ 91.8°; bp$_{20}$ 75.2°; bp$_{10}$ 61.1°; bp$_5$ 47.5°; bp$_{1.0}$ 10.3°. n_D^{20} 1.5490. Sol in alc, ether, benzene.

Caution: Irritants!

1430. 5-Bromouracil. *5-Bromo-2,4(1H,3H)-pyrimidinedione.* $C_4H_3BrN_2O_2$; mol wt 190.98. C 25.15%, H 1.58%, Br 41.84%, N 14.67%, O 16.76%. Major chemical mutagen; incorporates into DNA, altering base-pair sequencing by replacing thymine. Prepn: H. L. Wheeler, H. F. Merriam, *Am. Chem. J.* **29**, 478 (1903); P. A. Levene, F. B. LaForge, *Ber.* **45**, 608 (1912). UV effects on DNA containing 5-bromouracil: A. G. Skavronskaya *et al.*, *Mutat. Res.* **6**, 319 (1968); F. Hutchinson, *Quart. Rev. Biophys.* **6**, 201 (1973). Mechanisms of mutagenesis: I. Pietrzykowska, *Mutat. Res.* **19**, 1 (1973); E. M. Witkin, E. C. Parisi, *ibid.* **25**, 407 (1974); B. Rydberg, *Molec. Gen. Genet.* **152**, 19 (1977). Crystal structure and base stacking properties: H. Sternglanz, C. E. Bugg, *Biochim. Biophys. Acta* **378**, 1 (1975).

Prisms from water, mp 293°.
USE: Exptly as mutagen.

1431. Bromoxynil. *3,5-Dibromo-4-hydroxybenzonitrile;* 3,5-dibromo-4-hydroxyphenyl cyanide; 2,6-dibromo-4-cyanophenol; broxynil; ENT 20852; MB 10064; Brominil; Buctril. $C_7H_3Br_2NO$; mol wt 276.92. C 30.36%, H 1.09%, Br 57.71%, N 5.06%, O 5.78%. Selective contact herbicide. Prepn: K. Auwers, J. Reis, *Ber.* **29**, 2359 (1896); E. Müller *et al.*, *Ber.* **92**, 2278 (1959); K. Carpenter *et al.*, *Weed Res.* **4**, 175 (1964); Fr. pat. **1,375,311** (1964 to May & Baker), *C.A.* **62**, 3982h (1965). Herbicidal activity: K. Carpenter, B. J. Heywood, *Nature* **200**, 28 (1963). HPLC analysis: J. C.

Van Damme, M. Galoux, *J. Chromatogr.* **190**, 401 (1980). Persistance in soil: A. E. Smith, *Pestic. Sci.* **11**, 341 (1980).

Colorless solid, mp 194-195°. pKa 4.06. Subl at 135°/0.15 mm Hg. Slightly volatile in steam. Soly in (g/l) at 25°: water 0.13; methanol 90; acetone 170; tetrahydrofuran 410. LD_{50} orally in mice: 111 mg/kg, K. Carpenter *et al.*, *loc. cit.*

Octanoate ester, $C_{15}H_{17}Br_2NO_2$, *MB 10731, NPH 1320, RP 16272.* Waxy solid, mp 45-46°. Low volatility; subl at 90°/0.1 mm Hg. Insol in water.

USE: Herbicide.

1432. Bromperidol. *4-[4-(4-Bromophenyl)-4-hydroxy-1-piperidinyl]-1-(4-fluorophenyl)-1-butanone; 4-[4-(p-bromophenyl)-4-hydroxypiperidino]-4'-fluorobutyrophenone;* R 11,333; Azurene; Impromen; Tesoprel. $C_{21}H_{23}BrFNO_2$; mol wt 420.33. C 60.01%, H 5.51%, Br 19.01%, F 4.52%, N 3.33% O 7.61%. Bromine analog of haloperidol, *q.v.* Prepn (no data): P. A. J. Janssen, **Brit.** pat. **895,309** corresp to U.S. pat. **3,438,991** (1962, 1969 both to Janssen); C. J. E. Niemegeers, P. A. J. Janssen, *Arzneimittel-Forsch.* **24**, 45 (1974). Prepn of ^{82}Br bromperidol and preliminary tissue distribution studies: S. H. Vincent *et al.*, *J. Med. Chem.* **23**, 75 (1980). Pharmacological study: C. J. E. Niemegeers, P. A. J. Janssen, *Life Sci.* **24**, 2201 (1979). Preliminary clinical study: B. Woggon *et al.*, *Int. Pharmacopsychiatry* **14**, 213 (1979). Radioimmunoassay: E. Van Den Eeckhout *et al.*, *Eur. J. Drug Metab. Pharmacokinet.* **5**, 45 (1980).

Off-white amorphous or microcrystalline powder, mp 155-158°. uv max: 245 nm. pKa 8.6-8.7. Soly in water: 0.09 mg/ml; in 0.1*M* tartaric, lactic, citric and acetic acids: greater than or equal to 10 mg/ml.

THERAP CAT: Antipsychotic.

1433. Brompheniramine. *γ-(4-Bromophenyl)-N,N-dimethyl-2-pyridinepropanamine; 2-[p-bromo-α-(2-dimethylaminoethyl)benzyl]pyridine;* 1-(p-bromophenyl)-1-(2-pyridyl)-3-dimethylaminopropane; 3-(p-bromophenyl)-3-(2-pyridyl)-N,N-dimethylpropylamine; parabromdylamine. $C_{16}H_{19}BrN_2$; mol wt 319.26. C 60.19%, H 6.00%, Br 25.03%, N 8.78%. Prepn from α-(p-bromophenyl)-α-(β-dimethylaminoethyl)-2-pyridylacetonitrile: Sperber *et al.*, U.S. pats. **2,567,245** (1951) and **2,676,964** (1954 to Schering). Prepn of *d*-form: L. A. Walter, U.S. pat. **3,061,517** (1962 to Schering).

Oily liquid. Slightly yellow color. Characteristic amine-like odor. bp$_{0.5}$ 147-152°. Soluble in dilute acids.

Maleate, $C_{20}H_{23}BrN_2O_4$, *Dimegan, Dimetane, Dimotane, Ebalin, Ilvin, Nagemid, Symptom 3, Veltane.* Crystals, mp 132-134°. Sol in water, somewhat less sol in alcohol. pH of a 2% aq soln about 5.

d-Form, *dexbrompheniramine,* (+)-*parabromdylamine.* Oily liquid. $[\alpha]_D^{25}$ +42.7° (c = 1 in DMF).

d-Form maleate, $C_{20}H_{23}BrN_2O_4$, *Disomer, Ebalin.* Crystals, mp 103-113°.

THERAP CAT: Antihistaminic.

1434. Bromphenol Blue. *4,4'-(3H-2,1-Benzoxathiol-3-ylidene)bis[2,6-dibromophenol] S,S-dioxide;* α,α-bis(3,5-di-

bromo-4-hydroxyphenyl)-α-hydroxy-o-toluenesulfonic acid, *γ-sultone;* 3,3′,5,5′-tetrabromophenolsulfonphthalein; Albutest. $C_{19}H_{10}Br_4O_5S$; mol wt 670.02. C 34.06%, H 1.51%, Br 47.71%, O 11.94%, S 4.79%. Prepd by slow addition of excess bromine to a hot soln of phenolsulfonphthalein in glacial acetic acid: White, Acree, *J. Am. Chem. Soc.* **41**, 1205 (1919); Orndorff, Sherwood, *ibid.* **45**, 495 (1923). *See also* ref given *under* Bromcresol Green.

Elongated hexagonal prisms from acetic acid + acetone. Dec 279°. (An orange discoloration with formation of a green sublimate sets in at 210°). Soluble in water (about 0.4 g/100 ml); more sol in methyl and ethyl alcohol, and in benzene. Freely sol in NaOH solns with the formation of a water-sol sodium salt. pK = 4.0. To prepare a soln for use as pH indicator, dissolve 0.10 g in 7.45 ml $N/50$ NaOH and dil with water to 250 ml.

USE: As indicator, pH 3.0 yellow; pH 4.6 purple.

1435. Bromthymol Blue. *4,4′-(3H-2,1-Benzoxathiol-3-ylidene)bis[2-bromo-3-methyl-6-(1-methylethyl)phenol] S,S-dioxide; α,α-bis(6-bromo-5-hydroxycarvacryl)-α-hydroxy-o-toluenesulfonic acid, γ-sultone;* 3,3′-dibromothymolsulfonphthalein. $C_{27}H_{28}Br_2O_5S$; mol wt 624.39. C 51.93%, H 4.52%, Br 25.60%, O 12.81%, S 5.14%. Prepd by the action of bromine on thymol blue in glacial acetic acid: Clark, Lubs, *J. Wash. Acad. Sci.* **5**, 609 (1915); **6**, 482 (1916); *Chem. Zentr.* **1916**, I, 175; II, 1068. *See also* ref *under* Bromcresol Green.

Cream-colored crystals. Sparingly sol in water; sol in alcohol and in aq solns of alkalies. Also sol in ether. Less sol in benzene, toluene, xylene. Practically insol in petr ether. pK = 7.0. To prepare a soln for use as pH indicator, dissolve 0.10 g in 8.0 ml $N/50$ NaOH and dil with water to 250 ml. To prepare a soln for use as indicator in volumetric work, dissolve 0.1 g in 100 ml of 50% alc.

USE: As indicator, pH 6.0 yellow; pH 7.6 blue.

1436. p-Bromtripelennamine. *N-[(4-Bromophenyl)-methyl]-N′,N′-dimethyl-N-2-pyridinyl-1,2-ethanediamine; 2-[(p-bromobenzyl)(2-dimethylaminoethyl)amino]pyridine; N-(2-pyridyl)-N-(p-bromobenzyl)-N′,N′-dimethylethylenediamine; N,N-dimethyl-N′-(4-bromobenzyl)-N′-(2-pyridyl)ethylenediamine.* $C_{16}H_{20}BrN_3$; mol wt 334.28. C 57.49%, H 6.03%, Br 23.91%, N 12.57%. Prepn: Vaughan *et al., J. Org. Chem.* **14**, 228 (1949); Howard, U.S. pat. **2,572,-569** (1951 to Am. Cyanamid). Activity: Netter, Bodenschatz, *Biochem. Pharmacol.* **16**, 1627 (1967).

Yellow oil, $bp_{1.5}$ 184-190°.

Hydrochloride, $C_{16}H_{21}BrClN_3$, *Hibernon.* Crystals from ethyl acetate, mp 184-186.

THERAP CAT: Antihistaminic.

1437. Bronopol. *2-Bromo-2-nitro-1,3-propanediol; β-bromo-β-nitrotrimethyleneglycol;* Bronosol. $C_3H_6BrNO_4$; mol wt 200.00. C 18.02%, H 3.03%, Br 39.95%, N 7.00%, O 32.00%. $HOCH_2CBr(NO_2)CH_2OH$. Prepn: E. Schmidt, R. Wilkendorf, *Ber.* **52**, 389 (1919); R. Wessendorf, **Ger.** pats. **1,804,068, 1,954,173** corresp to U.S. pats. **3,658,921, 3,711,-561** (1970, 1971, 1972, 1973 all to Henkel & Cie). Antibacterial, antifungal and toxic properties: Z. Eckstein *et al., Bull. Acad. Polon. Sci. Ser. Sci. Chim.* **11**, 687 (1963); B. Croshaw *et al., J. Pharm. Pharmacol.* **16** (Suppl), 127T (1964); N. G. Clark *et al.,* **Brit.** pat. **1,057,131** corresp to U.S. pat. **3,558,788** (1967, 1971 both to Boots); R. J. Stretton, T. W. Mason, *J. Appl. Bacteriol.* **36**, 61 (1973). Biotransformation and distribution in rats: H. S. Buttar, R. H. Downie, *Toxicol. Lett.* **6**, 101 (1980). As a nitrosating agent for diethanolamine: I. Schmeltz, A. Wenger, *Food Cosmet. Toxicol.* **17**, 105 (1979); R. L. Elder, *J. Environ. Pathol. Toxicol.* **4**, 47 (1980). Review of bronopol and other preservatives: B. Croshaw, *J. Soc. Cosmet. Chem.* **28**, 3-16 (1977).

Odorless crystals from ethyl acetate-chloroform, mp 120-122°. Sol in water, alcohol, ethyl acetate. Slightly sol in chloroform, acetone, ether, benzene. Insol in ligroin. LD_{50} orally in mice, rats (mg/kg): 350, 400 (Croshaw).

USE: Preservative in cosmetics and toiletries. Antiseptic.

1438. Broparoestrol. *1-(2-Bromo-1,2-diphenylethenyl)-4-ethylbenzene; 1-bromo-2-(p-ethylphenyl)-1,2-diphenylethylene;* 1-bromo-2-(4-ethylphenyl)-1,2-diphenylethylene; α-bromo-α,β-diphenyl-β-(p-ethylphenyl)ethylene; 1-(p-ethylphenyl)-1,2-diphenyl-2-bromoethylene; B.D.P.E.; LN 107; Acnestrol; Longestrol. $C_{22}H_{19}Br$; mol wt 363.31. C 72.73%, H 5.27%, Br 22.00%. Prepn: Dvolaitsky, Jacques, *Bull. Soc. Chim. Biol.* **40**, 939 (1958).

cis-trans Mixture, crystals, mp 89°. (Isomer A, needles from ethanol, mp 112-113.5°; isomer B, mp 111.5-112°). The commercial medicinal product is the *cis-trans* mixture. Soluble in ether, benzene, chloroform; less sol in ethanol.

THERAP CAT: Estrogen. Used in dermatology.

1439. Brotizolam. *2-Bromo-4-(2-chlorophenyl)-9-methyl-6H-thieno[3,2-f][1,2,4]triazolo[4,3-a][1,4]diazepine;* 2-bromo-4-(o-chlorophenyl)-9-methyl-6H-thieno[3,2-f]-s-triazolo[4,3-a][1,4]diazepine; 8-bromo-6-(o-chlorophenyl)-1-methyl-4H-s-triazolo[3,4-c]thieno[2,3-e]-1,4-diazepine; WE 941-BS; Lendorm; Lendormin. $C_{15}H_{10}BrClN_4S$; mol wt 393.70. C 45.76%, H 2.56%, Br 20.30%, Cl 9.00%, N 14.23%, S 8.14%. One of a class of triazolo-1,4-thienodiazepines having psychotropic activity. Prepn: K. H. Weber *et al.,* **Ger.** pat. **2,410,030** corresp to U.S. pat. **4,094,984** (1975, 1978 to Boehringer, Ing.); *eidem, Ann.* **8**, 1257 (1978). Pharmacodynamics: J. Gruenberger *et al., Curr. Ther. Res.* **24**, 427 (1978). Bioavailability: B. Saletu *et al., Arzneimittel-Forsch.* **29**, 700 (1979). Comparative effects in monkeys: A. Nicholson, *Brit. J. Pharmacol.* **70**, 141P (1980). Effect on sleep and performance in man: *eidem, ibid.* 157P.

Pharmaco-EEG study: M. Fink, P. Irwin, *Clin. Pharmacol. Ther.* **30**, 336 (1981). Toxicity studies: C. Hewett *et al.,* *Arzneimittel-Forsch.* **36**, 592 (1986). Series of articles on pharmacology, pharmacokinetics, toxicology: *ibid.* 517-620.

Colorless cryst from ethanol, mp 212-214°. LD_{50} in mice, rats (mg/kg): >10000, >10000 orally; 920, 1000 i.p. (Hewett).

THERAP CAT: Sedative. Hypnotic.

1440. Brovincamine. *(3α,14β,16α)-11-Bromo-14,15-dihydro-14-hydroxyeburnamenine-14-carboxylic acid methyl ester; cis-11-bromovincamine.* $C_{21}H_{25}BrN_2O_3$; mol wt 433.35. C 58.21%, H 5.81%, Br 18.44%, N 6.46%, O 11.08%. Cerebrovascular agent, vincamine deriv. Prepn: P. Pfaeffli, **Ger.** pat. **2,458,164** corresp to **U.S.** pat. **4,146,643** (1975, 1979 both to Sandoz). HPLC determn in plasma: R. R. Brodie, L. F. Chasseaud, *J. Chromatog.* **228**, 413 (1982). Effect on autonomic nervous system in laboratory animals: K. Kushiku *et al., J. Pharmacobio.-Dyn.* **7**, 177 (1984). Cardiovascular effects, mechanistic studies: *eidem, Clin. Exp. Pharmacol. Physiol.* **12**, 121 (1985). Mode of action studies: T. Katsuragi *et al., Gen. Pharmacol.* **15**, 43 (1984). Clinical trials in patients with multi-infarct dementia: S. Hagstadius *et al., Psychopharmacology (Berlin)* **83**, 321 (1984).

Crystals from isopropanol, mp 214° (dec). $[\alpha]_D^{20}$ +8.7° (1% in $CHCl_3$).

Hydrogen fumarate, $C_{25}H_{29}BrN_2O_7$, *BV 26-723, Sabromin, Zabromin.* mp 144°. $[\alpha]_D^{20}$ +4.7 (0.388% in H_2O).

THERAP CAT: Peripheral vasodilator.

1441. Broxyquinoline. *5,7-Dibromo-8-quinolinol;* 5,7-dibromo-8-hydroxyquinoline; Brodiar; Broxykinolin; Colepur; Colipar; Fenilor; Intensopan; Paramibe. $C_9H_5Br_2NO$; mol wt 302.97. C 35.68%, H 1.66%, Br 52.76%, N 4.62%, O 5.28%. Prepn by bromination of 5-formyl-8-quinolinol: Matsumura, Ito, *J. Am. Chem. Soc.* **77**, 6671 (1955); by bromination of 8-hydroxyquinaldine: Irving, Pinnington, *J. Chem. Soc.* **1957**, 285; by bromination of 8-quinolinol: Luis, Palomo, *Afinidad* **28**, 163 (1951), *C.A.* **47**, 10533d (1953); Zinnei, Fiedler, *Arch. Pharm.* **291**, 493 (1958); Aristov, Kostina, *Zh. Obshch. Khim.* **34**, 3421 (1964). Crystal structure: Kashino, Haisa, *Bull. Chem. Soc. Japan* **46**, 1094 (1973). Metabolism: Rodriguez, Close, *Biochem. Pharmacol.* **17**, 1647 (1968).

Monoclinic needles from alc. d 2.189. mp 196°. Freely sol in chloroform, alcohol, benzene, acetic acid; slightly sol in ether; practically insol in water.

USE: In testing for Cu, Fe, Ti.

THERAP CAT: Antiseptic; disinfectant.

1442. Bruceantin. *15-[(3,4-Dimethyl-1-oxo-2-pentenyl)oxy]-13,20-epoxy-3,11,12-trihydroxy-2,16-dioxopicras-3-en-21-oic acid methyl ester;* NSC-165563. $C_{28}H_{36}O_{11}$; mol wt 548.60. C 61.30%, H 6.62%, O 32.08%. Antileukemic quassinoid from the simaroubaceous tree *Brucea antidysenterica* J. F. Mill. Isoln and structure: S. M. Kupchan, **Ger.** pat. **2,347,576** corresp to **U.S.** pat. **3,969,369** (1975, 1976 both to Research Corp.); S. M. Kupchan *et al., J. Org. Chem.* **40**, 648 (1975). Mode of action: L. L. Liao *et al., Mol. Pharmacol.* **12**, 167 (1976). *In vivo* study: R. K. John-

son *et al., Cancer Treat. Rep.* **62**, 1535 (1978). Pharmacology: S. M. Sieber *et al., ibid.* **60**, 1127 (1976); M. Fresno *et al., Biochim. Biophys. Acta* **518**, 104 (1978). Clinical study: A. Y. Bedikian *et al., Proc. Am. Assoc. Cancer Res.* **20**, 193 (1979). Toxicologic evaluation: T. R. Castles *et al., U.S. NTIS Rep.* **PB-257175** (1976) 348 pp. Synthetic studies: O. D. Dailey, P. L. Fuchs, *J. Org. Chem.* **45**, 216 (1980); R. J. Pariza, P. L. Fuchs, *J. Org. Chem.* **48**, 2306 (1983).

Crystals from ether, mp 225-226°. $[\alpha]_D^{25}$ −43° (c = 0.31 in pyridine). uv max (ethanol): 280, 221 nm (ϵ 8680, 18,000); (ethanol, NaOH): 328, 221 nm (ϵ 7290, 28,600). LD_{50} in male, female mice: 1.95, 2.58 mg/kg i.v., T. R. Castles *et al., loc. cit.*

1443. Brucine. *2,3-Dimethoxystrychnidin-10-one;* 10,11-dimethoxystrychnine. $C_{23}H_{26}N_2O_4$; mol wt 394.47. C 70.03%, H 6.64%, N 7.10%, O 16.22%. From Strychnos seeds *(Strychnos nux-vomica* L. and *S. ignatii* Berg., *Loganiaceae):* Hartwick, Geiger, *Arch. Pharm.* **239**, 491 (1901). Structure: Findlay, *J. Am. Chem. Soc.* **73**, 3008 (1951). Prepn from strychnine, *q.v.:* E.Tedeschi *et al., Tetrahedron* **24**, 4573 (1968); P. Rosenmund *et al., Ber.* **101**, 2754 (1968). ^{13}C-NMR study: Wenkert *et al., J. Org. Chem.* **43**, 1099 (1978). *Review:* H. L. Holmes in R. H. F. Manske, H. L. Holmes, *The Alkaloids* vol. I (Academic Press, New York, 1950) pp 377-420.

Needles from acetone + water, mp 178°. $[\alpha]_D$ −127° (chloroform), −85° (in abs alcohol). uv max (ethanol): 263, 301 nm (log ϵ 4.09, 3.93).

Tetrahydrate, monoclinic prisms. Also forms a dihydrate. Very bitter taste. Bitterness threshold 1:220,000. *Very poisonous!* Becomes anhydr at 100°. One gram dissolves in 0.8 ml methanol, 1.3 ml alcohol, 5 ml chloroform, 25 ml ethyl acetate, 36 ml glycerol, about 100 ml benzene, 187 ml ether, 1320 ml water, 750 ml boiling water. pH of satd water soln 9.5. pK_1 6.04, pK_2 11.7. *Caution:* Handle dry powder in hood only. LD_{50} orally in rats: 1 mg/kg, *RTECS* **Vol. 1**, R. J. Lewis, R. L. Tatken, Eds. (1979) p 324.

Hydrochloride, $C_{23}H_{26}N_2O_4$·HCl, crystals. Sol in water or alcohol. The solns are neutral or slightly acid.

Nitrate dihydrate, $C_{23}H_{26}N_2O_4$·HNO_3·$2H_2O$, prisms, dec 230°. Sol in water or alcohol.

Sulfate heptahydrate, $(C_{23}H_{26}N_2O_4)_2$·H_2SO_4·$7H_2O$, small, bitter crystals. One gram dissolves in 75 ml cold water, about 10 ml boiling water, 105 ml alcohol, 170 ml chloroform. Soly in water is decreased by H_2SO_4.

Human Toxicity: A highly toxic alkaloid resembling strychnine.

USE: Denaturing alcohol and oils; in analytical chemistry; for separating racemic mixtures. Has been patented as addition agent to lubricants.

THERAP CAT: Central stimulant.

1444. Bryonia. Bryony. Dried root of *Bryonia alba* L., or of *B. dioica* Jacq., *Cucurbitaceae. Habit.* Europe. *Constit.* Bryonin, bryonidin, bryonicine, bryoamarid glycoside, Δ⁷-

stigmastenol, volatile oil, resin. Isoln of constituents: Tunmann, Wienecke, *Arch. Pharm.* **293**, 195 (1960); Biglino, *Farmaco Ed. Sci.* **14**, 673 (1959). Bryogenine structure studies: Biglino *et al., Tetrahedron Letters* **1963**, 1651. *Poisonous!*

THERAP CAT: Cathartic.

THERAP CAT (VET): Formerly as a purgative.

1445. Bucetin. *N-(4-Ethoxyphenyl)-3-hydroxybutanamide; 3-hydroxy-p-butyrophenetidide; β-hydroxybutyric acid p-phenetidide; p-ethoxy-N-(β-hydroxybutyryl)aniline;* Betadid. $C_{12}H_{17}NO_3$; mol wt 223.27. C 64.55%, H 7.67%, N 6.27%, O 21.50%. Phenetidine derivative with analgesic and antipyretic activity. Prepn: G. Ehrhart *et al.,* **U.S.** pat. **2,830,087** (1958 to Hoechst); K. Kummerle *et al.,* **Belg.** pat. **852,340**; *eidem*, **U.S.** pat. **4,128,578** (1977, 1978 both to Hoechst). Analgesic activity and toxicity data: G. Doll, E. Hackenthal, *Arzneimittel-Forsch.* **13**, 68 (1963); H. Fujimura, *Folia Pharmacol. Japan.* **62**, 123 (1966). Prepn and pharmacology: G. Ehrhart *et al., ibid.* **15**, 727 (1965). Metabolism in rabbits: J. Shibasaki *et al., Chem. Pharm. Bull.* **16**, 1726, 2269 (1968). Synergistic analgesic and antipyretic effect with ethenzamide, *q.v.*: H. Fujimura, K. Shinozaki, **U.S.** pat. **3,284,298** (1966); H. Kojima *et al., Oyo Yakuri* **16**, 549 (1978), *C.A.* **90**, 81035y (1979); H. Kojimi, T. Sakurai, **Japan.** pat. **23,132** (1979 to Daiichi), *C.A.* **90**, 210147n (1979).

Crystals from isopropanol, mp 160°. Sparingly sol in water. LD_{50} in mice (mg/kg): 790 i.p., 2800 orally (Fujimura).

Note: Component in combination analgesics containing ethenzamide, caffeine and vitamin B_1, e.g. *Butylon, Bucetalon.*

THERAP CAT: Analgesic.

1446. Buchu. Bucco; bucku; buku. Dried leaves of *Barosma betulina* Bartl. & Wendl. (short buchu) or of *B. crenulata* (Linné) Hooker (oval buchu), or of *B. seratifolia* Willd. (long buchu), *Rutaceae. Habit.* Southern Africa (Cape of Good Hope). *Constit.* Diosphenol (barosma camphor), diosmin, bitter extractive, resin, *l*-menthone, mucilage, hesperidin, 1-2% volatile oil. *Review:* Feldman, Youngken, *J. Am. Pharm. Assoc.* **33**, 277 (1944).

THERAP CAT: Antiseptic (urinary).

THERAP CAT (VET): Has been used as a urinary antiseptic, diuretic.

1447. Bucillamine. *N-(2-Mercapto-2-methyl-1-oxopropyl)-L-cysteine; N-(2-mercapto-2-methylpropanoyl)-L-cysteine; N-(2-mercaptoisobutyryl)-L-cysteine; tiobutarit; DE-019; SA96; Rimatil.* $C_7H_{13}NO_3S_2$; mol wt 223.31. C 37.65%, H 5.87%, N 6.27%, O 21.49%, S 28.71%. Amino acid derivative structurally related to tiopronin, *q.v.,* which modulates the immune response. Prepn: T. Fujita *et al.,* **Ger.** pat. **2,709,820**; *eidem*, **U.S.** pat. **4,305,958** (1977, 1981 both to Santen); M. Oya *et al., Chem. Pharm. Bull.* **29**, 940 (1981). General pharmacology: T. Iso *et al., Oyo Yakuri* **25**, 123 (1983), *C.A.* **98**, 119191b (1983). Stimulation of lymphocyte proliferation *in vitro*: I. Yamamoto *et al., Int. J. Tissue React.* **4**, 1 (1982). Effect on macrophage function *in vitro* and *ex vivo*: M. Hayashi *et al., Int. J. Immunopharmacol.* **8**, 299 (1986). Preliminary clinical trial in rheumatoid arthritis: K. Ishikawa, M. Sakaguchi, *Scand. J. Rheumatol.* **15**, 85 (1986).

Crystals from ethyl acetate, mp 139-140°. $[\alpha]_D^{25}$ +32.3° (c = 1.0 in ethanol). LD_{50} in mice (mg/kg): 2285 i.p.; 989.6 i.v. (Fujita, 1981).

THERAP CAT: Immunomodulator.

1448. Bucladesine. *N-(1-Oxobutyl)adenosine cyclic 3',5'-(hydrogen phosphate) 2'-butanoate; N-(9-β-D-ribofuranosyl-9H-purin-6-yl)butyramide cyclic 3',5'-(hydrogen phosphate) 2'-butyrate; N6,2'-O-dibutyryladenosine 3',5'-cyclic monophosphate; N6,2'-O-dibutyryl cAMP; DBcAMP;* $C_{18}H_{24}N_5O_8P$; mol wt 469.39. C 46.06%, H 5.15%, N 14.92%, O 27.27%, P 6.60%. Vasodilating cyclic nucleotide derivative which can permeate the cell membrane. Mimics the action of endogenous cyclic AMP, *q.v.* Prepn: Th. Posternak *et al., Biochim. Biophys. Acta* **65**, 558 (1962); O. Nagase *et al.,* **Japan. Kokai** 76 113,896, *C.A.* **86**, 140409r (1977); T. Hirayama *et al.,* **Japan. Kokai** 77 39,698, 77 39,699, *C.A.* **87**, 136330m-136331n (1977) (1976, 1977, 1977 all to Daiichi Seiyaku). Pharmacology: H. S. White *et al., Eur. J. Pharmacol.* **57**, 107 (1979); J. D. Johnson *et al., ibid.* **91**, 343 (1983); N. Bondar *et al., J. Physiol.* **355**, 33 (1984). Clinical trial in congestive heart failure: S. Matsui *et al., Am. J. Cardiol.* **51**, 1364 (1983); after cardiopulmonary bypass surgery: T. Yoshitake *et al., Prog. Clin. Biol. Res.* **111**, 211 (1983).

Sodium salt, $C_{18}H_{23}N_5NaO_8P$, *DC-2797, Actosin.*
Barium salt, $C_{36}H_{46}BaN_{10}O_{16}P_2$. uv max (ethanol-0.5M ammonium acetate, 5:2) 270 nm.

THERAP CAT: Cardiostimulant.

1449. Buclizine. *1-[(4-Chlorophenyl)phenylmethyl]-4-[[4-(1,1-dimethylethyl)phenyl]methyl]piperazine; 1-(p-tert-butylbenzyl)-4-(p-chloro-α-phenylbenzyl)piperazine; 1-(p-chlorobenzhydryl)-4-(p-tert-butylbenzyl)diethylenediamine; 1-(p-tert-butylbenzyl)-4-(p-chlorodiphenylmethyl)piperazine; histabutyzine; histabutizine; UCB 4445; Buclifen; Longifene; Posdel; Postafen Tabl; Vibazine.* $C_{28}H_{33}ClN_2$; mol wt 433.04. C 77.66%, H 7.68%, Cl 8.19%, N 6.47%. Prepn: Morren, **U.S.** pat. **2,709,169** (1955 to U.C.B.), *see also* **Ger.** pat. **964,040**, Lui *et al., C.A.* **62**, 2776a (1965). Clinical evaluation as antiemetic in migraine: P. B. Jorgensen *et al., Curr. Ther. Res.* **16**, 1276 (1974); E. I. Adam, *J. Int. Med. Res.* **15**, 71 (1987).

$bp_{0.001}$ 217-220°.

Dihydrochloride, $C_{28}H_{35}Cl_3N_2$, *Aphilan R, Buclina, Softran.* mp 230-240°.

THERAP CAT: Antiemetic.

1450. Buclosamide. *N-Butyl-4-chloro-2-hydroxybenzamide; N-butyl-4-chlorosalicylamide; 4-chloro-2-hydroxybenzoic acid n-butylamide.* $C_{11}H_{14}ClNO_2$; mol wt 227.70. C 58.02%, H 6.20%, Cl 15.57%, N 6.15%, O 14.05%. Prepn: Ruschig *et al.,* **U.S.** pat. **2,923,737** (1960 to Hoechst).

Crystals from chloroform, mp 90-92°.

Marketed also in combination with salicylic acid, *Jadit* and with prednisolone.

THERAP CAT: Antifungal.

1451. Bucloxic Acid. *3-Chloro-4-cyclohexyl-α-oxobenzenebutanoic acid; 3-(3-chloro-4-cyclohexylbenzoyl)propionic acid;* 4-(4-cyclohexyl-3-chlorophenyl)-4-oxobutyric acid; bucloxonic acid. $C_{16}H_{19}ClO_3$; mol wt 294.78. C 65.19%, H 6.50%, Cl 12.03%, O 16.28%. Prepn: Krausz, Brunaud, **Ger.** pat. **2,021,445** (1970 to Clin-Byla), *C.A.* **75**, 5521d (1971) corresp to **Brit.** pat. **1,315,542**; Krausz *et al., Arzneimittel-Forsch* **24**, 1360, 1364 (1974). Review of pharmacology, metabolism, toxicity: *ibid.* 1368-1444.

Crystals from acetone, mp 163°. uv max (ethanol): 255 nm (ε 15,500). LD_{50} in mice, rats: 900, 120 mg/kg orally; 1100, 210 mg/kg i.p.

Calcium salt, $C_{32}H_{36}CaCl_2O_6$, *CB 804, Esfar.* LD_{50} in mice, rats: 1700, 175 mg/kg orally; 1700, 200 mg/kg i.p. THERAP CAT: Anti-inflammatory.

1452. Bucolome. *5-Butyl-1-cyclohexyl-2,4,6(1H,3H,5H)-pyrimidinetrione; 5-butyl-1-cyclohexylbarbituric acid;* 5-*n*-butyl-1-cyclohexyl-2,4,6-trioxoperhydropyrimidine; BCP; Paramidin. $C_{14}H_{22}N_2O_3$; mol wt 266.33. C 63.13%, H 8.33%, N 10.52%, O 18.02%. Prepn: Senda *et al.*, **U.S.** pats. **3,243,344** and **3,274,195** (both 1966 to Takeda). Metabolic studies in man: Yashiki *et al., Chem. Pharm. Bull.* **19**, 468, 869 (1971).

Needles from methanol, mp 84°. $bp_{0.8}$ 185-187°. THERAP CAT: Anti-inflammatory.

1453. Bucrylate. *2-Cyano-2-propenoic acid 2-methylpropyl ester; 2-cyanoacrylic acid isobutyl ester;* isobutyl cyanoacrylate; bucrilate; IBC; IBCA. $C_8H_{11}NO_2$; mol wt 153.18. C 62.72%, H 7.24%, N 9.15%, O 20.89%. Prepn: A. E. Ardis, **U.S.** pat. **2,467,926** (1949 to B. F. Goodrich). Use as tissue adhesive in surgery: H. W. Coover, **U.S.** pat. **2,794,788** (1957 to Eastman-Kodak); **Belg.** pat. **636,286** (1963 to Eastman-Kodak), *C.A.* **62**, 1811b (1965). Exptl use in aneurysms: P. H. Zanetti, F. E. Sherman, *J. Neurosurg.* **36**, 72 (1972); R. P. Leather *et al., Arch. Surg.* **114**, 1402 (1979); in cerebral arteriovenous malformations: L. D. Cromwell, A. B. Harris, *J. Neurosurg.* **52**, 705 (1980).

Liquid, $bp_{2.0}$ 170°. Polymerizes in <1 second on contact with ionic solns, e.g. saline or blood. USE: Surgical aid (tissue adhesive).

1454. Bucumolol. *8-[3-[(1,1-Dimethylethyl)amino]-2-hydroxypropoxy]-5-methyl-2H-1-benzopyran-2-one; (±)-8-[3-(t-butylamino)-2-hydroxypropoxy]-5-methylcoumarin; dl-8-(2-hydroxy-3-tert-butylaminopropoxy)coumarin.* $C_{17}H_{23}NO_4$; mol wt 305.38. C 66.86%, H 7.59%, N 4.59%, O 20.96%. β-Adrenergic blocker. Prepn: Y. Sato *et al.*, **Ger.** pat. **2,021,958**; *eidem,* **U.S.** pat. **3,663,570** (1970, 1972 both to Sankyo); *eidem, Chem. Pharm. Bull.* **20**, 905 (1972). Pharmacology: T. Oshima *et al., Japan. J. Pharmacol.* **23**, 497 (1973). Metabolic studies: R. Hayashi *et al., Chem. Pharm. Bull.* **23**, 1173 (1975). Antiarrhythmic activity in dogs: K. Nakayama *et al., Japan. J. Pharmacol.* **29**, 935

(1979). Early clinical study: H. Kishida *et al., Rinsho Yakuri* **11**, 99 (1980), *C.A.* **93**, 107380w (1980).

Hydrochloride, $C_{17}H_{24}ClNO_4$, *CS-359, Bucumarol.* Colorless crystals from isopropanol, mp 226-228° (dec). LD_{50} in male, female mice (mg/kg): 676, 692 orally; 33.1, 31.6 i.v. (Oshima).

THERAP CAT: Antianginal, antiarrhythmic.

1455. Budesonide. *16,17-Butylidenebis(oxy)-11,21-dihydroxypregna-1,4-diene-3,20-dione; (R,S)-11β,16α,17,21-tetrahydroxypregna-1,4-diene-3,20-dione cyclic 16,17-acetal with butyraldehyde;* S-1320; Budeson; Preferid; Pulmicort; Rhinocort; Spirocort. $C_{25}H_{34}O_6$; mol wt 430.55. C 69.74%, H 7.96%, O 22.30%. Non-halogenated glucocorticoid related to triamcinolone hexacetonide, *q.v.* with a high ratio of topical to systemic activity. Prepn: R. L. Brattsand *et al.*, **Ger.** pat. **2,323,215** corresp to **U.S.** pat. **3,929,768** (1973, 1975 both to Bofors). Synthesis and anti-inflammatory properties: A. Thalén, R. L. Brattsand, *Arzneimittel-Forsch.* **29**, 1787 (1979). Pharmacokinetic study: A. Ryrfeldt *et al., J. Steroid Biochem.* **10**, 317 (1979). HPLC determn of epimers, impurities, content: G. Roth *et al., J. Pharm. Sci.* **69**, 766 (1980). Review of pharmacodynamics and efficacy in asthma and rhinitis: S. P. Clissold, R. C. Heel, *Drugs* **28**, 485-518 (1984).

Crystals, mp 221-232° (dec). It is a mixture of two isomers; the content of the S-isomer in the mixture varies between 40-51%. $[\alpha]_D^{25}$ +98.9° (c = 0.28 in methylene chloride).

THERAP CAT: Anti-inflammatory.

1456. Budipine. *1-(1,1-Dimethylethyl)-4,4-diphenylpiperidine;* 1-*tert*-butyl-4,4-diphenylpiperidine. $C_{21}H_{27}N$; mol wt 293.46. C 85.95%, H 9.28%, N 4.77%. Centrally acting deriv of diphenylpiperidine. Prepn: H. G. Menge, J. Klosa, **Ger.** pat. **1,936,452**; *eidem,* pat. **4,016,280** (1969, 1977 both to Byk Gulden). Pharmacological studies: U. Brand, H. G. Menge, *Arzneimittel-Forsch.* **30**, 1242 (1980); M. Eltze, *ibid.* 1129; H. G. Menge, U. Brand, *ibid.* **32**, 85 (1982). Comparison to other drugs in treatment of dyskinesias: J. Siegfried, *Can. J. Neurol. Sci.* **6**, 89 (1979). Clinical evaluation with levodopa in Parkinson's disease: K. Jellinger, H. Bliesath, *J. Neurol.* **234**, 280 (1987).

Hydrochloride, $C_{21}H_{28}ClN$, *Parkinsan.* LD_{50} in male mice, rats (mg/kg): 120, 165 orally; 33, 28 i.v. (Menge, Brand).

THERAP CAT: Antiparkinsonian.

1457. Budralazine. *1(2H)-Phthalazinone (1,3-dimethyl-2-butenylidene)hydrazone;* 4-methyl-3-penten-2-one (1-

phthalazinyl)hydrazone; mesityl oxide (1-phthalazinyl)hydrazone; DJ-1461; Buterazine. $C_{14}H_{16}N_4$; mol wt 240.31. C 69.97%, H 6.71%, N 23.32%. Deriv of hydralazine, *q.v.* Prepn: K. Ueno *et al., **Ger. pat.** 2,145,359* corresp to *U.S. pat. 3,840,539* (1972, 1974 both to Daiichi Seiyaku); *eidem, Chem. Pharm. Bull.* **24**, 1068 (1976). Pharmacology: A. Akashi *et al., Eur. J. Pharmacol.* **29**, 161 (1974); *eidem, Arch. Int. Pharmacodyn. Ther.* **235**, 134 (1978). Metabolism: R. Moroi *et al., Chem. Pharm. Bull.* **24**, 2850 (1976); K. Ono *et al., Xenobiotica* **9**, 227 (1979). Reproduction studies: *Oyo Yakuri* **21**, 313-350 (1981), *C.A.* **95**, 73764y-7b (1981). Series of articles on cardiovascular action, α-adrenergic effects, antihypertensive and general pharmacological properties: *Arzneimittel-Forsch.* **31**, 1072-1087 (1981). Characterization of vasodilator action: T. Chiba *et al., ibid.* **33**, 112 (1983). Toxicity study: T. Onodera *et al., Toxicol. Appl. Pharmacol.* **44**, 431 (1978).

Cryst from isopropyl ether, mp 132-133°. uv max (methanol): 208, 240, 289, 357 nm (ϵ 2.7 × 10^4, 8.9 × 10^4, 2.0 × 10^4, 1.5 × 10^4). LD_{50} in mice, rats (mg/kg): 1820, 620 orally; 4020, 3570 i.p., T. Onodera *et al., loc. cit.*

THERAP CAT: Antihypertensive.

1458. Bufalin. *3,14-Dihydroxybufa-20,22-dienolide.* $C_{24}H_{34}O_4$; mol wt 386.53. C 74.58%, H 8.86%, O 16.56%. Cardiotonic steroid constituent of Ch'an Su (also called Senso), a galenical preparation of the dried venom of the Chinese toad, *Bufo bufo gargarizans.* Isoln and characterization: M. Kotake, K. Kuwada, *Sci. Papers Inst. Phys. Chem. Res. (Tokyo)* **36**, 106 (1939), *C.A.* **33**, 7304⁹ (1939); K. Meyer, *Helv. Chim. Acta* **32**, 1238 (1949). Structure: K. Kuwada, *J. Chem. Soc. Japan* **60**, 335 (1939), *C.A.* **35**, 51123⁹ (1941). Prepn from resibufogenin (*q.v.*): H. Kondo, S. Ohno, *U.S. pat. 3,134,772* (1964); from digitoxigenin (*q.v.*): G. R. Pettit *et al., U.S. pat. 3,687,944* (1972). Total synthesis: F. Sondheimer *et al., J. Am. Chem. Soc.* **91**, 1228 (1969). Additional syntheses: G. R. Pettit *et al., J. Org. Chem.* **35**, 2895 (1970); Y. Kamano, G. R. Pettit, *ibid.* **38**, 222 (1973); E. Yoshii *et al., Chem. Pharm. Bull.* **25**, 2249 (1977); K. Wiesner *et al., Helv. Chim. Acta* **66**, 2632 (1983). Pharmacology: S. Yoshida, T. Sakai, *Japan. J. Pharmacol.* **23**, 859 (1973); **24**, 97 (1974).

Needles from methanol/chloroform, mp 242-243°. $[\alpha]_D$ −20°. uv max (ethanol): 298 nm (log ϵ 3.77).
Sodium sulfate, $C_{24}H_{33}NaO_7S$, *sodium bufalin-3-sulfate.* Isoln from Japanese toad *Bufo vulgaris formosus:* K. Shimada *et al., Tetrahedron Letters* **1974**, 2767. Colorless amorphous solid, mp 165.5-166.5°. $[\alpha]_D^{23}$ −33.1° (c = 0.05 in methanol).

1459. Bufencarb. *3-(1-Ethylpropyl)phenol methylcarbamate mixture with 3-(1-methylbutyl)phenyl methylcarbamate (1:3); methylcarbamic acid m-(1-ethylpropyl)phenyl ester mixt. with m-(1-methylbutyl)phenyl ester;* metalkamate; Ortho 5353; ENT 27127; Bux. $C_{13}H_{19}NO_2$; mol wt 221.30. C 70.56%, H 8.65%, N 6.33%, O 14.46%. Prepn: Ospenson *et al., U.S. pats. 3,062,864* and *3,062,867* (1962 to Calif. Res. Corp.); *Brit. pat. 1,106,326* (1968 to Chevron). Acute toxic-

ity: T. B. Gaines, R. E. Linder, *Fundam. Appl. Toxicol.* **7**, 299 (1986).

R = —CH(CH₃)CH₂CH₂CH₃ and —CH(C₂H₅)₂

Yellow-amber solid, mp 26-39°. d^{26} 1.024. $bp_{0.04}$ 125°. Very sol in xylene, methanol. Soly in water < 50 ppm. Stable in neutral or acidic solns; increases in pH or temp increases rate of hydrolysis. Fairly rapidly degraded in the soil. LD_{50} in male, female rats (mg/kg): 97, 61 orally (Gaines, Linder).
Caution: Cholinesterase inhibitor. Absorbed through the skin: *Clinical Toxicology of Commercial Products,* R. E. Gosselin *et al.,* Eds. (Williams & Wilkins, Baltimore, 5th ed., 1984) Section II, p 304.

USE: Insecticide.

1460. Bufeniode. *4-Hydroxy-3,5-diiodo-α-[1-[(1-methyl-3-phenylpropyl)amino]ethyl]benzenemethanol;* 1-(4-hydroxy-3,5-diiodophenyl)-2-(1-methyl-3-phenylpropylamino)-propanol; diiodobuphenine; Diastal; Proclival. $C_{19}H_{23}I_2$-NO_2; mol wt 551.20. C 41.40%, H 4.20%, I 46.04%, N 2.54%, O 5.81%. Prepn: J. P. Fourneau, J. M. Delourme, *S. Afr. pat. 68 00,046* corresp to *U.S. pat. 3,542,870* (1968, 1970 to Lab. Houdé). Metabolism: Veyre *et al., C.R. Soc. Biol.* **163**, 136 (1969).

Crystals, mp (slow heating) 185° (dec), mp (fast heating) 212°. LD_{50} in mice: > 600 mg/kg i.p., > 2 g/kg orally (Fourneau, Delourme).

THERAP CAT: Antihypertensive; vasodilator.

1461. Bufetolol. *1-[(1,1-Dimethylethyl)amino]-3-[2-[(tetrahydro-2-furanyl)methoxy]phenoxy]-2-propanol; 1-(tert-butylamino)-3-[o-[(tetrahydrofurfuryl)oxy]phenoxy]-2-propanol.* $C_{18}H_{29}NO_4$; mol wt 323.44. C 66.84%, H 9.04%, N 4.33%, O 19.79%. β-Adrenergic blocker. Prepn: M. Nakanishi *et al., **Ger. pat.** 2,024,001; eidem, **U.S. pat.** 3,-723,476* (1970, 1973, both to Yoshitomi); *eidem, J. Med. Chem.* **15**, 45 (1972). Pharmacology: M. Nakanishi *et al., Yakugaku Zasshi* **91**, 1037 (1971); *ibid.* **92**, 375 (1972); Imamura *et al., ibid.* 1039. Metabolism: M. Nakanishi *et al., ibid.* 299; *Oyo Yakuri* **6**, 479 (1972), *C.A.* **78**, 66785b (1973). Toxicology: *eidem, ibid.* 485, 1267, *C.A.* **78**, 66945d; **79**, 337w (1973).

$bp_{0.07}$ 180-186°.
Hydrochloride, $C_{18}H_{30}ClNO_4$, Y-6124, Adobiol. White crystals from methanol-isopropanol, mp 153.5°-157°; also reported as mp 151-154°. (Diastereoisomer, mp 118°). Freely sol in water, methanol, glacial acetic acid. Very slightly sol in benzene. Practically insol in ether. Unstable on exposure to light. Average LD_{50} in mice, rats (mg/kg): 409, 1142 orally; 507, 1904 s.c., *Japan. Med. Gaz.* **11**(7), 4 (1974).

THERAP CAT: Antiarrhythmic; antianginal.

1462. Bufexamac. *4-Butoxy-N-hydroxybenzeneacetam-ide; 2-(p-butoxyphenyl)acetohydroxamic acid; 2-[p-(butyl-oxy)phenyl]acetohydroxamic acid; CP 1044 J3; Droxarol; Droxaryl; Feximac; Malipuran; Mofenar; Norfemac; Par-fenac; Parfenal.* $C_{12}H_{17}NO_3$; mol wt 223.28. C 64.55%, H 7.67%, N 6.27%, O 21.50%. Synthesis: Buu-Hoï *et al.*, *Compt. Rend.* **261**, 2259 (1965); *eidem,* **Belg.** pat. **661,226** corresp to U.S. pat. **3,479,396** (1965, 1969 to Madan). Pharmacology and toxicology: Lambelin *et al., Med. Phar-macol. Exp.* **15**, 545 (1966).

CH_3(CH_2)_3O—⟨benzene ring⟩—CH_2CNHOH ‖ O

Needles from acetone, mp 153-155°. Practically insol in water. LD_{50} orally in mice, rats: > 8, > 4 g/kg (Lambelin). THERAP CAT: Anti-inflammatory, analgesic, antipyretic.

1463. Buflomedil. *4-(1-Pyrrolidinyl)-1-(2,4,6-trimeth-oxyphenyl)-1-butanone; 2′,4′,6′-trimethoxy-4-(1-pyrroli-dinyl)butyrophenone; (2,4,6-trimethoxyphenyl) (3-pyrroli-dinopropyl) ketone.* $C_{17}H_{25}NO_4$; mol wt 307.40. C 66.42%, H 8.20%, N 4.56%, O 20.82%. Prepn: L. Lafon, **Ger.** pat. **2,122,144;** *idem,* U.S. pat. **3,895,030** (both 1971 to Orsy-monde). Pharmacology: J. Duteil *et al., Therapie* **30**, 207 (1975); C. Debray *et al., ibid.* 259. HPLC determn: J. A. Badmin *et al., J. Chromatog.* **172**, 319 (1979). Clinical phar-macokinetics: U. Gundert-Remy *et al., Eur. J. Clin. Phar-macol.* **20**, 459 (1981). Comparative safety and efficacy study: G. Rosas *et al., Angiology* **32**, 291 (1981). Review of pharmacology and therapeutic effects: S. P. Clissold *et al., Drugs* **33**, 430-460 (1987).

⟨structure: trimethoxyphenyl ketone with pyrrolidine⟩

Hydrochloride, $C_{17}H_{26}ClNO_4$, *LL 1656, Bufedil, Buflan, Fonzylane, Irrodan, Lofton, Loftyl, Provas.* White crystals from isopropanol, mp 192-193°. LD_{50} in mice (mg/kg): 80 ±4.6 i.v. (Lafon).
THERAP CAT: Peripheral vasodilator.

1464. Bufogenin B. *3,14,16-Trihydroxybufa-20,22-di-enolide; desacetylbufotalin.* $C_{24}H_{34}O_5$; mol wt 402.51. C 71.61%, H 8.51%, O 19.88%. Isolated from the Chinese drug Ch'an Su which is prepd from Chinese toads (*Bufo asiaticus* = *Bufo gargarizans* Cantor); also obtained from bufotalin, *q.v.*: Ruckstuhl, Meyer, *Helv. Chim. Acta* **40**, 1270 (1957).

⟨steroid structure⟩

Elongated prisms from methanol, begins to sinter at 195°, dec 210-223°. $[\alpha]_D^{19}$ +30° (c = 1.039 in dioxane). Very sparingly sol in chloroform, methanol, acetone.

1465. Buformin. *N-Butylimidodicarbonimidic diamide; 1-butylbiguanide; n-butylbiguanide; butyldiguanide; butfor-min; W 37.* $C_6H_{15}N_5$; mol wt 157.22. C 45.83%, H 9.62%, N 44.55%. Prepn: Shapiro *et al., J. Am. Chem. Soc.* **81**, 3728 (1959); Shapiro, Freedman, U.S. pat. **2,961,377** (1960 to USV).

CH_3CH_2CH_2CH_2NHCNHCNH_2 with ‖NH ‖NH above

Strong base. Very sol in water.
Hydrochloride, $C_6H_{16}ClN_5$, *Andere, Biforon, Bigunal, Bu-fonamin, Bulbonin, Diabrin, Dibetos, Gliporal, Insulamin, Krebon, Panformin, Silubin, Sindiatil, Tidemol, Ziavetine.* Crystals, mp 174-177°. Freely sol in water, alcohol. LD_{50} i.p. in mice: 380 mg/kg, Rikimaru *et al., J. Antibiot.* **18A**, 196 (1965).
Nitrate, $C_6H_{15}N_5 \cdot HNO_3$, crystals from ethanol, dec 125-126°.
THERAP CAT: Antidiabetic.

1466. Bufotalin. *16-(Acetyloxy)-3,14-dihydroxybufa-20,22-dienolide; 3β,14,16β-trihydroxy-5β-bufa-20,22-dien-olide 16-acetate.* $C_{26}H_{36}O_6$; mol wt 444.55. C 70.24%, H 8.16%, O 21.59%. One of the genins found in the venom of the common European toad (*Bufo vulgaris*). Isoln and structure: Faust, *Arch. Exp. Pathol. Pharmakol.* **47**, 279 (1902); **49**, 1 (1902); Wieland, Weil, *Ber.* **46**, 3315 (1913); Wieland, Weyland, *Sitzungsber. Bayr. Akad. Wiss.* (math.-physikal. Klasse) **1920**, 329; Wieland *et al., Ann.* **524**, 203 (1936); Meyer, *Helv. Chim. Acta* **32**, 1993 (1949); Pettit *et al., Chem. Commun.* **1970**, 1566. Partial synthesis: Kamano *et al., J. Org. Chem.* **39**, 3007 (1974).

⟨steroid structure with OCOCH_3⟩

Solvated crystals from alc. Sinters at 154°, dec 223°. Sub-limes in high vacuum at 225-230°. uv max: 300 nm. $[\alpha]_D^{20}$ +5.4° (c = 0.5 in chloroform). Sol in alcohol, chloroform.
3-Acetate, mp 269-272°.

1467. Bufotenine. *3-[2-(Dimethylamino)ethyl]-1H-in-dol-5-ol; 3-(2-dimethylaminoethyl)-5-indolol; 5-hydroxy-N,N-dimethyltryptamine; N,N-dimethylserotonin; 3-(β-di-methylaminoethyl)-5-hydroxyindole; mappine.* $C_{12}H_{16}N_2O$; mol wt 204.26. C 70.56%, H 7.90%, N 13.72%, O 7.83%. Isoln from toads: Wieland *et al., Ann.* **513**, 1 (1934); Wie-land, Wieland, *ibid.* **528**, 239 (1937); from toadstools: Wie-land, Motzel, *ibid.* **581**, 10 (1953). Isoln from *Piptadenia peregrina* Benth., *Leguminosae:* Stromberg, *J. Am. Chem. Soc.* **76**, 1707 (1954). Synthesis: Hoshino, Shimodaira, *Ann.* **520**, 19 (1935); Harley-Mason, Jackson, *Chem. & Ind.* (Lon-don) **1952**, 954; *see also* Serotonin; and Speetor, U.S. pat. **2,708,197** (1955 to Upjohn); Stoll *et al., Helv. Chim. Acta* **38**, 1452 (1955). Activity: Bhattacharya, Sanyal, *Indian J. Physiol. Pharmacol.* **15**, 133 (1971). Crystal and molecular structure: G. Falkenberg, *Acta Crystallogr.* **28B**, 3219 (1972).

⟨indole structure with dimethylaminoethyl⟩

Stout prisms from ethyl acetate, mp 146-147°. $bp_{0.1}$ 320°. uv max: 220, 265 nm (log ε 4.0, 3.7). Almost insol in water. Freely sol in alcohol, less sol in ether. Sol in dil acids and alkalies.
Methyliodide, $C_{13}H_{19}IN_2O$, stout prisms from methanol, dec 214-215°.
Monopicrate, yellow crystals which change to a red modi-fication at 120 to 140°, then mp 179-180°.
Dipicrate, red crystals from methanol, mp 176-177°.
Caution: This is a controlled substance (hallucinogen) listed in the U.S. Code of Federal Regulations, Title 21 Part 1308.11 (1985).

THERAP CAT: Hallucinogen.

1468. Bufotoxin. *16-(Acetyloxy)-3-[[8-[[4-[(aminoimi-nomethyl)amino]-1-carboxybutyl]amino]-1,8-dioxooctyl]-oxy]-14-hydroxybufa-20,22-dienolide;* vulgarobufotoxin; bufotalin 3-suberoylarginine ester. $C_{40}H_{60}N_4O_{10}$; mol wt 756.91. C 63.47%, H 7.99%, N 7.40%, O 21.14%. Principal toxin of the venom of the common European toad, *Bufo vulgaris.* Isoln and structure: Wieland, Alles, *Ber.* **55**, 1789 (1922); Wieland *et al., Ann.* **524**, 203 (1936); Meyer, *Helv. Chim. Acta* **32**, 1993 (1949); Urscheler *et al., ibid.* **38**, 883 (1955). Alternate structure: Kamano *et al., Tetrahedron Letters* **1968**, 5673; Linde-Tempel, *Helv. Chim. Acta* **53**, 2188 (1970). Partial synthesis: Pettit, Kamano, *Chem. Commun.* **1972**, 45.

Monohydrate, bitter needles from alc, dec 205°. uv max: 295 nm (log ϵ 3.74). Freely sol in methanol, pyridine, sparingly sol in abs alc. Practically insol in water, ether, acetone, chloroform, petr ether.

1469. Bufuralol. α-*[[(1,1-Dimethylethyl)amino]methyl]-7-ethyl-2-benzofuranmethanol;* α-[(*tert*-butylamino)methyl]-7-ethyl-2-benzofuranmethanol; 2-(2-*tert*-butylamino-1-hydroxyethyl)-7-ethylbenzofuran; 1-(7-ethylbenzofuran-2-yl)-2-*tert*-butylamino-1-hydroxyethane. $C_{16}H_{23}NO_2$; mol wt 261.36. C 73.53%, H 8.87%, N 5.36%, O 12.24%. β-Adrenergic blocker with peripheral vasodilating activity. Prepn: **Neth. pat. Appl. 6,606,441;** G. A. Fothergill *et al.,* **U.S. pat. 3,929,836** (1966, 1975 both to Hoffmann-La Roche). Prepn and resolution of isomers: *eidem, Experientia* **31**, 1322 (1975); *eidem, Arzneimittel-Forsch.* **27**, 981 (1977). Pharmacology: T. C. Hamilton, M. W. Parkes, *ibid.* 1410. HPLC determn in plasma: P. Haefelfinger, *J. Chromatog.* **221**, 327 (1980). Metabolism of isomers and racemate: R. J. Francis *et al., Eur. J. Clin. Pharmacol.* **23**, 529 (1982). Pharmacokinetics in hypertensive patients: M. Eckert *et al., ibid.* **24**, 479 (1983). Hemodynamic effects in patients with angina: M. Pfisterer *et al., J. Cardiovasc. Pharmacol.* **6**, 417 (1984).

Hydrochloride, $C_{16}H_{24}ClNO_2$, *Ro 3-4787, Angium.* Fine, white powder from acetone, mp 146°. LD_{50} in mice (mg/kg): 29.7 i.v.; 88.0 i.p.; 177 orally; in rats (mg/kg): 1400 s.c.; 750 orally (Hamilton, Parkes).

(+)-Hydrochloride, crystals from ethyl acetate-ether, mp 122-123°. $[\alpha]_{365}^{20}$ +135.0° (c = 1.0 in ethanol).

(−)-Hydrochloride, crystals from ethyl acetate-ether, mp 122-123°. $[\alpha]_{365}^{20}$ −136.0° (c = 1.0 in ethanol).

THERAP CAT: Antianginal; antihypertensive.

1470. Bulan®. *1,1'-(2-Nitrobutylidene)bis[4-chlorobenzene]; 1,1-bis(p-chlorophenyl)-2-nitrobutane;* 2-nitro-1,1-bis(p-chlorophenyl)butane. $C_{16}H_{15}Cl_2NO_2$; mol wt 324.22. C 59.27%, H 4.66%, Cl 21.87%, N 4.32%, O 9.87%. Prepn: Hass, Blickenstaff, **U.S. pat. 2,516,186** (1950 to Purdue Res. Found.).

LD_{50} orally in rats: 375 mg/kg, Lehman, *Assoc. Food Drug Officials U.S. Quart. Bull.* **14**, 82 (1951), *C.A.* **45**, 3517h (1951).

Combination with Prolan [CSC], *Dilan.*

USE: Insecticide.

1471. Bulbocapnine. *(S)-6,7,7a,8-Tetrahydro-11-methoxy-7-methyl-5H-benzo[g]-1,3-benzodioxolo[6,5,4-de]quinolin-12-ol; 10-methoxy-1,2-(methylenedioxy)-6aα-aporphin-11-ol.* $C_{19}H_{19}NO_4$; mol wt 325.35. C 70.14%, H 5.89%, N 4.31%, O 19.67%. From root of *Corydalis cava* (L.) Schweigg. & Körte *(C. tuberosa* DC), *Fumariaceae* and *Dicentra canadensis* Walp., *Papaveraceae.* Isoln: Freund, Josephi, *Ber.* **25**, 2411 (1892); Manske, *Can. J. Res.* **7**, 258 (1932). Structure: Gadamer, *Chem. Ztg.* **34**, 1004 (1910). Configuration: Ayer, Taylor, *J. Chem. Soc.* **1956**, 472; Corrodi, Hardegger, *Helv. Chim. Acta* **39**, 889 (1956). Synthesis: Kikkawa, *C.A.* **54**, 4649b (1960). Peripheral-dopamine-receptor blocking activity: R. G. Pendleton *et al., Arch. Pharmacol.* **289**, 171 (1975).

dl-Form, columns, mp 213-214°. Absorption max: Brustier, *C.A.* **49**, 12127b (1955).

l-Form, columns, mp 202-203°. $[\alpha]_D^{22}$ −239°.

d-Form, columns, mp 201-203°. $[\alpha]_D^{22}$ +231°. Practically insol in water, sol in alcohol, chloroform. LD_{50} s.c. in mice: 195 mg/kg.

Methyl ether, $C_{20}H_{21}NO_4$, crystals from petr ether, mp 129°. $[\alpha]_D^{20}$+260°. Absorption max: Girardet, *J. Chem. Soc.* **1931**, 2630.

1472. Bumadizon. *Butylpropanedioic acid mono(1,2-diphenylhydrazide); butylmalonic acid mono(1,2-diphenylhydrazide);* N-(2-carboxycaproyl)hydrazobenzene; α-carboxycaproyl-N,N'-diphenylhydrazine. $C_{19}H_{22}N_2O_3$; mol wt 326.40. C 69.92%, H 6.79%, N 8.58%, O 14.71%. Prepn: **Neth. pat. Appl. 6,406,412** corresp to Pfister *et al.,* **U.S. pat. 3,455,999** (1964, 1969 to Geigy); **Neth. pat. Appl. 6,600,685** (1966 to Byk-Gulden). Main product of the hydrolysis of phenylbutazone, *q.v.:* Schmid, *Helv. Chim. Acta* **53**, 2239 (1970). Chemical characterization: Pawelczyk, Wachowiak, *Acta Pol. Pharm.* **26**, 433 (1969), *C.A.* **72**, 136355p (1970). Series of articles on pharmacology: *Arzneimittel-Forsch.* **23**, 1226-1251, 1813-1822 (1973). Toxicity: R. Riedel, W. Schoetensack, *ibid.* 1215.

Crystals from ether-petr ether, mp 116-117°. Also reported as mp 77-79° (dependent on speed of crystallization). uv max (0.1N NaOH): 234, 264 nm (ϵ 16200, 3700).

Calcium salt hemihydrate, $C_{38}H_{42}CaN_4O_6\cdot\frac{1}{2}H_2O$, *Bumaflex, Eumotol, Rheumatol.* dec 154°. Sol in chloroform, alcohol, ether. Slightly sol in water. LD_{50} in mice, rats (mg/kg): 2500, 1250 orally; 258, 263 i.v. (Riedel, Schoetensack).

THERAP CAT: Analgesic; antipyretic; anti-inflammatory.

1473. Bumetanide. *3-(Aminosulfonyl)-5-(butylamino)-4-phenoxybenzoic acid; 3-(butylamino)-4-phenoxy-5-sulfamoylbenzoic acid;* PF 1593; Ro 10-6338; Bumex; Burinex; Fontego; Fordiuran; Lixil; Lunetoron; Segurex. $C_{17}H_{20}N_2O_5S$; mol wt 364.42. C 56.03%, H 5.53%, N 7.69%, O 21.95%, S 8.80%. Prepn: Feit, **Ger.** pats. **1,964,503** and **1,964,504** corresp to U.S. pat. **3,634,583** (1970, 1970, 1972 all to Leo Pharm.); Feit, *J. Med. Chem.* **14**, 432 (1971). Pharmacology: Oestergaard *et al., Arzneimittel-Forsch.* **22**, 66 (1972). Review of pharmacology and therapeutic use: A. Ward, R. C. Heel, *Drugs* **28**, 426-464 (1984).

Crystals from aq ethanol, mp 230-231°. LD_{50} i.v. in mice: 330 mg/kg (Oestergaard).
THERAP CAT: Diuretic.

1474. Bunaftine. *N-Butyl-N-[2-(diethylamino)ethyl]-1-naphthalenecarboxamide; N-butyl-N-[2-(diethylamino)ethyl]-1-naphthamide;* bunaphtide; bunaphtine. $C_{21}H_{30}N_2O$; mol wt 326.48. C 77.26%, H 9.26%, N 8.58%, O 4.90%. Prepn: Giannini, **Ger.** pat. **2,099,894** (1971 to Malesci), *C.A.* **75**, 20222a (1971); Giannini *et al., Farmaco Ed. Sci.* **28**, 429 (1973). Pharmacology: Ferroni, Monticelli, *Pharmacol. Res. Commun.* **5**, 151 (1973).

$bp_{0.1}$ 178°. LD_{50} i.p. in mice: 122 mg/kg.
Citrate, $C_{27}H_{38}N_2O_8$, *Meregon.*
THERAP CAT: Antiarrhythmic.

1475. Bunamidine. *N,N-Dibutyl-4-(hexyloxy)-1-naphthalenecarboximidamide; N,N-dibutyl-4-hexyloxy-1-naphthamidine.* $C_{25}H_{38}N_2O$; mol wt 382.57. C 78.48%, H 10.01%, N 7.32%, O 4.18%. Prepn: **Neth.** pat. **Appl. 6,403,016** (1964 to Wellcome Found.), *C.A.* **62**, 9083a (1965); **Neth.** pat. **Appl. 6,603,570** and **Fr.** pat. **M3766** (both 1966 to Wellcome Found.), *C.A.* **66**, 37701t, 68911r (1967); Harfenist, Baltzly, **U.S.** pat. **3,290,375** (1966 to Burroughs-Wellcome). Review of synthesis, chemistry and activity: Harfenist *et al., J. Med. Chem.* **14**, 97 (1971). Activity studies of salts: Hatton, *Vet. Rec.* **81**, 104 (1967).

Hydrochloride, $C_{25}H_{39}ClN_2O$, *Scolaban.* Crystals from ethanol + ethyl acetate + ether, mp 214-214.8°. LD_{50} orally in mice: 540 mg/kg (calcd as base).
3-Hydroxy-2-naphthoate, *Buban.* mp 169-170°.
THERAP CAT (VET): Anthelmintic (Cestodes).

1476. Bunamiodyl Sodium. *2-[[2,4,6-Triiodo-3-[(1-oxobutyl)amino]phenyl]methylene]butanoic acid monosodium salt; 3-butyramido-α-ethyl-2,4,6-triiodocinnamic acid sodium salt;* 3-(3-butyrylamino-2,4,6-triiodophenyl)-2-ethyl-acrylic acid sodium salt; α-ethyl-β-(2,4,6-triiodo-3-butyramidophenyl)acrylic acid sodium salt; α-(2,4,6-triiodo-3-butyrylaminobenzylidene)butyric acid sodium salt; sodium 3-butyramido-α-ethyl-2,4,6-triiodocinnamate; buniodyl;

Bunaiod; Orabilex; Orabilix. $C_{15}H_{15}I_3NNaO_3$; mol wt 661.00. C 27.25%, H 2.29%, I 57.60%, N 2.12%, Na 3.48%, O 7.26%. Prepn: Cassebaum, Dierbach, *Pharmazie* **16**, 389 (1961). Acute toxicity: J. O. Hoppe *et al., J. Med. Chem.* **13**, 997 (1970).

Crystals from water (free acid mp 105-120°). Slightly sol in water. LD_{50} in mice: 418 mg/kg i.v.; 2.78 g/kg orally (Hoppe).
THERAP CAT: Diagnostic aid (radiopaque medium).

1477. Bunazosin. *1-(4-Amino-6,7-dimethoxy-2-quinazolinyl)hexahydro-4-(1-oxobutyl)-1H-1,4-diazepine;* 1-(4-amino-6,7-dimethoxy-2-quinazolinyl)-4-butyrylhexahydro-1H-1,4-diazepine; 4-amino-6,7-dimethoxy-2-[4-(n-butyryl)homopiperazin-1-yl]quinazoline; DDQ. $C_{19}H_{27}N_5O_3$; mol wt 373.45. C 61.11%, H 7.29%, N 18.75%, O 12.85%. Aminoquinazoline derivative with potent hypotensive properties. Prepn: T. Takahashi, H. Sugimoto, **Belg.** pat. **806,626**; *eidem,* **U.S.** pat. **3,920,636** (1974, 1975 to Eisai); alternative prepn: *eidem,* **Japan. Kokai 75 140,474** (1975 to Eisai), *C.A.* **85**, 46769r (1976). *In vitro* and *in vivo* studies of adrenergic blocking action and antihypertensive properties: T. Shoji *et al., Japan. J. Pharmacol.* **30**, 763 (1980); K. Hoshi, S. Fujino, *ibid.* 427. Antihypertensive effect in volunteers: T. Kawasaki *et al., Eur. J. Clin. Pharmacol.* **20**, 399 (1981). Pharmacokinetics: C. Yamato *et al., Xenobiotica* **12**, 549 (1982).

Hydrochloride, $C_{18}H_{28}ClN_5O_3$, *E-643, Dentanol, Detantol.* Crystals from methanol/ethanol, mp 280-282°.
THERAP CAT: Antihypertensive.

1478. Bungarotoxins. Constituent proteins of the venom of the *Elapidae* snake, *Bungarus multicinctus* (Southeast Asian banded krait). Separation of the crude venom yields several fractions, the most important being α-bungarotoxin (α-Bgt) and β-bungarotoxin (β-BuTX): C. C. Chang, C. Y. Lee, *Arch. Int. Pharmacodyn. Ther.* **144**, 241 (1963). α-Bgt is a single polypeptide chain of mol wt about 8,000 containing 74 amino acid residues crosslinked by five S-S bridges: D. Mebo *et al., Biochem. Biophys. Res. Commun.* **44**, 711 (1971); *eidem, Z. Physiol. Chem.* **353**, 243 (1972). Purification, characterization, immunochemical studies: D. G. Clark *et al., Biochemistry* **11**, 1663 (1972). It is a postsynaptic neurotoxin with curare-like action that binds irreversibly to acetylcholine receptor sites, producing neuromuscular blockade and skeletal muscle paralysis. Activity and use as a probe for acetylcholine receptors: C. C. Chang, *Nature* **215**, 1177 (1967); J. P. Changeux *et al., Proc. Nat. Acad. Sci. USA* **67**, 1241 (1970); R. E. Oswald, J. A. Freeman, *Neuroscience* **6**, 1 (1981). β-Bungarotoxin has been shown to contain several components, the major protein being β_1- or β-bungarotoxin. It is composed of 2 subunits of mol wts of about 13,000 and 7,000, linked by disulfide bonds. The larger chain contains 120 amino acid residues including 13 half-cysteines; the smaller chain contains 60 residues including 7 half cysteines. Purification: T. Abe *et al., Eur. J. Biochem.* **80**, 1 (1977); *eidem, Proc. Roy. Soc. London B* **207**, 487 (1980). Chemical properties, amino acid sequence and composition of the two polypeptide chains: K. Kondo *et al., J. Biochem.* **83**, 91, 101 (1978).

Complete purification and characterization of its action on synaptosomal accumulation and release of acetylcholine: J. W. Spokes, J. O. Dolly, *Biochim. Biophys. Acta* **596**, 81 (1980). β-BuTX is a pre-synaptic neurotoxin that prevents acetylcholine release at skeletal neuromuscular junctions without affecting the sensitivity of the post-synaptic membrane. Its proposed phospholipase A_2 activity is believed to be responsible for some of its effects at motor nerve terminals. Activity studies: P. N. Strong *et al.*, *Proc. Nat. Acad. Sci. USA* **75**, 1029 (1976); T. Abe, R. Miledi, *Proc. Roy. Soc. London B* **200**, 225 (1978); M. T. Alderdice, R. L. Volle, *Arch. Pharmacol.* **316**, 126 (1981). *Review:* A. T. Tu, *Venoms: Chemistry and Molecular Biology* (Wiley, New York, 1977) pp 185-187, 240-251.

The crude venom of *Bungarus multicinctus* is quite toxic, having an LD_{50} value in mice of 0.16 μg/g s.c., C. C. Chang, C. Y. Lee, *loc. cit.*, also reported as 0.33 μg/g, D. Mebs *et al.*, *loc. cit.* (1972). LD_{50} in mice of α-Bgt: 0.21 μg/g s.c.; 0.15 μg/g i.p., D. Mebs *et al.*, *ibid.*; of β-BuTX: 0.019 μg/g i.p., K. Kondo *et al.*, *loc. cit.*

USE: As exptl tools in investigating neuromuscular processes.

1479. Bunitrolol. *2-[3-[(1,1-Dimethylethyl)amino]-2-hydroxypropoxy]benzonitrile; o-[3-(tert-butylamino)-2-hydroxypropoxy]benzonitrile;* 1-(2-cyanophenoxy)-2-hydroxy-3-*tert*-butylaminopropane; Kö 1366. $C_{14}H_{20}N_2O_2$; mol wt 248.32. C 67.71%, H 8.12%, N 11.28%, O 12.89%. β-Adrenergic blocker. Prepn: H. Koeppe *et al.*, **S. Afr. pat. 68 03,783;** *eidem,* **U.S. pats. 3,940,489** and **3,961,071** (1968, 1976, 1976 all to Boehringer, Ing.). Pharmacology: T. Kimura, *Experientia* **28**, 813 (1972). Studies on absorption, distribution, metabolism, excretion in rats and mice: T. Suzuki, T. Rikihisa, *Arzneimittel-Forsch.* **29**, 1707 (1979). Clinical studies: A. Reale *et al.*, *J. Int. Med. Res.* **4**, 338 (1976); J. J. C. Jonker *et al.*, *Arzneimittel-Forsch.* **31**, 1140 (1981). Toxicology study: M. Kanda *et al.*, *Oyo Yakuri* **9**, 457, 465, 499 (1975), *C.A.* **83**, 172616 (1975).

Hydrochloride, $C_{14}H_{21}ClN_2O_2$, *Betriol, Stresson.* Crystals from ethanol, mp 163-165°. LD_{50} in mice, rats (mg/kg): 1344-1440, 639-649 orally; 264-265, 222-225 i.p. (Kanda).

THERAP CAT: Antihypertensive, antiarrhythmic, antianginal.

1480. Bunte Salts. Water-sol salts of certain alkyl or aralkyl thiosulfuric acids of the general formula: $RSSO_2$-ONa. Prepn: Lecher, Hardy, U.S. pat. **2,712,547** (1955 to Am. Cyanamid); El-Heweki, Taeger, *J. Prakt. Chem.* **7**, 191 (1958); Tabushi *et al.*, *Kogyo Kagaku Zasshi* **67**, 478 (1964).

THERAP CAT (VET): Have been used as coccidiostats.

1481. Buparvaquone. *2-[[4-(1,1-Dimethylethyl)cyclohexyl]methyl]-3-hydroxy-1,4-naphthalenedione;* 2-[(4-*tert*-butylcyclohexyl)methyl]-3-hydroxy-1,4-naphthoquinone; 3-(4-*t*-butylcyclohexyl)methyl-2-hydroxy-1,4-naphthoquinone; BW 720C; Butalex. $C_{21}H_{26}O_3$; mol wt 326.44. C 77.27%, H 8.03%, O 14.70%. Analog of antimalarial hydroxynaphthoquinones; *t*-butyl deriv of parvaquone, *q.v.* Prepn: A. T. Hudson, A. W. Randall, **Eur. pat. Appl. 77,-550;** *eidem,* U.S. pat. **4,485,117** (1983, 1984 both to Wellcome). *In vivo* and *in vitro* antiprotozoal activity: A. T. Hudson *et al.*, *Parasitology* **90**, 45 (1985); N. McHardy *et al.*, *Res. Vet. Sci.* **39**, 29 (1985). Treatment of bovine theileriosis: S. Dhar *et al.*, *Vet. Rec.* **119**, 635 (1986). Efficacy as a chemoimmunoprophylactic of theileriosis: *eidem, ibid.* **120**, 375 (1987).

Crystals, mp 124-125°. LD_{50} orally in rats: >2 g/kg (Hudson).

THERAP CAT (VET): Antiprotozoal (Theileria).

1482. Buphanamine. *2,3-Didehydro-9-methoxycrinan-1-ol.* $C_{17}H_{19}NO_4$; mol wt 301.33. C 67.76%, H 6.36%, N 4.65%, O 21.24%. Isoln from bulb of *Buphane (Boöphane) disticha* Herb., Appleg. *(Haemanthus toxicarius* Herb.*) Amaryllidaceae:* Humber, Taylor, *Can. J. Chem.* **33**, 1268 (1955); Hauth, Stauffacher, *Helv. Chim. Acta* **44**, 491 (1961). From *B. fischeri* Baker: Renz *et al.*, *ibid.* **38**, 1209 (1955). Identity with "oily" haemanthine: Goosen, Warren, *J. Chem. Soc.* **1960**, 1094. Structure and stereochemistry: Fales, Wildman, *J. Org. Chem.* **26**, 881 (1961). Revised structure: Wildman in *The Alkaloids,* **vol. XI,** R. H. F. Manske, Ed. (Academic Press, New York, 1968) p 361. NMR of revised structure: Crain *et al.*, *J. Am. Chem. Soc.* **93**, 990 (1971). High-resolution mass spectrum: P. Longevialle *et al.*, *Org. Mass Spectrom.* **7**, 401 (1973).

Prisms from ethyl acetate, mp 183-185° (Fales); from acetone, mp 192-194° (Goosen). $[\alpha]_{589}^{24}$ −195°; $[\alpha]_{436}^{24}$ −408° (c = 0.97); $[\alpha]_D^{20}$ −205° (c = 0.69, 95% ethanol); $[\alpha]_D^{21}$ −194° (c = 0.247, chloroform). uv max: 287 nm (ε 1495).

Hydrochloride, $C_{17}H_{19}NO_4$·HCl, prisms from ethanol + ether, mp 180°.

1483. Buphanitine. *9-Methoxycrinan-1α,3α-diol;* hemanthine; nerbowdine. $C_{17}H_{21}NO_5$; mol wt 319.35. C 63.93%, H 6.63%, N 4.39%, O 25.05%. Major alkaloid from *Buphane (Boöphane) disticha* Herb., Appleg. *(Haemanthus toxicarius* Herb.), *Amaryllidaceae.* Isoln: Lewin, *Arch. Exp. Path. Pharmakol.* **68**, 333 (1912); Tutin, *ibid.* **69**, 314 (1912). Identity with hemanthine: Goosen, Warren, *Chem. Ind.* **1957**, 267; with nerbowdine: Lyle *et al.*, *J. Am. Chem. Soc.* **82**, 2620 (1960). Structure studies: Goosen, Warren, *J. Chem. Soc.* **1960**, 1097; Goosen *et al.*, *ibid.* **1961**, 4038. High-resolution mass spectrum: P. Longevialle *et al.*, *Org. Mass Spectrom.* **7**, 401 (1973).

Needles from chloroform + ether (changes crystalline form at 210°), mp 232°. Prisms from acetone. $[\alpha]_D^{20}$ −102° (c = 1 in chloroform).

Hydrochloride, $C_{17}H_{22}ClNO_5$, mp 265°.

Nitrate, $C_{17}H_{22}N_2O_8$, mp 222-224°.

1484. Bupirimate. *Dimethylsulfamic acid 5-butyl-2-(ethylamino)-6-methyl-4-pyrimidinyl ester;* 5-butyl-2-(ethylamino)-6-methyl-4-pyrimidinyl dimethylsulfamate; PP 588; Nimrod. $C_{13}H_{24}N_4O_3S$; mol wt 316.42. C 49.35%, H 7.64%, N 17.71%, O 15.17%, S 10.13%. Prepn: A. M. Cole *et al.*, **Ger. pat. 2,246,645** corresp to U.S. pat. **3,880,852**

(1973, 1975 to ICI). Activity: J. R. Finney *et al.*, *Proc. Brit. Insect. Fung. Conf.*, *8th* **2**, 667 (1975).

Pale tan waxy solid, mp 50-51°. Vapor press at 20°: 5×10^{-7} mm Hg. Soly in water at 25°: 22 mg/l. Sol in most organic solvents except paraffins. Easily hydrolyzed by dil acids. LD_{50} orally in rats: 4000 mg/kg, J. R. Finney *et al.*, *loc. cit.*

USE: Fungicide.

1485. Bupivacaine. *1-Butyl-N-(2,6-dimethylphenyl)-2-piperidinecarboxamide; dl-1-butyl-2',6'-pipecoloxylidide;* 1-*n*-butyl-2',6'-dimethyl-2-piperidinecarboxanilide; *dl-N-n*-butylpipecolic acid 2,6-xylidide; 1-butyl-2-(2,6-xylylcarbamoyl)piperidine; *dl*-1-*n*-butylpiperidine-2-carboxylic acid 2,6-dimethylanilide; Anekain; Marcaina. $C_{18}H_{28}N_2O$; mol wt 288.43. C 74.96%, H 9.78%, N 9.71%, O 5.54%. Prepn: **Brit. pat. 869,978** and Thuresson, Pettersson, **U.S. pat. 2,955,111** (1959, 1960 both to Bofors); Tullar, Bolen, **Brit. pat. 1,166,802** (1969 to Sterling Drug). Prepn, resolution of isomers and report on comparative activities: Luduena, Tullar, **S. Afr. pat. 68 02,611** and **Brit. pat. 1,180,712** (1968 to Sterling Drug), *C.A.* **73**, 25314a (1970); Tullar, *J. Med. Chem.* **14**, 891 (1971). Pharmacological and toxicological studies: Henn, Brattsand, *Acta Anaesthesiol. Scand. Suppl.* **21**, 9 (1966), *C.A.* **66**, 17863u (1967).

mp 107-108°.
Hydrochloride, $C_{18}H_{29}ClN_2O$, *AH-2250*, *LAC-43*, *Marcain*, *Marcaina*, *Carbostesin*. mp 255-256°. LD_{50} in mice (mg/kg): 7.8 i.v., 82 s.c. (Henn, Brattsand).

THERAP CAT: Local anesthetic.

1486. Bupranolol. *1-(2-Chloro-5-methylphenoxy)-3-[(1,1-dimethylethyl)amino]-2-propanol; 1-(tert-butylamino)-3-[(6-chloro-m-tolyl)oxy]-2-propanol;* 1-(6-chloro-3-methylphenoxy)-3-*tert*-butylaminopropan-2-ol; 1-*tert*-butylamino-3-(2-chloro-5-methylphenoxy)-2-propanol; bupranol; Ophtorenin. $C_{14}H_{22}ClNO_2$; mol wt 271.79. C 61.87%, H 8.16%, Cl 13.04%, N 5.15%, O 11.77%. β-Adrenergic blocker. Prepn: Kunz *et al.*, **Ger. pat. 1,236,523** (1967 to Sanol-Arzneimittel Dr. Schwarz), *C.A.* **67**, 64046k (1967) and **U.S. pat. 3,309,406** (1967). Pharmacology: Waterloh *et al.*, *Arzneimittel-Forsch.* **19**, 153, 330, 1710 (1969); Pendleton *et al.*, *Arch. Int. Pharmacodyn. Ther.* **187**, 75 (1970); P. Montastruc *et al.*, *Arch. Farmacol. Toxicol.* **3**, 93 (1977).

Hydrochloride, $C_{14}H_{23}Cl_2NO_2$, *KL-255*, *Betadran*, *Betadrenol*, *Looser*, *Panimit.* Crystals, mp 220-222°.
THERAP CAT: Antihypertensive, antianginal, antiarrhythmic, antiglaucoma agent.

1487. Buprenorphine. *[5α,7α(S)]-17-(Cyclopropylmethyl)-α-(1,1-dimethylethyl)-4,5-epoxy-18,19-dihydro-3-hydroxy-6-methoxy-α-methyl-6,14-ethenomorphinan-7-methanol; 21-cyclopropyl-7α-[(S)-1-hydroxy-1,2,2-trimethylpropyl]-6,14-endo-ethano-6,7,8,14-tetrahydrooripavine;* 21-cyclopropyl-7α-(2-hydroxy-3,3-dimethyl-2-butyl)-6,14-endo-ethano-6,7,8,14-tetrahydrooripavine; 6029-M; RX

6029-M. $C_{29}H_{41}NO_4$; mol wt 467.66. C 74.48%, H 8.84%, N 3.00%, O 13.68%. Analgesic that demonstrates narcotic agonist-antagonist properties. Prepn: K. W. Bentley *et al.*, **Brit. pat. 1,136,214** corresp to **U.S. pat. 3,433,791** (1968, 1969 to Reckitt & Sons). *See also:* K. W. Bentley, "The Morphine Alkaloids" in *The Alkaloids* **vol. 13**, R. F. Manske, Ed. (Academic Press, New York, 1971) pp 75-120. Pharmacology: A. Cowan in *Advan. Biochem. Psychopharmacol.* **vol. 8**, M. C. Braude *et al.*, Eds. (Raven Press, New York, 1974) pp 427-438; *idem, J. Pharm. Pharmacol.* **28**, 177 (1976); W. R. Martin *et al.*, *J. Pharmacol. Exp. Ther.* **197**, 517 (1976). Metabolism: M. J. Rance, J. S. Shillingford, *Biochem. Pharmacol.* **25**, 735 (1976). Clinical study in chronic pain: M. Kjaer *et al.*, *Brit. J. Clin. Pharmacol.* **13**, 487 (1982). *Review:* J. W. Lewis, "Ring C-Bridged Derivatives of Thebaine and Oripavine" in *Advan. Biochem. Psychopharmacol.* **vol. 8**, M. C. Braude *et al.*, Eds. (Raven Press, New York, 1974) pp 123-137. Review of pharmacology and therapeutic efficacy: R. C. Heel *et al.*, *Drugs* **17**, 81-110 (1979).

Crystals, mp 209°.
Hydrochloride, $C_{29}H_{42}ClNO_4$, *CL 112,302*, *NIH 8805*, *UM 952*, *Buprenex*, *Temgesic.*
Note: This is a controlled substance (narcotic) listed in the U.S. Code of Federal Regulations, Title 21 Part 1308.15 (1987).

THERAP CAT: Narcotic analgesic.

1488. Bupropion. *1-(3-Chlorophenyl)-2-[(1,1-dimethylethyl)amino]-1-propanone; (±)-2-(tert-butylamino)-3'-chloropropiophenone; m-chloro-α-(tert-butylamino)propiophenone;* amfebutamon(e). $C_{13}H_{18}ClNO$; mol wt 239.74. C 65.13%, H 7.57%, Cl 14.79%, N 5.84%, O 6.67%. Pharmacologically similar to the tricyclic antidepressants. Prepn: N. B. Mehta, D. A. Yeowell, **Ger. pat. 2,059,618** (1971 to Wellcome Found.) corresp to N. B. Mehta, **U.S. pats. 3, 819,706** and **3,885,046** (1974, 1975 to Burroughs Wellcome). Pharmacology: F. Soroko *et al.*, *J. Pharm. Pharmacol.* **29**, 767 (1977). Clinical evaluation in attention deficit hyperactivity disorder: T. H. Clay *et al.*, *Psychopharmacol. Bull.* **24**, 143 (1988). *Review:* R. A. Maxwell *et al.*, in *Pharmacological and Biochemical Properties of Drug Substances* **vol. 3**, M. E. Goldberg, Ed. (Am. Pharm. Assoc., Washington, DC, 1981) pp 1-55; S. G. Bryant *et al.*, *Clin. Pharm.* **2**, 525-537 (1983).

Pale yellow oil, $bp_{.005}$ 52°C. Sol in methanol, ethanol, acetone, ether, benzene. Very hygroscopic and susceptible to decompn.
Hydrochloride, $C_{13}H_{19}Cl_2NO$, *Wellbatrin*, *Wellbutrin.* Crystals from isopropanol and abs ethanol, mp 233-234°. Soly in mg/ml: water 312; alcohol 193; 0.1N HCl 333. LD_{50} in mice, rats (mg/kg): 230, 210 i.p.; 575, 600 orally (Soroko).

THERAP CAT: Antidepressant.

1489. Buquinolate. *4-Hydroxy-6,7-bis(2-methylpropoxy)-3-quinolinecarboxylic acid ethyl ester; 4-hydroxy-6,7-*

diisobutoxy-3-quinolinecarboxylic acid ethyl ester; ethyl 6,7-diisobutoxy-4-hydroxyquinoline-3-carboxylate; Bonaid. $C_{20}H_{27}NO_5$; mol wt 361.42. C 66.46%, H 7.53%, N 3.88%, O 22.13%. Prepn: Watson, **Belg. pat. 659,237** (1965 to Norwich), *C.A.* **64**, 2071c (1966); corresp to U.S. pat. **3,267,106** (1964).

Crystals, mp 288-291°.
THERAP CAT (VET): Coccidiostat.

1490. Buramate. *(Phenylmethyl)carbamic acid 2-hydroxyethyl ester; benzylcarbamic acid 2-hydroxyethyl ester;* 2-hydroxyethyl benzylcarbamate; glycol benzylcarbamate; AC-601; Hyamate. $C_{10}H_{13}NO_3$; mol wt 195.21. C 61.52%, H 6.71%, N 7.18%, O 24.59%. Prepn from glycol cyclic carbonate and benzylamine: Viard, **Brit. pat. 689,705** (1953 to Chauny & Cirey).

Crystals, mp 40°.
THERAP CAT: Anticonvulsant; antipsychotic.

1491. Burgundy Mixture. Soda Bordeaux. Variable compn, considered to be basic cupric carbonate and/or sulfate. Prepd by reaction of aq solns of $CuSO_4$ and Na_2CO_3: Frear, *Chemistry of the Pesticides* (Van Nostrand, 3rd ed., 1955) p 322; Mond, Haberlein, *J. Chem. Soc.* **115**, 908 (1919). *See also* Cupric Sulfate, Basic.

Blue to green colloidal ppt. Practically insol in water.
USE: Plant fungicide, more phytotoxic than Bordeaux mixture.

1492. Buserelin. *6-[O-(1,1-dimethylethyl)-D-serine]-9-(N-ethyl-L-prolinamide)-10-deglycinamideluteinizing hormone-releasing factor (pig);* HOE 766; Receptal; Suprecur. $C_{60}H_{86}N_{16}O_{13}$; mol wt 1239.44. C 58.14%, H 6.99%, N 18.08%, O 16.78%. Synthetic nonapeptide agonist analog of LH-RH, *q.v.* Synthesis: W. Konig *et al.*, **Ger. pat. 2,438,352**; *eidem*, **U.S. pat. 4,024,248** (1976, 1977 both to Hoechst); A. S. Dutta *et al.*, *J. Med. Chem.* **21**, 1018 (1978). Hormonal and pharmacological action: J. Sandow *et al.*, *Blue Book* (Hoechst AG) **29**, 363 (1979); A. Lemay *et al.*, *Fertil. Steril.* **37**, 193 (1982). Metabolism: J. Sandow, *J. Endocrinol.* **85**, 118 (1980). Use as contraceptive: M. von der Ohe *et al.*, **Ger. pat. 2,735,515**; *eidem*, **U.S. pat. 4,263,-282** (1977, 1981 both to Hoechst); C. Bergquist *et al.*, *Lancet* **2**, 215 (1979); W. Hardt, M. Schmidt-Gollwitzer, *Clin. Endocrinol.* **19**, 613 (1983). Veterinary use to increase conception rate: K. Moller, *N. Zealand Vet. J.* **29**, 214 (1981). Clinical efficacy in prostatic carcinoma: G. Tolis *et al.*, *Proc. Natl. Acad. Sci. USA* **79**, 1658 (1982); J. H. Waxman, *Brit. J. Urol.* **55**, 737 (1983). Clinical evaluation as ovulatory stimulant for *in vitro* fertilization: V. MacLachlan *et al.*, *N. Engl. J. Med.* **320**, 1233 (1989). Review of clinical applications: J. Sandow, *Clin. Endocrinol.* **18**, 571-592 (1983).

5-oxoPro-His-Trp-Ser-Tyr-D-Ser(*t*-Bu)-Leu-Arg-Pro-NHC$_2$H$_5$

$[\alpha]_D^{20}$ −40.4° (c = 1 in dimethylacetamide).
Monoacetate, $C_{62}H_{90}N_{16}O_{15}$, *Sprecur, Suprefact.*
THERAP CAT: Gonad-stimulating principle. Treatment of prostatic carcinoma.
THERAP CAT (VET): Gonad-stimulating principle.

1493. Buspirone. *8-[4-[4-(2-Pyrimidinyl)-1-piperazinyl]butyl]-8-azaspiro[4.5]decane-7,9-dione.* $C_{21}H_{31}N_5O_2$; mol wt 385.51. C 65.43%, H 8.11%, N 18.16%, O 8.30%. Nonbenzodiazepine anxiolytic; 5-hydroxytryptamine (5-HT$_1$) receptor agonist. Prepn: Y. H. Wu *et al.*, *J. Med. Chem.* **15**, 477 (1972); Y. H. Wu, J. W. Rayburn, **Ger. pat. 2,057,845** (1971 to Bristol-Myers); *eidem*, **U.S. pat. 3,717,634** (1973 to

Mead-Johnson). Pharmacology: L. E. Allen *et al.*, *Arzneimittel-Forsch.* **24**, 917 (1974). Comparison with diazepam in treatment of anxiety: H. L. Goldberg, R. J. Finnerty, *Am. J. Psychiat.* **136**, 1184 (1979); A. F. Jacobson *et al.*, *Pharmacotherapy* **5**, 290 (1985). Nonsynergistic effect with alcohol: T. Seppala *et al.*, *Clin. Pharmacol. Ther.* **32**, 201 (1982). Disposition and metabolism: S. Caccia *et al.*, *Xenobiotica* **13**, 147 (1983). Series of articles on chemistry, pharmacology, addictive potential, and clinical trials: *J. Clin. Psychiat.* **43**, pp 1-116 (1982); on pharmacology, safety and clinical comparison with clorazepate: *Am. J. Med.* **80**, Suppl. 3B, 1-51 (1986). Review of pharmacology and therapeutic efficacy: K. L. Goa, A. Ward, *Drugs* **32**, 114-129 (1986). *Review:* M. W. Jann, *Pharmacotherapy* **8**, 100-116 (1988); D. P. Taylor, *FASEB J.* **2**, 2445-2452 (1988).

Hydrochloride, $C_{21}H_{32}ClN_5O_2$, *Bespar, Buspar, Buspinol, Censpar, Lucelan, Travin.* Crystals from abs ethanol, mp 201.5-202.5°. LD_{50} i.p. in rats: 136 mg/kg (Allen).
THERAP CAT: Anxiolytic.

1494. Busulfan. *1,4-Butanediol dimethanesulfonate esters;* 1,4-bis(methanesulfonoxy)butane; 1,4-di(methanesulfonyloxy)butane; 1,4-di(methylsulfonoxy)butane; methanesulfonic acid tetramethylene ester; tetramethylene bis(methanesulfonate); busulphan; CB 2041; GT 41; Mielucin; Misulban; Mitosan; Myelosan; Myeleukon; Myeloleukon; Myleran; Sulfabutin. $C_6H_{14}O_6S_2$; mol wt 246.31. C 29.26%, H 5.73%, O 38.98%, S 26.04%. $CH_3SO_2O(CH_2)_4OSO_2CH_3$. Alkylating agent with antileukemic activity. Discovery: A. Haddow, G. M. Timmis, *Lancet* **1**, 207 (1953). Prepn: G. M. Timmis, **U.S. pat. 2,917,432** (1959 to Burroughs Wellcome). Chemosterilant effect in boll weevils: J. W. Haynes *et al.*, *J. Econ. Entomol.* **66**, 619 (1973); J. W. Haynes, J. E. Wright, *Southwest Entomol.* **7**, 56 (1982). Pharmacokinetics: H. Ehrsson *et al.*, *Clin. Pharmacol. Ther.* **34**, 86 (1983). HPLC determn in human plasma: W. D. Henner *et al.*, *J. Chromatog.* **416**, 426 (1987). Clinical pretreatment with cyclophosphamide, *q.v.*, for bone marrow transplants: G. W. Santos *et al.*, *N. Engl. J. Med.* **309**, 1347 (1983). Comparative clinical trial with mitobronitol, *q.v.*, in chronic myeloid leukemia: R. T. Silver *et al.*, *Cancer* **60**, 1442 (1987). Toxicity data: H. R. Scherf *et al.*, *Arzneimittel-Forsch.* **20**, 1467 (1970). Review of pharmacology: C. D. R. Dunn, *Exp. Hematol.* (Copenhagen) **2**, 101-117 (1974); of toxicology: J. B. Bishop, J. S. Wassom, *Mutat. Res.* **168**, 15-45 (1986). Comprehensive description: M. Tariq, A. A. Al Badr in *Analytical Profiles of Drug Substances* vol. 16, K. Florey, Ed. (Academic Press, New York, 1987) pp 53-83.

Crystals, mp 114-118°. Soly in acetone at 25°: 2.4 g/100 ml; in alcohol: 0.1 g/100 ml. Practically insol in water, but will dissolve slowly as hydrolysis takes place. LD_{50} i.v. in rats: 1.8 mg/kg (Scherf).
Note: This substance has been listed as a known carcinogen: *Fourth Annual Report on Carcinogens* (NTP 85-002, 1985) p 47.
USE: Insect sterilant.
THERAP CAT: Antineoplastic.

1495. Butabarbital Sodium. *5-Ethyl-5-(1-methylpropyl)-2,4,6(1H,3H,5H)-pyrimidinetrione sodium salt; 5-sec-butyl-5-ethylbarbituric acid sodium salt;* 5-ethyl-5-(1-methylpropyl)barbituric acid sodium salt; sodium 5-sec-butyl-5-ethylbarbiturate; sodium 5-ethyl-5-(1-methylpropyl)barbiturate; secbutobarbitone sodium; Asturidon; Bubarbital Sodium; Butabarbitone Sodium; Butabarpal sodium; Busodium; Busotran; Butabon; Butabar; Butak; Buta-Kay; Butrate; Butte; Buticaps; Butalan; Butalix; Butanotic; Butex; Butatran; Butazem; Carrbutabarb; Butased; Butisol Sodium; Loubarb; Neravan; Prelital; Sarisol. $C_{10}H_{15}N_2NaO_3$; mol wt 234.23. C 51.27%, H 6.46%, N 11.96%, Na 9.82%, O 20.49%. Prepn: Shonle, **U.S. pat. 1,856,792** (1932 to Lilly).

Bitter powder. (The free acid mp 165-168°.) One gram dissolves in 2 ml water, in about 7 ml alcohol. Practically insol in ether, benzene. The pH of a 10% aq soln is between 10.0 and 11.2: USP XXI (1985).

Caution: May be habit forming. This is a controlled substance (depressant) listed in the U.S. Code of Federal Regulations, Title 21, Parts 329.1 and 1308.13 (1987).

THERAP CAT: Sedative, hypnotic.

1496. Butacaine. *3-(Dibutylamino)-1-propanol 4-aminobenzoate;* 3-(*p*-aminobenzoxy)-1-di-*n*-butylaminopropane; dibutylaminopropyl-*p*-aminobenzoate; *p*-aminobenzoyldibutylaminopropanol; Butelline. $C_{18}H_{30}N_2O_2$; mol wt 306.44. C 70.55%, H 9.87%, N 9.14%, O 10.44%. Prepd from *p*-nitrobenzoyl chloride and γ-di-*n*-butylaminopropanol followed by reduction of the NO_2 group to NH_2: Kamm *et al.*, U.S. pat. 1,358,751 (1920); Adams, Volwiler; Weston, U.S. pats. 1,676,470 and 2,437,984 (1928 and 1948 to Abbott); Burnett *et al., J. Am. Chem. Soc.* **59**, 2248 (1937); Kaye, Roberts, *ibid.* **73**, 4762 (1951).

Liquid, $bp_{0.11}$ 178-182°.

Sulfate, $C_{36}H_{62}N_4O_8S$, *Butyn Sulfate.* Crystals from *n*-propanol, mp 138.5-139.5° (also reported as mp 100-103°). Produces numbness of tongue on tasting. Affected by light. One gram dissolves in somewhat less than 1 ml water, more rapidly on heating. Quite sol in warm alc, in acetone; slightly sol in chloroform; practically insol in ether. Aq soln is practically neutral to litmus and may be boiled for sterilization without dec. LD_{50} in mice: 12.4 mg/kg i.v., Schmidt *et al., Toxicol. Appl. Pharmacol.* **1**, 454 (1956). *Pharmaceutical Incompat.* Alkalies and alkaline-reacting substances liberate the free base as an oily liquid from solns. Bicarbonates produce a precipitate of butacaine carbonate. Iodine gives a brown precipitate. Chlorides form the almost insol butacaine chloride which may precipitate.

Hydrochloride, $C_{18}H_{30}N_2O_2$.HCl, crystals from ethanol, mp 157-158.5°.

THERAP CAT: Topical anesthetic.

THERAP CAT (VET): Topical anesthetic.

1497. Butacetin. *N-[4-(1,1-Dimethylethoxy)phenyl]acetamide; 4'-tert-butoxyacetanilide;* Tromal. $C_{12}H_{17}NO_2$; mol wt 207.26. C 69.54%, H 8.27%, N 6.76%, O 15.44%. Prepn: Bowden, Green, *J. Chem. Soc.* **1954**, 1795.

Plates from 66% aq alcohol, mp 130°.

THERAP CAT: Analgesic; antidepressant.

1498. Butachlor. *N-(Butoxymethyl)-2-chloro-N-(2,6-diethylphenyl)acetamide; N-(butoxymethyl)-2-chloro-2',6'-diethylacetanilide;* 2-chloro-2,6-diethyl-N-(butoxymethyl)-acetanilide; CP 53619; Machete; Butanex. $C_{17}H_{26}ClNO_2$; mol wt 311.86. C 65.47%, H 8.40%, Cl 11.37%, N 4.49%, O 10.26%. Selective pre-emergence and pre-plant herbicide. Prepn: J. F. Olin, U.S. pat. 3,442,945 (1969 to Monsanto). Photodecomposition: Y.-L. Chen, C.-C. Chen, *Nippon Noyaku Gakkaishi* **3**, 143 (1978); *C.A.* **89**, 174926j (1978). Degradation: Y.-L. Chen, T.-C. Wu, *ibid.* 411, *C.A.* **91**, 1258f (1979); Y.-L. Chen, *FFTC Book Ser.* **20**, 121 (1981).

Light yellow oil, $bp_{0.5}$ 196°. d_4^{30} 1.0695. Soly in water at 20°: 20 mg/l. Sol in most organic solvents. LD_{50} orally in rats: 1740 mg/kg, *RTECS* vol. I, R. J. Lewis, R. L. Tatken, Eds. (1980) p 25.

USE: Herbicide.

1499. Butaclamol. *3-(1,1-Dimethylethyl)-2,3,4,4a,8,9,-13b,14-octahydro-1H-benzo[6,7]cyclohepta[1,2,3-de]pyrido-[2,1-a]isoquinolin-3-ol;* (±)-3α-*tert*-butyl-2,3,4,4aβ-8,9,-13bα,14-octahydro-1H-benzo[6,7]cyclohepta[1,2,3-*de*]-pyrido[2,1-*a*]isoquinolin-3-ol. $C_{25}H_{31}NO$; mol wt 361.54. C 83.06%, H 8.64%, N 3.87%, O 4.43%. A highly potent neuroleptic agent; one of a novel class of psychopharmacologically active compounds. Prepn: F. T. Bruderlein, L. G. Humber, **Ger.** pat. 2,106,165 corresp to **U.S.** pat. 3,657,250 (1971, 1973 both to Ayerst); *eidem, J. Med. Chem.* **18**, 185 (1975). Crystallographic study and abs config: P. H. Bird *et al., Can. J. Chem.* **54**, 2715 (1976). Although the antipsychotic efficacy of the (±)-hydrochloride has been demonstrated in man, studies have shown that its pharmacological activity was due solely to the (+)-enantiomer: L. Humber *et al., Mol. Pharmacol.* **11**, 833 (1975); W. Lippmann *et al., Life Sci.* **16**, 213 (1975). Psychopharmacological profile: K. Voith, F. Herr, *Psychopharmacologia* **42**, 11 (1975). Clinical study of the (±)-hydrochloride in schizophrenia: D. H. Mielke *et al., Dis. Nerv. Syst.* **36**, 7 (1975). Butaclamol has been used in identifying stereospecific binding sites for neuroleptic agents: P. Seeman *et al., Proc. Nat. Acad. Sci. USA* **72**, 4376 (1975); L. G. Humber *et al., J. Med. Chem.* **22**, 761 (1979); A. H. Phillip *et al., ibid* 768.

(±)-Hydrochloride, $C_{25}H_{32}ClNO$, *AY-23,028.* Cryst from acetone, mp 282-284° (Bruderlein, Humber, 1973), also reported as mp 309-310° (Humber, 1975).

(+)-Hydrochloride, cryst, mp 304-307°. $[\alpha]_D^{25}$ +218.5° (methanol soln).

(−)-Hydrochloride, cryst, mp 305-307°. $[\alpha]_D^{25}$ −219.0° (methanol soln).

1500. 1,3-Butadiene. α,γ-Butadiene; bivinyl; divinyl; erythrene; vinylethylene; biethylene; pyrrolylene. C_4H_6; mol wt 54.09. C 88.82%, H 11.18%. CH_2=CHCH=CH_2. In the U.S.A. it is produced largely from petroleum gases, i.e., by catalytic dehydrogenation of butene or butene-butane mixtures. It can be obtained directly by cracking naphtha and light oil. Practical processes based on the conversion of ethyl alc were worked out during World War II. Manuf: Faith, Keyes & Clark's *Industrial Chemicals*, F. A. Lowenheim, M. K. Moran, Eds. (Wiley-Interscience, New York, 4th ed., 1975) pp 164-172. Toxicity: Carpenter *et al., J. Ind. Hyg. Toxicol.* **26**, 69 (1944). *Reviews:* Norton, *Chem. Rev.* **31**, 319 (1942); Alder, "The Diene Synthesis" in *Newer Methods of Preparative Organic Chemistry* (Interscience, New York, 1948); Konrad, *Angew. Chem.* **62**, 491 (1950); Hillyer, Stallings, *Petrol. Refiner* **35**(12), 157 (1956); A. S. Onishchenko, *Diene Synthesis* (New York, 1964); Bailey, "Butadiene" in *Vinyl and Diene Monomers* (part 2), E. C. Leonard, Ed. (Interscience, New York, 1971) pp 757-995; I. Kirshenbaum in Kirk-Othmer *Encyclopedia of Chemical Technology* vol. **4** (Wiley-Interscience, New York, 3rd ed., 1978) pp 313-337.

Colorless gas. mp −108.966°: Glasgow *et al., Anal.*

Chem. **22**, 1521 (1950). bp$_{760}$ −4.5°. d$_4^{-6}$ 0.650. Densities from −17.8° to 60°: *Ind. Eng. Chem., Anal. Ed.* **16**, 7 (1944). n$_D^{-6}$ 1.4223. bp (at pressures greater than 1 atm): 2 atm: 15.3°; 5 atm: 47.0°; 10 atm: 76.0°; 20 atm: 114.0°; 30 atm: 139.8°; 40 atm: 160.0°. Critical temperature 161.8°; critical pressure 42.6 atm. Infrared absorption spectrum: *ibid.* 422. Stability: *Ind. Eng. Chem.* **36**, 3 (1944). Stabilization with o-dihydroxybenzene: **Brit. pat. 569,412**; with aliphatic mercaptans: **U.S. pat. 2,373,754**. Polymerizes and copolymerizes easily, e.g. under the influence of sodium, thereby forming synthetic rubbers. Sol in organic solvents. Alcohol dissolves about 40 vols at room temp. LC for rabbits in air: 250,000 ppm (Carpenter).

USE: As polymer component in the manuf of synthetic rubber, *buna;* in the Diels-Alder condensation for the synthesis of many diverse compds. *Caution:* May be irritating to skin, mucous membranes, and narcotic in high concns.

1501. Butalamine. *N,N-Dibutyl-N'-(3-phenyl-1,2,4-oxadiazol-5-yl)-1,2-ethanediamine; 5-[[2-(dibutylamino)-ethyl]amino]-3-phenyl-1,2,4-oxadiazole;* 3-phenyl-5-(dibutylaminoethylamino)-1,2,4-oxadiazole. C$_{18}$H$_{28}$N$_4$O; mol wt 316.45. C 68.32%, H 8.92%, N 17.70%, O 5.06%. Prepn: **Fr. pat. M3334** corresp to Aron-Samuel, Sterne, **U.S. pat. 3,338,899** (1965, 1967, to Aron-Samuel). Pharmacology and toxicity studies: J. Sterne *et al., Therapie* **24**, 735 (1969); J. Sterne, *ibid.* 745; J. L. Fontaine, R. Fontaine, *ibid.* **25**, 961 (1970).

$$(C_4H_9)_2NCH_2CH_2NH \underset{N}{\overset{O-N}{\diagdown}} C_6H_5$$

Hydrochloride, C$_{18}$H$_{29}$ClN$_4$O, *LA 1221, Adrevil, Hemotrope, Surem, Surheme.* mp 145°. LD$_{50}$ in mice: 625 mg/kg orally; 2.5 g/kg s.c., J. Sterne, *loc. cit.*

THERAP CAT: Peripheral vasodilator.

1502. Butalbital. *5-(2-Methylpropyl)-5-(2-propenyl)-2,4,6(1H,3H,5H)-pyrimidinetrione; 5-allyl-5-isobutylbarbituric acid; 5-allyl-5-(2-methylpropyl)barbituric acid; 5-isobutyl-5-allylbarbituric acid; alisobumal; allylbarbital; itobarbital; tetrallobarbital; Sandoptal.* C$_{11}$H$_{16}$N$_2$O$_3$; mol wt 224.25. C 58.91%, H 7.19%, N 12.49%, O 21.40%. Prepn: Volwiler, *J. Am. Chem. Soc.* **47**, 2236 (1925).

$$\begin{array}{c} CH_2=CHCH_2 \\ (CH_3)_2CHCH_2 \end{array} \overset{}{\underset{}{\diagdown}}\ \text{(barbituric acid ring)}$$

Prisms, mp 138-139°. Slightly bitter taste. Practically insol in water and petr ether. Sol in alcohol, chloroform, ether, acetone, glacial acetic acid, also in solns of fixed alkali hydroxides. A satd aq soln is acid to litmus.

Caution: May be habit forming. This is a controlled substance (depressant) listed in the U.S. Code of Federal Regulations, Title 21, Parts 329.1 and 1308.13 (1987).

THERAP CAT: Sedative, hypnotic.

1503. Butallylonal. *5-(2-Bromo-2-propenyl)-5-(1-methylpropyl)-2,4,6(1H,3H,5H)pyrimidinetrione; 5-(2-bromoallyl)-5-sec-butylbarbituric acid;* sonbutal; Pernoston. C$_{11}$H$_{15}$BrN$_2$O$_3$; mol wt 303.16. C 43.58%, H 4.99%, Br 26.36%, N 9.24%, O 15.83%. Prepn: **U.S. pat. 1,739,662** (1929).

$$\begin{array}{c} CH_2=CBrCH_2 \\ CH_3CH_2CH \\ | \\ CH_3 \end{array} \overset{}{\underset{}{\diagdown}}\ \text{(barbituric acid ring)}$$

Crystals. Slightly bitter taste. mp 130-133°. Freely sol in

alcohol, ether, alkaline solns. Practically insol in water (acid to litmus), petr ether, aliphatic hydrocarbons.

Sodium salt, C$_{11}$H$_{14}$BrN$_2$NaO$_3$, *sodium 5-(2-bromoallyl)-5-sec-butylbarbiturate, Pernocton* (usually a 10% soln). Powder or crystals, bitter taste. Soluble in water, alcohol; slightly sol in ether, chloroform. A 10% aq soln (pH of about 9.5) is alkaline to litmus and phenolphthalein.

Caution: May be habit forming. This is a controlled substance (depressant) listed in the U.S. Code of Federal Regulations, Title 21 Parts 329.1 and 1308.13 (1987).

THERAP CAT: Sedative, hypnotic.

1504. Butamben. *4-Aminobenzoic acid butyl ester;* butyl aminobenzoate; *n*-butyl *p*-aminobenzoate; Butesin; Butoform; Planoform; Scuroforme. C$_{11}$H$_{15}$NO$_2$; mol wt 193.24. C 68.37%, H 7.82%, N 7.25%, O 16.56%. Prepn: Brill, *J. Am. Chem. Soc.* **43**, 1322 (1921); Adams, Volwiler, **U.S. pat. 1,440,652** (1923 to Abbott); **Brit. pat. 252,870**; *C.A.* **17**, 1243 (1923); *C.A.* **21**, 2478 (1927).

$$H_2N-\langle\text{benzene ring}\rangle-COOCH_2CH_2CH_3$$

Crystals from alc, mp 57-59°. bp$_8$ 174°. One gram dissolves in about 7 liters of water. Sol in dil acids, alcohol, chloroform, ether, and fatty oils. It is hydrolyzed slowly when boiled with water.

Picrate, C$_{28}$H$_{33}$N$_5$O$_{11}$, *Abbott 34842, Butesin Picrate. See* **U.S. pat. 1,596,259** (1926 to Abbott). Yellow powder, mp 109-110°. Sol in alc, chloroform, ether. Soly in water: 1 g/2000 ml.

THERAP CAT: Topical anesthetic.

THERAP CAT (VET): Topical anesthetic.

1505. Butamirate. *α-Ethylbenzeneacetic acid 2-[2-(diethylamino)ethoxy]ethyl ester; 2-phenylbutyric acid 2-[2-(diethylamino)ethoxy]ethyl ester;* 2-[2-(diethylamino)ethoxy]-ethyl 2-phenylbutyrate; butamyrate. C$_{18}$H$_{29}$NO$_3$; mol wt 307.44. C 70.32%, H 9.51%, N 4.56%, O 15.61%. Prepd by the esterification of phenylethylacetyl chloride with diethylaminoethoxyethanol: **Ger. pat. 1,151,515**, *C.A.* **60**, 462h (1964) and Heusser, **U.S. pat. 3,349,114** (1963, 1967, both to Hommel A.G.).

$$C_6H_5CHCOOCH_2CH_2OCH_2CH_2N(CH_2CH_3)_2 \\ | \\ CH_2CH_3$$

Practically colorless liquid with peculiar odor. bp$_1$ 140-155°. Practically insol in water. Very sol in alcohol, acetone, ether.

Citrate, C$_{24}$H$_{37}$NO$_{10}$, *Abbott 36581, HH-197, Acodeen, Panatus, Pertix, Sincodex, Sinecod.* White, hygroscopic crystals from acetone, mp 75°.

THERAP CAT: Antitussive.

1506. Butamisole. *2-Methyl-N-[3-(2,3,5,6-tetrahydro-imidazo[2,1-b]thiazol-6-yl)phenyl]propanamide; (−)-2-methyl-3'-(2,3,5,6-tetrahydroimidazo[2,1-b]thiazol-6-yl)-propionanilide.* C$_{15}$H$_{19}$N$_3$OS; mol wt 289.40. C 62.25%, H 6.62%, N 14.52%, O 5.53%, S 11.08%. Prepn: L. D. Spicer, J. J. Hand, **S. Afr. pat. 71 05,803** (1972 to Am. Cyanamid), *C.A.* **78**, 72151h (1973). Prepn and resolution of isomers: *eidem,* **Fr. pat. 2,199,979** corresp to **U.S. pat. 3,899,583** (1974, 1975 both to Am. Cyanamid). Efficacy and safety in dogs: B. T. Alford *et al., Vet. Med. Small Anim. Clin.* **74**, 487 (1979). Evaluation in horses and ponies: R. B. Grieve *et al., Am. J. Vet. Res.* **40**, 139 (1979).

$$(CH_3)_2CHCOHN-\langle\text{benzene ring}\rangle-\langle\text{thiazoline ring}\rangle$$

Hydrochloride, C$_{15}$H$_{20}$ClN$_3$OS, *CL 206214, Styquin.*

THERAP CAT (VET): Anthelmintic.

1507. Butane. *n*-Butane. C$_4$H$_{10}$; mol wt 58.12. C 82.66%, H 17.34%. CH$_3$CH$_2$CH$_2$CH$_3$. Occurrence: in petroleum, Mabery, *J. Am. Chem. Soc.* **30,** 143 (1908); in natural gas and in refinery cracking products. Prepd from C_2H_5I and sodium amalgam: Löwig, *Jahresber. Fortschr. Chem.* **1860,** 397, *Beilstein,* **vol. 1,** 118. Recovery of butanes from natural and refinery gases: Kirkbride, Bertelli, *Ind. Eng. Chem.* **35,** 1242 (1943); Walters, *ibid.* **47,** 2544 (1955); Gilmore, Bauer, *Oil Gas J.* **50,** 84, 90, 94, 119 (1951), *C.A.* **46,** 1743d (1952). Separation of butane and isobutane: Stone, *Petrol. Refiner* **25**(4), 164 (1946), *C.A.* **43,** 2414 (1949). Handbook: *Butane-Propane Gases,* L. C. Denny *et al.,* Eds. (Chilton Co., Los Angeles, 4th ed., 1962) 383 pp.

Flammable gas, flash pt $-138°$. bp $-0.50°$. d(gas) 2.046 (air = 1). One vol of water dissolves 0.15 vol and 1 vol of alcohol 18 vols of the gas at 17° and 770 mm; 1 vol of ether or chloroform at 17° dissolves 25 or 30 vols of the gas, resp.

USE: As producer gas; raw material for motor fuels, in the manuf of synthetic rubbers. *Caution:* May be narcotic in high concns. A simple asphyxiant.

1508. Butanilicaine. *2-(Butylamino)-N-(2-chloro-6-methylphenyl)acetamide; 2-(butylamino)-6′-chloro-o-acetotoluidide;* ω-*n*-butylaminoacetic acid 2-methyl-6-chloroanilide; 1-(butylaminoacetylamino)-2-chloro-6-methylbenzene; *N*-(butylaminoacetyl)-6-chloro-o-toluidine. $C_{13}H_{19}ClN_2O$; mol wt 254.77. C 61.29%, H 7.52%, Cl 13.92%, N 11.00%, O 6.28%. Prepn: Häussler, Ther, *Arzneimittel-Forsch.* **3,** 609 (1953); **Brit.** pats. **726,080; 759,744** (1955, 1956, both to Cilag); **Brit.** pat. **782,971** and Ehrhart *et al.,* **U.S.** pat. **2,912,-460** (1957, 1959, both to Hoechst); Epstein, Kaminsky, *J. Am. Chem. Soc.* **80,** 1892 (1958).

$$\text{Cl} \quad \text{NHCOCH}_2\text{NH(CH}_2\text{)}_3\text{CH}_3 \quad \text{CH}_3$$

Crystals, mp 45-46°. $bp_{0.001}$ 145°. $bp_{0.5}$ 166-167°.

Hydrochloride, $C_{13}H_{19}ClN_2O\cdot HCl$, crystals from ethanol, mp 232°, also reported as mp 236-239° (Epstein, Kaminsky, *loc. cit.*). Soluble in water. Aq soln is neutral. LD_{50} s.c. in mice: 700 mg/kg.

Phosphate, present in *Hostacaine.* Crystals, mp 126-127°.

THERAP CAT: Local anesthetic.

1509. Butaperazine. *1-[10-[3-(4-Methyl-1-piperazinyl)-propyl]-10H-phenothiazin-2-yl]-1-butanone;* 2-butyryl-10-[3-(4-methyl-1-piperazinyl)propyl]phenothiazine; *N*-[γ-(4′-methyl-1′-piperazinyl)propyl]-3-butyrylphenothiazine; butyrylperazine; Bayer 1362; Repoise; Tyrylen. $C_{24}H_{31}N_3OS$; mol wt 409.60. C 70.38%, H 7.63%, N 10.26%, O 3.91%, S 7.83%. Prepn: Hoerlein *et al.,* **Ger.** pat. **1,120,451** (1961 to Bayer), *C.A.* **57,** 4677c (1962).

$$\text{CH}_2\text{CH}_2\text{CH}_2-\text{N} \quad \text{N}-\text{CH}_3$$
$$\text{COCH}_2\text{CH}_2\text{CH}_3$$

$bp_{0.05}$ 270-280°.

Dimaleate, $C_{32}H_{39}N_3O_9S$, *Randolectil.*

Maleate, $C_{24}H_{31}N_3OS\cdot C_4H_4O_4$, crystals from carbon tetrachloride, mp 180-182°.

THERAP CAT: Antipsychotic.

1510. Butaverine. *β-Phenyl-1-piperidinepropanoic acid butyl ester; β-phenyl-1-piperidinepropionic acid butyl ester;* butyl β-phenyl-1-piperidinepropionate; butyl β-piperidinohydrocinnamate; *n*-butyl β-(*N*-piperidyl)-β-phenylpropionate; butamiverine. $C_{18}H_{27}NO_2$; mol wt 289.40. C 74.70%, H 9.40%, N 4.84%, O 11.06%. Prepn: Pollard, Mattson, *J. Am. Chem. Soc.* **78,** 4089 (1956); Pacheco *et al., Bull. Soc. Chim. France* **1962,** 1379.

$$\text{N}-\text{CHCH}_2\text{COOCH}_2\text{(CH}_2\text{)}_2\text{CH}_3 \quad \text{C}_6\text{H}_5$$

Hydrochoride, $C_{18}H_{28}ClNO_2$, *Espasmo-Gemora, Gemora.* Crystals, mp 170°.

THERAP CAT: Antispasmodic.

1511. Butazolamide. *N-[5-(Aminosulfonyl)-1,3,4-thia-diazol-2-yl]butanamide; N-(5-sulfamoyl-1,3,4-thiadiazol-2-yl)butyramide;* 5-butyramido-1,3,4-thiadiazole-2-sulfonamide; SKF 4965; Butamide. $C_6H_{10}N_4O_3S_2$; mol wt 250.31. C 28.79%, H 4.03%, N 22.38%, O 19.18%, S 25.62%. Prepn: Vaughan *et al., J. Org. Chem.* **21,** 700 (1956); **Brit.** pat. **769,-757** (1957 to Am. Cyanamid).

$$\text{C}_3\text{H}_7\text{CONH} \quad \text{S} \quad \text{SO}_2\text{NH}_2$$
$$\text{N}-\text{N}$$

Crystals, mp 260-262° (dec).

THERAP CAT: Carbonic anhydrase inhibitor, diuretic.

1512. Butedronic Acid. *(Diphosphonomethyl)butanedioic acid; (diphosphonomethyl)succinic acid;* 2,3-dicarboxypropane-1,1-diphosphonic acid; DPD. $C_5H_{10}O_{10}P_2$; mol wt 292.08. C 20.56%, H 3.45%, O 54.78%, P 21.21%. Prepn: A. Heins *et al.,* **Ger.** pat. **2,217,692** (1973 to Henkel); *eidem,* **U.S.** pat. **3,923,876** (1975 to Bayer A.G.); K. H. Worms, H. Blum, *Z. Anorg. Allgem. Chem.* **457,** 219 (1979); and physical-chemical properties: N. Vanlic-Razumenic *et al., J. Serb. Chem. Soc.* **51,** 63 (1986). Comparative biokinetics of the ^{99m}Tc complex in volunteers: C. Schümichen, H. Schmidt, *Nuklearmedizin,* Suppl. **19,** 930 (1982). Comparative evaluation in rabbits of ^{99m}Tc complex as skeletal imaging agent: G. Subramanian *et al., Radiology* **149,** 823 (1983); in humans: G. Godart *et al., Clin. Nucl. Med.* **11,** 92 (1986).

$$\text{HOOCCH}_2\text{CHCH} \quad \text{COOH} \quad \text{O} \quad \text{P} \quad \text{OH} \quad \text{OH} \quad \text{OH} \quad \text{OH} \quad \text{O} \quad \text{P}$$

Monohydrate, white crystalline powder from glacial acetic acid and water (15:1), mp 150°. uv max (water): 208 nm (ε 274).

Tetrasodium salt, $C_5H_6Na_4O_{10}P_2$, Tc 924 (DPD), Teceos (non-radioactive kit).

THERAP CAT: Diagnostic aid (radioactive imaging agent).

1513. 1-Butene. α-Butylene; ethylethylene. C_4H_8; mol wt 56.10. C 85.63%, H 14.37%. $CH_3CH_2CH=CH_2$. Occurs in oil and coal gas. Obtained by cracking of petr oils and by thermal decompn of butane or pentane or isopentane: Egloff *et al., Ind. Eng. Chem.* **28,** 1283 (1936); Calingaert, *J. Am. Chem. Soc.* **45,** 130 (1923); from butyl alcohol by treatment with conc H_2SO_4: *Compt. Rend.* **176,** 813 (1923).

Gas. Not solid at $-190°$. bp_{760} $-6.47°$. $d_4^{-6.47}$ 0.6255. Explodes in mixtures with oxygen. Description of properties: *Oil, Gas J.* **44,** 119 (1945), *C.A.* **39,** 3783; *J. Am. Chem. Soc.* **50,** 1427 (1928).

Caution: May be narcotic in high concns. A simple asphyxiant.

1514. 2-Butene. Pseudo-butylene; *sym*-dimethylethylene; β-butylene. C_4H_8; mol wt 56.10. C 85.63%, H 14.37%. $CH_3CH=CHCH_3$. Occurs in coal gas. Obtained by cracking of petroleum oils. From isobutanol by the action of hot $ZnCl_2$: LeBel, Greene, *Am. Chem. J.* **2,** 24 (1880). Configuration of the stereoisomeric forms: Kistiakowsky, *J. Am. Chem. Soc.* **57,** 879 (1935).

Flammable gas.

cis-Form, mp −139.3°; bp$_{760}$ +3.73°.
trans-Form, mp −105.8°; bp$_{744}$ +0.3-0.4°.

1515. Butethal. *5-Butyl-5-ethyl-2,4,6(1H,3H,5H)-pyrimidinetrione; 5-butyl-5-ethylbarbituric acid;* butobarbitone; Soneryl; Neonal; Butobarbital; Etoval. $C_{10}H_{16}N_2O_3$; mol wt 212.24. C 56.59%, H 7.60%, N 13.20%, O 22.61%. Prepn: U.S. pat. 1,609,520 (1926). Pharmacology and toxicity: G. A. Alles *et al., J. Pharmacol. Exp. Ther.* **89**, 356 (1947).

Crystals. Slightly bitter taste. mp 124-127°. One gram dissolves in about 5 ml alcohol, 10 ml ether. Practically insol in water; insol in petr ether, aliphatic hydrocarbons. LD$_{50}$ i.p. in mice: 1.506 mM/kg (Alles).
Caution: May be habit forming. This is a controlled substance (depressant) listed in the U.S. Code of Federal Regulations, Title 21 Parts 329.1 and 1308.13 (1987).
THERAP CAT: Sedative, hypnotic.

1516. Butethamate. *Benzeneacetic acid α-ethyl-2-(diethylamino)ethyl ester; 2-phenylbutyric acid 2-(diethylamino)ethyl ester; β-diethylaminoethyl ethylphenylacetate; 2-diethylaminoethyl 2-phenylbutyrate.* $C_{16}H_{25}NO_2$; mol wt 263.37. C 72.96%, H 9.57%, N 5.32%, O 12.15%. Prepn: Di Paco, Tauro, *Farmaco Ed. Sci.* **11**, 540 (1956); **Swiss** pat. 291,375 (1953 to AG Hommel's Haematogen) and **Swiss** pat. 292,596 (1953 to Chem. Fabrik "PARA"), *C.A.* **49**, 2505a,c (1955). Pharmacology: Jordan, *Arzneimittel-Forsch.* **8**, 716 (1958); Fleisch *et al., ibid.* **11**, 1119 (1961).

bp$_{11}$ 167-169°. n_D^{20} 1.4909.
Citrate, $C_{22}H_{36}NO_9$, *Abuphenine, Convenil, Hicoseen, Pertix, Phenesin, Phenetin.* Crystals from abs alcohol, mp 109-110°. Freely sol in water, alcohol.
THERAP CAT: Antitussive.

1517. Butethamine. *2-[(2-Methylpropyl)amino]ethanol 4-aminobenzoate (ester); 2-(isobutylamino)ethanol p-aminobenzoate (ester);* 2-(isobutylamino)ethyl *p*-aminobenzoate. $C_{13}H_{20}N_2O_2$; mol wt 236.30. C 66.07%, H 8.53%, N 11.86%, O 13.54%. Prepn of the hydrochloride: Goldberg, U.S. pat. 2,139,818 (1938 to Novocol Chem.); J. Büchi *et al., Arzneimittel-Forsch.* **14**, 161 (1964); **16**, 1657 (1966).

Formate, $C_{13}H_{20}N_2O_2.CH_2O_2$, mp 136-139°. Freely sol in water and alcohol. Slightly sol in chloroform, ether; very slightly sol in benzene. pH (1% aq soln): about 6.1.
Hydrochloride, $C_{13}H_{21}ClN_2O_2$, *Ibylcaine, Monocaine.* mp 192-196°. Sol in water, slightly sol in alcohol, chloroform, benzene. Practically insol in ether. pH (1% aq soln): about 4.7.
meta-Isomer hydrochloride, $C_{13}H_{21}ClN_2O_2$, *metabutethamine hydrochloride.* Bitter crystals, mp 181-184°. Sol in water. pH of 1:50 aq soln about 6.2. Slightly sol in alcohol, acetone, chloroform.
THERAP CAT: Anesthetic (local).
THERAP CAT (VET): Local anesthetic for nerve block.

1518. Buthalital Sodium. *Dihydro-5-(2-methylpropyl)-5-(2-propenyl)-2-thioxo-4,6(1H,5H)-pyrimidinedione mono-* *sodium salt; 5-allyl-5-isobutyl-2-thiobarbituric acid sodium salt;* sodium 5-allyl-5-isobutyl-2-thiobarbiturate; 5-allyl-5-(2-methylpropyl)-2-thiobarbituric acid sodium salt; thialbutone sodium; Baytinal; Buthalitone Sodium; Thialisobumalnatrium; Transithal; Ulbreval. $C_{11}H_{15}N_2NaO_2S$; mol wt 262.31. C 50.37%, H 5.76%, N 10.68%, O 12.20%, S 12.22%. Preparation: Miller *et al., J. Am. Chem. Soc.* **58**, 1090 (1936). Pharmacology: Schildt, Schildt, *Acta Pharmacol. Toxicol.* **19**, 377 (1962).

Pale yellow, hygroscopic powder. (The free acid, crystals, mp 147°.) Freely sol in water, yielding alkaline solns. Partly sol in ethanol. Insol in ether, benzene.
Note: This is a controlled substance (depressant) listed in the U.S. Code of Federal Regulations, Title 21 Part 1308.13 (1987).
THERAP CAT: Anesthetic (intravenous).

1519. Buthiazide. *6-Chloro-3,4-dihydro-3-(2-methylpropyl)-2H-1,2,4-benzothiadiazine-7-sulfonamide 1,1-dioxide; 6-chloro-3,4-dihydro-3-isobutyl-7-sulfamoyl-1,2,4-benzothiadiazine 1,1-dioxide; thiabutazide; butizide; isobutylhydrochlorothiazide;* Su 6187; S 3500; Eunephran; Saltucin. $C_{11}H_{16}ClN_3O_4S_2$; mol wt 353.86. C 37.34%, H 4.56%, Cl 10.02%, N 11.87%, O 18.09%, S 18.12%. Prepd from 5-chloro-2,4-disulfamoylaniline and isovaleraldehyde: Werner *et al., J. Am. Chem. Soc.* **82**, 1161 (1960); **Brit.** pats. 861,367; 885,078 (both 1961 to Ciba); Topliss *et al., J. Org. Chem.* **26**, 3842 (1961).

Crystals, mp 241-245° (Werner *et al., loc. cit.;* **Brit.** pats. *loc. cit.*), from methanol + chloroform, mp 228° (Topliss *et al., loc. cit.*).
An ingredient of *Modenol.*
THERAP CAT: Diuretic, antihypertensive.

1520. Buthiobate. *3-Pyridinylcarbonimidodithioic acid butyl [4-(1,1-dimethylethyl)phenyl]methyl ester;* butyl 4-*tert*-butylbenzyl *N*-(3-pyridyl)dithiocarbonimidate; S-1358; Denmert. $C_{21}H_{28}N_2S_2$; mol wt 372.58. C 67.70%, H 7.57%, N 7.52%, S 17.21%. Prepn: S. Tanaka *et al.,* **Ger.** pat. 2,119,174 corresp to U.S. pat. 3,832,351 (1971, 1974 to Sumitomo). Mode of action: T. Kato *et al., Agr. Biol. Chem.* **38**, 2377 (1974); **39**, 169 (1975); T. Kato, Y. Kawase, *ibid.* **40**, 2379 (1976). Metabolism: H. Ohkawa *et al., ibid.* **39**, 1605 (1975). Degradation: *eidem, ibid.* **40**, 943 (1976).

Yellowish, oily liquid, mp 31-33°. d$_{25}^{25}$ 1.0865; n$_D^{26.5}$ 1.596. Vapor press at 20°: 4.52 × 10^{-7} mm Hg. Practically insol in water; sol in most organic solvents. LD$_{50}$ orally in male, female rats: 4.9, 2.7 g/kg, H. Ohkawa *et al., loc. cit.* (1975).
USE: Fungicide.

1521. Butibufen. *α-Ethyl-4-(2-methylpropyl)benzeneacetic acid; 2-(4-isobutylphenyl)butyric acid;* Butilopan; Mijal. $C_{14}H_{20}O_2$; mol wt 220.31. C 76.32%, H 9.15%, O 14.52%. Prepn: L. Aparicio *et al.,* **Ger.** pat. 2,505,813 (1976 to Juste), *C.A.* **85**, 159695 (1976); J. M. Carretero *et al., Eur. J. Med. Chem.* **13**, 77 (1978). Pharmacology: L. Aparicio,

Arch. Int. Pharmacodyn. Ther. **227**, 130 (1977). Pharmacokinetics: R. Revilla De Granda *et al.*, *An. R. Acad. Farm.* **43**, 419 (1977), *C.A.* **88**, 83349 (1978).

Solid, mp 51-53°. LD_{50} orally in mice: 810 mg/kg (Carretero).

THERAP CAT: Anti-inflammatory.

1522. Butidrine Hydrochloride. *5,6,7,8-Tetrahydro-α-[[(1-methylpropyl)amino]methyl]-2-naphthalenemethanol hydrochloride; α-[(sec-butylamino)methyl]-5,6,7,8-tetrahydro-2-naphthalenemethanol hydrochloride;* butydrine hydrochloride; Betabloc; Recetan. $C_{16}H_{26}ClNO$; mol wt 283.86. C 67.70%, H 9.23%, Cl 12.49%, N 4.94%, O 5.64%. β-Adrenergic blocker. Prepn: Ferrari *et al., Boll. Chim. Farm.* **103**, 32 (1964), *C.A.* **61**, 5580d (1964); *Fr. pat.* **1,390,056** (1965 to Holding Ceresia), *C.A.* **62**, 16162d (1965). Pharmacology: R. Ferrini, *Arzneimittel-Forsch.* **18**, 48 (1968). Brief review of butidrine and other β-blockers: A. M. Karow *et al., Prog. Drug Res.* **15**, 103-122 (1971).

Crystals, mp 129-130°. LD_{50} in mice (mg/kg): 20.2 i.v.; 235 orally (Ferrini).

THERAP CAT: Antiarrhythmic.

1523. Butirosin. Ambutyrosin. $C_{21}H_{41}N_5O_{12}$; mol wt 555.60. C 45.40%, H 7.44%, N 12.60%, O 34.56%. Aminoglycosidic antibiotic complex obtained from fermentation filtrates of mucoid strains of *Bacillus circulans* (NRRL B-3312 and B-3313). Consists of two components, butirosin A (80-85%) and butirosin B (15-20%), isomers which differ only in the configuration at one carbon atom in the pentose moiety. Prepn: Woo *et al., Ger. pat.* **1,914,527** corresp to *U.S. pat.* **3,541,078** (1969, 1970 to Parke, Davis). Isoln and characterization: Dion *et al., Antimicrob. Ag. Chemother.* **2**, 84 (1972). Structures: Woo *et al., Tetrahedron Letters* **1971**, 2617, 2621, 2625. Activity: Howells *et al., Antimicrob. Ag. Chemother.* **2**, 79 (1972); Heifetz *et al., ibid.* 89. Synthesis of butirosin B: Ikeda *et al., J. Antibiot.* **25**, 741 (1972); Akita *et al., ibid.* **26**, 365 (1973). Proposed biosynthetic pathway to the butirosins: K. Takeda *et al., ibid.* **31**, 250 (1978).

Butirosin sulfate dihydrate, $C_{21}H_{41}N_5O_{12} \cdot 2H_2SO_4 \cdot 2H_2O$. No sharp mp, dec ~225°. $[\alpha]_D^{25}$ +29° (c = 2 in water). pKa' (water): 5.5, 7.2, 8.5, 9.4. Very sol in water; moderately sol in methanol; slightly sol in ethanol. LD_{50} i.v. in mice: 450-500 mg/kg, Howells *et al., loc. cit.*

Butirosin A. *(S)-O-2,6-Diamino-2,6-dideoxy-α-D-glucopyranosyl-(1→4)-O-[β-D-xylofuranosyl-(1→5)]-N'-(4-amino-2-hydroxy-1-oxobutyl)-2-deoxystreptamine.* R = OH, R' = H. White, amorphous solid, melts with dec over wide range, beginning at about 149°. $[\alpha]_D^{25}$ +26° (c = 1.46 in water). pKa' (water): 5.6, 7.3, 8.7, 9.8.

Butirosin B. *1-N-[(S)-4-Amino-2-(hydroxybutyryl)]ribostamycin.* R = H, R' = OH. Dihydrate, melts over wide range beginning at 146°. $[\alpha]_D^{25}$ + 33° (c = 1.5 in water). pKa' (water): 5.3, 7.1, 8.6, 9.8.

THERAP CAT: Antibacterial.

1524. Butobendine. *[S-(R*,R*)]-3,4,5-Trimethoxybenzoic acid 1,2-ethanediylbis[(methylimino)(2-ethyl-2,1-ethanediyl)] ester;* (2S,2'S)-N,N'-dimethyl-N,N'-bis-1-(3',4',5'-trimethoxybenzoyloxy)butyl-2-ethylenediamine; (+)-(S,S)-ethylenebis[(methylimino)(2-ethylethylene)]bis-(3,4,5-trimethoxybenzoate). $C_{32}H_{48}N_2O_{10}$; mol wt 620.74. C 61.92%, H 7.79%, N 4.51%, O 25.78%. Antiarrhythmic agent which increases cardiac blood flow. Prepn: M. Eckstein *et al., Ger. pat.* **2,435,380**; *eidem, U.S. pat.* **4,021,473** (1975, 1977 both to Polfa). Series of articles on pharmacology, pharmacokinetics, toxicology: *Pol. J. Pharmacol. Pharm.* **32**, 817-953 (1980). Physical properties: L. Krowczynski *et al., ibid.* 909. Toxicity data: J. Maj *et al., ibid.* 823.

Crystals, mp 60-62°. Also reported as mp 64-65° (Krowczynski). $[\alpha]_D^{20}$ +2.4° (c = 5 in ethanol). Partition coefficient (*n*-octanol/water) at 20°: 31.49.

Dihydrochloride, $C_{32}H_{50}Cl_2N_2O_{10}$, *M-71, Craviten.* White crystalline powder. Forms mono-, di- and trihydrates, melting range 83-113°. Also reported as crystals from isopropanol, mp 81-83° (Krowczynski). $[\alpha]_D^{20}$ −6.4° (c = 2.5 in ethanol). $[\alpha]_D^{20}$ −7.5° (c = 5 in H$_2$O). $[\alpha]_D^{20}$ −5.5° (c = 5 in pyridine). pH (8% aq soln): 2.20. Partition coefficient (*n*-octanol/water) at 20°: 4.49. Sol in water, chloroform, ethanol. Slightly sol in methanol. Practically insol in benzene, CCl$_4$, ether. LD_{50} in rats (mg/kg): 142.0 i.p., 15.8 i.v.; in mice (mg/kg): 550.0 i.p.; in rabbits (mg/kg): 5.1 i.v. (Maj).

THERAP CAT: Antiarrhythmic.

1525. Butoconazole. (±)-1-[4-(4-Chlorophenyl)-2-[(2,6-dichlorophenyl)thio]butyl]-1H-imidazole. $C_{19}H_{17}Cl_3$-N_2S; mol wt 411.78. C 55.42%, H 4.16%, Cl 25.83%, N 6.80%, S 7.79%. Imidazole derivative with antifungal properties. Prepn: K. A. M. Walker, *U.S. pat.* **4,078,071** (1978 to Syntex). Prepn, toxicity, activity vs *Candida albicans* in mice: K. A. M. Walker *et al., J. Med. Chem.* **21**, 840 (1978). *In vitro* comparison with other antifungal agents: F. C. Odds *et al., J. Antimicrob. Chemother.* **14**, 105 (1984). Clinical trials in treatment of vulvovaginal candidiasis: W. Droegemueller *et al., Obstet. Gynecol. (N.Y.)* **64**, 530 (1984); J. B. Jacobson *et al., Acta Obstet. Gynecol. Scand.* **64**, 241 (1985). Comparison with miconazole, *q.v.*: C. S. Bradbeer *et al., Genitourin. Med.* **61**, 270 (1985).

Crystals from cyclohexane, mp 68-70.5°.

Nitrate, $C_{19}H_{18}Cl_3N_3O_3S$, *RS-35887, Exelgyn, Femstat, Gynomyk.* Colorless blades from acetone/ethyl acetate, mp 162-163°. LD_{50} in mice, male, female rats (mg/kg): > 3200, > 3200, 1720 orally; > 1600, 940, 940 i.p. (Walker).

THERAP CAT: Topical antifungal.

1526. Butoctamide. *N-(2-Ethylhexyl)-3-hydroxybutanamide; N-(2-ethylhexyl)-3-hydroxybutyramide;* hexobutyramide. $C_{12}H_{25}NO_2$; mol wt 215.34. C 66.93%, H 11.70%, N 6.50%, O 14.86%. Prepn: A. Sakuma *et al.*, **Fr. pat. 8393**; *eidem*, **U.S. pat. 3,639,457**; *eidem*, **Japan. Kokai 79 24822**, *C.A.* **91**, 34947f (1979) (1971, 1972, 1979 all to Lion Denti-frice). Toxicology study: M. Kato *et al.*, *Oyo Yakuri* **17**, 287 (1979), *C.A.* **91**, 68665e (1979). Acute toxicity: R. Goshima *et al.*, *Toho Igakka Zasshi* **17**, 579 (1970), *C.A.* **74**, 138765n (1971). Clinical effect on rapid eye movement (REM) sleep: Y. Hayashi *et al.*, *Psychopharmacology* **77**, 367 (1982); in Down's syndrome: J. C. Grubar *et al.*, *ibid.* **90**, 119 (1986).

$$CH_3CHCH_2CONHCH_2CH(CH_2)_3CH_3$$
$$|\qquad\qquad\qquad |$$
$$OH\qquad\qquad\quad C_2H_5$$

Colorless viscous oil, $bp_{0.30\ mm}$ 149-150°.

Hydrogen succinate, $C_{16}H_{29}NO_5$, *butanedioic acid mono-[3-[(2-ethylhexyl)amino]-1-methyl-3-oxopropyl] ester; succinic acid monoester with N-(2-ethylhexyl)-3-hydroxybutyramide; N-2-ethylhexyl-β-oxybutyramide semisuccinate; BAHS; M-2H; Listomin S.* Colorless viscous oil; mp 46.5°.

Hydrogen succinate calcium salt, $C_{32}H_{58}CaN_2O_{10}$, white powder, slightly hygroscopic. uv max: 192 ($E_{1cm}^{1\%}$ 290). LD_{50} in mice (mg/kg): 476 i.p.; 2000 orally (Sakuma, 1971).

THERAP CAT: Sedative, hypnotic.

1527. Butofilolol. (±)-*1-[2-[3-[(1,1-Dimethylethyl)-amino]-2-hydroxypropoxy]-5-fluorophenyl]-1-butanone;* (±)-5-fluoro-2-(2-hydroxy-3-t-butylaminopropoxy)buty-rophenone; (±)-2'-[3-(tert-butylamino)-2-hydroxypropoxy]-5'-fluorobutyrophenone; CM 6805.* $C_{17}H_{26}FNO_3$; mol wt 311.41. C 65.57%, H 8.42%, F 6.10%, N 4.50%, O 15.41%. β-Adrenergic blocker. Prepn: H. Demarne le Florence, **Ger. pat. 2,528,147** corresp to **U.S. pat. 4,252,825** (1976, 1981 both to C. M. Industries). Pharmacological study: G. G. Re *et al.*, *Boll. Soc. Ital. Biol. Sper.* **56**, 1264 (1980), *C.A.* **93**, 197668y (1980). GC and HPLC determn in plasma and urine: J. P. Jeanniot *et al.*, *J. Chromatog.* **278**, 301 (1983). Pharmacokinetics in humans: G. Houin *et al.*, *Int. J. Clin. Pharm. Res.* **4**, 175 (1984).

Crystals from diisopropyl ether, mp 88-89°.

Maleate, $C_{21}H_{30}FNO_7$, *Cafide.*

THERAP CAT: Antihypertensive.

1528. Butonate. *Butanoic acid 2,2,2-trichloro-1-(dimethoxyphosphinyl)ethyl ester; butyric acid ester with dimethyl (2,2,2-trichloro-1-hydroxyethyl)phosphonate; O,O-dimethyl 2,2,2-trichloro-1-n-butyryloxyethylphosphonate; O,O-dimethyl-2,2,2-trichloro-1-phosphonoethyl butyrate; T-113; F-139.* $C_8H_{14}Cl_3O_5P$; mol wt 327.55. C 29.34%, H 4.31%, Cl 32.47%, O 24.42%, P 9.46%. Prepd by reacting Dipterex with *n*-butyric anhydride: Arthur, Casida, *J. Agr. Food Chem.* **6**, 360 (1958); from α-chloro-β,β,β-trichloroethyl butyrate and dimethyl phosphite: Casida, Arthur, **U.S. pat. 2,911,435** (1959 to Wisconsin Alumni Res. Found.). Mode of action studies: Knowles, Casida, *J. Agr. Food Chem.* **14**, 566 (1966).

Liquid. $bp_{0.5}$ 129°.

USE: Insecticide. *Caution:* A cholinesterase inhibitor.

THERAP CAT (VET): Anthelmintic for horses.

1529. Butopyronoxyl. *3,4-Dihydro-2,2-dimethyl-4-oxo-2H-pyran-6-carboxylic acid butyl ester;* butyl 3,4-dihydro-

2,2-dimethyl-4-oxo-2H-pyran-6-carboxylate; butyl mesityl oxide oxalate; α,α-dimethyl-α'-carboxydihydro-γ-pyrone butyl ester; Indalone. $C_{12}H_{18}O_4$; mol wt 226.26. C 63.70%, H 8.02%, O 28.28%. Prepd by condensation of mesityl oxide with dibutyl oxalate in the presence of sodium or sodium butoxide: Ford, **U.S. pat. 2,138,540** (1938 to Kilgore Dev.). Exists largely as the dihydropyrone in equilibrium with the open chain enol form: Hall *et al.*, *J. Am. Chem. Soc.* **67**, 1224 (1945).

Yellow to pale reddish-brown liquid. Aromatic odor. d_{25}^{25} 1.052-1.060. bp_{760} 256-270°. n_D^{25} 1.4745-1.4755. Practically insol in water. Misc with alcohol, chloroform, ether, glacial acetic acid. LD_{50} orally in mice, rats: 11.6, 7.4 ml/kg, J. H. Draize *et al.*, *J. Pharmacol. Exp. Ther.* **93**, 26 (1948).

USE: Insect repellent. *Toxicity:* Liver necrosis has been produced in exptl animals, N. I. Sax, *Dangerous Properties of Industrial Materials* (Reinhold, New York, 1968) p 496.

1530. Butorphanol. *17-(Cyclobutylmethyl)morphinan-3,14-diol;* (−)-*N-cyclobutylmethyl-3,14-dihydroxymorphinan; levo-BC 2627.* $C_{21}H_{29}NO_2$; mol wt 327.47. C 77.02%, H 8.93%, N 4.28%, O 9.77%. Mixed opioid agonist-antagonist. Prepn: I. Pachter *et al.*, **Ger. pat. 2,243,961** corresp to **U.S. pat. 3,819,635** (1973, 1974 to Bristol-Myers); I. Monkovic, T. Thomas, **U.S. pat. 3,775,414** (1973 to Bristol-Myers). Total synthesis and pharmacology: I. Monkovic *et al.*, *J. Am. Chem. Soc.* **95**, 7910 (1973); *eidem, Can. J. Chem.* **53**, 3094 (1975). Clinical study: F. Vargas-Arreola *et al.*, *Curr. Ther. Res., Clin. Exp.* **22**, 186 (1977). Antitussive effect: R. L. Cavanaugh *et al.*, *Arch. Int. Pharmacodyn. Ther.* **220**, 258 (1976). *Review:* R. C. Heel *et al.*, *Drugs* **16**, 473-505 (1978); F. S. Caruso *et al.*, in *Pharmacological and Biochemical Properties of Drug Substances* vol. 2, M. E. Goldberg, Ed. (Am. Pharm. Assoc., Washington, DC, 1979) pp 19-57.

Solid, mp 215-217°. $[\alpha]_D$ −70.0° (c = 0.1 in methanol).

Tartrate, $C_{25}H_{35}NO_8$, *Stadol, Torate, Torbugesic, Torbutrol.* Solid. mp 217-219°. $[\alpha]_D^{22}$ −64.0° (c = 0.4 in methanol). LD_{50} in mice, rats (mg/kg): 40-57, 17-20 i.v.; 395-527, 570-756 orally (Heel).

THERAP CAT: Narcotic analgesic.

THERAP CAT (VET): Narcotic analgesic; antitussive.

1531. Butoxycaine. *4-Butoxybenzoic acid 2-(diethylamino)ethyl ester;* 2-diethylaminoethyl *p-butoxybenzoate.* $C_{17}H_{27}NO_3$; mol wt 293.39. C 69.59%, H 9.28%, N 4.77%, O 16.36%. Prepd from p-butoxybenzoyl chloride and β-diethylaminoethanol: Christiansen, Harris, **U.S. pat. 2,412,966** (1946 to Squibb). Alternate prepn and activity: Reynaud *et al.*, *Chim. Ther.* **2**, 25 (1967). *See also* Büchi *et al.*, *Arzneimittel-Forsch.* **18**, 610 (1968).

Hydrochloride, $C_{17}H_{28}ClNO_3$, *Stadacain*, heavy crystals, mp 146°.

THERAP CAT: Local anesthetic.

1532. Butralin. *4-(1,1-Dimethylethyl)-N-(1-methylpro-*

pyl)-2,6-dinitrobenzenamine; N-sec-butyl-4-tert-butyl-2,6-dinitroaniline; dibutalin; Amchem 70-25; A-820; Amex; Tamex. $C_{14}H_{21}N_3O_4$; mol wt 295.34. C 56.93%, H 7.17%, N 14.23%, O 21.67%. Pre-emergence herbicide. Prepn: J. J. Damiano, **Ger.** pat. 2,058,201 corresp to **U.S.** pat. 3,672,866 (1971, 1972 to Amchem). Activity: S. R. Mc Lane *et al., Proc. S. Weed Sci. Soc.* 24, 58 (1971). Soil persistence and metabolism: P. C. Kearney *et al., J. Agr. Food Chem.* 22, 856 (1974). Photochemistry: J. R. Plimmer, U. I. Klinge-biel, *ibid.* 689.

Yellow-orange crystals, mp 60-61°, bp$_{0.5}$ 134-136°. Vapor press at 25°: 1.3×10^{-5} mm Hg. Flash pt, open cup: 97°F (36°C). Soly in water at 25°: 1 mg/l. Soly at 25° (kg/kg): methanol 0.125; acetone 4.48; benzene 2.7; xylene 3.88; butanone 9.55; carbon tetrachloride 1.46. LD$_{50}$ orally in rats: 2500 mg/kg, S. R. McLane *et al., loc. cit.*

USE: Herbicide.

1533. Butriptyline. *10,11-Dihydro-N,N,β-trimethyl-5H-dibenzo[a,d]cycloheptene-5-propanamine;* 5-(3-dimethyl-amino-2-methylpropyl)-10,11-dihydro-5H-dibenzo[a,d]-cycloheptene; 5-(2-methyl-3-dimethylaminopropyl)diben-zo[a,d][1,4]cycloheptadiene; butriptyline. $C_{21}H_{27}N$; mol wt 293.46. C 85.95%, H 9.28%, N 4.77%. Prepn: Winthrop, Davis, **Belg.** pat. 613,750 (1962 to Ayerst, McKenna & Harrison), *C.A.* 57, 15036c (1962); *eidem, J. Org. Chem.* 27, 230 (1962); Villani, **U.S.** pat. 3,409,640 (1968 to Schering). Review of pharmacology and clinical data: *see J. Med. (Basel)* 2, 249-343 (1971). Metabolic studies: Cameron *et al., Arzneimittel-Forsch* 24, 93 (1974). Toxicity: Voith, Herr, *Arch. Int. Pharmacodyn. Ther.* 182, 318 (1969).

Oil, bp$_1$ 180-185°.
Hydrochloride, $C_{21}H_{28}ClN$, *AY-62014, Evadene, Evadyne.* Crystals from isopropyl alcohol-ether, mp 188-190° (dec). uv max (methanol): 273, 270, 266 nm (ε 460, 441, 552). Freely sol in water; moderately sol in aliphatic alc, chloro-form. Insol in ether, paraffinic hydrocarbons. LD$_{50}$ in mice (mg/kg): 120 i.p.; 345 orally (Voith, Herr).

THERAP CAT: Antidepressant.

1534. Butropium Bromide. *[3(S)-endo]-8-[(4-Butoxy-phenyl)methyl]-3-(3-hydroxy-1-oxo-2-phenylpropoxy)-8-methyl-8-azoniabicyclo[3.2.1]octane bromide; 8-(p-butoxy-benzyl)-3α-hydroxy-1αH,5αH-tropanium bromide (−)-tropate; l-[1-(p-n-butoxybenzyl)hyoscyaminium] bromide;* BHB; Coliopan. $C_{28}H_{38}BrNO_4$; mol wt 532.53. C 63.15%, H 7.19%, Br 15.00%, N 2.63%, O 12.02%. Prepn: Tanaka, Hashimoto, **Ger.** pat. 1,950,378 (1970) corresp to **U.S.** pat. 3,696,110 (1972); **Japan.** pat. 22,715('72), *C.A.* 77, 79549g (1972), (all to Eisai); Tanaka *et al., J. Pharm. Soc. Japan* 92, 510 (1972). Prepn of the labelled compound: Fujita *et al., J. Label. Compounds* 9, 149, 555 (1972). Activity: Akutsu, Ichikawa, *Showa Igakkai Zasshi* 32, 494 (1972), *C.A.* 78, 119243g (1973). *Review: Japan. Med. Gaz.* 11(9), 10 (1974).

Crystals from ethanol-acetone, mp 166-168°; also report-ed as white needles from isopropanol, mp 158-160° (**U.S.** pat. 3,696,110). $[\alpha]_D^{20}$ −21.7° (c = 0.5 in water). Freely sol in glacial acetic acid; sol in chloroform, DMF. Sparingly sol in ethanol; slightly sol in water, 0.1N HCl, 0.1N NaOH. Practically insol in acetone, ether, benzene. LD$_{50}$ in male mice (mg/kg): 1500 orally; 660 s.c.; 12.0 i.v. (**U.S.** pat. 3,696,110).

THERAP CAT: Antispasmodic; anticholinergic.

1535. n-Butyl Acetate. *Acetic acid butyl ester.* $C_6H_{12}O_2$; mol wt 116.16. C 62.04%, H 10.41%, O 27.55%. CH$_3$COO-(CH$_2$)$_3$CH$_3$. Prepd from acetic acid and butyl alcohol: Leyes, Othmer, *Ind. Eng. Chem.* 37, 968 (1945); Vogel, *J. Chem. Soc.* 1948, 624; Zettlemoyer *et al.,* **U.S.** pat. 2,644,839 (1953 to FMC); Faith, Keyes & Clark's *Industrial Chemi-cals,* F. A. Lowenheim, M. K. Moran, Eds. (Wiley-Intersci-ence, New York, 4th ed., 1975) pp 171-177. Toxicity data: H. F. Smyth *et al., Arch. Ind. Hyg. Occup. Med.* 10, 61 (1954).

Liquid. d_{20}^{20} 0.8826. bp 125-126°. mp −77°. n_D^{20} 1.3951. Flash pt, closed cup: 72°F (22° C). Sol in about 120 parts water at 25°; misc with alcohol, ether; sol in most hydrocar-bons. LD$_{50}$ orally in rats: 14.13 g/kg (Smyth).

USE: Manuf lacquer, artificial leather, photographic films, plastics, safety glass. *Caution:* Irritating; may cause con-junctivitis. Narcotic in high concns: E. Browning, *Toxicity and Metabolism of Industrial Solvents* (Elsevier, New York, 1965) pp 529-532, 591-593.

1536. sec-Butyl Acetate. *Acetic acid 1-methylpropyl es-ter; acetic acid sec-butyl ester.* CH$_3$COOCH(CH$_3$)CH$_2$CH$_3$. Empirical formula, mol wt, etc.: *see n-butyl acetate above.* Prepd from *sec*-butanol and acetic anhydride: Altschul, *J. Am. Chem. Soc.* 68, 2605 (1946). Prepn of *d*-form: Kenyon *et al., J. Chem. Soc.* 1935, 1072; Bird, *Tetrahedron* 18, 1 (1962). Prepn of *l*- form: Kenyon *et al., loc. cit.* Manuf: Faith, Keyes & Clark's *Industrial Chemicals,* F. A. Lowen-heim, M. K. Moran, Eds. (Wiley-Interscience, New York, 4th ed., 1975) pp 171-177.

dl-Form, liquid. d$_4^{25}$ 0.865. bp 112-113°. n_D^{25} 1.3866. Flash pt, open cup: 88° F (31° C). Slightly sol in water; sol in alcohol, ether.

d-Form, liquid, bp 116-117°. d$_4^{19}$ 0.873. n_D^{18} 1.3899.

1537. tert-Butyl Acetate. *Acetic acid 1,1-dimethylethyl ester; acetic acid tert-butyl ester.* CH$_3$COOC(CH$_3$)$_3$. Empiri-cal formula, mol wt, etc.: *see n-butyl acetate above.* Prepn: Baker, Bordwell; Hauser *et al., Org. Syn.* **coll. vol. III,** 141, 142 (1955). Manuf from acetic acid and isobutylene: Young, Pare, **U.S.** pat. 3,031,495 (1962 to Sinclair); Wheeler *et al.,* **U.S.** pat. 3,102,905 (1963 to Celanese); Heisler *et al.,* **U.S.** pat. 3,096,365 (1963 to Texaco); Faith, Keyes & Clark's *Industrial Chemicals,* F. A. Lowenheim, M. K. Moran, Eds. (Wiley-Interscience, New York, 4th ed., 1975) pp 171-177.

Liquid, bp 97.8°. d$_4^{20}$ 0.8665, d$_4^{25}$ 0.8593. n_D^{20} 1.3870. Practically insol in water; miscible with alcohol, ether.

USE: As gasoline additive (Wheeler *et al., loc. cit.*).

1538. tert-Butylacetic Acid. *3,3-Dimethylbutanoic acid; 3,3-dimethylbutyric acid.* $C_6H_{12}O_2$; mol wt 116.16. C 62.04%, H 10.41%, O 27.55%. (CH$_3$)$_3$CCH$_2$COOH. Prepn: Homeyer *et al., J. Am. Chem. Soc.* 55, 4209 (1933); Botter-son, Shulman, *J. Org. Chem.* 27, 1059 (1962); A. Nilsson, R. Carlson, *Acta Chem. Scand.* B34, 621 (1980).

Liquid, bp$_{26}$ 96°, bp$_{739}$ 183.0-183.3°, bp 190°. mp 6-7°. n_D^{20}

1.4115 (Botterson, Shulman), also reported as 1.4096 (Homeyer *et al.*). d_4^{20} 0.9124.

1539. n-Butyl Acrylate. *2-Propenoic acid butyl ester;* acrylic acid *n*-butyl ester. $C_7H_{12}O_2$; mol wt 128.17. C 65.59%, H 9.44%, O 24.97%. CH_2=CHCOOC$_4$H$_9$. Prepn from *n*-butanol and methyl acrylate: Rehberg, *Org. Syn.* **coll. vol. III**, 146 (1955).

Liquid. d_4^{20} 0.8986; d_4^{15} 0.9110; d_4^{12} 0.9117; d_4^{0} 0.9202. bp$_{760}$ 145° (also reported 138°); bp$_{101}$ 84-86°; bp$_{25}$ 59°; bp$_{10}$ 39°; bp$_{8}$ 35°. n_D^{20} 1.4190; n_D^{11} 1.4254. Sp heat ($-60°$): 0.467 cal/g/°C; heat of vaporization 8.11 kcal/mol; heat of combustion 974.46 kcal/mol. Soly in water at 20°: 0.14 g/100 ml; at 40°: 0.12 g/100 ml. Soly of water in *n*-butyl acrylate at 20°: 0.8 ml/100 g. LD$_{50}$ orally in rats: 3.73 g/kg, H. F. Smyth *et al.*, *Arch. Ind. Hyg. Occup. Med.* **4**, 119 (1951).

Polymer, elastic, tacky substance. Brittle temp $-45°$.

USE: The monomer in the manuf of polymers and resins for textile and leather finishes, paint formulations, etc.

1540. n-Butyl Alcohol. *1-Butanol; butyl alcohol;* propyl carbinol. $C_4H_{10}O$; mol wt 74.12. C 64.81%, H 13.60%, O 21.59%. $CH_3CH_2CH_2CH_2OH$. Prepn by reduction of butyraldehyde with sodium borohydride: Chaikin, Brown, *J. Am. Chem. Soc.* **71**, 122 (1949). Manuf from ethylene oxide and triethylaluminum: Rudner, U.S. pat. **3,091,627** (1963 to Koppers); by oxidation of tributylborane: Mirviss, U.S. pat. **3,067,235** (1962 to Esso). Manuf by carbohydrate fermentation, by hydrogenation of butyraldehyde, from crotonaldehyde: Faith, Keyes & Clark's *Industrial Chemicals*, F. A. Lowenheim, M. K. Moran, Eds. (Wiley-Interscience, New York, 4th ed., 1975) pp 178-185. Purification and vapor pressure: Biddiscombe *et al.*, *J. Chem. Soc.* **1963**, 1954.

Highly refractive liquid; burns with a strongly luminous flame; leaves a transitory greasy spot on paper. Odor similar to that of fusel oil, but weaker. Its vapors irritate and cause cough. d_4^{20} 0.810. bp 117-118°. mp $-90°$. Flash pt 36-38°. n_D^{20} 1.3993. A mixture of 63% of the alcohol and 37% water forms a constant boiling mixture, boiling at 92°. Soly at 25°, 9.1 ml/100 ml H$_2$O: Booth, Everson, *Ind. Eng. Chem.* **40**, 1491 (1948). Miscible with alc, ether and many other organic solvents. LD$_{50}$ orally in rats: 4.36 g/kg, Smyth *et al.*, *Arch. Ind. Hyg. Occup. Med.* **4**, 119 (1951).

USE: As solvent for fats, waxes, resins, shellac, varnish, gums etc.; manuf lacquers, rayon, detergents, other butyl compds; in microscopy for preparing paraffin imbedding materials. *Caution:* May cause irritation of mucous membranes, contact dermatitis, headache, dizziness, drowsiness.

1541. sec-Butyl Alcohol. *2-Butanol;* butylene hydrate; 2-hydroxybutane; methyl ethyl carbinol. $C_4H_{10}O$; mol wt 74.12. C 64.81%, H 13.60%, O 21.59%. $CH_3CH_2CH(OH)$-CH_3. Prepn by reduction of 2-butanone: Chaikin, Brown, *J. Am. Chem. Soc.* **71**, 122 (1949); Nystrom *et al.*, *ibid.* 3245. Manuf by hydration of 2-butene or hydrocarbons contg butene: Dale *et al.*, *Ind. Engl. Chem.* **48**, 913 (1956); Archibald, Mottern, U.S. pat. **2,543,820** (1951 to Standard Oil); Limerick, Wylie, U.S. pat. **2,776,324** (1957 to Shell). Purification and vapor pressure: Biddiscombe *et al.*, *J. Chem. Soc.* **1963**, 1954. Prepn of *l*-form or *d*-form by hydroboration of *cis*-2-butene: Brown, U.S. pat. **3,078,313** (1963); Brown *et al.*, *J. Am. Chem. Soc.* **86**, 397 (1964). Resolution of *dl*-form: Kantor, Hauser, *ibid.* **75**, 1744 (1953). Absolute configuration of *l*-form: Brown *et al.*, *ibid.* **86**, 1071 (1964). Toxicity: H. F. Smyth *et al.*, *Arch. Ind. Hyg. Occup. Med.* **10**, 61 (1954).

dl-Form, liquid, bp 99.5°. mp $-114.7°$. d_4^{20} 0.808. n_D^{25} 1.3949. Flash pt, open cup: 88°F (31°C). Sol in 12 parts water; misc with alcohol, ether. LD$_{50}$ orally in rats: 6.48 g/kg (Smyth).

d-Form, liquid, d_4^{27} 0.8025. n_D^{20} 1.3954. $[\alpha]_D^{27}$ +13.52°. *Ref:* Leroux, Lucas, *J. Am. Chem. Soc.* **73**, 41 (1951).

l-Form, liquid, bp$_{744}$ 98°. d_4^{25} 0.8042. n_D^{20} 1.3970, n_D^{25} 1.3949. $[\alpha]_D^{25}$ $-13.51°$. *Ref:* Leroux, Lucas, *loc. cit.*

USE: In the synthesis of flotation agents, flavors, perfumes, dyestuffs, wetting agents. In industrial cleaners, paint removers. Solvent for many natural resins, linseed and castor oils.

1542. tert-Butyl Alcohol. *2-Methyl-2-propanol;* trimethyl carbinol. $C_4H_{10}O$; mol wt 74.12. C 64.81%, H

13.60%, O 21.59%. $(CH_3)_3COH$. Prepd from acetyl chloride and dimethylzinc: Butlerow, *Ann.* **144**, 1 (1867). Manuf by catalytic hydration of isobutylene: Kreps, Nachod, U.S. pat. **2,477,380** (1949 to Atlantic Refining); Serniuk, Vanderbilt, U.S. pat. **2,534,304** (1950 to Standard Oil); by reduction of *tert*-butyl hydroperoxide: Lorand, U.S. pat. **2,484,-841** (1949 to Hercules Powder); De Jong, U.S. pat. **2,853,-532** (1958 to Shell). Purification: Biddiscombe *et al.*, *J. Chem. Soc.* **1963**, 1954. Physical properties: *eidem, ibid.;* Dreisbach, Martin, *Ind. Eng. Chem.* **41**, 2875 (1949). Toxicity: Schaffarzick, Brown, *Science* **116**, 663 (1952).

Crystals, camphor-like odor, mp 25.6°. At 99.69 mol-% purity, bp 82.41°. mp 25.7°. d_4^{20} 0.78581, d_4^{25} 0.78086, d_4^{25} (calcd) 0.78080; n_D^{20} 1.38468, n_D^{25} 1.38231. Flash pt, closed cup: 52° F (11.1° C). Sol in water. Miscible with alcohol, ether. LD$_{50}$ orally in rats: 3.5 g/kg (Schaffarzick, Brown).

USE: Denaturant for ethanol, mfg flotation agents, flavors, perfumes; as solvent; in paint removers. Octane booster in gasoline.

1543. n-Butylamine. *1-Butanamine;* 1-aminobutane. $C_4H_{11}N$; mol wt 73.14. C 65.69%, H 15.16%, N 19.15%. $CH_3CH_2CH_2CH_2NH_2$. Prepn by reduction of butyraldoxime: Lycan *et al.*, *Org. Syn.* **coll. vol. II**, 319 (1943). Usually manuf by catalytic alkylation of ammonia with butyl alcohol: Davies *et al.*, U.S. pat. **2,609,394** (1952 to I.C.I.); Hindley, Fisher, U.S. pat. **2,782,237** (1957 to British Celanese); Lemon, Myerly, U.S. pat. **3,022,349** (1962 to Union Carbide); Shirley, Speranza, U.S. pat. **3,128,311** (1964 to Jefferson Chem.). Manuf from butyraldehyde and ammonia in the presence of Raney nickel: Brimer *et al.*, U.S. pat. **2,518,659** (1950 to Eastman Kodak).

Liquid, ammoniacal odor, bp 78°. mp $-50°$. d_4^{25} 0.7327. n_D^{20} 1.4010. Flash pt, open cup: 30°F ($-1°$C). Misc with water, alcohol, ether. LD$_{50}$ orally in rats: 500 mg/kg, C. H. Hine *et al.*, *Arch. Environ. Health* **1**, 343 (1960).

USE: Intermediate for pharmaceuticals, dyestuffs, rubber chemicals, emulsifying agents, insecticides, synthetic tanning agents. *Caution:* Potent skin, eye, mucous membrane irritant. Direct skin contact causes severe primary irritation and blistering.

1544. sec-Butylamine. *2-Butanamine;* 2-aminobutane; Frucote; Deccotane; Tutane. $CH_3CH_2CH(NH_2)CH_3$. Empirical formula, mol wt etc.: *see n*-butylamine. Prepn by reduction of ethyl methyl ketoxime: Lycan *et al.*, *Org. Syn.* **coll. vol. II**, 319 (1943). Manuf: Taylor *et al.*, U.S. pat. **2,636,902** (1953 to I.C.I.); Thurston, U.S. pat. **2,689,868** (1954 to Am. Cyanamid). Resolution of *dl*-form: Leithe, *Ber.* **63**, 800 (1930); Bruck *et al.*, *J. Chem. Soc.* **1956**, 921. Toxicity: Goldenthal, *Toxicol. Appl. Pharmacol.* **18**, 204 (1971).

dl-Form, liquid, bp 63°. mp $-104°$. d_4^{20} 0.724. n_D^{20} 1.394. Miscible with water, alc. LD$_{50}$ orally in rats: 380 mg/kg (Goldenthal).

d(S)-Form, liquid, bp 63°. d_4^{15} 0.7308. n_D^{15} 1.3963. α_D^{15} +7.80° (neat).

l-Form, liquid, bp 63°. d_4^{19} 0.728. $[\alpha]_D^{19}$ $-7.64°$.

USE: Fungistat. *Caution:* Irritating to skin and mucous membranes.

1545. tert-Butylamine. *2-Methyl-2-propanamine;* 2-aminoisobutane; 2-amino-2-methylpropane. $(CH_3)_3CNH_2$. Empirical formula, mol wt, etc. see *n*-butylamine. Prepn: Campbell *et al.*, *Org. Syn.* **coll. vol. III**, 148 (1955). Manuf: Gresham *et al.*, U.S. pat. **2,501,509** (1950 to du Pont); Albert, Kibler, U.S. pat. **2,773,097** (1956 to Firestone Tire & Rubber).

Liquid, bp 44-46°. mp $-72.65°$. d_4^{20} 0.6951, d_4^{25} 0.6867. n_D^{20} 1.37. Miscible with alcohol.

Hydrochloride, mp 310°. Sol in cold methanol, in boiling isopropyl alcohol.

1546. Butylate. *Bis(2-methylpropyl)carbamothioic acid S-ethyl ester; diisobutylthiocarbamic acid S-ethyl ester;* S-ethyl *N,N*-diisobutylthiocarbamate; butilate; diisocarb; R-1910; Sutan. $C_{11}H_{23}NOS$; mol wt 217.37. C 60.78%, H 10.66%, N 6.45%, O 7.36%, S 14.75%. Selective pre-planting herbicide. Prepn: H. Tilles, *J. Am. Chem. Soc.* **81**, 714 (1959); H. Tilles, J. Antognini, U.S. pat. **2,913,327** (1959 to

Stauffer). Metabolism: J. P. Hubbell, J. E. Casida, *J. Agr. Food Chem.* **25**, 404 (1977).

Clear liquid, bp_{21} 138°. n_D^{30} 1.4701; d 0.9417. Vapor press at 25°: 1.3×10^{-3} mm Hg. Soly in water at 25°: 45 mg/l. LD_{50} orally in rats: 4000 mg/kg, *RTECS* Vol. I, R. J. Lewis, Sr., R. L. Tatken, Eds. (1979) p 365.

USE: Herbicide.

1547. Butylated Hydroxyanisole. BHA; Antrancine 12; Embanox; Nipantiox 1-F; Sustane 1-F; Tenox BHA. $C_{11}H_{16}O_2$; mol wt 180.24. C 73.30%, H 8.95%, O 17.75%. A mixture of *2-tert-butyl-4-methoxyphenol* (also called *3-tert-butyl-4-hydroxyanisole*) and *3-tert-butyl-4-methoxyphenol* (also called *2-tert-butyl-4-hydroxyanisole*). Prepn from *p*-methoxyphenol and isobutene: Rosenwald, U.S. pats. **2,459,540** and **2,470,902** (1949 to Universal Oil Products).

Waxy solid, mp 48-55°. bp_{733} 264-270°. Insoluble in water. Soluble in petr ether (Skellysolve H), in 50% alcohol (or higher), in propylene glycol; alcohols. Sol in fats and oils. Exhibits antioxidant properties and synergism with acids, BHT, propyl gallate, hydroquinone, methionine, lecithin, thiodipropionic acid, etc. LD_{50} orally in mice, rats: 2000, 2200 mg/kg, A. J. Lehman *et al., Adv. Food Res.* **3**, 197 (1951).

USE: Antioxidant, esp in foods. The American Meat Institute Foundation has proposed an antioxidant mixture known as *AMIF-72* which contains 20% of butylated hydroxyanisole, 6% of propyl gallate, and 4% of citric acid in propylene glycol.

1548. Butylated Hydroxytoluene. *2,6-Bis(1,1-dimethylethyl)-4-methylphenol; 2,6-di-tert-butyl-p-cresol;* 2,6-di-*tert*-butyl-4-methylphenol; BHT; Antrancine 8; Tenox BHT; Ionol CP; Sustane; Dalpac; Impruvol; Vianol. $C_{15}H_{24}O$; mol wt 220.34. C 81.76%, H 10.98%, O 7.26%. Prepared from *p*-cresol and isobutylene: Stillson, U.S. pat. **2,428,745** (1947 to Gulf); McConnell, Davis, U.S. pat. **3,082,258** (1963 to Eastman Kodak). Inactivator of lipid-containing mammalian and bacterial viruses: Snipes *et al., Science* **188**, 64 (1975).

Crystals, mp 70°. d_4^{20} 1.048. bp 265°. Flash pt (open cup): 260°F (127°C). Insol in water. Freely sol in toluene, sol in methanol, ethanol, isopropanol, methyl ethyl ketone, acetone, Cellosolve, petr ether, benzene, most other hydrocarbon solvents. Soly in liquid petrolatum (white oil): 0.5% w/w. More sol in food oils and fats than butylated hydroxyanisole. Good soly in linseed oil. LD_{50} orally in mice: 1040 mg/kg, *J. Am. Pharm. Assoc.* **38**, 366 (1949).

USE: Antioxidant for food, animal feed, petrol products, synthetic rubbers, plastics, animal and vegetable oils, soaps. Antiskinning agent in paints and inks.

1549. *n*-Butylbenzene. 1-Phenylbutane. $C_{10}H_{14}$; mol wt 134.21. C 89.49%, H 10.51%. $C_6H_5(CH_2)_3CH_3$. Prepn: Radziszewski, *Ber.* **9**, 261 (1876); Balbiano, *Ber.* **10**, 296 (1877); Read, Foster, *J. Am. Chem. Soc.* **48**, 1606 (1926). Liquid. mp −88.5°. d_4^{20} 0.8604. bp_{760} 183.1°; bp_{400} 159.2°; bp_{200} 136.9°; bp_{100} 116.2°; bp_{60} 102.6°; bp_{40} 92.4°; bp_{20} 76.3°; bp_{10} 62.0°; bp_5 48.8°; $bp_{1.0}$ 22.7°. n_D^{20} 1.49040. Flash pt, open cup: 160° F (71° C). Insol in water; miscible with alcohol, ether, benzene.

1550. *sec*-Butylbenzene. *(1-Methylpropyl)benzene;* 2-phenylbutane. $C_{10}H_{14}$; mol wt 134.21. C 89.49%, H 10.51%. $C_6H_5CH(CH_3)CH_2CH_3$. Prepd from benzene and *n*-butyl chloride in presence of $AlCl_3$: Schramm, *Monatsh.* **9**, 621 (1888); by the action of sodium on γ-chloro-*sec*-butylbenzene: Braun *et al., Ber.* **46**, 1277 (1913); with other products by heating *n*- or *sec*-butyl alcohol with 80% H_2SO_4: Meyer, Bernhauer, *Monatsh.* **53**, 727 (1929). Liquid. mp −82.7°. d_4^{20} 0.8608. bp_{760} 173.5°; bp_{400} 150.3°; bp_{200} 128.8°; bp_{100} 109.5°; bp_{60} 96.0°; bp_{40} 86.2°; bp_{20} 70.6°; bp_{10} 57.0°; bp_5 44.2°; $bp_{1.0}$ 18.6°. n_D^{20} 1.48980. Flash pt, closed cup: 126° F (52° C). Insol in water; misc with alcohol, ether, benzene.

d-Form, $[\alpha]_D^{25}$ +26.6°: Bonner, Greenlee, *J. Am. Chem. Soc.* **81**, 3336 (1959).

l-Form, $[\alpha]_D^{25}$ −27.3°.

USE: Solvent; in organic syntheses.

1551. *tert*-Butylbenzene. *(1,1-Dimethylethyl)benzene;* 2-methyl-2-phenylpropane; trimethylphenylmethane; pseudobutylbenzene. $C_{10}H_{14}$; mol wt 134.21. C 89.49%, H 10.51%. $C_6H_5C(CH_3)_3$. Prepn: Konowalow, *Bull. Soc. Chim.* [3] **16**, 865 (1896); Shoesmith, Mackie, *J. Chem. Soc.* **1928**, 2336; Meyer, Bernhauer, *Monatsh.* **53**, 727 (1929); Wilt, Abegg, *J. Org. Chem.* **33**, 923 (1968). *See also* Groose, Ipatieff, *J. Am. Chem. Soc.* **57**, 2415 (1935); Ipatieff, Pines, *ibid.* **58**, 1056 (1936). Liquid. mp −58.1°. d_4^{20} 0.8669. bp_{760} 168.5°; bp_{400} 145.8°; bp_{200} 123.7°; bp_{100} 103.8°; bp_{60} 90.6°; bp_{40} 80.8°; bp_{20} 65.6°; bp_{10} 51.7°; bp_5 39.0°; $bp_{1.0}$ 13.0°. n_D^{20} 1.49235. Flash pt, open cup: 140°F (60°C). Insol in water; misc with alcohol, ether, benzene.

1552. *n*-Butyl Benzoate. *Benzoic acid butyl ester.* $C_{11}H_{14}O_2$; mol wt 178.22. C 74.13%, H 7.92%, O 17.95%. $C_6H_5COO(CH_2)_3CH_3$. Prepn: Newman, Fones, *J. Am. Chem. Soc.* **69**, 1046 (1947); Justoni, Brit. pat. **719,891** (1954 to Vismara). Thick, oily liquid. d 1.00. mp −22°. bp 250°. Practically insoluble in water; sol in alcohol or ether. LD_{50} orally in rats: 5.14 g/kg, Smyth *et al., Arch. Ind. Hyg. Occup. Med.* **10**, 61 (1954).

1553. *n*-Butyl Bromide. *1-Bromobutane.* C_4H_9Br; mol wt 137.03. C 35.06%, H 6.62%, Br 58.32%. $CH_3(CH_2)_3Br$. Prepd from *n*-butyl alc and a hydrobromic-sulfuric acid mixture: Kamm, Marvel, *Org. Syn.* vol. **1**, 5 (1921); Skau, McCullough, *J. Am. Chem. Soc.* **57**, 2440 (1935). Colorless liquid. d_4^{25} 1.2686. bp_{760} 101.3° (mp −112°). n_D^{20} 1.4398. Insol in water; sol in alcohol, ether.

1554. *sec*-Butyl Bromide. *2-Bromobutane;* methylethylbromomethane. C_4H_9Br; mol wt 137.03. C 35.06%, H 6.62%, Br 58.32%. $CH_3CH_2CHBrCH_3$. Prepn: Levene, Marker, *J. Biol. Chem.* **91**, 405 (1931); Kenyon *et al., J. Chem. Soc.* **1935**, 1080; Skau, McCullough, *J. Am. Chem. Soc.* **57**, 2440 (1935); Colson *et al., J. Chem. Soc.* **1965**, 2364. Prepn of optically pure isomers: Goodwin, Hudson, *J. Chem. Soc. (B)* **1968**, 1333.

dl-Form, colorless liquid, pleasant odor. d_4^{25} 1.2530. bp 91.2° (mp −112°). n_D^{25} 1.4344. Insol in water. Freely sol in alcohol, ether.

d-Form, n_D^{20} 1.4359-1.4362. α_D^{20} +42.64°.

l-Form, n_D^{20} 1.4368. α_D^{20} −43.7°.

Caution: Narcotic in high concns.

1555. *tert*-Butyl Bromide. *2-Bromo-2-methylpropane;* 2-bromoisobutane; trimethylbromomethane. C_4H_9Br; mol wt 137.03. C 35.06%, H 6.62%, Br 58.32%. $(CH_3)_3CBr$. Prepn: Brunel, *J. Am. Chem. Soc.* **39**, 1978 (1917); Bryce-Smith, Howlett, *J. Chem. Soc.* **1951**, 1141; Coe *et al., ibid.* **1954**, 2281.

Colorless liquid. d_4^{25} 1.2125. bp 73.3°. mp —16.3°. At 210° changes to isobutyl bromide. n_D^{25} 1.4249. Insol in water; miscible with organic solvents.

1556. n-Butyl n-Butyrate. *Butanoic acid butyl ester; butyric acid butyl ester.* $C_8H_{16}O_2$; mol wt 144.21. C 66.63%, H 11.18%, O 22.19%. $CH_3(CH_2)_2COO(CH_2)_3CH_3$. Prepn from butyl alcohol: Robertson, *Org. Syn.* **coll. vol. I**, 138 (1941); Horton, U.S. pat. **2,522,676** (1950 to Socony-Vacuum Oil).
Liquid, bp 165°. d_4^{20} 0.8692. n_D^{20} 1.4064. Practically insol in water; miscible with alcohol, ether.

1557. Butyl Carbitol®. *2-(2-Butoxyethoxy)ethanol;* diethylene glycol monobutyl ether; butyl digol; Butyl Diicinol. $C_8H_{18}O_3$; mol wt 162.22. C 59.23%, H 11.18%, O 29.59%. $HOCH_2CH_2OCH_2CH_2OC_4H_9$. Prepn: Riemschneider, Gross, *Monatsh.* **90**, 783 (1959). Purification: Miller, Yonan, *J. Am. Chem. Soc.* **79**, 5931 (1957); Ridley, Ridley, **Brit.** pat. **795,866** (1958 to Esso). Toxicity data: Smyth *et al.*, *J. Ind. Hyg. Toxicol.* **23**, 259 (1941).
Practically odorless liquid, bp 230.4°. mp —68.1°. d_{20}^{20} 0.9536. n_D^{27} 1.4258. Miscible in water, oils. Miscibility in other organic solvents: Jackson, Drury, *Ind. Eng. Chem.* **51**, 1491 (1959). Flash pt 110°. LD_{50} orally in rats, guinea pigs: 6.56, 2.00 g/kg (Smyth).
USE: Solvent.

1558. n-Butyl Carbonate. *Carbonic acid dibutyl ester;* dibutyl carbonate. $C_9H_{18}O_3$; mol wt 174.23. C 62.04%, H 10.41%, O 27.55%. $(C_4H_9O)_2CO$. Prepn from ethyl carbonate, butyl alcohol and ethylmagnesium bromide: Frank *et al., J. Am. Chem. Soc.* **66**, 1509 (1944); from butyl alcohol and CO in the presence of Pd and $CuCl_2$: Mador, Blackham, U.S. pat. **3,114,762** (1963 to National Distillers).
Liquid, bp 206.6°. d_4^{20} 0.9251, d_4^{25} 0.9388. n_D^{20} 1.4117. Practically insol in water. Miscible with ethanol, benzene, chloroform, acetone, ether and other organic solvents, *see:* Jackson, Drury, *Ind. Eng. Chem.* **51**, 1491 (1959).

1559. Butyl Cellosolve®. *2-Butoxyethanol;* ethylene glycol monobutyl ether. $C_6H_{14}O_2$; mol wt 118.17. C 60.98%, H 11.94%, O 27.08%. $HOCH_2CH_2OC_4H_9$. Prepn: L. H. Cretcher, W. H. Pittenger, *J. Am. Chem. Soc.* **46**, 1503 (1924); W. W. Carlson, U.S. pat. **2,448,767** (1948 to Mellon Inst. Ind. Res.); R. Riemschneider, P. Gross, *Monatsh.* **90**, 783 (1959). Toxicity: H. F. Smyth *et al., J. Ind. Hyg. Toxicol.* **23**, 259 (1941); C. P. Carpenter *et al., Arch. Ind. Health* **14**, 114 (1956).
Liquid, bp 171-172°. d_4^{20} 0.9012, d_{20}^{20} 0.9019. n_D^{20} 1.4196. Flash pt, closed cup: 141°F (60°C). Sol in water, mineral oil, most organic solvents. LD_{50} orally in rats: 1.48 g/kg (Smyth).
USE: Solvent for nitrocellulose, resins, grease, oil, albumin; dry cleaning. *Caution:* Toxic symptoms similar to those for Methyl Cellosolve.

1560. n-Butyl Chloride. *1-Chlorobutane;* n-propylcarbinyl chloride; butyl chloride. C_4H_9Cl; mol wt 92.57. C 51.90%, H 9.80%, Cl 38.30%. $CH_3CH_2CH_2CH_2Cl$. Prepd from n-butyl alcohol by heating with HCl and anhydr $ZnCl_2$: Whaley, Copenhaver, *J. Am. Chem. Soc.* **60**, 2497 (1938); *Org. Syn.* **coll. vol. I**, 142 (2nd ed., 1941).
Liquid. Highly flammable. d_4^{15} 0.89197; d_4^{20} 0.88648; d_4^{25} 0.88098. One gallon weighs 7.35 pounds. mp —123.1°. bp_{760} 78.5°. n_D^{20} 1.40223. Flash pt —6.7° (+20°F). Dipole moment: 1.95. Practically insol in water (0.066% at 12°). Misc with alcohol, ether. LD_{50} orally in rats: 2.67 g/kg, H. F. Smyth *et al., Arch. Ind. Hyg. Occup. Med.* **10**, 61 (1954).
USE: As butylating agent in organic synthesis, e.g., in the manuf of butyl cellulose.
THERAP CAT (VET): Anthelmintic.

1561. sec-Butyl Chloride. *2-Chlorobutane;* 2-chloro-3-methylpropane. C_4H_9Cl; mol wt 92.57. C 51.90%, H 9.80%, Cl 38.30%. $CH_3CH_2CHClCH_3$. Prepn from sec-butyl alcohol, hydrochloric acid and $ZnCl_2$: Norris, Taylor, *J. Am. Chem. Soc.* **46**, 756 (1924); Copenhaver, Whaley, *Org. Syn.* **coll. vol. I**, 143 (1941); from sec-butyl alcohol and PCl_3: Coulson *et al., J. Chem. Soc.* **1965**, 2364. Prepn of optically pure isomers: Goodwin, Hudson, *J. Chem Soc.* **(B) 1968**, 1333. Toxicity data: H. F. Smyth *et al., Am. Ind. Hyg. Assoc. J.* **30**, 470 (1969).

dl-Form, liquid; pleasant, ethereal odor. d_4^{20} 0.871. bp 68°. n_D^{20} 1.3960; n_D^{25} 1.3953. One gram dissolves in 1000 ml water at 25°; misc with alcohol, ether. LD_{50} orally in rats: 20.0 ml/kg (Smyth).
d-Form, n_D^{20} 1.3963. α_D^{20} +30.8°.
l-Form, n_D^{20} 1.3968. α_D^{20} —31.0°.

1562. tert-Butyl Chloride. *2-Chloro-2-methylpropane;* 2-chloroisobutane; trimethylchloromethane. C_4H_9Cl; mol wt 92.57. C 51.90%, H 9.80%, Cl 38.30%. $(CH_3)_3CCl$. Prepd by shaking tert-butyl alcohol with concd HCl and distilling: Norris, Olmsted, *Org. Syn.* **8**, 50 (1928). Prepn from tert-butyl alcohol and PCl_3: Gerrard *et al., J. Chem. Soc.* **1953**, 1920.
Liquid. d_4^{15} 0.847. n_D^{18} 1.38686. mp —26.5°. bp_{760} 51.0°; bp_{400} 32.6°; bp_{200} +14.6°; bp_{100} —1.0°; bp_{60} —11.4°; bp_{40} —19.0°. Sparingly sol in water, miscible with alcohol and ether. Boiling with water yields tert-butyl alcohol.

1563. tert-Butyl Chloroacetate. *Chloroacetic acid 1,1-dimethyl ester;* t-butyl chloroacetate; chloroacetic acid tert-butyl ester. $C_6H_{11}ClO_2$; mol wt 150.61. C 47.85%, H 7.36%, Cl 23.54%, O 21.25%. $ClCH_2COOC(CH_3)_3$. Prepd by reacting monochloroacetic acid and isobutylene in dioxane in the presence of sulfuric acid: Johnson *et al., J. Am. Chem. Soc.* **75**, 4995 (1953); from tert-butyl alcohol, chloroacetyl chloride and dimethylaniline: Baker, *Org. Syn.* **24**, 21 (1944).
Liquid, bp 155° (dec); bp_{11} 48-49°; bp_{16-17} 56-57°. n_D^{25} 1.4204-1.4210; n_D^{20} 1.4259-1.4260. Hydrolyzes to tert-butyl alcohol and chloroacetic acid.
USE: In the glycidic ester condensation.

1564. Butyl Citrate. *2-Hydroxy-1,2,3-propanetricarboxylic acid tributyl ester;* citric acid tributyl ester; n-butyl citrate; tributyl citrate. $C_{18}H_{32}O_7$; mol wt 360.44. C 59.98%, H 8.95%, O 31.07%. Prepn: Fodor, *Arch. Biochem.* **28**, 274 (1950). Synthesis from n-butyl alcohol and citric acid: Benedict, *Chemistry* **47**, 27 (1974).

$$CH_2COOC_4H_9$$
$$HOCCOOC_4H_9$$
$$CH_2COOC_4H_9$$

Colorless or pale yellow, odorless liquid. d_4^{20} 1.045. bp_{22} about 233°. mp —20°. Flash pt 185°. n_D^{20} 1.4460 (Benedict). Insol in water; miscible with most organic liquids.
USE: Plasticizer and solvent for nitrocellulose lacquers; in polishes, inks and similar prepns; also as anti-foam agent.

1565. α-Butylene Dibromide. *1,2-Dibromobutane.* $C_4H_8Br_2$; mol wt 215.92. C 22.25%, H 3.73%, Br 74.02%. $CH_3CH_2CHBrCH_2Br$. Prepn by bromination of butene: Wurtz, *Ann.* **152**, 21 (1869); of bromobutane: Kharasch *et al., J. Org. Chem.* **20**, 1430 (1955).
Yellowish liquid, bp 166°. mp —65°. d_4^{20} 1.7946. n_D^{20} 1.5144. Practically insol in water; miscible with alcohol.

1566. 1,3-Butylene Glycol. *1,3-Butanediol;* 1,3-dihydroxybutane; β-butyleneglycol; methyltrimethylene glycol; butane-1,3-diol. $C_4H_{10}O_2$; mol wt 90.12. C 53.31%, H 11.19%, O 35.51%. $CH_3CHOHCH_2CH_2OH$. Usually prepd by catalytic hydrogenation of aldol using Raney nickel: Hancock, Henson, *Ind. Eng. Chem.* **45**, 629 (1953); **Brit.** pat. **853,266** (1960 to Celanese); F. S. Wagner in Kirk-Othmer, *Encyclopedia of Chemical Technology* **vol. 11** (Wiley-Interscience, New York, 3rd ed., 1980) pp 956-962. Toxicity: Smyth *et al., Arch. Ind. Hyg. Occup. Med.* **4**, 119 (1951).
Viscous liquid. The pure compd is colorless. d_4^{20} 1.004-1.006. One gallon weighs 8.398 lbs at room temp. Viscosity in centistokes: 24.6 at 50°C, 96 at 25°, 590 at 0°, 3253 at —17.7°, 6059 at —23°. melts below 50°. Because of the high viscosity at low temps heating is necessary for pumping. Very hygroscopic: Will absorb 38.5 wt % of water within 144 hrs at 81% relative humidity. bp 207.5°. n_D^{20} 1.4401. Flash pt (tag open cup): 250°F (121°C). Dielectric constant 28.8 at 25°. Surface tension: 37.8 dynes/cm at 25°. Sol in water, acetone, methyl ethyl ketone, ethanol, dibutyl phthalate, castor oil. Practically insol in aliphatic hydrocarbons, benzene, toluene, carbon tetrachloride, ethanolamines, mineral oil, linseed oil. LD_{50} orally in rats: 22.8 g/kg (Smyth).

USE: Intermediate in the manufacture of polyester plasticizers, humectant for cellophane, tobacco. Has some mold inhibiting action.

1567. 2,3-Butylene Glycol. *2,3-Butanediol; 2,3*-dihydroxybutane; dimethylethylene glycol. $C_4H_{10}O_2$; mol wt 90.12. C 53.31%, H 11.19%, O 35.51%. $CH_3CH(OH)CH(OH)CH_3$. Occurs in 3 isomeric forms: *meso-* or *erythro-*, D(—)-*threo-*, and L(+)-*threo*-forms. The commercial product is usually either the *meso-* or the D(—)-form. Prepn of *meso*-form from *trans*-2,3-epoxybutane and of DL-form from *cis*-2,3-epoxybutane: Wilson, Lucas, *J. Am. Chem. Soc.* **58**, 2396 (1936). Prepn of D(—)- and L(+)-forms from corresponding D- and L-mannitols: Rubin et al., *ibid.* **74**, 425 (1952). Manuf of D(—)-form by fermentation of carbohydrate solns with organisms of the *Bacillus subtilis* group: Vergnaud, U.S. pat. **2,529,061** (1950 to Usines de Melle). Manuf from 2-butene: Cosby et al., U.S. pat. **2,808,429** (1957 to Allied Chem.); Keith et al., U.S. pat. **2,974,161** (1961 to Sinclair); from 2-butyne: Saegebarth, U.S. pat. **3,157,704** (1964 to du Pont). Configuration: Morell, Auernheimer, *J. Am. Chem. Soc.* **66**, 792 (1944); Leroux, Lucas, *ibid.* **73**, 41 (1951); Rubin et al., loc. cit.

meso-Form (*erythro*-form), hygroscopic crystals from dry diisopropyl ether, mp 34.4°. bp_{742} 181.7°, bp_{16} 89°. d_4^{25} 0.9939. n_D^{35} 1.4324. Moderately sol in diisopropyl ether.

DL-*threo*-Form, hygroscopic crystals from diisopropyl ether, mp 7.6°. bp_{742} 172.7°, bp_{16} 86°. n_D^{25} 1.4310. Very sol in diisopropyl ether.

D(—)-*threo*-Form, mp 19.7°. bp_{745} 179-180°, bp_{10} 77.5-78°. d_4^{25} 0.9869. n_D^{25} 1.4315. $[\alpha]_D^{25}$ —13.0° (neat).

L(+)-*threo*-Form, bp 179-182°. d^{25} 0.9872. n_D^{25} 1.4306.

1568. n-Butyl Ether. *1,1'-Oxybis[butane]; butyl ether; n*-dibutyl ether. $C_8H_{18}O$; mol wt 130.22. C 73.78%, H 13.93%, O 12.29%. $(CH_3CH_2CH_2CH_2)_2O$. Prepn: Smith, *J. Chem. Ed.* **39**, 212 (1962); d'Engenieres et al., *Bull. Soc. Chim. France* **1964**(10), 2471.

Liquid. d_{20}^{20} 0.769. mp —98°. bp 142-143°. Flash pt, closed cup: 100° F (37° C). Almost insol in water; misc with alcohol or ether. Tends to form explosive peroxides, especially when anhydr. LD_{50} orally in rats: 7.4 g/kg, Smith et al., *Arch. Ind. Hyg. Occup. Med.* **10**, 61 (1954), *C.A.* **48**, 13952a (1954).

1569. tert-Butyl Hydroperoxide. 1,1-Dimethylethylhydroperoxide. $C_4H_{10}O_2$; mol wt 90.12. C 53.31%, H 11.19%, O 35.51%. Prepn from *tert*-butyl alcohol and 30% H_2O_2: N. A. Milas, S. A. Harris, *J. Am. Chem. Soc.* **60**, 2434 (1938); N. A. Milas, Surgenor, *ibid.* **68**, 205 (1946); U.S. pat. **2,573,947** (1951 to Shell). By oxidation of *tert*-butyl-magnesium chloride: Walling, Buckler, *J. Am. Chem. Soc.* **75**, 4372 (1953); **77**, 6032 (1955).

$$CH_3 - \underset{\underset{CH_3}{|}}{\overset{\overset{CH_3}{|}}{C}} - O - O - H$$

Liquid. Stable to 75°. d_4^{20} 0.896. mp —8°. bp_{20} 35°. n_4^{20} 1.4007. Sol in organic solvents. Slow first-order decompn can be accelerated by the presence of 1 mol-% of Cu, Co and Mn salts.

USE: Catalyst in polymerization reactions. To introduce peroxy group into org molecules, in radical substitution reactions: Kharasch, Fono, *J. Org. Chem.* **23**, 325 (1948); see also Kharasch, Sosnovsky, *Tetrahedron* **3**, 97, 105 (1958). Caution: Flammable.

1570. tert-Butyl Hypochlorite. *Hypochlorous acid 1,1-dimethylethyl ester; t*-butyl hypochlorite. C_4H_9ClO; mol wt 108.57. C 44.25%, H 8.35%, Cl 32.66%, O 14.74%. $(CH_3)_3$-COCl. Prepd by adding aq *tert*-butyl alcohol to a cooled aq soln of NaOH and passing chlorine into the mixture: Teeter, Bell, *Org. Syn.* **32**, 20 (1952). By adding CO_2 to neutral aq *tert*-butyl alcohol and NaOCl satd with NaCl: Katz, U.S. pat. **2,694,722** (1954 to Bjorksten Res. Labs.).

Pale yellow liquid, irritating odor; attacks mucous membranes and eyes. d_{20}^{20} 0.910; d_4^{18} 0.9583. bp 77-78°. n_D^{20} 1.403. Hydrolyzes in presence of aq alkali but not aq acid. Dec

upon exposure to bright light to methyl chloride and acetone with considerable evolution of heat. Should be stored in an inert atmosphere in a dark, refrigerated place.

USE: Dehydration of alcohols; N- and C-chlorinations. Caution: Violent reaction occurs if exposed to rubber, strong light or overheating. Has far greater stability than the corresponding *n*-butyl and *sec*-butyl hypochlorites.

1571. Butylidene Chloride. *1,1-Dichlorobutane.* C_4H_8-Cl_2; mol wt 127.02. C 37.82%, H 6.35%, Cl 55.83%. CH_3-$CH_2CH_2CHCl_2$. Prepn from butyraldehyde and PCl_5: Henne et al., *J. Am. Chem. Soc.* **61**, 938 (1939).

Oily liquid. bp_{752} 114.8-115.1°. d^{25} 1.0797; n_D^{25} 1.4305: Brown, Ash, *J. Am. Chem. Soc.* **77**, 4019 (1955). Practically insol in water; sol in alcohol, chloroform, ether.

1572. n-Butyl Iodide. *1-Iodobutane.* C_4H_9I; mol wt 184.03. C 26.10%, H 4.93%, I 68.97%. $CH_3CH_2CH_2CH_2I$. Prepn: Franzen, *Ber.* **87**, 1148 (1954); Landauer, Rydon, *J. Chem. Soc.* **1954**, 2281; Stone, Shechter, *Org. Syn.* **coll. vol.** **IV**, 321 (1963). Manuf: Huber, Schenck, U.S. pats. **2,899,-471** and **3,053,901** (1959 and 1962, both to GAF).

Liquid, bp 130.4°. mp —103.0°. d_4^{20} 1.616. n_D^{20} 1.4998: Vogel, *J. Chem. Soc.* **1943**, 636. Practically insol in water; sol in alcohol, ether. *Protect from light.*

1573. sec-Butyl Iodide. *2-Iodobutane.* C_4H_9I; mol wt 184.03. C 26.10%, H 4.93%, I 68.97%. $CH_3CH_2CHICH_3$. Prepn of *dl*-form from 2-butanol: Kornblum et al., *J. Am. Chem. Soc.* **77**, 5528 (1955); from dibutyl ether: Long, Free-guard, *Chem. & Ind. (London)* **1965**, 223. Prepn of *d*-form from *l*-butanol: Kenyon et al., *J. Chem. Soc.* **1935**, 1072. Prepn of *l*-form from d-2-butanol: Levene, Rothen, *J. Biol. Chem.* **115**, 415 (1936); Kornblum et al., *J. Am. Chem. Soc.* **70**, 746 (1948).

dl-Form, liquid, rapidly turns brown on exposure to light. bp 120°. mp —104°. d_4^{20} 1.592, n_D^{20} 1.4991: Vogel, *J. Chem. Soc.* **1943**, 636. Practically insol in water; sol in alc, ether. *Protect from light.*

l-Form, liquid, bp 118°. d_4^{20} 1.596, n_D^{20} 1.495, $[\alpha]_D^{24}$ —15.9° (neat); also reported as $[\alpha]_D^{17}$ —31.98° (Kornblum et al.).

1574. n-Butylmalonic Acid. *Butylpropanedioic acid;* pentane-1,1-dicarboxylic acid. $C_7H_{12}O_4$; mol wt 160.17. C 52.49%, H 7.55%, O 39.96%. $CH_3(CH_2)_3CH(COOH)_2$. Prepd by the interaction of sodium *n*-amyl chloride and carbon dioxide under pressure with ligroin as solvent: Morton et al., *J. Am. Chem. Soc.* **58**, 754 (1936).

Prisms from water, mp 102°. K at 5° 1.1×10^{-3}. Sol in water: 100 g of a satd aq soln contain 11.6 g at 0°; 79.3 g at 50°. Sol in alc, ether. Heating to 150° yields caproic acid.

Diethyl ester, $C_{11}H_{20}O_4$. Prepared by the interaction of *n*-butyl bromide, diethyl malonate, and sodium alcoholate: Adams, Kamm, *Org. Syn.* **4**, 11 (1925). Liquid, bp_{760} 235-240°; bp_{40} 140-145°; bp_{20} 130-135°. n_D^{20} 1.425. Very sol in alcohol, ether.

1575. n-Butyl Mercaptan. *1-Butanethiol;* normal butyl thioalcohol; thiobutyl alcohol. $C_4H_{10}S$; mol wt 90.19. C 53.27%, H 11.18%, S 35.55%. $CH_3(CH_2)_2CH_2SH$. Formed by the action of yeast on a product obtained when hydrogen sulfide is made to react with butyraldehyde in alcoholic ammonia: Nord, *Ber.* **52**, 1209 (1919); by passing vapors of butanol and hydrogen sulfide over thorium oxide catalyst: Kramer, Reid, *J. Am. Chem. Soc.* **43**, 880 (1921); together with dibutyl sulfide in distilling an aq soln of sodium butyl sulfate and sodium sulfide: Gray, Gutekunst, *J. Am. Chem. Soc.* **42**, 858 (1920); by slightly warming dithiocarbamic butyl ester with aq KOH: v. Braun, Engelbertz, *Ber.* **56**, 1574 (1923); the sodium salt is formed when dibutyl disulfide is treated with sodium in ether or alcohol: Moses, Reid, *J. Am. Chem. Soc.* **48**, 777 (1926). Physical properties: Mathias, *ibid.* **72**, 1897 (1950); W. E. Haines et al., *J. Phys. Chem.* **60**, 549 (1956). Once cited to occur in "skunk" fluid, see Beckmann, *Pharm. Zentralhalle* **37**, 557 (1896); T. B. Aldrich, *J. Exp. Med.* **1**, 323 (1896); however, recent literature rebuts this: K. K. Anderson, D. T. Bernstein, *J. Chem. Ecol.* **1**, 493 (1975); eidem, *J. Chem. Ed.* **55**, 159 (1978).

Mobile liquid. Heavy skunk odor. mp —115.9°. bp_{766} 98.2°; bp_{760} 98.4°. Flammable. d_4^{25} 0.83679. n_D^{25} 1.44014. Slightly sol in water. Very sol in alcohol, ether, liquid hydrogen sulfide. Forms azeotropic mixtures with butyl alcohol (bp 97.8°; 85.16% butanethiol) and with butyl alcohol and water.

1576. sec-Butyl Mercaptan. *2-Butanethiol; sec*-butyl thioalcohol. $C_4H_{10}S$; mol wt 90.19. C 53.27%, H 11.18%, S 35.55%. $CH_3CH_2CH(SH)CH_3$. Prepd from 2-iodobutane by means of alcoholic KSH soln: Reymann, *Ber.* **7**, 1287 (1874); from 2-bromobutane in the same manner: Ellis, Reid, *J. Am. Chem. Soc.* **54**, 1674 (1932).

Mobile liq. Heavy skunk odor. mp −165°. bp 84-85°. Flammable. d^{17} 0.8299; d_4^{25} 0.8246. n_D^{25} 1.4338. Slightly sol in water. Very sol in alc, ether, liquid H_2S.

1577. tert-Butyl Mercaptan. *2-Methyl-2-propanethiol.* $C_4H_{10}S$; mol wt 90.19. C 53.27%, H 11.18%, S 35.55%. $(CH_3)_3CSH$. Prepd from *tert*-butyl iodide, ZnS and alcohol: Dobbin, *J. Chem. Soc.* **57**, 641 (1890). *See also* U.S. pats. **2,020,421**, *C.A.* **30**, 489 (1936) and **2,051,806**, *C.A.* **30**, 6760 (1936).

Mobile liquid. Heavy skunk odor. mp −0.5°. bp_{760} 63.7-64.2°. d_4^{25} 0.79426; n_D^{25} 1.41984: Mathias, *J. Am. Chem. Soc.* **72**, 1897 (1950). Flammable. Remarkably stable to oxidizing agents. Slightly sol in water. Very sol in alcohol, ether, liquid H_2S.

1578. n-Butylmercuric Chloride. *Butylchloromercury.* C_4H_9ClHg; mol wt 293.18. C 16.39%, H 3.09%, Cl 12.09%, Hg 68.43%. $CH_3CH_2CH_2CH_2HgCl$. Prepn: Slotta, Jacobi, *J. Prakt. Chem.* [2] **120**, 249 (1929); R. C. Larock, H. C. Brown, *J. Am. Chem. Soc.* **92**, 2467 (1970). Distribution in mice: M. Yonaha *et al.*, *Chem. Pharm. Bull.* **23**, 1718 (1975). For the prepn of the bromide *see* Slotta, Jacobi, *loc. cit.*; Marvel *et al.*, *J. Am. Chem. Soc.* **47**, 3009 (1925). Review and bibliography: Krause, von Grosse, *Die Chemie der Metallorganischen Verbindungen* (Berlin, 1937).

White needles from ethanol, sometimes leaflets, mp 130°. Soly (g/100 ml) in water at 18°: 1.4×10^{-4}; at 100°: 3.3×10^{-4}; soly (g/100 g) in ethanol at 18°: 1.5; at 78°: 9.0; in chloroform at 18°: 5.6.

1579. 1-Butyl-3-metanilylurea. *N-[(3-Aminophenyl)-sulfonyl]-N'-butylurea; N^1-metanilyl-N^2-n-butylurea; N^1-butyl-N^2-metanilylcarbamide; N^1-(3-aminobenzenesulfonyl)-N^2-n-butylurea;* SB-1; Sucrida Berna. $C_{11}H_{17}N_3O_3S$; mol wt 271.35. C 48.69%, H 6.32%, N 15.49%, O 17.69%, S 11.82%. Prepn: Haack, *Arzneimittel-Forsch.* **8**, 444 (1958); Zahler, U.S. pat. **3,015,673** (1962 to Swiss Serum and Vaccine Inst.).

Crystals, mp 117°.
THERAP CAT: Oral hypoglycemic.

1580. Butyl Methoxydibenzoylmethane. *1-[4-(1,1-Dimethylethyl)phenyl]-3-(4-methoxyphenyl)-1,3-propanedione;* 4-*tert*-butyl-4′-methoxydibenzoylmethane; Parsol 1789; Parsol A. $C_{20}H_{22}O_3$; mol wt 310.39. C 77.39%, H 7.14%, O 15.46%. UV-A blocker. Prepn: K.-F. De Polo, Ger. pat. **2,945,125**; *idem*, U.S. pat. **4,387,089** (1980, 1983 both to Givaudan). Clinical efficacy as sunscreen: R. W. Gange *et al.*, *J. Am. Acad. Dermatol.* **15**, 494 (1986); K. Kaidbey, R. W. Gange, *ibid.* **16**, 346 (1987); N. J. Lowe *et al.*, *ibid.* **17**, 224 (1987). Assessment of photostability: A. Deflandre, G. Lang, *Int. J. Cosmet. Sci.* **10**, 53 (1988). HPLC determn in cosmetic products: L. Gagliardi *et al.*, *J. Chromatog.* **408**, 409 (1987).

Crystals from methanol, mp 83.5°.
THERAP CAT: Ultraviolet screen.

1581. n-Butyl Nitrite. *Nitrous acid butyl ester.* C_4H_9-

NO_2; mol wt 103.12. C 46.59%, H 8.80%, N 13.58%, O 31.03%. $CH_3CH_2CH_2CH_2ONO$. Prepd by the action of nitrous acid on butyl alcohol: Noyes, *Org. Syn. coll. vol. II*, 108 (1943); Miller, Audrieth, *Inorg. Syn.* **2**, 139 (1946).

Oily liquid. Characteristic odor. Breathing of vapor causes headache and vasodilation. d_4^0 0.9114. bp_{760} 78.2° (some decompn). Miscible with alcohol, ether. Dec on storage. Polymerization products of butyraldehyde have been found in five-month-old samples.

USE: In the manuf of rare earth azides. *Caution:* Lowers blood pressure through vasodilation, causing headache, throbbing, weakness. Effects resemble those produced by amyl nitrate.

1582. tert-Butyl Nitrite. *Nitrous acid 1,1-dimethyl ethyl ester;* α,α-dimethylethyl nitrite; nitrous acid *tert*-butyl ester. $C_4H_9NO_2$; mol wt 103.12. C 46.59%, H 8.80%, N 13.58%, O 31.03%. $(CH_3)_3CONO$. Prepd by adding *t*-butyl chloride to a cooled slurry of silver nitrite and anhydr ether: Kornblum *et al.*, *J. Am. Chem. Soc.* **77**, 5528 (1955); by passing NO_2 or NO_2 + NO into *t*-butyl alcohol at 25-30°: Treacy, U.S. pat. **2,739,166** (1956); from *t*-butanol, sodium nitrite and sulfuric acid: Coe, Doumani, *J. Am. Chem. Soc.* **70**, 1516 (1948).

Yellow liquid, agreeable odor. d_4^0 0.8941; d_4^{20} 0.8671. bp_{760} 63°; bp_{250} 34°; n_D^{20} 1.3687. Very sol in alcohol, ether, chloroform, carbon disulfide; slightly sol in water; practically insol in glycerol.

USE: Jet propellant.

1583. Butylparaben. *4-Hydroxybenzoic acid butyl ester; n*-butyl *p*-hydroxybenzoate; Butoben; Butyl Chemosept; Butyl Parasept; Tegosept B. $C_{11}H_{14}O_3$; mol wt 194.22. C 68.02%, H 7.27%, O 24.71%. Prepn of analogous derivatives: *See* ethyl-, methyl- and propylparaben. Prepn of calcium and magnesium salts: Engels, Weijlard, U.S. pats. **2,046,324** and **2,056,176** (both 1936 to Merck & Co.).

Crystalline powder, mp 68-69°. Very slightly soluble in water (1:6500), glycerin; freely sol in acetone, alc, ether, chloroform, propylene glycol. Preserve in well-closed containers.
Calcium salt, $C_{22}H_{26}CaO_6$, cryst powder. Soly in water: approx 1:125.
Magnesium salt, $C_{22}H_{26}MgO_6$, cryst powder. Soly in water: approx 1:110.
USE: Pharmaceutic aid (antifungal). Preservative in foods.

1584. p-tert-Butylphenol. *4-(1,1-Dimethylethyl)phenol;* butylphen. $C_{10}H_{14}O$; mol wt 150.21. C 79.95%, H 9.40%, O 10.65%. Prepd by heating phenol with isobutanol in the presence of zinc chloride: Ger. pat. **17,311**; *Frdl.* **1**, 22; also from phenol, *tert*-butyl chloride and excess alkali in alcohol: Lewis, *J. Chem. Soc.* **83**, 329 (1903). *See also* Smith, *J. Am. Chem. Soc.* **55**, 3718 (1933); Natelson, *ibid.* **56**, 1583 (1934); Huston, Hsieh, *ibid.* **58**, 439 (1936); U.S. pat. **2,039,344** (1936 to Dow); Isagulyants, Bagryantseva, *C.A.* **33**, 8183 (1939).

Needles from water, mp 98°. d_4^{114} 0.9081. bp 237°. Volatile with steam. Practically insol in cold water. Sol in alcohol, ether. LD_{50} orally in rats: 3.25 ml/kg, H. F. Smyth *et al.*, *Am. Ind. Hyg. Assoc. J.* **30**, 470 (1969).

Sodium salt, $C_{10}H_{13}NaO$, deliquescent leaflets. Sol in water.

USE: Intermediate in the manuf of varnish and lacquer resins; as a soap antioxidant; ingredient in de-emulsifiers for oil field use; in motor oil additives.

1585. 4-*tert*-Butylphenyl Salicylate. *2-Hydroxybenzoic acid 4-(1,1-dimethylethyl)phenyl ester; salicylic acid p-tert-butylphenyl ester;* TBS. $C_{17}H_{18}O_3$; mol wt 270.31. C 75.53%, H 6.71%, O 17.76%. Prepd from salicylic acid and *p-tert*-butylphenol in the presence of $POCl_3$: Stoesser, Sommerfield, U.S. pat. 2,606,920 (1952 to Dow).

Crystals, slight odor resembling that of salol, mp 62-64°. Maximum light absorption at 290-330 nm. Soly (w/w): in water <0.1%; abs ethanol 79%; ethyl acetate 153%; methyl ethyl ketone 197%; toluene 158%; Stoddard solvent 39%. USE: Light absorber in plastic food wrappings.

1586. *n*-Butyl Phthalate. *1,2-Benzenedicarboxylic acid dibutyl ester; phthalic acid dibutyl ester;* dibutyl phthalate; DBP. $C_{16}H_{22}O_4$; mol wt 278.34. C 69.04%, H 7.97%, O 22.99%. $1,2-(C_4H_9OOC)_2C_6H_4$. Prepn: Farrar, Wienkauff, *Chem. & Ind. (London)* **1962**, 2144. Manuf: Bruno, U.S. pat. 2,628,249 (1953 to Pittsburgh Coke & Chemical).

Oily liquid. bp 340°. d^{20} 1.0459 and 1.0465, Kemppinen, Gokeen, *J. Phys. Chem.* **60**, 126 (1956). n_D^{20} 1.4900. Flash pt, open cup: 340°F (171°C). Sol in about 2500 parts water; very sol in alcohol, ether, acetone, benzene. Max single oral dose tolerated by rats: >8 g/kg body wt, Smith, *Arch. Ind. Hyg. Occup. Med.* **7**, 310 (1953), *C.A.* **47**, 12619h (1953). USE: Insect repellant for the impregnation of clothing.

1587. *n*-Butyl Propionate. *Propanoic acid butyl ester; propionic acid butyl ester.* $C_7H_{14}O_2$; mol wt 130.18. C 64.58%, H 10.84%, O 24.58%. $CH_3CH_2COOC_4H_9$. Prepn from 1-butanol and propionic acid: Vogel, *J. Chem. Soc.* **1948**, 616; Spindt, Stevens, U.S. pat. 2,470,876 (1949 to Gulf).

Liquid, bp 146.8°. mp −89°. d_4^{20} 0.8754. n_D^{20} 1.401. Very slightly sol in water; very sol in alcohol, ether.

1588. *N*-Butylscopolammonium Bromide. *[7(S)-(1α,-2β,4β,5α,7β)]-9-Butyl-7-(3-hydroxy-1-oxo-2-phenylpropoxy)-9-methyl-3-oxa-9-azoniatricyclo[3.3.1.0²,⁴]nonane bromide;* α-(hydroxymethyl)benzeneacetic acid 9-butyl-9-methyl-3-oxa-9-azoniatricyclo[3.3.1.0²,⁴]non-7-yl ester bromide; 8-butyl-6β,7β-epoxy-3α-hydroxy-1αH,5αH-tropanium bromide (−)-tropate; butylscopolamine bromide; hyoscine-*N*-butyl bromide; scopolamine-*N*-butyl bromide; scopolamine bromobutylate; Amisepan; Buscapina; Buscol; Buscolamin; Buscolysin; Buscopan; Butylmin; Butylscopolamine; Donopon; Monospan; Scobro; Scobron; Scobutil; Sparicon; Sporamin; Stibron; Tirantil. $C_{21}H_{30}BrNO_4$; mol wt 440.40. C 57.27%, H 6.87%, Br 18.15%, N 3.18%, O 14.53%. Ganglion blocking agent. Prepn: F. Adickes *et al.,* **Ger.** pat. 856,890; *eidem,* U.S. pat. 2,872,452 (1952, 1959 both to Boehringer Ing.). Biological activity studies: Bauer *et al., Arzneimittel-Forsch.* **18**, 1132 (1968). Toxicological studies: K. Stockhaus, H. Wick, *Arch. Int. Pharmacodyn. Ther.* **180**, 155 (1969). Use in endoscopic examinations: J. R. Lee, *Clin. Radiol.* **33**, 273 (1982); D. N. Hupscher, O. Dommerholt, *Diagn. Imag. Clin. Med.* 53, 77 (1984); A. C. Steger *et al., Am. J. Gastroenterol.* **81**, 615 (1986).

Crystals from methanol, mp 142-144°. $[\alpha]_D^{20}$ −20.8° (c = 3 in water). LD_{50} in mice (mg/kg): 15.6 i.v.; 74 i.p.; 570 s.c.; 3000 orally (Stockhaus, Wick).

THERAP CAT: Antispasmodic.

1589. Butyl Stearate. *Octadecanoic acid butyl ester.* $C_{22}H_{44}O_2$; mol wt 340.57. C 77.58%, H 13.02%, O 9.40%. $CH_3(CH_2)_{16}COOC_4H_9$. Prepn from silver stearate and *n*-butyl iodide: Whitby, *J. Chem. Soc.* **1926**, 1458; from stearic acid and *n*-butanol: Smith, *ibid.* **1931**, 802. Physical properties: *Beilstein* vol. 2, Suppl. 3, 1016; Smith, *loc. cit.; Ind. Eng. Chem.* **32**, 880 (1940).

Crystals from alcohol, propanol, or ether, mp 27°. Also reported as mp 16°. bp 343°. Closed-cup flash pt 160° (320°F). Open-cup flash pt 196° (385°F). d_{25}^{25} 0.855-0.875. Slightly sol in water; sol in alcohol, ether. USE: Solvent, spreading and softening agent in plastics, textiles, cosmetics, rubber industries.

1590. *n*-Butyl Sulfide. *1,1'-Thiobisbutane;* butylthiobutane; dibutyl sulfide. $C_8H_{18}S$; mol wt 146.29. C 65.68%, H 12.40%, S 21.92%. $(CH_3CH_2CH_2CH_2)_2S$. Prepd by refluxing a soln contg sodium sulfide and sodium *n*-butyl sulfate for several hrs: Gray, Gutekunst, *J. Am. Chem. Soc.* **42**, 856 (1920).

Liquid. mp −79.7°. d_0^{16} 0.839. d_4^0 0.852. bp 182°. Insol in water. Very sol in alcohol, ether.

1591. Butyraldehyde. *Butanal.* C_4H_8O; mol wt 72.10. C 66.62%, H 11.18%, O 22.19%. $CH_3CH_2CH_2CHO$. Prepn from butyryl chloride: Brown, Tsukamoto, *J. Am. Chem. Soc.* **83**, 2016 (1961); by reduction of corresponding nitrile: Gaiffe, Pallaud, *Compt. Rend.* **254**, 496 (1962); by alkali aluminum hydride reduction of methyl butyrate: Zakharkin *et al., Tetrahedron Letters* **1963**, 208; by oxidation of butanol: Harrison, *Proc. Chem. Soc.* **1964**, 110. Usually manuf by catalytic dehydrogenation of butanol, catalytic hydrogenation of crotonaldehyde, or by the oxo process from propene, *e.g., see:* Dunbar, Arnold, *J. Org. Chem.* **10**, 501 (1945); Horn, *Ind. Eng. Chem.* **51**, 655 (1959); W. L. Faith *et al., Industrial Chemicals* (John Wiley, New York, 3rd ed., 1965) pp 183, 304-305. *Review:* P. D. Sherman in Kirk-Othmer *Encyclopedia of Chemical Technology* vol. 4 (Wiley-Interscience, New York, 3rd ed., 1978) pp 376-386.

Flammable liquid. bp 74.8°. mp −99°. Closed-cup flash pt −6.67° (20°F). d_4^{20} 0.8016. n_D^{20} 1.379. Soly in water at 25°, 7.1 (wt-%): Smith, Bonner, *Ind. Eng. Chem.* **43**, 1169 (1951). Miscible with ethanol, ether, ethyl acetate, acetone, toluene, many other organic solvents and oils. Single-dose LD_{50} orally in rats: 5.89 g/kg, Smyth *et al., Arch. Ind. Hyg. Occup. Med.* **4**, 119 (1951). USE: Chiefly in the manuf of rubber accelerators, synthetic resins, solvents, plasticizers. *Caution:* May act as irritant, narcotic.

1592. *n*-Butyramide. *Butanamide.* C_4H_9NO; mol wt 87.12. C 55.14%, H 10.41%, N 16.08%, O 18.36%. $CH_3CH_2CH_2CONH_2$. Prepn by a modified Willgerodt reaction: Jelinek, U.S. pats. 2,572,809/10 (both 1951 to GAF); from butyryl chloride: Philbrook, *J. Org. Chem.* **19**, 623 (1954); from butyraldoxime: Huber, U.S. pat. 2,721,199 (1955 to du Pont); from butyric acid: Rahman, *Rec. Trav. Chim.* **79**, 188 (1960); from butyronitrile: Gilbert, Rumanowski, U.S. pat. 3,062,883 (1962 to Allied Chem.).

Crystals, mp 115-116°. bp 216°. Sol in water, alcohol; slightly sol in ether.

1593. Butyric Acid. *Butanoic acid; n-butyric acid; eth-ylacetic acid.* $C_4H_8O_2$; mol wt 88.10. C 54.53%, H 9.15%, O 36.32%. $CH_3CH_2CH_2COOH$. Discovered by Lieben and Rossi in 1869. Present in butter as an ester to the extent of

4-5%. Obtained by suitable fermentation of carbohydrates; prepn from n-propanol + CO at 200 atm in the presence of $Ni(CO)_4$ and NiI_2: Reppe *et al.*, *Ann.* **582**, 83 (1953); lab prepn from ethylmalonic acid: Gattermann-Wieland, *Praxis des Organischen Chemikers* (de Gruyter, Berlin, 40th ed., 1961) p 221.

Oily liq; unpleasant, rancid odor. d_4^{20} 0.959. bp 163.5°. mp $-7.9°$. n_D^{20} 1.3991. Flash pt, closed cup: 170° F (77° C). Neutralization value 636.79. Miscible with water, alcohol, ether. The calcium salt of this acid is less soluble in hot than in cold water (difference from isobutyric acid). LD_{50} orally in rats: 8.79 g/kg, Smyth *et al.*, *Arch. Ind. Hyg. Occup. Med.* **10**, 61 (1954).

Magnesium salt, $C_8H_{14}MgO_4$, *magnesium butyrate*. Deliquesc leaflets. Sol in water. *Keep well closed.*

USE: Manuf of esters, some of which serve as bases of artificial flavoring ingredients of certain liqueurs, soda-water syrups, candies; also for varnishes; as decalcifier of hides.

1594. Butyric Anhydride. *Butanoic acid anhydride;* butyryl oxide. $C_8H_{14}O_3$; mol wt 158.19. C 60.74%, H 8.92%, O 30.34%. $CH_3CH_2CH_2COOCOCH_2CH_2CH_3$. Prepn from butyric acid: Williams, Krynitsky, *Org. Syn.* **coll. vol. III**, 165 (1955); Kuwajima, Mukaiyama, *J. Org. Chem.* **29**, 1385 (1964). Manuf by catalytic hydrogenation of crotonic acid: Smith, Hunter, U.S. pat. **2,492,403** (1949 to Celanese); by catalytic carbonylation of the corresponding acid ester: Reppe, Friederich, U.S. pat. **2,730,546** (1956 to BASF).

Liq, bp 199.4-201.4°. mp $-75°$. d_4^{20} 0.9668. n_D^{20} 1.4070. Flash pt, open cup: 190° F (88° C). Sol in water and in alc with decompn; sol in ether.

1595. Butyroin. *5-Hydroxy-4-octanone;* 5-octanol-4-one. $C_8H_{16}O_2$; mol wt 144.21. C 66.62%, H 11.18%, O 22.19%. $CH_3(CH_2)_2CH(OH)CO(CH_2)_2CH_3$. Prepd by the reaction between an ether soln of ethyl butyrate and sodium or potassium: Bouveault, Locquin, *Bull. Soc. Chim.* [3] **35**, 629 (1906); Corson *et al.*, *J. Am. Chem. Soc.* **52**, 3988 (1930); Scheibler, Emden, *Ann.* **434**, 265 (1923); Snell, McElvain, *Org. Syn.* **coll. vol. II**, 114 (1943).

Liquid. d_4^0 0.9367; $d_4^{16.7}$ 0.91075. bp_{760} 180-190°; bp_{155} 150-154°; bp_{20} 95°; bp_{10} 85°. $n_4^{16.7}$ 1.43455. Reduces Fehling's soln.

Oxime, $C_8H_{17}NO_2$, viscous liquid, bp_{10} 143°.

1596. Butyrolactone. *Dihydro-2(3H)-furanone;* γ-butyrolactone; 1,2-butanolide; 1,4-butanolide; γ-hydroxybutyric acid lactone; 3-hydroxybutyric acid lactone; 4-hydroxybutanoic acid lactone. $C_4H_6O_2$; mol wt 86.09. C 55.80%, H 7.03%, O 37.17%. Prepd from acetylene and formaldehyde: Reppe, *Chem. Ing. Technik* **1950**, 365; *Chem. Eng.* **58**, no. 6, 176 (1951); also prepd from ethylene chlorohydrin, glutaric, ν-hydroxybutyric acid solns, tetrahydrofuran, or vinylacetic acid: F. C. Whitmore, *Organic Chemistry* (Van Nostrand, New York, 2nd ed., 1951). Alternate synthesis: Y. Ogata *et al.*, *J. Org. Chem.* **45**, 1320 (1980). Physical properties: McKinley, Copes *J. Am. Chem. Soc.* **72**, 5331 (1950). Toxicity: H. F. Smyth *et al.*, *Am. Ind. Hyg. Assoc. J.* **30**, 470 (1969).

Oily liquid. d_0^0 1.1441; d_0^{15} 1.1286. mp $-43.53°$. bp_{760} 204°. bp_{12} 89°. n_D^{25} 1.4348. Flash pt, open cup: 209°F (98°C). Volatile with steam. Misc with water. Sol in methanol, ethanol, acetone, ether, benzene. Hydrolyzed by hot alkaline solns. LD_{50} orally in rats: 17.2 ml/kg (Smyth).

USE: Intermediate in the synthesis of polyvinylpyrrolidone, DL-methionine, piperidine, phenylbutyric acid, thiobutyric acids. Solvent for polyacrylonitrile, cellulose acetate, methyl methacrylate polymers, polystyrene. Constituent of paint removers, textile aids, drilling oils.

1597. Butyronitrile. *Butanenitrile;* propyl cyanide; butyric acid nitrile. C_4H_7N; mol wt 69.10. C 69.52%, H 10.21%, N 20.27%. $CH_3CH_2CH_2CN$. Prepd from 1-butanol by controlled cyanation with NH_3 at 300° in the presence of $Ni-Al_2O_3$ catalysts: Popov, Shuikin, *Izvest. Akad. Nauk SSSR, Otdel. Khim. Nauk* **1958**, 713-718, *C.A.* **52**, 19924 (1958). Toxicity: H. F. Smyth *et al.*, *Am. Ind. Hyg. Assoc. J.* **23**, 95 (1962).

Liquid. d_4^0 0.8091; d_4^{15} 0.7954; d_4^{30} 0.7817. mp $-112°$. bp_{760} 117.5°; bp_{400} 96.8°; bp_{200} 76.7°; bp_{100} 59.0°; bp_{60} 47.3°; bp_{40} 38.4°; bp_{20} 25.7°; bp_{10} 13.4°; bp_5 +2.1°; $bp_{1.0}$ $-20.0°$. n_D^{20} 1.38385. Flash pt, open cup: 85°F (29°C). Viscosity ($\eta \times 10^5$) at 15° = 624; at 30° = 515. Dipole moment 3.5. Sparingly sol in water. Misc with alcohol, ether, dimethylformamide. LD_{50} orally in rats: 0.14 g/kg (Smyth).

Caution: Highly toxic.

1598. n-Butyryl Chloride. *Butanoyl chloride; butyryl chloride.* C_4H_7ClO; mol wt 106.55. C 45.09%, H 6.62%, Cl 33.28%, O 15.02%. $CH_3CH_2CH_2COCl$. Prepn from butyric acid: Helferich, Schaefer, *Org. Syn.* **coll. vol. I**, 147 (1941).

Liquid, bp 101-102°. mp $-89°$. $d_4^{20.6}$ 1.0263. n_D^{20} 1.412. Dissolves slowly with decompn in water, alcohol. Miscible with ether.

1599. Buzepide. *Hexahydro-α,α-diphenyl-1H-azepine-1-butanamide;* 2,2-diphenyl-4-hexamethyleneiminobutyramide; R 658. $C_{22}H_{28}N_2O$; mol wt 336.46. C 78.53%, H 8.39%, N 8.33%, O 4.76%. Prepn: Janssen *et al.*, *J. Med. Pharm. Chem.* **1**, 187 (1959); of methiodide as well as free base: Janssen, de Jongh, U.S. pat. **2,881,165** (1959 to N.V. Nederlandsche Combinatie Chem. Ind.).

Crystals from isopropanol, mp 141.5-143.5°.

Methiodide, $C_{23}H_{31}IN_2O$, *metazepium iodide, 1-(3-carbamoyl-3,3-diphenylpropyl)hexahydro-1-methylazepinium iodide, N-(3,3-diphenyl-3-carbamoylpropyl)-N-methylperhydroazepinium iodide, Spactin.* Crystals, dec 212-213°.

THERAP CAT: Anticholinergic.

C

1600. Cabenegrins. Orally active antidotes against snake venoms; isolated from the root of a South American plant called "Cabeca de Negra" and structurally related to pterocarpin, *q.v.* Isoln: L. L. Darko *et al., Eur.* pat. **Appl.** **89,229;** *eidem,* **U.S.** pat. **4,429,141** (1983, 1984 both to Richter, Budapest). Structure determn: M. Nakagawa *et al., Tetrahedron Letters* **1982,** 3855. Synthesis of (±)-cabenegrins A-I and A-II: M. Ishiguro *et al., ibid.* 3859.

cabenegrin A-I   cabenegrin A-II

Cabenegrin A-I, $C_{21}H_{20}O_6$, *[6aR-[4(E),6aα,12aα]]-6a,12a-dihydro-4-(4-hydroxy-3-methyl-2-butenyl)-6H-[1,3]dioxolo-[5,6]benzofuro[3,2-c][1]benzopyran-3-ol.* White crystalline solid, mp 167-168°. uv max (ethanol): 309 nm (ε 13000); uv max (methanol): 209, 233, 309 nm (ε 75000, 24000, 13000).
Cabenegrin A-II, $C_{21}H_{22}O_6$, *6a,12a-dihydro-3-hydroxy-β-methyl-6H-[1,3]dioxolo[5,6]benzofuro[3,2-c][1]benzopyran-2-butanol.* uv max (methanol): 204, 230, 292, 308 nm (ε 116000, 8000, 9400, 11800).

1601. Cacao Shell. Cocoa shells. Shells of the seed of *Theobroma cacao* L., *Sterculiaceae. Habit.* Brazil, Central America, Mexico, West Indies, and most tropical countries. *Constit.* Theobromine, caffeine, cacao red, protein, pentosans, pectic acid, starch. *Ref:* Dittmar, *Engenharia e quim* **5,** no. 1, 1 (1953), *C.A.* **48,** 2949e (1954). *Monographs:* E. M. Chatt, *Cocoa* (Interscience, New York, 1953); D. H. Urquhart, *Cocoa* (Longmans, London, 1961); Powell, Harris, *Chocolate and Cocoa* in Kirk-Othmer *Encyclopedia of Chemical Tehnology,* **vol. 5** (Interscience, New York, 2nd ed., 1964) pp 363-402.
Thin, papery, reddish-brown, concavo-convex shells; weak chocolate-like odor and taste.
USE: In the manuf of caffeine, theobromine. *Caution:* Occasionally causes allergic dermatitis from handling.

1602. Cacodyl. *Tetramethyldiarsine;* dicacodyl. C_4H_{12}-As_2; mol wt 209.96. C 22.88%, H 5.76%, As 71.36%. Prepn from cacodyl chloride by heating with zinc in an atm of carbon dioxide: Bunsen, *Ann.* **42,** 14 (1842); from dimethylarsine by the action of oxides of nitrogen, aqueous chromic acid, lead peroxide, cacodyl chloride or potassium ferricyanide: Dehn, Wilcox, *Am. Chem. J.* **35,** 1 (1906); by the action of free methyl on arsenic: Paneth, Loleit, *J. Chem. Soc.* **1935,** 366; by reduction of cacodyl oxide: Witten, **U.S.** pat. **2,531,487** (1950 to U.S.A.); Fuson, Shive, **U.S.** pat. **2,756,-245** (1956 to U.S.A.).

Oily liquid. Solidifies to large quadratic plates, mp −6°. Almost intolerable garlicky odor. Inflames spontaneously in dry air. bp_{760} 165°. Slightly sol in water. Controlled oxidation with moist air yields cacodyl oxide and cacodylic acid. Reduction with tin and HCl yields *Erytrarsin* $(CH_3As)_4As_2$-O_3.

1603. Cacodylic Acid. *Dimethylarsinic acid; hydroxydimethylarsine oxide;* Phytar. $C_2H_7AsO_2$; mol wt 137.99. C 17.41%, H 5.11%, As 54.29%, O 23.19%. $(CH_3)_2As(O)OH$. Prepn: Guinot, *J. Pharm. Chim.* **27,** 55 (1923), *C.A.* **17,**

2103[8] (1923); Inverni, *Boll. Chim. Farm.* **62,** 129 (1923), *C.A.* **17,** 2413[7] (1923); Challenger, Ellis, *J. Chem. Soc.* **1935,** 396; Fioretti, Portelli, *Ann. Chim. (Rome)* **53,** 1869 (1963), *C.A.* **60,** 10712c (1964). As herbicide: Sprague, **U.S.** pat. **3,056,-668** (1962 to Ansul). Toxicity: G. W. Bailey, J. L. White, *Residue Rev.* **10,** 97 (1965).
Crystals from alcohol + ether. mp 195-196°. Hygroscopic. *Poisonous!* Sol in 0.5 part water; very sol in alcohol; sol in acetic acid; practically insol in ether. *Keep well closed.* LD_{50} orally in rats: 1350 mg/kg (Bailey, White).
Mercury salt, $C_4H_{12}As_2HgO_4$, *mercuric cacodylate.* Hygroscopic, somewhat unstable cryst powder. *Poisonous!* Soluble in water, alcohol; practically insol in ether. *Keep well closed.*
USE: Herbicide.
THERAP CAT: Dermatologic.
THERAP CAT (VET): Has been used in chronic eczema, anemia and as a tonic.

1604. Cacotheline. *2,3-Dihydro-4-nitro-2,3-dioxo-9,10-secostrychnidin-10-oic acid.* $C_{21}H_{21}N_3O_7$; mol wt 427.40. C 59.01%, H 4.95%, N 9.83%, O 26.20%. A nitro derivative of brucine made by treating brucine with 10% nitric acid at 60-70°: Leuchs *et al., Ber.* **43,** 1042 (1910). Structure: Teuber, *ibid.* **86,** 232 (1953). Used as reversible redox indicator for Sn^{+2} titrations: P. Szarvas, J. Lantos, *Talanta* **10,** 477 (1963).

Yellow crystals; sparingly sol in water.
USE: Indicator.

1605. Cactinomycin. *Actinomycin C;* HBF 386; Sanamycin. Antibiotic complex produced by *Streptomyces chrysomallus:* Brockmann, Grubhofer, *Naturwiss.* **36,** 376 (1949); **37,** 494 (1950); *Ber.* **84,** 260 (1951); **U.S.** pat. **2,953,495** (1960 to Shenley Inds.); Lindenbein, *Arch. Mikrobiol.* **17,** 361 (1952). Mixture of actinomycins C_1 (dactinomycin *q.v.*), C_2 and C_3, 10%, 45% and 45%, resp: Brockmann, Pfennig, *Naturwiss.* **39,** 429 (1952); Brockmann, Gröne, *ibid.* **40,** 222 (1953). Description of other actinomycins: Waksman *et al., Proc. Nat. Acad. Sci. U.S.A.* **44,** 602 (1958). Structures: Brockmann *et al., Angew. Chem.* **68,** 70 (1956); Brockmann, Boldt, *Naturwiss.* **50,** 19 (1963). Synthesis of actinomycin C_3: Brockmann, Lackner, *ibid.* **47,** 230 (1960); **48,** 555 (1961); **51,** 407 (1964); Brockmann *et al., Ger.* pat. **1,172,680** (1964 to Bayer); Brockmann, Lackner, *Ber.* **100,** 353 (1967); **101,** 1312 (1968). Synthesis of actinomycin C_2: Brockmann, Lackner, *Tetrahedron Letters* **1964,** 3517. Comprehensive review: H. Brockmann, in *Fortschr. Chem. Org. Naturst.* **18,** 1-54 (1960).

Alizarin-red hexagonal bipyramids from ethyl acetate, mp 252°. $[α]_D^{25}$ −325 to −349° (c = 0.25 in ethanol). Sparingly sol in water; moderately sol in ethanol; sol in chloroform, ethyl acetate, benzene, acetone. *Protect from light.*

Actinomycin C₂. $C_{63}H_{88}N_{12}O_{16}$. R = D-valine; R' = D-alloisoleucine. Red bipyramids, prisms or needles from ethyl acetate, mp 237-239°. $[\alpha]_D^{21}$ −325° ± 10° (c = 0.23 in methanol). Abs max (methanol): 443 nm (ε 25,400).

Actinomycin C₃. $C_{64}H_{90}N_{12}O_{16}$. R = R' = D-alloisoleucine. Red hexagonal bipyramids from ethyl acetate or methanol, dec 235°. $[\alpha]_D^{17}$ −328° (c = 0.5 in ethanol). Absorption max (methanol): 443 nm (ε 24,100). Weak base.

THERAP CAT: Antineoplastic.

1606. Cactus Grandiflorus. Night-blooming cereus; large-flowered cereus. Fresh, succulent stems of *Selenicereus grandiflorus* (L.) Britt. and Rose, *Cactaceae. Habit.* Tropical America. *Constit.* Cactine(?), acrid resinous glucoside, fat, wax. Extract containing combined principles from freshly tinctured stems and petals is known as *cactoid.*

THERAP CAT: Cardiotonic.

THERAP CAT (VET): Has been used as circulatory stimulant.

1607. Cadalene. *1,6-Dimethyl-4-(1-methylethyl)naphthalene; 4-isopropyl-1,6-dimethylnaphthalene;* cadalin. $C_{15}H_{18}$; mol wt 198.29. C 90.85%, H 9.15%. Obtained from cadinene and other sesquiterpenes or sesquiterpene alcohols by dehydrogenation: Ruzicka, Meyer, *Helv. Chim. Acta* **4**, 505 (1921); Ruzicka, Stoll, *ibid.* **7**, 84 (1924). Synthesis: Ruzicka, Seidel, *ibid.* **5**, 369 (1922); Barnett, Cook, *J. Chem. Soc.* **1933**, 22; Johnson, Jones, *J. Am. Chem. Soc.* **69**, 792 (1947); Kohli *et al., Experientia* **28**, 131 (1972).

Liquid. d_4^{25} 0.9667. bp_{720} 291-292°; bp_{10} 149°. n_D^{25} 1.5785. uv max: 228, 232, 280, 284, 295, 310, 317, 325 nm. Insol in water; sol in fat solvents, oils.

1608. Cadaverine. *1,5-Pentanediamine;* pentamethylenediamine; animal coniine. $C_5H_{14}N_2$; mol wt 102.18. C 58.77%, H 13.81%, N 27.42%. $NH_2(CH_2)_5NH_2$. Biogenic polyamine and homolog of putrescine, q.v., produced by decarboxylation of lysine. Formed through action of comma bacillus on meat, fish, albumin, etc. Found in cholera discharge. Isoln: Bocklisch, *Ber.* **18**, 1922 (1885); Ackermann, *Z. Physiol. Chem.* **54**, 16 (1907). Prepn: Ladenburg, *Ber.* **18**, 2956 (1885). GC determn in foods: W. F. Staruszkiewicz, J. F. Bond, *J. Assoc. Offic. Anal. Chem.* **64**, 584 (1981). Metabolism study of ¹⁴C-cadaverine in rat brain: S. K. Salzman, M. Stepita-Klauco, *J. Neurochem.* **37**, 1308 (1981). Biosynthetic study in *S. ruminantium:* Y. Kamio *et al., J. Biol. Chem.* **257**, 3326 (1982).

Colorless, syrupy liquid; characteristic odor. Strong base, fumes and attracts CO_2 on exposure to air. pKa_1 10.25; pKa_2 9.13. *Poisonous!* d_4^{25} 0.873. mp 9°. bp 178-180°. n_D^{20} 1.463. Sol in water, alcohol; slightly sol in ether. *Keep well closed.*

Dihydrochloride, $C_5H_{14}N_2 \cdot 2HCl$, needles from water, mp 225-230°. Sol in water. Practically insol in abs alc.

Caution: Skin irritant and possible sensitizer.

1609. Cadexomer Iodine. Iodosorb. A hydrophilic modified starch polymer containing 0.9% (w/w) iodine within a helical matrix. Produced by the reaction of dextrin with epichlorohydrin coupled with ion exchange groups and iodine. Clinical use in venous ulcers: E. Skog *et al., Brit. J. Dermatol.* **109**, 77 (1983); M. C. Ormiston *et al., Brit. Med. J.* **291**, 308 (1985); L. Hillström, *Acta Chir. Scand. Suppl.* **544**, 53 (1988).

THERAP CAT: Vulnerary.

1610. Cadinenes. $C_{15}H_{24}$; mol wt 204.34. C 88.16%, H 11.84%. Sesquiterpenes occurring in essential oils from Juniper species and cedars (oil of cade). Nine possible isomers differing in stereochemistry and position of the double bonds, the principal isomer being β-cadinene: Sykora *et al., Chem. Listy* **52**, 1314 (1958). Prepn and structure: Campbell, Soffer, *J. Am. Chem. Soc.* **64**, 417 (1942); Campbell *et* al., *ibid.* 425; Rao *et al., Tetrahedron Letters* **1960**, 27; Herout *et al., Coll. Czech. Chem. Commun.* **31**, 3012 (1966). Synthesis: Soffer, Günay, *Tetrahedron Letters* **1965**, 1355. Of the nine possible isomers, all able to yield (−)-cadinene dihydrochloride, six are known: Herout, Sykora, *Tetrahedron* **4**, 246 (1958); Kartha *et al., ibid.* **19**, 241 (1963). Structure and configuration of isomers: Sykora *et al., Coll. Czech. Chem. Commun.* **23**, 2181 (1958). Isoln of α-*cadinene* from Japanese hop: Y. Naya, M. Kotake, *Bull. Chem. Soc. Japan* **42**, 1468 (1969). Synthesis: O. P. Vig, *Indian J. Chem.* **21B**, 145 (1982). Brief review in *Rodd's Chemistry of Carbon Compounds,* vol. 2, pt. C, S. Coffey, Ed. (Elsevier, New York, 1969) pp 268-270.

β-cadinene

β-Cadinene, oil. Slight pleasant odor. bp_9 124°. d_4^{20} 0.9239. n_D^{20} 1.5059. $[\alpha]_D^{20}$ −251°.

1611. Cadmium. Cd; at. wt 112.41; at. no. 48; valence 2. Group 2b element. Abundance in earth's crust: 0.1 to 0.2 ppm. Natural isotopes: 114 (28.86%); 112 (24.07%); 111 (12.75%); 110 (12.39%); 113 (12.26%); 116 (7.58%); 106 (1.22%); 108 (0.88%); known isotopes range in mass number from 103 to 121. Found in zinc ores; also as CdS, greenockite; $CdCO_3$, otavite. Obtained in vapor form when roasting zinc ores, as sludge from zinc sulfate purification. Lab prepns from $CdSO_4$: Treadwell, *Helv. Chim. Acta* **4**, 551 (1921). *Review:* Aylett "Group IIB" in *Comprehensive Inorganic Chemistry,* vol. 3, J. C. Bailar Jr. *et al.,* Eds. (Pergamon Press, Oxford, 1973) pp 187-328; M. L. Hollander, S. C. Carapella in Kirk-Othmer *Encyclopedia of Chemical Technology* vol. 4 (Wiley-Interscience, New York, 3rd ed., 1978) pp 387-396. Review of carcinogenicity studies: *IARC Monographs* **11**, 39-74 (1976).

Silver-white, blue-tinged, lustrous metal; distorted hexagonal close-packed structure; easily cut with a knife; available in the form of bars, sheets or wire or a gray, granular powder. mp 321°. bp 765°. d^{25} 8.65. Specific heat at constant pressure (25°) 6.22 cal/mole deg. Slowly oxidized by moist air to form CdO. E° (aq) Cd/Cd^{2+} 0.4025 V. Insol in water; reacts readily with dil HNO_3; reacts slowly with hot HCl; does not react with alkalies. Other reactions similar to those of zinc. Solns of cadmium salts and H_2S or Na_2S yield a yellow ppt insol in excess Na_2S. Cadmium and its salts are highly toxic.

Caution: Ingestion of metal and sol compounds causes increased salivation, choking, vomiting, abdominal pain, anemia, renal dysfunction, diarrhea, tenesmus. *Inhalation* (dust or fumes): throat dryness, cough, headache, vomiting, chest pain, extreme restlessness and irritability, pneumonitis, possibly bronchopneumonia. Implicated as causative agent in Itai-Itai disease in Japan. *See* E. Browning, *Toxicity of Industrial Metals* (Appleton-Century-Crofts, New York, 2nd ed., 1969) pp 98-108; Flick *et al., Environ. Res.* **4**, 71-85 (1971); Fassett, *Ann. Rev. Pharmacol.* **15**, 425-435 (1975). This substance and certain cadmium compods may reasonably be anticipated to be carcinogens: *Fourth Annual Report on Carcinogens* (NTP 85-002, 1985) p 48.

USE: A constituent of easily fusible alloys, e.g., Lichtenberg's, Abel's, Lipowitz', Newton's, and Wood's metal; soft solder and solder for aluminum; electroplating (major use), deoxidizer in Ni plating; process engraving, electrodes for cadmium vapor lamps, photoelectric cells; photometry of ultraviolet sun-rays; in Ni-Cd storage batteries. The powder is also used as an amalgam (1Cd:4Hg) in dentistry. To charge Jones reductors.

THERAP CAT (VET): Many cadmium salts, especially the oxide and anthranilate, are used or have been suggested as anthelmintics in swine and poultry.

1612. Cadmium Acetate. $C_4H_6CdO_4$; mol wt 230.49. C

20.84%, H 2.62%, Cd 48.77%, O 27.77%. $Cd(CH_3COO)_2$. Prepd by treating cadmium nitrate with acetic anhydride: Späth, *Monatsh.* **33**, 241 (1912); Wagenknecht, Juza, in *Handbook of Preparative Inorganic Chemistry,* vol. **2**, G. Brauer, Ed. (Academic Press, New York, 2nd ed., 1965) p 1105.

Dihydrate, crystals. Slight acetic acid odor; becomes anhydr at about 130°. d 2.01. d (anhydr) 2.341. mp (anhydr) 255°. Freely sol in water; sol in alcohol. pH of 0.2 molar aq soln 7.1.

USE: Producing iridescent effects on porcelains and pottery; as a reagent for determination of S, Se, and Te; in cadmium electroplating.

1613. Cadmium Bromide. Br_2Cd; mol wt 272.24. Br 58.71%, Cd 41.29%. $CdBr_2$. Usually prepd from the elements: Honigschmid, Schlee, *Z. Anorg. Allgem. Chem.* **227**, 184 (1936); Wagenknecht, Juza, in *Handbook of Inorganic Chemistry,* vol. **2**, G. Brauer, Ed. (Academic Press, New York, 2nd ed., 1965) p 1096.

Hexagonal, pearly flakes; highly hygroscopic. mp 566°; bp 963°; d 5.192. Freely sol in water, alc; moderately sol in acetone; slightly sol in ether. Crystallizes as the monohydrate below 36°, as the tetrahydrate above 36°.

USE: In photography, process engraving, and lithography.

1614. Cadmium Carbonate. $CCdO_3$; mol wt 172.42. C 6.97%, Cd 65.20%, O 27.84%. $CdCO_3$. Occurs in nature as the mineral, *otavite.* Prepn: de Schulten, *Bull. Soc. Chim. France* [3] **19**, 34 (1898); Biltz, *Z. Anorg. Allgem. Chem.* **220**, 312 (1934).

Powder or rhombohedral leaflets. d 4.26. Practically insol in water; sol in concd solns of ammonium salts, dil acids.

1615. Cadmium Chloride. Caddy; Vi-Cad. $CdCl_2$; mol wt 183.32. Cd 61.32%, Cl 38.68%. Prepn: *Gmelin's, Cadmium* (8th ed.) **33**, pp 82-83 (1925); supplement, pp 464-465 (1959); Pray, *Inorg. Syn.* **5**, 153 (1957).

Hygroscopic, rhombohedral crystals. mp 568°; bp 960°; d 4.05. Freely sol in water; sol in acetone; slightly sol in methanol, ethanol; practically insol in ether. LD_{50} orally in rats: 88 mg/kg, *Toxic Substances List,* H. E. Christensen *et al.,* Eds. (1974).

Hemipentahydrate, efflorescent granules or rhombohedral leaflets. d^{25} 3.33. Freely sol in water.

USE: Photography; dyeing and calico printing; in the vacuum tube industry; manuf of cadmium yellow; galvanoplasty; manuf of special mirrors; as ice-nucleating agent; as lubricant; in analysis of sulfides to absorb the H_2S; in testing for pyridine bases; as fungicide.

1616. Cadmium Cyanide. C_2CdN_2; mol wt 164.45. C 14.61%, Cd 68.36%, N 17.04%. $Cd(CN)_2$. Prepn from $Cd(OH)_2$ and HCN: Biltz, *Z. Anorg. Allgem. Chem.* **170**, 161 (1928); Wagenknecht, Juza in *Handbook of Preparative Inorganic Chemistry,* vol. **2**, G. Brauer, Ed. (Academic Press, New York, 2nd ed., 1965) p 1105.

Crystals or white powder. Turns brown on heating in air. *Very poisonous!* d 2.226. Soly in water (15°) 1.71 g/100 ml. Slightly sol in alc; sol in solns of alkali cyanides or hydroxides; not appreciably attacked by organic acids, but readily dec by dil mineral acids with evolution of hydrogen cyanide. Readily forms complex cyanides.

USE: In copper bright electroplating.

1617. Cadmium Fluoride. CdF_2; mol wt 150.41. Cd 74.74%, F 25.26%. Prepn from $CdCl_2$ and NH_4F: Kurtenacker *et al., Z. Anorg. Chem.* **211**, 89 (1933); from $CdCO_3$ and HF: Kwasnik in *Handbook of Preparative Inorganic Chemistry,* vol. **1**, G. Brauer, Ed. (Academic Press, New York, 2nd ed., 1963) p 243.

Cubic crystals. mp 1049; bp 1748°; d 6.33. *Poisonous!* Soly in water (25°) 4.3 g/100 ml. Sol in HF and other mineral acids; practically insol in alcohol and liquid ammonia.

USE: Manufacture of phosphors, glass; in nuclear reactor controls.

1618. Cadmium Hydroxide. Cadmium hydrate. CdH_2O_2; mol wt 146.43. Cd 76.77%, H 1.38%, O 21.85%. $Cd(OH)_2$. Usually prepd from a Cd salt by treatment with KOH: de Schulten, *Compt. Rend.* **101**, 72 (1885); Fricke,

Blaschke, *Z. Elektrochem.* **46**, 46 (1940); Scholder, Staufenbiel, *Z. Anorg. Allgem. Chem.* **247**, 271 (1941).

Powder or trigonal and hexagonal crystals. d 4.79. Dehydration starts at 130° and is complete at 200°. Absorbs CO_2 from the air. Practically insol in water; slightly sol in NaOH soln; sol in dil acids, in NH_4OH or NH_4Cl solns.

USE: In storage battery electrodes.

1619. Cadmium Iodide. CdI_2; mol wt 366.23. Cd 30.69%, I 69.31%. Prepd from the elements or from the sulfate and KI: Wagenknecht, Juza in *Handbook of Preparative Inorganic Chemistry,* vol. **2**, G. Brauer, Ed. (Academic Press, New York, 2nd ed., 1965) p 1096.

Hexagonal, lustrous, flake-like crystals; becomes yellow on long exposure to air and light. mp 388°; bp 787°; d 5.67. Sol in water, alc, ether, acetone.

USE: In electrodeposition of Cd; as nematocide; in manuf of phosphors; as lubricant; in photoconductors; in photography, process engraving; lithography; in analytical chemistry.

1620. Cadmium Nitrate. CdN_2O_6; mol wt 236.43. Cd 47.55%, N 11.85%, O 40.60%. $Cd(NO_3)_2$. Prepn: *Gmelin's Cadmium* (8th ed.) **33**, pp 76-78 (1925); suppl. p 446 (1959).

Tetrahydrate, hygroscopic, orthorhombic crystals. mp 59.5°. Sol in 0.6 part water; sol in alc, acetone, ethyl acetate; practically insol in concd HNO_3. *Keep well closed in a cool place.*

USE: In making other Cd salts; in photographic emulsions.

1621. Cadmium Oxide. CdO; mol wt 128.41. Cd 87.54%, O 12.46%. Prepn: *Gmelin's, Cadmium* (8th ed.) **33**, pp 69-70 (1925); supplement, pp 419-420 (1959).

Dark-brown, infusible powder or cubic crystals. d 8.15. Practically insol in water; sol in dil acids; slowly sol in ammonium salts. LC_{50} in rats, monkeys: 500, approx 15,000 mg/m^3, Barrett *et al., J. Ind. Hyg. Toxicol.* **29**, 279 (1947).

Active ingredient of *Aska-Rid,* a swine anthelmintic.

USE: In phosphors, semiconductors; manuf of silver alloys, glass; in storage battery electrodes; as nematocide; as catalyst for organic reactions, in cadmium electroplating; in ceramic glazes. *Caution:* Inhalation of fumes produces metal fume fever which may end in fatal pulmonary edema.

THERAP CAT (VET): As an ascaricide in swine.

1622. Cadmium Potassium Cyanide. *Potassium tetracyanocadmate.* $C_4CdK_2N_4$; mol wt 294.68. C 16.30%, Cd 38.15%, K 26.54%, N 19.01%. $K_2Cd(CN)_4$. Prepn: Dickinson, *J. Am. Chem. Soc.* **44**, 774 (1922); Biltz, *Z. Anorg. Allgem. Chem.* **170**, 161 (1928).

Highly refractive, cubic crystals. d 1.846. mp about 450°. When heated, melts to a colorless liquid solidifying to a gray, cryst mass on cooling. Sol in 3 parts cold, 1 part boiling water; slightly sol in alc.

1623. Cadmium Salicylate. $C_{14}H_{10}CdO_6$; mol wt 386.63. C 43.49%, H 2.61%, Cd 29.07%, O 24.83%. $Cd(C_7H_5O_3)_2$. Prepn: Prasad *et al., J. Indian Chem. Soc.* **35**, 267 (1958). Monohydrate, small needles or plates. mp 242° (dec). Slightly sol in cold water, freely sol in boiling water; very slightly sol in methanol, ethanol.

THERAP CAT: Antiseptic.

1624. Cadmium Selenate. CdO_4Se; mol wt 255.37. Cd 44.02%, O 25.06%, Se 30.92%. $CdSeO_4$. Prepn: *Gmelin's, Cadmium* (8th ed.) **33**, p 133 (1925); suppl. p 637 (1959). Dihydrate, orthorhombic crystals. Dec at 100°. d 3.632. Very sol in water.

1625. Cadmium Selenide. CdSe; mol wt 191.37. Cd 58.74%, Se 41.26%. Prepd by heating cadmium in a current of hydrogen selenide and subliming the product in hydrogen at a dull red heat: Margottet, *Recherches sur les Sulfures, les Séléniures, et les Tellures Métalliques,* Paris (1879); by passing hydrogen selenide over heated cadmium chloride and washing with warm water: Grzenkowsky, *Ueber Selenide und Erdalkaliferrocyanide,* (Danzig, 1925); from cadmium sulfate and hydrogen selenide: Wagenknecht, Juza in *Handbook of Preparative Inorganic Chemistry,* vol. **2**, G. Brauer, Ed. (Academic Press, New York, 2nd ed., 1965) p 1099. Prepn of high purity CdSe: Taylor, Conn, U.S. pat. **3,540,-859** (1970 to Merck & Co.).

White to brown cubic or hexagonal crystals; turns red in sunlight. d 5.8; mp 1350°. Dec in air or acids. Practically insol in water.

USE: In photoconductors, semiconductors, photoelectric cells, and rectifiers; in phosphors.

1626. Cadmium Succinate. *Succinic acid cadmium salt;* Cadminate. $C_4H_4CdO_4$; mol wt 228.47. C 21.03%, H 1.76%, Cd 49.20%, O 28.01%. $CdC_4H_4O_4$. Prepd from cadmium carbonate and succinic acid: Schiff, *Ann.* **104**, 325 (1857). Needles or plates. Soly in water at 40°, 0.367 g/100 ml. Practically insol in alc. LD_{50} orally in rats, mice: 660, 312 mg/kg, *Registry of Toxic Effects of Chemical Substances,* H. E. Christensen, T. T. Laginbyhl, Eds. (1975) p 1085.

USE: Plant fungicide.

1627. Cadmium Sulfate. CdO_4S; mol wt 208.47. Cd 53.92%, O 30.70%, S 15.38%. $CdSO_4$. Prepn: *Gmelin's, Cadmium* (8th ed.) **33**, p 121 (1925); suppl. pp 609-610. Hydrate, 8/3 moles water per mole cadmium sulfate, odorless, monoclinic crystals. On heating loses water above 40°, forming monohydrate by 80°. Does not become anhydrous on further heating. d 3.08. Freely sol in water; almost insol in alcohol, ethyl acetate. LD s.c. in dogs: 27 mg/kg, *Toxic Substances List,* H. E. Chistensen *et al.,* Eds. (1974) p 740.

USE: In electrodeposition of Cd, Cu, and Ni; in phosphors; manuf of standard cadmium elements; catalyst in the Marsh test for As, determining H_2S and detecting fumaric acid; as nematocide.

1628. Cadmium Sulfide. C.I. 77199; Capsebon. CdS; mol wt 144.47. Cd 77.81%, S 22.19%. Occurs in nature as the mineral *greenockite.* Prepd from $CdSO_4 + H_2S$: Milligan, *J. Phys. Chem.* **38**, 797 (1934); Frerichs, *Naturwiss.* **33**, 2181 (1946); prepn of single crystals: Grillot, *Compt. Rend.* **230**, 1280 (1950); Czyzak *et al., J. Appl. Phys.* **23**, 932 (1952); Wagenknecht, Juza in *Handbook of Preparative Inorganic Chemistry,* vol. II, G. Brauer, Ed. (Academic Press, New York, 2nd ed., 1965) pp 1098-1099. *See also Colour Index* vol. **4** (3rd ed., 1971) p 4659.

Light-yellow or orange-colored cubic or hexagonal crystals. Cubic structure: d 4.50; hexagonal structure: d 4.82. The light-yellow variety is also known as *Cadmium Yellow* or *Jaune Brilliant.* d 4.82. Sublimes at 980°. Soly in water (18°): 0.13 mg/100 g. Sol in concd or warm dil mineral acids with evolution of H_2S; readily dec and dissolved by moderately hot HNO_3.

USE: As a pigment being fast to light and not affected by H_2S; color for soaps; coloring glass yellow; coloring textiles, paper, rubber; in printing inks, ceramic glazes, fireworks; in phosphors and fluorescent screens; in scintillation counters, semiconductors, photoconductors.

THERAP CAT: Dermatologic.

1629. Cadmium Telluride. CdTe; mol wt 240.02. Cd 46.83%, Te 53.17%. Prepd by fusion of the elements or by the action of H_2Te on $CdCl_2$ solns: Dennis, Anderson, *J. Am. Chem. Soc.* **36**, 887 (1914); Kroger, de Nobel, *J. Electronics* **1**, 190 (1955); Kretschmar, Schilberg, *J. Appl. Phys.* **28**, 865 (1957); from cadmium telluride and a cadmium salt: Nitsche, U.S. pat. **2,767,049** (1956 to du Pont). Prepn of high purity CdTe: Taylor, Conn, U.S. pat. **3,540,859** (1970 to Merck & Co.); prepn of single crystals: Kyle, U.S. pat. **3,519,399** (1970 to Hughes Aircraft).

Brownish-black, cubic crystals by sublimation in hydrogen. d_4^{15} 6.2. mp 1041°. Oxidizes upon prolonged exposure to moist air. Practically insol in water and acids, except nitric, in which it is sol with decompn.

USE: In semiconductor research, in phosphors.

1630. Cadmium Tungstate(VI). CdO_4W; mol wt 360.33. Cd 31.20%, O 17.76%, W 51.04%. $CdWO_4$. Prepn: Karl, *Compt. Rend.* **196**, 1403 (1933); prepn of single crystals: Uitert, Soden, *J. Appl. Phys.* **31**, 328 (1960).

White or yellowish monoclinic crystals or powder. Practically insol in water or dil acids; sol in solns of alkali cyanides.

USE: In x-ray screens; in scintillation counters; in phosphors; as catalyst for organic reactions.

1631. Cadralazine. *2-[6-[Ethyl(2-hydroxypropyl)ami-*

no]-3-pyridazinyl]hydrazinecarboxylic acid ethyl ester; ethyl 6-[ethyl(2-hydroxypropyl)amino]-3-pyridazinecarbazate; 3-(2-carbethoxyhydrazino)-6-[N-(2-hydroxypropyl)ethylamino]pyridazine; DC 826; ISF 2469; Cadral; Cadraten; Cadrilan. $C_{12}H_{21}N_5O_3$; mol wt 283.33. C 50.87%, H 7.47%, N 24.72%, O 16.94%. Peripheral vasodilator similar to hydralazine, *q.v.* Prepn: C. Carpi *et al.,* **Belg.** pat. **811,847;** *eidem,* U.S. pats. **3,925,381; 4,002,753** (1974, 1975, 1977 all to ISF); F. Parravicini *et al., Farmaco Ed. Sci.* **34**, 299 (1979). Analytical profile: L. Citerio *et al., Boll. Chim. Farm.* **120**, 222 (1981). Pharmacology: C. Semeraro *et al., J. Cardiovasc. Pharmacol.* **3**, 455 (1981). HPLC determn in plasma and urine: T. Crolla *et al., J. Chromatog.* **310**, 139 (1984). Hemodynamic effects in dogs: L. Dorigotti *et al., Arzneimittel-Forsch.* **34**, 984 (1984); in humans: B. Persson *et al., Eur. J. Clin. Pharmacol.* **31**, 513 (1987). Pharmacokinetics and metabolism in humans: H. Schütz *et al., Eur. J. Drug Metab. Pharmacokinet.* **10**, 147 (1985); S. A. Hauffe *et al., ibid.* 217. Preliminary clinical evaluations: R. Buoninconti, M. Motolese, *Int. J. Clin. Pharmacol. Ther. Toxicol.* **23**, 613 (1985); A. Salvadeo *et al., Arzneimittel-Forsch.* **35**, 623 (1985).

Crystals from acetone, mp 160-162°. pKa 6.0. uv max: 248, 340 nm (ε 22100, 2250). Soly (mg/ml): water 1.3; HCl 235.0; DMSO 323.0; methanol 21.0; dioxane 18.6; chloroform 8.5; diethyl ether, benzene, cyclohexane <0.1. LD_{50} in rats, dogs (mg/kg): 269, approx 400 i.v.; 2060, >2000 orally (Semeraro); in mice (mg/kg): 700 i.p. (Parravicini).

THERAP CAT: Antihypertensive.

1632. Cafaminol. *3,7-Dihydro-8-[(2-hydroxyethyl)-methylamino]-1,3,7-trimethyl-1H-purine-2,6-dione;* 8-[(2-hydroxyethyl)methylamino]caffeine; 8-(β-oxyethyl)methyl-aminocaffeine; methylcoffanolamine; Rhinoptil; Rhinetten. $C_{11}H_{17}N_5O_3$; mol wt 267.30. C 49.43%, H 6.41%, N 26.20%, O 17.96%. Alkanolamine deriv of caffeine, *q.v.* Prepn: J. Klosa, Ger. pat. **1,085,530** (1958) corresp to U.S. pat. **3,094,531** (1963 to Delmar Chemicals Ltd.). Efficacy studies: E. Szirmai, *Praxis* **13**, 412 (1969); R. Leypoldt, *Therapiewoche* **26**, 3381 (1976). Bioavailability and absorption kinetics in humans: H. Walther, K. Kochler, *Pharmazie* **34**, 375 (1979).

Colorless cryst from ethanol, mp 162-164°. Soly in water is about 6%; pH of aq solns is 6.9. LD_{50} in male mice: 700 mg/kg s.c., J. Klosa, *loc. cit.*

THERAP CAT: Nasal decongestant.

1633. Cafestol. *[3bS-(3bα,5aβ,7β,8β-10aα,10bβ)]-3b,4,-5,6,7,8,9,10,10a,10b,11,12-Dodecahydro-7-hydroxy-10b-methyl-5a,8-methano-5aH-cyclohepta[5,6]naphtho[2,1-b]-furan-7-methanol;* cafesterol; $C_{20}H_{28}O_3$; mol wt 316.42. C 75.91%, H 8.92%, O 15.17%. Diterpenoid constituent of coffee. Isoln from green coffee oil: Slotta, Neisser, *Ber.* **71**, 1991, 2342 (1938); C. Djerassi *et al., J. Org. Chem.* **18**, 1449 (1953). Prepn and purification: R. Bertholet, U.S. pat. **4,692,534** (1987 to Nestec). Structure: C. Djerassi *et al., J. Am. Chem. Soc.* **81**, 2386 (1959); R. A. Finnegan, C. Djerassi, *ibid.* **82**, 4342 (1960). Stereochemical studies: R. A. Finnegan, *J. Org. Chem.* **26**, 3057 (1961); A. I. Scott *et al., J.*

Am. Chem. Soc. **84,** 3197 (1962); A. I. Scott et al., Tetrahedron **20,** 1339 (1964). Stereospecific total synthesis of (±)-form: E. J. Corey et al., J. Am. Chem. Soc. **109,** 4717 (1987).

vol. **15,** K. Florey, Ed. (Academic Press, New York, 1986) pp 71-150.

Crystals from hexane, mp 158-160°. $[\alpha]_D$ −101°. uv max: 222 nm (log ϵ 3.78).

Acetate, $C_{22}H_{30}O_4$, needles from petr ether, mp 167-168°. $[\alpha]_D$ −89°. uv max: 222 nm (log ϵ 3.80).

Tetrahydrocafestol, $C_{20}H_{32}O_3$, crystals from dil methanol, mp 154.5-157°.

1634. Caffeic Acid. *3-(3,4-Dihydroxyphenyl)-2-propenoic acid; 3,4-dihydroxycinnamic acid.* $C_9H_8O_4$; mol wt 180.15. C 60.00%, H 4.48%, O 35.52%. Constituent of plants, probably occurs in plants only in conjugated forms, e.g., chlorogenic acid. Isoln from green coffee: Wolfrom et al., J. Agr. Food Chem. **8,** 58 (1960); from roasted coffee: Krasemann, Arch. Pharm. **293,** 721 (1960). Formation by acid hydrolysis of chlorogenic acid: Fiedler, Arzneimittel-Forsch. **4,** 41 (1954); Whiting, Carr, Nature **180,** 1479 (1957); Guern, C.A. **61,** 9965h (1964). Synthesis: Hayduck, Ber. **36,** 2935 (1903); Posner, J. Prakt. Chem. **82,** 432 (1910); Mauthner, ibid. **142,** 33 (1935); Pandya et al., Proc. Indian Acad. Sci. **9A,** 511 (1939); Neish, Can. J. Biochem. Physiol. **37,** 1431 (1959). Review: Herrmann, Pharmazie **11,** 433 (1956).

Yellow crystals from concd aq solns. Monohydrate from dil solns. Dec 223-225° (softens at 194°). R_f values: Fiedler, loc. cit. Sparingly sol in cold water. Freely sol in hot water, cold alc. Alkaline solns turn from yellow to orange.

Methyl ester, $C_{10}H_{10}O_4$, colorless needles from water, mp 152-153°.

1635. Caffeine. *3,7-Dihydro-1,3,7-trimethyl-1H-purine-2,6-dione;* 1,3,7-trimethylxanthine; 1,3,7-trimethyl-2,6-dioxopurine; coffeine; thein; guaranine; methyltheobromine; No-Doz. $C_8H_{10}N_4O_2$; mol wt 194.19. C 49.48%, H 5.19%, N 28.85%, O 16.48%. Occurs in tea, coffee, maté leaves; also in guarana paste and cola nuts: Shuman, U.S. pat. **2,508,-545** (1950 to General Foods). Obtained as a by-product from the manuf of caffeine-free coffee: Barch, U.S. pat. **2,817,588** (1957 to Standard Brands); Nutting, U.S. pat. **2,802,739** (1957 to Hill Bros. Coffee); Adler, Earle, U.S. pat. **2,933,395** (1960 to General Foods). Crystal structure: Sutor, Acta Cryst. **11,** 453 (1958). Synthesis: Fischer, Ach, Ber. **28,** 2473, 3135 (1895); Gepner, Kreps, J. Gen. Chem. USSR **16,** 179 (1946); Bredereck et al., Ber. **83,** 201 (1950); Crippa, Crippa, Farmaco Ed. Sci. **10,** 616 (1955); Swidinsky, Baizer, U.S. pats. **2,785,162** and **2,785,163** (1957 to Quinine Chem. Works); Bredereck, Gotsmann, Ber. **95,** 1902 (1962). Reversed-phase HPLC study: J. W. Weyland et al., J. Chromatog. **247,** 221 (1982). Effect of pregnancy on the pharmacokinetics of caffeine: R. Knutti et al., Arch. Toxicol. **5,** Suppl., 187 (1982). Binding of caffeine on benzodiazepine receptors: V. Saano, M. M. Airaksinen, Acta Pharmacol. Toxicol. **51,** 300 (1982). Disposition of caffeine and its metabolites in man: D. D. Tang-Liu et al., J. Pharmacol. Exp. Ther. **224,** 180 (1983). Arrhythmogenic effects in humans: D. J. Dobmeyer et al., N. Engl. J. Med. **308,** 814 (1983). Teratogenicity study: P. E. Palm et al., Toxicol. Appl. Pharmacol. **44,** 1 (1978). Comprehensive description: M. U. Zubair et al. in Analytical Profiles of Drug Substances

Hexagonal prisms by sublimation, mp 238°. Sublimes 178°. Fast sublimation is obtained at 160-165° under 1 mm press. at 5 mm distance. d_4^{18} 1.23. pH of 1% soln 6.9. Aq solns of caffeine salts dissociate quickly. Absorption spectrum: Hartley, J. Chem. Soc. **87,** 1802 (1905). One gram dissolves in 46 ml water, 5.5 ml water at 80°, 1.5 ml boiling water, 66 ml alcohol, 22 ml alcohol at 60°, 50 ml acetone, 5.5 ml chloroform, 530 ml ether, 100 ml benzene, 22 ml boiling benzene. Freely sol in pyrrole; in tetrahydrofuran contg about 4% water; also sol in ethyl acetate; slightly in petr ether. Soly in water is increased by alkali benzoates, cinnamates, citrates or salicylates. LD_{50} orally in mice, hamsters, rats, rabbits (mg/kg): 127, 230, 355, 246 (males); 137, 249, 247, 224 (females) (Palm).

Monohydrate, felted needles, contg 8.5% H_2O. Effloresces in air; complete dehydration takes place at 80°.

Hydrochloride dihydrate, $C_8H_{11}ClN_4O_2.2H_2O$, crystals, dec 80-100° with loss of water and HCl. Sol in water and in alcohol with dec.

Mixture with citric acid, **citrated caffeine, "caffeine citrate".** White, crystalline powder; acid reaction. Sol in about 4 parts warm water.

THERAP CAT: CNS stimulant.

THERAP CAT (VET): Has been used as a cardiac and respiratory stimulant and as a diuretic.

1636. Calamine. Eczederm. Prepd calamine. Zinc oxide with about 0.5% ferric oxide.

Pink powder. Insol in water. Almost completely sol in mineral acids.

THERAP CAT: Topical protectant.

THERAP CAT (VET): Astringent. Skin protectant.

1637. Calamus. Sweet flag; calmus; sweet cane; sweet grass. Dried rhizome of Acorus calamus L., Araceae. Habit. Europe, North America, Western Asia; cultivated in Burma and Ceylon. Constit. Acorin, acoretin (choline), 1.5% volatile oil, 2.5% resins, 1.5% tannins; also reducing sugars and sterol bodies. Ref: Bose et al., J. Am. Pharm. Assoc. **49,** 32 (1960).

THERAP CAT: Carminative, anthelmintic.

1638. Calcifediol. (3β,5Z,7E)-9,10-Secocholesta-5,7,10-(19)-triene-3,25-diol; 25-hydroxyvitamin D_3; 25-hydroxycholecalciferol; 25-HCC; U-32070E; Calderol; Dedrogyl; Didrogyl; Hidroferol. $C_{27}H_{44}O_2$; mol wt 400.65. C 80.94%, H 11.07%, O 7.98%. The principal circulating form of vitamin D_3, formed in the liver by hydroxylation at C-25: Ponchon, DeLuca, J. Clin. Invest. **48,** 1273 (1969). It is the intermediate in the formation of 1α,25-dihydroxycholecalciferol, q.v., the biologically active form of vitamin D_3 in the intestine. Identification in rat as an active metabolite of vitamin D_3: Lund, DeLuca, J. Lipid Res. **7,** 739 (1966); Morii et al., Arch. Biochem. Biophys. **120,** 513 (1967). Evaluation of biological activity in comparison with vitamin D_3: Blunt et al., Proc. Nat. Acad. Sci. USA **61,** 717 (1968); ibid. 1503. Isoln from porcine plasma and establishment of structure: Blunt et al., Biochemistry **7,** 3317 (1968). Synthesis: Blunt, DeLuca, ibid. **8,** 671 (1969). Review of isoln, identification and synthesis: DeLuca, Am. J. Clin. Nutr. **22,** 412 (1969). Review of bioassays: J. G. Haddad Jr., Basic Clin. Nutr. **2,** 579-597 (1980).

uv max (ethanol): 265 nm (ε 18000) (Blunt, DeLuca).
THERAP CAT: Calcium regulator.

1639. Calcimycin. *6S-[6α(2S*,3S*),8β(R*),9β,11α]-5-(Methylamino)-2-[[3,9,11-trimethyl-8-[1-methyl-2-oxo-2-(1H-pyrrol-2-yl)ethyl]-1,7-dioxaspiro[5.5]undec-2-yl]methyl]-4-benzoxazolecarboxylic acid;* A 23187; antibiotic A23187. $C_{29}H_{37}N_3O_6$; mol wt 523.63. C 66.52%, H 7.12%, N 8.02%, O 18.33%. Polyether antibiotic produced by a strain of *Streptomyces chartreusensis* Calhoun and Johnson NRRL 3882. Activity as a divalent cation ionophore in isolated mitochondria: P. W. Reed, H. A. Lardy, *J. Biol. Chem.* **247**, 6970 (1972). Prepn and antimicrobial activity: R. M. Gale *et al.*, U.S. pat. **3,923,823** (1975 to Lilly). Elucidation of structure: M. O. Chaney *et al.*, *J. Am. Chem. Soc.* **96**, 1932 (1974). Spectral studies of ionophore and metal ion complexes: D. R. Pfeiffer *et al.*, *Biochemistry* **13**, 4007 (1974). Total synthesis and absolute configuration: D. A. Evans *et al.*, *J. Am. Chem. Soc.* **101**, 6789 (1979); P. A. Grieco *et al.*, *J. Org. Chem.* **45**, 3537 (1980). Stereospecific synthesis: G. R. Martinez *et al.*, *J. Am. Chem. Soc.* **104**, 1436 (1982); D. P. Negri, Y. Kishi, *Tetrahedron Letters* **28**, 1063 (1987). Review of cation binding and transport properties: D. R. Pfeiffer *et al.*, *Ann. N.Y. Acad. Sci.* **307**, 402-423 (1978). Use in model systems of calcium transport: M. Takamori *et al.*, *J. Neurol. Sci.* **50**, 89 (1981); M. Takamori *et al.*, *ibid.* **51**, 207 (1981); M. H. Freedman *et al.*, *Cell. Immunol.* **58**, 134 (1981); G. Thomas, *Eur. J. Pharmacol.* **81**, 35 (1982); V. L. Lew, J. Garcia-Sancho, *Cell Calcium* **6**, 15 (1985).

Crystalline solid, mp 181-182°. $[α]_D^{25}$ −56° (c = 1 in chloroform). uv max (ethanol): 204, 225, 278, 378 nm (E 28200, 26200, 18200, 8200). pKa_1 6.9 in 90% DMSO. Slightly sol in water, readily sol in ethyl acetate, chloroform, methanol, DMSO. Also reported as mp 184.5-186° (Evans, 1979). LD_{50} i.p. in mice: 10 mg/kg (Gale).
Mixed calcium-magnesium salt. Colorless crystalline solid, mp 230-250° (dec). uv max (ethanol, neutral): 202, 228, 303, 370 nm ($E_{1cm}^{1\%}$ 425, 490, 278, 109). Insol in water, pentane, hexane, heptane. Very slightly sol in methanol, DMSO. Very sol in methylene chloride, chloroform, acetone, methyl ethyl ketone, diethyl ketone, ethyl acetate.
USE: Biochemical tool used to study the role of divalent cations in various biological systems.

1640. Calcitonin. Thyrocalcitonin; TCA; TCT; Calsyn; Karil. Mol wt about 4,500. Calcium regulating hormone secreted from the mammalian thyroid gland and in non-mammalian species from the ultimobranchial gland. Postulation of a plasma-calcium lowering substance: Copp *et al.*, *Endocrinology* **70**, 638 (1962). Recognition as a hormone: Hirsch *et al.*, *ibid.* **73**, 244 (1963); of thyroid origin: Foster *et al.*, *Nature* **202**, 1303 (1964). Over-all action is to oppose

the bone and renal effects of parathyroid hormone, *q.v.*; inhibits bone resorption of Ca^{2+}, with accompanying hypocalcemia and hypophosphatemia and decreased urinary Ca^{2+} concentrations. Also abolishes the osteolytic effect of toxic doses of vitamins A and D. Calcitonin is highly active biologically, *e.g.* 50 mγ/min infused into a 100 g rat leads to a significant (1 mg/100 ml) decrease in the concn of the plasma calcium within 60 min (together with a corresponding fall in plasma phosphate). Activity is destroyed by trypsin, chymotrypsin, pepsin, polyphenol oxidase; also by hydrogen peroxide oxidation, photooxidation, and treatment with N-bromosuccinimide.
Calcitonin structures are single polypeptide chains containing 32 amino acid residues. Structure of porcine: Neher *et al.*, *Helv. Chim. Acta* **51**, 917 (1968); Potts *et al.*, *Proc. Nat. Acad. Sci. USA* **59**, 1321 (1968); Bell *et al.*, *J. Am. Chem. Soc.* **90**, 2704 (1968); *eidem, Biochemistry* **9**, 1665 (1970). Synthesis of porcine: Rittel *et al.*, *Helv. Chim. Acta* **51**, 924 (1968); Guttmann *et al.*, *ibid.* 1155. Isoln of human calcitonin from non-pathological thyroid glands: Haymovits, Rosen, *Endocrinology* **81**, 993 (1967); from medullary carcinoma of the thyroid: Neher *et al.*, *Nature* **220**, 984 (1968); *Helv. Chim. Acta* **51**, 1738 (1968); Neher, Riniker, **Ger.** pat. **1,929,957** (1970 to Ciba), *C.A.* **73**, 28902b (1970). Structure of human: Neher *et al.*, *Helv. Chim. Acta* **51**, 1900 (1968). Synthesis of human: Sieber *et al.*, *ibid.* 2057; J. Hirt *et al.*, *Rec. Trav. Chim.* **98**, 143 (1979). Biosynthetic studies: J. W. Jacobs *et al.*, *J. Biol. Chem.* **254**, 10600 (1979); S. G. Amara *et al.*, *ibid.* **255**, 2645 (1980). Amino acid sequence differs among mammalian species, salmon calcitonin showing a marked difference from that of the higher vertebrae as well as a more potent biological activity. Mechanism of action: E. M. Brown, G. D. Aurbach, *Vitam. Horm. (New York)* **38**, 236 (1980). Anorectic activity in rats: W. J. Freed *et al.*, *Science* **206**, 850 (1979). Growth inhibition of human breast cancer cells *in vitro*: Y. Iwasaki *et al.*, *Biochem. Biophys. Res. Commun.* **110**, 235 (1983). Review of early literature: Munson, Hirsch, *Clin. Orthop.* **49**, 209 (1966). Review of isoln, structure, synthesis: Behrens, Grinnan, *Ann. Rev. Biochem.* **38**, 83 (1969); Potts *et al.*, *Vitam. Horm. (New York)* **29**, 41 (1971). Comprehensive review: *Calcitonin, Proc. Symp. on Thyrocalcitonin and the C Cells,* S. Taylor, Ed. (Springer-Verlag, New York, 1968); Foster *et al.*, "Calcitonin" in *Clinics in Endocrinology and Metabolism,* I. MacIntyre, Ed. (W. B. Saunders, Philadelphia, 1972) pp 93-124. Review of pharmacology and therapeutic use: J. C. Stevenson, I. M. A. Evans, *Drugs* **21**, 257-272 (1981).
Calcitonin, salmon, *Calcimar, Salmotonin.*
Calcitonin, porcine, *Calcitar(e).*
Calcitonin, human synthetic, *Cibacalcin.*
Salcatonin, a synthetic polypeptide structurally similar to natural salmon calcitonin, is sold under the trademarks *Calsynar, Miacalcic* and *Tonocalcin. See also* Elcatonin.
THERAP CAT: Calcium regulator.

1641. Calcitriol. *(1α,3β,5Z,7E)-9,10-Secocholesta-5,7,10(19)-triene-1,3,25-triol;* 1α,25-dihydroxycholecalciferol; 1α,25-dihydroxyvitamin D_3; metabolite 4B; Peak IV; 1,25-DHCC; Ro 21-5535; Rocaltrol. $C_{27}H_{44}O_3$; mol wt 416.65. C 77.84%, H 10.64%, O 11.52%. The biologically active form of vitamin D_3 in intestinal calcium transport and bone calcium resorption: Haussler *et al.*, *Proc. Nat. Acad. Sci. USA* **68**, 177 (1971); Raisz *et al.*, *Science* **175**, 768 (1972). Formed by the sequential hydroxylation of vitamin D_3 at C-25 in the liver and at C-1 in the kidney: Blunt *et al.*, *Biochemistry* **7**, 3317 (1968); Fraser, Kodicek, *Nature* **228**, 764 (1970); Norman *et al.*, *Biochem. Biophys. Res. Commun.* **42**, 1082 (1971). Classification as a steroid hormone: Emtage *et al.*, *Nature* **246**, 100 (1973). First obtained from chick intestine and designated as metabolite 4B: Myrtle *et al.*, *J. Biol. Chem.* **245**, 1190 (1970); Myrtle, Norman, *Science* **171**, 79 (1971); Haussler *et al.*, *Proc. Nat. Acad. Sci. USA* **68**, 177 (1971). Isoln and identification: Holick *et al.*, *ibid.* **803**; Lawson *et al.*, *Nature* **230**, 228 (1971); Norman *et al.*, *Science* **173**, 51 (1971). Synthesis: Semmler *et al.*, *Tetrahedron Letters* **1972**, 4147; Barton *et al.*, *Chem. Commun.* **1974**, 203; H. E. Paaren *et al.*, *ibid.* **1977**, 890; T. Sato *et al.*, *Chem. Pharm. Bull.* **26**, 2933 (1978). *Review:* Suda, *Vitamins* **45**, 175-188 (1972), *C.A.* **77**, 69866u (1972); E. Kodicek, *Lancet* **1**, 325 (1974). Another polar metabolite of vitamin D$_3$, *21,25-dihydroxycholecalciferol,* C$_{27}$H$_{44}$O$_3$, shows marked action on mobilization of bone mineral and a small, but significant effect on intestinal calcium transport and has been isolated from pig plasma: Suda *et al., Biochemistry* **9**, 2917 (1970). *See also* DeLuca *et al., Proc. Nat. Acad. Sci. USA* **68**, 2131 (1971), which details the effect of calcium on *in vivo* synthesis of 1,25- and 21,25-dihydroxycholecalciferol. Comprehensive description: E. Debesis in *Analytical Profiles of Drug Substances* Vol. 8, K. Florey, Ed. (Academic Press, New York, 1979) pp 83-100.

White crystalline powder, mp 111-115°. uv max (abs ethanol): 264 nm (ϵ 19000). $[\alpha]_D^{25}$ +48° (methanol). Slightly sol in methanol, ethanol, ethyl acetate, THF. Air and light sensitive.

THERAP CAT: Calcium regulator.

1642. Calcium. Ca; at. wt 40.08; at. no. 20; valence 2. An alkaline earth metal. Occurrence in the earth's crust 3.64% (fifth element in order of abundance). Sea water contains about 400 g/ton. Isotopes: 40 (97%), 44 (2.06%), 42 (0.64%), 48 (0.18%), 43 (0.145%), 46 (0.003%). Found naturally only in the form of its compds, never uncombined. Principal commercial source is limestone, *q.v.* Essential constituent of bones, shells, teeth. First isolated by Davy in 1808. Produced by electrolysis of calcium chloride: Rathenau, Suter, **Ger. pat.** 155,433 (1903); *Z. Elektrochem.* **10**, 502 (1904); Goodwin, *J. Am. Chem. Soc.* **27**, 1403 (1905); also by thermal reduction of lime with aluminum, or with silicon. Prepn of the pure metal for laboratory use: Whaley, *Inorg. Syn.* **6**, 18 (1960). Purification of commercial material: Marshall, Whaley, *ibid.* 24. *Reviews:* Schaufler in *Ullmanns Encyklopädie der Tech. Chemie,* vol. 4 (Munich, 3rd ed., 1953) pp 830-836; Mantell in C. A. Hampel, *Rare Metals Handbook* (Reinhold, New York, 1954) p 17-29. Review of calcium and its compounds: Goodenough, Stenger, "Magnesium, Calcium, Strontium, Barium and Radium" in *Comprehensive Inorganic Chemistry,* Vol. 1, J. C. Bailar Jr. *et al.,* Eds. (Pergamon Press, Oxford, 1973) pp 591-664; C. J. Kunesh in Kirk-Othmer *Encyclopedia of Chemical Technology* vol. 4 (Wiley-Interscience, New York, 3rd ed., 1978) pp 412-421. Book: R. P. Rubin, *Calcium and Cellular Secretion* (Plenum, New York, 1982) 276 pp.

Lustrous, silver-white surface (when freshly cut); face-centered cubic structure below 300°C. Ignites in air when finely divided, then burns with crimson flame. Much harder than sodium, but softer than aluminum or magnesium. Acquires bluish-gray tarnish on exposure to moist air. d$_4^{20}$ 1.54. mp 850°. bp 1440°. Electrical resistivity at 20°: 3.5 μohm cm. Brinell hardness: 17. Heat of combustion 151.9 cal/g. sp ht (0-100°) 0.149 cal/g. Considerably less reactive than sodium. E° (aq) Ca^{2+}/Ca −2.87 V. Reacts with water, alcohols, and dil acids with evolution of hydrogen. Reacts with halogens. Dissolves in liquid ammonia to form a blue soln. Contact with alkali hydroxides or carbonates may cause detonation. Burns in air. Calcium salts impart brick red color to a flame. Insol in and inert towards benzene, kerosene.

USE: In metallurgy as deoxidizer for copper, beryllium, steel (together with silicon). To harden lead for bearings. Alloyed with cerium to make flints for cigarette and gas lighters. In manuf of electronic vacuum tubes as "getter" to fix residual gases as oxides, nitrides, hydrides of calcium.

1643. Calcium Acetate. C$_4$H$_6$CaO$_4$; mol wt 158.17. C

30.37%, H 3.82%, Ca 25.34%, O 40.46%. Ca(CH$_3$COO)$_2$. The technical product is known as *"brown acetate of lime"* or *"gray acetate of lime"*. Prepn: *Gmelin's, Calcium* (8th ed.) **28B**, 162-164, 982-991 (1958); electrolytic prepn: Schmidt, *Z. Anorg. Allgem. Chem.* **270**, 188 (1952).

Very hygroscopic, rod-shaped crystals. On heating above 160° dec to acetone and CaCO$_3$. d 1.50. Sol in water; slightly sol in methanol; practically insol in ethanol, acetone, benzene. *Keep well closed.*

Monohydrate, needles, granules or powder. Does not lose all its water below 150°. Sol in water; slightly sol in alcohol. pH of 0.2 molar aq soln 7.6. LD$_{50}$ orally in rats: 4.28 g/kg, Smyth *et al., Am. Ind. Hyg. Assoc. J.* **30**, 470 (1969).

Dihydrate, long, transparent needles. Loses water on standing in air, forming the monohydrate.

USE: Manuf of acetic acid, acetone; in dyeing, tanning, and curing skins; in lubricants; as food stabilizer; as corrosion inhibitor.

1644. Calcium Acetylsalicylate. *2-(Acetyloxy)benzoic acid calcium salt; salicylic acid acetate calcium salt;* acetylsalicylic acid calcium salt; calcium aspirin; soluble aspirin; Ascal; Cal-Aspirin; Dispril; Disprin; Kalmopyrin; Kalsetal; Solaspin; Solprin; Tylcalsin. C$_{18}$H$_{14}$CaO$_8$; mol wt 398.39. C 54.27%, H 3.54%, Ca 10.06%, O 32.13%. Ca(OOCC$_6$H$_4$O-COCH$_3$)$_2$. Prepd from calcium carbonate and acetylsalicylic acid: Lawrence, **U.S. pat.** 2,003,374 (1935 to Lee Labs.).

Dihydrate, amorphous, nonhygroscopic powder. One gram dissolves in 6 ml water, 80 ml alcohol.

Complex with urea, C$_{19}$H$_{18}$CaN$_2$O$_9$, *carbasalate calcium, acetylsalicylic acid calcium salt complex with urea, calcium acetylsalicylate carbamide, urea calcium acetylsalicylate, carbaspirin calcium, Alcacyl, Calurin, Iromin, Solupsan.* Amorphous powder, dec 243-245°. Soly in water at 37°: 231 mg/ml, pH 4.8. Description and hydrolytic stability: Parrott, *J. Pharm. Sci.* **51**, 897 (1962).

THERAP CAT: Analgesic; antipyretic; anti-inflammatory.

1645. Calcium Aluminosilicate. Aluminum calcium silicate. Many different forms of calcium aluminosilicate are known, the most common of which are CaAl$_2$Si$_2$O$_8$ and Ca$_2$Al$_2$SiO$_7$. It occurs in nature as the minerals: *anorthite, bavenite, clinozoisite, didymolite, epistilbite, gehlenite, gismondite, grossularite, heulandite, hibschite, laubanite, laumontite, lawsonite, levynite, margarite, meionite, plazolite, pumpellyite, scolecite, stellerite, vesuvianite, zoisite.* Prepn and properties: *Gmelin's, Aluminum* (8th ed.) **35B**, 576-586 (1934). Summary and references for minerals: Hey, *An Index of Mineral Species and Varieties* (British Museum, London, 2nd ed., 1962) pp 159-162.

USE: Constituent of cement; in refractories.

1646. Calcium Arsenate. Tricalcium arsenate; Pencal. As$_2$Ca$_3$O$_8$; mol wt 398.06. As 37.64%, Ca 30.21%, O 32.15%. Ca$_3$(AsO$_4$)$_2$. Prepn: Les Veaux, **U.S. pat.** 2,715,-562 (1955 to FMC). *Review:* Guerin, *Chim. & Ind. (Paris)* **77**, 1288 (1957).

Powder. *Poisonous!* Slightly sol in water; sol in dil acids. LD$_{50}$ orally in female rats: 298 mg/kg, T. B. Gaines, *Toxicol. Appl. Pharmacol.* **14**, 515 (1969).

USE: Insecticide; molluscicide.

1647. Calcium Arsenite. Variable composition. Prepd by passing steam over a dry mixture of CaO and As$_2$O$_3$: Altwegg, Dutel, **U.S. pat.** 1,700,756 (1929).

White, granular powder. *Poisonous!* Slightly sol in water; sol in acids.

USE: As insecticide, germicide, molluscicide.

1648. Calcium Ascorbate. *Ascorbic acid calcium salt.* C$_{12}$H$_{14}$CaO$_{12}$; mol wt 390.32. C 36.93%, H 3.61%, Ca 10.27%, O 49.19%. (C$_6$H$_7$O$_6$)$_2$Ca. Prepd from ascorbic acid and calcium carbonate by controlled precipitation in dilute acetone or alcohol: Ruskin, Merrill, *Science* **105**, 504 (1947); Ruskin, **U.S. pat.** 2,596,103 (1952); *idem,* **U.S. pat.** 2,631,-155 (1953 to Physiological Chemicals).

Dihydrate, triclinic crystals, $[\alpha]_D^{20}$ +95.6° (c = 2.4). Freely sol in water; practically insol in methanol, ethanol. Aq solns are neutral and oxidize quickly. After prolonged storage calcium oxalate may precipitate. The addition of various stabilizers such as cysteine, has been patented: Karrer, **U.S. pat. 2,442,461** (1948 to Hoffmann-La Roche).

THERAP CAT: Antiscorbutic vitamin.

1649. Calcium 3-Aurothio-2-propanol-1-sulfonate.
[2-Hydroxy-3-mercapto-1-propanesulfonato(2—)-O², S³]-aurate(1—) calcium (2:1); [(2-hydroxy-3-sulfopropyl)thio]-gold calcium salt; 2-hydroxy-3-mercapto-1-propanesulfonic acid S-gold deriv calcium salt; Chrisanol; Chrysanol; Kriza-nol. $C_6H_{12}Au_2CaO_8S_4$; mol wt 774.51. C 9.30%, H 1.56%, Au 50.90%, Ca 5.17%, O 16.53%, S 16.55%. (AuSCH₂-CHOHCH₂SO₃)₂Ca. Prepd by treating sodium thio-propanolsulfonate with gold chloride in the presence of SO_2 and converting the sodium salt to the calcium salt by means of calcium chloride: Lipovich, Mones, *J. Appl. Chem. (USSR)* **18**, 20 (1945); *cf.* Brit. pat. 265,777; Lumière, Perrin, *Compt. Rend.* **184**, 289 (1927).

Pale yellow powder. Has no definite mp. Soly in water about 2%. Practically insol in organic solvents. Marketed as *Oleochrysine* which is a suspension plus some Ca gluconate in peach kernel oil.

THERAP CAT: Antiarthritic.

1650. Calcium Bisulfite, Solution. $Ca(HSO_3)_2$ is known only in soln. Prepn from sulfite liquor: Arend, *Chem. Products* **10**, 53 (1947); Lougheed, *Pulp Paper Mag. Can.* **49**(3), 215 (1948); Schoeffel, **U.S. pat. 2,696,424** (1954 to Sterling Drug). The product here described is substantially a soln of calcium sulfite in an aq sulfur dioxide soln.

Colorless or slightly yellow liquid; strong SO_2 odor. On standing in the air, crystals of $CaSO_3.2H_2O$ form. d about 1.06. This soln corrodes metals.

USE: As germicide, preservative, and disinfectant; for washing (1:1000) casks in brewing to prevent souring and cloudiness of beer and to prevent secondary fermentation; as antichlor in bleaching fabrics; largely in manuf sulfite cellulose from wood for paper-making.

1651. Calcium Borate. Calcium pyroborate; calcium tetraborate. B_4CaO_7; mol wt 195.36. B 22.15%, Ca 20.52%, O 57.33%. CaB_4O_7. Prepn by direct fusion of B_2O_3 with $CaCO_3$: Griveau, *Compt. Rend.* **166**, 993 (1918). Various calcium borate minerals occur in nature. These include: *colemanite, ginorite, inyoite, meyerhofferite, pandermite, priceite.*

Hexahydrate, powder. Almost insol in cold, moderately sol in hot water; sol in dil acids.

USE: As flux in heavy-metal metallurgy; mfr of forsterite porcelain insulators; in glycol antifreeze; in fire-retardant paint.

1652. Calcium Borogluconate. *D-Gluconic acid cyclic 4,5-ester with boric acid calcium salt (2:1); calcium diboro-gluconate.* $C_{12}H_{20}B_2CaO_{16}$; mol wt 482.01. C 29.90%, H 4.18%, B 4.49%, Ca 8.32%, O 53.11%. Prepn and use in milk fever: H. Dryerre, J. R. Greig, *Vet. Rec.* **15**, 456 (1935); H. T. Macpherson, J. Stewart, *Biochem. J.* **32**, 76 (1938). Physical properties: L. Seekles, E. Havinga, *Norsk. Vet.-Tids.* **58**, No. 11, 433-445 (1946), *C.A.* **43**, 1907 (1949). Effect on plasma calcium concentration in sheep: D. A. H. Farningham, *Res. Vet. Sci.* **39**, 70 (1985). Spectrophotometric determn in plasma: D. J. Lyons, K. P. Spann, *J. Assoc. Off. Anal. Chem.* **68**, 160 (1985). Evaluation in bovine milk fever: P. A. Mullen, *Vet. Rec.* **97**, 87 (1975).

Crystals, freely sol in water. Soly in water 1:1 at 15°: 2.8:1 at 100°. A 20% aq soln has a pH of 3.5. If desired, the pH may be adjusted to 7.0 by the addition of CaO which is very sol in Ca borogluconate solns.

THERAP CAT (VET): In hypocalcemic states including bovine milk fever.

1653. Calcium Bromide. Br₂Ca; mol wt 199.91. Br

79.95%, Ca 20.05%. $CaBr_2$. Prepn: *Gmelin's, Calcium* (8th ed.) **28B**, 100-102, 584-599 (1958). The N.F. article is a hydrated salt, contg not less than 84% and not more than 94% $CaBr_2$.

Odorless, deliquesc granules or rhombic crystals; sharp, saline taste. Becomes yellow on long exposure to air. When anhydr mp 730°; d_2^{25} 3.353. When strongly heated in air, becomes alkaline due to loss of bromine and formation of lime. Very sol in water, methanol, ethanol; sol in acetone; practically insol in dioxane, chloroform, ether. The aq soln is neutral or only slightly alkaline to litmus. *Keep well closed.*

USE: In photography for making dry plates and light-sensitive papers; manuf mineral waters, NH₄Br, fire-extinguishing compositions.

THERAP CAT: Sedative, anticonvulsant.

THERAP CAT (VET): Has been used in hypocalcemic states such as canine eclampsia.

1654. Calcium Bromolactobionate. *D-Gluconic acid 4-O-β-D-galactopyranosyl calcium salt (2:1) compd with calcium bromide (CaBr₂) (1:1); lactobionic acid calcium salt, compound with calcium bromide;* calcium galactogluconate bromide; Brocalcin; Calabron; Calcibromin; Calcibronat; Calciobrom. $C_{24}H_{42}Br_2Ca_2O_{24}$; mol wt 954.59. C 30.20%, H 4.43%, Br 16.74%, Ca 8.40%, O 40.23%. $(C_{12}H_{21}O_{12})_2$-Ca.CaBr₂. Prepn: Isbell, *J. Res. Nat. Bur. Stand.* **17**, 331 (1936).

Hexahydrate, hexagonal prisms. Mild taste. Nonhygroscopic. Freely sol in water.

THERAP CAT: Sedative.

1655. Calcium N-Carbamoylaspartate. *N-(Aminocarbonyl)aspartic acid calcium salt (1:1); N-carbamoylaspartic acid calcium salt;* calcium ureidosuccinate. $C_5H_6CaN_2O_5$; mol wt 214.19. C 28.04%, H 2.82%, Ca 18.71%, N 13.08%, O 37.35%. Prepn of DL-form: Clemence, **Fr. pat. M6376** (1968 to Roussel-UCLAF), *C.A.* **74**, 76655r (1971); of optically active forms: Miyazaki *et al., Chem. Pharm. Bull.* **15**, 1604 (1967); Yoshimura *et al.,* **Japan. pat. 14,205**('68) (to Dainippon Pharm.), *C.A.* **71**, 112456a (1969).

DL-Form, *Pacilan.* mp > 350°. LD_{50} in rats: 1 g/kg.
L-Form, $[\alpha]_D^{20}$ +32.2° (c = 2.53).
D-Form, $[\alpha]_D^{20}$ −31.5° (c = 1.54).
THERAP CAT: Psychostimulant.

1656. Calcium Carbide. Acetylenogen. C_2Ca; mol wt 64.10. C 37.48%, Ca 62.53%. CaC_2. Prepn: Ehrlich in *Handbook of Preparative Inorganic Chemistry,* **Vol. 1,** G. Brauer, Ed. (Academic Press, New York, 2nd ed., 1963) p 943. *Review:* Brennan, *J. Electrochem. Soc.* **99**, 61c (1952).

Grayish-black, irregular lumps or orthorhombic crystals; dec by water with evolution of acetylene leaving a residue of lime. d 2.22; mp 2300°.

USE: Generating acetylene gas for lighting purposes (1 kg of carbide yields ~300 liters acetylene); as reducing agent, e.g., for direct reduction of copper sulfide to metallic copper; signal fires for marine service; manuf of calcium, iron, alloys, lampblack, cyanamide; welding and cutting metals.

1657. Calcium Carbonate. *Carbonic acid calcium salt (1:1);* Calcichew; Calcidia; Citrical. $CCaO_3$; mol wt 100.09. C 12.00%, Ca 40.04%, O 47.96%. $CaCO_3$. Exists in nature as the minerals *aragonite, calcite* and *vaterite.*

Odorless, tasteless powder or crystals. Two crystal forms are of commercial importance: Aragonite, orthorhombic, mp 825° (dec), d 2.83, formed at temps above 30°; Calcite, hexagonal-rhombohedral, mp 1339° (102.5 atm), $d^{25.2}$ 2.711, formed at temps below 30°. At about 825° is dec into CaO and CO_2. Practically insol in water. Sol in dil acids. *Incompat:* Acids, alum, ammonium salts.

Precipitated calcium carbonate, precipitated chalk, Aero-matt, Albacar, Purecal. Commercial $CaCO_3$ produced by chemical means. It is 98-99% pure. The byproduct process, the carbonation process, and the calcium chloride process of manuf from limestone are outlined in Kirk-Othmer *Encyclopedia of Chemical Technology,* **vol. 4** (Interscience, New

York, 2nd ed., 1964) pp 7-11. *Review:* Woerner in *Pigment Handbook* **vol. 1,** T. C. Patton, Ed. (John Wiley, New York, 1973) pp 119-128.

Prepared calcium carbonate, drop chalk, prepared chalk, whiting, English white, Paris white. Native $CaCO_3$ purified by elutriation.

USE: Manuf of paint, rubber, plastics, paper, dentifrices, ceramics, putty, polishes, insecticides, inks, shoe dressings; as a filler in production of adhesives, matches, pencils; crayons, linoleum, insulating compds, welding rods. In foods, cosmetics, pharmaceuticals, antibiotics; removing acidity of wines. In anal. chem for detecting and determining halogens in organic combinations; with NH_4Cl for decomposing silicates; preparing $CaCl_2$ soln for standardizing soap solns; for water analyses.

THERAP CAT: Antacid. Calcium supplement.

THERAP CAT (VET): Antacid, calcium supplement, antidiarrheal agent.

1658. Calcium Chlorate. $CaCl_2O_6$; mol wt 206.99. Ca 19.36%, Cl 34.26%, O 46.38%. $Ca(ClO_3)_2$. Prepn: Wilderman, **Brit.** pat. **183,671** (1921); Duveau, *Bull. Soc. Chim.* **10,** 374 (1943).

Dihydrate, monoclinic, hygroscopic crystals. d 2.711. mp 100° when rapidly heated. Sol in 0.6 part water; sol in alcohol. *Keep well closed.* LD_{50} orally in rats: 4.5 g/kg, *Handbook of Toxicology* vol. 1, W. S. Spector, Ed. (Saunders, Philadelphia, 1956) pp 58-59.

USE: Herbicide, insecticide, seed disinfectant.

1659. Calcium Chloride. Intergravin-orales. $CaCl_2$; mol wt 110.99. Ca 36.11%, Cl 63.89%. Forms mono-, di-, tetra- and hexahydrates. Obtained as a byproduct of the ammonia-soda (Solvay) process and as a joint product from natural salt brines: Faith, Keyes & Clark's *Industrial Chemicals,* F. A. Lowenheim, M. K. Moran, Eds. (Wiley-Interscience, New York, 4th ed., 1975) pp 186-190. Acute toxicity: I. B. Syed, F. Hosain, *Toxicol. Appl. Pharmacol.* **22,** 150 (1972).

Cubic crystals, granules or fused masses. Very hygroscopic. mp 772°. bp > 1600°; d_4^{15} 2.152. Freely sol in water (with liberation of much heat), alcohol. The commercial product is about 94-97% $CaCl_2$, the chief impurity being $Ca(OH)_2$. *Keep well closed.* LD_{50} i.v. in mice: 42.2 mg/kg (Syed, Hosain).

Dihydrate, hygroscopic granules, flakes or powder. Apparent (bulk) density: 0.835. Freely sol in water, alcohol. Commercial grades contain 73-80% $CaCl_2$. *Keep well closed.*

Hexahydrate, deliquesc trigonal crystals. mp 30°. d^{17} 1.68. Loses all H_2O at 200°. Extremely sol in water, alcohol. *Keep well closed.*

USE: The *anhydrous* form used as a drying and dehydrating agent for organic liquids and gases, and in desiccators. The *dihydrate* and *hexahydrate* forms are used for antifreeze and refrigerating solns, in fire extinguishers, etc. (a 40% soln freezes at −41°); to preserve wood, stone; manuf ice, glues, cements; fireproofing fabrics; automobile antifreeze mixtures; to melt ice and snow; as coagulant in rubber manuf, as size in admixture with starch paste; in concrete mixes to give quicker initial set and greater strength; freezeproofing of coal and ores; dust control on unpaved roads; sizing and finishing cotton fabrics; as brine for filling inflatable tires on tractors to increase traction.

THERAP CAT: Electrolyte replacement. Has been used as diuretic, urinary acidifier, antiallergic.

THERAP CAT (VET): May be used intravenously in hypocalcemic states such as milk fever.

1660. Calcium Chromate(VI). Calcium chrome yellow; gelbin; yellow ultramarine; C.I. 77223; C.I. Pigment Yellow 33. $CaCrO_4$; mol wt 156.09. Ca 25.68%, Cr 33.32%, O 41.00%. Prepn: Mylius, Wrochem, *Ber.* **33,** 3689 (1900); Udy, **U.S.** pats. **2,493,789; 2,494,215** (both 1950); Dunn, O'Brien, **U.S.** pats. **2,745,764/5** (both 1956 to Vanadium Corp. of America). Also occurs as hemihydrate, monohydrate, and dihydrate.

Yellow monoclinic or rhombic crystals. Sparingly sol in water; sol in dil acids; practically insol in alcohol.

USE: As a pigment, corrosion inhibitor; manuf of chromium; in oxidizing reactions; in battery depolarization.

1661. Calcium Citrate. *2-Hydroxy-1,2,3-propanetricarboxylic acid calcium salt* (2:3). $C_{12}H_{10}Ca_3O_{14}$; mol wt 498.44. C 28.91%, H 2.02%, Ca 24.12%, O 44.94%. $Ca_3(C_6H_5O_7)_2$. Prepn from citrus fruit: Cole, **U.S.** pat. **2,389,766** (1945 to California Fruit Growers Exchange). *Review:* Rudy, *Pharmazie* **4,** 393 (1949).

Tetrahydrate, odorless powder. Loses most of its water at 100° and all at 120°. Sol in 1050 parts cold water; somewhat more sol in hot water; insol in alcohol.

USE: In the production of citric acid and other citrates; improvement of baking properties of flour.

1662. Calcium Cyanamide. Calcium carbimide; "cyanamide"; nitrolime. $CCaN_2$; mol wt 80.11. C 14.99%, Ca 50.03%, N 34.99%. $N\equiv CN=Ca$. Prepn: Kastens, McBurney, *Ind. Eng. Chem.* **43,** 1020 (1951); Franck, Heimann, *Angew. Chem.* **44,** 372 (1931); Owen, *Trans. Faraday Soc.* **57,** 670 (1961); Dedman, Owen, *ibid.* 678.

Commercial grades may occur as grayish-black lumps of powder. While pure calcium cyanamide is nonvolatile and noncombustible, commercial grades may contain small amounts of calcium carbide which will produce acetylene in containers and processing vessels. Other contaminants are carbon, $Ca(OH)_2$, CaO, and $CaCO_3$. Pure calcium cyanamide occurs as glistening, hexagonal crystals belonging to the rhombohedral system. mp about 1340°; d_4^{20} 2.29. Sublimes at 1150-1200°. Heat of formn from $CaC_2 + N_2$: −69.0 kcal/mole (25°). Heat of fusion 1.29 cal/g. Essentially insol in water, but undergoes partial hydrolysis to the sol calcium hydrogen cyanamide, a source of cyanamide ions. No known solvent will bring about soln without decompn.

USE: As fertilizer, defoliant, herbicide, pesticide; manuf and refining of iron; manuf of calcium cyanide, melamine, dicyandiamide.

THERAP CAT (VET): Has been used as an anthelmintic.

1663. Calcium Cyanamide Citrated. Citrated calcium carbimide; CCC; carbimide; Colme; Dipsan; Abstem; Temposil. Contains citric acid in two parts by weight to one part of calcium cyanamide suitably purified for drug use. Used in treatment of alcoholism. Formulations: de Grunigen, Ferguson, **U.S.** pat. **2,998,350** (1961 to Cyanamid and Alcoholism Res. Found., Toronto). Acts by inhibiting aldehyde dehydrogenase: J. A. Smith *et al., J. Am. Med. Assoc.* **165,** 2181 (1957). Effect on cardiovascular system: J. E. Peachey *et al., Clin. Pharmacol. Ther.* **29,** 40 (1981); on liver cells: J. J. Vázquez, S. Cervera, *Lancet* **1,** 361 (1980). Comparison with disulfiram, *q.v.*: M. S. Levy *et al., Am. J. Psychiat.* **123,** 1018 (1967). Review of drug therapy for alcoholism: E. M. Sellers *et al., N. Engl. J. Med.* **305,** 1255 (1981).

THERAP CAT: Alcohol deterrent.

1664. Calcium Cyanide. Cyanogas. C_2CaN_2; mol wt 92.12. C 26.08%, Ca 43.52%, N 30.41%. $Ca(CN)_2$. Prepn: *Gmelin's, Calcium* (8th ed.) **28B,** 173-178, 958-960 (1958). Commercial prepns contain 40-50% $Ca(CN)_2$.

Rhombohedric crystals or powder; dec in moist air liberating hydrogen cyanide. *Very poisonous!* Sol in water with gradual liberation of HCN; even very weak acid (CO_2) liberates HCN; sol in alc. *Keep dry.* LD_{50} orally in rats: 39 mg/kg, H. F. Smyth *et al., Am. Ind. Hyg. Assoc. J.* **30,** 470 (1969).

USE: Fumigant; rodenticide; in stainless-steel manuf; in leaching ores of precious metals; stabilizer for cement.

1665. Calcium Cyclamate. *Cyclohexylsulfamic acid calcium salt; cyclohexanesulfamic acid calcium salt;* calcium cyclohexanesulfamate; calcium cyclohexylsulfamate; cyclamate calcium; Cyclan; Sucaryl Calcium. $C_{12}H_{24}CaN_2O_6S_2$; mol wt 396.54. C 36.34%, H 6.10%, Ca 10.11%, N 7.07%, O 24.21%, S 16.17%. $[C_6H_{11}NHSO_3^-]_2Ca^{2+}$. Prepn: Cummins, Johnson; McQuaid, **U.S.** pats. **2,799,700; 2,804,477** (both 1957 to du Pont); Freifelder, **U.S.** pat. **3,082,247** (1963 to Abbott); Birsten, Rosin, **U.S.** pats. **3,361,798; 3,366,670** (both 1968 to Baldwin-Montrose). Metabolism: Wallace *et al., J. Pharmacol. Exp. Ther.* **175,** 325 (1970); Prosky, O'Dell, *J. Pharm. Sci.* **60,** 1341 (1971); Renwick, Williams, *Biochem. J.* **129,** 869 (1972).

Dihydrate, crystals with pleasant, very sweet taste. Freely sol in water. Practically insol in alc, benzene, chloroform, ether. pH of 10% aq soln 5.5-7.5. Said to be more resistant to cooking temps than saccharin. *See* Cyclamic Acid.

Note: Consult latest Government regulations on use in foods.

USE: Non-nutritive sweetener.

1666. Calcium Dichromate(VI). Calcium bichromate. $CaCr_2O_7$; mol wt 256.10. Ca 15.65%, Cr 40.62%, O 43.73%. Prepn: Hartford *et al., J. Am. Chem. Soc.* **72**, 3353 (1950).

Trihydrate, bipyramidal orange-red crystals. Nonhygroscopic if pure. Dec on heating above 100° to $CaCrO_4$ and CrO_3. d_4^{30} 2.370. Very sol in water; insol in ether, CCl_4, hydrocarbons; dissolves in alc with immediate reduction of the dichromate and pptn of brown hydrous chromic chromate; dissolves in acetone with subsequent pptn of $CaCrO_4$.

USE: As catalyst; in manuf of $CrCl_3$ and CrO_3; corrosion inhibitor.

1667. Calcium 2-Ethylbutanoate. *2-Ethylbutanoic acid calcium salt; 2-ethylbutyric acid calcium salt;* diethylacetic acid calcium salt; Ethanion. $C_{12}H_{22}CaO_4$; mol wt 270.39. C 53.30%, H 8.20%, Ca 14.82%, O 23.67%. Prepn: *Beilstein* **vol. 2**, 333, 2nd suppl., 291-292, 3rd suppl., 752.

$$\left[\begin{array}{c} CH_3CH_2CHCOO \\ | \\ C_2H_5 \end{array} \right]_2 Ca$$

Monohydrate, prismatic twin crystals. Sol in water.

USE: As a stabilizer: Mack, **U.S. pat. 2,510,035** (1950 to Advance Solvents & Chem.).

THERAP CAT: Sedative.

1668. Calcium Ferrous Citrate. Ferrous calcium citrate; Ferrocal; Rarical. $C_{12}H_{10}Ca_2FeO_{14}$; mol wt 514.22. C 28.03%, H 1.96%, Ca 15.59%, Fe 10.86%, O 43.56%. Prepn: Opfermann, **U.S. pat. 2,691,666** (1954); Oroshnik, **U.S. pat. 2,812,344** (1956 to Ortho Pharm.). Probable structure:

$$\begin{array}{c} H_2C - COO \\ | \\ HO - C - COO \\ | \\ H_2C - COO \end{array} Ca \quad Ca \begin{array}{c} OOC - CH_2 \\ | \\ OOC - C - OH \\ | \\ OOC - CH_2 \end{array} Fe$$

Tetrahydrate, tasteless powder, very stable to air oxidation.

THERAP CAT: Hematinic.

1669. Calcium Fluoride. CaF_2; mol wt 78.08. Ca 51.33%, F 48.67%. Occurs in nature as the mineral *fluorite* or *fluorspar.* Prepd from $CaCO_3$ + HF: O. Ruff, *Die Chemie des Fluors* (Berlin, 1920) p 89; Emeleus in *Fluorine Chemistry,* **vol. I**, J. H. Simons, Ed. (Academic Press, 1950) p 36; Kwasnik in *Handbook of Preparative Inorganic Chemistry,* **Vol. 1**, G. Brauer, Ed. (Academic Press, New York, 2nd ed., 1963) p 233.

White powder or cubic crystals. When F ions are pptd with Ca^{2+} in the absence of CO_3 ions, a gel is obtained. For the crystals: d 3.18; mp 1403°; bp 2500°; Mohs' hardness: 4. Becomes luminous when heated. Practically insol in water (soly at 18°: 0.0015 g/100 ml); slightly sol in dil mineral acids; is dissolved by concd mineral acids with liberation of HF. LD orally in guinea pigs: > 5 g/kg.

USE: Fluorspar is the main primary source of fluorine and its compds. In ferrous metallurgy it is used as a flux to increase the fluidity of the slag. The steel industry is the largest consumer; the chemical industry, second and glass and ceramics, third. Synthetic fluorspar is used in the optical industry (transmits u.v. rays), and pure calcium fluoride is used as catalyst in dehydration and dehydrogenations. Used to fluoridate drinking water.

1670. Calcium Fluorophosphate. *Phosphorofluoridic acid calcium salt;* calcium monofluorophosphate. $CaFO_3P$; mol wt 138.06. Ca 29.03%, F 13.76%, O 34.77%, P 22.44%. $CaPO_3F$. Prepd from calcium chloride and sodium monofluorophosphate: Rowley, Stuckey, *J. Am. Chem. Soc.* **78**, 4262 (1956).

Dihydrate, monoclinic crystals with a tendency to form twins. Loses fluorine on heating. Soly in water (27°) 0.417

g/100 ml of soln. Practically insoluble in the usual organic solvents.

1671. Calcium Formate. $C_2H_2CaO_4$; mol wt 130.12. C 18.46%, H 1.55%, Ca 30.80%, O 49.19%. $Ca(HCOO)_2$. Prepn from $Ca(OH)_2$ and CO at high temp and pressure: Enderli, **U.S. pats. 1,920,851; 1,995,607** (1933, 1935 to Rudulf Koepp); Erasmus, Hamby, **U.S. pat. 2,913,318** (1959 to Union Carbide); from $CaCl_2$ and formic acid: Funk, Romer, *Z. Anorg. Allgem. Chem.* **239**, 288 (1938).

Orthorhombic crystals or cryst powder. Slight acetic acid-like odor. d 2.02. Sol in water, practically insol in alc.

USE: Preservative for food, silage; as binder for fine-ore briquets; in drilling fluids and lubricants.

1672. Calcium Gluconate. D-*Gluconic acid calcium salt (2:1);* Calciofon; Calglucon; Ebucin; Glucal; Glucobiogen. $C_{12}H_{22}CaO_{14}$; mol wt 430.38. C 33.49%, H 5.16%, Ca 9.31%, O 52.05%. $Ca[HOCH_2(CHOH)_4COO]_2$.

Odorless, tasteless crystals, granules, or powder. $[\alpha]_D^{20}$ about +6°. Does not lose its water on drying without some decomposition. Slowly sol in 30 parts cold, about 5 parts boiling water; insol in alc or other organic solvents. pH aq soln: 6-7. More concd (20 to 30%) aq solns are easily obtained by the addition of boric acid or similar complex-forming acids. The use of calcium D-saccharate for the prepn of supersatd injectable solns of calcium gluconate is described *under* Calcium D-Saccharate. Injectable solns of calcium gluconate contg sodium ascorbate are described in **Brit. pat. 495,675** and **Ger. pat. 702,185.** General directions and stability data: Siegrist, *Pharm. Acta Helv.* **24**, 430 (1949).

USE: In sewage purification; in coffee powders to prevent caking.

THERAP CAT: Calcium replenisher.

THERAP CAT (VET): In hypocalcemic states, including bovine milk fever.

1673. Calcium Glycerophosphate. *1,2,3-Propanetriol, mono(dihydrogen phosphate) calcium salt (1:1);* calcium glycerinophosphate; calcium phosphoglycerate; Neurosin. $C_3H_7CaO_6P$; mol wt 210.15. C 17.15%, H 3.36%, Ca 19.07%, O 45.69%, P 14.74%. Three isomers exist: The *β-glycerophosphoric acid calcium salt* ($(HOCH_2)_2CHOPO_3Ca$) and D(+)- and L(−)-*α-glycerophosphoric acid calcium salt* ($HOCH_2CH(OH)CH_2OPO_3Ca$). Commercial product is a mixture of calcium *β-* and DL-*α*-glycerophosphates: Toal, Phillips, *J. Pharm. Pharmacol.* **1**, 869 (1949). Prepn of the calcium salt of the *α-* and *β*-acids: King, Pyman, *J. Chem. Soc.* **105**, 1238 (1914); Toal, Phillips, *loc. cit.* Sepn of the *α*-acid from *β-* and polyglycerophosphoric acids via the *α*-acid calcium salt: Carrara, **Ital. pat. 460,219** (1950), *C.A.* **46**, 5077a (1952). Protective action against demineralization of dental enamel: T. H. Grenby, J. M. Bull, *Caries Res.* **14**, 210 (1980).

Commercial product, fine, odorless, almost tasteless, slightly hygroscopic powder; dec > 170°. Sol in about 50 parts water; almost insol in alc, boiling water. Soly in water is increased by citric or lactic acid. The aq soln is alkaline.

Mixture with calcium lactate, *Calphosan.*

USE: In dentifrices, baking powder, as food stabilizer.

THERAP CAT: Calcium and phosphorus source. Tonic.

THERAP CAT (VET): Has been used as dietary supplement.

1674. Calcium Hexafluorosilicate. Calcium fluosilicate; calcium silicofluoride. CaF_6Si; mol wt 182.17. Ca 22.00%, F 62.58%, Si 15.42%. $CaSiF_6$. Prepn from Ca salt and H_2SiF_6: Moller, Kreth, **Brit. pat. 263,780** (1925).

Dihydrate, powder. d 2.25. Almost insol in cold water; partially dec by hot water; practically insol in acetone. LD orally in guinea pigs: 250 mg/kg, *Handbook of Toxicology,* **vol. 1**, W. S. Spector, Ed. (Saunders, Philadelphia, 1956) pp 58-59.

USE: In wood, rubber, textile industries; flotation agent; insecticide.

1675. Calcium Hydride. CaH_2; mol wt 42.10. Ca 95.21%, H 4.79%. Prepd by direct combination of calcium and hydrogen at 300-400°: **Brit. pat. 597,055** (1948); by reduction of lime with magnesium in the presence of hydrogen: Gibb, *Trans. Electrochem. Soc.* **93**, 198-211 (1948); from $CaCl_2$ and hydrogen in the presence of sodium: Wade,

Alexander, U.S. pat. **2,702,740** (1955 to Metal Hydrides). *Reviews:* Halls, *Ind. Chem.* **22**, 680 (1946); Kilb, *USAEC* APEX-485, 57 pp (1959).

Orthorhombic crystals or powder; the commercial product is gray. d 1.7. mp 1.86. Decomposes with water, lower alcohols and carboxylic acids to form hydrogen; moderately powerful condensing agent with ketones and acid esters; more powerful reducing agent toward metal oxides than lithium or sodium hydrides.

USE: To prepare rare metals by reduction of their oxides; as a drying agent for liquids and gases; to generate hydrogen: 1 g of calcium hydride in water liberates 1 liter of hydrogen at STP; in organic syntheses.

1676. Calcium Hydroxide. Calcium hydrate; slaked lime. CaH_2O_2; mol wt 74.10. Ca 54.09%, H 2.72%, O 43.19%. $Ca(OH)_2$. Contains at least 95% $Ca(OH)_2$. Commercial prepn by hydration of lime. W. L. Faith *et al., Industrial Chemicals* (John Wiley, New York, 3rd ed., 1965) pp 483-484. Laboratory prepn by treating an aq soln of a calcium salt with alkali: Ehrlich in *Handbook of Preparative Inorganic Chemistry* vol. 1, G. Brauer (Academic Press, New York, 2nd ed., 1963) p 934.

Crystals or soft, odorless, granules or powder. Slightly bitter, alkaline taste. Readily absorbs CO_2 from air forming $CaCO_3$. Loses water when ignited; forms CaO. d 2.08-2.34. Slightly sol in water; sol in glycerol, sugar or NH_4Cl solns; sol in acids with evolution of much heat. pH of aq soln satd at 25°: 12.4. *Keep well closed.* LD_{50} orally in rats: 7.34 g/kg, Smyth *et al., Am. Ind. Hyg. Assoc. J.* **30**, 470 (1969).

USE: In mortar, plaster, cement and other building and paving materials; in lubricants, drilling fluids, pesticides, fireproofing coatings, water paints; as egg preservative; manuf of paper pulp; in SBR rubber vulcanization; in water treatment; dehairing hides.

THERAP CAT: Astringent.

1677. Calcium Hypochlorite. Losantin. $CaCl_2O_2$; mol wt 142.99. Ca 28.03%, Cl 49.59%, O 22.38%. $Ca(OCl)_2$. Pure product has not been prepd. Commercial product usually contain 50% or more $Ca(OCl)_2$. Preparation of solid product contg 90-94% $Ca(OCl)_2$: Cady, *Inorg. Syn.* **5**, 161 (1957). Impurities include: $Ca(ClO_4)_2$, $CaCl_2$, $CaCO_3$, $Ca(OH)_2$ and water.

USE: Algicide, bactericide, deodorant, disinfectant, fungicide; in sugar refining; oxidizing agent; bleaching agent. Ingredient of *Camporit* which also contains NaCl. Also ingredient of *HTH* and *Perchloron.*

1678. Calcium Hypophosphite. $CaH_4O_4P_2$; mol wt 170.07. Ca 23.57%, H 2.37%, O 37.63%, P 36.43%. $Ca(H_2PO_2)_2$. Prepn: *Gmelin's, Calcium* (8th ed.) **28B**, 1119-1121 (1958).

Monoclinic, prismatic crystals or granular powder. When heated above 300° it evolves spontaneously-inflammable phosphine. Sol in water; slightly sol in glycerol. Practically insol in alcohol. The aq soln is slightly acid. *Incompat.* KI, oxidizers.

USE: As corrosion inhibitor; in nickel plating. Pharmaceutic aid (retards oxidation of ferrous salts).

THERAP CAT: Calcium source.

THERAP CAT (VET): Has been used as a dietary supplement and also as a "nerve tonic".

1679. Calcium Iodate. Lautarite. CaI_2O_6; mol wt 389.90. Ca 10.28%, I 65.10%, O 24.62%. $Ca(IO_3)_2$. Prepd by passing chlorine into a hot soln of lime in which iodine has been dissolved: Bahl, Singh, *J. Indian Chem. Soc.* **17**, 397 (1940).

Nonhygroscopic, monoclinic-prismatic crystals. d_4^{15} 4.519. Stable up to 540°. Sensitive to reducing agents. Soly in water (g/100 ml): 0.10 (0°); 0.95 (100°). More sol in aq solns of iodides and in amino acid solns. Sol in nitric acid. Insol in alcohol.

Monohydrate, cubic crystals. Slightly sol in water.

Hexahydrate, orthorhombic crystals. Slightly sol in water.

USE: Nutritional source of iodine in foods and feedstuffs. More stable in table salts than iodides: *Food Field Reporter,* Aug. 8, 1956; Daum, *C.A.* **51**, 5324 (1957); to improve properties of yeast-leavened bakery products.

THERAP CAT: Antiseptic.

1680. Calcium Iodide. CaI_2; mol wt 293.90. Ca 13.64%, I 86.36%. Prepn: Farr, U.S. pat. **2,415,346** (1947 to Mallinckrodt); Chaigneau, *Bull. Soc. Chim. France* **1957**, 886; *Gmelin's, Calcium* (8th ed.) **28B**, 102, 610-622 (1958). The commercial product usually contains 16-20% water.

Very hygroscopic hexagonal lamella. Becomes yellow and completely insol on exposure to air due to liberation of I_2 and absorption of CO_2. mp 740°; bp 1100°. Very sol in water, methanol, ethanol, acetone; practically insol in ether, dioxane. The aq soln is neutral or slightly alkaline. *Keep tightly closed and protected from light.*

Hexahydrate, hexagonal, thick needles, or plates, or lumps, or powder. Very hygroscopic; becomes yellow in air, mp about 42°. Freely sol in water, alcohol. *Keep tightly closed and protected from light.*

THERAP CAT: Expectorant.

1681. Calcium Iodobehenate. *Iododocosanoic acid calcium salt;* Calioben; Saiodin; Sajodin. $C_{44}H_{84}CaI_2O_4$; mol wt 971.03. C 54.42%, H 8.72%, Ca 4.13%, I 26.14%, O 6.59%. $Ca(C_{21}H_{42}ICO_2)_2$. Prepn: *Beilstein,* vol. **2**, 392.

White or yellowish powder; odorless or with slight fat-like odor. Practically insol in water, alcohol, ether; freely sol in chloroform. *Protect from light.*

THERAP CAT: Internally for its iodine action.

THERAP CAT (VET): Has been used internally as a source of iodine.

1682. Calcium Iodostearate. *2-Iodooctadecanoic acid calcium salt;* stearodine. $C_{36}H_{68}CaI_2O_4$; mol wt 858.82. C 50.34%, H 7.98%, Ca 4.67%, I 29.56%, O 7.45%. $Ca(C_{17}H_{34}ICO_2)_2$. Prepn: *Beilstein,* vol. **2**, 1st suppl., 177.

Cream-colored, almost odorless powder. Practically insol in water or alcohol; sol in benzene, chloroform, ether.

THERAP CAT: Iodine source.

THERAP CAT (VET): Has been used internally as a source of iodine.

1683. Calcium Lactate. *2-Hydroxypropanoic acid calcium salt.* $C_6H_{10}CaO_6$; mol wt 218.22. C 33.02%, H 4.62%, Ca 18.37%, O 43.99%. $Ca[CH_3CH(OH)COO]_2$. Commercial prepn usually contains about 25% water, and on the anhydr basis it is at least 98% pure. Prepd commercially by neutralization of lactic acid, from fermentation of dextrose, molasses, starch, sugar or whey, with $CaCO_3$: Inskeep *et al., Ind. Eng. Chem.* **44**, 1955 (1952).

Pentahydrate, almost odorless, slightly efflorescent granules or powder. Becomes anhydr at 120°. pH: 6-7. Slowly sol in cold water, quickly sol in hot water; almost insol in alcohol.

USE: As a preservative in foods and beverages; in dentifrices.

THERAP CAT: Replenisher (calcium).

THERAP CAT (VET): May be used for hypocalcemic states.

1684. Calcium Levulinate. *4-Oxopentanoic acid calcium salt; levulinic acid calcium salt.* $C_{10}H_{14}CaO_6$; mol wt 270.30. C 44.43%, H 5.22%, Ca 14.83%, O 35.52%. $(CH_3COCH_2CH_2COO)_2Ca$. Prepn: Cox, Dodds, U.S. pat. **2,033,909** (1936 to Niacet Chemicals).

Dihydrate, crystals or granular powder. mp 125°. Loses 1 H_2O on drying *in vacuo* at room temp and all H_2O at 50°. Very sol in water; the aq soln is practically neutral.

THERAP CAT: Replenisher (calcium).

THERAP CAT (VET): May be used in hypocalcemic states.

1685. Calcium Mesoxalate. *Mesoxalic acid calcium salt;* calcium ketomalonate; calcium oxomalonate; ketomalonic acid calcium salt; oxomalonic acid calcium salt; Mesoxan. C_3CaO_5; mol wt 156.11. C 23.08%, Ca 25.67%, O 51.25%. Prepn: Scheiber, Hopfer, *Ber.* **53**, 908 (1920); Kobayashi, **Japan.** pat. **4157**('52), *C.A.* **48**, 5212a (1954); Yanagisawa, **Japan.** pat. **7463**('60), *C.A.* **55**, 5880a (1961).

Crystalline powder, dec 210-220°. Sparingly sol in glacial

acetic acid. *Note:* May also exist as calcium dihydroxymalonate.

THERAP CAT: Oral hypoglycemic.

1686. Calcium Methionate. *Methanedisulfonic acid calcium salt.* $CH_2CaO_6S_2$; mol wt 214.24. C 5.61%, H 0.94%, Ca 18.71%, O 44.81%, S 29.93%. $CaCH_2(SO_3)_2$. Prepn: *Beilstein vol. 1*, 1st suppl., 303; Jenkins, *J. Am. Pharm. Assoc.* **27**, 484 (1938).

Dihydrate, crystals or cryst powder, having a slight odor. Stable in air. Sol in about 2.5 parts water forming a neutral soln (pH 6-7); very slightly sol in alc.

1687. Calcium Molybdate(VI). $CaMoO_4$; mol wt 200.03. Ca 20.04%, Mo 47.98%, O 32.00%. Prepn from sodium molybdate and $CaSO_4$: Carosella, U.S. pat. **2,460,974** (1949 to U.S. Vanadium); by heating a stoichiometric mixture of CaO or $CaCO_3$ and molybdic acid: Kroger, *Nature* **159**, 674 (1947).

Tetragonal crystals. d. 4.35. Insol in water, alcohol; sol in concd mineral acids.

USE: In phosphors and luminescent materials.

1688. Calcium Nitrate. CaN_2O_6; mol wt 164.10. Ca 24.42%, N 17.07%, O 58.50%. $Ca(NO_3)_2$. Prepn: *Gmelin's Calcium* (8th ed.) **28B**, 59-69, 341-382 (1956).

Deliquesc granules, mp about 560°. Very sol in water, heat being evolved; freely sol in methanol, ethanol, acetone; almost insol in concd HNO_3. pH of 5% aq soln 6.0. *Keep well closed.*

Note: $Ca(NO_3)_2$ crystallizes also with $4H_2O$ (30.5%), melting at 45°. Technical flake usually contains 28.6% H_2O.

USE: In explosives, fertilizers, matches, pyrotechnics; manuf of incandescent mantles, radio tubes, HNO_3; corrosion inhibitor in diesel fuels.

1689. Calcium Nitrite. CaN_2O_4; mol wt 132.10. Ca 30.34%, N 21.21%, O 48.45%. $Ca(NO_2)_2$. Prepd by reaction of nitric oxide with a mixture of calcium ferrate(III) and calcium nitrate: Ray, Ogg, Jr., *J. Am. Chem. Soc.* **79**, 265 (1957).

White or yellowish, deliquesc, hexagonal crystals. d 2.23. Freely sol in water; slightly sol in alc. *Keep well closed.*

USE: Corrosion inhibitor in lubricants, concrete.

1690. Calcium Oleate. *9-Octadecenoic acid calcium salt; oleic acid calcium salt.* $C_{36}H_{66}CaO_4$; mol wt 602.97. C 71.71%, H 11.03%, Ca 6.65%, O 10.61%. $Ca(C_{18}H_{33}O_2)_2$. Prepn: Harrison, *Biochem. J.* **18**, 1222 (1924); Pink, *J. Chem. Soc.* **1939**, 619.

Pale-yellow transparent solid. Dec above 140°. Slowly absorbs moisture from the air to form the monohydrate. Practically insol in water, alcohol, ether, acetone, petr ether; sol in chloroform, benzene.

USE: Thickening lubricating grease; waterproofing concrete; emulsifier for benzene, kerosene, etc.; in modeling waxes to vary hardness.

1691. Calcium Oxalate. *Ethanedioic acid calcium salt.* C_2CaO_4; mol wt 128.10. C 18.75%, Ca 31.29%, O 49.96%. CaC_2O_4. Prepn from calcium formate: Bredt, U.S. pat. **1,622,991** (1927); from calcium cyanamide: Barsky, Buchanan, *J. Am. Chem. Soc.* **53**, 1270 (1931).

Monohydrate, cubic crystals. Loses all of its water at 200°. When ignited is converted into $CaCO_3$ or CaO without appreciable charring. d 2.2. Practically insol in water or acetic acid; sol in dil HCl or HNO_3.

USE: In ceramic glazes; as carrier for separation of rare earth metals; analysis for calcium: Ingols, Murray, *Anal. Chem.* **21**, 525 (1949).

1692. Calcium Oxide. Lime; burnt lime; calx; quicklime. CaO; mol wt 56.08. Ca 71.47%, O 28.53%. Properly stored lime of commerce contains 90-95% free CaO. Commercial production from limestone: W. L. Faith *et al., Industrial Chemicals* (John Wiley, New York, 3rd ed., 1965) pp 482-487. Lab prepn by ignition of $CaCO_3$: Ehrlich in *Handbook of Preparative Inorganic Chemistry* vol. 1, G. Brauer, Ed. (Academic Press, New York, 2nd ed., 1963) p 931. *Review:* R. S. Boynton in Kirk-Othmer *Encyclopedia of Chemical Technology* vol. **14** (Wiley-Interscience, New York, 3rd ed., 1981) pp 343-382.

Crystals, white or grayish-white lumps, or granular powder; commercial material sometimes has a yellowish or brownish tint, due to iron. mp 2572°; bp 2850°; d 3.32-3.35. Readily absorbs CO_2 and H_2O from air, becoming air-slaked. Sol in water forming $Ca(OH)_2$ and generating a large quantity of heat; sol in acids, glycerol, sugar soln; practically insol in alc. *Keep tightly closed and dry.*

USE: In bricks, plaster, mortar, stucco and other building and construction materials; manuf of steel, aluminum, magnesium, and flotation of non-ferrous ores; manuf of glass, paper, Na_2CO_3 (Solvay process), Ca salts and many other industrial chemicals; dehairing hides; clarification of cane and beet sugar juices; in fungicides, insecticides, drilling fluids, lubricants; water and sewage treatment; in laboratory to absorb CO_2 (the combination with NaOH is known as soda-lime, *q.v.*). *Caution:* A strong caustic. May cause severe irritation of skin, mucous membranes.

1693. Calcium Palmitate. *Hexadecanoic acid calcium salt; palmitic acid calcium salt.* $C_{32}H_{62}CaO_4$; mol wt 550.90. C 69.76%, H 11.34%, Ca 7.28%, O 11.62%. $Ca(C_{16}H_{31}O_2)_2$. Prepn: Harrison, *Biochem. J.* **18**, 1222 (1924).

Powder or rhombic crystals. Dec above 155°. Practically insol in water, alcohol, ether, acetone, petr ether; slightly sol in chloroform, benzene, acetic acid.

USE: Thickening lubricating oils; waterproofing fabrics and lubricating greases; as corrosion inhibitor in halohydrocarbons.

1694. Calcium Pantothenate. *N-(2,4-Dihydroxy-3,3-dimethyl-1-oxobutyl)-β-alanine calcium salt; pantothenic acid calcium salt;* calcium D(+)-N-(2,4-dihydroxy-3,3-dimethylbutyryl)-β-alaninate; Calpanate; Galamila; Pantholin. $C_{18}H_{32}CaN_2O_{10}$; mol wt 476.53. C 45.37%, H 6.77%, Ca 8.41%, N 5.88%, O 33.57%. $[HOCH_2C(CH_3)_2CHOH-CONHCH_2CH_2COO]_2Ca$. Prepn: Wehrmeister, U.S. pat. **2,780,645** (1957 to Commercial Solvents); Kagan, U.S. pat. **2,845,456** (1958 to Upjohn). Purification: Kapp, Griffith, U.S. pat. **2,935,528** (1960 to Nopco). Monograph: Greulich, Meiner, *Pharmazie* **12**, 643 (1957). *See also* Pantothenic Acid for ref relating to prepn. Only the *d*-isomer described here has vitamin activity.

Minute needles from CH_3OH. Sweetish taste with slightly bitter aftertaste. Dec 195-196°. Moderately hygroscopic. Reasonably stable to air and light. $[\alpha]_D^{25}$ +28.2° (c = 5). One gram dissolves in 2.8 ml H_2O. Sol in glycerol; slightly sol in alcohol, acetone. pH of aq soln (1 in 20): 7.2-8.0; pH in CO_2-free water: 8.7. Solns are most stable at pH 5-7. Rate of hydrolysis is a function of pH and is catalyzed by the presence of electrolytes. Solns are not stable to autoclaving, and sterilization by filtration is necessary. Stability data: Frost, McIntire, *J. Am. Chem. Soc.* **66**, 425 (1944).

THERAP CAT: Enzyme co-factor vitamin.

THERAP CAT (VET): As a dietary supplement (pantothenic acid source).

1695. Calcium Permanganate. $CaMn_2O_8$; mol wt 277.94. Ca 14.42%, Mn 39.53%, O 46.05%. $Ca(MnO_4)_2$. Prepn from $KMnO_4$ and $CaCl_2$: Brit. pat. **624,885** (1949 to Boots Pure Drug Co. and T. Hagyard); from $Al(MnO_4)_3$ and $Ca(OH)_2$: Jaskowiak, U.S. pat. **2,504,130** (1950 to Carus Chemical).

Violet or dark-purple, deliquesc crystals. Freely sol in water; dec in alcohol. *Keep tightly closed.*

USE: Antiseptic, disinfectant, deodorizer; with CaF_2 as binder for welding electrode coatings and fluxes.

1696. Calcium Peroxide. Calcium dioxide. CaO_2; mol wt 72.08. Ca 55.60%, O 44.40%. The commercial product usually contains about 60% CaO_2, water, and some $Ca(OH)_2$ and $CaCO_3$. Prepn: Young, U.S. pat. **2,533,660** (1950 to du Pont); Ehrlich in *Handbook of Preparative Inorganic Chemistry* vol. 1, G. Brauer, Ed. (Academic Press, New York, 2nd ed., 1963) p 936.

White or yellowish, odorless, almost tasteless powder. Dec in moist air. Slightly sol in water; sol in acids with formation of H_2O_2. *Keep well closed.*

USE: Stabilizer for rubber.

THERAP CAT: Antiseptic.

1697. Calcium Phenolsulfonate. *p-Hydroxybenzenesulfonic acid calcium salt;* calcium sulfocarbolate; calcium sulfophenolate. $C_{12}H_{10}CaO_8S_2$; mol wt 386.40. C 37.30%, H

2.61%, Ca 10.37%, O 33.12%, S 16.60%. $Ca[C_6H_4(OH)SO_3]_2$. Prepn: *Hagers Handb. Pharm. Praxis* **vol. 2**, 420 (Berlin, 1930).

Hydrate, odorless cryst powder. Sol in water or alcohol. The aq soln is neutral, and has a bitter, astringent taste.

THERAP CAT (VET): Has been used as an intestinal antiseptic, in dusting powders for ulcers and in ophthalmic solns.

1698. Calcium Phenoxide. Calcium carbolate; calcium phenate; calcium phenolate; calcium phenylate. $C_{12}H_{10}$- CaO_2; mol wt 226.28. C 63.69%, H 4.45%, Ca 17.72%, O 14.14%. $Ca(OC_6H_5)_2$. Prepn: Kluge, Drake, U.S. pat. 2,870,134 (1959 to Texas Co.).

Reddish powder. Dec in air. Slightly sol in water or alcohol. *Keep well closed.*

USE: Detergent; additive for motor oils.

1699. Calcium Phosphate, Dibasic. Calcium monohydrogen phosphate; dicalcium orthophosphate; secondary calcium phosphate. $CaHO_4P$; mol wt 136.06. Ca 29.46%, H 0.74%, O 47.04%, P 22.77%. $CaHPO_4$. Occurs in nature as the mineral *monetite*. Prepn from $CaCl_2$ and Na_2HPO_4: Jensen, Rathley, *Inorg. Syn.* **4**, 19, 20 (1953); from $Ca_3(PO_4)_2$ and H_3PO_4: Perloff, Posner, *ibid.* **6**, 16 (1960), where it is an intermediate in the preparation of hydroxyapatite.

Triclinic crystals. At red heat dehydrated to calcium pyrophosphate. Practically insol in water, alcohol.

Dihydrate, *brushite.* Monoclinic crystals. Loses water of crystn slowly below 100°. Dehydr at red heat to calcium pyrophosphate. d 2.31. Practically insol in water, alcohol; sol in dil HCl or HNO_3; slightly sol in dil acetic acid.

USE: Chiefly in animal feeds; mineral supplement in cereals and other foods; manuf of glass; in dental products, fertilizers (*see also* Calcium Phosphate, Monobasic).

THERAP CAT: Calcium replenisher.

THERAP CAT (VET): Has been used as a dietary supplement, and as an antacid.

1700. Calcium Phosphate, Monobasic. Acid calcium phosphate; calcium biphosphate; monocalcium orthophosphate; monocalcium phosphate; primary calcium phosphate; "calcium superphosphate". $CaH_4O_8P_2$; mol wt 234.06. Ca 17.12%, H 1.72%, O 54.69%, P 26.47%. $Ca(H_2PO_4)_2$. Commercial prepn for fertilizers by treating pulverized phosphate rock with H_2SO_4 or H_3PO_4: Faith, Keyes & Clark's *Industrial Chemicals*, F. A. Lowenheim, M. K. Moran, Eds. (Wiley-Interscience, New York, 4th ed., 1975) pp 191-200. Laboratory prepn from $CaCO_3$ and H_3PO_4: Jensen, Kathley, *Inorg. Syn.* **4**, 18 (1953).

Monohydrate, large, shining, triclinic plates, cryst powder or granules. Non-hygroscopic when pure, but traces of impurities such as H_3PO_4 cause material to be deliquesc. Strong acid taste. Loses H_2O at 100°, dec at 200°. d_4^{18} 2.220. Moderately sol in water; sol in dil HCl or HNO_3 or acetic acid.

Note: The products obtained from commercial processes are not pure monobasic calcium phosphate. The *superphosphate* obtained from the H_2SO_4 treatment is about 30% $CaH_4(PO_4)_2.H_2O$, 10% $CaHPO_4$, 45% $CaSO_4$, 10% iron oxide, silica, alumina, etc. and 5% water; it contains 18-21% available P_2O_5. The *triple superphosphate* obtained from the H_3PO_4 treatment contains from 43 to 50% available P_2O_5.

USE: Chiefly in fertilizers; as acidulant in baking powders and in wheat flours; mineral supplement for foods and feeds; in enameling.

1701. Calcium Phosphate, Tribasic. Tricalcium orthophosphate; tricalcium phosphate; tertiary calcium phosphate; Calcigenol Simple. $Ca_3O_8P_2$; mol wt 310.20. Ca 38.76%, O 41.26%, P 19.97%. $Ca_3(PO_4)_2$. It is about 96% pure, usually contg an excess of CaO. Occurs in nature as the minerals: *oxydapatit, voelicherite, whitlockite.* The technical product is also known as "*bone ash*". Commercial prepn from phosphate rock: Hignett, Hubbard, *Ind. Eng. Chem.* **38**, 1208 (1946); Elmore, U.S. pat. 2,474,831 (1949 to T.V.A.); Hollingsworth, U.S. pats. 2,556,541 and 2,562,718 (both 1951 to Coronet Phosphate); Brosheer, Hignett, *Chem. Eng. Rept.* no. 7, 143 pp (1953).

Amorphous, odorless, tasteless powder. mp 1670°. d 3.14. Practically insol in water, alcohol or acetic acid; sol in dil HCl or HNO_3.

USE: Manuf of fertilizers, H_3PO_4 and P compds; manuf milk-glass, polishing and dental powders, porcelains, pottery; enameling; clarifying sugar syrups; in animal feeds; as noncaking agent; in the textile industry.

THERAP CAT: Calcium replenisher.

THERAP CAT (VET): Has been used as a dietary supplement, and as an antacid.

1702. Calcium Phosphide. Photophor. Ca_3P_2; mol wt 182.20. Ca 65.99%, P 34.01%. Prepn: Ehrlich in *Handbook of Preparative Inorganic Chemistry*, **vol. 1**, G. Brauer, Ed. (Academic Press, New York, 2nd ed., 1963) p 943.

Red-brown cryst powder or gray lumps. Dec by moist air or water, evolving spontaneously-flammable phosphine. d 2.51; mp about 1600°. *Keep dry and tightly closed.*

USE: For signal fires; in purification of Cu and Cu alloys; as rodenticide.

1703. Calcium Phosphite. $CaHO_3P$; mol wt 120.07. Ca 33.38%, H 0.84%, O 39.98%, P 25.80%. $CaHPO_3$. Prepn: *Gmelin's, Calcium* (8th ed.) **28B**, 1121 (1958).

Monohydrate, crystals. Loses water at 200°; dec above 300°. Slightly sol in water; practically insol in alcohol.

USE: Fertilizers; polymerization catalyst.

1704. Calcium Polycarbophil. WL 140; Carbofil; Mitrolan; Quival; Sorboquel. Calcium salt of a synthetic loosely crosslinked hydrophilic resin of the polycarboxylic type. Review of pharmacology, toxicology, clinical efficacy, and adverse effects: I. E. Danhof, *Pharmacotherapy* **2**, 18-28 (1982).

THERAP CAT: Cathartic.

1705. Calcium Propionate. *Propionic acid calcium salt;* Mycoban. $C_6H_{10}CaO_4$; mol wt 186.22. C 38.70%, H 5.41%, Ca 21.52%, O 34.37%. $Ca(CH_3CH_2COO)_2$. Occurs as mono- or trihydrate. Prepn: *Beilstein* **vol. 2**, 238, 2nd suppl., 218, 3rd suppl., 516.

Powder or monoclinic crystals. Sol in water; slightly sol in methanol, ethanol; practically insol in acetone, benzene.

USE: As an inhibitor of molds and other microorganisms in foods, tobacco, pharmaceuticals; in butyl rubber to improve processability and scorching resistance.

THERAP CAT: Antifungal.

1706. Calcium Pyrophosphate. Calcium diphosphate. $Ca_2O_7P_2$; mol wt 254.12. Ca 31.54%, O 44.08%, P 24.38%. $Ca_2P_2O_7$. Prepn by ignition of $CaHPO_4$: St. Pierre, *J. Am. Chem. Soc.* **77**, 2197 (1955).

Polymorphous crystals or powder. d 3.09. mp 1353°. Practically insol in water; sol in dil HCl or HNO_3.

USE: Abrasive; fertilizer; feed supplement; in dentifrices, ceramic ware, china, glass, phosphors.

1707. Calcium D-Saccharate. *D-Glucaric acid calcium salt.* $C_6H_8CaO_8$; mol wt 248.21. C 29.03%, H 3.25%, Ca 16.15%, O 51.57%. $CaC_6H_8O_8$. The normal calcium salt of D-saccharic acid, a dicarboxylic sugar acid derived from the oxidation of D-gluconic acid. Calcium D-saccharate is a true chemical compd and should not be confused with saccharated lime, formerly called "calcium saccharate" and produced by the action of lime upon sugar. Prepn: *Beilstein,* **vol. 3**, 2nd suppl., 378; *Hagers Handb. Pharm. Praxis* **vol. 1**, 755 (Berlin, 1930).

Tetrahydrate, odorless, tasteless crystals or fine white powder. Stable to air. Becomes anhydr upon heating at 100° *in vacuo*. Practically insol in water, alcohol, ether. Sol in dil mineral acids and in calcium gluconate solns.

USE: Pharmaceutic aid (stabilizer for calcium gluconate solns). As plasticizer in cement, concrete, mortar.

1708. Calcium Selenide. CaSe; mol wt 119.04. Ca 33.67%, Se 66.33%. Prepd by reducing $CaSeO_4$ in a stream of H_2 at 400-500°: Ehrlich in *Handbook of Preparative Inorganic Chemistry*, vol. **1**, G. Brauer, Ed. (Academic Press, New York, 2nd ed, 1963) p 939.

White powder. In air may turn red within a few minutes and light brown in a few hours. d 3.82. Decomposed by water. Treatment with HCl produces H_2Se gas, and red Se separates.

USE: In electron emitters.

1709. Calcium Silicate. Many different forms of calcium silicate are known. Among the most common forms are $CaSiO_3$, Ca_2SiO_4 and Ca_3SiO_5. Usually occur in hydrated form contg various percentages of water of crystallization. Names of calcium silicate minerals are: *afwillite; akermanite; calcium pectolith; centrallasite; crestmoreite; eaklite; foshagite; foshallasite; gjellebaekite; grammite; gyrolite; hillebrandite; larnite; okenite; parawollastonite; pseudo-wollastonite; riversideite; table spate; tobermorite; wollastonite; xonaltite; xonotlite.* Commercial calcium silicate sold for industrial use, such as *Micro-Cell* and *Silene*, is prepared synthetically to control its absorbing power. The usual method of prepn is from lime and diatomaceous earth under carefully controlled conditions: Boss, *Chem. Eng. News* **27**, 677 (1949); Steinour, *Chem. Revs.* **40**, 391 (1947). The commercial product is described here.

White or slightly cream-colored, free-flowing powder. Approximate analysis: CaO 19%, SiO_2 67%, H_2O 6 to 8%. d^{25} 2.10. Bulk density: 15 to 16 lb/cu ft. Absorbs 1 to 2.5 times its weight of liquids and still remains a free-flowing powder. Total absorption power for water about 600%, for mineral oil about 500%. Available surface area: 95 to 175 m^2/g. Ultimate particle size: 0.02 to 0.07 μ. pH of aq slurry 8.0 to 10.0. Practically insol in water. Forms a siliceous gel with mineral acids.

USE: Constituent (produced *in situ*) of lime glass, portland cement; reinforcing filler in elastomers and plastics; absorbent for liquids, gases, vapors; as anti-caking agent, suspension agent, pigment and pigment extender; binder for refractory material; in chromatography; in road construction.

1710. Calcium Stearate. *Octadecanoic acid calcium salt; stearic acid calcium salt.* $C_{36}H_{70}CaO_4$; mol wt 607.00. C 71.23%, H 11.62%, Ca 6.60%, O 10.54%. $Ca(C_{18}H_{35}O_4)_2$. The commercial prepn also contains palmitate. Prepn: Harrison, *Biochem. J.* **18**, 1222 (1924); Kebrich, Petrot, U.S. pat. **2,650,932** (1953 to National Lead).

Granular, fatty powder. Bulk density about 20 lb/cu ft, mp 147-149° (determined by gradient bar). Practically insol in water, ether, chloroform, acetone, cold alcohol; slightly sol in hot alcohol, in hot vegetable and mineral oils; quite sol in hot pyridine.

USE: For waterproofing fabrics, cement, stucco, explosives; as a releasing agent for plastic molding powders; as a stabilizer for polyvinyl chloride resins; lubricant; in pencils and wax crayons. Food grade calcium stearate, derived from edible tallow, is used as a conditioning agent in certain food and pharmaceutical products.

1711. Calcium Stearyl-2 Lactylate. *Stearic acid, ester with lactate of lactic acid, calcium salt; calcium stelate; Verv-Ca.* $C_{48}H_{86}CaO_{12}$; mol wt 895.30. C 64.40%, H 9.68%, Ca 4.48%, O 21.45%. Use in improving the mixing characteristics of flour: Thompson, Buddemeyer, *Cereal Chem.* **31**, 296 (1954); in improving whipping and baking properties of dried egg whites: Gorman, Keith, U.S. pat. **2,919,992** (1960 to Seymour Foods). Metabolism: J. C. Phillips *et al.*, *Food Cosmet. Toxicol.* **19**, 7 (1981).

$$(C_{17}H_{35}COOCHCOOCHCOO)_2 Ca$$
$$\quad\quad\quad CH_3 \quad CH_3$$

Free flowing, nonhygroscopic powder. Sparingly sol in water. pH of a 2% aq suspension 4.7.

USE: Dough conditioner in yeast-leavened bakery products; emulsifier in cosmetic and pharmaceutical industry.

1712. Calcium Succinate. *Butanedioic acid calcium salt; succinic acid calcium salt;* Artume. $C_4H_4CaO_4$; mol wt 156.15. C 30.77%, H 2.58%, Ca 25.67%, O 40.98%. $CaC_4H_4O_4$. Prepn: *Beilstein* vol. **2**, 607, 2nd suppl., 548, 3rd suppl., 1657.

Trihydrate, needles or granules. Slightly sol in water; practically insol in alcohol; sol in dil acids.

THERAP CAT: Combined with salicylates for rheumatic fever and rheumatoid arthritis.

1713. Calcium Sulfate. CaO_4S; mol wt 136.14. Ca 29.44%, O 47.01%, S 23.55%. $CaSO_4$. *Review:* R. J. Wenk, P. L. Henkels in Kirk-Othmer *Encyclopedia of Chemical Technology*, vol. **4** (Wiley-Interscience, New York, 3rd ed., 1978) pp 437-448.

The natural form of anhydrous calcium sulfate is known as the mineral *anhydrite;* also as *karstenite, muriacite, anhydrous sulfate of lime, anhydrous gypsum.* Crystals are orthorhombic, color varies, e.g., white with blue, gray or reddish tinge, or brick red. d 2.96. Hardness 3-3.5 (Mohs'). Sol in water (18.75°) 0.2 pts/100 pts. *Insoluble anhydrite* or *dead-burned gypsum* which has the same crystal structure as the mineral is obtained upon complete dehydration of gypsum at above 650°. *Soluble anhydrite* is obtained in granular or powder form by complete dehydration of gypsum at below 300° in an electric oven. Estimated pore space is 38% by volume. Possesses high affinity for water and will absorb 6.6% of its weight of water forming the stable hemihydrate.

Hemihydrate, *dried calcium sulfate; dried gypsum; plaster of Paris; Annalin.* Fine, odorless, tasteless powder. When mixed with water, sets to a hard mass. *Keep well closed.*

Dihydrate, *native calcium sulfate; precipitated calcium sulfate; gypsum; alabaster; selenite; terra alba; satinite; mineral white; satin spar; light spar.* Lumps or powder. d 2.32. It loses only part of its water at 100-150°. Sol in water; very slowly sol in glycerol. Practically insol in most organic solvents.

USE: *Anhydrous:* Insol anhydrite is used in cement formulations and as a paper filler. Soluble anhydride, because of its strong tendency to absorb moisture, is useful as a drying agent for solids, organic liquids and gases; the desiccant used in laboratory and industry is known under the name *Drierite.* This material can be regenerated repeatedly and reused without noticeable decrease in its desiccating efficiency. The *hemihydrate* is used for wall plasters; wallboard; tiles and blocks for the building industry; moldings; statuary; in the paper industry. The *dihydrate* is used in the manuf of portland cement; in soil treatment to neutralize alkali carbonates and to prevent loss of volatile and dissolved nitrogenous compounds by volatilization and leaching; for the manuf of plaster of Paris, artificial marble; as a white pigment, filler or glaze in paints, enamels, pharmaceuticals, paper, insecticide dusts, yeast manuf, water treatment, polishing powders; in the manuf of sulfuric acid, CaC_2, $(NH_4)_2SO_4$, porous polymers. Pharmaceutic aid (in plaster casts).

1714. Calcium Sulfide. CaS; mol wt 72.14. Ca 55.56%, S 44.44%. Pure CaS prepd in the laboratory by heating pure $CaCO_3$ in a stream of $H_2S + H_2$ at 1000°: Ehrlich in *Handbook of Preparative Inorganic Chemistry* vol. **1**, G. Brauer, Ed. (Academic Press, New York, 2nd ed., 1963) p 938. Crude calcium sulfide, erroneously called *sulfurated lime, calcic liver of sulfur, liver of lime, hepar calcis,* made by igniting calcium sulfate with carbonaceous matter. Contains not less than 55% CaS; the balance is calcium sulfate, sulfite and carbonate, and the "ash" from the carbonaceous material. *See Mellor's* vol III, p 740 (1928). Luminous calcium sulfide or *Canton's phosphorus* made by igniting a mixture of $CaCO_3$ and S with very small quantities of Bi or Mn salts, etc.: Verneuil, *Compt. Rend.* **103**, 600 (1886); *Mellor's, loc. cit.*

White powder if pure; crude and luminous calcium sulfide may be yellowish to pale-gray. Odor of H_2S in moist air; unpleasant alkaline taste. Oxidizes in dry air and dec in moist air. mp > 2000°. d 2.59. Slightly sol in cold, more sol in hot water with partial decompn; freely sol in solns of ammonium salts; practically insol in alcohol; dec even by weak acids, evolving H_2S. *Keep well closed.*

USE: In phosphors; as lubricant additive. Pure CaS used in electron emitters. Luminous CaS used for making luminous paints or varnishes.

THERAP CAT (VET): Has been used in chronic suppurative lesions.

1715. Calcium Sulfite. CaO_3S; mol wt 120.14. Ca 33.36%, O 39.95%, S 26.69%. $CaSO_3$. Prepn: *Gmelin's, Calcium* (8th ed.) **28B**, 107-108, 660-674 (1958).

Dihydrate, crystals or powder. Slowly oxidizes in air to $CaSO_4$. Slightly sol in water, alcohol; sol in SO_2 solns, acids with liberation of SO_2.

USE: Preserving cider and other fruit juices; as disinfectant of brewing vats; antichlor in bleaching textiles; in sugar manuf; in paper pulp cooking; in cement.

1716. Calcium Tartrate. *2,3-Dihydroxybutanedioic acid calcium salt.* $C_4H_4CaO_6$; mol wt 188.15. C 25.53%, H 2.14%, Ca 21.30%, O 51.02%. $CaC_4H_4O_6$. A byproduct of the wine industry. Prepn from wine dregs: Dabul, U.S. pat. **3,114,770** (1963 to Orandi & Massera). *See also* the processes mentioned under L-tartaric acid.

Tetrahydrate, powder. Slightly sol in water (from about 0.04% at 10° to about 0.2% at 85°) or in alcohol; sol in dil HCl or HNO_3.

USE: As preservative for fruits, vegetables, seafoods; in deodorization of fish; as antacid.

1717. Calcium Thiocyanate. Calcium rhodanate; calcium sulfocyanate. $C_2CaN_2S_2$; mol wt 156.24. C 15.37%, Ca 25.65%, N 17.93%, S 41.04%. $Ca(SCN)_2$. Prepn: *Gmelin's, Calcium* (8th ed.) **28B**, 972-976 (1958).

Tetrahydrate, hygroscopic crystals or cryst powder. Dec on heating above 160°. Very sol in water; sol in methanol, ethanol, acetone. *Keep well closed.*

USE: In manuf of acrylonitrile polymers; for parchmentizing; for stiffening of textiles; soln as a solvent for textiles.

1718. Calcium Thioglycollate. *Mercaptoacetic acid calcium derivative;* Depil. $C_2H_2CaO_2S$; mol wt 130.19. C 18.45%, H 1.55%, Ca 30.79%, O 24.58%, S 24.63%. Prepn: Hoshall, *J. Assoc. Offic. Agr. Chem.* **23**, 727 (1940).

Trihydrate, prismatic rod crystals. Odorless or faint mercaptan odor; somewhat astringent and fetid taste. Slowly loses H_2O above 95°, darkens at 220° and partially fuses with decompn at 280-290°. Sol in water; very slightly sol in alcohol, chloroform; practically insol in ether, petr ether, benzene. Solns readily absorb CO_2 from air, pptg $CaCO_3$. *Keep well closed.*

USE: Depilatory; tanning leather; in hair-waving prepns.

1719. Calcium Thiosulfate. Calcium hyposulfite; Tecesal. CaS_2O_3; mol wt 152.20. Ca 26.33%, O 31.54%, S 42.13%. Prepn: Ballezo, Kaufmann, *Monatsh.* **80**, 220 (1949); Levenson, U.S. pat. **2,763,531** (1956 to Kodak).

Hexahydrate, triclinic crystals. When dry, dec on standing forming a yellow crust on surface; more stable if kept damp and stored below 0°. Spontaneously dec at 43-49°. d 1.87. Freely sol in water; practically insol in alcohol. *Keep well closed and in a cool place.*

THERAP CAT: In dermatitis and jaundice due to arsphenamine.

1720. Calcium Tungstate(VI). CaO_4W; mol wt 288.00. Ca 13.92%, O 22.23%, W 63.86%. $CaWO_4$. Occurs in nature as the mineral *scheelite.* The article of commerce is usually made by pptn: Boericke, Boericke, U.S. pat. **2,390,687** (1945); prepn by heating a stoichiometric mixture of CaO or $CaCO_3$ and tungstic acid: Kroger, *Nature* **159**, 674 (1947); prepn of single crystals: Uitert, Soden, *J. Appl. Phys.* **31**, 328 (1960).

Tetragonal crystals. d 6.06. Practically insol in water; dec by hot HCl or HNO_3.

USE: Very small crystals have been used for injection into malignant tumors, etc., thus affording by transillumination a means of x-ray treatment; for preparing screens for x-ray observations and photographs; in luminous paints; in scintillation counters.

1721. Caldariomycin. *(1S-trans)-2,2-Dichloro-1,3-cyclopentanediol;* 1,3-dihydroxy-2,2-dichlorocyclopentane. $C_5H_8Cl_2O_2$; mol wt 171.01. C 35.09%, H 4.71%, Cl 41.50%, O 18.70%. Mold metabolite from *Caldariomyces fumago* Woronichin. Isoln and structure: Clutterbuck *et al., Biochem. J.* **34**, 664 (1940). Biosynthesis: Shaw *et al., J. Biol. Chem.* **234**, 2560 (1959); Beckwith, *ibid.* **238**, 3086 (1963); Beckwith, Hager, *ibid.* 3091. Synthesis: *eidem, J. Org. Chem.* **26**, 5206 (1961); Burgstahler *et al., ibid.* **31**, 3516 (1966). Abs configuration: Johnson *et al., J. Am. Chem. Soc.* **90**, 136 (1968).

Fine, colorless needles, mp 121°. Decomposes at temps >180°; very stable to heat at mod. temps. $[\alpha]_{5461}^{20}$ +59.2° (c = 0.338 in H_2O). Readily sol in cold water, ethanol, ether and most other org solvents. Slightly sol in chloroform.

1722. C-Calebassine. C-Toxiferine II; C-strychnotoxine. $[C_{40}H_{48}N_4O_2]^{2+}$; mol wt 616.82. From calabash-curare: Wieland *et al., Ann.* **547**, 156 (1941); Karrer, Schmidt, *Helv. Chim. Acta* **29**, 1853 (1946); Zürcher *et al., J. Am. Chem. Soc.* **80**, 1500 (1958). Identity with C-toxiferine II and C-strychnotoxine: Wieland, Merz, *Ber.* **85**, 731 (1952). Structure: Hesse *et al., Helv. Chim. Acta* **44**, 2211 (1961); Fehlmann *et al., ibid.* **48**, 303 (1965). Synthesis from C-dihydrotoxiferine chloride: Bernauer *et al., ibid.* **40**, 1999 (1957); Grdinic *et al., J. Am. Chem. Soc.* **86**, 3357 (1964).

Dichloride, $C_{40}H_{48}Cl_2N_4O_2$, needles from methanol + ether. $[\alpha]_D^{20}$ +72.1° (c = 0.67 in water). uv max (water): 253, 302 nm (log ϵ 4.37, 3.77).

Dipicrate, $C_{52}H_{52}N_{10}O_{16}$, prisms from acetone + water, mp 215°. Slightly sol in acetone; practically insol in water, methanol, ethanol, dioxane.

THERAP CAT: Skeletal muscle relaxant.

1723. Calendula. Marigold; Mary-bud; gold-bloom; holligold. Dried, ligulate florets of *Calendula officinalis* L., *Compositae. Habit.* Southern Europe and Levant; cultivated everywhere in gardens. *Constit.* Volatile oil; calendulin, carotenoid pigments, a saponin on hydrolysis yields oleanolic acid, bitter principle (caledin). *Refs:* Zimmerman, *Helv. Chim. Acta* **29**, 445 (1946); Gedeon, *Pharmazie* **6**, 547 (1951); **9**, 922 (1954), Kasprzyk, *C.A.* **47**, 6918c (1953); Suchy, Herout, *Coll. Czech. Chem. Commun.* **26**, 890 (1961).

THERAP CAT: Topical anti-inflammatory.

1724. Californium. Cf; at. wt (most stable known isotope) 251; at. no. 98; valence 3, also 2. Man-made, radioactive element. First isotope, ^{245}Cf ($T_{1/2}$ 44 min), discovered in 1950: Thompson *et al., Phys. Rev.* **80**, 790 (1950). Obtained originally by bombarding ^{242}-curium with α-particles. Known isotopes (mass numbers): 240-255. ^{251}Cf ($T_{1/2}$ ~800 years, α-emitter) produced by neutron irradiation of uranium and higher actinides. Prepn of ^{249}Cf metal ($T_{1/2}$ 360 years, α-emitter) by reduction of Cf_2O_3 with lanthanum metal: Haire, Baybarz, *J. Inorg. Nucl. Chem.* **36**, 1295 (1974). Medical uses of ^{252}Cf (α-decay half-life 2.731 years; spontaneous fission half-life 85.5 years): Seaborg, *Handb. Exp. Pharmakol.* **36**, 929 (1973). Clinical trials in cervical cancer: Y. Maruyama *et al., Int. J. Radiat. Oncol. Biol. Phys.* **11**, 1475 (1985); Y. Maruyama *et al., Cancer* **59**, 1500 (1987). *Reviews:* Cunningham, *J. Chem. Ed.* **36**, 32-37 (1959); M. Haissinsky, J. P. Adloff, *Radiochemical Survey of the Elements* (Elsevier, New York, 1965) pp 28-29; C. Keller, *The Chemistry of the Transuranium Elements* (Verlag Chemie, Weinheim, English Ed., 1971) pp 567-581; Silva, "Trans-Curium Elements" in *MTP Int. Rev. Sci.: Inorg. Chem., Ser. One* vol. 8, A. G. Maddock, Ed. (University Park Press, Baltimore, 1972) pp 71-105; *Comprehensive Inorganic Chemistry* vol.5, J. C. Bailar, Jr. *et al.*, Eds. (Pergamon Press, Oxford, 1973) *passim;* several authors, *Handb. Exp. Pharmakol.* **36**, 689-928 (1973). Review of radiobiology and therapeutic applications: Y. Maruyama *et al., Oncology* **35**, 172 (1978).

Metal; two crystalline forms, face-centered cubic and hexagonal close-packed structures. mp 900±30°: Haire, Baybarz, *loc. cit.*

THERAP CAT: ^{252}Cf as antineoplastic (radiation source).

1725. Calmagite. *3-Hydroxy-4-[(2-hydroxy-5-methylphenyl)azo]-1-naphthalenesulfonic acid;* 1-(1-hydroxy-4-methyl-2-phenylazo)-2-naphthol-4-sulfonic acid. $C_{17}H_{14}-N_2O_5S$; mol wt 358.38. C 56.97%, H 3.94%, N 7.82%, O 22.32%, S 8.95%. Prepn from 1-amino-2-naphthol-4-sulfonic acid and *p*-cresol: Lindstrom, Diehl, *Anal. Chem.* **32**, 1123 (1960).

Red crystals from acetone. Sol in water. Absorption max (pH 10.10): 610 nm (ε 20,300). Functions as acid-base indicator: Aq solns are bright red at low pH, red at pH 7.1 to 9.1, blue at pH 9.1 to 11.4. The blue color at pH 10 is changed to red by the addition of calcium or magnesium.

USE: As indicator in titration of Ca or Mg with EDTA.

1726. Calmodulin. CaM; calcium-dependent regulator protein; CDR. A calcium-binding multifunctional regulatory protein, ubiquitously distributed in eukaryotic cells. It functions as an intracellular intermediary for calcium ions and activates a number of enzymes involved in fundamental cell processes, such as protein phosphorylation, contractile processes, and metabolism of cyclic nucleotides, of glycogen and of calcium, as well as in other metabolic reactions. Originally discovered as a protein activator of cyclic 3',5'-nucleotide phosphodiesterase: W. Y. Cheung, *Biochem. Biophys. Res. Commun.* **29**, 478 (1967); *idem, Biochim. Biophys. Acta* **191**, 303 (1969); *idem, Biochem. Biophys. Res. Commun.* **38**, 533 (1970); S. Kakiuchi *et al., ibid.* **41**, 1104 (1970). Calcium-binding activity: T. S. Teo, J. H. Wang, *J. Biol. Chem.* **248**, 5950 (1973). Calmodulin is a relatively small, acidic, stable monomer of mol wt 15,000-19,000, lacking cysteine, hydroxyproline, and tryptophan. It has a high content of acidic amino acids and low tyrosine content; almost all calmodulins isolated also contain a single, fully trimethylated lysyl residue. Purification from bovine brain: Y. M. Lin *et al., ibid.* **249**, 4943 (1974); *eidem, Methods Enzymol* **39**, Pt. C, 262 (1974). Amino acid sequence of CaM from bovine brain: T. C. Vanaman in *Calcium Binding Proteins and Calcium Function,* R. H. Wasserman *et al.*, Eds. (Elsevier, New York, 1977) pp 107-116; D. M. Watterson *et al., J. Biol. Chem.* **255**, 962 (1980); from rat testis: J. R. Dedman *et al., ibid.* **253**, 343 (1978); from sea invertebrate, *Renilla reniformis:* T. C. Vanaman, F. Sharief, *Fed. Proc.* **38**, 788 (1979). Radioimmunoassay: R. W. Wallace, W. Y. Cheung, *J. Biol. Chem.* **254**, 6564 (1979); J. C. Chafouleas *et al., ibid.* 10262. CaM has four Ca^{2+}-binding sites; binding

to any one of the sites results in a conformational change, which is required for calmodulin to regulate enzyme systems. Conformational transition study: C. B. Klee, *Biochemistry* **16**, 1017 (1977). An important feature of calmodulin's structure is the presence of four homologous internal amino acid sequences, called domains. Each of these domains (one for each bound calcium) are reportedly "E-F Hands", a structural concept described by R. H. Kretsinger in *Calcium Transport in Contraction and Secretions,* E. Carafoli *et al.*, Eds. (Elsevier, Amsterdam, 1975) pp 469-478. Calmodulins obtained from a wide range of phylogenetically different sources are similar in amino acid sequence and in physico-chemical and biological properties; hence, the protein lacks both species and tissue specificity and appears to be structurally and functionally conserved throughout evolution. *Reviews:* W. Y. Cheung, *Science* **207**, 19-27 (1980); A. R. Means, J. R. Dedman, *Nature* **285**, 73-77 (1980); C. B. Klee *et al., Ann. Rev. Biochem.* **49**, 489-515 (1980); A. R. Means, *Recent Progr. Horm. Res.* **37**, 333-367 (1981); Y. M. Lin, *Mol. Cell Biochem.* **45**, 101-112 (1982). Books: *Ann. N.Y. Acad. Sci.* **356**, entitled "Calmodulin and Cell Functions", D. M. Watterson, F. F. Vincenzi, Eds. (1980) 455 pp; *Calcium and Cell Function* vol. 1, W. Y. Cheung, Ed. (Academic Press, New York, 1980) 395 pp.

Isoelectric pt 3.9-4.3. $\epsilon^{1\%}_{276nm}$ 1.8; $\epsilon^{1\%}_{280nm}$ 2.1. Stable when subjected to heat at neutral or acidic pH.

1727. Calomelol. Colloidal calomel; sol calomel. Consists of 80% $HgCl$ and 20% proteins.

Whitish-gray, odorless, tasteless powder. Sol in water to an opalescent suspension; insol in alcohol, ether; pptd by acids, redissolved by alkalies. *Protect from light.*

THERAP CAT: Cathartic.

1728. Calotropin. *[2α(2S,3S,4S,6R),3β,5α]-14-Hydroxy-19-oxo-3,2-[(tetrahydro-3,4-dihydroxy-6-methyl-2H-pyran-2,3-diyl)bis(oxy)]card-20(22)-enolide.* $C_{29}H_{40}O_9$; mol wt 532.61. C 65.39%, H 7.57%, O 27.04%. African arrow poison isolated from milk sap of *Calotropis procera* Dryand., *Asclepiadaceae.* Isolation: Lewin, *Arch. Exp. Path. Pharmakol.* **71**, 142 (1913); Hesse *et al., Ann.* **526**, 252 (1936); **566**, 130 (1950); Rajagopalan *et al., Helv. Chim. Acta* **38**, 1809 (1955). Isoln from *Asclepias curassavica* L., *Asclepiadaceae:* S. M. Kupchan *et al., Science* **146**, 1685 (1964). Structure: G. Hesse, G. Lettenbauer, *Ann.* **623**, 142 (1959); Hesse *et al., ibid.* **625**, 157, 161 (1959); D. G. H. Crout *et al., Tetrahedron Letters* **1963**, 63; *J. Chem. Soc.* **1964**, 2187. Extraction from *C. procera* R.Br. and toxicity: F. Brüschweiler *et al., Helv. Chim. Acta* **52**, 2086 (1969). Revised structure: *eidem, ibid.* 2276. Sequestration by larvae of Monarch butterfly *Danaus plexippus* L.: J. N. Sieber *et al., J. Chem. Ecol.* **6**, 321 (1980). Quantitative analysis of cardenolides in latex and leaves of *C. procera: eidem, Phytochemistry* **21**, 2343 (1982). Biosynthesis of labelled compd: M. S. Lee, J. N. Sieber, *ibid.* **22**, 923 (1983).

Rectangular platelets from alcohol or ethyl acetate, mp 223° (dec). $[\alpha]_D^{18}$ +66.8° (in methanol). Sol in water, alc. Practically insol in ether. uv max: 217, 310 nm (log ε 4.21, 1.49). Lethal dose in cats: 0.12 mg/kg (Brüschweiler).

1729. Calumba. Colombo. Root of *Jatrorrhiza palmata* (DC.) Miers (*J. columba* Miers), *Menispermaceae.* Habit. Eastern Africa. *Constit.* Columbin, chasmanthin, palmarin (isomer of chasmanthin), jatrorrhizine, columbic acid, columbamine; contains no tannin: Barton, Elad, *J. Chem. Soc.*

1956, 2085, 2090. *Review:* Feist, *Arzneimittel-Forsch.* **1**, 418 (1951).

THERAP CAT (VET): Has been used as a stomachic.

1730. Calusterone. *17-Hydroxy-7,17-dimethylandrost-4-en-3-one; 7β,17α-dimethyltestosterone;* U -22550; Methosarb. $C_{21}H_{32}O_2$; mol wt 316.49. C 79.70%, H 10.19%, O 10.11%. Prepn: Campbell, Babcock, U.S. pats. **3,029,263; 3,341,557** (1962, 1967 both to Upjohn). Epimeric with bolasterone (*q.v.*). Clinical studies: Gordan *et al., J. Am. Med. Assoc.* **219**, 483 (1972).

Crystals from acetone, mp 127-129°. $[\alpha]_D$ +57° (CHCl$_3$). uv max (alcohol): 243 nm.

THERAP CAT: Antineoplastic.

1731. Calycanthine. $C_{22}H_{26}N_4$; mol wt 346.48. C 76.26%, H 7.56%, N 16.17%. In *Calycanthus floridus* L., *C. glaucus* Willd.; *Chimonanthus praecox* (L.) Link, *Calycanthaceae.* First isoln: Eccles, *Proc. Am. Pharm. Assoc.* **84**, 382 (1888). Extraction procedure: Manske-Marion, *Can. J. Res.* **17B**, 293 (1939). Structure: Woodward *et al., Proc. Chem. Soc.* **1960**, 76; Hamor *et al., ibid.* 78; Hamor, Robertson, *J. Chem. Soc.* **1962**, 194. Configuration: Clayton *et al., Tetrahedron* **18**, 1495 (1962). Synthesis of DL-form: Hendrickson *et al., ibid.* **20**, 565 (1964); Hall *et al., ibid.* **23**, 4131 (1967).

Crystals, mp 245° (evac tube). $[\alpha]_D^{18}$ +684° (c = 1.2 in abs alc). uv max (95% ethanol): 250, 309 nm (log ε 4.28, 3.80). Alkaline reaction to litmus. Freely sol in alc, chloroform; sol in ether, acetone, pyridine; slightly sol in water.

Monohydrate, orthorhombic bipyramidal crystals from water, mp 220°.

Caution: Highly toxic. Can cause violent convulsions, paralysis, cardiac depression.

1732. Camazepam. *Dimethylcarbamic acid 7-chloro-2,3-dihydro-1-methyl-2-oxo-5-phenyl-1H-1,4-benzodiazepin-3-yl ester; 7-chloro-1,3-dihydro-3-hydroxy-1-methyl-5-phenyl-2H-1,4-benzodiazepin-2-one dimethylcarbamate (ester); 7-chloro-1,3-dihydro-3-(N,N-dimethylcarbamoyl)-1-methyl-5-phenyl-2H-1,4-benzodiazepin-2-one;* SB-5833; Albego. $C_{19}H_{18}ClN_3O_3$; mol wt 371.82. C 61.38%, H 4.88%, Cl 9.53%, N 11.30%, O 12.91%. Prepn: G. Ferrari, C. Casagrande, **Ger.** pat. **2,142,181;** *eidem,* **U.S.** pat. **3,799,920** (1972, 1974 both to Siphar). Metabolism: F. Marcucci *et al., J. Pharm. Sci.* **67**, 1470 (1978). Pharmacology: L. Merlo *et al., Arzneimittel-Forsch.* **24**, 1759 (1974); R. Ferrini *et al., ibid.* 2029. Clinical studies: A. Tammaro *et al., ibid.* **27**, 2177 (1978); S. Carrara *et al., Eur. J. Clin. Pharmacol.* **13**, 335 (1978).

White crystalline powder from ethyl acetate, mp 173-174°. Sol in alcohol, moderately sol in water. LD_{50} orally in mice, rats: 970, >4000 mg/kg (Ferrini).

Note: This is a controlled substance (depressant) listed in the U.S. Code of Federal Regulations, Title 21 Part 1308.14 (1987).

THERAP CAT: Anxiolytic.

1733. Cambendazole. *[2-(4-Thiazolyl)-1H-benzimidazol-5-yl]carbamic acid 1-methylethyl ester; 2-(4-thiazolyl)-5-benzimidazolecarbamate isopropyl ester;* 5-isopropoxycarbonylamino -2 -(4 -thiazolyl)benzimidazole; 5-isopropoxycarbonylaminothiabendazole; isopropyl 2-(4 -thiazolyl)-5 -benzimidazolecarbamate; MK-905; Bonlam; Bovicam; Cambenzole; Cambet; Equiben; Novazole; Noviben. $C_{14}H_{14}N_4O_2S$; mol wt 302.35. C 55.61%, H 4.67%, N 18.53%, O 10.58%, S 10.60%. Prepn and activity studies: Hoff, Fisher, **S. Afr.** pat. **68 00,351** (1969 to Merck & Co.), *C.A.* **72**, 90461q (1970); Hoff *et al., Experientia* **26**, 550 (1970). Clinical studies with sheep, cattle and swine: Egerton *et al., Res. Vet. Sci.* **11**, 193, 495, 590 (1970).

Odorless, white crystalline solid, mp 238-240° (dec). Sol in alcohol, dimethylformamide; sparingly sol in acetone; slightly sol in benzene; very slightly sol in 0.1M HCl. Practically insol in isooctane and water (0.02 mg/ml). Stable in acid and base in range of pH 1 to 12. uv max (0.1N HCl): 319, 232 nm ($A_{1cm}^{1\%}$ 740, 670).

THERAP CAT (VET): Anthelmintic.

1734. Camostat. *4-[[4-[(Aminoiminomethyl)amino]benzoyl]oxy]benzeneacetic acid 2-(dimethylamino)-2-oxoethyl ester; N,N-dimethylcarbamoylmethyl -p-(p-guanidinobenzoyloxy)phenylacetate.* $C_{20}H_{22}N_4O_5$; mol wt 398.43. C 60.29%, H 5.57%, N 14.06%, O 20.08%. Orally active, non -peptide proteolytic enzyme inhibitor with anti-trypsin and anti-plasmin activities, related structurally to gabexate, *q.v.* Prepn: S. Fujii *et al.,* **Ger. pat. 2,548,886** corresp to **U.S.** pat. **4,021,472** (1976, 1977 both to Ono). Enzyme inhibition study: Y. Tamura *et al., Biochim. Biophys Acta* **484**, 417 (1977). Inhibition of exptl tumors in mice: M. Ohkoshi, *Gann* **72**, 959 (1981), *C.A.* **96**, 97304v (1982); M. Ohkoshi, S. Fujii, *ibid.* **73**, 108 (1982), *C.A.* **96**, 155182s (1982). Prevention of acute exptl pancreatitis: S. Takasugi *et al., Digestion* **24**, 36 (1982). Absorption and excretion: S. Hiraku *et al., Iyakuhin Kenkyu* **13**, 756 (1982), *C.A.* **97**, 120038t (1982).

Monomethanesulfonate, $C_{21}H_{26}N_4O_8S$, *Foy-305, camostat mesylate, Foypan.* Solid from methanol/ether, mp 150-155°. Sol in water.

THERAP CAT: Enzyme inhibitor (proteinase).

1735. Campesterol. *(24R)-Ergost-5-en-3β-ol.* $C_{28}H_{48}O$; mol wt 400.66. C 83.93%, H 12.08%, O 3.99%. Small amounts are found in rape-seed oil derived from *Brassica campestris* L., *Cruciferae*, in soybean oil, and in wheat germ oil. Isoln: Fernholz, MacPhillamy, *J. Am. Chem. Soc.* **63**, 1155 (1941). Structure: Fernholz, Ruigh, *ibid.* 1157. Synthesis: Tarzia *et al.*, *Gazz. Chim. Ital.* **97**, 102 (1967), *C.A.* **67**, 32883q (1967).

Crystals from acetone, mp 157-158°. $[\alpha]_D^{23}$ −33° (22.5 mg in 5 ml chloroform).
Acetate, $C_{30}H_{50}O_2$, crystals from alc, mp 137-138°. $[\alpha]_D^{23}$ −35° (28.8 mg in 1 ml chloroform).

1736. Camphene. *2,2-Dimethyl-3-methylenebicyclo-[2.2.1]heptane; 2,2-dimethyl-3-methylenenorbornane; 3,3-dimethyl-2-methylenenorcamphane.* $C_{10}H_{16}$; mol wt 136.23. C 88.16%, H 11.84%. Occurs in many essential oils, such as turpentine (*levo* and *dextro* forms), in cypress oil *(dextro* form), in camphor oil from species of *Lauraceae (dextro)*, in bergamot oil, in oil of citronella, neroli, ginger, valerian. Reviews on isolation, preparation and properties: J. L. Simonsen, *The Terpenes* vol. II (Cambridge Univ. Press, 1949) pp 280-322; E. Guenther, *The Essential Oils* vol. II (Van Nostrand, 1949) pp 66-70. Synthesis: G. W. Hana, H. Koch, *Ber.* **111**, 2527 (1968).

dl-Form, cubic crystals from alcohol. Large dodecahedra by slow sublimation. Volatilizes on exposure to air. Insipid odor. mp 51-52°. bp$_{760}$ 158.5-159.5°; bp$_{100}$ 92.4°; bp$_{16}$ 55-56°. d_4^{54} 0.8422. n_D^{54} 1.45514. Practically insol in water. Moderately sol in alcohol; sol in ether, cyclohexane, cyclohexene, dioxane, chloroform.
d-Form, mp 52°. $[\alpha]_D^{17}$ +103.5° (c = 9.67 in ether). d_4^{50} 0.8486. n_D^{50} 1.4605.
l-Form, mp 52°. $[\alpha]_D^{21}$ −119.11° (c = 2.33 in benzene). d_4^{54} 0.8422. n_D^{40} 1.4620.

1737. d-Camphocarboxylic Acid. *4,7,7-Trimethyl-3-oxobicyclo[2.2.1]heptane-2-carboxylic acid; d-2-oxo-3-bornanecarboxylic acid; d-3-camphorcarboxylic acid; d-2-oxo-3-camphanecarboxylic acid; d-3-carboxy-2-bornanone; d-3-carboxy-2-camphanone.* $C_{11}H_{16}O_3$; mol wt 196.24. C 67.32%, H 8.22%, O 24.46%. Prepd by carboxylation of *d*-camphor: Brühl, *Ber.* **24**, 3373 (1891).

Crystals from benzene, water, ether, or 50% alcohol. mp 127-128°. Sparingly sol in cold water, more sol in warm water; sol in alcohol, ether, chloroform, in about 2 parts boiling benzene. Sparingly sol in cold benzene. Practically insol in cold petr ether; very slightly sol in boiling petr ether.

For soln contg 0.38 g in 25 ml solvent: $[\alpha]_D$ +18° (benzene), +60° (alcohol), +73.3° (water).
Ammonium salt, $C_{11}H_{19}NO_3$, *camphydryl, Canfoxil, camphor solubilized.*
Basic bismuth salt, $C_{33}H_{46}Bi_2O_{11}$, *Bismo-Cymol, Angimuth, Camphobismol.* Prepn: Raiziss, Clemence, U.S. pat. **1,921,638** (1933 to Abbott). Powder. Odor of camphor. Practically insoluble in water; soluble in methanol, ether, benzene, oils.

THERAP CAT: Basic bismuth salt formerly as antisyphilitic.

1738. Camphor. *1,7,7-Trimethylbicyclo[2.2.1]heptan-2-one;* 2-bornanone; 2-camphanone; 2-keto-1,7,7-trimethylnorcamphane; gum camphor; Japan camphor; Formosa camphor; laurel camphor. $C_{10}H_{16}O$; mol wt 152.23. C 78.89%, H 10.60%, O 10.51%. Occurs in all parts of the camphor tree, *Cinnamomum camphora* T. Nees & Ebermeier, *Lauraceae. Habit:* Java, Sumatra, China (central provinces), Japan, Formosa, Brazil. Obtained by steam distillation from comminuted trees which should be at least 50 years old. Description of various indigenous processes: Gubelmann, Elley, *Ind. Eng. Chem.* **26**, 589 (1934); G. Etzel in Kirk-Othmer *Encyclopedia of Chemical Technology,* **vol. 4** (Wiley, New York, 2nd ed., 1964) pp 54-58. Modern processes start with vinyl chloride and cyclopentadiene to obtain the important intermediate dehydronorbornyl chloride. Review of syntheses: K. Alder in *New Methods of Preparative Organic Chemistry* (New York, 1948); A. F. Thomas in *The Total Synthesis of Natural Products* vol. 2, J. ApSimon, Ed. (Wiley-Interscience, New York, 1973) pp 149-154. More than three-fourths of the camphor sold in the U.S. is produced synthetically (usually from pinene), and most is sold in the racemic form, although the U.S.P. specifies the *d*-form. Configuration: Freudenberg *et al.*, *Ann.* **594**, 76 (1955). Toxicity: A. Smith, G. Margolis, *Am. J. Pathol.* **30**, 857 (1954).

Translucent mass with crystalline fracture. Rhombohedral crystals from alcohol. Cubic crystals by melting and chilling. Familiar fragrant and penetrating odor. Slightly bitter and cooling taste. d_4^{25} 0.992. mp 179.75° (corr., open capillary, 2 mm diam). bp 204°. Sublimes appreciably at room temp and press. *Keep in tight containers away from heat.* At 80° and 12 mm press 14% sublimes within 60 minutes. Very volatile in steam. $[\alpha]_D^{25}$ +41° to +43° (c = 10 in U.S.P. alcohol) according to U.S.P. specif. The water content of the ethanol influences the rotation considerably. $[\alpha]_D^{20}$ +43.8° (c = 7.5 in abs alcohol). uv max (CHCl$_3$): 292 nm. At 25° one gram dissolves in about 800 ml water (giving a colloidal soln), in 1 ml alcohol, 1 ml ether, 0.5 ml chloroform, 0.4 ml benzene, 0.4 ml acetone, 1.5 ml oil of turpentine, 0.5 ml glacial acetic acid. Sol in aniline, nitrobenzene, carbon disulfide, tetralin, decalin, methylhexalin, petr ether, in the higher alcohols, in fixed and volatile oils. Also sol in concd mineral acids in phenol, in liquid NH_3 and in liquid SO_2. Camphor has a peculiar tenacity and cannot be powdered in a mortar unless it is moistened with an organic solvent. Liquefies when triturated with chloral hydrate, menthol, resorcinol, salol, β-naphthol, thymol, phenol, urethan. LD_{50} i.p. in mice: 3000 mg/kg (Smith, Margolis).
Incompat: Incompatible with potassium permanganate; salts of any kind should not be added to camphor water.
Human Toxicity: Ingestion or injection may cause nausea, vomiting, vertigo, mental confusion, delirium, clonic convulsions, coma, respiratory failure, death, *Clinical Toxicology of Commercial Products,* R. E. Gosselin *et al.*, Eds. (Williams & Wilkins, Baltimore, 4th ed., 1976) Section III, pp 77-79.
USE: Excellent plasticizer for cellulose esters and ethers; used in manuf of plastics, esp celluloid; in lacquers and varnishes; in explosives; in pyrotechnics; as moth repellent; in embalming fluids; in manuf cymene; as preservative in phar-

maceuticals and cosmetics; in camphorated parachlorophenol and paregoric.

THERAP CAT: Topical anti-infective; topical antipruritic.

THERAP CAT (VET): Has been used internally as a stimulant and carminative; externally as an antipruritic, counterirritant and antiseptic.

1739. Camphoric Acid. *1,2,2-Trimethyl-1,3-cyclopentanedicarboxylic acid;* dextrocamphoric acid. $C_{10}H_{16}O_4$; mol wt 200.23. C 59.98%, H 8.06%, O 31.96%. By oxidation of camphor: Bredt, *Ber.* **26**, 3047 (1893). Synthesis: Perkin, Thorpe, *J. Chem. Soc.* **89**, 799 (1906); Toivonan, *Acta Chem. Scand.* **2**, 597 (1948).

Leaflets from water, monoclinic prisms from alc, mp 186-188°. d 1.186. $[\alpha]_D^{20}$ +47° to +48° (alc). One gram dissolves in 125 ml water, 10 ml boiling water, 1 ml alc, 20 ml glycerol; sol in chloroform, ether, fats, oils.

1740. d-Camphorsulfonic Acid. *7,7-Dimethyl-2-oxobicyclo[2.2.1]heptane-1-methanesulfonic acid; 2-oxo-10-bornanesulfonic acid;* 10-camphorsulfonic acid; camsylate; camphostyl; β-camphorsulfonic acid; Reychler's acid. $C_{10}H_{16}O_4S$; mol wt 232.31. C 51.70%, H 6.94%, O 27.55%, S 13.80%. Prepn from powdered camphor, concd H_2SO_4 and acetic anhydride: Reychler, *Bull. Soc. Chim. France* [3] **19**, 120 (1898); Armstrong, Lowry, *J. Chem. Soc.* **81**, 1447 (1902); Lipp, Knapp, *Ber.* **73B**, 915 (1940). Structure: Loudon, *J. Chem. Soc.* **1933**, 823; Komppa, *J. Prakt. Chem.* **162**, 19 (1943).

Prisms from glacial acetic acid or ethyl acetate, dec 193-195°. $[\alpha]_D^{20}$ +43.5° (c = 4.3 in alcohol); $[\alpha]_D^{20}$ +21.5° (c = 4.3 in water). Deliquesc in moist air. Practically insol in ether; slightly sol in glacial acetic acid, ethyl acetate.

Ammonium salt, $C_{10}H_{19}NO_4S$, needles from water. $[\alpha]_D^{16}$ +20.5° (c = 5 in water). Very sol in water.

Potassium salt, $C_{10}H_{15}KO_4S$, needles from alcohol. $[\alpha]_D^{16}$ +18.4° (c = 4.4 in water).

USE: Resolution of optically active isomers.

1741. Camphotamide. *3-[(Diethylamino)carbonyl]-1-methylpyridinium salt with 4,7,7-trimethyl-3-oxobicyclo-[2.2.1]heptane-2-sulfonic acid (1:1);* 3-(diethylcarbamoyl)-1-methylpyridinium 2-oxo-3-bornanesulfonate; camphosulfonyl-N-methylpyridine-β-diethylcarboxamide; camphetamide; camphramine; Tonicorine. $C_{21}H_{32}N_2O_5S$; mol wt 424.55. C 59.41%, H 7.60%, N 6.60%, O 18.84%, S 7.55%. Prepd by the condensation of equimolar amounts of camphorsulfonic acid and N-methylnikethamide: Lab. Lematte et Boinot in P. Lebeau, M. Janot, *Traité de Pharmacie Chimique* vol. 4, (Paris 1955-56) p 2471; by reacting nicotinic acid with methyl camphorsulfonate and treating with diethylamine: **Fr. pat. 812,032** (1937 to Soc. Franc. Recherches Biochim.), *C.A.* **32**, 1052 (1938).

Minute crystals, mp 174-175°. Slight camphor-like odor.

Bitter taste with sweet aftertaste. Soluble in water, alcohol, ether. Insol in benzene, other hydrocarbons.

THERAP CAT: Analeptic.

1742. Camptothecin. *4-Ethyl-4-hydroxy-1H-pyrano-[3',4':6,7]indolizino[1,2-b]quinoline-3,14(4H,12H)-dione.* $C_{20}H_{16}N_2O_4$; mol wt 348.34. C 68.96%, H 4.63%, N 8.04%, O 18.37%. Alkaloid exhibiting antileukemic and antitumor activities. Isoln from the stem wood of the Chinese tree, *Camptotheca acuminata* Decsne., *Nyssaceae*, and structure: Wall *et al., J. Am. Chem. Soc.* **88**, 3888 (1966). Approach to synthesis: Kepler *et al., J. Org. Chem.* **34**, 3853 (1969). Total synthesis: E. J. Corey *et al., ibid.* **40**, 2140 (1975). Total synthesis of racemate: Stork, Schultz, *J. Am. Chem. Soc.* **93**, 4074 (1971); Volkmann *et al., ibid.* 5576; Tang *et al., ibid.* **97**, 159 (1975); J. C. Bradley, G. Buchi, *J. Org. Chem.* **41**, 699 (1976); T. Kametani *et al., J. Chem. Soc. Perkin Trans. I* **1981**, 1563. Pharmacologic and clinical evaluation: Gottlieb *et al., Cancer Chemother. Rep.* **54**, 461 (1970); Gallo *et al., J. Nat. Cancer Inst.* **46**, 789 (1971); S. M. Sieber *et al., Cancer Treat. Rep.* **60**, 1127 (1976). Mechanism of action: J. W. Lown, H.-H. Chen, *Biochem. Pharmacol.* **29**, 905 (1980). *Reviews:* Horwitz in *Antibiotics*, **vol. 3**, J. W. Corcoran, F. E. Hahn, Eds. (Springer-Verlag, New York, 1975) pp 48-57; M. E. Wall, M. C. Wani, *Anticancer Agents Based on Natural Product Models*, J. M. Cassady, J. D. Douros, Eds. (Academic Press, New York, 1980) pp 417-436.

Pale yellow needles from methanol + acetonitrile, dec 264-267°. Also reported as mp 275-277° (Volkmann); 287-288° (Stork, Schultz). $[\alpha]_D^{25}$ +31.3° (in chloroform-methanol, 8:2). Exhibits intense blue fluorescence under uv light. uv max: 220, 254, 290, 370 nm (ε 37320, 29230, 4980, 19900). Does not form stable salts with acids.

Acetate, $C_{22}H_{18}N_2O_5$, crystals, dec 271-274°. uv max: 220, 254, 290, 360-370 nm (ε 39010, 28740, 6160, 22000).

Chloroacetate, $C_{22}H_{17}ClN_2O_5$, crystals, dec 245-248°.

1743. Camylofine. α-[[2-(Diethylamino)ethyl]amino]-benzeneacetic acid 3-methylbutyl ester; N-[2-(diethylamino)-ethyl]-2-phenylglycine isopentyl ester; isoamyl α-[N-(β-diethylaminoethyl)amino]phenylacetate; isoamyl N-(β-diethylaminoethyl)-α-aminophenylacetate; α-[(β-diethylaminoethyl)amino]phenylacetic acid isoamyl ester; acamylophenine; Adopon; Avadyl; Belosin; Navadyl; Novospasmin; Sintespasmil; Spasmocan. $C_{19}H_{32}N_2O_2$; mol wt 320.46. C 71.21%, H 10.07%, N 8.74%, O 9.99%. Prepn: Schmeisser *et al.,* **Ger. pat. 842,206** (1952 to Asta), *C.A.* **47**, 5445i (1953); **Brit. pat. 688,331** (1953); Chielmetti, *Farm. Sci. Tec. (Pavia)* **7**, 625 (1952), *C.A.* **47**, 11161 (1953); Edwards *et al., J. Pharm. Pharmacol.* **12**, 179 (1960).

Pale yellow oil, $bp_{1.5}$ 174-178°, bp_4 165-180°. Strongly alkaline reaction.

Dihydrochloride, $C_{19}H_{34}Cl_2N_2O_2$, *Avacan.* Crystals, mp 174-178°, also reported as mp 172° and 173°. Soluble in water. LD_{50} in mice: 760 mg/kg orally; 1.35 g/kg s.c.; 49.2 mg/kg i.v.

THERAP CAT: Anticholinergic.

1744. Canadine. *5,8,13,13a-Tetrahydro-9,10-dimethoxy-6H-benzo[g]-1,3-benzodioxolo[5,6-a]quinolizine; 9,10-dimethoxy-2,3-(methylenedioxy)berbine; 5,6,13,13a-tetrahydro-9,10-dimethoxy-2,3-(methylenedioxy)-8H-dibenzo-[a,g]quinolizine;* tetrahydroberberine; xanthopuccine. $C_{20}H_{21}NO_4$; mol wt 339.38. C 70.78%, H 6.24%, N 4.13%, O 18.86%. A protoberberine alkaloid from *Corydalis cava* (L.), Schweigg & Korte (*C. tuberosa* DC.), *Fumariaceae:* Späth,

Julian, *Ber.* **64**, 1131 (1931). Prepn by reduction of berberine: Bersch, Seufert, *ibid.* **70**, 1121 (1937); Awe, Hertel, *Arch. Pharm.* **288**, 516 (1955); Russell, *J. Am. Chem. Soc.* **78**, 3115 (1956). Configuration: Corrodi, Hardegger, *Helv. Chim. Acta* **39**, 889 (1956). Other syntheses: Kametani *et al., J. Chem. Soc.* (C) **1969**, 2036; **1971**, 2709; M. Cushman, F.W. Dekon, *J. Org. Chem.* **44**, 407 (1979); K. Iwasa *et al., ibid.* **46**, 4744 (1981); N. S. Narasimhan *et al., Tetrahedron Letters* **22**, 2797 (1981). Pharmacology: F. Sadritdinov, M. B. Sultanov, *C.A.* **66**, 74714v (1967). *Review:* R. H. F. Manske, H. L. Holmes, *The Alkaloids* **vol.** 4 (Academic Press, New York, 1954) pp 91-92.

d-Form, mp 132°. $[\alpha]_D^{15}$ +299° (chloroform).

l-Form, crystals from methanol, mp 135°. $[\alpha]_D^{22}$ −308°; −317° (c = 0.28; 0.4, both in methanol).

dl-Form, mp 172° (Späth, Julian), also reported as 163-165° (Cushman, Dekon). uv max (95% ethanol): 209, 284 nm (ϵ 28,300; 5200).

All three forms are sol in aq methanol. Racemic canadine is less sol in ether than the optically active forms. LD_{50} in mice (mg/kg): 940 orally; 790 s.c.; 100 i.v. (Sadritdinov, Sultanov).

1745. Canavanine. *O-[(Aminoiminomethyl)amino]homoserine; 2-amino-4-(guanidinooxy)butyric acid.* $C_5H_{12}N_4O_3$; mol wt 176.18. C 34.08%, H 6.87%, N 31.80%, O 27.24%. $H_2NC(NH)NHOCH_2CH_2CH(NH_2)COOH$. Basic amino acid isolated as L-canavanine from jack beans, *Canavalia ensiformis* (L.) DC., *Leguminosae:* Kitagawa, Tomiyama, *J. Biochem. (Tokyo)* **11**, 265 (1929). Constitutes about 1.5% of the dry weight of alfalfa seeds and sprouts: E. A. Bell *Biochem. J.* **75**, 618 (1960). Distribution in the plant kingdom: Turner, Harborne, *Phytochemistry* **6**, 863 (1967). Structure: Gulland, Morris, *J. Chem. Soc.* **1935**, 763; Kitagawa, Takani, *J. Biochem. (Tokyo)* **23**, 181 (1936). Configuration: Cadden, *Proc. Soc. Exp. Biol. Med.* **45**, 224 (1940). Synthesis of racemate: Nyberg, Christensen, *J. Am. Chem. Soc.* **79**, 1222 (1957); Frankel *et al., J. Chem. Soc.* **1963**, 3127. Alternate synthesis: Yamada *et al., Agr. Biol. Chem.* **37**, 2201 (1973). Bears resemblance to arginine, *q.v.* and is often used as substrate to some enzymes normally acting on arginine. Accordingly it is a potent growth inhibitor of many organisms: Pilcher *et al., Proc. Soc. Exp. Biol. Med.* **88**, 79 (1955). Toxic to mammals: B. Tschiersch, *Pharmazie* **17**, 621 (1962). L-Canavanine induces certain hematological and serologic abnormalities characteristic of systemic lupus erythematosus in monkeys: M. R. Malinow *et al., Science* **216**, 415 (1982).

Crystals from abs alcohol, mp 184°. $[\alpha]_D^{20}$ +7.9° (c = 3.2). Sulfate, $C_5H_{14}N_4O_7S$, crystals from dil alcohol, dec 172°. $[\alpha]_D^{17}$ +19.4° (c = 2). Freely sol in water.

DL-Form, crystals from ethanol, mp 180-182°. Soluble in water; practically insol in alcohol. Forms a monohydrochloride, mp 190°.

1746. Candelilla Wax. Obtained from the candelilla plants; *Euphorbia antisyphilitica* Zucc., *Euphorbiaceae* and *E. cerifera* Alcocer (which is only doubtfully distinct from *E. antisyphilitica)* are now the principal sources of candelilla wax. *Pedilanthus pavonis* Boiss. and *P. aphyllus* Boiss. are secondary sources, yielding a wax of lower melting range and lower saponification number. Most of the wax is produced in Mexico by immersing the plants in boiling water containing sulfuric acid and skimming off the wax which rises to the surface as described by Dickinson, *Am. J. Pharm.* **91**, 808 (1919); Hodge, Sineath, *Econ. Bot.* **10**, 134 (1956). The main constituent is the hydrocarbon hentriacontane. Brief review: C. S. Letcher in Kirk-Othmer *Encyclopedia of*

Chemical Technology **vol.** 24 (Wiley-Interscience, New York, 3rd ed., 1984) pp 468-469.

Brownish to yellowish-brown, hard, brittle, easily pulverizable lumps. d 0.950-0.990. mp 68-70°. Saponification number 50-65. Acid number 10-20. Iodine number 30-35. Practically insol in water; sparingly sol in alcohol; sol in acetone, benzene, carbon disulfide, decalin, hot petr ether, gasoline, oils, turpentine, hot chloroform, carbon tetrachloride.

USE: Manuf cosmetics, rubber substitutes, furniture and leather polishes, candles, sealing wax, phonograph records; for waterproofing boxes and fabrics; electric insulations; lithographic, printing, stamping and writing inks; molding compositions; sizing paper; hardening other waxes; protective coating for citrus fruits; formerly in chewing gum.

1747. Candicidin. Levorin; Candeptin; Candimon; Vanobid. Heptaene macrolide antifungal antibiotic complex composed of candicidins A, B, C and D (major component). Produced by a strain of *Streptomyces griseus* (Rutgers no. 3570): Lechevalier *et al., Mycologia* **45**, 155 (1953). Methods of isoln: Siminoff, U.S. pat. **2,872,373** (1959 to Penick); Waksman, Lechevalier, U.S. pat. **2,992,162** (1961 to Rutgers Res. & Ed. Found.). Activity: Kligman, Lewis, *Proc. Soc. Exp. Biol. Med.* **82**, 399 (1953); *Lancet* **1954**, 266, 507; Lechevalier, *Presse Med.* **61**, 1327 (1953). Structure of candicidin D: J. Zielinski *et al., Tetrahedron Letters* **1979**, 1791. Biosynthesis studies: Liu *et al., J. Antibiot.* **25**, 116 (1972). Anticholesteremic property and mode of action studies: I. H. Kwon, H. Fisher, *Nutr. Rep. Int.* **9**, 245 (1974); A. K. Singhal *et al., Lipids* **16**, 423 (1981). HPLC comparison of candicidin, hamycin, trichomycin, *q.q.v.:* P. Helboe *et al., J. Chromatog.* **189**, 249 (1980).

candicidin D

Small yellow needles or rosettes from aq tetrahydrofuran or pyridine/acetic acid/water soln (when pure). Absorption max: 403, 380 ($E_{1cm}^{1\%}$ 1150), 360 nm. Practically insol in water, alcohols, ketones, esters, ethers, hydrocarbons, and other lipophilic solvents. Sol in DMSO, DMF, and lower aliphatic acids. Very sol in 80% aq tetrahydrofuran soln. The addn of 5%-25% water to alcohols greatly increases soly. Forms sol salts in alkaline solns. Soly: Marsh, Weiss, *J. Assoc. Offic. Anal. Chem.* **50**, 457 (1967). LD_{50} i.p. in mice: 14 mg/kg, L. C. Vining *et al., Antibiot. Ann.* **1954-55**, 980.

Candicidin D, $C_{59}H_{89}N_2O_{18}$, *levorin* A_2. Identity with levorin A_2: Bosshardt, Bickel, *Experientia* **24**, 422 (1968); J. Zielinski *et al., loc. cit.*

THERAP CAT: Topical antifungal.

1748. Candidin. $C_{47}H_{71}NO_{17}$; mol wt 922.10. C 61.22%, H 7.76%, N 1.52%, O 29.50%. Main component of an antifungal antibiotic complex produced by the soil actinomyces *Streptomyces viridoflavus:* Taber *et al., Antibiot. & Chemother.* **4**, 455 (1954); Vining, Taber, *Can. J. Chem.* **34**, 1163 (1956). The two other active principles are designated as *candidinin* and *candidoin.* Prepn of pure crystalline candidin: Preud'hamme, Vuillemin, Fr. pat. **1,298,345** (1962 to Rhone-Poulenc), *C.A.* **58**, 420c (1963). Structure: Borowski *et al., Tetrahedron Letters* **1971**, 1987. Antifungal activity studies: Solotorovsky *et al., Antibiot. & Chemother.* **8**, 364 (1958).

Golden-yellow needles from methanol + chloroform + water. Does not melt, but darkens slowly above 180°. $[\alpha]_D^{27}$ +363° (c = 0.3 in DMF); $[\alpha]_D^{27}$ +205° (c = 0.3 in glacial acetic acid). uv max (methanol): 406, 383, 362, 347 nm. Practically insol in water and most organic solvents. Moderately sol in glacial acetic acid, pyridine, DMF. LD_{50} in mice: 1.5 mg/kg i.v.; 7-36 mg/kg i.p.; 30 mg/kg s.c.

1749. Canella. White cinnamon; wild cinnamon; false Winter's bark; Bahama white wood; wild canilla. Bark of *Canella alba* Murr. (*C. winterana* Gaertn.), *Canellaceae*. *Habit.* W. Indies and Florida. *Constit.* Eugenol, cineol, terpenes, caryophyllene, mannitol, resin, cancellin.
USE: As spice and as an addition to smoking tobacco.

1750. Cannabidiol. *2-[3-Methyl-6-(1-methylethenyl)-2-cyclohexen-1-yl]-5-pentyl-1,3-benzenediol; (3R,4R)-2-p-mentha-1,8-dien-3-yl-5-pentylresorcinol.* $C_{21}H_{30}O_2$; mol wt 314.47. C 80.21%, H 9.62%, O 10.17%. Occurs in *Cannabis sativa* L. (*C. sativa* var. *indica* Auth.), *Moraceae*. Isoln from the marihuana extract of Minnesota wild hemp: Adams *et al., J. Am. Chem. Soc.* **62**, 196, 2194 (1940); U.S. pat. **2,304,-669** (1942); Jacob, Todd, *J. Chem. Soc.* **1940**, 649; Russell, Todd, *ibid.* **1942**, 628; Schultz, Haffner, *Arch. Pharm.* **291**, 391 (1958). A physiologically inactive component: Mechoulam *et al., Science* **169**, 611 (1970). Structure: Mechoulam, Shvo, *Tetrahedron* **19**, 2073 (1963). Crystal and molecular structure: T. Ottersen *et al., Acta Chem. Scand. B* **31**, 807 (1977). Synthesis of (−)-form: T. Petrzilka *et al., Helv. Chim. Acta* **52**, 1102 (1969); H. J. Kurth *et al., Z. Naturforsch.* **36B**, 275 (1981); of (±)-form: R. Mechoulam, Y. Gaoni, *J. Am. Chem. Soc.* **87**, 3273 (1965). Abs config: *eidem, ibid.* **93**, 217 (1971).

Pale yellow resin or crystals, mp 66-67°. bp_2 187-190° (bath temp 220°). $bp_{0.001}$ 130°. d_4^{40} 1.040. n_D^{20} 1.5404. $[\alpha]_D^{27}$ −125° (0.066 g in 5 ml 95% ethanol). $[\alpha]_D^{18}$ −129° (c = 0.45 in ethanol). uv max (ethanol): 282, 274 nm (log ε 3.10, 3.12). Practically insol in water or 10% NaOH. Sol in ethanol, methanol, ether, benzene, chloroform, petr ether.

1751. Cannabinol. *6,6,9-Trimethyl-3-pentyl-6H-dibenzo[b,d]pyran-1-ol; 3-amyl-1-hydroxy-6,6,9-trimethyl-6H-dibenzo[b,d]pyran.* $C_{21}H_{26}O_2$; mol wt 310.42. C 81.25%, H 8.44%, O 10.31%. Constituent of the resinous exudate of the female flowers of *Cannabis sativa* L. (*C. sativa* var. *indica* Auth.), *Moraceae*. Physiologically inactive component of marihuana: Mechoulam *et al., Science* **169**, 611 (1970). Isoln: Wood *et al., J. Chem. Soc.* **69**, 539 (1896); Work *et al., Biochem. J.* **33**, 123 (1939). Structure: Cahn, *J. Chem. Soc.* **1930**, 986; **1931**, 630; **1932**, 1342; **1933**, 1400; Bergel, Vögele, *Ann.* **493**, 250 (1932). Crystal and molecular structure: T. Ottersen *et al., Acta Chem. Scand. B* **31**, 781 (1977). Synthesis: Adams *et al., J. Am. Chem. Soc.* **62**, 2204 (1940); P. C. Meltzer *et al., Synthesis* **1981**, 985; J. Novák, C. A. Salemink, *Tetrahedron Letters* **23**, 253 (1982).

Leaflets from petr ether, mp 76-77°. Sublimes at 4 mm with a bath temp of 180-190°. $bp_{0.05}$ 185°. Insol in water. Sol in methanol, ethanol, aq alkaline solns.

1752. Cannabis. Indian hemp; Indian cannabis; marihuana; marijuana; bhang; ganja; charas; kif; hasach; pot. Dried flowering tops of pistillate plants of *Cannabis sativa* L. (*C. sativa* var. *indica* Auth.), *Moraceae*. *Habit.* Persia, East India, U.S., Central America, cultivated in Europe. *Constit.* Isomeric tetrahydrocannabinols (the most active constituents), cannabinol, cannabidiol. Sepn of constituents by chromatographic methods: Parker *et al., Bull. Narcotics* **20**, 9 (1968). *Reviews:* Adams, *Science* **92**, 115 (1940); Todd, *Experientia* **2**, 55 (1946); Karbe, *Arzneimittel-Forsch.* **1**, 37 (1951). Status report, including many minor constituents: *Chem. & Eng. News,* July 6, **1970**, pp 30-33. Review of pharmacology: Paton, *Ann. Rev. Pharmacol.* **15**, 191 (1975). Review of health hazards and therapeutic potential: Council on Scientific Affairs, *J. Am. Med. Assoc.* **246**, 1823-1827 (1981). Brief summary of Nat. Acad. Sci. report: *Chem. & Eng. News,* March 8, 1982, pp 27-28. Book: *Marijuana: Chemistry, Pharmacology, Metabolism and Clinical Effects,* R. Mechoulam, Ed. (Academic Press, New York, 1973) pp 409. *See also* Hashish and Tetrahydrocannabinols.
Human Toxicity: When ingested or inhaled as smoke, may cause euphoria, delirium, hallucinations, weakness, hyporeflexia, drowsiness, *Clinical Toxicology of Commercial Products,* R. E. Gosselin *et al.,* Eds. (Williams & Wilkins, Baltimore, 4th ed., 1976) Section III, pp 212-219.
Note: This is a controlled substance (hallucinogen) listed in the U.S. Code of Federal Regulations, Title 21 Part 1308.11 (1985).
THERAP CAT (VET): Has been used as a sedative in equine colic.

1753. Canrenone. *(17α)-17-Hydroxy-3-oxopregna-4,6-diene-21-carboxylic acid γ-lactone; 17α-(2-carboxyethyl)-17β-hydroxyandrosta-4,6-dien-3-one lactone; 17α-(2-carboxyethyl)-17β-hydroxy-3-oxoandrosta-4,6-diene lactone; 6-dehydrotestosterone-17α-propionic acid γ-lactone; 3-(3-oxo-17β-hydroxy-4,6-androstadien-17α-yl)propionic acid γ-lactone; Phanurane.* $C_{22}H_{28}O_3$; mol wt 340.44. C 77.61%, H 8.29%, O 14.10%. Prepd by dehydrogenation of 17-hydroxy-3-oxo-17α-pregn-4-ene-21-carboxylic acid γ-lactone: Cella, Tweit, *J. Org. Chem.* **24**, 1109 (1959); Cella, U.S. pat. **2,900,383** (1959 to Searle).

Crystals from ethyl acetate, mp 149-151°, solidifies and remelts at 165°. $[\alpha]_D$ +24.5° (chloroform). uv max: 283 nm (ε 26,700).
Free acid potassium salt, $C_{22}H_{30}KO_4$, *potassium canrenoate, Kanrenol, Soldactone, Venactone.*
THERAP CAT: Aldosterone antagonist. Diuretic.

1754. Cantharides. Spanish fly; blistering fly; blistering beetle; *Cantharis vesicatoria. Habit.* Southern and Central Europe, mainly upon *Oleaceae* and *Caprifoliaceae. Constit.* 0.6-1% cantharidin, 10-15% fat; resinous substances, acetic and uric acids. Powerful irritant, vesicant, rubefacient. *Review:* Ude, Heeger, *Pharm. Zentralhalle* **82**, 193 (1941).

Human Toxicity: Highly toxic by ingestion or absorption from skin and mucous membranes; severe gastroenteritis, nephritis, collapse, death may occur, *Clinical Toxicology of Commercial Products,* R. E. Gosselin *et al.,* Eds. (Williams & Wilkins, Baltimore, 4th ed., 1976) Section II, p 175.

THERAP CAT: Vesicant.

THERAP CAT (VET): Vesicant, counterirritant.

1755. Cantharidin. *Hexahydro-3a,7a-dimethyl-4,7-epoxyisobenzofuran-1,3-dione; 2,3-dimethyl-7-oxabicyclo-[2.2.1]heptane-2,3-dicarboxylic anhydride; exo-1,2-cis-di-methyl-3,6-epoxyhexahydrophthalic anhydride;* cantharides camphor. $C_{10}H_{12}O_4$; mol wt 196.21. C 61.21%, H 6.17%, O 32.62%. Active principle of cantharides *(q.v.)* and other insects, in notorious "*Spanish Fly*" aphrodisiac. Synthesis and stereochemistry: Woodward, Loftfield, *J. Am. Chem. Soc.* **63,** 3167 (1941); Ziegler *et al., Ann.* **551,** 1 (1942). Stereospecific synthesis: Stork, Van Tamelen *et al., J. Am. Chem. Soc.* **75,** 384 (1953). Simple efficient synthesis: W. G. Dauben *et al., ibid.* **102,** 6893 (1980). Crystal structure: M. Zehnder, U. Thewalt, *Helv. Chim. Acta* **60,** 740 (1977).

Orthorhombic plates, scales, mp 218°. Sublimes at about 110° (12 mm Hg, 3-5 mm distance). Insol in cold water, somewhat sol in hot water. One gram dissolves in 40 ml acetone, 65 ml chloroform, 560 ml ether, 150 ml ethyl acetate. Sol in oils.

THERAP CAT: Vesicant.

THERAP CAT (VET): Rubefacient, vesicant, counterirritant.

1756. Canthaxanthin. *β,β-Carotene-4,4'-dione;* 4,4'-dioxo-β-carotene; Food Orange 8; C.I. 40850; Carophyll Red; Orobronze; Roxanthin Red 10; Carotaben plus. $C_{40}H_{52}O_2$; mol wt 564.82. C 85.05%, H 9.28%, O 5.67%. All *trans*-carotenoid pigment widely distributed in nature. Isoln from the edible mushroom *Cantharellus cinnabarinus* Adans. ex Fr., *Agaricaceae:* Haxo, *Bot. Gaz.* **112,** 228 (1950); also isolated from flamingo feathers. Structure and synthesis: Isler *et al., Verh. Naturforsch. Ges. Basel* **67,** 379 (1956); Zeller *et al., Helv. Chim. Acta* **42,** 841 (1959). Alternate syntheses: M. Akhtar, B. C. L. Weedon, *J. Chem. Soc.* **1959,** 4058; R. Rüegg, G. Saucy, U.S. pat. **2,983,752** and J. D. Surmatis, U.S. pat. **3,311,656** (1961, 1967 both to Hoffmann-La Roche); J. D. Surmatis *et al., Helv. Chim. Acta* **53,** 974 (1970); M. Rosenberger *et al., Pure Appl. Chem.* **51,** 871 (1979); *eidem, J. Org. Chem.* **47,** 2130 (1982).

Violet crystals from methylene chloride, dec 217°. Absorption max (cyclohexane): 470 nm ($E_{1cm}^{1\%}$ 2250). Sol in chloroform, oils. Oil solns are more red than those of β-carotene.

USE: Permissible color additive for food and drugs (exempt from certification): *Fed. Reg.* **34,** no. 5 (Jan. 8, 1969). Oral suntanning agent.

1757. Capobenic Acid. *6-[(3,4,5-Trimethoxybenzoyl)-amino]hexanoic acid; 6-(3,4,5-trimethoxybenzamido)hexanoic acid;* 3,4,5-trimethoxybenzoyl-ε-aminocaproic acid; ATBAC; TB-ACA; C-tre; C-3. $C_{16}H_{23}NO_6$; mol wt 325.37. C 59.07%, H 7.12%, N 4.31%, O 29.50%. Prepn: A. Garzia, Ger. pat. **2,034,192** corresp to U.S. pat. **3,697,563** (1971, 1972 to Ist. Chemioter. Ital.); Bottazzi *et al., Riv. Farmacol. Ter.* **11,** 215 (1971). Pharmacology: Razzaboni *et al., Boll. Soc. Ital. Biol. Sper.* **44,** 1783 (1968). Pharmacology, metabolism and clinical investigation of sodium salt: Greggia *et*

al., ibid. **45,** 1447 (1969); Fontanini *et al., Minerva Med.* **60,** 4857 (1969); several authors, *Riv. Farmacol. Ter.* **1** (1970); **2** (1971); **3** (1972).

White, odorless powder, mp 121-123°. uv max (ethanol): 214, 259 nm. Sol in ethanol, acetone, chloroform, alkaline solns. Practically insol in water, ether, carbon tetrachloride. LD_{50} i.p. in rats: 2.5 g/kg, A. Garzia, *loc. cit.*

Sodium salt, $C_{16}H_{22}NNaO_6$, *sodium capobenate, Capben.*

THERAP CAT: Cardiac depressant (anti-arrhythmic).

1758. Capreomycin. Capromycin; Caprolin; Capastat; Capostatin. Cyclic peptide antibiotic similar to viomycin; produced by *Streptomyces capreolus* NRRL 2773; Herr *et al.,* 140th Am. Chem. Soc. Meet. (Chicago, Sept. 1961), *Abstracts of Papers,* p 49C; *Chem. Eng. News* **39,** 57 (Sept. 18, 1961); **Brit.** pat. **920,563;** U.S. pat. **3,143,468** (1963, 1964, both to Lilly). Mixture of capreomycins IA, IB, IIA, and IIB in the approx percentages, 25%, 67%, 3%, 6%, resp.: Herr, Redstone, *Ann. N.Y. Acad. Sci.* **135,** 940 (1966). Activity studies: Sutton *et al., ibid.* 947; Lucchesi, *Antibiot. Chemother. (Basel)* **16,** 27 (1970). Proposed structure: Bycroft *et al., Nature* **231,** 301 (1971). Revised structure and total synthesis: T. Shiba *et al., Tetrahedron Letters* **1976,** 3907; S. Nomoto *et al., Tetrahedron* **34,** 921 (1978). Structure of IA and IB: *eidem, J. Antibiot.* **30,** 955 (1977).

The mixture is a white solid, sol in water; practically insol in most organic solvents. pKa in 66% aq DMF: 6.2, 8.2, 10.1, 13.3. Stable in aq soln at pH 4-8; unstable in strongly acidic or strongly basic solns.

Disulfate, *Caprocin, Ogostal.* Solubility data: March, Weiss, *J. Assoc. Offic. Anal. Chem.* **50,** 457 (1967). LD_{50} in mice, rats: 250, 325 mg/kg i.v.; 514, 1191 mg/kg s.c., Welles *et al., Ann. N.Y. Acad. Sci.* **135,** 960 (1966).

Capreomycin IA, $C_{25}H_{44}N_{14}O_8$, R = OH. mp 246-248° (dec). $[\alpha]_D^{22}$ −21.9° (c = 0.5 in water). uv max (0.1N HCl): 269 nm (ε 24,000); (H_2O): 268 nm (ε 23,900); (0.1N NaOH): 289 nm (ε 15,900) (all for $C_{25}H_{48}N_{14}O_8Cl_4 \cdot \frac{1}{2}C_2H_5OH$).

Capreomycin IB, $C_{25}H_{44}N_{14}O_7$, R = H. mp 253-255° (dec). $[\alpha]_D^{22}$ −44.6° (c = 0.5 in water). uv max (0.1N HCl): 268 nm (ε 22,700); (H_2O): 268 nm (ε 22,300); (0.1N NaOH): 290 nm (ε 14,400) (all for $C_{25}H_{48}N_{14}O_7Cl_4 \cdot \frac{1}{2}C_2H_5OH \cdot \frac{1}{2}H_2O$).

Capreomycins IIA, capreomycin IIB. Structures corresp to capreomycins IA and IB but lack β-lysine residues.

THERAP CAT: Antibacterial (tuberculostatic).

1759. *n*-Capric Acid. *Decanoic acid.* $C_{10}H_{20}O_2$; mol wt 172.26. C 69.72%, H 11.70%, O 18.58%. $CH_3(CH_2)_8COOH$. Prepn from octyl bromide: Shishido *et al., J. Am. Chem.*

Soc. **81**, 5817 (1959); Closson, De Pree, U.S. pat. **2,918,494** (1959 to Ethyl Corp.). Recovery from *Cuphea llavea* Llave et Lex, *Lythaceae* seed oil: Miwa *et al.*, U.S. pat. **2,964,546** (1960 to U.S.D.A.). *Review: Fatty Acids* Part I, K. S. Markley, Ed. (Interscience, New York, 2nd ed., 1960) pp 34, 39.

Cryst solid, mp 31.4°. Rancid odor. bp 270°. d_4^{50} 0.8782. n_D^{40} 1.4288. Practically insol in water (0.015 g/100 g at 20°); sol in ethanol, ether, chloroform, benzene, carbon disulfide. Also sol in dil HNO_3 (d 1.14) from which it precipitates unchanged by addition of water. LD_{50} i.v. in mice: 129 ±5.4 mg/kg, L. Orö, A. Wretlind, *Acta Pharmacol. Toxicol.* **18**, 141 (1961).

USE: Manuf of esters for artificial fruit flavors and perfumes; as an intermediate in other chemical syntheses.

1760. *n*-Caproic Acid. Hexanoic acid. $C_6H_{12}O_2$; mol wt 116.16. C 62.04%, H 10.41%, O 27.55%. $CH_3(CH_2)_4COOH$. Occurs in milk fats (about 2%), in coconut oil (< 1%), various palm and other oils. Prepn: Vliet *et al.*, *Org. Syn.* coll. vol. II, 417 (1943); Reid, Ruhoff, *ibid.*, 475. Manuf by catalytic reduction of corresponding β-lactone: Caldwell, U.S. pat. **2,484,486** (1949 to Kodak); from oleic acid: Follett, Murray, U.S. pat. **2,580,417** (1952 to Arthur D. Little); from castor oil or a ricinoleate: Steadman, Peterson, U.S. pat. **2,847,432** (1958 to National Res. Corp.); by ozonolysis of tall oil unsaturated fatty acids: Maggiolo, U.S. pat. **2,865,937** (1958 to Welsbach); from 1,3-butadiene and potassium acetate in presence of $NaNH_2$: Schmerling, Toekelt, U.S. pat. **3,075,010** (1963 to Universal Oil Prod.); from cyclohexanol: Bartlett, Lippincott, U.S. pat. **3,121,728** (1964 to Esso); by catalytic oxidation of *n*-hexanol: Hay, U.S. pat. **3,173,933** (1965 to General Electric). *Review: Fatty Acids*, Part **1**, K. S. Markley, Ed. (Interscience, New York, 2nd ed., 1960) pp 34, 37.

Oily liquid, bp 205°. Characteristic goat-like odor. mp −3.4°. d_4^{20} 0.9265. n_D^{20} 1.4163. Slightly soluble in water (1.082 g/100 g); readily soluble in ethanol, ether. LD_{50} orally in rats: 3.0 g/kg, H. F. Smyth, C. P. Carpenter, *J. Ind. Hyg. Toxicol.* **26**, 269 (1944).

USE: Manuf of esters for artificial flavors, and of hexyl derivatives, especially hexylphenols, hexylresorcinol, etc.

1761. Caproic Aldehyde. Hexanal; caproaldehyde; hexaldehyde. $C_6H_{12}O$; mol wt 100.16. C 71.94%, H 12.08%, O 15.97%. $CH_3(CH_2)_4CHO$. Prepn: Bagard, *Bull. Soc. Chim.* **1**, 307 (1907).

Liquid. d_4^{20} 0.8335. bp_{760} 131°; bp_{12} 28°. Autooxidizes and polymerizes, especially in the presence of traces of acid. LD_{50} orally in rats: 4.89 g/kg, Smyth *et al.*, *Arch. Ind. Hyg. Occup. Med.* **10**, 61 (1954).

1762. Caprolactam. Hexahydro-2H-azepin-2-one; ε-caprolactam; 2-oxohexamethylenimine; 2-ketohexamethylenimine; aminocaproic lactam. $C_6H_{11}NO$; mol wt 113.16. C 63.68%, H 9.80%, N 12.39%, O 14.14%. Prepn: Wallach, *Ann.* **312**, 187 (1900); **343**, 43 (1905); Ruzicka *et al.*, *Helv. Chim. Acta* **4**, 477 (1921); Eck, Marvel, *J. Biol. Chem.* **106**, 387 (1934); Marvel, Eck, *Org. Syn.* coll. vol. II, 371 (1943); Lazier, Rigby, U.S. pat. **2,234,566** (1941 to du Pont); Schlack, U.S. pat. **2,249,177** (1941 to I. G. Farben); Ger. pats. **739,953** (1943); **745,224** (1943); P. Smith, *J. Am. Chem. Soc.* **70**, 320 (1948); E. Schmitz *et al.*, *J. Prakt. Chem.* **319**, 274 (1977). Purification: Kampschmidt, U.S. pat. **2,786,052** (1957 to Stamicarbon N. V.). Stabilization with alkalies: Indest *et al.*, U.S. pat. **2,884,414** (1959 to Vereinigte Glanzstoff-Fabriken). *Reviews: CIOS Repts.* no. **22** and **31**, File XXXIII/Synthetic Fiber Developments in Germany, parts I & II; K. Kahr *et al.* in *Ullmann's Encyklopädie der Technischen Chemie* vol. **9**, E. Bartholome *et al.*, Eds. (Verlag Chemie, Weinheim, 4th ed., 1975) pp 96-114.

Hygroscopic leaflets from petr ether, mp 70°. d_4^{75} (liq) 1.02. bp_{50} 180°; bp_3 100°. Viscosity at 78° = 9 centipoises. Flash pt, open cup: 257°F (125°C). Freely sol in water, methanol, ethanol, ether, tetrahydrofurfuryl alcohol, di-

methylformamide. Also sol in chlorinated hydrocarbons, cyclohexene, petroleum fractions. A 70% aq soln has d_2^{25} 1.05; n_D^{31} 1.4965; n_D^{40} 1.4935. LD_{50} orally in rats: 2.14 g/kg, H. F. Smyth *et al.*, *Am. Ind. Hyg. Assoc. J.* **30**, 470 (1969).

USE: Manuf of synthetic fibers of the polyamide type (Perlon); solvent for high mol wt polymers.

1763. Caproyl Chloride. Hexanoyl chloride. $C_6H_{11}ClO$; mol wt 134.61. C 53.54%, H 8.24%, Cl 26.34%, O 11.89%. $CH_3(CH_2)_4COCl$. Prepn: Brown, *J. Am. Chem. Soc.* **60**, 1325 (1938). Manuf: Wygant, U.S. pat. **2,806,061** (1957 to Monsanto).

Liquid, bp 151-153°. mp −87.3°. d_4^{15} 0.9805. n_D^{15} 1.4286. Dec by water or alcohol. Sol in ether, chloroform.

1764. Caprylene. 1-Octene; octylene. C_8H_{16}; mol wt 112.21. C 85.63%, H 14.37%. $CH_3(CH_2)_5CH=CH_2$. Prepn from appropriate alkylmagnesium bromide and allyl bromide or chloride: Geisler, Pilz, *Ber.* **95**, 96 (1962); from formaldehyde or paraformaldehyde and triphenyl(phenylmethylene)phosphorane: Hauser *et al.*, *J. Org. Chem.* **28**, 372 (1963); by catalytic dehydration of 2-octanol: Lundeen, Hoozer, *J. Am. Chem. Soc.* **85**, 2180 (1963).

Liquid, bp 121°, bp_{100} 61.5-61.7°. mp −102°. d_4^{20} 0.7149, d_4^{25} 0.7109. n_D^{20} 1.4087, n_D^{25} 1.4062. Flash pt, open cup: 70°F (21°C). Practically insol in water; misc with alcohol, ether.

1765. Caprylic Acid. Octanoic acid. $C_8H_{16}O_2$; mol wt 144.21. C 66.63%, H 11.18%, O 22.19%. $CH_3(CH_2)_6COOH$. Prepn from 1-heptene: Dupont *et al.*, *Compt. Rend.* **240**, 628 (1955); by oxidation of octanol: Langenbeck, Richter, *Ber.* **89**, 202 (1956). Manuf: Alexander, U.S. pat. **2,821,534** (1958 to GAF); McAlister *et al.*, U.S. pat. **3,053,869** (1962 to Standard Oil Co., Indiana). Antifungal properties: O. Wyss *et al.*, *Arch. Biochem.* **7**, 418 (1945). Toxicity: P. M. Jenner *et al.*, *Food Cosmet. Toxicol.* **2**, 327 (1964). *Review: Fatty Acids*, Part **1**, K. S. Markley, Ed. (Interscience, New York, 2nd ed., 1960) pp 34, 38.

Oily liquid, bp 239.7°. Slightly unpleasant rancid taste. mp 16.7°. d_4^{20} 0.910. n_D^{20} 1.4280. Very slightly sol in water (0.068 g/100 g at 20°); freely sol in alcohol, chloroform, ether, carbon disulfide, petr ether, glacial acetic acid. LD_{50} orally in rats: 10,080 mg/kg (Jenner).

USE: An intermediate in manuf of esters used in perfumery; in manuf of dyes, etc.

1766. Caprylic Aldehyde. Octanal; caprylaldehyde; octaldehyde. $C_8H_{16}O$; mol wt 128.21. C 74.94%, H 12.58%, O 12.48%. $CH_3(CH_2)_6CHO$. Prepn: Stephen, *J. Chem. Soc.* **127**, 1874 (1925).

Liquid. d_4^{20} 0.821. bp_{760} 163.4°; bp_{20} 72°; bp_9 60°. n_D^{26} 1.41667. Slightly sol in water; misc with alc, ether.

1767. Capsaicin. (E)-N-[(4-Hydroxy-3-methoxyphenyl)methyl]-8-methyl-6-nonenamide; trans-8-methyl-N-vanillyl-6-nonenamide; N-(4-hydroxy-3-methoxybenzyl)-8-methylnon-trans-6-enamide; Mioton; Zostrix. $C_{18}H_{27}NO_3$; mol wt 305.40. C 70.78%, H 8.91%, N 4.59%, O 15.72%. Pungent principle in fruit of various species of *Capsicum*, *Solanaceae*. Isoln from paprika and cayenne: Thresh, *Pharm. J. and Trans.* **7**, 21 (1876); Micko, *Z. Nahr. Genussm.* **1**, 818 (1898). See *Beilstein* 13, suppl. I, 322. Early structure study: Nelson, *J. Am. Chem. Soc.* **42**, 597 (1920). Synthesis: Späth, Darling, *Ber.* **63**, 737 (1930); L. Crombie *et al.*, *J. Chem. Soc.* **1955**, 1025; O. P. Vig *et al.*, *Indian J. Chem.* **17B**, 558 (1979). Constitution and biosynthesis: D. J. Bennet, E. W. Kirby, *J. Chem. Soc. C* **1968**, 442. Pharmacology: Molnar *et al.*, *Acta Physiol.* **35**, 369 (1969). Capsaicin is a powerful irritant; initial administration causes intense pain. Prolonged treatment causes insensitivity to painful stimuli and induces selective degeneration of certain primary sensory neurons: G. Jancso *et al.*, *Nature* **270**, 741 (1977); R. Gamse, *Arch. Pharmacol.* **320**, 205 (1982); P. Holzer *et al.*, *Neurosci. Letters* **31**, 253 (1982). Neuronal depletion of substance P, q.v.: T. M. Jessell *et al.*, *Brain Res.* **152**, 183 (1978); T. L. Yaksh *et al.*, *Science* **206**, 481 (1979). Capsaicin pretreatment also induces long-lasting desensitization of airway mucosa to various mechanical and chemical irritants: J. M. Lundberg, A. Saria, *Nature* **302**, 251 (1983). Preliminary clinical evaluation in chronic postherpetic neuralgia: J. E. Bernstein *et al.*, *J. Am. Acad. Dermatol.* **17**, 93 (1987). *Reviews:* Molnar, *Arzneimittel-Forsch.* **15**,

718 (1965); Walker, Gavern, *Mfg. Chem. Aerosol News* **39** (6), 35 (1968); R. M. Virus, G. F. Gebhart, *Life Sci.* **25**, 1273 (1979); Y. Monsereenusorn *et al.*, *CRC Crit. Rev. Toxicol.* **10**, 321-339 (1982).

Monoclinic, rectangular plates, scales from petr ether, mp 65°. $bp_{0.01}$ 210-220° (air-bath temp). uv max: 227, 281 nm (ϵ 7000, 2500). Burning taste, one part in 100,000 can be detected by tasting. Practically insol in cold water. Freely sol in alc, ether, benzene, chloroform; slightly sol in CS_2.

USE: As a tool in neurobiological research.

THERAP CAT: Topical analgesic.

1768. Capsanthin. *(3R,3'S,5'R)-3,3'-Dihydroxy-β,k-caroten-6'-one.* $C_{40}H_{56}O_3$; mol wt 584.85. C 82.14%, H 9.65%, O 8.21%. Carotenoid pigment isolated from paprika *(Capsicum annuum L., Solanaceae)*: L. Zechmeister, L. Cholnoky, *Ann.* **454**, 54 (1927); L. Cholnoky *et al., Ann.* **606**, 194 (1957); Warren, Weedon, *J. Chem. Soc.* **1958**, 3972. Structure: Entschel, Karrer, *Helv. Chim. Acta* **43**, 89 (1960); Faigle, Karrer, *ibid.* **44**, 1257, 1904 (1961). Absolute configuration: Bartlett *et al., J. Chem. Soc. (C)* **1969**, 2527. Crystal structure: I. Ueda, W. Nowacki, *Verh. Schweiz. Naturforsch. Ges.* **1971**, 152, *C.A.* **77**, 106449h (1972).

Deep carmine-red needles from petr ether, mp 181-182°. Absorption max 483 nm ($\epsilon \times 10^{-3}$, 121). $[\alpha]_{Cd}$ +36° (chloroform). Freely sol in acetone, chloroform. Sol in methanol, ethanol, ether, benzene. Slightly sol in petr ether, CS_2.

Diacetate, $C_{44}H_{60}O_5$, red plates from methanol, mp 150°. Dipalmitate, $C_{72}H_{116}O_5$, bordeaux-red plates from benzene + methanol, mp 95°.

1769. Capsicum. Cayenne Pepper. The dried ripe fruit of *Capsicum frutescens L., Solanaceae,* known commercially as *African Chillies,* or of *Capsicum annuum L.,* var. *conoides* Irish, known as *Tabasco Pepper* or of *Capsicum annuum* var. *longum* Sendt, known as *Louisiana Long Pepper,* or of a hybrid between the Honka variety of Japanese Capsicum and the Old Louisiana Sport Capsicum known as *Louisiana Sport Pepper* (Fam. *Solanaceae). Constit.* Fixed oils; 0.1-1% capsaicin, capsanthin, *q.q.v.* Store in a cool place. Protect from light.

THERAP CAT (VET): Has been used externally as a counter-irritant; internally as a carminative and stomachic.

1770. Captafol. *3a,4,7,7a-Tetrahydro-2-[(1,1,2,2-tetra-chloroethyl)thio]-1H-isoindole-1,3(2H)-dione; N-(1,1,2,2-tetrachloroethylthio)-4-cyclohexene-1,2-dicarboximide; N-(1,1,2,2-tetrachloroethylmercapto)-4-cyclohexene-1,2-di-carboximide; N-(1,1,2,2-tetrachloroethylthio)-Δ^4-tetrahy-drophthalimide; N-(1,1,2,2-tetrachloroethylsulfenyl)-cis-4-cyclohexene-1,2-dicarboximide;* Difolatan; $C_{10}H_9Cl_4NO_2S$; mol wt 349.09. C 34.41%, H 2.60%, Cl 40.63%, N 4.01%, O 9.17%, S 9.19%. Prepn: Kohn, **Belg.** pat. **633,205** (1963 to California Res. Corp.), *C.A.* **60**, 15789cd (1964). Toxicology: R. Ben-Dyke *et al., World Rev. of Pest Control* **9**, 119 (1970); G. L. Kennedy, Jr. *et al., Food Cosmet. Toxicol.* **13**, 55 (1975).

Crystals, mp 160-161°. LD_{50} in rats, rabbits (mg/kg): 2500-6200 orally; 15400 dermally (Ben-Dyke *et al.*).

Note: The EPA has determined that captafol is oncogenic in mice and rats: *Fed. Reg.* **50**, 1103 (1985).

USE: Agricultural fungicide, especially for potatoes. *Compare:* Captan; Folpet.

1771. Captan. *3a,4,7,7a-Tetrahydro-2-[(trichloromethyl)thio]-1H-isoindole-1,3(2H)-dione; N-(trichloromethylthio)-4-cyclohexene-1,2-dicarboximide; N-trichloromethylthio-3a,4,7,7a-tetrahydrophthalimide; N-trichloromethylmercapto-4-cyclohexene-1,2-dicarboximide; N-(trichloromethylmercapto)-Δ^4-tetrahydrophthalimide;* ENT 26538; SR-406; Merpan; Orthocide-406; Vancide 89. $C_9H_8Cl_3$-NO_2S; mol wt 300.57. C 35.96%, H 2.68%, Cl 35.38%, N 4.66%, O 10.65%, S 10.67%. Prepn: A. R. Kittleson, **U.S.** pats. **2,553,771; 2,653,155** and **2,713,058** (1951, 1953, 1955, all to Standard Oil); *idem, Science* **115**, 84 (1952); *idem, J. Agr. Food Chem.* **1**, 677 (1953). Review of mutagenicity studies: B. A. Bridges, *Mutat. Res.* **32**, 3 (1975).

Odorless crystals from CCl_4, mp 178°; d 1.74. Practically insol in water. Soly at 26° in g/100 ml: chloroform 7.78; tetrachloroethane 8.15; cyclohexanone 4.96; dioxane 4.70; benzene 2.13; toluene 0.69; heptane 0.04; ethanol 0.29; ether 0.25. Colorimetric determination: A. R. Kittleson, *Anal. Chem.* **24**, 1173 (1952). LD_{50} orally in rats: 9000 mg/kg, B. A. Bridges, *loc. cit.*

USE: Fungicide; bacteriostat in soap. *Caution:* Ingestion of large quantities may cause vomiting, diarrhea.

1772. Captodiamine. *2-[[[4-(Butylthio)phenyl]phenyl-methyl]thio]-N,N-dimethylethanamine; 2-[p-(butylthio)-α-phenylbenzylthio]-N,N-dimethylethylamine; p-butylmercap-tobenzhydryl β-dimethylaminoethyl sulfide; p-butylthiodi-phenylmethyl 2-dimethylaminoethyl sulfide;* captodiam; captodramin. $C_{21}H_{29}NS_2$; mol wt 359.60. C 70.14%, H 8.13%, N 3.90%, S 17.83%. Prepn: Hübner, Petersen, **U.S.** pat. **2,830,088** (1958). Pharmacological studies: R. Kopf, I. Moller-Nielsen, *Arzneimittel-Forsch.* **8**, 154 (1958).

Hydrochloride, $C_{21}H_{30}ClNS_2$, Covatine, Covatix, Suvren. Crystals, mp 131-132°. LD_{50} (96 hr) in mice, rats (mg/kg): 180, 343 i.p.; 1630, 3800 orally (Kopf, Moller-Nielsen).

THERAP CAT: Anxiolytic.

1773. Captopril. *(S)-1-(3-Mercapto-2-methyl-1-oxo-propyl)-L-proline; (2S)-1-(3-mercapto-2-methylpropionyl)-*L-proline; D-2-methyl-3-mercaptopropanoyl-L-proline; SQ 14225; Acediur; Acepril; Aceplus; Alopresin; Acepress; Capoten; Captolane; Captoril; Cesplon; Dilabar; Garranil; Hypertil; Lopirin; Lopril; Tensobon; Tensoprel. C_9H_{15}-NO_3S; mol wt 217.28. C 49.75%, H 6.96%, N 6.45%, O 22.09%, S 14.75%. First orally active inhibitor of angioten-sin-converting enzyme (*see* angiotensin). Prepn: M. A. Ondetti, D. W. Cushman, **U.S.** pat. **4,046,889** (1977 to Squibb). Design and synthesis: M. A. Ondetti *et al., Science* **196**, 441 (1977); D. W. Cushman *et al., Biochemistry* **16**, 5484 (1977). Improved synthesis: D. H. Nam *et al., J. Pharm. Sci.* **73**, 1843 (1984). Pharmacology: B. Rubin *et al., Eur. J. Pharmacol.* **51**, 377 (1978); *eidem, Prog. Cardiovasc. Dis.* **21**, 183 (1978). Clinical studies: D. B. Case *et al., ibid.* **195**; H. R. Brunner *et al., Ann. Int. Med.* **90**, 19 (1979). Toxicology and metabolism: G. R. Keim in *Captopril and Hypertension,* D. B. Case, Ed. (Plenum, New York, 1980) p 137. GC/MS determn in biological fluids: T. Ito, Y. Matsuki, *J. Chromatog.* **417**, 79 (1987). Historical review and comprehensive bibliography: Z. P. Horovitz in *Pharmaco-logical and Biochemical Properties of Drug Substances* vol. 3,

M. E. Goldberg, Ed. (Am. Pharm. Assoc., Washington, DC, 1981) pp 148-175. Comprehensive description: H. Kadin in *Analytical Profiles of Drug Substances* **vol. 11**, K. Florey, Ed. (Academic Press, New York, 1982) pp 79-137. *Reviews: Am. Heart J.* **104**, pt. 2, 1125-1228 (1982); *Brit. J. Clin. Pharmacol.* **14**, Suppl. 2, 69S-252S (1982). Series of articles on pharmacology and therapeutic efficacy: *Postgrad. Med. J.* **62**, Suppl. 1, 1-191 (1986).

Crystals from ethyl acetate/hexane, mp 103-104° (Ondetti, Cushman). Generally regarded as polymorphic: stable form, mp 106°; unstable form, mp 86° (Florey); also reported as mp 87-88°, resolidifies, second mp 104-105° (Cushman). $[\alpha]_D^{22}$ −131.0° (c = 1.7 in ethanol). pK_1 3.7, pK_2 9.8. Freely sol in water, alc, chloroform, methylene chloride. LD_{50} in mice (mg/kg): 1040 i.v.; 6000 orally (Keim).

Combination with hydrochlorothiazide, *Acezide, Capozide, Captea, Ecazide.*

THERAP CAT: Antihypertensive.

1774. Capuride. *N-(Aminocarbonyl)-2-ethyl-3-methyl-pentanamide; (2-ethyl-3-methylvaleryl)urea; (ethyl-sec-butylacetyl)urea; (1-ethyl-2-methylpentanoyl)urea;* Pacinox. $C_9H_{18}N_2O_2$; mol wt 186.25. C 58.03%, H 9.74%, N 15.04%, O 17.18%. Prepn: Volwiler, Tabern, *J. Am. Chem. Soc.* **58**, 1352 (1936); Adams *et al.*, U.S. pat. **3,282,998** (1966 to Millmaster Onyx).

Crystals from xylene, mp 172°.

THERAP CAT: Hypnotic.

1775. Caramel. Burnt sugar coloring; burnt sugar. Made by heating sugar or glucose, adding small quantities of alkali, alkaline carbonate or a trace of mineral acid during the heating.

Dark-brown, thick liq; pleasant, bitter taste; odor of burnt sugar. d about 1.35. Sol in water, dil alc; insol in benzene, chloroform, ether, acetone, petr ether, oil turpentine.

USE: Coloring foods, confectionery, galenicals.

1776. Caramiphen Ethanedisulfonate. *1-Phenylcyclopentanecarboxylic acid 2-(diethylamino)ethyl ester 1,2-ethanedisulfonate(2:1); diethylaminoethyl 1-phenylcyclopentane-1-carboxylate ethanedisulfonate; bis-[1-(carbo-β-diethylaminoethoxy)-1-phenylcyclopentane] ethanedisulfonate; bis[1-(2-diethylaminoethoxycarbonyl)-1-phenylcyclopentane] ethanedisulfonate;* Alcopon; Taoryl; Toryn. $C_{38}H_{60}N_2O_{10}S_2$; mol wt 769.03. C 59.35%, H 7.86%, N 3.64%, O 20.81%, S 8.34%. One of the active ingredients in *Dondril.* For prepn see the references under caramiphen hydrochloride and **Swiss pat. 272,708** (1951 to Geigy); *Chem. Zentr.* **1952**, 1571; *cf.* U.S. pat. **2,404,588** (1946 to Geigy).

Crystals from acetone, mp 115-116°. More sol in water than the hydrochloride. Sol in alc, pharmaceutical syrups.

THERAP CAT: Antitussive.

1777. Caramiphen Hydrochloride. *1-Phenylcyclopentanecarboxylic acid 2-(diethylamino)ethyl ester hydrochloride;* diethylaminoethyl 1-phenylcyclopentane-1-carboxylate hydrochloride; Panparnit; Parpanit. $C_{18}H_{28}ClNO_2$; mol wt

325.87. C 66.34%, H 8.66%, Cl 10.88%, N 4.30%, O 9.82%. Prepn from 1-phenylcyclopentanecarboxylic acid chloride and diethylaminoethanol: **Swiss pat. 234,452** (1945 to Geigy); *C.A.* **43**, 6229 (1949).

Crystals, mp 145-146°. The free ester $bp_{0.05}$ 110-115°. Sol in alc. Slightly sol in water. LD_{50} i.p. in rats: 209 mg/kg, Kraatz *et al., J. Pharmacol. Exp. Ther.* **96**, 42 (1949).

THERAP CAT: Anticholinergic.

1778. Caraway. Dried ripe fruit of *Carum carvi* L., *Umbelliferae. Habit.* Europe, Central and Western Asia; cultivated in England, Russia, U.S. *Constit.* 3.5-7% volatile and fatty oils; resin, sugar, tannin, mucilage.

USE: As a spice in baking. Pharmaceutic aid (flavor).

1779. Carazolol. *1-(9H-Carbazol-4-yloxy)-3-[(1-methylethyl)amino]-2-propanol;* BM 51052; Conducton; Suacron. $C_{18}H_{22}N_2O_2$; mol wt 298.39. C 72.46%, H 7.43%, N 9.39%, O 10.72%. β-Adrenergic blocker. Prepn: H. Leinert *et al.,* **Ger. pat. 2,240,599** (1974 to Boehringer, Mann.), *C.A.* **80**, 133455a (1974). Comparative study of cardiac action: W. Bartsch *et al., Arzneimittel-Forsch.* **27**, 1022 (1977). Initial clinico-pharmacological study: E. Chorianopoulos *et al., Herz/Kreislauf* **9**, 965 (1977), *C.A.* **88**, 164460t (1978). Receptor binding studies: G. Kaiser *et al., Arch. Pharmacol.* **305**, 41 (1978); R. B. Innis *et al., Life Sci.* **24**, 2255 (1979). Use in treatment of stress syndrome in pigs: G. Ballarini, F. Guizzardi, *Tierärztl. Umschau* **36**, 171 (1981).

Hydrochloride, $C_{18}H_{23}ClN_2O_2$, crystals, mp 234-235°.

THERAP CAT: Antihypertensive, antianginal, antiarrhythmic.

THERAP CAT (VET): Treatment of stress in pigs.

1780. Carbachol. *2-[(Aminocarbonyl)oxy]-N,N,N-trimethylethanaminium chloride; (2-hydroxyethyl)trimethyl ammonium chloride carbamate;* choline chloride carbamate; carbamylcholine chloride; Carcholin; Miostat; Moryl; Doryl; Coletyl; Lentin. $C_6H_{15}ClN_2O_2$; mol wt 182.65. C 39.45%, H 8.28%, Cl 19.41%, N 15.34%, O 17.52%. $[NH_2COOCH_2CH_2N(CH_3)_3]^+Cl^-$. Prepd by the interaction of β-chloroethylcarbamate and trimethylamine: Kreitmair, *Arch. Exp. Path. Pharmakol.* **164**, 346 (1932); Merck, **Ger.** pats. **539,329; 553,148; 590,311.** Toxicity: Molitor, *J. Pharmacol. Exp. Ther.* **58**, 337 (1936).

Hard prismatic crystals. Hygroscopic. Odorless, but on standing in an open container develops a faint odor resembling that of an aliphatic amine. mp 200-203° (some decompn); mp 210-212°; mp 204-205° (Kreitmair). One gram dissolves in 1 ml water, 50 ml alcohol, 10 ml methanol. Almost insol in chloroform, ether. The aq soln is stable even when heated and is neutral to litmus. LD_{50} in mice (mg/kg): 15 orally, 0.3 i.v. (Molitor).

Note: The name "Doryl" is also used to designate a high temp insulating resin which has a diphenyl oxide structure linked with methylene groups: Micarta Div., Westinghouse Electric Corp. Ref: *Chem. Week*, Sept. 29, 1962, p 92.

THERAP CAT: Cholinergic, miotic.

THERAP CAT (VET): Parasympathomimetic, used chiefly in large animals, esp for simple colic in the horse.

1781. Carbacrylic Resins. A generic name for cross-linked polyacrylic polycarboxylic ion exchange resins.

Carbacylamine resins, Carbo-Resin. Combination of the

following ion exchange resins: carbacrylic resin, potassium carbacrylic resin, and polyaminemethylene resin.

THERAP CAT: Cation-exchange resin.

1782. Carbadox. *(2-Quinoxalinylmethylene)hydrazine-carboxylic acid methyl ester N,N'-dioxide; 3-(2-quinoxalin-ylmethylene)carbazic acid methyl ester N,N'-dioxide;* methyl 3-(2-quinoxalinylmethylene)carbazate N^1,N^4-dioxide; 2-formylquinoxaline-1,4-dioxide carbomethoxyhydrazone; GS-6244; Fortigro; Mecadox. $C_{11}H_{10}N_4O_4$; mol wt 262.23. C 50.38%, H 3.84%, N 21.37%, O 24.41%. Prepn: Johnston, **Belg.** pat. **669,353;** idem, **U.S.** pats. **3,371,090; 3,433,871** (1964, 1968, 1969, all to Pfizer). Animal studies: Thrasher et al., *J. Anim. Sci.* **26,** 911 (1967); Kornegay et al., ibid. **27,** 1134 (1968).

Minute yellow crystals, mp 239.5-240°. uv max (water): 236, 251, 303, 366, 373 nm (ε 11000, 10900, 36400, 16100, 16200). Practically insol in water.

THERAP CAT (VET): Antimicrobial.

1783. Carbamazepine. *5H-Dibenz[b,f]azepine-5-carbox-amide;* 5-carbamoyl-5H-dibenz[b,f]azepine; G 32883; Biston; Calepsin; Carbelan; Epitol; Finlepsin; Sirtal; Stazepine; Tegretal; Tegretol; Telesmin; Timonil. $C_{15}H_{12}N_2O$; mol wt 236.26. C 76.25%, H 5.12%, N 11.86%, O 6.77%. Prepn: Schindler, **U.S.** pat. **2,948,718** (1960 to Geigy). Metabolism: P. L. Morselli, A. Frigerio, *Drug Metab. Rev.* **4,** 97 (1975). Review of pharmacokinetics in man: L. Bertilsson, *Clin. Pharmacokinet.* **3,** 128-143 (1978); S. Pynnönen, *Ther. Drug Monit.* **1,** 409-431 (1979). Toxicity: E. G. Stenger, F. C. Roulet, *Med. Exp.* **11,** 191 (1964). Comprehensive description: H. Y. Aboul-Enein, A. A. Al-Badr in *Analytical Profiles of Drug Substances* vol. 9, K. Florey, Ed. (Academic Press, New York, 1980) pp 87-106.

Crystals from abs ethanol + benzene, mp 190-193°. Sol in alcohol, acetone, propylene glycol. Practically insol in water. LD_{50} orally in mice, rats: 3750, 4025 mg/kg (Stenger, Roulet).

THERAP CAT: Analgesic. Anticonvulsant.

1784. Carbamic Acid. Aminoformic acid; aminomethanoic acid. NH_2COOH. Not known in the free state. Its salts occur in blood and urine of mammals.

Its salts, e.g., ammonium carbamate, and its esters, such as the ethyl ester (urethan), are entered separately.

1785. Carbamyl Chloride. *Carbamic chloride;* chloroformamide. CH_2ClNO; mol wt 79.49. C 15.11%, H 2.54%, Cl 44.60%, N 17.62%, O 20.13%. H_2NCOCl. Prepd by passing HCl gas over heated cyanuric acid: Gattermann, Rossolymo, *Ber.* **23,** 1190 (1890); Gattermann, *Ber.* **32,** 1117 (1899). From ammonia and phosgene at 400°: Rupe, Labhard, *Ber.* **33,** 236 (1900), cf. Gattermann, *Ann.* **244,** 30 (1888).

Liquid. Acrid, offensive odor. Has been obtained cryst, mp about 50°, bp 61-62° (decompn). Reacts violently on contact with water, forming ammonium chloride and carbon dioxide. During storage it gives off HCl and slowly changes to cyanuric acid.

1786. Carbanilic Acid. *Phenylcarbamic acid; N-carboxyaniline.* $C_7H_7NO_2$; mol wt 137.14. $C_6H_5NHCOOH.$ Known only by its derivatives (esters, etc.).

1787. Carbanilide. *N,N'-Diphenylurea;* diphenylcarbamide; 1,3-diphenylurea; sym-diphenylurea. $C_{13}H_{12}N_2O$; mol wt 212.24. C 73.56%, H 5.70%, N 13.20%, O 7.54%. $C_6H_5NHCONHC_6H_5$. Obtained during the preparation of phenylurea from aniline hydrochloride and urea: Davis, Blanchard, *Org. Syn.* coll. vol. I, 453 (2nd ed., 1941). Crystal structure: W. Dannecker et al., *Cryst. Struct. Commun.* **8,** 429 (1979).

Orthorhombic prisms from alc. d 1.239. mp 238°. bp 260° (decompn). Sublimes in current of hydrogen at 220°. Sol in ether, glacial acetic acid. Sparingly sol in water (0.15 g/l), acetone, alcohol, chloroform. Moderately sol in pyridine (69.0 g/l).

1788. Carbarsone. *[4-[(Aminocarbonyl)amino]phenyl]-arsonic acid; N-carbamoylarsanilic acid; p-ureidobenzenearsonic acid; N-carbamylarsanilic acid; p-carbamidobenzenearsonic acid; 4-ureido-1-phenylarsonic acid; 4-carbamyl-aminophenylarsonic acid; p-arsonophenylurea;* Amabevan; Ameban; Amibiarson; Arsambide; Carb-O-Sep; Histocarb; Fenarsone; Leucarsone; Aminarsone; Amebarsone. $C_7H_9-AsN_2O_4$; mol wt 260.07. C 32.33%, H 3.49%, As 28.80%, N 10.77%, O 24.61%. Prepd from the sodium salt of arsanilic acid by treatment with potassium cyanate or cyanogen bromide: **Ger.** pat. **213,155;** Stickings, *J. Chem. Soc.* **1928,** 3131; from arsanilic acid by treatment with phosgene: Nakatsu, Kawase, *Ann. Rept. Takamine Lab.* **8,** 44-47 (1956); **Japan.** pat. **4418('58)** (to Sankyo).

White powder. mp 174°. Slightly sol in water, alcohol; sol in solns of alkali hydroxides and carbonates. The satd aq soln is acid to litmus. Nearly insol in ether or chloroform. LD_{50} orally in rats: 510 mg/kg.

THERAP CAT: Antiamebic.

THERAP CAT (VET): Antihistomonad in turkeys.

1789. Carbaryl. *1-Naphthalenol methylcarbamate; methyl carbamic acid 1-naphthyl ester;* 1-naphthyl N-methylcarbamate; ENT 23969; OMS 29; UC 7744; Arylam; Carylderm; Clinicide; Derbac; Dicarbam; Ravyon; Seffein; Sevin. $C_{12}H_{11}NO_2$; mol wt 201.22. C 71.62%, H 5.51%, N 6.96%, O 15.90%. Prepn and description: Haynes et al., *Contrib. Boyce Thompson Inst.* **18,** 507 (1957); Lambrech, **U.S.** pat. **2,903,478** (1959 to Union Carbide). Metabolism: W. E. Whitehurst et al., *J. Agr. Food Chem.* **11,** 167 (1963); J. B. Houston et al., *Xenobiotica* **5,** 637 (1975). Degradation: D. G. Crosby et al., *J. Agr. Food Chem.* **13,** 204 (1965); D. L. Heywood, *Environ. Qual. Saf.* **4,** 128 (1975). Toxicology: I. Nisbet, D. Miner, *Environment* **13,** 10 (1971). Toxicity: M. Vandekar et al., *Bull. World Health Org.* **44,** 241 (1971). Clinical trial in pediculosis: J. W. Maunder, *Clin. Exp. Dermatol.* **6,** 605 (1981). *Review: Carbamate Insecticides: Chemistry, Biochemistry and Toxicology,* R. J. Kuhr, H. W. Dorough, Eds. (CRC Press, Cleveland, 1976) 301 pp.

Crystals, mp 142°. d_{20}^{20} 1.232. Moderately sol in DMF, acetone, isophorone, cyclohexanone. Soly in water at 30°: 120 mg/l. Vapor pressure at 25°: $< 4 \times 10^{-5}$ mm Hg. Stable to heat, light, acids; hydrolyzed in alkalies; noncorrosive. LD_{50} orally in rats: 250 mg/kg (Vandekar).

Human Toxicity: Nausea, vomiting, diarrhea, bronchoconstriction, blurring vision, excessive salivation, muscle twitching, cyanosis, convulsions, coma, respiratory failure, *Clinical Toxicology of Commercial Products,* R. E. Gosselin *et al.*, Eds. (Williams & Wilkins, Baltimore, 4th ed., 1976) Section III, pp 79-82.

USE: Contact insecticide.

THERAP CAT: Pediculocide.

THERAP CAT (VET): Ectoparasiticide.

1790. Carbazochrome Salicylate. *2-Hydroxybenzoic acid monosodium salt compd with 2-(1,2,3,6-tetrahydro-3-hydroxy-1-methyl-6-oxo-5H-indol-5-ylidene)hydrazinecarboxamide (1:1); 3-hydroxy-1-methyl-1,5,6-indolinedione semicarbazone compd with sodium salicylate;* adrenochrome semicarbazone compd with sodium salicylate; adrenochrome monosemicarbazone sodium salicylate complex; Adenogen; Adrenosem; Adrenosem Salicylate; Adrestat-F; Statimo. $C_{17}H_{17}N_4NaO_6$; mol wt 396.35. C 51.52%, H 4.32%, N 14.14%, Na 5.80%, O 24.22%. $C_{10}H_{12}N_4O_3 \cdot C_7H_5NaO_3$. Prepd by refluxing adrenochrome semicarbazone and Na salicylate in 30% methanol: Iwao *et al.*, **Japan.** pat. **546('57)** (to Tanabe), *C.A.* **52,** 4693 (1958).

Orange-red powder, mp 196-197.5° (dec). Soly in water (25°): 0.61 mg/ml. Practically insol in ether, chloroform. pH of 10% aq soln 6.7 to 7.3. *See also* Adrenochrome.

THERAP CAT: Antihemorrhagic.

THERAP CAT (VET): Systemic hemostatic for capillary bleeding from increased capillary permeability.

1791. Carbazochrome Sodium Sulfonate. *5-[(Aminocarbonyl)hydrazono]-2,3,5,6-tetrahydro-1-methyl-6-oxo-1H-indole-1-sulfonic acid monosodium salt; 5,6-dihydro-1-methyl-5,6-dioxo-2-indolinesulfonic acid 5-semicarbazone sodium salt;* epinochrome-2-sulfonic acid 5-monosemicarbazone sodium salt; 1-methyl-5-semicarbazono-6-oxo-2,3,5,6-tetrahydroindole-2-sulfonic acid sodium salt; sodium 5,6-dihydro-1-methyl-5,6-dioxo-2-indolinesulfonate 5-semicarbazone; Ac-17; Adenaron; Adona; Adrechros; Carbazon; Donaseven; Emex; Odanon; Tazin. $C_{10}H_{11}N_4$-NaO_5S; mol wt 322.29. C 37.27%, H 3.44%, N 17.38%, Na 7.13%, O 24.82%, S 9.95%. Prepn: Iwao, *Pharm. Bull. (Tokyo)* **4,** 251 (1956); *C.A.* **51,** 6860f (1957); **Brit.** pat. **795,184** (1958 to Gohei Tanabe). Revision of structure from a 3- to a 2-sulfonic acid: Kawazu *et al.*, *J. Heterocycl. Chem.* **10,** 1059 (1973).

Yellow-orange needles from aq methanol, dec 227-228° (free acid dec 195°). Soluble in water.

THERAP CAT: Hemostatic.

1792. Carbazole. *9H-Carbazole;* 9-azafluorene; dibenzopyrrole; diphenylenimine. $C_{12}H_9N$; mol wt 167.20. C 86.19%, H 5.43%, N 8.38%. Prepn: Bunyan, Cadogan, *J. Chem. Soc.* **1963,** 42. Manuf from 2-biphenylamine: Conover, **U.S.** pat. **2,481,292** (1949 to Monsanto); Voltz *et al.*, **U.S.** pat. **2,891,965** (1959 to Houdry Process Corp.); Nevitt, Seelig, **U.S.** pat. **3,085,095** (1963 to Standard Oil, Indiana); from 2-nitrobiphenyl: Larrison, **U.S.** pat. **2,508,791** (1950 to Allied Chem.); by hydrogenolysis of coal-hydrogenation products: Murray *et al.*, **U.S.** pat. **2,913,397** (1959 to Union Carbide); from diphenylamine: Grotta, **U.S.** pat. **2,921,942** (1960 to American-Marietta); Bearse *et al.*, **U.S.** pat. **3,041,-349** (1962 to Martin-Marietta); Nevitt, Seelig, *loc. cit.* Purification: Rottschaefer, **U.S.** pat. **2,459,135** (1949 to GAF); Insinger, **U.S.** pat. **2,464,811** (1949 to Koppers Co.). Review of prepn and properties: W. C. Sumpter, F. M. Miller, *Heterocyclic Compounds* (Interscience, New York, 1954) pp 70-109.

Crystals from alcohol, benzene, toluene, glacial acetic acid, mp 245°. Sublimes. bp 355°, bp_{147} 200°. d_4^{18} 1.10. Ex-

hibits strong fluorescence and long phosphorescence on exposure to ultraviolet light. Extremely weak base. Insol in water. One gram dissolves in 3 ml quinoline, 6 ml pyridine, 9 ml acetone, 2 ml acetone at 50°, 35 ml ether, 120 ml benzene, 135 ml abs alcohol. Slightly sol in petr ether, chlorinated hydrocarbons, acetic acid. Dissolves in concd H_2SO_4 without dec. KOH fusion yields *N*-potassium salt. LD_{50} orally in rats: > 5 g/kg, Eagle, Carlson, *J. Pharmacol. Exp. Ther.* **99,** 450 (1950).

Picrate, red prisms, mp 186°.

USE: Important dye intermediate. Used in making photographic plates sensitive to ultraviolet light. Reagent for lignin, carbohydrates, and formaldehyde.

1793. 9-Carbazoleacetic Acid. Carbazyl-*N*-acetic acid; *N,N*-diphenyleneglycine. $C_{14}H_{11}NO_2$; mol wt 225.24. C 74.65%, H 4.92%, N 6.22%, O 14.21%. Prepd by condensing dry potassium carbazole with chloroethyl acetate and saponifying the resulting ester with 35% sodium hydroxide: **Ger.** pat. **255,304;** *Chem. Zentr.* **1913, I,** 350; *Frdl.* **11,** 171 (1912-14). *See also:* Seka, *Ber.* **57,** 1527 (1924).

Leaflets from ethyl acetate, mp 215°. Soluble in ether, glacial acetic acid, alcohol, chloroform, xylene. On heating to 240-250° it yields *N*-methylcarbazole.

Ethyl ester, $C_{16}H_{15}NO_2$, crystals from alcohol, mp 97°, sol in organic solvents.

USE: In the detection of nitrates.

1794. Carbendazim. *1H-Benzimidazol-2-ylcarbamic acid methyl ester; 2-benzimidazolecarbamic acid methyl ester;* 2-(methoxycarbonylamino)benzimidazole; methyl 2-benzimidazolecarbamate; carbendazole; BMC; MBC; BCM; BAS 3460; BAS 67054; CTR 6669; Hoe 17411; Bavistin; Derosal. $C_9H_9N_3O_2$; mol wt 191.18. C 56.54%, H 4.74%, N 21.98%, O 16.74%. Degradn product of benomyl, *q.v.* Prepn: H. M. Loux, **U.S.** pat. **3,010,968** (1961 to du Pont); H. A. Selling *et al.*, *Chem. & Ind. (London)* **1970,** 1625. Activity: G. P. Clemons, H. D. Sisler, *Pestic. Biochem. Physiol.* **1,** 32 (1971). Degradn in soil: A. Helweg, *Pestic. Sci.* **8,** 71 (1977).

Light gray powder, mp 302-307° (dec). pKa 4.48. Soly in water at 24°: 8 mg/l at pH 7; 29 mg/l at pH 4. Soly at 24° (mg/l): hexane 0.5; benzene 36; dichloromethane 68; ethanol 300. Slowly decomp in alkaline soln. LD_{50} orally in rats: 6400 mg/kg, *RTECS* **Vol. I,** R. J. Lewis, R. L. Tatken, Eds. (1979) p 229.

USE: Fungicide.

1795. Carbenes. Highly reactive, electron-deficient, divalent carbon intermediates having two unshared electrons, the simplest being *methylene*. Carbenes exist in two states: singlet and triplet. The unshared electrons of singlet carbenes reside in one orbital with paired spins, giving rise to nucleophilic and electrophilic properties. In triplet carbenes, these electrons are found in two orbitals with unpaired spins, and may be considered as diradicals. Carbenes are formed thermally or photochemically from ketenes or diazo compounds. They insert into carbon-hydrogen bonds, add to olefins to form cyclopropanes, dimerize and rearrange. Formation in interstellar space: D. K. Bohme, *Nature* **319,** 473 (1986). Nitrenes, *q.v.*, are nitrogen analogs of carbenes. Books: W. Kirmse, *Carbene Chemistry* (Academic Press, New York, 2nd ed., 1971); *Carbenes* vol. 1, M. Jones, R. A. Moss, Eds. (Wiley-Interscience, New York, 1973) 356 pp; **vol. 2** (1975) 373 pp. *Reviews:* R. A. Moss, M. Jones, *Reactive Intermediates* vol. 1, M. Jones, R. A. Moss, Eds. (Wiley, New York, 1978) pp 69-116; **vol. 2** (1981) pp 59-133.

1796. Carbenicillin. *6-[(Carboxyphenylacetyl)amino]-3,3-dimethyl-7-oxo-4-thia-1-azabicyclo[3.2.0]heptane-2-carboxylic acid; N-(2-carboxy-3,3-dimethyl-7-oxo-4-thia-1-azabicyclo[3.2.0]hept-6-yl)-2-phenylmalonamic acid;* α-carboxybenzylpenicillin; 6-(α-carboxyphenylacetamido)penicillanic acid; α-phenyl(carboxymethylpenicillin). $C_{17}H_{18}N_2O_6S$; mol wt 378.42. C 53.96%, H 4.79%, N 7.40%, O 25.37%, S 8.47%. Semi-synthetic antibiotic related to penicillin. Prepn of monopotassium salt: Hobbs, **U.S.** pat. **3,142,673** (1964 to Pfizer); of disodium salt: **Belg.** pat. **646,-991**; corresp to Brain, Nayler, **U.S.** pat. **3,282,926** (1964, 1966 to Beecham). Activity studies and pharmacology: Naumann, Kempf, *Arzneimittel-Forsch.* **19**, 1222 (1969). Chemistry and mode of action: Butler *et al., J. Infec. Dis.* **122**, *Suppl.*, 81 (1970). Clinical data: Hoffler *et al., ibid.* 1233; Gritz, Naumann, *ibid.* 1237. Toxicity data: E. I. Goldenthal, *Toxicol. Appl. Pharmacol.* **18**, 185 (1971).

Disodium salt, $C_{17}H_{16}N_2Na_2O_6S$, *BRL-2064, carbenicillin disodium, CP-15639-2, Anabactyl, Carbapen, Carbecin, Geopen, Hyoper, Microcillin, Pyocianil, Pyopen.* White powder. LD_{50} i.p. in rats: > 2000 mg/kg (Goldenthal).
THERAP CAT: Antibacterial.

1797. Carbenoxolone. *3-(3-Carboxy-1-oxopropoxy)-11-oxoolean-12-en-29-oic acid; 3β-hydroxy-11-oxoolean-12-en-30-oic acid hydrogen succinate;* 3-O-(β-carboxypropionyl)-11-oxo-18β-olean-12-en-30-oic acid; glycyrrhetic acid hydrogen succinate; 18β-glycyrrhetic acid hydrogen succinate; carbenoxalone. $C_{34}H_{50}O_7$; mol wt 570.74. C 71.55%, H 8.83%, O 19.62%. Anti-inflammatory glucocorticoid related to enoxolone, *q.v.* Prepn: Gottfried, Baxendale, **Brit.** pat. **843,133**; **U.S.** pat. **3,070,623** (1960, 1961 to Biorex). Monograph: *Carbenoxalone Sodium*, J. M. Robson, F. M. Sullivan, Eds. (Butterworths, London, 1969) 263 pp. Symposium on clinical efficacy: *Scand. J. Gastroenterol.* **15**, Suppl. 65, 1-121 (1980). Effect on gastric prostaglandin levels in humans: J. Rask-Madsen *et al., Eur. J. Clin. Invest.* **13**, 351 (1983); P. Minuz *et al., Pharmacol. Res. Commun.* **16**, 875 (1984).

Cream-colored crystals, mp 291-294°. $[\alpha]_D^{20}$ +128° (chloroform).
Disodium salt, $C_{34}H_{48}Na_2O_7$, *Biogastrone, Bioplex, Bioral, Duogastrone, Neogel, Pyrogastrone, Sanodin, Ulcus-Tablinen.* Creamy-white solid. Freely sol in water. LD_{50} in male mice (mg/kg): 198 i.v.; 120 i.p.; in male rats (mg/kg): 3200 orally (Robson, Sullivan).
THERAP CAT: Anti-ulcerative.

1798. Carbetapentane. *1-Phenylcyclopentanecarboxylic acid 2-(2-diethylaminoethoxy)ethyl ester;* 2-(diethylaminoethoxy)ethyl 1-phenyl-1-cyclopentanecarboxylate; 2-(diethylaminoethoxy)ethyl 1-phenylcyclopentyl-1-carboxylate; 1-phenylcyclopentane-1-carboxylic acid diethylaminoethoxyethyl ester; pentoxyverine; pentoxiverin; Atussil. $C_{20}H_{31}NO_3$; mol wt 333.46. C 72.03%, H 9.37%, N 4.20%, O 14.39%. Prepn: H. G. Morren, **Brit.** pat. **753,779** (1956), *C.A.* **51**, 7443d (1957). Antispasmodic activity: D. Wellens,

Arzneimittel-Forsch. **17**, 495 (1967). Clinical effect on lung function: E. Krieger, *ibid.* **22**, 389 (1972).

$bp_{0.01}$ 165-170°.
Citrate, $C_{26}H_{39}NO_{10}$, *UCB 2543, Antees, Aslos, Calnathal, Carbetane, Cossym, Fustpentane, Germapect, Pencal, Sedotussin, Toclase, Tosnone, Tuclase.* Crystals, mp 93°. Freely sol in water, chloroform; sol in alcohol, acetone, ethyl acetate. Practically insol in ether, petr ether, benzene.
THERAP CAT: Antitussive.

1799. Carbetidine. *1-[2-(2-Hydroxyethoxy)ethyl]-4-phenyl-4-piperidinecarboxylic acid ethyl ester; 1-[2-(2-hydroxyethoxy)ethyl]-4-phenylisonipecotic acid ethyl ester;* ethyl 1-[2-(2-hydroxyethoxy)ethyl]-4-phenylisonipecotate; 1-hydroxyethoxyethyl-4-phenyl-4-piperidine ethyl carboxylate; 1-(2-hydroxyethoxyethyl)-4-phenyl-4-carbethoxypiperidine; etoxeridine; Wy 2039; UCB 2073; Atenos. $C_{18}H_{27}NO_4$; mol wt 321.40. C 67.26%, H 8.47%, N 4.36%, O 19.91%. Prepn: Morren, **U.S.** pat. **2,858,316** (1958 to UCB).

Liquid. $bp_{0.02}$ 170°.
Hydrochloride, $C_{18}H_{27}NO_4 \cdot HCl$, mp 115°.
Caution: May be habit forming. This is a controlled substance (opiate) listed in the U.S. Code of Federal Regulations, Title 21 Part 1308.11 (1985).
THERAP CAT: Analgesic.

1800. Carbetocin. *1-Butanoic acid-2-(O-methyl-L-tyrosine)-1-carbaoxytocin;* 1-butyric acid-2-[3-(p-methoxyphenyl)-L-alanine]oxytocin; deamino-2-O-methyltyrosine-1-carbaoxytocin; 1-thia-4,7,10,13,16-pentaazacycloeicosane cyclic peptide deriv; 1-desamino-1-monocarba-[2-tyr-(OMe)]-OT; (2-O-methyltyrosine)deamino-1-carbaoxytocin; d(COMOT); Decomoton; Depotocin. $C_{45}H_{69}N_{11}O_{12}S$; mol wt 988.17. C 54.70%, H 7.04%, N 15.59%, O 19.43%, S 3.24%. Synthetic carba-analog of oxytocin, *q.v.* Prepn: I. Fric *et al., Coll. Czech. Chem. Commun.* **39**, 1290 (1974); J. H. Cort *et al.,* **Ger.** pat. **2,732,175** (1976 to Czech. Akad. Ved.). Chromatographic properties: M. Lebl, *ibid.* **45**, 2927 (1980). Uterotonic and galactogogic activity: T. Barth *et al., ibid.* 3045; T. Barth *et al., ibid.* **46**, 2441 (1981). Pharmacokinetics in lactating sows: N. Cort *et al., Am. J. Vet. Res.* **42**, 1804 (1981). Use in regulation of bovine labor: Z. Veznik *et al., ibid.* **40**, 425 (1979). Effect on milk let-down in sows: N. Cort *et al., ibid.* **43**, 1283 (1982).

Solid from methanol with ether $[\alpha]_D$ −69.0° (c = 0.25 in 1M acetic acid).
THERAP CAT (VET): Oxytocic, stimulates milk let-down.

1801. Carbic Anhydride. *3aα,4,7,7aα-Tetrahydro-4α,-7α-methanoisobenzofuran-1,3-dione; cis-endo-5-norborn-*

ene-2,3-dicarboxylic anhydride; endo-cis-bicyclo[2.2.1]hept-5-ene-2,3-dicarboxylic anhydride; 3,6-endomethylene-1,2,-3,6-tetrahydro-cis-phthalic anhydride; 3,6-endomethylene-Δ^4-tetrahydrophthalic anhydride; Nadic anhydride. $C_9H_8O_3$; mol wt 164.15. C 65.85%, H 4.91%, O 29.24%. Prepd by the reaction of maleic anhydride with cyclopentadiene in benzene: Diels, Alder, *Ann.* **460**, 98 (1928). Crystal structure: Destro *et al., Acta Crystallogr.* **25B**, 2465 (1969).

Shiny, orthorhombic crystals from petr ether, mp 164-165°. d 1.417. Converted to equilibrium mixtures with *exo-cis* isomers when heated above mp. Sol in benzene, toluene, acetone, carbon tetrachloride, chloroform, ethanol, ethyl acetate. Slightly sol in petr ether. Reacts with water to form the corresponding acid. Forms the γ-lactone of 5-hydroxy-2,3-norcamphanedicarboxylic acid in 50% H_2SO_4.

1802. Carbidopa. *S-α-Hydrazino-3,4-dihydroxy-α-methylbenzenepropanoic acid monohydrate;* (−)-L-α-hydrazino-3,4-dihydroxy-α-methylhydrocinnamic acid monohydrate; α-hydrazino-α-methyl-β-(3,4-dihydroxyphenyl)propionic acid monohydrate; L-α-(3,4-dihydroxybenzyl)-α-hydrazinopropionic acid monohydrate; α-methyldopahydrazine; HMD; MK-486; Lodosin; Lodosyn. $C_{10}H_{14}N_2O_4 \cdot H_2O$; mol wt 244.25. C 49.18%, H 6.60%, N 11.47%, O 32.75%. Peripheral decarboxylase inhibitor. Prepn of DL-form: Pfister, **Fr.** pat. M1553 (1962 to Merck & Co.), *C.A.* **59**, 12921e (1963); Sletzinger *et al., J. Med. Chem.* **6**, 101 (1963); **Brit.** pat. **940,596** corresp to Chemerda *et al.,* **U.S.** pat. **3,462,536** (1963, 1969 both to Merck & Co.). Synthesis of the L-form: Karady *et al.,* **Ger.** pats. **2,062,285**; **2,062,332** (both 1971 to Merck & Co.), *C.A.* **75**, 118122t, 118120r (1971); *eidem, J. Org. Chem.* **36**, 1946, 1949 (1971). Inhibition of dopa decarboxylase: Porter *et al., Biochem. Pharmacol.* **11**, 1067 (1962); Moran, Sourkes, *J. Pharmacol. Exp. Ther.* **148**, 252 (1962); Watanabe *et al., Clin. Pharmacol. Ther.* **11**, 740 (1970). Only the L-form is pharmacologically active: Lotti, Porter, *J. Pharmacol. Exp. Ther.* **172**, 406 (1970).

Crystals from hot water, mp 203-205° (dec). $[\alpha]_D$ −17.3° (methanol). Also reported mp 208°.
Combination with levodopa, *Isicom, Nacom, Sinemet.*
DL-Form, tan fluffy crystals, mp 206-208° (dec). uv max (methanol): 282.5 nm (ϵ 2940).
THERAP CAT: In combination with levodopa as antiparkinsonian.

1803. Carbimazole. *2,3-Dihydro-3-methyl-2-thioxo-1H-imidazole-1-carboxylic acid ethyl ester;* 1-ethoxycarbonyl-3-methyl-2-thio-4-imidazoline; ethyl 3-methyl-2-thioimidazoline-1-carboxylate; 1-methyl-3-carbethoxy-2-thioglyoxalone; athyromazole; Neo-mercazole; Neo-Thyreostat. $C_7H_{10}N_2O_2S$; mol wt 186.23. C 45.15%, H 5.41%, N 15.04%, O 17.18%, S 17.22%. Prepn: Rimington *et al.,* **U.S.** pat. **2,671,088** and Re-issue **24505; U.S.** pat. **2,815,349** (1954, 1958, 1957, all to Natl. Res. Dev. Corp.); Baker, *J. Chem. Soc.* **1958**, 2387.

Crystalline powder with characteristic odor; tasteless at first, followed by bitter taste; mp 122-125°. Sol (at 20°) in 500 parts of water; in 50 parts of ethanol; in 330 parts of

ether; in 3 parts of chloroform; in 17 parts of acetone. uv max in 0.1N HCl : water (1:8): 291 nm; in 0.1N H_2SO_4: 227 nm and 291 nm ($E_{1cm}^{1\%}$ 557).
THERAP CAT: Thyroid inhibitor.

1804. Carbinoxamine. *2-[(4-Chlorophenyl)-2-pyridinylmethoxy]-N,N-dimethylethanamine; 2-[p-chloro-α-(2-dimethylaminoethoxy)benzyl]pyridine;* paracarbinoxamine. $C_{16}H_{19}ClN_2O$; mol wt 290.80. C 66.09%, H 6.59%, Cl 12.19%, N 9.63%, O 5.50%. Prepn: Tilford, Shelton, **U.S.** pat. **2,606,195** (1952 to Wm. S. Merrell); Swain, **U.S.** pat. **2,800,485** (1957 to McNeil Labs.). Prepn of l-form: **Brit.** pat. **905,993** (1962 to McNeil Labs.), *C.A.* **58**, 5644a (1962). Abs config of l-form: V. Barouh *et al., J. Med. Chem.* **14**, 834 (1971). Pharmacology and toxicology: R. Cahen, *Ann. Pharm. Franc.* **20**, 463 (1962). GLC determn in serum: D. J. Hoffman *et al., J. Pharm. Sci.* **72**, 1342 (1983).

Liquid, $bp_{0.1}$ 158-162°.
Hydrochloride, $C_{16}H_{19}ClN_2O \cdot HCl$, crystals from isopropanol + ethyl acetate, dec 162-164°. Sol in water.
Maleate, $C_{20}H_{23}ClN_2O_5$, *Allergefon, Clistin, Ciberon, Hislosine, Lergefin, Polistin T-Caps.* Bitter crystals from ethyl acetate, mp 117-119°. Freely sol in water, alcohol, chloroform. Very slightly sol in ether. pH of 1% aq soln 4.6-5.1. LD_{50} in mice (mg/kg): 166 i.p. (Cahen).
l-Form, *levocarbinoxamine, rotoxamine, McN-R-73-Z.* $bp_{0.5}$ 143-144°. n_D^{20} 1.5522. $[\alpha]_D^{25}$ −6.8° (c = 2 in methanol).
l-Form *d*-tartrate, $C_{20}H_{25}ClN_2O_7$, *Twiston.* Crystals from isopropanol, mp 143-144.5°. $[\alpha]_D^{25}$ +37.2° (c = 20 in methanol).
THERAP CAT: Antihistaminic.

1805. Carbiphene. *α-Ethoxy-N-methyl-N-[2-[methyl(2-phenylethyl)amino]ethyl]-α-phenylbenzeneacetamide; 2-ethoxy-N-methyl-N-[2-(methylphenethylamino)ethyl]-2,2-diphenylacetamide;* etomide (rescinded USAN); etymide. $C_{28}H_{34}N_2O_2$; mol wt 430.57. C 78.10%, H 7.96%, N 6.51%, O 7.43%. Prepn: Krapcho, Turk, *J. Med. Chem.* **6**, 547 (1963).

Hydrochloride, $C_{28}H_{35}ClN_2O_2$, *SQ 10269, Jubalon, Bandol.* Crystals, mp 163-165°.
THERAP CAT: Analgesic.

1806. Carbitol®. *2-(2-Ethoxyethoxy)ethanol;* diethylene glycol monoethyl ether; ethyl digol. $C_6H_{14}O_3$; mol wt 134.17. C 53.71%, H 10.52%, O 35.77%. $CH_3CH_2OCH_2-CH_2OCH_2CH_2OH$. Prepn from ethylene oxide and 2-ethoxyethanol in the presence of SO_2: Britton, Sexton, **U.S.** pat. **2,807,651** (1957 to Dow). Toxicity data: Smyth, Carpenter, *J. Ind. Hyg. Toxicol.* **30**, 63 (1948).
Very hygroscopic liquid, bp 196°. d_4^{25} 0.9855, d_{20}^{20} 1.0273. n_D^{20} 1.4273. Flash pt 96° (Tag open cup). Miscible with acetone, benzene, chloroform, ethanol, ether, pyridine, etc.: Jackson, Drury, *Ind. Eng. Chem.* **51**, 1491 (1959). Also miscible with water. LD_{50} orally in rats: 8.69 g/kg (Smyth, Carpenter).
Acetate, $C_8H_{16}O_4$, liquid, bp 218.5°. mp −25°. d_{20}^{20} 1.0114. n_D^{20} 1.4213. Flash pt 110° (Tag open cup). Miscible with water, alc, ether, most oils. LD_{50} orally in rats: 11 g/kg (Smyth, Carpenter).
USE: As solvent for cellulose esters, in lacquer and thinner formulations, in quick-drying varnishes and enamels, for dyestuffs and wood stains. Acetate is used as a solvent and plasticizer for cellulose esters, gums, resins, etc.

1807. Carbobenzoxy Chloride. *Carbonochloridic acid phenylmethyl ester;* benzyl chloroformate; chloroformic acid benzyl ester; benzylcarbonyl chloride. $C_8H_7ClO_2$; mol wt 170.60. C 56.32%, H 4.14%, Cl 20.79%, O 18.76%. Prepn by action of phosgene absorbed in toluene on benzyl alcohol: Carter *et al.*, *Org. Syn.* **23**, 13 (1943); by reacting carbonyl chloride and benzyl alcohol at −20 to −30°: Farthing, *J. Chem. Soc.* **1950**, 3213.

CH₂OCOCl

Oily liquid; acrid odor; lacrimator. bp₂₀ 103°; bp₇ 85-87°. Dec to CO_2 and benzyl chloride upon heating at 100-155°.
USE: In peptide synthesis to block the amino group.

1808. Carbocloral. *(2,2,2-Trichloro-1-hydroxyethyl)carbamic acid ethyl ester;* ethyl (2,2,2-trichloro-1-hydroxyethyl)carbamate; chloral-urethane; CI-336; HY-185; Ural; Uraline; Uralium. $C_5H_8Cl_3NO_3$; mol wt 236.49. C 25.39%, H 3.41%, Cl 44.98%, N 5.93%, O 20.30%. $Cl_3CCH(OH)$-$NHCOOC_2H_5$. Prepd from chloral and urethane in the presence of HCl: Bischoff, *Ber.* **7**, 628 (1874).
Crystalline powder, mp about 103° with partial decompn. Practically insol in water; freely sol in alcohol, ether.
THERAP CAT: Hypnotic.

1809. Carbocysteine. *S-(Carboxymethyl)-L-cysteine;* 3-[(carboxymethyl)thio]alanine; S-carboxymethylcysteine; AHR 3053; LJ 206; Carbocit; Fluifort; Lisil; Lisomucil; Loviscol; Muciclar; Mucocis; Mucodyne; Mucolase; Mucolex; Mucopront; Mucotab; Mukinyl; Pectox; Pulmoclase; Reomucil; Rhinathiol; Siroxyl; Thiodril; Transbronchin. $C_5H_9NO_4S$; mol wt 179.21. C 33.51%, H 5.06%, N 7.82%, O 35.71%, S 17.90%. $HOOCCH_2SCH_2CH(NH_2)COOH$. Prepn: Armstrong, Lewis, *J. Org. Chem.* **16**, 749 (1951); Schöberl, Wagner, *Z. Physiol. Chem.* **304**, 97 (1956); Foye, Verderame, *J. Am. Pharm. Assoc.* **46**, 273 (1957); Goodman *et al.*, *J. Org. Chem.* **23**, 1251 (1958). Pharmacology: Huyen-Vu-Ngoc *et al.*, *C. R. Soc. Biol.* **160**, 1849 (1966); Quevauviller *et al.*, *Therapie* **22**, 485 (1967). Clinical trial in chronic bronchitis: M. Grillage, K. Bernard-Jones, *Brit. J. Clin. Pract.* **39**, 395 (1985). Review of pharmacology and clinical uses: D. T. Brown, *Drug. Intell. Clin. Pharm.* **22**, 603-608 (1988).
L-Form, mp 204-207. [α]²⁴_D 0.5° (1N HCl).
DL-Form, spherical aggregates of needles.
THERAP CAT: Mucolytic; expectorant.

1810. Carbofuran. *2,3-Dihydro-2,2-dimethyl-7-benzofuranol methylcarbamate; methyl carbamic acid 2,3-dihydro-2,2-dimethyl-7-benzofuranyl ester;* 2,2-dimethyl-2,3-dihydro-7-benzofuranyl-N-methylcarbamate; 2,2-dimethyl-7-coumaranyl N-methylcarbamate; BAY 70143; NIA 10242; Furadan. $C_{12}H_{15}NO_3$; mol wt 221.26. C 65.14%, H 6.83%, N 6.33%, O 21.69%. Prepn and use as insecticide: **Neth. pat. Appl.** 6,407,316 (1964 to Bayer), *C.A.* **63**, 583a (1965); **Neth. pat. Appl.** 6,500,340 corresp to W. G. Scharpf, **U.S. pats.** 3,474,170-1 (1965, 1969 to FMC); E. F. Orwoll, **U.S. pat.** 3,356,690 (1967 to FMC). Metabolism: H. W. Dorough, *J. Agr. Food Chem.* **16**, 319 (1968); J. B. Knaak, *ibid.* **18**, 832 (1970). Toxicity studies: J. S. Tobin, *J. Occup. Med.* **12**, 16 (1970); M. A. Fahmy *et al.*, *J. Agr. Food Chem.* **18**, 793 (1970). Teratogenicity study: K. D. Courtney *et al.*, *J. Environ. Sci. Health* **B20**, 373 (1985).

OOCNHCH₃

White crystalline solid, mp 150-153°. Soly in water at 25°: 700 ppm. Unstable in alk. LD₅₀ orally in mice: 2 mg/kg (Fahmy, 1970).
Caution: Cholinesterase inhibitor: *Clinical Toxicology of Commercial Products,* R. E. Gosselin *et al.*, Eds. (Williams & Wilkins, Baltimore, 5th ed., 1984) Section II, p 305.
USE: Systemic insecticide, acaricide, nematocide.

1811. Carbohydrazide. *Carbonic dihydrazide;* 1,3-diaminourea. CH_6N_4O; mol wt 90.09. C 13.33%, H 6.71%, N 62.20%, O 17.76%. $NH_2NHCONHNH_2$. Prepd by refluxing diethyl carbonate with hydrazine hydrate: Mohr *et al.*, *Inorg. Syn.* **4**, 32 (1953).
Crystals from water + ethanol, dec 153-154°. Freely sol in water. pH of 1% aq soln about 7.4. Practically insol in alcohol, ether, chloroform, benzene. Forms salts with acids. With nitrous acid it forms the highly explosive carbonyl azide $CO(N_3)_2$.

1812. γ-Carboline. *5H-Pyrido[4,3-b]indole;* 2H-pyrid-[4,3-b]indole; 5-carboline. $C_{11}H_8N_2$; mol wt 168.19. C 78.55%, H 4.79%, N 16.66%. Prepn: Robinson, Thornley, *J. Chem. Soc.* **125**, 2169 (1924). Prepn of derivs: Hörlein, *Ber.* **87**, 463 (1954); C. Ducrocq *et al.*, *J. Heterocycl. Chem.* **12**, 963 (1975). NMR studies: F. Balkau, M. L. Heffernan, *Aust. J. Chem.* **26**, 1501, 1523 (1973).

Monoclinic needles from water, mp 225°. d 1.352. Can be distilled at atmospheric pressure without dec. Strong base. Freely sol in methanol; somewhat less sol in ethanol. Slightly sol in benzene, water.
Picrate, yellow needles, mp 250°.

1813. Carbomycin. Magnamycin. Sixteen-membered-ring macrolide antibiotic complex similar to leucomycin, *q.v.* and erythromycin, *q.v.*, produced by *Streptomyces halstedii*. Isoln and antibacterial activity: F. W. Tanner *et al.*, *Antibiot. & Chemother.* **2**, 441 (1952). Two components have been isolated: Carbomycin A (major) and carbomycin B. Isoln of A: Friedman *et al.*, **U.S. pat.** 2,960,438 (1960 to Pfizer); of B: F. A. Hochstein, K. Murai, *J. Am. Chem. Soc.* **76**, 5080 (1954). Structure of A and B: R. B. Woodward, *Angew. Chem.* **69**, 50 (1957); revised structure: M. Kuehne, B. W. Benson, *J. Am. Chem. Soc.* **87**, 4660 (1965); R. B. Woodward *et al.*, *ibid.* 4662. Abs config of A and B: W. D. Celmer, *ibid.* **88**, 5028 (1966). Identity of A with deltamycin A₄: Y. Shimauchi *et al.*, *J. Antibiot.* **31**, 270 (1978). Synthesis of B: K. Tatsuta *et al.*, *J. Am. Chem. Soc.* **99**, 5826 (1977). Stereospecific total synthesis of B: *eidem, Tetrahedron Letters* **1980**, 2837. Retrosynthetic studies: K. C. Nicolaou *et al.*, *J. Am. Chem. Soc.* **103**, 1222 (1981). Reviews: D. Vazquez, in *Antibiotics* Vol. 1, D. Gottlieb, P. D. Shaw, Eds. (Springer-Verlag, New York, 1967) pp 366-377; W. Keller-Schierlein in *Fortschr. Chem. Org. Naturst.* **30**, 314-460 (1973).

Carbomycin A

Carbomycin A, $C_{42}H_{67}NO_{16}$, *(12S,13S)-9-deoxy-12,13-epoxy-12,13-dihydro-9-oxoleucomycin V 3-acetate 4^B-(3-methylbutanoate), M-4209, magnamycin A, deltamycin A₄.* Blunt needles from ethanol, mp 214°. [α]²⁵_D −58.6° (chloroform). uv max (abs ethanol): 238, 327 nm (E1%_{1cm} 185, 0.9). Carbomycin standard is the free base having a potency of 1080 units/mg. For stability of soln data see H. L. Martin, *Antibiot. & Chemother.* **3**, 865 (1953). Weak base, pKb 7.2. Solubilities determined by Weiss *et al.*, *ibid.* **7**, 374 (1957) in mg/ml at about 28°: water 0.295; methanol > 20; ethanol > 20. LD₅₀ i.v. in mice: 550 mg/kg (Tanner).
Carbomycin B, $C_{42}H_{67}NO_{15}$, *9-deoxy-9-oxoleucomycin V 3-acetate 4^B-(3-methylbutanoate), magnamycin B.* Colorless anisotropic plates from acetone/water, mp 141-144° (dec), softens at 138°. [α]²⁵_D −35° (c = 1 in chloroform). uv max

(abs ethanol): 278 nm ($E_{1cm}^{1\%}$ 276). pKb 7.56. Solubilities in mg/ml at 25°: ethanol 450; water 0.1-0.2.

THERAP CAT: Antibacterial.

THERAP CAT (VET): Antimicrobial.

1814. Carbon. C; at. wt 12.01115; at. no. 6; valence 4. Stable isotopes: 12 (98.892%); 13 (1.108%); radioactive isotopes: 9-11; 14-16. Abundance in earth's crust: approx 0.027%. Cosmic abundance: 6 atoms/atom Si. Occurs in 3 forms: (1) Diamond, *q.v.*; (2) Graphite, *q.v.* or black lead; (3) Amorphous carbon such as coal, lampblack, and the various forms of artificial carbon. Comprehensive reviews: P. L. Walker, *Am. Scientist* **50**, 259-293 (June 1962); Holliday *et al.* in *Comprehensive Inorganic Chemistry* vol. 1, J. C. Bailar, Jr. *et al.*, Eds. (Pergamon Press, Oxford, 1973) pp 1173-1294; several authors in Kirk-Othmer *Encyclopedia of Chemical Technology* vol. 4 (Wiley-Interscience, New York, 3rd ed., 1978) pp 556-709.

^{14}C isotope, continuously formed in the earth's atm by the bombardment of nitrogen with cosmic neutrons according to the reaction $^{14}_{7}N + ^{1}_{0}n \rightarrow ^{14}_{6}C + ^{1}_{1}H$. The ^{14}C is rapidly oxidized to CO_2, in this form it penetrates into animals and plants by photosynthesis and metabolism. The ^{14}C content of living matter is estimated at 15.3 disintegrations per minute and per gram of carbon, corresponding to the equilibrium reached between formation of ^{14}C and its exchange with ^{12}C. This equilibrium stops when the plant or animal dies, and the ^{14}C content begins to decrease, because the ^{14}C decays with a half-life of 5760 years. This fact can be used to date organic matter (not more than 40,000 years old) by comparison with the standard 15.3 disintegrations per min per gram: M. Haissinsky, J. P. Adloff, *Radiochemical Survey of the Elements* (Elsevier, New York, 1965) pp 30-32. Production of *buckminsterfullerene*, a stable cluster of 60 carbon atoms: H. W. Kroto *et al.*, *Nature* **318**, 162 (1985).

1815. Carbon, Amorphous. Carbon black; carbon, activated; carbon, decolorizing. A quasi-graphitic form of carbon of small particle size. By the term "carbon black" several forms of artificially prepared carbon or charcoal are designated, *e.g.*: (1) *Animal charcoal*, obtained by charring bones, meat, blood, etc.; (2) *Gas black; furnace black; channel black; C.I. 77266:* obtained by incomplete combustion of natural gas; (3) *Lamp black*, obtained by burning various fats, oils, resins, etc., under suitable conditions; (4) *Activated charcoal, e.g. Carbomix, Carboraffin, Medicoal, Norit, Opocarbyl, Ultracarbon*, prepd from wood and vegetables. Monograph: H. W. Davidson *et al.*, *Manufactured Carbon* (Pergamon Press, New York, 1968). *Reviews:* Cohan in *Science of Petroleum* vol. V, Pt 2, B. T. Brooks, A. E. Dunstan, Eds. (Oxford Univ. Press, 1953), pp 79-89; Smisek, Cerny, *Active Carbon* (Elsevier Publishing Co., Amsterdam, 1970).

USE: Number (4), *e.g.* Norit, Carboraffin, is used chiefly for clarifying, deodorizing, decolorizing and filtering. The others are used as a pigment for rubber tires; for printing, stenciling and drawing inks; for leather; stove polish, phonograph records, electrical insulating apparatus. Activated charcoal (from the destructive distillation of various organic materials) is used in medicine, *e.g.*, Opocarbyl; Norit; Ultracarbon. *Caution:* Carbon black obtained by the impingement or channel process, also known as gas black and channel black, has been banned by the FDA for use as a color additive in foods, drugs and cosmetics.

THERAP CAT: Activated charcoal as antidote; adsorptive.

THERAP CAT (VET): Internally as an adsorptive in diarrhea; externally in foul wounds.

1816. Carbon Dioxide. Carbonic acid gas; carbonic anhydride. CO_2; mol wt 44.01. C 27.29%, O 72.71%. Occurs in the atms of many planets. In our solar system, e.g., on Venus, the optical layer thickness due to CO_2 is 100,000 cm/atm, but only 220 cm/atm on Earth. Analyses of air in the temperate zones of the Earth show 0.027 to 0.036% (v/v) of CO_2: G. P. Kuiper, *The Atmospheres of the Earth and the Planets* (Univ. of Chicago Press, 1949); Landolt-Bornstein, *Zahlenwerte* vol. **III** (Springer-Verlag, 6th ed., 1952) pp 59 and 585. Constituent of carbonate type of minerals and products of animal metabolism. Necessary for the respiration cycle of plants and animals. Obtained industrially as a by-product in the manuf of lime during the "burning" of

limestone ($CaCO_3$). Also produced by burning coke or other carbonaceous material. In the U.S.A. large amounts are produced by fermentation (Backus process and Reich process). When glucose is fermented by yeast, the chief products are ethyl alcohol and CO_2. Prepd in the laboratory by dropping acid on a carbonate: E. H. Archibald, *The Preparation of Pure Inorganic Substances* (Wiley, New York, 1932) p 196; Loomis, Walters, *J. Am. Chem. Soc.* **48**, 3103 (1926). Purification: Glemser in *Handbook of Preparative Inorganic Chemistry*, G. Brauer, Ed. (Academic Press, New York, 2nd ed., 1963) p 647. Discovery of a second polymorph of dry ice: L.-G. Liu, *Nature* **303**, 508 (1983). *Reviews:* E. L. Quinn, *J. Chem. Ed.* **7**, 151-162 and 403-419 (1930); J. Kuprianoff, *Die feste Kohlensäure (Trockeneis)* (Enke, Stuttgart, 1939); E. L. Quinn, C. L. Jones, *Carbon Dioxide* (Reinhold, New York, 1947); W. R. Ballou, in Kirk-Othmer *Encyclopedia of Chemical Technology* vol. 4 (Interscience, New York, 3rd ed., 1978) pp 725-742.

Colorless, odorless, noncombustible gas. Faint acid taste. Usually a nonsupporter of combustion, athough burning magnesium continues to burn when transferred into a CO_2 atm. Usually marketed in steel cylinders (under sufficient pressure to keep it liquid) or in solid form as *Dry Ice* (compressed carbon dioxide snow, d 1.35). At atmospheric pressures the solid form changes into the gaseous phase without liquefaction. d (gas) 1.527 (air = 1); d (gas) 1.557 (N_2 = 1); abs d 0.1146 lb/cu ft at 25°; vol at 25°: 8.76 cu ft/lb. d (gas, 0°) 1.976 g/l at 760 mm; d (liq, 0°) 0.914 at 34.3 atm; d (solid, −56.6°) 1.512. Sublimes at −78.48° (760 mm). $mp_{5.2\,atm}$ −56.6°. The gas is not affected by heat until temp reaches about 2000°. Crit temp 31.3°; crit press 72.9 atm; crit density 0.464. Triple point −56.6° at 5.11 atm. Vapor press at −120°: 10.5 mm; at −100°: 104.2 mm; at −82°: 569.1 mm. Heat of formation 94.05 kcal/mol. Latent heat of vaporization 83.12 g cal/g. Specific heat 0.19 to 0.21 Btu/lb. Soly in water (ml CO_2/100 ml H_2O at 760 mm): 0° = 171; 20° = 88; 60° = 36. More sol at higher pressures. Less sol in alcohol, other neutral organic solvents. Absorbed by alkaline solns with the formation of carbonates.

Caution: When shipped in steel cylinders, CO_2 is in the form of gas over liquid and at 20° exerts a pressure of 830 psi. Humans cannot breathe air contg more than 10% CO_2 without losing consciousness. Use gloves when handling dry ice, as its temp is at least −78.5°; momentary skin contact with dry ice has caused serious frostbites and blisters.

USE: In the carbonation of beverages; manuf of carbonates; in fire prevention and extinction; for inerting flammable materials during manuf, handling and transfer; as propellant in aerosols; as dry ice for refrigeration; to produce harmless smoke or fumes on stage; as rice fumigant; as antiseptic in bacteriology and in the frozen food industry.

THERAP CAT: Respiratory stimulant.

THERAP CAT (VET): Respiratory stimulant (inhalant).

1817. Carbon Diselenide. Carbon selenide. CSe_2; mol wt 169.93. C 7.07%, Se 92.93%. Prepd by the action of methylene chloride vapor on heated selenium: Ives *et al.*, *J. Chem. Soc.* **1947**, 1080; or from a mixture of CCl_4 and H_2Se in a stream of N_2 at 500°: Grimm, Metzger, *Ber.* **69**, 1356 (1936); from the elements by electrical discharge on Se vapor in the presence of sugar charcoal: Steudel, *Z. Anorg. Allgem. Chem.* **361**, 195 (1968).

Light-sensitive, golden yellow, strongly refractive, liquid. Odor of rotten radishes. Turns brown to black on storage. d_0^{20} 2.6824; d_4^{25} 2.6626. mp −45.5°. bp 125-126°; $bp_{8.0}$ 10.0°. n_D^{20} 1.845. Heat of formation: 34 kcal/mol. Miscible with carbon tetrachloride, carbon disulfide, toluene, other organic solvents. Practically insol in water. Dec by alc, pyridine.

1818. Carbon Disulfide. Carbon bisulfide; dithiocarbonic anhydride. CS_2; mol wt 76.14. C 15.77%, S 84.23%. Minute amounts occur in coal tar and in crude petroleum. Prepd on an industrial scale by heating charcoal with vaporized sulfur; from sulfur and natural gas: Faith, Keyes & Clark's *Industrial Chemicals*, F. A. Lowenheim, M. K. Moran, Eds. (Wiley-Interscience, New York, 4th ed., 1975) pp 224-229. Laboratory purification: Glemser in *Handbook of Preparative Inorganic Chemistry* vol. 1, G. Brauer, Ed. (Academic Press, New York, 2nd ed., 1963) p 652. Review of production and uses: Bushell, *Chem. & Ind. (London)*

1961, 1465; R. W. Timmerman in Kirk-Othmer *Encyclopedia of Chemical Technology* vol. 4 (Wiley-Interscience, New York, 3rd ed., 1978) pp 742-757.

Highly refractive, mobile, very flammable liq. *Poisonous!* The purest distillates ever obtained are reported to have a sweet, pleasing, and ethereal odor, while the usual commercial and reagent grades are foul smelling. Dec on standing for a long time. Burns with a blue flame to CO_2 and SO_2. *Acute fire and explosion hazard*, can be ignited by hot steam pipes. Flash pt, closed cup: $-30°C$. Ignition pt: $100°$. Explosive range: 1 to 50% (v/v) in air. d_4^0 1.29272; d_4^{15} 1.27055; d_4^{20} 1.2632; d_4^{30} 1.24817. Vapors sink to the ground. Vapor density 2.67 (air = 1). mp: $-111.6°$. $bp_{1.0}$ $-73.8°$; bp_{10} $-44.7°$; bp_{100} $-5.1°$; bp_{400} $+28.0°$; bp_{760} $+46.5°$; $bp_{(2 atm)}$ $+69.1°$; $bp_{(5 atm)}$ $+104.8°$. Crit temp $280.0°$; crit press. 72.9 atm. n_D^{15} 1.63189; $n_D^{20.1}$ 1.62803; $n^{23.5}$ 1.62543. Surface tension at $20°$: 32.25. Coefficient of viscosity at $20°$: 0.363. Heat of vaporization at bp: 84.1 cal/g. Heat of fusion: 1.049 kcal/mole. Heat capacity at $24.3°$: 18.17 cal/mole/-deg: Brown, Manov, *J. Am. Chem. Soc.* **59**, 500 (1937). Ebullioscopic constant: $2.35°$. Dielectric constant at low frequencies: 2.641. Dipole moment: 0.0. Soly in water at $20°$: 0.294%. Soly of water in CS_2: < 0.005%. Azeotrope with water bp $42.6°$, contains 97.2% CS_2. Misc with anhydr methanol, ethanol, ether, benzene, chloroform, carbon tetrachloride, oils. Can be stored in iron, aluminum, glass, porcelain, Teflon.

Caution: Poisoning usually occurs from inhalation but also may be caused by ingestion and skin absorption. *Acute Toxicity:* euphoria, restlessness, mucous membrane irritation, nausea, vomiting, unconsciousness, terminal convulsions. *Chronic Toxicity:* marked psychic disturbances ranging from extreme irritability to mania with hallucinations, tremors, auditory and visual disturbances, weight loss, blood dyscrasias. Dermal contact with concd solns may cause burning pain, erythema, exfoliation. *See: Clinical Toxicology of Commercial Products*, R. E. Gosselin *et al.*, Eds. (Williams & Wilkins, Baltimore, 4th ed., 1976) Section III, pp 83-86.

USE: In the manuf of rayon, carbon tetrachloride, xanthogenates, soil disinfectants, electronic vacuum tubes. Solvent for phosphorus, sulfur, selenium, bromine, iodine, fats, resins, rubbers.

1819. Carbonic Anhydrase. *Carbonate dihydratase;* carbonate hydro-lyase. Mol wt approx 30,000. A small zinc-contg enzyme which catalyzes the hydration of CO_2. Found in higher concns in erythrocytes, renal cortex, and gastric mucosa of mammals; also found in other animal tissues, in plants and in some bacteria. Isoln from bovine erythrocytes: Lindskog, *Biochim. Biophys. Acta* **39**, 218 (1960); from human erythrocytes: Nyman, *ibid.* **52**, 1 (1961); from renal cortex: Höber, *Proc. Soc. Exp. Biol. Med.* **49**, 87 (1942); from gastric mucosa: Davenport, *Physiol. Rev.* **26**, 560 (1946). Human carbonic anhydrase consists of two isoenzymes with distinctly different amino acid sequences and specific activities. The high activity form is called carbonic anhydrase C; the low activity form is called B; modified forms of these two isoenzymes exist: Funakoshi, Deutsch, *J. Biol. Chem.* **243**, 6474 (1968); **244**, 3438 (1969). Amino acid sequence of carbonic anhydrase B: Anderson *et al.*, *Biochem. Biophys. Res. Commun.* **48**, 670 (1972); Lin, Deutsch, *J. Biol. Chem.* **248**, 1885 (1973); sequence of carbonic anhydrase C: Henderson, Henriksson, *Biochem. Biophys. Res. Commun.* **52**, 1388 (1973); Lin, Deutsch, *J. Biol. Chem.* **249**, 2329 (1974). Crystal structure of carbonic anhydrase C: Liljas *et al.*, *Nature New Biol.* **235**, 131 (1972). Catalyzes the reversible reaction of CO_2 and H_2O to HCO_3^- and H^+. Permits CO interchange between blood and tissues. In gastric mucosa, reaction rate is sufficient to neutralize the excess alkalinity produced by the ionization of water and secretion of hydrogen ions: Roughton, Clark in *The Enzymes* vol. 1, part 2, J. B. Sumner, K. Myrbäck, Eds. (Academic Press, New York, 1951) pp 1250-1265. In the kidney, participates in Na^+ transport. Review of physiology: Maren in *Oxygen Affinity of Hemoglobin and Red Cell Acid Base Status, Alfred Benzon Symposium IV*, P. Astrup, M. Roerth, Eds. (Academic Press, New York, 1972) pp 418-433. Review of metal ion function: Prince, Woolley, *Angew. Chem. Int. Ed.* **11**, 408-417 (1972). *Review:* Lindskog *et al.*, "Carbonic Anhy-

drase" in *The Enzymes*, **vol. 5**, P. D. Boyer, Ed. (Academic Press, New York, 1971) pp 587-665.

1820. Carbon Monoxide. CO; mol wt 28.01. C 42.88%, O 57.12%. Produced on an industrial scale by partial oxidation of hydrocarbon gases from natural gas or by the gasification of coal and coke. Conveniently prepd in the laboratory by heating calcium carbonate with Zn dust: Weinhouse, *J. Am. Chem. Soc.* **70**, 442 (1948); by dehydration of formic acid with H_2SO_4: Gilliland, Blanchard, *Inorg. Syn.* **2**, 81 (1946). Purification of carbon monoxide bought in steel cylinders: A. Klemenc, *Die Behandlung und Reindarstellung von Gasen* (Vienna, 1948) p 160; Glemser in *Handbook of Preparative Inorganic Chemistry* vol. 1, G. Brauer, Ed. (Academic Press, New York, 2nd ed., 1963) p 646. Review of toxic effects in humans: Stewart, *Ann. Rev. Biochem.* **15**, 409-423 (1975). *Review:* C. M. Bartish, G. M. Drissel in Kirk-Othmer *Encyclopedia of Chemical Technology* vol. 4 (Wiley-Interscience, New York, 3rd ed., 1978) pp 772-793.

Highly poisonous, odorless, colorless, tasteless gas. Very flammable, burns in air with a bright blue flame. Ignition pt in air: $700°$. mp $-205.0°$. bp $-191.5°$. d_4^{-195} (liq) 0.814. d (gas) 0.968 (air = 1.000). d_4^0 at 760 mm: 1.250 g/liter. The top pressure is 1500 psi. Flammable limits in air: 12 to 75 vol %. Crit press 35 atm, crit temp $-139°$. Heat capacity at $20°$: 6.95 cal/mole/°C. Heat value per m³: 3033 kcal. Heat of formation: -26.39 kcal/mol. Dec into carbon and carbon dioxide between 400 and 700°, at lower temp when in contact with catalytic surfaces. Above 800° the equilibrium reaction favors CO formation. Hopcalite, a mixture of the oxides of manganese and copper, catalyzes the decompn at room temp, as does Pd on silica gel. Sparingly sol in water: 3.3 ml/100 ml H_2O at 0°; 2.3 ml/100 ml H_2O at 20°; freely absorbed by a concd soln of cuprous chloride in HCl or in NH_4OH. Appreciably sol in some organic solvents, such as ethyl acetate, $CHCl_3$, acetic acid. The soly in methanol and ethanol is about 7 times as great as the soly in water.

Caution: Combines with the hemoglobin of the blood to form carboxyhemoglobin which is useless as an oxygen carrier. *Toxic symptoms:* Headache, mental dullness, dizziness, weakness, nausea, vomiting, loss of muscular control, increased then decreased pulse and respiratory rates, collapse, unconsciousness, death. *Antidote:* Oxygen. *See Patty's Industrial Hygiene and Toxicology* vol. 2C, G. D. Clayton, F. E. Clayton, Eds. (Wiley-Interscience, New York, 3rd ed., 1982) pp 4114-4124.

USE: As reducing agent in metallurgical operations especially in the Mond process for the recovery of nickel; in organic synthesis especially in the Fischer-Tropsch processes for petroleum-type products and in the oxo reaction; in the manuf of metal carbonyls.

1821. Carbon Suboxide. *1,2-Propadiene-1,3-dione;* tricarbon dioxide. C_3O_2; mol wt 68.03. C 52.96%, O 47.04%. $O=C=C=C=O$. Prepn by thermal decompn of malonic acid: Glemser in *Handbook of Preparative Inorganic Chemistry*, vol. 1, G. Brauer, Ed. (Academic Press, New York, 2nd ed., 1963) p 648. Reactions in organic synthesis: Dashkevich, Beilin, *Russ. Chem. Rev.* **36**, 391 (1967). Comprehensive reviews: Reyerson, Kobe, *Chem. Revs.* **7**, 479 (1930); Vol'kenshtein, *Uspeki Khim.* **4**, 610 (1935); Grauer, *Chimia* **14**, 11 (1960); T. Kappe, E. Ziegler, *Angew. Chem. Int. Ed.* **13**, 491-504 (1974), reprinted in *New Synthetic Methods* vol. 1 (Verlag Chemie, Weinheim, 1975) pp 29-69.

Colorless, highly refractive liquid or colorless gas which burns with a blue, sooty flame. Odor like acrolein and mustard oil. mp $-111.3°$. bp_{760} $6.8°$. d_4^0 1.114. n_D^0 1.45384; n_D^{-12} 1.46757. Vapor pressure at 0°: 587-589 mm. Explosive limits, 6 to 30 vol % in air. Dipole moment: 0.7D. Thermodynamic constants: Thompson, *Trans. Faraday Soc.* **37**, 249 (1941). The gas can be stored at pressures of up to 100 mm, but even at these pressures polymerization may occur, giving a red, water-sol product. This invariably occurs at higher pressure or in the liquid state. Polymerization facilitated by presence of P_2O_5. Dec when passed through heated glass tubes, forming a mirror surface. Difficultly sol in carbon disulfide, xylene. With water forms malonic acid quantitatively. Forms malonamide with ammonia.

USE: Prepn of malonates; improving dye affinity of fibers.
Caution: In small amounts acts as a lacrimator; in high

concns attacks eyes, nose, respiratory organs, producing a feeling of suffocation.

1822. Carbon Tetrachloride. *Tetrachloromethane; per-*chloromethane; Necatorina; Benzinoform. CCl_4; mol wt 153.84. C 7.81%, Cl 92.19%. Obtained from carbon disulfide and chlorine in presence of a catalyst, e.g., $SbCl_5$, Fe filings, or by the chlorination of hydrocarbons: Faith, Keyes & Clark's *Industrial Chemicals,* F. A. Lowenheim, M. K. Moran, Eds. (Wiley-Interscience, New York, 4th ed., 1975) pp 230-234; H. D. DeShon in Kirk-Othmer *Encyclopedia of Chemical Technology* vol. 5 (Wiley-Interscience, New York, 3rd ed., 1979) pp 704-714. Toxicity: Svirbely, *J. Ind. Hyg. Toxicol.* **29**, 382 (1947); E. Browning, *Toxicity and Metabolism of Industrial Solvents* (Elsevier, New York, 1965) pp 173-188. Review of carcinogenicity studies: *IARC Monographs* **20**, 371-399 (1979). Use in induction of experimental liver disease: P. Trivedi, A. P. Mowat, *Br. J. Exp. Pathol.* **64**, 25 (1983).
Colorless, clear, nonflammable, heavy liquid; characteristic odor. d_{25}^{25} 1.589. bp 76.7°. mp −23°. n_D^{20} 1.4607. One ml dissolves in 2000 ml water; misc with alcohol, benzene, chloroform, ether, carbon disulfide, petr ether, oils. LC_{50} for mice: 9528 ppm (Svirbely).
Human Toxicity: Poisoning by inhalation, ingestion or skin absorption. *Acute:* nausea, vomiting, diarrhea, headache, renal damage leading to anuria and azotemia, liver injury. Can be fatal. *Chronic:* primarily liver damage but kidney injury and visual disturbances also occur. Skin contact can lead to dermatitis through defatting action. *Caution:* Alcohol intensifies action. This substance may reasonably be anticipated to be a carcinogen: *Fourth Annual Report on Carcinogens* (NTP 85-002, 1985) p 50.
USE: As solvent for oils, fats, lacquers, varnishes, rubber waxes, resins; starting material in manuf of organic compds; grain fumigant. Pharmaceutic aid (solvent). Formerly used as dry cleaning agent and fire extinguisher. *Caution:* May form phosgene when used to put out electrical fires. *Use only when adequate ventilation is possible.*
THERAP CAT: Formerly as anthelmintic (Nematodes).
THERAP CAT (VET): Anthelmintic.

1823. Carbon Tetrafluoride. *Tetrafluoromethane;* Freon-14. CF_4; mol wt 88.01. C 13.65%, F 86.35%. Prepd from carbon or carbon monoxide and fluorine: Yost, *Inorg. Syn.* **1**, 34 (1939); Simons, Block, *J. Am. Chem. Soc.* **61**, 2962 (1939); Kwasnik in *Handbook of Preparative Inorganic Chemistry* vol 1, G. Brauer, Ed. (Academic Press, New York, 2nd ed., 1963) p 203. May also be prepd from SiC + F_2: Priest, *Inorg. Syn.* **3**, 178 (1950).
Colorless, odorless gas. Thermally stable. Chemically very inert. d (solid, −195°) 1.98. d (liq, −183°) 1.89. mp −183.6°. bp −127.8°. May be stored in steel cylinders.
USE: Low temp refrigerant; gaseous insulator. *Caution:* Narcotic in high concns.

1824. Carbon Tetraiodide. *Tetraiodomethane.* CI_4; mol wt 519.65. C 2.31%, I 97.69%. Prepd by the interaction of carbon tetrachloride and aluminum or calcium iodide: Gustavson, *Ann.* **172**, 173 (1874); boron iodide: Moissan, *Compt. Rend.* **113**, 19 (1891); lithium or calcium iodide: Lantenois, *ibid.* **156**, 1385 (1913); ethyl iodide in the presence of aluminum chloride: Walker, *J. Chem. Soc.* **85**, 1090 (1904); McArthur, Simons, *Inorg. Syn.* **3**, 37 (1950).
Red cubic crystals. Odor of iodine. Dec to iodine and tetraiodoethylene under the influence of light or heat. d_4^{20} 4.32. mp 171°. Sol in benzene, chloroform. Dec by hot alcohol. Practically insol in water, but hydrolyzes slowly in contact with water, forming iodoform and iodine.

1825. N,N′-Carbonyldiimidazole. *1,1′-Carbonylbis-*1H-imidazole. $C_7H_6N_4O$; mol wt 162.15. C 51.85%, H 3.73%, N 34.56%, O 9.87%. Prepd from phosgene and imidazole in dry tetrahydrofuran or dry benzene: Staab, *Ann.* **609**, 75 (1957); Anderson, Paul, *J. Am. Chem. Soc.* **80**, 4423 (1958).

Crystals from tetrahydrofuran or benzene, mp 115.5-116°. Should be handled under exclusion of atmospheric moisture. Hydrolyzed by water in a few sec with evolution of CO_2.
USE: In the synthesis of peptides. Reacts readily with carboxylic acids to form acyl imidazoles; subsequent reaction with amines to form amides goes smoothly.

1826. Carbonyl Fluoride. *Carbonic difluoride;* fluophosgene. CF_2O; mol wt 66.01. C 18.19%, F 57.57%, O 24.24%. COF_2. Prepd from CO and F_2 or BrF_3 and CO: Ruff, Miltschitzky, *Z. Anorg. Allgem. Chem.* **221**, 154 (1935); Kwasnik in *Handbook of Preparative Inorganic Chemistry,* **vol. 1,** G. Brauer, Ed. (Academic Press, New York, 2nd ed., 1963) p 206. Alternate route from CO + AgF_2: Farlow *et al., Inorg. Syn.* **6**, 155 (1960).
Pungent, very hygroscopic gas. d (solid, −190°): 1.388. d (liq, −114°): 1.139. mp −114.0°. bp −83.1°. Heat of formation: 166.6 kcal. Instantly hydrolyzed by water. *Caution:* A strong irritant to skin, eyes, mucous membranes, respiratory tract.

1827. Carbophenothion. *Phosphorodithioic acid S-[[(4-*chlorophenyl)thio]methyl] O,O-diethyl ester; S-[[(p-chlorophenyl)thio]methyl] O,O-diethyl phosphorodithioate; O,O-diethyl S-(p-chlorophenylthio)methyl phosphorodithioate; R 1303; Garrathion; Trithion. $C_{11}H_{16}ClO_2PS_3$; mol wt 342.85. C 38.53%, H 4.71%, Cl 10.36%, O 9.33%, P 9.03%, S 28.05%. Prepn: Fancher, U.S. pat. **2,793,224** (1957 to Stauffer).

Light amber liquid. $bp_{0.01}$ 82°. d_4^{25} 1.271. n_D^{25} 1.5970 (n_D^{26} 1.6198 in patent). Very low vapor pressure. Practically insol in water. Miscible with vegetable oils and most organic solvents. LD_{50} in female, male rats: 10, 30 mg/kg orally; 27, 54 mg/kg dermally, T. B. Gaines, *Toxicol. Appl. Pharmacol.* **14**, 515 (1969).
USE: Miticide; insecticide. *Caution:* Cholinesterase inhibitor.

1828. Carboplatin. *(SP-4-2)-Diammine[1,1-cyclobu-*tanedicarboxylato(2−)-O,O′]platinum; 1,1-cyclobutanedicarboxylic acid platinum complex; cis-diammine(1,1-cyclobutanedicarboxylato)platinum(II); CBDCA; JM8; NSC-241240; Paraplatin. $C_6H_{12}N_2O_4Pt$; mol wt 371.25. C 19.41%, H 3.26%, N 7.54%, O 17.24%, Pt 52.55%. Analog of cisplatin, *q.v.,* with reduced nephrotoxicity. Prepn and antitumor activity: Neth. pat. **Appl. 7,307,863;** M. J. Cleare *et al.,* U.S. pat. **4,140,707** (1973, 1979 both to Research Corp.). Improved prepn: R. C. Harrison *et al., Inorg. Chim. Acta* **46**, L15 (1980). Crystal structure: S. Neidle *et al., J. Inorg. Biochem.* **13**, 205 (1980). Comparison with other antitumor platinum complexes: M. J. Cleare *et al., Biochimie* **60**, 835 (1978). Early clinical studies: A. H. Calvert *et al., Cancer Chemother. Pharmacol.* **9**, 140 (1982). Clinical pharmacokinetics: S. J. Harland *et al., Cancer Res.* **44**, 1693 (1984); M. J. Egorin *et al., ibid.* 5432. Toxicity, activity in mice, rats, dogs: P. Lelieveld *et al., Eur. J. Cancer Clin. Oncol.* **20**, 1087 (1984). Comparison with cisplatin chemotherapy in advanced seminoma: M. J. Peckham *et al., Brit. J. Cancer* **52**, 7 (1985). In treatment of small cell lung cancer: I. E. Smith *et al., Cancer Treat. Rep.* **69**, 43 (1985).

White crystals, sol in water. LD_{50} in mice (mg/kg): 150 i.p., 140 i.v.; in rats (mg/kg): 85 i.v. (Lelieveld).
THERAP CAT: Antineoplastic.

1829. Carboprost. *9,11,15-Trihydroxy-15-methylprosta-*5,13-dien-1-oic acid; 7-[3,5-dihydroxy-2-(3-hydroxy-3-

methyl-1-octenyl)cyclopentyl]-5-heptenoic acid; (15S)-15-methyl PGF$_{2\alpha}$; U-32921; Prostin/15M. C$_{21}$H$_{36}$O$_5$; mol wt 368.52. C 68.44%, H 9.85%, O 21.71%. Analog of prostaglandin F$_{2\alpha}$, q.v. Prepn: G. L. Bundy et al., Ger. pat. 2,121,980; G. L. Bundy, U.S. pat. 3,728,382 (1971, 1973 both to Upjohn); eidem, Ann. N.Y. Acad. Sci. 180, 76 (1971); E. W. Yankee et al., J. Am. Chem. Soc. 96, 5865 (1974). Biological activity: J. R. Weeks et al., J. Pharmacol. Exp. Ther. 186, 67 (1973). Mechanism of action: A. I. Csapo, M. O. Pulkkinen, Prostaglandins 18, 479 (1979). Clinical studies: P. C. Schwallie, K. R. Lamborn, J. Reprod. Med. 23, 289 (1979); M. P. Mapa et al., Int. J. Gynaecol. Obstet. 20, 125 (1982). Teratological study: G. M. Szczech et al., Advan. Prostaglandin Thromboxane Res. 4, 157 (1978).

Tromethamine salt, C$_{25}$H$_{47}$NO$_8$, U-32921E, carboprost trometamol, Hemabate, Prostin/15M.

Methyl ester, C$_{22}$H$_{38}$O$_5$, U-36384, carboprost methyl. Crystals from ether/hexane, mp 55-56°. [α]$_D$ +24° (c = 0.81 in ethanol).

THERAP CAT: Oxytocic.

1830. Carboquone. 2-[2-[(Aminocarbonyl)oxy]-1-methoxyethyl]-3,6-bis(1-aziridinyl)-5-methyl-2,5-cyclohexadiene-1,4-dione; 2,5-bis(1-aziridinyl)-3-(2-hydroxy-1-methoxyethyl)-6-methyl-p-benzoquinone carbamate (ester); 2,5-bis(1-aziridinyl)-3-(2-carbamoyloxy-1-methoxyethyl)-6-methyl-1,4-benzoquinone; carbazilquinone; Esquinon. C$_{15}$H$_{19}$N$_3$O$_5$; mol wt 321.34. C 56.07%, H 5.96%, N 13.08%, O 24.89%. Prepn and antitumor activity: Nakao et al., Chem. Pharm. Bull. 20, 1968 (1972); Nakamura et al., Ger. pat. 1,905,224; Nakao et al., Japan. pat. 70 33,057 (1969, 1970 both to Sankyo). Effect on tumors in mice: Arakawa et al., Gann 61, 535 (1970). Acute toxicity: H. Masuda, Y. Suzuki, Oyo Yakuri 8, 501 (1974), C.A. 81, 163349g (1974).

Red to reddish-brown crystals, mp 202° (dec). Slightly sol in chloroform, acetone, abs alcohol. Practically insol in water. LD$_{50}$ in male mice, male rats (mg/kg): 6.09, 3.88 i.v.; 3.84, 3.16 i.p.; 30.8, 28.0 orally (Masuda, Suzuki).

THERAP CAT: Antineoplastic.

1831. Carbostyril. 2(1H)-Quinolinone; 2-hydroxyquinoline; 2-quinolinol; 2(1H)-quinolone; o-aminocinnamic acid lactam. C$_9$H$_7$NO; mol wt 145.15. C 74.47%, H 4.86%, N 9.65%, O 11.02%. Prepn from quinoline by heating with potassium hydroxide to 225° under anhydr conditions: Tschitschibabin, Ber. 56, 1883 (1923); Ger. pat. 406,208; Chem. Zentr. 1925, I, 1536; Frdl. 14, 515. Several other syntheses.

Prisms from methanol, mp 199-200°. Sublimes at atmospheric pressure without dec. A monohydrate has been obtained from a satd aq soln. Kb at 18° = 1.94 × 10^{-9}. Very sparingly sol in water: One gram dissolves in 950 ml H$_2$O at 22°. Sol in alcohol, ether, dil HCl. Forms easily hydrolyzed Na and K salts.

Compd with 1,3,5-trinitrobenzene, (C$_9$H$_7$NO)$_2$·C$_6$H$_3$N$_3$O$_6$, yellow needles, mp 178°.

1832. Carboxin. 5,6-Dihydro-2-methyl-N-phenyl-1,4-oxathiin-3-carboxamide; 2,3-dihydro-5-carboxanilido-6-methyl-1,4-oxathiin; DCMO; Vitavax. C$_{12}$H$_{13}$NO$_2$S; mol wt 235.31. C 61.25%, H 5.57%, N 5.95%, O 13.60%, S 13.63%. Systemic fungicide, effective against loose smut in cereals. Prepn: B. von Schmeling et al., U.S. pat. 3,249,499 (1966 to U.S. Rubber); M. Kulka et al., U.S. pat. 3,393,202 (1968 to Uniroyal). Fungicidal activity: B. von Schmeling, M. Kulka, Science 152, 659 (1966). Inhibitor of succinate oxidation: D. E. Mathre, Phytopathol. 60, 671 (1970); G. A. White, Biochem. Biophys. Res. Commun. 44, 1212 (1971); P. C. Mowery et al., ibid. 71, 354 (1976). Residue determn in crops: J. R. Lane, J. Agr. Food Chem. 18, 409 (1970); H. R. Sisken, J. E. Newell, ibid. 19, 738 (1971). Toxicity studies: T. V. Dyadicheva, Gig. Tr. Prof. Zabol. 1979(2), 55, C.A. 90, 146673b (1979).

Crystals, from ethanol or methanol, mp 93-95°. LD$_{50}$ in rats, mice (mg/kg): 430, 3200 orally (Dyadicheva).

USE: Systemic plant fungicide.

1833. β-Carboxyaspartic Acid. 2-Amino-1,1,2-ethanetricarboxylic acid; Asa. C$_5$H$_7$NO$_6$; mol wt 177.11. C 33.91%, H 3.98%, N 7.91%, O 54.20%. Amino acid found in ribosomal proteins of E. coli; homolog of γ-carboxyglutamic acid, q.v. Isoln and synthesis of DL-form: J. J. Van Buskirk, W. M. Kirsch, J. Am. Chem. Soc. 103, 3935 (1981). Synthesis of DL-form: E. B. Henson et al., Tetrahedron 37, 2561 (1981). Kinetics of decarboxylation to aspartic acid: P. V. Hauschka et al., Anal. Chem. 108, 57 (1980); M. R. Christy, T. H. Koch, J. Am. Chem. Soc. 104, 1771 (1982); N. E. Dixon, A. M. Sargeson, ibid. 6716. Crystal structure and pKa values of DL-zwitterion: B. Richey et al., Biochemistry 21, 4819 (1982).

DL-Form hydrochloride, C$_5$H$_8$ClNO$_6$, white crystalline powder. pK$_1$ 0.8 ±0.2, pK$_2$ 2.5 ±0.1, pK$_3$ 4.7 ±0.1, pK$_4$ 10.9 ±0.1.

1834. γ-Carboxyglutamic Acid. 3-Amino-1,1,3-propanetricarboxylic acid; γ-carboxy-L-glutamic acid; L-γ-carboxyglutamic acid; Gla. C$_6$H$_9$NO$_6$; mol wt 191.14. C 37.70%, H 4.75%, N 7.32%, O 50.22%. Amino acid found in blood coagulation proteins (prothrombin, Factor VII, Factor IX, Factor X, q.q.v.), plasma proteins and proteins from calcified tissue. Presence of Gla confers metal binding properties to the protein. Discovery in prothrombin: J. Stenflo, Proc. Nat. Acad. Sci. USA 71, 2730 (1974); G. L. Nelsestuen et al., J. Biol. Chem. 249, 6347 (1974). Synthesis: S. Danishefsky et al., J. Am. Chem. Soc. 101, 4385 (1979); of DL-form: H. R. Morris et al., Biochem. Biophys. Res. Commun. 62, 856 (1975); W. Märki, R. Schwyzer, Helv. Chim. Acta 58, 1471 (1975); B. Weinstein et al., J. Org. Chem. 41, 3634 (1976). Resolution of D- and L-forms: W. Märki et al., Helv. Chim. Acta 60, 798 (1977). Biosynthesis by vitamin K-dependent carboxylation of glutamic acid: C. T. Esmon et al., J. Biol. Chem. 250, 4744 (1975). Metal binding properties: G. L. Nelsestuen et al., Biochem. Biophys. Res. Commun. 65, 233 (1975); G. L. Nelsestuen, J. Biol. Chem. 251, 5648 (1976); R. Robertson et al., ibid. 253, 5880 (1978); S. P. Bajaj et al., ibid. 257, 3726 (1982). HPLC determn in proteins, bone, urine: M. Kuwada, K. Katayama, Anal. Biochem. 117, 259 (1981). Reviews: J. Stenflo, J. W. Suttie, Ann. Rev. Biochem. 46, 157-172 (1977); R. E. Olson, J. W. Suttie, Vitam. Horm. 35, 59-108 (1977); J. W. Suttie, CRC Crit. Rev. Biochem. 8, 191-223 (1980); J. P. Burnier et al., Mol. Cell. Biochem. 39, 191-207 (1981).

(HOOC)$_2$CHCH$_2$---C---COOH

Crystals, mp 167-167.5° (dec). [α]$_D^{20}$ +35.3° (c = 1 in 6N HCl).

DL-Form, white powder, mp 90-92°.

1835. Carboxymethylcellulose Sodium. *Carboxymethyl ether cellulose sodium salt;* CMC; sodium carboxymethylcellulose; sodium cellulose glycolate; Carmethose; Cel-O-Brandt; Cethylose; Glykocellon; Carbose D; Thylose; Xylo-Mucine; Tylose MGA; Cellolax; Polycell. R$_n$OCH$_2$COONa. Prepd by treating alkali cellulose with sodium chloroacetate: Faith, Keyes & Clark's *Industrial Chemicals*, F. A. Lowenheim, M. K. Moran, Eds. (Wiley-Interscience, New York, 4th ed., 1975) pp 235-238. Review and bibliography: Ott, *Cellulose and Cellulose Derivatives*, New York, 1946 (2nd ed., 1955).

White granules. Soly in water depends on degree of substitution. Water-soluble CMC is available in various viscosities (5-2000 centipoises in 1% soln), and the soly is equally good in hot and cold water (difference from methyl cellulose). Also the presence of metal salts has little effect on the viscosity. Solns are stable between pH 2 and 10. Below pH 2 precipitation of a solid occurs, above pH 10 the viscosity decreases rapidly. pKa 4.30. The free acid is obtained from aq soln at pH 2.5 and may be precipitated with alcohol.

USE: In drilling muds, in detergents as a soil-suspending agent, in resin emulsion paints, adhesives, printing inks, textile sizes, as protective colloid in general. As stabilizer in foods. Pharmaceutic aid (suspending agent; tablet excipient; viscosity-increasing agent).

1836. Carboxypolymethylene. *Carbomer; carbopol;* carboxyvinyl polymer. A vinyl polymer with active carboxyl groups. Description: *Chem. & Eng. News* **36**, 64 (Sept. 29, 1958).

White powder. Highly ionic and slightly acidic. Reacts with fatty amines to form thick and stable emulsions of oils in water.

USE: Thickening, suspending, dispersing, emulsifying agent. In the cosmetic and textile printing fields, in printing inks, in emulsion-based lubricants, in pharmaceuticals, polishes, waxes, paints, waterproof and oil-proof coatings, in industrial specialties.

1837. Carbromal. *N-(Aminocarbonyl)-2-bromo-2-ethylbutanamide;* (α-bromo-α-ethylbutyryl)urea; (α-bromo-α-ethylbutyryl)carbamide; bromodiethylacetylurea; bromodiethylacetylcarbamide; Tildin; Adalin; Planadalin; Diacid; Addisomnol; Bromadal; Uradal; Nyctal. C$_7$H$_{13}$BrN$_2$O$_2$; mol wt 237.11. C 35.46%, H 5.53%, Br 33.70%, N 11.82%, O 13.50%. (C$_2$H$_5$)$_2$CBrCONHCONH$_2$. Prepd by heating urea (to ~50°) with α-bromo-α-ethylbutyryl bromide (C$_2$H$_5$)$_2$-CBrCOBr, see **Ger. pat.** 225,710 (1910); *Frdl.* **10**, 1160; *Chem. Zentr.* **1910**, II, 1008. Large patent literature tabulated in Slotta, *Grundriss der modernen Arzneistoff-Synthese* (Stuttgart, 1931); H. P. Kaufmann, *Arzneimittel-Synthese* (Berlin, 1953).

Crystals, mp 116-119°. One gram dissolves in about 3000 ml water, 18 ml alcohol, 3 ml chloroform, 14 ml ether. It is very sol in boiling alcohol and dissolves in concd sulfuric, nitric and hydrochloric acids, from which it is precipitated on dilution with water. It is dissolved by solns of alkali hydroxides. LD orally in dogs: 450 mg/kg, *Handbook of Toxicology* **vol. 1**, W. S. Spector, Ed. (Saunders, Philadelphia, 1956) pp 14-15.

Caution: This substance may be habit forming and is listed in the U.S. Code of Federal Regulations, Title 21 Part 329.1 (1987).

THERAP CAT: Sedative, hypnotic.

1838. Carbubarb. *5-[2-[(Aminocarbonyl)oxy]ethyl]-5-butyl-2,4,6(1H,3H,5H)-pyrimidinetrione; carbamic acid ester with 5-butyl-5-(2-hydroxyethyl)barbituric acid;* 5-butyl-5-carbamoyloxyethylbarbituric acid; 5-butyl-5-(2-hydroxyethyl)barbituric acid carbamate; carbamic acid ester with 5-(β-hydroxyethyl)-5-butylmalonylurea; 5-(β-hydroxyethyl)-5-butylmalonylurea carbamate; Nogexan. C$_{11}$H$_{17}$-

N$_3$O$_5$; mol wt 271.27. C 48.70%, H 6.32%, N 15.49%, O 29.49%. Prepn: **Fr. pat.** M1059 (1962 to Consortium Mondial des Grandes Marques), *C.A.* **59**, 7539c (1963); Buzas, **U.S. pat.** 3,150,137 (1964). Purification and toxicity: **Fr. pat.** M2633 (1964 to Interco Fribourg), *C.A.* **62**, 1671e (1965).

Crystals from ethanol, mp 192-194°. LD$_{50}$ s.c. in mice: 1.4 g/kg (**Fr. pat.** M2633).

Note: This is a controlled substance (depressant) listed in the U.S. Code of Federal Regulations, Title 21 Part 1308.13 (1987).

THERAP CAT: Sedative, hypnotic.

1839. Carbutamide. *4-Amino-N-[(butylamino)carbonyl]benzenesulfonamide; 1-butyl-3-sulfanilylurea; N^1-(butylcarbamoyl)sulfanilamide; N^1-sulfanilyl-N^2-butylurea; N^1-sulfanilyl-N^2-butylcarbamide; N-(4-aminobenzenesulfonyl)-N'-butylurea;* aminophenurobutane; BZ 55; U 6987; Nadisan; Invenol; Emedan; Oranil; Orasulin; Glucofren; Bukarban; Bucarban; Cicloral; Glucidoral; Alentin; Norboral; Bucrol. C$_{11}$H$_{17}$N$_3$O$_3$S; mol wt 271.35. C 48.69%, H 6.32%, N 15.49%, O 17.69%, S 11.82%. *Ref:* Achelis, Hardebeck, *Deut. Med. Wochenschr.* **80**, 1452 (1955); Achelis *et al.*, *Arch. Exp. Pathol. Pharmakol.* **228**, 163 (1956). Prepn from butylurea and sulfanilamide: Samaniego, **Span. pat.** 229,696 (1956), *C.A.* **51**, 7413 (1957); E. Haack, A. Hagedorn, **U.S. pat.** 2,907,692 (1959 to Boehringer, Mann.).

H$_2$N⟨⟩SO$_2$NHCONH(CH$_2$)$_3$CH$_3$

Crystals, mp 144-145°. Sol in water at pH 5 to 8. LD$_{50}$ s.c. in mice: 3 g/kg (Haack, Hagedorn).

THERAP CAT: Antidiabetic.

1840. Carbuterol. *[5-[2-[(1,1-Dimethylethyl)amino]-1-hydroxyethyl]-2-hydroxyphenyl]urea; α-(t-butylaminomethyl)-4-hydroxy-3-ureidobenzyl alcohol.* C$_{13}$H$_{21}$N$_3$O$_3$; mol wt 267.33. C 58.41%, H 7.92%, N 15.71%, O 17.96%. A β-adrenergic agonist related to isoproterenol, *q.v.*, with selectivity for airway smooth muscle receptors. Prepn: C. Kaiser, S. T. Ross, **Ger. pat.** 2,106,620 corresp to **U.S. pat.** 3,763,232 (1971, 1973 both to SKF); C. Kaiser *et al.*, *J. Med. Chem.* **17**, 49 (1974). Pharmacology, mechanism of action, toxicity study: J. R. Wardell *et al.*, *J. Pharmacol. Exp. Ther.* **189**, 167 (1974). Analysis in aq soln: L. J. Ravin *et al.*, *J. Pharm. Sci.* **67**, 1523 (1978). Clinical study: T. D. James, H. A. Lyons, *J. Am. Med. Assoc.* **241**, 704 (1979).

HO⟨⟩CHCH$_2$NHC(CH$_3$)$_3$

Cryst, mp 174-176°. LD$_{50}$ in mice: 32.8 mg/kg i.v.; 3134.6 mg/kg orally; in rats: 77.2 mg/kg i.v., J. R. Wardell *et al.*, *loc. cit.*

Hydrochloride, C$_{13}$H$_{22}$ClN$_3$O$_3$, *SKF 40383, Bronsecur, Pirem.* Cryst from ethanol/ether, mp 205-207° (dec).

THERAP CAT: Bronchodilator.

1841. Cardamom Seed. Grains of paradise. Dried ripe seeds of *Elettaria cardamomum* Maton, *Zingiberaceae*. *Habit:* Malabar, cultivated in India, Ceylon, Guatemala. *Constit:* Resin; 2-8% essential oil, 1-2% fixed oil. The essential oil contains eucalyptol (cineol), sabinene, *d,α*-terpineol and acetate, borneol, limonene, terpinene, 1-terpinen-4-ol and its formate and acetate. The fixed oil consists of the glycerides of oleic, stearic, linoleic, palmitic, caprylic and caproic acids. The unsaponifiable matter from the fixed oil contains β-sitosterol. The seeds contain 1.3 γ of vitamin B_1/g. Traces of manganese have also been observed. The ash of the pod including the seeds consists of 24.8% SiO_2, 20.43% Na_2CO_3, 13.3% CaO, 2.54% Cl, 1.5% Al_2O_3 and 4.3% MnO. Other constituents of pods and seeds are pentosans and starch. Description of the cultivation, botany, and the processing of common and oriental cardamom varieties: Viehoever, Sung, *J. Am. Pharm. Assoc.* **26**, 872 (1937); *see also* E. Guenther, *The Essential Oils* vol. V (New York, 1952) pp 85-106.

USE: Flavoring baked goods, confectionery, curry powder. In the manufacture of oil of cardamom which is used for flavoring liqueurs. Pharmaceutic aid (flavor).

THERAP CAT: Adjuvant carminative.

THERAP CAT (VET): Has been used as a carminative.

1842. Cardiotoxin. One of the toxic principles from cobra venom. Isoln: Sarkar, *J. Indian Chem. Soc.* **21**, 227 (1947). A single peptide chain composed of 60 amino acid residues cross-linked by four disulfide bridges with leucine and asparagine at the N and C terminals respectively: Narita, Lee, *Biochem. Biophys. Res. Commun.* **41**, 339 (1970). Causes irreversible depolarization of cell membrane and contraction of skeletal and smooth muscle. Shows direct lytic and marked cardiotoxic activities, potentiated by phospholipase A and prevented by polyanions (heparin, RNA, gangliosides). Pharmacology: Lee *et al.*, *Arch. Exp. Pathol. Pharmakol.* **259**, 360 (1968).

1843. 3-Carene. *3,7,7-Trimethylbicyclo[4.1.0]hept-3-ene;* Δ^3-*carene*; 4,7,7-trimethyl-3-norcarene; isodiprene. $C_{10}H_{16}$; mol wt 136.23. C 88.16%, H 11.84%. Constituent of turpentine. The turpentine from *Pinus sylvestris* L. may contain as much as 42%, turpentine from *Pinus longifolia* Roxb., *Pinaceae* about 30%. Isoln and structure: Simonsen, *The Terpenes* vol. II (Cambridge, 1949) pp 64-72; Guenther, *The Essential Oils* vol. II (Van Nostrand, 1949) pp 49-51. Conformation: Acharya, *Tetrahedron Letters* **1966**, 4117. Absorption spectrum: Cole, *J. Chem. Soc.* **1954**, 3807.

d-Form, mobile liquid. Readily oxidized on exposure to air. Sweet and pungent odor, more agreeable than the odor of turpentine. d_{15}^{15} 0.8668; d_{90}^{30} 0.8586. bp_{705} 168-169°. bp_{200} 123-124°. $[\alpha]_D^{20}$ +7.69°. n_D^{30} 1.468. Practically insol in water. Miscible with fat solvents and oils.

d-Form nitrosate, $C_{10}H_{16}N_2O_4$, prepd from *d*-Δ^3-carene with amyl nitrite, acetic and nitric acid, prisms, dec 147.5°.

1844. Carfecillin Sodium. *6-[(1,3-Dioxo-3-phenoxy-2-phenylpropyl)amino]-3,3-dimethyl-7-oxo-4-thia-1-azabicyclo[3.2.0]heptane-2-carboxylic acid monosodium salt; N-(2-carboxy-3,3-dimethyl-7-oxo-4-thia-1-azabicyclo[3.2.0]hept-6-yl)-2-phenylmalonamic acid 1-phenyl ester sodium salt;* sodium α-phenoxycarbonylbenzylpenicillin; carbenicillin phenyl sodium; α-carboxybenzylpenicillin phenyl ester sodium salt; BRL 3475; Gripenin-O; Uticillin. $C_{23}H_{21}N_2NaO_6S$; mol wt 476.48. C 57.98%, H 4.44%, N 5.88%, Na 4.82%, O 20.15%, S 6.73%. Semi-synthetic antibiotic related to penicillin. Prepn: Hardy *et al.*, S. Afr. pat. 67 06,472 corresp to U.S. pats. 3,853,849 and 3,881,013 (1968, 1974, 1975, all to Beecham); Butler, S. Afr. pat. 69 00,060 (1969 to Pfizer), *C.A.* **72**, 111465m (1970). Prepn and activity studies: Clayton *et al.*, *J. Med. Chem.* **18**, 172 (1975).

Crystals from ethanol, $[\alpha]_D^{20}$ +216.2° (H_2O).

THERAP CAT: Antibacterial.

1845. Carfinate. α-*Ethynylbenzenemethanol carbamate;* α-*ethynylbenzyl carbamate; carbamic acid α-ethynylbenzyl ester;* α-ethynylbenzyl alcohol carbamate; phenylethynylcarbinol carbamate; Equilium; Nirvotin. $C_{10}H_9NO_2$; mol wt 175.18. C 68.56%, H 5.18%, N 8.00%, O 18.27%. Prepn: **Brit.** pat. 736,340 (1955 to Carlo Erba).

Crystals from alc, mp 86-87°.

THERAP CAT: Sedative; hypnotic.

1846. Cargutocin. *1-Butanoic acid-7-glycine-1,6-dicarbaoxytocin; 1-butyric acid-6-(L-2-aminobutyric acid)-7-glycineoxytocin;* 7-glycine-1,6-aminosuberic acid-oxytocin; [Gly^7, $Asu^{1,6}$]oxytocin; Y 5350; Statocin. $C_{42}H_{65}N_{11}O_{12}$; mol wt 916.06. C 55.07%, H 7.15%, N 16.82%, O 20.96%. Analog of oxytocin, *q.v.* Prepn: S. Sakakibara *et al.*, **Ger.** pat. 2,056,298 corresp to U.S. pat. 3,749,705 (1971, 1973 both to Yoshitomi). Synthesis and uterotonic activity: T. Yamanaka, S. Sakakibara, *Bull. Chem. Soc. Japan* **47**, 1228 (1974). Pharmacological and biochemical properties: T. Oka *et al.*, *Japan. J. Pharmacol.* **25**, 15 (1975). Comparison with prostaglandin $F_{2\alpha}$ as oxytocic: T. Hashimoto, *Iryo* **30**, 1121 (1976), *C.A.* **87**, 554e (1977). Distribution, metabolism, excretion: Y. Kato *et al.*, *Iyakuhin Kenkyu* **12**, 433 (1981), *C.A.* **95**, 126386k (1981). Teratogenicity studies: Y. Hamada *et al.*, *ibid.* **10**, 26, 41 (1979), *C.A.* **91**, 639u, 83856m (1979).

$[\alpha]_D^{25}$ −44.0° (c = 0.55 in water).

THERAP CAT: Oxytocic.

1847. Carindacillin. *6-[[3-[(2,3-Dihydro-1H-inden-5-yl)oxy]-1,3-dioxo-2-phenylpropyl]amino]-3,3-dimethyl-7-oxo-4-thia-1-azabicyclo[3.2.0]heptane-2-carboxylic acid; N-(2-carboxy-3,3-dimethyl-7-oxo-4-thia-1-azabicyclo[3.2.0]hept-6-yl)-2-phenylmalonamic acid 1-(5-indanyl) ester;* 6-[2-(5-indanyloxycarbonyl)phenylacetamido]-3,3-dimethyl-7-oxo-4-thia-1-azabicyclo[3.2.0]heptane-2-carboxylic acid; 1-(5-indanyl) N-(2-carboxy-3,3-dimethyl-7-oxo-4-thia-1-azabicyclo[3.2.0]hept-6-yl)-2-phenylmalonamate; 6-[2-phenyl-2-(5-indanyloxycarbonyl)acetamido]-penicillanic acid; 6-(2-carboxy-2-phenylacetamido)-3,3-dimethyl-7-oxo-4-thia-1-azabicyclo[3.2.0]heptane-2-carboxylic acid 6-(5-indanyl ester); α-(5-indanyloxycarbonyl)-benzylpenicillin; carbenicillin indanyl ester; CP 15464. $C_{26}H_{26}N_2O_6S$; mol wt 494.57. C 63.14%, H 5.30%, N 5.66%, O 19.41%, S 6.48%. Semi-synthetic antibiotic related to penicillin. Prepn: Butler, S. Afr. pat. 69 00,060; **Ger.** pat. 1,944,376 corresp to U.S. pats. 3,557,090 and 3,574,189 (1969, 1970, 1971, 1971); Nakanishi, **Ger.** pat. 1,959,569 (1970), *C.A.* **73**, 66565b (1970) (all to Pfizer). Activity studies: Butler, *Del. Med. J.* **43**, 366 (1971); English *et al.*, *Antimicrob. Ag. Chemother.* **1**, 185 (1972). Clinical evaluation: Wallace *et al.*, *Antimicrob. Ag. Chemother.* **1970**, 223; Bran *et al.*, *Clin. Pharmacol. Ther.* **12**, 525 (1971); *Indanyl Carbeni-*

cillin, H. Swarz, F. E. Storari, Eds. (Am. Elsevier, New York, 1974) 100 pp.

Sodium salt, $C_{26}H_{25}N_2NaO_6S$, *CP 15464-2, Carindapen, Geocillin, G.U.-Pen.*

THERAP CAT: Antibacterial.

1848. Carisoprodol. *(1-Methylethyl)carbamic acid 2-[[(aminocarbonyl)oxy]methyl]-2-methylpentyl ester; N-isopropyl-2-methyl-2-propyl-1,3-propanediol dicarbamate;* isopropyl meprobamate; isobamate; carisoprodate; Apesan; Arusal; Caprodat; Carisoma; Domarax; Flexal; Flexartal; Miolisodal; Mioril; Rela; Relasom; Sanoma; Soma; Somadril; Somalgit. $C_{12}H_{24}N_2O_4$; mol wt 260.33. C 55.36%, H 9.29%, N 10.76%, O 24.58%. Prepn: Berger, Ludwig, U.S. pat. **2,937,119** (1960 to Carter Prod.). Pharmacology: Berger *et al., J. Pharmacol. Exp. Ther.* **127**, 66 (1959).

Crystals, mp 92-93°. Slightly bitter taste. Very sparingly sol in water: 30 mg/100 ml at 25°; 140 mg/100 ml at 50°. Sol in many common organic solvents. Practically insol in vegetable oils. Stable in dil acids and alkalies. LD_{50} in mice, rats: 2340, 1320 mg/kg orally; 980, 450 mg/kg i.p., Berger *et al., loc. cit.*

THERAP CAT: Skeletal muscle relaxant.

1849. Carlsbad Salt Artificial. A mixture of 1 part potassium sulfate, 9 sodium chloride, 18 sodium bicarbonate and 22 anhydr sodium sulfate. One gram of the mixture in 200 ml water is approx equivalent to an equal vol of Carlsbad water (Sprudel). *Keep dry and well closed.*

Carlsbad salt artificial, effervescent. A mixture of 25 parts of Carlsbad salt artificial, 48 sodium bicarbonate, 17 tartaric acid and 25 citric acid.

THERAP CAT: Cathartic.

THERAP CAT (VET): Purgative.

1850. Carminic Acid. *7-α-D-Glucopyranosyl-9,10-dihydro-3,5,6,8-tetrahydroxy-1-methyl-9,10-dioxo-2-anthracenecarboxylic acid;* C.I. Natural Red 4; C.I. 75470. $C_{22}H_{20}O_{13}$; mol wt 492.38. C 53.66%, H 4.09%, O 42.24%. Glucosidal coloring matter from the scale insect *Coccus cacti* L., *Homoptera* (cochineal). The essential constituent of carmine. Isoln: Schunk, Marchlewski, *Ber.* **27**, 2979 (1894); Dimroth, Scheuer, *Ann.* **399**, 43 (1913). Structure: Dimroth, Kammerer, *Ber.* **53**, 471 (1920); Ali, Haynes, *J. Chem. Soc.* **1959**, 1033; revised structure: Bhatia, Venkataraman, *Indian J. Chem.* **3** (2), 92 (1965). *See also Colour Index* vol. 4 (3rd ed., 1971) p 4632.

Red prisms from alc, no distinct melting point, darkens at 120°. Has a deep red color in water and is yellow to violet in acid solns. uv max (water): 500 nm (ε 6800); (0.02N HCl): 490-500 nm (ε 5800); (0.0001N NaOH): 540 nm (ε 3450). $[α]_{654}^{15}$ +51.6° (water). Sol in water, alc, concd H_2SO_4; slightly sol in ether; practically insol in petr ether, benzene, chloroform.

Methyl tetra-*O*-methylcarminate, $C_{27}H_{30}O_{13}$, yellow needles from benzene + petr ether, mp 185-188°.

Aluminum calcium lake, *carmine.* Bright-red, light pieces; easily reduced to powder. Practically insol in cold water or dil acids; partly sol in hot water; sol in solns of alkali hydroxides or their carbonates giving deep red solns; also sol in borax.

USE: Free acid in color photography; pigment for artists' paints; as bacteriol. stain; rarely now as acid-base indicator or as oxidimetric indicator; as a reagent for aluminum; as a complexing agent for cations. Aluminum calcium lake as dye; in inks; coloring food products and galenicals; in microscopy for making various stains. Approved by the FDA for use in foods and drugs.

1851. Carmofur. *5-Fluoro-N-hexyl-3,4-dihydro-2,4-dioxo-1(2H)-pyrimidinecarboxamide;* 1-(*n*-hexylcarbamoyl)-5-fluorouracil; HCFU; Mifurol; Yamaful. $C_{11}H_{16}FN_3O_3$; mol wt 257.27. C 51.35%, H 6.27%, F 7.39%, N 16.33%, O 18.65%. Orally active cytostatic deriv of fluorouracil, *q.v.* Prepn: S. Ozaki *et al.*, **Japan. Kokai** 77 78886, *C.A.* **87**, 152265 (1977); *eidem*, U.S. pat. **4,071,519** (1977, 1978 both to Mitsui); *eidem, Bull. Chem. Soc. Japan* **50**, 2406 (1977). Anti-tumor activity: A. Hoshi *et al., Chem. Pharm. Bull.* **26**, 161 (1978). Determn in body fluids: S. Watanabe *et al., Chemotherapy (Tokyo)* **27**, 778 (1979); O. Nakajima *et al., J. Chromatog.* **225**, 91 (1981). General pharmacology: Z. Henmi *et al., Oyo Yakuri* **19**, 369 (1980), *C.A.* **93**, 125639s (1980). Pharmacokinetics: M. Iigo *et al., Cancer Chemother. Pharmacol.* **4**, 189 (1980). Metabolism: *eidem, Xenobiotica* **10**, 847 (1980); T. Kobari *et al., ibid.* **11**, 57 (1981). Phase I clinical study: Y. Koyama, Y. Koyama, *Cancer Treat. Rep.* **64**, 861 (1980). Mutagenicity study: Y. Seino *et al., Cancer Res.* **38**, 2148 (1978). *Review:* T. Taguchi, *Recent Results Cancer Res.* **70**, 125-132 (1980).

White cryst from ethanol, mp 110-111°. uv max (chloroform): 258 nm (ε 1.16 × 10⁴), S. Ozaki *et al., Bull. Chem. Soc. Japan* **50**, 2406 (1977). Also reported as mp 283° (dec), *eidem*, U.S. pat. **4,071,519.**

THERAP CAT: Antineoplastic.

1852. Carmustine. *N,N'-Bis(2-chloroethyl)-N-nitrosourea;* BCNU; BiCNU; NSC-409962; Becenun; Carmubris; Nitrumon. $C_5H_9Cl_2N_3O_2$; mol wt 214.04. C 28.05%, H 4.24%, Cl 33.12%, N 19.63%, O 14.95%. Chloroethylnitrosourea derivative with antitumor activity. Similar to chlorozotocin, lomustine, nimustine, ranimustine, *q.q.v.* Synthesis: Johnston *et al., J. Med. Chem.* **6**, 669 (1963). Properties: Loo *et al., J. Pharm. Sci.* **55**, 492 (1966). Decompn studies as related to antileukemic activity: Montgomery *et al., J. Med. Chem.* **10**, 668 (1967). Antifungal action: Hunt, Pittilo, *Antimicrob. Ag. Chemother.* **1965**, 710. Toxicology studies: Thompson, Larson, *Toxicol. Appl. Pharmacol.* **21**, 405 (1972).

Light yellow powder that melts to an oily liquid; mp 30-32°. Both powder and liquid are stable. Dec rapidly in acid and in soln above pH 7. Most stable in petroleum ether or aqueous soln at pH 4. Non-ionized at pH 7 with consequent high lipid solubility. Sol in water up to 4 mg/ml and in 50% ethanol up to 150 mg/ml: DeVita *et al., Cancer Res.* **25**, 1876 (1965). LD_{50} in mice (mg/kg): 19-25 orally, 26 i.p., 24 s.c.; in rats (mg/kg): 30-34 orally (Thompson, Larson).

Note: This substance may reasonably be anticipated to be a carcinogen: *Fourth Annual Report on Carcinogens* (NTP 85-002, 1985) p 45.

THERAP CAT: Antineoplastic.

1853. Carnauba Wax. Brazil wax. An exudate from the pores of the leaves of the Brazilian wax palm tree *Copernicia prunifera* (Muell.) H. E. Moore [*Copernicia cerifera* (Arruda da Camara) Mart.], *Palmae*. The botany of the tree and the native wax-collecting procedures are described adequately by A. H. Warth, *The Chemistry and Technology of Waxes* (Reinhold, New York, 1947). The hardness and high-polish capability of this important wax can be ascribed to the presence of esters of hydroxylated unsaturated fatty acids having about 12 carbon atoms in the acid chain. The usual names for the constituents, *i.e.* cerotic acid, melissyl cerotate, carnaubic acid etc., are meaningless. Brief review: C. S. Letcher in Kirk-Othmer *Encyclopedia of Chemical Technology* vol. 24 (Wiley-Interscience, New York, 3rd ed., 1984) p 469-470.

Hard greenish solid, cryst fracture. Sharp, characteristic, not unpleasant odor upon melting. mp 82-85.5°. d 0.990 to 0.999. Saponification number 78 to 89. Iodine number about 13. n_D^{90} 1.4500. Sparingly soluble in fat solvents at 25°, quite sol at 45°.

USE: Wherever a hard, high-polish wax is desired, *e.g.* in automobile waxes, floor wax emulsions, high quality shoe polishes, in the paper industry (especially for making carbon papers). As a plasticizer in dental impression compounds. To raise the melting point of other waxes; often used together with candelilla wax. The presence of the lower-melting ouricury wax is considered as an adulteration. Purified and bleached carnauba wax is used for cosmetic materials, such as depilatories and deodorant sticks. In pharmacy as the last stage in tablet coating. Skin sensitization or irritation by carnauba wax seems infrequent.

1854. Carnegine. *1,2,3,4-Tetrahydro-6,7-dimethoxy-1,2-dimethylisoquinoline*; 6,7-dimethoxy-1,2-dimethyl-1,2,3,4-tetrahydroisoquinoline; pectenine. $C_{13}H_{19}NO_2$; mol wt 221.29. C 70.55%, H 8.66%, N 6.33%, O 14.46%. In *Carnegiea gigantea* (Engelm.) Britt. & Rose *(Cereus giganteus* Engelm.), *Cactaceae*, a cactus of Arizona and Mexico. Extraction procedure: Heyl, *Arch. Pharm.* 266, 668 (1928). Synthesis: Späth, *Ber.* 62, 1021 (1929); Nakada, Nisgihara, *J. Pharm. Soc. Japan* 64, 74 (1944). Pharmacology: E. Santi-Soncin, M. Furlanut, *Fitoterapia* 43, 21 (1972), *C.A.* 80, 66667f (1974).

Viscous liquid, dec on standing. Distills at 1 mm pressure and 170° air bath temp. Sol in ether, alc, chloroform. LD_{50} i.p. in mice: 15.23 mg/kg, E. Santi-Soncin, M. Furlanut, *loc. cit.*

Hydrochloride monohydrate, $C_{13}H_{19}NO_2$·HCl·H_2O, clusters from dil alc, mp 207° (mp 211° when anhydr). Sol in water, slightly sol in alc.

Hydrobromide monohydrate, $C_{13}H_{19}NO_2$·HBr·H_2O, needles from alc, mp 228°.

Methyliodide, $C_{14}H_{22}NO_2I$, needles from methanol, mp 211° (evac tube) after drying at 100° and 10 mm.

1855. Carnidazole. *[2-(2-Methyl-5-nitro-1H-imidazol-1-yl)ethyl]carbamothioic acid O-methyl ester*; O-methyl [2-(2-methyl-5-nitroimidazol-1-yl)ethyl]thiocarbamate; R 25831; Spartrix. $C_8H_{12}N_4O_3S$; mol wt 244.27. C 39.33%, H 4.95%, N 22.94%, O 19.65%, S 13.12%. Prepn and antiprotozoal activity: J. Heeres *et al., Ger. pat.* 2,429,755; *eidem,* U.S. pat. 3,928,374 (both 1975 to Janssen); *eidem, Eur. J. Med. Chem.-Chim. Ther.* 11, 237 (1976). Crystal structure of monohydrate: N. M. Blaton *et al., Acta Crystallogr.* B35, 753 (1979). Clinical efficacy in vaginal trichomoniasis: A. Notowicz *et al., Brit. J. Vener. Dis.* 53, 129 (1977); P. Chaudhuri, A. C. Drogendijk, *Eur. J. Obstet. Gynecol. Reprod. Biol.* 10, 325 (1980).

Crystals from ethanol, mp 142.4° (dec).
THERAP CAT (VET): Antiprotozoal, trichomonacide.

1856. Carnitine. *3-Carboxy-2-hydroxy-N,N,N-trimethyl-1-propanaminium hydroxide, inner salt; (3-carboxy-2-hydroxypropyl)trimethylammonium hydroxide, inner salt;* γ-amino-β-hydroxybutyric acid trimethylbetaine; γ-trimethyl-β-hydroxybutyrobetaine; 3-hydroxy-4-(trimethylammonio)butanoate. $C_7H_{15}NO_3$; mol wt 161.20. C 52.15%, H 9.38%, N 8.69%, O 29.78%. $(CH_3)_3N^+CH_2CH(OH)CH_2$-$COO^-$. Essential cofactor of fatty acid metabolism; constituent of striated muscle and liver. Isoln from meat extract: W. Gulewitsch, R. Krimberg, *Z. Physiol. Chem.* 45, 326 (1905); S. P. Colowick, N. O. Kaplan, *Methods in Enzymology,* vol.III (Academic Press, New York, 1957) p 660. Isoln and characterization of *Vitamin B_T* as carnitine: H. E. Carter *et al., Arch. Biochem. Biophys.* 38, 405 (1952). Isoln of D-carnitine: S. Friedman *et al., Arch. Biochem. Biophys.* 66, 10 (1957). Synthesis of L-carnitine: M. Tomita, Y. Sendju, *Z. Physiol. Chem.* 169, 263 (1927); R. Voeffray *et al., Helv. Chim. Acta* 70, 2058 (1987). Synthesis of DL-carnitine: E. Strack *et al., Ber.* 86, 525 (1953); H. E. Carter, P. K. Bhattacharyya, *J. Am. Chem. Soc.* 75, 2503 (1953). Synthesis of DL-, D-, and L-carnitine from epichlorohydrin: E. Strack, I. Lorenz, *Z. Physiol. Chem.* 318, 129 (1960). Prepn by hydrolysis of γ-dimethylamino-β-hydroxybutyronitrile chloromethylate: Ger. pat. 1,090,676 (1960 to Labaz). Biosynthesis: G. Wolf, C. R. A. Berger, *Arch. Biochem. Biophys.* 92, 360 (1961). Absolute configuration: Kaneko, Yashida, *Bull. Chem. Soc. Japan* 35, 1153 (1962). Metabolism: M. E. Mitchell, *Am. J. Clin. Nutr.* 31, 293 (1978). Use of L-carnitine in hyperlipoproteinemia: C. Carazza, U.S. pat. 4,255,449 (1981); M. T. Ramacci, U.S. pat. 4,315,944 (1982 to Sigma-Tau). Effect on myocardial metabolism in coronary artery disease: R. Ferrari *et al., Clin. Trials J.* 21, 40 (1984). Clinical evaluation in congestive heart failure: O. Ghidini *et al., Int. J. Clin. Pharmacol. Ther. Toxicol.* 26, 217 (1988).

Monograph: Recent Research on Carnitine, G. Wolf, Ed. (MIT Press, Cambridge, Mass., 1965). Book: *Carnitine Biosynthesis, Metabolism, and Functions,* R. A. Frenkel, J. D. McGarry, Eds. (Academic Press, New York, 1980) 356 pp. *Reviews:* G. Fraenkel, S. Friedman, *Vitam. Horm.* XV, 73-118 (1957); S. L. De Felice, *Orphan Drugs,* F. E. Karch, Ed. (Marcel Dekker, New York, 1982) pp 33-56. Reviews of nutritional significance and therapeutic applications: C. J. Rebouche, D. J. Paulson, *Ann. Rev. Nutr.* 6, 41-66 (1986); of clinical pharmacology: J. J. Bahl, R. Bressler, *Ann. Rev. Pharmacol. Toxicol.* 27, 257-277 (1987); of pharmacokinetics, therapeutic use: K. L. Goa, R . N. Brogden, *Drugs* 34, 1-24 (1987); of metabolism and enzymology: L. L. Bieber, *Ann. Rev. Biochem.* 57, 261-283 (1988).

DL-Form, hygroscopic crystalline solid, dec 195-197°.

DL-Form hydrochloride, $C_7H_{16}ClNO_3$, *Bicarnesine, Flatistine.* Needles from ethanol, mp 196° (dec). Very sol in water; sol in hot ethanol; slightly sol in cold ethanol. Practically insol in acetone, ether.

D-Form, crystals, dec 210-212°. $[\alpha]_D$ +30.9°. Very sol in water and alcohol. Practically insol in acetone and ether.

D-Form hydrochloride, $C_7H_{16}ClNO_3$, crystals, dec 142°.

L-Form, *levocarnitine, vitamin B_T, Cardiogen, Carnitene, Carnicor, Carnitor, Carrier, Levocarnil, Metina.* Crystals from ethanol + acetone, dec 197-198°. Very hygroscopic solid. $[\alpha]_D^{30}$ −23.9° (c = 0.86 in water). Sol in water, hot alcohol. Practically insol in acetone, ether, benzene.

L-Form hydrochloride, $C_7H_{16}ClNO_3$, *Lefcar.* Crystals, dec 142°.

THERAP CAT: L-Form as antihyperlipoproteinemic; DL-hydrochloride as gastric and pancreatic secretion stimulant.

1857. Carnosine. *N-β-Alanyl-L-histidine;* ignotine. C_9-$H_{14}N_4O_3$; mol wt 226.23. C 47.78%, H 6.24%, N 24.76%, O 21.22%. Naturally occurring dipeptide found in muscle of man and of numerous animals, but some animals, e.g., pigeons and geese, have *N*-methylcarnosine (anserine, *q.v.*) in their muscles. Isoln: Gulewitsch, Amiradzibi, *Ber.* **33**, 1902 (1900); Wolff, Wilson, *J. Biol. Chem.* **95**, 495 (1932); **109**, 565 (1935). Synthesis from histidine and β-iodo- or β-nitropropionyl chloride: Baumann, Ingvaldsen, *ibid.* **35**, 271 (1918); Barger, Tutin, *Biochem. J.* **12**, 406 (1918). Later syntheses: Sifford, du Vigneaud, *J. Biol. Chem.* **108**, 753 (1935); R. A. Turner, *J. Am. Chem. Soc.* **75**, 2388 (1953); F. J. Vinick, S. Jung, *J. Org. Chem.* **48**, 392 (1983). Crystal structure: H. Itoh *et al., Acta Crystallogr.* **33B**, 2959 (1977). Studies on L-carnosine as a possible olfactory neurotransmitter; F. L. Margolis, *Science* **184**, 909 (1974). Possible role of L-carnosine in wound healing: D. E. Fischer *et al., Proc. Soc. Exp. Biol. Med.* **158**, 402 (1978).

L-Form, the natural form, crystals from aqueous ethanol, mp 262° (dec) (Vinick, Jung); also reported as 260° (capillary tube) and as 308–309° (Dennis bar) (Sifford, du Vigneaud). $[\alpha]_D^{25}$ +21.0° (c = 1.5 in water). pK$_1$ 2.64; pK$_2$ 6.83; pK$_3$ 9.51. Alkaline reaction. One gram dissolves in 3.1 ml water at 25°.

Nitrate, $C_9H_{15}N_5O_6$, crystals, dec 222°. $[\alpha]_D^{20}$ +24.1° (c = 1.5 in water). Very sol in water.

Hydrochloride, $C_9H_{15}ClN_4O_3$, crystals, dec 245°. Very sol in water.

D-Form, crystals, mp 260°. $[\alpha]_D^{28}$ −20.4° (c = 1.5).

1858. Caro's Acid. *Peroxymonosulfuric acid; sulfomonoperacid; persulfuric acid.* H_2O_5S; mol wt 114.09. H 1.77%, O 70.12%, S 28.11%. Dry reagent is prepd by stirring 10 g potassium persulfate into 11 g concd H_2SO_4 for 10 min and adding 30 g finely powdered potassium sulfate; liquid reagent is obtained by triturating potassium persulfate with three times as much (by weight) of H_2SO_4; dil reagent is prepd by stirring 10 g potassium persulfate into 11 g concd H_2SO_4 and adding 50 cc ice: Baeyer, Villiger, *Ber.* **32**, 3625 (1899); another prepn by reacting 90% H_2O_2 with chlorosulfonic acid at −40 to −50°: Ball, Edwards, *J. Am. Chem. Soc.* **78**, 1125 (1956).

The product is a sirupy liquid consisting of about equal amounts of Caro's acid and H_2SO_4. pK$_2$ of Caro's acid 9.4 ± 0.1. Oxygen is evolved at room temp; should be stored at dry ice temp.

USE: In prepn of dyes; oxidation of olefins to α-glycols; oxidation of ketones to lactones or esters; treating woolens to prevent felting and shrinking; in bleaching compositions. *Caution:* Can be dangerously unstable, like most peroxides. Description of explosion at Brown University: J. O. Edwards, *Chem. & Eng. News* **33**, 3336 (1955). Explosion at Sun Oil, *ibid.* **38**, 59 (Nov. 21, 1960). *Toxicity:* May be highly irritating to skin, eyes, mucous membranes.

1859. α-Carotene. $C_{40}H_{56}$; mol wt 536.85. C 89.49%, H 10.51%. About as widely distributed as its β-isomer, but in smaller amounts. Best sources for both the α- and β-isomers are carrots, palm oil, and green leaves of various species. As a provitamin A it is half as active as β-carotene. Found in the mother liquors after crystallizing β-carotene. Isoln by chromatography: Karrer, Walker, *Helv. Chim. Acta* **16**, 641 (1933). Structure: Kuhn, Lederer, *Ber.* **64**, 1349 (1931); Karrer *et al., Helv. Chim. Acta* **14**, 614 (1931); **16**, 975 (1933). Natural (+)-α-carotene has 6′R configuration: Eugster *et al., ibid.* **52**, 1729 (1969). Synthesis of dl-α-carotene: Eugster, Karrer, *ibid.* **38**, 610 (1955); Tscharner *et al., ibid.* **40**, 1676 (1957); Ruegg *et al., ibid.* **44**, 985 (1961).

Deep purple prisms, polyhedra from petr ether or from benzene + methanol, mp 187.50° (evacuated tube). $[\alpha]_{643}^{18}$ +385° (c = 0.08 in benzene). Absorption max (CHCl$_3$): 485, 454 nm. More sol than β-carotene. Freely sol in carbon disulfide, chloroform; sol in ether, benzene. Slightly sol in petr ether, alcohol. 100 ml hexane dissolves 294 mg at 0°. Practically insol in water, acids, alkalies. Absorbs oxygen from the air, giving rise to inactive colorless oxidation products. The oxidation in light is autocatalytic. *Store in darkness in sealed ampoules and at low temp (−20°C).*

THERAP CAT: Vitamin A precursor.

THERAP CAT (VET): Vitamin A precursor for all species except cats.

1860. β-Carotene. *β,β-Carotene;* Carotaben; Provatene; Solatene. $C_{40}H_{56}$; mol wt 536.85. C 89.49%, H 10.51%. Most important of the provitamins A. Widely distributed in the plant and animal kingdom. In plants it occurs almost always together with chlorophyll. Isoln from carrots: Willstätter, Escher, *Z. Physiol. Chem.* **64**, 47 (1910); Kuhn, Lederer, *Ber.* **64**, 1349 (1931); Barnett *et al.,* U.S. pat. **2,848,508** (1958). Chromatography: Karrer, Walker, *Helv. Chim. Acta* **16**, 641 (1933). Structure: Willstätter, Mieg, *Ann.* **355**, 1 (1907); Zechmeister *et al., Ber.* **61**, 566 (1928); **66**, 123 (1933); Karrer *et al., Helv. Chim. Acta* **12**, 1142 (1929); **13**, 1084 (1930); **14**, 1033 (1931); Kuhn, Brockmann, *Ber.* **65**, 894 (1932); **14**, 1319 (1933); **67**, 1408 (1934); *Ann.* **516**, 95 (1935). Crystal structure: Sterling, *Acta Cryst.* **17**, 1224 (1964). Synthesis: Milas *et al., J. Am. Chem. Soc.* **72**, 4844 (1950); Karrer, Eugster, *Compt. Rend.* **250**, 1920 (1950); Inhoffen *et al., Chem. Ztg.* **74**, 285, 309 (1950); Surmatis, Ofner, *J. Org. Chem.* **26**, 1171 (1961); Rüegg *et al., Helv. Chim. Acta* **44**, 985 (1961); Bestmann *et al., Ann.* **1973**, 760; Fischli, Mayer, *Helv. Chim. Acta* **58**, 1584 (1975). Industrial mfg procedure: Isler *et al., Helv. Chim. Acta* **39**, 249 (1956); Isler *et al.,* U.S. pat. **2,917,539** (1959 to Hoffmann-La Roche). Microbial production by *Choanephora trispora:* Zajic, U.S. pats. **2,959,521-2** and **3,128,236** (1960 and 1964 to Grain Processing); Miescher, U.S. pat. **3,001,-912** (1961 to C.S.C.). *Review:* Fleming, *Selected Organic Syntheses* (John Wiley, London, 1973) pp 70-74.

Deep-purple, hexagonal prisms from benzene + methanol. Red, rhombic, almost square leaflets from petr ether. mp 183° (evacuated tube). Absorption max (chloroform): 497, 466 nm. Less sol than α-carotene. Sol in CS$_2$, benzene, chloroform. Moderately sol in ether, petr ether, oils. 100 ml hexane dissolve 109 mg at 0°. Very sparingly sol in methanol and ethanol. Practically insol in water, acids, alkalies. Dil solns are yellow. Absorbs oxygen from the air, giving rise to inactive, colorless oxidation products. *Keep tightly closed and protected from light. Store at low temp (−20°C).* Commercial crystalline β-carotene has a vitamin A activity of 1.67 million U.S.P. units per gram. The I.U. of 0.6 μg β-carotene is almost exactly equivalent to 0.3 μg vitamin A.

USE: Yellow coloring agent for foods.

THERAP CAT: Vitamin A precursor. Ultraviolet screen.

THERAP CAT (VET): Vitamin A precursor for all species except cats.

1861. γ-Carotene. $C_{40}H_{56}$; mol wt 536.85. C 89.49%, H 10.51%. A rare carotenoid. Has provitamin A activity. Best source is *Penicillium sclerotiorum:* Mase *et al., Arch. Biochem. Biophys.* **68**, 150 (1957). Also present in small proportions in many plant materials, esp fruits containing β-carotene. Chromatographic isoln from crude carotenes: Kuhn, Brockmann, *Ber.* **66**, 407 (1933); Winterstein, *Z. Physiol. Chem.* **219**, 249 (1933). Structure: Kuhn, Brockmann, *loc. cit.; Naturwiss.* **21**, 44 (1933). Synthesis: Garbers *et al., Helv. Chim. Acta* **36**, 1783 (1953); Ruegg *et al., ibid.* **44**, 985 (1961).

Probably crystallizes in polymorphous forms. Synthetic form: red plates, mp 152-153.5°. Absorption max (petr ether): 437, 462, 494 nm ($E_{1cm}^{1\%}$ 2055, 3100, 2720). Natural form: minute, deep-red prisms with bluish luster from benzene + methanol. mp 177.5°. Absorption max (chloroform): 508.5, 475, 446 nm. Somewhat less sol than β-carotene. *Store in darkness in sealed ampoules at low temps (0°C).*

THERAP CAT: Vitamin A precursor.

THERAP CAT (VET): Vitamin A precursor for all species except cats.

1862. δ-Carotene. ε,ψ-*Carotene.* $C_{40}H_{56}$; mol wt 536.85. C 89.49%, H 10.51%. Extracted from the fruit of *Gonocaryum pyriforme* Mig., *Icacinaceae:* Winterstein, *Z. Physiol. Chem.* **219**, 249 (1933). Occurs also in carrots and certain varieties of tomatoes. Isoln from tomato mutants: Porter, Murphey, *Arch. Biochem. Biophys.* **32**, 21 (1951). Structure: Kargl, Quackenbush, *ibid.* **88**, 59 (1960). Synthesis: Manchand *et al., J. Chem. Soc.* **1965**, 2019. Absolute configuration: Buchecker, Eugster, *Helv. Chim. Acta* **54**, 327 (1971).

Long orange-red needles from CS_2 + hexane + ethanol, mp 140.5°. $[\alpha]_{Cd}$ +317°; $[\alpha]_D^{25}$ +352° ±16% (hexane). Absorption spectrum: Kargl, Quackenbush, *loc. cit.*

1863. Carotol. *[3R-(3α,3aα,8aα)]-2,3,4,5,8,8a-Hexahydro-6,8a-dimethyl-3-(1-methylethyl)-3a(1H)-azulenol; 2,3,-4,5,8,8a-hexahydro-3-isopropyl-6,8a-dimethyl-3a(1H)-azulenol.* $C_{15}H_{26}O$; mol wt 222.36. C 81.02%, H 11.79%, O 7.20%. From oil of carrot seeds, *Daucus carota* L., *Umbelliferae:* Asahina, Tsukamoto, *J. Pharm. Soc. Japan* **525**, 961 (1925), *C.A.* **20**, 2845 (1926); Sorm *et al., Coll. Czech. Chem. Commun.* **16**, 47 (1951); Pigulevskii, Kovaleva, *Zh. Priklad. Khim.* **32**, 2703 (1959). Structure: Sykora *et al., Coll. Czech. Chem. Commun.* **26**, 788 (1961); Zalkow *et al., J. Org. Chem.* **26**, 981 (1961). Stereochemistry: Levisalles, Rudler, *Bull. Soc. Chim. France* **1964**, 2020; **1967**, 2059. Synthesis of (+)-form: DeBroissia *et al., ibid.* **1972**, 4314; *eidem, Chem. Commun.* **1972**, 855.

(+)-carotol

Liquid. $bp_{2.5}$ 126°. $[\alpha]_D^{20}$ +30.4°. n_D^{20} 1.4964. d^{20} 0.9624.

1864. Caroverine. *1-[2-(Diethylamino)ethyl]-3-[(4-methoxyphenyl)methyl]-2(1H)-quinoxalinone; 1-(diethylaminoethyl)-3-(p-methoxybenzyl)dihydro-2-quinoxalone; Spasmium.* $C_{22}H_{27}N_3O_2$; mol wt 365.46. C 72.30%, H 7.45%, N 11.50%, O 8.76%. Prepn: Zellner *et al.,* U.S. pat. **3,028,384** (1962 to Donau-Pharm.). Polarographic study: P. Pflegel,

Pharmazie **24**, 667 (1969). Pharmacological study: F. Hahn *et al., Arch. Int. Pharmacodyn. Ther.* **199**, 108 (1972).

Crystals from isopropyl alcohol, mp 69°. $bp_{0.01}$ 202°. Hydrochloride, dec 188°.

THERAP CAT: Antispasmodic.

1865. Caroxazone. *2-Oxo-2H-1,3-benzoxazine-3(4H)-acetamide;* 3-acetamide-4H-1,3-benzoxazin-2-one; 4H-3-methylcarboxamide-3-benzoxazine-2-one; 4H-3-carbamoylmethyl-1,3-benzoxazine-2-one; FI 6654; Timostenil. $C_{10}H_{10}N_2O_3$; mol wt 206.20. C 58.25%, H 4.89%, N 13.58%, O 23.28%. Prepn and pharmacological properties: L. Bernardi *et al., Experientia* **24**, 774 (1968); L. Bernardi, S. Coda, S. Afr. pat. **74 02,435** (1974 to Farmitalia), *C.A.* **84**, 59497 (1976). Pharmacology: G. K. Suchowsky, L. Pegrassi, *Arzneimittel-Forsch.* **19**, 643 (1969); G. K. Suchowsky *et al., Eur. J. Pharmacol.* **6**, 327 (1969). Clinical study: S. Cecchini *et al., J. Int. Med. Res.* **6**, 388 (1978).

Solid, mp 203-205°. LD_{50} in mice: 728 mg/kg orally, *RTECS* **Vol. II**, E. J. Fairchild, Ed. (1977) p 211.

THERAP CAT: Antidepressant.

1866. Carpaine. $C_{28}H_{50}N_2O_4$; mol wt 478.70. C 70.25%, H 10.53%, N 5.85%, O 13.37%. Found in *Carica papaya* L. and in *Vasconcellosia hastata* Caruel, *Caricaceae.* Isoln from papaya leaves: Greshoff, *Mededeel. uit's Lands. Plant., Buitenzorg.* **No. 7**, 5 (1890); Rapoport, Baldridge, *J. Am. Chem. Soc.* **73**, 343 (1951). Reportedly causes bradycardia, CNS depression: Kakowski, *Arch. Int. Pharm.* **15**, 84 (1905). Structure and stereochemistry: Govindachari *et al., Tetrahedron Letters* **1965**, 1907. Absolute configuration: Coke, Rice, *J. Org. Chem.* **30**, 3420 (1965). Review of structural studies: Govindachari, *J. Indian Chem. Soc.* **45**, 945 (1968).

Monoclinic prisms from acetone, mp 119-120°. Sublimes at 120° under 0.05 mm pressure. $[\alpha]_D^{12}$ +24.7° (c = 1.07 in ethanol). Slightly sol in water. Sol in most organic solvents except petr ether.

1867. Carpetimycins. Carbapenem antibiotics related to thienamycin and olivanic acids, *q.q.v.* Carpetimycins A, B, C and D are known. Prodn of A and B by *Streptomyces* strain No. C-19393 (IFO 13866, ATCC 31486) now named *Streptomyces griseus* subsp. *cryophilus,* and antibacterial properties: **Neth. pat. Appl. 8,000,628;** A. Imada *et al.,* U.S. pat. **4,518,529** (1980, 1985 to Takeda); A. Imada *et al., J. Antibiot.* **33**, 1417 (1980); by *Streptomyces* strain KC-6643

(FERM-P 4467, ATCC 31493): M. Nakayama *et al., ibid.*
1388; M. Nakayama *et al.*, **Eur.** pat. **Appl. 47,611;** *eidem,*
U.S. pat. **4,415,497** (1982, 1983 to Kowa). Structure and
absolute configuration: S. Harada *et al., J. Antibiot.* **33,** 1425
(1980); M. Nakayama *et al., ibid.* **34,** 818 (1981). Mode of
action study: Y. Nozaki *et al., ibid.* 206. Antimicrobial and
β-lactamase inhibitory activity of A and B: F. Kobayashi *et
al., Antimicrob. Ag. Chemother.* **21,** 536 (1982). Total synthesis of (±)-carpetimycin A: H. Natsugari *et al., J. Chem.
Soc. Perkin Trans. I* **1983,** 403; M. Ihara *et al., Heterocycles*
20, 2182 (1983); J. D. Buynak, M. N. Rao, *J. Org. Chem.* **51,**
1571 (1986). Synthesis of (−)-carpetimycin A: T. Iimori *et
al., J. Am. Chem. Soc.* **105,** 1659 (1983); M. Aratani *et al.,
Tetrahedron Letters* **26,** 223 (1985); M. Shibasaki *et al., ibid.*
2217. Fermentation, isoln, physicochemical properties and
structure of carpetimycins C and D: M. Nakayama *et al., J.
Antibiot.* **36,** 943 (1983).

carpetimycin A: R = H
B: R = SO₃H

Carpetimycin A, C₁₄H₁₈N₂O₆S, *[5R-[3[R*(E)],5α,6β]]-3-
[[2-(acetylamino)ethenyl]sulfinyl]-6-(1-hydroxy-1-methyl-
ethyl)-7-oxo-1-azabicyclo[3.2.0]hept-2-ene-2-carboxylic
acid,* antibiotic Ab 651, antibiotic C 19393 H₂, antibiotic KA
6643-A, C-19393 H₂. Colorless solid, mp > 145° (dec). $[\alpha]_D^{24}$
−27° (c = 1.7 in water). uv max (water): 240, 288 nm ($E_{1cm}^{1\%}$
369, 300) (Nakamyama, 1980).
Carpetimycin A sodium salt, C₁₄H₁₇N₂NaO₆S. $[\alpha]_D^{24}$
−141° (c = 0.54 in H₂O). uv max (water): 244.5, 290 nm
(ε 14300, 11000) (Harada).
Carpetimycin B, C₁₄H₁₈N₂O₉S₂, *[5R-[3[R*(E)],5α,6β]]-3-
[[2-(acetylamino)ethenyl]sulfinyl]-6-[1-methyl-1-(sulfooxy)-
ethyl]-7-oxo-1-azabicyclo[3.2.0]hept-2-ene-2-carboxylic
acid,* antibiotic C-19393 S₂, antibiotic KA-6643-B, C-19393
S₂. Colorless solid melting above 130° (dec). $[\alpha]_D^{24}$ −145°
(c = 1 in water). uv max (water): 240, 285 nm ($E_{1cm}^{1\%}$ 357,
305) (Nakayama, 1980).
Carpetimycin B disodium salt, C₁₄H₁₆N₂Na₂O₉S₂. $[\alpha]_D^{24}$
−152° (c = 0.5 in water). uv max (H₂O): 243.5, 288 nm
(ε 15900, 13200) (Harada).

1868. Carphenazine. *1-[10-[3-[4-(2-Hydroxyethyl)-1-
piperazinyl]propyl]-10H-phenothiazin-2-yl]-1-propanone;*
10-[3-[4-(2-hydroxyethyl)-1-piperazinyl]propyl]-3-propi-
onylphenothiazine; 2-propionyl-10-[γ-(N′-β-hydroxyethyl-
piperazino)propyl]phenothiazine; Proketazine. C₂₄H₃₁N₃-
O₂S; mol wt 425.60. C 67.73%, H 7.34%, N 9.87%, O 7.52%,
S 7.54%. Prepn: Tislow *et al.*, **U.S.** pat. **3,023,146** (1962 to
Am. Home Products).

Dimaleate, C₂₄H₃₁N₃O₂S.2C₄H₄O₄, yellow crystals from
isopropanol, mp 175-177°.
THERAP CAT: Antipsychotic.

1869. Carpipramine. *1′-[3-(10,11-Dihydro-5H-dibenz-
[b,f]azepin-5-yl)propyl]-[1,4′-bipiperidine]-4′-carboxamide;*
5-[3-(4-piperidino-4-carbamoylpiperidino)propyl]-10,11-
dihydro-5H-dibenz[b,f]azepine; carbadipimidine; BAY b
4343 b. C₂₈H₃₈N₄O; mol wt 446.64. C 75.30%, H 8.57%, N
12.54%, O 3.58%. Prepn: Nakanishi, Munakata, **Japan.** pat.
6752('66) (to Yoshitomi); Nakanishi *et al., J. Med. Chem.*
13, 644 (1970). Pharmacology: Nakanishi *et al., Arzneimit-*

tel-Forsch. **18,** 1435 (1968). Metabolism: *eidem, J. Pharm.
Soc. Japan* **90,** 204 (1970).

Dihydrochloride monohydrate, C₂₈H₄₀Cl₂N₄O.H₂O, *Defekton, Prazinil.* Crystals, mp 260°. Sparingly sol in water.
LD₅₀ in mice, rats: 28.2, 37.0 mg/kg i.v.; 136.0, 76.0 mg/kg
i.p.; 2180, 1025 mg/kg orally.
THERAP CAT: Psychotropic.

1870. Carprofen. *6-Chloro-α-methyl-9H-carbazole-2-
acetic acid;* C 5720; Ro 20-5720/000; Imadyl; Rimadyl.
C₁₅H₁₂ClNO₂; mol wt 273.73. C 65.82%, H 4.42%, Cl
12.95%, N 5.12%, O 11.69%. Prepn: L. Berger, A. J. Cor-
raz, **Ger.** pat. **2,337,340;** *eidem,* **U.S.** pat. **3,896,145** (1974,
1975 both to Hoffmann-La Roche). Pharmacology: L. O.
Randall, H. Baruth, *Arch. Int. Pharmacodyn.* **220,** 94 (1976).
Stereospecific assay and disposition: J. M. Kemmerer *et al.,
J. Pharm. Sci.* **68,** 1274 (1979). Metabolism: F. Rubio *et al.,
ibid.* **69,** 1245 (1980). Pharmacokinetic and clinical studies
in gout: T. F. Yu, J. Perel, *J. Clin. Pharmacol.* **20,** 347
(1980). Use in treatment of rheumatoid arthritis: S. Jalava,
Scand. J. Rheumatol. **1983,** Suppl. 48, 5-12. Review of
pharmacology and clinical efficacy: W. M. O′Brien, G. F.
Bagby, *Pharmacotherapy* **7,** 16 (1987).

Cryst from chloroform, mp 197-198°. LD₅₀ orally in mice:
400 mg/kg (Berger, Corraz).
THERAP CAT: Anti-inflammatory.

1871. Carpronium Chloride. *3-Carboxy-N,N,N-trimeth-
yl-1-propanaminium chloride methyl ester; (3-carboxyprop-
yl)trimethylammonium chloride methyl ester;* γ-butyrobetaine chloride methyl ester; methyl N-trimethyl-γ-aminobutyrate chloride; Actinamin; Furozin. C₈H₁₈ClNO₂; mol wt
195.70. C 49.10%, H 9.27%, Cl 18.12%, N 7.16%, O 16.35%.
[CH₃OOCCH₂CH₂CH₂N(CH₃)₃]⁺Cl⁻. Prepn: Strack,
Försterling, *Ber.* **76B,** 14 (1943).
Cube-like, hygroscopic crystals from warm acetone, mp
126°. Very sol in water, methanol, ethanol; sol in acetone.
Practically insol in ether.
THERAP CAT: Cholinergic.

1872. Carrageenan. Carrageen; carrageenin. Structural
polysaccharide of the red seaweed *(Rhodophyceae).* Chief
sources are *Chondrus crispus* (L.) Stackhouse, and *Gigartina
stellata* (G. mammillosa) (Goodenough and Woodward) J.
Aghardh, *Gigartinaceae,* found most abundantly in North
Atlantic coastal regions from Norway to North Africa. The
name carrageenan is derived from the Irish coastal town of
Carragheen. Structure consists of alternating copolymers of
β-(1 → 3)-D-galactose and (1 → 4)-3,6-anhydro-D- or L-
galactose. Several members of the family are known; they
differ in the amount of sulfate ester and/or other substituent
groups they carry. *Iota*-carrageenan is the most highly sulfated member of the family; *kappa*-carrageenan contains
only one sulfate group in each disaccharide repeating unit.
Double helix conformation for both *iota*- and *kappa*-carra-
geenans has been determined: N. S. Anderson *et al., J. Mol.
Biol.* **45,** 85 (1969); S. Arnott, *ibid.* **90,** 253, 269 (1974).
Secondary and tertiary structure in solutions and gels: D.
A. Rees, E. J. Welsh, *Angew. Chem. Int. Ed.* **16,** 214 (1977).
Use in assay for anti-inflammatory drugs: C. A. Winter *et
al., Proc. Soc. Exp. Biol. Med.* **111,** 544 (1962). Reviews and
bibliographies: Towle in *Industrial Gums,* R. L. Whistler,
Ed. (Academic Press, New York, 2nd ed., 1973) pp 83-114;

I. W. Cottrell, J. K. Baird, "Gums" in Kirk-Othmer *Encyclopedia of Chemical Technology* vol. 12 (Wiley-Interscience, New York, 3rd ed., 1980) pp 45-66.

Completely soluble in hot water. Dilute aq solns are viscous, the viscosity increasing almost logarithmically with increased concns. Aq solns gel on addition of hydrophobic cations such as potassium and ammonium. Sol in anhydrous hydrazine, sparingly sol in formamide and methyl sulfoxide. Swells but does not dissolve in *N,N*-dimethylformamide. Generally insol in oils and org solvents. Precipitates proteins when pH of the soln is below the isoelectric point of the protein; at low concns causes agglomeration of milk proteins. Undergoes depolymerization in acid soln. Solutions at pH 9 are most stable.

USE: Gelling, emulsifying, and stabilizing agent and viscosity builder in foods and non-foods, but esp in milk or water systems. Demulcent. To induce experimental inflammation in laboratory animals.

1873. Carsalam. *2H-1,3-Benzoxazine-2,4(3H)-dione;* carbonylsalicylamide; 2,4-dioxodihydro-5,6-benzo-1,3-oxazine; CSA; Beaprine. $C_8H_5NO_3$; mol wt 163.13. C 58.90%, H 3.09%, N 8.59%, O 29.42%. Prepn: Shapiro *et al., J. Am. Chem. Soc.* **79**, 2811 (1957); D. N. Dhar, A. K. Bag, *Indian J. Chem.* **21B**, 366 (1982). Crystal and molecular structure: S. Kashino *et al., Acta Crystallogr.* **B34**, 2191 (1978).

Crystals from butanol, mp 229-230°.
THERAP CAT: Analgesic.

1874. Cartap. *Carbamothioc acid, S,S'-[2-(dimethylamino)-1,3-propanediyl] ester;* 1,3-bis(carbamoylthio)-2-*N,N*-(dimethylamino)propane. $C_7H_{15}N_3O_2S_2$; mol wt 205.28. C 40.96%, H 7.36%, N 20.47%, O 15.59%, S 15.62%. Insecticide modeled on a toxin isolated from the marine annelid *Lumbrineris heteropoda*. Prepn, insecticidal activity and toxicity: K. Konishi *et al.,* Fr. pat. **1,452,338** corresp to U.S. pat. **3,332,943** (1968, 1968 both to Takeda); K. Konishi, *Agr. Biol. Chem.* **34**, 935 (1970). Crystal structure of hydrochloride: C. J. Cheer, F. J. Pickles, *J. Chem. Soc. Perkin Trans. II* **1980**, 1805. Action against army worms: A. Pigatti *et al., Biologico (Sao Paulo)* **47**, 169 (1981); against chrysanthemum leaf miners: B. Gumey, N. W. Hussey, *Plant Pathol.* **23**, 127 (1974); against rice leaf beetles: S. Koyama *et al., Nippon Noyaku Gakkaishi* **8**, 183 (1983), *C.A.* **100**, 98291b (1984). Effect on soil enzyme activity: T. Endo *et al., ibid.* **7**, 101 (1982), *C.A.* **97**, 122045s (1982). Toxicity to carp: Y. Hashimoto, J. Fukami, *Bochu-Kagaku* **34**, 63 (1969), *C.A.* **71**, 90207r (1969); to aquatic organisms: S. Lakota *et al., Acta Hydrobiol.* **23**, 183 (1981). Chronic toxicity study in mice: Y. Tsubura *et al., Nara Igaku Zasshi* **26**, 368 (1975), *C.A.* **84**, 174837c (1976). *Review:* Y. Kono, *Japan. Pestic. Inf.* **34**, 22 (1978).

Colorless prisms from ethyl acetate, mp 130.5-131° (dec). Monohydrochloride, $C_7H_{16}ClN_3O_2S_2$, NTD 2, Padan. Colorless rods from methanol, mp 176° (dec). Sol in water. LD_{50} in mice, rats (mg/kg): 165, 250 orally (Konishi).
USE: Insecticide.

1875. Carteolol. *5-[3-[(1,1-Dimethylethyl)amino]-2-hydroxypropoxy]-3,4-dihydro-2(1H)-quinolinone;* 5-[3-(*tert*-butylamino)-2-hydroxypropoxy]-3,4-dihydrocarbostyril. $C_{16}H_{24}N_2O_3$; mol wt 292.38. C 65.73%, H 8.27%, N 9.58%, O 16.42%. β-Adrenergic blocker. Prepn: K. Nakagawa *et al.,* Ger. pat. **2,302,027** corresp to U.S. pat. **3,910,924** (1973, 1975 both to Otsuka); *eidem, J. Med. Chem.* **17**, 529 (1974). Pharmacological evaluation in rats: S. Morita *et al., Bio-*

chem. Pharmacol. **25**, 1836 (1976). HPLC determn in plasma and urine: S. Y. Chu, *J. Pharm. Sci.* **67**, 1623 (1978). *In vitro* study: S. Chiba, *Arzneimittel-Forsch.* **29**, 895 (1979). Carcinogenicity study: M. Kurosomi *et al., Farmaco Ed. Prat.* **34**, 202 (1979). Reproduction and toxicity studies: N. Tanaka *et al., J. Toxicol. Sci.* **4**, 47 (1979); M. Tamagawa *et al., ibid.* 59. Acute and subacute toxicity studies: N. Tanaka *et al., Oyo Yakuri* **11**, 165, 173 (1976), *C.A.* **88**, 115336y, 115337z (1978). Series of articles on pharmacology, toxicology and clinical studies: *Arzneimittel-Forsch.* **33**, 277-345 (1983). Acute toxicity data: W. Lang, *ibid.* 290.

Hydrochloride, $C_{16}H_{25}ClN_2O_3$, *Abbott 43326, OPC 1085, Arteoptic, Caltidren, Carteol, Cartrol, Endak, Mikelan, Optipress, Tenalet, Tenalin, Teoptic.* Crystals from ethanol, mp 278°. LD_{50} in male mice, rats (mg/kg): 810, 1380 orally; 54.5, 158 i.v.; 380, 400 i.p. (Lang).
THERAP CAT: Antihypertensive, antianginal, antiarrhythmic, antiglaucoma agent.

1876. Carthamin. *6-β-D-Glucopyranosyl-2-[[3-β-D-glucopyranosyl-2,3,4-trihydroxy-5-[3-(4-hydroxyphenyl)-1-oxo-2-propenyl]-6-oxo-1,4-cyclohexadien-1-yl]methylene]-5,6-dihydroxy-4-[3-(4-hydroxyphenyl)-1-oxo-2-propenyl]-4-cyclohexene-1,3-dione;* carthamic acid; safflor carmine; safflor red; C.I. Natural Red 26; C.I. 75140. $C_{43}H_{42}O_{22}$; mol wt 910.81. C 56.70%, H 4.65%, O 38.65%. Coloring principle from *Carthamus tinctorius* L., *Compositae* (safflower): Kametaka, Perkin, *J. Chem. Soc.* **97**, 1415 (1910). Extraction: Wada, *Japan.* pat. **8943** (1955). Structure: Kuroda, *J. Chem. Soc.* **1930**, 752, 765. Revised structure: H. Obara, J. Onodera, *Chem. Letters* **1979**, 201. Biogenesis: Shimokoriyama, Hattori, *Arch. Biochem. Biophys.* **54**, 93 (1955). *See also: Colour Index* vol. 4 (3rd ed., 1971) p 4625.

Dark red, granular powder with green luster. Slightly sol in water; sol in dil alkali carbonates, alcohol; practically insol in ether. Solns rapidly dec.
USE: As dye.

1877. Carthamus. Safflower; African saffron; thistle saffron; American saffron; false saffron; bastard saffron; Dyer's saffron. Florets of *Carthamus tinctorius* L. *Compositae. Habit.* Levant; Orient; cultivated extensively in Europe and America. *Constit.* Carthamin, safflor yellow.
USE: In dyeing; surrogate for Spanish saffron; coloring butter, liqueurs, confectionery, cosmetics.
THERAP CAT: Dietary supplement.

1878. Carticaine. *4-Methyl-3-[[1-oxo-2-(propylamino)-propyl]amino]-2-thiophenecarboxylic acid methyl ester;* 3-propylamino-α-propionylamino-2-carbomethoxy-4-methylthiophene; methyl 4-methyl-3-(2-propylaminopropionamido)thiophene-2-carboxylate. $C_{13}H_{20}N_2O_3S$; mol wt 284.37. C 54.91%, H 7.09%, N 9.85%, O 16.88%, S 11.27%. Prepn: H. Ruschig *et al.,* S. Afr. pat. **68 04,265** corresp to U.S. pat. **3,855,243** (1969, 1974, both to Farbwerke Hoechst). Pharmacological studies: R. Muschaweck, R. Rippel, *Prakt. Anaesth.* **9**, 135 (1974); H. Hofer *et al., ibid.* 157; A. Den Hertog, *Eur. J. Pharmacol.* **26**, 175 (1974). Toxicity studies: C. Baeder *et al., Prakt. Anaesth.* **9**, 147 (1974). Mechanism of action: U. Borchard, H. Drouin, *Eur. J. Pharmacol.* **62**, 73 (1980). *Review:* H. Schneider, *Schweiz. Monatsschr. Zahnheilkd.* **86**, 1188-1194 (1976).

bp$_{0.3}$ 162-167°.
Hydrochloride, C$_{13}$H$_{21}$ClN$_2$O$_3$S, *Hoe 045, Hoe 40045, Ul-tracain*. White, fine crystals, mp 177-178°. LD$_{50}$ i.v. in mice: 37 mg/kg, R. Muschaweck, R. Rippel, *loc. cit.*
THERAP CAT: Local anesthetic.

1879. Carubicin. *8-Acetyl-10-[(3-amino-2,3,6-trideoxy-α-L-lyxo-hexopyranosyl)oxy]-7,8,9,10-tetrahydro-1,6,8,11-tetrahydroxy-5,12-naphthacenedione; (1S,3S)-3-acetyl-1,2,-3,4,6,11-hexahydro-3,5,10,12-tetrahydroxy-6,11-dioxo-1-naphthacenyl 3-amino-2,3,6-trideoxy-α-L-lyxo-hexopyra-noside; 4-O-demethyldaunorubicin; carminomycin; carmi-nomycin I; karminomycin*. C$_{26}$H$_{27}$NO$_{10}$; mol wt 513.52. C 60.81%, H 5.30%, N 2.73%, O 31.16%. Anthracycline anti-tumor antibiotic, related to daunorubicin and doxorubicin, *q.q.v.* Isoln from *Actinomadura carminata:* G. F. Gauze *et al., Antibiotiki* **18,** 675 (1973); M. G. Brazhnikova *et al., ibid.* 678. Antitumor activity: V. A. Shorin *et al., ibid.* 681. Physico-chemical characteristics, structure: M. G. Brazhni-kova *et al., J. Antibiot.* **27,** 254 (1974). Pharmacokinetics, pharmacodynamics, toxicity study: L. E. Goldberg *et al., Antibiotiki* **19,** 57 (1974). Production: M. G. Brazhnikova *et al.,* U.S.S.R. pat. **508076** (1976 to Inst. Antibiot. Res., USSR), *C.A.* **86,** 15215 (1977). Stereochemistry: *eidem, J. Antibiot.* **29,** 469 (1976). Synthesis from daunomycinone: G. Cassinelli *et al., ibid.* **31,** 178 (1978). Molecular pharmacol-ogy: V. H. DuVernay *et al., Cancer Res.* **40,** 387 (1980). Analysis in human serum: S. E. Fandrich, K. A. Pittman, *J. Chromatog.* **223,** 155 (1981). Early clinical studies: L.H. Baker *et al., Cancer Treat. Rep.* **63,** 899 (1979). Embryotoxi-city and teratogenicity study: I. Damjanov, A. Celluzzi, *Res. Commun. Chem. Pathol. Pharmacol.* **28,** 497 (1980).

Hydrochloride, C$_{26}$H$_{28}$ClNO$_{10}$, *NSC-180024*. Cryst from ethanol/benzene. [α]$_D^{20}$ +289°. uv max (ethanol): 236, 255, 462, 478, 492 (E$_{1cm}^{1\%}$ 300), 510, 525 nm. pKa$_1$ 8.00; pKa$_2$ 10.16. Sol in water, methanol. Practically insol in other organic solvents. LD$_{50}$ in mice (mg/kg): 7.3 orally; 1.3 i.v.; 3.7 s.c., L. E. Goldberg *et al., loc. cit.*
THERAP CAT: Antineoplastic.

1880. Carumonam. *[2S-[2α,3α(Z)]]-[[[2-[[2-[[(Amino-carbonyl)oxy]methyl]-4-oxo-1-sulfo-3-azetidinyl]amino]-1-(2-amino-4-thiazolyl)-2-oxoethylidene]amino]oxy]acetic acid; (Z)-[[[(2-amino-4-thiazolyl)[[(2S,3S)-2-(hydroxy-methyl)-4-oxo-1-sulfo-3-azetidinyl]carbamoyl]methylene]-amino]oxy]acetic acid carbamate (ester); (3S,4S)-cis-3-[2-(2-amino-4-thiazolyl)-2-(Z)-carboxymethoxyiminoacet-amido]-4-carbamoyloxymethyl-2-azetidinone-1-sulfonic acid*. C$_{12}$H$_{14}$N$_6$O$_{10}$S$_2$; mol wt 466.40. C 30.90%, H 3.03%, N 18.02%, O 34.30%, S 13.75%. Synthetic monocyclic β-lactam (monobactam) antibiotic. Prepn: S. Kishimoto *et al.,* Eur. pat. Appl. **93,376;** T. Matsuo *et al.,* U.S. pat. **4,572,801** (1983, 1986 both to Takeda); M. Sendai *et al., J. Antibiot.* **38,** 346 (1985). Alternate synthesis: P. S. Man-chard *et al., J. Org. Chem.* **53,** 5507 (1988). Comparative *in vitro* antimicrobial activity: R. J. Fass, V. L. Helsel, *Antimi-crob. Ag. Chemother.* **28,** 834 (1985); B. R. Smith *et al., ibid.* **29,** 346 (1986); I. M. Hoepelman *et al., Chemotherapy (Ba-*

sel) **33,** 103 (1987). β-Lactamase stability: R. L. Then, *ibid.* **30,** 398 (1984). Pharmacokinetics in humans: E. Weide-kamm *et al., Antimicrob. Ag. Chemother.* **26,** 898 (1984); C. A. M. McNulty *et al., ibid.* **28,** 425 (1985).

Colorless powder. [α]$_D^{26}$ −45° (c = 1 in DMSO). Disodium salt, C$_{12}$H$_{12}$N$_6$Na$_2$O$_{10}$S$_2$, *AMA-1080, Ro 17-2301, Amasulin.*
THERAP CAT: Antibacterial.

1881. Carvacrol. *2-Methyl-5-(1-methylethyl)phenol;* 2-*p*-cymenol; 2-hydroxy-*p*-cymene; isopropyl-*o*-cresol; iso-thymol. C$_{10}$H$_{14}$O; mol wt 150.21. C 79.95%, H 9.40%, O 10.65%. Found in oil of origanum, thyme, marjoram, sum-mer savory: E. Guenther, *The Essential Oils* vol. **2** (Van Nostrand, New York, 1949) p 503; Carpenter, Easter, *J. Org. Chem.* **20,** 401 (1955). Prepn by chlorination of α-pinene with *tert*-butyl hypochlorite: J. Am. Chem. Soc. **72,** 2381 (1950); from 2-bromo-*p*-cymol: Strubell, Baumgartel, *Arch. Pharm.* **291,** 66 (1958). Toxicity data: Kochmann, *Arch. Exp. Pathol. Pharmakol.* **161,** 196 (1931).

Liquid; thymol odor. d$_4^{20}$ 0.976; d$_{25}^{25}$ 0.9751. bp$_{760}$ 237-238°; bp$_{18}$ 118-122°; bp$_3$ 93°. mp ~0°. n$_D^{20}$ 1.52295. uv max (95% ethanol): 277.5 nm (log ε 3.262). Volatile with steam. Practically insol in water. Freely soluble in alc or ether. LD orally in rabbits: 100 mg/kg (Kochmann).
USE: As disinfectant; in organic syntheses.
THERAP CAT: Has been used as anti-infective; anthelmintic (Nematodes).

1882. Carvedilol. *1-(9H-Carbazol-4-yloxy)-3-[[2-(2-methoxyphenoxy)ethyl]amino]-2-propanol; BM-14190; DQ-2466.* C$_{24}$H$_{26}$N$_2$O$_4$; mol wt 406.48. C 70.92%, H 6.45%, N 6.89%, O 15.74%. Nonselective β-adrenergic blocker with vasodilating activity. Prepn: F. Wiedemann *et al.,* Ger. pat. **2,815,926;** *eidem,* U.S. pat. **4,503,067** (1979, 1985 both to Boehringer Mannheim). Pharmacology: G. Sponer *et al., J. Cardiovasc. Pharmacol.* **9,** 317 (1987). Clinical pharmacolo-gy: E. von Möllendorff *et al., Clin. Pharm. Ther.* **39,** 677 (1986); L. X. Cubeddu *et al., ibid.* **41,** 31 (1987). HPLC determn in biological fluids: K. Reiff, *J. Chromatog.* **413,** 355 (1987). Hemodynamic effects as compared with pro-pranolol, *q.v.,* in hypertension: R. Eggertsen *et al., J. Hyper-tension* **2,** 529 (1984). Clinical trial in hypertension: M. E. Heber *et al., Am. J. Cardiol.* **59,** 400 (1987). Clinical efficacy as antianginal: J. C. Kaski *et al., ibid.* **56,** 35 (1985); E. A. Rodrigues *et al., ibid.* **58,** 916 (1986).

Colorless crystals from ethyl acetate, mp 114-115°.
THERAP CAT: Antihypertensive; antianginal.

1883. Carvone. *2-Methyl-5-(1-methylethenyl)-2-cyclo-hexene-1-one; p-mentha-6,8-dien-2-one; 1-methyl-4-iso-*

propenyl-Δ^6-cyclohexen-2-one; carvol. $C_{10}H_{14}O$; mol wt 150.21. C 79.95%, H 9.40%, O 10.65%. *d*-Carvone is found in caraway seed and dill seed oils: Schweizer, *J. Prakt. Chem.* **24**, 257 (1841); Gladstone, *J. Chem. Soc.* **25**, 1 (1872). Isoln of *d*-carvone from mandarin peel oil (*Citrus reticulata* Blanco, *Rutaceae):* Kugler, Kováts, *Helv. Chim. Acta* **46**, 1480 (1963). *l*-Carvone is found in spearmint and kuromoji oils: Kwaenick, *Ber.* **24**, 82 (1891). *dl*-Carvone is found in gingergrass oil: Walbaum, Hüthig, *J. Prakt. Chem.* **71**, 459 (1905). Preparation from α-pinene: Booth, Klein, U.S. pat. **2,796,428** (1957 to Glidden). Structure: Wagner, *Ber.* **27**, 2270 (1894); Tiemann, Semmler, *ibid.* **28**, 2147 (1895). Synthesis of *l*-carvone: Royals, Horne, *J. Am. Chem. Soc.* **73**, 5856 (1951); Shono *et al., Chem. Letters* **1975**, 4330. Synthesis of *dl*-carvone: Suga, *Bull. Chem. Soc. Japan* **31**, 569 (1958); I. Fleming, I. Patterson, *Synthesis* **1979**, 736. *Review:* J. L. Simonsen, *The Terpenes* vol. I (University Press, Cambridge, 2nd ed., 1947) pp 394-408; B. Singaram, J. Verghese, *Perfum. Flavor.* **2**, 47 (1977).

d-Form, liquid, bp$_{755}$ 230°; bp$_{5-6}$ 91°. d$_4^{20}$ 0.965. n$_D^{20}$ 1.4989. $[\alpha]_D^{20}$ +61.2°.
l-Form, liquid, bp$_{763}$ 230-231°. d$_{15}^{15}$ 0.9652. n$_D^{20}$ 1.4988. $[\alpha]_D^{20}$ −62.46°.
dl-Form, liquid, bp 230-231°. d$_{15}^{15}$ 0.9645. n$_D^{20}$ 1.5003. Practically insol in water, misc with alc. LD$_{50}$ orally in rats: 1640 mg/kg, P. M. Jenner *et al., Food Cosmet. Toxicol.* **2**, 327 (1964).

USE: As oil of caraway; also for flavoring liqueurs; in perfumery and soaps.

THERAP CAT: Carminative.

1884. Caryophyllene. *4,11,11-Trimethyl-8-methylenebicyclo[7.2.0]undec-4-ene; β-caryophyllene.* $C_{15}H_{24}$; mol wt 204.36. C 88.16%, H 11.84%. Sesquiterpenoid occurring in many essential oils and especially in clove oil, the oils from stems and flowers of *Syzygium aromaticum* (L.) Merrill & Perry (*Jambrosa caryophyllus* Niedenzu; *Eugenia caryophyllata* Thunb.), *Myrtaceae.* Occurs in nature as a mixture with isocaryophyllene and α-caryophyllene (humulene, *q.v.*). Isolation of mixture: Schreiner, Kremers, *Pharm. Arch.* **2**, 273, 293 (1899). Existence of isomers: Deussen, *Ann.* **356**, 1 (1907). Structure: Aebi *et al., J. Chem. Soc.* **1953**, 3124; Ramage, Whitehead, *ibid.* **1954**, 4336. Abs config: Barton, Nickon, *ibid.* **4665**. Total synthesis of *dl*-caryophyllene and *dl*-isocaryophyllene: Corey *et al., J. Am. Chem. Soc.* **86**, 485 (1964). Prepn of isocaryophyllene by rearrangement of caryophyllene: Rachlin, *Ger.* pat. **2,044,018** (1971 to I.F.F.), *C.A.* **75**, 49364j (1971). *Reviews:* Simonsen, *The Terpenes* vol. III (University Press, Cambridge, 1952) pp 39-71; Barton, de Mayo, *Quart. Rev. (London)* **11**, 197 (1957); Halsall, *ibid.* **16**, 101 (1962).

caryophyllene   isocaryophyllene

Liquid. Has a terpene odor about midway between odor of cloves and turpentine. bp$_{14}$ 129-130°; bp$_{9.7}$ 118-119°. $[\alpha]_D^{15}$ −5.2°. n$_D^{17}$ 1.5009; n$_D^{15}$ 1.5030. d$_4^{17}$ 0.9052.
Dihydrochloride, $C_{15}H_{24}\cdot2HCl$, mp 69-70°.
Nitrosochloride, $C_{15}H_{24}\cdot ClNO$. Crystals, dec 159°. $[\alpha]_D^{17}$ −98.07°.
Isocaryophyllene, γ-caryophyllene. Liquid. bp$_{19}$ 130°;

bp$_{14.5}$ 125-125.5°. $[\alpha]_D^{19}$ −26.17°. n$_D^{19}$ 1.4966. d^{19} 0.8995. When treated with hydrogen chloride gives β-caryophyllene dihydrochloride.
Nitrosochloride, $C_{15}H_{24}\cdot ClNO$, exists in two modifications: dec 122°, $[\alpha]_D^{20.5}$ +14.71°, and dec 146°, $[\alpha]_D^{18}$ −33.69°.

USE: In perfumery.

1885. Carzenide. *p-Sulfamoylbenzoic acid; p-sulfamylbenzoic acid; p-carboxybenzenesulfonamide; p-sulfonamidobenzoic acid; benzoic acid p-sulfamide;* Dirnate. $C_7H_7NO_4S$; mol wt 201.19. C 41.79%, H 3.51%, N 6.96%, O 31.81%, S 15.94%. May be obtained from crude saccharin: Fahlberg, *Ger.* pat. **64,624** (1891); from *p*-sulfamoyltoluene: Noyes, *Am. Chem. J.* **7**, 147 (1885); Amazu *et al., Japan.* pat. **180,-511** (1949 to Nissan). A metabolite of homosulfanilamide (*q.v.*): Kakemi *et al., J. Pharm. Soc. Japan* **82**, 1582 (1962). Has some inhibitory effect on carbonic anhydrase: Tanimukai *et al., Biochem. Pharmacol.* **14**, 961 (1965).

Flat, shiny prisms from water, dec 280°. K 3.05×10^{-4} at 18°. Practically insol in cold water, ether, benzene. Freely sol in alcohol.
Sodium salt, $C_7H_6NNaO_4S$, crystals. Neutral. Soly in water: ~20 g/100 ml.

1886. Carzinophilin. Cardinophyllin. Antitumor antibiotic from *Streptomyces sahachiroi.* Isoln: T. Hata *et al., J. Antibiot.* **7A**, 107 (1954); Hata, Sano, **Japan.** pat. **7590('56)** (to Kitasato Res. Inst. & Kyowa); **Brit.** pat. **777,287** (1957 to Kitasato Res. Inst.). Isoln of active component, carzinophilin A: H. Kamada *et al., J. Antibiot.* **8A**, 187 (1955). Purification: H. Kamada *et al.,* U.S. pat. **3,044,935** (1962 to Kyowa). Clinical studies: F. Koga, *J. Antibiot.* **7A**, 275 (1954); N. Shimada *et al., ibid.* **8A**, 67 (1955). Mechanism of action: J. W. Lown, K. C. Majumdar, *Can. J. Biochem.* **55**, 630 (1977). Structure studies of carzinophilin A: M. Onda *et al., J. Antibiot.* **22**, 42 (1969); *eidem, Chem Pharm. Bull.* **19**, 2013 (1971). Total structure of carzinophilin A: J. W. Lown, C. C. Hanstock, *J. Am. Chem. Soc.* **104**, 3214 (1982). Synthetic studies on carzinophilin A: M. Shibuya, *Tetrahedron Letters* **24**, 1175 (1983).

carzinophilin A

Crystals, mp 207° (dec). uv max (water): 217, 290 nm (E$_{1\%}^{1cm}$ 476, 190). Sol in water, alcohols, acetone, butyl acetate, chloroform, benzol; practically insol in petr ether. Activity is influenced by the change of pH, being most stable at pH 7.0, relatively stable at pH 8-9 and pH 6; loses biological activity quickly when pH is reduced to <5.0. Activity also influenced by temp change: heating in boiling water for 15 min at pH 7.0 causes activity to drop.
Carzinophilin A, $C_{50}H_{58}N_5O_{18}$, *CZ.* Crystals, mp 217-222° (dec). $[\alpha]_D^{28}$ +57.8.° in chloroform. uv max (methanol): 218, 250, 283 nm. Sol in acetone, chloroform, ethyl acetate, butyl acetate, benzene, dioxane, dil alk water; slightly sol in methanol, ethanol, ether, carbon tetrachloride, petr ether,

water. LD_{50} i.v. in mice: 150 γ/kg, H. Kamada et al., loc. cit. (1955).

THERAP CAT: Antineoplastic.

1887. Casanthranol. Cantralax; Peristim. Complex mixture of the purified anthranol glycosides derived from *Cascara sagrada (q.v.)*, practically devoid of free anthraquinones. Two active fractions have been identified as *casanthranol A* and *casanthranol B*. Upon oxidative hydrolysis, casanthranol A yields emodin, while casanthranol B under similar treatment yields the aglycone, aloe-emodin, and traces of chrysophanic acid. Prepn: Lee, Berger, U.S. pat. **2,552,896** (1951 to Hoffmann-La Roche).

Component of *Peri-Colace, Casakol*.

THERAP CAT: Cathartic.

1888. Cascara Amara. Honduras bark. Bark of *Picramnia antidesma* Sw., *Simaroubaceae* and related species. *Habit*. West Indies, Mexico.

1889. Cascara Sagrada. Sacred bark; Chittem bark; Chittim bark; Purshiana bark; Persian bark; bearberry bark; bearwood. Dried bark of *Rhamnus purshiana* DC., *Rhamnaceae*, from which a naturally occurring cathartic is extracted. *Habit*. Northern Idaho, west to Northern California. The cathartic properties are primarily due to the presence of *cascarosides*, anthraglycosides which are related to glycosides found in aloe *(q.v.)* and buckthorn. *See* A. Y. Leung, *Drug Cosmet. Ind.* **121**, 42 (December, 1977). Other constituents are aloins (C-glycosides), O-glycosides, and free anthraquinones: *Analyst* **93**, 749 (1968); *ibid.* **98**, 830 (1973) (Joint Committee Reports). Isoln of anthraquinone aglycones and glycosides: S. C. Yung Su, N. M. Ferguson, *J. Pharm. Sci.* **63**, 899 (1973). Structure of cascarosides A and B: J. W. Fairbairn et al., *J. Pharm. Pharmacol.* **15** (Suppl.) 292T, (1963); F. J. Evans et al., *ibid.* **27** (Suppl.), 91P (1975). Biological evaluation of cascara bark prepns: J. W. Fairbairn, G. E. D. H. Mahran, *ibid.* **5**, 827 (1953).

Commercial prepn, *Cas-Evac*.

THERAP CAT: Cathartic.

THERAP CAT (VET): Laxative.

1890. Cascarilla. Eleuthera bark; sweet-wood bark. Dried bark of *Croton eluteria* (L.) Sw., *Euphorbiaceae. Habit*. Bahama Islands, Cuba, Haiti. *Constit*. About 15% resin; 5.39-6.33% fat; 7.6-8.13% ash; theobromine, volatile oil, cascarillin, betaine, tannin. *Ref:* Farb, *C.A.* **44**, 8060e (1950).

USE: As an addition to smoking tobacco for flavoring.

THERAP CAT: Aromatic bitter.

1891. Cascarillin. 3-*(Acetyloxy)-1-[2-(3-furanyl)-2-hydroxyethyl]decahydro-5,6-dihydroxy-2,4a,5-trimethyl-1-naphthalenecarboxaldehyde.* $C_{22}H_{32}O_7$; mol wt 408.48. C 64.68%, H 7.90%, O 27.42%. From bark of *Croton eluteria* (L.) Sw., *Euphorbiaceae*. Isoln: Duval, *J. Pharm.* **8**, 91 (1845); Mylius, *Ber.* **6**, 1051 (1873); Naylor, Littlefield, *Pharm. J.* **57**, 95 (1896). Structure and stereochemistry: McEachan et al., *J. Chem. Soc. (B)* **1966**, 633.

Needles from alc, mp 205°. Very bitter taste. Freely sol in ether, hot alc; slightly sol in water, cold alc, chloroform.

THERAP CAT: Aromatic bitter.

1892. Casein. A mixture of related phosphoproteins occurring in milk and cheese. Present to the extent of 3% in bovine milk. One of the most nutritive milk proteins in that it contains all of the common amino acids and is rich in the essential ones. Produced in mammary tissue from amino acids supplied by the blood. Obtained from milk by removing the cream and acidifying the skimmed milk which causes casein to precipitate. In cheese manufacture, casein is precipitated by the lactic acid formed from the same milk by

fermentation. Precipitation by rennet is favored for casein intended for plastics manuf. Curdling by electricity has been described. The prepn of pure casein is described by Hammarsten, *Textbook of Physiological Chemistry* (New York, 7th ed., 1911) p 619, or in Abderhalden's *Handbuch* vol. **II** (Berlin, 1910) p 384. Alternate prepn: van Slyke, Baker, *J. Biol. Chem.* **35**, 127 (1918); *cf.* Cohn, Hendry, *Org. Syn.* coll. vol. **II**, 120 (1943). Prepn of casein free from vitamin B_{12} for nutrition experiments: Kissel, U.S. pat. **2,853,479** (1958 to Natl. Dairy Prods.). The major casein components may be distinguished by electrophoresis and are designated as α-, β-, γ- and K-caseins, in order of decreasing mobility at pH 7. The complete amino acid sequence of bovine β-casein is known and contains 209 residues with an approx. mol wt of 23,600: Ribadeaudumas et al., *Eur. J. Biochem.* **25**, 505 (1972). *Review:* McKenzie, *Advan. Protein Chem.* **22**, pp 75-135 (1967).

White, amorphous powder or granules without odor or taste. Very sparingly sol in water and in nonpolar organic solvents; sol in aqueous solns of alkalies, levorotatory. The isoelectric zone is around pH 4.7; sol in concd HCl with light violet color. Amphoteric; forms salts with both acids and bases. Present in bovine milk as neutral calcium caseinate and in human milk as potassium caseinate. Precipitated from solns satd with metallic salts. Forms a hard, insol plastic with formaldehyde.

Casamino acids, commercial acid-hydrolyzed casein. Hydrolysis is carried out until all the nitrogen in the casein is converted to amino acids or other compounds of relative chemical simplicity. Prepn: Mueller, Miller, *J. Immunol.* **40**, 21 (1941); Mueller, Johnson, *ibid.* 33. Typical analysis: N 10%, NaCl 14%, ash 20%, phosphorus as PO_4 2%, Fe 15 mg/3 g.

Note: Legumin, also known as *avenin* or *vegetable casein*, occurring in beans and nuts, is a globulin resembling casein. Isoln from *Avena sativa* L., *Graminea:* Sanson, *Compt. Rend.* **96**, 75 (1883). Structure: C. J. Bailey, D. Boulter, *Eur. J. Biochem.* **17**, 460 (1970).

USE: In the manuf of molded plastics, adhesives, paints, textile finishes, paper coatings, man-made fibers. Vitamin-free casein is used in diets of animals employed for the biological assay of vitamins. Medicinal grades are used in dietetic prepns and for determining the effectiveness of digestive enzyme prepns contg. pepsin, trypsin, papain. Casamino acids are used in microbial assays and in the prepn of microbiological media.

THERAP CAT: Nutrient.

1893. Casimiroedine. N-D-Glucoside of N-cinnamoyl-N-methylhistamine. $C_{21}H_{27}N_3O_6$; mol wt 417.47. C 60.42%, H 6.52%, N 10.07%, O 23.00%. From fruit, seed and bark of *Casimiroa edulis* Llave et Lex., *Rutaceae:* Power, Callan, *J. Chem. Soc.* **99**, 1993 (1911); Aebi, *Helv. Chim. Acta* **39**, 1495 (1956); Kincl et al., *J. Chem. Soc.* **1956**, 4163; Djerassi et al., *J. Org. Chem.* **21**, 1510 (1956). Structure work: Djerassi et al., *Tetrahedron* **2**, 168a (1958); Raman et al., *Tetrahedron Letters* **1962**, 357.

Needles from methanol, mp 226-228°. $[\alpha]_D^{20}$ −21.3° (c = 1.21 in 5% HCl); $[\alpha]_D$ −33° (aq 1% HCl). uv max (ethanol): 219, 280 nm (log ϵ 4.26, 4.30). Freely sol in dil HCl; dissolves slowly in 2N NaOH or saturated $NaHCO_3$; practically insol in water; sol in hot alc, in amyl alcohol, slightly sol in benzene, chloroform, ether.

Picrate, $C_{21}H_{27}N_3O_6 \cdot C_6H_3N_3O_7$, yellow crystals from methanol, mp 110-112°.

1894. Casimiroin.   *6-Methoxy-9-methyl-1,3-dioxolo-[4,5-h]quinolin-8(9H)-one;* 1-methyl-4-methoxy-7,8-methylenedioxycarbostyril; 4-methoxy-1-methyl-7,8-methylene-dioxy-2-quinolone. $C_{12}H_{11}NO_4$; mol wt 233.32. C 61.80%, H 4.75%, N 6.01%, O 27.44%. From the seed, trunk, root and bark of *Casimiroa edulis* Llave et Lex., *Rutaceae:* Kincl et al., *J. Chem. Soc.* **1956**, 4163; Iriarte et al., *ibid.* 4170. Structure: Meisels, Sondheimer, *J. Am. Chem. Soc.* **79**, 6328 (1957). Synthesis: Weinstein, Hylton, *Tetrahedron* **20**, 1725 (1964).

Crystals, mp 203°. uv max: 226 nm (log ε 4.50). Almost insol in water; sol in alcohol, ether, ethyl acetate.

Picrate, $C_{12}H_{11}NO_4 \cdot C_6H_3N_3O_7$, yellow needles from methanol, mp 193-194°.

1895. Cassaidine.   *(Dodecahydro-7β,10-dihydroxy-1α,-4bβ,8,8-tetramethyl-2(1H)-phenanthrenylidene)acetic acid 2-(dimethylamino)ethyl ester;* (E)-3β,7β-dihydroxy-14α-methylpodocarpane-Δ^{13,α}-acetic acid 2-(dimethylamino)ethyl ester. $C_{24}H_{41}NO_4$; mol wt 407.58. C 70.72%, H 10.14%, N 3.44%, O 15.70%. Cardiotonic principle from the bark of *Erythrophleum guineense* G. Don., *Leguminosae.* Isoln and structure: Ruzicka, Dalma, *Helv. Chim. Acta* **23**, 753 (1940); Engel, *ibid.* **42**, 1127 (1959). Synthesis: Turner et al., *J. Am. Chem. Soc.* **88**, 1766 (1966). Stereochemistry: Clarke et al., *ibid.* **88**, 5865 (1966).

Prisms from acetone + ether, mp 139.5°. $[\alpha]_D^{20} -98°$ (ethanol); $[\alpha]_D^{20} -104°$ (0.1N HCl). Slightly sol in methanol, ethanol, acetone, acetic, chloroform. Practically insol in ether, benzene.

Oxalate, $C_{24}H_{41}NO_4 \cdot (COOH)_2$, crystals from methanol + acetone, mp 198-201°. $[\alpha]_D^{17} -84°$ (c = 1.39 in water).

1896. Cassaine.   *(Dodecahydro-7β-hydroxy-1α,4bβ,8,8-tetramethyl-10-oxo-2(1H)-phenanthrenylidene)acetic acid 2-(dimethylamino)ethyl ester;* (E)-3β-hydroxy-14α-methyl-7-oxopodocarpane-Δ^{13,α}-acetic acid 2-(dimethylamino)ethyl ester. $C_{24}H_{39}NO_4$; mol wt 405.56. C 71.07%, H 9.69%, N 3.45%, O 15.78%. Cardiotonic principle from the bark of *Erythrophleum guineense* G. Don., *Leguminosae:* Dalma, *Helv. Chim. Acta* **22**, 1497 (1939). Structure: Turner et al., *Tetrahedron Letters* **1959**, 7; Gensler et al., *J. Am. Chem. Soc.* **81**, 5217 (1959). Absolute configuration: Hauth et al., *Helv. Chim. Acta* **48**, 1087 (1965); Clarke et al., *J. Am. Chem. Soc.* **88**, 5865 (1966). Synthesis: Turner et al., *ibid.* 1766.

Glossy flakes from ether, mp 142.5°. $[\alpha]_D^{23} -110.5°$ (alc); $[\alpha]_D^{23} -101°$ (0.1N HCl). Absorption spectrum: Ruzicka, Dalma, *Helv. Chim. Acta* **22**, 1516 (1939). Sol in methanol, ethanol, acetone, acetic acid, chloroform, ether, benzene.

Hydrochloride hydrate, $C_{24}H_{40}ClNO_4 \cdot H_2O$, minute crystals from alcohol + methyl ethyl ketone + ether (1:1:4), mp 220°.

1897. Cassamine.   *7-[2-[2-(Dimethylamino)ethoxy]-2-oxoethylidene]tetradecahydro-1,4a,8-trimethyl-9-oxo-1-phenanthrenecarboxylic acid methyl ester;* (E)-13-(carboxymethylene)-14α-methyl-7-oxopodocarpan-16-oic acid 13-[2-(dimethylamino)ethyl]methyl ester. $C_{25}H_{39}NO_5$; mol wt 433.57. C 69.25%, H 9.07%, N 3.23%, O 18.45%. Extracted from the bark of *Erythrophleum guineense* G. Don., *Leguminosae.* Isoln: Engel, Tondeur, *Helv. Chim. Acta* **32**, 2364 (1949). Structure: Arya, Engel, *ibid.* **44**, 1650 (1961).

Crystals from pentane, mp 86-87°. $[\alpha]_D^{20} -56°$ (ethanol). uv max: 225 nm (log ε 4.2). Soluble in petr ether.

Hydrochloride, $C_{25}H_{39}NO_5 \cdot HCl$, crystals from acetone, dec 214-217°. $[\alpha]_D^{17} -48°$ (c = 0.65 in water).

Sulfate, $C_{25}H_{39}NO_5 \cdot H_2SO_4$, crystals from alcohol + ether, mp 191-194°. $[\alpha]_D^{17} -49.5°$ (c = 0.47 in water).

1898. Cassella's Acid.   *7-Hydroxy-2-naphthalenesulfonic acid;* 2-naphthol-7-sulfonic acid; β-naphthol-δ-monosulfonic acid; β-naphtholsulfonic acid F; F Acid. $C_{10}H_8O_4S$; mol wt 224.23. C 53.56%, H 3.60%, O 28.54%, S 14.30%. Prepn: Bayer, Duisberger, *Ber.* **20**, 1426 (1887); Green, *J. Chem. Soc.* **55**, 33 (1889).

Needles from HCl, mp 89°. Readily sol in water, alcohol; practically insol in ether, benzene.

USE: Dyestuff intermediate.

1899. Cassella's Acid F.   *7-Amino-2-naphthalenesulfonic acid;* 2-naphthylamine-7-sulfonic acid; β-naphthylamine-δ-sulfonic acid. $C_{10}H_9NO_3S$; mol wt 223.26. C 53.80%, H 4.06%, N 6.28%, O 21.50%, S 14.36%. Prepn by sulfonation of β-naphthylamine and separation from the 6-amino isomer: Green, *J. Chem. Soc.* **55**, 33 (1889); from 7-hydroxy-2-naphthalenesulfonic acid and ammonia: Green, *loc. cit.;* Wait, U.S. pat. **1,492,497** (1924).

Monohydrate, crystals. Sol in 5040 parts cold water, 350 parts boiling water; sol in glacial acetic acid.

Copper salt, orange-yellow crystals. Sparingly sol in water. *Ref:* Green, Vakil, *J. Chem. Soc.* **113**, 35 (1918).

Note: *Bronner's acid* was first described as *6-amino-2-naphthalenesulfonic acid* or *2-naphthylamine-6-sulfonic acid.* However, this product obtained by sulfonation of β-naphthylamine, was subsequently shown to be a mixture of about equal parts of 6- and 7-amino-2-naphthalenesulfonic acids: Green, *loc. cit.*

USE: Both the title compd and its 6-amino isomer are used in the manuf of azo dyes, e.g., **Brit.** pat. **810,246** (1959 to Bayer).

1900. Cassia Fistula.   Cassia pods; drumstick; Indian laburnum; pudding-stick; pudding pipe; purging cassia. Dried fruit of *Cassia fistula* L. (*Cathartocarpus fistula* [L.] Pers.), *Leguminosae. Habit.* Upper Egypt, E. India; cultivated in tropical America and Africa. The pulp of the ripe fruit, *cassia pulp,* is an almost black, viscid mass with a sweetish taste. *Constit.* Hydroxymethylanthraquinones, gum, tannin, albuminoids, about 60% sugars.

THERAP CAT: Cathartic.

1901. Castanea. Chestnut. Leaves of *Castanea dentata* (Marsh.) Borkh., *Fagaceae,* collected in September and October. *Habit.* Southern Europe. There are hardly any chestnut trees left in the U.S. *Constit.* Tannin, gum, albumin, resin.

1902. Castanospermine. *[1S-(1α,6β,7α,8β,8aβ)]-Octahydro-1,6,7,8-indolizinetetrol;* 1,6,7,8-tetrahydroxyoctahydroindolizine; (1S,6S,7R,8R,8aR)-1,6,7,8-tetrahydroxyindolizidine. $C_8H_{15}NO_4$; mol wt 189.21. C 50.78%, H 7.99%, N 7.40%, O 33.82%. Polyhydroxy alkaloid isolated from the seeds of the Australian leguminous tree, *Castanospermum australe.* Inhibits enzymatic glycoside hydrolysis. Isoln of the naturally occurring (+)-form: L. D. Hohenschutz *et al., Phytochemistry* **20,** 811 (1981). Total synthesis and absolute configuration: R. C. Bernotas, B. Ganem, *Tetrahedron Letters* **25,** 165 (1984). Alternate synthesis: H. Hamana *et al., J. Org. Chem.* **52,** 5492 (1987). Inhibition of α- and β-glucosidases: R. Saul *et al., Arch. Biochem. Biophys.* **221,** 593 (1983); *eidem, ibid.* **230,** 668 (1984). Insect antifeedant activity: D. L. Dreyer *et al., J. Chem. Ecol.* **11,** 1045 (1985). Inhibition of HIV infectivity *in vitro:* B. D. Walker *et al., Proc. Nat. Acad. Sci. USA* **84,** 8120 (1987); R. A. Gruters *et al., Nature* **330,** 74 (1987).

Crystals from aq ethanol, mp 212-215° (dec). $[\alpha]_D^{25}$ +79.7° (c = 0.93 in water). pK 6.09.

1903. Castle's Intrinsic Factor. Intrinsic factor; IF. A thermolabile mucoprotein with mol wt about 60,000. Promotes vitamin B_{12} absorption by transporting it through the intestinal wall. Occurs in normal gastric juice, but is deficient in patients with pernicious anemia. Prepd from hog mucosa: Castle *et al., Am. J. Med. Sci.* **178,** 748, 764 (1929); **180,** 305 (1930); Glass *et al., Science* **115,** 101 (1952); Heinle *et al., Trans. Assoc. Am. Phys.* **65,** 214 (1952); Latner *et al., Biochem. J.* **55,** XXIII (1953); Callender *et al., Brit. Med. J.* **I,** 10 (1954); Latner *et al., Lancet* **I,** 497 (1954); Baum, Federman, U.S. pat. **2,912,360** (1959 to Lilly). Purification: Robbins, U.S. pat. **3,008,877** (1961 to Armour; Highley, Ellenbogen, U.S. pats. **3,434,927** and **3,591,678** (1969, 1971, both to Am. Cyanamid). In approx 30% of pernicious anemia patients, antibodies are produced in the serum which combine with IF, thus inhibiting its biological activity. In clinical tests diminished excretion of vitamin B_{12} in the feces is taken as evidence of intrinsic factor activity. Function in the metabolism of vitamin B_{12}: Glass, *Physiol. Rev.* **43,** 529 (1963). *Review:* Gräsbeck, *Progr. Hematol.* **6,** 233 (1969). Commercial prepns are usually marketed in combination with vitamin B_{12}, *e.g., Gastrhéma.*

THERAP CAT: Adjuvant in vitamin B_{12} utilization.

1904. Castor Oil. Ricinus oil; oil of Palma Christi; tangantangan oil; Neoloid. Fixed oil obtained by cold-pressing the seeds of *Ricinus communis* L., *Euphorbiaceae.* Triglyceride of fatty acids. Fatty acid composition is approx ricinoleic 87%, oleic 7%, linoleic 3%, palmitic 2%, stearic 1% and dihydroxystearic trace amounts: Binder *et al., J. Am. Oil Chem. Soc.* **39,** 513 (1962). Review and bibliography: Anderson, *J. Philippine Pharm. Assoc.* **42,** 5-16 (1955); Dominguez *et al., J. Chem. Ed.* **20,** 446 (1952); F. C. Naughton *et al.,* in Kirk-Othmer *Encyclopedia of Chemical Technology* **vol. 5** (Wiley-Interscience, New York, 3rd ed., 1979) pp 1-15.

Pale yellow, viscous oil. Slight somewhat characteristic odor. The crude oil tastes slightly acrid with a decidedly nauseating after-taste. Has excellent keeping qualities, does not turn rancid unless subjected to excessive heat. Dextrorotatory (undil. in sodium light). $d_{15.5}^{15.5}$ 0.961-0.963. Wt of tech grades: 8.1 to 8.9 lbs/gallon. n_D^{25} 1.473-1.477. n_D^{40} 1.466-1.473. Solidif −10° to −18°. Viscosity at 25°: 6-8 poises, also expressed as U ± ½ (Gardner-Holdt Scale).

Flash pt 445°F (230°C); ignition temp 840°F (449°C). Surface tension (dynes/cm): at 20°, 39.0; at 80°, 35.2. Acid value <4. Sapon no. 176-187. Iodine no. (Wijs') 81-91. Reichert-Meissl value <0.5. Polenske value <0.5. Acetyl value 144-150. Hydroxyl value 161-169. Miscible with abs ethanol, methanol, ether, chloroform, glacial acetic acid. Dissolves in its own vol of petr ether or 95% alcohol. Does not dissolve to any extent in mineral oil, unless mixed with another vegetable oil. When heated to 300° for several hours it polymerizes and becomes miscible with mineral oil.

USE: As an industrial raw material for the prepn of chemical derivs used in coatings, urethane derivs, surfactants and dispersants, cosmetics, lubricants; chief raw material for the production of sebacic acid, a basic ingredient in the production of synthetic resins and fibers; as lubricant in metal drawing, machine lubrication and 2-cycle engine fuels, in hydraulic fluids, rubber preservative and mold lubricants; constituent of embalming fluids; in soap manuf; to impart emollient and lubricant properties to cosmetic prepns; as Turkey-red oil (sulfated castor oil) for dyeing and finishing textiles; as dehydrated castor oil in alkyds, resinous copolymers, varnishes, oil-based paints, enamels, calks and putties; as blown oil (oxidized oil) for plasticizing oilcloth, artificial leather, coated fabrics, and lacquers; to plasticize rosin in the manuf of sticky fly-paper, for nitrocellulose and similar coating systems, hot melts, duplicating and stencil inks, adhesives and laminants; as release and anti-sticking agent in hard candy manuf.

THERAP CAT: Cathartic.

THERAP CAT (VET): Mild purgative, but considered unreliable in adult horses. Emollient.

1905. Castor Oil, Hydrogenated. Opalwax; Castorwax. Mol wt about 932. A hard, white wax, mp 86-88°. Iodine number (Wijs') <5.0. Extremely insol in water and in the more common organic solvents.

USE: In water-repellent coatings, candles, shoepolish, carbon paper, ointments and cosmetics; for impregnating paper, wood, cloth; for electrical condenser impregnation; as solid lubricant; as a pressure mold release agent in the manuf of formed plastics and rubber goods.

1906. Catalase. Caperase; Equilase; Optidase. Enzymes which promote reactions involving the decompn of hydrogen peroxide to water and oxygen. Although widely distributed among animals, plants, bacteria, and fungi, catalase in mammalian liver and blood has been most intensively studied. Catalase for commercial use obtained from animal liver, bacterial (*Micrococcus lysodeikticus),* and fungal (*Aspergillus niger)* sources. Isoln from mammalian livers and kidneys: Lolli, Cavanaugh, U.S. pat. **2,703,779** (1955 to Armour); Schroeder *et al., Biochim. Biophys. Acta* **58,** 611 (1962); Dan, U.S. pat. **2,992,167** (1961 to Chr. Hansen's Lab.); from *Aspergillus niger:* Faucett *et al.,* U.S. pat. **3,102,081** (1963 to Miles Labs.). All catalases so far isolated contain four tetrahedrally arranged subunits of equal size giving an approximate mol wt of 240,000. Each subunit consists of a single polypeptide chain associated with a single prosthetic group, ferric protoporphyrin IX. Amino acid sequence of bovine liver catalase subunit: Schroeder *et al., Arch. Biochem. Biophys.* **131,** 653 (1969). Bovine *hepatocatalase,* a term for catalase obtained from liver, was reported to lower serum cholesterol; the active form was found to contain a catalytic amount of zinc (0.32%): Azarnoff, Curran, *J. Lab. Clin. Med.* **60,** 856 (1962), *cf.* Laporte *et al., Biochem. Pharmacol.* **11,** 670 (1962). *Reviews:* Nicholls, Schonbaum, in *The Enzymes,* **vol. 8,** P. D. Boyer *et al.,* Eds. (Academic Press, New York, 1963) pp 147-225; Deisseroth, Daunce, *Physiol. Rev.* **50,** 319 (1970); Schonbaum, Chance, in *The Enzymes* **vol. 13** (part C), P. D. Boyer, Ed. (Academic Press, New York, 3rd ed., 1976) pp 363-408.

USE: In combination with glucose oxidase, for treatment of food wrappers to prevent oxidative deterioration of food: Sarett, Scott, U.S. pat. **2,765,233** (1956 to Ben L. Sarett). In the removal of traces of peroxide in the process of cold sterilization (preservation of milk and cheese by treatment with hydrogen peroxide). With glucose oxidase, *q.v.,* in food preservation.

1907. Catalposide. Catalpin. $C_{22}H_{26}O_{12}$; mol wt 482.45. C 54.77%, H 5.43%, O 39.80%. From unripe fruit of *Catalpa* *bignonioides* Walt. or *C. ovata* G. Don, *Bignoniaceae:* Claassen, *Am. Chem. J.* **10**, 228 (1888). Historical summary: Bobbitt *et al., J. Org. Chem.* **26**, 3090 (1961). Structure: Lunn *et al., Can. J. Chem.* **40**, 104 (1962). Stereochemistry: Bobbitt *et al., J. Org. Chem.* **31**, 500 (1966); **32**, 1459 (1967).

Crystals from methanol + ethyl acetate, mp 215-216.5°. $[\alpha]_D^{23.2}$ −184° (c = 0.87 in methanol). uv max (ethanol): 260 nm (log ϵ 4.27); in sodium hydroxide: 303 nm (log ϵ 4.35). Soluble in water, alcohol; slightly sol in benzene, chloroform, ether.

Hexaacetate, $C_{34}H_{38}O_{18}$, crystals from ethanol, mp 141.5°. $[\alpha]_D^{21.7}$ −106° (c = 0.75 in chloroform).

1908. Catechin. *2-(3,4-Dihydroxyphenyl)-3,4-dihydro-2H-1-benzopyran-3,5,7-triol; catechol; 3,3',4',5,7-flavan-pentol; catechinic acid; catechuic acid; cyanidol.* $C_{15}H_{14}O_6$; mol wt 290.28. C 62.06%, H 4.86%, O 33.07%. Flavonoid found primarily in higher woody plants as (+)-catechin along with (−)-*epicatechin* (*cis* form). From catechu (gambir and acacia), mahogany wood, etc.: Perkin, Yoshitake, *J. Chem. Soc.* **81**, 1160 (1902); Freudenberg *et al., Ber.* **55**, 1734 (1922). Structure: Freudenberg *et al., Ann.* **444**, 135 (1925). Stereochemistry: Clark-Lewis, *Chem. & Ind. (London)* **1955**, 1218; Hardegger *et al., Helv. Chim. Acta* **40**, 1819 (1957); Clark-Lewis, *J. Chem. Soc.* **1960**, 2433. Pharmacology: Van Cauwenberge *et al., C.R. Soc. Biol.* **165**, 1195 (1971). Metabolism: Das, Sothy, *Biochem. J.* **125**, 417 (1971). Biosynthesis of epicatechins: Zaprometov, Grisebach, *Z. Naturforsch.* **28c**, 113 (1973). Other catechins such as *afzelechin* and *gallocatechin* and catechin oligomers (*procyanidins*) also exist in nature: Thompson *et al., J. Chem. Soc. Perkin Trans. I* **1972**, 1287. *See also* Bioflavonoids.

OH

d-catechin

dl-Form, needles from water + acetic acid, mp 212-216°. Slightly sol in cold water, ether; sol in hot water, alcohol, glacial acetic acid, acetone. Practically insol in benzene, chloroform, petr ether.

d-Form, (+)-*cyanidanol-3, dexcyanidanol, Catergen. Pharmacol.* **20**, 3435 (1971).

Hydrated *d*-form, needles from water + acetic acid, mp 93-96°; 175-177° when anhydr. $[\alpha]_D^{18}$ +16° to +18.4°.

Hydrated *l*-form, needles from water + acetic acid, mp 93-96°; 175-177° when anhydr. $[\alpha]_D$ −16.8°.

Note: Catechin is also called catechol (flavan) to distinguish it from catechol (pyrocatechol, *q.v.*).

USE: In dyeing and tanning.

THERAP CAT: Astringent (diarrheal); *d*-form in treatment of hepatic diseases.

1909. Catechu Black. Cutch; cachou; pegu catechu; cashoo. Extract prepd from heartwood of *Acacia catechu* Willd., *Leguminosae. Habit.* India, Hindustan, Ceylon; naturalized in Jamaica. *Constit.* 25-35% catechutannic acid, 2-10% catechin; catechu red, quercetin, gum.

Incompat. Iron compds, gelatin, lime water, zinc sulfate. USE: Tanning, dyeing fabrics brown and black; staining wood; in toilet preparations.

THERAP CAT: Astringent (diarrheal).

THERAP CAT (VET): Has been used as intestinal astringent.

1910. Catharanthine. *3,4-Didehydroibogamine-18-carboxylic acid methyl ester; 7-ethyl-9,10,12,13-tetrahydro-6,9-methano-5H-pyrido[1',2':1,2]azepino[4,5-b]indole-6(6aH)-carboxylic acid methyl ester.* $C_{21}H_{24}N_2O_2$; mol wt 336.43. C 74.97%, H 7.19%, N 8.33%, O 9.51%. Precursor of vinblastine-type alkaloids. Isoln from *Vinca rosea* Linn. (*Catharanthus roseus* G. Don.) *Apocynaceae:* M. Gorman *et al., J. Am. Pharm. Assoc. Sci. Ed.* **48**, 256 (1959); G. H. Svoboda *et al., ibid.* 659. Structure: N. Neuss, M. Gorman, *Tetrahedron Letters* **1961**, 206. Abs config: K. Bláha *et al., ibid.* **1972**, 2763. Synthesis of (±)-form: A. A. Qureshi, A. I. Scott, *Chem. Commun.* **1968**, 947, 948; A. R. Battersby *et al., ibid.* 951; G. Büchi *et al., J. Am. Chem. Soc.* **92**, 999 (1970); J. P. Kutney, F. Bylsma, *Helv. Chim. Acta* **58**, 1672 (1975); B. M. Trost *et al., J. Org. Chem.* **44**, 2052 (1979); T. Imanishi *et al., Tetrahedron Letters* **21**, 3285 (1980).

Crystals from methanol, mp 126-128°. uv max (ethanol): 226, 284, 292 nm (log ϵ 4.56, 3.92, 3.88). $[\alpha]_D^{27}$ +29.8° (CHCl$_3$). pKa' 6.8.

1911. Cathepsins. Intracellular proteinases obtained from animal tissue extracts, the richest sources being liver, kidney and spleen. Located primarily in the lysosomal fraction within the cell. Part of the general enzymic apparatus of animal cells; in most cases they do not specialize in functions characteristic of individual tissues. Review of cathepsins A-C: J. S. Fruton in *The Enzymes* vol. 4, P. D. Boyer *et al.*, Eds. (Academic Press, New York, 1960) pp 233-241; of cathepsins A-E: M. J. Mycek, *Methods Enzymol.* **19**, 285-315 (1970); of cathepsins B, D and G: several authors, *Res. Monogr. Cell Tissue Physiol.* **2**, 57-89, 181-248 (1977); of cathepsins B, D, G, H, L, N and S: several authors, *Ciba Found. Symp.* **75**, 1-68 (1980).

Cathepsin C, dipeptidyl transferase, dipeptidyl amino peptidase I. Isoln from beef spleen: Tallan *et al., J. Biol. Chem.* **194**, 793 (1952); de la Haba *et al., ibid.* **234**, 316 (1959). Hydrolyzes dipeptidyl amides or esters bearing a free α-amino (or α-imino) group in the N-terminal position, esp. those containing an aromatic amino acid adjacent to the free α-amino group: Planta, Gruber, *Biochim. Biophys. Acta* **53**, 443 (1961); Wurz *et al., Biochemistry* **1**, 19 (1962). Enhances the proteolysis of prothrombin to thrombin and thus plays an important role in blood clotting: Purcell, Barnhart, *Biochim. Biophys. Acta* **78**, 800 (1963).

Cathepsin D. Isoln from bovine spleen: Press *et al., Biochem. J.* **74**, 501 (1960); Webb, *ibid.* **76**, 538 (1960); **84**, 455 (1962). Involved in the catabolism of cartilage and connective tissue: Weston *et al., Nature* **222**, 285 (1969).

Cathepsin G. Isoln from human spleen: G. Starkey *et al., Biochem. J.* **155**, 255 (1976). Chymotrypsin-like enzyme. Catalytic and immunological properties: P. M. Starkey, A. J. Barrett, *ibid.* 273.

1912. Cathinone. *(S)-2-Amino-1-phenyl-1-propanone; α-aminopropiophenone.* $C_9H_{11}NO$; mol wt 149.20. C 72.45%, H 7.43%, N 9.39%, O 10.72%. Pharmacologically active alkaloid extracted from the leaves of khat, *Catha edulis* Forsk., *Celastraceae.* Isoln and identification as active constituent of khat: D. W. Peterson *et al., Life Sci.* **27**, 2143

(1980). Prepn of racemic compd: L. Behr-Bergowski, *Ber.* **30**, 1515 (1897); S. Gabriel, *ibid.* **41**, 1127 (1908). Stereospecific synthesis of (−)-cathinone from norephedrine, characterization as HCl salt: B. D. Berrang *et al., J. Org. Chem.* **47**, 2643 (1982). Similarity of cardiovascular effects with those of (+)-amphetamine: J. D. Kohli, L. J. Goldberg, *J. Pharm. Pharmacol.* **34**, 338 (1982); P. Kalix, *Life Sci.* **32**, 801 (1983). Effect on dopamine metabolism: G. P. Mereu *et al., ibid.* 1383.

Sol in methanol (with racemization), methylene chloride. $[\alpha]_D$ −46.8 (c = 0.24 in methanol).
Hydrochloride, $C_9H_{12}ClNO$, crystals from isopropanol-THF, mp 189-190°.

1913. Catnep. Cataria; catnip; catmint. Herb of *Nepeta cataria* L., *Labiatae.* *Habit.* Europe, Asia; naturalized in U.S. *Constit.* Volatile oil, nepetalactone *(q.v.),* nepetalic acid and related compds, tannin: McElvain, Eisenbraun, *J. Am. Chem. Soc.* **77**, 1599 (1955). Its odor is very attractive to all members of the cat family.
THERAP CAT: Aromatic.

1914. Caulophylline. *1,2,3,4,5,6-Hexahydro-3-methyl-1,5-methano-8H-pyrido[1,2-a][1,5]diazocin-8-one; 12-methylcytisine; N-methylcytisine.* $C_{12}H_{16}N_2O$; mol wt 204.26. C 70.56%, H 7.89%, N 13.72%, O 7.83%. From *Caulophyllum thalictroides* Michx., *Berberidaceae:* Power, Salway, *J. Chem. Soc.* **103**, 191 (1913); from *Thermopsis rhombifolia* Richards., *Leguminosae:* Manske, Marion, *Can. J. Res.* **21B**, 144 (1943); from *Baptisia perfoliata* (L.) R. Br., *Leguminosae:* Marion, Turcotte, *J. Am. Chem. Soc.* **70**, 3253 (1948); from *Cytisus laburnum* L., *Leguminosae:* Pöhm, Galinovsy, *Monatsh.* **84**, 1197 (1953); from *Ormosia stipitata* Schery., *Leguminosae:* Lloyd, Horning, *J. Org. Chem.* **23**, 1074 (1958). Synthesis: Van Tamelen, Baran, *J. Am. Chem. Soc.* **77**, 4944 (1955).

Crystals from ethyl acetate + cyclohexane, mp 140-141°. $[\alpha]_{589}^{25}$ −223° (c = 0.905). Sol in 1.1 parts water, 0.9 part alcohol, 0.5 part methanol, 0.9 part chloroform, 2.8 parts benzene, 2.6 parts acetone. LD_{50} in mice (mg/kg): 21 i.v.; 51 i.p., R. B. Barlow, L. J. McLeod, *Brit. J. Pharmacol.* **35**, 161 (1969).

1915. Caulophyllum. Blue cohosh; papoose root; squaw root. Dried rhizome and roots of *Caulophyllum thalictroides* Michx., *Berberidaceae.* *Habit.* Canada to North Carolina, Missouri, and Nebraska; Japan. *Constit.* Leontin, caulophylline, saponin, resins.

1916. Ceanothic Acid. *2α-Carboxy-3β-hydroxy-A(1)-norlup-20(29)-en-28-oic acid;* emmolic acid. $C_{30}H_{46}O_5$; mol wt 486.67. C 74.03%, H 9.53%, O 16.45%. From *Ceanothus americanus* L., *Rhamnaceae* (New Jersey tea). Isoln: Julian *et al., J. Am. Chem. Soc.* **60**, 77 (1938). Identity of ceanothic acid and emmolic acid: Mechoulam, *Chem. & Ind. (London)* **1961**, 1835. Structure: de Mayo, Starratt, *Tetrahedron Letters* **1961**, 259; Mechoulam, *J. Org. Chem.* **27**, 4070 (1962). Stereochemistry: de Mayo, Starratt, *Can. J. Chem.* **40**, 788 (1962). Revised stereochemistry: Eade *et al., Aust. J. Chem.* **24**, 621 (1971).

Crystals from methanol + ether, mp 356-357°. $[\alpha]_D$ +38° (c = 1.20 in ethanol).
Dimethyl ester, $C_{32}H_{50}O_5$, crystals from ether + petr ether, mp 221-223°. $[\alpha]_D$ +41° (chloroform).

1917. Cecropins. A major class of non-specific antibacterial proteins induced in hemolymph of some insects as part of the immune response. First characterized in *Lepidoptera,* particularly *Hyalophora cecropia.* (*See also* attacins.) Small, basic polypeptides which possess similar amino acid sequences with relatively long hydrophobic regions, mol wt approx 4000 Da. Originally designated *protein P9,* three major peptides have been characterized, cecropins A, B, D; three minor peptides C, E, F are also known. Identification of cell free immune response in *Lepidoptera:* H. Boman *et al., Infect. Immun.* **10**, 136 (1974). Initial identification and isolation of immune response proteins in *H. cecropia:* I. Faye *et al., ibid.* **12**, 1426 (1975). Purification and preliminary characterization of cecropins A and B: D. Hultmark *et al., Eur. J. Biochem.* **106**, 7 (1980); antibacterial spectrum and amino acid sequence: H. Steiner *et al., Nature* **292**, 246 (1981). Partial synthesis of A: R. B. Merrifield *et al., Biochemistry* **21**, 5020 (1982). Total synthesis of A: D. Andreu *et al., Proc. Nat. Acad. Sci. USA* **80**, 6475 (1983); of B: P. van Hofsten *et al., ibid.* **82**, 2240 (1985). Isolation, characterization of cecropins C, D, E, F, and comparison with A and B: D. Hultmark *et al., Eur. J. Biochem.* **127**, 207 (1982); X. Qu *et al., ibid.* 219. Isoln of cecropin-like proteins from flesh fly larvae: M. Okada, S. Natori, *Biochem. J.* **211**, 727 (1983); from tsetse fly: G. P. Kaaya *et al., Insect. Biochem.* **17**, 309 (1987). Cecropin-like proteins designated as *lepidopterans* have been isolated from *Bombyx mori* silkworms: T. Teshima *et al., Tetrahedron Letters* **28**, 4705 (1987). Mechanism of action: H. Steiner *et al., Biochim. Biophys. Acta* **939**, 260 (1988); B. Christensen *et al., Proc. Nat. Acad. Sci. USA* **85**, 5072 (1988). Review of cecropins A and B: H. G. Boman, H. Steiner in *Current Topics in Microbiology and Immunology* **94/95**, W. Henle *et al.,* Eds. (Springer-Verlag, N.Y., 1981) pp 75-91.

1918. Cedrin. *3a,5,5a,9b-Tetrahydro-4,5-dihydroxy-9-methoxy-3,5a-dimethylnaphtho[1,2-b]furan-2,8(3H,4H)-dione.* $C_{15}H_{18}O_6$; mol wt 294.29. C 61.21%, H 6.17%, O 32.62%. From seeds of *Simaba cedron* Planch., *Simaroubaceae:* Lewy, *Compt. Rend.* **32**, 510 (1851); Stieren, *Arch. Pharm.* **23**, 398 (1885). Structure: Krebs, Ruber, *Arzneimittel-Forsch.* **10**, 500 (1960).

Rhombohedral crystals from methanol, mp 266°. uv max (methanol): 243 nm (log ε 3.5). Practically insol in petr ether, ether, acetone. Slightly sol in water, abs ethanol; more sol in dil ethanol, methanol, chloroform and phenol.

1919. Cedrol. *Octahydro-3,6,8,8-tetramethyl-1H-3a,7-methanoazulen-6-ol; 8βH-cedran-8-ol;* cedar camphor; cypress camphor; $C_{15}H_{26}O$; mol wt 222.36. C 81.02%, H 11.79%, O 7.20%. Occurs in cedar wood oil from *Juniperus* *virginiana* L. (*J. sabina* Kook), Cupressaceae, in cypress oil from *Cupressus sempervirens: Schimmel's Report* **1904, II,** 20; in the oil from *Juniperus chinensis:* Kondo, *J. Pharm. Soc. Japan* **1907,** 236; and from *Origanum smyrnaeum* L., Labiatae: *Schimmel's Report* **1906,** Oct., 72. Structure: Stork, Breslow, *J. Am. Chem. Soc.* **75,** 3291, 3292 (1953). Stereochemistry and total synthesis: Stork, Clarke, *ibid.* **83,** 3114 (1961). Synthesis of *dl*-form: Corey *et al., ibid.* **91,** 1557 (1969); *eidem, Tetrahedron Letters* **1973,** 3153; E. G. Breitholle, A. G. Fallis, *J. Org. Chem.* **43,** 1964 (1978).

Needles from dil methanol, mp 86-87°. $[\alpha]_D^{28}$ +9.9° (c = 5 in chloroform).
USE: In fragrances.

1920. Cefaclor. *7-[(Aminophenylacetyl)amino]-3-chloro-8-oxo-5-thia-1-azabicyclo[4.2.0]oct-2-ene-2-carboxylic acid monohydrate;* 7-(D-2-amino-2-phenylacetamido)-3-chloro-3-cephem-4-carboxylic acid monohydrate; 3-chloro-7-D-(2-phenylglycinamido)-3-cephem-4-carboxylic acid monohydrate; Compound 99638; Alfatil; Ceclor; Distaclor; Panacef; Panoral. $C_{15}H_{14}ClN_3O_4S.H_2O$; mol wt 385.83. C 46.70%, H 4.18%, Cl 9.19%, N 10.89%, O 20.73%, S 8.31%. Semi-synthetic cephalosporin antibiotic, related to cephalexin, *q.v.* Prepn: R. R. Chauvette, **Ger.** pat. 2,408,698 (1974 to Lilly), *C.A.* **82,** 4278n (1975); **U.S.** pat. **3,925,372** (1975 to Lilly); R. R. Chauvette, P. A. Pennington, *J. Med. Chem.* **18,** 403 (1975). *In vitro* studies: M. S. Silver *et al., Antimicrob. Ag. Chemother.* **12,** 591 (1977); R. N. Jones, *J. Antibiot.* **30,** 753 (1977); B. R. Meyers, S. Z. Hirschman, *J. Clin. Pharmacol.* **18,** 85 (1978). Metabolism: N. G. Waterman, L. F. Sharfenberger, *Antimicrob. Ag. Chemother.* **14,** 614 (1978). Human pharmacology: G. R. Hodges *et al., ibid.* 454; A. Glynne *et al., J. Antimicrob. Chemother.* **4,** 343 (1978). Clinical studies: J. D. Nelson *et al., Am. J. Dis. Child.* **132,** 992 (1978); B. M. Gray *et al., Antimicrob. Ag. Chemother.* **13,** 988 (1978). Comprehensive description: L. J. Lorenz in *Analytical Profiles of Drug Substances* **vol. 9,** K. Florey, Ed. (Academic Press, New York, 1980) pp 107-123.

Cryst solid. uv max (pH 7 buffer): 265 nm (ε 6800). Sol in water; practically insol in methanol, chloroform, benzene. Solns are stable at pH 2.5-4.5.
THERAP CAT: Antibacterial.

1921. Cefadroxil. *7-[[Amino-(4-hydroxyphenyl)acetyl]-amino]-3-methyl-8-oxo-5-thia-1-azabicyclo[4.2.0]oct-2-ene-2-carboxylic acid monohydrate;* 7-[D-(−)-α-amino-α-(4-hydroxyphenyl)acetamido]-3-methyl-3-cephem-4-carboxylic acid monohydrate; p-hydroxycephalexine monohydrate; BL-S 578; MJF 11567-3; Baxan; Bidocef; Cefa-Drops; Cefamox; Ceforal; Cephos; Duracef; Duricef; Kefroxil; Oracéfal; Sedral; Ultracef. $C_{16}H_{17}N_3O_5S.H_2O$; mol wt 381.41. C 50.39%, H 5.02%, N 11.02%, O 25.17%, S 8.40%. Semi-synthetic cephalosporin antibiotic. Prepn: **Neth.** Appl. 6,812,382; B. Crast, Jr., **Brit.** pat. **1,240,687** (1969, 1971 both to Bristol-Myers); T. Takahashi *et al.,* **Ger.** pat. **2,216,113;** *eidem,* **U.S.** pat. **3,816,253** (1972, 1974, both to Takeda); E. S. Granatek, J. E. Vogan, **Ger.** pat. **2,259,011** (1973 to Bristol-Myers), *C.A.* **79,** 66376q (1973); T. Ishimaru, Y. Kodama, **Ger.** pats. **2,163,514, 2,263,861,** *C.A.* **79,** 78826z, 78822v (1973); *eidem,* **U.S.** pat. **3,864,340** (1973, 1973, 1975, all to Toyama). Prepn of crystalline monohydrate: D. Bouzard *et al.,* **U.S.** pat. **4,504,657** (1985 to Bristol-Myers). Antimicrobial activity: R. E. Buck, K. E.

Price, *Antimicrob. Ag. Chemother.* **11,** 324 (1977). Pharmacology: M. Pfeffer *et al., ibid.* 331; A. I. Hartstein *et al., ibid.* **12,** 93 (1977). Review: *J. Antimicrob. Chemother.* **10,** Suppl. B, 1-162 (1982). Series of articles on clinical trials in respiratory tract infections: *Drugs* **32,** Suppl. 3, 1-56 (1986).

White crystals, mp 197° (dec).
THERAP CAT: Antibacterial.
THERAP CAT (VET): Antibacterial.

1922. Cefamandole. *7-[(Hydroxyphenylacetyl)amino]-3-[[(1-methyl-1H-tetrazol-5-yl)thio]methyl]-8-oxo-5-thia-1-azabicyclo[4.2.0]oct-2-ene-2-carboxylic acid; 7-mandelamido-3-[[(1-methyl-1H-tetrazol-5-yl)thio]methyl]-8-oxo-5-thia-1-azabicyclo[4.2.0]oct-2-ene-2-carboxylic acid;* 7-D-mandelamido-3-[[(1-methyl-1H-tetrazol-5-yl)thio]methyl]-3-cephem-4-carboxylic acid; 7-D-mandelamido-3-(1-methyl-1,2,3,4-tetrazole-5-thiomethyl)-Δ³-cephem-4-carboxylic acid; CMT; Compound 83405. $C_{18}H_{18}N_6O_5S_2$; mol wt 462.50. C 46.75%, H 3.92%, N 18.17%, O 17.30%, S 13.86%. Broad-spectrum semi-synthetic cephalosporin antibiotic. Prepn: C. W. Ryan, **Ger.** pat. **2,018,600** corresp to **U.S.** pat. **3,641,021** (1970, 1972 to Lilly); J. M. Greene, **Ger.** pat. **2,312,997** corresp to **U.S.** pat. **3,840,531** (1973, 1974 to Lilly). Biological properties: W. E. Wick, D. A. Preston, *Antimicrob. Ag. Chemother.* **1,** 221 (1972). Antibacterial activity: S. Eykyn *et al., ibid.* **3,** 657 (1973); H. C. Neu, *ibid.* **6,** 177 (1974); A. D. Russell, *J. Antimicrob. Chemother.* **1,** 97 (1975). Pharmacologic studies: B. R. Meyers *et al., Antimicrob. Ag. Chemother.* **9,** 140 (1976); R. S. Griffith *et al., ibid.* **10,** 814 (1976). Comprehensive description: R. H. Bishara, E. C. Rickard in *Analytical Profiles of Drug Substances* **vol. 9,** K. Florey, Ed. (Academic Press, New York, 1980) pp 125-154.

Nafate, $C_{19}H_{17}N_6NaO_6S_2$, *Bergacef, Cedol, Cefam, Cefiran, Cemado, Cemandil, Fado, Kefadol, Kefandol, Lampomandol, Mandokef, Mandol, Mandolsan, Neocefal, Pavecef.* White, odorless needles, mp 190° (dec). uv max (H₂O): 269 nm (ε 10800). pKa 2.6-3.0. Sol in water, methanol; practically insol in ether, chloroform, benzene, cyclohexane.
THERAP CAT: Antibacterial.

1923. Cefatrizine. *[6R-[6α,7β(R*)]]-7-[[Amino-(4-hydroxyphenyl)acetyl]amino]-8-oxo-3-[(1H-1,2,3-triazol-4-ylthio)methyl]-5-thia-1-azabicyclo[4.2.0]oct-2-ene-2-carboxylic acid;* 7-[D-α-amino-α-(4-hydroxyphenyl)acetamido]-3-(1,2,3-triazol-4(5)-ylthiomethyl)-3-cephem-4-carboxylic acid; BL-S640; SKF 60771; S 640P. $C_{18}H_{18}N_6O_5S_2$; mol wt 462.50. C 46.75%, H 3.92%, N 18.17%, O 17.30%, S 13.86%. Orally active semi-synthetic cephalosporin antibiotic. Prepn: G. L. Dunn, J. R. E. Hoover, **Ger.** pat. **2,316,866;** *eidem,* **U.S.** pat. **3,867,380** (1974, 1975 both to SK&F); D. Willner, L. B. Crast, **Ger.** pat **2,364,192** (1974 to Bristol-Myers), *C.A.* **81,** 105538 (1974); G. L. Dunn *et al., J. Antibiot.* **29,** 65 (1976). Prepn of the propylene glycolate: M. A. Kaplan, A. P. Granatek, **Ger.** pat. **2,500,385;** *eidem,* **U.S.** pat. **3,970,651** (1975, 1976 both to Bristol-Myers). *In vitro* and *in vivo* antibacterial activity and pharmacokinetic behavior: P. Actor *et al., ibid.* **28,** 594 (1975); F. Leitner *et al., Antimicrob. Ag. Chemother.* **7,** 298 (1975). General pharmacological properties: M. Matsuzaki *et al., Japan. J. Antibiot.*

29, 107 (1976), *C.A.* **85**, 311g (1976). Determn in serum and urine by HPLC: E. Crombez *et al., J. Chromatog.* **173**, 165 (1979). Disposition of ^{14}C-cefatrizine in man: R. C. Gaver, G. Deeb, *Drug Metab. Dispos.* **8**, 157 (1980). Activity and clinical use: A. J. Weinstein, *Drugs* **20**, 137 (1980). Acute toxicity: M. Matsuzaki *et al., Japan. J. Antibiot.* **29**, 612 (1976), *C.A.* **85**, 171724y (1976). Toxicological studies: *Japan. J. Antibiot.* **29**, 639-686 (1976), *C.A.* **85**, 171722w-171723x (1976). Series of articles on pharmacokinetics and clinical efficacy: *Drugs Exptl. Clin. Res.* **11**, 441-462 (1985).

Zwitterionic. LD$_{50}$ in male, female mice, male, female rats (mg/kg): 6880, 6410, 4325, 4325 i.p. (Matsuzaki).

Propylene glycolate, $C_{21}H_{26}N_4O_7S_2$, *Bricef, Cefaperos, Cepticol, Cefotrizin, Faretrizin, Ghimacef, Kefoxina, Tamyl, Zitrix.* Rod-like crystals. $[\alpha]_D^{23}$ +55.9° (c = 1% in 1N HCl). uv max: 227, 272 nm.

THERAP CAT: Antibacterial.

1924. Cefazedone. *7-[[(3,5-Dichloro-4-oxo-1(4H)-pyridinyl)acetyl]amino]-3-[[(5-methyl-1,3,4-thiadiazol-2-yl)-thio]methyl]-8-oxo-5-thia-1-azabicyclo[4.2.0]oct-2-ene-2-carboxylic acid;* 3-(5-methyl-1,3,4-thiadiazolyl-2-mercaptomethyl)-7-(3,5-dichloro-4-pyridon-1-ylacetamido)-3-cephem-4-carboxylic acid; EMD 30087. $C_{18}H_{15}Cl_2N_5O_5S_3$; mol wt 548.44. C 39.42%, H 2.75%, Cl 12.93%, N 12.77%, O 14.59%, S 17.54%. Semi-synthetic cephalosporin antibiotic. Prepn: R. Gerike *et al.*, Ger. pat. **2,345,402** (1975 to E. Merck, W. Ger.), *C.A.* **83**, 43359e (1975). *See also eidem,* **Ger.** pats. **2,621,011, 2,626,026, 2,627,204** (all 1977 to E. Merck, W. Ger.), *C.A.* **88**, 5093, 105383, 121213 (1978). Series of articles on prepn, *in vitro* and *in vivo* activity, kinetics, pharmacology, toxicology, and clinical trials: *Arzneimittel-Forsch.* **29**, 361-462 (1979). Toxicity data: M. von Eberstein *et al., ibid.* 424.

Crystals from acetonitrile.
Sodium salt, $C_{18}H_{14}Cl_2N_5NaO_5S_3$, *Refosporin.* LD$_{50}$ in mice, rats, rabbits, dogs (mg/kg): 6800, 4225, 3200, 3000 i.v. (von Eberstein).

THERAP CAT: Antibacterial.

1925. Cefazolin. *(6R-trans)-3[[(5-Methyl-1,3,4-thiadiazol-2-yl)thio]methyl]-8-oxo-7-[(1H-tetrazol-1-ylacetyl)-amino]-5-thia-1-azabicyclo[4.2.0]oct-2-ene-2-carboxylic acid;* 7-(1-(1H)-tetrazolylacetamido)-3-[2-(5-methyl-1,3,4-thiadiazolyl)thiomethyl]-Δ³-cephem-4-carboxylic acid; CEZ. $C_{14}H_{14}N_8O_4S_3$; mol wt 454.50. C 37.00%, H 3.10%, N 24.65%, O 14.08%, S 21.16%. Semi-synthetic antibiotic derived from 7-aminocephalosporanic acid, *q.v.* Prepn: Takano *et al.*, **S. Afr. pat. 68 04,513** corresp to U.S. pat. **3,516,997** (1969, 1970 to Fujisawa). Synthesis and properties: Kariyone *et al., J. Antibiot.* **23**, 131 (1970). Activity and clinical studies: Nishida *et al., ibid.* **137**, 184; Shibata, Fujii, *Antimicrob. Ag. Chemother.* **1970**, 467. Metabolic studies: Kozatani *et al., Chem. Pharm. Bull.* **20**, 1105 (1972). Toxicology: H. A. Birkhead *et al., J. Infec. Dis.* **128**, Suppl., 379 (1973). Comprehensive description: A. E. Zappala *et al.*, in *Analytical Profiles of Drug Substances* vol. **4**, K. Florey, Ed. (Academic Press, New York, 1975) pp 1-20. *Review:* H. Nakano in *Pharmacological and Biochemical Properties of Drug Substances* vol. **1**, M. E. Goldberg,

Ed. (Am. Pharm. Assoc., Washington, DC, 1977) pp 155-182.

Needles from aq acetone, mp 198~200° (dec). uv max (buffer pH 6.4): 272 nm (ε 13,150). Easily sol in DMF, pyridine; sol in aq acetone, aq dioxane, aq ethanol; slightly sol in methanol. Practically insol in chloroform, benzene, ether.

Sodium salt, $C_{14}H_{13}N_8NaO_4S_3$, *sodium CEZ, SKF 41558, Acef, Ancef, Atirin, Biazolina, Bor-Cefazol, Cefacidal, Cefamedin, Cefamezin, Cefazil, Cefazina, Elzogram, Firmacef, Gramaxin, Kefzol, Lampocef, Liviclina, Totacef, Zolicef.* White to yellowish-white, odorless crystalline powder with a bitter, salty taste. Crystallizes in α, β, and γ-forms, *see* Kariyone, *loc. cit.* Easily sol in water, slightly sol in methanol, ethanol. Practically insol in benzene, acetone, chloroform. LD$_{50}$ in mice, rats (mg/kg): 3.9, 3.18 i.v.; 6.2, 7.4 i.p. (Birkhead).

THERAP CAT: Antibacterial.

1926. Cefbuperazone. *7-[[2-[[(4-Ethyl-2,3-dioxo-1-piperazinyl)carbonyl]amino]-3-hydroxy-1-oxobutyl]amino]-7-methoxy-3-[[(1-methyl-1H-tetrazol-5-yl)thio]methyl]-8-oxo-5-thia-1-azabicyclo[4.2.0]oct-2-ene-2-carboxylic acid;* (6R,7S)-7-[(2R,3S)-2-(4-ethyl-2,3-dioxo-1-piperazinecarboxamido)-3-hydroxybutyramido]-7-methoxy-3-[[(1-methyl-1H-tetrazol-5-yl)thio]methyl]-8-oxo-5-thia-1-azabicyclo[4.2.0]oct-2-ene-2-carboxylic acid; 7β-[D-α-(4-ethyl-2,3-dioxo-1-piperazinecarboxamido)-β-(S)-hydroxybutanamido]-7α-methoxy-3-[5-(1-methyl-1,2,3,4-tetrazolyl)thiomethyl]-Δ³-cephem-4-carboxylic acid; T-1982. $C_{22}H_{29}N_9-O_9S_2$; mol wt 627.65. C 42.09%, H 4.66%, N 20.08%, O 22.94%, S 10.22%. Broad spectrum injectable cephamycin. Prepn: I. Saikawa *et al.*, **Belg. pat. 879,217** corresp to U.S. pat. **4,263,292** (1980, 1981 both to Toyama). Distribution of ^{14}C-cefbuperazone in mice: *eidem, Japan. J. Antibiot.* **35**, 2159 (1982), *C.A.* **98**, 11062w (1982). *In vitro* and *in vivo* antibacterial activity: M. Tai *et al., Antimicrob. Ag. Chemother.* **22**, 728 (1982). General pharmacology: A. Takai *et al., Japan. J. Antibiot.* **35**, 2139 (1982), *C.A.* **98**, 65142r (1982). Absorption, distribution, excretion: I. Saikawa *et al., ibid.* 2163, *C.A.* **98**, 100691t (1982). Series of articles on cefbuperazone: *Chemotherapy (Tokyo)* **30**, Suppl. 3 (1982) 986 pp. Stability and degradation studies: S. Takano *et al., Yakugaku Zasshi* **103**, 62 (1983).

Solid, mp 118-120° (dec).
Sodium salt, $C_{22}H_{28}N_9NaO_9S_2$, *T-1982, Keiperazon, Tomiporan.*

THERAP CAT: Antibacterial.

1927. Cefixime. *[6R-[6α,7β(Z)]]-7-[[(2-Amino-4-thiazolyl)[(carboxymethoxy)imino]acetyl]amino]-3-ethenyl-8-oxo-5-thia-1-azabicyclo[4.2.0]oct-2-ene-2-carboxylic acid;* 7-[2-(2-amino-4-thiazolyl)-2-(carboxymethoxyimino)acetamido]-3-vinyl-3-cephem-4-carboxylic acid; FK-027; FR 17027; CL 284635; Oroken; Cefspan; Suprax. $C_{16}H_{15}N_5-O_7S_2$; mol wt 453.44. C 42.38%, H 3.33%, N 15.44%, O 24.70%, S 14.14%. Orally active, third generation cephalosporin antibiotic. Prepn: T. Takaya *et al.*, **Eur. pat. Appl. 30,630**; *eidem*, U.S. pat. **4,409,214** (1981, 1983 both to Fujisawa); H. Yamanaka *et al., J. Antibiot.* **38**, 1738 (1985).

Synthesis and activity of *(E)-*isomer: K. Kawabata *et al.*, *Chem. Pharm. Bull.* **34**, 3458 (1986). Mechanism of action: Y. Shigi *et al., J. Antibiot.* **37**, 790 (1984). Comparative antibacterial spectrum *in vitro* and *in vivo*: T. Kamimura *et al., Antimicrob. Ag. Chemother.* **25**, 98 (1984). *In vitro* activity and β-lactamase stability: H. C. Neu *et al., ibid.* **26**, 174 (1984). HPLC determn in blood plasma and urine: Y. Tokuma *et al., J. Chromatog.* **311**, 339 (1984). Pharmacokinetics in humans: D. R. P. Guay *et al., Antimicrob. Ag. Chemother.* **30**, 485 (1986). Clinical trial in urinary tract infections: J. Levenstein *et al., S. Afr. Med. J.* **70**, 455 (1986); A. Irvani *et al., Am. J. Med.* **85**, Suppl. 3A, 17 (1988); in respiratory infections: R. Kiani *et al., ibid.* 6.

*(Z)-*Form disodium salt, $C_{16}H_{13}N_5Na_2O_7S_2$, mp > 250°. *(E)-*Form trihydrate, pale yellow solid, mp 218-225° (dec). THERAP CAT: Antibacterial.

1928. Cefmenoxime. *[6R-[6α,7β(Z)]]-7-[[(2-Amino-4-thiazolyl)(methoxyimino)acetyl]amino]-3-[[(1-methyl-1H-tetrazol-5-yl)thio]methyl]-8-oxo-5-thia-1-azabicyclo[4.2.0]-oct-2-ene-2-carboxylic acid; (6R,7R)-7-[2-(2-amino-4-thiazolyl)glyoxylamido]-3-[[(1-methyl-1H-tetrazol-5-yl)thio]-methyl]-8-oxo-5-thia-1-azabicyclo[4.2.0]oct-2-ene-2-carb-oxylic acid 7²-(Z)-(O-methyloxime); SCE-1365.* $C_{16}H_{17}N_9-O_5S_3$; mol wt 511.56. C 37.57%, H 3.35%, N 24.64%, O 15.64%, S 18.80%. Third generation cephalosporin antibiotic; related structurally to cefotaxime and ceftizoxime, *q.q.v.* The name cefmenoxime applies to the isomer having a *syn-*methoxyimino group. Prepn (unspecified stereochemistry): M. Ochiai *et al.*, **Ger.** pat. **2,556,736;** *eidem,* **U.S.** pat. **4,-098,888** (1976, 1978 both to Takeda); *syn-*isomer: R. Heymes, A. Lutz, **Ger.** pat. **2,713,272;** *eidem,* **U.S.** pat. **4,476,122** (1983, 1984 both to Roussel-Uclaf). Series of articles on antibacterial activity, absorption, excretion, metabolism, mechanism of action, clinical studies: *Chemotherapy (Tokyo)* **29**, Suppl. 1, 1-998 (1979). *In vitro* and *in vivo* study: K. Tsuchiya *et al., Antimicrob. Ag. Chemother.* **19**, 56 (1981). β-Lactamase stability: K. Okonogi *et al., ibid.* **20**, 171 (1981). Clinical pharmacokinetics study: D. Höffler, P. Koeppe, *Arzneimittel-Forsch.* **33**, 269 (1983). Symposium: *Am. J. Med.* **77**(6A), 1-59 (1984).

Hydrochloride (*syn-*isomer), $C_{32}H_{35}ClN_{18}O_{10}S_6$, *Abbott 50192, Bestcall, Cefmax, Cemix, Tacef.* Sodium salt (*syn-*isomer), $C_{16}H_{16}N_9NaO_5S_3$, $[\alpha]_D^{20}$ — 13.5° ±1 (c = 1 in water). THERAP CAT: Antibacterial.

1929. Cefmetazole. *7-[[[(Cyanomethyl)thio]acetyl]amino]-7-methoxy-3-[[(1-methyl-1H-tetrazol-5-yl)thio]methyl]-8-oxo-5-thia-1-azabicyclo[4.2.0]oct-2-ene-2-carboxylic acid;* CS 1170; SKF 83088. $C_{15}H_{17}N_7O_5S_3$; mol wt 471.54. C 38.21%, H 3.63%, N 20.79%, O 16.97%, S 20.40%. Semisynthetic antibiotic derived from cephamycin C, *q.v.* Prepn: H. Nakao *et al.*, **Ger.** pat. **2,455,884;** *eidem,* **U.S.** pat. **4,007,-177** (1973, 1977 both to Sankyo); *eidem, J. Antibiot.* **29**, 554 (1976). *In vitro* study: J. V. Uri *et al., ibid.* **31**, 82 (1978). Immunological study: M. Iwata, T. Matuhasi, *Japan. J. Exp. Med.* **48**, 401 (1978). Absorption, distribution, excretion, metabolism in various species: H. Shindo *et al., Chemother. (Tokyo)* **27**, Suppl. 1, 64 (1979). Pharmacology: S.

Kobayashi *et al., ibid.* **26**, Suppl. 5, 115 (1978). Toxicological studies: H. Masuda *et al., Sankyo Kenkyusho Nempo* **30**, 112 (1978), *C.A.* **90**, 180268h (1979). Comparative clinical study: M. Nishida *et al., J. Antibiot.* **32**, 1319 (1979).

Sodium salt, $C_{15}H_{16}N_7NaO_5S_3$, *Cefmetazon, Zefazone.* White solid. Very sol in water, methanol, sol in acetone. Practically insol in chloroform. LD_{50} i.v. in rats: > 5000 mg/kg (Masuda). THERAP CAT: Antibacterial.

1930. Cefminox. *[6R-[6α,7α,7(S*)]]-7-[[[(2-Amino-2-carboxyethyl)thio]acetyl]amino]-7-methoxy-3-[[(1-methyl-1H-tetrazol-5-yl)thio]methyl]-8-oxo-5-thia-1-azabicyclo-[4.2.0]oct-2-ene-2-carboxylic acid; 7β-(2-D-amino-2-carb-oxyethylthioacetamido)-7α-methoxy-3-[[(1-methyl-1H-tetrazol-5-yl)thio]methyl]-3-cephem-4-carboxylic acid.* $C_{16}H_{21}N_7O_7S_3$; mol wt 519.57. C 36.99%, H 4.07%, N 18.87%, O 21.56%, S 18.51%. Semisynthetic broad spectrum cephamycin antibiotic. Prepn: **Belg.** pat. **880,656,** K. Iwamatsu *et al.*, **U.S.** pat. **4,357,331** (1980, 1982 both to Meiji Seika). Synthesis, biological activity: *eidem, J. Antibiot.* **36**, 229 (1983). Structure-activity studies: S. Inouye *et al., ibid.* **37**, 1403 (1984). Proposed mechanism of action: T. Tsuruoka *et al., Eur. J. Biochem.* **151**, 209 (1985). *In vitro, in vivo* antibacterial activity compared with that of other cephalosporins: S. Inouye *et al., Antimicrob. Ag. Chemother.* **26**, 722 (1984). Activity against gram-negative infections in mice: T. Watanabe *et al., Drugs Exptl. Clin. Res.* **10**, 293 (1984). Toxicity studies: M. Kurebe *et al., Japan. J. Antibiot.* **37**, 847 (1984). Series of articles on pharmacokinetics, early clinical trials, treatment of urinary tract and respiratory infections: *Chemotherapy (Tokyo)* **32**, Suppl. 5 (1984) 561 pp.

Sodium salt heptahydrate, $C_{16}H_{20}N_7NaO_7S_3 \cdot 7H_2O$, *MT 141, Meicelin.* Crystals from ethanol-water, mp 90-91°. LD_{50} in male, female mice (mg/kg): 6100, 5200 i.v.; in male, female rats (mg/kg): 6600, 5700 i.v.; 8600, 8550 i.p.; > 15000 s.c. or orally (Kurebe). THERAP CAT: Antibacterial.

1931. Cefodizime. *[6R-(6α,7β(Z))]-7-[[(2-Amino-4-thiazolyl)(methoxyimino)acetyl]amino]-3-[[5-(carboxymeth-yl)-4-methyl-2-thiazolyl]thio]methyl]-8-oxo-5-thia-1-azabicyclo[4.2.0]oct-2-ene-2-carboxylic acid; (6R,7R)-7-[2-(2-amino-4-thiazolyl)glyoxylamido]-3-[[[5-(carboxymethyl)-4-methyl-2-thiazolyl]thio]methyl]-8-oxo-5-thia-1-azabicy-clo[4.2.0]oct-2-ene-2-carboxylic acid 7²(Z)-(O-methylox-ime).* $C_{20}H_{20}N_6O_7S_4$; mol wt 584.65. C 41.09%, H 3.45%, N 14.37%, O 19.16%, S 21.93%. Third generation cephalosporin antibiotic; derivative of cefotaxime, *q.v.* Prepn: **Belg.** pat. **865,632;** W. Dürckheimer *et al.*, **U.S.** pat. **4,278,793** (1978, 1981 both to Hoechst AG); and antibacterial activity: J. Blumbach *et al., J. Antibiot.* **40**, 29 (1987). *In vitro* comparative antibacterial spectrum: R. N. Jones *et al., Antimicrob. Ag. Chemother.* **20**, 760 (1981); and β-lactamase stability: M. Limbert *et al., J. Antibiot.* **37**, 892 (1984). *In vivo* antibacterial activity: N. Klesel *et al., ibid.* 1712. Enhancement of immune response *in vivo:* M. Limbert *et al., ibid.* 1719; and *ex vivo:* A. Fietta *et al., Chemotherapy (Basel)* **34**, 430 (1988). HPLC determn in biological fluids: T. Marunaka *et al., J. Chromatog.* **420**, 329 (1987). Pharmacokinetics: N. Klesel *et al., J. Antibiot.* **37**, 901 (1984); in humans: E. E. Dagrosa *et al., Clin. Ther.* **10**, 18 (1987). Clinical study in urogenital gonorrhea: A. H. van der Willigen *et al., Antimicrob. Ag. Chemother.* **32**, 426 (1988). Series of articles on pharmacology, pharmacokinetics and clinical studies: *Chemotherapy (Tokyo)* **36**, Suppl. 5, 1-980 (1988).

Disodium salt, $C_{20}H_{20}N_6Na_2O_7S_4$, *HR-221, THR-221.*
THERAP CAT: Antibacterial.

1932. Cefonicid. *7-[(Hydroxyphenylacetyl)amino]-8-oxo-3-[[[1-(sulfomethyl)-1H-tetrazol-5-yl]thio]methyl]-5-thia-1-azabicyclo[4.2.0]oct-2-ene-2-carboxylic acid; (6R,7R)-7-[(R)-mandelamido]-8-oxo-3-[[[1-(sulfomethyl)-1H-tetrazol-5-yl]thio]methyl]-5-thia-1-azabicyclo[4.2.0]oct-2-ene-2-carboxylic acid.* $C_{18}H_{18}N_6O_8S_3$; mol wt 542.56. C 39.85%, H 3.34%, N 15.49%, O 23.59%, S 17.73%. Injectable semi-synthetic cephalosporin antibiotic related to cefamandole, *q.v.* Exhibits long acting broad spectrum activity. Prepn: D. A. Berges, **Ger.** pat. **2,611,270,** *C.A.* **86,** 29854t (1977); **U.S.** pats. **4,093,723, 4,159,373** (1976, 1978, 1979 all to SmithKline). Antibacterial activity, pharmacokinetics: P. Actor *et al., Antimicrob. Ag. Chemother.* **13,** 784 (1978). Series of articles on stability, comparative *in vitro* activity, serum levels: *Curr. Chemother. Infect. Dis., Proc. 11th Int. Congr. Chemother.,* **vol. 1** (Am. Soc. Microbiol., Washington, D.C., 1979) pp 246-254. Stability towards β-lactamases: R. Mehta *et al., J. Antibiot.* **34,** 202 (1981). Kinetics and renal handling: D. Pitkin *et al., Clin. Pharmacol. Ther.* **30,** 587 (1981). *In vitro* evaluation vs Group B streptococci: A. S. Bayer *et al., Antimicrob. Ag. Chemother.* **21,** 344 (1982). Review of antibacterial activity, pharmacology, therapeutic use: E. Saltiel, R. N. Brogden, *Drugs* **32,** 222-259 (1986).

Disodium salt, $C_{18}H_{16}N_6Na_2O_8S_3$, *SKF 75073, Cefodie, Monocid, Monocidur.*
THERAP CAT: Antibacterial.

1933. Cefoperazone. *[6R-[6α,7β(R*)]]-7-[[[[(4-Ethyl-2,3-dioxo-1-piperazinyl)carbonyl]amino](4-hydroxyphenyl)-acetyl]amino]-3-[[(1-methyl-1H-tetrazol-5-yl)thio]methyl]-8-oxo-5-thia-1-azabicyclo[4.2.0]oct-2-ene-2-carboxylic acid; 7-[D-(−)-α-(4-ethyl-2,3-dioxo-1-piperazinecarboxamido)-α-(4-hydroxyphenyl)acetamido]-3-[[(1-methyl-1H-tetrazol-5-yl)thio]methyl]-3-cephem-4-carboxylic acid.* $C_{25}H_{27}N_9O_8S_2$; mol wt 645.68. C 46.51%, H 4.22%, N 19.52%, O 19.82%, S 9.93%. Broad spectrum third generation cephalosporin antibiotic. Prepn: I. Saikawa *et al.,* **Belg.** pat. **837,-682;** *eidem,* **U.S.** pat. **4,410,522** (1976, 1983 both to Toyama); *eidem, Yakugaku Zasshi* **99,** 929 (1979). Stability in aq soln: *eidem, ibid.* 1207. *In vitro* activity: M. V. Borobio *et al., Antimicrob. Ag. Chemother.* **17,** 129 (1980). Kinetics in rats: J. Fabre *et al., Schweiz. Med. Wochenschr.* **110,** 264 (1980); in humans: A. F. Allaz, *ibid.* **109,** 1999 (1979). Review of pharmacology and therapeutic efficacy: R. N. Brogden *et al., Drugs* **22,** 423-460 (1981). Symposium on clinical studies: *ibid.* Suppl. 1, 1-124.

Crystals from acetonitrile/water, mp 169-171° (hydrated). Stable at pH 4.0-7.0; slightly unstable in acid; highly unstable in alkaline soln.
Sodium salt, $C_{25}H_{26}N_9NaO_8S_2$, *CP-52,640-2, T-1551, Cefobid, Cefobine, Cefobis, Cefosint, Farecef, Tomabef.*
THERAP CAT: Antibacterial.

1934. Ceforanide. *7-[[[2-(Aminomethyl)phenyl]acetyl]-amino]-3-[[[1-(carboxymethyl)-1H-tetrazol-5-yl]thio]meth-yl]-8-oxo-5-thia-1-azabicyclo[4.2.0]oct-2-ene-2-carboxylic acid; (6R,7R)-7-[2-(α-amino-o-tolyl)acetamido]-3-[[[1-(carboxymethyl)-1H-tetrazol-5-yl]thio]methyl]-8-oxo-5-thia-1-azabicyclo[4.2.0]oct-2-ene-2-carboxylic acid; 7-[o-(aminomethyl)phenylacetamido]-3-[[[1-(carboxymethyl)-1H-tetrazol-5-yl]thio]methyl]-3-cephem-4-carboxylic acid; BL-S786.* $C_{20}H_{21}N_7O_6S_2$; mol wt 519.56. C 46.24%, H 4.07%, N 18.87%, O 18.48%, S 12.34%. Injectable, semi-synthetic cephalosporin antibiotic. Exhibits long-acting, broad-spectrum activity. Prepn: M. A. Kaplan *et al.,* **Ger.** pat. **2,538,804** corresp to **U.S.** pats. **4,172,196** and **4,182,863** (1976, 1979, 1980 all to Bristol-Myers); W. J. Gottstein *et al., J. Antibiot.* **29,** 1226 (1976). Laboratory evaluation: F. Leitner *et al., Antimicrob. Ag. Chemother.* **10,** 426 (1976). *In vitro* susceptibility comparisons: R. N. Jones *et al., J. Antibiot.* **30,** 576, 583 (1977). Comparative tissue distribution: F. H. Lee *et al., Antimicrob. Ag. Chemother.* **19,** 625 (1981). Pharmacokinetics: E. H. Estey *et al., Clin. Pharmacol. Ther.* **30,** 396 (1981); S. S. Hawkins *et al., ibid.* 468. Clinical study: R. J. Wallace *et al., Antimicrob. Ag. Chemother.* **20,** 648 (1981).

White solid, melts above 150° (dec).
Sodium salt, $C_{20}H_{21}N_7NaO_6S_2$; *Precef.*
THERAP CAT: Antibacterial.

1935. Cefotaxime. *[6R-[6α,7β(Z)]]-3-[(Acetyloxy)meth-yl]-7-[[(2-amino-4-thiazolyl)(methoxyimino)acetyl]amino]-8-oxo-5-thia-1-azabicyclo[4.2.0]oct-2-ene-2-carboxylic acid; (6R,7R)-7-[2-(2-amino-4-thiazolyl)glyoxylamido]-3-(hydr-oxymethyl)-8-oxo-5-thia-1-azabicyclo[4.2.0]oct-2-ene-2-carboxylate 7²-(Z)-(O-methyloxime) acetate; 7-[2-(2-ami-no-4-thiazolyl)-2-methoxyiminoacetamido]cephalosporanic acid.* $C_{16}H_{17}N_5O_7S_2$; mol wt 455.48. C 42.19%, H 3.76%, N 15.38%, O 24.59%, S 14.08%. Broad spectrum third generation cephalosporin antibiotic. The name cefotaxime applies to the isomer having a *syn*-methoxyimino group. Prepn (unspecified stereochemistry): M. Ochiai *et al.,* **Ger.** pat. **2,556,736;** *eidem,* **U.S.** pat. **4,098,888** (1976, 1978 both to Takeda); *syn*-isomer: R. Heymes, A. Lutz, **Ger.** pat. **2,702,-501;** *eidem,* **U.S.** pat. **4,152,432** (1977, 1979 both to Roussel-Uclaf). Antibacterial activity: R. Wise, *Antimicrob. Ag. Chemother.* **14,** 807 (1978); H. W. Van handuyt, M. Pycka-vet, *ibid.* **16,** 109 (1979). Human pharmacology: R. Lüthy *et al., ibid.* 127. Symposia: *Drug Therapy* **Suppl.,** 1-108 (Jan. 1981); *Rev. Infect. Dis.* **4,** Suppl., S281-S488 (1982); *Diagn. Microbiol. Infect. Dis.* **2**(3), Suppl., 1S-92S (1984); *Infection* **13,** Suppl. 1, S1-S162 (1985). Comprehensive description: F. J. Muhtadi, M. M. A. Hassan, in *Analytical Profiles of Drug Substances* vol. 11, K. Florey, Ed. (Academic Press, New York, 1982) pp 139-168.

Sodium salt (*syn*-isomer), $C_{16}H_{16}N_5NaO_7S_2$, **HR-756,** *Ru-24756, Cefotax, Chemcef, Claforan, Makrocef, Pretor, Tolycar.* $[\alpha]_D^{20}$ +55° ±2 (c = 0.8 in water).
THERAP CAT: Antibacterial.

1936. Cefotetan. *[6R-(6α,7α)]-7-[[[4-(2-Amino-1-carboxy-2-oxoethylidene)-1,3-dithietan-2-yl]carbonyl]-amino]-7-methoxy-3-[[(1-methyl-1H-tetrazol-5-yl)thio]-methyl]-8-oxo-5-thia-1-azabicyclo[4.2.0]oct-2-ene-2-carb-oxylic acid;* (6R,7S)-7-[4-(carbamoylcarboxymethylene)-1,3-dithietane-2-carboxamido]-7-methoxy-3-[[(1-methyl-1H-tetrazol-5-yl)thio]methyl]-8-oxo-5-thia-1-azabicyclo-[4.2.0]oct-2-ene-2-carboxylic acid. $C_{17}H_{17}N_7O_8S_4$; mol wt 575.62. C 35.47%, H 2.98%, N 17.03%, O 22.24%, S 22.28%. Broad-spectrum injectable semi-synthetic antibiotic derived from cephamycin C, *q.v.* Prepn: M. Iwanami *et al.*, **Ger. pat. 2,824,559;** *eidem,* **U.S. pat. 4,263,432** (1979, 1981 both to Yamanouchi); *eidem, Chem. Pharm. Bull.* **28,** 2629 (1980). *In vitro* comparison of antibacterial activity: B. Chatto-padhyay, J. C. Teli, *J. Antimicrob. Chemother.* **10,** 151 (1982). Toxicity study: K. Imamura *et al.*, *Chemotherapy (Tokyo)* **30,** Suppl. 1, 212 (1982). Series of articles on phar-macology, clinical studies, toxicology: *ibid.* 1-947; *J. Anti-microb. Chemother.* **11,** Suppl. A, 1-236 (1983). Review of antibacterial activity, pharmacokinetics, therapeutic use: A. Ward, D. M. Richards, *Drugs* **30,** 382-426 (1985).

Disodium salt, $C_{17}H_{15}N_7Na_2O_8S_4$, **ICI 156834,** *YM 09330, Apatef, Cefotan, Ceftenon, Cepan, Darvilen, Yamatetan.* LD_{50} in male mice, rats (g/kg): > 10 orally and s.c. both species; 6.35, 8.48 i.v.; 8.12, 8.37 i.p. (Imamura).
THERAP CAT: Antibacterial.

1937. Cefotiam. *7-[[(2-Amino-4-thiazolyl)acetyl]-amino]-3-[[[1-[2-(dimethylamino)ethyl]-1H-tetrazol-5-yl]-thio]methyl]-8-oxo-5-thia-1-azabicyclo[4.2.0]oct-2-ene-2-carboxylic acid;* 7β-[2-(aminothiazol-4-yl)acetamido]-3-[[[1-(2-dimethylaminoethyl)-1H-tetrazol-5-yl]thio]methyl]-ceph-3-em-4-carboxylic acid. $C_{18}H_{23}N_9O_4S_3$; mol wt 525.62. C 41.13%, H 4.41%, N 23.98%, O 12.18%, S 18.30%. Broad-spectrum, semi-synthetic cephalosporin antibiotic. Prepn: S. Tsushima *et al.*, **Ger. pat. 2,607,064** (1976 to Ta-keda), *C.A.* **86,** 72674 (1977). Synthesis: M. Numata *et al.*, *J. Antibiot.* **31,** 1262 (1978). Pharmacology: K. Tsuchiya *et al., ibid.* 1272. *In vitro* and *in vivo* activity: *eidem, Antimi-crob. Ag. Chemother.* **14,** 557 (1978). Absorption, distribu-tion, excretion: *eidem, J. Antibiot.* **31,** 1272 (1978). Metabo-lism: S. Tanayama *et al., ibid.* 703. Effect in exptl pneumo-nia: T. Nishi, K. Tsuchiya, *Antimicrob. Ag. Chemother.* **18,** 549 (1980). Review: H. C. Neu, *Ann. Rev. Pharmacol. Toxi-col.* **22,** 599-642 (1982). Series of articles on pharmacology, antibacterial activity and clinical use: *Chemotherapy (Tokyo)* **36,** Suppl. 6, 1-884 (1988).

Dihydrochloride, $C_{18}H_{25}Cl_2N_9O_4S_3$, *Abbott-48999, CGP-14221/E, SCE-963, Alospar, Ceradon, Halospor, Pansporin, Pansporine, Spizef, Sporidyn.* White to light yellow crystals. Sol in methanol. Slightly sol in ethanol.
THERAP CAT: Antibacterial.

1938. Cefoxitin. *3-[[(Aminocarbonyl)oxy]methyl]-7-methoxy-8-oxo-7-[(2-thienylacetyl)amino]-5-thia-1-azabi-cyclo[4.2.0]oct-2-ene-2-carboxylic acid;* 3-(hydroxymethyl)-7-methoxy-8-oxo-7-[2-(2-thienyl)acetamido]-5-thia-1-aza-bicyclo[4.2.0]oct-2-ene-2-carboxylic acid carbamate (ester); 3-carbamoyloxymethyl-7α-methoxy-7-[2-(2-thienyl)acet-amido]-3-cephem-4-carboxylic acid. $C_{16}H_{17}N_3O_7S_2$; mol wt 427.46. C 44.96%, H 4.01%, N 9.83%, O 26.20%, S 15.00%. Semi-synthetic antibiotic derived from cephamycin C, *q.v.*, possessing high resistance to β-lactamase inactiva-tion. Synthesis: Christensen *et al.*, **Ger. pats. 2,129,675; 2,203,653** corresp to **U.S. pat. 4,297,488** (1971, 1972, 1981 all to Merck & Co.); Karady *et al., J. Am. Chem. Soc.* **94,** 1410 (1972); Ratcliffe, Christensen, *Tetrahedron Letters* **1973,** 4653. Biological evaluation: Wallick, Hendlin, *Anti-microb. Ag. Chemother.* **5,** 25 (1974); Miller *et al., ibid.* 33; Onishi *et al., ibid.* 38; Hamilton, Miller *et al., J. Antibiot.* **27,** 42 (1974). Mode of action: Onishi *et al., Ann. N.Y. Acad. Sci.* **235,** 406 (1974). Toxicity: S. Takayama *et al., Chemo-therapy (Tokyo)* **26,** Suppl. 1, 150 (1978). Comprehensive reviews: *J. Antimicrob. Chemother.* **4,** Suppl. B, 1-256 (1978); R. N. Brogden *et al., Drugs* **17,** 1-37 (1979); E. O. Stapley, K. R. Brown, in *Pharmacological and Biochemical Properties of Drug Substances* vol. 3, M. E. Goldberg, Ed. (Am. Pharm. Assoc., Washington, DC, 1981) pp 262-290. Comprehensive description: G. S. Brenner in *Analytical Pro-files of Drug Substances* vol. 11, K. Florey, Ed. (Academic Press, New York, 1982) pp 169-195.

Crystals, mp 149-150° (dec). pKa 2.2. Very sol in ace-tone; sol in aq $NaHCO_3$; very slightly sol in water. Practi-cally insol in ether and chloroform.
Sodium salt, $C_{16}H_{16}N_3NaO_7S_2$, *MK-306, Betacef, Farmox-in, Mefoxin, Mefoxitin, Merxin, Cenomycin.* White crystals with characteristic odor. $[\alpha]_{589nm}^{25}$ +210° (c = 1 in metha-nol). Very sol in water; sol in methanol; sparingly sol in ethanol and acetone. Insol in aromatic and aliphatic hy-drocarbons. LD_{50} in mice, rats, dogs (g/kg): 5.10, 8.98, > 10.0 i.v. (Takayama).
THERAP CAT: Antibacterial.

1939. Cefpimizole. *[6R-[6α,7β(7R*)]]-1-[[2-Carboxy-7-[[[[(5-carboxy-1H-imidazol-4-yl)carbonyl]amino]phenyl-acetyl]amino]-8-oxo-5-thia-1-azabicyclo[4.2.0]oct-2-en-3-yl]methyl]-4-(2-sulfoethyl)pyridinium hydroxide, inner salt;* 1-[(6R,7R)-2-carboxy-7-[(R)-2-(5-carboxy-4-imidazolyl-carboxamido)-2-phenylacetamido]-8-oxo-5-thia-1-azabicy-clo[4.2.0]oct-2-en-3-ylmethyl]pyridino-4-ethylsulfonate; 7-β-[D-(—)-α-(4-carboxyimidazole-5-carboxamido)phenyl-acetamido]-3-(4-β-sulfoethylpyridinium)methyl-3-cephem-4-carboxylic acid; U-63196; AC 1370. $C_{28}H_{26}N_6O_{10}S_2$; mol wt 670.67. C 50.14%, H 3.91%, N 12.53%, O 23.86%, S 9.56%. Third generation injectable cephalosporin antibiotic. Prepn: N. Yasuda *et al.*, **Ger. pat. 2,826,546;** *eidem,* **U.S. pat. 4,217,450** (1979, 1980 both to Ajinomoto); N. Yasuda *et al., J. Antibiot.* **36,** 242 (1983). *In vitro* antibacterial activi-ty and β-lactamase stability: H. C. Neu, P. Labthavikul, *Antimicrob. Ag. Chemother.* **24,** 375 (1983). Potentiating effect on phagocyte functions: H. Ohnishi *et al., ibid.* **23,** 874 (1983). Toxicity study: S. Hashimoto *et al., Toxicol. Letters* **23,** 135 (1984). Therapeutic efficacy in mice: Y. Obana *et al., J. Antimicrob. Chemother.* **16,** 727 (1985). HPLC determn in human plasma and urine: D. B. Lakings, J. M. Wozniak, *J. Chromatog.* **308,** 261 (1984). Pharmacoki-netics in humans: D. B. Lakings *et al., Antimicrob. Ag.*

Chemother. **29**, 271 (1986). Efficacy and tolerance in gonorrhea in men: E. T. Sandberg *et al., ibid.* 849.

Monosodium salt, $C_{28}H_{25}N_6NaO_{10}S_2$, *U-63196E, Ajicef, Renilan.* $[\alpha]_D^{20}$ $-28.2°$ (c = 0.5 in water). uv max (water): 257 nm (ε 22400). Sol in water. LD_{50} in male, female mice, male, female rats (g/kg): 2.7, 2.9, 4.2, 3.5 i.v.; 8.2, 6.8, 12.2, 11.5 s.c.; all > 15.0 orally (Hashimoto).

THERAP CAT: Antibacterial.

1940. Cefpiramide. *[6R-[6α,7β(R*)]]-7-[[[[(4-Hydroxy-6-methyl-3-pyridinyl)carbonyl]amino](4-hydroxyphenyl)-acetyl]amino]-3-[[(1-methyl-1H-tetrazol-5-yl)thio]methyl]-8-oxo-5-thia-1-azabicyclo[4.2.0]oct-2-ene-2-carboxylic acid;* 7-[(R)-2-(4-hydroxy-6-methylnicotinamido)-2-(p-hydroxyphenyl)acetamido]-3-[[(1-methyl-1H-tetrazol-5-yl)-thio]methyl]-8-oxo-5-thia-1-azabicyclo[4.2.0]oct-2-ene-2-carboxylic acid; D-7-[[(4-hydroxy-6-methylnicotinamido)-4-hydroxyphenylacetamido]-3-(1-methyltetrazol-5-yl)thiomethylcephem-4-carboxylic acid. $C_{25}H_{24}N_8O_7S_2$; mol wt 612.63. C 49.01%, H 3.95%, N 18.29%, O 18.28%, S 10.47%. Broad spectrum semi-synthetic cephalosporin antibiotic. Prepn: H. Yamada *et al.,* **Belg.** pat. **833,063;** *eidem,* **U.S.** pats. **4,156,724; 4,160,087** (1975, 1979, 1979 all to Sumitomo); *eidem,* **Japan. Kokai 79 119,497** (1979 to Sumitomo), *C.A.* **92,** 11042j (1980); I. Isaka *et al.,* **Japan. Kokai 79 30,-197** (1979 to Yamanouchi), *C.A.* **91,** 57037a (1979); **Japan. Kokai 84 65,094** (1984 to Sumitomo), *C.A.* **101,** 110641w (1984). Prepn, NMR data, antibacterial activity: H. Yamada *et al., J. Antibiot.* **36,** 522, 543 (1983). Pharmacokinetics: H. Matsui *et al., Antimicrob. Ag. Chemother.* **22,** 213 (1982); K. Sata *et al., ibid.* **26,** 578 (1984). Metabolism: H. Imasaki *et al., ibid.* **24,** 42 (1983). In vitro activity: H. Wexler *et al., ibid.* **25,** 162 (1984); M. A. Pfaller *et al., ibid.* 368. Comparison with other cephalosporins: N. J. Khan *et al., ibid.* **26,** 585 (1984). Series of articles on *in vitro, in vivo* activity, determn in body fluids, pharmacokinetics, clinical studies: *Chemotherapy (Japan)* **31,** Suppl. 1, 1-842 (1983).

Yellow crystals, mp 213-215° (dec).

Sodium salt, $C_{25}H_{23}N_8NaO_7S_2$, *SM 1652, WY 44635, Cefpiran, Suncefal, Sepatren.*

THERAP CAT: Antibacterial.

1941. Cefpodoxime Proxetil. *[6R-[6α,7β(Z)]]-7-[[(2-Amino-4-thiazolyl)(methoxyimino)acetyl]amino]-3-(methoxymethyl)-8-oxo-5-thia-1-azabicyclo[4.2.0]oct-2-ene-2-carboxylic acid 1-[[(1-methylethoxy)carbonyl]oxy]ethyl ester;* 1-(isopropoxycarbonyloxy)ethyl (6R,7R)-7-[2-(2-amino-4-thiazolyl)-(Z)-2-(methoxyimino)acetamido]-3-methoxymethyl-3-cephem-4-carboxylate; CS-807; U-76252; Banan. $C_{21}H_{27}N_5O_9S_2$; mol wt 557.59. C 45.24%, H 4.88%, N 12.56%, O 25.82%, S 11.50%. Broad spectrum, orally absorbed third generation cephalosporin, ester prodrug of the active free acid metabolite, cefpodoxime. Prepn: H. Nakao *et al.,* **Eur.** pat. **Appl. 49,118;** *eidem,* **U.S.** pat. **4,486,425** (1982, 1984 both to Sankyo). Prepn, pharmacokinetics and NMR analysis: K. Fujimoto *et al., J. Antibiot.* **40,** 370 (1987). In vitro and in vivo antibacterial activity: Y. Utsui *et al., Antimicrob. Ag. Chemother.* **31,** 1085 (1987). In vitro antibacterial spectrum and susceptibility testing of free acid: R. N. Jones, A. L. Barry, *ibid.* **32,** 443 (1988). Series of ar-

ticles on antibacterial spectrum, pharmacokinetics, toxicology and clinical studies: *Chemotherapy (Tokyo)* **36,** Suppl. 1, 1-1126 (1988). Review of chemistry, antibacterial activity, toxicity and clinical studies: H. Nakao *et al., Sankyo Kenkyusho Nempo* **39,** 1-44 (1987), *C.A.* **109,** 27503x (1988).

LD_{50} in male, female mice, male, female rats (mg/kg): > 10000, > 10000, > 2000, > 2000 s.c., 3502, 2535, > 4000, > 4000 i.p.; > 8000, > 8000, > 4000, > 4000 orally (Nakao, 1988).

Free acid, $C_{15}H_{17}N_5O_6S_2$, *R-3763, U-76253, cefpodoxime.* Free acid, sodium salt, $C_{15}H_{16}N_5NaO_6S_2$, *R-3746, U-76253A.*

THERAP CAT: Antibacterial.

1942. Cefroxadine. *7-[(Amino-1,4-cyclohexadien-1-yl-acetyl)amino]-3-methoxy-8-oxo-5-thia-1-azabicyclo[4.2.0]-oct-2-ene-2-carboxylic acid;* 7-[D-2-amino-2-(1,4-cyclohexadienyl)acetamido]-3-methoxy-3-cephem-4-carboxylic acid; CGP-9000; Oraspor. $C_{16}H_{19}N_3O_5S$; mol wt 365.41. C 52.59%, H 5.24%, N 11.50%, O 21.89%, S 8.77%. Orally active cephalosporin deriv. Prepn: R. Scartazzini, H. Bickel, **Ger.** pat. **2,331,133** corresp to **U.S.** pat. **4,073,902** (1974, 1978 both to Ciba-Geigy). Antibiotic activity: O. Zak *et al., J. Antibiot.* **29,** 653 (1976). In vivo and in vitro microbiological evaluation: K. Yasuda *et al., Antimicrob. Ag. Chemother.* **18,** 105 (1980). Pharmacokinetics: T. Bergan, *Chemotherapy* **26,** 225 (1980). Series of articles on activity, toxicity, reproduction studies: *Chemotherapy (Tokyo)* **28,** Suppl 3, 1-335 (1980).

Internal salt, mp 170° (dec.). $[\alpha]_D^{20}$ +87° (c = 1.093 in 0.1N HCl). uv max (0.1N HCl): 267 nm (ε 6,100). LD_{50} in mice (mg/kg): > 6000 orally; 7,090 i.p., R. Scartazzini, H. Bickel, *loc. cit.*

THERAP CAT: Antibacterial.

1943. Cefsulodin. *[6R-[6α,7β(R*)]]-4-(Aminocarbonyl)-1-[[2-carboxy-8-oxo-7-[(phenylsulfoacetyl)amino]-5-thia-1-azabicyclo[4.2.0]oct-2-en-3-yl]methyl]pyridinium hydroxide inner salt;* 7-(α-sulphophenylacetamido)-3-(4'-carbamoylpyridinium)methyl-3-cephem-4-carboxylic acid. $C_{22}H_{20}N_4O_8S_2$; mol wt 532.55. C 49.62%, H 3.79%, N 10.52%, O 24.03%, S 12.04%. Third generation cephalosporin antibiotic. Prepns: S. Morimoto *et al.,* **Ger.** pat. **2,234,-280;** *eidem,* **U.S.** pat. **4,065,619** (1973, 1977 both to Takeda). Prepn and separation of isomers: H. Nomura *et al., J. Med. Chem.* **17,** 1312 (1974). In vitro and in vivo antibacterial activity: K. Tsuchiya *et al., Antimicrob. Ag. Chemother.* **13,** 137 (1978). Absorption, distribution, excretion in mice, rats, dogs: *eidem, J. Antibiot.* **31,** 593 (1978). Activity and susceptibility to β-lactamases: A. King *et al., Antimicrob. Ag. Chemother.* **17,** 165 (1980). In vitro comparison with other antibacterials: H. Grimm, *Arzneimittel-Forsch.* **30,** 1478 (1980). Clinical pharmacology: J. Fuellhaas *et al., Curr. Chemother., Proc. 10th Int. Congr. Chemother.* (Washington, D.C., 1978) **2,** 848-851. Review of activity: H. C. Neu, B. E. Scully, *Rev. Infec. Dis.* **6,** Suppl. 3, S667-S677 (1984). *Review:* A. Bryskier, *Lyon Pharm.* **34,** 343-355 (1983); D. B. Wright, *Drug Intell. Clin. Pharm.* **20,** 845-849 (1986).

Sodium salt, $C_{22}H_{19}N_4NaO_8S_2$, *Abbott-46811, CGP-7174/E, SCE-129, sulcephalosporin, Cefomonil, Monaspor, Pseudomonil, Pseudocef, Pyocefal, Takesulin, Tilmapor, Ulfaret.* Colorless needles from ethanol/water, mp 175° (dec). LD_{50} in mice (mg/kg): > 4000 i.p.; > 15000 orally (Bryskier).

d-Form sodium salt, $[\alpha]_D^{23}$ +16.5° (c = 1.08 in water). uv max (water): 263 nm (ϵ 14600).

l-Form sodium salt, $[\alpha]_D^{23}$ −16.8° (c = 1.01 in water).
THERAP CAT: Antibacterial.

1944. Ceftazidime. *[6R-[6α,7β(Z)]]-1-[[7-[[(2-Amino-4-thiazolyl)[(1-carboxy-1-methylethoxy)imino]acetyl]amino]-2-carboxy-8-oxo-5-thia-1-azabicyclo[4.2.0]oct-2-en-3-yl]methyl]pyridinium hydroxide, inner salt;* 1-[[(6R,7R)-7-[2-(2-amino-4-thiazolyl)glyoxylamido]-2-carboxy-8-oxo-5-thia-1-azabicyclo[4.2.0]oct-2-en-3-yl]methyl]pyridinium hydroxide, inner salt 7^2-(Z)-[O-(1-carboxy-1-methylethyl)-oxime]; GR-20263; Fortam; Kefamin; Modacin. $C_{22}H_{22}N_6O_7S_2$; mol wt 546.58. C 48.34%, H 4.06%, N 15.38%, O 20.49%, S 11.73%. Third generation cephalosporin antibiotic. Prepn: C. H. O'Callaghan *et al.,* **Ger. pat. 2,921,316;** *eidem,* **U.S. pat. 4,258,041** (1979, 1981 both to Glaxo). Prepn of crystalline pentahydrate: A. Brodie, L. A. Wetherill, **Ger. pat. 3,037,102;** *eidem,* **U.S. pat. 4,329,453** (1981, 1982 both to Glaxo). Chemical and antibacterial properties: *eidem, Antimicrob. Ag. Chemother.* **17,** 876 (1980). Activity vs. *Pseudomonas* and *Enterobacteriaceae:* L. Verbist, J. Verhaegen, *ibid.* 807. *In vitro* comparison with other β-lactams: R. Wise *et al., ibid.* 884; L. Verbist, *ibid.* **19,** 407 (1981). Symposium on clinical studies: *J. Antimicrob. Chemother.* **12,** Suppl. A, 1-414 (1983). Review of antibacterial activity, pharmacokinetics and therapeutic use: D. M. Richards, R. N. Brogden, *Drugs* **29,** 105-161 (1985).

Pentahydrate, $C_{22}H_{22}N_6O_7S_2 \cdot 5H_2O$, *Ceftim, Fortaz, Fortum, Glazidim, Kefazim, Panzid, Spectrum, Tazicef, Tazidime.* Crystalline solid. uv max (pH 6): 257 nm ($E_{1cm}^{1\%}$ 348).
THERAP CAT: Antibacterial.

1945. Cefteram. *[6R-[6α,7β(Z)]]-7-[[(2-Amino-4-thiazolyl)(methoxyimino)acetyl]amino]-3-[(5-methyl-2H-tetrazol-2-yl)methyl]-8-oxo-5-thia-1-azabicyclo[4.2.0]oct-2-ene-2-carboxylic acid;* (+)-(6R,7R)-7-[(Z)-2-(2-amino-4-thiazolyl)-2-(methoxyimino)acetamido]-3-[(5-methyl-2H-tetrazol-2-yl)methyl]-8-oxo-5-thia-1-azabicyclo[4.2.0]oct-2-ene-2-carboxylic acid; 7-[2-(2-aminothiazol-4-yl)-2-*syn*-methoxyiminoacetamido]-3-[2-(5-methyl-1,2,3,4-tetrazolyl)methyl]-Δ^3-cephem-4-carboxylic acid; (+)-(6R,7R)-7-[2-(2-amino-4-thiazolyl)glyoxylamido]-3-[(5-methyl-2H-tetrazol-2-yl)methyl]-8-oxo-5-thia-1-azabicyclo[4.2.0]oct-2-ene-2-carboxylic acid 7^2-(Z)-(O-methyloxime); ceftetrame; Ro 19-5247; T-2525. $C_{16}H_{17}N_9O_5S_2$; mol wt 479.49. C 40.08%, H 3.57%, N 26.29%, O 16.68%, S 13.37%. Third generation, orally active cephalosporin antibiotic. Prepn: **Belg. pat. 890,499;** H. Sadaki *et al.,* **U.S. pat. 4,489,072** (1982, 1984 both to Toyama). *In vitro* antibacterial activity: R. Wise *et al., Antimicrob. Ag. Chemother.* **29,** 1067 (1986); *in vitro* activity vs *Neisseria gonorrhoeae:* W. R. Bowie *et al., ibid.* **31,** 470 (1987). General pharmacology in animals: S. Hirai *et al., Japan. J. Antibiot.* **39,** 958 (1986), *C.A.* **105,**

183450a (1986). Pharmacokinetics and tissue distribution in animals: I. Saikawa *et al., ibid.* 979, *C.A.* **105,** 183338v (1986). Series of articles on activity, toxicology, pharmacology and clinical efficacy: *Chemotherapy (Tokyo)* **34,** Suppl. 2, 1-984 (1986). Acute toxicity: S. Sato *et al., ibid.* 166.

Crystals from acetone, mp > 200°.
Pivaloyloxymethyl ester, $C_{22}H_{27}N_9O_7S_2$, *cefteram pivoxil, T-2588, Tomiron.* Oral prodrug of cefteram. Crystals, mp 127-128° (dec). LD_{50} in male, female mice, rats (g/kg): > 6.00, 5.86, 5.63, 5.09 i.p.; in both species > 6.00 s.c.; in male mice, rats, dogs (g/kg): > 6.00, > 6.00, > 2.00 orally (Sato).
THERAP CAT: Antibacterial.

1946. Ceftezole. *8-Oxo-7-[(1H-tetrazol-1-ylacetyl)amino]-3-[(1,3,4-thiadiazol-2-ylthio)methyl]-5-thia-1-azabicyclo[4.2.0]oct-2-ene-2-carboxylic acid;* 7-(1H-tetrazol-1-yl)-acetamido-3-(1,3,4-thiadiazol-2-ylthio)methylceph-3-em-4-carboxylic acid; CG-B3Q; CTZ. $C_{13}H_{12}N_8O_4S_3$; mol wt 440.48. C 35.45%, H 2.75%, N 25.44%, O 14.53%, S 21.83%. Broad spectrum antibiotic similar to cefazolin, *q.v.* Prepn: T. Takano *et al.,* **S. Afr. pat. 68 04,513** corresp to **U.S. pat. 3,516,997** (1969, 1970 both to Fujisawa). *In vitro* and *in vivo* activity: T. Noto *et al., J. Antibiot.* **29,** 1058 (1976). Pharmacology: Y. Harada *et al., ibid.* 1071; M. Kakimoto *et al., Chemotherapy (Tokyo)* **24,** 722 (1976). Metabolism: K. Koyama *et al., ibid.* 619; Y. Tohira *et al., ibid.* 655. Clinical study: S. Ishiyama *et al., ibid.* 1006. Toxicity and teratogenicity studies: R. Niki *et al., ibid.* 671.

Solid, mp 155° (dec). uv max (phosphate buffer pH 6.4): 273 nm ($E_{1cm}^{1\%}$ 274).
Sodium salt, $C_{13}H_{11}N_8NaO_4S_3$, *FR 10123, Celoslin, Falomesin.* White needles, freely sol in water.
THERAP CAT: Antibacterial.

1947. Ceftibuten. *[6R-[6α,7β(Z)]]-7-[[2-(2-Amino-4-thiazolyl)-4-carboxy-1-oxo-2-butenyl]amino]-8-oxo-5-thia-1-azabicyclo[4.2.0]oct-2-ene-3-carboxylic acid;* 7β-[(Z)-2-(2-aminothiazol-4-yl)-4-carboxy-2-butenoylamino]-3-cephem-4-carboxylic acid; 7432-S; SCH 39720. $C_{15}H_{14}N_4O_6S_2$; mol wt 410.42. C 43.90%, H 3.44%, N 13.65%, O 23.39%, S 15.62%. Orally active third generation cephalosporin antibiotic. Prepn: Y. Hamashima, **Eur. pat. Appl. 136,721;** *idem,* **U.S. pat. 4,634,697** (1985, 1987 both to Shionogi); and *in vitro* antibacterial activity: Y. Hamashima *et al., J. Antibiot.* **40,** 1468 (1987). Industrial synthesis utilizing penicillin G: M. Yoshioka, *Pure Appl. Chem.* **59,** 1041 (1987). Comparative antibacterial spectrum: R. N. Jones, A. L. Barry, *Antimicrob. Ag. Chemother.* **32,** 1576 (1988). Pharmacokinetics in humans: M. Nakashima *et al., J. Clin. Pharmacol.* **28,** 246 (1988).

THERAP CAT: Antibacterial.

1948. Ceftiofur. *[6R-[6α,7β(Z)]]-7-[[(2-Amino-4-thi-azolyl)(methoxyimino)acetyl]amino]-3-[[(2-furanylcarbon-yl)thio]methyl]-8-oxo-5-thia-1-azabicyclo[4.2.0]oct-2-ene-2-carboxylic acid; (6R,7R)-7-[2-(2-amino-4-thiazolyl)gly-oxylamido]-3-(mercaptomethyl)-8-oxo-5-thia-1-azabicy-clo[4.2.0]oct-2-ene-2-carboxylic acid 7²-(Z)-(O-methylox-ime) 2-furoate (ester).* $C_{19}H_{17}N_5O_7S_3$; mol wt 523.55. C 43.59%, H 3.27%, N 13.38%, O 21.39%, S 18.37%. Broad spectrum, third generation cephalosporin antibiotic. Prepn: B. Labeeuw, A. Salhi, **Eur. pat. Appl. 36,812;** *eidem,* **U.S. pat. 4,464,367** (1981, 1984 both to Sanofi). *In vivo* and *in vitro* antibacterial activity and β-lactamase stability: R. J. Yancey *et al., Am. J. Vet. Res.* **48,** 1050 (1987).

Monosodium salt, $C_{19}H_{16}N_5NaO_7S_3$, *CM 31-916,* U 64279E, *Excenel, Naxcel.*
Monohydrochloride, $C_{19}H_{18}ClN_5O_7S_3$, *U 67279A.*
THERAP CAT (VET): Antibacterial.

1949. Ceftizoxime. *[6R-[6α,7β(Z)]]-7-[[(2-Amino-4-thiazolyl)(methoxyimino)acetyl]amino]-8-oxo-5-thia-1-aza-bicyclo[4.2.0]oct-2-ene-2-carboxylic acid; (6R,7R)-7-[2-(2-amino-4-thiazolyl)glyoxylamido]-8-oxo-5-thia-1-azabicy-clo[4.2.0]oct-2-ene-2-carboxylic acid 7²-(Z)-(O-methylox-ime).* $C_{13}H_{13}N_5O_5S_2$; mol wt 383.40. C 40.72%, H 3.42%, N 18.27%, O 20.87%, S 16.72%. Injectable third generation cephalosporin antibiotic related to cefotaxime, *q.v.* Exhibits broad spectrum activity and resistance to β-lactamase hydrolysis. Prepn: T. Takaya *et al.,* **Ger. pat. 2,810,922;** *eidem,* **U.S. pat. 4,427,674** (1978, 1984 both to Fujisawa). *In vitro* and *in vivo* antibacterial activity: T. Kamimura *et al., Antimicrob. Ag. Chemother.* **16,** 540 (1979). Comparison with other cephalosporins: M. Nishida *et al., J. Antibiot.* **32,** 1319 (1979). Pharmacokinetics: T. Murakawa *et al., Antimicrob. Ag. Chemother.* **17,** 157 (1980). Series of articles on metabolism, pharmacology, laboratory evaluation: *Arznei-mittel-Forsch.* **30,** 1662-1687 (1980). Toxicity and reproduction studies: K. Fukuhara *et al., ibid.* 1669. Phase I clinical study: N. Nakashima *et al., J. Clin. Pharmacol.* **21,** 388 (1981). *Reviews: J. Antimicrob. Chemother.* **10,** Suppl. C, 1-355 (1982); D. M. Richards, R. C. Heel, *Drugs* **29,** 281-329 (1985).

Crystals, mp 227° (dec).
Sodium salt, $C_{13}H_{12}N_5NaO_5S_2$, *FK-749, FR-13,479, SKF-88373, Cefizox, Ceftix, Epocelin, Eposerin.* LD_{50} in mice, rats (mg/kg): ~6000 i.v. (Fukuhara).
THERAP CAT: Antibacterial.

1950. Ceftriaxone. *[6R-[6α,7β(Z)]]-7-[[(2-Amino-4-thiazolyl)(methoxyimino)acetyl]amino]-8-oxo-3-[[(1,2,5,6-tetrahydro-2-methyl-5,6-dioxo-1,2,4-triazin-3-yl)thio]-methyl]-5-thia-1-azabicyclo[4.2.0]oct-2-ene-2-carboxylic acid; (6R,7R)-7-[2-(2-amino-4-thiazolyl)glyoxylamido]-3-[[(2,5-dihydro-6-hydroxy-2-methyl-5-oxo-as-triazin-3-yl)thio]methyl]-8-oxo-5-thia-1-azabicyclo[4.2.0]oct-2-ene-2-carboxylic acid 7²-(Z)-(O-methyloxime); cefatriaxone.* $C_{18}H_{18}N_8O_7S_3$; mol wt 554.58. C 38.98%, H 3.27%, N 20.21%, O 20.20%, S 17.34%. Parenteral third generation cephalosporin antibiotic. Prepn: M. Montavon, R. Reiner, **Brit. pat. Appl. 2,022,090;** *eidem,* **U.S. pat. 4,327,210** (1979,

1982 both to Hoffmann-La Roche); R. Reiner *et al., J. Antibiot.* **33,** 783 (1980). *In vitro* and *in vivo* studies: P. Angehrn *et al., Antimicrob. Ag. Chemother.* **18,** 913 (1980). Pharmacokinetics: M. Seddon *et al., ibid.* 240. Determn in plasma, urine, bile: K. H. Trautmann, P. Haefelfinger, *J. High Reso-lut. Chromatogr. Chromatogr. Commun.* **4,** 54 (1981), *C.A.* **94,** 202437 (1981). Mechanism of action study: R. B. Wright *et al., J. Antibiot.* **34,** 590 (1981). Kinetics in humans: K. Stoekel *et al., Clin. Pharmacol. Ther.* **29,** 650 (1981). Toxicity data: K. Teelman *et al.,* "Experimentelle Toxikologie von Ceftriaxon", in *Ceftriaxon ein neues paren-terales Cephalosporin,* Proc. Hahnenklee Symp., 1981, R. Grieshalber, Ed. (Editiones Roche, Basel, 1982) pp 91-111. Review of antibacterial activity, toxicology, pharmacology and clinical studies: D. M. Richards *et al., Drugs* **27,** 469 (1984); T. R. Beam, *Pharmacotherapy* **5,** 237-253 (1985).

Disodium salt hemiheptahydrate, $C_{18}H_{16}N_8Na_2O_7S_3$·3½$H_2O$, *Ro-13-9904/001, Rocefin, Rocephin(e).* White cryst powder, melts above 155° (dec). $[α]_D^{25}$ −165° (c = 1 in water) (calc for water-free substance). uv max (water): 242, 272 nm (ε 32300, 29530). pKa: ~3 (COOH), 3.2 (NH_3^+), 4.1 (enolic OH). Soly in water at 25°: approx 40 g/100 ml. LD_{50} in male, female, mice, rats (mg/kg): 3000, 2800, 2175, 2175 i.v.; >10000 all species orally; >5000 all species s.c. (Teelman).
THERAP CAT: Antibacterial.

1951. Cefuroxime. *[6R-[6α,7β(Z)]]-3-[[(Aminocarbon-yl)oxy]methyl]-7-[[2-furanyl(methoxyimino)acetyl]amino]-8-oxo-5-thia-1-azabicyclo[4.2.0]oct-2-ene-2-carboxylic acid; (6R,7R)-3-carbamoyloxymethyl-7-[2-(2-furyl)-2-(meth-oxyimino)acetamido]-8-oxo-5-thia-1-azabicyclo[4.2.0]oct-2-ene-2-carboxylic acid; (6R,7R)-3-carbamoyloxymethyl-7-[2-(2-furyl)-2-(methoxyimino)acetamido]ceph-3-em-4-carboxylic acid.* $C_{16}H_{16}N_4O_8S$; mol wt 424.40. C 45.28%, H 3.80%, N 13.20%, O 30.16%, S 7.55%. Prepn: M. C. Cook *et al.,* **Ger. pat. 2,439,880;** *eidem,* **U.S. pat. 3,974,153** (1973, 1976, both to Glaxo). *In vitro* studies: C. H. O'Cal-laghan *et al., Antimicrob. Ag. Chemother.* **9,** 511 (1976); R. N. Jones *et al., ibid.* **12,** 47 (1977). *In vitro* antibacterial activity, human pharmacokinetics: C. H. O'Callaghan *et al., J. Antibiot.* **29,** 29 (1976). Pharmacology: H. Freiesleben *et al., Proc. 10th Int. Congr. Chemother.,* Zürich, 1977 (Am. Soc. for Microbiol., Washington, 1978) **II,** pp 873-874. Pharmacokinetics: P. E. Gower, *ibid.* 877-878; J. Kosmidis *et al., ibid.* 875-876. Clinical studies: P. F. Wood *et al., ibid.* 1042-1044; R. Norrby *et al., J. Antimicrob. Chemother.* **3,** 355 (1977). Review of antibacterial activity, pharmacology and therapeutic efficacy: R. N. Brogden *et al., Drugs* **17,** 233-266 (1979).

White crystalline solid. $[α]_D^{20}$ +63.7° (c = 1.0 in 0.2M pH 7 phosphate buffer). uv max (pH 6 phosphate buffer): 274 nm (ε 17600).
Sodium salt, $C_{16}H_{15}N_4NaO_8S$, *Anaptivan, Biociclin, Bio-furex, Bioxima, Cefamar, Cefoprim, Cefossim, Cefumax, Cefurex, Cefurin, Curocef, Curoxim, Duxima, Gibicef, Ipacef, Kefurox, Kesint, Lampsporin, Medoxim, Polixima, Ultroxim, Zinacef.* White solid. $[α]_D^{20}$ +60° (c = 0.91 in water). uv max (water): 274 nm (ε 17400). Soly in water: 500 mg/2.5

ml. Solns are stable at room temp for 13 hrs; <10% decompn in 48 hrs at 25° (O'Callaghan, *J. Antibiot.*).

1-Acetoxyethyl ester, $C_{20}H_{22}N_4O_{10}S$, *cefuroxime axetil*, *CCI 15641, Axoril, Ceftin, Cepazine, Ximos, Zinnat*.

THERAP CAT: Antibacterial.

1952. Cefuzonam. *[6R-[6α,7β(Z)]]-7-[[(2-Amino-4-thiazolyl)(methoxyimino)acetyl]amino]-8-oxo-3-[(1,2,3-thiadiazol-5-ylthio)methyl]-5-thia-1-azabicyclo[4.2.0]oct-2-ene-2-carboxylic acid; (—)-(6R,7R)-7-[2-(2-amino-4-thiazolyl)-glyoxylamido]-8-oxo-3-[(1,2,3-thiadiazol-5-ylthio)methyl]-5-thia-1-azabicyclo[4.2.0]oct-2-ene-2-carboxylic acid 7²-(Z)-(O-methyloxime); 7β-[2-(2-aminothiazol-4-yl)-(Z)-2-methoxyiminoacetamido]-3-[(1,2,3-thiadiazol-5-yl)thio-methyl]ceph-3-em-4-carboxylic acid; CL-118523.* $C_{16}H_{15}N_7O_5S_4$; mol wt 513.58. C 37.42%, H 2.94%, N 19.09%, O 15.58%, S 24.97%. Semi-synthetic third generation cephalosporin. Prepn and bactericidal activity: W. V. Curran, A. S. Ross, **Fr. pat.** 2,488,258; *eidem*, **U.S. pat.** 4,399,132 (1982, 1983 both to Am. Cyanamid); *eidem, J. Antibiot.* **36**, 179 (1983). Improved prepn of sodium salt: K. Naito *et al.*, **Eur. pat.** 145,395 (1985 to Takeda). *In vitro* antibacterial spectrum: M. Hikada *et al., J. Antimicrob. Chemother.* **18**, 585 (1986). Toxicity study: S. Takita, *Chemotherapy (Tokyo)* **34**, Suppl. 3, 96 (1986). Conference proceedings: *Recent Adv. Chemother., Antimicrobial Sect. 2., Proc. 14th Int. Congr. Chemother., Kyoto 1985*, J. Ishigami, Ed. (Univ. Tokyo Press, Tokyo, 1985) pp 887-908. Series of articles: *Chemotherapy (Tokyo)* **34**, Suppl. 3, 1-732 (1986).

White amorphous solid. Absorption max (KBr): 5.62. LD_{50} in female, male mice, female, male rats (mg/kg): 4117, >4800, 4281, 4222 i.v.; 6424, 6783, >8000, >8000 i.p.; all >10000 orally (Takita).

Sodium salt dihydrate, $C_{16}H_{14}N_7NaO_5S_4 \cdot 2H_2O$, *L-105 (Lederle), CL-251931, LJC-10305, Cosmosin*. Yellowish crystalline powder. Sol in water.

THERAP CAT: Antibacterial.

1953. Celery Seed. Dried, ripe fruit of *Apium graveolens* L., *Umbelliferae. Habit.* Southern Europe; cultivated everywhere. *Constit.* Volatile and fixed oils; bitter extractive; resin.

1954. Celesticetin. *(S)-2-[(2-Hydroxybenzoyl)oxy]ethyl 6,8-dideoxy-7-O-methyl-6-[[(1-methyl-2-pyrrolidinyl)carbonyl]amino]-1-thio-D-erythro-α-D-galacto-octopyranoside; S-demethyl-3'-depropyl-S-[2-[(2-hydroxybenzoyl)oxy]ethyl]-7-O-methyllincomycin; 2-hydroxyethyl 6,8-dideoxy-7-O-methyl-6-(1-methyl-L-2-pyrrolidinecarboxamido)-1-thio-D-erythro-D-galacto-octopyranoside monosalicylate.* $C_{24}H_{36}N_2O_9S$; mol wt 528.64. C 54.53%, H 6.86%, N 5.30%, O 27.23%, S 6.07%. Antibiotic substance produced by *Streptomyces caelestis* from a soil obtained from Emigration Canyon near Salt Lake City: DeBoer *et al., Antibiotics Annual 1954-1955*, 831; Hoeksema *et al., ibid.* 837. Structure: Hoeksema, *J. Am. Chem. Soc.* **86**, 4224 (1964); **90**, 755 (1968).

Amphoteric, amorphous base. $[\alpha]_D^{24}$ +126.6° (c = 0.5 in chloroform). uv max (0.01N alcoholic H_2SO_4): 240, 310 nm ($E_{1cm}^{1\%}$ 183.7, 80.6). Soluble in acidic or strongly basic aq solns, practically insol in the pH range of 7 to 10. Soluble in the more polar organic solvents, but not in ether or light hydrocarbons. Active against gram-positive organisms and plant pathogens such as *B. stewartii, P. fascians, X. pruni* at 0.5 γ/ml. The antibacterial activity is rapidly destroyed in aq solns above pH 9 at 24°, but remains at pH 5 to 7 for at least 60 days. Celesticetin is more stable in acid than in basic solns, but is destroyed rapidly in 1N HCl at 100°. Cross resistance with erythromycin.

Salicylate, monoclinic tablets, mp 139°. $[\alpha]_D^{24}$ +90.2° (c = 0.5 in H_2O). Sol in water.

Oxalate, crystals, mp 149-154°. $[\alpha]_D^{24}$ +106.6° (c = 0.5 in H_2O). Sol in water.

1955. Celestin Blue. *1-(Aminocarbonyl)-7-(diethylamino)-3,4-dihydroxyphenoxazin-5-ium chloride;* 1-carbamoyl-7-(diethylamino)-3,4-dihydroxyphenazoxonium chloride; C.I. Mordant Blue 14; C.I. 51050; michrome no. 66; Corein 2R. $C_{17}H_{18}ClN_3O_4$; mol wt 363.80. C 56.12%, H 4.99%, Cl 9.75%, N 11.55%, O 17.59%. Prepd by the action of gallamide on *N,N*-diethyl-*p*-nitrosoaniline HCl: Gnehm, Bauer, *J. Prakt. Chem.* **72**, 249 (1905); Grandmougin, Bodmer, *ibid.* **77**, 502 (1908), *see also:* **U.S. pats.** 534,809 (1895) and 1,227,407 (1917); *Colour Index* vol. 4 (3rd ed., 1971). *Review:* H. J. Conn's *Biological Stains*, R. D. Lillie, Ed. (Williams & Wilkins, Baltimore, 9th ed., 1977) pp 407-408.

Greenish-black powder. With water it yields a reddish-purple soln, which appears blue on extreme diln. The alcoholic soln is blue in all concns. Approximate soly in water 2.0%, abs ethanol 1.5%, Cellosolve 2.25%, ethylene glycol 6.5%, xylene 0.005%.

USE: To dye fabrics a navy-blue color with faint reddish tinge. As nuclear and connective tissue stain. One of the ingredients of Picro-Mallory stain: Lendrum, McFarlane, *J. Pathol. Bacteriol.* **50**, 381 (1940).

1956. Celiprolol. *N'-[3-Acetyl-4-[3-[(1,1-dimethylethyl)amino]-2-hydroxypropoxy]phenyl]-N,N-diethylurea;* N-[3-acetyl-4-(3'-*tert*-butylamino-2'-hydroxy)propoxy]phenyl-*N'*-diethylurea; ST-1396. $C_{20}H_{33}N_3O_4$; mol wt 379.50. C 63.30%, H 8.76%, N 11.07%, O 16.86%. Cardioselective β_1-adrenergic blocker. Prepn: **Belg. pat.** 823,411; G. Zölss *et al.*, **U.S. pat.** 4,034,009 (1975, 1977 both to Chemie Linz). Hemodynamic effects: J. Bonelli *et al., Wien Klin. Wochenschr.* **90**, 350 (1978); H. Pittner, *Arch. Pharmacol.* **311**, Suppl., 180 (1980). Series of articles on determination in biological material, pharmacology, toxicology, clinical studies: *Arzneimittel-Forsch.* **33**, 1-79 (1983). Toxicity: W. Wendtlandt, H, Pittner, *ibid.* 41. Symposium on pharmacology, clinical efficacy and comparison with other β-blockers: *J. Cardiovasc. Pharmacol.* **8**, Suppl. 4, S1-S152 (1986).

Crystals, mp 110-112°.

Hydrochloride, $C_{20}H_{34}ClN_3O_4$, *Celectol, Selectol*. Crystals. Soly in water: >5%. LD_{50} in male mice, rats (mg/kg): 56.2, 68.3 i.v.; 1834, 3826 orally (Wendtlandt, Pittner).

THERAP CAT: Antihypertensive, antianginal.

1957. Cellobiose. *4-O-β-D-Glucopyranosyl-D-glucose;*

β-cellobiose; cellose; 4-(β-D-glucosido)-D-glucose. $C_{12}H_{22}O_{11}$; mol wt 342.30. C 42.10%, H 6.48%, O 51.42%. Unit of cellulose and lichenin. Does not occur free in nature, or as glucoside. Prepn from cotton: Braun, *Org. Syn.* coll. vol. II, 122, 124 (1943). Prepn from cell-free enzymatic hydrolyzate of cellulose: Whistler, Smart, *J. Am. Chem. Soc.* **75**, 1916 (1953). Structure: Haworth, Hirst, *J. Chem. Soc.* **119**, 193 (1923); Charlton *et al., ibid.* **1926**, 89; Zemplén, *Ber.* **59**, 1254 (1926); Haworth *et al., J. Chem. Soc.* **1927**, 2809; Peterson, Spencer, *J. Am. Chem. Soc.* **49**, 2822 (1927); Helferich *et al., Ber.* **63**, 992 (1930); Hess, Dziengel, *ibid.* **68**, 1594 (1935); Hassid, Ballou in *The Carbohydrates*, W. Pigman, Ed. (Academic Press, New York, 1957) p 490. Synthesis: Haskins *et al., J. Am. Chem. Soc.* **64**, 1289 (1942). *Review:* Pazur in *The Carbohydrates* vol. 2A, W. Pigman *et al.*, Eds. (Academic Press, New York, 2nd ed., 1970) pp 109-110; R. G. Edwards, *Dev. Food Carbohyd.* **2**, 229-273 (1980).

Minute crystals from dil alcohol which retain 0.25 to 0.50 mol water after drying in vacuo. Indifferent taste. Dec 225°. Shows mutarotation. $[\alpha]_D^{20}$ +14.2° → +34.6° (15 hrs, c = 8). One gram dissolves in 8 ml water, in 1.5 ml boiling water; almost insol in abs alc and ether. Reduces Fehling's soln. Hydrolysis with acid or emulsin yields 2 mols β-D-glucose. Not fermented by brewers' yeast, maltase, or invertase.

Octaacetyl-aldehydro-cellobiose, $C_{28}H_{38}O_{19}$, mp 105-110°. $[\alpha]_D^{20}$ +17.7° (c = 3 in chloroform).

Octaacetyl-α-cellobiose, $C_{28}H_{38}O_{19}$, mp 229°. $[\alpha]_D^{20}$ +41° (c = 6 in chloroform).

Octaacetyl-β-cellobiose, mp 202°. $[\alpha]_D^{20}$ −14.7° (c = 5 in chloroform).

1958. Cellocidin. *2-Butynediamide;* acetylenedicarboxamide; acetylenedicarboxylic acid diamide; aquamycin; lenamycin. $C_4H_4N_2O_2$; mol wt 112.09. C 42.86%, H 3.60%, N 24.99%, O 28.55%. $H_2NCOC≡CCONH_2$. Antibiotic substance with antibacterial activity. Produced by *Streptomyces chibaensis* from soil collected at Chiba City, Japan: Suzuki *et al., J. Antibiot.* **11**, 81 (1958). Synthesis from dimethyl acetylenedicarboxylate and concd ammonium hydroxide at −10°: Saggiomo, *J. Org. Chem.* **22**, 1171 (1957); Suzuki, Okuma, *J. Antibiot.* **11**, 84 (1958). Identity with lenamycin: Y. Sekizawa, *Meiji Seika Kenkyu Nempo* **1960**, 42, *C.A.* **56**, 14609a (1962). Biosynthesis: E. R. H. Jones, *J. Chem. Soc. Perkin Trans. I* **1973**, 148.

Crystals from dil methanol, dec. 216-218°. uv max (0.1N NaOH): 299 nm ($E_{1cm}^{1\%}$ 290). Sparingly sol in water, methanol, ethanol, acetone, chloroform, glacial acetic acid. Relatively stable in neutral or acid solns, showing no loss of activity at pH 2 to 7 when heated for 10 min at 100°. Unstable in alkaline soln evolving ammonia. LD_{50} i.v. in mice: 11 mg/kg (Suzuki).

1959. Cellophane. Francephane. Transparent, flexible cellulose sheeting made from viscose. The word "cellophane" is not a trademark in the U.S.

1960. Celluloid®. Pyralin; Zylonite. Prepd from nitrocellulose and camphor.

Colorless, amorphous mass. Flammable. Prone to spontaneous decompn which is retarded or prevented by the addition of urea, ZnO, $MgCO_3$, diphenylamine, etc. It is rendered less flammable by addition of ammonium phosphate. Softens in boiling water; sol in acetone.

USE: Plastic material for manuf of toilet articles, toys, photographic films; substitute for amber, ivory, ebonite, tortoise shell; also in surgery for bandages and in dentistry as substitute for rubber.

1961. Cellulose. $(C_6H_{10}O_5)_n$. Polysaccharide with the glucose units linked as in cellobiose. Chief constituent of the fiber of plants; cotton is the purest natural form, contg about 90%. Rayon is regenerated cellulose. Books: C. Dorée, *The Methods of Cellulose Chemistry* (Chapman & Hall, London, 1947); T. Lieser, *Kurzes Lehrbuch der Cellulosechemie* (Gebrüder Borntraeger, Berlin, 1953); S. D. Antonovskii, *Chemistry of Wood and Cellulose* (Vsesoyuz. Zaochnyi Lesotekh Instit., Leningrad, 1954); E. Ott *et al., Cellulose and Cellulose Derivatives,* **vols. 1-3** (Interscience, New York, 1954, 1955). *Reviews:* Several authors in *Encyclopedia of Polymer Science and Technology* vol. 3, N. M. Bikales, Ed. (Interscience, New York, 1965) pp 131-539; Shafizadeh, *Pure Appl. Chem.* **35**, 195-208 (1973); A. F. Turbak *et al.* in Kirk-Othmer *Encyclopedia of Chemical Technology* vol. 5 (Wiley-Interscience, New York, 3rd ed., 1979) pp 70-89. Comprehensive review on constitution, conformation, size of molecule, fine structure and superstructure: H. Krässig, *Papier (Darmstadt)* **33**, 9-20 (1979).

White substance. Practically insol in water or other usual solvents, but is dissolved by concd soln of zinc chloride, by ammoniacal copper hydroxide soln; also by caustic alkali with carbon disulfide.

Microcrystalline form, *Avicel.* Prepn and manuf of crystallite cellulosic aggregates: Battista, *Ind. Eng. Chem.* **42**, 502 (1950); Battista, Smith, U.S. pats. **2,978,446** and **3,141,-875** (1961 to Am. Viscose and 1964 to FMC). Non-fibrous powder. Particle shape: rigid rods. Refractive index: 1.55. Bulk density: 18-19 lb/cubic foot. Practically insol, but dispersible in water; partially sol with swelling in dil alkali; practically insol in and resistant to dil acid; practically insol and inert in organic acids.

USE: Fibrous form is the basic material for the textile and paper industries. Nitrated it yields nitrocellulose used for manuf of explosives, collodion, lacquers. Basic material also for cellulose acetate, cellulose xanthate. Also used in chromatography and as ion exchange material especially in the form of derivatives such as *DEAE-cellulose* (diethylaminoethyl cellulose) and ECTEOLA-cellulose, *q.v.* Microcrystalline forms of cellulose are used as combination binder-disintegrants in tableting, as separatory medium in thin-layer and column chromatography. Colloidal cellulose particles aid in stabilization and emulsification of liquid and foam systems. May be used as pure cellulose raw-material. Incorporation of cellulose crystallite aggregates in foods to reduce caloric content: Battista, U.S. pat. **3,023,104** (1962 to American Viscose); also used in food industry as stabilizer, thickener, texturizer.

1962. Cellulose Acetates. Partially acetylated cellulose, *q.v.* Several acetates of cellulose are known, which differ from one another only in the degree of acetylation. In triacetates, no less than 92% of the hydroxyl groups are acetylated. In characterizing the degree of acetylation, percent acetyl value and percent combined acetic acid are used. All cellulose acetates are obtained by treating cellulose with acetic anhydride at various temps for different lengths of time to produce amorphous white solid material in granular, flake, or powder form from which fibers may be formed by extrusion. In the plastics industry, it is usual to acetylate fully and then to lower the acetyl value to 52-56% by partial hydrolysis. Such material when compounded with suitable plasticizers gives a tough thermoplastic product. Manuf: Faith, Keyes & Clark's *Industrial Chemicals,* F. A. Lowenheim, M. K. Moran, Eds. (Wiley-Interscience, New York, 4th ed., 1975) pp 239-243. *Review:* G. A. Serad, J. R. Sanders, "Cellulose Acetate and Triacete Fibers" in Kirk-Othmer *Encyclopedia of Chemical Technology* Vol. 5 (Wiley-Interscience, New York, 3rd ed., 1979) pp 89-117.

Commercial products do not have sharp melting points. Solubility is affected by the acetyl value; the triacetate is insol in water, alcohol, ether, but sol in glacial acetic acid; the tetraacetate is insol in water, alcohol, ether, glacial acetic acid, methanol; the pentaacetate is insol in water, but sol in alcohol.

USE: Manuf rubber and celluloid substitutes, nonflammable photographic and cinema films, airplane dopes, varnishes and lacquers, filaments, phonograph records; waterproofing fabrics and rendering balloons gas-tight; sizing and finishing fabrics; coating skins; insulating electric wires; tow for cigarette smoke filters.

1963. Cellulose Ethyl Hydroxyethyl Ether. *Ethyl 2-*

hydroxyethyl ether cellulose; ethyl hydroxyethyl cellulose; Etulos. Prepn: Jullander, *C.A.* **48**, 6114g (1954); *idem*, *Ger. pat.* **1,000,367** (1957 to Mo och Domsjö Ab.), *C.A.* **54**, 5088f (1960). Use as laxative: Alm, *C.A.* **50**, 2122i (1956); *idem*, *Am. J. Dig. Dis.* [NS] **2**, 493 (1957).

R is H, —CH₂CH₃, or —CH₂CH₂OH

THERAP CAT: Cathartic.

1964. Centaurein. *7-(β-D-Glucopyranosyloxy)-5-hydroxy-2-(3-hydroxy-4-methoxyphenyl)-3,6-dimethoxy-4H-1-benzopyran-4-one;* 3',5,7-trihydroxy-3,4',6-trimethoxyflavone 7-β-D-glucoside. $C_{24}H_{26}O_{13}$; mol wt 522.47. C 55.17%, H 5.02%, O 39.81%. Isoln from root of *Centaurea jacea* L., *Compositae:* Bridel, Charaux, *Compt. Rend.* **175**, 833, 1168 (1922). Structure: Farkas *et al.*, *Ber.* **97**, 1666 (1964).

Monohydrate, yellow crystals, mp 208-209°. $[\alpha]_D^{20}$ —76.6° (c = 1.4 in methanol). uv max (methanol): 349, 258 nm (log ε 4.31, 4.30). Sol in hot water, hot alcohol, and hot acetone; practically insol in cold water, chloroform, ether.

1965. Centaury, American. Rose-pink; bitter-bloom; square-stem rose-gentian. Dried flowering plant of *Sabatia angularis* (L.) Pursh, *Gentianaceae*. *Habit.* Florida to Louisiana and Oklahoma, northern to southeastern New York; southern Ontario, Michigan, Wisconsin, Missouri. *Constit.* Gentiopicrin, amarogentin, gentisin: Korte, *Ber.* **87**, 1357 (1954); **88**, 704 (1955). Prepn of sedative extracts: Huneker, *Am. J. Pharm.* **43**, 207 (1871).

THERAP CAT: Bitter tonic.

1966. Centaury, Chilean. Chanchalagua; canchalagua. Dried flowering plant of *Centaurium cachanlahuen* (Mol.) Robins. *(Erythraea chilensis* (Willd.) Pers.), *Gentianaceae*. Formerly known as *Gentiana peruviana* Lam.; *Chironia chilensis* Willd. *Habit.* Chile, Peru. *Constit.* Gentiopicrin, amarogentin, gentisin: Korte, *Ber.* **87**, 1357 (1954); **88**, 704 (1955).

THERAP CAT: Bitter tonic.

1967. Centaury, Minor. Lesser centaury; bitter herb; bloodwort. Dried flowering plant of *Centaurium umbellatum* Gilib. *(Erythraea centaurium* Pers.), *Gentianaceae*. *Habit.* Europe, North America, Middle East, North Africa. *Constit.* Gentiopicrin, amarogentin, gentisin: Korte, *Ber.* **87**, 1357 (1954); **88**, 704 (1955).

THERAP CAT: Bitter tonic.

1968. Centaury, Spiked. Dried flowering plant of *Centaurium spicatum* (L.) Fern. *(Erythraea spicata* (L.) Pers.), *Gentianaceae*. *Habit.* Europe, Middle East, North Africa, North America. *Constit.* Gentiopicrin, amarogentin, gentisin: Korte, *Ber.* **87**, 1357 (1954); **88**, 704 (1955). Prepn of hypotensive extracts: Farrag, Sherif, *J. Pharm. Pharmacol.* **1**, 219 (1949).

1969. Cephacetrile Sodium. *3-[(Acetyloxy)methyl]-7-[(cyanoacetyl)amino]-8-oxo-5-thia-1-azabicyclo[4.2.0]oct-2-*ene-2-carboxylic acid monosodium salt; sodium 7-(2-cyanoacetamido)-3-(hydroxymethyl)-8-oxo-5-thia-1-azabicyclo[4.2.0]oct-2-ene-2-carboxylate acetate (ester); sodium 7-(2-cyanoacetamido)cephalosporanic acid; CIBA 36278-Ba; Celospor. $C_{13}H_{12}N_3NaO_6S$; mol wt 361.31. C 43.21%, H 3.35%, N 11.63%, Na 6.36%, O 26.57%, S 8.87%. Semisynthetic cephalosporin antibiotic. Prepn: **Neth. pat. Appl. 6,600,586** corresp to Bickel *et al.*, **U.S. pat. 3,483,197** (1966, 1969 both to Ciba). Activity studies: Knüsel *et al.*, *Antimicrob. Ag. Chemother.* **1970**, 140; Neu *et al.*, *J. Antibiot.* **25**, 400 (1972). Pharmacokinetics and clinical results: Kradolfer *et al.*, *Schweiz. Med. Wochenschr.* **103**, 711 (1973); Riess *et al.*, *ibid.* 718. Toxicology: Kradolfer *et al.*, *Antimicrob. Ag. Chemother.* **1970**, 150. Review: *Arzneimittel-Forsch.* **24**, 1446-1533 (1974).

White crystalline powder. LD_{50} in mice (mg/kg): 4500 ± 540 i.v.; 9100 ± 1500 s.c. (Kradolfer).

Free acid, needles from acetone + ether, mp 168-170° (dec). uv max (0.1N NaHCO₃): 260 nm (ε 9300).

THERAP CAT: Antibacterial.

1970. Cephaeline. *7',10,11-Trimethoxyemetan-6'-ol;* desmethylemetine; dihydropsychotrine. $C_{28}H_{38}N_2O_4$; mol wt 466.60. C 72.07%, H 8.21%, N 6.00%, O 13.72%. Next to emetine the most important alkaloid of ipecac, the ground roots of *Uragoga ipecacuanha* (Brot.) Baill. [*Cephaelis ipecacuanha* (Brot.) A. Rich.], *Rubiaceae*: Carr, Pyman, *J. Chem. Soc.* **105**, 1591 (1914); Hesse, *Ann.* **405**, 1 (1914). Structure: Pailer, Porschinski, *Monatsh.* **80**, 117 (1949). Stereochemistry: Van Tamelen *et al.*, *J. Am. Chem. Soc.* **79**, 4817 (1957). Partial synthesis: A. K. Garg, J. R. Gear, *Tetrahedron Letters* **50**, 4377 (1969). Biosynthetic studies: *eidem*, *ibid.* **1968**, 141. HPLC determn in biological fluids: D. J. Crouch *et al.*, *J. Anal. Toxicol.* **8**, 63 (1984). Review: M. Janot in Manske, Holmes, *The Alkaloids* vol. 3 (Academic Press, New York, 1953) pp 363-394.

Needles from ether. Faintly bitter taste. mp 115-116°. $[\alpha]_D^{20}$ —43.4° (c = 2 in chloroform). Practically insol in water. Freely sol in dil hydrochloric acid, dil sulfuric acid, acetic acid, methanol, ethanol, acetone, chloroform. Less sol in ether, petr ether.

Dihydrochloride heptahydrate, $C_{28}H_{40}Cl_2N_2O_4 \cdot 7H_2O$, prisms, sinters at 245°, slowly melts up to 270°. $[\alpha]_D^{20}$ +25.0° (c = 2). Sol in water. Solns turn yellow. Less sol in alc, acetone, chloroform. Practically insol in benzene, ligroin.

Dihydrobromide heptahydrate, $C_{28}H_{40}Br_2N_2O_4 \cdot 7H_2O$, prisms from dil hydrobromic acid, sinters at 266°, slowly melts up to 293°. Sol in water; moderately sol in alcohol, acetone. Practically insol in benzene.

Methyl ether, *see* Emetine.

THERAP CAT: Emetic; antiamebic.

THERAP CAT (VET): Has been used as an emetic and expectorant.

1971. Cephalexin. *7-[(Aminophenylacetyl)amino]-3-methyl-8-oxo-5-thia-1-azabicyclo[4.2.0]oct-2-ene-2-carboxylic acid;* 7-(D-α-aminophenylacetamido)desacetoxycephalosporanic acid; 7-(D-2-amino-2-phenylacetamido)-3-methyl-Δ^3-cephem-4-carboxylic acid; Ausocef; Cefadros; Cefaloto; Cefibacter; Ceporexine; Cex; Derantel; Efalexin; Farexin; Fergon 500; Garasin; Ibilex; Iwalexin; Larixin; Lexibiotico; Llonexina; Madlexin; Mamalexin; Mecilex; Neolexina; Ohlexin; Oracocin; Rinesal; Sencephalin; Syncl; Taicelexin; Tokiolexin; Xahl. $C_{16}H_{17}N_3O_4S$; mol wt 347.40. C 55.32%, H 4.93%, N 12.09%, O 18.42%, S 9.23%. Semi-synthetic cephalosporin antibiotic. Prepn: Ryan et al., J. Med. Chem. **12**, 310 (1969); R. B. Morin, B. G. Jackson, U.S. pat. 3,275,626 (1966 to Lilly). Pharmacology and toxicology: Muggleton et al., Antimicrob. Ag. Chemother. **1968**, 353; Kind et al., Welles et al., ibid. 361, 489. Comprehensive description: L. P. Marrelli in Analytical Profiles of Drug Substances vol. 4, K. Florey, Ed. (Academic Press, New York, 1975) pp 21-46. Clinical pharmacology, efficacy, adverse reactions: Postgrad. Med. J. **59**, Suppl. 5, 1-56 (1983).

Crystals. uv max: 260 nm (ϵ 7750). pKa 5.2, 7.3.
Monohydrate, *Cefa-Iskia, Ceporex, Keforal, Keflet, Keflex, Oracef, Ortisporina, Sartosona, Sintolexyn*. LD_{50} in mice, rats (g/kg): 1.6-4.5, >5.0 orally; 0.4-1.3, >3.7 i.p. (Welles).
Monohydrochloride monohydrate, $C_{16}H_{18}ClN_3O_4 \cdot H_2O$, *LY061188, Keftab*.
Sodium salt, $C_{16}H_{16}N_3NaO_4S$, *Alfaspoven*.
Pivalate ester, see pivcefalexin.
THERAP CAT: Antibacterial.
THERAP CAT (VET): Antibacterial.

1972. Cephalins. Kephalins; phosphatidylethanolamine. Group of phospholipids found in all living organisms. Significant constituent of nervous tissue and brain substance. Cephalins consist of glycerophosphoric acid in which the two free hydroxyls are esterified with long-chain fatty acid residues, and ethanolamine forms an ester linkage with the phosphate group. α-Isomers are derivatives of α-glycerophosphoric acid, q.v.; β-isomers are derivatives of β-glycerophosphoric acid, q.v. Natural products occur in the α-form, while the β-form is now recognized to be an artifact. Prepn from commercial "soybean lecithin": Scholfield, Dutton, Biochem. Prepn. **5**, 5 (1957); U.S. pat. 2,801,255 (1957 to USDA). Diagnostic use: C. R. Ratliff et al., Am. J. Gastroenterol. **55**, 589 (1971). Reviews on natural and synthetic cephalins: E. Baer in Progress in the Chemistry of Fats and Other Lipids vol. 6, (MacMillan, New York, 1963) pp 39-44; Van Deenen, de Haas in Advances in Lipid Research vol. 2 (Academic Press, New York, 1964) pp 183-189; Verkade, Bull. Soc. Chim. France **1963**, 1993.

R and R′ can be but are not necessarily identical fatty acids.

α-cephalin

Yellowish amorphous substances; characteristic odor and taste. Practically insol in water, acetone. Freely sol in chloroform, ether; slightly sol in ethanol.
USE: Clinical reagent (liver function test).
THERAP CAT: Local hemostatic.

1973. Cephaloglycin. *3-[(Acetyloxy)methyl]-7-[(aminophenylacetyl)amino]-8-oxo-5-thia-1-azabicyclo[4.2.0]oct-2-ene-2-carboxylic acid; 7-(2-amino-2-phenylacetamido)-3-(hydroxymethyl)-8-oxo-5-thia-1-azabicyclo[4.2.0]oct-2-ene-2-carboxylic acid acetate (ester);* 7-(D-2-amino-2-phenyl-acetamido)-3-acetoxymethyl-Δ^3-cephem-4-carboxylic acid;

7-(D-α-aminophenylacetamido)cephalosporanic acid; Kefglycin. $C_{18}H_{19}N_3O_5S$; mol wt 405.44. C 53.32%, H 4.72%, N 10.36%, O 23.68%, S 7.91%. Semi-synthetic cephalosporin antibiotic. Prepn: **Belg.** pat. 635,137 corresp to U.S. pat. 3,422,103 (1964, 1969 both to Glaxo); **Brit.** pat. 1,017,-624 (1966 to Merck & Co.); Spencer et al., J. Med. Chem. **9**, 746 (1966); of D- and DL-forms: **Brit.** pat. 985,747 (1965 to Lilly); Kurita et al., J. Antibiot. **19A**, 243 (1966). Metabolism and pharmacokinetics in man: J. Haginaka et al., J. Antibiot. **33**, 236 (1980).

Dihydrate, *Kafocin*. Crystalline powder, mp 223-250° (dec). uv max (2% DMF): 258 nm ($E_{1cm}^{1\%}$ 166). Isoelectric point and pH of maximum stability: 4.5.
THERAP CAT: Antibacterial.

1974. Cephalonic Acid. *8-Hydroxy-5-oxoophiobola-3,6,19-trien-25-oic acid;* ophiobolin D. $C_{25}H_{36}O_4$; mol wt 400.54. C 74.96%, H 9.06%, O 15.98%. Minor antibiotic metabolite produced by Cephalosporium caerulens. Exhibits weak activity against Staphylococcus aureus. Isolation, structure elucidation and physical data: Itai et al., Tetrahedron Letters **1967**, 4111; Nozoe et al., ibid. 4113. Crystal structure: Itai et al., Acta Cryst. **25B**, 872 (1969). Synthetic studies on related compounds: P. C. Dutta, T. K. Das, Syn. Commun. **6**, 253 (1976); W. G. Dauben, D. S. Hart, J. Org. Chem. **42**, 922 (1977); R. K. Boeckman et al., ibid. 3630.

Solid, mp 139°. $[\alpha]_D$ +76.2° (c = 0.54 in chloroform). uv max: 259 nm (ϵ 11700).

1975. Cephaloridine. *1-[[2-Carboxy-8-oxo-7-[(2-thienylacetyl)amino]-5-thia-1-azabicyclo[4.2.0]oct-2-en-3-yl]-methyl]pyridinium hydroxide inner salt; 1-[(7′-β-[2-(2-thienyl)acetamido]-8′-oxo-1′-aza-5′-thiabicyclo[4.2.0]oct-2′-en-3′-yl)methyl]pyridinium-2′-carboxylate; N-[7-[(2-thienyl)acetamido]ceph-3-em-3-ylmethyl]pyridinium-4-carboxylate; N-[7-(2′-thienylacetamidoceph-3-ylmethyl)]-pyridinium-2-carboxylate;* cefaloridin; Ceflorin; Cepaloridin; Cer; Faredina; Ceporan; Ceporin; Cilifor; Intrasporin; Keflodin; Keflordin (obsolete); Lloncefal; Sefacin; Cepalorin; Deflorin; Kefspor; Loridine; Aliporina; Ampligram; Floridin. $C_{19}H_{17}N_3O_4S_2$; mol wt 415.50. C 54.92%, H 4.13%, N 10.11%, O 15.40%, S 15.43%. Broad-spectrum semi-synthetic cephalosporin antibiotic. Prepn: **Fr.** pat. 1,384,197 (1965 to Glaxo), C.A. **63**, 11591c (1965). Toxicity studies: Atkinson et al., Toxicol. Appl. Pharmacol. **8**, 398 (1966).

Crystals. Soluble in water. Aq solns are slightly acid (pH 4.5-5); $[\alpha]_D$ +47.7° (c = 1.25 in water). Aq solns (20% w/v) are stable for four weeks at 4° in the dark. Exposure to sunlight causes discoloration. LD_{50} mice, rats (g/kg): >15, 2.5-4 orally; in monkeys (g/kg): >0.2 i.m. (Atkinson).
THERAP CAT: Antibacterial.
THERAP CAT (VET): Antibacterial.

1976. Cephalosporin C. *3-[(Acetyloxy)methyl]-7-[(5-amino-5-carboxy-1-oxopentyl)amino]-8-oxo-5-thia-1-azabicyclo[4.2.0]oct-2-ene-2-carboxylic acid; 7-(D-5-amino-5-carboxyvaleramido)-3-(hydroxymethyl)-8-oxo-5-thia-1-azabicyclo[4.2.0]oct-2-ene-2-carboxylic acid acetate.* $C_{16}H_{21}$-N_3O_8S; mol wt 415.44. C 46.26%, H 5.10%, N 10.11%, O 30.81%, S 7.72%. Antibiotic substance produced along with penicillin N by a *Cephalosporium* sp. cultivated from sea water near a sewage outlet on the coast of Sardinia: G. Brotzu, *Lav. Ist. Igiene Cagliari* (1948); Newton, Abraham, *Nature* **175**, 548 (1955); *eidem, Biochem. J.* **62**, 651 (1956); *eidem*, **Ger.** pat. **1,014,711** (1957 to Natl. Res. Dev. Corp.); **Brit.** pat. **810,196** (1959); Kelly *et al.*, U.S. pat. **3,082,155** (1963 to Natl. Res. Dev. Corp.); Demain, U.S. pats. **3,116,-216/7** (1963 to Merck & Co.). Biosynthesis: Demain, Newkirk, *Appl. Microbiol.* **10**, 321 (1962). Has the same side chain as penicillin N, but the nucleus is different and is named 7-aminocephalosporanic acid: Jago, Heatley, *Brit. J. Pharmacol.* **16**, 170 (1961). Structure: Abraham, Newton, *Biochem. J.* **79**, 377 (1961); Hodgkin, Maslen, *ibid.* 393. In *cephalosporin C_A* the acetoxy group is replaced by a pyridinium group: Hale *et al., ibid.* 403. Transformation to *cephalosporin C_C* for which greater activity is claimed: Abraham *et al.*, U.S. pat. **3,049,541** (1962 to Natl. Res. Dev. Corp.). Total synthesis: Woodward *et al., J. Am. Chem. Soc.* **88**, 852 (1966). Review of cephalosporin C and related compounds: Abraham, *Quart. Rev. Chem. Soc.* **21**, 231 (1967); Abraham, Loder, in *Cephalosporins and Penicillins*, E. H. Flynn, Ed. (Academic Press, New York, 1972) pp 1-26.

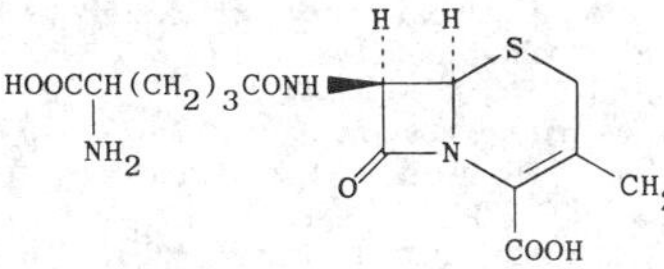

Sodium salt, dihydrate, $C_{16}H_{20}N_3NaO_8S.2H_2O$, monoclinic crystals from dil alc. $[\alpha]_D^{20}$ +103° (H_2O). uv max: 260 nm ($E_{1cm}^{1\%}$ 200). pKa values: < 2.6, 3.1, 9.8. Sol in water. Practically insol in ethanol, ether. Aq solns are stable at pH 2.5 to 8. Low toxicity. Shows *in vitro* activity against penicillin-resistant staphylococci.

THERAP CAT: Antibacterial.

1977. Cephalosporin P₁. *6α,16β-Bis(acetyloxy)-3α,7β-dihydroxy-29-nordammara-17(20),24-dien-21-oic acid.* $C_{33}H_{50}O_8$; mol wt 574.73. C 68.96%, H 8.77%, O 22.27%. Antibiotic substance produced by a *Cephalosporium* sp. cultivated from the sea near Sardinia. Crude cephalosporin P contains at least 5 components: P_1, P_2, P_3, P_4, P_5, the major active substance being cephalosporin P_1. Isoln: Burton, Abraham, *Biochem. J.* **50**, 168 (1951). Structure: Baird *et al., Proc. Chem. Soc.* **1961**, 257; Halsall *et al., ibid.* **1963**, 16; Melera, *Experientia* **19**, 565 (1963); Halsall *et al., Chem. Commun.* **1966**, 685.

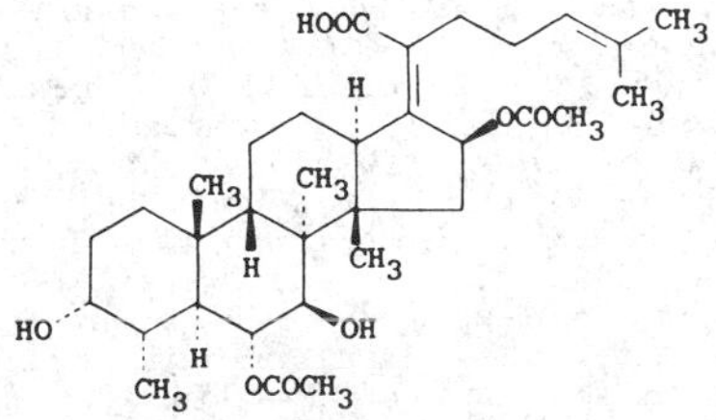

Orthorhombic crystals from 50% aq ethanol, $C_{33}H_{50}O_8$.-$H_2O.\frac{1}{2}C_2H_5OH$. mp 147°. $[\alpha]_D^{20}$ +28° (c = 2.7 in chloroform). uv max: 211 nm (ε 9140). Readily sol in chloroform and ethanol; sparingly sol in hexane and water.

1978. Cephalothin. *3-[(Acetyloxy)methyl]-8-oxo-7-[(2-thienylacetyl)amino]-5-thia-1-azabicyclo[4.2.0]oct-2-ene-2-carboxylic acid; 3-(hydroxymethyl)-8-oxo-7-[2-(2-thienyl)-acetamido]-5-thia-1-azabicyclo[4.2.0]oct-2-ene-2-carboxylic acid acetate;* 7-(2-thienylacetamido)cephalosporanic acid; 7-(thiophene-2-acetamido)cephalosporanic acid. $C_{16}H_{16}N_2$-

O_6S_2; mol wt 396.44. C 48.48%, H 4.07%, N 7.06%, O 24.22%, S 16.17%. Semi-synthetic cephalosporin antibiotic. Prepn: Chauvette *et al., J. Am. Chem. Soc.* **84**, 3401 (1962); **Fr.** pat. **1,384,197** (1965 to Glaxo), *C.A.* **63**, 11591c (1965). Bacteriology and pharmacology: Lee, Anderson, *Antimicrob. Ag. Chemother.* **1962**, 695; Walters *et al., ibid.* 706; Naumann, *Arzneimittel-Forsch.* **16**, 1099 (1966). Acute toxicity: M. Kuramoto *et al., Japan. J. Antibiot.* **27**, 746 (1974), *C.A.* **83**, 71972t(1975). Comprehensive description: R. J. Simmons in *Analytical Profiles of Drug Substances* vol. 1, K. Florey, Ed. (Academic Press, New York, 1972) pp 319-341.

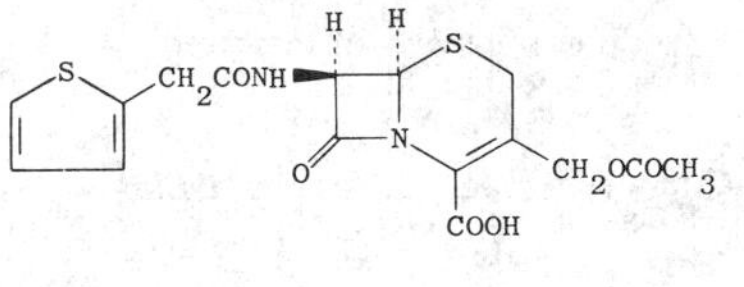

mp 160-160.5°. $[\alpha]_D^{20}$ +50° (c = 1.03 in acetonitrile), *see* R. B. Woodward *et al., J. Am. Chem. Soc.* **88**, 852 (1966).

Sodium salt, $C_{16}H_{15}N_2NaO_6S_2$, *Averon-1, Cefalotin, Cephation, Ceporacin, Cepovenin, Chephalotin, Coaxin, Keflin, Lospoven, Microtin, Synclotin, Toricelocin.* mp 204-205°. LD_{50} in mice, rats (mg/kg): > 20000, > 10000 orally; 5670, 7716 i.p. (Kuramoto).

THERAP CAT: Antibacterial.

1979. Cephamycins. A family of β-lactam antibiotics produced by various *Streptomyces* species. Detection and production: Stapley *et al., Antimicrob. Ag. Chemother.* **2**, 122 (1972). Chemical characterization: Miller *et al., ibid.* 132. Cephamycins A, B, and C have been isolated, the latter being identical to a *Streptomyces clavuligerus* metabolite: Nagarajan *et al., J. Am. Chem. Soc.* **93**, 2308 (1971). Structures: Albers-Schoenberg *et al., Tetrahedron Letters* **1972**, 2911. Antibacterial activity studies: Miller *et al., Antimicrob. Ag. Chemother.* **2**, 281, 287 (1972); Daoust *et al., ibid.* **3**, 254 (1973). Review of syntheses: T. Hiraoka *et al., Heterocycles* **8**, 719 (1977).

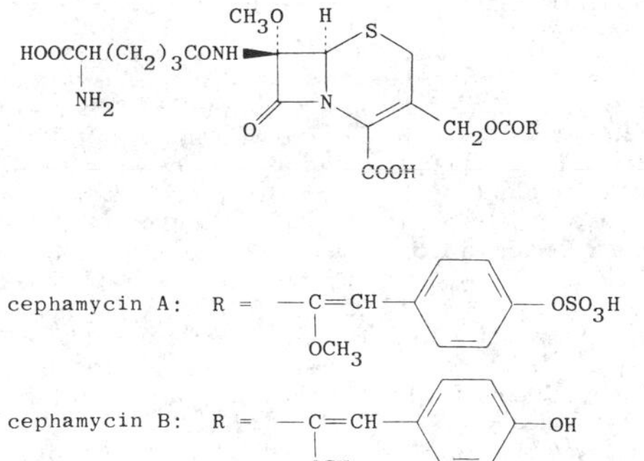

cephamycin A:

cephamycin B:

cephamycin C:

1980. Cephapirin Sodium. *3-[(Acetyloxy)methyl]-8-oxo-7-[[(4-pyridinylthio)acetyl]amino]-5-thia-1-azabicyclo-[4.2.0]oct-2-ene-2-carboxylic acid monosodium salt; 3-(hydroxymethyl)-8-oxo-7-[2-(4-pyridylthio)acetamido]-5-thia-1-azabicyclo[4.2.0]oct-2-ene-2-carboxylic acid acetate monosodium salt;* 7-[α-(4-pyridylthio)acetamido]cephalosporanic acid sodium salt; sodium 7-(pyrid-4-ylthioacetamido)cephalosporanate; BL-P 1322; Ambrocef; Brisfirina; Bristocef; Cefadyl; Cefa-Lak; Cefatrexyl; Piricef; ToDay. $C_{17}H_{16}N_3$-NaO_6S_2; mol wt 445.45. C 45.84%, H 3.62%, N 9.43%, Na 5.16%, O 21.55%, S 14.39%. A cephalosporin C antibiotic effective against gram-pos. and gram-neg. bacteria including *Staphylococcus aureus, Escherichia coli, Klebsiella pneumoniae* and *Proteus mirabilis*. Prepn: Crast, **S. Afr.** pat. **67 07783** corresp to **U.S.** pat. **3,422,100**; Silvestri, Johnson; Havranek, Crast, **U.S.** pats. **3,503,967**; **3,578,661** (1968, 1969, 1970, 1971, all four to Bristol-Myers); Crast *et al., J. Med. Chem.*

16, 1413 (1973). Activity studies: Chisholm *et al., Antimicrob. Ag. Chemother.* **1969,** 244; Axelrod *et al., Appl. Microbiol.* **22,** 904 (1971); Bran *et al., Antimicrob. Ag. Chemother.* **1,** 35 (1972); Wiesner *et al., ibid.* 303.

White crystalline powder, sol in water.
THERAP CAT: Antibacterial.
THERAP CAT (VET): Antibacterial.

1981. Cepharanthine. *6',12'-Dimethoxy-2,2'-dimethyl-6,7-[methylenebis(oxy)]oxyacanthan.* $C_{37}H_{38}N_2O_6$; mol wt 606.69. C 73.24%, H 6.31%, N 4.62%, O 15.82%. From tubers of *Stephania cephalantha* Hayata, and *Stephania sasahii* Hayata, *Menispermaceae:* Kondo *et al.,* U.S. pat. **2,206,407** (1940); Kondo, Hasegawa, U.S. pat. **2,248,241** (1941). Structure: Kondo, Keimatsu, *Ber.* **71,** 2553 (1938); Tomita, Sasaki, *Pharm. Bull. (Japan)* **2,** 89, 375 (1954). Total synthesis (*dl*-form): Tomita *et al., Tetrahedron Letters* **1967,** 1201.

Yellow powder, mp 145-155°. Obtained by drying solvated needles from acetone + benzene. $[\alpha]_D^{20}$ +277° (c = 2 in chloroform). Soluble in the usual organic solvents except petr ether.

1982. Cephradine. *[6R-[6α,7β(R*)]]-7-[(Amino-1,4-cyclohexadien-1-yl-acetyl)amino]-3-methyl-8-oxo-5-thia-1-azabicyclo[4.2.0]oct-2-ene-2-carboxylic acid;* 7-[D-2-amino-2-(1,4-cyclohexadienyl)acetamido]desacetoxycephalosporanic acid; cefradin; SQ 11436; Anspor; Cefradex; Cefrag; Cefro; Celex; Cesporan; Dimacef; Ecosporina; Eskacef; Lenzacef; Lisacef; Medicef; Megacef; Samedrin; Sefril; Velocef; Velosef. $C_{16}H_{19}N_3O_4S$; mol wt 349.41. C 55.00%, H 5.48%, N 12.03%, O 18.32%, S 9.17%. A semisynthetic cephalosporin antibiotic. Prepn and activity data: Weisenborn *et al.,* U.S. pat. **3,485,819** (1969 to Squibb); Dolfini *et al., J. Med. Chem.* **14,** 117 (1971). Clinical trials: Limson *et al., Curr. Ther. Res.* **14,** 101 (1972); Landa, *ibid.* 496. Comprehensive description: K. Florey, Ed. in *Analytical Profiles of Drug Substances* vol. 5 (Academic Press, New York, 1976) pp 21-59. *Review:* W. E. Brown, J. R. Knill, in *Pharmacological and Biochemical Properties of Drug Substances* vol. 2, M. E. Goldberg, Ed. (Am. Pharm. Assoc., Washington, DC, 1979) pp 279-304.

Monohydrate, *Forticef.* Small, colorless crystals. mp 140-142° (dec). pK_1 2.63; pK_2 7.27. Sol in propylene glycol; slightly sol in acetone, ethanol. Insol in ether, chloroform, benzene, hexane.
Dihydrate, *SQ 22022.* mp 183-185°.
THERAP CAT: Antibacterial.

1983. Cerberoside. *3-[(O-β-D-Glucopyranosyl-(1 → 6)-O-D-glucopyranosyl-(1 → 4)-6-deoxy-3-O-methyl-α-L-glucopyranosyl)oxy]-14-hydroxycard-20(22)-enolide;* Thevetin B; Thevanil. $C_{42}H_{66}O_{18}$; mol wt 858.95. C 58.73%, H 7.75%, O 33.53%. Glycoside isolated from *Cerbera odollam* Gaertn. and *Thevetia neriifolia* Juss., *Apocynaceae:* Chen, Steldt, *J. Pharmacol. Exp. Ther.* **76,** 167 (1942); Bloch *et al., Helv. Chim. Acta* **43,** 652 (1960). Structure: K. Tori *et al., Tetrahedron Letters* **1977,** 717.

O-thevetose-gentiobiose

Crystals from water. mp 197-201°. $[\alpha]_D^{24}$ −61.4° ±1.5° (c = 1.50 in methanol).
Acetyl deriv, $C_{59}H_{82}O_{26}$, crystals from methanol + ether, mp 145-149°. $[\alpha]_D^{24}$ −56.9° ±2° (c = 1.37 in chloroform).
THERAP CAT: Cardiotonic.

1984. Ceresin. Purified ozokerite; earth wax; mineral wax; cerosin; cerin. A mixture of hydrocarbons of complex composition purified by treatment with concd H_2SO_4 and filtration through bone-black. Found in Ukraine, Lake Baikal, also in Utah, Texas.
White or yellow, tasteless, waxy cakes. The white is odorless, the yellow has a slight odor. Fracture is very much like that of white wax. d 0.91-0.92. mp 61-78°. It is very stable toward oxidizing agents. Insol in water; sol in 30 parts abs alcohol; sol in benzene, chloroform, petr ether, hot oils.
USE: Substitute for beeswax; for making candles, wax figures; for waxed paper and cloth; in polishes, electrical insulators; waterproofing fabrics; for bottles for hydrofluoric acid; in dentistry for impression and inlay waxes and modeling compounds.

1985. Ceric Fluoride. CeF_4; mol wt 216.13. Ce 64.84%, F 35.16%. Prepd from CeF_3 and F_2: Klemm, Henkel, Z. *Anorg. Allgem. Chem.* **220,** 181 (1934); von Wartenberg, *ibid.* **244,** 343 (1940); Kwasnik in *Handbook of Preparative Inorganic Chemistry,* vol. 1, G. Brauer, Ed. (Academic Press, New York, 2nd ed., 1963) pp 247-248; from CeO_2 and F_2: Asker, Wylie, *Aust. J. Chem.* **18,** 959 (1965).
Minute crystals, d 4.77. mp above 650°. Practically insol in water; very slowly hydrolyzed by cold water. Thermally stable below 550°. May be reduced to CeF_3 by H_2 at 300°.
Monohydrate, white powder. Sol in acids.
USE: Fluorinating agent.

1986. Ceric Oxide. Ceria. CeO_2; mol wt 172.13. Ce 81.41%, O 18.59%. Usually prepd by ignition of Ce salts Duval, *Anal. Chim. Acta* **1,** 341 (1947); other prepns: Warf, U.S. pat. **2,564,241** (1951 to USAEC); Wilke, *Z. Anorg. Allgem. Chem.* **330,** 164 (1964). Purification: Wetzel in *Handbook of Preparative Inorganic Chemistry,* vol. 2, G. Brauer, Ed. (Academic Press, New York, 2nd ed., 1965) pp 1132-1135. *Review:* Davis, Wayman, *Can. Chem. Process Ind.* **29,** 230 (1945).
Heavy powder or cubic crystals; white if pure, but usually pale yellow; traces of lanthanum, praseodymium, etc., give a pink to reddish-brown color. Practically insol in water, acids. LD_{50} orally in rats: ∼1 g/kg, *Toxic Substances List,* H. E. Christensen, Ed. (1972) p 129.
USE: Polishing and decolorizing glass; as opacifier for vitreous enamels; in heat-resistant alloy coatings; in coatings for infrared filters to prevent reflection; in analysis for Ce and oxidimetry; as catalyst for organic reactions.

1987. Ceric Sulfate. CeO_8S_2; mol wt 332.25. Ce 42.18%, O 38.53%, S 19.30%. $Ce(SO_4)_2$. Prepd by heating CeO_2 with concd H_2SO_4: Vanino, *Handbuch der Präparativen Chemie,* vol. 1 (Stuttgart, 3rd ed., 1925) p 755. *Review:* Smith, *Ceric Sulfate* (G. Frederick Smith Chem. Co., Columbus, Ohio,

2nd ed., 1935) 51 pp. The commercial salt usually contains small amounts of associated rare earths.

Tetrahydrate, yellow to orange powder or orthorhombic crystals. On heating loses water, becoming anhydr at 180-200°; dec above 350°, forming $CeOSO_4$. Sol in a small quantity of water, but dec with much water with separation of a basic salt. Slowly sol in cold, more readily sol in hot mineral acids; sol in dil H_2SO_4.

USE: Oxidizing agent; analytical reagent; in radiation dosimeters.

1988. Cerium. Ce; at. wt 140.12; at. no. 58; valences 2, 3, 4. A lanthanide. Four naturally occurring isotopes: 140 (88.48%); 142 (11.07%); 138 (0.25%); 136 (0.193%); [142]Ce is radioactive, $T_{1/2} > 5 \times 10^{15}$ years. Artificial radioactive isotopes: 132-135; 137; 139; 141; 143-148. The heavier of these are fission products, the most important of which is [144]Ce ($T_{1/2}$ 284.5 days; β^- decay to [144]Pr); one of the most abundant products (5.3%) of the fission bomb. Estimated abundance of cerium in the earth's crust: 20-46 ppm. Found in the minerals monazite, bastaesite and cerite. Discovered by Klaproth, Hisinger and Berzelius in 1803. Can be separated from other rare earths by selective precipitation of ceric (4+) salts from buffered solns (pH 3-4); also by ion exchange techniques. Prepn of metal by reduction of trichloride with calcium: Spedding et al., Ind. Eng. Chem. **44**, 553 (1952). Review of prepn, properties and compds of cerium and other lanthanides: The Rare Earths, F. H. Spedding, A. H. Daane, Eds. (Krieger, Huntington, N.Y., 1971, reprint of 1961 ed.) 641 pp; Hulet, Bode, "Separation Chemistry of the Lanthanides and Transplutonium Actinides" in MTP Int. Rev. Sci.: Inorg. Chem., Ser. One, vol. 7, K. W. Bagnall, Ed. (University Park Press, Baltimore, 1972) pp 1-45; Moeller, "The Lanthanides" in Comprehensive Inorganic Chemistry, vol. 4, J. C. Bailar, Jr. et al., Eds. (Pergamon Press, Oxford, 1973) pp 1-101; W. L. Silvernail in Kirk-Othmer Encyclopedia of Chemical Technology vol. 5 (Wiley-Intersciene, New York, 3rd ed., 1979) pp 315-327.

Iron-gray, ductile, malleable metal; face-centered cubic structure at room temp. Stable in dry air, but superficially oxidized in moist air; when finely divided may ignite spontaneously. d 6.77. mp 795°. E^0(aq) Ce^{3+}/Ce −2.48 V (calc). Slowly dec by cold, rapidly by hot water; sol in dil mineral acids. Ceric salts usually are yellow to orange-red in color and liberate iodine from KI. Cerous salts are usually white and give a white ppt with alkali hydroxides or sulfides, insol in excess of reagent; they also are pptd by ammonium oxalate from cold dil acid sols.

USE: Manuf ferro-cerium (spark metal) for gas lighters; as catalyst in manuf of ammonia. Most of the cerium is used in the form of hydrated ceric oxide for decolorizing glass. The radioactive isotope [144]Ce as tracer and in atomic warfare. Cerium chelates are under study as antiknock compounds: Eisentraut et al., U.S. pat. **3,794,473** (1974 to U.S. Air Force).

1989. Cerous Bromide. Br_3Ce; mol wt 379.98. Br 63.11%, Ce 36.89%. $CeBr_3$. Prepn: Jantsch, Wein, Monatsh. **69**, 161 (1936).

Heptahydrate, colorless, deliquesc needles. mp 732°. Sol in water, alcohol. Keep well closed.

1990. Cerous Carbonate. $C_3Ce_2O_9$; mol wt 460.29. C 7.83%, Ce 60.89%, O 31.28%. $Ce_2(CO_3)_3$. Prepn: Head, Holley, Jr., USAEC LADC-5579, 8 pp (1962).

Pentahydrate, powder, or microcryst prisms. Almost insol in water; sol in dil mineral acids and in soln of ammonium salts.

1991. Cerous Chloride. $CeCl_3$; mol wt 246.50. Ce 56.85%, Cl 43.15%. Prepn: Kleinheksel, Kremers, J. Am. Chem. Soc. **50**, 959 (1928); Didtschenko, U.S. pat. **2,932,553** (1960 to Union Carbide); Harmon, Wickers, U.S. pat. **2,982,603** (1961 to USAEC); Taylor, Carter, J. Inorg. Nucl. Chem. **24**, 387 (1962).

Very fine powder, mp 822°. d^{25} 3.97. Sol in water, alcohol (exothermic). LD_{50} in mice, guinea pigs: 352, 110 mg/kg i.p., Graca et al., Arch. Environ. Health **5**, 437 (1962).

Heptahydrate, colorless to yellow deliquesc orthorhombic crystals. Begins to lose water above 90°, becomes anhyd by 230°. Very sol in H_2O, alcohol. Keep well closed.

USE: In manuf of Ce and Ce salts; with Al or Mg salts as catalyst in the polymerization of olefins.

1992. Cerous Fluoride. CeF_3; mol wt 197.13. Ce 71.09%, F 28.91%. Prepd by the action of HF on CeO_2: von Wartenberg, Z. Anorg. Allgem. Chem. **244**, 343 (1940); by the action of F_2 or concd HF on $CeSi_2$: Sterba, Ann. Chim. Phys. [8] **2**, 193 (1904); from CCl_2F_2 and CeO_2: Pausewang, Rüdorff, Z. Anorg. Allgem. Chem. **369**, 89 (1969).

Hexagonal crystals or powder. d 6.157. mp 1460°. Practically insol in, but slowly hydrolyzed by water.

1993. Cerous Iodide. CeI_3; mol wt 520.86. Ce 26.90%, I 73.10%. Prepn from Ce and HgI_2: Asprey et al., Inorg. Chem. **3**, 1137 (1964); from the oxide and HI in the presence of NH_4I: Taylor, Curtis, J. Inorg. Nucl. Chem. **24**, 387 (1962).

Bright yellow, orthorhombic crystals. mp 752°. Dec in moist air. Sol in water.

Nonahydrate, white to reddish-white crystals. Very sol in water, the soln readily decomposing with separation of iodine; sol in alc. Keep well closed and protected from light.

1994. Cerous Nitrate. CeN_3O_9; mol wt 326.15. Ce 42.96%, N 12.88%, O 44.15%. $Ce(NO_3)_3$. Prepn from Ce sulfate and $Ca(NO_3)_2$: Blumenfeld, U.S. pat. **2,166,702** (1939 to Soc. de Produits Chimiques des Terres Rares); in separation of cerium from other rare earths: Pierce, Butler, Inorg. Syn. **2**, 51 (1946); Scargill et al., J. Inorg. Nucl. Chem. **4**, 304 (1957). For physical and chemical properties see USAEC IDO-14504, 126 pp (1961).

Hexahydrate. LD_{50} in rats: 290 mg/kg i.p.; 4.2 g/kg orally, Bruce et al., Toxicol. Appl. Pharmacol. **5**, 750 (1963).

USE: Separation of Ce from other rare earths; catalyst in hydrolysis of phosphoric acid esters; mixture with $Th(NO_3)_4$ (1:99) formerly used in manuf incandescent mantles.

1995. Cerous Oxalate. Sedemesis. $C_6Ce_2O_{12}$; mol wt 544.32. C 13.24%, Ce 51.49%, O 35.27%. $Ce_2(C_2O_4)_3$. Prepn from $Ce(NO_3)_3.6H_2O$ and oxalic acid: Wylie, J. Chem. Soc. **1947**, 1687. The article of commerce contains variable amounts of the oxalates of cerium, lanthanum, and other associated elements.

Nonahydrate, white or slightly pink, tasteless powder. Practically insol in water; sol in hot, moderately dil H_2SO_4 or HCl.

THERAP CAT: Antiemetic.

1996. Cerous Sulfate. $Ce_2O_{12}S_3$; mol wt 568.44. Ce 49.30%, O 33.78%, S 16.92%. $Ce_2(SO_4)_3$. Prepn: Blandin, Rerat, Compt. Rend. **239**, 1055 (1954); Wendlandt, J. Inorg. Nucl. Chem. **7**, 51 (1958).

Octahydrate, orthorhombic octahedral crystals. d. 2.87. Loses water on heating, becoming anhyd by 250°; loses SO_3 to form a basic salt above 650°. Soluble in water.

USE: Developing aniline black; said to be superior to vanadium for this purpose.

1997. Cerulenin. *3-(1-Oxo-4,7-nonadienyl)oxiranecarboxamide; (2R,3S)-2,3-epoxy-4-oxo-7E,10E-dodecadienamide;* 2,3-epoxy-4-oxo-7,10-dodecadienoylamide; helicocerin. $C_{12}H_{17}NO_3$; mol wt 223.28. C 64.55%, H 7.68%, N 6.27%, O 21.50%. Antifungal antibiotic isolated from Cephalosporium caerulens; Acryocylindrium oryzae; Helicoceras oryzae: T. Hata et al., Japan. J. Bacteriol. **15**, 1075 (1960); Matsumae et al., J. Antibiot. **16A**, 236, 239 (1963); Sano et al., ibid. **20A**, 344 (1967); Furuya, Shirasaka, Japan. pat. **21,638**('70) (to Sankyo), C.A. **73**, 108271k (1970). Biological characteristics: Matsumae et al., J. Antibiot. **17A**, 1 (1964). Structure: Omura et al., ibid. **20A**, 349 (1967). Abs config: eidem, Chem. Pharm. Bull. **17**, 2361 (1969); Arison, Omura, J. Antibiot. **27**, 28 (1974). Stereoselective synthesis of (+)- and (−)-*tetrahydrocerulenin* and corrected abs config of (+)-cerulenin: H. Ohrui, S. Emoto, Tetrahedron Letters **1978**, 2095; J.-R. Pougny, P. Sinay, ibid. 3301. Stereoselective synthesis of (+)-form from D-glucose: N. Sueda et al., ibid. **1979**, 2039; M. Pietraszkiewicz, P. Sinay, ibid. 4741. Interrupts yeast-type fungi growth by inhibiting the biosynthesis of sterols and fatty acids. Mechanism of action studies: Nomura et al., J. Biochem. **71**, 783 (1972); eidem, J. Antibiot. **25**, 365 (1972); see also D. Vance, Biochem. Biophys. Res. Commun. **48**, 649 (1972). Total synthesis of (±)-cerulenin: R. K. Boeckman, Jr., E.W. Thomas, *J. Am. Chem. Soc.* **99**, 2805 (1977); A. A. Jakubowski *et al., Tetrahedron Letters* **1977**, 2399; E. J. Corey, D. R. Williams, *ibid.* 3847; K. Mikami *et al., Chem. Letters* **1981**, 1721; A. A. Jakubowski *et al., J. Org. Chem.* **47**, 1221 (1982).

White needles from benzene, mp 93-94°. bp 120° (10⁻⁸ mm). $[\alpha]_D^{16} +63°$ (c = 2 in methanol). Stable in neutral and acidic solns. Sol in ethanol, acetone, benzene and most common solvents. Slightly sol in water. Practically insol in petr ether. LD_{50} in mice (mg/kg): 154 i.v.; 211 i.p.; 547 orally (Matsumae 1964).
DL-Form, mp 40-43°.
USE: Biochemical tool.

1998. Ceruletide. *Caerulein;* cerulein; FI 6934. $C_{58}H_{73}$-$N_{13}O_{21}S_2$; mol wt 1352.43. C 51.51%, H 5.44%, N 13.46%, O 24.84%, S 4.74%. Cholecystokinin analog. Decapeptide discovered in the skins of Australian amphibians: V. Erspamer *et al., Nature* **212**, 204 (1966). Isoln from *Hyla caerulea* and structure: A. Anastasi *et al., Experientia* **23**, 699 (1967); *eidem, Arch. Biochem. Biophys.* **125**, 57 (1968). Synthesis: L. Bernardi *et al., Experientia* **23**, 700 (1967); *eidem*, **S. Afr.** pat. **67 04,716;** *eidem*, **U.S.** pat. **3,472,832** (1968, 1969 to Soc. Farm. Italia). Shows hypotensive activity, stimulates smooth muscle and increases gastric, pancreatic, and biliary secretions: V. Erspamer *et al., Experientia* **23**, 702 (1967). Series of articles on metabolism: *Eur. J. Biochem.* **91**, 21-48 (1978). Pharmacodynamics: N. Iwatsuki, O. H. Petersen, *J. Cell. Biol.* **79**, 533 (1978); T. Fujita *et al., Adv. Exp. Med. Biol.* **106**, 147 (1978). Diagnostic use: R. Fujita *et al., Acta Gastroenterol. Belg.* **40**, 167 (1977); C. Monti *et al., Radiology* **129**, 611 (1978). Toxicology: T. Chieli *et al., Toxicol. Appl. Pharmacol.* **23**, 480 (1972). Reviews of pharmacology and clinical applications: G. Bertaccini, *Pharmacol. Rev.* **28**, 127 (1976); and toxicology: M. E. Vincent *et al., Pharmacotherapy* **2**, 223 (1982).

```
5-oxo-Pro-Gln-Asp-Tyr-Thr-Gly-Trp-Met-Asp-Phe-NH₂
                   |
                  SO₃H
```

mp 224-226° (dec). $[\alpha]_D^{20} -26°$ (c = 1 in DMF). Diethylamine salt, *Ceosunin, Cerulen, Takus, Tymtran.* Off-white hygroscopic powder. Sol in DMF and DMSO. Insol in acetone, diethyl ether. LD_{50} i.v. in mice: 1012 mg/kg (Chieli).
THERAP CAT: Stimulant (gastric secretory). Diagnostic aid (pancreatic function; cholecystokinetic agent in cholecystography).

1999. Ceruloplasmin. Caeruloplasmin; ferroxidase. Intensely blue colored copper-containing glycoprotein of the α_2-globulin fraction of mammalian blood; it is the principal copper transport protein and is believed to play an important role in iron mobilization via its ferroxidase activity. First reported by C. G. Holmberg, *Acta Physiol. Scand.* **8**, 227 (1944). Isoln, purification and description of properties of porcine and human ceruloplasmin: C. G. Holmberg, C. B. Laurell, *Acta Chem. Scand.* **2**, 550 (1948). Ceruloplasmin accounts for 90-95% of the circulating copper in normal mammals. Its concentration increases by a factor of 2 to 3 during pregnancy and varies significantly in several diseases and hormonal conditions. Prepd by Cohn cold ethanol fractionation: Cohn *et al., J. Am. Chem. Soc.* **68**, 459 (1946); Steinbuch, Quentin, *Nature* **183**, 323 (1959); and further purified from fraction IV: Sanders *et al., Arch. Biochem. Biophys.* **84**, 60 (1959); **U.S.** pat. **3,003,918** (1961 to Merck & Co.). Different mol wts have been reported for human ceruloplasmin, ranging from 126,000 to 160,000; the most generally accepted is 134,000 ±3000, cf. L. Ryder, I. Björk, *Biochemistry* **15**, 3411 (1976). Chemical and structural studies of porcine ceruloplasmin: Osaki *et al., J. Biochem. (Tokyo)* **48**, 190 (1960); **50**, 24, 29 (1961); Mukasa *et al., Biochim. Biophys. Acta* **168**, 132 (1968); Matsunaga, Nosoh, *ibid.* **215**,

280 (1970); of human: Kasper, Deutsch, *J. Biol. Chem.* **238**, 2325 (1963); Jamieson, *ibid.* **240**, 2019 (1965); Poillon, Bearn, *Biochim. Biophys. Acta* **127**, 407 (1966); Simons, Bearn, *ibid.* **175**, 260 (1969); Ryden, *Eur. J. Biochem.* **26**, 380 (1972); T. G. Samsonidze *et al., Int. J. Peptide Protein Res.* **14**, 161 (1979); V. N. Zaitsev *et al., Kristallografiya* **25**, 174 (1980), *C.A.* **92**, 210415q (1980). Human metabolism: Kekki *et al., Nature* **209**, 1252 (1966). Reviews of biological role: E. Frieden, H. S. Hsieh, *Adv. Exp. Med. Biol.* **74**, 505-529 (1976); J. M. C. Gutteridge, *Ann. Clin. Chem.* **15**, 293-296 (1978). Comprehensive review: S. H. Laurie, E. S. Mohammed, *Coord. Chem. Rev.* **33**, 279-312 (1980).
Absorption max: 280, 610 nm ($E_{1cm}^{1\%}$ 14.9, 0.68). Electrophoretic mobility (cm/volt/sec): -5.05×10^{-5} at pH 7.0 (0.1M phosphate buffer); -5.32×10^{-5} at pH 8.6 (0.1M barbital sodium buffer).

2000. Cervicarcin. *1,2,3,4-Tetrahydro-1,3,4,5,10-pentahydroxy-2-methyl-3-[(3-methyloxiranyl)carbonyl]-4a,9a-epoxyanthracen-9(10H)-one; 4a,9a-epoxy-3-(2,3-epoxybutyryl)-1,2,3,4,4a,9a-hexahydro-1,3,4,5,10-pentahydroxy-2-methylanthrone.* $C_{19}H_{20}O_9$; mol wt 392.37. C 58.16%, H 5.14%, O 36.70%. Antineoplastic antibiotic produced by *Streptomyces ogaensis*: Okuma *et al., J. Antibiot.* **15A**, 152, 247 (1962). Prepn: Sumiki *et al.,* **Japan.** pat. **7400('64)** (to Inst. Phys. & Chem. Res.). Structure: Marumo *et al., J. Am. Chem. Soc.* **86**, 4507 (1964); *eidem, Agr. Biol. Chem.* **32**, 209 (1968). Stereochemistry: *eidem, ibid.* **35**, 1931 (1971). Activity studies: Itakura *et al., J. Antibiot.* **16A**, 231 (1963).

Needles, mp 203-205° (dec). $[\alpha]_D^{20} -144°$ (in ethanol). uv max: 227, 264, 323 nm (ε 14,200, 7,800, 2,600). pKa 9.0. Sol in acetone, lower alcohols, acetic acid, pyridine. Moderately sol in ethyl acetate, ethyl ether, chloroform, carbon tetrachloride. Slightly sol in water, benzene. Insol in petr ether, ligroin. Stable in neutral or acidic soln; unstable in alkaline soln. LD_{50} i.p. in mice: 48.5 mg/kg.

2001. Cesium. Caesium. Cs; at. wt 132.9054; at. no. 55; valence 1. Alkali metal. Occurrence in earth's crust: one ppm. One natural isotope: ¹³³Cs; artificial isotopes (mass nos.): 123, 125-132, 134-144. Occurs in nature in the aluminosilicates, *pollucite* and lepidolite, and in the borate, *rhodizite.* Discovered by Bunsen and Kirchhoff in 1860. Prepn from pollucite: *Inorg. Syn.* **4**, 5 (1953). ¹³⁷Cs ($T_{1/2}$ 30 years; β^- 0.514, 1.176 MeV) is an important constituent of radioactive fallout; decays to and reaches radioactive equilibrium with ¹³⁷ᵐBa ($T_{1/2}$ 2.554 min; γ 0.662 MeV). Review of cesium and its compounds: Whaley, "Sodium, Potassium, Rubidium, Cesium, and Francium" in *Comprehensive Inorganic Chemistry*, vol. 1, J. C. Bailar, Jr. *et al.,* Eds. (Pergamon Press, Oxford, 1973) pp 369-529; C. T. Williams in Kirk-Othmer *Encyclopedia of Chemical Technology* vol. 5 (Wiley-Interscience, New York, 3rd ed., 1979) pp 327-338.
Silver-white, ductile metal; body-centered cubic structure. Oxidizes rapidly in air; in moist air may ignite spontaneously; colors nonluminous flame reddish-violet. d 1873. mp 28.5°. bp 705°. Sp heat (25°) 0.057 cal/g/°C. Electrical resistivity 36.6 micro-ohm cm at 30°. Reacts with water to form hydroxide with evolution of hydrogen which ignites spontaneously; sol in liquid ammonia. *Keep immersed in mineral oil.*
USE: In photoelectric cells, as a "getter" in vacuum tubes. The construction and operation of one type of atomic clock is based on the vibrational frequency (9,192.76 megacycles/sec) of the cesium atom. Although ⁶⁰Co is by far the most commonly used radioisotope for encapsulated energy sources, ¹³⁷Cs offers competition because as an abundant fission product it is available in large quantities from the reprocessing of fuel elements used in nuclear reactors. *Caution:* Low toxicity in animal expts. Can act as an analog of potassium and thus be potentially harmful. For radioactive ¹³⁷Cs: max permissible body burden: 30 μcurie; max per-

missible concn in air: 2×10^{-8} μcurie; in water: 2×10^{-4} μcurie: *U.S. Bureau of Standards, Handbook* **69**, 58 (1959).

2002. Cesium Bromide. BrCs; mol wt 212.83. Br 37.55%, Cs 62.45%. CsBr. Prepn from pollucite: Thomas, Steton, *Compt. Rend.* **241**, 56 (1955).
Crystals. d 4.44. mp 636°. bp about 1300°. Very sol in water; sol in alc; practically insol in acetone. LD_{50} i.p. in rats: 1.4 g/kg, Cochran *et al., Arch. Ind. Hyg.* **1**, 637 (1950).
USE: In x-ray fluorescent screens, spectrometer prisms, absorption-cell windows.

2003. Cesium Carbonate. CCs_2O_3; mol wt 325.83. C 3.69%, Cs 81.58%, O 14.73%. Cs_2CO_3. Prepn: Suhrmann, Clusius, *Z. Anorg. Allgem. Chem.* **152**, 52 (1926); Thomas, *Ann. Chim.* **6**, 367 (1957); Bernard, *Compt. Rend.* **243**, 1528 (1956).
Very deliquesc crystals; melts at red heat. Extremely sol in water, alcohol; sol in ether.
USE: Catalyst in ethylene oxide polymerization; in coating for spatter-free welding of steel in CO_2; in oxide cathode.

2004. Cesium Chloride. ClCs; mol wt 168.37. Cl 21.06%, Cs 78.94%. CsCl. Prepn from carnallite and from pollucite: Donges in *Handbook of Preparative Inorganic Chemistry* vol. **1**, G. Brauer, Ed. (Academic Press, New York, 2nd ed., 1963) pp 951, 955.
Deliquesc cubic crystals. d 3.99. mp 646°. bp 1303°. Very sol in water; sol in alc. *Keep well closed.* LD_{50} i.p. in rats: 1.5 g/kg, Cochran *et al., Arch. Ind. Hyg.* **1**, 637 (1950).
Note: Cescan-131 is ^{131}CsCl.
USE: In the final evacuation of radio and television vacuum tubes; in x-ray fluorescent screens; as radiographic contrast medium; in manuf of cesium.

2005. Cesium Hydroxide. Cesium hydrate. CsHO; mol wt 149.92. Cs 88.65%, H 0.67%, O 10.67%. CsOH. Prepd by electrolysis of Cs salts: Winslow *et al., J. Phys. Colloid Chem.* **51**, 967 (1947); Jolibois, Berger, *Compt. Rend.* **224**, 78 (1947). Alternate prepn: H. Jacobs, B. Harbrecht, *Z. Naturforsch.* **B36**, 270 (1981).
White or yellowish, fused, very deliquescent, cryst mass; strongly alkaline reaction. mp 272°. d 3.68. *Corrosive!* Sol in about 0.25 part water, much heat being evolved; sol in alcohol. *Keep tightly closed.* LD_{50} i.p. in rats: 100 mg/kg, Cochran *et al., Arch. Ind. Hyg.* **1**, 637 (1950).
USE: In storage-battery electrolytes; as catalyst in the polymerization of cyclic siloxanes.

2006. Cesium Iodide. CsI; mol wt 259.82. Cs 51.15%, I 48.85%. Prepn: *Gmelin's, Cesium* (8th ed.) **25**, pp 188-204 (1938).
Deliquesc crystals or cryst powder. d 4.5. mp 621°. bp about 1280°. Very sol in water; sol in ethanol; slightly sol in methanol; practically insol in acetone. *Keep well closed.* LD_{50} i.p. in rats: 1.4 g/kg, Cochran *et al., Arch. Ind. Hyg.* **1**, 637 (1950).
USE: Prisms for infrared spectroscopy; in x-ray fluorescent screens, scintillation counters.

2007. Cesium Nitrate. $CsNO_3$; mol wt 194.92. Cs 68.19%, N 7.19%, O 24.62%. Prepn from pollucite: Watt, *Inorg. Syn.* **4**, 6 (1953).
White, lustrous hexagonal or cubic prisms; saltpeter taste. d_4^{20} 3.64-3.68. mp 414°; dec at higher temp. Sol in 5 parts cold, 0.5 part boiling water; sol in acetone; very slightly sol in alc. LD_{50} i.p. in rats: 1.2 g/kg, Cochran *et al., Arch. Ind. Hyg.* **1**, 637 (1950).
USE: Prepn of other Cs salts.

2008. Cesium Sulfate. Cs_2O_4S; mol wt 361.89. Cs 73.45%, O 17.69%, S 8.86%. Cs_2SO_4. Prepn: *Gmelin's, Cesium* (8th ed.) **25**, pp 218-225 (1938).
Orthorhombic or hexagonal prisms. d 4.24. mp 1019°. Very sol in water; practically insol in alc, acetone, pyridine.
USE: With V or V_2O_5 as catalyst for SO_2 oxidation.

2009. Cetalkonium Chloride. *N-Hexadecyl-N,N-dimethylbenzenemethanaminium chloride; benzylhexadecyldimethylammonium chloride;* cetyldimethylbenzylammonium chloride; hexadecyldimethylbenzylammonium chloride; Banicol; Acetoquat CDAC; Acquat CDAC; Ammonyx G; Zettyn; Ammonyx T; Cetol. $C_{25}H_{46}ClN$; mol wt 396.12. C 75.81%,

H 11.71%, Cl 8.95%, N 3.54%. Prepn: **Fr. pat. 771,746** (1934 to I. G. Farben); Piggott, **U.S. pat. 2,075,958** (1937 to I.C.I.); Westphal, Jerchel, *Ber.* **73**, 1011 (1940). *Review* of this type of quaternary ammonium compounds: J. P. Sisley, P. J. Wood, *Encyclopedia of Surface-Active Agents* (Chem. Publishing Co., New York, 1952) p 96 sqq.

Leaflets from ethyl acetate + petr ether, mp 59°. Sol in water, alc, acetone, ethyl acetate, propylene glycol, sorbitol solns, glycerol, ether, CCl_4. pH of aq solns 7.2.
USE: Cationic quaternary ammonium surfactant germicide and fungicide. Used in leather processing, textile dyeing. A mildew preventive in silicone-based water repellents. Compatible with many non-ionic detergents. Active in moderately alkaline solns.
THERAP CAT: Topical anti-infective.

2010. Cetamolol. *2-[2-[3-[[(1,1-Dimethylethyl)amino]-2-hydroxypropoxy]phenoxy]-N-methylacetamide; 2-[o-[3-(t-butylamino)-2-hydroxypropoxy]phenoxy]-N-methylacetamide; (±)-1-t-butylamino-3-(o-N-methylcarbamoylmethoxyphenoxy)propan-2-ol;* $C_{16}H_{26}N_2O_4$; mol wt 310.40. C 61.91%, H 8.44%, N 9.02%, O 20.62%. Cardioselective β_1-adrenergic blocker. Prepn: **Belg. pat. 767,781;** D. J. Lecount, C. J. Squire, **U.S. pat. 4,059,622** (1971, 1977 both to ICI). Cardiovascular effects in dogs: G. Beaulieu *et al., Can. J. Physiol. Pharmacol.* **63**, 610 (1984). Determn in human serum: M. D. Stern, *Clin. Biochem.* **17**, 162 (1984). Pharmacology and pharmacodynamics in humans: M. A. Klausner *et al., Curr. Ther. Res.* **36**, 379 (1984); J. Coelho *et al., Brit. J. Clin. Pharmacol.* **19**, 411 (1985). Evaluation in hypertensive patients: W. B. White *et al., Clin. Pharmacol. Ther.* **39**, 664 (1986).

Crystals from benzene-petr ether, mp 96-97°.
Hydrochloride, $C_{16}H_{27}ClN_2O_4$, *AI 27303, ICI 72222, Betacor.*
THERAP CAT: Antihypertensive.

2011. Cethexonium Bromide. *N-Hexadecyl-2-hydroxy-N,N-dimethylcyclohexanaminium bromide; hexadecyl(2-hydroxycyclohexyl)dimethylammonium bromide;* cetyldimethyl-(2-hydroxycyclohexyl)ammonium bromide; dimethylcetyl-2-cyclohexanolammonium bromide; *N-cetyl-N,N-dimethyl-2-cyclohexanolammonium bromide;* Biocidan. $C_{24}H_{50}Br$-NO; mol wt 448.59. C 64.26%, H 11.24%, Br 17.82%, N 3.12%, O 3.57%. Prepd by the action of cetyl bromide upon 2-dimethylaminocyclohexanol: Mousseron *et al., Bull. Soc. Chim. Biol.* **33**, 369 (1951).

White powder, mp 75°. Sol in water, alcohol, chloroform. Practically insol in petr ether.
THERAP CAT: Antiseptic.

2012. Cetiedil. *α-Cyclohexyl-3-thiopheneacetic acid*

2-(hexahydro-1H-azepin-1-yl)ethyl ester; α-cyclohexyl-α-(3-thienyl)acetic acid 2-hexamethyleneiminoethyl ester. $C_{20}H_{31}NO_2S$; mol wt 349.54. C 68.72%, H 8.94%, N 4.01%, O 9.16%, S 9.17%. Prepn: Pons, Robba, **Fr.** pat. **1,460,571** and Pons *et al.*, **Fr.** pat. **M5504** (1966, 1967, both to Innothera), *C.A.* **68,** 59429d (1968); **71,** 91286c (1969). Prepn and activity: Robba, LeGuen, *Chim. Ther.* **2,** 120 (1967). Antisickling effect: T. Asakura *et al.*, *Proc. Nat. Acad. Sci. USA* **77,** 2955 (1980); L. R. Berkowitz, E. P. Orringer, *J. Clin. Invest.* **68,** 1215 (1981).

Citrate, $C_{26}H_{39}NO_9S$, *Stratene, Vasocet.* Crystals from ethanol-ether, mp 115°.

Hydrochloride, $C_{20}H_{32}ClNO_2S$, crystals from acetonitrile, mp 152° (Robba, LeGuen); also mp 143° (Pons, Robba).

THERAP CAT: Vasodilator (peripheral).

2013. Cetirizine. *[2-[4-[(4-Chlorophenyl)phenylmethyl]-1-piperazinyl]ethoxy]acetic acid; [2-[4-(p-chloro-α-phenylbenzyl)-1-piperazinyl]ethoxy]acetic acid.* $C_{21}H_{25}$-ClN_2O_3; mol wt 388.89. C 64.86%, H 6.48%, Cl 9.12%, N 7.20%, O 12.34%. Nonsedating type histamine H_1-receptor antagonist; major metabolite of hydroxyzine, *q.v.* Prepn: E. Baltes *et al.*, **Eur.** pat. **Appl. 58,146;** *eidem,* **U.S.** pat. **4,525,-358** (1982, 1985 both to UCB). Pharmacology: C. De Vos *et al.*, *Ann. Allergy* **59,** 278 (1987); L. Juhlin *et al.*, *J. Allergy Clin. Immunol.* **80,** 599 (1987). Clinical evaluation in asthma: A. Brik *et al., ibid.* 51. Mode of action by eosinophil inhibition: R. Fadel *et al.*, *Clin. Allergy* **17,** 373 (1987). Clinical evaluation of antihistaminic and psychomotor effects: F. M. Gengo *et al.*, *Clin. Pharmacol. Ther.* **42,** 265 (1987).

Crystals from ethanol, mp 110-115°.

Dihydrochloride, $C_{21}H_{27}Cl_3N_2O_3$, *P071, Virlix, Zirtek, Zyrtec.* Crystals from isopropanol, mp 225°.

THERAP CAT: Antihistaminic.

2014. Cetotiamine. *Thiocarbonic acid O-ethyl ester, S-ester with N-[(4-amino-2-methyl-5-pyrimidinyl)methyl]-N-(4-hydroxy-2-mercapto-1-methyl-1-butenyl)formamide ethyl carbonate (ester); O,S-bis(ethoxycarbonyl)thiamine; O,S-dicarbethoxythiamine; DCET.* $C_{18}H_{26}N_4O_6S$; mol wt 426.51. C 50.69%, H 6.15%, N 13.14%, O 22.51%, S 7.52%. Prepn: Takamizawa, Hirai, *Chem. Pharm. Bull.* **10,** 1102 (1962); Takamizawa *et al., ibid.* 1107; Yamamoto *et al., Bitamin* **25,** 472 (1962), *C.A.* **60,** 9773e (1964); **Brit.** pat. **944,641** (1963 to Shionogi).

Prisms from ethyl acetate + petr ether, mp 113.5-114.5°.

Hydrochloride monohydrate, $C_{18}H_{27}ClN_4O_6S.H_2O$, *dicethiamin, Dicetamin.* Crystals from ethyl acetate, dec 122-124°. Sol in water, methanol. Practically insol in ether, benzene.

THERAP CAT: Vitamin B_1 source.

2015. Cetoxime. *N-Hydroxy-2-[phenyl(phenylmethyl)-amino]ethanimidamide; 2-(N-benzylanilino)acetamidoxime; α-(N-benzyl-N-phenylamino)acetamidoxime.* $C_{15}H_{17}N_3O$; mol wt 255.31. C 70.56%, H 6.71%, N 16.46%, O 6.27%. Prepd from (N-benzylanilino)acetonitrile via its thioamide: Benn *et al.*, **Brit.** pat. **895,495** (1962 to Boots Pure Drug).

Crystals, mp 107-108°.

Hydrochloride, $C_{15}H_{18}ClN_3O$, *Febramine.* Crystals from abs alcohol + ether, mp 164-165°.

THERAP CAT: Antihistaminic.

2016. Cetraric Acid. *9-(Ethoxymethyl)-4-formyl-3,8-dihydroxy-1,6-dimethyl-11-oxo-11H-dibenzo[b,e][1,4]dioxepin-7-carboxylic acid;* cetrarin. $C_{20}H_{18}O_9$; mol wt 402.34. C 59.70%, H 4.51%, O 35.79%. From Iceland moss, *Cetraria islandica (L.)* Ach., *Parmeliaceae.* Isoln: Schnedermann, Knopp, *Ann.* **55,** 144 (1845). Structure: Asahina, Asano, *Ber.* **66,** 893 (1933).

Very bitter prisms from alcohol or acetic acid. Bitterness threshold 1:50,000. Practically insol in hot water, petr ether, benzene, ether, or in cold methanol, alc, acetone and acetic acid. Sol in aq solns of alkalies or their carbonates forming a yellow soln that turns brown on standing.

2017. Cetraxate. *4-[[[4-(Aminomethyl)cyclohexyl]carbonyl]oxy]benzenepropanoic acid; p-hydroxyhydrocinnamic acid trans-(4-aminomethyl)cyclohexanecarboxylate; tranexamic acid p-(2-carboxyethyl)phenyl ester.* $C_{17}H_{23}NO_4$; mol wt 305.38. C 66.86%, H 7.59%, N 4.59%, O 20.96%. Deriv of tranexamic acid, *q.v.* Prepn: O. Atsuji *et al.*, *J. Med. Chem.* **15,** 247 (1972); S. Kitahara, **Japan. Kokai 73 75547** (1973 to Daiichi), *C.A.* **80,** 59727x (1974). Mechanism of action: Y. Suzuki *et al.*, *Japan. J. Pharmacol.* **29,** 829 (1979), *C.A.* **92,** 88029 (1980). Anti-ulcer effects in rats: T. Hashizume *et al.*, *Arch. Int. Pharmacodyn. Ther.* **240,** 314 (1979). Clinical study: A. Ishimori *et al.*, *Arzneimittel-Forsch.* **29,** 1625 (1979); S. Yamagata, K. Miura, *ibid.* **33,** 1191 (1983).

Crystals from methanol, melts over a range of 200-280°.

Hydrochloride, $C_{17}H_{24}ClNO_4$, *DV-1006, Neuer.* Crystals from methanol/ether, mp 238-240°.

THERAP CAT: Anti-ulcerative.

2018. Cetrimonium Bromide. *N,N,N-Trimethyl-1-hexadecanaminium bromide; hexadecyltrimethylammonium bromide;* cetyltrimethylammonium bromide; Bromat; Cetab; Cetavlon; Cetylamine; C.T.A.B.; Lissolamine V; Micol; Quamonium. $C_{19}H_{42}BrN$; mol wt 364.48. C 62.61%, H 11.62%, Br 21.93%, N 3.84%. $[CH_3(CH_2)_{15}N(CH_3)_3]Br$. Prepd from cetyl bromide and trimethylamine: Shelton *et al.*, *J. Am. Chem. Soc.* **68,** 753 (1946). Toxicity and pharmacology: B. Isomaa, K. Bjondahl, *Acta Pharmacol. Toxicol.* **47,** 17 (1980).

Crystals, mp 237-243°. Soluble in about 10 parts water. Freely sol in alc; sparingly sol in acetone. Practically insol in ether, benzene. Stable in acid soln. LD_{50} i.v. in mice, rats: 32.0, 44.0 mg/kg, B. Isomaa, K. Bjondahl, *loc. cit.*

p-Toluenesulfonate analog, $C_{26}H_{49}NO_3S$, *Cetats*.

Note: Cetrimide is a mixture consisting chiefly of tetradecyltrimethylammonium bromide together with smaller amounts of dodecyltrimethylammonium bromide and cetrimonium bromide.

USE: As cationic detergent and antiseptic; as laboratory reagent.

THERAP CAT: Topical antiseptic.

THERAP CAT (VET): Antiseptic, cleansing agent.

2019. Cetrimonium Stearate. *N,N,N-Trimethyl-1-hexadecanaminium octadecanoate; hexadecyltrimethylammonium stearate;* cetyltrimethylammonium stearate; trimethylhexadecylammonium stearate; Arquad 16 stearate; Dynafac. $C_{37}H_{77}NO_2$; mol wt 568.04. C 78.24%, H 13.66%, N 2.47%, O 5.63%. $[CH_3(CH_2)_{16}COO][CH_3(CH_2)_{15}N(CH_3)_3]$. Prepn: Gautier *et al., Bull. Soc. Chim. France* **1955**, 634.

Solid, mp 142-143°. Practically insol in water, alcohol.

Note: The commercial product, a waxy solid, also contains other alkyltrimethylammonium stearates, since the hexadecyl chain is derived from soybean fatty acids.

2020. Cetyl Alcohol. *1-Hexadecanol;* ethal; ethol; palmityl alcohol. $C_{16}H_{34}O$; mol wt 242.43. C 79.26%, H 14.-14%, O 6.60%. $CH_3(CH_2)_{14}CH_2OH$. Discovered by Chevreul in 1813. Obtained from spermaceti by saponification: Spada, Gavioli, *Farm. Sci. e Tec.* (Pavia) **7**, 435 (1952), *C.A.* **47**, 891c (1953). Prepn from palmitoyl chloride + $NaBH_4$: Caikin, Brown, *J. Am. Chem. Soc.* **71**, 122 (1949); from methylthiopalmitate + Raney Ni: Ruzicka, Prelog, U.S. pat. **2,509,171** (1950 to Ciba); from hexadecyl bromide: Levine, Clippinger, U.S. pat. **3,018,308** (1962 to California Res. Corp.).

White crystals. d 0.811. mp 49°. bp 344°; bp_{15} 190°. n_D^{79} 1.4283. Practically insol in water. Sol in alcohol, chloroform, ether.

Note: The *hexadecyl alcohol* developed by Esso Res. & Eng. Co. for cosmetics is a liquid, primary, branched chain, C_{16} alcohol, made up of an array of isomeric compds maintained in constant proportion by a complex manufacturing process (not from spermaceti): Edman, Lowden, *Drug Cosmet. Ind.* **93**, 631 (Nov. 1963). Liquid, d_{20}^{20} 0.842. bp_{50} 195-205°. Freezes at < −60°. Miscible with most alcohols, glycols, esters, ketones, cosmetic oils and aromatics. Immiscible with water.

USE: In cosmetics as emollient, emulsion modifier, coupling agent. Pharmaceutic aid (emulsifying and stiffening agent).

2021. Cetyldimethylethylammonium Bromide. *N-Ethyl-N,N-dimethyl-1-hexadecanaminium bromide; ethylhexadecyldimethylammonium bromide;* ethyl cetab; CDA; Ammonyx DME; Bretol. $C_{20}H_{44}BrN$; mol wt 378.49. C 63.47%, H 11.72%, Br 21.11%, N 3.70%. Cationic germicidal detergent. Prepn and antibacterial activity: R. S. Shelton *et al., J. Am. Chem. Soc.* **68**, 753 (1946).

$$\left[(CH_3)_2 - \overset{+}{\underset{\underset{CH_2CH_3}{|}}{N}} - (CH_2)_{15}CH_3 \right] \ Br^-$$

White powder, mp 178-186°. Sol in water, alcohol; slightly sol in chloroform, benzene, ether. LD_{50} orally in rats: 500 mg/kg, *RTECS* Vol. 1, R. J. Lewis, R. L. Tatken, Eds. (1979) p 107.

USE: Disinfectant; laboratory reagent.

THERAP CAT: Topical antiseptic.

THERAP CAT (VET): Topical antiseptic.

2022. Cetyl Lactate. *2-Hydroxypropanoic acid hexadecyl ester; 1-hexadecanol lactate;* lactic acid cetyl ester; lactic acid hexadecyl ester; Ceraphyl 28. $C_{19}H_{38}O_3$; mol wt 314.49. C 72.56%, H 12.18%, O 15.26%. $CH_3CHOHCOOC_{16}H_{33}$. Preparation: Rehberg, Marion, *J. Am. Chem. Soc.* **72**, 1918 (1950).

Waxy solid. mp 41°. $bp_{0.1}$ 132°; bp_1 170°; bp_{10} 219°. n_D^{40} 1.4410; n_D^{50} 1.4370.

USE: Non-ionic emollient. To improve feel and texture of cosmetic and pharmaceutical prepns.

2023. Cetyl Palmitate. *Hexadecanoic acid hexadecyl ester; palmitic acid hexadecyl ester;* hexadecyl palmitate. $C_{32}H_{64}O_2$; mol wt 480.83. C 79.93%, H 13.41%, O 6.65%. $CH_3(CH_2)_{14}COOCH_2(CH_2)_{14}CH_3$. Prepn from palmitoyl chloride and cetyl alcohol in the presence of Mg: Paquot, Bouquet, *Bull. Soc. Chim. France* **1947**, 321; by CrO_3-H_2SO_4 oxidation of cetyl alcohol: Cymerman-Craig, Horning, *J. Org. Chem.* **25**, 2098 (1960). Biosynthesis using inoculum of *Nocardia salmonicolor:* Davis, U.S. pat. **3,169,099** (1965 to Socony Mobil Oil).

Monoclinic leaflets, mp 54°. d^{20} 0.989. n_D^{70} 1.4398. Practically insol in water; sol in abs alc, ether.

2024. Cetylpyridinium Chloride. *1-Hexadecylpyridinium chloride;* Ceepryn; Cepacol; Cetamium; Dobendan; Medilave; Merocet; Pristacin; Pyrisept. $C_{21}H_{38}ClN$; mol wt 339.99. C 74.19%, H 11.26%, Cl 10.43%, N 4.12%. Pharmacology and toxicology: *J. Pharmacol. Exp. Ther.* **74**, 401 (1942). Review of early literature: C. L. Huyck, *Am. J. Pharm.* **116**, 50 (1944). Toxicity data: J. W. Nelson, S. C. Lyster, *J. Am. Pharm. Assoc.* **35**, 89 (1946).

$$\left[\underset{\text{(pyridinium ring)}}{\overset{+}{N}-(CH_2)_{15}CH_3} \right] \ Cl^-$$

Monohydrate, white powder. mp 77-83°. Freely sol in water, alcohol, chloroform; very slightly sol in benzene, ether. pH (1% aq soln): 6.0 to 7.0. Surface tension (25°): 43 dyn/cm (0.1% aq soln); 41 dyn/cm (1.0%); 38 dyn/cm (10%). LD_{50} in rats (mg/kg): 250 s.c.; 6 i.p.; 30 i.v.; 200 orally (Nelson, Lyster).

USE: Pharmaceutic aid (preservative).

THERAP CAT: Topical anti-infective.

THERAP CAT (VET): Topical antiseptic; disinfectant.

2025. Cevadine. *(Z)-4α,9-Epoxycevane-3β,4,12,14,16β,-17,20-heptol 3-(2-methyl-2-butenoate);* veratrine. $C_{32}H_{49}NO_9$; mol wt 591.72. C 64.95%, H 8.35%, N 2.37%, O 24.33%. From seeds of *Schoenocaulon officinale* (Schlecht & Cham.) A. Gray *(Sabadilla officinarum* Brandt), *Liliaceae:* Poetsch *et al., J. Am. Pharm. Assoc.* **38**, 525 (1949); Ringel, *ibid.* **45**, 433 (1956). Structure: Kupchan, Alfonso, *ibid.* **49**, 242 (1960). *Review:* Wintersteiner in Graff, *Essays in Biochemistry* (Wiley, New York, 1956) pp 308-321.

Flat needles from ether, decomp 213-214.5°. $[\alpha]_D^{20}$ +12.8° (c = 3.2 in alc). One gram dissolves in about 15 ml alc or ether; slightly sol in water. LD_{50} i.p. in mice: 3.5 mg/kg, Swiss, Bauer, *Proc. Soc. Exp. Biol. Med.* **76**, 847 (1951).

Aurichloride, fine yellow needles from alc, dec 190°.

Mercurichloride, $C_{32}H_{49}NO_9.HCl.HgCl_2$, silvery scales, dec 172°.

USE: Evaluation as insecticide: Ikawa, Link *et al., J. Biol. Chem.* **159**, 517 (1945). *Caution:* Extremely irritating locally, particularly to mucous membranes. Serious poisoning has resulted from local application. Caution must be used in handling. *See also* Veratrine (Mixture).

2026. Cevine. *4,9-Epoxycevane-3α,4β,12,14,16β,17,20-heptol.* $C_{27}H_{43}NO_8$; mol wt 509.65. C 63.63%, H 8.51%, N 2.74%, O 25.12%. By hydrolysis of cevadine. Structure and stereochemistry: Barton *et al., Experientia* **10**, 81 (1954); Kupchan *et al., J. Am. Chem. Soc.* **80**, 1769 (1958); Kupchan *et al., Tetrahedron* **7**, 47 (1959); Eeles, *Tetrahedron Letters* no. 7, 24 (1960).

Hemiheptahydrate, $C_{27}H_{43}NO_8.3\frac{1}{2}H_2O$, triclinic prisms from dil alc. After drying at 110° it sinters at 165° and becomes a transparent resin at 165-170°: Ikawa, Link *et al., J. Biol. Chem.* **159**, 517 (1945). $[\alpha]_D^{17}$ −17.5° (aq alc). Sol in water, alc; slightly sol in ether. LD_{50} i.v. in mice: 87 mg/kg. Hydrochloride, mp 247°.

2027. Chalcomycin. $C_{35}H_{56}O_{14}$; mol wt 700.80. C 59.98%, H 8.05%, O 31.96%. Antibiotic substance produced from *Streptomyces bikiniensis* NRRL 2737: Frohardt *et al.,* **Ger.** pat. 1,109,835 (1960 to Parke, Davis). Structure: Woo *et al., J. Am. Chem. Soc.* **86**, 2726 (1964). *Review:* Jordan, "Chalcomycin" in *Antibiotics* I, D. Gottlieb, P. Shaw, Eds. (Springer-Verlag, New York, 1967) pp 446-450.

Crystals from butanol, mp 121-123°. $[\alpha]_D^{27}$ −43.5° (ethanol). uv max: 218 nm (ϵ 22,770). Sol in ethyl acetate, methanol, ethanol, benzene, toluene, chloroform; sparingly sol in carbon tetrachloride, ether, water. Practically insol in petr ether.

2028. Chalcone. *1,3-Diphenyl-2-propen-1-one;* chalkone; benzylideneacetophenone; benzalacetophenone; phenyl styryl ketone. $C_{15}H_{12}O$; mol wt 208.25. C 86.51%, H 5.81%, O 7.68%. $C_6H_5CH=CHCOC_6H_5$. Prepd by the action of NaOH on an alcoholic soln of benzaldehyde and acetophenone: E. P. Kohler, H. M. Chadwell, *Org. Syn. coll. vol. I,* 78 (1941).

Pale-yellow orthorhombic prisms from petr ether. mp 57-58°. d_4^{62} 1.0712. bp_{760} 345-348° (slight decompn). bp_{25} 208°. n_D^{62} 1.6458. Freely sol in ether, chloroform, carbon

bisulfide, benzene; slightly in alcohol; very sparingly in cold petr ether. *Caution:* Skin irritant.

Picrate, $C_{15}H_{12}O.2C_6H_3N_3O_7$, yellow precipitate, mp 93-97°. Easily dec by water.

Note: Exists in other forms (polymorphic) with different mps. Irradiation produces four forms of dimers.

2029. Chalcopyrite. Cupric ferrous sulfide; copper pyrites; yellow copper; Chalkopyrite; Kupferkies. $CuFeS_2$; mol wt 183.51. Cu 34.62%, Fe 30.43%, S 34.94%. Occurs as a mineral. It is an important copper ore found in Canada, Montana, Utah, Arizona, Tennessee, Chile, Peru, Bolivia, Europe, Japan. Prepd synthetically by treating $KFeS_2$ with an ammoniacal soln of Cu^+ ions: Schneider, *J. Prakt. Chem.* [2] **38**, 569 (1888); **56**, 415 (1897); Boon, *Rec. Trav. Chim.* **63**, 69 (1954).

Yellow brass-colored or bronze-colored crystals. Metallic, greenish sheen. Tetragonal crystal system. d 4.1-4.3. mp 950°. Specific heat 0.1291. Hardness 3.5-4.0. Soluble in nitric acid, but dissolves faster in aqua regia. Practically insol in hydrochloric acid.

Note: Copper iron sulfide of the formula $FeS.2Cu_2S.CuS$ occurs as the mineral *bornite.*

USE: The ore as source of copper. The synthetic material in semiconductor research.

2030. D-Chalcose. *4,6-Dideoxy-3-O-methyl-D-xylo-hexose;* 4,6-dideoxy-3-O-methyl-D-glucose; lankavose. $C_7H_{14}O_4$; mol wt 162.18. C 51.84%, H 8.70%, O 39.46%. Component of chalcomycin and lankomycin, *q.v.* Prepn from chalcomycin and structure: Woo *et al., J. Am. Chem. Soc.* **83**, 3352 (1961). Stereochemistry: Woo *et al., ibid.* **84**, 1066 (1962); Foster *et al., J. Chem. Soc.* **1965**, 2318. Synthesis: Kochetkov, Usov, *Tetrahedron Letters* **1963**, 519; McNally, Overend, *Chem. & Ind. (London)* **1964**, 2021; Lawton *et al., Can. J. Chem.* **47**, 2899 (1969). Synthesis of DL-form: R. M. Srivastava, R. K. Brown, *Can. J. Chem.* **48**, 830 (1970); K. Torssell, M. P. Tyagi, *Acta Chem. Scand. B* **31**, 1 (1977); S. Danishefsky, J. F. Kerwin, *J. Org. Chem.* **47**, 1597 (1982).

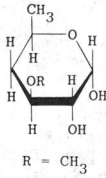

R = CH_3

Crystals, mp 96-99°. $[\alpha]_D^{24}$ +120° (2 min) → +97° (10 min) → +76° (3 hr and 26 hr) (c = 1.5 in water).

Methyl chalcoside, $C_8H_{16}O_4$, crystals, mp 101.5-102°. $[\alpha]_D^{27}$ −21° (c = 2.04 in chloroform).

2031. Chamazulene. *7-Ethyl-1,4-dimethylazulene;* 1,4-dimethyl-7-ethylazulene; dimethulene. $C_{14}H_{16}$; mol wt 184.27. C 91.25%, H 8.75%. Anti-inflammatory principle obtained from chamazulenogenic compds found in chamomile (*Matricaria chamomilla* L.), wormwood (*Artemisia absinthium* L.), and yarrow (*Achillea millefolium* L., Compositae). Isoln: Ruzicka, Rudolph, *Helv. Chim. Acta* **9**, 118 (1926); Ruzicka, Haagen-Smit, *ibid.* **14**, 1104 (1931). Structure: Sorm *et al., Coll. Czech. Chem. Commun.* **18**, 527 (1953); Meisels, Weizmann, *J. Am. Chem. Soc.* **75**, 3865 (1953); Suchy *et al., Coll. Czech. Chem. Commun.* **21**, 477 (1956). Synthesis: Mangoni, Bandiera, *Gazz. Chim. Ital.* **90**, 947 (1960); White, Winter, *Tetrahedron Letters* **1963**, 137; D. Mukherjee *et al., J. Am. Chem. Soc.* **101**, 251 (1979). Pharmacology: I. G. Boldina, *Farmakol. Toksikol.* **29**, 672 (1966); I. G. Boldina, M. V. Nazarenko, *C.A.* **67**, 42441h (1967).

Blue oil. bp_{12} 161°; bp_{11} 145°. d_4^{20} 0.9883. uv max: 370 nm (log ϵ 3.7). LD_{50} i.m. in white mice: 3 g/kg (Boldina, Nazarenko).

Trinitrobenzene derivative, $C_{20}H_{19}N_3O_6$, dark violet needles from abs ethanol, mp 131.5-132.5°.

2032. Chamomile. Camomile; Roman chamomile; garden chamomile; ground apple. Dried flower heads of *Anthemis nobilis* L., *Compositae*, from cultivated plants. *Habit.* Southern and Western Europe. Cultivated in Germany, Great Britain, France, Belgium; somewhat naturalized in the U.S. *Constit.* Bitter glucoside (anthemic acid), anthesterol, anthemene (anthemidin), 0.8-1% volatile oil (which contains tiglic acid esters); resin, tannin. *See also* Matricaria. Active principles: Verzar-Petri, Korbely, *Gyogyszereszet* **13**, 174 (1969), *C.A.* **71**, 73939p (1969).

THERAP CAT: Aromatic bitter.

2033. Chanoclavine. *2-Methyl-3-[1,3,4,5-tetrahydro-4-(methylamino)benz[cd]indol-5-yl]-2-propen-1-ol; (E)-8,9-didehydro-6-methyl-6,7-secoergoline-8-methanol;* chanoclavin-I; secaclavine. $C_{16}H_{20}N_2O$; mol wt 256.34. C 74.96%, H 7.87%, N 10.93%, O 6.24%. Precursor of the tetracyclic ergolines, agroclavine, elymoclavine, and lysergic acid amide, *q.q.v.*: Gröger *et al., Z. Naturforsch.* **21b**, 827 (1966); Floss *et al., Chem. Commun.* **1967**, 105. Occurs in sclerotia of *Claviceps purpurea* (Fries) Tul., *Hypocreaceae*, in *Ipomea tricolor* Cav., *Convolvulaceae* and is one of the active principles of the ancient Aztec drug "Ololiuqui," *Rivea coryrubosa* (L.) Hall. f., *Convolvulaceae:* Hofmann, Tscherter, *Experientia* **16**, 414 (1960). Isoln and structure: Hofmann *et al., Helv. Chim. Acta* **40**, 1358 (1957). Isoln of stereoisomers: Stauffacher, Tscherter, *ibid.* **47**, 2186 (1964). Stereochemistry: Acklin *et al., Chem. Commun.* **1966**, 799. Biosynthetic studies: Floss *et al., J. Am. Chem. Soc.* **90**, 6500 (1968). Total synthesis of (±)-chanoclavine: H. Plieninger *et al., Tetrahedron Letters* **1975**, 1827; H. Plieninger, D. Schmalz, *Ber.* **109**, 2140 (1976); M. Natsume, H. Muratake, *Heterocycles* **16**, 375 (1981).

Thick prisms and polyhedra from acetone or methanol, mp 220-222°. $[\alpha]_D^{20}$ −240° (pyridine), $[\alpha]_D^{20}$ −205° (c = 0.75 in alcohol). uv max: 225, 284, 293 nm (ϵ 4.44, 3.82, 3.72). One gram dissolves in 25 ml boiling methanol, 140 ml boiling acetone, 170 ml boiling ethyl acetate or 350 ml boiling chloroform. Practically insol in water.

Epimer at position 10, *chanoclavine II.*

2034. CHAPS. *N,N-Dimethyl-N-(3-sulfopropyl)-3-[[(3α,5β,7α,12α)-3,7,12-trihydroxy-24-oxocholan-24-yl]amino]-1-propanaminium hydroxide inner salt;* 3-[(3-cholamidopropyl)dimethylammonio]-1-propanesulfonate. $C_{32}H_{58}N_2O_7S$; mol wt 614.88. C 62.51%, H 9.51%, N 4.56%, O 18.21%, S 5.21%. Sulfobetaine zwitterionic derivative of cholic acid, *q.v.* Prepn: L. M. Hjelmeland, *Proc. Nat. Acad. Sci. USA* **77**, 6368 (1980); *idem*, U.S. pat. **4,372,888** (1983 to U.S. Secy. Health and Human Services); and physical properties: L. M. Hjelmeland *et al., Anal. Biochem.* **130**, 72 (1983). Physical characterization: R. E. Stark *et al., J. Phys. Chem.* **88**, 6063 (1984). Micellar properties: M. A. Partearroyo *et al., Biochem. Int.* **16**, 259 (1988). Solubilization of

membrane proteins: A. J. Bitonti *et al., Biochemistry* **21**, 3650 (1982); W. Wouters *et al., Eur. J. Pharmacol.* **115**, 1 (1985); Ll. Bancells *et al., Biochem. Pharmacol.* **36**, 2539 (1987); of opioid receptors: J. Simon *et al., J. Neurochem.* **46**, 695 (1986).

Crystals from absolute methanol at 0°.

Hydroxy analog, $C_{32}H_{58}N_2O_8S$, *CHAPSO.* White granular detergent.

USE: Nondenaturing biological detergent.

2035. Chartreusin. *10-[[6-Deoxy-2-O-(6-deoxy-3-O-methyl-α-D-galactopyranosyl)-β-D-galactopyranosyl]oxy]-6-hydroxy-1-methylbenzo[h][1]benzopyrano[5,4,3-cde][1]benzopyran-5,12-dione;* antibiotic X-465A; lambdamycin; NSC 5159. $C_{32}H_{32}O_{14}$; mol wt 640.58. C 60.00%, H 5.04%, O 34.96%. Antibiotic substance produced by *Streptomyces chartreusis* from African soil, also by another *Streptomyces* sp. from North American soil: B. E. Leach *et al., J. Am. Chem. Soc.* **75**, 4011 (1953). Identity with antibiotic X-465A: J. Berger *et al., ibid.* **80**, 1636 (1958). Structure: L. H. Sternbach *et al., ibid.* 1639; E. Simonitsch *et al., Helv. Chim. Acta* **47**, 1459 (1964); W. Eisenhuth *et al., ibid.* 1475. Anticancer activity: J. P. McGovern *et al., Cancer Res.* **37**, 1666 (1977). Synthesis of *chartarin*, the chartreusin aglycone: T. R. Kelly *et al., J. Am. Chem. Soc.* **102**, 798 (1980); F. M. Hauser, D.W. Combs, *J. Org. Chem.* **45**, 4071 (1980).

Yellow plates from acetone or methylene chloride + ethanol, mp 184-186°. Weak acid. $[\alpha]_D^{25}$ +132.5° (c = 0.2 in pyridine); $[\alpha]_D^{25}$ −33° (c = 0.3 in glacial acetic acid). uv max: 237, 262, 332, 382, 405, 422 nm. Practically insol in water; sol in acetone.

Sodium salt, gold-colored needles or plates from water. Soly in water: at least 20 mg/ml (pH 9.5). LD_{50} i.v. in mice: 250 mg/kg, B. E. Leach *et al., loc. cit.*

2036. Chaulmoogra Oil. Hydnocarpus oil; gynocardia oil. Fixed oil expressed from chaulmoogra, the seeds of *Taraktogenos kurzii* King, *Bixaceae; Hydnocarpus wightiana* Blume or *H. anthelmintica* Pierre, *Flacourtiaceae.* The seeds contain about 50% chaulmoogra oil. *Constit.* Glycerides of chaulmoogric and hydnocarpic acids with small amounts of glycerides of palmitic acid, etc.

Yellow or brownish-yellow oil. Below 25° it is a soft solid; characteristic odor. About d_{25}^{25} 0.95; d_{15}^{45} 0.940. $[\alpha]_D^{20}$ +48° to +60°. Sol in benzene, chloroform, ether, petr ether, slightly in cold alcohol; almost entirely sol in hot alcohol, carbon disulfide.

THERAP CAT: Antibacterial (leprostatic).

2037. Chaulmoogric Acid. *2-Cyclopentene-1-tridecanoic*

acid; D-*13-(2-cyclopenten-1-yl)tridecanoic acid;* hydnocarp-ylacetic acid. $C_{18}H_{32}O_2$; mol wt 280.44. C 77.09%, H 11.50%, O 11.41%. Isoln from chaulmoogra oil and charac-terization: Power, Gornall, *J. Chem. Soc.* **85**, 838, 851 (1904); Barrowcliff, Power, *ibid.* **91**, 557 (1907); Hashimoto, *J. Am. Chem. Soc.* **47**, 2325 (1925); Cole, Cardoso, *ibid.* **61**, 2349, 2351, 3442 (1939). Structure: Shriner, Adams, *ibid.* **47**, 2727 (1925). Synthesis: Noller, Adams, *ibid.* **48**, 1080 (1926); Perkins, Cruz, *ibid.* **49**, 1070 (1927). Synthesis and stereochemistry: Mislow, Steinberg, *ibid.* **77**, 3807 (1955). Activity against *Mycobacterium leprae:* L. Levy, *Am. Rev. Respir. Dis.* **111**, 703 (1975). Biosynthesis: T. Kaneda, *Biochem. Biophys. Res. Commun.* **99**, 1226 (1981).

$$(CH_2)_{12}COOH$$

Shiny leaflets from petr ether or alcohol. mp 68.5°. bp_{20} 247-248°. $[\alpha]_D^{20}$ +60.3° (c = 4 in chloroform). Iodine value 90.5. Freely sol in ether, chloroform, ethyl acetate; sol in other organic fat solvents.

Methyl ester, $C_{19}H_{34}O_2$, prepd by passing HCl gas into a methanol soln of the acid. Needles, mp 22°. bp_{20} 227°. d_{25}^{25} 0.9119. $[\alpha]_D^{15}$ +50° (c = 5 in chloroform).

Ethyl ester, *chaulmestrol, Moogrol.* Essentially a mixt of the ethyl esters of the unsaturated fatty acids (chaulmoogric and hydnocarpic) of chaulmoogra oil. Pale yellow, clear liquid; slight fruity odor; not unpleasant taste. Iodine no. 90-100. d_{25}^{25} about 0.904. $[\alpha]_D^{15}$ +44.5° (chloroform). Insol in water; misc with alc, chloroform.

THERAP CAT: Ethyl ester in treatment of leprosy, sarcoid-osis.

2038. Chavicine. *1-[5-(1,3-Benzodioxol-5-yl)-1-oxo-2,4-pentadienyl]piperidine; (Z,Z)-1-piperoylpiperidine.* C_{17}-$H_{19}NO_3$; mol wt 285.33. C 71.56%, H 6.71%, N 4.91%, O 16.82%. One of the most active constituents of black pep-per: Buchheim, *Arch. Exp. Pathol. Pharmakol.* **5**, 455 (1876); Ott, Eichler, *Ber.* **55**, 2653 (1922); Ott, Lüdemann, *Ber.* **57**, 214 (1924). Stereoisomer of piperine: Lohaus, Gall, *Ann.* **517**, 282 (1935). Loss of pungency of ground pepper during storage attributed to gradual isomerization of chavicine into piperine *(q.v.):* Newman, *Chem. Prod.* **16**, 379 (1953). Syn-thesis of chavicine and questioning of its existence as con-stituent of black pepper: R. Grewe *et al., Ber.* **103**, 3752 (1970). Spectroscopic structural elucidation and prepn: R. DeCleyn, M. Verzele, *Bull. Soc. Chim. Belg.* **84**, 435 (1975).

Yellowish, oily mass; sharp, peppery taste. Sol in alc, ether, petr ether. uv max (methanol): 318 nm (ε 16,200).

2039. Chavicol. *4-(2-Propenyl)phenol; p-allylphenol; γ-(p-hydroxyphenyl)-α-propylene.* $C_9H_{10}O$; mol wt 134.17. C 80.56%, H 7.51%, O 11.92%. Found together with ter-penes in volatile betel oils (from leaves of *Piper betle* L., *Pip-eraceae):* Eijkman, *Ber.* **22**, 2739 (1889). Prepd from a mix-ture of estragole, ethyl bromide and magnesium in benzene: Grignard, *Compt. Rend.* **151**, 323 (1910); from 4-allylphen-ylmagnesium bromide in ether by the action of oxygen: Quelet, *Bull. Soc. Chim. France* [4] **45**, 265 (1929).

Liquid. mp 15.8°. d_4^{15} 1.0203; d_4^{20} 1.0175. bp_{760} 238°; bp_{16}

123°. n_D^{18} 1.5441. Miscible with alcohol, ether, chloroform, petr ether.

Methyl ether, *see* Estragole.

2040. Cheirolin. *1-Isothiocyanato-3-(methylsulfonyl)-propane; isothiocyanic acid 3-(methylsulfonyl)propyl ester; γ-isothiocyanopropyl methyl sulfone; methylpropylsulfone mustard oil.* $C_5H_9NO_2S_2$; mol wt 179.26. C 33.50%, H 5.06%, N 7.81%, O 17.85%, S 35.78%. $CH_3SO_2(CH_2)_3NCS$. From seed of *Cheiranthus cheiri* L., *Cruciferae:* Wagner, *Chem. Ztg.* **32**, 76 (1908); Schneider, *Ber.* **41**, 4466 (1909); **42**, 3416 (1910). Structure: *idem., Ann.* **375**, 207 (1910). Synthesis: Kjaer *et al., Acta Chem. Scand.* **7**, 1370 (1953). Prepn by enzymatic hydrolysis of glucocheirolin: Kjaer, Jensen, *ibid.* **12**, 1746 (1958).

Crystals from ether, mp 47-48°. bp_3 200°. Slightly sol in water, ether; freely sol in alcohol, chloroform, ethyl acetate.

2041. Chelerythrine. *1,2-Dimethoxy-12-methyl[1,3]ben-zodioxolo[5,6-c]phenanthridinium;* toddaline. $[C_{21}H_{18}$-$NO_4]^+$. From root of *Chelidonium majus* L., *Papaveraceae:* Kratzmann, *Pharm. Monatsh.* **5**, 161 (1924), *C.A.* **18**, 3406[2] (1924); Platonova *et al., J. Gen. Chem. USSR* **26**, 181 (1956). Structure: Späth, Kuffner, *Ber.* **64**, 1123 (1931); N. Decau-dain *et al., Ann. Pharm. Franc.* **35**, 521 (1977). Identity with toddaline: Govindachari, Thyagarajan, *J. Chem. Soc.* **1956**, 769. Synthesis of chelerythrine chloride: Bailey, Worthing, *ibid.* 4535; S. V. Kessar *et al., J. Org. Chem.* **53**, 1708 (1988). Pharmacology: Chelombit'o, Murav'eva, *Aktual Prob. Farmakol Farm., Vses. Nauch. Konf.* **1971**, 183, *C.A.* **76**, 112g (1972).

Crystals from chloroform + methanol, mp 207° [Manske, *Can. J. Res.* **21B**, 140 (1943)]. The free base is colorless, but its quaternary salts are yellow. Aq solns of base and salts show violet fluorescence. uv spectrum of chelerythrine chlo-ride: Hruban *et al., Coll. Czech. Chem. Commun.* **35**, 3420 (1970).

2042. Chelidonic Acid. *4-Oxo-4H-pyran-2,6-dicarboxyl-ic acid;* 4-oxo-1,4-pyran-2,6-dicarboxylic acid; Jerva acid; jervasic acid. $C_7H_4O_6$; mol wt 184.10. C 45.67%, H 2.19%, O 52.14%. Presence in various plants: Stransky, *Arch. Pharm.* **258**, 56 (1920); Ramstad, *Pharm. Acta Helv.* **28**, 45 (1953). Structure: Verkade, *Rec. Trav. Chim.* **43**, 879 (1924). Synthesis: Riegel, Reinhard, *J. Am. Chem. Soc.* **48**, 1334 (1926); Riegel, Zwilgmeyer, *Org. Syn. coll. vol. II*, 126 (1943); Toomey, Riegel, *J. Org. Chem.* **17**, 1492 (1952).

Crystals, dec 257°. uv max (water): 270 nm (ε 11,500). One gram dissolves in 65 ml water, 26 ml boiling water. Sparingly sol in alc.

Monohydrate, monoclinic crystals from hot alcohol + water. The water of crystn is given up when heated at 102° and then at 160° to constant weight.

Sodium salt, sparingly sol in water.

2043. Chelidonine. 5b,6,7,12b,13,14-Hexahydro-13-methyl[1,3]benzodioxolo[5,6-c]-1,3-dioxolo[4,5-i]phenan-thridin-6-ol; stylophorin. $C_{20}H_{19}NO_5$; mol wt 353.36. C 67.98%, H 5.42%, N 3.96%, O 22.64%. Hexahydrobenzo-phenanthridine alkaloid. Occurs in nature in (+)-form (depicted below), (-)-form and racemic form. Isoln of (+)-form from root of *Chelidonium majus* L., *Stylophorum diphyllum* (Michx.) Nutt., and *Dicranostigma franchetianum* (Prain) Fedde, *Papaveraceae:* J. M. Probst, *Ann.* **29**, 113

(1839); E. Schmidt, F. Selle, *Arch. Pharm.* **228**, 441 (1890); F. Selle, *ibid.* 96; Manske, *Can. J. Res.* **20B**, 53 (1942); J. Slavik, *Coll. Czech. Chem. Commun.* **20**, 198 (1955); from *Symphoricarpos albus* L., Blake, *Caprifoliaceae:* M. Szaufer *et al., Phytochemistry* **17**, 1446 (1978). Isoln of (−)-form from *Glaucium corniculatum* Curt., *Papaveraceae:* J. Slavik, L. Slavikova, *Coll. Czech. Chem. Commun.* **22**, 279 (1957). Structure: F. von Bruchhausen, H. W. Bersch, *Ber.* **63**, 2520 (1930); E. Späth, F. Kuffner, *ibid.* **64**, 370 (1931); H. W. Bersch, *Arch. Pharm.* **291**, 491 (1958). Identity of (±)-form with diphylline: J. Slavik, *Coll. Czech. Chem. Commun.* **26**, 2933 (1961). Absolute configuration of (+)-*p*-bromobenzoate: N. Takao *et al., Tetrahedron Letters* **1979**, 495. Conformation: M. Cushman, T.-C. Choong, *Heterocycles* **14**, 1935 (1980); M. Sugiura *et al., J. Chem. Soc. Perkin Trans.* II, 175 (1986). Chiroptic properties and absolute configuration of (+)-chelidonine: N. Takao *et al., Arch. Pharm. (Weinheim)* **317**, 223 (1984). Biosynthesis: E. Leete, *J. Am. Chem. Soc.* **85**, 473 (1963). Total synthesis of (±)-form: W. Oppolzer, K. Keller, *ibid.* **93**, 3836 (1971); M. Cushman *et al., J. Org. Chem.* **45**, 5067 (1980); W. Oppolzer, C. Rabbiani, *Helv. Chim. Acta* **66**, 1119 (1983); M. Hanaoka *et al., Chem. Lett.* **1986**, 736. Effect on smooth muscle: P. J. Hanzlik: *J. Pharm. Exp. Ther.* **7**, 99 (1915). Inhibition of reverse transcriptase activity: M. L. Sethi, *Can. J. Pharm. Sci.* **16**, 29 (1981). Cytotoxic effects: M. Cushman *et al., loc. cit.* Review of pharmacological effects: V. Preininger, "The Biology of Papaveraceae Alkaloids" in *The Alkaloids* vol. **15**, R. H. F. Manske, Ed. (Academic Press, Orlando, 1975) pp 241-242. General review: V. Simanek, "Benzophenanthridine Alkaloids" *ibid.* vol. **26**, A. Brossi, Ed. (1985) pp 185-240. Toxicity data: R. C. Anderson, K. K. Chen, *Fed. Proc.* **5**, 163 (1946).

(+)-Form: Monoclinic prisms from methanol, ethanol or ethanol + chloroform, mp 135-136°. bp$_{0.002}$ 220° (air-bath temp). $[\alpha]_D^{22}$ +115° ±3° (ethanol); $[\alpha]_D^{20}$ +117° (c = 3 in CHCl$_3$). uv max (methanol): 289, 239, 208 nm (log ε 3.89, 3.88, 4.69). Sol in alc, chloroform, ether, amyl alc. Practically insol in water. LD$_{50}$ in mice (mg/kg): 34.6 ±2.44 i.v. (Anderson, Chen).

(+)-*O*-Acetylchelidonine, C$_{22}$H$_{21}$NO$_6$, crystals from chloroform, mp 184-186°. $[\alpha]_D$ +110°.

(+)-Benzoylchelidonine, C$_{27}$H$_{23}$NO$_6$, crystals from chloroform, mp 210-211°.

(−)-Form: Crystals from aqueous ethanol, mp 136°. $[\alpha]_D^{22}$ −112° ±3° (c = 0.47 in ethanol).

(±)-Form: *Diphylline.* Crystals from ethanol, mp 215-216°.

2044. Chenodiol. (3α,5β,7α)-3,7-*Dihydroxycholan-24-oic acid; 3α,7α-dihydroxy-5β-cholanic acid;* anthropodesoxycholic acid; gallodesoxycholic acid; 17β-(1-methyl-3-carboxypropyl)etiocholane-3α,7α-diol; chenic acid; chenodeoxycholic acid; CDC; Chendol; Chenix; Chenocedon; Chenocol; Chenodex; Chenofalk; Chenossil; Chenosäure; Cholanorm; Fluibil; Kebilis; Ulmenide. C$_{24}$H$_{40}$O$_4$; mol wt 392.56. C 73.43%, H 10.27%, O 16.30%. A major bile acid in many vertebrates, occurring as the *N*-glycine and/or *N*-taurine conjugate. With other bile acids, forms mixed micelles with lecithin in bile which solubilize cholesterol and thus facilitates its excretion. Facilitates fat absorption in the small intestine by micellar solubilization of fatty acids and monoglycerides. Has cathartic properties since it induces fluid secretion from large intestine. Main constituent of the bile of hens, geese and other fowl; occurs in appreciable amounts in the bile of hamster, hog, guinea pig, bear and man. Epimeric with ursodiol, *q.v.* Isoln: Windhaus *et al., Z. Physiol. Chem.* **140**, 177 (1924); Wieland, Reveney, *ibid.* 186. Configuration: Lettré, *Ber.* **68**, 766 (1935). Prepn

from cholic acid: Fieser, Rajagopalan, *J. Am. Chem. Soc.* **72**, 5530 (1950); Hauser *et al., Helv. Chim. Acta* **43**, 1595 (1960); Hofmann, *Acta Chem. Scand.* **17**, 173 (1963). Alternate prepns: Sato, Ikekawa, *J. Org. Chem.* **24**, 1367 (1959); T. Iida, F. C. Chang, *ibid.* **46**, 2786 (1981). Stereoselective total synthesis: T. Kametani *et al., J. Am. Chem. Soc.* **103**, 2890 (1981). Asymmetric total synthesis of (+)-form: *eidem, J. Org. Chem.* **47**, 2331 (1982). Dissolution of cholesterol gallstones: Danzinger *et al., N. Engl. J. Med.* **286**, 1 (1972); Bell *et al., Lancet* **II**, 1213 (1972). Use in long-term treatment of cerebrotendinous xanthomatosis: V. M. Berginer *et al., N. Engl. J. Med.* **311**, 1649 (1984). Monograph on bile acids: *The Bile Acids*, 2 vols., P. P. Nair, D. Kritchevsky, Eds. (Plenum Press, New York, 1971, 1973). Review of pharmacology and therapeutic use of chenodeoxycholic acid: J. H. Iser, A. Sali, *Drugs* **21**, 90-119 (1981). Effect on cholesterol and bile acid metabolism: G. S. Tint *et al., Gastroenterology* **91**, 1007 (1986).

Needles from ethyl acetate + heptane, mp 119°. $[\alpha]_D^{20}$ +11.5° (dioxane). Freely sol in methanol, alc, acetone, acetic acid; more sol in ether and ethyl acetate than deoxycholic acid. Practically insol in water, petr ether, benzene. High solvent power for alkali soaps, but does not form "choleic" acid addition compds as does deoxycholic acid. Forms beautiful cryst salts of Na, K and Ba. While the acid is tasteless, the Na salt tastes slightly sweet at first, then bitter.

Diformate, C$_{25}$H$_{40}$O$_6$, clusters of needles from alc; mp with slight effervescence at 137°, upon further heating solidifies again, and finally melts around 172°.

Methyl ester, C$_{25}$H$_{42}$O$_4$, fine needles from benzene + heptane, mp 90-91°. $[\alpha]_D^{25}$ +20°.

THERAP CAT: Anticholelithogenic.

2045. Chicle. The partially evaporated, milky juice from *Manilkara zapotilla* (Jacq.) Gilly (*Achras sapota* L.), *Sapotaceae. Habit.* West Indies, Mexico and Central America. USE: In the chewing gum industry.

2046. Chimaphila. Pipsissewa; Prince's pine; bitter wintergreen; rheumatism weed; ground holly; pyrola; pine tulip. Dried leaves of *Chimaphila umbellata* (L.) Nutt., *Ericaceae. Habit.* Europe, Asia, North America. *Constit.* Chimaphilin, arbutin, ericolin, urson, tannin, resin.

THERAP CAT: Urinary antiseptic.

2047. Chimaphilin. 2,7-Dimethyl-1,4-naphthalenedione; 2,7-dimethyl-1,4-naphthoquinone. C$_{12}$H$_{10}$O$_2$; mol wt 186.20. C 77.40%, H 5.41%, O 17.19%. From *Chimaphila carymbosa* Pursh. and *Pyrola incarnata* (Fisch.) Fernald, *Ericaceae:* DiModica *et al., Gazz. Chim. Ital.* **83**, 393 (1953); Inouye, *J. Pharm. Soc. Japan* **76**, 976 (1956). Structure: DiModica, Tira, *Gazz. Chim. Ital.* **86**, 234 (1956).

Yellow needles, mp 113.5-114.5°. uv max (alc): 233, 256, 339 nm. Polymerizes in sunlight to needles, mp 232-233°. Diacetate, crystals from ethanol + water, mp 90.5-91.5°.

2048. Chimonanthine. 1-Demethylcalycanthidine. C$_{22}$H$_{26}$N$_4$; mol wt 346.46. C 76.26%, H 7.56%, N 16.17%. From leaves of *Chimonanthus fragrans* Lindle, *Calicanthaceae:* Hodson *et al., Proc. Chem. Soc.* **1961**, 465. Structure:

Clayton *et al.*, *Tetrahedron* **18**, 1495 (1962). Synthesis: Hendrickson *et al.*, *Proc Chem. Soc.* **1962**, 383; Scott *et al.*, *J. Am. Chem. Soc.* **86**, 302 (1964); Hall *et al.*, *Tetrahedron* **23**, 4131 (1967).

Crystals, mp 188-189°. Weak diacidic base of equivalent weight 173. uv max: 246, 304 nm.
Dimethiodide, crystals, dec 235-236°.

2049. Chimyl Alcohol. (+)-3-(*Hexadecyloxy*)-1,2-pro-panediol; α-hexadecylglyceryl ether; testriol. $C_{19}H_{40}O_3$; mol wt 316.51. C 72.10%, H 12.74%, O 15.16%. $CH_3(CH_2)_{15}O$-$CH_2CHOHCH_2OH$. Isoln from bull and boar testes: Hirano, *J. Pharm. Soc. Japan* **56**, 122 (1936); Prelog *et al.*, *Helv. Chim. Acta* **27**, 674 (1944). Synthesis: Baer, Fischer, *J. Biol. Chem.* **140**, 397 (1941).
Leaflets from hexane, mp 64°. $bp_{0.005}$ 120°. $[\alpha]_D^{20}$ +3.0° (c = 1.16 in chloroform). Sol in acetone, hexane, chloroform.
Bis(phenylurethan), $C_{33}H_{50}N_2O_5$, leaflets from petr ether, mp 97.5-98.5°.

2050. Chinese Wax. The excretion of an insect, *Coccus ceriferus* Fabr., or *C. pela* Westwood, deposited on the twigs and branches of a species of ash tree in Western China. Chief constituent is ceryl cerotate.
White to yellowish-white, practically odorless, tasteless solid. d about 0.93. mp about 92°. Solidif 80-81°. Sapon no. 80-92. Iodine no. 1.4. Acid no. 63. Insol in water; freely sol in benzene, slightly in alcohol or ether.
USE: Manuf candles, leather and furniture polish; treating silk and cotton fabrics; sizing and glazing papers.

2051. Chirata. Chiretta; chirayita; bitter stick; East Indian balmony. Dried plant, *Swertia (Ophelia) chirata* (Roxb.) Buch.-Ham., *Gentianaceae*. *Habit.* India (Himalaya). *Constit.* Chiratin, ophelic acid.
THERAP CAT: Bitter tonic, antimalarial.

2052. Chitin. $(C_8H_{13}NO_5)_n$. C 47.29%, H 6.45%, N 6.89%, O 39.37%. Cellulose-like biopolymer consisting predominantly of unbranched chains of β-(1 → 4)-2-acetamido-2-deoxy-D-glucose (also named *N*-acetyl-D-glucosamine) residues. Found in fungi, yeasts, marine invertebrates and arthropods, where it is a principal component in the exoskeletons. May be regarded as a derivative of cellulose, in which the C-2 hydroxyl groups have been replaced by acetamido residues. Occurrence in lower animals: A. G. Richards, *The Integument of Arthropods* (University of Minnesota Press, Minneapolis, 1951). Occurrence in the plant kingdom: F. von Wettstein, *Handbuch der systematischen Botanik* (F. Deuticke, Leipzig and Vienna, 4th ed., 1933). Typical isoln: Hackman, *Aust. J. Biol. Sci.* **7**, 168 (1954); Horowitz *et al.*, *J. Am. Chem. Soc.* **79**, 5046 (1957). Occurrence in the anthozoan *Stylobates aeneus* Dall, the first sea anemone proved capable of synthesizing chitin: D. F. Dunn, M. H. Liberman, *Science* **221**, 157 (1983). Structure: Dweltz, *Biochim. Biophys. Acta* **44**, 416 (1960); **51**, 283 (1961); Carlstrom, *ibid.* **59**, 361 (1962); Ramachadran, Ramakrishman, *ibid.* **63**, 307 (1962). *Review:* Foster, Webber, *Advan. Carbohyd. Chem.* **15**, 371-393 (1960); C. Jeuniaux, "Chitinous Structures" in *Comprehensive Biochemistry* vol. **26c**, M. Florkin, E. H. Stotz, Eds. (Elsevier, New York, 1971) pp 595-632. Book: R. A. A. Muzzarelli, *Chitin* (Pergamon Press, New York, 1977). Review of properties and possible novel applications: P. R. Austin *et al.*, *Science* **212**, 749-753 (1981).

Amorphous solid. Practically insol in water, dil acids, dil and concd alkalies, alcohol and other organic solvents; sol in concd HCl, H_2SO_4, 78-97% H_3PO_4, anhydr HCOOH. There are substantial variations in solubility, molecular weight, acetyl values, specific rotation among chitins of different origins and prepared by different methods.
Acetate, $(C_{14}H_{19}NO_{11})_n$. Prepn: Shorigin, Hait, *Ber.* **68**, 971 (1935). Sol in HCOOH, 50% resorcinol, concd HCl, H_2SO_4, HNO_3; practically insol in organic solvents.
Sulfate. Process for sulfating chitin: Cushing, Kratovil, U.S. pat. 2,755,275 (1956 to Abbott). Sulfated chitin has been found to possess anticoagulant properties in laboratory animals: Roth *et al.*, *Am. J. Physiol.* **171**, 761 (1952); *Proc. Soc. Exp. Biol. Med.* **86**, 315 (1954).
USE: Deacylated chitin, *chitosan*, used in water treatment; in photographic emulsions; in improving the dyeability of synthetic fibers and fabrics.
THERAP CAT: Vulnerary.

2053. Chlophedianol. 2-*Chloro*-α-*[2-(dimethylamino)-ethyl]-α-phenylbenzenemethanol;* 2-*chloro*-α-*(2-dimethyl-aminoethyl)benzhydrol;* 1-*o*-chlorophenyl-1-phenyl-3-di-methylamino-1-propanol; 1-phenyl-1-(*o*-chlorophenyl)-3-dimethylaminopropanol; α-(dimethylaminoethyl)-*o*-chloro-benzhydrol; clofedanol; Tussistop. $C_{17}H_{20}ClNO$; mol wt 289.80. C 70.45%, H 6.96%, Cl 12.24%, N 4.83%, O 5.52%. Prepn: Henecka, Lorenz, *Ger.* pat. 1,083,277 (1960); Lorenz *et al.*, U.S. pat. 3,031,377 (1962, both to Bayer).

mp 120°. LD_{50} i.v. in mice: 70 mg/kg (Lorenz).
Hydrochloride, $C_{17}H_{21}Cl_2NO$, Coldrin, Detigon, Pectoli-tan, Refugal, Ulone, ULO. Crystals, mp 190-191°. Freely sol in water, methanol, ethanol. Sparingly sol in ether, benzene, ethyl acetate. LD_{50} orally in rats: 350 mg/kg; s.c. in mice: 95 mg/kg (Lorenz).
THERAP CAT: Antitussive.

2054. Chloracetyl Chloride. $C_2H_2Cl_2O$; mol wt 112.95. C 21.27%, H 1.79%, Cl 62.78%, O 14.17%. $ClCH_2COCl$. Prepn from chloroacetic acid and benzoyl chloride: Brown, *J. Am. Chem. Soc.* **60**, 1325 (1938); by chlorination of ketene: Erickson, Prill, *J. Org. Chem.* **23**, 141 (1958); Prill, U.S. pat. 2,889,365 (1959 to Monsanto); from chloroacetic acid and pyrocatechylphosphorus trichloride: Gross, Gloede, *Ber.* **96**, 1387 (1963). Manuf from chloroacetic acid and $POCl_2$: MacKenzie, Morris, U.S. pat. 2,848,491 (1958 to Dow); from ketene: Prill, U.S. pat. 2,889,365 (1959 to Monsanto). Physical data: McDonald *et al.*, *J. Chem. Eng. Data* **4**, 311 (1959).
Liquid, very pungent odor. bp 106°. mp −21.77°. d_4^{20} 1.4202. n_D^{20} 1.4541. Dec by water. *Protect from moisture.*
Caution: Irritating to eyes, mucous membranes.
USE: In the synthesis of organic compounds.

2055. Chloracizine. 2-*Chloro*-10-*[3-(diethylamino)-1-oxopropyl]-10H-phenothiazine;* 2-*chloro*-10-(*N,N-diethyl-β-alanyl)phenothiazine;* 10-(β-diethylaminopropionyl)-2-chlorophenothiazine; 2-chloro-10-(β-diethylaminopropion-yl)phenothiazine; chloracysin; chlorocyzine; chlorocizin; khloratsizin; G-020. $C_{19}H_{21}ClN_2OS$; mol wt 360.93. C 63.23%, H 5.87%, Cl 9.82%, N 7.76%, O 4.43%, S 8.89%. Prepn: Horclois, Metivier, *Fr.* pat. 1,060,715 (1954 to Rhône Poulenc); *Brit.* pat. 740,932 (1955); Zhuravlev, Gritsenko, *Zhur. Obshch. Khim.* **26**, 3385 (1956), *C.A.* **51**, 9623h (1957). For other prepns in Russian journals *see C.A.* **55**, 6484b, 9425a (1961).

O=CCH₂CH₂N(C₂H₅)₂ structure

Hydrochloride, $C_{19}H_{21}ClN_2OS \cdot HCl$, crystals from isopropanol, mp 168-169°. Sol in water.

THERAP CAT: Coronary vasodilator.

2056. Chloral Alcoholate. *2,2,2-Trichloro-1-ethoxyethanol;* chloral ethylalcoholate; trichloroacetaldehyde monoethylacetal. $C_4H_7Cl_3O_2$; mol wt 193.47. C 24.83%, H 3.65%, Cl 54.98%, O 16.54%. Prepd by refluxing trichloroacetaldehyde or chloral hydrate with alcohol: Personne, *Compt. Rend.* **69**, 1363 (1869); Magnani, McElvain, *J. Am. Chem. Soc.* **60**, 2212 (1938); Post, *J. Org. Chem.* **6**, 830 (1941).

Cl₃C—CH—OC₂H₅ structure with OH

Crystals, d 1.143. mp 47.5°. bp₇₆₀ 116°. Less sol in water than chloral hydrate. Sol in organic solvents. With concd H_2SO_4 it forms trichloroacetaldehyde and ethyl sulfate.

2057. Chloral Ammonia. *1-Amino-2,2,2-trichloroethanol.* $C_2H_4Cl_3NO$; mol wt 164.43. C 14.61%, H 2.45%, Cl 64.69%, N 8.52%, O 9.73%. $Cl_3CCH(OH)NH_2$. Prepd from chloral and NH_3: Schiff, *Ber.* **10**, 165 (1877); Aschan, *ibid.* **48**, 874 (1915). Believed to be a bimolecular compd, $Cl_3CCH(OH)NHCH(NH_2)CCl_3 \cdot H_2O$ or $C_4H_8Cl_6N_2O_2$: Aschan, *loc. cit.*

Crystals, mp 62-64° (Schiff, *loc. cit.*), mp 72-74° (Aschan, *loc. cit.*). Dissolves slowly in cold water with evolution of NH_3. Sol in ether, benzene.

2058. Chloralantipyrine. Antipyrine chloral hydrate; Hypnal. $C_{13}H_{15}Cl_3N_2O_3$; mol wt 353.64. C 44.15%, H 4.28%, Cl 30.09%, N 7.92%, O 13.57%. $CCl_3CH(OH)_2 \cdot C_{11}H_{12}N_2O$. Prepn: Pfeiffer, Seidel, *Z. Physiol. Chem.* **178**, 97 (1928).

Crystals, large octahedra, from water, mp 68°. One gram dissolves in 10-12 ml water, 4 ml alcohol.

Note: See also Dichloralantipyrine.

THERAP CAT: Hypnotic, analgesic.

2059. Chloral Betaine. *1-Carboxy-N,N,N-trimethylmethanaminium hydroxide inner salt compd with 2,2,2-trichloro-1,1-ethanediol (1:1);* Beta-Chlor; Somilan. $C_7H_{14}Cl_3NO_4$; mol wt 282.57. C 29.75%, H 4.99%, Cl 37.64%, N 4.96%, O 22.65%. Prepd by warming betaine hydrate and chloral hydrate: Petrow *et al., Brit.* pat. **874,246** (1959 to British Drug Houses).

Prisms from water, mp 122.5-124.5°.

Caution: May be habit forming. This is a controlled substance (depressant) listed in the U.S. Code of Federal Regulations, Title 21 Part 1308.14 (1987).

THERAP CAT: Sedative, hypnotic.

2060. Chloral Formamide. *N-(2,2,2-Trichloro-1-hydroxyethyl)formamide;* chloralamide; chloramide. $C_3H_4Cl_3NO_2$; mol wt 192.44. C 18.72%, H 2.10%, Cl 55.27%, N 7.28%, O 16.63%. $HCONHCH(OH)CCl_3$. Prepn or manuf from chloral and formamide: Bennett, Campbell, *Quart. J. Pharm. Pharmacol.* **8**, 398 (1955); Reuter, Sehring, *Ger.* pat. **1,168,161** (1964 to Boehringer, Ing.), *C.A.* **62**, 1024g (1965). Has herbicidal activity (Reuter, Sehring, *loc. cit.*).

Crystals from water of 30% alc, mp 124-126°, also reported as 115-116° (*see* Bennett, Campbell, *loc. cit.*). Soluble in water; readily sol in alcohol, ether, glycerol, acetone, ethyl acetate. Dec by hot solvents. *Incompat:* Alkalies.

Caution: This substance may be habit forming and is listed in the U.S. Code of Federal Regulations, Title 21 Part 329.1 (1987).

THERAP CAT: Has been used as sedative, hypnotic.

2061. Chloral Hydrate. *2,2,2-Trichloro-1,1-ethanediol;* trichloroacetaldehyde monohydrate; Escre; Noctec; Nycton; Somnos; Lorinal; Chloraldurat. $C_2H_3Cl_3O_2$; mol wt 165.42. C 14.52%, H 1.83%, Cl 64.30%, O 19.35%. Made by adding the required amount of water to trichloroacetaldehyde, *q.v.* First synthesized by Liebig in 1832. Introduced as hypnotic by Liebreich in 1869. Comprehensive description: J. E. Fairbrother in *Analytical Profiles of Drug Substances* **vol. 2**, K. Florey, Ed. (Academic Press, New York, 1973) pp 85-143. Toxicity data: E. Goldenthal, *Toxicol. Appl. Pharmacol.* **18**, 185 (1971).

Cl₃C—CHOH structure with OH

Monohydrate, large, monoclinic plates. Perfect 001 cleavage. Skin irritant. Aromatic, penetrating and sightly acrid odor, slightly bitter, caustic taste. Slowly volatilizes on exposure to air. d 1.91. mp 57° when heated in an open vessel. bp 98° with dissociation into chloral and water. Freely sol in water. One ml of water dissolves the following amounts of chloral hydrate: 2.4 g at 0°; 8.3 g at 25°, 14.3 g at 40°. One gram of chlorate hydrate dissolves in 1.3 ml alcohol, in 2 ml chloroform, in 1.5 ml ether, in 1.4 ml olive oil, in 0.5 glycerol, in 68 g carbon disulfide. Freely sol in acetone, methyl ethyl ketone. Moderately or sparingly sol in turpentine, petr ether, carbon tetrachloride, benzene, toluene. Dissociation in water: Piguet *et al., Helv. Chim. Acta* **46**, 406 (1963). An alcoholic soln of chloral hydrate (1 in 20) does not at once redden moistened blue litmus paper. Dec by NaOH into chloroform. Reduces ammoniacal $AgNO_3$. Liquefies when triturated with an equal quantity of menthol, camphor or thymol. LD_{50} orally in rats: 479 mg/kg (Goldenthal).

Pharmaceutical incompat. Alcohol, iodide, cyanide, permanganate, borax, alkali hydroxides and carbonates, lead acetate, monobromated camphor, Diuretin, acetophenetidin, quinine sulfate, salol, theobromine sodiosalicylate, sodium phosphate, urea, urethane.

Caution: May be habit forming. This is a controlled substance (depressant) listed in the U.S. Code of Federal Regulations, Title 21 Parts 329.1 and 1308.14 (1987).

USE: Manuf DDT.

THERAP CAT: Hypnotic, sedative.

THERAP CAT (VET): Narcotic, anesthetic, sedative.

2062. α-Chloralose. *(R)-1,2-O-(2,2,2-Trichloroethylidene)-α-D-glucofuranose;* α-D-glucochloralose; glucochloral; anhydroglucochloral; chloralosane; Alphakil; Dorcalm; Somio. $C_8H_{11}Cl_3O_6$; mol wt 309.54. C 31.04%, H 3.58%, Cl 34.36%, O 31.01%. Prepd from anhydr glucose and trichloroacetaldehyde (anhydr chloral); β-chloralose is also formed. Prepn: M. Hanroit, C. Richet, *C.R. Acad. Sci.* **116**, 63 (1893); and structure: Pictet, Reichel, *Helv. Chim. Acta* **6**, 621 (1923); Freudenberg, Vajda, *J. Am. Chem. Soc.* **59**, 1955 (1937); *cf.* Coles *et al., ibid.* **51**, 519 (1929); White, Hixon, *ibid.* **55**, 2438 (1933). Crystal structure: T. Taga *et al., Acta Crystallogr.* **B38**, 1874 (1982). Immobilizing agent in control of depredating birds: Redpath *et al., Ann. Appl. Biol.* **49**, 77 (1961). Anesthetic for dogs in experimental work: B. G. Bass, N. M. Buckley, *Am. J. Physiol.* **210**, 854 (1966); R. H. Cox, *ibid.* **223**, 660 (1972). CNS depressant activity: J. L. Barker, *Nature* **252**, 52 (1974). GLC determn in rodenticides: J. Theobald, *J. Chromatog.* **129**, 444 (1976); in tissue: E. Odam *et al., Analyst* **109**, 1335 (1984). Acute toxicity: E. W. Schafer, *Toxicol. Appl. Pharmacol.* **21**, 315 (1972). Review of pharmacology: G. U. Balis, R. R. Monroe, *Psychopharmacologia* **6**, 1-30 (1964); and toxicology: P. Lees, *Vet. Rec.* **91**, 330-333 (1972). Review as canine anes-

glucofuranose ring structure with CH₂OH, HOCH, OH, H, O, CCl₃

thetic: H. Holzgrefe *et al., Lab. Animal Sci.* **37**, 587-595 (1987).

Needles from alcohol or ether, mp 187°. $[\alpha]_D^{22}$ +19° (c = 5 in 98% alc). One gram dissolves in 225 ml water at 15°, in 120 ml at 37°, in 15 ml alc at 25°. Sol in ether, glacial acetic acid; slightly in chloroform. Almost insol in petr ether. LD_{50} orally in mice: 400 mg/kg (Schafer).

β-*Chloralose*, mp 227-230°, is much less sol in water, alcohol, and ether.

Caution: This substance may be habit forming and is listed in the U.S. Code of Federal Regulations, Title 21 Part 329.1 (1987).

USE: Rodenticide. Control of avian pests.

THERAP CAT: Sedative, hypnotic.

THERAP CAT (VET): Surgical anesthetic for laboratory animals.

2063. Chloramben. *3-Amino-2,5-dichlorobenzoic acid.* $C_7H_5Cl_2NO_2$; mol wt 206.02. C 40.81%, H 2.45%, Cl 34.41%, N 6.80%, O 15.53%. Selective pre-emergence herbicide. Prepn: S. R. McLane *et al.,* U.S. pat. **3,014,063** (1961 to Amchem). Activity: A. J. Tafuro, *Proc. Northeast Weed Control Conf.* **1959**, 423. Degradn in soil: R. E. Wildung *et al., Weed Res.* **8**, 213 (1968). Carcinogenicity study: J. F. Robens, *Vet. Human Toxicol.* **22**, 328 (1980).

Cryst, mp 200-201°. Vapor press at 100°: 7×10^{-3} mm Hg. Soly in water at 25°: 700 mg/l. Soly in (g/100 ml): acetone 23.3; methanol 22.3; ethanol 17.3; chloroform 0.09; benzene 0.02.

Ammonium salt, $C_7H_8Cl_2N_2O_2$, *ACPM-629, Amiben.* mp 194-197° (dec). Soly in (g/100 gm) at 25°: water 29.6; ethanol 6.9; isopropanol 1.7. Soly in water at 20°: 120 mg/l.

Methyl ester, $C_8H_7Cl_2NO_2$, *Amchem-65-81B, Vegiben.* mp 63-64°. Soly in water at 20°: 120 mg/l.

USE: Herbicide.

2064. Chlorambucil. *4-[Bis(2-chloroethyl)amino]benzenebutanoic acid; 4-[p-[bis(2-chloroethyl)amino]phenyl]-butyric acid;* γ-[p-di(2-chloroethyl)aminophenyl]butyric acid; *N,N-di-2-chloroethyl-γ-p-aminophenylbutyric acid;* chloraminophene; chloroambucil; CB 1348; Amboclorin; Leukeran. $C_{14}H_{19}Cl_2NO_2$; mol wt 304.23. C 55.27%, H 6.30%, Cl 23.31%, N 4.60%, O 10.52%. Nitrogen mustard deriv. Prepn: Everett *et al., J. Chem. Soc.* **1953**, 2390. Improved process: Phillips, Mentha, U.S. pat. **3,046,301** (1962 to Burroughs Wellcome). Chemotherapeutic activity: Van Putten *et al., Eur. J. Cancer* **7**, 11 (1971). Toxicity: W. C. J. Ross, *Biochem. Pharmacol.* **13**, 969 (1964). Review of carcinogenicity studies: *IARC Monographs* **9**, 125-134 (1975). Comprehensive description: M. Tariq, A. A. Al-Badr in *Analytical Profiles of Drug Substances* vol. **16**, K. Florey, Ed. (Academic Press, New York, 1987) pp 85-118.

Flattened needles from petr ether, mp 64-66°. Soluble in ether. Soluble at 20° in 1.5 parts alcohol, in 2.5 parts chloroform, in 2 parts acetone. Practically insol in water. LD_{50} i.p. in rats: 58.2 μmole/kg (Ross).

Note: This substance has been listed as a known carcinogen: *Fourth Annual Report on Carcinogens* (NTP 85-002, 1985) p 54.

THERAP CAT: Antineoplastic.

THERAP CAT (VET): Antineoplastic agent, esp for leukemia.

2065. Chloramine-B. *N-Chlorobenzenesulfonamide sodium salt; (N-chlorobenzenesulfonamido)sodium;* sodium benzenesulfochloramine; Neomagnol. $C_6H_5ClNNaO_2S$; mol wt 213.64. C 33.73%, H 2.36%, Cl 16.60%, N 6.56%, Na 10.76%, O 14.98%, S 15.01%. $C_6H_5SO_2NNaCl.$ Prepd via

benzenesulfonamide: Chattaway, *J. Chem. Soc.* **87**, 145 (1905); Cuiban *et al., Pharmazie* **13**, 407 (1958).

Trihydrate, prisms. Soluble in 20 parts of water, more sol in hot water. Sol in 25 parts of ethanol, yielding a turbid soln. Very sparingly sol in ether, chloroform. An aq soln first turns red litmus paper blue, then gradually bleaches it. Gives a red color wth phenolphthalein.

THERAP CAT: Antibacterial.

THERAP CAT (VET): Topical antiseptic.

2066. Chloramine-T. *N-Chloro-4-methylbenzenesulfonamide sodium salt; (N-chloro-p-toluenesulfonamido)sodium;* sodium *p*-toluenesulfonchloramide; chloramine; Aktiven; Chloraseptine; Chlorazene; Chlorazone; Clorina; Euclorina; Gansil; Gyneclorina; Halamid; Mianine; Tochlorine; Tolamine. $C_7H_7ClNNaO_2S$; mol wt 227.67. C 36.93%, H 3.10%, Cl 15.58%, N 6.15%, Na 10.10%, O 14.06%, S 14.09%. *p*-$CH_3C_6H_4SO_2NNaCl$. Prepd via *p*-toluenesulfonamide: Chattaway, *J. Chem. Soc.* **87**, 145 (1905); Inglis, *J. Soc. Chem. Ind. (London)* **37**, 288 (1918); F. J. Welcher, *Organic Analytical Reagents* vol. **4** (Van Nostrand, New York, 1948) pp 316-320.

Trihydrate, prisms. Loses water on drying. Dec slowly on exposure to air. Keep well closed. Fairly sol in water; practically insol in benzene, chloroform, ether. Dec by alc.

USE: Detection of bromate and halogens, F. J. Welcher, *loc. cit.*

THERAP CAT: Antibacterial.

THERAP CAT (VET): Topical antiseptic.

2067. Chloraminophenamide. *4-Amino-6-chloro-1,3-benzenedisulfonamide; 4-amino-6-chloro-m-benzenedisulfonamide;* 5-chloro-2,4-bis(sulfonamido)aniline; 6-amino-4-chlorobenzene-1,3-disulfonamide; 5-chloro-2,4-disulfamylaniline; Salamid; Idorese. $C_6H_8ClN_3O_4S_2$; mol wt 285.74. C 25.22%, H 2.82%, Cl 12.41%, N 14.71%, O 22.40%, S 22.44%. Prepn: Novello, Sprague, *J. Am. Chem. Soc.* **79**, 2028 (1957); Novello, U.S. pats. **2,809,194** and **2,965,655-6** (1957 and 1960 to Merck & Co.); Novello *et al., J. Org. Chem.* **25**, 965 (1960).

Crystals from aq ethanol, mp 251-252°. uv max (ethanol): 223.5-224.5, 265-266, 312-314 nm (ε 41,776, 18,633, 3874). Slightly sol in water; more freely sol in alkalies.

THERAP CAT: Diuretic.

2068. Chloramphenicol. *2,2-Dichloro-N-[2-hydroxy-1-(hydroxymethyl)-2-(4-nitrophenyl)ethyl]acetamide;* D-*threo*-N-dichloroacetyl-1-*p*-nitrophenyl-2-amino-1,3-propanediol; D(−)-*threo*-2-dichloroacetamido-1-*p*-nitrophenyl-1,3-propanediol; D-*threo*-N-(1,1'-dihydroxy-1-*p*-nitrophenylisopropyl)dichloroacetamide; Ak-Chlor; Alficetyn; Amphicol; Anacetin; Aquamycetin; Austracol; C.A.F.; Chemicetina; Chlomycol; Chloramex; Chloramfilin; Chloramsaar; Chlorasol; Chloricol; Chlorocaps; Chlorocid; Chloromycetin; Chloronitrin; Chloroptic; Cidocetine; Ciplamycetin; Cloramfen; Cloramficin; Cloramicol; Clorocyn; Cloromisan; Cylphenicol; Duphenicol; Embacetin; Enicol (capsules); Enteromycetin; Farmicetina; Fenicol; Globenicol; Interomycetin; Intramycetin; Juvamycetin; Kamaver; Kemicetine; Klorita; Leukomycin; Levomicetina; Levomycetin; Loromisin; Mastiphen; Medichol; Micloretin; Micoclorina; Microcetina; Mychel; Mycinol; Novomycetin; Opclor; Ophtho-chlor; Pantovernil; Paraxin; Quemicetina; Ronfenil; Septicol; Sintomicetina; Sno Phenicol; Stanomycetin; Synthomycetine; Tega-Cetin; Tevcocin; Tifomycine; Treomicetina; Unimycetin; Veticol; Viceton. $C_{11}H_{12}Cl_2N_2O_5$; mol wt 323.14. C 40.88%, H 3.74%, Cl 21.95%, N 8.67%, O 24.76%. Broad spectrum antibiotic obtained from cultures of the soil bacterium *Streptomyces venezuelae:* Bartz, *J. Biol. Chem.* **172**, 445 (1948); Gottlieb *et al., J. Bact.* **55**, 409 (1948); Ehrlich *et al., ibid.* **56**, 467 (1948). Isoln from the moon snail, *Lunatia* *heros:* C. A. Price *et al., J. Antibiot.* **34,** 118 (1981). Structure: Rebstock *et al., J. Am. Chem. Soc.* **71,** 2458 (1949). Synthesis: Controulis *et al., ibid.* 2463; Long, Troutman, *ibid.* 2469, 2473. *See also* U.S. pats. **2,483,871; 2,483,884; 2,483,892.** Alternate synthesis: Ehrhart *et al., Ber.* **90,** 2088 (1957); **Brit.** pats. **795,131; 796,901,** *C.A.* **53,** 2161 (1959) (both to Chinoin); U.S. pat. **2,839,577** (1958 to Chinoin). Review and evaluation of toxicity studies: *IARC Monographs* **10,** 85-98 (1976). Review of pharmacology and clinical efficacy: I. Shalit, M. I. Marks, *Drugs* **28,** 281-291 (1984). Comprehensive description: D. Szulczewski, F. Eng, in *Analytical Profiles of Drug Substances* vol. 4, K. Florey, Ed. (Academic Press, New York, 1975) pp 47-90; A. A. Al-Badr, H. A. El-Obeid, *ibid.* **vol. 15,** pp 701-760 (1986). *Reviews:* Hahn in *Antibiotics,* vol. 1, D. Gottlieb, P. D. Shaw, Eds. (Springer-Verlag, New York, 1967) pp 308-330; Pestka, *ibid.* **vol. 3,** J. W. Corcoran, F. E. Hahn, Eds. (1975) pp 370-395; O. Pongs, *ibid.* **vol. 5,** pt. 1, F. E. Hahn, Ed. (1979) pp 26-42.

Needles or elongated plates from water or ethylene dichloride. mp 150.5-151.5°. Sublimes in high vacuum. $[\alpha]_D^{27}$ +18.6° (c = 4.86 in ethanol). $[\alpha]_D^{25}$ −25.5° (ethyl acetate). uv max: 278 nm ($E_{1cm}^{1\%}$ 298). Soly (25°) in water: 2.5 mg/ml; in propylene glycol: 150.8 mg/ml. Very sol in methanol, ethanol, butanol, ethyl acetate, acetone. Fairly sol in ether; insol in benzene, petr ether, vegetable oils. Soly in 50% acetamide soln about 5%. Additional soly data: Weiss *et al., Antibiot. & Chemother.* **7,** 374 (1957). Aq solns are neutral. Neutral and acid solns are stable on heating.

Monosuccinate sodium salt, *Protophenicol.* Freely sol in water (about 50% w/w).

Palmitate, *see* separate entry.

Monosuccinate arginine salt, $C_{21}H_{30}Cl_2N_6O_{10}$, *chloramphenicol arginine succinate, Paraxin Succinate A.* mp 135-145° (dec.). *See Japan. Med. Gaz.* 7(10), 15 (1970).

Cinnamate, *Cloromilen.*

THERAP CAT: Antibacterial. Antirickettsial.

THERAP CAT (VET): Antimicrobial.

2069. Chloramphenicol Palmitate. *[R-(R*,R*)]-Hexadecanoic acid 2-[(dichloroacetyl)amino]-3-hydroxy-3-(4-nitrophenyl)propyl ester;* Chloromycetin palmitate; Chloropal; Clorolifarina. $C_{27}H_{42}Cl_2N_2O_6$; mol wt 561.54. C 57.75%, H 7.54%, Cl 12.63%, N 4.99%, O 17.10%. Semisynthetic antibiotic prepd by the action of palmitoyl chloride on chloramphenicol in pyridine: Edgerton, U.S. pat. **2,662,- 906** (1953 to Parke, Davis). Structure: Edgerton *et al., J. Am. Chem. Soc.* **77,** 27 (1955). Description: Glazko *et al., Antibiot. & Chemother.* **2,** 234 (1952).

Crystals from benzene, mp 90°. Practically tasteless. $[\alpha]_D^{26}$ +24.6° (c = 5 in ethanol). uv max (ethanol): 271 nm ($E_{1cm}^{1\%}$ 179). Very slightly sol in water (1.05 mg/ml at 28°); petr ether (0.225 mg/ml). Freely sol in methanol, ethanol, chlo-

roform, ether, benzene. Additional soly data: Weiss *et al., Antibiot. & Chemother.* **7,** 374 (1957).

USE: In pharmaceutical prepns to circumvent the bitter taste of chloramphenicol. The inert palmitate is hydrolyzed to free, biologically active chloramphenicol by intestinal and pancreatic lipases.

THERAP CAT: Antibacterial; antirickettsial.

THERAP CAT (VET): Antimicrobial.

2070. Chloramphenicol Pantothenate. *[2R-[(1R*),2R*,- 3R*]]-N-(2,4-Dihydroxy-3,3-dimethyl-1-oxobutyl)-β-alanine 2-[(dichloroacetyl)amino]-3-hydroxy-3-(4-nitrophenyl)propyl ester; pantothenic acid ester with 2,2-dichloro-N-[β-hydroxy-α-(hydroxymethyl)-p-nitrophenethyl]acetamide;* D(−)-*threo*-2-(dichloroacetamido)-1-(*p*-nitrophenyl)-3-pantothenyloxy-1-propanol; pantothenic acid ester with chloramphenicol. $C_{20}H_{27}Cl_2N_3O_9$; mol wt 524.37. C 45.81%, H 5.19%, Cl 13.52%, N 8.01%, O 27.46%. Semisynthetic antibiotic. Prepn of title compd and the complexes, *chloramphenicol sodium pantothenate* (1:1 and 2:1) and *chloramphenicol calcium pantothenate* (2:1 and 4:1): Vilax, **Brit.** pats. **866,787, 866,788** and **866,789** (all 1961), U.S. pat. **3,078,300** (1963).

Hygroscopic, sinters at 93°.

Chloramphenicol calcium pantothenate complex (4:1), $C_{62}H_{80}CaCl_8N_{10}O_{30}$, *chloramphenicol pantothenate complex, Pantofenicol.*

THERAP CAT: Antibacterial; antirickettsial.

2071. Chloranil. *2,3,5,6-Tetrachloro-2,5-cyclohexadiene-1,4-dione; 2,3,5,6-tetrachloro-p-benzoquinone;* tetrachloroquinone; Spergon; Vulklor. $C_6O_2Cl_4$; mol wt 245.89. C 29.31%, Cl 57.68%, O 13.01%. Prepd from *p*-phenylenediamine or phenol by treating with $KClO_3$ and HCl. Because of its great resistance to further oxidation, chloranil is formed as the final product of the chlorate-HCl oxidation of many aromatic compds. For comprehensive list *see* Huntress, *Organic Chlorine Compounds* (New York, 1948). Laboratory procedure starting with phenol or *p*-chlorophenol: Fierz-David, Blangey, *Grundlegende Operationen der Farbenchemie* (Vienna, 5th ed., 1943) p 140.

Golden-yellow platelets from acetic acid or acetone. Monoclinic prisms from benzene or toluene or by sublimation *in vacuo,* mp 290°. Absorption spectrum in chloroform: Lifschitz *et al., Rec. Trav. Chim.* **43,** 276, 658 (1924). Insol in water; almost insol in cold petr ether, cold alcohol; sol in ether; sparingly sol in chloroform, carbon tetrachloride, carbon disulfide. Solubility data: Dimroth, Bamberger, *Ann.* **438,** 103, 106 (1924). LD_{50} orally in rats: 4.0 g/kg, *Toxic Substances List,* H. E. Christensen, Ed (1973) p 909.

USE: Fungicide. In manuf of chloranil electrodes for pH measurements. As reagent for pamaquine (Plasmochin) in urine (blue color): Schulemann *et al., Chem. Zentr.* **1928, I,** 2193. Two of the chlorine atoms in *para* position are easily substituted, and compds such as 2,5-dianilino-3,6-dichloroquinone are much used in the dye industry. *Caution:* May be irritating to skin, mucous membranes.

2072. Chloranilic Acid. *2,5-Dichloro-3,6-dihydroxy-2,5-cyclohexadiene-1,4-dione; 2,5-dichloro-3,6-dihydroxy-p-benzoquinone.* $C_6H_2Cl_2O_4$; mol wt 208.99. C 34.48%, H 0.96%, Cl 33.93%, O 30.62%. Prepd from chloranil by alkaline hydrolysis: Conant, Fieser, *J. Am. Chem. Soc.* **46,** 1866 (1924).

Red crystals, mp 283-284°. Tendency to sublime. Relatively strong dibasic acid. Absorption max (pH 4 in water): 290-340 nm, 520-555 nm.

USE: Reacts with metal cations to form stable salts. Used in spectrophotometry: Hart, *Org. Chem. Bull.* **33**, no. 3 (1961).

2073. Chlorazanil. *N-(4-Chlorophenyl)-1,3,5-triazine-2,4-diamine; 2-amino-4-(p-chloroanilino)-s-triazine; N-(p-chlorophenyl)-2,4-diamino-s-triazine; 2-(p-chloroanilino)-4-amino-s-triazine; chlorazinil; ASA-226; Diurazine; Triazurol; Orpizin; Daquin; Neo-Urofort; Neurofort.* C_9H_8-ClN_5; mol wt 221.65. C 48.77%, H 3.64%, Cl 16.00%, N 31.60%. Prepn: Clauder, Bulcsu, *Magyar Kém. Folyoirat* **57**, 68 (1951), *C.A.* **46**, 4023g (1952); **Austrian** pat. **168,063** (1951) and **Brit.** pat. **676,024** (1952), *C.A.* **47**, 8097g, 3887i (1953). Synthesis and diuretic activity: Shah *et al., J. Med. Chem.* **11**, 1167 (1968). *See also* Takagi, Ueda, *Chem. Pharm. Bull.* **17**, 1061 (1969).

Crystals, mp 233-234°. Also reported as crystals from ethanol, mp 256-258° (Shah *et al., loc. cit.*).

Hydrochloride, $C_9H_9Cl_2N_5$, *Doclizid-T, Orpidan.* mp 277-278°.

THERAP CAT: Diuretic.

2074. Chlorbenside. *1-Chloro-4-[[(4-chlorophenyl)methyl]thio]benzene; p-chlorobenzyl p-chlorophenyl sulfide; chlorocide; Chlorparacide; Chlorsulphacide; Mitox.* $C_{13}H_{10}$-Cl_2S; mol wt 269.20. C 58.00%, H 3.74%, Cl 26.34%, S 11.91%. Prepn of analogous compd, p-chlorobenzyl p-nitrobenzyl sulfide, from p-chlorophenylthiol and p-nitrobenzyl chloride: Stevenson *et al.,* **Brit.** pat. **738,170** (1955 to Boots Pure Drug). Chemical and biological properties: Cranham *et al., Chem. & Ind. (London)* **1953**, 1206.

Crystals, mp 75-76°. Technical grades may have almond-like odor. d_4^{25} 1.4210. Vapor tension at 30°: 1.21×10^{-5} mm. uv max (95% ethanol): 262 nm: Watson, *Chem. & Ind. (London)* **1956**, 349. Soly in water < 1:5000, in ethanol about 2.9%, in kerosene 5-7.5%. Sol in acetone, benzene, toluene, xylene, petr ether. Resistant to acid and alkaline hydrolysis. Strong oxidizing agents convert it to the corresponding sulfoxide and sulfone.

USE: Acaricide, esp for the control of eggs and larvae of red spider mites. *Caution:* May be irritating to skin. Kidney and liver injuries have been produced in exptl animals.

2075. Chlorbenzoxamine. *1-[2-[(2-Chlorophenyl)phenylmethoxy]ethyl]-4-[(2-methylphenyl)methyl]piperazine; 1-[2-(o-chloro-α-phenylbenzyloxy)ethyl]-4-o-methylbenzyl-piperazine; 1-[2'-(o-chlorobenzhydryloxy)ethyl]-4-(o-methylbenzyl)diethylenediamine; 1-(o-chlorobenzhydryloxyethyl)-4-(o-methylbenzyl)piperazine; chlorbenzoxyethamine; chlorbenzoxamine.* $C_{27}H_{31}ClN_2O$; mol wt 434.99. C 74.54%, H 7.18%, Cl 8.15%, N 6.44%, O 3.68%. Prepn: Morren *et al., Ind. Chim. Belge* **22**, 409 (1957); Morren, **Brit.** pat. **837,986** (1960).

bp$_{0.01}$ 234-236°; bp$_{0.005}$ 235°.

Dihydrochloride, $C_{27}H_{33}Cl_3N_2O$, *UCB 1474, Libratar, Antiulcera Master, Gastomax.* Bitter crystals, mp 197-200°. Freely sol in methanol; sol in ethanol, chloroform, acetic acid; slightly sol in water, acetone. Practically insol in acetonitrile, ether, benzene. LD$_{50}$ s.c., i.v. in rats: 4000, 66 mg/kg; orally in mice: 1400 mg/kg, Levis *et al., Arch. Int. Pharmacodyn. Ther.* **118**, 167 (1959).

THERAP CAT: Anticholinergic.

2076. Chlorbetamide. *2,2-Dichloro-N-[(2,4-dichlorophenyl)methyl]-N-(2-hydroxyethyl)acetamide; 2,2-dichloro-N-(2,4-dichlorobenzyl)-N-(2-hydroxyethyl)acetamide; N-(2,4-dichlorobenzyl)-N-(2-hydroxyethyl)dichloroacetamide; Win 5047; Mantomide; Pontalin.* $C_{11}H_{11}Cl_4NO_2$; mol wt 331.05. C 39.91%, H 3.35%, Cl 42.84%, N 4.23%, O 9.67%. Prepn: Surrey, *J. Am. Chem. Soc.* **76**, 2214 (1954); **U.S.** pat. **2,732,402** (1956 to Sterling Drug).

Crystals from benzene + Skellysolve A, mp 112.4-113.4°. Slightly bitter taste. Very slightly sol in water; sol in 95% ethanol to at least 5%.

THERAP CAT: Antiamebic.

2077. Chlorbicyclen. *5,6-Bis(chloromethyl)-1,2,3,4,7,7-hexachlorobicyclo[2.2.1]hept-2-ene; 1,2,3,4,7,7-hexachloro-5,6-bis(chloromethyl)-2-norbornene; 5,6-bis(chloromethyl)-1,2,3,4,7,7-hexachloro-2-norbornene; 1,4,5,6,7,7-hexachloro-2,3-bis(chloromethyl)bicyclo[2.2.1]hept-5-ene; Alodan (obsolete).* $C_9H_6Cl_8$; mol wt 397.74. C 27.18%, H 1.52%, Cl 71.30%. Prepn: Fensch, Finkenbrink, **Ger.** pat. **1,002,341** (1957 to Hoechst); Hoch, Clegg, *J. Am. Chem. Soc.* **81**, 5413 (1959); **U.S.** pat. **2,951,098** (1960 to Hooker). Synthesis of the *trans*-isomer: Schuphan, Tölg, *Tetrahedron Letters* **1969**, 3389.

Solid, mp 104-106°. bp$_2$ 172-176°. mp of *trans*-isomer, 74-76°.

USE: Formerly as an insecticide and acaricide.

2078. Chlorcyclizine. *1-[(4-Chlorophenyl)phenylmethyl]-4-methylpiperazine; 1-(p-chloro-α-phenylbenzyl)-4-methylpiperazine; 1'-(4-chlorobenzhydryl)-4-methylpiperazine; N-methyl-N'-(4-chlorobenzhydryl)piperazine; chlorocyclizine; compd 47-282; Alergicide; Perazyl; Trihistan.* $C_{18}H_{21}ClN_2$; mol wt 300.85. C 71.86%, H 7.04%, Cl 11.79%, N 9.31%. Prepn: Baltzly *et al., J. Org. Chem.* **14**, 775 (1949); Murfitt, Dewing, **Brit.** pat. **656,043** (1951 to Wellcome Found.); **U.S.** pat. **2,630,435** (1953 to Burroughs Wellcome).

Oil, bp$_{0.1-0.15}$ 137-145°.

Hydrochloride, $C_{18}H_{22}Cl_2N_2$, *AH-289, Di-Paralene, Perazil, Histantin* (Burroughs Wellcome). Cryst powder, mp 226-227°. One gram dissolves in about 2 ml water, in 11 ml alc, in about 4 ml chloroform. Practically insol in ether, benzene.

Dihydrochloride, $C_{18}H_{23}Cl_3N_2$, prisms from alc + ether, mp 216-216.5°. Freely sol in water; sol in alc. Aq soln is acid to litmus. LD$_{50}$ i.p. in mice: 137 mg/kg, Castillo *et al.*, *J. Pharmacol. Exp. Ther.* **96**, 388 (1949).

Di[(*tert*-butyl)naphthalene sodium sulfonate], $C_{46}H_{51}Cl$-$N_2Na_2O_6S_2$, *Bexedan*.

THERAP CAT: Antihistaminic.

THERAP CAT (VET): The hydrochloride as antihistaminic.

2079. Chlordan(e). *1,2,4,5,6,7,8,8-Octachloro-2,3,3a,4,7,-7a-hexahydro-4,7-methano-1H-indene; 1,2,4,5,6,7,8,8-octachloro-3a,4,7,7a-tetrahydro-4,7-methanoindan;* 1,2,4,5,6,7,-8,8-octachloro-4,7-methane-3a,4,7,7a-tetrahydroindane; CD-68; Velsicol 1068; Toxichlor; Niran; Octachlor; Ortho-Klor; Synklor; Corodane; Belt. $C_{10}H_6Cl_8$; mol wt 409.80. C 29.31%, H 1.48%, Cl 69.22%. The commercial product is a mixture containing 60 to 75% of the pure compound and 25 to 40% of related compds. Chlorine content: 64-67%. Prepn: Hyman, **Brit.** pat. **618,432** (1949); Kleiman, **U.S.** pat. **2,598,561** (1952 to Velsicol).

Viscous, amber-colored liquid. Viscosity 69 poises at 25° (about that of 95% glycerol). Viscosity reduced considerably by heating to 120-140°F when it may be sprayed directly. d^{25} 1.59-1.63. n_D^{25} 1.56-1.57. Insol in water; miscible with aliphatic and aromatic hydrocarbon solvents, including deodorized kerosene. Loses its chlorine in presence of alkaline reagents, and should not be formulated with any solvent, carrier, diluent or emulsifier, which has an alkaline reaction. LD$_{50}$ i.p. in male rats: 343 mg/kg, R. B. Harbison, *Toxicol. Appl. Pharmacol.* **32**, 443 (1975).

Note: The EPA has cancelled registrations of pesticides containing this compound with the exception of its use through subsurface ground insertion for termite control and the dipping of roots or tops of non-food plants: *Fed. Reg.* **vol. 40**, p 28850 (July 9, 1975).

USE: Insecticide: Stomach poison, contact poison, fumigant. See Bussart, Schor, *Soap Sanit. Chemicals* **24**, 126 (1948). *Human Toxicity:* Moderately irritating to skin. Poisoning may occur by ingestion, inhalation, or percutaneous absorption. *Acute Toxicity:* Irritability, convulsions, deep depression. Continued ingestion may cause degenerative changes in liver.

THERAP CAT (VET): Insecticide, acaricide.

2080. Chlordantoin. *5-(1-Ethylpentyl)-3-[(trichloromethyl)thio]-2,4-imidazolidinedione; 5-(1-ethylpentyl)-3-(trichloromethylthio)hydantoin;* 5-(1-ethylamyl)-3-(trichloromethylthio)hydantoin; 3-(trichloromethylthio)-5-(1-ethylpentyl)hydantoin; 3-(trichloromethylthio)-5-(1-ethylamyl)hydantoin. $C_{11}H_{17}Cl_3N_2O_2S$; mol wt 347.72. C 38.00%, H 4.93%, Cl 30.59%, N 8.06%, O 9.20%, S 9.22%. Prepn of analogous compounds: Kittleson, *Science* **115**, 84 (1952); *cf.* Kupferberg, Doscher, *Antibiot. & Chemother.* **11**, 73 (1961).

Component of *Sporostacin.*

THERAP CAT: Antifungal.

2081. Chlordecone. *1,1a,3,3a,4,5,5,5a,5b,6-Decachloro-* octahydro-1,3,4-metheno-2H-cyclobuta[cd]pentalen-2-one; GC-1189; Kepone. $C_{10}Cl_{10}O$; mol wt 490.68. C 24.48%, Cl 72.26%, O 3.26%. Prepn: Gilbert, Giolito, U.S. pats. **2,616,825** and **2,616,928** (1952 to Allied Chem., reissue **24,435**; 1958). Structure: McBee *et al., J. Am. Chem. Soc.* **78**, 1511 (1956). *Review:* Ungnade, McBee, *Chem. Rev.* **58**, 249-320 (1958). Carcinogenicity study: *Am. Ind. Hyg. Assoc. J.* **37**, 680 (1976). Toxicity data: Gaines, *Toxicol. Appl. Pharmacol.* **14**, 515 (1969). Comparative toxicity of chlordecone in man and animals: P. S. Guzelian, *Ann. Rev. Pharmacol. Toxicol.* **22**, 89-113 (1982).

Crystals, dec 350°. Slightly sol in water and hydrocarbon solvents. Sol in alcohols, ketones, acetic acid. LD$_{50}$ orally in rats: 125 mg/kg (Gaines).

Caution: Has caused tremors in factory workers exposed to this substance: *Chem. & Eng. News* **54**, 19 (Feb. 2, 1976). This substance may reasonably be anticipated to be a carcinogen: *Fourth Annual Report on Carcinogens* (NTP 85-002, 1985) p 120.

USE: Insecticide, fungicide.

2082. Chlordiazepoxide. *7-Chloro-N-methyl-5-phenyl-3H-1,4-benzodiazepin-2-amine 4-oxide; 7-chloro-2-methyl-amino-5-phenyl-3H-1,4-benzodiazepine 4-oxide;* metaminodiazepoxide; methaminodiazepoxide; clopoxide; Decacil; Disarim; Eden-psich; Helogaphen; Libritabs; Menrium; Mesural; Multum; O.C.M.; Risolid; Silibrin; Sonimen; Tropium; Zeisin; Zetran. $C_{16}H_{14}ClN_3O$; mol wt 299.75. C 64.11%, H 4.71%, Cl 11.83%, N 14.02%, O 5.34%. Prepn: Sternbach, U.S. pat. **2,893,992** (1959 to Hoffmann-La Roche); Sternbach, Reeder, *J. Org. Chem.* **26**, 1111 (1961). Comprehensive description: A. MacDonald *et al.*, in *Analytical Profiles of Drug Substances* **vol. 1**, K. Florey, Ed. (Academic Press, New York, 1972) pp 15-51.

Light yellow plates from ethanol, mp 236-236.5°.

Hydrochloride, $C_{16}H_{15}Cl_2N_3O$, *Ansiacal, Bent, Benzodiapin, Calmoden, Cebrum, Chlordiazachel, Corax, Diazachel (obsolete), Droxol, Elenium, Equibral, J-Liberty, Kalmocaps, Labican, Lentotran, Librium, Mildmen, Napoton, Novosed, Psichial, Psicosan, Psicoterina, Reliberan, Sophiamin, Seren Vita, SK-Lygen, Tensinyl, Timosin, Trakipeal, Viansin, Viopsicol.* Crystals from methanol, mp 213°. Mixture with pentaerythritol tetranitrate: *Pentrium;* mixture with clidinium bromide: *Librax.*

Note: This is a controlled substance (depressant) listed in the U.S. Code of Federal Regulations, Title 21 Part 1308.14 (1987).

THERAP CAT: Anxiolytic.

THERAP CAT (VET): Tranquilizer.

2083. Chlordimeform. *N'-(4-Chloro-2-methylphenyl)-N,N-dimethylmethanimidamide; N'-(4-chloro-o-tolyl)-N,N-dimethylformamidine;* chlorophenamidine; chlorphenamidine; spanon; CDM; Ciba-8514; Schering 36,268; Fundal; Galecron. $C_{10}H_{13}ClN_2$; mol wt 196.67. C 61.07%, H 6.66%, Cl 18.03%, N 14.24%. Member of a class of insecticides possessing the formamidine moiety, effective vs many organophosphate and carbamate resistant pests. Prepn: Arndt, Steinhausen, **Ger.** pat. **1,172,081** corresp to **U.S.** pats.

3,378,437 and **3,502,720** (1964, 1968, 1970, all to Schering AG). Acts by interfering with amine regulatory mechanisms: Beeman, Matsumura, *Nature* **242**, 273 (1973); Aziz, Knowles, *ibid.* 417. Review of metabolism: Knowles, *J. Agr. Food Chem.* **18**, 1038 (1970). Degradation products: Witkonton, Ercegovich, *ibid.* **20**, 569 (1972). Toxicity data: C. P. Robinson, P. W. Smith, *J. Toxicol. Environ. Health* **3**, 565 (1977). Toxicological studies: D. S. Folland *et al.*, *J. Am. Med. Assoc.* **13**, 1052 (1978); K. T. Maddy *et al.*, *Toxicol. Letters* **33**, 37 (1986). Review of carcinogenicity studies: *IARC Monographs* **30**, 61-72 (1983).

Colorless crystals, mp 35°. $bp_{0.4}$ 156-157°. n_D^{25} 1.5885. d_4^{25} 1.105. Vapor pressure at 20°: 3.5×10^{-4} mm. Slightly sol in water; easily sol in organic solvents. LD_{50} i.p. in rats: 238 mg/kg (Robinson, Smith).

USE: Acaricide, insecticide.

2084. Chlorendic Anhydride. *4,5,6,7,8,8-Hexachloro-3a,4,7,7a-tetrahydro-4,7-methanoisobenzofuran-1,3-dione; 1,4,5,6,7,7-hexachloro-endo-5-norbornene-2,3-dicarboxylic anhydride;* hexachloroendomethylenetetrahydrophthalic anhydride; 1,4,5,6,7,7-hexachloro-*endo*-bicyclo[2.2.1]hept-5-ene-2,3-dicarboxylic anhydride. $C_9H_2Cl_6O_3$; mol wt 370.86. C 29.15%, H 0.54%, Cl 57.36%, O 12.94%. Prepd from hexachlorocyclopentadiene and maleic anhydride: Herzfeld *et al.*, U.S. pat. **2,606,910** (1952 to Velsicol); Baranauckas *et al.*, U.S. pat. **2,903,463** (1959 to Hooker Chem.). Purification: Zimmer *et al.*, Ger. pat. **1,113,451** (1961 to Hooker Chem.), *C.A.* **57**, 697a (1962).

Crystals, mp 231-235°.

USE: In prepn of polyester resins.

2085. Chlorfenac. *2,3,6-Trichlorobenzeneacetic acid; 2,3,6-trichlorophenylacetic acid;* Fenac; Tri-Fene. $C_8H_5Cl_3O_2$; mol wt 239.50. C 40.12%, H 2.10%, Cl 44.41%, O 13.36%. Prepd by conversion of 2,3,6-trichlorotoluene to the benzyl chloride, the benzyl cyanide, then hydrolysis: Brit. pat. **860,310** (1961 to Hooker).

Crystals from benzene, mp 161°. Sol in acetone, alcohol, ether; practically insol in water. LD_{50} orally in rats: 3000 mg/kg, G. W. Bailey, J. L. White, *Residue Rev.* **10**, 97 (1965).

USE: Herbicide.

2086. Chlorfenethol. *4-Chloro-α-(4-chlorophenyl)-α-methylbenzenemethanol; 4,4'-dichloro-α-methylbenzhydrol; p,p'-dichlorodiphenylmethyl carbinol;* 1,1-bis(p-chlorophenyl)ethanol; 1,1-bis(p-chlorophenyl)methyl carbinol; DMC; DCPC; Dimite. $C_{14}H_{12}Cl_2O$; mol wt 267.16. C 62.94%, H 4.53%, Cl 26.54%, O 5.99%. Prepn: Ruthruff *et al.*, U.S. pat **2,430,586** (1947 to Sherwin-Williams); Brit. pat. **831,421** (1960 to Metal & Thermit). Activity: O. Grummitt, *Science* **111**, 361 (1950).

Crystals from Skellysolve C, mp 69-69.5°. Sol in organic solvents. Practically insol in water. LD_{50} orally in rats: 500 mg/kg, *RTECS* **Vol. I**, R. J. Lewis, R. L. Tatken, Eds. (1979) p 225.

USE: Acaricide. *Caution:* Effects similar to DDT, *q.v.*

2087. Chlorfenvinphos. *Phosphoric acid 2-chloro-1-(2,4-dichlorophenyl)ethenyl diethyl ester; O,O*-diethyl *O*-[2-chloro-1-(2,4-dichlorophenyl)vinyl] phosphate; 2,4-dichloro-α-(chloromethylene)benzyl alcohol diethyl phosphate; CVP; SD 7859; Compound 4072; Birlane; Dermaton; Sapecron; Steladone; Supona. $C_{12}H_{14}Cl_3O_4P$; mol wt 359.56. C 40.08%, H 3.92%, Cl 29.58%, O 17.80%, P 8.61%. Commercial prepn contains at least 90% of the active Z-isomer. Prepn: E. E. Gilbert *et al.*, U.S. pat. **3,003,916** (1961 to Allied Chemical); Phillips *et al.*, U.S. pat. **3,102,842** (1963 to Shell); Whetstone, Phillips *et al.*, *J. Agr. Food Chem.* **14**, 352 (1966). Metabolism: W. F. Chamberlain, D. E. Hopkins, *J. Econ. Entomol.* **55**, 86 (1962); D. H. Hotson *et al.*, *Biochem. J.* **102**, 133 (1967). Toxicity studies: A. M. Ambrose *et al.*, *Toxicol. Appl. Pharmacol.* **17**, 323 (1970); Bunyan *et al.*, *Pestic. Sci.* **2**, 148 (1971).

Amber liquid, mild odor, $bp_{0.001}$ 120°, $bp_{0.5}$ 167-170°. Vapor press at 25°C: 7.5×10^{-6} mm Hg. n_D^{25} 1.5272. Soly in water at 23°: 145 ppm. Miscible with acetone, ethanol, propylene glycol. Slowly hydrolyzed by water. Corrosive to metal. Toxic to fish. LD_{50} in rats: 6.6 mg/kg i.v.; 9.66 mg/kg orally, A. M. Ambrose *et al.*, *loc. cit.*

USE: Insecticide; acaricide.

2088. Chlorguanide. *N-(4-Chlorophenyl)-N'-(1-methylethyl)imidodicarbonimidic diamide; 1-(p-chlorophenyl)-5-isopropylbiguanide; N^1-p*-chlorophenyl-*N^5*-isopropyldiguanide; chloroguanide; M 4888; RP 3359; SN 12837; Diguanyl; Drinupal; Guanatol; Palusil; Proguanil; Tirian. $C_{11}H_{16}ClN_5$; mol wt 253.75. C 52.07%, H 6.36%, Cl 13.97%, N 27.60%. Prepn: Curd, Rose, *J. Chem. Soc.* **1946**, 729; Curd *et al.*, *ibid.* **1948**, 1630; Ainley *et al.*, *ibid.* **1949**, 98. Manuf of acetate: Gailliot, Fr. pat. **1,001,548** (1952 to Rhône-Poulenc), *C.A.* **51**, 7411e (1957).

Rectangular plates from toluene, mp 129°.

Acetate, $C_{13}H_{20}ClN_5O_2$, crystals from acetone, mp 189-190°.

Hydrochloride, $C_{11}H_{17}Cl_2N_5$, *Paludrine*. Crystals from water or aq ethanol, mp 243-244°. uv max (alc): 259 nm. Sol in alc; slightly sol in water; practically insol in chloroform and ether. pH of satd aq soln 5.8-6.3. LD_{50} orally in rats: 200 mg/kg, Schmidt *et al.*, *J. Pharmacol. Exp. Ther.* **90**, 233 (1947).

THERAP CAT: Hydrochloride as antimalarial.

2089. Chlorhexadol. *2-Methyl-4-(2,2,2-trichloro-1-hydroxyethoxy)-2-pentanol;* 2-methyl-2-hydroxy-4-(2,2,2-chloro-1-hydroxyethoxy)pentane; 2-hydroxy-2-methyl-4-(2,2,2-trichloro-1-hydroxyethoxy)pentane; Chloralodol; Lora (formerly); Mecoral; Medodorm. $C_8H_{15}Cl_3O_3$; mol wt 265.58. C 36.18%, H 5.69%, Cl 40.05%, O 18.07%. Prepd from chloral and 2-methyl-2,4-pentanediol: Christensen, U.S. pat. **2,931,838** (1960 to Danske Medicinal).

$$CH_3CH_2CH\underset{OH}{|}CH\underset{OCHCCl_3}{|}CH_3$$
$$\underset{OH}{|}$$

Crystals, slightly bitter taste, mp 102-104°. Readily sol in alcohol, chloroform; moderately sol in ether; slightly sol in CCl_4. Hydrolyzes in aq soln.

Caution: May be habit forming. This is a controlled substance (depressant) listed in the U.S. Code of Federal Regulations, Title 21 Part 1308.13 (1985).

THERAP CAT: Hypnotic.

2090. Chlorhexidine. *N,N''-Bis(4-chlorophenyl)-3,12-diimino-2,4,11,13-tetraazatetradecanediimidamide; 1,1'-hexamethylenebis[5-(p-chlorophenyl)biguanide];* 1,6-bis[N'-(p-chlorophenyl)-N⁵-biguanido]hexane; 1,6-bis(N⁵-p-chlorophenyl-N'-diguanido)hexane; 1,6-di(4'-chlorophenyldiguanido)hexane; 10040; Nolvasan; Sterilon. $C_{22}H_{30}Cl_2N_{10}$; mol wt 505.48. C 52.28%, H 5.98%, Cl 14.03%, N 27.71%. Bisbiguanide with bacteriostatic activity. Prepn: Rose, Swain, *J. Chem. Soc.* **1956**, 4422; *eidem,* U.S. pat. **2,684,924** (1954 to I.C.I.). Antibacterial activity and acute toxicity: G. E. Davies *et al., Brit. J. Pharmacol.* **9,** 192 (1954). Review of toxicology and clinical uses: D. M. Foulkes, *J. Periodont. Res.* **8,** Suppl. 12, 55-60 (1973). Series of articles on clinical efficacy in gingivitis and plaque control: *ibid.* **21,** Suppl. 16, 1-89 (1986).

Cl—⟨ ⟩—NHCNHCNH(CH₂)₆NHCNHCNH—⟨ ⟩—Cl
NH NH   NH NH

Crystals from methanol, mp 134°. Strong alkaline reaction. Soly in water at 20°: 0.08% (w/v).

Dihydrochloride, $C_{22}H_{32}Cl_4N_{10}$, *Lisium.* Crystals, dec 260-262°. Soly in water at 20°: 0.06 g/100 ml.

Diacetate, $C_{26}H_{38}Cl_2N_{10}O_4$, *Chlorasept 2000.* Crystals, mp 154-155°. Neutral reaction. Soly in water at 20°: 1.9 g/100 ml. Aq solns dec when heated above 70°. Soluble in alcohol, glycerol, propylene glycol, polyethylene glycols. LD_{50} orally in mice: 2 g/kg (Davies).

Digluconate, $C_{34}H_{54}Cl_2N_{10}O_{14}$, *Bacticlens, Corsodyl, Hibiclens, Hibidil, Hibiscrub, Hibitane, Orahexal, Peridex, pHiso-Med, Plac Out, Plurexid, Rotersept, Unisept.* Soly in water at 20°: >50% (w/v). LD_{50} in mice (mg/kg): 22 i.v.; 1800 orally (Foulkes).

THERAP CAT: Topical antibacterial; disinfectant.

THERAP CAT (VET): Topical and uterine antiseptic; disinfectant.

2091. Chloric Acid. $ClHO_3$; mol wt 84.46. Cl 41.98%, H 1.19%, O 56.83%. $HClO_3$. Prepd from barium chlorate and sulfuric acid: Lamb *et al., J. Am. Chem. Soc.* **42,** 1643 (1920); from sodium chlorate using ion-exchange resins: Klement, *Z. Anorg. Allgem. Chem.* **260,** 271 (1949).

Known in aq soln only. Aq solns are stable if pure and protected from light. 1% aq soln d_4^{18} 1.0044; 6% soln d_4^{18} 1.0344; 10% soln d_4^{18} 1.0594; 16% soln d_4^{18} 1.0991; 20% soln d_4^{18} 1.1273; 24% soln d_4^{18} 1.1568. A 40% soln corresponds to $HClO_3 \cdot 7H_2O$, d_4^{20} 1.282. If higher concns are attempted by evaporation the soln begins to dec with evolution of chlorine and oxygen and formation of perchloric acid. The salts of chloric acid are known as chlorates.

USE: Oxidizing agent; with H_2SO_4 as catalyst in acrylonitrile polymerization. *Caution:* Strongly irritating to skin, mucous membranes.

2092. Chlorimuron Ethyl. *2-[[[[(4-Chloro-6-methoxy-2-pyrimidinyl)amino]carbonyl]amino]sulfonyl]benzoic acid ethyl ester;* ethyl 2-[[[[(4-chloro-6-methoxypyrimidin-2-yl)amino]carbonyl]amino]sulfonyl]benzoate; DPX-F6025; Classic. $C_{15}H_{15}ClN_4O_6S$; mol wt 414.82. C 43.43%, H 3.64%, Cl 8.55%, N 13.51%, O 23.14%, S 7.73%. Selective sulfonylurea herbicide. Prepn: A. D. Wolf, **Austrian** pat. **8,316,181;** *idem,* U.S. pat. **4,547,215** (1984, 1985 both to Du Pont). Effect on plant growth and pigment synthesis: R. M. Devlin, Z. K. Koszanski, *Proc. Ann. Meet. Northeast.*

Weed Sci. Soc. **40,** 115 (1986). Metabolism by plants: H. M. Brown, S. M. Neighbors, *Pestic. Biochem. Physiol.* **29,** 112 (1987). Field trial in soybeans: G. N. Rhodes *et al., Tenn. Farm Home Sci.* **142,** 21 (1987). HPLC determn in crops: J. L. Prince, R. A. Guinivan, *J. Agr. Food Chem.* **36,** 63 (1988). Brief review: J. S. Claus, *Weed Technol.* **1,** 114-115 (1987).

[Structure of chlorimuron ethyl: benzene ring with COOC₂H₅ and SO₂NHCONH group linked to pyrimidine ring bearing Cl and OCH₃]

Crystals from butyl chloride, mp 198-201°. Soly (ppm): acetone 71000, acetonitrile 31000, benzene 8000, methylene chloride 153000, water (pH 7) 1200, (pH 6.5) 450, (pH 5) 11. LD_{50} in male, female rats (mg/kg): 4102, 4236 orally (Claus).

USE: Herbicide.

2093. Chlorinated Lime. Bleaching powder. Improperly called *"chloride of lime"* or *"calcium oxychloride".* A relatively unstable chlorine carrier in solid form; a complex chemical compd of indefinite composition, presumably consisting of varying proportions of $Ca(OCl)_2$, $CaCl_2$, $Ca(OH)_2$ and H_2O in its molecular structure. Maximum available chlorine content approaches 39%. Commercial products usually range between 24% and 37% of available chlorine.

White or grayish-white powder; strong odor of chlorine. On exposure to air it becomes moist and rapidly decomposes. Most of it dissolves in water or alcohol. *Keep dry and tightly closed.*

Human Toxicity: Strong solns irritate skin. Inhalation of fumes may cause laryngeal and pulmonary irritation, pulmonary edema, death. Ingestion may produce severe oral, esophageal, gastric irritation.

USE: Bleaching of wood pulp, linen, cotton, straw, oils, soaps, and in laundering; oxidizer in calico printing to obtain white designs on a colored ground; destroying caterpillars; disinfecting drinking water, sewage, etc.; as a decontaminant for mustard gas and similar substances.

THERAP CAT: Disinfectant.

THERAP CAT (VET): Disinfectant for premises. Has been used as a topical antiseptic for superficial wounds.

2094. Chlorindanol. *7-Chloro-2,3-dihydro-1H-inden-4-ol; 7-chloro-4-indanol.* C_9H_9ClO; mol wt 168.63. C 64.10%, H 5.38%, Cl 21.03%, O 9.49%. Prepn: Buck *et al., J. Am. Chem. Soc.* **79,** 3559 (1957); Buck, U.S. pat. **2,990,-324** (1961 to Esta Med. Labs.).

[Structure of chlorindanol: indanol ring with Cl and OH substituents]

Needles from petr ether, mp 91-93°. Absorption spectra: Buck *et al., loc. cit.* Ingredient of *Lanesta.*

USE: Spermaticide.

2095. Chlorine. Cl; at. wt 35.453; at. no. 17; valences 1 to 7; elemental state: Cl_2. A halogen. Abundance in igneous rock (95% of earth's crust): 0.031% by wt; in seawater: 1.9% by wt (primarily as NaCl). Natural isotopes: 35 (75.53%); 37 (24.47%); seven radioactive isotopes and two isomers are known; radioactive tracer elements: ³⁶Cl ($T_{1/2}$ 3.08 × 10⁵ yrs; β⁻, EC); ³⁸Cl ($T_{1/2}$ 37.29 min; β⁻; formed in atm by bombardment with cosmic rays. Discovered in 1774 by Scheele; recognized as an element in 1810 by Davy. Produced on a large scale by electrolysis from fused chlorides. The industrial product is about 99.3% pure. Contaminants are traces of bromide, hexachloroethane, hexachlorobenzene, and water. Purification: Fye, Beaver, *J. Am. Chem. Soc.* **63,** 1268 (1941); A. Klemenc, *Die Behandlung und Reindarstellung von Gasen* (Vienna, 2nd ed., 1948) p

153. Lab prepn from MnO_2 and HCl: Schmeisser in *Handbook of Preparative Inorganic Chemistry* vol. 1, G. Brauer, Ed. (Academic Press, New York, 2nd ed., 1963) p 272. Manuf: Faith, Keyes & Clark's *Industrial Chemicals,* F. A. Lowenheim, M. K. Moran, Eds. (Wiley-Interscience, New York, 4th ed., 1975) pp 244-253. *Reviews: Ciba Review* vol. 12, no. 139 (Aug. 1960); *Chlorine,* J. S. Sconce, Ed., A.C.S. Monograph Series, no. 154 (Reinhold, New York, 1962) 901 pp; *MTP Int. Rev. Sci.: Inorg. Chem., Ser. One,* vol. 3, V. Gutmann, Ed. (Butterworths, London, 1972); Downs, Adams, "Chlorine, Bromine, Iodine and Astatine" in *Comprehensive Inorganic Chemistry,* vol. 2, J. C. Bailar, Jr. *et al.,* Eds. (Pergamon Press, Oxford, 1973) pp 1107-1594.

Greenish-yellow, diatomic gas; suffocating odor. mp $-101.00°$. bp $-34.05°$, d^{20} at 6.864 atm 1.4085 (liq.); d^{-35} at 0.9949 atm 1.5649 (liq.). Critical temp 144°; critical pressure 76.1 atm. Heat capacity at constant pressure (gas, 25°) 8.11 cal/mole/°C. Marketed in the form of gas over liquid compressed into steel cylinders. *Caution: Never heat cylinders.* Vapor pressure data: Giauque, Powell, *J. Am. Chem. Soc.* 61, 1970 (1939). Sol in water (25°) with formation of aqueous Cl_2 (0.062 moles/l), HOCl (0.030 moles/l) and Cl^- (0.030 moles/l); total soly: 0.092 moles/l. More sol in alkalies. *See* Chlorinated Lime and Sodium Hypochlorite Soln. Acts as an electron-acceptor in forming complexes with many donor species: Bent, *Chem. Rev.* 68, 587 (1968). Very reactive; E^0 (aq) $\frac{1}{2}Cl_2/Cl^-$ 1.356 V; dissociation energy (25°): 57.978 kcal; combines readily with all elements except the rare gases (xenon excluded) and nitrogen. Forms explosive mixtures with hydrogen; many finely divided metals will burn in an atm of chlorine. Oxides are strong oxidizing agents and explosive. Monatomic chlorine is unstable under ordinary conditions, however, it can be formed as a result of thermal or optical dissociation, by an electrical discharge or as an intermediate during chemical reactions. *Dangerous to inhale:* K. C. Back *et al., Reclassification of Materials Listed as Transportation Health Hazards* (TSA-20-72-3; PB 214-270). LC_{50} (one hr), inhalation by rats, mice: 293 ppm, 137 ppm.

Caution: A powerful irritant. Can cause fatal pulmonary edema. Threshold odor detection: 0.2-0.4 ppm, *cf.* Patty's *Industrial Hygiene and Toxicology* vol. 2B, G. D. Clayton, F. E. Clayton, Eds. (Wiley-Interscience, New York, 3rd ed., 1981) pp 2954-2965.

USE: Largely for manuf chlorinated lime which is used in bleaching all kinds of fabrics; for purifying water; disinfecting; detinning and dezincing iron; manuf synthetic rubber and plastics, chlorinated hydrocarbons, and a large number of other chemicals. It is an indispensable reagent in synthetic chemistry. Has been used as a military poison gas under the name *bertholite.* ^{36}Cl is considered an extinct nuclide and provides a method of determining the geological age of meteors.

2096. Chlorine Dioxide. Chlorine peroxide. ClO_2; mol wt 67.46. Cl 52.56%, O 47.44%. Prepn from chlorine and sodium chlorite: Derby, Hutchinson, *Inorg. Syn.* 4, 152 (1953); from potassium chlorate and sulfuric acid: Bodenstein *et al., Z. Anorg. Allgem. Chem.* 147, 233 (1925); by passing NO_2 through a column of sodium chlorate: Hutchinson, Derby, *Cereal Chem.* 24, 372 (1947); alternate methods of prepn: Schmeisser in *Handbook of Preparative Inorganic Chemistry* vol. 1, G. Brauer, Ed. (Academic Press, New York, 2nd ed., 1963) p 301. *Review:* Bedumeau, *Rev. Prod. Chim.* 57, 173-177, 257-261 (1954).

Strongly oxidizing, yellow to reddish-yellow gas at room temp. Unpleasant odor similar to that of chlorine and reminiscent of that of nitric acid. Unstable in light; stable in dark if pure, but chlorides catalyze its decompn even in the dark. *Reacts violently with organic materials.* In concns in excess of 10% at atm pressure easily detonated by sunlight, heat, contact with mercury or carbon monoxide. mp $-59°$; bp 11°; d^0(liq) 1.642: Cheesman, *J. Chem. Soc.* 1930, 35. Sol in water (3.01 g/l at 25° and 34.5 mm Hg), with slight hydrolysis to chlorous and chloric acid; sol in alkaline and sulfuric acid solns. Solid ClO_2 is yellowish-red cryst mass; liquid ClO_2 is reddish-brown.

USE: Bleaching cellulose, paper-pulp, flour, leather, fats and oils, textiles, beeswax; purification of water; taste and odor control of water; cleaning and detanning leather; manuf

of chlorite salts; oxidizing agent; bactericide and antiseptic. *Caution:* May be highly irritating to skin and mucous membranes of respiratory tract. May cause pulmonary edema.

2097. Chlorine Heptoxide. Dichlorine heptoxide; perchloric anhydride. Cl_2O_7; mol wt 182.91. Cl 38.77%, O 61.23%. Prepd by dehydration of perchloric acid with P_2O_5: Michael, Conn, *Am. Chem. J.* 23, 445 (1900); 25, 92 (1901); Meyer, Kessler, *Ber.* 54, 566 (1921); Goodeve, Powney, *J. Chem. Soc.* 1932, 2078.

Colorless, very volatile oily liquid. d_4^0 1.86. mp $-91.5°$; bp 82°; $bp_{23.7}$ 0°. Trouton constant 23.4. The most stable oxide of chlorine. *Explodes violently upon concussion or on contact with a flame or iodine.* Dipole moment in CCl_4 at 20° is 0.72. Does not attack wood or paper. Slowly hydrolyzed by water, forming perchloric acid.

USE: Catalyst in cellulose esterification. *Caution:* May be irritating to skin, mucous membranes.

2098. Chlorine Monofluoride. Chlorine fluoride. ClF; mol wt 54.46. Cl 65.11%, F 34.89%. Prepd from Cl_2 and F_2 at 400°: Ruff *et al., Z. Anorg. Allgem. Chem.* 176, 256 (1928); Kwasnik in *Handbook of Preparative Inorganic Chemistry* vol. 1, G. Brauer, Ed. (Academic Press, New York, 2nd ed., 1963) p 153.

Colorless gas. Slightly yellow when liquid. mp $-155.6°$. bp $-100.1°$. Crit temp $-14°$. d (liq; $-108°$) 1.67. *Extremely reactive.* Destroys glass instantly, attacks quartz readily in presence of moisture; organic matter bursts into flame on contact; violent reaction with water.

Caution: Extremely corrosive and irritating to skin, eyes, mucous membranes, respiratory tract.

2099. Chlorine Monoxide. Dichlorine monoxide; dichloromonoxide; dichloroxide; hypochlorous anhydride. Cl_2O; mol wt 86.91. Cl 81.59%, O 18.41%. First preparation by J. L. Gay-Lussac in 1842 from yellow mercuric oxide and chlorine. See also: Cady, *Inorg. Syn.* 5, 156 (1957); Schmeisser in *Handbook of Preparative Inorganic Chemistry* vol. 1, G. Brauer, Ed. (Academic Press, New York, 2nd ed., 1963) p 299. Use as a powerful and selective chlorinating agent: F. D. Marsh *et al., J. Am. Chem. Soc.* 104, 4680 (1982).

Yellowish-brown gas. Disagreeable, penetrating odor. *Explodes on contact with organic matter.* Can also be caused to explode by a spark or by heating. Dec at moderate rate at room temp. mp $-120.6°$. bp $+2.2°$. Trouton constant: 22.5. Suffers photochemical and thermal decompn: at 100-140° there is an induction period, followed by a second order reaction, see N. V. Sidgwick, *The Chemical Elements and Their Compounds* vol. II (Oxford, 1950) p 1201. One vol of water at 0° dissolves more than 100 vols Cl_2O with formation of HClO; sol in CCl_4. Stored as a liquid or solid at temps below $-80°$.

USE: Chlorinating agent. *Caution:* Intensely irritating to eyes, skin, mucous membranes, respiratory tract.

2100. Chlorine Trifluoride. ClF_3; mol wt 92.46. Cl 38.35%, F 61.65%. Prepd from F_2 and Cl_2 or ClF. May be purified by distillation in suitable steel apparatus: Ruff, Krug, *Z. Anorg. Allgem. Chem.* 190, 270 (1930); Grisard *et al., J. Am. Chem. Soc.* 73, 5724 (1951); Kwasnik in *Handbook of Preparative Inorganic Chemistry* vol. 1, G. Brauer, Ed. (Academic Press, New York, 2nd ed., 1963) p 156. *Reviews:* Kemmitt, Sharp, *Advan. Fluorine Chem.* 4, 240-241 (1965); Stein, "Physical and Chemical Properties of Halogen Fluorides" in *Halogen Chemistry* vol. 1, V. Gutmann, Ed. (Academic Press, New York, 1967) pp 133-224; Meinert, *Z. Chem.* 7, 41-57 (1967).

Corrosive, colorless gas. Somewhat sweet, suffocating odor. mp $-76.34°$. bp 11.75°. *Extremely reactive.* Glass wool and organic matter burst into flames on contact even with dil vapors. Violently hydrolyzed by water. Attacks quartz if traces of moisture are present. Liquid is yellow-green in color and solid is white.

USE: Fluorinating agent; in nuclear reactor fuel processing; incendiary; igniter and propellant for rockets; pyrolysis inhibitor for fluorocarbon polymers. *Caution:* Highly irritating to skin, eyes, mucous membranes.

2101. Chlorisondamine Chloride. *4,5,6,7-Tetrachloro-2,3-dihydro-2-methyl-2-[2-(trimethylammonio)ethyl]-2H-*

isoindolium dichloride; 4,5,6,7-tetrachloro-2-(2-dimethyl-aminoethyl)-2-methylisoindolinium chloride methochloride; 4,5,6,7-tetrachloro-2-(2-dimethylaminoethyl)isoindoline dimethylchloride; *N*-[(2-dimethylammonium)ethyl]-4,5,6,7-tetrachloroisoindolinium dimethochloride; chlorisondamine dimethochloride; SU 3088; Ecolid; Ecolid chloride. $C_{14}H_{20}$-Cl_6N_2; mol wt 429.07. C 39.19%, H 4.70%, Cl 49.58%, N 6.53%. Prepn: Allen, Ocampo, *J. Electrochem. Soc.* **103**, 452, 682 (1956); Huebner, U.S. pat. **3,025,294** (1962 to Ciba).

Non-hygroscopic crystals contg ethanol of crystallization, mp 258-265° (dec). Sol in water, methanol, ethanol. pH of aq solns 4.7 to 6.2. Forms yellow, chloroform-soluble complex with bromcresol green, absorption max 420 nm.
THERAP CAT: Antihypertensive.

2102. Chlormadinone Acetate. *17-(Acetyloxy)-6-chloro-pregna-4,6-diene-3,20-dione; 6-chloro-17-hydroxypregna-4,6-diene-3,20-dione acetate;* 6-chloro-6-dehydro-17α-hydroxyprogesterone; 6-chloro-6-dehydro-17α-acetoxyprogesterone; 17α-acetoxy-6-chloro-6,7-dehydroprogesterone; Cero; Gestafortin; Lormin; Luteran; Lutoral; Matrol; Normenon; Menstridyl; Prostal; Traslan; Verton. $C_{23}H_{29}ClO_4$; mol wt 404.94. C 68.22%, H 7.22%, Cl 8.76%, O 15.80%. Orally active progestogen with antiandrogenic activity; has been used in combinations as an oral contraceptive. Prepn: Brückner, **Ger. pat. 1,075,114** (1960 to E. Merck, AG); Brückner *et al., Ber.* **94**, 1225 (1961); Sciaky, *Gazz. Chim. Ital.* **91**, 545 (1961). Endocrinological activities: D. M. Brennan, R. J. Kraay, *Acta Endocrinol.* **44**, 367 (1963). Metabolism: S. Honma *et al., Chem. Pharm. Bull.* **25**, 2019 (1977). Clinical evaluation in prostatic carcinoma: R. Nishimura, K. Shida, *Prostate,* Suppl. 1, 27 (1981); in benign prostatic hypertrophy: T. Usui *et al., Acta Urol. Japan* **27**, 327 (1981), *B.A.* **73**, 27225 (1982). Review of carcinogenicity studies: *IARC Monographs* **21**, 365-375 (1979).

Crystals from methanol or ether, mp 212-214°. $[\alpha]_D$ +6° (c = 1 in CHCl₃). uv max: 283.5, 286 nm (ε 23400, 22100). Mixture with ethinyl estradiol, *q.v., Amenyl, Lutestral, Menova.* Mixture with mestranol, *q.v., C-Quens, Gestamestrol, Sequens.*
THERAP CAT: Progestogen. Antineoplastic (hormonal).
THERAP CAT (VET): Progestogen. Estrus regulator.

2103. Chlormequat Chloride. *2-Chloro-N,N,N-trimeth-ylethanaminium chloride; (2-chloroethyl)trimethylammonium chloride;* chlorocholine chloride; choline dichloride; AC 38555; CCC; Cycocel; Cycogan. $C_5H_{13}Cl_2N$; mol wt 158.07. C 37.99%, H 8.29%, Cl 44.86%, N 8.86%. [ClCH₂CH₂N⁺-(CH₃)₃]Cl⁻. Prepn: Kauffmann, Vorlander, *Ber.* **43**, 2740 (1910); Freiss, Carville, *J. Am. Chem. Soc.* **76**, 2260 (1954); H. Linser *et al.,* **Austrian** pat. **222,145** (1962 to OSSW), *C.A.* **57**, 7660f (1962). Activity: N. E. Tolbert, *J. Biol. Chem.* **235**, 475 (1960); J. Namokar, *Pesticides* **11**, 53 (1977). Metabolism: R. C. Blinn, *J. Agr. Food Chem.* **15**, 984 (1967).
White cryst solid, mp 245° (dec). Fish-like odor; very hygroscopic. Sol in water, lower alcs. Insol in ether, hydro-

carbons. Aq solns corrosive to metal. LD_{50} in mice: 7 mg/kg i.v., 54 mg/kg orally, G. Hennighausen, B. Wiegershausen, *Acta Biol. Med. Ger.* **27**, 663 (1971).
USE: Plant growth regulator.

2104. Chlormerodrin. *[3-[(Aminocarbonyl)amino]-2-methoxypropyl]chloromercury; 1-[3-(chloromercuri)-2-meth-oxypropyl]urea;* chlormeroprin; Katonil; Mercoral; Diurone; Mercloran; Percapyl; Merilid; Neohydrin; Oricur. $C_5H_{11}Cl$-HgN_2O_2; mol wt 367.22. C 16.35%, H 3.02%, Cl 9.66%, Hg 54.63%, N 7.63%, O 8.71%. Prepn: Rowland *et al., J. Am. Chem. Soc.* **72**, 3595 (1950); Foreman, U.S. pat. **2,635,983** (1953 to Lakeside).

$$ClHgCH_2CHCH_2NHCNH_2$$
$$OCH_3 \quad O$$

Crystals from ethanol, mp 152-153°. Bitter, metallic taste. Stable to light and air. Soly in water, methanol: both 1.1 g/100 ml; in ethanol: 0.56 g/100 ml. Sparingly sol in chloroform. Freely sol in alkaline solns. pH of a 0.5% aq soln 4.3-5.0. LD_{50} orally in rats: ~82 mg/kg, E. I. Goldenthal, *Toxicol. Appl. Pharmacol.* **18**, 185 (1971).
THERAP CAT: Diuretic.
THERAP CAT (VET): Mercurial diuretic.

2105. Chlormezanone. *2-(4-Chlorophenyl)tetrahydro-3-methyl-4H-1,3-thiazin-4-one 1,1-dioxide;* 2-(p-chlorophenyl)perhydro-3-methyl-1,3-thiazin-4-one 1,1-dioxide; 2-(4-chlorophenyl)-3-methyl-4-metathiazanone 1,1-dioxide; dichloromethazanone; chlormethazanone; Alinam; Banabin-Sintyal; Fenarol; Lobak; Mio-Sed; Rexan; Rilansyl; Rilaquil; Rilassol; Supotran; Suprotan; Tanafol; Trancopal; Trancote; Transanate. $C_{11}H_{12}ClNO_3S$; mol wt 273.75. C 48.26%, H 4.42%, Cl 12.95%, N 5.12%, O 17.53%, S 11.72%. Prepn starting with 4-chlorobenzylidenemethylamine: Surrey *et al., J. Am. Chem. Soc.* **80**, 3469 (1958); **Brit.** pat. **815,-203** (1959 to Sterling Drug).

Crystals, mp 116.2-118.2°. Soly in water at 25° less than 0.25% (w/v); in 95% ethanol at 25° less than 1.0% (w/v).
THERAP CAT: Anxiolytic; muscle relaxant (skeletal).

2106. Chlormidazole. *1-[(4-Chlorophenyl)methyl]-2-methyl-1H-benzimidazole; 1-(p-chlorobenzyl)-2-methylbenz-imidazole;* 2-methyl-1-(p-chlorobenzyl)benzimidazole; clomidazole. $C_{15}H_{13}ClN_2$; mol wt 256.74. C 70.18%, H 5.10%, Cl 13.81%, N 10.91%. Prepn: Herrling *et al.,* U.S. pat. **2,876,233** (1959 to Gruenenthal).

bp₁₂ 240-242°.
Monohydrate, mp 67-68°.
Hydrochloride, $C_{15}H_{14}Cl_2N_2$, *Diamyceline, Futrican, H-115.* mp 227-228°.
THERAP CAT: Antifungal.

2107. Chlornaphazine. *N,N-Bis(2-chloroethyl)-2-naph-thylamine;* dichloroethyl-β-naphthylamine; β-naphthyldi-(2-chloroethyl)amine; β-naphthylbis(β-chloroethyl)amine; di(2-chloroethyl)-β-naphthylamine; CB 1048; R48; Cloronaftina; Erysan. $C_{14}H_{15}Cl_2N$; mol wt 268.20. C 62.70%, H 5.64%, Cl 26.44%, N 5.22%. Prepd from 2-$C_{10}H_7N(C_2H_4$-

OH)$_2$ by treatment with POCl$_3$: Ross, *J. Chem. Soc.* **1949**, 183; Ghielmetti, *Farmaco Ed. Sci.* **5**, 275 (1950); **11**, 603 (1956). Mechanism of action studies: Jeney *et al., Kiserl. Orvostud.* **20**, 369 (1968), *C.A.* **70**, 18766j (1969). Review and evaluation of studies of carcinogenicity in laboratory animals and in humans: *IARC Monographs* **4**, 119-124 (1974).

Platelets from petr ether, mp 54-56°. bp$_5$ 210°. Very sparingly sol in water, glycerol. More sol (in ascending degree) in petr ether, ethanol, olive oil, ether, acetone, benzene.

Note: This substance has been listed as a known carcinogen: *Fourth Annual Report on Carcinogens* (NTP 85-002, 1985) p 44.

THERAP CAT: Antineoplastic.

2108. Chloroacetaldehyde. 2-Chloro-1-ethanal; monochloroacetaldehyde. C$_2$H$_3$ClO; mol wt 78.50. C 30.60%, H 3.85%, Cl 45.17%, O 20.38%. ClCH$_2$CHO. Prepd industrially by carefully controlled chlorination of acetaldehyde: Söll, *Ger. pat.* **844,595** (1943 to Knapsack-Griesheim); Shawinigan Chem. Ltd., *Brit. pat.* **644,914** (1947); Guinot, Tabuteau, *Compt. Rend.* **231**, 234 (1950); *Fr. pat.* **1,012,991** (1950 to Comp. des Prod. Chim. et Electrometallurg.).

Liquid. Acrid, penetrating odor. *Very corrosive to mucous membranes.* bp$_{760}$ 85-86°. The anhydr substance polymerizes on standing but reverts to the monomer on distillation.

Hemihydrate, probably ClCH$_2$CH(OH)OCH(OH)CH$_2$Cl, platelets from water, mp 43-50°, bp 85.5° with decompn into water and chloraldehyde. Sol in water, alcohol, ether.

USE: In the manufacture of 2-aminothiazole; to facilitate bark removal from tree trunks. *Caution:* Intensely irritating to eyes, skin, mucous membranes, respiratory tract.

2109. Chloroacetamide. 2-Chloroacetamide. C$_2$H$_4$ClNO; mol wt 93.51. C 25.68%, H 4.31%, Cl 37.91%, N 14.98%, O 17.11%. ClCH$_2$CONH$_2$. Prepn from ethyl chloroacetate and ammonia: Jacobs, Heidelberger, *Org. Syn. coll. vol.* **I**, 153 (2nd ed., 1941); from chloroacetyl chloride and ammonium acetate: Finan, Fothergill, *J. Chem. Soc.* **1962**, 2824.

Crystals from water, mp 119-120°. bp about 225° (dec). Sol in 10 parts water, 10 parts abs alc; very slightly sol in ether. Two isomorphous crystalline modifications: α-form, "stable form", obtained by sublimation and by crystallization from nonpolar solvents; β-form, "unstable form", obtained by quenching the melt or by crystallization from polar solvents, *see* Katayama, *Acta Crystallogr.* **9**, 986 (1956); **10**, 468 (1957); B. Kalyanaraman *et al., J. Cryst. Mol. Struct.* **8**, 175 (1978).

2110. Chloroacetanilide. C$_8$H$_8$ClNO; mol wt 169.61. C 56.65%, H 4.75%, Cl 20.91%, N 8.26%, O 9.43%. Prepn of *m-*, *o-* and *p-*isomers: Roberts *et al., J. Org. Chem.* **24**, 654 (1959). Sepn of *o-* and *p-*isomers: Orton, Bradford, *J. Chem. Soc.* **1927**, 986.

m-Chloroacetanilide, N-(3-chlorophenyl)acetamide, 3'-chloroacetanilide. Needles from 50% glacial acetic acid, mp 77-78°. Readily sol in alcohol, benzene, carbon disulfide; very slightly sol in ligroin. uv max (95% ethanol): 245 nm (log ε 4.19).

o-Chloroacetanilide. Needles from dil glacial acetic acid, mp 87-88°. Sublimes at about 50-60°. Practically insol in water, alkalies; sol in alc; more sol in benzene than corresponding *p-*isomer. uv max (95% ethanol): 240 nm (log ε 4.02).

p-Chloroacetanilide. Orthorhombic crystals from aq glacial acetic acid, alc, or acetone. mp 178-179°. d$_4^{22}$ 1.385.

Practically insol in water; readily sol in alc, ether, carbon disulfide; slightly sol in CCl$_4$, benzene. uv max (95% ethanol): 249 nm (log ε 4.25).

2111. Chloroacetic Acid. Chloroethanoic acid; monochloroacetic acid; MCA. C$_2$H$_3$ClO$_2$; mol wt 94.50. C 25.42%, H 3.20%, Cl 37.52%, O 33.86%. ClCH$_2$COOH. Made by chlorination of glacial acetic acid in presence of small amount of sulfur or iodine; also by hydrolysis of trichlorethylene with 90% H$_2$SO$_4$. Lab prepn: Gattermann-Wieland, *Praxis des organischen Chemikers* (de Gruyter, Berlin, 40th ed., 1961) pp 109, 110. Manuf: Faith, Keyes & Clark's *Industrial Chemicals*, F. A. Lowenheim, M. K. Moran, Eds. (Wiley-Interscience, New York, 4th ed., 1975) pp 254-257. Toxicity: Woodward *et al., J. Ind. Hyg. Toxicol.* **23**, 78 (1941).

Colorless or white, deliquesc crystals. d 1.580. Exists in α, β and γ forms having mp 63°, 55-56° and 50° respectively. mp for acid of commerce is 61-63°. bp for all forms is 189°. Very sol in water; sol in alcohol, benzene, chloroform, ether. *Keep well closed and in a cool place.*

Sodium salt, C$_2$H$_2$ClNaO$_2$, *Monoxone.* White crystalline solid. Soly in water at 20°: 85 g/100 ml. LD$_{50}$ in rats, mice, guinea pigs (mg/kg): 76, 255, 80 orally (Woodward).

USE: Herbicide. Manuf various dyes and other organic chemicals. *Caution:* Irritating to skin, mucous membranes.

2112. Chloroacetic Anhydride. *Chloroacetic acid anhydride;* monochloroacetic acid anhydride; *sym-*dichloroacetic anhydride. C$_4$H$_4$Cl$_2$O$_3$; mol wt 170.98. C 28.10%, H 2.36%, Cl 41.47%, O 28.07%. (ClCH$_2$CO)$_2$O. Prepd by treating monochloroacetic acid in tetrahydrofuran with methoxyacetylene at 15° and distilling the mixture: Eglinton *et al., J. Chem. Soc.* **1954**, 1860; by mixing monochloroacetic acid and HCN with 1,4-dioxane satd with HCl and fractionating the mixture: Krieble, Smellie, *U.S. pat.* **2,390,106** (1945); by heating chloroacetyl chloride with monochloroacetic acid in the presence of AlCl$_3$: Strosacker, Schwegler, *U.S. pat.* **1,713,104** (1929 to Dow).

Prisms from benzene; d^{20} 1.5494; mp 46°. bp$_{760}$ 203°; bp$_{116}$ 163°; bp$_{62}$ 149°; bp$_{24}$ 126°; bp$_{10}$ 109-110°; bp$_{0.05}$ 118-120°. Freely sol in ether, chloroform; slightly sol in benzene, practically insol in cold ligroin.

USE: In *N-*acetylation of amino acids in alkaline soln; prepn of cellulose chloroacetates.

2113. Chloroacetone. *1-Chloro-2-propanone;* chloracetone; monochloroacetone; monochloracetone; acetonyl chloride; chloropropanone; 1-chloro-2-ketopropane; 1-chloro-2-oxopropane. C$_3$H$_5$ClO; mol wt 92.53. C 38.94%, H 5.45%, Cl 38.32%, O 17.29%. ClCH$_2$COCH$_3$. Prepn by the action of chlorine upon diketene: A. B. Boese, Jr., *U.S. pat.* **2,209,683** (1940 to Carbide and Carbon Chem.); by chlorination of acetone: E. J. Rahrs, *U.S. pat.* **2,235,562** (1941 to Eastman Kodak); G. H. Morey, *U.S. pat.* **2,243,484** (1941 to Commercial Solvents Corp.). Stabilization: *idem, U.S. pat.* **2,229,651** (1941 to Commercial Solvents Corp.); E. J. Rahrs, *U.S. pat.* **2,263,010** (1941 to Eastman Kodak). Forms binary azeotropes with many organic liquids: L. H. Horsley, *Azeotropic Data*, Advances in Chemistry Series No. 6 (Washington, 1952) p 73. Reacts with aryl Grignard reagents to form stilbenes: Huang, *J. Chem. Soc.* **1954**, 2539. Toxicology: E. V. Sargent *et al., Am. Ind. Hyg. Assoc. J.* **47**, 375 (1986).

Liquid. Pungent odor. Lacrimator. Turns dark and resinifies on prolonged exposure to light. May be stabilized by addition of 0.1% H$_2$O or 1.0% CaCO$_3$. d$_4^{25}$ 1.123; d$_4^{15}$ 1.135. mp −44.5°. bp 119.7°; bp$_{50}$ 61°, bp$_{12}$ 20°. Volatile with steam. Surface tension at 20° = 35.27 dyn/cm. Dipole moment in hexane, 2.36D. Sol in 10 parts water (w/w). Miscible with alcohol, ether, chloroform. LD$_{50}$ (14 day) in mice, rats (mg/kg): 127, 100 orally; LC$_{50}$ (1 hr) in rats (ppm): 262 by inhalation (Sargent).

2,4-Dinitrophenylhydrazone, yellow needles from alc, mp 124°.

USE: Has been proposed as tear gas component for police and military use; in the manuf of couplers for color photography; as enzyme inactivator; intermediate in the manuf of perfumes, antioxidants, drugs; in insecticide formulations; in photopolymerization of vinyl compds. Proposed as catalyst in tetraethyllead production, as selective solvent for separating diolefins. *Caution:* Intensely irritating to eyes, skin, mucous membranes.

2114. *p*-Chloroacetophenone. *1-(4-Chlorophenyl)ethanone; 4′-chloroacetophenone.* C_8H_7ClO; mol wt 154.59. C 62.15%, H 4.56%, Cl 22.94%, O 10.34%. Prepd from chlorobenzene and acetic anhydride in the presence of $AlCl_3$: Adams, Noller, *Org. Syn.* coll. vol. I, 109 (1941). Physical properties: *eidem, ibid.*; Dreisbach, Martin, *Ind. Eng. Chem.* **41**, 2875 (1949).

Cl —⟨ ⟩— $COCH_3$

Liquid, bp_{24} 124-126°, bp 237°. mp 20-21° (Adams, Noller); also reported as 18.4° (Dreisbach, Martin). d_4^{20} 1.192, d_4^{25} 1.188. n_D^{20} 1.555, n_D^{25} 1.553. Practically insol in water; miscible with alcohol, ether.
Caution: Highly irritating to eyes, mucous membranes.

2115. ω-Chloroacetophenone. *2-Chloro-1-phenylethanone; 2-chloroacetophenone; α-chloroacetophenone; phenacyl chloride; Chemical Mace; CN.* C_8H_7ClO; mol wt 154.59. C 62.15%, H 4.56%, Cl 22.94%, O 10.35%. $C_6H_5COCH_2Cl$. Chemical warfare agent with lacrimatory properties. Prepn: Scholl, Korten, *Ber.* **34**, 1902 (1901); Schaefer, Sonnenberg, *J. Org. Chem.* **28**, 1128 (1963). Purification: Lofton, U.S. pat. **2,414,418** (1947 to Pennsylvania Coal Prod.). Melting point determination and review of prepn and mp: Macy, *J. Chem. Ed.* **24**, 222 (1947). Comparative toxicity of CN and *o*-chlorobenzylidenemalononitrile, *q.v.*: B. Ballantyne, D. W. Swanston, *Arch. Toxicol.* **40**, 75 (1978).
Crystals from dil alcohol, carbon tetrachloride, or light petr. mp 58-59°, also reported as 54°, Macy, *loc. cit.*, and 56.5°, Rheinboldt, Perrier, *J. Am. Chem. Soc.* **69**, 3148 (1947). bp 244-245°. d^{15} 1.324. Vapor press. at 20°: 5.4 × 10^{-3} mm Hg. Practically insol in water; freely sol in alcohol, ether, benzene. LD_{50} in rats: 41 mg/kg i.v., 36 mg/kg i.p., 127 mg/kg orally; LC_{50} in rats: 8,750 mg/min/m³, B. Ballantyne, D.W. Swanston, *loc. cit.*
USE: Riot control agent. *Caution:* Potent eye, throat and skin irritant; possible corneal damage; may be a sensitizer in human beings. *See Fed. Proc.* **30**, 84-99 (1971).

2116. Chloroacetyl Isocyanate. Chloroacetic acid anhydride with isocyanic acid. $C_3H_2ClNO_2$; mol wt 119.52. C 30.15%, H 1.69%, Cl 29.67%, N 11.72%, O 26.77%. $ClCH_2$-CONCO. Obtained by the reaction of chloroacetamide with oxalyl chloride: Speziale, Smith, *J. Org. Chem.* **28**, 1808 (1963).
Liquid. bp_{20} 50-55°. $n_D^{21.5}$ 1.4580.

2117. *N*-(3-Chloroallyl)hexaminium Chloride. *1-(3-Chloro-2-propenyl)-3,5,7-triaza-1-azoniatricyclo[3.3.1.1^{3,7}]-decane chloride; 1-(3-chloroallyl)-3,5,7-triaza-1-azoniaadamantane chloride;* Dowicide Q. $C_9H_{16}Cl_2N_4$; mol wt 251.17. C 43.04%, H 6.42%, Cl 28.23%, N 22.31%. Prepn: Scott, Wolf, *Appl. Microbiol.* **10**, 211 (1962).

Cream-colored powder. Freely sol in water (up to 25% w/w). Concd solns are yellow.
USE: Bactericide.

2118. Chloroaniline. C_6H_6ClN; mol wt 127.57. C 56.49%, H 4.74%, Cl 27.79%, N 10.99%. Prepn of *m*-, *o*- and *p*-isomers: Sidgwick, Rubie, *J. Chem. Soc.* **119**, 1013 (1921). Manufacture of *m*- and *p*-isomers by catalytic hydrogenation of chloronitrobenzene: Pray, Trager, U.S. pat. **2,791,613** (1957 to Columbia-Southern Chem.); by

reduction of chloronitrobenzene with NaHS: Latourette *et al.*, U.S. pat. **2,894,035** (1959 to Food Machinery & Chem.). Properties of *o*-isomer: Dreisbach, Martin, *Ind. Eng. Chem.* **41**, 2875 (1949). Toxicity data: H. F. Smyth *et al.*, *Am. Ind. Hyg. Assoc. J.* **23**, 95 (1962).

m-Chloroaniline, *3-chlorobenzenamine.* Liquid, bp 230.5°. mp −10.4°. d_4^{22} 1.2150. n_D^{20} 1.5931. Practically insol in water. Sol in most common organic solvents.
o-Chloroaniline, liquid. At 99.61 mol-% purity, bp 208.84°; mp −1.94°. d_4^{22} 1.2114. n_D^{20} 1.5895. Practically insol in water. Sol in acid and in most organic solvents.
p-Chloroaniline, orthorhombic crystals from alcohol or petr ether, mp 72.5°. bp 232°. d_4^{77} 1.169. Sol in hot water; freely sol in alcohol, ether, acetone, carbon disulfide. LD_{50} orally in rats: 0.31 g/kg (Smyth).

2119. Chloroarsenol. *(2-Chloro-1-heptenyl)arsonic acid; 2-chloro-1-heptene-1-arsonic acid.* $C_7H_{14}AsClO_3$; mol wt 256.56. C 32.77%, H 5.50%, As 29.20%, Cl 13.82%, O 18.71%. Prepn: Ger. pat. **296,915** (1917 to Bayer).

$CH_3(CH_2)_4CCl$
‖
$CH—AsO(OH)_2$

Needles from water, mp 115°.
Ammonium salt, $C_7H_{17}AsClNO_3$, *ammonium chloroheptenearsonate.* Freely sol in water. Supplied as a 1% soln made isotonic with 0.7% NaCl under the names: *Solarson; Arsion.*

2120. Chloroazodin. *N,N″-Dichlorodiazenedicarboximidamide; α,α′-azobis[chloroformamidine]; chlorazodin; dichloroazodicarbonamidine; Azochloramide.* $C_2H_4Cl_2N_6$; mol wt 183.01. C 13.12%, H 2.20%, Cl 38.75%, N 45.93%. Prepd by treating an AcOH-NaOAc soln of guanidine nitrate with sodium hypochlorite in the cold: Braz *et al.*, *Appl. Chem. USSR* **17**, 565 (1944), *C.A.* **40**, 2267 (1946). Structure and uv spectra: Kumler, *J. Am. Chem. Soc.* **75**, 3092 (1953).

H_2N NH_2
    $CN=NC$
ClN NCl

Bright yellow needles, plates, flakes. Faint chlorine odor, burning taste. Dec explosively at about 155°. Decompn is accelerated by contact with metals. Very slightly sol in water; sparingly sol in alcohol; slightly sol in glycerol, glyceryl triacetate (1:100), vegetable oils, ether, chloroform. Practically insol in carbon tetrachloride and liquid petrolatum. Oil solns are generally made by diluting a triacetin (glyceryl triacetate) soln of chloroazodin with oil. Solns of chloroazodin in glycerol and in alcohol dec rapidly on warming; all solns dec on exposure to light.
THERAP CAT: Antibacterial.
THERAP CAT (VET): Topical anesthetic.

2121. Chlorobenzene. Monochlorobenzene; benzene chloride. C_6H_5Cl; mol wt 112.56. C 64.02%, H 4.48%, Cl 31.50%. Produced by the chlorination of benzene in the presence of a catalyst. Review of manuf, properties, and use: Faith, Keyes & Clark's *Industrial Chemicals*, F. A. Lowenheim, M. K. Moran, Eds. (Wiley-Interscience, New York, 4th ed., 1975) pp 258-265.
Colorless, very refractive liquid; faint, not unpleasant odor. d_4^{20} 1.107. bp 131-132°. Solidif −55°. mp −45°. Flash pt 28°. n_D^{20} 1.5248. Insol in water; freely sol in alcohol, benzene, chloroform, ether.
USE: Manuf phenol, aniline, DDT; solvent for paints; heat transfer medium.

2122. *p*-Chlorobenzenesulfonic Acid. *4-Chlorobenzene-*

sulfonic acid; closylate. $C_6H_5ClO_3S$; mol wt 192.62. C 37.41%, H 2.62%, Cl 18.41%, O 24.92%, S 16.64%. Prepd by heating chlorobenzene with concd sulfuric acid or oleum under dehydrating conditions: Meyer, *Ann.* **433**, 327, 333 (1923).

Monohydrate, crystals from water, mp 67°. The anhydr material is usually obtained as a syrup, bp_{25} 148°. Sol in water, alc. Practically insol in ether, benzene.

Sodium salt monohydrate, $C_6H_4ClNaO_3.H_2O$, cubic crystals from water. Sol in water.

2123. Chlorobenzilate. *4-Chloro-α-(4-chlorophenyl)-α-hydroxybenzeneacetic acid ethyl ester; 4,4'-dichlorobenzilic acid ethyl ester;* ethyl 4,4'-dichlorobenzilate; ethyl 2-hydroxy-2,2-bis(4-chlorophenyl)acetate; chlorbenzilat; G 23922; compd 338; Acaraben; Akar; Folbex. $C_{16}H_{14}Cl_2O_3$; mol wt 325.20. C 59.09%, H 4.34%, Cl 21.81%, O 14.76%. Prepn: Häfliger, **U.S.** pat. **2,745,780** (1956 to Geigy); **Brit.** pat. **831,421** (1960 to Metal & Thermit).

Viscous liquid. The commercial product is yellow. $bp_{0.04}$ 146-148°. n_D^{20} 1.5727. Vapor press. at 20°: 2.2×10^{-6} mm. Slightly sol in water. Sol in most organic solvents. Incompatible with lime. LD_{50} orally in male, female rats: 1040, 1220 mg/kg, T. B. Gaines, *Toxicol. Appl. Pharmacol.* **14**, 515 (1969).

USE: Acaricide in spider-mite control; synergist for DDT. *Human Toxicity:* Symptoms similar to DDT, *q.v.*

2124. m-Chlorobenzoic Acid. *3-Chlorobenzoic acid.* $C_7H_5ClO_2$; mol wt 156.57. C 53.69%, H 3.22%, Cl 22.65%, O 20.44%. Prepn by catalytic oxidation of 1-chloro-3-ethylbenzene: Emerson *et al., J. Am. Chem. Soc.* **71**, 1742 (1949); by chlorination of benzoic acid: Gorvin, *Chem. & Ind. (London)* **1951**, 910; by the von Richter reaction from 1-chloro-4-nitrobenzene and alcoholic KCN: Samuel, *J. Chem. Soc.* **1960**, 1318.

Crystals, mp 158°. d_4^{25} 1.496. Sol in 2850 parts cold water; more sol in hot water; freely sol in alcohol, ether.

Methyl ester, $C_8H_7ClO_2$, *methyl m-chlorobenzoate.* mp 21°, bp 231°.

2125. o-Chlorobenzoic Acid. *2-Chlorobenzoic acid.* Empirical formula, etc.: *see m-isomer above.* Prepd by $KMnO_4$ oxidation of o-chlorotoluene: Clarke, Taylor, *Org. Syn. coll. vol.* **II**, 135 (1943). Crystal structure: Ferguson, Sim, *Acta Cryst.* **12**, 941 (1959), *C.A.* **55**, 24188i (1961).

Monoclinic crystals, mp 142°. d_4^{20} 1.544. Sol in 900 parts cold water; more sol in hot water, freely sol in alc, ether.

USE: Preservative for glues, paints. Intermediate in the manufacture of fungicides and dyes.

2126. p-Chlorobenzoic Acid. Chlorodracylic acid. Empirical formula etc.: *see corresp m-isomer above.* Prepn by catalytic oxidation of 1-chloro-4-ethylbenzene: Emerson *et al., J. Am. Chem. Soc.* **71**, 1742 (1949). Manuf by oxidation of p-chlorobenzaldehyde: Shipman, **U.S.** pat. **3,124,611** (1964 to ICI). Crystal structure: Toussaint, *Acta Cryst.* **4**,

71 (1951), *C.A.* **45**, 6604e (1951); Pollock, Woodward, *ibid.* **7**, 605 (1954), *C.A.* **48**, 13331e (1954).

Triclinic crystals, mp 243°. Sol in 5290 parts water; freely sol in alcohol, ether.

Methyl ester, $C_8H_7ClO_2$, *methyl p-chlorobenzoate,* mp 44°.

Sodium salt, $C_7H_4ClNaO_2$, *sodium p-chlorobenzoate.* White, odorless, crystalline powder, freely sol in water. Ingredient of *Microbin* which also contains the *ortho* isomer.

USE: Sodium salt as a preservative.

2127. o-Chlorobenzylidenemalononitrile. *[(2-Chlorophenyl)methylene]propanedinitrile;* o-chlorobenzalmalononitrile; *β,β*-dicyano-*o*-chlorostyrene; CS. $C_{10}H_5ClN_2$; mol wt 188.61. C 63.68%, H 2.67%, Cl 18.80%, N 14.85%. Lacrimatory chemical warfare agent. Prepn: B.B. Corson, R. W. Stoughton, *J. Am. Chem. Soc.* **50**, 2825 (1928). Pharmacology: R. W. Brimblecombe *et al., Brit. J. Pharmacol.* **44**, 561 (1972). Toxicology: C. L. Punte *et al., Toxicol Appl. Pharmacol.* **4**, 656 (1962); J. R. Gaskins *et al., Arch. Environ. Health* **24**, 449 (1972); B. Ballantyne, S. Callaway, *Med. Sci. Law* **12**, 43 (1972). Comparative toxicity of CS and *ω*-chloroacetophenone, *q.v.*: B. Ballantyne, D. W. Swanston, *Arch. Toxicol.* **40**, 75 (1978).

White crystalline solid, mp 95-96°, bp 310-315°. Vapor press at 20°: 3.4×10^{-5} mm Hg. Sparingly sol in water; sol in acetone, dioxane, methylene chloride, ethyl acetate, benzene. LD_{50} in rats: 28 mg/kg i.v., 48 mg/kg i.p.; LC_{50} in rats: 88,480 mg/min/m³, B. Ballantyne, D. W. Swanston, *loc. cit.*

USE: Riot control agent; a more potent irritant than *ω*-chloroacetophenone but less incapacitating. *Caution:* Potent eye, throat, and skin irritant.

2128. p-Chlorobenzylpseudothiuronium Chloride. *Carbamimidothioic acid (4-chlorophenyl)methyl ester monohydrochloride.* $C_8H_{10}Cl_2N_2S$; mol wt 237.16. C 40.51%, H 4.25%, Cl 29.90%, N 11.81%, S 13.52%. Prepd by refluxing a mixture of p-chlorobenzyl chloride and thiourea in ethanol: Dewey, Sperry, *J. Am. Chem. Soc.* **61**, 3251 (1939).

Crystals from concd HCl + water, mp 197°.
USE: Identification of organic acids.

2129. Chlorobutanol. *1,1,1-Trichloro-2-methyl-2-propanol; β,β,β*-trichloro-*tert*-butyl alcohol; acetone chloroform; chlorbutol; Chloretone; Coliquifilm; Methaform; Sedaform. $C_4H_7Cl_3O$; mol wt 177.47. C 27.07%, H 3.98%, Cl 59.94%, O 9.02%. Chlorobutanol may be anhydrous or it may contain up to about one-half molecule of water. Prepn: Fishburn, Watson, *J. Am. Pharm. Assoc.* **28**, 491 (1939); E. Bergman, **U.S.** pat. **2,446,453** (1948 to Polymerisable Products Ltd.); P. Riegger, H. Richtzenbain, **Ger.** pat. **1,271,697** (1968 to Dynamit Nobel A.-G.), *C.A.* **69**, 95967g (1968); M. Saljoughian, *Monatsh.* **114**, 813 (1983). Toxicity in fish, rabbits, dogs: E. Impens, *Arch. Int. Pharmacodyn. Ther.* **8**, 77 (1901).

Crystals. Camphor odor and taste. A crystal of chlorobutanol placed on a water surface "dances" like a camphor crystal. Sublimes easily. The anhydrous form mp 97° (the hemihydrate mp 78°). bp$_{760}$ 167°; bp$_{246}$ 135°. Easily sol in hot water; one gram dissolves also in 1 ml alcohol, 10 ml glycerol; sol in chloroform, ether, acetone, petr ether, glacial acetic acid, oils. MLD orally in dogs, rabbits (mg/kg): 238, 213 (Impens).

Compd with chloral hydrate, mp 65°. Ref: **Ger. pat. 151,188** (1904).

Caution: This substance may be habit forming and is listed in the U.S. Code of Federal Regulations, Title 21 Part 329.1 (1987).

USE: Plasticizer for cellulose esters and ethers. Preservative for biological fluids, hypodermic solns, and solns of alkaloids. Pharmaceutic aid (antimicrobial).

THERAP CAT: Dental analgesic.

THERAP CAT (VET): Has been used as a mild sedative, antipruritic, and antiseptic.

2130. 1-Chloro-2-butene. α-Chloro-β-butylene; crotyl chloride; γ-methallyl chloride; γ-methylallyl chloride. C$_4$-H$_7$Cl; mol wt 90.55. C 53.05%, H 7.79%, Cl 39.16%. CH$_3$-CH=CHCH$_2$Cl. Prepn of *cis*- and *trans*-isomers: Hatch, Nesbitt, *J. Am. Chem. Soc.* **72**, 727 (1950). Manuf: Carlson, **U.S. pat. 2,494,034** (1950 to Shell); Montagna, Hess, **U.S. pat. 3,055,954** (1962 to Union Carbide). Equilibrium constant for isomerization of 3-chloro-1-butene to 1-chloro-2-butene: Dittmer, Marcantonio, *J. Org. Chem.* **29**, 3473 (1964).

trans-Isomer, liquid, bp$_{752}$ 84.8°. n_D^{20} 1.4350, n_D^{25} 1.4327. d$_4^{20}$ 0.9295.

cis-Isomer, liquid, bp$_{758}$ 84.1°. n_D^{20} 1.4390. d$_4^{20}$ 0.9426. *Caution:* Irritates eyes, respiratory passages.

2131. 3-Chloro-1-butene. γ-Chloro-α-butylene; α-methallyl chloride; α-methylallyl chloride. CH$_2$=CH-CHClCH$_3$. Empirical formula, etc.: *see* 1-Chloro-2-butene. Prepn of DL-form: Böhme, *Ber.* **71**, 2372 (1938); Curtin, Gerber, *J. Am. Chem. Soc.* **74**, 4052 (1952); of D(−)-form: Böhme, *loc. cit.;* of L(+)-form: Young, Caserio, *J. Org. Chem.* **26**, 245 (1961). Absolute configuration: Young, Caserio, *loc. cit.* Equilibrium constant for isomerization to 1-chloro-2-butene: Dittmer, Marcantonio, *J. Org. Chem.* **29**, 3473 (1964).

DL-Form, liquid, bp 63.9-64.2°, bp$_{26}$ −5°. d$_4^{20}$ 0.9001. n_D^{20} 1.4150.

D(−)-Form, liquid, bp$_{26}$ −5°. α$_D^{20}$ −2.52° (neat).

L(+)-Form, liquid, α$_D^{25}$ +5.87 (neat).

Caution: Irritates eyes, respiratory passages.

2132. 3-Chloro-*d*-camphor. *3-Chloro-1,7,7-trimethylbicyclo[2.2.1]heptan-2-one; 3-chloro-d-2-bornanone.* C$_{10}$-H$_{15}$ClO; mol wt 186.68. C 64.33%, H 8.10%, Cl 18.99%, O 8.57%. Of the two isomers, the *endo*-form is more stable than the *exo*-form: Cookson, *J. Chem. Soc.* **1954**, 282. Prepn of *endo*-form: Kipping, Pope, *ibid.* **63**, 548 (1893). Prepn of *exo*-form by isomerization of *endo*-form: Lowry, Steele, *ibid.* **107**, 1382 (1915). uv spectra: Lowry, Owen, *J. Chem. Soc.* **129**, 606 (1926). Crystal structure of *endo*-form: Wiebenga, Krom, *Rec. Trav. Chim.* **65**, 663 (1946). Configuration: Cookson, *loc. cit.;* Kumler *et al., J. Am. Chem. Soc.* **83**, 2711 (1961). *Review: The Terpenes* vol. II, J. L. Simonsen, Ed. (University Press, Cambridge, 2nd ed., 1949) p 397.

endo-form   *exo*-form

endo-Form, *α-chloro-d-camphor, 3α-chloro-d-camphor, camphor monochlorated.* Monoclinic prisms from alc, mp 94°. Volatile with steam. [α]$_D$ +96.2° (c = 5 in alc). Practically insol in water. Sol in alcohol, chloroform, ether. uv max (cyclohexane): 305 nm (log ε 1.72).

exo-Form, *α'-chloro-d-camphor, 3β-chloro-d-camphor.* Crystals from alc, mp 117°. [α]$_D$ +35° (c = 5 in alc). Readily sol in all ordinary solvents, except water and formamide.

Much more sol in 96% alc than *endo*-form. On standing, loses HCl. uv max (cyclohexane): 306 nm (log ε 1.75).

2133. 4-Chloro-*m*-cresol. *4-Chloro-3-methylphenol;* 3-methyl-4-chlorophenol; chlorocresol; parachlorometacresol; 6-chloro-*m*-cresol; 6-chloro-3-hydroxytoluene; 2-chloro-5-hydroxytoluene. C$_7$H$_7$ClO; mol wt 142.58. C 58.96%, H 4.95%, Cl 24.88%, O 11.22%. Prepd by chlorination of *m*-cresol: **Ger. pat. 90,847** (1897 to Kalle), *Frdl.* **4**, 94; Laschinger, **U.S. pat. 1,847,566** (1932), *C.A.* **26**, 2471 (1932); Sah, Anderson, *J. Am. Chem. Soc.* **63**, 3165 (1941).

Dimorphous crystals, mp 55.5° and mp 66° (ligroin). Said to be odorless when very pure, but usually a phenolic odor persists. bp 235°. Volatile with steam. One gram dissolves in 260 ml water at 20°, more sol in hot water. Freely sol in alcohol, benzene, chloroform, ether, acetone, petr ether, fixed oils, terpenes, aq alkaline solns. Aq solns turn yellow on exposure to light and air.

USE: Antiseptic, disinfectant, more active in acid than in alkaline solns.

THERAP CAT: Topical antiseptic.

THERAP CAT (VET): Has been used as a topical antiseptic.

2134. Chlorocyanohydrin. *3,3,3-Trichloro-2-hydroxypropanenitrile; 3,3,3-trichlorolactonitrile;* 3,3,3-trichloro-2-hydroxypropionitrile; chloral hydrocyanide. C$_3$H$_2$Cl$_3$NO; mol wt 174.43. C 20.66%, H 1.16%, Cl 60.98%, N 8.03%, O 9.17%. Cl$_3$CCH(OH)CN. Prepd from chloral hydrate and HCN: Chwala, Wassmuth, *Monatsh.* **81**, 843 (1950).

Crystals, mp 61°. Odor of HCN and chloral. Sublimes easily. bp 220°. Freely sol in water, alcohol, ether, and most other common organic solvents; moderately sol in carbon tetrachloride; practically insol in petr ether.

USE: Instead of cherry-laurel or bitter-almond water; is superior to these because HCN content is uniform and permanent, hence exact dose is possible.

2135. 1-Chloro-2,4-dinitrobenzene. 2,4-Dinitro-1-chlorobenzene; 4-chloro-1,3-dinitrobenzene; 6-chloro-1,3-dinitrobenzene. C$_6$H$_3$ClN$_2$O$_4$; mol wt 202.56. C 35.57%, H 1.49%, Cl 17.50%, N 13.83%, O 31.60%. Obtained by nitrating *o*-nitrochlorobenzene: Borsche, Rantscheff, *Ann.* **379**, 152 (1911); Hodgson, Dodgson, *J. Chem. Soc.* **1948**, 1006. The mixture of 2,4- and 2,6-dinitrochlorobenzenes (of which the 2,4-compd is present in largest amount) can be separated with an alkaline ethanol-water soln: Molard, Vaganay, *Mém. Poudres* **39**, 111 (1957).

Yellow crystals. *Skin irritant.* d about 1.7. mp 52-54°. bp 315°. Practically insol in water; sparingly sol in cold, freely sol in hot alc; sol in ether, benzene, CS$_2$. LD$_{50}$ orally in rats: 1.07 g/kg, H. F. Smyth *et al., Am. Ind. Hyg. Assoc. J.* **23**, 95 (1962).

USE: As a reagent for the detection and determination of nicotinic acid, nicotinamide, and other pyridine compds. *Caution:* May cause dermatitis of both primary and allergic types.

2136. 2-Chloro-1,3-dinitrobenzene. 1-Chloro-2,6-dinitrobenzene; 2,6-dinitro-1-chlorobenzene. C$_6$H$_3$ClN$_2$O$_4$; mol wt 202.56. C 35.57%, H 1.49%, Cl 17.50%, N 13.83%, O 31.60%. Prepd from 2,6-dinitroaniline by the Sandmeyer reaction: Gunstone, Tucker, *Org. Syn.,* **coll. vol. IV,** 160 (1963).

Yellow crystals from benzene + petr ether or from acetic acid. *Skin irritant.* mp 86-87°. $d_4^{16.5}$ 1.6867. bp 315°. Practically insol in water. Sol in alc. Moderately sol in ether, benzene.
Caution: See 1-Chloro-2,4-dinitrobenzene.

2137. β-Chloroethyl Acetate. *2-Chloroethanol acetate.* $C_4H_7ClO_2$; mol wt 122.55. C 39.20%, H 5.76%, Cl 28.94%, O 26.11%. $CH_3COOCH_2CH_2Cl$. Prepn from ethylene oxide and acetyl chloride: Gustus, Stevens, *J. Am. Chem. Soc.* **55**, 378 (1933).
Liquid, bp 145°. d^{20} 1.1460. n_D^{20} 1.4234. Flash pt 53.89° (closed cup): *Ind. Eng. Chem.* **32**, 880 (1940). Practically insol in water; sol in alcohol.
USE: Gasoline additive, Henderson, U.S. pat. **3,179,506** (1965 to Shell).

2138. N-(2-Chloroethyl)dibenzylamine Hydrochloride. *N-(2-Chloroethyl)-N-(phenylmethyl)benzenemethanamine hydrochloride; N,N-dibenzyl-β-chloroethylamine hydrochloride; N,N-dibenzylaminoethyl chloride hydrochloride; Dibenamine hydrochloride.* $C_{16}H_{19}Cl_2N$; mol wt 296.23. C 64.87%, H 6.46%, Cl 23.94%, N 4.73%. $(C_6H_5CH_2)_2NCH_2-CH_2Cl.HCl$. α-Adrenergic blocker. Prepn: Rabald, Dimroth, Ger. pat. **824,208** (1951 to Boehringer, Mann.), *C.A.* **47**, 2206 (1953). Pharmacology and toxicity: M. Nickerson, G. M. Nomaguchi, *J. Pharmacol. Exp. Ther.* **101**, 379 (1951). Use in adrenergic receptor differentiation: R. F. Furchgott, *ibid.* **111**, 265 (1954); in specific receptor labelling: R. D. Green *et al., ibid.* **187**, 524 (1973). *In vivo* studies of protection against hepatotoxic agents: H. M. Maling *et al., Toxicol. Appl. Pharmacol.* **27**, 380 (1974); E. K. Weisburger *et al., ibid.* **28**, 477 (1974); H. M. Maling *et al., Biochem. Pharmacol.* **23**, 1479 (1974).
Crystals, mp 192°, also given as 180-181° (the free base is an oily liquid). Practically insol in water near neutrality. Sol in dil acids (2% at pH 2.1, 1% at pH 2.4 and 0.5% at pH 2.7), in 95% alcohol and in propylene glycol. Stable in acid soln, but rapidly loses activity in neutral or alkaline solns. LD_{50} s.c. in mice: 800 mg/kg (Nickerson, Nomaguchi).

2139. 2-Chloroethyl Vinyl Ether. *(2-Chloroethoxy)ethene.* C_4H_7ClO; mol wt 106.55. C 45.08%, H 6.62%, Cl 33.28%, O 15.02%. $ClCH_2CH_2OCH=CH_2$. May be prepd by the action of solid NaOH + triethanolamine upon β,β'-dichlorodiethyl ether: Chitwood, Perkins, U.S. pat. **2,104,-717** (1938 to Carbide and Carbon Chem.); *cf.* Cretcher *et al., J. Am. Chem. Soc.* **47**, 1175 (1926).
Liquid. d_{15}^{15} 1.0525. bp_{740} 109°. Quite stable to NaOH solns. Even dil acids produce hydrolysis to acetaldehyde and ethylene chlorohydrin [2-chloroethanol]. LD_{50} orally in rats: 250 mg/kg, Smyth *et al., J. Ind. Hyg. Toxicol.* **31**, 60 (1949).
USE: Mfg anesthetics, sedatives, cellulose ethers.

2140. Chlorofluorocarbons. CFCs; FCCs. Chemically stable series of chlorinated and fluorinated compounds usually with methane or ethane skeleton, marketed under general names such as **Arcton, Freon, Frigen, Genetron**. Known collectively as CFCs, individually identified by a CFC code based on the "rule of 90". (To derive the chemical formula for CFC-12, e.g., add "90" to 12; the resulting number "102" indicates 1 carbon, 0 hydrogen, 2 fluorine yielding the formula CCl_2F_2.) Initial report on suitability as refrigerants: T. Midgley, Jr., A. L. Henne, *Ind. Eng. Chem.* **22**, 542 (1930). Review including physical and chemical properties of various CFCs: E. Heiskel, *Aerosol Rep.* **22**, 403-415 (1983). CFCs do not decompose in the lower atmosphere. Photodecomposition occurs in the stratosphere via absorption of uv radiation and subsequent release of atomic chlorine which can catalyze ozone breakdown. CFC-ozone depletion hypothesis: M. J. Molina, F. S. Rowland, *Nature* **249**, 810 (1974). Reviews focusing on atmospheric chemistry of CFCs, uses, potential hydrogen-substituted replace-ments, environmental and regulatory issues: J. P. Cohn, *BioScience* **37**, 647-650 (1987); R. Pool, *Science* **242**, 666-668 (1988); F. S. Rowland, *Environ. Conserv.* **15**, 101-115 (1988); L. B. Weisfeld, *Plast. Comp.* **1988**, 15-22, 40-43; F. S. Rowland, *Am. Scientist* **77**, 36-45 (1989). For prepn information *see* dichlorodifluoromethane (CFC-12), trichlorofluoromethane (CFC-11), cryofluorane (CFC-114).
Colorless, essentially odorless, nonflammable, noncorrosive. Miscible with aliphatic, alicyclic and aromatic hydrocarbons, halogenated hydrocarbons, monovalent low molecular alcohols.
Note: Consult latest Government regulations on use as aerosol propellant.
USE: In aerosol propellants (CFC-11, 12, 113); air conditioning; refrigeration (CFC-12); blowing agents for making foam (CFC-11, 12); cleaning fluids (CFC-113); solvents for the electronics industry, bedding and packaging.

2141. Chloroform. *Trichloromethane.* $CHCl_3$; mol wt 119.39. C 10.05%, H 0.84%, Cl 89.10%. Improperly called "formyl trichloride". Made from acetone and bleaching powder by addition of sulfuric acid: $2CH_3COCH_3 + 6CaOCl_2.H_2O \rightarrow 2CHCl_3 + (CH_3COO)_2Ca + 2Ca(OH)_2 + 3CaCl_2 + 6H_2O$. May also be prepd by carefully controlled chlorination of methane: Faith, Keyes & Clark's *Industrial Chemicals*, F. A. Lowenheim, M. K. Moran, Eds. (Wiley-Interscience, New York, 4th ed., 1975) pp 266-269. Has been used as an anesthetic and in pharmaceutical preparations. Toxicity data: H. F. Smyth *et al., Am. Ind. Hyg. Assoc. J.* **23**, 95 (1962); E. T. Kimura *et al., Toxicol. Appl. Pharmacol.* **19**, 699 (1971). Review of toxicology: L. R. Pohl, *Rev. Biochem. Toxicol.* **1**, 79-108 (1979). Review of carcinogenicity studies: *IARC Monographs* **20**, 401-427 (1979). *Review:* H. D. DeShon in Kirk-Othmer *Encyclopedia of Chemical Technology* vol. 5 (Wiley-Interscience, New York, 3rd ed., 1979) pp 693-703.
Highly refractive, nonflammable, heavy, very volatile, sweet-tasting liquid; characteristic odor. d_{20}^{20} 1.484. bp 61-62°. mp −63.5°. n_D^{20} 1.4476. Forms a constant boiling mixture with 7% alc, boiling at 59°. d 1.474-1.478 for U.S.P. chloroform contg 0.5-1% ethanol as stabilizer. One ml dissolves in about 200 ml water at 25°. Misc with alcohol, benzene, ether, petr ether, carbon tetrachloride, carbon disulfide, oils. Pure chloroform is light sensitive and reagent grade chloroform usually contains 0.75% ethanol as stabilizer. *Protect from light and keep cool.* LD_{50} (14 day) orally in rats: 2.18 ml/kg (Smyth); 0.9 ml/kg (Kimura).
Human Toxicity: Inhalation of large doses may cause hypotension, respiratory and myocardial depression and death. Banned by FDA from use in drug, cosmetic and food packaging products in 1976. This substance may reasonably be anticipated to be a carcinogen: *Fourth Annual Report on Carcinogens* (NTP 85-002, 1985) p 56.
USE: As a solvent for fats, oils, rubber, alkaloids, waxes, gutta-percha, resins; as cleansing agent; in fire extinguishers to lower the freezing temp of carbon tetrachloride; in the rubber industry.

2142. Chlorogenic Acid. *3-[[3-(3,4-Dihydroxyphenyl)-1-oxo-2-propenyl]oxy]-1,4,5-trihydroxycyclohexanecarboxylic acid; 1,3,4,5-tetrahydroxycyclohexanecarboxylic acid 3-(3,4-dihydroxycinnamate);* 3-caffeoylquinic acid; 3-(3,4-dihydroxycinnamoyl)quinic acid. $C_{16}H_{18}O_9$; mol wt 354.30. C 54.24%, H 5.12%, O 40.64%. Important factor in plant metabolism. Isoln from green coffee beans: Freudenberg, *Ber.* **53**, 237 (1920). Chlorogenic acid and its isomers isochlorogenic acid and neochlorogenic acid occur also in fruit, leaves and other tissues of dicotyledenous plants: Sondheimer, *Arch. Pharm.* **293**, 721 (1960). Forms caffeic acid on hydrolysis: Fiedler, *Arzneimittel-Forsch.* **4**, 41 (1954). Structure:

Fischer, Dangschat, *Ber.* **65**, 1037 (1932); Barnes *et al., J. Am. Chem. Soc.* **72**, 4178 (1950); Corse *et al., Tetrahedron* **18**, 1207 (1962). Synthesis: Panizzi *et al., Gazz. Chim. Ital.* **86**, 913 (1956).

Hemihydrate, needles from water. Becomes anhydr at 110°. mp 208°. $[\alpha]_D^{26}$ $-35.2°$ (c = 2.8). K at 27° = 2.2 × 10^{-3}. R_f values: Fiedler, *loc. cit.* Soly in water at 25° about 4%, much more sol in hot water. Alkaline solns acquire an orange color. Freely sol in alcohol, acetone. Very slightly sol in ethyl acetate. Heating with dil HCl yields caffeic acid. Forms a black compd with iron, said to be responsible for the blackening of cut and cooked potatoes: *Chem. & Ind. (London)* **1958**, 627.

3'-Methyl ether, $C_{17}H_{20}O_9$, *3-feruloylquinic acid.* Crystals from ethyl acetate + petr ether, mp 196-197°. $[\alpha]_D^{25}$ $-42.8°$ (ethanol). uv max (ethanol): 325 nm (ϵ 19,200).

2143. Chlorogenin. *(25R)-5α-Spirostan-3β,6α-diol.* $C_{27}H_{44}O_4$; mol wt 432.62. C 74.96%, H 10.25%, O 14.79%. Isoln from bulbs of the California soap plant, amole: *Chlorogalum pomeridianum* (DC.) Kunth, *Liliaceae:* Liang, Noller, *J. Am. Chem. Soc.* **57**, 525 (1935). Chlorogenin occurs in amole as a saponin which kills or stuns fish without rendering them inedible. Structure: Marker, Rohrmann, *ibid.* **61**, 947, 3479 (1939); Marker *et al., ibid.* **62**, 2537, 3006 (1940). On hydrogenation the 3β,6β-isomer (β-chlorogenin) is produced.

Needles from methanol, mp 273-276°. $[\alpha]_{546}^{24}$ $-52°$ (chloroform or isopropanol). Less sol in methanol, more sol in isopropanol than tigogenin.

Diacetate, crystals from dil methanol, mp 154-155°.

Dibenzoate, crystals from methanol + chloroform, mp 200.5-204.5°, $[\alpha]_{546}^{26}$ $+9.5°$ (chloroform).

2144. 1-Chlorohexane. *n*-Hexyl chloride. $C_6H_{13}Cl$; mol wt 120.62. C 59.74%, H 10.86%, Cl 29.40%. $CH_3(CH_2)_4$-CH_2Cl. Prepd from 1-hexanol by treatment with fuming HCl: Henry, *Chem. Zentr.* **1905**, II, 214; with excess $SOCl_2$ or with PCl_5 + $ZnCl_2$: Clark, Streight, *Trans. Roy. Soc. Can.* [3] **23**, III, 77 (1929).

Mobile liquid. d_4^{20} 0.8780. bp_{760} 134°. n_D^{20} 1.4236 (Clark, Streight, *loc. cit.*); n_D^{20} 1.4195 (Mumford, Phillips, *J. Chem. Soc.* **1950**, 75). Insol in water. Refluxing with 10% aq NaOH decomposes 1-chlorohexane to 1-hexanol.

2145. α-Chlorohydrin. *3-Chloro-1,2-propanediol;* 3-chloro-1,2-dihydroxypropane; α-monochlorohydrin; β,β'-dihydroxyisopropyl chloride; glycerol α-monochlorohydrin; 3-chloropropylene glycol; Epibloc; $C_3H_7ClO_2$; mol wt 110.54. C 32.59%, H 6.38%, Cl 32.08%, O 28.95%. CH_2Cl-$CHOHCH_2OH$. Prepd from glycerol and HCl gas: Conant, Quayle, *Org. Syn.* **coll. vol. I**, 294 (1941).

Liquid. Sweetish taste. Tendency to turn straw color. d_4^{20} 1.3218. n_D^{20} 1.4831. bp_{760} 213° (dec); bp_{14} 114-120°; bp_{11} 115-117°. Sol in water, alcohol, ether. LD_{50} orally in mice, rats: 0.16, 0.15 g/kg, C. H. Hine *et al., Arch. Ind. Health* **14**, 250 (1956).

USE: To lower the freezing point of dynamite; in the manuf of dye intermediates. As rodent chemosterilant.

2146. Chloromethyl Methyl Ether. *Chloromethoxymethane;* methyl chloromethyl ether; monochloromethyl ether; chlorodimethyl ether; CMME; C_2H_5ClO; mol wt 80.51. C 29.84%, H 6.26%, Cl 44.03%, O 19.87%. CH_3-OCH_2Cl. Prepd by passing HCl through a mixture of formalin and methanol: C. S. Marvel, P. K. Porter, *Org. Syn.*

coll. vol. I, 377 (1941). *See also Beilstein* **1**, 580 (1918) and supplements. Commercial product usually contaminated by *sym*-dichloromethyl ether, *q.v.* Review and evaluation of studies of carcinogenicity in laboratory animals: *IARC Monographs* **4**, 239-245 (1974).

Colorless liquid, bp 59°. d_4^{20} 1.0605. n_D^{20} 1.39737.

Note: The technical grade has been listed as a known carcinogen: *Fourth Annual Report on Carcinogens* (NTP 85-002, 1985) p 46.

USE: In synthesis of chloromethylated compounds.

2147. 1-Chloro-2-methylpropene. α-Chloroisobutylene; β,β-dimethylvinyl chloride; isocrotyl chloride. C_4H_7Cl; mol wt 90.55. C 53.05%, H 7.79%, Cl 39.16%. Prepn from isobutyraldehyde: Kirrmann, *Bull. Soc. Chim. France* **1948**, 163; by isomerization of 3-chloro-2-methylpropene with 80% H_2SO_4: Backhurst *et al., J. Chem. Soc.* **1959**, 2742.

$$\underset{CH_3}{\overset{\overset{\displaystyle CH_3}{|}}{C}}=CHCl$$

Liquid, bp 68.1°. d_4^{20} 0.9186. n_D^{20} 1.4221.

USE: In organic syntheses. *Caution:* Irritates eyes, respiratory passages; has anesthetic properties.

2148. 3-Chloro-2-methylpropene. γ-Chloroisobutylene; methallylchloride; β-methallyl chloride; β-methylallylchloride. Empirical formula, etc: *see* 1-Chloro-2-methylpropene above. Prepn by chlorination of isobutylene: Burgin *et al., Ind. Eng. Chem.* **31**, 1413 (1939). Manuf: Carlson, U.S. pat. **2,494,034**, Cherniavsky, Brown, U.S. pat. **2,612,530**, and Cheney *et al.,* U.S. pat. **2,642,464** (1950, 1952, and 1953, all to Shell).

$$\underset{ClCH_2}{\overset{\overset{\displaystyle CH_3}{|}}{C}}=CH_2$$

Liquid, bp 71-72°. d_4^{15} 0.9210, d_4^{20} 0.9165 (commercial grade d_{20}^{20} 0.926-0.930). n_D^{15} 1.4318, n_D^{20} 1.4274.

USE: Insecticide, fumigant; in organic syntheses. *Caution:* Irritates eyes, respiratory passages.

2149. 1-Chloronaphthalene. α-Chloronaphthalene. C_{10}-H_7Cl; mol wt 162.61. C 73.86%, H 4.34%, Cl 21.80%. Prepd by passing chlorine into boiling naphthalene with or without solvent such as chlorobenzene and with or without catalyst such as I_2: DeWitt, Ekeley, *Univ. Colo. Stud.* **18**, 119 (1931), *C.A.* **26**, 2974 (1932). From α-naphthylamine by diazotization and Sandmeyer CuCl reaction: von Auwers, Frühling, *Ann.* **422**, 194, 200, 202 (1921); Hampson, Weissberger, *J. Chem. Soc.* **1936**, 394.

Oily liquid. Volatile with steam. d_4^{20} 1.19382. mp $-2.5°$. bp_{760} 259.3°; bp_{400} 230.8°; bp_{200} 204.2°; bp_{100} 180.4°; bp_{60} 165.6°; bp_{40} 153.2°; bp_{20} 134.4°; bp_{10} 118.6°; bp_5 104.8°; $bp_{1.0}$ 80.6°. n_D^{20} 1.63321. Sol in benzene, petr ether, alcohol.

Picrate, yellow needles, mp 137°.

Compound with 2,4,6-trinitro-*m*-cresol, $C_{17}H_{12}ClN_3O_7$, mp 78°.

Styphnate, $C_{16}H_{10}ClN_3O_8$, yellow needles from alcohol, mp 112°.

USE: As immersion liquid in the (microscopic) determn of the refractive index of crystals. Solvent for oils, fat, DDT.

2150. 2-Chloronaphthalene. β-Chloronaphthalene. C_{10}-H_7Cl; mol wt 162.61. C 73.86%, H 4.34%, Cl 21.80%. Prepd from β-naphthylamine by diazotization and Sandmeyer CuCl reaction: Scheid, *Ber.* **34**, 1813 (1901); van der Kam, *Rec. Trav. Chim.* **45**, 568 (1926); Hampson, Weissberger, *J. Chem. Soc.* **1936**, 394. By chlorination of gaseous naphthalene at 530° in the presence of iodine, ~50% of the product being 1-chloronaphthalene which is sepd by fractional freezing: Britton, Reed, U.S. pat. **1,917,822** (1933).

Monoclinic plates, leaflets from dil alcohol. mp 59.5°. bp_{760} 256°; bp_{60} 161°; bp_{20} 132.6°; bp_{11} 119.7°. Volatile with steam. $n_D^{70.7}$ 1.60787. Sol in alcohol, ether, benzene, chloroform, carbon disulfide.

Picrate, yellow needles, mp 81.5°.

2151. Chloronitrobenzene. $C_6H_4ClNO_2$; mol wt 157.56. C 45.74%, H 2.56%, Cl 22.50%, N 8.89%, O 20.31%. Prepn of *m*-isomer: W. W. Hartman, M. R. Brethen, *Org. Syn.* **coll. vol. I** (2nd ed., 1964) p 162.

m-Chloronitrobenzene, 1-chloro-3-nitrobenzene, m-nitro-chlorobenzene. Pale-yellow orthorhombic prisms from alcohol. d_4^{20} 1.534. mp 46°. bp_{760} 236°; bp_{12} 117°. Insol in water; sparingly sol in cold alcohol; freely sol in hot alcohol, chloroform, ether, carbon disulfide, glacial acetic acid.

o-Chloronitrobenzene. Yellow crystals. d 1.305. mp 32-33°. bp 245-246°. Insol in water; sol in alcohol, benzene, ether.

p-Chloronitrobenzene. Yellow crystals. d 1.520. mp 82-84°. bp 242°. Flash pt 127°. Insol in water; sparingly sol in cold alcohol, freely in boiling alcohol, ether, carbon disulfide.

USE: In dye chemistry. *Caution:* Severe poison!

2152. Chlorophacinone. *2-[(4-Chlorophenyl)phenylacetyl]-1H-indene-1,3(2H)-dione; 2-[(p-chlorophenyl)phenylacetyl]-1,3-indandione;* LM 91; Caid; Drat; Liphadione; Quick; Raviac; Rozol. $C_{23}H_{15}ClO_3$; mol wt 374.82. C 73.70%, H 4.03%, Cl 9.46%, O 12.80%. Prepn: Molho *et al.,* U.S. pat. **3,153,612** (1964 to Lipha).

Crystals, mp 140°. Absorption band at 325 nm in acetone soln. Sol in organic solvents; very sparingly sol in water. LD_{50} orally in rats: 20.5 mg/kg.

USE: Anticoagulant rodenticide.

2153. p-Chlorophenacyl Bromide. *2-Bromo-1-(4-chlorophenyl)ethanone; 2-bromo-4'-chloroacetophenone;* 4-chloro-ω-bromoacetophenone; α-bromo-p-chloroacetophenone. C_8H_6BrClO; mol wt 233.51. C 41.15%, H 2.59%, Br 34.23%, Cl 15.19%, O 6.85%. Prepn from bromoacetyl chloride, chlorobenzene, and $AlCl_3$: Collet, *Compt. Rend.* **125**, 718 (1897); by bromination of *p*-chloroacetophenone: *idem, Bull. Soc. Chim. France* [3] **21**, 69 (1899).

Needles, mp 96-96.5°. LD_{50} orally in mice: > 2000 mg/kg, Dat-Xuong *et al., Med. Exp.* **11**(3), 137 (1964).

USE: In the prepn of quaternary salts of methenamine and of chlorophenol glyoximes.

2154. Chlorophenol. C_6H_5ClO; mol wt 128.56. C 56.05%, H 3.92%, Cl 27.58%, O 12.44%. Prepn of *o*-form: Holleman, Rinkes, *Rec. Trav. Chim.* **30**, 48 (1911); Huston, Neely, *J. Am. Chem. Soc.* **57**, 2176 (1935). *cis-trans* equilibria in *o*-halophenols: Baker, *ibid.* **80**, 3598 (1958). Prepn of *p*-form: Crawford, Willson, U.S. pat. **1,910,679** (1927 to Monsanto); Neu, *Ber.* **72**, 1505 (1939); Hodgson, Norris, *J.*

Chem. Soc. **1949,** Suppl. 1, S181. Manuf: Britton, Keil, U.S. pat. **2,725,402** (1955 to Dow). Prepn of *m*-form: Acheson, Taylor, *J. Chem. Soc.* **1956**, 4727. Manuf: Stoesser, Gentry; Barnard, Meyer, U.S. pats. **2,835,707** and **2,852,567** (both 1958 to Dow). Toxicity data: Deichmann, *Fed. Proc.* **2**, 76 (1943).

m-Chlorophenol, needles, mp 33.5°. bp 214°. d_4^{45} 1.245, d_4^{78} 1.214. n_D^{40} 1.5565. Dipole moment in benzene, 2.10D. Slightly sol in cold water; sol in alcohol, ether, caustic alkali solns. LD_{50} orally in rats: 0.57 g/kg (Deichmann).

o-Chlorophenol, liquid, bp 175°. mp 9.3°. d_4^{23} 1.2573. n_D^{25} 1.5565; n 1.5473. Dipole moment in benzene, 1.33D. Slightly sol in water; freely sol in alcohol, ether, caustic alkali solns. LD_{50} orally in rats: 0.67 g/kg (Deichmann).

p-Chlorophenol, parachlorophenol. Crystals with characteristic phenolic odor, mp 43.2-43.7°. bp 220°. d_4^{78} 1.2238. n_D^{55} 1.5419; n_D^{40} 1.5579. Dipole moment in benzene, 2.22D. Sparingly sol in water, liquid petrolatum; very sol in alcohol, glycerin, ether, chloroform, fixed and volatile oils: *U.S.P.* **XXI**, 1467. LD_{50} orally in rats: 0.67 g/kg (Deichmann).

Caution: May cause tremors, convulsions, dyspnea, coma. Irritating to skin. *See:* Patty's *Industrial Hygiene and Toxicology* Vol. **2A**, G. D. Clayton, F. E. Clayton, Eds. (Wiley-Interscience, New York, 3rd ed., 1981) pp 2612-2615.

THERAP CAT: Parachlorophenol as antiseptic.

2155. Chlorophyll. The green pigment of plants. Higher plants and green algae contain chlorophyll a and chlorophyll b in the approx ratio of 3:1. Chlorophyll c is found together with chlorophyll a in many types of marine algae: Jeffrey, *Biochem. J.* **86**, 313 (1963). Red algae *(Rhodophyta)* contain principally chlorophyll a and also chlorophyll d: Manning, Strain, *J. Biol. Chem.* **151**, 1 (1943). Isoln by chromatography: Tswett, *Ber. Deutsch. Bot. Ges.* **24**, 316, 385 (1906); Willstätter, Stoll, *Investigations on Chlorophyll* (transl by Schertz and Merz: Lancaster, 1928); Schertz, *Ind. Eng. Chem.* **30**, 1073 (1938); Fischer-Orth-Stern, *Die Chemie des Pyrrols* Vol. **II**, part 2 (Leipzig, 1940); Zechmeister, Cholnoky, *Principles and Practice of Chromatography* (New York, 1943). Industrial large-scale isoln processes: Judah *et al., Ind. Eng. Chem.* **46**, 2262 (1954). Structure: Fischer-Orth-Stern, *loc. cit.;* Ficken *et al., J. Chem. Soc.* **1956**, 2273. Total synthesis of chlorophyll a: Woodward *et al., J. Am. Chem. Soc.* **82**, 3800 (1960); *Angew. Chem.* **72**, 651 (1960); Strell *et al., ibid.* 169; Woodward, *Pure Appl. Chem.* **2**, 383 (1963). Abs config of chlorophylls a and b: Brockmann, *Ann.* **754**, 139 (1971); Brockmann, Bode, *Ann.* **1974** (7), 1017. ^{13}C-NMR study of chlorophyll a: S. Lötjönen, P. H.

Hynninen, *Org. Mag. Reson.* **16**, 304 (1981); of chlorophyll b: N. Risch, H. Brockmann, *Tetrahedron Letters* **24**, 173 (1983). Review of syntheses: Johnson, *Sci. Progr.* **49**, 77 (1961). Comprehensive reviews with bibliography: Stoll, Wiedemann, *Fortschr. Chem. Org. Naturst.* **1**, 159-254 (1938); *Fortschr. Chem. Forsch.* **2**, 538 (1952); *The Chlorophylls*, L. P. Vernon, G. R. Seely, Eds. (Academic Press, NewYork, 1966) 679 pp; Inhoffen *et al.*, *Fortschr. Chem. Org. Naturst.* **26**, 284-298 (1968); Inhoffen, *Pure Appl. Chem.* **17**, 443-460 (1968).

Chlorophyll a, $C_{55}H_{72}MgN_4O_5$. R = CH₃. Sepn and purification: Anderson, Calvin, *Nature* **194**, 285 (1962). Waxy blue-black microcrystals, usually aggregates of thin, lancet-like leaflets, mp 117-120°. $[\alpha]_D^{20}$ —262° (acetone). Absorption max (ether): 660, 613, 577, 531, 498, 429, 409 nm. Freely sol in ether, ethanol, acetone, chloroform, carbon disulfide, benzene. Sparingly sol in cold methanol. Practically insol in petr ether. The alcoholic soln is blue-green with a deep-red fluorescence.

Chlorophyll b, $C_{55}H_{70}MgN_4O_6$. R = CHO. Waxy blue-black microcrystals. Sinters between 86° and 92°, becomes a viscous liquid at 120-130° and then begins to puff up in large bubbles. $[\alpha]_D^{20}$ —267° (acetone-methanol). Absorption max (ether): 642, 593, 565, 545, 453, 427 nm Sparingly sol in petr ether, ligroin, cold methanol. Freely sol in abs alcohol, ether. The ether soln has a brilliant green color. Solns with other organic solvents are usually green to yellowish-green with red fluorescence.

Chlorophyll c. Structure: Dougherty *et al.*, *J. Am. Chem. Soc.* **88**, 5037 (1966). Reddish-black hexagonal bipyramids or four-sided plates from dil ethanol. Absorption max (acetone): 628, 580, 442 nm ($E_{1cm}^{0.1\%}$ 15.8, 10.7, 115.9). Sol in methanol, ethanol, ethyl acetate; practically insol in ether, acetone.

Chlorophyll d, $C_{54}H_{70}MgN_4O_6$. Chlorophyll a with the 9-vinyl group replaced by a formyl group: Holt, Morley, *Can. J. Chem.* **37**, 507 (1959); Holt, *Can. J. Bot.* **39**, 327 (1961). Absorption max (ether): 686, 445 nm.

The chlorophyll of commerce is an intensely dark-green, aq, alc, or oil soln. Careful alkaline hydrolysis of chlorophyll opens the cyclopentanone ring and replaces the methyl and phytyl ester groups with Na or K; the resulting salts are called *chlorophyllins* and are water sol, e.g. sodium magnesium chlorophyllin, $C_{34}H_{31}N_4Na_3MgO_6$. Acid treatment of chlorophyll removes the Mg replacing it with H_2 which can be replaced with other metals, e.g. iron pheophytin, $C_{55}H_{72}$-FeN_4O_5, sol in oil. Comprehensive review with 89 refs: "Chlorophyll Derivatives, Their Chemistry, Commercial Preparation and Uses" by A. C. Kephart in *Econ. Bot.* **9**, 3-38 (1955). Exhaustive analysis and description of commercial chlorophyllin prepns: Strell *et al.*, *Arzneimittel-Forsch.* **5**, 640 (1955); **6**, 8 (1956). Trademarks: *Biophyll, Chloresium, Chlorofolin, Darotol, Ennds, Exodor-Grun, Nullo Chlorophyll, Stozzon-Chlorophyll*.

USE: To color soaps, oils, fats, waxes, confectionery, preserves, liquors, cosmetics, perfumes. Source of phytol. For dyeing leather. As sensitizer for color film. Has been used as antiknock agent in gasoline; as accelerator in the vulcanizing of rubber; in deodorizers.

THERAP CAT: Deodorant.

THERAP CAT (VET): Has been used orally to reduce odors, and topically to promote healing of skin lesions.

2156. Chloropicrin. *Trichloronitromethane;* acquinite; nitrochloroform; Larvacide 100; Picfume. CCl_3NO_2; mol wt 164.39. C 7.31%, Cl 64.71%, N 8.52%, O 19.47%. First prepd in 1848 by Stenhouse from picric acid and bleach powder. Review of prepn, properties, physiological action, and uses: Jackson, *Chem. Rev.* **14**, 251 (1934). Manuf from nitromethane and alkaline hypochlorite: Wilhelm, U.S. pat. **3,106,588** (1963).

Slightly oily liquid, intense odor; bp_{757} 112°. mp —64°; mp —69.2° (corr). d_4^{20} 1.6558, d_4^{25} 1.6483. n_D^{20} 1.4611, n_D^{25} 1.4596. Dipole moment in heptane or benzene, 1.80D. Practically insol in water (0.2272 g and 0.1621 g/100 ml at 0° and 25°); miscible with benzene, abs alc, carbon disulfide; sol in ether.

USE: Disinfecting cereals and grains; in synthesis, esp in manuf of methyl violet; fumigant; soil insecticide; war gas. *Caution:* Produces severe sensory irritation in upper respiratory passages. Has strong lacrimatory properties, produces

increased sensitivity after frequent exposures. Orally causes severe nausea, vomiting, colic, diarrhea. Potent skin irritant: Fairhall, *Industrial Toxicology* (Hafner, New York, 1969) p 196.

2157. Chloroprednisone. *6α-Chloro-17,21-dihydroxy-pregna-1,4-diene-3,11,20-trione;* 6α-chloro-1,4-pregnadiene-17α,21-diol-3,11,20-trione. $C_{21}H_{25}ClO_5$; mol wt 392.89. C 64.20%, H 6.41%, Cl 9.03%, O 20.36%. Prepn: Batres *et al.*, Ger. pat. **1,079,042** (1960 to Syntex). Prepn of the 21-acetate by selenium dioxide dehydrogenation of 6α-chlorocortisone 21-acetate: Ringold *et al.*, *J. Am. Chem. Soc.* **80**, 6464 (1958); Ringold, Rosenkranz, U.S. pat. **2,957,895** (1960 to Syntex).

21-Acetate, $C_{23}H_{27}ClO_6$, *Topilan.* Crystals from acetone + hexane, mp 217-219°. uv max (ethanol): 237 nm (log ε 4.19). $[\alpha]_D$ +144° (chloroform).

THERAP CAT: Topical glucocorticoid.

2158. 2-Chloroprocaine Hydrochloride. *4-Amino-2-chlorobenzoic acid 2-(diethylamino)ethyl ester monohydrochloride;* 2-chloro-4-aminobenzoic acid diethylaminoethyl ester hydrochloride; Nesacaine. $C_{13}H_{20}Cl_2N_2O_2$; mol wt 307.22. C 50.82%, H 6.56%, N 9.12%, Cl 23.08%, O 10.42%. Description: Hädicke, *Pharm. Zentralh.* **94**, 384 (1955).

Crystals, mp 176-178° (microstage). Bitter taste. Slowly sol in water. Soly in water at 20° about one gram in 22 ml. Aq solns are just acid to litmus and turn yellow on standing. Soly in 95% ethanol about one gram in 100 ml.

THERAP CAT: Local anesthetic.

2159. β-Chloropropionic Acid. *3-Chloropropanoic acid.* $C_3H_5ClO_2$; mol wt 108.53. C 33.20%, H 4.64%, Cl 32.67%, O 29.48%. CH_2ClCH_2COOH. Prepd by the hydrolysis of ethylene cyanohydrin with hydrochloric acid; by the oxidation of β-chloropropionaldehyde or of trimethylene chlorohydrin by nitric acid: Moureu, Chaux, *Org. Syn.* **8**, 54 (1928); Powell, *ibid.* 58; Paal, Lobeck, *Ber.* **64**, 2142 (1931).

Leaflets from ligroin. Somewhat hygroscopic. mp 41°. bp_{765} 200°; bp_{35} 127°; bp_{25} 124°; bp_{12} 108°. K at 25° = 1.01 × 10⁻⁴. Freely sol in water, alc, chloroform, slightly less in ether.

Methyl ester, $C_4H_7ClO_2$, from β-chloropropionic acid with methanol and HCl or from acrylic acid chloride and methanol. d_4^0 1.198. bp 155-157°.

2160. β-Chloropropionitrile. *3-Chloropropanenitrile;* 3-chloropropanonitrile. C_3H_4ClN; mol wt 89.53. C 40.25%, H 4.50%, Cl 39.60%, N 15.65%. $ClCH_2CH_2CN$. Prepd from acrylonitrile and hydrogen chloride or bromide: Moureu, Clarke, *Bull. Soc. Chim. France* [4] **27**, 905 (1920); Stewart, Clarke, *J. Am. Chem. Soc.* **69**, 714 (1947); Shirley, *Preparation of Organic Intermediates* (Wiley, New York, 1951) p 82.

Liquid. Acrid, characteristic odor. *Poisonous!* mp —51°. d_4^{25} 1.1363. bp_{760} 176° (dec); bp_{200} 132° (dec); bp_{50} 95.2°; bp_5 46.0°. Flash pt 168°F (75.5°C). n_D^{25} 1.4341. Begins to dec when heated above 130° evolving HCl. Absorbs strongly in the infrared. Transparent to uv above 220 nm. Soly in water at 25°: 4.5 g/100 ml. Soly of water in β-chloropropionitrile at 25°: 2.2 ml/100 g. Miscible with ethanol, ether, acetone, benzene, carbon tetrachloride. LD_{50} orally in mice, rats: 9, 100 mg/kg, Fassett in *Industrial Hygiene and Toxicology* vol. 2, F. A. Patty, Ed. (Interscience, New York, 2nd ed., 1962) pp 2025-2026.

USE: In pharmaceutical and polymer synthesis. Combines the reactivity of a nitrile and an alkyl halide. Because of the cyano group the chlorine atom is more reactive than in ordinary alkyl halides. *Caution:* Exposure by any route should be avoided. Somewhat less hazardous than acrylonitrile because of lower vapor pressure. Readily penetrates skin to produce systemic cyanide poisoning, death.

2161. 6-Chloropurine. $C_5H_3ClN_4$; mol wt 154.56. C 38.86%, H 1.94%, Cl 22.95%, N 36.26%. Prepd by the action of phosphorus oxychloride on hypoxanthine in *N,N*-dimethylaniline: Bendich *et al., J. Am. Chem. Soc.* **76,** 6073 (1954). Antineoplastic activity *in vivo:* A. C. Sartorelli, B. A. Booth, *Experientia* **21,** 457 (1965).

Blunt needles from water, dec 175-177° (hot stage preheated to 170°). Also reported as not melted at 290°. uv max (pH 5.2): 265 nm (ε 9120); (pH 13): 274 nm (ε 8790). Soly in water: 0.5% at 20°, also reported as 1 g/182 ml at 24°. Sol in ether, dimethylformamide.

2162. Chloropyramine. *N-[(4-Chlorophenyl)methyl]-N',N'-dimethyl-N-2-pyridinyl-1,2-ethanediamine;* 2-[(*p*-chlorobenzyl)[2-(dimethylamino)ethyl]amino]pyridine; *N*-dimethylaminoethyl-*N*-*p*-chlorobenzyl-α-aminopyridine; *N,N*-dimethyl-*N'*-(*p*-chlorobenzyl)-*N'*-(2-pyridyl)ethylenediamine; halopyramine; Chloropyribenzamine; Synpen. $C_{16}H_{20}ClN_3$; mol wt 289.82. C 66.31%, H 6.96%, Cl 12.23%, N 14.50%. Prepn: Phillips, Cates, U.S. pat. **2,607,778** (1952 to Merck & Co.); Howard, U.S. pat. **2,569,314** (1951 to Am. Cyanamid); **Swiss** pats. **264,754; 266,234; 266,235** (1950); **Brit.** pat. **651,596** (1951) (all to Geigy).

Light yellow, viscous, oily liquid. Pungent odor. $bp_{0.2}$ 154-155°.

Hydrochloride, $C_{16}H_{21}Cl_2N_3$, *Alegan S, Synopen.* Crystals from acetone, mp 172-174°.

THERAP CAT: Antihistaminic.

2163. Chloroquine. *N⁴-(7-Chloro-4-quinolinyl)-N¹,N¹-diethyl-1,4-pentanediamine;* 7-chloro-4-(4-diethylamino-1-methylbutylamino)quinoline; SN 7618; RP 3377; Aralen; Capquin; Nivaquine B; Sanoquin; Artrichin; Bipiquin; Reumachlor; Bemaphate; Résoquine. $C_{18}H_{26}ClN_3$; mol wt 319.89. C 67.59%, H 8.19%, Cl 11.08%, N 13.14%. Prepd by the condensation of 4,7-dichloroquinoline with 1-diethylamino-4-aminopentane: **Ger.** pat. **683,692** (1939); H. Andersag *et al.,* U.S. pat. **2,233,970** (1941 to Winthrop); Surrey, Hammer, *J. Am. Chem. Soc.* **68,** 113 (1946). *Review:* Hahn in *Antibiotics* vol. 3, J. W. Corcoran, F. E. Hahn, Eds. (Springer-Verlag, New York, 1975) pp 58-78. Comprehensive description: D. D. Hong in *Analytical Profiles of Drug Substances* vol. 5, K. Florey, Ed. (Academic Press, New York, 1976) pp 61-85. Comparative clinical trial with dapsone in rheumatoid arthritis: P. D. Fowler *et al., Ann. Rheum. Dis.* **43,** 200 (1984); with penicillamine: T. Gibson *et al., Brit. J. Rheumatol.* **26,** 279 (1987).

mp 87°.

Diphosphate, $C_{18}H_{32}ClN_3O_8P_2$, *Arechin, Avloclor, Imagon, Malaquin, Resochin, Silbesan, Tresochin.* Bitter, colorless crystals. Dimorphic. One modification, mp 193-195°; the other, mp 215-218°. Freely sol in water; pH of 1% soln about 4.5; less sol at neutral and alkaline pH. Stable to heat in solns of pH 4.0 to 6.5. Practically insol in alcohol, benzene, chloroform, ether.

Sulfate, $C_{18}H_{28}ClN_3O_4S$, *Nivaquine.*

THERAP CAT: Antimalarial; antiamebic. Antirheumatic. Lupus erythematosus suppressant.

2164. Chloroselenic Acid. $ClHO_3Se$; mol wt 163.43. Cl 21.70%, H 0.62%, O 29.37%, Se 48.32%. $HClSeO_3$. Prepd by the reaction of hydrogen chloride and selenium trioxide: Le Geyt *et al., J. Chem. Soc.* **123,** 2870 (1923).

Liquid. Fumes in air. d 2.26. mp −46°. Dec by heat. Sol in water and selenium oxychloride. Dec in alc. Insol in ether, benzene, chloroform, carbon tetrachloride.

2165. *N*-Chlorosuccinimide. *1-Chloro-2,5-pyrrolidinedione;* succinchlorimide. $C_4H_4ClNO_2$; mol wt 133.54. C 35.97%, H 3.02%, Cl 26.55%, N 10.49%, O 23.96%. Prepn from succinimide: Hirst, Macbeth, *J. Chem. Soc.* **121,** 2169 (1922); Zimmer, Audrieth, *J. Am. Chem. Soc.* **76,** 3856 (1954). Crystal structure: Brown, *Acta Cryst.* **9,** 193 (1956); **14,** 711 (1961).

Orthorhombic crystals, mp 150-151°. Odor of chlorine. Acid to litmus (1:50 aq soln). One gram dissolves in about 70 ml water, 150 ml alcohol, 50 ml benzene. Sparingly sol in ether, chloroform, carbon tetrachloride. Liberates iodine from potassium iodide solns, and bromine from sodium bromide solns. MLD orally in rats: 2.7 g/kg, Stohlman, Smith, *U.S. Publ. Health Repts.* **59,** 541 (1944).

USE: Chlorinating agent.

2166. Chlorosulfonic Acid. Sulfuric chlorohydrin; chlorosulfuric acid. $ClHO_3S$; mol wt 116.53. Cl 30.43%, H 0.87%, O 41.19%, S 27.52%. $SO_2(OH)Cl$. Prepd by the reaction of HCl gas with SO_3: Simon, Kratsch, *Z. Anorg. Allgem. Chem.* **242,** 369 (1939); Briggs, U.S. pat. **1,442,335** (1922 to General Chem.). Purification: Kaplar, Shechter, *Inorg. Syn.* **4,** 52 (1953). Review: H. O. Burrus in Kirk-Othmer *Encyclopedia of Chemical Technology* vol. 5 (Wiley-Interscience, New York, 3rd ed., 1979) pp 873-880.

Colorless or slightly yellow, very corrosive liquid, causes severe burns; fumes in air; pungent odor. d_{20}^{20} 1.76-1.77; d_4^0 1.784; d_4^{20} 1.753. mp −80°. bp_{755} 151-152°; bp_{19} 74-75°; $bp_{2.4}$ 60-64°, n_D^{14} 1.437. On dropping into water dec with explosive violence. *Keep tightly closed.* When used for the prepn of sulfate esters, the common solvent is pyridine. Other solvents are liquid sulfur dioxide and dichloroethane.

USE: Manuf sulfone compds, saccharin. As chlorosulfonating and condensing agent in organic syntheses. *Caution:* Highly irritating and corrosive to eyes, skin, mucous membranes.

2167. Chlorothalonil. *2,4,5,6-Tetrachloro-1,3-benzenedicarbonitrile; tetrachloroisophthalonitrile; m-tetrachlorophthalodinitrile;* 2,4,5,6-tetrachloro-1,3-dicyanobenzene; 1,3-dicyano-2,4,5,6-tetrachlorobenzene; chlorthalonil; DAC-2787; Daconil 2787; Bravo; Forturf; Exotherm Termil; Termil. $C_8Cl_4N_2$; mol wt 265.89. C 36.14%, Cl 53.33%, N 10.53%. Fungicidal properties first described by N. J. Turner *et al., Contrib. Boyce Thompson Inst.* **22,** 303 (1964).

Prepn: R. D. Battershell, H. Bluestone, U.S. pat. 3,290,353 (1966 to Diamond Alkali); R. M. Bimber, U.S. pat. 3,652,-637 (1972 to Diamond Shamrock). Toxicity studies in mice: H. Yoshikawa, K. Kawai, *Ind. Health* **4**, 11 (1966).

Crystals. d_4^{25} 1.7. mp 250-251°. b_{760} 350°. Vapor press < 0.01 at 40°. Practically insol in water (soly at room temp reported as 0.6 ppm). Soly in organic solvents at 25° (w/w): xylene 8%, cyclohexane 3%, acetone 2%, kerosine < 1.0%. LD_{50} orally in rats > 10.0 g/kg, Turner *et al., loc. cit.*

USE: Fungicide, bactericide, nematocide. Agricultural and horticultural fungicide.

2168. Chlorothen. *N-[(5-Chloro-2-thienyl)methyl]-N',N'-dimethyl-N-2-pyridinyl-1,2-ethanediamine; 2-[(5-chloro-2-thenyl)(2-dimethylaminoethyl)amino]pyridine; N,N-dimethyl-N'-(2-pyridyl)-N'-(5-chloro-2-thenyl)ethylenediamine; N,N-dimethyl-N'-(α-pyridyl)-N'-(2-methyl-5-chlorothienyl)ethylenediamine; N-5-chloro-2-thienylmethyl-N',N'-dimethyl-N-2-pyridylethylenediamine; chloropyrilene; chloromethapyrilene; chlorothenylpyramine.* $C_{14}H_{18}ClN_3S$; mol wt 295.85. C 56.84%, H 6.13%, Cl 11.98%, N 14.20%, S 10.84%. Prepd by the condensation of 5-chloro-2-thenyl chloride and N,N-dimethyl-N'-(2-pyridyl)ethylenediamine in the presence of sodium or potassium amide: Clapp *et al., J. Am. Chem. Soc.* **69**, 1549 (1947).

Liquid. $bp_{1.0}$ 155-156°. Strong base.

Hydrochloride, $C_{14}H_{19}Cl_2N_3S$, *Thenclor.* Crystals, mp 106-108°. Freely sol in water.

Citrate, $C_{20}H_{26}ClN_3O_7S$, *Tagathen.* Crystals, mp 112-116°. On further heating gradually solidifies and remelts 125-140° (dec). uv max: 240 nm ($E_{1cm}^{1\%}$ 390-410). One gram dissolves in 35 ml water, in about 65 ml alc. Practically insol in ether, chloroform, benzene. pH of 1% aq soln 3.9 to 4.1. LD_{50} i.p. in mice: 105 mg/kg, Castillo *et al., J. Pharmacol. Exp. Ther.* **96**, 388 (1949).

THERAP CAT: Antihistaminic.

2169. Chlorothiazide. *6-Chloro-2H-1,2,4-benzothiadiazine-7-sulfonamide 1,1-dioxide; 6-chloro-7-sulfamoyl-1H-1,2,4-benzothiadiazine 1,1-dioxide; 6-chloro-7-sulfamyl-1,2,4-benzothiadiazine 1,1-dioxide; Alurene; Chlorosal; Chlorurit; Chlotride; Clotride; Diuresal; Diuril; Diurilix; Diurite; Exuril; Urinex; Flumen; Minzil; Neo-Dema; Ro-Chlorozide; Salisan; Salunil; Saluretil; Saluric; Warduzide; Yadulan.* $C_7H_6ClN_3O_4S_2$; mol wt 295.72. C 28.43%, H 2.04%, Cl 11.99%, N 14.21%, O 21.64%, S 21.68%. Synthesis: Novello, Sprague, *J. Am. Chem. Soc.* **79**, 2028 (1957); Novello, Hinkley, U.S. pats. **2,809,194; 2,937,169** (1957, 1960, both to Merck & Co.).

Crystals, dec 342.5-343°. Soly in water: about 0.400 g/l (pH 4); 0.65 g/l (pH 7). Soluble in alkaline aq solns with decompn upon standing or heating.

Sodium salt, $C_7H_5ClN_3NaO_4S_2$, *6-chloro-7-sulfamoyl-2H-1,2,4-benzothiadiazin-2-ylsodium 1,1-dioxide, Diuril Lyovac, Lyovac Diuril.*

THERAP CAT: Diuretic, antihypertensive.
THERAP CAT (VET): Diuretic.

2170. Chlorothricin. $C_{50}H_{63}ClO_{16}$; mol wt 955.50. C 62.85%, H 6.65%, Cl 3.71%, O 26.79%. Macrolide antibiotic active against gram-positive bacteria. Isolated together with its dechloro analog from Tü 99, a strain of *Streptomyces antibioticus:* Keller-Schierlein *et al., Helv. Chim. Acta* **52**, 127 (1969). Degradation products studies: Muntwyler *et al., ibid.* **53**, 1544 (1970). Structure: Muntwyler, Keller-Schierlein, *ibid.* **55**, 2071 (1972). Non-competitive inhibitor of pyruvate carboxylase: P. W. Schindler, H. Zaehner, *Arch. Mikrobiol.* **82**, 66, (1972); P. W. Schindler, M. C. Scrutton, *ibid.* **55**, 543 (1975). Biosynthetic studies: O. Mascaretti *et al., J. Nat. Prod.* **42**, 455 (1979); *eidem, Biochemistry* **20**, 919 (1981). Partial synthesis of the aglycone, *chlorothricolide:* R. E. Ireland, W. J. Thompson, *J. Org. Chem.* **44**, 3041 (1979); R. E. Ireland *et al., ibid.* **46**, 4863 (1981); W. R. Roush, S. E. Hall, *J. Am. Chem. Soc.* **103**, 5200 (1981). Alternate synthetic approach: R. E. Ireland, M. D. Varney, *J. Org. Chem.* **51**, 635 (1986). *Review:* Keller-Schierlein, *Fortschr. Chem. Org. Naturst.* **30**, 394-396 (1973); H. G. Floss, C.-J. Chang, *Antibiotics* **vol. IV**, J. W. Corcoran, Ed. (Springer-Verlag, New York, 1981) pp 193-214.

Colorless crystals from methylene chloride-methyl acetate, mp 206-207°. uv max (alc): 222, 260 nm (log ∈ 4.20, 3.81); (0.01N alc KOH): 221, 259 nm (log ∈ 4.09, 3.95). Dibasic acid; pK: 5.01, 7.91. (The above data applies to a 5:1 mixture with dechlorothricin, Muntwyler, Keller-Schierlein, *loc. cit.*). Slightly sol in water; readily sol in organic solvents of medium polarity.

2171. Chlorothymol. *4-Chloro-5-methyl-2-(1-methylethyl)phenol; 6-chlorothymol; 4-chlorothymol; 1-methyl-3-hydroxy-4-isopropyl-6-chlorobenzene; 6-chloro-4-isopropyl-1-methyl-3-phenol.* $C_{10}H_{13}ClO$; mol wt 184.66. C 65.04%, H 7.10%, Cl 19.20%, O 8.66%. Made by the action of sulfuryl chloride on thymol in CCl_4: Satriana *et al., J. Am. Pharm. Assoc.* **39**, 135 (1950); H. Pahlicke, **Ger.** pat. **905,738** (1954 to Diwag), *C.A.* **52**, 16294i (1958). Antifungal properties: M. Iannarone, *Drug Standards* **25**, 190 (1957).

Crystals. mp 62-64° (59-61° *N.F.* **XII**). One gram dissolves in about 1000 ml water, 0.5 ml alcohol, 2 ml benzene,

2 ml chloroform, 1.5 ml ether, about 10 ml petr ether; also sol in dil aq NaOH.

2172. Chlorotoluene. C_7H_7Cl; mol wt 126.58. C 66.42%, H 5.57%, Cl 28.01%. Prepn from diazotized toluidine: Neogi, Mitra, *J. Chem. Soc.* **1928**, 1332; Marvel, McElvain, *Org. Syn. coll. vol.* **I**, 170 (1941). Manuf of *o*- and *p*-isomers by catalytic chlorination of toluene: Di Bella, U.S. pat. **3,000,975** (1959 to Heyden Newport Chem.). Absorption spectrum of *m*- and *p*-isomers: Baly, *J. Chem. Soc.* **99**, 1704 (1911): of *m*- and *o*-isomers: Purvis, *ibid.* 1704. Metabolism of *o*-isomer by bacteria: P. A. Vandenbergh *et al., Appl. Environ. Microbiol.* **42**, 737 (1981); by rats: G. B. Quistad *et al., J. Agr. Food Chem.* **31**, 1158 (1983). Brief review: S. Gelfand in Kirk-Othmer *Encyclopedia of Chemical Technology* vol. **5** (Wiley-Interscience, New York, 3rd ed., 1979) pp 819-827.

m-Chlorotoluene, 1-chloro-3-methylbenzene. Liquid, bp 161.75°. $d_4^{18.7}$ 1.0760. mp −47.8°. n_D^{20} 1.5218.

o-Chlorotoluene, 1-chloro-2-methylbenzene. Liquid, bp 158.97°. Vapor harmful. d_4^{20} 1.0826. mp −35.59°. n_D^{20} 1.5258. Volatile with steam. Slightly sol in water; freely sol in alcohol, benzene, chloroform, ether.

p-Chlorotoluene, 1-chloro-4-methylbenzene. Liquid, bp 162.4°. d_4^{20} 1.0697. mp 7.5°. n_D^{20} 1.5211. Slightly sol in water; sol in alcohol, benzene, chloroform, ether.

USE: Solvent; dyestuff intermediate; in organic syntheses.

2173. Chlorotrianisene. *1,1′,1″-(1-Chloro-1-ethenyl-2-ylidene)tris[4-methoxybenzene]; chlorotris(p-methoxyphenyl)ethylene;* tri-*p*-anisylchloroethylene; tris(*p*-methoxyphenyl)chloroethylene; Hormonisene; Merbentul; Tace. $C_{23}H_{21}$-ClO_3; mol wt 380.86. C 72.53%, H 5.56%, Cl 9.31%, O 12.60%. Prepd by reacting tri-*p*-anisylethylene or tri-*p*-anisylethanol with Cl in an inert solvent: Basford and I.C.I., Brit. pat. **561,508** (1944). Synthesis from *p*-(*p*-anisoyl)anisole: Shelton, Van Campen, Jr., U.S. pat. **2,430,891** (1947 to Merrell).

Crystals from methanol, mp 114-116°. Softens at 108°. uv max (chloroform): 310 nm, ($E_{1cm}^{1\%}$ 423); min 278 nm. Practically insol in water. Soly in alcohol: 0.28 g/100 ml; in ether: 3.6 g/100 ml. Also sol in glacial acetic acid, acetone, chloroform, carbon tetrachloride, benzene, vegetable oils.

THERAP CAT: Estrogen.

2174. Chlorotris(triphenylphosphine)rhodium. Wilkinson's catalyst. $C_{54}H_{45}ClP_3Rh$; mol wt 925.22. C 70.10%, H 4.90%, Cl 3.83%, P 10.04%, Rh 11.12%. $ClRh[P(C_6H_5)_3]_3$. Prepd from excess triphenylphosphine and $RhCl_3.3H_2O$ in ethanol: Young *et al., Chem. Commun.* **1965**, 131; Bennett, Longstaff, *Chem. & Ind. (London)* **1965**, 846; Osborn *et al., J. Chem. Soc. (A)* **1966**, 1711; Osborn, Wilkinson, *Inorg. Syn.* **10**, 67 (1967). Crystal structure: M. J. Bennett, P. B. Donaldson, *Inorg. Chem.* **16**, 655 (1977).

Dark, burgundy-red crystals from ethanol. Orange crystals from concd ethanol soln refluxed for five minutes; converted to red form on further refluxing. mp 157-158° (Osborn *et al.*); 138° (dec) (Bennett, Longstaff). Soly in chloroform, dichloromethane (25°): ~20 g/l; in benzene, toluene (25°): ~2 g/l; much less sol in acetic acid, acetone and other ketones, methanol, lower aliphatic alcohols. Insol in paraf-

fins, cyclohexane. Reacts with donor solvents such as pyridine, dimethyl sulfoxide, acetonitrile.

USE: Homogeneous catalyst.

2175. Chloroxine. *5,7-Dichloro-8-quinolinol;* 5,7-dichloro-8-hydroxyquinoline; Capitrol. $C_9H_5Cl_2NO$; mol wt 214.06. C 50.50%, H 2.35%, Cl 33.13%, N 6.54%, O 7.47%. Prepd by chlorinating 8-quinolinol: Hebebrand, *Ber.* **21**, 2977 (1888), F. J. Welcher, *Organic Analytical Reagents* vol. **I** (Van Nostrand, 1947) pp 328-329.

Crystals from alc, mp 179-180°. Soluble in benzene, acetone; slightly sol in cold alcohol, acetic acid; readily sol in sodium and potassium hydroxides and in acids, forming yellow solns.

Mixture with 5-chloro-8-quinolinol and 7-chloro-8-quinolinol, *see* Halquinol.

USE: As an analytical reagent.

THERAP CAT: Antiseborrheic. In mixtures as topical antiinfective.

2176. Chloroxylenol. *4-Chloro-3,5-dimethylphenol;* *p*-chloro-*m*-xylenol; 4-chloro-3,5-xylenol; parachlorometaxylenol; 2-chloro-*m*-xylenol; 2-chloro-5-hydroxy-*m*-xylene; 2-chloro-5-hydroxy-1,3-dimethylbenzene; Benzytol; Dettol. C_8H_9ClO; mol wt 156.61. C 61.35%, H 5.79%, Cl 22.64%, O 10.22%. Prepd by treating 3,5-dimethylphenol with Cl_2 or SO_2Cl_2: Lesser, Gad, *Ber.* **56**, 974, 976 (1923); von Auwers *et al., Chem. Zentr.* **1924**, II, 2267, *C.A.* **19**, 2339 (1925); Gladden, Cocker, U.S. pat. **2,350,677** (1944).

Crystals from benzene, mp 115.5°. Phenolic odor. Volatile with steam. bp 246°. One gram dissolves in 3 liters of water at 20°. More stable in hot water. Soluble in 1 part of 95% alcohol, ether, benzene, terpenes, fixed oils, in solns of alkali hydroxides.

USE: Antiseptic and germicide; for mildew prevention. Claimed to be about 60 times as potent as phenol.

THERAP CAT: Antibacterial, topical and urinary antiseptic.

THERAP CAT (VET): Topical antiseptic.

2177. Chlorozotocin. *2-[[[(2-Chloroethyl)nitrosoamino]carbonyl]amino]-2-deoxy-D-glucose;* 2-[3-(2-chloroethyl)-3-nitrosoureido]-2-deoxy-D-glucopyranose; 1-(2-chloroethyl)-1-nitroso-3-(D-glucos-2-yl)urea; DCNU; NSC-178248. $C_9H_{16}ClN_3O_7$; mol wt 313.69. C 34.46%, H 5.14%, Cl 11.30%, N 13.40%, O 35.70%. Chloroethylnitrosourea derivative with antitumor activity. Similar to carmustine, lomustine, nimustine, ranimustine, *q.q.v*; 2-chloroethyl analog of streptozotocin, *q.v.* Synthesis: H. D. Burns *et al., Org. Prep. Proced. Int.* **6**, 259 (1974); T. P. Johnston *et al., J. Med. Chem.* **18**, 104 (1975). Pharmacology: T. Anderson *et al., Cancer Res.* **35**, 761 (1975); P.S. Schein *et al., Cancer Treatment Rep.* **60**, 801 (1976). Decomposition in aqueous media: J. A. Montgomery *et al., J. Med. Chem.* **18**, 568 (1975).

Ivory colored crystals, mp 147-148° (dec with the evolution of gas), H. D. Burns, N. D. Heindel, *loc. cit.* Also reported as mp 140-141° (dec), T. P. Johnston *et al., loc. cit.* Sol in water.

THERAP CAT: Antineoplastic.

2178. Chlorphenesin. *3-(4-Chlorophenoxy)-1,2-propanediol;* p-chlorophenyl α-glyceryl ether; Adermykon; Mycil. $C_9H_{11}ClO_3$; mol wt 202.64. C 53.34%, H 5.47%, Cl 17.50%, O 23.69%. Prepd by condensing equimol amts of p-chlorophenol and glycidol in the presence of a tertiary amine or a quaternary ammonium salt as catalyst: Bradley, Forrest, **Brit.** pat. **628,497** (1949 to British Drug Houses).

Crystals, mp 77-79°. Soly in water is less than 1%, may be increased by the addition of solubilizers such as ethylurea or propylene glycol: Berger *et al.,* **U. S.** pat. **2,468,423** (1949 to British Drug Houses).

THERAP CAT: Topical antifungal.

2179. Chlorphenesin Carbamate. *3-(4-Chlorophenoxy)-1,2-propanediol-1-carbamate; carbamic acid 3-(p-chlorophenoxy)-2-hydroxypropyl ester;* 3-(p-chlorophenoxy)-2-hydroxypropyl carbamate; 1,2-propanediol-3-(p-chlorophenoxy)-1-carbamate; Maolate; Rinlaxer. $C_{10}H_{12}ClNO_4$; mol wt 245.68. C 48.89%, H 4.93%, Cl 14.43%, N 5.70%, O 26.05%. Prepn: Collins, Matthews; Parker, **U.S.** pats. **3,161,567; 3,214,336** (1964, 1965, both to Upjohn).

Crystals from benzene + toluene, mp 89-91°. Readily sol in 95% ethanol, acetone, ethyl acetate; fairly readily sol in dioxane. Almost insol in cold water, benzene, cyclohexane. LD_{50} orally in rats: 748 mg/kg, *J. New Drugs* **2,** 366 (1962).

THERAP CAT: Relaxant (skeletal muscle).

THERAP CAT (VET): Relaxant (skeletal muscle).

2180. Chlorpheniramine. *γ-(4-Chlorophenyl)-N,N-dimethyl-2-pyridinepropanamine; 2-[p-chloro-α-(2-dimethylaminoethyl)benzyl]pyridine;* 1-(p-chlorophenyl)-1-(2-pyridyl)-3-dimethylaminopropane; 1-(p-chlorophenyl)-1-(2-pyridyl)-3-N,N-dimethylpropylamine; 3-(p-chlorophenyl)-3-(2-pyridyl)-N,N-dimethylpropylamine; γ-(4-chlorophenyl)-γ-(2-pyridyl)propyldimethylamine; chlorprophenpyridamine; chlorphenamine; Haynon. $C_{16}H_{19}ClN_2$; mol wt 274.80. C 69.93%, H 6.97%, Cl 12.90%, N 10.20%. Synthesis: Sperber *et al.,* **U.S.** pats. **2,567,245, 2,676,964** (1951, 1954, both to Schering). Prepn of d-form: L. A. Walter, **U.S.** pat. **3,061,517** (1962 to Schering). Solutions: Foley, Ilavsky, **U.S.** pat. **2,766,174** (1956 to Schering). Pharmacology: F. E. Roth, W. M. Govier, *J. Pharmacol. Exp. Ther.* **124,** 347 (1958). Toxicity data: R. B. Smith *et al., Toxicol. Appl. Pharmacol.* **28,** 240 (1974). Comprehensive description: C. G. Eckhart, T. McCorkle in *Analytical Profiles of Drug Substances* vol. 7, K. Florey, Ed. (Academic Press, New York, 1978) pp 43-80.

Oily liquid, $bp_{1.0}$ 142°.

Maleate, $C_{20}H_{23}ClN_2O_4$, *Allerclor, Allergisan, Alunex, Antagonate, Chlormene, Chlor-Trimeton, Chlor-Tripolon, Cloropiril, C-Meton, Histadur, Histalen, Histaspan, Lorphen, M.P. Chlorcaps T.D., Piriton, Polaronil (Germany), Pyridamal-100, Teldrin.* Crystals, mp 130-135°. uv max (water): 261 nm (ε 5760). Soly in mg/ml at 25°: ethanol 330; chloroform 240; water 160; methanol 130. Slightly sol in benz-

ene, ether. pH of a 2% aq soln about 5. LD_{50} orally in mice: 162 mg/kg (Smith).

d-Form, *dexchlorpheniramine, d-chlorpheniramine.* Oily liquid. $[\alpha]_D^{25}$ +49.8° (c = 1 in DMF).

d-Form maleate, $C_{20}H_{23}ClN_2O_4$, *Chlo-amine, Fortamine, Isomerine, Phenamin, Phendextro, Polaramine, Polaronil, Sensidyn.* Crystals from ethyl acetate, mp 113-115°. $[\alpha]_D^{25}$ +44.3° (c = 1 in dimethylformamide). pH of 1% soln 4-5.

THERAP CAT: Antihistaminic.

THERAP CAT (VET): Antihistaminic.

2181. Chlorphenoxamide. *2,2-Dichloro-N-(2-hydroxyethyl)-N-[[4-(4-nitrophenoxy)phenyl]methyl]acetamide; N-(β-hydroxyethyl)-N-[p-(4-nitrophenoxy)benzyl]dichloroacetamide; N-(β-hydroxyethyl)-N-[p-phenoxy-(4'-nitro)-benzyl]dichloroacetamide; dichloro-N-(β-hydroxyethyl)-N-[p-(4'-nitrophenoxy)benzyl]acetamide; clefamide; chlorphenoxamide; Mebinol.* $C_{17}H_{16}Cl_2N_2O_5$; mol wt 399.25. C 51.14%, H 4.04%, Cl 17.76%, N 7.02%, O 20.04%. Prepn: Logemann *et al., Farmaco Ed. Sci.* **13,** 139 (1958); **U.S.** pat. **2,824,894** (1958 to Carlo Erba).

Crystals from 95% ethanol, mp 136-137°. Practically insol in water (soly ~3 γ/ml). Sol in ethanol, acetone, dioxane. LD_{50} orally in mice: > 5 g/kg, Carneri, *C.A.* **53,** 5519c (1959).

Note: Do not confuse chlorphenoxamide with chlorphenoxamine, *q.v.*

THERAP CAT: Antiamebic.

2182. Chlorphenoxamine. *2-[1-(4-Chlorophenyl)-1-phenylethoxy]-N,N-dimethylethanamine; 2-[(p-chloro-α-methyl-α-phenylbenzyl)oxy]-N,N-dimethylethylamine;* β-dimethylaminoethyl (p-chloro-α-methylbenzhydryl) ether; [1-(p-chlorophenyl)-1-phenyl]ethyl (β-dimethylaminoethyl) ether; Systral. $C_{18}H_{22}ClNO$; mol wt 303.84. C 71.15%, H 7.30%, Cl 11.67%, N 4.61%, O 5.27%. Prepn: Arnold *et al.,* **U.S.** pat. **2,785,202** (1957 to Asta-Werke). Synthesis: G. Cahiez *et al., Tetrahedron Letters* **29,** 3659 (1988). Pharmacology: *eidem, Arzneimittel-Forsch.* **4,** 189 (1954); Brock *et al., ibid.* 262. Toxicity studies: Kerley *et al., Toxicol. Appl. Pharmacol.* **4,** 638 (1962). Metabolism: C. Koppel *et al., Arzneimittel-Forsch.* **37,** 1062 (1987). GC-MS determn in urine: H. Maurer, K. Pfleger, *J. Chromatog.* **428,** 43 (1988).

$bp_{0.05}$ 150-155°.

Hydrochloride, $C_{18}H_{23}Cl_2NO$, *1766, Clorevan, Contristamine, Phenoxene, Phenoxine, Systral.* Needles, mp 128°. Soluble in water; aq solns are stable.

Note: Do not confuse chlorphenoxamine with chlorphenoxamide, *q.v.*

THERAP CAT: Anticholinergic.

2183. Chlorphentermine. *4-Chloro-α,α-dimethylbenzeneethanamine; 4-chloro-α,α-dimethylphenethylamine; α,α-dimethyl-p-chlorophenethylamine;* 1-(p-chlorophenyl)-2-methyl-2-propylamine; 1-(p-chlorophenyl)-2-methyl-2-aminopropane; clorfentermina; Avicol; Lucofen; Teramine. $C_{10}H_{14}ClN$; mol wt 183.69. C 65.39%, H 7.68%, Cl 19.30%, N 7.63%. Prepn: Bachman *et al., J. Am. Chem. Soc.* **76,** 3972 (1954); Ferrari, *Farmaco Ed. Sci.* **15,** 337 (1960); **Fr.** pats. **M1299** and **1,296,132** (1962 to Simes), *C.A.* **58,** 4467e, 3352g (1963); **Brit.** pat. **906,331** (1962 to Kefalas), *C.A.* **58,** 6654c (1963). Pharmacology: Jun *et al., Can. J. Pharm. Sci.* **4,** 27 (1969); Ciborska *et al., Acta Pol. Pharm.* **26,** 595 (1969); Moeller-Nielsen, Dubnick, *Proc. Int. Symp. Amphetamines Relat. Compounds 1969,* E. Costa, Ed. (Raven Press, New York, 1970) pp 63-73.

Liquid. bp_2 100-102°.

Hydrochloride, $C_{10}H_{15}Cl_2N$, *Pre-Sate.* Crystals from alcohol + ether, mp 234°. Soly in water: > 20%. pH of 1% aq soln ~5.5. LD_{50} in mice (mg/kg): 270 orally; 267 s.c. (Ciborska).

Note: This is a controlled substance (stimulant) listed in the U.S. Code of Federal Regulations, Title 21 Part 1308.13 (1987).

THERAP CAT: Anorexic.

2184. Chlorproethazine. *2-Chloro-N,N-diethyl-10H-phenothiazine-10-propanamine; 2-chloro-10-(3-diethylaminopropyl)phenothiazine;* 3-chloro-10-(3-diethylaminopropyl)phenothiazine; RP 4909. $C_{19}H_{23}ClN_2S$; mol wt 346.92. C 65.78%, H 6.68%, Cl 10.22%, N 8.08%, S 9.24%. Prepn: Buisson *et al.,* U.S. pat. **2,769,002** (1956 to Rhône-Poulenc).

Hydrochloride, $C_{19}H_{24}Cl_2N_2S$, *Neuriplege.* Crystals, mp 178° (free base bp_1 225-240°). Sensitive to light. Soly in water about 1.0 g/60 ml, ethanol about 1.0 g/300 ml, chloroform 1.0 g/5 ml. Practically insol in acetone, ether, benzene. pH of 1% aq soln 4.8.

THERAP CAT: Muscle relaxant (skeletal); antipsychotic.

2185. Chlorproguanil. *N-(3,4-Dichlorophenyl)-N'-(1-methylethyl)imidodicarbonimidic diamide; 1-(3,4-dichlorophenyl)-5-isopropylbiguanide;* N^1-3,4-dichlorophenyl-N^5-isopropylbiguanide; N^1-3,4-dichlorophenyl-N^5-isopropyldiguanide; M5943. $C_{11}H_{15}Cl_2N_5$; mol wt 288.18. C 45.84%, H 5.25%, Cl 24.61%, N 24.30%. Method of prepn: Crowther *et al., J. Chem. Soc.* **1951**, 1780; Curd *et al.,* U.S. pat. **2,544,827** (1951); Crowther *et al.,* Brit. pat. **667,116** (1952) (both to I.C.I.).

Hydrochloride, $C_{11}H_{16}Cl_3N_5$, *Lapudrine.* Crystals, mp 246-247°. Soly in water: 1 g/100 ml. Solns may be boiled without dec.

THERAP CAT: Antimalarial.

2186. Chlorpromazine. *2-Chloro-N,N-dimethyl-10H-phenothiazine-10-propanamine; 2-chloro-10-(3-dimethylaminopropyl)phenothiazine;* 3-chloro-10-(3-dimethylaminopropyl)phenothiazine; N-(3-dimethylaminopropyl)-3-chlorophenothiazine; 2601 A; HL 5746; RP 4560; SKF 2601-A; Chlorderazin; Chlorpromados; Contomin; Esmind; Fenactil; Novomazina; Promactil; Prozil; Plegomazin; Sanopron; Aminazine; Ampliactil; Amplictil; Promazil; Chlor-Promanyl; Proma; Elmarin; Wintermin. $C_{17}H_{19}ClN_2S$; mol wt 318.88. C 64.03%, H 6.01%, N 8.79%, Cl 11.12%, S 10.06%. Prepn: Charpentier *et al., Compt. Rend.* **235**, 59 (1952); Charpentier, U.S. pat. **2,645,640** (1953 to Rhône-Poulenc). Review of pharmacology: Crismon, *Psychopharmacol. Bull.* **4**, 151 pp (Oct. 1967).

Oily liq. Amine odor. Alkaline reaction. $bp_{0.8}$ 200-205°.

Maleate, *Clordelazin.*

Hydrochloride, $C_{17}H_{20}Cl_2N_2S$, *Hebanil, Hibanil, Hibernal, Klorpromex, Largactil, Largaktyl, Megaphen, Promacid, Chloractil, Chlorazin, Sonazine, Marazine, Propaphenin, Taroctyl, Thorazine, Torazina.* Crystals, dec 179-180° (capillary); 194-196° (microblock). uv curve: Neuhoff, Auterhoff, *Arch. Pharm.* **288**, 400 (1955). pH of 5% aq soln 4.0-5.5. One gram dissolves in 2.5 ml water. Sol in methanol, ethanol, chloroform. Practically insol in ether, benzene. Slightly acid to litmus. LD_{50} orally in rats: 225 mg/kg, E. I. Goldenthal, *Toxicol. Appl. Pharmacol.* **18**, 185 (1971).

THERAP CAT: Anti-emetic; antipsychotic.

THERAP CAT (VET): Anti-emetic; tranquilizer; sedative. Slight antihistaminic and antiadrenaline actions. Peripheral vasodilator.

2187. Chlorpropamide. *4-Chloro-N-[(propylamino)carbonyl]benzenesulfonamide; 1-(p-chlorophenylsulfonyl)-3-propylurea;* 1-(p-chlorobenzenesulfonyl)-3-propylurea; N-propyl-N'-(p-chlorobenzenesulfonyl)urea; P-607; Adiaben; Asucrol; Catanil; Chloronase; Diabechlor; Diabenal; Diabetoral; Diabinese; Melitase; Millinese; Oradian; Stabinol. $C_{10}H_{13}ClN_2O_3S$; mol wt 276.75. C 43.40%, H 4.73%, Cl 12.81%, N 10.12%, O 17.34%, S 11.59%. Prepn: Marshall, Sigal, *J. Org. Chem.* **23**, 927 (1958); Brit. pat. **853,555** corresp to W. M. McLamore, U.S. pat. **3,349,124** (1960, 1967 to Pfizer); Bauer *et al., J. Org. Chem.* **31**, 3440 (1960). Pharmacology and metabolism: Khurana *et al., Indian J. Med.* **55**, 1084 (1967); Brotherton *et al., Clin. Pharmacol. Ther.* **10**, 505 (1969); Madsen *et al., Eur. J. Pharmacol.* **13**, 374 (1971).

Crystals from dil ethanol, mp 127-129°. uv max (0.01N HCl): 232.5 nm (ϵ 16,500). Soly in water at pH 6: 2.2 mg/ml. Practically insol at pH 7.3. Sol in alc; moderately sol in chloroform; sparingly sol in ether, benzene. LD_{50} i.p. in rats: 580 mg/kg, E. I. Goldenthal, *Toxicol. Appl. Pharmacol.* **18**, 185 (1971).

THERAP CAT: Antidiabetic.

2188. Chlorpropham. *(3-Chlorophenyl)carbamic acid 1-methylethyl ester; m-chlorocarbanilic acid isopropyl ester;* isopropyl-m-chlorocarbanilate; isopropyl N-(3-chlorophenyl)carbamate; chloro-IPC; chloropropham; CIPC; Chlor-IFC; Furloe; Sprout-Nip. $C_{10}H_{12}ClNO_2$; mol wt 213.68. C 56.21%, H 5.66%, Cl 16.59%, N 6.56%, O 14.98%. Prepn: E. D. Witman, U.S. pat. **2,695,225** and Strain, U.S. pat. **2,734,911** (1954, 1956 to Columbia-Southern Chem.); Brockway, U.S. pat. **2,806,051** (1957 to B. F. Goodrich). Toxicology: E. M. Boyd, E. Carsky, *Arch. Environ. Health* **19**, 621 (1969).

Solid, mp 40.7-41.1°. bp_2 149°. n_D^{20} 1.5388. Commercial product is a liquid. Slightly sol in water; miscible with most oils and organic solvents. LD_{50} orally in rats: 1.2 g/kg, E. M. Boyd, E. Carsky, *loc. cit.*

USE: Herbicide; plant growth regulator.

2189. Chlorprothixene. *3-(2-Chloro-9H-thioxanthen-9-ylidene)-N,N-dimethyl-1-propanamine; 2-chloro-N,N-dimethylthioxanthene-$\Delta^{9,\gamma}$-propylamine;* 2-chloro-N,N-dimethyl-3-thioxanthen-9-ylidenepropylamine; 2-chloro-9-(3'-dimethylaminopropylidene)thioxanthene; α-2-chloro-10-(3-dimethylaminopropylidene)thioxanthene; N-714; Taractan; Truxal; Truxaletten; Tarasan. $C_{18}H_{18}ClNS$; mol wt 315.86. C 68.44%, H 5.74%, Cl 11.23%, N 4.44%, S 10.15%. Prepn: Brit. pat. **829,763** and Sprague, Engelhardt, U.S. pat. **2,951,082** (both 1960 to Merck & Co.); Brit. pat. **834,143** (1960 to Am. Cyanamid). Comprehensive description: B. C. Rudy, B. Z. Senkowski, in *Analytical Profiles of Drug*

Substances vol. **2**, K. Florey, Ed. (Academic Press, New York, 1973) pp 63-84.

Pale yellow crystals, mp 97-98°. Practically insol in water; sol in alcohol, ether, chloroform. Incompatible with acids, alkalies, phenobarbital, thiopental sodium, mepazine. LD$_{50}$ orally in rats: 380 mg/kg, *RTECS* Vol. II, E. J. Fairchild, Ed. (1977) p 909.

Hydrochloride, $C_{18}H_{18}ClNS.HCl$, crystals, mp 221°. Freely sol in water at pH 6 to 6.5.

THERAP CAT: Antipsychotic.

2190. Chlorpyrifos. *Phosphorothioic acid O,O-diethyl O-(3,5,6-trichloro-2-pyridinyl) ester; O,O-diethyl O-3,5,6-trichloro-2-pyridyl phosphorothioate; chlorpyrifos-ethyl; Dowco 179; ENT 27311; Dursban; Lorsban; Pyrinex.* $C_9H_{11}Cl_3NO_3PS$; mol wt 350.57. C 30.83%, H 3.16%, Cl 30.34%, N 4.00%, O 13.69%, P 8.83%, S 9.15%. Prepn: Rigterink, **Fr.** pat. **1,360,901** corresp to **U.S.** pat. **3,244,586** (1964, 1966, both to Dow); Rigterink, Kenaga, *J. Agr. Food Chem.* **14**, 394 (1966). Activity: E. E. Kenaga *et al., J. Econ. Entomol.* **58**, 1043 (1965). Metabolism: G. N. Smith *et al., J. Agr. Food Chem.* **15**, 132 (1967); W. H. Gutenmann *et al., ibid.* **16**, 45 (1968).

White granular crystals, mp 41-42°. Vapor press at 25°: 1.87 × 10^{-5} mm Hg. Soly at 25°: water 2 ppm; isooctane 79% w/w; methanol 43% w/w. Readily sol in other org solvents. LD$_{50}$ orally in rats: 145 mg/kg, Schafer, *Toxicol. Appl. Pharmacol.* **21**, 315 (1972).

O,O-Dimethyl analog, $C_7H_7Cl_3NO_3PS$, *chlorpyrifos-methyl, Dowco 214, ENT 27520, Reldan.* Crystals, mp 45.5-46.5°. Vapor press at 25°: 4.22 × 10^{-5} mm Hg. Soly in water at 25°: 5 mg/l.

USE: Insecticide; acaricide.

2191. Chlorquinaldol. *5,7-Dichloro-2-methyl-8-quinolinol; 5,7-dichloro-8-quinaldinol; 5,7-dichloro-8-hydroxyquinaldine; 5,7-dichloro-2-methyl-8-hydroxyquinoline; hydroxydichloroquinaldine; chloroquinaldol; Afungil; Quesil; Siogène; Gyno-Sterosan; Gynotherax; Saprosan; Sterosan; Steroxin; Siosteran.* $C_{10}H_7Cl_2NO$; mol wt 228.08. C 52.66%, H 3.09%, Cl 31.09%, N 6.14%, O 7.01%. Prepd by chlorination of 8-hydroxyquinaldine with or without formic acid as solvent: Senn, **U.S.** pat. **2,411,670** (1946 to Geigy).

Yellow needles from alc, mp 114-115° (slight decompn). Medicinal odor. uv max (ethanol): 316 nm (A$_{1cm}^{1\%}$ 170); min 280 nm. Practically insol in water. Soly (25°) in ethanol 1.0 g/100 ml of soln; chloroform 5.0 g; acetone 4.0 g; ether 3.0 g; 0.1N NaOH 1.4 g. Also sol in benzene, glacial acetic acid.

THERAP CAT: Antibacterial.

THERAP CAT (VET): Topical antibacterial, fungistat.

2192. Chlorsulfuron. *2-Chloro-N-[[(4-methoxy-6-methyl-1,3,5-triazin-2-yl)amino]carbonyl]benzenesulfonamide; DPX 4189;* Glean; Telar. $C_{12}H_{12}ClN_5O_4S$; mol wt 357.78. C 40.28%, H 3.38%, Cl 9.91%, N 19.58%, O 17.89%, S 8.96%. Selective pre- and post-emergence herbicide. Prepn: G.

Levitt, **Ger.** pat. **2,715,786;** *idem,* **U.S.** pat. **4,127,405** (1977, 1978 both to du Pont). Activity: G. Levitt *et al., J. Agr. Food Chem.* **29**, 416 (1981). Mode of action: T. B. Ray, *Pestic. Biochem. Physiol.* **17**, 10 (1982).

Crystals from ether, mp 174-178°. Soly in water: 125 ppm. Moderately sol in methylene chloride; less sol in acetone, acetonitrile. Low soly in hydrocarbon solvents. LD$_{50}$ in male, female rats: 5545, 6293 mg/kg (Levitt).

USE: Herbicide.

2193. Chlortetracycline. *7-Chloro-4-dimethylamino-1,4,4a,5,5a,6,11,12a-octahydro-3,6,10,12,12a-pentahydroxy-6-methyl-1,11-dioxo-2-naphthacenecarboxamide;* 7-chlorotetracycline; Acronize; Aureocina; Aureomycin; Biomitsin; Biomycin; Chrysomykine. $C_{22}H_{23}ClN_2O_8$; mol wt 478.88. C 55.17%, H 4.84%, Cl 7.40%, N 5.85%, O 26.73%. Antibiotic substance isolated from the substrate of *Streptomyces aureofaciens:* Duggan, *Ann. N.Y. Acad. Sci.* **51**, 177 (1948); **U.S.** pat. **2,482,055** (1949 to Am. Cyanamid); Broschard *et al., Science* **109**, 199 (1949). Structure: Stephens *et al., J. Am. Chem. Soc.* **74**, 4976 (1952); **76**, 3568 (1954). Crystal structure: Donohue *et al., ibid.* **85**, 851 (1963). Absolute configuration: Dobrynin *et al., Tetrahedron Letters* **1962**, 901. Purification: Winterbottom *et al.,* **U.S.** pat. **2,899,422** (1959 to Am. Cyanamid). Improved process: Miller *et al.;* Goodman, **U.S.** pats. **2,987,449; 3,050,446** (1961, 1962 to Am. Cyanamid). Toxicity: E. I. Goldenthal, *Toxicol. Appl. Pharmacol.* **18**, 185 (1971). Comprehensive description: G. Schwartzman *et al.,* in *Analytical Profiles of Drug Substances* vol. **8**, K. Florey, Ed. (Academic Press, New York, 1979) pp 101-137.

Golden-yellow crystals, mp 168-169°. [α]$_D^{23}$ −275.0° (methanol). uv max (0.1N HCl): 230, 262.5, 367.5 nm; (0.1N NaOH): 255, 285, 345 nm. Soly in water: 0.5-0.6 mg/ml. Very sol in aq solns above pH 8.5. Freely sol in the Cellosolves, dioxane, and Carbitol. Slightly sol in methanol, ethanol, butanol, acetone, ethyl acetate, benzene. Practically insol in ether, petr ether.

Crystal salt, white powder. Insol in water. Marketed as suspension preserved with methyl and propyl *p*-hydroxybenzoate; stable and fully active for at least one year at room temps.

Hydrochloride, $C_{22}H_{24}Cl_2N_2O_8$, *Aureociclina, Isphamycin.* Bitter, yellow rhomboid crystals. Dec above 210°. [α]$_D^{23}$ −240°. Soly at about 28° (mg/ml): water 8.6; methanol 17.4; ethanol 1.7: Weiss *et al., Antibiot. Chemother.* **7**, 374 (1957). Sol in solns of alkali hydroxides and carbonates. Practically insol in acetone, ether, chloroform, dioxane. pH of satd aq soln 2.8-2.9. LD$_{50}$ orally in rats: 10300 mg/kg (Goldenthal).

THERAP CAT: Antibacterial; antiamebic.

THERAP CAT (VET): Antimicrobial.

2194. Chlorthalidone. *2-Chloro-5-(2,3-dihydro-1-hydroxy-3-oxo-1H-isoindol-1-yl)benzenesulfonamide; 2-chloro-5-(1-hydroxy-3-oxo-1-isoindolinyl)benzenesulfonamide; 3-hydroxy-3-(4-chloro-3-sulfamylphenyl)phthalimidine; 2-chloro-5-(3-hydroxy-1-oxoisoindolin-3-yl)benzenesulfonamide; 1-oxo-3-(3-sulfamyl-4-chlorophenyl)-3-hydroxyisoindoline; 3-(4'-chloro-3'-sulfamoylphenyl)-3-hydroxyphthalimidine; 1-keto-3-(3'-sulfamyl-4'-chlorophenyl)-3-hydroxyisoindoline; chlorphthalidolone; phthalamu-*

dine; phthalamodine; G 33182; Hydro-long; Hydroton; Hygroton; Thalitone. $C_{14}H_{11}ClN_2O_4S$; mol wt 338.78. C 49.63%, H 3.27%, Cl 10.47%, N 8.27%, O 18.89%, S 9.47%. Prepn: Graf *et al.*, *Helv. Chim. Acta* **42**, 1085 (1959); U.S. pat. **3,055,904** (1962 to Geigy). Activity and side effects: Holtmeier *et al.*, *Med. Welt* **1967**, 1384; Zsoter *et al.*, *J. Pharmacol. Exp. Ther.* **180**, 723 (1972). Metabolism: Beisenherz *et al.*, *Arch. Int. Pharmacodyn. Ther.* **161**, 76 (1966). Comprehensive description: J. M. Singer *et al.*, in *Analytical Profiles of Drug Substances* vol. 14, K. Florey, Ed. (Academic Press, New York, 1985) pp 1-36.

Crystals from 50% acetic acid, dec 224-226°. mp range may extend from 218° to 264° on slow heating. Can form a monohydrate. uv max (methanol): < 220 nm. Soly in water: 12 mg/100 ml (20°); 27 mg/100 ml (37°); in 0.1N Na_2CO_3: 577 mg/100 ml (20°); 990 mg/100 ml (37°). More sol in aq solns of NaOH. Soluble in warm ethanol; slightly sol in ether.

THERAP CAT: Diuretic, antihypertensive.

THERAP CAT (VET): Diuretic.

2195. Chlorthenoxazin(e). *2-(2-Chloroethyl)-2,3-dihydro-4H-1,3-benzoxazin-4-one;* 2-(β-chloroethyl)-2,3-dihydro-4-oxo-(benzo-1,3-oxazine); 4-oxo-2-(β-chloroethyl)-2,3-dihydrobenzo-1,3-oxazine; 2-(β-chloroethyl)-2,3-dihydro-4-keto-(benzo-1,3-oxazine); Ap 67; Valtorin; Piroxina; Ossazone; Reumagrip; Ossipirina; Apirazin; Valmorin. $C_{10}H_{10}ClNO_2$; mol wt 211.65. C 56.74%, H 4.76%, Cl 16.75%, N 6.62%, O 15.12%. Prepn: Ohnacker, U.S. pat. **2,943,087** (1960 to Thomae).

Crystals from ethanol, mp 146-147° (dec). Sol in chloroform. uv max (chloroform): 297.5 nm. Ingredient of *Trigatan, Fiobrol.*

THERAP CAT: Antipyretic; analgesic.

2196. Chlorthion®. *Phosphorothioic acid O-(3-chloro-4-nitrophenyl) O,O-dimethyl ester; O,O-dimethyl O-(3-chloro-4-nitrophenyl) thionophosphate; p-nitro-m-chlorophenyl dimethyl thionophosphate;* compd 22/190; Chlorothion. $C_8H_9ClNO_5PS$; mol wt 297.68. C 32.28%, H 3.05%, Cl 11.91%, N 4.71%, O 26.87%, P 10.41%, S 10.77%. Prepn: Schrader, *Angew. Chem.* **66**, 265 (1954); U.S. pat. **2,701,259** (1955 to Bayer).

The pure compound as yellow crystals, mp 21°. d_4^{20} 1.437. n_D^{20} 1.5661. Commercial product is a yellow oil, $bp_{0.1}$ 125°. Miscible in benzene, alcohol, ether; practically insol in water. Hydrolyzed by alkali. LD_{50} orally in male, female rats: 880, 980 mg/kg, T. B. Gaines, *Toxicol. Appl. Pharmacol.* **14**, 515 (1969).

USE: Insecticide. *Caution:* A cholinesterase inhibitor. Symptoms similar to Parathion.

2197. Chlorzoxazone. *5-Chloro-2(3H)-benzoxazolone; 5-chloro-2-benzoxazolol;* 5-chloro-2-hydroxybenzoxazole; 2-hydroxy-5-chlorobenzoxazole; 5-chlorbenzoxazolin-2-

one; 5-chlorobenzoxazolidone; Paraflex; Biomioran; Solaxin. $C_7H_4ClNO_2$; mol wt 169.58. C 49.58%, H 2.38%, Cl 20.91%, N 8.26%, O 18.87%. Prepn from 2-amino-5-chlorobenzoxazole: Marsh, U.S. pat. **2,895,877** (1959 to McNeil Labs.). LC determn in plasma: I. L. Honigsberg *et al.*, *J. Pharm. Sci.* **68**, 253 (1979). Fluorometric determn in tablets and biological fluids: J. T. Stewart, C. W. Chan, *ibid.* 910. Clinical studies: R. Herman, *Curr. Ther. Res.* **9**, 537 (1967); J. J. Scheiner, *ibid.* **19**, 51 (1976). Metabolism study: R. Twele, G. Spiteller, *Arzneimittel-Forsch.* **32**, 759 (1982). Toxicity: G. Hofrichter *et al.*, *Arzneimittel-Forsch.* **17**, 242 (1967). Review of clinical studies: J. K. Elenbaas, *Am. J. Hosp. Pharm.* **37**, 1313 (1980). Comprehensive description: J. T. Stewart, C. A. Janicki in *Analytical Profiles of Drug Substances* vol. 16, K. Florey, Ed. (Academic Press, New York, 1987) pp 119-144.

Crystals from acetone, mp 191-191.5°. Sparingly sol in water; sol in methanol, ethanol, isopropanol. Freely sol in aq solns of alkali hydroxides and ammonia. LD_{50} in mice (mg/kg): 3650 orally, 380 i.p. (suspensions); 440 orally, 183 i.p. (solns of Na salt) (Hofrichter).

THERAP CAT: Skeletal muscle relaxant.

2198. Cholane. *5β-Cholane.* $C_{24}H_{42}$; mol wt 330.58. C 87.19%, H 12.81%. Prepn from bisnorcholyl methyl ketone: Wieland *et al.*, *Z. Physiol. Chem.* **161**, 80, 109 (1926); from potassium cholanate: Kazuno *et al.*, *Proc. Japan Acad.* **28**, 416 (1952).

Stout prisms from alc, mp 90°; $bp_{0.001}$ 190°. Sparingly sol in methanol.

2199. Cholanic Acid. *5β-Cholan-24-oic acid;* ursocholanic acid; 17β-(1-methyl-3-carboxypropyl)etiocholane. $C_{24}H_{40}O_2$; mol wt 360.56. C 79.94%, H 11.18%, O 8.87%. Steroidal acid probably formed by the dehydration and hydrogenation of certain bile acids commonly found in animals. Considered to be a chemical trademark certifying the prehistoric presence of some type of animal, *see* Seifert, *Pure Appl. Chem.* **34**, 633 (1973). This status as a natural product of exclusive animal origins now questioned by its isolation from the embryo of the jequirity bean, *Abrus precatorius, Leguminosae:* Mandava *et al.*, *Steroids* **23**, 357 (1974). Prepn from coprostane, cholic acid, lithocholic acid, desoxy- and chenodesoxycholic acid: Wieland, Weil, *Z. Physiol. Chem.* **80**, 287 (1912); Wieland, Boersch, *ibid.* **106**, 193 (1919); Windaus, Neukirchen, *Ber.* **52**, 1915 (1919); Wieland, Vocke, *Z. Physiol. Chem.* **191**, 69 (1930). From 3-ketocholanic acid diethyl thioacetal by hydrogenolysis: Bernstein, Dorfman, U.S. pat. **2,440,660** (1948 to Am. Cyanamid). 5β-Cholanic acid differs from the thermodynamically more stable 5α-isomer, by being *cis* at the A/B steroid ring function rather than *trans.*

Needles from alc, mp 163-164°. $[\alpha]_D^{20}$ +21.7° (chloroform). Sol in chloroform, alcohol, acetic acid. Forms a molecular compd with allocholanic acid, mp 163.5°.

Methyl ester, $C_{25}H_{42}O_2$, mp 86-87°.

Ethyl ester, $C_{26}H_{44}O_2$, needles from 80% alc, mp 93-94°; bp$_{12}$ 273°. $[\alpha]_D^{19}$ +21° (chloroform).

Note: The name cholanic acid was formerly applied to desoxybilianic acid, $C_{24}H_{36}O_7$.

USE: In chemotaxonomical classification.

2200. Cholanthrene. *1,2-Dihydrobenz[j]aceanthrylene.* $C_{20}H_{14}$; mol wt 254.31. C 94.45%, H 5.55%. The numbering system is that of Chem. Abstracts. The structural formula may also be arranged and numbered like a sterol (*see* 20-Methylcholanthrene). Synthesis from 1-β-naphthylhydrindene: Cook *et al.*, *J. Chem. Soc.* **1935**, 667; from 4-indanyl-MgBr and 1-naphthoyl chloride: Fieser, Seligman, *J. Am. Chem. Soc.* **57**, 2174 (1935).

Faintly yellow plates from benzene + ether, mp 173°. Sublimes 210-215° at 0.2 mm Hg. Absorption max: Fieser, Hershberg, *J. Am. Chem. Soc.* **60**, 940 (1938). Sol in benzene, xylene, toluene. Slightly sol in methanol. Insol in water. May be solubilized by aq solns of Na desoxycholate.

Picrate, $C_{26}H_{17}N_3O_7$, dark purple needles from benzene, mp 169-170°.

Molecular complex with 2,4,7-trinitrofluorenone, olive green crystals, mp 245-246°: Orchin, Woolfolk, *J. Am. Chem. Soc.* **68**, 1727 (1946).

2201. Cholecystokinin. Cholecystokinin-pancreozymin; Pancreozymin; CCK-PZ. Polypeptide hormone found in the mammalian gastrointestinal tract and brain. Stimulates pancreatic exocrine secretion and growth. May also play a role in appetite satiation, pain perception, and neuronal transmission. First shown to cause gallbladder contraction: Ivy, Oldberg, *Am. J. Physiol.* **86**, 599 (1928). Discovery of a substance, designated as pancreozymin, which promotes secretion of digestive enzymes by the pancreas: Harper, Raper, *J. Physiol. (London)* **102**, 115 (1943). Identity with pancreozymin: Jorpes *et al.*, *Acta Chem. Scand.* **18**, 2408 (1964). The C-terminal pentapeptide has been shown to be identical to that of gastrin and caerulein: V. Mutt, J. E. Jorpes, *Eur. J. Biochem.* **6**, 156 (1968); eidem, *Biochem. J.* **125**, 57P (1971). Various biologically active, amino-truncated forms have been identified. Cholecystokinin consisting of 33 amino acids (CCK-33) is the predominant gastrointestinal form; CCK-39 and CCK-58 have also been identified. CCK-8 is the predominant CNS form. Identification of CCK in brain: J. J. Vanderhaeghen *et al.*, *Nature* **257**, 604 (1975); G. J. Dockray, *ibid.* **264**, 568 (1976). Distribution and molecular heterogeneity: J. F. Rehfeld, *J. Biol. Chem.* **253**, 4022 (1978). Synthesis of the C-terminal dodecapeptide: M. A. Ondetti *et al.*, *J. Am. Chem. Soc.* **92**, 195 (1970). Synthesis of the N-terminal hexapeptide of porcine CCK-33: Bodanszky *et al.*, *J. Org. Chem.* **37**, 2303 (1972). Cloning and nucleotide sequence of the human cholecystokinin gene: Y. Takahashi *et al.*, *Proc. Nat. Acad. Sci. USA* **82**, 1931 (1985). Total synthesis of porcine CCK-33: Y. Kurano, *Chem. Commun.* **1987**, 323; of human CCK-33: N. Fujii *et*

al., ibid. **1988**, 324. Proposed role in suppression of food intake: M. A. Della-Fera, C. A. Baile, *Science* **206**, 471 (1979); C. J. Savory, M. J. Gentle, *Experientia* **36**, 1191 (1980); M. A. Della-Fera *et al.*, *Science* **212**, 687 (1981); in regulation of hypothalamic peptides: S. Itoh *et al.*, *Life Sci.* **25**, 1725 (1979); in modulation of catecholaminergic activity: K. Fuxe *et al.*, *Eur. J. Pharmacol.* **67**, 329 (1980). There is also evidence that CCK acts as a specific antagonist of opiate analgesia: P. L. Faris *et al.*, *Science* **219**, 310 (1983). *Reviews:* E. Straus, R. S. Yalow, *Fed. Proc.* **38**, 2320-2324 (1979); V. Mutt, *Biochem. Soc. Trans.* **8**, 11-14 (1980); idem, *Vitam. Horm.* **39**, 231-426 (1982). Review of physiology: G. J. Dockray, *Brit. Med. Bull.* **38**, 253-258 (1982); of role in appetite satiation and pain perception: G. Stacher, *Psychoneuroendocrinology* **11**, 39-48 (1986). Symposium on neuronal CCK: *Ann. N.Y. Acad. Sci.* **448**, 1-697 (1985).

Lys-Ala-Pro-Ser-Gly-Arg-Met-Ser-Ile-Val-Lys-Asn-Leu-Gln-Asn-Leu-Asp-

Pro-Ser-His-Arg-Ile-Ser-Asp-Arg-Asp-Tyr-Met-Gly-Trp-Met-Asp-Phe-NH$_2$
                |
               SO$_3$H

Human CCK-33

C-Terminal octapeptide, *see* Sincalide.

2202. Cholestane. $C_{27}H_{48}$; mol wt 372.65. C 87.02%, H 12.98%. The *trans*-decalin homolog of coprostane, *q.v.* Prepd from cholesteryl chloride: Diels, Linn, *Ber.* **41**, 548 (1908); Windaus, *Ber.* **50**, 136 (1917); Ruzicka *et al.*, *Helv. Chim. Acta* **16**, 327 (1933). Crystal structure: Haner, Norton, *Acta Cryst.* **20**, 930 (1966).

Scales from ether + alcohol. mp 80°. $[\alpha]_D^{20}$ +24.4° or +30.2° (c = 2 in chloroform). n_D^{88} 1.4887. Freely sol in chloroform, ether, benzene, slightly sol in abs alcohol.

2203. Cholestanol. Dihydrocholesterol; 3β-hydroxycholestane; β-cholestanol. $C_{27}H_{48}O$; mol wt 388.65. C 83.43%, H 12.45%, O 4.12%. Occurs in human feces, in gallstones, in eggs. Prepn by reduction of cholesterol: Willstätter, Mayer, *Ber.* **41**, 2199 (1908); Ellis, Gardner, *Biochem. J.* **12**, 72 (1918). From coprostenone: Diels, Abderhalden, *Ber.* **39**, 884 (1906). *See also* Bruce, Ralls, *Org. Syn.* coll. vol. **II**, 191 (1943).

Monohydrate, scales from alc, mp 141.5-142°. $[\alpha]_D^{22}$ +24.2° (c = 1.3 in chloroform). One gram dissolves in about 100 ml alcohol, in 200 ml dry methanol. Freely sol in hot alc, ether, chloroform. Pptd by digitonin.

Methyl ether, $C_{28}H_{50}O$, needles from acetone, mp 82.5-83°. $[\alpha]_D^{20}$ +20.0°.

Acetate, $C_{29}H_{50}O_2$, prisms from ethyl acetate + methanol, mp 111°. $[\alpha]_D^{20}$ +13.3° (c = 2 in chloroform).

2204. Cholesterol. *Cholest-5-en-3β-ol;* cholesterin. $C_{27}H_{46}O$; mol wt 386.64. C 83.87%, H 11.99%, O 4.14%. Principal sterol of the higher animals. Found in all body tissues, esp in the brain, spinal cord, and in animal fats or oils.

Main constituent of gallstones. Prepd commercially from the spinal cord of cattle by petr ether extraction of the non-saponifiable matter. Also produced from wool grease. Cholesterol from animal organs always contains cholestanol (dihydrocholesterol) and other satd sterols. Purification by repeated bromination: Schoenheimer, *J. Biol. Chem.* **105**, 355 (1934); Fieser, *Org. Syn.*, **coll. vol. IV**, 195 (1963). Laboratory procedure for isoln from gallstones: L. F. Fieser, *Organic Experiments* (Heath, Boston, 3rd ed., 1964) p 70. Total synthesis: Keana, Johnson, *Steroids* **4**, 457 (1964). Reviews and bibliographies: Fieser, Fieser, *Steroids* (Reinhold, New York, Chapman & Hall, London, 1959); Lettré *et al.*, *Ueber Sterine, Gallensäuren und verwandte Naturstoffe* (Stuttgart, 2nd ed., 1955); R. P. Cook, *Cholesterol (Chemistry, Biochemistry and Pathology)* (Academic Press, New York, 1958) 542 pp; J. T. Gwynne, J. F. Strauss, *Endocrine Rev.* **3**, 299-329 (1982).

Monohydrate, pearly leaflets or plates from dil alcohol. Becomes anhydr at 70-80°. When anhydr mp 148.5°. Has been sublimed as orthorhombic needles. $bp_{0.5}$ 233°; bp_{760} 360° (some decompn). d 1.03 (monohydrate); d_4^{19} 1.052 (anhydr). $[\alpha]_D^{20}$ −31.5° (c = 2 in ether); $[\alpha]_D^{20}$ −39.5° (c = 2 in chloroform). Absorption spectrum: Heilbron *et al.*, *J. Chem. Soc.* **1928**, 47. Practically insol in water (about 0.2 mg/100 ml H_2O). Slightly sol in alc (1.29% w/w at 20°), more sol in hot alc (100 g of satd 96% alcoholic soln contains 28 g at 80°). One gram dissolves in 2.8 ml ether, in 4.5 ml chloroform, in 1.5 ml pyridine. Also sol in benzene, petr ether, oils, fats. Soly in aq solns of bile salts: Rosin, *Z. Physiol. Chem.* **124**, 282 (1923). Solubilization: Gemant, *Life Sci.* **1**, 233 (June 1962). Is pptd by digitonin. Gives intense red color with rosaniline in chloroform soln.

Methyl ether, $C_{28}H_{48}O$, crystals from acetone, mp 84°. $[\alpha]_D^{20}$ −45.8° (c = 1.2 in chloroform).

Acetate, $C_{29}H_{48}O_2$, needles from acetone, mp 115-116°. $[\alpha]_D^{20}$ −47.4° (c = 2 in chloroform).

Benzoate, $C_{34}H_{50}O_2$, mp 145.5° (the melt becomes clear at 180°). $[\alpha]_D^{25}$ −13.7° (c = 0.9 in chloroform).

USE: Pharmaceutic aid (emulsifying agent).

2205. Cholestyramine Resin.

Colestyramin; Dowex 1-X2-Cl; MK-135; Cholybar; Colybar; Cuemid; Quantalan; Questran. A synthetic, strongly basic anion exchange resin contg quaternary ammonium functional groups which are attached to a styrene-divinylbenzene copolymer. Main constituent: Polystyrene trimethylbenzylammmonium as Cl⁻ anion, also contains divinylbenzene (~2%) water (~43%). Cross linkage %: 1-10. Particle size: 50-100 mesh. Percent volume increase of exhausted resin (Cl⁻ to OH⁻): 20%. Stable at temperatures up to 150°. Capacity: 3.5 meq/g dry, 1.33 meq/ml wet. Pharmacology review: H.

typified structure of main polymeric groups

R. Casdorph in *Lipid Pharmacology* Vol. 2(2), R. Paoletti, C. J. Glueck, Eds. (Academic Press, New York, 1976) pp 222-256. Aids detoxification in kepone poisoning: J. J. Boylan *et al.*, *Science* **199**, 893 (1978).

White to buff-colored, hygroscopic, fine powder. Odorless or has not more than a slight amine-like odor. Insol in water, alcohol, benzene, chloroform, ether.

THERAP CAT: Ion-exchange resin (bile salts); antihyperlipoproteinemic.

2206. Cholic Acid.

$3\alpha,7\alpha,12\alpha$-*Trihydroxy-5β-cholan-24-oic acid*; $3\alpha,7\alpha,12\alpha$-*trihydroxy-5β-cholanic acid*; cholalic acid; 17β-(1-methyl-3-carboxypropyl)etiocholane-$3\alpha,7\alpha$,-12α-triol; Colalin. $C_{24}H_{40}O_5$; mol wt 408.56. C 70.55%, H 9.87%, O 19.58%. Occurs in conjugation with glycine or taurine in bile of most vertebrates. Extraction from beef bile: Wieland, Siebert, *Z. Physiol. Chem.* **262**, 1 (1939); laboratory procedure: Gattermann-Wieland, *Praxis des Organischen Chemikers* (de Gruyter, Berlin, 40th ed., 1961) p 360. Structure and synthesis: Fieser, Fieser, *Steroids* (Reinhold, New York, 1959).

Monohydrate, plates from dil acetic acid. Bitter taste with sweetish aftertaste. When anhydr mp 198°. $[\alpha]_D^{20}$ +37° (c = 0.6 in alcohol). pK = 6.4. Not precipitated by digitonin. Soly at 15° in water 0.28 g/l; in alcohol 30.56 g/l; in ether 1.22 g/l; in chloroform 5.08 g/l; in benzene 0.36 g/l; in acetone 28.24 g/l; in glacial acetic acid 152.12 g/l. Sol in solns of alkali hydroxides or carbonates.

Methyl ester, $C_{25}H_{42}O_5$, crystals from 95% alcohol +water, mp 155-156°.

Ethyl ester, $C_{26}H_{44}O_5$, crystals from ethyl acetate +petr ether, mp 162-163°.

Sodium salt, $C_{24}H_{39}NaO_5$, *sodium cholate.* Crystals. Soly in water at 15°: >568.9 g/l. *Caution:* The names "sodium cholate" and "sodium choleate" are sometimes used for mixtures of bile salts. The term "sodium choleate" is to be preferred for bile salts, *see* Ox Bile Extract.

Note: The property of forming molecular compds is common to bile acids. For instance a blue molecular compd $(C_{24}H_{40}O_5)_4 \cdot KI \cdot H_2O$, may be prepd by mixing an alcoholic soln of cholic acid and a soln of iodine in aq potassium iodide: Barger, Field, *J. Chem. Soc.* **101**, 1404 (1912).

THERAP CAT: Choleretic.

2207. Choline.

2-Hydroxy-N,N,N-trimethylethanaminium; *(β-hydroxyethyl)trimethylammonium*; bilineurine. $[C_5H_{14}NO]^+$; mol wt 104.17. $[HOCH_2CH_2N^+(CH_3)_3]$. Basic constituent of lecithin. Found in many plants and animal organs, *e.g.* bile, brain, yolk of eggs, hops, belladonna, strophanthus. Usually made synthetically from trimethylamine and ethylene chlorohydrin or ethylene oxide: Blackett, Soliday, U.S. pat. **2,774,759** (1956 to Am. Cyanamid). *Review:* T. H. Jukes in Kirk-Othmer *Encyclopedia of Chemical Technology* vol. 6 (Wiley-Interscience, New York, 3rd ed., 1979) pp 19-28.

Hydroxide, $C_5H_{15}NO_2$, *bursine, fagine, gossypine, luridine, sincaline, vidine.* Viscid, strongly alkaline liq, has been crystallized; absorbs CO_2 from the air. Very sol in water, alcohol; insol in ether. *Keep tightly closed.*

2208. Choline Chloride.

2-Hydroxy-N,N,N-trimethylethanaminium chloride; Biocolina; Hepacholine; Lipotril. $C_5H_{14}ClNO$; mol wt 139.63. C 43.01%, H 10.11%, Cl 25.39%, N 10.03%, O 11.46%. $[HOCH_2CH_2N^+(CH_3)_3]Cl^-$. Prepn: Klein, Kapp, U.S. pat. **2,623,901** (1952 to Nopco Chem.). Polymorphism: Collin, *J. Am. Chem. Soc.* **79**, 6086 (1957). Enhancement of human serial learning: N. Sitaram *et al.*, *Life Sci.* **22**, 1555 (1978); *eidem*, *Science* **201**, 274

(1978). Toxicity data: Hartung, Cornish, *Toxicol. Appl. Pharmacol.* **12**, 486 (1968).

Deliquesc crystals. Very sol in water or alcohol. The aq soln is practically neutral. *Keep tightly closed.* LD$_{50}$ in rats (g/kg): 0.400 i.p.; 6.64 orally (Hartung, Cornish).

THERAP CAT: Lipotropic.

THERAP CAT (VET): Nutritional factor. Dietary source of choline (lipotropic factor) in poultry.

2209. Choline Dehydrocholate. Dehydrocholic acid salt of choline; Biscolan. C$_{29}$H$_{48}$NO$_6$; mol wt 506.71. C 68.74%, H 9.55%, N 2.76%, O 18.95%. Prepn: Giovambattista, *Rev. Asoc. Bioquim. Arg.* **19**, 180 (1954), *C.A.* **49**, 13275 (1955). Crystals from methanol+ ether, mp 196-198°.

THERAP CAT: Lipotropic.

2210. Choline Dihydrogen Citrate. (2-Hydroxyethyl)trimethylammonium citrate; Chothyn; Cirrocolina; Citracholine; Neurotropan. C$_{11}$H$_{21}$NO$_8$; mol wt 295.29. C 44.74%, H 7.17%, N 4.74%, O 43.35%. [HOCH$_2$CH$_2$N$^+$(CH$_3$)$_3$]-[C$_6$H$_7$O$_7^-$]. Prepn: Klein *et al.*, U.S. pat. **2,870,198** (1959). Granules, mp 105-107.5°. Acrid taste. Freely sol in water; very slightly sol in alcohol. Practically insol in benzene, chloroform, ether. The pH of a 25% aq soln is 4.25. May be formulated with calcium citrate for better taste and easier tabletting.

THERAP CAT: Lipotropic.

THERAP CAT (VET): Dietary source of choline (lipotropic factor) in poultry.

2211. Choline Esterase. There are at least two choline esterases. *Acetylcholinesterase* is a "specific" choline esterase, hydrolyzing predominantly choline esters, and characterized by high concns in brain, nerve and red blood cells. The other type, called *butyrylcholinesterase*, is a nonspecific choline esterase ("pseudo" choline esterase), hydrolyzing other esters as well as choline esters, and found in blood serum, pancreas and liver, *cf.* Sumner and Somers, *Chemistry and Methods of Enzymes* (New York, 1947). Prepn from dog pancreas: Mendel, Mundell, *Biochem. J.* **37**, 64 (1943); from horse serum: Strelitz, *ibid.* **38**, 86 (1944). *Reviews:* Davies, Green, *Advan. Enzymol.* **20**, 283 (1958); Froede, Wilson, "Acetylcholinesterase" in *The Enzymes*, P. D. Boyer, Ed. (Academic Press, N. Y., 3rd ed., 1971) vol. **5**, pp 87-114.

Specific choline esterase develops its max activity at pH 7 and at low levels of acetylcholine (less than 2.5 mg %). Both enzymes are inhibited by very small quantities of physostigmine (eserine). Acetylcholinesterase is inhibited by phosphorus-containing insecticides and nerve gases.

2212. Choline Salicylate. *2-Hydroxy-N,N,N-trimethylethanaminium salt with 2-hydroxybenzoic acid (1:1); (2-hydroxyethyl)trimethylammonium salicylate;* choline salicylic acid salt; salicylic acid choline salt; Actasal; Arret; Arthropan; Artrobione; Audax; Mundisal. C$_{12}$H$_{19}$NO$_4$; mol wt 241.28. C 59.73%, H 7.94%, N 5.81%, O 26.53%. Prepn from choline chloride and sodium salicylate: Broh-Kahn, *Intern. Rec. Med.* **173**, 219 (Apr. 1960); *cf.* Johnson, **Brit. pat. 8031** (1919); Broh-Kahn *et al.*, U.S. pat. **3,069,321** (1962 to Labs. for Pharmaceut. Dev.). Variations of process: Belg. pat. **583,513** (1960 to Mundipharma, AG).

Extremely hygroscopic solid, mp 49.5-50.0°. Very freely sol in water. Also sol in alcohol, acetone, other hydrophilic solvents. Practically insol in ether, petr ether, benzene, oils. Aq solns are stable, they contain the compd in the form of its dissociated choline and salicylate ions. pH of 10% aq soln 6.5. Aq solns are easily discolored by minute traces of iron. The addition of acid to aq solns immediately precipitates free salicylic acid, while choline base, readily recognized by its fishy odor, is liberated upon the addition of alkali.

THERAP CAT: Analgesic; antipyretic.

2213. Choline Theophyllinate. *2-Hydroxy-N,N,N-trimethylethanaminium salt with 3,7-dihydro-1,3-dimethyl-1H-purine-2,6-dione;* theophylline cholinate; theophylline salt of choline; oxtriphylline; oxytrimethylline; Sabidal; Teokolin; Teofilcolina; Filoral; Cholinophylline; Choledyl; Theoxylline; Soliphylline. Prepn: Ladenburg *et al.*, U.S. pat. **2,776,287** (1957 to Nepera Chem.).

White granules (contains about 60% of anhydr theophylline). One gram dissolves in 1 ml water. Also sold as Cholinophylline Magnesium Glycinate which contains 43% of anhydr theophylline.

THERAP CAT: Bronchodilator.

2214. Chondrillasterol. *Stigmasta-7,22-dien-3-ol.* C$_{29}$H$_{48}$O; mol wt 412.67. C 84.40%, H 11.72%, O 3.88%. Stereoisomeric with α-spinasterol, *q.v.* Isolated from the green alga *Scenedesmus obliquus* (Turpin) Kuetz., *Scenedesmaceae*: Bergmann, Feeney, *J. Org. Chem.* **15**, 812 (1950). Synthesis: W. Sucrow *et al.*, *Phytochemistry* **15**, 1533 (1976); M. Anastasia *et al.*, *J. Chem. Soc. Perkin Trans. I* **1981**, 2561.

Crystals from chloroform-methanol, mp 168-169°. [α]$_D^{24}$ −2° (chloroform).

Acetate, C$_{31}$H$_{50}$O$_2$, crystals from chloroform-methanol, mp 174.5-175.5°. [α]$_D^{24}$ −0.7° (chloroform).

Benzoate, C$_{36}$H$_{52}$O$_2$, crystals from dioxane, mp 194-195°. [α]$_D^{24}$ +3.9° (chloroform).

2215. Chondrocurine. *6,6′-Dimethoxy-2,2′-dimethyltubocuraran-7′,12′-diol;* d-chondocurine; d-tubocurine. C$_{36}$H$_{38}$N$_2$O$_6$; mol wt 594.68. C 72.70%, H 6.44%, N 4.71%, O 16.14%. Isoln from *Chondodendron tomentosum* Ruiz & Pav., *Menispermaceae*: Wintersteiner, Dutcher, *Science* **97**, 467 (1943); King, *J. Chem. Soc.* **1948**, 1945; Bick, Clezy, *ibid.* **1960**, 2402; Boissier *et al.*, *Lloydia* **28**, 191 (1965), *C.A.* **64**, 948b (1966). Structure: Dutcher, *J. Am. Chem. Soc.* **68**, 419 (1946); Hultin, *Acta Chem. Scand.* **15**, 1130 (1961). Configuration: Bick, Clezy, *J. Chem. Soc.* **1953**, 3893; Hultin, *Acta Chem. Scand.* **17**, 753 (1963). Revised structure: Everett *et al.*, *Chem. Commun.* **1970**, 1020. ^{13}C-NMR: L. Koike *et al.*, *J. Org. Chem.* **46**, 2385 (1981).

Slender needles from methanol, mp 232-234°; also reported as 218-220° (Boissier). [α]$_D^{24}$ +200° (c = 0.5 in 0.1N HCl). [α]$_D^{24}$ +105° (c = 0.9 in pyridine).

Sulfate tetrahydrate, C$_{36}$H$_{38}$N$_2$O$_6$.H$_2$SO$_4$.4H$_2$O, rectangular plates from water. Contains 9.43% H$_2$O. After drying, mp 263-265° (dec). [α]$_D^{24}$ +184° (c = 0.375 in methanol).

Methiodide, mp 270-275°.

Dimethiodide, C$_{38}$H$_{44}$I$_2$N$_2$O$_6$, pale amorphous precipitate, mp 275° (dec). [α]$_D^{24}$ +184° (c = 0.375 in methanol). uv max (water): 225, 280 nm (ε 62,000, 7030).

USE: Methiodides have curarizing power.

2216. Chondrofoline. *6,6′,7′-Trimethoxy-2,2′-dimeth-*

yltubocuraran-12'-ol; (R,R)-7-O-methylbebeerine; *7-O*-methylcurine. $C_{37}H_{40}N_2O_6$; mol wt 608.74. C 73.00%, H 6.62%, N 4.60%, O 15.77%. Alkaloid isolated from the leaves of *Chondrodendron platyphyllum* (A. St. Hil.) Miers, *Menispermaceae:* King, *J. Chem. Soc.* **1940**, 737. Structure, stereochemistry, NMR and mass spectrum: Baldas *et al.*, *Chem. Commun.* **1971**, 132. ^{13}C-NMR: L. Koike *et al.*, *J. Org. Chem.* **46**, 2385 (1981).

Solvated plates from methanol, mp about 135°. $[\alpha]_{5461}^{20}$ −281° (0.1*N* HCl).

Nitrate hydrate, $C_{37}H_{40}N_2O_6 \cdot 2HNO_3 \cdot H_2O$, needles from water, dec 225°.

2217. Chondroitin Sulfate. Chondroitinsulfuric acid; Chonsurid; Structum. Mol wt estimated at 50,000 depending on source and method of prepn: Schubert, *Fed. Proc.* **17**, 1099 (1958). High viscosity mucopolysaccharides (glycosaminoglycans) with *N*-acetylchondrosine as a repeating unit and with one sulfate group per disaccharide unit. These biological polymers act as the flexible connecting matrix between the tough protein filaments in cartilage to form a polymeric system similar to reinforced rubber. Chondroitin 4-sulfate and chondroitin 6-sulfate are the most abundant mucolysaccharides in the body and occur both in skeletal and soft connective tissue. Dermatan sulfate, formerly called chondroitin sulfate B, is present in soft connective tissue and is abundant in skin, arterial walls and heart valves. (It differs from the chondroitins by containing iduronic acid in place of glucuronic acid, its epimer, at carbon atom 5). Isoln: Bray *et al.*, *Biochem. J.* **38**, 142 (1944); Patat, Elias, *Z. Physiol. Chem.* **316**, 1 (1959); Kasavina *et al.*, **USSR** pat. **157,466** (1962); Wheat, Davidson, *Biochem. Prepns.* **10**, 52 (1963); Haneno, **Japan.** pat. **7650('64)** (to Yasushi Hano). Structure: Davidson, Meyer, *J. Am. Chem. Soc.* **77**, 4796 (1955). Absorption spectrum of A: Orr, *Biochim. Biophys. Acta* **14**, 173 (1954); of B + C: Mathews, *Nature* **181**, 421 (1958). Clinical trials in atherosclerosis: K. Nakazawa, K. Murata, *J. Int. Med. Res.* **6**, 217 (1978); eidem, *Z. Alternsforsch.* **34**, 153 (1979). *Reviews:* K. Meyer, "Chondroitin Sulfates" in *Polysaccharides in Biology*, Trans. 4th Conf. 1958, G. F. Springer, Ed. (Josiah Macy Jr. Foundn., New York, 1959) p 11; Muir, *Am. J. Med.* **47**, 673-690 (1969); Rodén, *Pure Appl. Chem.* **35**, 181-193 (1973). *See also* Chondrosine.

Chondroitin 4-sulfate, chondroitin sulfate A, CSA, Atheroitin. $R_1 = SO_3H$, $R_2 = H$. $[\alpha]_D$ −28 to −32°.
Chondroitin 4-sulfate disodium salt, Condrosulf, Lacrypos P.O.S.
Chondroitin 6-sulfate, chondroitin sulfate C. $R_1 = H$, $R_2 = SO_3H$. $[\alpha]_D$ −12° to −18°.
Dermatan sulfate, chondroitin sulfate B, β-heparin. $R_1 = SO_3H$, $R_2 = H$, C-5' epimer. $[\alpha]_D$ −60° to −70°.
Note: Mucopolysaccharidepolysulfuric acid esters now being marketed include: *Arteparon, Ateroid, Eleparon.*
THERAP CAT: Antihyperlipoproteinemic.

2218. Chondrosine. *2-Amino-2-deoxy-3-O-β-*D*-glucopyranurosyl-*D*-galactose;* β-D-glucopyranosyluronic acid 2-deoxy-2-amino-D-galactose. $C_{12}H_{21}NO_{11}$; mol wt 355.31. C 40.56%, H 5.96%, N 3.94%, O 49.53%. Disaccharide unit of chondroitins 4-sulfate and 6-sulfate. Isoln: Hebting, *Biochem. Z.* **63**, 353 (1914); Levene, *J. Biol. Chem.* **140**, 267 (1941); Wolfrom *et al.*, *J. Am. Chem. Soc.* **74**, 1491 (1952); Davidson, Meyer, *ibid.* **76**, 5686 (1954). Structure: *eidem, loc. cit.* **77**, 4796 (1955). Synthesis: Takanashi *et al.*, *ibid.* **84**, 3029 (1962). Chondrosine yields on acid hydrolysis chondrosamine (D-galactosamine, *q.v.*). Prepn and derivatives: Stacey, *J. Chem. Soc.* **1944**, 272; Wolfrom, Onodera, *J. Am. Chem. Soc.* **79**, 4737 (1957).

D-glucuronic acid   chondrosamine

Crystals from aq ethanol. $[\alpha]_D^{24}$ +40° (0.05 HCl). $[\alpha]_D^{20}$ +39° (water). Dec on heating.

Methyl ester hydrochloride, $C_{13}H_{24}ClNO_{11}$, crystals from hot ethanol, mp 159-161°. $[\alpha]_D^{23}$ +42° (c = 2 in water).

2219. Chonemorphine. *(3β,5α,20S)-N²⁰,N²⁰-Dimethylpregnane-3,20-diamine;* 3β-amino-20α-dimethylamino-5α-pregnane. $C_{23}H_{42}N_2$; mol wt 346.58. C 79.70%, H 12.22%, N 8.08%. From bark and leaves of *Chonemorpha macrophylla* G. Don, *Ch. fragrans* Alston, and *Ch. penangensis* Ridl., *Apocynaceae:* Greshoff, *Ber.* **23**, 3545 (1890); Das, Pillay, *J. Sci. Ind. Res. (India)* **13B**, 602, 701 (1954); Chatterjee, Das, *Chem. & Ind. (London)* **1959**, 1445. Structure and synthesis: *eidem, ibid.* **1960**, 290; Janot *et al.*, *Bull. Soc. Chim. France* **1962**, 111. Stereochemistry: Chien *et al.*, *J. Org. Chem.* **29**, 315 (1964).

Bitter crystals, mp 149°. $[\alpha]_D$ +25° (chloroform). Sol in chloroform, ether.

N-Benzilidenechonemorphine, $C_{30}H_{46}N_2$, crystals, mp 182-183°. $[\alpha]_D$ +36°. uv max: 248, 277, 287 nm.

N-Acetylchonemorphine, $C_{25}H_{44}N_2O$, crystals, mp 262-265°. $[\alpha]_D$ +17°.

2220. Chorismic Acid. *(3R-trans)-3-[(1-Carboxyethenyl)oxy]-4-hydroxy-1,5-cyclohexadiene-1-carboxylic acid;* 3-enolpyruvic ether of *trans*-3,4-dihydroxycyclohexa-1,5-diene carboxylic acid; α-(5-carboxy-1,2-dihydro-2-hydroxyphenoxy)acrylic acid. $C_{10}H_{10}O_6$; mol wt 226.19. C 53.10%, H 4.46%, O 42.44%. The first branch point intermediate in the biosynthesis of aromatic amino acids via the shikimate pathway in bacteria, fungi, and higher plants; naturally occurring as the (−)-form. Its existence was predicted, then discovered during development of a mutant of *A. aerogenes:* M. I. Gibson *et al.*, *Nature* **195**, 1173 (1962); M. I. Gibson, F. Gibson, *Biochim. Biophys. Acta* **65**, 160 (1962). NMR and preliminary structure study: F. Gibson, M. Jackman, *Nature* **198**, 388 (1963). Isoln and metabolism study: M. Gibson, F. Gibson, *Biochem. J.* **90**, 248 (1964). Prepn and characterization of the barium salt: F. Gibson, *ibid.* 256. Structure, relative and abs config, prepn of the monohydrate: J. M. Edwards, L. M. Jackman, *Aust. J. Chem.* **18**, 1227 (1965). Total synthesis of racemic chorismic

acid: D. A. McGowan, G. A. Berchtold, *J. Am. Chem. Soc.*
104, 1153 (1982); B. Ganem *et al., ibid.* 6787; improved synthesis: G. A. Berchtold *et al., ibid.* **105**, 6265 (1983); short
formal synthesis: G. H. Posner *et al., J. Org. Chem.* **52**, 4836
(1987). Total synthesis of (−)-form: J. L. Pawlak, G. A.
Berchtold, *ibid.* 1765. Potential use in development of herbicides: S. Stinson, *Chem. & Eng. News* **60**, 31 (Dec. 6,
1982). Reviews of chorismic acid in biosynthesis of aromatic
amino acids: F. Gibson, J. Pittard, *Bacteriol. Rev.* **32**, 465-
492 (1968); R. J. Ife *et al., J. Chem. Soc. Perkin Trans. I*
1976, 1776-1783; U. Weiss, J. M. Edwards, *The Biosynthesis
of Organic Compounds* (Wiley, New York, 1980) pp 134-143.
See also shikimic acid.

Crystals may be obtained but show marked tendency toward solvent retention, mp 105-108° (dec). $[\alpha]_D^{21}$ −274°
(c = 0.16 in water).
Barium salt trihydrate, $C_{10}H_8BaO_6.3H_2O$. Unstable. uv
max (aq soln): 272 nm (ε 2700).
Monohydrate, colorless crystals from ethyl acetate+ light
petroleum, mp 148-149° (dec); from ether + light petroleum, mp 112° (dec); from ethyl acetate + carbon tetrachloride, mp 115° (dec). $[\alpha]_{5890}^{25}$ −295.5° (c = 0.2 in water). uv
max (water): 275 nm (ε 2630).
(±)-Form, crystals from ethyl acetate/hexane, mp 139.4-
141° (dec).

2221. Chromic Acetate. $C_6H_9CrO_6$; mol wt 229.14. C
31.45%, H 3.96%, Cr 22.70%, O 41.90%. $Cr(CH_3COO)_3$.
The commercial material, usually sold as a concd soln of the
basic acetate, $Cr(OH)(C_2H_3O_2)_2$, contains Na acetate or
Na_2SO_4 impurities. Industrial prepn: Stover; Drew, U.S.
pats. **2,650,239**; **2,678,328** (1953, 1954 both to Socony-
Vacuum Oil). Prepn of hexahydrate: Hein, Herzog in
Handbook of Preparative Inorganic Chemistry vol. 2, G.
Brauer, Ed. (Academic Press, New York, 2nd ed., 1965)
p 1371. Review: *Chromium* vol. 1, M. J. Udy, Ed., A.C.S.
Monograph Series no. **132** (Reinhold, New York, 1956)
pp 229-233.
Hydrate, approx $Cr(C_2H_3O_2)_3.H_2O$, gray-green powder or
violet plates. Slightly sol in water; practically insol in alc.
MLD i.v. in mice: 2.29 g/kg, *Handbook of Toxicology* vol. 1,
W. S. Spector, Ed. (Saunders, Philadelphia, 1956) pp 70-71.
Hexahydrate, *hexaaquochromium triacetate*. Blue-violet
needles. Readily sol in water, with partial hydrolysis, giving
a soln which is blue under the incident light and red under
transmitted light; solvolyzed by alc.
Basic, $Cr(OH)(C_2H_3O_2)_2$, violet cryst powder. Freely sol
in water.
Many other Werner complexes of chromic acetate are
known: *See* Udy, *loc. cit.*
USE: As a mordant in dyeing; in tanning; in hardening
photographic emulsions; to improve light stability and dye
affinity of textiles and polymers; in catalyst for polymerization of olefins; as an oxidation catalyst.

2222. Chromic Bromide. Br_3Cr; mol wt 291.76. Br
82.17%, Cr 17.83%. $CrBr_3$. Prepd by passing Br_2 vapor over
Cr powder at 1000°: Hein, Herzog, in *Handbook of Preparative Inorganic Chemistry* vol. 2, G. Brauer, Ed. (Academic
Press, New York, 2nd ed., 1965) p 1341.
Black, lustrous, hexagonal crystals; green in transmitted
and reddish in reflected light. Sol in cold water only upon
addition of chromous salts; sol in boiling water.
Hexahydrate, at least two isomeric forms exist. *Dibromotetraaquochromium bromide dihydrate*, $[CrBr_2(H_2O)_4]Br.$-
$2H_2O$; green, deliquesc crystals; sol in water, alc. *Hexaaquochromium tribromide*, $[Cr(H_2O)_6]Br_3$; violet, deliquesc crystals; sol in water; insol in alc. Both forms practically insol in
ether.
USE: In catalysts for polymerization of olefins.

2223. Chromic Carbonate. $Cr_2O_3.xCO_2.yH_2O$. A basic
carbonate of indefinite composition, contg. 50-55% Cr.
Obtained by pptg a chromic salt with a sol carbonate or
bicarbonate.
Blue-green amorphous powder. Practically insol in water;
slightly sol in CO_2-contg water; sol in mineral acids.
USE: Prepn of chromic salts.

2224. Chromic Chloride. Cl_3Cr; mol wt 158.38. Cl
67.16%, Cr 32.84%. $CrCl_3$. Prepn: Heisig *et al., Inorg. Syn.*
2, 193 (1946); Pray, *ibid.* **5**, 153 (1953); Vavoulis *et al., ibid.*
6, 129 (1960); Hein, Herzog in *Handbook of Preparative
Inorganic Chemistry* vol. 2, G. Brauer, Ed. (Academic Press,
New York, 2nd ed., 1965) pp 1338-1340; prepn of hexahydrates: *eidem, ibid.,* pp 1348-1350.
Violet, lustrous, hexagonal, cryst scales. Greasy feel. mp
1152°; dissociates above 1300°. d^{25} 2.87. The rate of soln in
water, acids, organic solvents is extremely slow. Addition of
a trace of $CrCl_2$ or wetting agent aid in rapid soln in water,
alcohol. Also exists in hygroscopic, sol, peach-blossom
colored form: Pray, *loc. cit. Keep tightly closed.* MLD i.v. in
mice: 801 mg/kg: *Handbook of Toxicology* vol. 1, W. S.
Spector, Ed. (Saunders, Philadelphia, 1956) pp 70-71.
Hexahydrates, several known isomers. The commercially
available, dark green salt is *dichlorotetraaquochromium
chloride dihydrate*, [*trans*-$[CrCl_2(H_2O)_4]Cl.2H_2O$]; monoclinic crystals; d 1.849: Freeman, Dance, *Inorg. Chem.* **4**, 1555
(1965); Morosin, *Acta Cryst.* **21**, 280 (1966). The violet,
rhombohedral hydrate is *hexaaquochromium trichloride*,
[$[Cr(H_2O)_6]Cl_3$]. A light green isomer is *chloropentaaquo-
chromium dichloride monohydrate*, [$[CrCl(H_2O)_5]Cl_2.H_2O$].
All are deliquesc in air. Sol in water; dil aq solns are violet,
concd aq solns are green. pH of 0.2 molar aq soln 2.4. Sol
in alcohol; slightly sol in acetone; practically insol in ether.
Keep well closed.
USE: In chromizing; manuf of Cr metal and compds; as
catalyst for polymerization of olefins and other organic reactions; as textile mordant; in tanning; in corrosion inhibitors;
as waterproofing agent.

2225. Chromic Fluoride. CrF_3; mol wt 109.01. Cr
47.71%, F 52.29%. Prepd by heating $CrCl_3$ in a current of
HF: Kwasnik in *Handbook of Preparative Inorganic Chemistry* vol. 1, G. Brauer, Ed. (Academic Press, New York, 2nd
ed., 1963) p 257. Review of chromium halides: Fergusson
in *Halogen Chemistry* vol. 3, V. Gutmann, Ed. (Academic
Press, New York, 1967) pp 227-302.
Dark green needles, d 3.8. mp 1100°. Sublimes at 1100-
1200°. Practically insol in water, alcohol; sol in HCl with
violet color.
Trihydrate, *triaquochromium trifluoride*. Green crystals
from solns of Cr or $Cr(OH)_3$ in hydrofluoric acid. Sparingly
sol in water.
Other hydrates or Werner complexes are discussed in
Chromium vol. 1, M. J. Udy, Ed., A.C.S. Monograph Series
no. **132** (Reinhold, New York, 1956) p 183.
USE: The hydrates in printing and dyeing woolens; coloring and hardening marble; mothproofing woolen fabrics;
treating silk; polishing metals; halogenation catalyst.

2226. Chromic Formate. $C_3H_3CrO_6$; mol wt 187.06. C
19.26%, H 1.61%, Cr 27.80%, O 51.32%. $Cr(HCOO)_3$.
Prepd from $Cr(OH)_3$ and formic acid: Akhmedli, Negretov,
J. Gen. Chem. USSR **20**, 2045 (1950).
Hexahydrate, fine green crystals or gray-green powder.
Dec above 300°, evolving CO and CO_2. Sol in water; concd
aq soln is blue under incident light, red under transmitted
light; dil aq soln is green.
USE: In printing cotton in skeins; in leather tanning and
waterproofing.

2227. Chromic Hydroxide. *Chromium hydroxide;* chromic oxide gel; chromic oxide, hydrous. CrH_3O_3; mol wt
103.03. Cr 50.48%, H 2.93%, O 46.59%. $Cr(OH)_3$. Occurs
only as hydrates. Prepn: Ruthroff, *Inorg. Syn.* **2**, 190
(1946); Hein, Herzog in *Handbook of Preparative Inorganic
Chemistry* vol. 2, G. Brauer, Ed. (Academic Press, New
York, 2nd ed., 1965) pp 1345-1346.
Trihydrate, $Cr(OH)_3.3H_2O$, or $Cr_2O_3.9H_2O$, blue-green
powder. Practically insol in water; sol in dil mineral acids

when freshly prepd, giving blue or green soln; becomes insol in acids on aging.

$Cr(OH)_3 . nH_2O$, shining, vitrous, jet-black particles. Useful as catalyst in dehydrogenation of alcohols and paraffins, hydrogenation of olefins.

USE: As pigment; in tanning industry; as mordant; as catalyst for organic reactions.

2228. Chromic Nitrate. CrN_3O_9; mol wt 238.03. Cr 21.85%, N 17.65%, O 60.50%. $Cr(NO_3)_3$. Prepn of anhydr salt from N_2O_5 and $Cr(CO)_6$: Addison, Chapma, *J. Chem. Soc.* **1964**, 539; prepn of nonahydrate from $Cr(OH)_3$ and dil HNO_3 or by reducing CrO_3 in the presence of HNO_3: *Chromium* vol. 1, M. J. Udy, Ed., A.C.S. Monograph Series, no. **132** (Reinhold, New York, 1956) pp 203-204.

Pale green, extremely deliquesc powder. Non-volatile. Dec above 60°. Sol in water, ethyl acetate, DMSO; practically insol in benzene, CCl_4, $CHCl_3$; reacts with ether, sometimes vigorously.

Nonahydrate, deep violet, rhombic, monoclinic crystals. mp about 60°; dec above 100°. Sol in water, alcohol. Aq soln slowly becomes green on heating and rapidly recovers the reddish-violet color on cooling. LD_{50} in rats: 3.25 g/kg orally, Smyth *et al., Am. Ind. Hyg. Assoc. J.* **30**, 470 (1969).

USE: Prepn of Cr catalyst; in textile printing; corrosion inhibitor.

2229. Chromic Oxide. Anadonis green; chrome green; chrome ocher; chrome oxide green; chromia; chromium sesquioxide; green cinnabar; green oxide of chromium; green rouge; leaf green; oil green; ultramarine green; C.I. Pigment Green 17; C.I. 77288. Cr_2O_3; mol wt 152.02. Cr 68.43%, O 31.57%. Prepd by reaction of sodium dichromate or chromate with sulfur: Copson in *Chromium* vol. 1, M. J. Udy, Ed., ACS Monograph Series no. **132** (Reinhold, New York, 1956) pp 277-278. *Review:* Wiesburg, *Paint Ind. Mag.* **71**(2), 11 (1956). *See also Colour Index* vol. 4 (3rd ed., 1971) p 4662. Use as catalyst: R. Uma, J. C. Kuriacose, *Indian Chem. Mfr.* **8**, 11 (1970). Cytotoxic effects: V. Bianchi *et al., Toxicology* **17**, 219 (1980).

Light to dark green, fine, hexagonal crystals. mp about 2435°. bp about 3000°. d^{25} 5.22. Turns brown on heating but reverts to green color on cooling. Cryst Cr_2O_3 is extremely hard; will scratch quartz, topaz, zircon. Practically insol in water, alc, acetone. Slightly sol in acids, alkalies.

Caution: Trivalent chromium may cause skin irritation: S. Fregert, H. Rorsman, *Arch. Derm.* **90**, 4 (1964).

USE: In abrasives, refractory materials, electric semiconductors; as pigment, particularly in coloring glass; in alloys; printing fabrics and banknotes; as catalyst for organic and inorganic reactions.

2230. Chromic Phosphate. CrO_4P; mol wt 146.99. Cr 35.38%, O 43.54%, P 21.07%. $CrPO_4$. Prepn: Ness *et al., J. Am. Chem. Soc.* **74**, 4685 (1952); Eickhoff, Kebrich, U.S. pat. **2,749,214** (1956 to National Lead); Wegenknecht, Ger. pats. **1,046,597** (1958); **1,056,104** (1959).

Gray-brown to black crystals of amorphous solid. Does not melt below 1800°. $d^{32.5}$ 2.94. Partially oxidizes to CrO_3 on heating in air. Practically insol in water, acetic acid, HCl, aqua regia.

Hemiheptahydrate, *Arnaudon's green, Plessy's green.* Blue-green powder. d 2.15. Practically insol in water; sol in acids.

Hexahydrate, violet crystals. Loses water gradually on heating, becoming anhyd after one hour at 800° or 3-4 hrs at 500°. d^{14} 2.121. Practically insol in water; slightly sol in acetic acid solns; readily sol in mineral acids, alkalis, oxalic acid solns.

Radioactive form, $Cr^{32}PO_4$, *Phosphocol.*

USE: Green pigment; in wash primers; in catalysts for dehydrogenation of hydrocarbons or polymerization of olefins.

THERAP CAT: $Cr^{32}PO_4$ as radioactive agent.

2231. Chromic Potassium Oxalate. *Tripotassium tris-(ethanedioato)chromate(3−); tripotassium tris(oxalato)chromate(3−);* potassium trioxalatochromate(III); potassium chromic oxalate. $C_6CrK_3O_{12}$; mol wt 433.38. C 16.63%, Cr 12.00%, K 27.06%, O 44.30%. $K_3[Cr(C_2O_4)_3]$. Prepd by treatment of oxalic acid and $K_2C_2O_4$ with $K_2Cr_2O_7$: Bailar, Jr., Jones, *Inorg. Syn.* **1**, 37 (1939); Hein, Herzog in Hand-

book of Preparative Inorganic Chemistry vol. 2, G. Brauer, Ed. (Academic Press, New York, 2nd ed., 1965) p 1372.

Trihydrate, *potassium trioxalatotriaquochromate(III).* Black-green, monoclinic scales with transparent blue edges. Freely sol in water; practically insol in alcohol.

USE: In tanning industry; dyeing chromate colors on wool.

2232. Chromic Potassium Sulfate. Potassium chromic sulfate; potassium disulfatochromate(III). $CrKO_8S_2$; mol wt 283.23. Cr 18.36%, K 13.81%, O 45.19%, S 22.64%. $KCr(SO_4)_2$. Dodecahydrate produced by reduction of $K_2Cr_2O_7$ with SO_2: Copson in *Chromium* vol. 1, M. J. Udy, Ed., A.C.S. Monograph Series no. **132** (Reinhold, New York, 1956) p 281; electrolytic manuf: Nishihara *et al.,* **Japan.** pat. **2164('60)**, *C.A.* **55**, 5200e (1961).

Dodecahydrate, *chrome alum*, $K[Cr(SO_6H_4)_2(H_2O)_2]$.-$6H_2O$: Duval, *Chim. Anal. (Paris)* **44**, 102 (1962), *C.A.* **57**, 9479d (1962). Large, violet-red to black, octahedral, cubic crystals; ruby-red under transmitted light. d^{25} 1.83. mp 89°; at 400° loses all its H_2O. Sol in 4 parts cold, 2 parts boiling water; practically insol in alcohol. The aq soln is violet when cold, green when hot. The violet color returns after a few weeks at room temp.

USE: Mordant for dyeing fabrics uniformly; tanning leather; printing calico; rendering glue and gum insol; manuf ink, other chromium salts; waterproofing fabrics; hardening photographic emulsions.

2233. Chromic Sulfate. $Cr_2O_{12}S_3$; mol wt 392.20. Cr 26.52%, O 48.95%, S 24.52%. $Cr_2(SO_4)_3$. Prepn of anhydr salt by dehydration of hydrated forms: Rollinson, Bailar, Jr., *Inorg. Syn.* **2**, 197 (1946).

Peach-colored solid. d 3.012. Practically insol in water and acids. MLD i.v. in mice: 247 mg/kg, *Handbook of Toxicology* vol. 1, W. S. Spector, Ed. (Saunders, Philadelphia, 1956) pp 70-71.

Hydrates are known in both green and violet modifications, and have several degrees of hydration up to $18H_2O$: Lukaszewski, Redfern, *Nature* **190**, 805 (1961); Udy in *Chromium* vol. 1, M. J. Udy, Ed., A.C.S. Monograph Series no. **132** (Reinhold, New York, 1956) pp 213-217, 288. The technical product comes in the form of a finely granular, dark-green flake or powder approximating the formula $Cr_2(SO_4)_3 . 10H_2O$. Readily sol in water; almost insol in alc.

Basic chromic sulfates of the type $Cr(OH)SO_4 . nH_2O$ are of importance in the tanning industry: Udy, *loc. cit.* and pp 278-280, 305-308. Technical grades are available in two degrees of basicity, one-third and one-half, as finely granular dark-green flakes or powder contg about 25% Cr_2O_3. Readily sol in water.

USE: Insolubilization of gelatin; in catalyst prepn; as mordant in textile industry; in tanning of leather; in chrome plating; in manuf of Cr, CrO_3, and Cr alloys; to improve dispersibility of vinyl polymers in water; in manuf of green varnishes, paints, inks, glazes for porcelain.

2234. Chromium. Cr; at. wt 51.996; at. no. 24; valences 1-6. Four naturally occurring isotopes: 50 (4.35%); 52 (83.79%); 53 (9.50%); 54 (2.36%); artificial radioactive isotopes: 46-49; 51; 55; 56; longest-lived isotope is ^{51}Cr ($T_{1/2}$ 27.8 days) prepd by (n,γ) reaction from ^{50}Cr. Reported abundance in earth's crust varies from 100 to 300 ppm. Discovered by 1797 by Vauquelin. Obtained from chrome ore, *chromite* ($FeCr_2O_4$), by a silicothermic or aluminothermic process. Reviews of chromium, its alloys and compds: *Chromium*, M. J. Udy, Ed., A.C.S. Monograph Series, no. **132** (Reinhold, New York, 1956) vol. 1, 433 pp; vol. 2, 402 pp; Rollinson, "Chromium, Molybdenum and Tungsten" in *Comprehensive Inorganic Chemistry* vol. 3, J. C. Bailar, Jr. *et al.,* Eds. (Pergamon Press, Oxford, 1973) pp 623-700; J. H. Westbrook in Kirk-Othmer *Encyclopedia of Chemical Technology* vol. 6 (Wiley-Interscience, New York, 3rd ed., 1979) pp 54-82. Important trace element. Review of biological function of the chromium(III) ion: Mertz, *Physiol. Rev.* **49**, 163-239 (1969). Review of carcinogenicity studies of chromium and chromium compds: *IARC Monographs* **2**, 100-125 (1973); *ibid.* **23**, 205-323 (1980). Book: *Chromium: Metabolism and Toxicity*, D. Burrows, Ed. (CRC Press, Boca Raton, 1983) 172 pp.

Steel-gray, lustrous metal; body-centered cubic structure; hard as corundum and less fusible than platinum. Takes a high polish. d 7.14. mp 1900°. bp 2642°. Heat capacity (25°): 5.58 cal/g-atom deg. Latent heat of fusion: approx 3.5 kcal/g-atom; latent heat of vaporization: approx 81.7 kcal/g-atom. Reacts with dil HCl, H_2SO_4; not with HNO_3; attacked by caustic alkalies and alkali carbonates. Not oxidized by air, even in presence of much moisture.

Caution: Chromic acid or chromate salts constitute industrial hazards. Irritant effects on the skin and respiratory passages lead to ulceration. Oral ingestion may lead to severe irritation of the gastrointestinal tract, circulatory shock and renal damage. Chromium(III) compounds show little or no toxicity. This substance and certain chromium compounds have been listed as known carcinogens: *Fourth Annual Report on Carcinogens* (NTP 85-002, 1985) p 58.

USE: In manuf of chrome-steel or chrome-nickel-steel alloys (stainless steel); for greatly increasing resistance and durability of metals; for chromeplating of other metals. The man-made ^{51}Cr isotope as tracer in various blood diseases and in the determination of blood volume (as the chloride or as Na chromate).

2235. Chromium Carbonyl. Chromium hexacarbonyl. C_6CrO_6; mol wt 220.07. C 32.75%, Cr 23.63%, O 43.62%. $Cr(CO)_6$. Prepn from Cr salts and CO in the presence of a Grignard reagent: Owen *et al., Inorg. Syn.* **3**, 156 (1950); in the presence of Mg and ether: Wender, U.S. pat. **3,012,858** (1961 to Diamond Alkali); in the presence of Na and diglyme: Podall *et al., J. Am. Chem. Soc.* **83**, 2057 (1961); in the presence of Na and an aromatic hydrocarbon: Pruett, Wyman, U.S. pat. **3,053,629** (1962 to Union Carbide); in the presence of I_2 and a nitrile: Wotiz, U.S. pat. **3,100,687** (1963 to Diamond Alkali).

Orthorhombic, highly refractive crystals. Sublimes at room temp. Sinters at 90°; dec at 130°; explodes at 210°. Burns with a luminous flame. d^{18} 1.77. Vapor pressure (mm): 0.04 (0°); 1.0 (48°); 66.5 (100°). Almost insol in water, ethanol, methanol; sol in ether, $CHCl_3$, other organic solvents. Solns or impure solid dec by light. LD_{50} in mice: 100 mg/kg i.v., Strohmeier, *Z. Naturforsch.* **19b**, 540 (1964).

USE: In catalysts for olefin polymerization and isomerization; gasoline additive to increase octane number; prepn of chromous oxide, CrO.

2236. Chromium Dioxide. CrO_2; mol wt 84.00. Cr 61.90%, O 38.10%. Prepn: Wöhler, *Ann.* **111**, 117 (1859); Thamer *et al., J. Am. Chem. Soc.* **79**, 547 (1957); Swoboda *et al., J. Appl. Phys.* **32**, Suppl. no. 3, 374 (1961); Arthur; Arthur, Ingraham, U.S. pats. **2,959,955; 3,117,093** (1960, 1964 both to du Pont). *Reviews:* Hund, *Farbe + Lack* **78**, 11-16 (1972); Rollinson in *Comprehensive Inorganic Chemistry* vol **3**, J. C. Bailar, Jr. *et al.,* Eds. (Pergamon Press, Oxford, 1973) pp 689-690.

Black, ferromagnetic crystals; rutile structure. d 4.89. Metastable in air; various temperatures (250-500°) reported for decompn to Cr_2O_3.

USE: In magnetic recording tapes; as catalyst.

2237. Chromium Tetrafluoride. CrF_4; mol wt 128.01. Cr 40.63%, F 59.37%. Prepd by reaction of F_2 with Cr or $CrCl_3$: von Wartenberg, *Z. Anorg. Allgem. Chem.* **247**, 136 (1941); Clark, Sadana, *Can. J. Chem.* **43**, 50 (1964).

Very dark greenish-black, amorphous solid; on exposure to moist air becomes brown on surface due to hydrolysis. mp (estimated) 277°: Fergusson in *Halogen Chemistry* **Vol. 3**, V. Gutmann, Ed. (Academic Press, New York, 1967) p 242. bp about 400° evolving steel-blue vapor; sublimes *in vacuo* above 100°; d 2.89. Less reactive than CrF_3, does not readily form Werner complexes; does not react with NH_3, SO_2, SO_3, pyridine. Sol in water with rapid hydrolysis; practically insol in most organic solvents. Can be stored indefinitely *in vacuo* in well dried Pyrex or silica vessels.

Caution: A strong irritant.

2238. Chromium Trioxide. Chromic acid; chromic anhydride. CrO_3; mol wt 100.01. Cr 52.00%, O 48.00%. Produced commercially by the action of concd H_2SO_4 on a soln of chromate or dichromate: Faith, Keyes & Clark, *Industrial Chemicals,* F. A. Lowenheim, M. K. Moran, Eds. (Wiley-Interscience, New York, 4th ed., 1975) pp 270-274.

Dark red, deliquesc bipyramidal prismatic crystals, flakes or granular power. d 2.70. mp 197°. Dec at 250° to Cr_2O_3

and O_2. Very sol in water; sol in H_2SO_4. Powerful oxidizer; oxidizes alcohol and most other organic substances, sometimes with dangerous violence. *Contact with combustible material may cause fire. Pharmaceutical Incompat:* Alcohol, ether, glycerol, spirit nitrous ether and almost every organic substance; bromides, chlorides, iodides, hypophosphites, sulfites, sulfides.

Caution: Dermal contact can cause primary irritation and ulceraton as well as allergic eczema. Inhalation can cause nasal irritation, septal perforation. Pulmonary irritation, bronchogenic carcinoma may result from breathing chromate dust. Ingestion causes violent G.I. irritation with vomiting, diarrhea. Renal injury has been reported in exptl animals: E. Browning, *Toxicity of Industrial Metals* (Appleton-Century-Crofts, New York, 2nd ed., 1969) pp 119-131; *Clinical Toxicology of Commercial Products,* R. E. Gosselin *et al.,* Eds. (Williams & Wilkins, Baltimore, 4th ed., 1976) Section II, p 75.

USE: Chromium plating; copper stripping; aluminum anodizing; corrosion inhibitor; photography; purifying oil and acetylene; hardening microscopical prepns; oxidizing agent in organic chemistry.

THERAP CAT (VET): Has been used in solution as a topical antiseptic and astringent.

2239. Chromocarb. *4-Oxo-4H-1-benzopyran-2-carboxylic acid;* 2-chromonecarboxylic acid; 4-oxo-4H-chromene-2-carboxylic acid; benzo-γ-pyronecarboxylic acid. $C_{10}H_6O_4$; mol wt 190.16. C 63.16%, H 3.18%, O 33.66%. Prepn: S. Ruhemann, H. E. Stapleton, *J. Chem. Soc.* **77**, 1179 (1900); J. Schmutz *et al., Helv. Chim. Acta* **34**, 767 (1951); G. Pifferi *et al., J. Heterocycl. Chem.* **14**, 1257 (1977). Prepn of diethylamine salt: P. A. Tronche, **S. Afr.** pat. **68 7352;** *eidem,* **U.S.** pat. **3,816,470** (1969, 1974 both to Ferlux). Pharmacology of diethylamine salt in animals: J. Couquelet *et al., C.R. Soc. Biol.* **164**, 329 (1970); P. Conquet *et al., ibid.* 800. Clinical comparison with dipyridamole of effect on platelet function: A. Vittoria *et al., Curr. Ther. Res.* **35**, 1033 (1984). Clinical trial in diabetes with vascular disease: N. Ciavarella *et al., ibid.* **36**, 293 (1984). Bioavailability in humans: J.-M. Aiache *et al., Biopharm. Drug Dispos.* **7**, 301 (1986).

Colorless needles from alcohol, mp 250-251° (dec) (Ruhemann, Stapleton); also reported as mp 255-256° (Pifferi). uv max: 230, 305 nm (ε 20220, 8075). Sol in alcohol, ammonia. Sparingly sol in water.

Diethylamine, $C_{14}H_{17}NO_4$, *Angiophtal, Campel, Fludarene.* Microcrystalline powder from alcohol + acetone, mp 138°. Sol in water. LD_{50} in mice: ~800 mg/kg i.v.; >5 g/kg orally (Tronche, 1974).

THERAP CAT: Diethylamine salt as capillary protectant.

2240. Chromomycins. An antibiotic complex produced by *Streptomyces griseus;* composed of chromomycins group A, group B, and C, with chromomycin A_3 as the major component. Isoln of complex and antibacterial activity: M. Shibata *et al., J. Antibiot.* **13B**, 1 (1960), *C.A.* **54**, 22835g (1960). Separation of components: Mizuno and Tatsuoka *et al., ibid.* **13B**, 329, 332 (1960), *C.A.* **56**, 1745gi (1962); Miyake *et al.,* **Japan.** pat. **7842('60)** and **Ger.** pat. **1,072,775** (both 1960 to Takeda), *C.A.* **55**, 7758b, 26361f (1961); Mizuno, *J. Antibiot.* **16A**, 22 (1963). Structure of A_3: Miyamoto *et al., Tetrahedron Letters* **1964**, 2367, 2371; Tatsuoka *et al., Proc. Japan Acad.* **40**(3), 236 (1964), *C.A.* **61**, 10767h (1964); Miyamoto *et al., Tetrahedron* **23**, 421 (1967). Abs config of A_3: Harada *et al., J. Am. Chem. Soc.* **91**, 5896 (1969). Revised structure of A_3: J. Thiem, B. Meyer, *J. Chem. Soc. Perkin Trans. II* **1979**, 1331. Series of articles on partial syntheses: *Ber.* **113**, 3039-3074 (1980); J. H. Dodd *et al., J. Org. Chem.* **47**, 4045 (1982). Pharmacology and toxicity: M. Slavik, S. K. Carter, *Adv. Pharmacol. Chemother.* **12**, 1 (1975). *Review:* J. D. Skarbek, M. K. Speedie in *Antitumor*

Compounds of Natural Origin vol. **1**, A. Aszalos, Ed. (CRC Press, Boca Raton, 1981) pp 191-235.

chromomycin A₃

Chromomycin A₃, $C_{57}H_{82}O_{26}$, *3^B-O-(4-O-acetyl-2,6-dideoxy-3-C-methyl-α-L-arabino-hexopyranosyl)-7-methylolivomycin D, aburamycin B, Toyomycin.* Yellow powder, dec. 185°. $[\alpha]_D^{23} -55°$ (ethanol). uv max (ethanol): 230, 281, 304, 318, 330, 412 nm (log ε 4.39, 4.72, 3.85, 3.92, 3.84, 4.07). LD_{50} in mice (mg/kg): 1.85 i.v.(Slavik, Carter); 1.7 i.p. (Shibata).

THERAP CAT: Antineoplastic.

2241. Chromonar. *[[3-[2-(Diethylamino)ethyl]-4-methyl-2-oxo-2H-1-benzopyran-7-yl]oxy]acetic acid ethyl ester; 3-(β-diethylaminoethyl)-4-methyl-7-(carbethoxymethoxy)-coumarin; 3-(β-diethylaminoethyl)-4-methyl-7-(carbethoxymethoxy)-2H-1-benzopyran-2-one; ethyl [[3-[2-(diethylamino)ethyl]-4-methyl-2-oxo-2H-1-benzopyran-7-yl]oxy]acetate; carbochromen; carbocromen.* $C_{20}H_{27}NO_5$; mol wt 361.42. C 66.46%, H 7.53%, N 3.88%, O 22.13%. Prepn: Belg. pat. **621,327** corresp to Ritter *et al.*, U.S. pat. **3,282,-938** (1963, 1966 both to Casella Farbwerke Mainkur). Series of articles on pharmacology: *Arzneimittel-Forsch.* **13**, 243-268 (1963); on distribution: *ibid.* **22**, 479-511 (1972). Metabolic studies: Schraven *et al.*, *ibid.* **20**, 1905 (1970).

Practically insol in water.

Hydrochloride, $C_{20}H_{28}ClNO_5$, *Cassella 4489, Antiangor, Intenkordin, Intensain.* Cryst powder, mp 159-160°. Freely sol in water, alc, methylene chloride, chloroform. Sparingly sol in acetone, methyl ethyl ketone, benzene, ether. Aq solns show blue fluorescence. LD_{50} in mice (g/kg): 6.3 orally; 0.528 i.p.; 0.0355 i.v., R. E. Nitz, E. Potzch, *Arzneimittel-Forsch.* **13**, 243 (1963).

THERAP CAT: Coronary vasodilator.

2242. Chromotrope 2B. *4,5-Dihydroxy-3-[(4-nitrophenyl)azo]-2,7-naphthalenedisulfonic acid disodium salt; C.I. Acid Red 176; p-nitrobenzeneazochromotropic acid sodium salt; C.I.* 16575. $C_{16}H_9N_3Na_2O_{10}S_2$; mol wt 513.38. C 37.43%, H 1.77%, N 8.19%, Na 8.96%, O 31.17%, S 12.49%. Prepn from diazotized p-nitroaniline and chromotropic acid, *see: Colour Index* vol. **4** (3rd ed., 1971) p 4097.

Reddish-brown powder. Sol in water with a yellowish-red color. Insol in alcohol.

USE: As a dye; as a reagent for boric acid or borates.

2243. Chromotropic Acid. *4,5-Dihydroxy-2,7-naphthalenedisulfonic acid;* 1,8-dihydroxynaphthalene-3,6-disulfonic acid. $C_{10}H_8O_8S_2$; mol wt 320.29. C 37.50%, H 2.52%, O 39.96%, S 20.02%. Prepn from 4-chloro-5-hydroxy-2,7-naphthalenedisulfonic acid: **Ger. pat. 147,852** (1904 to BASF); alternate methods: *Frdl.* **3**, 460-466.

Dihydrate, white needles, soluble in water.

Disodium salt dihydrate, $C_{10}H_6Na_2O_8S_2.2H_2O$, white needles or leaflets. Very sol in water.

USE: Analytical reagent; intermediate for azo dyes.

2244. Chromous Acetate. $C_4H_6CrO_4$; mol wt 170.10. C 28.24%, H 3.56%, Cr 30.58%, O 37.62%. $Cr(C_2H_3O_2)_2$. Prepn of monohydrate: Balthis, Jr., Bailar, Jr., *Inorg. Syn.* **1**, 122 (1939); Hatfield, *ibid.* **3**, 148 (1950); Kranz, Witkowska, *ibid.* **6**, 144 (1960); Ocone, Block, *ibid.* **8**, 125 (1966).

Monohydrate, deep red powder, or monoclinic crystals; composed of dimeric units. Easily oxidized, especially when moist, to chromic acetate. Loses H_2O when dried over P_2O_5 at 100°, changing color to brown. d 1.79. Slightly sol in cold water; readily sol in hot water; sol in and reacts with most acids; slightly sol in alcohol; practically insol in ether. *Keep tightly closed.* LD_{50} orally in rats: 11.26 g/kg, H. F. Smyth *et al.*, *Am. Ind. Hyg. Assoc. J.* **30**, 470 (1969).

USE: Prepn of other chromous salts; as O_2 absorber in gas analysis.

2245. Chromous Bromide. Br_2Cr; mol wt 211.84. Br 75.45%, Cr 24.55%. $CrBr_2$. Prepn: Hein, Herzog in *Handbook of Preparative Inorganic Chemistry*, Vol. 2, G. Brauer, Ed. (Academic Press, New York, 2nd ed. 1965) p 1340. Review of chromium halides: Fergusson in *Halogen Chemistry*, Vol. 3, V. Gutmann, Ed. (Academic Press, New York, 1967) pp 227-302.

White monoclinic crystals, becoming yellow when heated. mp 842°; d_4^{25} 4.236. Stable in dry air, oxidizes in moist air. Sol in water (exothermic) giving blue soln; sol in alcohol. *Keep well closed.*

USE: In chromizing.

2246. Chromous Chloride. Cl_2Cr; mol wt 122.92. Cl 57.69%, Cr 42.31%. $CrCl_2$. Prepn: Burg, *Inorg. Syn.* **3**, 150 (1950); Hein, Herzog, in *Handbook of Preparative Inorganic Chemistry*, Vol. 2, G. Brauer, Ed. (Academic Press, New York, 2nd ed. 1965) pp 1336-1338. Prepn of tetrahydrate: Balthis, Jr., Bailar, Jr., *Inorg. Syn.* **1**, 125 (1939); Holah, Fackler, *ibid.* **10**, 26 (1967).

Lustrous needles or fused, fibrous mass. mp 824°; d_4^{14} 2.751. Very hygroscopic; stable in dry air but oxidizes rapidly if moist. Powerful reducing agent. Soluble in water giving a blue soln. *Keep well closed.* LD_{50} orally in rats: 1.87 g/kg, H. F. Smyth *et al.*, *Am. Ind. Hyg. Assoc. J.* **30**, 470 (1969).

Tetrahydrate, *tetraaquochromium dichloride.* Bright blue, hygroscopic crystals. Above 38° changes to isomeric green modification. Loses $1H_2O$, forming trihydrate, at 51°. Absorbs O_2 even when dry, forming a greenish-black oxychloride. Sol in water; almost insol in concd HCl. On standing in soln it is oxidized by the water with liberation of H_2. *Keep well closed.*

USE: In chromizing; in prepn of Cr metal; in catalysts for organic reactions; as O_2 absorbent; in analysis.

2247. Chromous Fluoride. CrF_2; mol wt 90.01. Cr

57.78%, F 42.22%. Prepd from $CrCl_2$ + HF: Poulenc, *Compt. Rend.* **116**, 254 (1893); Kwasnik in *Handbook of Preparative Inorganic Chemistry*, **Vol. 1**, G. Brauer, Ed. (Academic Press, New York, 2nd ed., 1963) pp 256-257. Review of other preparative methods: Sturm, *Inorg. Chem.* **1**, 665-672 (1962).

Blue-green, monoclinic crystals with iridescent sheen. d 3.79. mp 894°. Sparingly sol in water; practically insol in alcohol; sol in boiling HCl; not attacked by hot dil H_2SO_4 or HNO_3. Changes to Cr_2O_3 when heated in air.

USE: In chromizing; in catalytic cracking of hydrocarbons; alkylation catalyst; in nuclear reactor fuels. *Caution*: A strong irritant.

2248. Chromous Formate. $C_2H_2CrO_4$; mol wt 142.05. C 16.91%, H 1.42%, Cr 36.61%, O 45.05%. $Cr(HCOO)_2$. Prepd from $CrCl_2$ and Na formate: Lux, Illman, *Ber.* **91**, 2143 (1958); Earnshaw *et al.*, *Proc. Chem. Soc.* **1963**, 281. Prepn of blue and red anhydrous forms from hydrated formates: Herzog, Kalies, *Z. Chem.* **5**, 273 (1965).

Hemihydrate, violet to blue crystals. Sol in water.

Monohydrate, red needles. Sol in water to give blue soln.

Hemipentahydrate, large, dark-red cubic crystals. Sol in water, alcohol, ether, acetone.

USE: In baths for Cr electroplating; in catalysts for organic reactions.

2249. Chromous Oxalate. C_2CrO_4; mol wt 140.02. C 17.15%, Cr 37.14%, O 45.71%. CrC_2O_4. Prepn from chromous acetate and oxalic acid: *Chromium*, **Vol. 1**, M. J. Udy, Ed., A.C.S. Monograph Ser., no. **132** (Reinhold, New York, 1956) p 227; from chromous sulfate and sodium oxalate: Lux, Illmann, *Ber.* **91**, 2143 (1958). Reported as monohydrate (Udy, *loc. cit.*) or dihydrate (Lux, Illman, *loc. cit.*).

Hydrate, yellow to yellowish-green crystalline powder. d 2.468. Most stable of chromous salts. Not appreciably oxidized by moist air. Practically insol in cold water, alcohol; sol in hot water, dil acids.

2250. Chromous Sulfate. CrO_4S; mol wt 148.07. Cr 35.13%, O 43.22%, S 21.65%. $CrSO_4$. Prepn of pentahydrate: Lux, Illman, *Ber.* **91**, 2143 (1958); Holah, Fackler, *Inorg. Syn.* **10**, 26 (1967). The common hydrated form, long considered to be the heptahydrate, has been reported to be the pentahydrate: Lux, Illman, *loc. cit.*; Lux *et al.*, *Ber.* **97**, 503 (1964). Prepn of standard soln for use as an analytical reagent: Lingane, Pecsok, *Anal. Chem.* **20**, 425 (1948).

Pentahydrate, blue crystals. Stable in air if dry. Sol in water; slightly sol in alcohol; practically insol in acetone. Soluble in dil, dec by concd H_2SO_4. Solns are rapidly oxidized by atmospheric oxygen.

USE: Analytical reagent; absorption of O_2 from gas mixtures; dehydrohalogenating and reducing agent.

2251. Chromyl Chloride. *Dichlorodioxochromium;* chromium dioxychloride. Cl_2CrO_2; mol wt 154.92. Cl 45.77%, Cr 33.57%, O 20.66%. CrO_2Cl_2. Prepn from CrO_3 + HCl: Sisler, *Inorg. Syn.* **2**, 205 (1946); Hein, Herzog in *Handbook of Preparative Inorganic Chemistry*, **Vol. 2**, G. Brauer, Ed. (Academic Press, New York, 2nd ed., 1965) p 1384; and $AlCl_3$: Flesch, Svec, *J. Am. Chem. Soc.* **80**, 3189 (1958); from Cr_2O_3 and $TiCl_4$: Braos, Cohen, U.S. pat. **3,111,380** (1963 to Natl. Distillers & Chem.). Review of chromium halides: Fergusson in *Halogen Chemistry*, **Vol. 3**, V. Gutmann, Ed. (Academic Press, New York, 1967) pp 227-302.

Deep red liq; appears black under reflected light. Fumes in moist air. mp −96.5°; bp 117°; d_4^{25} 1.91. Slightly less viscous than water. Nonconductor of electricity. Strong oxidizing agent; *can react explosively with combustible organic and inorganic substances*. Hydrolyzes vigorously on contact with water. Reacts vigorously with liq or gaseous ammonia. Sol in CCl_4, CS_2, benzene, nitrobenzene, $CHCl_3$, $POCl_3$. Its soln in CCl_4 is fairly stable. Liquid CrO_2Cl_2 is stable indefinitely in glass, aluminum, stainless steel containers if protected from light and moisture. CrO_2Cl_2 dissolves CrO_3, yielding a powerful oxidant. LD_{50} s.c. in mice: 5.45 mg/kg (approx): *Toxic Substances List*, H. E. Christensen, Ed. (1972) p 145.

Caution: Burns and blisters the skin. *Handle only in well-ventilated hood.*

USE: Catalyst for polymerization of olefins; oxidation of

hydrocarbons; in the Etard reaction for production of aldehydes and ketones; in the prepn of various coordination complexes of Cr.

2252. Chromyl Fluoride. *Difluorodioxochromium;* chromium oxyfluoride. CrF_2O_2; mol wt 122.01. Cr 42.63%, F 31.14%, O 26.23%. CrO_2F_2. Prepn of impure product from CrO_2Cl_2 and F_2: von Wartenberg, *Z. Anorg. Allgem. Chem.* **247**, 140 (1941). Prepn from CrO_3 and SeF_4: Bartlett, Robinson, *J. Chem. Soc.* **1961**, 3549; from $K_2Cr_2O_7$ and HF or from Cr, KNO_3 and HF: Wiechert, *Z. Anorg. Allgem. Chem.* **261**, 315 (1950). Review of earlier prepns and prepn from CrO_3 and ClF, COF_2, MoF_6 or WF_6: P. J. Green, G. L.Gard, *Inorg. Chem.* **16**, 1243 (1977). Review of chromium halides: Fergusson in *Halogen Chemistry*, **vol. 3**, V. Gutmann, Ed. (Academic Press, New York, 1967) pp 227-302.

Red-violet to black rhombic or monoclinic crystals; described as a gas or a liq in earlier work with impure samples. Sublimation temp 29.6°; mp 31.6° (885 mm). *Extremely reactive;* etches glass, quartz. Ignites hydrocarbons such as methane or butane at elevated temps; mixtures burn with a bright flame producing fumes of Cr_2O_3 and CrF_3. Can be stored in a sealed aluminum phosphate glass tube or in a Kel-F tube. Stable indefinitely at room temp in the dark. Polymerizes slowly if exposed to visible, ultraviolet or infrared light to a gray-white solid, mp 200°.

USE: Fluorination catalyst; to increase olefin-polymer receptivity for dyes.

2253. Chrysamminic Acid. *1,8-Dihydroxy-2,4,5,7-tetranitro-9,10-anthracenedione;* 1,8-dihydroxy-2,4,5,7-tetranitroanthraquinone; 2,4,5,7-tetranitrochrysazin; chrysammic acid. $C_{14}H_4N_4O_{12}$; mol wt 420.20. C 40.01%, H 0.96%, N 13.33%, O 45.69%. Prepn by heating chrysazin with fuming nitric acid: Liebermann, *Ann.* **183**, 193 (1876). Structure: Robinson, Simonsen, *J. Chem. Soc.* **95**, 1088 (1909).

Golden-yellow, lustrous, bitter leaflets. Almost insol in water; sol in alcohol with deep-red and in ether with yellow color. *Caution*: The acid and its salts explode when ignited or heated rapidly in air.

2254. Chrysanthemaxanthin. *5,8-Epoxy-5,8-dihydro-β,ε-carotene-3,3'-diol.* $C_{40}H_{56}O_3$; mol wt 584.85. C 82.14%, H 9.65%, O 8.21%. Carotenoid pigment. Occurs together with flavoxanthin. Has same structural formula, mp, rotation, and absorption as flavoxanthin, *q.v.*, but differs sterically. For absolute configuration see: H. Cadosch *et al.*, *Helv. Chim. Acta* **61**, 783 (1978). Can be separated chromatographically: In a zinc carbonate column the chrysanthemaxanthin zone locates itself below flavoxanthin and above lutein epoxide. Isoln from asters: Karrer, Jucker, *ibid.* **26**, 626 (1943). Partial synthesis from lutein, *eidem, ibid.* **28**, 300 (1945).

Golden yellow leaflets, mp 184-185°. $[\alpha]_C^{20}$ +180° to +190° (c = 0.04 in benzene). Absorption max ($CHCl_3$): 459, 430 nm. Freely sol in chloroform, benzene, acetone. Less sol in methanol, ethanol. Almost insol in petr ether.

Note: Chrysanthemaxanthin has no vitamin A activity.

2255. Chrysanthemic Acid. *2,2-Dimethyl-3-(2-methyl-1-propenyl)cyclopropanecarboxylic acid;* chrysanthemummonocarboxylic acid; chrysanthemumic acid. $C_{10}H_{16}O_2$; mol wt 168.23. C 71.39%, H 9.59%, O 19.02%. Occurs as esters in pyrethrum flowers, see Pyrethrin I and Cinerin I, also Allethrin (a synthetic product). Isoln and structure: Staudinger, Ruzicka, *Helv. Chim. Acta* **7**, 177, 201 (1924). Synthesis: Staudinger *et al.*, *ibid.* 390; Campbell, Harper, *J. Chem. Soc.* **1945**, 283; M. J. Devos, A. Krief, *Tetrahedron Letters* **1978**, 1845; **1979**, 1891; O. A. Nesmeyanova *et al.*, *Synthesis* **1982**, 296. Asymmetric synthesis: T. Aratani *et al.*, *ibid.* **1977**, 2599. Stereoselective synthesis of *dl-trans*-chrysanthemic acid: Mills *et al.*, *Chem. Commun.* **1971**, 555; *eidem, J. Chem. Soc. Perkin Trans. I* **1973**, 133; S. C. Welch, T. A. Valdes, *J. Org. Chem.* **42**, 2108 (1977); J. P. Gehet *et al.*, *Tetrahedron Letters* **21**, 3183 (1980); of *dl-cis*-form: J. Mann, A. Thomas, *ibid.* **27**, 3533 (1986). Synthesis of (−)-*cis*- and (+)-*trans*-forms: Gopichand *et al.*, *Indian J. Chem.* **13**, 433 (1975).

dl-cis-Form, cubic prisms from ethyl acetate, mp 115-116°.

dl-trans-Form, long prisms, mp 54°. Very sol in ethyl acetate. Exhibits a marked negative heat of soln in ethyl acetate or in methanol.

l-trans-Form, elongated prisms, mp 17-21°. $[\alpha]_D^{25}$ −14.01° (c = 1.535 in abs alc).

d-trans-Form, elongated prisms, mp 17-21°. $[\alpha]_D^{25}$ +14.16° (c = 1.554 in abs alc).

2256. Chrysanthenone. *2,7,7-Trimethylbicyclo[3.1.1]heptan-6-one; 2-pinen-7-one.* $C_{10}H_{14}O$; mol wt 150.22. C 79.96%, H 9.39%, O 10.65%. A constituent of the essential oil of *Chrysanthemum indicium* L., also extracted from *C. sinense* Sabin, *Compositae.* Prepared by photochemical rearrangement of *l-* or *d-*verbenone (*q.v.*): Hurst, Whitham, *J. Chem. Soc.* **1960**, 2864; Erman, *J. Am. Chem. Soc.* **89**, 3828 (1967); Schuster, Widman, *Tetrahedron Letters* **1971**, 3571. Structure: Kotake, Nonaka, *Ann.* **607**, 153 (1957); Blanchard, *Chem. & Ind. (London)* **1958**, 293. Chemistry: Erman, *J. Am. Chem. Soc.* **91**, 779 (1969).

d-Form, herbaceous odor. bp_{12} 88-89°. n_D^{22} 1.4720. $[\alpha]_D$ +37° (c = 2.1 in chloroform). uv max: 290 nm (ε 120). Insoluble in water; sol in alcohol.

2257. Chrysarobin. Purified Goa powder; purified araroba. Improperly called *"medicinal chrysophanic acid".* The name chrysarobin today has two meanings: (1) A commercial product consisting of 70-85% of different anthraquinone derivs extracted from Goa powder (Araroba); (2) A pure substance, (*1,8-dihydroxy-3-methyl-9-anthrone; 3-methyl-1,8,9-anthracenetriol*), reduction product of chrysophanic acid, *q.v.* This pure substance constitutes about 30% of commercial chrysarobin. Isoln of the commercial product from the wood of *Andira araroba* Aguiar [*Vouacapoua araroba* (Aguiar) Lyons], *Leguminosae:* Liebermann, Siedler, *Ann.* **212**, 29 (1882); Hesse; *ibid.* **309**, 53 (1899). Prepn of the pure substance by reduction of chrysophanic acid: Naylor, Gardner, *J. Am. Chem. Soc.* **53**, 4114 (1931).

Commercial product, brownish to orange-yellow, micro-

crystalline, odorless, tasteless powder; irritating to mucous membranes. *Caution: Causes dangerous inflammation if it enters the eye!* Very slightly sol in water; 1 g dissolves in 385 ml alcohol, 30 ml benzene, 15 ml chloroform, 160 ml ether, 180 ml carbon disulfide; also sol in fats. Dissolves in alkali hydroxides or in H_2SO_4 with a red color and is pptd from the H_2SO_4 soln by diluting with H_2O.

Pure substance, $C_{15}H_{12}O_3$, yellow needles from glacial acetic acid, mp 203.4-204°.

Human Toxicity: Ingestion may cause renal damage, severe gastroenteritis. Application to large areas of skin may cause renal irritation by percutaneous absorption. Inadvertent conjunctival contact frequently produces conjunctivitis. Allergic reaction of skin has been reported.

THERAP CAT: Has been used as antipsoriatic.

THERAP CAT (VET): Has been used in ringworm, and noninfectious diseases of the skin.

2258. 6-Chrysenamine. 6-Chrysenylamine; 6-aminochrysene; 6-chrysylamine; CP 1001; Chrysenex. $C_{18}H_{13}N$; mol wt 243.29. C 88.86%, H 5.39%, N 5.76%. Prepd by reduction of 6-nitrochrysene: Newman, Cathcart, *J. Org. Chem.* **5**, 620 (1940).

Leaflets from alc, mp 210-211°. Slightly sol in alcohol, benzene, ethyl acetate.

USE: In biochemical research to produce leukopenia.

2259. Chrysene. 1,2-Benzphenanthrene. $C_{18}H_{12}$; mol wt 228.28. C 94.70%, H 5.30%. Occurs in coal tar. Is formed during distillation of coal, in very small amount during distillation or pyrolysis of many fats and oils. Isoln from coal tar: Liebermann, *Ann.* **158**, 299, 307 (1871). Purification by chromatography: Winterstein, Schön, *Z. Physiol. Chem.* **230**, 146 (1934); Winterstein *et al.*, *ibid.* 158. Synthesis by heating H_2 and acetylene to 800°: Meyer, *Ber.* **45**, 1633 (1912). From cholesterol on heating with palladium charcoal or activated charcoal: Schmid, Zentner, *Monatsh.* **49**, 96 (1928).

Orthorhombic bipyramidal plates from benzene. d_4^{20} 1.274. mp 254°. Sublimes easily *in vacuo.* bp 448°. Strong fluorescence under ultraviolet light. Absorption spectrum: Marchlewski, Moroz, *Bull. Soc. Chim. France* [4] **33**, 1406 (1923). Slightly sol in alc, ether, carbon bisulfide, glacial acetic acid. At 25° one gram dissolves in 1300 ml abs alc, 480 ml toluene. About 5% is sol in toluene at 100°. Moderately sol in boiling benzene. Insol in water. Chrysene is generally only slightly sol in cold organic solvents, but fairly sol in these solvents when hot, including glacial acetic acid.

2260. Chrysergonic Acid. $C_{32}H_{30}O_{14}$; mol wt 638.59. C 60.19%, H 4.74%, O 35.08%. Yellow pigment from ergot. Diastereoisomer of secalonic acids, *q.v.* For isoln and structure *see* secalonic acids.

Pale yellow needles from chloroform, mp 268-270°; pale yellow plates from glacial acetic acid, mp 250-257°. $[\alpha]_D^{20}$ −3° → +34° (c = 1.11 in pyridine). uv max (glacial acetic acid): 335 nm (log ε 4.57). One part sol in 1000 parts boiling chloroform; in 800-1000 parts boiling glacial acetic acid.

2261. Chrysin. *5,7-Dihydroxy-2-phenyl-4H-1-benzopyran-4-one; 5,7-dihydroxyflavone;* chrysidenon 1438. $C_{15}H_{10}O_4$; mol wt 254.23. C 70.86%, H 3.96%, O 25.17%. From heartwood of *Pinus monticola* Dougl., *P. excelsa* Wall., and *P. aristata* Engelm., *Pinaceae:* Linstedt, *Acta Chem. Scand.* **3**, 1147, 1375 (1949); **4**, 55 (1950); from bark of *Dolichandrone falcata* Seem., *Bisnomiaceae:* Kincl, *Naturwiss.* **42**, 646 (1955). Synthesis: Seka, Prosche, *Monatsh.* **69**, 284 (1936); Hutchins, Wheeler, *J. Chem. Soc.* **1939**, 91; Teoule *et al., Bull. Soc. Chim. France* **1961**, 546.

Light yellow prisms from methanol, mp 285°. uv max: 270, 329 nm (log ε 4.40, 3.90). Practically insol in water; sol in alkali hydroxide solns; slightly sol in alcohol, chloroform, ether.

Diacetoxychrysin, $C_{19}H_{14}O_6$, crystals from ethanol, mp 194-195°.

Methylchrysin, $C_{16}H_{12}O_4$, *tectochrysin*. Yellow needles, mp 163°. Sol in alcohol, benzene, chloroform. It is present as such or in the form of a glucoside in buds of *Populus* spp., *Salicaceae*. Use of tectochrysin as diuretic: Perrault, U.S. pat. **3,155,579** (1964 to Laroche Navarron).

2262. Chrysoidine. *4-(Phenylazo)-1,3-benzenediamine monohydrochloride;* 4-phenylazo-*m*-phenylenediamine hydrochloride; 2,4-diaminoazobenzene hydrochloride; chrysoidine orange; chrysoidine Y; C.I. 11270; C.I. Basic Orange 2. $C_{12}H_{13}ClN_4$; mol wt 248.71. C 57.95%, H 5.27%, Cl 14.26%, N 22.53%. Prepn: Hofmann, *Ber.* **10**, 213 (1877); Maximoff, U.S. pat. **2,053,095** (1935 to Azodal Co.). Prepn of salt: **Ger.** pat. **562,392** (1933 to Chem.-Pharm. Fabrik Hubold & Bartsch). *See also Colour Index* vol. **4** (3rd ed., 1971) p 4019; *H. J. Conn's Biological Stains*, R. D. Lillie, Ed. (Williams & Wilkins, Baltimore, 9th ed., 1977) p 87.

Reddish-brown crystalline powder, mp 118-118.5°. Solubility at 15°: water 5.5%; abs ethanol 4.75%; cellosolve 6.0%; anhydr ethylene glycol 9.5%; xylene 0.005%; slightly sol in acetone. Practically insol in benzene. Gives yellow soln in concd H_2SO_4, orange soln in dil H_2SO_4 and in HNO_3.

Citrate, $C_{18}H_{21}ClN_4O_7$, *4-phenylazo-m-phenylenediamine hydrochloride citrate, 2,4-diaminoazobenzene hydrochloride citrate, Azoangin, Azohel.* Sol in water (up to 4%), alcohol. Free base, $C_{12}H_{12}N_4$, *C.I. Solvent Orange 3*.

USE: Dyeing silk and cotton. Biological stain.

THERAP CAT: Citrate as antiseptic.

2263. Chrysophanic Acid. *1,8-Dihydroxy-3-methyl-9,10-anthracenedione; 1,8-dihydroxy-3-methylanthraquinone;* 3-methylchrysazin; chrysophanol. $C_{15}H_{10}O_4$; mol wt 254.23. C 70.86%, H 3.96%, O 25.17%. Occurs in the free state and as glucoside in cascara sagrada, senna and various species of *Rumex* and *Rheum* (rhubarb). Isoln from rhubarb root: Tutin, Clewer, *J. Chem. Soc.* **99**, 946 (1911); Siesto, Bartoli, *Farmaco Ed. Prat.* **12**, 517 (1957); Carelli, Giuliano, *ibid.* 184; from *Penicillium islandicum* Sopp.: Howard, Raistrick, *Biochem. J.* **46**, 49 (1950); from *Chaetonium affine* Corda: Arkley *et al., Croat. Chem. Acta* **29**, 141 (1957), *C.A.* **53**, 1287h (1959). Synthesis: Eder, Widmer, *Helv. Chim. Acta* **5**, 3 (1922); **6**, 419 (1923); Ayyangar *et al., J. Sci. Ind. Res. (India)* **20B**, 493 (1961). Total synthesis: M. E. Jung, J. A. Lowe, *Chem. Commun.* **1978**, 95.

Hexagonal or monoclinic crystals from alcohol or benzene, mp 196°. Sublimes. Absorption max: 226, 256, 278, 288, 436 nm (ε × 10^{-3} 41, 28, 14, 14, 11.8). Practically insol in water. Slightly sol in cold, freely in boiling alc; sol in benzene, chloroform, ether, glacial acetic acid, acetone, solns of alkali hydrides, and in hot solns of alkali carbonates; very slightly sol in petr ether.

Glucoside, $C_{21}H_{20}O_9$, *chrysophanein, chrysophaniin*. Fine yellow needles from alc, mp 248-249°. Slightly sol in hot water; sol in pyridine. Practically insol in cold water, chloroform, ether.

2264. Chymopapain. BAX 1526; Chymodiactin; Discase. Proteolytic enzyme which is the major component of the crude latex of *Carica papaya, Caricaceae*. A sulfhydryl enzyme similar to papain, *q.v.*, with respect to substrate specificities, but differing in electrophoretic mobility, stability and solubility. Original crystallization and partial characterization: Jansen, Balls, *J. Biol. Chem.* **137**, 459 (1941); *eidem*, U.S. pat. **2,313,875** (1943). Consists of four components, two of which have molecular wts of about 35,000 and have been isolated and studied. Chymopapain A: Erbata, Yasunobu, *J. Biol. Chem.* **237**, 1086 (1962); chymopapain B: Kunimitsu, Yasunobu, *Biochim. Biophys. Acta* **139**, 405 (1966); Isunoda, Yasunobu, *J. Biol. Chem.* **241**, 4610 (1966). Purification: Stern, U.S. pat. **3,558,433** (1971 to Baxter Labs.). Specificity studies: Ebata, Takahashi, *Biochim. Biophys. Acta* **118**, 201 (1966). *Reviews:* Kunimitsu, Yasunobu, *Methods Enzymol.* vol. **XIX**, G. E. Perlmann, L. Lorand, Eds. (Academic Press, New York, 1970) pp 244-252; Glazer, Smith, *The Enzymes* vol. **III**, P. D. Boyer, Ed. (Academic Press, New York, 3rd ed., 1971) pp 537-538. Review of use in herniated disk treatment: M. J. David, *J. Am. Med. Assoc.* **243**, 2043 (1980).

Powder, more sol in aq soln than papain. Extremely stable at pHs as low as 1.8 and sol in satd NaCl above pH 3. Most active at pHs 2.5 to 4.0. uv max (pH 7): 280 nm ($A_{1cm}^{1\%}$ 18.7 (A); 18.4 (B)). The ratio of milk-clotting to proteolytic activity is twice that of papain although the clotting capacities are the same.

USE: In meat tenderizer.

THERAP CAT: Proteolytic enzyme (used in chemonucleolysis).

2265. Chymotrypsins. Avazyme; Catarase; Chymar; Chymetin; Chymolase; Enzeon; Kymo-trypure; Quimar; Quimoral; Quimotrase; Zolyse. A group of major proteolytic enzymes in the pancreatic juice. Produced in the form of inactive chymotrypsinogen by the acinous cells of pancreas, and carried as such by the pancreatic juice into the duodenum where they are activated by trypsin. They split secondary amide or peptide bonds, carboxylic or phenolic ester bonds and even carbon-carbon bonds. Their main function is to hydrolyze peptide bonds during the intestinal digestion of proteins. Cattle pancreatic juice contains nearly equal quantities of two chymotrypsinogens—A which is cationic at pH 8 and B which is anionic. Chymotrypsinogen A is converted by trypsin into π-chymotrypsin. In the presence of large amounts of trypsin, π-chymotrypsin is degraded to δ-chymotrypsin. In the presence of small amounts of trypsin, π-chymotrypsin is autolyzed to α-chymotrypsin. After long standing α-chymotrypsin gives rise to β- and γ-chymotrypsin which differ from it in their crystal habits. Prepn of π- and δ-chymotrypsins: Bettelheim, Neurath, *J. Biol. Chem.* **212**, 241 (1955). Prepn of α-, β-, γ- and δ-chymotrypsins (and chymotrypsin B): Laskowski, *Methods Enzymol.* **2**, 8 (1955). Amino acid sequence of chymotrypsinogen A: Hartley, *Nature* **201**, 1284 (1964); Melorin *et al., Biochim. Biophys. Acta* **130**, 543 (1966); of chymotrypsinogen B: Hartley *et al., Nature* **207**, 1157 (1965); Smillie *et al., ibid.* **218**, 343 (1968). *Reviews:* Bender *et al., J. Polymer Sci.* **49**, 75 (1961); Desnuelle, *The Enzymes* vol. **4**, P. D. Boyer *et al.*, Eds. (Academic Press, New York, 2nd ed., 1960) pp 93-118; Kraut; Blow; Hess, *ibid.* vol. **3** (3rd ed., 1971) 165, 185, 213.

α-Chymotrypsin, *Alpha-Chymocutan, Chymotase (tabl.), Chymozym, Alfapsin, Kimopsin, Kimoral, Impral.* Arrangement of molecules in monoclinic crystal form: Blow *et al., J. Mol. Biol.* **8**, 65 (1964). Structure: Matthews *et al., Nature* **214**, 652 (1967); Blow *et al., ibid.* **221**, 337 (1969). *Review:* Niemann, *Science* **143**, 1287 (1964).

THERAP CAT: Enzyme (proteolytic).
THERAP CAT (VET): Topically for enzymatic debridement.

2266. Ciafos. *Phosphorothioic acid O-(4-cyanophenyl) O,O-dimethyl ester; phosphorothioic acid O,O-dimethyl ester, O-ester with p-hydroxybenzonitrile; O,O-dimethyl O-(4-cyanophenyl) phosphorothioate; O,O-dimethyl O-(4-cyanophenyl) thionophosphate; cyanophos; BAY 34727; S 4084; Sumitomo S 4084; Cyanox.* $C_9H_{10}NO_3PS$; mol wt 243.21. C 44.44%, H 4.14%, N 5.76%, O 19.74%, P 12.73%, S 13.18%. Prepn: Nishizawa, Mizutani, **Japan.** pat. **62 18,184;** Kuramoto *et al.,* **U.S.** pat. **3,150,040** (1962, 1964 both to Sumitomo); Sakamoto, Nishizawa, *Agr. Biol. Chem.* **26,** 252 (1962); Bernhart, **Ger.** pat. **2,162,344** corresp to U.S. pat. **3,792,132** (1972, 1974 to Stauffer). Pharmacology and metabolism: H. Yamamoto *et al., Oyo Yakuri* **5,** 75 (1971), *C.A.* **76,** 68907a (1972); Wakimura, Miyamoto, *Agr. Biol. Chem.* **35,** 410 (1971). Evaluation as an insecticide: Nishiyawa *et al., ibid.* **25,** 597 (1961); Tamura *et al., ibid.* 773. Evaluation of neurotoxicity: H. Ohkawa *et al., Biochem. Pharmacol.* **29,** 2721 (1980).

CH3O–P(=S)(CH3O)–O–⟨benzene⟩–CN

Yellow to reddish-yellow transparent liquid. $bp_{0.09}$ 119-120° (dec). mp 14-15°. $n_D^{32.5}$ 1.5404. Very sol in methanol, ethanol, acetone, chloroform. Sparingly sol in *n*-hexane, kerosene; slightly sol in water. Rapidly dec under alkaline conditions and upon exposure to light, *Japan. Med. Gaz.* **11**(7), 7 (1974). LD_{50} in mice (mg/kg): 1000 orally; 880 i.p. (Yamamoto).
USE: Insecticide. *Caution:* Cholinesterase inhibitor.

2267. Cichoriin. *7-(β-D-Glucopyranosyloxy)-6-hydroxy-2H-1-benzopyran-2-one; 6,7-dihydroxycoumarin-7-glucoside.* $C_{15}H_{16}O_9$; mol wt 340.28. C 52.94%, H 4.74%, O 42.32%. Isomeric with esculin. In flowers of the chicory plant (*Cichorium intybus* L., *Compositae*). Extraction procedure: Merz, *Arch. Pharm.* **270,** 476 (1932). Synthesis: Head, Robertson, *J. Chem. Soc.* **1939,** 1266; also Merz, *loc. cit.* Review: Sethna, Shah, *Chem. Rev.* **36,** 1 (1945).

glucose-O–⟨benzopyranone⟩–O=O, HO

Needles with $2H_2O$, mp 213-215° (after drying in desiccator). $[\alpha]_D^{18}$ −105° (c = 3 in 50% dioxane). Sol in hot water, alc, glacial acetic acid. Insol in ether, petr ether. Sol in dil alkalies with yellow color, but no fluorescence (difference from esculin). Absorption spectrum of soln: Merz, *loc. cit.*

2268. Cicletanine. *(±)-3-(4-Chlorophenyl)-1,3-dihydro-6-methylfuro[3,4-c]pyridin-7-ol; 1,3-dihydro-3-(4-chlorophenyl)-7-hydroxy-6-methylfuro[3,4-c]pyridine; cicletanide; cycletanide.* $C_{14}H_{12}ClNO_2$; mol wt 261.71. C 64.25%, H 4.62%, Cl 13.55%, N 5.35%, O 12.23%. Furopyridine derivative with antihypertensive and diuretic properties. Prepn: A. Esanu, **Belg.** pat. **891,797;** *idem,* **U.S.** pat. **4,383,998** (1982, 1983 both to Soc. Conseils Recher. Appl. Sci.). Studies on mechanism of action: P. Braquet *et al., Lancet* **1,** 1218 (1983); G. R. Elliott *et al., Thromb. Res.* **33,** 549 (1984). Cardiovascular pharmacology in dogs: R. Jouve *et al., J. Cardiovasc. Pharmacol.* **8,** 208 (1986). *In vitro* comparison of antihistaminic activity of isomers: P. Schoeffter *et al., Eur. J. Pharmacol.* **136,** 235 (1987). HPLC determn in biological fluids: G. Cuisinaud *et al., J. Chromatog.* **341,** 97 (1985). Effect on human prostaglandin metabolism: P. Guinot, J. C. Frölich, *Arzneimittel-Forsch.* **35,** 1714 (1985). Effect on potassium ion transport in hypertension patients: R. Garay *et al., J. Hypertension* **4,** Suppl. 5, S208 (1986). Pharmacokinetics in humans: J. M. Lize *et al., Therapie* **42,** 399 (1987).

⟨furopyridine structure with OH, H_3C, N, O, Cl-phenyl, H⟩

Hydrochloride, $C_{14}H_{13}Cl_2NO_2$, *BN 1270, Coverine, Secletan, Tenstaten.* White crystals, mp 219-228°. Insol in water.
THERAP CAT: Antihypertensive.

2269. Ciclonicate. *3-Pyridinecarboxylic acid 3,3,5-trimethylcyclohexyl ester; trans-3,3,5-trimethylcyclohexyl nicotinate; cyclonicate; P-350;* Bled. $C_{15}H_{21}NO_2$; mol wt 247.35. C 72.84%, H 8.56%, N 5.66%, O 12.94%. Deriv of nicotinic acid, *q.v.* Prepn: K. Matsuda *et al.,* **Ger.** pat. **1,910,481** (1969 to Takeda), *C.A.* **72,** 12769g (1970); G. Massaroli, **Ger.** pat. **2,406,849** (1974 to Poli), *C.A.* **82,** 4129 (1975). Chromatographic study: G. Sekules *et al., Boll. Chim. Farm.* **119,** 521 (1980). Antilipolytic activity in rats: D. Faini *et al., Farmaco Ed. Prat.* **36,** 478 (1981). *In vitro* effect on lipolysis: F. Tessari *et al., Farmaco Ed. Sci.* **36,** 1029 (1981). Clinical study: C. Fossati, *Gazz. Med. Ital.* **139,** 43 (1980).

⟨pyridine-COO-cyclohexyl structure with CH_3, H_3C, H⟩

Liq, $bp_{0.6}$ 127-128°.
THERAP CAT: Vasodilator.

2270. Ciclopirox. *6-Cyclohexyl-1-hydroxy-4-methyl-2(1H)-pyridinone.* $C_{12}H_{17}NO_2$; mol wt 207.28. C 69.53%, H 8.27%, N 6.76%, O 15.44%. Broad spectrum antimycotic agent with some antibacterial activity. Prepn: G. Lohaus, W. Dittmar, **S. Afr.** pat. **69 06039;** *eidem,* **U.S.** pat. **3,883,-545** (1970, 1975 both to Hoechst). *In vitro* study: *eidem, Arzneimittel-Forsch.* **23,** 670 (1973). Series of articles on pharmacokinetics, pharmacology, teratology, toxicity studies: *Oyo Yakuri* **9,** 57-115 (1975), *C.A.* **83,** 53159d, 53538b, 53539c, 71844c, 90833q (1975). Series of articles on chemistry, mechanism of action, toxicology, clinical trials: *Arzneimittel-Forsch.* **31,** 1309-1386 (1981). Toxicity data: H. G. Alpermann, E. Schutz, *ibid.* 1328. Review: S. G. Jue *et al., Drugs* **29,** 330-341 (1985).

⟨pyridinone structure with OH, cyclohexyl, O, N, CH_3⟩

Solid, mp 144°.
Ethanolamine salt (1:1), $C_{14}H_{24}N_2O_3$, *HOE 296, ciclopirox olamine, Batrafen, Brumixol, Cicloche, Dafnegin, Loprox, Micoxolamina, Mycoster, Terit.* LD_{50} in mice, rats (mg/kg): 2898, 3290 orally (Alpermann, Schutz).
THERAP CAT: Antifungal.

2271. Ciclosidomine. *N-(Cyclohexylcarbonyl)-3-(4-morpholinyl)sydnone imine.* $C_{13}H_{20}N_4O_3$; mol wt 280.33. C 55.70%, H 7.19%, N 19.98%, O 17.12%. Peripheral vasodilator similar to molsidomine, *q.v.* Prepn: **Fr.** pat. **1,551,-013; Brit.** pat. **1,198,283** (1968, 1970 both to Boehringer, Ing.). Clinical evaluation: J. Vos, E. J. D. Mees, *Brit. J. Clin. Pharmacol.* **8,** 155 (1979). Effect on blood pressure and heart rate: R. G. Shanks *et al., ibid.* **18,** 232 (1984). Tissue distribution and excretion: M. R. Matteo *et al., Arzneimittel-Forsch.* **35,** 329 (1985).

Hydrochloride, $C_{13}H_{21}ClN_4O_3$, *PR-G 138-Cl, Neopres.* Crystalline powder, mp 187°.

THERAP CAT: Antihypertensive.

2272. Cicrotoic Acid. *β-Methylcyclohexaneacrylic acid;* β-cyclohexyl-β-methacrylic acid; 3-cyclohexyl-2-butenoic acid; β-methylcyclohexanepropenoic acid; AD-106; Accroibile. $C_{10}H_{16}O_2$; mol wt 168.24. C 71.39%, H 9.59%, O 19.02%. Prepn: Young *et al., J. Org. Chem.* **28**, 928 (1963); Redel, Raymond, **Fr.** pat. **M4665** (1967 to Soc. Chim. Org. Biol.), *C.A.* **72**, 42949h (1970).

Colorless prisms from methanol-water, mp 85-86°. uv max: 219 nm (ε 12,000). LD$_{50}$ orally in mice, rats: 1.925, 2.900 g/kg.

THERAP CAT: Choleretic.

2273. Cicutoxin. *(E,E,E)-(−)-8,10,12-Heptadecatriene-4,6-diyne-1,14-diol.* mol wt 258.35. C 79.03%, H 8.58%, O 12.39%. Poisonous principle of the water hemlock *(Cicuta virosa* L., *Umbelliferae)*; naturally occurring as the (−)-form. Isoln: Jacobson, *J. Am. Chem. Soc.* **37**, 916 (1915). Structure: Anet *et al., J. Chem. Soc.* **1953**, 309. Synthesis of (±)-form: Hill *et al., ibid.* **1955**, 1770. Antileukemic activity: T. Konoshima, K.-H. Lee, *J. Nat. Prod.* **49**, 1117 (1986). Simple analytical method: R. A. Smith, D. Lewis, *Vet. Hum. Toxicol.* **29**, 240 (1987).

$$HOCH_2(CH_2)_2-(C\equiv C)_2-(CH=CH)_3-\underset{|}{\overset{OH}{CH}}CH_2CH_2CH_3$$

Prisms from ether + petr ether, transformed in air and light to yellow oily resin. mp 54°. $[\alpha]_D^{15}$ −14.5° (c = 1.7 in ethanol). Sol in alc, chloroform, ether, hot water, alkali hydroxides. Practically insol in petr ether. Very toxic.

(±)-Form, crystals from ether + petr ether, mp 67°. uv max (alc): 242, 252, 318.5, 335.5 nm (ε × 10^{-3} 14.6, 21.6, 50.6, 60.3).

2274. Cifenline. *2-(2,2-Diphenylcyclopropyl)-4,5-dihydro-1H-imidazole;* (±)-2-(2,2-diphenylcyclopropyl)-2-imidazoline; 1-(2-Δ²-imidazolinyl)-2,2-diphenylcyclopropane; cibenzoline; Ro 22-7796; UP 33-901. $C_{18}H_{18}N_2$; mol wt 262.35. C 82.41%, H 6.91%, N 10.68%. Prepn: **Belg.** pat. **807,630**; J. C. Cognaco, **U.S.** pat. **3,903,104** (1974, 1975 both to Hexachemie). Activity and tolerance: D. Herpin *et al., Acta Cardiol.* **36**, 131 (1981). Electrophysiological effects in man: J. F. Thizy *et al., Lyon Med.* **245**, 119 (1981). Clinical trial: D. Herpin *et al., Curr. Ther. Res.* **30**, 742 (1981).

Crystals from petr ether, mp 103-104°. LD$_{100}$ in rats (mg/kg): 64 i.v. (Cognaco).

Succinate, $C_{22}H_{24}N_2O_4$, *Cipralan, Ritmalan.* Crystals from ethanol + ether, mp 165°.

THERAP CAT: Anti-arrhythmic.

2275. Cilastatin. *[R-[R*,S*(Z)]]-7-[(2-Amino-2-carboxyethyl)thio]-2-[[(2,2-dimethylcyclopropyl)carbonyl]amino]-2-heptenoic acid;* MK-791. $C_{16}H_{26}N_2O_5S$; mol wt 358.46. C 53.61%, H 7.31%, N 7.81%, O 22.32%, S 8.94%. Prevents renal metabolism of penem and carbapenem antibiotics by specific and reversible dehydropeptidase I inhibition. Synthesis and combination with thienamycins: D. W. Graham *et al.,* **Eur.** pat. **Appl. 48,301**; H. Kropp *et al.,* **Eur.**

pat. **Appl. 48,025** (both 1982 to Merck & Co.), *C.A.* **97**, 145271b, 145270a (1982). Combination with penems: F. M. Kahan, H. Kropp, **Eur.** pat. **Appl. 72,014** (1983 to Merck & Co.), *C.A.* **99**, 70272h (1983). The articles cited below discuss the activity of cilastatin alone and in combination with imipenem, *q.v.* Dipeptidase inhibition, pharmacokinetics: S. R. Norrby *et al., Antimicrob. Ag. Chemother.* **23**, 300 (1983). Stimulation of granulocyte function: H. Gnarpe *et al., ibid.* **25**, 179 (1984). HPLC determn in serum: C. M. Myers, J. L. Blumer, *ibid.* **26**, 78 (1984). Enhances intrathecal and ocular penetration of imipenem: A. W. Chow *et al., ibid.* **23**, 634 (1983) and K. R. Finlay *et al., Invest. Ophthalmol. Visual Sci.* **24**, 1147 (1983), respectively. In experimental meningitis: D. E. Washburn *et al., J. Antimicrob. Chemother.* **12**, 39 (1983). Series of articles on pharmacokinetics, safety and tolerance and efficacy of cilastatin/imipenem: *ibid.* **12**, Suppl. D, 1-155 (1983); *Infection* **14**, Suppl. 2, S111-S180 (1986).

Sodium salt, $C_{16}H_{25}N_2NaO_5S$, *cilastatin sodium.* Off-white to yellowish-white hygroscopic, amorphous solid. pKa$_1$ 2.0; pKa$_2$ 4.4; pKa$_3$ 9.2. Very sol in water, methanol. Combination with imipenem: *Primaxin, Zienam.*

THERAP CAT: Dipeptidase inhibitor (in combination with imipenem).

2276. Cilazapril. *[1S-[1α,9α(R*)]]-9-[[1-(Ethoxycarbonyl)-3-phenylpropyl]amino]octahydro-10-oxo-6H-pyridazino[1,2-a][1,2]diazepine-1-carboxylic acid monohydrate;* (1S,9S)-9-[[(S)-1-carboxy-3-phenylpropyl]amino]octahydro-10-oxo-6H-pyridazino[1,2-a][1,2]diazepine-1-carboxylic acid 9-ethyl ester monohydrate; Ro 31-2848; Inhibace. $C_{22}H_{31}N_3O_5 \cdot H_2O$; mol wt 435.52. C 60.67%, H 7.64%, N 9.65%, O 22.04%. Angiotensin-converting enzyme inhibitor. Prepn: M. R. Attwood *et al.,* **Ger.** pat. **3,317,290**; *eidem,* **U.S.** pat. **4,512,924** (1983, 1985 both to Hoffmann-La Roche); *eidem, J. Chem. Soc. Perkin Trans. I* **1986**, 1011. Pharmacology of cilazapril and its active diacid metabolite, cilazaprilat: I. L. Natoff *et al., J. Cardiovasc. Pharmacol.* **7**, 569 (1985). Cardiovascular effects in dogs: M. Holck *et al., ibid.* **8**, 99 (1986). Series of articles on clinical pharmacology and pharmacokinetics: *ibid.* **9**, 26-44 (1987). Enzyme immunoassay in human plasma: H. Tanaka *et al., J. Pharm. Sci.* **76**, 224 (1987). Preliminary evaluation in essential hypertension: A. A. Ajayi *et al., Brit. J. Clin. Pharmacol.* **22**, 167 (1986).

Crystals from aq ethanol, mp 95-97°. $[\alpha]_D^{20}$ −62.51° (c = 1% in ethanol).

Diacid, $C_{20}H_{27}N_3O_5$, *cilazaprilat, Ro 31-3113.* Crystals from water, mp 242°. $[\alpha]_D^{20}$ −74.7° (c = 0.5 in 1M NaOH).

THERAP CAT: Antihypertensive.

2277. Cilostazol. *6-[4-(1-Cyclohexyl-1H-tetrazol-5-yl)butoxy]-3,4-dihydro-2(1H)-quinolinone;* 6-[4-(1-cyclohexyl-1H-tetrazol-5-yl)butoxy]-3,4-dihydrocarbostyril; 6-[4-(1-cyclohexyl-5-tetrazolyl)butoxy]-1,2,3,4-tetrahydro-2-oxoquinoline; OPC-13013; Pletaal. $C_{20}H_{27}N_5O_2$; mol wt 369.52. C 65.01%, H 7.38%, N 18.95%, O 8.66%. Prepn: **Belg.** pat. **878,548**; T. Nishi, K. Nakagawa, **U.S.** pat. **4,277,-479** (1979, 1981 both to Otsuka); T. Nishi *et al., Chem. Pharm. Bull.* **31**, 1151 (1983). Physical properties: T. Shimizu *et al., Arzneimittel-Forsch.* **35**, 1117 (1985). Series of articles on pharmacology, metabolism, mechanism of action and clinical evaluations: *ibid.* 1125-1208. HPLC determn in plasma: H. Akiyama *et al., J. Chromatog.* **338**, 456 (1985). Series of articles on toxicology: *Iyakuhin Kenkyu* **16**, 1200-1324 (1985), *C.A.* **104**, 81784z-81787c, 102207f-102209h (1986).

Colorless needle-like crystals from methanol, mp 159.4-160.3°. uv max (methanol): 257 nm (ε 15200). Freely sol in acetic acid, chloroform, *n*-methyl-2-pyrrolidone, DMSO. Practically insol in ether, water, 0.1*N* HCl, 0.1*N* NaOH.
THERAP CAT: Antithrombotic.

2278. Cimaterol. *2-Amino-5-[1-hydroxy-2-[(1-methylethyl)amino]ethyl]benzonitrile;* (±)-5-[1-hydroxy-2-(isopropylamino)ethyl]anthranilonitrile; 1-(4-amino-3-cyanophenyl)-2-isopropylaminoethanol; AB-A 663; CL 263780; AC 263780. $C_{12}H_{17}N_3O$; mol wt 219.29. C 65.73%, H 7.81%, N 19.16%, O 7.30%. β-Adrenergic agonist, related to clenbuterol and mabuterol, *q.q.v.* Prepn: G. Engelhardt *et al.,* Ger. pat. 2,261,914 (1974 to Thomae). Pharmacology: G. Engelhardt, *Arzneimittel-Forsch.* **34**, 1625 (1984). Effect on lipid metabolism in sheep: C. Y. Hu *et al., J. Anim. Sci.* **66**, 1393 (1988). Efficacy as a repartitioning agent in livestock: R. W. Jones *et al., ibid.* **61**, 905 (1985); D. H. Beermann *et al., ibid.* **62**, 370 (1986). *Review:* M. Lafontan *et al., Reprod. Nutr. Develop.* **28**, 61-84 (1988).

Crystals, mp 159-161°.
THERAP CAT (VET): Repartitioning agent.

2279. Cimetidine. *N-Cyano-N'-methyl-N''-[2-[[(5-methyl-1H-imidazol-4-yl)methyl]thio]ethyl]guanidine;* SKF 92334; Acibilin; Acinil; Cimal; Cimetag; Cimetum; Edalene; Dyspamet; Eureceptor; Gastromet; Metracin; Peptol; Tagamet; Tametin; Tratul; Ulcedin; Ulcedine; Ulcerfen; Ulcimet; Ulcofalk; Ulcomedina; Ulcomet; Ulhys; Valmagen; Venopex. $C_{10}H_{16}N_6S$; mol wt 252.34. C 47.60%, H 6.39%, N 33.30%, S 12.71%. Competitive histamine H_2-receptor antagonist which inhibits gastric acid secretion and reduces pepsin output. Prepn: G. J. Durant *et al.,* Belg. pat. 804,-144; *eidem,* U.S. pat. 3,950,333 (1974, 1976 both to SK&F); P. Kairisalo, E. Honkanen, *Arch. Pharm. (Weinheim)* **316**, 688 (1983). Chemistry, pharmacology and toxicology: R. W. Brimblecombe *et al., J. Int. Med. Res.* **3**, 86 (1975); *eidem, Brit. J. Pharmacol.* **53**, 435p (1975). Use in combination with other drugs for cancer treatment: R. D. Thornes *et al., Lancet* **2**, 328 (1982); S. Borgström *et al., N. Engl. J. Med.* **307**, 1080 (1982); **308**, 591 (1983). Controlled clinical study in treatment of acute upper gastrointestinal tract bleeding: D. Barer *et al., ibid.* 1571; for prevention of recurrent duodenal ulcer: S. Sontag *et al., ibid.* **311**, 689 (1984). Immunoregulatory effect: J. L. Jorizzo *et al., Ann. Int. Med.* **92**, 192 (1980); W. B. White, M. Ballow, *N. Engl. J. Med.* **312**, 198 (1985). Review of effect on endocrine secretion: C. Scarpignato, G. Bertaccini, *Drugs Exp. Clin. Res.* **5**(4-5), 129-140 (1979). Symposium on clinical efficacy, cytoprotection and antifibrinolytic effects: *Scand. J. Gastroenterol.* **21**, Suppl. 121, 1-62 (1986).

Crystals, mp 141-143°. Soly in water at 37°: 1.14%. Soly increased by dil HCl. LD_{50} in mice, rats (mg/kg): 2600, 5000 orally; 150, 106 i.v.; 470, 650 i.p. (Brimblecombe).
Hydrochloride, $C_{10}H_{17}ClN_6S$, *Aciloc, Biomag, Brumetidina, Cimet, Notul.*
THERAP CAT: Anti-ulcerative.

2280. Cimetropium Bromide. *[7(S)-(1α,2β,4β,5α,7β)]-9-(Cyclopropylmethyl)-7-(3-hydroxy-1-oxo-2-phenylpropoxy)-9-methyl-3-oxa-9-azoniatricyclo[3.3.1.0^{2,4}]nonane bromide;* 8-(cyclopropylmethyl)-6β,7β-epoxy-3α-hydroxy-1αH,5αH-tropanium bromide (−)-(S)-tropate; N-cyclopropylmethylscopolamine bromide; DA-3177; Alginor. $C_{21}H_{28}BrNO_4$; mol wt 438.36. C 57.54%, H 6.44%, Br 18.23%, N 3.19%, O 14.60%. Spasmolytic agent with affinity for intestinal muscarinic receptors. Prepn: S. Casadio, A. Donetti, Ger. pat. 2,316,728; *eidem,* U.S. pat. 3,853,886 (1973, 1974 both to De Angeli). Crystal and molecular structure: G. Giuseppetti *et al., Farmaco, Ed. Sci.* **35**, 231 (1980). Molecular conformation in solution: A. Gallazzi *et al., ibid.* 913. Pharmacology: A. Schiavone *et al., Arzneimittel-Forsch.* **35**, 796 (1985); C. Scarpignato, G. Bianchi Porro, *Int. J. Clin. Pharm. Res.* **5**, 467 (1985). Clinical trials in treatment of irritable bowel syndrome: G. Piai *et al., Clin. Trials J.* **23**, 6 (1986); A. Ferrari *et al., Clin. Ther.* **8**, 320 (1986).

Crystals from acetonitrile, mp 174°. $[\alpha]_D^{20}$ −18.3° (c = 3).
THERAP CAT: Antispasmodic.

2281. Cimicifuga. Black cohosh; black snake root; Actaea; bugbane; bugwort. Dried rhizome and roots of *Cimicifuga racemosa* (L.) Nutt., *Ranunculaceae.* *Habit.* U.S., Canada. *Constit.* Isoferulic acid, 15-20% cimicifugin, tannin, volatile oil, resin, a sugar.
THERAP CAT: Astringent bitter.

2282. Cimigenol. *16,23:16,24-Diepoxy-9,19-cyclo-9β-lanostane-3β,15α,25-triol;* cimicifugol. $C_{30}H_{48}O_5$; mol wt 488.68. C 73.73%, H 9.90%, O 16.37%. From the resin of *Cimicifuga racemosa* Nutt. (*Actaea racemosa* L.), *Ranunculaceae:* Corsano, Spano, *Atti Accad. Nazl. Lincei, Rend., Classe Sci. Fiz. Mat. Nat.* **32**, 674 (1962), *C.A.* **58**, 11408e (1963). Structure: Corsano *et al., Chem. Commun.* **1965**, 185; Corsano, Piancatelli, *Ric. Sci.* **37**, 366 (1967); *eidem, Gazz. Chim. Ital.* **99**, 1140 (1969). Identity with cimicifugol: *J. Pharm. Soc. Japan* **87**, 1569 (1967).

Needles from methanol, mp 227.5-228.5°. $[\alpha]_D$ +38° (c = 0.86 in chloroform).
Diacetate, $C_{34}H_{52}O_7$, needles from hexane, mp 202-204°. $[\alpha]_D$ +46.3° (c = 1.07 in chloroform).
Triacetate, $C_{36}H_{54}O_8$, needles from hexane, mp 149.5-150.5°. $[\alpha]_D$ +25.4° (chloroform).

2283. Cinametic Acid. *3-[4-(2-Hydroxyethoxy)-3-meth-oxyphenyl]-2-propenoic acid;* 4-(2-hydroxyethoxy)-3-meth-oxycinnamic acid; Transoddi. $C_{12}H_{14}O_5$; mol wt 238.24. C 60.50%, H 5.92%, O 33.58%. Prepn: **Fr. pat. M6380** (1968 to Anphar), *C.A.* **76,** 140232w (1972).

CH=CHCOOH
OCH$_3$
OCH$_2$CH$_2$OH

THERAP CAT: Hepato-biliary regulator.

2284. Cinamiodyl. *2-[(3-Amino-2,4,6-triiodophenyl)-methylene]butanoic acid;* 3-amino-α-ethyl-2,4,6-triiodocin-namic acid; α-ethyl-β-(3-amino-2,4,6-triiodophenyl)acrylic acid; 3-(3-amino-2,4,6-triiodophenyl)-2-ethylacrylic acid; cinamiodil; dehydroiopanoic acid; AG 34-55. $C_{11}H_{10}I_3NO_2$; mol wt 568.91. C 23.22%, H 1.77%, I 66.92%, N 2.46%, O 5.62%. Prepn: Chao, Hu, *C.A.* **52,** 15467f (1958).

C$_2$H$_5$
CH=CCOOH
I
I
NH$_2$
I

Crystals from ethanol, mp 251-253°.
Sodium salt, $C_{11}H_9I_3NNaO_2$, *Triodan.*
THERAP CAT: Diagnostic aid (x-ray contrast medium).

2285. Cinchomeronic Acid. *3,4-Pyridinedicarboxylic acid.* $C_7H_5NO_4$; mol wt 167.12. C 50.31%, H 3.02%, N 8.38%, O 38.29%. Prepd in 42% yield by boiling quinine with nitric and fuming nitric acid: Ternájgo, *Monatsh.* **21,** 448 (1900); *cf.* Kirpal, *ibid.* **23,** 248 (1902); Kaas, *ibid.* 252; by selenium dioxide oxidation of isoquinoline: Mueller, U.S. pat. **2,436,660** (1948 to Allied Chem.).

N
COOH
COOH

Crystals from acidulated water, mp 256°. On heating, a small portion sublimes without decompn. Strong acid. Sparingly sol in hot alc, ether, benzene. Insol in chloroform.
Disodium salt dihydrate, $C_7H_3NNa_2O_4.2H_2O$, plates, moderately sol in water, alcohol.
Dimethyl ester, $C_9H_9NO_4$, crystals, mp 141°.

2286. Cinchona. Calisaya bark; Peruvian bark; Cinchona bark; Jesuit's bark. Dried bark of stem or root of various species of *Cinchona,* fam. *Rubiaceae,* largely of *C. officinalis* L. (*C. ledgeriana* Moens) cultivated mostly in Java, of *C. officinalis* L. (*C. calisaya* Wedd.) from Bolivia, of *C. officinalis* L. and *C. micrantha* R. & P. from Peru, of *C. pubescens* Vahl. (*C. succirubra* Pav.) from Ecuador and *C. pitayensis* Wedd. from Colombia. *Habit.* South America, cultivated mostly in Java, also in India. *Constit.* About 35 alkaloids; cinchotannic, quinic and quinovic acids; cinchona red; vola-tile oils. The alkaloid content varies according to the source of the bark. The cultivated bark contains 7-10% total alka-loids of which about 70% is quinine. *Cinchona succirubra* Pav. and some other varieties contain more cinchonidine and cinchonine and sometimes quinidine than the cultivated. The standardized method for the assay of cinchona, named "Brussels 1949" is published in the journal *De Belgische Chemische Industrie (Ind. Chim. Belge)* **15,** 328-338 (1950).
On heating a small portion of the bark in a test tube a characteristic purple vapor is evolved. Dil H_2SO_4 extracts have a blue fluorescence.

THERAP CAT: Antimalarial.

2287. Cinchonamine. *2-(5-Ethenyl-1-azabicyclo[2.2.2]-oct-2-yl)-1H-indole-3-ethanol;* 3-(β-hydroxyethyl)-2-(5-vinyl-2-quinuclidyl)indole. $C_{19}H_{24}N_2O$; mol wt 296.40. C 76.99%, H 8.16%, N 9.45%, O 5.40%. From the bark of *Remijia purdieana* Wedd., *Rubiaceae:* Arnaud, *Compt. Rend.* **93,** 593 (1881); Hesse, *Ann.* **225,** 211 (1884). Prepn by reduction of quinamine with lithium aluminum hydride and structure: Goutarel *et al., Helv. Chim. Acta.* **33,** 150 (1950). Synthesis: Chen *et al., C.A.* **53,** 7219e (1959). Total synthe-sis: G. Grethe *et al., Helv. Chim. Acta* **59,** 2271 (1976). Stereochemistry: Wenkert, Bringi, *J. Am. Chem. Soc.* **80,** 3484 (1958); Augustine, *Chem. & Ind. (London)* **1959,** 1071; Sawa, Matsumura, *Tetrahedron* **26,** 2923 (1970). Biosynthe-tic studies: Battersby, Parry, *Chem. Commun.* **1971,** 31. Conversion of chinchona alkaloids of the quinoline series to those of the indole series: Ochiai *et al., C.A.* **59,** 14040h (1963).

OH
N
H
H
CH$_2$

Triboluminescent, orthorhombic prisms from methanol, also reported as mp 194°. $[\alpha]_D^{20}$ +123° (c = 0.66 in ethanol). pK in 80% methyl Cellosolve: 8.28. uv max (methanol): 223, 292 nm (log ε 4.60, 3.88). One gram dis-solves in about 35 ml alcohol, 105 ml ether. Sol in benzene, chloroform, petr ether, CS_2; practically insol in water.
Hydrochloride monohydrate, $C_{19}H_{24}N_2O.HCl.H_2O$, cubic crystals. One gram dissolves in 200 ml water. Sol in alc.

2288. Cinchonidine. *Cinchonan-9-ol;* cinchovatine; α-quinidine. $C_{19}H_{22}N_2O$; mol wt 294.38. C 77.51%, H 7.53%, N 9.52%, O 5.43%. Occurs in most varieties of cin-chona bark, especially in bark of *Cinchona pubescens* Vahl. (*C. succirubra* Pav.) and *C. pitayensis* Wedd., *Rubiaceae.* Isoln: Leers, *Ann.* **82,** 147 (1852). Structure: Rabe, *ibid.* **365,** 359 (1909). Stereoisomeric with cinchonine: Koenigs, *ibid.* **347,** 182 (1906). Configuration: Prelog, Zalán, *Helv. Chim. Acta* **27,** 535 (1944); Prelog, Häfliger, *ibid.* **33,** 2021 (1950); Roth, *Pharmazie* **16,** 257 (1961); Lyle, Keefer, *Tetra-hedron* **23,** 3253 (1967). Biosynthetic studies: Battersby, Parry, *Chem. Commun.* **1971,** 31.

CH$_2$=CH
H
H
N
HO
H
N

Orthorhombic plates, prisms from alcohol, mp 210°. $[\alpha]_D^{20}$ −109.2° (alc). uv absorption data: Kamath *et al., Indian J. Chem.* **6,** 510 (1968). Sol in alcohol and chloroform; mod-erately sol in ether; practically insol in water. pK$_1$ 5.80, pK$_2$ 10.03. *Protect from light.* LD$_{50}$ i.p. in rats: 206 mg/kg.
Epicinchonidine, mp 104°, $[\alpha]_D^{20}$ +63° (c = 0.804 in alc): Rabe *et al., Ann.* **492,** 253 (1932).
Dihydrochloride, $C_{19}H_{22}N_2O.2HCl$, white or slightly yel-low crystals or powder. Freely sol in water or alcohol. *Protect from light.*
Hydrochloride dihydrate, $C_{19}H_{22}N_2O.HCl.2H_2O$, cryst powder; loses all of its H_2O at 120°. $[\alpha]_D^{20}$ −117.5° (water). Sol in 25 parts cold water, more sol in boiling water; sol in alcohol, chloroform, slightly in ether. The aq soln is prac-tically neutral. *Protect from light.*
Sulfate trihydrate, $(C_{19}H_{22}N_2O)_2.H_2SO_4.3H_2O$, silky, acic-ular crystals; effloresce on exposure to air and darkens in light. mp when anhydr, at about 240°, with decompn. One gram dissolves in 70 ml water, 20 ml hot water, 90 ml alco-

hol, 40 ml hot alc, 620 ml chloroform; practically insol in ether. The aq soln is practically neutral. *Protect from light.*

THERAP CAT: Antimalarial.

2289. Cinchonine. *Cinchonan-9-ol.* $C_{19}H_{22}N_2O$; mol wt 294.38. C 77.51%, H 7.53%, N 9.52%, O 5.43%. Occurs in most varieties of cinchona bark, especially in bark of *Cinchona micrantha* R. & P., *Rubiaceae.* Isoln and structure: Rabe, *Ber.* **41**, 63 (1908). Stereoisomeric with cinchonidine: Koenigs, *Ann.* **347**, 182 (1906). Configuration: Prelog, Zalán, *Helv. Chim. Acta* **27**, 535 (1944); Prelog, Häfliger, *ibid.* **33**, 2021 (1950); Roth, *Pharmazie* **16**, 257 (1961); Lyle, Keefer, *Tetrahedron* **23**, 3253 (1967). Crystal and molecular structure: B. Oleksyn *et al., Acta Crystallogr.* **B35**, 440 (1979). Biosynthetic studies: Battersby, Parry, *Chem. Commun.* **1971**, 31.

Prisms, needles from alcohol or ether. mp about 265°; begins to sublime at 220°. $[\alpha]_D$ +229° (alcohol). uv absorption data: Kamath *et al., Indian J. Chem.* **6**, 510 (1968). One gram dissolves in 60 ml alcohol, 25 ml boiling alcohol, 110 ml chloroform, 500 ml ether. Practically insol in water. pK_1 5.85, pK_2 9.92. *Protect from light.* LD_{50} i.p. in rats: 152 mg/kg, C. C. Johnson, C. F. Poe, *Acta Pharmacol. Toxicol.* **4**, 265 (1949).

Epicinchonine, mp 83°, $[\alpha]_D^{22}$ +120.3° (c = 0.806 in alc), Rabe *et al., Ann.* **492**, 253 (1932).

Dihydrochloride, $C_{19}H_{22}N_2O \cdot 2HCl$, white or faintly yellow crystals or cryst powder. Freely sol in water or alcohol. *Protect from light.*

Hydrochloride dihydrate, $C_{19}H_{22}N_2O \cdot HCl \cdot 2H_2O$, fine crystals. mp when anhydr, about 215° with decompn. One gram dissolves in 20 ml water, 3.5 ml boiling water, 1.5 ml alcohol, 20 ml chloroform; slightly sol in ether. The aq soln is practically neutral. *Protect from light.*

Sulfate dihydrate, $(C_{19}H_{22}N_2O)_2 \cdot H_2SO_4 \cdot 2H_2O$, lustrous, very bitter crystals. mp when anhydr about 198°. One gram dissolves in 65 ml water, 30 ml hot water, 12.5 ml alc, 7 ml hot alc, 47 ml chloroform; slightly sol in ether. The aq soln is practically neutral.

THERAP CAT: Antimalarial.

2290. Cinchophen. *2-Phenyl-4-quinolinecarboxylic acid; 2-phenylcinchoninic acid;* Atophan; Quinophan; Phenoquin; Agotan; Artam; Alutyl; Cinconal; Vantyl; Viophan; Atocin; Mylofanol; Rhematan; Rheumin; Tophol; Polyphlogin; Quinofen; Tophosan. $C_{16}H_{11}NO_2$; mol wt 249.26. C 77.09%, H 4.45%, N 5.62%, O 12.84%. Quinoline derivative formerly used in treatment of chronic gout: A. B. Gutman, *Arthritis Rheum.* **16**, 431 (1973). Prepd by heating pyruvic acid with aniline and benzaldehyde or with benzylidene aniline in abs alcohol: Doebner, Gieseke, *Ann.* **242**, 290 (1887). Also from acetophenone and isatinic acid in alcoholic KOH: Pfitzinger, *J. Prakt. Chem.* **38**, 582 (1882); **56**, 293 (1897). Cinchophen as ulcerogenic agent: T. P. Churchill, F. H. van Wagoner, *Arch. Path.* **13**, 850 (1932); N. Umeda *et al., Toxicol. Appl. Pharmacol.* **18**, 102 (1971); T. H. Stewart *et al., J. Pathol.* **131**, 363 (1980). Mitochondrial toxicity of cinchophen and derivatives: H. Vainio *et al., Biochem. Pharmacol.* **20**, 1589 (1971).

Needles, mp 213-216°. Stable to air, but turns yellow under the influence of light. Slightly bitter taste. Practically insol in water. One gram dissolves in about 400 ml chloroform, in about 100 ml ether and in about 120 ml alcohol.

Sodium salt, water soluble. In combination with sodium salicylate as *Atophanyl, Arthrosan, Cincosal.*

Hydrochloride, $C_{16}H_{12}ClNO_2$, *Chloroxyl.* Yellow cryst powder, slightly bitter, astringent taste; mp about 223°. Practically insol in water; slightly sol in alcohol; sol in chloroform, ether.

Lithium salt octahydrate, $C_{16}H_{10}LiNO_2 \cdot 8H_2O$, *Ektophanol.* Crystals, rapidly losing water of crystn on exposure to air. Dec 345°. Sol in water, alc; practically insol in ether.

Allyl ester, $C_{19}H_{15}NO_2$, *allyl 2-phenylcinchoninate, Atoquinol.* Prepn: Gams, U.S. pat. **1,336,952** (1920). Long needles from alc, mp 30°. bp_{15} 260°. Practically insol in water; readily sol in alcohol, ether, acetone, oils. Tasteless.

USE: Experimentally to induce ulcers.

THERAP CAT: Formerly as analgesic.

2291. Cinchotoxine. *(cis)-3-(3-Ethenyl-4-piperidinyl)-1-(4-quinolinyl)-1-propanone; 9-deoxy-9-oxo-1,8-secocinchonine;* cinchonicine. $C_{19}H_{22}N_2O$; mol wt 294.38. C 77.51%, H 7.53%, N 9.52%, O 5.43%. A rearrangement product of cinchonine or cinchonidine obtained by boiling with acetic acid: Pasteur, *Compt. Rend.* **37**, 110 (1853). Miller, Rohde, *Ber.* **33**, 3214 (1900). Conversion to cinchonidinone and cinchonine: Rabe, *Ber.* **44**, 2088 (1911); **41**, 62 (1908). Prepn from cinchonine: Prostenik, Prelog, *Helv. Chim. Acta* **26**, 1965 (1943).

Bitter needles. mp 58-60°. $[\alpha]_D^{15}$ +47° in alcohol. Sparingly sol in water; sol in alcohol, ether, chloroform.

2292. Cinepazet Maleate. *4-[1-Oxo-3-(3,4,5-trimethoxyphenyl)-2-propenyl]-1-piperazineacetic acid ethyl ester (Z)-2-butenedioate(1:1); 4-(3,4,5-trimethoxycinnamoyl)-1-piperazineacetic acid ethyl ester maleate;* ethyl 4-(3,4,5-trimethoxycinnamoyl)-1-piperazineacetate maleate; [4-(3',4',5'-trimethoxycinnamoyl)piperazinyl]-2-ethyl acetate maleate; cinepazic acid ethyl ester maleate; ethyl cinepazate maleate; MD 6753; Vascoril. $C_{24}H_{32}N_2O_{10}$; mol wt 508.53. C 56.68%, H 6.34%, N 5.51%, O 31.46%. $C_{20}H_{28}N_2O_6 \cdot C_4H_4$-$O_4$. Prepn: Fauran *et al.,* **Brit.** pat. **1,168,108** corresp to **U.S.** pat. **3,590,034** (1969, 1971 to Delalande). Synthesis and structure-activity studies of the hydrochloride: Fauran, Turin, *Chim. Ther.* **4**, 290 (1969); Huguet *et al., ibid.* 293. Pharmacology and toxicology: Pourrias *et al., Therapie* **26**, 845, 859 (1971). Metabolic data: Chasseaud *et al., Arzneimittel-Forsch.* **22**, 2003 (1972).

Crystals, mp 130°.

Free base, $C_{20}H_{28}N_2O_6$, mp 96° (from isopropyl ether).

Hydrochloride, mp 200° (dec). LD_{50} in mice (mg/kg): 300 i.v., 1300 orally, Fauran *et al., loc. cit.*

THERAP CAT: Anti-anginal.

2293. Cinepazide. *1-[2-Oxo-2-(1-pyrrolidinyl)ethyl]-4-[1-oxo-3-(3,4,5-trimethoxyphenyl)-2-propenyl]piperazine; 1-[(1-pyrrolidinylcarbonyl)methyl]-4-(3,4,5-trimethoxycinnamoyl)piperazine;* 1-[4-[2-oxo-2-(1-pyrrolidinyl)ethyl]-1-piperazinyl]-3-(3,4,5-trimethoxyphenyl)-2-propen-1-one; 1-[4-[(3',4',5'-trimethoxycinnamoyl)-1-piperazinyl]acetyl]pyrrolidine. $C_{22}H_{31}N_3O_5$; mol wt 417.51. C 63.29%, H

7.48%, N 10.06%, O 19.16%. Prepn: **Belg. pat. 730,345** (1969 to Delalande), *C.A.* **72**, 132784e (1970); Fauran, Turin, *Chim. Ther.* **4**, 290 (1969). Pharmacology: Huguet *et al., ibid.* 293; Marmo *et al., Farmaco Ed. Prat.* **28**, 132 (1973); Pourrias *et al., Therapie* **29**, 29, 43 (1974).

Maleate, $C_{26}H_{35}N_3O_9$, **MD 67350, Brendil, Vasodistal.** Crystals from abs ethanol, mp 135°. LD_{50} in mice, rats: 1000, 1310 mg/kg orally; 617, 414 mg/kg i.v., Pourrias *et al., loc. cit.*

THERAP CAT: Peripheral vasodilator.

2294. Cinerins. Active insecticidal constiutents of pyrethrum flowers. Isoln: Ward, *Chem. & Ind. (London)* **1953**, 586; Stephenson, *Pyrethrum Post* **5**, no. 4, 22 (1960); *C.A.* **55**, 13753g (1961). Structure: Schechter *et al., J. Am. Chem. Soc.* **71**, 3165 (1949); Godin *et al., J. Chem. Soc. (C)* **1966**, 332. Stereochemistry: Begley *et al., Chem. Commun.* **1972**, 1276. *Review:* Crombie, Elliott, *Fortschr. Chem. Org. Naturst.* **19**, 120-164 (1961).

Cinerin I, $C_{20}H_{28}O_3$, *2,2-dimethyl-3-(2-methyl-1-propenyl)cyclopropanecarboxylic acid 3-(2-butenyl)-2-methyl-4-oxo-2-cyclopenten-1-yl-ester.* R = CH_3. Viscous liquid. Oxidizes rapidly and becomes inactive in presence of air. $bp_{0.008}$ 136-138°. n_D^{20} 1.5064. $[\alpha]_D^{20}$ −22° (hexane). uv max: 222 nm (ε 21,400). Practically insol in water; sol in alcohol, petr ether, kerosene, carbon tetrachloride, ethylene dichloride, nitromethane.

Cinerin II, $C_{21}H_{28}O_5$, *3-(3-methoxy-2-methyl-3-oxo-1-propenyl)-2,2-dimethylcyclopropanecarboxylic acid 3-(2-butenyl)-2-methyl-4-oxo-2-cyclopenten-1-yl ester.* R = COO-CH_3. Viscous liquid. Oxidizes rapidly and becomes inactive in presence of air. $bp_{0.001}$ 182-184°. n_D^{20} 1.5183. $[\alpha]_D^{16}$ +16° (isooctane). uv max: 229 nm (ε 28,700). Practically insol in water; sol in alcohol, petr ether (less sol than Cinerin I), kerosene, carbon tetrachloride, ethylene dichloride, nitromethane.

USE: Insecticide. *Caution:* Diarrhea, convulsions, prostration, death from respiratory paralysis; injury to liver, kidneys.

2295. Cinmetacin. *5-Methoxy-2-methyl-1-(1-oxo-3-phenyl-2-propenyl)-1H-indole-3-acetic acid;* 1-cinnamoyl-5-methoxy-2-methylindole-3-acetic acid; 1-cinnamoyl-2-methyl-5-methoxy-3-indolylacetic acid; Cindomet; Indolacin. $C_{21}H_{19}NO_4$; mol wt 349.39. C 72.19%, H 5.48%, N 4.01%, O 18.32%. Prepn: Yamamoto, Nakao, **S. Afr. pat. 67 02,683** corresp to **U.S. pat. 3,576,800** (1968, 1971 both to Sumitomo); *eidem, J. Med. Chem.* **12**, 176 (1969). Pharmacology: T. Komatsu *et al., Arzneimittel-Forsch.* **23**, 1690 (1973).

Yellow crystals from acetone-water, mp 170-172°; also

reported as mp 164-165°. LD_{50} in mice, rats: 360, 590 mg/kg i.p.; 750, 1020 mg/kg orally, T. Komatsu *et al., loc. cit.*

THERAP CAT: Anti-inflammatory.

2296. Cinmethylin. *exo-(±)-1-Methyl-4-(1-methylethyl)-2-[(2-methylphenyl)methoxy]-7-oxabicyclo[2.2.1]heptane;* (±)-2-exo-(2-methylbenzyloxy)-1-methyl-4-isopropyl-7-oxabicyclo[2.2.1]heptane; SD 95481; Cinch. $C_{18}H_{26}O_2$; mol wt 274.40. C 78.79%, H 9.55%, O 11.66%. Pre-emergence grass herbicide. Member of the cineole (eucalyptol, *q.v.*) family. Prepn: G. B. Payne *et al., Eur. pat.* **Appl. 81,893** (1983 to Shell); *eidem,* **U.S. pat. 4,670,041** (1987 to Du Pont). Physical and chemical properties: B. T. Grayson *et al., Pestic. Sci.* **21**, 143 (1987). Mechanism of action study: M. H. El-Deek, F. D. Hess, *Weed Sci.* **34**, 684 (1986). Metabolism in rats: P. W. Lee *et al., J. Agric. Food Chem.* **34**, 162 (1986). Mobility in soil: D. R. Wendt *et al., Proc. South. Weed Sci. Soc.* **1987**, 391; and herbicidal activity: P. C. Lolas, A. Galopoulos, *Zizaniology* **1**, 221 (1985). Field studies in tobacco: P. C. Lolas, *Brit. Crop Prot. Conf. - Weeds* **1985**, 841; in soybeans: P. C. Bhowmik, *Weed Sci.* **36**, 678 (1988).

Colorless liquid, bp 313° ±2. d^{20} 1015 kg/m³. Viscosity (20°): 70-90 mPa. Vapor pressure (20°): 7.6 × 10⁻⁵ mm Hg. Moderately volatile. Partition coefficient (*n*-octanol/-water): 6850. Miscible with most organic solvents. Slightly sol in water. Soly in water (20°): 63 ± 2 mg/l. LD_{50} orally in rats: 4.5 g/kg; LD_{50} dermally in rabbits: > 2 g/kg (Lee).

USE: Herbicide.

2297. Cinnabarine. *2-Amino-9-hydroxymethyl-3-oxo-3H-phenoxazine-1-carboxylic acid;* 1-carboxy-2-amino-9-hydroxymethylphenoxazin-3-one; polystictine. $C_{14}H_{10}N_2O_5$; mol wt 286.24. C 58.74%, H 3.52%, N 9.79%, O 27.95%. Antibiotic substance produced by the fungus *Trametes cinnabarina* Jacq.: Gripenberg, *Acta Chem. Scand.* **5**, 590 (1951); from *Coriolus sanguineus* (Fr.) (*Polystictus cinnabarinus* Jacq.): Cavill *et al., J. Chem. Soc.* **1953**, 525. Identity with polystictine: Gripenberg *et al., Acta Chem. Scand.* **11**, 1485 (1957). Structure: Gripenberg, *ibid.* **12**, 603 (1958); **13**, 1305 (1959); Cavill *et al., Tetrahedron* **5**, 275 (1959). Syntheses: Weinstein, Brattesani, *J. Heterocyclic Chem.* **4**, 151 (1967); Schäfer, Schlude, *Tetrahedron Letters* **1968**, 2161.

Red needles, dec 320°. Absorption max: 234, 430, 455 nm (log ε 4.2, 4.0, 4.0). Soly in acetone: 0.04% (20°); in dioxane: 0.1% (100°); practically insol in ethanol, chloroform and dimethyl sulfoxide. Sol in cold concd HCl.

O-Methyl ether, $C_{15}H_{12}N_2O_5$, orange red prisms from ethyl acetate + petr ether, mp 200-202°.

2298. Cinnamaldehyde. *3-Phenyl-2-propenal;* cinnamic aldehyde; phenylacrolein; cinnamal. C_9H_8O; mol wt 132.15. C 81.79%, H 6.10%, O 12.11%. $C_6H_5CH=CHCHO$. Found in Ceylon and Chinese cinnamon oils. Prepn by condensation of benzaldehyde and acetaldehyde: Peine, *Ber.* **17**, 2117 (1884); **Japan. pat. 163,097** (1944 to Ogawa Chem. Ind.); from 2-chloroallylbenzene: Bert, Dorier, *Compt. Rend.* **191**, 332 (1930); Bert, Annequin, *ibid.* **192**, 1315 (1931); by condensation of styrene with formylmethylaniline: **Brit. pat. 504,125** (1939 to I. G. Farben); by oxidization of cinnamyl alc: Holum, *J. Org. Chem.* **26**, 4814 (1961); Traynelis, Hergenrother, *J. Am. Chem. Soc.* **86**, 298 (1964). Isoln from woodrotting fungus, *Stereum subpileatum* Berk. & Curt.: Birkinshaw *et al.*, *Biochem. J.* **66**, 188 (1957). Toxicity data: P. M. Jenner *et al.*, *Food Cosmet. Toxicol.* **2**, 327 (1964).

Yellowish oily liquid, strong odor of cinnamon. d_{25}^{25} 1.048-1.052. Volatile with steam. mp −7.5°. $bp_{1.0}$ 76.1°; bp_5 105.8°; bp_{10} 120.0°; bp_{20} 135.7°; bp_{40} 152.2°; bp_{100} 163.7°; bp_{100} 177.7°; bp_{200} 199.3°; bp_{400} 222.4°; bp_{760} 246.0° (some dec). n_D^{20} 1.618-1.623. Dissolves in about 700 parts water, in about 7 vols of 60% alc. Misc with alcohol, ether, chloroform, oils. LD_{50} in rats (mg/kg): 2220 orally (Jenner).

USE: In the flavor and perfume industry.

2299. Cinnamedrine. α-[1-[Methyl(3-phenyl-2-propenyl)amino]ethyl]benzenemethanol; *N*-cinnamylephedrine; α-[1-(*N*-cinnamyl-*N*-methylamino)ethyl]benzyl alcohol. $C_{19}H_{23}NO$; mol wt 281.38. C 81.10%, H 8.24%, N 4.98%, O 5.69%. Smooth muscle relaxant. Prepn: Stolz, Flaecher, U.S. pat. **1,959,392** (1934 to Winthrop Chem.); Welsh, Keenan, *J. Am. Pharm. Assoc., Sci. Ed.* **30**, 123 (1941).

dl-Hydrochloride, $C_{19}H_{24}ClNO$, bitter crystals from ethyl acetate + chloroform, mp 180-185° (base, mp 72-78°). Sol in about 80 parts water and 10 parts alc at 20°, more easily sol at higher temps; sol in 50 parts chloroform; sparingly sol in ether, benzene.

Component of *Midol*.

THERAP CAT: Antispasmodic.

2300. Cinnamic Acid. *3-Phenyl-2-propenoic acid;* β-phenylacrylic acid. $C_9H_8O_2$; mol wt 148.15. C 72.96%, H 5.44%, O 21.60%. $C_6H_5CH=CHCOOH$. The ordinary synthetic cinnamic acid which is described here is the *trans*-isomer. Occurs free and partly esterified in storax, balsam Peru or Tolu, oil of cinnamon, coca leaves. Isoln: Beilstein, Kuhlberg, *Ann.* **163**, 123 (1872); von Miller, *Ann.* **188**, 196 (1877). Synthesis (Perkin reaction) from benzaldehyde, acetic anhydride, and potassium acetate: *Org. Reactions* I, 248 (1942); from oxalyl bromide + styrene: Treibs *et al.*, *Naturwiss.* **45**, 85 (1958); from acetylene + benzaldehyde: Herbetz, *Ber.* **92**, 541 (1959). Prepn of *cis*- and *trans*-isomers: Comte *et al.*, *Compt. Rend.* **245**, 1144 (1957). Isoln from wood-rotting fungus, *Stereum subpileatum* Berk. & Curt.: Birkinshaw *et al.*, *Biochem. J.* **66**, 188 (1957).

Monoclinic crystals, d_4^4 1.2475. mp 133°; bp 300°. Distilling at 146° causes decarboxylation to styrene. K at 25° = 3.5×10^{-5}. uv max (alc): 273 nm. One gram dissolves in about 2000 ml water at 25° (more sol in hot water), in 6 ml alc, 5 ml methanol, 12 ml chloroform. Freely sol in benzene, ether, acetone, glacial acetic acid, carbon disulfide, oils. The alkali salts are sol in water.

Methyl ester, $C_{10}H_{10}O_2$, crystals, odor fruity and balsamic, reminiscent of strawberries. mp 36°. d_0^6 1.042. bp 261.9°; bp_{15} 132.5-134°. n_D^{21} 1.5766. Freely sol in alcohol, ether. Practically insol in water. Clearly sol in 80% alc.

Ethyl ester, $C_{11}H_{12}O_2$, *ethyl cinnamate, ethyl phenylacrylate*. Almost colorless, oily liquid, fruity and balsamic odor, reminiscent of cinnamon with an amber note. d_{25}^{25} 1.045-1.048. d_4^{20} 1.049. bp 271°. mp 6-10°. n_D^{20} 1.559-1.561. Miscible with alcohol, ether. Insol in water. Soluble in 3 vols of 70% alc.

n-Butyl ester, $C_{13}H_{16}O_2$, *butyl cinnamate, Eliminoxy*. Liquid, agreeable ethereal odor when pure. bp_{13} 145°. d_4^{18} 1.012. Very sparingly sol in water (<0.5%). Sol in 95% alc, ether, acetone, chloroform, benzene. Incompatible with alkalies; stable to light, air, and storage temps.

Anhydride, $C_{18}H_{14}O_3$, *cinnamic anhydride*. Crystals, mp 136°. Practically insol in water. Freely sol in warm benzene; slightly sol in alc.

USE: The main use of cinnamic acid is in the manuf of the methyl, ethyl, and benzyl esters for the perfume industry.

Ethyl ester used in glass prisms, lenses. Some esters, such as chaulmoogryl and other derivatives, are used in medicine.

2301. Cinnamon, Ceylon. Dried bark of cultivated trees of *Cinnamomum zeylanicum* Nees, *Lauraceae*. *Habit.* Ceylon, Sumatra, Borneo; cultivated in tropical Africa, America and Asia. *Constit.* Cinnamic aldehyde, tannin, 1-2% volatile oil, sugar.

USE: Pharmaceutical aid (flavor).

THERAP CAT: Carminative.

THERAP CAT (VET): Has been used as a carminative and antidiarrheal.

2302. Cinnamon, Saigon. Dried bark of *Cinnamomum loureirii* Nees, *Lauraceae*. *Habit.* Anam in Cochin China; cultivated in Java, Sumatra, S. America.

USE: Pharmaceutical aid (flavor).

THERAP CAT: Carminative.

THERAP CAT (VET): Has been used as a carminative and antidiarrheal.

2303. Cinnamoyl Chloride. *3-Phenyl-2-propenoyl chloride.* C_9H_7ClO; mol wt 166.60. C 64.87%, H 4.24%, Cl 21.28%, O 9.60%. $C_6H_5CH=CHCOCl$. Prepn by the action of oxalyl chloride on sodium cinnamate: Adams, Ulich, *J. Am. Chem. Soc.* **42**, 605 (1920).

Yellowish crystals. mp 35-36°; bp_{58} 170-171°; bp_{25} 154°; bp_{16} 147°; bp_2 101°; $d_4^{45.3}$ 1.1617; $n_4^{42.5}$ 1.614.

USE: Titrimetric determination of small amts of water.

2304. Cinnamoylcocaine. *[1R-(exo,exo)]-8-Methyl-3-[(1-oxo-3-phenyl-2-propenyl)oxy]-8-azabicyclo[3.2.1]octane-2-carboxylic acid methyl ester; ecgonine cinnamate methyl ester;* cinnamoylecgonine methyl ester; cinnamylcocaine; cinnamoylmethylecgonine. $C_{19}H_{23}NO_4$; mol wt 329.38. C 69.28%, H 7.04%, N 4.25%, O 19.43%. From leaves of *Erythroxylon coca* Lam., *Erythroxylaceae*, particularly from the Javanese leaves: Liebermann, *Ber.* **22**, 2661 (1889); de Jong, *Rec. Trav. Chim.* **67**, 484 (1948); Hegnauer, Fikenscher, *Pharm. Acta Helv.* **35**, 43 (1960); from *E. monogynum* Roxb.: Chopra, Gosh, *Arch. Pharm.* **276**, 340 (1938).

Needles, mp 121°. $[\alpha]_D$ −4.7° (chloroform). Almost insol in water; sol in alcohol, chloroform, ether.

2305. Cinnamyl Alcohol. *3-Phenyl-2-propen-1-ol;* cinnamic alcohol; styryl carbinol; γ-phenylallyl alcohol. $C_9H_{10}O$; mol wt 134.17. C 80.56%, H 7.51%, O 11.92%. $C_6H_5CH=CHCH_2OH$. Occurs (in the esterified form) in storax and in balsam Peru, cinnamon leaves, hyacinth oil. Obtained by the alkaline hydrolysis of storax. Prepd synthetically by reducing cinnamal diacetate with iron filings and acetic acid; from cinnamaldehyde by Meerwein-Ponndorf reduction with aluminum isopropoxide: Meerwein, Schmidt, *Ann.* **444**, 221 (1925).

Needles or cryst mass. Odor of hyacinth. mp 33°. d_{35}^{35} 1.0397. $bp_{1.0}$ 72.6°; bp_5 102.5°; bp_{10} 117.8°; bp_{20} 133.7°; bp_{40} 151.0°; bp_{60} 162.0°; bp_{100} 177.8°; bp_{200} 199.8°; bp_{400} 224.6°; bp_{760} 250.0°. n_D^{20} 1.58190; n_D^{33} 1.57580. When small amounts of impurities are present as in the natural article (cinnamyl alcohol from storax), it remains fluid at lower temps than the melting point. Minimum congealing points specified by the Essential Oil Assn are: 33.0° for cinnamic alcohol pure; 28.0° for cinnamic alcohol prime; 20.0° for cinnamic alcohol from storax. Is oxidized slowly on exposure to heat, light and air. Sol in water, glycerol. Clearly sol in 3 vols 50% alc. Freely sol in alc, ether, other common organic solvents.

USE: In perfumery; as deodorant in 12.5% soln in glycerol.

2306. Cinnamyl Anthranilate. *2-Aminobenzoic acid 3-phenyl-2-propenyl ester; anthranilic acid cinnamyl ester;* 3-phenyl-2-propenyl anthranilate; cinnamyl *o*-aminobenzoate. $C_{16}H_{15}NO_2$; mol wt 253.31. C 75.87%, H 5.97%, N 5.53%, O 12.63%. Synthetic imitation grape or cherry flavoring agent. Prepn: A. Seldner, *Am. Perfumer Essent. Oil Rev.* **54**, 295 (1949); R. P. Staiger, E. B. Miller, *J. Org.*

Chem. **24**, 1214 (1959). Carcinogenicity studies: G. D. Stoner *et al.*, *Cancer Res.* **33**, 3069 (1973); *Fed. Reg.* **45**, 85832 (1980). *Review:* D. L. Opdyke, *Food Cosmet. Toxicol.* **13**, Suppl., 751-752 (1975).

Cryst, mp 61-61.5°.
USE: Flavoring agent in food; fragrance in soaps and perfumes.

2307. Cinnamyl Cinnamate. *3-Phenyl-2-propenoic acid 3-phenyl-2-propenyl ester; cinnamic acid cinnamyl ester;* cinnyl cinnamate; styracin. $C_{18}H_{16}O_2$; mol wt 264.31. C 81.79%, H 6.10%, O 12.11%. $C_6H_5CH=CHCOOCH_2$-$CH=CHC_6H_5$. From buds of *Populus balsamifer* L., *Salicaceae*: Goris, Canal, *Bull. Soc. Chim. France* **3**, 1982 (1936); from *Lavanga scandens* Buch.-Ham., *Lavangalata*: Baslas, Deshapande, *J. Indian Chem. Soc.* **27**, 379 (1950). Synthesis of *trans*-cinnamyl *trans*-cinnamate: Klemm *et al.*, *Tetrahedron* **20**, 871 (1964).
trans-trans-Form, needles from abs ethanol, mp 44°. uv max (95% ethanol): 216, 223 nm (log ε 3.45, 3.25). Practically insol in water. 1 g dissolves in 22 ml cold, 3 ml boiling alcohol, 3 ml ether.

2308. Cinnarizine. *1-(Diphenylmethyl)-4-(3-phenyl-2-propenyl)piperazine; 1-cinnamyl-4-diphenylmethylpiperazine;* N-benzhydryl-*N'-trans*-cinnamylpiperazine; 1-*trans*-cinnamyl-4-diphenylmethylpiperazine; 1-cinnamyl-4-benzhydrylpiperazine; 1-diphenylmethyl-4-*trans*-cinnamylpiperazine; cinnipirine; 516 MD; Aplactan; Aplexal; Apotomin; Artate; Carecin; Cerebolan; Cerepar; Cinaperazine; Cinazyn; Cinnacet; Cinnageron; Corathiem; Denapol; Dimitron; Eglen; Folcodal; Giganten; Glanil; Hilactan; Ixertol; Izaberizin; Katoseran; Labyrin; Midronal; Mitronal; Olamin; Processine; Sedatromin; Sepan; Siptazin; Spaderizine; Stugeron; Stutgeron; Stutgin; Toliman. $C_{26}H_{28}N_2$; mol wt 368.50. C 84.74%, H 7.66%, N 7.60%. Calcium channel blocker with anti-allergic and anti-vasoconstricting activity. Prepn: Janssen, U.S. pat. 2,882,271 (1959, Janssen). Pharmacology: Van Nueten, Janssen, *Arch. Int. Pharmacodyn. Ther.* **204**, 37 (1973). Metabolism: Soudijn, van Wijngaarden, *Life Sci.* **7**, 231 (1968). Clinical study in treatment of intermittent claudication: J. H. Barber *et al.*, *Pharmatherapeutica* **2**, 401 (1980). Assessment of calcium antagonist effects: M. Spedding, *Arch. Pharmacol.* **318**, 234 (1982). HPLC determn in blood and plasma: M. Puttemans *et al.*, *J. Liq. Chromatog.* **7**, 2237 (1984). Comparison of clinical pharmacokinetics with other antihistamines: D. M. Paton, D. R. Webster, *Clin. Pharmacol.* **10**, 477-497 (1985).

Hydrochloride, $C_{26}H_{29}ClN_2$, mp 192° (dec). Soly in water: 2 mg/100 ml.
Mixture with vitamin B_6, *Emesazine*.
THERAP CAT: Antihistaminic. Peripheral and cerebral vasodilator.

2309. Cinnoline. 1,2-Benzodiazine; benzo[c]pyridazine; 1,2-diazanaphthalene; α-phenodiazine. $C_8H_6N_2$; mol wt 130.14. C 73.83%, H 4.65%, N 21.53%. First prepared by reduction of 4-chlorocinnoline: Busch, Rast, *Ber.* **30**, 521, 524 (1897). Synthesis from methyl anthranilate: Jacobs *et al.*, *J. Am. Chem. Soc.* **68**, 1310 (1946); from cinnoline-4-carboxylic acid: Morley, *J. Chem. Soc.* **1951**, 1971. Crystal and molecular structure: C. Huiszoon *et al.*, *Acta Crystallogr.* **33B**, 1867 (1977).

Pale yellow clusters of crystals from ligroin, mp 38°; also reported as mp 40-41° (Morley, *loc. cit*). Solvated needles from ether (contg 1 mol ether), mp 24-25°. $bp_{0.35}$ 114°. uv spectrum: Hearn *et al.*, *J. Chem. Soc.* **1951**, 3318. Geranium-like odor. *Poisonous!* Bitter taste resembling the taste of quinine. Must be kept under nitrogen. On exposure to air it has a tendency to liquefy and turn green. Strong base: *ibid.* **1949**, 1356. Freely sol in the usual solvents.
Hydrochloride, $C_8H_6N_2$.HCl, brownish needles from alcohol + ether, mp 156-160°. Volatilizes around 100° on slow heating. Freely sol in water, alcohol.
Picrate, $C_8H_6N_2.C_6H_3N_3O_7$, amber-colored prisms from alc, dec 196-196.5°.

2310. Cinobufotalin. *16-(Acetyloxy)-14,15-epoxy-3,5-dihydroxybufa-20,22-dienolide; 14,15β-epoxy-3β,5,16β-trihydroxy-5β-bufa-20,22-dienolide 16-acetate.* $C_{26}H_{34}O_7$; mol wt 458.53. C 68.10%, H 7.47%, O 24.43%. Isolated from the Chinese drug Ch'an Su which is prepd from Chinese toads (*Bufo asiaticus = bufo gargarizans* Cantor). Isoln: Kotake, Kuwada, *C.A.* **31**, 7065[2] (1937); Ruckstubl, Meyer, *Helv. Chim. Acta* **40**, 1270 (1957). Structure: Bernoulli *et al., ibid.* **45**, 240 (1962).

Octahedrons from acetone, mp 259-262°. $[\alpha]_D^{20}$ +11°. uv max: 295 nm (log ε 3.72).

2311. Cinoxacin. *1-Ethyl-1,4-dihydro-4-oxo[1,3]dioxolo[4,5-g]cinnoline-3-carboxylic acid;* 1-ethyl-6,7-methylenedioxy-4(1H)-oxocinnoline-3-carboxylic acid; Compound 64716; Cinobac; Noxigram; Uronorm. $C_{12}H_{10}N_2O_5$; mol wt 262.22. C 54.97%, H 3.84%, N 10.68%, O 30.51%. Quinolone antibacterial; analog of oxolinic acid, *q.v.* Prepn: W. A. White, Ger. pat. 2,005,104; *idem*, U.S. pat. 3,669,965 (1970, 1972 both to Lilly). *In vitro* and *in vivo* study: W. E. Wick *et al.*, *Antimicrob. Ag. Chemother.* **4**, 415 (1973). Pharmacology: J. J. Szwed *et al.*, *J. Antimicrob. Chemother.* **4**, 451 (1978); K. S. Israel *et al.*, *J. Clin. Pharmacol.* **18**, 491 (1978). Metabolism: R. L. Wolen *et al.* in *Stable Isotopes*, T. A. Baillie, Ed. (Univ. Park Press, Baltimore, 1978) pp 113-125. Clinical studies: S. Colleen *et al.*, *J. Antimicrob. Chemother.* **3**, 579 (1977); S. N. Rous, *J. Urol.* **120**, 196 (1978). Toxicity: I. Narama *et al.*, *Chemotherapy (Tokyo)* **28**, Suppl. 4, 406 (1980), *C.A.* **94**, 25007 (1981). Pharmacokinetics: M. Ohkawa *et al.*, *J. Antimicrob. Chemother.* **8**, 447 (1981); R. Barbhaiya *et al.*, *Antimicrob. Ag. Chemother.* **21**, 472 (1982). *Review:* J. M. Scavone *et al.*, *Pharmacotherapy* **2**, 266-271 (1982). Review of pharmacology and therapeutic efficacy: T. S. Sisca *et al.*, *Drugs* **25**, 544-569 (1983).

Light tan crystals, mp 261-262° (dec). Sol in most polar organic solvents. LD_{50} in rats (mg/kg): 4160 orally; 900 i.v. (Narama).

Sodium salt, $C_{12}H_9N_2NaO_5$, white crystalline solid. Sol in aqueous solvents.

THERAP CAT: Antibacterial.

2312. Cinoxate. *3-(4-Methoxyphenyl)-2-propenoic acid 2-ethoxyethyl ester; p-methoxycinnamic acid 2-ethoxyethyl ester;* 2-ethoxyethyl p-methoxycinnamate; Give-Tan; Sun-Dare. $C_{14}H_{18}O_4$; mol wt 250.28. C 67.18%, H 7.25%, O 25.57%. Prepn: **Brit. pat. 856,411** (1960 to Givaudan).

$$CH_3O-\!\!\!\bigcirc\!\!\!-CH=CHCOOCH_2CH_2OCH_2CH_3$$

Viscous liq; bp_2 184-187°. May have a slightly yellow tinge. Practically odorless. Solidifies below −25°. d_{25}^{25} 1.102. n_D^{20} 1.5670. Sapon no. 225.5. Practically insol in water (soly ~0.05%). Soly in glycerol 0.5%, in propylene glycol 5.0%. Miscible with alcohols, esters, vegetable oils.

USE: Sunscreen agent.

2313. Cinromide. *(E)-3-(3-Bromophenyl)-N-ethyl-2-propenamide; trans-3-bromo-N-ethylcinnamamide;* BW 122u; Vumide. $C_{11}H_{12}BrNO$; mol wt 254.12. C 51.99%, H 4.76%, Br 31.44%, N 5.51%, O 6.30%. Cinnamic acid derivative with anti-epileptic activity. Prepn: E. M. Grivsky, **Ger. pat. 2,535,599** corresp to **U.S. pat. 4,041,071** (1976, 1977 to Burroughs Wellcome). Pharmacology: F. E. Soroko *et al., J. Pharm. Pharmacol.* **33,** 741 (1981); G. H. Fromm *et al., Epilepsia* **24,** 394 (1983). Pharmacokinetics and metabolism: R. R. Maddox, B. B. Wannamaker, *Curr. Ther. Res.* **32,** 165 (1982). Mass spec determn in plasma and urine: R. J. Perchalski *et al., Anal. Chem.* **54,** 1466 (1982). Controlled clinical trials: J. W. Allen *et al., Epilepsia* **24,** 422 (1983); B. Spilker *et al., Curr. Ther. Res.* **34,** 7 (1983).

Crystals from ethanol-water, mp 89-90°. LD_{50} in mice (mg/kg): 660 ± 28 i.p.; 2277 ± 250 orally (Soroko).

THERAP CAT: Anticonvulsant.

2314. Ciprofibrate. *2-[4-(2,2-Dichlorocyclopropyl)phenoxy]-2-methylpropanoic acid;* 2-[4-(2,2-dichlorocyclopropyl)phenoxy]isobutyric acid; Win 35833; Ciprol; Lipanor. $C_{13}H_{14}Cl_2O_3$; mol wt 289.15. C 54.00%, H 4.88%, Cl 24.52%, O 16.60%. Hypolipemic agent, related structurally to clofibrate, *q.v.* Prepn: D. K. Phillips, **Ger. pat. 2,343,606** corresp to **U.S. pat. 3,948,973** (1974, 1976 both to Sterling). Pharmacokinetics in animals and man: C. Davison *et al., Drug Metab. Dispos.* **3,** 520 (1975). Inhibition of cholesterol biosynthesis in rats: A. Arnold *et al., Atherosclerosis* **32,** 155 (1979). Blood levels, distribution, duration of action: J. Edelson *et al., ibid.* **33,** 351 (1979). Metabolic effects: A. Arnold *et al., J. Pharm. Sci.* **68,** 1557 (1979). Teratology study: H. Tuchman-Duplessis *et al., Toxicology* **12,** 1 (1979).

Pale cream solid from hexane, mp 114-116°.

THERAP CAT: Antihyperlipoproteinemic.

2315. Ciprofloxacin. *1-Cyclopropyl-6-fluoro-1,4-dihydro-4-oxo-7-(1-piperazinyl)-3-quinolinecarboxylic acid;* Bay o 9867; Ciflox; Ciprobay; Ciproxan; Ciproxin; Velmonit. $C_{17}H_{18}FN_3O_3$; mol wt 331.35. C 61.62%, H 5.48%, F 5.73%, N 12.68%, O 14.49%. Fluorinated quinolone antibacterial. Prepn: K. Grohe *et al.,* **Ger. pat. 3,142,854;** *eidem,* **U.S. pat. 4,670,444** (1983, 1987 both to Bayer AG); K. Grohe, H. Heitzer, *Ann.* **1987,** 29. Use as plant fungicide: K. Grohe *et al.,* **Ger. pat. 3,248,507;** *eidem,* **U.S. pat. 4,563,459** (1984,

1986 both to Bayer AG). *In vitro* activity against anaerobic bacteria: B. Watt, F. V. Brown, *J. Antimicrob. Chemother.* **17,** 605 (1986). Comparison of activity with other nalidixic acid analogs: C. M. Bassey *et al., ibid.* 623. HPLC determn in biological fluids: W. Gau *et al., J. Liq. Chromatog.* **8,** 485 (1985). Pharmacokinetics: G. Hoffken *et al., Antimicrob. Ag. Chemother.* **27,** 375 (1985). Clinical trial: C. A. Ramirez *et al., ibid.* **28,** 128 (1985); in *Pseudomonas aeruginosa* infections: H. Giamarellou *et al., Drugs Exp. Clin. Res.* **11,** 351 (1985); B. E. Scully *et al., Lancet* **1,** 819 (1986). Series of articles on antibacterial activity, pharmacokinetics, clinical efficacy and safety: *Chemotherapy (Tokyo)* **33,** Suppl. 7, 1-1024 (1985). Symposium on antibacterial spectrum and clinical use: *Am. J. Med.* **82,** Suppl. 4A, 1-404 (1987).

Dec 255-257°.

Monohydrochloride monohydrate, $C_{17}H_{19}ClFN_3O_3 \cdot H_2O$, *Cipro.*

THERAP CAT: Antibacterial.

2316. Ciramadol. *3-[(Dimethylamino)(2-hydroxycyclohexyl)methyl]phenol;* (−)-cis-2-(α-dimethylamino-m-hydroxybenzyl)cyclohexanol; WY-15705. $C_{15}H_{23}NO_2$; mol wt 249.36. C 72.25%, H 9.30%, N 5.62%, O 12.83%. Novel analgesic with narcotic agonist-antagonist properties. Prepn: J. P. Yardley, P. B. Russel, **Ger. pat. 2,317,183,** *C.A.* **80,** 47637c (1974) and **U.S. pats. 3,928,626, 4,017,637** (1973, 1975 and 1977 all to Am. Home). Synthesis, stereochemistry, biological activity: J. P. Yardley *et al., Experientia* **34,** 1124 (1978). Determn in pharmaceutical dosage forms: H. K. Chan, A. G. Fogg, *Anal. Chim. Acta* **105,** 423 (1979). Metabolism: C. O.Tio *et al., Fed. Proc.* **38,** 2717 (1979). Respiratory effects in man: H. P. Wuest, *Arch. Int. Pharmacodyn. Ther.* **240,** 45 (1979). Analgesic effect in chronic pain: M. J. Staquet, *Curr. Med. Res. Opin.* **6,** 475 (1980).

Cryst from acetone/hexane, mp 191-193°. $[\alpha]_D^{25}$ −46.92° (c = 1.061 in methanol).

Hydrochloride, $C_{15}H_{24}ClNO_2$, *Ciradol.* Cryst from acetone/methanol, mp 255-257° (effervescence). $[\alpha]_D^{25}$ −15.31° (c = 1.067 in methanol).

THERAP CAT: Analgesic.

2317. Circulins. Polypeptide antibiotics related to polymyxins. Produced by *Bacillus circulans,* a microorganism found in soil and dust: Murray, Tetrault, *Proc. Soc. Am. Bact.* **1,** 20 (1948); McLeod, *J. Bacteriol.* **56,** 749 (1948); Murray *et al., ibid.* **57,** 305 (1949). Isoln yields circulins A and B, circulin A being the major component: Tetrault, **U.S. pat. 2,676,133** (1954 to Purdue Res. Found.). *Review:* Vogner, Studer, *Experientia* **22,** 345 (1966).

DAB = α,γ-diaminobutyric acid

Circulin A, $C_{53}H_{100}N_{16}O_{13}$, R = (+)-6-methyloctanoyl. Structure: Fujikawa *et al.*, *Experientia* **21,** 307 (1965). Synthesis: Studer *et al.*, *Helv. Chim. Acta* **49,** 974 (1966).

Sulfate, $C_{53}H_{100}N_{16}O_{13}.2\frac{1}{2}H_2SO_4$, crystals or amorphous solid, dec 226-228°. $[\alpha]_D^{25}$ −61.6° (c = 1.25). Sol in water; less sol in the lower alcs. Practically insol in acetone and water-immiscible solvents.

Circulin B, $C_{52}H_{98}N_{16}O_{13}$, R = isooctanoyl. Structure: Hayashi *et al.*, *Experientia* **24,** 656 (1968).

2318. Cisapride. *cis-4-Amino-5-chloro-N-[1-[3-(4-fluorophenoxy)propyl]-3-methoxy-4-piperidinyl]-2-methoxybenzamide; cis*-4-amino-5-chloro-*N*-[1-[3-(*p*-fluorophenoxy)propyl]-3-methoxy-4-piperidinyl]-*o*-anisamide. $C_{23}H_{29}ClFN_3O_4$; mol wt 465.95. C 59.29%, H 6.27%, Cl 7.61%, F 4.08%, N 9.02%, O 13.73%. Gastrointestinal prokinetic agent. Prepn: G. Van Daele, **Eur. pat. Appl. 76,530** (1983 to Janssen), *C.A.* **99,** 194812d (1983); and pharmacology: G. Van Daele *et al.*, *Drug Dev. Res.* **8,** 225 (1986). Pharmacokinetics: J. A. Barone *et al.*, *Clin. Pharm.* **6,** 640 (1987). Review of pharmacology and clinical evaluations: A. Reyntjens *et al.*, *Drug. Dev. Res.* **8,** 251 (1986). Series of articles on pharmacology, mechanism of action and clinical efficacy: *Curr. Ther. Res.* **36,** 1029-1070 (1984).

Monohydrate, $C_{23}H_{29}ClFN_3O_4.H_2O$, *R 51619, Acenalin, Prepulsid, Risamal.* Crystals from 2-propanol, mp 109.8°.
THERAP CAT: Peristaltic stimulant.

2319. Cisplatin. *(SP-4-2)-Diamminedichloroplatinum; cis-diamminedichloroplatinum; cis*-platinum II; *cis*-DDP; CACP; CPDC; DDP; NSC-119875; Briplatin; Cismaplat; Cisplatyl; Citoplatino; Lederplatin; Neoplatin; Platamine; Platinex; Platiblastin; Platinol; Platinoxan; Platistin; Platosin; Randa. $Cl_2H_6N_2Pt$; mol wt 300.05. Cl 23.63%, H 2.02%, N 9.33%, Pt 65.02%. Prepn: M. Peyrone, *Ann.* **51,** 1 (1845); G. B. Kauffman, D. O. Cowan, *Inorg. Syn.* **7,** 239 (1963); S. C. Dhara, *Indian J. Chem.* **8,** 193 (1970). Originally known as *Peyrone's salt* or *Peyrone's chloride*, it was primarily of interest in the development of coordination theory. Early structural studies: R. Werner, *Z. Anorg. Chem.* **3,** 267 (1893); H. D. K. Drew *et al.*, *J. Chem. Soc.* **1932,** 988. Discovery of anti-tumor activity: B. Rosenberg *et al.*, *Nature* **205,** 698 (1965); **222,** 385 (1972). Use as neoplasm inhibitor: M. L. Tobe *et al.*, **Ger. pat. 2,318,020** (1972 to Rustenburg Platinum Mines Ltd.), *C.A.* **80,** 55897e (1974); M. J. Cleare *et al.*, **Ger. pat. 2,329,485** (1972 to Research Corp.), *C.A.* **81,** 21172v (1974). Mechanism of action: J. J. Roberts, J. M. Pascoe, *Nature* **235,** 282 (1972); A. Khan, J. M. Hill, *Transplant* **13,** 55 (1972); G. L. Cohen *et al.*, *Science* **203,** 1014 (1979). X-ray structure of cisplatin-DNA adduct: S. E. Sherman *et al.*, *Science* **230,** 412 (1985). Inhibition of *in vitro* DNA synthesis: A. L. Pinto, S. J. Lippard, *Proc. Natl. Acad. Sci. USA* **82,** 4616 (1985). Pharmacology: A. Sirica *et al.*, *Proc. Am. Assoc. Cancer Res.* **12,** 4 (1971); C. L. Litterst *et al.*, *Cancer Res.* **36,** 2340 (1976); N. P. Johnson *et al.*, *Chem. Biol. Interact.* **23,** 267 (1978). Metabolism: R. C. Lange *et al.*, *J. Nucl. Med.* **14,** 191 (1973). Clinical studies: J. J. Ochs *et al.*, *Cancer Treat. Rep.* **62,** 239 (1978); H. M. Pinedo *et al.*, *Eur. J. Cancer* **14,** 1149 (1978). Cisplatin has been shown to have trypanocidal effects: M. S. Wysor *et al.*, *Science* **217,** 454 (1982). Toxicology: R. L. Dixon, *Proc. 7th Int. Congr. Chemother. Vol. 2*

(University Park Press, Baltimore, 1972) pp 241-243; R. W. Fleishman *et al.*, *Toxicol. Appl. Pharmacol.* **33,** 320 (1975). *Reviews:* D. D. Von Hoff, M. Rozencweig in *Advances in Pharmacology and Chemotherapy* vol. 16, S. Garattini *et al.*, Eds. (Academic Press, New York, 1979) pp 273-294; S. J. Lippard, *Science* **218,** 1075-1082 (1982). Review of carcinogenicity studies: *IARC Monographs* **26,** 154-161 (1981). Comprehensive description: C. M. Riley, L. M. Sternson in *Analytical Profiles of Drug Substances* vol. **14,** K. Florey, Ed. (Academic Press, New York, 1985) pp 77-105. Book: *Cisplatin, Current Status and New Developments,* A. W. Prestayko *et al.*, Eds. (Academic Press, New York, 1980) 527 pp.

Deep yellow solid, dec 270°. Soly in water 0.253 g/100 g at 25°; slowly changes to *trans*-form in aq soln. Insol in most common solvents. Sol in DMF. LD_{50} in guinea pigs: 9.7 mg/kg i.p. (Fleishman).
THERAP CAT: Antineoplastic.

2320. Citalopram. *1-[3-(Dimethylamino)propyl]-1-(4-fluorophenyl)-1,3-dihydro-5-isobenzofurancarbonitrile;* 1-[3-(dimethylamino)propyl]-1-(4-fluorophenyl)-5-phthalancarbonitrile; nitalapram; Lu 10-171. $C_{20}H_{21}FN_2O$; mol wt 324.40. C 74.05%, H 6.52%, F 5.86%, N 8.64%, O 4.93%. Serotonin uptake inhibitor. Prepn: K. P. Boegesoe, A. S. Toft, **Ger. pat. 2,657,013;** *eidem,* **U.S. pat. 4,136,193** (1977, 1979 both to Kefalas); A. J. Bigler *et al.*, *Eur. J. Med. Chem. Chim. Ther.* **12,** 289 (1977). Pharmacology: A. V. Christensen *et al.*, *Eur. J. Pharmacol.* **41,** 153 (1977). Pharmacokinetics: K. F. Overo, *Eur. J. Clin. Pharmacol.* **14,** 69 (1978). HPLC determn in plasma and urine: E. Oyehaug *et al.*, *J. Chromatog.* **308,** 199 (1984). Clinical evaluations in depression: H. Dufour *et al.*, *Int. Clin. Psychopharmacol.* **2,** 225 (1987); L. Timmerman *et al., ibid.* 239.

$bp_{0.03}$ 175-181°.
Hydrobromide, $C_{20}H_{22}BrFN_2O$, crystals from isopropanol, mp 182-183°.
THERAP CAT: Antidepressant.

2321. Citicoline. *Cytidine 5'-(trihydrogen diphosphate) mono[2-(trimethylammonio)ethyl] ester hydroxide inner salt;* choline cytidine 5'-pyrophosphate (ester); cytidine diphosphate choline ester; CDP-choline; Audes; Cereb; Citifar; Colite; Corenalin; Cyscholin; Difosfocin; Emicholin; Ensign; Haocolin; Hornbest; Neucolis; Nicholin; Nicolin; Niticolin; Reagin; Recognan; Rexort; Sintoclar; Somazina; Suncholin. $C_{14}H_{26}N_4O_{11}P_2$; mol wt 488.33. C 34.43%, H 5.37%, N 11.47%, O 36.04%, P 12.68%. Naturally occurring nucleotide; coenzyme in one of the biosynthetic pathways of lecithin synthesis. Isoln from liver and yeast: Kennedy, Weiss, *J. Am. Chem. Soc.* **77,** 250 (1955); Lieberman *et al.*, *Science* **124,** 81 (1956). Molecular structure: M. A. Viswamitra *et al.*, *Nature* **258,** 497 (1975). Synthesis: Kennedy, *J. Biol. Chem.* **222,** 185 (1956); **Japan. pats. 6541('64)** and **1384('67)** (both to Takeda), *C.A.* **61,** 13406c (1964) and *C.A.* **66,** 85992k (1967); Kikugawa *et al.*, *Chem. Pharm. Bull.* **19,** 1011, 2466 (1971); N. H. Phuong *et al.*, *Bull. Soc. Chim. France* **1979**(pt 2), 518. Biosynthesis: Ewing, Finamore, *Biochim. Biophys. Acta* **218,** 463 (1970). *Review:* B. Nilsson, *Clin. Pharmacol. Drug Epidemiol.* **2,** 273-277 (1979). Series of articles on pharmacology, toxicity, distribution, pharmacokinetics and clinical studies: *Arzneimittel-Forsch.* **33,** 1009-1080 (1983). Toxicity: T. Grau *et al., ibid.* 1033.

Amorphous, somewhat hygroscopic powder. $[\alpha]_D^{20}$ +17.2° (H_2O); $[\alpha]_D^{25}$ +19.0° (c = 0.86 in H_2O). uv max (pH 1): 280 nm (ϵ 12800). Dissolves readily in water to form acidic soln. Practically insol in most organic solvents. pKa 4.4. LD_{50} in mice, rats (mg/kg): 4600 ±335, 4150 ±370 i.v.; both species 8 g/kg orally (Grau).

Sodium salt, $C_{14}H_{25}N_4NaO_{11}P_2$, *Brassel, Cebroton, Cidifos, Gerolin, Logan, Neuroton, Sinkron*.

THERAP CAT: Cerebral vasodilator.

2322. Citiolone. *2-Acetamido-4-mercaptobutyric acid γ-thiolactone;* N-(tetrahydro-2-oxo-3-thienyl)acetamide; N-acetylhomocysteinethiolactone; α-acetamido-γ-thiobutyrolactone; AHCTL; BO 714; Citiolase; Thioxidrene. C_6H_9-NO_2S; mol wt 159.20. C 45.26%, H 5.70%, N 8.80%, O 20.10%, S 20.14%. Prepn: Wagner *et al.,* Ger. pat. **1,134,683** (1962 to Degussa, vorm Roessler), *C.A.* **58,** 4648c (1963); Nakanishi *et al.,* **Japan.** pat. **16,712('62)** (to Yoshitomi), *C.A.* **59,** 11660h (1963); Takayananagi *et al.,* and Ogino, Yasui, **Japan.** pats. **3420('64)** and **1376('65)** (both to Sumitomo), *C.A.* **61,** 3193h (1964) and **62,** 14822c (1965). Chemistry: Benesch, Benesch, *J. Am. Chem. Soc.* **78,** 1597 (1956); Laliberté *et al., J. Chem. Soc.* **1963,** 2756. Pharmacology: Kirnberger *et al., Arzneimittel-Forsch.* **8,** 72 (1958); Varga *et al., ibid.* **13,** 1867 (1963).

Needles from toluene, mp 111.5-112.5°. uv max: 238 nm (ϵ 4400). LD_{50}: 1.20 g/kg i.v. in mice; 1.95 g/kg i.p. in rats.

USE: Photographic antifogging agent: Dersch, **U.S.** pat. **3,068,100** (1962); Weber, **Ger.** pat. **1,164,828** (1964 to Adox Fotowerke Schleussner GmbH), *C.A.* **60,** 14050f (1964); protector against radiation: Langendorff, Koch, *Strahlentherapie* **106,** 451 (1958); Braun *et al., ibid.* **108,** 262 (1959), *C.A.* **52,** 18841h (1958); **53,** 17325e (1959).

THERAP CAT: In treatment of hepatic disorders.

2323. Citraconic Acid. *2-Methyl-2-butenedioic acid;* methylmaleic acid. $C_5H_6O_4$; mol wt 130.10. C 46.16%, H 4.65%, O 49.19%. Obtained by carefully heating citric acid at about 175°. Production of citraconic anhydride from itaconic acid: Humphrey, **U.S.** pat. **2,966,498** (1960 to Pfizer).

Hygroscopic monoclinic crystals; characteristic odor. d 1.62. mp about 90° with decompn. Freely sol in water, alc, ether, slightly in chloroform; insol in benzene or petr ether. LD_{50} orally in rats, mice: 1320, 2260 mg/kg, P. M. Jenner *et al., Food Cosmet. Toxicol.* **2,** 327 (1964).

2324. Citral. *3,7-Dimethyl-2,6-octadienal.* $C_{10}H_{16}O$; mol wt 152.23. C 78.89%, H 10.59%, O 10.51%. Constituent (75 to 85%) of oil of lemon grass, the volatile oil of *Cymbopogon citratus* (DC.) Stapf, or of *Cymbopogon flexuosus* (Nees) Stapf, *Gramineae.* Also present to a limited extent in oils of verbena, lemon, and orange. Isoln: Tiemann, Semuler, *Ber.* **26,** 2708 (1893); Tiemann, *ibid.* **31,** 3310, 3317 (1898); Hibbert, Cannon, *J. Am. Chem. Soc.* **46,** 119 (1924). Synthesis based on acetylene chemistry: Kimel, Sax, **U.S.** pat. **2,661,368** (1953 to Hoffmann-La Roche). Synthesis

from isoprene: Leets *et al., J. Gen. Chem. USSR* **27,** 1584 (1957), *C.A.* **53,** 4336e (1959). Alternate synthesis: T. Mandai, *Tetrahedron Letters* **22,** 763 (1981); K. Takabe *et al., Syn. Commun.* **13,** 297 (1983). Citral from natural sources is a mixture of two geometric isomers *geranial* and *neral.* Separation of isomers: Naves, *Bull. Soc. Chim. France* **1952,** 521. Comprehensive reviews: J. L. Simonsen, *The Terpenes,* vol. **I,** 83-100 (1947); E. Guenther, *The Essential Oils,* vol. **II,** 326-336 (1949); Gildemeister-Hoffmann, *Die Aetherischen Oele,* vol. **IV,** 307-356 (4th ed., 1956).

geranial (citral a)   neral (citral b)

Geranial, light oily liquid. Strong lemon odor. $bp_{2.6}$ 92-93°. d_4^{20} 0.8888. n_D^{20} 1.48982. Practically insol in water. Miscible with alc, ether, benzyl benzoate, diethyl phthalate, glycerol, propylene glycol, mineral oil, essential oils.

Neral, light oily liquid. Lemon odor not as intense but sweeter than gerianal. $bp_{2.6}$ 91-92°. d_4^{20} 0.8869. n_D^{20} 1.48690. Solubilities same as gerianal.

USE: In the synthesis of vitamin A, ionone and methylionone. As a flavor, for fortifying lemon oil. In perfumery for its citrus effect in lemon and verbena scents, in cologne odors, in perfumes for colored soaps. Not stable to alkalies and strong acids. Will cause discoloration of white soaps and alkaline cosmetics.

2325. Citramalic Acid. *2-Hydroxy-2-methylbutanedioic acid;* 2-methylmalic acid; 2-hydroxy-2-methylsuccinic acid; α-hydroxypyrotartaric acid; *trans*-methylbutanedioic acid. $C_5H_8O_5$; mol wt 148.11. C 40.54%, H 5.44%, O 54.01%. Enzymatic synthesis: Barker, Blair, *Biochem. Prepns.* **9,** 21 (1962). Chemical synthesis: Barker, *ibid.* 25; J. B. Wilkes, R. G. Wall, *J. Org. Chem.* **45,** 247 (1980). Stereoselective synthesis: E. G. J. Staring *et al., Rec. Trav. Chim.* **105,** 374 (1986).

dl-Form, deliquescent monoclinic prisms from ethyl acetate + petr ether, mp 117°. Sublimes. Freely sol in water, acetone. Sol in ethyl acetate, ether. Practically insol in petr ether, benzene.

d-Form, crystals, mp 112.2-112.8°. $[\alpha]_D^{22}$ +23.6° (c = 3 in H_2O).

l-Form, crystals, mp 112-113°. $[\alpha]_D^{20}$ -23.4° (c = 3 in H_2O).

2326. β-Citraurin. *3-Hydroxy-8'-apo-β,ψ-carotenal;* citraurin. $C_{30}H_{40}O_2$; mol wt 432.62. C 83.28%, H 9.32%, O 7.40%. Carotenoid pigment found only in orange peel. Isoln by chromatography: Zechmeister, Tuzson, *Ber.* **69,** 1878 (1936); **70,** 1966 (1937). The peels from 100 kilos of oranges yield about 35 mg. Structure: Zechmeister, Tuzson, *loc. cit.;* Karrer, Solmssen, *Helv. Chim. Acta* **20,** 682 (1937); Karrer *et al., ibid.* 1020; Zechmeister, v. Cholnoky, *Ann.* **530,** 291 (1937); Karrer *et al., Helv. Chim. Acta* **21,** 445 (1938). Abs config: Bartlett *et al., J. Chem. Soc.* (C) **1969,** 2527. Synthesis: H. Pfander *et al., Helv. Chim. Acta* **63,** 1377 (1980).

Thin orange or yellow-colored plates from benzene + petr ether, mp 147°. Absorption max (benzene): 497, 467 nm. Freely sol in acetone, ethanol, ether, benzene, and carbon disulfide. Sparingly sol in petr ether.

2327. Citrazinic Acid. *1,2-Dihydro-6-hydroxy-2-oxo-4-pyridinecarboxylic acid; 2,6-dihydroxyisonicotinic acid; 2,6-dihydroxy-4-pyridinecarboxylic acid.* $C_6H_5NO_4$; mol wt 155.11. C 46.46%, H 3.25%, N 9.03%, O 41.26%. Prepn from citric acid with aq NH_3 at 140-160° under pressure: Bavley, Hamilton, U.S. pat. 2,729,647 (1956 to Pfizer). Purification: Bavley et al., U.S. pat. 2,738,352 (1956 to Pfizer).

Yellowish powder with a greenish tinge; carbonizes above 300° without melting. Ultrapure material which is white or colorless, has been prepared. Almost insol in water; slightly sol in hot HCl; sol in alkali hydroxide or carbonate solns. Alkaline solns turn blue on standing.

2328. Citric Acid. *2-Hydroxy-1,2,3-propanetricarboxylic acid; β-hydroxytricarballylic acid.* $C_6H_8O_7$; mol wt 192.12. C 37.51%, H 4.20%, O 58.29%. Widely distributed in plants and in animal tissues and fluids. Produced by mycological fermentation on an industrial scale using crude sugar solns, such as molasses and strains of *Aspergillus niger*: See review by Von Loesecke, *Chem. & Eng. News* **23**, 1952 (1945); Schweiger, U.S. pat. 2,970,084 (1961 to Miles Labs.); Faith, Keyes & Clark's *Industrial Chemicals*, F. A. Lowenheim, M. K. Moran, Eds. (Wiley-Interscience, New York, 4th ed., 1975) pp 275-279. Also extracted from citrus fruits (lemon juice contains 5 to 8%) and from pineapple waste. *Reviews:* Wilson, *Chem. & Met. Eng.* **29**, 787 (1923); Browne, *Ind. Eng. Chem.* **13**, 81 (1921); Warneford, Hardy, *ibid.* **17**, 1283 (1925); E. F. Bouchard, E. G. Merritt in Kirk-Othmer *Encyclopedia of Chemical Technology* vol. 6 (Wiley-Interscience, New York, 3rd ed., 1979) pp 150-179. Toxicity: Gruber, Halbeisen, *J. Pharmacol. Exp. Ther.* **94**, 65 (1948).

Anhydr form, mp 153°. Crystals are monoclinic holohedra and crystallize from hot concd aq soln. d 1.665. At 25°, pK_1 3.128; pK_2 4.761; pK_3 6.396, Bates, Pinching, *J. Am. Chem. Soc.* **71**, 1274 (1949). Soly in water: 54.0% w/w at 10°; 59.2% at 20°; 64.3% at 30°; 68.6% at 40°; 70.9% at 50°; 73.5% at 60°; 76.2% at 70°; 78.8% at 80°; 81.4% at 90°; 84.0% at 100°.
Monohydrate, orthorhombic crystals from cold aq solns. Pleasant, sour taste. d 1.542. Monohydrate crystals lose water of crystn in dry air or when heated at about 40 to 50°, slightly deliquesce in moist air. Softens at 75°. mp ~100°. pH of 0.1N soln = 2.2. Densities of aq soln (15°/15°): 10% = 1.0392; 20% = 1.0805; 30% = 1.1244; 40% = 1.1709; 50% = 1.2204; 60% = 1.2738. Soly in g/100 g satd soln: ether 2.17; chloroform 0.007; amyl alcohol 15.43; amyl acetate 5.98; ethyl acetate 5.98. Soly at 19° in g/100 g solvent: methanol 197; propanol 62.8. LD_{50} i.p. in rats: 975 mg/kg (Gruber, Halbeisen).
Pharmaceutical Incompatibilities: Potassium tartrate, alkali and alkaline earth carbonates and bicarbonates, acetates, sulfides. Dilute aq solns may ferment on standing.
Barium salt heptahydrate, $C_{12}H_{10}Ba_3O_{14}.7H_2O$, *barium citrate.* Powder. Loses all H_2O at 150°. Sol in 1750 parts water; freely sol in dil HCl or HNO_3; practically insol in alcohol.
Ethyl ester, $C_{12}H_{20}O_7$, *ethyl citrate, triethyl citrate.* Bitter, oily liq. d^{20} 1.137. bp_{760} 294°; $bp_{1.0}$ 127°. Viscosity at 25°: 35.2 cps. Pour pt ~10°. n_D^{20} 1.4455. Soly: water ~6.9%; peanut oil 0.8%. Misc with alc, ether.

USE: Acidulant in beverages, confectionery, effervescent salts, in pharmaceutical syrups, elixirs, in effervescent powders and tablets, to adjust the pH of foods and as synergistic antioxidant, in processing cheese. Used in beverages, jellies, jams, preserves and candy to provide tartness. In the manuf of alkyd resins; in esterified form as plasticizer, foam inhibitor. In the manuf of citric acid salts. As sequestering agent to remove trace metals. As mordant to brighten colors; in electroplating; in special inks; in analytical chemistry for determining citrate-soluble P_2O_5; as reagent for albumin, mucin, glucose, bile pigments.
THERAP CAT: Component of anticoagulant citrate solns (citrate dextrose soln; citrate phosphate dextrose soln; citric acid syrup).

2329. Citrinin. *(3R-trans)-4,6-Dihydro-8-hydroxy-3,4,5-trimethyl-6-oxo-3H-2-benzopyran-7-carboxylic acid;* Antimycin. $C_{13}H_{14}O_5$; mol wt 250.24. C 62.39%, H 5.64%, O 31.97%. Antibiotic substance produced by a white spore aspergillus which has been placed under the species name *Aspergillus niveus* (Thorn and Raper). Also produced in small quantities by *Penicillium citrinum:* Hetherington, Raistrick, *Trans. Roy. Soc. London* **B220**, 269 (1931); Raistrick, Smith, *Chem. & Ind (London)* **60**, 828 (1941); Timonin, *Science* **96**, 494 (1942); Timonin, Rovatt, *Can. J. Pub. Health* **35**, 80 (1944). Identity with antimycin: Haese, *Arch. Pharm.* **296**, 227 (1963). Structure: Brown et al., *J. Chem. Soc.* **1949**, 867; Warren et al., *J. Am. Chem. Soc.* **79**, 3812 (1957); Kovac et al., *Nature* **190**, 1104 (1961). Synthesis: Cartwright et al., *J. Chem. Soc.* **1949**, 1563; J. A. Barber et al., *J. Chem. Soc. Perkin Trans. I* **1986**, 2101. Stereochemistry: Cram, *J. Am. Chem. Soc.* **72**, 1001 (1950); Mehta, Whalley, *J. Chem. Soc.* **1963**, 3777; Mathieson, Whalley, *ibid.* **1964**, 4640. Physical characteristics and toxicity: Nagai et al., *Chem. Zentr.* **1958**, 8088, *C.A.* **55**, 1914 (1961). Crystal and molecular structure: Rodig, *Chem. Commun.* **1971**, 1553. Biosynthesis: J. Barber et al., *J. Chem. Soc. Perkin Trans. I* **1981**, 2577; L. Colombo et al., *ibid.* 2594. Toxicology: A. M. Ambrose, F. De Eds, *J. Pharmacol. Exp. Ther.* **88**, 173 (1946). *Review:* Saito et al., "Yellowed Rice Toxins" in *Microbial Toxins*, A. Ciegler, S. Kadis, A. Ajl, Eds. (Academic Press, New York, 1971) **vol. VI**, pp 357-367.

Lemon-yellow needles from alcohol, dec 175°. $[α]_D^{18}$ −37.4° (c = 1.15 in alc.). uv max: 250, 331 nm ($E_{1cm}^{1\%}$ 370, 418). Strong acid. Practically insol in water. Sol in alcohol, dioxane, dilute alkali. Solns change color with changes in pH, from lemon-yellow at pH 4.6 to cherry-red at pH 9.9. *Poisonous!* LD_{50} i.p. in mice, rats: 35, 67 mg/kg, A. M. Ambrose, F. De Eds, *loc. cit.*
Methyl citrinin, $C_{14}H_{16}O_5$, plates from benzene, dec 139°. $[α]_D^{18}$ +217.1° (c = 0.38 in acetone). uv max: 260, 334 nm ($E_{1cm}^{1\%}$ 520, 151.6). Sol in hot alcohol; moderately sol in chloroform. Practically insol in petr ether.
Dihydrocitrinin, $C_{13}H_{16}O_5$, prisms from benzene, dec 171°. $[α]_D^{18}$ −18.8° (c = 4.148 in chloroform). uv max: 260, 330 nm ($E_{1cm}^{1\%}$ 400, 100). Sol in alcohol, acetone, chloroform; sparingly sol in benzene, petr ether.

2330. Citromycetin. *8,9-Dihydroxy-2-methyl-4-oxo-4H,5H-pyrano[3,2-c][1]benzopyran-10-carboxylic acid;* frequentic acid. $C_{14}H_{10}O_7$; mol wt 290.22. C 57.94%, H 3.47%, O 38.59%. Antibiotic substance produced by *Penicillium frequentans* Westling and *P. vesiculosum* Bainier and by *Citromyces* spp: Hetherington, Raistrick, *Phil. Trans. Roy. Soc. London, Ser. B*, **220**, 209 (1931); Grove, Brian, *Nature* **167**, 995 (1951). Structure: Robertson et al., *J. Chem. Soc.* **1951**, 2013. Biosynthesis: Birch et al., *ibid.* **1958**, 4576; Money, *Nature* **199**, 592 (1963). Total synthesis: M. Yamauchi et al., *J. Chem. Soc. Perkin Trans. I* **1987**, 395.

Dihydrate, yellow crystals, effervescence at 155°, dec 290-300° (considerable antecedent blackening). Freely sol in ethanol; readily sol in aq sodium carbonate soln; sparingly sol in water, chloroform. Insol in benzene, hexane. Stable to acid and alkali at 100°.

2331. Citronellal. *3,7-Dimethyl-6-octenal.* $C_{10}H_{18}O$; mol wt 154.24. C 77.86%, H 11.76%, O 10.37%. Chief constituent of citronella oil; also found in many other volatile oils, such as lemon, lemon grass, melissa: Tiemann, *Ber.* **32**, 834 (1899); Spoon, *Chem. Weekbl.* **54**, 236 (1958). Structure: Naves, *Bull. Soc. Chim. France* **1951**, 505; Eschinazi, *J. Org. Chem.* **26**, 3072 (1961).

Liquid. bp_1 47°. n_D^{20} 1.4460. $[\alpha]_D^{25}$ +11.50°. d 0.848-0.856. Soluble in alcohols; very slightly sol in water.
USE: In soap perfumes; insect repellent.

2332. β-Citronellol. *3,7-Dimethyl-6-octen-1-ol;* 2,6-dimethyl-2-octen-8-ol; citronellol; cephrol. $C_{10}H_{20}O$; mol wt 156.26. C 76.86%, H 12.90%, O 10.24%. *l*-Form as a constituent of rose and geranium oils. *d*-Form occurs in Ceylon and Java citronella oils. History: J. L. Simonsen, L. N. Owen, *The Terpenes* vol. I (University Press, Cambridge, 2nd ed, 1947). Prepn of (±)-form: Adams, Garvey, *J. Am. Chem. Soc.* **48**, 477 (1926); Ofner *et al., Helv. Chim. Acta* **42**, 2577 (1959). Prepn of (+)-form: Rienäcker, Ohloff, *Angew. Chem.* **73**, 240 (1961); Naves, Tullen, *Helv. Chim. Acta* **44**, 1867 (1961); Eschinazi, *J. Org. Chem.* **26**, 3072 (1961); Rienäcker, *Chimia* **27**, 97 (1973); C. G. Overberger, J. L. Weise, *J. Am. Chem. Soc.* **90**, 3525 (1968); T. Sato *et al., Tetrahedron Letters* **1980**, 3377. Prepn of (−)-form: Ohloff, *loc. cit.;* Rienäcker, *loc. cit.;* Shono *et al., Tetrahedron Letters* **1974**, 1295; K. Mori, T. Sugai, *Synthesis* **1982**, 752. Synthesis of (+) or (−)-form from isoprene: Hidai *et al., Chem. Commun.* **1975**, 170. Stereospecific prepn via microbiological (*Saccharomyces cerevisiae*) reduction: P. Gramatica *et al., Experientia* **38**, 775 (1982). Manuf: Woroch *et al.;* Bain; Webb, U.S. pats. **2,990,422; 3,005,845; 3,028,431** (1961, 1961, 1962, all to Glidden); Eschinasi, U.S. pat. **3,052,730** (1962 to Givaudan). Abs config of the (+)-form: Freudenberg, Hohmann, *Ann.* **584**, 54 (1953); Freudenberg, Lwowski, *ibid.* **587**, 213 (1954). NMR, HPLC determn of *R/S* enantiomer ratios: D. Valentine *et al., J. Org. Chem.* **41**, 62 (1976).

R-(+)-β-citronellol

(+)-Form, oily liquid, bp 224.5°, bp_{10} 108.4°. d_4^{20} 0.8550. n_D^{20} 1.4559. $[\alpha]_D^{20}$ +5.22°. Very slightly sol in water, miscible with with alcohol, ether.

(−)-Form, *β-rhodinol, Levocitrol.* bp_{10} 108-109°. d_4^{18} 1.4576. $[\alpha]_D^{20}$ −4.76°.
(±)-Form, *dihydrogeraniol.* $d_4^{23.5}$ 0.851. $n_D^{23.5}$ 1.454.
USE: In perfumery. *See also* Rhodinol.

2333. Citrulline. N^5-*(Aminocarbonyl)-*L-*ornithine;* δ-ureidonorvaline; α-amino-δ-ureidovaleric acid; $N^δ$-carbamylornithine. $C_6H_{13}N_3O_3$; mol wt 175.19. C 41.13%, H 7.48%, N 23.99%, O 27.40%. $H_2NCONH(CH_2)_3CH(NH_2)$-COOH. An amino acid, first isolated from the juice of watermelon, *Citrullus vulgaris* Schrad., *Cucurbitaceae:* Wada, *Biochem. Z.* **224**, 420 (1930); isoln from casein: Wada, *ibid.* **257**, 1 (1933). Synthesis from ornithine through copper complexes: Kurtz, *J. Biol. Chem.* **122**, 477 (1938); by alkaline hydrolysis of arginine: Fox, *ibid.* **123**, 687 (1938); from cyclopentanone oxime: Fox *et al., J. Org. Chem.* **6**, 410 (1941). Crystallization: Matsuda *et al.,* **Japan.** pat. **00,174('71)** (to Ajinomoto), *C.A.* **74**, 126056u (1971). Crystal and molecular structure: Naganathan, Venkatesan, *Acta Crystallogr.* **27B**, 1079 (1971); Ashida *et al., ibid.* **28B**, 1367 (1972). Use in asthenia and hepatic insufficiency: Fr. pat. **2,198,739** (1974 to Hublot & Vallet), *C.A.* **82**, 144952c (1975). Clinical trial in treatment of lysinuric protein intolerance: J. Rajantie *et al., J. Pediatr.* **97**, 927 (1980); T. O. Carpenter *et al., N. Engl. J. Med.* **312**, 290 (1985). Prisms from methanol + water, mp 222°. $[\alpha]_D^{20}$ +3.7° (c = 2). pK_1 2.43; pK_2 9.41. Sol in water. Insol in methanol, ethanol.
Hydrochloride, $C_6H_{13}N_3O_3$·HCl, crystals, dec 185°. $[\alpha]_D^{22}$ +17.9° (c = 2).
Malate (salt), $C_{10}H_{19}N_3O_8$, *Stimol.*
THERAP CAT: Treatment of asthenia.

2334. Citrullol. $C_{22}H_{38}O_4$; mol wt 366.52. C 72.09%, H 10.45%, O 17.46%. From fruit pulp of *Citrullus colocynthis* Schrad., *Cucurbitaceae:* Power, Moore, *J. Chem. Soc.* **97**, 99 (1910); Power, Salway, *ibid.* **103**, 399, 1022 (1913); Khadem, Rahman, *Tetrahedron Letters* **1962**, 1137.
Crystals, mp 282-283°. uv max: 242, 272, 282 nm (log ε 2.85, 2.68, 2.68). Sol in pyridine; practically insol in usual organic solvents.
Diacetate, $C_{26}H_{42}O_6$, crystals, mp 162°.

2335. Citrus Red 2. *1-[(2,5-Dimethoxyphenyl)azo]-2-naphthalenol; C.I. Solvent Red 80; C.I.* 12156. $C_{18}H_{16}N_2O_3$; mol wt 308.34. C 70.12%, H 5.23%, N 9.08%, O 15.57%. Prepn: H. W. Elley, H. W. Daudt, U.S. pat. **2,224,904** (1940 to du Pont). Metabolism: J. L. Radomski, *J. Pharmacol. Exp. Ther.* **134**, 100 (1961); **136**, 378 (1962). Toxicology: M. Sharratt *et al., Food Cosmet. Toxicol.* **4**, 493 (1966). Review of carcinogenicity studies: *IARC Monographs* **8**, 101-106. *See also Colour Index* vol. 4 (3rd ed., 1971) p 4033.

Crystals, mp 155-157°. Slightly sol in water; partially sol in ethanol and vegetable oils.
USE: To color orange skins.

2336. Civet. Zibeth. Unctuous secretion from receptacles between the anus and genitalia of both male and female civet cat. *Constit.* Civetone and similar compds.
Semi-solid, yellowish to brown unctuous substance; unpleasant, subacrid, bitter taste; fusible and burns without leaving much residue. Insol in water; partly sol in hot alcohol or in ether.
USE: As a fixative in perfumery.

2337. Civetone. *9-Cycloheptadecen-1-one.* $C_{17}H_{30}O$; mol wt 250.41. C 81.53%, H 12.08%, O 6.39%. 17-Membered macrocyclic musk, constituent of civet: Ruzicka, *Helv. Chim. Acta* **9**, 230 (1926); Ruzicka *et al., ibid.* **10**, 695 (1927). Occurs in nature as *cis*-form. Synthesis of *cis*-civetone: Stoll *et al., ibid.* **31**, 543 (1948); J. Tsuji, T. Mondai, *Tetra-*

hedron Letters **1977**, 3285; E. Seoane *et al., Chem. & Ind. (London)* **1978**, 165. Synthesis of *trans*-form: H. Hunsdiecker, *Ber.* **77**, 185 (1944); H. H. Mathur, S. C. Bhattacharyya, *J. Chem. Soc.* **1968**, 114. Crystal and molecular structure of *cis*-civetone: G. Bernardinelli, R. Gerdil, *Helv. Chim. Acta* **65**, 558 (1982).

Crystals, mp 31-32°. Musky odor becoming pleasant in extreme dilns. d_4^{33} 0.917. bp_{742} 342°; bp_2 59°. $n_D^{33.4}$ 1.4830.
USE: In perfumery.

2338. Clanobutin. *4-[(4-Chlorobenzoyl)(4-methoxyphenyl)amino]butanoic acid; 4-[p-chloro-N-(p-methoxyphenyl)benzamido]butyric acid; N-(p-chlorobenzoyl)-γ-(p-*anisidino)butyric acid; Bykahepar. $C_{18}H_{18}ClNO_4$; mol wt 347.80. C 62.16%, H 5.22%, Cl 10.19%, N 4.03%, O 18.40%. Prepn: K. Klemm *et al.,* Ger. pat. **1,917,036** corresp to U.S. pat. **3,780,095** (1971, 1973 both to Byk-Gulden). Series of articles on synthesis, physical and pharmacological properties: *Arzneimittel-Forsch.* **29**, 1-15 (1979). *In vitro* biochemical study: H. Wolf *et al., Biochem. Pharmacol.* **29**, 1649 (1980). Effect on bile excretion in rats, dogs: P. Berchtold *et al., Arzneimittel-Forsch.* **30**, 1878 (1980).

Cryst from ethyl acetate, mp 115-116°. pKa 5.04. Soly in water at 37°: 4.02 × 10^{-2} mol/l at pH 7. LD_{50} in rats (mg/kg): > 2000 orally; 570 i.v. (Klemm).
Sodium salt, $C_{18}H_{17}ClNNaO_4$, *Bykehepar.*
THERAP CAT: Choleretic.
THERAP CAT (VET): Choleretic; in treatment of piroplasmosis and anaplasmosis.

2339. Clarase®. Diastatic and proteolytic enzymes. Contains mainly amylase and in smaller amounts maltase, protease, peptidase, lipase and other enzymes: Dzialoszynski, *Nature* **160**, 464 (1947). In determination of thiamin and riboflavin in cowpea: J. K. Edijala, *Analyst (London)* **104**, 637 (1979).
USE: In analytical chemistry for digesting food samples (protein solubilizing, amino acid liberating); in industry for starch conversions, removal of starches, manuf sizing materials, paste making.

2340. Clarithromycin. *6-O-Methylerythromycin;* A 56268; TE-031; Clathromycin. $C_{38}H_{69}NO_{13}$; mol wt 747.96. C 61.02%, H 9.30%, N 1.87%, O 27.81%. Semisynthetic antibiotic; derivative of erythromycin, *q.v.* Prepn: Y. Watanabe *et al.,* Eur. pat. **Appl. 41,355**; *eidem,* U. S. pat. **4,331,-803** (1981, 1982 both to Taisho); and *in vitro* antibacterial activity: S. Morimoto *et al., J. Antibiot.* **37**, 187 (1984). *In vitro* and *in vivo* antibacterial activity: P. B. Fernandes *et al., Antimicrob. Ag. Chemother.* **30**, 865 (1986). Comparative antibacterial spectrum *in vitro:* C. Benson *et al., Eur. J. Clin. Microbiol.* **6**, 173 (1987); H. M. Wexler, S. M. Finegold, *ibid.* 492. HPLC determn in biological fluids: D. Croteau *et al., J. Chromatog.* **419**, 205 (1987). Acute toxicity study: S. Abe *et al., Chemotherapy* **36**, Suppl. 3, 274 (1988). Series of articles on pharmacology, pharmacokinetics and clinical studies: *ibid.,* 1-1113.

Colorless needles from chloroform + diisopropyl ether (1:2), mp 217-220° (dec). Also reported as crystals from ethanol, mp 222-225° (Morimoto). uv max ($CHCl_3$): 288 nm (ε 27.9). $[\alpha]_D^{24}$ −90.4° (c = 1 in $CHCl_3$). Stable at acidic pH. LD_{50} in male, female mice, male, female rats (mg/kg): 2740, 2700, 3470, 2700 orally, 1030, 850, 669, 753 i.p., > 5000 all s.c. (Abe).
THERAP CAT: Antibacterial.

2341. Clathrates. Compounds that are capable of trapping other substances within their own crystal lattices. The cavities of the host molecules are classified as cages, tunnels, or layered types, depending on the way they include guest molecules. The geometry of the cavities limits the guest molecules by size and shape, rather than by chemical similarity with the host molecules. Among common clathrates are *molecular sieves, cyclotriphosphazenes,* and *Dianin's compound,* as well as hydroquinone, cyclodextrins, *o*-thymotide, and deoxycholic acid, *q.q.v. Cavitands* are organic hosts with enforced (rigid) cavities: D. J. Cram, *Science* **219**, 1177 (1983); R. C. Helgeson *et al., Chem. Commun.* **1983**, 101. Comprehensive book: *Clathrate Compounds,* V. M. Bhatnagar, Ed. (Chemical Pub. Co., New York, 1970) 244 pp. *Reviews:* D. D. MacNicol *et al., Chem. Soc. Rev.* **7**, 65-87 (1978); E. C. Makin, "Clathration" in Kirk-Othmer *Encyclopedia of Chemical Technology* Vol. 6 (Wiley-Interscience, New York, 3rd ed., 1979) pp 178-189.
USE: As complexing agent; stabilizing agent. In analytical separations.

2342. Clavulanic Acid. *3-(2-Hydroxyethylidene)-7-oxo-4-oxa-1-azabicyclo[3.2.0]heptane-2-carboxylic acid;* MM 14151. $C_8H_9NO_5$; mol wt 199.16. C 48.24%, H 4.55%, N 7.03%, O 40.17%. β-Lactamase inhibitor. Antibiotic produced by *Streptomyces clavuligerus;* first reported naturally occurring fused β-lactam containing oxygen. Isoln: M. Cole *et al.,* Ger. pat. **2,517,316** (1975 to Beecham), *C.A.* **84**, 72635t (1976); A. G. Brown *et al., J. Antibiot.* **29**, 668 (1976). Structure, x-ray crystallography: T. T. Howarth *et al., Chem. Commun.* **1976**, 266. Total synthesis of (±)-form: P. H. Bentley *et al., ibid.* **1977**, 748, 905; *eidem, Tetrahedron Letters* **1979**, 1889. β-Lactamase inhibition and antibacterial spectrum: C. Reading, M. Cole, *Antimicrob. Ag. Chemother.* **11**, 852 (1977). Mechanism of action: B. G. Spratt *et al., ibid.* **12**, 406 (1977). Antibacterial activity, pharmacology and clinical efficacy of combination with amoxicillin: A. P. Ball *et al., Lancet* **1**, 620 (1980); R. N. Brogden *et al., Drugs* **22**, 337-362 (1981). *In vitro* and *in vivo* synergism with ticarcillin: R. Sutherland *et al., Am. J. Med.* **79**, Suppl. 5B, 13 (1985).

Combination of potassium salt with amoxicillin trihydrate, *Amoksiklav, Augmentin, Stacillin;* with ticarcillin disodium, *Betabactyl, Timentin.*
Methyl ester, $C_9H_{11}NO_5$, oil. $[\alpha]_D^{22}$ +38°.
p-Nitrobenzyl ester, $C_{15}H_{14}N_2O_7$, monoclinic crystals, mp 117.5-118°.

THERAP CAT: Combination with β-lactam antibiotics as antibacterial.

2343. Clazuril. *2-Chloro-α-(4-chlorophenyl)-4-[4,5-dihydro-3,5-dioxo-1,2,4-triazin-2(3H)-yl]benzeneacetonitrile;* (±)-[2-chloro-4-(4,5-dihydro-3,5-dioxo-*as*-triazin-2(3H)-yl)phenyl](p-chlorophenyl)acetonitrile; Appertex. $C_{17}H_{10}Cl_2N_4O_2$; mol wt 373.20. C 54.71%, H 2.70%, Cl 19.00%, N 15.01%, O 8.57%. Prepn: G. M. Boeckx *et al.,* **Eur. pat. Appl. 170,316;** *eidem,* U.S. pat. **4,631,278** (both 1986 to Janssen). Anticoccidial effect in pigeons: W. Coussement *et al., Res. Vet. Sci.* **45,** 117 (1988).

mp 196.8°.
THERAP CAT (VET): Anticoccidial.

2344. Clebopride. *4-Amino-5-chloro-2-methoxy-N-[1-(phenylmethyl)-4-piperidinyl]benzamide;* 4-amino-N-(1-benzyl-4-piperidyl)-5-chloro-*o*-anisamide; N-(1'-benzyl-4'-piperidyl)-2-methoxy-4-amino-5-chlorobenzamide. $C_{20}H_{24}ClN_3O_2$; mol wt 389.88. C 61.61%, H 6.20%, Cl 9.09%, N 10.78%, O 12.31%. Dopamine receptor antagonist related to metoclopramide, *q.v.* Prepn: R. G. Spickett *et al.,* **Ger. pat. 2,513,136** corresp to A. V. Noverola *et al.,* U.S. pat. **4,138,492** (1975, 1979 both to Anphar); J. Prieto *et al., J. Pharm. Pharmacol.* **29,** 147 (1977). Pharmacological study: J. L. Masso, D. J. Roberts, *ibid.* **32,** 727 (1980). *In vitro* metabolism studies: G. Huizing *et al., Xenobiotica* **10,** 211 (1980); G. Huizing, A. H. Beckett, *ibid.* 593. Pharmacokinetics: J. Segura *et al., J. Pharm. Pharmacol.* **33,** 214 (1981). Comparative study of anti-emetic and gastrointestinal effects: P. Berga *et al., Arch. Farmacol. Toxicol.* **7,** 189 (1981). Efficacy in healing exptl ulcers: Y. Matsuo *et al., Oyo Yakuri* **24,** 251 (1982); S. Okabe *et al., ibid.* 261, *C.A.* **97,** 208081n, 208082p (1982). Determn in urine and plasma by radioimmunoassay: M. Yano *et al., Chem. Pharm. Bull.* **32,** 1491 (1984). Clinical effect in postoperative nausea: D. F. Duarte *et al., Clin. Ther.* **7,** 365 (1985).

Cryst from methanol, mp 194-195°.
Hydrochloride monohydrate, $C_{20}H_{25}Cl_2N_3O_2 \cdot H_2O$, cryst, mp 217-219°. Approx oral LD_{50} in male Swiss mice: >1000 mg/kg (Prieto).
Malate, $C_{24}H_{30}ClN_3O_7$, *Amicos, Clanzol, Clast, Cleboril, Cleprid, Motilex.*
THERAP CAT: Anti-emetic. Antispasmodic.

2345. Clemastine. *2-[2-[1-(4-Chlorophenyl)-1-phenylethoxy]ethyl]-1-methylpyrrolidine;* (+)-2-[2-[(p-chloro-α-methyl-α-phenylbenzyl)oxy]ethyl]-1-methylpyrrolidine; 1-methyl-2-[2-(α-methyl-p-chlorobenzhydryloxy)ethyl]-pyrrolidine; 1-methyl-2-[2-(methyl-p-chlorodiphenylmethyloxy)ethyl]pyrrolidine; meclastine. $C_{21}H_{26}ClNO$; mol wt 343.90. C 73.34%, H 7.62%, Cl 10.31%, N 4.07%, O 4.65%. Prepn: **Brit. pat. 942,152** (1963 to Sandoz). Synthesis and abs config: A. Ebnöther, H. P. Weber, *Helv. Chim. Acta* **59,** 2462 (1976). Pharmacology and toxicology of the hydrogen fumarate: Weidmann *et al., Boll. Chim. Farm.* **106,** 467 (1967).

Free base, $bp_{0.02}$ 154°. n_D^{22} 1.5582. $[\alpha]_D^{20}$ +33.6° (ethanol). Hydrogen fumarate, $C_{25}H_{30}ClNO_5$, HS-592, *Aloginan, Alphamin, Anhistan, Clemanil, Fuluminol, Inbestan, Kinotomin, Lacretin, Lecasol, Maikohis, Mallermin-F, Marsthine, Masleline, Piloral, Reconin, Tavegil, Tavegyl, Tavist, Telgin-G, Trabest, Xolamin.* mp 177-178°. $[\alpha]_D^{21}$ +16.9° (methanol). LD_{50} in mice, rats: 730, 3550 mg/kg orally; 43, 82 mg/kg i.v.
THERAP CAT: The hydrogen fumarate as antihistaminic.

2346. Clemizole. *1-[(4-Chlorophenyl)methyl]-2-(1-pyrrolidinylmethyl)-1H-benzimidazole;* 1-(p-chlorobenzyl)-2-(1-pyrrolidinylmethyl)benzimidazole; 1-(p-chlorobenzyl)-2-pyrrolidylmethylenebenzimidazole; Allercur; Histacuran; Klemidox; Reactrol. $C_{19}H_{20}ClN_3$; mol wt 325.85. C 70.03%, H 6.19%, Cl 10.88%, N 12.90%. Prepn: Jerchel *et al., Ann.* **575,** 173 (1952).

Crystals from dil alc, mp 167°. Forms a water-soluble hydrochloride and sulfate. Used to form a repository form of penicillin G (as salt of benzylpenicillinic acid), trademarks: *Neopenyl, Lergopenin, Depocural.* Ref: *Arzneimittel-Forsch.* **4,** 487 (1954).
THERAP CAT: Antihistaminic.

2347. Clenbuterol. *4-Amino-3,5-dichloro-α-[[(1,1-dimethylethyl)amino]methyl]benzenemethanol;* 4-amino-α-[(tert-butylamino)methyl]-3,5-dichlorobenzyl alcohol; NAB 365; Monores. $C_{12}H_{18}Cl_2N_2O$; mol wt 277.18. C 52.00%, H 6.54%, Cl 25.58%, N 10.11%, O 5.77%. Substituted phenylethanolamine with β_2 sympathomimetic activity. Prepn: J. Keck *et al.,* **S. Afr. pat. 67 05692** corresp to U.S. pat. **3,536,712** (1968, 1970 both to Thomae); *eidem, Arzneimittel-Forsch.* **22,** 861 (1972). Series of articles on pharmacology, reproductive toxicology, pharmacokinetics and metabolism: *ibid.* **26,** 1403-1419, 1427-1460 (1976). Toxicity: M. Ueberberg *et al., ibid.* 1420. Clinical studies: M. Tschan *et al., Eur. J. Clin. Pharmacol.* **15,** 159 (1979); C. Pasotti *et al., Int. J. Clin. Pharmacol. Biopharm.* **17,** 176 (1979).

Hydrochloride, $C_{12}H_{19}Cl_3N_2O$, NAB 365Cl, *Spiropent, Ventipulmin.* Colorless microcrystalline powder from isopropyl alc, mp 174-175.5°. Very sol in water, methanol, ethanol, slightly sol in chloroform; insol in benzene. LD_{50} in mice, rats, guinea pigs (mg/kg): 176, 315, 67.1 orally; 27.6, 35.3, 12.6 i.v. (Ueberberg).
THERAP CAT: Anti-asthmatic.
THERAP CAT (VET): Bronchodilator; tocolytic.

2348. 1,6-Cleve's Acid. *5-Amino-2-naphthalenesulfonic acid;* 1-naphthylamine-6-sulfonic acid. $C_{10}H_9NO_3S$; mol wt

223.25. C 53.80%, H 4.06%, N 6.27%, O 21.50%, S 14.36%. Prepn: Erdmann, *Ann.* **275**, 262 (1893).

Crystals. Sol in 1000 parts water. Practically insol in alc, ether.

USE: In manuf of azo dyes.

2349. 1,7-Cleve's Acid. *8-Amino-2-naphthalenesulfonic acid;* 1-naphthylamine-7-sulfonic acid. Empirical formula etc.: *see* 1,6-Cleve's Acid above. Prepn: Erdmann, *Ann.* **275**, 262 (1893); Roos *et al.,* U.S. pat. **2,875,243** (1959 to Bayer). Industrial synthesis: Nakahara *et al., J. Syn. Org. Chem. Japan.* **29**, 1129 (1971).

Needles or prisms from water contg $1H_2O$. Sol in 220 parts water; very slightly sol in alcohol, ether.

USE: In manuf of azo dyes.

2350. Clidanac. *6-Chloro-5-cyclohexyl-2,3-dihydro-1H-indene-1-carboxylic acid;* (±)-6-chloro-5-cyclohexyl-indan-1-carboxylic acid; TAI-284; Britai; Indanal. $C_{16}H_{19}$-ClO_2; mol wt 278.78. C 68.93%, H 6.87%, Cl 12.72%, O 11.48%. Prepn: P. F. Juby *et al.,* Ger. pat. **2,004,038** corresp to U.S. pat. **3,565,943** (1970, 1971 both to Bristol-Myers); S. Noguchi *et al., Chem. Pharm. Bull.* **19**, 646 (1971). Synthesis and anti-inflammatory activity: *eidem, ibid.* **22**, 529 (1974); G. R. Allen *et al., J. Med. Chem.* **15**, 934 (1972); P. F. Juby *et al., ibid.* 1297. Pharmacology: S. Kuzuna *et al., Japan. J. Antibiot.* **24**, 695 (1974), *C.A.* **82**, 80297h (1976). X-ray crystallographic study: K. Kamiya *et al., Chem. Pharm. Bull.* **23**, 1589 (1975). Metabolism: S. Tanayama, Y. Kanai, *Xenobiotica* **7**, 145 (1977). Immuno-genicity study: S. Kuzuna *et al., Oyo Yakuri* **19**, 951 (1980), *C.A.* **94**, 76682 (1981). Inhibition of prostaglandin biosyn-thesis: S. Tamura *et al., J. Pharm. Pharmacol.* **33**, 29 (1981). Antipyretic effect in cancer patients: H. Furue, *Curr. Ther. Res.* **30**, 71 (1981).

Colorless cryst from petr ether, mp 150.5-152.5°. LD_{50} in rats: 41 mg/kg orally, P. F. Juby *et al., loc. cit.* (1972). (S)-(+)-Form, colorless needles from petr ether, mp 135-136°. $[\alpha]_D^{25}$ −28.7 and $[\alpha]_{365}^{25}$ +87.7 (c = 2 in ethanol). LD_{50} in rats: 35 mg/kg orally, P. F. Juby *et al., loc. cit.* (1972). (R)-(-)-Form, colorless cryst from petr ether, mp 134-135°, $[\alpha]_D^{25}$ −28.2 and $[\alpha]_{365}^{25}$ −87.5° (c = 2 in ethanol).

THERAP CAT: Anti-inflammatory, antipyretic.

2351. Clidinium Bromide. *3-[(Hydroxydiphenylacetyl)-oxy]-1-methyl-1-azoniabicyclo[2.2.2]octane bromide; 3-hydr-oxy-1-methylquinuclidinium bromide benzilate;* 1-methyl-3-benziloyloxyquinuclidinium bromide; 3-benziloyloxy-1-aza-bicyclo[2.2.2]octane methobromide; Ro 2-3773; Quarzan. $C_{22}H_{26}BrNO_3$; mol wt 432.38. C 61.11%, H 6.06%, Br 18.48%, N 3.24%, O 11.10%. Prepn: Sternbach, U.S. pat. **2,648,667** (1953 to Hoffmann-La Roche). Comprehensive description: B. C. Rudy, B. Z. Senkowski, in *Analytical Profiles of Drug Substances* vol. 2, K. Florey, Ed. (Academic Press, New York, 1973) pp 145-161.

Crystals from methanol + acetone + ether, mp 240-241°.

THERAP CAT: Anticholinergic.

2352. Clindamycin. *(2S-trans)-Methyl 7-chloro-6,7,8-trideoxy-6-[[(1-methyl-4-propyl-2-pyrrolidinyl)carbonyl]-amino]-1-thio-L-threo-α-D-galacto-octopyranoside;* 7(S)-chloro-7-deoxylincomycin; 7-deoxy-7(S)-chlorolincomycin; clinimycin (rescinded); U-21251; Antirobe; Cleocin; Dalacin C; Klimicin; Sobelin. $C_{18}H_{33}ClN_2O_5S$; mol wt 424.98. C 50.87%, H 7.83%, Cl 8.34%, N 6.59%, O 18.82%, S 7.54%. Semi-synthetic antibiotic prepd from lincomycin, *q.v.* Prepn: Magerlein *et al., Antimicrob. Ag. Chemother.* **1966**, 727; Birkenmeyer, U.S. pat. **3,475,407**; Kagan, Magerlein, U.S. pats. **3,509,127** and **3,544,551**; Birkenmeyer, Kagan, U.S. pat. **3,513,155** (1969, 1970, 1970, 1970 all to Upjohn). Synthesis and structure: Birkenmeyer, Kagan, *J. Med. Chem.* **13**, 616 (1970). Stability studies: Oesterling, *J. Pharm. Sci.* **59**, 63 (1970). Activity studies: McGehee *et al., Am. J. Med. Sci.* **256**, 279 (1968); D. A. Leigh, *J. Antimicrob. Chemother.* **7**(suppl. A), 3 (1981). Toxicology: J. E. Gray *et al., Toxicol. Appl. Pharmacol.* **21**, 516 (1972). Comprehen-sive description of the hydrochloride: L. W. Brown, W. F. Beyer in *Analytical Profiles of Drug Substances* vol. 10, K. Florey, Ed. (Academic Press, New York, 1981) pp 75-91.

Yellow, amorphous solid. $[\alpha]_D$ +214° (chloroform).

Hydrochloride monohydrate, $C_{18}H_{34}Cl_2N_2O_5S \cdot H_2O$, *Dalactine.* White crystals from ethanol-ethyl acetate, mp 141-143°. $[\alpha]_D$ +144° (H_2O). pKa 7.6. Sol in water, pyri-dine, ethanol, DMF. LD_{50} in mice (mg/kg): 245 i.v.; 361 i.p.; 2618 orally (Gray).

2-Dihydrogen phosphate, $C_{18}H_{34}ClN_2O_8PS$, *U-28508, Dalacin T.*

THERAP CAT: Antibacterial.

2353. Clinofibrate. *2,2'-[Cyclohexylidenebis(4,1-phen-yleneoxy)]bis[2-methylbutanoic acid];* 2,2'-(4,4'-cyclohexyl-idenediphenoxy)-2,2'-dimethyldibutyric acid; S-8527; Lipo-clin. $C_{28}H_{36}O_6$; mol wt 468.60. C 71.77%, H 7.74%, O 20.49%. Anti-atherosclerosis agent, related structurally to clofibrate, *q.v.* Prepn: Y. Nakamura *et al.,* Ger. pat. **2,017,-331** corresp to U.S. pat. **3,716,583** (1970, 1973 both to Sumitomo). Hypolipidemic activity: K. Toki *et al., Athero-sclerosis* **18**, 101 (1973); K. Suzuki *et al., Japan. J. Pharma-col.* **24**, 407 (1974), *C.A.* **82**, 25821z (1975). Effect on cho-lesterol and lipoprotein metabolism in rats: K. Suzuki, *Bio-chem. Pharmacol.* **25**, 325 (1976). General pharmacology: H. Nakatani *et al., Oyo Yakuri* **16**, 687 (1978), *C.A.* **90**, 145808 (1979).

Off-white powder, mp 143-146° (dec). Sol in methanol, ethanol, acetone, chloroform, glacial acetic acid. Slightly sol in CCl_4. Practically insol in water. LD_{50} in male mice, rats (mg/kg): 1800, >4000 orally; 255, 205 i.p.; 410, 2200 s.c., *Japan. Med. Gaz.* **18**(8), 7 (1981).

THERAP CAT: Antihyperlipoproteinemic.

2354. Clioxanide. *2-(Acetyloxy)-N-(4-chlorophenyl)-3,5-diiodobenzamide;* 4'-chloro-3,5-diiodosalicylanilide acetate; acetoxy-4'-chloro-3,5-diiodosalicylanilide; CI-633; SYD-230; Tremerad. $C_{15}H_{10}ClI_2NO_3$; mol wt 541.54. C 33.27%, H 1.86%, Cl 6.55%, I 46.87%, N 2.59%, O 8.86%. Prepn from 4'-chloro-3,5-diiodosalicylanilide: Campbell *et al., Experientia* **23**, 992 (1967); **Neth.** pat. **Appl. 6,604,303** (1966 to Parke, Davis), *C.A.* **66**, 55247d (1967), corresp to **Brit.** pat. **1,070,516.** Anthelmintic activity: Presidente *et al., Am. J. Vet. Res.* **33**, 1593 (1972).

Needles from aq DMF or acetone, mp 215-216°. uv max (methanol): 262 nm. LD_{50} by rumenal route in sheep: approx 420 mg/kg; by abomasal route: 1600 mg/kg.

THERAP CAT (VET): Anthelmintic (Trematodes).

2355. Clobazam. *7-Chloro-1-methyl-5-phenyl-1H-1,5-benzodiazepine-2,4(3H,5H)-dione;* 1-phenyl-5-methyl-8-chloro-1,2,4,5-tetrahydro-2,4-dioxo-3H-1,5-benzodiazepine; H-4723; HR 376; LM 2717; Frisium; Urbadan; Urbanyl. $C_{16}H_{13}ClN_2O_2$; mol wt 300.74. C 63.90%, H 4.36%, Cl 11.79%, N 9.31%, O 10.64%. Benzodiazepine psychotherapeutic agent in which the nitrogens in the heterocyclic ring are in the 1,5- rather than in the more common 1,4-positions. Prepn: Hauptmann *et al.,* **S. Afr.** pat. **68 00,803** corresp to K.-H. Weber *et al.,* **U.S.** pat. **3,684,798** (1968, 1972 to Boehringer, Ing.); Rossi *et al., Chim. Ind. (Milan)* **51**, 479 (1969); Weber *et al., Ann.* **756**, 128 (1972). Pharmacology and toxicology: Barzaghi *et al., Arzneimittel-Forsch.* **23**, 683 (1973). Review of pharmacology: R. N. Brogden *et al., Drugs* **20**, 161-178 (1980); *Drug Dev. Res.* **1982**, Suppl. 1, 1-186.

Crystals, mp 180-182°. LD_{50} orally in mice, rats: 840, >2000 mg/kg; i.p. in mice: 510 mg/kg (Barzaghi).

Note: This is a controlled substance (depressant) listed in the U.S. Code of Federal Regulations, Title 21 Part 1308.14 (1987).

THERAP CAT: Anxiolytic.

2356. Clobenfurol. *α-(4-Chlorophenyl)-2-benzofuranmethanol;* 2-benzofuryl-p-chlorophenylcarbinol; Menacor. $C_{15}H_{11}ClO_2$; mol wt 258.71. C 69.64%, H 4.29%, Cl 13.70%, O 12.37%. Prepn: **Belg.** pat. **644,176;** Ghelardoni, Russo, **Brit.** pat. **1,160,925** (1964, 1969 both to Menarini), *C.A.* **63**, 8319h (1965); **71**, 124212f (1969); Ghelardoni, Russo, *Boll. Chim. Farm.* **109**, 48 (1970). Pharmacology of the hemisuc-

cinate deriv: Pisanti, Volterra, *Farmaco Ed. Sci.* **26**, 312 (1971).

White, odorless powder, mp 48-49°.
Hemisuccinate, $C_{19}H_{15}ClO_5$, mp 95-97° from petr ether. LD_{50} i.p. in mice, rats: 321, 300 mg/kg (Pisanti, Volterra).

THERAP CAT: Coronary vasodilator.

2357. Clobenoside. *Ethyl 5,6-bis-O-[(chlorophenyl)-methyl]-3-O-propyl-D-glucofuranoside;* ethyl 3-O-n-propyl-5,6-di-O-(4-chlorobenzyl)-D-glucofuranoside; ZY-15028; Arvigol; Floganol. $C_{25}H_{32}Cl_2O_6$; mol wt 499.43. C 60.12%, H 6.46% Cl 14.20%, O 19.22%. Congener of tribenoside, *q.v..* Prepn: **Neth.** pat. **Appl. 6,812,906;** A. Rossi, **U.S.** pat. **3,542,761** (1969, 1970 both to Ciba). Pharmacology and bioavailability in cats: G. Hennings *et al., Arzneimittel-Forsch.* **35**, 498 (1985). Effect on glycoprotein biosynthesis *in vitro:* W. Reutter, C. Bauer, *Adv. Enzyme Regul.* **24**, 405 (1985). Mechanism of action study: D. Araujo *et al., Arch. Int. Pharmacodyn.* **277**, 192 (1985).

Faintly yellowish oil, $[\alpha]_D^{20}$ −17° ± 1 (c = 1 in chloroform). Poorly sol in water.

THERAP CAT: Vasoprotectant.

2358. Clobenzepam. *7-Chloro-10-[2-(dimethylamino)-ethyl]-5,10-dihydro-11H-dibenzo[b,e][1,4]diazepin-11-one;* 7-chloro-10-(2-dimethylaminoethyl)dibenzo[b,e][1,4]diazepin-11(10H)-one; 7-chloro-10-(β-dimethylaminoethyl)-10,-11-dihydro-11-oxo-5H-dibenzo[b,e][1,4]diazepine. $C_{17}H_{18}-ClN_3O$; mol wt 315.80. C 64.66%, H 5.74%, Cl 11.22%, N 13.31%, O 5.07%. Prepn: **Belg.** pat. **611,926** (1962 to Parke, Davis), *C.A.* **57**, 13785a (1962); **Brit.** pat. **961,106** corresp. to Schmutz, Hunziker, **U.S.** pat. **3,419,547** (1964, 1968 both to Wander). Pharmacology: Ackermann *et al., Med. Exp.* **6**, 205 (1962); Hunziker *et al., Arzneimittel-Forsch.* **13**, 324 (1963).

Colorless prisms from acetone-ether, mp 165-166°. uv max (alcohol): 230 nm (32,740).
Hydrochloride, $C_{17}H_{19}Cl_2N_3O$, *Tarpan.* Colorless prisms from alcohol-ether, mp 225-233°. LD_{50} orally in mice: 330 mg/kg.

THERAP CAT: Antihistaminic.

2359. Clobenzorex. *N-[(2-Chlorophenyl)methyl]-α-methylbenzeneethanamine;* (+)-N-(o-chlorobenzyl)-α-methylphenethylamine; d-N-(1-phenyl-2-propyl)-2-chlorobenzylamine. $C_{16}H_{18}ClN$; mol wt 259.78. C 73.98%, H 6.98%, Cl 13.65%, N 5.39%. Prepn and pharmacology: Boissier, Tatouis, **Fr.** pat. **1,429,306** (1966 to Soc. Ind. Fabric. Antibiot.), *C.A.* **66**, 46195h (1967). Synthesis of labelled compound: *J. Label. Compounds* **6**, 289 (1970). Metabolic studies: Thomasset *et al.*, *Experientia* **26**, 692 (1970); Glasson *et al.*, *Arzneimittel-Forsch.* **21**, 1985 (1971).

bp$_{0.1}$ 132-134°.
Hydrochloride, $C_{16}H_{19}Cl_2N$, *BA 7205, SD 271-12, Dinintel, Rexigen.* Crystals, mp 182-183° (isopropanol). $[\alpha]_D^{20}$ +26.3° (water). Sol in water, ethanol. Slightly sol in methanol, chloroform. LD$_{50}$ in rats: 103 mg/kg i.p.
THERAP CAT: Anorexic.

2360. Clobenztropine. *3-[(4-Chlorophenyl)phenylmethoxy]-8-methyl-8-azabicyclo[3.2.1]octane; 3-[(p-chloro-α-phenylbenzyl)oxy]tropane;* tropine 4-chlorobenzhydryl ether; Teprin. $C_{21}H_{24}ClNO$; mol wt 341.89. C 73.78%, H 7.08%, Cl 10.37%, N 4.10%, O 4.68%. Prepn: Fromer, U.S. pat. **2,799,680** (1957); Nield, Bosch, U.S. pat. **2,782,200** (1957 to Schenley Labs.).

Hydrochloride, $C_{21}H_{24}ClNO \cdot HCl$, crystals from isopropanol, mp 215-217°. Sol in water, ethanol; practically insol in acetone, benzene, ether.
Hydrobromide, $C_{21}H_{24}ClNO \cdot HBr$, mp 197-200°. Sol in hot water, hot ethanol; practically insol in acetone, benzene, ether.
Methobromide, $C_{21}H_{24}ClNO \cdot CH_3Br$, mp 245-248°. Sol in hot water, in ethanol; practically insol in acetone, benzene, ether.
Methochloride, $C_{21}H_{24}ClNO \cdot CH_3Cl$, mp 261-263°. Sol in water, ethanol; practically insol in acetone, benzene, ether.
THERAP CAT: Antihistaminic.

2361. Clobetasol. *(11β,16β)-21-Chloro-9-fluoro-11,17-dihydroxy-16-methylpregna-1,4-diene-3,20-dione.* $C_{22}H_{28}$-$ClFO_4$; mol wt 410.91. C 64.31%, H 6.87%, Cl 8.63%, F 4.62%, O 15.57%. Prepn: Elks *et al.*, **Ger.** pat. **1,902,340**; *eidem*, **U.S.** pat. **3,721,687** (1969, 1973 both to Glaxo).

17-Propionate, $C_{25}H_{32}ClFO_5$, *GR 2/925, Clobesol, Dermoval, Dermovate, Dermoxin, Dermoxinale, Temovate.* Crystals, mp 195.5-197°. $[\alpha]_D$ +103.8° (c = 1.04 in dioxane). uv max (ethanol): 237 nm (ε 15000).
THERAP CAT: Glucocorticoid; anti-inflammatory.

2362. Clobetasone. *(16β)-21-Chloro-9-fluoro-17-hydroxy-16-methylpregna-1,4-diene-3,11,20-trione.* 21-chloro-11-dehydrobetamethasone. $C_{22}H_{26}ClFO_4$; mol wt 408.90. C 64.62%, H 6.40%, Cl 8.66%, F 4.64%, O 15.68%. Prepn: Elks *et al.*, **Ger.** pat. **1,902,340**; *eidem*, **U.S.** pat. **3,721,687** (1969, 1973, both to Glaxo). Activity: Munro, Wilson, *Brit. Med. J.* **3**, 626 (1975). Radioimmunoassay: W. A. Webb, R. V. Brooks, *Acta Endocrinol. Suppl.* **212**, 203 (1977). Toxicology: J. Tamura *et al.*, *J. Toxicol. Sci.* **5**, 45, 177 (1980). Comparative clinical studies: A. Lassus, *Curr. Med. Res. Opin.* **6**, 165 (1979); T. Fredriksson, K. Nordin, *ibid.* 322. Clinical evaluation of ophthalmic form: D. Lloyd-Jones *et*

al., Brit. J. Ophthalmol. **65**, 641 (1981); L. A. Eilon, S. R. Walker, *ibid.* 644; K. R. Wilhelmus *et al., ibid.* 699.

17-Butyrate, $C_{26}H_{32}ClFO_5$, *GR/1214, Emovate, Eumovate, Molivate* (obsolete). Crystals from methanol, mp 90-100°.
THERAP CAT: Glucocorticoid; anti-inflammatory.

2363. Clobutinol. *4-Chloro-α-[2-(dimethylamino)-1-methylethyl]-α-methylbenzeneethanol; p-chloro-α-[2-(dimethylamino)-1-methylethyl]-α-methylphenethyl alcohol;* 2-(p-chlorobenzyl)-3-dimethylaminomethyl-2-butanol; 1-p-chlorophenyl-2,3-dimethyl-4-dimethylamino-2-butanol. $C_{14}H_{22}ClNO$; mol wt 255.79. C 65.74%, H 8.67%, Cl 13.86%, N 5.48%, O 6.25%. Prepn: **Brit.** pat. **898,010** corresponds to Berg, **U.S.** pat. **3,121,087** (1962, 1964, both to Thomae). Pharmacology of hydrochloride: Engelhorn, *Arzneimittel-Forsch.* **10**, 785 (1960). Metabolism of hydrochloride: Beisenherz *et al., ibid.* **19**, 79 (1969).

bp$_{12}$ 179-180°.
Hydrochloride, $C_{14}H_{23}Cl_2NO$, *KAT 256, Biotertussin, Pertoxil, Silomat.* Minute crystals, mp 169-170°. Slightly bitter, acidic taste. Numbs the tongue. Sol in water. LD$_{50}$ in mice: 600 mg/kg orally; 130 mg/kg i.p., Engelhorn, *loc. cit.*
THERAP CAT: Antitussive.

2364. Clobuzarit. *2-[[4'-Chloro(1,1'-biphenyl)-4-yl]methoxy]-2-methylpropanoic acid; 2-[[p-(p-chlorophenyl)-benzyl]oxy]-2-methylpropionic acid;* ICI 55897; Clozic. $C_{17}H_{17}ClO_3$; mol wt 304.78. C 67.00%, H 5.62%, Cl 11.63%, O 15.75%. Analog of clofibrate, *q.v.* Prepn: T. Leigh, L. A. McArdle, **Brit.** pat. **1,140,748**; *eidem*, **U.S.** pat. **3,549,690** (1969, 1970 both to ICI). Pharmacological studies: M. E. J. Billingham, *Brit. J. Pharmacol.* **72**, 523 (1981); M. E. J. Billingham *et al., ibid.* 551. Effects on platelet function in atherosclerosis: J. R. O'Brien *et al., Thromb. Haemostasis* **40**, 75 (1978). Comparative clinical evaluation with penicillamine: D. Y. Bulgen *et al., Ann. Rheum. Dis.* **38**, 567 (1979); H. A. Bird *et al., Clin. Exp. Rheumatol.* **1**, 93 (1983).

Crystals from benzene, mp 154-155°.
THERAP CAT: Antirheumatic.

2365. Clocapramine. *1'-[3-(3-Chloro-10,11-dihydro-5H-dibenz[b,f]azepin-5-yl)propyl][1,4'-bipiperidine]-4'-carboxamide;* 3-chloro-5-[3-(4-piperidino-4-carbamoyl-piperidino)propyl]-10,11-dihydro-5H-dibenz[b,f]azepine; 3-chlorocarpipramine; clocarpramine. $C_{28}H_{37}ClN_4O$; mol wt 481.09. C 69.91%, H 7.75%, Cl 7.37%, N 11.65%, O 3.32%. Prepn: Nakanishi, Tashiro, **Ger.** pat. **1,905,765** corresp to **U.S.** pat. **3,668,210** (1969, 1972, both to Yoshitomi); M. Nakanishi *et al., J. Med. Chem.* **13**, 644 (1970). Pharmacology and toxicity: M. Nakanishi *et al., Arzneimittel-Forsch.* **21**, 391 (1971). Effect on metabolism of serotonin and catecholamines: M. Nakanishi, M. Setoguchi, *ibid.* **23**, 806 (1973). Metabolism: M. Nakanishi *et al., J. Pharm. Soc. Japan* **91**, 1042 (1971). *Review:* Japan. *Med. Gaz.* **10**(7), 5 (1973).

Dihydrochloride monohydrate, $C_{28}H_{39}Cl_3N_4O.H_2O$, *Y-4153*, *Clofekton*. White, bitter tasting, crystalline powder, mp 267°. Soly in water: < 3.5 g/100 ml. Freely sol in glacial acetic acid. Practically insol in ether, acetone. LD_{50} in mice, rats (mg/kg): 160, 125 i.p.; 2550, 6800 orally (Nakanishi).

THERAP CAT: Neuroleptic.

2366. Clocinizine. *1-[(4-Chlorophenyl)phenylmethyl]-4-(3-phenyl-2-propenyl)piperazine; 1-(p-chloro-α-phenylbenzyl)-4-cinnamylpiperazine;* 1-cinnamyl-4-(p-chloro-α-phenylbenzyl)piperazine; 1-cinnamyl-4-(4-chlorobenzhydryl)-piperazine; cliocinizine. $C_{26}H_{27}ClN_2$; mol wt 402.99. C 77.49%, H 6.75%, Cl 8.80%, N 6.95%. Prepn: **Brit. pat. 809,760** (1959 to Janssen).

Hydrochloride, crystals, mp 200-201°. Component of *Denoral*.

THERAP CAT: Antihistaminic.

2367. Cloconazole. *1-[1-[2-[(3-Chlorophenyl)methoxy]-phenyl]ethenyl]-1H-imidazole;* 1-[1-[2-[(3-chlorobenzyl)-oxy]phenyl]vinyl]-1H-imidazole; croconazole. $C_{18}H_{15}ClN_2$-O; mol wt 310.78. C 69.57%, H 4.86%, Cl 11.41%, N 9.01%, O 5.15%. Antimycotic vinylimidazole. Prepn and antifungal properties: M. Ogata *et al.*, Belg. pat. **883,665;** *eidem*, U.S. pat. **4,328,348** (1980, 1982 both to Shionogi); *eidem*, J. Med. Chem. **26**, 768 (1983). Pharmacology: K. Yamamoto *et al.*, Oyo Yakuri **27**, 533 (1984), C.A. **101**, 16881c (1984). Series of articles on metabolism in rats and rabbits: K. Mizojiri *et al.*, Iyakuhin Kenkyu **16**, 177-206 (1985), C.A. **103**, 47788y; 47789z; 47790t (1985). Mechanism of action study: T. Hiratani, H. Yamaguchi, Chemotherapy (Tokyo) **33**, 579 (1985).

mp 72-73°. Sol in ethyl acetate.
Monohydrochloride, $C_{18}H_{16}Cl_2N_2O$, *710674-S*, *Pilzcin*. Crystals from ethyl acetate + acetonitrile, mp 148.5-150°. LD_{50} in rats (mg/kg): 7000 s.c.; 2500 orally (Ogata, 1983).

THERAP CAT: Topical antifungal.

2368. Clocortolone. *9-Chloro-6α-fluoro-11β,21-dihydroxy-16α-methylpregna-1,4-diene-3,20-dione;* 9-chloro-6α-fluoro-16α-methyl-1,4-pregnadiene-11β,21-diol-3,20-dione. $C_{22}H_{28}ClFO_4$; mol wt 410.91. C 64.31%, H 6.87%, Cl 8.63%, F 4.62%, O 15.57%. Prepn: **Neth. pat. Appl. 6,412,-708** (1965); Kaspar, Philippson, Ger. pat. **2,011,559** corresp to U.S. pat. **3,729,495** (1971, 1973) (all to Schering AG). Clinical trials: Wortmann, Wien. Med. Wochenschr. **122**, 701 (1972); Baumann, Praxis **61**, 536 (1972).

Crystals from methanol-methylene chloride, mp 254° (dec).

21-Acetate, $C_{24}H_{30}ClFO_5$, *SH 818*. Crystals from methanol-methylene chloride, mp 252° (dec).

21-Pivalate, $C_{27}H_{36}ClFO_5$, *CL 68*, *SH 863*, *Cloderm*, *Purantix*.

THERAP CAT: Glucocorticoid.

2369. Clodronic Acid. *(Dichloromethylene)bisphosphonic acid; (dichloromethylene)diphosphonic acid;* dichloromethanediphosphonic acid; Cl_2MDP; DMDP. $CH_4Cl_2O_6P_2$; mol wt 244.89. C 4.90%, H 1.65%, Cl 28.95%, O 39.20%, P 25.30%. Bone resorption inhibitor structurally related to inorganic pyrophosphoric acid and to etidronic acid, *q.q.v.* Prepn and use as detergent additive: **Belg. pat. 672,205** (1966 to Procter and Gamble), *C.A.* **67**, 4040u (1967). Prepn: O. T. Quimby *et al.*, J. Org. Chem. **32**, 4111 (1967). Effect on hydroxyapatite (durapatite, *q.v.*) crystal aggregation: N. M. Hansen *et al.*, Biochim. Biophys. Acta **451**, 549 (1976); S. Bisaz *et al.*, ibid. 560. Effect on osteoclast morphology in rats: D. J. Rowe, E. Hausmann, Calcif. Tissue Res. **20**, 53 (1976); S. C. Miller, W. S. S. Jee, Anat. Rec. **193**, 439 (1979). Effect (with piroxicam, *q.v.*) on osteolytic and metastatic spread of rat prostate adenocarcinoma cells: M. Pollard, P. H. Luckert, Prostate **8**, 81 (1986). Study of macrophage function with liposome-encapsulated DMDP: E. Claassen, N. Van Rooijen, J. Microencapsulation **3**, 109 (1986). Clinical study in Paget's disease: D. L. Douglas *et al.*, Lancet **1**, 1043 (1980); in metastatic bone disease: I. Elomaa *et al.*, ibid. **1**, 146 (1983); A. Jung *et al.*, N. Engl. J. Med. **308**, 1499 (1983); in hypercalcemia of cancer: R. C. Percival *et al.*, Brit. J. Cancer **51**, 665 (1985). Symposium on therapeutic efficacy in neoplastic bone disease: Bone **8**, Suppl. 1, S1-S86 (1987).

mp 249-251°.
Disodium salt, $CH_2Cl_2Na_2O_6P_2$, *clodronate disodium*, *DClMDP*, *Bonefos*, *Clasteon*, *Difosfonal*, *Ossiten*, *Ostac*.

THERAP CAT: Calcium regulator.

2370. Clofazimine. *N,5-Bis(4-chlorophenyl)-3,5-dihydro-3-[(1-methylethyl)imino]-2-phenazinamine; 3-(p-chloroanilino)-10-(p-chlorophenyl)-2,10-dihydro-2-(isopropylimino)phenazine;* 2-(4-chloroanilino)-3-isopropylimino-5-(4-chlorophenyl)-3,5-dihydrophenazine; 2-p-chloroanilino-5-p-chlorophenyl-3,5-dihydro-3-isopropyliminophenazine; G 30320; B 663; Lampren(e). $C_{27}H_{22}Cl_2N_4$; mol wt 473.41. C 68.50%, H 4.68%, Cl 14.98%, N 11.83%. Prepn: Barry *et al.*, Nature **179**, 1013 (1957); Barry *et al.*, J. Chem. Soc. **1958**, 859; Belton *et al.*, Proc. Roy. Irish Acad. Sect. B **62**, 9 (1961), C.A. **58**, 4556d (1963). Activity studies: Vischer, Arzneimittel-Forsch. **18**, 1529 (1968). Toxicity data: Stenger *et al.*, ibid. **20**, 794 (1970). Clinical studies: Mathies, ibid. 1838; Karat *et al.*, Brit. Med. J. **1**, 198 (1970). Brief review: ibid. **3**, 175 (1971).

Dark-red water-insol crystals, mp 210-212°. LD_{50} orally in mice, rats, and guinea pigs: > 4 g/kg (Stenger).
THERAP CAT: Antibacterial (tuberculostatic, leprostatic).

2371. Clofenamide. *4-Chloro-1,3-benzenedisulfonamide; 4-chloro-m-benzenedisulfonamide;* 1-chlorobenzene-2,4-disulfonamide; chlorphenamide; monochlorphenamide; Salco; Saltron; Soluran; Aquedux; Haflutan. $C_6H_7ClN_2O_4S_2$; mol wt 270.73. C 26.62%, H 2.61%, Cl 13.10%, N 10.35%, O 23.64%, S 23.69%. Prepn: Olivier, *Rec. Trav. Chim.* **37**, 307 (1918); Davies, Wood, *J. Chem. Soc.* **1928**, 1125.

Needles, mp 206-207° (Davies), mp 217-219° (Olivier). Sol in hot alcohol, water; slightly sol in cold solvents.
THERAP CAT: Diuretic.

2372. Clofenciclan. *2-[[1-(4-Chlorophenyl)cyclohexyl]oxy]-N,N-diethylethanamine; 2-[[1-(p-chlorophenyl)cyclohexyl]oxy]triethylamine;* 1-(p-chlorophenyl)cyclohexyl β-diethylaminoethyl ether; chlorphencyclan. $C_{18}H_{28}ClNO$; mol wt 309.89. C 69.77%, H 9.11%, Cl 11.44%, N 4.52%, O 5.16%. Prepd from 1-(p-chlorophenyl)cyclohexanol and β-diethylaminoethyl chloride in the presence of $NaNH_2$: Winter, Stach, **Ger. pat. 1,096,347** (1961 to Boehringer, Mann.); Stach, Winter, *Arzneimittel-Forsch.* **12**, 25 (1962).

Free base, $bp_{0.8}$ 145-152°.
Hydrochloride, $C_{18}H_{29}Cl_2NO$, crystals, mp 145-146°. Ingredient of *Vesitan; Tonoquil* (also contg thiopropazate).
THERAP CAT: CNS stimulant.

2373. Clofentezine. *3,6-Bis(2-chlorophenyl)-1,2,4,5-tetrazine;* bisclofentezin; NC 21314; Apollo. $C_{14}H_8Cl_2N_4$; mol wt 303.15. C 55.48%, H 2.66%, Cl 23.38%, N 18.48%. Contact acaricide active against eggs and early motile stages of phytophagous mites. Prepn: J. H. Parsons, **Eur. pat. Appl. 5912**; idem, **U.S. pat. 4,237,127** (1979, 1980 both to Fisons). Physical properties, toxicity and bioactivity: K. M. G. Bryan et al., *Proc. Brit. Crop Prot. Conf.-Pests Dis.* **1981**, 67. Field trial on apple trees: A. J. Read, *Proc. 36th N.Z. Weed Pest Control Conf.*, 261 (1983). *Review:* F. Rauch, *Def. Veg.* **39**, 11-17 (1985).

Magenta solid. Crystals from ethyl acetate, mp 179-182°. Soly: 50 g/l in chloroform; 2.5 g/l in benzene; < 1 g/l in hexane; < 1 mg/l in water. LD_{50} in rats, mice (mg/kg):

> 3200 orally; LC_{50} (96 hr) in rainbow trout: 100 mg/l (Bryan).
USE: Pesticide.

2374. Clofibrate. *2-(4-Chlorophenoxy)-2-methylpropanoic acid ethyl ester;* ethyl 2-(p-chlorophenoxy)-2-methylpropionate; ethyl p-chlorophenoxyisobutyrate; Amotril; Anparton; Apolan; Artevil; Ateculon; Ateriosan; Atheropront; Atromidin; Atromid-S; Bioscleran; Claripex; Cloberat; Clobren-SF; Clofinit; CPIB; Hyclorate; Lipavlon; Liprinal; Neo-Atromid; Normet; Normolipol; Recolip; Regelan; Serotinex; Sklerolip; Skleromexe; Sklero-Tablinen; Ticlobran; Xyduril. $C_{12}H_{15}ClO_3$; mol wt 242.71. C 59.38%, H 6.23%, Cl 14.61%, O 19.78%. Prepn: Jones et al., **Brit. pat. 860,303** corresp to **U.S. pat. 3,262,850** (1961, 1966 to I.C.I.). Clinical trials in the primary prevention of ischaemic heart disease: *Brit. Heart J.* **40**, 1069 (1978). Follow-up of the clinical trials: *Lancet* **2**, 379 (1980). Review of pharmacology: M. Chevais, *Therapie* **35**, 5-22 (1980). Review of metabolism: M. N. Cayen, *Rev. Drug Metab. Drug Interact.* **3**, 77-103 (1980). Review of toxicity studies: *IARC Monographs* **24**, 39-58 (1980). Comprehensive description: M. M. A. Hassan, A. A. Elazzouny, in *Analytical Profiles of Drug Substances* vol. **11**, K. Florey, Ed. (Academic Press, New York, 1982) pp 197-224.

Oil, bp_{20} 148-150°. Practically insol in water; misc with ethanol, acetone, chloroform, ether. LD_{50} orally in mice, rats: 1.28, 1.65 g/kg, G. Metz et al., *Arzneimittel-Forsch.* **27**, 1173 (1977).
Mixture with androsterone, *Atromid.*
THERAP CAT: Antihyperlipoproteinemic.

2375. Clofibric Acid. *2-(4-Chlorophenoxy)-2-methylpropanoic acid; 2-(p-chlorophenoxy)-2-methylpropionic acid;* α-(p-chlorophenoxy)isobutyric acid; chlorophibrinic acid; Arteriohom; Dimetrop; Regulipid. $C_{10}H_{11}ClO_3$; mol wt 214.64. C 55.95%, H 5.17%, Cl 16.52%, O 22.36%. Prepn: Galimberti, Defranceschi, *Gazz. Chim. Ital.* **77**, 431 (1947); Gilman, Wilder, *J. Am. Chem. Soc.* **77**, 6644 (1955); Jones et al., **Brit. pat. 860,303** (1961 to I.C.I.). Prepn and resolution of isomers: Witiak et al., *J. Med. Chem.* **11**, 1086 (1968). Activity thought to be due to its displacement of thyroxin and its inhibition of cholesterol biosynthesis. *See* Chang et al., *Biochem. Pharmacol.* **16**, 2053 (1967); Walsh et al., *Arch. Biochem. Biophys.* **130**, 7 (1969); Witiak et al., *J. Med. Chem.* **14**, 754 (1971). Metabolism: Almirante et al., *Boll. Chim. Farm.* **108**, 292 (1969).

Crystals from water or methanol, mp 118-119°.
Basic aluminum salt, $C_{20}H_{21}AlCl_2O_7$, Alufibrate, Atherolip, Atherolipin.
Magnesium salt, $C_{20}H_{20}Cl_2MgO_6$, *magnesium clofibrate*, UR-112, Clomag. Ingredient of *Davistar.* White powder, mp 326-328°. Soly in g/100 mg: water 4.5; abs ethanol 0.7; chloroform 0.02.
Pyridoxine salt, $C_{18}H_{22}ClNO_6$, *pyridoxine p-chlorophenoxyisobutyrate, Claresan.* Prepn: Sarbach et al., **Ger. pat. 1,915,497** (1970 to Inst. Rech. Sci.), *C.A.* **74**, 13011g (1971).
Ethyl ester, *see* clofibrate.
3-Pyridylmethyl ester, *see* clofenpyride.
Etofylline ester, *see* theofibrate.
THERAP CAT: Antihyperlipoproteinemic.

2376. Cloflucarban. *N-(4-Chlorophenyl)-N'-[4-chloro-3-(trifluoromethyl)phenyl]urea; 4,4'-dichloro-3-(trifluoromethyl)carbanilide;* Irgasan CF3; Irgosan CF3. $C_{14}H_9Cl_2F_3N_2O$; mol wt 349.15. C 48.16%, H 2.60%, Cl 20.31%, F 16.33%, N 8.03%, O 4.58%. Prepn: Schetty et al., U.S. pat.

2,745,874 (1956 to Geigy); Sydor, **Belg.** pat. **617,117** (1962 to Am. Cyanamid), compd included but not specifically described in abstract, *C.A.* **59,** 1542a (1963).

White crystalline solid, mp 214-215°. Insol in water; dissolves well in organic solvents.

THERAP CAT: Disinfectant.

2377. Clofoctol. *2-[(2,4-Dichlorophenyl)methyl]-4-(1,1,3,3-tetramethylbutyl)phenol;* α-(2,4-dichlorophenyl)-4-(1,1,3,3-tetramethylbutyl)-*o*-cresol; Gramplus; Octofène. $C_{21}H_{26}Cl_2O$; mol wt 365.34. C 69.04%, H 7.17%, Cl 19.41%, O 4.38%. Prepn: J. Debat, U.S. pat. **3,830,852** (1974 to Inst. Recherches Chim. Biol. Appl.). Activity against gram positive and gram negative bacteria: A. Buogo, *Drugs Exp. Clin. Res.* **10,** 321 (1984). Pharmacokinetics in rats: M. Del Tacca *et al.* in *Recent Advances in Chemotherapy, Antimicrobial Section 3,* J. Ishigami, Ed. (University of Tokyo Press, 1985) pp 1927-28; in man: *eidem, J. Antimicrob. Chemother.* **19,** 679 (1987). Mechanism of action: J. Combe *et al., J. Pharmacol.* **11,** 411 (1980); F. Yablonsky, G. Simonet, *ibid.* **13,** 515 (1982). Clinical trials in respiratory infections: J. Vialatte, *Diagnostics* **207,** 65 (1978); in infectious bronchopulmonary diseases: R. Danesi, M. Del Tacca, *Int. J. Clin. Pharm. Res.* **5,** 175 (1985); in ear, nose and throat infections: P. L. Ghilardi, A. Casani, *Drugs Exp. Clin. Res.* **11,** 815 (1985).

Crystals from petr ether, mp 78°. LD_{50} orally in male rats: > 4 g/kg (Debat).

THERAP CAT: Antibacterial.

2378. Cloforex. *[2-(4-Chlorophenyl)-1,1-dimethylethyl]-carbamic acid ethyl ester;* (p-chloro-α,α-dimethylphenethyl)-*carbamic acid ethyl ester;* ethyl (p-chloro-α,α-dimethylphenethyl)carbamate; ethyl *N*-[α,α-dimethyl-β-(*p*-chlorophenyl)ethyl]carbamate; Avicol SL; Frenapyl; Oberex. $C_{13}H_{18}$-$ClNO_2$; mol wt 255.75. C 61.05%, H 7.09%, Cl 13.87%, N 5.48%, O 12.51%. Prepn: **Belg.** pat. **619,886** (1963 to Troponwerke Dinklage), *C.A.* **59,** 11359c (1963), corresp to **Brit.** pat. **970,565.** Metabolism: Ryrfeldt, *Acta Pharmacol. Toxicol.* **28,** 391 (1970); Buelow *et al., Arzneimittel-Forsch.* **21,** 86 (1971).

Crystals, mp 52.5-53°. $bp_{0.005}$ 88-90°.

THERAP CAT: Anorexic.

2379. Clomacran. *2-Chloro-9,10-dihydro-N,N-dimethyl-9-acridinepropanamine;* 2-chloro-9-[3-(dimethylamino)-propyl]acridan; SKF 14336D. $C_{18}H_{21}ClN_2$; mol wt 300.83. C 71.87%, H 7.04%, Cl 11.78%, N 9.31%. Prepn: C. L. Zirkle, **U.S.** pat. **3,131,190** (1964 to SKF). Effect on *in vivo* histamine metabolism: R. Schayer, M. A. Reilly, *Arch. Int. Pharmacodyn. Ther.* **203,** 123 (1973). Clinical potency in relation to dopamine-binding: P. Seeman *et al., Nature* **261,** 717 (1976). Pharmacology: P. Fowler *et al., Arzneimittel-Forsch.* **27,** 866 (1977). Photooxidation study: W. R. Knappe, *J. Pharm. Sci.* **67,** 318 (1978).

Phosphate, $C_{18}H_{24}ClN_2OP_4$, *SKF 14336, Devryl, Olaxin.* LD_{50} in rats, mice (mg/kg): 350, 222 orally (Fowler).

THERAP CAT: Antipsychotic.

2380. Clomestrone. *16-Chloro-3-methoxyestra-1,3,5-(10)-trien-17-one;* 16α-chloroestrone 3-methyl ether; Arterolo; Atheran; Iposclerone. $C_{19}H_{23}ClO_2$; mol wt 318.85. C 71.57%, H 7.27%, Cl 11.12%, O 10.04%. Prepn: Mueller *et al., J. Am. Chem. Soc.* **80,** 1769 (1958); Mueller, **U.S.** pat. **2,855,411** (1958 to Searle); Ravizza, **Swiss** pat. **343,396** (1959). Alternate route: Labs. Sobio, **Fr.** pat. **1,350,100** (1964).

Crystals from chloroform, mp 179-180°. $[\alpha]_D$ +161° (chloroform).

THERAP CAT: Antihyperlipoproteinemic.

2381. Clometacin. *3-(4-Chlorobenzoyl)-6-methoxy-2-methyl-1H-indole-1-acetic acid;* [2-methyl-3-(*p*-chlorobenzoyl)-6-methoxyindol-1-yl]acetic acid; clometazin; mindolic acid; R 3959; C 1656; Dupéran. $C_{19}H_{16}ClNO_4$; mol wt 357.81. C 63.78%, H 4.51%, Cl 9.91%, N 3.91%, O 17.89%. Prepn: Allais, Nomine, **Ger.** pat. **1,901,167** (1969 to Roussel-UCLAF), *C.A.* **72,** 43440x (1970).

White powder, mp 242°.

THERAP CAT: Analgesic.

2382. Clomethiazole. *5-(2-Chloroethyl)-4-methylthiazole;* 4-methyl-5-(β-chloroethyl)thiazole; chlorethiazol; chlormethiazole; S.C.T.Z. C_6H_8ClNS; mol wt 161.66. C 44.58%, H 4.99%, Cl 21.93%, N 8.67%, S 19.84%. Prepd by condensation of thioformamide with chloroacetopropyl alcohol: Buchman, *J. Am. Chem. Soc.* **58,** 1803 (1936); by chlorination of 4-methyl-5-(β-ethoxyethyl)thiazole: Clarke, Gurin, *ibid.* **57,** 1876 (1935); by treating 5-(2-hydroxyethyl)-4-methylthiazole with thionyl chloride: Sawa, Ishida, *J. Pharm. Soc. Japan* **76,** 337 (1956); by H_2O_2 oxidation of 2-mercapto-4-methyl-5-(β-chloroethyl)thiazole: **Swiss** pat. **200,248** (1938 to Hoffmann-La Roche). Prepn of methane and ethanedisulfonate: Charonnat *et al.*, **U.S.** pat. **3,031,457** (1962).

Oily, viscous liquid. d_4^{25} 1.233. Characteristic disagreeable odor of thiazole compds. bp_7 92°.

Hydrochloride, $C_6H_9Cl_2NS$, crystals from abs ethanol + ether, mp 130°. Sol in water, alcohol.

Methanedisulfonate, $C_{13}H_{20}Cl_2N_2O_6S_4$, crystals from methanol + ether, mp 120°.

Ethanedisulfonate, $C_{14}H_{22}Cl_2N_2O_6S_4$, *Distraneurin, Hemi-*

neurin, Heminevrin. Crystals from methanol + ether, mp 124°.

USE: Intermediate in some processes of vitamin B_1 manufacture.

THERAP CAT: Sedative, hypnotic, anticonvulsant.

2383. Clometocillin. *6-[[(3,4-Dichlorophenyl)methoxyacetyl]amino]-3,3-dimethyl-7-oxo-4-thia-1-azabicyclo-[3.2.0]heptane-2-carboxylic acid;* 3,4-dichloro-α-methoxybenzylpenicillin; 6-(α-methoxy-3,4-dichlorophenylacetamido)penicillanic acid; clometacillin; clomethacillin; no. 356; penicillin 356. $C_{17}H_{18}Cl_2N_2O_5S$; mol wt 433.33. C 47.12%, H 4.19%, Cl 16.36%, N 6.47%, O 18.46%, S 7.40%. Semisynthetic antibiotic related to penicillin. Prepn: Vanderhaeghe *et al., Antimicrob. Ag. Chemother.* **1961**, 581; U.S. pat. **3,007,920** (1961 to Recherche et Ind. Therap.). Activity studies: van Dijck *et al., Antibiot. Chemother.* **12,** 192 (1962).

Sodium salt (racemic mixture), $C_{17}H_{17}Cl_2N_2NaO_5S$, *Rixapen.* $[\alpha]_D$ 210-220°.

Pure epimers were prepared by fractional crystn. First crop: Crystals, dec 232-235°. $[\alpha]_D^{20}$ +182° (water). Second crop: $[\alpha]_D^{20}$ +207°. Third crop: Crystals, dec 180-183°. $[\alpha]_D^{20}$ +261° (water).

The free acid yields two diastereoisomers: $[\alpha]_D$ +177° and $[\alpha]_D$ +257°.

THERAP CAT: Antibacterial.

2384. Clomiphene. *2-[4-(2-Chloro-1,2-diphenylethenyl)phenoxy]-N,N-diethylethanamine; 2-[p-(2-chloro-1,2-diphenylvinyl)phenoxy]triethylamine;* 2-[p-(β-chloro-α-phenylstyryl)phenoxy]triethylamine; 1-[p-(β-diethylaminoethoxy)phenyl]-1,2-diphenylchloroethylene; clomifene; chloramiphene; MRL-41. $C_{26}H_{28}ClNO$; mol wt 405.98. C 76.92%, H 6.95%, Cl 8.73%, N 3.45%, O 3.94%. Synthetic estrogen agonist-antagonist. Prepn: Allen *et al.,* U.S. pat. **2,914,563** (1959 to Merrell). Stereochemistry of the geometric isomers: Ernst *et al., J. Pharm. Sci.* **65,** 148 (1976). Induction of ovulation: R. B. Greenblatt, *Fertil. Steril.* **12,** 402 (1961). Clinical trial in anovulatory women: J. Garcia *et al., ibid.* **28,** 707 (1977). Use in male infertility: P. J. Sorbie, R. Perez-Marrero, *J. Urol.* **131,** 425 (1984). HPLC determn of isomers in human plasma: C. L. Baustian, T. J. Mikkelson, *J. Pharm. Biomed. Anal.* **4,** 237 (1986).

Dihydrogen citrate, $C_{32}H_{36}ClNO_8$, *Clomid, Clomphid, Clomivid, Clostilbegyt, Dyneric, Ikaclomine, Pergotime, Serophene.* Crystals, mp 116.5-118°. Slightly sol in water, chloroform; freely sol in methanol; sparingly sol in alcohol. Insol in ether.

*cis-*Form, *zuclomiphene.*
*trans-*Form, *enclomiphene.*

THERAP CAT: Gonad-stimulating principle.

2385. Clomipramine. *3-Chloro-10,11-dihydro-N,N-dimethyl-5H-dibenz[b,f]azepine-5-propanamine; 3-chloro-5-[3-(dimethylamino)propyl]-10,11-dihydro-5H-dibenz[b,f]azepine;* 3-chloro-10,11-dihydro-5-(3-dimethylaminopropyl)-5H-dibenz[b,f]azepine; 5-(γ-dimethylaminopropyl)-3-chloroiminodibenzyl; chlorimipramine; G 34586. $C_{19}H_{23}-ClN_2$; mol wt 314.87. C 72.48%, H 7.36%, Cl 11.26%, N 8.90%. Prepn: Craig *et al., J. Org. Chem.* **26,** 135 (1961); Schindler, Dietrich, **Swiss.** pat. **371,799** (1963 to Geigy), *C.A.* **60,** 11995c (1964).

$bp_{0.3}$ 160-170°.

Hydrochloride, $C_{19}H_{24}Cl_2N_2$, *Anafranil.* Crystals from acetone + ether or methanol + ether, mp 189-190° (Craig *et al., loc. cit.*), 191-192° (Schindler, Dietrich, *loc. cit.*). Crystal and molecular structure: M. L. Post, A. S. Horn, *Acta Crystallogr.* **33B,** 2590 (1977).

THERAP CAT: Antidepressant.

2386. Clomocycline. *7-Chloro-4-(dimethylamino)-1,4,-4a,5,5a,6,11,12a-octahydro-3,6,10,12,12a-pentahydroxy-N-(hydroxymethyl)-6-methyl-1,11-dioxo-2-naphthacenecarboxamide;* chlormethylenecycline; N^2-(hydroxymethyl)chlortetracycline; N-methylol-7-chlortetracycline; Megaclor. $C_{23}-H_{25}ClN_2O_9$; mol wt 508.93. C 54.28%, H 4.95%, Cl 6.97%, N 5.51%, O 28.29%. Semi-synthetic broad spectrum antibiotic related to tetracycline, *q.v.* Prepn: Banci, Tubaro, *Boll. Chim. Farm.* **102,** 471 (1963); **Belg.** pat. **628,142** (1963 to AB Leo), *C.A.* **60,** 15805f (1964).

Yellow material. Dec 145-170°. Sensitive to light and air. pH of 2% aq soln 6.3-6.5. Sol in citrate-phosphate buffer, pH 6-8.

THERAP CAT: Antibacterial.

2387. Clonazepam. *5-(2-Chlorophenyl)-1,3-dihydro-7-nitro-2H-1,4-benzodiazepin-2-one;* 7-nitro-5-(2-chlorophenyl)-3H-1,4-benzodiazepin-2(1H)-one; Ro 5-4023; Clonopin; Iktorivil; Klonopin; Landsen; Rivotril. $C_{15}H_{10}ClN_3-O_3$; mol wt 315.72. C 57.07%, H 3.19%, Cl 11.23%, N 13.31%, O 15.20%. Antiepileptic agent with anxiolytic and antimanic properties. Prepn: L. H. Sternbach *et al., J. Med. Chem.* **6,** 261 (1963); O. Keller *et al.,* U.S. pat. **3,121,076** (1964 to Hoffmann-La Roche). Prepd but not claimed: J. Kariss, H. L. Newmark, U.S. pat. **3,116,203** (1963 to Hoffmann-La Roche). Alternate process: A. Focella, A. I. Rachlin, U.S. pat. **3,335,181** (1967 to Hoffmann-La Roche). Pharmacology: Guerrero-Figueroa *et al., Curr. Ther. Res. Clin. Exp.* **11,** 40 (1969); Lechat *et al., Therapie* **25,** 893 (1970); D'Armagnac *et al., ibid.* **26,** 439 (1971). Toxicology: Blum *et al., Arzneimittel-Forsch.* **23,** 377 (1971). Studies on the detection of clonazepam and its main metabolites: S. Ebel, H. Schütz, *ibid.* **27,** 325 (1977). Clinical efficacy in acute mania: G. Chouinard, *Psychosomatics* **26**(12), Suppl., 7 (1985); in panic disorders and agitation: R. Fontaine, *ibid.* 13. Comprehensive description: W. C. Winslow in *Analytical Profiles of Drug Substances* vol. 6, K. Florey, Ed. (Academic Press, New York, 1977) pp 61-81.

White crystals from ethanol-methylene chloride, mp 236.5-238.5°. uv max (7.5% methanol in isopropanol): 248, 310 nm (ε 14500, 11600). Soly in mg/ml at 25°: acetone 31; chloroform 15; methanol 8.6; ether 0.7; benzene 0.5; water <0.1. pK_1 1.5, pK_2 10.5. LD_{50} orally in mice: >4000 mg/kg (Blum).

Consult the cross index before using this section.

Caution: May be habit forming. This is a controlled substance (depressant) listed in the U.S. Code of Federal Regulations, Title 21 Part 1308.14 (1985).

THERAP CAT: Anticonvulsant.

2388. Clonidine. *2,6-Dichloro-N-2-imidazolidinylidenebenzenamine; 2-(2,6-dichloroanilino)-2-imidazoline; 2-(2,6-dichloroanilino)-1,3-diazacyclopentene-(2); 2-[(2,6-dichlorophenyl)amino]-2-imidazoline.* $C_9H_9Cl_2N_3$; mol wt 230.10. C 46.98%, H 3.94%, Cl 30.82%, N 18.26%. α_2-Adrenergic agonist. Prepn: Zeile *et al.*, U.S. pat. 3,202,660 (1965 to Boehringer, Ing.). Use in shaving soap formulations: *eidem,* U.S. pat. 3,190,802 (1965 to Boehringer, Ing.). Pharmacology: Bolme, Fuxe, *Eur. J. Pharmacol.* **13**, 168 (1971). Revised structure: L. M. Jackman, T. Jen, *J. Am. Chem. Soc.* **97**, 2811 (1975). GC determn in plasma: P. O. Edlund, *J. Chromatog.* **187**, 161 (1980). Preliminary studies on potential antidepressant activity: D. C. Jimerson *et al., Biol. Psychiatry* **15**, 45 (1980); J. B. Malick, *Prog. Clin. Biol. Res.* **71**, 165 (1981). Exptl use in drug rehabilitation: M. S. Gold, A. C. Pottash, *Ann. N.Y. Acad. Sci.* **362**, 191-202 (1981). Activity as α-adrenoceptor agonist: A. G. Roach *et al., J. Pharmacol. Exp. Ther.* **227**, 421 (1983). Symposium on cardiovascular and psychotropic pharmacology and clinical experience: *Central Blood Pressure Regulation: The Role of α_2-Receptor Stimulation,* K. Hayduk, K. B. Bock, Eds. (Steinkopff Verlag, Darmstadt, 1983) 284 pp. Effects in acute smoking withdrawal syndrome: A. H. Glassman *et al., Science* **226**, 864 (1984); in alcoholism withdrawal: Z. Jraidi *et al., Therapie* **42**, 21 (1987). Clinical trial in Tourette's syndrome: J. F. Leckman *et al., Neurology* **35**, 343 (1985); in cigarette smoking cessation: R. Davison *et al., Clin. Pharmcol. Ther.* **44**, 265 (1988). *Reviews:* H. Schmitt, *Handb. Exp. Pharmacol.* **39**, 299-396 (1977); A. Walland in *Pharmacological and Biochemical Properties of Drug Substances* vol. **1**, M. E. Goldberg, Ed. (Am. Pharm. Assoc., Washington, D.C., 1977) pp 67-107; M. C. Houston, *Prog. Cardiovasc. Dis.* **23**, 337-350 (1981).

Crystals, mp 130°.

Hydrochloride, $C_9H_{10}Cl_3N_3$, *St 155, Catapres, Catapresan, Clonistada, Dixarit, Ipotensium, Isoglaucon, Tenso-Timelets.* Crystals, mp 305°. LD_{50} in mice, rats (mg/kg): 328, 270 orally; 18, 29 i.v. (Walland).

USE: In shaving soaps.

THERAP CAT: Antihypertensive.

2389. Clonitazene. *2-[(4-Chlorophenyl)methyl]-N,N-diethyl-5-nitro-1H-benzimidazole-1-ethanamine; 2-p-chlorobenzyl-1-(2-diethylaminoethyl)-5-nitrobenzimidazole; 1-(β-diethylaminoethyl)-2-p-chlorobenzyl-5-nitrobenzimidazole.* $C_{20}H_{23}ClN_4O_2$; mol wt 386.89. C 62.09%, H 5.99%, Cl 9.17%, N 14.48%, O 8.27%. Prepn: Hunger *et al., Experientia* **13**, 400 (1957); Hoffmann *et al.,* U.S. pat. 2,935,514 (1960 to Ciba).

Crystals, mp 75-76°.

Hydrochloride, $C_{20}H_{24}Cl_2N_4O_2$, mp 238-240°.

Methanesulfonate, $C_{21}H_{27}ClN_4O_5S$, mp 163-166°.

Caution: May be habit forming. This is a controlled substance (opiate) listed in the U.S. Code of Federal Regulations, Title 21 Part 1308.11 (1987).

THERAP CAT: Narcotic analgesic.

2390. Clonitrate. *3-Chloro-1,2-propanediol dinitrate;*

3-chloropropyleneglycol dinitrate; dinitrochlorohydrin; Dylate. $C_3H_5ClN_2O_6$; mol wt 200.53. C 17.97%, H 2.51%, Cl 17.68%, N 13.97%, O 47.87%. Prepn from 3-chloro-1,2-propanediol: Henry, *Ann.* **155**, 165 (1870).

Slightly yellow liq, bp_{760} 190-195° (some decompn); bp_{15} 121-123°; bp_{10} 117.5°. d^9 1.5112; d^{15} 1.5408. Sol in alc, acetone, chloroform; practically insol in water, acids.

THERAP CAT: Vasodilator (coronary).

2391. Clopamide. *(cis)-3-(Aminosulfonyl)-4-chloro-N-(2,6-dimethyl-1-piperidinyl)benzamide; 1-(4-chloro-3-sulfamoylbenzamido)-2,6-dimethylpiperidine; N-(2',6'-dimethyl-1'-piperidyl)-3-sulfamoyl-4-chlorobenzamide; 4-chloro-N-(2',6'-dimethyl-1'-piperidyl)-3-sulfamoylbenzamide; chlosudimeprimyl; DT-327; Adurix; Aquex; Brinaldix.* $C_{14}H_{20}ClN_3O_3S$; mol wt 345.86. C 48.62%, H 5.83%, Cl 10.25%, N 12.15%, O 13.88%, S 9.27%. Prepn (stereochemistry unspecified): Belg. pat. 610,039; E. Jucker, A. J. Lindenmann, U.S. pat. 3,459,756 (1962, 1969 both to Sandoz). Prepn of cis- and trans-forms: *eidem, Helv. Chim. Acta* **45**, 2316 (1962). HPLC determn: R. T. Sane *et al., J. Chromatog.* **356**, 468 (1986). Pharmacokinetics and diuretic effect in humans: J. J. McNeil *et al., Clin. Pharmacol. Ther.* **42**, 299 (1987). Clinical trial of combination with pindolol, *q.v.,* in essential hypertension: D. Crowder, E. G. M. Cameron, *Curr. Med. Res. Opin.* **6**, 342 (1979).

Mixture with pindolol, *Viskaldix.*

Hydrazine deriv., crystals from methanol + diisopropyl ether, mp 244-246°.

THERAP CAT: Antihypertensive; diuretic.

2392. Clopenthixol. *4-[3-(2-Chloro-9H-thioxanthen-9-ylidene)propyl]-1-piperazineethanol; 2-chloro-9-[3-[4-(2-hydroxyethyl)-1-piperazinyl]propylidene]thioxanthene.* $C_{22}H_{25}ClN_2OS$; mol wt 400.99. C 65.90%, H 6.28%, Cl 8.84%, N 6.99%, O 3.99%, S 8.00%. Thioxanthene neuroleptic. Prepn (configuration not specified): Belg. pat. 585,338; P. V. Petersen *et al.,* U.S. pat. 3,116,291 (1960, 1963 both to Kefalas A/S). Prepn of the pharmacologically active cis-isomer: Belg. pat. 816,855; N. Lassen, U.S. pat. 3,996,211 (1974, 1976 both to Kefalas A/S). Pharmacology: Cazzullo, Andreola, *Acta Neurol. (Naples)* **20**, 162 (1965); Weissman, *Mod. Probl. Pharmacopsychiat.* **2**, 15 (1969); Moeller Nielsen, *ibid.* 23. Metabolism: Khan, *Acta Pharmacol. Toxicol.* **27**, 202 (1969). HPLC determn of isomers in serum: T. Aaes-Jorgensen, *J. Chromatog.* **188**, 239 (1980). Series of articles on pharmacology and clinical studies: *Acta Psychiatr. Scand.* **64**, Suppl. 294, 1-77 (1981).

Colorless syrup. Sparingly sol in ether. Readily sol in methanol.

Dihydrochloride, $C_{22}H_{27}Cl_3N_2OS$, *AY 62021, N-746, Ciatyl, Sordenac, Sordinol.* Crystals from ethanol, mp 250-260° (dec). Freely sol in water; sparingly sol in alcohol. Practically insol in other organic solvents. LD_{50} in male mice (mg/kg): 111 i.v. (Lassen).

cis(Z)-Form, α-*clopenthixol, zuclopenthixol.* Crystals, mp 84-85°.

cis(Z)-Form dihydrochloride, *Cisordinol, Clopixol.* Crystals, mp 250-260° (dec). LD_{50} in male mice (mg/kg): 105 i.v. (Lassen).

THERAP CAT: Antipsychotic.

2393. Cloperastine. *1-[2-[(4-Chlorophenyl)phenylmethoxy]ethyl]piperidine; 1-[2-[(p-chloro-α-phenylbenzyl)oxy]ethyl]piperidine;* p-chlorobenzhydryl 2-(1-piperidyl)ethyl ether. $C_{20}H_{24}ClNO$; mol wt 329.88. C 72.82%, H 7.33%, Cl 10.75%, N 4.25%, O 4.85%. Prepn: **Brit. pat.** **670,622** (1952 to Parke, Davis); Fujie, *Yakugaku Zasshi* **81,** 693 (1961), *C.A.* **55,** 25848f (1961). Prepn of salts: **Brit. pat. 1,179,945** (1970 to Yoshitomi Pharm.), *C.A.* **72,** 132549g (1970). Pharmacology: Takagi *et al., Yakugaku Zasshi* **85,** 550 (1965); **87,** 907 (1967), *C.A.* **63,** 8932f (1965); **67,** 107101u (1967). Metabolism: Kato, Furuta, *Oyo Yakuri* **5,** 735 (1972).

$bp_{0.06}$ 172-174°, $bp_{0.15}$ 178-180°.

Hydrochloride, $C_{20}H_{25}ClN_2O$, *Hustazol, Nitossil, Novotusil, Sekin.* Crystals, mp 147.9°.

Methiodide, $C_{21}H_{27}ClINO$, crystals, mp 140°.

THERAP CAT: Antitussive.

2394. Clopidol. *3,5-Dichloro-2,6-dimethyl-4-pyridinol;* meticlorpindol; clopindol; Coyden. $C_7H_7Cl_2NO$; mol wt 192.06. C 43.78%, H 3.68%, Cl 36.92%, N 7.29%, O 8.33%. Prepn: Stevenson, U.S. pat. **3,206,358** (1965 to Dow). Metabolism studies in chickens: Smith, *Poultry Sci.* **48,** 420 (1969).

Solid, mp > 320°. Practically insol in water. LD_{50} orally in rats: 18 g/kg, Plisek *et al., Vet. Spofa* **12,** 111 (1970), *C.A.* **74,** 138780p (1971).

THERAP CAT (VET): Coccidiostat.

2395. Clopirac. *1-(4-Chlorophenyl)-2,5-dimethyl-1H-pyrrole-3-acetic acid;* BRL 13856; CP 172AP; Clopiran. $C_{14}H_{14}ClNO_2$; mol wt 263.72. C 63.76%, H 5.35%, Cl 13.44%, N 5.31%, O 12.13%. Prepn: Lambelin *et al.,* **Ger. pat. 2,261,965** (1973 to Continental Pharma), *C.A.* **79,** 78604a (1973).

THERAP CAT: Anti-inflammatory.

2396. Cloprednol. *(11β)-6-Chloro-11,17,21-trihydroxy-pregna-1,4,6-triene-3,20-dione;* 6-chloro-Δ^{1,4,6}-pregnatriene-11β,17α,21-triol-3,20-dione; RS-4691; Cloradryn; Syntestan. $C_{21}H_{25}ClO_5$; mol wt 392.88. C 64.20%, H 6.41%, Cl 9.02%, O 20.36%. Prepn: H. J. Ringold, G. Rosenkranz, **Fr. pat. 1,271,981,** *C.A.* **58,** 1148a (1963); *eidem,* U.S. pat. **3,232,965** (1962, 1966 both to Syntex). Metabolic effects: E. Ortega *et al., J. Clin. Pharmacol.* **16,** 122 (1976). Effects on hypothalamic-pituitary-adrenal axis function: *eidem, J. Int. Med. Res.* **4,** 326 (1976). Bioavailability in humans: E.

Mroszczak *et al., J. Pharm. Sci.* **67,** 920 (1978). HPLC determn in plasma: C. Lee *et al., ibid.* **70,** 669 (1981).

Cryst from acetone/hexane.

THERAP CAT: Glucocorticoid.

2397. Cloprostenol. *7-[2-[4-(3-Chlorophenoxy)-3-hydroxy-1-butenyl]-3,5-dihydroxycyclopentyl]-5-heptenoic acid.* $C_{22}H_{29}ClO_6$; mol wt 424.92. C 62.19%, H 6.88%, Cl 8.34%, O 22.59%. Aryloxymethyl analog of prostaglandin $F_{2α}$, *q.v.* Prepn: J. Bowler, **Ger. pat. 2,223,365** (1972 to ICI), *C.A.* **78,** 110692v (1973); D. Binder *et al., Prostaglandins* **6,** 87 (1974). Synthesis and biological activity: *eidem, ibid.* **15,** 773 (1978). Short synthesis: N. R. A. Beeley *et al., Tetrahedron* **37,** Suppl. 9, 411 (1981). Disposition in the rat and marmoset: G. R. Bourne *et al., Xenobiotica* **9,** 623 (1979). Effect on fertility in cows: N. Baishya *et al., Brit. Vet. J.* **136,** 227 (1980); on superovulation, fertilization, egg transport in ewes: D. Whyman, R. W. Moore, *J. Reprod. Fertil.* **60,** 267 (1980).

Sodium salt, $C_{22}H_{28}ClNaO_6$, *ICI 80996, Estrumate, Planate.*

THERAP CAT (VET): In treatment of infertility in sows, gilts. In synchronization of estrus.

2398. Clopyralid. *3,6-Dichloro-2-pyridinecarboxylic acid;* 3,6-dichloropicolinic acid; 3,6-DCP; Dowco 290; Lontrel; Shield; Reclaim. $C_6H_3Cl_2NO_2$; mol wt 192.00. C 37.53%, H 1.57%, Cl 36.93%, N 7.30%, O 16.67%. Systemic post-emergence herbicide for use in food crops and mesquite. Prepn: H. Johnston, **Belg. pat. 644,105;** *idem,* U.S. pat. **3,317,549** (1964, 1967 both to Dow). Alternate process: S. D. McGregor, U.S. pat. **4,087,431** (1978 to Dow). Physicochemical properties, toxicity and herbicidal activity: J. G. Brown, S. D. Uprichard, *Proc. Brit. Crop Prot. Conf. - Weeds* **1976,** 119. GLC determn: A. J. Pik, G. W. Hodgson, *J. Assoc. Offic. Anal. Chem.* **59,** 264 (1976). Persistence in soil: A. J. Pik *et al., J. Agr. Food Chem.* **25,** 1054 (1977). Field trial in mesquite control: R. W. Bovey, R. E. Meyer, *Weed Sci.* **33,** 349 (1985).

White, odorless crystalline solid, mp 151-152°. Vapor pressure at 25°: 1.2×10^{-5} mm Hg. Soly at 25°: approx 1000 ppm in water, > 25% w/w in methanol, acetone, xylene. LD_{50} in male, female rats (mg/kg): > 5000, 4300 orally; LC_{50} (96 hr) to rainbow trout: 103.5 mg/l (Brown, Uprichard).

USE: Herbicide.

2399. Cloranolol. *1-(2,5-Dichlorophenoxy)-3-[(1,1-dimethylethyl)amino]-2-propanol;* 1-(tert-butylamino)-3-(2,5-dichlorophenoxy)-2-propanol. $C_{13}H_{19}Cl_2NO_2$, mol wt 292.20. C 53.44%, H 6.55%, Cl 24.26%, N 4.79%, O 10.95%. β-Adrenergic blocker. Prepn: G. Richter, **Ger. pat. 2,213,-**

044 (1972 to Res. Inst. Pharm. Chem. Hungary), *C.A.* **78**, 15780p (1973). Effect on plasma renin and blood pressure: I. Tenyi, *Curr. Ther. Res.* **21**, 823 (1977). Pharmacology: M. I. K. Fekete *et al., Arch. Int. Pharmacodyn.* **248**, 190 (1980). Electrophysiological characteristics: F. Solti, E. Czako, *Int. J. Clin. Pharmacol. Ther. Toxicol.* **18**, 229 (1980). Clinical-pharmacological evaluation: E. Török *et al., ibid.* 200. GC-MS determn in plasma: E. Tomori, E. Elekes, *J. Chromatog.* **204**, 355 (1981). Multicenter clinical trial: E. Török *et al., Int. J. Clin. Pharmacol. Ther. Toxicol.* **23**, 650 (1985).

Crystals, mp 82-83°.
Hydrochloride, $C_{13}H_{20}Cl_3NO_2$, *GYKI-41099, Tobanum.* Crystals, mp 210-212°.
THERAP CAT: Antiarrhythmic.

2400. Clorazepate. *7-Chloro-2,3-dihydro-2,2-dihydroxy-5-phenyl-1H-1,4-benzodiazepine-3-carboxylic acid.* $C_{16}H_{13}ClN_2O_4$; mol wt 332.74. C 57.75%, H 3.94%, Cl 10.65%, N 8.42%, O 19.23%. Prepn: J. Schmitt, **Fr. pat. AD 91,403** (1968 to Clin-Byla), *C.A.* **71**, 91548q (1969). *See also* prepn of the intermediate 2-oxo-compd: **Neth. pat. Appl. 6,507,637** corresp to J. Schmitt, **U.S. pat. 3,516,988** (1965, 1970 both to Clin-Byla). Synthesis and activity of the dipotassium salt: Schmitt *et al., Chim. Ther.* **4**, 239 (1969). Metabolism and kinetics: Gros, Raveux; Raveux, Briot, *ibid.* 312; 303 resp. Pharmacology and toxicology: Brunaud *et al., Arzneimittel-Forsch.* **20**, 123 and sqq (1970). Comprehensive description: J. A. Raihle, V. E. Papendick, in *Analytical Profiles of Drug Substances* vol. 4, K. Florey, Ed. (Academic Press, New York, 1975) pp 91-112.

Dipotassium salt, $C_{16}H_{11}ClK_2N_2O_4$, *clorazepate dipotassium, CB 4306, Belseren, Mendon, Tranxilène, Tranxilium, Transene, Tranxène.* White powder, freely sol in water. Very poorly sol in ethanol. Practically insol in ether, chloroform. Aq solns are alkaline to phenolphthalein. LD_{50} in mice (mg/kg): 700 orally; 290 i.p. (Brunaud).
Monopotassium salt, $C_{16}H_{12}ClKN_2O_4$, *Azene.*
Caution: May be habit forming. This is a controlled substance (depressant) listed in the U.S. Code of Federal Regulations, Title 21 Part 1308.14 (1987).
THERAP CAT: Anxiolytic.

2401. Clorexolone. *6-Chloro-2-cyclohexyl-2,3-dihydro-3-oxo-1H-isoindole-5-sulfonamide; 6-chloro-2-cyclohexyl-3-oxo-5-isoindolinesulfonamide;* 5-chloro-2-cyclohexyl-1-oxo-6-sulfamoylisoindoline; 5-chloro-2-cyclohexyl-6-sulfamoylisoindolin-1-one; 5-chloro-2-cyclohexylphthalimidine-6-sulfonamide; M & B 8430; RP 12833; Flonatril; Nefrolan. $C_{14}H_{17}ClN_2O_3S$; mol wt 328.84. C 51.13%, H 5.21%, Cl 10.78%, N 8.52%, O 14.60%, S 9.75%. Prepn: **Belg. pat. 620,654** corresp to G. E. Lee, W. R. Wragg, **U.S. pat. 3,183,243** (1963, 1965 to May & Baker).

Prisms, mp 266-268°. Soly in water at 25°: 16 mg/l.
THERAP CAT: Diuretic.

2402. Clorindione. *2-(4-Chlorophenyl)-1H-indene-1,3(2H)-dione; 2-(p-chlorophenyl)-1,3-indandione;* 2-(4-chlorophenyl)indan-1,3-dione; chlophenadione; G 25766; MG 2552; Indaliton. $C_{15}H_9ClO_2$; mol wt 256.70. C 70.19%, H 3.53%, Cl 13.81%, O 12.47%. Prepn: Cavallini *et al., Farmaco Ed. Sci.* **10**, 710 (1955); **Brit. pat. 748,251** (1956 to Geigy). Pharmacokinetics: Danek, Pogonowska-Wala, *Pol. J. Pharmacol. Pharm.* **25**, 307 (1973), *C.A.* **79**, 111540b (1973).

Dark red needles from alcohol, pale yellow needles from dil acetic acid, mp 145-146°. Practically insol in water; sol in organic solvents except petr ether. LD_{50} orally in mice: 1220 mg/kg, Fontaine *et al., Med. Pharmacol. Exp.* **17**, 497 (1967).
THERAP CAT: Anticoagulant.

2403. Clorophene. *4-Chloro-2-(phenylmethyl)phenol; 4-chloro-α-phenyl-o-cresol;* 2-benzyl-4-chlorophenol; 5-chloro-2-hydroxydiphenylmethane; o-benzyl-p-chlorophenol; Santophen 1; Septiphene. $C_{13}H_{11}ClO$; mol wt 218.69. C 71.40%, H 5.07%, Cl 16.21%, O 7.32%. Prepn: Klarmann *et al., J. Am. Chem. Soc.* **54**, 3315 (1932); Klarmann, Gate, **U.S. pat. 1,967,825** (to Lehn & Fink); from Na or K phenoxide and benzyl chloride followed by chlorination of the resulting o-benzylphenol: Kaiser, **Ger. pat. 703,955** (1941 to Deutsche Hydrierwerke).

Crystals, mp 48.5°. $bp_{3.5}$ 160-162°. $d_{15.5}^{55}$ 1.186-1.190.
THERAP CAT: Disinfectant.

2404. Clorprenaline. *2-Chloro-α-[[(1-methylethyl)amino]methyl]benzenemethanol; o-chloro-α-(isopropylaminomethyl)benzyl alcohol;* 1-(o-chlorophenyl)-2-isopropylaminoethanol; N-(β-o-chlorophenyl-β-hydroxyethyl)isopropylamine; isophenamine; isoprofenamine. $C_{11}H_{16}ClNO$; mol wt 213.72. C 61.82%, H 7.55%, Cl 16.59%, N 6.56%, O 7.49%. β-Adrenergic agonist. Prepn: Mills, **U.S. pat. 2,816,059** (1957 to Lilly); Nash, **U.S. pat. 2,887,509** (1959 to Lilly).

Hydrochloride monohydrate, $C_{11}H_{17}Cl_2NO.H_2O$, *Broncon, Clopinerin, Conselt, Fusca, Kalutein, Pentadoll, Restanolon, Vortel* (mixt). mp 163-164°.
THERAP CAT: Bronchodilator.

2405. Clorsulon. *4-Amino-6-(trichloroethenyl)-1,3-benzenedisulfonamide;* 4-amino-6-(trichlorovinyl)-m-benzenedisulfonamide; MK-401; Curatrem. $C_8H_8Cl_3N_3O_4S_2$; mol wt 380.65. C 25.24%, H 2.12%, Cl 27.94%, N 11.04%, O 16.81%, S 16.84%. Benzenedisulfonamide deriv with fasciolicidal activity. Prepn: H. H. Mrozik, **Ger. pat. 2,556,122** (1976 to Merck & Co.), *C.A.* **85**, 94100n (1976); H. Mrozik *et al., J. Med. Chem.* **20**, 1225 (1977). GLC determn in biological fluids: W. J. A. Vanden Heuvel *et al., J. Agr. Food*

Chem. **25,** 389 (1977). Efficacy against liver fluke, *Fasciola hepatica:* D. A. Ostlind *et al., Brit. Vet. J.* **133,** 211 (1977); M. D. Schulman *et al., J. Parasitol.* **65,** 555 (1979). *In vitro* inhibition of glycolysis: M. D. Schulman, D. Valentino, *Exp. Parasitol.* **49,** 206 (1980). Mechanism of action: M. D. Schulman *et al., Mol. Biochem. Parasitol.* **5,** 133 (1982). Pharmacokinetics: *eidem, J. Parasitol.* **68,** 603 (1982).

Crystals from ether, mp 194-203°. Another crystal form from water, mp 203-205°. uv max (CH_3OH): 325, 267, 227 nm (ϵ 4530, 17395, 36310). LD_{50} in mice (mg/kg): 761 i.p.; >10,000 orally (Ostlind).

THERAP CAT (VET): Anthelmintic (Trematodes).

2406. Clortermine. *2-Chloro-α,α-dimethylbenzeneethanamine; o-chloro-α,α-dimethylphenethylamine;* 1-(*o*-chlorophenyl)-2-methyl-2-propylamine; 1-(*o*-chlorophenyl)-2-methyl-2-aminopropane. $C_{10}H_{14}ClN$; mol wt 183.68. C 65.39%, H 7.68%, Cl 19.30%, N 7.62%. Prepn and use as anorexic agent: **Belg. pat. 665,244; Brit. pat. 1,066,616;** Finocchio, Huebner, **U.S. pat. 3,415,937** (1965, 1967, 1968, all to Ciba). Toxicity data: E. I. Goldenthal, *Toxicol. Appl. Pharmacol.* **18,** 185 (1971).

bp$_{16}$ 116-118°.
Hydrochloride, $C_{10}H_{15}Cl_2N$, *Su-10568, Voranil.* Crystals from ethanol, mp 245-246°. LD_{50} orally in rats: 332 ±23 mg/kg (Goldenthal).

Caution: Excessive use may lead to tolerance and physical dependence. This is a controlled substance (stimulant) listed in the U.S. Code of Federal Regulations, Title 21 Part 1308.13 (1985).

THERAP CAT: Anorexic.

2407. Closantel. *N-[5-Chloro-4-[(4-chlorophenyl)cyanomethyl]-2-methylphenyl]-2-hydroxy-3,5-diiodobenzamide;* R 31520; Flukiver; Seponver. $C_{22}H_{14}Cl_2I_2N_2O_2$; mol wt 663.07. C 39.85%, H 2.13%, Cl 10.69%, I 38.28%, N 4.22%, O 4.83%. Salicylanilide derivative. Prepn: M. A. C. Janssen, V. K. Sipido, **Belg. pat. 839,481;** *eidem,* **U.S. pat. 4,005,218** (1976, 1977 both to Janssen). Effectiveness against *Taenia pisiformis* in rabbits: R. A. F. Chevis *et al., Vet. Parasitol.* **7,** 333 (1980); against *Ancylostoma caninum:* J. Guerrero *et al., J. Parasitol.* **68,** 616 (1983); against *Fasciola hepatica* in sheep: B. E. Stromberg *et al., ibid.* **70,** 446 (1984). Prolonged effect on *Haemonchus contortus* in sheep: C. A. Hall *et al., Res. Vet. Sci.* **31,** 104 (1981). Acts by uncoupling oxidative phosphorylation: H. Van den Bossche *et al., Arch. Int. Physiol. Biochim.* **87,** 851 (1979); H. J. Kane *et al., Mol. Biochem. Parasitol.* **1,** 347 (1980).

Crystals from methanol, mp 217.8°.
THERAP CAT (VET): Anthelmintic.

2408. Clospirazine. *8-[3-(2-Chloro-10H-phenothiazin-10-yl)propyl]-1-thia-4,8-diazaspiro[4.5]decan-3-one;* 8-[3-(2-chloro-10-phenothiazinyl)propyl]-3-oxo-1-thia-4,8-diazaspiro[4.5]decane; spiclomazine. $C_{22}H_{24}ClN_3OS_2$; mol wt 446.02. C 59.24%, H 5.42%, Cl 7.95%, N 9.42%, O 3.59%, S 14.37%. Prepn: Nakanishi *et al.,* **Japan. pat. 23,091('69), U.S. pat. 3,574,204** (1969, 1971 both to Yoshitomi). Pharmacological and metabolic studies: Nakanishi *et al., J. Pharm. Soc. Japan* **90,** 800, 808 (1970); Imamura *et al., ibid.* 813. *Review: Japan. Med. Gaz.* **8**(11), 10 (1971).

mp 154-156° (acetone).
Hydrochloride, *APY-606, Diceplon, Disepron.* White to brownish, tasteless, odorless crystals or powder, mp 262-264° (dec). Sparingly sol in methanol; slightly sol in chloroform; very slightly sol in water. Decomposes slowly when exposed to light. LD_{50} in mice, rats (mg/kg): 3400, 2950 i.p.; >5490, >5490 s.c.; 3800, 4000 orally (Nakanishi, 1970).

THERAP CAT: Antipsychotic.

2409. Clostebol. *4-Chloro-17β-hydroxyandrost-4-en-3-one;* 4-chlorotestosterone. $C_{19}H_{27}ClO_2$; mol wt 322.89. C 70.68%, H 8.43%, Cl 10.98%, O 9.91%. Prepn: Camerino *et al., Farmaco Ed. Sci.* **11,** 586 (1956); *eidem, J. Am. Chem. Soc.* **78,** 3540 (1956); Ringold *et al., J. Org. Chem.* **21,** 1432 (1956); Camerino, **U.S. pat. 2,953,582** (1960 to Farmitalia); Julian, Printy, **U.S. pat. 2,933,510** (1960 to Julian Labs.). Clinical pharmacology of the acetate: Molinatti *et al., Folia Endocrinol.* **14,** 528 (1961); Krueskemper, Morgner, *Int. Z. Klin. Pharmakol. Ther. Toxikol.* **1,** 455 (1968).

Crystals from acetone + hexane, mp 188-190°. $[\alpha]_D^{20}$ +148° ($CHCl_3$). uv max (95% ethanol): 256 nm (log ϵ 4.13).

Acetate, $C_{21}H_{29}ClO_3$, *Macrobin, Steranabol, Turinabol.* Crystals from methanol, mp 228-230°. $[\alpha]_D$ +118° ±4° ($CHCl_3$). uv max: 255 nm (ϵ 13,300). Sol in ethanol.

THERAP CAT: Anabolic.

2410. Clothiapine. *2-Chloro-11-(4-methyl-1-piperazinyl)dibenzo[b,f][1,4]thiazepine;* Entumine; Etumine. $C_{18}H_{18}ClN_3S$; mol wt 343.89. C 62.87%, H 5.27%, Cl 10.31%, N 12.22%, S 9.33%. Prepn: **Fr. pat CAM 51** (1964 to Dr. A. Wander), *C.A.* **61,** 8328h (1964).

Crystals from ether + petr ether, mp 118-120°.
THERAP CAT: Antipsychotic.

2411. Clotiazepam. *5-(2-Chlorophenyl)-7-ethyl-1,3-dihydro-1-methyl-2H-thieno[2,3-e]-1,4-diazepin-2-one;* Y 6047; Clozan; Rise; Rize; Rizen; Tienor; Trecalmo; Veratran. $C_{16}H_{15}ClN_2OS$; mol wt 318.82. C 60.28%, H 4.74%, Cl 11.12%, N 8.78%, O 5.02%, S 10.06%. A thienodiazepine tranquilizer with biological activity similar to the benzodiazepines. Prepn: M. Nakanishi *et al.*, **Ger.** pat. **2,107,356**; *eidem*, **U.S.** pat. **3,849,405** (1971, 1974 both to Yoshitomi); *eidem*, *J. Med. Chem.* **16**, 214 (1973). Pharmacology: *eidem*, *Arzneimittel-Forsch.* **22**, 1905 (1972). Effects on biogenic amine metabolism: *eidem*, *ibid.* 1914. Metabolism: *eidem*, *Yakugaku Zasshi* **93**, 311 (1973), *C.A.* **79**, 13397r (1973). Clinical studies: S. Sieberns, *Fortschr. Med.* **97**, 1705 (1979).

Crystals from hexane, mp 105-106°. LD$_{50}$ in mice (mg/kg): 440 i.p., 636 orally (Nakanishi).
Note: This is a controlled substance (depressant) listed in the U.S. Code of Federal Regulations, Title 21 Part 1308.14 (1985).

THERAP CAT: Anxiolytic.

2412. Clotrimazole. *1-[(2-Chlorophenyl)diphenylmethyl]-1H-imidazole; 1-(o-chloro-α,α-diphenylbenzyl)imidazole; 1-[α-(2-chlorophenyl)benzhydryl]imidazole; 1-[(o-chlorophenyl)diphenylmethyl]imidazole; diphenyl-(2-chlorophenyl)-1-imidazolylmethane; 1-(o-chlorotrityl)imidazole*; FB 5097; BAY b 5097; Canesten; Canifug; Empecid; Gyne-Lotrimin; Lotrimin; Mono-Baycuten; Mycelex-G; Mycofug; Mycosporin; Pedisafe; Rimazole; Tibatin; Trimysten. $C_{22}H_{17}ClN_2$; mol wt 344.84. C 76.63%, H 4.97%, Cl 10.28%, N 8.12%. Prepn: Buechel *et al.*, **S. Afr.** pats. **68 05,392** and **69 00,039** (both 1969 to Bayer), *C.A.* **71**, 91473m (1969); **72**, 66939f (1970). Pharmacology: Plempel *et al.*, *Antimicrob. Ag. Chemother.* **1969**, 271; *eidem*, *Deut. Med. Wochenschr.* **94**, 1356 (1969). Clinical findings: Oberste-Lehn *et al.*, *ibid.* 1365. Series of articles on prepn, toxicology, pharmacokinetics, clinical studies: *Arzneimittel-Forsch.* **22**, 1260-1272, 1276-1299 (1972). Toxicity: D. Tettenborn, *ibid.* 1276. Comprehensive description: J. G. Hoogerheide, B. E. Wyka in *Analytical Profiles of Drug Substances* vol. **11**, K. Florey, Ed. (Academic Press, New York, 1982) pp 225-255.

Crystals, mp 147-149°. A weak base, slightly sol in water, benzene, toluene; sol in acetone, chloroform, ethyl acetate, DMF. Hydrolyzes rapidly upon heating in aq acids. LD$_{50}$ in male mice, rats (mg/kg): 923, 708 orally (Tettenborn). Hydrochloride, $C_{22}H_{18}Cl_2N_2$, mp 159°.

THERAP CAT: Antifungal.
THERAP CAT (VET): Antifungal.

2413. Clove. Caryophyllus. Dried flower-buds of *Eugenia caryophyllata* Thunb. *(Caryophyllus aromaticus* L.), *Myrtaceae. Habit.* Molucca Islands, Zanzibar, Sumatra, S. America, W. Indies. *Constit.* 15-18% eugenol, caryophyllin, tannin, gum, resin.
USE: Manuf oil of clove, eugenol; in baking; confections. Pharmaceutic aid (flavor).

THERAP CAT: Analgesic (dental).

2414. Cloxacillin. *6-[[[3-(2-Chlorophenyl)-5-methyl-4-isoxazolyl]carbonyl]amino]-3,3-dimethyl-7-oxo-4-thia-1-azabicyclo[3.2.0]heptane-2-carboxylic acid;* [3-(o-chloro-

phenyl)-5-methyl-4-isoxazolyl]penicillin; [5-methyl-3-(o-chlorophenyl)-4-isoxazolyl]penicillin; 6-[3-(o-chlorophenyl)-5-methyl-4-isoxazolecarboxamido]penicillanic acid. $C_{19}H_{18}ClN_3O_5S$; mol wt 435.88. C 52.36%, H 4.16%, Cl 8.13%, N 9.64%, O 18.35%, S 7.35%. Semi-synthetic antibiotic related to penicillin; chlorinated deriv of oxacillin, *q.v.* Prepn: Doyle *et al.*, *J. Chem. Soc.* **1963**, 5838. Manuf: *Ind. Chem.* **39**, 513 (1963), *C.A.* **60**, 1543a (1964). Properties and pharmacology: Naylor *et al.*, *Nature* **195**, 1264 (1962). Toxicity data: E. I. Goldenthal, *Toxicol. Appl. Pharmacol.* **18**, 185 (1971). Comprehensive description: D. L. Mays in *Analytical Profiles of Drug Substances* vol. **4**, K. Florey, Ed. (Academic Press, New York, 1975) pp 113-136.

Sodium monohydrate, $C_{19}H_{17}ClN_3NaO_5S.H_2O$, *sodium cloxacillin, BRL-1621, Bactopen, Cloxapen, Cloxypen, Ekvacillin, Gelstaph, Orbenin, Methocillin-S, Prostaphlin-A, Staphobristol-250, Staphybiotic, Tegopen, Tepogen.* Microcryst powder, dec 170°. $[\alpha]_D^{20}$ +163°. pH 6.0-7.5. Sol in water, methanol, ethanol, pyridine, ethylene glycol. LD$_{50}$ in rats, mice (mg/kg): 1630 ±112, 1280 ±50 i.p. (Goldenthal).
Benzathine salt, $C_{54}H_{56}Cl_2N_8O_{10}S_2$, *Boviclox, Dry-Clox, Noroclox DC, Opticlox, Orbenin Dry Cow, Triclox.*

THERAP CAT: Antibacterial.
THERAP CAT (VET): Antibacterial.

2415. Cloxazolam. *10-Chloro-11b-(2-chlorophenyl)-2,3,7,11b-tetrahydrooxazolo[3,2-d][1,4]benzodiazepin-6(5H)-one;* 7-chloro-5-(2-chlorophenyl)tetrahydrooxazolo[5,4-b]-2,3,4,5-tetrahydro-1H-1,4-benzodiazepin-2-one; 10-chloro-11b-(2-chlorophenyl)-2,3,5,6,7,11b-hexahydrobenzo[6,7]-1,4-diazepino[5,4-b]oxazol-6-one; 10-chloro-11b-(2-chlorophenyl)-6-oxo-2,3,5,6,7,11b-hexahydrooxazolo[3,2-d]-[1,4]benzodiazepine; CS-370; Enadel; Lubalix; Olcadil; Sepazon; Tolestan. $C_{17}H_{14}Cl_2N_2O_2$; mol wt 349.21. C 58.47%, H 4.04%, Cl 20.30%, N 8.02%, O 9.16%. Prepn: Tachikawa *et al.*, **Ger.** pats. **1,812,252** and **1,952,201** corresp to **U.S.** pats. **3,772,371** and **3,696,094** (1969, 1970, 1973, 1972, all to Sankyo); Miyadera *et al.*, *J. Med. Chem.* **14**, 520 (1971). Pharmacology: Kamioka *et al.*, *Arzneimittel-Forsch.* **22**, 884 (1972). Metabolism: Murata *et al.*, *Chem. Pharm. Bull.* **21**, 404 (1973). Multicenter trials and complementary studies: K. A. Fischer-Cornelssen, *Arzneimittel-Forsch.* **31**, 1757 (1981).

Crystals, mp 202-204° (dec). Freely sol in glacial acetic acid; sparingly sol in chloroform; slightly sol in acetone, dehydrated ethanol, ethyl acetate, benzene. Practically insol in water. LD$_{50}$ in mice (g/kg): 3.3 orally; >2.0 i.p. (Kamioka).
Note: This is a controlled substance (depressant) listed in the U.S. Code of Federal Regulations, Title 21 Part 1308.14 (1987).

THERAP CAT: Anxiolytic.

2416. Cloxyquin. *5-Chloro-8-quinolinol;* 5-chloro-8-hydroxyquinoline; 5-chloro-8-oxychinolin; cloxiquine. C_9H_6ClNO; mol wt 179.62. C 60.18%, H 3.37%, Cl 19.74%, N 7.80%, O 8.91%. Prepn: Bratz, Niementowski, *Ber.* **52**, 189 (1919); Weizmann, Bograchov, *J. Am. Chem. Soc.* **69**, 1222

(1947); Manske *et al.*, *Can. J. Res.* **27F**, 359 (1949). Crystal structure: T. Banerjee, N. N. Saha, *Acta Crystallog.* **C42**, 1408 (1986).

Crystals from ethanol, mp 130°. Sparingly sol in cold, dil HCl.

Hydrochloride, $C_9H_7Cl_2NO$, yellow needles. mp 256-258°.

Acetate (ester), $C_{11}H_8ClNO_2$, *5-chloro-8-acetoxyquinoline*, *Silital.*

THERAP CAT: Antibacterial; antifungal.

2417. Clozapine. *8-Chloro-11-(4-methyl-1-piperazin-yl)-5H-dibenzo[b,e][1,4]diazepine;* HF 1854; Clorazil; Clozaril; Leponex; Lepotex. $C_{18}H_{19}ClN_4$; mol wt 326.83. C 66.15%, H 5.86%, Cl 10.85%, N 17.14%. Prepn: **Fr. pat. 1,334,944** (1963 to Wander); Schmutz, Hunziker, **U.S. pat. 3,539,573** (1970); **Neth. pat. Appl. 293,201** (1965 to Wander), *C.A.* **64**, 8221a (1966); Hunziker *et al.*, *Helv. Chim. Acta* **50**, 1588 (1967). Structure-activity studies: Schmutz *et al.*, *Chim. Ther.* **2**, 424 (1967). Pharmacology: Stille *et al.*, *Farmaco, Ed. Prat.* **26**, 603 (1971). Metabolism: Gauch, Michaelis, *ibid.* 667. Toxicology: Lindt *et al.*, *ibid.* 585. Clinical studies: De Maio, *Arzneimittel-Forsch.* **22**, 919 (1972). Clinical efficacy in treatment-resistant schizophrenia: J. M. Hane *et al.*, *Psychopharmacol. Bull.* **24**, 62 (1988). Review: A. C. Sayers, H. A. Amsler, in *Pharmacological and Biochemical Properties of Drug Substances* **vol. 1**, M. E. Goldberg, Ed. (Am. Pharm. Assoc., Washington, DC, 1977) pp 1-31.

Yellow crystals from acetone-petr ether, mp 183-184°. uv max (ethanol): 215, 230, 261, 297 nm (ϵ 27400, 25800, 16800, 10500). LD_{50} in mice, rats (mg/kg): 61, 58 i.v.; 199, 260 orally (Lindt).

THERAP CAT: Antipsychotic.

2418. Clupeine. Protamine found in herring (*Clupea pallasii*) sperm. Isoln from herring testes contg ripe sperm: Kossel, *The Protamines and Histones* (London, 1928); Rasmussen, Z. *Physiol. Chem.* **224**, 97 (1934); Felix, Mager, *ibid.* **249**, 111 (1937); Block *et al.*, *Proc. Soc. Exp. Biol. Med.* **70**, 494 (1949). Separated into two main fractions, Y and Z, and fraction Y separated into Y_I and Y_{II}: Ando, Sawada, *J. Biochem. (Tokyo)* **49**, 252 (1961). Chemical structure of fraction Z: Ando *et al.*, *Biochem. Biophys. Acta* **56**, 628 (1962); Felix, Hashimoto, Z. *Physiol. Chem.* **330**, 205 (1963). Complete amino acid sequence of Z component: Iwai *et al.*, *J. Biochem. (Tokyo)* **69**, 493 (1971); of Y_{II} component: Suzuki, Ando, *ibid.* **72**, 1419 (1972); of Y_I component: *eidem, ibid.* 1433. Solid-phase synthesis of clupeine Z: Yonezawa *et al.*, *C.A.* **79**, 19093k (1973).

White powder, strongly alkaline reaction. pKa 7.4-8.0; pKb 2.9-3.3.

Usually isolated as the sulfate B-2H$_2$SO$_4$; white powder, $[\alpha]_D^{22}$ -85.49° (satd aq soln). One gram dissolves in 80 ml water at room temp. Freely sol in hot water, separates from the supersatd soln on cooling as a clear, colorless oil contg 50% H_2O, n_D^{20} 1.4435. Clupeine is split by protaminase, ac-

tive trypsin and by chymotrypsin. Compds of clupeine with nucleic acids are described by Kossel, *loc. cit.*

2419. Cnicin. *3,4-Dihydroxy-2-methylenebutanoic acid 2,3,3a,4,5,8,9,11a-octahydro-10-(hydroxymethyl)-6-methyl-3-methylene-2-oxocyclodeca[b]furan-4-yl ester; 6α,8α,15-trihydroxygermacra-1(10),4,11(13)-trien-12-oic acid 12,6-lactone 8-(3,4-dihydroxy-2-methylenebutyrate);* cynisin; centaurin. $C_{20}H_{26}O_7$; mol wt 378.41. C 63.48%, H 6.93%, O 29.60%. Bitter principle of *Cnicus benedictus* L., *Compositae.* Isolation and review: Korte, Bechmann, *Naturwiss.* **45**, 390 (1958). Structure: Suchy *et al.*, *Tetrahedron Letters* no. 10, 5 (1969); *Ber.* **93**, 2449 (1960). Revised structure: Samek *et al.*, *Tetrahedron Letters* **1969**, 2931. Stereochemistry: Tori *et al.*, *J. Chem. Soc. (B)* **1971**, 1084.

Crystals, mp 143°. $[\alpha]_D^{20}$ +158° (c = 2.3 in ethanol). uv max: 220 nm (log ϵ 4.34). Slightly soluble in water, freely in alcohol; practically insol in ether.

2420. Coal Tar. Clinitar; Psorigel; T/Gel. A by-product in the destructive distillation of coal. *Constit.* Benzene, toluene, naphthalene, anthracene, xylene, and other aromatic hydrocarbons; phenol, cresol, and other phenol bodies; ammonia, pyridine, and some other organic bases; thiophene. Monographs: *Coal Tar Data Book,* 2nd ed., 1965, compiled and published by The Coal Tar Res. Assoc., Gomersal (Leeds), England; H. G. Franck, G. Collin, *Steinkohlenteer* (Springer Verlag, Heidelberg-New York, 1968) 255 pp. Review of carcinogenicity studies: *IARC Monographs* **35**, 83-159 (1985).

Almost black, thick liq or semisolid; characteristic odor. A small portion of coal tar dissolves in water; all or almost all dissolves in benzene or nitrobenzene; partly dissolves in alcohol, ether, chloroform, carbon disulfide, methanol, acetone, petr ether, or sodium hydroxide soln. Practically insol in water. Sol in 20 parts alcohol; miscible with abs alcohol, acetone, petrolatum, oils and fats.

Note: Purified colorless coal tar, also known as *pixalbol* is a light-yellow, thin, oily liq. Occupational exposure to soots, tars and certain mineral oils is known to be carcinogenic: *Fourth Annual Report on Carcinogens* (NTP 85-002, 1985) pp 205-207.

THERAP CAT: Topical anti-eczematic.

THERAP CAT (VET): Topically in skin disorders.

2421. Cobalt. Co; at. wt 58.9332; at. no. 27; valences 1, 2, 3; rarely 4, 5. One naturally occurring isotope: ^{59}Co; artificial, radioactive isotopes: 54-58; 60-64. Widely distributed in nature; abundance in earth's crust 0.001-0.002%. Principle ores include *cobaltite* ($CoS_2 \cdot CoAs_2$), *linnaeite* (Co_3S_4), *smaltite* ($CoAs_2$) and erythrite ($3CoO \cdot As_2O_5 \cdot 8H_2O$). Metal first isolated in 1735 by Brandt. Reviews of prepn: Whittemore in *Rare Metals Handbook*, C. A. Hampel, Ed. (Reinhold, New York, 1956) pp 105-146; Houot, *Ann. Mines* **1969** (April), 9-36. Prepn of high purity metal: Ware in *Ultrapurification of Semiconductor Materials*, M. S. Brooks, J. K. Kennedy, Eds. (Macmillan, New York, 1962) pp 192-204. Cobalt appears to be essential to life. Plays an important part in animal nutrition; the absence of cobalt-contg vitamin B_{12} causes pernicious anemia. The reactor-produced ^{60}Co ($T_{1/2}$ 5.263 years; β 0.314 Mev; γ 1.173, 1.332 Mev) is a widely used source of radioactivity: Centre d'Information du Cobalt, *Cobalt Monograph* (Brussels, 1960) 515 pp. Comprehensive reviews of cobalt and its compds: *Cobalt, Its Chemistry, Metallurgy and Uses,* R. S. Young, Ed., A.C.S. Monograph Series, no. **149** (Reinhold, New York, 1960) 424 pp; Nicholls in *Comprehensive Inorganic Chemistry* **vol. 3**, J. C. Bailar, Jr. *et al*, Eds. (Pergamon Press, Oxford, 1973) pp 1053-1107; F. Planinsek, J. B. Newkirk in Kirk-Othmer *Encyclopedia of Chemical Technol-*

ogy **vol. 6** (Wiley-Interscience, New York, 3rd ed., 1979) pp 481-494.

Gray, hard, magnetic, ductile, somewhat malleable metal. Exists in two allotropic forms. At room temp the hexagonal form is more stable than the cubic form; both forms can exist at room temperature. Stable in air or toward water at ordinary temp. d 8.92. mp 1493°. bp about 3100°. Brinell hardness: 125. Latent heat of fusion 62 cal/g, latent heat of vaporization 1500 cal/g. Specific heat (15-100°): 0.1056 cal/g/°C. Readily sol in dil HNO_3; very slowly attacked by HCl or cold H_2SO_4. The hydrated salts of cobalt are red, and the sol salts form red solns which become blue on adding concd HCl.

Caution: Inhalation of the dust may cause pulmonary symptoms. Powder may cause dermatitis. Ingestion of sol salts produces nausea and vomiting by local irritation, E. Browning, *Toxicity of Industrial Metals* (Appleton-Century-Crofts, New York, 2nd ed., 1969) pp 132-142.

USE: For alloys; manuf cobalt salts; in nuclear technology. Since ^{60}Co can be encapsulated compactly, it has replaced radium in exptl medicine and cancer research. Cobalt is also used in the cobalt bomb, a hydrogen bomb surrounded by a cobalt metal shell. When the nuclear explosion occurs ^{60}Co is formed from [59]Co by neutron capture. Considered a "dirty bomb" because of long half-life and intense β^- and γ radiation. Max permissible concentration of ^{60}Co in air: $10^{-7} \mu Ci/cc$, *National Bureau of Standards Handbook* **69**, 31 (1959).

THERAP CAT: Trace mineral; ^{60}Co as radiation source (antineoplastic).

2422. Cobaltic Acetate. $C_6H_9CoO_6$; mol wt 236.07. C 30.52%, H 3.84%, Co 24.97%, O 40.66%. $Co(C_2H_3O_2)_3$. Prepd by electrolytic oxidation of $Co(C_2H_3O_2)_2.4H_2O$ in glacial acetic acid contg 2% (v/v) water: Sharp, White, *J. Chem. Soc.* **1952**, 110; by oxidation of solns of cobaltous salts by alkaline persulfates in the presence of acetic acid: Peschanski, Wormser, *Compt. Rend.* **252**, 1607 (1961). Generally considered to be a hydroxo-bridged binuclear complex.

Dark-green, very hygroscopic powder or green octahedral crystals. Becomes black and dec on heating to 100°. Sol in water, acetic acid, alcohol, *n*-butanol; dec by mineral acids. Aq solns hydrolyze slowly at room temp, rapidly at 60-70°; reduced by light or $FeSO_4$.

USE: Catalyst for cumene hydroperoxide decompn.

2423. Cobaltic-Cobaltous Oxide. Cobalto-cobaltic oxide; cobaltosic oxide; tricobalt tetroxide. Co_3O_4; mol wt 240.82. Co 73.43%, O 26.58%. Prepn: *Gmelin's, Cobalt* (8th ed.) **58** (Part A), 231-235 (1932) and supplement, 202, 491-496 (1961). *Review:* de Bie, Doyen, *Cobalt* **15**, 3-13; **16**, 3-15 (1962).

Black or grey octahedral crystals of cubic system. d 6.11. Commercial material is black and contains about 71% Co. Above 900° loses O_2 to form CoO; absorbs O_2 at lower temps but cryst structure is unchanged. Absorbs water but no definite hydrate has been identified. Readily reduced to Co metal by C, CO, or H_2. Practically insol in water; sol in acids, alkalies.

USE: In enamels; in semiconductors; in grinding wheels.

2424. Cobaltic Fluoride. Cobalt trifluoride. CoF_3; mol wt 115.93. Co 50.83%, F 49.17%. Prepn from F_2 and CoF_2, $CoCl_2$, or Co_2O_3: Priest, *Inorg. Syn.* **3**, 175 (1950); Kwasnik in *Handbook of Preparative Inorganic Chemistry* **vol. 1**, G. Brauer, Ed. (Academic Press, New York, 2nd ed., 1963) p 268; from ClF_3 and $CoCl_2$: Rochow, Kukin, *J. Am. Chem. Soc.* **74**, 1615 (1952).

Minute, light brown, hexagonal crystals, d 3.88. Discolors rapidly on exposure to moist air. Reacts with water giving off O_2. Comparatively stable to heat, at 600° the fluorine pressure over the solid is less than 0.1 atm. Volatilizes in an F_2 stream at 600-700°. May be stored in hermetically sealed glass, quartz, or metal containers.

USE: Important fluorinating agent, particularly for complete fluorination of hydrocarbons by the Fowler process.

2425. Cobaltic Oxide Monohydrate. *Cobalt hydroxide oxide*; cobaltic hydroxide. $CoHO_2$; mol wt 91.94. Co 64.10%, H 1.10%, O 34.80%. $CoO(OH)$. Alternate formula

$Co_2O_3.H_2O$. Prepn: Glemser in *Handbook of Preparative Inorganic Chemistry* **vol 2**, G. Brauer, Ed. (Academic Press, New York, 2nd ed., 1965) pp 1520-1521; Schrader, Petzold, *Z. Anorg. Allgem. Chem.* **353**, 186 (1967). Existence of anhydrous cobaltic oxide (Co_2O_3) has not been established: Pagel *et al.*, *J. Am. Chem. Soc.* **57**, 2552 (1935); Nicholls in *Comprehensive Inorganic Chemistry* **vol 3**, J. C. Bailar, Jr. *et al.*, Eds. (Pergamon Press, Oxford, 1973) p 1096. *See also* de Bie, Doyen, *Cobalt* **15**, 3 (1962). Structure: Schrader, Petzold, *loc. cit.*; Delaplane *et al.*, *J. Chem. Phys.* **50**, 1920 (1969).

Dark-brown to black powder; hexagonal crystal structure. Converted to Co_3O_4 on heating to 148-150° in a vacuum. Practically insol in water; sol in HCl, evolving Cl_2; sol in HNO_3, H_2SO_4.

USE: Oxidation catalysts, in separation of cobalt from nickel.

2426. Cobaltic Potassium Nitrite. *Hexakis(nitrito-N)cobaltate(3−) tripotassium;* potassium hexanitrocobaltate(III); potassium cobaltinitrite; potassium nitrocobaltate(III); cobalt yellow; Fischer's yellow; C.I. Pigment Yellow 40; C.I. 77357. $CoK_3N_6O_{12}$; mol wt 452.29. Co 13.03%, K 25.93%, N 18.58%, O 42.45%. $K_3Co(NO_2)_6$. Incorrectly called "*Indian yellow*". Prepd by addition of KNO_2 to a solution of a Co salt: Salyer, Sweet, *Anal. Chem.* **32**, 548 (1962).

Sesquihydrate, yellow, octahedral, cubic crystals. Very slightly sol in water, dil acetic acid; practically insol in alcohol; dec by mineral acids.

USE: As an oil- and water-color pigment; in painting on glass and porcelain; in coloring rubber; in separation of Co from Ni; in Co analysis.

2427. Cobaltous Acetate. $C_4H_6CoO_4$; mol wt 177.03. C 27.14%, H 3.42%, Co 33.29%, O 36.15%. $Co(C_2H_3O_2)_2$. Prepd commercially from cobaltous hydroxide or carbonate and an excess of dil acetic acid: Morral in Kirk-Othmer *Encyclopedia of Chemical Technology* **vol. 5** (Interscience, New York, 2nd ed., 1964) p 737; prepn from powdered Co and acetic acid: Hahl, U.S. pat. **3,133,942** (1964 to BASF); prepn by oxidation of Co carbonyls in the presence of acetic acid: Gwynn *et al.*, U.S. pat. **3,246,024** (1966 to Gulf). *Review:* de Bie, Doyen, *Cobalt* **15**, 3-13; **16**, 3-15 (1962).

Light-pink crystals. Readily sol in water.

Tetrahydrate, *bis(acetato)tetraaquocobalt*. Intense red, monoclinic, prismatic crystals. d 1.705. On heating becomes anhydrous by 140°. Sol in water, alcohols, dil acids, pentyl acetate. pH of 0.2 molar aq soln 6.8.

USE: Bleaching agent and drier for lacquers, varnishes; in anodizing; catalyst for oxidation and esterification; foam stabilizer for malt beverages.

2428. Cobaltous Arsenate. C.I. 77350. $As_2Co_3O_8$; mol wt 454.64. As 32.95%, Co 38.89%, O 28.15%. $Co_3(AsO_4)_2$. Octahydrate occurs in nature as *erythrite* or *cobalt bloom*. Prepn: *Gmelin's, Cobalt* (8th ed.) **58**, (part A), 305 (1932) and supplement, 752 (1961); Charles-Messance *et al.*, *Bull. Soc. Chim. France* **1962**, 574. *See Colour Index* **vol. 4** (3rd ed., 1971) p 4664.

Octahydrate, pink to blood-red, monoclinic, fine needles. On heating becomes anhydr by 400°. Dec by 1000° to Co_6-As_2O_{11}. d 2.9-3.1. Practically insol in water; sol in dil mineral acids, in NH_4OH.

USE: Painting on glass and porcelain.

2429. Cobaltous Bromide. Cobalt dibromide. Br_2Co; mol wt 218.77. Br 73.06%, Co 26.94%. $CoBr_2$. Prepn of hexahydrate from $CoCO_3$ and HBr: Clark, Buchner, *J. Am. Chem. Soc.* **44**, 230 (1922). Prepn of anhydr: *eidem, loc. cit.;* Watt *et al.*, *ibid.* **77**, 2752 (1955); Wydeven, Gregory, *J. Phys. Chem.* **68**, 3249 (1964).

Bright green solid or lustrous green cryst leaflets. mp 678° (under HBr and N_2); d_4^{25} 4.909. Hygroscopic, forms hexahydrate in air. Readily sol in water, methanol, ethanol, acetone, methyl acetate.

Hexahydrate, red to reddish-purple, deliquesc, prismatic crystals. mp 47-48°. d_4^{25} 2.46. Loses $4H_2O$ at 100° giving the purple dihydrate, and all H_2O by 130°. Sol in water giving red or blue soln depending on concn and temp, in methanol giving red soln, in ethanol, acetone, ether, methyl acetate giving blue solns. *Keep well closed.*

USE: Chiefly in hygrometers; also in catalysts for organic reactions.

2430. Cobaltous Carbonate. $CCoO_3$; mol wt 118.95. C 10.10%, Co 49.55%, O 40.35%. $CoCO_3$. Occurs in nature as the mineral *cobalt spar* or *sphaerocobaltite*. Prepd by heating a soln of a cobaltous salt with Na_2CO_3: Schlessinger, *Inorg. Syn.* **6**, 189 (1963) where it is the starting material for the prepn of trinitrotriamminecobalt. *Review:* de Bie, Doyen, *Cobalt* **15**, 3-13; **16**, 3-15 (1962).

Red powder or rhombohedral crystals. d 4.13. Almost insol in water, alcohol, methyl acetate. Does not react with cold concd HNO_3 or HCl; when heated, dissolves with evolution of CO_2. Oxidized by air or weak oxidizing agents to cobaltic carbonate.

Hexahydrate, pink to violet-red cryst needles. Pptd when excess CO_2 is present during prepn. On heating becomes anhyd by 140°. Stable in air.

Cobaltous carbonate basic, $C_2H_6Co_5O_{12}$, *cobalt carbonate hydroxide.* $Co_5(OH)_6(CO_3)_2$. Pale-red powder, usually containing some water. Practically insol in water; sol in dilute acids and ammonia.

USE: In ceramics; manuf of Co pigments; prepn of Co compds.

THERAP CAT (VET): Nutritional factor. Used in cobalt deficiency in ruminants.

2431. Cobaltous Chloride. Cobalt dichloride. Cl_2Co; mol wt 129.85. Cl 54.61%, Co 45.39%. $CoCl_2$. Prepn of anhydr from Co powder and Cl_2: Osthoff, West, *J. Am. Chem. Soc.* **76**, 4732 (1954); from the acetate and acetyl chloride: Watt *et al., ibid.* **77**, 2752 (1955); by dehydration of the hexahydrate with $SOCl_2$: Hecht, *Z. Anorg. Chem.* **254**, 51 (1947); prepn of the hexahydrate by treating an aqueous soln of a cobaltous salt with HCl: *Cobalt—Its Chemistry, Metallurgy, and Uses,* R. S. Young, Ed., A.C.S. Monograph Series no. **149** (Reinhold, New York, 1960) p 76. *Review:* de Bie, Doyen, *Cobalt* **15**, 3-13; **16**, 3-15 (1962).

Pale-blue hygroscopic leaflets; colorless in very thin layers; turns pink on exposure to moist air. mp 735°; bp 1049°; d_4^{25} 3.367. Dec 400° on long heating in air. Sublimes at 500° in HCl gas, forming iridescent, fluffy, colorless cryst. Sol in water, alcohols, acetone, ether, glycerol, pyridine. LD_{50} in mice, guinea pigs: 80, 55 mg/kg orally, Krasovskii, Fridlyand, *Gig. Sanit.* **36**, 95 (1971), *C.A.* **74**, 138803y (1971).

Hexahydrate, monoclinic crystals. Structure is reported to be $[CoCl_2(H_2O)_4]\cdot2H_2O$: Mizuno *et al., J. Phys. Soc. Japan* **14**, 383 (1959), *C.A.* **53**, 14630i (1959). Pink to red, slightly deliquesc, monoclinic, prismatic crystals. mp 87°; d^{20} 1.924. On heating loses $4H_2O$ at 52-56° forming the dihydrate, violet or blue crystals, d_4^{25} 2.477, stable unless exposed directly to moisture. Loses another H_2O by 100°, giving monohydrate, violet, hygroscopic, amorphous solid or needles. Remaining H_2O lost at 120-140°. Sol in water, alcohols, acetone, ether, glycerol. pH of 0.2 molar aq soln 4.6. The aq soln is pink to red, but turns blue when heated or when HCl or H_2SO_4 is added. *Keep well closed.* MLD s.c. in rabbits: 200 mg/kg, *Handbook of Toxicology* **vol. 1**, W. S. Spector, Ed. (Saunders, Philadelphia, 1956) pp 72-73.

Human Toxicity: Large amounts of $CoCl_2$ depress erythrocyte production. May lead to death in children. Other effects include cutaneous flushing, chest pains, dermatitides, tinnitus, nausea and vomiting, nerve deafness, thyroid hyperplasia, myxedema, congestive heart failure. *See* E. Beutler *et al., Clinical Disorders of Iron Metabolism* (Grune & Stratton, New York, 1963) pp 175-178.

USE: Invisible ink; humidity and water indicator; in hygrometers; temp indicator in grinding; in electroplating; for painting on glass and porcelain; prepn of catalysts; fertilizer and feed additive; foam stabilizer in beer; as absorbent for military poison gas and ammonia; in manuf of vitamin B_{12}. Radioactive cobalt chloride, $^{57}CoCl_2$ (half-life 270 days, pure gamma emitter) used in Mössbauer effect (nuclear clock).

THERAP CAT: Hematinic.

THERAP CAT (VET): Nutritional factor. Used in cobalt deficiency in ruminants.

2432. Cobaltous Chromate(III). Cobalt chromite. $CoCr_2O_4$; mol wt 226.92. Co 25.97%, Cr 45.83%, O 28.20%. Prepn: *Gmelin's, Cobalt* (8th ed.) **58**, (part A), 479 (1932) and supplement, 874-876 (1961).

Brilliant greenish-blue powder having a cubic spinel structure. Almost insol in concd HCl and HNO_3.

USE: Green pigment for ceramics.

2433. Cobaltous Cyanide. *Cobalt cyanide.* C_2CoN_2; mol wt 110.98. C 21.64%, Co 53.11%, N 25.25%. Prepn: Ray, Sahu, *J. Indian Chem. Soc.* **23**, 161 (1946); *Gmelin's, Cobalt* (8th ed.) **58**, (part A), 364 (1932) and supplement, 712 (1961). Structure reported as $Co_3[Co(CN)_6]_2$: P. S. Poskozim *et al., J. Inorg. Nucl. Chem.* **35**, 687 (1973). Prepn and structure as $Co(CN)_2$: D. M. S. Mosha, D. Nicholls, *Inorg. Chim. Acta* **38**, 127 (1980).

Deep-blue, very hygroscopic powder. d_4^{25} 1.872. Di- to trihydrate, pink to reddish-brown powder or needles. Practically insol in water, acids, methyl acetate; sol in alkali cyanide solns.

USE: In cobalt catalysts.

2434. Cobaltous Fluoride. Cobalt difluoride. CoF_2; mol wt 96.94. Co 60.80%, F 39.20%. Prepd by the action of HF on anhydr $CoCl_2$: Kwasnik in *Handbook of Preparative Inorganic Chemistry,* **vol. 1**, G. Brauer, Ed. (Academic Press, New York, 2nd ed., 1963) p 267; on $CoCO_3$: Clark, Buchner, *J. Am. Chem. Soc.* **44**, 230 (1922); on Co: Muetterties, Castle, *J. Inorg. Nucl. Chem.* **18**, 148 (1961).

Rosy-red tetragonal crystals. mp 1100-1200°, forming a red liq. Volatilizes at about 1400°. d 4.43. Sparingly sol in water; readily sol in warm mineral acids. Forms di-, tri-, and tetrahydrates, all sol in water; their aq solns are dec by boiling, forming the oxyfluoride $CoF_2\cdot CoO\cdot H_2O$.

USE: Catalyst for organic reactions.

2435. Cobaltous Formate. $C_2H_2CoO_4$; mol wt 148.98. C 16.12%, H 1.35%, Co 39.56%, O 42.96%. $Co(HCOO)_2$. Prepn: *Gmelin's, Cobalt* (8th ed.) **58** (part A), 350 (1932) and supplement, 702 (1961).

Dihydrate, red, cryst powder. d_4^{22} 2.13. Sol in water; almost insol in alcohol. Becomes anhydr at 140°.

USE: In prepn of Co catalysts.

2436. Cobaltous Hydroxide. CoH_2O_2; mol wt 92.95. Co 63.40%, H 2.17%, O 34.43%. $Co(OH)_2$. Prepd from a solution of a cobaltous salt and an alkali hydroxide: Glemser in *Handbook of Preparative Inorganic Chemistry,* **vol. 2**, G. Brauer, Ed. (Academic Press, New York, 2nd ed., 1965) p 1521; Weiser, Milligan, *J. Phys. Chem.* **36**, 722 (1932). *Review:* de Bie, Doyen, *Cobalt* **15**, 3-13; **16**, 3-15 (1962).

Blue-green or rose-red powder or microscopic rhombohedral crystals; red form is the more stable of the two. d_4^{15} 3.597. Easily oxidized by air or weak oxidizing agents to $Co(OH)_3$. Amphoteric. Loses water on heating, forming CoO at 168° *in vacuo.* Very slightly sol in water; readily sol in acids; practically insol in dil alkalies; sol in ammonia.

USE: Manuf of Co compds; drier for paints; in enhancing drying properties of lithographic printing inks; in storage battery electrode impregnating solns.

2437. Cobaltous Iodide. Cobalt diiodide. CoI_2; mol wt 312.76. Co 18.84%, I 81.16%. Prepn: Clark, Buchner, *J. Am. Chem. Soc.* **44**, 230 (1922); Chaigneau, *Bull. Soc. Chim. France* **1957**, 886; Chaigneau, Chastagnier, *ibid.* **1958**, 1192; Glemser in *Handbook of Preparative Inorganic Chemistry,* **vol. 2**, G. Brauer, Ed. (Academic Press, New York, 2nd ed., 1965) p 1518.

The anhydr salt exists in two isomorphous forms. α-CoI_2: black, graphite-like solid. mp 515-520° (in high vacuum). d_4^{25} 5.584. Very hygroscopic, becomes blackish-green in air. Sol in water to give pink to red soln. β-CoI_2: ochre-yellow powder. Blackens at 400° and converts to α-form. d_4^{25} 5.45. Very hygroscopic; deliquesc in moist air forming green droplets. Sol in water to give colorless soln which becomes pink on heating.

Hexahydrate, dark red hexagonal prisms. Loses H_2O on heating becoming anhydr by 130°. d 2.90. Loses I_2 on exposure to air and light. Sol in water to give soln which is red below 20°, olive green at 20 to 40°, and green at higher temps. Sol in ethanol (blue soln), ether (blue to green soln), chloroform (blue soln), acetone.

USE: Indicator for moisture and humidity; determination of water in organic solvents; catalyst for organic reactions.

2438. Cobaltous Nitrate. CoN_2O_6; mol wt 182.96. Co

32.22%, N 15.31%, O 52.47%. $Co(NO_3)_2$. Prepn: *Gmelin's, Cobalt* (8th ed.) **58** (part A), 252-262 (1932) and supplement, 515-521 (1961); Weigel *et al., Bull. Soc. Chim. France* **1964**, 836; Addison, Sutton, *J. Chem. Soc.* **1964**, 5553. Toxicity: *Handbook of Toxicology* vol. **1**, W. S. Spector, Ed. (Saunders, Philadelphia, 1956) pp 72-73. *Review:* de Bie, Doyen, *Cobalt* **15**, 3-13; **16**, 3-15 (1962).

Pale red powder. Dec at 100-105°. d 2.49. Sol in water. LD orally in rabbits: 250 mg/kg (Spector).

Hexahydrate, red, deliquesc, monoclinic crystals. mp ~55°. Red liq becomes green and dec to the oxide above 74°. d 1.88. Very sol in water, alcohol, most organic solvents. *Keep well closed in a cool place.*

USE: Manuf of cobalt pigments and invisible inks; decorating stoneware and porcelain; prepn of catalysts; production of vitamin B_{12} supplements.

2439. Cobaltous Oxalate. C_2CoO_4; mol wt 146.96. C 16.34%, Co 40.11%, O 43.55%. CoC_2O_4. Prepn: Robin, *Bull. Soc. Chim. France* **1953**, 1078; *Gmelin's, Cobalt* (8th ed.) **58** (part A) p 355 (1932) and supplement, pp 704-706 (1961). *Review:* de Bie, Doyen, *Cobalt* **15**, 3-13; **16**, 3-15 (1962).

d_4^{25} 3.021. Readily absorbs moisture from air to form hydrates.

Dihydrate, light pink microcryst powder or needles. Almost insol in water; slightly sol in acids; almost insol in aq oxalic acid; freely sol in aq ammonia. Dec on heating with aq KOH or Na_2CO_3 soln.

Tetrahydrate, yellowish-pink amorphous powder. Effloresces on exposure to air. Loses water on heating to 100° giving the dihydrate. Very slightly sol in water; slightly sol in acids; readily sol in aq ammonia.

USE: Prepn of Co catalysts, Co metal powder for powder-metallurgical applications; stabilizer for HCN; temperature indicator.

2440. Cobaltous Oxide. CoO; mol wt 74.94. Co 78.65%, O 21.35%. Prepn: Amiel *et al., Compt. Rend.* **259**, 3512 (1964); Wilke, *Z. Anorg. Allgem. Chem.* **330**, 164 (1964). *Review:* de Bie, Doyen, *Cobalt* **15**, 3-13; **16**, 3-15 (1962).

Powder, or cubic or hexagonal crystals. Color varies from olive green to red, depending on the particle size, but the commercial material is usually dark grey and contains about 76% Co. mp about 1935°. d 5.7 to 6.7, depending on method of prepn. Readily absorbs O_2 even at room temp. Practically insol in water, sol in acids or alkalies. Easily reduced to Co by C or CO. Reacts at high temperatures with silica, alumina, zinc oxide to form pigments. LD_{50} orally in rats: 1.70 g/kg, Smyth *et al., Am. Ind. Hyg. Assoc. J.* **30**, 470 (1969).

Note: The commercial oxides are usually not definite chemical compds but mixtures of the cobalt oxides.

USE: In pigments for ceramics; glass coloring and decolorization; oxidation catalyst for drying oils, fast-drying paints and varnishes; prepn of cobalt-metal catalysts, Co powder for binder in sintered tungsten carbide; in semiconductors.

2441. Cobaltous Phosphate. C.I. Pigment Violet 14; C.I. 77360. $Co_3O_8P_2$; mol wt 366.77. Co 48.21%, O 34.90%, P 16.89%. $Co_3(PO_4)_2$. Prepn from $CoCl_2$ and $(NH_4)_2HPO_4$: Klement, Haselbeck, *Z. Anorg. Allgem. Chem.* **334**, 27 (1964); from $Ca(H_2PO_4)_2$: Vickery, U.S. pat. **2,914,380** (1959 to Horizons). *Review:* de Bie, Doyen, *Cobalt* **15**, 3-13; **16**, 3-15 (1962). *See also Colour Index* vol. **4** (3rd ed., 1971) p 4665.

Octahydrate, pink to lavender amorph powder. d 2.77. Practically insol in water; sol in mineral acids.

USE: In ceramic pigments; in artists' colors, plastic resins.

2442. Cobaltous Potassium Sulfate. Potassium cobaltosulfate; potassium cobaltous sulfate; potassium disulfato-cobaltate(II). $CoK_2O_8S_2$; mol wt 329.26. Co 17.90%, K 23.75%, O 38.88%, S 19.47%. $K_2Co(SO_4)_2$. Prepn: *Gmelin's, Cobalt* (8th ed.) **58** (part A), 414 (1932) and supplement, 782-786 (1961).

Hexahydrate, red, monoclinic, prismatic crystals. Loses water on heating above 75°, forming the reddish-brown dihydrate which is stable from 120 to 150°; becomes anhyd (violet solid) at about 200°. d_4^{20} 2.22. Freely sol in water.

2443. Cobaltous Sulfate. CoO_4S; mol wt 155.00. Co

38.03%, O 41.29%, S 20.68%. $CoSO_4$. Hexahydrate occurs in nature as the mineral *bieberite*. Prepn: Clark *et al., J. Am. Chem. Soc.* **42**, 2483 (1920); Hammel, *Ann. Chim.* **11**, 247 (1939); *Gmelin's, Cobalt* (8th ed.) **58**, (part A) 324-336 (1932) and supplement, 628-647 (1961). *Review:* de Bie, Doyen, *Cobalt* **15**, 3-13; **16**, 3-15 (1962).

Red to lavender dimorphic, orthorhombic crystals. d_4^{25} 3.71. Stable to 708°. Dissolves slowly in boiling water.

Monohydrate, rose-colored, monoclinic crystals. Structure reported to be $Co(H_2SO_5)$. d_4^{25} 3.08. Dissolves slowly in boiling water.

Heptahydrate, structure reported to be $[Co(H_2O)_6]\cdot[H_2SO_5]$. Pink to red monoclinic, prismatic crystals. On heating dehydrates to the hexahydrate (monoclinic, prismatic crystals) at 41.5°, and to the monohydrate at 71°. d_4^{25} 2.03. Sol in water; slightly sol in methanol, ethanol.

USE: Usual source of water-soluble cobalt since it is the most economical and it shows less tendency to deliquesc or dehydrate than the chloride or nitrate. Used in storage batteries; in Co-electroplating baths; as drier for lithographic inks, varnishes; in ceramics, enamels, glazes to prevent discoloring; in Co pigments for decorating porcelain.

2444. Cobaltous Sulfide. CoS; mol wt 91.01. Co 64.76%, S 35.23%. Prepn: Glemser in *Handbook of Preparative Inorganic Chemistry*, vol. **2**, G. Brauer, Ed. (Academic Press, New York, 2nd ed., 1965) p 1523.

Exists in two forms. α-CoS: black, amorphous powder. Forms Co(OH)S in air. Sol in HCl. β-CoS: grey powder or reddish-silver octahedral crystals. mp above 1100°; d 5.45. Practically insol in water; sol in acids.

USE: Catalyst for hydrogenation or hydrodesulfurization.

2445. Cobaltous Thiocyanate. Cobaltous rhodanide; cobaltous sulfocyanate. $C_2CoN_2S_2$; mol wt 175.10. C 13.72%, Co 33.66%, N 16.00%, S 36.62%. $Co(SCN)_2$. Prepn: *Gmelin's, Cobalt* (8th ed.) **58** (part A), 380 (1932) and supplement, 720-722 (1961); Schlessinger, *Inorganic Laboratory Preparations* (Chemical Publishing, New York, 1962) p 44.

Yellow-brown powder. Soluble in water to give a rose-colored soln; sol in ethanol, methanol, ether, acetone, $CHCl_3$ to give blue solns.

Trihydrate, violet to violet-brown rhombic crystals; red in transmitted light. Sol in water to give blue soln which becomes pink on dilution; sol in ethanol, ether, acetone to give blue solns.

USE: As humidity indicator.

2446. Cobamamide. *Cobinamide, 3'-ester with 5,6-dimethyl-1-α-D-ribofuranosylbenzimidazole, Co-(5'-deoxyadenosine-5') deriv., hydroxide, dihydrogen phosphate (ester), inner salt;* cobamamidum; coenzyme B_{12}; DBC; adenosyl-B_{12}; 5'-deoxyadenosyl-B_{12}; 5'-deoxyadenosylcobalamine; dibencozide; dibenzcozamide; dimebenzcozamide; 5,6-dimethylbenzimidazolylcobamide coenzyme; 5,6-dimethylbenzimidazolylcobamide 5'-deoxyadenosine; vitamin B_{12} coenzyme; Actimide; Ademide; Anabasi; Calomide; Cobalion; Cobaltamin S; Cobanzyme; Cobazymase; Dolonevran; Enzicoba; Héraclène; Hi-Fresmin; Hycobal; Indusil; Ripresil; Sabalamin; Xobaline. $C_{72}H_{100}CoN_{18}O_{17}P$; mol wt 1579.57. The coenzyme is the metabolically active form of Vitamin B_{12}: Barker *et al., Proc. Nat. Acad. Sci. USA* **44**, 1093 (1958); Weissbach, Taylor, *Vitam. Horm. (New York)* **26**, 395 (1968). Isoln from a culture of *Propionibacterium shermanii:* Barker *et al., J. Biol. Chem.* **235**, 181, 480 (1960); *Biochem. Prepn.* **10**, 27 (1964). Prepn of aquocobalamin and 2',3'-O-isopropylidene-5'-O-p-tolylsulfonyladenosine: Hogenkamp, Pailes, *ibid.* **12**, 124 (1968); from treatment of a cobalamin-thiol complex: Murakami *et al.,* U.S. pat. **3,461,-114** (1969 to Yamanouchi). The coenzyme differs from Vitamin B_{12} by the presence of a 5'-deoxyadenosyl group in the axial ligand occupied by the cyanide in the vitamin. Structure: Lenhert, Hodgkin, *Nature* **192**, 937 (1961). The nucleoside is linked to the cobalt via the 5'-carbon atom of its deoxyribosyl moiety and the biological and chemical reactivity reside in this C to Co bond: Hogenkamp *et al., J. Biol. Chem.* **240**, 3641 (1965) and *Fed. Proc.* **25**, 1623 (1966). *Review:* Smith, *Vitamin B_{12}* (Methuen & Co., London, 3rd ed., 1965).

Yellow-orange six-faced crystals which become deep red upon exposure to air. Absorption max (H_2O): 260, 375, 522 nm (A $\times$ 10^{-6} 34.7, 10.9, 8.0 $cm^2/mole$). Soly in water (24°): 16.4 mmol. Solns of pH < 3.5 are yellow, > 3.5, red. Sol in ethanol, phenol; practically insol in acetone, ether, dichloroethylene, dioxane. pKa 3.5. Stability studies: Collado, Nieto, *Ann. Pharm. Fr.* **27**, 427 (1969). Highly sensitive to light, to cyanide and moderately sensitive to acid. Solns are most stable at pH 6-7 stored in the dark. Heating of acid or alkaline solns produces slow inactivation.

THERAP CAT: Hematopoietic vitamin.

2447. Cobrotoxin. The main toxic protein in cobra venom. Crystallization and properties: Yang, *J. Biol. Chem.* **240**, 1616 (1965). Composed of a single peptide chain having 62 amino acid residues intramolecularly cross-linked by four disulfide bonds. Amino acid sequence: Yang *et al., Biochem. Biophys. Acta* **188**, 65 (1969). Complete structure: *eidem, ibid.* **214**, 355 (1970). Synthetic study: H. Aoyagi *et al., ibid.* **263**, 823 (1972). *Review:* C. C. Yang, "Biochemical Studies on the Toxic Nature of Snake Venom Cobrotoxin from Formosan Cobra Venom" in *Toxins of Animal and Plant Origin* I, A. de Vries, E. Kochva, Eds. (Gordon and Breach Science Publishers, New York, 1971) pp 205-236. *See also* A. T. Tu, *Venoms: Chemistry and Molecular Biology* (John Wiley and Sons, New York, 1977) pp 174-189.

2448. Coca. Erythroxylon; cuca; hayo; ipado. Dried leaves of *Erythroxylon coca* Lam., *Erythroxylaceae*. *Habit.* Bolivia, Brazil, Peru, cultivated in Java. *Constit.* 0.5-1% alkaloids in the South American, 1.5-2.5% in the Javanese leaves. In the former, the major alkaloid is cocaine; in the latter, there is very little cocaine, the alkaloids consisting chiefly of ecgonine derivatives such as benzoyl ecgonine, methyl ecgonine, etc. The Java leaves also contain small quantities of tropococaine which is apparently absent in the South American leaves. Other "cocaine" alkaloids present are truxillococaine, isatropylcocaine or cocamine, cocaicine.
Note: This is a controlled substance listed in the U.S. Code of Federal Regulations, Title 21 Part 1308.12 (1987).

THERAP CAT: CNS stimulant.

2449. Cocaethylene. *[1R-(exo,exo)]-3-(Benzoyloxy)-8-methyl-8-azabicyclo[3.2.1]octane-2-carboxylic acid ethyl*

ester; ecgonine ethyl ester benzoate (ester); *O*-benzoyl-*l*-ecgonine ethyl ester; ethylbenzoylecgonine; Homocaine. $C_{18}H_{23}NO_4$; mol wt 317.37. C 68.11%, H 7.31%, N 4.41%, O 20.16%. Homolog of cocaine. Prepn: Merck, *Ber.* **18**, 2952 (1885); Einhorn, *ibid.* **21**, 47 (1888).

Prisms from alcohol, mp 109°. Almost insol in water; sol in alcohol, ether.

THERAP CAT: Local anesthetic.

2450. Cocaine. *[1R-(exo,exo)]-3-(Benzoyloxy)-8-methyl-8-azabicyclo[3.2.1]octane-2-carboxylic acid methyl ester; 3β-hydroxy-1αH,5αH-tropane-2β-carboxylic acid methyl ester benzoate;* 2β-carbomethoxy-3β-benzoxytropane; ecgonine methyl ester benzoate; *l*-cocaine; β-cocaine; benzoylmethylecgonine. $C_{17}H_{21}NO_4$; mol wt 303.35. C 67.31%, H 6.98%, N 4.62%, O 21.10%. From the leaves of *Erythroxylon coca* Lam. and other species of *Erythroxylon, Erythroxylaceae* or by synthesis. Extraction procedure: Squibb, *Pharm. J.* [3] **15**, 775, 796; **16**, 67 (1885); Emde in *Ullmann's Enzyklopädie der Technischen Chemie;* Schwyzer, *Die Fabrikation pharmazeutischer und chemisch-technischer Produkte* (Berlin, 1931). Configuration: Findlay, *J. Am. Chem. Soc.* **76**, 2855 (1954); O. Kovacs *et al., Helv. Chim. Acta* **37**, 892 (1954). Synthesis: R. Willstätter *et al., Ann.* **434**, 111 (1923). Stereospecific synthesis of *dl*-form: J. J. Tufariello *et al., Tetrahedron Letters* **1978**, 1733; *eidem, J. Am. Chem. Soc.* **101**, 2435 (1979). Biosynthesis: E. Leete, *Chem. Commun.* **1980**, 1170. Absorption spectrum: J. J. Dobbie, J. J. Fox, *J. Chem. Soc.* **103**, 1193 (1913); Fischer, *Arch. Exp. Pathol. Pharmakol.* **170**, 610 (1933). Toxicity: C. L. Rose *et al., J. Lab. Clin. Med.* **15**, 731 (1930). Vapor pressure studies: A. H. Lawrence *et al., Can. J. Chem.* **62**, 1886 (1984). Comprehensive description: F. J. Muhtadi, A. A. Al-Badr in *Analytical Profiles of Drug Substances* vol. 15, K. Florey, Ed. (Academic Press, New York, 1986) pp 151-231.

Monoclinic tablets from alcohol, mp 98°. Volatile, esp above 90°, but the sublimate is not crystalline. $bp_{0.1}$ 187-188°. $[\alpha]_D^{18}$ −35° (50% alcohol); $[\alpha]_D^{20}$ −16° (c = 4 in chloroform). Aq solns are alkaline to litmus. pKa at 15° = 5.59. One gram dissolves in 600 ml water, 270 ml water at 80°, 6.5 ml alcohol, 0.7 ml chloroform, 3.5 ml ether, 12 ml oil turpentine, 12 ml olive oil, 30-50 ml liquid petrolatum. Also sol in acetone, ethyl acetate, carbon disulfide. LD_{50} i.v. in rats: 17.5 mg/kg (Rose).

Hydrochloride, $C_{17}H_{22}ClNO_4$, *cocaine muriate.* Crystals, granules, or powder; saline, slightly bitter taste; numbs tongue and lips. mp about 195°. $[\alpha]_D$ −72° (c = 2 in aq soln pH 4.5). One gram dissolves in 0.4 ml water; 3.2 ml cold, 2 ml hot alcohol; 12.5 ml chloroform. Also sol in glycerol, acetone. Insol in ether or oils. Avoid heat in preparing soln as it decomposes. Preserve in well-closed, light-resistant containers.

Nitrate dihydrate, $C_{17}H_{22}N_2O_7 \cdot 2H_2O$, crystals, mp 58-63°. Freely sol in water or alcohol; slightly sol in ether.

Sulfate, $C_{17}H_{23}NO_8S$, white, granular, crystalline powder. Sol in water or alcohol.

Caution: May be habit forming. Cocaine and its derivatives are controlled substances listed in the U.S. Code of Federal Regulations, Title 21 Parts 329.1 and 1308.12 (1987).

THERAP CAT: Topical anesthetic.
THERAP CAT (VET): Topical anesthetic (ophthalmic).

2451. Cocarboxylase. *3-[(4-Amino-2-methyl-5-pyrimidinyl)methyl] 4-methyl-5-(4,6,6-trihydroxy-3,5-dioxa-4,6-*

diphosphahex-1-yl)thiazolium chloride, P,P'-dioxide; 3-[(4-amino-2-methyl-5-pyrimidinyl)methyl]-5-[2-[[hydroxy-(phosphonooxy)phosphinyl]oxy]ethyl]-4-methylthiazolium chloride; thiamine pyrophosphoric acid ester chloride; thiamine diphosphate ester chloride; Bioxilasi; Bivitasi; Cocalose; Cocarbina; Berolase; Biosyth. $C_{12}H_{19}ClN_4O_7P_2S$; mol wt 460.76. C 31.28%, H 4.16%, Cl 7.69%, N 12.16%, O 24.31%, P 13.44%, S 6.96%. The coenzyme or prosthetic group of the yeast enzyme carboxylase which is composed of a protein, apocarboxylase, and cocarboxylase. Cocarboxylase is the key substance in biochemical decarboxylation, it catalyzes the decarboxylation of many α-oxo acids. Enzymatic synthesis: Lohmann, Schuster, *Biochem. Z.* **294**, 183 (1937); Tauber, *Enzymologia* **2**, 171 (1937). Enzymatic synthesis stops when the apoenzyme is satd and is useless for preparative purposes. Chemical synthesis: Weijlard,Tauber, *J. Am. Chem. Soc.* **60**, 2263 (1938); Weil-Malherbe, *Biochem. J.* **34**, 980 (1940); Weijlard, *J. Am. Chem. Soc.* **63**, 1160 (1941); Karrer, Viscontini, *Helv. Chim. Acta* **29**, 711 (1946); Galamon, Filipowicz, *C.A.* **69**, 19108n (1968). Review of enzyme activity: Ullrich *et al., Vitam. Horm. (New York)* **28**, 365 (1970).

Monohydrate, crystals from alc contg some HCl, dec 240-244°. mp 238-240° from abs ethanol. uv max: 242 nm. Soluble in water. pH of 0.3% soln 2.23. The dry substance is very stable. Aq solns are somewhat less stable than solns of thiamine chloride. The free ester forms a stable tetrahydrate, $C_{12}H_{18}N_4O_7P_2S.4H_2O$, dec 220-225°. Prepn: Wenz, Göttmann, Koop, U.S. pat. **2,991,284** (1961 to E. Merck).

2452. Cocculus. Fish-berry; Indian berry; *Cocculus indicus;* oriental berry. Dried fruit of *Anamirta cocculus* (L.) Wight & Arn., *Menispermaceae. Habit.* East Indies, Malay Archipelago. *Constit.* Menispermine, paramenispermine, about 1% picrotoxin, picrotoxic acid, cocculine alkaloid, about 50% fat. *Poisonous!*
THERAP CAT: Central and respiratory stimulant.

2453. Cochineal. The dried female insect, *Coccus cacti* L., enclosing the young larvae. *Habit.* Mexico, Central America; cultivated in West Indies, Canary Islands, Algiers, and Southern Spain. About 70,000 insects to 1 lb. *Constit.* About 10% carminic acid, about 2% coccerin (a wax), about 10% fat. The coloring matter—alkali carminate—is contained only in the fatty parts of the insect and in the yolk of the eggs, to the extent of 10-14%.
USE: Coloring food products and toilet preparations; the source of carmine and carminic acid for manuf red and pink inks and lakes.

2454. Cocillana. Dried bark of *Guarea rusbyi* (Britt.) Rusby, *Meliaceae. Habit.* Bolivia. *Constit.* Rusbyine, about 2.5% resins, about 2.5% fat, tannin.
THERAP CAT: Expectorant.
THERAP CAT (VET): Has been used as an expectorant.

2455. Coclaurine. *(S)-1,2,3,4-Tetrahydro-1-[(4-hydroxyphenyl)methyl]-6-methoxy-7-isoquinolinol;* 1-(p-hydroxybenzyl)-6-methoxy-7-hydroxy-1,2,3,4-tetrahydroisoquinoline; machiline. $C_{17}H_{19}NO_3$; mol wt 285.33. C 71.56%, H 6.71%, N 4.91%, O 16.82%. Isolated as the racemate from species of *Machilus (Lauraceae)* and *Cocculus (Menispermaceae).* First isoln from *C. laurifolius* D.C. believed to be of the d-form: Kondo, Kondo, *J. Pharm. Soc. Japan* no. **524**, 876 (1925), *C.A.* **20**, 604⁷ (1926); *see also* Johns *et al., Aust. J. Chem.* **20**, 1729 (1967). Structure: Kondo, Kondo, *J. Pharm. Soc. Japan* **48**, 1156 (1928); Tomita, Kusuda, *ibid.* **72**, 280 (1952). Synthesis: Kratzl, Billek, *Monatsh.* **82**, 568 (1951); Finkelstein, *J. Am. Chem. Soc.* **73**, 550 (1951). Identity with machiline: Tomita *et al., J. Pharm. Soc. Japan* **83**, 218 (1963), *C.A.* **59**, 2874a (1963). Crystal structure and

absolute configuration: Fridrichsons, Mathieson, *Tetrahedron* **24**, 5785 (1968).

Plates, tablets from alc, mp 220-221°. Sol in hot alc, hot acetone; slightly sol in water, alc, chloroform, ether, acetone; practically insol in benzene, petr ether.
Hydrochloride, $C_{17}H_{19}NO_3.HCl$, crystals, mp 263-264°.

2456. Cocoa. A powder prepd from the roasted and cured kernels of ripe seeds of *Theobroma cacao* L. and other species of *Theobroma, Sterculiaceae.* For bibliography *see* Cacao Shell.
Brownish powder of chocolate odor and taste.
USE: In nutrient beverages; as flavoring.

2457. Coconut Oil. Copra oil. Expressed oil from kernels of *Cocos nucifera* L., *Palmae. Constit.* Trimyristin, trilaurin, tripalmitin, tristearin; also various other glycerides.
White, semisolid, lard-like fat; stable to air. Remains bland and edible for several years under ordinary storage conditions. d_0^0 0.903. mp 21-25°. n_D^{40} 1.4485-1.4495. Sapon. no. 255-258. Iodine no. 8-9.5. Acid no. not over 6. Surface tension (20°): 33.4 dyn/cm; (80°): 28.4 dyn/cm. Practically insol in water, 95% alc, more sol in abs alc; very sol in chloroform, ether, carbon disulfide. Soly data: Rao, Arnold, *J. Am. Oil Chem. Soc.* **33**, 389 (1956).
USE: Manuf soap, edible fats, chocolate, candies; in baking instead of lard; in candles and night lights; in dyeing cotton; as an ointment base; in hair dressing; in massage.

2458. Codamine. *(S)-1-[(3,4-Dimethoxyphenyl)methyl]-1,2,3,4-tetrahydro-6-methoxy-2-methyl-7-isoquinolinol; 1,2,3,4-tetrahydro-6-methoxy-2-methyl-1-veratryl-7-isoquinolinol.* $C_{20}H_{25}NO_4$; mol wt 343.41. C 69.95%, H 7.34%, N 4.08%, O 18.64%. Minor opium alkaloid. Constitutes about 0.003% of Turkish opium. Isoln: Hesse, *Ann.* **282**, 213 (1894). Structure: Späth, Epstein, *Ber.* **59B**, 2791 (1926). Synthesis: Schöpf, Thierfelder, *Ann.* **537**, 143 (1939); Billek, *Monatsh.* **87**, 106 (1956).

dl-Form, large, six-sided prisms from ether, mp 127°. Very sol in alcohol, chloroform. Somewhat sol in boiling water. In soln codamine reacts strongly basic. The salts are bitter in taste, whereas the base is said to be tasteless.

2459. Codeine. *(5α,6α)-7,8-Didehydro-4,5-epoxy-3-methoxy-17-methylmorphinan-6-ol;* methylmorphine; morphine monomethyl ether; morphine 3-methyl ether; Codicept. $C_{18}H_{21}NO_3$; mol wt 299.36. C 72.22%, H 7.07%, N 4.68%, O 16.03%. Present in opium from 0.7 to 2.5%, depending on the source, but mostly prepd by methylation of morphine, *q.v.* Discussion of structure and bibliography: Small, Lutz, "Chemistry of the Opium Alkaloids," *U.S. Public Health Reports,* Suppl. No. 103, Washington (1932). Prepn of (+)-codeine and racemate: Goto, Yamamoto, *Proc. Japan Acad.* **30**, 769 (1954), *C.A.* **50**, 1052h (1956); of (−)-form: E. J. Bijsterveld, H. J. Sinnige, *Rec. Trav. Chim.* **95**, 24 (1976); H. C. Beyerman *et al., ibid.* **97**, 127 (1978).

Manuf from morphine: W. R. Heumann, *Bulletin on Narcotics* **X**, 15 (1958). Facile synthesis from thebaine, *q.v.:* W. G. Dauben *et al., J. Org. Chem.* **44**, 1567 (1979). Toxicity of the hydrochloride: Eddy, Sumwalt, *J. Pharmacol. Exp. Ther.* **67**, 127 (1939). Comprehensive description of codeine and codeine phosphate, *q.v.:* F. J. Muhtadi, M. M. A. Hassan in *Analytical Profiles of Drug Substances* **vol. 10**, F. Florey, Ed. (Academic, New York, 1981) pp 93-138.

Monohydrate, orthorhombic sphenoidal rods or tablets (octahedra) from water or dil alcohol, mp 154-156° (after drying at 80°). Sublimes (when anhyd) at 140-145° under 1.5 mm pressure. Melts to oily drops when heated in an amount of water insufficient for complete soln, crystallizes on cooling. d_4^{20} 1.32. $[\alpha]_D^{15}$ −136° (c = 2 in alcohol), $[\alpha]_D^{15}$ −112° (c = 2 in chloroform). pK at 15°: 6.05; pH of satd aq soln 9.8. One gram dissolves in 120 ml water, 60 ml water at 80°, 2 ml alcohol, 1.2 ml hot alcohol, 13 ml benzene, 18 ml ether, 0.5 ml chloroform; freely sol in amyl alcohol, methanol, dil acids. Almost insol in petr ether or in solns of alkali hydroxides.

Acetate, $C_{20}H_{25}NO_5$. Dihydrate, crystals; acetic acid odor. Sol in water, alc. Loses acetic acid on keeping, then becomes incompletely sol in water. *Keep tightly closed.*

Hydrobromide, $C_{18}H_{22}BrNO_3$. Dihydrate, crystals. Anhyd, mp 190-192°. $[\alpha]_D^{22}$ −96.6°. One gram dissolves in 60 ml water, 110 ml alcohol. pH about 5.

Hydrochloride, $C_{18}H_{22}ClNO_3$. Dihydrate, small needles, mp at about 280° with some decompn. $[\alpha]_D^{0}$ −108°. One gram dissolves in 20 ml water, 1 ml boiling water, 180 ml alcohol. pH about 5. LD_{50} s.c. in mice: 300 mg/kg (Eddy, Sumwalt).

Salicylate, $C_{25}H_{27}NO_6$, white, crystalline powder. Slightly sol in water; freely sol in alcohol or ether.

Caution: May be habit forming. This is a controlled substance (opiate) listed in the U.S. Code of Federal Regulations, Title 21 Parts 329.1 and 1308.12 (1987).

THERAP CAT: Narcotic analgesic; antitussive.

THERAP CAT (VET): Narcotic analgesic; antitussive.

2460. Codeine Methyl Bromide. Eucodin. $C_{19}H_{24}Br$-NO_3; mol wt 394.32. C 57.87%, H 6.13%, Br 20.27%, N 3.55%, O 12.17%. $C_{18}H_{21}NO_3.CH_3Br$.

Crystals, mp about 260°. Sol in 2-3 parts water, in hot methanol, sparingly in alc. Insol in chloroform, ether.

Caution: May be habit forming. This is a controlled substance (opium derivative) listed in the U.S. Code of Federal Regulations, Title 21 Parts 329.1, 1308.11 (1987).

THERAP CAT: Narcotic analgesic; antitussive.

2461. Codeine *N*-Oxide. Genocodeine; genkodein; Codeigene. $C_{18}H_{21}NO_4$; mol wt 315.36. C 68.55%, H 6.71%, N 4.44%, O 20.29%. Prepn from codeine and 30% hydrogen peroxide: Freund, Speyer, *Ber.* **43**, 3313 (1910); Kelentey *et al., Arzneimittel-Forsch.* **7**, 594 (1957).

Platelets from water, mp 231-232°.

Monohydrate, crystals from alc, mp 215°. $[\alpha]_D^{18}$ −97.1° (c = 2.1 in water).

Hydrochloride monohydrate, $C_{18}H_{21}NO_4.HCl.H_2O$, crystals, loses crystal water at 110°, mp 219-220°. $[\alpha]_D^{20}$ −105.8° (c = 2 in water). One gram dissolves in 9.5 ml water.

Caution: May be habit forming. This is a controlled substance (opium derivative) listed in the U.S. Code of Federal Regulations, Title 21 Part 1308.11 (1987).

THERAP CAT: Antitussive.

2462. Codeine Phosphate. Galcodine. $C_{18}H_{24}NO_7P$; mol wt 397.37. C 54.41%, H 6.09%, N 3.53%, O 28.18%, P 7.79%.

Hemihydrate, (U.S.P.), fine, white, needle-shaped crystals or cryst powder. Odorless; affected by light. Solns acid to litmus. Freely sol in water; very sol in hot water; slightly sol in alcohol; more sol in boiling alcohol.

Sesquihydrate, very efflorescent, small crystals or cryst powder. One gram dissolves in 2.3 ml water, 0.5 ml water at 80°, 325 ml alcohol, 125 ml boiling alcohol, 4500 ml chloroform, 1875 ml ether. pH of a 2% aq soln: 4.6. *Keep well closed.*

Caution: May be habit forming. This is a controlled substance (opiate) listed in the U.S. Code of Federal Regulations, Title 21 Part 1308.12 (1987).

THERAP CAT: Narcotic analgesic; antitussive.

THERAP CAT (VET): Narcotic analgesic; antitussive.

2463. Codeine Sulfate. $C_{36}H_{44}N_2O_{10}S$; mol wt 696.82. C 62.05%, H 6.37%, N 4.02%, O 22.96%, S 4.60%.

Trihydrate, crystals or cryst powder. One gram dissolves in 30 ml water, 6.5 ml water at 80°, 1300 ml in chloroform or ether. pH: 5.0. Store in airtight containers; protect from light. *Incompat.* of codeine prepns in general: Alkalies, alkaloidal precipitants, ammonium chloride, bromide, valerate; salts of copper, iron or lead.

Caution: May be habit forming. This is a controlled substance (opiate) listed in the U.S. Code of Federal Regulations, Title 21 Part 1308.12 (1987).

THERAP CAT: Narcotic analgesic; antitussive.

THERAP CAT (VET): Narcotic analgesic; antitussive

2464. Cod Liver Oil. Gaduol; Tunol. The partially destearinated fixed oil expressed from fresh livers of *Gadus morrhua* L., and other species of *Gadidae*. *Constit.* Most important are vitamins A and D, each gram containing at least 850 U.S.P. units vitamin A (255 µg) and at least 85 U.S.P. units vitamin D (2.125 µg); glycerides of palmitic, stearic, etc. acids (≈19% saturated fatty acids; remainder unsaturated); cholesterol, batyl alcohol esters. Source of omega 3 fatty acid: N. Haagsma *et al., J. Am. Oil Chem. Soc.* **59**, 117 (1982). Brief description: D. Hilditch, P. Williams, *The Chemical Constitution of Natural Fats* (Wiley-Interscience, New York, 4th ed., 1964) p 43; *Bailey's Industrial Oils & Fat Products* Vol. 1, D. Swern (Wiley-Interscience, New York, 4th ed., 1979) pp 451-453.

Pale-yellow, thin liq; bland, slightly fishy taste and odor. Becomes yellow, acquires a somewhat disagreeable odor on exposure to air and light. d 0.918-0.927. n_D^{40} 1.4705-1.4745. Sapon no. 180-190. Iodine no. 145-180. Acid no. not over 1.2. Slightly sol in alcohol; freely sol in chloroform, ether, carbon disulfide, ethyl acetate, petr ether.

THERAP CAT: Vitamins A and D source.

THERAP CAT (VET): Source of vitamins A and D. Locally to promote healing.

2465. Coenzyme A. CoA. $C_{21}H_{36}N_7O_{16}P_3S$; mol wt 767.55. C 32.86%, H 4.73%, N 12.78%, O 33.35%, P 12.11%, S 4.18%. A co-factor in enzymatic acetyl transfer reactions. The molecule is built up from pantetheine [*Lactobacillus bulgaricus* factor consisting of pantothenic acid and cysteamine (thioethylamine; β-mercaptoethylamine; decarboxylated cysteine)], adenosine, and phosphoric acid. Isoln from animal sources: Lipmann *et al., J. Biol. Chem.* **167**, 869 (1947); **186**, 235 (1950). Many microorganisms contain large amounts of the coenzyme. Isoln from *Streptomyces fradiae*: Kaplan, Lipmann, *ibid.* **174**, 37 (1948). Purifications: De Vries *et al., J. Am. Chem. Soc.* **72**, 4838 (1950); Gregory *et al., ibid.* **74**, 854 (1952). A pure prepn contains 413 units/mg. Structure: Baddiley *et al., Nature* **171**, 76 (1953). Total synthesis: Moffatt, Khorana, *J. Am. Chem. Soc.* **81**, 1265 (1959); **83**, 663 (1961); Shimizu *et al., Chem. Pharm. Bull.* **13**, 1142 (1965). Review and bibliography: Lipmann, *Bacteriol. Rev.* **17**, 1-16 (1953); Baddiley, *Advan. Enzymol.* **16**, 1 (1955); Jaenicke, Lynen in *The Enzymes* vol. **3**, P. D. Boyer *et al.,* Eds. (Academic Press, New York, 2nd ed., 1960) pp 3-103.

White powder. Characteristic thiol odor. May be dried *in vacuo* over phosphorus pentoxide at 34°. uv max: 259.5 nm (ϵ 16,800). Fairly strong acid. pK 9.6 (thiol); pK 6.4 (secondary phosphate); pK 4.0 (adenine NH_3^+). Soluble in water. Practically insol in ethanol, ether, acetone. Strength of solns expressed in Lipmann units (that amount of CoA which gives half-saturation of the sulfanilamide acetylation system). One unit of CoA weighs 2.43 γ and contains 0.7 γ pantothenic acid; 1.0 mg CoA equals 413 Lipmann units. The pure dry coenzyme is best stored in evacuated ampuls at room temp. Readily oxidized by air to the catalytically inactive disulfide.

2466. Coffee, Green. Coffee beans. Dry, unroasted seeds of *Coffea arabica* L., *Rubiaceae. Habit.* Tropical Africa, especially Abyssinia; cultivated in many tropical countries, *e.g.,* Java, West Indies, Brazil, etc. *Constit.* Caffeine 1-2%, coffee oil 10-15%, sucrose and other sugars about 8%, proteins about 11%, ash about 5%, chlorogenic and caffeic acids about 6%. Other constituents include cellulose, hemicelluloses, trigonelline, tannic acid, volatile oils. Monograph: M. Sivetz, H. E. Foote, *Coffee Processing Technology,* 2 vols (Avi Publ., 1963).

2467. Coffee Oil. Coffee bean oil. A fatty oil found in coffee beans to the extent of about 15% (dry basis). Byproduct in the manufacture of instant coffees. Obtained in 2.7% yield by petr ether extraction of green Santos beans after previous dewaxing with tetrachloroethane: Schuette *et al., J. Am. Chem. Soc.* **56**, 2085 (1934). Coffee oil seems devoid of linolenic acid, but positive tests for linoleic (25.66%) and oleic (12.36%) acids were obtained.

Greenish-brown oil. Odor characteristic of green coffee beans. d_{25}^{25} 0.9653; n_D^{25} 1.4790. Iodine no. (Wijs) 100.72. Saponification no. 195.53; Reichert-Meissl no. 0.36. Saturated acids 33.60%. Unsaturated fatty acids 38.02%. Unsaponifiable matter 12.63%.

2468. Coherin. A peptide factor present in the bovine neurohypophysis which may be involved in the normal physiological regulation of intestinal motility in mammals. Appears to stimulate the coordinated or "coherent" contraction of the intestine needed to keep food moving down the tract. Mol wt ~4000. Amino acid analysis (residue/mole): lysine 0.59; aspartic acid 3.9; threonine 0.11; serine 0.22; glutamic acid 4.0; proline 5.1; glycine 4.7; alanine 0.90; half-cystine 4.9; isoleucine 3.3; leucine 3.1; tyrosine 2.9; phenylalanine 0.88. Isoln from bovine posterior pituitary and preliminary data: Goodman, Hiatt, *Science* **178**, 419 (1972). Localization of formation site: I. Galasinska-Pomykol, M. Chilimoniuk, *Endokrynol. Pol.* **24**, 429 (1973), *C.A.* **81**, 102499j (1974). Effect on electrical rhythm of dog ileum *in vivo:* R. B. Hiatt *et al., Am. J. Dig. Dis.* **22**, 108 (1977). Effects on idiopathic delayed gastric emptying in humans: E. D. Davidson *et al., Am. J. Gastroenterol.* **74**, 419 (1980).

Relatively thermostable, losing only 10% of its activity when heated in a boiling water bath at pH 3 for 1 hour. Isoelectric point: pH 6.0.

2469. Colchiceine. (*S*)-*N*-(5,6,7,9-*Tetrahydro*-10-*hydroxy*-1,2,3-*trimethoxy*-9-*oxobenzo[a]heptalen*-7-*yl*)*acetamide;* 7-acetamido-10-hydroxy-1,2,3-trimethoxy-6,7-dihydrobenzo[a]heptalen-9(5*H*)-one. $C_{21}H_{23}NO_6$; mol wt 385.40. C 65.44%, H 6.02%, N 3.63%, O 24.91%. Deriv of colchicine, *q.v.,* isolated from *Colchicum autumnale* L., *Liliaceae:* Santavy, Macák, *Coll. Czech. Chem. Commun.* **19**, 805 (1954). Structure and synthesis: Nakamura, *Chem. Pharm. Bull.* **10**,

299 (1962). Circular dichroism: J. Hrbek *et al., Coll. Czech. Chem. Commun.* **47**, 2258 (1982). Effect on antibody response induced *in vitro:* J. Sterzl *et al., Folia Microbiol.* **27**, 256 (1982).

Yellow crystals from dioxane + ether, mp 178-179°. $[\alpha]_D^{25}$ −255.1° (c = 1 in chloroform). uv max (95% ethanol): 351, 244 nm (log ϵ 4.28, 4.51). Slightly sol in water; freely sol in alcohol, chloroform; almost insol in benzene, ether.

Ethyl ether, $C_{23}H_{27}NO_6$, needles from ethyl acetate + ether, mp 135-139°. $[\alpha]_D^{25}$ −129.4° (c = 1 in chloroform). uv max (95% ethanol): 351, 243.5 nm (log ϵ 4.21, 4.45). LD_{50} in mice: 84 mg/kg i.p., *RTECS* vol. 1, R. J. Lewis, R. L. Tatken, Eds. (1980) p 534.

2470. Colchicine. *N*-(5,6,7,9-*Tetrahydro*-1,2,3,10-*tetramethoxy*-9-*oxobenzo[a]heptalen*-7-*yl*)*acetamide.* $C_{22}H_{25}NO_6$; mol wt 399.43. C 66.15%, H 6.31%, N 3.51%, O 24.03%. A major alkaloid of *Colchicum autumnale* L., *Liliaceae.* Extraction procedure: Chemnitius, *J. Prakt. Chem.* [II] **118**, 29 (1928); F. E. Hamerslag, *Technology and Chemistry of Alkaloids* (New York, 1950) pp 66-80. Structure: Dewar, *Nature* **155**, 141 (1945); King *et al., Acta Cryst.* **5**, 437 (1952); Horowitz, Ullyot, *J. Am. Chem. Soc.* **74**, 487 (1952). Crystal structure: L. Lessinger, T. N. Margulis, *Acta Crystallogr.* **B34**, 578 (1978). Total synthesis: Schreiber *et al., Helv. Chim. Acta* **44**, 540 (1961); Van Tamelen *et al., Tetrahedron* **14**, 8 (1961); Nakamura, *Chem. Pharm. Bull.* **8**, 843 (1960); Sunagawa *et al., ibid.* **9**, 81 (1961); **10**, 281 (1962); Scott *et al., Tetrahedron* **21**, 3605 (1965); Woodward, *Harvey Lectures,* Ser. **59** (Academic Press, New York, 1965) p 31; Kotani *et al., Chem. Commun.* **1974**, 300; D. A. Evans *et al., J. Am. Chem. Soc.* **103**, 5813 (1981). Biosynthesis: Leete, *Tetrahedron Letters* **1965**, 333; Battersby *et al., J. Chem. Soc.* **1964**, 4257; Hill, Unrau, *Can. J. Chem.* **43**, 709 (1965). Tubulin-binding activity: J. M. Andreu, S. N. Timasheff, *Proc. Nat. Acad. Sci. USA* **79**, 6753 (1982). Toxicity: S. J. Rosenbloom, F. C. Ferguson, *Toxicol. Appl. Pharmacol.* **13**, 50 (1968); R. P. Beliles, *ibid.* **23**, 537 (1972). Clinical evaluations in cirrhosis of the liver: M. M. Kaplan *et al., N. Engl. J. Med.* **315**, 1448 (1986); D. Kershenobich *et al., ibid.* **318**, 1709 (1988). Bibliography of early literature: Eigsti, *Lloydia* **10**, 65 (1947). Monograph: O. J. Eigsti, P. Dustin, Jr., *Colchicine in Agriculture, Medicine, Biology and Chemistry* (Iowa State College Press, Ames, Iowa, 1955). *Reviews:* Fleming, *Selected Organic Syntheses* (John Wiley, London, 1973) pp 183-207; G. Lagrue *et al., Ann. Med. Interne* **132**, 496-500 (1981); F. D. Malkinson, *Arch. Dermatol.* **118**, 453-457 (1982). Comprehensive description: D. K. Wyatt *et al.,* in *Analytical Profiles of Drug Substances* vol. 10, K. Florey, Ed. (Academic, New York, 1981) pp 139-182.

Pale yellow scales or powder, mp 142-150°. Darkens on exposure to light. Has been crystallized from ethyl acetate, pale yellow needles, mp 157°. $[\alpha]_D^{17}$ −429° (c = 1.72). $[\alpha]_D^{17}$ −121° (c = 0.9 in chloroform). pK at 20°: 12.35; pH of 0.5% soln: 5.9. uv max (95% ethanol): 350.5, 243 nm (log ϵ 4.22; 4.47). One gram dissolves in 22 ml water, 220 ml ether, 100 ml benzene; freely sol in alcohol or chloroform. Practically insol in petr ether. Forms two cryst compds with chloroform, B.CHCl$_3$ or B.2CHCl$_3$, which do not give up their chloroform unless heated between 60 and 70° for considerable time. LD$_{50}$ in rats (mg/kg): 1.6 i.v. (Rosenbloom, Ferguson); in mice (mg/kg): 4.13 i.v. (Beliles).

USE: In research in plant genetics (for doubling chromosomes).

THERAP CAT: Gout suppressant. Treatment of Familial Mediterranean Fever.

THERAP CAT (VET): Has been used as an antineoplastic.

2471. Colchicum Corm. Meadow saffron; autumn crocus; wild saffron; meadow crocus. Dried corm of *Colchicum autumnale* L., *Liliaceae*. *Habit.* Central and Southern Europe, North Africa. *Constit.* 0.3-0.5% colchicine, colchicein, colchicoresin, starch, etc. Detn of colchicine in colchicine corm: Self, Corfield, *Quart. J. Pharm. Pharmacol.* **5**, 347 (1932).

THERAP CAT: Gout suppressant.

2472. Colestipol. *N-(2-Aminoethyl)-N'-[2-[(2-aminoethyl)amino]ethyl]-1,2-ethanediamine polymer with (chloromethyl)oxirane.* A basic anion exchange resin described as a high molecular wt copolymer of diethylenetriamine and 1-chloro-2,3-epoxypropane (hydrochloride), with approx 1 out of 5 amine nitrogens protonated, for which no structural formula has been assigned and for which no specific mol wt information is available, due to the highly cross-linked and insoluble nature of the material. Prepn: Nelson, Van den Berg, **Ger.** pat. **1,927,336**; *eidem,* **U.S.** pat. **3,692,895** (1969, 1971 to Upjohn); Lednicer, Peery, **Ger.** pat. **2,053,585**; *eidem,* **U.S.** pat. **3,803,237** (1971, 1974 to Upjohn). Activity studies: Parkinson *et al., Atherosclerosis* **11**, 531 (1970); **17**, 167 (1973); Marmo *et al., G. Arterioscler.* **8**, 229 (1973); Goodman *et al., J. Clin. Invest.* **52**, 2646 (1973). Toxicology: Webster, Bollert, *Toxicol. Appl. Pharmacol.* **28**, 57 (1974). Review of pharmacology and therapeutic efficacy: R. C. Heel *et al., Drugs* **19**, 161-180 (1980).

Hydrochloride, *U-26597A, Cholestabyl, Colestid, Lestid.* LD$_{50}$ in rats (mg/kg): >4000 i.p.; >1000 orally (Webster, Bollert).

THERAP CAT: Antihyperlipoproteinemic.

2473. Colforsin. *[3R-(3α,4aβ,5β,6β,6aα,10α,10aβ,-10bα)]-5-(Acetyloxy)-3-ethenyldodecahydro-6,10,10b-trihydroxy-3,4a,7,7,10a-pentamethyl-1H-naphtho[2,1-b]pyran-1-one; 7β-acetoxy-8,13-epoxy-1α,6β,9α-trihydroxylabd-14-en-11-one;* forskolin; boforsin (obsolete); HL-362. C$_{22}$H$_{34}$O$_7$; mol wt 410.50. C 64.37%, H 8.35%, O 27.28%. Diterpene isolated from *Coleus forskohlii*, Briq. *Labiatae*, possessing vasodilating and cardiostimulatory properties. Isoln and characterization: S. V. Bhat *et al., Tetrahedron Letters* **1977**, 1669; *eidem,* **Ger.** pat. **2,557,784**; *eidem,* **U.S.** pat. **4,088,659** (1977, 1978 both to Hoechst). Synthesis by hydroxylation of 9-deoxyforskolin: N. J. Hrib, *Tetrahedron Letters* **28**, 19 (1987); *see also* F. E. Ziegler *et al., J. Am. Chem. Soc.* **109**, 8115 (1987). Stereocontrolled synthesis of the 3-ring system: S. Hashimoto *et al., Chem. Commun.* **1987**, 24. Total synthesis of (±)-forskolin: S. Hashimoto *et al., J. Am Chem. Soc.* **110**, 3670 (1988); E. J. Corey *et al., ibid.* 3672. Positive inotropic and blood-pressure lowering activity: E. Lindner *et al., Arzneimittel-Forsch.* **28**, 284 (1978). Activates adenylate cyclase: H. Metzger, E. Lindner, *ibid.* **31**, 1248 (1981). Binds to catalytic site of adenylate cyclase: T. Pfeuffer, H. Metzger, *FEBS Letters* **146**, 369 (1982). Lowers intraocular pressure: J. Capriolli, M. Sears, *Lancet* **1**, 958 (1983). In treatment of glaucoma: *eidem,* **U.S.** pat. **4,476,-140** (1984 to Yale University). *Reviews:* K. B. Seaman, J. W. Daly, *J. Cyclic Nucl. Res.* **1981**, 201; K. B. Seaman, "Forskolin and Adenylate Cyclase: New Opportunities in Drug Design" in *Ann. Rep. Med. Chem.* **19**, D. M. Bailey, Ed. (Academic Press, New York, 1984) pp 293-302.

Colorless crystals from ethyl acetate/petr ether, mp 230-232°. uv max 210, 305 nm (ϵ 1000, 50). $[\alpha]_D^{25}$ −26.19° (c = 1.68 in CHCl$_3$).

USE: In purification of adenylate cyclase.

2474. Colicins. Colicine. Antibiotic substances, or complexes of antibiotic substances, which are highly specific, are produced by certain strains of intestinal bacteria, and act upon other related strains: Fredericq, *Ann. Rev. Microbiol.* **11**, 7 (1957). Colicins produced by various strains may differ in many characteristics, the most conspicuous of which are activity spectra and the specificity of resistant mutants. They give the general reactions of proteins, are antigenic, and their activity is destroyed by proteolytic enzymes: Fredericq, *J. Theoret. Biol.* **4**, 159 (1963). Production of colicin A: Barry *et al., Nature* **198**, 211 (1963); of colicin V: Hutton, Goebel, *Proc. Nat. Acad. Sci. USA* **47**, 1498 (1961); of colicins E$_4$, N, P,V$_2$, V$_3$, V$_4$ and V$_5$: Hamon, Péron, *Ann. Inst. Pasteur* **107**, 44 (1964). Mechanism of action of colicins: Nomura, *Proc. Nat. Acad. Sci. USA* **52**, 1514 (1964). *Reviews:* idem in *Antibiotics* vol. 1, D. Gottlieb, P. D. Shaw, Eds. (Springer-Verlag, New York, 1967) pp 696-704; Wendt, *ibid.* vol. 3, J. W. Corcoran, F. E. Hahn, Eds. (1975) pp 588-605.

2475. Colistin. Colimycin; Coly-Mycin; Totazina; Polymyxin E; Colisticina. Cyclopolypeptide antibiotic produced by *Bacillus colistinus (Aerobacillus colistinus)* isolated from Japanese soil: Koyama, **Japan.** pat. **1546('52).** Composed of colistins A, B and C: Suzuki *et al., J. Biochem. (Tokyo)* **54**, 25 (1963). Structure of colistin A: *eidem, ibid.* 173, 412. Separation into polymyxins E$_1$ and E$_2$ and identity of colistin A with polymyxin E$_1$: Wilkinson, Lowe, *J. Chem. Soc.* **1964**, 4107. Synthesis of colistin A: Studer *et al., Helv. Chim. Acta* **48**, 1371 (1965). *Reviews:* Vogler, Studer, *Experientia* **22**, 345 (1966); *Medicamenta* **61** (509), 177-234 (1973).

DAB = α,γ-diaminobutyric acid

Colistin sodium methanesulfonate, C$_{58}$H$_{105}$N$_{16}$Na$_5$O$_{28}$S$_5$, *colistimethate sodium, Alficetin, Methacolimycin.* The injectable form. Sol in water and stable in the dry form. LD$_{50}$ i.v. in mice: >550 mg/kg, Barnett *et al., Brit. J. Pharmacol. Chemother.* **23**, 552 (1964).

Colistin formaldehyde-sodium bisulfite, crystals, dec 290-295°. Sol in water, slightly sol in methanol, ethanol, acetone, ether, Koyama *et al., Japan.* pat. **4898('57).**

Sulfate, *Multimycine.*

Acetylcolistin, Koyama, **Japan.** pat. **9313('60)** (to Kayaku Antibiotics Res.).

Colistin A, C$_{53}$H$_{100}$N$_{16}$O$_{13}$, *polymyxin E$_1$.* R = (+)-6-methyloctanoyl. $[\alpha]_{5461}^{22}$ −93.3° (2% acetic acid).

Polymyxin E$_2$, C$_{52}$H$_{98}$N$_{16}$O$_{13}$. R = 6-methylheptanoyl. $[\alpha]_{5461}^{22}$ −94.5° (2% acetic acid).

THERAP CAT: Antibacterial.

2476. Collagen. Ossein. Mol wt about 130,000. Polypeptide substance comprising one third of the total protein in mammalian organisms; main constituent of skin, connective tissue, and the organic substance of bones and teeth. Prepd from bones by dissolving the mineral part of the bones with phosphoric acid: Sciallano, **Fr.** pat. **688,104** (1929), *C.A.* **25**, 786 (1931). Isoln of sol collagens from corium tissue: Reizo, Sakata, *Kumamoto Med. J.* **13**, 27 (1960). Purification: Bloch, Oneson, **U.S.** pat. **2,973,302** (1961 to Ethicon). Commercial extraction process: Highberger, **U.S.** pat. **2,979,438** (1961 to United Shoe Machinery). Collagen production in the body is preceded by the formation of a much larger molecule, the biosynthetic precursor *procollagen*, which is degraded by specific enzymes to make collagen. Different types of collagens exist. They are all composed of molecules containing three polypeptide chains, α-chains, arranged in a triple helical conformation. The amino acid sequence of the α-chain is mostly a repeating structure with glycine in every third position and proline or 4-hydroxyproline frequently preceding the glycine residues. Slight differences in the primary structure (amino acid sequence) establish differences between types. Review of structural studies: Tanzer, *Science* **180**, 561-566 (1973); P. Bornstein, H. Sage, *Ann. Rev. Biochem.* **49**, 959 (1980). Collagen is differentiated from the accompanying fibrous proteins (elastin, *q.v.*, and reticulin) by (*1*) its content of proline, hydroxyproline and hydroxylysine, by (*2*) the absence of tryptophan and its low tyrosine and sulfur content, but particularly by (*3*) its high content of polar groups originating from the difunctional amino acids. The polar groups are responsible for the swelling properties leading eventually to dispersion of collagen in dil acid. Denaturation of collagen is the conversion of the rigidly coiled helix to a random coil called gelatin, *q.v.* Description of the native and denatured states of sol collagen: Boedtker, Doty, *J. Am. Chem. Soc.* **78**, 4267 (1956). Biosynthesized in fibroblastic cells. Review of biosynthesis: Bornstein, *Ann. Rev. Biochem.* **43**, 567-603 (1974). *Reviews:* Gross, *Sci. Am.* **204**, 121 (May 1961); Harrington, von Hippel, *Advan. Protein Chem.* **16**, 1-138 (1961); Grassmann *et al.*, *Fortschr. Chem. Org. Naturst.* **23**, 195-314 (1965); P. Bornstein, W.Traub, "The Chemistry and Biology of Collagen" in *The Proteins*, **vol. IV**, H. Neurath, R. L. Hill, Eds. (Academic Press, New York, 1979) pp 411-632; D. R. Eyre, *Science* **207**, 1315-1322 (1980); *Methods in Enzymology* **vol. 82**, L. W. Cunningham, D. W. Frederiksen, Eds. (Academic Press, New York, 1982) pp 3-555. Summary of recent studies of the role of collagen in disease: *Chem. & Eng. News* **60**, 32 (Jan. 25, 1982). Books: A. Veis, *The Macromolecular Chemistry of Gelatin* (Academic Press, New York, 1964) 433 pp; *Biochemistry of Collagen*, G. N. Ramachandran *et al.*, Eds. (Plenum, New York, 1976) 536 pp; P. P. Fietzek in *Collagen in Health and Disease*, M. I. V. Jayson, J. B. Weiss, Eds. (Churchill-Livingstone, New York, 1982).

USE: As fibers in sutures, in leather substitutes; as a gel in photographic emulsions, in coatings; in food casings.

2477. Collagenase. Clostridiopeptidase A; Santyl. Rare proteolytic enzyme capable of digesting native undenatured collagen, *q.v.*, found in certain *Clostridia* bacteria culture filtrates. Crude prepn from *C. histolyticum:* Mandl *et al.*, *J. Clin. Invest.* **32**, 1323 (1953). Purified prepn: Keller, Mandl, *Arch. Biochem. Biophys.* **101**, 81 (1963). Also isolated from culture media of human wound tissue, skin, bone, leukocytes, gingiva, cornea, rheumatoid synovium; from involuting rat uterus and tadpole tailfin tissue. Sepn of fractions A and B having the respective mol wts 105,000 and 57,400: Harper *et al.*, *Biochem. Biophys. Res. Commun.* **18**, 627 (1965). Amino acid composition: Mandl *et al.*, *Biochemistry* **3**, 1737 (1964); Yoshida, Noda, *Biochim. Biophys. Acta* **105**, 562 (1965). Potential use in treatment of herniated discs: Sussman, *Fr. pat.* **2,008,611** (1970 to Worthington Biochemical). Review of early studies: Mandl, *Advan. Enzymol.* **23**, 163-264 (1961). *Reviews: idem, Science* **169**, 1234-1238 (1970); Nordwig, *Advan. Enzymol. Relat. Areas Mol. Biol.* **34**, 155-205 (1971); Lazarus, *Brit. J. Dermatol.* **86**, 193-199 (1972). Book: *Collagenase*, I. Mandl, Ed. (Gordon and Breach, N.Y., 1972) 215 pp.

USE: In investigation of the structure and biosynthesis of collagen; in dispersion of cells for tissue culture studies.

THERAP CAT: Debriding agent.

2478. Collinomycin. 6-[2-(4,9-Dihydro-8-hydroxy-5,7-dimethoxy-4,9-dioxonaphtho[2,3-b]furan-2-yl)ethyl]-7,8-dihydroxy-1-oxo-1H-2-benzopyran-3-carboxylic acid methyl ester; α-rubromycin. $C_{27}H_{20}O_{12}$; mol wt 536.46. C 60.45%, H 3.76%, O 35.79%. Antibiotic substance produced by *Streptomyces collinus:* Brockmann, Renneberg, *Naturwiss.* **40**, 166 (1953); **Ger. pat. 918,162** (1954 to Bayer). Identity

with α-rubromycin: Brockmann *et al.*, *Tetrahedron Letters* **1966**, 3525. Structure: Brockmann, Zeeck, *Ber.* **103**, 1709 (1970). Shows considerable *in vitro*-activity against *Staph. aureus*.

Orange prisms from chloroform-methanol, mp 280-282°. Moderately sol in chloroform, acetone, dioxane; sparingly sol in ether, low-molecular alcohols. Practically insol in petr ether, water, aq sodium bicarbonate soln.

2479. Collinsonia. Stone-root; knob root; horse balm; richweed. Root of *Collinsonia canadensis* L., *Labiatae. Habit.* North America, from Ontario to Florida and west to Kansas. *Constit.* Resin, saponin, tannin, mucilage.

2480. Collodion. A soln of 4 g pyroxylin (chiefly nitrocellulose) in 100 ml of a mixture of 25 ml alcohol and 75 ml ether. Contains about 70% ether and 24% abs alc by vol.

Colorless, or slightly yellow, clear or slightly opalescent, syrupy liquid. Has the odor of ether. d_{25}^{25} 0.765-0.775. Exposed in thin layers, it evaporates leaving a tough, colorless film. On the addition of water the pyroxylin ppts.

Flexible collodion, mixture of simple collodion with 2% camphor and 3% castor oil (by wt). Yellow syrupy liquid; contains about 67% ether and about 22% abs alcohol by volume.

Styptic collodion, mixture of flexible collodion with 18% tannic acid by wt. Contains about 61% ether and about 21% abs alcohol by volume.

Caution: Highly flammable! Keep tightly closed in a cool place and away from flame!

USE: Collodion in photography; manuf lacquers, patent and artificial leathers, artificial pearls; process engraving; in cements.

THERAP CAT: Collodion and flexible collodion as topical protectant; styptic collodion as hemostatic.

THERAP CAT (VET): Skin protectant.

2481. Collodion, Cantharidal. Blistering collodion; vesicating collodion. One hundred parts represent 60 parts cantharides. Contains about 57% ether and 22% abs alc by vol.

Olive-green, syrupy liquid.

THERAP CAT: Rubefacient.

2482. Colocynth. Bitter apple; bitter cucumber; bitter gourd. Dried pulp of fruit of *Citrullus colocynthis* Schrad., *Cucurbitaceae. Habit.* Mediterranean region, Africa, Syria. *Constit.* Colocynthin, colocynthidin, pectin, albuminoids.

THERAP CAT: Cathartic.

THERAP CAT (VET): Has been used as a purgative.

2483. Colocynthin. 25-(Acetyloxy)-2-(β-D-glucopyranosyloxy)-16,20-dihydroxy-9-methyl-19-norlanosta-1,5,23-triene-3,11,22-trione; 2-O-β-D-glucopyranosylcucurbitacin E. $C_{38}H_{54}O_{13}$; mol wt 718.81. C 63.49%, H 7.57%, O 28.94%. Glucoside from fruit of *Citrullus colocynthis* Schrad., *Cucurbitaceae:* Walz, *Neues Jahrb. Pharm.* **9**, 16, 225 (1858); **16**, 10 (1861); Henke, *Arch. Pharm.* **221**, 200 (1883); Naylas, Chappel, *Pharm. J.* **25**, 117 (1907); Lavie *et al.*, *Acta Univ. Int. Contra Cancerum* **15**, 177 (1959); Khadem, Rahman, *Tetrahedron Letters* **1962**, 1137. Other isolns: H. Ripperger, K. Seifert, *Tetrahedron* **31**, 1561 (1975); H. Ripperger, *ibid.* **32**, 1567 (1976). Structure: Khadem, Rahman, *J. Chem. Soc.* **1963**, 4991.

Yellow crystals, mp 158-160°. $[\alpha]_D$ +50° (c = 0.4 in ethanol). uv max: 234-236 nm (log ε 4.11). Sol in ethanol, acetone, chloroform; slightly sol in ether, water.

THERAP CAT: Cathartic.

2484. Colostrokinin. A substance occurring in the colostrum from which it is liberated by kallikrein. Colostrokinin lowers blood pressure and contracts the uterus and intestine. Isoln from cow colostrum: Werle, Trautschold, *Z. Biol.* **112**, 169-180 (1960), *C.A.* **55**, 1725 (1961); Yamazaki, Moriya, *Biochem. Pharmacol.* **18**, 2303 (1969). Properties: *eidem, ibid.* 2313.

2485. Colpormon. *3,16α-Bis(acetyloxy)estra-1,3,5(10)-trien-17-one; 3,16α-dihydroxyestra-1,3,5(10)-trien-17-one diacetate;* 16α-hydroxyestrone diacetate; 3,16α-diacetoxy-Δ^1,3,5-estratrien-17-one; Colpogynon. $C_{22}H_{26}O_5$; mol wt 370.43. C 71.33%, H 7.08%, O 21.60%. Prepn: Leeds *et al., J. Am. Chem. Soc.* **76**, 2943 (1954); Fr. pat. **M1390** (1962 to Lab. Albert Rolland), *C.A.* **58**, 3484c (1963).

Crystals, mp 179-180°. $[\alpha]_D^{28}$ +122° (chloroform).
THERAP CAT: Estrogen.

2486. Coltsfoot. Coughwort. Dried leaves of *Tussilago farfara* L., *Compositae. Habit.* Northern Europe and Asia; naturalized in Northeastern U.S. *Constit.* Pectin, bitter glucoside, tannin, volatile oil, resin, saponin, caoutchouc.

2487. Colubrines. $C_{22}H_{24}N_2O_3$; mol wt 364.43. C 72.50%, H 6.64%, N 7.69%, O 13.17%. Minor alkaloids found in *Strychnos nux vomica* L.; naturally occurring in two chemically distinct forms, α and β. Isolation from mother liquor of commercial strychnine extraction: K. Warnat, *Helv. Chim. Acta* **14**, 997 (1931). Isolation and characterization from *S. nux vomica* L.: G. B. Marini-Bettolo *et al., J. Assoc. Offic. Anal. Chem.* **51**, 185 (1968). Structure of α- and β-colubrine: S. P. Findlay, *J. Am. Chem. Soc.* **73**, 3008 (1951); of β-colubrine: P. Rosenmund, *Ber.* **95**, 2639 (1962); of α-colubrine: J. M. Culver, M. Sainsbury, *J. Chem. Res. (S)* **1987**, 304. Synthesis from strychnine: P. Rosenmund, H. Franke, *Ber.* **97**, 1677 (1964). Determn by GLC: N. G. Bisset, P. Fouché, *J. Chromatog.* **37**, 172 (1968); by HPLC: G. M. Iskander *et al., J. Liq. Chromatog.* **5**, 1481 (1982). Chiroptical properties of β-colubrine: J. W. Snow, T. M. Hooker, Jr., *Can. J. Chem.* **56**, 1422c (1978).

α-colubrine: $R_1 = OCH_3$; $R_2 = H$

β-colubrine: $R_1 = H$; $R_2 = OCH_3$

α-Colubrine, 3-methoxystrychnicin-10-one, 11-methoxystrychnine. Pyramids from ethyl acetate, mp 189-193°. $[\alpha]_D^{20}$ −72.4° (c = 0.9 in alc). uv max (ethanol): 255, 297 nm (log ε 4.03, 3.77). Sol in alc, benzene, chloroform.
β-Colubrine, 2-methoxystrychnicin-10-one, 10-methoxystrychnine. Crystals from dil alcohol, mp 222°. Very bitter. $[\alpha]_D^{19}$ −107.7° (c = 2.5 in 80% alcohol). $[\alpha]_D^{21}$ −156° (c = 0.042 in chloroform). uv max (ethanol): 262, 297 nm (log ε 4.40, 3.80). Sol in alcohol, benzene, chloroform.

2488. Columbamine. *5,6-Dihydro-2-hydroxy-3,9,10-trimethoxydibenzo[a,g]quinolizinium; 7,8,13,13α-tetradehydro-2-hydroxy-3,9,10-trimethoxyberbinium.* $[C_{20}H_{20}NO_4]^+$. Found in *Jatrorrhiza palmata* (DC.) Miers (*J. columba*

Miers), *Menispermaceae:* Breslau, *Arch. Pharm.* **245**, 586 (1908); Späth, Burger, *Ber.* **59**, 1486 (1926); Reed, *Diss. Abstr.* **23**, 3638 (1963); from *Berberis lambertii* R. N. Parker, *Berberidaceae:* Chatterjee, Banerjee, *J. Indian Chem. Soc.* **30**, 705 (1953). Synthesis from berberine: Cava, Reed, *J. Org. Chem.* **32**, 1640 (1967).

The free base has not been described, but its quaternary salts are known.
Chloride hemipentahydrate, $C_{20}H_{20}ClNO_4 \cdot 2\frac{1}{2}H_2O$, orange-yellow needles, mp 194°. Sol in water, alcohol.
Chloride tetrahydrate, $C_{20}H_{20}ClNO_4 \cdot 4H_2O$, orange-yellow prisms, mp 184°.
Iodide, orange-yellow needles, mp 224°. Sol in water, alcohol.

2489. Columbin. *9-(3-Furanyl)decahydro-4-hydroxy-4a,10a-dimethyl-1,4-etheno-3H,7H-benzo[1,2-c:3,4-c']dipyran-3,7-dione; 15,16-epoxy-1β,4,12-trihydroxy-5,9-dimethyl-17,18-dinor-8βH,9βH,10α-labda-2,13(16),14-triene-19,20-dioic acid 19,1:20,12-dilactone.* $C_{20}H_{22}O_6$; mol wt 358.38. C 67.02%, H 6.19%, O 26.79%. Major bitter principle from the root of *Jatrorrhiza palmata* (DC.) Miers (*J. columba* Miers), *Menispermaceae.* Isoln: Wittstock, Poggendorff's *Ann.* **19**, 298 (1830). Structure: Barton, Elad, *J. Chem. Soc.* **1956**, 2090. Stereochemistry: Cava *et al., Tetrahedron Letters* **1959**, 1; Overton *et al., J. Chem. Soc. (C)* **1966**, 1482; Cheung *et al., ibid. (B)* **1966**, 853.

Very bitter needles from methanol, mp 195°. $[\alpha]_D$ +52.5° (pyridine). uv max (ethanol): 209 nm (log ε 3.78). Practically insol in water. One gram dissolves in 30 ml acetone, 50 ml ethyl acetate, 75 ml methanol; sol also in chloroform, methylene chloride. Bitterness threshold 1:60,000. Easily isomerized to isocolumbin.
Isocolumbin, $C_{20}H_{22}O_6$, needles from methanol, dec 190°. $[\alpha]_D^{17}$ +0.17° (alk alc). uv max (alc): 209 nm (log ε 3.80).

2490. Concanavalin A. ConA. The most extensively investigated member of the lectin family of plant proteins. Unlike most lectins, it lacks covalently bound carbohydrate and therefore is not a glycoprotein. Isolated from jack bean, *Canavalia ensiformis, Papilionatae:* J. B. Sumner, S. F. Howell, *J. Bacteriol.* **32**, 227 (1936). Its function in *C. ensiformis* is unknown, but it agglutinates a variety of somatic and germ line cells through specific interaction with saccharide-containing cell surface receptors and restores the growth pattern of virus-transformed fibroblasts in tissue culture to that of normal cells, cf. M. M. Burger, K. D. Noonan, *Nature* **228**, 512 (1970); G. M. Edelman, C. F. Millette, *Proc. Nat. Acad. Sci. USA* **68**, 2436 (1971). Differential toxicity on normal and transformed cells *in vitro* and inhibition of tumor development *in vivo* have also been reported: J. Shoham *et al., Nature* **227**, 1244 (1970). The molecule consists of identical polypeptide subunits of mol wt about 27,000, existing as dimers in soln at pH <6 and as tetramers at physiologic pH. The proposed amino acid sequence contains 238 residues; con A has also been shown to have binding site for transition metal ions and calcium ions in addition to saccharide binding sites: G. M. Edelman *et al., Proc. Nat. Acad. Sci. USA* **69**, 2580 (1972). The transition ion, usually Mn^{+2} or Ca^{+2}, apparently stabilizes the formation of the specific saccharide binding site: M. Shoham *et al., Biochemistry* **12**, 1914 (1973). Circular dichroism-NMR study of metal binding sites: A. R. Palmer *et al., ibid.* **19**, 5063 (1980). Oligosaccharide binding study: A. Vanlands *et al., Eur. J. Biochem.* **103**, 307 (1980). Use of con A to study immunoregulation of human T cells: D. M. Dwyer, C. Johnson, *Clin. Exp. Immunol.* **46**, 237 (1981); E. L. Larson *et al., Immunobiology* **161**, 5 (1982). For general refs, *see* Lectins.

USE: As a reagent in analytical and preparative biochemistry; as a probe in studies of cell surface membrane dynamics and cell division.

2491. Condurangin. Kondurangin. Bitter principle from the bark of the Condurango vine, *Marsdenia condurango* Reichb. f., *Asclepiadaceae,* also known as *eagle vine, mataperro, condor vine. Habit.* Ecuador, Peru. Isoln: Korte, Korte, *Z. Naturforsch.* **10b**, 223 (1955); Zechner, Zölss, *Sci. Pharm.* **24**, 107 (1956). Separation of fractions: Cellarius, Zechner, *ibid.* **34**, 10 (1966). Proposed structure of the aglycone, *condurangogenin A:* Tschesche *et al., Tetrahedron* **21**, 1777, 1797 (1965); **23**, 1461 (1967).

Crystals from dil methanol, mp 186-188°. uv max (methanol): 277 nm. Soluble in chloroform, methanol; slightly sol in water. Practically insol in ether, petr ether. Bitterness threshold 1:20,000.

THERAP CAT: Astringent.

2492. Conessine. 3β-*(Dimethylamino)con-5-enine; nerine; roquessine; wrightine.* $C_{24}H_{40}N_2$; mol wt 356.58. C 80.84%, H 11.31%, N 7.86%. Antiamebic principle in Kurchi bark, the bark of *Holarrhena anti-dysenterica* Wall., *Apocynaceae,* native to India; also found in African species, such as *H. africana* A.DC.; *H. congolensis* Stapf; *H. wulfsbergii* Stapf, and *H. febrifuga* Klotzsch, *Apocynaceae.* Isoln: Haines, *Trans. Med. Soc. Bombay* **4**, 28 (1858); Warnecke, *Ber.* **19**, 60 (1886); Bertho, *Arch. Pharm.* **277**, 237 (1939); Siddiqui, Pillay, *J. Indian Chem. Soc.* **9**, 553 (1932). Structure: Haworth *et al., J. Chem. Soc.* **1953**, 1102; Haworth, McKenna, *Chem. & Ind. (London)* **1957**, 1510; Bertho, Götz, *Ann.* **619**, 96 (1958). Discussion of structure in Fieser, Fieser, *Steroids* (New York, 1959), p 858. Synthesis from Δ^5-pregnene-3β,20β-diol: Barton, Morgan, *Proc. Chem. Soc.* **1961**, 206; *J. Chem. Soc.* **1962**, 622. Total synthesis: Marshall, Johnson, *J. Am. Chem. Soc.* **84**, 1485 (1962); Stork *et al., ibid.* 2018; Nagata *et al., Tetrahedron Letters* **1963**, 869.

Leaflets or broad plates from acetone, mp 127-128.5°. $[\alpha]_D^{20}$ +25.3° (c = 0.7 in abs alc). Sparingly sol in water. Dihydrochloride monohydrate, $C_{24}H_{42}Cl_2N_2 \cdot H_2O$, crystals, sol in water, mp >340°. $[\alpha]_D^{20}$ +9.3° (c = 0.9 in H_2O).
Dihydrobromide, $C_{24}H_{42}Br_2N_2$, crystals, dec 340°. Very bitter taste. $[\alpha]_D^{20}$ +7.0° (c = 5 in H_2O). Sol in water. Slightly sol in 95% ethanol. Very slightly sol in ether. Practically insol in petr ether.

2493. Congo Red. *3,3′-[[1,1′Biphenyl]-4,4′-diylbis-(azo)]bis[4-amino-1-naphthalenesulfonic acid] disodium salt; C.I. Direct Red 28;* sodium diphenyldiazo-bis-α-naphthylaminesulfonate; *C.I.* 22120. $C_{32}H_{22}N_6Na_2O_6S_2$; mol wt 696.67. C 55.17%, H 3.18%, N 12.06%, Na 6.60%, O 13.78%, S 9.21%. Prepn: Böttiger, **DRP 28753** (1884); *Frdl.*

1, 470; Whitehead, *Chem. Trade J.* **77**, 386 (1925); Kline, *J. Chem. Ed.* **15**, 129 (1938). Purification: Bedaux *et al., Pharm. Weekbl.* **98**, 189 (1963). *See also Colour Index* vol. 4 (3rd ed., 1971) p 4166; *H. J. Conn's Biol. Stains,* R. D. Lillie, Ed. (Williams & Wilkins., Baltimore, 9th ed., 1977) p 148. Toxicity: Richardson, Dillon, *Am. J. Med. Sci.* **198**, 73 (1939). Clinical evaluation as diagnostic aid: E. Ouchi *et al., Tohoku J. Exp. Med.* **118**, Suppl., 191 (1976). Use in combination with Sirius red as histological stain: I. R. Hinds, *Lab. Med.* **16**, 366 (1985).

Brownish-red powder. Absorption max (pH 7.3): 488 nm ($E_{1cm}^{1\%}$, 595). Sol in water (yellowish-red) and in ethanol (orange); very slightly sol in acetone. Practically insol in ether. LD_{50} i.v. in rats: 190 mg/kg (Richardson, Dillon).
Congo Red Paper, *Riegel's paper, Herzberg's paper.* Paper charged with a 0.1% aq soln of Congo Red.
USE: As indicator, usually in 0.1% aq soln for estimating free mineral acids, particularly in presence of organic acids. pH: 3.0 blue-violet, 5.0 red. Detecting and estimating free HCl in gastric contents; detecting acidity of papers. As addition to culture media; reagent for bitter-almond water; dye; biological stain.

THERAP CAT: Diagnostic aid (amyloidosis).

2494. Congressane. *Decahydro-3,5,1,7-[1,2,3,4]butane-tetraylnaphthalene;* pentacyclo[7.3.1.1^{4,12}.0^{2,7}.0^{6,11}]tetradecane; diamantane. $C_{14}H_{20}$; mol wt 188.30. C 89.29%, H 10.71%. Isoln from petroleum: Hala *et al., Angew. Chem. Int. Ed.* **5**, 1045 (1966). Synthesis: Cupas *et al., J. Am. Chem. Soc.* **87**, 917 (1965); Gund *et al., J. Org. Chem.* **39**, 2979 (1974). Structure: Karle, Karle, *J. Am. Chem. Soc.* **87**, 918 (1965).

Crystals, mp 236-237°.
Note: Structure is congress emblem of the International Congress of Pure and Applied Chemistry.

2495. Conhydrine. *α-Ethyl-2-piperidinemethanol; 2-(α-hydroxypropyl)piperidine.* $C_8H_{17}NO$; mol wt 143.22. C 67.08%, H 11.96%, N 9.78%, O 11.17%. From seeds of *Conium maculatum* L., *Umbelliferae.* Isoln: Späth, Adler, *Monatsh.* **63**, 127 (1933). Stereochemistry: Hill, *J. Am. Chem. Soc.* **80**, 1609 (1958); Sicher, Ticky, *Coll. Czech. Chem. Commun.* **23**, 2081 (1958).

Crystals from ether. mp 121°. bp 226°. $[\alpha]_D$ +10°. Slightly sol in water; sol in alc, chloroform, ether.

2496. β-Coniceine. *2-(1-Propenyl)piperidine; α-allyl-piperidine.* $C_8H_{15}N$; mol wt 125.21. C 76.74%, H 12.08%, N 11.19%. Prepn from conhydrine: Löffler, Friederick, *Ber.* **42**, 107 (1909); Löffler, Tschunke, *ibid.* 929.

dl-Form, liquid, solidifies around 8°. bp 168°. d_4^{15} 0.8716.

dl-Form hydrochloride, $C_8H_{15}N.HCl$, mp 206°.
d-Form, needles, mp 39°, bp 168°. $[\alpha]_D^{45}$ +50°.
d-Form hydrochloride, mp 182°.
l-Form, crystals, mp 40°, bp 168°. $[\alpha]_D^{45}$ −50°.
l-Form hydrochloride, mp 183°.
All 3 hydrochlorides are sol in water. The free bases are sol in alcohol and ether; very slightly in water.

2497. γ-Coniceine. *2,3,4,5-Tetrahydro-6-propylpyridine;* 2-*n*-propyl-3,4,5,6-tetrahydropyridine; 2*n*-propyl-Δ^1-piperidine. $C_8H_{15}N$; mol wt 125.21. C 76.74%, H 12.08%, N 11.19%. Easily reduced to *dl*-coniine *(q.v.)*. From seeds of *Conium maculatum* L., *Umbelliferae:* Cromwell, *Biochem. J.* **64**, 259 (1956); Fairbairn, Challen, *ibid.* **72**, 556 (1959); Fairbairn, Sewal, *Phytochemistry* **1**, 38 (1961). Structure: Beyerman *et al., Rec. Trav. Chim.* **80**, 513 (1961). Synthesis: Lukes *et al., Coll. Czech. Chem. Commun.* **12**, 641 (1947); Cervinka, *Chem. Listy* **52**, 1145 (1958).

Alkaline liquid, mousy odor. bp 171°; bp_{15} 63°. Volatile with steam. d_4^{15} 0.8753. n_D^{16} 1.4661. Slightly sol in water; freely sol in alcohol, chloroform, ether.
Hydrochloride, $C_8H_{15}N.HCl$, hygroscopic crystals from ether, mp 143°.
Picrate, $C_{14}H_{18}N_4O_7$, crystals from abs ethanol, mp 75°.

2498. Coniferin. *4-(3-Hydroxy-1-propenyl)-2-methoxy-phenyl* β-D-*glucopyranoside;* 4-hydroxy-3-methoxy-1-(γ-hydroxypropenyl)benzene-4-D-glucoside; abietin; laricin. $C_{16}H_{22}O_8$; mol wt 342.35. C 56.13%, H 6.48%, O 37.39%. Principal glucoside of the conifers. Also in comfrey root, sugar beet, and asparagus. Extraction from the cambium layer of fir: Solntsev, *C.A.* **38**, 3780 (1944). Synthesis: Pauley, Feuerstein, *Ber.* **60**, 1031 (1927). Hydrolysis by emulsin yields coniferyl alcohol and D-glucose. Yields lignin-like material by enzymatic dehydrogenation and polymerization: Freudenberg *et al., Ber.* **85**, 641 (1952); Freudenberg, Rasenack, *Ber.* **86**, 756 (1953).

Dihydrate, crystals from water. Anhydr after 4 hrs at 100°. Anhydr coniferin, mp 186°. $[\alpha]_D^{20}$ −68° (c = 0.5). Absorption spectrum: Patterson, Hibbert, *J. Am. Chem. Soc.* **65**, 1862 (1943). One gram dissolves in 200 ml water, freely sol in boiling water; sol in pyridine; sparingly sol in alc; practically insol in ether.

2499. Coniferyl Alcohol. *4-(3-Hydroxy-1-propenyl)-2-methoxyphenol;* 3-(4-hydroxy-3-methoxyphenyl)-2-propen-1-ol; 4-hydroxy-3-methoxycinnamic alcohol; γ-hydroxyisoeugenol. $C_{10}H_{12}O_3$; mol wt 180.20. C 66.65%, H 6.71%, O 26.64%. In benzoin, especially Siam benzoin, as coniferyl benzoate (up to 78%): Reinitzer, *Arch. Pharm.* **264**, 131 (1926). From coniferin by hydrolysis with emulsin. From ferula aldehyde (coniferyl aldehyde) by the action of fermenting yeast. Small amounts of coniferyl alcohol are found in brandy from beet molasses.

Prisms, mp 74°. Freely sol in ether; moderately sol in alc; almost insol in water; sol in alkalies. Polymerized by dil acids; converted to an amorphous gum. Absorption spectrum: Herzog, Hillmer, *Ber.* **64B**, 1288 (1931).

2500. Coniine. *2-Propylpiperidine;* cicutine; conicine. $C_8H_{17}N$; mol wt 127.22. C 75.52%, H 13.47%, N 11.01%. Toxic principle of poison hemlock, *Conium maculatum* L., *Umbelliferae:* Ladenburg, *Ber.* **19**, 439 (1886); Koller, *Monatsh.* **47**, 393 (1926). Occurs naturally as the (*S*)-(+)-isomer. Review of early literature: Marion in *The Alkaloids,* **vol. I**, R. H. F. Manske, H. L. Holmes, Eds. (Academic Press, New York, 1950) pp 211-217. Isoln from the pitcher plant, *Sarracenia flava* and insect paralyzing properties: N. V. Mody *et al. Experientia* **32**, 829 (1976). Absolute configuration: J. C. Craig, S. K. Roy, *Tetrahedron* **21**, 401 (1965). Resolution of (±)-form: J. C. Craig, A. R. Pinder, *J. Org. Chem.* **36**, 3648 (1971). Synthesis of (*S*)-(+)-form: K. Aketa *et al., Chem. Pharm. Bull.* **24**, 621 (1976); of (*R*)-(−)-form: D. Lathbury, T. Gallagher, *Chem. Commun.* **1986**, 114; of (±)-form: T. Nagasaka *et al., Heterocycles* **27**, 1685 (1988). Enantiospecific synthesis of (+)- and (−)-forms: L. Guerrier *et al., J. Am. Chem. Soc.* **105**, 7754 (1983).

Colorless alkaline liquid, darkens and polymerizes on exposure to light and air. Mousy odor. mp ~ −2°. bp 166°-166.5°; bp_{20} 65-66°. Volatile with steam. d_4^{20} 0.844-0.848. n_D^{23} 1.4505. $[\alpha]_D^{25}$ +8.4° (c = 4.0 in $CHCl_3$); $[\alpha]_D^{23}$ +14.6° (neat). pK 3.1. One ml dissolves in 90 ml water, less sol in hot water. The base dissolves about 25% water at room temp. Sol in alcohol, ether, acetone, benzene, amyl alcohol; slightly sol in chloroform.
(*R*)-(−)-Form, liquid, bp_{756} 165°. $[\alpha]_D^{25}$ −8.1° (c = 4.0 in $CHCl_3$); $[\alpha]_D^{23}$ −14.2° (neat).
(±)-Form, bp 200-210°.
Caution: Ingestion causes weakness, drowsiness, nausea, vomiting, labored respiration, paralysis, asphyxia, death.

2501. Coniine Hydrobromide. $C_8H_{18}BrN$; mol wt 208.15. C 46.16%, H 8.72%, Br 38.39%, N 6.73%.
Prisms, mp 211°. One gram dissolves in 2 ml water, 3 ml alcohol; sol in chloroform, ether. *Very poisonous! Incompat.* Albumin; salts of Al, Cu, Fe, Mn, and Zn.
THERAP CAT: Antispasmodic.

2502. Coniine Hydrochloride. $C_8H_{18}ClN$; mol wt 163.69. C 58.70%, H 11.08%, Cl 21.66%, N 8.56%.
Rhomboids, mp 221°. Freely sol in water, alcohol, chloroform. *Very poisonous!*
THERAP CAT: Antispasmodic.

2503. Conium Fruit. Hemlock; poison hemlock; spotted hemlock; poison parsley; spotted cowbane. Full-grown, but unripe, carefully dried fruit of *Conium maculatum* L., *Umbelliferae. Habit.* Europe, Asia, naturalized in U.S. *Constit.* 0.5-1.5% coniine, conhydrine, pseudoconhydrine, methylconiine, ethylpiperidine, coniic acid, volatile and fixed oil. Has antispasmodic activity.

2504. Conjugated Estrogenic Hormones. Amnestrogen; Conestron; Estrifol; Genisis; Premarin. An amorphous preparation contg water-soluble, conjugated forms of mixed estrogens obtained from urine of pregnant mares. The principal estrogen present is sodium estrone sulfate. The total estrogenic potency of the preparation is expressed in terms of an equiv quantity of sodium estrone sulfate. Prepn: Bates, Cohen, U.S. pat. 2,565,115 (1951 to Squibb); Stiller, O'Keefe, U.S. pat. 2,720,483 (1955 to Olin Mathieson).

Note: This substance has been listed as a known carcinogen: *Fourth Annual Report on Carcinogens* (NTP 85-002, 1985) p 61.

THERAP CAT: Estrogen.

2505. Conquinamine. $C_{19}H_{24}N_2O_2$; mol wt 312.40. C 73.04%, H 7.74%, N 8.97%, O 10.24%. Stereoisomer of quinamine (*q.v.*) from which it differs in configuration at one or more of the asymmetric centers. From bark of *Cinchona pubescens* Vahl. (*C. succirubra* Pav.), and *C. rosulenta* How., *Rubiaceae.* Isoln: Hesse, *Ber.* **10**, 2158 (1877); *Ann.* **209**, 62 (1881); Oudemans, *ibid.* **38**. Identity with epiquinamine: Culvenor *et al., J. Chem. Soc.* **1950**, 1485.

Prisms. mp 123°. $[\alpha]_D^{15}$ +200° (alcohol). Slightly sol in water; freely sol in alcohol, chloroform, ether.

2506. Convallamarogenin. $C_{27}H_{42}O_4$; mol wt 430.61. C 75.31%, H 9.83%, O 14.86%. Main genin of the glycoside, *convallamarin.* From root of *Convallaria majalis* L., *Liliaceae.* Isoln and structure: Tschesche *et al., Ber.* **94**, 1699 (1961). From *Reineckia carnea* Kunth., *Liliaceae:* Takeda *et al., Tetrahedron* **19**, 759 (1963).

Prisms from chloroform + methanol, mp 259-261°. $[\alpha]_D^{27}$ −79.1° (c = 1.02 in $CHCl_3$).
Diacetate, $C_{31}H_{46}O_6$, prisms, mp 214-215°. $[\alpha]_D^{27}$ −88.2° (c = 1.10 in $CHCl_3$).

2507. Convallaria. Lily of the valley; May lily; Park lily; May blossom. Dried flowers of *Convallaria majalis* L., *Liliaceae. Habit.* U.S., Europe, Northern Asia; cultivated in gardens. *Constit.* of dried flowers: Convallatoxin, convallarin, volatile oil; of dried rhizome and roots: Glycosides of convallamarin, convallatoxin, convallarin. Extraction and identification of glycosides: Tschesche, Seehofer, *Ber.* **87**, 1108 (1954); Erbring, Patt, *Arzneimittel-Forsch.* **8**, 554 (1958).

Convallaria glycosides, Convacard, Convallan, Convallen, Convalyt, Convasid. An aqueous extract containing the total glycosides of Convallaria.

THERAP CAT: Cardiotonic.

THERAP CAT (VET): Has been used as a cardiotonic and diuretic.

2508. Convallatoxin. *3β-[(6-Deoxy-α-L-mannopyranosyl)oxy]-5,14-dihydroxy-19-oxo-5β-card-20(22)-enolide;* strophanthidin α-L-rhamnoside; Convallaton; Corglykon; Korglykon. $C_{29}H_{42}O_{10}$; mol wt 550.63. C 63.25%, H 7.69%, O 29.06%. From blossoms of lily of the valley (*Convallaria majalis* L., *Liliaceae*): Karrer, *Helv. Chim. Acta* **12**, 506 (1929); from *Ornithogalum umbellatum* L., *Liliaceae:* Mrozik *et al., ibid.* **42**, 683 (1959); from *Antiaris toxicaria* Lesch, *Moraceae:* Juslen *et al., ibid.* **46**, 117 (1963). Structure: Tschesche, Haupt, *Ber.* **69**, 459 (1936); Fieser, Jacobsen, *J. Am. Chem. Soc.* **59**, 2335 (1937). With the aid of the method of Mannich and Siewert, *Ber.* **75**, 737 (1942), using HCl in cold acetone, the cleavage into L-rhamnose and strophanthidin has been accomplished with good yields: Reichstein, Katz, *Pharm. Acta Helv.* **18**, 521 (1943); *see also C.A.* **38**, 2046. Synthesis from strophanthidin and acetobromrhamnose: Reyle *et al., Helv. Chim. Acta* **33**, 1541 (1950); Haede *et al., Ger.* pat. **1,933,090** (1971 to Hoechst).

Prisms from methanol + ether. mp 235-242°. $[\alpha]_D^{22}$ −1.7° ± 3° (c = 0.65 in methanol); $[\alpha]_D^{25}$ −9.4° ± 3° (c = 0.72 in dioxane). Sol in alcohol, acetone; slightly sol in chloroform, ethyl acetate, and water (1:2000); practically insol in ether, petr ether. Gives Legal's reaction. MLD i.v. in frogs: 0.3 mg/kg.

Tri-*O*-acetyl-convallatoxin, $C_{35}H_{48}O_{13}$, needles from acetone + ether, mp 215-238°. $[\alpha]_D^{25}$ −5.5° ± 2° (c = 0.962 in chloroform).

THERAP CAT: Cardiotonic.

2509. Convicine. *6-Amino-5-(β-D-glucopyranosyloxy)-2,4(1H,3H)-pyrimidinedione.* $C_{10}H_{15}N_3O_8$; mol wt 305.24. C 39.35%, H 4.95%, N 13.77%, O 41.93%. From seeds of *Vicia sativa* L., *Leguminosae:* Ritthausen, *J. Prakt. Chem.* [2] **24**, 202 (1881); *Ber.* **29**, 894 (1896). Structure: Johnson, *J. Am. Chem. Soc.* **36**, 337 (1914); Fisher, Johnson, *ibid.* **54**, 2038 (1932); Bendick, Clements, *Biochim. Biophys. Acta* **12**, 462 (1953); Bien *et al., J. Chem. Soc.* (C) **1968**, 496. Synthesis: *eidem, J. Chem. Soc. Perkin Trans.* I **1973**, 1089.

Leaflets from boiling water, dec without melting at 287°. uv max (pH 7): 245, 271 nm (log ε 3.43, 4.16). Sol in hot water, dil NaOH; slightly sol in cold water, alcohol; practically insol in chloroform, glacial acetic acid.

2510. Copaene. *[1R-(1α,2α,6α,7α,8α)]-1,3-Dimethyl-8-(1-methylethyl)tricyclo[4.4.0.0²,⁷]dec-3-ene; (1R,2S,6S,7S,8S)-(−)-8-isopropyl-1,3-dimethyltricyclo[4.4.0.0²,⁷]dec-3-ene;* α-copaene. $C_{15}H_{24}$; mol wt 204.34. C 88.16%, H 11.84%. Tricyclic sesquiterpene occurring in African copaiba balsam oil; in the so-called supa oil from *Sindora supa* Merr. (*S. wallichii* F.-Vill.), *Leguminosae:* Henderson *et al., J. Chem. Soc.* **1926**, 3077. Also in the essential oil of *Phyllocladus trichomanoides* D. Don, *Podocarpaceae:* Briggs, Sutherland, *J. Org. Chem.* **13**, 1 (1948); from *Cyperus articulatus* Michx., *Cyperaceae:* Couchman *et al., Tetrahedron* **20**, 2037 (1964). Structural studies: Kapadia *et al., ibid.* **21**, 607 (1965). Absolute stereostructure: deMao *et al., ibid.* 619. Synthesis: Heathcock, *J. Am. Chem. Soc.* **88**, 4110 (1966); Heathcock *et al., ibid.* **89**, 4133 (1967); Corey, Watt, *ibid.* **95**, 2303 (1973). *Review:* J. Simonsen, D. H. R. Barton, *The Terpenes* vol. III (University Press, Cambridge, 1952) pp 88-91. *See also* Ylangene.

Oily liquid, bp 246-251°; bp_{10} 119-120°; d_{15}^{15} 0.9077; n_D^{20} 1.4894. $[\alpha]_D^{22}$ −6.3° (c = 1.20 in chloroform).

2511. Copaiba. Balsam copaiba; balsam capivi; Jesuit's balsam. Oleoresin from South American species of *Copaifera (Copaiba), Leguminosae.* *Habit.* Brazil, Venezuela, Colombia, especially the Amazon valley and banks of Orinoco. *Constit.* Volatile oil, resin; illuric and metacopaivic acid (in Maracaibo balsam); copaivic and oxycopaivic acids (in Para balsam). Brief description of balsam copaiba and its constituents: J. A. Wenninger *et al., J. Assoc. Offic. Anal. Chem.* **50,** 1304 (1967); D. L. J. Opdyke, *Food Cosmet. Toxicol.* **14,** 687 (1976); of copaiba oil: *idem, ibid.* **11,** 1075 (1973).

Transparent, viscid, pale yellow to brownish-yellow liq; peculiar odor; bitter, acrid, nauseating taste. d 0.930-0.995. Acid no. 28-95. Insol in water. Sol in benzene, chloroform, ether, oils, carbon disulfide, abs alcohol, petr ether, partly in 95% alcohol. *Incompat.* Mineral acids, magnesia, water.

USE: In varnishes; for removing old oil varnish from oil paintings; manuf photographic paper.

2512. Copal. Resin copal; gum copal; anime (soft copal); kaurie; cowrie. A resin found as a fossil in Zaire, or exuding from various species of *Trachylobium, Hymenaea courbaril* L., etc., *Leguminosae.* Occurs as hard or soft copals. *Habit.* Zanzibar, Mozambique, also S. America, Australia, Philippine Islands and West Indies. *Constit.* Zanzibar copal contains 80% trachylolic acid, 4% isotrachylolic acid, 6% resene and volatile oil. Kaurie copal contains dammaric acid, dammaran and a resin.

Yellowish to yellowish-brown pieces of various sizes; conchoidal fracture; odorless and tasteless. Hard copals are almost insol in usual solvents; soft copals are partly sol in alcohol, chloroform or glacial acetic acid; both copals after having been fused are sol in oil turpentine and linseed oil.

USE: In varnishes and cements; as substitute for amber; manuf oil cloths and linoleum. In dentistry, for modeling compds and cavity varnishes.

2513. Coparaffinate. *Isopar.* A mixture of water-insol isoparaffinic acids partially neutralized with hydroxybenzyl-dialiphatic amines. The water-insol isoparaffinic acids are obtained by oxidation of petroleum hydrocarbons by the passage of a current of oxygen under pressure at an elevated temp in the presence of a metal catalyst. The water-insol mono- and dicarboxylic acids with from 6 to 16 carbon atoms are separated and purified by fractional distillation. The hydroxybenzyldialiphatic amines are combined directly with the isoparaffinic acids or in suitable solvent. The latter is then removed by distillation: C. E. Earle, U.S. pat. **2,262,720** (1941).

Dark-brown, viscous, oily liq. Odor of burnt petroleum. d 0.970-0.980. Immiscible with water. Freely sol in alc, volatile and fixed oils. Marketed as a 17% ointment, which also contains 4% titanium dioxide in an ointment base consisting of beeswax, cetyl alcohol, lanolin, and petrolatum.

THERAP CAT: Topical antifungal.

2514. Copper. Cu; at. wt 63.546; at. no. 29; valences 1, 2. Occurrence in the earth's crust: 70 ppm; also present in seawater: 0.001-0.02 ppm. Two naturally occurring isotopes: 63 (69.09%), 65 (30.91%); nine artificial isotopes: 58-62, 64, 66-68. One of the earliest known metals. Found in nature in its native state; also in combined form in several minerals including chalcopyrite, chalcocite, bornite, *q.q.v.,* **tetrahedrite** ($Cu_{12}Sb_4S_{13}$), *enargite* (Cu_3AsS_4), antlerite. Extraction from ores: Clark-Hawley, *Encyclopedia of Chemistry* (Reinhold, New York, 2nd ed., 1966) p 288. Metallurgy of copper and its alloys: *Metals Reference Book,* vols. 1, 2, C. J. Smithells, Ed. (Butterworth's, London, 3rd ed., 1962). A trace element essential to many plants and animals. Occurs in biological complexes such as **pheophytin** (analog of chlorophyll), hemocyanin, tyrosinase and ceruloplasmin, *q.q.v.* Reviews of copper and copper compounds: *Copper,* A. Butts, Ed., A.C.S. Monograph Series, no. **122** (Reinhold, New York, 1954) 936 pp; Massey, "Copper" in *Comprehensive Inorganic Chemistry,* vol. **3,** J. C. Bailar, Jr. *et al.,* Eds. (Pergamon Press, Oxford, 1973) pp 1-78; W. M. Tuddenham, P. A. Dougall, Kirk-Othmer, *Encyclopedia of Chemical Technology* vol. **6** (Wiley-Interscience, New York, 3rd ed.,

1979) pp 819-869. Book: *Inflammatory Diseases and Copper,* J. R. J. Sorenson, Ed. (Humana Press, Clifton, N J, 1982) 622 pp.

Reddish, lustrous, ductile, malleable metal; face-centered cubic structure; commercially available in the form of ingots, sheets, wire or powder. Becomes dull when exposed to air. In moist air gradually becomes coated with green basic carbonate. d 8.94. mp 1083°. bp 2595°. Mohs' hardness 3.0. Resistivity 1.673 microohm-cm. Heat of fusion 48.9 cal/g; heat of vaporization 1150 cal/g. Heat capacity at constant pressure (solid) 0.092 cal/g/°C (20°), (liq) 0.112 cal/g/°C. E^0 (aq) Cu^+/Cu +0.521 V; E^0 (aq) Cu^{2+}/Cu +0.337 V. Very slowly attacked by cold hydrochloric or dil sulfuric acid; readily by dil nitric acid, and by both hot concd H_2SO_4 and HBr. It is also attacked by acetic and other organic acids. Slowly sol in ammonia water. Water-soluble cupric salts yield with sodium hydroxide a bluish-green precipitate of cupric hydroxide which is changed to black cupric oxide on warming. Potassium ferrocyanide produces a brownish-red precipitate of copper ferrocyanide. Hydrogen sulfide produces in acid solns a black precipitate of cupric sulfide which is sol in soln of sodium cyanide. Aluminum, iron or zinc precipitate metallic copper from its solns.

USE: Manuf bronzes, brass, other copper alloys, electrical conductors, ammunition, copper salts, works of art. *Toxicity:* Copper itself probably has little or no toxicity, although there are conflicting reports in the literature. Soluble salts, notably copper sulfate, are strong irritants to skin, mucous membranes. Copper oxide fumes can cause metal fume fever. A relationship between copper and hemochromatosis has been reported. See E. Browning, *Toxicity of Industrial Metals* (Appleton-Century-Crofts, New York, 2nd ed., 1969) pp 145-152.

2515. Copper Phthalocyanine. $(SP-4-1)$-$[29H,31H$-*Phthalocyaninato*$(2-)$-$N^{29},N^{30},N^{31},N^{32}]$*copper;* C.I. Pigment Blue 15; C.I. 74160. $C_{32}H_{16}CuN_8$; mol wt 576.08. C 66.72%, H 2.80%, Cu 11.03%, N 19.45%. Discoverers: Dandridge *et al.,* **Brit.** pat. **322,169** (1928 to Scottish Dyes). Prepn: Baumann *et al., Angew. Chem.* **68,** 133 (1956); Sanielevici *et al., Rec. Chim. (Bucharest)* **12,** 281 (1961); Raab, Hoernle, **Brit.** pat. **930,150** (1963 to Bayer). Purification: Heinle, Mau, **Ger.** pat. **1,167,307** (1964 to G. Siegle); Sanielevici *et al.,* **Belg.** pat. **657,307** (1965 to Romania, Ministry of Petroleum and Chemical Industry); H. Tomoda *et al., Chem. Letters* **1980,** 1277. Properties: Dent, Linstead, *J. Chem. Soc.* **1934,** 1027; Easton, Smith, *J. Oil Colour Chem. Assoc.* **49,** 614 (1966). *See also Colour Index,* vol. **4** (3rd ed., 1971) p 4618.

Bright blue microcrystals with purple lustre; sol in 98% H_2SO_4 from which it can be almost quantitatively pptd by dilution with water. Practically insol in water, alcohol, and hydrocarbons. Stable toward heat (crystalline and analytically pure sublimate obtained at about 580° in low pressure atmosphere of nitrogen or carbon dioxide), alkalies, dilute acids. Dec by hot nitric acid, or dilute acid permanganate to yield phthalimide.

Exists in two forms: the thermodynamically less stable, redder α-form is a better pigment than the more stable greener β-form. In the presence of aromatic solvents, heat, high shear, etc., α-form is converted to β-form.

USE: In inks and paints. Approved by FDA for use in polypropylene sutures.

2516. Coproergostane. 5β-*Ergostane;* pseudoergostane.

$C_{28}H_{50}$; mol wt 386.68. C 86.97%, H 13.03%. Prepn from cholanic acid: Fernholz, *Ber.* **69**, 1792 (1936).

Needles from acetone. mp 64°. $[\alpha]_D^{19}$ +25.3° (c = 2 in chloroform).

2517. Coprogen. $C_{35}H_{53}FeN_6O_{13}$; mol wt 821.70. C 51.16%, H 6.50%, Fe 6.80%, N 10.23%, O 25.31%. Iron-contg, growth-promoting complex. Isoln from *Penicillium* spp: Hesseltine *et al., J. Am. Chem. Soc.* **74**, 1362 (1952). Pidacks *et al., ibid.* **75**, 6064 (1953); Gaeumann, Vischer, U.S. pat. 3,297,526 (1967 to Ciba). Isoln from *Aspergillaceae:* Zaehner *et al., Arch. Mikrobiol.* **45**, 119 (1963). Structure: Keller-Schierlein, Diekmann, *Helv. Chim. Acta* **53**, 2035 (1970).

Clusters of brick-red needles from ethanol. Absorption max (ethanol): 217, 250, 440 nm (log ∈ 4.46, 4.22, 3.47). Sol in water, methanol, ethanol (amorphous coprogen only), propanol, benzyl alcohol, Cellosolves; practically insol in ether, ethyl acetate, chloroform, benzene, Cellosolve esters. *Desferricoprogen,* iron-free coprogen, may be obtained by treating an aq solution with 8-quinolinol yielding a light brown powder.

USE: Coprogen as growth-promoting agent in various organisms. Desferricoprogen to bring about excretion of iron.

2518. Coprostane. *5β-Cholestane;* pseudocholestane. $C_{27}H_{48}$; mol wt 372.65. C 87.02%, H 12.98%. The *cis*-decalin homolog of cholestane. Prepn: Wieland, Jacobi, *Ber.* **59**, 2064 (1926); Young *et al., J. Am. Chem. Soc.* **81**, 1452 (1959); Dart, Henbest, *J. Chem. Soc.* **1960**, 3563; Nickon, Bagli, *J. Am. Chem. Soc.* **83**, 1498 (1961); Caglioti, Grasselli, *Chem. & Ind. (London)* **1964**, 153.

Needles from alcohol, mp 72°. $[\alpha]_D^{11}$ +25.1° (c = 2 in $CHCl_3$). n_D^{88} 1.4884. Freely sol in ether, chloroform; slightly in abs alcohol.

2519. Coprosterol. *5β-Cholestan-3β-ol; 3β-*coprostanol; stercorin. $C_{27}H_{48}O$; mol wt 388.65. C 83.43%, H 12.45%, O 4.12%. Found in feces of man and of carnivorous animals: Bondzynski, Humnicki, *Z. Physiol. Chem.* **22**, 396 (1896); Wells *et al., Arch. Biochem. Biophys.* **57**, 437 (1955); Samuel *et al., J. Chromatog.* **14**, 508 (1964). Prepn from coprostenone: Ruzicka *et al., Helv. Chim. Acta* **17**, 1407 (1934); Shoppee, Summers, *J. Chem. Soc.* **1950**, 687. From coprostenol: Schönheimer, Evans, *J. Biol. Chem.* **114**, 567 (1936); Ruzicka *et al., Helv. Chim. Acta* **21**, 498 (1938); Agashe, Summers, *J. Chem. Soc.* **1957**, 3107; Dart, Henbest, *ibid.* **1960**, 3563. Prepn from cholesterol: Rosenfeld, Gallagher, *Steroids* **4**, 515 (1964).

Needles from methanol, mp 101°. $[\alpha]_D^{18}$ +28° (c = 1.8 in $CHCl_3$). Freely sol in ether, chloroform, benzene; slightly sol in methanol (one gram dissolves in 145 ml MeOH); insol in water.

2520. Coptine. From root of *Coptis trifolia* Salisb., *Ranunculaceae:* Gross, *Am. J. Pharm.* **45**[4], 193 (1873); Mollett, Christensen, *J. Am. Pharm. Assoc.* **23**, 310 (1934); from *C. teeta* Wall.: Chatterjee *et al., J. Indian Chem. Soc.* **28**, 97 (1951); from *C. chinensis* Wils.: Schramm, Tang, *Pharmazie* **14**, 405 (1959).

Crystals. Practically insol in water, ammonia, alkali hydroxides; sol in hot alcohol, dil acids.

2521. Coptis. Goldthread. Dried plant of *Coptis trifolia* Salisb., *Ranunculaceae. Habit.* U.S., northern and middle Atlantic States; Canada. *Constit.* Berberine, coptine.

THERAP CAT: Bitter tonic.

2522. Coptisine. *6,7-Dihydrobis[1,3]benzodioxolo[5,6-a: 4',5'-g]quinolizinium; 7,8,13,13a-tetrahydro-2,3:9,10-bis-methylenedioxyberbinium;* bis[methylenedioxy]protoberberine. $[C_{19}H_{14}NO_4]^+$; mol wt 320.33. C 71.24%, H 4.40%, N 4.37%, O 19.98%. From root of *Coptis japonica* Makino, *Ranunculaceae.* Extraction procedure: Kitasato, *C.A.* **21**, 2700 (1927). Structure: Späth, Posega, *Ber.* **62**, 1029 (1929); *cf.* Huang-Minlon, *Ber.* **69**, 1744 (1936). Prepn from rhoeageninediol: Klásek *et al., Tetrahedron Letters* **1968**, 4549.

Hydroxide, $C_{19}H_{15}NO_5$, yellowish needles from alc, mp 218°. Absorption spectrum: Kitasato, *Acta Phytochim.* **3**,

175 (1927). Very slightly sol in water; sparingly in alcohol; sol in alkalies.

Chloride, orange prisms, not melted at 300°.

Iodide, yellow needles, dec above 280°.

Sulfate, yellow crystals, insol in water and alcohol (enables separation from berberine and worenine).

2523. Cord Factor. Trehalose 6,6-dimycolate; 6,6'-di-O-mycolyl-α,α-trehalose; (6-O-mycolyl-α-D-glucopyranosyl)-6-O-mycolyl-α-D-glucopyranoside. Toxic glycolipids responsible for the cord formation and the leukotoxic effect of virulent bacilli. The term cord factor is widely used for the natural mixture of trehalose dimycolates produced by virulent *Mycobacteria, Nocardia, Corynebacteria* and attenuated BCG, *q.v.* For precise designation, the strain from which the preparation was isolated must be mentioned. First isolated from *Mycobacterium tuberculosis:* H. Bloch, *J. Exp. Med.* **91**, 197 (1950). Structure: Y. Asselineau, E. Lederer, *Biochim. Biophys. Acta* **17**, 161 (1955); H. Noll *et al., ibid.* **20**, 299 (1956). Synthesis of *Mycobacterium tuberculosis* cord factor and of trehalose diesters with synthetic mycolic acids: T. Gendre, E. Lederer, *Bull. Soc. Chim.* **1956**, 1478; J. Polonsky *et al., Carbohydr. Res.* **65**, 295 (1978). Antitumor activity: A. Bekierkunst *et al., Science* **174**, 1240 (1971); M. V. Pimm *et al., Int. J. Cancer* **24**, 780 (1979). *Review:* E. Lederer, *Springer Semin. Immunopathol.* **2**, 133-148 (1979).

Colorless wax. Yields trehalose and mycolic acid upon alkaline hydrolysis.

2524. Cordycepin. *3'-Deoxyadenosine;* 9-cordyceposido-adenine. $C_{10}H_{13}N_5O_3$; mol wt 251.24. C 47.80%, H 5.22%, N 27.88%, O 19.11%. First reported nucleoside antibiotic. Isoln from culture fluids of *Cordyceps militaris* (Linn.) Link: K. G. Cunningham *et al., J. Chem. Soc.* **1951**, 2299; N. M. Kredich, A. J. Guarino, *Biochim. Biophys. Acta* **41**, 363 (1960). Proposed structure: H. R. Bentley *et al., J. Chem. Soc.* **1951**, 2301. Identity with 3'-deoxyadenosine and revised structure: E. A. Kaczka *et al., Biochem. Biophys. Res. Commun.* **14**, 456 (1964). Biosynthesis: R. Suhadolnik *et al., J. Am. Chem. Soc.* **86**, 948 (1964). Synthesis: A. R. Todd, T. L. Ulbricht, *J. Chem. Soc.* **1960**, 3275; W. W. Lee *et al., J. Am. Chem. Soc.* **83**, 1906 (1961); E. Walton *et al., ibid.* **86**, 2952 (1964); Y. Ito *et al., ibid.* **103**, 6739 (1981). Cordycepin and cordycepin triphosphate have been used extensively in the study of messenger RNA transcription, *see* H. T. Shigeura, G. E. Boxer, *Biochim. Biophys. Res. Commun.* **17**, 758 (1964); S. Penman *et al., Proc. Nat. Acad. Sci. USA* **67**, 1878 (1970). *Reviews:* J. J. Fox. *et al.,* "Nucleoside Antibiotics" in *Progr. Nucleic Acid Res. Mol. Biol.* **5**, 258-262 (1966); A. J. Guarino, "Cordycepin" in *Antibiotics* I, D.

Gottlieb, P. Shaw, Eds. (Springer-Verlag, New York, 1967) pp 468-480.

Needles from ethanol, *n*-butanol, *n*-propanol or water. mp 225-226°. $[\alpha]_D^{20}$ −47°. $[\alpha]_D^{27}$ −42°. uv max (ethanol): 260 nm (ε 14,600). pH aq soln: 7.1.

Triphosphate, $C_{10}H_{13}N_5O_{12}P_3$, *cordycepin-5'-triphosphate, 3'-deoxy ATP, 3'-deoxyadenosine-5'-(tetrahydrogen triphosphate)*. Inhibits the final step of RNA biosynthesis by termination of the ribonucleotide chain due to the absence of the 3'-hydroxyl group. Formation by conversion of 3'-deoxyadenosine in Ehrlich ascites tumor: H. Klenow, *Biochim. Biophys. Acta* **76**, 347 (1963). Metabolism in KB cells: H. Shigeura, S. Sampson, *ibid.* **138**, 26 (1967). Synthesis: J. J. Novak, F. Sorm, *Coll. Czech. Chem. Commun.* **38**, 113 (1973); M. Blandin, *J. Carbohyd. Nucl., Nucl.* **3**(5/6), 341 (1976).

2525. Coriamyrtin. *[1aS-(1aα,1bβ,2β,5β,6aβ,7β,7aα,-8S*)]-Hexahydro-1b-hydroxy-6a-methyl-8-(1-methylethenyl)spiro[2,5-methano-7H-oxireno[3,4]cyclopent[1,2-d]oxepin-7,2'-oxiran]-3(2H)-one.* $C_{15}H_{18}O_5$; mol wt 278.29. C 64.73%, H 6.52%, O 28.75%. Sesquiterpene of the picrotoxane group. Main toxic principle from leaves and fruit of *Coraria myrtifolia* L., *Coriariaceae:* M. J. Riban, *C. R. Acad. Sci.* **57**, 798 (1863) and *C. japonica* A. Gray, *Coriariaceae:* Kariyone, Sato, *J. Pharm. Soc. Japan* **50**, 106 (1930); Okuda, *Pharm. Bull. (Japan)* **2**, 185 (1954). Series of articles on structure, absolute configuration, stereochemistry and derivatives: T. Okuda, T. Yoshida, *Chem. Pharm. Bull.* **9**, 379, 404 (1961); **15**, 1687, 1691 (1967). Structural relationship to tutin, *q.v.: eidem, Tetrahedron Letters* **1965**, 439. Biosynthetic studies: M. Biollaz, D. Arigoni, *Chem. Commun.* **1969**, 633. Synthetic studies: K. Tanaka *et al., Chem. Pharm. Bull.* **31**, 1943, 1958, 1972 (1983). Pharmacology: E. E. Swanson, K. K. Chen, *J. Pharmacol. Exp. Ther.* **57**, 410 (1936). Structure-activity relationship: C. H. Jarboe *et al., J. Med. Chem.* **11**, 729 (1968).

Bitter, monoclinic prisms, mp 229-230°. $[\alpha]_D^{14}$ +79°. Slightly sol in water, cold alc; freely sol in hot alc, in ether. LD_{50} i.p. in mice: 3 mg/kg (Jarboe).

2526. Coriander. Dried ripe fruit of *Coriandrum sativum* L., *Umbelliferae. Habit.* Asia, Europe. *Constit.* About 1% volatile oil; fixed oils, malic acid, tannin, mucilage.

THERAP CAT: Carminative, aromatic.

2527. Cork. The bark of the cork oak tree, *Quercus suber* L., *Fagaceae* and to a lesser degree of *Quercus occidentalis* Gray, *Fagaceae,* indigenous to the African and European shores of the western Mediterranean. *Constit.* 35-60% suberin, 30-33% cellulose, 27-32% lignin; small amounts of cerin (a wax), fats, inorganic manganese compds, decacrylic acid. Suberin contains phellonic acid, suberic acid, phloionic acid, phloionolic acid, suberolcarboxylic acid, eicosadicarboxylic acid, stearic acid, cortic acid. Used to seal wine casks since antiquity: Theophrastus, *Historia Plantarum* **3**, 16 and 17. Plinius (major), *Historia Naturalis* **16**, 8, 13. *Reviews and monographs:* A. Klauber, *Monographie des Korkes* (Weber, Berlin, 1920); W. Herrmann, *Kork* in *Ullmanns Encyklopädie der Technischen Chemie* vol. **10**, (Urban & Schwarzenberg, Munich, 3rd ed., 1958) pp 634-637; G. B. Cooke, *Cork and the Cork Tree* (Pergamon Press, New York, 1961).

Pale tan, elastic mass. Floats on water. Specific gravity about 0.10 to 0.25 g/ml. Excellent insulator against heat and sound. Stable to heat up to about 100°. Unaffected by water, brine, dil acids, alcohols, bland oils. Attacked by concd mineral acids, strong alkali and ammonia, free halo-gens, hydrogen peroxide, ozone.

USE: Bottle stoppers, sound deadeners, heat insulators, life preservers, gaskets, linoleum manufacture, cork tile, bulletin boards, dart boards, inlays for shoes.

2528. Corn Oil. Maize oil; Maydol; Mazola. Obtained as a byproduct by wet milling the grain of *Zea mays* L., *Gramineae* for the manuf of corn starch, corn syrup, glucose, dextrins, etc.: E. W. Eckey, *Vegetable Fats and Oils* (Rein-hold, New York, 1954). *Constit.* Glycerides of the following fatty acids: Myristic 0.1-1.7%, palmitic 8-12%, stearic 2.5-4.5%, hexadecenoic 0.2-1.6%, oleic 19-49%, linoleic 34-62%. Unsaponifiable fraction: 1-3% (γ-tocopherol 0.1%, the rest is mostly isomeric sitosterols and wax such as myricyl and ceryl alcohols). The crude oil may contain up to 2% phos-pholipids (vegetable lecithin, inositol esters). The following constants are for the refined product.

Yellow oil. Faint characteristic odor and taste. d_{25}^{25} 0.916-0.921. mp -18 to $-10°$. Titer 14-20°. Flash pt 610°F (321°C). Ignition pt 740°F (393°C). n_D^{25} 1.470-1.474; n_D^{40} 1.464-1.468. Acid value 2-6. Saponification value 187-196. Iodine value 109-133. Thiocyanogen value 71-77. Hydroxyl value 8-12. Reichert-Meissl value <0.5. Polen-ske value <0.5. Hehner value 92-96. Classed as a semidry-ing oil. On prolonged exposure to air it thickens and be-comes rancid. Miscible with chloroform, ether, benzene, petr ether. Slightly sol in alc.

USE: As salad and cooking oil; in prepn of margarine. As pharmaceutic aid (solvent). Some use in the preparation of non-yellowing enamel paint.

2529. Corn Steep Liquor. Corn steep water. Cleaned corn grain is washed thoroughly and then steeped for 36 to 40 hours in approx twice its volume of water contg 0.2% SO_2 at a temp of 46-50°C. As the steep water is drawn from the corn it contains between 6 and 9 lbs of solids per 100 lbs of water. The corn should be soft to the touch, but must not be slimy or smeary—in the latter case it was steeped either too hot or too long. The steep liquor is then evaporated [under 25 mm Hg pressure] in a "Swenson" evaporator to what is called heavy steep liquor with a gravity of 16 to 20° Bé and contg 50-60% of solids. In this form it is sold to manufacturers of antibiotics who may have their own speci-fications as to age, gravity and SO_2 content. *See* Bowden, Peterson, *Arch. Biochem.* **9,** 387-399 (April 1946); Graefe, *Stärke* **4,** 275-282 (1952).

USE: In production of penicillin, *meso*-inositol; source of phytin.

2530. Cornus. Dogwood; flowering dogwood. Dried root bark of *Cornus florida* L., *Cornaceae*. *Habit.* Eastern U.S. and Ontario. *Constit.* Cornine.

THERAP CAT: Astringent bitter.

2531. Coroxon. *Phosphoric acid 3-chloro-4-methyl-2-oxo-2H-1-benzopyran-7-yl diethyl ester; 3-chloro-7-hydr-oxy-4-methylcoumarin diethyl phosphate;* 3-chloro-4-meth-ylumbelliferone diethyl phosphate; *O,O*-diethyl *O*-(3-chlo-ro-4-methyl-7-coumarinyl) phosphate; *O,O*-diethyl *O*-7-(3-chloro-4-methylumbelliferone) phosphate; coumaphos oxygen analog; Coralox. $C_{14}H_{16}ClO_6P$; mol wt 346.70. C 48.50%, H 4.65%, Cl 10.23%, O 27.69%, P 8.93%. Metabo-lite of coumaphos, *q.v.* Prepn: Fusco *et al.,* U.S. pat. **2,951,-851** (1960 to Montecatini); G. C. Amin, C. M. Christian, *Indian Chem. Manuf.* **12,** 22 (1974). Crystal and molecular structure: M. R. Gifkins, R. A. Jacobson, *J. Agr. Food Chem.* **24,** 232 (1976). Determn in eggs and milk: K. D. White *et al., J. Assoc. Offic. Anal. Chem.* **66,** 1358 (1983). Ophthalmic cholinesterase inhibitor: M. Kadin, *Am. J. Ophthalmol.* **50,** 622 (1960).

Practically insol in water. Somewhat sol in acetone, chlo-roform, corn oil.

Mixture with phenothiazine, *Coopex.*

THERAP CAT (VET): Anthelmintic.

2532. Corticosterone. *(11β)-11,21-Dihydroxypregn-4-ene-3,20-dione;* 4-pregnene-11β,21-diol-3,20-dione; com-pound B; Kendall's compound B; Reichstein's substance H. $C_{21}H_{30}O_4$; mol wt 346.45. C 72.80%, H 8.73%, O 18.47%. Isoln from adrenal cortex: Mason *et al., J. Biol. Chem.* **114,** 613 (1936); Reichstein, von Euw, *Helv. Chim. Acta* **21,** 1197 (1938); Jeanloz *et al.,* U.S. pat. **2,676,904** (1954 to Searle); from adrenal extract: Reichstein, U.S. pat. **2,166,877** (1939 to Roche-Organon). Prepn from desoxy-cholic acid: Wallis, Chakravorty, U.S. pat. **2,341,250** (1944 to Research Corp.); from 11-deoxycorticosterone: Zaffa-roni, U.S. pat. **2,671,752** (1954 to Syntex); from cortisone: Oliveto *et al.,* U.S. pat. **2,927,108** (1960 to Schering).

Trigonal plates from acetone, mp 180-182°. $[\alpha]_D^{15}$ +223° (c = 1.1 in alc). uv max: 240 nm. Insol in water. Sol in usual organic solvents. Upon moistening with concd H_2SO_4 it gives an orange-yellow soln which has a strong fluores-cence.

21-Acetate, $C_{23}H_{32}O_5$, clusters of needles from acetone + ether, mp 145°. $[\alpha]_D^{20}$ +195° (c = 0.62 in acetone).

THERAP CAT: Glucocorticoid.

2533. Cortisone. *17α,21-Dihydroxy-4-pregnene-3,11,20-trione;* 17-hydroxy-11-dehydrocorticosterone; 11-dehydro-17-hydroxycorticosterone; Δ^4-pregnene-17α,21-diol-3,11,-20-trione; Kendall's "compound E"; Wintersteiner's "com-pound F"; Reichstein's "substance Fa"; KE; Incortin; Cor-tone; Cortadren; Scheroson; Corlin; Cortogen; Adreson. $C_{21}H_{28}O_5$; mol wt 360.46. C 69.98%, H 7.83%, O 22.19%. Isoln from suprarenal glands: Pfiffner *et al., J. Biol. Chem.* **111,** 585 (1935); **116,** 291 (1936); Mason *et al., ibid.* **114,** 613 (1936); Reichstein, *Helv. Chim. Acta* **19,** 1107 (1936); Kui-zenga, Cartland, *Endocrinology* **24,** 526 (1939). Synthesis of the monoacetate from desoxycholic acid: Sarett, *J. Biol. Chem.* **162,** 601 (1946). Further development of prepn methods: Meystre, Wettstein, *Experientia* **3,** 185 (1947); *Helv. Chim. Acta* **30,** 1037, 1256 (1947); Reichstein *et al., ibid.* **26,** 562, 705, 721 (1943); **27,** 821 (1944); Reichstein, U.S. pat. **2,403,683** (1946); Kendall *et al., J. Biol. Chem.* **166,** 345 (1946); Gallagher, U.S. pat. **2,447,325** (1948); Peterson, Murray, *J. Am. Chem. Soc.* **74,** 1871 (1952); Perl-man, *ibid.* 2126; Sarett, *ibid.* **70,** 1454 (1948); **71,** 2443 (1949); Mattox, Kendall, *ibid.* **70,** 882 (1948). Stereospecific total synthesis: Sarett *et al., ibid.* **74,** 4974 (1952).

Rhombohedral platelets from 95% alcohol, mp 220-224° (some decompn) when heated in evac capillary. $[\alpha]_D^{25}$ +209° (c = 1.2 in 95% alcohol); $[\alpha]_{546}^{25}$ +269° (c = 0.125 in benz-ene); $[\alpha]_{546}^{25}$ +248° (c = 0.1 to 0.2 in alcohol). uv max: 237 nm (ϵ 1.4 × 10⁴) *see* Mason *et al., J. Biol. Chem.* **116,** 267 (1936); Wintersteiner, Pfiffner, *ibid.* 291. Fairly sol in cold methanol, ethanol, acetone; much less sol in ether, benzene, chloroform; slightly sol in water (28 mg/100 ml at 25°). The water soln is neutral. Gives orange-red soln with intense green fluorescence in concd H_2SO_4. Reduces Benedict's soln on heating.

Monoacetate (C_{21}-acetate), $C_{23}H_{30}O_6$, *cortisone acetate, Cortistab, Cortelan, Cortisyl Artriona.* Flat needles from acetone; clusters of radiating rods from chloroform. Becomes opaque at 70-100°, mp 235-238° with slight sintering at 230°. $[\alpha]_D^{25}$ +164° (c = 0.5 in acetone), $[\alpha]_D^{25}$ +208 to +217° (dioxane). uv max: 238 nm (ϵ 1.58 × 10⁴), *see* Sarett, *J. Biol. Chem.* **162**, 630 (1946). Soly at 25° in water: 2.2 mg/100 ml; in propylene glycol 44 mg/100 ml; in chloroform 182 mg/g. Reduces ammoniacal silver nitrate soln at room temp. Sol in sulfuric acid giving a yellow soln without fluorescence (difference from hydrocortisone acetate).

THERAP CAT: Glucocorticoid.

THERAP CAT (VET): Glucocorticoid, anti-inflammatory agent.

2534. Cortisone, 21β-Cyclopentanepropionate. Cortisone, 21-cyclopentylpropionate; 4-pregnene-3,11,20-trione-17α,21-diol, 21β-cyclopentylpropionate. $C_{29}H_{40}O_6$; mol wt 484.61. C 71.87%, H 8.32%, O 19.81%. Prepd by the action of β-cyclopentylpropionyl chloride on cortisone in pyridine: Ott, U.S. pat. **2,746,978** (1956 to Upjohn).

Needles from diisopropyl ether, mp 158-161°. $[\alpha]_D^{20}$ +190° (chloroform). uv max (ethanol): 239 nm (ϵ 16,350). Sol in ether, glycols, vegetable oils, especially sesame, peanut, and corn oils.

2535. Cortisone Phosphate. Cortisone 21-dihydrogen phosphate; 21-cortisonephosphoric acid. $C_{21}H_{29}O_8P$; mol wt 440.40. C 57.27%, H 6.64%, O 29.06%, P 7.03%. Prepn from 17α-hydroxy-21-iodo-4-pregnene-3,11,20-trione by metathesis with silver dihydrogen phosphate in boiling acetonitrile: Poos *et al.*, *Chem. & Ind. (London)* **1958**, 1260; Cutler *et al.*, *J. Am. Chem. Soc.* **80**, 6300 (1958). Alternate methods: Conbere, Pfister, U.S. pat. **2,870,177** (1959); Christensen *et al.*, U.S. pat. **2,932,657** (1960 to Merck & Co.).

Crystals, decomp 198-204° (Poos); decomp 190-193.5° (Cutler). uv max (methanol): 238 nm (ϵ 15,200); (water): 244 nm (ϵ 15,900).

Monosodium salt, $C_{21}H_{28}NaO_8P$, dec 166-170°. Sol in water.

Disodium salt, $C_{21}H_{27}Na_2O_8P$, dec 290-295°. Sol in water. Dimethyl ester, $C_{23}H_{33}O_8P$, crystals, dec 228.5-230.5°. uv max (methanol): 238 nm (ϵ 15,600). Practically insol in water.

2536. Cortivazol. *(11β,16α)-21-(Acetyloxy)-11,17-dihydroxy-6,16-dimethyl-2'-phenyl-2'H-pregna-2,4,6-trieno-[3,2-c]pyrazol-20-one;* 1,2,3,3a,3b,7,10,10a,10b,11,12,12a-dodecahydro-1,11-dihydroxy-2,5,10a,12a-tetramethyl-7-

phenylcyclopenta[7,8]phenanthro[2,3-c]pyrazol-1-yl hydroxymethyl ketone acetate; 6,16α-dimethyl-11β,17α,21-trihydroxy-2'-phenyl[3,2-c]pyrazolo-4,6-pregnadien-20-one 21-acetate; MK-650; H 3625; Altim; Diaster; Dilaster; Idaltim. $C_{32}H_{38}N_2O_5$; mol wt 530.66. C 72.43%, H 7.21%, N 5.28%, O 15.07%. Prepn: Fried *et al.*, *J. Am. Chem. Soc.* **85**, 236 (1963); Tishler *et al.*, U. S. pats. **3,067,194** and **3,300,-483** (1962, 1967 to Merck & Co.).

Double mp 160-165° and 229-230°; $[\alpha]_D^{23}$ +14° ($CHCl_3$). uv max (methanol): 283, 315 nm (ϵ 15,700, 19,000).

THERAP CAT: Glucocorticoid.

2537. Cortol. *(3α,5β,11β,20S)-Pregnane-3,11,17,20,21-pentol.* $C_{21}H_{36}O_5$; mol wt 368.50. C 68.44%, H 9.85%, O 21.71%. Isolated from human urine after administration of hydrocortisone or ACTH: Fukushima *et al.*, *J. Biol. Chem.* **212**, 449 (1955).

α-Cortol; 5β-pregnane-3α,11β,17,20α,21-pentol. Crystals from ethyl acetate, mp 250.5-254°. $[\alpha]_D^{26}$ +23.7° (ethanol).

α-Cortol 3,20,21-triacetate, $C_{27}H_{42}O_8$, crystals from benzene + cyclohexane, mp 168-170°.

β-Cortol; 5β-pregnane-3α,11β,17,20β,21-pentol. Crystals from methanol, mp 262-264.5°. $[\alpha]_D^{25}$ +33.3° (ethanol).

2538. Cortolone. *3α,17α,20α,21-Tetrahydroxypregnan-11-one;* 3α,17α,20α,21-pregnanetetrol-11-one. $C_{21}H_{34}O_5$; mol wt 366.48. C 68.82%, H 9.35%, O 21.83%. Isolated from human urine after administration of cortisone, hydrocortisone or ACTH: Fukushima *et al.*, *J. Biol. Chem.* **212**, 449 (1955). Synthesis: Sarett, *J. Chem. Soc.* **71**, 1169 (1949); Soloway *et al.*, *ibid.* **76**, 2941 (1954); Fukushima *et al., loc. cit.*

Crystals from acetone, mp 208-209°. $[\alpha]_D^{25}$ +44° (c = 0.5 in ethanol); also reported as $[\alpha]_D^{28}$ +34.2° (ethanol).

3,20,21-Triacetate, $C_{27}H_{40}O_8$, crystals from methanol, mp 214-216°. $[\alpha]_D^{28}$ +28° (acetone?).

2539. Corybulbine. *5,8,13,13a-Tetrahydro-2,9,10-trimethoxy-13-methyl-6H-dibenzo[a,g]quinolizin-3-ol; 2,9,10-trimethoxy-13-methyl-13a-berbin-3-ol;* 3-hydroxy-13-methyl-2,9,10-trimethoxyberbine; Corydalis-G. $C_{21}H_{25}$-NO_4; mol wt 355.42. C 70.96%, H 7.09%, N 3.94%, O 18.01%. In *Corydalis cava* (L.) Schweigg & Korte (*C. tuber-*

osa DC.*);* *Fumariaceae:* Freund, Josephi, *Ann.* **277**, 1 (1893); in *C. platycarpa* Makino, *Fumariaceae:* Manske, *Can. J. Res.* **21B**, 13 (1943). Structure: Späth, Dobrowsky, *Ber.* **58**, 1274 (1925).

Needles from methanol, mp 242°. $[\alpha]_D^{20}$ +303° (c = 1.4 in CHCl$_3$). Unstable in light. Weakly basic. Sol in acetone, chloroform, hot benzene; sparingly sol in water, alc, ether. Hydrochloride, C$_{21}$H$_{25}$NO$_4$·HCl, prisms, dec 245-250°.

2540. Corycavamine. (+)-4,6,7,14-Tetrahydro-5,14-dimethylbis[1,3]benzodioxolo[4,5-c:5',6'-g]azecin-13(5H)-one; 7,13-dimethyl-2,3:9,10-bis(methylenedioxy)-7,13a-secoberbin-13a-one. C$_{21}$H$_{21}$NO$_5$; mol wt 367.39. C 68.65%, H 5.76%, N 3.81%, O 21.77%. In *Corydalis cava* (L.) Schweigg & Korte (*C. tuberosa* DC.), *Fumariaceae:* Gadamer *et al.*, *Arch. Pharm.* **240**, 81 (1902). Structure: von Bruchhausen, *ibid.* **263**, 570 (1925). Corycavamine and corycavine originally depicted as the keto and enol tautomers, resp.

Rhombic columns from ether and alcohol, mp 149°. Is converted to corycavine at the mp. $[\alpha]_D^{20}$ +167° (chloroform). Sol in alcohol and chloroform.

dl-Form, *corycavine.* Orthorhombic plates from alc, mp 218°, mp 222° in vac. Very slightly sol in water; sol in abs alcohol and chloroform; unstable to light.

2541. Corycavidine. (+)-5,7,8,15-Tetrahydro-3,4-dimethoxy-6,15-dimethylbenzo[e][1,3]dioxolo[4,5-k][3]benzazecin-14(6H)-one. C$_{22}$H$_{25}$NO$_5$; mol wt 383.43. C 68.91%, H 6.57%, N 3.65%, O 20.86%. In *Corydalis cava* (L.) Schweigg & Korte (*C. tuberosa* DC.), *Fumariaceae:* Gadamer, *Arch. Pharm.* **249**, 30 (1911); von Bruchhausen, *ibid.* **263**, 570 (1925). Synthesis of *dl*-form: Govindachari, Rajadurai, *J. Sci. Ind. Res. (New Delhi)* **16B**, 506 (1957).

Crystals from methanol, mp 212°. $[\alpha]_D^{20}$ +211°. Unstable to light; racemizes at the mp: *dl*-form, prisms from ether, mp 194°.

Hydrochloride, C$_{22}$H$_{25}$NO$_5$·HCl, crystals, easily sol in water and alcohol.

2542. Corydaldine. 3,4-Dihydro-6,7-dimethoxy-1(2H)-isoquinolinone; 3,4-dihydro-6,7-dimethoxyisocarbostyril; 1,2,3,4-tetrahydro-6,7-dimethoxy-1(2H)-1-isoquinolone. C$_{11}$H$_{13}$NO$_3$; mol wt 207.22. C 63.75%, H 6.32%, N 6.76%, O 23.16%. Synthesis: Späth, Dobrowsky, *Ber.* **58**, 1274 (1925); Mohunta, Ray, *J. Chem. Soc.* **1934**, 1263; Wiesner *et*

al., J. Am. Chem. Soc. **77**, 675 (1955); Brossi *et al.*, *Helv. Chim. Acta* **43**, 1459 (1960); Mahuzier, Hamon, *Bull. Soc. Chim. France* **1969**, 684.

Monoclinic prisms from water or alcohol, mp 175°. Sol in water, alcohol, ether, benzene, chloroform. Practically insol in petr ether.

2543. Corydaline. 5,8,13,13a-Tetrahydro-2,3,9,10-tetramethoxy-13-methyl-6H-dibenzo[a,g]quinolizine; 2,3,9,10-tetramethoxy-13α-methyl-13aβ-berbine. C$_{22}$H$_{27}$NO$_4$; mol wt 369.44. C 71.52%, H 7.37%, N 3.79%, O 17.32%. Alkaloid isolated from many species of *Corydalis*. Discovery in *C. tuberosa*: Wackenroder, *Kustner's Arch.* **8**, 423 (1826), as cited in: C. Wehmer, *Die Pflanzenstoffe* I (G. Fischer, Jena, 1929) p 389. Isoln from *Corydalis aurea* Willd. and *C. solida* (L.) Swartz, *Fumariaceae:* Manske, *Can. J. Res.* **16B**, 81 (1938); *Can. J. Chem.* **34**, 1 (1956). Structure: von Bruchhausen, Stippler, *Arch. Pharm.* **265**, 152 (1927). Synthesis: Späth, Kruta, *Ber.* **62**, 1024 (1929); T. Kametani *et al.*, *J. Chem. Soc. Perkin I* **1977**, 1151; M. Cushman, F. W. Dekon, *Tetrahedron* **1978**, 1435; Z. Kiparissides *et al.*, *Can. J. Chem.* **58**, 2770 (1980). Stereochemistry: Bersch, *Arch. Pharm.* **291**, 595 (1958); Jeffs, *Experientia* **21**, 690 (1965). Pharmacology: Berezhinskaya, *C.A.* **78**, 119087j (1973).

d-Form, prisms from alc, mp 135°. $[\alpha]_D^{20}$ +311° (c = 0.8 in alc). uv max: 396 nm. Sol in chloroform; moderately sol in ether; sparingly sol in methanol, ethanol. Practically insol in water. LD$_{50}$ i.v. in mice: 135.5 ± 12.8 mg/kg, R. C. Anderson, K. K. Chen, *Fed. Proc.* **5**, 163 (1946).

Hydrochloride hydrate, C$_{22}$H$_{27}$NO$_4$·HCl.H$_2$O, four-sided columns, mp 230-240°.

dl-Form, mp 135°. Slightly sol in water.

dl-Mesocorydaline, mp 158°. The (13R-cis)-analog of corydaline. Slightly sol in water.

d-Mesocorydaline, rhombic crystals, mp 152°. $[\alpha]_D^{20}$ +82° (c = 1.4); +180° (c = 3 in chloroform).

l-Mesocorydaline, rhombic crystals, mp 152°. $[\alpha]_D^{20}$ −85° (c = 1.4); −181° (c = 3 in chloroform).

2544. Corydalis. Squirrel corn; turkey corn. Dried tuber of *Dicentra (Bicuculla) cucullaria* (L.) Bernh., or of *Dicentra canadensis* (DC.) Walp., *Fumariaceae*. *Habit*. Ontario to Kentucky and Missouri. *Constit*. Corydaline, bulbocapnine, isocorydine, corytuberine, corycavine, corybulbine, corydine, fumaric acid, acrid resin, protopine.

2545. Corydine. 5,6,6a,7-Tetrahydro-2,10,11-trimethoxy-6-methyl-4H-dibenzo[de,g]quinolin-1-ol; 2,10,11-trimethoxy-6aα-aporphin-1-ol; 1-hydroxy-2,10,11-trimethoxyaporphine. C$_{20}$H$_{23}$NO$_4$; mol wt 341.39. C 70.36%, H 6.79%, N 4.10%, O 18.75%. The methyl ether of corytuberine. Occurs in *Corydalis cava* (L.) Schweigg & Körte (*C. tuberosa* DC), *Fumariaceae*, in the dextrorotatory form. Isoln: Gadamer, Ziegenbein, *Arch. Pharm.* **240**, 94 (1902). Structure: Späth, Berger, *Ber.* **64**, 2038 (1931). Synthesis: Hey, Palluel, *J. Chem. Soc.* **1957**, 2926; Arumugam *et al.*, *Ber.* **91**, 40 (1958); Jackson, Martin, *Chem. Commun.* **1965**, 142, 420; *eidem*, *J. Chem. Soc. (C)* **1966**, 2222. uv spectrum: Shamma, Yao, *J. Org. Chem.* **36**, 3253 (1971).

Tetragonal prisms from ether, mp 149°. $[\alpha]_D^{20}$ +204° (c = 1.6 in chloroform). Freely sol in chloroform, alcohol, ethyl acetate; moderately sol in ether.

dl-Hydrochloride monohydrate, $C_{20}H_{23}NO_4.HCl.H_2O$, crystals from ethanol + ether, sinters at 205°, dec 228°. Sparingly sol in water.

2546. Corynantheine. *(16E)-16,17,18,19-Tetradehydro-17-methoxycorynan-16-carboxylic acid methyl ester; (E)-16,-17,18,19-tetradehydro-17-methoxy-17,18-secoyohimban-16-carboxylic acid methyl ester.* $C_{22}H_{26}N_2O_3$; mol wt 366.44. C 72.10%, H 7.15%, N 7.65%, O 13.10%. Isolated from *Corynanthe johimbe* K. Schum., and *Pseudocinchona africana* A. Cheval., *Rubiaceae:* Karrer *et al., Helv. Chim. Acta* **9**, 1059 (1926); **35**, 851 (1952). Structure: Janot *et al., ibid.* **34**, 1207 (1951). Stereochemistry: Van Tamelen *et al., J. Am. Chem. Soc.* **79**, 6426 (1957). Total synthesis of *dl-* form: Van Tamelen, Wright, *Tetrahedron Letters* **1964**, 295; *eidem, J. Am. Chem. Soc.* **91**, 7349 (1969). Total synthesis of *d*-form: Autrey, Scullard, *ibid.* **90**, 4917 (1968).

Crystals: α-form mp 103-107°; β-form mp 165-166°. $[\alpha]_D$ +28° (c = 1 in methanol). uv max (methanol): 227, 280, 291 nm (log ε 4.64, 3.82, 3.80).

Hydrochloride, $C_{22}H_{26}N_2O_3.HCl$, needles from alcohol + ether, mp 194-206°. Also reported as mp 176-179°. $[\alpha]_D^{20}$ +12° (water). Sol in alcohol, sparingly sol in water.

Dihydrocorynantheine, $C_{22}H_{28}N_2O_3$. Total synthesis of *dl*-form: Van Tamelen, Hester, *J. Am. Chem. Soc.* **81**, 3805 (1959); *eidem, ibid.* **91**, 7342 (1969). Plates from alcohol + water, mp 177-177.5°. $[\alpha]_D$ +28°.

Hydrochloride, $C_{22}H_{28}N_2O_3.HCl$, mp 212-213°. Also reported as crystals from 95% ethanol-ethyl acetate, mp 242.2-243.3° (dec) (sealed tube), Van Tamelen, Hester, *loc. cit.* (1969).

2547. Corynanthine. *17α-Hydroxyyohimban-16β-carboxylic acid methyl ester;* rauhimbine. $C_{21}H_{26}N_2O_3$; mol wt 354.43. C 71.16%, H 7.39%, N 7.90%, O 13.54%. From bark of *Pseudocinchona africana* Chev., *Corynanthe johimbe* K. Schum., *Rubiaceae* and *Rauwolfia serpentina* (L.) Benth., *Apocynaceae:* Raymond-Hamet, *Compt. Rend.* **212**, 305 (1941); Jorio, *Ann. Chim. Farm.* **1939**, 50, *C.A.* **33**, 9306[9] (1939); Le Hir *et al., Ann. Pharm. Franc.* **11**, 546 (1953); Hofmann, *Helv. Chim. Acta* **37**, 314 (1954). Identity with rauhimbine: *idem, ibid.* 849. Structure and stereochemistry: Janot *et al., Bull. Soc. Chim. France* **1952**, 1085; **1961**, 637.

Stout prisms from acetone, dec 225-226°. $[\alpha]_D^{19}$ −85° (c = 0.5 in pyridine). uv max (methanol): 226, 283, 290 nm (log ε 4.56, 3.87, 3.79). Practically insol in water or petr ether. Sol in 40 parts of boiling chloroform, in 60 parts of boiling benzene, in 20 parts of boiling ethyl acetate, in 5 parts of boiling alcohol.

O,N-Dibutyrylcorynanthine hydrochloride, crystals from benzene, mp 208-210°. Prepn: Reiser *et al.,* U.S. pat. **2,-975,183** (1961 to Chemische Werke Albert).

O,N-Dipropionylcorynanthine hydrochloride, crystals from isopropanol, mp 236-237°. Prepn: Reiser *et al., loc. cit.*

2548. Corypalmine. *5,8,13,13a-Tetrahydro-2,9,10-trimethoxy-6H-dibenzo[a,g]quinolizin-3-ol; 2,9,10-trimethoxyberbin-3-ol;* 3-hydroxy-2,9,10-trimethoxyberbine; tetrahydrojatrorrhizine. $C_{20}H_{23}NO_4$; mol wt 341.39. C 70.36%, H 6.79%, N 4.10%, O 18.75%. In *Corydalis cava* (L.) Schweigg & Körte *(C. tuberosa* D.C.) and other varieties of *Corydalis, Fumariaceae.* Synthesis of *dl*-corypalmine: Govindachari *et al., Ber.* **92**, 1654 (1959).

d-Form, small crystals, mp 236°. $[\alpha]_D^{16}$ +280° (chloroform). Insol in water; sol in alcohol and chloroform: Späth *et al., Ber.* **56**, 878 (1923); **58**, 2133 (1925); **60**, 383 (1927).

l-Form, crystals from chloroform-methanol, mp 246° (in vac): Manske, *Can. J. Res.* **20B**, 57 (1942); **21B**, 111 (1943).

dl-Form, crystals from methanol, mp 207°.

2549. Corytuberine. *5,6,6a,7-Tetrahydroxy-2,10-dimethoxy-6-methyl-4H-dibenzo[de,g]quinoline-1,11-diol; 2,10-dimethoxy-6aα-aporphine-1,11-diol;* 1,11-dihydroxy-2,10-dimethoxyaporphine. $C_{19}H_{21}NO_4$; mol wt 327.37. C 69.70%, H 6.47%, N 4.28%, O 19.55%. In *Corydalis cava* (L.) Schweigg & Körte *(C. tuberosa* DC), in *Dicentra formosa* (Andr.) Walp., *Fumariaceae.* Isoln: Späth, Berger, *Ber.* **64**, 2038 (1931); Manske, *Can. J. Res.* **10**, 521 (1934). Synthesis of (±)-form: Tomita, Kikkawa, **Japan.** pat. **6466('58)** (to Shionogi), *C.A.* **54**, 1584i (1960); T. Kametani, M. Ihara, *J. Chem. Soc. Perkin Trans. I* **1980**, 629. Biosynthesis studies: Blaschke, *Biochem. Physiol. Alkaloide, Int. Symp., 4th,* K. Mothes, Ed. (Akad.-Verlag, Berlin, 1972) pp 283-286. Biomimetric total synthesis: T. Kametani *et al., Heterocycles* **5**, 175 (1976).

Pentahydrate, leaflets, plates, turning gray on exposure to light. mp 240° dec (dry). uv max (methanol): 227, 272, 311.5 nm. Sol in alc, hot water; slightly sol in ether, chloroform, ethyl acetate. $[\alpha]_D^{20}$ +283° (alc). The dried crystals are hygroscopic and quickly attract 1 mol H_2O from the air.

Hydrochloride, $C_{19}H_{21}NO_4$.HCl, crystals from alcohol-ether, dec above 250°; $[\alpha]_D^{20}$ +168°; sparingly sol in water.

Methyl ether, *see* Corydine.

2550. Cosyntropin. α^{1-24}-*Corticotropin;* β^{1-24}-corticotropin; tetracosactide; tetracosactrin; Actholain; Cortrosinta; Cortrosyn; Synacthen. $C_{136}H_{210}N_{40}O_{31}S$; mol wt 2,933.57. C 55.68%, H 7.22%, N 19.10%, O 16.91%, S 1.09%. Ser-Tyr-Ser-Met-Glu-His-Phe-Arg-Trp-Gly-Lys-Pro-Val-Gly-Lys-Lys-Arg-Arg-Pro-Val-Lys-Val-Tyr-Pro. Structure and synthesis: Kappeler, Schwyzer, *Helv. Chim. Acta* **44**, 1136 (1961); Schwyzer, Kappeler, *ibid.* **46**, 1550 (1963).

Hexaacetate tetradecahydrate, $C_{148}H_{228}N_{40}O_{43}S.14H_2O$. $[\alpha]_D^{22}$ −88 ± 2° (c = 0.511 in 1% acetic acid).

THERAP CAT: Adrenocorticotropic hormone.

2551. Cotarnine. *5,6,7,8-Tetrahydro-4-methoxy-6-methyl-1,3-dioxolo[4,5-g]isoquinolin-5-ol.* $C_{12}H_{15}NO_4$; mol wt 237.25. C 60.75%, H 6.37%, N 5.90%, O 26.97%. Prepd by the oxidation of narcotine with dil nitric acid. Synthesis: Salway, *J. Chem. Soc.* **97**, 1208 (1910). Is tautomeric. Structure studies: Small, Lutz, "Chemistry of the Opium Alkaloids," Supplement No. 103, *Public Health Reports* (Washington, 1932) p 49; Schneider, Müller, *Ann.* **615**, 34 (1958). Absorption spectrum: Csokán, *Z. Anal. Chem.* **124**, 344 (1942).

Small needles from benzene, dec 132-133°. Sol in alcohol, chloroform, ether, benzene; slightly sol in water; sol in dil acids, in ammonia or sodium carbonate soln, but only slightly in potassium hydroxide soln. Aq or alcoholic solns are yellow.

Hydrochloride, $C_{12}H_{16}ClNO_4$, *Secalysat.*

Chloride, $C_{12}H_{14}ClNO_3$, *7,8-dihydro-4-methoxy-6-methyl-1,3-dioxolo[4,5-g]isoquinolinium chloride, cotarnine chloride, cotarninium chloride, Stypticin.* Dihydrate, light-yellow powder; deliquesc in moist air. Sol in about 1 part water, 4 parts alc. *Keep well closed.*

Phthalate, $C_{32}H_{32}N_2O_{10}$, *Styptol.*

THERAP CAT: Hemostatic.

2552. Cotinine. *1-Methyl-5-(3-pyridinyl)-2-pyrrolidinone;* *N-methyl-2-(3-pyridyl)-5-pyrrolidone.* $C_{10}H_{12}N_2O$; mol wt 176.21. C 68.16%, H 6.86%, N 9.08%. Nicotine metabolite first described by Pinner, *Arch. Pharm.* **231**, 378 (1893). Isolated from autoxidized nicotine, nicotine treated with hydrogen peroxide, from nicotine irradiated with ultraviolet light: Frankenburg, Vaitekunas, *J. Am. Chem. Soc.* **79**, 149 (1957).

Viscous oil, bp_6 210-211°. Absorption spectra: Frankenburg, Vaitekunas, *loc. cit.*

Fumarate, $C_{24}H_{28}N_4O_6$, *Scotine.*

THERAP CAT: Antidepressant.

2553. Cotoin. *(2,6-Dihydroxy-4-methoxyphenyl)phenylmethanone;* *2,6-dihydroxy-4-methoxybenzophenone.* $C_{14}H_{12}O_4$; mol wt 244.25. C 68.84%, H 4.95%, O 26.20%. Crystalline principle from true coto bark, the bark of *Aniba coto* (Rusby) Kost, *Lauraceae.* Extraction procedure: Jobst, Hesse, *Ann.* **199**, 17 (1879). Isoln from South American rosewood trees, *Aniba rosaeodora* Ducke and *A. duckei* Kost. *Lauraceae:* Gottlieb, Mors, *J. Am. Chem. Soc.* **80**, 2263 (1958). Structure: Karrer, Lichtenstein, *Helv. Chim. Acta* **11**, 789 (1928).

Yellow needles from benzene, mp 131-132°. Sharp taste. Irritates mucous membranes. Almost insol in water; sol in alcohol, methanol, amyl alcohol, benzene, chloroform, ether, carbon bisulfide, acetone, solns of alkali hydroxides and carbonates. Solns are yellow. Slowly reduces Fehling's soln or ammoniacal $AgNO_3$ in the cold. LD s.c. in frogs: 8 mg/kg, Jodbauer, Kurz, *Biochem. Z.* **74**, 340 (1916).

2554. Cotton-root Bark. Dried root-bark of one or more of the cultivated species of *Gossypium herbaceum* L. and of other species of *Gossypium, Malvaceae. Habit.* Asia (India, China, Arabia), Egypt, U.S., West Indies, S. America, Australia, Spain, etc. *Constit.* Yellow chromogen about 8% of a pale yellow resin, fixed oil, sugar.

THERAP CAT: Oxytocic.

2555. Cottonseed Oil. Fixed oil from seeds of cultivated varieties of *Gossypium herbaceum* L. or of other species of *Gossypium.*

Pale yellow, oily, practically odorless liq. d_{25}^{25} 0.915-0.921. Solidif 0° to −5°. n_D^{40} 1.4645-1.4655. Sapon no.: 190-198. Iodine no.: 105-114. Surface tension (20°) = 35.4 dyn/cm; (80°) = 31.3 dyn/cm. Slightly sol in alcohol; miscible with chloroform, ether, carbon disulfide, petr ether.

USE: Manuf soaps, oleomargarine, hydrogenated fats, lard substitute, glycerol, leather dressings, lubricants, cosmetics; emollient; also as salad and cooking oil; packing fish. Pharmaceutic aid (solvent).

THERAP CAT (VET): Pediculicide, acaricide, laxative, emollient.

2556. Coumachlor. *3-[1-(4-Chlorophenyl)-3-oxobutyl]-4-hydroxy-2H-1-benzopyran-2-one; 3-(α-acetonyl-p-chlorobenzyl)-4-hydroxycoumarin; 3-(α-p-chlorophenyl-β-acetylethyl)-4-hydroxycoumarin;* Geigy rodenticide exp. 332; Ratilan; Tomorin. $C_{19}H_{15}ClO_4$; mol wt 342.79. C 66.57%, H 4.41%, Cl 10.34%, O 18.67%. Prepn: F. Litvan, W. Stoll, U.S. pat. **2,648,682** (1953 to J. R. Geigy). Anticoagulant activity: M. Reiff, R. Weismann, *Acta Tropica* **8**, 97 (1951), *C.A.* **50**, 10976d (1956). Comparative toxicology of coumachlor and warfarin, *q.v.:* H. Wanntorp, *Acta Pharmacol. Toxicol.* **16**(suppl. 2), 123 pp (1959).

Crystals, mp 169-171°. Sol in alc, acetone, chloroform. Slightly sol in benzene, ether; practically insol in water. Highly toxic to dogs and pigs. LD_{50} orally in rats: 900 mg/kg, *RTECS Vol. I*, R. J. Lewis, R. L. Tatken, Eds. (1979) p 449.

USE: Rodenticide. *Caution:* Similar to warfarin, with delayed actions on prothrombin level and blood clotting, resulting in death by hemorrhage. *Antidote:* Vitamin K_1.

2557. Coumafuryl. *3-[1-(2-Furanyl)-3-oxobutyl]-4-hydroxy-2H-1-benzopyran-2-one; 3-(α-acetonylfurfuryl)-4-hydroxycoumarin; 3-[α-(2-furyl)-β-acetylethyl]-4-hydroxycoumarin;* Fumarin. $C_{17}H_{14}O_5$; mol wt 298.28. C 68.45%, H 4.73%, O 26.82%. Prepn: **Brit.** pat. **734,142** (1955 to Norddeutsche Affinerie and C. F. Spress & Sohn).

Crystals, mp 124°.

USE: Rodenticide.

2558. Coumalic Acid. *2-Oxo-1,2H-pyran-5-carboxylic acid; α-pyrone-5-carboxylic acid.* $C_6H_4O_4$; mol wt 140.09. C 51.44%, H 2.88%, O 45.68%. Prepd by heating anhydr malic acid with oleum: v. Pechmann, *Ann.* **264**, 262, 272 (1891); Wiley, Smith, *Org. Syn. coll. vol. IV*, 201 (1963).

Bright yellow prisms from methanol, mp 205-210° (partial decompn). bp$_{120}$ 218°; sublimes partially. Sparingly sol in cold water. Dec by boiling water. Sol in alcohol, glacial acetic acid. Slightly sol in ether, acetone, ethyl acetate. Insol in chloroform, benzene, ligroin.
Methyl ester, $C_7H_6O_4$, leaflets fom ligroin, mp 73-74°. bp$_{60}$ 178-180°. bp$_{760}$ 250-260°.
Ethyl ester, $C_8H_8O_4$, crystals, mp 36°. bp$_{760}$ 262-265°. Distills without decompn.

2559. Coumaphos. *Phosphorothioic acid O-(3-chloro-4-methyl-2-oxo-2H-1-benzopyran-7-yl) O,O-diethyl ester; 3-chloro-7-hydroxy-4-methylcoumarin O-ester with O,O-diethyl phosphorothioate;* 3-chloro-4-methylumbelliferone, *O-ester with O,O-diethyl phosphorothioate; O,O-diethyl O-(3-chloro-4-methyl-7-coumarinyl) phosphorothioate; O,O-diethyl O-(3-chloro-4-methylumbelliferone) thiophosphate; coumafos;* Bayer 21/199; Asuntol; Baymix; Co-ral; Meldane; Muscatox; Perizin; Resitox. $C_{14}H_{16}ClO_5PS$; mol wt 362.78. C 46.35%, H 4.45%, Cl 9.77%, O 22.05%, P 8.54%, S 8.84%. Prepn: Schrader, *Ger. pat.* **881,194**; U.S. pat. **2,748,146** (1951, 1956, both to Bayer); Krueger *et al., J. Agr. Food Chem.* **7**, 183 (1959). Toxicity: T. B. Gaines, *Toxicol. Appl. Pharmacol.* **14**, 515 (1969).

Crystals, mp 91°. The commercial product may be slightly brownish. Practically insol in water. Somewhat sol in acetone, chloroform, corn oil. Stable in water. LD$_{50}$ in female, male rats (mg/kg): 16, 41 orally (Gaines).
Note: Minor ingredient of Neguvon A, *see* under Trichlorfon.
USE: Insecticide, nematocide.
THERAP CAT (VET): Anthelmintic.

2560. Coumaran. *2,3-Dihydrobenzofuran;* cumaran; dihydrocoumarone. C_8H_8O; mol wt 120.14. C 79.97%, H 6.71%, O 13.32%. Synthesis: Bennett, Mahmoud Hafez, *J. Chem. Soc.* **1941**, 287; *cf.* Hurd, Hoffmann, *J. Org. Chem.* **5**, 212 (1940); Rindfusz, *J. Am. Chem. Soc.* **41**, 669 (1919); J. M. Bakke, H. M. Roholdt, *Acta Chem. Scand.* **B34**, 73 (1980).

Oily liq, bp 188-189°, bp$_{13}$ 74-75°. d$_4^{25}$ 1.058, n_D^{20} 1.5426. Sol in alcohol, ether, chloroform, carbon disulfide.

2561. p-Coumaric Acid. *3-(4-Hydroxyphenyl)-2-propenoic acid; p-hydroxycinnamic acid; β-[4-hydroxyphenyl]-acrylic acid.* $C_9H_8O_3$; mol wt 164.15. C 65.85%, H 4.91%, O 29.24%. Isoln: Hlasiwetz, *Ann.* **136**, 31 (1865); Bamberger, *Monatsh.* **12**, 459 (1891). Prepn: Eigel, *Ber.* **20**, 2528 (1887); Konek, Pacsu, *Ber.* **51**, 856 (1918). uv data: Wheeler, Covarrubias, *J. Org. Chem.* **28**, 2015 (1963).

Needles, mp 210-213° (reported in early lit as 206°). Crystallizes in anhydrous form from conc hot aq soln, but as monohydrate from dil aq soln on slow cooling. Slightly sol in cold water; sol in hot water and alc, ether. Practically insol in benzene and ligroin. uv max (95% ethanol): 223, 286 nm (ε 14,450, 19,000).

2562. Coumarilic Acid. *2-Benzofurancarboxylic acid;* coumarone-2-carboxylic acid. $C_9H_6O_3$; mol wt 162.14. C 66.66%, H 3.73%, O 29.60%. Synthesis from coumarin: Fuson *et al., Org. Syn. coll. vol. III*, 209 (1955). *Review:* Sethna, Shah, *Chem. Rev.* **36**, 1 (1945).

Needles from water, bitter taste. mp 192-193°. bp 310-315° with slight decompn. Sol in boiling water, in alcohol; slightly sol in chloroform, carbon disulfide.

2563. Coumarin. *2H-1-Benzopyran-2-one; 1,2-benzopyrone; cis-o-coumarinic acid lactone;* cumarin; coumarinic anhydride; tonka bean camphor. $C_9H_6O_2$; mol wt 146.14. C 73.96%, H 4.14%, O 21.90%. In tonka beans, lavender oil, woodruff (*Asperula* species), in sweet clover (*Melilotus*). Crystal and molecular structure: *Chem. Commun., Univ. Stockholm* **1976**, 21. Toxicity: Jenner *et al., Food Cosmet. Toxicol.* **2**, 327 (1964). *Review:* Sethna, Shah, *Chem. Rev.* **36**, 1 (1945); W. C. Meuly in Kirk-Othmer *Encyclopedia of Chemical Technology vol. 7* (Wiley-Interscience, New York, 3rd ed., 1979) pp 196-206.

Orthorhombic, rectangular plates. Pleasant, fragrant odor resembling that of vanilla beans, burning taste. mp 68-70°. bp 297-299°. bp$_5$ 139°. One gram dissolves in 400 ml cold, 50 ml boiling water. Freely sol in alc, chloroform, ether, oils; also sol in alkali hydroxide solns. LD$_{50}$ orally in rats, guinea pigs: 680, 202 mg/kg (Jenner).
USE: Pharmaceutic aid (flavor).

2564. Coumarin-3-carboxylic Acid. *2-Oxo-2H-1-benzopyran-3-carboxylic acid.* $C_{10}H_6O_4$; mol wt 190.15. C 63.16%, H 3.18%, O 33.66%. By heating salicylaldehyde with malonic acid in glacial acetic acid. *Review:* Sethna, Shah, *Chem. Rev.* **36**, 1 (1945).

Needles from water, mp 188° (dec). Slightly sol in water; sol in alc, alkalies; insol in ether, benzene, petr ether.

2565. Coumestrol. *3,9-Dihydroxy-6H-benzofuro[3,2-c]-[1]benzopyran-6-one; 2-(2,4-dihydroxyphenyl)-6-hydroxy-3-benzofurancarboxylic acid δ-lactone;* 7',6-dihydroxycoumarino(3',4',3,2)coumarone. $C_{15}H_8O_5$; mol wt 268.21. C 67.17%, H 3.01%, O 29.83%. An estrogenic factor occurring naturally in forage crops, esp in ladino clover (*Trifolium repens* L.), strawberry clover (*T. fragiferum* L.) and alfalfa (*Medicago sativo* L., *Leguminosae*). Isoln: Bickoff *et al.*, *J. Agr. Food Chem.* **6**, 536 (1958); Bickoff, Booth, U.S. pat. **2,890,116** (1959 to U.S.A.). Structure: Bickoff *et al.*, *J. Am. Chem. Soc.* **80**, 3969 (1958). Synthesis: Emerson, Bickoff, *ibid.* 4381; U.S. pat. **2,884,427** (1959 to U.S.A.); Jurd, *Tetrahedron Letters* **1963**, 1151; Kappe, Brandner, *Z. Naturforsch.* **29B**, 292 (1974). Biosynthesis: Grisebach, Barz, *Chem. & Ind. (London)* **1963**, 690.

Crystals, mp 385°. Sublimes at 325°; sublimes in high vacuum at about 175°. uv max (methanol): 208, 243, 343 nm. Exhibits bright blue fluorescence in neutral or acid soln, greenish-yellow fluorescence in strong alkali. Absorption and fluorescence spectra: O. S. Wolfbeis, K. Schaffner, *Photochem. Photobiol.* **32**, 143 (1980). Practically insol in water at acid and neutral pH, in petr ether. Sparingly sol in water at alkaline pH (pH 11-12); slightly sol in methanol, chloroform, ether; very slightly sol in carbon tetrachloride, benzene.

Diacetate, $C_{19}H_{12}O_7$, crystals from acetic acid, mp 237°.

Dimethyl ether, $C_{17}H_{12}O_5$, crystals from methanol, mp 198°.

2566. Coumetarol. *3,3'-(2-Methoxyethylidene)bis[4-hydroxy-2H-1-benzopyran-2-one]; 3,3'-(2-methoxyethylidene)-bis[4-hydroxycoumarin]*; 1-methoxy-2,2-bis[4-hydroxy-3-coumarinyl]ethane; 4,4'-dihydroxy-3,3'-(2-methoxyethylidene)dicoumarin; 2,2-bis[4'-hydroxy-3'-coumarinyl]ethyl methyl ether; cumetharol; cumethoxaethane; Ph 137; Dicoumoxyl; Dicumoxane. $C_{21}H_{16}O_7$; mol wt 380.34. C 66.31%, H 4.24%, O 29.45%. Prepn: Veldstra *et al.*, *Rec. Trav. Chim.* **72**, 358 (1953); **Brit. pat. 736,388** and **Neth. pat. 80,005** (both 1955 to N. V. Amsterdamsche Chininefabriek), *C.A.* **50**, 10795a, 16874b (1956). Stereodynamics: Laruelle, *Tetrahedron Letters* **1970**, 2235. Has antivitamin K action. Activity studies: de Jongh, Kok, *Arch. Int. Pharmacodyn.* **94**, 470 (1953); Gilgenkrantz *et al.*, *Ann. Med. Nancy* **3**, 115 (1964). Series of articles on clinical studies: Hutinel, *Lyon Med.* **23**, 1569-1604 (1964).

Crystals from ethanol or methanol, mp 156-157°.

THERAP CAT: Anticoagulant.

2567. Coumingine. *3-Hydroxy-3-methylbutanoic acid 7-[2-[2-(dimethylamino)ethoxy]-2-oxoethylidene]tetradecahydro-1,1,4a,8-tetramethyl-9-oxo-2-phenanthrenyl ester.* $C_{29}H_{47}NO_6$; mol wt 505.67. C 68.88%, H 9.37%, N 2.77%, O 18.98%. Cardiotonic principle from the bark of *Erythrophleum couminga* Baillon, *Leguminosae*. Isoln: Ruzicka *et al.*, *Helv. Chim. Acta* **24**, 63 (1941). Structure: Ruzicka *et al.*, *ibid.* 1449.

Thin shiny needles from ether, mp 142°. $[\alpha]_D^{20}$ −70°. Sol in methanol, ethanol, acetone, acetic acid, benzene, chloroform. Practically insol in water, hexane, ether.

Hydrochloride, $C_{29}H_{48}ClNO_6$, crystals from ethanol + ether, mp 195°. Sol in ethanol, water.

2568. Coumithoate. *Phosphorothioic acid O,O-diethyl O-(7,8,9,10-tetrahydro-6-oxo-6H-dibenzo[b,d]pyran-3-yl) ester; 2-(2,4-dihydroxyphenyl)-1-cyclohexene-1-carboxylic acid δ-lactone O,O-diethylphosphorothioate; O,O-diethyl O-7-hydroxy-3,4-tetramethylenecoumarinyl phosphorothioate; O,O-diethylthiophosphoric ester of 3,4-tetramethyleneumbelliferone; 3,4-tetramethyleneumbelliferone O,O-diethylthiophosphoric ester; 7-hydroxy-3,4-tetramethylenecoumarin O,O-diethylthiophosphate; Dithion; Dition.* $C_{17}H_{21}O_5PS$; mol wt 368.40. C 55.43%, H 5.75%, O 21.72%, P 8.41%, S 8.70%. Prepn: Fusco *et al.*, **U.S. pat. 2,860,085** (1958 to Montecatini).

Needles from benzene + petr ether, mp 88-88.5°. Practically insol in water. Has only limited solubility in organic solvents. LD_{50} orally in rats: 67 mg/kg, *World Rev. Pest Contr.* **9**, 119 (1970).

USE: Insecticide. *Caution:* A cholinesterase inhibitor. Symptoms similar to parathion.

2569. Crataegus. Hawthorn; English hawthorn; haws; haw apple; aubépine. Berries, flowers, and leaves of *Crataegus oxyacantha* L., *Rosaceae*. Habit. All temperate zones, abundant in Europe, a garden escape in North America. *Constit.* Triterpene acids (oleanolic, ursolic, crataegolic); purines; anthocyanin type pigments and flavone deriv (pelargonin, quercitrin); choline; acetylcholine; trimethylamine; chlorogenic acid; caffeic acid; ascorbic acid; a growth hormone for caterpillars; a substance named RN 30/9: Hahn *et al.*, *Arzneimittel-Forsch.* **10**, 825 (1960). Some commercial extracts are: *Curtacrat; Crataegus-Kreussler; Esbericard.* Review and bibliography: F. Berger, *Handbuch der Drogenkunde* vol. 3 (Vienna, 1952) pp 202-211.

THERAP CAT: Cardiotonic, coronary vasodilator.

2570. Creatine. *N-(Aminoiminomethyl)-N-methylglycine; N-amidinosarcosine; (α-methylguanido)acetic acid; N-methyl-N-guanylglycine; methylglycocyamine.* $C_4H_9N_3O_2$; mol wt 131.14. C 36.63%, H 6.92%, N 32.05%, O 24.40%. Present in muscular tissue of many vertebrates. Commercially isolated from meat extracts. Small amounts occur in the blood, but it is not found in normal urine from adults. The greater part of creatine in muscle is combined with phosphoric acid as phosphocreatine. Produced by liver and kidneys by the transfer of the guanidine moiety of arginine to glycine which is then methylated to give creatine. *In vitro* synthesis by liver and kidney tissues: Borsook, Dubnoff, *J. Biol. Chem.* **134**, 635 (1940). Review and bibliography: Peters, Van Slyke, *Quantitative Clinical Chemistry* Vol. I, *Interpretations* (Baltimore, 2nd ed., 1946). Synthesis by heating cyanamide with sarcosine: Strecker, *Jahresber. Chem.* **1868**, 686; *cf.* Volhard, *Z. Chem.* **5**, 318 (1869); Paulmann, *Arch. Pharm.* **232**, 638 (1894); Bergmann, Zervas, *Z. Physiol. Chem.* **173**, 80 (1928); King *J. Chem. Soc.* **1930**, 2374.

Monohydrate, monoclinic prisms from water. Becomes anhydr at 100°; dec 303°. Neutral reaction to litmus. Kb at 20° = 9.6 × 10⁻¹². Adsorption on various chromatographic agents: Grettie, Williams, *J. Am. Chem. Soc.* **50**, 671 (1928). Absorption spectrum: Abderhalden, Haas, *Z. Physiol. Chem.* **164**, 7 (1927). One gram of the monohydrate dissolves in 75 ml water, in about 9 liters alcohol. Insol in ether. In aq soln creatinine is formed, and aq and alkaline solns contain an equilibrium mixture of creatine and creatinine, while in acid solns the formation of creatinine is complete: Cannan, Shore, *Biochem. J.* **22**, 924 (1928). The ratio of the molar concn of creatine to creatinine in water and in various buffer solns is given by Edgar and Shiver, *J. Am. Chem. Soc.* **47**, 1179 (1925).

Picrate, yellow needles from water, mp 218-220°.

2571. Creatinine. *2-Amino-1,5-dihydro-1-methyl-4H-imidazol-4-one; 2-amino-1-methyl-4-imidazolidinone;* 1-methylhydantoin-2-imide; 1-methylglycocyamidine. C_4H_7-N_3O; mol wt 113.12. C 42.47%, H 6.24%, N 37.15%, O 14.14%. The end product of creatine catabolism. Normal constituent of urine; daily output about 25 mg per kg of body weight. Also found together with creatine in muscle tissues and blood. Occurs in all soils and in grain seeds and other vegetable matter. Has been found in certain fish and in crab meat extract. Isoln from urine: Maly, *Ann.* **159**, 279 (1871); Folin, *J. Biol. Chem.* **17**, 463 (1914); Benedict, *ibid.* **18**, 183 (1914). The isoln from urine is tedious, and creatinine is usually prepd from commercial creatine by treatment with HCl: Edgar, Hinegardner, *ibid.* **56**, 881 (1923); *Org. Syn.* **4**, 15 (1925).

Monoclinic plates. Leaflets from water. Dec about 300°. Kb at 40° = 3.57 × 10⁻¹¹. Sol in 12 parts water; slightly sol in alc; practically insol in acetone, ether, chloroform.

Picrate, pale-yellow nedles from water, mp 220-221°.

2572. Creolin®. Creolin-Pearson. Prepd from refined coal-tar oils. Approx composition: Tar acids and oils 75-77%; emulsifying soaps 15-17%; water 8-10%.

Dark brown liquid, characteristic odor resembling that of phenol. Phenol coefficient (against *B. typhosus*) about 10. d 1.02-1.04. Forms stable milky emulsions when dil with much water. Miscible with a small amount of water; also miscible with alcohol, ether, chloroform.

USE: As a general industrial and household disinfectant and deodorant in 1-3% emulsion in water.

THERAP CAT (VET): Antiseptic, parasiticide. Do not use on cats.

2573. Creosol. *2-Methoxy-4-methylphenol; 2-methoxy-p-cresol;* 4-methylguaiacol; 3-methoxy-4-hydroxytoluene; 4-hydroxy-3-methoxy-1-methylbenzene. $C_8H_{10}O_2$; mol wt 138.16. C 69.54%, H 7.29%, O 23.16%. Occurs in beech-wood tar. It is one of the active constituents of creosote. Obtained by the Clemmensen reduction of vanillin using amalgamated zinc and toluene as auxiliary solvent: Fletcher, Tarbell, *J. Am. Chem. Soc.* **65**, 1431 (1943); Schwarz, Hering, *Org. Syn., coll. vol. IV,* 203 (1963).

Colorless to yellowish, strongly refractive, aromatic liq. d_4^{25} 1.092. bp_{760} 220°; bp_{15} 105°; bp_4 79°. mp 5.5°. n_D^{25} 1.5353. Slightly sol in water; miscible with alcohol, benzene, chloroform, ether, glacial acetic acid.

Note: Not to be confused with cresol.

2574. Creosote, Coal Tar. Coal tar creosote. A distillate of coal tar produced by high temp carbonization of bituminous coal. *Constit.* Liq and solid aromatic hydrocar-

bons, tar acids (up to 3%) and tar bases. History and composition: Roche, *J. Forest Prod. Res. Soc.* **2**, 75 (1952). Characterization by GLC: F. H. M. Nestler, *Anal. Chem.* **46**, 46 (1974). GC-MS analysis in treated railroad ties: W. Rotard, W. Mailahn, *ibid.* **59**, 65 (1987). Toxicity: W. B. Deichmann, M. L. Keplinger in Patty's *Industrial Hygiene and Toxicology* Vol. 2A, G. D. Clayton, F. E. Clayton, Eds. (Wiley-Interscience, New York, 3rd ed., 1981) pp 2601-2604. Review of constituents, uses and carcinogenicity studies: *IARC Monographs* **35**, 83-159 (1985).

Translucent brown to black, oily liq. Characteristic sharp odor. Heavier than water. *Typical specification:* $d_{15.5}^{38.0}$ 1.06. Distillation ranges: Up to 210° not >5%; up to 235° not >25% nor <5%; up to 270° not <20%; up to 355° not >85% nor <60%. Flash pt 165°F (75°C); ignition temp 637°F (335°C). Practically insol in water.

Note: Occupational exposure to soots, tars and certain mineral oils is known to be carcinogenic: *Fourth Annual Report on Carcinogens* (NTP 85-002, 1985) pp 205-207.

USE: Wood preservative. Disinfectant, insecticide.

THERAP CAT (VET): Has been used as an anthelmintic.

2575. Creosote, Wood. Wood creosote; beechwood creosote; Creasote. Liquid obtained from wood tars by distillation; composed chiefly of guaiacol and creosol, *q.q.v.* Acute toxicity: T. Miyazato *et al., Oyo Yakuri* **21**, 899 (1981), *C.A.* **96**, 28311h (1982). Chronic toxicity and carcinogenicity studies: *eidem, ibid.* **28**, 909, 925 (1984), *C.A.* **102**, 41238b, 57380c (1985). *See also: Clinical Toxicology of Commercial Products,* R. E. Gosselin *et al.,* Eds. (Williams & Wilkins, Baltimore, 5th ed., 1984) Section II, p 192.

Almost colorless or yellowish, very refractive, oily liq; characteristic smoky odor; caustic, burning taste. d_{25}^{25} not below 1.076. Begins to boil at about 203° and at least 90% by vol distills between 203-220°. Does not solidify at −20°. Sol in 150-200 parts water, in glycerol, glacial acetic acid, fixed alkali hydroxide solns; miscible with alcohol, chloroform, ether, oils. *Incompat:* Acacia, albumin; cupric, ferric, gold and silver salts; oxidizers.

Calcium deriv, *calcium creosotate, Calcreose.* A mixture of calcium compds of creosote constituents. Dark-brown powder, creosote odor, sharp phenolic taste. Partly sol in water; soln deposits $Ca(OH)_2$ and $CaCO_3$ on standing.

Benzoate, yellowish liquid. Insol in water. Freely sol in alcohol or ether.

Carbonate, colorless to yellowish, clear, viscid, oily liquid; odorless and tasteless or slight odor and taste of creosote. d_{25}^{25} not below 1.145. Insol in water. Freely sol in alcohol; sol in petr ether, fixed oils. Miscible with benzene, chloroform.

Oleate, *Oleocreosote.* Yellowish, oily liquid. d 0.950. Insol in water. Sol in benzene, chloroform, ether.

Phosphate, *Phosote.* Mixture of the phosphoric acid esters of creosote. Yellowish, almost odorless, viscid mass. d_{15}^{15} 1.19. bp 230-235° with decompn. Insol in water. Sol in alcohol.

Valerate, *Eosote.* Colorless to yellowish liquid. bp about 240°. Insol in water. Sol in alcohol, ether.

THERAP CAT: Antiseptic; expectorant.

THERAP CAT (VET): Antiseptic, parasiticide, deodorant; has been used as expectorant, gastric sedative and gastrointestinal antiseptic. Do not use in cats.

2576. Creosotic Acid. Mixture of the isomeric cresotic acids.

White or reddish-white powder. Slightly sol in water; sol in alcohol, ether, alkali hydroxide solns.

USE: Disinfectant generally and in veterinary medicine in 1:500 aq or soap soln; manuf dyes and artificial tannin.

2577. Creslan®. Fiber X-54; Exlan. A copolymer of acrylonitrile possibly with acrylamide or a substituted acrylamide. *Ref:* R. W. Moncrieff, *Man-made Fibres* (John Wiley, New York, 1963) p 483.

Off-white fiber. Specific gravity 1.17. Sticks at 210°. Shrinkage in boiling water, 1%. Sunlight has negligible effect. Resistance to acids: good except to mineral acids. Resistance to alkalies: fair to dilute, poor to concd. Good resistance to cleaning solvents. Superior to its forerunner, *fiber X-51,* in the ease with which it can be dyed.

USE: Recommended for those uses that acrylics generally fill. It is claimed that it can be durably pleated.

2578. Cresol(s). Cresylic acid; cresylol; tricresol. C_7-H_8O; mol wt 108.13. C 77.75%, H 7.46%, O 14.80%. $HOC_6H_4CH_3$. Mixture of the three isomeric cresols, in which the *m*-isomer predominates. Obtained from coal tar: Paulsen, U.S. pat. 2,998,457 (1962 to Ashland Oil & Ref.). Usually contains a few per cent phenol. Prepn by sulfonation of toluene: Englund *et al., Ind. Eng. Chem.* **45**, 189 (1953); by oxidation of toluene: Braunwarth, Winsted, U.S. pat. 2,994,722 (1961 to Pure Oil).

Colorless, yellowish, brownish-yellow or pinkish liq; phenolic odor; becomes darker with age and on exposure to light. *Poisonous!* d_{25}^{25} 1.030-1.038. Not less than 90% by vol distils between 195-205°. Soluble in about 50 parts water; miscible with alcohol, benzene, ether, glycerol, petr ether; also sol in solns of fixed alkali hydroxides. A soln in water is neutral to bromocresol purple. *Protect from light.*

Human Toxicity: Orally 8 g or more produces rapid circulatory collapse, death. Chronic poisoning from oral or percutaneous absorption may produce digestive disturbances, nervous disorders with faintness, vertigo, mental changes, skin eruptions, jaundice, oliguria, uremia. *Caution:* General protoplasmic poison. *See also* Phenol.

USE: For making synthetic resins.

THERAP CAT: Disinfectant.

THERAP CAT (VET): Local antiseptic, parasiticide, disinfectant; has been used as an intestinal antiseptic.

2579. m-Cresol. *3-Methylphenol.* C_7H_8O; mol wt 108.13. C 77.75%, H 7.46%, O 14.80%. Obtained from coal tar: Maesawa, Kurakano, **Japan.** pat. 8929('55) (to Osaka Gas), *C.A.* **52**, 1231d (1958); Macak, Rehak, *Brennstoff Chem.* **43**, 80 (1962). Prepn from toluene: Toland, U.S. pat. 2,760,991 (1956 to California Res. Corp.); by oxidation of *o*- or *p*-toluic acid: Kaeding *et al., Ind. Eng. Chem.* **53**, 805 (1961). Review of manuf processes: Faith, Keyes & Clark's *Industrial Chemicals*, F. A. Lowenheim, M. K. Moran, Eds. (Wiley-Interscience, New York, 4th ed., 1975) pp 285-293.

Colorless or yellowish liquid; phenolic odor. d_4^{20} 1.034. bp 202°. mp 11-12°. Flash pt, closed cup: 187°F (86°C). n_D^{20} 1.5398. Sol in about 40 parts water, in solns of fixed alkali hydroxides; miscible with alc, chloroform, ether. LD_{50} orally in rats: 2.02 g/kg, Deichmann, Witterup, *J. Pharmacol. Exp. Ther.* **80**, 233 (1944).

USE: In disinfectants and fumigants; in photographic developers, explosives. *Caution:* See Phenol.

2580. o-Cresol. *2-Methylphenol; o-cresylic acid; o-hydroxytoluene.* Prepn from *m*-toluic acid: Toland, U.S. pat. 2,766,294 (1956 to California Res. Corp.); Barnard, Meyer, U.S. pat. 2,852,567 (1958 to Dow). Review of manuf processes: Faith, Keyes & Clark's *Industrial Chemicals*, F. A. Lowenheim, M. K. Moran, Eds. (Wiley-Interscience, New York, 4th ed., 1975) pp 285-293.

Crystals or liq, becoming dark with age and exposure to air and light; phenolic odor. d_4^{20} 1.047. bp 191-192°. mp 30°. Flash pt 81-83°. n_D^{20} 1.553. Sol in about 40 parts water, in solns of the fixed alkali hydroxides; miscible with alcohol, chloroform, ether. *Protect from light.* LD_{50} orally in rats: 1.35 g/kg, Deichmann, Witterup, *J. Pharmacol. Exp. Ther.* **80**, 233 (1944).

USE: As disinfectant like phenol; also as a solvent. *Caution:* See Phenol.

2581. p-Cresol. *4-Methylphenol.* Obtained from coal tar. Lab prepn from *p*-toluenesulfonic acid by fusion with potassium hydroxide: W. W. Hartman, *Org. Syn.* **coll. vol. I**, 175 (2nd ed., 1941); from toluene: Braunwarth, U.S. pat. 3,046,305 (1962 to Pure Oil). Review of manuf processes: Faith, Keyes & Clark's *Industrial Chemicals*, F. A. Lowenheim, M. K. Moran, Eds. (Wiley-Interscience, New York, 4th ed., 1975) pp 285-293.

Crystals. Phenolic odor. d_4^{20} 1.0341. mp 35.5°. bp$_{760}$ 201.8°; bp$_{200}$ 179.4°; bp$_{100}$ 140.0°; bp$_{40}$ 117.7°; bp$_{20}$ 102.3°; bp$_{10}$ 88.6°; bp$_5$ 76.5°; bp$_{1.0}$ 53.0°. Flash pt (closed cup) 86°C (187°F). Volatile in steam. n_D^{20} 1.5395. 100 ml water dissolves about 2.5 g at 50°, about 5 g at 100°. Sol in aq solns of alkali hydroxides; in the usual organic solvents. LD_{50} orally in rats: 1.8 g/kg, Deichmann, Witterup, *J. Pharmacol. Exp. Ther.* **80**, 233 (1944).

THERAP CAT: Disinfectant.

2582. o-Cresolphthalein. *3,3-Bis(4-hydroxy-3-methyl-phenyl)-1(3H)-isobenzofuranone; 3,3'-dimethylphenolphthalein.* $C_{22}H_{18}O_4$; mol wt 346.36. C 76.28%, H 5.24%, O 18.48%. Prepn: F. J. Welcher, *Organic Analytical Reagents* vol. 4 (Van Nostrand, New York, 1948) p 482; Hubacher *et al., J. Am. Pharm. Assoc.* **42**, 23 (1953).

Crystals from alc, mp 223°. Slightly sol in water; sol in alcohol. pK 9.4.

Note: The sodium salt exists in the quinone form. See Phenolphthalein Sodium.

USE: As indicator: pH range, 8.2 colorless, 9.8 red. Analytical reagent, Welcher, *loc. cit.*

2583. Cresol Red. *4,4'-(3H-2,1-Benzoxathiol-3-ylidene)bis(2-methylphenol) S,S-dioxide; α-hydroxy-α,α-bis(4-hydroxy-m-tolyl)-o-toluenesulfonic acid γ-sultone; o-cresolsulfonphthalein.* $C_{21}H_{18}O_5S$; mol wt 382.42. C 65.95%, H 4.74%, O 20.92%, S 8.39%. Prepn: Sohon, *Am. Chem. J.* **20**, 265 (1898), *Beilstein* **vol. 19**, 91; 1st suppl., 650; Lubs, Clark, *J. Wash. Acad. Sci.* **5**, 609 (1915), **6**, 481 (1916). Tautomerism: Ramart-Lucas, *Bull. Soc. Chim. France* [5] **10**, 282 (1943).

Reddish-brown cryst powder. Can be recrystallized from glacial acetic acid; sol in alc, water, dil acid (yellow soln), dil alkalies (purple soln): Lubs, Acree, *J. Am. Chem. Soc.* **38**, 2772 (1916). pK 8.3: Sager *et al., ibid.* **70**, 732 (1948).

USE: As indicator. pH range: 7.2 yellow, 8.8 red, 2-3 orange to amber.

2584. m-Cresotic Acid. *2-Hydroxy-4-methylbenzoic acid; 2,4-cresotic acid; m-homosalicylic acid; m-cresotinic acid; 2-hydroxy-p-toluic acid; γ-cresotic acid.* $C_8H_8O_3$; mol wt 152.14. C 63.15%, H 5.30%, O 31.55%. Prepn: Prelog *et al., Helv. Chim. Acta* **30**, 675 (1947); Baine *et al., J. Org. Chem.* **19**, 510 (1954); Hauptschein *et al., J. Am. Chem. Soc.* **77**, 2284 (1955).

Crystals or leaflets, mp 177°; volatile with steam. Soly as of the *o*-acid.

USE: In manuf of dyes. *Caution:* Toxicity is similar to salicylic acid, *q.v.*

2585. *o*-Cresotic Acid. *2-Hydroxy-3-methylbenzoic acid; 2,3-cresotic acid; o*-homosalicylic acid; *o*-cresotinic acid; 2-hydroxy-*m*-toluic acid; 3-methylsalicylic acid. Prepn: Baine *et al., J. Org. Chem.* **19,** 510 (1954); Jones, *Chem. & Ind. (London)* **1958,** 228; Blicke, McCarty, *J. Org. Chem.* **24,** 1061 (1959); Wessely *et al., Ber.* **93,** 2840 (1960).

White to slightly reddish, odorless cystals, mp 165-166°; volatile with steam. Slightly sol in cold, more sol in hot water; sol in chloroform, alcohol, ether, alkali hydroxides.

USE: In manuf of dyes. *Caution:* Toxicity is similar to salicylic acid.

2586. *p*-Cresotic Acid. *2-Hydroxy-5-methylbenzoic acid; 2,5-cresotic acid; p*-homosalicylic acid; *p*-cresotinic acid; 6-hydroxy-*m*-toluic acid. Prepn: Cameron *et al., J. Org. Chem.* **15,** 233 (1950); Baine *et al., ibid.* **19,** 510 (1954); Thomas *et al., J. Am. Chem. Soc.* **80,** 5864 (1958); Blicke, McCarty, *J. Org. Chem.* **24,** 1061 (1959).

White or slightly reddish, almost odorless crystals, mp 151°. Volatile with steam. Sublimes with partial decompn. Soly as of the *o*-acid.

Caution: Toxicity is similar to salicylic acid.

2587. *m*-Cresyl Acetate. *Acetic acid 3-methylphenyl ester; m*-tolyl acetate; *m*-cresol acetic acid ester; acetic acid *m*-cresol ester; acetylmetacresol; metacresol acetate; Cresatin; Cresatin-Sulzberger; Cresatin Metacresylacetate; Metacresylacetate-Sulzberger; Kresatin. $C_9H_{10}O_2$; mol wt 150.17. C 71.98%, H 6.71%, O 21.31%. Prepd by refluxing *m*-cresol with acetic anhydride: Claus, Hirsch, *J. Prakt. Chem.* [2] **39,** 62 (1889); from *m*-cresol and acetyl chloride: Eijkman, *Chem. Weekbl.* **1,** 453 (1904); *see also* **U.S.** pat. **1,031,971** (1912 to N. Sulzberger). Purification: Lebeau-Janot, *Traité Pharm. Chim.* **II,** 882 (Paris, 1955).

Oily liq. Volatile with steam. Characteristic phenolic odor with a reminiscence of acetone. d_4^{26} 1.048. bp 212°; bp_{13} 99°. uv max (methanol): 262.5, 269.5 nm. Practically insol in water, glycerol. Miscible with alc, ether, chloroform, petr ether, benzene. Soly in liq petrolatum about 5%. Also sol in cottonseed oil.

THERAP CAT: Topical antiseptic, antifungal.

THERAP CAT (VET): Topical antiseptic, fungicide.

2588. *o*-Cresyl Acetate. *Acetic acid 2-methylphenyl ester; acetic acid o-tolyl ester; o*-cresol acetate; acetyl-*o*-cresol; *o*-cresylic acetate; *o*-tolyl acetate. $C_9H_{10}O_2$; mol wt 150.17. C 71.98%, H 6.71%, O 21.31%. $CH_3C_6H_4OCOCH_3$.

Liquid. bp about 208°. Almost insol in cold, sol in hot water; freely sol in usual organic solvents, in oils.

2589. CRF. *Corticotropin-releasing factor;* CRH; corticoliberin(e); corticotropin-releasing hormone. Hypothalamic substance that stimulates secretion of corticotropin

(ACTH, *q.v.*) and β-endorphin from the pituitary. First direct evidence for its presence in hypothalami: R. Guillemin, B. Rosenberg, *Endocrinology* **57,** 599 (1955); M. Saffran, A. V. Schally, *Can. J. Biochem. Physiol.* **33,** 408 (1955). Purification and characterization of a 41-residue ovine hypothalamic CRF that is highly potent in stimulating secretion of corticotropin and β-endorphin-like immunoactivities: W. Vale *et al., Science* **213,** 1394 (1981). Primary structure: J. Spiess *et al., Proc. Nat. Acad. Sci. USA* **78,** 6517 (1981). When centrally administered, CRF activates the sympathetic nervous system and may therefore function as a key hormone in stress mobilization of the organism, *cf.* W. Vale *et al., loc. cit.;* M. Brown *et al., Life Sci.* **30,** 207 (1982). CRF also stimulates α-MSH secretion and cyclic AMP accumulation in rat pars intermedia cells: H. Meunier *et al., ibid.* **31,** 2129 (1982). A functional relationship between CRF and dynorphin-related opioid peptides has also been suggested: K. A. Roth *et al., Science* **219,** 189 (1983). *Reviews:* A. V. Schally *et al., Recent Progr. Horm. Res.* **24,** 497-588 (1968); R. Burgus, R. Guillemin, *Ann. Rev. Biochem.* **39,** 499-526 (1970); M. Saffran in *Hypothalamic Peptide Hormones and Pituitary Regulation,* J. C. Porter, Ed. (Plenum Press, New York, 1977) pp 225-235; W.Vale *et al.,* in *The Role of Peptides in Neuronal Function,* J. L. Barker, J. G. Smith, Eds. (Dekker, New York, 1980) pp 432-454; several authors in *Polypeptide Hormones,* R. F. Beers, E. G. Bassett, Eds. (Raven Press, New York, 1980) pp 165-271; B. Lutz-Bucher *et al., J. Physiol.* **77,** 939-950 (1981); N. Yasuda *et al., Endocrine Rev.* **3,** 123-140 (1982); E. Stark, G. B. Makara in *Hormonally Active Brain Peptides: Structure and Function,* K. W. McKerns, V. Pantic, Eds. (Plenum Press, New York, 1982) pp 157-179.

Activity is destroyed by trypsin; unaffected by thioglycolate, pepsin, chymotrypsin.

2590. Crimidine. *2-Chloro-N,N,6-trimethyl-4-pyrimidinamine; 2-chloro-4-(dimethylamino)-6-methylpyrimidine;* W-491; Castrix. $C_7H_{10}ClN_3$; mol wt 171.64. C 48.99%, H 5.87%, Cl 20.66%, N 24.48%. Prepd from an appropriate 2,4-dihalopyrimidine and dimethylamine: Westphal, **U.S.** pat. **2,219,858** (1940 to Winthrop).

Commercial product is a brown waxy solid, mp 87°. Practically insol in water; sol in most organic solvents. *Poisonous!* LD_{50} orally in female mice: 1.2 mg/kg. Pyridoxine was found to be an excellent antidote for mice poisoned with crimidine: Karlog, Knudsen, *Nature* **200,** 790 (1963).

USE: Rodenticide. *Caution:* May cause serious CNS damage leading to fatal convulsions.

2591. Croceic Acid. *7-Hydroxy-1-naphthalenesulfonic acid;* 2-naphthol-8-sulfonic acid; Bayer's acid. $C_{10}H_8O_4S$; mol wt 224.23. C 53.56%, H 3.60%, O 28.54%, S 14.30%. Prepn: Nietzki, Zübelen, *Ber.* **22,** 453 (1889); Forster, Keyworth, *J. Soc. Chem. Ind. (London)* **46,** 26T (1927).

On evaporation of an aq soln, the free acid dec into sulfuric acid and β-naphthol. However, very concd solns of the acid can be obtained. The normal sodium salt is sol in water, and readily sol in alcohol, crystallizing with 2 mols of alcohol. *Ref:* Forster, Keyworth, *loc. cit.*

2592. Crocetin. *8,8′-Diapocarotenedioic acid.* $C_{20}H_{24}O_4$; mol wt 328.39. C 73.14%, H 7.37%, O 19.49%. Carotenoid dicarboxylic acid isolated from *Crocus sativus* L.; *C. albiflorus* var. *neapolitanus* Hort.; *C. luteus* Lam., Iridaceae. Extraction procedure and structure: Jucker, Karrer, *Carotinoide* (Basel, 1948) p 282.

trans-Form, brick-red rhombs from acetic anhydride, mp 285°. Absorption max (pyridine): 464, 436, 411 nm. Sol in pyridine, in very dil NaOH solns. Very sparingly sol in water, and in the usual organic solvents except pyridine and similar organic bases. Forms a solid dipyridyl salt.

Dimethyl ester, $C_{22}H_{28}O_4$, brick-red elongated leaflets, mp 222.5°. Total synthesis: Buchta, Andree, *Naturwiss.* **46**, 74 (1959).

2593. Crocin. *8,8'-Diapo-ψ,ψ-carotenedioic acid bis(6-O-β-D-glucopyranosyl-β-D-glucopyranosyl) ester;* α-*crocin;* di-gentiobiose ester of crocetin. $C_{44}H_{64}O_{24}$; mol wt 977.00. C 54.09%, H 6.60%, O 39.30%. The coloring principle of saffron. Occurs in crocus and gardenia. Isoln and structure: Karrer, Salomon, *Helv. Chim. Acta.* **10**, 397 (1927); **11**, 513, 711 (1928); Karrer *et al., ibid.* **12**, 985 (1929); **13**, 392 (1930); Kuhn, Winterstein, *Ber.* **67**, 344 (1934); Reichstein, *Angew. Chem.* **74**, 887 (1962). Crocin and crocetin dimethyl ester were shown to play a role in sex processes of algae of the *Chlamydomonas* group: Kuhn, *ibid.* **53**, 1 (1940).

Hydrated brownish-red needles from methanol, mp 186° (effervescence). Absorption max (methanol): 464, 434 nm. Freely sol in hot water giving an orange-colored soln. Sparingly sol in abs alcohol, ether, other organic solvents.

2594. Cromolyn. *5,5'-[(2-Hydroxy-1,3-propanediyl)bis(oxy)]bis[4-oxo-4H-1-benzopyran-2-carboxylic acid]; 5,5'-[(2-hydroxytrimethylene)dioxy]bis(4-oxo-4H-1-benzopyran-2-carboxylic acid); 5,5'-(2-hydroxytrimethylenedioxy)bis(4-oxochromene-2-carboxylic acid); 1,3-bis(2-carboxychromon-5-yloxy)-2-hydroxypropane; 1,3-di(2-carboxy-4-oxochromen-5-yloxy)propan-2-ol; cromoglycic acid; Duracroman.* $C_{23}H_{16}O_{11}$; mol wt 468.38. C 58.98%, H 3.44%, O 37.58%. Chromone complex which blocks mast cell degranulation. Prepn: Fitzmaurice, Lee, **Brit. pat. 1,144,906** (1969 to Fisons). Metabolism: M. J. Ashton *et al., Toxicol. Appl. Pharmacol.* **26**, 319 (1973). Mechanism of action: T. C. Theoharides *et al., Science* **207**, 80 (1980); R. G. Alvarez *et al., Agents Actions* **11**, 94 (1981). Toxicology in primates: J. E. Beach *et al., Toxicol. Appl. Pharmacol.* **57**, 367 (1981). Clinical study in allergic conjunctivitis: G. A. Friday *et al., Am. J. Ophthalmol.* **95**, 169 (1983). Review of pharmacology and clinical use: G. G. Shapiro, P. Konig, *Pharmacotherapy* **5**, 156-170 (1985). *Review:* J. S. G. Cox in *Pharmacological and Biochemical Properties of Drug Substances* vol. 1, M. E. Goldberg, Ed. (Am. Pharm. Assoc., Washington, DC, 1977) pp 277-310.

Disodium salt, $C_{23}H_{14}Na_2O_{11}$, *cromolyn sodium, disodium cromoglycate, DSCG, FPL 670, Aarane, Aararre, Alercrom, Alerion, Allergocrom, Colimune, Fivent, Frenasma, Gastrofrenal, Inostral, Intal, Introl, Irtan, Lomudal, Lomudas, Lomupren, Lomuspray, Nalcrom, Nalcron, Nasalcrom, Nasmil, Opticrom, Opticron, Rynacrom, Vividrin.* White or creamy-white powder, mp 241-242° (dec). Freely sol in

water (100 mg/ml at 20°). Practically insol in chloroform and alcohol.

THERAP CAT: Prophylactic anti-asthmatic. Anti-allergic.

2595. Cropropamide. *N-[1-[(Dimethylamino)carbonyl]-propyl]-N-propyl-2-butenamide; N-(1-dimethylcarbamoylpropyl)-N-propylcrotonamide;* α-*(N'*-crotonyl-*N'*-propyl)-amino-N,N-dimethylbutyramide.* $C_{13}H_{24}N_2O_2$; mol wt 240.34. C 64.96%, H 10.07%, N 11.66%, O 13.31%. Prepn: Martin, Gysin, **U.S. pat. 2,447,587** (1948 to Geigy).

Liquid, $bp_{0.25}$ 128-130°. Easily sol in water, ether. Combination with crotethamide, *prethcamide, Micoren.*

THERAP CAT: Analgesic. Combination as respiratory stimulant.

2596. Crotamine. One of the poisonous principles found in the venom of Brazilian rattlesnakes, *Crotalus terrificus crotaminicus* Goncalves, *Crotalidae.* A polypeptide having mol wt 10,000-15,000 and isoelectric point at pH 10.3. Purification and properties: Goncalves in *Venoms,* E. E. Buckley, N. Porges, Eds. (Publ. no. **44** of the Am. Assoc. Adv. Science, Washington, D.C., 1956) pp 261-274. Pharmacology: Moussatché *et al., ibid.* pp 275-279.

2597. Crotamiton. *N-Ethyl-N-(2-methylphenyl)-2-butenamide; N-ethyl-o-crotonotoluidide;* crotonyl-N-ethyl-o-toluidine; Crotamitex; Eurax; Euraxil; Veteusan. $C_{13}H_{17}NO$; mol wt 203.27. C 76.81%, H 8.43%, N 6.89%, O 7.87%. Prepd by treating crotonic acid o-toluidide with diethyl sulfate: **Swiss pats. 253,472-3; Brit. pat. 615,137** (1949 to Geigy).

Yellowish oil, bp_{13} 153-155°. Sol in methanol, ethanol.
USE: Fungicide, insecticide.
THERAP CAT: Scabicide.
THERAP CAT (VET): Scabicide; antipruritic.

2598. Crotethamide. *N-[1-[(Dimethylamino)carbonyl]-propyl]-N-ethyl-2-butenamide; N-(1-dimethylcarbamoylpropyl)-N-ethylcrotonamide;* α-*(N'*-crotonyl-*N'*-ethyl)amino-N,N-dimethylbutyramide.* $C_{12}H_{22}N_2O_2$; mol wt 226.31. C 63.68%, H 9.80%, N 12.38%, O 14.14%. Prepn: Martin, Gysin, **U.S. pat. 2,447,587** (1948 to Geigy).

Liquid, $bp_{0.03}$ 132-134°. Easily sol in water, ether. Combination with cropropamide, *prethcamide, Micoren.*

THERAP CAT: Analgesic. Combination as respiratory stimulant.

2599. Crotonaldehyde. *2-Butenal;* crotonic aldehyde; β-methylacrolein. C_4H_6O; mol wt 70.09. C 68.54%, H 8.63%, O 22.83%. $CH_3CH=CHCHO$. The crotonaldehyde of commerce has the *trans*-configuration: Young, *J. Am. Chem. Soc.* **54**, 2498 (1932). Obtained by adding aldol to a boiling dil acid soln and removing the crotonaldehyde by distillation: v. Auwers, Eisenlohr, *J. Prakt. Chem.* **82**, 115 (1910); Hibbert, *J. Am. Chem. Soc.* **37**, 1759 (1915). Alternate synthesis: J. Smidt *et al., Angew. Chem.* **71**, 176 (1959). *Review:* W. F. Baxter in Kirk-Othmer *Encyclopedia of Chemical Technology* vol. 7 (Wiley-Interscience, New York, 3rd ed., 1979) pp 207-218.

Flammable liq. *Lacrimator!* Vapors extremely irritating. mp $-76.5°$; bp $104.0°$; d_{20}^{20} 0.853; $n_D^{17.3}$ 1.4384. Flash pt when anhydr: $13°$ ($55°F$) (open cup). Explosive limits in air 2.95-15.5% v/v. Vapor density 2.41 (air = 1). Heat capacity: 0.7 cal/g/°C; heat of vaporization: 8.62 cal/mole. Commercial crotonaldehyde (93%) is stabilized with water (a solid phase separates out at $-5°$). Readily resinifies to a dimer when pure, slowly oxidizes to crotonic acid. Soly in water (g/100 g) at 20°: 18.1; at 5°: 19.2. Soly of water in crotonaldehyde (g/100 g) at 20°: 9.5; at 5°: 8.0. LD_{50} orally in rats: 0.3 g/k, H. F. Smyth, C. P. Carpenter, *J. Ind. Hyg. Toxicol.* **26**, 269 (1944).

Caution: Keep away from heat and open flame. Keep container closed. Use with adequate ventilation. Highly irritating to eyes, skin, mucous membranes. Do not breathe vapor. In case of contact, immediately flush skin or eyes with plenty of water for at least 15 min; for eyes get medical attention. Remove and wash clothing before re-use. When necessary the lacrimatory effect of the vapors may be counteracted by ammonia fumes.

Note: *cis-Crotonaldehyde* has also been prepared: M. Perrier, F. Rouessac, *Nouv. J. Chim.* **1**, 367 (1977).

USE: Manuf of butyl alcohol, butyraldehyde, quinaldine. As warning agent in fuel gases, in locating breaks and leaks in pipes. Minor amounts are used in the manuf of maleic acid, crotyl alcohol, butyl chloral hydrate, and in rubber accelerators. In organic syntheses; as solvent in purification of mineral oils, manuf of resins, rubber antioxidants, insecticides. In chemical warfare.

2600. Crotonic Acid. *trans-2-Butenoic acid; β-methylacrylic acid; α-crotonic acid; solid crotonic acid.* $C_4H_6O_2$; mol wt 86.09. C 55.80%, H 7.03%, O 37.17%. Has been found in clay soil in Texas; formed during the dry distillation of wood. Obtained on a commercial scale exclusively by oxygen- or air-oxidation of crotonaldehyde: Kennedy, U.S. pat. 2,413,235 (1946 to Shawinigan Chem.); Leupold, PB report 70249 (1942); Matthews, B.I.O.S. report 758 (1946). Laboratory procedure using alkaline silver oxide: Young, *J. Am. Chem. Soc.* **54**, 2498 (1932); from acetaldehyde and malonic acid in pyridine: v. Auwers, *Ann.* **432**, 58 (1923); Backer, Bloemen, *Rec. Trav. Chim.* **45**, 102 (1926); Florence, *Bull. Soc. Chim France* [4] **41**, 440 (1927); Letch, Linstead, *J. Chem. Soc.* **1932**, 454. *Review:* W. F. Baxter in Kirk-Othmer *Encyclopedia of Chemical Technology* **vol. 7** (Wiley-Interscience, New York, 3rd ed., 1979) pp 218-226. *Caution:* Explosions have occurred.

Monoclinic needles, prisms from water or ligroin. d_4^{15} 1.018; d_4^{80} 0.964. mp 71.6°. bp_{10} 80.0°; bp_{20} 93.0°; bp_{40} 107.8°; bp_{60} 116.7°; bp_{100} 128.0°; bp_{200} 146.0°; bp_{400} 165.5°; bp_{760} 185.0°. n_D^{80} 1.4228. K at 25° = 2.0 × 10^{-5}. Soly in water (g/l) at 0°: 41.5; at 10°: 54.6; at 20°: 76.1; at 25°: 94; at 30°: 122; at 40°: 656. Soly in ethanol at 25°: 52.5% w/w; acetone: 53.0% w/w; toluene: 37.5% w/w. LD_{50} orally in rats: 1.0 g/kg, Smyth, Carpenter, *J. Ind. Hyg. Toxicol.* **26**, 269 (1944).

Methyl ester, $C_5H_8O_2$, liq, bp 121°; d_4^{20} 0.9444; n_D^{20} 1.4242. Ethyl ester, $C_6H_{10}O_2$, liq, bp 138°; d_4^{20} 0.9175; n_D^{20} 1.4245. Vinyl ester, $C_6H_8O_2$, liq, bp 133°; d_4^{20} 0.9410; n_D^{20} 1.450.

USE: In the manuf of copolymers with vinyl acetate used in lacquers and paper sizing; in the manuf of softening agents for synthetic rubber. In medicinal chemistry, e.g., in the manuf of DL-threonine, vitamin A.

2601. Croton Oil. Fixed oil expressed from seeds of *Croton tiglium* L., *Euphorbiaceae.* *Constit.* Croton resin—a powerful vesicant; glycerides of stearic, palmitic, myristic, lauric, tiglic, etc., acids; crotin; phorbol (q.v.) esters. *Review:* Hecker, Schmidt, *Fortschr. Chem. Org. Naturst.* **31**, 377 (1974).

Pale-yellow or brownish-yellow, rather viscid liq; slight disagreeable odor. *Poisonous!* d_{25}^{25} 0.935-0.950. n_D^{40} 1.470-1.473. Sapon no. 205-220. Iodine no. 102-118. Slightly sol in alcohol, increasing with age. Freely sol in chloroform,

ether, oils, petr ether, carbon disulfide, glacial acetic acid. *Keep well closed and protected from light.*

USE: Has been used as an irritant and cocarcinogen in cancer research.

THERAP CAT: Cathartic, counterirritant.

THERAP CAT (VET): Formerly used as a counterirritant and a purgative.

2602. Crotoxin. Neurotoxin. One of the toxic principles isolated from rattlesnake venom: Slotta, Fraenkel-Conrat, *Ber.* **71**, 1076 (1938). A polypeptide complex of probably two components, one acidic and one basic, having mol wts of ~9000 and 12,000 resp. The subunits do not retain the toxicity of crotoxin but show synergistic action in combination: Hendon, Fraenkel-Conrat, *Proc. Nat. Acad. Sci. USA* **68**, 1560 (1971). Amino acid analyses and fractionation studies: Fraenkel-Conrat, Singer, *Arch. Biochem. Biophys.* **60**, 64 (1956); Rübsamen *et al.*, *Arch. Pharmacol.* **270**, 274 (1971); Horst *et al.*, *Biochem. Biophys. Res. Commun.* **46**, 1042 (1972). Review of early chemical studies: Fraenkel-Conrat, Singer, *Publ. Am. Assoc. Advan. Sci.* **44**, 259 (1956). *Monograph:* Behringwerk-Mitteilungen Sonderband, *Die Giftschlangen der Erde* (N. G. Elwert, Marburg, 1963) 464 pp. Pharmacology: Brazil, Excell, *J. Physiol. (London)* **212**, 34P (1970); Brazil *et al.*, *ibid.* **234**, 63P (1973); Habermann *et al.*, *Arch. Pharmacol.* **273**, 313 (1972). *See also* Crotamine.

Human Toxicity: When injected by snake bite, local pain, redness, hemorrhage and necrosis result. Systemic effects are dizziness, sensory and motor depression, collapse, shock. May be fatal. *Antidote:* Rattlesnake antivenin.

2603. Crotoxyphos. *3-[(Dimethoxyphosphinyl)oxy]-2-butenoic acid 1-phenylethyl ester; 3-hydroxycrotonic acid α-methylbenzyl ester dimethyl phosphate; O,O-dimethyl O-[1-methyl-2-(1-phenylcarbethoxy)vinyl] phosphate; dimethyl 2-(α-methylbenzyloxycarbonyl)-1-methylvinyl phosphate;* ENT 24717; SD 4294; Ciodrin. $C_{14}H_{19}O_6P$; mol wt 314.28. C 53.50%, H 6.09%, O 30.55%, P 9.86%. Prepn: **Brit.** pat. 855,238 (1960 to Bataafse Petroleum Maatschappij N.V.); Whetstone, Stiles, U.S. pat. 2,982,686 (1961 to Shell). Activity: C. P. Weidenback, R. L. Younger, *J. Econ. Entomol.* **55**, 793 (1962). Metabolism and degradn: K. I. Beynon *et al.*, *Residue Rev.* **47**, 55 (1973).

Light straw-colored liq, $bp_{0.03}$ 135°. n_D^{25} 1.4988. d^{25} 1.19. Vapor press. at 20°: 1.4 × 10^{-5} mm Hg. Soly in water at room temp: 0.1%. Slightly sol in kerosine, saturated hydrocarbons. Sol in acetone, chloroform, ethanol, highly chlorinated hydrocarbons. LD_{50} orally in female, male rats: 74, 110 mg/kg, T. B. Gaines, *Toxicol. Appl. Pharmacol.* **14**, 515 (1969).

USE: Insecticide for external use on livestock. *Caution:* A cholinesterase inhibitor.

2604. Crotyl Alcohol. *2-Buten-1-ol;* crotonyl alcohol. C_4H_8O; mol wt 72.10. C 66.63%, H 11.18%, O 22.19%. $CH_3CH=CHCH_2OH$. Prepn by reduction of crotonaldehyde: Nystrom, Brown, *J. Am. Chem. Soc.* **69**, 1197 (1947); Olivier, Young, *ibid.* **81**, 5811 (1959); Smith, Holm; Finch, Furman, U.S. pats. 2,761,883; 2,763,696 (both 1956 to Shell); Foreman, U.S. pat. 3,109,865 (1963 to Standard Oil Co., Ohio). Prepn of cis- and trans-forms: Hatch, Nesbitt, *J. Am. Chem. Soc.* **72**, 727 (1950); Hiskey *et al.*, *J. Org. Chem.* **21**, 429 (1956); of cis-form only: Clarke, Crombie, *Chem. & Ind. (London)* **1957**, 143.

Liquid, bp 122°. d_4^{20} 0.8532. n_D^{20} 1.4285. Sol in 6 parts water; misc with alc. LD_{50} orally in rats: 0.93 ml/kg, H. F. Smyth *et al.*, *Am. Ind. Hyg. Assoc. J.* **23**, 95 (1962).

*cis-*Form, liquid, bp 123.6°. mp $-90.15°$. d_4^{20} 0.8662. n_D^{20} 1.4342.

trans-Form, liquid, bp 121.2°. d_4^{25} 0.8454. n_D^{20} 1.4289, n_D^{25} 1.4270.

2605. Crown Ethers. Crown compounds. Macrocyclic polyethers with the repeating unit $(-CH_2-CH_2-O-)_n$, where *n* is greater than 2. Crown compounds with other heteroatoms (N, S, P) are known. Prepd by C. J. Pedersen, *J. Am. Chem. Soc.* **89**, 2495, 7017 (1967). Described as "crown" ethers due to appearance of space-filling models and ability to "crown" cations. Proposed nomenclature lists non-ring substituents, number of atoms in ring, the class (crown), and the number of heteroatoms in the ring; e.g. dibenzo-18-crown-6. Bicyclic crowns are called *cryptates*. Crowns complex with cations and solubilize inorganic reagents in organic solvents. Selectivity results from the definite size of the crown cavity which admits only cations of corresponding ionic radii. The stability of complexes depends on the size of the cation and the size of the polyether ring. Chiral crowns are used in optical resolution of enantiomers. *Reviews:* D. J. Cram, J. M. Cram, *Science* **183**, 803-809 (1974); *eidem, Accounts Chem. Res.* **11**, 8-14 (1978); J. J. Christensen *et al., Chem. Rev.* **74**, 351-384 (1974); G. W. Gokel, H. D. Durst, *Synthesis* **1976**, 168-184; A. C. Knipe, *J. Chem. Ed.* **53**, 618-622 (1976); V. Prelog, *Pure Appl. Chem.* **50**, 893-904 (1978). Historical overview: C. J. Pedersen, *Science* **241**, 536-540 (1988).

Dibenzo-18-crown-6

18-Crown-6, $C_{12}H_{24}O_6$, *1,4,7,10,13,16-hexaoxacyclooctadecane.* Prepn: G. W. Gokel *et al., Org. Syn.* **57**, 30 (1977). Crystals from acetonitrile, mp 38-39.5°.

Dibenzo-18-crown-6, $C_{20}H_{24}O_6$, *2,3,11,12-dibenzo-1,4,7,-10,13,16-hexaoxacyclooctadeca-2,11-diene.* The first crown ether. Prepn: C. J. Pedersen, U.S. pat. **3,687,978** (1972 to Du Pont). White crystals from acetone, mp 164°, bp$_{679}$ 380-384°.

Dicyclohexano-18-crown-6, $C_{20}H_{36}O_6$, *"dicyclohexyl-18-crown-6".* Causes eye, skin irritation in test animals. Approximate lethal dose in rats (mg/kg): 300 orally; 130 skin absorption (Pedersen).

USE: In organic synthesis as phase transfer reagents, catalysts and in sepn of enantiomers.

2606. CRP. C-Reactive Protein. Homogeneous acute phase protein, found in man and most other animals. Discovered as a result of its precipitation with C-polysaccharide (CPS) in sera of patients with infections and inflammatory disease: W. S. Tillet, T. Francis, *J. Exp. Med.* **52**, 561 (1930). Characterization as a protein and identification of its calcium ion requirement for interaction with CPS: T. J. Abernathy, O. T. Avery, *ibid.* **73**, 173 (1941). CRP is a trace serum protein consisting of single subunits of mol wt about 21,000. In serum and in the purified state, the subunits aggregate as cyclic pentamers having mol wts of 110,000-144,000. The concentrations of C-reactive protein increase up to 1000-fold in inflammatory conditions and in tissue necrosis. In addition, it can initiate reactions of agglutination, precipitation, and opsonization for phagocytosis and can activate the complement system. Other biological activities with platelets and lymphocytes have also been described, but the full biological functions of CRP have not yet been completely elucidated. Review of CRP and the acute phase response: H. Gewurz *et al., Advan. Int. Med.* **27**, 345-372 (1982). Review of structure and function: M. B. Pepys, *Eur. J. Rheumatol. Inflammation* **5**, 386-397 (1982). Book: *Ann. N.Y. Acad. Sci.* **389**, entitled "C-Reactive Protein and the Plasma Protein Response to Tissue Injury", I. Kushner *et al.,* Eds. (1982) 482 pp.

2607. Crufomate. *Methylphosphoramidic acid 2-chloro-4-(1,1-dimethylethyl)phenyl methyl ester; methylphosphoramidic acid 4-tert-butyl-2-chlorophenyl methyl ester; 4-tert-butyl-2-chlorophenyl N-methyl O-methylphosphoramidate;* Dowco 132; Montrel; Ruelene. $C_{12}H_{19}ClNO_3P$; mol wt 291.71. C 49.41%, H 6.56%, Cl 12.16%, N 4.80%, O 16.45%, P 10.61%. Prepn: Wasco *et al.,* U.S. pat. **2,929,762** (1960 to Dow).

Crystals from petr ether, mp 60-60.5°. Relatively stable in aq dispersions at pH 7 or below. Commercial product is a yellow oil, bp$_{0.01}$ 117-118°. n_D^{20} 1.5142. Sol in acetone, benzene, carbon tetrachloride; practically insol in water, light petroleum. LD$_{50}$ orally in male, female rats: 635, 460 mg/kg, T. B. Gaines, *Toxicol. Appl. Pharmacol.* **14**, 515 (1969).

USE: Insecticide.

THERAP CAT (VET): Anthelmintic.

2608. Cryofluorane. *1,2-Dichloro-1,1,2,2-tetrafluoroethane;* Freon 114; Frigen 114; Arcton 114. $C_2Cl_2F_4$; mol wt 170.93. C 14.05%, Cl 41.49%, F 44.46%. ClCF$_2$CClF$_2$. Prepn: Henne, *Organic Reactions* **2**, 49 (1944).

Colorless, practically odorless, noncorrosive, nonirritating, nonflammable gas. Faint, ether-like odor in high concentrations. d_{liq}^0 1.5312. mp $-94°$. bp$_{760}$ +4.1°. n_D^0 1.3092. Critical temp 145.7°; crit pressure 474 lb/sq in. abs. Practically insol in water. Sol in alcohol, ether. Absorbs less than 0.0025% water. Has little, if any, anesthetic or toxic effect on humans, but toxic substances may be formed on contact with a flame or hot metal surface.

Note: Consult latest Government regulations on use as aerosol propellant.

USE: Refrigerant, aerosol propellant.

2609. Cryolite. Kryolith; ice spar; sodium aluminum fluoride. AlF_6Na_3; mol wt 209.97. Al 12.85%, F 54.29%, Na 32.86%. 3NaF.AlF$_3$. A mineral; large deposits exist in Greenland and in the Urals. Now prepd synthetically for the aluminum industry. Aluminum oxide is sol and dissociable in molten cryolite, and electrodeposition of aluminum metal is thus possible. Synthetic cryolite is usually made from NaAlO$_2$, NaHCO$_3$ and NaF. *See* the review by Vogel, *Chemiker-Ztg.* **75**, 603 (1951).

Snow-white, semi-opaque masses, vitreous fracture. Seldom monoclinic crystals with cubic habit, perfect cleavage (001) (110) (101). The natural form may be colored reddish or brown or even black, but loses this discoloration on heating. d 2.95. Mohs' hardness 2.5-3. Fuses fairly easily, mp 1000°. Sol in concd H$_2$SO$_4$. Dec by boiling with aq alkali hydroxides or aq calcium hydroxide. LD$_{50}$ orally in rats: 200 mg/kg, *RTECS* Vol. II, R. J. Lewis, R. L. Tatken, Eds. (1979) p 531.

USE: In the aluminum and fluorine industry. Deville, *Ann. Chim. Phys.* [3] **61**, 329 (1861), found that fused cryolite is dissociated by the passage of electric current, with deposition of aluminum metal. Although most aluminum is prepd from bauxite, the presence of cryolite is necessary, since the process operates as the principle mentioned.

2610. Cryptenamine Tannates. Unitensen Tannate. A prepn of ester alkaloids from *Veratrum viride* Ait., Liliaceae, obtained by non-aqueous benzene-triethylamine extraction process: Cavallito, U.S. pat. **2,789,977** (1957 to Irwin, Neisler). Contains hypotensive alkaloids protoveratrines A and B, germitrine, neogermitrine, germerine, germidine; also jervine, rubijervine, isorubijervine, germbudine: Kupchan, Gruenfeld, *J. Am. Pharm. Assoc.* **48**, 727 (1959).

Cryptamine tannate is a tan, amorphous powder. Sol in alcohol and slightly sol in water.

Cryptenamine itself is a white, amorphous powder which induces sneezing. Freely sol in alc, practically insol in water. Readily sol in dil acid soln. $[\alpha]_D^{25}$ -25 to $-40°$ (ethanol).

THERAP CAT: Antihypertensive.

2611. Cryptopine. *4,6,7,13-Tetrahydro-9,10-dimethoxy-5-methylbenzo[e]-1,3-dioxolo[4,5-l][2]benzazecin-12(5H)-one;* cryptocavine. $C_{21}H_{23}NO_5$; mol wt 369.40. C 68.28%, H 6.28%, N 3.79%, O 21.66%. Occurs in opium (0.003-0.03%). Has been found in *Corydalis sempervirens* (L.) Pers., and in *Dicentra* spp., *Fumariaceae:* Manske, *Can. J. Res.* **8**, 407 (1933); **7**, 265 (1932); **15B**, 274 (1937). Structure: Perkin, *J. Chem. Soc.* **115**, 713 (1919). Synthesis: Haworth, Perkin, *ibid.* **1926**, 1769.

Crystallizes as six-sided prisms or plates from benzene, mp 220-221°. d 1.35. Almost insol in water and ether. Soluble in chloroform, acetic acid. One part dissolves in 455 parts alc or 80 parts boiling alc. Sparingly sol in most other organic solvents. May be crystallized from hot alc, benzene, petr ether, methyl ethyl ketone, isoamyl alcohol, acetophenone, pyridine, or an alcohol-pyridine mixture. The salts tend to oil out, but can be crystallized by warming, redissolving, and slow cooling.

2612. Cryptoxanthin. *(3R)-β,β-Caroten-3-ol;* cryptoxanthol; caricaxanthin; hydroxy-β-carotene; β-caroten-3-ol. $C_{40}H_{56}O$; mol wt 552.85. C 86.90%, H 10.21%, O 2.89%. Carotenoid pigment with vitamin A activity, best isolated from petals and berries of *Physalis* species: Kuhn, Grundmann, *Ber.* **66**, 1746 (1933). Widely distributed, also was isolated from orange rind, *Carica papaya* L., *Caricaceae, Cucurbita pepo* L., *Cucurbitaceae, Zea mays* L., *Gramineae,* egg yolk, butter, bovine blood serum. Structure: Kuhn, Grundmann, *Ber.* **66**, 1746 (1933); Karrer, Schlientz, *Helv. Chim. Acta* **17**, 55 (1934); Cholnoky *et al., Ann.* **616**, 207 (1958). Total synthesis: Isler *et al., Helv. Chim. Acta* **40**, 456 (1957); Loeber *et al., J. Chem. Soc. (C)* **1971**, 404.

All-*trans*-cryptoxanthin, red plates with metallic luster from ether + methanol. mp 158-159° (racemate); 169° (natural). Absorption max: 452, 480 nm ($E^{1\%}_{1cm}$ 2380, 2080). Freely sol in chloroform, benzene, pyridine, carbon disulfide. Less sol in ligroin, petr ether, alcohol, methanol. Monoacetate, $C_{42}H_{58}O_2$; deep red leaflets, mp 117.5°.

2613. Crystallins. Major water-soluble structural proteins found in fiber cells of vertebrate eye lenses; account for the transparency of the lens. Heterogeneous family composed of four groups α, β, γ, δ which have been separated on the basis of size, charge, immunological properties and source. α, β, δ crystallins occur in avian and reptilian lenses while α, β, γ crystallins occur in all other lenses. Several other minor forms have also been identified. Identification and isolation of α, β, γ: C. T. Mörner, *Z. Physiol. Chem.* **18,** 61 (1894); of δ: M. Rabaey, *Exp. Eye Res.* **1**, 310 (1962). Review of early isolation procedures: A. Spector, *Invest. Ophthalmol.* **3**, 182-193 (1964). Structural studies of α: A. F. Van Dam, G. Ten Cates, *Biochim. Biophys. Acta* **121**, 183 (1966); H. A. Kramps *et al., Eur. J. Biochem.* **50**, 503 (1975); of β: H. P. C. Driessen *et al., ibid.* **121**, 83 (1981); G. Wistow *et al., FEBS Letters* **133**, 9 (1981); of γ: T. Blundell *et al., Nature* **289**, 771 (1981); L. Summers *et al., Peptide Protein Rev.* **3**, 147 (1984); of δ: J. Horwitz, J. Piatigorsky, *Biochim. Biophys. Acta* **624**, 21 (1980); L. A. Williams *et al., ibid.* **708**, 49 (1982). Short-range spatial ordering is essential

for lens transparency: M. Delaye, A. Tardieu, *Nature* **302,** 415 (1983). Photoinduced aggregation; implications for cataract formation: U. P. Andley *et al., Photochem. Photobiol.* **40**, 343 (1984); K. Mandal *et al., ibid.* **47**, 583 (1988); M. Kono *et al., ibid.* 593. Role of crystallin aging in cataract formation: R. J. Siezen *et al., Proc. Nat. Acad. Sci. USA* **82**, 1701 (1985); H. J. Hoenders, G. J. H. Bessems, *Lens Res.* **3**, 281 (1986). Evolution and expression of crystallin genes: J. Piatigorsky, *Cell* **38**, 620 (1984); W. W. de Jong, W. Hendriks, *J. Mol. Evol.* **24**, 121 (1986). *Reviews:* A. Spector, *Invest. Ophthalmol.* **4**, 579-591 (1965); W. W. de Jong, *Co-ord. Regul. Gene Expression - Proc. 2nd Int. Workshop,* R. M. Clayton, Ed. (Plenum Press, New York, 1986) pp 281-291; G. J. Wistow, J. Piatigorsky, *Ann. Rev. Biochem.* **57**, 479-504 (1988). Review of δ-crystallin: J. Piatigorsky, *Mol. Cell. Biochem.* **59**, 33-56 (1984). Book: *Molecular and Cellular Biology of the Eye Lens,* H. Bloemendal, Ed. (John Wiley & Sons, New York, 1981).

α-Crystallin. High molecular wt multimeric (35-50 subunits) polypeptide for which two major subunit forms exist $αA_1$, $αA_2$; M_r unit 20 kDa. Most acidic crystallin, pI 4.8-5.0.

β-Crystallin. Dioctamer for which several (6-7) subunit forms exist; M_r unit 23-25 kDa. pI 5.7-7.0.

γ-Crystallin. Monomer, M_r 20kDa. pI 7.1-8.1.

δ-Crystallin. Tetramer for which two possible subunit forms exist; M_r (monomer) 48-50 kDa. pI 4.9-5.3.

2614. Cubane. *Pentacyclo[4.2.0.0^{2,5}.0^{3,8}.0^{4,7}]octane.* C_8H_8; mol wt 104.14. C 92.26%, H 7.74%. Prepn from bromocyclopentadienone dimer: Eaton, Cole, *J. Am. Chem. Soc.* **86**, 3157 (1964). Structure: Fleischer, *ibid.* 3889.

Rhombic crystals from methanol followed by sublimation at just above room temperature and atmospheric pressure, mp 130-131° (sealed capillary). Thermally unstable. Dec at 200°.

2615. Cubeb. Tailed pepper; Java pepper. Dried, unripe, nearly full-grown fruit of *Piper cubeba* L.f., *Piperaceae. Habit.* Southern Asia, Java, Borneo, Sumatra; cultivated in W. Indies and Ceylon. Berries are globular, 4-5 mm in diameter; blackish-gray; intern. whitish and hollow; strong, spicy odor; aromatic, pungent taste. *Constit.* 10-18% volatile oil, cubebin, about 1% cubebic acid, resin, fat, wax. Total ether extractive about 20%.

2616. Cubebin. *3,4-Bis(1,3-benzodioxol-5-ylmethyl)tetrahydro-2-furanol; tetrahydro-3,4-dipiperonyl-2-furanol;* 5-hydroxy-3,4-bis(3,4-methylenedioxybenzyl)tetrahydrofuran. $C_{20}H_{20}O_6$; mol wt 356.36. C 67.40%, H 5.66%, O 26.94%. From cubeb. Isoln: Capitaine, Soubeiran, *Ann.* **31**, 190 (1839). Structure: Haworth, Kelly, *Chem. & Ind. (London)* **1936**, 901; Haensel *et al., Arch. Pharm.* **300**, 559 (1967). Synthesis: Batterbee *et al., J. Chem. Soc. (C)* **1969**, 2470.

Slender prisms from methanol, mp 131-132°. $[α]_D^{14}$ −17° (acetone). Practically insol in water. Sol in alcohol, chloroform, ether.

Note: "Cubebine" is the French designation for ethereal extract of cubeb.

2617. Cucurbitacins. A group of tetracyclic triterpenes, commonly referred to as "bitter principles of cucurbits", which have antineoplastic and anti-gibberellin activity. They are isolated from various spp of cucurbitaceous plants known since antiquity for their beneficial and toxic properties. The plants have been used as vermifuges, emetics, narcotics, and antimalarials and have been implicated in sporadic livestock poisoning in S. Africa. Seventeen cucurbitacins have been isolated, most from plants of the *Cucurbitaceae* family, but also from *Begoniaceae, Cruciferae, Datisceae, Euphorbiaceae,* and *Scrophulariaceae.* Cucurbitacins B and E are the most commonly identified. Isoln of cucurbitacins A, B, C, D, F: P. R. Enslin, *J. Sci. Food Agr.* **5,** 410 (1954); of G, H, J, K, L: *idem, ibid.* **8,** 673 (1957); of E and I: D. Lavie, S. J. Szinai, *J. Am. Chem. Soc.* **80,** 707 (1958); of O, P, Q and anti-tumor activity: S. Kupchan *et al., J. Org. Chem.* **35,** 2891 (1970). Structures of A, B, C, D, E, I: W. T. DeKock *et al., J. Chem. Soc.* **1963,** 3828; of B, D, F: D. Lavie *et al., Chem. & Ind. (London)* **1959,** 951; of G, H: C. W. Holzapfel, P. R. Enslin, *J. S. Afr. Chem. Inst.* **17,** 142 (1964), *C.A.* **62,** 10467c (1965); of J, K, L: P. R. Enslin, K. B. Norton, *J. Chem. Soc.* **1964,** 529. Stereochemistry of B, D, E, F, I: D. Lavie *et al., J. Org. Chem.* **28,** 1790 (1963). ^{13}C-NMR study of cucurbitacins: V. V. Velde, D. Lavie, *Tetrahedron* **39,** 317 (1983). Use as plant growth regulators: J. Guha, S. P. Sen, *Nature (New Biol.)* **244,** 273 (1973). *Reviews:* D. Lavie, E. Glotter, *Fortschr. Chem. Org. Naturst.* **29,** 308-357 (1971); J. Guha, S. P. Sen, *Plant Biochem. J.* **2,** 12-28 (1975); A. Shrotria, *Botanica* **26,** 28-31 (1976).

cucurbitacin B

Cucurbitacin B, $C_{32}H_{46}O_8$, *25-(acetyloxy)-2β,16α,20-trihydroxy-9β-methyl-19-nor-10α-lanosta-5,23E-diene-3,11,-22-trione, 1,2-dihydro-α-elaterin.* Cryst from abs ethanol, mp 184-186°. $[\alpha]_D^{25}$ +88° (c = 1.55 in ethanol). LD_{10} in mice: 5 mg/kg orally, J. LeMen *et al., Chim. Ther.* **4,** 459 (1969).

Cucurbitacin E, $C_{32}H_{44}O_8$, *25-(acetyloxy)-2β,16α,20-trihydroxy-9β-methyl-19-nor-10α-lanosta-1,5,23E-triene-3,-11,22-trione, α-elaterin.* White hexagonal plates from methanol, mp 232-233° (dec). $[\alpha]_D$ −59° (c = 0.7 in chloroform). uv max (chloroform): 234, 267 nm (ε 11,700, 8,350). LD_{50} in mice: 340 mg/kg orally, O. Albert *et al., Chim. Ther.* **5,** 205 (1970).

2618. Cudbear. Crottle. Common names for the lichen *Ochrolechia tartarea* L., *Lecanoraceae* and for the coloring matter from this and other lichens, especially *Lecanoraceae* and *Roccellaceae.* A source of litmus, *q.v.* Acids in lichens (e.g. lecanoric acid) hydrolyze to orcinol, *q.v.*, which, in the presence of ammonia can be oxidized to the dye orcein, *q.v.* The dyes *French Purple, Persio, Orchil,* and *Orseilles* derive from the salts of orcein: *Colour Index* Vol. 3 (3rd ed., 1971) p 3241.

Purplish-red powder. Imparts fine purplish-red color to alcohol, acid and neutral liquids. The purple tinge may be covered by adding a few drops of burnt-sugar coloring.

USE: Color for syrups, elixirs etc.

2619. Cumene. *(1-Methylethyl)benzene;* cumol; isopropylbenzene. C_9H_{12}; mol wt 120.19. C 89.94%, H 10.06%. Manuf: Faith, Keyes & Clark's *Industrial Chemicals,* F. A. Lowenheim, M. K. Moran, Eds. (Wiley-Interscience, New York, 4th ed., 1975) pp 294-297. *Review:* D. J. Ward in Kirk-Othmer *Encyclopedia of Chemical Technology* vol. 7 (Wiley-Interscience, New York, 3rd ed., 1979) pp 286-290.

Colorless liquid. d_4^{20} 0.862. bp 152-153°. Flash pt, closed cup: 102°F (39°C). n_D^{20} 1.4914: Hirschler, Faulconer, *J. Am. Chem. Soc.* **68,** 210 (1946). Insol in water; sol in alcohol and many other organic solvents. LD_{50} orally in rats: 2.91 g/kg, Smith *et al., Arch. Ind. Hyg. Occup. Med.* **4,** 119 (1951).

Caution: Narcotic in high concns.

USE: In manuf of phenol, acetone, acetophenone, α-methylstyrene.

2620. Cumic Acid. *4-(1-Methylethyl)benzoic acid;* p-isopropylbenzoic acid; cuminic acid. $C_{10}H_{12}O_2$; mol wt 164.20. C 73.14%, H 7.37%, O 19.49%.

Triclinic crystals. d^4 1.162. mp 115-117°. Sparingly sol in water; sol in alcohol, ether; also sol in concd H_2SO_4 without decompn, and pptd unchanged by adding water.

2621. Cumic Alcohol. *4-(1-Methylethyl)benzenemethanol;* cuminol; cuminyl alcohol; p-isopropylbenzyl alcohol. $C_{10}H_{14}O$; mol wt 150.21. C 79.95%, H 9.39%, O 10.65%. Found in caraway seed. Prepd by the reduction of cuminal; also by treating a suspension of cuminylmagnesium chloride in ether with dry oxygen.

Liquid; intense, persistent, caraway-like odor and burning, aromatic taste. d^{15} 0.981. bp 248.4°. n_D^{24} 1.522. Insol in water; miscible with alcohol, ether.

2622. Cumidine. *4-(1-Methylethyl)benzenamine;* 4-aminocumene; p-isopropylaniline; 4-amino-1-isopropylbenzene; β-(4-aminophenyl)propane. $C_9H_{13}N$; mol wt 135.20. C 79.95%, H 9.69%, N 10.36%. Prepn from aniline and isopropyl alcohol: Louis, *Ber.* **16,** 111 (1883); by nitration and subsequent reduction of cumene: Constam, Goldschmidt, *Ber.* **21,** 1157 (1898); Vavon, Callier, *Bull. Soc. Chim. France* [4] **41,** 677 (1927).

Liquid. bp_{745} 226-227°. d 0.9526. Insol in water. Hydrochloride, $C_9H_{13}N.HCl$, prisms from water or alc. Sulfate, $C_9H_{13}N.\frac{1}{2}H_2SO_4$, leaflets, mp 205°. Sol in hot water or alcohol, less sol in cold water, ether.

USE: In the determination of tungsten.

2623. Cuminaldehyde. *4-(1-Methylethyl)benzaldehyde;* p-isopropylbenzaldehyde; cuminal; cumaldehyde. $C_{10}H_{12}O$; mol wt 148.20. C 81.04%, H 8.16%, O 10.80%. $(CH_3)_2$-CHC_6H_4CHO. Constituent of eucalyptus, myrrh, cassia, cumin, and other essential oils; prepared synthetically from p-isopropylbenzoyl chloride. Alternate synthesis by formylation of isopropylbenzene under pressure: Crounse, *J. Am. Chem. Soc.* **71**, 1263 (1949).

Colorless to yellowish, oily liq; strong persistent odor; acrid, burning taste. d^{20} 0.978. bp_{760} 235-236°; bp_{35} 131-135°. n_D^{20} 1.5301. Practically insol in water; sol in alc, ether. LD_{50} orally in rats: 1390 mg/kg, P. M. Jenner *et al., Food. Cosmet. Toxicol.* **2**, 327 (1964).

USE: In perfumery; in the synthesis of its thiosemicarbazone, *q.v.*

2624. Cuminaldehyde Thiosemicarbazone. *p-Isopropylbenzaldehyde thiosemicarbazone;* cuminal thiosemicarbazone; *p*-isopropylbenzylidene thiosemicarbazone; cumaldehyde thiosemicarbazone; SHCH-58; Cuthizone; Cutizon; Kuthison; Kutizon. $C_{11}H_{15}N_3S$; mol wt 221.32. C 59.70%, H 6.83%, N 18.99%, S 14.48%. Prepn by the reaction of thiosemicarbazide with *p*-isopropylbenzaldehyde in warm, dil acetic acid: Sah, Daniels, *Rec. Trav. Chim.* **69**, 1545 (1950); Bernstein *et al., J. Am. Chem. Soc.* **73**, 906 (1951); Behnisch *et al.,* Ger. pat. **885,705** (1953 to Bayer).

Stout platelets from methanol, mp 147°. Practically insol in water. Moderately sol in alc.

THERAP CAT: Antiviral.

2625. Cupferron. *N-Hydroxy-N-nitrosobenzenamine ammonium salt; N*-nitrosophenylhydroxylamine ammonium salt. $C_6H_9N_3O_2$; mol wt 155.16. C 46.44%, H 5.85%, N 27.08%, O 20.62%. Prepd from phenylhydroxylamine by treating with $NaNO_2$ at 0° in presence of HCl, filtering, dissolving in ether and treating with NH_3, or by treating an ether soln of phenylhydroxylamine with butyl or amyl nitrite and NH_3 in the cold: Marvel, Kamm, *Org. Syn.* **4**, 19 (1925).

Crystals, mp 163-164°. Freely sol in water or alc. *Keep in well-closed containers to which a piece of ammonium carbonate has been added.*

Note: This substance may reasonably be anticipated to be a carcinogen: *Fourth Annual Report on Carcinogens* (NTP 85-002, 1985) p 63.

USE: As a reagent for separating Cu and Fe from other metals. It ppts iron quantitatively from strongly acid soln; as a quantitative reagent for vanadates with which it gives a dark-red ppt sol in alkali soln, and for Ti with which it forms a yellow ppt; also suitable for the colorimetric estimation of Al.

2626. Cupreine. *Cinchonan-6',9-diol;* hydroxycinchonidine; ultraquinine. $C_{19}H_{22}N_2O_2$; mol wt 310.38. C 73.52%, H 7.14%, N 9.03%, O 10.31%. In bark and seeds of *Remijia pedunculata* Flueck., *Rubiaceae.* Isoln: Howard, Hodgkin, *J. Chem. Soc.* **41**, 16 (1882). Separated from homoquinine, *q.v.*, by treatment with NaOH: Hesse, *Ann.* **230**, 55 (1885). Reduction to *hydrocupreine:* Giemsa, Halberkann, *Ber.* **51**, 1325 (1918). Conversion into cinchonidine: King, *J. Chem. Soc.* **1946**, 523.

Monoclinic plates from alcohol, mp 202°. $[\alpha]_D^{17}$ −176°, (methanol). pK_1 6.57. Sol in alcohol, NaOH soln, but not in NH_4OH. Sparingly sol in water, ether, chloroform, benzene, petr ether. The neutral salts give yellow aq solns; the acid salts remain colorless in aq soln.

Dihydrate, monoclinic prisms, pseudo-orthorhombic by twinning.

Hydrochloride monohydrate, $C_{19}H_{22}N_2O_2 \cdot HCl \cdot H_2O$, needles, $[\alpha]_D^{17}$ −157°.

Dihydrochloride dihydrate, $C_{19}H_{22}N_2O_2 \cdot 2HCl \cdot 2H_2O$, crystals, $[\alpha]_D^{17}$ −211°.

Hydrogen sulfate hexahydrate, $(C_{19}H_{22}N_2O_2)_2 \cdot H_2SO_4 \cdot 6H_2O$, plates, tablets, when anhydr mp 257° dec, $[\alpha]_D^{22}$ −198°. Sol in 815 parts water.

Diacetate, $C_{23}H_{26}N_2O_4$, scales, plates, mp 88°.

Monomethyl ether, *see* Quinine.

2627. Cupric Acetate. Crystallized verdigris; neutralized verdigris. $C_4H_6CuO_4$; mol wt 181.63. C 26.45%, H 3.33%, Cu 34.98%, O 35.24%. $Cu(CH_3COO)_2$. Prepd by the action of acetic acid on CuO or $CuCO_3$: Winter *et al.,* in Kirk-Othmer *Encyclopedia of Chemical Technology,* vol. 6 (Interscience, New York, 2nd ed., 1965) p 267.

Monohydrate, dark green, monoclinic crystals; exists in dimeric units. mp 115°; dec at 240°; d 1.882. Sol in water, alcohol; slightly sol in ether, glycerol. LD_{50} orally in rats: 0.71 g/kg, H. F. Smyth *et al., Am. Ind. Hyg. Assoc. J.* **30**, 470 (1969).

USE: Intermediate in manuf of Paris green; as fungicide; as catalyst for organic reactions, including rubber aging; in textile dyeing; as pigment for ceramics.

2628. Cupric Acetate, Basic. Cupric subacetate. Several complexes are known with different cupric acetate:cupric hydroxide:water ratios. *Common verdigris* or *green verdigris* is a 2:1:5 complex; *blue verdigris* or *French verdigris* a 1:1:5 complex; other complexes exist with ratios 1:2:0 and 1:3:2. Review of prepn and properties: Gauthier, *Compt. Rend.* **242**, 644-647 (1956); *idem, Rev. Gen. Sci. Pures Appl. Bull. Assoc. Franc. Avance. Sci.* **66**, 67-78 (1957); Schweizer, Muehlethaler, *Farbe + Lack* **74**, 1159-1173 (1968).

The complexes exist as blue crystals or blue to green powders. Slightly sol in water, alc; sol in dil acids, ammonia.

USE: In manuf of Paris green and other pigments and insecticides; in dyeing and printing fabrics; as fungicide and mold-preventative; preparing gilder's wax in fire gilding; in artificial flowers; as water- or oil-color pigment.

2629. Cupric Acetoarsenite. *(Acetato)trimetaarsenitodicopper; Paris green;* copper acetate arsenite; emerald green; French green; imperial green; mineral green; Mitis green; parrot green; Schweinfurt green; Vienna green; C.I. Pigment Green 21; C.I. 77410. $C_4H_6As_6Cu_4O_{16}$; mol wt 1013.71. C 4.74%, H 0.60%, As 44.34%, Cu 25.07%, O 25.25%. $Cu(C_2H_3O_2)_2 \cdot 3Cu(AsO_2)_2$. Usually contains some water. Prepn: Serciron, U.S. pat. **2,159,864** (1939); Krefft, U.S. pat. **2,268,123** (1944 to Chemical Marketing); Glemser, Sauer in *Handbook of Preparative Inorganic Chemistry,* vol. 2, G. Brauer, Ed. (Academic Press, New York, 2nd ed., 1965) p 1027. *See also Colour Index* vol. 4 (3rd ed., 1971) p 4667.

Emerald green, cryst powder. *Poisonous!* Stable to air, light. Practically insol in water; dec on prolonged heating in water. Unstable in acids, bases, and toward H_2S. LD_{50} orally in female rats: 100 mg/kg, T. B. Gaines, *Toxicol. Appl. Pharmacol.* **14**, 515 (1969).

USE: Insecticide; wood preservative; as pigment, particularly for ships and submarines. *Caution:* Toxicity primarily result of arsenic content. Gastric disturbance, tremors or muscular cramps, and peripheral neuritis. Local effects on the skin, mucous membranes and conjunctiva.

2630. Cupric Arsenite. *Arsonic acid copper(2+) salt (1:1); arsenious acid copper(2+) salt (1:1);* Scheele's green. Compound of variable composition. Usually considered to be $CuHAsO_3$; mol wt 187.46. Also reported as $Cu(AsO_2)_2$: Bhadraver, *Bull. Chem. Soc. Japan* **35**, 1770 (1962). Prepd by reaction of a cupric salt with an alkaline or ammoniacal arsenite: R. N. Kust in Kirk-Othmer *Encyclopedia of Chemical Technology,* vol. 7 (Interscience, New York, 3rd ed., 1979) p 102. *See also: Colour Index* vol. 4 (3rd ed., 1971) p 4667.

Yellowish-green powder. *Poisonous!* Practically insol in water, alcohol; sol in acids, ammonia.

USE: As pigment, wood preservative, insecticide, fungicide, rodenticide.

2631. Cupric Borate. A material of indefinite composition prepd by the reaction of a borax soln with $CuSO_4$: Barnard, Jr. in *Copper*, A. Butts, Ed., ACS Monograph, **no. 122** (Reinhold, New York, 1954) p 794.

Blue to blue-green solid. Extremely hard material.

USE: As ceramic color, pigment in oil paint; in dehydrogenation catalyst; as insulating and sealing medium for lead wires in lamps and mercury rectifiers, as fireproofing agent and preservative for wood; as fungicide, insecticide, parasiticide. *Caution:* Symptoms of acute toxicity include vomiting, diarrhea, tremors, convulsions, shock, dermatitis; renal, hepatic and CNS degeneration. Symptoms of chronic ingestion include skin disorders, kidney lesions.

2632. Cupric Bromide. Br_2Cu; mol wt 223.37. Br 71.55%, Cu 28.45%. $CuBr_2$. Prepn: Watt *et al.*, *J. Am. Chem. Soc.* **77**, 2752 (1955); Glemser, Sauer in *Handbook of Preparative Inorganic Chemistry*, **vol. 2**, G. Brauer, Ed. (Academic Press, New York, 2nd ed., 1965) p 1009.

Almost black, iodine-like, deliquesc monoclinic crystals or cryst powder. mp 498°; bp 900°; d_4^{20} 4.710. Very sol in water; sol in alcohol, acetone, ammonia; practically insol in benzene, ether, concd H_2SO_4. *Keep well closed.*

USE: As intensifier in photography; as brominating agent in organic synthesis; as humidity indicator; as wood preservative; in solid-electrolyte battery; as stabilizer for acetylated polyformaldehyde.

2633. Cupric Butyrate. $C_8H_{14}CuO_4$; mol wt 237.73. C 40.42%, H 5.94%, Cu 26.73%, O 26.92%. $Cu(CH_3CH_2CH_2-COO)_2$. Prepn from a cupric salt and butyric acid or Na butyrate: Graddon, *J. Inorg. Nucl. Chem.* **17**, 222 (1955); Martin, Waterman, *J. Chem. Soc.* **1957**, 2545; from powdered Cu and butyric acid: Hahl, U.S. pat. **3,133,942** (1964 to BASF).

Monohydrate, large, dark green, monoclinic, hexagonal plates. Becomes dull and disintegrates after several days exposure to air. Sol in water, dioxane, benzene; slightly sol in chloroform, alcohol.

USE: In catalysts; in lubricants.

2634. Cupric Carbonate, Basic. *Copper carbonate hydroxide;* cupric subcarbonate; Bremen blue; Bremen green. $CH_2Cu_2O_5$; mol wt 221.11. C 5.43%, H 0.91%, Cu 57.47%, O 36.18%. $CuCO_3 \cdot Cu(OH)_2$. Occurs in nature as the mineral *malachite* (dark green monoclinic crystals). Prepd by adding $CuSO_4$ soln to Na_2CO_3 soln: Winter *et al.*, in Kirk-Othmer *Encyclopedia of Chemical Technology*, **vol. 6** (Interscience, New York, 2nd ed., 1965) p 270. The commercial product usually contains 50-55% Cu.

Green to blue amorphous powder or dark-green monoclinic crystals. Practically insol in water, alcohol; sol in dil acids, ammonia.

Note: Another form of basic copper carbonate, $2CuCO_3 \cdot Cu(OH)_2$, occurs in nature as the mineral *azurite* or *chessylite.*

See also Burgundy Mixture.

USE: As seed treatment fungicide; in pyrotechnics; as paint and varnish pigment; in animal and poultry feeds; in sweetening of petrol sour crude stock; in manuf of other Cu salts.

THERAP CAT (VET): Nutritional factor. Used in copper deficiency in ruminants.

2635. Cupric Chlorate. Cl_2CuO_6; mol wt 230.45. Cl 30.77%, Cu 27.57%, O 41.66%. $Cu(ClO_3)_2$. Prepn: *Gmelin's, Copper* (8th ed.) **60B**, 332-334 (1958).

Hexahydrate, blue to green, deliquesc, octahedral crystals. mp 65°. Dec at 100°. Very sol in water, alcohol. *Keep well closed and out of contact with organic matter.*

USE: Mordant in dyeing and printing of textiles.

2636. Cupric Chloride. Cl_2Cu; mol wt 134.45. Cl 52.74%, Cu 47.26%. $CuCl_2$. Prepn: Pray, *Inorg. Syn.* **5**, 153 (1957); *Gmelin's, Copper* (8th ed.) **60B**, 253-295 (1958); Glemser, Sauer in *Handbook of Preparative Inorganic Chemistry*, **vol. 2**, G. Brauer, Ed. (Academic Press, New York, 2nd ed., 1965) p 1008.

Yellow to brown, deliquesc, microcryst powder. Partially dec above 300° to CuCl + Cl_2. mp (extrapolated) 630°. mp of 498° usually reported for this compd is the melting pt of a mixture of $CuCl_2$ and CuCl. d_4^{25} 3.39. Forms the dihydrate in moist air. Sol in water, alc, acetone. *Keep well closed.*

Dihydrate, green to blue powder or orthorhombic, bipyramidal crystals. Deliquesc in moist air, effloresce in dry air. Water loss occurs from 70-200°: Bell, Coultard, *J. Chem. Soc. (A)* **1968**, 1417. mp about 100°; d 2.51. Freely sol in water, methanol, ethanol; moderately sol in acetone, ethyl acetate; slightly sol in ether. The aq soln is acid to litmus; pH 0.2 molar soln 3.6.

USE: As catalyst for organic and inorganic reactions; in petroleum industry as deodorizing, desulfurizing, and purifying agent; as mordant for dyeing and printing textiles; as oxidizing agent for aniline dyestuffs; in indelible, invisible and laundry-marking inks; in metallurgy in wet process for recovering mercury from ores, in refining Cu, Ag, Au; in tinting-baths for Fe, Sn; in electrotyping baths for plating Cu on Al; in photography as a fixer, desensitizer and re-agent; in producing color in pyrotechnic compositions; in manuf of acrylonitrile, fast black (melanin); in pigments for glass, ceramics; as feed-additive, wood preservative, disinfectant. *Caution:* Irritating to skin, mucous membranes.

2637. Cupric Chromate(VI). Neutral cupric chromate. $CrCuO_4$; mol wt 179.55. Cr 28.97%, Cu 35.39%, O 35.64%. $CuCrO_4$. Prepn from $CuCO_3$, Na_2CrO_4 and CrO_3: Briggs, *J. Chem. Soc.* **1929**, 242; from CuO and CrO_3: Campbell, Lemaire, *Can. J. Res.* **25B**, 243 (1947); from $Cu(OH)_2$ and $K_2Cr_2O_7$: Winter *et al.*, in Kirk-Othmer *Encyclopedia of Chemical Technology* **vol. 6** (Interscience, New York, 2nd ed., 1965) p 270.

Reddish-brown crystals. Gradually dec to $CuCr_2O_4$ above 400°. Practically insol in water; sol in acids.

Basic cupric chromates. Several cupric chromate-cupric hydroxide compounds are known. $CuCrO_4 \cdot Cu(OH)_2$ forms yellow, copper-red or chocolate-brown to lilac crystals; $CuCrO_4 \cdot 2Cu(OH)_2$, a light brown powder; $2CuCrO_4 \cdot 3Cu(OH)_2$, yellow to yellowish-brown crystals. Prepn: Campbell, Lemaire, *loc. cit.;* Winter *et al.*, *loc. cit.* All are practically insol in water.

USE: In fungicides, seed protectants, and wood preservatives; as mordant in dyeing textiles; in protecting textiles against insects and microorganisms.

2638. Cupric Chromite. Cupric chromate(III). Cr_2CuO_4; mol wt 231.56. Cr 44.92%, Cu 27.44%, O 27.64%. $CuCr_2O_4$. Prepd by heating $CuCrO_4$ at 400°: Stroupe, *J. Am. Chem. Soc.* **71**, 569 (1949); or by decomp of $(NH_4)_2Cu(CrO_4)_2 \cdot 2NH_3$ at 700°: Whipple, Wold, *J. Inorg. Nucl. Chem.* **24**, 23 (1962).

Grayish-black to black tetragonal crystals. Dec to $CuCrO_2$ and CrO_3 above 900°. Practically insol in water, dil acids, concd HCl.

Copper-chromium oxide. A mixture of $CuCr_2O_4$ and CuO formed on decompn at temps below 400°, of orange copper ammonium chromate complex prepd from $Cu(NO_3)_2$, $Na_2-Cr_2O_7$ and NH_4OH: Adkins *et al.*, *J. Am. Chem. Soc.* **72**, 2626 (1950); Wagner in *Handbook of Preparative Inorganic Chemistry* **vol. 2**, G. Brauer, Ed. (Academic Press, New York, 2nd ed., 1965) p 1672. Fine, brownish-black to black powder. Stable in atmospheric O_2 and moisture. Practically insol in water, dil acids.

USE: Selective hydrogenation catalysts.

2639. Cupric Citrate. *2-Hydroxy-1,2,3-propanetricarboxylic acid copper salt (1:2);* Cuprocitrol. $C_6H_4Cu_2O_7$; mol wt 315.18. C 22.86%, H 1.28%, Cu 40.32%, O 35.53%. Prepd by the interaction of hot aq solns of $CuSO_4$ and Na citrate: U.S.D. 25th ed., p 1653.

Hemipentahydrate, green or bluish-green, odorless, cryst powder. Loses its water at about 100°. Slightly sol in water; sol in ammonia, dil acids; slightly sol in cold, freely in hot solns of alkali citrates.

THERAP CAT: Astringent, antiseptic.

2640. Cupric Ferrocyanide. *Hexakis(cyano-C)ferrate-(4−) copper(2+) (1:2); hexacyanoferrate(4−) dicopper(2+);* cupric hexacyanoferrate(II); Hatchett's brown; C.I. Pigment Brown 9. $C_6Cu_2FeN_6$; mol wt 339.04. C 21.25%, Cu

37.48%, Fe 16.47%, N 24.79%. $Cu_2Fe(CN)_6$. Prepd by addition of $K_4Fe(CN)_6$ to an aqueous soln of a sol cupric salt: Weiser *et al.*, *J. Phys. Chem.* **42**, 945 (1938); **46**, 99 (1942).

Reddish-brown powder or cubic crystals; ppts as a colloid or gel. Practically insol in water, dil acids, most organic solvents; sol in NH_4OH, solns of alkali cyanides.

USE: As pigment; in photographic toning baths; to lower electric resistance of soil and electrode-to-soil contacts.

2641. Cupric Fluoride. CuF_2; mol wt 101.54. Cu 62.58%, F 37.42%. Prepd by fluorinating Cu or a Cu salt with F_2: von Wartenberg, *Z. Anorg. Allgem. Chem.* **241**, 381 (1939); Jache, Cady, *J. Phys. Chem.* **56**, 1106 (1952); Haendler *et al.*, *J. Am. Chem. Soc.* **76**, 2178 (1954); Ritter, Smith, *J. Phys. Chem.* **70**, 805 (1966); **71**, 2036 (1971).

Monoclinic crystals; turns blue in moist air due to formation of dihydrate. mp about 785° (N_2 atm), 950° (HF atm); d 4.85. Soly in water (20°) 4.7 g/100 ml; hydrolyzed to CuFOH in hot water. May be stored in sealed glass ampuls.

Dihydrate, blue monoclinic crystals. Dec above 130°. d_4^{25} 2.934. Slightly sol in cold water, hydrolyzed to oxyfluoride in hot water.

USE: Anhydr in cathodes in nonaq galvanic cells; high temperature fluorinating agent; dihydrate as flux in casting gray iron.

2642. Cupric Formate. Tubercuprose. $C_2H_2CuO_4$; mol wt 153.58. C 15.64%, H 1.31%, Cu 41.37%, O 41.67%. $Cu(HCOO)_2$. Prepd from cupric carbonate and formic acid: Martin, Waterman, *J. Chem. Soc.* **1959**, 1359.

Three forms of anhydrous compd exist: powder-blue, turquoise, or royal blue crystals. Sol in water; practically insol in most organic solvents.

Dihydrate, very pale blue, monoclinic needles. Loses $2H_2O$ on standing in air. Turquoise, anhydrous modification formed by dehydration at 100° *in vacuo* over phosphoric oxide. Sol in water.

Tetrahydrate, large, light-blue, monoclinic, holohedral prisms. Powder-blue modification formed by dehydration over $CaCl_2$ under reduced press. Sol in water; very slightly sol in alc; practically insol in most organic solvents.

USE: As antibacterial agent in the treatment of cellulose.

2643. Cupric Gluconate. $C_{12}H_{22}CuO_{14}$; mol wt 453.85. C 31.76%, H 4.89%, Cu 14.00%, O 49.36%. $Cu[CH_2OH-(CHOH)_4CO_2]_2$. Prepd from gluconic acid and basic cupric carbonate: Suzuki *et al.*, **Japan.** pat. **2889**('63) (to Dainippon), *C.A.* **59**, 11264c (1963).

Hydrate, light blue to bluish-green, odorless crystals or cryst powder. Astringent taste. Soly in water 30 g/100 ml at 25°; slightly sol in alc; practically insol in most other organic solvents.

USE: In dietary supplements as a readily assimilable form of copper; as oral deodorant.

2644. Cupric Glycinate. *Bis(glycinato)copper;* cupric aminoacetate; glycine copper complex; glycocoll-copper. $C_4H_8CuN_2O_4$; mol wt 211.66. C 22.70%, H 3.81%, Cu 30.02%, N 13.24%, O 30.24%. $(H_2NCH_2COO)_2Cu$. Prepn from glycine and a cupric salt: Tomita, *Bull. Chem. Soc. Japan* **34**, 280 (1960).

Monohydrate, long, deep-blue, rhombic needles. Loses H_2O at 123°, chars at 213°, and dec with gas evolution at 228°. Sol in water; slightly sol in alcohol.

Dihydrate, light blue, powdery crystals. Loses one H_2O at 103°, remaining H_2O at about 140°. Dec with gas evolution at about 225°. Sol in water.

USE: In photometric analysis for copper.

THERAP CAT (VET): Nutritional factor. Given parenterally for copper deficiency in ruminants.

2645. Cupric Hexafluorosilicate. Cupric fluosilicate; cupric silicofluoride. CuF_6Si; mol wt 205.64. Cu 30.90%, F 55.44%, Si 13.66%. $CuSiF_6$. Prepn: Worthington, Haring, *Ind. Eng. Chem., Anal. Ed.* **3**, 7 (1931).

Tetrahydrate, blue, monoclinic, efflorescent crystals. d 2.56: Clark *et al.*, *Can. J. Chem.* **47**, 3859 (1969). Readily sol in water. *Keep well closed.*

USE: Dyeing and hardening white marble; treating plant diseases.

2646. Cupric Hydroxide. Copper hydrate; hydrated cupric oxide. CuH_2O_2; mol wt 97.56. Cu 65.13%, H 2.07%, O 32.80%. $Cu(OH)_2$. Commercial prepn: Furness, U.S. pat. **1,800,828** (1931 to Cellosilk); *idem,* U.S. reissue pat. **24,324** (1957 to Copper Research); Rowe, U.S. pat. **2,536,-096** (1951 to Mountain Copper); laboratory prepn: Weiser *et al.*, *J. Am. Chem. Soc.* **64**, 503 (1942); Gauthier, *Bull. Soc. Chim. France* **1960**, 353.

Blue to blue-green gel or light blue cryst powder. Stability is dependent on the method of prepn; may dec to black CuO on standing a few days or on heating. d 3.37. Practically insol in water; sol in concd alkali when freshly pptd; sol in acids, NH_4OH.

USE: In manuf of rayon, battery electrodes, other Cu salts; as mordant in dyeing; as pigment; in fungicides, insecticides; as feed additive; in treating and staining paper; in prepn of Schweitzer's reagent; in catalysts.

2647. Cupric Nitrate. CuN_2O_6; mol wt 187.56. Cu 33.88%, N 14.94%, O 51.18%. $Cu(NO_3)_2$. Prepn: *Gmelin's, Copper* (8th ed.) **60B**, 164-179 (1958); Addison, Hathaway, *J. Chem. Soc.* **1958**, 3099. Toxicity: H. F. Smyth *et al.*, *Am. Ind. Hyg. Assoc. J.* **30**, 470 (1969).

Large, blue-green, deliquesc, orthorhombic crystals. Sublimes at 150-225°. mp 255-256°. Sol in water, ethyl acetate, dioxane; dissolves in and reacts vigorously with ether. *Keep well closed.*

Trihydrate, *gerhardite*. Blue, deliquesc, rhombic plates. mp 114.5°; d 2.05. Freely sol in water, alcohol; practically insol in ethyl acetate. pH of 0.2 molar aq soln 4.0. *Keep well closed.* LD_{50} orally in rats: 0.94 g/kg (Smyth).

Hexahydrate, blue, deliquesc prismatic crystals. Decomposes to trihydrate at 26.4°. d 2.07. Freely sol in water; sol in alc. *Keep well closed.*

USE: In light-sensitive reproductive papers; as ceramic color; as mordant and oxidizing agent in textile dyeing and printing; as reagent for burnishing iron, for giving a black "antique" finish to copper, for coloring zinc brown; in nickel-plating baths; in aluminum brighteners; in wood-preservatives, fungicides, herbicides; in pyrotechnic compositions; as catalyst component in solid rocket fuel; as nitrating agent for aromatic organosilicon compds; as catalyst for organic reactions. *Caution:* Irritating to skin, mucous membranes.

2648. Cupric Oleate. *9-Octadecenoic acid copper salt.* $C_{36}H_{66}CuO_4$; mol wt 626.43. C 69.02%, H 10.62%, Cu 10.14%, O 10.22%. $(C_{17}H_{33}COO)_2Cu$. Prepd from $CuSO_4$ and K oleate: Nelson, Pink, *J. Chem. Soc.* **1954**, 4412.

Blue to green solid. Practically insol in water; slightly sol in alcohol; sol in ether.

USE: In antifouling compositions; as emulsifier and dispersing agent; as antioxidant in lubricating oils; as combustion-improver in fuel oils; as stabilizer for amide polymers; as catalyst.

Note: Ingredient of *Cuprex,* a hydrocarbon mixture used as a topical pediculicide.

2649. Cupric Oxalate. *Ethanedioic acid copper salt.* C_2CuO_4; mol wt 151.57. C 15.85%, Cu 41.92%, O 42.23%. CuC_2O_4. Prepd by reaction of $CuSO_4$ with oxalic acid: David, *Bull. Soc. Chim. France* **1960**, 719. Usually contains some water.

Blue-white powder. Loses any hydrated water by 200°; dec in air at 310° to CuO. Practically insol in water, alcohol, ether, acetic acid; sol in NH_4OH.

USE: As catalyst for organic reactions; as stabilizer for acetylated polyformaldehyde; in anticaries compositions; in seed treatments to repel birds and rodents.

2650. Cupric Oxide. Black copper oxide. CuO; mol wt 79.54. Cu 79.88%, O 20.12%. Occurs in nature as the minerals *tenorite* (triclinic crystals) and *paramelaconite* (tetrahedral, cubic crystals). Prepn: Glemser, Sauer in *Handbook of Preparative Inorganic Chemistry,* **vol. 2**, G. Brauer, Ed. (Academic Press, New York, 2nd ed., 1965) p 1012.

Black to brownish-black amorphous or cryst powder or granules. d_4^{14} 6.315. Practically insol in water, alc. Soly study: Hayward *et al.*, *Nature* **215**, 730 (1967). Sol in dil acids, alkali cyanides, $(NH_4)_2CO_3$ soln; slowly sol in NH_3.

USE: As pigment in glass, ceramics, enamels, porcelain glazes, artificial gems; in manuf of rayon, other Cu compds;

in sweetening petr gases; in galvanic electrodes; as flux in Cu metallurgy; in correcting Cu deficiencies in soil; as optical-glass polishing agent; to impart flux-and abrasion-resistance to glass fibers; in antifouling paints, pyrotechnic compositions; welding fluxes for bronze; as exciter in phosphor mixtures; as catalyst for organic reactions.

2651. Cupric Perchlorate. Cl_2CuO_8; mol wt 262.45. Cl 27.02%, Cu 24.21%, O 48.77%. $Cu(ClO_4)_2$. Prepd from $Cu(NO_3)_2$ and perchloric acid or nitrosyl perchlorate: Caven, Bryce, *J. Chem. Soc.* **1934**, 514; Hathaway, Underhill, *ibid.* **1960**, 648; **1961**, 3091; Hathaway, *Proc. Chem. Soc.* **1958**, 344.
Very pale green, hygroscopic crystals. Volatilizes on heating. Thermally stable to 130°; above 130° dec to a basic perchlorate. mp about 230-240°. Soluble in water, ether, dioxane, ethyl acetate. Practically insol in benzene, CCl_4, hexane.
Hexahydrate, deep blue, monoclinic crystals. Freely sol in water, methanol, ethanol, acetic acid, acetic anhydride, acetone; slightly sol in ether, ethyl acetate.
USE: Analytical reagent: Kolb, *Ind. Eng. Chem., Anal. Ed.* **11**, 197 (1939); **16**, 38 (1944); Serjeant, *Nature* **186**, 963 (1960). Also in copper electrodeposition; in catalysts for combustion and propellants.

2652. Cupric p-Phenolsulfonate. *p-Hydroxybenzenesulfonic acid copper salt;* cupric sulfocarbolate; Cupriaseptol. $C_{12}H_{10}CuO_8S_2$; mol wt 409.86. C 35.16%, H 2.46%, Cu 15.50%, O 31.23%, S 15.65%. $[C_6H_4(OH)SO_3]_2Cu$. Prepd by double decomposition of $CuSO_4$ and Ba p-phenolsulfonate: Lagenbeck, Mahrwald, *Ann.* **605**, 111 (1957).
Hexahydrate, blue to green, rhombic prisms or needles. Soluble in water, alcohol.
USE: In baths for Cu electroplating; esterification catalyst.

2653. Cupric Phosphate. $Cu_3O_8P_2$; mol wt 380.57. Cu 50.09%, O 33.63%, P 16.28%. $Cu_3(PO_4)_2$. Basic salts occur in nature as the minerals: *cornetite, dihydrite, libethenite, phosphorochalcite, pseudolibethenite, pseudomalachite, tagilite.* Prepn from $CuSO_4$ and $(NH_4)_2HPO_4$: Klement, Haselbeck, *Z. Anorg. Allgem. Chem.* **334**, 27 (1964).
Trihydrate, blue or olive orthorhombic crystals or blue to bluish-green powder. Dec on heating. Practically insol in cold water; slightly sol in hot water; sol in acids, NH_4OH.
USE: Fungicide; catalyst for organic reactions, fertilizer; emulsifier; corrosion inhibitor for H_3PO_4; protectant for metal surfaces against oxidation.

2654. Cupric Salicylate. *2-Hydroxybenzoic acid copper salt.* $C_{14}H_{10}CuO_6$; mol wt 337.76. C 49.78%, H 2.98%, Cu 18.81%, O 28.42%. $Cu[C_6H_4(OH)COO]_2$. Prepd from reaction of $CuSO_4$ with Na salicylate: Hanic, Michalov, *Acta Cryst.* **13**, 299 (1960); Inoue *et al., Inorg. Chem.* **3**, 239 (1964).
Brown, blue-green or green powder. Absorbs water in moist air to form the tetrahydrate.
Tetrahydrate, pale-blue needles or blue-green plates. Blue-green modification loses water on standing in air. Sol in water, alcohol, NH_4OH.
USE: In the separation of monoolefinic from di- or polyolefinic hydrocarbons.

2655. Cupric Selenate. CuO_4Se; mol wt 206.50. Cu 30.77%, O 30.99%, Se 38.24%. $CuSeO_4$. Prepd by the action of selenic acid on cupric carbonate: Klein, *Ann. Chim.* **14**, 263 (1940).
Pentahydrate, light-blue triclinic crystals. Loses water above 80°, forming the monohydrate at 150-220° and becoming anhyd by 265°; dec to the selenite and basic selenite at ~480° and to CuO at ~700°. d 2.56. Sol in water; very slightly sol in acetone; practically insol in alcohol.
USE: In coloring Cu or Cu alloys black.

2656. Cupric Selenide. CuSe; mol wt 142.50. Cu 44.59%, Se 55.41%. Occurs in nature as the mineral *klockmannite.* Prepd by reduction of cupric selenite with hydrazine: Benzing *et al., J. Am. Chem. Soc.* **80**, 2657 (1958); Kulifay, *J. Inorg. Nucl. Chem.* **25**, 75 (1965).
Blue-black to greenish-black prismatic needles or hexagonal plates. Dec at a dull red heat. d 6.0-6.6. Sol in HCl

with H_2Se evolved, in H_2SO_4 with SO_2 evolved; oxidized to $CuSeO_3$ by HNO_3.
USE: Catalyst in Kjeldahl digestions; in semiconductors.

2657. Cupric Selenite. CuO_3Se; mol wt 190.50. Cu 33.35%, O 25.20%, Se 41.45%. $CuSeO_3$. Occurs in nature as the dihydrate, *chalcomenite.* Prepd by the action of selenious acid or an alkaline selenite on a cupric salt: Rocchiccioli, *Compt. Rend.* **247**, 1108 (1958).
Dihydrate, blue, orthorhombic or monoclinic crystals. Becomes anhyd by 265°. Dec to $CuSeO_3 \cdot CuO$ above 460° and to CuO above 660°. d 3.31. Practically insol in water, H_2SeO_3; sol in acids, NH_4OH.

2658. Cupric Stearate. *Octadecanoic acid copper salt.* $C_{36}H_{70}CuO_4$; mol wt 630.46. C 68.58%, H 11.19%, Cu 10.08%, O 10.15%. $(C_{17}H_{35}COO)_2Cu$. Prepd by metathesis of alcohol cupric acetate with an alcohol soln of stearic acid: Martin, Waterman, *J. Chem. Soc.* **1957**, 2545; Rai, Mehrotra, *J. Inorg. Nucl. Chem.* **21**, 311 (1961).
Pale-blue to blue-green amorphous powder. Considered to be dimeric. mp about 250°: Grant, *Can. J. Chem.* **42**, 951 (1964). Practically insol in water, methanol, ethanol, acetone, ether; sol in pyridine, dioxane, acetic anhydride; sol in hot but practically insol in cold benzene, toluene, CCl_4, $CHCl_3$, ethyl acetate.
USE: In antifouling paints; in textile and wood preservatives; in rubber aging; as catalyst.

2659. Cupric Sulfate. CuO_4S; mol wt 159.61. Cu 39.81%, O 40.10%, S 20.09%. $CuSO_4$. Occurs in nature as the mineral *hydrocyanite.* Commercial preparation of pentahydrate: Faith, Keyes & Clark's *Industrial Chemicals,* F. A. Lowenheim, M. K. Moran, Eds. (Wiley-Interscience, New York, 4th ed., 1975) pp 280-284; other prepns: *Gmelin's, Copper* (8th ed.) **60B**, 491-560 (1958). Toxicity: Smyth *et al., Am. Ind. Hyg. Assoc. J.* **30**, 470 (1969).
Grayish-white to greenish-white rhombic crystals or amorphous powder. On heating dec above 560°. d 3.6. Hygroscopic. Sol in water; practically insol in alcohol. *Keep tightly closed.*
Monohydrate, *dried cupric sulfate.* Hygroscopic, almost white powder. Sol in water; practically insol in alcohol. *Keep tightly closed.*
Pentahydrate, *bluestone, blue vitriol, Roman vitriol, Salzburg vitriol.* Occurs in nature as the mineral *chalcanthite.* Large, blue or ultramarine, triclinic crystals or blue granules or light-blue powder. Slowly efflorescent in air. Loses $2H_2O$ at 30°, 2 more H_2O at 110°; becomes anhyd by 250°. $d_4^{15.6}$ 2.286. Very sol in water; sol in methanol, glycerol; slightly sol in ethanol. pH of 0.2 molar aq soln 4.0. LD_{50} orally in rats: 960 mg/kg (Smyth).
Human Toxicity: A strong irritant. *See also* Copper.
USE: Anhyd salt for detecting and removing trace amounts of water from alcohols and other organic compds; as fungicide. Pentahydrate as agricultural fungicide, algicide, bactericide, herbicide; food and fertilizer additive; in insecticide mixtures; in manuf of other Cu salts; as mordant in textile dyeing; in prepn of azo dyes; in preserving hides; in tanning leather; in preserving wood; in electroplating solns; as battery electrolyte; in laundry and metal-marking inks; in petroleum refining; as flotation agent; pigment in paints, varnishes and other materials; in mordant baths for intensifying photographic negatives; in pyrotechnic compositions; in water-resistant adhesives for wood; in metal coloring and tinting baths; in antirusting compositions for radiator and heating systems; as reagent toner in photography and photoengraving; etc.
THERAP CAT: Antidote to phosphorus; topical antifungal.
THERAP CAT (VET): Nutritional factor. Used in copper deficiency of ruminants. Has also been used as an anthelmintic, emetic, fungicide.

2660. Cupric Sulfate, Basic. *Copper hydroxide sulfate;* cupric subsulfate; Basi-Cop; Cuproxat. Salts of variable compositions of cupric sulfate and cupric hydroxide or oxide. Occurs in nature as the mineral *dolerophane,* $CuSO_4 \cdot CuO$. *Reviews:* Frear, *Chemistry of the Pesticides* (Van Nostrand, New York, 3rd ed., 1955) pp 316-323; *idem, Agricultural Chemistry,* **vol. 2** (Van Nostrand, New York,

1951) pp 524-525; *Gmelin's, Copper* (8th ed.) **60B**, 579-589 (1958).

Copper sulfate tribasic, $Cu_4H_6O_{10}S$. Occurs in nature as the mineral *brochantite;* monohydrate as the mineral *langite.* Initial product formed during preparation of *Bordeaux mixture* from $CuSO_4$ and $Ca(OH)_2$: Frear, *loc. cit.* Also prepd from $CuSO_4$ and $Cu(OH)_2$: Denk, Leschhorn, *Z. Anorg. Allgem. Chem.* **336**, 58 (1965). Very finely divided, light blue, gelatinous particles. Practically insol in water. Sol in plant and mineral acids, NH_4OH.

Copper sulfate dibasic, $Cu_3H_4O_8S$. Occurs in nature as the mineral *antlerite.* Blue-green, rhombic, bipyramidal crystals. Practically insol in water.

See also Burgundy Mixture.

USE: Fungicide for plants, seed treatment.

2661. Cupric Sulfide. CuS; mol wt 95.61. Cu 66.46%, S 33.54%. Occurs in nature as the mineral *covellite,* or *indigo copper* (blue, hexagonal or monoclinic crystals). Prepn: Glemser, Sauer in *Handbook of Preparative Inorganic Chemistry,* vol. 2, G. Brauer, Ed. (Academic Press, New York, 2nd ed., 1965) p 1017.

Black powder. Stable in air when dry, oxidized to $CuSO_4$ by moist air. Practically insol in water, alcohol, dil acids, alkalies; sol in KCN soln, NH_4OH, hot HNO_3.

USE: In antifouling paints; in prepn of mixed catalysts; in development of aniline black dye in textile printing.

2662. Cupric Tartrate. *2,3-Dihydroxybutanedioic acid copper salt.* $C_4H_4CuO_6$; mol wt 211.61. C 22.70%, H 1.91%, Cu 30.03%, O 45.37%. Prepd by reaction of K tartrate with $Cu(NO_3)_2$: Kirschner, Kiesling, *J. Am. Chem. Soc.* **82**, 4174 (1960).

Trihydrate, green to blue, odorless powder. Slightly sol in water; sol in acids, alkali solns.

USE: In baths for Cu electroplating.

2663. Cupric Tungstate(VI). Cupric wolframate. CuO_4W; mol wt 311.46. Cu 20.40%, O 20.55%, W 59.05%. $CuWO_4$. Prepn: Kosek *et al., Coll. Czech. Chem. Commun.* **24**, 2034 (1959).

Dihydrate, light green powder. Becomes brown to greyish-yellow on heating, with loss of H_2O. Practically insol in water; slightly sol in acetic acid; sol in NH_3, H_3PO_4; dec by mineral acids.

USE: In semiconductors, nuclear reactors; as catalyst for polyester formation.

2664. Cuprobam. *(Dimethyldithiocarbamato)copper, compound with copper chloride;* dimethyldithiocarbamic acid copper salt, complex with copper chloride; cuprous dimethyldithiocarbamate, cuprous chloride complex; cuprobamé. Prepd as monocuprous chloride complex $[(CH_3)_2NCSSCu.CuCl]$: Couillard, Racine, **U.S. pat. 3,037,041** (1962 to Roussel-UCLAF). Prepd as dicuprous chloride complex $[(CH_3)_2NCSSCu.2CuCl]$: **Brit. pat. 825,901** (1959 to UCLAF). Both forms are practically insol in water.

USE: Fungicide, molluscicide.

2665. Cuprous Acetate. *Acetic acid copper(1+) salt.* $C_2H_3CuO_2$; mol wt 122.58. C 19.59%, H 2.47%, Cu 51.83%, O 26.10%. CH_3COOCu. Obtained as a sublimate by heating cupric acetate *in vacuo* to temps above 220°: Angel, Harcourt, *J. Chem. Soc.* **81**, 1385 (1902); prepn from cuprous oxide and acetic acid-acetic anhydride: Shimizu, Weller, *J. Am. Chem. Soc.* **74**, 4469 (1952). Crystal structure: T. Ogura *et al., J. Am. Chem. Soc.* **95**, 949 (1973).

Transparent, leafy crystals. Volatilizes on heating; dec on strong heating. Rapidly hydrolyzed by water with the formation of yellow Cu_2O.

2666. Cuprous Bromide. BrCu; mol wt 143.46. Br 55.71%, Cu 44.29%. CuBr. Prepn: Briggs, *J. Chem. Soc.* **127**, 496 (1925); Keller, Wycoff, *Inorg. Syn.* **2**, 3 (1946); Wagner, Wagner, *J. Chem. Phys.* **26**, 1597 (1957).

White powder or cubic crystals, (zinc blende structure); turns green to dark-blue on exposure to sunlight. mp 504°; bp 1345°; d_4^{25} 4.72. Slightly sol in cold water, dec in hot water, HNO_3. Sol in HCl, HBr, NH_4OH with formation of complexes. Practically insol in acetone, concd H_2SO_4. *Keep tightly closed in a dark place.*

USE: As catalyst for organic reactions.

2667. Cuprous Chloride. Cu-lyt. ClCu; mol wt 99.00. Cl 35.82%, Cu 64.18%. CuCl. Occurs in nature as the mineral *nantokite* (colorless to grey cubic crystals). Prepn: Keller, Wycoff, *Inorg. Syn.* **2**, 1 (1946); Glemser, Sauer in *Handbook of Preparative Inorganic Chemistry,* vol. 2, G. Brauer, Ed. (Academic Press, New York, 2nd ed., 1965) p 1005.

White cryst powder or cubic crystals (zinc-blende structure); stable to air and light if dry, but in presence of moisture turns green on exposure to air and blue to brown on exposure to light. mp 430°; d_4^{25} 4.14. Sparingly sol in water with partial decompn; practically insol in alcohol, acetone; sol in concd HCl, concd NH_4OH with formation of complexes. Solns oxidize rapidly in air. *Keep tightly closed and protected from light.*

USE: As catalyst for organic reactions; catalyst, decolorizer and desulfuring agent in petr industry; in denitration of cellulose; as condensing agent for soaps, fats and oils; in gas analysis to absorb carbon monoxide.

2668. Cuprous Cyanide. Cupricin. CCuN; mol wt 89.56. C 13.41%, Cu 70.95%, N 15.64%. CuCN. Commercial prepns: Winter *et al.,* Kirk-Othmer *Encyclopedia of Chemical Technology,* vol. 6 (Interscience, New York, 2nd ed., 1965) p 271; laboratory prepns: Barber, *J. Chem. Soc.* **1943**, 79; Norberg, Jacobson, *Acta Chem. Scand.* **3**, 174 (1949); Vaughan, McCane, *J. Am. Chem. Soc.* **76**, 2504 (1954).

White to cream-colored powder, colorless or dark green orthorhombic crystals, or dark red monoclinic crystals. *Poisonous!* mp 474°. Practically insol in water, alcohol, cold dil acids; sol in NH_4OH; dec by HNO_3, boiling dil HCl. Sol in alkali cyanide solns because of formation of stable cyanocuprate(I) ions.

USE: In electroplating Cu or Fe; as insecticide, fungicide; as antifouling agent in marine paints; as polymerization catalyst.

2669. Cuprous Iodide. Hydro-Giene. CuI; mol wt 190.46. Cu 33.36%, I 66.64%. Occurs in nature as the mineral *marshite* (red-brown crystals). Prepn: Kaufman, Pinnel, *Inorg. Syn.* **6**, 3 (1960); Glemser, Sauer in *Handbook of Preparative Inorganic Chemistry,* vol 2, G. Brauer, Ed. (Academic Press, New York, 2nd ed., 1965) p 1007.

Dense powder or cubic crystals (zinc-blende structure). Less photosensitive than CuBr or CuCl. mp 588-606°; bp about 1290°; d_4^{25} 5.63. Extremely insol in water; practically insol in dil acids, alcohol; sol in aq solns of NH_3, alkali cyanides, thiosulfates, iodides; dec by concd H_2SO_4 and HNO_3.

USE: As catalyst in organic reactions; as ice-nucleating chemical; as coating in cathode-ray tubes; as source of iodine in animal feeds; with HgI_2 as temp indicator; bactericide.

2670. Cuprous Mercuric Iodide. *Dicopper tetraiodomercurate(2−);* cuprous tetraiodomercurate(II); mercuric cuprous iodide. Cu_2HgI_4; mol wt 835.37. Cu 15.21%, Hg 24.01%, I 60.77%.

Deep-red, thermochromic cryst powder. *Poison!* Practically insol in water, alcohol.

USE: For detecting overheating of machine bearings, etc., the red color changes to brownish-black at 60-70° and again becomes red on cooling.

2671. Cuprous Oxide. Red copper oxide; C.I. 77402; Perenex; Yellow Cuprocide; Copper-Sandoz; Caocobre. Cu_2O; mol wt 143.08. Cu 88.82%, O 11.18%. Occurs in nature as the mineral *cuprite* (red to reddish-brown octahedral or cubic crystals). Prepd commercially by furnace reduction of mixtures of copper oxides with Cu: Drapeau, Johnson, **U.S. pats. 2,758,014; 2,891,842** (1956, 1959 to Glidden); by decompn of copper ammonium carbonate: Rowe, **U.S. pat. 2,474,497**; Klein, **U.S. pat. 2,474,533** (both 1949 to Lake Chemical); Rowe, **U.S. pat. 2,536,096**, Munn, **U.S. pat. 2,670,273** (1951, 1954 to Mountain Copper); by treatment of $Cu(OH)_2$ with SO_2: Rowe, **U.S. pat. 2,665,192** (1954 to Mountain Chemical); or by electrolysis of an aq soln of NaCl between Cu electrodes: Arend, *Paint Technology* **13**, 265 (1948). Laboratory prepns: Glemser, Sauer in *Handbook of Preparative Inorganic Chemistry,* vol. 2, G.

Brauer, Ed. (Academic Press, New York, 2nd ed., 1965) p 1011.

Cubic crystals or microcryst powder. Color may be yellow, red, or brown depending on the method of prepn and the particle size. Stable in dry air; gradually oxidizes in moist air to CuO. mp 1232°; d_4^{25} 6.0. Practically insol in water; sol in NH_4OH; in HCl forming CuCl which dissolves in excess HCl. With dil H_2SO_4 or dil HNO_3, the cupric salt is formed and half the copper is pptd as the metal. LD_{50} orally in rats: 0.47 g/kg, H. F. Smyth et al., Am. Ind. Hygiene Assoc. J. 30, 470 (1969).

USE: Fungicide; antiseptic for fishnets; in antifouling paints for marine use; in photoelectric cells; as red pigment for glass, ceramic glazes; in brazing pastes; in rectifiers; as catalyst.

2672. Cuprous Potassium Cyanide. Potassium bis(cyano-C)cuprate(1−); potassium dicyanocuprate(I); potassio-cuprous cyanide; potassium cyanocuprate(I). C_2CuKN_2; mol wt 154.67. C 15.53%, Cu 41.08%, K 25.28%, N 18.11%. $KCu(CN)_2$. Prepd by evaporation of an aq soln of CuCN and KCN: Staritzky, Walker, Anal. Chem. 28, 419 (1956).

Monoclinic, prismatic crystals. d 2.38. Practically insol in water; dec to CuCN by cold dil H_2SO_4 or heating in water; sol in DMSO.

USE: In electroplating copper and brass.

2673. Cuprous Selenide. Berzeline; selenkupfer. Cu_2Se; mol wt 206.04. Cu 61.68%, Se 38.32%. Occurs in nature as the mineral berzelianite which in addition to copper and selenium contains 4.73-8.50% silver, 0.35-0.54% iron, up to 0.38% thallium and traces of lead and magnesium. Prepd by reduction of an ammoniacal soln of a cupric salt and H_2SeO_3 with hydrazine: Benzing et al., J. Am. Chem. Soc. 80, 2657 (1958); Kulifay, J. Inorg. Nucl. Chem. 25, 75 (1963); alternate methods: Glemser, Sauer in Handbook of Preparative Inorganic Chemistry, vol. 2, G. Brauer, Ed. (Academic Press, New York, 2nd ed., 1965) p 1019.

Blue-black to black, tetragonal or cubic crystals with a metallic luster. mp 1113°. d_4^{21} 6.84. Sol in HCl with evolution of H_2Se, in H_2SO_4 with evolution of SO_2, in KCN soln; oxidized by HNO_3 to $CuSeO_3$.

USE: In semiconductors.

2674. Cuprous Sulfide. Cu_2S; mol wt 159.15. Cu 79.85%, S 20.15%. Occurs in nature as the mineral chalcocite also called copper glance (grey, black, green, blue, or violet rhombic crystals). Prepn: Glemser, Sauer in Handbook of Preparative Inorganic Chemistry, vol. 2, G. Brauer, Ed. (Academic Press, New York, 2nd ed., 1965) p 1016.

Blue to grayish-black lustrous powder, granules, or orthorhombic crystals. mp about 1100°. d_4^{20} 5.6. On heating in the absence of air it forms Cu and CuS; in the presence of air it forms CuO, $CuSO_4$ and SO_2. Practically insol in water, acetic acid; very sparingly sol in HCl; dec by HNO_3, concd H_2SO_4. Partially sol in NH_4OH, readily in cyanide solns due to complex ion formation.

USE: In luminous paints; in electrodes for thermoelements; in preparation of $CuSO_4$; in solid-lubricant compositions; as catalyst.

2675. Cuprous Sulfite. Cu_2O_3S; mol wt 207.14. Cu 61.35%, O 23.17%, S 15.48%. Cu_2SO_3. Prepn and review: Dasent, Morrison, J. Inorg. Nucl. Chem. 26, 1122 (1964).

Hemihydrate, Etard's salt. White to pale-yellow hexagonal crystals. Slightly sol in water; sol in HCl, NH_4OH, alkali solns; practically insol in ether, alcohol.

Cupro-cupric sulfate. $Cu_3O_6S_2$, Chevreul's salt. Dihydrate, red microcryst powder or prismatic crystals. Practically insol in water, alcohol; sol in NH_4OH, HCl.

Rogojski's salt. Formerly considered to be $Cu_2SO_3 \cdot H_2O$; recently shown to be an equimolar mixture of Chevreul's salt and metallic copper (Dasent, Morrison, loc. cit.). Brick-red solid.

USE: As fungicide for grape vines; in dyeing polyacrylic fibers; polymerization catalyst. Chevreul's salt is a selective molluscicide.

2676. Cuprous Thiocyanate. Cuprous sulfocyanate; cuprous sulfocyanide. CCuNS; mol wt 121.62. C 9.87%, Cu 52.24%, N 11.52%, S 26.36%. CuSCN. Prepn: Demmerle et

al., Ind. Eng. Chem. 42, 2 (1950); Newman, Analyst 88, 500 (1963).

White to yellow amorphous powder. d 2.85. Practically insol in water, dil acids, alc, acetone; sol in NH_4OH, ether, solns of alkali thiocyanates; dec by concd mineral acids.

USE: In marine antifouling paints; in primer compositions for explosives industry.

2677. Cuproxoline. Tetrahydrogen bis[8-hydroxy-5,7-quinolinedisulfonato(3−)-N^1,O^8]cuprate(4−) compd with N-ethylethanamine (1:4); 8-hydroxy-5,7-quinolinedisulfonic acid copper derivative compound with diethylamine; cupric bis[8-hydroxyquinoline di(diethylammonium sulfonate)]; copper DOS; Dicuprene; Cujec; Cuprimyl. $C_{34}H_{56}CuN_6O_{14}$-S_4; mol wt 964.67. C 42.33%, H 5.85%, Cu 6.59%, N 8.71%, O 23.22%, S 13.30%. Clinical efficacy in treatment of arthritis: W. C. Kuzell et al., Ann. Rheum. Dis. 10, 328 (1951). Veterinary use as parenteral copper supplement: I. J. Cunningham, N. Z. Vet. J. 7, 15 (1959); N. F. Suttle, Vet. Rec. 109, 304 (1981); G. J. Judson et al., Aust. Vet. J. 61, 40 (1984). Review of use of organic copper complexes in treatment of rheumatoid and degenerative diseases: J. R. J. Sorenson, W. Hangarter, Inflammation 2, 217-238 (1977). Review of veterinary use: W. M. Allen, C. B. Mallinson, Vet. Rec. 114, 451-454 (1984).

Green platelets. Sol in water, giving a dark-green soln. A 10% aq soln is almost neutral and can be sterilized by autoclaving. LD_{50} i.m. in rats: 126 mg/kg (Kuzell).

THERAP CAT: Antirheumatic.

THERAP CAT (VET): Copper supplement.

2678. Curare. Ourari; urari; woorari; woorali; wourara. Rendering of an Indian name given to the unstandardized extracts derived mainly from the bark of various spp. of Strychnos and Chondodendron, prepared for use as arrow poisons by Indians in the Amazon and Orinoco valleys, and in the Guianas; the physiologically active principle of which is (+)-tubocurarine chloride (q.v.). Three kinds of curare have appeared in commerce, distinguished by the kind of containers in which they were packed: Tube cucare or bamboo curare, pot curare, and gourd curare or calabash curare. Listing of members of the family Menispermaceae—botanical components of tube or bamboo curare and of pot curare: Krukoff, Moldenke, Studies of American Menispermaceae, with special reference to species used in the preparation of arrow poisons in Brittonia 3, 1-74 (1938); also suppl. no. 1-5. (Note that the curare available for medical use under the name Intocostrin is a physiologically standardized extract from Chondodendron tomentosum R. & P., Menispermaceae). Listing of Strychnos spp. Loganiaceae—botanical components of calabash curare (and of pot curare): Krukoff, Monachino, The American Species of Strychnos in Brittonia 4, 248-322 (1942), also suppl. no. 1-6. Alkaloids from calabash curare and bark of Strychnos spp.: P. Karrer, J. Pharm. Pharmacol. 8, 161-164 (1956); Schmid, Karrer, Helv. Chim. Acta 29, 1853 (1946); 30, 1162 (1947); Marino-Bettolo, Festschrift Arthur Stoll (Birkhäuser-Verlag, Basel 1957) pp 257-280. Review of history, chemistry, and use of arrow-poison curare: McIntyre, Curare (Chicago, 1947); Bovet et al., Curare and Curare-like Agents (Van Nostrand, Princeton, 1959).

Curare is sol in water and in dil alcohol. Stable aq solns (in ampuls) are standardized to contain 20 units per ml.

THERAP CAT: Skeletal muscle relaxant.

THERAP CAT (VET): Muscle relaxant.

2679. C-Curarine I. $[C_{40}H_{44}N_4O]^{2+}$; mol wt 596.78. Isoln from calabash-curare: Wieland, Pistor, Ann. 536, 68

(1938); Karrer, Schmid, *Helv. Chim. Acta* **29**, 1853 (1946); Zürcher *et al., J. Am. Chem. Soc.* **80**, 1500 (1958). Structure: Nagyváry *et al., Tetrahedron* **14**, 138 (1961); Grdinic *et al., J. Am. Chem. Soc.* **86**, 3357 (1964); Jones, Nowacki, *Chem. Commun.* **1972**, 805. Prepn of C-dihydrotoxiferine chloride: Bernauer *et al., Helv. Chim. Acta* **40**, 1999 (1957).

Dichloride, $C_{40}H_{44}Cl_2N_4O$, needles from methanol + ether, mp over 350°. $[\alpha]_D^{20}$ +73.6° (c = 1 in water). uv max (95% alc): 260, 296 nm (log ε 4.41, 4.07). Sol in water, alc; practically insol in ether, acetone.

2680. C-Curarine III. *C*-Fluorocurarine. $[C_{20}H_{23}\text{-}N_2O]^+$; mol wt 307.42. From calabash-curare: Wieland *et al., Ann.* **547**, 140 (1941); Schmid *et al., Helv. Chim. Acta* **35**, 1864 (1952); Zürcher *et al., J. Am. Chem. Soc.* **80**, 1500 (1958). Identity with C-fluorocurarine: Philipsborn *et al., Helv. Chim. Acta* **41**, 1257 (1958). Structure: Philipsborn *et al., ibid.* **42**, 461 (1959). Synthesis from C-curarine I: Boekelheide *et al., J. Chem. Soc.* **81**, 2256 (1959); from strychnine: Fritz *et al., Ann.* **663**, 150 (1963).

Chloride, $C_{20}H_{23}ClN_2O$, rods from abs alcohol, dec 270-274°. $[\alpha]_D^{20}$ −937° (water). uv max (methanol): 240, 300, 360 nm (log ε 4.00, 3.60, 4.23).
Picrate, $C_{26}H_{25}N_5O_8$, plates from water, mp 189°.

2681. Curcumin. *1,7-Bis(4-hydroxy-3-methoxyphenyl)-1,6-heptadiene-3,5-dione;* turmeric yellow; diferuloylmethane. $C_{21}H_{20}O_6$; mol wt 368.37. C 68.47%, H 5.47%, O 26.06%. Coloring matter from root of *Curcuma longa* L., Zingiberaceae. Isoln: Vogel, *Ann.* **44**, 297 (1842); Perkin, Phipps, *J. Chem. Soc.* (Trans.) **85, I,** 64 (1904); Rao, Shintre, *J. Soc. Chem. Ind.* **47**, 54T (1928). Synthesis: Lampe, *Ber.* **51**, 1347 (1918). Production: Stieglitz, Horn, **Ger.** pat. 859,145 (1952 to Hoechst). Biosynthesis studies: Roughley, Whiting, *Tetrahedron Letters* **1971**, 3741. Chromatography: Srinivasan, *J. Pharm. Pharmacol.* **5**, 448 (1953). Has anti-inflammatory activity. Pharmacology: Srimal, Dhawan, *ibid.* **25**, 447 (1973).

Orange-yellow, cryst powder, mp 183°. Insol in water, ether. Sol in alcohol, glacial acetic acid. Gives a brownish-red color with alkali; a light-yellow color with acids.

USE: For preparing curcuma paper, pH range 8-9. In the detection of boron.

2682. Curine. *(1β)-6,6'-Dimethoxy-2,2'-dimethyltubocuraran-7',12'-diol; l*-bebeerine. $C_{36}H_{38}N_2O_6$; mol wt 594.72. C 72.70%, H 6.44%, N 4.71%, O 16.14%. From tubocurare *(Chondodendron tomentosum* R. & P., Menispermaceae): Boehm, *Arch. Pharm.* **235**, 660 (1897); Späth *et al., Ber.* **61**, 1698 (1928). Structure: Späth, Kuffner, *Ber.* **67**, 55 (1934); Faltis *et al., Ber.* **69**, 1269 (1936); King, *J. Chem. Soc.* **1939**, 1157. Configuration: Bick, Clezy, *ibid.* **1953**, 3893; Hultin, *Acta Chem. Scand.* **17**, 753 (1963). *See* Bebeerine for the structural formula.

Efflorescent crystals from methanol. mp 213°; mp 221° *in vacuo;* four sided prisms from benzene containing one mol benzene, mp 161°. $[\alpha]_D^{20}$ −328° (pyridine). Sol in benzene, chloroform, pyridine.
Hydrochloride, $C_{36}H_{38}N_2O_6$.HCl, crystals, dec 273°, sol in water and alcohol.
Dihydrochloride, $C_{36}H_{38}N_2O_6$.2HCl, crystals.
Methiodide, $C_{36}H_{38}N_2O_6$.CH$_3$I, crystals, mp 253°.

2683. Curium. Cm; at. wt (most stable isotope) 247; at. no. 96; valence 3; also 4. ^{242}Cm, first prepared in 1944 by Seaborg, James and Ghiorsi. Formed by an (α,n)-process from ^{239}Pu or by spontaneous β-emission from ^{242}Am. ^{242}Cm is an α-emitter (T$_{1/2}$ 163 days); decays to ^{238}Pu. Metallic ^{242}Cm first prepd by reduction of CmF$_3$ by barium: Wolfmann *et al., J. Am. Chem. Soc.* **73**, 493 (1951). Thirteen curium isotopes have been produced; mass numbers 238-250. Most important isotopes: 244 (T$_{1/2}$ 17.6 yrs); 245 (T$_{1/2}$ 9.3 × 10³ yrs); 246 (T$_{1/2}$ 5.5 × 10³ yrs); 247 (T$_{1/2}$ 1.6 × 10⁷ yrs); 248 (T$_{1/2}$ 4.7 × 10⁵ yrs); decay by α-emission; available in subkilogram quantities. ^{247}Cm (half-life approx the age of the earth); considered an extinct nuclide. The radioactive actino-uranium family may have been produced from it. ^{250}Cm (T$_{1/2}$ 1.7 × 10⁴ yrs; spontaneous fission), probably generated during supernova star explosions; subsequent spontaneous fission produces the comparatively high abundances of certain isotopes of xenon and tellurium. *See* M. Haissinsky, J.-P. Adloff, *Radiochemical Survey of the Elements* (Elsevier, New York, 1965) pp 41-43. *Reviews:* J. J. Katz, G. T. Seaborg, *The Chemistry of the Actinide Elements* (John Wiley, New York, 1957) pp 373-385; Keenan, *J. Chem. Ed.* **36**, 27-31 (1959); C. Keller, *The Chemistry of the Transuranium Elements* (Verlag Chemie, Weinheim, English Ed., 1971) pp 529-551; *Comprehensive Inorganic Chemistry* **vol 5,** J. C. Bailar, Jr. *et al.,* Eds. (Pergamon Press, Oxford, 1973) *passim;* several authors, *Handb. Exp. Pharmakol.* **36**, 689-928 (1973).

Silvery, hard, brittle metal. Two forms reported: α-form, double hexagonal, close-packed structure, d (calc; ^{244}Cm) 13.51, Cunningham, Wallman, *J. Inorg. Nucl. Chem.* **26**, 271 (1964). β-form, face-centered cubic structure, d (calc; ^{244}Cm) 12.9, Smith *et al., J. Chem. Phys.* **50**, 5066 (1969). mp 1350 ± 50°. Oxidized rapidly in the presence of traces of oxygen. Chemistry of trivalent state similar to that of trivalent lanthanides.
Caution: Intensely radioactive. Max allowable concn of sol ^{244}Cm in air: 3 × 10⁻¹² μCi/cc; of insol ^{244}Cm: 3 × 10⁻¹¹ μCi/cc, *National Bureau of Standards Handbook* **69**, 89 (1959).

2684. Curvularin. *4,5,6,7,8,9-Hexahydro-11,13-dihydroxy-4-methyl-2H-3-benzoxacyclododecin-2,10(1H)-dione.* $C_{16}H_{20}O_5$; mol wt 292.32. C 65.74%, H 6.90%, O 27.37%. Mold metabolite from a species of *Convalaria:* Musgrave, *J. Chem. Soc.* **1956**, 4301; from *Penicillium steckii:* Fennell *et al., Chem. & Ind.* (London) **1959**, 1382. Structure: Birch *et al., J. Chem. Soc.* **1959**, 3146. Synthesis: H. Gerlach, *Helv. Chim. Acta* **60**, 3039 (1977); of *dl-O,O-*dimethyl ether: Baker *et al., J. Chem. Soc.* (C) **1967**, 1913; T. Takahashi *et* *al., Tetrahedron Letters* **21**, 3885 (1980); H. H. Wasserman, R. J. Gambale, *ibid.* **22**, 4849 (1981).

Plates from hot benzene + methanol, mp 206-206.5°. $[\alpha]_D^{18}$ − 36.3° (c = 3.8 in ethanol). uv max (ethanol): 223, 272, 304.5 nm (ε 11,300; 6350; 5100). Sol in ethanol, methanol, dioxane, acetone, pyridine, concd H_2SO_4; moderately sol in acetic acid, ether; sparingly sol in benzene, petr ether, chloroform, water. Gives yellow solns with aq ammonia, sodium carbonate, and sodium hydroxide. Alkaline solns darken rapidly in air, eventually becoming purple.

Dibenzoate, $C_{30}H_{28}O_7$, needles from benzene + petr ether, mp 133-134°. $[\alpha]_D^{18}$ − 10.8° (c = 1.9 in chloroform). uv max (ethanol): 236 nm (ε 36,700).

O,O-Dimethyl ether, $C_{18}H_{24}O_5$, short rods from aq ethanol, mp 72°. $[\alpha]_D^{18}$ − 2.9° (c = 2.7 in chloroform). uv max (ethanol): 223, 267.5 nm (ε 10,500; 5100).

2685. Cuscohygrine. *1,3-Bis(1-methyl-2-pyrrolidinyl)-2-propanone;* cuskhygrine; bellaradine. $C_{13}H_{24}N_2O$; mol wt 224.34. C 69.60%, H 10.78%, N 12.49%, O 7.13%. In coca leaves of various origin. Found in crude hygrine. Readily converted to hygrine by acids and bases. Isoln: Liebermann, *Ber.* **22**, 679 (1898); Liebermann, Cybulski, *ibid.* **28**, 578 (1895). Identity with bellaradine: Steinegger, Phokas, *Pharm. Acta Helv.* **30**, 441 (1955). Structure: Hess, Fink, *Ber.* **53**, 794 (1920); Sohl, Shriner, *J. Am. Chem. Soc.* **55**, 3828 (1933); Rapoport, Jorgensen, *J. Org. Chem.* **14**, 664 (1949). Synthesis: Späth, Tuppy, *Monatsh.* **79**, 119 (1948); Galinovsky *et al., ibid.* **82**, 551 (1951). Enzymatic synthesis: Tuppy, Faltaous, *ibid.* **91**, 167 (1960). Stereochemistry: Galinovsky, Zuber, *ibid.* **84**, 798 (1953). Biosynthesis: E. Leete, *Chem. Commun.* **1980**, 1170.

Oily liquid, bp_{23} 169-170°; bp_{14} 152°; bp_2 118-125°. d_4^{20} 0.9733. n_D^{20} 1.4832. Miscible with water. Sol in alcohol, ether, benzene.

Hemiheptahydrate, needles, mp 40°.

Hydrobromide, $C_{13}H_{24}N_2O$.2HBr, prisms from alc, mp 234°.

2686. Cuspareine. *2-[2-(3,4-Dimethoxyphenyl)ethyl]-1,2,3,4-tetrahydro-1-methylquinoline.* $C_{20}H_{25}NO_2$; mol wt 311.41. C 77.13%, H 8.09%, N 4.50%, O 10.28%. From Angostura bark *(Cusparia trifoliata* (Willd.) Engl., *Rutaceae)*: Tröger, Runne, *Arch. Pharm.* **249**, 174 (1911); Späth, Pikl, *Monatsh.* **55**, 352 (1930). Structure: Schläger, Leeb, *ibid.* **81**, 714 (1950). Synthesis: Stanek, *ibid.* **88**, 250 (1957).

Crystals from ether, mp 56°. $[\alpha]_D^{20}$ − 20.4° (c = 6.8 in abs ethanol).

Methyliodide, $C_{20}H_{25}NO_2$.CH_3I, crystals, mp 156°.

2687. Cusparine. *2-[2-(1,3-Benzodioxol-5-yl)ethyl]-4-methoxyquinoline.* $C_{19}H_{17}NO_3$; mol wt 307.33. C 74.25%, H 5.58%, N 4.56%, O 15.62%. From Angostura bark: Tröger, Runne, *Arch. Pharm.* **249**, 174 (1911); Späth, Pikl, *Monatsh.* **55**, 352 (1930). Synthesis from 2-methyl-4-methoxyquinoline and piperonal: Späth, Brunner, *Ber.* **57**, 1243 (1924).

Apparently trimorphous. White or yellow needles from

petr ether, mp 92°, or amber prisms, mp 110-122°. Freely sol in organic solvents; practically insol in water.

2688. Cyacetacide. *Cyanoacetic acid hydrazide;* malononitrile hydrazide; cyanoacetohydrazide; cyanazide; cyanizide; cyanacethydrazine; cyanacethydrazide; cyanoethydrazide; cyanacetylhydrazide; Dictyzide; Mackreazid; Armazal; Reacid; Reazide; Hidacian; Leandin; Neohydrazid; Dictycide. $C_3H_5N_3O$; mol wt 99.09. C 36.36%, H 5.09%, N 42.41%, O 16.15%. $NCCH_2CONHNH_2$. Prepd by boiling ethyl or methyl cyanoacetate with hydrazine hydrate in alc: Rothenburg, *Ber.* **27**, 687 (1894); Darapsky, Hillers, *J. Prakt. Chem.* **92**[2], 313 (1915); Klosa, *Arch. Pharm.* **288**, 453 (1955). Prepn of dosage forms: Muset, U.S. pat. **2,849,369** (1958 to Labs OM).

Stout prisms from alc, mp 114.5-115° (evac tube). Rapidly dec by heat. Very slightly acid to litmus. Very sol in water. Sol in alcohol. Practically insol in ether. Aq solns become discolored after a few days and the pH becomes alkaline.

Hydrochloride, $C_3H_6ClN_3O$, crystals, mp 145°. Freely sol in water. Acid reaction. Aq solns are more stable than those of the free hydrazide.

THERAP CAT: Antibacterial (tuberculostatic).

THERAP CAT (VET): Anthelmintic (lungworms).

2689. Cyamelide. *1,3,5-Trioxane-2,4,6-triimine;* insoluble cyanuric acid. $C_3H_3N_3O_3$; mol wt 129.08. C 27.91%, H 2.34%, N 32.56%, O 37.19%. Formed by polymerization of cyanic acid (gaseous or liquid). Prepd by digesting equal parts of potassium cyanate and oxalic acid: Liebig, Wöhler, *Berzelius Jahresber.* **11**, 85. Crude cyamelide contains cyanuric acid which is removed by washing with water: Senier, Walsh, *J. Chem. Soc.* **81**, 290 (1902). Mechanism of polymerization: Werner, Fearon, *ibid.* **117**, 1358 (1920).

White amorphous powder. Practically insol in water. Insol in organic solvents. Slightly sol in ammonia water, in concd acids, in concd NaOH with salt formation.

2690. Cyamemazine. *10-[3-(Dimethylamino)-2-methylpropyl]-10H-phenothiazine-2-carbonitrile;* 2-cyano-10-(3-dimethylamino-2-methylpropyl)phenothiazine; 10-(3-dimethylamino-2-methylpropyl)-3-cyanophenothiazine; cyamepromazine; RP 7204; TH 2602; Kyamepromazine; Cianatil; Tercian. $C_{19}H_{21}N_3S$; mol wt 323.47. C 70.55%, H 6.54%, N 12.99%, S 9.91%. Prepn: Jacob, Robert, U.S. pat. **2,877,224** (1959 to Rhône-Poulenc). Synthesis: Craig *et al., J. Org. Chem.* **26**, 1138 (1961). Biotransformation studies: Robinson, *J. Pharm. Pharmacol.* **18**, 19 (1966). Chemistry: Kiger, Kiger, *Ann. Pharm. Franc.* **23**, 489 (1965).

Yellow oil, $bp_{0.2-0.5}$ 205-220°; also reported as yellow powder, mp 89-96° (Kiger). Practically insol in water. Sol in ethanol and in organic solvents.

Maleate, $C_{23}H_{25}N_3O_4S$, pale yellow plates from methanol-ethanol, mp 196-197° (Craig).

THERAP CAT: Antipsychotic.

2691. Cyanamide. Carbodiimide; hydrogen cyanamide; carbimide; cyanogenamide; amidocyanogen. CH_2N_2; mol wt 42.04. C 28.57%, H 4.80%, N 66.64%. $H_2NC{\equiv}N$. Prepd commercially by continuous carbonation of calcium cyanamide in water according to the equations: $2CaCN +$ $2H_2O \rightarrow Ca(HNCN)_2 + Ca(OH)_2$; $Ca(HNCN)_2 + CO_2 +$

$H_2O \rightarrow 2H_2NCN + CaCO_3$. *Ref.* Brochure "Cyanamide," edited by Process Chemicals Dept., Am. Cyanamid (New York, Aug. 1959) p 19.

Deliquescent, orthorhombic, elongated, six-sided tablets from dimethyl phthalate. d_4^{20} 1.282. mp 45-46°. $bp_{0.5}$ 83°. Formn of dicyandiamide begins at 122°. Dipole moment in benzene at 20°: 3.8. Cryoscopic constant (water) 26.8-28.4. Sp ht 0.547 cal/g/°C between 0° and 39°. Heat of formation 14.05 kcal/mole (25°); heat of combustion −176.4 kcal/mole (25°); heat of fusion 2.1 kcal/mole. Heat of vaporization 16.4 kcal/mole. Soly (g/100 g soln) in water at 15°: 77.5, at 43°: 100; in butanol at 20°: 28.8, in methyl ethyl ketone: 50.5, in ethyl acetate: 42.4. Sol in alcohols, phenols, amines, ethers, ketones. Very sparingly sol in benzene, halogenated hydrocarbons. Practically insol in cyclohexane. Solid cyanamide should be stored in a cool, dry place. Polymerizes at 122°. Optimum pH for storage of solns is ∼4. Attacks various metals. Solns can be stored in glass provided they are stabilized with phosphoric, acetic, sulfuric, or boric acid. LD_{50} in rats: 125 mg/kg orally.

Note: Term "cyanamide" is also used to designate calcium cyanamide.

Caution: Very irritating and caustic; produces severe dermatitis on moist skin. Inhalation may cause irritation of mucous membranes. Ingestion or inhalation: transitory intense redness of face, headache, vertigo, increased respirations, tachycardia, hypotension. Does not produce "cyanide" effect.

2692. Cyanazine. *2-[[4-Chloro-6-(ethylamino)-1,3,5-triazin-2-yl]amino]-2-methylpropanenitrile; 2-[[4-chloro-6-(ethylamino)-s-triazin-2-yl]amino]-2-methylpropionitrile;* SD 15418; WL 19805; DW 3418; Bladex; Fortrol. $C_9H_{13}ClN_6$; mol wt 240.68. C 44.91%, H 5.44%, Cl 14.73%, N 34.92%. Selective pre-emergence herbicide. Prepn: **Brit. pat. 1,132,306** corresp to W. Schwarze, **U.S. pat. 3,505,325** (1968, 1970 to Degussa). Activity: T. Chapman *et al., Proc. Brit. Weed Control Conf.,* 9th **2,** 1018 (1968). Metabolism: J. V. Crayford, D. H. Hutson, *Pestic. Biochem. Physiol.* **2,** 295 (1975). Persistence in soil: J. T. Majka, T. L. Lavy, *Weed Sci.* **25,** 401 (1977).

White cryst, mp 167.5-169°. Vapor press at 20°: 1.6 × 10^{-9} mmHg. Soly in water at 25°: 171 mg/l. Soly at 25° (g/l): benzene 15, chloroform 210, ethanol 45, hexane 15. LD_{50} orally in rats, mice: 182, 380 mg/kg, T. Chapman *et al., loc. cit.*

USE: Herbicide.

2693. Cyanic Acid. Hydrogen cyanate. CHNO; mol wt 43.03. C 27.91%, H 2.34%, N 32.56%, O 37.19%. N≡COH. Best prepd in the laboratory by dry distillation of cyanuric acid: Linhard, *Z. Anorg. Allgem. Chem.* **236,** 200 (1938).

Liquid. *Acrid odor, strong lacrimator and vesicant.* d_4^{20} 1.140°; d_4^{-20} 1.156. mp −86°. bp_{760} 23.5°. Heat of formation −36.5 kcal/mol. Dipole moment 1.592. K at 20° = 2.2 × 10^{-4}. Rapid heating of the liq may result in an explosion. Polymerizes on standing forming cyamelide and cyanuric acid. Sol in water with decompn to carbon dioxide and ammonia. Dil solns in ice water may be kept for several hours. Dil solns in ether, benzene, toluene can be kept for several weeks. *See also* Isocyanic Acid.

USE: In formation of some cyanates. *Caution:* Highly irritating to eyes, skin, mucous membranes, respiratory tract.

2694. Cyanidin Chloride. *2-(3,4-Dihydroxyphenyl)-3,5,7-trihydroxy-1-benzopyrylium chloride; 3,3',4',5,7-pentahydroxyflavylium chloride; 3,3',4',5,7-pentahydroxy-2-phenylbenzopyrylium chloride.* $C_{15}H_{11}ClO_6$; mol wt 322.70. C 55.83%, H 3.44%, Cl 10.99%, O 29.75%. Prepd by acid hydrolysis of cyanin chloride: Willstätter, Everest, *Ann.* **401,** 189 (1913). Isoln from bananas: Simmonds, *Nature* **173,** 402 (1954). Prepn by reduction of quercetin: Bauer *et al., Chem. & Ind. (London)* **1954,** 433; King, White, *J. Chem.*

Soc. **1957,** 3901. Structure: Willstätter, Mallison, *Ann.* **408,** 147 (1915). Synthesis: Willstätter *et al., Ber.* **57,** 1938 (1924); Robertson, Robinson, *J. Chem. Soc.* **1928,** 1528. Biosynthesis: Fritsch *et al., Z. Naturforsch.* **26b,** 581 (1971). *See also* Bioflavonoids.

Metallic brownish-red needles (monohydrate) from dil alcoholic HCl. The anhydr compd does not melt below 300°. Absorption max (methanolic HCl): 535 nm. Freely sol in alcohol and in amyl alcohol giving a violet soln. Sol in sodium carbonate soln with blue color. Sparingly sol in dil HCl or H_2SO_4.

3-Glucoside, $C_{21}H_{21}ClO_{11}$, chrysanthemin, asterin, kuromanin. From winter aster (*Chrysanthemum indicum* L., *Compositae):* Willstätter, Bolton, *Ann.* **412,** 136 (1917); from wild strawberries (*Fragaria vesca* L., *Rosaceae):* Sondheimer, Karash, *Nature* **178,** 648 (1956); from sweet cherries (*Prunus avium* L., *Rosaceae):* Li, Wagenknecht *ibid.* **182,** 657 (1958). Identity with asterin: Robinson, Willstätter, *Ber.* **61,** 2503 (1928). Structure and synthesis: Murakami *et al., J. Chem. Soc.* **1931,** 2665. Reddish-brown plates or prisms with a metallic shine from dil alcoholic HCl, dec 205° without melting. Absorption max (methanolic HCl): 525 nm. Sol in alc with strong fluorescence and in sodium carbonate with violet-blue color.

3-Rhamnoglucoside, $C_{27}H_{31}ClO_{15}$, keracyanin, antirrhinin, sambucin, Meralop, Meralops. From skin of sweet cherries (*Prunus avium* L,. *Rosaceae):* Willstätter, Zollinger, *Ann.* **412,** 164 (1917); from sour cherries (*Prunus cerasus* L., *Rosaceae):* Li, Wagenknecht, *J. Am. Chem. Soc.* **78,** 979 (1956). Structure: Robertson, Robinson, *J. Chem. Soc.* **1927,** 2196. Prepn by reduction of quercetin-3-rhamnoglucoside: Bauer *et al., Chem. & Ind. (London)* **1954,** 433. Reduces time to adjust to darkness: F. Trimarchi *et al., Min. Oftalmol.* **18,** 143 (1977). Mechanism of action: F. Trimarchi *et al., Ann. Ottalmol. Clin. Ocul.* **105,** 111 (1979). Red needles from dil HCl or dark prisms from dil methanolic HCl. Absorption max (ethanolic HCl): 532, 333, 282 nm. Sol in hot water, alcohol.

3-Galactoside, $C_{21}H_{21}ClO_{11}$, idaein, idein. From cranberries (*Vaccinium vitis idaea* L., *Ericaceae):* Willstätter, Mallison, *Ann.* **408,** 15 (1915); from Winesap apple (*Pyrus malus* Linn., *Rosaceae):* Duncan, Dustman, *J. Am. Chem. Soc.* **58,** 1511 (1936). Structure and synthesis: Grove, Robinson, *J. Chem. Soc.* **1931,** 2722. Red needles with bronze metallic luster from dil ethanolic HCl, dec 210-212°. Sol in water, ethanol, methanol, dil HCl. Practically insol in 7% HCl.

3,5-Diglucoside, $C_{27}H_{31}ClO_{16}$, cyanin, shisonin A. From cornflower (*Centaurea cyanus* L., *Compositae):* Willstätter, Everest, *loc. cit.* Structure: Léon, Robinson, *J. Chem. Soc.* **1932,** 221. Synthesis: Robinson, Todd, *ibid.* 2488. Plates with a metallic luster from dil alcoholic HCl, mp 203-204°. $[\alpha]_D$ −258° (in 0.05% HCl). Absorption max (methanolic HCl): 522 nm.

3-Sophoroside, $C_{27}H_{31}ClO_{16}$, mecocyanin. From the flowers of *Papaver rhoeas* L., *Papaveraceae:* Willstätter, Weil, *Ann.* **412,** 231 (1917); from sour cherries: Li, Wagenknecht, *J. Am. Chem. Soc., loc. cit.* Structure: Harborne, *Experientia* **19,** 7 (1963); *Phytochem.* **2,** 85 (1963). (Alternate structure: Grove *et al., J. Chem. Soc.* **1934,** 1608). Dark-red crystals from dil HCl + HOAc or dark-red needles from 2% alcoholic HCl. Absorption max (methanolic HCl): 523 nm. Sol in water. Pptd by glacial acetic acid or acetone.

THERAP CAT: 3-Rhamnoglucoside in treatment of night blindness.

2695. Cyanoacetamide. Malonamide nitrile. $C_3H_4N_2O$; mol wt 84.08. C 42.85%, H 4.80%, N 33.32%, O 19.03%.

$CNCH_2CONH_2$. Prepd by the action of aq or alcoholic ammonia on cyanoacetic ester: Hesse, *Am. Chem. J.* **18**, 724 (1896); Thole, Thorpe, *J. Chem. Soc.* **99**, 429 (1911); Ott, Löpmann, *Ber.* **55**, 1258 (1922); Corson *et al.*, *Org. Syn.* **coll. vol. I** (2nd ed., 1941) p 179.

Needles from alc, mp 119.5°. One gram dissolves in 6.5 ml of cold water. Soly in 100 ml of 95% alc: 1.3 g at 0°; 3.1 g at 26°; 7.0 g at 44°; 14.0 g at 62°; 21.5 g at 71°.

USE: In organic syntheses (*e.g.*, starting material for vita-min B_6 synthesis).

2696. Cyanoacetic Acid. Malonic mononitrile. C_3H_3-NO_2; mol wt 85.06. C 42.36%, H 3.56%, N 16.47%, O 37.62%. $N\equiv CCH_2COOH$. Prepd from chloroacetic acid and NaCN: Ruggli, Businger, *Helv. Chim. Acta* **25**, 35 (1942); Lapworth, Baker, *Org. Syn.* **coll. vol. I**, 181 (1941); **Brit. pat. 824,640** (1959); Eaker, **U.S. pat. 2,539,238** (1951 to Monsanto).

Hygroscopic crystals, mp 66°; dec at 160° into CO_2 and acetonitrile. bp_{15} 108°. Sol in water, alcohol, ether, slightly in benzene, chloroform. *Keep well closed.*

USE: Synthesis of intermediates; manuf barbital.

2697. Cyanofenphos. *Phenylphosphonothioic acid O-(4-cyanophenyl) O-ethyl ester; phenylphosphonothioic acid O-ethyl ester O-ester with p-hydroxybenzonitrile;* CYP; S-4087; Surecide. $C_{15}H_{14}NO_2PS$; mol wt 303.32. C 59.40%, H 4.65%, N 4.62%, O 10.55%, P 10.21%, S 10.57%. Prepn: S. Kuramoto *et al.*, **Japan. pat. 2870('63),** *C.A.* **60**, 455c (1964), **Brit. pat. 929,738,** *C.A.* **59**, 14025a (1963) (both to Sumitomo); **Belg. pat. 628,298,** *C.A.* **60**, 12055h (1964), (1963 to Stauffer). Insecticidal activity: S. Tamura *et al.*, *Agr. Biol. Chem.* **25**, 773 (1961).

Crystalline solid, mp 83°. n_D^{25} 1.5839. Soly in water at 30°: 6 ppm. Moderately sol in ketones and aromatic solvents. Vapor press at 25°: 1.32×10^{-5} mm Hg. Toxic to fish and bees. LD_{50} orally in mice: 43.7 mg/kg.

USE: Insecticide.

2698. Cyanogen. *Ethanedinitrile;* dicyan; oxalic acid dinitrile. C_2N_2; mol wt 52.04. C 46.16%, N 53.84%. $N\equiv C-C\equiv N$. Usually prepd by adding an aq soln of sodi-um or potassium cyanide to an aq soln of copper(II) sulfate or chloride. Procedure: Janz, *Inorg. Syn.* **5**, 43 (1957). Alternate prepn from HCN by the use of CuO: Fierce, Mil-likan, **U.S. pat. 2,841,472** (1958 to Pure Oil); from HCN and NO_2: Fierce, Sander, *Ind. Eng. Chem.* **53**, 985 (1961). Re-view: Brotherton, Lynn, *Chem. Rev.* **59**, 841-883 (1959).

Highly poisonous gas. Almond-like odor. Acrid and pungent when in lethal concns. Burns with pink flame having a bluish border. mp −27.9° (also reported as −34.4°). bp −21.17°. $d_4^{-21.17}$ 0.9537. Heat of vaporization (liquid) 5.778 kcal/mole. Above 500° polymerizes to insol paracyanogen $(CN)_n$. One vol of water dissolves about 4 vols of cyanogen gas. Also sol in alcohol, ether. Slowly hydrolyzed in aq soln giving oxalic acid and ammonia.

Caution: Toxic effects similar to those of hydrogen cya-nide (*q.v.*).

2699. Cyanogen Azide. Carbon pernitride. CN_4; mol wt 68.04. C 17.65%, N 82.34%. $N^-=N^+=N-C\equiv N$. Prepd by suspending NaN_3 in dry acetonitrile and distilling cyano-gen chloride into the cooled suspension: Marsh, Hermes, *J. Am. Chem. Soc.* **86**, 4506 (1964); *see also Chem. & Eng. News* **42**, 51 (Oct. 26, 1964); Marsh, **U.S. pat. 3,410,658** (1968); Marsh, *J. Org. Chem.* **37**, 2966 (1972). Spectrum, structure, dipole moment: Costain, Kroto, *Can. J. Phys.* **50**, 1453 (1972); Almenningen *et al.*, *Acta Chem. Scand.* **27**, 1531 (1973).

Clear, colorless, oily liquid. *Caution: The pure azide deto-nates violently upon thermal, electrical or mechanical shock!* Can be handled relatively safely in solvents. Half-life of a

27% soln in acetonitrile is 15 days at room temp, more stable at lower temps.

USE: In organic synthesis. Has a versatility and scope of chemical reactivity that is very broad and useful, *e.g.* reacts with alkanes to give primary alkylcyanamides: Anastassiou *et al.*, *J. Am. Chem. Soc.* **87**, 2296 (1965).

2700. Cyanogen Bromide. Bromine cyanide. CBrN; mol wt 105.93. C 11.34%, Br 75.44%, N 13.22%. $BrC\equiv N$. Prepd from potassium or sodium cyanide and bromine: Hartman, Dreger, *Org Syn.* **11**, 30 (1931). Industrial prepn from sodium bromide, sodium chlorate, and sodium cyanide in 30% H_2SO_4: Ewan, *Chem. Zentr.* **1907, I**, 591; Göpner, *ibid.* **1908, I**, 1807; Grignard, Crouzier, *Bull. Soc. Chim. France* **[4] 29**, 214 (1921).

Needles, cubes; volatile at ordinary temps. *Vapors are highly irritant and very poisonous!* d_4^{20} 2.015. mp 52°. bp 61-62°. Freely sol in water, alcohol, ether. Aqueous solns of alkalies dec it to alkali cyanide and alkali bromide. Pure cyanogen bromide, completely dried by distillation over sodium, may be stored in a desiccator for several months. Impure material dec rapidly and tends to explode.

Caution: Toxic effects similar to those of hydrogen cya-nide (*q.v.*).

2701. Cyanogen Chloride. CClN; mol wt 61.48. C 19.54%, Cl 57.68%, N 22.79%. CNCl. By action of Cl on HCN. Toxicity data: *Handbook of Toxicology* vol. 1, W. S. Spector, Ed. (Saunders, Philadelphia, 1956) pp 82-83.

Liquid. *Vapors are highly irritant and very poisonous!* d 1.186. bp 13.8°. mp −6°. Sol in water, alc, ether. LD s.c. in rabbits: 20 mg/kg (Spector).

USE: In chemical synthesis. Military poison gas. *Caution:* Toxic effects similar to those of hydrogen cyanide, *q.v.*

2702. Cyanogen Iodide. *Iodine cyanide.* CIN; mol wt 152.94. C 7.85%, I 82.99%, N 9.16%. CNI. Prepd by the action of iodine on sodium cyanide: Bak, Hillebert, *Org. Syn.* **coll. vol. IV**, 207 (1963).

White needles; very pungent odor; acrid taste. *Very poi-sonous!* mp 146-147°. Sol in water, alcohol, ether, volatile oils. LD orally in cats: 18 mg/kg, *Handbook of Toxicology* vol. 1, W. S. Spector, Ed. (Saunders, Philadelphia, 1956) pp 82-83.

USE: Generally for destroying all lower forms of life. In taxidermy for preserving insects, butterflies, etc. *Caution:* Causes convulsions, paralysis, death from respiratory failure. Lacks inhalation hazard.

2703. Cyanopsin. Photoreceptor protein found in the retinal cone cells of fresh water and migratory fish, lam-preys, and certain amphibians; corresponds to the rod pig-ment, porphyropsin, *q.v.* Absorption maximum approxi-mately 620 nm. Composed of the chromophore 11-*cis* 3-dehydroretinal, *q.v.*, bound to a photopsin, the specific pro-tein component of cone cells (*see* Opsins). Proposed exis-tence as a visual pigment and *in vitro* synthesis from 3-dehy-droretinal and chicken photopsin: G. Wald, *Science* **118**, 505 (1953). Demonstration of natural occurrence by micro-spectrophotometry: P. Liebman, G. Entine, *Nature* **216**, 501 (1967). Methods for prepn and assay: R. Hubbard *et al.*, "Methodology of Vitamin A and Visual Pigments" in *Meth-ods in Enzymology* Vol. 18, D. B. McCormick, L. D. Wright, Eds. (Academic Press, New York, 1971) pp 615-653. Pho-tochemistry and biological activity: G. Wald, *Science* **162**, 230 (1968). A trichromatic cone system utilizing 3-dehy-droretinal as chromophore has been noted to occur in cer-tain amphibians and fish. Three photochemically distinct pigments, corresponding to those of retinal-based color vi-sion, have been identified: P. A. Liebman in *Handbook of Sensory Physiology* Vol. VII(1), H. J. A. Dartnall, Ed. (Springer-Verlag, New York, 1972) pp 481-528. A visual system is generally based on one type of chromophore com-bined with various opsins. However, pigments utilizing both types of chromophore have been found to co-exist in the retina of some of these species. Demonstration of paired-pigment cone systems: E. R. Loew, H. J. A. Dartnall, *Vision Res.* **11**, 551 (1976); A. T. C. Tsin *et al.*, *ibid.* **21**, 943 (1981). Absorption spectrum: J. E. M. Mooij, T. J. T. P. van den Berg, *ibid.* **23** (1983).

2704. Cyanuric Acid. *1,3,5-Triazine-2,4,6(1H,3H,5H)-*

trione; normal cyanuric acid; _sym_-triazinetriol; 2,4,6-trihydroxy-1,3,5-triazine; tricyanic acid; trihydroxycyanidine. $C_3H_3N_3O_3$; mol wt 129.08. C 27.91%, H 2.34%, N 32.56%, O 37.19%. Formed on heating urea. Prepd from carbonyldiurea or from carbonyldiubiuret by heating, by boiling with alkalies or by heating in excess phosgene at 150°: Schmidt, _J. Prakt. Chem._ [2] **5**, 41-52 (1872). From allantoin by oxidation with H_2O_2 in slightly alkaline soln at 80°: Venable, Moore, _J. Am. Chem. Soc._ **39**, 1752 (1917). Review of prepn, properties and uses: _The Chemistry of Heterocyclic Compounds,_ A. Weissberger, Ed., **vol. 13**, entitled _s-Triazines and Derivatives_ by E. M. Smolin, L. Rapoport (Interscience, New York, 1959) pp 17-48.

HO—[triazine structure]—OH / OH

Dihydrate, efflorescent monoclinic prisms (octahedra) from water. The water of crystn is lost on exposure to air. Anhydr crystals from concd HCl or H_2SO_4. When anhydr d_4^{20} 2.500. Does not melt, evolves cyanic acid on heating. Ka_1 at 25° = 6.31×10^{-8}; $Ka_2 = 7.94 \times 10^{-12}$. One gram dissolves in about 200 ml water, tends to form supersaturated solns. Much more sol in hot waters. Sol in hot alcohols, pyridine, concd HCl and H_2SO_4 without decompn. Also sol in aq solns of NaOH and KOH. Insol in cold methanol, ether, acetone, benzene, chloroform.

USE: Convenient lab source of cyanic acid gas. In prepn of melamine, sponge rubber. Selective herbicide. Very toxic to certain types of barley and radishes.

2705. Cycasin. _(Methyl-ONN-azoxy)methyl β-D-gluco-pyranoside;_ methylazoxymethanol β-D-glucoside; β-D-glucosyloxyazoxymethane. $C_8H_{16}N_2O_7$; mol wt 252.22. C 38.09%, H 6.39%, N 11.11%, O 44.41%. Toxic substance from seeds of _Cycas revoluta_ Thumb. and _C. circinalis_ L., Cycadaceae from Guam: Nishida _et al., Bull. Agr. Chem. Soc. Japan_ **19**, 77 (1955), _C.A._ **50**, 13756g (1956); Riggs, _Chem. & Ind. (London)_ **1956**, 926; Matsumoto, Strong, _Arch. Biochem. Biophys._ **101**, 299 (1963). Biological effects, metabolism and mechanism of action of cycasin and its aglycone, _methylazoxymethanol:_ Spatz, _Ann. N.Y. Acad. Sci._ **163**, 848 (1969). Structure: Korsch, Riggs, _Tetrahedron Letters_ **1964**, 523. Carcinogenic activity studies: Hirono _et al., J. Natl. Cancer Inst._ **40**, 1003 (1968); Hirono _et al., Cancer Res._ **31**, 283 (1971). Toxicity data: I. Hirono, _Fed. Proc._ **31**, 1493 (1972). Review and evaluation of carcinogenicity studies of cycasin and methylazoxymethanol acetate in laboratory animals: _IARC Monographs_ **10**, 121-138 (1976). Toxicology: Laqueur, Spatz, _Cancer Res._ **28**, 2262 (1968).

[chemical structure]

Needles from water + acetone + ether, dec 154°. $[\alpha]_D^{18}$ −44° (c = 0.62). uv max (0.4M H_2SO_4): 217 nm. LD_{50} in rats: 562 mg/kg (Hirono).

Tetraacetate, $C_{16}H_{24}N_2O_{11}$, plates from acetone + petr ether, mp 137°. $[\alpha]_D^{18}$ −27° (c = 0.98 in $CHCl_3$).

Note: This substance may reasonably be anticipated to be a carcinogen: _Fourth Annual Report on Carcinogens_ (NTP 85-002, 1985) p 64.

2706. Cyclacillin. _6-[[(1-Aminocyclohexyl)carbonyl]-amino]-3,3-dimethyl-7-oxo-4-thia-1-azabicyclo[3.2.0]hept-ane-2-carboxylic acid;_ 6-(1-aminocyclohexanecarboxamido)penicillanic acid; (1-aminocyclohexyl)penicillin; ciclacillin; Wy 4508; Calthor; Citosarin; Cyclapen; Syngacillin; Ultracillin; Vastcillin; Vatracin; Vipicil; Wyvital. $C_{15}H_{23}$-N_3O_4S; mol wt 341.43. C 52.77%, H 6.79%, N 12.31%, O 18.74%, S 9.39%. Semi-synthetic antibiotic related to penicillin. Prepn: Alburn _et al.,_ **U.S. pat. 3,194,802**; Robinson,

Nescio, **Ger. pat. 1,800,698** corresp to **U.S. pat. 3,478,018** (1965; 1969, all to Am. Home Prods.). Activity studies: Rosenman _et al., Antimicrob. Ag. Chemother._ **1967**, 590; Hopper _et al., ibid._ 597; Yurchenco _et al., ibid._ 602; _Chemotherapy_ **15**, 209 (1970). Metabolic studies: Poole, _J. Pharm. Sci._ **59**, 1255 (1970); Tucker _et al., Toxicol. Appl. Pharmacol._ **19**, 361 (1971). Physicochemical data: Alburn _et al., Antimicrob. Ag. Chemother._ **1967**, 586; Hou, Poole, _J. Pharm. Sci._ **58**, 1510 (1969); Poole, Bahal, _ibid._ **59**, 1265 (1970).

[chemical structure with NH2, CONH, S, CH3, CH3, COOH]

Crystals, mp 182-183° (anhydrate) (Hou, Poole); 156-158° (dec) (Alburn _et al.). $[\alpha]_D^{25}$ 268° (water). pK_1, pK_2 in water: 2.68, 7.50; in 50% dioxane: 4.16, 7.04. Stable in acid. Soly at 38°: about 29 mg/ml.

THERAP CAT: Antibacterial.

2707. Cyclamic Acid. _Cyclohexylsulfamic acid; cyclo-hexanesulfamic acid;_ Hexamic Acid. $C_6H_{13}NO_3S$; mol wt 179.24. C 40.20%, H 7.31%, N 7.82%, O 26.78%, S 17.89%. Prepn: Audrieth, Sveda, **U.S. pat. 2,275,125** (1942 to du Pont); _J. Org. Chem._ **9**, 89 (1944); Thompson, **U.S. pat. 2,800,501** (1957 to du Pont); Shah, Bernsen, **U.S. pat. 3,361,799** (1968 to Abbott). Prepn of the Na salt: Robinson, **U.S. pat. 2,383,617** (1945 to du Pont). Other prepns and metabolism: _See_ Calcium Cyclamate. Sweetness of the sodium salt discovered by Michael Sveda at the University of Illinois in 1937. Toxicity: Taylor _et al., Food Cosmet. Toxicol._ **6**, 313 (1968). Review of long-term toxicity and carcinogenicity of sodium cyclamate in mice: Brantom _et al., Food Cosmet. Toxicol._ **11**, 735 (1973).

[chemical structure with NHSO3H]

Sweet-sour crystals. mp 169-170°. Fairly strong acid. Very sparingly soluble in water. Slowly hydrolyzed by hot water.

Sodium salt, $C_6H_{12}NNaO_3S$, _sodium cyclohexylsulfamate, cyclamate sodium, sodium cyclamate,_ Assugrin, Sucaryl Sodium, Sucrosa. Crystals. Pleasantly sweet to the taste. Freely sol in water. About 30 times as sweet as refined cane sugar. Sweetness is still easily perceptible at a dilution of 1:10,000 (sugar 1:140; saccharin 1:50,000). pH of 10% aq soln between 5.5 and 7.5. Practically insol in alcohol, ether, benzene, chloroform. LD_{50} orally in mice, rats: 17.0, 15.25 g/kg (Taylor).

Note: Consult latest Government regulations on use in food.

USE: Non-nutritive sweetener.

2708. Cyclandelate. _α-Hydroxybenzeneacetic acid 3,3,5-trimethylcyclohexyl ester; mandelic acid 3,3,5-trimethylcyclo-hexyl ester;_ 3,3,5-trimethylcyclohexyl mandelate; 3,5,5-tri-methylcyclohexyl amygdalate; 3,3,5-trimethylcyclohexanol α-phenyl-α-hydroxyacetate; BS 572; Cyclergine; Cyclobral; Cyclolyt; Cyclomandol; Cyclospasmol; Natil; Novodil; Perebral; Spasmocyclon. $C_{17}H_{24}O_3$; mol wt 276.36. C 73.88%, H 8.75%, O 17.37%. Prepn: Brock _et al., Arzneimit-tel-Forsch._ **2**, 165 (1952); Funcke _et al., ibid._ **3**, 503 (1953); van Sluis, _Chem. Products_ **17**, 375 (1954); **Brit. pat. 707,-227** (1954 to Brocades-Stheeman). Purification: Flitter, **U.S. pat. 3,663,597** (1972 to Am. Home Prods.). Pharmacokinetics: A. Orr, J. R. Whittier, _Int. J. Nucl. Med. Biol._ **1**, 205 (1974). GLC determn in human plasma: G. Andermann, M. Dietz, _J. Chromatog._ **223**, 365 (1981). Clinical trial in chronic cerebrovascular disease: S. Bassi _et al., Brit. J. Clin. Prac._ **38**, 344 (1984). Review of efficacy as cerebral and peripheral vasodilator: C. B. Blakemore, _ibid._ **34**,

Suppl., 3-9 (1984). Symposium on calcium modulation and clinical effects: *Drugs* **33**, Suppl. 2, 1-141 (1987).

Crystals, mp 50-53°. bp$_{14}$ 192-194°. Practically insol in water. Sol in lipoids and their solvents.
THERAP CAT: Vasodilator (peripheral, cerebral).

2709. Cyclarbamate. *1,1-Cyclopentanedimethanol bis-(phenylcarbamate); 1,1-cyclopentanedimethanol dicarbanilate;* cyclopentane-1,1-dimethanol bis(phenylcarbamate); 1,1-dihydroxymethylcyclopentane *N,N'*-diphenylcarbamate; cyclopentaphene; BSM 906M; Calmalone; Casmalon. $C_{21}H_{24}N_2O_4$; mol wt 368.44. C 68.46%, H 6.57%, N 7.60%, O 17.37%. Prepn: Bourdais, *Bull. Soc. Chim. France* **1962**, 266; E. Rosenberg, **Brit.** pat. **904,410** corresp to **U.S.** pat. **3,067,240** (both 1962 to Lab. Cassenne).

Needles from methanol, mp 151-152°. Practically insol in water. Slightly sol in alcohol, glycerol, propylene glycol.
THERAP CAT: Anxiolytic; muscle relaxant (skeletal).

2710. Cyclazocine. *3-(Cyclopropylmethyl)-1,2,3,4,5,6-hexahydro-6,11-dimethyl-2,6-methano-3-benzazocin-8-ol;* WIN 20740; NSC-107429. $C_{18}H_{25}NO$; mol wt 271.39. C 79.66%, H 9.29%, N 5.16%, O 5.90%. Analgesic with mixed narcotic agonist-antagonist properties. Prepn: Archer, **Belg.** pat. **611,000** (1962 to Sterling Drug), *C.A.* **58**, 2439c (1963). Crystal structure: I. L. Karle *et al., Acta Crystallogr.* **B25**, 1469 (1969). Supraspinal analgesic effects: S. Sasson, C. Kornetsky, *Life Sci.* **38**, 21 (1986). Evaluation in anorexia and motor disruption in rats: J. W. Henck *et al., Pharmacol. Biochem. Behav.* **22**, 671 (1985). Determn by HPLC: I. Jane, A. McKinnon, *J. Chromatog.* **323**, 191 (1985). Review of use in treatment of opiate addiction: M. J. Goldstein, *Int. J. Addict.* **15**, 939 (1980).

Crystals from methanol, mp 201-204°.
THERAP CAT: Narcotic antagonist.

2711. Cyclethrin. *2,2-Dimethyl-3-(2-methyl-1-propenyl)cyclopropanecarboxylic acid 3-(2-cyclopenten-1-yl)-2-methyl-4-oxo-2-cyclopenten-1-yl ester;* 3-(2-cyclopentenyl)-2-methyl-4-oxo-2-cyclopentenyl ester of chrysanthemum-monocarboxylic acid; chrysanthemummonocarboxylic acid ester with 3-(2-cyclopenten-1-yl)-2-methyl-4-oxo-2-cyclopenten-1-ol. $C_{21}H_{28}O_3$; mol wt 328.44. C 76.79%, H 8.59%, O 14.61%. Prepn and insecticidal activity: H. L. Haynes *et al., Contrib. Boyce Thompson Inst.* **18**, 1 (1954); H. R. Guest, H. A. Stansbury, **U.S.** pat. **2,891,888** (1959 to Union Carbide). Product is a mixture of isomers including four racemic forms or eight optical and geometric isomers which have not yet been isolated and evaluated. Toxicology: C. P. Carpenter *et al., Arch. Ind. Hyg. Occup. Med.* **10**, 162 (1954).

Liquid. n_D^{30} 1.5120. d_{20}^{20} 1.033. LD$_{50}$ orally in rats: 1.4-2.8 g/kg (Carpenter).
USE: Insecticide for flies, roaches, and grain pests.

2712. Cyclexanone. *2-(1-Cyclopenten-1-yl)-2-[2-(4-morpholinyl)ethyl]cyclopentanone;* 2-(Δ-1'-cyclopentenyl)-2-(β-tetrahydroparoxazinoethyl)cyclopentanone; 1-(β-*N*-morpholinoethyl)-1-cyclopentenylcyclopentane-2-one; pentethylcyclanone. $C_{16}H_{25}NO_2$; mol wt 263.37. C 72.96%, H 9.57%, N 5.32%, O 12.15%. Prepd from 2-cyclopentylidenecyclopentanone and *N*-(β-chloroethyl)morpholine: Ueberwasser, **Ger.** pat. **1,059,901**; **Brit.** pat. **878,677** (1959, 1961 both to Ciba).

Liquid, bp$_{0.035}$ 112°.
Hydrochloride, $C_{16}H_{26}ClNO_2$, *Exopan, Exopon.* Crystals from aq ethanol, mp 209°.
Methiodide, $C_{16}H_{25}NO_2$.CH$_3$I, crystals, mp 106°.
THERAP CAT: Antitussive.

2713. Cyclexedrine. *N,β-Dimethylcyclohexaneethanamine;* cyclohexylisopropylmethylamine; 2-cyclohexyl-*N*-methylpropylamine; 1-cyclohexyl-1-methyl-2-methylaminoethane; ethylhexedrine; isopropylhexedrine. $C_{10}H_{21}N$; mol wt 155.28. C 77.34%, H 13.63%, N 9.02%. Prepn: **Ger.** pat. **970,480** (1958 to Knoll).

Hydrochloride, $C_{10}H_{22}ClN$, *Eventin.* Crystalline powder, bitter taste. mp 138-140°. Soluble in water, ethanol, chloroform; slightly sol in ether. LD$_{50}$ orally in mice: 304 mg/kg.
THERAP CAT: Sympathomimetic, anorexic.

2714. Cyclic AMP. *Adenosine cyclic 3',5'-(hydrogen phosphate);* adenosine 3',5'-cyclic monophosphate; adenosine 3',5'-cyclic phosphate; adenosine 3',5'-monophosphate; adenosine 3',5'-phosphate; cyclic adenosine 3',5'-monophosphate; acrasin; 3',5'-AMP; cAMP. $C_{10}H_{12}N_5O_6P$; mol wt 329.22. C 36.48%, H 3.68%, N 21.27%, O 29.16%, P 9.41%. Key intracellular regulator of a number of cellular processes; found in most animal cells, in bacteria, and in some higher plants. First isoln and identification: Rall *et al., J. Biol. Chem.* **224**, 463 (1957); Sutherland, Rall, *ibid.* **232**, 1077 (1958). Molecular structure and conformation: Cook *et al., J. Am. Chem. Soc.* **79**, 3607 (1957); Lipkin *et al., ibid.* **81**, 6198 (1959); Watenpaugh *et al., Science* **159**, 206 (1968). Syntheses: Smith *et al., J. Am. Chem. Soc.* **83**, 698 (1961); Borden, Smith, *J. Org. Chem.* **31**, 3247 (1966). Physical data: D. Lipkin *et al., J. Am. Chem. Soc.* **81**, 6075 (1959). Functions as a mediator of hormone-action for a variety of hormones such as epinephrine, glucagon and ACTH (*q.v.*). Activates phosphorylation of proteins by protein kinases. Defined as a "second messenger" because of its response to hormones ("first messengers"). Converted from adenosine triphosphate (ATP) by the enzyme *adenylate cyclase.* Deactivated by *cyclic nucleotide phosphodiesterases* which convert it to 5'-adenylic acid. Reviews of biochemical model: Sutherland *et al., J. Biol. Chem.* **237**, 1220-1243 (1962); Robison *et al., Ann. Rev. Biochem.* **37**, 149 (1968); Jost, Rickenberg, *ibid.* **40**, 741 (1971); Sutherland, *J. Am. Med. Assoc.* **214**, 1281 (1970); Pastan, Perlman, *Nature New Biol.* **229**, 5 (1971); G. A. Robison *et al.*, Eds., *Ann. N.Y. Acad. Sci.* **185** (1971); Losert, *Pharmazie* **28**, 351 (1973). See *also* Cyclic GMP. Identity with acrasin from cellular slime molds (*Dictyostelium* species), where it acts as a "first" rather than a "second messenger": Konijn *et al., Proc. Nat. Acad. Sci. USA* **58**, 1152 (1967). Review of quantitative methods: *Methods in Molecular Biology*, vol. **3** entitled "Methods in Cyclic Nucleotide Research", Mark Chasin, Ed. (Marcel Dekker, New York, 1972). Books: G. A. Robison *et al., Cyclic AMP* (Academic Press, New York, 1971); *Advan. Cyclic Nucleotide Res.* **vols. 1, 2, 3**, P. Greengard, G. A. Robison, Eds. (Raven Press, New York, 1972, 1973); *Handb. Exp. Pharmacol.* **58** entitled "Cyclic Nucleotides", Pt I: Biochemistry and Pt II: Physiology & Pharmacology, J. A. Nathanson, J. W. Kobabian, Eds. (Springer-Verlag, New York, 1982) 736 and 1000 pp resp.

Hydrate crystallizes from water as platelets with pearly luster, mp 219-220° (with gas evolution). $[\alpha]_D -51.3°$ (c = 0.67). uv max (pH 2): 256 nm (ϵ 14500); (pH 7): 258 nm (ϵ 14650). Heat stable; resistant to inactivation by acid or alkali. Found in concns of 10^{-7} to 10^{-6} moles/kg in living cells.

2715. Cyclic GMP. *Guanosine cyclic 3',5'-(hydrogen phosphate);* cyclic guanosine 3',5'-monophosphate; guanosine 3',5'-cyclic monophosphate; guanosine 3',5'-cyclic phosphate; guanosine 3',5'-monophosphate; 3',5'-GMP; cGMP. $C_{10}H_{12}N_5O_7P$; mol wt 345.22. C 34.79%, H 3.51%, N 20.29%, O 32.44%, P 8.97%. A cellular regulatory agent which has been described as a "second messenger". *See* Cyclic AMP. First synthesized because of the interest in cyclic nucleotides generated by cyclic AMP: Smith *et al., J. Am. Chem. Soc.* **83**, 698 (1961); Borden, Smith, *J. Org. Chem.* **31**, 3247 (1966). Has subsequently been found in animal and bacterial cells in concentrations of 10^{-8} to 10^{-6} moles/kg. First isoln from rat urine: Ashman *et al., Biochem. Biophys. Res. Commun.* **11**, 330 (1963). Structure and conformation: Chwang, Sundaralingam, *Nature, New Biol.* **244**, 136 (1973). Proposed as an antagonist of cyclic AMP in bidirectional systems such as contraction-relaxation or glycogen synthesis-glycogen breakdown. Cyclic GMP levels increase in response to a variety of hormones including acetylcholine, insulin and oxytocin. Formed by conversion of guanosine triphosphate by the enzyme *guanylate cyclase* and hydrolyzed by cyclic nucleotide phosphodiesterases. Found to activate specific protein kinases. Reviews of biochemical model: Goldberg *et al.,* in *Pharmacology and the Future of Man* vol. **5**, R. A. Maxwell, G. H. Acheson, Eds. (Karger, New York, 1973) pp 146-169; *eidem* in *Advan. Cyclic Nucleotide Res.* vol. **3**, P. Greengard, G. A. Robison, Eds. (Raven

Press, New York, 1973) pp 155-223; Kolata, *Science* **182**, 149-151 (1973); *Nature* **246**, 186 (1973); N. D. Goldberg, M. K. Haddox, *Ann. Rev. Biochem.* **46**, 823-896 (1977). Book: *Handb. Exp. Pharmacol.* **58**, entitled "Cyclic Nucleotides", Pt. I: Biochemistry and Pt. II: Physiology & Pharmacology, J. A. Nathanson, J. W. Kobabian, Eds. (Springer-Verlag, New York, 1982) 736 and 1000 pp resp.

Calcium salt, $C_{20}H_{22}CaN_{10}O_{14}P_2$, decahydrate. uv max (pH 1): 256.5 nm (ϵ 11,350); (pH 7): 254 nm (ϵ 12,950); (pH 12): 262 nm (ϵ 12,400).

2716. Cyclizine. *1-Diphenylmethyl-4-methylpiperazine; N*-benzhydryl-*N'*-methylpiperazine; *N*-methyl-*N'*-benzhydrylpiperazine; (*N*-benzhydryl)(*N'*-methyl)diethylenediamine; Compd 47-83; Wellcome prepn 47-83; Marzine; Marezine; Nautazine; Neo-Devomit. $C_{18}H_{22}N_2$; mol wt 266.37. C 81.16%, H 8.33%, N 10.52%. Prepd by the action of benzhydryl chloride on *N*-methyl piperazine: Baltzly *et al., J. Org. Chem.* **14**, 775 (1949); U.S. pat. **2,630,435** (1953 to Burroughs Wellcome). Metabolism: R. Kuntzman *et al., Ann. N.Y. Acad. Sci.* **226**, 131 (1973). Comprehensive description: S. A. Benezra in *Analytical Profiles of Drug Substances* vol. **6**, K. Florey, Ed. (Academic Press, New York, 1977) pp 83-97. Clinical antiemetic efficacy: G. Rowlands, W. J. Currie, *Brit. J. Clin. Pract.* **30**, 197 (1976); W. N. Chestnutt *et al., Eur. J. Anaesthesiol.* **3**, 27 (1986).

Light-sensitive crystals from petr ether, mp 105.5-107.5°. uv max (0.1N HCl): 269, 263, 258, 225 (ϵ 540, 742, 694, 11300). Soly in g/ml at 25°: chloroform 1.1; ether 0.17; ethanol 0.17; water < 0.001. LD_{50} orally in mice: 147 mg/kg, *RTECS* vol. **II**, E. J. Fairchild, Ed. (1977) p 716.

Hydrochloride, $C_{18}H_{23}ClN_2$, *Valoid.* Crystals, moderately sol in water. *Protect from light.*

THERAP CAT: Antiemetic.
THERAP CAT (VET): Antiemetic.

2717. Cyclobarbital. *5-(1-Cyclohexen-1-yl)-5-ethyl-2,4,6(1H,3H,5H)-pyrimidinetrione; 5-(1-cyclohexen-1-yl)-5-ethylbarbituric acid;* cyclobarbitone; hexemal; tetrahydrophenobarbital; Cavonyl; Cyclodorm; Cyklodorm; Fanodormo; Irifan; Namuron; Palinum; Phanodorm; Phanodorn; Philodorm; Prälumin; Pro-Sonil; Sonaform. $C_{12}H_{16}N_2O_3$; mol wt 236.26. C 61.00%, H 6.83%, N 11.86%, O 20.32%. Prepn: Brit. pat. **231,150** (1924 to Bayer); Schulemann, Meisenburg, U.S. pat. **1,690,796** (1929 to Winthrop); Eckstein, *Przemysl Chem.* **9**, 390 (1953), *C.A.* **49**, 11668c (1955). Toxicity data: G. Hofrichter *et al., Arzneimittel-Forsch.* **17**, 242 (1967).

Shiny crystals, insipid bitter taste, mp 171-174°. Very slightly sol in cold water, appreciably sol in boiling water. One gram dissolves in 5 ml alcohol, 20 ml ether. LD_{50} in mice, rats (mg/kg): 350, 290 i.p. (Hofrichter).

Calcium salt, $C_{24}H_{30}CaN_4O_6$, *hexemal calcium, Itridal, Kollerdormfix, Pronox.* Minute crystals. Appreciably bitter taste. Slightly basic reaction. Soly in water about 1 part in 70 parts (w/w), in 95% alc about 1 part in 500 parts. Practically insol in ether, chloroform.

Caution: May be habit forming. This is a controlled substance (depressant) listed in the U.S. Code of Federal Regulations, Title 21 Parts 329.1 and 1308.13 (1987).

THERAP CAT: Sedative, hypnotic.
THERAP CAT (VET): Has been used as an anesthetic and sedative.

2718. Cyclobendazole. *[5-(Cyclopropylcarbonyl)-1H-benzimidazol-2-yl]carbamic acid methyl ester;* ciclobendazole; CC-2481; R-17147; Haptocil. $C_{13}H_{13}N_3O_3$; mol wt 259.26. C 60.23%, H 5.05%, N 16.21%, O 18.51%. Prepn: J. L. Van Gelder *et al.*, **Ger.** pat. **2,029,637** corresp to **U.S.** pat. **3,657,267** (1971, 1972 both to Janssen). Synthesis and anthelmintic activity: A. H. M. Raeymaekers *et al.*, *Arzneimittel-Forsch.* **28**, 586 (1978). Embryotoxic and antimytotic properties: P. Delatour, Y. Richard, *Therapie* **31**, 505 (1976). Pharmacokinetics: R. R. Brodie *et al.*, *Arzneimittel-Forsch.* **27**, 593 (1977). Biotransformation: B. C. Mayo *et al.*, *Drug Metab. Dispos.* **6**, 518 (1978). Clinical evaluation: A. Degrémont, E. Stahel, *Schweiz. Med. Wochenschr.* **108**, 1430 (1978).

Cryst from acetic acid, mp 250.5°.

THERAP CAT: Anthelmintic (Nematodes).

2719. Cyclobenzaprine. *3-(5H-Dibenzo[a,d]cyclohepten-5-ylidene)-N,N-dimethyl-1-propanamine; N,N-dimethyl-5H-dibenzo[a,d]cyclohepten-Δ5,γ-propylamine;* 5-(3-dimethyl-aminopropylidene)dibenzo[a,e]cycloheptatriene; 1-(3-dimethylaminopropylidene)-2,3:6,7-dibenzo-4-suberene; proheptatriene; MK-130; Ro 4-1577; RP 9715. $C_{20}H_{21}N$; mol wt 275.38. C 87.22%, H 7.69%, N 5.09%. Prepn: **Brit.** pat. **858,187** (1961 to Hoffmann-La Roche); Villani *et al.*, *J. Med. Pharm. Chem.* **5**, 373 (1962); Winthrop *et al.*, *J. Org. Chem.* **27**, 230 (1962). Pharmacology: C. D. Barnes, W. L. Adams, *Neuropharmacology* **17**, 445 (1978); N. N. Share, *ibid.* 721; and toxicology: J. Metysova *et al.*, *Arch. Int. Pharmacodyn. Ther.* **144**, 481 (1963). Metabolism: G. Belvedere *et al.*, *Biomed. Mass Spectrom.* **1**, 329 (1974); H. B. Hucker *et al.*, *Drug Metab. Dispos.* **6**, 184 (1978). Bioavailability: *eidem, J. Clin. Pharmacol.* **17**, 719 (1977). Clinical studies: J. V. Basmajian, *Arch. Phys. Med. Rehabil.* **5**, 58 (1978); B. R. Brown, J. Womble, *J. Am. Med. Assoc.* **240**, 1151 (1978). Comprehensive description: M. L. Cotton, G. R. B. Down in *Analytical Profiles of Drug Substances* vol. 17, K. Florey, Ed. (Academic Press, New York, 1988) pp 41-72.

bp$_1$ 175-180°. uv max: 224, 289 nm (log ε 4.57, 4.02), Villani *et al.*, *loc. cit.*

Hydrochloride, $C_{20}H_{22}ClN$, *Flexeril, Flexiban*. Use as muscle relaxant: N. N. Share, **Fr.** pat. **2,100,873** (1972 to Frosst), *C.A.* **78**, 47801n (1973). Crystals from isopropanol, mp 216-218°. Soly in water: > 20 g/100 ml. Freely sol in water, methanol, ethanol; sparingly sol in isopropanol; slightly sol in chloroform, methylene chloride. Practically insol in hydrocarbons. uv max: 226, 295 nm (ε 52,300, 12,000). LD$_{50}$ in mice (mg/kg): 35 i.v., 250 orally (Metysova).

THERAP CAT: Muscle relaxant (skeletal).

2720. Cyclobutane. Tetramethylene. C_4H_8; mol wt 56.10. C 85.63%, H 14.37%. Prepd by hydrogenaton of cyclobutene in the presence of nickel at 100°: Willstätter, Bruce, *Ber.* **40**, 3988 (1907); *see also* Heisig, *J. Am. Chem. Soc.* **63**, 1698 (1941); by pyrolysis of ether, along with other products: Peytral, *Bull. Soc. Chim. France* [4] **35**, 964 (1924). In 39% yield from cyclobutanecarboxylic acid: Cason, Way, *J. Org. Chem.* **14**, 31 (1949), for the prepn of this acid *see* Heisig, Stodola, *Org. Syn.* **coll. vol. III**, 213 (1955).

Gas. Burns with a luminous flame. mp −80°. bp$_{741}$

+13.08°. d^0 0.7038; d^{-5} 0.7185. n$_D^0$ 1.37520. Insol in water; freely sol in alcohol, acetone.

2721. Cyclobutyrol. *α-Ethyl-1-hydroxycyclohexaneacetic acid;* 1-cyclohexanol-α-butyric acid; α-(1-hydroxycyclohexyl)butyric acid; 1-hydroxy-α-ethylcyclohexylacetic acid. $C_{10}H_{18}O_3$; mol wt 186.24. C 64.49%, H 9.74%, O 25.77%. Prepd by hydrolysis of the ethyl ester obtained by the Reformatski reaction: Kon, Naravanan, *J. Chem. Soc.* **1927**, 1536; Kandiah, Linstead, *ibid.* **1929**, 2150; Maillard *et al.*, *Bull. Soc. Chim. France* **1958**, 244; *eidem,* **Ger.** pat. **1,094,-254** corresp to **U.S.** pat. **3,065,134** (1959, 1962 to Lab. Jacques Logeais). Activity studies: Roquet, Jousse, *Arch. Int. Pharmacodyn.* **125**, 172 (1960).

Colorless crystals from ether-petr ether, mp 81-82°. Formerly reported as viscous oil. d$_4^{18.8}$ 1.0010. bp$_{24}$ 164°; bp$_{16}$ 167-170°; n$_D^{18.8}$ 1.4680 (Kon, Naravanan, *loc. cit.*). Slightly sol in water, petr ether. Very sol in alcohols, acetone, dioxane, chloroform, ether. Completely sol in aq alkalies at pH 7.5-8.

Sodium salt, $C_{10}H_{17}NaO_3$, *Bicol, Bilimix, Bis-Bil, Colepan, Dimene, Epa-Bon, Hebucol, Maricolene, Tachicol, Tri-Bil, Tribilina, Viobilina*. White crystalline, slightly hygroscopic powder, mp 299-300°.

THERAP CAT: Choleretic.

2722. Cyclobuxine. *14-Methyl-3β,20α-bis(methylamino)-4-methylene-9,19-cyclo-5α,9β-pregnan-16α-ol.* $C_{25}H_{42}N_2O$; mol wt 386.60. C 77.66%, H 10.95%, N 7.25%, O 4.14%. From *Buxus sempervirens* L., *Buxaceae*: Schlittler *et al.*, *Helv. Chim. Acta* **32**, 2209 (1949). Probably identical with Alkaloid A: Heusler, Schlittler, *ibid.* 2226. Structure: Brown, Kupchan, *J. Am. Chem. Soc.* **84**, 4590 (1962). Stereochemistry: *eidem, ibid.* **86**, 4424 (1964).

Crystals, dec 245-247°. [α]$_D^{23}$ +98° (chloroform).

Dihydrobromide, $C_{25}H_{42}N_2O·2HBr$, crystals, dec 288-292°.

N,N′-Dimethylcyclobuxine, $C_{27}H_{46}N_2O$, crystals, dec 204-205°. [α]$_D^{25}$ +99° (chloroform).

2723. Cyclocumarol. *3,4-Dihydro-2-methoxy-2-methyl-4-phenyl-2H,5H-pyrano[3,2-c][1]benzopyran-5-one;* 2-methyl-2-methoxy-4-phenyl-5-oxodihydropyrano[3,2-c][1]benzopyran; 3,4-(2′-methyl-2′-methoxy-4′-phenyl)dihydropyranocoumarin; 4-hydroxycoumarin anticoagulant no 63; BL-5; Cumopyran; Cumopyrin. $C_{20}H_{18}O_4$; mol wt 322.34. C 74.52%, H 5.63%, O 19.85%. Prepn: Ikawa *et al.*, *J. Am. Chem. Soc.* **66**, 902 (1944).

Crystals from ethanol, mp 166°. One gram dissolves in 100 ml water. Slightly sol in alcohol, vegetable oil.

THERAP CAT: Anticoagulant.

2724. Cyclodextrins. Cycloamyloses; cycloglucans; Schardinger dextrins. Naturally occurring clathrates, *q.v.* Obtained from the action of *Bacillus macerans* amylase on starch to form homogeneous cyclic α-(1 → 4) linked D-glucopyranose units. α-, β- and γ-cyclodextrins are composed of six, seven and eight units, resp. Isoln of β-cyclodextrin: A. Villiers, *Compt. Rend.* **112**, 536 (1891); of α- and β-cyclodextrin: F. Schardinger, *Z. Untersuch. Nahr. Genussm.* **6**, 865 (1903); of γ-cyclodextrin: K. Freudenberg, R. Jacobi, *Ann.* **518**, 102 (1935). X-ray mol wt determn: D. French, R. E. Rundle, *J. Am. Chem. Soc.* **64**, 1651 (1942); D. French *et al., ibid.* **72**, 5150 (1950). Have hydrophobic cavities; form inclusion compds with organic substances, salts and halogens in the solid state or in aq solns. Model for enzyme action: I. Tabushi, *Acc. Chem. Res.* **15**, 66 (1982). *Reviews:* D. French, *Adv. Carbohyd. Chem.* **12**, 189-260 (1957); J. A. Thoma, L. Stewart, *Starch: Chemistry and Technology* vol. **1**, R. L. Whistler, E. F. Paschall, Eds. (Academic Press, New York, 1965) pp 209-249; F. Cramer, H. Hettler, *Naturwiss.* **154**, 625-632 (1967). Book: M. L. Bender, M. Komiyama, *Cyclodextrin Chemistry* (Springer-Verlag, New York, 1978) 96 pp.

α-cyclodextrin

α-*Cyclodextrin*, $C_{36}H_{60}O_{30}$, *cyclohexaamylose.* Hexagonal plates or blade-shaped needles. $[\alpha]_D$ +150.5°.

β-*Cyclodextrin*, $C_{42}H_{70}O_{35}$, *cycloheptaamylose.* Parallelogram shaped crystals. $[\alpha]_D$ +162.0°.

γ-*Cyclodextrin*, $C_{48}H_{80}O_{40}$, *cyclooctaamylose.* Square plates or rectangular rods. $[\alpha]_D$ +177.4°.

USE: As complexing agent; in study of enzyme action.

2725. Cyclodrine. 1-Hydroxy-α-phenylcyclopentaneacetic acid 2-diethylaminoethyl ester; β-diethylaminoethyl (1-hydroxycyclopentyl)phenylacetate; 2-phenyl-2-(1-hydroxycyclopentyl)acetic acid β-(diethylamino)ethyl ester. $C_{19}H_{29}NO_3$; mol wt 319.45. C 71.44%, H 9.15%, N 4.38%, O 15.03%. Prepn: Blicke, U.S. pat. 2,922,795 (1960 to Univ. of Michigan). Activity studies: Lands, Luduena, *J. Pharmacol. Exp. Ther.* **117**, 331 (1956); Karczman, Long, *ibid.* **123**, 230 (1958).

Hydrochloride, $C_{19}H_{30}ClNO_3$, *GT 92, Cyclopent.* Crystals from isopropanol-isopropyl ether, mp 133-135°.

THERAP CAT: Anticholinergic.

2726. Cyclofenil. 4-[[4-(Acetyloxy)phenyl]cyclohexylidenemethyl]phenol acetate; α-cyclohexylidene-α-(p-hydroxyphenyl)-p-cresol diacetate; bis(p-acetoxyphenyl)cyclohexylidenemethane; 4,4'-diacetoxybenzhydrylidenecyclohexane; F 6066; H 3452; ICI-48213; Fertodur; Neoclym; Ondonid; Ondogyne; Rehibin; Sanocrisin; Sexadieno; Sexovid. $C_{23}H_{24}O_4$; mol wt 364.44. C 75.80%, H 6.63%, O 17.56%. Prepn: Miquel *et al., J. Med. Chem.* **6**, 774 (1963); Olsson *et al.,* U.S. pat. 3,287,397 (1966). Pharmacology: Hiramatsu *et al., Oyo Yakuri* **6**, 1045 (1972), *C.A.* **78**, 131963z (1973).

Crystals from ethanol, mp 135-136°. uv max (ethanol): 247 nm (ε 17000).

Free diol, $C_{19}H_{20}O_2$, *F 6060.* mp 235-236°.

THERAP CAT: Gonad-stimulating principle.

2727. Cycloguanil. 1-(4-Chlorophenyl)-1,6-dihydro-6,6-dimethyl-1,3,5-triazine-2,4-diamine; 4,6-diamino-1-(p-chlorophenyl)-1,2-dihydro-2,2-dimethyl-s-triazine; 2,4-diamino-1-p-chlorophenyl-1,6-dihydro-6,6-dimethyl-1,3,5-triazine; 1-p-chlorophenyl-2,4-diamino-6,6-dimethyl-1,6-dihydro-1,3,5-triazine; chlorazine (Russian); chlorguanide triazine; TCl; 10,580; M 10580; D-20. $C_{11}H_{14}ClN_5$; mol wt 251.73. C 52.48%, H 5.61%, Cl 14.09%, N 27.82%. Metabolic product formed from the antimalarial drug chlorguanide: Carrington *et al., Nature* **168**, 1080 (1951). Prepn: Modest *et al., J. Am. Chem. Soc.* **74**, 855 (1952); Loo, *ibid.* **76**, 5096 (1954); Carrington *et al., J. Chem. Soc.* **1954**, 1017; Bami, *J. Sci. Ind. Res. (India)* **14C**, 231 (1955); Modest, *J. Org. Chem.* **21**, 1 (1956); U.S. pat. 2,900,385 (1959 to Children's Cancer Res. Found.); Elslager, Worth, U.S. pat. 3,074,947 (1963 to Parke, Davis).

Prisms from chloroform + ether, mp 146°. uv max (water): 241 nm (log ε 4.11).

Hydrochloride, $C_{11}H_{14}ClN_5$.HCl, prisms from water, mp 210-215°. uv max (water): 241 nm (log ε 4.12).

Dihydrochloride, $C_{11}H_{14}ClN_5$.2HCl, crystals from acetone, mp 190-196°.

Pamoate, $C_{45}H_{44}Cl_2N_{10}O_6$, *cycloguanil embonate, CI-501, Camolar.* Yellow crystals, mp 231-234°. Soly in water: 0.003%.

THERAP CAT: Antimalarial.

2728. Cycloheptanone. Suberone; ketoheptamethylene; ketocycloheptane. $C_7H_{12}O$; mol wt 112.17. C 74.95%, H 10.78%, O 14.26%. Prepn from cyclohexanol with 1-(aminomethyl)cyclohexanol as the intermediate: Blicke *et al., J. Am. Chem. Soc.* **74**, 2924 (1952). Other methods: *Org. Syn. coll. vol. IV*, 221-228 (1963).

Liquid. d_4^{20} 0.9490. bp_{760} 179-181°. bp_{16} 66-70°. n_D^{20} 1.4608. Dipole moment 2.98. Practically insol in water. Freely sol in alcohol. Sol in ether.

Oxime, $C_7H_{13}NO$, *suberoxime.* mp 23°, bp_{20} 152°.

2729. Cyclohexane. Hexahydrobenzene; hexamethylene; hexanaphthene. C_6H_{12}; mol wt 84.16. C 85.63%, H 14.37%. Occurs in petr (0.5-1.0%). Obtained in the distillation of petr or by hydrogenation of benzene. In the distillation of petr the C_4-400°F boiling range naphthas are fractionated to obtain a C_5-200°F naphtha contg 10-14% cyclohexane which on superfractionation yields an 85% concentrate (which is sold as such); further purification necessitates isomerization of pentanes to cyclohexane, heat cracking for removing open chain hydrocarbons and sulfuric acid treatment to remove aromatic compds. The hydrogenation of benzene is done in the liq phase at 150° using Raney nickel catalyst and at least 10 atm H_2 pressure: Sabatier, *Ind. Eng. Chem.* **18**, 1005 (1926). Review and bibliography: Sachanen, *Chemical Constituents of Petroleum* (New York, 1945). Prepn of high purity cyclohexane: Seyer *et al.*, *Ind. Eng. Chem.* **31**, 759 (1939). Cyclohexane can exist in two interconvertible conformations, the boat and the chair. In the chair form its 12 extracyclic bonds fall into two classes: six lie parallel to the main axis of symmetry and are designated "axial", while six extend radially outward at ±109.5° angles to the axis and are designated as "equatorial", Barton *et al.*, *Nature* **172**, 1096 (1954); *Science* **119**, 49 (1954). Solubility: F. P. Schwarz, *Anal. Chem.* **52**, 10 (1980). Toxicity: Lazarew, *Arch. Exp. Pathol. Pharmacol.* **143**, 223 (1929). Physical properties and methods of purification: L. Scheflan, M. B. Jacobs, *The Handbook of Organic Solvents* (Van Nostrand, 1953) p 233; *Techniques of Chemistry*, A. Weissberger, Ed., **vol. II**, 3rd ed., entitled *"Organic Solvents"* by J. D. Riddick, W. B. Bunger (Wiley-Interscience, New York, 1970) p 592. Review: M. L. Campbell in Kirk-Othmer *Encyclopedia of Chemical Technology*, **vol. 12** (Wiley-Interscience, New York, 3rd ed., 1980) pp 931-937.

boat   chair

Flammable liq. Solvent odor. Pungent when impure. d_4^{20} 0.7781; d_4^{80} 0.7206. mp +6.47°. bp_{760} 80.7°; bp_{400} 60.8°; bp_{200} 42.0°; bp_{100} 25.5°; bp_{60} 14.7°; bp_{40} 6.7°. n_D^{20} 1.4264. Flash pt, closed cup: 1°F (−18°C). Flammability limits in air 1.3-8.4% v/v. Soly in water at 23.5°C (w/w): 0.0052%. 100 ml of methanol dissolves 57 grams at 20°C; miscible with ethanol, ethyl ether, acetone, benzene, carbon tetrachloride. LC in mice: ~60-70 mg/l air (Lazarew).

Caution: High concns may act as narcotic, skin irritant: E. Browning, *Toxicity and Metabolism of Industrial Solvents* (Elsevier, New York, 1965) pp 130-134.

USE: Solvent for lacquers and resins. Paint and varnish remover. In the extraction of essential oils. In analytical chemistry for mol wt determinations (cryoscopic constant 20.3). In the manuf of adipic acid, benzene, cyclohexyl chloride, nitrocyclohexane, cyclohexanol and cyclohexanone. In the manuf of solid fuel for camp stoves. In fungicidal formulations (possesses slight fungicidal action). In the industrial recrystn of steroids.

2730. Cyclohexanecarboxylic Acid. Hexahydrobenzoic acid. $C_7H_{12}O_2$; mol wt 128.17. C 65.59%, H 9.44%, O 24.97%. Prepn from anisic acid: Lumsden, *J. Chem. Soc.* **87**, 90 (1905); from 2-chlorocycloheptanone: Gutsche, *J. Am. Chem. Soc.* **71**, 3513 (1949); by carbonation of cyclohexylmagnesium chloride: Wagner, Moore, ibid. **72**, 974 (1950); from cyclohexane + KI + active Ni: Reppe *et al.*, *Ann.* **582**, 38 (1953); by oxidation of cycloheptanone: Payne, Smith, *J. Org. Chem.* **22**, 1680 (1957); from cyclohexane + HCOOH or CO_2: McKursick *et al.*, *J. Am. Chem. Soc.* **82**, 723 (1960); McKursick, U.S. pat. **2,940,913** (1960 to du Pont).

COOH

Liquid. bp 232.5°; bp_{20} 131°; bp_8 110°; $bp_{<1}$ 63-67°. Crystallizes, on cooling, in monoclinic prisms, mp 29°. Odorless but when liq or in soln has a valerian odor. d_4^{15} 1.0480. n_D^{20} 1.4530. Soly in 100 g water at 15°: 0.201 g. Sol in most organic solvents.

Methyl ester, $C_8H_{14}O_2$, fragrant liq, bp 183°. d_4^{15} 0.9954.

USE: Solubilizer for vulcanized rubber; clarifier for mineral oil; in insecticide formulations.

2731. Cyclohexanol. Hexalin; hexahydrophenol. $C_6H_{12}O$; mol wt 100.16. C 71.95%, H 12.08%, O 15.97%. Obtained by oxidation of cyclohexane or hydrogenation of phenol: Faith, Keyes & Clark's *Industrial Chemicals*, F. A. Lowenheim, M. K. Moran, Eds. (Wiley-Interscience, New York, 4th ed., 1975) pp 304-309.

OH

Hygroscopic crystals; camphor odor. d^{20} 0.962. mp 23-25°. bp 161°. Flash pt, closed cup: 154°F (68°C). n_D^{22} 1.465. At 20° soly in water: 3.6% (w/w); soly of water in cyclohexanol: 11% (w/w). Miscible with ethanol, ethyl acetate, linseed oil, petr solvents, aromatic hydrocarbons. LD_{50} orally in rats: 2.06 g/kg, H. F. Smyth *et al.*, *Am. Ind. Hyg. Assoc. J.* **23**, 95 (1962).

USE: Solvent for alkyd resins, alcohol-sol phenolic resins, ethyl cellulose. Manuf celluloid; finishing textiles; insecticides. *Caution:* Narcotic-like action. Has caused liver, kidney, vascular injury in exptl animals: E. Browning, *Toxicity and Metabolism of Industrial Solvents* (Elsevier, New York, 1965) pp 385-388.

2732. Cyclohexanone. Ketohexamethylene; pimelic ketone; Hytrol O; Anone; Nadone. $C_6H_{10}O$; mol wt 98.14. C 73.43%, H 10.27%, O 16.30%. Obtained from cyclohexanol by catalytic dehydrogenation or by oxidation (which yields cyclohexanone and adipic acid) or from cyclohexane by oxidation (yielding cyclohexanone and cyclohexanol): **Brit.** pat. **310,055** (1928 to Schering-Kahlbaum); **U.S.** pats. **2,223,493; 2,223,494; 2,285,914** (1940, 1940, 1942 to du Pont). Review of manuf processes: Faith, Keyes & Clark's *Industrial Chemicals*, F. A. Lowenheim, M. K. Moran, Eds. (Wiley-Interscience, New York, 4th ed., 1975) pp 304-309. Toxicity: Smyth *et al.*, *Am.. Ind. Hyg. Assoc. J.* **30**, 470 (1969).

Oily liq. Odor reminiscent of peppermint and acetone. *Caution:* Vapor harmful. d_4^{20} 0.9478; d_4^{25} 0.9421. mp −32.1°. bp_{760} 155.6°; bp_{400} 132.5°; bp_{200} 110.3°; bp_{100} 90.4°; bp_{60} 77.5°; bp_{40} 67.8°; bp_{20} 52.5°; bp_{10} 38.7°; bp_5 26.4°; $bp_{1.0}$ 1.4°. n_D^{20} 1.4507. Flash pt closed cup: 147°F (63°C). Soly in water: 150 g/l at 10°; 50 g/l at 30°. Soly of water in cyclohexanone: 87g/l at 20°. Sol in alcohol, ether and in other common organic solvents. LD_{50} orally in rats: 1.62 ml/kg (Smyth).

2,4-Dinitrophenylhydrazone, golden yellow needles, mp 160°. uv max ($CHCl_3$): 262 nm (ϵ 22500); (0.01N NaOH): 435 nm (ϵ 19000).

USE: Solvent for cellulose acetate, nitrocellulose, natural resins, vinyl resins, crude rubber, waxes, fats, shellac, DDT. In the production of adipic acid for nylon. In the prepn of cyclohexanone resins. 2,4-Dinitrophenylhydrazone deriv as standard in elemental analysis.

2733. Cyclohexene. 1,2,3,4-Tetrahydrobenzene. C_6H_{10}; mol wt 82.14. C 87.73%, H 12.27%. Occurs in coal tar. Prepd by dehydration of cyclohexanol at high temps over various catalysts. Lab prepn using H_2SO_4 as dehydrating agent: Coleman, Johnstone, *Org. Syn.* **5**, 33 (1925); Wagner, *J. Chem. Ed.* **10**, 113 (1933); by distn of cyclohexanol over silica gel or alumina: Hershberg, Ruhoff, *Org. Syn.* **17**, 25 (1937).

Liquid. d_4^{20} 0.8098; d^{50} 0.7823; d^{100} 0.7355. mp −103.5°. bp_{760} 83°. n_D^{20} 1.4465; n_D^{27} 1.4428. Absorption spectrum: Purvis, *Proc. Cambridge Phil. Soc.* **23**, 588 (1927); *Chem. Zentr.* **1927**, II, 379; cf. Hartley, Dobbie, *J. Chem. Soc.* **77**, 846 (1900).

USE: Alkylation component. In the manuf of adipic acid, maleic acid, hexahydrobenzoic acid and aldehyde. To prepare butadiene in the laboratory. Has been suggested for the synthesis of maleic acid and as stabilizer for high octane gasoline.

2734. Cycloheximide. *4-[2-(3,5-Dimethyl-2-oxocyclohexyl)-2-hydroxyethyl]-2,6-piperidinedione; 3-[2-(3,5-dimethyl-2-oxocyclohexyl)-2-hydroxyethyl]glutarimide;* naramycin A; NSC 185; U-4527; Actidione. $C_{15}H_{23}NO_4$; mol wt 281.34. C 64.03%, H 8.24%, N 4.98%, O 22.75%. Antibiotic substance isolated from the beers of streptomycin-producing strains of *Streptomyces griseus:* Leach *et al., J. Am. Chem. Soc.* **69**, 474 (1947); Ford, Leach, *ibid.* **70**, 1223 (1948); Whiffen *et al.*, U.S. pat. **2,574,519** (1951 to Upjohn). Prodn, assay and antibiotic activity: Whiffen, *J. Bacteriol.* **56**, 283 (1948). Improved prodn method: Kominek, U.S. pats. **3,915,802-3** (both 1975 to Upjohn). Structure: Kornfeld *et al., J. Am. Chem. Soc.* **71**, 150 (1949). Abs config: Eisenbraun *et al., ibid.* **80**, 1261 (1958); Johnson, Starkovsky, *Tetrahedron Letters* **1962**, 1173; Johnson *et al., J. Am. Chem. Soc.* **87**, 4612 (1965). Synthesis of *dl-* and *l*-forms: Johnson *et al., ibid.* **88**, 149 (1966).

Plates from amyl acetate or water or 30% methanol. mp 119.5-121° (U.S. pat. **2,574,519**); mp 115-116° (Johnson *et al.*). $[\alpha]_D^{29}$ −3.38° (c = 9.47 in methanol); $[\alpha]_D^{25}$ +6.8° (c = 2 in H_2O). Soly at 2°: water 2.1 g/100 ml, amyl acetate 7 g/100 ml. Also sol in chloroform, ether, acetone, methanol, ethanol, other common organic solvents except satd hydrocarbons. Relatively heat-stable, acid-stable, destroyed by boiling in aq soln at pH 7 for 1 hr, but shows no loss of activity after 15 min boiling. At pH 2 it is not destroyed by boiling for 1 hr. Rapidly inactivated at room temp by dil alkali with the formation of a volatile, fragrant ketone, 2,4-dimethylcyclohexanone. Extremely repellant to rats. LD_{50} i.v. in mice: 150 mg/kg, Leach *et al., loc. cit.*

Acetate, $C_{17}H_{25}NO_5$, glistening plates from methanol, mp 148-149°. $[\alpha]_D^{25}$ +22° (c = 2.3 in methanol).

USE: Fungicide; plant growth regulator. As protein synthesis inhibitor.

2735. Cyclohexylamine. *Cyclohexanamine;* aminocyclohexane; hexahydroaniline. $C_6H_{13}N$; mol wt 99.17. C 72.66%, H 13.21%, N 14.12%. Prepd by the catalytic hydrogenation of aniline at elevated temps and pressures. Fractionation of the crude reaction product yields cyclohexylamine, unchanged aniline, and a high-boiling residue contg N-phenylcyclohexylamine (cyclohexylaniline) and dicyclohexylamine. Review and bibliography: Carswell, Morrill, *Ind. Eng. Chem.* **29**, 1247 (1937). Toxicity data: H. F. Smyth *et al., Am. Ind. Hyg. Assoc. J.* **30**, 470 (1969).

Liquid. Strong, fishy, amine odor. d_{25}^{25} 0.8647. mp

−17.7°. bp_{760} 134.5°; bp_{500} 118.9°; bp_{300} 102.5°; bp_{100} 72.0°; bp_{50} 56.0°; bp_{30} 45.1°; bp_{25} 41.3°; bp_{20} 36.4°; bp_{15} 30.5°. n_D^{25} 1.4565. Strong base. Completely misc with water and with common organic solvents, including alcohols, ethers, ketones, esters, aliphatic hydrocarbons, aromatic hydrocarbons and their chlorinated derivatives. On distillation with water, cyclohexylamine forms an azeotropic mixture contg 44.2% cyclohexylamine by weight, bp_{760} 96.4°. Reacts with excess ammonia and zinc chloride at 350° to produce α-picoline: Nordt, *Conversion of Hexahydroaniline into Pyridine Bases,* Off. Pub. Bd., Report PB 704 (1941). LD_{50} orally in rats: 0.71 ml/kg (Smyth).

USE: In organic synthesis, manuf insecticides, plasticizers, corrosion inhibitors, rubber chemicals, dyestuffs, emulsifying agents, dry-cleaning soaps, acid gas absorbents. *Caution:* Can cause irritation and sensitization. High concs cause nausea and narcotic effects.

2736. Cyclohexyl Bromide. *Bromocyclohexane.* C_6H_{11}-Br; mol wt 163.06. C 44.19%, H 6.80%, Br 49.01%.

Liquid; penetrating odor. d_{15}^{15} 1.329. bp 163-165°. n_D^{15} 1.4956. Insol in water; sol in alcohol, etc.

2737. Cyclohexylcarbinol. *Cyclohexanemethanol;* cyclohexanecarbinol; hydroxymethylcyclohexane; hexahydrobenzyl alcohol. $C_7H_{14}O$; mol wt 114.18. C 73.63%, H 12.36%, O 14.01%. Prepd from cyclohexylmagnesium bromide and formaldehyde in ether: Gilman, Catlin, *Org. Syn.* coll. vol. I (2nd ed., 1941) p 188; Hiers, Adams, *J. Am. Chem. Soc.* **48**, 2385 (1926); Marvel *et al., ibid.* **50**, 2810 (1928). For prepn of cyclohexylmagnesium bromide see *Org. Syn.,* loc. cit.; Gilman, Zoellner, *J. Am. Chem. Soc.* **53**, 1945 (1931).

Liquid. Slight odor of camphor. d_4^{25} 0.9215. bp_{784} 184-186°; bp_{23} 91-92°. n_D^{25} 1.4640. Sol in alc, ether.

Acetate, bp_{740} 199-201°.

2738. Cyclohexyl Chloride. *Chlorocyclohexane.* C_6H_{11}-Cl; mol wt 118.61. C 60.75%, H 9.35%, Cl 29.89%.

Colorless liquid; suffocating odor. d_4^{20} 1.000. mp −44°. bp 142°. n_D^{20} 1.4626. Insol in water; sol in alcohol, etc.

2739. 2-Cyclohexyl-4,6-dinitrophenol. 2,4-Dinitro-6-cyclohexylphenol; dinitro-*o*-cyclohexylphenol; DNOCHP; SN 46. $C_{12}H_{14}N_2O_5$; mol wt 266.25. C 54.13%, H 5.30%, N 10.52%, O 30.05%. Prepd from phenol, H_2SO_4, and cyclohexene or cyclohexanol: **Brit.** pat. **620,026** (1949 to Pest Control); from *o*-cyclohexylphenol by nitration in chloroform: Baroni, Kleinan, *Monatsh.* **68**, 251 (1936).

Crystals, mp 106.5-107.5°. Soly in petr oils 2.5 to 6%. Sol in benzene, DMF; very slightly sol in water.

Sodium salt, $C_{12}H_{13}N_2NaO_5$, *2-cyclohexyl-4,6-dinitrophenate, Anobesina.* Crystals, sol in water, alcohol.

USE: Insecticide, esp in control of citrus red mite.

2740. Cycloleucine. *1-Aminocyclopentanecarboxylic acid;* ACPC; NSC-1026. $C_6H_{11}NO_2$; mol wt 129.16. C 55.79%, H 8.58%, N 10.85%, O 24.77%. Synthetic amino acid thought to act as a valine antagonist. Prepn: Zelinsky, Stadnikoff, *Z. Physiol. Chem.* **75**, 350 (1911); Connors, Ross, *J. Chem. Soc.* **1960**, 2119; Cremlyn, *ibid.* **1962**, 3977; Sudo, Ichihara, *Bull. Chem. Soc. Japan* **36**, 34 (1963). Manuf: **Neth.** pat. Appl. **6,607,754** (1966 to Rohm & Haas), *C.A.* **67**, 73159b (1967). *Review:* Ross *et al., J. Med. Pharm. Chem.* **3**, 1 (1961). Immunopharmacology: Rosenthale *et al., J.*

Pharmacol. Exp. Ther. **180,** 501 (1972); Brambilla *et al.,* *Cancer Chemother. Rep., Part 1,* **56,** 579 (1972).

Crystals from ethanol-water, mp 330° (dec). Soly in water approx 5 g/100 ml. Forms stable metal salts.

Hydrochloride, prismic crystals from water, mp 274° (dec). LD_{50} orally in mice, rats: 309, 260-290 mg/kg.

USE: Biological testing material for immunosuppressive properties in mice; for amino acid transport studies.

2741. Cyclomethycaine. *4-(Cyclohexyloxy)benzoic acid 3-(2-methyl-1-piperidinyl)propyl ester; p-cyclohexyloxybenzoic acid ester of N-(3-hydroxypropyl)pipecoline; 3-(2-methylpiperidino)propyl p-cyclohexyloxybenzoate;* Surfacaine; Surfathesin; Topocaine. $C_{22}H_{33}NO_3$; mol wt 359.51. C 73.50%, H 9.25%, N 3.90%, O 13.35%. Prepn: McElvain, Carney, *J. Am. Chem. Soc.* **68,** 2592 (1946), *eidem,* U.S. pat. **2,439,818** (1948). Toxicity data: Schmidt *et al., Toxicol. Appl. Pharmacol.* **1,** 454 (1959).

Hydrochloride, $C_{22}H_{34}ClNO_3$, crystals, dec 178-180°. Generally used as the hydrochloride. Soly in water: slightly >1 g/100 ml. Also used as the sulfate, $[(C_{22}H_{33}NO_3)_2.H_2SO_4]$; crystals, mod sol in water.

THERAP CAT: Topical anesthetic.

THERAP CAT (VET): Surface anesthetic.

2742. Cyclonite. *Hexahydro-1,3,5-trinitro-1,3,5-triazine; sym-trimethylenetrinitramine;* cyclotrimethylenetrinitramine; 1,3,5-trinitrohexahydro-*s*-triazine; T_4; RDX; Hexogen. $C_3H_6N_6O_6$; mol wt 222.26. C 16.22%, H 2.72%, N 37.83%, O 43.22%. Prepd by treating methenamine with fuming nitric acid: Henning, Ger. pat. **104,280** (1899); Hale, *J. Am. Chem. Soc.* **47,** 2754 (1925). Two moles of cyclonite can be obtained from one mole of methenamine, if ammonium nitrate and acetic anhydride are added: Bachmann, Sheehan, *J. Am. Chem. Soc.* **71,** 1842 (1949). Manuf by nitrolysis of hexamethylenetetramine: Ruth, U.S. pat. **3,049,-543** (1962 to Olin Mathieson). Structure: Wood, *Proc. Int. Meet. Mol. Spectrosc., 4th, Bologna, 1959,* **2,** 955 (1962), *C.A.* **59,** 4681b (1963).

Orthorhombic crystals from acetone, mp 205-206°. d_4^{20} 1.82. One gram dissolves in 25 ml acetone. Slightly sol in methanol, ether, ethyl acetate, glacial acetic acid. Practically insol in water, carbon tetrachloride, carbon disulfide.

USE: High explosive; rat poison. For physical data relating to detonation, explosion, and blasting action *see* M. A. Cook, *Science of High Explosives,* ACS Monograph no. 139 (Reinhold, New York, 1958). *Caution:* Can cause epileptiform seizures, Kaplan *et al., Arch. Environ. Health* **10,** 877 (1965).

2743. Cyclonium Iodide. *1-[(2-Cyclohexyl-2-phenyl-1,3-dioxolan-4-yl)methyl]-1-methylpiperidinium iodide; N-methyl-N-[(2-cyclohexyl-2-phenyl-1,3-dioxolan-4-yl)methyl]piperidinium iodide; 1-(2-phenyl-2-cyclohexyl-1,3-dioxolan-4-yl)methyl-1-methylpiperidinium iodide;* ciclonium iodide; oxapium iodide; SH-100; Esperan. $C_{22}H_{34}INO_2$; mol wt 471.42. C 56.05%, H 7.27%, I 26.92%, N 2.97%, O 6.79%. Prepn and separation of isomers: Kimura *et al.,* **Japan.** pat. **6585('66),** *C.A.* **65,** 7191 (1966) and Saikawa, **Japan.** pat. **16,475('70),** *C.A.* **73,** 45355t (1970) (both to Toyama). Stability studies: Shiuchi, Miyazaki, *Bunseki*

Kagaku (Japan Analyst) **18,** 455 (1969), *C.A.* **71,** 74097z (1969). Incompatibility studies: Kunita *et al.,Yakuzaigaku (Pharmacology)* **28,** 84 (1968), *C.A.* **69,** 99335x (1968). Anticholinergic activity and acute toxicity: J. Faff *et al., Pol. J. Pharmacol. Pharm.* **30,** 493 (1978). *Review: Japan. Med. Gaz.* **8**(9), 11 (1971).

α-Form, white crystals from monochlorobenzene or isopropyl alcohol, mp 195-197°. Insol in trichloroethylene.

β-Form, white crystals from monochlorobenzene, mp 150-152°. Sol in hot trichloroethylene.

Both sol in methanol, ethanol, chloroform and tetrachlorethane. Hardly sol in benzene, toluene, xylene and water. LD_{50} i.p. in mice: 106 mg/kg (Faff).

THERAP CAT: Antispasmodic.

2744. Cyclopentadiene. *1,3-Cyclopentadiene.* C_5H_6; mol wt 66.10. C 90.85%, H 9.15%. Obtained from the distillates produced in carbonization of coal, esp from the foreruns of coke-oven light oil: Ward, U.S. pat. **2,211,038** (1940); process involving the liquefaction of coke-oven gas: Horclois, *Chim. Ind. (Paris),* special no. April 1934, pp 357-363. Also obtained during the cracking of petr hydrocarbons: Tropsch *et al., Ind. Eng. Chem.* **30,** 169 (1938); *cf.* Dedussenko, *J. Gen. Chem. USSR* **7,** 1467 (1937); *Chem. Zentr.* **109, I,** 793 (1938). Synthesis by passing vaporized cyclopentane over activated alumina and oxides of molybdenum, chromium, or vanadium: Grosse *et al., Ind. Eng. Chem.* **32,** 309 (1940); U.S. pats. **2,157,202-3; 2,157,939.** Lab prepn by depolymerization of *dicyclopentadiene:* Moffett, *Org. Syn.* **coll. vol. IV,** 238 (1963). Review and discussion of structure: Wilson, Wells, *Chem. Rev.* **34,** 1 (1944). *Review:* M. Fefer, A. B. Small in Kirk-Othmer *Encyclopedia of Chemical Technology* **vol. 7** (Wiley-Interscience, New York, 3rd ed., 1979) pp 417-429.

Liquid. d_4^0 0.8235; d_4^{10} 0.8131; d_4^{20} 0.8021; d_4^{25} 0.7966; d_4^{30} 0.7914. mp −85°. bp_{760} 41.5-42.0°. n_D^{16} 1.44632. Absorption spectum: Pickett *et al., J. Am. Chem. Soc.* **63,** 1073 (1941). Insol in water. Miscible with alc, ether, benzene, carbon tetrachloride. Sol in carbon disulfide, aniline, acetic acid, liquid petrolatum. Cyclopentadiene polymerizes to dicyclopentadiene on standing. Polymerization is accelerated by the presence of peroxides or trichloroacetic acid. The dimer is cryst, mp 32.5°, with a camphor-like odor, having the structure of a partially hydrogenated indene contg a bridged methylene group. It is a more convenient form in which to handle cyclopentadiene, and is easily depolymerized by distilling at atmospheric pressure. LD_{50} of dimer in rats: 0.82 g/kg orally, Smyth *et al., Arch. Ind. Hyg. Occup. Med.* **10,** 61 (1954).

USE: Manuf resins; in organic synthesis as the diene in the Diels-Alder reaction producing sesquiterpenes, synthetic alkaloids, camphors.

2745. Cyclopentamine. *N,α-Dimethylcyclopentaneethanamine; N,α-dimethylcyclopentaneethylamine; 1-cyclopentyl-2-methylaminopropane; β-(methylamino)propylcyclopentane;* cyclopentadrine; Sinos; Cyklosal; Cyclonarol; Clopane. $C_9H_{19}N$; mol wt 141.25. C 76.52%, H 13.56%, N 9.92%. Prepn: Rohrmann, U.S. pat. **2,520,015** (1950 to Lilly).

Hydrochloride, $C_9H_{19}N·HCl$, crystals, mp 113-115° (base bp_{30} 83-86°, n_D^{25} 1.4500). Bitter taste. Freely sol in water.

THERAP CAT: Adrenergic (vasoconstrictor); nasal decongestant.

2746. Cyclopentane. Pentamethylene. C_5H_{10}; mol wt 70.13. C 85.63%, H 14.37%. Occurs in petroleum. Found in petr ether fractions. Prepd by cracking cyclohexane in the presence of alumina at high temps and pressure: Haensel, Ipatieff, *Ind. Eng. Chem.* **35**, 632 (1943); by reduction of cyclopentadiene: David *et al.*, *Bull. Soc. Chim. France* [5] **11**, 561 (1944).

Mobile, flammable liq. mp −94.4°. bp 49.3°. d_4^{20} 0.7460. n_D^{20} 1.4068. Insol in water. Miscible with other hydrocarbon solvents, alcohol, ether. Lethal concn for mice in air: 38,000 ppm, *Handbook of Toxicology* vol. **1**, W. S. Spector, Ed. (Saunders, Philadelphia, 1956) pp 330-331.

2747. Cyclopentanol. Cyclopentyl alcohol; hydroxycyclopentane. $C_5H_{10}O$; mol wt 86.13. C 69.72%, H 11.70%, O 18.58%. Prepd by modified Meerwein-Pondorf-Verley reduction of cyclopentanone in the presence of aluminum isopropoxide and sodium hydroxide: Truett, Moulton, *J. Am. Chem. Soc.* **73**, 5913 (1951); by catalytic hydrogenation of pure cyclopentanone with copper chromite at 150° and 150 atm: Kögl, Ultée, *Rec. Trav. Chim.* **69**, 1576 (1950); by catalytic hydrogenation of cyclopentanone with platinum oxide and platinum black at 2-3 atm: Noller, Adams, *J. Am. Chem. Soc.* **48**, 1084 (1926); by hydration of cyclopentene in aq H_2SO_4: Hepp, U.S. pat. **2,414,646** (1947 to Phillips Petr.); by reduction of cyclopentanone with $LiAlH_4$ in ether at room temp: Nystrom, Brown, *J. Am. Chem. Soc.* **69**, 1197 (1947).

Liquid, odor of amyl alcohol. d_4^0 0.96253; d_4^{15} 0.95078; d_4^{20} 0.9488; d_4^{30} 0.93908. mp −19°. bp 140.85°. Flash pt 124°F. n_D^{15} 1.45512; n_D^{20} 1.4520. Sparingly sol in water. Sol in ethanol.

2748. Cyclopentanone. Ketocyclopentane; ketopentamethylene; adipic ketone. C_5H_8O; mol wt 84.11. C 71.39%, H 9.59%, O 19.02%. Prepd by heating adipic acid to 285-295° in the presence of barium hydroxide, distilling, extracting with ether and fractionating: Thorpe, Kon, *Org. Syn.* coll. vol. **I**, 192 (2nd ed., 1941).

Liquid. Agreeable odor, somewhat like peppermint. d_4^{18} 0.9509. mp −58.2°. bp_{760} 130.6°; bp_{10} 23-24°. Flash pt 85°F. n_D^{20} 1.4366. Slightly sol in water. Miscible with alc, ether. Polymerizes easily, esp in presence of acids.

2749. 1,2-Cyclopentenophenanthrene. $C_{17}H_{14}$; mol wt 218.28. C 93.53%, H 6.47%. Prepn: Ruzicka *et al.*, *Helv. Chim. Acta* **16**, 838 (1933); *cf.* Kon, *J. Chem. Soc.* **1933**, 1081; Cook, Hewett, *Chem. & Ind. (London)* **52**, 451 (1933); Hawthorne, Robinson, *J. Chem. Soc.* **1936**, 763; Bachmann, Kloetzel, *J. Am. Chem. Soc.* **59**, 2207 (1937).

Needles from alc or petr ether, mp 135-136°. Absorption

spectrum: Mayneord, Roe, *Proc. Roy. Soc. London* **A152**, 299 (1935).

Picrate, $C_{23}H_{17}N_3O_7$, orange needles from benzene, mp 133-134°.

2750. Cyclopenthiazide. *6-Chloro-3-(cyclopentylmethyl)-3,4-dihydro-2H-1,2,4-benzothiadiazine-7-sulfonamide 1,1-dioxide;* 6-chloro-3-cyclopentylmethyl-3,4-dihydro-7-sulfamoyl-2H-1,2,4-benzothiadiazine 1,1-dioxide; 3-cyclopentylmethyl-6-chloro-7-sulfamyl-3,4-dihydro-1,2,4-benzothiadiazine 1,1-dioxide; cyclomethiazide; tsiklometiazid; Su 8341; Navidrex; Navidrix; Salimid. $C_{13}H_{18}ClN_3O_4S_2$; mol wt 379.89. C 41.10%, H 4.78%, Cl 9.33%, N 11.06%, O 16.85%, S 16.88%. Prepn: Whitehead *et al.*, *J. Org. Chem.* **26**, 2814 (1961); **Belg. pat. 587,225** (1960 to Ciba). Pharmacology: Barrett *et al.*, *Arch. Int. Pharmacodyn.* **131**, 325 (1961).

Crystals from dil alc, mp 230°. LD_{50} i.v. in rats, mice: 142, 232 mg/kg.

THERAP CAT: Antihypertensive.

2751. Cyclopentobarbital. *5-(2-Cyclopenten-1-yl)-5-(2-propenyl)-2,4,6(1H,3H,5H)-pyrimidinetrione); 5-allyl-5-(2-cyclopenten-1-yl)barbituric acid;* Cyclopal. $C_{12}H_{14}N_2O_3$; mol wt 234.25. C 61.53%, H 6.02%, N 11.96%, O 20.49%. Prepd by condensing the ethyl ester of cyclopentenylmalonic acid with urea: Compagnie de Bethune: **Ger. pat. 589,947** (1930), *Frdl.* **19**, 1198; **Brit. pat. 349,455**; **Fr. pat. 38,680**.

Crystals, mp 139-140°. Bitter taste. Slightly sol in cold water; moderately sol in hot water. Freely sol in alcohol, organic solvents.

Sodium salt, $C_{12}H_{13}N_2NaO_3$, *Cyclopal Sodium.* White powder, very sol in water.

Caution: May be habit forming. This is a controlled substance (depressant) listed in the U.S. Code of Federal Regulations, Title 21 Parts 329.1 and 1308.13 (1987).

THERAP CAT: Sedative, hypnotic.

2752. Cyclopentolate. *α-(1-Hydroxycyclopentyl)benzeneacetic acid 2-(dimethylamino)ethyl ester; 1-hydroxy-α-phenylcyclopentaneacetic acid 2-(dimethylamino)ethyl ester;* 2-dimethylaminoethyl 1-hydroxy-α-phenylcyclopentaneacetate; β-dimethylaminoethyl (1-hydroxycyclopentyl)phenylacetate; 2-phenyl-2-(1-hydroxycyclopentyl)ethanoic acid, β-(dimethylamino)ethyl ester. $C_{17}H_{25}NO_3$; mol wt 291.38. C 70.07%, H 8.65%, N 4.81%, O 16.47%. Ophthalmic anticholinergic. Prepn: Treves, U.S. pat. **2,554,511** (1951 to Schieffelin).

Hydrochloride, $C_{17}H_{26}ClNO_3$, *Ak-Pentolate, Mydplegic, Cyclogyl, Mydrilate, Zyklolat.* Crystals from ethyl acetate, mp 137-141°. pH of 1% aq soln: 5.0-5.4. Freely sol in water, alcohol. Practically insol in ether.

THERAP CAT: Mydriatic.

2753. Cyclophosphamide. *N,N-Bis(2-chloroethyl)tetra-*

hydro-2H-1,3,2-oxazaphosphorin-2-amine 2-oxide; 2-[bis(2-chloroethyl)amino]tetrahydro-2H-1,3,2-oxazophosphorine 2-oxide; 1-bis(2-chloroethyl)amino-1-oxo-2-aza-5-oxa-phosphoridin; bis(2-chloroethyl)phosphamide cyclic propanolamide ester; bis(2-chloroethyl)phosphoramide cyclic propanolamide ester; N,N-bis(β-chloroethyl)-N',O-propylenephosphoric acid ester diamide; N,N- bis(β-chloroethyl)-N',O-trimethylenephosphoric acid ester diamide; cyclophosphane; cytophosphane; B 518; Cycloblastin; Cyclostin; Cytoxan; Endoxan; Procytox; Sendoxan. $C_7H_{15}Cl_2N_2O_2P$; mol wt 261.10. C 32.20%, H 5.79%, Cl 27.16%, N 10.73%, O 12.26%, P 11.86%. Prepn: H. Arnold *et al., Naturwiss.* **45**, 64 (1957); *eidem, Nature* **181**, 931 (1958); H. Arnold, F. Bourseaux, *Angew. Chem.* **70**, 539 (1958). Abs config of enantiomers: I. L. Karle *et al., J. Am. Chem. Soc.* **99**, 4803 (1977); D. A. Adamaik *et al., Angew. Chem. Int. Ed.* **16**, 330 (1977). Optical resolution study: T. Kawashima *et al., J. Org. Chem.* **43**, 1111 (1978). Asymmetric synthesis of enantiomers: T. Sato *et al., ibid.* **48**, 98 (1983). Metabolism: Connors *et al., Biochem. Pharmacol.* **21**, 1373 (1972); Cox *et al., ibid.* **25**, 993 (1976). Use in combination with azathioprine and hydroxychloroquine, *qqv.*, for the treatment of rheumatoid arthritis: D. J. McCarty, G. F. Carrera, *J. Am. Med. Assoc.* **248**, 1718 (1982). Clinical study with ACTH in multiple sclerosis patients: S. L. Hauser *et al., N. Engl. J. Med.* **308**, 173 (1983). Toxicology: A. G. Wheeler *et al., Toxicol. Appl. Pharmacol.* **4**, 324 (1962). Review of carcinogenicity studies: *IARC Monographs* **9**, 135-156 (1975). Review of field trials as defleecing agent for sheep: M. H. Fahmy, Y. Moride, *Anim. Breed. Abstr.* **52**, 7-19 (1984).

Monohydrate, mp 41-45°. Soly in water: 40 g/l. Slightly sol in alcohol, benzene, ethylene glycol, carbon tetrachloride, dioxane; sparingly sol in ether and acetone. LD_{50} orally in mice, rats: 350, 94 mg/kg (Wheeler).
Note: This substance has been listed as a known carcinogen: *Fourth Annual Report on Carcinogens* (NTP 85-002, 1985) p 64.
USE: Proposed as defleecing agent for sheep.
THERAP CAT: Antineoplastic.

2754. Cyclopregnol. *6β-Hydroxy-3,5-cyclopregnan-20-one;* Neurosterone. $C_{21}H_{32}O_2$; mol wt 316.47. C 79.70%, H 10.19%, O 10.11%. Prepn: Patel *et al., J. Chem. Soc.* **1957**, 665; Petrow *et al.,* U.S. pat. **2,816,901** (1957) and Kirk *et al.,* U.S. pat. **2,823,213** (1958 to Brit. Drug Houses).

Needles from acetone, mp 181°. $[\alpha]_D^{24}$ +123° (c = 0.663 in chloroform).
THERAP CAT: Psychotropic.

2755. Cyclopropane. Trimethylene. C_3H_6; mol wt 42.08. C 85.63%, H 14.37%. Prepd by reduction of 1,2-dibromocyclopropane with zinc and alcohol: Lott, Christiansen, *J. Am. Pharm. Assoc.* **19**, 341 (1930); Schlatter, *J. Am. Chem. Soc.* **63**, 1733 (1941); from 1,3-dibromopropane with zinc and alcohol in absence of water: Ashdown *et al., ibid.* **58**, 850 (1936); by the action of sodium vapor on 1,3-dibromopropane: Bawn, Hunter, *Trans. Faraday Soc.* **34**, 608 (1938); from 1,3-dichloropropane heated with an excess of zinc dust, iodine and ethanol in 80% yield: Hass *et al., Ind. Eng. Chem.* **28**, 1178 (1936); from 1,3-dichloropropane with zinc in presence of sodium iodide: U.S. pat. **2,102,556** (1937); cf.

U.S. pat. **2,098,239**; Brit. pat. **498,225**; U.S. pats. **2,211,787**; **2,235,679**; **2,235,762**; **2,240,513**; **2,240,514**; **2,242,235**. From ethylene by the reaction with methylene iodide and a zinc-copper couple in 29% yield: Simmons, Smith, *J. Am. Chem. Soc.* **80**, 5323 (1958); *Chem. & Eng. News* **36**, 40 (Dec 8, 1958).

Flammable gas. Characteristic odor resembling that of petr ether. mp −127°. bp −33°. Liquefies at 4-6 atms. One liter of cyclopropane (at 1 atm, 0°) weighs 1.879 g. One vol of cyclopropane dissolves in about 2.7 vols of water at 15°. Freely sol in alcohol, ether. Sol in fixed oils. Concd H_2SO_4 absorbs the gas readily.
Caution: Mixture of cyclopropane with oxygen or air may explode when brought in contact with a flame or other causes of ignition. Explosive limits (% by vol in air), lower: 2.41; upper: 10.3. The explosibility is greater than that of other anesthetic-oxygen mixtures because of the comparatively larger amounts of oxygen that are compatible with cyclopropane anesthesia. Rich oxygen mixtures are therefore to be avoided.
THERAP CAT: Anesthetic (inhalation).
THERAP CAT (VET): Inhalation anesthetic.

2756. Cyclopropyl Methyl Ether. *Methoxycyclopropane;* cyprome ether. C_4H_8O; mol wt 72.10. C 66.63%, H 11.18%, O 22.19%. Prepd from 1,3-dibromo-2-methoxypropane by reduction with Zn: Olson *et al., J. Am. Chem. Soc.* **69**, 2454 (1947); *see also* Krantz, Jr. *et al., J. Pharmacol.* **69**, 207 (1940); Krantz, Jr., Drake, U.S. pat. **2,330,979** (1944); Haller, U.S. pat. **2,424,029** (1947).

Mobile, flammable liq. Odor similar to that of cyclopropane. d_4^{25} 0.786. mp −119°; bp 44.7°; n_D^{20} 1.3802. 100 ml of water dissolve 5.5 g of cyprome ether at 25°. Oil-water coefficient: 6.7 (ethyl ether = 4.5). Vapor press. at 26° (measured in nitrometer): 414 mm (ethyl ether = 555 mm). Explosive mixture, lower limit for air or oxygen: 2.0% v/v (ethyl ether vapor = 2.5 v/v).
USE: Insecticide. *Caution:* Narcotic in high concns.

2757. Cyclorphan. *17-(Cyclopropylmethyl)morphinan-3-ol;* (−)-3-hydroxy-N-cyclopropylmethylmorphinan. $C_{20}H_{27}NO$; mol wt 297.42. C 80.76%, H 9.15%, N 4.71%, O 5.38%. Prepn: Gates, Montzka, *J. Med. Chem.* **7**, 127 (1964); Gates, U.S. pat. **3,285,922** (1966 to Res. Corp.). Pharmacology: Harris *et al., Arch. Int. Pharmacodyn. Ther.* **165**, 112 (1967).

Crystals from ethyl acetate, mp 187.5-189°. $[\alpha]_D^{30}$ −120° (c = 2.26). Rings B/C *cis.* LD_{50} i.v. in mice, rats: 24, 23 mg/kg, Gates, *loc. cit.*

2758. Cycloserine. D-4-*Amino-3-isoxazolidinone;* D-4-amino-3-isoxazolidone; orientomycin; PA-94; 106-7; Closina; Farmiserina; Micoserina; Oxamycin; Seromycin. $C_3H_6N_2O_2$; mol wt 102.09. C 35.29%, H 5.92%, N 27.44%, O 31.34%. Antibiotic substance produced by *Streptomyces garyphalus* sive *orchidaceus*: Kuehl, Jr., *et al., J. Am. Chem. Soc.* **77**, 2344 (1955); Hidy *et al., ibid.* 2345; Shull, Sardinas, *Antibiot. & Chemother.* **5**, 398 (1955); Shull *et al.,* U.S. pat.

2,773,878 (1956 to Pfizer); Harned, U.S. pat. **2,789,983** (1957 to Commercial Solvents); **Brit.** pat. **768,007** (1957 to Commercial Solvents), *C.A.* **51**, 10847e (1957); U.S. pat. **3,124,590** (1964 to Commercial Solvents); Howe, U.S. pat. **2,845,433** (1958 to Merck & Co.). Synthesis: Stammer *et al., J. Am. Chem. Soc.* **77**, 2346 (1955); Peck, U.S. pat. **2,772,280** (1956 to Merck & Co.); Plattner *et al., Helv. Chim. Acta* **40**, 1531 (1957); Holly, Stammer, U.S. pat. **2,840,565** (1958 to Merck & Co.). Comprehensive description: J. W. Lamb in *Analytical Profiles of Drug Substances* vol. 1, K. Florey, Ed. (Academic Press, New York, 1972) pp 53-64.

Crystals, decomp 155-156°. $[\alpha]_D^{23}$ +116° (c = 1.17); $[\alpha]_{546}^{25}$ +137° (c = 5 in 2*N* NaOH). uv max: 226 nm ($E_{1cm}^{1\%}$ 402). Sol in water. Aq solns have a pH around 6. Forms salts with acids and bases. Neutral or acid solns are unstable. Aq solns buffered to pH 10 with sodium carbonate can be stored without loss for one week at refrigerator temps. Slightly sol in methanol, propylene glycol. Prepn of crystalline calcium and magnesium salts: Harris *et al.*, U.S. pat. **2,832,788** (1958 to Merck & Co.).

THERAP CAT: Antibacterial (tuberculostatic).

2759. Cyclosporins. A group of nonpolar cyclic oligopeptides with immunosuppressant activity; produced by *Tolypocladium inflatum Gams* (formerly designated as *Trichoderma polysporum* [Link ex Pers.] Rifai) and other fungi imperfecti. The major component, cyclosporin A, has been identified along with several other minor metabolites, cyclosporins B through I. A number of synthetic analogs have also been prepared. Isoln of A and C, structure of A: A. Rüegger *et al., Helv. Chim. Acta* **59**, 1075 (1976); M. Dreyfuss *et al., J. Appl. Microbiol.* **3**, 125 (1976). Crystal and molecular structure of the iodo deriv of A: T. J. Petcher *et al., Helv. Chim. Acta* **59**, 1480 (1976). Structure of C: R. Traber *et al., ibid.* **60**, 1247 (1977). Production of A and C: E. Härri *et al.*, U.S. pat. **4,117,118** (1978 to Sandoz). Isoln, characterization and antifungal activity of B, D, E, structures of A through D: R. Traber *et al., Helv. Chim. Acta* **60**, 1568 (1977). Isoln and structures of E, F, G, H, I: *eidem, ibid.* **65**, 1655 (1982). Total synthesis of cyclosporin A and analogs: R. Wenger, *Transplant. Proc.* **15**(4), Suppl. 1, 2230 (1983). Biosynthetic studies: H. Kobel *et al., Experientia* **39**, 873 (1983); R. Zocher *et al., Phytochemistry* **29**, 549 (1984). *In vivo* immunosuppressant activity: J. F. Borel *et al., Agents Actions* **6**, 468 (1976); *eidem, Immunology* **32**, 1017 (1977); R. Y. Calne, *Clin. Exp. Immunol.* **35**, 1 (1979). Selective effect on T-cells: D. J. White *et al., Transplantation* **27**, 55 (1979); M. Y. Gordon, J. W. Singer, *Nature* **279**, 433 (1979). Inhibition of γ-interferon synthesis: G. H. Reem *et al., Science* **221**, 63 (1983). Immunosuppressive profile of cyclosporin G: P. C. Hiestand *et al., Transplant. Proc.* **17**, 1362 (1985); *eidem, Immunology* **55**, 249 (1985). Comparative inhibition of cytokine production by A and G: R. M. McKenna *et al., Transplantation* **47**, 343 (1989). HPLC determn of A and G in human serum: P. E. Wallemacq, M. Lesne, *J. Chromatog.* **413**, 131 (1987). Clinical studies of A in graft-versus-host disease: R. Y. Calne *et al., Lancet* **2**, 1323 (1978); R. L. Powles *et al., ibid.* 1327; R. L. Powles *et al., ibid.* **1**, 327 (1980); P. J. Tutschka *et al., Blood* **61**, 318 (1983); in myasthenia gravis: R. S. A. Tindall *et al., N. Engl. J. Med.* **316**, 719 (1987). Toxicological evaluation of cyclosporin A: B. Ryffel *et al., Arch. Toxicol.* **53**, 107 (1983). Comparative toxicological study of A and G: J. I. Duncan *et al., Transplantation* **42**, 395 (1986). Review of cyclosporin A: *Progr. Allergy* vol. 38, K. Ishizaka *et al.*, Eds., entitled "Ciclosporin" by J. F. Borel, Ed. (Karger, Basel, 1986) 474 pp. Comprehensive description: M. M. Hassan, M. A. Al-Yahya in *Analytical Profiles of Drug Substances* vol. 16, K. Florey, Ed. (Academic Press, New York, 1987) pp 145-206. Review of mechanism of action studies: A. Aszalos, *J. Med.* **19**, 297-316 (1988); of use in rheumatoid arthritis: W. B. Harrison, B. von Graffenried, *Agents Actions* **24**, Suppl. 236-253 (1988).

Cyclosporin A, $C_{62}H_{111}N_{11}O_{12}$, *27-400, cyclosporine, ciclosporin, Sandimmun(e)*. R = CH_2CH_3. White prismatic needles from acetone at -15°. mp 148-151°. $[\alpha]_D^{20}$ -244° (c = 0.6 in chloroform); $[\alpha]_D^{20}$ -189° (c = 0.5 in methanol). Sol in methanol, ethanol, acetone, ether, chloroform; slightly sol in water and saturated hydrocarbons. LD_{50} in mice, rats, rabbits (mg/kg): 107, 25, >10 i.v.; 2329, 1480, >1000 orally (Ryffel).

Cyclosporin B, $C_{61}H_{109}N_{11}O_{12}$, *Ala²-cyclosporine*. R = CH_3. Amorphous white powder, mp 149-152°. $[\alpha]_D^{20}$ -238° (c = 0.62 in chloroform). $[\alpha]_D^{20}$ -168° (c = 0.56 in methanol).

Cyclosporin C, $C_{62}H_{111}N_{11}O_{13}$, *Thr²-cyclosporine*. R = CH(OH)—CH_3. Colorless prismatic needles from acetone at -15°. mp 152-155°. $[\alpha]_D^{20}$ -255° (c = 0.5 in chloroform). $[\alpha]_D^{20}$ -182° (c = 0.5 in methanol). Solubility similar to cyclosporin A.

Cyclosporin D, $C_{63}H_{113}N_{11}O_{12}$, *Val²-cyclosporine*. R = CH(CH_3)$_2$. Colorless prisms from acetone at -15°. mp 148-151°. $[\alpha]_D^{20}$ -245° (c = 0.52 in CHCl$_3$). $[\alpha]_D^{20}$ -211° (c = 0.51 in methanol).

Cyclosporin G, $C_{63}H_{113}N_{11}O_{12}$, *7-L-norvaline cyclosporin A, Nva²-cyclosporine*. R = $CH_2CH_2CH_3$. Colorless, polyhedric crystals from ether + petroleum ether, mp 196-197°. $[\alpha]_D^{20}$ -245° (c = 1.0 in chloroform). $[\alpha]_D^{20}$ -191° (c = 1.04 in methanol).

THERAP CAT: Cyclosporin A as immunosuppressant.

2760. Cyclothiazide. *3-Bicyclo[2.2.1]hept-5-en-2-yl-6-chloro-3,4-dihydro-2H-1,2,4-benzothiadiazine-7-sulfonamide 1,1-dioxide; 6-chloro-3,4-dihydro-3-(2-norbornen-5-yl)-2H-1,2,4-benzothiadiazine-7-sulfonamide 1,1-dioxide; 6-chloro-3,4-dihydro-3-(2-norbornen-5-yl)-7-sulfamoyl-1,2,4-benzothiadiazine 1,1-dioxide; 6-chloro-3-(2-norbornen-5-yl)-7-sulfamyl-3,4-dihydro-1,2,4-benzothiadiazine 1,1-dioxide*; Lilly 35483; Aquirel; Anhydron; Doburil; Fluidil. $C_{14}H_{16}ClN_3O_4S_2$; mol wt 389.91. C 43.13%, H 4.14%, Cl 9.09%, N 10.78%, O 16.41%, S 16.45%. Prepn: Whitehead *et al., J. Org. Chem.* **26**, 2814 (1961). Comprehensive description: C. D. Wentling in *Analytical Profiles of Drug Substances* vol. 1, K. Florey, Ed. (Academic Press, New York, 1972) pp 65-77.

Crystals from dil alc, mp 234°.
THERAP CAT: Diuretic, antihypertensive.
THERAP CAT (VET): Diuretic.

2761. Cyclovalone. *2,6-Divanillylidenecyclohexanone; 2,6-bis(4-hydroxy-3-methoxybenzylidene)cyclohexanone; divanillalcyclohexanone*; Beveno; Divanil; Divanon (obsolete); DVC; Flavugal; Vanilone. $C_{22}H_{22}O_5$; mol wt 366.40. C 72.11%, H 6.05%, O 21.83%. Prepn: Vorländer, Koch, *Ber.* **62**, 534 (1929); Rumpel, **Austrian** pat. **180,258** (1954 to Chem-Pharm. Fabrik Waldheim).

Consult the cross index before using this section.

Crystals from acetic acid, mp 178-179°. Soluble in alcohol, water.

THERAP CAT: Choleretic, cholagogue.

2762. Cycothiamin(e). *N-[(4-Amino-2-methyl-5-pyrimidinyl)methyl]-N-[1-(2-oxo-1,3-oxathian-4-ylidene)ethyl]-formamide;* carbothiamine; cyclocarbothiamine; cycotiamine; CCT; Cometamine; Commetamin. $C_{13}H_{16}N_4O_3S$; mol wt 308.36. C 50.64%, H 5.23%, N 18.17%, O 15.57%, S 10.40%. Prepn: **Neth.** pat. **Appl. 6,511,842** corresp. to Murakami *et al.,* **U.S.** pat. **3,324,124** (1966, 1967 to Yamanouchi Seiyaku).

Crystals from water, dec 175.5°. LD_{50} orally in mice: 13,390 mg/kg.

Hydrochloride, $C_{13}H_{17}ClN_4O_3S$, crystals from methanol + ethyl acetate, or from ethylene dichloride, dec 179-180°.

THERAP CAT: Oral thiamine therapy.

2763. Cycrimine Hydrochloride. *α-Cyclopentyl-α-phenyl-1-piperidinepropanol hydrochloride;* 1-phenyl-1-cyclopentyl-3-piperidino-1-propanol hydrochloride; compd 08958; Pagitane hydrochloride. $C_{19}H_{30}ClNO$; mol wt 323.90. C 70.45%, H 9.34%, N 4.32%, Cl 10.95%, O 4.94%. Prepd by the addition of a Grignard reagent to the corresponding β-aminoketone: Denton *et al., J. Am. Chem. Soc.* **72,** 3795 (1950). *See also* Trihexyphenidyl Hydrochloride.

Crystals, dec 241-244°. Very bitter taste. Soly (g/100 ml) at 25° in water: 0.6; in ethanol: 2.0; in chloroform: 3.0. pH of a 0.5% aq soln: 4.9-5.4.

THERAP CAT: Anticholinergic.

2764. Cyfluthrin. *3-(2,2-Dichloroethenyl)-2,2-dimethyl-cyclopropanecarboxylic acid cyano(4-fluoro-3-phenoxyphenyl)methyl ester; (R,S)-α-cyano-4-fluoro-3-phenoxybenzyl-(1R,S)-cis,trans-3-(2,2-dichlorovinyl)-2,2-dimethylcyclo-propanecarboxylate;* cyfoxylate; FCR 1272; BAY-FCR 1272; Baythroid. $C_{22}H_{18}Cl_2FNO_3$; mol wt 434.29. C 60.84%, H 4.18%, Cl 16.33%, F 4.37%, N 3.23%, O 11.05%. Synthetic pyrethroid insecticide. Commercial product is mixture of 8 isomers, the *(1R)*-isomers primarily responsible for the bioactivity. Prepn of racemic mixture: R. Fuchs *et al.,* **Ger.** pat. **2,709,264** corresp to **U.S.** pat. **4,218,469** (1978, 1980 both to Bayer AG). Prepn of stereoisomers: *eidem,* **Eur.** pat. **Appl. 22,970** corresp to **U.S.** pat. **4,287,208** (both 1981 to Bayer AG). Chiral phase HPLC separation of enantiomers: R. A. Chapman, *J. Chromatog.* **258,** 175 (1983). Metabolism by cell suspension cultures: U. Preiss *et al., Chemosphere* **13,** 861 (1984). Field evaluation against cotton insect pests: J. A. Durant, *J. Agric. Entomol.* **1,** 201 (1984). Review of chemistry, bioactivity and field studies: I. Hammann, R. Fuchs, *Pflanzenshutz-Nachr.* **34,** 121-151 (1981). Review of formulations and potential uses: W. Behrenz *et al., ibid.* **36,** 127-176 (1983).

Yellowish-brown oil, n_D^{23} 1.5511. Soly in water (20°): 1-2 $\times 10^{-6}$ g/l. LD_{50} in male, female rats, mice (mg/kg): 500-800, 1200, 300, 600 orally in Lutrol. LC_{50} (96 hr) in rainbow trout: 0.0006 mg/l (Hammann).

(1R)(3R)(αR)-Form, colorless oil. $[\alpha]_D^{20}$ −15.0° (c = 1.0 in $CHCl_3$).

(1R)(3R)(αS)-Form, crystals, mp 50-52°. $[\alpha]_D^{20}$ +24.5° (c = 1.0 in $CHCl_3$).

(1R)(3S)(αS)-Form, crystals from *m*-hexane mp 68-69°. $[\alpha]_D^{20}$ −2.1° (c = 1.0 in $CHCl_3$).

USE: Agricultural insecticide.

2765. Cyhalothrin. *3-(2-Chloro-3,3,3-trifluoro-1-propenyl)-2,2-dimethylcyclopropanecarboxylic acid cyano(3-phenoxyphenyl)methyl ester;* Grenade. $C_{23}H_{19}ClF_3NO_3$; mol wt 449.86. C 61.41%, H 4.26%, Cl 7.88%, F 12.67%, N 3.11%, O 10.67%. Synthetic pyrethroid insecticide. Prepn: R. K. Huff, **Ger.** pat. **2,802,962** corresp to **U.S.** pat. **4,183,-948** (1977, 1980 to ICI). Activity: P. D. Bently *et al., Pestic. Sci.* **11,** 156 (1980); V. K. Stubbs *et al., Austr. Vet. J.* **59,** 152 (1982).

USE: Insecticide; acaricide.

2766. Cyheptamide. *10,11-Dihydro-5H-dibenzo[a,d]-cycloheptene-5-carboxamide;* dibenzo[a,d][1,4]cycloheptadiene-5-carboxamide; AY 8682. $C_{16}H_{15}NO$; mol wt 237.29. C 80.98%, H 6.37%, N 5.90%, O 6.74%. Prepn: M. A. Davis *et al.,* **Fr.** pat. **1,355,829** (1964 to Ayerst, McKenna & Harrison); M. A. Davis, S. O. Winthrop, **U.S.** pat. **3,242,212** (1966 to Am. Home Prods.). Prepn and anticonvulsant activity: M. A. Davis *et al., J. Med. Chem.* **7,** 88 (1964). Physical characteristics: H. J. Doorenbos *et al., Pharm. Weekblad* **104,** 732 (1969). Pharmacology: A. B. H. Funcke *et al., Arch. Int. Pharmacodyn.* **187,** 174 (1970). Toxicological studies: C. J. van Eeken *et al., ibid.* **188,** 79 (1970). Metabolism in animals and man: M. Kraml *et al., Biochem. Pharmacol.* **20,** 2327 (1971). Comparison of potency with other anticonvulsants: G. L. Jones *et al., J. Pharm. Sci.* **70,** 618 (1981). Elucidation of structure by x-ray diffraction, mode of action: P. W. Codding *et al., J. Med. Chem.* **27,** 649 (1984).

Long needles from acetonitrile, mp 193-194°. Sol in chloroform; sparingly sol in methanol, acetone; slightly sol in ethanol, ether. Practically insol in water. LD_{50} in mice (g/kg): 4.2-5.2 orally; 2.4-2.6 i.p. (Funcke).

2767. Cyhexatin. *Tricyclohexylhydroxystannane;* TCTH; tricyclohexylstannol; tricyclohexyltin hydroxide; ENT 27395; Dowco 213; Plictran. $C_{18}H_{34}OSn$; mol wt 385.16. C 56.13%, H 8.90%, O 4.15%, Sn 30.82%. Prepn: E. E. Kenaga, **U.S.** pat. **3,264,177** (1966 to Dow). Activity studies: W. E. Allison *et al., J. Econ. Entomol.* **61,** 1254 (1968); L. R. Jeppson *et al., ibid.,* 1502. Biochemical mode of action: S. Ahmad, C. O. Knowles, *Comp. Gen. Pharmacol.* **3,** 125 (1972). Biodegradation: E. H. Blair, *Environ. Qual. Safety*

Suppl. **3**, 406 (1975). Toxicity data: T. B. Gaines, R. E. Linder, *Fundam. Appl. Toxicol.* **7**, 299 (1986). Acute toxicity studies in livestock: J. H. Johnson *et al., Toxicol. Appl. Pharmacol.* **31**, 66 (1975).

Whitish crystalline powder, mp 195-198°. Insol in water. Slightly sol in most organic solvents. Degraded when exposed in thin layers to uv light. Non-toxic to bees, toxic to fish. LD₅₀ in adult male, female rats (mg/kg): 779, 826 orally (Gaines, Linder).

Caution: Irritating to eyes.

USE: Acaricide.

2768. Cymarin. *3β-[(2,6-Dideoxy-3-O-methyl-β-D-ribo-hexopyranosyl)oxy]-5β,14-dihydroxy-19-oxocard-20(22)-enolide;* K-strophanthin-α; Alvonal MR. C₃₀H₄₄O₉; mol wt 548.65. C 65.67%, H 8.08%, O 26.24%. A glycoside of cymarose, the aglucon being strophanthidin. Isoln from *Strophanthus kombé* Oliv., *Apocynaceae:* Jacobs, Hoffmann, *J. Biol. Chem.* **67**, 609 (1926); Stoll *et al., Helv. Chim. Acta* **20**, 1484 (1937); from *Adonis vernalis* L., and *A. chrysocyathus* Hook., *Ranunculaceae:* Reichstein, Rosenmund, *Pharm. Acta Helv.* **15**, 150 (1940); Pitra, Cekan, *Coll. Czech. Chem. Commun.* **26**, 1551 (1961); Abubakirov, Yamatova, *Zh. Obshch. Khim.* **31**, 2424 (1961); from *Pentopetia androsaemifolia* Decne., *Asclepiadaceae:* Wyss *et al., Helv. Chim. Acta* **43**, 664 (1960); from *Castilloa elastica* Cerv., *Moraceae:* Adams, Wilkinson, *J. Pharm. Pharmacol.* **13**, 279 (1961). Structure: Kochetkov *et al., Doklady Akad. Nauk SSSR* **136**, 613 (1961). Toxicity data: V. G. Vogel, E. Kluge, *Arzneimittel-Forsch.* **11**, 848 (1961).

Needles from methanol, mp 148°. [α]²⁰_D +39.2° (methanol); [α]²²_D +39.0° (c = 1.7 in chloroform). Sol in methanol, chloroform. Practically insol in water. LD₅₀ i.v. in rats: 24.8±1.8 mg/kg (Vogel, Kluge). Sesquihydrate, hexagonal plates from dil ethanol, mp 184-185°.

Monoacetylcymarin, C₃₂H₄₆O₁₀, needles from dil methanol, mp 175-176°. [α]²²_D +45.1° (ethanol). Sol in chloroform. Practically insol in water.

THERAP CAT: Cardiotonic.

2769. Cymarose. *2,6-Dideoxy-3-O-methyl-ribo-hexose;* 3-methyldigitoxose. C₇H₁₄O₄; mol wt 162.18. C 51.84%, H 8.70%, O 39.46%. By hydrolysis of a number of cardiac glycosides from the *Apocynaceae* family. From *Apocynum cannabinum* L., *A. androsaemifolium* L., *A. venetum* L.: Windaus, Hermanns, *Ber.* **48**, 979 (1915); Trabert, *Arzneimittel-Forsch.* **10**, 197 (1960); from *Strophanthus kombé* Oliv., *Periploca graeca* L.: Jacobs, Hoffmann, *J. Biol. Chem.* **67**, 609 (1926); **79**, 519 (1928); from *S. emini* Aschers & Pax: Lamb, Smith, *J. Chem. Soc.* **1936** 442; also from *Castilla elastica* Cerv., *Moraceae:* Adams, Wilkinson, *J. Pharm. Pharmacol.* **13**, 279 (1961). Structure: S. F. Dyke, *The Carbohydrates* (Interscience, New York, 1960) p 104. Synthesis: Bollinger, Ulrich, *Helv. Chim. Acta* **35**, 93 (1952). Review: R. C. Elderfield, "The Carbohydrate Components of Cardiac Glycosides" in W. W. Pigman, M. L. Wolfrom, *Advances in Carbohydrate Chemistry* vol. I (Academic Press, New York, 1945) pp 147-173.

Prisms from ether + petr ether, mp 93-94°. [α]²⁰_D +54.7° (c = 3.2 in water after 24 hrs). Sol in water, alcohol, acetone; sparingly sol in ether, petr ether, chloroform, benzene. Reduces Fehling soln. Does not form an osazone.

2770. Cymene. isopropyltoluene; methylisopropylbenzene. C₁₀H₁₄; mol wt 134.21. C 89.49%, H 10.51%. Usually prepd by alkylation of toluene; *m-, o-* and *p-*isomers obtained: Allen, Yats, *J. Am. Chem. Soc.* **83**, 2799 (1961). Purification and properties of the three isomers: Streiff *et al., Anal. Chem.* **27**, 411 (1955). Separation of the three isomers by gas chromatography: Rihani, Froment, *J. Chromatog.* **18**, 150 (1965).

m-Cymene, methyl(3-methylethyl)benzene. Liquid, bp 175.14°. mp −63.75°. d²⁰_4 0.8610, d²⁵_4 0.8570. n²⁰_D 1.4930, n²⁵_D 1.4906. Practically insol in water; miscible with alcohol, ether.

o-Cymene, liquid, bp 178.15°. mp −71.54° (also −75.24° and −81.53° for two unstable solid forms, *see* Streiff, *loc. cit.*). d²⁰_4 0.8766, d²⁵_4 0.8726. n²⁰_D 1.5006, n²⁵_D 1.4982. Practically insol in water; miscible with organic solvents.

p-Cymene, Dolcymene. Occurs in a number of essential oils. Liquid, bp 177.10°. mp −67.94°. d²⁰_4 0.8573, d²⁵_4 0.8533. n²⁰_D 1.4909, n²⁵_D 1.4885. Flash pt, closed cup: 117°F (47°C). Practically insol in water; misc with alcohol, ether. LD₅₀ orally in rats: 4750 mg/kg, P. M. Jenner *et al., Food Cosmet. Toxicol.* **2**, 327 (1964).

2771. Cymiazole. *2,4-Dimethyl-N-(3-methyl-2(3H)-thiazolylidene)benzenamine;* 2-(2',4'-dimethylphenylimino)-3-methylthiazoline; *N-*(3-methyl-4-thiazolin-2-ylidene)-2,4-xylidene; xymiazole; CGA 50439; Tifatol. C₁₂H₁₄N₂S; mol wt 218.31. C 66.02%, H 6.46%, N 12.83%, S 14.69%. Prepn: D. Duerr, W. D. Traber, Ger. pat. **2,619,724** (1976 to Ciba-Geigy), *C.A.* **86**, 72627h (1977). Activity: R. M. Immler *et al., Proc. Brit. Crop Prot. Conf.-Pests Dis.* **1977**, 383.

USE: Acaricide.

THERAP CAT (VET): Ectoparasiticide.

2772. Cynanchogenin. *12-[(3,4-Dimethyl-1-oxo-2-pentenyl)oxy]-3,8,14-trihydroxypregn-5-en-20-one.* C₂₈H₄₂O₆; mol wt 474.62. C 70.85%, H 8.92%, O 20.23%. From root of *Cynanchum caudatum* Max., *Asclepiadaceae:* Mitsuhashi, Shimizu, *Chem. Pharm. Bull.* **8**, 313, 318 (1960). Proposed structure: *eidem, Steroids* **2**, 373 (1963). Revised structure: *eidem, Tetrahedron* **24**, 4143 (1968). ¹³C-NMR studies: T. Yamagishi *et al., Tetrahedron Letters* **1973**, 3527, 3531.

Fine needles from ether + petr ether, mp 165-167°. [α]²²_D −39.5° (c = 1.24 in 90% ethanol). uv max (ethanol): 218 nm (log ε 4).

Monoacetate, C₃₀H₄₄O₇, crystals from petr ether + benzene, mp 150°.

2773. Cynarin(e). *(1α,3α,4α,5β)-1,3-Bis[[3-(3,4-dihydroxyphenyl)-1-oxo-2-propenyl]oxy]-4,5-dihydroxycyclohexanecarboxylic acid;* 1,3-dicaffeoylquinic acid; 1,5-dicaffeylquinic acid; 3,4-dihydroxycinnamic acid 1-carboxy-4,5-dihydroxy-1,3-cyclohexylene ester; caffeic acid 1-carboxy-4,5-dihydroxy-1,3-cyclohexylene ester; 1-carboxy-4,5-dihydroxy-1,3-cyclohexylenebis-(3,4-dihydroxycinnamate); quinic acid 1,5-dicaffeic ester; Cinarine; Listrocol; Plemocil. $C_{25}H_{24}O_{12}$; mol wt 516.44. C 58.14%, H 4.68%, O 37.18%. Active principle of artichoke, Cynara scolymus L., Compositae: L. Panizzi, M. L. Scarpati, Gazz. Chim. Ital. **84,** 792 (1954). Synthesis of originally proposed structure: L. Panizzi et al., ibid. 806; L. Panizzi et al., U.S. pats. **2,863,909; 3,100,224** (1958, 1963 both to Farmitalia); Alberti et al., U.S. pat. **2,918,477** (1959). Revised structure: L. Panizzi, M. L. Scarpati, Gazz. Chim. Ital. **95,** 71 (1965). NMR confirmation of structure: I. Horman et al., J. Agr. Food Chem. **32,** 538 (1984). HPLC determn in pharmaceutics: A. Bettero, Boll. Chim. Farm. **120,** 49 (1981). Inhibition of fatty acid mobilization: P. Dorigo, G. Fassina, Pharmacol. Res. Commun. **2,** 109 (1970). Reduction of cholesterol levels in rats: J. Wojcicki, Drug Alcohol Depend. **3,** 143 (1978). Clinical trial in hyperlipemic syndrome: M. Montini et al., Arzneimittel-Forsch. **25,** 1311 (1975).

Crystals from dil acetic acid. Sweet taste. mp 225-227°. $[\alpha]_D^{25}$ −59° (c = 2 in methanol). uv max (methanol): 326 nm ($E_{1cm}^{1\%}$ 616). Sparingly sol in cold, more in boiling water; sol in glacial acetic acid, alcohols.

THERAP CAT: Choleretic.

2774. Cyoctol. Hexahydro-4-(5-methoxyheptyl)-2(1H)-pentalenone; 6-(5-methoxy-1-heptyl)bicyclo[3.3.0]octan-3-one; 2-(5-methoxyhept-1-yl)bicyclo[3.3.0]octan-7-one; X-Andron; CPC 10997; Exandron. $C_{16}H_{28}O_2$; mol wt 252.40. C 76.14%, H 11.18%, O 12.68%. Topically active nonsteroidal androgen-receptor blocker. Prepn: W. J. Kasha, PCT Int. pat Appl. 83 04,019 (1983 to CBD Corp.), C.A. **101,** 23037k (1984). Prepn (not claimed) and use: idem, U.S. pat. **4,689,345** (1987 to CBD Corp.). Antiandrogenic and antineoplastic activity in vitro: L. C. Ford et al., Chemotherapy (Basel) **31,** 362 (1985). Series of articles on inhibition of human and microbial dihydrotestosterone-receptor binding in vitro: Rec. Adv. Chemother., Proc. 14th Int. Congr. Chemother. Antimicrob. Sect. 1 (Univ. Tokyo Press, Tokyo, 1985) pp 261-270; 273-274.

Clear, colorless oil.

THERAP CAT: Antiacne; antialopecia agent.

2775. Cypermethrin. 3-(2,2-Dichloroethenyl)-2,2-dimethylcyclopropanecarboxylic acid cyano(3-phenoxyphenyl)-methyl ester; (±)-α-cyano-3-phenoxybenzyl-(±)-cis,trans-3-(2,2-dichlorovinyl)-2,2-dimethylcyclopropane carboxylate; NRDC 149; FMC 30980; PP 383; Agrothrin; Ammo; Arrivo; Barricade; Cymbush; Cypercare; Cyperkill; Cypersect; Demon; Dysect; Fastac; Flectron; Nurelle; Polytrin; Ripcord; Rycopel; Sherpa; Topclip Parasol. $C_{22}H_{19}Cl_2NO_3$; mol wt 416.30. C 63.47%, H 4.60%, Cl 17.03%, N 3.37%, O 11.53%. Potent synthetic pyrethroid insecticide. Prepn of racemic mixture: M. Elliot et al., Ger. pat. **2,326,077;** eidem, U.S. pat. **4,024,163** (1974, 1977 to NRDC). Activity: eidem, Pestic. Sci. **6,** 537 (1975); M. H. Breese, ibid. **8,** 264 (1977). Soil degradation: T. R. Roberts, M. E. Standen, ibid. 305; D. D. Kaufman et al., J. Agr. Food Chem. **29,** 239

(1981). Metabolism: T. Shono et al., ibid. **27,** 316 (1979); M. J. Crawford et al., ibid. **29,** 130 (1981). Repellent activity against flies on cattle: J. A. Shemanchuk, Pestic. Sci. **12,** 412 (1981); J. E. Hillerton et al., Brit. Vet. J. **141,** 160 (1985).

Commercial product is a mixture of eight isomers. Viscous semi-solid, mp 60-80°. Insol in water. Sol in methanol, acetone, xylene, methylene dichloride.

USE: Insecticide.

THERAP CAT (VET): Ectoparasiticide.

2776. Cyphenothrin. 2,2-Dimethyl-3-(2-methyl-1-propenyl)cyclopropanecarboxylic acid cyano(3-phenoxyphenyl)-methyl ester; (RS)-α-cyano-3-phenoxybenzyl (1R)-cis,trans-chrysanthemate; α-cyano-m-phenoxybenzyl 2,2-dimethyl-3-(2-methylpropenyl)cyclopropanecarboxylate; 3-phenoxy-α-cyanobenzylchrysanthemate; S-2703; S-2703 Forte; Gokilaht. $C_{24}H_{25}NO_3$; mol wt 375.47. C 76.77%, H 6.71%, N 3.73%, 12.78%. Synthetic pyrethroid. Prepn: T. Matsuo et al., Ger. pat. **2,231,312;** eidem, U.S. pat. **3,835,176** (1973, 1974 to Sumitomo). Insecticidal activity: M. Elliot et al., Pestic. Sci. **13,** 407 (1982); T. Matsunaga et al., Pesticide Chemistry: Human Welfare and the Environment **2,** J. Miyamoto et al., Eds (Pergamon Press, Oxford, 1983) pp 231-238; T. Itoh et al., J. Am. Mosquito Cont. **2,** 503 (1986). Mechanism of action: T. Narahashi, Neurobehav. Toxicol. Teratol. **4,** 753 (1982). Series of articles on stereoselective syntheses of labelled cis and trans forms: H. Kanamaru et al., Radioisotopes **35,** 103, 109, 242 (1986).

(1R)-trans form

Yellowish viscous liquid. d_{25}^{25} 1.083. Vapor pressure at 30°: 3.11 × 10⁻⁶ mm Hg. Viscosity at 30°: 808.8 cP. Note: Commercial product is a mixture (35:65) of cis and trans forms.

USE: Insecticide.

2777. Cyprenorphine. 17-(Cyclopropylmethyl)-4,5-epoxy-3-hydroxy-6-methoxy-α,α-dimethyl-6,14-ethenomorphinan-7-methanol; N-(cyclopropylmethyl)-7,8-dihydro-7α-(1-hydroxy-1-methylethyl)-O⁶-methyl-6,14-endo-ethenonormorphine; N-(cyclopropylmethyl)-6,14-endo-etheno-7α-(2-hydroxy-2-propyl)tetrahydronororipavine; N-(cyclopropylmethyl)-6,14-endo-ethenotetrahydronororipavine; $C_{26}H_{33}NO_4$; mol wt 423.53. C 73.73%, H 7.85%, N 3.31%, O 15.11%. Narcotic antagonist closely related to diprenorphine, q.v. Prepn (7α- or 7β-linkage unspecified): Belg. pat. **629,070** (1963 to J. F. Macfarlan), C.A. **61,** 13364g (1964); K. W. Bentley, U.S. pat. **3,474,101** (1969 to Reckitt & Sons); of 7α: K. W. Bentley, D. G. Hardy, J. Am. Chem. Soc. **89,** 3281 (1967). Activity in rats: K. W. Bentley et al., Nature **206,** 102 (1965). Use in large animals: M. R. Jainudeen, Vet. Rec. **89,** 686 (1971); K. R. Presnell et al., J. Wildl. Dis. **9,** 336 (1973). Binding to neuroleptic receptors: A. Czlonkowski et al., Life. Sci. **22,** 953 (1978). Determn by HPLC: I. Jane, A. McKinnon, J. Chromatog. **323,** 191 (1985).

Prisms from methanol, mp 234° (7α- or 7β-linkage unspecified).

Hydrochloride, $C_{26}H_{34}ClNO_4$, M 285. Crystals from methanol, mp 248°.

Caution: May be habit forming. This is a controlled substance (opium derivative) listed in the U.S. Code of Federal Regulations, Title 21 Part 1308.11 (1985).

THERAP CAT (VET): Etorphine antagonist.

2778. Cypripedium. Lady's slipper; American valerian; nerve root; yellow moccasin flower; Noah's ark. Dried rhizome and roots of *Calypso bulbosa* (L.) Oakes (*Cypripedium bulbosum* L.), *Cypripedium pubescens* Willd., or of *Cypripedium parviflorum* Salisb., *Orchidaceae*. Habit. Nova Scotia south to Alabama, west to Nebraska and Missouri. Constit. Volatile oil, volatile acid, tannin, resins.

THERAP CAT: Sedative.

2779. Cyproheptadine. *4-(5H-Dibenzo[a,d]cyclohepten-5-ylidene)-1-methylpiperidine;* 1-methyl-4-(5H-dibenzo-[a,d]cycloheptenylidene)piperidine; 5-(1-methylpiperidylidene)-5H-dibenzo[a,d]cycloheptene; 1-methyl-4-(5-di-benzo[a,e]cycloheptatrienylidene)piperidine; Periactinol. $C_{21}H_{21}N$; mol wt 287.39. C 87.76%, H 7.37%, N 4.87%. Prepn: Engelhardt, U.S. pat. **3,014,911** (1961 to Merck & Co.); Engelhardt *et al., J. Med. Chem.* **8,** 829 (1965). Human metabolism: C. C. Porter *et al., Drug Metab. Dispos.* **3,** 189 (1975). Comprehensive description: H. Y. Aboul-Enein, A. A. Al-Badr in *Analytical Profiles of Drug Substances* vol. 9, K. Florey, Ed. (Academic Press, New York, 1980) pp 155-179.

Crystals from dil ethanol, mp 112.3-113.3°.

Hydrochloride sesquihydrate, $C_{21}H_{22}ClN·1½H_2O$, *Anarexol, Antegan, Ifrasarl, Nuran, Periactin, Vimicon, Ciprac-tin, Peritol.* Crystals from abs ethanol + ether, dec 252.6-253.6°. uv max $(0.1N\ H_2SO_4)$: 224, 285 nm $(E_{1cm}^{1\%}$ 1656, 355). 1.0 g is sol in 1.5 ml methanol, 16 ml chloroform, 35 ml alc, 275 ml H_2O. Practically insol in ether. LD_{50} orally in mice: 74.2 mg/kg, J. J. Loux *et al., Arzneimittel-Forsch.* **28,** 1644 (1978).

Hydrochloride monohydrate, crystals, mp 214-216°. Soly in water: about 5 mg/ml.

THERAP CAT: Antihistaminic, antipruritic.

2780. Cyproquinate. *6,7-Bis(cyclopropylmethoxy)-4-hy-droxy-3-quinolinecarboxylic acid ethyl ester;* ethyl 6,7-bis-(cyclopropylmethoxy)-4-hydroxyquinoline-3-carboxylate; cyproquinidate; cyproxyquine; Su 18137; Coxytrol. $C_{20}H_{23}-NO_5$; mol wt 357.41. C 67.21%, H 6.48%, N 3.92%, O 22.38%. Prepn by ring closure of corresp dialkyl (alkoxy-phenylamino)methylenemalonates: Mizzoni *et al., Experientia* **24,** 1188 (1968); Mizzoni, DeStevens, S. Afr. pat. 67 05,655 (1968 to Ciba), *C.A.* **70,** 47316t (1969). Review of structure and biology: Mizzoni *et al., J. Med. Chem.* **13,** 870 (1970).

Crystals from DMF, mp 288-288.5° (dec).

THERAP CAT (VET): Coccidiostat.

2781. Cyproterone. *6-Chloro-1,2-dihydro-17-hydroxy-3'H-cyclopropa[1,2]pregna-1,4,6-triene-3,20-dione; 6-chlo-ro-17-hydroxy-1α,2α-methylenepregna-4,6-diene-3,20-dione;* 6-chloro-6-dehydro-17α-hydroxy-1,2α-methyleneproges-terone; 6-chloro-1,2α-methylene-4,6-pregnadien-17α-ol-3,20-dione; SH 881; SH 80881. $C_{22}H_{27}ClO_3$; mol wt 374.92. C 70.48%, H 7.26%, Cl 9.46%, O 12.80%. Prepn of free alcohol: Wiechert, Neumann, **Ger. pat. 1,189,991,** *C.A.* **63,** 1842e (1965); Wiechert, U.S. pat. **3,234,093** (1965, 1966 both to Schering AG). Biodynamics in man: Gerhards *et al., Arzneimittel-Forsch.* **23,** 1550 (1973). The free alcohol is an anti-androgen; the acetate is both an anti-androgen and a progestogen. Effect on hormone secretion and on spermatogenesis in man: L. Moltz *et al., Contraception* **21,** 393 (1980). Review of pharmacology and clinical studies (acetate) on acne and hirsutism in women: J. Hammerstein *et al., J. Steroid Biochem.* **19,** 591 (1983).

Crystals from ethyl acetate, mp 237.5-240°.

Acetate, $C_{24}H_{29}ClO_4$, *SH 714, CPA, Androcur, Cyprostat.* Crystals from diisopropyl ether, mp 200-201°. uv max (methanol): 281 nm (ε 17280). Mixture with ethinyl estradiol, *Dianette, Diane 35.*

THERAP CAT: The acetate as anti-androgen; combination with estrogen in treatment of acne.

2782. Cyromazine. *N-Cyclopropyl-1,3,5-triazine-2,4,6-triamine;* 2-cyclopropylamino-4,6-diamino-s-triazine; CGA 72662; Larvadex; Vetrazin. $C_6H_{10}N_6$; mol wt 166.18. C 43.36%, H 6.07%, N 50.57%. Insect growth regulator. Prepn: H. U. Brechbuhler *et al.,* Ger. pat. **2,736,876** corresp to U.S. pat. **4,225,598** (1978, 1980 to Ciba-Geigy). Activity: R. D. Hall, M. C. Foehse, *J. Econ. Entomol.* **73,** 564 (1980); R. J. Hart *et al., Aust. Vet. J.* **59,** 104 (1982).

Crystals, mp 219-222°.

USE: Insecticide.

THERAP CAT (VET): Ectoparasiticide.

2783. Cystamine. *2,2'-Dithiobisethanamine; 2,2'-di-thiobis[ethylamine];* β,β'-diaminodiethyl disulfide; bis-[β-aminoethyl]disulfide; decarboxycystine. $C_4H_{12}N_2S_2$; mol wt 152.29. C 31.55%, H 7.94%, N 18.40%, S 42.11%. $(H_2NCH_2CH_2)_2S_2$. Forms when cystine is distilled: Neuberg, Ascher, *Chem. Zentr.* **1907,** II, 1156. Prepd by H_2O_2 oxidation of $H_2NCH_2CH_2SH$: Mills, Jr., Bogert, *J. Am. Chem. Soc.* **62,** 1173 (1940); Barnett, *J. Chem. Soc.* **1944,** 5.

Viscous oil. Cannot be distilled without decompn (even in high vacuum). Freely sol in water. Sol in alc.

Dihydrochloride, $C_4H_{14}Cl_2N_2S_2$, needles from methanol, dec 203-214°. Sol in water.

Picrate, $C_4H_{12}N_2S_2.2C_6H_3N_3O_7$, yellow crystals, mp 204°.

2784. L-Cystathionine. *(R)-S-(2-Amino-2-carboxyethyl)-L-homocysteine*; L-*2-amino-4-[(2-amino-2-carboxyethyl)thio]butyric acid.* $C_7H_{14}N_2O_4S$; mol wt 222.28. C 37.83%, H 6.35%, N 12.61%, O 28.79%, S 14.43%. An amino acid whose homocysteine and serine moieties are both L in configuration. Intermediate in transsulfuration whereby the mammal transfers the sulfur of methionine via homocysteine to cysteine: Hope, *Proc. Intern. Congr. Biochem. 4th, Vienna, 1958*, 13, 63 (1960). Preparation: du Vigneaud *et al.*, *J. Biol. Chem.* **143**, 59 (1942); Weiss, Stekol, *J. Am. Chem. Soc.* **73**, 2497 (1951); Schöberl, Täuber, *Ann.* **599**, 23 (1956). Isolation from human brain: Tallau *et al.*, *J. Biol. Chem.* **230**, 707 (1958). *Review:* J. P. Greenstein, M. Winitz, *Chemistry of the Amino Acids* vol. 3 (John Wiley, N. Y., 1961) p 2682.

$$
\begin{array}{c}
\text{COOH} \\
|\\
\text{H}_2\text{NCH} \\
|\\
\text{HCH} \\
|\\
\text{HCH} \\
|\\
\text{S} \\
|\\
\text{HCH} \\
|\\
\text{HCNH}_2 \\
|\\
\text{COOH}
\end{array}
$$

Crystals (rectangular parallelepipeds), dec 312° (darkening at 270°). $[\alpha]_D^{20}$ +23.7° (1N HCl).

2785. Cysteamine. *2-Aminoethanethiol;* mercaptamine; β-mercaptoethylamine; 2-aminoethyl mercaptan; thioethanolamine; decarboxycysteine; MEA; mercamine; L-1573; Becaptan; Lambratene (formerly). C_2H_7NS; mol wt 77.15. C 31.14%, H 9.15%, N 18.16%, S 41.56%. $HSCH_2CH_2NH_2$. A sulfhydryl compound with a variety of biological effects. Prepn: Gabriel, Leupold, *Ber.* **31**, 2837 (1898); Knorr, Rössler, *ibid.* **36**, 1281 (1903); Mills, Jr., Bogart, *J. Am. Chem. Soc.* **62**, 1173 (1940); Wenker, *ibid.* **57**, 2328 (1935); D. A. Shirley, *Preparation of Organic Intermediates* (Wiley, New York, 1951) p 189. Use in treatment of paracetamol (acetaminophen) poisoning: L. F. Prescott *et al.*, *Lancet* **2**, 109 (1976); A. L. Harris, *Brit. Med. J.* **284**, 825 (1982). Effects in nephropathic cystinosis: M. Yudkoff *et al.*, *N. Engl. J. Med.* **304**, 141 (1981). Radioprotective effects: R. P. Bird, *Radiat. Res.* **72**, 290 (1980); C. J. Koch, R. L. Howell, *ibid.* **87**, 265 (1981). Cysteamine has been shown to be a duodenal ulcerogen in rats: H. Selye, S. Szabo, *Nature* **244**, 458 (1973); S. Szabo, *Am. J. Path.* **93**, 273 (1978); P. Kirkegaard *et al.*, *Scand. J. Gastroenterol.* **15**, 621 (1980). *Review:* S. Szabo, *Lab. Invest.* **51**, 121 (1984). It has also been found to deplete somatostatin concentration: S. Szabo, S. Reichlein, *Endocrinology* **109**, 2255 (1981); S. M. Sagar *et al.*, *J. Neuroscience* **2**, 225 (1982). In pituitary tissue, cysteamine is a potent depletor of prolactin concentrations *in vivo* and *in vitro*: W. J. Millard *et al.*, *Science* **217**, 452 (1982). Toxicity studies: E. Beccari *et al.*, *Arzneimittel-Forsch.* **5**, 421 (1955); D. L. Klayman *et al.*, *J. Med. Chem.* **12**, 510 (1969); P. K. Srivastava, L. Field, *ibid.* **18**, 798 (1975).

Crystals by sublimation *in vacuo*. Disagreeable odor. mp 97-98.5°. Oxidizes to cystamine on standing in air. Freely sol in water, alkaline reaction. LD_{50} in mice (mg/kg): 625 orally; 250 i.p. (Klayman); 250 i.p. (Srivastava, Field).

Hydrochloride, $C_2H_7NS.HCl$, crystals from alc, mp 70.2-70.7°. Sol in water, alcohol. LD_{50} (cg/kg): 23.19 i.p. in rats; 14.95 i.v. in rabbits (Beccari).

USE: Experimentally as a radioprotective agent and to produce acute and chronic duodenal ulcers in rats.

THERAP CAT: Antidote to acetaminophen.

2786. Cysteic Acid. *3-Sulfoalanine;* α-amino-β-sulfopropionic acid. $C_3H_7NO_5S$; mol wt 169.16. C 21.30%, H 4.17%, N 8.28%, O 47.29%, S 18.96%. $HOOCCH(NH_2)-CH_2SO_3H$. Has been isolated from human hair oxidized with permanganate: Lissizin, *Z. Physiol. Chem.* **173**, 309

(1928). Occurs normally in the outer part of the sheep's fleece, where the wool is exposed to light and weather: Martin, Synge, *Advances in Protein Chemistry* **2**, 3 (1945). Prepd from cystine or cysteine by oxidation with bromine in water: Friedmann, *Beiträge zur Chemischen Physiologie und Pathologie* **3**, 25, 38; Gortner, Hoffman, *J. Biol. Chem.* **72**, 435 (1927).

L-Form, octahedra or needles from dil alc (also forms a monohydrate, prismatic needles). When anhydr, dec 260°. $[\alpha]_D^{20}$ +8.66° (1.85 g in 25 ml). $Ka_1 = 1.3 \times 10^{-2}$ at 25°; $Ka_2 = 2 \times 10^{-9}$; Kb about 2×10^{-13}. Soluble in water; insol in alcohol.

DL-Form, crystals, dec 245°.

2787. Cysteine. L-*Cysteine*; cys (IUPAC abbrev.); β-mercaptoalanine; 2-amino-3-mercaptopropanoic acid; 2-amino-3-mercaptopropionic acid; α-amino-β-thiolpropionic acid. $C_3H_7NO_2S$; mol wt 121.16. C 29.74%, H 5.82%, N 11.56%, O 26.41%, S 26.47%. $HSCH_2CH(NH_2)COOH$. An amino acid. A primary cleavage product of proteins; converted to L-cystine upon prolonged hydrolysis in the presence of air. Prepn from proteins by hydrolysis: Okuda, *Pr. Acad. Tokyo* **2**, 277; *Chem. Zentr.* **1926**, II, 2728; by electrolytic reduction of cystine: Koperina, Gavrilov, *J. Gen. Chem. USSR* **17**, 1651 (1947), *C.A.* **42**, 3725h (1948). Alternate prepns: Rapkine; Farlow; Holloway, Young, U.S. pats. **2,376,186; 2,406,362; 2,414,303** (1945, 1946, 1947). Simple synthesis of racemic cysteine: V. J. Martens *et al.*, *Angew. Chem. Int. Ed.* **20**, 668 (1981).

Crystals. $[\alpha]_D^{25}$ +6.5° (5N HCl); $[\alpha]_D^{25}$ +13.0° (glacial acetic acid). pK_1 1.71; pK_2 8.33; pK_3 10.78. Absorption spectrum: Abderhalden, Rossner, *Z. Physiol. Chem.* **178**, 160 (1928). Freely sol in water, alcohol, acetic acid, ammonia water. Insol in ether, acetone, ethyl acetate, benzene, carbon disulfide, carbon tetrachloride. In neutral or slightly alkaline aq solns it is oxidized to cystine by air or by filtration through charcoal. Acidic solns can be stored for some days, esp if kept under nitrogen. The pure product is much more stable than a prepn which is contaminated by traces of metals. Copper and iron are the worst contaminants.

Hydrochloride, $C_3H_7NO_2S.HCl$, crystals, dec 175-178°. $[\alpha]_D^{25}$ +5.0° (5N HCl); $[\alpha]_D^{25}$ +10.0° (glacial acetic acid). Sol in water, alcohol, acetone; the aq soln is acid. *Keep tightly closed.* Decomposes and oxidizes slowly; hygroscopic.

USE: As dough conditioner.

THERAP CAT (VET): Has been used as a detoxicant.

2788. Cystine. 3,3′-Dithiobis(2-aminopropanoic acid); dicysteine; β,β'-dithiodialanine; α-diamino-β-dithiolactic acid; β,β'-diamino-β,β'-dicarboxydiethyl disulfide; bis(β-amino-β-carboxyethyl) disulfide. $C_6H_{12}N_2O_4S_2$; mol wt 240.30. C 29.99%, H 5.03%, N 11.66%, O 26.63%, S 26.69%. An amino acid classified as nonessential with respect to its growth effect in rats. Formed by air oxidation in alkaline water solns of cysteine. Found in urine, especially after ingestion of cysteine; present in large quantities in cystine calculi; constituent of hair. Usually obtained from horse hair which contains about 8%. Several syntheses, e.g., the old Fischer synthesis (1908) starting with serine methyl ester hydrochloride and starting with benzylthiomethyl chloride and sodiophthalimidomalonic ester: Wood, duVigneaud, *J. Biol. Chem.* **131**, 267 (1939). Prepn by hydrolysis of keratin (hair): Gortner, Hoffman, *Org. Syn.* **5**, 39 (1925). Sepn from tyrosine: Vassel, U.S. pat. **2,929,840** (1960 to Intl. Minerals & Chem.). Synthesis: Gasparini *et al.*, *Bull. Soc. Chim. France* **1965**, 794.

$$
\begin{array}{cc}
\text{NH}_2 & \text{NH}_2 \\
| & | \\
\text{HOOCCHCH}_2 - \text{S} - \text{S} - \text{CH}_2\text{CHCOOH} &
\end{array}
$$

L-Form, *Gelucystine*. Naturally occurring levorotatory form. Hexagonal tablets from water, dec 260-261° (sealed tube). $[\alpha]_D^{20}$ −223.4° (1.0N HCl). pK_1 1; pK_2 2.1; pK_3 8.02; pK_4 8.71 at 35°. Soly in water (g/l) at 25°: 0.112; at 50°: 0.239; at 75°: 0.523; at 100°: 1.142. Quite sol in aq solns below pH 2 or above pH 8. Soly curves: Sano, *Biochem. Z.* **168**, 14 (1926). Insol in alc. Absorption spectrum: Marchlewski, Nowotonowna, *Bull. Soc. Chim. France* [4] **39**, 163, 166 (1926).

D-Form, crystals. $[\alpha]_D^{20}$ +223° (1.0N HCl). Soly in water at 25°: 0.057 g/l.

DL-Form, crystals. Soly in water at 25°: 0.057 g/l. *meso*-Cystine, crystals. Soly in water: 0.056 g/l.

THERAP CAT: Adjuvant in treatment of alopecia, seborrhea, brittleness of nails.

2789. Cystine S-Dioxides. $C_6H_{12}N_2O_6S_2$; mol wt 272.30. C 26.46%, H 4.44%, N 10.29%, O 35.25%, S 23.55%. So-called *cystine disulfoxide* as prepd by Lavine, Toennies, *J. Biol. Chem.* **113**, 571 (1936) and Emilozzi, Pichat, *Bull. Soc. Chim. France* **1959**, 1887 has been shown to be a mixture of two isomers, *cystine S,S-dioxide* and *cystine S,S'-dioxide.* The isomers can be separated by adding 0.25N HCl, filtering of the undissolved S,S-dioxide, and pptg the dissolved S,S'-dioxide with 8N NH₄OH. The pH should never exceed 6, since both isomers start to dec at pH 6-7. *Ref:* Utzinger, *Experientia* **17**, 374 (1961).

I   II

The S,S-dioxide (structure I), *2-amino-2-carboxyethyl 2-amino-2-carboxyethanethiosulfonate;* L-*cystine thiosulfonate.* Main product of oxidation of cystine hydrochloride with a mixture of hydrogen peroxide and formic acid (Utzinger, *loc. cit.*). Dec on heating, becoming partly viscous at about 205°. Practically insol in water.

The S,S'-dioxide (structure II), *sym*-L-*cystine disulfoxide, 3,3'-disulfinyldialanine.* Main product of oxidation of cystine perchlorate in acetonitrile with perbenzoic acid in chloroform (Utzinger, *loc. cit.*). Dec on heating, remaining dry and darkening slowly up to 240°. Slightly more sol in water than the S,S-dioxide.

2790. Cytarabine. *4-Amino-1-β-*D-*arabinofuranosyl-2(1H)-pyrimidinone; 1-β-*D-*arabinofuranosylcytosine; β-*cytosine arabinoside; CHX-3311; U-19920; Alexan; Arabitin; Aracytidine; Aracytine; Ara-C; Cytosar; Erpalfa; Iretin; Udicil.* $C_9H_{13}N_3O_5$; mol wt 243.22. C 44.44%, H 5.39%, N 17.28%, O 32.89%. Prepn: Walwick *et al., Proc. Chem. Soc.* **1959**, 84; Hunter, U.S. pat. **3,116,282** (1963 to Upjohn); Shen *et al., J. Org. Chem.* **30**, 835 (1965); Fromageot, Reese, *Tetrahedron Letters* **1966**, 3499; Roberts, Dekker, *J. Org. Chem.* **32**, 816 (1967); prebiosynthetic route: *Chem. & Eng. News* **47**, 30 (Oct. 6, 1969). Conversion of cytidylic acid into aracytidine-3'-phosphate: Nagyvary, Tapiero, *Tetrahedron Letters* **1969**, 3481. Efficient synthesis of cytarabine: Hessler, *J. Org. Chem.* **41**, 1828 (1976). Mechanism of action studies: Eridani, *Haematologica* **57**, 341 (1972); Gray, Renis, *J. Theor. Biol.* **39**, 623 (1973). Clinical pharmacology and toxicology: R. C. Donehower *et al., Cancer Treat. Rep.* **70**, 1059 (1986). Symposium on clinical pharmacology, pharmacokinetics and efficacy in leukemia: *Scand. J. Haematol.* **36**, Suppl. 44, 1-74 (1986).

Prisms from 50% ethanol, mp 212-213°. $[\alpha]_D^{23}$ +158°; $[\alpha]_D^{24}$ +153° (c = 0.5 in water). uv max at pH 2: 281.0, 212.5 nm (ε 13171, 10230); at pH 12: 272.5 nm (ε 9259).

THERAP CAT: Antineoplastic. Antiviral.

2791. Cythioate. *Phosphorothioic acid O-[4-(aminosulfonyl)phenyl] O,O-dimethyl ester; phosphorothioic acid O,O-dimethyl ester O-ester with p-hydroxybenzenesulfonamide; O,O-dimethyl O-p-sulfamoylphenylphosphorothioate;* CL 26691; ENT 25640; Proban. $C_8H_{12}NO_5PS_2$; mol wt 297.28. C 32.32%, H 4.07%, N 4.71%, O 26.91%, P 10.42%, S 21.57%. Prepn: G. Berkelhammer, U.S. pat. **3,005,004** and R. I. Hewitt, G. Berkelhammer, U.S. pat. **3,179,560** (1961, 1965 to Am. Cyanamid). Pharmacodynamics: H. G. Smith, R. L. Goulding, *J. Econ. Entomol.* **63**, 1640 (1970). Efficacy as ectoparasiticide: C. P. Doval, I. Gupta, *Indian Vet. J.* **55**, 890 (1978); P. M. Bowen, N. J. Caldwell, *Vet. Med. Small Anim. Clinic.* **77**, 79 (1982). Toxicity data: E. E. Kenaga, W. E. Allison, *Bull. Entomol. Soc. Am.* **15**, 85 (1969).

Crystals, mp 70-71°. n_D^{25} 1.5346. LD₅₀ orally in rats: 160 mg/kg (Kenaga, Allison).

USE: Insecticide.

THERAP CAT (VET): Ectoparasiticide.

2792. Cytidine. *4-Amino-1-β-*D-*ribofuranosyl-2-(1H)-pyrimidinone; cytosine riboside; 1-β-*D-*ribofuranosylcytosine.* $C_9H_{13}N_3O_5$; mol wt 243.22. C 44.44%, H 5.39%, N 17.28%, O 32.89%. Constituent of nucleic acids. Isoln from yeast nucleic acid: Levene, Jacobs, *Ber.* **43**, 3154 (1910); Levene, La Forge, *ibid.* **45**, 608 (1912). Sepn from other nucleosides by ion-exchange chromatography: Cohn in Chargaff-Davidson, *The Nucleic Acids* vol. I (New York, 1955) p 211. Synthesis: Howard *et al., J. Chem. Soc.* **1947**, 1052. Crystal structure: Furberg *et al., Acta Cryst.* **18**, 313 (1965). *Review: Basic Principles in Nucleic Acid Chemistry* vol. 1, P. O. P. Ts'o, Ed. (Academic Press, New York, 1974) *passim. See also* Nucleic Acids.

Long needles from 90% ethanol, dec 220-230°. $[\alpha]_D^{25}$ +31° (c = 0.7 in water). Freely sol in water, less sol in alcohol. pK (amino, cationic) 4.22; pK (sugar, anionic) 12.5. uv max (pH 8.2): 271 nm (ε 9100); (pH 2.2): 280 nm (ε 13,400), Voet *et al., Biopolymers* **1**, 193 (1963).

Sulfate, $(C_9H_{13}N_3O_5)_2 \cdot H_2SO_4$, long prismatic needles, mp 224-225° (dec with effervescence). $[\alpha]_{589}^{25}$ +34°; $[\alpha]_{546}^{25}$ +43°.

Picrate, $C_9H_{13}N_3O_5 \cdot C_6H_3N_3O_7$, rosettes of microcrystals from hot alc, mp 187°.

2793. 2'-Cytidylic Acid. *Cytidine-2'-monophosphate; cytidylic acid a; 2'-cytidinephosphoric acid; cytidine-2'-phosphate; 2'-cytosylic acid; 2'-CMP.* $C_9H_{14}N_3O_8P$; mol wt 323.19. C 33.44%, H 4.37%, N 13.00%, O 39.61%, P 9.58%. Ribonuclease inhibitor. Prepn from yeast ribonucleic acid: Cohn, Carter, *J. Am. Chem. Soc.* **72**, 2606 (1950); Loring *et al., J. Biol. Chem.* **196**, 807 (1952); by phosphorylation of $N^6,O^{3'},O^{5'}$-tribenzoylcytidine: Rammler, Khorana, *J. Am. Chem. Soc.* **84**, 3112 (1962). Crystal structure and conformation of the trihydrate: Kartha *et al., Science* **179**, 495 (1973). *Reviews: See* Cytidine; Nucleic Acids.

Crystals, dec 238-240°. $[\alpha]_D^{20}$ +20.7° (water); $[\alpha]_D^{25}$ —3° (aq NaOH adjusted to pH 10). uv max at pH 2: 278 nm (ϵ 12.7); at pH 12: 272 nm (ϵ 8.6). $pKa_1 = 0.8$; $pKa_2 = 4.36$; $pKa_3 = 6.17$. Sparingly sol in water. Less sol than 3'-cytidylic acid.

2794. 3'-Cytidylic Acid. Cytidine-3'-monophosphate; cytidylic acid b; 3'-cytidinephosphoric acid; cytidine-3'-phosphate; 3'-cytosylic acid; 3'-CMP. $C_9H_{14}N_3O_8P$; mol wt 323.19. C 33.44%, H 4.37%, N 13.00%, O 39.61%, P 9.58%. Ribonuclease inhibitor. Prepn from yeast ribonucleic acid: *see* refs under 2'-Cytidylic Acid. Prepn by pancreatic-ribonuclease-catalyzed ring opening of dicyclohexylguanidinium cytidine 2',3'-cyclic phosphate: Lohrmann, Khorana, *J. Am. Chem. Soc.* **86**, 4188 (1964). X-ray analysis studies of orthorhombic form: Alver, Furberg, *Acta Chem. Scand.* **13**, 910 (1959); Sundaralingam, Jensen, *J. Mol. Biol.* **13**, 914 (1965); of monoclinic form: Bugg, Marsh, *ibid.* **25**, 67 (1967). *Reviews: See* Cytidine; Nucleic Acids.

Crystals, dec 232-234°. $[\alpha]_D^{20}$ +49.4° (c = 1 in water). $[\alpha]_D^{25}$ +50.0° (c = 1 in aq NaOH adjusted to pH 10). uv max at pH 2: 279 nm (ϵ 13.0); at pH 12: 272 nm (ϵ 8.90). $pKa_1 = 0.8$; $pKa_2 = 4.28$; $pKa_3 = 6.0$. Moderately sol in water; more sol than 2'-cytidylic acid.

2795. Cytisine. *1,2,3,4,5,6-Hexahydro-1,5-methano-8H-pyrido-[1,2-a][1,5]diazocin-8-one;* baptitoxine; sophorine; ulexine; Cytiton. $C_{11}H_{14}N_2O$; mol wt 190.24. C 69.44%, H 7.42%, N 14.73%, O 8.41%. Toxic principle in seed of *Laburnum anagyroides* Medik. and other *Leguminosae*. Extraction: Ing, *J. Chem. Soc.* **1931**, 2200; Späth, Galinovsky, *Ber.* **65**, 1526 (1932); **66**, 1338 (1933); Lecoq, *Bull. Soc. Chim. France* **10**, 153 (1943). Structure: Ing, *J. Chem. Soc.* **1932**, 2778. Synthesis: Bohlmann et al., *Angew. Chem.* **67**, 708 (1955); Van Tamelen, Baran, *J. Am. Chem. Soc.* **77**, 4944 (1955). Abs config: Okuda et al., *Chem. & Ind. (London)* **1961**, 1751. Pharmacological properties: R. B. Barlow, L. J. McLeod, *Brit. J. Pharmacol.* **35**, 161 (1969).

Orthorhombic prisms from acetone, mp 152-153°. Sublimes. bp_2 218°. $[\alpha]_D^{17}$ —120°. pK_1 6.11; pK_2 13.08. Sol in 1.3 parts water, 13 parts acetone, 1.3 parts methanol, 3.5 parts alcohol, 30 parts benzene, 10 parts ethyl acetate, 2.0 parts chloroform. Practically insol in petr ether. LD_{50} in mice (mg/kg): 1.73 i.v.; 9.4 i.p.; 101 orally, R. B. Barlow, L. J. McLeod, *loc. cit.*

Hydrochloride, $C_{11}H_{14}N_2O \cdot HCl$, deliquescent crystals, sol in water and alcohol. pH of 0.1 molar aq soln 4.3.

Nitrate monohydrate, $C_{11}H_{14}N_2O \cdot HNO_3 \cdot H_2O$, needles or leaflets, $[\alpha]_D$ —81.5°. Sol in water, slightly sol in alcohol, practically insol in ether.

2796. Cytochalasins. A class of mold metabolites exhibiting a number of unusual and varied effects on animal cells. More than twenty cytochalasins are known, isolated from several different mold spp. All are characterized by a highly substituted hydrogenated isoindole ring of known configuration to which is fused a macrocyclic ring. This ring may vary from 11 to 14 atoms in size and may be either a carbocycle or a lactone. Isoln and structure of cytochalasins A, B, C, D: Aldridge et al., *J. Chem. Soc. (C)* **1967**, 1667; eidem, *Chem. Commun.* **1967**, 26; of E and F: Aldridge et al., *ibid.* **1972**, 148. Revised structures of cytochalasins E and F: Aldridge et al., *ibid.* **1973**, 551; Büchi et al., *J. Am. Chem. Soc.* **95**, 5423 (1973). Isoln of H, also known as *paspaline-P* or *kodocytochalasin-1:* G. S. Pendse, *Experientia* **30**, 107 (1974). Structure of H: S. A Patwardhan et al., *Phytochemistry* **13**, 1985 (1974); M. A. Beno et al., *J. Am. Chem. Soc.* **99**, 4123 (1977); x-ray crystal and molecular structure: J. A. McMillan et al., *Chem. Commun.* **1977**, 105. Isoln of K, L, M from *Chalara microspora* and proposed structures: T. Fex, *Tetrahedron Letters* **22**, 2703 (1981). Isoln of E and K from *Aspergillus clavatus* and proposed alternate structure of K: P. S. Steyn et al., *J. Chem. Soc. Perkins Trans.* **I 1982**, 541. Partial synthesis of A and B: S. Masamune et al., *J. Am. Chem. Soc.* **99**, 6756 (1977). Total synthesis of B: G. Stork et al., *ibid.* **100**, 7775 (1978). Major biological effects are the blockage of cytoplasmic cleavage by blocking formation of contractile microfilament structues, resulting in multinucleate cell formation, the reversible inhibition of cell movement, and the induction of nuclear extrusion: Carter, *Nature* **213**, 261 (1967); Krishan, *J. Cell. Biol.* **54**, 657 (1972); E. D. Korn, *Physiol. Rev.* **62**, 703 (1982). Correlation between effects of cytochalasins on cellular structures and cellular events and those on actin *in vitro:* I. Yahara et al., *J. Cell Biol.* **92**, 69 (1982). Other reported effects include the inhibition of glucose transport, of thyroid secretion, of growth hormone release, of phagocytosis, and of platelet aggregation and clot contraction. *See* D. A. Hume et al., *Nature* **272**, 359 (1978). Nomenclature: M. Binder et al., *J. Chem. Soc. Perkin Trans.* **I 1973**, 1146. *Reviews:* M. Binder, C.Tamm, *Angew. Chem. Int. Ed.* **12**, 370 (1973); R. B. Herbert in *The Alkaloids*, vol. 7, J. E. Saxton, Ed. (The Chemical Society, London, 1977) pp 29-30; W. G. Thilly et al., *Front. Biol.* **46**, 53-64 (1978); L. V. Domnina et al., *Proc. Nat. Acad. Sci. USA* **79**, 7754-7757 (1982); W. Siess et al., *ibid.* 7709-7713.

cytochalasin B

Cytochalasin B. $C_{29}H_{37}NO_5$. *7,20-Dihydroxy-16-methyl-10-phenyl-24-oxo[14]cytochalasa-6(12),13,21-triene-1,23-dione; (E,E)-16-benzyl-6,7,8,9,10,12a,13,14,15,15a,16,17-dodecahydro-5,13-dihydroxy-9,15-dimethyl-14-methylene-2H-oxacyclotetradec[2,3-d]isoindole-2,18(5H)-dione; phomin.* The most important and biologically studied of the cytochalasins. Formerly isolated from cultures of a *Phoma* sp. and called phomin: Rothweiler, Tamm, *Experientia* **22**,

750 (1966). Felted needles from acetone, mp 218-221°. Completely stable under normal conditions. Solutions in DMSO show no decrease in potency when stored at 4° for three years (Aldrich data sheet). Solubility (mg/ml at 24°): acetone 10.3; ethanol 35.4; DMSO 371; DMF 492. Insol in water.

USE: As tools in cytological research and in characterization of polymerization properties of actin, q.v.

2797. Cytochrome c. Myohematin; hematin-protein; Cromoci; Cytorest; Landrax. Hemoprotein in which the catalytically active prosthetic group is a derivative of iron protoporphyrin IX. Occurs in the cells of all aerobic organisms. Found in animal cells in the mitochondrial protein-lipid complex. Can be readily converted to corresponding hemochromogen by replacement of the protein with an organic base, usually pyridine. Because of their ready fluctuation within the cell between the ferrous and ferric states, they are all efficient biological electron-transporters and play a vital role in cellular oxidations in both plants and animals. Generally regarded as universal catalysts of respiration, forming the essential electron-bridge between the respirable substrates and oxygen: Slater, Nord, *Advan. Enzymol. Relat. Subj. Biochem.* **20**, 147 (1958); James, Leech, *Endeavour* **19**, 108 (1960). Isoln of cytochrome c from bakers' yeast: Keilin, *Proc. Roy. Soc. (London) Ser. B* **106**, 418 (1930); Hagikara et al., *Nature* **178**, 629 (1956); from ox heart: Kirby et al., *Acta Chem. Scand.* **10**, 148 (1956); from beef heart: Paleus, *ibid.* **14**, 1743 (1960); Morrison et al., *Biochim. Biophys. Acta* **41**, 334 (1960); from horse heart: Nozaki et al., *J. Biochem. (Tokyo)* **44**, 453 (1957); from fish heart: Hagikara et al., *Nature* **179**, 249 (1957); from wheat germ: Hagikara et al., *ibid.* **181**, 1950 (1958); from human heart: Matsuba, Vasunobu, *J. Biol. Chem.* **236**, 1701 (1961). Isoln of cytochrome c_1 from submitochondrial particles: Green et al., *Biochim. Biophys. Acta* **31**, 34 (1955); from beef heart: Sekugu, *J. Biochem. (Tokyo)* **48**, 214 (1960). Prepn of cytochrome c_2 from *Rhodospirillum rubrum* and cytochrome c_3 from *Desulfovibrio desulfuricans:* Hario, Kamen, *Biochim. Biophys. Acta* **48**, 266 (1961). Crystn of cytochrome c: Butt, Keilin, *Proc. Roy. Soc. (London) Ser. B*, **156**, 429 (1962). Improved prepn from mammalian sources: Margoliash, Walasek, U.S. pat. **3,342,796** (1967 to Abbott). Structure consists of a single polypeptide chain of 104 or more amino acid residues with the heme group attached through cysteine residues at positions 14 and 17. Sequence varies among species although a band of 11 residues in positions 70 through 80 are identical in all plants and animals studied. First elucidation of amino acid sequence was that of horse heart cytochrome c: Margoliash et al., *Nature* **192**, 1125 (1961); *J. Biol. Chem.* **237**, 2161 (1962); of human heart cytochrome c: Matsubara, Smith, *ibid.* **237**, 3575 (1962). Sequences are known for a great number of mammalian species, see review by Margoliash, Schejter, *Advan. Protein Chem.* **21**, 113 (1966). Homologies of bacterial and mammalian cytochrome c: Dickerson, *J. Mol. Biol.* **57**, 1 (1971). Synthetic studies: Moroder et al., *Biopolymers* **12**, 493-534, 693-750 (1973). *Reviews:* Lemberg, Legge, *Hematin Compounds and Bile Pigments* (New York, 1949); Morton, *Rev. Pure Appl. Chem.* **8**, 161 (1958); Morrison, *Ann. Rev. Biochem.* **30**, 11 (1960); Margoliash, *ibid.*, 549; Margoliash, Schejter, *loc. cit.;* several authors in *Structure and Function of Cytochromes,* Okunuki et al., Eds. (University Park Press, Baltimore, 1968); Harbury, Marks, in *Inorganic Biochemistry* vol. 2, G. L. Eichhorn, Ed. (Elsevier, New York, 1973) pp 904-954.

Cytochrome c. Reduced form crystallizes as separate needles; oxidized form as rosettes. Mol wt about 13,000. Extinction ratio: $E_{550\,red}/E_{280\,ox} = 1.15$-1.25. $E_{1cm}^{1\%}$ 550 nm: 22.6-23.1. $E_0' = E_0'$ (quinone) -0.06 log [Ferricyt/Ferrocyt] $= +0.251$ v. Absorption max of reduced form, *Ferrocytochrome c:* 416, 520, 550 nm; oxidized form, *Ferricytochrome c:* 280, 361, 410, 529 nm. Contains one heme per mole weight of 12,000.

Cytochrome c_1. Identity with *cytochrome e:* Kellin, Hartree, *Nature* **176**, 200 (1955). Extinction ratio: $E_{418\,red}/E_{411\,ox} = 1.27$. Extinction coefficient, $E_{553\,nm\,red} = 15.3 \times 10^3$ M^{-1} cm^{-1}. $E_0' = E_0'$ (quinone) -0.06 log [Ferricyt/Ferrocyt] $= \pm 0.223$ v. Acidic. Non-autoxidizable. Isoelectric pt: pH 3.6. Absorption max of reduced form: 418, 523, 553 nm; oxidized form: 278, 411, 523 nm. Reduced by

sodium dithionate, potassium borohydride, ascorbic acid, sodium thioglycolate, cysteine. Stable at 0°. Denatured and coagulated by heating at 50° for 5 min.

Cytochrome c_2. Needles changing to squares. Mol wt about 13,000. Extinction ratio: $E_{550\,red}/E_{275\,ox} = 1.14$. Absorption max of reduced form: 272, 316, 415, 521, 550 nm; oxidized form: 275, 357, 410, 520-530 nm.

Cytochrome c_3. Needles. Mol wt 11,300. Subject to rapid autoxidation. Total extinction: $A_{552\,nm}^{red}$34.2, $A_{570\,nm}^{red}$1.90, $A_{280\,nm}^{ox}$11.62. Absorption max of reduced form: 418, 522, 552 nm; oxidized form: 349, 409 nm. Contains 2 hemes per mole weight of 12,000.

2798. Cytochromes P_{450}. P_{450}. Hemoproteins involved in enzyme hydroxylation, demethylation, and N-oxidation; present in a number of mammalian tissues but in highest concentrations in adrenal and liver microsomes. Has also been found in insects, yeast, plants, and bacteria. Existence of mitochondrial P_{450} first reported by Harding et al., *Biochim. Biophys. Acta* **92**, 415 (1964). Extraction and purification: Mitani, Horie, *J. Biochem.* **65**, 269 (1969). Acts as a monooxygenase in catalyzing metabolic functions; influences the rate of drug metabolism. Name derived from the characteristic absorption at 450 nm. *Reviews:* several authors in *Oxidases and Redox Systems* vol. 2, King et al., Eds. (University Park Press, Baltimore, 1971); J. T. Groves, *Advances in Inorganic Biochemistry*, vol. 1, G. L. Eichhorn, L. G. Marzilla, Eds. (Elsevier/North-Holland, New York, 1979) pp 119-145; several authors in *Xenobiotica* **12**, 671-800 (1982). Books: *Cytochromes P-450 & B5—Structure, Function & Interaction,* D. Y. Cooper et al., Eds. (Plenum, New York, 1975) 554 pp; *Cytochrome P-450,* R. Sato, T. Omura, Eds. (Academic Press, New York, 1978) 233 pp; *Biochemistry, Biophysics and Regulation of Cytochrome P-450,* J. A. Gustafsson et al., Eds. (Elsevier/North-Holland, New York, 1980) 626 pp.

2799. Cytohemin. *Chloro[7-ethenyl-17-formyl-12-(1-hydroxy-5,9,13-trimethyl-4,8,12-tetradecatrienyl)-3,8,13-trimethyl-21H,23H-porphine-2,18-dipropanoato(4—)-N²¹,N²²,-N²³,N²⁴]ferrate(2—) dihydrogen; chloro[dihydrogen-3-formyl-8-(1-hydroxy-5,9,13-trimethyltetradecyl)-7,12,17-trimethyl-13-vinyl-2,18-porphinedipropionato(2—)]iron.* $C_{49}H_{62}$-$ClFeN_4O_6$; mol wt 894.37. C 65.81%, H 6.99%, Cl 3.96%, Fe 6.24%, N 6.26%, O 10.73%. Isoln from heart muscle: Warburg, Gewitz, *Z. Physiol. Chem.* **288**, 1 (1951); Lemberg, *Biochem. Z.* **338**, 97 (1963). Structure: Grassl et al., *ibid.* **337**, 35 (1963); **338**, 771 (1963).

Amorphous powder.

2800. Cytolipin H. *1-O-(4-O-β-D-Galactopyranosyl-β-D-glucopyranosyl)ceramide;* lactosylceramide; ceramide-β-lactoside; cerebronylsphingosylglucosidogalactoside; Gal β1→4Glcβ1→1cer. A glycosphingolipid containing lactose and a ceramide (N-acyl fatty acid deriv of a sphingosine). Fatty acid composition is variable but is primarily palmitic, behenic and lignoceric acids. First isolated from human epidermoid carcinoma cells, later found to be one of the major neutrophil glycolipids serving as a neutrophil-differentiation marker. Isoln: M. M. Rapport et al., *Cancer* **12**, 438 (1959). Simplified prepn from ox spleen: M. M. Rapport *et al., J. Biol. Chem.* **237**, 1056 (1962). Structure elucidation: A. C. Schram *et al., Nature* **197**, 1074 (1963). Synthesis: D. Shapiro, E. S. Rachaman, *ibid.* **201**, 878 (1964); K. C. Nicolaou *et al., J. Am. Chem. Soc.* **110**, 7910 (1988). Serological activity: M. M. Rapport, L. Graf, *Nature* **201**, 879 (1964); and effect of fatty acid chain length: L. Graf, M. M. Rapport, *Chem. Phys. Lipids* **13**, 367 (1974). Characterization and distribution of glycolipids in leukemic and nonleukemic blood cells: J. Hildebrand *et al., J. Lipid Res.* **12**, 361 (1971); in neutrophils: B. A. Macher, J. C. Klock, *J. Biol. Chem.* **255**, 2092 (1980). Role as neutrophil cell surface marker: F. W. Symington *et al., J. Immunol.* **134**, 2498 (1985). Intracellular localization: *eidem, J. Biol. Chem.* **262**, 11356 (1987). TLC determn: K. Ogawa *et al., J. Chromatog.* **426**, 188 (1988). Clinical implications in inflammatory bowel disease: C. R. Stevens *et al., Gut* **29**, 580 (1988). Review of structure, organization and role of glycolipids in the cell surface membrane: S. Hakomori, *Ann. Rev. Biochem.* **50**, 733-764 (1981).

Coral-like clusters from pyridine + acetone, mp 230-240° with sintering at 180-190°. $[\alpha]_D^{24}$ −10.8° (c = 2 in pyridine). Practically insol in water, cold methanol, acetone, cold acetic acid, ether, acetonitrile. Soluble in pyridine, chloroform, hot methanol, hot acetic acid. Dissolves readily in chloroform + methanol mixtures of high chloroform content.

2801. Cytosine. *4-Amino-2(1H)-pyrimidinone;* 4-amino-2-oxo-1,2-dihydropyrimidine; 4-amino-2-pyrimidinol; 4-amino-2-hydroxypyrimidine. $C_4H_5N_3O$; mol wt 111.10. C 43.24%, H 4.54%, N 37.82%, O 14.40%. Widely distributed in nature; constituent of nucleic acids. Isoln by hydroly-

sis of thymus nucleic acids: Levene, Bass, *Nucleic Acids* (New York, 1931) p 57. Syntheses: Hilbert, Johnson, *J. Am. Chem. Soc.* **52**, 1152 (1930); Hilbert *et al., ibid.* **57**, 552 (1935); *cf.* Hunter, Hlynka, *Biochem. J.* **31**, 486 (1937); Ballweg, *Tetrahedron Letters* **1968**, 2171; David, Lubineau, *Bull. Soc. Chim. France* **1969**, 816. Synthesis from carbon dioxide and ammonia on kaolinite: G. R. Harvey *et al., Naturwiss.* **58**, 624 (1971). Exists in keto form. Structure: Barker, Marsh, *Acta Crystallogr.* **17**, 1581 (1964); of monohydrate: Jeffrey, Kinoshita, *ibid.* **16**, 20 (1963); R. J. McClure, B. M. Craven, *ibid.* **B29**, 1234 (1973). Tautomerism: Y. P. Wong, *J. Am. Chem. Soc.* **95**, 3511 (1973). *Review:* Ts'o, "Bases, Nucleosides and Nucleotides" in *Basic Principles of Nucleic Acid Chemistry* **vol. 1**, P. O. P. Ts'o, Ed. (Academic Press, New York, 1974) pp 453-584. *See also* Nucleic Acids.

Monohydrate, lustrous monoclinic or triclinic platelets from water. Anhydr at 100°; brown around 300°; dec 320-325°. uv max (pH 8.8): 196.5, 267 nm ($\epsilon \times 10^{-3}$ 22.5, 6.1). pK_1 = 4.60; pK_2 = 12.16. One gram dissolves in 130 ml water. Slightly sol in alcohol, insol in ether. Gives red color when dissolved in soln of sodium hypochlorite to which a drop of NH_4OH is added. Forms salts with acids.

D

2802. 2,4-D. *(2,4-Dichlorophenoxy)acetic acid;* Hedonal; Trinoxol. $C_8H_6Cl_2O_3$; mol wt 221.04. C 43.47%, H 2.74%, Cl 32.08%, O 21.71%. Prepd from 2,4-dichlorophenol and monochloroacetic acid in aq NaOH: Pokorny, *J. Am. Chem. Soc.* **63**, 1768 (1941); Foster, **Brit.** pat. **573,476** (1945); by chlorination of molten phenoxyacetic acid: Manske, **U.S.** pat. **2,471,575** (1949 to U.S. Rubber); from 2,4-dichlorophenol, sodium, and ethyl chloroacetate followed by hydrolysis of the ester: Haskelberg, *J. Org. Chem.* **12**, 426 (1947). Toxicology: J. M. Way, *Residue Rev.* **26**, 37 (1969).

Crystals from benzene, mp 138°. $bp_{0.4}$ 160°. Almost insol in water; sol in organic solvents. LD_{50} orally in mice, rats: 368, 375 mg/kg, V. K. Rowe, T. A. Hymas, *Am. J. Vet. Res.* **15**, 622 (1954).

Sodium salt, $C_8H_5Cl_2NaO_3$, needles from alc, dec 215°. Soly in water: 3.5%.

Isopropyl ester, $C_{11}H_{12}Cl_2O_3$, *Esteron 44.* Liquid, d_{25}^{25} 1.255-1.270. Solidifies at +5°. Sol in oil.

Butyl ester, $C_{12}H_{14}Cl_2O_3$, *Lironox, Weedone Aero Concentrate.* Prepn: Nagel, **Ger.** pat. **1,144,288** (1963 to Bayer), *C.A.* **59**, 6315e (1963).

Amine salts. Sol in water. The following amines form useful salts: Dimethylamine, isopropylamine, triethylamine, diethanolamine, triethanolamine. These salts are much more sol than the sodium salt. A small amount of sequestering agent, such as ethylenediaminetetraacetic acid, is generally added to prevent complex formation in hard water.

Note: The name Hedonal was formerly a synonym for carbamic acid 1-methylbutyl ester.

USE: Herbicide. To increase latex output of old rubber trees. *Caution:* Causes irritation of eyes, G.I. disturbances.

2803. Dacarbazine. *5-(3,3-Dimethyl-1-triazenyl)-1H-imidazole-4-carboxamide;* 5(or 4)-(dimethyltriazeno)imidazole-4(or 5)-carboxamide; DIC; DTIC; NSC-45388; Dacatic; DTIC-Dome; Deticene. $C_6H_{10}N_6O$; mol wt 182.18. C 39.55%, H 5.53%, N 46.13%, O 8.78%. First synthesized in 1959 at the Southern Research Institute. Prepn: Shealy *et al., J. Org. Chem.* **27**, 2150 (1962); Hano *et al., Gann* **59**, 207 (1968), *C.A.* **69**, 42527g (1968). Antitumor activity: Shealy *et al., Biochem. Pharmacol.* **11**, 674 (1962); Hano *et al., Gann* **56**, 417 (1965), *C.A.* **63**, 18856g (1965). Mechanism of action studies: Saunders, Schultz, *Biochem. Pharmacol.* **19**, 911 (1970). Metabolism: Skibba *et al., ibid.* 2043; Mizuno, Humphrey, *Cancer Chemother. Rep., Part 1* **56**, 465 (1972). Review: Carter, Friedman, *Europ. J. Cancer* **8**, 85-92 (1972), *see also* the series of articles on history, activity, mechanism of action and clinical studies: *Cancer Treatment Rep.* **60**, 123-214 (1976).

Ivory microcryst substance; explosive decomp 250-255°. Also reported as mp 205°, Hano *et al., loc. cit.* (1965). uv max (0.1N HCl): 223 nm (7500); (pH 7): 237 nm (11,200); both solns protected from light. Stable in neutral soln in absence of light.

THERAP CAT: Antineoplastic.

2804. Dactinomycin. *Actinomycin D;* actinomycin-[thr-val-pro-sar-meval]; meractinomycin; actinomycin A_{IV}; actinomycin IV; actinomycin C_1; actinomycin I_1; actinomycin X_1; NSC 3053; Cosmegen. $C_{62}H_{86}N_{12}O_{16}$; mol wt 1255.47. C 59.31%, H 6.90%, N 13.39%, O 20.39%. Antibiotic sub-

stance belonging to the actinomycin complex, produced by several *Streptomyces* spp. Historical background of actinomycins and chemistry, toxicology, pharmacology of actinomycin D: *Ann. N.Y. Acad. Sci.* **89**, 285-485 (1960). Isoln from broth cultures of *S. parvulus:* Manaker *et al., Antibiot. Ann.* 1954/55, 853. Structure: Bullock, Johnson, *J. Chem. Soc.* **1957**, 3280. Synthesis: Brockmann, Manegold, *Naturwiss.* **51**, 383 (1964); Brockmann, Lackner, *ibid.* 384, 435 (1964); Brockmann *et al., Ger.* pat. **1,172,680** (1964 to Bayer); *eidem, Ber.* **101**, 1312 (1968); Meienhofer, *J. Am. Chem. Soc.* **92**, 3771 (1970); T. Tanaka *et al., Bull. Chem. Soc. Japan* **53**, 1352 (1980); K. Nakajima *et al., Pept. Chem.* **19**, 143-148 (1981). Conformation from NMR: Conti, DeSantis, *Nature* **227**, 1239 (1970); Lackner, *Ber.* **104**, 3653 (1971). Toxicity data: F. S. Philips *et al., Ann. N.Y. Acad. Sci.* **89**, 348 (1960). Review and evaluation of studies of carcinogenic action in laboratory animals: *IARC Monographs* **12**, 29-41 (1976).

Trihydrate, bright red, rhomboid prisms from abs alc, dec 241.5-243°. $[\alpha]_D^{28}$ −315° (c = 0.25 in methanol). Abs max (methanol): 244, 441 nm ($A_{1cm}^{1\%}$ 281, 206). Abs solns are very sensitive to light. Sol in alc, propylene glycol, water + glycol mixtures. LD_{50} orally in mice, rats: 13.0, 7.2 mg/kg (Philips).

THERAP CAT: Antineoplastic.

2805. Daidzein. *7-Hydroxy-3-(4-hydroxyphenyl)-4H-1-benzopyran-4-one;* 4′,7-dihydroxyisoflavone. $C_{15}H_{10}O_4$; mol wt 254.23. C 70.86%, H 3.96%, O 25.17%. The aglucon of daidzin, prepd by hydrolysis with HCl in methanol: Walz, *Ann.* **489**, 118 (1931). Isoln from red clover: Wong, *J. Sci. Food Agr.* **13**, 304 (1962); from the mold *Micromonospora halophytica:* Ganguly, Sarre, *Chem. & Ind. (London)* **1970**, 201. Synthesis: Baker *et al., J. Chem. Soc.* **1933**, 274; Wessely *et al., Ber.* **66**, 685 (1933); Mahal *et al., J. Chem. Soc.* **1934**, 1769; Baker *et al., ibid.* **1953**, 1852; Farkas, *Ber.* **90**, 2940 (1957). Biosynthetic study: Grisebach, Zilg, *Z. Naturforsch.* **23b**, 494 (1968).

Pale-yellow prisms from dil alc, dec 315-323°. uv max: 250 nm (log ε 4.44). Sol in alcohol, ether.

7-Glucoside, $C_{21}H_{20}O_9$, *daidzin.* Isoln from soybean meal *(Soja hispida):* Walz, *loc. cit.* Synthesis: Farkas, Várady, *Ber.* **92**, 819 (1959). Monohydrate, needles from water, dec 234-236°. $[\alpha]_D^{20}$ −36.4° (0.02N NaOH). Sol in water and aq alcohol.

2806. Dalapon. *2,2-Dichloropropanoic acid;* α,α-dichloropropionic acid. $C_3H_4Cl_2O_2$; mol wt 142.97. C 25.20%, H 2.82%, Cl 49.60%, O 22.38%. Prepn: **Brit.** pat. **752,761** (1956 to Dow); Kowolik, Fisher, **Brit.** pat. **892,584** (1962 to British Celanese); of free acid and sodium salt: Smeykal, Pallutz, *Chem. Tech. (Leipzig)* **15** (11), 654 (1963), *C.A.* **60**, 6741c (1964); of sodium salt: Barrons, **U.S.** pat. **2,642,354** and Maylott, Meyer, **U.S.** pat. **3,007,964** (1953, 1961, both to Dow). Acute toxicity: T. B. Gaines, R. E. Linder, *Fundam. Appl. Toxicol.* **7**, 299 (1986). Toxicity data for the sodium salt: G. W. Bailey, J. L. White, *Residue Rev.* **10,** 97 (1965).

$$CH_3 \overset{Cl}{\underset{Cl}{\overset{|}{C}}} COOH$$

Liquid, bp$_{20}$ 98-99°. d^{20} 1.4014. n$_D^{20}$ 1.4551. LD$_{50}$ in male, female rats (mg/kg): 7126, 6936 orally (Gaines, Linder). Sodium salt, $C_3H_3Cl_2NaO_2$, *Dowpon, Radapon, Basfapon B.* Crystals, dec 174-176°. Solubility in water at 25°, 45 g/100 ml; aq solns hydrolyze above 70°. Corrosive to iron.
USE: Sodium salt as herbicide.

2807. Daltroban. *4-[2-[[(4-Chlorophenyl)sulfonyl]-amino]ethyl]benzeneacetic acid; [p-[2-(p-chlorobenzenesul-fonamido)ethyl]phenyl]acetic acid;* BM 13505; SK&F 96148. $C_{16}H_{16}ClNO_4S$; mol wt 353.82. C 54.31%, H 4.56%, Cl 10.02%, N 3.96%, O 18.09%, S 9.06%. Thromboxane A$_2$ receptor antagonist. Prepn: E. C. Witte *et al.,* **Ger.** pat. **3,000,377;** *eidem,* **U.S.** pat. **4,443,477** (1981, 1984 both to Boehringer, Mann.). Receptor binding study: A. Yanagi-sawa *et al., Eur. J. Pharmacol.* **133,** 89 (1987). Pharmacological evaluations in animals: D. J. Lefer *et al., Arch. Int. Pharmacodyn.* **287,** 89 (1987); E. F. Smith, III, J. McDonald, *Pharmacology* **36,** 340 (1988); P. Löbel *et al., Biomed. Biochem. Acta* **47,** S 86 (1988). GC determn in biological fluids: V. Uebis, *J. Chromatog.* **419,** 345 (1987).

$$Cl \longrightarrow SO_2NHCH_2CH_2 \longrightarrow CH_2COOH$$

THERAP CAT: Antithrombotic.

2808. Damar. Gum Damar; Dammar; resin Damar. Resinous exudate from a species of *Shorea, Dipterocarpaceae. Habit.* East Indies, Philippines. *Constit.* Volatile oil, resins, bitter substance.
Yellowish-white, roundish, or stalactite shaped, friable masses; semi-transparent, conchoidal fracture; varying degrees of hardness. d 1.04-1.12. mp about 120°. Insol in water; sol in alcohol, chloroform, ether, carbon disulfide, oil rosemary; partly sol in oil turpentine.
USE: In plasters, varnishes, lacquers, etc. A soln of the purified material in chloroform or xylene is used for preserving animal and vegetable specimens for microscopy.

2809. Damascenine. *3-Methoxy-2-(methylamino)benzoic acid methyl ester; 2-(methylamino)-m-anisic acid methyl ester;* methyl 2-(methylamino)-3-methoxybenzoate; methyl-damascenine; nigelline. $C_{10}H_{13}NO_3$; mol wt 195.21. C 61.52%, H 6.71%, N 7.18%, O 24.59%. Odoriferous principle of oil of Nigella from seeds of *Nigella damascena* L. and *Nigella arvensis* L., *Ranunculaceae.* Early syntheses: Ewins, *J. Chem. Soc.* **101,** 544 (1912); Kaufman, Rothlin, *Ber.* **49,** 578 (1916); Sornet, *Manuf. Chem.* **5,** 87 (1924); Keller, Schulze, *Arch. Pharm.* **263,** 481 (1925). Improved synthesis: Mutschler, *ibid.* **298,** 861 (1965); M. Thoinet *et al., Ann. Pharm. Franc.* **36,** 337 (1978).

$$\underset{OCH_3}{\overset{COOCH_3}{\bigcirc}} NHCH_3$$

Prisms from abs. alc. Nutmeg-like odor, mp 27-29°. bp 270° (slight dec), bp$_{10}$ 147°. Volatile with steam. Insol in water. Freely sol in alc, ether, chloroform, petr ether, oils.
Hydrochloride, deliquesc prisms, mp 156°. (Hydrate, mp 122°.) LD$_{50}$ in mice: 1800 mg/kg orally (Bekemeier *et al., Arch. Int. Pharmacodyn. Ther.* **168,** 199 (1967)).

2810. Daminozide. *Butanedioic acid mono(2,2-dimethylhydrazide); N-(dimethylamino)succinamic acid;* succinic acid 2,2-dimethylhydrazide; B-9; B-995; Alar; B-Nine; Kylar. $C_6H_{12}N_2O_3$; mol wt 160.17. C 44.99%, H 7.55%, N

17.49%, O 29.97%. $HOOCCH_2CH_2CONHN(CH_3)_2$. Prepn: **Belg.** pat. **613,799** *C.A.* **57,** 10281d (1962); H. A. Hageman, W. L. Hubbard, **U.S.** pat. **3,257,414** (1962, 1966 both to U.S. Rubber). Effect in controlling growth in apples: F. W. Southwick *et al., Proc. Am. Soc. Hort. Sci.* **92,** 71 (1968); D. W. Greene *et al., Fruit Var. J.* **40,** 41 (1986). *In vitro* and *in vivo* mechanism of action studies: K. Ryugo, R. M. Sachs, *J. Am. Soc. Hort. Sci.* **94,** 529 (1969). Hydrolysis to 1,1-dimethylhydrazine, *q.v.,* and determn in food products: W. H. Newsome, *J. Agr. Food Chem.* **28,** 319 (1980); M. K. Conditt, J. R. Baumgardner, *J. Assoc. Offic. Anal. Chem.* **71,** 735 (1988). Discussion of toxicological evaluation: D. Campt, *EPA Journal* **13,** 32 (1987).
Crystals, mp 154-155°. Soly 10% in water, 2.5% in acetone, 5% in methanol.
USE: Plant growth regulator.

2811. Danazol. *Pregna-2,4-dien-20-yno[2,3-d]isoxazol-17-ol; 17α-pregn-4-en-20-yno[2,3-d]isoxazol-17-ol;* 1-ethynyl-2,3,3a,3b,4,5,10,10a,10b,11,12,12a-dodecahydro-10a,-12a-dimethyl-1H-cyclopenta[7,8]phenanthro[3,2-d]isoxazol-1-ol; 17α-ethynyl-17β-hydroxy-4-androsteno[2,3-d]-isoxazole; WIN 17757; Bonzol; Chronogyn; Cyclomen; Danocrine; Danol; Danoval; Ladogal; Winobanin. $C_{22}H_{27}$-NO_2; mol wt 337.47. C 78.30%, H 8.07%, N 4.15%, O 9.48%. Anterior pituitary supressant. Anabolic steroid deriv of ethisterone, *q.v.,* with mild androgenic side effects (an impeded androgen). Prepn: **Brit.** pat. **905,844** (1962 to Sterling Drug), *C.A.* **58,** 6895c (1963); Manson *et al., J. Med. Chem.* **6,** 1 (1963); Clinton, Manson, **U.S.** pat. **3,135,743** (1964 to Sterling Drug). Activity studies: Sherins *et al., J. Clin. Endocrinol. Metab.* **32,** 522 (1971); Dmowski *et al., Fert. Steril.* **22,** 9 (1971). Clinical studies in endometriosis and other endocrine disorders: R. B. Greenblatt *et al., ibid.* 102. Series of articles on pharmacology, pharmacokinetics and clinical use: *Drugs* **19,** 321-372 (1980). Use in idiopathic thrombocytopenic purpura: Y. S. Ahn *et al., N. Engl. J. Med.* **308,** 1396 (1983); in hemophilia: H. R. Gralnick *et al., ibid.* 1393.

Crystals from acetone, mp 224.4-226.8°. [α]$_D^{25}$ +7.5° (ethanol); [α]$_D^{25}$ +21.9° (chloroform). uv max (ethanol): 286 nm (ε 11300).
THERAP CAT: Antigonadotropin.

2812. Dansyl Chloride. *5-(Dimethylamino)-1-naphthalenesulfonyl chloride.* $C_{12}H_{12}ClNO_2S$; mol wt 269.74. C 53.43%, H 4.48%, Cl 13.14%, N 5.19%, O 11.86%, S 11.89%. Reagent for fluorescent labelling of amines, amino acids and proteins. Synthesis: G. Weber, *Biochem. J.* **51,** 155 (1952); A. Mendel, *J. Chem. Eng. Data* **15,** 340 (1970). In amino acid analysis: J. Airhart *et al., Anal. Biochem.* **53,** 132 (1973). In HPLC analysis of phenols: R. M. Cassidy, D. S. LeGay, *J. Chromatog. Sci.* **12,** 85 (1974). In protein microassays: T. Kinoshita *et al., Anal. Biochem.* **66,** 104 (1975). In measurement of desquamation: R. Marks *et al., Brit. J. Dermatol.* **111,** 265 (1984).

Yellow-orange crystals from hexane, mp 66.5-68°. Sol in acetone, pyridine, benzene, dioxane. Insol in water.
USE: In fluorescent labelling, column chromatography, HPLC.

2813. Danthron. *1,8-Dihydroxy-9,10-anthracenedione;* *1,8-dihydroxyanthraquinone;* chrysazin; dantron; Altan; Antrapurol; Diaquone; Dorbane; Duolax; Istin; Istizin; Modane. $C_{14}H_8O_4$; mol wt 240.20. C 70.00%, H 3.36%, O 26.64%. Prepd from 1,8-anthraquinone potassium disulfonate: Fierz-David, Blangey, *Farbenchemie* (Vienna, 5th ed., 1943) pp 224-225; Kozlov, *Doklady Akad. Nauk SSSR* **61**, 281 (1948). Mutation study: J. P. Brown, R. J. Brown, *Mutat. Res.* **40**, 203 (1976).

Orange needles from alc, mp 193-197°. Sublimes. Absorption max: 430, 250 nm (log ε 4.35, 4.60). Almost insol in water (6.5 × 10^{-6} mols/l at 25°), in alcohol (1:2000); moderately sol in ether (1:500), in chloroform; sol in 10 parts hot glacial acetic acid. Very slightly sol in aq solns of alkali hydroxides: about 0.8 g dissolves in 100 ml 0.5N NaOH. (Disodium salt is reported to have a water soly of 0.05%.) LD_{50} in mice: 500 mg/kg i.p., *RTECS Vol. I*, R. J. Lewis, R. L. Tatken, Eds. (1979) p 142.

USE: Important intermediate in the manuf of alizarin and indanthrene dyestuffs. Forms insol Ca, Ba, Pb lakes and has been proposed as analytical reagent.

THERAP CAT: Cathartic.

THERAP CAT (VET): Purgative.

2814. Dantrolene. *1-[[[5-(4-Nitrophenyl)-2-furanyl]-methylene]amino]-2,4-imidazolidinedione;* *1-[[5-(p-nitro-phenyl)furfurylidene]amino]hydantoin.* $C_{14}H_{10}N_4O_5$; mol wt 314.26. C 53.51%, H 3.21%, N 17.83%, O 25.45%. Prepn: Neth. pat. **Appl. 6,612,588** corresp to Davis, Snyder, U.S. pat. **3,415,821** (1967, 1968, both to Norwich Pharmacal); Snyder *et al.*, *J. Med. Chem.* **10**, 807 (1967); Frimm *et al.*, *Chem. Zvesti* **23**, 916 (1969). Pharmacology and mechanism of action: Ellis, Carpenter, *Arch. Pharmacol.* **275**, 83 (1972); Ellis *et al.*, *J. Pharm. Sci.* **62**, 948 (1973). Metabolic studies: Cox *et al.*, *ibid.* **58**, 987 (1969). Review of pharmacology and use in malignant hypothermia, neuroleptic malignant syndrome, muscle spasticity: A. Ward *et al.*, *Drugs* **32**, 130-168 (1986).

Crystals from aqueous DMF, mp 279-280°.

Sodium salt hemiheptahydrate, $C_{14}H_9N_4NaO_5 \cdot 3\frac{1}{2}H_2O$, *F-440, Dantamacrin, Dantrium.* Orange powder. Slightly soluble in water; more sol in alkaline soln.

Sodium salt tetrahydrate, $C_{14}H_9N_4NaO_5 \cdot 4H_2O$, *Dantrix.*

THERAP CAT: Skeletal muscle relaxant.

2815. Daphnandrine. *6,6',12'-Trimethoxy-2-methyloxy-acanthan-7-ol.* $C_{36}H_{38}N_2O_6$; mol wt 594.68. C 72.70%, H 6.44%, N 4.71%, O 16.14%. From bark of *Daphnandra micrantha* Benth., *Monimiaceae*. Isoln: Pyman, *J. Chem. Soc.* **105**, 1679 (1914); Bick *et al.*, *ibid.* **1953**, 695. Structure: Bick *et al.*, *ibid.* **1949**, 2767; **1960**, 4928.

Needles from methanol, dec 270°. $[\alpha]_D^{16}$ +480° (c = 1.2 in chloroform). uv max (methanol): 284 nm (log ε 3.92). Practically insol in hot water, ethyl acetate, acetone, ether, petr ether; sparingly sol in hot methanol, ethanol, xylene, cold chloroform; sol in hot chloroform.

Hydrochloride hemihydrate, $C_{36}H_{38}N_2O_6 \cdot 2HCl \cdot \frac{1}{2}H_2O$, prisms from water, dec 282°. $[\alpha]_D$ +400° (c = 1.12). Very sol in hot water; sparingly sol in cold water.

Hydrobromide monohydrate, $C_{36}H_{38}N_2O_6 \cdot 2HBr \cdot H_2O$, prisms from water, dec 291°. $[\alpha]_D$ +386° (c = 3.8). Very sol in hot water; sparingly sol in cold water.

2816. Daphnetin. *7,8-Dihydroxy-2H-1-benzopyran-2-one;* *7,8-dihydroxycoumarin.* $C_9H_6O_4$; mol wt 178.14. C 60.68%, H 3.40%, O 35.93%. The aglucon of daphnin. By boiling daphnin with dil mineral acids; by enzymatic hydrolysis of daphnin; by sublimation from daphnin; by synthesis: v. Pechmann, *Ber.* **17**, 934 (1884); Gatterman, Köbner, *Ber.* **32**, 287 (1899); Baker, Savage, *J. Chem. Soc.* **1938**, 1602; Späth, Galinovsky, *Ber.* **70B**, 235 (1937); Sethna, Shah, *Chem. Rev.* **36**, 1 (1945).

Crystals from dil alc, mp 256° (dec). Sublimes. Sol in boiling water, hot dil alcohol, hot glacial acetic acid; very sparingly sol in ether, in carbon bisulfide, in chloroform, in benzene. Sol in alkali and alkali carbonate solns giving a yellow color. Aq solns give a green color with ferric chloride which turns red on addition of sodium carbonate.

2817. Daphnin. *7-β-D-Glucopyranosyloxy-8-hydroxy-2H-1-benzopyran-2-one;* 7,8-dihydroxycoumarin 7-β-D-glucoside; daphnetin 7-β-D-glucoside; 7-glucosido-8-hydroxycoumarin. $C_{15}H_{16}O_9$; mol wt 340.28. C 52.94%, H 4.74%, O 42.32%. Extraction from fresh bark of *D. mezereum* L., *Thymelaeaceae*: Beilstein **XXXI**, 248. Structure: Wessely, Sturm, *Ber.* **63**, 1299 (1930); Hattori, *Chem. Zentr.* **1930** II, 2138. Synthesis: Gandini, *Gazz. Chim. Ital.* **70**, 611 (1940), *C.A.* **35**, 1394 (1941).

Monohydrate, prisms from water. Anhydr after drying at 80-90° at 2 mm Hg. When anhydrous dec 215° (Wessely, Sturm); dec 224° (Hattori); dec 229° (older data). Optical rotations of monohydrate: $[\alpha]_D^{22}$ −124° (15 mg in 4 ml MeOH); after 2 recrystns from water: $[\alpha]_D^{22}$ −115° (abs MeOH) (Wessely, Sturm); $[\alpha]_D^{15}$ −84.2° (MeOH) (Hattori). Sol in hot water, alcohol; insol in ether; sol in aq solns of alkalies and alkali carbonates with yellow color. Is hydrolyzed by dilute acid yielding 7,8-dihydroxycoumarin and D-glucose. Also split by emulsin.

2818. Daphnoline. *6,6'-Dimethoxy-2-methyloxyacanthan-7,12'-diol;* trilobamine. $C_{35}H_{36}N_2O_6$; mol wt 580.65. C 72.39%, H 6.25%, N 4.82%, O 16.53%. From bark of *Daphnandra micrantha* Benth., *Monimiaceae*. Isoln: Pyman, *J. Chem. Soc.* **105**, 1679 (1914). Identity with trilobamine:

Bick *et al.*, *ibid.* **1949**, 2767. Structure: Bick *et al.*, *ibid.* **1960**, 4928.

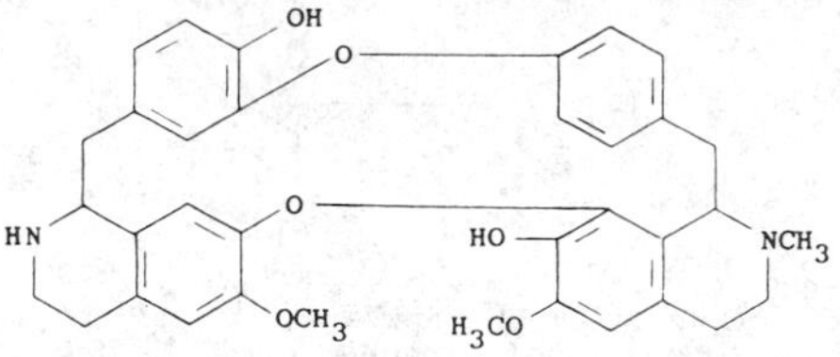

Crystals from chloroform, mp 194-196°. $[\alpha]_D$ +459° (c = 0.3 in chloroform). uv max (methanol): 285 nm (log ϵ 3.9). Practically insol in water, ethyl acetate, acetone, ether, petr ether; sparingly sol in methanol, ethanol, xylene, hot chloroform sol in dil acids, cold aq 5% sodium hydroxide.

Hydrochloride, $C_{35}H_{36}N_2O_6 \cdot 2HCl$, crystals from aq ethanol, dec 290°. $[\alpha]_D$ +355° (c = 2.1 in water). Sol in cold water, hot alcohol.

Hydrobromide monohydrate, $C_{35}H_{36}N_2O_6 \cdot 2HBr \cdot H_2O$, microscopic needles from water, dec 286°. Very sol in hot water; sparingly sol in cold water.

2819. Dapiprazole. *5,6,7,8-Tetrahydro-3-[2-[4-(2-methylphenyl)-1-piperazinyl]ethyl]-1,2,4-triazolo[4,3-a]pyridine; 5,6,7,8-tetrahydro-3-[2-(4-o-tolyl-1-piperazinyl)ethyl]-s-triazolo[4,3-a]pyridine.* $C_{19}H_{27}N_5$; mol wt 325.46. C 70.12%, H 8.36%, N 21.52%. α-Adrenergic blocker. Prepn: B. Silvestrini, L. Baiocchi, **Ger.** pat. **2,915,318**; *eidem*, **U.S.** pat. **4,252,721** (1979, 1981 both to Angelini Francesco). Psychopharmacological profile and toxicity: B. Silvestrini *et al.*, *Arzneimittel-Forsch.* **32**, 668 (1982). Antihistaminic and adrenolytic effects: R. Lisciani *et al.*, *ibid.* 674. Effect on intraocular pressure and pupillary size in rabbits: B. Silvestrini *et al.*, *ibid.* 678. Distribution and excretion in rats: P. Valeri *et al.*, *Pharmacol. Res. Commun.* **17**, 417 (1985). Topical miotic and ocular hypotensive effects in humans: A. Reibaldi *et al.*, *Acta Ther.* **10**, 381 (1984); C. Malpassi *et al.*, *ibid* **12**, 55 (1986).

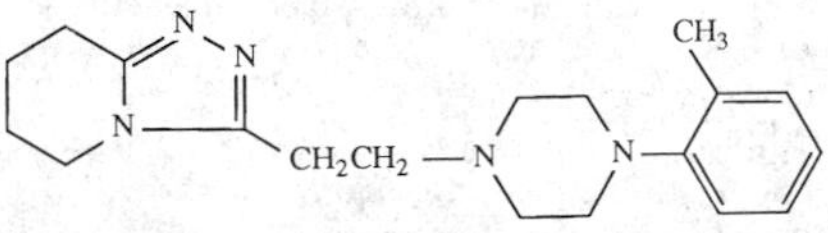

mp 158-160°.

Monohydrochloride, $C_{19}H_{28}ClN_5$, *AF 2139*, *Glamidolo, Reversil.* Crystals from absolute ethanol, mp 206-207°. LD_{50} in mice (mg/kg): 260 i.p. (Silvestrini).

THERAP CAT: Antiglaucoma agent.

2820. Dapsone. *4,4'-Sulfonylbisbenzeneamine; 4,4'-sulfonyldianiline;* bis(4-aminophenyl)sulfone; 4,4'-diaminodiphenyl sulfone; DDS; diaphenylsulfone; DADPS; 1358F; Dumitone; Avlosulfone; Sulfona-Mae; Croysulfone; Croysulphone; Disulone; Sulphadione; Avlosulfon; Eporal; Diphone; Novophone; Diphenasone; Udolac. $C_{12}H_{12}N_2O_2S$; mol wt 248.30. C 58.04%, H 4.87%, N 11.28%, S 12.91%, O 12.89%. Prepn: **Fr.** pat. **829,926** (1938 to I. G. Farbenind.), *C.A.* **33**, 1761 (1939); Buckles, *J. Chem. Ed.* **31**, 36 (1954); Ferry *et al.*, *Org. Syn.* **coll. vol. III**, 239 (1955). Mechanism of toxic action: Wu, DuBois, *Arch. Int. Pharmacodyn. Ther.* **183**, 36 (1970). Comprehensive description: C. E. Orzech *et al.*, in *Analytical Profiles of Drug Substances* **vol. 5**, K. Florey, Ed. (Academic Press, New York, 1976) pp 87-114.

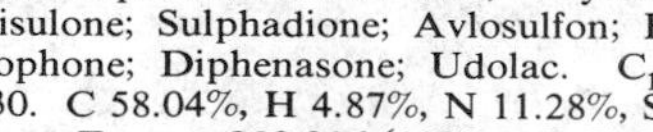

Crystals from 95% ethanol. mp 175-176° (also a higher melting form, mp 180.5°). pKb 13.0. Soluble in alcohol, methanol, acetone, dil hydrochloric acid. Practically insol in water.

USE: Hardening agent in the curing of epoxy resins.

THERAP CAT: Antibacterial (leprostatic); dermatitis herpetiformis suppressant.

THERAP CAT (VET): Antibacterial, antiprotozoan.

2821. Darvan®. Vinylidene cyanide-vinyl acetate copolymer; Darlan; Travis. *Methylenemalononitrile copolymer with vinyl acetate* contg about equal parts of each component. Polymerization of vinylidene cyanide with vinyl acetate: Gilbert *et al.*, *J. Am. Chem. Soc.* **78**, 1669 (1956); Gilbert, Miller, **U.S.** pat. **2,615,866** and Sayre, **U.S.** pat. **2,975,-158** (1952, 1961, both to B. F. Goodrich). *Review:* R. W. Moncrieff, *Man-Made Fibres* (John Wiley, New York, 4th ed., 1963) pp 488-494. Approximate structure:

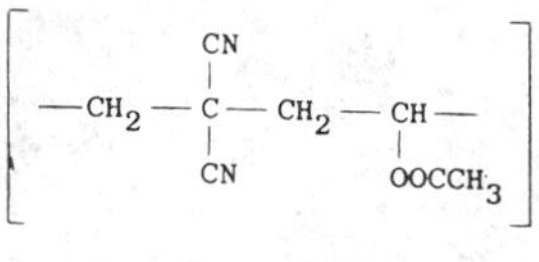

A wool-like fiber, sp gr 1.18. Softens at about 175°. Flammability similar to cotton and viscose. Insol in acetone, methylene chloride. Good resistance to sunlight, clothes-moth and carpet-beetle attack, and mildew; moderate resistance to 10% H_2SO_4 at 100° for hrs, and to 0.5% caustic soda at 43° for 7 days. Shrinks about 1% when immersed in boiling water for 3 minutes. Difficult to dye. Relatively resistant to pilling.

USE: Knitted garments, pile coats, and in blends for tropical suitings and men's shirtings.

2822. Datiscetin. *3,5,7-Trihydroxy-2-(2-hydroxyphenyl)-4H-1-benzopyran-4-one; 2',3,5,7-tetrahydroxyflavone; 2'-hydroxycrysidenolon 1493.* $C_{15}H_{10}O_6$; mol wt 286.23. C 62.94%, H 3.52%, O 33.54%. Cleavage product of datiscin: Stenhouse, *Ann.* **98**, 166 (1856). From roots of *Datisca cannabina* L., *Datiscaceae:* Marchlewski, *Biochem. Z.* **3**, 286 (1907). Structure: Leskiewicz, Marchlewski, *Ber.* **47**, 1599 (1914). Synthesis: Kalf, Robinson, *J. Chem. Soc.* **127**, 1968 (1925); Simpson, Wally, *ibid.* **1955**, 166.

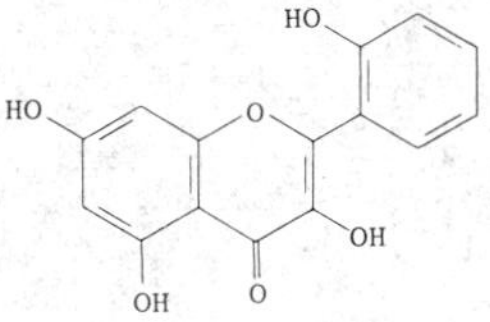

Pale yellow needles from alc, mp 271°. uv max (96% alc): 264.0, 375.0 nm (log ϵ 4.265, 4.005). Sol in alcohol, ether, other organic solvents; practically insol in water. Sol in alkaline solns with yellow color.

2823. Datura. Dried leaves and dried seeds of *Datura fastuosa* L., var. *alba* Nees (also leaves from *D. metel* L.), *Solanaceae. Habit.* India. *Constit.* About 0.5% hyoscine, with traces of hyoscyamine and atropine; the seeds contain also resin and fixed oil. Comprehensive monograph: A. F. Blakeslee, *The Genus Datura* (Ronald Press, New York, 1959) 289 pp.

THERAP CAT: Anticholinergic.

2824. Daucol. *Octahydro-6,8a-dimethyl-3-(1-methylethyl)-1H-3a,6-epoxyazulen-7-ol; 2,3,4,5,6,7a,8,8a-octahydro-3-isopropyl-6,8a-dimethyl-1H-3a,6-epoxyazulen-7-ol.* $C_{15}H_{26}O_2$; mol wt 238.36. C 75.58%, H 11.00%, O 13.42%. From oil of carrot seeds, *Daucus carota* L., *Umbelliferae:* Richter, *Arch. Pharm.* **247**, 391 (1909). Structure: Sykora *et al.*, *Coll. Czech. Chem. Commun.* **26**, 788 (1961); Zalkow *et al.*, *J. Org. Chem.* **26**, 981 (1961). Stereochemistry: Levisalles, Rudler, *Bull. Soc. Chim. France* **1964**, 2020; **1967**, 2059. Synthesis of (−)-form: DeBroissia *et al.*, *Chem. Commun.* **1972**, 855; *eidem*, *Bull. Soc. Chim. France* **1972**, 4314.

(−)-daucol

Crystals (from petr ether at −30°), mp 113-115°. bp$_2$ 124-132°. [α]$_D^{20}$ −16.9° (c = 2.76 in ethanol).

2825. Daunorubicin. (8S-cis)-8-Acetyl-10-[(3-amino-2,3,6-trideoxy-α-L-lyxo-hexopyranosyl)oxy]-7,8,9,10-tetrahydro-6,8,11-trihydroxy-1-methoxy-5,12-naphthacenedione; daunomycin; leukaemomycin C; rubidomycin; Cerubidin. C$_{27}$H$_{29}$NO$_{10}$; mol wt 527.51. C 61.47%, H 5.54%, N 2.66%, O 30.33%. Anthracycline antibiotic related to the rhodomycins, q.v. Isolated from fermentation broths of Streptomyces peucetius: G. Cassinelli, P. Orezzi, Giorn. Microbiol. 11, 167 (1963), C.A. 62, 9482b (1965); A. Di Marco et al., Nature 201, 706 (1964); eidem, Belg. pat. 639,897; eidem, U.S. pat. 4,012,284 (1964, 1977 both to Soc. Farmaceut. Italia). Daunorubicin is a glycoside formed by a tetracyclic aglycone, daunomycinone, (C$_{21}$H$_{18}$O$_8$) and an amino sugar, daunosamine, (C$_6$H$_{13}$NO$_3$), 3-amino-2,3,6-trideoxy-L-lyxohexose: F. Arcamone et al., J. Am. Chem. Soc. 86, 5334, 5335 (1964); R. H. Iwamoto et al., Tetrahedron Letters 1968, 3891. Absolute stereochemistry: F. Arcamone et al., Gazz. Chim. Ital. 100, 949-989 (1970). Identity with rubidomycin: G. L. Tong et al., J. Pharm. Sci. 56, 1691 (1967). Synthesis of daunosamine: J. P. Marsh et al., Chem. Commun. 1967, 973; T. Yamaguchi, M. Kojimo, Carbohyd. Res. 59, 343 (1977); P. M. Wovkulich, M. R. Uskokovic, J. Am. Chem. Soc. 103, 3956 (1981); of daunomycinone: C. M. Wong et al., Can. J. Chem. 51, 466 (1973); J. S. Swenton, P. W. Reynolds, J. Am. Chem. Soc. 100, 6188 (1978); K. Krohn, K. Tolkiehn, Ber. 112, 3453 (1979); F. M. Hauser, S. Prasanna, J. Am. Chem. Soc. 103, 6378 (1981). Total synthesis of daunorubicin: E. M. Acton et al., J. Med. Chem. 17, 659 (1974). Purification: E. Oppici et al., Belg. pat. 898,506; eidem, Brit. pat. Appl. 2,133,005 (both 1984 to Farmitalia). Toxicity data: A. Di Marco et al., Cancer Chemother. Rep. (part 1) 53, 33 (1969). Review of properties, biosynthesis, fermentation: R. J. White, R. M. Stroshane, Drugs Pharm. Sci. 22, 569-594 (1984); of carcinogenic action in laboratory animals: IARC Monographs 10, 145-152 (1976); of toxicology: R. J. Maral et al., Cancer Treat. Rep. 65, Suppl. 4, 9-18 (1981); of use in treatment of solid tumors: R. B. Weiss et al., ibid. 25-28; of interactions with nucleic acids: S. Neidle, M. R. Sanderson, in Molecular Aspects of Anti-cancer Drug Action, S. Neidle, M. J. Waring, Eds. (Verlag-Chemie, Florida, 1983) pp 35-55; of mechanism of cytotoxicity: H. S. Schwartz, ibid. pp 93-125; of metabolism and clinical pharmacokinetics: C. E. Riggs, Jr., Sem. Oncol. 11, Suppl. 3, 2-11 (1984). Review: A. DiMarco et al., Antibiotics vol. 3, J. W. Corcoran, F. E. Hahn, Eds. (Springer Verlag, New York, 1975) pp 101-128.

daunomycinone

daunosamine

mp 208-209°. LD$_{50}$ in mice, rats (mg/kg): 20, 13 i.v.; 5, 8 i.p. (DiMarco, 1977).

Hydrochloride, C$_{27}$H$_{30}$ClNO$_{10}$, Cérubidine, Daunoblastina, Ondena. Thin red needles, dec 188-190°. [α]$_D^{20}$ +248 ± 5°

(c = 0.05-0.1 in methanol). Sol in water, methanol, aq alcohols. Practically insol in chloroform, ether, benzene. Color of aq soln changes from pink at acid pH to blue at alkaline pH. Absorption max (methanol): 234, 252, 290, 480, 495, and 532 nm (E$_{1cm}^{1\%}$ 665, 462, 153, 214, 218, and 112). LD$_{50}$ in mice (mg/kg): 26 i.v. (DiMarco, 1969).
THERAP CAT: Antineoplastic.

2826. Dauricine. 4-[(1,2,3,4-Tetrahydro-6,7-dimethoxy-2-methyl-1-isoquinolinyl)methyl]-2-[4-[(1,2,3,4-tetrahydro-6,7-dimethoxy-2-methyl-1-isoquinolinyl)methyl]phenoxy]-phenol. C$_{38}$H$_{44}$N$_2$O$_6$; mol wt 624.75. C 73.05%, H 7.10%, N 4.48%, O 15.37%. From Menispermum dauricum DC., and M. canadense L., Menispermaceae: Kondo, Narita, J. Pharm. Soc. Japan no. 542, 279 (1927); C.A. 21, 2700 (1927); Manske, Can. J. Res. 21B, 17 (1943). Structure: Kondo et al., Ber. 68, 519 (1935); Kondo, Tomita, Arch. Pharm. 274, 65 (1936); Inubushi, Niwa, J. Pharm. Soc. Japan 72, 762 (1952); C.A. 47, 6430e (1953). Total synthesis: Kametani, Fukumoto, J. Chem. Soc. 1965, Suppl. 2, 6141.

Slightly yellow amorphous base, mp 115°. [α]$_D^{11}$ −139° in methanol. Sol in alc, acetone, benzene; slightly sol in ether. Dimethiodide, C$_{38}$H$_{44}$N$_2$O$_6$·2CH$_3$I, needles, mp 204°. [α]$_D^{20}$ −114°.

2827. Dazomet. Tetrahydro-3,5-dimethyl-2H-1,3,5-thiadiazine-2-thione; 2-thio-3,5-dimethyltetrahydro-1,3,5-thiadiazine; 3,5-dimethyl-2-thionotetrahydro-1,3,5-thiadiazine; dimethylformocarbothialdine; DMTT; Crag 974; Mylone. C$_5$H$_{10}$N$_2$S$_2$; mol wt 162.27. C 37.01%, H 6.21%, N 17.27%, S 39.52%. Prepd by the action of carbon disulfide on trimethyltrimethylenetriimine: Delépine, Bull. Soc. Chim. 15, 891 (1897); from formaldehyde and methylammonium methyldithiocarbamate: Bodendorf, J. Prakt. Chem. 126, 233 (1930); Ainley et al., J. Chem. Soc. 1944, 147. Prepd commercially by the reaction of carbon disulfide, methylamine and caustic soda: Chem. Week 79, no. 18, 82-83 (1956).

Needles from benzene, mp 106-107°. Soluble in alc with decompn. Dec by water, dil acids. uv max (cyclohexane): 242, 289 nm (ε 7150, 9900). LD$_{50}$ orally in rats: 363 mg/kg, Toxic Substances List, H. E. Christensen, Ed. (1974) p 757.
USE: Soil fungicide, nematocide, weed killer.

2828. 2,4-DB. 4-(2,4-Dichlorophenoxy)butanoic acid; 4-(2,4-dichlorophenoxy)butyric acid; Butyrac; Legumex D. C$_{10}$H$_{10}$Cl$_2$O$_3$; mol wt 249.09. C 48.22%, H 4.05%, Cl 28.46%, O 19.27%. Selective post-emergence, translocated herbicide. First described as a plant growth regulator: M. E. Synerholm and P. W. Zimmerman, Contrib. Boyce Thompson Inst. 14, 369 (1947). Prepn: G. W. Kitchingman, A. C. Tucker, Brit. pats. 804,565 and 883,255 (1958, 1961, both to ICI). Similar in activity to 2,4-D, q.v., but more selective. Mechanism of selectivity studies: D. L. King, D. E. Bayer, Proc. West Soc. Weed Sci. 25, 37 (1972). Toxicity data: J. W. Bailey, J. L. White, Residue Rev. 10, 97 (1965).

White crystals, mp 117-119°. Soly in water at 25°: 46

ppm. Sol in acetone, ethanol, diethyl ether. Slightly sol in benzene, toluene, kerosene.

Sodium salt, $C_{10}H_9Cl_2NaO_3$, *Embutox*. Water soluble; hard water will show precipitate as the calcium and magnesium salts.

Dimethylamine salt, $C_{12}H_{17}Cl_2NO_3$, *Butoxone*. Water sol; calcium and magnesium salts precipitate in hard water.

USE: Herbicide.

2829. DBMC. *2,4-Bis(1,1-dimethylethyl)-5-methylphenol; 4,6-di-tert-butyl-m-cresol;* 3-methyl-4,6-di-*tert*-butylphenol. $C_{15}H_{24}O$; mol wt 220.34. C 81.76%, H 10.98%, O 7.26%. Prepn: Weinrich, *Ind. Eng. Chem.* **35**, 264 (1943).

Crystals, mp 62.1°. bp 282°, bp_{100} 211°, bp_{20} 167°. d_4^{80} 0.912. Practically insol in water, ethylene glycol; sol in alcohol, benzene, carbon tetrachloride, ether, acetone.

USE: Intermediate in the production of rubber chemicals modified phenolic resins, synthetic musks of the ambrette type.

2830. DCPA. *2,3,5,6-Tetrachloro-1,4-benzenedicarboxylic acid dimethyl ester; 2,3,5,6-tetrachloroterephthalic acid dimethyl ester;* dimethyl 2,3,5,6-tetrachloroterephthalate; chlorthal-methyl; Dacthal; Rid. $C_{10}H_6Cl_4O_4$; mol wt 331.99. C 36.18%, H 1.82%, Cl 42.72%, O 19.28%. Prepn: Lindemann, U.S. pat. **2,923,634** (1960 to Diamond Alkali). Toxicity data: G. W. Bailey, J. L. White, *Residue Rev.* **10**, 97 (1965).

Crystals from methanol, mp 155-156°. Solubility: < 5% in water; > 5% in acetone, cyclohexanone, xylene. LD_{50} orally in rats: > 3000 mg/kg (Bailey, White).

USE: Pre-emergent herbicide.

2831. DDD (Analytical). *6,6'-Dithiobis-2-naphthalenol; 6,6'-dithiodi-2-naphthol;* bis(6-hydroxy-2-naphthyl) disulfide; 2,2'-dihydroxy-6,6'-dinaphthyl disulfide; 6,6'-dithiobis(2-naphthol). $C_{20}H_{14}O_2S_2$; mol wt 350.46. C 68.54%, H 4.03%, O 9.13%, S 18.30%. Prepn: Zincke, Dereser, *Ber.* **51**, 352 (1918).

Leaflets, mp 221-222°.

Note: Not to be confused with the insecticide, p,p'-DDD [1,1-dichloro-2,2-bis(p-chlorophenyl)ethane].

USE: For determination of protein-bound sulfhydryl groups.

2832. DDT. *1,1'-(2,2,2-Trichloroethylidene)bis[4-chlorobenzene]; 1,1,1-trichloro-2,2-bis(p-chlorophenyl)ethane; α,α-bis(p-chlorophenyl)-β,β,β-trichlorethane;* dichlorodiphenyltrichloroethane; chlorophenothane; clofenotane; dicophane; pentachlorin; p,p'-DDT; Agritan; Gesapon; Gesarex; Gesarol; Guesapon; Neocid. $C_{14}H_9Cl_5$; mol wt 354.50. C 47.43%, H 2.56%, Cl 50.01%. Polychlorinated nondegradable pesticide. Prepn: Zeidler, *Ber.* **7**, 1180 (1874); P. Müller, U.S. pat. **2,329,074** (1944 to Geigy); Rueggeberg, Torrans, *Ind. Eng. Chem.* **38**, 211 (1946); Cook et al., ibid. **39**, 868, 1683 (1947). Convenient lab procedures: Bailes, *J.*

Chem. Ed. **22**, 122 (1945); Ginsburg, *Science* **108**, 339 (1948). Large scale production: Mosher et al., *Ind. Eng. Chem.* **38**, 916 (1946). Chemical composition of technical grade: H. L Haller et al., *J. Am. Chem. Soc.* **67**, 1591 (1945). Activity: P. Müller, *Helv Chim. Acta* **29**, 1560 (1946). Comprehensive monograph (in English and German): *DDT*, P. Müller, Ed., 3 vols (Birkhäuser Verlag, Basel and Stuttgart, 1955). Toxicity data: Gaines, *Toxicol. Appl. Pharmacol.* **2**, 88 (1960); **14**, 515 (1969).

Biaxial elongated tablets, needles from 95% alc. mp 108.5-109°. uv max (95% alc): 236 nm. Vapor pressure at 20° = 1.5×10^{-7} mm Hg. Practically insol in water, dil acids, alkalies. Soly (g/100 ml): acetone 58; benzene 78; benzyl benzoate 42; carbon tetrachloride 45; chlorobenzene 74; cyclohexanone 116; 95% alc 2; ethyl ether 28; gasoline 10; isopropanol 3; kerosene 8-10; morpholine 75; peanut oil 11; pine oil 10-16; tetralin 61; tributyl phosphate 50; freely sol in pyridine, dioxane. The soly in organic solvents increases sharply with a rise in temp. Resistant to destruction by light and oxidation. Its unusual stability has resulted in difficulties in residue removal from water, soil and foodstuffs. DDT should not be kept in iron containers and should not be mixed with iron and aluminum salts nor with alkaline substances. High storage temps should also be avoided. Setting point of technical grade: 88.6-91.4°. LD_{50} in male, female rats: 113, 118 orally (Gaines, 1960).

Human Toxicity: Poisoning may occur by ingestion or by absorption through skin or respiratory tract. *Acute:* tremors of head and neck muscles, tonic and clonic convulsions, cardiac or respiratory failure, death. Estimated oral fatal dose 500 mg/kg body wt of the solid material. Solvents such as kerosene increase toxicity. Death occurs in 2 to 24 hrs. *Chronic:* Hepatic damage, CNS degeneration, agranulocytosis, dermatitis, weakness, convulsions, coma, death, cf. *Clinical Toxicology of Commercial Products,* R. E. Gosselin et al., Eds. (Williams & Wilkins, Baltimore, 4th ed., 1976) Section III, pp 116-119. This substance may reasonably be anticipated to be a carcinogen: *Fourth Annual Report on Carcinogens* (NTP-85-002, 1985) p 66.

USE: Contact insecticide.

THERAP CAT: Ectoparasiticide; pediculicide.

2833. Deaminooxytocin. *1-(3-Mercaptopropanoic acid)-oxytocin;* demoxytocin; desaminooxytocin; ODA-914; Sandopart. $C_{43}H_{65}N_{11}O_{12}S_2$; mol wt 992.19. C 52.05%, H 6.60%, N 15.53%, O 19.35%, S 6.46%. Highly potent analog of the posterior pituitary hormone oxytocin q.v., differing in structure at the 1 position where the free amino group in the half-cystine residue is replaced by hydrogen. Synthesis and biological activity: du Vigneaud et al., *J. Biol. Chem.* **235**, P.C. 64 (1960); Hope et al., ibid. **237**, 1563 (1962); Takashima et al., *J. Am. Chem. Soc.* **90**, 1323 (1968). Isoln and purifn: Ferrier et al., *J. Biol. Chem.* **240**, 4264 (1965). Structure-activity correlation: Jarvis, du Vigneaud, *Science*

143, 545 (1964). Synthesis of D-isomer: G. Flouret, V. du Vigneaud, *J. Med. Chem.* **14**, 556 (1971). Crystal structure analysis: S. P. Wood *et al.*, *Science* **232**, 633 (1986). Pharmacology and comparison with oxytocin: Chan, du Vigneaud, *Endocrinology* **71**, 977 (1962). *Review: Advan. Exp. Med. Biol.* vol. 2, 53-104 (1968).

L-Isomer, white plate-like crystals from water, mp 179° (Ferrier); 182-183° (Takashima). For amorphous powder; $[\alpha]_D^{20}$ −88.3° (Ferrier); $[\alpha]_D^{21}$ −107° (Hope); $[\alpha]_D^{25}$ −95.1° (Takashima); all (c = 0.5 in 1N acetic acid). Anhydrous form is very hygroscopic.

D-Isomer, white, fluffy powder, $[\alpha]_D^{20}$ +104° (c = 0.5 in 1N acetic acid).

THERAP CAT: Oxytocic.

2834. Deanol. 2-*(Dimethylamino)ethanol*; β-dimethyl-aminoethyl alcohol; *N,N*-dimethyl-2-hydroxyethylamine; Liparon. $C_4H_{11}NO$; mol wt 89.14. C 53.89%, H 12.44%, N 15.71%, O 17.95%. $(CH_3)_2NCH_2CH_2OH$. Prepd from equimolar amounts of ethylene oxide and dimethylamine: Knorr, *Ber.* **37**, 3508 (1904); Hanhart, Ingold, *J. Chem. Soc.* **1927**, 1012.

Liquid. d_4^{20} 0.8866. bp_{758} 135°. n_D^{20} 1.43. Miscible with water, alcohol, ether.

Bitartrate, *Atrol, Paxanol, Dimethaen*. Crystals, sol in water.

Dipropylacetate, $C_{12}H_{27}NO_3$, *Brainine*.

Hemisuccinate, $C_8H_{17}NO_5$, *Tonibral*. White, hygroscopic crystalline powder. Very soluble in water, ethanol; sparingly sol in chloroform, benzene.

THERAP CAT: Antidepressant.

2835. Deanol Aceglumate. *N-Acetyl-L-glutamic acid compd with 2-(dimethylamino)ethanol (1:1)*; dimethylamino-ethanol acetyl-L-glutamate; demanol aceglumate; Cleregil; Otrun; Risatarun. $C_{11}H_{22}N_2O_6$; mol wt 278.32. C 47.47%, H 7.97%, N 10.07%, O 34.49%. Prepn: **Fr. pat. M2487** (1964 to Interco Fribourg), *C.A.* **61**, 12085c (1964).

$$HOOCCH_2CH_2CHCOOH \quad . \quad HOCH_2CH_2N(CH_3)_2$$
$$| $$
$$NHCOCH_3$$

Soluble in water. LD_{50} s.c. in mice: 4 mg/kg. LD_{50} i.p. in rats: 3100 mg/kg, *Toxic Substances List*, H. E. Christensen, Ed. (1974) p 382.

THERAP CAT: Antidepressant.

2836. Deanol Acetamidobenzoate. 4-*(Acetylamino)benzoic acid compd with 2-(dimethylamino)ethanol (1:1)*; 2-(dimethylamino)ethanol *p*-acetamidobenzoate; DMAE *p*-acetamidobenzoate; Deaner; Diforene. $C_{13}H_{20}N_2O_4$; mol wt 268.31. C 58.19%, H 7.51%, N 10.44%, O 23.85%. Prepn: **Brit. pat. 879,259** (1957 to Riker Labs); Lasslo *et al.*, *J. Am. Pharm. Assoc. Sci. Ed.* **48**, 345 (1959).

$$COOH$$
$$·HOCH_2CH_2N(CH_3)_2$$
$$NHCOCH_3$$

Crystals from absolute ethanol + ethyl acetate, mp 159-161.5°. Soluble in water. LD_{50} i.p. in mice: 1020 mg/kg (calc'd as base), Healy *et al.*, *Fed. Proc.* **19**, 23 (1960).

THERAP CAT: Antidepressant.

2837. Debrisoquin. 3,4-*Dihydro-2(1H)-isoquinolinecarboximidamide*; 3,4-*dihydro-2(1H)-isoquinolinecarboxamidine*; 2-amidino-1,2,3,4-tetrahydroisoquinoline; isocaramidine. $C_{10}H_{13}N_3$; mol wt 175.23. C 68.54%, H 7.48%, N 23.98%. Preparation of derivs: Wenner, **Belg. pat. 629,007** corresp to **U.S. pat. 3,157,573** (1963, 1964, both to Hoffmann-La Roche); *idem, J. Med. Chem.* **8**, 125 (1965). Pharmacology: Moe *et al.*, *Curr. Ther. Res. Clin. Exp.* **6**, 299 (1964); Abrams *et al.*, *J. New Drugs* **4**, 268 (1964). Possible mechanism of action: Medina *et al.*, *Biochem. Pharmacol.*

18, 891 (1969); Pocelinko *et al.*, *Int. Z. Klin. Pharmakol. Ther. Toxicol.* **2**, 13 (1969).

Sulfate, $C_{20}H_{28}N_6O_4S$, *Ro 5-3307/1, Declinax, Tendor.* Crystals, mp 278-280°, 284-285° or 266-268° (H_2O), depending on humidity. Freely sol in water. LD_{50} orally in rats (mg/kg): 88±18 (neonates), 1580±163 (adults), E. J. Goldenthal, *Toxicol. Appl. Pharmacol.* **18**, 185 (1971).

THERAP CAT: Antihypertensive.

2838. Decaborane(14). Decaboron tetradecahydride. $B_{10}H_{14}$; mol wt 122.31. B 88.46%, H 11.54%. Prepd by the pyrolysis of diborane(6). Reviews of chemistry: Griffo, *Diss. Abstr.* **22**, 2976 (1962); Stanko *et al.*, *Usp. Khim.* **34**(6), 1011-1039 (1965). Review of structure and properties: Campbell, Jr., in *Progress in Boron Chemistry*, Steinberg, McCloskey, Eds., vol. 1 (Macmillan, New York, 1964) pp 173-188. Reviews of toxicity: Levinskas, "Toxicology of Boron Compounds" in *Boron, Metallo-Boron Compounds and Boranes*, R. M. Adams, Ed. (Interscience, New York, 1964) pp 693-737; E. Browning, *Toxicity of Industrial Metals* (Appleton-Century-Crofts, New York, 2nd ed., 1969) pp 92-97. General reviews: Stock, *Hydrides of Boron and Silicon* (Cornell Univ. Press, Ithaca, 1933) *passim;* Siegel, Mack, *J. Chem. Ed.* **34**, 314-317 (1957); Major, *Chem. Eng. Progr.* **54**(3), 49-54 (1958); Lipscomb, *Boron Hydrides* (Benjamin, New York, 1963) *passim;* Adams in *Boron, Metallo-Boron Compounds and Boranes, loc. cit.*, pp 647-663; Hawthorn in *Advan. Inorg. Chem. Radiochem.* **5**, 307-345 (1963); Greenwood in *Comprehensive Inorganic Chemistry*, vol. 1, J. C. Bailar, Jr. *et al.*, Eds. (Pergamon Press, Oxford, 1973) pp 818-837.

Orthorhombic crystals. mp 99.6-99.7°; bp 213°; bp_{19} 100°. d_4^{25} 0.94; d_4^{100} (liq) 0.78. Stable indefinitely at room temp; dec slowly into B + H_2 at 300°. Heat of fusion: 7.8 kcal/mole; of sublimation: 18.33 kcal/mole; of vaporization: 11.6 kcal/mole. Slightly sol in cold water; hydrolyzes in hot water. Sol in ethyl acetate, 1-bromopropane, ethyl silicate, carbon disulfide, benzene, alcohol, acetic anhydride, acetic acid, ethyl borate, carbon tetrachloride. Reacts with amides, acetone, butyraldehyde, acetonitrile at room temp. Decaborane mixtures with carbon tetrachloride are dangerously shock sensitive.

USE: In rocket propellants; as catalyst in olefin polymerization. *Caution:* Dizziness, nausea, vomiting, muscular tremors and evidence of liver injury have been reported in man.

2839. Decalin®. *Decahydronaphthalene;* perhydronaphthalene; bicyclo[4.4.0]decane; naphthalane; naphthane; Dec; DeKalin. $C_{10}H_{18}$; mol wt 138.24. C 86.88%, H 13.12%. Occurs in two forms: *cis* and *trans*. Hydrogenation of naphthalene in glacial acetic acid in the presence of platinum catalyst at 25° and 130 atm yields a mixture of 77% *cis*-Decalin and 23% *trans*-Decalin. Hydrogenation of Tetralin under the same conditions yields almost only *cis*-Decalin: Baker, Schuetz, *J.Am. Chem. Soc.* **69**, 1250 (1947). Hydrogenation of Δ$^{4a(8a)}$-octalin in ethanol in the presence of platinum yields *trans*-Decalin as the main component: Linstead *et al., J. Chem. Soc.* **1937**, 1136. Separation of *trans*- and *cis*-forms: Seyer, Walker, *J. Am. Chem. Soc.* **60**, 2125 (1938). Prepn of *trans*-Decalin from *cis*-Decalin: Zelinsky, *Ber.* **65**, 1299 (1932). Conformation studies: Barton, *J. Chem. Soc.* **1948**, 340; Moniz, Dixon, *J. Am. Chem. Soc.* **83**, 1671 (1961).

trans

cis

Liquid. Slight odor resembling menthol. Pure Decalin does not smell of naphthalene. Volatile with steam. The commercial product may be practically all *trans*-Decalin, or a mixture contg up to 60% *cis*-Decalin. The commercial mixture has a flash pt (closed cup) of about 136°F (58°C). Autoignition temp 504°F. Insol in water; very sol in alcohol, methanol, ether, chloroform. Miscible with propyl and iso-propyl alcohol; miscible with most ketones and esters. LD_{50} orally in rats: 4.2 g/kg. Lethal concn for rats in air: 500 ppm, Smyth *et al.*, *Arch. Ind. Hyg. Occup. Med.* **4**, 119 (1951).

cis-Form, mp −43.26°. bp 195.7°; bp_9 67.0°. d_4^{20} 0.8963. n_D^{20} 1.48113.

trans-Form, mp −30.4°. bp 187.25°; bp_9 62.0°. d_4^{20} 0.8700. n_D^{20} 1.46968.

USE: Solvent for naphthalene, fats, resins, oils, waxes; used instead of turpentine in lacquers, shoe polishes, floor waxes. In motor fuel and lubricants. As patent fuel in stoves.

2840. Decamethonium Bromide. *N,N,N,N',N',N'-Hexamethyl-1,10-decanediaminium dibromide; decamethyl-enebis[trimethylammonium bromide];* C-10; Syncurine. C_{16}-$H_{38}Br_2N_2$; mol wt 418.36. C 45.94%, H 9.16%, Br 38.21%, N 6.70%. $[(CH_3)_3N^+(CH_2)_{10}N^+(CH_3)_3].2Br^-$. Prepn: Blomquist *et al.*, *J. Am. Chem. Soc.* **81**, 678 (1959).

Crystals from methanol + acetone, dec 268-270°. Freely sol in water, alcohol; very slightly sol in chloroform. Practically insol in ether. Aq solns are stable and may be sterilized by autoclaving.

THERAP CAT: Skeletal muscle relaxant.

2841. Decamethylcyclopentasiloxane. $C_{10}H_{30}O_5Si_5$; mol wt 370.80. C 32.40%, H 8.16%, O 21.58%, Si 37.85%. Isolated from the hydrolysis product of dimethyldichlorosilane: Patnode, Wilcock, *J. Am. Chem. Soc.* **68**, 358 (1946).

Oily liq. mp −38°. bp 210°; bp_{20} 101°. d 0.9593. n_D^{20} 1.3982.

2842. Decamethylene Glycol. *1,10-Decanediol.* $C_{10}H_{22}$-O_2; mol wt 174.28. C 68.91%, H 12.72%, O 18.36%. HO-$(CH_2)_{10}OH$. Prepd by the reduction of dimethyl or diethyl sebacate with sodium and ethyl alcohol: Bouveault, Blanc, *Compt. Rend.* **137**, 329 (1903); *Bull. Soc. Chim.* [3] **31**, 1205 (1904); **Ger.** pat. **164,294;** *Frdl.* **8**, 1260 (1905); Chuit, *Helv. Chim. Acta* **9**, 264 (1926); Manske, *Org. Syn.* **14**, 20 (1934). By catalytic hydrogenation of sebacic esters: Folkers, Adkins, *J. Am. Chem. Soc.* **54**, 1146 (1932).

Needles from water or dil alcohol. mp 74°. bp_{20} 192°; bp_{11} 179°; bp_8 170°. Freely sol in alcohol, warm ether; almost insol in petr ether and water.

Diethyl ether, $C_{14}H_{30}O_2$, *1,10-diethoxydecane.* d_0^0 0.850; bp 260°.

2843. Decamethyltetrasiloxane. $C_{10}H_{30}O_3Si_4$; mol wt 310.71. C 38.67%, H 9.74%, O 15.45%, Si 36.14%. Prepn: Patnode, Wilcock, *J. Am. Chem. Soc.* **68**, 358 (1946).

Liquid. bp 194°. d 0.8536. n_D^{20} 1.3895. mp ∼−70°.

Stable. Inert to most chemical reagents and rubber. Maintains about the same viscosity over a wide temp range. Sol in benzene and the lighter hydrocarbons; slightly sol in alc and heavy hydrocarbons.

USE: As a basis for silicone oils or fluids designed to withstand extremes of temp; as a foam suppressant in petr lubricating oil.

2844. Dechlorane® Plus. *1,2,3,4,7,8,9,10,13,13,14,14-Dodecachloro-1,4,4a,5,6,6a,7,10,10a,11,12,12a-dodecahydro-1,4:7,10-dimethanodibenzo[a,e]cyclooctene.* $C_{18}H_{12}Cl_{12}$; mol wt 653.73. C 33.07%, H 1.85%, Cl 65.08%. Polychlorinated hydrocarbon. Prepn by Diels-Alder reaction between cyclo-octadiene and hexachlorocyclopentadiene: K. Ziegler, H. Froitzheim-Kühlhorn, *Ann.* **589**, 157 (1954). Use as fire-retardant for plastics: E. V. Gouinlock, Jr., **U.S.** pat. **3,382,204** (1965 to Hooker). Accumulation in juvenile Atlantic salmon: V. Zitko, *Chemosphere* **9**, 73 (1980). Brief review of fire retardant props: T. J. Machmer, *Life Prop. Protect., Fire Retard. Chem. Assoc., Semi-Annu. Meet.* **1977**, 156-174.

Colorless crystals, mp > 325°. Soluble in *o*-dichlorobenzene.

USE: Fire retardant for plastics.

2845. Decimemide. *4-(Decyloxy)-3,5-dimethoxybenzam-ide;* EGYT 1050; Denegyt. $C_{19}H_{31}NO_4$; mol wt 337.47. C 67.62%, H 9.26%, N 4.15%, O 18.96%. Prepn: **Neth.** pat. **Appl. 6,605,460** corresp to E. Kasztreiner *et al.*, **U.S.** pat. **3,432,549** (1966, 1969 both to EGYT). Pharmacology: E. Kasztreiner, *Magy. Kém. Lapja* **27**, 370 (1972), *C.A.* **78**, 66697z (1973); I. Kiraly *et al.*, *Acta. Physiol. Acad. Sci. Hung.* **47**, 200 (1976). Metabolism study: G. Molnar *et al.*, *Acta Pharm. Hung.* **48**, Suppl., 58 (1978).

Cryst from methanol, mp 121-122°. LD_{50} in mice, rats: 2950, 1650 mg/kg orally, *Psychotropic Drugs and Related Compounds*, E. Usdin, D. H. Efron, Eds. (NIH, Washington, D. C., 2nd Ed., 1972) p 377.

THERAP CAT: Anticonvulsant.

2846. Decoquinate. *6-Decyloxy-7-ethoxy-4-hydroxy-3-quinolinecarboxylic acid ethyl ester;* ethyl 6-(*n*-decyloxy)-7-ethoxy-4-hydroxyquinoline-3-carboxylate; M & B 15497; Deccox. $C_{24}H_{35}NO_5$; mol wt 417.53. C 69.03%, H 8.45%, N 3.35%, O 19.16%. Description of compd: Ball *et al.*, *Chem. & Ind.* (*London*) **1968**, 56; **Belg.** pat. **698,305** (1966 to May & Baker). Metabolism and residue studies: Craine *et al.*, *J. Agr. Food Chem.* **19**, 1228 (1971); Kouba *et al.*, *ibid.* 1234.

THERAP CAT (VET): Coccidiostat.

2847. *n*-Decyl Alcohol. *1-Decanol;* nonylcarbinol. C_{10}-$H_{22}O$; mol wt 158.28. C 75.88%, H 14.01%, O 10.11%. $CH_3(CH_2)_8CH_2OH$. Prepn from caprinaldehyde: Krafft, *Ber.* **16**, 1717 (1883); from capric acid methyl ester: Bouveault, Blanc, *Bull. Soc. Chim.* [3] **31**, 674 (1904); from 1-

chlorodecane: Schultz, *Ber.* **42,** 3611 (1909); from nonyl-magnesium bromide and formaldehyde: Yohe, Adams, *J. Am. Chem. Soc.* **50,** 1507 (1928). Manuf: Faith, Keyes & Clark's *Industrial Chemicals,* F. A. Lowenheim, M. K. Moran, Eds. (Wiley-Interscience, New York, 4th ed., 1975) pp 310-317.

Moderately viscous, strongly refractive liq. Solidifies forming rectangular plates or leaflets; mp 6.4°. bp_{760} 232.9°; bp_{15} 115-120°; bp_8 109.5°. d_4^{20} 0.8297. n_D^{20} 1.43587. Insol in water; sol in alcohol, ether.

USE: In the manuf of plasticizers, synthetic lubricants, petroleum additives, herbicides, surface active agents, solvents. Has moderate antifoaming capacity.

2848. Deet. *N,N-Diethyl-3-methylbenzamide; N,N-diethyl-m-toluamide;* M-Det; *m*-DETA; ENT 20218; Autan; *m*-Delphene; Detamide; Dieltamid; Flypel; Metadelphene; Off; Repel. $C_{12}H_{17}NO$; mol wt 191.26. C 75.35%, H 8.96%, N 7.32%, O 8.37%. Prepn from *m*-toluoyl chloride and diethylamine in benzene or ether: Maxim, *Bull. Soc. Chim. Romania* **11,** 29 (1929); *Chem. Zentr.* **1929,** II, 2324, *C.A.* **24,** 94 (1930); *Beilstein* **9,** II, 325; McCabe *et al., J. Org. Chem.* **19,** 493, 496 (1954). Activity: I. H. Gilbert *et al., J. Econ. Entomol.* **48,** 741 (1955). Metabolism: L. Blomquist, W. Thorsell, *Acta Pharmacol Toxicol.* **41,** 235 (1977).

CON$(C_2H_5)_2$

CH$_3$

Liquid, d_4^{20} 0.996. bp_{19} 160°; $bp_{1.0}$ 111°. n_D^{25} 1.5206. Practically insol in water. Freely sol in alcohol, ether, benzene. Sparingly sol in petr ether. LD_{50} orally in rats about 2 g/kg, Ambrose *et al., Toxicol. Appl. Pharmacol.* **1,** 97 (1959).

USE: Insect repellent. *Caution:* Irritant to eyes, mucous membranes, but not to skin. Ingestion can cause CNS disturbances.

2849. Defensins. A class of low molecular weight cationic peptides isolated from mammalian neutrophils which have *in vitro* microbiocidal activity against various bacteria, fungi, and viruses. Possess similar arginine-rich, cysteine-rich structures of 29-34 amino acids with a conserved core of 8-11 residues. Abbreviated NP (neutrophil peptide) where NP is for rabbit, HNP for human, GPNP for guinea pig etc.; defensins represent a major protein component of granulocytes. Initial identification studies: H. I. Zeya, J. K. Spitznagel, *Science* **142,** 1085 (1961); *eidem, J. Bacteriol.* **91,** 750 (1966). Isoln, antimicrobial activities and amino acid sequence of six NPs: M. E. Selsted *et al., Infect. Immun.* **45,** 150 (1984); M. E. Selsted *et al., J. Biol. Chem.* **260,** 4579 (1985); of three HNPs: T. Ganz *et al., J. Clin. Invest.* **76,** 1427 (1985); M. E. Selsted *et al., ibid.* 1436. Direct inactivation of enveloped viruses: K. A. Daher *et al., J. Virol.* **60,** 1068 (1986). *In vitro* cytolysis of tumor cells: A. Lichtenstein *et al., Blood* **68,** 1407 (1986). Mechanism of cytolysis: *eidem, J. Immunol.* **140,** 2686 (1988).

2850. Deferoxamine. *N'-[5-[[4-[[5-(Acetylhydroxyamino)pentyl]amino]-1,4-dioxobutyl]hydroxyamino]pentyl]-N-(5-aminopentyl)-N-hydroxybutanediamide; N-[5-[3-[[5-aminopentyl)hydroxycarbamoyl]propionamido]pentyl]-3-[[5-(N-hydroxyacetamido)pentyl]carbamoyl]propionohydroxamic acid;* 1-amino-6,17-dihydroxy-7,10,18,21-tetraoxo-27-(N-acetylhydroxylamino)-6,11,17,22-tetraazaheptaeicosane; desferrioxamine B. $C_{25}H_{48}N_6O_8$; mol wt 560.71. C 53.55%, H 8.63%, N 14.99%, O 22.83%. Natural product forming iron complexes, isolated from *Streptomyces pilosus* Ettlinger *et al.* Isoln: Bickel *et al., Helv. Chim. Acta* **43,** 2118 (1960). Prepn from ferrioxamine B and structure: Bickel *et al., ibid.* 2129. Syntheses: Prelog, Walser, *ibid.* **45,** 631 (1962). Derivatives: Bickel *et al., ibid.* **46,** 1385 (1963). Clinical efficacy in thallasemia major (Cooley's anemia): B. Modell *et al., Brit. Med. J.* **284,** 1081 (1982). Treatment of hemodialysis-induced aluminum accumulation in brain: P. Ackril *et al, Lancet* **2,** 692 (1980); in bone: H. H. Malluche *et al., N. Engl. J. Med.* **311,** 140 (1984). Neurotoxicity study: N. F. Olivieri *et al., ibid.* **314,** 869 (1986). Review of clinical ex-

perience: R. Propper, D. Nathan, *Ann. Rev. Med.* **33,** 509-519 (1982).

$$NH_2(CH_2)_5N-C(CH_2)_2CNH(CH_2)_5N-C(CH_2)_2CNH(CH_2)_5N-CCH_3$$
$$\quad\quad HO\ \ O \quad\quad O \quad\quad HO\ \ O \quad\quad O \quad\quad HO\ \ O$$

Monohydrate, crystals from dil alcohol, mp 138-140°. Sol in water at 20°: 1.2%.

Hydrochloride, $C_{25}H_{49}ClN_6O_8$, Ba-29837. Crystals from slightly acidic methanol, mp 172-175°.

Methanesulfonate, $C_{26}H_{52}N_6O_{11}S$, *desferrioxamine mesylate, DFOM, Desferal.* Crystals from dil alcohol, mp 148-149°. Sol in water at 20°: > 20%.

N-Acetyl derivative, $C_{27}H_{50}N_6O_8$, *N-acetyldeferoxamine, N-acetyldesferrioxamine B.* Crystals from *n*-propanol, mp 180-182°.

Iron complex, $C_{25}H_{45}FeN_6O_8$, *Ferrioxamine B.* A natural microbial growth factor. Isoln: Gaeumann *et al.,* U.S. pats. **3,118,823** and **3,153,621** (1964 to Ciba).

Note: Reviews on iron-containing metabolites with growth stimulating properties, *sideramines* or with antibiotic properties, *sideromycins, ferrimycins:* Bickel *et al., Experientia* **16,** 129-133 (1960); Prelog, *Pure Appl. Chem.* **6,** 327-338 (1963); Knüsel, Nüesch, *Nature* **206,** 674-676 (1965); Emery, *Advan. Enzymol. Relat. Areas Mol. Biol.* **35,** 135-185 (1971); Knüsel, Zimmermann, in *Antibiotics,* **vol. 3,** J. W. Corcoran, F. E. Hahn, Eds. (Springer-Verlag, New York, 1975) pp 653-667.

THERAP CAT: Parenteral chelating agent (iron and aluminum). Methanesulfonate as antidote to iron poisoning.

2851. Defibrotide. Fraction P; defibrinotide; Noravid; Prociclide. Sodium salt of polydeoxyribonucleotide extract from animal organs, mol wt 45,000-50,000 daltons. Prepn: A. Butti *et al.,* Ger. pat. **2,154,278;** *eidem,* U.S. pat. **3,899,-481** (1972, 1975 both to Crinos). Antithrombotic activity: R. Niada *et al., Thromb. Res.* **23,** 233 (1981). Pharmacokinetics and profibrinolytic activity in rabbits: R. Pescador *et al., ibid.* **30,** 1 (1983). Clinical trial as antithrombotic: S. Coccheri *et al., Int. J. Clin. Pharm. Res.* **2,** 227 (1982). Clinical evaluation in thrombotic microangiopathy: V. Bonomini *et al., Nephron* **40,** 195 (1985). Symposium on pharmacology and clinical efficacy: O. N. Ulutin, Ed., *Haemostasis* **16,** Suppl. 1, 1-68 (1986).

THERAP CAT: Antithrombotic, fibrinolytic.

2852. Deflazacort. *(11β,16β)-21-(Acetyloxy)-11-hydroxy-2'-methyl-5'H-pregna-1,4-dieno[17,16-d]oxazole-3,20-dione; 11β,21-dihydroxy-2'-methyl-5'βH-pregna-1,4-dieno[17,16-d]oxazole-3,20-dione 2l-acetate;* pregna-1,4-diene-11β,21-diol-3,20-dione[17α,16α-d]-2'-methyloxazoline 21-acetate; oxazacort; azacort; DL-458-IT; L-5458; Calcort; Deflan; Lantadin. $C_{25}H_{31}NO_6$; mol wt 441.52. C 68.01%, H 7.08%, N 3.17%, O 21.74%. Systemic corticosteroid; oxazoline derivative of prednisolone, *q.v.* Prepn: G. Nathansohn, G. Winters, **Belg.** pat. **679,820;** *eidem,* **Brit.** pat. **1,077,393;** *eidem,* **U.S.** pat. **3,436,389** (1966, 1967, 1969 all to Lepetit); G. Nathansohn *et al., J. Med. Chem.* **10,** 799 (1967). Pharmacology: P. Schiatti *et al., Arzneimittel-Forsch.* **30,** 1543 (1980). Pharmacokinetics: A. Assandri *et al., Eur. J. Drug Metab. Pharmacokinet.* **5,** 207 (1980); *eidem, Adv. Exp. Med. Biol.* **171,** 9 (1984). Immunosuppressive activity in humans: B. H. Hahn *et al., J. Rheumatol.* **8,** 783 (1981). Effect on human mineral metabolism: T. J. Hahn *et al., Calcif. Tissue Int.* **31,** 109 (1980); on human glucose metabolism: P. Cavallo-Perin *et al., Eur. J. Clin. Pharmacol.* **26,** 357 (1984). Clinical comparison with prednisone, *q.v.,* in

CH_2OOCCH_3
CO
CH_3
N
CH_3
HO
O
CH_3
O

rheumatoid arthritis and lupus: B. Imbimbo *et al., Adv. Exp. Med. Biol.* **171**, 241 (1984).

Crystals from acetone-hexane, mp 255-256.5°. $[\alpha]_D$ +62.3° (c = 0.5 in chloroform). uv max (methanol): 241-242 nm ($E_{1cm}^{1\%}$ 352.5). LD$_{50}$ orally in mice: 5200 mg/kg (Schiatti).

THERAP CAT: Anti-inflammatory. Glucocorticoid.

2853. Defosfamide. *N,N-Bis(2-chloroethyl)-N'-(3-hydroxypropyl)phosphorodiamidic acid 2-chloroethyl ester;* 2-chloroethyl *N,N-bis(2-chloroethyl)-N'-(3-hydroxypropyl)phosphorodiamidate; N,N,O-tris(2-chloroethyl)-N'-(3-hydroxypropyl)phosphoric acid diamide;* desmofosfamide; B 612; Mitarson. $C_9H_{20}Cl_3N_2O_3P$; mol wt 341.63. C 31.64%, H 5.90%, Cl 31.14%, N 8.20%, O 14.05%, P 9.07%. Prepn: Arnold *et al, Ger.* pat. **1,061,780** (1959 to Asta-Werke).

Oily liquid. Sp gr 1.3675. Practically insol in water. Sol in organic solvents.

THERAP CAT: Antineoplastic.

2854. Deguelin. *13,13a-Dihydro-9,10-dimethoxy-3,3-dimethyl-3H-bis[1]benzopyrano[3,4-b:6',5'-e]pyran-7(7aH)-one.* $C_{23}H_{22}O_6$; mol wt 394.41. C 70.04%, H 5.62%, O 24.34%. From roots of *Tephrosia toxicaria* Pers., leaves of *Tephrosia vogelii* Hook f., *Leguminosae,* and from derris and cubé roots: H. L. Haller, F. B. LaForge, *J. Am. Chem. Soc.* **56**, 2415 (1934); E. Brierly, H. J. Smith, *J. Pharm. Pharmacol.* **20**, 840 (1968). Structure: Clark, *J. Am. Chem. Soc.* **54**, 3000 (1932). Stereoselective synthesis: P. B. Anzeveno, *J. Org. Chem.* **44**, 2578 (1979). Total synthesis of racemate: Fukami *et al., Agr. Biol. Chem. (Tokyo)* **25**, 252 (1961); Omokawa, Yamashita, *ibid.* **38**, 1731 (1974). Stereochemical studies: Djerassi *et al., J. Chem. Soc.* **1961**, 1448.

Yellow oil, $[\alpha]_D^{27}$ −97.2° (c = 0.2 in benzene).
(±)-Form, pale green crystals, mp 171°. Practically insol in water, sol in alc.

USE: Insecticide. *Caution:* Irritates skin. Inhalation may cause fatal pulmonary damage.

2855. Dehydroacetic Acid. *3-Acetyl-6-methyl-2H-pyran-2,4(3H)-dione;* 2-acetyl-5-hydroxy-3-oxo-4-hexenoic acid δ-lactone; methylacetopyronone; DHA. $C_8H_8O_4$; mol wt 168.14. C 57.14%, H 4.80%, O 38.06%. Polymerization product of ketene: Steele *et al., J. Chem. Soc.* **14**, 460 (1949). From ethyl acetoacetate: Arndt, *Org. Syn. coll. vol.* **III**, 231 (1955). Series of articles on toxicity studies, pharmacology, mechanism of action, absorption, distribution, renal action: *J. Pharmacol. Exp. Ther.* **99**, 57-111 (1950).

White to cream cryst powder, mp 109-111° (sublimes). bp 269.9°. Soly (w/w at 25°): in acetone 22%; benzene 18%; methanol 5%; carbon tetrachloride 3%; U.S.P. ethanol 3%; ether 5%; glycerol <0.1%; *n*-heptane 0.7%; olive oil 1.6%; propylene glycol 1.7%; water <0.1%. LD$_{50}$ orally in rats:

1000 mg/kg, H. C. Spencer *et al., J. Pharmacol. Exp. Ther.* **99**, 57 (1950).

Sodium salt hydrate, $C_8H_7NaO_4 \cdot H_2O$, *DHA-S.* Tasteless white powder. Soly (w/w at 25°): in water 33%; propylene glycol 48%; olive oil <0.1%; methanol 14%; U.S.P. ethanol 1%; *n*-heptane <0.1%; glycerol 15%; ether <0.1%; carbon tetrachloride <0.1%; benzene <0.1%; acetone 0.2%. LD$_{50}$ orally in rats: 570 mg/kg, H. C. Spencer *et al., loc. cit.*

USE: In organic syntheses; as plasticizer, compatible with nitrocellulose, polystyrene, methacrylate, vinylite resins; as fungicide and bactericide; in antienzyme toothpastes; to reduce pickle bloating.

2856. Dehydroascorbic Acid. L-*threo-2,3-Hexodiulosonic acid γ-lactone.* $C_6H_6O_6$; mol wt 174.11. C 41.39%, H 3.47%, O 55.14%. The reversibly oxidized form of ascorbic acid. Prepd by the action of benzoquinone on ascorbic acid: Ohle, Erlbach, *Ber.* **67**, 555 (1934); Moll, Wieters, *E. Merck's Jahresber.* **50**, 65 (1936); by the action of iodine: Herbert *et al., J. Chem. Soc.* **1933**, 1270; Kenyon, Munro, *ibid.* **1948**, 158; by oxidn with *peri*-naphthindan-2,3,4-trione hydrate: Moubasher, *J. Biol. Chem.* **176**, 533 (1948); *see also* Müller-Mulot, *Z. Physiol. Chem.* **351**, 52 (1970). Isomerization and formn of derivs: Egge, *Tetrahedron Letters* **1969**, 801. Structure studies: Teichmann, Ziebarth, *Z. Prakt. Chem.* **33**, 124 (1966). Toxicity studies: Gaudiano *et al., Boll. Soc. Ital. Biol. Sper.* **43**, 674 (1967).

Fine needles, dec 225°. Sol in water at 60°. In soln the two carbonyl groups (in position 2 and 3) assume the hydrated form $-C(OH)_2-C(OH)_2-$. Practically neutral reaction. pKa: 3.90. $[\alpha]_D^{20}$ +56°. Aq solns are much less stable than those of ascorbic acid. Detailed stability data: Bogdanski, Bogdanska, *Bull. Acad. Polonaise Sci.* (IIe classe) **3**, 41 (1955). *See also* Velisek *et al., Coll. Czech. Chem. Commun.* **37**, 1465 (1972). Undecomposed dehydroascorbic acid in soln is easily converted to ascorbic acid by reduction with sulfurous acid. Has same antiscorbutic activity in humans as ascorbic acid (upon oral ingestion).

2857. 7-Dehydrocholesterol. *Cholesta-5,7-dien-3-ol;* provitamin D$_3$. $C_{27}H_{44}O$; mol wt 384.62. C 84.31%, H 11.53%, O 4.16%. Occurs in higher animals and in man. Isoln from the horned snail, *Buccinum undatum:* Windaus *et al., Z. Physiol. Chem.* **241**, 100 (1936); from pigskin: Windaus, Bock, *ibid.* **245**, 168 (1937); Boer *et al., Proc. Koninkl. Akad. Wetenschappen Amsterdam* **39**, 622 (1936). Synthesis: Windaus *et al., Ann.* **520**, 98 (1935); Rosenberg, Tinker, U.S. pat. **2,215,727**; Buisman *et al., Rec. Trav. Chim.* **66**, 83 (1947); P. N. Confalone *et al., J. Org. Chem.* **46**, 1030 (1981). Irradiation with ultraviolet light produces vitamin D$_3$; also lumisterol$_3$ and tachysterol$_3$: Windaus *et al., Ann.* **533**, 118 (1938).

Hydrated plates from ether-methanol. The water of crystn is held tenaciously: Schenck *et al., Ber.* **69**, 2705 (1936). When anhydr mp 150-151°. $[\alpha]_D^{20}$ −113.6° (c = 1 in chloroform), Koch, Koch, *J. Biol. Chem.* **116**, 757 (1936); $[\alpha]_D^{20}$ −127.1° in benzene (Boer, *loc. cit.*). All provitamins D have the same uv maxima at 260, 270, 281, 293.5 nm. Insol in water, sol in the usual organic solvents.

Acetate, $C_{29}H_{46}O_2$, crystals from methanol, mp 129-130°; $[\alpha]_D^{20}$ −85.3° (c = 1.2 in benzene).

Benzoate, $C_{34}H_{48}O_2$, plates from chloroform-acetone, mp 139-140°; $[\alpha]_D^{20}$ — 53.2° in chloroform.

2858. Dehydrocholic Acid. *3,7,12-Trioxocholan-24-oic acid;* 3,7,12-triketocholanic acid; Acolen; Bilidren; Bilostat; Cholan-DH; Cholagon; Cholepatin; Chologon; Decholin; Dehychol; Deidrocolico Vita; Didrocolo; Dilabil; Erebile; Felacrinos; Oxycholin; Procholon. $C_{24}H_{34}O_5$; mol wt 402.51. C 71.61%, H 8.51%, O 19.87%. Prepd from cholic acid by oxidation with CrO_3 in glacial acetic acid: Hammarsten in *Abderhalden's Handbuch der Biol. Arbeitsmethoden,* Abt. **I,** Teil 6, p 238 (1925); by using chlorine instead of CrO_3: Hinkley, Singleton, **U.S.** pat. **2,966,499** (1960 to Merck & Co.).

Bitter crystals from acetone, mp 237°. $[\alpha]_D^{20}$ +26° (c = 1.4 in alc). Soly at 15° (g/l): water 0.18; alcohol 3.3; ether 0.46; chloroform 9.04; benzene 1.04; acetone 7.76; ethyl acetate 7.4; glacial acetic acid 7.42.

Sodium salt, $C_{24}H_{33}NaO_5$, *sodium dehydrocholate, Decholin Sodium, Biliton, Carachol, Dycholium, Suprachol, Procholon Sodium.* Very sol in water. pH of aq soln about 9.0.

THERAP CAT: Choleretic.

2859. 11-Dehydrocorticosterone. *21-Hydroxypregn-4-ene-3,11,20-trione;* Δ^4-pregnen-21-ol-3,11,20-trione; 17-(1-keto-2-hydroxyethyl)-Δ^4-androsten-3,11-dione; Kendall's compound A. $C_{21}H_{28}O_4$; mol wt 344.43. C 73.22%, H 8.19%, O 18.58%. Found in adrenal cortex. 1000 lbs of beef glands yield 333 mg: Kendall, *Cold Spring Harbor Symposia Quant. Biol.* **5,** 299 (1937). Isoln procedure: Mason *et al., J. Biol. Chem.* **114,** 613 (1936); *see also* Kendall *et al., Proc. Staff Meet. Mayo Clinic* **12,** 136 (1937). Prepn from corticosterone 21-acetate: Reichstein, *Helv. Chim. Acta* **20,** 953 (1937); from desoxycholic acid: Lardon, Reichstein, *ibid.* **26,** 747 (1943); Gallagher, *Recent Progress in Hormone Research* (New York, 1946) p 83; from 3α-acetoxy-11-ketobisnorcholanic acid: Sarett, *J. Biol. Chem.* **162,** 601 (1946); *J. Am. Chem. Soc.* **68,** 2478 (1946); from 3α-acetyl-11-ketolithocholic acid methyl ester: Wettstein, Meystre, *Helv. Chim. Acta* **30,** 1262 (1947). Manuf: Wettstein, **U.S.** pat. **2,778,-776** (1957 to Ciba).

Large prisms from aq acetone, mp 178-180°. Can be distilled in high vacuum. $[\alpha]_{546}^{25}$ +299°, also given as +347° (c = 0.23 in benzene); $[\alpha]_D^{25}$ +258° (alc). Relatively good soly in benzene.

Acetate, $C_{23}H_{30}O_5$, needles from alc, mp 179-181°. $[\alpha]_{546}^{18}$ +285° (dioxane); $[\alpha]_D^{18}$ +233.7° (dioxane).

2860. Dehydroemetine. *2,3-Didehydro-6',7',10,11-tetramethoxyemetan;* 2,3-dehydroemetine; 2-dehydroemetine. $C_{29}H_{38}N_2O_4$; mol wt 478.61. C 72.77%, H 8.00%, N 5.85%, O 13.37%. Synthetic analog of emetine, *q.v.*; the (−)-form is therapeutically active. Synthesis of racemic 2,3-dehydroemetine: Brossi *et al., Helv. Chim. Acta* **42,** 772 (1959). Stereospecific synthesis of (±)-form and (±)-*2,3-dehydroisoemetine,* the (1'β)-isomer: Clark *et al., J. Chem. Soc.* **1962,** 2479. Separation of the four possible isomers: Brossi, Burkhardt, *Experientia* **18,** 211 (1962). Absolute configura-

tion and bioactivity of isomers: A. Brossi *et al., Helv. Chim. Acta* **45,** 2219 (1962). Prepn of pure (−)-form: A. Brossi, Belg. pat. **629,898;** *eidem,* **U.S.** pat. **3,311,633** (1963, 1967 both to Hoffmann-La Roche); N. Whittaker *J. Chem. Soc. (C)* **1969,** 94. Antiprotozoal activity *in vitro:* G. H. Al-Khateeb *et al., Chemotherapy (Basel)* **23,** 267 (1977). Review of use in amebiasis: G. Woolfe, *Prog. Drug. Res.* **8,** 12 (1965).

Crystals from isopropyl ether, mp 94-96°. $[\alpha]_D$ —183°. (−)-Form dihydrobromide, $C_{29}H_{40}Br_2N_2O_4$, crystals, mp 243-245°. $[\alpha]_D$ —97° (methanol). uv max (alc): 282 nm (E 7300).

(+)-Form dihydrobromide, $C_{29}H_{40}Br_2N_2O_4$, crystals, mp 241-243°. $[\alpha]_D$ +95° (methanol). uv max (alc): 282 nm (E 7350).

(±)-Form dihydrochloride, $C_{29}H_{40}Cl_2N_2O_4$, *Dametin, Mebadin.* Crystals from ethanol + ether, mp 235°.

(±)-2,3-Dehydroisoemetine hydrochloride, crystals from ethanol + ether, mp 220-225°.

(−)-2,3-Dehydroisoemetine dihydrobromide, $C_{29}H_{40}Br_2-N_2O_4$, crystals, mp 257-260°. $[\alpha]_D$ —107° (methanol). uv max (alc): 285 nm (ε 7400).

(+)-2,3-Dehydroisoemetine dihydrobromide, $C_{29}H_{40}Br_2-N_2O_4$, crystals, mp 257-258°. $[\alpha]_D$ +109° (methanol). uv max (alc): 285 nm (ε 7450).

THERAP CAT: Anti-amebic.

2861. Dehydroergosterol. *Ergosta-5,7,9(11),22-tetraen-3β-ol.* $C_{28}H_{42}O$; mol wt 394.62. C 85.22%, H 10.73%, O 4.05%. Prepd from ergosterol: Windhaus, Linsert, *Ann.* **465,** 148 (1928); Callow, Rosenheim, *J. Chem. Soc.* **1933,** 387. From isopyrocalciferol: Windhaus, Dimroth, *Ber.* **70,** 376 (1937).

Solvated plates from alcohol, needles from ether. When dry, mp 146°. bp$_{0.5}$ 230°. $[\alpha]_D^{15}$ +149.2° (c = 1.9 in chloroform). Absorption spectrum: Morton, de Gouveia, *J. Chem. Soc.* **1934,** 916. One gram dissolves in 800 ml methanol. Freely sol in ether, chloroform, benzene.

Acetate, $C_{30}H_{44}O_2$, mp 146°. $[\alpha]_D^{16}$ +204° (c = 1.1 in chloroform).

Methyl ether, $C_{29}H_{44}O$, mp 106°. $[\alpha]_D^{20}$ +166° (c = 2 in chloroform).

2862. 3-Dehydroretinal. *3,4-Didehydroretinal;* 3,7-dimethyl-9-(2,6,6-trimethyl-1,3-cyclohexadien-1-yl)-2,4,6,8-nonatetraenal; retinal 2; retinene 2; vitamin A$_2$ aldehyde. $C_{20}H_{26}O$; mol wt 282.42. C 85.06%, H 9.28%, O 5.66%. Carotenoid component of the visual pigments of fresh-water and migratory fish, lampreys and certain amphibians. Of the 16 possible stereoisomers, 7 have been synthesized. The 11-*cis*-isomer is the chromophore of porphyropsin and cyanopsin, *q.q.v.* Isoln from retinas of fresh-water fish: G. Wald, *Nature* **139,** 1017 (1937). Recognition as vitamin A$_2$ aldehyde: R. A. Morton, *ibid.* **153,** 69 (1944). Prepn by the oxidation of vitamin A_2: G. Wald, *Biochim. Biophys. Acta* **4**, 215 (1950). Synthesis of *trans*-isomer: K. R. Farrar *et al.*, *J. Chem. Soc.* **1952**, 1414; of stereoisomers: U. Schwieter *et al.*, *Helv. Chim. Acta* **45**, 517, 528, 541 (1962); R. S. H. Liu *et al.*, *J. Am. Chem. Soc.* **99**, 8095 (1977). HPLC separation: K. Tsukida *et al.*, *J. Chromatog.* **192**, 395 (1980); T. Suzuki, M. Makino-Tasaka, *Anal. Biochem.* **129**, 111 (1983). Biological activity is thought to be analogous with that of retinal, *q.v.*: G. Wald, *J. Gen. Physiol.* **22**, 775 (1939); *idem*, *Fed. Proc.* **12**, 606 (1953). *Reviews:* G. Wald, *Science* **162**, 230-239 (1968); R. Hubbard *et al.*, "Methodology of Vitamin A and Visual Pigments" in *Methods of Enzymology* **18**, D. B. McCormick, L. D. Wright, Eds. (Academic Press, New York, 1971) pp 615-653. Book: *The Retinoids*, Vol. **1-2**, M. B. Sporn *et al.*, Eds. (Academic Press, New York, 1984).

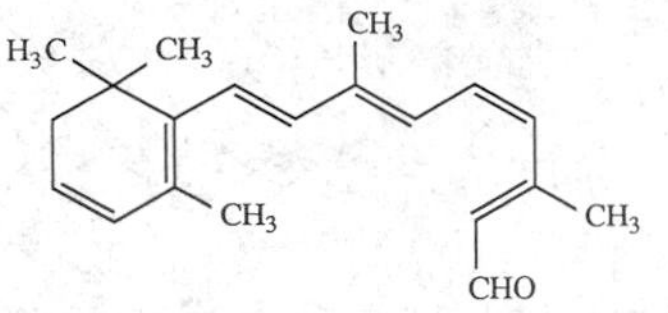

11-*cis* isomer

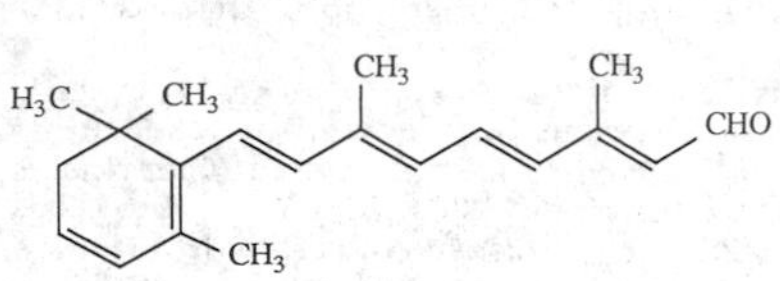

All-*trans*

trans-Isomer, orange-red prisms from pentane, mp 77-78. uv max (ethanol): 401 nm ($E^{1\%}_{1cm}$ 1470).

11-*cis*-Isomer, oil from petr ether-ether. uv max (ethanol): 393 nm ($E^{1\%}_{1cm}$ 882).

USE: As tool in biological energy transduction research.

2863. 7-Dehydrositosterol. *Stigmasta-5,7-dien-3β-ol.* $C_{29}H_{48}O$; mol wt 412.67. C 84.40%, H 11.72%, O 3.88%. From soy bean oil β-sitosterol: Wunderlich, *Z. Physiol. Chem.* **241**, 116 (1936). Starting with cholesterol: U.S. pat. **2,386,635** (1945).

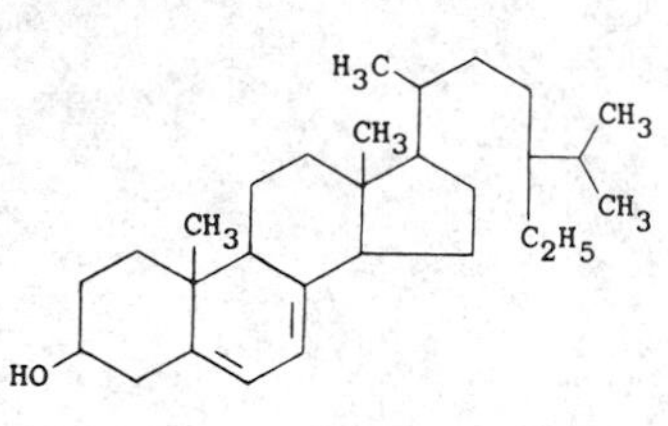

Platelets from alc, browns on contact with air, mp 144-145°. $[\alpha]^{21}_D -116°$ (c = 2 in chloroform). Sparingly sol in methanol, somewhat more in alcohol, freely sol in the other usual organic solvents. Insol in water.

Acetate, $C_{31}H_{50}O_2$, mp 151-152°. $[\alpha]^{21}_D -71°$ (c = 2 in chloroform).

2864. Delapril. *(S)-N-(2,3-Dihydro-1H-inden-2-yl)-N-[N-[1-(ethoxycarbonyl)-3-phenylpropyl]-L-alanyl]glycine; N-[N-[(S)-1-(ethoxycarbonyl)-3-phenylpropyl]-L-alanyl]-N-(indan-2-yl)glycine; ethyl (S)-2-[[(S)-1-[(carboxymethyl)-2-indanylcarbamoyl]ethyl]amino]-4-phenylbutyrate;* alindapril; indalapril; $C_{26}H_{32}N_2O_5$; mol wt 452.52. C 69.01%, H 7.13%, N 6.19%, O 17.67%. Angiotensin-converting enzyme inhibitor. Prepn: Y. Oka *et al.*, *Eur.* pat. *Appl.* **51,391**; *eidem*, U.S. pat. **4,385,051** (1982, 1983 both to Takeda); J. T. Suh *et al.*, *Eur. J. Med. Chem. - Chim. Ther.* **20**, 563 (1985); A. Miyake *et al.*, *Chem. Pharm. Bull.* **34**, 2852 (1986). Pharmacology in animals: Y. Inada *et al.*,

Japan. J. Pharmacol. **42**, 1 (1986); *eidem, ibid.*, 99. Pharmacology and metabolism in humans: T. Ogihara *et al.*, *Curr. Ther. Res.* **41**, 809 (1987). Pharmacokinetics and evaluation in hypertension: H. Shionoiri *et al.*, *Clin. Pharmacol. Ther.* **41**, 74 (1987).

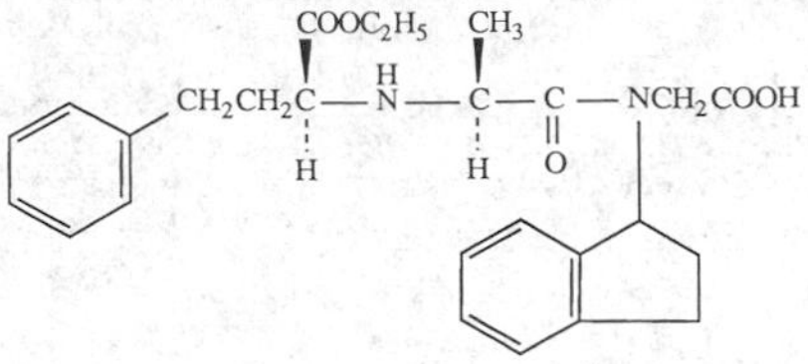

Hydrochloride, $C_{26}H_{33}ClN_2O_5$, *CV-3317, REV-6000A, Adecut.* Colorless plates from acetone + hydrochloric acid, mp 166-170° (dec). $[\alpha]^{22}_D +18.5°$ (c = 1 in methanol).

THERAP CAT: Antihypertensive.

2865. Delmadinone Acetate. *17-(Acetyloxy)-6-chloropregna-1,4,6-triene-3,20-dione; 6-chloro-17-hydroxypregna-1,4,6-triene-3,20-dione acetate;* 1,6-bisdehydro-6-chloro-17α-acetoxyprogesterone; Δ^1-chlormadinone acetate; RS-1301; Estrex; Tardastrex; Zenadrex. $C_{23}H_{27}ClO_4$; mol wt 402.92. C 68.56%, H 6.76%, Cl 8.80%, O 15.88%. Prepn: Ringold *et al.*, *J. Am. Chem. Soc.* **81**, 3485 (1959); **Brit.** pat. **890,315** (1962 to Upjohn), *C.A.* **57**, 5996c (1962) and **76**, 14813y (1972); Campbell, Babcock, **Ger.** pat. **1,243,682** (1967 to Upjohn), *C.A.* **67**, 100343r (1967). Activity studies: Dorfman, Kincl, *Steroids* **1**, 185 (1963); Weichert *et al.*, *Arzneimittel-Forsch.* **17**, 1103 (1967); Gerber *et al.*, *J. Small Animal Pract.* **14**, 151 (1973).

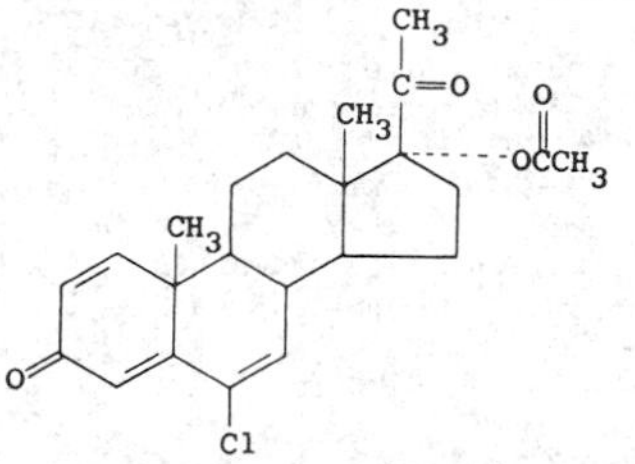

Crystals, mp 168-170°. $[\alpha]_D -83°$ (chloroform). uv max (ethanol): 229, 258, 297 nm (log ∈ 4.00, 4.00, 4.03).

THERAP CAT: Progestogen; anti-androgen; anti-estrogen.

THERAP CAT (VET): Progestogen; anti-androgen; anti-estrogen.

2866. Delphinidin. *3,5,7-Trihydroxy-2-(3,4,5-trihydroxyphenyl)-1-benzopyrylium chloride; 3,3',4',5,5',7-hexahydroxyflavylium chloride; 3,3',4',5,5',7-hexahydroxy-2-phenylbenzopyrylium chloride;* delphinidol. $C_{15}H_{11}ClO_7$; mol wt 338.70. C 53.19%, H 3.27%, Cl 10.47%, O 33.07%. The aglucone of delphinin: Willstätter, Mieg, *Ann.* **408**, 61 (1915); the aglucone of violanin: Willstätter, Weil, *ibid.* **412**, 178 (1917). Occurrence in plants: Acheson, *Can. J. Biochem. Physiol.* **37**, 1 (1959). Synthesis: Pratt, Robinson, *J. Chem. Soc.* **127**, 166 (1925); Bradley *et al., ibid.* **1930**, 793.

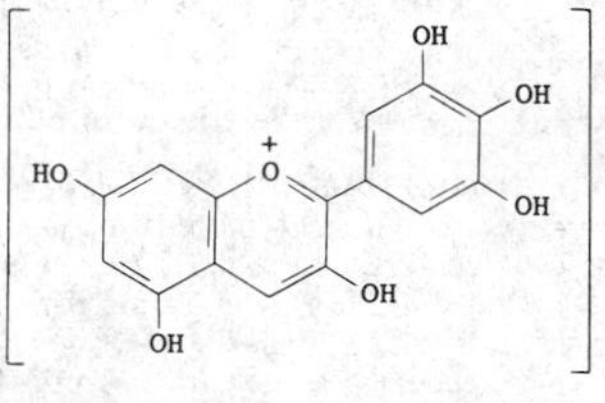

Chocolate-brown prisms or needles with metallic luster from 5% HCl. Crystallizes from water with 1, 2 or 4 mols H_2O. When anhydr does not melt below 350°. Absorption

max (methanolic HCl): 544 nm. Sol in methanol, ethanol, ethyl acetate.

3-Glucoside, $C_{21}H_{21}ClO_{12}$, *myrtillin-a.* From the whortleberry *(Vaccinium myrtillus* L., *Ericaceae):* Willstätter, Zollinger, *Ann.* **408,** 83 (1915); **412,** 195 (1916); from the pansy *(Viola tricolor* L., *Violaceae):* Karrer, de Meuron, *Helv. Chim. Acta* **16,** 292 (1933). Structure and synthesis: Reynolds, Robinson, *J. Chem. Soc.* **1934,** 1039. Deep purple crystals from dil HCl. Absorption max (methanolic HCl): 535 nm.

3,5-Diglucoside, $C_{27}H_{31}ClO_{17}$, *3,5-bis(glucosyloxy)-3',4',- 5',7-tetrahydroxyflavylium chloride; delphin; delphoside; hyacin.* From flowers of *Salvia patens* Cav., *Labiatae,* structure and synthesis: Reynolds *et al., J. Chem. Soc.* **1934,** 1235; from flowers of *Verbena hybrida* Hart., *Verbenaceae:* Scott-Moncrieff, Sturgess, *Biochem. J.* **34,** 268 (1940); from grape skins: Bockian *et al., J. Agr. Food Chem.* **3,** 695 (1955). Identity with hyacin: Saito *et al., Proc. Japan. Acad.* **36,** 340 (1960). Leaflets with bronze luster from dil HCl. Absorption max (methanolic HCl): 534 nm. Practically insol in cold water, alcohol, dil acids; sol in hot dil HCl.

Compound with glucose + hydroxybenzoic acid, $C_{41}H_{39}$- ClO_{21}, *delphinin.* From *Delphinium consolida* L., *Ranunculaceae:* Willstätter, Mieg, *loc. cit.* Dark red-brown plates or prisms from 3% HCl. dec 200-203°. Sol in water with decompn, dil alcohol, acetone, boiling alcohol.

Compound with glucose + rhamnose + *p*-hydroxycinnamic acid, $C_{36}H_{37}ClO_{18}$, *violanin.* From flowers of *Viola tricolor* L., *Violaceae:* Willstätter, Weil, *loc. cit.* Structure: Karrer, de Meuron, *loc. cit.* Bluish-violet plates with green metallic luster from methanolic HCl + ether. Sol in amyl alcohol, in 0.15-0.5% HCl; less sol in 1% HCl; slightly sol in water; practically insol in 5-12% HCl.

2867. Delphinine. *(1α,6α,14α,16β)-1,6,16-Trimethoxy-4-(methoxymethyl)-20-methylaconitane-8,13,14-triol 8-acetate 14-benzoate.* $C_{33}H_{45}NO_9$; mol wt 599.70. C 66.09%, H 7.56%, N 2.34%, O 24.01%. A toxic alkaloid from seeds of *Delphinium staphisagria* L., *Ranunculaceae.* Isoln: Markwood, *J. Am. Pharm. Assoc.* **16,** 928 (1927). Purification: Schneider, *Arch. Pharm.* **283,** 283 (1950). Structure: Jacobs, Pelletier, *J. Am. Chem. Soc.* **78,** 3542 (1956); Wiesner *et al., Can. J. Chem.* **47,** 2734 (1969). Configuration: Wiesner *et al., Tetrahedron Letters* **1959,** no. 3, 12; **1960,** no. 15, 23; Gilman, Marion, *ibid.* **1962,** 923; Birnbaum *et al., ibid.* **1971,** 867. Synthesis: Wiesner *et al., Can. J. Chem.* **50,** 1925 (1972).

Orthorhombic, six-sided plates from alc, mp 197.5-199°. $[\alpha]_D^{20}$ +25° (ethanol). Practically insol in water. Sol in 25 parts alcohol, 20 parts chloroform, 10 parts ether.

Hydrochloride, $C_{33}H_{45}NO_9 \cdot HCl$, crystals, dec 214°.

2868. Delsoline. *20-Ethyl-6,14,16-trimethoxy-4-(methoxymethyl)aconitane-1,7,8-triol.* $C_{25}H_{41}NO_7$; mol wt 467.59. C 64.21%, H 8.84%, N 3.00%, O 23.95%. R = CH_3. From seeds of *Dephinium consolida* L., *Ranunculaceae:* Markwood, *J. Am. Pharm. Assoc.* **13,** 696 (1924); Cionga, Iliescu, *Ber.* **74,** 1031 (1941); Marion, Edwards, *J. Am. Chem. Soc.* **69,** 2010 (1947). Structure: Anet *et al., Can. J. Chem.* **35,** 397 (1957); Skaric, Marion, *J. Am. Chem. Soc.* **80,** 4434 (1958); *eidem, Can. J. Chem.* **38,** 2433 (1960); **39,** 1579 (1961).

Prisms from methanol, mp 213-216.5° (when immersed at 205°). $[\alpha]_D^{22}$ +53.4° (c = 2.04 in chloroform). Slightly sol in water, sol in alcohol or chloroform.

Delcosine, $C_{24}H_{39}NO_7$, R = H. Alkaloid also found in *D. consolida.* mp 203-204°, $[\alpha]_D^{25}$ +56.8° (c = 2.01 in chloroform). Quite sol in water.

USE: Glycosidal dyestuff.

2869. Deltamethrin. *3-(2,2-Dibromoethenyl)-2,2-dimethylcyclopropanecarboxylic acid cyano(3-phenoxyphenyl)-methyl ester;* (S)-α-cyano-3-phenoxybenzyl-(1R)-cis-3-(2,2-dibromovinyl)-2,2-dimethylcyclopropane carboxylate; decamethrin; esbecythrin; FMC 45498; NRDC 161; RU 22974; Butox; Decis; K-Othrine. $C_{22}H_{19}Br_2NO_3$; mol wt 505.22. C 52.30%, H 3.79%, Br 31.63%, N 2.77%, O 9.50%. Potent synthetic pyrethroid insecticide. Prepn of racemic mixture: M. Elliot *et al., Ger.* pat. **2,439,177** (1975 to NRDC), *C.A.* **83,** 73519z (1975); of decamethrin and isomers: *eidem, Pestic. Sci.* **9,** 105 (1978). Activity: *eidem, Nature* **248,** 710 (1974); *eidem, Pestic. Sci.* **9,** 112 (1978). Abs config: J. D. Owen, *J. Chem. Soc. Perkin Trans. I* **1975,** 1865. Photochemistry: L. O. Ruzo *et al., J. Agr. Food Chem.* **25,** 1385 (1977). Metabolism: *eidem, ibid.* **26,** 918 (1978). Toxicology: R. Kavlock *et al., J. Environ. Pathol. Toxicol.* **2,** 751 (1979). Pharmacological effects on central nervous system: P. H. Chanh *et al., Arzneimittel-Forsch.* **34,** 175 (1984).

Crystals, mp 98-101°. Sol in ethanol, acetone, dioxane. Insol in water. LD_{50} in female rats (mg/kg): 31 orally; 4 i.v. (Kavlock).

USE: Insecticide.

2870. Demanyl Phosphate. *Mono[2-(dimethylamino)-ethyl]phosphoric acid ester; 2-dimethylaminoethanol dihydrogen phosphate;* phosphoryldimethylaminoethanol; phosphoryldimethylcolamine; phosphoryldimethylethanolamine; P-DMEA; Panclar. $C_4H_{12}NO_4P$; mol wt 169.12. C 28.41%, H 7.15%, N 8.28%, O 37.84%, P 18.31%. $(CH_3)_2NCH_2CH_2$-OPO_3H_2. Prepn: Aron-Samuel *et al., Fr.* pat. **992,802** (1951), *C.A.* **51,** 458e (1957); Cherbuliez, Rabinowitz, *Helv. Chim. Acta* **41,** 1168 (1958); **Brit.** pat. **861,463;** U.S. pat. **3,098,072** (1961, 1963 to Ciba). The phosphate ester of a possible choline precursor. Has been isolated from a choline-requiring mutant of *Neurospora crassa,* strain 47904: Wolf, Nyc, *J. Biol. Chem.* **234,** 1068 (1959). Biosynthetic studies: Ansell, Chojnacki, *Nature* **196,** 545 (1962); *eidem, Biochem. J.* **98,** 303 (1966).

Crystals, mp 175-176°. Monohydrate, mp 78-81°.

THERAP CAT: Psychotonic.

2871. Demecarium Bromide. *3,3'-[1,10-Decanediylbis-[(methylimino)carbonyloxy]]bis[N,N,N-trimethylbenzenamin-ium]dibromide; (m-hydroxyphenyl)trimethylammonium bromide, decamethylenebis(methylcarbamate); N,N'-bis[3-trimethylammoniumphenoxycarbonyl]-N,N'-dimethyldecamethylenediamine dibromide; decamethylenebis[m-dimethylaminophenyl N-methylcarbamate] dimethobromide; decamethylenebis[N-methylcarbamic acid m-dimethylaminophenyl ester] bromomethylate; BC 48; Tosmilen; Humorsol.* $C_{32}H_{52}Br_2N_4O_4$; mol wt 716.61. C 53.63%, H 7.31%, Br

22.30%, N 7.82%, O 8.93%. Preparation: Schmid, **U.S.** pat. **2,789,981** (1957 to Oesterreichische Stickstoffwerke).

White, slightly hygroscopic powder, dec 162-167°. Freely sol in water, alc. Sparingly sol in acetone. Insol in ether. Aq solns are neutral, stable, and may be sterilized by heat.
THERAP CAT: Cholinergic (ophthalmic).

2872. Demeclocycline. *7-Chloro-4-dimethylamino-1,4,- 4a,5,5a,6,11,12a-octahydro-3,6,10,12,12a-pentahydroxy- 1,11-dioxo-2-naphthacenecarboxamide; 7-chloro-6-demeth- yltetracycline;* demethylchlortetracycline (obsolete); RP 10192; Bioterciclin; Declomycin; Deganol; Ledermycin; Periciclina. $C_{21}H_{21}ClN_2O_8$; mol wt 464.88. C 54.26%, H 4.55%, Cl 7.63%, N 6.03%, O 27.53%. Antibiotic related to tetracycline produced by *Streptomyces aureofaciens.* Prepn: McCormick *et al., J. Am. Chem. Soc.* **79,** 4561 (1957); **U.S.** pat. **2,878,289** (1959 to Am. Cyanamid). Improved fermen- tation processes: Szumski; Goodman, Matrishin; Good- man, **U.S.** pats. **3,012,946** and **3,019,172; 3,050,446** (1961 and 1962 to Am. Cyanamid); **Fr.** pat. **1,344,645** (1963 to Merck & Co.); Neidleman, **U.S.** pat. **3,154,476** (1964 to Olin Mathieson). Abs config: Dobrynin *et al., Tetrahedron Let- ters* **1962,** 901. Toxicity data (hydrochloride): E. I. Golden- thal, *Toxicol. Appl. Pharmacol.* **18,** 185 (1971).

Sesquihydrate, mp 174-178° (dec). $[\alpha]_D^{25}$ −258° (c = 0.5 in $0.1N$ H_2SO_4). Solubilities: Marsh, Weiss, *J. Assoc. Offic. Anal. Chem.* **50,** 457 (1967).
Hydrochloride, *Clortetrin, Demetraciclina, Detravis, Meci- clin, Mexocine.* LD_{50} orally in rats: 2372 mg/kg (Golden- thal).
THERAP CAT: Antibacterial.
THERAP CAT (VET): Antibacterial.

2873. Demecolcine. *6,7-Dihydro-1,2,3,10-tetramethoxy- 7-(methylamino)benzo[a]heptalen-9(5H)-one; N-deacetyl-N- methylcolchicine; N-methyl-N-desacetylcolchicine;* colcha- mine; Santavy's substance F; Colcemid; Omaine. $C_{21}H_{25}$- NO_5; mol wt 371.42. C 67.90%, H 6.78%, N 3.77%, O 21.54%. Isoln from *Colchicum autumnale* L., *Liliaceae:* Santavy, *Pharm. Acta Helv.* **25,** 248 (1950); Santavy, Reich- stein, *Helv. Chim. Acta* **33,** 1606 (1950); Schlittler, Uffer, **Ger.** pat. **936,268** (1955 to Ciba), *C.A.* **53,** 1396 (1959). Structure: Santavy *et al., Helv. Chim. Acta* **36,** 1319 (1953). Synthesis: Uffer *et al., ibid.* **37,** 18 (1954); H. G. Capraro, A. Brossi, *ibid.* **62,** 965 (1979).

Pale yellow prisms from ethyl acetate + ether, mp 186°. $[\alpha]_D^{20}$ −129.0° (c = 1 in chloroform). uv max (alc): 245, 355

nm (log ε 4.55, 4.24). Basic reaction. Soluble in acidified water, in alcohol, ether, chloroform, benzene.
THERAP CAT: Antineoplastic.

2874. Demegestone. *17-Methyl-19-norpregna-4,9-di- ene-3,20-dione;* 17α-methyl-19-nor-$\Delta^{4,9}$-pregnadiene-3,20- dione; 17α-methyl-Δ^9-19-norprogesterone; R 2453; Lutio- nex. $C_{21}H_{28}O_2$; mol wt 312.45. C 80.73%, H 9.03%, O 10.24%. Prepn: **Neth.** pat. **Appl. 6,517,141** corresp to Vig- nau *et al.,* and Joly *et al.,* **U.S.** pats. **3,453,267** and **3,547,959** (1966, 1969, 1970, all to Roussel-UCLAF); Velluz *et al., Tetrahedron,* **Suppl 8,** part II, 495 (1966); Joly *et al., Bull. Soc. Chim. France* **1973,** 2694.

Crystals from ether, mp 106°. $[\alpha]_D$ −275° (c = 0.5 in eth- anol). uv max (ethanol): 214, 302 nm (ε 6350, 21000).
THERAP CAT: Progestogen.

2875. Demeton. *Phosphorothioic acid O,O-diethyl O-[2- (ethylthio)ethyl] ester mixture with O,O-diethyl S-[2-(ethyl- thio)ethyl]phosphorothioate;* mercaptophos; Bayer 8169; E-1059; Systox. $C_8H_{19}O_3PS_2$; mol wt 258.34. C 37.19%, H 7.41%, O 18.58%, P 11.99%, S 24.82%. Isomeric mixture consisting of demeton-O and demeton-S (O,O-diethyl O(and S)-ethylmercaptoethyl thiophosphates). Another name for demeton-S is *Isosystox.* Prepn: Schrader, **U.S.** pats. **2,571,- 989** and **2,597,534** (1951, 1952 both to Bayer); Gardner, Heath, *Anal. Chem.* **25,** 1849 (1953).

Oily liq. Faint odor. bp_2 134°. Virtually insol in water. Sol in ethanol, propylene glycol, toluene and similar hydro- carbons. Absorbed by plants, rendering the foliage and plant fluids toxic to insects. LD_{50} in female, male rats: 2.5, 6.2 mg/kg orally; 8.2, 14 mg/kg dermally, T. B. Gaines, *Toxicol. Appl. Pharmacol.* **14,** 515 (1969).
USE: Insecticide. *Caution:* Cholinesterase inhibitor. Readily absorbed through skin.

2876. Demexiptiline. *5H-Dibenzo[a,d]cyclohepten-5-one O-[2-(methylamino)ethyl]oxime.* $C_{18}H_{18}N_2O$; mol wt 278.35. C 77.67%, H 6.52%, N 10.06%, O 5.75%. Prepn: S. Schütz *et al.,* **Ger.** pat. **1,247,302** corresp to **U.S.** pat. **3,963,778** (1967, 1976 both to Bayer). Synthesis and pharmacological properties: G. Aichinger *et al., Arzneimittel-Forsch.* **19,** 838 (1969); S. Rossi *et al., Farmaco Ed. Sci.* **24,** 685 (1969).

Colorless oil.
Hydrochloride, $C_{18}H_{19}ClN_2O$, *LM 2909, Deparon.* Cryst from isopropanol, mp 232-233°. LD_{50} in rats (mg/kg): 330 orally, 25 i.v., S. Rossi *et al., loc. cit.*
THERAP CAT: Antidepressant.

2877. Denatonium Benzoate. *N-[2-[(2,6-Dimethylphen- yl)amino]-2-oxoethyl]-N,N-diethylbenzenemethanaminium benzoate;* benzyldiethyl[(2,6-xylylcarbamoyl)methyl]ammo- nium benzoate; lignocaine benzyl benzoate; Bitrex. $C_{28}H_{34}$- N_2O_3; mol wt 446.57. C 75.30%, H 7.67%, N 6.27%, O 10.75%. Prepn: J. E. Hay, **U.S.** pat. **3,080,327** (1963 to

T. & H. Smith Ltd.); *idem*, **U.S.** pat. **3,268,575** (1966 to Edinburgh Pharm. Ind.). HPLC determn in rapeseed oil: C. E. Damon, B. C. Pettitt, Jr., *J. Chromatog.* **195,** 243 (1980). Brief review of substance history and uses: H. A. S. Payne, *Chem. & Ind. (London)* **22,** 721-723 (1988).

Crystals from isopropyl alcohol + ethyl acetate, mp 166-170°. Among the most bitter substances known to man. Soluble in water, alcohol; sparingly sol in acetone. Practically insol in ether.

USE: Pharmaceutic aid (alcohol denaturant; flavor). Added to toxic substances as a deterrent to accidental ingestion. Can replace brucine or quassin as denaturant for ethyl alcohol.

2878. Denopamine. *(R)-α-[[[2-(3,4-Dimethoxyphenyl)ethyl]amino]methyl]-4-hydroxybenzenemethanol;* (−)-*(R)-α-[[(3,4-dimethoxyphenethyl)amino]methyl]-p-hydroxybenzyl alcohol;* (−)-*(R)-1-(4-hydroxyphenyl)-2-(3,4-dimethoxyphenethylamino)ethanol;* TA-064; Carguto; Kalgut. $C_{18}H_{23}NO_4$; mol wt 317.38. C 68.12%, H 7.30%, N 4.41%, O 20.16%. Selective β_1-adrenoceptor agonist with positive inotropic activity. Prepn: M. Ikezaki *et al.,* **Belg.** pat. **833,-731;** *eidem,* **U.S.** pat. **4,032,575** (1976, 1977 both to Tanabe Seiyaku); N. Umino *et al., Chem. Pharm. Bull.* **27,** 1479 (1979). Effect on carbohydrate and lipid metabolism in rats: M. Inamasu *et al., Biochem. Pharmacol.* **33,** 2171 (1984). Cardiovascular pharmacology in animals: T. Nagao *et al., Japan. J. Pharmacol.* **35,** 415 (1984); in humans: M. Kino *et al., Am. J. Cardiol.* **51,** 802 (1983). Metabolism in humans: T. Suzuki *et al., Drug Metabol. Dispos.* **11,** 377 (1983). GC-MS determn in human urine: *eidem, Chem. Pharm. Bull.* **33,** 2549 (1985). Binding affinity and selectivity: K. Naito *et al., Japan. J. Pharmacol.* **38,** 235 (1985). Clinical efficacy in congestive cardiomyopathy: J. Thormann *et al., Am. Heart J.* **110,** 426 (1985). Series of articles on pharmacokinetics, metabolism and disposition in animals: *Arzneimittel-Forsch.* **36,** 643-667 (1986).

l-Form hydrochloride, $C_{18}H_{24}ClNO_4$, crystals from isopropanol, mp 138-139.5°. $[\alpha]_D^{25}$ −38.0° (c = 1 in methanol). *dl*-Form hydrochloride, crystals from isopropanol + isopropyl ether, mp 164-167°.

THERAP CAT: Cardiotonic.

2879. Denopterin. *N-[4-[[1-(2-Amino-1,4-dihydro-4-oxo-6-pteridinyl)ethyl]methylamino]benzoyl]-L-glutamic acid;* 9,10-dimethylpteroylglutamic acid; N^{10},9-dimethylfolic acid; Dimetfol. $C_{21}H_{23}N_7O_6$; mol wt 469.47. C 53.73%, H 4.94%, N 20.89%, O 20.45%. Prepn: Hultquist *et al., J. Am. Chem. Soc.* **71,** 619 (1949); Hultquist, Smith, **Brit.** pat. **667,-098** (1952 to Am. Cyanamid).

Yellow-orange microcrystals. uv max (0.1*N* NaOH): 253, 305, 365 nm; (0.1*N* HCl): 315 nm.

THERAP CAT: Antineoplastic.

2880. 6-Deoxy-L-ascorbic Acid. $C_6H_8O_5$; mol wt 160.12. C 45.00%, H 5.04%, O 49.96%. Prepn starting with 2,3-monoacetone-L-sorbomethylose: Müller, Reichstein, *Helv. Chim. Acta* **21,** 273 (1938). Alternative procedure: **Swiss.** pats. **203,549** and **209,587** (1939 and 1940, both to Hoffmann-La Roche).

Stout prisms from ethyl acetate, mp 168°. Sublimes at 10^{-3} mm and 160°. $[\alpha]_D^{22}$ +36.7° (0.1*N* HCl). Freely sol in water; sol in acetone, alcohol; sparingly sol in ethyl acetate. Practically insol in ether.

2881. Deoxycholic Acid. *(3α,5β,12α)-3,12-Dihydroxy-5-cholan-24-oic acid;* 17β-(1-methyl-3-carboxypropyl)etiocholane-3α,12α-diol; desoxycholic acid. $C_{24}H_{40}O_4$; mol wt 392.56. C 73.43%, H 10.27%, O 16.30%. Lacks the C-7 hydroxy group of cholic acid. Occurs in bile of man, ox, sheep, dog, goat, and rabbit. Isoln: Wieland, Siebert, *Z. Physiol. Chem.* **262,** 1 (1939). From cholic acid: Sifferd, **U.S.** pat. **2,765,316** (1956 to Armour). Structure and synthesis: L. F. Fieser, M. Fieser, *Steroids* (Reinhold, New York, 1959). Laboratory prepn from ox bile: Gattermann-Wieland, *Praxis des organischen Chemikers* (de Gruyter, Berlin, 40th ed., 1961) p 361. Forms molecular coordination compds (so-called *choleic acids*) with many substances. Complexes wth fatty acids have been studies extensively: Sobotka, Goldberg, *Biochem. J.* **26,** 555 (1932); Sobotka, *Chem. Rev.* **15,** 362 (1934). Surface activity: *idem, Chemistry of the Steroids,* New York (1938).

Crystals from alc, mp 176-178°. $[\alpha]_D^{20}$ +55° (alc). pK = 6.58. Not precipitated by digitonin. Soly at 15° in water 0.24 g/l; in alcohol 220.7 g/l; in ether 1.16 g/l; in chloroform 2.94 g/l; in benzene 0.12 g/l; in acetone 10.46 g/l; in glacial acetic acid 9.06 g/l. Sol in solns of alkali hydroxides or carbonates.

Sodium salt, $C_{24}H_{39}NaO_4$, *sodium deoxycholate.* Soly in water at 15°: > 333 g/l.

An adduct with menthol is marketed under the name *Degalol.*

THERAP CAT: Choleretic.

2882. Deoxycorticosterone. *21-Hydroxypregn-4-ene-3,20-dione;* 4-pregnen-21-ol-3,20-dione; 21-hydroxyprogesterone; desoxycorticosterone; 11-deoxycorticosterone; cortexone; desoxycortone; Kendall's desoxy compound B; Reichstein's substance Q. $C_{21}H_{30}O_3$; mol wt 330.45. C 76.32%, H 9.15%, O 14.52%. Occurs in adrenal cortex: Reichstein, von Euw, *Helv. Chim. Acta* **21,** 1197 (1938); Steiger, Reichstein, *ibid.* **20,** 1164 (1937). Numerous prepns from other steroids: Schindler *et al., ibid.* **24,** 371 (1941); Reichstein, **Ger.** pat. **875,353** (1953 to Schering); Bockmühl *et al.,* **Ger.** pat. **871,153** (1953 to Hoechst); Wettstein *et al.,* **U.S.** pat. **2,778,776** (1957 to Ciba); **Dutch** pat. **89,575** (1958 to Organon); Kaspar *et al.,* **Ger.** pat. **1,028,572** (1958 to Schering). Isoln from the prothoracal glands of the water beetle, *Dytiscus marginalis:* Schildknecht *et al., Angew. Chem.* **78,** 392 (1966).

Plates from ether, mp 141-142°. $[\alpha]_D^{22}$ +178° (alc). uv max: 240 nm. Freely sol in alcohol, acetone.

Acetate, *see* Deoxycortiscosterone Acetate.

Tetraacetyl-β-D-glucoside, $C_{35}H_{48}O_{12}$, clusters of needles from 50% alc, mp 176-176.5°. $[\alpha]_D^{24}$ +80° (c = 0.515 in chloroform). Believed to occur in nature in this glycosidic form: Johnson, *J. Am. Chem. Soc.* **63**, 3238 (1941).

THERAP CAT: Mineralocorticoid.

2883. Deoxycorticosterone Acetate. *21-Acetyloxypregn-4-ene-3,20-dione;* 21-hydroxypregn-4-ene-3,20-dione 21-acetate; 11-deoxycorticosterone acetate; cortexone acetate; desoxycorticosterone acetate; deoxycortone acetate; desoxycortone acetate; DCA; Cortate; Cortenil; Cortesan; Cortifar; Cortigen; Cortiron; Cortivis; Cortixyl; Decortin; Decosterone; Deoxycostone Acetate; Descorterone; Descotone; Doca; Dorcostrin; Ocriten; Percorten; Percotol; Primocort; Primocortan; Sincortex; Syncort; Syncorta; Syncortyl; Unidocan. $C_{23}H_{32}O_4$; mol wt 372.49. C 74.16%, H 8.66%, O 17.18%.

Orthorhombic needles from alcohol, stable in air, mp 154-160°. Sublimes in high vacuum. $[\alpha]_D^{20}$ +168 to +176° (dioxane). Slightly sol in alcohol, methanol, acetone, ether, dioxane, propylene glycol (10 mg/ml), vegetable oils. Practically insol in water.

THERAP CAT: Mineralocorticoid.

THERAP CAT (VET): Adrenocortical steroid, mineralocorticoid.

2884. Deoxydihydrostreptomycin. *O-2-Deoxy-2-(methylamino)-α-L-glucopyranosyl-(1→2)-O-3,5-dideoxy-3-(hydroxymethyl)-α-L-arabinofuranosyl-(1→4)-N,N'-bis-(aminoiminomethyl)-D-streptamine;* dihydrodeoxystreptomycin; Desoxymycin; Mervastrept. $C_{21}H_{41}N_7O_{11}$; mol wt 567.62. C 44.44%, H 7.28%, N 17.27%, O 31.01%. Reduc-

tion product of streptomycin. Prepn and structure: Ikeda, *C.A.* **50**, 13765g (1956); Yabuta *et al.*, U.S. pat. **2,837,510** (1958); Gailliot, Baget, Fr. pat. **1,198,042** (1959 to Rhône-Poulenc). TLC separation: H. Heding, *Acta Chem. Scand.* **24**, 3086 (1970). Comparative antibacterial activity of streptomycin analogs: H. Heding, O. Lützen, *J. Antibiot.* **25**, 287 (1972).

Needles, dec 190-195°. $[\alpha]_D^{20}$ −102.5° (water).

Sulfate, $[\alpha]_D^{20}$ −78° (c = 2).

THERAP CAT: Antibacterial (tuberculostatic).

2885. Deoxyepinephrine. *4-[2-(Methylamino)ethyl]-1,2-benzenediol; 4-[2-(methylamino)ethyl]pyrocatechol;* 3,4-dihydroxyphenylethylamine; methyl[β-(3,4-dihydroxyphenyl)ethyl]amine; desoxyepinephrine; Epinine. $C_9H_{13}NO_2$; mol wt 167.20. C 64.65%, H 7.84%, N 8.38%, O 19.14%. Prepd by heating HCl and 1-keto-6,7-dimethoxy-2-methyltetrahydroisoquinoline obtained from laudanosine or papaverine: Pyman, *J. Chem. Soc.* **95**, 1266, 1610 (1909). Synthesis from veratrole: Kindler, Peschke, *Arch. Pharm.* **270**, 340 (1932); Kindler, Hesse, *ibid.* **271**, 439 (1933). From methylhomoveratrylamine: Buck, *J. Am. Chem. Soc.* **52**, 4119 (1930); Bretschneider, *Monatsh.* **76**, 335 (1947).

Crystals, mp 188-189°.

Hydrochloride, $C_9H_{13}NO_2 \cdot HCl$, crystals. Sol in water, alcohol.

THERAP CAT: The hydrochloride as adrenergic; vasoconstrictor.

2886. 2-Deoxy-D-glucose. *2-Deoxy-D-arabino-hexose;* D-arabino-2-desoxyhexose; 2-deoxyglucose; 2-DG; Ba 2758. $C_6H_{12}O_5$; mol wt 164.16. C 43.90%, H 7.37%, O 48.73%. Antimetabolite of glucose, *q.v.*, with antiviral activity. Synthesis: M. Bergmann *et al.*, *Ber.* **55**, 158 (1922); **56**, 1052 (1923); J. C. Sowden, H. O. L. Fischer, *J. Am. Chem. Soc.* **69**, 1048 (1947); H. R. Bolliger, *Helv. Chim. Acta* **34**, 989 (1954); H. R. Bolliger, M. D. Schmid, *ibid.* 1597, 1671; H. R. Bolliger, "2-Deoxy-D-arabino-hexose (2-Deoxy-D-glucose)" in *Methods in Carbohydrate Chemistry* vol. I, R. L. Whistler, M. L. Wolfrom, Eds. (Academic Press, New York, 1962) pp 186-189. Inhibition of influenza virus multiplication: E. D. Kilbourne, *Nature* **183**, 271 (1959). Effects on herpes simplex virus: R. J. Courtney *et al.*, *Virology* **52**, 447 (1973). Mechanism of action studies: M. R. Steiner *et al.*, *Biochem. Biophys. Res. Commun.* **61**, 745 (1974); E. K. Ray *et al.*, *Virology* **58**, 118 (1978). Use in human genital herpes infections: H. A. Blough, R. L. Giuntoli, *J. Am. Med. Assoc.* **241**, 2798 (1979); L. Corey, K. K. Holmes, *ibid.* **243**, 29 (1980). Effect vs respiratory syncytial viral infections in calves: S. B. Mohanty *et al.*, *Am. J. Vet. Res.* **42**, 336 (1981).

α-2-deoxy-D-glucose

Cryst from acetone or butanone, mp 142-144°. $[\alpha]_D^{17.5}$ +38.3° (35 min) → +45.9° (c = 0.52 in water); +22.8° (24 hrs) → +80.8° (c = 0.57 in pyridine).

α-Form, cryst from isopropanol, mp 134-136°. $[\alpha]_D^{26}$ +156° → +103° (c = 0.9 in pyridine).

USE: Exptlly as an antiviral agent.

2887. 1-Deoxynojirimycin. *[2R-(2α,3β,4α,5β)]-2-(Hydroxymethyl)-3,4,5-piperidinetriol;* D-5-amino-1,5-dideoxyglucopyranose; moranoline; 1,5-dideoxy-1,5-imino-D-glucitol; (2R,3R,4R,5S)-2-hydroxymethyl-3,4,5-trihydroxypiperidine; BAY n 5595; S-GI. $C_6H_{13}NO_4$; mol wt 163.17. C 44.17%, H 8.03%, N 8.58%?, O 39.22%. α-Glucosidase inhibitor. Prepd by reduction of nojirimycin: S. Inouye et al., J. Antibiot. **19A**, 290 (1966); S. Inouye et al., Tetrahedron **23**, 2125 (1968). Isoln from mulberry, Morus spp. (Moraceae): M. Yagi et al., Nippon Nogei Kagaku Kaishi **50**, 571 (1976), C.A. **86**, 167851r (1977). Production by Bacillus subtilis DSM704: D. C. Stein et al., Appl. Environ. Microbiol. **48**, 280 (1984). Total synthesis: H. Setoi et al., Chem. Pharm. Bull. **34**, 2642 (1986); H. J. G. Broxterman et al., Rec. Trav. Chim. **106**, 571 (1987). HPLC determn: M. D. Cole et al., J. Chromatog. **445**, 295 (1988). Inhibition of α-glucosidase: Y. Yoshikuni, Agr. Biol. Chem. **52**, 121 (1988). Antiviral activity: P. S. Sunkara et al., Biochem. Biophys. Res. Commun. **148**, 206 (1987). Inhibition of glyco-protein synthesis and secretion: N. Peyrieras et al., EMBO J. **2**, 823 (1983); V. Gross et al., Biochem. J. **236**, 853 (1986); M. Bollen et al., Biochem. Pharmacol. **37**, 905 (1988).

Prisms from water/ethanol, mp 195°. $[\alpha]_D^{21}$ +47° (c = 1.045 in water); $[\alpha]_D^{20}$ +46.7° (c = 0.2 in water); $[\alpha]_D^{20}$ +36° (c = 0.02 in water). pKa 6.6.

2888. Deoxyribonuclease (Pancreatic). Desoxyribonuclease (pancreatic); pancreatic dornase; pancreatic desoxyribonuclease; DNase I; Deanase; Dinase; Dornavac. Mol wt about 31,000. Enzyme capable of hydrolyzing highly polymerized DNA by splitting phosphodiester linkages, preferentially adjacent to a pyrimidine nucleotide, yielding 5'-phosphate-terminated nucleotides: Matsuda, Ogoshi, J. Biochem. **59**, 230 (1966). Prepn from beef pancreas: Kunitz, Science **108**, 19 (1948); Baumgarten, Johnson, U.S. pat. 2,801,956 (1957 to Merck & Co.). Prepn of a stabilized form: eidem, U.S. pat. 3,042,587 (1962 to Merck & Co.). Mol wt and amino acid composition: Lindberg, Biochemistry **6**, 335 (1967). Properties of purified DNase I: Price et al., J. Biol. Chem. **244**, 917 (1969). Reviews: McDonald, Methods Enzymol. **2**, 437 (1955); Laskowski in The Enzymes, P. D. Boyer, Ed. (Academic Press, New York, 3rd ed., 1971) pp 289-311.

THERAP CAT: In resolution of clots and exudates following trauma and inflammation.

THERAP CAT (VET): For enzymatic debridement.

2889. Deoxyribonucleic Acid. Desoxyribonucleic acid; DNA; thymus nucleic acid; Desoxiribon; Eucytol. Polynucleotide; essential component of chromosomes in cell nuclei. In its role as the carrier of genetic information, DNA must have two functions: be exactly reproducible in order to transmit its genetic information to future generations; contain information, in chemical code, to direct the development of the cell according to its inheritance. Reviews of biological function: Hotchkiss in The Nucleic Acids vol. 2, E. Chargaff, J. N. Davidson, Eds. (Academic Press, New York, 1955) pp 435-473; Crick, Nature **227**, 561 (1970); J. N.

Davidson, The Biochemistry of Nucleic Acids (Academic Press, New York, 7th ed., 1972) pp 6-28. Description of DNA components: see Nucleic Acids. The purine and pyrimidine bases of the nucleosides are primarily adenine, guanine, cytosine and thymine; the sugar is D-2-deoxyribose. The nucleosides are linked together by phosphates in diester linkage from the 3'-hydroxyl of one sugar to the 5'-hydroxyl of the next. The repeating sugar-phosphate linkage forms the backbone of the single polynucleotide strand which is the primary structure of DNA.

Chemical analyses of DNA from different species show that the purine content is equal to the pyrimidine content; adenine content equal to thymine; guanine equal to cytosine: Chargaff, Experientia **6**, 201 (1950); idem, Fed. Proc. **10**, 654 (1951). In the Watson-Crick model of its secondary structure (based on chemical analysis and x-ray studies), DNA consists of two polynucleotide chains forming right-handed helices coiled around the same axis with the sequence of atoms in the two sugar-phosphate backbones running in opposite direction. Two major families of right-handed helix were proposed. A-DNA and B-DNA, each having its own intrinsic restrictions on chain-folding and structure. B-DNA is believed to be the predominant form in biological systems. The purine and pyrimidine bases are inside the helical structure formed by the sugar phosphate backbones; those on one chain form hydrogen bonds to those on the other. Adenine in one chain is always bonded to thymine in the complementary chain by hydrogen bonds; similarly guanine is bonded to cytosine. The linear sequence of bases in one strand completely determines the sequence in the complementary strand. Thus each strand can serve as a template for the replication of the original DNA molecule: Watson, Crick, Nature **171**, 737, 964 (1953). X-ray studies: Wilkins et al., ibid. 738; Marvin et al., J. Mol. Biol. **3**, 547 (1961); Fuller et al., ibid. **12**, 60 (1965). DNA also acts as a template in the formation of ribonucleic acids, q.v., which play a fundamental role in the synthesis of proteins in the cell. Another form of DNA, termed Z-DNA, is also known. Its structure is an antiparallel double helix with Watson-Crick base pairing, but it is a left-handed helix with the ribose-phosphate backbone following a zig-zag course. Molecular structure, atomic resolution x-ray crystallographic analysis: A. H.-J. Wang et al., Nature **282**, 680 (1979). First identification of Z-DNA in material of biological origin: A. Nordheim et al., ibid. **294**, 417 (1981). Studies of B- and Z-DNA: D. J. Patel et al., Proc. Nat. Acad. Sci. USA **79**, 1413 (1982). Comparison of A-, B-, and Z-DNA: R. E. Dickerson et al., Science **216**, 475 (1982). Demonstration of Z-DNA immunoreactivity in rat tissues: G. Morgenegg et al., Nature **309**, 540 (1983). Comprehensive reviews and monographs: see Nucleic Acids.

2890. D-2-Deoxyribose. 2-Deoxy-D-erythro-pentose; desoxyribose; D-2-deoxyarabinose; D-2-ribodesose; D-erythro-2-deoxypentose; thyminose. $C_5H_{10}O_4$; mol wt 134.13. C 44.77%, H 7.52%, O 47.71%. Isoln from deoxyribonucleic acid by acidic hydrolysis of purine deoxyribonucleosides which have been isolated by ion-exchange resin chromatography: Laland, Overend, Acta Chem. Scand. **8**, 192 (1954). Synthesis: Felton, Freudenberg, J. Am. Chem. Soc. **57**, 1637 (1935); Deriaz et al., J. Chem. Soc. **1949**, 1879, 2836; Hough, Chem. & Ind. (London) **1951**, 406; Sowden, Biochem. Prepn. **5**, 75 (1957); I. Ziderman, E. Dimant, J. Org. Chem. **32**, 1267 (1967); J. R. Hauske, H. Rapoport, ibid. **44**, 2472 (1979); T. Harada, T. Mukaiyama, Chem. Letters **1981**, 1109. Review: Overend, Stacey, in Chargaff-Davidson, Nucleic Acids vol. 1, E. Chargaff, N. J. Davidson, Eds. (Academic Press, New York, 1955) pp 1-80. See also Nucleic Acids.

Crystals from isopropanol, mp 91°. Shows mutarotation.

Final $[\alpha]_D^{22}$ $-56.2°$ (H_2O). Sol in water, pyridine. Slightly sol in alc.

1,3,4-Triacetate, $C_{11}H_{16}O_7$, needles from methanol, mp 98°. $[\alpha]_D^{23}$ $-171.8°$ (c = 0.56 in chloroform): Allerton, Overend, *J. Chem. Soc.* **1951**, 1480.

3,4,5-Triacetate, oily liq, $bp_{0.001}$ 105°. $[\alpha]_D^{21}$ $+3.4°$ (c = 4.57 in pyridine): Zinner *et al.*, *Ber.* **90**, 2696 (1957).

1,3,4-Tribenzoate, small white nodules from ethanol, mp 127°. $[\alpha]_D^{15}$ $-65°$ (c = 1.02 in chloroform): Allerton, Overend, *loc. cit.* Probably a mixture of the two anomeric 2-deoxy-D-ribopyranose tribenzoates: Pedersen *et al.*, *J. Am. Chem. Soc.* **82**, 3425 (1960).

3,4,5-Tribenzoate, fine needles from ethyl acetate + petr ether, mp 118-119°. $[\alpha]_D^{18}$ $-2.8°$ (c = 1.44 in pyridine): Zinner *et al.*, *loc. cit.*

2891. 2-Deoxystreptamine. 2-deoxy-1,3-*myo*-inosadiamine; 1,3-diamino-4,5,6-trihydroxycyclohexane. $C_6H_{14}N_2O_3$; mol wt 162.18. C 44.43%, H 8.70%, N 17.27%, O 29.60%. Important component of several aminocyclitol antibiotics. Production by acid hydrolysis of neamine, *q.v.* and structure: F. A. Kuehl *et al.*, *J. Am. Chem. Soc.* **73**, 881 (1952). Also obtained from hydrolysis products of kanamycins, paromomycins, gentamicins, *q.q.v.* and other aminoglycoside antibiotics. See K. L. Rinehart, *J. Infec. Dis.* **119**, 345 (1969); S. Hanessian, T. H. Haskell, *The Carbohydrates*, vol. IIA, W. Pigman, D. Horton, Eds. (Academic Press, New York, 2nd ed., 1970) pp 159-172. Configuration: J. Daly *et al.*, *J. Am. Chem. Soc.* **82**, 5928 (1960); H. E. Carter *et al.*, *ibid.* **83**, 3723 (1961); R. U. Lemieux, R. J. Cushley, *Can. J. Chem.* **41**, 858 (1963). Synthesis: M. Nakajima *et al.*, *Tetrahedron Letters* **1964**, 967; T. Suami *et al.*, *ibid.* **1967**, 2671; S. Ogawa *et al.*, *J. Org. Chem.* **39**, 812 (1974); H. Prinzbach *et al.*, *Angew. Chem. Int. Ed.* **14**, 225 (1975); S. Ogawa *et al.*, *J. Org. Chem.* **42**, 3083 (1977).

Colorless crystals from ethanol, mp 225-228°.

Dihydrochloride, $C_6H_{16}Cl_2N_2O_3$. Crystals, mp 325° (dec), H. Hitomi *et al.*, *Chem. Pharm. Bull.* **9**, 340 (1961).

2892. Deoxyuridine. 2'-Deoxyuridine; 1-(2-deoxy-β-D-erythro-pentofuranosyl)uracil; 1-(2-deoxy-β-D-ribofuranosyl)uracil; uracil deoxyriboside. $C_9H_{12}N_2O_5$; mol wt 228.20. C 47.37%, H 5.30%, N 12.28%, O 35.06%. Prepn: Dekker, Todd, *Nature* **166**, 557 (1950); Brown *et al.*, *J. Chem. Soc.* **1958**, 3035; Prystas *et al.*, *Coll. Czech. Chem. Commun.* **28**, 3140 (1963); Smejkal *et al.*, *ibid.* **31**, 291 (1966). Conformation: Rahman, Wilson, *Nature* **232**, 333 (1971).

Needles from abs alc or 95% alc, mp 163° (Dekker, Todd) and 167° (Brown *et al.*). $[\alpha]_D^{22}$ $+50°$ (c = 1.1 in N NaOH).

Note: α-Anomer, 1-(2-deoxy-α-D-erythro-pentofuranosyl)uracil; 1-(2-deoxy-α-D-ribofuranosyl)uracil. Prepn: Prystas *et al.*, *loc. cit.*

2893. Deprenyl. N,α-Dimethyl-N-2-propynylbenzeneethanamine; N,α-dimethyl-N-2-propynylphenethylamine; phenylisopropylmethylpropynylamine; deprenalin; deprenil. $C_{13}H_{17}N$; mol wt 187.29. C 83.37%, H 9.15%, N 7.48%. Monoamine oxidase inhibitor related structurally to pargy-

line, *q.v.* Prepn of the racemic mixture: **Fr. pat. M 2635** (1964 to Chinoin), *C.A.* **63**, 2927e (1965); of the (−)-form: **Neth. pat. Appl. 6,605,956** (1966 to Chinoin), *C.A.* **67**, 21611y (1967); J. S. Fowler, *J. Org. Chem.* **42**, 2637 (1977). Effects on central nervous system: J. Knoll *et al.*, *Arch. Int. Pharmacodyn.* **155**, 154 (1965). Comparative pharmacology of the optical isomers: K. Magyar *et al.*, *Acta Physiol. Acad. Sci. Hung.* **32**, 377 (1967). Effects in Parkinson's disease: G. P. Reynolds *et al.*, *Biochem. Soc. Trans.* **7**, 143 (1979); P. Reiderer, G. P. Reynolds, *Brit. J. Clin. Pharmacol.* **9**, 98 (1980). Preliminary study in Alzheimer's disease: P. N. Tariot *et al.*, *Arch. Gen. Psychiat.* **44**, 427 (1987).

Oil, bp_5 103-110°. n_D^{20} 1.5224. LD_{50} in rats (mg/kg): 75 i.v., 218 s.c. (Magyar).

(−)-Form, *selegiline, Eldéprine, Movergan.* Oil, $bp_{0.8}$ 92-93°. n_D^{20} 1.5180. $[\alpha]_D^{20}$ $-11.2°$. LD_{50} in rats (mg/kg): 81 i.v., 280 s.c. (Magyar).

(−)-Form hydrochloride, $C_{13}H_{18}ClN$, *E 250, Eldepryl, Jumex.* Crystals, mp 141-142°. $[\alpha]_D^{25}$ $-10.8°$ (c = 6.48 in water). LD_{50} of the racemic hydrochloride in rats (mg/kg): 63 i.v., 126 s.c., 385 orally (Knoll).

THERAP CAT: (−)-Form as antiparkinsonian.

2894. Deptropine. endo-3-[(10,11-Dihydro-5H-dibenzo-[a,d]cyclohepten-5-yl)oxy]-8-methyl-8-azabicyclo[3.2.1]octane; 3α-[(10,11-dihydro-5H-dibenzo[a,d]cyclohepten-5-yl)oxy]-1αH,5αH-tropane; dibenzheptropine. $C_{23}H_{27}NO$; mol wt 333.45. C 82.84%, H 8.16%, N 4.20%, O 4.80%. Prepn: Van der Stelt *et al.*, *J. Med. Pharm. Chem.* **4**, 335 (1961). Pharmacology: Funcke *et al.*, *Arch. Int. Pharmacodyn.* **148**, 135 (1964); Timmerman *et al.*, *ibid.* **187**, 291 (1970). Metabolism: Hespe *et al.*, *ibid.* **164**, 397 (1966).

Citrate, $C_{29}H_{35}NO_8$, *BS 6987, Brontine.* LD_{50} in mice (mg/kg): 32 i.v.; 300 orally (Timmerman).

Maleate, $C_{23}H_{27}NO.C_4H_4O_4$, crystals from ethanol, acetone, ethyl acetate or chloroform, mp 133-136°.

THERAP CAT: Anticholinergic, antihistaminic.

2895. Dequalinium Acetate. *1,1'-(1,10-Decanediyl)bis-[4-amino-2-methylquinolinium diacetate]; 1,1'-decamethylenebis[4-aminoquinaldinium acetate];* Dequadin acetate. $C_{34}H_{46}N_4O_4$; mol wt 574.74. C 71.05%, H 8.07%, N 9.75%, O 11.14%. Ref: Babbs *et al.*, *J. Pharm. Pharmacol.* **8**, 110 (1956). See also Dequalinium Chloride.

Soly in water at 25°: 3 parts in 4 parts.

Ingredient of *Micrin, Gargilon.*

THERAP CAT: Antiseptic, disinfectant.

2896. Dequalinium Chloride. *1,1'-(1,10-Decanediyl)bis-[4-amino-2-methylquinolinium chloride]; 1,1'-decamethylenebis[4-aminoquinaldinium chloride];* BAQD 10; decamine; dekamin; Dekadin; Dequadin Chloride; Dequafungan; Dequavet; Dequavagyn; Eriosept; Evazol; Grocreme; Labosept; Optipect; Phylletten; Polycidine; Sorot. $C_{30}H_{40}Cl_2N_4$; mol wt 527.60. C 68.30%, H 7.64%, Cl 13.44%, N 10.62%. Prepn: Taylor *et al.*, **Brit. pat. 745,956** (1956 to Allen & Hanburys), *C.A.* **50**, 16878e (1956); Austin *et al.*, *J. Chem. Soc.* **1958**, 1489.

Crystals from ethanol, mp 326° (dec). Soly in water (25°): about one g/200 ml.

USE: Bacteriostat. Ingredient of *Efisol, Gargilon, Gramipan, Hexalyse, Micrin.*

THERAP CAT: Antiseptic, disinfectant.

THERAP CAT (VET): Antimicrobial.

2897. Dermostatin. Viridofulvin; Dermastatin. Antifungal polyene antibiotic mixture produced by *Streptomyces viridogriseus* Thirum. Comprised of about 43% *Dermostatin A*, $C_{40}H_{64}O_{11}$, and 57% *Dermostatin B*, $C_{41}H_{66}O_{11}$, the first hexaenes to have their structures determined. Preliminary isoln and physico-chemical studies: Bhate *et al., Hindustan Antibiot. Bull.* **4**, 159 (1962), *C.A.* **57**, 13013d (1962); Thirumalachar, Bhate, **Indian** pat. **76,253** (1963 to Hindustan Antibiotics), *C.A.* **59**, 8095b (1963). Purifn: Narasimhachari *et al., Hindustan Antibiot. Bull.* **8**, 111 (1966), *C.A.* **65**, 5301c (1966). Structural studies: Narasimhachari, Swami, *Chemotherapy* **13**, 181 (1968). Revised structure: Pandey *et al., J. Antibiot.* **26**, 475 (1973). Antifungal activity: Menon, *Hindustan Antibiot. Bull.* **4**, 106 (1962), *C.A.* **57**, 5121c (1962); Gordee, Butler, *J. Antibiot.* **24**, 561 (1971).

Dermostatin A   R = CH_3
Dermostatin B   R = CH_2CH_3

Golden-yellow needles, sinters at 180° and darkens at 200°. $[\alpha]_D$ $-82°$ (c = 0.2 in methanol). uv max: 383, 282, 223 nm ($E_{1cm}^{1\%}$ 1000, 212, 130). Sol in aq methanol.
Acetate, prisms from methanol, mp 146-147°. $[\alpha]_D^{23}$ $-59.8°$ (c = 1.37 in chloroform).

THERAP CAT: Antifungal.

2898. Derris Root. Tuba root; Deguelia root. The roots of plants belonging to the genus *Derris,* of which more than 80 species have been described. *Derris elliptica* (Wall.) Benth. and *D. malaccensis* Prain, *Leguminosae* are cultivated in British Malaya and the Dutch East Indies. Toxicity data: Haag *et al., Proc. Soc. Exp. Biol. Med.* **54**, 140 (1943).
LD$_{50}$ orally in mice: 350 mg/kg (as the dry powdered root (Haag)).

USE: As insecticide; as a source of rotenone and rotenoid compds.

2899. Desaspidin. *2-[[2,4-Dihydroxy-6-methoxy-3-(1-oxobutyl)phenyl]methyl]-3,5-dihydroxy-4,4-dimethyl-6-(1-oxobutyl)-2,5-cyclohexadien-1-one; 3'-[(5-butyryl-2,4-dihydroxy-3,3-dimethyl-6-oxo-1,4-cyclohexadien-1-yl)methyl]-2',6'-dihydroxy-4'-methoxybutyrophenone;* rosapin. $C_{24}H_{30}O_8$; mol wt 446.48. C 64.56%, H 6.77%, O 28.67%. Isoln from the rhizome of *Dryopteris austriaca* (Jacq.) Wojnar, *Polypodiaceae:* Aebi *et al., Helv. Chim. Acta* **40**, 266 (1957); from *D. caucasica* (A. Br.) Fraser-Jenkins et Corley: Widén *et al., ibid.* **56**, 831 (1973). Structure: Aebi *et al., ibid.* **40**, 572 (1957).

Crystals from ether + petr ether, mp 150-150.5°. uv max: 230, 274 nm (ϵ 23,000, 16,800) in cyclohexane. Freely sol in ether, benzene, acetone; practically insol in methanol, ethanol, petr ether. LD$_{50}$ orally in mice: 340 mg/kg, Airaksinan *et al., Acta Pharmacol. Toxicol.* **25**, 33 (1967).

2900. Desatrine. *4α,9-Epoxycevane-3β,4,6α,7α,14,15α,-16β,20-octol 6,7-diacetate 15-(2-methylbutanoate);* protoverine 6,7-diacetate 15-(2-methylbutyrate); protoverine 6,7-diacetate 15(l)-2'-methylbutyrate. $C_{36}H_{55}NO_{12}$; mol wt 693.85. C 62.32%, H 7.99%, N 2.02%, O 27.67%. Prepd from protoveratrine B: Kupchan, Ayres, *J. Am. Chem. Soc.* **82**, 2252 (1960); Kupchan, U.S. pats. **3,009,917** and **3,066,-144** (1962 to Wisconsin Alumni Res. Found.). Structure: Kupchan, Ayres, *loc. cit.*

Needles from acetone + petr ether, dec 232-233°. $[\alpha]_D^{26}$ $-46°$ (c = 0.95 in pyridine).

USE: Esters, *e.g., desatrine-3-(N,N-diethylaminoacetate),* used as insecticides.

2901. Deserpidine. *17α-Methoxy-18β-[(3,4,5-trimethoxybenzoyl)oxy]-3β,20α-yohimban-16β-carboxylic acid methyl ester;* 11-desmethoxyreserpine; canescine; recanescine; raunormine; Harmonyl. $C_{32}H_{38}N_2O_8$; mol wt 578.64. C 66.43%, H 6.62%, N 4.84%, O 22.12%. Isoln from roots of *Rauwolfia canescens* L., *Apocynaceae:* Stoll, Hofmann, *J. Am. Chem. Soc.* **77**, 820 (1955); Klohs *et al., ibid.* 4084; Neuss *et al., ibid.* 4087; MacPhillamy *et al., ibid.* 4335; Schlittler *et al., Experientia* **11**, 64 (1955); **Brit.** pat. **791,241** (1958 to Penick); **Brit.** pat. **809,912** (1959 to Ciba). Stereochemistry: Huebner *et al., Experientia* **11**, 303 (1955); Aldrich *et al., J. Am. Chem. Soc.* **81**, 2481 (1959). Synthesis: Bláha *et al., Coll. Czech. Chem. Commun.* **25**, 237 (1960); **Can.** pat. **678,216** (1964 to Roussel-UCLAF).

Exists in three cryst forms from methanol: α-form, mp 228-232°; β-form, mp 230-232°; and γ-form, mp 138° and 226-232° with resolidification at 175°. $[\alpha]_D^{20}$ $-163°$ (c = 0.5 in pyridine). pKa 6.68 in 40% methanol. uv max (ethanol): 218, 272, 290 nm (log ϵ 4.79, 4.26, 4.07).
Hydrochloride, $C_{32}H_{38}N_2O_8$·HCl, thin rectangular plates from acetone, dec 253-256°.
Nitrate, $C_{32}H_{38}N_2O_8$·HNO$_3$, crystals, dec 254-260°.
Oxalate, $C_{32}H_{38}N_2O_8$·H$_2$C$_2$O$_4$, crystals, dec 239-243°.

10-Methoxy deriv, $C_{33}H_{40}N_2O_9$, *10-methoxydeserpidine*, *methoserpidine*, *Decaserpyl*.

THERAP CAT: Antihypertensive.

2902. Desipramine. *10,11-Dihydro-N-methyl-5H-dibenz[b,f]azepine-5-propanamine; 10,11-dihydro-5-[3-(methylamino)propyl]-5H-dibenz[b,f]azepine;* 5-(γ-methylaminopropyl)iminodibenzyl; *N*-(3-methylaminopropyl)iminobibenzyl; desmethylimipramine; norimipramine. $C_{18}H_{22}N_2$; mol wt 266.37. C 81.16%, H 8.33%, N 10.52%. Prepn of hydrochloride: **Belg.** pat. **614,616,** *C.A.* **58,** 11338c (1963); of free base and hydrochloride: **Brit.** pat. **908,788** (both 1962 to Geigy). Chemistry, pharmacology: E. Eriksoo, O. Rohte, *Arzneimittel-Forsch.* **20,** 1561 (1970). Clinical response, plasma levels, pharmacokinetics: P. D. Hrdina, Y. D. Lapierre, *Prog. Neuropsychopharmacol.* **4,** 591 (1980). Mechanism of action study: S. A. Checkley *et al.*, *Brit. J. Psychiatry* **138,** 248 (1981). Efficacy in depression: W. Z. Potter *et al.*, *Psychopharmacol. Bull.* **17,** 26 (1981); J. W. Stewart *et al.*, *ibid.* 136. Teratological study: L. Aeppli, *Arzneimittel-Forsch.* **19,** 1617 (1969). Carcinogenicity and mutagenicity study: H. Kubinski *et al.*, *Mutat. Res.* **89,** 95 (1981). Has been found to be one of the most effective cmpds described for *in vitro* and *in vivo* reversal of chloroquine, *q.v.*, resistance in *Plasmodium falciparum:* A. J. Bitonti *et al.*, *Science* **242,** 1301 (1988).

$bp_{0.02}$ 172-174°. uv max: 213, 252 nm (log ε 4.39, 3.93). Hydrochloride, $C_{18}H_{23}ClN_2$, *G-35020, JB-8181, NSC-114901, Irene, Norpramin, Nortimil, Pertofran, Pertofrane.* Water-sol crystals, mp 214-218° (**Brit.** pat., *loc. cit.*), 206-208° (**Belg.** pat., *loc. cit.*). LD_{50} in mice, rats (mg/kg): 500, 385 orally; 94, 48 i.p.; 420, 183 s.c. (Eriksoo, Rohte).

THERAP CAT: Antidepressant.

2903. Deslanoside. *(3β,5β,12β)-3-[(O-β-D-Glucopyranosyl-(1 → 4)-O-2,6-dideoxy-β-D-ribo-hexopyranosyl-(1 → 4)-O-2,6-dideoxy-β-D-ribo-hexopyranosyl-(1 → 4)-2,6-dideoxy-β-D-ribo-hexopyranosyl)oxy]-12,14-dihydroxycard-20(22)-enolide; deacetyllanatoside C;* desacetyldigilanide C; Purpurea glycoside C; Cedilanid-D; Desace; Desaci; Lanimerck (ampuls). $C_{47}H_{74}O_{19}$; mol wt 943.11. C 59.86%, H 7.91%, O 32.23%. From leaves of *Digitalis lanata* Ehrh., *Scrophulariaceae:* Stoll, Kreis, *Helv. Chim. Acta* **16,** 1049, 1390 (1933). Prepn by alkaline degradation of lanatoside C: Kroszczynski *et al.*, *Acta Polon. Pharm.* **21,** 357 (1964), *C.A.* **62,** 10292b (1965).

Crystals from methanol, dec 265-268°. $[\alpha]_D^{20}$ +12° (c = 1.084 in 75% alc). One part dissolves in 5000 parts water, 200 parts methanol and 2500 parts ethanol; very slightly sol in chloroform. Practically insol in ether.

THERAP CAT: Cardiotonic.

2904. Desmopressin. *1-(3-Mercaptopropanoic acid)-8-D-arginine vasopressin; 8-D-arginine-1-(3-mercaptopropanoic acid) vasopressin;* 1-desamino-8-D-arginine vasopressin; DDAVP; Adiuretin SD; DAV Ritter; Desmospray; Minirin. $C_{46}H_{64}N_{14}O_{12}S_2$; mol wt 1069.24. C 51.67%, H 6.03%, N 18.34%, O 17.96%, S 6.00%. An analog of vasopressin possessing high antidiuretic activity. Syntheses: Huguenin,

Boissonnas, *Helv. Chim. Acta* **49,** 695 (1966); Zaoral *et al.*, *Coll. Czech. Chem. Commun.* **32,** 1250 (1967). Clinical studies: Vavra *et al.*, *Lancet* **1,** 948 (1968); Andersson, Arner, *Acta Med. Scand.* **192,** 21 (1972). Antidiuretic action in rats: Vavra *et al.*, *J. Pharmacol. Exp. Ther.* **188,** 241 (1974).

$[\alpha]_D^{25}$ +85.5° ± 2° (calculated for the free peptide). Monoacetate, $C_{48}H_{68}N_{14}O_{14}S_2$, *Octostim.*

THERAP CAT: Antidiuretic.

2905. Desmosterol. *3β-Cholesta-5,24-dien-3-ol;* 24-dehydrocholesterol; desmesterol. $C_{27}H_{44}O$; mol wt 384.62. C 84.31%, H 11.53%, O 4.16%. Has been isolated from chick embryos and the skin of rats: Stokes *et al.*, *J. Biol. Chem.* **220,** 415 (1956). From barnacles: Fagerlund, Idler, *J. Am. Chem. Soc.* **79,** 6473 (1957); from red algae: Idler *et al.*, *Steroids* **11,** 465 (1968). Synthesis: Dasgupta *et al.*, *J. Org. Chem.* **39,** 1658 (1974); M. A. Apfel, *ibid.* **43,** 2284 (1978). Stereospecific synthesis: M. Koreeda *et al.*, *ibid.* **45,** 1172 (1980); S. Takano *et al.*, *Chem. Commun.* **1983,** 760.

Platelets from methanol, mp 121.5°, also reported as 119-119.5°. $[\alpha]_D^{27}$ −41.0° (c = 1 in chloroform).

2906. Desogestrel. *(17α)-13-Ethyl-11-methylene-18,19-dinorpregn-4-en-20-yn-17-ol;* 17α-ethynyl-18-methyl-11-methylene-Δ⁴-estren-17β-ol; Org-2969. $C_{22}H_{30}O$; mol wt 310.48. C 85.11%, H 9.74%, O 5.15%. Progestogen with low androgenic potency. Prepn: A. J. van den Broek, **Ger.** pat. **2,361,120;** *idem,* **U.S.** pat. **3,927,046** (1974, 1975 both to Akzo); A. J. van den Broek *et al.*, *Rec. Trav. Chim.* **94,** 36 (1975). Biological effects: L. Viinikka *et al.*, *Acta Endocrinol.* **83,** 429 (1976). Endocrinological studies in animals: J. van der Vies, J. de Visser, *Arzneimittel-Forsch.* **33,** 231 (1983). Radioimmunoassay: L. Viinikka, *J. Steroid Biochem.* **9,** 979 (1978). Metabolism, pharmacokinetics in humans: L. Viinikka *et al.*, *Eur. J. Clin. Pharmacol.* **15,** 349 (1979). Receptor binding study: E. W. Bergink *et al.*, *J. Steroid Biochem.* **14,** 175 (1981). Clinical study: M. J. Weijers, *Clin. Ther.* **4,** 359 (1982).

Crystals, mp 109-110°. $[\alpha]_D^{20}$ +55° (chloroform). Mixture with ethinyl estradiol, *q.v.*, *Cyclosa, Dicromil, Marvelon 150/30, Mercilon, Oviol, Varnoline.*

THERAP CAT: Progestogen. In combination with estrogen as oral contraceptive.

2907. Desomorphine. *4,5a-Epoxy-17-methylmorphinan-3-ol;* dihydrodesoxymorphine-D; Permonid. $C_{17}H_{21}NO_2$; mol wt 271.35. C 75.24%, H 7.80%, N 5.16%, O 11.79%. Synthesis, physical data: L. F. Small *et al.*, *J. Am. Chem. Soc.* **55,** 3863 (1933).

Rectangular plates from acetone-water. Sublimes in high vacuum between 140° and 170°. (Anhydr base mp 189°.) $[\alpha]_D^{28}$ −77° (c = 1.6 in methanol). Sol in acetone, ethyl acetate. Alcoholic solns darken and acquire a reddish color.

Hemihydrate, crystals from ethyl acetate, mp 162-164°. $[\alpha]_D^{28}$ −79° (c = 1.08 in ethyl acetate).

Hydrochloride, $C_{17}H_{21}NO_2 \cdot HCl$, crystals from alcohol, $[\alpha]_D^{27}$ −67° (c = 0.90). Very sol in water.

Hydrogen sulfate dihydrate, $(C_{17}H_{21}NO_2)_2 \cdot H_2SO_4 \cdot 2H_2O$, scales, plates from water, brown at 210°, unmelted at 230°. Sol in 40 parts water; freely sol in hot water. Soln is stable to boiling. $[\alpha]_D^{29}$ −60° (c = 1.43).

Caution: May be habit forming. This is a controlled substance (opium derivative) listed in the U.S. Code of Federal Regulations, Title 1 Part 1308.11 (1985).

THERAP CAT: Narcotic analgesic.

2908. Desonide. *11,21-Dihydroxy-16,17-[(1-methylethylidene)bis(oxy)]pregna-1,4-diene-3,20-dione; 11β,16α,17,-21-tetrahydroxypregna-1,4-diene-3,20-dione cyclic 16,17-acetal with acetone;* 16α-hydroxyprednisolone-16α,17-acetonide; 11β,21-dihydroxy-16α,17-isopropylidenedioxy-1,4-pregnadiene-3,20-dione; 16α-hydroxy-Δ¹-hydrocortisone-16,17α-acetonide; 16α,17α-isopropylidenedioxyprednisolone; prednacinolone; D-2083; Locapred; Sterax; Steroderm; Topifug; Tridesilon. $C_{24}H_{32}O_6$; mol wt 416.52. C 69.21%, H 7.74%, O 23.05%. Prepn and activity: Bernstein *et al., J. Am. Chem. Soc.* **81**, 4573 (1959); Bernstein, Allen, U.S. pat. **2,990,401** (1961 to Am. Cyanamid); Lee *et al.* and Diassi, Principe, U.S. pats. **3,536,586** and **3,549,498** (both 1970 to Squibb). Pharmacological and toxicological studies: Mascitelli-Coriandoli, Fraia, *Arzneimittel-Forsch.* **20**, 11 (1970); Phillips *et al., Toxicol. Appl. Pharmacol.* **20**, 522 (1971). Structure-activity studies: Ringler *et al., Proc. Soc. Exp. Biol. Med.* **107**, 451 (1961). Chemical and physical properties: Mantica *et al., Arzneimittel-Forsch.* **20**, 109 (1970).

Small plates or white to off-white odorless powder, mp 274-275° from methanol (Mantica); also reported as mp 263-266° from ethyl acetate-petr ether (Bernstein *et al.*). $[\alpha]_D^{25}$ +123° (c = 0.5 in DMF). uv max: 242 nm ($E_{1cm}^{1\%}$ 356). LD_{50} in rats (mg/kg): 93 s.c. (Phillips).

THERAP CAT: Anti-inflammatory.

2909. Desosamine. *3,4,6-Trideoxy-3-(dimethylamino)-D-xylo-hexose;* 4-dimethylaminotetrahydro-6-methylpyran-2,3-diol; picrocine. $C_8H_{17}NO_3$; mol wt 175.22. C 54.83%, H 9.78%, N 7.99%, O 27.39%. Sugar component of several macrolide antibiotics, such as the erythromycins. Isoln: Clark, *Antibiot. & Chemother.* **3**, 663 (1953). Identity with picrocine: Brockmann *et al., Ber.* **87**, 856 (1954). Stereochemistry: Woo *et al., Tetrahedron Letters* **1962**, 735. Configuration: Bolton *et al., Chem. & Ind. (London)* **1962**, 1945; Foster *et al., J. Chem. Soc.* **1965**, 2318. Synthesis of DL-form: Korte *et al., Tetrahedron* **18**, 657 (1962); Newman, *Chem. & Ind. (London)* **1963**, 372; *J. Org. Chem.* **29**, 1461 (1964). Stereospecific synthesis of the natural D-form: Richardson, *J. Chem. Soc.* **1964**, 5364; of the L-form: Baer, Chiu, *Can. J. Chem.* **52**, 122 (1974).

Crystals, mp 86-87°.

Hydrochloride, $C_8H_{18}ClNO_3$, needles from ethanol + acetone, dec 191-193°. $[\alpha]_D^{20}$ +49.5° (c = 10.0); $[\alpha]_D^{20}$ +53.4° (c = 2.1 in ethanol).

2910. Desoximetasone. *9-Fluoro-11,21-dihydroxy-16-methylpregna-1,4-diene-3,20-dione;* 9α-fluoro-16α-methyl-17-desoxyprednisolone; 9α-fluoro-16α-methyl-Δ¹-corticosterone; desoxymethasone; A 41-304; R-2113; HOE 304; Esperson; Stiedex; Topicorte; Topiderm; Topisolon. $C_{22}H_{29}FO_4$; mol wt 376.47. C 70.19%, H 7.76%, F 5.05%, O 17.00%. Prepn: R. Joly *et al.,* **Fr. pat. 1,296,544;** *eidem,* **U.S. pat. 3,099,654** (1962, 1963 to Roussel-UCLAF); **Belg. pat. 614,196;** Kieslich *et al.,* **U.S. pat. 3,232,839** (1962, 1966 to Schering AG); R. Joly *et al., Arzneimittel-Forsch.* **24**, 1 (1974). Activity studies: Branceni *et al., Steroids* **6**, 451 (1965); Schröder *et al., Arzneimittel-Forsch.* **24**, 3 (1974). NMR data: Lukacs *et al., C. R. Acad. Sci. Ser. C* **274**, 1458 (1972). Review of pharmacology and therapeutic efficacy in dermatoses: R. C. Heel *et al., Drugs* **16**, 302-321 (1978).

Crystals from ethyl acetate, mp 217°. $[\alpha]_D$ +109° (chloroform). uv max: 238 nm (ϵ 15750). Sol in alcohol, acetone, chloroform, hot ethyl acetate; slightly sol in ether, benzene. Insol in water, dil aq acids and alkalies.

THERAP CAT: Anti-inflammatory; glucocorticoid.

2911. 6-Desoxy-D-glucosamine. *2-Amino-2,6-dideoxy-D-glucose;* 6-deoxy-D-glucosamine; 2,6-didesoxy-2-amino-D-glucose. $C_6H_{13}NO_4$; mol wt 163.17. C 44.16%, H 8.03%, N 8.58%, O 39.22%. May be prepd by tosylation of the primary alcohol group of a suitable derivative of D-glucosamine, followed by acetylation and replacement of the tosyl group by iodine, and catalytic cleavage of the halide with Raney nickel in the presence of triethylamine: Morel, *Helv. Chim. Acta* **41**, 1501 (1958).

Hydrochloride, $C_6H_{14}ClNO_4$, crystals from abs ethanol + ether, dec 172-173°. Shows mutarotation. Final value: $[\alpha]_D^{20}$ +55° (H_2O).

1,3,4-Tri-O-acetyl-N-acetyl-6-desoxy-β-D-glucosamine, $C_{14}H_{21}NO_8$. Crystals from alc, mp 209-210°. $[\alpha]_D^{20}$ +17.5° (chloroform).

USE: In the investigation of amino acid metabolism.

2912. 11-Desoxy-17-hydroxycorticosterone. *17,21-Dihydroxypregn-4-ene-3,20-dione;* 17-hydroxydesoxycorticosterone; cortexolone; 4-pregnene-17α,21-diol-3,20-dione; 17-(1-keto-2-hydroxyethyl)-4-androsten-17α-ol-3-one; Reichstein's "Substance S"; 11-desoxycortisone. $C_{21}H_{30}O_4$; mol wt 346.45. C 72.80%, H 8.73%, O 18.47%. Isoln and partial synthesis: Reichstein, von Euw, *Helv. Chim. Acta* **21**, 1197 (1938); Reichstein, *ibid.* 1490; Reichstein, von Euw, *ibid.* **23**, 1258 (1940). Partial synthesis involving the chromic acid oxidation of 4-pregnene-17α,20,21-triol-3-one 21-monoacetate: Sarett, *J. Biol. Chem.* **162**, 627 (1946). Prepn from 3α-formoxy-17α-hydroxypregnan-20-one: Gallagher *et al.*, *J. Am. Chem. Soc.* **71**, 3262 (1949); from 16,17-oxido-5-pregnen-3β-ol-20-one acetate: Julian *et al.*, *ibid.* 3574; from 5-pregnen-3β-ol-20-one: Julian *et al.*, *ibid.* **72**, 5145 (1950).

Fine, glistening plates from ether. mp 212.8-216.8°; gives no depression of melting point when mixed with cortisone. Very sparingly sol in water, ether. Sol in acetone, methanol, alcohol. Gives a carmine-red fluorescence reaction with concd. H_2SO_4. Reduces ammoniacal silver nitrate soln at room temp. uv max: 242 nm ($E_{1cm}^{1\%}$ 500). Oxidation with chromic acid in glacial acetic acid yields 4-androstene-3,17-dione.

Acetate, $C_{23}H_{32}O_5$, mp 237.2-240.2° (sinters at 230°); $[\alpha]_D^{24}$ +116° (acetone); uv max (methanol): 242 nm ($E_{1cm}^{1\%}$ 448). When dissolved in concd sulfuric acid produces a typical scarlet color.

2913. 4-Desoxypyridoxine Hydrochloride. *5-Hydroxy-4,6-dimethyl-3-pyridinemethanol hydrochloride;* 2,4-dimethyl-3-hydroxy-5-hydroxymethylpyridine hydrochloride; 3-hydroxy-5-hydroxymethyl-2,4-dimethylpyridine hydrochloride; 4-desoxyadermin hydrochloride. $C_8H_{12}ClNO_2$; mol wt 189.64. C 50.66%, H 6.38%, Cl 18.70%, N 7.39%, O 16.87%. Synthesis starting with 3-cyano-4,6-dimethyl-2-pyridone: Zima, Jung, **Ger.** pat. **707,266** (1941 to E. Merck); *Chem. Zentr.* **1941**, II, 2353; van Wagtendonk, Wibaut, *Rec. Trav. Chim.* **61**, 728 (1942); from 5-acetoxy-3-cyano-2-hydroxy-4,6-dimethylpyridine which on treatment with PCl_5 gives the 2-chloro analog which on catalytic hydrogenation, diazotization, and hydrolysis gives 4-desoxypyridoxine: Coover, Bowman, **U.S.** pat. **2,481,573** (1949 to Merck & Co.); the quickest and best prepn is by catalytic reduction of pyridoxine: Harris, *J. Am. Chem. Soc.* **62**, 3203 (1940). Inhibits vitamin B₆ in chick, rat, mouse, dog, monkey, and man. Readily produces symptoms of vitamin B₆ deficiency: Mueller, Vilter, *J. Clin. Invest.* **29**, 193 (1950).

Crystals from alcohol + ether + acetone, mp 257°. Sol in water, alcohol.

2914. Desthiobiotin. *(4R-cis)-5-Methyl-2-oxo-4-imidazolidinehexanoic acid; 5-methyl-2-oxo-4-imidazolidinecaproic acid;* ε-(4-methyl-5-imidazolidone-2)-caproic acid; 4-methyl-5-(ω-carboxyamyl)imidazolidone-2. $C_{10}H_{18}N_2O_3$; mol wt 214.26. C 56.05%, H 8.47%, N 13.08%, O 22.40%. Prepd from biotin by hydrogenolysis of the sulfide linkage: du Vigneaud *et al.*, *J. Biol. Chem.* **146**, 475 (1942). Improved prepn procedure: Melville *et al.*, *Science* **98**, 497 (1943). Synthesis: Melville, *J. Am. Chem. Soc.* **66**, 1422 (1944); Wood, du Vigneaud, *ibid.* **67**, 210 (1945); Duschinsky, Dolan, *ibid.* 2079; Bourquin *et al.*, *Helv. Chim. Acta* **28**, 528 (1954).

Long needles from water. mp 156-158°. $[\alpha]_D^{21}$ +10.7° (c = 2). Sol in water.

Methyl ester, $C_{11}H_{20}N_2O_3$, crystals from methanol. Can be sublimed at 10^{-5} mm and 100°. mp 69-70°. $[\alpha]_D^{28}$ +2.6° (c = 2 in chloroform). Readily sol in chloroform.

2915. Destomycin A. *O-6-Amino-6-deoxy-L-glycero-D-galacto-heptopyranosylidene-(1 → 2-3)-O-β-D-talopyranosyl-(1 → 5)-2-deoxy-N′-methyl-D-streptamine;* 5-O-[2,3-O-[6-(1-amino-2-hydroxyethyl)tetrahydro-3,4,5-trihydroxy-2H-pyran-2-ylidene]-β-D-talopyranosyl]-2-deoxy-N³-methyl-D-streptamine; Destonate 20. $C_{20}H_{37}N_3O_{13}$; mol wt 527.54. C 45.53%, H 7.07%, N 7.96%, O 39.43%. Aminoglycoside antibiotic, member of the orthosomycin family. Destomycins A, B and C are known; A is the major component. Isoln of A and B from culture broth of *Streptomyces rimofaciens*: S.-I. Kondo *et al.*, *J. Antibiot.* **18A**, 38 (1965); of C: *eidem, ibid.* **28**, 83 (1975). Prepn: *eidem,* **Japan.** pat. **67 07598** (1967 to Meiji Seika Kaisha). Structure of A: *eidem, J. Antibiot.* **19A**, 139 (1966); of A and B: eidem, ibid. **28**, 79 (1975). Stereochemical study: S. Horito *et al.*, *Bull. Chem. Soc. Japan* **54**, 2147 (1981). Synthesis of destomycin C: J. Yoshimura *et al.*, *Chem. Letters* **1985**, 1335; J.-I. Tamura *et al.*, *Carbohydr. Res.* **174**, 181 (1988). Anthelmintic activity in poultry: I. Sawada, *Kiseichugaku Zasshi* **21**, 45 (1972), *C.A.* **77**, 135668j (1972). HPLC determn in pork samples: K. Nakaya *et al.*, *Shokuhin Eiseigaku Zasshi* **28**, 487 (1987), *C.A.* **109**, 5359c (1988).

White powder, mp 180~190° (dec). $[\alpha]_D^{22}$ +7° (c = 2). Freely sol in water, lower alcohols. Insol or poorly sol in most organic solvents. LD_{50} in mice (delayed toxicity) (mg/kg): 5~10 i.v.; 50~100 orally (Kondo, 1965).

THERAP CAT (VET): Broad spectrum antimicrobial; anthelmintic.

2916. Detaxtran. *Dextran 2-(diethylamino)ethyl ether;* diethylaminoethyldextran; DEAE-dextran; basic dextran. Mol wt ~500,000. Cationic derivative of dextran, *q.v.*, with antilipemic activity. Prepn: W. McKernan, C. R. Ricketts, *Chem. & Ind. (London)* **1959**, 1490; E. Antonini *et al.*, *Giorn. Biochem.* **14**, 88 (1965); *C.A.* **63**, 4375h (1965). Effect on platelet aggregation: R. Brossmer, Th. Pfleiderer, *Naturwiss.* **53**, 464 (1966); A. Larcon *et al.*, *Experientia* **28**, 1096 (1972). Use to lower hypercholesterolemia: T. M. Parkinson, **U.S.** pat. **3,627,872** (1971 to Upjohn); to reduce blood lipid levels: F. Kuzuya *et al.*, **U.S.** pat. **3,851,057** (1974 to Meito Sangyo). Clinical trial in hyperlipoproteinemia: F. Pupita, A. Barone, *Int. J. Clin. Pharm. Res.* **3**, 287 (1983).

Powder, soluble in water, saline.

Hydrochloride, *Dexide; Pulsar; Lipalt.*

THERAP CAT: Antihyperlipoproteinemic.

2917. Detomidine. *4-[(2,3-Dimethylphenyl)methyl]-1H-imidazole;* 4-(2′,3′-dimethylbenzyl)imidazole. $C_{12}H_{14}N_2$; mol wt 186.26. C 77.38%, H 7.58%, N 15.04%. α₂-Adrenoceptor agonist with sedative and analgesic activity. Prepn: A. J. Karjalayne, K. O. A. Kurkela, **Eur.** pat. **Appl. 24,829**; *eidem,* **U.S.** pat. **4,443,466** (1981, 1984 both to Farmos). Physical studies: E. Laine *et al.*, *Acta Pharm. Suec.* **20**, 451 (1983). Crystal structure: L. H. J. Lajunen *et al.*, ibid. **21**, 163 (1984). Pharmacology: R. Virtanen, L. Nyman, *Eur. J. Pharmacol.* **108**, 163 (1985); R. Virtanen, E. MacDonald, ibid. **115**, 277 (1985). Mechanism of action: *eidem, J. Vet. Pharmacol. Ther.* **8**, 30 (1985).

Crystals from acetone, mp 114-116°. LD_{50} i.v. in mice: 35 mg/kg (Karjalayne, Kurkela).

Hydrochloride, $C_{12}H_{15}ClN_2$, *Domosedan*. Crystals, mp 160°. Converts reversibly to monohydrate at room temp, 80% humidity.

THERAP CAT (VET): Sedative.

2918. Detoxin Complex. DX.C. A group of selective antagonists of blasticidin S, *q.v.*, produced by *Streptomyces caespitosus* var. *detoxicus* 7072 GC₁. The complex counteracts the inhibitory action of blasticidin S (BS) against *Bacillus cereus* but not against *Piricularia oryzal*, the causative agent of rice blast disease. DX.C also depresses the phytotoxicity of BS to rice plants, and a combination of the two showed less eye irritation in animals than BS alone. Isoln, production, biological properties: H. Yonehara *et al., J. Antibiot.* **21**, 369 (1968); *eidem, Agr. Biol. Chem.* **37**, 2771 (1973). Separation of the complex yields eight groups of detoxins designated A through H, the major components being C and D: N. Otake *et al., J. Antibiot.* **21**, 371 (1968); *eidem, Agr. Biol. Chem.* **37**, 2777 (1973). Detoxin D_1, the major component of the D group, has been shown to have the highest specific activity. It is a unique depsipeptide containing the amino acid *detoxinine (β,3-dihydroxy-2-pyrrolidinepropanoic acid)*. Structure of D_1: K. Kakenuma *et al., Tetrahedron Letters* **1972**, 2509. ¹³C-NMR: T. Ogita *et al., Agr. Biol. Chem.* **42**, 2403 (1978). Structures of the minor components of the D group: N. Otake *et al., J. Antibiot.* **27**, 484 (1974). Structures of detoxins A_1, B_1, B_3, C_1, C_2, C_3, E_1: N. Otake *et al., Experientia* **37**, 926 (1981); T. Ogita *et al., Agr. Biol. Chem.* **45**, 2605 (1981). Effect of D on BS uptake in *B. cereus*: A. Shimazu *et al., Experientia* **37**, 365 (1981).

Light colored fine powder.

Detoxin D_1, $C_{28}H_{41}N_3O_8$, R=CH(CH₃)CH₂CH₃, *N-(2-methyl-1-oxobutyl)-L-phenylalanine 1-[3-(acetyloxy)-1-(2-amino-3-methyl-1-oxobutyl)-2-pyrrolidinyl]-2-carboxyethyl ester*. Fine cryst powder, mp 156-158°. $[\alpha]_D^{25}$ −16° (c = 1 in methanol). uv max (methanol): 253, 258, 265, 268 nm ($E_{1cm}^{1\%}$ 3.1, 3.58, 2.77, 1.85). Amphoteric. pKa 4.0, 8.0.

Detoxin C_1, $C_{25}H_{35}N_3O_8$, R=CH₃, *detoxin C_{al}, N-acetyl-L-phenylalanine 1-[3-(acetyloxy)-1-(2-amino-3-methyl-1-oxobutyl)-2-pyrrolidinyl]-2-carboxyethyl ester*. Microneedles from water, mp 142-144°. $[\alpha]_D^{25}$ −23° (c = 1 in methanol). uv max (methanol): 248, 253, 259, 265, 269 nm ($E_{1cm}^{1\%}$ 3.7, 3.18, 2.65, 2.8, 1.85). Amphoteric. pKa 8.0, 3.9.

USE: In mixtures with blasticidin S, to decrease phytotoxicity to rice plants.

2919. Deuterium. Heavy hydrogen. ²H or D. Exists in the diatomic state; mol wt 4.028. A stable, non-radioactive isotope of hydrogen used in nuclear work. May be obtained by electrolysis of heavy water (obtained by the H_2S/H_2O exchange process) or by fractional distillation of liq hydrogen: Urey *et al., Phys. Rev.* **39**, 164 (1932); Spevack, U.S. pat. **2,787,526** (1957 to USAEC). *Review:* Mackay, Dove, in *Comprehensive Inorganic Chemistry,* vol. **1**, J. C. Bailar, Jr. *et al.,* Eds. (Pergamon Press, Oxford, 1973) pp 77-116; J. J. Katz in Kirk-Othmer *Encyclopedia of Chemical Technology* vol. **7** (Wiley-Interscience, New York, 3rd ed., 1979) pp 539-553.

Colorless, odorless, flammable gas (*see* Hydrogen). d (liq, 20.4°K) 0.169. bp −249.49° (23.67°K). mp −254.43°

(18.73°K) at 128.5 mm (triple point). Crit temp −234.75°. Crit press. 16.432 atm. Flammable limits in air: 5-75%. Calculation of vapor press. from the triple point to the crit point: Friedman *et al., J. Am. Chem. Soc.* **73**, 1310 (1951).

Human Toxicity: Considered non-toxic (except for its asphyxiating and flammable properties) as used in laboratories and nuclear power plants. Stunts growth of mammals when drunk regularly as heavy water: J. F. Thomson, *Biological Effects of Deuterium* (Pergamon Press, 1964).

USE: Used extensively in small amts as tracer in the establishment of rates and kinetics of chemical reactions: Wiberg, *Chem. Revs.* **55**, 713-743 (1955). The hydrogen bomb contains lithium deuteride (LiD) as explosive and plutonium (Pu) as initiator. After detonation by Pu the following reaction sequence takes place (nuclear fusion): ⁶Li + D = 2⁴He; ⁶Li + n = ⁴He + T; ⁶Li + T = 2⁴He + n (n = neutrons, coming from Pu in the second reaction). Since the explosion does not start by itself, there is no critical mass and no limit to the size of the bomb. If controlled thermonuclear fusions can be achieved, deuterium gas will become increasingly important.

2920. Deuterium Oxide. Heavy water; water-d_2. D_2O; mol wt 20.028. D 20.11%, O 79.89%.

More associated than H_2O. mp 3.81°; triple point temp 3.82°. bp 101.42°. Critical temp 371.5°. d^{25} 1.1044. Temp of max density 11.23°; $d^{11.23}$ 1.1059. Sp heat of liq (4-25°) 1.028 cal/g/°C. Heat of fusion 1.501 kcal/mole: Long, Kemp, *J. Am. Chem. Soc.* **58**, 1829 (1936). Heat of evapn 9.917 kcal/mole. Dielectric const (25°): 78.06, Vidulich *et al., J. Phys. Chem.* **71**, 656 (1966). Dipole moment in benzene (25°) 1.78, in dioxane 1.87. The surface tension is very slightly smaller and the viscosity at 25° is 1.23 times as great as that of water. On mixing with water, heat is evolved. Reacts with hydrogen at 100° in the presence of 0.2 to 1N NaOH forming DOH + HD + OD⁻. Ionization const 1.95 × 10⁻¹⁵: W. F. K. Wynne-Jones, *Trans. Faraday Soc.* **32**, 1397 (1936). pK at 25° 14.955 (molarity scale); 16.653 (mole fraction scale): A. K. Covington *et al., J. Phys. Chem.* **70**, 3820 (1966). *See also* Water.

USE: To study chemical reaction rates and mechanisms. The cross section of deuterium for the capture of thermal neutrons is very low which makes it useful, in the form of heavy water, as a neutron moderator in nuclear reactors. Produces a considerable decrease in neutron energy per collision.

2921. Devarda's Metal. Devarda's alloy. Consists of 50 parts Cu, 45 Al and 5 Zn.

Gray powder. Partly sol in HCl with residue of Cu.

USE: Reducing agent for determination of nitrogen in nitrates and nitrites.

2922. Dexamethasone. *(11β,16α)-9-Fluoro-11,17,21-trihydroxy-16-methylpregna-1,4-diene-3,20-dione;* 9α-fluoro-16α-methylprednisolone; 16α-methyl-9α-fluoro-1,4-pregnadiene-11β,17α,21-triol-3,20-dione; 16α-methyl-9α-fluoroprednisolone; 1-dehydro-16α-methyl-9α-fluorohydrocortisone; 16α-methyl-9α-fluoro-Δ¹-hydrocortisone; hexadecadrol; Calonat; Decasone; Decacortin; Deronil; Decadron; Fluormone; Dekacort; Dexa-Cortisyl; Dinormon; Millicorten; Fortecortin; Anaflogistico; Spoloven; Luxazone; Dectancyl; Dexameth; Dexapos; Dexasone; Dexinoral; Gammacorten; Dextelan; Policort; Dexacortal; Dexa-Mamallet; Decalix; Dexacortin; Deseronil; Corson; Dergramin; Hexadrol; Deltafluorene; Oradexon; Dexa-Cortidelt; Aeroseb-D; Dexinolon; Decaderm; Dexafarma; Cortisumman; Dexa-sine; Lokalison F; Loverine; Maxidex; Dexa-Scheroson; Pet Derm III; Isopto-Dex. $C_{22}H_{29}FO_5$; mol wt 392.45. C 67.32%, H 7.45%, F 4.84%, O 20.38%. Prepn: Arth *et al., J. Am. Chem. Soc.* **80**, 3161 (1958); Oliveto *et al., ibid.* 4431; Muller *et al.,* U.S. pat. **3,007,923** (1961 to Lab. Franc. Chimiothér.); Arth *et al.,* Ger. pat. **1,113,690** (1961 to Merck & Co.); **Brit.** pat. **869,511** (to Upjohn). Pharmacology in guinea pigs: L. Zicha *et al., Arzneimittel-Forsch.* **10**, 831 (1960). Clinical trial of anti-emetic efficacy in chemotherapy-induced nausea: M. Markman *et al., N. Engl. J. Med.* **311**, 549 (1984). Use as diagnostic aid in depression: B. J. Carroll *et al., Arch. Gen. Psychiat.* **38**, 15 (1981); P. P. Hubain *et al., Neuropsychobiology* **16**, 57 (1986); M. Ansseau *et al., ibid.* 68. Comprehensive description: E.

M. Cohen in *Analytical Profiles of Drug Substances* vol. 2, K. Florey, Ed. (Academic Press, New York, 1973) pp 163-197. Reviews of diagnostic use in Cushing's Syndrome: L. Crapo, *Metabolism* **28**, 955 (1979); in depression: L. Braddock, *Brit. J. Psychiat.* **148**, 363 (1986).

Crystals from ether, mp 262-264°; mp 268-271° (Arth, 1961). $[\alpha]_D^{25}$ +77.5° (dioxane). Soly in water (25°): 10 mg/100 ml. Sol in acetone, ethanol, chloroform.

21-Acetate, $C_{24}H_{31}FO_6$, *Decadron-LA, Panasone*. Crystals, mp 215-221° (Arth, 1958); mp 229-231° (Oliveto); mp 238-240° (Arth, 1961). $[\alpha]_D^{25}$ +73° (chloroform) (Arth, 1958); $[\alpha]_D$ +77.6° (Oliveto). uv max: 239 nm (ϵ 14900).

21-(3,3-Dimethylbutyrate), $C_{28}H_{39}FO_6$, *dexamethasone tert-butylacetate, Decadron TBA*.

21-Phosphate, $C_{22}H_{30}FO_8P$, *Wymesone*.

21-Phosphate disodium salt, $C_{22}H_{28}FNa_2O_8P$, *dexamethasone 21-(dihydrogen phosphate) disodium salt, Ak-Dex, Baldex, Dalaron, Dexabene, Dezone, Solu-Decadron, Turbinaire, Orgadrone, Colvasone, Soldesam*. Crystals, mp 233-235°. $[\alpha]_D$ +74° ±4° (calcd on water-free and alcohol-free basis, concn of 10 mg/ml): *USP* **XIX**, p 124. uv max (ethanol): 238-239 nm (ϵ 14000). Sol in water. Used as an injectable form of dexamethasone. Prepn: Chemerda *et al.*, U.S. pat. 2,939,873 (1960 to Merck & Co.); Irmscher, *Chem. & Ind. (London)* **1961**, 1035.

Tetrahydrophthalate, $C_{30}H_{33}O_8FNa$, *Millicorten*.

21-Diethylaminoacetate, $C_{28}H_{41}FNO_6$, *Solu-Forte-Cortin*.

21-Isonicotinate, $C_{28}H_{32}FNO_6$, *dexamethasone 21-(4-pyridinecarboxylate), Auxiloson, Auxisone, Voren*. Crystals, mp 250-252°. $[\alpha]_D^{27}$ +183.5° (dioxane).

17,21-Dipropionate, $C_{28}H_{37}FO_7$, *THS-101, Methaderm*.

21-Palmitate, $C_{38}H_{59}FO_6$, *Limethasone*.

THERAP CAT: Glucocorticoid; anti-inflammatory. Diagnostic aid (Cushing's Syndrome, depression).

THERAP CAT (VET): Glucocorticoid; anti-inflammatory.

2923. Dexetimide. *3-Phenyl-1'-(phenylmethyl)-[3,4'-bipiperidine]-2,6-dione; (S)-(+)-2-(1-benzyl-4-piperidyl)-2-phenylglutarimide; (+)-3-(1-benzyl-4-piperidyl)-3-phenylpiperidine-2,6-dione; (+)-1-benzyl-4-(2,6-dioxo-3-phenyl-3-piperidyl)piperidine; dextrobenzetimide; dexbenzetimide*. $C_{23}H_{26}N_2O_2$; mol wt 362.45. C 76.21%, H 7.23%, N 7.73%, O 8.83%. The dextroenantiomer responsible for the pharmacological activity of racemic benzetimide, *q.v.* Resolution of isomers and comparative pharmacology: Janssen *et al.*, *Arzneimittel-Forsch.* **21**, 1365 (1971). Abs config studies: van Wijngaarden *et al.*, *Life Sci.* **9**, part 1, 1289 (1970); Spek *et al.*, *Nature* **232**, 575 (1971). Clinical trials: De Smedt *et al.*, *J. Clin. Pharmacol.* **10**, 207 (1970).

Crystals, mp 181-183°. $[\alpha]_D^{20}$ +125° (chloroform).

Hydrochloride, $C_{23}H_{27}ClN_2O_2$, *R 16470, Tremblex*. Crystals, mp 270-275°. $[\alpha]_D^{20}$ +125° (methanol). LD_{50} in rats: 45 mg/kg i.v., Janssen *et al.*, *loc. cit.*

THERAP CAT: Anticholinergic; antiparkinsonian.

2924. Dexpanthenol. *2,4-Dihydroxy-N-(3-hydroxypropyl)-3,3-dimethylbutanamide; D(+)-α,γ-dihydroxy-N-(3-hydroxypropyl)-β,β-dimethylbutyramide; pantothenylol;*

N-pantoyl-3-propanolamine; pantothenol; pantothenyl alcohol; Alcopan-250; Intrapan; Pantenyl; Panthoderm; Pantonyl; Motilyn; Bepanthen; Cozyme; Ilopan; Urupan. $C_9H_{19}NO_4$; mol wt 205.25. C 52.66%, H 9.33%, N 6.82%, O 31.18%. Prepd by the addition of propanolamine to optically active α,γ-dihydroxy-β,β-dimethylbutyrolactone: Schnider, *Jubilee vol. Emil Barell* **1946**, 85; Swiss pat. **227,706** (1943); **Brit.** pat. **582,156** (1946); U.S. pat. **2,413,077** (1946 to Hoffmann-La Roche). Only the D(+)-form has vitamin activity.

Viscous, somewhat hygroscopic liq. Slightly bitter taste. d_{20}^{20} 1.2. $bp_{0.02}$ 118-120°. Easily dec on distn. $[\alpha]_D^{20}$ +29.5° (c = 5). n_D^{20} 1.497. Freely sol in water, alcohol, methanol. Slightly sol in ether. Natural pH about 9.5. Reasonably stable to usual sterilization time and temp in aq solns adjusted to pH 3.0-4.0, but long heating causes racemization. Hydrolyzed by alkali and strong acid. Usually more stable than salts of pantothenic acid if pH can be adjusted between 3 and 5. For add'l stability data *see* Rubin, *J. Am. Pharm. Assoc., Sci. Ed.* **37**, 502 (1948). Aq solns can be stabilized with pantolactone: U.S. pat. **2,898,373** (1959).

dl-Form, *panthenol*.

THERAP CAT: Cholinergic; *dl*-form as vitamin.

THERAP CAT (VET): Nutritional factor. Dietary source of pantothenic acid.

2925. Dextran. *Dextraven; Expandex; Gentran; Hemodex; Intradex; Macrose; Onkotin; Plavolex; Polyglucin; Promit*. A term applied to polysaccharides produced by bacteria growing on a sucrose substrate, contg a backbone of D-glucose units linked predominantly α-D(1 → 6). Several organisms produce dextrans but only *Leuconostoc mesenteroides* and *L. dextranicum* (*Lactobacteriaceae*) have been used commercially. Chemical and physical properties of the dextrans vary with the methods of production. Native dextrans usually have high mol wt; lower mol wt clinical dextrans usually prepared by depolymerization of native dextrans or by synthesis. All dextrans are composed exclusively of α-D-glucopyranosyl units, differing only in degree of branching and chain length. Prepn: Tarr, Hibbert, *Can. J. Res.* **5**, 414 (1931); Novak, Stoycos, U.S. pat. **2,841,578** (1958 to Commonwealth Eng. of Ohio). Enzymic synthesis: Sugg, Hehre, *J. Immunol.* **43**, 119 (1942); Behrens, Ringpfeil, U.S. pat. **3,044,940** (1962 to Serum Werk Bernburg). The crude dextran may be isolated from the culture by precipitation with methanol. Continuous dialysis process: Shurter, U.S. pat. **2,717,853** (1955 to C.S.C.). Elimination of pyrogens: Levi, Lozinski, U.S. pat. **2,762,727** (1956 to Frosst). Method of producing clinical dextran: Novak, Witt, U.S. pat. **2,972,567** (1961). Structure studies: Fowler *et al.*, *Can. J. Res.* **B15**, 486 (1937); Fairhead *et al.*, *ibid.* **B16**, 151 (1938); Peat *et al.*, *J. Chem. Soc.* **1939**, 581; Goldstein, Whelan, *ibid.* **1962**, 170, 176. ^{13}C-NMR structure study: F. R. Seymour *et al.*, *Carbohyd. Res.* **51**, 179 (1976). Reviews: Evans, Hibbert, *Advan. Carbohyd. Chem.* **2**, 204 (1946); Neely, *ibid.* **15**, 341 (1960); Ricketts, *Progr. Org. Chem.* **5**, 73 (1961); Murphy, Whistler, in *Industrial Gums*, R. L. Whistler, Ed. (Academic Press, New York, 2nd ed., 1973) pp 513-542.

Dextran 40, LMD, LMWD, LVD, Eudextran, Gentran 40, Rheomacrodex, Rheotran. Produced by action of *L. mesenteroides* on sucrose; average mol wt: 40,000.

Dextran 70, Hyskon, Macrodex. Average mol wt: 70,000.

Dextran 75, Gentran 75. Average mol wt: 75,000.

USE: In soft center confections, as a partial substitute for barley malt. Mixed ethers and esters of dextran can be used in lacquers.

THERAP CAT: Plasma volume expander; Dextran 40 also as blood flow adjuvant.

THERAP CAT (VET): Plasma extender.

2926. Dextranase. α-1,6-Glucan 6-glucanohydrolase; E.C. 3.2.1.11. Enzyme which hydrolyzes the α-1 → 6 glucosidic linkages of the bacterial polysaccharide dextran. *Endodextranases* (dextranases which preferentially split glucosidic linkages remote from end groups) are secreted by various molds and a few bacteria. *Exodextranases* occur predominantly in mammalian tissues. Prepn: from *Penicillium lilacinum, P. funiculosum,* and *Verticillium coccorum,* Nordström, Hultin, *Svensk. Kem. Tid.* **60,** 283 (1948), *C.A.* **43,** 3050i (1949); from *Aspergillus,* Carlson, Carlson, *Science* **115,** 43 (1952), *eidem,* **U.S.** pat. **2,709,150** (1955 to Enzymatic Chemicals), *eidem,* **U.S.** pat. **2,716,084,** and Novak, Stoycos, **U.S.** pat. **2,841,578** (1955 and 1958 both to Commonwealth Eng. of Ohio); from *P. lilacinum, P. funiculosum, P. verruculosum,* and *Spicaria violacea,* Tsuchiya *et al.,* **U.S.** pat. **2,742,399** and Corman, Tsuchiya, **U.S.** pat. **2,776,925** (1956 and 1957 both to U.S. Secy. Agr.); from *P. lilacinum,* Charles, Farrell, *Can. J. Microbiol.* **3,** 239 (1957); from *Lactobacillus bifidus,* Bailey, Clarke, *Biochem. J.* **72,** 49 (1959). Tested as dental caries-control agent in hamsters: Fitzgerald *et al., J. Am. Dental Assoc.* **76,** 301 (1968). Review: E. H. Fischer, E. A. Stein, "Cleavage of *O*- and *S*-Glycosidic Bonds (Survey)" in *The Enzymes,* vol. 4, P. D. Boyer *et al.,* Eds. (Academic Press, New York, 2nd ed., 1960) pp 304-307.

USE: In prepn of dextran for clinical use; in dentifrices.

2927. Dextran Iron Complex. Iron-dextran complex; Fenate; Imferon; Ironorm. Prepn: London, Twigg, **U.S.** pat. **2,820,740** (1958 to Benger); **U.S.** reissue **24,642** (1959); Herb, **U.S.** pat. **2,885,393** (1959 to Laros). Marketed as a soln contg 2-5% elemental iron. Acute toxicity: Beliles, *Toxicol. Appl. Pharmacol.* **23,** 537 (1972). Review of carcinogenicity studies: *IARC Monographs* **2,** 161-178 (1973). LD_{50} i.v. in mice: 2.24 g Fe/kg (Beliles).

Note: This substance may reasonably be anticipated to be a carcinogen: *Fourth Annual Report on Carcinogens* (NTP 85-002, 1985) p 119.

THERAP CAT: Hematinic.

THERAP CAT (VET): Nutritional factor (parenteral). Used in iron deficiency anemia, chiefly in pigs.

2928. Dextranomer. *Dextran 2,3-dihydroxypropyl 2-hydroxy-1,3-propanediyl ethers;* Debrisan; Debrisorb. Three-dimensional hydrophilic network of a dextran polymer, linked by cross-chains of epichlorohydrin, *q.v.;* it absorbs moisture and small molecules from suppurating wounds. Chronic tissue response to implantation: J. Falk, G. Tollerz, *Clin. Ther.* **1**(3), 185 (1977). Potential allergic contact sensi-

tization in guinea pigs: G. Jonsson, *ibid.* **1**(4), 260 (1978). Efficacy in treatment of ulcers and wounds: J. Soul, *Brit. J. Clin. Pract.* **32,** 172 (1978); P. N. Sawyer *et al., Surgery* **85,** 201 (1979); S. Di Mascio, *Am. J. Nurs.* **79,** 684 (1979). Review: R. C. Heel *et al., Drugs* **18,** 89-102 (1979).

Insol in all solvents; stable in water, salt solns and in alkaline and weakly acidic soln (Falk, Tollerz, *loc. cit.*).

THERAP CAT: Vulnerary.

2929. Dextran Sulfate Sodium. Dextran sulfuric acid ester sodium salt; Asuro; Colyonal; Dexulate; Dextrarine; MDS. Heparin-like polysaccharide containing approx. 17% S with up to three sulfate groups per glucose molecule. Mol wt ranges from 4,000-500,000 Da; variations in mol wt are associated with differences in biological activity. Prepn, properties and anticoagulant activity: A. Grönwall *et al., Upsala Lakarefören. Förh.* **50,** 397 (1945); C. R. Ricketts, *Biochem. J.* **51,** 129 (1952). Evaluation of toxicity as a function of mol wt: K. W. Walton, *Brit. J. Pharmacol.* **9,** 1 (1954); of carcinogenicity: I. Hirono *et al., Cancer Letters* **18,** 29 (1983). Use in serum HDL cholesterol determn: P. R. Finley *et al., Clin. Chem.* **24,** 931 (1978); G. R. Warnick *et al., ibid.* **28,** 1379 (1982). Antiscrapie effect: B. Ehlers, H. Diringer, *J. Gen. Virol.* **65,** 1325 (1984); R. H. Kimberlin, C. A. Walker, *Antimicrob. Ag. Chemother.* **30,** 409 (1986). Anti-HIV-1 activity *in vitro:* H. Mitsuya *et al., Science* **240,** 646 (1988); M. Baba *et al., Proc. Nat. Acad. Sci. USA* **85,** 6132 (1988).

White powder from alcohol + ether. Freely sol in water. Activity about 17 international heparin units/mg. Aq solns must be buffered (*e.g.,* with sodium bicarbonate) to prevent dec during autoclaving.

USE: Clinical reagent (HDL cholesterol determn).

THERAP CAT: Anticoagulant; antihyperlipoproteinemic.

2930. Dextri-Maltose®. Maltose and dextrins obtained by enzymic action of barley malt on corn flour.

Light, amorphous powder. Readily sol in water or milk. One leveled tablespoonful (8 grams) supplies 27 calories.

USE: As carbohydrate modifier for use with milk and milk products in infants' formulas.

2931. Dextrin. Pyrodextrin; torrefaction dextrin. $(C_6-H_{10}O_5)_n$·xH_2O. Produced by the dry heating of unmodified starches. The term also includes products resulting from enzyme or acid-catalyzed hydrolysis of wet starch. *Review:* R. W. Satterthwaite, D. J. Iwinski, in *Industrial Gums,* R. L. Whistler, Ed. (Academic Press, New York, 2nd ed., 1973) pp 577-599.

British gum, starch gum. Produced at high temp in the absence of acid. Dark brown color, odorous. High viscosity; very sol in cold water. Does not reduce Fehling's soln; gives reddish-brown color with iodine.

Canary dextrin, yellow dextrin. Hydrolyzed at high temp for long period of time in the presence of small amts of acid. Light brown to yellow color, slight odor. Low viscosity; very sol in cold water.

White dextrin. Hydrolyzed at low temp for short period of time in the presence of large amts of acid. White color, odorless. Slightly sol in cold water giving a red color with iodine. Very sol in hot water giving a blue color with iodine.

USE: Excipient for dry extracts and pills; for preparing emulsions and dry bandages; for thickening dye pastes and mordants used in printing fabrics in fast colors; sizing paper and fabrics; printing tapestries; preparing felt; manuf printer's inks, glues and mucilage; polishing cereals; in matches, fireworks, and explosives.

2932. Dextroamphetamine Sulfate. *(S)-α-Methylbenzeneethanamine sulfate; d-α-methylphenethylamine sulfate; d*-amphetamine sulfate; *d*-1-phenyl-2-aminopropane sulfate; *d*-β-phenylisopropylamine sulfate; Dexampex; Dexedrine Sulfate; Afatin; Dexamphetamine; *d*-Amfetasul; Domafate; Obesedrin; Dexten; Maxiton; Sympamin; Simpamina-D; Albemap; Dadex; Ardex; Dexalone; Amsustain; Betafedrina; *d*-Betaphedrine; Diocurb. $C_{18}H_{28}N_2O_4S$; mol wt 368.49. C 58.67%, H 7.66%, N 7.60%, S 8.70%, O 17.37%. $[C_6H_5CH_2CH(NH_2)CH_3]_2$·$H_2SO_4$. *See* Amphetamine for structural formula. Prepn by resolution of amphetamine with *d*-tartaric acid followed by treatment with 10% H_2SO_4: Temmler, **Brit.** pat. **508,757** (1939); Nabenhauer, U.S. pat. **2,276,508** (1942 to SK&F); Magidson, Garkusha, *J. Gen. Chem. (USSR)* **11,** 339 (1941); from D-phenylalanine: D. B. Repke *et al., J. Pharm. Sci.* **67,** 1167 (1978). Toxicity data: E. J. Warawa *et al., J. Med. Chem.* **18,** 71 (1975).

Plates, rods [different from amphetamine sulfate crystals, *see* Keenan, *J. Am. Pharm. Assoc.* **37,** 519 (1948)]. Slightly bitter taste followed by a sensation of numbness. $[\alpha]_D^{20}$ +21.8° (c = 2). mp above 300°. One part dissolves in approx 10 parts of water, in 500 parts of 95% alcohol. pH 5% aq soln: 5.0 to 6.0. LD_{50} orally in mice: 10 mg/kg (Warawa).

Compounds formed from dextroamphetamine and carboxymethylcellulose are marketed as *Carboxyphen* (mol wt 568,000) and *Bontril* (about 40 mg = 15 mg of dextroamphetamine sulfate).

Note: This is a controlled substance (stimulant) listed in the U.S. Code of Federal Regulations, Title 21 Part 1308.12 (1987).

THERAP CAT: CNS stimulant; anorexic.
THERAP CAT (VET): Sympathomimetic, CNS stimulant.

2933. Dextromoramide. *1-[3-Methyl-4-(4-morpholinyl)-1-oxo-2,2-diphenylbutyl]pyrrolidine; 1-(3-methyl-4-morpholino-2,2-diphenylbutyryl)pyrrolidine;* 4-[2-methyl-4-oxo-3,3-diphenyl-4-(1-pyrrolidinyl)butyl]morpholine; d-morpholylmethyldiphenylbutyrylpyrrolidine; (+)-1-(3-methyl-4-morpholino-2,2-diphenylbutyryl)pyrrolidine; d-2,2-diphenyl-3-methyl-4-morpholinobutyrylpyrrolidine; d-3-methyl-2,2-diphenyl-4-morpholinobutyrylpyrrolidine; pyrrolamidol; R-875; SKF 5137; *D*-Moramide; Palfium; Palphium; Jetrium; Dimorlin. $C_{25}H_{32}N_2O_2$; mol wt 392.52. C 76.49%, H 8.22%, N 7.14%, O 8.15%. Synthesis: Janssen, *J. Am. Chem. Soc.* **78,** 3862 (1956); **Brit. pat. 822,055** (1959 to Janssen).

Crystals, mp 180-184°. $[\alpha]_D^{20}$ +25.5° (c = 5 in benzene). uv max (0.01N isopropanol-HCl): 254, 260, 264 nm. Practically insol in water. Soly in 0.1N HCl: 1:25 (w/v). Soly (g/100 ml) in ethanol 50; in methanol 40; in acetone 50; in ethyl acetate 40; in benzene 5; in chloroform 5. Sol in ether.

Bitartrate, $C_{25}H_{32}N_2O_2 \cdot C_4H_6O_6$, minute crystals, bitter taste. Dec 189-192°. Soly (w/v) at 25°: Water 20%, chloroform 30%, methanol 40%, ethanol 100%, acetone 100%.

Caution: May be habit forming. This is a controlled substance (opiate) listed in the U.S. Code of Federal Regulations, Title 21 Part 1308.11 (1985).

THERAP CAT: Narcotic analgesic.

2934. Dezocine. *[5R-(5α,11α,13S*)]-13-Amino-5,6,7,8,9,10,11,12-octahydro-5-methyl-5,11-methanobenzocyclodecen-3-ol;* WY-16225; Dalgan. $C_{16}H_{23}NO$; mol wt 245.36. C 78.32%, H 9.45%, N 5.71%, O 6.52%. Synthetic opiate agonist-antagonist. Prepn: M. E. Freed *et al.,* **Ger. pat. 2,159,324** (1972 to Am. Home Products), *C.A.* **79,** 53094w (1973); *eidem, J. Med. Chem.* **16,** 595 (1973). Pharmacology: J. L. Malis *et al., J. Pharmacol. Exp. Ther.* **194,** 488 (1975). Metabolism: S. F. Sisenwine, C. O. Tio, *Drug Metab. Dispos.* **9,** 37 (1981). Pharmacokinetics: S. F. Sisenwine *et al., ibid.* **10,** 366 (1982). Clinical studies: W. Oosterlinck, A. Verbaeys, *Curr. Med. Res. Opin.* **6,** 472 (1980); J. W. Downing *et al., Brit. J. Anaesth.* **53,** 59 (1981).

Hydrobromide salt, $C_{16}H_{24}BrNO$, crystalline powder, mp 269-270°. pH (2% aq soln): 4.6. Soly in water: >20 mg/ml. LD_{50} in female mice, rats (mg/kg): 313, 232 orally; 129, 270 i.m. (Malis).

THERAP CAT: Narcotic analgesic.

2935. DFDD. *1,1'-(2,2-Dichloroethylidene)bis[4-fluorobenzene]; 1,1-dichloro-2,2-bis(p-fluorophenyl)ethane;* difluorodiphenyldichloroethane. $C_{14}H_{10}Cl_2F_2$; mol wt 287.14. C 58.56%, H 3.51%, Cl 24.70%, F 13.23%. Prepn: Müller, *Helv. Chim. Acta* **9,** 1560 (1946).

Crystals from petr ether, mp 77°.
USE: Contact insecticide.

2936. DFDT. *1,1'-(2,2,2-Trichloroethylidene)bis[4-fluorobenzene]; 1,1,1-trichloro-2,2-bis(p-fluorophenyl)ethane;* difluorodiphenyltrichloroethane; fluorogesarol; HO-2,474; Gix. $C_{14}H_9Cl_3F_2$; mol wt 321.59. C 52.29%, H 2.82%, Cl 33.08%, F 11.82%. Prepn: Sumerford, *J. Am. Pharm. Assoc.* **36,** 127 (1947); Cross, **U.S. pat. 2,581,174** (1952). Toxicity: Piekarski, Holz, *Arch. Exp. Pathol. Pharmakol.* **210,** 71 (1950).

Crystals from aq ethanol, mp 44-45°. Practically insol in water; very sol in oils, most organic solvents.
USE: Contact insecticide. *Caution:* Effects similar to DDT [1,1,1-trichloro-2,2-bis(p-chlorophenyl)ethane], *q.v.*

2937. Dhurrin. *(S)-α-(β-D-Glucopyranosyloxy)-4-hydroxybenzeneacetonitrile;* p-hydroxymandelonitrile-β-D-glucoside; β-D-glucopyranosyloxy-L-p-hydroxymandelonitrile. $C_{14}H_{17}NO_7$; mol wt 311.28. C 54.02%, H 5.51%, N 4.50%, O 35.98%. From the fruit of young Egyptian plants of *Sorghum vulgare* Pers., *Gramineae:* Dunstan, Henry, *Chem. News* **85,** 301 (1902). Biosynthesis: Koukol *et al., J. Biol. Chem.* **237,** 3223 (1962). Absolute configuration: Towers *et al., Tetrahedron* **20,** 71 (1964). Allied to amygdalin (*q.v.*) and the mandelonitrile glucosides.

Leaflets from water, prisms from alcohol, dec 200°. $[\alpha]_D^{20}$ −65° (alc). uv max: 228 nm (ε 13,000). Sol in water, alcohol. Alkali labile.

Pentaacetate, $C_{24}H_{27}NO_{12}$, needles from ethyl acetate + petr ether, mp 132-132.5°. $[\alpha]_D^{20}$ −50.5° (c = 0.24 in alc).
Caution: Yields HCN on hydrolysis.

2938. Diaboline. *1-Acetyl-19,20-didehydro-17,18-epoxycuran-17-ol;* N-acetyl-Wieland-Gumlich aldehyde. $C_{21}H_{24}N_2O_3$; mol wt 352.42. C 71.57%, H 6.86%, N 7.95%, O 13.62%. From *Strychnos diaboli* Sandw., *Loganiaceae:* King, *J. Chem. Soc.* **1949,** 955; Bader *et al., Helv. Chim. Acta* **36,** 1256 (1953); Casinovi *et al., Nature* **193,** 1178 (1962). Structure: Battersby, Hodson, *Proc. Chem. Soc.* **1959,** 123. Synthesis: Deyrup *et al., Helv. Chim Acta* **45,** 2266 (1962). ^{13}C-NMR study: E. Wenkert *et al., J. Org. Chem.* **43,** 1099 (1978).

Needles from ether, mp 187°. $[\alpha]_D^{22}$ +37.8° (c = 1.72 in chloroform). uv max (ethanol): 249 nm (log ϵ 4.06).

Hydrochloride monohydrate, $C_{21}H_{24}N_2O_3 \cdot HCl \cdot H_2O$, needles from alcohol + ethyl acetate, mp 300°. For anhydr, $[\alpha]_{5461}^{20}$ +184° (c = 0.57 in water).

2939. Diacerein. *4,5-Bis(acetyloxy)-9,10-dihydro-9,10-dioxo-2-anthracenecarboxylic acid; 9,10-dihydro-4,5-dihydroxy-9,10-dioxo-2-anthroic acid diacetate;* 1,8-diacetoxy-3-carboxyanthraquinone; diacerhein; diacetylrhein; DAR; SF-277; Artrodar; Fisiodar. $C_{19}H_{12}O_8$; mol wt 368.30. C 61.96%, H 3.28%, O 34.75%. Diacetyl derivative of rhein, *q.v.* Demonstrates anti-arthritic activity without inhibiting prostaglandin synthesis. Prepn: A. Tschirch, K. Heuberger, *Arch. Pharm.* **240**, 596 (1902); V. K. Murty *et al., Tetrahedron* **23**, 515 (1967). Effect on prostaglandin release: P. Pomarelli *et al., Farmaco Ed. Sci.* **35**, 836 (1980). Inhibition of proteases: L. Raimondi *et al., Pharmacol. Res. Commun.* **14**, 103 (1982). Use in arthritis: C. A. Friedmann, *Ger. pat.* 2,711,493; *idem, U.S. pat.* 4,244,968 (1977, 1981 both to Proter). Preliminary clinical study in osteoarthritis: A. G. Kay *et al., Curr. Med. Res. Opin.* **6**, 548 (1980). Use in multiple sclerosis, amyotrophic lateral sclerosis: C. A. Friedmann, *U.S. pat.* 4,346,103 (1982); *idem, Japan. Kokai 83 225,015* (1983 to Proter), *C.A.* **100**, 144991e (1984). Clinical trial in osteoarthritis: S. Adami *et al., Clin. Ter.* **112**, 439 (1985).

Yellow plates from acetic acid, mp 217-218°.
THERAP CAT: Antiarthritic.

2940. Diacetazotol. *N-Acetyl-N-[2-methyl-4-[(2-methylphenyl)azo]phenyl]acetamide; 4''-(o-tolylazo)-o-diacetotoluidide;* diacetylaminoazotoluene; *N,N*-diacetyl-o-tolyazo-o-toluidine; 2,3'-dimethyl-4'-(diacetylamino)azobenzene; Dermagan; Diacetotoluide; Dimazon; Pellidol. $C_{18}H_{19}N_3O_2$; mol wt 309.36. C 69.88%, H 6.19%, N 13.58%, O 10.34%. Prepd by treating aminoazotoluene with acetic anhydride or with acetyl chloride in the presence of sodium acetate: *Ger. pat.* 253,884 (1912 to Kalle); Ettel, Hebky, *Coll. Czech. Chem. Commun.* **13**, 161 (1948). Biological properties: Rerábek *et al., C.A.* **46**, 7234i (1952). Allergenic properties: Jirasek *et al., Cesk. Dermatol.* **41**, 38 (1966).

Crystallizes in 2 modifications: Brick-red needles, mp 65°, or stout, red prisms, mp 75° (resembling crystals of potassium dichromate). The German Pharmacopoeia recognizes only a material which melts at 74-76° (yellowish-red powder). Insol in water; sol in alcohol, benzene, chloroform, ether, acetone, fixed oils, fats, petrolatum.

Mixture with albumin and tetraiodopyrrole, *Azodolen.*
THERAP CAT: Topical dermatologic.
THERAP CAT (VET): Has been used in ointment or dusting powder to stimulate wound epitheliazation.

2941. Diacetin. *1,2,3-Propanetriol diacetate;* glycerol diacetate; glyceryl diacetate. $C_7H_{12}O_5$; mol wt 176.17. C 47.72%, H 6.87%, O 45.41%. Commercial material is probably a mixture of glycerol 1,2- and 1,3-diacetates. Prepn of the 1,2- and 1,3-compounds: Wegscheider, Zmerzlikar, *Monatsh.* **34**, 1061 (1913); of the 1,2-compd: Golendeev, *J. Gen. Chem. (USSR)* **6**, 1841 (1936); of the 1,3-compd: Lagenbeck, Bollow, *Naturwiss.* **42**, 389 (1955).

1,2-compd   1,3-compd

Hygroscopic liquid, bp 259°. d_4^{16} 1.184. n_D^{20} 1.44. Sol in water, alc, ether, benzene; practically insol in CS_2.

dl-1,2-Compd, *1,2-diacetin.* bp_{12} 140-142°, bp_{40} 172-173.5°. d_4^{15} 1.1173.

1,3-Compd, *1,3-diacetin.* bp_{40} 172-174°. d^{15} 1.179. n_D^{20} 1.4395.

USE: Technical grade as a plasticizer and softening agent, and as a solvent.

2942. Diacetonamine. *4-Amino-4-methyl-2-pentanone.* $C_6H_{13}NO$; mol wt 115.17. C 62.57%, H 11.38%, N 12.16%, O 13.89%. $CH_3COCH_2C(CH_3)_2NH_2$. Prepd from mesityl oxide with aq NH_3: Haeseler, *J. Am. Chem. Soc.* **47**, 1195 (1925); Orthner, *Ann.* **456**, 245 (1927); Haeseler, *Org. Syn.* **coll. vol. I** (2nd ed., 1941) p 196. From mesityl oxide with liquid ammonia: Smith, Adkins, *J. Am. Chem. Soc.* **60**, 408 (1938).

Liq, $bp_{0.2}$ 25°. Amine odor. Lighter than water. Strongly alkaline. Sol in water, but aq solns prepd in the cold become turbid on heating; miscible with alc, ether. Dec easily forming mesityl oxide and NH_3. Decompn is prone to take place in aq solns or during distillation.

Acid oxalate monohydrate, $C_6H_{13}NO \cdot C_2H_2O_4 \cdot H_2O$, spear-shaped crystals, mp 126-127°.

2943. Diacetone Acrylamide. *N-(1,1-Dimethyl-3-oxobutyl)-2-propenamide; N-(1,1-dimethyl-3-oxobutyl)acrylamide; N-[2-(2-methyl-4-oxopentyl)]acrylamide.* $C_9H_{15}NO_2$; mol wt 169.22. C 63.88%, H 8.93%, N 8.28%, O 18.91%. Highly reactive vinyl monomer. Prepd by the reaction of acrylonitrile with acetone in presence of conc sulfuric acid: Coleman *et al., J. Polym. Sci.* **3A**, 1601 (1965); Coleman, *U.S. pat.* 3,277,056 (1966 to Lubrizol). Toxicity data: K. Hashimoto *et al., Arch. Toxicol.* **47**, 179 (1981). *Review: idem,* "N-Oxoalkylacrylamides" in *Encyclopedia of Polymer Science and Technology,* vol. 15, N. M. Bikales, Exec. Ed. (Wiley-Interscience, New York, 1971) pp 353-364.

White crystalline solid, mp 57-58°. $bp_{0.1-0.3}$ 93-100°; bp_8 120°. Hygroscopic. Has long shelf life; shows no polymerization on prolonged storage. A 50% neutral water soln is stable at pH 7.5-7.7 for up to six months. Similar to acrylic esters in polymerization behavior. Forms high mol wt polymers in bulk, solution, or emulsion. Esp suitable for electropolymerization. LD_{50} orally in mice: 7.7 mmol/kg (Hashimoto).

USE: In the manufacture of coatings, laminates, sealers, adhesives, lubricating oils.

2944. Diacetone Alcohol. *4-Hydroxy-4-methyl-2-pentanone;* Pyranton. $C_6H_{12}O_2$; mol wt 116.16. C 62.04%, H 10.41%, O 27.55%. $(CH_3)_2C(OH)CH_2COCH_3$. Prepd by the action of barium hydroxide or of calcium hydroxide on acetone: Conant, Tuttle, *Org. Syn.* **vol. 1**, p 45 (1921); **coll. vol. I**, 193; *cf.* Jacquemain, *Compt. Rend.* **196**, 1622 (1933). Industrial process using KOH and acetone: Schmitt, Disteldorf, *Ger. pat.* **1,052,970** (1959 to Hibernia).

Flammable liq; faint, pleasant odor. d_4^{25} 0.9306 (0.940 for tech grades). bp_{760} 167.9°. bp_{100} 108.2°. bp_{20} 72.0°. bp_{10} 58.8°. $bp_{1.0}$ 22.0°. mp —44°; n_D^{20} 1.4232: Lantz, *J. Am. Chem. Soc.* **62**, 3260 (1940). Flash pt reagent grade: 66° (151°F). Flash pt commercial grades: 8° (48°F) (closed cup); 13° (55°F) (open cup). Vapor density 4.00 (air = 1.00). Miscible with water, alcohol, ether, and other solvents. Is dec by prolonged exposure to alkalies and by distillation at atmospheric pressure. LD_{50} orally in rats: 4.0 g/kg, Smyth, Carpenter, *J. Ind. Hyg. Toxicol.* **30**, 63 (1948).

USE: Solvent for cellulose acetate, nitrocellulose, celluloid, fats, oils, waxes, resins. As a preservative in pharmaceutical prepns. In some antifreeze solns and in hydraulic fluids. *Caution:* Narcotic in high concns. Irritating to mucous membranes. Kidney and liver injury as well as anemia have been produced in exptl animals: E. Browning, *Toxicity and Metabolism of Industrial Solvents* (Elsevier, New York, 1965) pp 389-392.

2945. Diacetoneglucose. *1,2:5,6-Bis-O-(1-methylethyl-idene)-D-glucofuranose; 1,2:5,6-diisopropylidene-D-glucose;* 1,2:5,6-diisopropylidene-D-glucofuranose. $C_{12}H_{20}O_6$; mol wt 260.28. C 55.37%, H 7.75%, O 36.88%. Prepd by shaking glucose wih acetone contg 1% HCl: Fischer, Rund, *Ber.* **49**, 90, 93 (1916). Improved procedure (90% yield) using 85% H_3PO_4 and anhyd $ZnCl_2$ in a mechanically stirred suspension of glucose in acetone: Glen *et al.*, **U.S. pat. 2,715,-121** (1955 to Am. Home Prod.).

Needles from ether or petr ether. Bitter taste. mp 110-111°. Sublimes. $[\alpha]_D^{20}$ —18.5° (c = 5). Reduces Fehling's soln. One part dissolves in about 7 parts of boiling water and in about 200 parts of boiling petr ether. Freely sol in chloroform, alc, acetone, warm ether. Precipitated from aq soln by NaOH soln. Is not split by yeast or emulsin.

2946. Diacetyl. *2,3-Butanedione;* biacetyl; dimethyl diketone; dimethylglyoxal; 2,3-diketobutane. $C_4H_6O_2$; mol wt 86.09. C 55.80%, H 7.03%, O 37.17%. $CH_3COCOCH_3$. Found in bay and other oils; also in butter. Made from methyl ethyl ketone by converting to the isonitroso compd and then decomposing to diacetyl by hydrolysis with HCl; by special fermentation of glucose *via* methylacetylcarbinol: Suomalainen, Jännes, *Nature* **157**, 336 (1946).

Yellowish-green liq; quinone odor; the vapors have a Cl-like odor. d_{15}^{15} 0.990. bp 88°. n_D^{18} 1.3933. Sol in about 4 parts water; misc with alcohol, ether. LD_{50} orally in rats: 1580 mg/kg, P. M. Jenner *et al.*, *Food Cosmet. Toxicol.* **2**, 327 (1964).

USE: Carrier of aroma of butter, vinegar, coffee, and other foods.

2947. Diacetyldihydromorphine. *4,5-Epoxy-17-methyl-morphinan-3,6-diol diacetate (ester);* dihydroheroin; paralaudin. $C_{21}H_{25}NO_5$; mol wt 371.42. C 67.90%, H 6.78%, N 3.77%, O 21.54%. Prepn from dihydromorphine: Eddy, Howes, *J. Pharmacol.* **53**, 430 (1935); Small, Mallonee, *J. Org. Chem.* **12**, 558-566 (1947); K. W. Bentley, *The Chemistry of the Morphine Alkaloids* (Oxford, 1954).

Needles, mp 165-167°. Slightly sol in water; sol in chloroform, alcohol, ether.

Hydrochloride, $C_{21}H_{25}NO_5$·HCl, flakes, scales. mp 215-218°. $[\alpha]_D^{29}$ —59° (c = 1.2). Very sol in water.

2948. Diacetylmorphine. *(5α,6α)-7,8-Didehydro-4,5-epoxy-17-methylmorphinan-3,6-diol diacetate (ester);* heroin; diamorphine; acetomorphine. $C_{21}H_{23}NO_5$; mol wt 369.40. C 68.28%, H 6.28%, N 3.79%, O 21.66%. Narcotic analgesic prepared from morphine and acetic anhydride: O. Hesse, *Ann.* **220**, 203 (1883); from morphine and acetyl chloride: Small, Lutz, *Chemistry of the Opium Alkaloids,* Supplement No. 103, Public Health Reports, Washington (1932); K. W. Bentley, *The Chemistry of the Morphine Alkaloids* (Oxford, 1954). Comparison with acetylmorphine: N. B. Eddy, H. A. Howes, *J. Pharmacol. Exp. Ther.* **53**, 430 (1935). Pharmacodynamics: J. G. Umans, C. E. Inturrisi, *ibid.* **218**, 409 (1981). Toxicity: *eidem, Eur. J. Pharmacol.* **85**, 317 (1982). Vapor pressure studies: A. H. Lawrence *et al.*, *Can. J. Chem.* **42**, 1886 (1984). Comprehensive description: D. K. Wyatt, L. T. Grady in *Analytical Profiles of Drug Substances* vol. **10**, K. Florey, Ed. (Academic Press, New York, 1981) pp 357-403.

Orthorhombic plates, tablets from ethyl acetate. mp 173°. bp_{12} 272-274°. $[\alpha]_D^{15}$ —166° (c = 1.49 in methanol). One gram dissolves in 1.5 ml chloroform, 31 ml alcohol, 100 ml ether, 1700 ml water. Slightly sol in ammonia or sodium carbonate soln, sol in alkalies, dec by boiling with water. Turns pink and emits acetic odor on prolonged exposure to air. LD_{50} i.v. in mice: 59 μmol/kg (Umans, Inturrisi).

Hydrochloride monohydrate, $C_{21}H_{24}ClNO_5$·H_2O, fine crystals, mp 243-244°. $[\alpha]_D^{24}$ —156° (c = 1.044). Sol in 2 parts water, 11 parts alcohol. Insol in ether.

Methyl iodide, $C_{22}H_{26}INO_5$, needles, mp 252°, $[\alpha]_D^{15}$ —107° (c = 0.896).

Note: "**China White**" has been used as a term for very pure Southeast Asia heroin. This term has also been erroneously used to refer to 3-methyl and α-methylfentanyl, *q.v.*, which are potent derivs of fentanyl, *q.v.*

Caution: May be habit forming. This is a controlled substance (opium derivative) listed in the U.S. Code of Federal Regulations, Title 21 Parts 329.1, 1308.11 (1987). Because of its addiction liability, the importation or manuf of diacetylmorphine and its salts is now forbidden in the USA by Federal statute.

2949. Dialifor. *Phosphorodithioic acid S-[2-chloro-1-(1,3-dihydro-1,3-dioxo-2H-isoindol-2-yl)ethyl] O,O-diethyl ester; phosphorodithoic acid O,O-diethyl ester S-ester with N-(2-chloro-1-mercaptoethyl)phthalimide;* O,O-diethyl S-(2-chloro-1-phthalimidoethyl)phosphorodithioate; dialifos; Hercules 14503; Torak. $C_{14}H_{17}ClNO_4PS_2$; mol wt 393.84. C 42.70%, H 4.35%, Cl 9.00%, N 3.56%, O 16.25%, P 7.86%, S 16.28%. Prepn: Jamison, **U.S. pat. 3,355,353** (1967 to Hercules). Acute toxicity: T. B. Gaines, R. E. Linder, *Fundam. Appl. Toxicol.* **7**, 299 (1986).

White crystalline solid, mp 67-69°. Also reported as oil; when recrystallized from toluene + hexane, mp 62-64°. Insoluble in water. Generally sol in aromatic hydrocarbons, ethers, esters, ketones. LD_{50} in adult male, female rats (mg/kg): 24, 6 orally (Gaines, Linder).

USE: Insecticide; acaricide.

2950. Diallate. *Bis(1-methylethyl)carbamothioic acid S-(2,3-dichloro-2-propenyl) ester; diisopropylthiocarbamic acid S-2,3-dichloroallyl ester;* S-2,3-dichloroallyl diisopropylthiocarbamate; DATC; CP 15336; Avadex. $C_{10}H_{17}Cl_2$-NOS; mol wt 270.24. C 44.45%, H 6.34%, Cl 26.24%, N 5.18%, O 5.92%, S 11.87%. Prepn: **Brit. pat. 882,111;** M. W. Harman, J. J. D'Amico, **U.S. pat. 3,330,823** (1961, 1967 both to Monsanto). Toxicity data: Bailey, White, *Residue Rev.* **10**, 97 (1965). Mutagenicity studies: F. DeLorenzo *et al., Cancer Res.* **38**, 13 (1978); I. Schuphan *et al., Science* **205**, 1013 (1979). Herbicidal activity: R. Grover *et al., Weed Res.* **19**, 363 (1979). GLC determn in milk and plant tissue: L. W. Cook *et al., J. Assoc. Offic. Anal. Chem.* **65**, 215 (1982).

Brown liquid, bp$_9$ 150°. Very slightly sol in water (40 ppm); sol in acetone, benzene, chloroform, ether, heptane. LD$_{50}$ orally in rats: 395 mg/kg (Bailey, White).

USE: Herbicide.

2951. Diallylamine. *N-2-Propenyl-2-propen-1-amine;* di-2-propenylamine. $C_6H_{11}N$; mol wt 97.16. C 74.17%, H 11.41%, N 14.42%. $(CH_2=CHCH_2)_2NH$. Prepd from allylamine and allyl bromide: Ladenburg, *Ber.* **14**, 1879 (1881); from allylamine and allyl chloride: Liebermann, Hagen, *Ber.* **16**, 1641 (1883); from diallylcyanamide by refluxing with dil H_2SO_4: Vliet, *J. Am. Chem. Soc.* **46**, 1307 (1924); *Org. Syn.* coll. vol. I (2nd ed.,1941) p 201.

Liquid. bp 112°. LD$_{50}$ orally in rats: 0.65 g/kg, H. F. Smyth *et al., Am. Ind. Hyg. Assoc. J.* **23**, 95 (1962).

2952. Diallylcyanamide. *Di-2-propenylcyanamide; N-cyanodiallylamine.* $C_7H_{10}N_2$; mol wt 122.17. C 68.82%, H 8.25%, N 22.93%. $(CH_2=CHCH_2)_2NCN$. Prepd by the action of allyl bromide on disodium cyanamide: Vliet, *J. Am. Chem. Soc.* **46**, 1307 (1924); *Org. Syn.* coll. vol. I, 203 (1941); North, **U.S. pat. 1,659,793** (1929).

Liquid. bp$_{90}$ 140-145°; bp$_{57}$ 128-133°; bp$_{18}$ 105-110°. Insol in water, sol in the usual organic solvents.

2953. Diamfenetide. *N,N'-[Oxybis(2,1-ethanediyloxy-4,1-phenylene)]bisacetamide; β,β'-oxybis[p-acetopheneti-dide]; β,β'-bis(4-acetamidophenyloxy)ethyl ether;* diamphenethide; Coriban. $C_{20}H_{24}N_2O_5$; mol wt 372.42. C 64.50%, H 6.50%, N 7.52%, O 21.48%. Prepn: Harfenist, **Ger. pat. 2,143,570** (1972), *C.A.* **77**, 19404w (1972). Activity: Dickerson *et al., Brit. Vet. J.* **127**, xl (1971); Kingsbury, Rowlands, *ibid.* **128**, 235 (1972); Armour, Corba, *Vet. Rec.* **91**, 211 (1972). *Review:* Rowlands, *Pestic. Sci.* **4**, 883 (1973).

THERAP CAT (VET): Anthelmintic (Trematodes).

2954. *p*-Diaminoazobenzene. *4,4'-Azobisbenzenamine; 4,4'-azodianiline; p-azoaniline;* 4,4'-diaminoazobenzene. $C_{12}H_{12}N_4$; mol wt 212.25. C 67.90%, H 5.70%, N 26.40%. Prepn: Witt, Kopetschini, *Ber.* **45**, 1136 (1912); Brode *et al., J. Am. Chem. Soc.* **77**, 2762 (1955). Lab procedure starting with *p*-aminoacetanilide: Santurri *et al., Org. Syn.* **40**, 18 (1960).

Golden-yellow needles, dec 238-241°. Absorption max (ethanol): 400 nm. Slightly soluble in water, benzene, petr ether; freely sol in alcohol.

2955. 3,5-Diaminobenzoic Acid. $C_7H_8N_2O_2$; mol wt

152.14. C 55.25%, H 5.30%, N 18.41%, O 21.03%. Prepn from 3,5-dinitrobenzoic acid: Griess, *Ann.* **154**, 327 (1870).

Monohydrate, needles. Loses its water of crystn at about 110°, mp 228°. Slightly sol in water; the aqueous soln dec on standing. Sol in alcohol or ether.

Ethyl ester, $C_9H_{12}N_2O_2$, prisms from ether, mp 84°. Moderately sol in hot water. Sol in alcohol, benzene, other organic solvents.

USE: Detection and determination of nitrites.

2956. 2,6-Diamino-2'-butyloxy-3,5'-azopyridine. *3-[(6-Butoxy-3-pyridinyl)azo]-2,6-pyridinediamine; 2',6'-di-amino-2-butyloxy-5,5'-azopyridine; 2',6'-diamino-2-bu-toxy-5,5'-azopyridine; 2,6-diamino-6'-butoxy-3,3'-azo-pyridine; 2,6-diamino-2'-butoxy-3,5'-azopyridine; 2,6-di-amino-3-(6-butoxy-3-pyridylazo)pyridine;* Niazo; Neotropin; Savapyrin. $C_{14}H_{18}N_6O$; mol wt 286.33. C 58.72%, H 6.34%, N 29.35%, O 5.59%. Prepn: Dohrn, Diedrich, **U.S. pat. 1,862,361** (1932 to Schering-Kahlbaum, A.G.).

Orange-red crystals. mp 129°. Slightly sol in water, freely in organic solvents.

THERAP CAT: Urinary antiseptic.

2957. 4,4'-Diaminodiphenylamine. *N-(4-Aminophenyl)-1,4-benzenediamine; p-p'-diaminodiphenylamine.* $C_{12}H_{13}$-N_3; mol wt 199.25. C 72.33%, H 6.58%, N 21.09%. Prepn from *p*-phenylenediamine and aniline hydrochloride: Nietzki, *Ber.* **16**, 474 (1883); from *p*-aminoazobenzene: Barbier, Sisley, *Bull. Soc. Chim. France* [3] **33**, 1232 (1905); by reducing bis(*p*-nitrophenyl)nitroxide with tin chloride and hydrochloric acid: Wieland, Roth, *Ber.* **53**, 213, 226 (1920).

Leaflets from water, mp 158°. Oxidized to bromanil and bromonitromethane by bromine and concd nitric acid.

USE: Dyeing fur: **Ger. pats. 367,690; 371,231;** manuf dyes. In the detection of hydrogen cyanide.

2958. *p,p'*-Diaminodiphenylmethane. *4,4'-Methylene-bis[benzenamine]; 4,4'-methylenedianiline.* $C_{13}H_{14}N_2$; mol wt 198.26. C 78.75%, H 7.12%, N 14.13%. Prepn from aniline and formaldehyde: Scanlon, *J. Am. Chem. Soc.* **57**, 887 (1935); by hydrogenolysis of *p,p'*-diaminobenzophenone with LiAlH$_4$: Conover, Tarbell, *ibid.* **72**, 3586 (1950).

Crystals from water or benzene. mp 91.5-92°. bp$_{768}$ 398-399°; bp$_{18}$ 257°; bp$_{15}$ 249-253°; bp$_9$ 232°. Slightly sol in cold water. Very sol in alcohol, benzene, ether.

Caution: Hepatotoxic. Causative agent in "Epping Jaundice"; symptoms include jaundice, weakness, abdominal pain, nausea and/or vomiting, anorexia, fever, chills: H. Kopelman *et al., Brit. Med. J.* **1**, 514 (1966); D. B. McGill, J. D. Motto, *N. Engl. J. Med.* **291**, 278 (1974). This substance and its dihydrochloride may reasonably be anticipated to be carcinogens: *Fourth Annual Report on Carcinogens* (NTP 85-002, 1985) p 131.

USE: As curing agent for epoxy resins and urethane elastomers; intermediate in prepn of polyurethanes, Spandex fibers; in production of polyamides; in the determination of tungsten and sulfates; in prepn of azo dyes; as corrosion inhibitor.

2959. 2,4-Diamino-6-hydroxypyrimidine. *2,6-Diamino-4(1H)-pyrimidinone.* $C_4H_6N_4O$; mol wt 126.12. C 38.09%, H 4.80%, N 44.43%, O 12.69%. Prepn from guanidine hydrochloride and cyanoacetic ester: Traube, *Ber.* **33**, 1371 (1900).

Crystallizes from water as the monohydrate. When anhydrous, dec 286°.
USE: In the detection of nitrites and nitrates.

2960. 2,3-Diaminophenazine. *2,3-Phenazinediamine.* $C_{12}H_{10}N_4$; mol wt 210.23. C 68.55%, H 4.79%, N 26.65%. Prepn from *o*-phenylenediamine: Ullmann, Mauthner, *Ber.* **35**, 4304 (1902); Knoevenagel, *J. Prakt. Chem.* [2] **89**, 25 (1914); Richter *Ber.* **44**, 3469 (1911).

Brown to yellow needles. On careful heating sublimes forming yellow leaflets. mp 264°. Sol in alcohol, benzene. Yields 2-aminophenazine and small amounts of phenazine when heated with zinc dust.
USE: Detection of bismuth, cadmium, lead, copper, and mercury.

2961. 2,4-Diaminophenol. $C_6H_8N_2O$; mol wt 124.14. C 58.05%, H 6.50%, N 22.57%, O 12.89%. Prepn by reduction of 2,4-dinitrophenol: Braude *et al., J. Chem. Soc.* **1954**, 3586; Neilson *et al., ibid.* **1962**, 371; from *m*-nitroaniline: Bean, U.S. pat. **2,525,515** (1950 to Eastman Kodak).

Crystals, dec 78-80°. Somewhat sol in alcohol, acetone; sparingly sol in ether, chloroform, petr ether; readily sol in acids, alkalies.
Dihydrochloride. $C_6H_{10}ClN_2O$, *Acrol, Amidol.* Crystals, mp 205°. Soly in water at 15°: 27.5 g/100 ml. Slightly sol in alcohol.
USE: Dihydrochloride as photographic developer, in fur and hair dyeing, in test for formaldehyde and ammonia.

2962. 2,3-Diaminopropionic Acid. *3-Aminoalanine;* 2,3-diaminopropanoic acid; *α,β*-diaminopropionic acid. C_3H_8-N_2O_2; mol wt 104.11. C 34.61%, H 7.75%, N 26.91%, O 30.74%. Prepd by treating 2,3-dibromopropionic acid with ammonia: Klebs, *Z. Physiol. Chem.* **19**, 314 (1894); Winterstein, Küng, *ibid.* **59**, 146 (1909); Neuberg, Silbermann, *Ber.* **37**, 341 (1904); Frankland, *J. Chem. Soc.* **97**, 1318 (1910); by treating 2-amino-3-chloropropionic acid with ammonia under pressure: Fisher *et al., Ber.* **40**, 3717 (1907); by the reduction of cyanooximoacetic acid esters: Godefroi, U.S. pat. **2,738,363** (1956 to Parke, Davis). Configuration: Karrer *et al., Helv. Chim. Acta* **9**, 314 (1926). *See also* Bergmann, Grafe, *Z. Physiol. Chem.* **187**, 187 (1930).

$$CH_2 - CHCOOH$$
$$\ \ |\qquad\ |$$
$$NH_2\ \ \ NH_2$$

DL-Form, radially arranged cryst masses. Hygroscopic,

also absorbs CO_2 from air. Begins to melt at 97°, completely liq at 110-120°. Sol in water, yielding alkaline solns. Practically insol in alcohol, ether.
Hydrochloride, $C_3H_8N_2O_2$·HCl, needles or leaflets, mp 236-237°. One gram dissolves in 12 ml water, practically insol in alcohol.
Sulfate hemihydrate, $(C_3H_8N_2O_2)_2$·H_2SO_4·½H_2O, six-sided leaflets from water, dec 233-234°. One gram dissolves in 31 ml water at 20°.
Methyl ester dihydrochloride, $C_4H_{10}N_2O_2$·2HCl, crystals, dec 170°. Freely sol in water.
Ethyl ester dihydrochloride, $C_5H_{12}N_2O_2$·2HCl, crystals, mp 164.5-165° (some decompn). Freely sol in water.
L-Form hydrochloride, $C_3H_8N_2O_2$·HCl, needles from water, $[\alpha]_D^{20}$ +25.0° (c = 5 in 1.0N HCl).
D-Form hydrochloride, $C_3H_8N_2O_2$·HCl, needles from water, $[\alpha]_D^{20}$ −25.3° (c = 5 in 1.0N HCl).
USE: The salts as growth inhibitors for microorganisms.

2963. 2,6-Diaminopurine. *1H-Purine-2,6-diamine;* 2,6-diamino-9H-purine. $C_5H_6N_6$; mol wt 150.15. C 39.99%, H 4.03%, N 55.98%. Antagonist of naturally occurring purines: Hitchings, Elion in *Metabolic Inhibitors,* **vol. I**, R. M. Hochster, J. H. Quastel, Eds. (Academic Press, New York, 1963) pp 216-232. Molecular and electronic structure: Veillard, Pullman, *J. Theoret. Biol.* **4**, 37 (1963). First synthesized by Traube, *Ber.* **37**, 4547 (1904) by cyclization of 2,4,5,6-tetraaminopyrimidine. Prepn from guanidine carbonate in 70% yield and refs to earlier prepns: Taylor *et al., J. Am. Chem. Soc.* **81**, 2442 (1959).

Crystals from ethanol + water, mp 302°. uv max (pH 1.9): 241, 282 nm (log ε 3.98, 4.00).

2964. Diamond. A cryst modification of carbon. Mined as a mineral, principally in South Africa. (Non-commercial) synthesis from other carbon compds (e.g., lignin) by means of elevated temperatures (about 2700°) and pressures (about 800,000 lbs/sq inch): Desch, *Nature* **152**, 148 (1943); Neuhaus, *Angew. Chem.* **66**, 525 (1954); Hall, *Chem. Eng. News* **33**, 718 (1955); Bridgman, *Sci. Amer.* **1955**, 46; Hall, *J. Chem. Ed.* **38**, 484 (1961); Bundy, *Ann. N.Y. Acad. Sci.* **vol. 105**, art 17, pp 951-982 (1964). Books: S. Tolansky, *History and Use of Diamond* (London, 1962) 166 pp; R. Berman, *Physical Properties of Diamond* (Oxford, 1965) 442 pp.
Face-centered cubic crystal lattice. Burns when heated with a hot enough flame (over 800°, oxygen torch). d_4^{25} 3.513. n_D^{20} 2.4173. Hardness = 10 (Mohs' scale). Sp heat at 100°K: 0.606 cal/g-atom/°K. Entropy at 298.16°K: 0.5684 cal/g-atom/°K. Band gap energy: 6.7 ev. Dielectric constant 5.7. Electron mobility: ~1800 cm²/v-sec. Hole mobility: 1200 cm²/v-sec. Can be pulverized in a steel mortar. Attacked by laboratory-type cleaning soln (potassium dichromate + concd H_2SO_4). In the jewelry trade the unit of weight for diamonds is one carat = 200 mg. *Ref: Wall Street J.* **164**, no. 36, p 10 (Aug 19, 1964).
USE: Jewelry. Polishing, grinding, cutting glass, bearings for delicate instruments; manuf dies for tungsten wire and similar hard wires; making styli for recorder heads, long-lasting phonograph needles. In semiconductor research.

2965. Diamond Ink. Etching ink. A mixture of HF, $BaSO_4$ and fluorides.
Milky-white liq with a heavy sediment. Shake well before using and warm gently in a lead dish. *Keep in plastic, hard-rubber or intern. paraffin-coated bottles.*
USE: For etching glass.

2966. Diampromide. *N-[2-[Methyl(2-phenylethyl)amino]propyl]-N-phenylpropanamide; N-[2-(methylphenethylamino)propyl]propionanilide.* $C_{21}H_{28}N_2O$; mol wt 324.45. C 77.73%, H 8.70%, N 8.63%, O 4.93%. Synthesis: Wright *et al., J. Am. Chem. Soc.* **81**, 1518 (1959); U.S. pat. **2,944,081** (1960 to American Cyanamid).

Liquid, $bp_{0.5}$ 174-178°. n_D^{26} 1.546.
Sulfate, $C_{21}H_{30}N_2O_5S$, crystals from ethanol + ether, mp 110-111°.
Caution: May be habit forming. This is a controlled substance (opiate) listed in the U.S. Code of Federal Regulations, Title 21 Part 1308.11 (1987).
THERAP CAT: Narcotic analgesic.

2967. Diamthazole Dihydrochloride. *6-[2-(Diethylamino)ethoxy]-N,N-dimethyl-2-benzothiazolamine dihydrochloride; 6-[2-(diethylamino)ethoxy]-2-dimethylaminobenzothiazole dihydrochloride;* 2-dimethylamino-6-(β-diethylaminoethoxy)benzothiazole dihydrochloride; dimazole dihydrochloride; Ro 2-2453; Asterol Dihydrochloride; Atelor. $C_{15}H_{25}Cl_2N_3OS$; mol wt 366.35. C 49.17%, H 6.88%, Cl 19.36%, N 11.47%, O 4.37%, S 8.75%. Prepn: Steiger, Keller, **U.S.** pat. **2,578,757** (1951 to Hoffmann-La Roche).

Crystals, dec 269° (patent, mp 240-243°). Freely sol in water, methanol, ethanol. A 5% aq soln has a pH of $\sim$ 2.
THERAP CAT: Antifungal.

2968. Diamyl Sodium Sulfosuccinate. *Sulfobutanedioic acid 1,4-dipentyl ester sodium salt; sulfosuccinic acid dipentyl ester sodium salt;* Aerosol AY; Alphasol AY. $C_{14}H_{25}NaO_7S$; mol wt 360.41. C 46.65%, H 6.99%, Na 6.38%, O 31.08%, S 8.90%. The amyl or 1-methylbutyl diester of the monosodium salt of sulfosuccinic acid or a mixture of both. Wetting agent prepd by the action of the appropriate alcohols on maleic anhydride followed by addtion of sodium bisulfite: Jaeger, **U.S.** pats. **2,028,091** and **2,176,423** (1936, 1939 to Am. Cyanamid).

Available as a mixture of white, hard pellets and powder. Soly in water at 25° = 392 g/liter; at 70° = 502 g/liter. Maximum concn of electrolyte soln in which 1% of the wetting agent is sol: 3% NaCl; 2-4% NH_4Cl (turbid); 3-6% $(NH_4)_2HPO_4$ (turbid); 4% $NaNO_3$ (slightly turbid); 15% Na_2SO_4 (very slightly turbid). Also sol in pine oil, oleic acid, acetone, hot kerosene, carbon tetrachloride, glycerol, hot olive oil; insol in liq petrolatum. Surface tension in water: 0.001% = 69.4 dyn/cm; 0.02% = 68.3 dyn/cm; 0.1% = 50.2 dyn/cm; 0.25% = 41.6 dyn/cm; 1% = 29.2 dyn/cm. Interfacial tension 1% in water *vs.* liquid petrolatum: 5 seconds = 7.55 dyn/cm; 30 seconds = 7.37 dyn/cm; 15 minutes = 7.03 dyn/cm. Interfacial tension 0.1% in water *vs.* liquid petrolatum: 5 seconds = 29.5 dyn/cm; 30 seconds = 28.6 dyn/cm; 15 minutes = 27.5 dyn/cm. Stable in acid and neutral solns, hydrolyzes in alkaline solns.
USE: As emulsifier in emulsion polymerization and as a wetting agent.

2969. 1,2-Dianilinoethane. *N,N'-Diphenyl-1,2-ethanediamine; N,N'-diphenylethylenediamine; N,N'-diphenyl-α,ω-diaminoethane; sym-diphenylethylenediamine.* $C_{14}H_{16}N_2$; mol wt 212.28. C 79.21%, H 7.60%, N 13.20%. Prepd by heating aniline with dichloroethane: Büchi et al., *Helv. Chim. Acta* **39**, 950 (1956); by heating aniline with dibromoethane: Wanzlick, Löchel, *Ber.* **86**, 1463 (1953).

Crystals from dil ethanol, mp 67.5° (also reported as the monohydrate). bp_{12} 228-230°. Very sol in alc, ether.

USE: Identification of aldehydes; stabilizer for resins and rubber. Intermediate in the manufacture of antihistamines.

2970. Dianisidine. *3,3'-Dimethoxy-[1,1'-biphenyl]-4,4'-diamine; 3,3'-dimethoxybenzidine; 3,3'-dimethoxy-4,4'-diaminobiphenyl.* $C_{14}H_{16}N_2O_2$; mol wt 244.28. C 68.83%, H 6.60%, N 11.47%, O 13.10%. Prepn: Meier, Böhler, *Ber.* **89**, 2301 (1956); Sogn, **U.S.** pat. **2,794,047** (1957 to Allied Chem.). Manuf of stabilized product: Cashion, **U.S.** pat. **2,966,519** (1960 to Allied Chem.). Review and evaluation of studies of carcinogenicity in laboratory animals: *IARC Monographs* **4**, 41-47 (1974).

Crystals, becoming violet, mp 137-138°. Practically insol in water. Sol in alcohol, benzene, ether.
Note: This substance may reasonably be anticipated to be a carcinogen: *Fourth Annual Report on Carcinogens* (NTP 85-002, 1985) p 87.
USE: Manuf of azo dyes.

2971. Diastase of Malt. Maltin. A mixture contg amylolytic enzymes obtained from malt. Converts at least 50 times its wt of potato starch into sugars (dextrin and maltose) in 30 minutes.
Yellowish-white, amorphous powder or translucent scales. Loses amylolytic power on keeping, on heating its soln above 85°, or on adding much acid. Sol in water with some turbidity; almost insol in alc. *Keep well closed.*
USE: Manuf of sol starch; converting starch into sugar; removing starch from fabrics.

2972. Diathymosulfone. *4,4'-[Sulfonylbis(4,1-phenyleneazo)]bis[5-methyl-2-(1-methylethyl)]phenol; 6,6'-[sulfonylbis(p-phenyleneazo)]dithymol;* bis[p-(4-hydroxy-2-methyl-5-isopropylphenylazo)phenyl] sulfone; 4,4'-bis(p-isopropyl-m-cresylazo)diphenyl sulfone; thymol sulfone; Diatox. $C_{32}H_{34}N_4O_4S$; mol wt 570.72. C 67.34%, H 6.00%, N 9.82%, O 11.21%, S 5.62%. Prepn: **Brit.** pat. **758,744** (1956 to Lab. Laborec); Chiarlo et al., *Arch. Ital. Sci. Farmacol.* **7**, 167 (1957), *C.A.* **51**, 16949c (1957).

Red-brown microcrystalline powder from ethanol + water, mp 222-224°. Absorption max (alcohol): 400 nm. Practically insol in water. Sol in alkali solns with red color, in dioxane, acetone; less sol in alcohol, ether.
Di-silver salt, $C_{32}H_{32}Ag_2N_4O_4S$, *thymolated silver sulfone.*
THERAP CAT: Antibacterial (leprostatic).

2973. Diatretyne I. *(E)-8-Amino-8-oxo-2-octene-4,6-diynoic acid; trans-7-carbamoyl-2-heptene-4,6-diynoic acid; 7-carbamidohepta-2-ene-4,6-diynoic acid; diatretyne amide.* $C_8H_5NO_3$; mol wt 163.13. C 58.90%, H 3.09%, N 8.59%, O 29.42%. HOOC—CH=CHC≡CC≡CCONH₂. Antibiotic substance produced by the fungus *Clitocybe diatreta:* Anchel, *J. Am. Chem. Soc.* **75**, 4621 (1953); *Arch. Biochem. Biophys.* **78**, 100 (1958). Structure: Bu'Lock et al., *Chem. & Ind. (London)* **1954**, 990. Synthesis: Ashworth et al., *J. Chem. Soc.* **1958**, 950; Prevost et al., *Bull. Soc. Chim. France* **1961**, 2171.
Crystals from methanol, explodes 198°. uv max (95% ethanol): 224, 274, 291, 309.5 nm ($\epsilon \times 10^{-3}$ 37.7, 16.5, 23, 19).

2974. Diatretyne II. *7-Cyano-2-heptene-4,6-diynoic acid;* diatretyne nitrile; nudic acid B. $C_8H_3NO_2$; mol wt

145.12. C 66.21%, H 2.08%, N 9.65%, O 22.05%. HOOC-CH=CHC≡CC≡CCN. Antibiotic substance produced by the basidiomycete *Tricholoma nudum* and by the fungus *Clitocybe diatreta*. Isoln: Anchel, *J. Am. Chem. Soc.* **74**, 1588 (1952). Structure: Anchel, *Science* **121**, 607 (1955). Identity of nudic acid B and diatretyne II: Heatly, Stephenson, *Nature* **179**, 1078 (1957). Synthesis: Ashworth *et al.*, *J. Chem. Soc.* **1958**, 950.

Short needles from ether + light petr, dec 179-180°. Darkens and turns black within 2 hrs on exposure to sunlight. Sol in the usual organic solvents except hexane; dissolves slowly in sodium carbonate soln; practically insol in water and dil acids.

2975. Diatrizoate Sodium. *3,5-Bis(acetylamino)-2,4,6-triiodobenzoic acid sodium salt; 3,5-diacetamido-2,4,6-triiodobenzoic acid sodium salt;* sodium 3,5-diacetamido-2,4,6-triiodobenzoate; urografic acid sodium salt; sodium amidotrizoate; sodium diatrizoate; Hypaque Sodium; Triognost. $C_{11}H_8I_3N_2NaO_4$; mol wt 635.92. C 20.78%, H 1.27%, I 59.87%, N 4.41%, Na 3.62%, O 10.06%. Prepn of the acid: Larsen *et al.*, *J. Am. Chem. Soc.* **78**, 3210 (1956). Description: Langecker *et al.*, *Arch. Exp. Pathol. Pharmakol.* **222**, 584 (1954). Use in separation of leukocytes from whole blood: R. J. Perper *et al.*, *J. Lab. Clin. Med.* **72**, 842 (1968); L. P. Bignold, A. Ferrante, *J. Immunol. Meth.* **96**, 29 (1987); J. R. Kalmar *et al.*, *ibid.* **110**, 275 (1988). Comprehensive description: H. H. Lerner in *Analytical Profiles of Drug Substances* **vol. 4**, K. Florey, Ed. (Academic Press, New York, 1975) pp 137-167.

Rhombic needles, slightly salty taste, mp 261-262° (dec). Soly in water at 20°: 60 g/100 ml. pH of a 50% aq soln 7.0 to 7.5. On chilling solns may become cloudy or form a precipitate which disappears at room temp. Solns may be sterilized by autoclaving. Should be protected from light. LD_{50} i.v. in rats: 14.7 g/kg (Langecker).

Free acid, crystals from dil DMF, mp >300°.

Note: See also Meglumine Diatrizoate.

USE: Density gradient reagent for blood cell separation.

THERAP CAT: Diagnostic aid (radiopaque medium).

2976. Diaveridine. *5-[(3,4-Dimethoxyphenyl)methyl]-2,4-pyrimidinediamine; 2,4-diamino-5-veratrylpyrimidine; 2,4-diamino-5-(3′,4′-dimethoxybenzyl)pyrimidine.* $C_{13}H_{16}N_4O_2$; mol wt 260.29. C 59.98%, H 6.20%, N 21.53%, O 12.29%. Prepn: Falco *et al.*, *J. Am. Chem. Soc.* **73**, 3758 (1951). Hitchings, Falco, U.S. pat. **2,624,732** (1953 to Burroughs Wellcome); Stenbuck *et al.*, *J. Org. Chem.* **28**, 1983 (1963); Hoffer, U.S. pat. **3,341,541** (1967 to Hoffmann-La Roche); M. Hoffer *et al.*, *J. Med. Chem.* **14**, 462 (1971). Crystal structure: T. F. Koetzle, G. Williams, *Acta Crystallogr.* **B34**, 323 (1978).

Crystals, mp 233°.

Ingredient of *Darvisul* which also contains sulfaquinoxaline.

THERAP CAT (VET): Antiprotozoan (in combination with sulfaquinoxaline).

2977. Diazepam. *7-Chloro-1,3-dihydro-1-methyl-5-phenyl-2H-1,4-benzodiazepin-2-one;* 7-chloro-1-methyl-5-phenyl-3H-1,4-benzodiazepin-2(1H)-one; methyl diazepinone; diacepin; La III; Ro 5-2807; Wy 3467; NSC-77518; Alupram; Ansiolin; Apaurin; Apozepam; Atensine; Atilen;

Bialzepam; Calmpose; Ceregulart; Dialar; Diazemuls; Di-pam; Eridan; Eurosan; Evacalm; Faustan; Gewacalm; Lamra; Lembrol; Levium; Mandrozep; Morosan; Neurolytril; Noan; Novazam; Paceum; Pacitran; Paxate; Paxel; Pro-Pam; Q-Pam; Relanium; Sedapam; Seduxen; Setonil; Solis; Stesolid; Stesolin; Tranimul; Tranquase; Tranquo-Puren; Tranquo-Tablinen; Unisedil; Valaxona; Valiquid; Valium; Valrelease; Vival; Vivol. $C_{16}H_{13}ClN_2O$; mol wt 284.76. C 67.49%, H 4.60%, Cl 12.45%, N 9.84%, O 5.62%. Prepn: Sternbach, Reeder, *J. Org. Chem.* **26**, 4936 (1961); Reeder, Sternbach, U.S. pat. **3,371,085** (1968 to Hoffmann-La Roche); prepd but not claimed: *eidem*, U.S. pats. **3,109,-843** and **3,136,815** (1963, 1964 to Hoffmann-La Roche); M. Gates, *J. Org. Chem.* **45**, 1675 (1980); M. Ishikura *et al.*, *ibid.* **47**, 2456 (1982). Purification: Chase, U.S. pat. **3,102,116** (1963 to Hoffmann-La Roche). Pharmacology: Hudson, Wolpert, *Arch. Int. Pharmacodyn. Ther.* **186**, 388 (1970). Metabolism: Randall *et al.*, *Curr. Ther. Res. Clin. Exp.* **7**, 590 (1965); Van der Kleijn, *Ann. N.Y. Acad. Sci.* **179**, 115 (1971). Toxicity: E. I. Goldenthal, *Toxicol. Appl. Pharmacol.* **18**, 185 (1971). Binding study in rat brain: R. F. Squires, C. Braestrup, *Nature* **266**, 732 (1977). Comprehensive description: A. MacDonald *et al.*, in *Analytical Profiles of Drug Substances* **vol. 1**, K. Florey, Ed. (Academic Press, New York, 1972) pp 79-99. Clinical pharmacokinetics: M. Mandelli *et al.*, *Clin. Pharmacokinet.* **3**, 72 (1978).

Plates from acetone + petr ether, mp 125-126°. pK_a 3.4. Sol in chloroform, DMF, benzene, acetone, alc; slightly sol in water. LD_{50} orally in rats: 710 mg/kg (Goldenthal).

Caution: May be habit forming. This is a controlled substance (depressant) listed in the U.S. Code of Federal Regulations, Title 21 Part 1308.14 (1987).

THERAP CAT: Anxiolytic; muscle relaxant (skeletal).

2978. Diazinon. *Phosphorothioic acid O,O-diethyl O-[6-methyl-2-(1-methylethyl)-4-pyrimidinyl] ester;* thiophosphoric acid 2-isopropyl-4-methyl-6-pyrimidyl diethyl ester; *O,O-diethyl O-2-isopropyl-4-methyl-6-pyrimidyl thiophosphate;* diethyl 2-isopropyl-4-methyl-6-pyrimidyl thionophosphate; dimpylate; G-24480; Basudin; Diazol; Garden Tox; Neocidol; Sarolex; Spectracide. $C_{12}H_{21}N_2O_3$-PS; mol wt 304.36. C 47.36%, H 6.95%, N 9.20%, O 15.77%, P 10.18%, S 10.54%. Prepn: Gysin, Margot, U.S. pat. **2,754,243** (1956 to Geigy). Insecticidal properties: Gasser, *Z. Naturforsch.* **8b**, 225 (1953). Toxicity data: T. B. Gaines, *Toxicol. Appl. Pharmacol.* **14**, 515 (1969).

Liquid. Faint ester-like odor. d_4^{20} 1.116-1.118. $bp_{0.002}$ 83-84°. n_D^{20} 1.4978-1.4981. Vapor pressure at 20°: 1.4×10^{-4}; at 40°: 1.1×10^{-3} (about 5 times vapor pressure of parathion). Volatility at 20°: 2.4 mg/cubic meter; at 40°: 17.6 mg/cubic meter. Decomposes above 120°. Soly in water at 20°: 0.004%. Miscible with alcohol, ether, petr ether, cyclohexane, benzene, and similar hydrocarbons. More stable in alkaline formulations, then when at neutral or acid pH. LD_{50} in male, female rats (mg/kg): 250, 285 orally (Gaines).

Caution: Cholinesterase inhibitor: *Clinical Toxicology of Commercial Products*, R. E. Gosselin *et al.*, Eds. (Williams & Wilkins, Baltimore, 5th ed., 1984) Section II, p 294.

USE: Insecticide.

2979. Diaziquone. *[2,5-Bis(1-aziridinyl)-3,6-dioxo-1,4-cyclohexadiene-1,4-diyl]biscarbamic acid diethyl ester; 2,5-bis(1-aziridinyl)-3,6-dioxo-1,4-cyclohexadiene-1,4-dicarbamic acid diethyl ester;* 2,5-diaziridinyl-3,6-bis(ethoxycarbonylamino)-1,4-benzoquinone; aziridinylbenzoquinone; AZQ; CI-904; NSC-182986. $C_{16}H_{20}N_4O_6$; mol wt 364.36. C 52.74%, H 5.53%, N 15.38%, O 26.35%. Quinone-containing lipophilic alkylating agent. Prepn as intermediate in dyestuffs: S. Petersen *et al.*, U.S. pat. **2,913,453** (1959 to Schenley Ind.). Synthesis and antitumor activity: A. H. Khan, J. S. Driscoll, *J. Med. Chem.* **19**, 313 (1976); J. S. Driscoll *et al.*, U.S. pat. **4,146,622** (1979 to U.S. Gov't). *In vivo* antitumor studies: J. S. Driscoll *et al.*, *J. Pharm. Sci.* **68**, 185 (1979). Intracerebral penetration and tissue distribution in humans: N. Savaraj *et al.*, *J. Neuro-Oncol.* **1**, 15 (1983). LC determn: B. A. Allen *et al.*, *J. Chromatog.* **222**, 146 (1981). Pharmacokinetics: S. Zimm *et al.*, *Cancer Res.* **44**, 1698 (1984). Clinical studies in treatment of CNS neoplasms: R. T. Eagan *et al.*, *J. Neuro-Oncol.* **5**, 309 (1987); E. Tapazoglou *et al.*, *Am. J. Clin. Oncol.* **11**, 474 (1988). *Review:* J. F. Bender *et al.*, *Invest. New Drugs* **1**, 71-84 (1983).

Orange needles from ethanol, mp 230° (dec); also reported as yellowish-brown crystals from ethanol decomposing at temperatures above 250° (Petersen). uv max (methanol): 340 nm (log ε 4.17). Soly in water: 0.5 mg/ml LD_{50} in mice (mg/m²): 30.9 i.v. (Bender).

THERAP CAT: Antineoplastic.

2980. Diazoacetic Ester. Ethyl diazoacetate; ethyl diazoethanoate. $C_4H_6N_2O_2$; mol wt 114.10. C 42.10%, H 5.30%, N 24.55%, O 28.04%. $^-N=N^+=CHCOOC_2H_5$. Prepd by the action of sodium nitrite on glycine ethyl ester hydrochloride: Curtius *J. Prakt. Chem.* [2] **38**, 396 (1888); Silberrad, *J. Chem. Soc.* **81**, 600 (1902); Womack, Nelson, *Org. Syn.* **24**, 56 (1944).

Yellow oil. Pungent odor. Very volatile. Explosive. *Distillation, even under reduced pressure, is dangerous.* mp $-22°$. $d_4^{17.6}$ 1.0852. bp_5 42°; bp_{12} 45°; bp_{88} 85-86°; bp_{720} 140-141°. $n_D^{17.6}$ 1.4588. Slightly sol in water. Neutral reaction. Volatile with steam, ether, and benzene vapors. Miscible with alcohol, benzene, ligroin, ether. Explodes on contact with concd H_2SO_4.

2981. Diazoaminobenzene. *1,3-Diphenyl-1-triazene;* anilinoazobenzene; benzeneazoaniline. $C_{12}H_{11}N_3$; mol wt 197.11. C 73.07%, H 5.62%, N 21.31%. Made by diazotizing aniline dissolved in HCl with $NaNO_2$ and then adding a concd soln of sodium acetate: Hartman, Dickey, *Org. Syn.* coll. vol. II, 163 (1943).

Golden-yellow, small crystals. mp 98°; explodes when heated to 150°. Insol in water; freely sol in benzene, ether, hot alc.

2982. p-Diazobenzenesulfonic Acid. *4-Sulfobenzenediazonium hydroxide inner salt;* sulfanilic acid diazide. $C_6H_4N_2O_3S$; mol wt 184.17. C 39.13%, H 2.19%, N 15.21%, O 26.06%, S 17.41%. Prepn by diazotization of sulfanilic acid and structure: Whetsel *et al.*, *J. Am. Chem. Soc.* **78**, 3360 (1956); Bugg *et al.*, *Acta Cryst.* **17**, 767 (1964).

White or slightly red crystals; when dry, easily explodes by heat, a blow, or on rubbing; hence marketed as a paste. Slightly sol in cold water, alcohol; freely sol in hot water, dil alkalies, HCl.

USE: As reagent for phenol, fecal matter in water, glucose, aldehydes, and albumin; also in manuf of azo and other dyes.

2983. Diazomethane. Azimethylene. CH_2N_2; mol wt 42.04. C 28.57%, H 4.80%, N 66.64%. $CH_2=N^+=N^-$. Prepd from chloroform and hydrazine by reaction with potassium hydroxide: Staudinger, Kupfer, *Ber.* **45**, 501 (1912); McManus *et al.*, *J. Org. Chem.* **33**, 4272 (1968); from KOH and nitrosomethylurea: Dessaux, Durand, *Bull. Soc. Chim. France* **1963**, 41. These methods yield gaseous diazomethane. The following procedures yield ether solns of diazomethane. From *N*-nitroso-β-methylaminoisobutyl methyl ketone in ether and isopropanol by reaction with sodium isopropoxide or from the same ketone in ether by reaction with sodium cyclohexoxide: Redemann *et al.*, *Org. Syn.* **25**, 28 (1945). By KOH saponification of nitrosomethylurea in ether: Arndt, *Org. Syn.* coll. vol. II, 165 (1943), or of nitrosomethylurethan in ether: von Pechmann, *Ber.* **27**, 1888 (1894); **28**, 855 (1895); Meerwein, Burneleit, *Ber.* **61**, 1845 (1928). In the laboratory diazomethane may be prepd most simply by the action of alkali on the commercially available *N*-methyl-*N*-nitroso-*N'*-nitroguanidine, cf. McKay, *J. Am. Chem. Soc.* **70**, 1974 (1948); McKay *et al.*, *Can. J. Res.* **28**, 683 (1950); Fieser and Fieser, *Organic Chemistry* (Reinhold, New York, 3rd ed., 1956) p 176. Alternate intermediate: *p*-tolylsulfonylmethylnitrosamide ("Diazald" Aldrich Chemical Co.), *see* De Boer, Backer, *Org. Syn.* coll. vol. IV, 250 (1963).

Very toxic yellow gas. mp $-145°$; bp $-23°$. Undil liq and concd solns may explode violently, especially if impurities are present. Gaseous diazomethane may explode on heating to 100° or on rough glass surfaces. Ground glass apparatus and glass stirrers with glass sleeve bearings where grinding may occur, should not be used. Alkali metals also produce explosions with diazomethane. Sol in ether, dioxane. Such solns dec only slowly at low temps. Decompn is more rapid if alcohols or water are present. Copper powder causes active decompn with the evolution of nitrogen and the formation of insol white flakes of polymethylene $(CH_2)_x$. Solid calcium chloride or boiling stones have the same effect. This phenomenon appears to occur always during the action of diazomethane on solid substances.

USE: Powerful methylating agent for acidic compds such as carboxylic acids, phenols, enols. For syntheses with diazomethane *see* the reviews by Smith, *Chem. Revs.* **23**, 193 (1938); Eistert, *Z. Angew. Chem.* **54**, 99, 124 (1941) translated by Spangler in *Newer Methods of Preparative Organic Chemistry* (New York, 1948) p 513; J. S. Pizey, *Synthetic Reagents* vol. 2 (John Wiley, New York, 1974) pp 65-142. *Caution:* Explosive (use safety screen), insidious poison (a well-ventilated hood is absolutely necessary), avoid vapor. Strong irritant. Does not cause discernible reaction at the time of contact, but later, even in minute amounts, produces an inflammatory reaction. Hypersensitivity results which makes it impossible to work with diazomethane without attacks of asthma and associated symptoms.

2984. 6-Diazo-5-oxo-L-norleucine. DON. $C_6H_9N_3O_3$; mol wt 171.16. C 42.10%, H 5.30%, N 24.55%, O 28.05%. $N_2CHCOCH_2CH_2CH(NH_2)COOH$. Antitumor antibiotic produced by an unidentified species of *Streptomyces* from Peruvian soil: Dion *et al.*, *J. Am. Chem. Soc.* **78**, 3075 (1956); Ehrlich *et al.*, *Antibiot. & Chemother.* **6**, 487 (1956). Several methods of synthesis: De Wald, Moore, *ibid.* **80**, 3941 (1958); *eidem*, U.S. pat. **2,965,634** (1960 to Parke, Davis). *Review:* Pittillo, Hunt, *Antibiotics* vol 1, D. Gottlieb, P. D. Shaw, Eds. (Springer-Verlag, New York, 1967) pp 481-493.

Pale yellow crystals from dil ethanol, dec 145-155°. $[\alpha]_D^{26}$ +21° (c = 5.4). uv max (phosphate buffer): 244, 275 nm ($E_{1cm}^{1\%}$ 376, 683). Freely sol in water. The pH of aq solns should be maintained between 4.5 and 6.5. Also sol in aq solns of methanol, ethanol, acetone. Sparingly sol in absolute alcs.

THERAP CAT: Antineoplastic.

2985. 5-Diazouracil. *5-Diazo-2,4(1H,3H)-pyrimidinedione;* DU; 2,4-dioxo-5-diazopyrimidine; 5-diazo-2,4-dioxopyrimidine. $C_4H_2N_4O_2$; mol wt 138.09. C 34.79%, H 1.46%, N 40.58%, O 23.17%. Prepd by diazotization of 5-aminouracil: Johnson et al., *Ber.* **64**, 2629 (1931). Revised structure: Thurber, Townsend, *J. Heterocycl. Chem.* **9**, 629 (1972). Has significant activity against gram-positive and gram-negative bacteria *in vivo:* Hunt, Pittillo, *Appl. Microbiol.* **16**, 1792 (1968).

Crystals, conflagrates at 198°. The stout prisms obtained from ice water should be white, not yellow or red. Sensitive to light, temp and air. Best stored in evacuated, refrigerated ampuls. Acid reactions. IR spectrum: Band at 4.57 μ. Forms alcoholates, also a red monohydrate, $C_4H_4N_4O_3$, from which a potassium salt, $KC_4H_3N_4O_3$, is obtained which is slightly sol in water with neutral reaction.

USE: In cancer research.

2986. Diazoxide. *7-Chloro-3-methyl-2H-1,2,4-benzothiadiazine 1,1-dioxide;* 3-methyl-7-chloro-1,2,4-benzothiadiazine 1,1-dioxide; SRG 95213; Eudemine Injection; Proglicem; Hyperstat; Hypertonalum; Mutabase; Proglycem. $C_8H_7ClN_2O_2S$; mol wt 230.70. C 41.65%, H 3.06%, Cl 15.37%, N 12.15%, O 13.87%, S 13.90%. Prepn: Rubin et al., *Science* **133**, 2067 (1961); Topliss et al., U.S. pat. **2,986,-573** (1961 to Schering); Raffa, Monzani, *Farmaco Ed. Sci.* **17**, 244 (1962). Crystal and molecular structure: G. Bandoli, M. Nicolini, *J. Cryst. Mol. Struct.* **7**, 229 (1978).

Crystals from dil alc, mp 330-331°. uv max (methanol): 268 nm (ε 11,300). Sol in alcohol and alkaline solns. Insol in water.

THERAP CAT: Antihypertensive.

2987. Dibekacin. *O-3-Amino-3-deoxy-α-D-glucopyranosyl-(1 → 6)-O-[2,6-diamino-2,3,4,6-tetradeoxy-α-D-erythrohexopyranosyl-(1 → 4)]-2-deoxy-D-streptamine;* DKB; 3',4'-dideoxykanamycin B; debecacin; Icacine; Kappati; Orbicin; Panamicin. $C_{18}H_{37}N_5O_8$; mol wt 451.54. C 47.88%, H 8.26%, N 15.51%, O 28.35%. Semisynthetic analog of kanamycin, *q.v.*, effective against kanamycin-resistant bacteria. Prepn: Umezawa et al., *J. Antibiot.* **24**, 485 (1971); *eidem,* *Bull. Chem. Soc. Japan* **45**, 3624 (1972); Umezawa et al., Ger. pat. **2,135,191** (1972 to Microbiochem. Res. Found.). Improved synthesis: T. Yoneta et al., *Bull. Chem. Soc. Japan* **52**, 1131 (1979). Antibacterial activity: A. G. Paradelis et al., *Antimicrob. Ag Chemother.* **14**, 514 (1978). Metabolic studies: Shimizu, *Japan. J. Antibiot.* **26**, 522 (1973). Toxicology: Koeda et al., *ibid.* 221. Pharmacokinetics and acute toxicity: I. Komiya et al., *J. Pharmacobio.-Dyn.* **4**, 356 (1981). HPLC determn in serum: H. Kubo et al., *Antimicrob. Ag. Chemother.* **28**, 521 (1985). Review: P. Noone, *Drugs* **27**, 548-578 (1984).

$[\alpha]_D^{20}$ +132° (c = 0.65). LD_{50} in mice (mg/kg): 61.0-68.0 i.v., 373.0-380.0 i.m. (Komiya).

Sulfate, *Débékacyl, Panimycin, Tokocin.* White or yellowish-white powder. Slightly bitter taste. Sol in water. Practically insol in ethanol, acetone, other organic solvents.

THERAP CAT: Antibacterial.

2988. Dibenzalacetone. *1,5-Diphenyl-1,4-pentadien-3-one;* dibenzylidene acetone; distyryl ketone. $C_{17}H_{14}O$; mol wt 234.28. C 87.15%, H 6.02%, O 6.83%. $C_6H_5CH=CH-COCH=CHC_6H_5$. Prepn from benzaldehyde + acetone: Conrad, Dolliver, *Org. Syn.* **12**, 22 (1939); Haslam, U.S. pat. **2,719,863** (1955 to du Pont); Tokár et al., *Acta Chim. Acad. Sci. Hung.* **19**, 83 (1959). Prepn of geometrical isomers: Dinwiddie et al., *J. Org. Chem.* **27**, 327 (1962).

trans-trans-Form, crystals from hot ethyl acetate, mp 110-111°. uv max: 330 nm (ε 34,300). Practically insol in water. Slightly sol in alc, ether; sol in acetone, chloroform.

cis-trans-Form, light yellow needles from ethanol, mp 60°. uv max 295 nm (ε 20,000).

cis-cis-Form, yellow oil, $bp_{0.02}$ 130°. uv max: 287 nm (ε 11,000).

USE: In sun protection preparations.

2989. 1,2:5,6-Dibenzanthracene. *Dibenz[a,h]anthracene.* $C_{22}H_{14}$; mol wt 278.33. C 94.93%, H 5.07%. From methyl dinaphthyl ketone: Clar, *Ber.* **62**, 350, 1378 (1929); Fieser, Dietz, *ibid.* 1827; Bachmann, *J. Org. Chem.* **1**, 347 (1937). Separation by chromatography: Winterstein, Schön, *Z. Physiol. Chem.* **230**, 146 (1934). Absorption spectrum: Mayneord, Roe, *Proc. Roy. Soc. London* **A152**, 299 (1935). *Review:* E. Clar, *Polycyclic Hydrocarbons,* 2 vols. (Academic Press, New York, 1964). Review of carcinogenicity studies: *IARC Monographs* **3**, 178-196 (1973).

Plates, leaflets from acetic acid. Crystals may be monoclinic or orthorhombic. Sublimes. mp 266°. Sol in petr ether, benzene, toluene, xylene, oils, other organic solvents; slightly sol in alcohol and ether. Insol in water.

Note: This substance may reasonably be anticipated to be a carcinogen: *Fourth Annual Report on Carcinogens* (NTP 85-002, 1985) p 72.

2990. Dibenzepin. *10-[2-(Dimethylamino)ethyl]-5,10-dihydro-5-methyl-11H-dibenzo[b,e][1,4]diazepin-11-one;* 5-methyl-10β-dimethylaminoethyl-10,11-dihydro-11-oxodibenzo[b,e][1,4]diazepine; HF 1927. $C_{18}H_{21}N_3O$; mol wt 295.37. C 73.19%, H 7.17%, N 14.23%, O 5.42%. Prepn: F. Hunziker et al., *Arzneimittel-Forsch.* **13**, 324 (1963); **Brit. pat. 961,106** corresp to J. Schmutz, F. Hunziker, U.S. pats. **3,312,689** and **3,419,547** (1964, 1967, 1968 all to Wander). Metabolism: H. Lehner et al., *Arzneimittel-Forsch.* **17**, 185 (1967). Comprehensive description: A. Egli, W. R. Michaelis in *Analytical Profiles of Drug Substances* vol. 9, K. Florey, Ed. (Academic Press, New York, 1980) pp 181-206.

mp 116-117°. bp$_{0.01}$ 185°.

Hydrochloride, $C_{18}H_{22}ClN_3O$, *Neodalit, Noveril.* Cryst, mp 238°. uv max (0.1N HCl): 204, 220 nm (log ε 4.530, 4.458). pKa 8.25. Sol in water, alcohol, chloroform. LD$_{50}$ orally in mice: 215 mg/kg, F. Hunziker *et al., loc. cit.*

THERAP CAT: Antidepressant.

2991. Dibenzoylmethane. *1,3-Diphenyl-1,3-propanedione;* phenyl-α-hydroxystyryl ketone; ω-benzoylacetophenone; phenyl phenacyl ketone; γ-hydroxychalcone. $C_{15}H_{12}O_2$; mol wt 224.25. C 80.34%, H 5.39%, O 14.27%. $C_6H_5CO\text{-}CH_2COC_6H_5$. Prepd by the action of sodium methylate on benzylideneacetophenone dibromide: Allen *et al., Org. Syn.* coll. vol. I (2nd ed., 1941) p 205.

Orthorhombic plates, tablets from petr ether, mp 80°. bp$_{18}$ 219-221°. 100 parts alc dissolve 4.43 parts at 19°. Sol in ether, chloroform, aq NaOH. Equilibrium mixtures in methanol, alcohol, and glacial acetic acid contain 96-98% of the enol form.

Monoxime, $C_{15}H_{13}NO_2$, prepd by adding 4 mols hydroxylamine to 1 mol dibenzoylmethane in alcohol. Prisms from ether, mp 165°. Sparingly sol in cold alcohol.

2992. 2,3:6,7-Dibenzphenanthrene. *Pentaphene;* dibenzo[*b,h*]phenanthrene; β,β'-dibenzphenanthrene; naphtho-2',3'-1,2-anthracene. $C_{22}H_{14}$; mol wt 278.33. C 94.93%, H 5.07%. Prepn by pyrolysis of 1,2- or 1,4-di-*o*-toluylbenzene: Clar *et al., Ber.* **62**, 947 (1924); Clar, John, *Ber.* **64**, 986 (1931); Winterstein, Schön, *Z. Physiol. Chem.* **230**, 146 (1934); from phthalic anhydride + *o*-tolylmagnesium bromide: Clar, Stewart, *J. Chem. Soc.* **1951**, 3215; from 2-benzoylbenzoic acid: Marsili, Isola, *Tetrahedron Letters* **1965**, 3023; Franck, Zander, *Ber.* **99**, 396 (1966).

Yellowish-green needles or leaflets from xylene, mp 257°. Absorption max (alc): 423.5, 412, 399, 379, 356, 345, 329, 314.5, 302, 289.5, 257.5, 245 nm. Sol in benzene, xylene; sparingly sol in alcohol, ether. Solns exhibit blue fluorescence in daylight.

Dipicrate, $C_{22}H_{14}\cdot 2C_6H_3N_3O_7$, orange needles, mp 184°.

2993. Dibenzylamine. *N-(Phenylmethyl)benzenemethanamine.* $C_{14}H_{15}N$; mol wt 197.27. C 85.23%, H 7.66%, N 7.10%. $C_6H_5CH_2NHCH_2C_6H_5$. Prepd by the action of alcoholic ammonia on benzyl chloride: Mason, *J. Chem. Soc.* **63**, 1312 (1893).

Oil; ammoniacal odor. bp 300° with partial decompn. Practically insol in water; sol in alcohol or ether.

USE: Detection of cobalt, cyanate and iron.

2994. Dibenzyl Chlorophosphonate. *Phosphorochloridic acid bisphenylmethyl ester; benzyl phosphorochloridate;* dibenzylchlorophosphate; DBPCl; dibenzylphosphoryl chloride. $C_{14}H_{14}ClO_3P$; mol wt 296.69. C 56.67%, H 4.76%, Cl 11.95%, O 16.18%, P 10.44%. Prepd by adding sulfuryl chloride to dibenzyl phosphite at 17-19° in CCl$_4$ diluent: Atherton *et al., J. Chem. Soc.* **1948**, 1106; by adding chlorine to dibenzyl phosphite in CCl$_4$ at below 0°: Atherton *et al.,* U.S. pat. 2,490,573 (1949 to Hoffmann-La Roche).

Thick oil, dec on standing or distillation. Identified by its reaction with ammonia or amines to form corresponding dibenzyl aminophosphonates.

USE: In solns with inert solvents such as chloroform, carbon tetrachloride, benzene and ether for the phosphorylation of nucleosides and amino acids.

2995. Dibenzyl Disulfide. *Bis(phenylmethyl) disulfide;* benzyl disulfide; α-(benzyldithio)toluene; di(phenylmethyl)-disulfide. $C_{14}H_{14}S_2$; mol wt 246.37. C 68.25%, H 5.73%, S 26.03%. $C_6H_5CH_2SSCH_2C_6H_5$. Prepd by the reaction of benzyl chloride with sodium disulfide or polysulfide: Blanksma, *Rec. Trav. Chim.* **20**, 137 (1901); Moran, Crandall, U.S. pat. **2,113,092** (1938); Wojcik, U.S. pat. **2,185,007** (1939).

Leaflets from alcohol, mp 71-72°. Another modification mp 69-70°. Dec > 270°. Practically insol in water. Sol in ether, benzene, hot methanol, hot ethanol.

USE: Antioxidant in rubber compounding, stabilizer for petr fractions, additive to silicone oils. The soly in oils is increased by the presence of benzyl alcohol.

2996. Dibenzyl Phosphite. *Phosphonic acid bis(phenylmethyl) ester;* dibenzyl hydrogen phosphite; benzyl phosphite. $C_{14}H_{15}O_3P$; mol wt 262.24. C 64.12%, H 5.77%, O 18.30%, P 11.81%. $(HO)P(OCH_2C_6H_5)_2$. Prepd by adding dropwise a mixture of dimethylaniline and benzyl alc to a soln of phosphorus trichloride in benzene: Atherton *et al., J. Chem. Soc.* **1945**, 382; U.S. pat. 2,490,573 (1949 to Hoffmann-La Roche).

Liquid. mp −5 to +5°. bp$_{0.1}$ 160-164°; (decompn); bp$_{0.01}$ 110-120°; also reported as bp$_{0.001}$ 120-130°. n$_D^{18}$ 1.5521.

USE: In preparation of *N*-phosphorylated amines.

2997. Diborane(6). Boroethane; diboron hexahydride. B_2H_6; mol wt 27.69. B 78.16%, H 21.84%. Review of methods of prepn: Adams in *Borax to Boranes, Advances Chem. Ser. No. 32* (American Chemical Society, 1961) pp 60-68. Review of structure and properties: Campell, Jr. in *Progress in Boron Chemistry,* Steinberg, McCloskey, Eds., vol. 1 (Macmillan, New York, 1964) pp 167-184. Reviews of reaction chemistry: Schenker, *Angew. Chem.* **73**, 81-107 (1961); Long in *Advan. Inorg. Chem. Radiochem.* **16**, 201-296 (1974). Review of toxicity: see Decaborane(14). General reviews: Stock, *Hydrides of Boron and Silicon* (Cornell Univ. Press, Ithaca, 1933), *passim;* Siegel, Mack, *J. Chem. Ed.* **34**, 314-317 (1957); Major, *Chem. Eng. Progr.* **54**(3), 49-54 (1958); Mikhailov, *Usp. Khim.* **31**, 417-451 (1962); *Russian Chem. Rev.* (Eng. Ed.) **31**, 224-235 (1962); Adams, *Boron, Metalloboron Compounds and Boranes* (Interscience, New York, 1964) pp 555-605; Lipscomb, *Boron Hydrides* (Benjamin, New York, 1963) *passim;* Long, *Progr. Inorg. Chem.* **15**, 1-99 (1972); Greenwood in *Comprehensive Inorganic Chemistry* vol. 1, J. C. Bailar, Jr., *et al.,* Eds. (Pergamon Press, Oxford, 1973) pp 763-783.

Colorless, flammable gas; repulsive, sickly-sweet odor. mp −165°; bp −92.5°. d^{-112} 0.447; d$^{-29.6}$ 0.33; d$^{15.0}$ 0.210. Critical temp 16.7°; critical press. 39.5 atm; Cp at 25°: 13.60 cal/mole/°C. Dec at red heat to B + H$_2$, at lower temps to H$_2$ and other boron hydrides. Spontaneous ignition temp in air about 40-50°; presence of contaminants may lower the temp limit so that ignition or detonation of diborane(6)-air mixtures may occur at or below room temp. Hydrolyzes in water to H$_2$ + H$_3$BO$_3$. Sol in CS$_2$. Reacts with NH$_3$ to form diborane diammoniate; reacts slowly with Br$_2$ and explosively with Cl$_2$ to form boron halides; reacts with hydrocarbons or organoboron compds to give alkyl- or arylboron compds; reacts with metal alkyls to form metal borohydrides; reacts with strong electron pair donors to form borane addn compds.

USE: As catalyst for olefin polymerization; as rubber vulcanizer; as reducing agent; as flame-speed accelerator; in rocket propellants; as intermediate in prepn of the boron hydrides; in conversion of olefins to trialkylboranes and primary alcohols; as a doping gas. *Caution:* Inhalation may produce irritation of lungs, pulmonary edema. High exposures produce symptoms resembling metal fume fever. *See also* Decaborane.

2998. Diboron Tetrachloride. *Tetrachlorodiborane(4); boron chloride.* B_2Cl_4; mol wt 163.47. B 13.24%, Cl 86.76%. Prepd by passing gaseous boron trichloride through flow discharge between mercury or copper electrodes: Urry *et al., J. Am. Chem. Soc.* **76**, 5293 (1954); Wartik *et al., Inorg. Syn.* **10**, 118 (1967); by passing boron trichloride vapor over boron monoxide at 200°: McCloskey *et al., J. Am. Chem. Soc.* **83**, 4750 (1961). Infrared and Raman spectra: Linevsky *et al., J. Am. Chem. Soc.* **75**, 3287 (1953). Review of boron halides: Massey, *Advan. Inorg. Chem. Radiochem.* **10**, 1-152 (1967).

Liquid, ignites in air. mp $-92.6°$. bp_1 $-63.5°$; bp_{45} 0°; bp (calc) 65.5°.

2999. Diboron Tetrahydroxide. *Tetrahydroxydiborane-(4); hypoboric acid;* sub-boric acid. $B_2H_4O_4$; mol wt 89.67. B 24.13%, H 4.49%, O 71.37%. $B_2(OH)_4$. Prepd by action of steam on diboron tetrachloride: Wartik, Apple, *J. Am. Chem. Soc.* **77**, 6400 (1955); in quantitative yield by aq hydrolysis of tetraethoxydiboron at 0°: McCloskey *et al., ibid.* **83**, 4750 (1961).

Minute crystals, starts to lose water at about 90° *in vacuo*.

3000. 9,10-Dibromoanthracene. $C_{14}H_8Br_2$; mol wt 336.04. C 50.04%, H 2.40%, Br 47.56%. Made by brominating a suspension of anthracene in CCl_4: Heilbron, Heaton, *Org. Syn.* **3**, 41 (1923).

Yellow needles from xylene, mp 226°. Sublimes. Absorption max: Conrad-Billroth, *Z. Physik. Chem. (Leipzig)* **33B**, 133 (1936). Insol in water; slightly sol in alcohol, ether, cold benzene; sol in hot benzene, hot toluene. Oxidation gives anthraquinone.

Addn compd with *sym*-trinitrobenzene, $C_{20}H_{11}Br_2N_3O_6$, mp 179°.

3001. p-Dibromobenzene. $C_6H_4Br_2$; mol wt 235.92. C 30.54%, H 1.71%, Br 67.75%. Prepn from diazotized *p*-bromoaniline: Fry, Grote, *J. Am. Chem. Soc.* **48**, 710 (1926); by catalytic bromination of bromobenzene: Ferguson *et al., ibid.* **76**, 1250 (1954).

Crystals, mp 87.31°. bp 220.40°. $d^{99.6}$ 0.9641. $n_D^{99.3}$ 1.5743. Practically insol in water; sol in about 70 parts alcohol; sol in benzene, chloroform; very sol in ether.

3002. α,α'-Dibromo-d-camphor. *d-3,3-Dibromo-1,7,7-trimethylbicyclo[2.2.1]heptan-2-one; d-3,3-dibromo-2-bornanone;* dibromated camphor. $C_{10}H_{14}Br_2O$; mol wt 310.04. C 38.74%, H 4.55%, Br 51.55%, O 5.16%. Prepd by brominating α-bromocamphor: Lowry, *J. Chem. Soc.* **73**, 569 (1898). Conformation: Brutcher *et al., J. Am. Chem. Soc.* **81**, 4915 (1959). Review: J. L. Simonsen, L. N. Owen, *The Terpenes,* **vol. 2** (Cambridge University Press, London, 2nd ed., 1949) pp 401, 405.

Crystals from alcohol, benzene, or petr ether, mp 61°. Not very stable and readily assumes yellow color. Volatile with steam. $d_4^{21.6}$ 1.854. $[\alpha]_D^{20}$ $+39.6°$ (c = 17 in benzene). Practically insol in water; sol in about 5 parts abs alcohol; readily sol in ether, benzene, ethyl acetate, petr ether. uv max (cyclohexane): 323 nm (log ϵ 1.88): Lowry, Owen, *J. Chem. Soc.* **1926**, 606.

3003. Dibromochloropropane. *1,2-Dibromo-3-chloropropane;* 3-chloro-1,2-dibromopropane; DBCP; OS1897; Fumazone; Nemafume; Nemagon. $C_3H_5Br_2Cl$; mol wt 236.36. C 15.24%, H 2.13%, Br 67.62%, Cl 15.00%. $ClCH_2$-$CHBrCH_2Br$. Prepn: Darmstädter, *Ann.* **152**, 320 (1869). Activity: C. W. McBeth, G. B. Bergeson, *Plant Dis. Rep.* **39**, 223 (1955). Toxicity study: T. R. Torkelson *et al., Toxicol. Appl. Pharmacol.* **3**, 545 (1961). Carcinogenicity studies: W. A. Olsen *et al., J. Nat. Cancer Inst.* **51**, 1993 (1973); E. K. Weisburger, *Environ. Health Perspect.* **21**, 7 (1977). Mutagenicity study: H. S. Rosenkranz, *Bull. Environ. Contam. Toxicol.* **14**, 8 (1975).

Brown liq; pungent odor. bp 196°; bp_{16} 78°; $bp_{0.8}$ 21°. d^{14} 2.093. n_D^{14} 1.553. Vapor pressure at 21°: 0.8 mm Hg. Slightly sol in water; misc with oils, dichloropropane, isopropyl alcohol. LD_{50} in rats, mice (g/kg): 0.17, 0.26 orally (Torkelson).

Caution: May be irritating to skin, mucous membranes; may cause CNS depression: *Clinical Toxicology of Commercial Products,* R. E. Gosselin *et al.,* Eds. (Williams & Wilkins, Baltimore, 5th ed., 1984) Section II, pp 167-168. This substance may reasonably be anticipated to be a carcinogen: *Fourth Annual Report on Carcinogens* (NTP 85-002, 1985) p 76.

USE: Soil fumigant; nematocide.

3004. 1,2-Dibromo-2,4-dicyanobutane. *2-Bromo-2-(bromomethyl)pentanedinitrile;* 2-bromo-2-(bromomethyl)-glutaronitrile; Tektamer 38. $C_6H_6Br_2N_2$; mol wt 265.94. C 27.10%, H 2.27%, Br 60.10%, N 10.53%. Prepn: N. Grier, S. J. Lederer, **Ger.** pat. **2,164,723** corresp to **U.S.** pat. **3,833,731** (1972, 1974 to Merck & Co.).

Crystals from ethanol, mp 51.2-52.5°. Mildly pungent odor. Very sol in DMF, acetone, chloroform, ethyl acetate, benzene. Sol in methanol, ethanol, ether. Insol in water.

USE: Preservative for latex paints, adhesives, latex emulsions, dispersed pigments, joint cements, metal working fluids.

3005. 4',5'-Dibromofluorescein. *4',5'-Dibromo-3',6'-dihydroxyspiro[isobenzofuran-1(3H),9'-[9H]xanthen]-3-one;* 4,5-dibromo-3,6-fluorandiol; D & C Orange No. 5; C.I. Solvent Red 72; C.I. 45370:1. $C_{20}H_{10}Br_2O_5$; mol wt 490.12. C 49.01%, H 2.06%, Br 32.61%, O 16.32%. Prepn from fluorescein, *q.v.*: A. Baeyer, *Ann.* **183**, 1 (1876); M. A. Phillips, *J. Chem. Soc.* **1932**, 724. Metabolism: J. M. Webb *et al., J. Pharmacol. Exp. Ther.* **137**, 141 (1962). *See also: Colour Index* **vol. 4** (3rd ed., 1971) p 4425.

Red plates, mp 285°. Slightly sol in water (orange with faint yellow fluorescence); sol in ethanol (orange with greenish-yellow fluorescence) and in acetone (pink with yellow fluorescence). Red-yellow soln in conc H_2SO_4, turning yellow-brown with orange ppt on dilution. Absorption spectra: R. C. Gibbs, C. V. Shapiro, *J. Am. Chem. Soc.* **51**, 1769 (1929).

Disodium salt, $C_{20}H_8Br_2Na_2O_5$, *C.I. Acid Orange 11.*

USE: D & C Orange No. 5 permitted for use in lipsticks, mouthwashes and dentifrices: *Fed. Regist.* **47,** 49632 (1982).

3006. Dibromogallic Acid. *2,6-Dibromo-3,4,5-trihydroxybenzoic acid; 2,6-dibromogallic acid;* gallobromol. C_7H_4-Br_2O_5; mol wt 327.93. C 25.64%, H 1.23%, Br 48.74%, O 24.40%. Prepd by bromination of gallic acid: Etti, *Ber.* **11,** 1879 (1878).

Light-brown powder or crystals; crystallizes from water with one H_2O which it loses at 120°. mp 150°, also stated as about 140° with decompn. Soluble in 8 parts cold water, 0.5 part boiling water; freely sol in alcohol, ether; practically insol in chloroform.

3007. 3,5-Dibromo-4-hydroxybenzenesulfonic Acid. 2,6-Dibromophenol-4-sulfonic acid; *p*-hydroxy-3,5-dibromobenzenesulfonic acid. $C_6H_4Br_2O_4S$; mol wt 332.00. C 21.71%, H 1.21%, Br 48.14%, O 19.28%, S 9.66%. Prepn: Senhofer, *Ann.* **156,** 103 (1870); Armstrong, Brown, *J. Chem. Soc.* **25,** 858 (1872). Prepn of the sodium salt: Cannell, *J. Am. Chem. Soc.* **79,** 2927 (1957).

Rectangular platelets. Freely sol in water, alcohol; less sol in ether.

Sodium salt dihydrate, $C_6H_3Br_2NaO_4S.2H_2O$, *Dibromol.* Needles. Freely sol in hot water, less sol in cold water.

THERAP CAT: Topical disinfectant.

3008. Dibromopropamidine. *4,4'-[1,3-Propanediylbis-(oxy)]bis(3-bromobenzenecarboximidamide); 4,4'-(trimethylenedioxy)bis(3-bromobenzamidine); 2',2''-dibromo-4',4''-diamidino-1,3-diphenoxypropane.* $C_{17}H_{18}Br_2N_4O_2$; mol wt 470.16. C 43.43%, H 3.86%, Br 33.99%, N 11.92%, O 6.80%. Prepn: S. S. Berg, G. Newbery, Brit. pat. **598,911** (1948 to May & Baker); *eidem, J. Chem. Soc.* **1949,** 642. Antibacterial activity: R. Wien *et al., Lancet* **254,** 711 (1948); A. D. Russell, J. R. Furr, *Int. J. Pharm.* **34,** 115 (1986). Mode of action: W. Woodside, *Microbios* **8,** 23 (1973). Clinical use in *Acanthamoeba* keratitis: J. J. Wiens, W. B. Jackson, *Can. J. Ophthalmol.* **23,** 107 (1988). HPLC determn in cosmetics: B. Wyhowski de Bukanski, M. O. Masse, *Int. J. Cosmet. Sci.* **6,** 283 (1984).

Isethionate, $C_{21}H_{30}Br_2N_4O_{10}S_2$, *Brolene, Broline Ointment, Brulidine.* Prismatic needles from ethanol, mp 226°. Freely sol in water. One gram dissolves in 2 ml of water at 20°, in 60 g of 95% alcohol at 20°. Sol in glycerol. Practically insol in ether, chloroform, fixed oils, liquid petrolatum. Aq solns are very slightly acidic. Solns may be sterilized by heating at 100° for 30 min, but they should not be stored for more than a few days, and then only in neutral glass containers since some hydrolysis occurs with the formation of sparingly sol urea derivs. *Incompat.* with chlorides, sulfates, and many organic anions, all of which form sparingly sol salts.

USE: Preservative in cosmetics.

THERAP CAT: Antiseptic; antiamebic.

THERAP CAT (VET): Antiseptic, antimicrobial.

3009. 2,3-Dibromopropene. 2,3-Dibromopropylene; 2-bromoallyl bromide; *α*-bromoallyl bromide; *α*-epidibromohydrin. $C_3H_4Br_2$; mol wt 199.89. C 18.02%, H 2.02%, Br 79.96%. $BrCH_2CBr=CH_2$. Prepd by the action of sodium hydroxide on 1,2,3-tribromopropane: Lespieau, Bourguel, *Org. Syn.* **5,** 49 (1925); *cf.* Tapley, Giesy, *J. Am. Pharm. Assoc.* **15,** 173 (1926).

Liquid. d_4^{20} 1.9336. bp_{760} 140-143°; bp_{75} 75-76°; bp_{18} 42-43°. n_D^{20} 1.5157. May be distilled under atmospheric pressure with very slight decompn, but becomes highly colored on standing in glass-stoppered bottle.

3010. 2,6-Dibromoquinone-4-chlorimide. *2,6-Dibromo-4-(chloroimino)-2,5-cyclohexadien-1-one; 2,6-dibromo-N-chloro-p-benzoquinoneimine;* 2,6-dibromo-*N*-chloroquinonimine; BQC reagent. $C_6H_2Br_2ClNO$; mol wt 299.37. C 24.07%, H 0.67%, Br 53.39%, Cl 11.84%, N 4.68%, O 5.34%. Prepn: Hartman *et al., Org. Syn.* **coll. vol. II,** 175 (1943).

Yellow prisms from alc or glacial acetic acid, mp 83°. Sol in about 17,000 parts water; moderately sol in hot alc, hot glacial acetic acid. Gives a blue color with alk phenol soln.

USE: As a reagent for phenol and phosphatases: Ljunggren, *Proc. 13th Int. Dairy Congr.* **3,** 1319 (1953).

3011. 3,5-Dibromosalicylaldehyde. *3,5-Dibromo-2-hydroxybenzaldehyde;* Dalyde. $C_7H_4Br_2O_2$; mol wt 279.93. C 30.03%, H 1.44%, Br 57.10%, O 11.43%. Prepd by bromination of salicylaldehyde in glacial acetic acid with cooling: Wentworth, Brady, *J. Chem. Soc.* **117,** 1043 (1920); Brewster, *J. Am. Chem. Soc.* **46,** 2464 (1924); by bromination in the presence of sodium acetate in glacial acetic acid at 50°: Lindemann, Forth, *Ann.* **435,** 223 (1924).

Pale yellow prisms, mp 86°. Sublimes forming leaflets and needles. Volatile with steam. Sparingly sol in water. Aq solns are yellow. Readily sol in ether, benzene, chloroform, hot petr ether, alcohol, glacial acetic acid.

THERAP CAT: Antibacterial.

3012. 3,5-Dibromosalicylic Acid. *3,5-Dibromo-2-hydroxybenzoic acid.* $C_7H_4Br_2O_3$; mol wt 295.94. C 28.41%, H 1.36%, Br 54.01%, O 16.22%. Prepd from salicylic acid and Br_2: Earle, Jackson, *J. Am. Chem. Soc.* **28,** 111 (1906); Robertson, *J. Chem. Soc.* **82,** 1481 (1902); and HOBr: Leulier, Pinet, *Bull. Soc. Chim. France* [4] **41,** 1362 (1927).

Needles, mp 223°. Sparingly sol in H_2O; sol in alc, ether.

3013. 2,3-Dibromosuccinic Acid. *2,3-Dibromobutanedioic acid; α,α'-dibromosuccinic acid; sym-dibromosuccinic acid.* $C_4H_4Br_2O_4$; mol wt 275.90. C 17.41%, H 1.46%, Br 57.93%, O 23.20%. Prepn of *dl*-form by bromination of maleic acid: McKenzie, *J. Chem. Soc.* **101,** 1196 (1912); Young *et al., J. Am. Chem. Soc.* **61,** 1640 (1939); of *meso*-form by bromination of fumaric acid: *eidem, ibid.;* Rhinesmith, *Org. Syn.* **coll. vol. II,** 177 (1943). Resolution of *dl*-form: McKenzie, *loc. cit.*

BrCHCOOH
|
BrCHCOOH

dl-Form, **threo-2,3-dibromosuccinic acid**. Crystals, mp 167°. Very sol in cold water.

d-Form, crystalline powder from ethyl acetate + CCl$_4$, mp 157-158°. $[\alpha]_D^{18}$ +64.4° (c = 5 in water), +135.8° (c = 5 in alc), +147.8° (c = 5 in ethyl acetate).

l-Form, needles from benzene, decomp 157-158°. $[\alpha]_D^{13}$ -148.0° (c = 5.8 in ethyl acetate). Very sol in cold water; sol in ethyl acetate, acetone, methanol, ethanol; sparingly sol in chloroform, petr ether, CCl$_4$.

meso-Form, **erythro-2,3-dibromosuccinic acid**. Cryst from water, dec 255-256°, also reported as mp 270-273°. Sol in 50 parts cold water, more sol in hot water; sol in alcohol, ether; sparingly sol in chloroform.

3014. 3,5-Dibromo-L-tyrosine. β-(3,5-dibromo-4-hydroxyphenyl)alanine; Biotiren; Bromotiren. C$_9$H$_9$Br$_2$NO$_3$; mol wt 339.00. C 31.88%, H 2.68%, Br 47.15%, N 4.13%, O 14.16%. Obtained upon saponification of gorgonin, a substance isolated from coral *(Primnoa lepadifera)* stems: Mörner, *Z. Physiol. Chem.* **88**, 139, 152 (1913). Prepd by the action of bromine vapor on tyrosine: Gorup-Besanez, *Ann.* **125**, 281 (1863). By treating tyrosine in aq HBr with Br: Aloy, Rabaut, *Bull. Soc. Chim. France* [4], **3**, 392 (1908); Zeynek, *Z. Physiol. Chem.* **114**, 275 (1921). Clinical evaluation in thyroid function: A. Isidori, *Int. J. Clin. Pharmacol. Biopharm.* **16**, 180 (1978).

L-Form dihydrate, efflorescent needles, plates (orthorhombic) from water. When anhydr, dec 245°. $[\alpha]_D^{20}$ +1.3° (c = 5 in 4% HCl). pK$_1$ 2.17; pK$_2$ 6.45; pK$_3$ 7.60. One gram dissolves in 250 ml water at 25°, in 30 ml at boiling temp. Slightly sol in alcohol; insol in ether. Freely sol in alkalies and in dil mineral acids. Forms HBr and HCl salts; stable in boiling water.

DL-Form, efflorescent prisms, platelets from water probably contg 1 mol water of crystn. Dec at about 245°. Aq solns are acid to litmus. Behaves as a monobasic acid when titrated with NaOH and phenolphthalein. One gram dissolves in about 590 ml water at 20°.

THERAP CAT: Thyroid inhibitor.

3015. Dibromsalicil. *Bis(5-bromo-2-hydroxyphenyl)ethanedione; 5,5'-dibromosalicil;* DBS; 5,5'-dibromo-2,2'-dihydroxybenzil; 5,5'-dibromo-2,2'-dihydroxybibenzoyl; Dibrosal. C$_{14}$H$_8$Br$_2$O$_4$; mol wt 400.04. C 42.03%, H 2.02%, Br 39.95%, O 16.00%. Prepd by heating dibromodimethoxybenzil in nitrobenzene in the presence of aluminum chloride: Kuhn *et al.*, *Ber.* **76**, 900 (1943).

Flexible, lemon-colored needles or yellow powder from glacial acetic acid. Odorless, tasteless. mp 212-213°. Practically insol in water. Freely sol in alcohol, benzene. Sol in acetone, ether. Sol in dil NaOH yielding solns of an intense deep yellow color.

USE: Antiseptic, germicide.

3016. Dibucaine Hydrochloride. *2-Butoxy-N-[2-(diethylamino)ethyl]-4-quinolinecarboxamide monohydrochloride; 2-butoxy-N-(2-diethylaminoethyl)cinchoninamide hydrochloride;* benzolin; Nupercaine Hydrochloride; Percaine; Cincaine; Sovcaine; Cinchocaine. C$_{20}$H$_{30}$ClN$_3$O$_2$; mol wt 379.92. C 63.22%, H 7.96%, Cl 9.33%, N 11.06%, O 8.42%. Prepn: Miescher, U.S. pat. **1,825,623** (1931 to Ciba). Com-

prehensive description: G. R. Padmanabhan in *Analytical Profiles of Drug Substances* vol. **12**, K. Florey, Ed. (Academic Press, New York, 1983) pp 105-134.

Hygroscopic crystals. Dec 90-98°. (The base mp 64°.) uv max (1N HCl): 247, 320 nm (ε 24700, 8810). Sol in 0.5 part water; freely sol in alcohol, acetone, chloroform; slightly sol in benzene, ethyl acetate and toluene (on warming); insol in ether and in oils. The aq soln, about 1 in 20, is faintly alkaline to litmus.

THERAP CAT: Local anesthetic.

3017. Dibunate Sodium. *3,6-Bis(1,1-dimethylethyl)-1-naphthalenesulfonic acid sodium salt mixt. with 3,7-bis(1,1-dimethylethyl)-1-naphthalenesulfonic acid sodium salt; 3,6-(and 3,7-)di-tert-butyl-1-naphthalenesulfonic acid sodium salt; 2,7-(and 2,6-)di-tert-butylnaphthalene-4-sulfonic acid sodium salt; sodium dibunate;* L 1633; 1633 Labaz; Becantal; Becantex; Dibunafon; Keuten; Linctussal. Although most sources refer to this compound as sodium 2,6-di-*tert*-butylnaphthalene sulfonate, it is a mixture of at least two isomers. Prepn and separation of isomers: Menard *et al.*, *Can. J. Chem.* **39**, 729 (1961).

Slightly hygroscopic crystals, dec > 300°. Slightly sol in cold water (0.5-1.0%); freely sol in hot water; sol in methanol, less sol in ethanol. Aq solns are stable to boiling.

THERAP CAT: Antitussive.

3018. Dibutoline Sulfate. *2-[[(Dibutylamino)carbonyl]oxy]-N-ethyl-N,N-dimethylethanaminium sulfate (2:1); bis[ethyl(2-hydroxyethyl)dimethylammonium] sulfate bis(dibutylcarbamate); (2-dibutyl-carbamyloxyethyl)dimethylethylammonium sulfate; dimethylethyl-β-hydroxyethylammonium sulfate dibutylurethan; Dibuline Sulfate.* C$_{30}$H$_{66}$N$_4$O$_8$S; mol wt 642.93%. C 56.04%, H 10.35%, N 8.72%, O 19.91%, S 4.99%. Prepd by treating dibutoline iodide with Ag$_2$SO$_4$: Swan, White, U.S. pats. **2,408,893** (1946); **2,432,049** (1947).

Extremely hygroscopic powder. Dec 166°. Usually furnished as a colorless aqueous soln. Also sol in benzene. A surface-active agent; a 5% aq soln (pH 6.75-7.50) has an air-water interfacial tension of 42 dynes/cm at 25°; also reported as 47 dynes/cm: Swan, White, *Am. J. Ophthalmol.* **27**, 933 (1944). Ampuled aq solns may be stored at room temp for at least one year without precipitation, darkening, change of pH, or loss of potency. Heating of aq solns at 100° causes decompn. Sterilization by filtration is recommended.

THERAP CAT: Anticholinergic.

3019. *n*-Dibutylamine. *N-Butyl-1-butanamine.* C$_8$H$_{19}$N; mol wt 129.24. C 74.34%, H 14.82%, N 10.84%. (C$_4$H$_9$)$_2$NH. Prepn from butyl bromide and ammonia with separation of the mono-, di-, and tributylamines formed: Werner, *J. Chem. Soc.* **115**, 1010 (1919). Manuf: Engel, Hoog, U.S. pat. **2,574,693** (1951 to Shell); Davies *et al.*, U.S. pat. **2,609,394** (1952 to I.C.I.); Lemon, Myerly, U.S. pat. **3,022,349** (1962 to Union Carbide); Brois, Rutkowski, U.S. pat. **3,147,310** (1964 to Esso).

Liquid, bp 159-160°. mp −60° to −59°. d_4^{20} 0.7601. n_D^{20} 1.4177. Flash pt, open cup: 135°F (57°C). Sol in water, alc. LD_{50} orally in rats: 550 mg/kg, Smyth *et al., Arch. Ind. Hyg. Occup. Med.* **10,** 61 (1954).

3020. Di-*tert*-butyl Ether. *2,2'-Oxybis[2-methylpropane].* $C_8H_{18}O$; mol. wt. 130.22. C 73.78%, H 13.93%, O 12.29%. $[(CH_3)_3C]_2O$. Prepn: R. J. Moore, G. J. O'Donnell, **Brit. pat. 652,809** (1951 to Bataafsche Petrol.); *C.A.* **46,** 1023d (1952). Synthesis from *t*-butyl chloride and silver carbonate: J. Erickson, W. M. Ashton, *J. Am. Chem. Soc.* **63,** 1769 (1941); from *t*-butyl perbenzoates and Grignard reagents: S. O. Lawesson, N. C. Yang, *ibid.* **81,** 4230 (1959); by alkylation of *t*-butyl alcohol with trimethylcarbenium fluoroantimonate: G. A. Olah *et al., Synthesis* **1975,** 315; from lithium *tert*-butoxide with aromatic sulfonyl chlorides: H. Masada *et al., Tetrahedron Letters* **1979,** 1315.

Clear liquid with camphorous odor. bp 106.5-107.0°. n_D^{20} 1.3949; d_4^{20} 0.7658. Decomposes in the presence of acids.

USE: Gasoline additive.

3021. Di-*tert*-butyl Malonate. *Propanedioic acid bis(1,1-dimethylethyl) ester;* malonic acid di-*tert*-butyl ester. $C_{11}H_{20}O_4$; mol wt 216.27. C 61.09%, H 9.32%, O 29.59%. $(CH_3)_3COOCCH_2COOC(CH_3)_3$. Prepd from malonic acid, liq isobutylene and sulfuric acid: McCloskey *et al., Org. Syn.* **34,** 26 (1954).

Liquid. mp −6.1 to −5.9°. bp_{31} 112-115°; bp_{10} 93°; bp_1 65-67°. n_D^{20} 1.4184; $n_D^{24.2}$ 1.4161; n_D^{25} 1.4158-1.4161.

USE: In the prepn of ketones.

3022. 2,6-Di-*tert*-butylpyridine. *2,6-Bis(1,1-dimethylethyl)pyridine.* $C_{13}H_{21}N$; mol wt 191.31. C 81.61%, H 11.06%, N 7.32%. Prepd by reacting *tert*-butyllithium with 2-*tert*-butylpyridine: Brown, Kanner, *J. Am. Chem. Soc.* **75,** 3865 (1953); U.S. pat. **2,780,626** (1957 to Research Corp.).

Liquid, mp 2.2°. bp_{23} 100-101°. n_D^{20} 1.5733. pKa 3.58. The base neutralizes HCl but does not react with alkyl halides or boron trifluoride. Undergoes nuclear sulfonation with sulfur trioxide forming a sulfonic acid, $C_{13}H_{21}NSO_3$, mp 310° (decompn).

Chloroaurate, $C_{13}H_{22}NAuCl_4$, mp 184.2-184.5°.

USE: Has been proposed as an additive for lubricating oil, gasoline and for stabilizing Cl-containing polymers.

3023. Dibutyl Succinate. *Butanedioic acid dibutyl ester; succinic acid dibutyl ester;* di-*n*-butyl succinate; Tabutrex; Tabatrex. $C_{12}H_{22}O_4$; mol wt 230.30. C 62.58%, H 9.63%, O 27.79%. $CH_3(CH_2)_3OOCCH_2CH_2COO(CH_2)_3CH_3$. Prepn from butyl alcohol and succinic acid: Vogel, *J. Chem. Soc.* **1948,** 624.

Liquid, bp 274.5°, bp_3 121°. mp −29.0°. d_4^{20} 0.9768. n_D^{20} 1.4299. LD_{50} orally in rats: 8 g/kg.

USE: Insect repellent esp against biting flies of cattle and household pests and roaches.

3024. Di-*tert*-butyl Succinate. *Butanedioic acid bis(1,1-dimethylethyl) ester;* succinic acid di-*tert*-butyl ester. $C_{12}H_{22}O_4$; mol wt 230.30. C 62.58%, H 9.63%, O 27.79%. $(CH_3)_3COOCCH_2COOC(CH_3)_3$. Prepd from succinic acid, liq isobutylene and sulfuric acid in dioxane: McCloskey *et al., Org. Syn.* coll. **vol. IV,** 263 (1963).

Liquid, mp 36-37°. bp_9 109-110°; bp_7 105-107°.

USE: In Stobbe condensation of ketones.

3025. Dibutyltin Dilaurate. *Dibutylbis[(1-oxododecyl)oxy]stannane; dibutylbis(lauroyloxy)tin;* Butynorate; Davainex; Tinostat. $C_{32}H_{64}O_4Sn$; mol wt 631.55. C 60.86%, H 10.21%, O 10.13%, Sn 18.79%. $(C_4H_9)_2Sn[OOC(CH_2)_{10}CH_3]_2$. Prepn: Sheverdina *et al., Khim. Prom.* **1962** (10), 707, *C.A.* **59,** 8776f (1963); **Fr. pat. 1,320,473** (1963 to Noury & van der Lande), *C.A.* **59,** 8789b (1963), corresp to **Brit. pat. 975,369.** Compd claimed but not described: Eberly, U.S. pat. **2,560,034** (1951 to Firestone).

Soft crystals or yellow liq, mp 22-24° (Sheverdina *et al., loc. cit.).* n_D^{20} 1.4683, (**Fr. pat. 1,320,473**). Practically insol in water, methanol. Soluble in petr ether, benzene, acetone, carbon tetrachloride, ether, organic esters.

USE: Stabilizer for polyvinyl chloride resins. Catalyst for curing certain silicones.

THERAP CAT (VET): Anthelmintic (vs. chicken tapeworms).

3026. Dicamba. *3,6-Dichloro-2-methoxybenzoic acid; 3,6-dichloro-o-anisic acid;* 2-methoxy-3,6-dichlorobenzoic acid; dianat; Velsicol 58-CS-11; Banvel D; Mediben. $C_8H_6Cl_2O_3$; mol wt 221.04. C 43.47%, H 2.74%, Cl 32.08%, O 21.71%. Sometimes misnamed *methoxydichlorobenzoate.* Prepn: S. B. Richter, U.S. pat. **3,013,054** (1961 to Velsicol). Metabolism: D. D. Oehler, G. W. Ivie, *J. Agr. Food Chem.* **28,** 685 (1980). Toxicity: Bailey, White, *Residue Rev.* **10,** 97 (1965).

Crystals from pentane, mp 114-116°. Vapor pressure at 100°: 3.75 × 10^{-3} mm Hg. Sol in ethanol and acetone. Very slightly sol in water. LD_{50} orally in rats: 1040 mg/kg (Bailey, White). One of the ingredients of **Banlene** and of **Cambilene.**

USE: Herbicide.

3027. Dicapthon. *Phosphorothioic acid O-(2-chloro-4-nitrophenyl) O,O-dimethyl ester; O,O-*dimethyl *O-*(2-chloro-4-nitrophenyl)phosphorothioate; O,O-*dimethyl-*O-*(2-chloro-4-nitrophenyl)thionophosphate; *p*-nitro-*o*-chlorophenyl dimethyl thionophosphate; isochlorothion; American Cyanamid 4124; insecticide ACC 4124; Dicaptan. $C_8H_9ClNO_5PS$; mol wt 297.68. C 32.28%, H 3.05%, Cl 11.91%, N 4.71%, O 26.87%, P 10.41%, S 10.77%. Prepn: Fletcher and Geoghegan, McPherson, U.S. pats. **2,664,437** and **2,784,207** (1953, 1957, both to Am. Cyanamid); Schrader, *Angew. Chem.* **66,** 265 (1954); *idem, Die Entwicklung neuer insektizider Phosphorsäure-Ester* (Verlag Chemie, 3rd ed., 1963) pp 289-291.

Crystals from methanol, mp 53°. Sol in acetone, cyclohexanone, ethyl acetate, toluene, xylene, ethylene glycol, propylene glycol, some oils. Practically insol in water. LD_{50} in male, female rats: 400, 330 mg/kg orally; 790, 1250 mg/kg dermally, T. B. Gaines, *Toxicol. Appl. Pharmacol.* **14,** 515 (1969).

USE: Insecticide. *Caution:* Cholinesterase inhibitor.

3028. Dicentrine. *6,7,7a,8-Tetrahydro-10,11-dimethoxy-7-methyl-5H-benzo[g]-1,3-benzodioxolo[6,5,4-de]quinoline; 9,10-dimethoxy-1,2-(methylenedioxy)aporphine;* 1,2-methylenedioxy-9,10-dimethoxyaporphine. $C_{20}H_{21}NO_4$; mol wt 339.38. C 70.78%, H 6.24%, N 4.13%, O 18.86%. *d*-Form prevalent in nature. Found in *Dicentra pusilla* Sieb. & Zucc., *Fumariaceae* and several other *Dicentra* species. Related to actinodaphnine, *q.v.,* and laurotetanine, *q.v.* Isoln: Y. Asahina, *Arch. Pharm.* **247,** 201 (1909). Isoln of *l*-form from *Duguetia* A. St. Hil., *Annonaceae:* Casagrande, Ferrari, *Farmaco Ed. Sci.* **25,** 442 (1970). Synthesis of *dl*-form: Haworth *et al., J. Chem. Soc.* **127,** 2018 (1925); Cava *et al., Tetrahedron* **29,** 2245 (1973). Resolution of isomers: Haworth *et al., J. Chem. Soc.* **129,** 29 (1926).

d-dicentrine

dl-Form, prisms from methanol, mp 178-179°. Freely sol in chloroform, ethyl acetate, acetone, benzene, hot alcohol. Moderately sol in ether, cold alcohol. Absorption max: Girardet, *J. Chem. Soc.* **1931**, 2630.

d-Form, long prisms from ether. mp 169°. $[\alpha]_D^{17}$ +64.1° (c = 1.433 in chloroform). Freely sol in alcohol, ethyl acetate, benzene.

l-Form, long prisms from ether. mp 169°. $[\alpha]_D^{17}$ −63.5° (c = 1.70 in chloroform).

dl-Hydrochloride, $C_{20}H_{21}NO_4 \cdot HCl$, small needles from water, dec 263-265°.

dl-Methiodide monohydrate, $C_{20}H_{21}NO_4 \cdot CH_3I \cdot H_2O$, plates, mp 228-229°.

3029. Dichlobenil. *2,6-Dichlorobenzonitrile;* H 133; Niagara 5006; Casoron; Casoron-133. $C_7H_3Cl_2N$; mol wt 172.02. C 48.88%, H 1.76%, Cl 41.22%, N 8.14%. Prepn: Reich, *Bull. Soc. Chim. France* [4] **21**, 217 (1917); Norris, Klemka, *J. Am. Chem. Soc.* **62**, 1432 (1940); Chang *et al.*, *C.A.* **53**, 6134d (1959); Koopman, *Rec. Trav. Chim.* **80**, 1075 (1961); Hackmann, ten Haken **Brit.** pat. **861,899;** Higson, **Brit.** pat. **862,937** (both 1961 to Shell). Use as herbicide: Koopman, Daams, **U.S.** pat. **3,027,248** (1962 to N. A. Phillips). Fate in crops, soil and animals: Beynon, Wright, *Residue Rev.* **43**, 23 (1972); Veroop, *ibid.* 55. Toxicity data: G. W. Bailey, J. L. White, *ibid.* **10**, 97 (1965).

Crystals from petr ether, mp 144-145°. Soly in water at 25°: 25 ppm; at 20°: 18 ppm. Vapor pressure at 20°: 3 × 10^{-6} mm Hg; at 25°: 5 × 10^{-4} mm. Absorption spectrum see Koopman, *loc. cit.* LD_{50} in rats, mice (mg/kg): 2710, 6800 orally (Bailey, White).

USE: Herbicide.

3030. Dichlofenthion. *Phosphorothioic acid O-2,4-dichlorophenyl O,O-diethyl ester; O-2,4-dichlorophenyl O,O-*diethyl phosphorothioate; *O,O-*diethyl *O-*2,4-dichlorophenyl phosphorothioate; VC-13 Nemacide; Bromex; Nemacide VC-13; VC-13. $C_{10}H_{13}Cl_2O_3PS$; mol wt 315.17. C 38.11%, H 4.16%, Cl 22.50%, O 15.23%, P 9.83%, S 10.18%. Prepd from *O,O-*diethyl phosphorochloridothioate, 2,4-dichlorophenol and NaOH: Smithey, Jr., **U.S.** pat. **3,004,054** (1960 to Virginia-Carolina Chemical).

Liquid, $bp_{0.1}$ 164-169°. n_D^{25} 1.5291. d_{20} 1.300. Slightly sol in water; miscible with most organic solvents. LD_{50} orally in rats: 270 mg/kg., *Toxic Substances List,* H. E. Christensen, Ed. (1972) p 180.

USE: Nematocide, insecticide. *Caution:* Cholinesterase inhibitor.

3031. Dichlofluanid. *1,1-Dichloro-N-[(dimethylamino)-sulfonyl]-1-fluoro-N-phenylmethanesulfenamide; N-[(di-chlorofluoromethyl)thio]-N',N'-dimethyl-N-phenylsulf-*

amide; Bayer 47531; KUE 13032c; Elvaron; Euparen(e). $C_9H_{11}Cl_2FN_2O_2S_2$; mol wt 333.21. C 32.44%, H 3.33%, Cl 21.28%, F 5.70%, N 8.41%, O 9.60%, S 19.24%. Prepn: E. Klauke *et al.*, **Belg.** pat. **609,868** corresp to **U.S.** pats. **3,-285,929, 3,341,403** (1962, 1966, 1967, all to Bayer). Activity: E. Kuhle *et al.*, *Angew. Chem.* **76**, 807 (1964); F. Grewe, *Phytiat.-Phytopharm.* **17**, 47 (1968). UV degradation: T. Clark, D. Watkins, *Pestic. Sci.* **9**, 225 (1978).

White powder, mp 105.0-105.6°. Vapor press at 20°: 1 × 10^{-6} mm Hg. Insol in water; sol in acetone, methanol, xylene. Light sensitive; decomp by strong alkaline media. Toxic to fish, harmless to bees. LD_{50} orally in mice: 1250 mg/kg, I. Petersone, D. Berzina, *C.A.* **85**, 117680t (1976).

USE: Fungicide.

3032. Dichlone. *2,3-Dichloro-1,4-naphthalenedione; 2,3-dichloro-1,4-naphthoquinone;* USR 604; Phygon; Phygon Paste; Phygon XL. $C_{10}H_4Cl_2O_2$; mol wt 227.06. C 52.90%, H 1.78%, Cl 31.23%, O 14.09%. Prepd by chlorination of 1,4-naphthoquinone: Bertheim, *Ber.* **34**, 1554 (1901); Brass, Köhler, *Ber.* **55**, 2554 (1922); Sjöstrand, **U.S.** pat. **2,975,196** (1961 to Svenska Oljeslageri Aktiebolaget); by oxidation of 2,3-dichloro-5,8-dihydro-1,4-naphthohydroquinone: Gaertner, **U.S.** pat. **2,750,427** (1956 to Monsanto); by oxidation of 2,3-dichloro-*p*-toluquinone: Grinev *et al.*, *Zh. Obshch. Khim.* **29**, 90 (1959). Toxicity data: G. W. Bailey, J. L. White, *Residue Rev.* **10**, 97 (1965). *Review:* Ter Horst, *Ind. Eng. Chem.* **35**, 1255 (1943).

Golden yellow needles or leaflets from alc, mp 193°. Sublimes. Practically insol in water (soly about 1 part in 10 million parts H_2O). Soly in xylene and *o*-dichlorobenzene about 4%. Moderately sol in acetone, ether, benzene, dioxane. LD_{50} orally in rats: 1300 mg/kg (Bailey, White).

Caution: Irritating to skin, mucous membranes; CNS depressant: *Clinical Toxicology of Commercial Products,* R. E. Gosselin *et al.*, Eds. (Williams & Wilkins, Baltimore, 5th ed., 1984) Section II, p 318.

USE: Fungicide for agriculture and textiles; herbicide.

3033. Dichloralphenazone. *1,2-Dihydro-1,5-dimethyl-2-phenyl-3H-pyrazol-3-one compd with 2,2,2-trichloro-1,1-ethanediol(1:2); Dichloralantipyrine.* Bihypnal; Bonadorm; Dormwell; Sedor; Sominat; Welldorm. $C_{15}H_{18}Cl_6N_2O_5$; mol wt 519.07. C 34.71%, H 3.49%, Cl 40.98%, N 5.40%, O 15.41%. $[CCl_3CH(OH)_2]_2 \cdot C_{11}H_{12}N_2O$. Prepn: Pfeiffer, Seidel, *Z. Physiol. Chem.* **178**, 97 (1928).

Small prismatic needles from water, mp 68°.

Note: See also Chloralantipyrine.

THERAP CAT: Sedative, hypnotic.

3034. Dichloramine T. *N,N-Dichloro-4-methylbenzene-sulfonamide; N,N-dichloro-p-toluenesulfonamide; p-toluene-*sulfonic acid dichloramide. $C_7H_7Cl_2NO_2S$; mol wt 240.11. C 35.01%, H 2.94%, Cl 29.55%, N 5.83%, O 13.33%, S 13.35%. Prepn from *p*-toluenesulfonic acid: Soper, *J. Chem. Soc.* **125**, 1899 (1924); van Andel, **U.S.** pat. **2,495,489** (1950 to Shell).

Prisms from chloroform + petr ether, mp 83°. Strong chlorine odor; dec on exposure to air with loss of chlorine. mp about 80°. Almost insol in water, dec by alcohol when warmed; one gram dissolves in about 1 ml benzene, 1 ml chloroform, 2.5 ml carbon tetrachloride; sol in eucalyptol, chlorinated paraffin hydrocarbons, glacial acetic acid; slightly sol in petr ether. *Keep well closed and protected from light.*
USE: Germicide.
THERAP CAT: Antibacterial.

3035. Dichlorisone. *9,11β-Dichloro-17,21-dihydroxypregna-1,4-diene-3,20-dione;* 9α,11β-dichloro-1,4-pregnadiene-17α,21-diol-3,20-dione; 9α,11β-dichloro analog of prednisolone; Diloderm; Disoderm. $C_{21}H_{26}Cl_2O_4$; mol wt 413.35. C 61.02%, H 6.34%, Cl 17.15%, O 15.48%. Prepn: Robinson *et al., J. Am. Chem. Soc.* **81**, 2191 (1959); Gould *et al.,* U.S. pat. **2,894,963** (1959 to Schering).

Crystals from acetone, dec 238-241°. $[\alpha]_D^{20}$ +134° (pyridine). uv max (methanol): 237 nm (ϵ 15,400).
21-Acetate, $C_{23}H_{28}Cl_2O_5$, *Astroderm.* Crystals from acetone, dec 246-253°. $[\alpha]_D^{25}$ +162° (dioxane). uv max (methanol): 237 nm (ϵ 15,000).
THERAP CAT: Topical antipruritic.

3036. Dichlorisoproterenol. *3,4-Dichloro-α-[[(1-methylethyl)amino]methyl]benzenemethanol; 3,4-dichloro-α-(isopropylaminomethyl)benzyl alcohol;* 1-(3,4-dichlorophenyl)-2-isopropylaminoethanol; β-hydroxy-N-isopropyl-3,4-dichlorophenethylamine; N-[β-(3,4-dichlorophenyl)-β-hydroxyethyl]isopropylamine; dichloroisoproterenol; DCI. $C_{11}H_{15}Cl_2NO$; mol wt 248.16. C 53.24%, H 6.09%, Cl 28.57%, N 5.65%, O 6.45%. Prepn: Mills, U.S. pat. **2,938,921** (1960 to Lilly). NMR studies: A. Balsamo *et al., Eur. J. Med. Chem.* **17**, 471 (1982). Adrenergic receptor affinity study: T. Solmajer *et al., J. Med. Chem.* **25**, 1413 (1982).

bp$_{0.01}$ 125-135°. LD$_{50}$ in mice (mg/kg): 165 orally; 39 i.v., R. Ferrini, *Arzneimittel-Forsch.* **18**, 48 (1968).
Hydrochloride, $C_{11}H_{15}Cl_2NO.HCl$, mp 153-154°.
USE: Exptl adrenergic agent.

3037. Dichloroacetic Acid. Bichloracetic acid; dichlorethanoic acid; DCA. $C_2H_2Cl_2O_2$; mol wt 128.95. C 18.63%, H 1.56%, Cl 54.99%, O 24.82%. $CHCl_2COOH$. Transformation of chloral to dichloroacetic acid: Wallach, *Ann.* **173**, 288 (1874); Frantzen, Fikentscher, *ibid.* **623**, 68 (1959); Rosenblum *et al., Chem. & Ind. (London)* **1960**, 718. Toxicity: Smyth *et al., Arch. Ind. Hyg. Occup. Med.* **4**, 119 (1911); P. W. Stacpoole *et al., N. Engl. J. Med.* **300**, 372 (1979). Mutagenicity studies: V. Herbert *et al., Am. J. Clin. Nutr.* **33**, 1179 (1980). Use in treatment of lactic acidosis: P. W. Stacpoole *et al., N. Engl. J. Med.* **309**, 390 (1983).
Liquid; pungent odor. d$_4^{20}$ 1.563. bp 193-194°. Apparently occurs in two cryst forms, mp 9.7° and −4°. n_D 1.4659. Miscible with water, alc, ether. LD$_{50}$ orally in rats: 2.82 g/kg (Smyth).
Ethyl ester, $C_4H_6Cl_2O_2$, liquid, bp 158.3-158.7°. d$_4^{20}$ 1.282. n$_D^{20}$ 1.4386. Slightly sol in water; miscible with alcohol, ether.

THERAP CAT: Caustic; keratolytic; topical astringent.

3038. 1,1-Dichloroacetone. *1,1-Dichloro-2-propanone;* α,α-dichloroacetone; uns-dichloroacetone; dichloromethyl methyl ketone. $C_3H_4Cl_2O$; mol wt 126.98. C 28.38%, H 3.18%, Cl 55.85%, O 12.60%. $CH_3COCHCl_2$. Prepn: Borsche, Fittig, *Ann.* **133**, 113 (1865).
Oily liq. d$_{15}^{18}$ 1.305. bp 120°. Slightly sol in water; sol in alcohol; miscible with ether. Boiling with Na_2CO_3 gives acrylic acid.

3039. 1,3-Dichloroacetone. *1,3-Dichloro-2-propanone;* α,γ-dichloroacetone; sym-dichloroacetone; bis(chloromethyl) ketone. $C_3H_4Cl_2O$; mol wt 126.98. C 28.38%, H 3.18%, Cl 55.85%, O 12.60%. $CH_2ClCOCH_2Cl$. Prepd by the oxidation of dichlorohydrin with sodium dichromate: Conant, Quayle, *Org. Syn.* **2**, 13 (1922).
Plates, needles on distillation. mp 45°. bp 173°. d$_4^{46}$ 1.3826. n$_D^{46}$ 1.47144. Sol in water, very sol in alc, ether.
Caution: Lacrimator, vesicant.

3040. 2,2-Dichloroacetyl Chloride. Dichloroethanoyl chloride. C_2HCl_3O; mol wt 147.40. C 16.30%, H 0.68%, Cl 72.17%, O 10.85%. $Cl_2CHCOCl$. Prepd from pentachloroethane: Ott, **Ger.** pat. **362,748** (1922 to Weiler ter Meer); from chloroform and carbon monoxide in presence of aluminum chloride at high pressure (21% yield): Frank *et al., Ind. Eng. Chem.* **41**, 2061 (1949).
Liquid. Fumes in air; acrid, penetrating odor. d$_4^{16}$ 1.5315. bp$_{760}$ 107-108°; bp$_{739}$ 106-107°. n$_D^{16}$ 1.4638. Dec upon contact with water, alcohol. Miscible with ether. LD$_{50}$ orally in rats: 2.46 g/kg, Smyth *et al., Arch. Ind. Hyg. Occup. Med.* **4**, 119 (1951).
Caution: Irritating to eyes, mucous membranes.

3041. 3,4-Dichloroaniline. *3,4-Dichlorobenzeneamine.* $C_6H_5Cl_2N$; mol wt 162.03. C 44.48%, H 3.11%, Cl 43.76%, N 8.65%. Prepd by chlorination of *p*-chloroaniline with $AlCl_3$—HCl: Suthers *et al., J. Org. Chem.* **27**, 447 (1962). Manuf by catalytic reduction of 3,4-dichloro-1-nitrobenzene: Dietzler, Keil, U.S. pat. **3,067,253** (1962 to Dow).

Crystals, mp 71-72°. bp 272°. Practically insol in water; very sol in alcohol, ether; slightly sol in benzene.

3042. Dichlorobenzalkonium Chloride. Dichloran; Tetrosan. A mixture of alkyl-3,4-dichlorodimethylbenzylammonium chlorides. Prepn of this type of compound: **Fr.** pat. **806,662**, *C.A.* **31**, 4992¹ (1937) and **Ger.** pat. **685,321**, *C.A.* **34**, 4078⁸ (1940) (1936, 1939 to I. G. Farben).

Very bitter crystals. Cationic. Soluble in water, alc.
Dodecyldimethyl-3,4-dichlorobenzylammonium chloride, $C_{21}H_{36}Cl_3N$, *Riseptin, Dynaltone, Dynium Chloride.* R = $C_{12}H_{25}$.
Note: The name dichloran also applies to the fungicide, 2,6-dichloro-4-nitrobenzenamine; spelled also dicloran.
USE: Antiseptic, germicide, algicide, sanitizer, deodorant.
Caution: Irritating to skin.

3043. *m*-Dichlorobenzene. *1,3-Dichlorobenzene.* $C_6H_4Cl_2$; mol wt 147.01. C 49.02%, H 2.74%, Cl 48.24%. Prepn by Sandmeyer procedure from the appropriate chloroaniline, and, along with *o*- and *p*-dichlorobenzenes, by chlorination of chlorobenzene: Engelsma *et al.*, *Rec. Trav. Chim.* **76**, 325 (1957). Separation of mixture contg *m*-, *o*-, and *p*-dichlorobenzenes by distillation and crystn: Mueller, Wolz, Fr. pat. **1,374,863** (1964 to Bayer), *C.A.* **62**, 4936e (1965), corresp to Brit. pat. **999,845**.

Liquid, bp 173°. mp −24.76°. d_4^{20} 1.2884, d_4^{25} 1.2828. n_D^{20} 1.5459. Practically insol in water; sol in alcohol, ether.

3044. *o*-Dichlorobenzene. *1,2-Dichlorobenzene;* ortho-dichlorobenzene. Empirical formula, prepn, and separation from *m*- and *p*-dichlorobenzenes, *see m*-isomer above. Manuf: Faith, Keyes & Clark's *Industrial Chemicals*, F. A. Lowenheim, M. K. Moran, Eds. (Wiley-Interscience, New York, 4th ed., 1975) pp 258-265.

Liquid, bp 180.5°. mp −17.03°. d_4^{20} 1.3059, d_4^{25} 1.3003. n_D^{20} 1.5515, n_D^{25} 1.5491. Flash pt, closed cup: 151°F (66°C). Practically insol in water; miscible with alc, ether, benzene.

USE: Solvent for waxes, gums, resins, tars, rubbers, oils, asphalts; insecticide for termites and locust borers; fumigant; removing sulfur from illuminating gas; as degreasing agent for metals, leather, wool; as ingredient of metal polishes; as heat transfer medium; as intermediate in the manuf of dyes. *Caution:* Can cause injury to liver, kidneys. High concns cause CNS depression.

3045. *p*-Dichlorobenzene. Paracide; PDB; Paradichlorobenzene; Para-zene; Di-chloricide; Paramoth. Empirical formula, prepn, and separation from *m*- and *o*-dichlorobenzene, *see m*-isomer above. Crystal structure of triclinic form (β-modification): Housty, Clastre, *Acta Cryst.* **10**, 695 (1957); of monoclinic form (α-modification) and its transformation to triclinic form: Panatoni *et al.*, *Gazz. Chim. Ital.* **93**, 813 (1963). Manuf: Faith, Keyes & Clark's *Industrial Chemicals*, F. A.Lowenheim, M. K. Moran, Eds. (Wiley-Interscience, New Yok, 4th ed., 1975) pp 258-265.

Volatile crystals with a characteristic penetrating odor. Sublimes at ordinary temps. mp 53.5° (α-modification), 54° (β-modification). bp 174.12°. n_D^{60} 1.5285. Flash pt (closed cup) 150°F. Practically insol in water; sol in alcohol, ether, benzene, chloroform, carbon disulfide. Non-corrosive; non-staining. LD_{50} orally in rats: 500 mg/kg, *Toxic Substances List*, H. E. Christensen, Ed. (1973) p 321.

USE: Insecticidal fumigant. Popular for domestic use against clothes moths. *Caution:* Vapors may cause irritation to skin, throat, and eyes. Prolonged exposure to high concns may show weakness, dizziness, loss of weight; liver injury may develop.

3046. 2,2′-Dichlorobenzidine. *2,2′-Dichloro[1,1′-biphenyl]-4,4′-diamine.* $C_{12}H_{10}Cl_2N_2$; mol wt 253.13. C 56.94%, H 3.98%, Cl 28.01%, N 11.07%. Prepn from *m*-chloronitrobenzene: Laubenheimer, *Ber.* **8**, 1625 (1875); Cain, May, *J. Chem. Soc.* **97**, 723 (1910).

Needles from water or prisms from alc, mp 165°. Almost insol in water; moderately sol in alc; readily sol in ether. Hydrochloride, $C_{12}H_{10}Cl_2N_2 \cdot 2HCl$, leaflets from water, moderately sol in water.

USE: Manuf azo dyes.

3047. 3,3′-Dichlorobenzidine. *3,3-Dichloro-(1,1′-biphenyl)-4,4′-diamine;* 3,3′-dichloro-4,4′-biphenyldiamine; DCB. $C_{12}H_{10}Cl_2N_2$; mol wt 253.13. C 56.94%, H 3.98%, Cl 28.01%, N 11.07%. Prepn from o-chloronitrobenzene: Cohn, *Ber.* **33**, 3552 (1900). Review and evaluation of studies of carcinogenicity in laboratory animals and in humans: *IARC Monographs* **4**, 49-55 (1974).

Needles from alc or benzene, mp 132-133°. Almost insol in water. Readily sol in alc, benzene, glacial acetic acid. Hydrochloride, $C_{12}H_{10}Cl_2N_2 \cdot 2HCl$, needles. Slightly sol in water, readily in alc.

Note: This substance may reasonably be anticipated to be a carcinogen: *Fourth Annual Report on Carcinogens* (NTP 85-002, 1985) p 79.

USE: Manuf azo dyes; as an intermediate for the Benzidine Yellow pigments.

3048. Dichlorobenzyl Alcohol. *2,4-Dichlorobenzene-methanol;* Dybenal. $C_7H_6Cl_2O$; mol wt 177.04. C 47.49%, H 3.42%, Cl 40.05%, O 9.04%. Prepn: Van de Lande, *Rec. Trav. Chim.* **51**, 98 (1932); Metayer, Dat-Xuong, *Bull. Soc. Chim. France* **1954**, 615.

Crystals, mp 59.5°.

THERAP CAT: Antiseptic.

3049. 1,1-Dichloro-2,2-bis(*p*-chlorophenyl)ethane. *1,-1′-(2,2-Dichloroethylidene)bis[4-chlorobenzene];* tetrachlorodiphenylethane; TDE; dichlorodiphenyldichloroethane; DDD; *p,p′*-DDD; *p,p′*-TDE; Rhothane. $C_{14}H_{10}Cl_4$; mol wt 320.05. C 52.54%, H 3.15%, Cl 44.31%. Non-degradable pesticide; a component of technical grade DDT. Prepd by condensing dichloroacetaldehyde with chlorobenzene: Haller *et al.*, *J. Am. Chem. Soc.* **67**, 1596, 1600 (1945).

Crystals, mp 109-110°. The chemical properties and solys are similar to those of DDT. LD_{50} orally in rats: > 4000 mg/kg, T. B. Gaines, *Toxicol. Appl. Pharmacol.* **14**, 515 (1969).

USE: Insecticide. *Caution:* Slightly irritating to skin. *Acute toxicity symptoms:* lethargy but no convulsions. Estimated fatal oral dose 5 g/kg body wt. *Chronic toxicity symptoms:* atrophy of adrenal cortex, liver damage. Symptoms similar to DDT, *q.v.*

3050. 1,1-Dichloro-2,2-bis(*p*-ethylphenyl)ethane. *1,1′-(2,2-Dichloroethylidene)bis[4-ethylbenzene];* 2,2-bis(*p*-ethylphenyl)-1,1-dichloroethane; di(*p*-ethylphenyl)dichloroethane; Perthane; Q-137. $C_{18}H_{20}Cl_2$; mol wt 307.27. C 70.36%, H 6.56%, Cl 23.08%. Prepn: McKeever, Némec, U.S. pat. **2,917,553** (1955 to Rohm & Haas).

Crystals from ethanol mp 56-57°. Practically insol in water; sol in acetone, kerosene, diesel fuel. Of moderate persistence in the soil. LD_{50} orally in rats: > 4000 mg/kg, T. B. Gaines, *Toxicol. Appl. Pharmacol.* **14**, 515 (1969).

USE: Insecticide. *Caution:* Symptoms similar to DDT.

3051. Dichloro(2-chlorovinyl)arsine. *(2-Chloroethenyl)-arsonous dichloride;* 2-chlorovinyldichloroarsine; chlorovinylarsine dichloride; Lewisite. $C_2H_2AsCl_3$; mol wt 207.32. C 11.59%, H 0.97%, As 36.13%, Cl 51.31%. ClCH=CH-$AsCl_2$. Obtained together with bis(2-chlorovinyl)chloroarsine and tris(2-chlorovinyl)arsine upon passing acetylene into a mixture of arsenic trichloride and aluminum chloride and fractionating the product in a current of HCl gas: Wieland, *Ann.* **431**, 38 (1923); Lewis, Perkins, *Ind. Eng. Chem.* **15**, 290 (1923); Lewis, Stiegler, *J. Am. Chem. Soc.* **47**, 2546 (1925); Gibson, Johnson, *J. Chem. Soc.* **1931**, 754; Jarman in *Advances in Chemistry Series* **23**, 328-337 (1959).

Liquid. Faint odor of geranium. *Vesicant. War gas. Respiratory and systemic poison.* Solidifies at −13°. mp 0.1°. $bp_{12.5}$ 76-77°; bp_{26} 93°; bp_{760} 190° (dec). d_4^{20} 1.888. Vapor pressure at 0°: 0.087 mm Hg; at 20°: 0.395 mm Hg. Sol in the ordinary organic solvents; insol in water, dil mineral acids. Hydrolyzed by alkalies. Neutralized and inactivated by bleaching powder, sodium hypochlorite.

Caution: Extremely toxic! Produces severe vesication, even through rubber. If left on skin, as little as 0.5 ml may give rise to sufficient absorption to produce severe systemic effects; 2 ml may cause death. *Antidote:* British Anti-Lewisite (BAL, dimercaptopropanol).

3052. 2,3-Dichloro-5,6-dicyanobenzoquinone. *4,5-Dichloro-3,6-dioxo-1,4-cyclohexadiene-1,2-dicarbonitrile;* DDQ. $C_8Cl_2N_2O_2$; mol wt 227.02. C 42.32%, Cl 31.23%, N 12.34%, O 14.10%. Prepn: Thiele, Günther, *Ann.* **349**, 45 (1906); Walker, Waugh, *J. Org. Chem.* **30**, 3240 (1965).

Yellow to orange crystals, mp 213.5-215°. Dec in water. Sol in benzene, dioxane, acetic acid. Slightly sol in chloroform, methylene chloride.
USE: Oxidizing agent, especially in steroid synthesis.

3053. Dichlorodifluoromethane. Difluorodichloromethane; Arcton 12; Freon 12; Frigen 12; Genetron 12; Halon; Isotron 2. CCl_2F_2; mol wt 120.92. C 9.93%, Cl 58.65%, F 31.43%. Prepn: Henne, *Organic Reactions* **2**, 49 (1944); Swarts, *Chem. Zentr.* **1907, II**, 581. Manuf: Faith, Keyes & Clark's *Industrial Chemicals,* F. A. Lowenheim, M. K. Moran, Eds. (Wiley-Interscience, New York, 4th ed., 1975) pp 325-330.

Colorless, practically odorless, noncorrosive, nonirritating, nonflammable gas. Faint, ether-like odor in high concentrations. Stable up to 550°. $d_{liq}^{-29.8}$ 1.486. mp −158°. bp_{760} −29.8°; bp_{2atm} −12.2°; bp_{5atm} +16.1°; bp_{10atm} +42.4°; bp_{20atm} +74.0°; bp_{30atm} +95.6°. Critical temp 111.5°; critical pressure 39.6 atm (582 lb/sq inch). Dipole moment 0.51. Insol in water. Sol in alcohol, ether. Absorbs less than 0.0025% water. Has little, if any, anesthetic or toxic action, but toxic substances may be formed on contact with a flame or hot metal surface.
Note: Consult latest Government regulations on use as aerosol propellant.
USE: Refrigerant, aerosol propellant.

3054. 1,3-Dichloro-5,5-dimethylhydantoin. *1,3-Dichloro-5,5-dimethyl-2,4-imidazolidinedione;* Dactin; Halane. $C_5H_6Cl_2N_2O_2$; mol wt 197.03. C 30.48%, H 3.07%, Cl 35.99%, N 14.22%, O 16.24%. Prepd by passing chlorine through an aq soln of 5,5-dimethylhydantoin: Biltz, Slotta, *J. Prakt. Chem.* **113**, 248 (1926); Orazi, Orio, *Annales Asoc. Quim. Argentina* **41**, 153 (1953), *C.A.* **48**, 13634 (1954). Structure: Bogaert-Verhoogen, Martin, *Bull. Soc. Chim. Belges* **58**, 567 (1949). Purification by dissolving in concd H_2SO_4 and diluting with ice-water: Lorenz, U.S. pat. **2,828,308** (1958 to Purex).

Four-sided, pointed prisms from chloroform, d_{20}^{20} 1.5. mp 132°. Sublimes at 100°. Turns brown and conflagrates at 212° (after melting at 132°). The dry crystals [combined

available chlorine 77.6% (theory)] may be stored without much loss of available chlorine. After 14 weeks at 60° the loss was 1.5% Cl compared with a loss of 37.5% suffered by 70% calcium hypochlorite. On contact with water and esp hot water hypochlorous acid is liberated. At pH 9 nitrogen chloride is formed. Soly in water at 25° 0.21%; at 60° 0.60%. pH of aq soln about 4.4. Freely sol in chlorinated and highly polar solvents at 25°: Chloroform 14%, methylene chloride 30.0%, carbon tetrachloride 12.5%, ethylene dichloride 32.0%, *sym*-tetrachlorethane 17.0%, benzene 9.2%.
USE: Chlorinating agent, disinfectant, industrial deodorant. In water treatment. Active ingredient of powder laundry bleaches such as Sage's Dry Bleach, Colgate's Pruf. Intermediate for amino acids, drugs, insecticides. Stabilizer for vinyl chloride polymers. Polymerization catalyst.

3055. *sym*-Dichloroethyl Ether. *1,1'-Oxybis[2-chloroethane];* bis(2-chloroethyl) ether; *β,β'*-dichloroethyl ether; DCEE; Chlorex. $C_4H_8Cl_2O$; mol wt 143.02. C 33.59%, H 5.64%, Cl 49.60%, O 11.19%. $[ClCH_2CH_2]_2O$. Prepn: O. Kamm, J. H. Waldo, *J. Am. Chem. Soc.* **43**, 2223 (1921). Toxicity: H. F. Smyth, Jr., C. P. Carpenter, *J. Ind. Hyg. Toxicol.* **30**, 66 (1948). Carcinogenicity study: *IARC Monographs* **9**, 117 (1975).

Colorless, clear liq; pungent odor. Vapor harmful. d_{20}^{20} 1.22. bp 178°. mp −50°. n_D^{20} 1.457. Flash pt, closed cup: 145°F (63°C), Buckman Laboratories, Inc. Literature; also reported as 131°F (55°C), *Ind. Eng. Chem.* **32**, 880 (1940). Insol in water; sol in most organic solvents. It dissolves oils, fats, greases, etc. LD_{50} orally in rats: 75 mg/kg (Smyth, Carpenter).
USE: Reagent for organic synthesis; solvent. Has been used as a scouring agent for textiles; as soil fumigant. *Toxicity:* Strongly irritating to skin, eyes, mucous membranes.

3056. 4',5'-Dichlorofluorescein. *4',5'-Dichloro-3',6'-dihydroxyspiro[isobenzofuran-1(3H),9'-[9H]xanthen]-3-one;* D & C Orange no. 8; 4,5-dichloro-3,6-fluorandiol. $C_{20}H_{10}Cl_2O_5$; mol wt 401.19. C 59.87%, H 2.51%, Cl 17.68%, O 19.94%. Prepd by condensation of 2-chlororesorcinol with phthalic anhydride: *Colour Index* vol. 4 (3rd ed., 1971) p 4424.

Orange powder, insol in water and in dil acids. Sol in alcohol and in dil alkali yielding solns which are orange with a greenish-yellow fluorescence. Slightly sol in glycerol and the glycols. Practically insol in oils, fats and waxes. If a soln of the dye in strong alkali is heated, a violet color is produced. *Caution:* Delisted by FDA for use in drugs and cosmetics.

Disodium salt, $C_{20}H_8Cl_2Na_2O_5$, *C.I. Solvent Orange 32, C.I.* **45365.**
Note: Structure depicted as the open tautomer.

3057. 2,6-Dichloroindophenol Sodium. *2,6-Dichloro-4-[(4-hydroxyphenyl)imino]-2,5-cyclohexadien-1-one sodium salt;* sodium 2,6-dichloroindophenol; 2,6-dichloro-*N*-(*p*-hydroxyphenyl)-*p*-benzoquinone imine sodium; sodium 2,6-dichloro-*N*-(*p*-hydroxyphenyl)-*p*-benzoquinone imine; 2,6-dichloro-1,4-benzoquinone-4-(4-hydroxyanil) sodium; 2,6-dichlorophenol-indophenol sodium; Tillman's reagent. $C_{12}H_6Cl_2NNaO_2$; mol wt 290.09. C 49.68%, H 2.08%, Cl 24.45%, N 4.83%, Na 7.93%, O 11.03%. Prepd from 2,6-dichloro-*p*-quinone-4-chlorimide and phenol in alkaline soln: Gibbs *et al.*, *Publ. Health Repts.* **40**, 650 (1925); Tillmans *et al.*, *Z. Unters. Lebensm.* **56**, 273 (1932).

Dark green powder. May contain up to 2 mols H_2O. Freely sol in water, alc. The aq soln is deep blue, changed to red by acids. It liberates iodine from KI in acid solns.

USE: Analytical reagent in the determination of ascorbic acid (vitamin C) which reduces the dye to a colorless hydroxy compd.

3058. *sym*-Dichloromethyl Ether. *Oxybis[chloromethane]; bis(chloromethyl) ether;* BCME. $C_2H_4Cl_2O$; mol wt 114.97. C 20.89%, H 3.51%, Cl 61.68%, O 13.92%. $(CH_2Cl)_2O$. Review and evaluation of studies of carcinogenicity in laboratory animals and in humans: *IARC Monographs* 4, 231-238 (1974).

Colorless liquid; suffocating odor. d_4^{20} 1.315. bp 106°. n_D^{20} 1.4346. Unstable in moist air; dec by water into HCl and formaldehyde. *Caution:* Strong irritant to eyes, respiratory tract.

Note: This substance has been listed as a known carcinogen: *Fourth Annual Report on Carcinogens* (NTP 85-002, 1985) p 46.

3059. Dichlorophen(e). *2,2'-Methylenebis[4-chlorophenol];* 2,2'-dihydroxy-5,5'-dichlorodiphenylmethane; 5,5'-dichloro-2,2'-dihydroxydiphenylmethane; bis[5-chloro-2-hydroxyphenyl]methane; di[5-chloro-2-hydroxyphenyl]-methane; Anthiphen; Dicestal; Didroxane; Di-phenthane-70; G-4; Hyosan; Parabis; Plath-Lyse; Preventol G-D; Teniathane; Teniatol; Wespuril. $C_{13}H_{10}Cl_2O_2$; mol wt 269.12. C 58.02%, H 3.75%, Cl 26.35%, O 11.89%. Prepn by adding a CH_2O yielding reagent, e.g. aq HCHO, to *p*-chlorophenol: Gump, Luthy, **U.S. pat. 2,334,408** (1944 to Burton T. Bush). Acute toxicity: T. B. Gaines, R. E. Linder, *Fundam. Appl. Toxicol.* **7**, 299 (1986).

Crystals from toluene, mp 177-178°. Practically insol in water. Sparingly sol in toluene. One gram dissolves in 1 g of 95% ethanol, in less than 1 g of ether. Also sol in methanol, isopropyl ether, petr ether. Sol (with decompn) in alkaline aq solns. LD_{50} in adult male, female rats (mg/kg): 1506, 1683 orally (Gaines, Linder).

USE: Agricultural fungicide; antimicrobial; germicide in soaps, shampoos, etc.

THERAP CAT: Anthelmintic (Cestodes).

THERAP CAT (VET): Anthelmintic, antiprotozoan, fungicide.

3060. Dichlorophenarsine Hydrochloride. *(3-Amino-4-hydroxyphenyl)arsonous dichloride monohydrochloride; (3-amino-4-hydroxyphenyl)dichloroarsine hydrochloride;* RP 2591; Arseclor; Chlorarsen; Chlorarsol; Clorarsen; Dichloro-Mapharsen; Filarsen; Fontarsol; Halarsol. $C_6H_7AsCl_3$-NO; mol wt 290.41. C 24.81%, H 2.43%, As 25.80%, Cl 36.63%, N 4.82%, O 5.51%. Prepd by the action of arsenic trichloride on *o*-aminophenol, or from the corresp arsine oxide and an excess of HCl: **Ger. pat. 281,101;** from arsphenamine by treatment with mercuric chloride and HCl gas: Binz, Bauer, *Z. Angew. Chem.* **34**, 261 (1921). Toxicity: Beck, *Proc. Soc. Exp. Biol. Med.* **78**, 392 (1951).

White hygroscopic powder. mp 200°. Readily sol in wa-

ter yielding a soln of oxophenarsine (Mapharsen). Usually marketed mixed with 3.33 times its weight of sodium citrate-carbonate as buffer to make a soln of pH 6.2 isotonic with human blood. LD_{50} i.p. in mice: 41 mg/kg (Beck)

THERAP CAT: Formerly as antisyphilitic.

THERAP CAT (VET): Has been used as a filaricide.

3061. 2,4-Dichlorophenol. $C_6H_4Cl_2O$; mol wt 163.01. C 44.21%, H 2.47%, Cl 43.50%, O 9.82%. Key intermediate in synthesis of the herbicide 2,4-D, *q.v.* Prepn by chlorination of phenol: M. Kohn, S. Sussmann, *Monatsh.* **46**, 575 (1926); B. O. Pray, D. N. Sukow, **U.S. pat. 2,759,981** (1956 to Columbia-Southern Chem. Corp.); by chlorination of *o*- or *p*-chlorophenol: L. G. Groves *et al., J. Chem. Soc.* **1929**, 512; from 2,4-dinitrophenol: A. Ghosh, *J. Indian Chem. Soc.* **28**, 155 (1951). ¹H NMR study: J. B. Rowbotham, T. Schaefer, *Can. J. Chem.* **52**, 3037 (1974).

Needle-like crystals, mp 45°. bp 209-211°. Volatile with steam. Sol in CCl_4.

3062. 2,6-Dichlorophenol. $C_6H_4Cl_2O$; mol wt 163.01. C 44.21%, H 2.47%, Cl 43.50%, O 9.82%. Synthesis from ethyl 3,5-dichloro-4-hydroxybenzoate: D. S. Tarbell *et al., Org. Syn. coll. vol. III*, 267. Crystal structure: C. Bavoux, P. Michel, *Acta Crystallogr.* **30B**, 2043 (1974). Isoln and identification as sex pheromone of lone star tick: R. S. Berger, *Science* **177**, 704 (1972).

White crystals from petr ether, mp 64.5-65.5°.

3063. 1,3-Dichloro-2-propanol. α-Dichlorohydrin; *sym*-glycerol dichlorohydrin; glycerol α,γ-dichlorohydrin; *sym*-dichloroisopropyl alcohol. $C_3H_6Cl_2O$; mol wt 128.99. C 27.93%, H 4.69%, Cl 54.98%, O 12.40%. $CH_2ClCHOH$-CH_2Cl. Prepared from glycerol, acetic acid and HCl gas: Conant, Quayle, *Org. Syn.* **2**, 29 (1922); **coll. vol. I** (2nd ed.) p 292.

Liquid. Ethereal odor. d_4^{17} 1.3506. mp −4°. bp_{760} 174.3°; bp_{100} 114.8°; bp_{40} 93°; bp_{20} 78°; bp_5 52°; $bp_{1.0}$ 28°. n_D^{17} 1.480245. Sol in 10 parts water; miscible with alcohol, ether. LD_{50} orally in rats: 110 mg/kg, Smyth *et al., Am. Ind. Hyg. Assoc. J.* **23**, 95 (1962).

USE: Solvent for hard resins and nitrocellulose; manuf photographic and Zapon lacquer; cement for celluloid; binder for water colors.

3064. 1,3-Dichloropropene. α,γ-Dichloropropylene; 1,3-dichloropropylene; γ-chloroallyl chloride; Telone II. $C_3H_4Cl_2$; mol wt 110.98. C 32.47%, H 3.63%, Cl 63.90%. $ClCH_2CH=CHCl$. Prepd from 1,2,3-trichloropropane with NaOH: Friedel, Silva, *Compt. Rend.* **75**, 81 (1872); Reboul, *Ann. Chim.* [3] **60**, 37 (1860); from 3-chloro-2-propen-1-ol with PCl_3: Kirrmann *et al., Bull. Soc. Chim. France* [5] **1**, 860 (1934); from acrolein with PCl_5: van Romburgh, *ibid.* [2] **36**, 549 (1881); from 3,3-dichloropropene by isomerization with conc HCl: *idem, ibid.* The usual industrial prepn is from 1,3-dichloro-2-propanol by dehydration with $POCl_3$ or with P_2O_5 in benzene: Hurd, Webb, *J. Am. Chem. Soc.* **58**, 2191 (1936). Prepn of stereoisomers: L. F. Hatch, A. C. Moore, *ibid.* **66**, 285 (1944). Configuration: H. A. Smith, W. H. King, *ibid.* **70**, 3528 (1948); L. F. Hatch, R. H. Perry, *ibid.* **71**, 3262 (1949). Persistence in soil: I. J. Thomason, M. V. McKenry, *Hilgardia* **42**, 393 (1974). GC determn of isomers in rat blood: P. E. Kastl, E. A. Hermann, *J. Chro-*

matog. **265,** 277 (1983). Toxicity study: T. R. Torkelson, F. Oyen, *Am. Ind. Hyg. Assoc. J.* **38,** 217 (1977).

Liquid with chloroform odor. d^{25} 1.220. bp 108°. n_D^{22} 1.4735. Technical product is mixture of *cis-* and *trans-*isomers. LD_{50} (92% technical product) in male, female rats (mg/kg): 713, 470 orally (10% soln in corn oil); in rabbits (mg/kg): 504 dermally (Torkelson, Oyen). Component of *D-D, Vidden D,* and *Telone* which also contain 1,2-dichloropropane.

Cis-Form, bp 104.3°. d_4^{20} 1.224. n_4^{20} 1.4682.
Trans-Form, bp 112.0°. d_4^{20} 1.217. n_4^{20} 1.4730.

Caution: Irritating to skin, eyes, and mucous membranes; liver and kidney injury have been produced in exptl animals, *c.f. Clinical Toxicology of Commercial Products,* R. E. Gosselin *et al.,* Eds. (Williams & Wilkins, Baltimore, 5th Ed., 1984) Section III, p 141.

USE: Soil fumigant. Nematocide.

3065. Dichlororiboflavin. *7,8-Dichloro-7,8-didemethyl-riboflavin; 7,8-dichloro-10-(D-ribo-2,3,4,5-tetrahydroxypentyl)isoalloxazine; 7,8-dichloro-10-ribitylisoalloxazine; 6,7-dichloro-9-ribitylisoalloxazine; dichloroflavin.* $C_{15}H_{14}Cl_2-N_4O_6$; mol wt 417.21. C 43.18%, H 3.38%, Cl 17.00%, N 13.43%, O 23.01%. Riboflavin antagonist prepd by hydrogenation of 1,2-dichloro-4-nitro-5-D-ribitylaminobenzene, followed by condensation with alloxan: Kuhn *et al., Ber.* **76,** 1044 (1943); Fujita *et al., Nippon Kagaku Zasshi* **77,** 1344 (1956), *C.A.* **53,** 5270d (1959).

Yellow needles from alcohol, mp 273-275°. Sparingly sol in water.

3066. Dichloroxylenol. *2,4-Dichloro-3,5-dimethylphenol; 2,4-dichloro-3,5-xylenol;* DCMX. $C_8H_8Cl_2O$; mol wt 191.06. C 50.29%, H 4.22%, Cl 37.11%, O 8.37%. Prepn: Lesser, Gad, *Ber.* **56,** 963 (1923); Jones, *J. Chem. Soc.* **1941,** 267; Gemmell, *Mfg. Chemist* **23,** 63 (1952).

Crystals from petr ether, mp 95-96°. Sublimes. Volatile with steam. Soly in water: one part in 5000. Soly at 15° in 100 parts of solvent: benzene, 14; toluene, 15; acetone, 73; diethyl ketone, 59; petr ether, 4; chloroform, 25; CCl$_4$, 10.

USE: Bacteriostat in soaps; mold inhibitor, preservative.

3067. Dichlorphenamide. *4,5-Dichloro-1,3-benzenedisulfonamide; 1,3-disulfamyl-4,5-dichlorobenzene; 4,5-dichloro-1,3-disulfamoylbenzene; Antidrasi; Daranide; Oratrol.* $C_6H_6Cl_2N_2O_4S_2$; mol wt 305.16. C 23.61%, H 1.98%, Cl 23.24%, N 9.18%, O 20.97%, S 21.02%. Prepn: Schultz, U.S. pat. 2,835,702 (1958 to Merck & Co.).

Needles from DMSO + water, mp 239-241° (patent, mp 228.5-229°). Practically insol in water. Sol in alkaline solns.

THERAP CAT: Carbonic anhydrase inhibitor.

3068. Dichlorprop. *2-(2,4-Dichlorophenoxy)propanoic acid; 2-(2,4-dichlorophenoxy)propionic acid;* 2,4-DP; dichloroprop; Cornox RK; Hedonal DP. $C_9H_8Cl_2O_3$; mol wt 235.05. C 45.99%, H 3.43%, Cl 30.16%, O 20.42%. Selective pre- and post-emergence herbicide. First prepd and described as a plant growth regulator: M. E. Synerholm, P. W. Zimmerman, *Contrib. Boyce Thompson Inst.* **14,** 91 (1945). Commercial prepn: H. A. Stevenson, R. F. Brookes, **Brit.** pat. 822,199 (1959 to Boots). Exists in two optically active forms; only the (+)-isomer shows biological activity. Commercial products contain approx equal amounts of the two isomers.

Colorless crystals, mp 117-118°. Soly in water at 20°: 350 ppm. Readily sol in organic solvents. Corrosive to metals in the presence of water. Commercial prepns may contain the isooctyl or butyl esters or aq solns of the potassium salt. LD_{50} orally in rats and mice: 800, 400 mg/kg, *RTECS Vol. II,* E. J. Fairchild, Ed. (1977) p 770.

USE: Herbicide.

3069. Dichlorvos. *Phosphoric acid 2,2-dichloroethenyl dimethyl ester; phosphoric acid 2,2-dichlorovinyl dimethyl ester; O,O-dimethyl O-(2,2-dichlorovinyl) phosphate;* dichlorophos; dichlorovos; DDVP; SD 1750; Astrobot; Atgard; Canogard; Dedevap; Dichlorman; Divipan; Equigard; Equigel; Estrosol; Herkol; Nogos; Nuvan; Task; Vapona; Verdisol. $C_4H_7Cl_2O_4P$; mol wt 220.98. C 21.74%, H 3.19%, Cl 32.09%, O 28.96%, P 14.01%. Prepn: Whetstone, Harman, U.S. pat. 2,956,073 (1960 to Shell). Toxicity: T. B. Gaines, *Toxicol. Appl. Pharmacol.* **14,** 515 (1969).

Liquid. Practically non-flammable. d_4^{25} 1.415. bp$_{20}$ 140°; bp$_{1.0}$ 84°; bp$_{0.5}$ 72°; bp$_{0.01}$ 30°. n_D^{25} 1.451. Vapor pressure at 20°: 1.2×10^{-2} mm Hg. Misc with alcohol and most nonpolar solvents. Soly in water: about 1 g/100 ml; in glycerol: about 0.5 g/100 ml. LD_{50} orally in male, female rats: 80, 56 mg/kg (Gaines).

Caution: Cholinesterase inhibitor.

USE: Insecticide.

THERAP CAT (VET): Anthelmintic, insecticide.

3070. Diclobutrazol. *(R*,R*)-(±)-β-[(2,4-Dichlorophenyl)methyl]-α-(1,1-dimethylethyl)-1H-1,2,4-triazole-1-ethanol;* 1-(2,4-dichlorophenyl)-4,4-dimethyl-2-(1,2,4-triazol-1-yl)pentan-3-ol; 1-t-butyl-2-(1,2,4-triazol-1-yl)-2-(2',4'-dichlorobenzyl)ethanol; dichlobutrazol; PP 296; Vigil. $C_{15}H_{19}Cl_2N_3O$; mol wt 328.24. C 54.89%, H 5.83%, Cl 21.60%, N 12.80%, O 4.87%. Prepn: **Belg.** pat. 857,836; S. Balasubramanyan, M. C. Shephard, U.S. pat. 4,243,405 (1978, 1981 both to ICI). Activity, physical properties, and toxicity: K. J. Bent, A. M. Skidmore, *Proc. Brit. Crop Prot. Conf. - Pests Dis.* **1979,** 477. Fungicidal activity: B. Ram *et al., Test Agrochem. Cultiv.* **3,** 38 (1982). Regulation of plant growth: J. B. Shanks, *Proc. Plant Growth Reg. Soc. Am.* **9,** 68 (1982). Mechanism of action: P. Gadher *et al., Pest. Biochem. Physiolog.* **19,** 1 (1983); T. E. Wiggins, B. C. Baldwin, *Pestic. Sci.* **15,** 206 (1984). Resolution of enantiomers by GC: R. S. Burden *et al., J. Chromatog.* **391,** 273 (1987).

Odorless, colorless, crystalline solid, mp 147-149°. Vapor pressure at 20°: 1×10^{-8} to 2×10^{-8} mm Hg. Soly in water at 20°: 9 mg/l. Sol up to 50 g/l in methanol, ethanol, acetone, and chloroform. LD_{50} in rats (mg/kg): about 4000 orally; >1000 dermally; in mallard ducks: >9000 orally (Bent, Skidmore).

USE: Agricultural fungicide.

3071. Diclofenac Sodium. *2-[(2,6-Dichlorophenyl)amino]benzeneacetic acid monosodium salt; [o-(2,6-dichloroanilino)phenyl]acetic acid sodium salt;* sodium [o-[(2,6-dichlorophenyl)amino]phenyl]acetate; GP 45840; Allvoran; Assaren; Benfofen; Delphimix; Dichronic; Diclobenin; Diclo-Phlogont; Diclo-Puren; Diclord; Dicloreum; Dolobasan; Duravolten; Ecofenac; Effekton; Evinopon; Kriplex; Neriodin; Novapirina; Prophenatin; Rhumalgan; Tsudohmin; Urigon; Valetan; Voldal; Voltaren; Voltarol; Voveran; Xenid. $C_{14}H_{10}Cl_2NNaO_2$; mol wt 318.13. C 52.85%, H 3.17%, Cl 22.29%, N 4.40%, Na 7.23%, O 10.06%. Prepn: **Neth. pat. Appl. 6,604,752;** A. Sallmann, R. Pfister, **U.S. pat. 3,558,690** (1966, 1971 both to Geigy). Pharmacology: Renaud, Lecompte, *Thromb. Diath. Haemorrh.* **24**, 577 (1970), *C.A.* **74**, 86215m (1971); Krupp *et al.*, *Experientia* **29**, 450 (1973). Symposium on pharmacology and clinical experience: *Semin. Arthritis Rheum.* **15**, Suppl. 1, 57-110 (1985); on pharmacology, efficacy and safety: *Am. J. Med.* **80**, Suppl. 4B, 1-87 (1986).

Crystals from water, mp 283-285°. LD_{50} in mice, rats (mg/kg): ~390, 150 orally (Krupp).

Free acid, $C_{14}H_{11}ClNO_2$, crystals from ether-petr ether, mp 156-158°.

THERAP CAT: Anti-inflammatory.

3072. Diclofop-Methyl. *2-[4-(2,4-Dichlorophenoxy)phenoxy]propanoic acid methyl ester;* methyl 2-[4-(2,4-dichlorophenoxy)phenoxy]propionate; HOE 23408; Hoelon; Hoegrass; Il(1)oxan. $C_{16}H_{14}Cl_2O_4$; mol wt 341.19. C 56.33%, H 4.13%, Cl 20.78%, O 18.76%. Selective post-emergence herbicide. Prepn: W. Becker *et al.*, **Ger. pat. 2,223,894** corresp to **U.S. pat. 3,954,442** (1973, 1976 to Hoechst). Soil degradn: A. E. Smith, *J. Agr. Food Chem.* **25**, 893 (1977); **27**, 1145 (1979).

mp 39-41°. $bp_{0.1}$ 175-177°. Soly in water: 0.3 mg/100 ml. Soly (g/100 ml): acetone 249; ethanol 11; xylene 253.

USE: Herbicide.

3073. Dicloxacillin. *6-[[[3-(2,6-Dichlorophenyl)-5-methyl-4-isoxazolyl]carbonyl]amino]-3,3-dimethyl-7-oxo-4-thia-1-azabicyclo[3.2.0]heptane-2-carboxylic acid;* 6-[3-(2,6-dichlorophenyl)-5-methyl-4-isoxazolylcarboxamido]penicillanic acid; 3-(2,6-dichlorophenyl)-5-methyl-4-isoxazolylpenicillin; BRL-1702; Maclicine. $C_{19}H_{17}Cl_2N_3O_5S$; mol wt 470.33. C 48.52%, H 3.64%, Cl 15.07%, N 8.93%, O 17.01%, S 6.82%. Semi-synthetic antibiotic related to penicillin. Prepn: Naylor, **Brit. pat. 978,299** corresp to **U.S. pat. 3,239,507** (1965, 1969 to Beecham). Toxicity data: C. Gloxhuber *et al.*, *Arzneimittel-Forsch.* **15**, 322 (1965). Series

of articles on chemistry, pharmacology and toxicology: *ibid.* 322-348; Matsuzaki, *Japan J. Antibiot.* **21**, 274, 284 (1968), *C.A.* **70**, 95267z, 95265x (1969).

Sodium salt monohydrate, $C_{19}H_{16}Cl_2N_3NaO_5S.H_2O$, *sodium dicloxacillin monohydrate, P-1011, Brispen, Constaphyl, Dichlor-Stapenor, Diclocil, Dycill, Dynapen, Noxaben, Pathocil, Pen-Sint, Stampen, Syntarpen, Veracillin.* Crystals, dec 222-225°. $[\alpha]_D^{20}$ 127.2° (water). Sol in water, methanol; less sol in butanol; slightly sol in acetone and the usual organic solvents. LD_{50} in mice (g/kg): 0.9 i.v.; in rats (g/kg): 0.63 i.p.; >5 orally (Gloxhuber).

THERAP CAT: Antibacterial.

THERAP CAT (VET): Antibacterial.

3074. Dicobalt Octacarbonyl. *Di-μ-carbonylhexacarbonyldicobalt;* octacarbonyldicobalt; cobalt tetracarbonyl; cobalt octacarbonyl. $C_8Co_2O_8$; mol wt 341.94. C 28.10%, Co 34.47%, O 37.43%. Prepn: Mond *et al.*, *J. Chem. Soc.* **97**, 798 (1910); Gilmont, Blanchard, *Inorg. Syn.* **2**, 238 (1946); Wender *et al.*, *ibid.* **5**, 190 (1957); Chini *et al.*, *Chim. Ind. (Milan)* **55**, 120 (1973). Crystal structure: Sumner *et al.*, *Acta Cryst.* **17**, 732 (1964). Two isomeric forms exist in soln: Noack, *Spectrochim. Acta* **19**, 1925 (1963); Bor, *ibid.* 2065; Noack, *Helv. Chim. Acta* **47**, 1064, 1555 (1964). Toxicity studies: Kincaid *et al.*, *Arch. Ind. Hyg. Occup. Med.* **10**, 210 (1954); Brief *et al.*, *Arch. Environ. Health* **23**, 373 (1971); Kalekin, *C.A.* **78**, 67751z (1973); Spiridonova, Shabalina, *Gig. Sanit.* **1973**, 73, *C.A.* **78**, 119835b (1973). Reviews of prepn, properties and chemistry of cobalt carbonyls: I. Wender *et al.*, *U.S. Bur. Mines, Bull.* **600**, 83 pp (1960); *Organic Syntheses via Metal Carbonyls* vol. 1, I. Wender, P. Pino, Eds. (Interscience, New York, 1968) *passim;* Chalk, Harrod, "Catalysis by Cobalt Carbonyls" in *Advances in Organometallic Chemistry* **6**, 119-170 (1968); Nicholls in *Comprehensive Inorganic Chemistry* vol. 3, J. C. Bailar, Jr. *et al.*, Eds. (Pergamon Press, Oxford, 1973) pp 1059-1064.

Orange platelets obtained by vacuum sublimation. d 1.87. mp 51°; dec above 52°. Stable in an atm of H_2 and CO; dec on exposure to air. Insol in water. Sol in organic solvents such as ether, alcohol, CS_2, naphtha. Slowly attacked by HCl, H_2SO_4, more rapidly by HNO_3, Br_2. LD_{50} in mice, rats (mg/kg): 377.7, 753.8 by gavage (Spiridonova, Shabalina).

Cobalt carbonyl hydride, C_4HCoO_4, *cobalt hydrocarbonyl, tetracarbonylhydridocobalt, tetracarbonylhydrocobalt.* Unstable catalytic intermediate; active species in some reactions catalyzed by $Co_2(CO)_8$. Prepn: Gilmont, Blanchard, *loc. cit.*, Wender *et al.*, *loc. cit.* See also reviews above. Yellow, toxic, foul-smelling liquid. mp -26°. Dec at room temp to $Co_2(CO)_8$. Sparingly sol in water (3×10^{-3} moles/l); behaves as a strong acid. Readily oxidized in air.

USE: Catalyst for hydroformylation, hydrogenation, hydrosilation, isomerization, carboxylation, carbonylation and polymerization reactions.

3075. Dicofol. *4-Chloro-α-(4-chlorophenyl)-α-(trichloromethyl)benzenemethanol; 4,4'-dichloro-α-(trichloromethyl)benzhydrol;* 1,1-bis(p-chlorophenyl)-2,2,2-trichloroethanol; di(p-chlorophenyl)trichloromethylcarbinol; DTMC; ENT 23648; FW 293; Acarin; Kelthane; Mitigan. $C_{14}H_9Cl_5O$; mol wt 370.47. C 45.39%, H 2.45%, Cl 47.84%, O 4.32%. Prepn: Wilson, Craig; Wilson, Wolffe, **U.S. pats. 2,812,280; 2,812,362** (both 1957 to Rohm & Haas); Bergmann, Kaluszyner, *J. Org. Chem.* **23**, 1306 (1958); **Brit.** pat. **831,421** (1960 to Metal & Thermit Corp.). Miticidal activity: J. S. Baker, F. B. Maughan, *J. Econ. Entomol.* **49**, 458 (1956). Toxicology: R. B. Smith *et al.*, *Toxicol. Appl. Pharmacol.* **1**, 119 (1959). Metabolism: J. R. Brown *et al.*, *ibid.* **15**, 30 (1969); K. Tabata *et al.*, *Appl. Entomol. Zool.* **14**, 490 (1979).

Crystals from petr ether, mp 77-78°. uv max (ethanol): 226, 258, 266, 276 nm (4.43, 2.82, 2.85, 2.60). Insol in water; sol in most aliphatic and aromatic solvents. LD_{50} in rats: 1495 mg/kg orally; 1150 mg/kg i.p., J. R. Brown *et al.*, *loc. cit.*

USE: Acaricide.

3076. Dicolinium Iodide. *2-[[2-(Diethylmethylammonio)ethoxy]carbonyl]-1,1,6-trimethylpiperidinium diiodide; 2-carboxy-1,1,6-trimethylpiperidinium iodide diethyl(2-hydroxyethyl)methylammonium iodide ester;* 1,6-dimethylpipecolic acid diethylaminoethyl ester diiodomethylate; dicoline; Dikolin. $C_{16}H_{34}I_2N_2O_2$; mol wt 540.29. C 35.57%, H 6.34%, I 46.98%, N 5.19%, O 5.92%. Prepn: Rubstov *et al.*, *Zh. Obshch. Khim.* **26**, 130 (1956). Pharmacology: T. E. Vyshinskaya, *Sov. Med.* **30**(4), 123 (1967). Ganglion blocking agent.

Crystals, mp 201-202°. Sol in water, alcohol.

3077. Dicrotophos. *Phosphoric acid 3-(dimethylamino)-1-methyl-3-oxo-1-propenyl dimethyl ester; phosphoric acid dimethyl ester, ester with cis-3-hydroxy-N,N-dimethylcrotonamide;* 3-(dimethoxyphosphinyloxy)-*N,N*-dimethyl-*cis*-crotonamide; dimethyl 2-dimethylcarbamoyl-1-methylvinyl phosphate; dimethyl 1-dimethylcarbamoyl-1-propen-2-yl phosphate; C 709; ENT 24482; SD 3562; Bidrin; Carbicron; Ektafos. $C_8H_{16}NO_5P$; mol wt 237.21. C 40.51%, H 6.80%, N 5.91%, O 33.73%, P 13.06%. Compound included in general patent claim, prepn of corresp diethyl ester described: Whetstone, Stiles, U.S. pat. **2,802,855** (1957 to Shell). Activity: R. A. Corey, *J. Econ. Entomol.* **58**, 112 (1965). Metabolism and degradn: R. I. Beynon *et al.*, *Residue Rev.* **47**, 55 (1973).

Liquid, commercial grade is brown. bp_{760} 400°. d_{15}^{15} 1.216. n_D^{23} 1.468. Misc with water, ethanol, xylene; somewhat sol in kerosene. Dec at 90° after 7 days; at 75° after 31 days. LD_{50} in female, male rats: 16, 21 mg/kg orally; 42, 43 mg/kg dermally, T. B. Gaines, *Toxicol. Appl. Pharmacol.* **14**, 515 (1969).

USE: Insecticide. *Caution:* Highly toxic. A cholinesterase inhibitor.

3078. Dicryl. *N-(3,4-Dichlorophenyl)-2-methyl-2-propenamide; 3',4'-dichloro-2-methylacrylanilide; N-(3,4-dichlorophenyl)methacrylamide;* Niagara 4556. $C_{10}H_9Cl_2NO$; mol wt 230.10. C 52.20%, H 3.94%, Cl 30.82%, N 6.09%, O 6.95%. Post-emergence herbicide. Prepn from 3,4-dichloroaniline with methacryloyl chloride in the presence of tri-

ethylamine. Prepn described but not claimed by Thompson, U.S. pat. **3,169,850** (1965).

Crystals from ethanol + petr ether, mp 128°. Practically insol in water. Soluble in acetone, alcohol, isophorone, dimethyl sulfoxide. LD_{50} orally in rats: 3160 mg/kg, G. W. Bailey, J. L. White, *Residue Rev.* **10**, 97 (1965).

USE: Herbicide.

3079. Dictamnine. *4-Methoxyfuro[2,3-b]quinoline;* dictamine. $C_{12}H_9NO_2$; mol wt 199.20. C 72.35%, H 4.55%, N 7.03%, O 16.06%. Alkaloid from root of *Dictamnus albus* Linn., *Skimmia repens* Nakai, *Aegle marmelos* Correa, *Zanthoxylum alatum* Roxb., *Rutaceae:* Thoms, *Ber.* **33**, 68 (1923); Asahina *et al.*, *ibid.* **63**, 2045 (1930); Chatterjee, Roy, *J. Indian Chem. Soc.* **36**, 267 (1959); Deb *et al.*, *ibid.* **39**, 493 (1962). Structure: M. F. Brundon, N. J. McCorkindale, *Chem. & Ind. (London)* **1956**, 1091. Synthesis: Tuppy, Bohm, *Angew. Chem.* **68**, 388 (1956); M. Sato *et al.*, *Chem. Pharm. Bull. (Tokyo)* **35**, 1319 (1987).

Prisms from alcohol, ethyl acetate, benzene + ethyl acetate, mp 133°. Sol in hot alcohol, chloroform; slightly sol in ether. Practically insol in water.

Hydrochloride, $C_{12}H_{10}ClNO_2$, needles from alcohol, dec 170°.

3080. Dicumarol. *3,3'-Methylenebis[4-hydroxy-2H-1-benzopyran-2-one]; 3,3'-methylenebis[4-hydroxycoumarin];* bishydroxycoumarin (rescinded); Dicoumarol; Dicoumarin; Dicumol; Dufalone; Melitoxin. $C_{19}H_{12}O_6$; mol wt 336.29. C 67.86%, H 3.60%, O 28.55%. Originally isolated from spoiled sweet clover (improperly cured Melilotus hay): Link *et al.*, *J. Biol. Chem.* **138**, 21, 513, 529 (1941); **142**, 941 (1942); now prepd synthetically by the action of formaldehyde on 4-hydroxycoumarin (obtained as the Na derivative by the action of sodium metal on methyl acetylsalicylate): Link, *Fed. Proc.* **4**, 176 (1945), and patents (The Wisconsin Alumni Res. Found.). Toxicity data: Rose *et al.*, *Proc. Soc. Exp. Biol. Med.* **50**, 228 (1942).

Minute crystals. Slight pleasant odor. Has a slightly bitter taste. mp 287-293°. Sol in aq alkaline solns, in pyridine and similar organic bases. Slightly sol in benzene and chloroform. Practically insol in water, alcohol, ether. LD_{50} orally in rats: 541.6 mg/kg (Rose).

THERAP CAT: Anticoagulant.

3081. Dicyanine. *1-Ethyl-2-[3-(1-ethyl-2-methyl-4-(1H)-quinolinylidene)-1-propenyl]-4-methylquinolinium iodide;* 1-ethyl-2-[3-(1-ethyl-4(1H)-quinaldylidene)propenyl]lepidinium iodide; 1-ethyl-2-[3-(1-ethyl-2-methyl-4-(1H)-quinolylidene)propenyl]-4-methylquinolinium iodide; 2',4-dimethyl-1,1'-diethyl-2,4'-carbocyanine iodide. $C_{27}H_{29}IN_2$; mol wt 508.44. C 63.78%, H 5.75%, I 24.96%, N 5.51%. Prepn by J. A. Aeschlimann as reported by Mills, Odams, *J. Chem. Soc.* **125**, 1913 (1924). See also *Beilstein,* 2nd suppl. **vol. 23**, p 284.

Olive-green crystals with metallic sheen from methanol. Crystals are usually solvated. Drying at 110° expels methanol of crystn. Decomp 244-252°. Absorption max (methanol): 603.5, 655.5 nm ($A_{1cm}^{1\%}$ 63, 218). Soly in methanol: about 2%. Methanol solns appear deep blue when viewed under incident daylight. Soly in water about 0.2%. Aq solns are dichroic and may appear brownish green. Also sol in abs ethanol (0.5%), in ethylene glycol (1.5%), and in Cellosolve (2.0%). Practically insol in benzene, xylene.
USE: In color photography (sensitizer for extreme reds). Has photoconductive properties that increase sharply with a rise in temp and is classed as an O_2-photoconductor: Meier, *Z. Phys. Chem. (Leipzig)* 208, 325 (1958). Gives a color reaction with magnesium ions: Babenko, *Zhur. Anal. Khim.* 13, 496 (1958). Diagnostic stain for cytodiagnosis of ruptured fetal membranes.

3082. Dicyanodiamide. *Cyanoguanidine.* $C_2H_4N_4$; mol wt 84.08. C 28.57%, H 4.80%, N 66.64%. Prepared by controlled polymerization of cyanamide in water in the presence of ammonia, alkaline earth hydroxides, or other suitable bases: Grube, Nitsche, *Z. Angew. Chem.* 27, 368 (1914); Pinck, Hetherington, *Ind. Eng. Chem.* 27, 834 (1935); Pinck, *Inorg. Syn.* 3, 43 (1950).

$$NH$$
$$\|$$
$$NH_2CNHCN$$

Monoclinic prismatic crystals from water or alcohol, mp 209.5°. d_4^{25} 1.400. Eutectic with cyanamide at 35.6° (15% dicyanodiamide). Specific heat 0.456 at 0-204°. Neutral reaction. Soly in water at 13°: 2.26%, more sol in hot water. Solns above 80° dec slowly, yielding ammonia. Soly in abs ethanol at 13°: 1.26%, in ether: 0.01%. Insol in benzene, chloroform. Sol in liq ammonia.
USE: In the plastics industry (manuf of melamine). In the pharmaceutical industry (barbiturates, guanidine derivs).

3083. Dicyanodiamidine Sulfate. (*Aminoiminomethyl)-urea sulfate (2:1)*; biuretamidine sulfate; carbamylguanidine sulfate; guanylurea sulfate. $C_4H_{14}N_8O_6S$; mol wt 204.20. C 15.89%, H 4.67%, N 37.07%, O 31.76%, S 10.61%. Prepd by heating dicyanodiamide with H_2SO_4 and H_2O.

$$\left[\begin{array}{c} NH \\ \| \\ NH_2CNHCONH_2 \end{array} \right]_2 \cdot H_2SO_4$$

Dihydrate, white needles. At 110° loses its water. Sol in about 20 parts cold, 3 parts boiling water; slightly sol in alc.
USE: For detecting and determining Ni and its separation from Co and other metals.

3084. 9-Dicyanomethylene-2,4,7-trinitrofluorene. (*2,4,7-trinitro-9H-fluoren-9-ylidene)propanedinitrile*; DTF. $C_{16}H_5N_5O_6$; mol wt 363.26. C 52.90%, H 1.39%, N 19.28%, O 26.43%. Prepn from 2,4,7-trinitrofluorenone: Mukherjee, Levasseur, *J. Org. Chem.* 30, 644 (1965).

Yellow crystals from acetonitrile, mp 266-268°. uv max (dichloromethane): 365 nm (log ϵ 4.38).
USE: Forms charge-transfer complexes with aromatic hydrocarbons and amines.

3085. Dicyclohexylamine. *N-Cyclohexylcyclohexanamine*; dodecahydrodiphenylamine. $C_{12}H_{23}N$; mol wt 181.31. C 79.49%, H 12.79%, N 7.73%. Prepn by hydrogenation of equimolar amounts of cyclohexanone and cyclohexylamine: C. F. Winans, H. Adkins, *J. Org. Chem.* 54, 306 (1932); R. M. Robinson, W. C. Braaten, U.S. pat. 3,154,580 (1964 to Abbott). Forms crystalline derivatives with N-protected amino acids: E. Klieger *et al.*, *Ann.* 640, 157 (1961). *Review:* T. S. Carswell, H. L. Morrill, *Ind. Eng. Chem.* 29, 1265 (1937).

Liquid. Faint fishy odor. d_{25}^{25} 0.9104. mp $-0.1°$. pKa 10.4. bp_{760} 255.8°; bp_{300} 214.5°; bp_{200} 199.0°; bp_{100} 174.4°; bp_{50} 154.3°; bp_{25} 135.4°; bp_{11} 121°; bp_4 99.3°; $bp_{1.0}$ 83°. n_D^{25} 1.4823. Flash pt = 110°. Strong base. Sparingly sol in water; sol in the usual organic solvents. Miscible with cyclohexylamine. Readily forms adducts with solvents.
USE: Industrial solvent; corrosion inhibitor. *Caution:* A skin irritant and possible sensitizer.

3086. Dicyclohexylcarbodiimide. *N,N'-Methanetetrayl-biscyclohexanamine*; carbodicyclohexylimide; DCC; DCCI. $C_{13}H_{22}N_2$; mol wt 206.32. C 75.67%, H 10.75%, N 13.58%. Coupling agent in peptide synthesis: Sheehan, Hess, *J. Am. Chem. Soc.* 77, 1067 (1955); Khorana, *Chem. & Ind. (London)* 1955, 1087. Prepn: Schmidt *et al.*, *Ber.* 71, 1933 (1938); Schmidt, Schnegg, U.S. pat. 2,656,383 (1953 to Bayer); Stevens *et al.*, *J. Org. Chem.* 32, 2895 (1967); Bestmann *et al.*, *Ann.* 718, 24 (1968).

Cryst mass, mp 35-36°. bp_{11} 154-156°; $bp_{0.5}$ 98-100°. *Caution:* Contact allergen.
USE: In the synthesis of peptides.

3087. Dicyclomine Hydrochloride. *[1,1'-Bicyclohexyl]-1-carboxylic acid 2-(diethylamino)ethyl ester hydrochloride*; diethylaminocarbethoxybicyclohexyl hydrochloride; β-diethylaminoethyl-1-cyclohexylhexahydrobenzoate hydrochloride; β-diethylaminoethyl 1-cyclohexylcyclohexanecarboxylate hydrochloride; bis(cyclohexyl)carboxylic acid diethylaminoethyl ester hydrochloride; dicycloverin hydrochloride; Atumin; Benacol; Bentomine; Bentyl Hydrochloride; Diocyl Hydrochloride; Di-Syntramine; Dyspas; Mamiesan; Merbentyl; Procyclomin; Wyovin Hydrochloride. $C_{19}H_{36}ClNO_2$; mol wt 345.94. C 65.96%, H 10.49%, Cl 10.25%, N 4.05%, O 9.25%. Prepd from 1-cyclohexylcyclohexyl cyanide by alcoholysis and esterification: Tilford *et al.*, *J. Am. Chem. Soc.* 69, 2903 (1947).

Crystals from butanone, mp 164-166°. Solubility in water about 25%.
THERAP CAT: Anticholinergic.

3088. Didecyldimethylammonium Chloride. *N-Decyl-N,N-dimethyl-1-decanaminium chloride*; dimethyldidecylammonium chloride; Arquad 10; Bardac 2250/2280; BTC 1010; Dodigen 1881; Querton 210CL. $C_{22}H_{48}ClN$; mol wt 362.08. C 72.98%, H 13.36%, Cl 9.79%, N 3.87%. Quaternary alkyl ammonium compound with fungicidal and bactericidal activity. Prepn: A. W. Ralston *et al.*, *J. Org. Chem.* **13**, 186 (1948). Conductivity of aqueous soln: *eidem, J. Am. Chem. Soc.* **70**, 977 (1948). Mothproofing agent: A. M. Schwartz, C. A. Rader, U.S. pat. **3,746,767** (1971 to USDA). In carpet dyeing: F. F. Bartsch, R. Feigin, U.S. pat. **3,758,269** (1971 to Sybron). Germicidal activity: R. J. Wright, Ger. pat. **2,810,998**; *eidem,* U.S. pat. **4,272,395** (1978, 1981 both to Unilever). Ion-pair extracting agent in carbapenem isoln: J. D. Hood *et al.*, *J. Antibiot.* **32**, 295 (1979). In froth-flotation recovery of metals: V. Petrovich, U.S. pat. **4,225,428** (1980). Eliminates ice-nucleating effect of bacteria in plants: E. A. Youngman, R. C. Schnell, Eur. pat. Appl. **37,593**; *eidem,* U.S. pat. **4,311,517** (1981, 1982 both to Shell). Breaks oil-in-water emulsions created during oil-spill recovery procedures: J. Newcombe, U.S. pat. **4,374,734** (1983 to Cities Service). Fungidical wood preservative: A. F. Preston: *J. Am. Oil Chem. Soc.* **60**, 567 (1983).

$$\left[\begin{array}{c} CH_3(CH_2)_9 \quad CH_3 \\ N^+ \\ CH_3(CH_2)_9 \quad CH_3 \end{array} \right]^+ \quad Cl^-$$

Extremely hygroscopic. Sol in acetone, extremely sol in benzene. Insol in hexane.

Caution: Causes eye damage and skin irritation. Harmful if swallowed.

USE: General purpose disinfectant, sanitizer; mildew preventative in commercial laundries; water treatment in cooling towers and oil field flood waters; wood preservative.

3089. Didemnins. Biologically active depsipeptides extracted from a Caribbean tunicate (sea squirt) family *Didemnidae,* genus *Trididemnum.* The three components, didemnins A, B and C are weakly basic compds with antiviral, antitumor activity. Didemnin A is the most abundant, whereas didemnin B is generally the most active. Extraction and purification: K. L. Rinehart, Jr., Eur. pat. Appl. **48,149** and U.S. pat. **4,493,796** (1982, 1985 both to Univ. Illinois). Structure determn by NMR, MS: K. L. Rinehart, Jr. *et al.*, *J. Am. Chem. Soc.* **103**, 1857 (1981); revised structure and total synthesis: U. Schmidt *et al.*, *Tetrahedron Letters* **29**, 3507 (1988); *eidem, ibid.* 4407. Crystal structure of B: M. B. Hossain *et al.*, *Proc. Nat. Acad. Sci. USA* **85**, 4118 (1988). Efficient total synthesis of A and B: Y. Hamada *et al.*, *J. Am. Chem. Soc.* **111**, 669 (1989). *In vitro* antiviral activity: K. L. Rinehart, Jr. *et al.*, *Science* **212**, 933 (1981); *in vivo* activity of A and B: P. G. Canonico *et al.*, *Antimicrob. Ag.*

Didemnin A: R = H

Didemnin B: R = CH₃CHOHCO—N———CO—

Didemnin C: R = CH₃CHOHCO—

Chemother. **22**, 696 (1982); S. D. Weed, D. A. Stringfellow, *Antiviral Res.* **3**, 269 (1983); K. L. Rinehart, Jr. *et al.*, *Fed. Proc.* **42**, 87 (1983). *In vitro* antitumor activity of didemnin B: T. L. Jiang *et al.*, *Cancer Chemother. Pharmacol.* **11**, 1 (1983). *In vitro* cytotoxicity: S. L. Crampton *et al.*, *Cancer Res.* **44**, 1796 (1984). Inhibition of protein synthesis: L. H. Li *et al.*, *Cancer Letters* **23**, 279 (1984). *In vitro* and *in vivo* immunosuppressive activity: D. W. Montgomery, C. F. Zukowski, *Transplantation* **40**, 49 (1985). HPLC determn in biological fluids: J. N. Hartshorn *et al.*, *J. Liq. Chromatog.* **9**, 1489 (1986). Review of activity and toxicology: H. G. Chun *et al.*, *Invest. New Drugs* **4**, 279 (1986).

Didemnin A, C₄₉H₇₈N₆O₁₂. Greenish-white solid. Sol in methanol, ethanol, dioxane, chloroform. Insol in water.

Didemnin B, C₅₇H₈₉N₇O₁₅, NSC 325519. Yellow-white amorphous solid. Sol in methanol, ethanol, dioxane, ethyl acetate, chloroform. Insol in water.

Didemnin C, C₅₂H₈₁N₆O₁₄, oil. Sol in methanol, ethanol, isopropanol, dioxane, ethyl acetate and chloroform. Sparingly sol in toluene. Insol in water.

3090. Dideoxyadenosine. *2′,3′-Dideoxyadenosine;* 6-amino-9-(2′,3′-dideoxy-β-D-*glycero*-pentofuranosyl)purine; DDA; ddAdo. C₁₀H₁₃N₅O₂; mol wt 235.25. C 51.06%, H 5.57%, N 29.77%, O 13.60%. Purine nucleoside analog with antiviral (retrovirus) activity. Prepn: M. J. Robins, R. K. Robins, *J. Am. Chem. Soc.* **86**, 3585 (1964); J. R. McCarthy *et al.*, *J. Am. Chem. Soc.* **88**, 1549 (1966). *In vitro* inhibition of bacterial DNA synthesis: L. Toji, S. S. Cohen, *Proc. Nat. Acad. Sci. USA* **63**, 871 (1969). Antibacterial activity in mice: G. Beskid *et al.*, *Antimicrob. Ag. Chemother.* **19**, 424 (1981). Inhibition of *in vitro* infectivity of HIV-1 (HTLV-III/LAV) virus: H. Mitsuya, S. Broder, *Proc. Nat. Acad. Sci. USA* **83**, 1911 (1986). HPLC determn in biological fluids: P. A. Blau *et al.*, *J. Chromatog.* **420**, 1 (1987). Metabolism by cloned human T-cells: D. A. Cooney *et al.*, *Biochem. Pharmacol.* **36**, 1765 (1987).

Crystals from ethanol, mp 184-186°. [α]$_D^{25}$ −25.2° (c = 1.01 in water). uv max in methanol: 259.5 nm (ε 14800).

THERAP CAT: Antiviral.

3091. Dideoxycytidine. *2′,3′-Dideoxycytidine;* ddCyd. C₉H₁₃N₃O₃; mol wt 211.22. C 51.18%, H 6.20%, N 19.89%, O 22.72%. Pyrimidine nucleoside analog with antiviral activity. Prepn: J. P. Horwitz *et al.*, *J. Org. Chem.* **32**, 817 (1967); R. Marumoto, M. Honjo, *Chem. Pharm. Bull.* **22**, 128 (1974). Prepn and structure-activity study: T.-S. Lin *et al.*, *J. Med. Chem.* **30**, 440 (1987). *In vitro* inhibition of viral DNA synthesis: P. C. van der Vliet, M. M. Kwant, *Biochemistry* **20**, 2628 (1981); of HIV-1 (HTLV-III/LAV) virus replication: H. Mitsuya, S. Broder, *Proc. Nat. Acad. Sci. USA* **83**, 1911 (1986). Phosphorylation by human cells: D. A. Cooney *et al.*, *Biochem. Pharmacol.* **35**, 2065 (1986). Metabolism by human cells: M. C. Starnes, Y. Cheng, *J. Biol. Chem.* **262**, 988 (1987). Potential application in acquired immunodeficiency syndrome (AIDS): E. De Clercq, *J. Med. Chem.* **29**, 1561 (1986).

Crystals from ethanol + benzene, mp 215-217° (Horwitz); also reported as mp 209-210° (Lin). $[\alpha]_D^{25}$ +81° (c = 0.635 in water). uv max in 0.1N HCl: 280 nm (ϵ 17720); in 0.1N NaOH: 270 nm (ϵ 8410).

THERAP CAT: Antiviral.

3092. Dideoxyinosine. *2',3'-Dideoxyinosine;* DDI; ddIno; NSC 612049. $C_{10}H_{12}N_4O_3$; mol wt 236.23. C 50.84%, H 5.12%, N 23.72%, O 20.32%. Hypoxanthine nucleoside with antiviral activity; metabolic product of dideoxyadenosine, *q.v.* Enzymatic prepn from dideoxyadenosine: W. Plunkett, S. S. Cohen, *Cancer Res.* **35**, 1547 (1975). Synthesis: G. W. Koszalka, T. A. Krenitsky, *Eur.* pat. *Appl.* **206,497** (1986 to Wellcome Found.); R. R. Webb *et al.*, *Nucleosides Nucleotides* **7**, 147 (1988). *In vitro* inhibition of HIV-I: H. Mitsuya, S. Broder, *Proc. Natl. Acad. Sci. USA* **83**, 1911 (1986); of HIV-II: D. D. Richman, *Antimicrob. Ag. Chemother.* **31**, 1879 (1987). Antiretroviral spectrum *in vitro:* J. E. Dahlberg *et al.*, *Proc. Natl. Acad. Sci. USA* **84**, 2469 (1987). Cellular pharmacology in cloned human T-cells: G. Ahluwalia *et al.*, *Biochem. Pharmacol.* **36**, 3797 (1987). HPLC determn in human plasma: G. Ray, E. Murrill, *Anal. Letters* **20**, 1815 (1987).

White solid, mp 160-163°. uv max 248 nm (pH 2); 254 nm (pH 12).

THERAP CAT: Antiviral.

3093. Dieldrin. *(1aα,2β,2aα,3β,6β,6aα,7β,7aα)-3,4,5,6,- 9,9-Hexachloro-1a,2,2a,3,6,6a,7,7a-octahydro-2,7:3,6-dimethanonaphth[2,3-b]oxirene; 1,2,3,4,10,10-hexachloro-6,7-epoxy-1,4,4a,5,6,7,8,8a-octahydro-endo,exo-1,4:5,8-dimethanonaphthalene;* compound 497; insecticide no. 497; HEOD; ENT 16225; Octalox. $C_{12}H_8Cl_6O$; mol wt 380.93. C 37.84%, H 2.12%, Cl 55.85%, O 4.20%. Activity: C. W. Kearns *et al., J. Econ. Entomol.* **42**, 127 (1949). Prepn from aldrin, *q.v.*: Soloway, U.S. pat. **2,676,131** (1954 to Shell); Payne, Smith, U.S pat. **2,776,301** (1957 to Shell). Synthesis: Korte, Rechmeier, *Ann.* **656**, 131 (1962). Metabolism: M. K. Baldwin *et al., Chem. & Ind. (London)* **1970**, 595; C. T. Bedford, D. H. Hutson, *ibid.* **1976**, 440. Toxicity data: Gaines, *Toxicol. Appl. Pharmacol.* **14**, 515 (1969). A stereoisomer of endrin, *q.v.*

Crystals, mp 176-177°. Vapor press at 20°: 3.1×10^{-6} mm Hg. Practically insol in water. Moderately sol in common organic solvents except aliphatic petr solvents and methyl alcohol. Stable in org and inorg alkalies and acids commonly used in agriculture. Affected by strong mineral acids. LD_{50} orally in rats: 46 mg/kg (Gaines).

Human Toxicity: Readily absorbed through skin. Causes malaise, headache, nausea, vomiting, dizziness, tremors, clonic and tonic convulsions, coma, respiratory failure, *cf. Clinical Toxicology of Commercial Products,* R. E. Gosselin *et al.,* Eds. (Williams & Wilkins, Baltimore, 4th ed., 1976) Section III, pp 121-124.

USE: Formerly as insecticide; manuf and use has been discontinued in the U.S.

3094. Dienestrol. *4,4'-(1,2-Diethylidene-1,2-ethanediyl)bisphenol; 4,4'-(diethylideneethylene)diphenol;* 3,4-bis(p-hydroxyphenyl)-2,4-hexadiene; 4,4'-dihydroxy-γ,δ-diphenyl-β,δ-hexadiene; di(p-oxyphenyl)-2,4-hexadiene; dienoestrol; estrodienol; Cycladiene; Dienol; Dinovex; DV; Estroral; Gynefollin; Hormofemin; Oestrasid; Oestrodiene; Oestroral; Restrol; Retalon; Synestrol. $C_{18}H_{18}O_2$; mol wt 266.32. C 81.17%, H 6.81%, O 12.01%. Synthesis: Dodds *et al., Proc. Roy. Soc.* **127B**, 162 (1939); Hobday, Short, *J. Chem. Soc.* **1943**, 609; Short, Hobday, U.S. pat. **2,464,203** (1949 to Boots); Adler, U.S. pat. **2,465,505** (1949 to Hoffmann-La Roche). Configuration: Koch, *Nature* **161**, 309 (1948); Lane, Spialter, *J. Am. Chem. Soc.* **73**, 4408 (1951).

Minute needles from dil alcohol, mp 227-228°. Sublimes at 130° and 1 mm Hg. The sublimate mp 231-234°. Freely sol in alcohol, methanol, ether, acetone, propylene glycol; sol in chloroform, aq solns of alkali hydroxides; sol in vegetable oils after warming, but crystallizes out on standing. Practically insol in water, dil acids.

Diacetate, $C_{22}H_{22}O_4$, *Farmacyrol, Lipamone, Retalon-Oral.* Prisms from alc, mp 119-120°.

Note: Isodienestrol, mp 179°, has no estrogenic activity.

THERAP CAT: Estrogen.

THERAP CAT (VET): Estrogenic hormone therapy.

3095. Dienochlor. *1,1',2,2',3,3',4,4',5,5'-Decachlorobi-2,4-cyclopentadien-1-yl;* bis(pentachloro-2,4-cylopentadien-1-yl); decachlor; HRS-16; Pentac. $C_{10}Cl_{10}$; mol wt 474.64. C 25.31%, Cl 74.69%. Acaricide to control mites on ornamental plants. Prepn from hexachlorocyclopentadiene, characterization: E. T. McBee *et al., J. Am. Chem. Soc.* **77**, 4375 (1955). Prepn and rubber vulcanizing props: E. C. Ladd, U.S. pat. **2,732,409** (1956 to U.S. Rubber). Prepn and insecticidal props: J. T. Rucker, U.S. pat. **2,934,470** (1960 to Hooker). Effectiveness as miticide: W. W. Allen *et al., J. Econ. Entomol.* **57**, 187 (1964). Crystal structure: G. Smith *et al., J. Chem. Soc. Perkin Trans. II* **1976**, 796. Photodegradation: G. B. Quistad, K. M. Mulholland, *J. Agr. Food Chem.* **31**, 621 (1983). Effect on oogenesis of Mexican bean beetle: P. R. Hughes, M. A. Penton, *J. Econ. Entomol.* **76**, 1156 (1983).

Yellow prisms from petr ether, mp 121.5-122°. uv max: 330 nm (ϵ 2950). Stable to alkali.

USE: Miticide.

3096. Diethadione. *5,5-Diethyldihydro-2H-1,3-oxazine-2,4(3H)-dione;* 5,5-diethyl-2,3,5,6-tetrahydro-4H-1,3-oxazine-2,4-dione; Dioxone; Ledosten; Persisten; Tocèn. C_8-

$H_{13}NO_3$; mol wt 171.19. C 56.12%, H 7.65%, N 8.18%, O 28.04%. Prepn: Testa *et al., J. Org. Chem.* **24**, 1928 (1959); Brit. pat. **855,244** (1960 to Lepetit, SpA).

C₂H₅ structure (6-membered ring with two C=O, NH)

Crystals from ether, mp 97-98°. LD_{50} in mice, rats: 52, 31.6 mg/kg i.p.; 130, 70.5 mg/kg orally, Maffii *et al., J. Pharm. Pharmacol.* **13**, 244 (1961).

THERAP CAT: Analeptic.

3097. Diethanolamine. *2,2'-Iminobisethanol; 2,2'-iminodiethanol;* diethylolamine; bis(hydroxyethyl)amine; 2,2'-dihydroxydiethylamine. $C_4H_{11}NO_2$; mol wt 105.14. C 45.69%, H 10.55%, N 13.32%, O 30.44%. $(HOCH_2CH_2)_2$-NH. Produced along with mono- and triethanolamine by ammonolysis of ethylene oxide. *See refs under Ethanolamine.*

Deliquescent prisms, mp 28°. Usually offered as a viscous liquid. Mild ammoniacal odor. d_4^{30} 1.0881; d_4^{60} 1.0693. One U.S. gallon weighs 9.09 lbs at 30°. Viscosity at 30°: 351.9 cp; at 60°: 53.85 cp. bp_{760} 268.8°; bp_{150} 217°; $bp_{0.01}$ 20°. Strong base: pH of $0.1N$ aq soln: 11.0. n_D^{30} 1.4753. Dipole moment 2.81. Flash pt 300°F. Miscible with water, methanol, acetone. Soly at 25° in benzene: 4.2%, in ether: 0.8%, in carbon tetrachloride: <0.1%, in *n*-heptane: <0.1%. LD_{50} orally in rats: 12.76 g/kg, H. F. Smyth *et al., J. Ind. Hyg. Toxicol.* **23**, 259 (1941).

USE: To scrub gases as indicated under ethanolamine. Diethanolamine can be used with cracking gases and coal or oil gases which contain carbonyl sulfide that would react with monoethanolamine. As rubber chemicals intermediate. In the manuf of surface active agents used in textile specialties, herbicides, petr demulsifiers. As emulsifier and dispersing agent in various agricultural chemicals, cosmetics, and pharmaceuticals. In the production of lubricants for the textile industry. As humectant and softening agent. In organic syntheses.

3098. Diethazine. *N,N-Diethyl-10H-phenothiazine-10-ethanamine;* 10-(2-diethylaminoethyl)phenothiazine; N-(diethylaminoethyl)thiodiphenylamine; N-(2'-diethylaminoethyl)dibenzoparathiazine; RP 2987; Deparkin; Dinezin; Dolisina; Eazaminum; Ethylemin; Parkazin. $C_{18}H_{22}N_2S$; mol wt 298.44. C 72.44%, H 7.43%, N 9.39%, S 10.74%. Prepd by reacting 10-phenothiazineethyl chloride with diethylamine in presence of copper powder or by reacting diethylaminoethyl chloride with phenothiazine: Charpentier, *Compt. Rend.* **225**, 306 (1947); Huttrer, *Enzymologia* **12**, 293 (1948); Charpentier, U.S. pat. **2,530,451** (1950 to Rhône-Poulenc); Berg, Ashley, U.S. pat. **2,607,773** (1952 to Rhône-Poulenc).

CH₂CH₂N(C₂H₅)₂ phenothiazine structure

Oily liquid. $bp_{4.5}$ 195-208°; $bp_{0.4-0.5}$ 167-175°. Hydrochloride, $C_{18}H_{23}ClN_2S$, *Antipar, Aparkazin, Casantin, Diparcol, Latibon, Thiantan, Thiontan.* Crystals, mp 184-186°. Burning taste, producing a temporary numbness of the tongue. One part dissolves in about 5 parts water, 6 parts ethanol, 5 parts chloroform. Practically insoluble in ether. pH of 10% aq soln 5.0-5.3. LD_{50} orally in mice: 450 mg/kg, Bovet *et al., Therapie* **2**, 115 (1947).

THERAP CAT: Anticholinergic, antiparkinsonian.

3099. Diethylacetic Acid. *2-Ethylbutanoic acid.* C_6H_{12}-O_2; mol wt 116.16. C 62.04%, H 10.41%, O 27.55%. $(CH_3$-$CH_2)_2CHCOOH$.
Colorless liq; odor somewhat resembling that of caproic acid. d_4^{18} 0.920. bp 194-195°, also stated as 190°. mp about

−15°. n_D^{10} 1.4179. Slightly sol in water; freely sol in alcohol or ether.

3100. Diethylamine. *N-Ethylethanamine.* $C_4H_{11}N$; mol wt 73.14. C 65.68%, H 15.16%, N 19.15%. $(C_2H_5)_2NH$. Prepn from ethyl iodide and NH_3 with separation of mono-, di-, and triethylamines formed: Watt, Otto, *J. Am. Chem. Soc.* **69**, 836 (1947). Manuf from ethanol and NH_3, obtained along with mono- and triethylamines: Davies *et al.,* U.S. pat. **2,609,394** (1952 to ICI); Lemon, Myerly, U.S. pat. **3,022,349** (1962 to Union Carbide). Toxicity data: Smyth *et al., Arch. Ind. Hyg. Occup. Med.* **4**, 119 (1951).

Flammable, strongly alkaline liq, bp 55.5°. mp −50°. d_4^{20} 0.7074. n_D^{20} 1.3864. Flash pt <20°F. Forms a hydrate, $B_2.H_2O$, mp −19°. Miscible with water, alc. It is usually supplied as a soln. *Keep well closed.* LD_{50} orally in rats: 540 mg/kg (Smyth).

Hydrochloride, $C_4H_{12}ClN$, crystals from alcohol + ether, mp 226°. Hygroscopic. bp 320-330°. d_4^{21} 1.048. Sol in water, alcohol, chloroform. Practically insol in ether.

Caution: May be irritating to skin, mucous membranes: Patty's *Industrial Hygiene and Toxicology* Vol. **2B**, G. D. Clayton, F. E. Clayton, Eds. (Wiley-Interscience, New York, 3rd ed., 1981) p 3149.

USE: In the rubber and petroleum industry; in flotation agents; in resins; dyes, pharmaceuticals.

3101. 2-Diethylaminoethanol. *β-Diethylaminoethyl alcohol;* 2-hydroxytriethylamine. $C_6H_{15}NO$; mol wt 117.19. C 61.49%, H 12.90%, N 11.95%, O 13.65%. $(C_2H_5)_2NCH_2$-CH_2OH. Prepd by the action of ethylene chlorohydrin on diethylamine: Ladenburg, *Ber.* **14**, 1878 (1881); Soderman, Johnson, *J. Am. Chem. Soc.* **47**, 1394 (1925); Hartman, *Org. Syn. coll. vol.* **II**, 183 (1943); by the action of ethylene oxide on diethylamine: Horne, Shriner, *J. Am. Chem. Soc.* **54**, 2928 (1932); Headlee *et al., ibid.* **55**, 1066 (1933).

Liquid. d^{25} 0.8800. bp_{760} 163°; bp_{80} 100°; bp_{10} 55°. n_D^{25} 1.4389. Sol in water, alcohol, ether, benzene.

p-Nitrophenylurethan, crystals, mp 59-60°.

3102. Diethylaniline. *N,N-Diethylbenzenamine.* C_{10}-$H_{15}N$; mol wt 149.24. C 80.48%, H 10.13%, N 9.39%. Prepd by ethylation of aniline: Whitman, U.S. pat. **2,501,-556** (1950 to du Pont); Voltz, *J. Org. Chem.* **22**, 48 (1957); Closson *et al., ibid.* **22**, 646 (1957); Horyna, Cerny, *Coll. Czech. Chem. Commun.* **21**, 906 (1956); from bromobenzene, sodium amide + diethylamine: Bunnett, Brotherton, *J. Org. Chem.* **22**, 832 (1957).

N(C₂H₅)₂ benzene ring structure

Colorless to yellow liq. d_4^{25} 0.9302. bp 215-216°; bp_3 62-66°. mp −38°. d^{24} 1.5394. Volatile with steam. uv max (isooctane): 303, 259 nm ($\epsilon \times 10^{-3}$ 2.37, 16.7). Slightly sol in alcohol, chloroform, ether. One gram dissolves in 70 ml water at 12°.

USE: As dyestuff intermediate, in organic syntheses.

3103. *N,N*-Diethylbenzhydrylamine. *N,N-Diethyl-α-phenylbenzenemethanamine;* diethylaminodiphenylmethane. $C_{17}H_{21}N$; mol wt 239.35. C 85.30%, H 8.84%, N 5.85%. Prepd by addn of an excess of phenylmagnesium bromide to *N,N*-diethylformamide and treating the reaction mixture with ammonium chloride: Maxim, Mavrodineanu, *Bull. Soc. Chim. France* [5] **3**, 1084 (1936).

C₆H₅—CHN(C₂H₅)₂—C₆H₅ structure

Viscous liquid, bp_{17} 170°. Solidifies on deep cooling, then mp 56° (Maxim, Mavrodineanu); mp 58-59° [Sommelet, *Compt. Rend.* **175**, 1149 (1922)]; mp 61° [Titov, *C.A.* **43**, 4217 (1949)].

USE: In the detection of nitrates. Proposed as an antihistaminic: Capraro, *C.A.* **41**, 6989 (1947).

3104. Diethylberyllium. Beryllium diethyl. $C_4H_{10}Be$; mol wt 67.13. C 71.56%, H 15.02%, Be 13.43%. $Be(C_2H_5)_2$. Prepd from $BeCl_2$ and a Grignard reagent: Gilman, Schulze, *J. Chem. Soc.* **1927**, 2663; Goubeau, Rodewald, *Z. Anorg. Chem.* **258**, 162 (1949).

Liquid; bp_3 63°; dec above 85°. Spontaneously flammable in air; burns with luminous flame evolving dense white fumes of BeO. Reacts violently with water, HCl, alcohols, producing ethane. Sol in ether, benzene.

3105. Diethylbromoacetamide. *2-Bromo-2-ethylbutanamide; 2-bromo-2,2-diethylacetamide;* carbromide. C_6H_{12}BrNO; mol wt 194.08. C 37.13%, H 6.23%, Br 41.18%, N 7.22%, O 8.24%. $(C_2H_5)_2CBrCONH_2$. Prepn: *Ger.* pat. **158,220** (1905 to Kalle).

Crystalline powder; camphor odor; bitter, cooling taste. mp 67°. Sol in 115 parts cold water; sol in alc, benzene, ether, oils. Freely sol in warm water.

Note: Formerly marketed as *Neuronal.*

Caution: This substance may be habit forming and is listed in the U.S. Code of Federal Regulations, Title 21 Part 329.1 (1987).

THERAP CAT: Sedative, hypnotic.

3106. Diethylcarbamazine. *N,N-Diethyl-4-methyl-1-piperazinecarboxamide;* 1-diethylcarbamoyl-4-methylpiperazine; carbamazine; 1-diethylcarbamyl-4-methylpiperazine; 84L; RP 3799; Carbilazine; Caricide; Cypip; Ethodryl; Notézine; Spatonin. $C_{10}H_{21}N_3O$; mol wt 199.29. C 60.26%, H 10.62%, N 21.08%, O 8.03%. Prepn: Kushner *et al., J. Org. Chem.* **13**, 151 (1948); Kushner, Brancone, U.S. pats. **2,467,893; 2,467,895** (1949 to Am. Cyanamid). Pharmacology and toxicology: Harned *et al., Ann. N.Y. Acad. Sci.* **50**, 141 (1948). GC determn in blood: S. Nene *et al., J. Chromatog.* **308**, 334 (1984). Mode of action: J.-Y. Cesbron *et al., Nature* **325**, 533 (1987). Review of pharmacology, mechanisms of action and clinical uses: C. D. MacKenzie, M. A. Kron, *Trop. Dis. Bull.* **82**(10), R1-R37 (1985).

Crystals, mp 47-49°. bp_3 108.5-111°.

Hydrochloride, $C_{10}H_{22}ClN_3O$, crystals from acetone, mp 156.5-157°. Very sol in water; sol in chloroform, dioxane.

Citrate (hydrogen citrate), $C_{16}H_{29}N_3O_8$, *Banocide, Franocide, Filazine, Loxuran, Longicid, Dirocide, Filaribits, Hetrazan.* Crystals, mp 141-143°. Freely sol in water (>75% at 20°). Sparingly sol in cold alc; freely sol in hot alc. Practically insol in benzene, acetone, ether, chloroform. LD_{50} orally in rats: 1.38 g/kg (Harned).

Phosphate, *Ditrazin.* Crystals, freely sol in water.

THERAP CAT: Anthelmintic (Nematodes).

THERAP CAT (VET): Citrate as an anthelmintic.

3107. N,N'-Diethylcarbanilide. *N,N'-Diethyl-N,N'-diphenylurea;* sym-diethyldiphenylurea. $C_{17}H_{20}N_2O$; mol wt 268.35. C 76.08%, H 7.51%, N 10.44%, O 5.96%. $C_2H_5$$(C_6H_5)NCON(C_6H_5)C_2H_5$. Prepn from ethylaniline and phosgene: Michler, *Ber.* **9**, 712 (1876).

Crystals from alc, mp 79°. Insol in water.

USE: Proposed for use in retarding the aging of vulcanized rubber.

3108. Diethyl Carbitol®. *1,1'-Oxybis(2-ethoxy)ethane;* diethylene glycol diethyl ether. $C_8H_{18}O_3$; mol wt 162.22. C 59.23%, H 11.18%, O 29.59%. $(C_2H_5OCH_2CH_2)_2O$. Toxicity data: H. F. Smyth *et al., J. Ind. Hyg. Toxicol.* **23**, 259 (1941).

Liquid. d_4^{20} 0.907. bp 188°. Very sol in water, alcohol, and other organic solvents. LD_{50} orally in rats: 4.97 g/kg (Smyth).

USE: Solvent; high boiling reaction medium.

3109. Diethylene Glycol. *2,2'-Oxybisethanol;* 2,2'-oxydiethanol. $C_4H_{10}O_3$; mol wt 106.12. C 45.27%, H 9.50%, O 45.23%. $HOCH_2CH_2OCH_2CH_2OH$. Manuf from ethylene oxide and glycol: Faith, Keyes & Clark's *Industrial Chemicals,* F. A. Lowenheim, M. K. Moran, Eds. (Wiley-Interscience, New York, 4th ed., 1975) pp 397-402.

Colorless, hygroscopic, practically odorless liq; sharply sweetish taste. d_{20}^{20} 1.118. Solidifies at −10.45° (when pure). mp −6.5°. bp 244-245°. n_D^{20} 1.4475. Flash pt, open cup: 290°F (143°C). Misc with water, alc, ether, acetone, ethylene glycol; insol in benzene, carbon tetrachloride. LD_{50} orally in rats, guinea pigs: 20.76, 13.21 g/kg, Smyth *et al., J. Ind. Hyg. Toxicol.* **23**, 259 (1941).

Human Toxicity: Symptoms on ingestion similar to ethylene glycol, *q.v.* Fatal poisoning resulted from its use as a solvent in an elixir. *See* E. Browning, *Toxicity and Metabolism of Industrial Solvents* (Elsevier, New York, 1965) pp 624-628, 686-690; Patty's *Industrial Hygiene and Toxicology* vol. **2C**, G. D. Clayton, F. E. Clayton, Eds. (Wiley-Interscience, New York, 3rd ed., 1982) pp 3832-3838.

USE: In antifreeze soln for sprinkler systems, water seals for gas tanks, etc. (water with 40% diethylene glycol freezes at −18°; with 50% at −28°); as lubricating and finishing agent for wool, worsted, cotton, rayon, and silk; as solvent for vat dyes; in composition corks, glues, gelatin, casein, and pastes to prevent drying out.

3110. Diethylene Glycol Monolaurate. *Dodecanoic acid 2-(2-hydroxyethoxy)ethyl ester;* diethylene glycol laurate; diglycol laurate; Glaurin. $C_{16}H_{32}O_4$; mol wt 288.42. C 66.63%, H 11.18%, O 22.19%. Prepared by controlled esterification of diethylene glycol with lauric acid.

Oily liq. d_{25}^{25} 0.9572; d_{20}^{20} 0.963-0.968. mp 17-18°. bp ~270° (some dec). May be heated to 250° without decompn. Practically insol in water. Sol in methanol, ethanol, benzene, toluene, chlorinated hydrocarbons, acetone, ethyl acetate, cottonseed oil. Slightly sol in petr naphtha.

USE: Dispersing agent, emulsifier, plasticizer.

3111. Diethyl Ketone. *3-Pentanone;* dimethylacetone; propione; methacetone. $C_5H_{10}O$; mol wt 86.13. C 69.72%, H 11.70%, O 18.58%. $CH_3CH_2COCH_2CH_3$.

Liquid, acetone odor. d_4^{19} 0.816. bp 101.5°. mp −42°. n_D^{25} 1.3905. Sol in about 25 parts water; miscible with alcohol, ether. LD_{50} orally in rats: 2.1 g/kg, Smyth *et al., Arch. Ind. Hyg. Occup. Med.* **10**, 61 (1954).

3112. Diethylmagnesium. Magnesium diethyl. C_4H_{10}Mg; mol wt 82.44. C 58.27%, H 12.23%, Mg 29.50%. Mg$(C_2H_5)_2$. Prepd by the action of magnesium metal on mercury diethyl in ether: Schlenk, Jr., *Ber.* **64**, 734, 736 (1931).

Solvated crystals from ether, $MgEt_2.Et_2O$, plates, rods. mp 0°. Liquid at room temp. Spontaneously flammable in air. Violent explosion on contact with water. Loses its ether of crystn on heating *in vacuo.* Not volatile in high vac up to 250° when decompn sets in. Sol in ether. Dec by alc and ammonia. Will glow and catch fire even in CO_2. Must be handled in high vac, or under dry nitrogen or hydrogen.

3113. Diethyl Maleate. *(Z)-2-Butenedioic acid, diethyl ester; maleic acid, diethyl ester;* ethyl maleate. $C_8H_{12}O_4$; mol wt 172.18. C 55.81%, H 7.02%, O 37.17%. Prepn from maleic acid and ethanol: V. M. Mitchovitch, *Bull. Soc. Chim. France* **4**, 1667 (1937). Physical properties: G. H. Jeffrey, A. I. Vogel, *J. Chem. Soc.* **1948**, 658. Copolymerization with styrene: F. M. Lewis *et al., J. Am. Chem. Soc.* **70**, 1519, 1529 (1948). In Diels-Alder reactions: N. L. Bauld *et al., Tetrahedron Letters* **1972**, 2443. In Michael additions: G. Arsenault *et al., Chem. Commun.* **1983**, 437. Conjugation with glutathione, *q.v.:* E. Boyland, L. F. Chasseaud, *Biochem. J.* **104**, 95 (1967); *eidem, ibid.* **109**, 651 (1968). Toxicity: H. F. Smyth *et al., J. Ind. Hyg. Toxicol.* **31**, 60 (1949).

$$\underset{C_2H_5OOC}{\overset{H}{\diagdown}}C=C\underset{COOC_2H_5}{\overset{H}{\diagup}}$$

Oil, bp_{758} 219.5°. d_4^{20} 1.0674. n_D^{20} 1.4402. Insol in water. Soluble in ethanol, ether. LD_{50} orally in rats: 0.30 g/kg (Smyth).

USE: In organic synthesis.

3114. Diethylmalonic Acid. *Diethylpropanedioic acid;* 3,3-pentanedicarboxylic acid. $C_7H_{12}O_4$; mol wt 160.17. C 52.49%, H 7.55%, O 39.96%. Prepn: Daimler, *Ann.* **249**, 173 (1888); Speck, *J. Am. Chem. Soc.* **74**, 2876 (1952). Spectral studies: J. L. Delarbre *et al., J. Raman. Spectrosc.* **16**, 11 (1985). Crystal structure: A. Dubourg *et al., Acta Crystallogr.* **C44**, 1987 (1988).

$$\underset{C_2H_5}{\overset{C_2H_5}{\diagdown}}C\underset{COOH}{\overset{COOH}{\diagup}}$$

Crystals, mp 127°. Dec at 170-180° with liberation of CO_2 and formation of diethylacetic acid. Very sol in water; freely sol in alcohol, ether, slightly in chloroform.

3115. Diethyl Oxalate. *Ethanedioic acid diethyl ester;* ethyl oxalate; diethyl ethanedioate; oxalic acid diethyl ester. $C_6H_{10}O_4$; mol wt 146.14. C 49.31%, H 6.90%, O 43.79%. $C_2H_5OOCCOOC_2H_5$. Prepn: Clarke, Davis, *Org. Syn.;* **coll. vol. I**, 256 (2nd ed., 1941); Kenyon, *ibid.* 261.

Liquid. d_4^{20} 1.0785. mp -40.6°; bp_{760} 185.7°; bp_{100} 130.8°; bp_{20} 96.8°; $bp_{1.0}$ 47°. n_D^{20} 1.41011. Flash pt, open cup: 168° F (75° C). Sparingly sol in water which dec it gradually; miscible with alcohol, ether, and other usual org solvents.

USE: Manuf phenobarbital, ethylbenzyl malonate, triethylamine, and similar chemicals, plastics, dyestuff intermediates. Solvent for cellulose esters, perfumes.

3116. Diethylpropion. *2-(Diethylamino)-1-phenyl-1-propanone; 2-diethylaminopropiophenone;* α-benzoyltriethylamine; amfepramone. $C_{13}H_{19}NO$; mol wt 205.30. C 76.05%, H 9.33%, N 6.82%, O 7.79%. Prepn: Hyde *et al., J. Am. Chem. Soc.* **50**, 2287 (1928); Schutte, U.S. pat. 3,001,-910 (1961 to Temmler-Werke). Pharmacology: Cahen *et al., Therapie* **17**, 373-412 (1962).

$$\underset{\underset{CH_3}{|}}{C_6H_5COCHN(C_2H_5)_2}$$

Hydrochloride, $C_{13}H_{20}ClNO$, *Anfamon, Anorex, Danylen, Dobesin, Frekentine, Keramik, Magrene, Modulor, Moderatan, Parabolin, Prefamone, Regenon, Tenuate, Tenuate Dospan, Tepanil, Tylinal.* Crystals, dec 168°.

Note: Regenon and Danylen also contain trace elements. Component of *Derfon.*

Note: This is a controlled substance (stimulant) listed in the U.S. Code of Federal Regulations, Title 21 Part 1308.14 (1987).

THERAP CAT: Anorexic.

3117. Diethylsilane. $C_4H_{12}Si$; mol wt 88.23. C 54.45%, H 13.71%, Si 31.84%. $(C_2H_5)_2SiH_2$. Prepd from ethylene and silane: Hurd, U.S. pat. 2,537,763 (1951); Fritz, *Z. Naturforsch.* **7b**, 207 (1952); White, Rochow, *J. Am. Chem. Soc.* **76**, 3897 (1954); Clasen, *Angew. Chem.* **70**, 180 (1958).

Liquid. Stable to air when pure. bp_{741} 56°; d_4^{20} 0.6843; n_D^{20} 1.3921. Quickly oxidized by Ag_2O and HgO. Oxidation can be slowed by soln in petr ether or by cooling to -80°.

3118. Diethylstilbestrol. *4,4'-(1,2-Diethyl-1,2-ethenediyl)bisphenol;* α,α'-*diethylstilbenediol;* stilbestrol; stilboestrol; 3,4-bis(p-hydroxyphenyl)-3-hexene; 4,4'-dihydroxy-α,β-diethylstilbene; DES; Antigestil; Bio-des; Bufon; Cyren A; Distilbene; Domestrol; Estrobene; Estrosyn; Fonatol; Grafestrol; Hi-Bestrol; Makarol; Micrest; Milestrol; Neo-Oestranol I; Oestrogenine; Oestromenin; Oestromensyl; Oestromon; Palestrol; Percutacrine Oestrogénique Iscovesco;

Serral; Sexocretin; Sibol; Stilbetin; Stilboefral; Stilboestroform; Stilkap; Synestrin (tablets); Synthoestrin; Vagestrol. $C_{18}H_{20}O_2$; mol wt 268.34. C 80.56%, H 7.51%, O 11.92%. Prepn: E. C. Dodds *et al., Nature* **141**, 247 (1938); *eidem, Proc. Roy. Soc. London, Ser. B* **127**, 140 (1939). Comprehensive review of the early literature on synthetic estrogens: U. V. Solmssen, *Chem. Rev.* **37**, 481-598 (1945). Review of metabolism: P. W. Aschbacher, *J. Toxicol. Environ. Health,* Suppl. 1, 45-59 (1976); M. Metzler, *Crit. Rev. Biochem.* **10**, 171-212 (1981). Review of the toxic and carcinogenic effects of DES attributed to its use in the prevention of miscarriage in humans, and as a growth promotant in livestock: K. E. McMartin *et al., J. Environ. Pathol. Toxicol.* **1**, 279-313 (1978). Book: *Developmental Effects of Diethylstilbestrol (DES) in Pregnancy,* A. L. Herbst, H. A. Bern, Eds. (Thieme-Verlag, New York, 1981) 203 pp.

$$HO-\text{benzene ring}-\underset{\underset{CH_3CH_2}{|}}{\overset{\overset{CH_2CH_3}{|}}{C}}=C-\text{benzene ring}-OH$$

Small plates from benzene, mp 169-172°. Almost insol in water. Sol in alcohol, ether, chloroform, fatty oils, dil hydroxides.

Dipalmitate, $C_{50}H_{80}O_4$, *stilpalmitate.* Plates from hot alcohol, mp 77-78°.

Diphosphate, *see* Fosfestrol.

Dipropionate, *see* separate entry.

Dimethyl ether, *see* Dimestrol.

Disulfate, $C_{18}H_{20}O_8S_2$, *Idroestril.* Voluminous white cryst powder from dil HCl. Sparingly sol in water.

Monobenzyl ether, $C_{25}H_{26}O_2$, *Monozol.*

Monomethyl ether, *see* Mestilbol.

Caution: Banned by the FDA in 1979 as a growth promotant in livestock. Diethylstilbestrol has been listed as a known carcinogen: *Fourth Annual Report on Carcinogens* (NTP 85-002, 1985) p 85.

THERAP CAT: Estrogen.

THERAP CAT (VET): Formerly in estrogenic hormone therapy.

3119. Diethylstilbestrol Dipropionate. *(E)-4,4'-(1,2-Diethyl-1,2-ethenediyl)bisphenol dipropanoate;* α,α'-diethyl-4,4'-stilbenediol dipropionyl ester; Clinestrol; Cyren B; Dibestil (formerly); Estilben; Estrobene DP; Euvestin; Neo-Oestranol II; Orestol; Pabestrol D; sin-estrol; Stilbestronate; Stilboestrol DP; Stilbofax; Stilronate; Synestrin (ampuls); Syntestrine; Willestrol. $C_{24}H_{28}O_4$; mol wt 380.41. C 75.77%, H 7.42%, O 16.82%. Prepn: Dodds *et al., Proc. Roy. Soc. London, Ser. B* **127**, 140 (1939).

$$C_2H_5COO-\text{benzene ring}-\underset{\underset{CH_3CH_2}{|}}{\overset{\overset{CH_2CH_3}{|}}{C}}=C-\text{benzene ring}-OOCC_2H_5$$

Plates from methanol, mp 104°. Sol in organic solvents and in vegetable oils.

THERAP CAT: Estrogen.

THERAP CAT (VET): Estrogenic hormone therapy.

3120. Diethyl Sulfate. *Sulfuric acid diethyl ester.* $C_4H_{10}O_4S$; mol wt 154.19. C 31.16%, H 6.54%, O 41.50%, S 20.80%. $(C_2H_5)_2SO_4$. Prepd from ethanol + sulfuric acid; by absorption of ethylene in sulfuric acid; from diethyl ether and fuming sulfuric acid. Review of prepn and uses: R. Page, J. A. John in *Ethylene and its Industrial Derivatives,* S. A. Miller, Ed. (Benn, London, 1969) pp 774-787. Toxicity data: H. F. Smyth *et al., J. Ind. Hyg. Toxicol.* **31**, 60 (1949). Review and evaluation of studies of carcinogenicity in laboratory animals: *IARC Monographs* **4**, 277-281 (1974). Reviews: C. M. Suter, *The Organic Chemistry of Sulfur* (John Wiley, 1944) pp 62-65; E. E. Gilbert, *Sulfonation and Related Reactions* (Interscience, New York, 1965).

Colorless, oily liquid; peppermint odor. Darkens with

age. bp 209.5° with decompn; bp_{15} 96°; bp_5 75°. mp $-25°$. d_4^{23} 1.1774. n_D^{20} 1.40037, A. I. Vogel, D. M. Cowan, *J. Chem. Soc.* **1943**, 16. Flash pt: 220°F. Vapor density (air = 1) 5.31. Vapor pressure at 47° = 1 mm. Practically insol in water and gradually dec by it. Rapidly dec by hot water into monoethyl sulfate and alcohol. Miscible with alcohol, ether. LD_{50} orally in rats: 0.88 g/kg (Smyth).

See also Ethyl Sulfate, $C_2H_5OSO_2OH$.

Note: Diethyl sulfate may reasonably be anticipated to be a carcinogen: *Fourth Annual Report on Carcinogens* (NTP 85-002, 1985) p 86.

USE: Chiefly as an ethylating agent; as an accelerator in the sulfation of ethylene; in some sulfonations.

3121. Diethylzinc. Zinc diethyl. $C_4H_{10}Zn$; mol wt 123.50. C 38.90%, H 8.16%, Zn 52.94%. $Zn(C_2H_5)_2$. Prepd by the interaction of zinc and ethyl iodide: Simonowitch, *Chem. Zentr.* **1899**, I, 1066; Lachmann, *Am. Chem. J.* **24**, 32 (1900); Dennis, Hance, *J. Am. Chem. Soc.* **47**, 370 (1925); from zinc and ethyl bromide, ethyl iodide: C. R. Noller, *Org. Syn.* **coll. vol. II**, 184. Used in synthesis of cyclopropanes: J. Furukawa *et al., Tetrahedron* **24**, 53 (1968); *eidem, Tetrahedron Letters* **1968**, 3495; in synthesis of ketocarbenes: L. T. Scott, W. D. Cotton, *J. Am. Chem. Soc.* **95**, 2708 (1973); in ring expansion of arenes: S. Miyano, H. Hashimoto, *Chem. Commun.* **1973**, 216; in preservation of papers: J. C. Williams, G. B. Kelly Jr., U.S. pat. **3,969,549** (1976 to U.S.A.); *eidem, Advan. Chem. Ser.* **193**, 109 (1981).

Mobile liq. Stable in sealed tube and carbon dioxide. d^{20} 1.2065; d_4^8 1.245; mp $-28°$; bp_{760} 118°; bp_{30} 27°; $n_{H\alpha}^{20}$ 1.4936. *Ignites in air,* burns with a blue flame, giving off a peculiar, garlic-like odor. Miscible with ether, petr ether, benzene, other hydrocarbons.

USE: In organic synthesis; in preservation on archival papers.

3122. Difemerine. α-*Hydroxy-α-phenylbenzeneacetic acid 2-(dimethylamino)-2-methylpropyl ester;* 2-(dimethylamino)-1,1-dimethylethyl benzilate. $C_{20}H_{25}NO_3$; mol wt 327.41. C 73.36%, H 7.70%, N 4.28%, O 14.66%. Prepn: S. G. Kuznetsov, A. G. El'tsov, *Zh. Obshch. Khim.* **32**, 511 (1962), *C.A.* **58**, 470d (1963); C. Hoffmann, Fr. pat. **M3406** (1965 to Labs. U.P.S.A.), *C.A.* **63**, 13163ab (1965). Toxicity: S. N. Golikov, S. G. Kuznetsov, *C.A.* **61**, 8790b (1964). Comparison of side effects with those of atropine: L. Moser, P. V. Lundt, *Med. Welt* **31**, 1795 (1980). Crystal structure: A. Carpy *et al., Acta Crystallogr.* **B35**, 882 (1979).

$$C_6H_5\overset{OH}{\underset{C_6H_5}{\overset{|}{\underset{|}{C}}}}—COO—\overset{CH_3}{\underset{CH_3}{\overset{|}{\underset{|}{C}}}}CH_2N(CH_3)_2$$

Solid, mp 78-78.3°.

Hydrochloride, $C_{20}H_{26}ClNO_3$, *Luostyl.* Crystals from isopropanol, mp 182°.

THERAP CAT: Anticholinergic; antispasmodic.

3123. Difenamizole. 2-*(Dimethylamino)-N-(1,3-diphenyl-1H-pyrazol-5-yl)propanamide;* 1,3-diphenyl-5-(2-dimethylaminopropionamido)pyrazole; AP-14; Pasalin. $C_{20}H_{22}N_4O$; mol wt 334.43. C 71.83%, H 6.63%, N 16.75%, O 4.78%. Prepn: Kameyama, Niwa, **Japan. pat. 68 06,621** (1968 to Takeda), *C.A.* **69**, 106704x (1968). Physicochemical properties and stability: Uda, Yashiki, *Takeda Kenkyusho Ho* **29**, 42 (1970), *C.A.* **73**, 28847n (1970). Pharmacology: Fukuda *et al., ibid.* **30**, 389 (1971); Saji *et al., ibid.* **406**; Nagawa *et al., ibid.* **427**, *C.A.* **76**, 108025v, 108021r, 108023t (1972), resp. Toxicity studies: Tanaka *et al., Oyo Yakuri (Pharmacometrics)* **6**, 279 (1972). Mechanism of action studies: Yasuhara *et al., ibid.* **313**. Acute toxicity: M. Hayashi *et al., Gifu Daigaku Igakubu Kiyo* **20**, 360 (1972), *C.A.* **78**, 92505e (1973). *Review:* Japan. *Med. Gaz.* **10**(11), 12 (1973).

White to pale yellow, odorless powder, mp 123-128°. Also reported as mp 120-122°. Freely sol in acetone, chloroform, benzene. Practically insol in water. Stable to heat, humidity and sunlight for 3 months. LD_{50} in mice (mg/kg): 103 i.v.; 186 i.p.; 525 s.c.; 560 orally (Hayashi).

THERAP CAT: Analgesic, anti-inflammatory.

3124. Difenoxin. 1-*(3-Cyano-3,3-diphenylpropyl)-4-phenyl-4-piperidinecarboxylic acid;* 1-(3-cyano-3,3-diphenylpropyl)-4-phenylisonipecotic acid; difenoxylic acid; difenoxilic acid; McN-JR-15403-11; Lyspafen. $C_{28}H_{28}N_2O_2$; mol wt 424.54. C 79.22%, H 6.64%, N 6.60%, O 7.54%. Active metabolite of diphenoxylate, *q.v.*, from which it may be prepared by hydrolysis of the ethyl ester: Soudijn, van Wijngaarden, Ger. pat. **1,953,342** corresp to U.S. pat. **3,646,207** (1970, 1972 to Janssen). Series of articles on pharmacology, toxicology, metabolism: *Arzneimittel-Forsch.* **22**, 513-531 (1972). Toxicity data: Niemegeers *et al., ibid.* 516.

Hydrochloride, $C_{28}H_{29}ClN_2O_2$, *R 15403.* White amorphous power, mp 290°. Very slightly sol in water (0.023%); sparingly sol in chloroform, tetrahydrofuran, dimethylacetamide, DMSO. Stable, can be stored for several years under normal conditions; not hygroscopic; not affected by light. LD_{50} orally in rats: 149 mg/kg (Niemegeers).

Note: This is a controlled substance (opiate) listed in the U.S. Code of Federal Regulations, Title 21 Part 1308.11 (1987).

THERAP CAT: Antiperistaltic; antidiarrheal.

3125. Difenpiramide. N-2-*Pyridinyl-[1,1'-biphenyl]-4-acetamide;* diphenpyramide; Z 876; Difenax. $C_{19}H_{16}N_2O$; mol wt 288.5. C 79.14%, H 5.59%, N 9.71%, O 5.55%. Prepn: F. Tenconi, R. M. Tagliabue, Ger. pat. **2,325,309** (1973 to Zambeletti) corresp to L. Molteni *et al.,* U.S. pat. **3,868,380** (1975). Pharmacological properties: S. Caliari *et al., Arzneimittel-Forsch.* **27**, 2086 (1977). Physicochemical profile: A. Trebbi, G. Filippi, *Boll. Chim. Farm.* **118**, 729 (1979). Therapeutic use: C. Ortolani *et al., Clin. Ter.* **89**, 391 (1979).

Cryst, mp 122-124°. LD_{50} in male mice, rats (mg/kg): 2590, 2075 orally; 1421, 1396 i.p., S. Caliari *et al., loc. cit.*

THERAP CAT: Anti-inflammatory.

3126. Diflorasone. (6α,11β,16β)-6,9-*Difluoro-11,17,21-trihydroxy-16-methylpregna-1,4-diene-3,20-dione;* 6α,9α-difluoro-16β-methyl-$\Delta^{1,4}$-pregnadiene-11β,17α,21-triol-3,20-dione; 6α,9α-difluoro-16β-methylprednisolone. $C_{22}H_{28}F_2O_5$; mol wt 410.46. C 64.38%, H 6.87%, F 9.26%, O 19.49%. The 16β-analog of flumethasone, *q.v.* Prepn of free alcohol and 21-acetate: Brit. pat. **881,334** (1961 to Pfizer), *C.A.* **56**, 15586c (1962); Brit. pat. **898,293**; F. H. Lincoln *et al.,* U.S. pat. **3,557,158** (1962, 1971 both to Upjohn); Brit. pat. **912,015** (1962 to Merck & Co.). Prepn of the 17,21-diacetate: D. E. Ayer *et al.,* Ger. pat. **2,308,731**; *eidem,* U.S. pat. **3,980,778** (1973, 1976 both to Upjohn). Proposed mechanism of action: S. Hammarstrom *et al., Science* **197**, 994 (1977). Pharmacology: S. Wickrema *et al., J. Invest. Dermatol.* **71**, 372 (1978). Clinical study: S. M. Bluefarb *et al., J. Int. Med. Res.* **4**, 454 (1976).

Diacetate, $C_{26}H_{32}F_2O_7$, *U 34865, Dermaflor, Difral, Florone, Maxiflor, Soriflor.* Crystals from ethyl acetate-Skellysolve C and acetone-methanol, mp 221-223° (dec). uv max (alc): 238 nm (ε 17250). $[\alpha]_D$ +61° (chloroform).

THERAP CAT: Topical anti-inflammatory; glucocorticoid.

3127. Difloxacin. *6-Fluoro-1-(4-fluorophenyl)-1,4-dihydro-7-(4-methyl-1-piperazinyl)-4-oxo-3-quinolinecarboxylic acid.* $C_{21}H_{19}F_2N_3O_3$; mol wt 399.40. C 63.15%, H 4.79%, F 9.51%, N 10.52%, O 12.02%. Fluoroquinolone antibacterial structurally related to norfloxacin, *q.v.* Prepn: D. T. W. Chu, **Eur.** pat. **Appl. 131,839;** *idem,* **U.S.** pat. **4,730,000** (1985, 1988 both to Abbott); D. T. W. Chu *et al., J. Med. Chem.* **28,** 1558 (1985); H. Narita *et al.,* **Japan. Kokai 85 237,069** (1985 to Toyama). Antibacterial spectrum *in vitro:* J. M. Stamm *et al., Antimicrob. Ag. Chemother.* **29,** 193 (1986); *in vivo:* P. B. Fernandes *et al., ibid.* 201. Activity vs anaerobic bacteria: *eidem, J. Antimicrob. Chemother.* **18,** 693 (1986). HPLC determn in biological fluids: G. R. Granneman, L. T. Sennello, *J. Chromatog.* **413,** 199 (1987). Pharmacokinetics in humans: G. R. Granneman *et al., Antimicrob. Ag. Chemother.* **30,** 689 (1986).

Monohydrochloride, $C_{21}H_{20}ClF_2N_3O_3$, *Abbott-56619, A-56619.* Crystals, mp > 275°.

THERAP CAT: Antibacterial.

3128. Diflubenzuron. *N-[[(4-Chlorophenyl)amino]carbonyl]-2,6-difluorobenzamide;* 1-(4-chlorophenyl)-3-(2,6-difluorobenzoyl)urea; difluron; DU 112307; PH 60-40; TH-6040; ENT 29054; OMS 1804; Dimilin; Duphacid; Micromite. $C_{14}H_9ClF_2N_2O_2$; mol wt 310.68. C 54.12%, H 2.92%, Cl 11.41%, F 12.23%, N 9.02%, O 10.30%. Prepn: K. Wellinga, R. Mulder, **Ger.** pat. **2,123,236;** *eidem,* **U.S.** pats. **3,748,356, 3,933,908** and **3,989,842** (1971, 1973, 1976, 1976 all to Philips). Insecticidal activity studies: K. Wellinga *et al., J. Agr. Food Chem.* **21,** 348, 993 (1973); Mulder, Gijswijt, *Pestic. Sci.* **4,** 737 (1973). Inhibits chitin incorporation in insect cuticle: Hunter, Vincent, *Experientia* **30,** 1432 (1974). *Review:* J. L. Marx, *Science* **197,** 1170, 1172 (1977).

mp 239°. Soly in water ∼ 0.3 ppm. LD_{50} orally in mice, rats (formulation with 50% kaolin): 4.64 g/kg, > 10 g/kg (Mulder, Gijswijt).

USE: Insecticide (larvicide).

3129. Diflucortolone. *6,9-Difluoro-11,21-dihydroxy-16-methylpregna-1,4-diene-3,20-dione;* 6α,9α-difluoro-16α-methyl-1,4-pregnadiene-11β,21-diol-3,20-dione; 6α,9α-difluoro-16α-methyl-1-dehydrocorticosterone. $C_{22}H_{28}F_2O_4$; mol wt 394.46. C 66.99%, H 7.15%, F 9.63%, O 16.23%. The 9α-fluoro deriv of fluocortolone, *q.v.* Prepn: **Belg.** pat. **639,708** corresp to K. Kieslich *et al.,* **U.S.** pat. **3,426,128** (1964, 1969, both to Schering AG); K. Kieslich *et al., Arzneimittel-Forsch.* **26,** 1462 (1976). Toxicity: P. Gunzel *et al., ibid.* 1476. Series of articles on pharmacology, toxicity, metabolism and clinical studies: *ibid.* **26,** 1463-1513 (1976).

Crystals, mp 248-249°. $[\alpha]_D^{22}$ +111° (methanol). Also reported as crystals from ethyl acetate-ether, mp 242-244°. uv max: 237 nm (ε 16,600) (U.S. pat. **3,426,128**)

21-Valerate, $C_{27}H_{36}F_2O_5$, *DFV, Nerisona, Nerisone, Temetex, Texmeten.* Crystals, mp 195-195.5°. $[\alpha]_D^{22}$ +100.8° (dioxane). Approx LD_{50} in mice, rats: > 4, 3.1 g/kg orally; 180, 13 mg/kg s.c.; 450, 98 mg/kg i.p. (Gunzel).

THERAP CAT: Anti-inflammatory.

3130. Diflunisal. *2',4'-Difluoro-4-hydroxy-[1,1'-biphenyl]-3-carboxylic acid; 2',4'-difluoro-4-hydroxy-3-biphenylcarboxylic acid;* 2',4'-difluoro-4-hydroxy-[1',1-diphenyl]-3-carboxylic acid; 2-(hydroxy)-5-(2,4-difluorophenyl)benzoic acid; 5-(2,4-difluorophenyl)salicylic acid; MK-647; Adomal; Dolisal; Dolobid; Dolobis; Flovacil; Fluniget; Fluodonil; Flustar; Reuflos. $C_{13}H_8F_2O_3$; mol wt 250.20. C 62.41%, H 3.22%, F 15.19%, O 19.18%. Prepn: Ruyle *et al.,* **S. Afr.** pat. **67 01,021** (1968 to Merck & Co.), *C.A.* **70,** 106209k (1969); *eidem,* **Fr.** pat. **1,522,570** corresp to U.S. pat. **3,714,226** (1968, 1973, both to Merck & Co.); J. Hannah *et al., J. Med. Chem.* **21,** 1093 (1978). Metabolism: P. J. DeSchepper *et al., Brit. J. Clin. Pharmacol.* **4,** 645P (1977). Clinical studies: J. A. Wojtulewski *et al., Curr. Med. Res. Opin.* **5,** 562 (1978); J. A. Hicklin, *ibid.* 572; in sickle cell anemia: E. A. Oyewo, *Clin. Trials J.* **24,** 249 (1987). Review of chemistry, pharmacology, clinical pharmacology: S. L. Steelman *et al., ibid.* 506-514. Comprehensive description: M. Cotton, R. A. Hux in *Analytical Profiles of Drug Substances,* vol. 14, K. Florey, Ed. (Academic Press, New York, 1985) pp 491-526. *Review:* Brit. J. Clin. Pharmacol. **4,** Suppl. 1, 1S-52S (1977); C. A. Winter *et al.,* in *Pharmacological and Biochemical Properties of Drug Substances* vol. **3,** M. E. Goldberg, Ed. (Am. Pharm. Assoc., Washington, DC, 1981) pp 291-323. Comprehensive review of pharmacological properties and therapeutic use in pain: R. N. Brogden *et al., Drugs* **19,** 84-106 (1980); *Pharmacotherapy* **3,** no. 2, pt. 2, 1S-82S (1983).

Crystals from toluene, mp 210-211°. Sparingly sol in water. LD_{50} orally in female mice: 439 mg/kg (Stone).

THERAP CAT: Analgesic; anti-inflammatory.

3131. 2,4-Difluoroaniline. *2,4-Difluorobenzenamine.* $C_6H_5F_2N$; mol wt 129.11. C 55.81%, H 3.90%, F 29.43%, N 10.85%. Prepn: Swarts, *Rec. Trav. Chim.* **35,** 164 (1916); Kutepov, Rozanova, *Zh. Obshch. Khim.* **27,** 2848 (1957), *C.A.* **52,** 8067 (1958); Nad *et al., C.A.* **53,** 14976i (1959).

mp −7.5°. bp$_{753}$ 169.5°. n_D^{25} 1.5043. Specific gravity at 25°: 10.72 lb/gal. Flash pt 158°F (70°C).
USE: In organic syntheses.

3132. p-Difluorobenzene. $C_6H_4F_2$; mol wt 114.09. C 63.16%, H 3.53%, F 33.31%. Prepd by diazotizing *p*-fluoroaniline in hydrofluoric acid and heating the diazonium salt to 120°: Swarts, *Chem. Zentr.* **1913,** II, 760.

Liquid, pungent, aromatic odor. mp −23.7°. d^{20} 1.17006. bp 88.82°. $n_D^{18.9}$ 1.44219.

3133. 4,4′-Difluorodiphenyl. *4,4′-Difluoro-1,1′-biphenyl.* $C_{12}H_8F_2$; mol wt 190.18. C 75.78%, H 4.24%, F 19.98%. NMR spectrum: T. K. Halstead *et al., J. Chem. Soc. Faraday Trans. 2,* **1981,** 1817.

Crystalline powder; aromatic odor. d 1.04. mp 92-95°. bp 254-255°. Insol in water. Freely sol in alc, chloroform, ether, oils.

3134. Difluprednate. *21-(Acetyloxy)-6,9-difluoro-11-hydroxy-17-(1-oxobutoxy)pregna-1,4-diene-3,20-dione; 6α,9-difluoro-11β,17,21-trihydroxypregna-1,4-diene-3,20-dione 21-acetate 17-butyrate; 6α,9α-difluoroprednisolone-21-acetate-17-butyrate;* CM-9155; W-6309; Epitopic; Myser. $C_{27}H_{34}F_2O_7$; mol wt 508.57. C 63.77%, H 6.74%, F 7.47%, O 22.02%. Prepn: A. Ercoli, R. Gardi, S. Afr. pat. 68 03686 corresp to U.S. pat. 3,780,177 (1968, 1973 both to Warner-Lambert). Anti-inflammatory properties: G. Di Pasquale *et al., Steroids* **16,** 663, 679 (1970); R. Gardi *et al., J. Med. Chem.* **15,** 556 (1972). Pharmacological study: J. Navarro *et al., Arzneimittel-Forsch.* **28,** 2302 (1978). Vasoconstrictive activity in man: A. Barbier *et al., Therapie* **33,** 607 (1978). Bioavailability study: C. Lafille *et al., Dermatologica* **159,** 277 (1979).

Crystals from methylene chloride/ether/petr ether, mp 191-194°. [α]$_D^{22}$ +31.7° (c = 0.5 in dioxane). uv max (ethanol): 237-238 nm (E$_{1cm}^{1\%}$ 320).
THERAP CAT: Anti-inflammatory.

3135. Digallic Acid. *3,4-Dihydroxy-5-[(3,4,5-trihydroxybenzoyl)oxy]benzoic acid;* gallic acid 5,6-dihydroxy-3-carboxyphenyl ester; 4,5-dihydroxybenzoic acid monogallate; gallic acid 3-monogallate; *m*-digallic acid; *m*-galloylgallic acid. $C_{14}H_{10}O_9$; mol wt 322.22. C 52.18%, H 3.13%, O 44.69%. Isoln from Aleppo gallotannin and Chinese gallotannin: Nierensten, *Ber.* **43,** 628 (1910). Synthesis: Fischer, Freudenberg, *Ber.* **46,** 1128 (1913). Structure: Nierenstein *et al., J. Am. Chem. Soc.* **47,** 846 (1925). Determn by HPLC:

P. Delahaye, M. Verzele, *J. Chromatog.* **265,** 363 (1983); in tannic acids: M. Verzele *et al., Bull. Soc. Chim. Belg.* **92,** 181 (1983). Role in corrosion inhibition: A. M. Beccaria, E. D. Mor, *Brit. Corrosion J.* **11,** 156 (1976).

Hydrated needles from alc + water, anhydr at 110°; dec 280°. Sol in 1900 parts water at 25°; in 50-60 parts boiling water; sol in methanol, ethanol, acetone; sparingly sol in ether, glacial acetic acid. Upon addn of KCN to an aq soln a transitory pink color appears, which returns on shaking.
Note: In pharmaceutical literature the name digallic acid is frequently confused with tannic acid.

3136. Digalogenin. *(25R)-5α-Spirostan-3β,15β-diol.* $C_{27}H_{44}O_4$; mol wt 432.62. C 74.95%, H 10.25%, O 14.79%. Aglycon of digalonin. Isoln from saponin mixture and partial synthesis from digitogenin: Tschesche, Wulff, *Ber.* **94,** 2019 (1961).

Crystals from methanol, mp 218.5-220.5°. [α]$_D^{21}$ −75° (CHCl$_3$). IR spectrum: Tschesche, Wulff *loc. cit.* Sol in chloroform.

3137. Diginatigenin. *3β,12β,14,16β-Tetrahydroxy-5β-card-20(22)-enolide;* 12-hydroxygitoxigenin; 16-hydroxydigoxigenin. $C_{23}H_{34}O_6$; mol wt 406.50. C 67.95%, H 8.43%, O 23.62%. Aglycone of diginatin from *Digitalis lanata* Ehrk., *Scrophulariaceae:* Murphy, *J. Am. Pharm. Assoc. Sci. Ed.* **44,** 719 (1955). Aglycone of lanatoside D: Angliker *et al., Ann.* **607,** 131 (1957). Obtained by microbiological conversion of gitoxigenin: Tamm, Gubler, *Helv. Chim. Acta* **41,** 1762 (1958); Okado *et al., Chem. Pharm. Bull.* **8,** 530 (1960). Structure: Linde *et al., Helv. Chim. Acta* **42,** 2040 (1959); Okado *et al., Chem. Pharm. Bull.* **8,** 535 (1960).

Needles from water, mp 157°. [α]$_D^{20}$ +34° (methanol). Absorption max (ethanol): 318 nm (log ε 4.18); (98% w/w H$_2$SO$_4$): 230, 310, 390, 490 nm (E$_{1cm}^{1\%}$ 160, 130, 210, 85).

3138. Diginatin. *(3β,5β,12β,16β)-3-[(O-2,6-Dideoxy-β-D-ribo-hexopyranosyl-(1 → 4)-O-2,6-dideoxy-β-D-ribo-hexopyranosyl-(1 → 4)-2,6-dideoxy-β-D-ribo-hexopyranosyl)oxy]-12,14,16-trihydroxy-card-20(22)-enolide.* $C_{41}H_{64}O_{15}$; mol wt 796.96. C 61.79%, H 8.09%, O 30.11%. Cardiac glycoside from *Digitalis lanata* Ehrk., *Scrophulariaceae:* Murphy, *J. Am. Pharm. Assoc. Sci. Ed.* **44,** 719 (1955); Angliker *et al., Ann.* **607,** 131 (1957); von Wartburg *et al., Experientia* **14,** 439 (1958). Acid hydrolysis yields 1 mol diginatigenin + 3 mols D-digitoxose. The sugar residue is attached to the hydroxyl group at C-3 of the aglycon.

Prisms from acetone, mp 251-253°. Somewhat hygroscopic. $[\alpha]_D^{20}$ +20.5° ± 2°; $[\alpha]_{5463}^{20}$ +26.3° ± 2° (c = 0.6 in methanol). Absorption max (98% w/w H_2SO_4): 230, 310, 390, 480, 560 nm ($E_{1cm}^{1\%}$ 185, 175, 315, 230, 120). Sol in alcohol, dil alcohol, dioxane; slightly sol in acetone. One part sol in 1000 parts water and in 2000 parts chloroform at 25°.

Acetyldiginatin-α, $C_{43}H_{66}O_{16}$, flat plates from methanol + chloroform + ether, mp 199-202°. $[\alpha]_D^{20}$ +30.1° (c = 0.6 in pyridine). uv max: 218 nm (log ϵ 4.1).

3139. Diginin. (3β,12α,14β,17α,20S)-3-[(2,6-Dideoxy-3-O-methyl-D-lyxohexopyranosyl)oxy]-12,20-epoxypregn-5-ene-11,15-dione;3β-(diginosyloxy)-12α,20α-epoxy-14β,17α-pregn-5-ene-11,15-dione. $C_{28}H_{40}O_7$; mol wt 488.60. C 68.83%, H 8.25%, O 22.92%. Isoln from leaves of Digitalis purpurea L., Scrophulariaceae: Karrer in E. Barell Festschrift (1936) p 238; Chem. Zentr. 1936, II, 2727. Isoln: Shoppee, Reichstein, Helv. Chilm. Acta 23, 975 (1940); Shoppee, ibid. 27, 426 (1944). Structure: Shoppee et al., J. Chem. Soc. 1962, 3610; Tschesche, Brügmann, Tetrahedron 20, 1469 (1964). Mild hydrolysis yields diginigenin and diginose. Attempted partial synthesis of diginigenin: Tschesche, Schwinum, Ber. 100, 464 (1967); Tschesche, Müller-Albrecht, Ber. 103, 350 (1970).

Stout prisms from dil alc. Indistinct melting range: 155-183°. $[\alpha]_D^{14}$ −223° (c = 2.3 in chloroform). uv max (ethanol): 309 nm (log ϵ 1.94). Freely sol in chloroform; slightly sol in ether, acetone, ethyl acetate, carbon tetrachloride. Practically insol in water. Positive Legal's test. The Keller reaction produces a brilliant yellow band with diginigenin.

3140. Digitalin. (3β,5β,16β)-3-[(6-Deoxy-4-O-β-D-glucopyranosyl-3-O-methyl-β-D-galactopyranosyl)oxy]-14,16-dihydroxy-card-20(22)-enolide; digitalinum verum; Schmiedeberg's digitalin; Diginorgin. $C_{36}H_{56}O_{14}$; mol wt 712.81. C 60.66%, H 7.92%, O 31.42%. Obtained from seeds of Digitalis purpurea L., Scrophulariaceae, and from the roots of Adenium honghel A. DC., Apocynaceae: Schmiedeberg, Arch. Exp. Path. Pharmakol. 3, 16 (1874); Windaus, Haack, Ber. 62, 475 (1929); Hunger, Reichstein,

Helv. Chim. Acta 33, 76 (1950); Sasakawa, J. Pharm. Soc. Japan 74, 474 (1954); Miyatake et al., Pharm. Bull. (Tokyo) 5, 157 (1957).

Crystals from methanol + ether and methanol + water, mp 240-243°. $[\alpha]_D^{20}$ −1.1° (c = 0.894 in methanol). Slightly sol in water, chloroform or ether; sol in alcohol.

16-Acetyldigitalinum verum, $C_{38}H_{58}O_{15}$, amorphous powder. Prepn: Miyatake et al., U.S. pat. 3,023,147 (1962). Freely sol in water, alcohol, acetone. Practically insol in benzene, ether. uv max: 217 nm (log ϵ 4.16). $[\alpha]_D^{26}$ −21.1° (methanol).

THERAP CAT: Cardiotonic.

3141. Digitalis. Foxglove; purple foxglove; fairy gloves; Digifortis; Digitora; Neodigitalis; Pil-Digis. Dried leaves of Digitalis purpurea L., Scrophulariaceae. Habit. Southern and central Europe, cultivated in the U.S. Constit. Digitoxin (0.2-0.4%), digitonin, digitalin, antirhinic acid, digitalosmin, digitoflavone, inositol, pectin. One U.S.P. digitalis unit represents the potency of 0.1 g of the U.S.P. Digitalis Reference Standard. Toxicology: A. P. Somlyo, Am. J. Cardiol. 5, 523 (1960). Review of mechanism of action and clinical use: T. W. Smith, N. Engl. J. Med. 318, 358-365 (1988). Book: C. Fisch, B. Surawicz, Digitalis (Grune & Stratton, New York, 1969) 244 pp.

Human Toxicity: The therapeutic dose is close to the toxic dose which causes anorexia, nausea, salivation, vomiting, diarrhea, headache, drowsiness, disorientation, delirium, hallucinations; death may result, Clinical Toxicology of Commercial Products, R. E. Gosselin et al., Eds. (Williams & Wilkins, Baltimore, 4th ed., 1976) Section III, pp 124-133.

THERAP CAT: Cardiotonic.

THERAP CAT (VET): Cardiotonic.

3142. Digitalose. 6-Deoxy-3-O-methylgalactose; 3-methyl-D-fucose. $C_7H_{14}O_5$; mol wt 178.18. C 47.18%, H 7.92%, O 44.90%. First obtained by Kiliani, Ber. 25, 2116 (1892), by hydrolysis of Digitalinum verum, a glycoside of gitoxigenin with glucose and digitalose. By hydrolysis of a glycoside from the seeds of Strophanthus eminii Aschers & Pax., Apocynaceae: Lamb, Smith, J. Chem. Soc. 1936, 442. Configuration: Schmidt et al., Naturwiss. 31, 247 (1943); Ann. 555, 26 (1944). Elderfield in Advances in Carbohydrate Chemistry vol. I (New York, 1945) p 150. Prepn by isoln and hydrolysis of emicymarin from Strophanthus eminii seeds: Elderfield, ibid. 156. Synthesis: Reber, Reichstein, Helv. Chim. Acta 29, 343 (1946). Schmidt, Wernicke, Ann. 558, 70 (1947).

Rosettes of needles from ethyl acetate. mp 106° when freshly prepd, mp 115° after 3 days, mp 119° after 4 mos. $[\alpha]_{5461}^{127}$ +109° (15 min c = 1.7) → $[\alpha]_{5461}^{122}$ +126° or $[\alpha]_D^{122}$ +106° (17 hrs). By the Bertrand micro method, 1 mg digitalose has the same reducing power as 9.32 mg anhydr glucose. Soluble in water.

3143. Digitogenin. (25R)-5α-Spirostan-2α,3β,15β-triol. $C_{27}H_{44}O_5$; mol wt 448.62. C 72.28%, H 9.89%, O 17.83%. The aglycon of digitonin: Tschesche, Ber. 68, 1090 (1935); Tschesche, Hagedorn, Ber. 69, 797 (1936). Structure and stereochemistry: Marker et al., J. Am. Chem. Soc. 64, 1843 (1942); Meystre et al., Helv. Chim. Acta 38, 381 (1955). Klass et al., J. Am. Chem. Soc. 77, 3829 (1955); and IR spectra: Djerassi et al., ibid. 78, 3166 (1956).

Consult the cross index before using this section.

Needles from alc, dec 296°. $[\alpha]_D^{19}$ −81° (c = 1.4 in chloroform). Practically insol in water. Sol in 30 parts chloroform, 35 parts boiling alc, 100 parts alc at 20°.

2,3-Diacetate, crystals from methanol-chloroform, mp 241.5-242°. $[\alpha]_D$ −104°.

Triacetate, minute needles from ether, mp 190°.

3144. Digitonin. Digitin. $C_{56}H_{92}O_{29}$; mol wt 1229.30. C 54.71%, H 7.54%, O 37.74%. Obtained from the seeds of *Digitalis purpurea* L., Scrophulariaceae. Extraction procedure: Gisvold, *J. Am. Pharm. Assoc.* **23**, 664 (1934). Purification of commercial digitonin and its separation into two fractions: G. Ruhenstroth-Bauer, P. M. Breitenfeld, *Z. Physiol. Chem.* **302**, 111 (1955). Structure: Tschesche, Wulff, *Tetrahedron* **19**, 621 (1963). Use as clinical reagent: H. H. Leffler, *Am. J. Clin. Pathol.* **31**, 310 (1959).

R = digitogenin

Crystals from alc, sinters 225°. Indistinct mp 235-240°. $[\alpha]_D^{20}$ −54° (0.45 g in 15.8 ml methanol). One gram dissolves in 57 ml abs alc, in 220 ml 95% alc. Practically insol in water, forming a soapy suspension. Also practically insol in chloroform, ether.

USE: Clinical reagent (cholesterol determination).

3145. Digitoxigenin. *3,14-Dihydroxycard-20(22)-enolide; $\Delta^{20:22}$-3,14,21-trihydroxynorcholenic acid lactone;* cerberigenin; echujetin; evonogenin; Thevetigenin. $C_{23}H_{34}O_4$; mol wt 374.50. C 73.76%, H 9.15%, O 17.09%. The aglycon of digitoxin, thevetin, cerberin, echujin, evomonosid. Prepn by refluxing digitoxin in a mixture of water + alcohol + HCl: Cloetta, *Arch. Exp. Pathol. Pharmakol.* **88**, 113 (1920); Stoll, Kreiss, *Helv. Chim. Acta* **17**, 592 (1934). Structure: Jacobs, Hoffmann, *J. Biol. Chem.* **67**, 333 (1926); Jacobs, Gustus, *ibid.* **78**, 573; **79**, 533 (1928); **82**, 402 (1929); Jacobs, Elderfield, *ibid.* **108**, 497 (1935); Elderfield, *Chem. Rev.* **17**, 187 (1935). Stereochemistry: Meyer, *Helv. Chim. Acta* **30**, 1976 (1947). Synthesis: Danieli *et al., Tetrahedron* **22**, 3189 (1966); W. Fritsch *et al., Ann.* **1974**, 621; S. F. Donovan *et al., Tetrahedron Letters* **1979**, 3287; R. Marinibettolo *et al., Can. J. Chem.* **59**, 1403 (1981); T. Milkova *et al., Tetrahedron Letters* **23**, 413 (1982). Biosynthesis from neriifolin, *q.v.*: A. Cruz *et al., J. Org. Chem.* **42**, 3580 (1977).

Stout prisms from 40% alc, mp 253°. $[\alpha]_D^{17}$ +19.1° (c = 1.36 in methanol). Sol in alc, chloroform, acetone; slightly sol in ethyl acetate; very sparingly sol in ether, water.

3-Acetyldigitoxigenin, $C_{25}H_{36}O_5$, hexagonal plates from acetone + ether, mp 222-227°. $[\alpha]_D^{17}$ +21.4° (c = 1.02 in chloroform).

3146. Digitoxin. *(3β,5β)-3-[(O-2,6-Dideoxy-β-D-ribohexopyranosyl-(1→4)-O-2,6-dideoxy-β-D-ribo-hexopyranosyl-(1→4)-2,6-dideoxy-β-D-ribo-hexopyranosyl)oxy]-14-hydroxycard-20(22)-enolide;* digitalin, crystalline; digitophyllin; Asthenthilo; Cardigin; Carditoxin; Coramedan; Cristapurat; Crystodigin; Digicor; Digilong; Digimerck; Digimed; Digipural; Ditaven; Digisidin; Digitaline Nativelle; Lanatoxin; Mono-Glycocard; Myodigin; Purodigin; Purpurid; Tardigal; Unidigin. $C_{41}H_{64}O_{13}$; mol wt 764.92. C 64.37%, H 8.43%, O 27.19%. Secondary glycoside from *Digitalis purpurea* L., Scrophulariaceae. Extracted from the dried leaves with 50% alc. Extraction procedure: Cloetta, *Arch. Exp. Pathol. Pharmakol.* **112**, 261 (1926). Purification: Windaus, Freese, *Ber.* **58**, 2503 (1925). Ten kilo leaves yield about 6 grams pure digitoxin. Identity with digitophyllin: Cloetta, *Arch. Exp. Pathol. Pharmakol.* **88**, 113 (1920). Structure: See reviews by Elderfield in *Chem. Rev.* **17**, 187 (1935), and Stoll, *The Cardiac Glycosides, London* (1937); also Shoppee in *Ann. Rev. Biochem.* **11**, 103 (1942). Acid hydrolysis yields 1 mol digitoxigenin + 3 mol digitoxose. The sugar residue is attached to the hydroxyl group at C-3 of the aglycon. Structure of the sugar moiety: Lichti *et al., Helv. Chim. Acta* **45**, 868 (1962). Digitoxin U.S.P. is either pure digitoxin or a mixture of cardioactive glycosides obtained from *Digitalis purpurea* and consisting chiefly of digitoxin. The potency of digitoxin U.S.P. corresp to the potency of an equal wt of U.S.P. Digitoxin Reference Standard. A deviation of 20% is permitted. The physical characteristics given below are those of the pure compd. Toxicity data: Foerster *et al., Arch. Int. Pharmacodyn. Ther.* **159**, 1 (1966). Comprehensive description: I. M. Jakovljevic in *Analytical Profiles of Drug Substances* vol. 3, K. Florey, Ed. (Academic Press, New York, 1974) pp 149-172.

Very small elongated, rectangular plates from dil alc. May contain ½ or 1 mol H_2O or EtOH which is given up at 118° *in vacuo*. When anhydrous, mp 256-257°. $[\alpha]_D^{20}$ +4.8° (c = 1.2 in dioxane). One gram dissolves in about 40 ml chloroform, in about 60 ml alcohol, and in about 400 ml ethyl acetate. Also sol in acetone, amyl alcohol, pyridine, sparingly sol in ether, petr ether, water (1 g/100 liter at 20°). LD_{50} orally in guinea pigs, cats (mg/kg): 60.0, 0.18 (Foerster).

Acetyl derivatives, see Acetyldigitoxin.

THERAP CAT: Cardiotonic.

THERAP CAT (VET): Cardiac tonic.

3147. Digitoxose. *2,6-Dideoxy-D-ribo-hexose;* 2-desoxy-D-altromethylose; 2,6-desoxy-D-allose. $C_6H_{12}O_4$; mol wt 148.16. C 48.64%, H 8.16%, O 43.20%. Obtained by mild acid hydrolysis of the glycosides digitoxin, gitoxin and digoxin: Cloetta, *Arch. Exp. Pathol. Pharmakol.* **88**, 113 (1920); **112**, 261 (1926); Windaus, Stein, *Ber.* **61**, 2436 (1928); Kraft, *Arch. Pharm.* **250**, 118 (1912); Mannich *et al., ibid.* **268**, 453 (1930); Smith, *J. Chem. Soc.* **1930**, 508; **1931**, 23. Configuration: Micheel, *Ber.* **63**, 347 (1930). Structure: S. F. Dyke, *The Carbohydrates* (Interscience, New York, 1960) p 104. Synthesis: Gut, Prins, *Helv. Chim. Acta* **30**, 1223 (1947); Bolliger, Ulrich, *ibid.* **35**, 93 (1952). Stereochemical study: S. Tsukamoto *et al., J. Chem. Soc. Perkin*

Trans. I **1988,** 2621. *Review:* R. C. Elderfield in W. W. Pigman, M. L. Wolfrom, *Advances in Carbohydrate Chemistry* vol. I (Academic Press, New York, 1945) pp 159-164.

Crystals from methanol + ether, from ethyl acetate or from acetone + ether, mp 112°. $[\alpha]_D^{17}$ +46.3° (in water); $[\alpha]_D^{20}$ +39.1° (in methanol); $[\alpha]_D^{18}$+27.9° → +43.3° (after 24 hrs in pyridine). Freely sol in water; sol in acetone, ethanol. Practically insol in ether.

3148. Diglyme. *1,1'-Oxybis[2-methoxyethane]; bis(2-methoxyethyl) ether;* diethylene glycol dimethyl ether. $C_6H_{14}O_3$; mol wt 134.17. C 53.71%, H 10.52%, O 35.77%. $(CH_3OCH_2CH_2)_2O$. Prepn: Cretcher, Pittenger, *J. Am. Chem. Soc.* **47,** 163 (1925); Gallaugher, Hibbert, *ibid.* **58,** 813 (1936).

Liquid, bp_{760} 162°; bp_{200} 116°; bp_{35} 75°; bp_3 20°. d_4^{20} 0.9451. mp −68°. Flash pt (open cup) 158°F (70°C). n_D^{20} 1.4097. Miscible with water, alcohol, ether, hydrocarbon solvents.

USE: Solvent; reaction medium for Grignard and similar syntheses.

3149. Digoxigenin. *3,12,14-Trihydroxycard-20(22)-enolide;* $\Delta^{20:22}$-3β,12β,14,21-tetrahydroxynorcholenic acid lactone; lanadigenin. $C_{23}H_{34}O_5$; mol wt 390.53. C 70.74%, H 8.78%, O 20.49%. The aglycone of digoxin. By hydrolysis of digoxin: Smith, *J. Chem. Soc.* **1930,** 508. From *Digitalis orientalis* L. and *D. lanata* Ehrh., *Scrophulariaceae:* Mannick, Schneider, *Arch. Pharm.* **279,** 223 (1941); Pataki *et al., Helv. Chim. Acta* **36,** 1295 (1953). Structure: Meyer, Reichstein, *Experientia* **9,** 253 (1953); Cardwell, Smith, *ibid.* 367; *eidem, J. Chem. Soc.* **1954,** 2012. Synthesis: P. Welzel, H. Stein, *Tetrahedron Letters* **22,** 3385 (1981).

Dihydrate, prismatic rods from dil alc. Anhyd as stout prisms from ethyl acetate, mp 222°. $[\alpha]_{546}^{20}$ +27.0° (c = 1.77 in methanol). Although it is a 3β-alcohol, it is not precipitated by digitonin: Pataki *et al., loc. cit.*

3,12-Diacetyldigoxigenin, prisms from dil methanol, mp 222-223°. $[\alpha]_{546}^{20}$ +61.3° (c = 2 in methanol).

3150. Digoxin. (3β,5β,12β)-3-[(O-2,6-*Dideoxy-β-D-ribo-hexopyranosyl-(1 → 4)-O-2,6-dideoxy-β-D-ribo-hexopyranosyl-(1 → 4)-2,6-dideoxy-β-D-ribo-hexopyranosyl]oxy]-12,14-dihydroxycard-20(22)-enolide;* Cordioxil; Davoxin; Digacin; Dilanacin; Dixina; Dokim; Dynamos; Lanacordin; Lanicor; Lanoxin; LenoxiCaps; Lenoxin; Longdigox; Neo-Dioxanin; Rougoxin; Stillacor; Vanoxin. $C_{41}H_{64}O_{14}$; mol wt 780.92. C 63.06%, H 8.26%, O 28.69%. Secondary glycoside from *Digitalis lanata* Ehrh., or *D. orientalis* Lam., *Scrophulariaceae:* Smith, *J. Chem. Soc.* **1930,** 508; Stoll, *The Cardiac Glycosides* (London, 1937); Dhar *et al.,* **Indian** pat. **62,497** (1958 to Council Sci. Indust. Res.), *C.A.* **53,** 653b (1959). *See also* ref under Digoxigenin. Acid hydrolysis of digoxin yields 1 mol digoxigenin + 3 mols digitoxose. The sugar residue is attached to the hydroxyl group at C-3 of the aglycon. Clinical pharmacokinetics: J. K. Aronson, *Clin. Pharmacokinet.* **5,** 137 (1980). Comprehensive description: P. R. B. Foss, S. A. Benezra in *Analytical Profiles of Drug Sub-*

stances vol. **9,** K. Florey, Ed. (Academic Press, New York, 1980) pp 207-243.

Radially arranged, four- and five-sided triclinic plates from dil alcohol or dil pyridine, decomp 230-265°. $[\alpha]_{Hg}^{25}$ +13.4° to 13.8° (c = 10 in pyridine). uv max (ethanol): 220 nm (ε 12800). Sol in dil alcohol, pyridine, or mixt of chloroform and alcohol. Almost insol in ether, acetone, ethyl acetate, chloroform, water. More sol in hot 80% alcohol than gitoxin.

β-Methyldigoxin, $C_{42}H_{66}O_{14}$, *medigoxin, metildigoxin, 4'''-O-methyldigoxin, 3β,12β,14β-trihydroxy-5β-card-20(22)-enolide-3-(4'''-O-methyltridigitoxoside), Cardiolan, Lanirapid, Lanitop.* Obtained by the *O*-methylation of digoxin: Kaiser *et al.,* **S. Afr.** pat. **68 06,079** (1969 to Boehringer, Mann.), *C.A.* **72,** 13002p (1970). Crystals, mp 227-231°. Pharmacology and toxicity studies: Schaumann, Wegerle, *Arzneimittel-Forsch.* **21,** 225 (1971); Czerwek *et al., ibid.* 231. LD_{50} in rats, mice (mg/kg): 4.8, 4.9 i.v.; 6.2, 4.8 i.p.; 8.3, 7.8 orally (Czerwek).

α-Acetyldigoxin, $C_{43}H_{66}O_{15}$, *Lanatilin, Decardil, Dioxanin, Sandolanid.* Obtained by enzymatic hydrolysis of digilanide. Prisms from methanol + chloroform, dec 225°. $[\alpha]_D^{20}$ +18.9° (pyridine). Very sparingly sol in ethyl acetate.

β-Acetyldigoxin, *Cor-Puren, Novodigal, Agolanid, Kardiamed.* Needles from alcohol + chloroform, dec 240°. $[\alpha]_D^{20}$ +30.4° (c = 1.2 in alc). More sol in ethyl acetate than the α-form.

THERAP CAT: Cardiotonic.

THERAP CAT (VET): Cardiotonic.

3151. Dihexyverine. *[1,1'-Bicyclohexyl]-1-carboxylic acid 2-(1-piperidinyl)ethyl ester;* 1-cyclohexylcyclohexane-carboxylic acid β-piperidinoethyl ester; dicyclohexylacetic acid 2-piperidinoethyl ester; dihexiverine. $C_{20}H_{35}NO_2$; mol wt 321.49. C 74.71%, H 10.97%, N 4.36%, O 9.95%. Prepn: Kopp, Tchoubar, *Bull. Soc. Chim. France* **1952,** 84. Pharmacology: Buchel *et al., Arch. Sci. Physiol.* **16,** 57 (1962); **21,** 537 (1967); Guyonneau, *Med. Exp.* **6,** 245 (1962).

Hydrochloride, $C_{20}H_{36}ClNO_2$, *JL 1078, Metaspas, Diverine, Olimplex, Seclin, Dispas, Neospasmina, Spasmodex, Spasmalex.* Crystals, mp 175° (copper block). mp 200° (capillary). LD_{50} i.p. in mice: 212 mg/kg (Buchel).

THERAP CAT: Anticholinergic.

3152. Dihydralazine. *2,3-Dihydro-1,4-phthalazinedione dihydrazone; 1,4-dihydrazinophthalazine.* $C_8H_{10}N_6$; mol wt 190.21. C 50.51%, H 5.30%, N 44.19%. Vasodilator. Prepn: J. Druey, **U.S.** pat. **2,484,785** (1949 to Ciba); J. Druey, B. H. Ringier, *Helv. Chim. Acta* **34,** 195 (1951); Zerweck, Kunze, **U.S.** pat. **2,786,839** (1957 to Cassella Farbw.). Pharmacology: P. A. Van Zwieten, *Arzneimittel-Forsch.* **18,** 79 (1968). Mechanism of action: I. W. Reimann *et al., Clin. Sci.* **61,** 319s (1981); *eidem, Clin. Exp. Pharmacol. Physiol.* **12,** 79 (1985). Cerebrovascular effects in rats: L. M. Auer *et al., Acta Med. Scand.* **Suppl. 678,** 73 (1983); D. I. Barry *et al., Stroke* **15,** 102 (1984). Hemodynamic responses in man: G. G. Belz, *Clin. Pharmacol. Ther.* **37,** 48 (1985). Clinical studies: A. Salvadeo *et al., Int. J. Clin. Pharmacol. Ther. Toxicol.* **19,** 372 (1981); P. I. Salmela *et al., Ann. Clin. Res.* **13,** 433 (1981). Toxicity data: G. Steiner *et al., J. Med. Chem.* **24,** 59 (1981).

Orange needles from water, dec about 180°. LD_{50} i.p. in rats: 1084 μmol/kg (Steiner).

Hydrogen sulfate, $C_8H_{12}N_6SO_4$, *Nepresol (tabl)*, *Népressol*, *Pressunic*, *Depressan (tabl)*. Needles, dec 233°.

Hydrogen sulfate hemipentahydrate, $C_8H_{12}N_6SO_4\cdot$ $.2\frac{1}{2}H_2O$, *Dihyzin*.

Mesylate, $C_9H_{14}N_6SO_4$, *Nepresol (amp)*, *Depressan (amp)*.
THERAP CAT: Antihypertensive.

3153. Dihydrazine Sulfate. $H_{10}N_4O_4S$; mol wt 162.18. H 6.22%, N 34.55%, O 39.46%, S 19.77%. $(N_2H_4)_2.H_2SO_4$. White crystalline flakes. Apparent density 55 lbs/cu ft. mp about 104°, decomposes about 180°. Soluble in water (25°): 202 g/100 ml H_2O. Relatively insol in most organic solvents. *See also* Hydrazine Sulfate.
Caution: Irritating to eyes, skin, mucous membranes.

3154. Dihydrocodeine. *4,5-Epoxy-3-methoxy-17-methylmorphinan-6-ol*; 6-hydroxy-3-methoxy-N-methyl-4,5-epoxymorphinan; dihydroneopine; drocode; Codhydrine; Dehacodin; DF 118; DH-codeine; Didrate; Dihydrin; Hydrocodin; Nadeine; Novicodin; Paracodin; Parzone; Rapacodin. $C_{18}H_{23}NO_3$; mol wt 301.37. C 71.73%, H 7.69%, N 4.65%, O 15.93%. Prepd by reduction of codeine or neopine: Skita, Franck, *Ber.* **44**, 2862 (1911); Wieland, Koralek, *Ann.* **433**, 269 (1923); Stein, *Pharmazie* **10**, 180 (1955).

Crystals from methanol + water, mp 112-113°; bp_{15} 248°.
Bitartrate (acid tartrate), $C_{22}H_{29}NO_9$, *Fortuss, DHC, Dico.* Crystals from methanol, contains 66.8% dihydrocodeine when completely anhydr. mp 192-193° (Stein). The commercial medicinal grade usually melts at 186-190°. $[\alpha]_D^{25}$ −72° to −75° (c = 1 in H_2O). One gram dissolves in 4.5 ml water. Sparingly sol in alcohol. Insol in ether.
Caution: May be habit forming. This is a controlled substance (opiate) listed in the U.S. Code of Federal Regulations, Title 21 Part 1308.12 (1985).
THERAP CAT: Narcotic analgesic. Antitussive.

3155. Dihydrocodeinone Enol Acetate. *6,7-Didehydro-4,5-epoxy-3-methoxy-17-methylmorphinan-6-ol acetate (ester)*; demethyldihydrothebaine acetate (ester); acetyldemethyldihydrothebaine; acetyldihydrocodeinone; thebacon. $C_{20}H_{23}NO_4$; mol wt 341.39. C 70.36%, H 6.79%, N 4.10%, O 18.75%. Prepn by refluxing dihydrocodeinone with acetic anhydride and anhydr sodium acetate: Small *et al.*, *J. Org. Chem.* **3**, 204 (1938).

Needles from methanol, mp 154°. Practically insol in water. Soluble in most organic solvents.
Hydrochloride, $C_{20}H_{24}ClNO_4$, *Acedicon.* Crystals, mp 132-135° (dec), very sol in water, stable in boiling water.
Caution: May be habit forming. This is a controlled substance (opium derivative) listed in the U.S. Code of Federal Regulations, Title 21 Part 1308.11 (1985).
THERAP CAT: Narcotic analgesic. Antitussive.

3156. Dihydroequilin. *Estra-1,3,5(10),7-tetraene-3,17-diol.* $C_{18}H_{22}O_2$; mol wt 270.36. C 79.96%, H 8.20%, O 11.84%. Occurs in the 17α- and 17β-forms. Prepn of β-form by reduction of equilin with sodium in alc: David, *Acta Brev. Nederland* **4**, 63 (1934); Serini *et al.*, *U.S. pat.*

2,221,340 (1940 to Schering). Isoln of β-form from pregnant mares' urine: Gaudry, Glen, *Ind. Chim. Belge*, **Suppl. 2**, 435 (1959). Prepn of α-form by reduction of equilin with aluminum isopropoxide: Carol *et al.*, *J. Biol. Chem.* **185**, 267 (1950). Isoln of α-form from pregnant mares' urine: Glen *et al.*, *Nature* **177**, 753 (1956). α-Form was formerly called β-dihydroequilin and vice versa: Banes *et al.*, *J. Biol. Chem.* **187**, 557 (1950).

β-Form

17β-Form, crystals from acetone, mp 174.5-174.6°. $[\alpha]_D^{20}$ +220° (dioxane).
17α-Form, plates from 30% ethanol, mp 205.5-205.6°. $[\alpha]_D^{20}$ +213° (ethanol).

3157. Dihydroergotamine. *9,10-Dihydro-12'-hydroxy-2'-methyl-5'-(phenylmethyl)ergotaman-3',6',18-trione.* $C_{33}H_{37}N_5O_5$; mol wt 583.67. C 67.90%, H 6.39%, N 12.00%, O 13.71%. Prepd by the catalytic hydrogenation of ergotamine: Stoll, Hofmann, *Helv. Chim. Acta* **26**, 2070 (1943).
Strongly refractive prisms from dil acetone, contg 2 mols acetone and 2 mols water of crystn. mp 239°. $[\alpha]_D^{20}$ −64°; $[\alpha]_{546}^{20}$ −79° (c = 0.5 in pyridine). Insol in water. Sparingly sol in methanol, ethanol, chloroform, benzene.
Hydrochloride, $C_{33}H_{38}ClN_5O_5$, fine needles from methanol, dec 220-225°.
Tartrate, $C_{70}H_{80}N_{10}O_{16}$, *Divegal.* Six-sided plates from methanol, dec 210-215°.
Methanesulfonate, $C_{34}H_{41}N_5O_8S$, *dihydroergotamine mesylate, DHE-45, DET-MS, Agit, Angionorm, Dergotamine, Dirgotarl, Diergotan, Dihydergot, Dihytam, D-Tamin retard L.U.T., Endophleban, Ergomimet, Ergont, Ergotex, Ergotonin, Hydro-Tamin, Ikaran, Kidira, Morena, Orstanorm, Seglor, Tonopres, Verladyn.* Large prisms from 95% alc. mp 230-235°. Moderately sol in water.
THERAP CAT: Vasoconstrictor (specific in migraine).

3158. Dihydro-β-erythroidine. *(3β)-1,6-Didehydro-14,17-dihydro-3-methoxy-16(15H)-oxaerythrinan-15-one; 12,13-didehydro-2,7,13,14-tetrahydro-α-erythroidine.* $C_{16}H_{21}NO_3$; mol wt 275.34. C 69.79%, H 7.69%, N 5.09%, O 17.43%. The more active of the two isomers that constitute the principal alkaloidal fraction of seeds from several *Erythrina* spp. In contrast to curare, *q.v.*, it retains its ability to block neuromuscular transmission even when administered orally: K. Unna, J. G. Greslin, *J. Pharmacol.* **80**, 53 (1944); K. Unna *et al.*, *ibid.* 39. Prepd by catalytic hydrogenation of β-erythroidine or a salt of β-erythroidine: Folkers, Koniuszy, U.S. pat. 2,370,651 (1945 to Merck & Co.). Structure: Boekelheide *et al.*, *J. Am. Chem. Soc.* **75**, 2550 (1953). Configuration: Hanson, *Proc. Chem. Soc.* **1963**, 52.

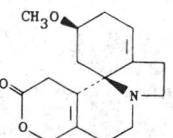

Crystals from anhydr ethyl ether, dec 85-86°. $[\alpha]_D^{25}$ +102.5°. Soluble in ethanol. Alkaline hydrolysis yields sodium dihydro-β-erythroidinate.
Hydrochloride, crystals from abs ethanol, mp 238°. $[\alpha]_D^{25}$ +124.7°.
Hydrobromide, crystals from abs ethanol or abs ethanol + abs ether, dec 242°. Bitter taste. $[\alpha]_D^{25}$ +106-107.5°. Soly in water 0.8 g/ml; in ethanol 0.5 g/100 ml.

3159. Dihydroisocodeine. *4,5α-Epoxy-3-methoxy-17-methylmorphinan-6β-ol*; DHIC. $C_{18}H_{23}NO_3$; mol wt 301.37.

C 71.73%, H 7.69%, N 4.65%, O 15.93%. Differs from dihydrocodeine only by the spatial arrangement of the —CHOH— group. Prepn by catalytic reduction of isocodeine: Speyer, Krauss, *Ann.* **432**, 233 (1923); Lutz, Small, *J. Am. Chem. Soc.* **54**, 4724 (1932); Rapoport, Payne, *J. Org. Chem.* **15**, 1097 (1950); by epimerization of dihydrocodeine: Baizer, U.S. pat. **2,774,762** (1956 to N.Y. Quinine and Chem. Works).

Prisms from ethanol, mp 199-200°.

Acid tartrate, $C_{18}H_{23}NO_3 \cdot C_4H_6O_6$, crystals, mp 192°, $[\alpha]_D^{29}$ −65.3°.

Acid tartrate trihydrate, $C_{18}H_{23}NO_3 \cdot C_4H_6O_6 \cdot 3H_2O$, crystals from water, mp about 180°. $[\alpha]_D^{26}$ −62.4° (c = 1.94). Soly at 24° = 4.5 g/100 ml H_2O.

Methiodide, $C_{18}H_{23}NO_3 \cdot CH_3I$, crystals from ethanol, dec 269-272°.

Picrate, crystals, mp 235-237°.

3160. Dihydromorphine. *(5α,6α)-4,5-Epoxy-17-methylmorphinan-3,6-diol.* $C_{17}H_{21}NO_3$; mol wt 287.35. C 71.05%, H 7.37%, N 4.87%, O 16.70%. Prepd by hydrogenation of morphine or opium; by demethylation of tetrahydrothebaine; from dihydrocodeine. *Ref:* Small, Lutz, *Chemistry of the Opium Alkaloids,* Suppl. No. 103, Public Health Reports, Washington (1932); Eddy, Reid *J. Pharmacol.* **52**, 468 (1934); K. W. Bentley, *The Chemistry of the Morphine Alkaloids* (Oxford, 1954).

Monohydrate, prisms from chloroform, mp (dry) 157° (dec). Insoluble in water. Sol in acetone, alcohol, chloroform, dil acids.

Hydrochloride, $C_{17}H_{22}ClNO_3$, *paramorphan, paramorfan.* Prismatic crystals, unmelted at 280° *in vacuo.* $[\alpha]_D^{25}$ −112° (c = 1.6). Very sol in water; sparingly sol in abs alcohol.

Hydriodide, $C_{17}H_{22}INO_3$, crystals, soften at 270°. mp 275°.

Caution: May be habit forming. This is a controlled substance (opium derivative) listed in the U.S. Code of Federal Regulations, Title 21 Parts 329.1, 1308.11 (1987).

THERAP CAT: Narcotic analgesic.

3161. Dihydroresorcinol. *1,3-Cyclohexanedione;* hydroresorcinol. $C_6H_8O_2$; mol wt 112.12. C 64.27%, H 7.19%, O 28.54%.

Crystals, mp about 105° with decompn. Sol in water, alcohol, chloroform, acetone, boiling benzene; slightly sol in ether, carbon disulfide, petr ether.

3162. Dihydrostreptomycin. *O-2-Deoxy-2-(methylamino)-α-L-glucopyranosyl-(1 → 2)-O-5-deoxy-3-C-(hydroxymethyl)-α-L-lyxofuranosyl-(1 → 4)-N,N'-bis(aminoiminomethyl)-D-streptamine;* DHSM; DST; Abiocine; Vibriomycin. $C_{21}H_{41}N_7O_{12}$; mol wt 583.62. C 43.22%, H 7.08%, N 16.80%, O 32.90%. Semi-synthetic antibiotic prepd by reduction of streptomycin: Bartz *et al., J. Am. Chem. Soc.* **68**, 2163 (1946); Fried, Wintersteiner, *ibid.* **69**, 79 (1947); Peck, U.S. pat. **2,498,574** (1950 to Merck & Co.); Carboni, Regna, U.S. pat. **2,522,858** (1950 to Pfizer); Levy, U.S. pat. **2,663,-685** (1953 to Schenley); Dolliver, Semenoff, U.S. pat. **2,717,-236** (1955 to Olin Mathieson); Kaplan, U.S. pat. **2,790,792** (1957 to Bristol); Sokol, Popino, U.S. pat. **2,784,181** (1957 to Cyanamid); Jurist, U.S. pat. **2,945,850** (1960 to Olin Mathieson). Isoln from fermentation broth of *Streptomyces*

humidus: Nakazawa *et al.,* and Tatsuoka *et al.,* U.S. pats. **2,931,756** and **2,950,277** (both 1960 to Takeda). Crystn of free base (obtained from the sulfate): Rhodehamel *et al., Science* **111**, 233 (1950). Crystn of the hydrochloride: Wolf *et al., ibid.* **109**, 515 (1949); Wolf, U.S. pats. **2,590,139; 2,594,245** (both 1952 to Merck & Co.). Crystn of the sulfate: Solomons, Regna, *Science* **109**, 515 (1949); Wolf *et al., ibid.* 515; Wolf, U.S. pats. **2,590,140; 2,590,141** (both 1952 to Merck & Co.); R. B. Peet, U.S. pat. **2,640,054** (1953 to Heyden Chem.); Katz, U.S. pat. **2,744,892** (1956 to Schenley). Total synthesis: Umezawa *et al., J. Am. Chem. Soc.* **96**, 920 (1974); *eidem, Bull. Chem. Soc. Japan* **48**, 563 (1975); T. Yamasaki *et al., J. Antibiot.* **31**, 1233 (1978).

R = CH_3
R' = CH_2OH

Hydrate, crystals from aq acetone. Chars at 240°, turning black up to 300° without melting.

Trihydrochloride, $C_{21}H_{44}Cl_3N_7O_{12}$, amorphous solid, dec 190-195°, or crystals from methanol. Soly in methanol at 25°: 45 mg/ml for crystalline form; > 1 g/ml for amorphous form. Crystals contain methanol of crystn which is lost on heating at 100°C. $[\alpha]_D^{25}$ −95° (1% soln).

Sesquisulfate, $C_{42}H_{88}N_{14}O_{36}S_3$, *Didromycine, Doublemycin, Sol-Mycin, Streptomagma.* Amorphous solid; crystalline form from water + (methanol, methyl ethyl ketone, or other low boiling solvent). Crystals, dec 255-265°, also reported as dec 250°. $[\alpha]_D^{25}$ −88.5° (1% soln). Crystals show very little hygroscopicity in contrast to amorphous form. Both forms are very sol in water. Solubility in 50% methanol + water: 0.8 mg/ml for crystals; 100 mg/ml for amorphous form. Soly at 28° (mg/ml): > 20 in water; 0.35 in methanol; 0.10 in ethanol: Weiss *et al., Antibiot. Chemother.* **7**, 374 (1957). Each milligram contains 800 micrograms of dihydrostreptomycin base on a potency basis.

Pantothenate, $C_{30}H_{58}N_8O_{17}$, *Didrothenat, Pantostrep.*

THERAP CAT: Antibacterial (tuberculostatic).

THERAP CAT (VET): Antibacterial.

3163. Dihydrotachysterol. *9,10-Secoergosta-5,7,22-trien-3β-ol;* dichystrolum; anti-tetany substance 10; AT 10; Antitanil; Calcamine; Dygratyl; Dihydral; Hytakerol; Parterol; Tachyrol. $C_{28}H_{46}O$; mol wt 398.65. C 84.35%, H 11.63%, O 4.01%. Prepn by reduction of tachysterol: Windaus *et al., Ann.* **499**, 198 (1932); v. Werder, U.S. pat. **2,228,491** (1941 to Winthrop), *cf. Z. Physiol. Chem.* **260**, 119 (1939). Activity similar to parathyroid extract.

Needles from 90% methanol, mp 125-127°. $[\alpha]_D^{22} +97.5°$ (chloroform). uv max: 242, 251, 261 nm (E$_{1cm}^{1\%}$ 870, 1010, 650). Insol in water. Easily sol in organic solvents.

THERAP CAT: Regulator (calcium).

3164. Dihydrothebaine. *6,7-Didehydro-4,5α-epoxy-3,6-dimethoxy-17-methylmorphinan.* $C_{19}H_{23}NO_3$; mol wt 313.38. C 72.82%, H 7.40%, N 4.47%, O 15.32%. Prepd by reduction of thebaine: Small *et al., J. Am. Chem. Soc.* **58,** 1457 (1936); K. W. Bentley, *The Chemistry of the Morphine Alkaloids* (Oxford, 1954) p 204.

Prisms from ethyl acetate, mp 162-163°. $[\alpha]_D^{20} -267°$ (c = 1.02 in benzene). Sol in alcohol, benzene, ethyl acetate; insol in water, alkalies.

Methyl iodide, $C_{20}H_{23}NO_4 \cdot CH_3I$, rods from alc, mp 257°.

3165. Dihydrovitamin K$_1$. *2-Methyl-3-(3,7,11,15-tetramethyl-2-hexadecenyl)-1,4-naphthalenediol; 2-methyl-3-phytyl-1,4-naphthohydroquinone;* phytonadiol; vitamin K$_1$ hydroquinone; α-phyllohydroquinone. $C_{31}H_{48}O_2$; mol wt 452.69. C 82.24%, H 10.69%, O 7.07%. Prepn: Fieser, *J. Am. Chem. Soc.* **61,** 2559, 3467 (1939); improved method: Wendler, U.S. pat. **2,831,899** (1958 to Merck & Co.); U.S. pat. **2,906,780** (1959). Prepn of esters: Hirschmann, U.S. pat. **3,076,004** (1963 to Merck & Co.).

Waxy mass. Freely sol in ether. Sparingly sol in petr ether, more sol in hot naphtha. Insol in water.

Diphosphate (O^1,O^4-diphosphoric acid) disodium salt, $C_{31}H_{48}Na_2O_8P_2$, *Kayhydrin, Mephyton DK.* mp 138°. Sol in water and methanol. Also used as the tetrasodium salt.

THERAP CAT: Prothrombogenic.

3166. Dihydroxyacetone. *1,3-Dihydroxy-2-propanone;* 1,3-dihydroxydimethyl ketone; Protosol; Ketochromin. $C_3H_6O_3$; mol wt 90.08. C 40.00%, H 6.71%, O 53.29%. HOCH$_2$COCH$_2$OH. Produced from glycerol by *Acetobacter* sp. under aerobic conditions: Bernhauer, Schoen, *Z. Physiol. Chem.* **177,** 107 (1928); Rutten, *Rec. Trav. Chim.* **70,** 449 (1951); Bousfield *et al., J. Inst. Brewing* **53,** 258 (1947); U.S. pat. **2,948,658** (1960 to Baxter). *Review: Chem. & Eng. News* **38,** 62 (Feb. 22, 1960), 54 (Aug. 15, 1960).

Cryst powder; fairly hygroscopic; characteristic odor; sweet, cooling taste. mp about 75-80°. The normal form is a dimer, slowly sol in 1 part water, 15 parts alcohol. When freshly prepd reverts rapidly to monomer in soln. The monomer is very sol in water, alcohol, ether, acetone.

USE: Artificial tanning agent; active ingredient in *Man-Tan; Oxatone; Tanorama; Q.T. (Quick Tan); Tan Tone; Magic Tan.* Formulations: Barmak, *Am. Perfum.,* March 1960, p 57; Andreadis, Miklean, U.S. pat. **2,949,403** (1960).

3167. Dihydroxyaluminum Acetylsalicylate. *Dihydroxy-(acetylsalicylato)aluminum;* dihydroxyaluminum aspirin. $C_9H_9AlO_6$; mol wt 240.15. C 45.01%, H 3.78%, Al 11.23%, O 39.98%. Prepn: Beekman, U.S. pat. **2,698,332** (1954 to Reheis); Schenck *et al.,* U.S. pat. **2,918,485** (1959 to Keystone Chemurgic).

May contain one or more molecules of Al(OH)$_3$. Stable in aq suspension at neutral pH, dec below pH 4.

THERAP CAT: Analgesic, antipyretic.

3168. Dihydroxyaluminum Aminoacetate. *(Glycinato-N,O)dihydroxyaluminum;* dihydroxy(glycinato)aluminum; aluminum aminoacetate (basic); aluminum dihydroxyaminoacetate; aluminum glycinate; glycine, aluminum salt; Ada; Alamine; Alglyn; Alminate; Alubasine; Alzinox; Aspogen; Dimothyn; Doraxamin; Elcosal; Robalate. $C_2H_6AlNO_4$; mol wt 135.05. C 17.79%, H 4.48%, Al 19.97%, N 10.37%, O 47.39%. NH$_2$CH$_2$COOAl(OH)$_2$. Prepd by adding a soln of aluminum isopropoxide in isopropanol to an aq soln of glycine: Krantz *et al., J. Pharmacol.* **82,** 247 (1944).

Very fine powder. Bland taste. Insol in water. Mixes easily with water forming suspensions which do not separate readily. At 25° the pH of the aq suspension (1 in 25) is 7.4. One gram consumes 200 ml of 0.1N HCl under conditions of the pharmacopoeial test for aluminum hydroxide gel.

THERAP CAT: Antacid.

3169. Dihydroxyaluminum Sodium Carbonate. *[Carbonato(1—)-O]dihydroxyaluminum monosodium salt;* aluminum sodium carbonate hydroxide; Kompensan; Minicid. CH$_2$AlNaO$_5$; mol wt 144.00. C 8.34%, H 1.40%, Al 18.74%, Na 15.97%, O 55.55%. (HO)$_2$AlOCO$_2$Na. Prepd by the reaction between an aluminum alkoxide and NaHCO$_3$ in water: Grote, U.S. pat. **2,783,179** (1957 to Chattanooga Medicine).

Amorphous powder or poorly formed crystals. May contain 10 to 11% H$_2$O. d 2.144. pH of water suspension: 9.7. Acid consuming power: 230 ml 0.1N HCl/gram.

THERAP CAT: Antacid.

3170. Dihydroxymaleic Acid. *(Z)-2,3-Dihydroxy-2-butenedioic acid;* 1,2-dihydroxyethylenedicarboxylic acid. $C_4H_4O_6$; mol wt 148.07. C 32.44%, H 2.72%, O 64.83%. Prepn from tartaric acid: Fenton, *J. Chem. Soc.* **87,** 811 (1905); Nef, *Ann.* **357,** 291 (1907).

Plates from water, usually crystallizes as the dihydrate. The anhyd acid dec 155° without melting. Slightly sol in cold water, ether, acetic acid; more sol in alcohol. Aq solns are unstable.

USE: In the detection of fluorides and of titanium. Has been proposed as antioxidant for frozen foods.

3171. 9,10-Dihydroxystearic Acid. *9,10-Dihydroxyoctadecanoic acid.* $C_{18}H_{36}O_4$; mol wt 316.47. C 68.31%, H 11.47%, O 20.22%. CH$_3$(CH$_2$)$_7$CH(OH)CH(OH)(CH$_2$)$_7$-COOH.

White, odorless, tasteless, lustrous crystals; fatty feel. mp 132-136°. Insol in water. Sol in hot alc or acetone, slightly in ether.

USE: Manuf cosmetic and toilet preparations.

3172. Dihydroxytartaric Acid. *Tetrahydroxybutanedioic acid;* diketosuccinic acid. $C_4H_6O_8$; mol wt 182.09. C 26.38%, H 3.32%, O 70.29%. HOOCC(OH)$_2$C(OH)$_2$COOH.

White, cryst powder. mp about 114-115° with decompn. Very sol in water; the aq soln dec on heating.

USE: As a reagent for the determination of sodium.

3173. 2,4-Diiodoaniline. *2,4-Diiodobenzenamine.* $C_6H_5I_2N$; mol wt 344.95. C 20.89%, H 1.46%, I 73.59%, N 4.07%.

Brown crystals. d 2.75. mp 95-96°. Slightly sol in cold water, moderately in hot water or in alcohol; freely sol in chloroform, ether, acetone, carbon disulfide, boiling alc.

3174. 4',5'-Diiodofluorescein. *3',6'-Dihydroxy-4',5'-diiodospiro[isobenzofuran-1(3H),9'-[9H]xanthen]-3-one;* hydroxydiiodo-o-carboxyphenylfluorone; D & C Orange No. 10; C.I. Solvent Red 73; C.I. 45425:1. $C_{20}H_{10}I_2O_5$; mol wt 584.12. C 41.12%, H 1.73%, I 43.46%, O 13.70%. Prepd by reaction of fluorescein with I_2 and HIO_4 or ICl in alkali: *Colour Index* vol. **4** (3rd ed., 1971) p 4427.

Orange-red powder. Slightly sol in water; sol in alcohol, in alkali hydroxide solns.

Sodium salt, $C_{20}H_8I_2Na_2O_5$, *Erythrosine Extra Yellowish, D & C Orange No. 11, C.I. Acid Red 95, C.I. 45425.*

Note: Structure depicted as the open tautomer.

USE: As an adsorption indicator in the determn of iodides in the presence of chlorides and bromides. The salts in dyeing and printing cotton, in printing half-silk, in dyeing jute, straw, etc.

3175. 3,5-Diiodosalicylic Acid. *2-Hydroxy-3,5-diiodobenzoic acid.* $C_7H_4I_2O_3$; mol wt 389.94. C 21.56%, H 1.03%, I 65.09%, O 12.31%. Prepn from salicylic acid and ICl: Woollett, Johnson, *Org. Syn. coll.* vol. **II**, 343 (1943); and iodine + H_2O_2: Jurd, *Aust. J. Sci. Res.* **3A**, 587 (1950).

Colorless, odorless needles or slightly yellow cryst powder. Sweetish, bitter taste. mp 235-236° dec. Sol in 5200 parts water at 25°; freely sol in alcohol, ether and most other organic solvents. Practically insol in $CHCl_3$, benzene.

Bismuth salt, *Bijosal:* Pollano, *Minerva Med.* **1930 I**, 786; *C.A.* **24**, 4854[7] (1930).

Ethyl ester, $C_9H_8I_2O_3$, *ethyl 3,5-diiodosalicylate.* Crystals, mp 133°. Practically insol in H_2O. Sol in alc, ether, benzene.

USE: As I_2 source in foods; in animal feeds as growth promotant.

3176. 3,5-Diiodothyronine. *O-(4-Hydroxyphenyl)-3,5-diiodotyrosine; 3-[4-(p-hydroxyphenoxy)-3,5-diiodophenyl]-alanine.* $C_{15}H_{13}I_2NO_4$; mol wt 525.10. C 34.31%, H 2.50%, I 48.34%, N 2.67%, O 12.19%. Prepn of DL-form: Harington, McCartney, *Biochem. J.* **21**, 852 (1927); Borrows *et al., J. Chem. Soc.* **1949**, Suppl. Issue No. 1, S199, S204; Siedel *et al.,* U.S. pat. **2,894,977** (1959 to Hoechst). Prepn of L(+)-form: Harington, *Biochem. J.* **22**, 1429 (1928); Chambers *et al., J. Chem. Soc.* **1949**, 3424; Hillmann, U.S. pat. **2,886,592** (1959); as the hydrochloride: Anthony, U.S. pat. **2,950,315** (1960 to Baxter); Meltzer, U.S. pat. **3,102,136** (1963 to Warner-Lambert). Prepn of D(−)-form: Harington, *loc. cit.;* Elks, Waller, *J. Chem. Soc.* **1952**, 2366.

DL-Form, plates, dec 256-257°.

L(+)-Form, crystals, dec 256°. $[\alpha]_D^{22}$ +26° (c = 1.06 in a 1:2 mixture of $1N$ HCl + alcohol).

L(+)-Form hydrochloride, crystals, mp 235-240°.

D(−)-Form, crystals, dec 256° (Harington), 265° (Elks, Waller). $[\alpha]_D^{20}$ −27.1° (c = 1 in $1N$ alc HCl:alc, 1:2 by vol).

USE: An intermediate in manuf of thyroxine.

3177. 3,5-Diiodotyrosine. *3,5-Diiodo-4-hydroxy-β-phenylalanine;* iodogorgoic acid; Agontan. $C_9H_9I_2NO_3$; mol wt 432.97. C 24.96%, H 2.09%, I 58.62%, N 3.24%, O 11.09%. Found in the skeletal proteins of corals, sponges, and other marine organisms. Isoln from *Gorgonia cavollini* as the DL acid: Drechsel, *Z. Biologie* **33**, 99 ; *Jahresber. Tierchemie* **1896**, 574; Henze, *Z. Physiol. Chem.* **38**, 71 (1903); isoln from the common sponge: Wheeler, Mendel, *J. Biol. Chem.* **7**, 1 (1909-10); Low, *J. Marine Res.* **10**, 239 (1951); from marine algae: Coulson, *Chem. & Ind. (London)* **1953**, 997; prepn from L- or DL-tyrosine: Henze, *Z. Physiol. Chem.* **51**, 67 (1907); Borrows *et al., J. Chem. Soc.* **1949**, Issue no. 1, S185; Jurd, *J. Am. Chem. Soc.* **77**, 5747 (1955); Boyle, Zlatkis, U.S. pat. **2,835,700** (1958 to Basic, Inc.).

L-Form, bunches of needles from water or 70% alcohol, dec 213°. $[\alpha]_D^{20}$ +2.89° (0.246 g in 5 g 4% HCl); $[\alpha]_D^{20}$ +2.27° (0.227 g in 5 g 25% NH_3). pK_1 2.12; pK_2 6.48; pK_3 7.82. Soly in water (g/l): at 0° = 0.204; at 25° = 0.617; at 50° = 1.862; at 75° = 5.62; at 100° = 17.00. On boiling with dil alc the crystals swell, and after prolonged boiling a gelatinous precipitate is formed.

DL-Form, double wedge crystals from 70% alc, rectangular plates from water. Dec about 200°. Soly in water (g/l): at 0° = 0.149; at 25° = 0.340; at 50° = 0.773 . On cooling a boiled hydroalcoholic soln, no gelatinous precip is formed.

THERAP CAT: Thyroid inhibitor.

3178. Diisoamylamine. *3-Methyl-N-(3-methylbutyl)-1-butanamine;* isodiamylamine. $C_{10}H_{23}N$; mol wt 157.29. C 76.36%, H 14.74%, N 8.91%. $(CH_3)_2CHCH_2CH_2NHCH_2-CH_2CH(CH_3)_2$.

Liquid. d_4^{20} 0.767. bp 188°. mp −44°. n_D^{21} 1.4229. Slightly sol in water; sol in alcohol, chloroform, ether.

Hydrochloride, $C_{10}H_{23}N\cdot HCl$. Vitreous mass, mp 276°. Freely sol in water, alcohol, ether.

Caution: Irritating to skin, mucous membranes. Has pressor effect.

3179. Diisobutyl Sodium Sulfosuccinate. *Sulfosuccinic acid diisobutyl ester S-sodium salt;* Aerosol IB; Alphasol IB. $C_{12}H_{21}NaO_7S$; mol wt 332.35. C 43.36%, H 6.37%, Na 6.91%, O 33.70%, S 9.67%. The isobutyl or butyl or 1-methylpropyl diester of the monosodium salt of sulfosuccinic acid or a mixture of all three esters. Wetting agent prepd by the action of the appropriate alcohols on maleic anhydrid followed by addition of sodium bisulfite: Jaeger, U.S. pats. **2,028,091; 2,176,423; Brit.** pat. **446,568; Fr.** pat. **776,495.**

The mixture of the three esters is available as a white, powder-like, easily grindable material. Soly in water at 25°: 760 g/l; at 60°: 804 g/l. Maximum concn of electrolyte soln in which 1% of the wetting agent is sol: 20% NaCl; 20% NH_4Cl; 30% $NaNO_3$; 20% Na_2SO_4. Also sol in glycerol, pine oil, oleic acid. Insol in acetone, kerosene, liq petrolatum, carbon tetrachloride, 2B ethanol, benzene, olive oil. Surface tension (dynes/cm) in water: 0.001% = 72.2; 0.1% = 67.5; 0.5% = 53.4; 1% = 49.1. Interfacial tension 1% in water vs. liq petrolatum: 10 seconds = 32.5 dynes/cm; 30 seconds = 32.3 dynes/cm; 15 minutes = 31.2 dynes/cm. Stable in acid and neutral solns; hydrolyzes in alkaline solns.

Caution: Irritating to eyes, mucous membranes.

USE: Wetting agent.

3180. Diisopromine. *N,N-Bis(1-methylethyl)-γ-phenyl-benzenepropanamine; N,N-diisopropyl-3,3-diphenylpropyl-*

amine; 1,1-diphenyl-3-diisopropylaminopropane; disopromine. $C_{21}H_{29}N$; mol wt 295.45. C 85.36%, H 9.89%, N 4.74%. Prepn: **Brit.** pat. **808,158** (1959 to N. V. Nederland. Comb. voor Chem. Ind. and Janssen, N. V.).

Hydrochloride, $C_{21}H_{30}ClN$, *Agofell, Bilagol.* Crystals, mp 175-176°. uv max: 254, 259.5, 268.8 nm. Soluble in water, methanol, ethanol, chloroform; practically insol in ether, aq solns of alkalies and ammonia.

THERAP CAT: Antispasmodic.

3181. Diisopropylamine. *N-(1-Methylethyl)-2-propanamine;* DIPA. $C_6H_{15}N$; mol wt 101.19. C 71.21%, H 14.94%, N 13.84%. $[(CH_3)_2CH]_2NH$. Prepn: A. Siersch, *Ann.* **148,** 263 (1868); M. Van der Zande, *Rec. Trav. Chim.* **8,** 202 (1889).

Liquid; characteristic odor; strongly alkaline. d^{22} 0.722. bp 84°. Flash pt, open cup: 21°F (-6°C). Sol in water, alc. LD_{50} orally in rats: 0.77 g/kg, H. F. Smyth *et al., Arch. Ind. Hyg. Occup. Med.* **10,** 61 (1954).

Lithium salt, *LDA, lithiodiisopropylamine, lithium diisopropylamide.* Review of chemistry: *Reagents for Organic Synthesis* vol. **8,** M. Fieser, Ed. (Wiley-Interscience, New York, 1980) pp 292-299. Powder, melts with decompn. Air and moisture sensitive.

Caution: Irritating to skin, mucous membranes.
USE: In organic synthesis.

3182. Diisopropylamine Dichloroacetate. *Dichloroacetic acid compd with N-(1-methylethyl)-2-propanamine (1:1);* dichloroacetic acid diisopropylammonium salt; diisopropylammonium dichloroacetate; diisopropylamine dichloroethanoate; DADA; DIPA-DCA; DIEDI; IS 401; Dapocel; Dedyl; Disotat; Kalodil; Oxypangam; Tensicor. $C_8H_{17}Cl_2NO_2$; mol wt 230.15. C 41.75%, H 7.45%, Cl 30.81%, N 6.09%, O 13.90%. Prepn: **Brit.** pat. **862,248** (1961 to Italseber). Pharmacology: V. A. E. Kraushaar *et al., Arzneimittel-Forsch.* **13,** 109 (1963); P. W. Stacpoole, *J. Clin. Pharmacol. J. New Drugs* **9,** 282 (1969). Mutagenicity study: M. D. Gelernt, V. Herbert, *Nutr. Cancer* **3,** 129 (1982). Constituent of some products sold in the U.S. as "pamganic acid".

Crystals, mp 119-121°. Soly in water: > 50%. LD_{50} orally in mice: 1700 mg/kg, V. A. E. Kraushaar *et al., loc. cit.*

THERAP CAT: Vasodilator, hypotensive.

3183. Diisopropyl Paraoxon. *Bis(1-methylethyl)phosphoric acid 4-nitrophenyl ester;* diisopropyl p-nitrophenyl phosphate; Mioticol; Propicol. $C_{12}H_{18}NO_6P$; mol wt 303.26. C 47.53%, H 5.98%, N 4.62%, O 31.66%, P 10.22%. Prepn: **Brit.** pat. **687,596** (1953 to Albright & Wilson).

THERAP CAT: Cholinergic (ophthalmic).

3184. Dikegulac. *2,3:4,6-Bis-O-(methylethylidene)-α-L-xylo-2-hexulofuranosonic acid; α-2,3:4,6-di-O-isopropylidene-L-xylo-hexulofuranosonic acid;* di-O-isopropylidene-2-keto-L-gulonic acid; diacetone-2-ketogulonic acid; diacetone-2-oxo-L-gulonic acid; oxogulonic acid diacetonide. $C_{12}H_{18}O_7$; mol wt 274.27. C 52.55%, H 6.61%, O 40.84%. Intermediate in the manuf of ascorbic acid, *q.v.* Manufacturing process: G. M. Jaffe, E. J. Pleven, **Ger.** pat. **2,123,-621;** *eidem,* U.S. pat **3,832,355** (1970, 1974 both to Hoffmann-La Roche). Use as a plant growth regulator: W.

Szkrybalo, **Ger.** pat. **2,339,239;** *eidem,* U.S. pat. **4,337,080** (1974, 1982 both to Hoffmann-La Roche); P. Bocian *et al., Nature* **258,** 142 (1975). Physicochemical properties, toxicity and growth retardant effect: W. H. de Silva *et al., Proc. Brit. Crop. Prot. Conf.-Weeds* **1976,** 349. Activity: S. S. Purohit, *Comp. Physiol. Ecol.* **4,** 264 (1979); **6,** 261 (1981).

Sodium salt, $C_{12}H_{17}NaO_7$, *dikegulac-sodium, Ro 7-6145, Atrinal.* Powder, mp > 300°. Vapor pressure at 25°: < 10^{-10} mm Hg. Soly at 20° (g/l): water 590; methanol 390; ethanol 230; chloroform 60; acetone < 10; hexane < 10; cyclohexanone < 10. LD_{50} in mice, male, female rats (mg/kg): 19500, 31000, 18000 orally; LC_{50} (96 hr) in bluegill sunfish, rainbow trout (ppm): > 10000, > 5000 (de Silva).

USE: Plant growth regulator; herbicide.

3185. Dilazep. *3,4,5-Trimethoxybenzoic acid (tetrahydro-1H-1,4-diazepine-1,4(5H)-diyl)di-3,1-propanediyl ester; 3,4,5-trimethoxybenzoic acid diester with tetrahydro-1H-1,4-diazepine-1,4(5H)-dipropanol;* 1,4-bis[3-(3,4,5-trimethoxybenzoyloxy)propyl]perhydro-1,4-diazepine; *N,N'*-bis[3-(3,4,5-trimethoxybenzoyloxy)propyl]homopiperazine; *N,N'*-(bis-ω-hydroxypropyl)homopiperazine 3,4,5-trimethoxybenzoate (diester). $C_{31}H_{44}N_2O_{10}$; mol wt 604.70. C 61.57%, H 7.33%, N 4.63%, O 26.46%. Prepn: **Brit.** pat. **1,107,470** corresp to Arnold *et al.,* U.S. pat. **3,532,685** (1968 and 1970 to Asta-Werke). Series of articles on pharmacology and metabolism: *Arzneimittel-Forsch.* **22,** 639-666 (1972). Toxicology: H. H. Abel *et al., ibid.* 667. Clinical results: Messerich, *Med. Welt.* **1972,** 563.

Dihydrochloride, $C_{31}H_{46}Cl_2N_2O_{10}$, *Asta C 4898, Comelian, Cormelian, Labitan.* Crystals from ethanol, mp 194-198° (monohydrate). LD_{50} in male mice, male rats (mg/kg): 26.6, 19.1 i.v.; 161, 90.1 i.p.; 3740, > 2150 orally (Abel).

THERAP CAT: Coronary vasodilator.

3186. Dilevalol. *[R-(R*,R*)]-2-Hydroxy-5-[1-hydroxy-2-[(1-methyl-3-phenylpropyl)amino]ethyl]benzamide; (R,R)-*labetalol; SCH 19927; Unicard. $C_{19}H_{24}N_2O_3$; mol wt 328.41. C 69.49%, H 7.37%, N 8.53%, O 14.61%. Non-selective β-adrenergic blocker with vasodilating activity; active isomer of labetalol, *q.v.* For prepn of racemate see labetalol. Synthesis and preliminary pharmacology: J. E. Clifton *et al., J. Med. Chem.* **25,** 670 (1982); E. H. Gold *et al., ibid.* 1363. Absolute configuration: P. Murray-Rust *et al., Acta Crystallogr.* **C40,** 825 (1984). Adrenoceptor blocking properties in comparison with labetalol: E. J. Sybertz *et al., J. Pharmacol. Exp. Ther.* **218,** 435 (1981). Antihypertensive and hemodynamic effects in rats and dogs: T. Baum *et al., ibid.* 444; and cardiac effects: J. J. Lynch *et al., ibid.* **239,** 719 (1986). HPLC determn in biological fluids: K. B. Alton *et al., J. Chromatog.* **425,** 363 (1988). Clinical evaluation in hypertension: J. Soberman *et al., J. Clin. Hypertens.* **3,** 271 (1987); J. D. Wallin *et al., Arch. Intern. Med.* **148,** 534 (1988). Symposium on pharmacology and clinical efficacy: *J. Cardiovasc. Pharmacol.* **11,** Suppl. 2, S1-S45 (1988).

Gum. $[\alpha]_D$ —21.7°.

Hydrochloride, $C_{19}H_{25}ClN_2O_3$, polymorphic crystals from ethanol, mp 133-134° (dec); mp 192-193.5° (dec). $[\alpha]_D^{26}$ —30.6° (c = 1.0 in ethanol).

THERAP CAT: Antihypertensive.

3187. Diloxanide. *2,2-Dichloro-N-(4-hydroxyphenyl)-N-methylacetamide; 2,2-dichloro-4'-hydroxy-N-methylacet-anilide; N-dichloroacet-4-hydroxy-N-methylanilide; dichlo-roacet-4-hydroxy-N-methylanilide; 4-hydroxy-N-methyldi-chloroacetanilide; Entamide; Ame-Boots.* $C_9H_9Cl_2NO_2$; mol wt 234.09. C 46.18%, H 3.87%, Cl 30.29%, N 5.98%, O 13.67%. Prepn: Oxley *et al.*, U.S. pat. **2,912,438** (1959 to Boots Pure Drug).

Crystals from ethyl acetate, mp 175°.

Furoate, *Furamide, Histomibal, Miforon, diloxanide 2-fu-roic acid ester.* Do not confuse with the amide of furoic acid.

THERAP CAT: Antiamebic.

3188. Diltiazem. *(2S-cis)-3-(Acetyloxy)-5-[2-(dimeth-ylamino)ethyl]-2,3-dihydro-2-(4-methoxyphenyl)-1,5-benzo-thiazepin-4(5H)-one; (+)-cis-5-[2-(dimethylamino)ethyl]-2,3-dihydro-3-hydroxy-2-(p-methoxyphenyl)-1,5-benzothi-azepin-4(5H)-one acetate (ester).* $C_{22}H_{26}N_2O_4S$; mol wt 414.52. C 63.75%, H 6.32%, N 6.76%, O 15.44%, S 7.73%. Calcium channel blocker with coronary vasodilating activity. Prepn: H. Kugita *et al.*, Ger. pat. **1,805,714**; *eidem*, U.S. pat. **3,562,257** (1969, 1971 to Tanabe Seiyaku); *eidem*, *Chem. Pharm. Bull.* **19**, 595 (1971). Resolution of optical isomers: H. Inoue *et al.*, *Yakugaku Zasshi* **93**, 729 (1973), *C.A.* **79**, 66331w (1974); **Japan. Kokai 59 20,273** (1984 to Tanabe Seiyaku), *C.A.* **101**, 38486e (1984). Stereospecific synthesis: K. Igarashi, T. Honma, Ger. pat. **3,415,035**; *eidem*, U.S. pat. **4,552,695** (1984, 1985 both to Shionogi). Vasodilating action is stereospecific for the *d-cis*-isomer. Structure-activity studies: Sato *et al.*, *Arzneimittel-Forsch.* **21**, 1338 (1971); T. Nagao *et al.*, *Chem. Pharm. Bull.* **21**, 92 (1973). Pharmacology and toxicity: *eidem*, *Japan. J. Phar-macol.* **22**, 467 (1972). Metabolism: Meshi *et al.*, *Chem. Pharm. Bull.* **19**, 1546 (1971). Comparative study in variant angina: D. D. Waters *et al.*, *Am. J. Cardiol.* **47**, 179 (1981). Clinical effect in reinfarction: R. S. Gibson *et al.*, *N. Engl. J. Med.* **315**, 423 (1986). Review of pharmacology and effica-cy: M. Chaffman, R. N. Brogden, *Drugs* **29**, 387-454 (1985). Symposium: *Acta Pharmacol. Toxicol.* **57**, Suppl. II, 1-73 (1985).

d-cis Form hydrochloride, $C_{22}H_{27}ClN_2O_4S$, **CRD-401**, *Altiazem, Anginyl, Angizem, Britiazim, Bruzem, Calcicard, Cardizem, Cardiem, Dilpral, Dilzem, Dilzene, Herbesser, Masdil, Tildiem.* Fine needles from ethanol-isopropanol, mp 207.5-212°. $[\alpha]_D^{24}$ +98.3 ± 1.4° (c = 1.002 in methanol). Freely sol in water, methanol, chloroform; slightly sol in abs ethanol. Practically insol in benzene. LD_{50} in male, female mice, male, female rats (mg/kg): 61, 58, 38, 39 i.v.; 260, 280, 520, 550 s.c.; 740, 640, 560, 610 orally (Nagao, 1972).

THERAP CAT: Anti-anginal.

3189. Dimecrotic Acid. *3-(2,4-Dimethoxyphenyl)-2-bu-tenoic acid; 2,4-dimethoxy-β-methylcinnamic acid; 3-(2,4-*

dimethoxyphenyl)crotonic acid. $C_{12}H_{14}O_4$; mol wt 222.24. C 64.85%, H 6.35%, O 28.80%. Prepn: **Ger. pat. 1,915,023** (1969 to Unicler), *C.A.* **72**, 66630y (1970).

mp 149°.

Magnesium salt, $C_{24}H_{26}MgO_8$, *Hepadial.* mp 135°. LD_{50} in mice, rats (mg/kg): 1.3, 1.0 i.p. (**Ger. pat.**).

THERAP CAT: Choleretic.

3190. Dimefline. *8-[(Dimethylamino)methyl]-7-meth-oxy-3-methyl-2-phenyl-4H-1-benzopyran-4-one; 8-[(di-methylamino)methyl]-7-methoxy-3-methylflavone; 8-di-methylaminomethyl-7-methoxy-3-methyl-2-phenylchro-mone.* $C_{20}H_{21}NO_3$; mol wt 323.38. C 74.28%, H 6.55%, N 4.33%, O 14.84%. Prepn: Da Re *et al.*, *Arzneimittel-Forsch.* **10**, 800 (1960); **Brit.** pat. **882,537** corresp to Da Re, U.S. pat. **3,147,258** (1961, 1964 both to Recordati). Pharmacolo-gy: Setnikar *et al.*, *J. Pharmacol. Exp. Ther.* **128**, 176 (1960); *eidem*, *J. Med. Pharm. Chem.* **3**, 471 (1961).

Hydrochloride, $C_{20}H_{22}ClNO_3$, *Rec 7-0267, Remeflin.* Crystals from alcohol + ether, dec 213-214°.

THERAP CAT: Respiratory stimulant.

3191. Dimefox. *Tetramethylphosphorodiamidic fluoride;* tetramethyldiamidophosphoric fluoride; bis(dimethylami-do)phosphoryl fluoride; fluophosphoric acid di(dimethylam-ide); bisdimethylaminofluorophosphine oxide; bis(dimethyl-amido)fluorophosphate; Pestox XIV (obsolete); Terrasytam. $C_4H_{12}FN_2OP$; mol wt 154.13. C 31.17%, H 7.85%, F 12.33%, N 18.17%, O 10.38%, P 20.10%. Prepd by fluorina-tion of bis(dimethylamido)phosphoryl chloride: Schrader, *Angew. Chem.*, suppl. **62**, 18 (1951); *B.I.O.S. Final Rept.* no. **714** (1947); Holmsted, *Acta Physiol. Scand.*, suppl. **90**, 33 (1951); Heap, Saunders, *J. Chem. Soc.* **1948**, 1313; **Brit.** pat. **692,446** (1953); McCombie *et al.*, U.S. pat. **2,489,917** (1949). Toxicity data: A. J. Okinaka *et al.*, *J. Pharmacol. Exp. Ther.* **112**, 231 (1954). *Review:* Kilby, *Chem. & Ind. (London)* **1953**, 856.

Liquid. Fishy odor. d_4^{20} 1.1151. $bp_{4.0}$ 67°; bp_{15} 86°. n_D^{20} 1.4267. Freely sol in water, ether, benzene. Aq solns are stable. LD_{50} in rats (mg/kg): 5.0 i.p.; 7.5 orally (Okinaka).

Caution: A highly toxic cholinesterase inhibitor. Symp-toms similar to parathion, *q.v.*: *Clinical Toxicology of Com-mercial Products*, R. E. Gosselin *et al.*, Eds. (Williams & Wilkins, Baltimore, 5th ed., 1984) Section II, p 299.

USE: Insecticide; acaricide.

3192. Dimemorfan. *(9α,13α,14α)-3,17-Dimethylmor-phinan; d-3-methyl-N-methylmorphinan; AT-17.* $C_{18}H_{25}N$; mol wt 255.41. C 84.65%, H 9.87%, N 5.48%. Prepn: Murakami *et al.*, Ger. pat. **2,128,607**; *eidem*, U.S. pat. **3,-786,054** (1971, 1974 both to Yamanouchi); *eidem*, *Chem. Pharm. Bull.* **20**, 1706 (1972). Pharmacology: Ida, Fujii, *Oyo Yakuri* **6**, 1207 (1972); Y. Kasé *et al.*, *Arzneimittel-Forsch.* **26**, 353 (1976).

Pale yellow oil, $bp_{0.3}$ 130-136° or white crystals from acetone, mp 90-93°.

Phosphate, $C_{18}H_{28}NO_4P$, *Astomin*. White to yellowish-white, odorless, bitter crystals, mp 267-269°. $[\alpha]_D^{23}$ +25.7° (c = 0.5 in methanol). Freely sol in glacial acetic acid; sparingly sol in water, methanol. Practically insol in ethanol, acetone, chloroform, benzene, ether. LD_{50} in mice (mg/kg): 223 s.c.; 475 orally (Kasé).

THERAP CAT: Antitussive.

3193. Dimenhydrinate. *8-Chloro-3,7-dihydro-1,3-dimethyl-1H-purine-2,6-dione compd with 2-(diphenylmethoxy)-N,N-dimethylethanamine (1:1); 2-(benzhydryloxy)-N,N-dimethylethylamine 8-chlorotheophyllinate; β-dimethylaminoethyl benzhydryl ether 1,3-dimethyl-8-chloroxanthine; diphenhydramine 8-chlorotheophyllinate; N,N-dimethyl-2-diphenylmethoxyethylamine 8-chlorotheophyllinate; O-benzhydryldimethylaminoethanol 8-chlorotheophyllinate; chloranautine; Amosyt; Anautine; Andramine; Antemin; Diamarin; Dimate; Dramamine; Dramarin; Dramocen; Dramyl; Emedyl; Emes; Epha; Gravol; Menhydrinate; Novamin; Reidamine; Removine; Travel-Gum; Travelin; Travelmin; Vomex A; Xamamina; Faston.* $C_{24}H_{28}ClN_5O_3$; mol wt 469.96. C 61.33%, H 6.01%, Cl 7.54%, N 14.90%, O 10.21%. Prepn: Cusic, *Science* **109**, 574 (1949); Cusic, U.S. pats. **2,499,058** and **2,534,813** (1950 to Searle).

Crystals, mp 102-107°. Freely sol in alcohol, chloroform. Soluble in benzene. Soly in water: about 3 mg/ml. Almost insol in ether. A satd aq soln has a pH between 6.8 and 7.3.

THERAP CAT: Anti-emetic.

THERAP CAT (VET): Has been used as an anti-emetic.

3194. Dimenoxadol. *α-Ethoxy-α-phenylbenzeneacetic acid 2-(dimethylamino)ethyl ester; ethoxydiphenylacetic acid 2-dimethylaminoethyl ester; α,α-diphenyl-α-ethoxyacetic acid β-dimethylaminoethyl ester; 2-dimethylaminoethyl ethoxydiphenylacetate; Lokarin.* $C_{20}H_{25}NO_3$; mol wt 327.41. C 73.36%, H 7.70%, N 4.28%, O 14.66%. Prepd by the reaction of diphenylethoxyacetic acid with dimethylchloroethylamine in the presence of isopropanol: Klosa, *Arch. Pharm.* **288**, 42 (1955). Improved prepn: C. N. Yakhontov *et al.*, *Pharm. Chem. J.* **8**, 189 (1974). HPLC determn: I. Jane, A. McKinnon, *J. Chromatog.* **323**, 191 (1985).

Hydrochloride, $C_{20}H_{26}ClNO_3$, long needles, mp 170-172°. Sol in water and alcohols; only slightly sol in chloroform and acetone. Insol in ether, ethyl acetate, benzene. pH (2% aq soln) 4.0-5.5. LD_{50} i.v. in rats: 66 mg/kg (Yakhontov).

Caution: May be habit forming. This is a controlled substance (opiate) listed in the U.S. Code of Federal Regulations, Title 21 Part 1308.11 (1987).

THERAP CAT: Narcotic analgesic.

3195. Dimepheptanol. *β-[2-(Dimethylamino)propyl]-α-ethyl-β-phenylbenzeneethanol; 6-(dimethylamino)-4,4-di-*

phenyl-3-heptanol; 2-dimethylamino-4,4-diphenyl-5-heptanol; 3-hydroxy-4,4-diphenyl-6-dimethylaminoheptane; bimethadol; Methadol; Pangerin. $C_{21}H_{29}NO$; mol wt 311.47. C 80.98%, H 9.39%, N 4.50%, O 5.14%. Prepn of α-isomers by catalytic or lithium aluminum hydride reduction of methadone: Speeter *et al.*, *J. Am. Chem. Soc.* **71**, 57 (1949); Pohland *et al.*, *ibid.* 460. Prepn of β-isomers by sodium propanol reduction of methadone: Eddy *et al.*, *J. Org. Chem.* **17**, 321 (1952); Clark, U.S. pats. **2,565,592** and **2,668,814** (1951 and 1954 to Merck & Co.); Speeter, U.S. pat. **2,649,445** (1953 to Bristol). Stereochemistry: Portoghese, Williams, *152nd Am. Chem. Soc. Meet.* (New York, Sept. 1966) Abstracts of Papers, p 5.

β-dl-Form, mp 127-128°. Hydrochloride, rods from acetone, mp 210-212°

β-d-Form, prisms from alc + water, mp 106-107°; $[\alpha]_D^{20}$ +178° (c = 0.63 in alc). Hydrochloride, plates from acetone + ether, mp 206-208°. $[\alpha]_D^{20}$ +73.9° (c = 0.69).

β-l-Form, prisms, mp 105-107°. $[\alpha]_D^{20}$ -178° (c = 1.04 in alc). Hydrochloride, mp 206-208°. $[\alpha]_D^{20}$ -74° (c = 0.94).

α-dl-Form hydrochloride, mp 192-193°.

α-d-Form hydrochloride, mp 169-171°. $[\alpha]_D^{25}$ +34° (c = 0.26).

α-l-Form hydrochloride, mp 169-171°. $[\alpha]_D^{25}$ -35° (c = 0.30).

Caution: May be habit forming. This is a controlled substance (opiate) listed in the U.S. Code of Federal Regulations, Title 21 Part 1308.11 (1987).

THERAP CAT: Narcotic analgesic.

3196. Dimercaprol. *2,3-Dimercapto-1-propanol;* 1,2-dithioglycerol; British Anti-Lewisite; BAL; Dicaptol; Sulfactin. $C_3H_8OS_2$; mol wt 124.21. C 29.00%, H 6.49%, O 12.88%, S 51.62%. $CH_2SHCHSHCH_2OH$. Developed as an anti-gas warfare agent. Effective in stopping the toxic action of Lewisite upon the pyruvate oxidase system in the brain: Peters *et al.*, *Nature* **156**, 616 (1945); Waters, Stock, *Science* **102**, 601 (1945). Prepd by the bromination of allyl alcohol to glycerol dibromohydrin followed by reaction with sodium hydrosulfide under pressure. Also prepd by hydrogenation of hydroxypropylene trisulfide: U.S. pat. **2,402,665** (1946 to du Pont). Stereospecific synthesis: Anisuzzaman, Owen, *J. Chem. Soc. (C)* **1967**, 1021.

Viscous oily liq. Pungent offensive odor of mercaptans. d_4^{25} 1.2385; $bp_{0.2}$ 60°; $bp_{5.6}$ 100°; bp_{15} 120°; bp_{25} 130°; bp_{40} 140°. n_D^{25} 1.5720. 8.7 grams dissolve in 100 ml water (dec). Sol in vegetable oils. Marketed as a 10% soln in peanut oil with 20% benzyl benzoate (stabilizer). LD_{50} i.m. in rats: 86.7 mg/kg, P. Zvirblis, R. I. Ellin, *Toxicol. Appl. Pharmacol.* **36**, 297 (1976).

THERAP CAT: Antidote (to arsenic, gold and mercury poisoning).

THERAP CAT (VET): Chelating agent. Detoxicant for heavy metal poisoning.

3197. 2,3-Dimercapto-1-propanesulfonic Acid. 2,3-Dithiolpropanesulfonic acid. $C_3H_8O_3S_3$; mol wt 188.27. C 19.14%, H 4.28%, O 25.49%, S 51.09%. Chelating agent related to BAL (dimercaprol, *q.v.*). Prepn: N. S. Johary, L. N. Owen, *J. Chem. Soc.* **1955**, 1307; V. E. Petrun'kin, *Ukrain. Khim. Zh.* **22**, 603 (1956), *C.A.* **51**, 5692h (1957). Distribution and excretion in rats: B. Gabard, *Arch. Toxicol.* **39**, 289 (1978). Metabolism study: B. Gabard, R. Walser, *J. Toxicol. Environ. Health* **5**, 759 (1979). Use in removal of internally deposited gold: B. Gabard, *Brit. J. Pharmacol.* **68**, 607 (1980). Protection of mice vs lethal effects of sodium arsenite: H. V. Aposhian *et al.*, *Toxicol. Appl. Pharmacol.* **61**, 385 (1981). Pharmacokinetics in dogs: P. Wiedemann *et al.*, *Biopharm. Drug Dispos.* **3**, 267 (1982). Anti-lewisite activity and stability: H. V. Aposhian *et al.*, *Life Sci.* **31**, 2149 (1982). Toxicological studies: F. Planas-Bohne *et al.*, *Arzneimittel-Forsch.* **30**, 1291 (1980).

HSCH$_2$—CH—CH$_2$SO$_3$H
|
SH

Sodium salt, C$_3$H$_7$NaO$_3$S$_3$, *DMPS, unitiol, Dimaval.* Leaflets, mp 235°. LD$_{50}$ in mice: 5.22 mmol/kg i.p., H. V. Aposhian *et al., loc. cit.* (1981).

THERAP CAT: Antidote (to heavy metal poisoning).

3198. Dimestrol. *(E)-1,1'-(1,2-Diethyl-1,2-ethenediyl)-bis[4-methoxybenzene]; α,α'-diethyl-4,4'-dimethoxystilbene;* 4,4'-dimethoxy-α,β-diethylstilbene; stilbestrol dimethyl ether; 3,4-bis(p-methoxyphenyl)-3-hexene; 3,4-dianisyl-3-hexene; Depot-Oestromenine; Depot-Oestromon; Synthila. C$_{20}$H$_{24}$O$_2$; mol wt 296.39. C 81.04%, H 8.16%, O 10.80%. Dimethyl ether of diethylstilbestrol, *q.v.* Prepd from 3,3-dianisyl-4-hexanone: Sisido, Nozaki, *J. Am. Chem. Soc.* **70**, 777 (1948). Alternate prepns: Dodds *et al., Nature* **142**, 211, 247 (1938); Reid, Wilson, *J. Am. Chem. Soc.* **64**, 1625 (1942). Improved stereospecific prepn: T. Hiyama, H. Nozaki, *Bull. Chem. Soc. Japan* **46**, 2248 (1973). Crystal structure: G. Ruban, P. Luger, *Acta Crystallogr.* **B31**, 2658 (1975). Biological activities: Y. Inamori *et al., Chem. Pharm. Bull.* **33**, 4478 (1985). Review and bibliography: Solmssen, *Chem. Rev.* **37**, 481 (1945). *See also* Mestilbol.

Crystals from petr ether, mp 124°. Less sol than the monomethyl ether, mestilbol. Practically insol in water. Sol in alcohol; freely sol in acetone, ether; also sol in dil aq or alcoholic solns of alkali hydroxides and in vegetable oils.

THERAP CAT: Estrogen.

3199. Dimetacrine. *N,N,9,9-Tetramethyl-10(9H)-acridinepropanamine; 10-[3-(dimethylamino)propyl]-9,9-dimethylacridan;* 5,5-dimethyl-10-dimethylaminopropylacridan; dimethacrine. C$_{20}$H$_{26}$N$_2$; mol wt 294.42. C 81.58%, H 8.90%, N 9.52%. Prepn: Holm, **Brit.** pat. **933,875** (1963 to Kefalas), *C.A.* **60**, 510a (1964); Molnar, Wagner-Jauregg, *Helv. Chim. Acta* **48**, 1782 (1965); Fr. pat. **1,438,357** (1966 to Siegfried), *C.A.* **66**, 10856k (1967). Pharmacology and metabolism: Schatz *et al., Arzneimittel-Forsch.* **18**, 862 (1968); Ishitani *et al., Japan. J. Pharmacol.* **20**, 432 (1970).

Free base, bp$_1$ 200°.
Hydrochloride, C$_{20}$H$_{26}$N$_2$·HCl, crystals from abs alc + acetone, mp 151-154°.
Tartrate, SD 709, *Istonil, Linostil.*
THERAP CAT: Tartrate as antidepressant.

3200. Dimetan®. *Dimethylcarbamic acid 5,5-dimethyl-3-oxo-1-cyclohexen-1-yl ester;* 5,5-dimethyldihydroresorcinol dimethylcarbamate; G-19258. C$_{11}$H$_{17}$NO$_3$; mol wt 211.25. C 62.54%, H 8.11%, N 6.63%, O 22.72%. Prepd by the action of dimethylcarbamic acid chloride on 5,5-dimethyldihydroresorcinol: **Swiss.** pat. **272,772** (1951); Gysin, **U.S.** pat. **2,592,890** (1952 to Geigy). As insecticide: Wiesmann *et al., Experientia* **7**, 117 (1951); Ferguson, Alexander, *J. Agr. Food Chem.* **1**, 888 (1953); Müller, Spindler, *Experientia* **10**, 91 (1954).

Crystals from cyclohexane, mp 45-46°. bp$_{11}$ 170-180°; bp$_{0.3}$ 122-124°. Soly in water at 20° = 3.15 g/100 ml. Freely sol in alc, acetone, ether. Moderately sol in petr ether, gasoline, cyclohexane.

USE: Insecticide (aphicide). *Caution:* A cholinesterase inhibitor. Similar in action to physostigmine.

3201. Dimethadione. *5,5-Dimethyl-2,4-oxazolidinedione;* DMO; AC 1198; BAX 1400Z; NSC 30152; Eupractone. C$_5$H$_7$NO$_3$; mol wt 129.11. C 46.51%, H 5.47%, N 10.85%, O 37.18%. Active metabolite of trimethadione, *q.v.* Prepn: F. Urech, *Ber.* **13**, 485 (1880); R. W. Stoughton, *J. Am. Chem. Soc.* **63**, 2376 (1941). Anticonvulsant activity: C. D. Withrow *et al., J. Pharmacol. Exp. Ther.* **161**, 335 (1968); in comparison with trimethadione: H. Ferngren, *Acta Pharmacol. Toxicol.* **26**, 177 (1968). Use in measurement of intracellular pH and cellular pH gradients: S. Addanki *et al., J. Biol. Chem.* **243**, 2337 (1968); V. Ehrhardt, *Biochim. Biophys. Acta* **775**, 182 (1984); J. B. Arnold *et al., J. Cereb. Blood Flow Metab.* **5**, 369 (1985). GLC determn: W. Gazdzik, W. Kmiotek, *J. Chromatog.* **378**, 482 (1986); LC determn in serum: E. Tanaka, S. Misawa, *J. Chromatog.* **413**, 376 (1987).

Crystals, mp 76-77°. Weak organic acid, pKa (37°) 6.13. LD$_{50}$ i.v. in mice: 450 mg/kg (Stoughton).
USE: *In vivo* measurement of intracellular pH.
THERAP CAT: Anticonvulsant.

3202. Dimethazan. *7-[2-(Dimethylamino)ethyl]-3,7-dihydro-1,3-dimethyl-1H-purine-2,6-dione; 7-(2-dimethylaminoethyl)theophylline;* 1,3-dimethyl-7-(2-dimethylaminoethyl)xanthine; Elidin. C$_{11}$H$_{17}$N$_5$O$_2$; mol wt 251.29. C 52.57%, H 6.82%, N 27.87%, O 12.73%. Prepn from theophylline and dimethyl-2-chloroethylamine: Moussalli *et al.,* **Brit.** pat. **669,070** (1952), *C.A.* **47**, 5435i (1953); Klosa, *Arch. Pharm.* **288**, 301 (1955). Pharmacology: Balatre, Merlen, *Therapie* **11**, 1146 (1956); Yago, *Japan. Circ. J.* **26**, 407 (1962).

Solid, mp 95°.
Hydrochloride, C$_{11}$H$_{17}$N$_5$O$_2$·HCl, mp 260°.
THERAP CAT: Antidepressant.

3203. Dimethazone. *2-[(2-Chlorophenyl)methyl]-4,4-dimethyl-3-isoxazolidinone;* FMC 57020; Command. C$_{12}$H$_{14}$ClNO$_2$; mol wt 239.70. C 60.13%, H 5.89%, Cl 14.79%, N 5.84%, O 13.35%. Pre-emergence broad-spectrum herbicide. Prepn and phytotoxic activity: J. H. Chang, **Belg.** pat. **889,040**; *idem,* **U.S.** pat. **4,405,357** (1982, 1983 both to FMC). Use on soybeans: M. P. Mascianica, *Proc. Ann. Meet. Northeast Weed Sci. Soc.* **39**, 25 (1985); on white potatoes: C. C. Kupatt *et al., ibid.* 166. Comparison with alachlor, metolachlor, *q.q.v.:* G. R. Marion *et al., ibid.* 147.

Oil.
USE: Pre-emergent herbicide.

3204. Dimethindene. *N,N-Dimethyl-3-[1-(2-pyridinyl)-ethyl]-1H-indene-2-ethanamine; 2-[1-[2-[2-(dimethylami-*

no)ethyl]inden-3-yl]ethyl]pyridine; 3-[α-(2'-pyridyl)ethyl]-2-(β-dimethylaminoethyl)indene; dimethylpyrindene. $C_{20}H_{24}N_2$; mol wt 292.41. C 82.14%, H 8.27%, N 9.58%. Synthesis: Huebner *et al., J. Am. Chem. Soc.* **82**, 2077 (1960); Huebner, U.S. pat. **2,970,149** (1961 to Ciba). Pharmacology: W. E. Barrett *et al., Toxicol. Appl. Pharmacol.* **3**, 534 (1961).

Maleate, $C_{24}H_{28}N_2O_4$, *Fenistil, Fenostil, Forhistal.* Crystals, mp 159-161°. LD_{50} in rats (mg/kg): 26.8 i.v.; 618.2 orally (Barrett).

THERAP CAT: Antihistaminic.

3205. Dimethiodal Sodium. *Diiodomethanesulfonic acid sodium salt;* sodium diiodomethanesulfonate; Intron; Tenebryl. CHI_2NaO_3S; mol wt 369.90. C 3.25%, H 0.27%, I 68.62%, Na 6.22%, O 12.98%, S 8.67%. CHI_2SO_3Na. Prepd from acetonesulfonic acid and iodine-yielding agent: Allardt, Ger. pat. **575,678** (1933 to Schering-Kahlbaum, AG). Toxicity data: A. Binz, H. Maier-Bode, *Biochem. Z.* **252**, 16 (1932).

LD i.v. in mice: 3.33 mg/g (Binz, Maier-Bode).

THERAP CAT: Diagnostic aid (radiopaque medium).

3206. Dimethirimol. *5-Butyl-2-(dimethylamino)-6-methyl-4(1H)-pyrimidinone; 5-butyl-2-(dimethylamino)-6-methyl-4-pyrimidinol;* 2-dimethylamino-4-methyl-5-*n*-butyl-6-hydroxypyrimidine; PP 675; Milcurb. $C_{11}H_{19}N_3O$; mol wt 209.29. C 63.13%, H 9.15%, N 20.08%, O 7.64%. Prepn: B. K. Snell *et al.,* **S. Afr.** pat. **67 01,373** corresp to U.S. pat. **3,980,781** (1968, 1976 to ICI). Systemic fungicidal activity: R. S. Elias, *Nature* **219**, 1160 (1968); K. J. Bent, *Ann. Appl. Biol.* **66**, 103 (1970). Metabolism and mode of action: A. Calderbank, *Acta Phytopathol.* **6**, 355 (1971). Metabolism: H. Bratt *et al., Food Cosmet. Toxicol.* **10**, 489 (1972).

Needles from ethanol, mp 102°. Vapor press at 30°: 1.1 × 10⁻⁵ mm Hg. uv max (methanol): 229, 304 nm (ε 15500, 7700). Soly in (g/l) at 25°: water 1.2; acetone 45; chloroform 1200; ethanol 65; xylene 360. LD_{50} in female rats: 200-400 mg/kg i.p., >4000 mg/kg orally, R. S. Elias *et al., loc. cit.*

USE: Fungicide.

3207. Dimethisoquin. *2-[(3-Butyl-1-isoquinolinyl)oxy]-N,N-dimethylethanamine; 3-butyl-1-[2-(dimethylamino)ethoxy]isoquinoline;* 1-(β-dimethylaminoethoxy)-3-*n*-butylisoquinoline; quinisocaine. $C_{17}H_{24}N_2O$; mol wt 272.38. C 74.96%, H 8.88%, N 10.29%, O 5.87%. Prepn: Wilson *et al., J. Am. Chem. Soc.* **71**, 937 (1949); Anderson *et al., J. Am. Pharm. Assoc. Sci. Ed.* **41**, 643 (1952); Ullyot, U.S. pat. **2,-612,503** (1952 to SK & F). Clinical studies: Robinson: *Southern Med. J.* **50**, 367 (1957).

Liquid, bp₃ 155-157°. n_D^{20} 1.5486.

Hydrochloride, $C_{17}H_{25}ClN_2O$, *Isochinol, Pruralgan, Pruralgin, Quotane.* Crystals, sol in water, alcohol, chloroform. Practically insol in ether. pH of 1% aq soln: ~4.2. LD_{50} i.p. in rats: 45-50 mg/kg, Fellows, Macko, *J. Pharmacol. Exp. Ther.* **103**, 306 (1951).

THERAP CAT: Topical anesthetic.
THERAP CAT (VET): Topical anesthetic.

3208. Dimethisterone. *17β-Hydroxy-6-methyl-17-(1-propynyl)androst-4-en-3-one; 6α-methyl-17-(1-propynyl)testosterone;* 6α,21-dimethyl-17β-hydroxy-17α-pregn-4-en-20-yn-3-one; 6α,21-dimethylethisterone; 17α-ethynyl-6α,-21-dimethyltestosterone; 17α-ethynyl-17-hydroxy-6α,21-dimethylandrost-4-en-3-one; Secrosteron. $C_{23}H_{32}O_2$; mol wt 340.49. C 81.13%, H 9.47%, O 9.40%. Orally active progestogen; formerly used in combinations as oral contraceptive. Prepn: S. P. Barton *et al., J. Chem. Soc.* **1959**, 1957; *eidem,* U.S. pat. **2,939,819** (1960 to Brit. Drug Houses). Androgenic effects: H. F. L. Scholer, A. M. de Wachter, *Acta Endocrinol.* **38**, 128 (1961). Progestational activity: G. K. Suchowsky, G. Baldratti, *J. Endocrinol.* **30**, 159 (1964). Evaluation of risk in endometrial cancer: N. S. Weiss *et al., N. Engl. J. Med.* **302**, 551 (1980). Review of carcinogenicity studies: *IARC Monographs* **21**, 377-385 (1979).

Crystals, mp 102°. $[α]_D^{20}$ +10° (c = 1 in chloroform). uv max (isopropanol): 240 nm ($E_{1cm}^{1\%}$ 450). Practically insol in water. Sol in ethanol. Slightly sol in acetone, chloroform. Mixture with ethinyl estradiol, *q.v., Oracon, Ovin, Tova.*

THERAP CAT: Progestogen.

3209. Dimethoate. *Phosphorodithioic acid O,O-dimethyl S-[2-(methylamino)-2-oxoethyl] ester; phosphorodithioic acid O,O-dimethyl ester, ester with 2-mercapto-N-methylacetamide; O,O-dimethyl S-methylcarbamoylmethyl phosphorodithioate;* American Cyanamid 12880; Cygon; Fostion MM; Perfekthion; Rogor; Roxion. $C_5H_{12}NO_3PS_2$; mol wt 229.28. C 26.19%, H 5.28%, N 6.11%, O 20.94%, P 13.51%, S 27.97%. Prepn: Cassaday *et al.,* Young, U.S. pats. **2,494,-283** and **2,996,531** (1950, 1961, both to Am. Cyanamid); **Brit.** pat. **791,824** (1958 to Montecatini), *C.A.* **52**, 18222 (1958). Toxicity data: E. W. Schafer, *Toxicol. Appl. Pharmacol.* **21**, 315 (1972). Teratogenicity study: K. D. Courtney *et al., J. Environ. Sci. Health* **B20**, 373 (1985).

Crystals, mp 52-52.5°. d⁶⁵ 1.277. n_D^{65} 1.5334. Burns readily on contact with flame. Very slightly sol in water. Freely sol in most organic solvents, except saturated hydrocarbons. Stable in aq soln; hydrolyzed by aq alkali. LD_{50} orally in rats: 250 mg/kg (Schafer).

Caution: Cholinesterase inhibitor: *Clinical Toxicology of Commercial Products,* R. E. Gosselin *et al.,* Eds. (Williams & Wilkins, Baltimore, 5th ed., 1984) Section II, p 297.

USE: Systemic and contact insecticide.

3210. Dimethocaine. *3-(Diethylamino)-2,2-dimethyl-1-propanol p-aminobenzoate;* 3-diethylamino-2,2-dimethylpropyl p-aminobenzoate; 1-aminobenzoyl-2,2-dimethyl-3-diethylaminopropanol; p-aminobenzoate of diethylaminoneopentyl alcohol; Larocaine. $C_{16}H_{26}N_2O_2$; mol wt 278.38. C 69.03%, H 9.41%, N 10.06%, O 11.50%. Prepn: Mannich, Wilder, *Ber.* **65**, 378 (1932); Mannich, U.S. pat. **1,889,678** (1933).

CH3
|
COOCH2CCH2N(C2H5)2
|
CH3

NH2

Hydrochloride, $C_{16}H_{26}N_2O_2 \cdot HCl$, minute crystals or powder, sometimes fine leaflets. Bitter taste. mp 196-197°. One part dissolves in 3 parts water at 20°, more sol in hot water. One part dissolves also in 10 parts of cold alcohol, in 5 parts of boiling alcohol. Practically insol in ether, oils, fats. Aq solns are slightly acid to litmus, and may be sterilized by heating at 105° for 10 minutes. LD s.c. in mice: 300 mg/kg.

THERAP CAT: Local anesthetic.

3211. Dimethoxanate. *10H-Phenothiazine-10-carboxylic acid 2-[2-(dimethylamino)ethoxy]ethyl ester; β-dimethylaminoethoxyethyl phenothiazine-10-carboxylate; Tussidin; Cotrane.* $C_{19}H_{22}N_2O_3S$; mol wt 358.48. C 63.66%, H 6.19%, N 7.82%, O 13.39%, S 8.95%. Prepn: von Seeman, U.S. pat. 2,778,824 (1957 to American Home Prod.).

COOCH2CH2OCH2CH2N(CH3)2
N
S

Hydrochloride, $C_{19}H_{22}N_2O_3S \cdot HCl$, crystals from anhydr acetone or methanol + ether, mp 161-163° (dec).

Ingredient of *Cothera Syrup*.

THERAP CAT: Antitussive.

3212. Dimethoxane. *2,6-Dimethyl-1,3-dioxan-4-ol acetate; 6-acetoxy-2,4-dimethyl-m-dioxane; 2,6-dimethyl-m-dioxan-4-yl acetate; acetomethoxane; Dioxin (obsolete); Giv-Gard DXN.* $C_8H_{14}O_4$; mol wt 174.19. C 55.16%, H 8.10%, O 36.74%. Prepn: C. S. Marvel *et al., J. Org. Chem.* **4**, 252 (1939); Späth *et al., Ber.* **76**, 57 (1943). Description: *Am. Perfum. Cosmet.* **77**, no. 12, 32-34 (1962), *C.A.* **58**, 8848c (1963).

CH3   O   CH3
O
OOCCH3

Liquid, bp_6 74-75°. n_D^{20} 1.4310; d_4^{20} 1.0655. Mustard-like odor. Miscible with water, many organic solvents.

USE: Preservative for cutting oils, resin emulsions, water-based paints, cosmetics, inks. Effective range of concn: 0.03-0.1%. Gasoline additive: Chafetz *et al.*, U.S. pat. **3,036,904** (1962 to Texaco).

3213. 1,2-Dimethoxyethane. Ethylene glycol dimethyl ether; monoglyme; α,β-dimethoxyethane; glyme; Dimethyl Cellosolve. $C_4H_{10}O_2$; mol wt 90.12. C 53.31%, H 11.19%, O 35.51%. $CH_3OCH_2CH_2OCH_3$. Prepd from ethylene glycol monomethyl ether, methyl sulfate and metallic sodium: Kranzfelder, Vogt, *J. Am. Chem. Soc.* **60**, 1714 (1938); from ethylene glycol monomethyl ether, methyl chloride and sodium: Capinjola, *ibid.* **67**, 1615 (1945); from chloromethyl methyl ether in the presence of sodium by Wurtz reaction: Geist, Mason, *ibid.* **76**, 3728 (1954).

Liquid. Sharp ethereal odor. d_4^{15} 0.86877; d_4^{20} 0.86285; d_4^{33} 0.8602; d_{20}^{20} 0.8692. mp −58° (also reported as mp −71°). bp_{760} 82-83°; $bp_{61.2}$ 20°; bp_{50} 16°; bp_{10} −14°. n_D^{24} 1.3739; n_D^{20} 1.3813. Flash pt 4.5°C (40°F). Miscible with water, alcohol. Sol in hydrocarbon solvents.

USE: Solvent, to facilitate the formation of alkali metal-hydrocarbon adducts; in the Reformatsky reaction with methyl γ-bromocrotonate.

3214. 2,6-Dimethoxyquinone. *2,6-Dimethoxy-2,5-cyclo-*

hexadiene-1,4-dione; 2,6-dimethoxybenzoquinone. $C_8H_8O_4$; mol wt 168.14. C 57.14%, H 4.80%, O 38.06%. Obtained from pyrogallol-1,3-dimethyl ether-2-acetate upon oxidation with potassium dichromate: Hofmann, *Ber.* **11**, 337 (1878); or by heating with nitric acid in alc soln: Graebe, Hess, *Ann.* **340**, 237 (1905); by treating pyrogallol trimethyl ether with nitric acid: Will, *Ber.* **21**, 608 (1888); from syringa acid: Graebe, Martz, *ibid.* 221; from sinapic acid: Gadamer, *Ber.* **30**, 2333 (1897); from antiarole by oxidation with ferric chloride: Baker, *J. Chem. Soc.* **1928**, 1029.

O
CH3O   OCH3
O

Monoclinic golden-yellow prisms from acetic acid, mp 256°. Sublimes easily. Volatile with steam. Slightly sol in hot water, in ether, in alcohol; freely sol in hot glacial acetic acid and in alkaline solns.

3215. Dimethylacetal. *1,1-Dimethoxyethane; acetaldehyde dimethyl acetal; ethylidene dimethyl ether.* $C_4H_{10}O_2$; mol wt 90.12. C 53.31%, H 11.19%, O 35.51%. $CH_3CH(OCH_3)_2$. Prepn from acetaldehyde and methanol: Meadows, Darwent, *Can. J. Chem.* **30**, 501 (1952); Frevel, Hedelund, U.S. pat. 2,691,684 (1954 to Dow).

Liquid, bp 64.5°. d_4^{20} 0.8516. n_D^{20} 1.3665. Miscible with water, alcohol, chloroform, ether. LD_{50} orally in rats: 6.5 g/kg; lethal concn for rats in air: 16,000 ppm, H. F. Smyth *et al., J. Ind. Hyg. Toxicol.* **31**, 60 (1949).

USE: As Mering's mixture which is 2 vol dimethylacetal and 1 vol chloroform.

3216. N,N-Dimethylacetamide. Acetic acid dimethylamide; DMAC. C_4H_9NO; mol wt 87.12. C 55.14%, H 10.41%, N 16.08%, O 18.37%. $CH_3CON(CH_3)_2$. Prepn from tris(dimethylamido)phosphate and acetic anhydride: Dye, U.S. pat. 2,667,510 (1954 to Chemstrand); from acetic anhydride and dimethylformamide: Coppinger, *J. Am. Chem. Soc.* **76**, 1372 (1954).

Liquid. d_4^{25} 0.9366; d_4^{20} 0.9429; d_4^0 0.9599. bp_{760} 163-165°; bp_{80} 96°; bp_{33} 85-87°; bp_{26} 74-74.5°; bp_{15} 66-67°; bp_{12} 62-63°. n_D^{20} 1.4373 (also reported as n_D^{20} 1.4230); n_D^{25} 1.4358. Flash pt 66° (151°). Miscible with water and most organic solvents. LD_{50} orally in rats: 5.4 ml/kg, W. Bartsch *et al., Arzneimittel-Forsch.* **26**, 1581 (1976).

USE: Solvent for many organic reactions and in industrial applications. *Caution:* On decompn can emit fumes highly irritating to eyes, mucous membranes.

3217. Dimethylamine. *N-Methylmethanamine;* C_2H_7N; mol wt 45.08. C 53.28%, H 15.65%, N 31.09%. $(CH_3)_2NH$. Prepn from methanol + ammonia: Smith, U.S. pat. 2,456,599 (1948 to Commercial Solvents); Serban, *Rev. Chim. (Bucharest)* **14**, 451 (1963), *C.A.* **60**, 5097b (1964); by catalytic hydrogenation of nitrosodimethylamine: Livering, Maury, **Brit.** pat. 797,483 (1958 to Hercules Powder).

A gas at ordinary temp; characteristic odor. d_4^0 of liq 0.680. bp +7°, mp −96°. Very sol in water forming a very strong alkaline soln; sol in alcohol or ether. It is usually marketed in compressed liq form in tubes or as a 33% aq soln. LD i.v. in rabbits: 4.0 g/kg.

Hydrochloride, $C_2H_7N \cdot HCl$, deliquesc leaflets, mp 171°. Very sol in water; sol in alcohol, chloroform; practically insol in ether. *Keep well closed.*

USE: As accelerator in vulcanizing rubber, tanning, manuf detergent soaps, or attracting boll weevils to exterminate them. As reagent for Mg. *Caution:* Irritating to skin, mucous membranes.

3218. p-Dimethylaminoazobenzene. *N,N-Dimethyl-4-(phenylazo)benzenamine; butter yellow; methyl yellow; C. I. Solvent Yellow 2; C.I. 11020.* $C_{14}H_{15}N_3$; mol wt 225.28. C 74.64%, H 6.71%, N 18.65%. Prepn: *Colour Index vol. 4* (3rd ed., 1971) p 4014. Review of carcinogenicity studies: *IARC Monographs* **8**, 125-146 (1975).

Yellow cryst leaflets, mp 114-117°. Insol in water. Sol in alc, benzene, CHCl₃, ether, petr ether, mineral acids, oils. *Note:* This substance may reasonably be anticipated to be a carcinogen: *Fourth Annual Report on Carcinogens* (NTP 85-002, 1985) p 89.

USE: For determination of free HCl in gastric juice; spot test identification of peroxidized fats; pH indicator (red 2.9, yellow 4.0).

3219. *p*-Dimethylaminobenzaldehyde. 4-Dimethylaminobenzenecarbonal; Ehrlich's reagent. $C_9H_{11}NO$; mol wt 149.19. C 72.45%, H 7.43%, N 9.39%, O 10.72%. Prepd by formylation of dimethylaniline with dimethylformamide: Campaigne, Archer, *Org. Syn.*, **coll. vol. IV**, 331 (1963).

Small granular crystals or leaflets from alcohol + water. Unless extremely pure, the crystals are lemon-colored and may turn pink on exposure to light. mp 74°. bp₁₇ 176-177°. Slightly sol in water; sol in alcohol, ether, chloroform, acetic acid, many other organic solvents.

Hydrochloride, crystals, mp 109°.

USE: Manuf dyes; reagent for arsphenamine, anthranilic acid, antipyrine, indole, skatole, indican, tryptophan, albumin, ergot alkaloids, colon bacteria, typhoid coli; for differentiating between serum eruptions and true scarlet fever.

3220. *p*-Dimethylaminobenzalrhodanine. 5-[[4-(Dimethylamino)phenyl]methylene]-2-thioxo-4-thiazolidinone; 5-[p-(dimethylamino)benzylidene]rhodanine. $C_{12}H_{12}N_2OS_2$; mol wt 264.36. C 54.52%, H 4.58%, N 10.60%, O 6.05%, S 24.26%. Prepn: Mackie, Misra, *J. Chem. Soc.* **1954**, 3919.

Deep red needles from xylene, dec 270°, also reported as dec 246°. Sintering about 260° and melting about 280°. Practically insol in water. Very slightly sol in chloroform, ether, benzene; slightly sol in boiling alc; moderately sol in acetone; sol in strong acids with a yellow color.

USE: As 0.03% soln in acetone for detection of silver, mercury, copper, gold, platinum and palladium ions.

3221. 4-(Dimethylamino)benzoic Acid. $C_9H_{11}NO_2$; mol wt 165.19. C 65.43%, H 6.71%, N 8.48%, O 19.37%. Prepn: Willstätter, Kahn, *Ber.* **37**, 411 (1904); Morton, Stevens, *J. Am. Chem. Soc.* **53**, 4028, 4031 (1931); Gilman, Banner, *ibid.* **62**, 344 (1940); Bowman, Stroud, *J. Chem. Soc.* **1950**, 1342. Use of esters as sunscreening agents: Kreps, Ohlsson, **U.S.** pat. **3,403,207** (1968); Worthington, **Brit.** pat. **1,162,337** and **Fr.** pat. **1,566,396** (both 1969 to Van Dyk).

Crystals from water, mp 242.5-243.5°. Soluble in alcohol, HCl and KOH solns, sparingly sol in ether. Practically insol in acetic acid. pKa = 6.027; pKb = 11.488.

Pentyl ester, $C_{14}H_{21}NO_2$, *padimate A, Escalol 506.*

2-Ethylhexyl ester, $C_{17}H_{27}NO_2$, *padimate O, Escalol 507.*

Glyceryl ester, *see* Glyceryl *p*-Aminobenzoate.

THERAP CAT: Esters as ultraviolet screen.

3222. *p*-Dimethylaminobenzophenone. *[4-(Dimethylamino)phenyl]phenylmethanone.* $C_{15}H_{15}NO$; mol wt 225.28. C 79.97%, H 6.71%, N 6.22%, O 7.10%. p-(CH₃)₂NC₆H₄-COC₆H₅. Prepd from benzanilide, dimethylaniline, and phosphorus oxychloride: **Ger.** pat. **41,751,** *Frdl.* **1**, 44 (1887); Meisenheimer *et al., Ann.* **423**, 84 (1921); Hurd, Webb, *Org. Syn.* **7**, 24 (1927).

Leaflets from alc, mp 92-93°. Insol in water. Slightly sol in cold alcohol; freely sol in hot alcohol and in ether. Very weak base, sol in concd mineral acids from which it is precipitated by water.

syn-Oxime, $C_{15}H_{16}N_2O$, prisms from H_2O + alc, mp 163°.

anti-Oxime, mp 176°.

3223. *N,N*-Dimethylaniline. *N,N-Dimethylbenzenamine;* dimethylphenylamine. $C_8H_{11}N$; mol wt 121.18. C 79.29%, H 9.15%, N 11.56%. Made by heating aniline, methyl alcohol, and H_2SO_4 under pressure, the sulfate formed being converted by NaOH to the free base. Toxicity data: H. F. Smyth *et al., Am. Ind. Hyg. Assoc. J.* **23**, 95 (1962).

Oily liq. *Poisonous!* d_4^{20} 0.956. bp 192-194°. mp 2°. Flash pt 61°. n_D^{20} 1.5582. Insol in water. Freely sol in alcohol, chloroform, ether. LD_{50} orally in rats: 1.41 ml/kg (Smyth).

USE: Solvent; manuf vanillin, Michler's ketone, methyl violet and other dyes. As reagent for methanol, methyl furfural, H_2O_2, nitrate, alcohol, formaldehyde.

3224. 9,10-Dimethyl-1,2-benzanthracene. *7,12-Dimethylbenz[a]anthracene;* 1,4-dimethyl-2,3-benzphenanthrene. $C_{20}H_{16}$; mol wt 256.33. C 93.71%, H 6.29%. Synthesis from o-(α-methyl-α-1-naphthyl)toluic acid: Newman, *J. Am. Chem. Soc.* **60**, 1141 (1938). From 9,10-dimethyl-9,10-dihydroxy-9,10-dihydro-1,2-benzanthracene: Bachmann, Chemerda, *J. Am. Chem. Soc.* **60**, 1023 (1938). From 9,10-dimethyl-9,10-dimethoxy-9,10-dihydro-1,2-benzanthracene: *eidem, ibid.* **61**, 2358 (1939). From 9-methyl-1,2-benzanthracen-10-one by a Reformatsky reaction: Mikhailov, Chernova *J. Gen. Chem. (USSR)* **9**, 2171 (1939), *C.A.* **34**, 4068 (1940).

Plates, leaflets from acetone-alc, faint greenish-yellow tinge. mp 122-123°. Maximum fluorescence at 440 nm: *C.A.* **38**, 1276 (1944). Freely sol in benzene; moderately sol in acetone; slightly sol in alcohol. Insol in water. May be solubilized in water by purines such as caffeine, tetramethyluric acid. The nucleosides, adenosine, and guanosine also show a solvent action.

Monopicrate, $C_{20}H_{16}.C_6H_3N_3O_7$, black needles from abs alc, mp 112-113°.

Dipicrate, $C_{20}H_{16}.2C_6H_3N_3O_7$, reddish-brown needles from abs alc, mp 103-106°.

3225. 5,6-Dimethylbenzimidazole. $C_9H_{10}N_2$; mol wt 146.19. C 73.94%, H 6.90%, N 19.16%. Obtained by acid hydrolysis of vitamin B₁₂: Brink, Folkers, *J. Am. Chem. Soc.* **71**, 2951 (1949); **72**, 4442 (1950). Synthesis by refluxing 4,5-diamino-1,2-dimethylbenzene and formic acid in 4N HCl: Hobrecker, *Ber.* **5**, 921 (1872); Phillips, *J. Chem. Soc.* **1928**, 2393; Brink, Folkers, *loc. cit.*

Crystals from ether, mp 205-206°. Sublimes at 140° and 3 mm. uv max (95% ethanol acidified with HCl to be 0.01N): 274.5, 284 nm. Alkaline to litmus. Soluble in water, chloroform, ether; freely sol in dil acids.

3226. p,α-Dimethylbenzyl Alcohol. α,4-Dimethylbenzenemethanol; p-tolylmethylcarbinol; methyl-p-tolylcarbinol; 4-(α-hydroxyethyl)toluene; 4-methyl-α-phenethyl alcohol; 1-p-tolyl-1-ethanol. $C_9H_{12}O$; mol wt 136.19. C 79.37%, H 8.88%, O 11.75%. Constituent of the essential oil from Curcuma longa L., Zingiberaceae and related plants: Dieterle, Kaiser, Arch. Pharm. **271**, 337 (1933). Prepd from p-tolylmagnesium bromide and acetaldehyde in ether: v. Auwers, Kolligs, Ber. **55**, 42 (1922); Eisenlohr, Schulz, Ber. **57**, 1816 (1924); from 4-methylacetophenone: Gastaldi, Cherchi, Gazz. Chim. Ital. **45**, II, 274 (1915).

Viscous liq. Odor somewhat like that of menthol. The natural product is probably levorotatory. Constants for the synthetic dl-form: $d_4^{15.5}$ 0.9668. bp_{756} 219°; bp_{14} 134°; bp_{11} 115-116°. Very sparingly sol in water. Miscible with abs alcohol, ether. Also sol in isopropanol, liq petrolatum.
Phenylurethan, $C_{16}H_{17}NO_2$, crystals, mp 97.5°.

3227. Dimethylberyllium. Beryllium dimethyl. C_2H_6Be; mol wt 39.08. C 61.46%, H 15.48%, Be 23.06%. $Be(CH_3)_2$. Prepn from a beryllium halide and a Grignard reagent: Gilman, Schulze, J. Chem. Soc. **1927**, 2663; from beryllium and dimethylmercury: Rabideau et al., U.S.A.E.C. **LA-1687**, 11 pp (1954).
Orthorhombic crystals. Sublimes at 200° without melting. Spontaneously inflammable in air; burns with a luminous flame evolving dense white fumes of BeO. Reacts violently with water, HCl, producing methane; sol in ether; slightly sol in benzene.

3228. 2,3-Dimethyl-1,3-butadiene. Diisopropenyl; β,γ-dimethyl-Δ-α,γ-butadiene. C_6H_{10}; mol wt 82.14. C 87.73%, H 12.27%. Occurs in the 66-70° fraction of Puertollano shale oil. Prepd by slow distillation of anhydr pinacol in the presence of hydrobromic acid: Allen, Bell, Org. Syn. **coll. vol. III**, 312; by rapid distillation of anhydr pinacol over alumina at 420-470°: Newton, Coburn, ibid. 313.

Liquid. d_{20}^{20} 0.7273; d_4^{20} 0.7267; d_4^{25} 0.7222. mp −76°. bp_{769} 69.2°; bp_{760} 68.8°. n_D^{25} 1.4362.
USE: In the manuf of synthetic rubber and polymers.

3229. Dimethylcadmium. Cadmium dimethyl. C_2H_6Cd; mol wt 142.48. C 16.86%, H 4.24%, Cd 78.90%. $Cd(CH_3)_2$. Prepn from a Grignard reagent and a cadmium halide: Krause, Ber. **50**, 1813 (1917); Gilman, Nelson, Rec. Trav. Chim. **55**, 518 (1936); Anderson, Taylor, J. Phys. Chem. **56**, 161 (1952).
Liquid. Disagreeable odor. May be kept in a sealed tube. mp −4.5°; bp_{758} 105.5°; $d_4^{17.9}$ 1.9846; n_D 1.5488. Dec with explosive violence when heated above 150°. Catches fire when dropped on filter paper and produces dense clouds of first white, then brown cadmium oxide smoke. When thrown into water, it sinks to the bottom in large drops, which dec in a series of sudden explosive jerks, with crackling sounds. Sol in hydrocarbons.
USE: In organic synthesis; as polymerization catalyst.

3230. Dimethyl Carbate. (endo,endo)-Bicyclo[2.2.1]hept-5-ene-2,3-dicarboxylic acid dimethyl ester; cis-5-norbornene-2,3-dicarboxylic acid dimethyl ester; cis-3,6-endomethylene-

Δ⁴-tetrahydrophthalic acid dimethyl ester; dimalone. $C_{11}H_{14}O_4$; mol wt 210.22. C 62.84%, H 6.71%, O 30.44%. Prepd by a Diels-Alder condensation of cyclopentadiene and maleic anhydride, followed by esterification of resulting anhydride: Bode, Ber. **70**, 1167 (1937); Morgan et al., J. Am. Chem. Soc. **66**, 404 (1944).

Crystals, mp 38° when very pure. The usual product is a viscous, syrupy liquid. $bp_{12.5}$ 137°; bp_9 130°; $bp_{1.5}$ 115°. d_4^{21} 1.164. n_D^{20} 1.4852. Practically insol in water. Soluble in the usual organic solvents. Solubility in kerosene and mineral oils ∼6%. LD_{50} orally in mice, rats: 1.4, 1.0 g/kg, Toxic Substances List, H. E. Christensen, Ed. (1974) p 539.
USE: Insect repellent. Caution: Can cause CNS excitation followed by depression.

3231. 5,5-Dimethyl-1,3-cyclohexanedione. 1,1-Dimethyl-3,5-diketocyclohexane; 1,1-dimethyl-3,5-cyclohexanedione; dimethyldihydroresorcinol; dimedone; methone. $C_8H_{12}O_2$; mol wt 140.18. C 68.54%, H 8.63%, O 22.83%. Prepd from mesityl oxide and diethyl malonate: Shriner, Todd, Org. Syn. **coll. vol. II**, 200 (1943).

Needles from water, prisms from alcohol + ether. Dec 148-150°. Slightly volatile with steam (50 ml of a distillate contained 0.016 g). The dry crystals may be stored in a brown bottle for several years at room temp. Aq solns oxidize on exposure to air and light or upon prolonged storage in the dark. 100 ml of a satd aq soln contains 0.401 g at 19°; 0.416 g at 25°; 1.185 g at 50°; 3.020 g at 80°; 3.837 g at 90°. Monobasic acid in water. K at 25° = 0.71 × 10⁻⁵. Dipole moment 3.46. Also sol in methanol, ethanol, chloroform, benzene, acetic acid, and in 50% ethanol-water mixture.
USE: For the separation and identification of aldehydes. Gives insol condensation products with aldehydes, but not with ketones. The use of piperidine as a catalyst is recommended.

3232. N,N-Dimethylformamide. DMF; DMFA. C_3H_7NO; mol wt 73.09. C 49.30%, H 9.65%, N 19.17%, O 21.89%. $HCON(CH_3)_2$. Prepd from dimethylamine and formic acid: Mitchell, Reid, J. Am. Chem. Soc. **53**, 1879 (1931); Brown, J. Appl. Chem. (London) **1**, Suppl. Issue no. 2, S159 (1951); Campbell, U.S. pat. **3,015,674** (1962 to Commercial Solvents); Surman, U.S. pat **3,072,725** (1963 to du Pont); from dimethylamine + HCN: Benneville et al., J. Org. Chem. **21**, 772 (1956); from HCN + methanol: Fukuoka, Kominami, Chem. Tech. **1972** (Nov.), 640. Reviews of chemical uses: R. S. Kittila, Dimethylformamide Chemical Uses (du Pont, Wilmington, 1967) 264 pp and Suppl. (1973) 148 pp; J. S. Pizey, Synthetic Reagents Vol. **1** (John Wiley, New York, 1974) pp 4-99; C. L. Eberling in Kirk-Othmer Encyclopedia of Chemical Technology vol. **11** (Wiley-Interscience, New York, 3rd ed., 1980) pp 263-268.
Colorless to very slightly yellow liquid. Faint amine odor. mp −61°. bp_{760} 153°; bp_{39} 76°; $bp_{3.7}$ 25°. d_4^{25} 0.9445. n_D^{25} 1.42803. Flash pt, open cup: 153°F (67°C). Misc with water and most common organic solvents. pH of 0.5 molar soln in H_2O = 6.7. LD_{50} in mice, rats: 6.8, 7.6 ml/kg orally; 6.2, 4.7 ml/kg i.p., W. Bartsch et al., Arzneimittel-Forsch. **26**, 1581 (1976).
Human Toxicity: Vapor harmful. Irritating to skin, eyes and mucous membranes. Liver injury has been produced in exptl animals by prolonged inhalation of 100 ppm: Kittila, loc. cit. pp 221-224.
USE: Solvent for liqs and gases. In the synthesis of organic compounds. Solvent for Orlon and similar polyacrylic fibers. Wherever a solvent with a slow rate of evaporation is required. Has been termed the universal organic solvent.

3233. N,N-Dimethylglycine. (Dimethylamino)acetic acid; N-methylsarcosine; DMG. $C_4H_9NO_2$; mol wt 103.12. C 46.59%, H 8.79%, N 13.59%, O 31.03%. $(CH_3)_2NCH_2$-COOH. Prepn: H. T. Clarke et al., J. Am. Chem. Soc. 55, 4571 (1933); D. E. Pearson, J. D. Bruton, ibid. 73, 864 (1951). Mutagenicity study: N. Colman et al., Proc. Soc. Exp. Biol. Med. 164, 9 (1980). Constituent of some products sold in the U.S. as "pangamic acid".
Hygroscopic crystals, mp 157-160°.
Hydrochloride, $C_4H_{10}ClNO_2$, mp 189-190°. Sol in water, alc. Insol in chloroform, acetone.

3234. N,N-Dimethylglycine Hydrazide Hydrochloride. Girard's reagent D. $C_4H_{12}ClN_3O$; mol wt 153.62. C 31.27%, H 7.88%, Cl 23.08%, N 27.35%, O 10.42%. $(CH_3)_2$-NCH$_2$CONHNH$_2$.HCl. Prepd from N,N-dimethylglycine ethyl ester and hydrazine hydrate in abs alc: Viscontini, Meier, Helv. Chim. Acta 33, 1773 (1950).
Crystals from alc, mp 181°.
Dihydrochloride, $C_4H_{13}Cl_2N_3O$, mp 214.5°.
USE: Reagent for aldehydes and ketones.

3235. Dimethylglyoxime. 2,3-Butanedionedioxime; diacetyldioxime. $C_4H_8N_2O_2$; mol wt 116.12. C 41.37%, H 6.95%, N 24.13%, O 27.56%. Prepn: Semon, Damerell, Org. Syn. coll. vol. II, 204 (1943). Manuf: Kamlet, U.S. pat. 2,732,404 (1956 to National Distillers Product Corp.). Crystal structure: Anal. Chem. 21, 1428 (1949); Merritt, Lanterman, Acta Cryst. 5, 811 (1952).

$$CH_3C=NOH$$
$$|$$
$$CH_3C=NOH$$

Triclinic crystals from alc + water, mp 238-240°, also reported as melting 242° and 246°. Practically insol in water. Sol in alc, ether, pyridine, acetone.
USE: Detection and determination of Ni and its separation from Co and many other metals. Forms a scarlet red ppt with Ni even in dil solns. Separation of Pd from Sn, Au, Rh, and Ir, also to detect Bi with which it forms a bright yellow color and ppt.

3236. 1,1-Dimethylhydrazine. unsym-Dimethylhydrazine; asym-dimethylhydrazine; N,N-dimethylhydrazine; UDMH; Dimazine. $C_2H_8N_2$; mol wt 60.10. C 39.97%, H 13.42%, N 46.62%. $(CH_3)_2NNH_2$. Prepd industrially by the reaction of dimethylamine and chloramine; by reduction of nitrosodimethylamine (obtained by treating a dimethylamine salt with sodium nitrite): Hatt, Org. Syn. coll. vol. II, 211 (1943). Subacute toxicity study: Cornish, Hartung, Toxicol. Appl. Pharmacol. 15, 62 (1969). Acute toxicity data: Witkin, Arch. Ind. Health 13, 34 (1956). Review and evaluation of studies of carcinogenicity in laboratory animals: IARC Monographs 4, 137-143 (1974).
Flammable, hygroscopic, mobile liq. Fumes in air and gradually turns yellow. Corrosive to skin! Characteristic ammoniacal odor of aliphatic hydrazines. d_4^{22} 0.791; d_{25}^{25} 0.782. mp −58°. bp$_{760}$ 63.9°. $n_D^{22.3}$ 1.40753. Miscible with water with evolution of heat. Also miscible with alcohol, ether, dimethylformamide, hydrocarbons. LD_{50} in mice, rats (mg/kg): 265, 122 orally; 250, 119 i.v. (Witkin).
Hydrochloride, $C_2H_9ClN_2$, hygroscopic crystals from abs ethanol, mp 83°. Sol in water, ethanol. Practically insol in ether.
Caution: Irritating to skin, eyes, mucous membranes; may cause CNS stimulation and convulsions: Patty's Industrial Hygiene and Toxicology vol. 2A, G. D. Clayton, F. E. Clayton, Eds. (Wiley-Interscience, New York, 3rd ed., 1981) pp 2801-2803. This substance may reasonably be anticipated to be a carcinogen: Fourth Annual Report on Carcinogens (NTP 85-002, 1985) p 92.
USE: The base in rocket fuel formulations.

3237. 1,2-Dimethylhydrazine. N,N'-Dimethylhydrazine; sym-dimethylhydrazine; SDMH. $C_2H_8N_2$; mol wt 60.10. C 39.97%, H 13.42%, N 46.62%. CH$_3$NHNHCH$_3$. Prepn from dibenzoylhydrazine: Folpmers, Rec. Trav. Chim. 34, 34 (1915); Hatt, Org. Syn. coll. vol. II, 208 (1943). Electrosynthesis from nitromethane: Iversen, Ber. 104, 2195

(1971). Review and evaluation of studies of carcinogenicity in laboratory animals: IARC Monographs 4, 145-150 (1974).
Flammable, hygroscopic, mobile liquid. Fumes in air and gradually turns yellow. Corrosive to skin! Characteristic ammoniacal odor of aliphatic hydrazines. d_4^{20} 0.8274. bp$_{753}$ 81°. n_D^{20} 1.4209. Miscible with water with much evolution of heat. Also miscible with alcohol, ether, dimethylformamide, hydrocarbons. LD_{50} in mice, rats: 36, 160 mg/kg orally; 29, 175 mg/kg i.v., Witkin, Arch. Ind. Health 13, 34 (1956).
Dihydrochloride, $C_2H_8N_2$.2HCl, extremely hygroscopic prisms, dec 168°. Freely sol in water, ethanol.
Picrate, mp 150°.
Caution: For toxic symptoms see 1,1-Dimethylhydrazine.

3238. Dimethylmercury. Mercury, dimethyl; methyl mercury. C_2H_6Hg; mol wt 230.66. C 10.41%, H 2.62%, Hg 86.96%. $(CH_3)_2Hg$. Environmental contaminant found together with monomethyl mercury compounds in fish and birds. Prepns: Buckton, Ann. 108, 103 (1858); 109, 219 (1859); Jones, Werner, J. Am. Chem. Soc. 40, 1257 (1918); Gilman, Brown, ibid. 52, 3314 (1930); Wade, U.S. pat. 3,-636,020 (1972 to Ventron); H. H. Sisler et al., J. Org. Chem. 45, 1329 (1980). Synthesized in bottom sediments: Jensen, Jernelöv, Nature 223, 753 (1969). Physical properties: Wilde, J. Chem. Soc. 1949, 72. Formation and degradation studies: Bertilsson, Neujahr, Biochemistry 10, 2805 (1971); DeSimone et al., Biochim. Biophys. Acta 304, 851 (1973); Spangler et al., Science 180, 192 (1973); eidem, Appl. Microbiol. 25, 488 (1973). Neurotoxicity studies: Dales, Am. J. Med. 53, 219 (1972); Skerfving, Food Cosmet. Toxicol. 10, 545 (1972); Kojima, Fujita, Toxicology 1, 43 (1973).
Colorless, volatile, toxic liquid; easily inflammable. bp$_{740}$ 92°. d^{20} 3.1874. n_D^{20} 1.5452. Easily sol in ether, alc. Insol in H_2O.
USE: As inorganic reagent.

3239. Dimethyl(methylene)ammonium Iodide. N-Methyl-N-methylenemethanaminium iodide; Eschenmoser's salt. C_3H_8IN; mol wt 185.00. C 19.48%, H 4.36%, I 68.59%, N 7.57%. $H_2C=N(CH_3)_2I$. Mannich type intermediate originally developed to introduce methyl groups into corrin chromophore. Prepn from trimethylamine and diiodomethane: J. Schreiber et al., Angw. Chem. Int. Ed. 10, 330 (1971); from N,N,N',N'-tetramethylmethylenediamine: T. A. Bryson et al., J. Org. Chem. 45, 524 (1980). Used in prepn of Mannich bases: J. Hooz, J. N. Bridson, J. Am. Chem. Soc. 95, 602 (1973). In functionalization of indoles: A. P. Kozikowski, H. Isida, Heterocycles 14, 55 (1980). In prepn of α-methylene carbonyls: J. L. Roberts et al., Tetrahedron Letters 1977, 1621; of terminal olefins: eidem, ibid. 1299.
Colorless crystals from tetrahydrothiophene dioxide, dec ~240°. Sublimes at 120° at 0.05 torr.
USE: In organic synthesis.

3240. N,N-Dimethyl-1-naphthylamine. N,N-Dimethyl-1-naphthalenamine; 1-dimethylaminonaphthalene. C_{12}-H$_{13}$N; mol wt 171.23. C 84.17%, H 7.65%, N 8.18%. Prepn by heating in a closed tube 1-naphthylamine and methyl iodide in methyl alcohol: Landshoff, Ber. 11, 643 (1878); by heating 1-naphthylamine with dimethyl sulfate in sodium hydroxide in presence of pyridine: Germuth, J. Am. Chem. Soc. 51, 1556 (1929). Use: F. G. Germuth, Ind. Eng. Chem. Anal. Ed. 1, 28 (1929).

[structure: N(CH$_3$)$_2$]

Oil; aromatic odor. bp$_{711}$ 274.5°, bp$_{90}$ 193°, bp$_{69}$ 184.5°, bp$_{13}$ 139-140°; d_4^4 1.0522, d_{15}^{15} 1.0446, d_{25}^{25} 1.0391; n_D^{20} 1.622. Readily sol in alcohol or ether.
USE: Detection and determination of nitrites.

3241. Dimethylolpropionic Acid. 3-Hydroxy-2-(hydroxymethyl)-2-methylpropanoic acid; 2,2-bis(hydroxymethyl)-propionic acid; dihydroxypivalic acid; DMPA. $C_5H_{10}O_4$; mol wt 134.13. C 44.77%, H 7.52%, O 47.71%. Prepn: H. Koch, T. Zerner, Monatsh. 22, 447, 450 (1891); K. E. Wilzbach et al., J. Am. Chem. Soc. 70, 4069 (1948); R. Riemschneider *et al., Monatsh.* **88,** 1099 (1957); H. Vieregge, J. F. Arens, *Rec. Trav. Chim.* **78,** 921 (1959).

$$CH_3C(CH_2OH)(COOH)(CH_2OH)$$

Free-flowing granular powder, mp 181-185°. Sol in water, methanol. Slightly sol in acetone. Insol in benzene.
Note: Do not confuse with the herbicide DMPA, *q.v.*
USE: In prepn of water-sol alkyl resins.

3242. Dimethyl-*p*-phenylenediamine. *N,N-Dimethyl-1,4-benzenediamine; p*-aminodimethylaniline. $C_8H_{12}N_2$; mol wt 136.19. C 70.55%, H 8.88%, N 20.58%. Prepn from *p*-nitrosodimethylaniline which is obtained from dimethylaniline: Gattermann-Wieland, *Praxis des Organischen Chemikers* (de Gruyter, Berlin, 40th ed., 1961) p 273. By reduction of methyl orange: A. I. Vogel, *Practical Organic Chemistry* (Longmans, London, 3rd ed., 1959) p 624.

$$H_2N-C_6H_4-N(CH_3)_2$$

Reddish-violet crystals. mp 53°, also stated as 41°. bp 262°. Sol in water, alcohol, chloroform, ether. *Keep tightly closed and protected from light.*
Dihydrochloride, $C_8H_{12}N_2 \cdot 2HCl$, white to grayish-white, hygroscopic crystalline powder. Freely sol in water; sol in alcohol.
USE: Dihydrochloride in microscopy and in tests for acetone, uric acid, thallic salts, oxydases, lignin, ozone, H_2O_2, H_2S, Br.

3243. Dimethyl Phthalate. *1,2-Benzenedicarboxylic acid dimethyl ester; phthalic acid dimethyl ester;* methyl phthalate; dimethyl-1,2-benzenedicarboxylate; DMP; Palatinol M; Fermine; Avolin; Mipax. $C_{10}H_{10}O_4$; mol wt 194.19. C 61.85%, H 5.19%, O 32.96%. Prepd industrially from phthalic anhydride and methanol: Faith, Keyes & Clark's *Industrial Chemicals,* F. A. Lowenheim, M. K. Moran, Eds. (Wiley-Interscience, New York, 4th ed., 1975) pp 318-324. Metabolite of *Gibberella fujikuroi:* Cross *et al., J. Chem. Soc.* **1963,** 2937. Toxicity data: J. H. Draize *et al., J. Pharmacol. Exp. Ther.* **93,** 26 (1948).

$$C_6H_4(COOCH_3)_2$$

Oily liq. Slight aromatic odor. $d^{15.6}_{15.6}$ 1.196; d^{20}_{20} 1.1940; d^{25}_{25} 1.189. One gallon weighs 9.93 lbs. Flash pt 295°F (146°C). mp 5.5° (the commercial product freezes around 0°). bp_{760} 283.7°; bp_{400} 257.8°; bp_{200} 232.7°; bp_{100} 210.0°; bp_{60} 194.0°; bp_{40} 182.8°; bp_{20} 164.0°; bp_{10} 147.6°; bp_5 131.8°; $bp_{1.0}$ 100.3°. Vapor pressure at 20° <0.01 mm. The vapor is heavy, d = 6.69 (air = 1). n^{20}_D 1.5168. Viscosity at 25° = 17.2 cps. Heat of vaporization: 93.1 g-cal/g; heat of combustion: 119.7 kg-cal/mole. uv max (ethanol): 277 nm ($E^{1\%}_{1cm}$ 57.7). Miscible with alcohol, ether, chloroform. Practically insol in water (0.43 g/100 ml), petr ether, and other paraffin hydrocarbons. Soly in mineral oil at 20°: 0.34 g/100 g. LD_{50} in mice, rats, guinea pigs (ml/kg): 7.2, 6.9, 2.4 orally (Draize).
Caution: Irritating to mucous membranes, eyes; can cause CNS depression when ingested. Not irritating to or absorbed through skin: *Clinical Toxicology of Commercial Products,* R. E. Gosselin *et al.,* Eds. (Williams & Wilkins, Baltimore, 5th ed., 1984) Section II, p 205.
USE: Solvent and plasticizer for cellulose acetate and cellulose acetate-butyrate compositions. Insect repellant for personal protection against biting insects.

3244. Dimethyl Sulfate. *Sulfuric acid dimethyl ester;* DMS. $C_2H_6O_4S$; mol wt 126.13. C 19.04%, H 4.80%, O 50.74%, S 25.42%. $(CH_3)_2SO_4$. Prepn by distillation of a mixture of oleum and methanol: Guyot, Simon, *Compt.*

Rend. **169,** 796 (1919). Technical production from dimethyl ether and SO_3: *BIOS Final Report* No. 986, p 176. *Review:* C. M. Suter, *The Organic Chemistry of Sulfur* (John Wiley, New York, 1944) p 49-61. Use as methylating agent: L. Fieser, M. Fieser, *Reagents for Organic Synthesis* (John Wiley, New York, 1967) pp 293-295. Toxicity data: Smyth *et al., Arch. Ind. Hyg. Occup. Med.* **4,** 119 (1951). Review of carcinogenicity studies: *IARC Monographs* **4,** 271-276 (1974). Review of mutagenicity studies: G. R. Hoffmann, *Mutat. Res.* **75,** 63-129 (1980).
Colorless oily liq. bp ~188° (with dec); bp_{15} 76°. mp −27°. d^{20}_4 1.3322. n^{20}_D 1.3874. Flash pt 182°F (83°C). Soly in water 2.8 g/100 ml at 18°. Hydrolysis is rapid at or above this temp. Vapor density 4.35. Sol in ether, dioxane, acetone, aromatic hydrocarbons. Sparingly sol in carbon disulfide, aliphatic hydrocarbons. LD_{50} orally in rats: 440 mg/kg (Smyth).
See also Methyl Sulfate, CH_3OSO_2OH.
Caution: Extremely hazardous. No warning characteristics (e.g. odor, irritation). Delayed appearance of symptoms may permit unnoticed exposure to lethal quantities. Liquid produces severe blistering, necrosis of skin. Sufficient skin absorption can occur to give serious poisoning. Vapors, after relatively asymptomatic latent period, cause severe inflammation and necrosis of eyes, mouth, respiratory tract. Severe and fatal pulmonary damage may result. Systematically causes prostration, convulsions, delirium, paralysis, coma, delayed damage to kidneys, liver, heart with ensuing death in severe cases: E. Browning, *Toxicity and Metabolism of Industrial Solvents* (Elsevier, New York, 1965) pp 713-721. This substance may reasonably be anticipated to be a carcinogen: *Fourth Annual Report on Carcinogens* (NTP 85-002, 1985) p 93.
USE: Methylating agent in the manuf of many organic chemicals. War gas.

3245. 2,4-Dimethylsulfolane. *Tetrahydro-2,4-dimethylthiophene 1,1-dioxide;* 2,4-dimethylthiacyclopentane 1,1-dioxide; 2,4-dimethyltetramethylene sulfone; 2,4-dimethylcyclotetramethylene sulfone. $C_6H_{12}O_2S$; mol wt 148.24. C 48.62%, H 8.16%, O 21.59%, S 21.63%. Prepd by catalytic hydrogenation of 2,4-dimethyl-3-sulfolene: Morris, Melchior, U.S. pat. **2,451,298** (1948 to Shell).

Colorless to yellow liquid. d^{20}_4 1.1362. n^{20}_D 1.4733. bp_5 123.3; bp 280-281° (some decompn). Miscible with lower aromatic hydrocarbons; partially miscible with naphthenes, olefins, paraffins. Limited miscibility with water.
USE: Solvent for liquid-liquid and vapor-liquid extraction processes.

3246. Dimethyl Sulfone. *Sulfonylbismethane;* $DMSO_2$; methyl sulfone; methylsulfonylmethane. $C_2H_6O_2S$; mol wt 94.33. C 25.52%, H 6.43%, O 34.00%, S 34.06%. CH_3SO_2-CH_3. Has been found in primitive plants such as *Equisetum arvense* L., *Equisetaceae* and in the adrenal cortex of cattle: Pfiffner, North, *J. Biol. Chem.* **134,** 781 (1940). Easily prepd by oxidation of dimethyl sulfide: Douglas, *J. Am. Chem. Soc.* **68,** 1072 (1946); McAllan *et al., ibid.* **73,** 3627 (1951). Crystals, mp 109°. bp_{760} 238°. Sublimes at 13 mm and 90° to 100°. Infrared absorption (solid) 7.6-8.7 μ. Dipole moment 4.44 (vapor). Freely sol in water, methanol, ethanol, acetone. Sparingly sol in ether.
USE: High temp solvent for many inorganic and organic substances.

3247. Dimethyl Sulfoxide. *Sulfinylbismethane; methyl sulfoxide;* DMSO; SQ 9453; DMS-70; DMS-90; Deltan; Demasorb; Demavet; Demeso; Dermasorb; Dolicur; Domoso; Dromisol; Gamasol 90; Hyadur; Infiltrina; Kemsol; Rimso-50; Sclerosol; Somipront; Syntexan; Topsym (rescinded). C_2H_6OS; mol wt 78.13. C 30.74%, H 7.74%, O 20.48%, S 41.03%. $(CH_3)_2SO$. Prepd by air-oxidation of dimethyl sulfide in the presence of nitrogen oxides: Smedslund, U.S. pat. **2,581,050** (1952 to Aktiebolaget Central-laboratorium); Coma, Gerttula, U.S. pat. **3,045,051** (1962 to Crown Zellerbach). Usually obtained as a by-product of wood pulp manuf for the paper and allied industries: Robbins, *Chem. Eng.* **68**, No. 13, 100 (1961). Purification: Traynelis *et al., J. Org. Chem.* **27**, 2377 (1962). Forms stable coordination complexes with metals: Meek *et al., J. Am. Chem. Soc.* **82**, 6013 (1960). Toxicity data: Brown *et al., J. Pharm. Pharmacol.* **15**, 688 (1963); Willson *et al., Toxicol. Appl. Pharmacol.* **7**, 104 (1965); W. Bartsch *et al., Arzneimittel-Forsch.* **26**, 1581 (1976). Reviews of pharmacology and toxicology: Jacob, Wood, *Arzneimittel-Forsch* **17**, 1553-1560 (1967); David, *Ann. Rev. Pharmacol.* **12**, 353-374 (1972); of use in organic chemistry: Agami, *Bull. Soc. Chim. France* **1965**, 1021-1039. *Reviews:* Ranky, Nelson, "Dimethyl Sulfoxide" in *Organic Sulfur Compounds,* vol. 1, N. Kharasch, Ed. (Pergamon Press, New York, 1961) pp 170-182; *Ann. N.Y. Acad. Sci.* **411**, entitled "Biological Actions and Medical Applications of Dimethyl Sulfoxide", J. C. de la Torre, Ed. (1983) pp 1-402.

Very hygroscopic liquid. Practically no odor or color. Slightly bitter taste with sweet after-taste. d_4^{20} 1.100. mp 18.45° (supercools easily). bp 189°; bp_{17} 83°; $bp_{5.11}$ 56.6°; $bp_{2.82}$ 47.4°; $bp_{0.79}$ 30°; $bp_{0.37}$ 20°. Flash pt, open cup: 203°F (95°C). n_D^{20} 1.4795; n_D^{21} 1.4787. Viscosity at 27° = 1.1 cp. Specific heat 0.7 cal/g (liq). Dielectric constant: 45. Sol in water, ethanol, acetone, ether, benzene, chloroform. LD_{50} orally in rats: 17.9 ml/kg (Bartsch).

Human Toxicity: Skin contact results in primary irritation with redness, itching and sometimes scaling. Urticarial wheals are not uncommon. Corneal opacities have been produced in exptl animals.

USE: Solvent for acetylene, sulfur dioxide and other gases. In organic reactions. As antifreeze or hydraulic fluid when mixed with water. Solvent for Orlon. As paint and varnish remover. Dissolves some hydrocarbons more than others, *see* data sheets issued by Crown Zellerbach Corp., Camas, Wash.

THERAP CAT: Topical anti-inflammatory.

THERAP CAT (VET): Proposed as analgesic, anti-inflammatory agent, and penetrant carrier to enhance absorption.

3248. Dimethylthiambutene. *N,N-Dimethyl-4,4-di-2-thienyl-3-buten-2-amine; N,N,1-trimethyl-3,3-di-2-thienyl-allylamine;* 3-dimethylamino-1,1-bis(2-thienyl)-1-butene; 3-dimethylamino-1,1-di(2′-thienyl)but-1-ene; NIH-4542; 338C48; Ohton; Aminobutene; Dimethibutin; Kobaton; Takaton. $C_{14}H_{17}NS_2$; mol wt 263.42. C 63.84%, H 6.51%, N 5.32%, S 24.35%. Prepn: Grignard reaction of ethyl β-dimethylaminobutyrate with 2-thienyllithium and dehydration of the resulting 3-dimethylamino-1,1-di-2-thienylbutanol with HCl: Adamson, *J. Chem. Soc.* **1950**, 885; U.S. pat. **2,561,899** (1951 to Burroughs Wellcome); Brit. pat. **657,301** (1951 to Wellcome Foundation). Toxicity data: Y. Kase *et al., Chem. Pharm. Bull.* **7**, 372 (1959).

Viscous, dark brown or yellow oil, $bp_{0.05}$ 123-125°; bp_3 157-158°. Sol in chloroform, ether. LD_{50} in mice (mg/10 kg): 1.21 s.c.; in dogs (mg/kg): 20.3 i.v. (Kase).

Hydrochloride, $C_{14}H_{18}ClNS_2$, may be recrystallized from mixt of ethyl acetate and ethanol, mp 168-169°. Sol in water, chloroform. LD_{50} in mice (mg/10 kg): 1.37 s.c.; in dogs (mg/kg): 20.3 i.v. (Kase).

Picrate, mp 169-170°.

Caution: May be habit forming. This is a controlled substance (opiate) listed in the U.S. Code of Federal Regulations, Title 21 Part 1308.11 (1987).

THERAP CAT: Narcotic analgesic.

3249. 2,4-Dimethylthiazole. C_5H_7NS; mol wt 113.18. C 53.06%, H 6.23%, N 12.38%, S 28.33%. Prepd from chloroacetone and thioacetamide: Hantzsch, *Ann.* **250**, 265 (1889);

Merck, Ger. pat. **670,131**, *C.A.* **33**, 2909 (1939); Schwarz, *Org. Syn.* **25**, 35 (1945).

Hygroscopic liquid. Penetrating odor. d_4^{15} 1.0601. bp_{719} 144-145.5°. Miscible with ice-cold water, but separation occurs upon warming. Sol in alcohol, ether.

Picrate, prisms, mp 137-138°.

Methyl iodide, crystals, dec 260°.

3250. N,N′-Dimethylthiourea. Dimethylthiocarbamide. $C_3H_8N_2S$; mol wt 104.18. C 34.58%, H 7.74%, N 26.89%, S 30.78%. $CH_3NHCSNHCH_3$.

Colorless, exceedingly deliquesc crystals. mp 60-62°. Very sol in water, alcohol, acetone; sparingly sol in benzene, ether, carbon disulfide, very slightly in petr ether. *Keep tightly closed.*

3251. N,N-Dimethyltryptamine. *N,N-Dimethyl-1H-indole-3-ethanamine; 3-[2-(dimethylamino)ethyl]indole;* DMT. $C_{12}H_{16}N_2$; mol wt 188.26. C 76.55%, H 8.57%, N 14.88%. Occurs naturally in plants with hallucinogenic properties. Isoln from the leaves of *Prestonia amazonica* (Benth.) Macbride (*Haemadictyon amazonicum* Spruce & Benth.), *Apocynaceae:* Hockstein, Paradies, *J. Am. Chem. Soc.* **79**, 5735 (1957). Synthesis: Szára, *Experientia* **12**, 441 (1956) using the method of Speeter, Anthony, *J. Am. Chem. Soc.* **76**, 6209 (1954). Relationship between hallucinogenic activity and electronic configuration: Snyder, Merril, *Proc. Nat. Acad. Sci. USA* **54**, 258 (1965).

Crystals, mp 44.6-46.8°. pKa 8.68 (ethanol-water). Freely sol in dil acetic and dil mineral acids.

Picrate, mp 169.5-170°.

Methiodide, mp 216-217°.

Caution: This is a controlled substance (hallucinogen) listed in the U.S. Code of Federal Regulations, Title 21 Part 1308.11 (1985).

3252. Dimethylzinc. Zinc dimethyl. C_2H_6Zn; mol wt 95.45. C 25.17%, H 6.34%, Zn 68.50%. $Zn(CH_3)_2$. Prepd by the interaction of zinc and methyl iodide: Simonowitch, *Chem. Zentr.* **1899, I**, 1066; Bamford *et al., J. Chem. Soc.* **1946**, 468.

Mobile liquid. Stable in sealed tube and under CO_2. $d_4^{10.5}$ 1.386; mp −40°; bp 46°. *Ignites in air*, burns with a blue flame, giving off a peculiar, garlic-like odor. Very slow oxidation with traces of air produces methylzinc methylate, CH_3ZnOCH_3. Sol in ether, miscible with hydrocarbons.

3253. Dimetilan. *Dimethylcarbamic acid 1-[(dimethylamino)carbonyl]-5-methyl-1H-pyrazol-3-yl ester; dimethylcarbamic acid ester with 3-hydroxy-N,N,5-trimethylpyrazole-1-carboxamide;* 2-dimethylcarbamoyl-3-methyl-5-pyrazolyl dimethylcarbamate; G 22870; GS-13332; Snip. $C_{10}H_{16}N_4O_3$; mol wt 240.27. C 49.99%, H 6.71%, N 23.32%, O 19.98%. Prepn: T. Grauer, H. Urwyler, U.S. pat. **3,452,043** (1969 to Geigy). Toxicity data: E. F. Edson, *Pharm. J.* **185**, 361 (1960). Insecticidal activity: B. H. Wilson, *J. Econ. Entomol.* **61**, 1764 (1968). Persistence in water: O. M. Aly *et al., Water Res.* **5**, 1191 (1971). HPLC determn: G. Blaicher *et al., Chromatographia* **13**, 438 (1980).

Colorless solid, mp 68-71° (tech: yellow to reddish-brown

solid, mp 55-65°). Sol in water, chloroform, DMF, ethanol, acetone, xylene, other org solvents. Boiling range 200-210° (at 13 mm Hg). Hydrolyzed by acid and alkali. LD_{50} in rats (mg/kg): 25 orally; 600-700 dermally (Edson).
USE: Insecticide.

3254. Dimetofrine. *4-Hydroxy-3,5-dimethoxy-α-[(methylamino)methyl]benzenemethanol;* 1-(3,5-dimethoxy-4-hydroxyphenyl)-2-(methylamino)ethanol; dimethophrine; dimetrophine. $C_{11}H_{17}NO_4$; mol wt 227.27. C 58.13%, H 7.54%, N 6.16%, O 28.16%. Selective $α_1$-adrenergic receptor agonist. Prepn: **Brit.** pat. **1,145,637**; M. Giani *et al.,* **U.S.** pat. **3,646,144** (1969, 1972 both to Zambeletti). Pharmacology: Scrollini *et al., Atti Accad. Med. Lomb.* **25,** 193, 203 (1970), *C.A.* **76,** 54270a, 54425e (1972). Pharmacokinetics and metabolism: P. Boselli *et al., Arzneimittel-Forsch.* **26,** 2038 (1976); M. S. Benedetti *et al., ibid.* **27,** 158 (1977). TLC determn: G. Musumarra *et al., J. Chromatog.* **350,** 151 (1985). Double blind clinical evaluations: V. Baldrighi *et al., Curr. Med. Res. Opin.* **9,** 78 (1984); U. Marini *et al., ibid.* 265.

CH₃O

HO⟩⟨CHCH₂NHCH₃
        OH
CH₃O

mp 178° (dec). pK 9.71.
Hydrochloride, $C_{11}H_{18}ClNO_4$, *Pressamina.* mp 171-173°.
THERAP CAT: Antihypotensive.

3255. Dimetridazole. *1,2-Dimethyl-5-nitro-1H-imidazole;* RP 8595; Emtryl; Emtrylvet; Unizole. $C_5H_7N_3O_2$; mol wt 141.13. C 42.55%, H 5.00%, N 29.78%, O 22.67%. Prepn: Bhagwat, Pyman, *J. Chem. Soc.* **127,** 1832 (1925).

         CH₃
         |
O₂N⟩ N   CH₃
      ⟨ ⟩
         N

Needles from water, mp 138-139°. Freely sol in alcohol; sparingly sol in cold water, ether.
Hydrochloride, $C_5H_7N_3O_2$.HCl, prisms from dil HCl, mp 195°. Freely sol in water, alcohol; sparingly sol in acetone. Also used as the dihydrogen phosphate, freely sol in water.
THERAP CAT (VET): Antiprotozoal *(Histomonas).*

3256. Diminazene Aceturate. *N-Acetylglycine compd. with 4,4'-(1-triazene-1,3-diyl)bis(benzenecarboximidamide) (2:1); N-acetylglycine compound with 4,4'-(diazoamino)dibenzamidine;* diminazene diaceturate; 4,4'-(diazoamino)dibenzamidine diaceturate; 1,3-bis(p-amidinophenyl)triazene bis(N-acetylglycinate); 1,3-bis[4-guanylphenyl]triazene diaceturate; 4,4'-diamidinodiazoaminobenzene diaceturate; *p,p'-*diguanyldiazoaminobenzene diaceturate; Azidin; Ganasag; Berenil. $C_{22}H_{29}N_9O_6$; mol wt 515.54. C 51.25%, H 5.67%, N 24.45%, O 18.62%. Prepn: Brodersen *et al.,* **U.S.** pat. **2,838,485** (1958 to Hoechst); Stavrovskaya, Drusvyatskaya, *C.A.* **59,** 15199a (1963). Mode of action: Festy *et al., C.R. Acad. Sci. Ser. D* **271,** 730 (1970). Chemotherapeutic characteristics: Schmulevich *et al., Veterinariya (Moscow)* **38,** 23 (1961), *C.A.* **56,** 13512b (1962). *Review:* Newton in *Antibiotics* vol. **3,** J. W. Corcoran, F. E. Hahn, Eds. (Springer-Verlag, New York, 1975) pp 34-47.

H₂N—C⟩⟨NHN=N⟩⟨C—NH₂ · 2 (HOOCCH₂NHC—CH₃)
    ‖   ‖   ‖
    NH   NH   O

Yellow solid, dec 217°. Sol in 14 parts water (20°); slightly sol in alcohol; very slightly sol in ether, chloroform.
Dilactate, $C_{20}H_{27}N_7O_6$, *Babesin.*
THERAP CAT (VET): Antiprotozoal (Trypanosoma, Babesia).

3257. Dimorpholamine. *N,N'-1,2-Ethanediylbis[N-butyl-4-morpholinecarboxamide]; N,N'-dibutyl-N,N'-dicarb-*

oxymorpholideethylenediamine; *N,N'-*dibutyl-*N,N'-*dicarboxyethylene diaminemorpholide; Th 1064; Amipan T; Théraleptique; Theraptique. $C_{20}H_{38}N_4O_4$; mol wt 398.54. C 60.27%, H 9.61%, N 14.06%, O 16.06%. Prepn: Boon, *J. Chem. Soc.* **1947,** 307; *idem,* **U.S.** pats. **2,398,283** and **2,409,-829** (both 1946 to I.C.I.). Toxicity data: Y. Kasé *et al., Arch. Int. Pharmacodyn. Ther.* **163,** 133 (1966).

         C₄H₉   C₄H₉
          | |
O⟩N—CONCH₂CH₂NCO—N⟨O

Crystals from petr ether, mp 41-42°. $bp_{0.4}$ 229°. Solubility in water: 50%. LD_{50} in mice (mg/10 g): 0.54 i.v.; 0.80 i.p.; 1.04 s.c.; 3.80 orally (Kasé).
THERAP CAT: Analeptic.

3258. Dimoxyline. *1-(4-Ethoxy-3-methoxybenzyl)-6,7-dimethoxy-3-methylisoquinoline;* 6,7-dimethoxy-3-methyl-1-(3'-methoxy-4'-ethoxybenzyl)isoquinoline; 6,7-dimethoxy-1-(4'-ethoxy-3'-methoxybenzyl)-3-methylisoquinoline; dioxyline; Paveril. $C_{22}H_{25}NO_4$; mol wt 367.43. C 71.91%, H 6.86%, N 3.81%, O 17.42%. Prepn: Shepard, **U.S.** pat. **2,728,769** (1955 to Lilly).

           OCH₃
           |
           ⟩⟨OC₂H₅
CH₃O⟩ |
  ⟨ ⟩⟨ ⟩N
CH₃O⟨ ⟩   CH₃

Crystals from dil ethanol, mp 124-125°.
Hydrochloride, $C_{22}H_{25}NO_4$.HCl, crystals, dec 196-208°.
Phosphate, $C_{22}H_{25}NO_4$.H_3PO_4, crystals, dec 197-199°. More sol than the hydrochloride. LD_{50} i.v. in mice: 112.7 mg/kg.
THERAP CAT: Vasodilator.

3259. Dimsyl. *Sulfinylbismethane ion (1−).* A name for *methylsulfinylcarbanion.*

         O
         ‖
CH₃S⟩⟨
       CH₂   ⊖

Sodium dimsyl is easily prepd by the reaction of sodium hydride or sodium metal with dimethyl sulfoxide: Corey, Chaykovsky, *J. Am. Chem. Soc.* **84,** 866 (1962); *Proc. Chem. Soc. (London)* **1964,** 108.

3260. β,β'-Dinaphthylamine. *N-2-Naphthalenyl-2-naphthalenamine.* $C_{20}H_{15}N$; mol wt 269.33. C 89.19%, H 5.61%, N 5.20%. Prepn from β-naphthol: Merz, Weith, *Ber.* **13,** 1300 (1880), Benz, *Ber.* **16,** 15 (1883); from β-naphthylamine: Klopsch, *Ber.* **18,** 1586 (1885); Liebermann, Jacobson, *Ann.* **211,** 43 (1882).

       H
       |
⟩⟨⟩ N ⟩⟨⟩

Glistening silvery leaves from benzene, mp 170.5°. Slightly sol in boiling alc; readily sol in boiling glacial acetic acid.
USE: Detection of nitrites, nitrates, and chlorates: Sa, *Anales Farm. Bioquim. (Buenos Aires)* **5,** 111 (1934), *C.A.* **30,** 6672 (1936).

3261. Diniconazole. *(E)-(±)-β-[(2,4-Dichlorophenyl)-methylene]-α-(1,1-dimethylethyl)-1H-1,2,4-triazole-1-ethanol; (E)-1-(2,4-dichlorophenyl)-4,4-dimethyl-2-(1,2,4-triazol-1-yl)-1-penten-3-ol;* S-3308-10; S-3308L; XE-779L;

Ortho Spotless; Spotless; Sumi-8. $C_{15}H_{17}Cl_2N_3O$; mol wt 326.23. C 55.23%, H 5.25%, Cl 21.74%, N 12.88%, O 4.90%. Prepn: Y. Funaki *et al.*, Ger. pat. **3,010,560**; *eidem*, **U.S.** pat. **4,554,007**; resolution of enantiomers: Y. Funaki *et al.*, **Eur.** pat. **Appl. 54,431**; *eidem*, **U.S.** pat. **4,435,203** (1980, 1985, 1982, 1984 all to Sumitomo). Metabolism of racemate in rats: N. Isobe *et al.*, *Nippon Noyaku Gakkaishi* **10**, 475 (1985), *C.A.* **104**, 103703b (1986); of enantiomers: *eidem*, *ibid.* **12**, 421 (1987); *C.A.* **108**, 1715x (1988). Mechanism of sterol inhibition: Y. Yoshida *et al.*, *Biochem. Biophys. Res. Commun.* **137**, 513 (1986); T. Katagi, *J. Agr. Food Chem.* **36**, 344 (1988). Fungicidal activity of the (−)-form and plant growth regulating activity of the (+)-form: C. S. Kvien *et al.*, *J. Plant Growth Regul.* **6**, 233 (1987); D. J. Daigle *et al.*, *Proc. 14th Ann. Mtg. Plant Growth Regul. Soc. Am. 1987*, p 108. Field trials as foliar fungicide: A. R. Biggs, J. Warner, *Can. J. Plant Pathol.* **9**, 41 (1987); A. K. Hagan *et al.*, *J. Environ. Hort.* **6**, 67 (1988); against soil-borne diseases: A. S. Csinos *et al.*, *Appl. Agr. Res.* **2**, 113 (1987). Comprehensive description: H. Takano, *Japan. Pest. Info.* **49**, 18 (1986).

Crystals from isopropanol, mp 148-149°. Vapor pressure at 20°: 2.2×10^{-5} mm Hg. Soly at 25° (% v/w): water 4.01; at 23°: acetone 9.5; methanol 9.5; xylene 1.4. Stable to heat, sunlight and moisture. LD_{50} in male, female rats (mg/kg): 639, 474 orally; >5000, >5000 dermally (Takano).

(−)-Form, crystals from carbon tetrachloride/*n*-hexane, mp 160-161°. $[\alpha]_D^{24}$ −31.7° (c = 1 in CHCl₃).

(+)-Form, crystals from carbon tetrachloride/*n*-hexane, mp 160-161°. $[\alpha]_D^{24}$ +26° (c = 1 in CHCl₃).

USE: Agricultural fungicide.

3262. Dinitolmide. *2-Methyl-3,5-dinitrobenzamide*; *3,5-dinitro-o-toluamide*; 3,5-dinitro-2-methylbenzamide; Zoalene; Zoamix. $C_8H_7N_3O_5$; mol wt 225.16. C 42.67%, H 3.13%, N 18.66%, O 35.53%. Prepn: McGookin *et al.*, *J. Soc. Chem. Ind.* **59**, 92 (1940); Harris *et al.*, U.S. pat. **2,937,-204** (1960 to Dow). Activity studies: Peterson, *Poultry Sci.* **39**, 739 (1960); Joyner, *Res. Vet. Sci.* **1**, 363 (1960). Metabolism: Smith *et al.*, *J. Agr. Food Chem.* **11**, 253 (1963); Smith, *Anal. Biochem.* **7**, 461 (1964).

Crystals from dil alc, mp 181°.
THERAP CAT (VET): Coccidiostat.

3263. 2,4-Dinitroaniline. *2,4-Dinitrobenzenamine.* $C_6-H_5N_3O_4$; mol wt 183.12. C 39.35%, H 2.75%, N 22.95%, O 34.95%. Prepn: Wells, Allen, *Org. Syn. coll. vol. II*, 221 (1943). Liquid chromatographic determn: A. L. Scher, *J. Assoc. Off. Anal. Chem.* **68**, 474 (1985). Metabolism and disposition in rats: H. B. Matthews *et al.*, *Xenobiotica* **16**, 1 (1986). Toxicity study: E. G. Feldmann, W. O. Foye, *J. Am. Pharm. Assoc. Sci. Ed.* **48**, 419 (1959).

Yellow needles from dil acetone, greenish-yellow plates from alcohol. mp 187.5-188°. pKa 18.46. Practically insol in cold water, Very sparingly sol in boiling water. 5.8 parts dissolve in 1000 parts of 88% alc at 18°; one part dissolves in 132.6 parts of 95% alc at 21°.
Caution: May be irritating to skin, mucous membranes.
USE: In prepn of azo dyes.

3264. 2,6-Dinitroaniline. *2,6-Dinitrobenzenamine.* $C_6-H_5N_3O_4$; mol wt 183.12. C 39.35%, H 2.75%, N 22.95%, O 34.95%. Synthesis starting with chlorobenzene, fuming sulfuric acid, and potassium nitrate: Schultz, *Org. Syn.* **coll. vol. IV**, 364 (1963).

Light orange needles from alc, mp 139-140°. Soly in 95% ethanol about 0.4 g/100 ml. Also sol in ether, hot benzene. Practically insol in water, petr ether. Dipole moment: 1.9.

3265. 2,4-Dinitrobenzaldehyde. $C_7H_4N_2O_5$; mol wt 196.12. C 42.87%, H 2.06%, N 14.29%, O 40.79%. Prepn: *p*-Nitrosodimethylaniline hydrochloride, obtained by reacting dimethylaniline and sodium nitrite, is condensed with 2,4-dinitrotoluene in the presence of sodium carbonate. The product, dinitrobenzylidene-*p*-aminodimethylaniline is split under steam agitation with HCl: Bennett, Bell, *Org. Syn.* **coll. vol. II**, 223 (1943).

Yellow to light brown crystals, mp 72°. bp_{10-20} 190-210°. Freely sol in alc, ether, benzene, slightly sol in petr ether, water.
USE: In prepn of Schiff bases.

3266. Dinitrobenzene. $C_6H_4N_2O_4$; mol wt 168.11. C 42.86%, H 2.40%, N 16.67%, O 38.07%.

m-Dinitrobenzene. Yellowish crystals. mp 89-90°. bp 300-303°. Volatile with steam. One gram dissolves in 2000 ml cold water, 320 ml boiling water, 37 ml alc, 20 ml boiling alcohol; freely sol in benzene, chloroform, ethyl acetate.

o-Dinitrobenzene. White crystals. d 1.57. mp 118°. bp 319°. Volatile with steam. One gram dissolves in 6600 ml cold water, 2700 ml boiling water, about 60 ml alc, 3 ml boiling alc, 20 ml benzene; freely sol in chloroform, ethyl acetate.

p-Dinitrobenzene. White crystals; sublimable. d 1.63. mp 173-174°. bp 299°. Volatile with steam. One gram dissolves in 12,500 ml cold water, 555 ml boiling water, 300 ml alcohol; sparingly soluble in benzene, chloroform, ethyl acetate. LD orally in cats: 29.4 mg/kg, *Handbook of Toxicology* vol. **1**, W. S. Spector, Ed. (Saunders, Philadelphia, 1956) pp 116-117.

3267. 2,4-Dinitrobenzenesulfenyl Chloride. $C_6H_3ClN_2-O_4S$; mol wt 234.62. C 30.71%, H 1.29%, Cl 15.11%, N 11.94%, O 27.28%, S 13.67%. Prepd from bis(2,4-dinitrophenyl)disulfide and chlorine in ethylene bromide: Kharasch *et al.*, *J. Am. Chem. Soc.* **69**, 1612 (1947); **72**, 1796 (1950); *J. Chem. Ed.* **33**, 585 (1956); from 2,4-dinitrothiophenol and chlorine in benzene: Perold, Snyman, *J. Am. Chem. Soc.* **73**, 2379 (1951); from 2,4-dinitrophenylbenzyl sulfide and sulfuryl chloride in ethylene chloride: Kharasch, Langford, *Org. Syn.* **44**, 47 (1964). *Review:* Kharasch, *Organic Sulfur Compounds* vol. **1** (Pergamon Press, New York, 1961) pp 375-396.

Yellow crystals from carbon tetrachloride, mp 96°. Sol in glacial acetic acid, methylene chloride, ethylene chloride, trichlorethylene, benzene, xylene. Somewhat less sol in carbon tetrachloride. Insol in ether. Reacts with alc (even in the cold) to produce ethyl 2,4-dinitrobenzenesulfenate.

USE: Analytical reagent for the characterization of organic compds: Kharasch, *J. Chem. Ed.* **33**. 585 (1956); Langford, Lawson, *ibid.* **34**, 510 (1957).

3268. 3,4-Dinitrobenzoic Acid. $C_7H_4N_2O_6$; mol wt 212.12. C 39.63%, H 1.90%, N 13.21%, O 45.26%. Prepd by treating 3-nitro-4-aminotoluene with Caro's acid and oxidizing the 3-nitro-4-nitrosotoluene with potassium dichromate: Langley, *Org. Syn.* **22**, 47 (1942); by oxidation of 3,4-dinitrotoluene: Goldstein, Voegeli, *Helv. Chim. Acta.* **26**, 475 (1943).

Crystals from water + alc. Bitter taste. mp 166°; (mp 165.5-166.5°, Goldstein). 0.673 g dissolves in 100 parts water at 25°. Freely sol in alc, ether and hot water. Sublimes.

USE: In quantitative sugar analysis.

3269. 3,5-Dinitrobenzoic Acid. $C_7H_4N_2O_6$; mol wt 212.12. C 39.63%, H 1.90%, N 13.21%, O 45.26%. Prepd by nitration of benzoic acid with fuming nitric acid: Brewster *et al.*, *Org. Syn.* **22**, 48 (1942); Saunders *et al.*, *Biochem. J.* **36**, 368 (1942). Purification: Jensen *et al.*, *Chem. Weekbl.* **43**, 731 (1947), *C.A.* **42**, 2307d (1948).

Monoclinic prismatic crystals from alc. mp 205-207°, (mp 206.5-207.2° purified). One gram dissolves in 53 parts of boiling water; much less in cold water. Very sol in alc and glacial acetic acid; sparingly sol in ether, carbon disulfide and benzene. Sublimes.

USE: Identification of alcohols, alkyl halides; chromatographic determination of the essential oil constituents.

3270. 3,5-Dinitrobenzoyl Chloride. $C_7H_3ClN_2O_5$; mol wt 230.57. C 36.46%, H 1.31%, Cl 15.38%, N 12.15%, O 34.70%. Prepd from 3,5-dinitrobenzoic acid and PCl_5 at 120-130°: Saunders *et al.*, *Biochem. J.* **36**, 368 (1942).

Needles from light petr (bp 40-60°), mp 69.5°. bp_{10-12} 196°.

USE: Identification of amino acids; characterization of the

alcohol part of acetals and ketals without prior hydrolysis and separation of the alcohol. *Caution:* A strong irritant.

3271. 4,4'-Dinitrocarbanilide. *N,N'-Bis(4-nitrophenyl)-urea;* DNC; 4,4'-dinitrodiphenylurea. $C_{13}H_{10}N_4O_5$; mol wt 302.24. C 51.66%, H 3.34%, N 18.54%, O 26.47%. Prepd by heating 4-nitroaniline and diphenyl carbonate for 3 hrs at 200°: Vittenet, *Bull. Soc. Chim. France* [3] **21**, 149 (1899); by nitration of diphenylurea with 68.28% HNO_3 in concd H_2SO_4 at 0°: Kogan, Kutepov, *J. Gen. Chem. USSR* **21**, 1297 (1951), *C.A.* **46**, 2003d (1952); by the action of dil HNO_3 on diphenylurea at up to 100°: *eidem*, *U.S.S.R.* pat. **78,379** (1949), *C.A.* **48**, 7056i (1954).

Yellow needles from alc, mp 312° (dec). Sublimes above 310°. Moderately sol in boiling nitrobenzene; sparingly sol in boiling alc, somewhat more sol in a mixture of glacial acetic acid and nitric acid. Practically insol in acetone, chloroform, benzene, dioxane, acetic acid, ether and linseed oil. Solubilities in g/100 ml solvent at 25°: water 2×10^{-6}; ethanol 0.007; ethyl acetate 0.015; petr ether 0.0; xylene < 0.01; methyl Cellosolve 0.1; dimethyl acetamide 0.14; dimethyl sulfoxide 0.47.

3272. Dinitrocresol. *2-Methyl-4,6-dinitrophenol; 4,6-dinitro-o-cresol;* 3,5-dinitro-2-hydroxytoluene; 3,5-dinitro-*o*-cresol; DN; DNC; DNOC; Antinonnin; Detal; Dinitrol; Elgetol; K III; K IV; Ditrosol; Prokarbol; Effusan; Lipan; Selinon; Sinox; Dekrysil. $C_7H_6N_2O_5$; mol wt 198.13. C 42.43%, H 3.05%, N 14.14%, O 40.38%. Prepd by sulfonation of *o*-cresol followed by controlled nitration: Noelting, de Salis, *Ber.* **14**, 987 (1881); Bovini, *Chem. Zentr.* **1928**, II, 112; Bures, *C.A.* **22**, 63 (1928); Datta, Varma, *J. Indian Chem. Soc.* **4**, 321 (1927); Monti, Cianetti, *Gazz. Chim. Ital.* **67**, 628 (1937).

Yellow prisms from alc, mp 87.5°. Moderately volatile with steam. Sparingly sol in water; readily sol in alkaline aq solns, in ether, acetone, alcohol (about 10%); sparingly sol in petr ether. LD_{50} orally in rats: 30 mg/kg, *Toxic Substances List*, H. E. Christensen, Ed. (1973) p 718.

Sodium salt, $C_7H_5N_2O_5Na$, red powder. Hydrate, yellow needles very freely sol in water.

The NH_4, K, Ca salts also are sol in water.

USE: Selective herbicide, insecticide (ovicidal spray for dormant fruit trees). *Caution:* Skin contact may lead to local necrosis and dangerous systemic effects. Symptoms similar to dinitrophenol, q.v. but is more toxic. Acts as a cumulative poison.

3273. 3,7-Dinitro-5-oxophenothiazine. *3,7-Dinitro-10H-phenothiazine 5-oxide;* dinitrodiphenylamine sulfoxide. $C_{12}H_7N_3O_5S$; mol wt 305.27. C 47.21%, H 2.31%, N 13.77%, O 26.21%, S 10.50%. Prepd from thiodiphenylamine and nitric acid: Buchanan, Schryver, *Brit. Food J.* **11**, 101 (1908); *C.A.* **3**, 2329 (1909); Kehrmann, Nossenko, *Ber.* **46**, 2818 (1913).

Yellowish leaflets from glacial acetic acid, dec 260°. Insol in water and in the usual organic solvents. Sol in ammonia water and in aq solns of alkalies, forming bluish-red solns. USE: In the determination of tin.

3274. 2,4-Dinitrophenol. α-Dinitrophenol; Aldifen. C_6-$H_4N_2O_5$; mol wt 184.11. C 39.14%, H 2.19%, N 15.22%, O 43.45%. Prepd by the action of NaOH on 1-chloro-2,4-dinitrobenzene. Crystal and molecular structure: F. Iwasaki, Y. Kawano, *Acta Crystallogr.* **33B**, 2455 (1977).

Yellowish to yellow orthorhombic crystals. d 1.683. mp 112-114°. Sublimes when carefully heated. Volatile with steam. Very sparingly sol in cold water; soly in water (g/100 g of satd soln): at 54.5° = 0.137; at 75.8° = 0.301; at 87.4° = 0.587; at 96.2° = 1.22. Soly at 15° (g/100 g soln): Ethyl acetate 15.55; acetone 35.90; chloroform 5.39; pyridine 20.08; carbon tetrachloride 0.423; toluene 6.36. Also sol in alcohol and benzene. Sol in aq alkaline solns. Forms crystalline sodium salt which is sol in water. LD_{50} orally in rats: 30 mg/kg, E. W. Schafer, *Toxicol. Appl. Pharmacol.* **21**, 315 (1972).

Human Toxicity: Highly toxic material. Readily absorbed through intact skin. Vapors absorbed through respiratory tract. Produces marked increase in metabolism and temp, profuse sweating, nausea, vomiting, collapse, death. May cause dermatitis, cataracts, wt loss, granulocytopenia, polyneuropathy, exfoliative dermatitis, *Clinical Toxicology of Commercial Products*, R. E. Gosselin *et al.*, Eds. (Williams & Wilkins, Baltimore, 4th ed., 1976) Section III, pp 134-137.

USE: Manuf dyes, diaminophenol, etc; wood preservative; insecticide; also as indicator. pH range: 2.6 colorless, 4.4 yellow; as a reagent for the detection of potassium and ammonium ions.

3275. 2,5-Dinitrophenol. Yellow crystals. mp 108°; also stated as 104°. Slightly sol in water or cold alcohol; sol in hot alcohol, ether, fixed alkali hydroxides.

USE: Manuf dyes and organic chemicals, as indicator. pH range: 4.0 colorless, 5.4 yellow.

3276. 2,6-Dinitrophenol. Light yellow crystals. mp 63-64°. Slightly sol in cold water or cold alc; freely sol in chloroform, ether, boiling alc, fixed alkali hydroxide solns.

USE: As of the preceding. pH range: 2.0 colorless, 4.0 yellow.

3277. 2,4-Dinitrophenylhydrazine. $C_6H_6N_4O_4$; mol wt 198.14. C 36.37%, H 3.05%, N 28.28%, O 32.30%. C_6H_3-$(NO_2)_2NHNH_2$.

Red, cryst powder. mp about 200°. Slightly sol in water or alcohol; sol in moderately dil inorganic acids; readily sol in diglyme.

USE: For the determination of aldehydes and ketones.

3278. 2,4-Dinitroresorcinol. *2,4-Dinitro-1,3-benzenediol.* $C_6H_4N_2O_6$; mol wt 200.11. C 36.01%, H 2.01%, N 14.00%, O 47.97%.

Yellow crystals, mp 146-148°. Explodes when strongly heated. Sublimes partially undec. Very slightly sol in water or cold alcohol; sol in solns of fixed alkali hydroxides.

USE: For dyeing fabrics mordanted with iron a green color. As a reagent for Co (brown-red ppt) and for Fe (olive-green color).

3279. 5,5'-Dinitrosalicil. *Bis(2-hydroxy-5-nitrophenyl)-*

ethanedione; 2,2'-dihydroxy-5,5'-dinitrobenzil; DNS 55. $C_{14}H_8N_2O_8$; mol wt 332.22. C 50.61%, H 2.43%, N 8.43%, O 38.53%. Prepd by oxidation of 2,2'-dimethoxybenzoin, followed by nitration and demethylation of the resulting 2,2'-dimethoxybenzil: Moureu *et al.*, *Bull. Soc. Chim. France* **1951 M**, 741; **1955 M**, 1155.

Pale yellow needles from acetone, dec 304°. Soly in water about 0.1 mg/ml H_2O. Sparingly sol in the usual organic solvents. Sol in aq 10% NaOH solns with decompn to sodium 5-nitrosalicylate.

3280. Dinobuton. *Carbonic acid 1-methylethyl 2-(1-methylpropyl)-4,6-dinitrophenyl ester; carbonic acid 2-sec-butyl-4,6-dinitrophenyl isopropyl ester;* 2-sec-butyl-4,6-dinitrophenyl isopropyl carbonate; dinitro-sec-butylphenyl isopropyl carbonate; isopropyl 2,4-dinitro-6-sec-butylphenyl carbonate; Acrex; Dessin; Sytasol. $C_{14}H_{18}N_2O_7$; mol wt 326.30. C 51.53%, H 5.56%, N 8.59%, O 34.32%. Prepn: Pianka, Polton, U.S. pats. **3,234,082; 3,234,260** (both 1966 to Murphy Chem.). Metabolism studies: Bandal, Casida, *J. Agr. Food Chem.* **20**, 1235 (1972).

Crystals from methanol or petr ether, mp 56-57°. LD_{50} orally in male, female rats: 59, 71 mg/kg, T. B. Gaines, *Toxicol. Appl. Pharmacol.* **14**, 515 (1969).

USE: Miticide.

3281. Dinocap. DNOCP; CR 1639; ENT 24727; Karathane; Arathane; Mildex; Isocothane; Crotothane. $C_{18}H_{24}$-N_2O_6; mol wt 364.39. C 59.33%, H 6.64%, N 7.69%, O 26.35%. Originally thought to be 2-(1-methylheptyl)-4,6-dinitrophenyl butenoate, it is actually a mixture of 2(or 4)-octyl-4,6(or 2,6)-dinitrophenyl butenoate in which the octyl is a mixture of 1-methylheptyl, 1-ethylhexyl, and 1-propylpentyl. Prepn: W. F. Hester *et al.*, U.S. pat. **2,526,660** (1950 to Rohm & Haas). Fungicidal activity and chemical constitution: A. H. M. Kirby *et al.*, *Ann. Appl. Biol.* **57**, 211 (1966); R. J. W. Byrde *et al.*, *ibid.* 223. Miticidal activity: N. D. Rishi, A. Q. Rather, *J. Entomol. Res.* **7**, 39 (1983). Degradation in soil: W. Mittelstaedt, F. Fuhr, *J. Agr. Food Chem.* **32**, 1151 (1984). Persistence on crops: B. D. Ripley *et al.*, *Can. J. Plant Sci.* **65**, 229 (1985). HPLC residue analysis in crops: D. Liang *et al.*, *J. Chromatog.* **387**, 385 (1987). Acute and chronic toxicity: P. S. Larson *et al.*, *Arch. Int. Pharmacodyn. Ther.* **119**, 31 (1959). Evaluation of developmental toxicity in rodents of technical grade and selected isomers: L. E. Gray, Jr. *et al.*, *Teratogen. Carcinogen. Mutagen.* **6**, 33 (1986); J. M. Rogers *et al.*, *ibid.* **7**, 341 (1987).

R = ── CH(CH$_2$)$_5$CH$_3$ or   ── CH(CH$_2$)$_4$CH$_3$ or   ── CH(CH$_2$)$_3$CH$_3$
          | | |
          CH$_3$ CH$_2$CH$_3$ (CH$_2$)$_2$CH$_3$

Dark brown liquid. $bp_{0.05}$ 138-140°. Incompatible with oil, oil based sprays, and lime-sulfur mixtures. LD_{50} in male rats (mg/kg): 23 i.v.; 980 orally (Larson).

Caution: Skin irritant: *Clinical Toxicology of Commercial Products*, R. E. Gosselin *et al.*, Eds. (Williams & Wilkins, Baltimore, 5th ed., 1984) Section II, p 197. The EPA has determined that dinocap is teratogenic in rabbits: *Fed. Reg.* **50**, 1119 (1985).

USE: Acaricide; fungicide.

3282. Dinoseb. *2-(1-Methylpropyl)-4,6-dinitrophenol; 2-sec-butyl-4,6-dinitrophenol;* DNBP; ENT 1122; WSX-8365; Chemox PE; Dow General; Premerge; Subitex; Caldon; Basanite. $C_{10}H_{12}N_2O_5$; mol wt 240.22. C 50.00%, H 5.04%, N 11.66%, O 33.30%. Prepn: L. E. Mills, B. L. Fayerweather, U.S. pat. **2,192,197** (1940 to Dow). Activity: A. S. Crafts, *Science* **101**, 417 (1945). Metabolism: S. K. Bandal, J. E. Casida, *J. Agr. Food Chem.* **20**, 1235 (1972); K. Ingebrigtsen, A. Froeslie, *Acta Pharmacol. Toxicol.* **46**, 326 (1980). Toxicology: R. G. Bough *et al.*, *Toxicol. Appl. Pharmacol.* **7**, 353 (1965); T. B. Gaines, R. E. Linder, *Fundam. Appl. Toxicol.* **7**, 299 (1986).

Orange-brown viscous liquid, mp 38-42°. pKa 4.62. Sol in most organic solvents. LD_{50} in adult male, female rats (mg/kg): 27, 28 orally (Gaines, Linder).

Acetate, $C_{12}H_{14}N_2O_6$, *Hoe 2904, Aretit, Ivosit.* Brown oil, vinegar-like odor, mp 26-27°. Vapor pressure at 20°: 6×10^{-4} mm Hg. Sol in aromatics.

Ammonium salt, $C_{10}H_{15}N_3O_5$, *Chemox Selective, Dow Selective, Sinox W.*

Triethanolamine salt, $C_{16}H_{27}N_3O_8$, *DN 289, Chemox DN, Gebutox.*

Methacrylate, *see* Binapacryl.

USE: Herbicide; insecticide; miticide.

3283. Dinosterol. *4,23-Dimethylergost-22-en-3-ol; 4α-methyl-5α(H)-Δ²²-23,24-dimethylcholesten-3β-ol;* Black Sea sterol. $C_{30}H_{52}O$; mol wt 428.75. C 84.04%, H 12.23%, O 3.73%. Biogenetically important marine sterol isolated from the toxic dinoflagellate, *Gonyaulax tamarensis.* Isoln and structure determn: Y. Shimizu *et al.*, *J. Am. Chem. Soc.* **98**, 1059 (1976). Stereochemistry: J. Finer *et al.*, *J. Org. Chem.* **43**, 1990 (1978). Identity with the Black Sea sterol: J. J. Boon *et al.*, *Nature* **277**, 125 (1979). Stereospecific synthesis: A. Y. L. Shu, C. Djerassi, *Tetrahedron Letters* **22**, 4627 (1981).

Needles from methanol-chloroform, mp 220-222°. $[\alpha]_D$ ±5° (c = 0.6 in CHCl₃).

3284. Dinsed. *N,N'-1,2-Ethanediylbis[3-nitrobenzenesulfonamide]; N,N'-ethylenebis[3-nitrobenzenesulfonamide]; N,N'-bis[3-nitrobenzenesulfonyl]ethylenediamine; di-m-nitrobenzenesulfonylethylenediamine.* $C_{14}H_{14}N_4O_8S_2$; mol wt 430.43. C

39.07%, H 3.28%, N 13.02%, O 29.74%, S 14.90%. Prepn: Chattaway, *J. Chem. Soc.* **87**, 381 (1905).

Faintly yellow platelets from glacial acetic acid, mp 189-191° (some decompn).

THERAP CAT (VET): Coccidiostat.

3285. Diopterin. *N-[N-[4-[[(2-Amino-1,4-dihydro-4-oxo-6-pteridinyl)methyl]amino]benzoyl]-L-α-glutamyl]-L-glutamic acid;* pteroyl-α-glutamylglutamic acid; pteroyldiglutamic acid; PDGA. $C_{24}H_{26}N_8O_9$; mol wt 570.51. C 50.52%, H 4.59%, N 19.64%, O 25.24%. Folic acid analog. Synthesis: Mowat *et al.*, *J. Am. Chem. Soc.* **70**, 1096 (1948); Hutchings, and Geraci, U.S. pats. **2,500,825** and **2,501,168** (both 1950 to Am. Cyanamid); Geraci, U.S. pat. **2,766,240** (1956 to Aries Labs.).

Precipitate from hot water. Sol in dil NaOH solns.

3286. Dioscin. *(25R)-Spirost-5-en-3β-yl O-6-deoxy-α-L-mannopyranosyl-(1 → 2)-O-[6-deoxy-α-L-mannopyranosyl-(1 → 4)]-β-D-glucopyranoside;* diosgenin bis-α-L-rhamnopyranosyl-(1→2 and 1→4)-β-D-glucopyranoside. $C_{45}H_{72}O_{16}$; mol wt 869.08. C 62.19%, H 8.35%, O 29.46%. From *Dioscorea tokoro* Mal., *Dioscoreaceae.* Isoln: Tsukamoto *et al.*, *Pharm. Bull. (Tokyo)* **4**, 35 (1956); Heitz, *Compt. Rend.* **248**, 283 (1958). Structure: Kawasaki, Yamauchi, *Chem. Pharm. Bull.* **10**, 703 (1962).

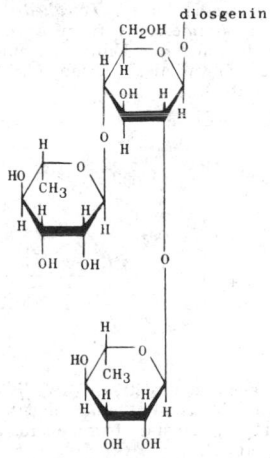

diosgenin

Crystals, decomp 275-277°. $[\alpha]_D^{13}$ −115° (c = 0.373 in ethanol).

3287. Dioscorea. Wild yam; colic root; rheumatism root. Dried rhizome of *Dioscorea villosa* L., *Dioscoreaceae.* Habit. North America. *Constit.* Saponin, acrid resin. Account of the nature, origins, cultivation and utilization of the useful members of the *Dioscoreaceae*: D. G. Coursey, *Yams* (London, Longmans, 1967) 230 pp. *See also* Yam, Mexican.

3288. Dioscorine. *[1R-(1α,4α,5α)]-2,4'-Dimethylspiro-[2-azabicyclo[2.2.2]octane-5,2'-[2H]pyran]-6'(3'H)-one.* $C_{13}H_{19}NO_2$; mol wt 221.29. C 70.55%, H 8.65%, N 6.33%, O 14.46%. Found in the tubers of *Dioscorea hirsuta*

Blume and *D. hispida* Dennst., *Dioscoreaceae:* H. W. Schutte, *Chem. Zentr.* **68, II,** 130 (1897); M. K. Gorter, *Rec. Trav. Chim.* **30,** 161 (1911); A. R. Pinder, *Nature* **168,** 1090 (1951); *idem, J. Chem. Soc.* **1952,** 2236. Structure: W. A. M. Davies *et al., Chem. & Ind. (London)* **1961,** 1410; *eidem, Tetrahedron* **18,** 405 (1962); Morris, A. R. Pinder, *J. Chem. Soc.* **1963,** 1841. Synthesis: Page, A. R. Pinder, *ibid.* **1964,** 4811. Absolute configuration: A. F. Beecham *et al., Tetrahedron Letters* **1969,** 3745. Biosynthetic studies: E. Leete, A. R. Pinder, *Chem. Commun.* **1971,** 1499; E. Leete, *J. Am. Chem. Soc.* **99,** 648 (1977). Pharmacology: A. R. Pinder, *J. Chem. Soc.* **1953,** 1826. Evaluation of toxic components: J. Webster *et al., J. Agr. Food Chem.* **32,** 1087 (1984).

Greenish-yellow prisms from ether, mp 54-55°. $[\alpha]_D^{18}$ −35.0° (c = 3.4 in chloroform). uv max (methanol): 215 nm (ϵ 10,160). Distills unchanged *in vacuo*. Sol in water, alcohol, acetone, chloroform; slightly sol in ether, benzene, petr ether.

Hydrochloride, $C_{13}H_{20}ClNO_2$, needles from alcohol + ether, dec 210-211°.

3289. Diosgenin. *(25R)-Spirost-5-en-3β-ol;* nitogenin. $C_{27}H_{42}O_3$; mol wt 414.61. C 78.21%, H 10.21%, O 11.58%. Aglycone of saponin dioscin. From *Dioscorea tokoro* Makino, *Dioscoreaceae:* Tsukamoto, *J. Pharm. Soc. Japan* **56,** 135 (1936); from rhizomes of *Trillium erectum* L., *Liliaceae* and *D. villosa* L., *Dioscoreaceae:* Marker *et al., J. Am. Chem. Soc.* **62,** 2542 (1940). Plant sources for diosgenin: Marker *et al., ibid.* **65,** 1199 (1943); Wall *et al., J. Am. Pharm. Assoc. Sci. Ed.* **46,** 653 (1957). Identity with nitogenin: Marker *et al., J. Am. Chem. Soc.* **65,** 1248 (1943). Structure: Marker *et al., ibid.* **62,** 2525 (1940). Configuration at C_{25}: James, *J. Chem. Soc.* **1955,** 637. Synthesis: Mazur *et al., J. Am. Chem. Soc.* **82,** 5889 (1960); Kessar *et al., Tetrahedron Letters* **1966,** 4319. Obtained commercially from *Dioscorea composita* Hemsl. and *D. terpinapensis* Uline (see under Barbasco). Isoln from barbasco varieties: Julian, U.S. pat. **3,019,220** (1962 to Julian Labs.).

Crystals from acetone, mp 204-207°. $[\alpha]_D^{25}$ − 129° (c = 1.4 in CHCl₃). Sol in the usual organic solvents, in acetic acid. Acetate, $C_{29}H_{44}O_4$, crystals from acetic acid, mp 198°. $[\alpha]_D^{20}$ − 119° (pyridine).

USE: Can be converted to pregnenolone and progesterone: Marker *et al., J. Am. Chem. Soc.* **69,** 2167 (1947).

3290. Diosmetin. *5,7-Dihydroxy-2-(3-hydroxy-4-methoxyphenyl)-4H-1-benzopyran-4-one; 3',5,7-trihydroxy-4'-methoxyflavone;* cyanidenon-4'-methyl ether 1479; luteolin-4'-methyl ether. $C_{16}H_{12}O_6$; mol wt 300.26. C 64.00%, H 4.03%, O 31.97%. Aglycone of diosmin, *q.v.* Prepn from diosmin isolated from various plant sources: O. A. Oesterle, G. Wander, *Helv. Chim. Acta* **8,** 519 (1925); isoln from lemons *(Citrus limon* Linn., *Rutaceae):* R. M. Horowitz, *J. Org. Chem.* **21,** 1184 (1956). Synthesis and structural elucidation: G. Zemplén, R. Bognár, *Ber.* **76,** 452 (1943). Synthesis: A. Lovecy *et al., J. Chem. Soc.* **1930,** 817; N. B. Lorette *et al., J. Org. Chem.* **16,** 930 (1951); J. H. Looker, M.

J. Holm, *ibid.* **24,** 1019 (1959). HPLC determn in biological fluids: D. Baylocq *et al., Ann. Pharmaceut. Franc.* **41,** 115 (1983).

Hemimethanolate, $C_{16}H_{12}O_6 \cdot \frac{1}{2}CH_3OH$, yellow needles from alcohol/ethyl acetate, sinters at 248°. mp 253-254°. Also reported as small yellow needles from methanol, mp 258-259° (Horowitz). uv max: 345, 268, 253 nm (log ϵ 4.32, 4.25, 4.28).

Triacetate, $C_{22}H_{18}O_9$, colorless needles from methanol, mp 195-196°.

3291. Diosmin. *7-[[6-O-6-Deoxy-α-L-mannopyranosyl)-β-D-glucopyranosyl]oxy]-5-hydroxy-2-(3-hydroxy-4-methoxyphenyl)-4H-1-benzopyran-4-one; 3',5,7-trihydroxy-4'-methoxyflavone-7-rutinoside; 5-hydroxy-2-(3-hydroxy-4-methoxyphenyl)-7-(O⁶-α-L-rhamnopyranosyl-β-D-glucopyranosyloxy)chromen-4-one; 5-hydroxy-2-(3-hydroxy-4-methoxyphenyl)-7-β-rutinosyloxy-4H-chromen-4-one;* diosmetin 7-β-rutinoside; barosmin; buchu resin; Daflon; Diosmil; Diovenor; Flebopex; Flebosmil; Flebosten; Flebotropin; Hemerven; Insuven; Tovene; Varinon; Ven-Detrex; Venex; Veno-V; Venosmine. $C_{28}H_{32}O_{15}$; mol wt 608.55. C 55.26%, H 5.30%, O 39.44%. Naturally occurring flavonic glycoside; rhamnoglycoside of diosmetin, *q.v.* Isolation from various plant sources: O. A. Oesterle, G. Wander, *Helv. Chim. Acta* **8,** 519 (1925). Elucidation of structure: G. Zemplén, R. Bognár, *Ber.* **76,** 452 (1943). Prepn from hesperidin, *q.v.*: *eidem, ibid.*; N. B. Lorette *et al., J. Org. Chem.* **16,** 930 (1951). Isoln from lemon peel *(Citrus limon* Linn. *Rutaceae):* R. M. Horowitz, *J. Org. Chem.* **21,** 1184 (1956); from *Zanthoxylum avicennae, Rutaceae:* H. R. Arthur *et al., J. Chem. Soc.* **1956,** 632; H. R. Arthur *et al., ibid.* **1959,** 4007; from flowers of *Sophora microphylla* Ait. *Leguminosae:* L. H. Briggs *et al., ibid.* **1960,** 1955. Toxicology studies: H. Heusser, W. Osswald, *Arch. Farmacol. Toxicol.* **3,** 33 (1977). NMR spectrum: J. L. Nieto, A. M. Gutierrez, *Spectroscop. Letters* **19,** 427 (1986). Pharmacology: C. Boudet, L. Peyrin, *Arch. Int. Pharmacodyn.* **283,** 312 (1986). Efficacy in animal models in high protein edema: J. R. Caseley-Smith, J. R. Caseley-Smith, *Agents Actions* **17,** 1 (1985); in inflammatory granuloma: M. Damon *et al., Arzneimittel-Forsch.* **37,** 1149 (1987). HPLC determn in biological fluids: D. Baylocq *et al., Ann. Pharmaceut. Franc.* **41,** 115 (1983). Clinical study in post-phlebitic ulcers: M. C. Nguyen, K. Morere, *Gaz. Med.* **92,** 71 (1985); in acute hemorrhoids: A. Tajana *et al., Minerva Med.* **79,** 387 (1988). Clinical trial in chronic venous insufficiency: R. Laurent *et al., Int. Angiol.* **7,** Suppl. 2, 39 (1988).

Monohydrate, $C_{28}H_{32}O_{15} \cdot H_2O$, mp 275-277° (dec) (Zemplén). Also reported as fine needles from aq pyridine or aq DMF, mp 283° (dec) (Briggs). uv max (ethanol): 255, 268, 345 nm (log ϵ 4.28, 4.25, 4.30). Practically insol in water, alcohol.

Combination with hesperidin (9:1), *S-5682, Daflon-500 mg.* Moderately hygroscopic. Practically insol in water. Sol in DMSO.

THERAP CAT: Capillary protectant.

3292. Diosphenol. *2-Hydroxy-3-methyl-6-(1-methyleth-yl)-2-cyclohexen-1-one; 1-methyl-4-isopropyl-1-cyclohexen-2-ol-3-one;* 2-hydroxypiperitone; 1-*p*-menthen-2-ol-3-one; Buchu camphor; Barosma camphor. $C_{10}H_{16}O_2$; mol wt 168.23. C 71.39%, H 9.59%, O 19.02%. The crystalline portion of an oil obtained from Buchu leaves which come from various species of *Barosma*, such as *B. betulina* Bartl. & Wendl., *B. serratifolia* (Curt.) Willd. and *B. crenulata* (L.) Hook., *Rutaceae:* Flückiger, *Pharm. J.* **11**, 174, 219 (1880); Spica, *Gazz. Chim. Ital.* **15**, 195 (1885); Shimoyama, *Arch. Pharm.* **226**, 403 (1888); Kondakov *et al., J. Prakt. Chem.* [II] **54**, 433 (1896); [II] **63**, 49 (1901). Structure: Semmler, McKenzie, *Ber.* **39**, 1160 (1906).

Crystals, mp 83°. Sublimes. bp_{760} 233° (dec); bp_{10} 109°. $d_4^{99.2}$ 0.9542. $n_D^{99.8}$ 1.4607. Absorption spectrum: Lowry, Lishmund, *J. Chem. Soc.* **1935**, 1313; Gillan *et al., ibid.* **1941**, 62. Sparingly sol in water; moderately sol in alcohol; sol in ether, chloroform, carbon disulfide.

3293. Dioxadrol. *2-(2,2-Diphenyl-1,3-dioxolan-4-yl)pi-peridine;* 2,2-diphenyl-4-(2-piperidyl)-1,3-dioxolane. $C_{20}H_{23}NO_2$; mol wt 309.39. C 77.64%, H 7.49% N 4.53%, O 10.34%. Prepn of *dl*-forms and resolution of α-racemates: Hardie, Halverstadt, **Belg. pat.** 613,262; **U.S. pat.** 3,262,938 (1962, 1966 both to Cutter Labs.); W. R. Hardie *et al., J. Med. Chem.* **9**, 127 (1966).

Oily liquid.
Hydrochloride, $C_{20}H_{24}ClNO_2$, *Rydar*. Crystals from methanol, mp 256-260°. LD_{50} orally in mice: 240 mg/kg (Hardie).
d-Form hydrochloride, *dexoxadrol hydrochloride, Relane*. Crystals, dec 254°. $[\alpha]_D^{20}$ +34° (c = 2 in methanol). LD_{50} orally in mice: 340 mg/kg (Hardie).
l-Form hydrochloride, *levoxadrol hydrochloride, Levoxan*. Crystals, mp 248-254°. $[\alpha]_D^{20}$ −34.5° (c = 2 in methanol). LD_{50} orally in mice: 230 mg/kg (Hardie).
THERAP CAT: Base as antidepressant; *d*-form hydrochloride as stimulant (central), analgesic; *l*-form hydrochloride as anesthetic (local), relaxant (smooth muscle).

3294. Dioxane. 1,4-Diethylene dioxide. $C_4H_8O_2$; mol wt 88.10. C 54.53%, H 9.15%, O 36.32%. Prepd by distilling ethylene glycol with dil H_2SO_4. Monograph: W. Stumpf, *Chemie und Anwendungen des 1,4-Dioxans* (Verlag Chemie, 1956). Toxicity data: E. P. Laug *et al., J. Ind. Hyg. Toxicol.* **21**, 173 (1939). Carcinogenicity studies: M. F. Argus *et al., J. Nat. Cancer Inst.* **35**, 949 (1965); R. J. Kociba *et al., Toxicol. Appl. Pharmacol.* **30**, 275 (1974).

Flammable liq; faint pleasant odor. Vapor harmful. d_4^{20} 1.0329. mp 11.80°. bp_{760} 101.1°; bp_{400} 81.8°; bp_{200} 62.3°; bp_{100} 45.1°; bp_{60} 33.8°; bp_{40} 25.2°; bp_{20} 12°. Cryoscopic constant 4.83. Trouton constant 21.90. Heat of combustion: 581 kcal/mol. Heat of fusion: 2.98 kcal/mol. Specific heat at 20° = 0.0370 kcal/mol/°C. Viscosity at 25° = 0.0120 poise. Crit temp 312°. Crit press 50.7 atm. Flash pt 5-18°C. n_D^{20}

1.4175. Dipole moment: zero. Azeotropic mixture with water: 81.6% dioxane, bp 87.8°. Azeotropic mixture with ethanol: 9.3% dioxane, bp 78.1°. Sol in water and the usual organic solvents. Tends to form explosive peroxides, especially when anhydr. LD_{50} in mice, rats (mg/kg): 5.7, 5.2 orally (Laug).
Caution: May cause CNS depression, necrosis of liver and kidneys. May be irritating to skin, lungs, mucous membranes. This substance may reasonably be anticipated to be a carcinogen: *Fourth Annual Report on Carcinogens* (NTP 85-002, 1985) p 95.
USE: Solvent for cellulose acetate, ethyl cellulose, benzyl cellulose, resins, oils, waxes, oil and spirit-sol dyes, and many other organic as well as some inorganic compds.

3295. Dioxaphetyl Butyrate. α,α-*Diphenyl-4-morpho-linebutanoic acid ethyl ester;* ethyl 4-morpholino-2,2-di-phenylbutyrate; 4-morpholino-2,2-diphenylbutyric acid ethyl ester; Amidalgon; Spasmoxal(e). $C_{22}H_{27}NO_3$; mol wt 353.44. C 74.75%, H 7.70%, N 3.96%, O 13.58%. Prepn: Specter *et al., J. Am. Chem. Soc.* **71**, 57 (1949); Dupré *et al., J. Chem. Soc.* **1949**, 500; Bockmühl, Ehrhart, **U.S. pat.** 2,230,774 (1940 to Winthrop).

Hydrochloride, $C_{22}H_{28}ClNO_3$, mp 168-169°.
Caution: May be habit forming. This is a controlled substance (opiate) listed in the U.S. Code of Federal Regulations, Title 21 Part 1308.11 (1987).
THERAP CAT: Narcotic analgesic. Antispasmodic.

3296. Dioxathion. *Phosphorodithioic acid S,S'-1,4-diox-ane-2,3-diyl O,O,O',O'-tetraethyl ester;* 2,3-*p*-dioxanedithiol S,S-bis(O,O-diethyl phosphorodithioate); AC 528; ENT 22879; Hercules 528; Delnav; Navadel. $C_{12}H_{26}O_6P_2S_4$; mol wt 456.54. C 31.57%, H 5.74%, O 21.03%, P 13.57%, S 28.09%. Prepn from 2,3-dichloro-*p*-dioxane and O,O-diethyl hydrogen phosphorodithioate; or from *p*-dioxene and bis(diethoxyphosphinothioyl)disulfide: Diveley, Lohr; Speck, **U.S. pats.** 2,725,328 and 2,815,350 (1955, 1957 both to Hercules); Diveley *et al., J. Am. Chem. Soc.* **81**, 139 (1959).

Tan liquid, d_4^{26} 1.257. mp −20°. n_D^{20} 1.5420. Practically insol in water, partly sol in hexane. Hydrolyzed by alkali and by heating. LD_{50} in female, male rats: 23, 43 mg/kg orally; 63, 235 mg/kg dermally, T. B. Gaines, *Toxicol. Appl. Pharmacol.* **14**, 515 (1969).
USE: Insecticide; acaricide.

3297. Dioxethedrine. *4-[2-(Ethylamino)-1-hydroxy-propyl]-1,2-benzenediol;* N-ethyl-3,4-dihydroxynorephedrine; α-(1-ethylaminoethyl)protocatechuyl alcohol; 2-ethylami-no-1-(3',4'-dihydroxyphenyl)-1-propanol; 1-(3',4'-dihy-droxyphenyl)-2-ethylamino-1-propanol; C 247. $C_{11}H_{17}NO_3$; mol wt 211.25. C 62.54%, H 8.11%, N 6.63%, O 22.72%. β-Adrenergic agonist. Prepd by catalytic hydrogenation of the corresponding aminoketone: Lespagnol, Cuignet, *Ann. Pharm. Franc.* **18**, 445 (1960). Only one of the two possible isomers, believed to be the "erythro" form, was isolated.

Hydrochloride, $C_{11}H_{18}ClNO_3$, crystals from methanol + ether, mp 212-214°.

One of the ingredients of *Bexol*.

THERAP CAT: Bronchodilator.

3298. Dioxybenzone. *(2-Hydroxy-4-methoxyphenyl)(2-hydroxyphenyl)methanone; 2,2'-dihydroxy-4-methoxybenzo-phenone;* 4-methoxy-2,2'-dihydroxybenzophenone; benzo-phenone-8; Cyasorb UV 24 (obsolete); Spectra-Sorb UV 24. $C_{14}H_{12}O_4$; mol wt 244.24. C 68.84%, H 4.95%, O 26.20%. Prepn: Hardy *et al.*, **U.S. pat.** 2,853,521 (1958 to Am. Cyanamid).

Yellow powder, mp 68°. Soly in g/100 ml at 25°: ethanol 21.8; isopropanol 17; propylene glycol 6.2; ethylene glycol 3.0; *n*-hexane 1.5.

THERAP CAT: Ultraviolet screen.

3299. Dioxypyramidon. *(Dimethylamino)oxoacetic acid 2-acetyl-2-methyl-1-phenylhydrazide; 1-acetyl-1,5,5-tri-methyl-2-phenylsemioxamazide;* α-(dimethylamidooxalyl)-β,β-methylacetylphenylhydrazine; dioxyaminopyrine; dioxo-aminopyrine. $C_{13}H_{17}N_3O_3$; mol wt 263.29. C 59.30%, H 6.51%, N 15.96%, O 18.23%. Prepd from aminopyrine by oxidation with 30% hydrogen peroxide: Charonnat, Delaby, *Compt. Rend.* **189**, 850 (1929); Valyashko, Depeshko, *Zh. Obshch. Khim.* **20**, 1667 (1950).

Orthorhombic, translucent prisms from water. Somewhat bitter taste. mp 105.5° (softens at 96°). bp_2 194-201°. uv max (hexane): 248.5 nm (ε 16,000). Soly in water at 20° = 7.69 g/100 ml; at 37° = 48.2 g/100 ml. Also sol in alcohol. Hydrolyzed by alkali.

3300. 2,5-Di-*tert*-pentylhydroquinone. *2,5-Bis(1,1-di-methylpropyl)-1,4-benzenediol;* 2,5-di-*tert*-amylhydroquin-one; 2,5-bis(1,1-dimethylpropyl)hydroquinone; Santovar A. $C_{16}H_{26}O_2$; mol wt 250.37. C 76.75%, H 10.47%, O 12.78%. Prepn: Erickson, **Brit. pat.** 596,461 (1948 to Mathieson Alkali Works).

Crystals, mp 179.4-180.4°.

USE: As a staining protector in rubber.

3301. Diperodon Hydrochloride. *3-(1-Piperidinyl)-1,2-propanediol bis(phenylcarbamate) monohydrochloride; 3-pi-peridino-1,2-propanediol dicarbanilate hydrochloride;* 3-(1-piperidyl)-1,2-propanediol dicarbanilate hydrochloride; di-(phenylurethan) of 1-piperidinepropane-2,3-diol hydrochlo-ride; Diothane Hydrochloride; Proctodon. $C_{22}H_{28}ClN_3O_4$;

mol wt 433.93. C 60.89%, H 6.50%, Cl 8.17%, N 9.68%, O 14.75%. Prepn: T. H. Rider, *J. Am. Chem. Soc.* **52**, 2115 (1930); *idem,* **U.S. pat.** 2,004,132 (1935). Stability of solns: E. S. Cook, T. H. Rider, *J. Am. Pharm. Assoc.* **26**, 222 (1937). Pharmacology: *eidem, J. Pharmacol.* **64**, 1 (1938). X-ray crystallographic study: R. C. Sullivan, K. P. O'Brien, *Bull. Narcotics* **20**, 31 (1968). IR spectra: *eidem, ibid.* **22**, 35 (1970).

Crystals. Bitter taste, followed by a sense of numbness. Stable in air at ordinary temps. Dec 195-200°. Sol in alco-hol. Slightly sol in water (1:100), acetone and ethyl acetate. The aq soln is faintly acid to litmus. Extemporaneously prepd aq solns should be used promptly to avoid precipita-tion of the free base by traces of alkali. The soly in water is increased by the addn of sodium chloride. Insol in benzene or ether.

THERAP CAT: Anesthetic (local).

3302. Diphemanil Methylsulfate. *4-(Diphenylmethyl-ene)-1,1-dimethylpiperidinium methyl sulfate; p-(α-phenyl-benzylidene)-1,1-dimethylpiperidinium methylsulfate; N,N-*dimethyl-4-piperidylidene-1,1-diphenylmethane methylsul-fate; Demotil; Diphenatil; Nivelona; Prantal; Variton. $C_{21}H_{27}NO_4S$; mol wt 389.50. C 64.75%, H 6.99%, N 3.60%, O 16.43%, S 8.23%. Prepd from *N*-methyl-4-piperidyldiphen-ylcarbinol which is obtained from methyl *N*-methylisonipe-cotate: Sperber *et al., J. Am. Chem. Soc.* **73**, 5010 (1951). Pharmacological properties: S. Margolin *et al., Proc. Soc. Exp. Biol. Med.* **78**, 576 (1951).

Crystals, mp 194-195°. Soluble in water. LD_{50} in rats, mice, guinea pigs (mg/kg): 1107, 64, 404 orally, S. Margolin *et al., loc. cit.*

THERAP CAT: Anticholinergic.

THERAP CAT (VET): Anticholinergic.

3303. Diphemethoxidine. *2-(Diphenylmethyl)-1-piperi-dineethanol;* 2-benzhydryl-1-piperidineethanol; 1-(β-hydr-oxyethyl)-2-(diphenylmethyl)piperidine; Ba 18189; Cleofil. $C_{20}H_{25}NO$; mol wt 295.41. C 81.31%, H 8.53%, N 4.74%, O 5.42%. Prepn: **Brit. pat.** 861,815 (1961 to Ciba), *C.A.* **55**, 15517h (1961).

mp 106-107°. $bp_{0.1}$ 180-181°.

Hydrochloride, mp 166-167°.

THERAP CAT: Anorexic.

3304. Diphenadione. *2-(Diphenylacetyl)-1H-indene-1,3(2H)-dione;* 2-diphenylacetyl-1,3-diketohydrindene; diphacinone; diphacin (Turkey); U-1363; Dipaxin; Oragu-lant; Solvan; Didandin; Diphacin. $C_{23}H_{16}O_3$; mol wt 340.36. C 81.16%, H 4.74%, O 14.10%. Prepd by the condensation of dimethyl phthalate with diphenylacetone in an inert sol-vent in the presence of sodium methoxide: Thomas, **U.S. pat.** 2,672,483 (1954 to Upjohn).

Pale yellow crystals from ethanol, mp 146-147°. Acid reaction. Practically insol in water. Slightly sol in benzene, hot ethanol. Sol in acetone, acetic acid. Forms a sodium salt which is sparingly sol in water.

USE: Rodenticide.

THERAP CAT: Anticoagulant.

3305. Diphenamid. *N,N-Dimethyl-α-phenylbenzeneacetamide; N,N-dimethyl-2,2-diphenylacetamide; N,N-dimethyl-α,α-diphenyl acetamide; 2,2-diphenyl-N,N-dimethylacetamide; L 34314; Dymid; Enide.* $C_{16}H_{17}NO$; mol wt 239.30. C 80.30%, H 7.16%, N 5.85%, O 6.69%. Selective pre-emergence herbicide. Prepn: Cheney *et al., J. Org. Chem.* **17**, 770 (1952).

Crystals from ethyl acetate, mp 134.5-135.5°. Soluble in water, acetone, dimethylformamide, xylene, phenyl Cellosolve. LD_{50} orally in rats: 700 mg/kg, G. W. Bailey, J. L. White, *Residue Rev.* **10**, 97 (1965).

USE: Herbicide.

3306. Diphenane. *4-(Phenylmethyl)phenol carbamate; carbamic acid α-phenyltolyl ester; p-benzylphenyl carbamate; p-benzylphenylurethan; carbamic acid p-benzylphenyl ester; p-hydroxydiphenylmethane carbamate; α-phenyl-p-cresyl carbamate; diphenan; Butolan; Butolen; Carphenol; Oxylan; Palafuge; Parabencil.* $C_{14}H_{13}NO_2$; mol wt 227.25. C 73.99%, H 5.77%, N 6.16%, O 14.08%. Prepd from α-phenyl-p-cresol, $COCl_2$, and NH_4OH: Kropp, U.S. pat. **1,252,452** (1818).

Crystals from alcohol, mp 147-150°. Practically insol in water. Sparingly sol in 90% ethanol; sol in methanol, abs alcohol, chloroform, ether, benzene.

THERAP CAT: Anthelmintic (Nematodes).

3307. Diphenazoline. *2-[(Diphenylmethoxy)methyl]-4,5-dihydro-1H-imidazole; 2-[(diphenylmethoxy)methyl]-2-imidazoline; 2-benzhydryloxymethyl-2-imidazoline; 2-imidazolinylmethyl benzhydryl ether.* $C_{17}H_{18}N_2O$; mol wt 266.33. C 76.66%, H 6.81%, N 10.52%, O 6.01%. Prepn: Djerassi, Scholz, U.S. pat. **2,516,108** (1950 to Ciba).

mp 102-103°.

Hydrochloride, $C_{17}H_{19}ClN_2O$, *MG 322, Antadril.* mp 205.5-207°.

THERAP CAT: Antihistaminic.

3308. Diphenhydramine. *2-Diphenylmethoxy-N,N-dimethylethanamine; 2-(benzhydryloxy)-N,N-dimethylethylamine; β-dimethylaminoethyl benzhydryl ether; O-benzhydryldimethylaminoethanol; β-dimethylaminoethanol diphenylmethyl ether; α-(2-dimethylaminoethoxy)diphenylmethane; benzhydramine; Alledryl; Allergin; Amidryl; Bagodryl; Benodin; Benylan; Benzantin; Dibendrin; Dibondrin; Dihidral; Diphantine; S 8; Sekundal-D; Syntedril.* $C_{17}H_{21}NO$; mol wt 255.35. C 79.96%, H 8.29%, N 5.49%, O 6.27%. Prepn: Rieveschl, Jr., U.S. pats. **2,421,714; 2,427,-878** (1947 to Parke, Davis), *cf.* U.S. pat. **2,397,799** (1946).

Toxicity: Gruhzit, Fisken, *J. Pharmacol. Exp. Ther.* **89**, 227 (1947). Comprehensive description of the hydrochloride: I. J. Holcomb, S. A. Fusari in *Analytical Profiles of Drug Substances* **vol. 3**, K. Florey, Ed. (Academic Press, New York, 1974) pp 173-232.

$bp_{2.0}$ 150-165°.

Hydrochloride, $C_{17}H_{22}ClNO$, *BAX, Benadryl, Benocten, Benzehist, Dabylen, Dolestan, Felben, Fenylhist, Halbmond, Rohydra, Sedopretten, Valdrene, Wehydryl.* Crystals from abs alcohol + ether, mp 166-170°. Bitter taste. Slowly darkens on exposure to light. Stable under ordinary conditions. One gram dissolves in 1 ml water, 2 ml alcohol, 2 ml chloroform, 50 ml acetone. Very slightly sol in benzene, ether. pH of 1% aq soln about 5.5. The aq soln forms a pink precipitate with satd Reinecke's salt soln. LD_{50} orally in rats: 500 mg/kg (Gruhzit, Fisken).

Ascorbate, *Antamin* (German), *Cetamin.*

Compd with p-sulfonamide benzoate, *Neobenadol.* See **Japan. pats. 203,405** and **243/64** (to Taisho).

THERAP CAT: Antihistaminic.

THERAP CAT (VET): The HCl as an antihistaminic. Also in motion sickness.

3309. Diphenic Acid. *[1,1'-Biphenyl]-2,2'-dicarboxylic acid; o,o'-bibenzoic acid.* $C_{14}H_{10}O_4$; mol wt 242.22. C 69.42%, H 4.16%, O 26.42%. Prepd from diazotized anthranilic acid by treatment with a cuproammonia-sulfite reducing agent: Vorländer, Meyer, *Ann.* **320**, 122 (1902); Atkinson, Lawler, *Org. Syn.* **coll. vol. I** (2nd ed, 1941) p 222. By chromic acid oxidation of phenanthrenequinone: Roberts, Johnson, *J. Am. Chem. Soc.* **47**, 1399 (1925). By heating potassium o-bromobenzoate with copper powder: Hurtley, *J. Chem. Soc.* **1929**, 1870.

Monoclinic prismatic rods upon slow cooling from water, leaflets from hot water, needles by careful sublimation. mp 228-229°. The satd aq soln is 0.0052N at 25°; soluble in the usual organic solvents.

Dimethyl ester, $C_{16}H_{14}O_4$, monoclinic prismatic plates, tablets from methanol, mp 73.5°; bp_{14} 204-206°.

Diethyl ester, $C_{18}H_{18}O_4$, cubes from alcoholic HCl, mp 42°.

3310. Diphenicillin Sodium. *6-[([1,1'-Biphenyl]-2-ylcarbonyl)amino]-3,3-dimethyl-7-oxo-4-thia-1-azabicyclo-[3.2.0]heptane-2-carboxylic acid sodium salt; sodium (2-biphenylylcarboxamido)penicillanate; (2-biphenylylcarboxamido) penicillanic acid sodium salt; sodium biphenylylpenicillin; biphenylylpenicillin sodium; 6-(2-phenylbenzamido)-penicillanic acid sodium salt; sodium diphenicillin; Ancillin.* $C_{21}H_{19}N_2NaO_4S$; mol wt 418.46. C 60.28%, H 4.58%, N 6.70%, Na 5.49%, O 15.29%, S 7.66%. Semi-synthetic antibiotic related to penicillin. Prepn: Doyle *et al.,* U.S. pat. **2,951,839** (1960); Dolan, Belg. pat. **615,855** (1962 to SK & F), *C.A.* **57**, 16762h (1962). Toxicity data: E. I. Goldenthal, *Toxicol. Appl. Pharmacol.* **18**, 185 (1971).

LD_{50} s.c. in rats: > 2500 mg/kg (Goldenthal).

THERAP CAT: Antibacterial.

3311. Diphenidol. *α,α-Diphenyl-1-piperidinebutanol;*

1,1-diphenyl-4-piperidino-1-butanol; diphenyl-[3-(1-piperidyl)propyl]carbinol; defenidol; Vontrol. $C_{21}H_{27}NO$; mol wt 309.43. C 81.51%, H 8.80%, N 4.53%, O 5.17%. Prepn: Miescher, Marxer, U.S. pat. **2,411,664** (1946 to Ciba); Barrett, Wilkinson, **Brit.** pat. **683,950** (1952 to Wellcome Foundation), *C.A.* **48**, 2112e (1954). Structure-activity studies: Gautier *et al., Med. Pharmacol. Exp.* **13**, 325 (1965). Clinical studies: Cutt *et al., Aerosp. Med.* **39**, 682 (1968); Benson, *ibid.* **40**, 589 (1969). Acute toxicity: E. I. Goldenthal, *Toxicol. Appl. Pharmacol.* **18**, 185 (1971); *Japan Med. Gaz.* **10**(7), 6 (1973).

Needles from petr ether, mp 104-105°. LD_{50} s.c. in rats: 50 mg/kg (Goldenthal).

Hydrochloride, $C_{21}H_{28}ClNO$, *Ansmin, Cefadol, Celmidol, Difenidolin, Maniol, Mecalmin, Pineroro, Satanolon, Tenesdol, Wansar, Yesdol.* Crystals from chloroform + ethyl acetate, mp 212-214°. Freely sol in methanol; sol in water, chloroform. Practically insol in ether, benzene, petr ether. LD_{50} in mice, rats (mg/kg): 430, 515 orally; 105, 82 i.p.; 37, 29 i.v. *(Japan Med. Gaz.).*

Note: Vontrol is also available as the hydrochloride and pamoate.

THERAP CAT: Antiemetic.

3312. Diphenolic Acid. *4-Hydroxy-γ-(4-hydroxyphenyl)-γ-methylbenzenebutanoic acid;* 4,4-bis[4'-hydroxyphenyl]pentanoic acid; γ,γ-bis-(*p*-hydroxyphenyl)valeric acid; DPA. $C_{17}H_{18}O_4$; mol wt 286.31. C 71.31%, H 6.34%, O 22.35%. Prepd by condensing 2.25-4.0 moles of phenol with one mole of levulinic acid in the presence of HCl: **Brit.** pat. **768,206** (1957 to S. C. Johnson & Son); from one mole of phenol, 0.5 mole levulinic acid and HCl, H_2SO_4 or H_3PO_4: Bader, Kantowicz, *J. Am. Chem. Soc.* **76**, 4465 (1954); Bader, U.S. pat. **2,933,520** (1960 to S. C. Johnson & Son).

Crystals from hot water. Higher melting modification, mp 171-172°. Appreciably sol in hot water; sol in acetone, acetic acid, ethanol, isopropanol, methyl ethyl ketone.

USE: Intermediate for surface coatings, lubricating oil additives, cosmetics, surfactants, plasticizers, textile chemicals.

3313. Diphenoxylate. *1-(3-Cyano-3,3-diphenylpropyl)-4-phenyl-4-piperidinecarboxylic acid ethyl ester;* 1-(3-cyano-3,3-diphenylpropyl)-4-phenylisonipecotic acid ethyl ester; ethyl 1-(3-cyano-3,3-diphenylpropyl)-4-phenylisonipecotate; ethyl 1-(3-cyano-3,3-diphenylpropyl)-4-phenyl-4-piperidinecarboxylate; 2,2-diphenyl-4-carbethoxy-4-phenylpiperidino)butyronitrile; R-1132. $C_{30}H_{32}N_2O_2$; mol wt 452.57. C 79.61%, H 7.13%, N 6.19%, O 7.07%. Prepn: Janssen, U.S. pat. **2,898,340** (1959). Pharmacokinetics and metabolism: Karim *et al., J. Pharmacol. Exp. Ther.* **177**, 546 (1971); *Clin. Pharmacol. Ther.* **13**, 407 (1972). *See also* Difenoxin, the active metabolite of diphenoxylate. Comprehensive description: D. D. Hung in *Analytical Profiles of Drug Substances* vol. 7, K. Florey, Ed. (Academic Press, N.Y., 1978) pp 149-169.

Hydrochloride, $C_{30}H_{32}N_2O_2 \cdot HCl$, crystals, mp 220.5-222°. uv max (methanol): 252, 258, 264 nm. Soly in mg/ml at 25°: acetic acid 500; DMF 500; chloroform 360; methanol > 50; ethanol 3; water 0.8; hexane 0.5.

An ingredient of *Lomotil, Diarsed,* and *Reasec,* which also contain atropine sulfate.

Caution: May be habit forming. This is a controlled substance (opiate) listed in the U.S. Code of Federal Regulations, Title 21 Part 1308.12 (1987).

THERAP CAT: Antiperistaltic; antidiarrheal.

3314. Diphenyl. *1,1'-Biphenyl;* bibenzene; phenylbenzene. $C_{12}H_{10}$; mol wt 154.20. C 93.46%, H 6.54%. Toxicity data: Deichmann *et al., J. Ind. Hyg. Toxicol.* **29**, 1 (1947). *Review:* W. C. Weaver *et al.,* in Kirk-Othmer *Encyclopedia of Chemical Technology* vol. 7 (Wiley-Interscience, New York, 3rd ed., 1979) pp 782-793.

Colorless leaflets; pleasant, peculiar odor. d 1.041. mp 69-71°. bp 254-255°. n_D^{77} 1.588. Insol in water. Sol in alc, ether. LD_{50} orally in rats: 3280 mg/kg (Deichmann).

Toxicity: CNS depression, paralysis, convulsions have been observed in exptl animals (Deichmann).

USE: As heat transfer agent; fungistat for oranges (applied to inside of shipping container or wrappers); in organic syntheses.

3315. Diphenylacetamide. Acetyldiphenylamine. $C_{14}H_{13}NO$; mol wt 211.25. C 79.59%, H 6.20%, N 6.63%, O 7.57%. $CH_3CON(C_6H_5)_2$.

White, cryst powder. mp 103°. Sublimes without decomposition. Slightly sol in water; sol in alcohol, ether.

3316. Diphenylacetic Acid. α-*Phenylbenzeneacetic acid;* diphenylmethane-α-carboxylic acid. $C_{14}H_{12}O_2$; mol wt 212.24. C 79.22%, H 5.70%, O 15.08%. $(C_6H_5)_2CHCOOH$. Prepd by the reduction of benzilic acid with hydriodic acid and red phosphorus: C. S. Marvel *et al., Org. Syn.* **coll. vol. I**, 224 (2nd ed., 1941). Ecologically and economically improved synthesis from benzilic acid: P. Strazzolini *et al., Syn. Commun.* **17**, 1919 (1987).

Small plates from alcohol, needles from water, mp 148°. Sublimes. Sol in hot water, alcohol, ether, chloroform.

Methyl ester, $C_{15}H_{14}O_2$, crystals from dil alcohol, mp 60°.

Ethyl ester, $C_{16}H_{16}O_2$, crystals from alcohol, mp 58°. bp_{15} 178°.

Amide, $C_{14}H_{13}NO$, mp 167-168°.

Nitrile, $C_{14}H_{11}N$, *diphenylacetonitrile,* α-*cyanodiphenylmethane, diphenatrile, Dipan.* mp 76°. bp_{12} 181°. Sol in alcohol, ether, propylene glycol. LD_{50} orally in rats: 3500 mg/kg, G. W. Bailey, J. L. White, *Residue Rev.* **10**, 97 (1965).

Chloride, $C_{14}H_{11}ClO$, crystals from petr ether, mp 57°; bp_{15} 178°. Very sol in benzene. Sol in hot ligroin, in hot isopropyl ether.

Anhydride, $C_{28}H_{22}O_3$, mp 98°; bp_{15} 220°.

USE: Nitrile as herbicide.

3317. Diphenylamine. *N-Phenylbenzeneamine.* $C_{12}H_{11}N$; mol wt 169.22. C 85.17%, H 6.55%, N 8.28%. $C_6H_5NH-C_6H_5$. Made by heating aniline with aniline hydrochloride.

Crystals; floral odor. d 1.16. mp 53-54°. bp 302°. Flash pt 153°. Discolors in light. One gram dissolves in 2.2 ml alcohol, 4.5 ml propyl alcohol; freely sol in benzene, ether, glacial acetic acid, carbon disulfide. Insol in water. Forms salts with strong acids. *Keep protected from light.*

Hydrochloride, $C_{12}H_{12}ClN$, crystals, turn blue in air. Freely sol in water, alcohol.

Sulfate, $C_{12}H_{13}NO_4S$, white to yellowish powder, mp 123-125°. Practically insol in water; sol in alc, in H_2SO_4.

USE: Manuf dyes; stabilizing nitrocellulose explosives and celluloid. In anal. chem for the detection of NO_3, ClO_3 and other oxidizing substances with which, in the presence of H_2SO_4, it gives a deep-blue color. *Caution:* May be irritating to mucous membranes. Methemoglobinemia has been produced experimentally. Symptoms similar to aniline, but diphenylamine is less toxic.

THERAP CAT (VET): Topically in anti-screwworm mixtures. In tests for nitrate or nitrite poisoning.

3318. Diphenylamine-2,2'-dicarboxylic Acid. *2,2'-Imi-*

nobis[benzoic acid]. $C_{14}H_{11}NO_4$; mol wt 257.24. C 65.37%, H 4.31%, N 5.45%, O 24.88%. Prepn from the alkali salts of anthranilic acid and of 2-chlorobenzoic acid in presence of copper: Ullmann, *Ann.* **355,** 352 (1907).

Yellow crystals from alcohol, mp 296-297° (dec). Insoluble in water; very slightly sol in alcohol, ether, chloroform, glacial acetic acid.

USE: Instead of diphenylamine in reactions with oxidizing agents. May be used as an oxidation-reduction indicator in strongly acidic solns.

3319. N,N'-Diphenylbenzidine. *N,N'-Diphenyl-[1,1'-biphenyl]-4,4'-diamine.* $C_{24}H_{20}N_2$; mol wt 336.42. C 85.68%, H 5.99%, N 8.33%.

Leaflets or plates, mp 242°. Insol in water; freely sol in boiling toluene or ethyl acetate, slightly in alc, acetone. *Keep protected from light.*

USE: Like diphenylamine; is more sensitive to oxidizing substances than the former.

3320. 1,4-Diphenyl-1,3-butadiene. *1,1'-(1,3-Butadiene-1,4-diyl)bisbenzene;* distyryl; bistyryl; 1,4-diphenylerythrene. $C_{16}H_{14}$; mol wt 206.27. C 93.16%, H 6.84%. C_6H_5-CH=CHCH=CHC$_6$H$_5$. Prepd by the condensation of phenylacetic acid and cinnamic aldehyde: Kuhn, Winterstein, *Helv. Chim. Acta* **11,** 103 (1928); Corson, *Org. Syn.* **16,** 28 (1936). This method yields the *trans,trans*-form.

cis,cis-Form, leaflets, needles, mp 70.5°. Changes to the *trans,trans*-form under the influence of light. $d_4^{100.6}$ 0.9697. $n_{H\alpha}^{100.6}$ 1.61831; $n_{H\beta}^{100.6}$ 1.66748; $n_{He}^{100.6}$ 1.63473. Soluble in ether, chloroform, benzene, petr ether, hot glacial acetic acid, slightly sol in alcohol.

cis,trans-Form, obtained only in the absence of light. Oily liquid or crystals, mp 88°. $bp_{0.1}$ 133-135°. d_4^{22} 0.9974. $n_{H\alpha}^{22.2}$ 1.59679; $n_{H\beta}^{22.2}$ 1.62830; $n_{He}^{22.2}$ 1.60532.

trans,trans-Form, crystals, mp 149.7°. bp_{720} 350°. Soluble in alcohol; sparingly sol in ether.

3321. sym-Diphenylcarbazide. *2,2'-Diphenylcarbonic dihydrazide.* $C_{13}H_{14}N_4O$; mol wt 242.27. C 64.44%, H 5.82%, N 23.13%, O 6.60%. C_6H_5NHNHCONHNHC$_6$H$_5$. White, cryst powder, gradually becomes pink. mp 168-171°. Very slightly sol in water; sol in hot alcohol, acetone, glacial acetic acid. *Keep protected from light.*

USE: As indicator in titrating Fe; for the colorimetric determination of Cr, detection of Cd, Hg, Mg, aldehydes, emetine.

3322. Diphenylcarbazone. *Phenyldiazenecarboxylic acid 2-phenylhydrazide.* $C_{13}H_{12}N_4O$; mol wt 240.26. C 64.99%, H 5.03%, N 23.32%, O 6.66%. C_6H_5N=NCONHNHC$_6$H$_5$. Orange-red needles. mp about 157° with decompn. Insol in water; sol in alcohol, chloroform, benzene.

USE: As a sensitive reagent for Hg, with which it gives a blue color.

3323. 1,1-Diphenylethene. *1,1'-Ethenylidenebis[benzene];* unsym-diphenylethylene; α,α-diphenylethylene; α-methylene-diphenylmethane. $C_{14}H_{12}$; mol wt 180.24. C 93.29%, H 6.71%. $(C_6H_5)_2$C=CH$_2$. Prepd by the action of phenylmagnesium bromide on ethyl acetate in ether, followed by treatment of the reaction product with ammonium chloride soln: Allen, Converse, *Org. Syn.* **coll. vol. I** (2nd ed., 1941) p 226.

Liquid, mp 8.2°. d_4^{20} 1.0232. bp_{760} 277.0°; bp_{400} 249.8°; bp_{200} 222.8°; bp_{100} 198.6°; bp_{60} 183.4°; bp_{40} 170.8°; bp_{20} 151.8°; bp_{10} 135.0°; bp_5 119.6°; $bp_{1.0}$ 87.4°. n_D^{20} 1.60849. Absorption spectrum: Lardy, *J. Chim. Phys.* **21,** 361 (1924).

3324. N-(1,2-Diphenylethyl)nicotinamide. *N-(1,2-Diphenylethyl)-3-pyridinecarboxamide;* 1-nicotinylamino-1,2-diphenylethane; C-1065; Lyspamin Forte. $C_{20}H_{18}N_2O$; mol wt 302.36. C 79.44%, H 6.00%, N 9.27%, O 5.29%. Prepn: Suter, U.S. pat. **2,483,250** (1949 to Cilag).

Needles from ethanol, mp 159°. Readily sol in 5N HCl; moderately sol in methanol, acetone, ether, benzene; slightly sol in ethanol; practically insol in water and alkalies.

Mixture with phenobarbital, *Lyspamin.*

THERAP CAT: Anticholinergic.

3325. 1,3-Diphenylguanidine. *N,N'-Diphenylguanidine;* sym-diphenylguanidine; Melaniline; Vulkazit. $C_{13}H_{13}N_3$; mol wt 211.26. C 73.90%, H 6.20%, N 19.89%. Prepn: Naunton, *J. Soc. Chem. Ind. (London)* **45,** 376T (1926); Macholdt-Erdniss, *Ber.* **91,** 1992 (1958); Ferris, Schutz, *J. Org. Chem.* **28,** 71 (1963). Pharmacology: P. Valade *et al.,* *C.R. Soc. Biol.* **143,** 815 (1949).

Crystals from ether, mp 150°. Dec at about 170°. d 1.13. Sparingly sol in water; sol in alcohol, chloroform, hot benzene, hot toluene; readily sol in dil mineral acids. Aq soln is strongly alkaline. MLD s.c. in guinea pigs: 200 mg/kg; i.v. in dogs: 25 mg/kg (Valade).

Phthalate, $C_{34}H_{32}N_6O_4$, *Guantal.* Obtained as the hemihydrate, blue-white to light gray powder. mp 178°. d_4^{25} 1.20. Sol in alcohol. Practically insol in benzene, gasoline.

USE: Recommended as a primary material for standardizing acids. Free base and phthalate as accelerators for vulcanization of rubber.

3326. 1,1-Diphenylhydrazine. $C_{12}H_{12}N_2$; mol wt 184.23. C 78.23%, H 6.57%, N 15.21%. $(C_6H_5)_2$NNH$_2$. Yellow crystals. d 1.19. mp 34.5°; also stated as 44°. bp_{40} 220°. Insol in water; freely sol in alc, ether.

Hydrochloride, $C_{12}H_{12}N_2$.HCl, white to grayish white, crystalline powder. Slightly sol in water; freely sol in alc.

USE: Hydrochloride as reagent for arabinose and lactose.

3327. Diphenylketene. *Diphenylethenone.* $C_{14}H_{10}O$; mol wt 194.22. C 86.57%, H 5.19%, O 8.24%. $(C_6H_5)_2$C=C=O. Prepd from benzil monohydrazone: Smith, Hoehn, *Org. Syn.* **coll. vol. III,** 356 (1955).

Reddish-yellow liquid. Best stored under nitrogen. The addition of hydroquinone helps to retard polymerization. Solidifies in refrigerator. $d_4^{13.7}$ 1.1107. bp_{760} 265-270° (dec); bp_{12} 146°; $bp_{3.5}$ 119-121°. Should be distilled under reduced pressure (3 to 5 mm) in an atm of nitrogen. $n_D^{14.1}$ 1.615. Dipole moment in benzene at 25° = −1.9.

3328. Diphenylmagnesium. Magnesium diphenyl. $C_{12}H_{10}Mg$; mol wt 178.52. C 80.73%, H 5.65%, Mg 13.62%. $Mg(C_6H_5)_2$. Prepd by the action of magnesium on mercury diphenyl: Schlenk, Jr., *Ber.* **64,** 736 (1931); Gilman, Brown, *Rec. Trav. Chim.* **49,** 202 (1930).

Solvated, feathery crystals from ether. Loses its ether *in vacuo* at around 37°. Dec 280°, dissociating into magnesium and diphenyl. Extremely reactive; catches fire in moist, although not in dry, air. Violently dec by water. The etherate is sol in benzene, but the dry compd is not.

3329. Diphenylmethane. *1,1'-Methylenebis[benzene];* benzylbenzene; ditan. $C_{13}H_{12}$; mol wt 168.23. C 92.81%, H 7.19%. $(C_6H_5)_2$CH$_2$. Prepd from methylene chloride and benzene with aluminum chloride as catalyst: Friedel, Crafts, *Bull. Soc. Chim.* [2] **41,** 324 (1884); from benzyl chloride and benzene: Hartman, Phillips, *Org. Syn.* **14,** 34 (1934); L. F. Fieser, *Experiments in Organic Chemistry* (Boston, 3rd ed., 1955) p 157.

Orthorhombic needles. Odor of oranges. d_4^{10} 1.3421 (solid). mp 25.9°. d_4^{26} 1.0008 (liq). bp_{760} 264.5°; bp_{400} 237.5°;

bp_{200} 210.7°; bp_{100} 186.3°; bp_{60} 170.2°; bp_{40} 157.8°; bp_{20} 139.8°; bp_{10} 122.8°; bp_5 107.4°; $bp_{1.0}$ 76.0°. n_D^{20} 1.57683. Absorption spectrum in hexane: Castille, *Bull. Soc. Chim. Belg.* **36,** 296; *Chem. Zentr.* **1927,** I, 1126; in alcohol: Ondorff, *J. Am. Chem. Soc.* **49,** 1541 (1927). Freely sol in alcohol, ether, chloroform, hexane, benzene; insol in liq ammonia.

3330. Diphenylmethane-4,4'-disulfonamide. *4,4'-Methylenebis[benzenesulfonamide];* Nirexon. $C_{13}H_{14}N_2O_4S_2$; mol wt 326.38. C 47.84%, H 4.32%, N 8.58%, O 19.61%, S 19.65%. Prepd from the corresp bis[sulfonyl chloride] deriv by treatment with ammonia: Schraufstätter *et al., Medizin und Chemie* V, (Verlag Chemie, Weinheim, 1956) p 107; **Brit. pat. 764,226** (1956 to Bayer), *C.A.* **52,** 2078d (1958).

$$H_2NSO_2-\!\!\bigcirc\!\!-CH_2-\!\!\bigcirc\!\!-SO_2NH_2$$

Crystals from 25% methanol, mp 184°. Slightly sol in water. The soly is increased as the pH is increased.
THERAP CAT: Carbonic anhydrase inhibitor, diuretic.

3331. N,N'-Diphenyl-p-phenylenediamine. *N,N-Diphenyl-1,4-benzenediamine;* 1,4-dianilinobenzene; DPPD. $C_{18}H_{16}N_2$; mol wt 260.32. C 83.04%, H 6.20%, N 10.76%. Prepd by condensing hydroquinone or p-aminophenol with aniline: Calm, *Ber.* **16,** 2805 (1883); Clemens, Magoffin, **U.S. pat. 2,503,712** (1950 to Kodak). Purification: Pecherer, **U.S. pat. 2,833,824** (1958 to Hoffmann-La Roche).

$$C_6H_5-NH-\!\!\bigcirc\!\!-NH-C_6H_5$$

Colorless leaflets from alcohol. Commercial grades are greenish-brown. d 1.20. mp 150-151° (uncorr., Pecherer, *loc. cit.*). $bp_{0.5}$ 220-225°. Sol in monochlorobenzene, benzene, DMF, ether, chloroform, acetone, ethyl acetate, isopropyl acetate, glacial acetic acid. Slightly sol in alcohol. Almost insol in petr ether, water.
USE: Polymerization inhibitor. Antioxidant for rubber, petr oils, feedstuffs.
THERAP CAT (VET): Antioxidant for feedstuffs.

3332. 1,1-Diphenyl-2-picrylhydrazyl (Free Radical). *2,2-Diphenyl-1-(2,4,6-trinitrophenyl)hydrazyl;* DPPH. Prepn: Goldschmidt, Renn, *Ber.* **55B,** 628 (1922); Lyons, Watson, *J. Polymer Sci.* **18,** 141 (1955); Arbuzov, Valitova, *Zh. Obshch. Khim.* **27,** 2354 (1957). Structure: Poirier *et al., J. Org. Chem.* **17,** 1437 (1952).

$$\begin{array}{c} \cdot N-N(C_6H_5)_2 \\ O_2N-\!\!\bigcirc\!\!-NO_2 \\ NO_2 \end{array}$$

Large, dark violet prisms from benzene + petr ether, mp 127-129° (dec). mp also reported as 132-133°.
USE: Analytical reagent for reducing substances: Schenck *et al., Tetrahedron Letters* **1967,** 193.

3333. α,α-Diphenyl-2-piperidinepropanol. *β-Methyl-α,α-diphenyl-1-piperidineethanol;* 1,1-diphenyl-2-piperidino-1-propanol; α,α-diphenyl-β-methyl-β-piperidineethanol; diphepanol; Hoechst 10682. $C_{20}H_{25}NO$; mol wt 295.41. C 81.31%, H 8.53%, N 4.74%, O 5.42%. Description: Lindner, Stein, *Arzneimittel-Forsch.* **9,** 94 (1959). Prepn: Stein, Lindner, **U.S. pat. 2,827,460** (1958 to Hoechst).

$$\begin{array}{c} C_6H_5 \quad OH \\ \quad | \\ CCH-N\!\!\bigcirc \\ \quad | \\ C_6H_5 \quad CH_3 \end{array}$$

bp_2 182°.

Hydrochloride, $C_{20}H_{25}NO.HCl$, crystals, dec 196-197°. Sol in water, alc. Ingredient of *Tussukal.*
Guaiacolsulfonate, *Viatussin.*
THERAP CAT: Antitussive.

3334. Diphenylpyraline. *4-(Diphenylmethoxy)-1-methylpiperidine;* diphenylpyrilene; 4-(benzhydryloxy)-1-methylpiperidine; 4-benzhydryloxy-N-methylpiperidine; 1-methyl-4-piperidyl benzhydryl ether; 1-methyl-4-hydroxypiperidine benzhydryl ether; P 253; Allergen; Antivom; Belfene; Dayfen; Diafen; Hispril; Histryl; Histyn; Hystryl; Lyssipoll; Mepiben; Neargal. $C_{19}H_{23}NO$; mol wt 281.38. C 81.10%, H 8.24%, N 4.98%, O 5.69%. Prepd by refluxing 1-methyl-4-piperidinol and benzhydryl bromide in xylene: Knox, Kapp, **U.S. pat. 2,479,843** (1949 to Nopco). Alternate prepn: Phillips, **U.S. pat. 2,595,405** (1952 to Merck & Co.).

$$CH_3-N\!\!\bigcirc\!\!-OCH\!\!\begin{array}{c} C_6H_5 \\ \\ C_6H_5 \end{array}$$

Hydrochloride, $C_{19}H_{24}ClNO$, *Anginosan, Kolton (jelly), Lergoban.* Crystals from isopropanol + ether, mp 206°. Sol in water, ethanol, isopropanol. Practically insol in ether, benzene.
Hydrobromide, $C_{19}H_{24}BrNO$, crystals, mp 201-202°. Soluble in water, ethanol.
8-Chlorotheophyllinate, *see* Piprinhydrinate.
THERAP CAT: Antihistaminic.

3335. sym-Diphenylpyrophosphorodiamidic Acid. *N,N'-Diphenyl-P,P'-diamidodiphosphoric acid.* $C_{12}H_{14}N_2O_5P_2$; mol wt 328.20. C 43.90%, H 4.30%, N 8.54%, O 24.38%, P 18.88%. Prepd from O-benzylphenylphosphoroamidic acid: Boger, Friedman, *J. Am. Chem. Soc.* **80,** 2583 (1958).

$$\begin{array}{ccccc} & O & & O & \\ & \| & & \| & \\ C_6H_5-NHP & -O- & PNH-C_6H_5 \\ & | & & | & \\ & OH & & OH & \end{array}$$

Dicyclohexylammonium salt, $C_{24}H_{40}N_4O_5P_2$, mp 223-224°. Sol in water. No spontaneous hydrolysis at pH 4.0 to pH 11.5 after two hours at 37°.
Dipotassium salt, $C_{12}H_{12}K_2N_2O_5P_2$, crystals.
USE: Colorimetric estimation of enzymes.

3336. Diphenyl Sulfone. *1,1'-Sulfonylbisbenzene; phenyl sulfone;* sulfobenzide. $C_{12}H_{10}SO_2$; mol wt 218.27. C 66.03%, H 4.62%, S 14.69%, O 14.66%. $(C_6H_5)_2SO_2$. Prepd by sulfonation of benzene with sulfuric acid; formed as a by-product on production of benzenesulfonic acid and benzenesulfonyl chloride.
White monoclinic prisms or leaflets, mp 128-129°. bp 378-379°. Insoluble in cold water; slightly sol in boiling water; sol in hot alcohol, in benzene. LD_{50} orally in rats: > 2 g/kg, *Residue Revs.* **36,** 240 (1971).
USE: Ovicide (esp vs eggs and larval stages of mites).

3337. sym-Diphenylthiourea. *N,N'-Diphenylthiourea; thiocarbanilide;* sulfocarbanilide. $C_{13}H_{12}N_2S$; mol wt 228.31. C 68.39%, H 5.30%, N 12.27%, S 14.04%. $C_6H_5NHCSNH-C_6H_5$. Prepn from aniline and carbon disulfide: Fry, *J. Am. Chem. Soc.* **35,** 1539 (1913); Stasse, **U.S. pat. 2,435,295** (1948 to Allied Chem. & Dye); from potassium ethylxanthate and aniline: Aravindakshan *et al., Indian J. Chem.* **1**(9), 395 (1963). Physical properties and toxicity data: P. J. Hanzlik, A. Irvine, *J. Pharmacol. Exp. Ther.* **17,** 349 (1921). White crystalline solid, mp 153-154°. d 1.32. Distinct bitter taste. Practically insol in water. Sol in alcohol, ether, chloroform. MLD orally in rabbits: 1.5 g/kg (Hanzlik, Irvine).
USE: Vulcanizing accelerator; sulfur dyes.

3338. Diphetarsone. *[1,2-Ethanediylbis(imino-4,1-phenylene)]bis[arsonic acid] disodium salt; N,N'-ethylenediarsanilic acid disodium salt;* bis[p-arsonophenylamino]-1,2-ethane disodium; N,N'-bis(p-arsonophenyl)ethylenediamine disodium; RP 4763; Amebarsin; Bémarsal; Rodameb. $C_{14}H_{16}$-

As$_2$N$_2$Na$_2$O$_6$; mol wt 504.10. C 33.36%, H 3.20%, As 29.72%, N 5.56%, Na 9.12%, O 19.04%. Prepn: Hamilton, *J. Am. Chem. Soc.* **45**, 2751 (1923).

HO—O=As—⟨ ⟩—NHCH$_2$CH$_2$NH—⟨ ⟩—As=O—OH
NaO   ONa

Decahydrate, powder, no definite melting range. Freely sol in water; slightly sol in boiling 95% ethanol; practically insol in abs ethanol, acetone, chloroform. pH of a 2% aq soln: 6 to 8. Precipitated from aq solns by acids.

THERAP CAT: Antiamebic.

3339. Diphosgene. *Carbonochloridic acid trichloromethyl ester; chloroformic acid trichloromethyl ester;* trichloromethylchloroformate; trichloromethylcarbonochloridate. C$_2$Cl$_4$O$_2$; mol wt 197.83. C 12.14%, Cl 71.68%, O 16.17%. ClCOOCCl$_3$. Used in synthesis as substitute for phosgene, *q.v.* Prepn by photochlorination of methyl formate: W. Hentschel, *J. Prakt. Chem.* **36**, 209 (1887); F. Grignard *et al., C.R. Acad. Sci.* **169**, 1074 (1919); *eidem, ibid.* 1143; by photochlorination of methyl chloroformate: A. Kling *et al., ibid.* 1046; K. Kurita, Y. Iwakura, *Org. Syn.* **59**, 195 (1980). Effect on air-blood barrier in rabbits, toxicity: E. Klika, A. Mysliveckova, *Folia Morphol.* **19**, 5 (1971), *C.A.* **75**, 3433c (1971). Use in synthesis of isocyanides: G. Skorna, J. Ugi, *Angew. Chem. Int. Ed.* **16**, 259 (1977).

Suffocating liquid. bp 128°. bp$_{50}$ 49°. d$_{15}$ 1.664. n$_D^{22}$ 1.45664. Stable at room temp, decomposes to phosgene at ~300°C. LC$_{100}$ in rabbits (10-20 minutes exposure to vapor): 0.9 mg/l air (Klika, Mysliveckova).

USE: In organic synthesis; as war gas.

3340. Dipicrylamine. *2,4,6-Trinitro-N-(2,4,6-trinitrophenyl)benzenamine;* 2,4,6,2′,4′,6′-hexanitrodiphenylamine. C$_{12}$H$_5$O$_{12}$N$_7$; mol wt 439.22. C 32.81%, H 1.15%, N 22.33%, O 43.71%. *Ref:* Winkel, Maas, *Angew. Chem.* **49**, 827 (1936).

Yellow prisms. Has a tendency to explode. mp about 238° with dec. Insol in water, acetone, alcohol or ether; sol in alkalies, glacial acetic acid.

USE: Reagent for gravimetric determination of potassium.

3341. Dipin. *1,4-Bis[bis(1-aziridinyl)phosphinyl]piperazine; 1,4-piperazinediylbis[bis(1-aziridinyl)phosphine oxide];* piperazine-1,4-bis(N,N′-diethylenephosphonediamide); tetraethyleneimidopiperazine-N,N′-diphosphoric acid. C$_{12}$H$_{24}$N$_6$O$_2$P$_2$; mol wt 346.32. C 41.62%, H 6.98%, N 24.27%, O 9.24%, P 17.89%. Alkylating agent; exptl antineoplastic. Prepn: Kropacheva *et al., Zh. Obshch. Khim.* **30**, 3584 (1960), *C.A.* **55**, 18695c (1961). Spectrophotometric determn: L. Kh. Kartashova, E. M. Salomatin, *Farmatsiya (Moscow)* **33**, 72 (1983), *C.A.* **102**, 119736j (1985). Metabolism in rats: V. V. Chistyakov *et al., Khim.-Farm. Zh.* **21**, 398 (1987), *C.A.* **107**, 146734s (1987).

Crystals, mp 187°.

3342. Dipipanone. *4,4-Diphenyl-6-(1-piperidinyl)-3-heptanone; dl-4,4-diphenyl-6-piperidinoheptan-3-one;* 6-piperidino-4,4-diphenylheptan-3-one; 2-(1-piperidino)-4,4-diphenyl-5-heptanone; phenylpiperone; 378C48; Hoechst 10805; Piperidyl Amidone; Pamedone(e); Fenpidon; Pipadone. C$_{24}$H$_{31}$NO; mol wt 349.50. C 82.47%, H 8.94%, N 4.01%, O 4.58%. Prepn: Ofner, Walton, *J. Chem. Soc.* **1950,**

2158; **Brit.** pat. **654,975** (1951 to Wellcome Foundation). Variations of the synthesis using phenyllithium as condensing agent: Kazuhiko, Kubota, *Kumamoto Med. J.* **12**, 304-307 (1960).

CH$_3$ C$_6$H$_5$
N—CHCH$_2$C—COCH$_2$CH$_3$
        C$_6$H$_5$

Hydrochloride, C$_{24}$H$_{32}$ClNO, minute prisms from wet alcohol-ether, mp 123-126°. Also reported as 126-127°.

Hydrobromide, C$_{24}$H$_{32}$BrNO, crystals from water, mp 103-106°.

Caution: May be habit forming. This is a controlled substance (opiate) listed in the U.S. Code of Federal Regulations, Title 21 Part 1308.11 (1987).

THERAP CAT: Narcotic analgesic.

3343. Dipiproverine. *α-Phenyl-1-piperidineacetic acid 2-(1-piperidinyl)ethyl ester;* piperidinoethyl α-piperidinophenylacetate. C$_{20}$H$_{30}$N$_2$O$_2$; mol wt 330.46. C 72.69%, H 9.15%, N 8.48%, O 9.68%. Prepn: Klosa, *Pharmazie* **8**, 723 (1953); Wunderlich, *ibid.* **11**, 201 (1956); Najer *et al., Bull. Soc. Chim. France* **1958**, 355. Pharmacology: Vaccari *et al., Clin. Ter.* **32**, 483 (1965).

C$_6$H$_5$
N—CHCOOCH$_2$CH$_2$—N

bp$_1$ 188-190°.

Dihydrochloride, C$_{20}$H$_{32}$Cl$_2$N$_2$O$_2$, *LD 935, Lévospasme, Levospasmol, Spasmonal.* Crystals from isopropanol, mp 212-214° (Wunderlich), 238° (Najer). Freely sol in water. Soly in water at 20°, 100°: 184, 323 g/100 ml. Soly in isopropanol at 20°, 50°, 80°: 5, 12.5, 27 g/100 g. Initial pH of a 2.5% soln: 3.35.

THERAP CAT: Anticholinergic.

3344. Dipivefrin. *2,2-Dimethylpropanoic acid 4-[1-hydroxy-2-(methylamino)ethyl]-1,2-phenylene ester;* (±)-3,4-dihydroxy-α-[(methylamino)methyl]benzyl alcohol 3,4-dipivalate; 1-(3′,4′-dipivaloyloxyphenyl)-2-methylamino-1-ethanol; dipivalyl epinephrine; DPE. C$_{19}$H$_{29}$NO$_5$; mol wt 351.45. C 64.93%, H 8.32%, N 3.99%, O 22.76%. Dipivalyl ester of epinephrine, *q.v.* Prepn: D. Henschler *et al., Ger. pat.* **2,152,058** corresp to U.S. pat. **4,085,270** (1973, 1978 both to Klinge); A. Hussain, J. E. Truelove, **Ger.** pat. **2,343,657** corresp to U.S. pats. **3,809,714** and **3,839,584** (all 1974 to Interx). *In vitro* study: A. H. Neufeld, E. D. Page, *Invest. Ophthalmol. Vis. Sci.* **16**, 1118 (1977). Pharmacology: B. C. Wang *et al., J. Pharmacol. Exp. Ther.* **203**, 442 (1977). Effects on intraocular pressure in dogs: R. M. Gwin *et al., Am. J. Vet. Res.* **39**, 83 (1978). Metabolism: I. Abramovsky, J. S. Mindel, *Arch. Ophthalmol.* **97**, 1937 (1979). Clinical study: M. A. Kass *et al., ibid.* 1865. General pharmacology, toxicology and clinical experience in glaucoma: D. A. McClure, *ACS Symposium Series* **14** (1975) pp 224-235.

(CH$_3$)$_3$CCOO—⟨ ⟩—CHCH$_2$NHCH$_3$
                          OH
(CH$_3$)$_3$CCOO

Crystals from ether, mp 146-147°.

Hydrochloride, C$_{19}$H$_{30}$ClNO$_5$, *Diopine, D-Epifrin, Pivalephrine, Propine, Vistapin.* Crystals from ethyl acetate, mp 158-159°. Sol in water and ethanol.

THERAP CAT: Adrenergic (ophthalmic); antiglaucoma.

3345. Diploicin. *2,4,7,9-Tetrachloro-3-hydroxy-8-methoxy-1,6-dimethyl-11H-dibenzo[b,e][1,4]dioxepin-11-one;* 3,5-dichloro-6-[(3,5-dichloro-6-hydroxy-4-methoxy-o-tolyl)oxy]-4,2-cresotic acid ε-lactone; 5,6′-dimethyl-2′,3-dihydroxy-4′-methoxy-2,3′,4,5′-tetrachloro-6-carboxydiphenyl ether 2′,6-lactone. C$_{16}$H$_{10}$Cl$_4$O$_5$; mol wt 424.06. C

45.32%, H 2.38%, Cl 33.44%, O 18.86%. Antibiotic isolated from the lichen *Buellia canescens* (Dicks.) De Not. [*Diploicia canescens* (Dicks.) Massal.], *Lecideaceae:* Zopf, *Ann.* **336**, 58 (1904); Nolan, *Sci. Proc. Roy. Dublin Soc.* **21**, 67 (1934); Nolan *et al., Chem. & Ind. (London)* **1935**, 577; Barry, *Nature* **158**, 131 (1946). Structure: Nolan *et al., Sci. Proc. Roy. Dublin Soc.* **24**, 319 (1948). Synthesis: Ollis, *ibid.* **27**, 161 (1956); Brown *et al., Proc. Chem. Soc.* **1960**, 393; Hendrickson, Ramsay, *Chem. Commun.* **1968**, 1101; Hendrickson *et al., J. Am. Chem. Soc.* **94**, 6834 (1972); P. D. Djura *et al., J. Chem. Soc., Perkin Trans I,* **1976**, 147.

White needles from methanol, mp 233-234°. uv max: 270 nm (log ε 3.79).

Acetate, $C_{18}H_{12}Cl_4O_6$, mp 234-235°.

3346. Diponium Bromide. *2-[(Dicyclopentylacetyl)oxy]-N,N,N-triethylethanaminium bromide; triethyl(2-hydroxyethyl)ammonium bromide dicyclopentylacetate;* diethylaminoethyl α,α-dicyclopentylacetate ethobromide; dicyclopentylacetic acid β-diethylaminoethyl ester ethobromide; dipenine bromide; Sa 267; Unospaston. $C_{20}H_{38}BrNO_2$; mol wt 404.45. C 59.39%, H 9.47%, Br 19.76%, N 3.46%, O 7.91%. Prepn: Barrelet *et al., Ind. Chim. Belg. Suppl.* **2**, 428 (1959); *eidem,* **Swiss** pat. **344,054** (1960 to Siegfried), *C.A.* **55**, 418b (1961). Pharmacology: G. Stille, *Arzneimittel-Forsch.* **10**, 911 (1960); G. Häusler, U. Jahn, *ibid.* **15**, 878 (1965).

Crystals, mp 185-186°. Freely sol in water. LD_{50} in mice (mg/kg): 570 orally; 88 i.p.; 6.2 i.v. (Stille); in rats (mg/kg): 6.6 i.v.; 780 orally (Häusler, Jahn).

THERAP CAT: Antispasmodic.

3347. Diprenorphine. *(5α,7α)-17-(Cyclopropylmethyl)-4,5-epoxy-3,19-dihydro-3-hydroxy-α,α-dimethyl-6,14-ethenomorphinan-7-methanol;* 21-cyclopropyl-6,7,-8,14-tetrahydro-7α-(1-hydroxy-1-methylethyl)-6,14-*endo*-ethanooripavine; *N*-(cyclopropylmethyl)-19-methylnororvinol; M5050; RX 5050M. $C_{26}H_{35}NO_4$; mol wt 425.57. C 73.38%, H 8.28%, N 3.29%, O 15.04%. Closely related to cyprenorphine, *q.v.* Prepn: K. W. Bentley, D. G. Hardy, *J. Am. Chem. Soc.* **89**, 3281 (1967). Activity in rats: G. F. Blane, *J. Pharm. Pharmacol.* **19**, 367 (1967); G. F. Blane, D. Dugdall, *ibid.* **20**, 547 (1968). Use as etorphine antagonist in dogs: M. Grange *et al., Rev. Med. Vet.* **124**, 899 (1973); in large animals: B. T. Alford *et al., J. Am. Vet. Med. Assoc.* **164**, 702 (1974). Binding to opiate receptors: C. B. Pert *et al., Life Sci.* **16**, 1849 (1975); J. Pearson *et al., ibid.* **26**, 1047 (1980); to μ opiate receptors: J. J. Frost *et al., ibid.* **38**, 1597 (1986). Treatment of experimental stroke in cats: D. S. Baskin *et al., Neuropeptides* **5**, 307 (1985); *eidem, J. Neurosurg.* **64**, 99 (1986). Determn by HPLC: I. Jane, A. McKinnon, *J. Chromatog.* **323**, 191 (1985). Toxicity data: N. S. Duggett *et al., Toxicol. Appl. Pharmacol.* **31**, 141 (1977).

Crystals from methanol, mp 185°.

Hydrochloride, $C_{26}H_{36}ClNO_4$, *Revivon.* LD_{50} s.c. in mice: 316.0 ± 20 mg/kg (Duggett).

THERAP CAT (VET): Narcotic antagonist.

3348. Dipropalin®. *4-Methyl-2,6-dinitro-N,N-dipropylbenzenamine; 2,6-dinitro-N,N-dipropyl-p-toluidine; N,N-*dipropyl-2,6-dinitro-4-methylaniline; 2,6-dinitro-*N,N*-dipropyl-4-methylaniline; 3,5-dinitro-4-dipropylaminotoluene; L-35355. $C_{13}H_{19}N_3O_4$; mol wt 281.31. C 55.50%, H 6.81%, N 14.94%, O 22.75%. Prepd from 2,6-dinitro-*p*-cresol, *p*-toluenesulfonate and dipropylamine: Hantzsch, *Ber.* **43**, 1662 (1910).

Yellow crystals, mp 80°.

USE: Herbicide.

3349. Dipropetryn. *6-(Ethylthio)-N,N'-bis(1-methylethyl)-1,3,5-triazine-2,4-diamine; 2-(ethylthio)-4,6-bis(isopropylamino)-s-triazine;* GS 16068; Cotofor; Sancap. $C_{11}H_{21}N_5S$; mol wt 255.39. C 51.73%, H 8.29%, N 27.42%, S 12.55%. Pre-emergence herbicide. Prepn: **Neth. pat. Appl.** **6,414,460** corresp to H. Yamamoto, T. Namekawa, **U.S.** pat. **3,326,912** (1965, 1967 to Nippon Kayaku). Absorption and translocation in plants: E. Basler *et al., Weed Sci.* **26**, 358 (1978).

Powder, mp 104-106°. Vapor pressure at 20°: 7.3×10^{-7} mm Hg. Soly in water at 20°: 16 mg/l. Soluble in organic solvents.

USE: Herbicide.

3350. *n*-Dipropylamine. *N-Propyl-1-propanamine.* $C_6H_{15}N$; mol wt 101.19. C 71.21%, H 14.94%, N 13.84%. $(CH_3CH_2CH_2)_2NH$.

Colorless liq; odor of ammonia. d_4^{20} 0.738. bp 110°. mp -63°. n_D^{20} 1.40455. Freely sol in water or alcohol. Forms a hydrate with H_2O. LD_{50} orally in rats: 0.93 g/kg, H. F. Smyth *et al., Am. Ind. Hyg. Assoc. J.* **23**, 95 (1962).

3351. Dipropyl Ketone. *4-Heptanone;* butyrone. $C_7H_{14}O$; mol wt 114.18. C 73.62%, H 12.36%, O 14.01%. $CH_3CH_2CH_2COCH_2CH_2CH_3$. Made by passing butyric acid over precipitated $CaCO_3$ at 450°.

Colorless, very refractive liquid; penetrating odor; burning taste. d_4^{15} 0.821. bp 144°. mp -32.6°. n_D^{22} 1.4073. Insol in water; miscible with alcohol, ether.

3352. 5,5-Dipropyl-2,4-oxazolidinedione. Propazone. $C_9H_{15}NO_3$; mol wt 185.22. C 58.36%, H 8.16%, N 7.56%, O 25.91%. Prepd by refluxing dipropylglycolamide and ethylchlorocarbonate with toluene: Stoughton, *J. Am. Chem. Soc.*

63, 2376 (1941); *see also* Altwegg, Ebin, **U.S.** pat. **1,375,949** (1921).

CH₃CH₂CH₂ / CH₃CH₂CH₂ structure

Crystals, mp 42-43°. bp₃ 141-143°.
Sodium salt, white powder, very sol in water.

3353. Diprotrizoate Sodium. *2,4,6-Triiodo-3,5-bis[(1-oxopropyl)amino]benzoic acid monosodium salt; 2,4,6-triiodo-3,5-dipropionamidobenzoic acid sodium salt;* sodium 3,5-dipropionamido-2,4,6-triiodobenzoate; sodium diprotrizoate; Miokon Sodium. $C_{13}H_{12}I_3N_2NaO_4$; mol wt 663.98. C 23.52%, H 1.82%, I 57.34%, N 4.22%, Na 3.46%, O 9.64%. Prepn: Larsen *et al., J. Am. Chem. Soc.* **78,** 3210 (1956).

Free acid, crystals from dilute DMF, dec > 300°. LD₅₀ i.v. in mice: 11.80 mg/kg (Larsen).
THERAP CAT: Diagnostic aid (radiopaque medium).

3354. Dipyridamole. *2,2′,2″,2‴-[(4,8-Di-1-piperidinylpyrimido[5,4-d]pyrimidine-2,6-diyl)dinitrilo]tetrakisethanol;* 2,6-bis(diethanolamino)-4,8-dipiperidinopyrimido-[5,4-d]pyrimidine; NSC-515776; RA-8; Anginal; Cardoxin; Cleridium 150; Coronarine; Curantyl; Dipyridan; Gulliostin; Natyl; Peridamol; Persantine; Piroan; Prandiol; Protangix. $C_{24}H_{40}N_8O_4$; mol wt 504.62. C 57.12%, H 7.99%, N 22.21%, O 12.68%. Prepn: **Brit.** pat. **807,826** (1959 to Thomae). Activity studies: Saraf, Seth, *Indian J. Physiol. Pharmacol.* **15,** 135 (1971). Toxicological study: Takenaka *et al., Arzneimittel-Forsch.* **22,** 892 (1972).

Deep yellow needles from ethyl acetate, mp 163°. Bitter taste. Slightly sol in H_2O; sol in dil acid having a pH of 3.3 or below; very sol in methanol, ethanol, chloroform; not too sol in acetone, benzene, ethyl acetate. Solns are yellow and show strong blue-green fluorescence. LD₅₀ in rats: 8.4 g/kg orally; 208 mg/kg i.v. (Takenaka).
THERAP CAT: Coronary vasodilator.

3355. α,α′-Dipyridyl. *2,2′-Bipyridine;* 2,2′-dipyridyl; CI-588. $C_{10}H_8N_2$; mol wt 156.18. C 76.90%, H 5.16%, N 17.94%. Prepn from pyridine: Smith, *J. Am. Chem. Soc.* **46,** 414 (1924). Manuf: Freeman, Ghosh, **U.S.** pat. **2,962,502** (1960 to I.C.I.). Absorption spectrum: P. Krumholz, *J. Am. Chem. Soc.* **73,** 3487 (1951). Pharmacologic study: Bass *et al., J. Pharmacol. Exp. Ther.* **152,** 104 (1966). Toxicity data: R. W. Grady *et al., ibid.* **196,** 478 (1976).

Crystals from dil alcohol, mp 69.7°. bp 272-273°. Sol in about 200 parts water; very sol in alcohol, ether, benzene, chloroform, petr ether. LD₅₀ i.p. in mice: 200 mg/kg (Grady).
USE: As a reagent for the determination of iron.

3356. γ,γ′-Dipyridyl. *4,4′-Bipyridine;* 4,4′-dipyridyl. $C_{10}H_8N_2$; mol wt 156.18. C 76.90%, H 5.16%, N 17.94%. Prepn: Dimroth, Frister, *Ber.* **55,** 3695 (1922); Smith, *J. Am. Chem. Soc.* **46,** 414 (1924).

Dihydrate, bitter needles from water, mp 73°. Anhydr form, mp 111-112°. bp₇₆₀ 304.8°. Freely sol in alc, benzene, chloroform, ether; slightly sol in water. Absorption spectrum: Krumholz, *J. Am. Chem. Soc.* **73,** 3487 (1951).
Dihydrochloride, $C_{10}H_8N_2 \cdot 2HCl$, prisms from water. Soluble in water; practically insol in ether.

3357. Dipyrocetyl. *2,3-Bis(acetyloxy)benzoic acid;* o-pyrocatechuic acid diacetate; diacetylpyrocatechol-3-carboxylic acid; 2,3-diacetoxybenzoic acid; Movirene; Artromialgina. $C_{11}H_{10}O_6$; mol wt 238.19. C 55.46%, H 4.23%, O 40.30%. Prepn: Rietz, **U.S.** pat. **1,140,716** (1914); Simokoriyama, *Bull. Chem. Soc. Japan* **16,** 284 (1941).

mp 148-150° (Rietz); mp 146-170° (Simokoriyama). Sol in organic solvents. Practically insol in water.
THERAP CAT: Analgesic, antipyretic.

3358. Dipyrone. *[(2,3-Dihydro-1,5-dimethyl-3-oxo-2-phenyl-1H-pyrazol-4-yl)methylamino]methanesulfonic acid sodium salt monohydrate; (antipyrinylmethylamino)methanesulfonic acid sodium salt;* 1-phenyl-2,3-dimethyl-5-pyrazolone-4-methylaminomethanesulfonate sodium; noraminopyrine methanesulfonate sodium; 4-methylamino-1,5-dimethyl-2-phenyl-3-pyrazolone sodium methanesulfonate; sodium methylaminoantipyrine methanesulfonate; methylmelubrin; methampyrone; metamizol; analgin; Alginodia; Algocalmin; Bonpyrin; Conmel; D-Pron; Divarine; Dya-Tron; Espyre; Farmolisina; Feverall; Fevonil; Keypyrone; Metilon; Narone; Nartate; Nevralgina; Novacid; Novaldin; Novalgin; Novemina; Novil; Paralgin; Pydirone; Pyralgin; Pyril; Pyrilgin; Pyrojec; Sulpyrin; Tega-Pyrone; Temp; Unagen. $C_{13}H_{16}N_3NaO_4S \cdot H_2O$; mol wt 351.35. C 44.44%, H 5.16%, N 11.96%, Na 6.54%, O 22.77%, S 9.13%. Prepd by methylating the amino group of sulfamidpyrine, *q.v.,* then treating with formaldehyde sodium bisulfite soln: **Ger.** pats. **254,-711; 259,503; 259,577** (1911 to Hoechst).

Minute crystals from alc. Sol in water (1 g/1.5 ml), methanol. Less sol in ethanol. Practically insol in ether, acetone, benzene, chloroform. Aq solns are neutral, may acquire a yellow discoloration without apparent potency loss. One of the ingredients of *Palerol (Pelerol).*
Magnesium salt, $C_{26}H_{32}MgN_6O_8S_2$, *Magnopyrol.*
THERAP CAT: Analgesic, antipyretic.
THERAP CAT (VET): Analgesic, antispasmodic.

3359. Diquat Dibromide. *6,7-Dihydrodipyrido[1,2-a:2′,1′-c]pyrazinediium dibromide;* 1,1′-ethylene-2,2′-dipyridylium dibromide; FB/2; Aquacide; Reglone. $C_{12}H_{12}$-

Br$_2$N$_2$; mol wt 344.07. C 41.89%, H 3.52%, Br 46.45%, N 8.14%. Prepn: Fielden *et al.*, U.S. pat. **2,823,987** (1958 to ICI). Herbicidal properties: Brian *et al.*, *Nature* **181**, 446 (1958). Fate in the environment and toxicity: Simsiman *et al.*, *Residue Rev.* **62**, 131-174 (1976). *Review:* Akhavein, Linscott, *ibid.* **23**, 97-145 (1968).

Monohydrate, pale yellow crystals from water, mp below 320° (dec). Also reported as mp 335-340°. uv max: 308.31 nm (ε 18,000). Soly in water at 20°: 70%. Insoluble in organic solvents; slightly sol in alcohol. Stable in acid or neutral soln. LD$_{50}$ orally in rats, mice, rabbits, guinea pigs: 231, 125, 101, 100 mg/kg, *Toxic Substances List,* H. E. Christensen, Ed. (1973) p 381.
USE: Contact herbicide used also to produce desiccation and defoliation. *Toxicity:* Minimal absorption from gastrointestinal tract although distension and irritation occur after large doses in exptl animals: Dalgaard-Mikkelsen, Poulsen, *Pharmacol. Rev.* **14**, 225 (1962).

3360. 1,3-Di-6-quinolylurea. *N,N'-Di-6-quinolinylurea; sym*-di-(6-quinolyl)urea; bis(6-quinolyl)urea; 6,6'-diquinolylurea; Babesan. C$_{19}$H$_{14}$N$_4$O; mol wt 314.33. C 72.60%, H 4.49%, N 17.83%, O 5.09%. Synthesis starting with 6-aminoquinoline: Schönhöfer, Henecka, **Ger.** pat. **583,207** (1933 to I. G. Farben); Haskelberg, *J. Org. Chem.* **12**, 434 (1947); Reuter, *Aust. Chem. Inst. J. & Proc.* **16**, 164 (1949).

Crystals from pyridine, mp 262°. Sol in dil acid.
Bismethosulfate, C$_{23}$H$_{26}$N$_4$O$_9$S$_2$, *6,6'-ureylenebis[1,1'-dimethylquinolinium] sulfate, quinuronium sulfate, 6,6'-ureylenebis(1-methylquinolinium)bis(methosulfate), dimethyl quinolyl methylsulfate urea, SN 5870, Acaprin, Zothelone, Baburan, Pirevan, Pyroplasmin, Atral.* Yellow crystals from methanol, dec 237°. Freely sol in water.
THERAP CAT (VET): Antiprotozoal *(Babesia).*

3361. Diresorcinol. *[1,1'-Biphenyl]-3,3',5,5'-tetrol; 3,5,3',5'-tetrahydroxydiphenyl.* C$_{12}$H$_{10}$O$_4$; mol wt 218.21. C 66.05%, H 4.61%, O 29.33%.

White to slightly yellowish cryst powder. mp 310°; becomes anhyd at 100°. Sparingly sol in cold water; less sol in hot water or alcohol.

3362. Disilane. Disilicoethane; disilicon hexahydride; silicoethane; disilicane. H$_6$Si$_2$; mol wt 62.23. H 9.73%, Si 90.27%. Si$_2$H$_6$. Obtained by separation of mixed silanes prepd from magnesium silicide and hydrochloric acid: Moissan, Smiles, *Compt. Rend.* **134**, 569, 1549 (1902); Stock, Somiesky, *Ber.* **49**, 111 (1916); **54B**, 524 (1921); **56B**, 247 (1923); Culbertson, U.S. pat. **2,551,571** (1951 to Union Carbide); prepd by conversion of silane to higher silanes in an ozonizer type electric discharge: Spanier, MacDiarmid, *Inorg. Chem.* **1**, 432 (1962).
Gas; repulsive odor. d$_4^{-25}$ 0.686. mp −132.5°. bp −14.5°. Dec 300°. Ignites spontaneously in air. Slowly dec in water. Explodes on contact with sulfur hexafluoride; reacts vigorously with carbon tetrachloride and chloroform. Sol in carbon disulfide, ethyl alcohol, benzene, and ethyl silicate. Potassium hydroxide liberates hydrogen.

Caution: A powerful irritant.

3363. Disodium Dihydrogen Hypophosphate. Disodium dihydrogen subphosphate; sodium dihydrogen hypophosphate; sodium acid hypophosphate. H$_2$Na$_2$O$_6$P$_2$; mol wt 205.96. H 0.98%, Na 22.33%, O 46.61%, P 30.08%. Na$_2$H$_2$-P$_2$O$_6$. Prepd by the oxidation of red phosphorus with sodium chlorite in water: Leininger, Chulski, *Inorg. Syn.* **4**, 68 (1953).
Hexahydrate, monoclinic plates from water, d 1.849. Becomes anhyd at 110°. mp 250°. Soly of the hexahydrate in water at 25°: 2.0 g/100 ml; at 100°: 25.0 g/100 ml. Insol in alc. Aq solns have an acid reaction.

3364. Disodium Phenyl Phosphate. C$_6$H$_5$Na$_2$O$_4$P; mol wt 218.07. C 33.04%, H 2.31%, Na 21.09%, O 29.35%, P 14.21%. C$_6$H$_5$PO$_4$Na$_2$.
White, cryst powder. Freely sol in water, sparingly in alcohol; insol in acetone or ether.
USE: Reagent in the testing of milk for proper pasteurization and for the presence of unpasteurized milk.

3365. Disophenol. *2,6-Diiodo-4-nitrophenol;* DNP; Ancylol. C$_6$H$_3$I$_2$NO$_3$; mol wt 390.93. C 18.43%, H 0.77%, I 64.93%, N 3.58%, O 12.28%. Prepd by iodination of *p*-nitrophenol with iodine—potassium iodide soln in aq ammonia: Datta, Prosad, *J. Am. Chem. Soc.* **39**, 446 (1917). As an anthelmintic: Thorson *et al.*, U.S. pat. **3,081,224** (1963 to Am. Cyanamid). Toxicity studies: J. A. Kaiser, *Toxicol. Appl. Pharmacol.* **6**, 232 (1964); Fowler, *Brit. Vet. J.* **127**, 304 (1971). Comparative field trial in sheep: C. A. Hall *et al.*, *Res. Vet. Sci.* **31**, 104 (1981).

Light yellow, feathery crystals from glacial acetic acid, mp 157°. Freely sol in alcohol, very sparingly sol in water. LD$_{50}$ in rats, mice (mg/kg): 170, 212 orally; 105, 88 i.v.; 105, 107 i.p.; 122, 110 s.c. (Kaiser).
THERAP CAT (VET): Anthelmintic (hookworm).

3366. Disopyramide. α-[2-[Bis(1-methylethyl)amino]ethyl]-α-phenyl-2-pyridineacetamide; α-[2-(diisopropylamino)ethyl]-α-phenyl-2-pyridineacetamide; 4-(diisopropylamino)-2-phenyl-2-(2-pyridyl)butyramide; H 3292; SC 7031; Dicorantil; Lispine; Ritmodan; Rythmodan. C$_{21}$H$_{29}$N$_3$O; mol wt 339.47. C 74.30%, H 8.61%, N 12.38%, O 4.71%. Prepn: Cusic, Sause, **Belg.** pat. **617,730**, *C.A.* **58**, 12522e (1963) and U.S. pat. **3,225,054** (1962 and 1965, both to Searle). Synthesis and biological activity: Adelstein, *J. Med. Chem.* **16**, 309 (1973). Pharmacology: *J. Am. Med. Assoc.* **213**, 697 (1970); Mathur, *Am. Heart J.* **84**, 764 (1972); B. Befeler, R. Lazzara, *Heart Lung* **9**, 475 (1980). Conformation-activity study: J. L. Czeisler, R. M. El-Rashidy, *J. Pharm. Sci.* **74**, 750 (1985). Toxicity: P. C. Ruenitz, C. M. Mokler, *J. Med. Chem.* **22**, 1142 (1979). *Reviews:* R. R. Dean *et al.*, in *Pharmacological and Biochemical Properties of Drug Substances* vol. 2, M. E. Goldberg, Ed. (Am. Pharm. Assoc., Washington, DC, 1979) pp 165-185; E. H. Taylor, A. A. Pappas, *Ann. Clin. Lab. Sci.* **16**, 289-295 (1986).

Crystals from hexane, mp 94.5-95.0°. pKa 10.2; also reported as 10.45. LD$_{50}$ i.p. in mice: 517 μmol/kg (Ruenitz, Mokler).
Phosphate, C$_{21}$H$_{32}$N$_3$O$_5$P, *SC 13957, Dirythmin SA, Disco-Duriles, Norpace, Rythmodul.*

THERAP CAT: Cardiac depressant (anti-arrhythmic).

3367. Disparlure. *2-Decyl-3-(5-methylhexyl)oxirane; cis-7,8-epoxy-2-methyloctadecane;* ENT 34886; Disrupt. $C_{19}H_{38}O$; mol wt 282.51. C 80.78%, H 13.56%, O 5.66%. The potent sex pheromone of the gypsy moth *Lymantria dispar* (formerly *Porthetria dispar*). Isoln and characterization: B. A. Bierl *et al., Science* **170,** 87 (1970); *eidem, J. Econ. Entomol.* **65,** 659 (1972). Synthesis: S. Iwaki *et al., J. Am. Chem. Soc.* **96,** 7842 (1974); D. G. Farnum *et al., Tetrahedron Letters* **1977,** 4009; K. Mori *et al., Tetrahedron* **35,** 833 (1979); B. E. Rossiter *et al., J. Am. Chem. Soc.* **103,** 464 (1981); K. Mori, T. Ebata, *Tetrahedron Letters* **22,** 4281 (1981). Synthesis of (±)-form: K. Eiter, *Angew. Chem. Int. Ed.* **11,** 60 (1972); *idem,* Ger. pat. **2,145,454** corresp to U.S. pat. **3,975,409** (1973, 1976 to Bayer); H. J. Bestmann *et al., Ber.* **109,** 3375 (1976). Activity of (+)- and (±)-forms: R. T. Cardé, R. P. Webster, *J. Chem. Ecol.* **5,** 935 (1979).

Viscous colorless oil, $bp_{0.25}$ 146-148°. $[\alpha]_D^{23}$ +0.8° ±0.2° (c = 10 in CCl_4). n_D^{23} 1.4450. LD_{50} orally in rats > 34,600 mg/kg, M. Beroza *et al., Toxicol. Appl. Pharmacol.* **31,** 421 (1975).

USE: Insect sex attractant.

3368. Distigmine Bromide. *3,3'-[1,6-Hexanediylbis-[(methylimino)carbonyl]oxy]bis[1-methylpyridinium]dibromide; 3-hydroxy-1-methylpyridinium bromide hexamethylenebis[methylcarbamate];* hexamethylenebis[methylcarbamic acid] ester of 3-hydroxy-1-methylpyridinium bromide; hexamethylenebis[N-methylcarbamic acid ester bromomethylate]; hexamethylenebis[N-methylcarbaminoyl-1-methyl-3-hydroxypyridinium bromide]; hexamarium; BC 51; Ubretid. $C_{22}H_{32}Br_2N_4O_4$; mol wt 576.36. C 45.85%, H 5.60%, Br 27.73%, N 9.72%, O 11.10%. Prepn: Schmid, U.S. pat. **2,789,981** (1957 to Oesterr. Stickstoffwerke). Pharmacology: Hertting *et al., Arzneimittel-Forsch.* **18,** 479 (1968); Hohenegger, Lindner, *Wien. Klin. Wochenschr.* **81,** 823 (1969). Stability studies: Suzuki, Tanimura, *Yakugaku Zasshi* **90,** 762 (1970).

Crystals, dec 149°.
THERAP CAT: Cholinesterase inhibitor.

3369. Disulfamide. *4-Chloro-6-methyl-1,3-benzenedisulfonamide; 5-chlorotoluene-2,4-disulfonamide;* 2-chloro-4-methylbenzene-1,5-disulfonamide; 5-chloro-2,4-disulfamyltoluene; disulphamide; Disamide; Natirene 25. $C_7H_9-ClN_2O_4S_2$; mol wt 284.75. C 29.53%, H 3.19%, Cl 12.45%, N 9.84%, O 22.48%, S 22.52%. Prepn: **Brit.** pat. **851,287** (1960 to Brit. Drug Houses). Pharmacology: David, Fellowes, *J. Pharm. Pharmacol.* **12,** 65 (1960).

Non-hygroscopic cryst powder, mp 260°. uv max (ethanol): 285 nm (ϵ 805). Practically insol in water; slightly sol in ethanol. Soly in abs ethanol at 25° = 1.89% w/w; in 90% ethanol = 2.23% w/w; in isopropanol = 0.35%; in chloroform = 0.001%. Sol in cold NaOH solns.

THERAP CAT: Diuretic.

3370. Disulfiram. *Tetraethylthioperoxydicarbonic diamide; bis(diethylthiocarbamoyl) disulfide;* tetraethylthiuram disulfide; bis(diethylthiocarbamyl) disulfide; teturamin; TTD; Cronetal; Abstensil; Stopetyl; Contralin; Antadix; Antietanol; Exhoran; Ethyl Thiurad; Antabuse; Etabus; Ro-Sulfiram; Abstinyl; Thiuranide; Esperal; Tetradine; Noxal; Tetraetil. $C_{10}H_{20}N_2S_4$; mol wt 296.54. C 40.50%, H 6.80%, N 9.45%, S 43.25%. Prepn: Bailey, U.S. pat. **1,796,-977** (1931 to Roessler and Hasslacher); Adams, Newser, U.S. pat. **1,782,111** (1931 to Naugatuck); *cf.* Cummings, Simmons, *Ind. Eng. Chem.* **20,** 1173 (1928). Comprehensive description: N. G. Nash, R. D. Daley in *Analytical Profiles of Drug Substances* vol. 4, K. Florey, Ed. (Academic Press, New York, 1975) pp 168-191.

Crystals, mp 70°. d 1.30. Practically insol in water (0.02 g/100 ml). Sol in alcohol (3.82 g/100 ml), in ether (7.14 g/100 ml), also sol in acetone, benzene, chloroform, carbon disulfide. LD_{50} orally in rats: 8.6 g/kg, Child, Cramp, *Acta Pharmacol. Toxicol.* **8,** 305 (1952). *Caution:* Therapy should not be instituted for at least 12 hours after ingestion of alcohol.

Human Toxicity: Ingestion of alc after disulfiram administration causes intense vasodilatation of face and neck, tachycardia and tachypnea followed by nausea, vomiting, pallor, hypotension. Occasionally convulsions, cardiac arrhythmias, myocardial infarction may occur. High doses of alc may cause dizziness, headache, dyspnea, unconsciousness, death, *cf. Clinical Toxicology of Commercial Products,* R. E. Gosselin *et al.,* Eds. (Williams & Wilkins, Baltimore, 4th ed., 1976) Section III, pp 137-140.

USE: Rubber accelerator; vulcanizer; seed disinfectant; fungicide.
THERAP CAT: Alcohol deterrent.

3371. Disulfoton. *Phosphorodithioic acid O,O-diethyl S-[2-(ethylthio)ethyl] ester; O,O-diethyl-S-ethylmercaptoethyl dithiophosphate;* thiodemeton; dithiodemeton; BAY 19639; ENT 23347; Dithiosystox; Di-Syston; Frumin AL; Frumin G; Solvirex. $C_8H_{19}O_2PS_3$; mol wt 274.38. C 35.02%, H 6.98%, O 11.66%, P 11.29%, S 35.05%. Prepn: Schrader, Lorenz, U.S. pat. **2,759,010** and **Brit.** pat. **797,307** (1956, 1958, both to Bayer). Homolog of phorate, *q.v.*

Colorless oil, $bp_{0.01}$ 108°, $bp_{1.5}$ 132-133°. d_4^{20} 1.144. n_D^{20} 1.5348. Vapor press. at 20°: 1.8×10^{-4} mm Hg. Insol in water. LD_{50} in female, male rats: 2.3, 6.8 mg/kg orally; 6, 15 mg/kg dermally, T. B. Gaines, *Toxicol. Appl. Pharmacol.* **14,** 515 (1969).

USE: Insecticide.

3372. Disul-sodium. *2-(2,4-Dichlorophenoxy)ethanol hydrogen sulfate sodium salt;* sodium 2,4-dichlorophenoxyethyl sulfate; 2,4-dichlorophenoxyethyl hydrogen sulfate sodium salt; 2,4-DES-Na; SES; sesone; Crag Herbicide-1; Crag Sesone. $C_8H_7Cl_2NaO_5S$; mol wt 309.13. C 31.08%, H 2.28%, Cl 22.94%, Na 7.44%, O 25.88%, S 10.37%. Prepn: J. A. Lambrech, U.S. pat. **2,573,769** (1951 to Union Carbide).

Crystals, mp 245° (dec). Soly in water (25°): 25.5%. Insol in most organic solvents except methanol. LD_{50} orally in

rats: 730 mg/kg, G. W. Bailey, J. L. White, *Residue Rev.* **10,** 97 (1965).
USE: Herbicide.

3373. Dita Bark. Alstonia cortex; Australian fever bark. The dried bark of *Alstonia scholaris* (L.) R. Br. and of *A. constricta* F. Muell., *Apocynaceae. Habit.* India, Ceylon, Philippine Islands, Australia. *Constit.* The alkaloids echit-amine, echitenine, alsonine, porphyrine; echicerin—a non-nitrogenous substance; echitin, alstonidine and echitein.

3374. Ditazol. *2,2'-[(4,5-Diphenyl-2-oxazolyl)imino]-bisethanol;* N-(4,5-diphenyloxazol-2-yl)diethanolamine; 2-[bis(β-hydroxyethyl)amino]-4,5-diphenyloxazole; 2,2'-di-hydroxy-N-(4,5-diphenyloxazol-2-yl)diethylamine; dieth-amphenazol; S 222; Ageroplas. $C_{19}H_{20}N_2O_3$; mol wt 324.38. C 70.35%, H 6.22%, N 8.63%, O 14.80%. Prepn: Marchetti *et al., J. Med. Chem.* **11,** 1092 (1968); **Fr. pat.** 1,538,009 corresp to Marchetti, **U.S. pat.** 3,557,135 (1968, 1971 to Ist. Farmacol. Serono). Series of articles on pharmacology and toxicology: L. Caprino *et al., Arzneimittel-Forsch.* **23,** 1272-1291 (1973). Metabolism: Marchetti *et al., ibid.* 1291.

Monohydrate, crystals from ethyl ether-petr ether, mp 96-98°. LD_{50} in mice, rats (mg/kg): 9621, 11380 orally; 3390, 7770 i.p. (Caprino).
THERAP CAT: Anti-inflammatory.

3375. Dithianone. *5,10-Dihydro-5,10-dioxonaphtho-[2,3-b]-1,4-dithiin-2,3-dicarbonitrile;* 1,4-dithiaanthraqui-none-2,3-dicarbonitrile; 2,3-dicyano-1,4-dithiaanthraqui-none; Delan. $C_{14}H_4N_2O_2S_2$; mol wt 296.33. C 56.74%, H 1.36%, N 9.46%, O 10.80%, S 21.64%. Prepn: van Schoor *et al.,* **U.S. pat.** 2,976,296 (1961 to E. Merck).

Gray-brown needles from acetone, mp 220°. Insoluble in water; sol in dioxane, chlorobenzene, chloroform. LD_{50} orally in rats, mice, guinea pigs: 1015, 1140, 115 mg/kg. USE: Fungicide.

3376. Dithiazanine Iodide. *3-Ethyl-2-[5-(3-ethyl-2-(3H)-benzothiazolylidene)-1,3-pentadienyl]benzothiazolium iodide;* 3,3'-diethylthiadicarbocyanine iodide; Abminthic; Anelmid; Anguifugan; Delvex; Dejo; Déselmine; Dilombrin; Dizan; Nectocyd; Partel; Telmicid; Telmid. $C_{23}H_{23}IN_2S_2$; mol wt 518.47. C 53.28%, H 4.47%, I 24.48%, N 5.40%, S 12.37%. Prepn from 1-methylbenzothiazole ethiodide and β-(ethylmercapto)acrolein diethyl acetal: Kendall, Edwards, **U.S. pat.** 2,412,815 (1946 to Ilford).

Green needles from methanol, dec 248°. Practically insol in water. Can be solubilized with polyvinylpyrrolidone: **Can. pat.** 676,636 (1963 to GAF).
USE: Sensitizer for photographic emulsions.
THERAP CAT: Anthelmintic (Nematodes).
THERAP CAT (VET): Anthelmintic.

3377. 2,2'-Dithiobis[benzothiazole]. MBTS; 2,2'-di-benzothiazyl disulfide; benzothiazyl disulfide; dibenzthiazyl disulfide; mercaptobenzthiazyl ether; Thiofide. $C_{14}H_8N_2S_4$;

mol wt 332.46. C 50.57%, H 2.43%, N 8.43%, S 38.58%. Formed by oxidation of 2-mercaptobenzothiazole.

Pale yellow needles from benzene. d 1.50. mp 180° (also reported as 186°, the commercial product mp 168° min). Insoluble in water. Solubility at 25° (g/100 ml): in alcohol <0.2 ; in acetone <0.5; in benzene <0.5; in carbon tetra-chloride <0.2 (somewhat more in chloroform); in ether <0.2; in naphtha <0.5.
USE: As accelerator in the rubber industry.

3378. 2,4-Dithiobiuret. *Thioimidodicarbonic diamide.* $C_2H_5N_3S_2$; mol wt 135.20. C 17.77%, H 3.73%, N 31.08%, S 47.42%. $NH_2CSNHCSNH_2$. Prepd by the interaction of dicyandiamide and hydrogen sulfide in inert solvents under pressure: Sperry, **U.S. pat.** 2,371,112 (1945 to Am. Cyan-amid). *Review:* Kurzer, *Chem. Rev.* **56,** 138-144, 179-181 (1956).
Monoclinic or triclinic crystals. *Poisonous when ingested!* Apparent bulk density 1.522 g/ml at 30°. Dec 181°. uv max: 225, 280 nm (log ϵ 2.0, 2.2). Solubility at 27° = 0.27 g/100 ml water; 2.2 g/100 g ethanol; 16 g/100 g acetone; about 34 g/100 g Cellosolve. Solubility in boiling water about 8.0%. pH of satd aq soln at 30° = 5.8. Soluble in alkalies with formation of water-soluble salts. Solubility in 1% NaOH = 3.6 g/100 g soln; in 5% NaOH = 16 g/100 g; in 10% NaOH = 29 g/100 g.
USE: Plasticizer, rubber accelerator, intermediate in resin manuf, in making insecticides and rodenticides. Can be used to delay the wilting of flowers. *Toxicity:* Limited animal expts suggest high toxicity due to respiratory paralysis.

3379. 4,4'-Dithiodimorpholine. *4,4'-Dithiobis[morpho-line];* morpholine, N,N'-disulfide; dimorpholine N,N'-disul-fide. $C_8H_{16}N_2O_2S_2$; mol wt 236.37. C 40.65%, H 6.82%, N 11.85%, O 13.54%, S 27.13%. Prepn: Blake, *J. Am. Chem. Soc.* **65,** 1267 (1943); *idem,* **U.S. pat.** 2,343,524 (1944 to Monsanto); Harman, **U.S. pat.** 2,766,236 and **Brit. pat.** 774,570 (1956, 1957, both to Monsanto). Lambrech *et al.,* **U.S. pat.** 2,902,402 (1959 to Pfizer).

Crystals, mp 124-125°.
USE: As staining protector in rubber; in vulcanization of rubber. As fungicide: Ladd, **U.S. pat.** 2,429,097 (1949 to U.S. Rubber).

3380. 3,3-Dithiodipyridine Dihydrochloride. *3,3'-Di-thiobis[pyridine] dihydrochloride;* 3,3-dipyridyl disulfide di-hydrochloride. $C_{10}H_{10}Cl_2N_2S_2$; mol wt 293.24. C 40.96%, H 3.44%, Cl 24.18%, N 9.55%, S 21.87%. Prepd from 3-pyri-dinesulfonyl chloride and Na_2SO_3: Gibbs, Penny, **Brit. pat.** 582,638 (1946).

Crystals from abs alc, mp 183°. Sol in water.
Picrate, yellow needles from alc, mp 185°.

3381. Dithiosalicylic Acid. *2-Hydroxybenzenecarbodithi-oic acid.* $C_7H_6OS_2$; mol wt 170.25. C 49.38%, H 3.55%, O 9.40%, S 37.67%. Prepn from salicylaldehyde and ammoni-um polysulfide: Bruni, Levi, *Atti Accad. Naz. Lincei* **32,** i, 5 (1923), *C.A.* **18,** 2694 (1924); *Beilstein,* 2nd suppl, **vol. 10,** 78.

CSSH

OH

Orange-yellow needles, mp 48-50°. Moderately sol in water; sol in alcohol, ether, benzene, methanol.

3382. 1,4-Dithiothreitol. *(R*,R*)-1,4-Dimercapto-2,3-butanediol;* Cleland's reagent; *threo*-2,3-dihydroxy-1,4-dithiolbutane. $C_4H_{10}O_2S_2$; mol wt 154.24. C 31.14%, H 6.53%, O 20.75%, S 41.57%. $HSCH_2(CHOH)_2CH_2SH$. Prepn: Evans *et al., J. Chem. Soc.* **1949**, 253.

Slightly hygroscopic needles from ether, mp 42-43°. Can be sublimed at 37° and 0.005 mm press; bp_2 125-130°. Redox potential: −0.33 volts at pH 7. Freely sol in water, ethanol, acetone, ethyl acetate, chloroform, ether.

USE: Protective agent for SH groups: Cleland, *Biochemistry* **3**, 480 (1964), *see also* "Cleland's Reagent" a detailed brochure and bibliography published by Calbiochem, Los Angeles.

3383. Dithizone. *Phenyldiazenecarbothioic acid 2-phenylhydrazide; (phenylazo)thioformic acid 2-phenylhydrazide;* diphenylthiocarbazone. $C_{13}H_{12}N_4S$; mol wt 256.32. C 60.91%, H 4.72%, N 21.86%, S 12.51%. $C_6H_5N=NCSNH-NHC_6H_5$. Prepn: E. Fischer, E. Besthorn, *Ann.* **212**, 316 (1882). Use as an analytical reagent: A. Stock, E. Pohland, *Angew. Chem.* **39**, 791 (1926); H. Fischer, *ibid.* **46**, 442 (1933); **50**, 919 (1937); *Organic Reagents for Metals* (Hopkins & Williams, 1934). Crystal structure: M. Laing, *J. Chem. Soc. Perkin Trans.* 2 **1977**, 1248. Conformation in soln: A. T. Hutton, H. M. Irving, *Chem. Commun.* **1981**, 735. Monograph: *Analytical Sciences Monograph No. 5* entitled "Dithizone", H. M. Irving, Ed. (Chemical Society, London, 1977) 112 pp. Review: *idem, CRC Crit. Rev. Anal. Chem.* **8**, 321-405 (1980).

Bluish-black cryst powder, mp 168° (dec). Insol in water; sparingly sol in alc, freely in carbon tetrachloride, chloroform. Solns are not stable, but may be preserved by a layer of aq SO_2.

USE: Sensitive reagent for several heavy metals, Co, Cu, Pb, and Hg; esp for the estimation of minute amounts of Pb. *Toxicity:* Has caused glycosuria and ocular injury in exptl animals.

3384. Ditiocarb Sodium. *Diethylcarbamodithioic acid sodium salt; diethyldithiocarbamic acid sodium salt;* diethyldithiocarbamate sodium; DTC; DDC; DEDC; DDTC; DeDTC; dithiocarb; Imuthiol. $C_5H_{10}NNaS_2$; mol wt 171.27. C 35.06%, H 5.89%, N 8.18%, Na 13.43%, S 37.45%. $(C_2H_5)_2NCS_2Na$. Chelating agent with strong affinity for Hg, Cu, Ni and Zn. Also active as T-cell specific immunostimulant. Prepn: A. M. Clifford, J. G. Lichty, *J. Am. Chem. Soc.* **54**, 1163 (1932); A. L. Klebanskii, L. P. Fomina, *Zh. Obshch. Khim.* **30**, 794 (1960). Absorption spectrum: H. P. Koch, *J. Chem. Soc.* **1949**, 401. Inhibition of superoxide dismutase in mice: R. E. Heikkila *et al., J. Biol. Chem.* **251**, 2182 (1976); of cisplatin nephrotoxicity in rats: R. F. Borch *et al., Proc. Nat. Acad. Sci. USA* **77**, 5441 (1980). Clinical use in acute nickel carbonyl poisoning: F. W. Sunderman, F. W. Sunderman Jr., *Am. J. Med. Sci.* **236**, 26 (1958); F. W. Sunderman, *Ann. Clin. Lab. Sci.* **9**, 1 (1979); in acute cadmium poisoning: G. R. Gale *et al., ibid.* **11**, 476 (1981) and **13**, 207 (1983). Specific effects on T-cell regulation: A. Pompidou *et al., Int. J. Immunopharmacol.* **7**, 561 (1985). Clinical evaluation in T-cell deficient diseases: E. Lemarie *et al., Meth. Find. Exp. Clin. Pharmacol.* **8**, 51 (1986); in AIDS-related complex: A. Pompidou *et al., Comp. Immun. Microbiol. Infect. Dis.* **9**, 263 (1986). Review of pharmacology, toxicity and clinical uses: G. Renoux, *J. Pharmacol.* **13**, Suppl. 1, 95-134 (1982); of immunopharmacology and use in cancer immunotherapy: G. Renoux *et al., Advan. Exp. Med. Biol.* **166**, 223-239 (1983).

Trihydrate, $C_5H_{10}NNaS_2.3H_2O$, thin, irregular plate-like crystals from acetone, mp 94-102°. Also reported as 90-92° (Sunderman, Sunderman). Freely sol in water; sol in ethanol, methanol, acetone. Insol in ether, benzene. The aq soln is alkaline to litmus and phenolphthalein and slowly dec. (pH of 10% aq soln is 11.6 at room temp). The addition of an acid to the aq soln produces a white turbidity due to the liberation of carbon disulfide. uv max (ethanol): 257, 290 nm (ε 1200, 13000). LD_{50} orally in rats, mice: 2830, 1870 mg/kg; i.v. in mice: > 1000 mg/kg (Renoux, 1982).

USE: For colorimetric determination of small quantities of copper and for its separation from other metals.

THERAP CAT: Immunomodulator. Chelating agent (Wilson's disease). Antidote (nickel, cadmium poisoning).

3385. 1,1-Di-p-tolylethane. *1,1'-Ethylidenebis[4-methylbenzene];* α,α-di-p-tolylethane; *asym*-di-p-tolylethane; 4,4'-α-trimethyldiphenylmethane. $C_{16}H_{18}$; mol wt 210.30. C 91.37%, H 8.63%. Prepd by passing acetylene into a mixture of toluene and concd H_2SO_4 in presence of mercuric sulfate: Reichert, Nieuwland, *Org. Syn.* **IV**, 23 (1925).

CH₃ CH CH₃

Oily, highly refractive liquid. Aromatic odor. d_4^{20} 0.974. Not solid at −20°. bp 295-300°; bp_{12} 155-157°. Soluble in acetic acid.

3386. 1,2-Di-p-tolylethane. *1,1'-(1,2-Ethanediyl)bis[4-methylbenzene];* α,β-di-p-tolylethane; *sym*-di-p-tolylethane; 4,4'-dimethyldibenzyl. $C_{16}H_{18}$; mol wt 210.30. C 91.37%, H 8.63%. Prepd by passing p-xylene vapor over red hot platinum wires: Meyer, Hofmann, *Monatsh.* **37**, 690 (1916).

CH₃ CH₂CH₂ CH₃

Leaflets from methanol; plates from ligroin. mp 82°; bp_{730} 296-298°; bp_{18} 178°. Sol in benzene, moderately sol in alcohol, petr ether.

3387. p-Ditolylmercury. *Bis(4-methylphenyl)mercury.* $C_{14}H_{14}Hg$; mol wt 382.86. C 43.91%, H 3.69%, Hg 52.40%. $(CH_3C_6H_4)_2Hg$. Prepd by the action of sodium iodide on p-tolylmercuric chloride in ethanol: Whitmore *et al., Org. Syn.* **3**, 65 (1923); by boiling p-bromotoluene with sodium amalgam in xylene in the presence of ethyl acetate as a catalyst: Dreher, Otto, *Ann.* **154**, 171 (1870); LaCoste, Michaelis, *ibid.* **201**, 246 (1880); *Compt. Rend.* **68**, 1298 (1869).

Needles from xylene. mp 238°. Insol in water; slightly sol in cold alcohol; sol in hot benzene, xylene, chloroform, carbon disulfide.

3388. Diuron. *N'-(3,4-Dichlorophenyl)-N,N-dimethylurea;* 1,1-dimethyl-3-(3,4-dichlorophenyl)urea; Diurex; Karmex; Telvar; Urox D. $C_9H_{10}Cl_2N_2O$; mol wt 233.10. C 46.37%, H 4.32%, Cl 30.42%, N 12.02%, O 6.86%. Prepn: Jones, U.S. pat. **2,768,971** (1956 to I.C.I.). Toxicity data: E. M. Boyd, V. Krupa, *J. Agr. Food Chem.* **18**, 1104 (1970).

Cl NHCON(CH₃)₂

Cl

Crystals, mp 158-159°. Soly in water at 25°: 42 ppm. Very low soly in hydrocarbon solvents. Vapor pressure at 50°: 3.1×10^{-6} mm Hg. LD_{50} orally in rats: 437 mg/kg (Boyd, Krupa).

Caution: Repeated doses produce anemia in rats and methemoglobinemia if the compound is hydrolyzed *in vivo* to dichloroaniline: du Pont, *Condensed Technical Information,* April 1961.

USE: Pre-emergent herbicide.

3389. Divicine. *2,6-Diamino-1,6-dihydro-4,5-pyrimidinedione; 2,6-diamino-4,5-pyrimidinediol;* 2,6-diamine-5-hydroxy-4(3H)-pyrimidinone; 2,4-diamino-5,6-dihydroxypyrimidine. $C_4H_6N_4O_2$; mol wt 142.12. C 33.80%, H 4.26%, N 39.43%, O 22.52%. From vicine by heating with dil H_2SO_4: Ritthausen, *Ber.* **29**, 2108 (1896); Levene, Senior, *J. Biol. Chem.* **25**, 607 (1916). Structure: Bendich, Clements, *Biochim. Biophys. Acta* **12**, 462 (1953). Synthesis:

Davall, Laney, *J. Chem. Soc.* **1956**, 2124; Chesterfield *et al.*, *ibid.* **1964**, 1001.

Brownish needles, dec above 280°. One gram dissolves in 100 ml boiling water, in about 350 ml cold water. Soluble in 10% KOH.

Diacetate, $C_8H_{10}N_4O_4$, crystals from water, dec 309-312°.

3390. Dixanthogen. *Thioperoxydicarbonic acid diethyl ester; dithiobis[thioformic acid] O,O-diethyl ester; O,O-diethyl dithiobis[thioformate]; bisethylxanthogen; diethyl xanthogenate; ethylxanthic disulfide; preparation K; EXD; Auligen; Aulinogen; Bexide; Herbisan; Lenisarin; Sulfasan.* $C_6H_{10}O_2S_4$; mol wt 242.40. C 29.73%, H 4.16%, O 13.20%, S 52.91%. Prepn: Losse, Wottgen, *J. Prakt. Chem.* **13**, 260 (1961). Metabolic studies: Gutenmann, Lisk, *J. Agr. Food Chem.* **19**, 200 (1971). Toxicity: G. W. Bailey, J. L. White, *Residue Rev.* **10**, 97 (1965).

Yellow needles, mp 28-32°. Onion-like odor. Soly in alcohol: 2 g/100 ml. Freely sol in benzene, ether, petr ether, oils. Almost insol in water. LD_{50} orally in rats: 480 mg/kg (Bailey, White).

USE: Insecticide formulations; herbicide.

THERAP CAT: Ectoparasiticide.

3391. Dixyrazine. *2-[2-[4-[2-Methyl-3-(10H-phenothiazin-10-yl)propyl]-1-piperazinyl]ethoxy]ethanol; 10-[2-methyl-3-(4-hydroxyethoxyethyl-1-piperazinyl)propyl]-phenothiazine; 1-[2-(2-hydroxyethoxy)ethyl]-4-[2-methyl-3-(10-phenothiazinyl)propyl]piperazine; UCB 3412; Esocalm; Esucos.* $C_{24}H_{33}N_3O_2S$; mol wt 427.60. C 67.41%, H 7.78%, N 9.83%, O 7.48%, S 7.50%. Prepn: Morren, **Brit. pat. 861,420** (1961).

Dihydrochloride, $C_{24}H_{33}N_3O_2S.2HCl$, crystals from isopropanol, mp 192°.

THERAP CAT: Antipsychotic.

3392. Dizocilpine. *10,11-Dihydro-5-methyl-5H-dibenzo[a,d]cyclohepten-5,10-imine; (+)-5-methyl-10,11-dihydro-5H-dibenzo[a,d]cyclohepten-5,10-imine.* $C_{16}H_{15}N$; mol wt 221.30. C 86.84%, H 6.83%, N 6.33%. Non-competitive NMDA, *q.v.*, receptor antagonist. Prepn of racemate: M. E. Christy *et al.*, *J. Org. Chem.* **44**, 3117 (1979); T. R. Lamanec *et al.*, *ibid.* **53**, 1768 (1988); and of enantiomers: **Belg. pat. 882,361;** P. Anderson *et al.*, **U.S. pat. 4,399,141** (1980, 1983 both to Merck & Co.). Series of papers on pharmacological activities: B. V. Clineschmidt *et al.*, *Drug. Devel. Res.* **2**, 123-163 (1982). NMDA receptor binding studies: E. H. F. Wong *et al.*, *Proc. Nat. Acad. Sci. USA* **83**, 7104 (1986); A. C. Foster, E. H. F. Wong, *Brit. J. Pharmacol.* **91**, 403 (1987); J. E. Huettner, B. P. Bean, *Proc. Nat. Acad. Sci. USA* **85**, 1307 (1988). Metabolism in animals: H. B. Hucker *et al.*, *Drug. Metab. Dispos.* **11**, 54 (1983). Pharmacokinetics: A. S. Troupin *et al.*, *Curr. Probl. Epilepsy* **4**, 191 (1986). Protective effects against excitotoxic amino acid-induced neuronal degeneration: G. N. Woodruff *et al.*, *Neuropharmacology* **26**, 903 (1987); A. C. Foster *et al.*, *Neuroscience Letters* **76**, 307 (1987); A. C. Foster *et al.*, *J. Neurosci.* **8**, 4745 (1988); specifically in ischemia: R. Gill *et al.*, *Neuroscience* **25**, 847 (1988); E. Ozyurt *et al.*, *J. Cereb. Blood Flow Metab.* **8**, 138 (1988). Anticonvulsant activity in models of epilepsy: K. Sato *et al.*, *Brain Res.* **463**, 12 (1988); M. E. Gilbert, *ibid.* 90. Clinical evaluation in attention deficit disorder: F. W. Reimherr *et al.*, *Psychopharmacol. Bull.* **22**, 237 (1986).

White solid from cyclohexane, mp 68.5-69°. $[\alpha]_{589}^{20}$ +161.4° (c = 0.038 g/2 ml ethanol).

Maleate, $C_{20}H_{19}NO_4$, **MK-801, Neurogard.** Crystalline solid, mp 208.5-210° $[\alpha]_D^{20}$ +114° (c = 0.0128 g/2 ml ethanol).

(−)-Form, white solid from cyclohexane, mp 68.5-69.5°. $[\alpha]_{589}$ −160.8° (c = 0.032 g/2 ml ethanol).

THERAP CAT: Neuroprotective.

3393. Djenkolic Acid. *S,S'-Methylenebis-L-cysteine; 3,3'-(methylenedithio)dialanine; 3,3'-methylenedithiobis(2-aminopropanoic acid);* L-cysteine thioacetal of formaldehyde. $C_7H_{14}N_2O_4S_2$; mol wt 254.33. C 33.06%, H 5.55%, N 11.02%, O 25.16%, S 25.22%. $CH_2[SCH_2CH(NH_2)COOH]_2$. An amino acid isolated from the djenkol bean (*Pithecolobium lobatum* Benth., *Leguminosae*): van Veen, Hyman, *Rec. Trav. Chim.* **54**, 493 (1935). Synthesis from 1 mol formaldehyde and 2 mols L-cysteine in strong HCl solution: Armstrong, du Vigneaud, *J. Biol. Chem.* **168**, 373 (1947); *cf.* du Vigneaud, Patterson, *ibid.* **114**, 633 (1936); Middlebrook, Phillips, *Biochem. J.* **41**, 218 (1947). Synthesis of homologs: Frankel, Gertner, *J. Chem. Soc.* **1960**, 898.

Rosettes of needles of various lengths. Gradually dec between 300 and 350°. $[\alpha]_D^{20.5}$ −65.0° (1.0N HCl); $[\alpha]_D^{25}$ −47.5° (c = 2 in 1.0N HCl). Very sparingly sol in cold water. Soly in boiling water about 1 in 200. Readily sol in aq solns of alkalies or acids.

Monohydrochloride, $C_7H_{15}ClN_2O_4S_2$, slender prisms, dec 250-300°, sol in water.

Dibenzoyldjenkolic acid, $C_{21}H_{22}N_2O_6S_2$, crystals, mp 87.5-89°.

3394. DMPA. *(1-Methylethyl)phosphoramidothioic acid O-(2,4-dichlorophenyl) O-methyl ester; isopropylphosphoramidothioic acid O-2,4-(dichlorophenyl) O-methyl ester; O-(2,4-dichlorophenyl) O-methyl isopropylphosphoramidothioate; K-22023; Dow 1329; ENT 25647; OMS 115; Dowco 118; Zytron.* $C_{10}H_{14}Cl_2NO_2PS$; mol wt 314.18. C 38.23%, H 4.49%, Cl 22.57%, N 4.46%, O 10.19%, P 9.86%, S 10.21%. Prepn: Blair *et al.*, *J. Agr. Food Chem.* **11**, 237 (1963). Synthesis of the optical isomers: Seiber, Tolkmith, *Tetrahedron* **25**, 381 (1969). Use as a herbicide: Leasure, **U.S. pat. 3,074,790** (1963 to Dow); as a plant growth regulator: Holmsen, *Weed Sci.* **17**, 187 (1969). Neurotoxicity in chickens: B. M. Francis *et al.*, *J. Environ. Sci. Health* **B15**, 313 (1980).

Solid, mp 51.4°. Vapor pressure at 150°: 2 mm. Slightly sol in water (5 ppm); freely sol in acetone, benzene, carbon tetrachloride. LD_{50} orally in rats: 270-360 mg/kg, E. W. Schafer, *Toxicol. Appl. Pharmacol.* **21**, 315 (1972).

Note: DMPA® is dimethylolpropionic acid, *q.v.*

USE: Herbicide; plant growth regulator.

3395. Dobesilate Calcium. *2,5-Dihydroxybenzenesulfonic acid calcium salt;* calcium dobesilate; hydroquinone calcium sulfonate; Dexium; Doxium. $C_{12}H_{10}CaO_{10}S_2$; mol wt 418.41. C 34.45%, H 2.41%, Ca 9.58%, O 38.24%, S 15.32%. Prepn: **Span. pat. 335,945** (1967 to Labs. Esteve), *C.A.* **69**, 106253z (1968); **Fr. pat. M6163** corresp to Esteve-Subirana, **U.S. pat. 3,509,207** (1968 and 1970 to Labs. OM). Clinical trials:

Berson, *Praxis* **59**, 1305 (1970); **61**, 52 (1972). Metabolism: A. Benakis *et al.*, *Therapie* **29**, 211 (1974).

White, powdery crystals from water, mp > 300° (dec). Color deepens to pink upon exposure to air. Very soluble in water and alcohol. Practically insol in ether, benzene, chloroform. LD_{50} in mice: 700 mg/kg, Esteve, Subirana, *loc. cit.*

THERAP CAT: Vasotropic.

3396. Dobutamine. (±)-*4-[2-[[3-(4-Hydroxyphenyl)-1-methylpropyl]amino]ethyl]-1,2-benzenediol;* (±)-*4-[2-[[3-(p-hydroxyphenyl)-1-methylpropyl]amino]ethyl]pyrocatechol;* 3,4-*dihydroxy-N-[3-(4-hydroxyphenyl)-1-methylpropyl]-β-phenylethylamine;* Compound 81929. $C_{18}H_{23}NO_3$; mol wt 301.39. C 71.73%, H 7.69%, N 4.65%, O 15.93%. $β_1$-Adrenergic agonist; derivative of dopamine, *q.v.* Prepn: Tuttle, Mills, *Ger. pat.* **2,317,710** corresp to *U.S. pat.* **3,987,200** (1973, 1976 to Lilly). Pharmacology: R. Weber, R. R. Tuttle in *Pharmacological and Biochemical Properties of Drug Substances* vol. 1, M. E. Goldberg, Ed. (Am. Pharm. Assoc., Washington, DC, 1977) pp 109-124; E. H. Sonnenblick *et al.*, *N. Engl. J. Med.* **300**, 17 (1979). Comprehensive description: R. H. Bishara, H. B. Long in *Analytical Profiles of Drug Substances* vol. 8, K. Florey, Ed. (Academic Press, New York, 1979) pp 139-158.

Hydrochloride, $C_{18}H_{24}ClNO_3$, *Inotrex* (obsolete), *Dobutrex.* Crystals, mp 184-186°. uv max (methanol): 281, 223 nm (ε 4768, 14400). pKa 9.45. Rapidly oxidized at pH 11-13. LD_{50} i.v. in mice: ~73 mg/kg (Weber, Tuttle).

THERAP CAT: Cardiotonic.

3397. Docusate Calcium. *Sulfobutanedioic acid 1,4-bis-(2-ethylhexyl)ester calcium salt; bis[2-ethylhexyl]calcium sulfosuccinate;* calcium dioctyl sulfosuccinate; dioctyl calcium sulfosuccinate; *Surfak.* $C_{40}H_{74}CaO_{14}S_2$; mol wt 883.24. C 54.40%, H 8.45%, Ca 4.54%, O 25.36%, S 7.26%. Prepd from dioctyl sodium sulfosuccinate dissolved in isopropanol and from calcium chloride dissolved in methanol: Klotz, *U.S. pat.* **3,035,973** (1962 to Lloyd Brothers).

White precipitate. Sol in mineral and vegetable oils, liq polyethylene glycol. Practically insol in glycerol. Claimed to have greater surface-active wetting properties than the sodium salt.

Ingredient of *Doxidan* which also contains danthron.

THERAP CAT: Stool softener.

3398. Docusate Sodium. *Sulfobutanedioic acid 1,4-bis-(2-ethylhexyl) ester sodium salt; sulfosuccinic acid 1,4-bis(2-ethylhexyl) ester S-sodium salt;* bis(2-ethylhexyl)sodium sulfosuccinate; dioctyl sodium sulfosuccinate; sodium dioctyl sulfosuccinate; DSS; Aerosol OT; Alphasol OT; Colace; Comfolax; Complemix; Coprol; Dioctylal; Dioctyl-Medo Forte; Diotilan; Diovac; Disonate; Doxinate; Doxol; Dulsivac; Molatoc; Molcer; Molofac; Nevax; Regutol; Softil; Soliwax; Solusol; Sulfimel DOS; Valsol OT; Velmol; Waxsol; Yal. $C_{20}H_{37}NaO_7S$; mol wt 444.56. C 54.03%, H 8.39%, Na

5.17%, O 25.19%, S 7.21%. Prepn: Jaeger, *U.S. pats.* **2,028,-091; 2,176,423** (1936, 1939, both to Am. Cyanamid). Structure and wetting power: Caryl, *Ind. Eng. Chem.* **33**, 731 (1941). Comprehensive description: S. Ahuja, J. Cohen, in *Analytical Profiles of Drug Substances* vol. 2, K. Florey, Ed. (Academic Press, New York, 1973) pp 199-219; vol. 12 (1983) pp 713-720.

Available as wax-like solid, usually in rolls of tissue-thin material; also as 50-75% solns in various solvents. Soly in water (g/l): 15 (25°), 23 (40°), 30 (50°), 55 (70°). Sol in CCl_4, petr ether, naphtha, xylene, dibutyl phthalate, liq petrolatum, acetone, alcohol, vegetable oils. Very sol in water + alcohol, water + water-miscible organic solvents. Stable in acid and neutral solns; hydrolyzes in alkaline solns.

Note: Ingredient of the laxative, *Jamylene,* which also contains danthron, and of *Peri-Colace* which also contains casanthranol.

Potassium salt, $C_{20}H_{37}KO_7S$, *docusate potassium, Rectalad Enema.*

USE: Sodium salt as pharmaceutic aid (surfactant); as wetting agent in industrial, pharmaceutical, cosmetic and food applications; dispersing and solubilizing agent in foods; adjuvant in tablet formation.

THERAP CAT: Stool softener.

3399. Dodecahedrane. *Hexadecahydro-5,2,1,6,3,4-[2,3]butanediyl[1,4]diylidenedipentaleno[2,1,6-cde:2',1',6'-gha]pentalene;* pentagonal dodecahedrane. $C_{20}H_{20}$; mol wt 260.38. C 92.26%, H 7.74%. Classical uniform convex polyhedrane. Theoretical studies: O. Ermer, *Angew. Chem. Int. Ed.* **16**, 411 (1977); R. L. Disch, J. M. Schulman, *J. Am. Chem. Soc.* **103**, 3297 (1981). Review of synthetic studies: P. E. Eaton, *Tetrahedron* **35**, 2189-2223 (1979). Total synthesis: R. J. Ternansky *et al., J. Am. Chem. Soc.* **104**, 4503 (1982).

Crystals from benzene, mp > 450°.

3400. Dodecamethylcyclohexasiloxane. $C_{12}H_{36}O_6Si_6$; mol wt 444.96. C 32.40%, H 8.16%, O 21.58%, Si 37.85%. Isolated from the hydrolysis product of dimethyldichlorosilane: Patnode, Wilcock, *J. Am. Chem. Soc.* **68**, 358 (1946).

Oily liquid. mp −3°. bp 245°. bp_{20} 128°. d 0.9762. n_D^{20} 1.4015.

3401. Dodecamethylpentasiloxane. $C_{12}H_{36}O_4Si_5$; mol wt

384.87. C 37.46%, H 9.43%, O 16.64%, Si 36.47%. Prepn: Patnode, Wilcock, *J. Am. Chem. Soc.* **68**, 358 (1946).

$$(CH_3)_3Si-O-\underset{\underset{CH_3}{|}}{\overset{\overset{CH_3}{|}}{Si}}-O-\underset{}{Si}(CH_3)_3$$

Liquid. bp$_{710}$ 229°. d 0.8755. n_D^{20} 1.3925. mp ~ −80°. Stable. Inert to most chemical reagents and rubber. Maintains about the same viscosity over a wide temperature range. Sol in benzene and the lighter hydrocarbons; slightly sol in alcohol and the heavy hydrocarbons.

USE: As a basis for silicone oils or fluids designed to withstand extremes of temperature; as a foam suppressant in petr lubricating oil.

3402. 1-Dodecanol. Dodecyl alcohol; lauryl alcohol. $C_{12}H_{26}O$; mol wt 186.33. C 77.35%, H 14.07%, O 8.59%. $CH_3(CH_2)_{11}OH$. Prepd by the reduction of esters of lauric acid with sodium and abs alcohol: Levene, Allen, *J. Biol. Chem.* **27**, 443 (1916); Ford, Marvel, *Org. Syn.* **10**, 62 (1930), or by high pressure hydrogenation of the esters using copper chromite catalyst: Adkins, Folkers, *J. Am. Chem. Soc.* **53**, 1095 (1931).

Leaflets from dil alc, mp 24°. d_4^{24} 0.8309 (liq); d_4^{40} 0.8201; d_4^{99} 0.7781. bp$_{760}$ 259°; bp$_{400}$ 235.7°; bp$_{200}$ 213°; bp$_{100}$ 192°; bp$_{60}$ 177.8°; bp$_{40}$ 167.2°; bp$_{20}$ 150°; bp$_{10}$ 134.7°; bp$_5$ 120.2°; bp$_{1.0}$ 91.0°. Insol in water; sol in alcohol and ether.

Phenylurethan, $C_{19}H_{31}NO_2$, needles from dilute methanol, mp 84°.

USE: Manuf of sulfuric acid esters which are used as wetting agents.

3403. 3-(1,3,5,7,9-Dodecapentaenyloxy)-1,2-propanediol. $C_{15}H_{22}O_3$; mol wt 250.35. C 71.97%, H 8.86%, O 19.17%. A potent mutagen detected in human feces: W. R. Bruce *et al.*, *Cold Spring Harbor Conf. Cell Proliferation* **3**, 1641 (1977). Isoln and characterization: T. D. Wilkins *et al.*, *Am. J. Clin. Nutr.* **33**, 2513 (1980). Structure: N. Hirai *et al.*, *J. Am. Chem. Soc.* **104**, 6149 (1982). Implicated in human colon cancer: *Chem. & Eng. News* **60**, 22 (Sept. 27, 1982); T. H. Maugh, *Science* **218**, 363 (1982).

$$HO-\overset{H}{\underset{HO}{\diagdown}}O\diagup\diagdown\diagup\diagdown\diagup\diagdown\diagup CH_3$$

Unstable, decomp rapidly in presence of air or acid. uv max: 325, 345, 365 nm. Sol in chloroform, benzene, ether. Insol in water.

3404. Dodecarbonium Chloride. *N-[2-(Dodecylamino)-2-oxoethyl]-N,N-dimethylbenzenemethanaminium chloride; benzyl[(dodecylcarbamoylmethyl)dimethyl]ammonium chloride; N,N-dimethyl-N-benzyl-N-chloro-N'-dodecylglycinamide;* (dodecylcarbamoylmethyl)benzyldimethylammonium chloride; Straminol; Urolocide. $C_{23}H_{41}ClN_2O$; mol wt 397.03. C 69.57%, H 10.41%, Cl 8.93%, N 7.05%, O 4.03%. Preparation: Hentrich *et al.*, U.S. pat. 2,295,655 (1942 to Patchem, AG).

$$\left[\underset{}{\overset{}{C_6H_4}}-CH_2-\underset{\underset{CH_3}{|}}{\overset{\overset{CH_3}{|}}{N^+}}-CH_2CONH(CH_2)_{11}CH_3\right]Cl^-$$

Bitter crystals, mp 147-148°. Sol in water, alcohol. Insol in ether, acetone, benzene. LD$_{50}$ orally in rats: 100 mg/kg.
THERAP CAT: Antiseptic, disinfectant.

3405. Dodemorph. *4-Cyclododecyl-2,6-dimethylmorpholine.* $C_{18}H_{35}NO$; mol wt 281.49. C 76.81%, H 12.53%, N 4.98%, O 5.68%. Prepn: W. Sanne *et al.*, Belg. pat. 614,214 corresp to U.S. pat. 3,686,399 (1962, 1972 to BASF); K.-H. König *et al.*, *Angew. Chem. Int. Ed.* **4**, 336 (1965). Activity: J. Kradel, E. H. Pommer, *Proc. Brit. Insectic. Fungic. Conf., 4th* **1967**, 170.

$$(CH_2)_{11}-CH-N\overset{\diagup}{\diagdown}\underset{}{\overset{CH_3}{\underset{CH_3}{O}}}$$

Oil, bp$_{1.5}$ 161-162°. n_D^{25} 1.4907. Acetate, $C_{20}H_{39}NO_3$, *cyclomorph,* **BAS 238F, Meltatox, Milban.** Yellow liquid, d 0.93. Misc with water. LD$_{50}$ i.p. in mice: 100 mg/kg, D. Marchand, G. Serra, *Def. Veg.* **27**, 144 (1973), *C.A.* **79**, 133569v (1973).

USE: Fungicide.

3406. Dodine. *Dodecylguanidine monoacetate;* AC 5223; Carpene; Cyprex; Melprex. $C_{15}H_{33}N_3O_2$; mol wt 287.44. C 62.67%, H 11.57%, N 14.62%, O 11.13%. Prepn: G. Lamb, U.S. pats. 2,867,562; 2,921,881 (1959, 1960, both to Am. Cyanamid).

$$CH_3(CH_2)_{11}NHC\overset{\overset{NH}{\|}}{}NH_2 \cdot CH_3COOH$$

Slightly waxy solid, mp 136°. Sol in alcohol, hot water; slightly sol in other solvents. LD$_{50}$ orally in rats: 566 mg/kg, *RTECS* **Vol. I**, R. J. Lewis, R. L. Tatken, Eds. (1979) p 726.

USE: Fungicide. *Caution:* Strong solns irritating to skin, mucous membranes. Ingestion causes vomiting, diarrhea.

3407. Doisynoestrol. *dl-cis-1-Ethyl-1,2,3,4-tetrahydro-7-methoxy-2-methyl-2-phenanthrenecarboxylic acid;* 1-ethyl-2-methyl-7-methoxy-1,2,3,4-tetrahydrophenanthryl-1-carboxylic acid; tetradehydrodoisynolic acid methyl ether; α-bisdehydrodoisynolic acid methyl ether; dehydrofolliculinic acid; fénocycline; Fenocylin; Surestrine; Surestryl. $C_{19}H_{22}O_3$; mol wt 298.37. C 76.48%, H 7.43%, O 16.09%. Prepn: Heer *et al.*, *Helv. Chim. Acta* **28**, 1342 (1945); Miescher *et al.*, U.S. pat. 2,621,208 (1952 to Ciba); Gastambide-Odier *et al.*, *Bull. Soc. Chim. France* **1963**, 1777. *Review:* Miescher, *Chem. Rev.* **43**, 367 (1948). *See also* Doisynolic Acid.

$$\underset{CH_3O}{\overset{CH_3}{\diagup}}\overset{\overset{CH_3}{\diagdown}COOH}{\underset{\underset{H}{\diagup}CH_2CH_3}{}}$$

Crystals from acetone, mp 228-230°. Sol in aq solns of sodium bicarbonate. The isomeric *trans*-isomer is inactive.

3408. Doisynolic Acid. *1-Ethyl-1,2,3,4,4a,9,10,10a-octahydro-7-hydroxy-2-methyl-2-phenanthrenecarboxylic acid;* 1-ethyl-7-hydroxy-2-methyl-1,2,3,4,4a,9,10,10a-octahydrophenanthrene-2-carboxylic acid; 3-hydroxy-16,17-seco-estra-1,3,5(10)-trien-17-oic acid. $C_{18}H_{24}O_3$; mol wt 288.37. C 74.97%, H 8.39%, O 16.64%. Prepd by alkali fusion of estrone: MacCorquodale *et al.*, *J. Biol. Chem.* **99**, 327 (1933); **101**, 753 (1933); Heer, Miescher, *Helv. Chim. Acta* **28**, 156 (1945). Sixteen isomers are possible. If the hydrogen atom at position 4a is α- and the hydrogen atom at position 10a is β-, the acid belongs to the A series. If the hydrogen at 4a is β- and the hydrogen at 10a is α-, the acid belongs to the B series. If both hydrogens are α-, the acid belongs to the C series. If both hydrogens are β-, the acid belongs to the D series: Miescher, *Chem. Rev.* **43**, 367 (1948). If the hydrogen atom at position 1 and the methyl group at position 2 have either the α,α- or β,β-orientation, the acid is designated *cis;* if they are α,β- the acid is designated *trans.* L. F. Fieser, M. Fieser, *Steroids* (Reinhold, New York, 1959) pp 487-495. Synthesis of *dl-cis* and *dl-trans* of the A series and *dl-cis* of the B series: Anner, Miescher, *Helv. Chim. Acta* **29**, 1889 (1946); **30**, 1422 (1947). Synthesis of *dl-cis* of the C series: Jilck, Protiva, *Coll. Czech. Chem. Commun.* **23**, 692 (1958). Stereochemistry of *d-cis*-form: Iriarte, Crabbe, *Chem.*

Commun. 1972, 1110. *Reviews:* Shoppee, *Annual Reports on the Progress of Chemistry* **44,** 190 (1948); Miescher in *Recent Progress in Hormone Research* **vol. 3,** G. Pincus Ed. (Academic Press, New York, 1948) pp 47-69.

dl-cis Acid of the A series: Crystals from methanol, mp 181-182°. Appears to have the highest estrogenic potency.

dl-trans Acid of the A series: Plates from methanol, mp 175-177°.

dl-cis Acid of the B series: Crystals from methanol, mp 212-214°.

dl-cis Acid of the C series: Crystals from methanol, mp 113-117°.

3409. Domesticine. *(S)-5,6,6a,7-Tetrahydro-2-methoxy-6-methyl-4H-benzo[de][1,3]benzodioxolo[5,6-g]quinolin-1-ol; 2-methoxy-9,10-(methylenedioxy)-6aα-aporphin-1-ol;* 1-hydroxy-2-methoxy-9,10-methylenedioxyaporphine. $C_{19}H_{19}NO_4$; mol wt 325.35. C 70.14%, H 5.89%, N 4.31%, O 19.67%. In *Nandina domestica* Thunb., *Berberidaceae.* Isoln: Kitasato, Shishido, *Ann.* **527,** 176 (1937). Syntheses: Govindachari *et al., Indian J. Chem.* **7,** 841 (1969); Kessar *et al., ibid.* **8,** 468 (1970); Kametani *et al., J. Chem. Soc. (C)* **1971,** 2446, 2712; *eidem, Chem. Pharm. Bull.* **21,** 766 (1973); Horii *et al., ibid.* **22,** 583 (1974); Hoshino *et al., ibid.* **23,** 2048 (1975).

Natural base from methanol + water, mp 115-116°; from abs methanol or benzene, mp 84-85°; dried at 60° over P_2O_5, mp 152-153°. *(dl-Form,* needles from methanol, mp 185-186° (dec), Kametani *et al., loc. cit.)* Easily oxidized in air. uv max (ethanol): 221, 283, 310 nm (log ε 4.56, 4.01, 4.17). Very sol in chloroform, sol in hot alc, ethyl acetate, acetic acid, alkalies; slightly sol in ether. Practically insol in water. Methyl ether, *nantenine, domestine, epidicentrine.* mp 139°. $[\alpha]_D^{18}$ +102° (c = 0.528 in chloroform).

3410. Domiodol. *2-(Iodomethyl)-1,3-dioxolane-4-methanol;* 4-hydroxymethyl-2-iodomethyl-1,3-dioxolane; MG 13608; Mucolitico. $C_5H_9IO_3$; mol wt 244.02. C 24.61%, H 3.72%, I 52.00%, O 19.67%. Organic iodide mixture of *cis* and *trans* isomers. Prepn: M. Carissimi *et al.,* Ger. pat. **2,610,704;** *eidem,* U.S. pat. **4,085,223** (1976, 1978 both to Maggioni). Prepn, expectorant and mucolytic activities: G. Cantarelli *et al., Farmaco Ed. Prat.* **34,** 393 (1979). Pharmacology, toxicity: K. Kogi *et al., Arznemittel-Forsch.* **33,** 1281 (1983). Separation of isomers and comparison of their pharmacological activity: M. Riva *et al., ibid.* 1091. Metabolism and tissue distribution in rats: T. Ohtsuki *et al., Farmaco Ed. Prat.* **39,** 291 (1984). Comparative study with S-carboxymethylcysteine, *q.v.,* in chronic obstructive lung disease: L. Casali *et al., Int. J. Clin. Pharmacol. Ther. Toxicol.* **20,** 554 (1982).

LD_{50} in male, female mice (mg/kg): 79, 89 i.p.; 140, 145 orally (Riva).

cis-Form: $bp_{0.2}$ 106-108°. LD_{50} orally in male mice: 135 mg/kg (Riva).

trans-Form: $bp_{0.2}$ 114-116°. LD_{50} orally in male mice: 150 mg/kg (Riva).

THERAP CAT: Mucolytic.

3411. Domiphen Bromide. *N,N-Dimethyl-N-(2-phenoxyethyl)-1-dodecanaminium bromide; dodecyldimethyl(2-phenoxyethyl)ammonium bromide; (β-phenoxyethyl)dimethyldodecylammonium bromide;* PDDB; phenododecinium bromide; NSC-39415; Bradosol Bromide; Oradol; Modicare. $C_{22}H_{40}BrNO$; mol wt 414.46. C 63.75%, H 9.73%, Br 19.28%, N 3.38%, O 3.86%. Prepd by heating phenoxyethyl-dimethylamine with dodecyl bromide: Hartmann, Bosshard, U.S. pat. **2,581,336** (1952 to Ciba).

Crystals, mild characteristic odor, bitter taste, mp 112-113°. Freely sol in water (100 g/100 ml), much less sol at low temps. Sol in ethanol, acetone, ethyl acetate, chloroform; very slightly sol in benzene. Aq solns are clear, colorless, and foam profusely on shaking. pH of 10% soln: 6.42; of 1% commercial product at 25°: 5.5; of 0.1%: 6.8. Surface tension values at 25° (by the capillary rise method) range from 26.75 dynes/cm (10% soln) to 22.08 dynes/cm (0.1% soln). Incompatible with soap.

THERAP CAT: Topical anti-infective.

3412. Domperidone. *5-Chloro-1-[1-[3-(2,3-dihydro-2-oxo-1H-benzimidazol-1-yl)propyl]-4-piperidinyl]-1,3-dihydro-2H-benzimidazol-2-one;* 5-chloro-1-[1-[3-(2-oxo-1-benzimidazolinyl)propyl]-4-piperidyl]-2-benzimidazolinone; R 33,812; Euciton; Evoxin; Mod; Motilium; Nauzelin; Peridon; Peridys. $C_{22}H_{24}ClN_5O_2$; mol wt 425.92. C 62.04%, H 5.68%, Cl 8.32%, N 16.44%, O 7.51%. A novel *in vitro* dopamine antagonist with antinauseant properties. Prepn: J. Vandenberk *et al.,* Ger. pat. **2,632,870** corresp to U.S. pat. **4,066,772** (1977, 1978 both to Janssen). Pharmacology: C. Ennis *et al., J. Pharm. Pharmacol.* **31**(Suppl.), 14P (1979). Gastrokinetic properties: J. M. Van Neuten *et al., Life Sci.* **23,** 453 (1978). ³H-domperidone studies: M. P. Martres *et al., ibid.* 1781; M. Baudry *et al., Arch. Pharmacol.* **308,** 231 (1979). Clinical studies: A. J. Reyntjens *et al., Arzneimittel-Forsch.* **28,** 1194 (1978); D. B. Wilson, J. W. Dundee, *Anaesthesia* **34,** 765 (1979). Review of pharmacology, pharmacokinetics and therapeutic efficacy: R. N. Brogden *et al., Drugs* **24,** 360-400 (1982).

Crystals from DMF/water, mp 242.5°.

THERAP CAT: Anti-emetic.

3413. Donovan's Solution. An arsenic and mercuric iodides solution prepd from 1 g each of AsI_3 and HgI_2 plus 0.9 g sodium bicarbonate in water to make 100 ml.

Clear, colorless or pale yellow, becoming darker with age. Miscible with water, alcohol.

Incompat. Alkalies, alkaloids or their salts.

THERAP CAT (VET): Has been used in chronic diseases of the skin.

3414. Dopa. *3-Hydroxytyrosine; 3-(3,4-dihydroxyphenyl)alanine; β-(3,4-dihydroxyphenyl)-α-alanine; 2-amino-3-*

(3,4-dihydroxyphenyl)propanoic acid. $C_9H_{11}NO_4$; mol wt 197.19. C 54.82%, H 5.62%, N 7.10%, O 32.46%. An amino acid found in seedlings, pods and beans of *Vicia faba* L. (broad beans) *Stizolobium deeringianum* L. (velvet beans) *Leguminosae:* Torquati, *Arch. farm. sper.* **15**, 213, 308 (1913), *C.A.* **7**, 2774 (1913); Sealock, *Biochemical Preparations* vol. **1**, 25 (1949). Absolute configuration: Guggenheim, *Z. Physiol. Chem.* **88**, 276 (1913). Prepn of DL-, D-, and L-dopa: Yamada *et al., Chem. Pharm. Bull.* **10**, 693 (1962).

HO—[benzene ring]—CH_2CHCO_2H with NH_2, HO

DL-Form, prisms from water or from aq $NaHSO_3$ soln, dec 270-272°. Soly in water: 144 mg/40 ml. Readily sol in dil acids and alkalies; slightly sol in benzene, carbon disulfide; practically insol in abs alcohol, ether, glacial acetic acid, petr ether, chloroform. Oxidizes readily.

D-Form, needles from water, dec 276-278°. $[\alpha]_D^{11} +13.0°$ (c = 5.27 in 1N HCl). Soly in water: 66 mg/40 ml.

L-Form, *see* Levodopa.

3415. Dopamine. *4-(2-Aminoethyl)-1,2-benzenediol; 4-(2-aminoethyl)pyrocatechol;* 3-hydroxytyramine; 3,4-dihydroxyphenethylamine; α-(3,4-dihydroxyphenyl)-β-aminoethane. $C_8H_{11}NO_2$; mol wt 153.18. C 62.72%, H 7.24%, N 9.14%, O 20.89%. Endogenous catecholamine with α and β-adrenergic activity. Isoln from *Hermidium alipes* (S. Watson) *Nyctaginaceae:* Buelow, Gisvold, *J. Am. Pharm. Assoc.* **33**, 270 (1944). Prepn from aminotyramine: Waser, Sommer, *Helv. Chim. Acta* **6**, 61 (1923). From homoveratrylamine: Schöpf, Bayeler, *Ann.* **513**, 196 (1934); Hahn, Stiehl, *Ber.* **69**, 2640 (1936). Comprehensive description of the hydrochloride: J. E. Carter *et al.,* in *Analytical Profiles of Drug Substances* vol. **11**, K. Florey, Ed. (Academic Press, New York, 1982) pp 257-272. Review of pharmacology and clinical efficacy in oliguria: J. F. Dasta, M. G. Kirby, *Pharmacotherapy* **6**, 304 (1986).

HO—[benzene ring]—$CH_2CH_2NH_2$, HO

Free base, stout prisms, highly sensitive to oxygen; discolors quickly.

Hydrochloride, $C_8H_{12}ClNO_2$, *ASL-279, Cardiosteril, Dopastat, Dynatra, Inovan, Intropin.* Rosettes of needles from water, dec 241°; may be recrystallized from methanol + ether. Freely sol in water; sol in methanol, in hot 95% ethanol; in aq solns of alkali hydroxides. Practically insol in ether, petr ether, chloroform, benzene, toluene.

Hydrobromide, $C_8H_{11}NO_2 \cdot HBr$, crystals, dec 210-214°.

THERAP CAT: Cardiotonic; antihypotensive.

3416. Dopan. *5-[Bis(2-chloroethyl)amino]-6-methyl-2,4(1H,3H)-pyrimidinedione; 5-[bis(2-chloroethyl)amino]-6-methyluracil;* 6-methyl-5-[bis(2-chloroethyl)amino]uracil; 4-methyl-5-[bis(β-chloroethyl)amino]uracil; 2,6-dihydroxy-4-methyl-5-bis[2-chloroethyl]aminopyrimidine; Elderfield pyrimidine mustard; NSC-23436. $C_9H_{13}Cl_2N_3O_2$; mol wt 266.12. C 40.62%, H 4.92%, Cl 26.65%, N 15.79%, O 12.02%. Description: Larionova, Platonova, *Voprosy Onkologii* **1**, no. 5, 36 (1955); *C.A.* **51**, 6862 (1957). Prepn: Nemets *et al.,* U.S.S.R. pat. **116,912** (1959), *C.A.* **53**, 17438i (1959). Outline of synthesis of the demethyl compd: Peter ing, Lyttle, *Chem. & Eng. News* **36**, 47 (Sept. 22, 1958). Monograph: L. F. Larionov, *Cancer Chemotherapy* (Pergamon Press, New York, 1965).

H_3C—[pyrimidinedione ring structure with N, H, O, NH], $(ClCH_2CH_2)_2N$

Snow-white crystals, dec 178-179°. Practically insol in cold water, acetone and benzene. Slightly sol in alc.

USE: In anticancer research: Louis, *J. Chronic Diseases* **15**, 273-281 (March 1962).

3417. Dopastin. *N-[2-(Hydroxynitrosoamino)-3-methylbutyl]-2-butenamide;* N-(2-nitrosohydroxylamino-3-methylbutyl)crotonamide. $C_9H_{17}N_3O_3$; mol wt 215.25. C 50.22%, H 7.96%, N 19.52%, O 22.30%. Dopamine β-hydroxylase inhibitor isolated from a culture filtrate of a bacterium of the *Pseudomonas* genus. Prodn, isoln and characterization: Iinuma *et al., Agr. Biol. Chem.* **38**, 2093 (1974). Structure and synthesis: Iinuma *et al., ibid.* 2099. Biochemical and biological studies: Iinuma *et al., ibid.* 2107.

CH_3, $CH_3CHCHCH_2NHCOCH=CHCH_3$, N—OH, NO

Crystals from acetone + *n*-hexane or from ethyl ether, mp 116-119°. Monobasic acid, pKa 5.1. $[\alpha]_D^{21} -246°$ (c = 0.5 in alcohol). uv max (0.01N HCl): 215 nm (ε 20,900).

Copper salt, $C_9H_{16}N_3O_3 \cdot \frac{1}{2}Cu$, dark blue needles from ethanol, mp 142-144°.

3418. Dopexamine. *4-[2-[[6-[(2-Phenylethyl)amino]hexyl]amino]ethyl]-1,2-benzenediol;* 4-[2-[[6-(phenethylamino)hexyl]amino]ethyl]pyrocatechol; FPL 60278; Dopacard. $C_{22}H_{32}N_2O_2$; mol wt 356.51. C 74.12%, H 9.05%, N 7.86%, O 8.97%. Dopamine receptor and β_2-adrenoreceptor agonist. Prepn: J. B. Farmer *et al.,* Eur. pat. Appl. **72,061** (1983 to Fisons). Unlike dopamine, dopexamine has little or no activity at α- and β_1-adrenoceptors: R. A. Brown *et al., Brit. J. Pharmacol.* **85**, 599 (1985). Cardiovascular activity in dogs: *eidem, ibid.* 609. Hemodynamic effects in chronic congestive heart failure: J. R. Dawson *et al., Brit. Heart J.* **54**, 313 (1985); G. Svensson *et al., Eur. Heart J.* **7**, 697 (1986). Human renovascular effects: F. Magrini *et al., Eur. J. Clin. Pharmacol.* **32**, 1 (1987). Symposium on pharmacology and clinical efficacy: *Am. J. Cardiol.* **62**, Suppl., 1C-88C (1988).

[structure: OH, OH on benzene ring, CH₂CH₂NH(CH₂)₆NHCH₂CH₂, second benzene ring] $CH_2CH_2NH(CH_2)_6NHCH_2CH_2$

Dihydrochloride, $C_{22}H_{34}Cl_2N_2O_2$, *dopexamine hydrochloride, FPL 60278AR.*

Dihydrobromide, $C_{22}H_{34}Br_2N_2O_2$, crystals from ethanol, mp 227-228°.

THERAP CAT: Cardiotonic.

3419. Dothiepin. *3-Dibenzo[b,e]thiepin-11(6H)-ylidene-N,N-dimethyl-1-propanamine; N,N-dimethyldibenzo[b,e]-thiepin-Δ11(6H),γ-propylamine;* 11-(3-dimethylaminopropylidene)-6,11-dihydrodibenzo[b,e]thiepin; dosulepin. $C_{19}H_{21}NS$; mol wt 295.45. C 77.24%, H 7.16%, N 4.74%, S 10.85%. Prepn: Protiva *et al., Experientia* **18**, 326 (1962); Gadient *et al., Helv. Chim. Acta* **45**, 1860 (1962); Stach, Spingler, *Monatsh.* **93**, 896 (1962); Belg. pat. **618,591** (1962 to SPOFA), *C.A.* **58**, 9036g (1963); Rajsner, Protiva, *Cesk. Farm.* **11**, 404 (1962), *C.A.* **59**, 2772g (1963); Protiva *et al., Coll. Czech. Chem. Commun.* **29**, 2161 (1964). Synthesis of isomers: Rajsner *et al., ibid.* **34**, 1963 (1969). Pharmacology: Metysova-Stramkova *et al., Arzneimittel-Forsch.* **13**, 1039 (1963); Benesova *et al., ibid.* **14**, 100 (1964). Metabolism: Horesovsky *et al., Biochem. Pharmacol.* **16**, 2421 (1967).

CHCH$_2$CH$_2$N(CH$_3$)$_2$

bp$_{0.05}$ 171-172°. mp 55-57°.
Hydrochloride, C$_{19}$H$_{22}$ClNS, *Altapin, Depresym, Prothiaden*. Crystals from ethanol-ether, mp 218-221°. uv max (methanol): 232, 260, 309 nm (log ε 4.41, 3.97, 3.53), Rajsner, Protiva, *loc. cit.*
THERAP CAT: Antidepressant.

3420. Dowicide 9. *4-Chloro-2-cyclopentylphenol.* C$_{11}$-H$_{13}$ClO; mol wt 196.67. C 67.18%, H 6.66%, Cl 18.03%, O 8.14%. Prepn: **Ger.** pat. **615,448** (1935 to Hoffmann-La Roche), *C.A.* **29**, 6248[1] (1935); Pajeau, Begue, *Bull. Soc. Chim. France* **1962**, 1923. Bactericidal properties: **Neth.** pat. **Appl. 6,513,777** corresp to Lorah, **U.S.** pat. **3,323,988** (1966 and 1967, both to Dow).

bp$_{18}$ 181-185°; bp$_{11}$ 160-162°.
USE: Germicide.

3421. Doxapram. *1-Ethyl-4-[2-(4-morpholinyl)ethyl]-3,3-diphenyl-2-pyrrolidinone.* C$_{24}$H$_{30}$N$_2$O$_2$; mol wt 378.50. C 76.15%, H 7.99%, N 7.40%, O 8.45%. Prepn: C. D. Lunsford, A. D. Cale, **Belg.** pat. **613,734**; *eidem*, **U.S.** pat. **3,192,-206** (1962, 1965 both to A. H. Robins); Lunsford *et al., J. Med. Chem.* **7**, 302 (1964). Clinical comparison with mepixanox in chronic bronchopulmonary disease: M. Parziale *et al., G. Ital. Mal Torace* **38**, 323 (1984). Toxicity data: E. I. Goldenthal, *Toxicol. Appl. Pharmacol.* **18**, 185 (1971).

Hydrochloride monohydrate, C$_{24}$H$_{31}$ClN$_2$O$_2$.H$_2$O, *AHR-619, Dopram, Stimulexin*. Crystals from isopropyl ether, mp 217-219°. Bitter taste. Sol in water; sparingly sol in alc, slightly sol in chloroform. LD$_{50}$ orally in rats: 261 mg/kg (Goldenthal).
Benzoate, C$_{31}$H$_{36}$N$_2$O$_4$, crystals, mp 123-124°.
THERAP CAT: Respiratory stimulant.
THERAP CAT (VET): Respiratory stimulant.

3422. Doxazosin. *1-(4-Amino-6,7-dimethoxy-2-quinazolinyl)-4-[(2,3-dihydro-1,4-benzodioxin-2-yl)carbonyl]-piperazine;* 4-amino-2-[4-(1,4-benzodioxan-2-carbonyl)-piperazin-1-yl]-6,7-dimethoxyquinazoline; UK-33274. C$_{23}$-H$_{25}$N$_5$O$_5$; mol wt 451.48. C 61.19%, H 5.58%, N 15.51%, O 17.72%. Selective α$_1$-adrenergic blocker related to prazosin, *q.v.* Prepn: S. F. Campbell, **Ger.** pat. **2,847,623**; *idem,* **U.S.** pat. **4,188,390** (1979, 1980 both to Pfizer). Cardiovascular pharmacology in animals: P. B. Timmermans *et al., Arch. Int. Pharmacodyn. Ther.* **245**, 218 (1980). HPLC determn in plasma: M. G. Cowlishaw, J. R. Sharman, *J. Chromatog.* **344**, 403 (1985). Pharmacokinetics in humans: H. L. Elliott *et al., Brit. J. Clin. Pharmacol.* **13**, 699 (1982). Comparison with prazosin in hypertension: P. W. de Leeuw *et al., Eur. J. Clin. Pharmacol.* **23**, 397 (1982). *In vitro* effect on low density lipoprotein-receptor activity: T. P. Leren, *Acta Pharmacol. Toxicol.* **56**, 269 (1985). Symposium on pharmacology and clinical efficacy: *Brit. J. Clin. Pharmacol.* **21**, Suppl. 1, 1S-92S (1986); *Am. J. Cardiol.* **59**, 1G-104G (1987).

Monohydrochloride, C$_{23}$H$_{26}$ClN$_5$O$_5$, crystals, mp 289-290°.
Monomethanesulfonate, C$_{24}$H$_{29}$N$_5$O$_8$S, *doxazosin mesylate, UK-33274-27, Cardura, Carduran*.
THERAP CAT: Antihypertensive.

3423. Doxefazepam. *7-Chloro-5-(2-fluorophenyl)-1,3-dihydro-3-hydroxy-1-(2-hydroxyethyl)-2H-1,4-benzodiazepin-2-one;* N-1-hydroxyethyl-3-hydroxyflurazepam; SAS-643; Doxans. C$_{17}$H$_{14}$ClFN$_2$O$_3$; mol wt 358.76. C 58.55%, H 4.05%, Cl 10.16%, F 5.45%, N 8.03%, O 13.76%. Prepn by condensation of the benzodiazepine N-oxide with 2-bromoethanol or 2-bromoethyl acetate: G. F. Tamagnone *et al., Arzneimittel-Forsch.* **25**, 720 (1975); *eidem, J. Pharm. Pharmacol.* **26**, 566 (1974); F. De Marchi, G. F. Tamagnone, **Ger.** pat. **2,338,058** (1974 to Schiapparelli), *C.A.* **80**, 121016v (1974). Comparison with flurazepam, *q.v.*, and pharmacology, toxicity studies: M. Babbini *et al., Arzneimittel-Forsch.* **25**, 1294 (1975). Structure-activity studies: *eidem, Pharmacol. Res. Commun.* **7**, 337 (1975). GC determn in blood and tissue: F. Marcucci *et al., J. Chromatog.* **198**, 180 (1980); in plasma, urine: S. Mardente *et al., Therap. Drug Monit.* **3**, 351 (1981). Quantitative EEG, behavioral effects: W. G. Sanita *et al., ibid.* 341. Effect on sleep: G. Rodriguez *et al., Neuropsychobiology* **11**, 133 (1984).

CH$_2$CH$_2$OH

Crystals from methylene chloride/petr ether, mp 138-140°. LD$_{50}$ in rats, mice (mg/kg): 2550, > 1500 orally; 586, 774 i.p. (Babbini).
THERAP CAT: Sedative, hypnotic.

3424. Doxenitoin. *5,5-Diphenyl-4-imidazolidinone;* 5,5-diphenyltetrahydroglyoxalin-4-one; tetrahydro-4-oxo-5,5-diphenylglyoxaline; SKF 2599; Glior. C$_{15}$H$_{14}$N$_2$O; mol wt 238.28. C 75.60%, H 5.92%, N 11.76%, O 6.71%. Prepd by the reduction of diphenylthiohydantoin with sodium: Biltz, Seydel, *Ann.* **391**, 218 (1912); with Raney nickel: Whalley *et al., J. Am. Chem. Soc.* **77**, 745 (1955); Goodman, **U.S.** pat. **2,744,852** (1956).

Stout plates from methanol, mp 183° (Goodman); crystals from alc, mp 185.5-186.5° (Biltz). Moderately sol in glacial acetic acid; less sol in alc, ethyl acetate, benzene, chloroform. Practically insol in water, ligroin.
Hydrochloride, C$_{15}$H$_{15}$ClN$_2$O, dec 205-206°.
Picrate, C$_{21}$H$_{17}$N$_5$O$_8$, prisms with 1H$_2$O or needles with 2C$_6$H$_6$. mp 158°. Very sol in glacial acetic acid, ethyl acetate; sol in alc, hot benzene, chloroform.
THERAP CAT: Anticonvulsant.

3425. Doxepin. *3-Dibenz[b,e]oxepin-11(6H)-ylidene-N,-N-dimethyl-1-propanamine; N,N-dimethyldibenz[b,e]oxepin-Δ$^{11(6H),\gamma}$-propylamine;* 11-(3-dimethylaminopropylidene)-6,11-dihydrodibenz[b,e]oxepin; P 3693A. C$_{19}$H$_{21}$NO;

mol wt 279.37. C 81.68%, H 7.58%, N 5.01%, O 5.73%. Prepn of mixture of cis- and trans-isomers: K. Stach, F. Bickelhaupt, *Monatsh.* **93**, 896 (1962); F. Bickelhaupt *et al., ibid.* **95**, 485 (1964); **Neth. pat. Appl. 6,407,758;** K. Stach, **U.S. pat. 3,438,981** (1965, 1969 both to Boehringer Mann.); and separation and activity of isomers: B. M. Bloom, J. R. Tretter, **Belg. pat. 641,498;** *eidem,* **U.S. pat. 3,420,851** (1964, 1969 both to Pfizer). Pharmacology: A. Ribbentrop, W. Schaumann, *Arzneimittel-Forsch.* **15**, 863 (1965). Metabolism in animals: D. C. Hobbs, *Biochem. Pharmacol.* **18**, 1941 (1969). Determn in plasma by GC/MS: T. P. Davis *et al., J. Chromatog.* **273**, 436 (1983); by HPLC: T. Emm, L. J. Lesko, *ibid.* **419**, 445 (1987). Clinical study in depression: K. Rickels *et al., Arch. Gen. Psychiat.* **42**, 134 (1985). Comparative clinical trial with cimetidine, *q.v.,* in treatment of ulcer: R. K. Shrivastava *et al., Clin. Ther.* **7**, 181 (1985). Review of pharmacology and therapeutic efficacy: R. M. Pinder *et al., Drugs* **13**, 161 (1977).

Oily liq consisting of a mixture of cis- and trans-isomers. $bp_{0.03}$ 154-157°, $bp_{0.2}$ 260-270°. LD_{50} in mice, rats (mg/kg): 26, 16 i.v.; 79, 182 i.p.; 135, 147 orally (Ribbentrop, Schaumann).

Hydrochloride, $C_{19}H_{22}ClNO$, *Adapin, Aponal, Curatin, Novoxapin, Quitaxon, Sinequan.* Crystals, mp 184-186°, 188-189°. (A *cis-trans* mixture of approx 1:5).

Maleate, crystals, mp 161-164°, 168-169°.

trans-Form hydrochloride, mp 192-193°.

cis-Form hydrochloride, *P-4599, cidoxepin.* Crystals, mp 209-210°.5.

THERAP CAT: Antidepressant.

THERAP CAT (VET): Antipruritic.

3426. Doxifluridine. *5'-Deoxy-5-fluorouridine;* 1-(β-D-5'-deoxyribofuranosyl)-5-fluorouracil; 5'-DFUR; 5'-dFUrd; Ro 21-9738; Flutron; Furtulon. $C_9H_{11}FN_2O_5$; mol wt 246.20. C 43.91%, H 4.50%, F 7.72%, N 11.38%, O 32.49%. Fluorinated pyrimidine nucleoside with cytostatic activity. Prepn: A. F. Cook, **U.S. pat. 4,071,680** (1978 to Hoffmann-La Roche); H. Hrebabecky, J. Beranek, *Nucleic Acids Res.* **5**, 1029 (1978); A. F. Cook *et al., J. Med. Chem.* **22**, 1330 (1979). Stereospecific synthesis: J. Kiss *et al., Helv. Chim. Acta* **65**, 1522 (1982). Mechanism of action studies: H.-R. Hartmann, A. Matter, *Cancer Res.* **42**, 2412 (1982); R. D. Armstrong *et al., Cancer Chemother. Pharmacol.* **11**, 102 (1983). Kinetics and metabolism in humans: J.-P. Sommadossi *et al., Cancer Res.* **43**, 930 (1983). Clinical trials in colorectal carcinoma: R. Abele *et al., J. Clin. Oncol.* **1**, 750 (1983); S. D. Fossa *et al., Cancer Chemother. Pharmacol.* **15**, 161 (1985). Series of articles on animal toxicology: *Yakuri to Chiryo* **13**, Suppl. 2, 221-430 (1985); acute toxicity: M. Shimizu *et al., ibid.* **209**, *C.A.* **104**, 14673z-14678e (1986). Evaluation of neurotoxicity in humans: M. S. Heier, S. D. Fossa, *Acta Neurol. Scand.* **73**, 449 (1986).

Crystals from ethyl acetate, mp 189-190° (Cook). Also reported as crystals from 2-propanol, mp 186-188° (Hrebabecky, Beranek); needles from methanol + ethyl acetate, mp 192-193° (Kiss). pKa 7.4. $[\alpha]_D^{25}$ +18.4° (c = 0.419 in water). uv max (in methanol): 268-269 nm (ϵ 8550). LD_{50} (14

day) in mice or rats (mg/kg): > 1000 i.v.: > 2000 s.c.; in male, female mice, male, female rats (mg/kg): > 5000, > 5000, 3471, 3390 orally (Shimizu).

THERAP CAT: Antineoplastic.

3427. Doxofylline. *7-(1,3-Dioxolan-2-ylmethyl)-3,7-dihydro-1,3-dimethyl-1H-purine-2,6-dione; 7-(1,3-dioxolan-2-ylmethyl)theophylline;* 2-(7'-theophyllinemethyl)-1,3-dioxolane; doxophylline; dioxyfilline; ABC 12/3; Ansimar; Ventax. $C_{11}H_{14}N_4O_4$; mol wt 266.26. C 49.62%, H 5.30%, N 21.04%, O 24.04%. Prepn: U. Avico *et al., Farmaco Ed. Sci.* **17**, 73 (1962). Use as bronchodilator: **Belg. pat. 868,556;** J. S. Franzone, T. Tamietto, **U.S. pat. 4,187,308** (1978, 1980 to Istituto Biologico Chemioterapico ABC). Pharmacology: J. S. Franzone *et al., Farmaco Ed. Sci.* **36**, 201 (1981). Pharmacodynamics and toxicity in rats: J. S. Franzone *et al., ibid.* 220. HPLC determn in pharmaceutical compositions: C. Badini *et al., Farmaco Ed. Prat.* **37**, 320 (1982). Clinical trial in obstructive pneumopathy: C. Bucca *et al., Int. J. Clin. Pharm. Res.* **11**, Suppl 1, 101 (1982).

Crystals, mp 144-145.5°. Sol in water, acetone, ethyl acetate, benzene, chloroform, dioxane, hot methanol or hot ethanol. Practically insol in ethyl ether or petr ether. LD_{50} in mice (mg/kg): 841 orally; 215.6 i.v.; in rats: 1022.4 orally, 445 i.p. (Franzone).

THERAP CAT: Bronchodilator.

3428. Doxorubicin. *(8S-cis)-10-[(3-Amino-2,3,6-trideoxy-α-L-lyxo-hexopyranosyl)oxy]-7,8,9,10-tetrahydro-6,8,11-trihydroxy-8-(hydroxyacetyl)-1-methoxy-5,12-naphthacenedione;* 14-hydroxydaunomycin; adriamycin (former generic name); NSC-123127; FI 106; Adriablastina. $C_{27}H_{29}NO_{11}$; mol wt 543.54. C 59.66%, H 5.38%, N 2.57%, O 32.38%. Anthracycline antibiotic isolated from *Streptomyces peucetius* var *caesius:* F. Arcamone *et al.,* **S. Afr. pat. 68 02378** and **U.S. pat. 3,590,028** (1968 and 1971 to Farmitalia); *eidem, Biotechnol. Bioeng.* **11**, 1101 (1969). Synthesis of derivs: F. Arcamone *et al.,* **Ger. pat. 1,917,874** (1969 to Farmitalia), *C.A.* **73**, 45799r (1970). Structural studies: F. Arcamone *et al., Tetrahedron Letters* **1969**, 1007. Synthesis from daunomycin, *q.v.:* *eidem, Chim. Ind. (Milan)* **51**, 834 (1969); *see also:* E. M. Acton *et al., J. Med. Chem.* **17**, 659 (1974); from 7-deoxydaunomycinone: T. H. Smith *et al.,* **U.S. pat. 4,012,448** (1977 to Stanford Res. Inst.). Biochemical comparison with daunomycin: Wang *et al., Proc. Am. Assoc. Cancer Res.* **12**, No. 62, 77 (1971). In acid environment doxorubicin (adriamycin) breaks up into a water-insol aglycone, *adriamycinone* ($C_{21}H_{18}O_9$), and a water-sol basic, reducing aminosugar, *daunosamine* ($C_6H_3NO_3$), 3-amino-2,3,6-trideoxy-L-lyxohexose: A. Di Marco *et al., Cancer Chemother. Rep.* (part 1) **53**, 33 (1969). Total synthesis of adriamycinone: F. Suzuki *et al., J. Am. Chem. Soc.* **100**, 2272 (1978); regiospecific synthesis: J. S. Swenton, P. W. Reynolds, *ibid.* 6188; of daunosamine: P. M. Wovkulich, M. R. Uskonovic, *Tetrahedron* **41**, 3455 (1985). Pharmacokinetic and chemotherapeutic studies: E. Arena *et al., Arzneimittel-Forsch.* **21**, 1258 (1971). Purification: E. Oppici *et al.,* **Belg. pat. 898,506;** *eidem,* **Brit. pat. Appl. 2,133,005** (both 1984 to Farmitalia). As modulator of immune response in mice: E. Mihich, M. J. Ehrke, *Transplant. Proc.* **16**, 499 (1984). Doxorubicin's cytotoxicity appears to be due to its ability to intercalate with DNA, interact with plasma membranes and take part in oxidation-reduction reactions: T. R. Tritton, G. Yee, *Science* **217**, 248 (1982); H. Simpkins *et al., Cancer Res.* **44**, 613 (1984); R. S. Youngman, E. F. Elstner, *Arch. Biochem. Biophys.* **231**, 424 (1984). In treatment of cancer of the bladder: M. Pavone-Macaluso *et al., Urology* **23**, 40 (1984); breast: D. C. Tormey *et al., Am. J.*

Clin. Oncol. **7**, 231 (1984); prostate: H. Scher *et al., J. Urol.* **131**, 1099 (1984). Toxicology: C. Bertazzoli *et al., Experientia* **26**, 389 (1970); *eidem, Toxicol. Appl. Pharmacol.* **21**, 287 (1972); R. D. Olson *et al., Life Sci.* **29**, 1393 (1981). Review of properties, biosynthesis, fermentation: R. J. White, R. M. Stroshane, *Drugs Pharm. Sci.* **22**, 569-594 (1984); of efficacy: H. L. Davis, T. E. Davis, *Cancer Treat. Rep.* **63**, 809-815 (1979). *Review:* R. H. Blum, S. K. Carter, *Ann. Int. Med.* **80**, 249-259 (1974); G. Aubel-Sadron, B. Londos-Gagliardi, *Biochimie* **66**, 333-352 (1984). Comprehensive description: A. Vigevani, M. J. Williamson in *Analytical Profiles of Drug Substances* vol. 9, K. Florey, Ed. (Academic Press, New York, 1980) pp 245-274. Book: *Doxorubicin,* F. Arcamone, Ed. (Academic Press, New York, 1981) 354 pp.

adriamycinone

daunosamine

mp 229-231°.

Hydrochloride, $C_{27}H_{30}ClNO_{11}$, *Adriacin, Adriablastina, Adriamycin.* Orange-red colored thin needles, mp 204-205° (dec). $[\alpha]_D^{20}$ +248° (c = 0.1 in methanol). Absorption max (methanol): 233, 252, 288, 479, 496, 529 nm. Sol in water, methanol, aq alcohols. Practically insol in acetone, benzene, chloroform, ethyl ether and petroleum ether. Aq solns are yellow-orange at acid pHs, orange-red at neutral pHs and violet-blue at pH > 9. Aq solns unchanged after one month at 5° but unstable at higher temperatures or at either acid or alkaline pHs. LD_{50} i.v. in mice: 21.1 mg/kg (Bertazzoli, 1970).

Note: This substance may reasonably be anticipated to be a carcinogen: *Fourth Annual Report on Carcinogens* (NTP 85-002, 1985) p 17.

THERAP CAT: Antineoplastic.

3429. Doxycycline. *4-(Dimethylamino)-1,4,4a,5,5a,6,-11,12a-octahydro-3,5,10,12,12a-pentahydroxy-6-methyl-1,11-dioxo-2-naphthacenecarboxamide monohydrate;* α-6-deoxy-5-hydroxytetracycline monohydrate; α-6-deoxyoxytetracycline monohydrate; 5-hydroxy-α-6-deoxytetracycline monohydrate; GS-3065; Azudoxat; Doxitard; Doxy-Puren; Investin; Liviatin; Nordox; Spanor; Vibramycin; Vibravenös. $C_{22}H_{24}N_2O_8 \cdot H_2O$; mol wt 462.46. C 57.14%, H 5.67%, N 6.05%, O 31.14%. Prepn of 6-deoxytetracyclines: Wittenau *et al., J. Am. Chem. Soc.* **84**, 2645 (1962); Stephens *et al., ibid.* **85**, 2643 (1963); Blackwood *et al.,* U.S. pat. **3,200,149** (1965 to Pfizer). Biological properties: English, *Proc. Soc. Exp. Biol. Med.* **122**, 1107 (1966). Pharmacology: Fabre, *Chemotherapia* **11**, 73 (1966); Gibaldi, *ibid.* **12**, 265 (1967). Prepn, separation and configuration of 6α- and 6β-epimers: Wittenau *et al., J. Am. Chem. Soc.* **84**, 2645 (1962); Stephens *et al., ibid.* **85**, 2643 (1963). ¹H-NMR study: Wittenau, Blackwood, *J. Org. Chem.* **31**, 613 (1966). Toxicity of hydrochloride: Goldenthal, *Toxicol. Appl. Pharmacol.* **18**, 185 (1971). Clinical trial in prophylaxis of leptospirosis: E. T. Takafuji *et al., N. Engl. J. Med.* **310**, 497 (1984). *Review:* C. Edwards in *Pharmacological and Biochemical Properties of Drug Substances* vol. 2, M. E. Goldberg, Ed. (Am. Pharm. Assoc., Washington, DC, 1979) pp 305-332.

·H_2O

Hydrochloride, $C_{22}H_{25}ClN_2O_8$, *doxycycline hyclate, Diocimex, Doryx, Doxatet, Doxigalumicina, Doxy-II (caps), Doxylar, Doxy-Tablinen, Doxytem, duradoxal, Ecodox, Granudoxy, Hydramycin, Liomycin, Mespafin, Midoxin, Nivocilin, Novadox, Retens, Roximycin, Samecin, Sigadoxin, Tanamicin, Tecacin, Tetradox, Vibradox, Vibramycin Hyclate, Vibra-Tabs, Zadorin.* Light yellow powder which crystallizes from ethanol + HCl as the hemihydrate hemialcoholate. Chars without melting at about 201°. $[\alpha]_D^{25}$ −110° (c = 1 in 0.01N methanolic HCl). uv max (0.01N methanolic HCl): 267, 351 nm (log ε 4.24, 4.12). Sol in water. The alcohol and water of crystallization are lost by drying at 100° under reduced pressure. More active biologically than the corresponding 6β-epimer hydrochloride, Wittenau *et al., loc. cit.* LD_{50} i.p. in rats: 262 mg/kg (Goldenthal).

THERAP CAT: Antibacterial.

3430. Doxylamine. *N,N-Dimethyl-2-[1-phenyl-1-(2-pyridinyl)ethoxy]ethanamine; 2-[α-(2-dimethylaminoethoxy)-α-methylbenzyl]pyridine;* phenyl-2-pyridylmethyl-β-N,N-dimethylaminoethyl ether; 2-dimethylaminoethoxyphenyl-methyl-2-picoline. $C_{17}H_{22}N_2O$; mol wt 270.38. C 75.52%, H 8.20%, N 10.36%, O 5.92%. Prepd from phenyl-2-pyridylmethylcarbinol and β-N,N-dimethylaminoethyl chloride in the presence of sodamide in xylene: Sperber *et al., J. Am. Chem. Soc.* **71**, 887 (1949). GC determn: H. C. Thompson *et al., J. Chromatog. Sci.* **20**, 373 (1982). Pharmacology, antihistaminic activity: B. B. Brown, H. Werner, *J. Lab. Clin. Med.* **33**, 325 (1948). Hypnotic efficacy: F. Sjöqvist, L. Lasagna, *Clin. Pharmacol. Ther.* **8**, 48 (1967). *Review:* T. J. Haley, *Dangerous Prop. Ind. Mater. Rep.* **2**, 17 (1982).

Liquid, $bp_{0.5}$ 137-141°. Sol in acids. Slightly volatile, darkens on exposure to light.

Succinate, $C_{21}H_{28}N_2O_5$, *mereprine, Alsadorm, Decapryn succinate, Gittalun, Hoggar N, Sedaplus, Unisom.* Crystals, mp 100-104°, sol in water. One gram dissolves in 1 ml water, 2 ml alcohol, 2 ml chloroform. Slightly sol in benzene and ether. pH (1% aq soln): 4.9 to 5.1. LD_{50} in mice, rabbits (mg/kg): 470, 250 orally; 62, 49 i.v.; in mice, male rats, female rats (mg/kg): 460, 440, 445 s.c. (Brown, Werner).

Note: A combination with pyridoxine hydrochloride, *q.v.,* has been marketed as *Bendectin* for nausea of pregnancy. Prior to 1976, Bendectin also contained dicyclomine, *q.v.* Discussion of Bendectin and the issue of teratogenicity: J. F. Cordero *et al., J. Am. Med. Assoc.* **245**, 2307 (1981); corr. *ibid.* **247**, 2234 (1982); L. B. Holmes, *Teratology* **27**, 277 (1983); L. J. Sheffield, R. Batagol, *Med. J. Aust.* **143**, 143 (1985).

THERAP CAT: Antihistaminic. Hypnotic.
THERAP CAT (VET): Antihistaminic.

3431. Dragon's Blood. A resinous secretion found on the fruits of *Daemonorops propinquus* Becc., *D. draco* Blume, and probably other species of *Daemonorops, Palmae* (Rattan palms). *Habit.* Sumatra, Borneo, India. *Constit.* About 55% of a red resin contg about 12-15% of bright-yellow, amorphous dracoresene; 2-3% white amorphous dracoalban. Isoln of the main coloring matter, dracorubin: Brockmann, Haase, *Ber.* **69**, 1950 (1936). Chemical studies of resin pigments: Olaniyi *et al., J. Chem. Soc. Perkin Trans. I* **1973**, 179.

Red sticks, pieces, or cakes; vitreous fracture; makes a bright-crimson powder; odorless and almost tasteless. mp at about 120° with sublimation of some benzoic acid. Insol in water; sol in alcohol.

USE: For coloring lacquers and varnishes; occasionally for coloring plasters; in photoengraving on zinc to protect metal parts against etching.

3432. Drazoxolon. *3-Methyl-4-[(2-chlorophenyl)hydrazone]-4,5-isoxazoledione;* 4-(2-chlorophenylhydrazono)-3-methyl-5(4H)-isoxazolone; 3-methyl-4-(o-chlorophenylhydrazono)-5-isoxazolone; PP 781; Ganocide; Mil-Col; Saisan.

$C_{10}H_8ClN_3O_2$; mol wt 237.65.   C 50.54%, H 3.39%, Cl 14.92%, N 17.68%, O 13.47%.   Prepn: Green, Roberts, **Brit. pat. 1,049,103** (1966 to ICI).   Structure: Summers, Shields, *Chem. & Ind. (London)* **1964**, 1264.   Soil degradation studies: Anderson, Horsgood, *Soil Biol. Biochem.* **3**, 271 (1971).   Hydrolysis: Lehtonen *et al., Pestic. Sci.* **3**, 357 (1972).   Toxicology: D. G. Clark, T. F. McElligott, *Food Cosmet. Toxicol.* **7**, 481 (1969).   *Review:* Yuen, *Anal. Methods Pestic. Plant Growth Regul.* **7**, 665-673 (1973).

Yellow crystals, possessing faint odor, from methanol-benzene, mp 168°.   Practically insol in water, acids, and aliphatic hydrocarbons.   Sol in alkali, aromatic hydrocarbons (4%), chloroform (about 10%), ethanol (1%), ketones (5%).   Stable to dilute acids and forms stable salts with alkalies. LD_{50} orally in female rats, mice: 126, 129 mg/kg (Clark, McElligott).

USE: Fungicide.

3433. Drimenin.   *5,5a,6,7,8,9,9a,9b-Octahydro-6,6,9a-trimethylnaphtho[1,2-c]furan-1(3H)-one.* $C_{15}H_{22}O_2$; mol wt 234.33.   C 76.88%, H 9.46%, O 13.66%.   From bark of *Drimys winteri* Forst., *Magnoliaceae:* Appel, Dohr, *Scientia* **25**, 137 (1958); *C.A.* **54**, 4663f (1960).   Structure and stereochemistry: Appel *et al., J. Chem. Soc.* **1960**, 4685.   Synthesis: Wenkert, Strike, *J. Am. Chem. Soc.* **86**, 2044 (1964); Yamagawa, *et al., Synthesis* **1970**, 257; M. Jallali-Naini *et al., Tetrahedron Letters* **22**, 2995 (1981).

Crystals from methanol and sublimed at 110°/0.1 mm, mp 133°. $[\alpha]_D$ −42° (c = 0.76 in benzene); $[\alpha]_D^{25}$ −35.8° (chloroform).   Soluble in organic solvents; practically insol in water, acids, bases.

3434. Drocarbil.   *1,2,5,6-Tetrahydro-1-methyl-3-pyridinecarboxylic acid methyl ester, mono[[3-(acetylamino)-4-hydroxyphenyl]arsonate]; 1,2,5,6-tetrahydro-1-methylnicotinic acid methyl ester, compound with N-acetyl-4-hydroxy-m-arsanilic acid;* arecoline-acetarsone; arecoline-acetarsol; acetarsone arecoline salt; 3-acetamido-4-hydroxyphenylarsonic acid arecoline salt; Cestarsol; Larumen; Nemural; Tenoban. $C_{16}H_{23}AsN_2O_7$; mol wt 430.29.   C 44.66%, H 5.39%, As 17.41%, N 6.51%, O 26.03%.   The commercial product contains 34 to 36.5% arecoline and 64 to 67% acetarsone.

White to pale-yellow powder.   Stable when kept dry. Freely sol in water.

THERAP CAT (VET): Anthelmintic.

3435. Drofenine.   *α-Cyclohexylbenzeneacetic acid 2-(diethylamino)ethyl ester; α-phenylcyclohexaneacetic acid 2-(diethylamino)ethyl ester;* 2-diethylaminoethyl α-phenylcyclohexaneacetate; hexahydroadiphenine. $C_{20}H_{31}NO_2$; mol wt 317.48.   C 75.66%, H 9.84%, N 4.41%, O 10.08%.   Prepn: Miescher, Hoffman, **U.S. pats. 2,265,184-5** (both 1941 to

Ciba).   Pharmacology: Fleisch *et al., Arzneimittel-Forsch.* **11**, 1119 (1961).

$bp_{0.15}$ 158°.

Hydrochloride, $C_{20}H_{32}ClNO_2$, **Trasentine-A, Trasentine 6-H.**   Crystals from alc + petr ether, mp 145-147°.   Freely sol in water; very sparingly sol in alc, ether.   A 5% aq soln is neutral to litmus. LD_{50} in mice: 65.6 mg/kg i.v., Fleisch *et al., loc. cit.*

THERAP CAT: Antispasmodic.

3436. Dromostanolone Propionate.   *2α-Methyl-17β-(1-oxopropoxy)-5α-androstan-3-one; 17β-hydroxy-2α-methylandrostan-3-one propionate;* 2α-methylandrostan-17β-ol-3-one propionate; 2α-methyldihydrotestosterone propionate; drostanolone propionate; Drolban; Emdisterone; Masterid; Masteril; Masterone; Permastril. $C_{23}H_{36}O_3$; mol wt 360.52.   C 76.62%, H 10.07%, O 13.31%.   Prepn: Ringold *et al., J. Am. Chem. Soc.* **81**, 427 (1959); Ringold, Rosenkranz, **U.S. pat. 2,908,693** (1959 to Syntex).

Crystals from hexane, mp 126-130°. $[\alpha]_D$ +24°.

THERAP CAT: Antineoplastic.

3437. Droperidol.   *1-[1-[4-(4-Fluorophenyl)-4-oxobutyl]-1,2,3,6-tetrahydro-4-pyridinyl]-1,3-dihydro-2H-benzimidazol-2-one;* 1-[1-[3-(p-fluorobenzoyl)propyl]-1,2,3,6-tetrahydro-4-pyridyl]-2-benzimidazolinone; 1-[1-[4-(p-fluorophenyl)-4-oxobutyl]-1,2,3,6-tetrahydro-4-pyridyl]-2-benzimidazolinone; dehydrobenzperidol; R 4749; Dridol; Droleptan; Inapsine. $C_{22}H_{22}FN_3O_2$; mol wt 379.44.   C 69.64%, H 5.85%, F 5.01%, N 11.07%, O 8.43%.   Prepn: Janssen, Gardocki, **U.S. pat. 3,141,823** (1964 to Janssen).   Pharmacology: J. Yelnosky *et al., Toxicol. Appl. Pharmacol.* **6**, 37 (1964).   Comprehensive description: C. A. Janicki, R. K. Gilpin in *Analytical Profiles of Drug Substances* vol. 7, K. Florey, Ed. (Academic Press, NewYork, 1978) pp 171-192.

Hydrate, crystals, mp 145-146.5°.   uv max (9:1, 0.1M HCl:methanol): 245, 280 nm (ε 15600, 7500).   Soly at 25° (g/100 ml): chloroform 40; DMF 17; benzene 0.55; methanol 0.41; ethanol 0.34; ether 0.24; 0.1M HCl 0.15; water <0.001.   pKa 7.64.   Heat and light sensitive. LD_{50} in mice (mg/kg): 125 s.c.; 43 i.v., J. Yelnosky *et al., loc. cit.*

Note: Ingredient of *Thalamonal, Innovar.*

THERAP CAT: Antipsychotic.

3438. Droprenilamine.   *N-(2-Cyclohexyl-1-methylethyl)-γ-phenylbenzenepropanamine; (±)-N-(3,3-diphenylpropyl)-α-methylcyclohexaneethylamine;* droprenilamine. $C_{24}H_{33}N$;

mol wt 335.53. C 85.91%, H 9.91%, N 4.18%. Vasodilator with antiarrhythmic and hypotensive activity. Prepn: M. Carissimi *et al.*, **Ger.** pat. **2,521,113** (1976 to Maggioni), *C.A.* **84**, 164388t (1976). Synthesis and pharmacological study: G. Carenini *et al.*, *Arzneimittel-Forsch.* **26**, 2127 (1976). Hemodynamic and metabolic effects: R. J. Marshall, J. R. Parratt, *Brit. J. Pharmacol.* **59**, 311 (1977). Clinical evaluation: F. Rengo *et al.*, *G. Ital. Cardiol.* **10**, 473 (1980).

Hydrochloride, $C_{24}H_{34}ClN$, *MG-8926, Valcor.* Cryst from isopropanol, mp 175-176°. LD_{50} in mice (mg/kg): 2850 orally; 68 i.p., G. Carenini *et al., loc. cit.*

THERAP CAT: Vasodilator (coronary).

3439. Dropropizine. *3-(4-Phenyl-1-piperazinyl)-1,2-propanediol;* 1-phenyl-4-(2,3-dihydroxypropyl)piperazine; 1-(2,3-dihydroxypropyl)-4-phenylpiperazine; 1-phenyl-4-(2,3-dihydroxypropyl)diethylenediamine; UCB 1967; Ribex. $C_{13}H_{20}N_2O_2$; mol wt 236.31. C 66.07%, H 8.53%, N 11.85%, O 13.54%. Cough suppressive phenylpiperazine derivative. Prepn: H. G. Morren, **Belg.** pat. **601,394;** *eidem,* **U.S.** pat. **3,163,649** (1961, 1964 to UCB); H. Howell *et al., J. Org. Chem.* **27**, 1709 (1962); J. Bourdais, *Bull. Soc. Chim. France* **1968**, 3246. Prepn of enantiomers: M. Borsa *et al.*, **Eur.** pat. **Appl. 147,847;** *eidem,* **U.S.** pat. **4,699,911** (1985, 1987 both to Dompé). Pharmacology: K. Cartwright, J. L. Paterson, *J. Pharm. Pharmacol.* **23**, Suppl., 2475 (1971). Controlled clinical trials: G. C. Moreo *et al., Gazz. Med. Ital.* **140**, 409 (1981); A. Ravetta, M. Ravetta, *ibid.* **141**, 531 (1982). Chronic oral toxicity: P. R. B. Noel, *Arzneimittel-Forsch.* **19**, 1246 (1969).

Crystals from benzene, mp 105° (Morren); also reported as mp 108° (Bourdais). LD_{50} in rats (mg/kg): 200 i.v., 750 orally (Morren).

S-Form, *levodropropizine, Danka, Levotuss.* White solid from acetone, mp 98-100°. $[\alpha]_D^{25} -10°$ (c = 1.0 in ethanol). LD_{50} in mice, rats (mg/kg): 1287.2, 886.6 orally; 408.0, 401.3 i.p. (Borsa).

R-Form, crystals from acetone, mp 104-105°. $[\alpha]_D^{25} +9.7°$ (c = 1.0 in ethanol). LD_{50} in mice, rats (mg/kg): 871.7, 721.3 orally; 319.2, 363.4 i.p. (Borsa).

THERAP CAT: Antitussive.

3440. Drosera. Common sundew; round-leafed sundew; youthwort. Air-dried, flowering plant, *Drosera rotundifolia* L., frequently mixed with closely allied species *D. anglica* Hudson and *D. longifolia* L., *Droseraceae. Habit.* Europe, Asia, North America, south to Florida. *Constit.* Malic and citric acids, resin, tannin.

3441. Drosophilin A. *2,3,5,6-Tetrachloro-4-methoxyphenol; p*-methoxytetrachlorophenol. $C_7H_4Cl_4O_2$; mol wt 261.93. C 32.10%, H 1.54%, Cl 54.15%, O 12.22%. Antibiotic substance isolated from the basidiomycete *Drosophila subatrata* (Batsch ex Fr.) Quel., grown in a corn-steep medium: Kavanagh, Hervey, *Proc. Nat. Acad. Sci. USA* **38**, 555 (1952). Prepn from tetrachloro-*p*-anisidine: Bures, Hutter, *Casopis Ceskoslov. Lekarnictva* **11**, 29, 57 (1931); *C.A.* **25**, 5153; *Chem. Zentr.* **1931**, II, 225; from 1,4-dimethoxytetrachlorobenzene: Anchel, *J. Am. Chem. Soc.* **74**, 2943 (1952).

Crystals from hexane or dil alcohol, mp 114°. uv max:

301 nm. Sparingly sol in water; sol in hexane, benzene, acetone, alcohol, chloroform, ether.

3442. Drotebanol. *3,4-Dimethoxy-17-methylmorphinan-6,14-diol;* 6β,14-dihydroxy-3,4-dimethoxy-N-methyl-morphinan; 14-hydroxydihydro-6β-thebainol 4-methyl ether; dihydro-14-hydroxy-4-*O*-methyl-6β-thebainol; dihydro-14-hydroxy-6β-thebainol 4-methyl ether; oxymethebanol; Metebanyl. $C_{19}H_{27}NO_4$; mol wt 333.41. C 68.44%, H 8.16%, N 4.20%, O 19.20%. Prepn from thebaine: **Japan.** pat. **26,726('64)** (to Sankyo), *C.A.* **62**, 11869b (1965). Metabolism: H. Shindo *et al., Yakugaku Zasshi* **90**, 36 (1970), *C.A.* **72**, 88520q (1970). Pharmacology: S. Kobayashi *et al., Arzneimittel-Forsch.* **20**, 43 (1970). Dependence potential in rhesus monkeys: T. Yanagita *et al., Bull. Narc.* **29**, 33 (1977).

Minute crystals, very slightly hygroscopic, slightly bitter taste. mp 165.5°-166.5°. Slightly sol in water. Freely sol in ethanol, chloroform. Sol in acetone, benzene, toluene. Very slightly sol in ether. LD_{50} in mice (mg/kg): 1300 orally, 1150 s.c., 91 i.v. (Kobayashi).

Caution: May be habit forming. This is a controlled substance (opium derivative) listed in the U.S. Code of Federal Regulations, Title 21 Part 1308.11 (1985).

THERAP CAT: Antitussive.

3443. Droxicam. *5-Methyl-3-(2-pyridinyl)-2H,5H-1,3-oxazino[5,6-c][1,2]benzothiazine-2,4(3H)-dione 6,6-dioxide;* E 3128. $C_{16}H_{11}N_3O_5S$; mol wt 357.34. C 53.78%, H 3.10%, N 11.76%, O 22.39%, S 8.97%. Prodrug of piroxicam, *q.v.* Prepn: J. E. Soler, **Fr.** pat. **2,528,433;** *idem,* **U.S.** pat. **4,563,452** (1983, 1986 both to Provesan SA). Spectroscopic structural analysis: J. Frigola, *J. Chem. Soc. Perkins Trans. II,* **1988**, 241. Pharmacology: A. J. Farré *et al., Methods Find. Exp. Clin. Pharmacol.* **8**, 407 (1986). Metabolism and pharmacokinetics in animals: A. Esteve *et al., ibid.* 423. Inhibition of platelet aggregation: J. Esteve *et al., ibid.* **9**, 209 (1987). Evaluation of ulcerogenic effects in comparison with piroxicam: G. Palacios *et al., ibid.* 353. Clinical pharmacology: L. Martinez *et al., ibid.* **10**, 729 (1988). Brief review of pharmacology: J. Esteve *et al., Gen. Pharmacol.* **19**, 49-54 (1988).

Crystals from acetone. mp 259-261°. LD_{50} orally in male, female mice, and male, female rats: 6192, 8841, 1434, 1994 mg/kg (Soler, 1986).

THERAP CAT: Anti-inflammatory.

3444. Droxidopa. *threo-β,3-Dihydroxy-L-tyrosine;* L-*threo*-3-(3,4-dihydroxyphenyl)serine; (−)-(2S,3R)-2-amino-3-hydroxy-3-(3,4-dihydroxyphenyl)propionic acid; *threo*-dopaserine; L-*threo*-DOPS; L-DOPS; SM-5688; Dops. $C_9H_{11}NO_5$; mol wt 213.19. C 50.71%, H 5.20%, N 6.57%, O 37.52%. Synthetic amino acid precursor of norepinephrine, *q.v.* Prepn of racemate: K. W. Rosenmund, H. Dornsaft, *Ber.* **52B**, 1734 (1919). Separation and resolution of diastereomers: B. Hegedüs *et al., Helv. Chim. Acta* **58**, 147 (1975); B. Hegedüs, A. Krasso, **U.S.** pat. **3,920,728** (1975 to Hoffmann-La Roche). Improved process for production: N.

Ohashi *et al.*, U.S. pat. **4,319,040** (1982 to Sumitomo). Pharmacology of stereoisomers: G. Bartholini *et al.*, *J. Pharmacol. Exp. Ther.* **193**, 523 (1975). Clinical pharmacology of L-*threo*-form and clinical evaluation in familial amyloid polyneuropathy (FAP): T. Suzuki *et al.*, *Eur. J. Clin. Pharmacol.* **17**, 429 (1980). Reversed-phase chromatography determn in plasma and urine: F. Boomsma *et al.*, *J. Chromatog.* **427**, 219 (1988). Pharmacokinetics in FAP: T. Suzuki *et al.*, *Eur. J. Clin. Pharmacol.* **23**, 463 (1982); in parkinsonism: T. Suzuki *et al.*, *Neurology* **34**, 1446 (1984). Metabolism to norepinephrine: T. Suzuki *et al.*, *Life Sci.* **36**, 435 (1985). Clinical studies in Parkinson's disease: N. Ogawa *et al.*, *J. Med.* **16**, 525 (1985); H. Narabayashi *et al.*, *Advan. Neurol.* **45**, 593 (1986).

Crystals from ethanol and ether, mp 232-235° (dec). $[\alpha]_D^{20}$ −39° (c = 1 in 1*N* aq HCl). Also cited as crystals from water and L-ascorbic acid, mp 229-232° (dec) (Ohashi). $[\alpha]_D^{20}$ −42.0° (c = 1 in 1*N* aq HCl).

THERAP CAT: Antiparkinsonian.

3445. DSIP. *Delta sleep-inducing peptide (rabbit);* delta sleep peptide; delta sleep factor. $C_{35}H_{48}N_{10}O_{15}$; mol wt 848.84. C 49.53%, H 5.70%, N 16.50%, O 28.27%. A nonapeptide that shows enhancement and induction of delta (slow-wave) and spindle EEG patterns. Its occurrence was suspected during dialysis of cerebral venous blood of rabbits during sleep induced by electrical stimulation of the thalamus: M. Monnier, L. Hösli, *Science* **146**, 796 (1964). Initial isoln: *eidem*, *Pflügers Arch.* **282**, 60 (1965). Isoln, characterization: G. A. Schoenberger *et al.*, *Experientia* **28**, 919 (1972). Amino acid sequence, synthesis of DSIP and analogs: G. A. Schoenberger, M. Monnier, *Proc. Nat. Acad. Sci. USA* **74**, 1282 (1977). Solid phase synthesis: Y. P. Shvachkin *et al.*, *Zh. Obshch. Khim.* **51**, 719 (1981), *C.A.* **95**, 43644s (1981). Rapid liquid phase synthesis: S. Nozaki, I. Muramatsu, *Bull. Chem. Soc. Japan* **55**, 2165 (1982). HPLC separation: M. Dizaroglu *et al.*, *J. Chromatog.* **237**, 417 (1982). Effect on human sleep: D. Schneider-Helmert *et al.*, *Lancet* **1**, 1256 (1981); *eidem*, *Int. J. Clin. Pharmacol. Ther. Toxicol.* **19**, 341 (1981); D. Schneider-Helmert, G. A. Schoenberger, *Experientia* **37**, 913 (1981).

```
Trp-Ala-Gly-Gly-Asp-Ala-Ser-Gly-Glu
```

3446. DTBP. *Bis(1,1-Dimethylethyl) peroxide;* di-*tert*-butyl peroxide. $C_8H_{18}O_2$; mol wt 146.22. C 65.71%, H 12.41%, O 21.88%. $(CH_3)_3COOC(CH_3)_3$.
Flammable liq; d_4^{20} 0.7940; mp −40°; bp_{284} 80°; n_D^{20} 1.3890. Flash pt (Tag open cup) 65°F (18°C). Soluble in organic solvents, in most resin monomers and in partial polymers. Soly in water about 0.01%.

USE: As polymerization catalyst.

3447. Dulcamara. Bittersweet; woody nightshade; scarlet berry. Dried stems of *Solanum dulcamara* L., *Solanaceae*. *Habit.* Europe, Western Asia, Northern Africa, natural in U.S. *Constit.* Solaniceine (about 1%), dulcamarin, dulcamaric and dulcamaretic acids.

3448. Dulcin. *(4-Ethoxyphenyl)urea;* p-phenetolcarbamide; p-phenetylurea; Sucrol; Valzin. $C_9H_{12}N_2O_2$; mol wt 180.20. C 59.98%, H 6.71%, N 15.55%, O 17.76%. Made by treating p-phenetidine with phosgene and then with ammonia: Berlinerblau, *J. Prakt. Chem.* **30**, 103 (1883); from p-phenetidine and urea: Kurzer, *Org. Syn.* coll. vol. IV, 52 (1963).

Lustrous needles; very sweet taste—about 250 times as sweet as cane sugar. mp 173-174°. Sol in 800 parts cold water, 50 parts boiling water, 25 parts alcohol.

USE: Non-nutritive sweetener.

3449. Durapatite. *Hydroxylapatite;* calcium phosphate hydroxide; calcium orthophosphate, basic; hydroxyapatite; Alveograf; Ossopan; Periograf. $3Ca_3(PO_4)_2 \cdot Ca(OH)_2$ or $Ca_{10}(PO_4)_6(OH)_2$. Also considered as pentacalcium monohydroxyorthophosphate $Ca_5(OH)(PO_4)_3$. Calcd as $Ca_{10}H_2$-$O_{26}P_6$; Ca 39.89%, H 0.20%, O 41.41%, P 18.50%. Occurs as a mineral in phosphate rock. Constitutes the mineral portion of bone. Prepn from $Ca(NO_3)_2$ and KH_2PO_4: Warington, *J. Chem. Soc.* **26**, 983 (1873); Rathje, *Ber.* **74**, 342 (1941); Hayek in *Handbook of Preparative Inorganic Chemistry*, G. Brauer, Ed. (Academic Press, 2nd ed., 1963) p 545; from calcium phosphate, dibasic: Perloff, Posner, *Inorg. Syn.* **6**, 16 (1960); from $Ca(NO_3)_2 \cdot 4H_2O$ and $(NH_4)_2PO_4$ plus NH_4OH: Hayek, Newesely, *ibid.* **7**, 63 (1963). Formation and structure of synthetic bone hydroxyapatites: A. S. Posner *et al.*, *Prog. Cryst. Growth Charact.* **3**, 3 (1980).
Hexagonal needles arranged in rosettes. Dec above 1100°. Practically insol in water, even when freshly prepd. Crystallographic data: a_0 9.425; C_0 6.935; C_0/a_0 0.736.

USE: Prosthetic aid (artificial bone and teeth).

THERAP CAT: Calcium and phosphorus supplement.

3450. Durene. *1,2,4,5-Tetramethylbenzene;* Durol. C_{10}-H_{14}; mol wt 134.21. C 89.49%, H 10.51%. Occurs in coal tar. Usually prepd from xylene and methyl chloride in the presence of $AlCl_3$: Smith, *Org. Syn.* vol. 10, 32 (1930); *cf.* Smith, Dobrovolny, *J. Am. Chem. Soc.* **48**, 1413 (1926).

Scales with camphor-like odor from alcohol. d_4^{81} 0.84. mp 80°. bp 191-193°. Sublimes and is volatile with steam. Insol in water; freely sol in alcohol, ether, benzene.

3451. Durohydroquinone. *2,3,5,6-Tetramethyl-1,4-benzenediol;* tetramethyl-*p*-hydroquinone; dihydroxydurene. $C_{10}H_{14}O_2$; mol wt 166.21. C 72.26%, H 8.49%, O 19.25%. For prepn *see* refs under Duroquinone.

Needles from alcohol. mp 233°. Begins to sinter at 220°. Sparingly sol in ether. Treatment with ferric chloride yields duroquinone.
Diacetyldurohydroquinone, needles from alc, mp 207°.

3452. Duroquinone. *2,3,5,6-Tetramethyl-2,5-cyclohexadiene-1,4-dione;* tetramethyl-*p*-benzoquinone. $C_{10}H_{12}O_2$; mol wt 164.20. C 73.14%, H 7.36%, O 19.49%. Prepn by reduction of dinitrodurene: Smith, *Org. Syn.* vol. 10, 40 (1930); Smith, Dobrovolny, *J. Am. Chem. Soc.* **48**, 1420 (1926); by condensation of 2,3-diketopentane with itself in presence of alkalies: von Pechmann, *Ber.* **21**, 1420 (1888); by the action of alkalies on 3,3-dichloropentan-2-one: Faworsky, *J. Prakt. Chem.* [2] **51**, 538 (1895).

Yellow needles from alc, mp 111-112°. Sublimes. Volatile with steam. Insol in water; sol in alcohol, benzene, ether, hot petr ether.

3453. Dyclonine. *1-(4-Butoxyphenyl)-3-(1-piperidinyl)-1-propanone; 3-piperidino-4'-butoxypropiophenone; β-piperidinoethyl-4-butoxyphenyl ketone; 4-butoxy-β-piperidinopropiophenone; 4-n-butoxy-β-(1-piperidyl)propiophenone; 4-butoxyphenyl piperidineethyl ketone; 2-(1-piperidinyl)ethyl p-butoxyphenyl ketone.* $C_{18}H_{27}NO_2$; mol wt 289.43. C 74.70%, H 9.40%, N 4.84%, O 11.05%. Prepd from p-butoxyacetophenone by condensation with formaldehyde and piperidine hydrochloride: Pofft, *Chem. Tech. (Berlin)* **4**, 241 (1952), *C.A.* **47**, 10531 (1953). *Compare* Falicain.

Hydrochloride, $C_{18}H_{28}ClNO_2$, *Dyclone, Tanaclone.* Crystals, mp 175-176°. Sol in water, alc, acetone. Phenol coefficient 3.6.

THERAP CAT: Anesthetic (topical).

3454. Dydrogesterone. *(9β,10α)-Pregna-4,6-diene-3,20-dione; 10α-pregna-4,6-diene-3,20-dione; 6-dehydro-retro-progesterone; 10α-isopregnenone; Dufaston; Duphaston; Gestatron; Gynorest; Prodel; Retrone.* $C_{21}H_{28}O_2$; mol wt 312.44. C 80.73%, H 9.03%, O 10.24%. Prepn: Westerhof, Reerink, *Rec. Trav. Chim.* **79**, 771 (1960); Rappoldt, Westerhof, *ibid.* **80**, 43 (1961).

Crystals from acetone + hexane, mp 169-170°. $[\alpha]_D^{25}$ −484.5° (chloroform). uv max: 286.5 nm (ε 26,400).
THERAP CAT: Progestogen.

3455. Dymanthine. *N,N-Dimethyl-1-octadecanamine; N,N-dimethyloctadecylamine; dimantine.* $C_{20}H_{43}N$; mol wt 297.55. C 80.73%, H 14.57%, N 4.71%. $CH_3(CH_2)_{17}N-(CH_3)_2$. Prepd from octadecylamine and formaldehyde: Reck *et al., J. Org. Chem.* **12**, 517 (1947).

mp 22.89°.

Hydrochloride, $C_{20}H_{44}ClN$, *GS 1339, NSC 5547, Thelmesan.*

THERAP CAT: Anthelmintic (Nematodes).
THERAP CAT (VET): Anthelmintic (Nematodes).

3456. Dymixal®. Aldrich Dye Mixture. A mixture of three dyes contg 46.1% or 1.5 parts of crystal violet (hexamethyl p-rosaniline HCl), 30.8% or 1.0 part of brilliant green (diaminotetraethyltriphenyl carbinol anhydride sulfate), 23.1% or 0.75 part of acriflavine neutral. It may be prepd by mechanical mixing of the three dyes in their solid state.

Sol in water, forming a neutral greenish-blue soln which shows a yellow fluorescence under ultraviolet light.

THERAP CAT: Topical for burns.
THERAP CAT (VET): Topically for burns (in soln or jelly).

3457. Dynel®. Staple fiber copolymer of 40% acrylonitrile and 60% vinyl chloride wet spun from acetone, stretched hot as much as 1300% and then annealed by heat treatment: E. E. Stout, *Introduction to Textiles* (John Wiley,

New York, 1960) pp 198-201; R. W. Moncrieff, *Man-made Fibres* (John Wiley, New York, 1963) pp 411-420; Kennedy, "Modacrylic Fibers" in *Encyclopedia of Polymer Science and Technology* **vol. 8**, N. M. Bikales, Ed. (Interscience, New York, 1968) pp 812-839.

Light cream fiber which can be bleached nearly white. Specific gravity 1.31. Tenacity ranges from 2.0 to 3.5 g/denier. Elongation is 30-40%; hygroscopicity is 0.4% under standard conditions. Has extremely good chemical resistance. Acetone is the best solvent; cyclohexanone and dimethylformamide also have some solvent action. Acetic anhydride, acetaldehyde, aniline, ethylene dichloride, and methyl ethyl ketone all plasticize or swell dynel. Resistant to clothes moths' larvae, to carpet beetles, and to mildew and fungus. Will burn in a flame, but if the flame is removed it is self-extinguishing,. Resistant to water and non-felting and non-shrinking below the boil. Hot water delusters dynel. Must be ironed with the lowest iron setting and a dry cotton cover over the fabric. May be heat-set in permanent pleats. Can be dyed readily and may be solution-dyed. Resistant to perspiration and to salt-water deterioration.

USE: In apparel and household furnishings; simulated fur coats; chemically resistant clothing. In making wigs and doll hair; the hair can be washed, combed, set, and in some instances, redyed.

3458. Dynorphin. Extremely potent, widely distributed neuropeptide that has 17 amino acid residues and contains leu[5]-enkephalin as its NH_2-terminal sequence. Its name is derived from "dynamis", the Greek word for power, and endorphin, the name applied to the group of opioid peptides to which it belongs. Initially isolated from porcine pituitaries and termed *slow-reversing endorphin:* L. I. Lowney *et al. Life Sci.* **24**, 2377 (1979). Purification, description of properties and amino acid sequence of the first 13 residues: A. Goldstein *et al., Proc. Nat. Acad. Sci. USA* **76**, 6666 (1979). Complete amino acid sequence of the heptadecapeptide from porcine pituitary: *eidem, ibid.* **78**, 7219 (1981). Isoln from porcine duodenum and identity with pituitary dynorphin: S. Tachibana *et al., Nature* **295**, 339 (1982). Synthesis of porcine dynorphin$_{1-13}$: M. Wakimasu *et al., Chem. Pharm. Bull.* **29**, 2592 (1981). Soln conformation: R. Maroun, W. L. Mattice, *Biochem. Biophys. Res. Commun.* **103**, 442 (1981). Radioimmunoassay: V. E. Ghazarossian *et al., Life Sci.* **27**, 75 (1980). Comparison of distribution of dynorphin and enkephalin systems in brain: S. J. Watson *et al., Science* **218**, 1134 (1982). Behavioral effects of dynorphin$_{1-13}$ in mice and rats: J. M. Walker *et al., Peptides* **1**, 341 (1980); H. Zwiers *et al., Life Sci.* **28**, 2545 (1981). Opiate activity and receptor selectivity: M. Wuester *et al., Neurosci. Letters* **20**, 79 (1980); C. Chavkin, A. Goldstein *Nature* **291**, 591 (1981). In the guinea pig ileum bioassay it has been shown to be 700 times more potent than leu[5]-enkephalin and its agonist effects are 1/13th as sensitive to naloxone antagonism: *eidem, Proc. Nat. Acad. Sci. USA* **78**, 6543 (1981). Dynorphin$_{1-13}$ has been proposed as the specific endogenous ligand of the *kappa* opioid receptor (*cf.* endorphins): C. Chavkin *et al., Science* **215**, 413 (1982). It has also been suggested that dynorphin$_{1-8}$ or dynorphin$_{1-9}$ may be transmitters or modulators at the kappa binding site and dynorphins 1-13 and 1-17 may act at a distance from the release site: A. D. Corbett *et al., Nature* **299**, 79 (1982). A possible regulatory role of dynorphin on morphine and β-endorphin-induced analgesia has been proposed: F. Tulunay *et al., J. Pharmacol. Exp. Ther.* **219**, 296 (1981); E. C. Petrie *et al., Peptides* **3**, 41 (1982). Several non-opiate effects have also been described: J. M. Walker *et al., Science* **218**, 1136 (1982); R. Przewlocki *et al., ibid.* **219**, 71 (1983).

```
1                          8                13
Tyr-Gly-Gly-Phe-Leu-Arg-Arg-Ile-Arg-Pro-Lys-Leu-Lys-

14          17
Trp-Asp-Asn-Gln

            porcine dynorphin
```

Dynorphin$_{1-13}$, $C_{75}H_{126}N_{24}O_{15}$, *1-13-dynorphin (pig).* $[\alpha]_D^{23}$ −62.9° (c = 0.5 in 1% acetic acid).

3459. Dyphylline. *7-(2,3-Dihydroxypropyl)-3,7-dihy-*

dro-1,3-dimethyl-1H-purine-2,6-dione; 7-(2,3-dihydroxy-propyl)theophylline; (1,2-dihydroxy-3-propyl)theophylline; glyphylline; glyfyllin; diprophylline; AFI-phyllin; Astmamasit; Asthmolysin; Astrophyllin; Circair; Coronarin; Cor-Theophyllin; Dilor; Hiphyllin; Hyphylline; Lufyllin; Neostenovasan; Neothylline; Neotilina; Neo-Vasophylline; Neutrafil; Neutraphylline; Silbephylline; Solufilin; Solufyllin; Theal ampules; Thefylan. $C_{10}H_{14}N_4O_4$; mol wt 254.25. C 47.24%, H 5.55%, N 22.04%, O 25.17%. Prepn: Jones, Maney, U.S. pat. 2,575,344 (1951 to State Univ. of Iowa); Roth, Arch. Pharm. 292, 234 (1959). Toxicity data: R. A. Al'tshuler, Med. Prom. SSSR 16, 57 (1962), C.A. 61, 6230g (1964).

Extremely bitter crystals from alcohol, mp 158°. uv max (0.001% in H_2O): 273 nm ($A_{1cm}^{1\%}$ 361). Freely sol in water; one gram dissolves in 3 ml H_2O at 25°. Soly (g/100 ml): alc 2; chloroform 1. pH (1% aq soln) 6.6 to 7.3. LD_{50} in mice (mg/kg): 3400 orally; 1430 s.c. (Al'tshuler).

THERAP CAT: Bronchodilator.

3460. Dypnone. 1,3-Diphenyl-2-buten-1-one; β-methyl-chalcone. $C_{16}H_{14}O$; mol wt 222.27. C 86.45%, H 6.35%, O 7.20%. Prepd by self-condensation of acetophenone in the presence of aluminum tert-butoxide: Adkins, Cox, J. Am. Chem. Soc. 60, 1151 (1938); W. Wayne, H. Adkins, Org. Syn. coll. vol. III, 367 (1955).

Liquid. d_4^{15} 1.1080. n_D^{20} 1.6343. bp_{760} 340-345° (partial decompn); bp_{22} 225°; $bp_{1.0}$ 150-155°. Insol in water. Sol in alcohol, ether. LD_{50} orally in rats: 3.6 g/kg, H. F. Smyth et al., J. Ind. Hyg. Toxicol. 31, 60 (1949).

USE: Sunscreen: Ind. Eng. Chem. 46, 15A (July 1954).

3461. Dysprosium. Dy; at. wt 162.50; at. no. 66; valence 3. Rare earth metal; yttrium group. Seven naturally occurring isotopes: 156 (0.052%); 158 (0.090%); 160 (2.294%); 161 (18.88%); 162 (25.53%); 163 (24.97%); 164 (28.18%); artificial radioactive isotopes: 149-155; 157; 159; 165-167. Abundance in earth's crust: 4.5-7.5 ppm. Occurs in gadolinite, xenotime, samarskite and other rare earth minerals. Discovered in 1886 by Lecoq de Boisbaudran. Sepn: Urbain, Compt. Rend. 139, 736 (1904); 141, 521 (1905); 142, 785 (1906); Spedding et al., J. Am. Chem. Soc. 69, 2812 (1947); 76, 2557 (1954). Prepn in form of a salt: Engle,

Balke, ibid. 39, 53 (1917); prepn of a Dy-Al alloy: Schumacher, Harris, ibid. 48, 3108 (1926). Reviews of prepn, properties and compds of dysprosium: The Rare Earths, F. H. Spedding, A. H. Daane, Eds. (Krieger, Huntington, N.Y., 1971, reprint of 1961 ed.) 641 pp; Hulet, Bode, "Separation Chemistry of the Lanthanides and Transplutonium Actinides" in MTP Int. Rev. Sci.: Inorg. Chem., Ser. One, vol. 7, K. W. Bagnall, Ed. (Univ. Park Press, Baltimore, 1972) pp 1-45; Moeller, "The Lanthanides" in Comprehensive Inorganic Chemistry, vol. 4, J. C. Bailar, Jr. et al., Eds. (Pergamon Press, Oxford, 1973) pp 1-101.

Silver metal; tarnishes in moist air. Hexagonal close-packed structure. d 8.559. mp 1407°. Forms greenish-yellow salts.

Oxide, Dy_2O_3, dysprosia. White substance. Prepd by heating the oxalate or sulfate. d^{27} 7.81.

Hydroxide, $Dy(OH)_3$, a gelatinous precipitate. Prepd by adding ammonia to an aq soln of a dysprosium salt; forms a blue colloidal soln.

Chloride, $DyCl_3$, obtained by passing S_2Cl_2 over heated dysprosia. A hexahydrate is formed from the aq soln. The anhydr chloride, yellow shining crystals, d 3.67; mp 680°. LD_{50} in mice: 585 mg/kg i.p.; 7.65 g/kg orally, Haley et al., Toxicol. Appl. Pharmacol. 8, 37 (1966).

Sulfate, $Dy_2(SO_4)_3$, octahydrate, yellow crystals. Prepd by dissolving the oxide in sulfuric acid and precipitating with alcohol; stable in air at 110°, dehydrated at 360°.

Nitrate, $Dy(NO_3)_3$, pentahydrate, melts at 88.6° in its water of crystn. Sol in water. LD_{50} (hexahydrate) in rats: 295 mg/kg i.p.; 3.1 g/kg orally, Bruce et al., ibid. 5, 750 (1962).

USE: Oxide used in control rods of some nuclear power reactors.

3462. Dystrophin. Protein product of the human Duchenne muscular dystrophy (DMD) gene. Mol wt approx 400,000 daltons. Present in very small amounts in normal muscle (approx 0.002% of total muscle protein) but either absent or abnormal in muscular dystrophy patients. Thought to be an intracellular structural component of the plasma membrane system in normal muscle fibers. Complete cloning of DMD cDNA: M. Koenig et al., Cell 50, 509 (1987). Isoln of dystrophin from mouse and human muscle tissue: E. P. Hoffman et al., ibid. 51, 919 (1987). Subcellular localization studies: E. P. Hoffman et al., Nature 330, 754 (1987); C. M. Knudson et al., J. Biol. Chem. 263, 8480 (1988); E. E. Zubrzycka-Gaarn et al., Nature 333, 466 (1988); S. C. Watkins et al., ibid. 863. Amino acid sequence and structural similarity to α-actinin and to the cytoskeletal protein, spectrin: M. Koenig et al., Cell 53, 219 (1988). Characterization of DMD gene expression in normal and diseased human muscle: M. Oronzi Scott et al., Science 239, 1418 (1988); in animal muscle and brain: U. Nudel et al., Nature 331, 635 (1988). Differentiation of muscle and brain DMD mRNA: U. Nudel et al., ibid. 337, 76 (1988). Correlation of clinical phenotype with dystrophin abnormalities: E. P. Hoffman et al., N. Engl. J. Med. 318, 1363 (1988). Use of dystrophin cDNA for prenatal diagnosis and carrier detection in muscular dystrophy: B. T. Darras et al., Am. J. Med. Genetics 29, 713 (1988).

E

3463. Ebimar®. C 16; polysaccharide C 16. Degraded carrageenan, *q.v.* Sulfated polysaccharide, extracted from *Chondrus crispus* (L.) Stackhouse, *Gigartinaceae.* Prepn: Anderson, *J. Pharm. Pharmacol.* **13,** 139 (1961); Anderson, Hargreaves, **Brit. pat.** 840,623 (1960 to Evans Medical).

Off-white powder containing 28-30% unchanged bound sulfate. Typical data: S 9.4%, Ca 0.2%, Na 7%, K 1.3%. $[\alpha]_D$ +34°. Freely sol in water and acid. Aq soln less viscous than the aq soln of the parent carrageenan.

THERAP CAT: Antacid.

3464. Eburnamonine. *Eburnamenin-14(15H)-one.* C_{19}-$H_{22}N_2O$; mol wt 294.40. C 77.52%, H 7.53%, N 9.51%, O 5.43%. One of the *Vinca* alkaloids, naturally occurring as the (+)- and (±)-forms. Isoln of the (+)-form from *Hunteria eburnea* Pinchon, *Apocynaceae:* M. F. Bartlett et al., *Compt. Rend.* **249,** 1259 (1959); of the (±)-form from *Vinca minor* L., *Apocynaceae:* J. Mokry et al., *Experientia* **17,** 354 (1961). The (−)-form is obtained by acid hydrolysis of vincamine, *q.v.:* J. Trajanek et al., *Tetrahedron Letters* **1961,** 702; O. Clauder et al., *ibid.* **1962,** 1147; eidem, **Hung. pat.** 151,295 (1964 to Gedeon Richter), *C.A.* **60,** 14558e (1964). Structure and synthesis of the (±)-form: M. F. Bartlett, W. I. Taylor, *Tetrahedron Letters* **20,** 20 (1959); eidem, *J. Am. Chem. Soc.* **82,** 5941 (1960); E. Wenkert, B. Wickberg, *ibid.* **87,** 1580 (1965). Short synthesis of the (±)-form: E. Wenkert et al., *ibid.* **100,** 4893 (1978); high yield total synthesis: J. L. Hermann et al., *ibid.* **101,** 1540 (1979); T. Imanishi et al., *Chem. Pharm. Bull.* **30,** 1521 (1982). Synthesis of the (−)-form: D. Cartier et al., *Bull. Soc. Chim. France* **1976,** 1961. Structural and biogenetic relationship to vincamine: J. Mokry et al., *Tetrahedron Letters* **1962,** 433. Pharmacology of the (−)-form: P. Lacroix et al., *Arzneimittel-Forsch.* **29,** 1094 (1979). Series of articles on the pharmacodynamics, metabolism, and therapeutic use of the (−)-form: *Eur. Neurol.* **17,** Suppl. 1, 1-172 (1978). *Reviews:* W. I. Taylor in *The Alkaloids* vol. 8, R. H. F. Manske, Ed. (Academic Press, NewYork, 1965) pp 253-259; idem, *ibid.* vol. 11, pp 108-110 (1968).

Cryst, mp 174°. $[\alpha]_D^{25}$ +89° (chloroform). uv max: 241, 268, 296, 302 nm (ε 19800, 10200, 4800, 4800).

(±)-Form, $C_{19}H_{22}N_2O$, (±)-*eburnamenin-14(15H)-one,* *vincanorine.* Crystals from methanol, mp 201-202.5°. uv max (ethanol): 227, 287, 294 nm (log ε 4.49, 3.89, 3.87).

(−)-Form, $C_{19}H_{22}N_2O$, 3α,16α-*eburnamenin-14(15H)-one,* 16-oxoeburnane, vincamone, CH 846, Cervoxan. Solid, mp 168-170°; also reported as mp 177-178° (Cartier). $[\alpha]_D^{25}$ −102° (chloroform) (Clauder); $[\alpha]_D$ −100° (c = 0.783 in chloroform) (Cartier). uv max: 205, 240, 265, 290, 300 nm (log ε 4.28, 4.16, 3.90, 3.59, 3.57).

(−)-Form phosphate, *Eburnal.*

THERAP CAT: The (−)-form as vasodilator.

3465. Ecdysones. Insect molting hormones controlling the pupation of insects. The terms *zooecdysones* and *phytoecdysones* used to distinguish ecdysones isolated from insects and crustaceae from those of plant origin. About 30 structurally known ecdysones. All have poor crystalline properties and are characterized by intense absorption max in ethanol: 243 nm. α-Ecdysone is the major component of insect extractions; β-ecdysone, of plant extractions. Configuration: Koreeda et al., *J. Am. Chem. Soc.* **93,** 4084 (1971). Chromosomal action: M. Ashburner, *Nature* **285,** 435

(1980). *Reviews:* Kilby, *Discovery* **18,** 13 (1957); P. Karlson, *Angew. Chem. Int. Ed.* **2,** 175-182 (1963); K. Nakanishi, *Pure Appl. Chem.* **25,** 167-195 (1971); M. Koreeda, B. A. Teicher, *Anal. Biochem. Insects* **1977,** 207-240; P. Karlson, *Dev. Endocrinol.* **7,** 1-11 (1980).

α-Ecdysone. $C_{27}H_{44}O_6$. 2,3,14,22,25-*Pentahydroxycholest-7-en-6-one.* R = H. First isoln from silkworm moths, *Bombyx mori:* Butenandt, Karlson, *Z. Naturforsch.* **9b,** 389 (1954). Isoln from *Polypodium vulgare* L.; *Pteridinium aquilinum:* Heinrich, Hoffmeister, *Experientia* **23,** 995 (1967); Kaplanis et al., *Science* **157,** 1436 (1967). Structure: Karlson et al., *Ber.* **98,** 2394 (1965). Configuration: Huber, Hoppe, *ibid.* 2403. Synthesis: Kerb et al., *Helv. Chim. Acta* **49,** 1601 (1966); Siddall et al., *J. Am. Chem. Soc.* **88,** 379, 862 (1966); Mori et al., *Chem. Pharm. Bull.* **16,** 563 (1968). mp 238-239° (Mori). $[\alpha]_{578}^{20}$ +62°. uv max: 243 nm (ε 11,600).

β-Ecdysone. $C_{27}H_{44}O_7$. 2,3,14,20,22,25-*Hexahydroxycholest-7-en-6-one,* 20-*hydroxyecdysone, ecdysterone, crustecdysone, isoinokosterone.* R = OH. Isoln from *B. mori; Jasus lalandei:* Hocks, Wiechert, *Tetrahedron Letters* **1966,** 2989; Hampshire, Horn, *Chem. Commun.* **1966,** 37. The first isolated phytoecdysone, from *Achyranthes fauriei:* Takemoto et al., *Yakugaku Zasshi* **87,** 325 (1967). From *P. elatus:* Galbraith, Horn, *Chem. Commun.* **1966,** 905. Configuration: Dammeier, Hoppe, *Ber.* **104,** 1660 (1971). Synthesis: Hüppi, Siddall, *J. Am. Chem. Soc.* **89,** 6790 (1967); Kerb et al., *Tetrahedron Letters* **1968,** 4277; Mori, Shibata, *Chem. Pharm. Bull.* **17,** 1970 (1969). Total synthesis: T. Kametani et al., *Tetrahedron Letters* **21,** 4855 (1980). mp 240-242° from methanol-ethyl acetate (Hüppi, Siddall). uv max: 243 nm (ε 10,300). Unstable in alkaline soln.

3466. Ecgonidine. 8-*Methyl-8-azabicyclo[3.2.1]oct-2-ene-2-carboxylic acid;* 2-*tropidinecarboxylic acid;* anhydro-ecgonin. $C_9H_{13}NO_2$; mol wt 167.20. C 64.65%, H 7.84%, N 8.38%, O 19.14%. Prepn from ecgonine: Findlay, *J. Am. Chem. Soc.* **75,** 1033 (1953). Prepn and formation of *l*-form: de Jong, *Rec. Trav. Chim.* **42,** 980 (1923), **66,** 99 (1947). Synthesis of *dl*-form: Grundmann, Ottmann, *Ann.* **605,** 24 (1957); **U.S. pat.** 2,783,235 (1957 to Olin Mathieson). Structure: Findlay, *loc. cit.*

dl-Form, crystals from alcohol + ether, dec 235-236°. Soluble in water; sparingly sol in alcohol.

l-Form, crystals from abs alcohol, mp 235°. $[\alpha]_D^{14}$ −84.6° (c = 1.7).

THERAP CAT: Topical anesthetic.

3467. Ecgonine. [1R-(exo,exo)]-3-*Hydroxy-8-methyl-8-azabicyclo[3.2.1]octane-2-carboxylic acid;* 3β-*hydroxy-1αH,-5αH-tropane-2β-carboxylic acid.* $C_9H_{15}NO_3$; mol wt 185.22. C 58.36%, H 8.16%, N 7.56%, O 25.91%. The principal part of the cocaine molecule. Obtained by hydrolysis of cocaine: Willstätter et al., *Ann.* **434,** 111 (1923); Bell, Archer, *J. Am. Chem. Soc.* **82,** 4642 (1960). From coca leaves: de Jong, *Rec. Trav. Chim.* **59,** 687 (1940). Structure: Gadamer, John, *Arch. Pharm.* **259,** 227 (1921). Stereochemistry: Fodor, *Nature* **170,** 278 (1952); Fodor, Kovács, *J. Chem. Soc.* **1953,** 724. Synthesis: Willstätter, Bommer, *Ann.* **422,** 15 (1920). *Review:* Stoll, Jucker, *Angew. Chem.* **66,** 376 (1954).

Consult the cross index before using this section.

l-Form monohydrate, triboluminescent, monoclinic prisms from alc, mp 198° (anhydr, dec 205°). $[\alpha]_D^{15}$ −45° (c = 5). Neutral to litmus. pKa 11.11; pKb 11.22. One gram dissolves in 5 ml water, 67 ml alc, 20 ml methanol, 75 ml ethyl acetate. Sparingly sol in acetone, ether, benzene, chloroform, petr ether.

Hydrochloride, $C_9H_{16}ClNO_3$, triclinic plates from water, mp 246°. $[\alpha]_D^{15}$ −59° (c = 10). Sol in water; slightly in alc.

dl-Form trihydrate, plates from 90% alcohol, mp 93-118° (anhydr dec 212°).

Note: This is a controlled substance listed in the U.S. Code of Federal Regulations, Title 21 Part 1308.12 (1987).

THERAP CAT: Topical anesthetic.

3468. Echinacea. Cone flower. Dried rhizome and roots of *Echinacea pallida* (Nutt.) Britt. or *E. angustifolia* DC. (*Brauneria pallida* Nutt.) Britt., *Compositae. Habit.* Saskatchewan to Texas and Alabama. *Constit.* Inulin, sucrose, betaine, two isomeric 2-methyltetradecadienes, echinacein (neoherculin, α-sanshool), echinacoside (a caffeic acid glycoside), resins, fatty acids. *Refs:* Stoll, *Helv. Chim. Acta* **33**, 1877 (1950); Jacobson, *Science* **120**, 1028 (1954); Crombie, Taylor, *J. Chem. Soc.* **1957**, 2760.

Note: A commercial extract of *Echinaceae* spp, **echinacin**, has been used experimentally as a hyaluronidase antagonist: Koch, Haase, *Arzneimittel-Forsch.* **2**, 464 (1952); Büsing, *ibid.* **5**, 320 (1955).

3469. Echinenone. β,β-*Caroten-4-one;* 4-oxo-β-carotene; 4-keto-β-carotene; aphanin; myoxanthin. $C_{40}H_{54}O$; mol wt 550.83. C 87.21%, H 9.88%, O 2.90%. Carotenoid pigment occurring in algae. Isoln from *Aphanizomenon flosaqua:* Tischer, *Z. Physiol. Chem.* **251**, 109 (1938); **260**, 257 (1939); from *Paracentrotus lividus:* Lederer, *Compt. Rend.* **201**, 300 (1935); from *Oscillatoria rubrescens:* Heilbron, Lythgoe, *J. Chem. Soc.* **1936**, 1376. Structure: Goodwin, Taha, *Biochem. J.* **47**, 244 (1950); **48**, 513 (1951); Goodwin, *ibid.* **63**, 481 (1956); Ganguly *et al., Arch. Biochem. Biophys.* **60**, 345 (1956). Synthesis: Akhtar, Weedon, *J. Chem. Soc.* **1958**, 3986; **1959**, 4058.

Orange-red crystals from benzene + methanol, mp 178-180°. Absorption max (chloroform): 472-478 nm. Freely sol in carbon disulfide, chloroform, benzene. Slightly sol in pyridine, ether. Practically insol in methanol. Provitamin A activity 54% of that of all-*trans*-β-carotene.

Oxime, $C_{40}H_{55}NO$, red crystals from benzene, mp 208°. Absorption max (chloroform): 464-468 nm.

3470. Echinochrome A. 2-*Ethyl-3,5,6,7,8-pentahydroxy-1,4-naphthalenedione;* 2-ethyl-3,5,6,7,8-pentahydroxy-1,4-naphthoquinone. $C_{12}H_{10}O_7$; mol wt 266.20. C 54.14%, H 3.78%, O 42.07%. The red pigment of the eggs of the sea urchin (*Arbacia pustulosa*). Isoln and structure: Kuhn, Wallenfels, *Ber.* **72**, 1407 (1939). Synthesis from 2-ethyl-1,3,4-trimethoxybenzene and dibenzoyloxymaleic anhydride: Wallenfels, Gauhe, *Ber.* **76**, 325 (1943).

Deep-red needles from dioxane-water, mp 220° (some

decompn). Sublimes at 10^{-4} mm at 120°. Absorption max (chloroform): 533, 497, 462 nm. Very sparingly sol in cold water, yet more than sufficiently sol for chemotaxic effects. Readily sol in carbon disulfide, ether, chloroform, benzene, concd H_2SO_4.

3471. Echinomycin. Quinomycin A. $C_{51}H_{64}N_{12}O_{12}S_2$; mol wt 1101.27. C 55.62%, H 5.86%, N 15.26%, O 17.43%, S 5.82%. A *"quinoxaline antibiotic"* similar to the triostins, *q.v.:* Kuroya, Ishida, *J. Antibiot.* **14A**, 324 (1961). Powerful, selective inhibitor of nucleic acid synthesis *in vitro.* Produced by *Streptomyces echinatus* from soil of Cuanza (Angola): Corbaz *et al., Helv. Chim. Acta* **40**, 199 (1957). Structure: Keller-Schierlein *et al., ibid.* **42**, 305 (1959). Revised structure: Dell *et al., J. Am. Chem. Soc.* **97**, 2497 (1975); D. G. Martin *et al., J. Antibiot.* **28**, 332 (1975). Identity with quinomycin A: Katagiri, Sugiura, *Antimicrob. Ag. Chemother.* **1961**, 162. Biosynthesis: Arif *et al., Indian J. Biochem.* **7**, 193 (1970). Isoln and properties of the other quinomycins: Yoshida, Katagiri, *J. Antibiot.* **14A**, 330 (1961); Otsuka, Shoji, *ibid.* **19A**, 128 (1966); *eidem, Tetrahedron* **23**, 1535 (1967). Conformation in soln: H. T. Cheung *et al., J. Am. Chem. Soc.* **100**, 46 (1978). Review of chemistry and biochemistry of echinomycin: M. J. Waring in *Antibiotics,* **vol. 5**(pt. 2), F. E. Hahn, Ed. (Springer-Verlag, New York, 1979) pp 173-194.

Slightly hygroscopic crystals, mp 217-218°. $[\alpha]_D^{20}$ −310° (c = 0.86 in chloroform). uv max (methanol): 243, 320 nm ($E_{1cm}^{1\%}$ 622, 100). Has lipophile solubilities. Easily sol in chloroform, dioxane. Insol in water, petr ether, hexane.

3472. Echinopsine. 1-*Methyl-4(1H)-quinolinone;* 1-*methyl-4(1H)-quinolone;* 1,4-dihydro-1-methyl-4-oxoquinoline; N-methyl-4-quinolone. $C_{10}H_9NO$; mol wt 159.18. C 75.45%, H 5.69%, N 8.80%, O 10.05%. From *Echinops ritro* L. and other spp of *Echinops, Compositae:* Greshoff, *Rec. Trav. Chim.* **19**, 360 (1900); Ban'kovskii *et al., Dokl. Akad. Nauk SSSR* **148**, 1073 (1963). Structure: Späth, Kolbe, *Monatsh.* **43**, 469 (1923). Synthesis: Kondo, Ikawa, *J. Pharm. Soc. Japan* **51**, 702 (1931); Allison *et al., J. Chem. Soc.* **1954**, 403; King, Abramo, *J. Org. Chem.* **23**, 1609 (1958); Kamiya, *Chem. Pharm. Bull.* **10**, 669 (1962). Simple synthesis: J. R. Merchant, V. Shankaranarayan, *Chem. & Ind. (London)* **1979**, 320.

Needles from benzene, mp 152°. One gram dissolves in about 60 ml water, 6 ml boiling water. Soluble in alcohol, chloroform, hot benzene; slightly sol in ether. LD_{100} in mice: 600 mg/kg.

Hydrochloride, $C_{10}H_9NO \cdot HCl$, crystals, mp 185-186°.

3473. Echinuline. *(3S-cis)-3-[[2-(1,1-Dimethyl-2-pro-penyl)-5,7-bis(3-methyl-2-butenyl)-1H-indol-3-yl]methyl]-6-methyl-2,5-piperazinedione.* $C_{29}H_{39}N_3O_2$; mol wt 461.63. C 75.45%, H 8.52%, N 9.10%, O 6.93%. From the dry my-celia of *Aspergillus echinulatus:* Quilico, Panizzi, *Ber.* **76**, 348 (1943). Structure: Casnati *et al., Gazz. Chim. Ital.* **92**, 105 (1962); Romanet *et al., Bull. Soc. Chim. France* **1963**, 1048. Absolute configuration: Nakashima, Slater, *Tetrahedron Letters* **1967**, 4433; **1971**, 2649. Synthetic studies: Hough-ton, Saxton, *J. Chem. Soc. (C)* **1969**, 595, 1003; Takamatsu *et al., Tetrahedron Letters* **1971**, 4661. Total synthesis of optically active *cis* isomer: *eidem, ibid.* 4665; S. Inoue *et al., J. Pharm. Soc. Japan* **97**, 558 (1977). Epimerization of echi-nuline with triethylamine in ethanol gives the *trans* isomer, *epi-echinuline*, Westley *et al., Anal. Chem.* **40**, 1888 (1968).

Needles from butanol, mp 242-243°. $[\alpha]_D^{20}$ —26.0° (chloro-form). uv max (ethanol): 230, 279, 286 nm (log ϵ 4.60, 3.98, 3.96). Sol in glacial acetic acid, chloroform, pyridine, diox-ane; less sol in warm alcohol, warm butanol; slightly sol in benzene, ether, petr ether, carbon tetrachloride, acetone, cold alcohol.

3474. Echitamine. *3,17-Dihydroxy-16-(methoxycarbon-yl)-4-methyl-2,4(1H)-cyclo-3,4-secoakuammilanium;* di-taine. $[C_{22}H_{29}N_2O_4]^+$; mol wt 385.48. From the bark of *Alstonia scholaris* (L.) R.Br., *A. congensis* Engl., *Apocynaceae; A. neriifolia:* Hesse, *Ann.* **203**, 150 (1880); Goodson, Henry, *J. Chem. Soc.* **127**, 1640 (1925); Goodson, *ibid.* **1932**, 2626; Chakravarti *et al., Bull. Calcutta Sch. Trop. Med.* **16**, 81 (1968), *C.A.* **71**, 3529f (1969); Roy, Chatterjee, *J. Ind. Chem. Soc.* **45**, 21 (1968). Structure: Hamilton *et al., Proc. Chem. Soc.* **1961**, 63; *J. Chem. Soc.* **1962**, 5061. Abs config: Ma-nohar, Ramaseshan, *Tetrahedron Letters* **1961**, 814. Synthe-tic studies: Fritz, Fischer, *Tetrahedron* **20**, 1737 (1964); Dolby, Esfandiari, *J. Org. Chem.* **37**, 43 (1972). *Review:* Govindachari, *J. Ind. Chem. Soc.* **45**, 945-957 (1968).

Hydroxide, $C_{22}H_{30}N_2O_5$, white crystals, mp 206°. $[\alpha]_D^{20}$ —29° (alcohol). Sol in water, alcohol, chloroform, ether. Chloride, $C_{22}H_{29}O_4N_2{}^+Cl^-$, long needles from water, mp 295°. $[\alpha]_D^{15}$ —58°. uv max (ethanol): 235, 295 nm (log ϵ 3.93, 3.63).

3475. Echothiophate Iodide. *2-[(Diethoxyphosphinyl)-thio]-N,N,N-trimethylethanaminium iodide; (2-mercaptoeth-yl)trimethylammonium iodide O,O-diethyl phosphorothioate;* diethoxyphosphinylthiocholine iodide; 2-diethoxyphosphin-ylthioethyltrimethylammonium iodide; *O,O-diethyl S-(2-tri-methylammoniumethyl)phosphorothioate iodide; S-β-di-methylaminoethyl-O,O-diethylthionophosphate methiodide;* ecothiopate iodide; 217 MI; Phospholine Iodide. $C_9H_{23}I$-NO_3PS; mol wt 383.22. C 28.21%, H 6.05%, I 33.12%, N 3.66%, O 12.53%, P 8.08%, S 8.37%. Prepn: Tammelin, *Acta Chem. Scand.* **11**, 1340 (1957); Fitch, U.S. pat. **2,911,430** (1959 to Campbell Pharmaceuticals). Comprehensive de-scription: R. D. Daley in *Analytical Profiles of Drug Sub-stances* **vol. 3,** K. Florey, Ed. (Academic Press, New York, 1974) pp 233-251.

Crystals, mp 138° (Tammelin); mp 124-124.5° (Fitch). Sol in water, alcohol, chloroform.

THERAP CAT: Cholinergic (ophthalmic).

3476. Econazole. *1-[2-[(4-Chlorophenyl)methoxy]-2-(2,4-dichlorophenyl)ethyl]-1H-imidazole; 1-[2,4-dichloro-β-[(p-chlorobenzyl)oxy]phenethyl]imidazole; SQ 13050.* C_{18}-$H_{15}Cl_3N_2O$; mol wt 381.68. C 56.64%, H 3.96%, Cl 27.86%, N 7.34%, O 4.19%. Prepn: Godefroi *et al., J. Med. Chem.* **12**, 784 (1969); Godefroi, Heeres, Ger. pat. **1,940,388** cor-resp to U.S. pat. **3,717,655** (1970, 1973 both to Janssen). Biological and toxicological properties: Thienpont *et al., Arzneimittel-Forsch.* **25**, 224 (1975). Review of antifungal activity and therapeutic efficacy: R. C. Heel *et al., Drugs* **16**, 177-201 (1978).

mp 86.8°.

Nitrate, $C_{18}H_{16}ClN_3O_4$, *R 14827, Ecostatin, Epi-Pevaryl, Gyno-Pevaryl, Ifenec, Micofugal, Micogin, Palavale, Pargin, Pevaryl, Spectazole.* Crystals from a mixture of 2-propanol, methanol and diisopropyl ether, mp 162°. Very slightly sol in water; very slightly to slightly sol in most common organ-ic solvents. LD_{50} in mice, rats (mg/kg): 462.7, 667.7 orally (Thienpont).

THERAP CAT: Antifungal.

3477. ECTEOLA-Cellulose. Cellulose-ECTEOLA. Anion-exchange material prepd from cellulose, epichlorohy-drin, and triethanolamine: Peterson, Sober, *J. Am. Chem. Soc.* **78**, 751 (1956); Veder, *J. Chromatog.* **10**, 507 (1963).

3478. Ectylurea. *(Z)-N-(Aminocarbonyl)-2-ethyl-2-butenamide; (2-ethylcrotonoyl)urea; (α-ethylcrotonyl)urea; (α-ethylcrotonyl)carbamide;* Astyn; Ektyl; Levanil; Nostal; Nostyn; Cronil; Neuroprocin; Pacetyn. $C_7H_{12}N_2O_2$; mol wt 156.18. C 53.83%, H 7.74%, N 17.94%, O 20.49%. Prepd by treating (2-bromo-2-ethylbutyryl)urea with dil boiling NaOH: Newbery, *J. Chem. Soc.* **127**, 295 (1925). Prepn from carbromal and isopropanol by refluxing with silver oxide: Fancher, U.S. pats. **2,854,379, 2,931,832** (1958, 1960 to Miles Labs.). Pharmacology and toxicity: M. H. Pindell *et al., Fed. Proc.* **12**, 357 (1953).

Needles from benzene, mp 198°, also reported as mp 191-193° (lower melting form, mp 158°). Very slightly sol in ether, hot alcohol, other organic solvents. Sol in concd aq solns of alkalies and acids. LD_{50} in rats (mg/kg): 2500 orally; 900 i.p. (Pindell).

THERAP CAT: Sedative, hypnotic.

3479. Edestin. Globular protein originally obtained from hemp seed (*Cannabis sativa*): Osborne, *Am. Chem. J.* **14**, 662 (1892); Stockwell *et al., Proc. Seed Protein Conf., New Orleans* **1963**, 56, *C.A.* **60**, 3090d (1964). Approx mol wt 300,000. Since the passage of the Marihuana Tax Act in the U.S.A. commercial hemp seed must be heat-treated to destroy its narcotic properties, a process which also destroys the edestin. Very closely related proteins may be obtained from seeds of *Cucurbita pepo* L. (pumpkin), *C. moschata* Duchesne, and *C. maxima* Duchesne (squash), *Citrullus* *vulgaris* Schrader (watermelon), *Cucumis melo* L. (cantaloupe), and *Cucumis sativus* L., *Cucurbitaceae* (cucumber). Isoln from *Cucurbita pepo:* Vickery *et al., Biochem. Prepn.* **2,** 5 (1952). Extraction from hemp seed, prepn of crystalline form and x-ray data: J. Drenth, E. W. Wiebenga, *Rec. Trav. Chim.* **74,** 813 (1955). The amino acid compositions of these globulins are slightly different from hemp seed globulin and from each other: Smith *et al., J. Biol. Chem.* **164,** 159 (1946); Smith, Greene, *ibid.* **167,** 833 (1947); **172,** 111 (1948). Structure studies: Hall, *ibid.* **185,** 45 (1959); Cleemann, Kratky, Z. *Naturforsch.* **15b,** 526 (1960); Dlouhá *et al., Coll. Czech. Chem. Commun.* **28,** 2779 (1963); **29,** 1835 (1964). Electron microscopy and optical diffraction: A. M. H. Schepman *et al., Biochim. Biophys. Acta* **271,** 279 (1972). Electron spin resonance studies: L. J. Dimmey, W. Gordy, *Proc. Nat. Acad. Sci. USA* **77,** 343 (1980).

Octahedral crystals. Completely sol to a clear soln in 10% salt soln. Sol in dil mineral acids. Forms a water-soluble hydrochloride.

3480. Edetate Calcium Disodium.

[[N,N'-1,2-Ethanediylbis[N-(carboxymethyl)glycinato]](4−)-N,N',O,O',O^N,-O^N]calciate(2−)disodium; [(ethylenedinitrilo)tetraacetato]-calciate(2−) disodium; ethylenediaminetetraacetic acid calcium disodium chelate; calcium disodium (ethylenedinitrilo)-tetraacetate; calcium disodium ethylenediaminetetraacetate; EDTA calcium; edathamil calcium disodium; calcium disodium edetate; edetic acid calcium disodium salt; sodium calciumedetate; Calcitetracemate Disodium; Calcium Disodium Versenate; Ledclair; Mosatil; Antallin; Sormetal; Versene CA. $C_{10}H_{12}CaN_2Na_2O_8$; mol wt 374.28. C 32.09%, H 3.23%, Ca 10.71%, N 7.49%, Na 12.29%, O 34.20%. Prepn: Astakhov, Kiseleva, *Zh. Obshch. Khim.* **20,** 1780 (1950), *C.A.* **45,** 2409 (1951).

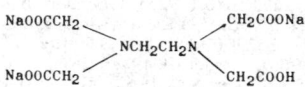

Tetrahydrate, powder. Sol in water: at 30° a 0.1M soln can be prepd (pH ∼7). Practically insol in organic solvents. Exchanges its calcium for lead or other heavy metal ions forming water-sol complexes of the heavy metals.

USE: Color retention agent in foods, flavoring agent.

THERAP CAT: Chelating agent (metal).

THERAP CAT (VET): Chelating agent in lead poisoning.

3481. Edetate Disodium.

N,N'-1,2-Ethanediylbis[N-(carboxymethyl)glycine] disodium salt; (ethylenedinitrilo)-tetraacetic acid disodium salt; ethylenediaminetetraacetic acid disodium salt; ethylenebis(iminodiacetic acid) disodium salt; edetic acid disodium salt; edathamil disodium; disodium edathamil; EDTA disodium; tetracemate disodium; disodium ethylenediaminetetraacetate; disodium edetate; Cheladrate; Chelaplex III; Endrate disodium; Sequestrene NA 2; Sodium Versenate; Titriplex III; Versene disodium salt. $C_{10}H_{14}N_2Na_2O_8$; mol wt 336.21. C 35.72%, H 4.20%, N 8.33%, Na 13.68%, O 38.07%. Prepn: Bersworth, U.S. pat. **2,407,645** (1946 to Martin Dennis); Bersworth, U.S. pat. **2,461,519** (1949). Toxicity: J. E. Wynn *et al., Toxicol. Appl. Pharmacol.* **16,** 807 (1970). *Review:* Biermans, Henrard, *Ind. Chim.* **39,** 6 (1952).

Dihydrate, crystals, mp 252° (dec). pH about 5.3. Sol in water. Has characteristics of weak acid, displacing CO_2 from carbonates and reacting with metals to form hydrogen. LD_{50} orally in rats: 2 g/kg (Wynn).

USE: Sequestering agent. Pharmaceutic aid (chelating agent).

THERAP CAT: Chelating agent (metal).

THERAP CAT (VET): Anticoagulant.

3482. Edetate Sodium.

N,N'-1,2-Ethanediylbis[N-(carboxymethyl)glycine] tetrasodium salt; (ethylenedinitrilo)tetraacetic acid tetrasodium salt; ethylenediaminetetraacetic acid tetrasodium salt; sodium edetate; tetrasodium ethylenediaminetetraacetate; ethylenebis(iminodiacetic acid) tetrasodium salt; tetrasodium ethylenebis(iminodiacetate); EDTA tetrasodium; edetic acid tetrasodium salt; tetracemate tetrasodium; tetrasodium edetate; tetracemin; Endrate Tetrasodium; Questex; Versene; Sequestrene; Tetrine; Kalex; Trilon B; Komplexon; Nullapon; Aquamollin; Complexone; Distol 8; Irgalon; Calsol; Syntes 12a; Tyclarosol; Nervanaid B. C_{10}-$H_{12}N_2Na_4O_8$; mol wt 380.20. C 31.59%, H 3.18%, N 7.37%, Na 24.19%, O 33.67%. $(NaOOCCH_2)_2NCH_2CH_2N(CH_2-COONa)_2$. Prepn: Bersworth, U.S. pat. **2,407,645** (1946 to Martin Dennis Co.); *idem.,* U.S. pat. **2,461,519** (1949). For bibliography and applications *see* "Keys to Chelation" issued by Dow, Midland, Mich.

Powder, mp > 300°. Apparent density: 6.9 lb/gallon. Very sol in water (about 103 g/100 ml). pH of 1% soln 11.3. Reacts with most divalent and trivalent metallic ions forming sol metal chelates. Less sol in alcohol than the potassium salt.

USE: Sequestering agent: one gram complexes 215 mg $CaCO_3$. Usually added to pharmaceuticals in the form of the calcium disodium salt to prevent calcium-depleting action in the body.

THERAP CAT: Chelating agent.

3483. Edetate Trisodium.

N,N'-1,2-Ethanediylbis[N-(carboxymethyl)glycine] trisodium salt; (ethylenedinitrilo)tetraacetic acid trisodium salt; EDTA trisodium; ethylenediaminetetraacetic acid trisodium salt; trisodium ethylenediaminetetraacetate; trisodium edetate; edetic acid trisodium salt; Limclair; Versene-9; Sequestrene NA 3. $C_{10}H_{13}N_2Na_3$-O_8; mol wt 358.20. C 33.53%, H 3.66%, N 7.82%, Na 19.25%, O 35.73%. Prepn: Bersworth, U.S. pat. **2,461,519** (1949). *Review:* Biermans, Henrard, *Ind. Chim.* **39,** 6 (1952).

Monohydrate, crystals from water, mp > 300°. More sol in water than the corresp disodium salt or free acid. pH of 1% aq soln 9.3. One gram complexes at least 242 mg $CaCO_3$.

THERAP CAT: Chelating agent.

3484. Edetic Acid.

N,N'-1,2-Ethanediylbis[N-(carboxymethyl)glycine]; (ethylenedinitrilo)tetraacetic acid; ethylenediaminetetraacetic acid; edathamil; EDTA; Havidote; Versene Acid. $C_{10}H_{16}N_2O_8$; mol wt 292.24. C 41.10%, H 5.52%, N 9.59%, O 43.80%. $(HOOCCH_2)_2NCH_2CH_2N(CH_2-COOH)_2$. Prepn: Münz, U.S. pat. **2,130,505** (1938 to General Aniline); Bersworth, U.S. pat. **2,407,645** (1946 to Martin Dennis Co.); Bersworth, U.S. pat. **2,461,519** (1949); Smith *et al., J. Org. Chem.* **14,** 355 (1949). Prepn of α- and β-form of crystals: Le Blanc, Spell, *J. Phys. Chem.* **64,** 949 (1960). *Review:* Biermans, Henrard, *Ind. Chim.* **39,** 6 (1952).

Crystals from water, dec 220°. Soly in water at 25°: 0.50 g/l, Palei, Udal'tsova, *C.A.* **56,** 13620c (1962). The free acid is less stable than its salts, and tends to decarboxylate when heated to temps of 150°. Stable on storage and on boiling in aq soln.

Cobalt salt, **Kelocyanor.**

Dipotassium salt dihydrate, $C_{10}H_{14}K_2N_2O_8 \cdot 2H_2O$, *edetate dipotassium, EDTA dipotassium.*

USE: As antioxidant in foods. Pharmaceutic aid (chelating agent).

THERAP CAT (VET): Pharmaceutic aid (chelating agent) in lead and heavy metal poisoning of farm animals.

3485. Edifenphos.

Phosphorodithioic acid O-ethyl S,S-diphenyl ester; O-ethyl S,S-diphenyl phosphorodithioate; EDDP; ediphenphos; Bayer 78418; Hinosan. $C_{14}H_{15}O_2PS_2$; mol wt 310.36. C 54.18%, H 4.87%, O 10.31%, P 9.98%, S

20.66%. Prepn: **Neth. pat. Appl. 6,611,860** corresp to G. Schrader *et al.,* **U.S. pat. 3,499,951** (1967, 1970 to Bayer). Degradn in soil: C. Tomizawa *et al., J. Environ. Sci. Health* **B11,** 231 (1976). Metabolism: I. Ueyama *et al., Agr. Biol. Chem.* **42,** 885 (1978). Mechanism of action: O. Kodama *et al., ibid.* **44,** 1015 (1980). Toxicology: T. S. Chen *et al., Toxicol. Appl. Pharmacol.* **23,** 519 (1972).

Clear yellow to light-brown liquid, bp$_{0.01}$ 154°. d$_4^{20}$ 1.23; n$_D^{22}$ 1.61. Practically insol in water; sol in acetone, xylene. LD$_{50}$ i.p. in female, male rats: 25.5, 66.5 mg/kg, T. S. Chen *et al., loc. cit.*
USE: Fungicide.

3486. Edoxudine. *2'-Deoxy-5-ethyluridine;* 5-ethyl-2'-deoxyuridine; 5-ethyl-3-(2'-deoxyribosyl)uracil; ORF 15817; EDU; EUDR; Aedurid; Edurid. C$_{11}$H$_{16}$N$_2$O$_5$; mol wt 256.26. C 51.56%, H 6.29%, N 10.93%, O 31.22%. Substituted uracil with anti-herpes activity. Prepn and properties: K. K. Gauri, **Brit. pat. 1,170,565;** *eidem,* **U.S. pat. 3,553,192** (1968, 1971 both to Robugen). Synthesis via organopalladium intermediates: D. E. Bergstrom, J. L. Ruth, *J. Am. Chem. Soc.* **98,** 1587 (1976); D. E. Bergstrom, M. K. Ogawa, *ibid.* **100,** 8106 (1978). Physical properties: M. Swierkowski, D. Shugar, *J. Med. Chem.* **12,** 533 (1969). Antiviral activity: E. De Clercq, D. Shugar, *Biochem. Pharmacol.* **24,** 1073 (1975). Preferential inhibition of herpes simplex virus type 2: C.-Z. Teh, S. L. Sacks, *Antimicrob. Ag. Chemother.* **23,** 637 (1983). Synergistic effect with interferon-β, *q.v.,* on herpes simplex virus: C. Janz, R. Wigand, *Arch. Virol.* **73,** 135 (1982). Inhibitory effect on growth of murine leukemia cells: J. Balzarini *et al., Invest. New Drugs* **2,** 35 (1984).

Long clear needles from acetone, mp 152-153°. uv max: 267 nm (ε 9610) at pH 2, 267 nm (ε 7280) at pH 1. pKa = 9.98.
THERAP CAT: Antiviral (herpes simplex).

3487. Edrophonium Chloride. *N-Ethyl-3-hydroxy-N,N-dimethylbenzenaminium chloride;* ethyl(m-hydroxyphenyl)-dimethylammonium chloride; (3-hydroxyphenyl)dimethyl-ethylammonium chloride; 3-hydroxy-N,N-dimethyl-N-eth-ylanilinium chloride; Antirex; Enlon; Reversol; Tensilon. C$_{10}$H$_{16}$ClNO; mol wt 201.70. C 59.55%, H 8.00%, Cl 17.58%, N 6.94%, O 7.93%. Cholinesterase inhibitor. Prepn: J. A. Aeschlimann, A. Stempel, **U.S. pat. 2,647,924** (1953 to Hoffmann-La Roche). LC determn in biological fluids: M. G. M. DeRuyter *et al., J. Chromatog.* **183,** 193 (1980). Clinical evaluation in reversal of neuromuscular blockade: D. M. Fisher *et al., Anesthesiology* **61,** 428 (1984). Clinical study in supraventricular tachycardia: J. Frieden *et al., Am. J. Cardiol.* **27,** 294 (1971); J. D. Cantwell *et al., Arch. Int. Med.* **130,** 221 (1972). Diagnostic use in myasthenia gravis: G. A. Nicholson *et al., Clin. Exp. Neurol.* **19,** 45 (1983); I. M. Williams *et al., ibid.* **22,** 1 (1986); in esophageal chest pain: J. E. Richter *et al., Ann. Int. Med.* **103,** 14 (1985); C. A. Lee *et al., Dig. Dis. Sci.* **32,** 682 (1987).

Crystals from isopropanol, dec 162-163°. pH of 1% aq soln 4-5. Very sol in water; freely sol in alcohol. Insol in chloroform, ether.
Bromide, C$_{10}$H$_{16}$BrNO, *edrophone bromide, Ro 2-3198.* Crystals from ethanol + ether, dec 151-152°. Bitter taste. Sol in water (more than 10%). Moderately sol in alcohol. Practically insol in ether. Solns are stable.
THERAP CAT: Cholinergic; antidote to curare principles. Diagnostic aid (myasthenia gravis; esophageal chest pain).

3488. EEDQ. *2-Ethoxy-1(2H)-quinolinecarboxylic acid ethyl ester;* N-carbethoxy-2-ethoxy-1,2-dihydroquinoline; N-ethoxycarbonyl-2-ethoxy-1,2-dihydroquinoline; BC-681. C$_{14}$H$_{17}$NO$_3$; mol wt 247.30. C 67.99%, H 6.93%, N 5.67%, O 19.41%. Coupling agent used in the synthesis of peptides: Belleau, Malek, *J. Am. Chem. Soc.* **90,** 1651 (1968); Yajima, Kawatani, *Chem. Pharm. Bull.* **19,** 1905 (1971); Sipos, Gaston, *Synthesis* **1971,** 321. Preparation: Weinberg, **U.S. pats. 3,389,142** and **3,452,140** (1968, 1969, to Bristol-Myers). Has CNS depressant activity: Belleau, *J. Am. Chem. Soc.* **90,** 823 (1968). Pharmacological studies: Martel *et al., Can. J. Physiol. Pharmacol.* **47,** 909 (1969); Chang *et al., Pharmacol. Res. Commun.* **2,** 63 (1970); Weissman, Muren, *J. Med. Chem.* **14,** 49 (1971).

mp 56-57°. bp$_{0.1}$ 125-128°.
USE: In the synthesis of peptides.

3489. Eflornithine. *2-(Difluoromethyl)-DL-ornithine;* α-difluoromethylornithine; DFMO; RMI 71782. C$_6$H$_{12}$F$_2$-N$_2$O$_2$; mol wt 182.18. C 39.56%, H 6.64%, F 20.86%, N 15.37%, O 17.57%. Irreversible inhibitor of ornithine decarboxylase. Prepn: B. W. Metcalf *et al., J. Am. Chem. Soc.* **100,** 2551 (1978); P. Bey *et al., J. Org. Chem.* **44,** 2732 (1979). Antiproliferative effects on cultured tumor cells: P. S. Mamont *et al., Biochem. Biophys. Res. Commun.* **81,** 58 (1978); on Ehrlich ascites cells in rats: L. Alhonen-Hongisto *et al., Acta Chem. Scand.* **B33,** 559 (1979). Inhibition of polyamine biosynthesis: E. Hölttä *et al., Biochem. J.* **178,** 109 (1979). Antitrypanosomal activity in mice: C. J. Bacchi *et al., Science* **210,** 332 (1980). Pharmacokinetics in humans: K. D. Haegele *et al., Clin. Pharmacol. Ther.* **30,** 210 (1981). Clinical evaluations in *Pneumocystis carinii* pneumonia: J. A. Golden *et al., West. J. Med.* **141,** 613 (1984); in trypanosomiasis: S. Van Nieuwenhove *et al., Trans. Royal Soc. Trop. Med. Hyg.* **79,** 692 (1985); in cancer chemotherapy: M. D. Abeloff *et al., Cancer Treat. Rep.* **70,** 843 (1986); V. A. Levin *et al., ibid.* **71,** 459 (1987).

Hydrochloride monohydrate, C$_6$H$_{13}$ClF$_2$N$_2$O$_2$·H$_2$O, *Ornidyl.* Crystals from ethanol/water, mp 183°.
THERAP CAT: Antineoplastic; antipneumocystis; antiprotozoal (Trypanosoma).

3490. Efloxate. *[(4-Oxo-2-phenyl-4H-1-benzopyran-7-yl)oxy]acetic acid ethyl ester;* 7-flavone ethyl hydroxyacetate; ethyl flavon-7-yloxyacetate; ethyl 7-flavonoxyacetate; 7-flavonoxyacetic acid ethyl ester; oxyflavil; Re 1-0185; Recordil.

$C_{19}H_{16}O_5$; mol wt 324.32. C 70.36%, H 4.98%, O 24.66%. Prepn: Colleoni, Setnikar, *Farmaco Ed. Sci.* **13**, 561 (1958); **Brit. pats.** 803,372, 824,547 (1958, 1959 to Recordati); Da Re, Colleoni, *Ann. Chim. (Rome)* **49**, 1632 (1959).

Crystals from 50% ethanol, mp 123-124°. Soluble in the usual organic solvents; slightly sol in water. LD_{50} i.p. in rats: 3200 mg/kg.

THERAP CAT: Coronary vasodilator.

3491. Efrotomycin. *31-O-[6-Deoxy-4-O-(6-deoxy-2,4-di-O-methyl-α-L-mannopyranosyl)-3-O-methyl-β-D-allopyranosyl]-1-methylmocimycin; 31-O-[6-deoxy-4-O-(6-deoxy-2,4-di-O-methylhexopyranosyl)-3-O-methylhexopyranosyl]-1-methylmocimycin;* FR-02A; MK-621; Producil. $C_{59}H_{88}$-N_2O_{20}; mol wt 1145.36. C 61.87%, H 7.74%, N 2.45%, O 27.94%. Antibiotic produced by *Streptomyces lactamdurans* NRRL 3802: R. G. Wax, W. M. Maiese, Ger. pat. **2,450,-813** (1975 to Merck & Co.), *C.A.* **83**, 145755y (1975); R. G. Wax *et al.*, *J. Antibiot.* **29**, 670 (1976). *In vitro* and *in vivo* activity: B. M. Frost *et al.*, ibid. 1083; **32**, 626 (1979). Production and growth promoting activity: W. M. Maiese, R. G. Wax, U.S. pat. 4,024,251 (1977 to Merck & Co.). Synergism with bottromycin, q.v.: B. M. Frost *et al.*, *J. Antibiot.* **32**, 1046 (1979). Structure: R. S. Dewey *et al.*, ibid. **38**, 1691 (1985). Stereospecific total synthesis: R. E. Dolle, K. E. Nicolaou, *J. Am. Chem. Soc.* **107**, 1691, 1695 (1985). HPLC determn in feeds: J. D. Strong, *Analyst* **111**, 853 (1986). Effect on gain and feed efficiency in swine: A. G. Foster *et al.*, *J. Anim. Sci.* **65**, 877 (1987).

Pale yellow solid. uv max (pH 7): 232, 327 nm ($E_{1cm}^{1\%}$ 464, 216). LD_{50} in mice (g/kg): >4 orally; >2 s.c. (Frost).

THERAP CAT (VET): Growth stimulant.

3492. EGF-Urogastrone. EGF-URO. Related polypeptides that are both potent stimulators of cellular proliferation and inhibitors of gastric acid secretion. *Urogastrone* was originally detected as an antisecretory agent during experiments on human urine: J. S. Gray *et al.*, *Science* **89**, 489 (1939); M. H. F. Friedman *et al.*, *Proc. Soc. Exp. Biol. Med.* **41**, 509 (1935). Isoln: J. S. Gray *et al.*, *Endocrinol.* **30**, 129 (1942); R. A. Gregory, *J. Physiol.* **129**, 528 (1955). Improved procedures led to the isoln and amino acid sequence determn of two polypeptides, β-*urogastrone* and γ-*urogastrone:* H. Gregory, *Nature* **257**, 325 (1975). These two peptides contain 3 disulfide bonds and consist of 53 and 52 amino acid residues, respectively. The only difference between them is the absence of a C-terminal arginine residue in the γ-peptide. *Epidermal growth factor*, or *EGF*, was isolated from submaxillary glands of male mice after first being detected during purification of a nerve growth promoting protein: S. Cohen, *Proc. Nat. Acad. Sci. USA* **46**, 302 (1960); idem, *J. Biol. Chem.* **237**, 1555 (1962). The primary structure of mouse EGF (mEGF) was found to be a 53 amino acid polypeptide containing 3 disulfide bonds: C. R. Savage *et al.*, ibid. **247**, 7612 (1972); **248**, 7669 (1973). Sequence of an mEGF cDNA clone that predicts the synthesis of EGF as a large protein precursor of 1,168 amino acids: A. Gray *et al.*, *Nature* **303**, 722 (1983). mEGF causes premature eye opening and stimulation of epithelial cell tissue growth when injected daily into newborn mice: S. Cohen, *loc. cit.* (1962); S. Cohen, J. M. Taylor, *Recent Progr. Horm. Res.* **30**, 533 (1974). The structural similarities between EGF and urogastrone led to experiments that established the gastric acid-inhibitory activity of mEGF and the proliferative activity of urogastrone, thus showing that both polypeptides have the same intrinsic biological activities, *cf.* H. Gregory, *loc. cit.* It has also been shown that both mEGF and urogastrone can share the same receptor sites in human tissue with almost equal affinities: M. D. Hollenberg, H. Gregory, *Life Sci.* **20**, 267 (1976). Human EGF has been isolated from urine: S. Cohen, G. Carpenter, *Proc. Nat. Acad. Sci. USA* **72**, 1317 (1975). The similarity of its amino acid composition and physicochemical properties to urogastrone has suggested the identity of these polypeptides. Although the mouse and human polypeptides are chemically distinct (16 differences in amino acid sequence), their identical intrinsic biological activities and their ability to share receptor sites with similar affinities has led to the use of the combined term EGF-urogastrone. *Reviews:* S. Cohen, R. Savage, *Recent Progr. Horm. Res.* **30**, 551-574 (1974); M. D. Hollenberg, *Vitam. Horm.* **37**, 69-110 (1979); D. Gospodarowicz, *Ann. Rev. Physiol.* **43**, 251-263 (1981); P. Walker, *J. Endocrinol. Invest.* **5**, 183-196 (1982); M. Das, *Int. Rev. Cytol.* **78**, 233-256 (1982).

Mouse EGF-URO. Mol wt 6041. uv max: 280 nm ($E_{1cm}^{1\%}$ 30.9). Isoelectric pt 4.60. Heat stable and non-dialyzable. Biological activity stable in boiling water but destroyed by heating in dil acid or alkali. Incubation with chymotrypsin or a bacterial protease also destroys biological activity.

Human EGF-URO, anthelone, anthelone U, uroanthelone, uroenterone. Mol wt 6201. Isoelectric pt 4.5. Very sol in water. Sol in methanol, ethylene gycol.

3493. Egg Oil. Oil of egg yolk. Obtained from fresh egg yolks by extraction with ethylene dichloride. Contains fatty glycerides, cholesterol, lecithin. The glyceride fraction is a mixture of the glycerides of satd and unsatd fatty acids. Palmitic, stearic, oleic, linoleic and clupanodonic acids have been isolated from both the glyceride and lecithin fractions. Preparation: Levin, Lerman, *J. Am. Oil Chem. Soc.* **28**, 441 (1951); U.S. pat. 2,503,312 (to Vio-Bin Corp.).

Dark oil. d 0.95. n_D^{20} 1.4790. Sol in the common organic solvents. Miscible with other oils. Insol in water, but disperses readily when shaken with it to form emulsions.

USE: In the formulation of hydrophilic ointment bases for medicinal ointments and cosmetic creams: Bandelin, Tuschhoff, *J. Am. Pharm. Assoc. [Pract. Pharm. Ed.]* **14**, 106, 120 (1953); *Schimmel-Briefs*, no. 235 (Oct. 1954).

3494. Eicosamethylnonasiloxane. $C_{20}H_{60}O_8Si_9$; mol wt 681.49. C 35.25%, H 8.87%, O 18.78%, Si 37.10%. Prepd by the reaction of hexamethyldisiloxane with octamethylcyclotetrasiloxane and sulfuric acid; by varying the proportions of these reactants, other methylpolysiloxanes of a desired average molecular weight may be prepd: Patnode, Wilcock, *J. Am. Chem. Soc.* **68**, 362 (1946).

Liquid. $bp_{4.9}$ 173°. d 0.918. n_D^{20} 1.3980. Stable. Inert to most chemical reagents and rubber. Maintains about the same viscosity over a wide temp range. Sol in benzene and the lighter hydrocarbons; slightly sol in alc and the heavy hydrocarbons.

USE: As a basis for silicone oils or fluids designed to withstand extremes of temp; as a foam suppressant in petr lubricating oil.

3495. 5,8,11,14,17-Eicosapentaenoic Acid. *all-cis-5,8,-11,14,17-eicosapentaenoic acid; all-cis*-fatty acid 20:5

omega-3; EPA. $C_{20}H_{30}O_2$; mol wt 302.46. C 79.42%, H 10.00%, O 10.58%. Important polyunsaturated fatty acid of the marine food chain that serves as a precursor for the prostaglandin-3 and thromboxane-3 families. It differs from arachidonic acid, *q.v.* (the eicosatetraenoic acid that is a precursor for the prostaglandin and thromboxane-2 families) by the extra double bond between the third and fourth carbons from the "methyl end" of the molecule. Isoln from cod liver oil: E. Klenk, D. Eberhagen, *Z. Physiol. Chem.* **307**, 42 (1957). Enzymatic conversion to prostaglandin E_3: S. Bergström *et al., J. Biol. Chem.* **239**, PC 4006 (1964). Effects on role of platelets in thrombosis: K. C. Srivastava *et al., Biochem. Exp. Biol.* **16**, 317 (1980). Effects on prostacyclin-like material in human umbilical vasculature: J. Dyerberg, K. A. Jorgensen, *Artery* **8**, 12 (1980). A possible relationship between diets rich in EPA in marine oils and low rates of ischemic heart disease has been proposed: H. O. Bang *et al., Am. J. Clin. Nutr.* **33**, 2657 (1980); J. Dyerberg, *Philos. Trans. Roy. Soc. London [Biol.]* **294**, 373 (1981). Clinical evaluation of lipid lowering effect: Y. Nagakawa *et al., Atherosclerosis* **47**, 71 (1983); of use in rheumatoid arthritis: J. M. Kremer *et al., Ann. Int. Med.* **106**, 497 (1987); of use in Raynaud's phenomenon: R. A. DiGiacomo *et al., Am. J. Med.* **86**, 158 (1989).

Colorless oil. n_D^{20} 1.49865.
Ethyl ester, $C_{22}H_{34}O_2$, *Eskima.*
THERAP CAT: Antihyperlipoproteinemic.

3496. Einsteinium. Es; at. wt (most stable known isotope) 252; at. no. 99; valence 3, also 2. Man-made radioactive element. ^{253}Es, ($T_{1/2}$ 20.47 days, α-emitter), originally discovered in debris from a thermonuclear explosion in Nov. 1952. Reviews of discovery: Ghiorso *et al., Phys. Rev.* **99**, 1048-1049 (1955); Ghiorso, Seaborg, *Sci. Am.* **195** (6), 67-80 (1956). Known isotopes range in mass number from 243-256. ^{254}Es ($T_{1/2}$ 276 days, α-emitter), longest-lived isotope; best prepd by irradiation of plutonium (or other elements of high at. no.) for several years in a high flux reactor: Seaborg, *J. Chem. Ed.* **36**, 38 (1959). *Reviews:* C. Keller, *The Chemistry of the Transuranium Elements* (Verlag Chemie, Weinheim, English Ed., 1971) pp 583-589. Silva, "Trans-Curium Elements" in *MTP Int. Rev. Sci.: Inorg. Chem., Ser. One* vol. **8**, A. G. Maddock, Ed. (University Park Press, Baltimore, 1972) pp 71-105; *Comprehensive Inorganic Chemistry* vol. **5**, J. C. Bailar, Jr. *et al.,* Eds. (Pergamon Press, Oxford, 1973) *passim;* several authors in *Handb. Exp. Pharmakol.* **36**, 689-928 (1973).

3497. Elaidic Acid. *9-Octadecenoic acid.* $C_{18}H_{34}O_2$; mol wt 282.45. C 76.54%, H 12.13%, O 11.33%. A stereoisomer of oleic acid, *q.v.* Prepd by the action of HNO_2 on oleic acid. ^{13}C-NMR: J. G. Batchelor *et al., J. Org. Chem.* **39**, 1698 (1974).

White leaflets. d^{79} 0.851. mp 44-45°; also stated as 51°. bp_{100} 288°; bp_{15} 234°. n_D^{100} 1.4308.

3498. Elaiomycin. *4-Methoxy-3-(1-octenyl-ONN-azoxy)-2-butanol;* (2S,3S)-4-methoxy-3-(1'-cis-octenyl-cis-azoxy)-2-butanol. D-*threo*-4-methoxy-3-(1-octenylazoxy)-2-butanol. $C_{13}H_{26}N_2O_3$; mol wt 258.35. C 60.43%, H 10.14%, N 10.84%, O 18.58%. Antibiotic substance produced by *Streptomyces hepaticus:* Haskell *et al., Antibiot. & Chemother.* **4**, 141 (1954); Ehrlich *et al.,* ibid. 338; Anderson *et al.,* ibid. **6**, 100 (1956). Inhibits the growth of *Mycobacterium tuberculosis* var. *hominis in vitro:* Karlson, ibid. **12**, 446 (1962). Reported to have carcinogenic activity: Schoental, *Nature* **221**, 765 (1969). Structure: Stevens *et al., J. Am. Chem. Soc.* **78**, 3229 (1956); **80**, 6088 (1958). Configuration: Stevens *et al.,* ibid. **81**, 1435 (1959); McGahren, Kunstmann,

ibid. **92**, 1587 (1970). Total synthesis: R. A. Moss, M. Matsuo, *ibid* **99**, 1643 (1977). Biosynthesis: R. J. Parry *et al., ibid.* **104**, 339 (1982).

Pale yellow oil. Stable to air. $[\alpha]_D^{26}$ +38.4° (c = 2.8 in abs ethanol). n_D^{25} 1.4798. uv max: 237.5 nm ($E_{1cm}^{1\%}$ 428). Sparingly sol in water; sol in practically all of the common organic solvents. Stable in neutral or slightly acidic aq solns. Dec into a yellow product when dissolved in 0.1*N* NaOH.
Acetate, $C_{15}H_{28}N_2O_4$, oily liquid, $bp_{0.5}$ 84-90°. $[\alpha]_D^{27}$ +25.3° (c = 3 in alc). uv max (alc): 237.5 nm (ϵ 11,000).

3499. Elastase. Pancreatopeptidase E; Elaszym. A protease of broad specificity, obtained from dried pancreas. Mol wt about 25,000. Capable of hydrolyzing proteins at N-terminal peptide bonds of aliphatic residues. Breaks down elastin, the specific protein of elastic fibers. Digests other proteins, such as fibrin, hemoglobin, albumin, soybean proteins and casein. Isoln by an adsorption method: Bagdy, Banga, *Acta Physiol. Acad. Sci. Hung.* **11**, 371 (1957); *cf.* Lewis *et al., J. Biol. Chem.* **222**, 705 (1956); Miller, Blum, *ibid.* **218**, 131 (1956). Porcine elastase is a single polypeptide chain of 240 amino acid residues: Shotton, Hartley, *Biochem. J.* **131**, 643 (1973).

White, lyophilized powder of slightly yellowish shade. Isoelectric point 9.5. Optimum pH between 8.1 and 8.8 according to the buffer employed. At refrigerator temps (4 to 6°) the thawed enzyme remains stable for several days at pH values between 5 and 10. Rapidly inactivated in highly acid media.

3500. Elastin. Elastic, load-bearing protein fibers of animal connective tissue, particularly the ligaments of the vertebrae and the walls of the large arteries. Elastin is an insoluble, highly cross-linked hydrophobic protein, rich in non-polar amino acid residues, such as valine, leucine, isoleucine, and phenylalanine. On the average, about every third residue is glycine and about every ninth residue is a prolyl residue. Unlike collagen, *q.v.,* which is rich in hydroxyproline, elastin contains only modest amounts. At least two types of elastin are distinguishable: type I elastin isolated from *ligamentum nuchae,* aorta or skin and type II elastin isolated from cartilage. The most ready source of high purity elastin is the *ligamentum nuchae* of the larger ruminants: Partridge *et al., Biochem. J.* **61**, 11 (1955). Amino acid composition studies: Petruska, Sandberg, *Biochem. Biophys. Res. Commun.* **33**, 222 (1968). Its elastic properties are brought about by a cross-linked structure. Identification of new cross-linking amino acids in elastin: Partridge *et al., Biochem. J.* **93**, 30C (1964); Franzblau *et al., Biochem. Biophys. Res. Commun.* **21**, 575 (1965); Starcher *et al., Biochemistry* **6**, 2425 (1967). Molecular model: Gray *et al., Nature* **246**, 461 (1973). *Review:* Partridge, *Advan. Protein Chem.* **17**, 227-302 (1962); C. Franzblau, "Elastin" in *Comprehensive Biochemistry* vol. **26c**, M. Florkin, E. H. Stotz, Eds. (Elsevier, New York, 1971) pp 659-712; Ross, Bornstein, *Sci. Amer.* **224**, 44 (June, 1971). Book (in English): I. Banga, *Structure and Function of Elastin and Collagen* (Akademiai Kiado, Budapest, 1967). Series of articles on structure, biosynthesis, and immunology: *Methods in Enzymology* vol. **82**, L. W. Cunningham, D. W. Frederiksen, Eds. (Academic Press, New York, 1982) pp 559-765.

Purified elastin has a pale yellow color and a bluish fluorescence in uv light. Resists acid and alkaline hydrolysis. Practically insol in a wide range of hydrogen-bond-breaking solvents at temps up to 100° and swells, but does not dissolve, in phenolic solvents. It appears practically impossible to bring elastin into soln except by hydrolytic reagents capable of rupturing peptide bonds. Elastin is one of only a few polymeric substances which, in the presence of water, exist in a form with rubber-like extensibility and low modulus of elasticity. Enzymes which dissolve fibers of the insol protein elastin undamaged and free from contamination are called elastases.

3501. Elcatonin. *1-Butanoic acid-26-L-aspartic acid-27-L-valine-29-L-alanine-1,7-dicarbacalcitonin (salmon); 1,4,7,10,13,16-hexaazacyclotricosane cyclic peptide deriv;* [aminosuberic acid 1,7]-eel calcitonin; [ASU^{1,7}]-E-CT; carbocalcitonin; HC-58; Calcinil; Carbicalcin; Elcitonin; Turbocalcin. $C_{148}H_{244}N_{42}O_{47}$; mol wt 3363.82. C 52.85%, H 7.31%, N 17.49%, O 22.35%. Synthetic analog of eel calcitonin; amino acid sequence also similar to that of salmon calcitonin. Polypeptide hormone effective in the reduction of plasma calcium. More stable than natural calcitonins because of the absence of disulfide bridges. Prepn: S. Sakakibara *et al.*, **Ger. pat. 2,616,399;** *eidem*, **U.S. pat. 4,086,221** (1976, 1978 both to Toyo Jozo); T. Morikawa *et al.*, *Experientia* **32**, 1104 (1976). *In vitro* growth inhibition of human breast cancer cells: Y. Iwasaki *et al.*, *Biochem. Biophys. Res. Commun.* **110**, 235 (1983). Clinical pharmacology: T. Ishioka, *Pharmatherapeutica* **1**, 625 (1977). Clinical study in Paget's disease: A. Caniggia *et al.*, *Minerva Med.* **74**, 993 (1983); in osteoporosis: Italian Osteoporosis Network, *Curr. Ther. Res.* **45**, 502 (1989).

O
‖
C——— (CH$_2$)$_4$ ——— CH$_2$
| |
| O
| ‖
Ser-Asp-Leu-Ser-Thr-NH-CH-C-Val-Leu-Gly-Lys-Leu-Ser-Glu-Glu-Leu-
| |
NH$_2$ NH$_2$

His-Lys-Leu-Glu-Thr-Tyr-Pro-Arg-Thr-Asp-Val-Gly-Ala-Gly-Thr-Pro-NH$_2$
|
NH$_2$

THERAP CAT: Calcium regulator.   Treatment of Paget's disease.

3502. Eledoisin. $C_{54}H_{85}N_{13}O_{15}S$; mol wt 1188.44. C 54.57%, H 7.21%, N 15.32%, O 20.19%, S 2.70%. A hendecapeptide from the posterior salivary glands of eledone spp (small octopus spp): Anastasi, Erspamer, *Brit. J. Pharmacol.* **19**, 326 (1962); *eidem, Experientia* **18**, 58 (1962). Synthesis: Sandrin, Boissonnas, *ibid.* 59; *eidem, Helv. Chim. Acta* **47**, 1294 (1964); Lubke *et al., Ann.* **679**, 195 (1964); **Brit. pat. 984,810** (1965 to Farmitalia); solid-phase synthesis: P. G. Pietta *et al., J. Org. Chem.* **39**, 44 (1974). Its physiologic action resembles that of the other tachykinins, substance P, *q.v.*, and physalaemin, *q.v.* Stimulates extravascular smooth muscle, acts as a potent vasodilator and hypotensive agent; in certain species causes salivation, and increases capillary permeability. Pharmacology: Erspamer, Erspamer, *Brit. J. Pharmacol.* **19**, 337 (1962); Sicuteri *et al., Experientia* **19**, 44 (1963); G. Bertaccini, *Pharmacol. Rev.* **28**, 127 (1976). Review: Erdös, *Advan. Pharmacol.* **4**, 64-72 (1966).

5-oxo-Pro-Pro-Ser-Lys-Asp-Ala-Phe-Ile-Gly-Leu-Met-NH$_2$

Sesquihydrate, powder, dec 230°. $[\alpha]_D^{22}$ −44° (c = 1 in 95% acetic acid). Slowly loses activity when incubated in blood.

Trifluoroacetate, $C_{56}H_{86}F_3O_{17}S$, Eloisin.

THERAP CAT: Hypotensive; stimulator of lacrimal secretion.

3503. Element 104. *Unnilquadium;* kurchatovium; rutherfordium; dubnium. Proposed symbols: Unq, Ku, Rf. Initial production of element 104 claimed by two groups. Prepn of isotope ^{260}104 (T$_{1/2}$ 0.1 seconds; spontaneous fission) by bombardment of ^{242}Pu with ^{22}Ne ions reported by Flerov *et al., At. Energ. (USSR)* **17**, 310 (1964), *C.A.* **62**, 189d (1965); *eidem, Phys. Letters* **13**, 73 (1964). Preparation of two new, α-emitting isotopes, ^{257}104 (T$_{1/2}$ 4.5 seconds) and ^{259}104 (T$_{1/2}$ 3 seconds) by bombardment of ^{249}Cf with ^{12}C and ^{13}C: Ghiorso *et al., Phys. Rev. Letters* **22**, 1317 (1969). Discussions of conflicting claims: Ghiorso; Zvara and discussion in *Proceedings of the Robert A. Welch Foundation Conferences on Chemical Research, XIII. The Transuranium Elements—The Mendeleev Centennial,* W. O. Milligan, Ed.

(Houston, Texas, 1970) pp 107-241; Zvara *et al., Inorg. Nucl. Chem. Letters* **7**, 1109 (1971); Ghiorso *et al., ibid.* 1117; *Nature* **229**, 603 (1971). Five isotopes reported; mass numbers 257-261. Longest-lived known isotope, ^{261}104, α-emitter, T$_{1/2}$ 65 ± 10 seconds; produced by bombardment of ^{248}Cm with ^{18}O ions: Ghiorso *et al., Phys. Letters B* **32**, 95 (1970). Chemical properties similar to hafnium and zirconium: Zvara *et al., At. Energ. (USSR)* **21** (2), 83 (1966), *C.A.* **65**, 16357h (1966); Silva *et al., Inorg. Nucl. Chem. Letters* **6**, 871 (1970). Reviews of history, prepn and properties: C. Keller, *The Chemistry of the Transuranium Elements* (Verlag Chemie, Weinheim, English Ed., 1971) pp 613-618; Silva, "Trans-Curium Elements" in *MTP Int. Rev. Sci.: Inorg. Chem., Ser. One* vol. 8, A. G. Maddock, Ed. (University Park Press, Baltimore, 1972) pp 71-105; Ghiorso, *Handb. Exp. Pharmakol.* **36**, 691-715 (1973).

3504. Element 105. *Unnilpentium;* hahnium; nielsbohrium; eka-tantalum; bohrium. Proposed symbols: Unp, Ha. Prepn of α-emitting isotope, ^{260}Ha (T$_{1/2}$ 1.6 ± 0.3 sec), by bombardment of ^{249}Cf with ^{15}N ions: Ghiorso *et al., Phys. Rev. Letters* **24**, 1498 (1970). Prepn of isotope, most probably ^{261}105, by bombardment of ^{243}Am with ^{22}Ne ions; decay by spontaneous fission measured, T$_{1/2}$ ~2 sec: Flerov *et al., Joint Inst. Nucl. Invest. Reps.* **JINR-P7-5108** and **JINR-P7-5164** (1970), *C.A.* **74**, 118785r (1971); *C.A.* **75**, 28639n (1971); *eidem, At. Energy (USSR)* **29** (4), 243 (1970); *eidem, Nucl. Phys. A* **160**, 181 (1971); *eidem, Proc. Int. Conf. Heavy Ion Phys.* **1971** pp 125-143, *C.A.* **80**, 139758y (1974). Prepn of two additional α-emitting isotopes: ^{261}Ha (T$_{1/2}$ 1.7 sec) from ^{249}Bk and ^{16}O ions or from ^{250}Cf and ^{15}N ions; ^{262}Ha (T$_{1/2}$ 40 sec) from ^{249}Bk and ^{18}O ions: Ghiorso *et al., Phys. Rev. C* **4**, 1850 (1971). Discussion of conflicting claims to discovery of element 105: Holcomb, *Science* **168**, 810 (1970); Flerov, *ibid.* **170**, 15 (1970); Ghiorso, *ibid.* **171**, 127 (1971). Reviews of history and prepn: C. Keller, *The Chemistry of the Transuranium Elements* (Verlag Chemie, Weinheim, English Ed., 1971) pp 619-622; Silva, "Trans-Curium Elements" in *MTP Int. Rev. Sci.: Inorg. Chem., Ser. One* vol. 8, A. G. Maddock, Ed. (University Park Press, Baltimore, 1972) pp 71-105; Ghiorso, *Handb. Exp. Pharmakol.* **36**, 691-715 (1973).

3505. Element 106. *Unnilhexium.* Proposed symbol: Unh. Prepn of isotope (mass no. 263) by bombardment of ^{249}Cf with ^{18}O ions: Ghiorso *et al., Phys. Rev. Letters* **33**, 1490 (1974). ^{263}106 is an α-emitter; T$_{1/2}$ 0.9 ± 0.2 seconds. Lighter isotopes prepd by Flerov and co-workers by bombardment of ^{207}Pb and ^{208}Pb with ^{54}Cr; decay by spontaneous fission; half-lives 4-10 milliseconds: *Chem. Eng. News* **52**, 4 (Sept. 16, 1974); Maugh, *Science* **186**, 42 (1974); Ghiorso *et al., loc. cit.*

3506. Element 107. *Unnilseptium.* Proposed symbol: Uns. Prepn of α-emitting isotope ^{261}107 by bombardment of ^{209}Bi with ^{54}Cr ions; decay by spontaneous fission, T$_{1/2}$ 1-2 msec: Yu. Ts. Organessian *et al., Nucl. Phys. A* **273**, 505 (1976); G. N. Flerov, *C.A.* **87**, 173830v (1977); A. S. Iljinov *et al., Report* **JINR-E-7-9686** (1976), *C.A.* **90**, 62354k (1979). Detection of ^{262}107 by correlated α-decays: G. Muenzenberg, *Prax. Naturwiss., Chem.* **31**, 174 (1982).

3507. Elenolide. *4-(1-Formyl-1-propenyl)-3,4-dihydro-2-oxo-2H-pyran-5-carboxylic acid methyl ester.* $C_{11}H_{12}O_5$; mol wt 224.21. C 58.92%, H 5.40%, O 35.68%. Hypotensive lactone from fruits, bark and leaves of the olive tree, *Olea europaea* L., Oleaceae: Veer *et al., Rec. Trav. Chim.* **76**, 839 (1957); **Brit. pat. 789,427** (1958 to Organon); **U.S. pat. 3,033,877** (1962 to Organon). Structure: Panizzi *et al., Gazz. Chim. Ital.* **90**, 1464 (1960); Beyerman *et al., Bull. Soc. Chim. France* **1961**, 1812. Synthesis from secologanin: L. F. Tietze, H. C. Uzar, *Angew. Chem. Int. Ed.* **18**, 539 (1979).

Needles from hot alc, mp 155.5°. $[\alpha]_D^{20}$ +369° (chloroform). uv max (ethanol): 225, 317 nm (log ϵ 4.29, 1.75). *Compare* Oleuropein.

3508. Ellagic Acid. *2,3,7,8-Tetrahydroxy[1]benzopyrano-[5,4,3-cde][1]benzopyran-5,10-dione; 4,4',5,5',6,6'-hexahydrodiphenic acid 2,6,2',6'-dilactone; benzoaric acid; Lagistase.* $C_{14}H_6O_8$; mol wt 302.19. C 55.64%, H 2.00%, O 42.36%. Occurs free or combined in galls. Isoln from the kino of *Eucalyptus maculata* Hook and *E. hemipholia* F. Muell., *Myrtaceae:* Gell et al., *Aust. J. Chem.* **11**, 372 (1958); Hills, Carle, *ibid.* **16**, 147 (1963). Prepd by sodium persulfate oxidation of gallic acid or by acid hydrolysis of crude tannin from walnuts: Perkin, Nierenstein, *J. Chem. Soc.* **87**, 1412 (1905); Jurd, *J. Am. Chem. Soc.* **78**, 3445 (1956); **79**, 6043 (1957). Purification: **Fr.** pat. **1,478,523** (1967 to Prod. Chim. Celluloses Rey), *C.A.* **68**, 78267r (1968). Physicochemical properties: Press, Hardcastle, *J. Appl. Chem.* **19**, 247 (1969). Pharmacology: Bhargava et al., *J. Pharm. Sci.* **57**, 1728 (1968). Ellagic acid is a potent antagonist of the mutagenicity of bay-region diol epoxides of several aromatic hydrocarbons. Inhibition of the mutagenicity of the ultimate carcinogenic metabolite of benzo[*a*]pyrene: A. W. Wood et al., *Proc. Nat. Acad. Sci. USA* **79**, 5513 (1982). Study of the reaction between this metabolite and ellagic acid: J. M. Sayer et al., *J. Am. Chem. Soc.* **104**, 5562 (1982). Brief discussion of ellagic acid as a prototype of a new class of cancer-preventing drugs: J. Fox, *Chem. & Eng. News* **60**, 26 (Oct. 25, 1982).

Cream colored needles from pyridine, mp above 360°. uv max (ethanol): 366, 255 nm (log ϵ 3.93, 4.60). Slightly sol in water or alcohol; sol in alkalies, in pyridine. Practically insol in ether.

Tetraacetate, $C_{22}H_{14}O_{12}$, needles from acetic anhydride, mp 340°.

THERAP CAT: Hemostatic.

3509. Ellipticine. *5,11-Dimethyl-6H-pyrido[4,3-b]carbazole.* $C_{17}H_{14}N_2$; mol wt 246.30. C 82.90%, H 5.73%, N 11.37%. Antitumor alkaloid isolated from *Ochrosia elliptica* Labill., *O. sandwicensis* A. DC., *O. viellardii, O. silvatica, Apocynaceae:* Goodwin et al., *J. Am. Chem. Soc.* **81**, 1903 (1959); Kan Fan et al., *Phytochemistry* **9**, 1351 (1970); Cosson, Schmid, *ibid.* 1353. Structure and synthesis: Woodward et al., *J. Am. Chem. Soc.* **81**, 4434 (1959). Recent syntheses: J. Bergman, R. Carlsson, *Tetrahedron Letters* **1977**, 4663; D. A. Taylor et al., *J. Chem. Res. (M)* **1979**, 4801; S. Kano et al., *J. Org. Chem.* **46**, 2979 (1981); R. Besselievre, H. P. Husson, *Tetrahedron* **37**, 241 (1981). Review of syntheses: M. Sainsbury, *Synthesis* **1977**, 437-448; R. Barone, M. Chanon, *Heterocycles* **16**, 1357-1365 (1981). Review of antineoplastic activity: K. W. Kohn et al., *Antibiotics* **vol. 5**(pt. 2), F. E. Hahn, Ed. (Springer-Verlag, New York, 1979) pp 195-213.

Bright yellow needles from ethyl acetate, mp 311-315° (dec). uv max: 239, 277, 286, 294, 332, 382, 400 nm (log ϵ 4.23, 4.61, 4.76, 4.74, 3.65, 3.61, 3.53). LD_{50} in mice: 19.5-22.4 mg/kg i.v.; 178-204 mg/kg orally, Rakieten et al., *U.S. Govt. Res. Develop. Rep.* **67**, 38 (1967), *C.A.* **68**, 20827g (1968).

3510. Elliptinium Acetate. *9-Hydroxy-2,5,11-trimethyl-6H-pyrido[4,3-b]carbazolium acetate (salt); 9-hydroxy-2-methylellipticinium acetate; HME; NSC-264137; Celiptium.* $C_{20}H_{20}N_2O_3$; mol wt 336.39. C 71.41%, H 5.99%, N 8.33%, O 14.27%. Deriv of ellipticine, *q.v.*, with anticancer activity. Prepn: J. LePecq et al., **Ger.** pat. **2,618,223** (1976 to Agence Nat. Valorisation), *C.A.* **86**, 55620h (1977). Activity vs L 1210 mouse leukemia: *eidem, Compt. Rend. Ser. D.* **281**, 1365 (1975). HPLC determn: G. Muzard, J. B. LePecq, *J. Chromatog.* **169**, 446 (1979). Metabolism, disposition: N. Van-Bac et al., *Cancer Treat. Rep.* **64**, 879 (1980). Antitumor activity, pharmacology, toxicity study: C. Paoletti et al., *Recent Results Cancer Res.* **74**, 107 (1980). Use in treatment of breast cancers: J. Rouesse et al., *Bull. Cancer* **68**, 437 (1981). Cytochemical and autoradiographic study: N. Sales, E. Puvion, *Eur. J. Cancer Clin. Oncol.* **18**, 291 (1982).

THERAP CAT: Antineoplastic.

3511. Elliptone. *12,12a-Dihydro-8,9-dimethoxy[1]benzopyrano[3,4-b]furo[2,3-h][1]benzopyran-6(6aH)-one;* $C_{20}H_{16}O_6$; mol wt 352.33. C 68.17%, H 4.58%, O 27.25%. Found in derris resins of low rotenone content. Isoln of naturally occurring (−)-form from *Derris elliptica* (Wall.) Benth., *Leguminosae:* Buckley, *J. Soc. Chem. Ind.* **55**, 285T (1936); Harper, *Chem. & Ind. (London)* **57**, 1059 (1938); **58**, 292 (1939); *J. Chem. Soc.* **1939**, 1099; see also Meyer, Koolhaas, *Rec. Trav. Chim.* **58**, 207 (1939). Structure: Harper, *J. Chem. Soc.* **1939**, 1424; **1942**, 593. Comparison with similar substances: Seiferle, Frear, *Ind. Eng. Chem.* **40**, 683 (1948). Total synthesis of (±)-form: Fukami et al., *Agr. Biol. Chem.* **29**, 82 (1965), *C.A.* **62**, 13115e (1963). Synthesis of (−)-form from rotenone *(q.v.):* P. B. Anzeveno, *J. Heterocycl. Chem.* **16**, 1643 (1979).

Needles from ethanol, mp 160° (Harper), also reported as mp 177-178° (Anzeveno). $[\alpha]_D^{20}$ −18° (benzene); $[\alpha]_D^{20}$ +55° (acetone).

3512. Elymoclavine. *8,9-Didehydro-6-methylergoline-8-methanol.* $C_{16}H_{18}N_2O$; mol wt 254.32. C 75.56%, H 7.13%, N 11.02%, O 6.29%. An ergot alkaloid obtained from cultures of fungi parasitic on *Elymus mollis* Trin.: Abe et al., **U.S.** pat. **2,835,675** (1958). Found in fungi parasitic on *Pennisetum typhoideum* Rich.: Stoll et al., *Helv. Chim. Acta* **37**, 1815 (1954). Biosyntheses utilizing *Claviceps* cultures: Naidoo et al., *Chem. Commun.* **1970**, 472; Ogunlana et al., *ibid.* 775; Seiler et al., *ibid.* 1394; Cavender, Anderson, *Biochim. Biophys. Acta* **208**, 345 (1970). Structure and stereochemistry: Schreier, *Helv. Chim. Acta* **41**, 1984 (1958).

Monoclinic prisms from methanol, 248-252° (dec). $[\alpha]_D^{20}$ −59° (c = 0.1 in ethanol) (Abe); $[\alpha]_D^{20}$ −152° (c = 0.9 in pyridine) (Stoll). uv max: 227, 283, 293 nm (log ϵ 4.31, 3.84, 3.76). Fairly sol in water with alkaline reaction. Sol in pyridine; very slightly sol in organic solvents.

3513. Embelin. *2,5-Dihydroxy-3-undecyl-2,5-cyclo-hexadiene-1,4-dione; 2,5-dihydroxy-3-undecyl-p-benzoquinone;* "Embelic acid". $C_{17}H_{26}O_4$; mol wt 294.39. C 69.36%, H 8.90%, O 21.74%. From fruit of *Embelia ribes* Burm., *Myrsinaceae*. Isoln: Heffter, Feuerstein, *Arch. Pharm.* **238**, 15 (1900). Structure and synthesis: Fieser, Chamberlin, *J. Am. Chem. Soc.* **70**, 71 (1948).

Glistening orange plates from alcohol, benzene or acetic acid, mp 142-143°. Soluble in the usual hot organic solvents or in alkali hydroxide solns; very slightly sol in petr ether. Practically insol in water.

Diammonium salt, $C_{17}H_{34}N_2O_4$, *ammonium embelate.* Grayish-violet powder. Soluble in water, dil alcohol. *Caution: Handle with care; irritates mucous membranes and may cause prolonged and violent sneezing!*

Disemicarbazone, $C_{19}H_{32}N_6O_4$, pale-brown needles from dil alcohol, dec 236°.

Dioxime, $C_{17}H_{28}N_2O_4$, pale-yellow needles from acetic acid, mp 278°.

THERAP CAT: Ammonium embelate as an anthelmintic (Cestodes).

3514. Embramine. *2-[1-(4-Bromophenyl)-1-phenyleth-oxy]-N,N-dimethylethanamine; 2-[(p-bromo-α-methyl-α-phenylbenzyl)oxy]-N,N-dimethylethylamine;* [2-(1-p-bromo-phenyl-1-phenylethoxy)ethyl]dimethylamine; 2-[1-(4-bro-modiphenyl)ethoxy]-N,N-dimethylethylamine; p-bromo-α-methylbenzhydryl 2-dimethylaminoethyl ether; 1-(p-bromo-phenyl)-1-phenyl-1-(2-dimethylaminoethoxy)ethane; β-di-methylaminoethyl p-bromo-α-methylbenzhydryl ether. $C_{18}H_{22}BrNO$; mol wt 348.30. C 62.07%, H 6.37%, Br 22.95%, N 4.02%, O 4.59%. Prepn: Novak, Protiva, *Coll. Czech. Chem. Commun.* **24**, 3966 (1959); *eidem*, *Ger.* pat. **1,240,877** (1967 to SPOFA), *C.A.* **67**, 73338j (1967).

$bp_{0.3}$ 135-140°. bp_1 163-170°.

Hydrochloride, $C_{18}H_{23}BrClNO$, *mebrophenhydramine, Bromadryl, Mebryl.* Crystals from acetone + ether, mp 152-155°. LD_{50} in white mice: 80 mg/kg i.v.; 330 mg/kg orally. THERAP CAT: Antihistaminic.

3515. Emepronium Bromide. *N-Ethyl-N,N,α-trimethyl-γ-phenylbenzenepropanaminium bromide; ethyldimethyl(1-methyl-3,3-diphenylpropyl)ammonium bromide;* ethyl(3,3-diphenyl-1-methylpropyl)dimethylammonium bromide; (1-methyl-3,3-diphenylpropyl)dimethylethylammonium bromide; Cetiprin; Hexanium; Restenacht; Ripirin; Uro-Ripirin. $C_{20}H_{28}BrN$; mol wt 362.37. C 66.29%, H 7.79%, Br 22.05%, N 3.87%. Prepn: Carlsson, **Swedish** pat. **136,606** (1952 to Aktiebolaget Recip.), *C.A.* **48**, 729a (1954).

Crystals, mp 204°.
THERAP CAT: Antispasmodic.

3516. Emetamine. *1',2',3',4'-Tetrahydro-6',7',10,-11-tetramethoxyemetan.* $C_{29}H_{36}N_2O_4$; mol wt 476.60. C 73.08%, H 7.61%, N 5.88%, O 13.43%. Occurs in small amounts in ipecac. Isoln: Pyman, *J. Chem. Soc.* **111**, 419 (1917). Structure: Battersby, *Spec. Publ. Chem. Soc. (London)* **3**, 36 (1955); Battersby *et al., J. Chem. Soc.* **1959**, 1744. Synthesis: Evstigneeva *et al., Proc. Acad. Sci. USSR* **117**, 227 (1957); Battersby *et al., J. Chem. Soc.* **1961**, 3899.

Crystals from ether, mp 142-143°. uv max (ethanol): 236, 283 nm (log ϵ 4.85, 3.86).

3517. Emetine. *6',7',10,11-Tetramethoxyemetan;* cephaeline methyl ether. $C_{29}H_{40}N_2O_4$; mol wt 480.63. C 72.47%, H 8.39%, N 5.83%, O 13.32%. Principal alkaloid of ipecac, the ground roots of *Uragoga ipecacuanha* (Brot.) Baill. *Rubiaceae.* Extraction procedures: E. Merck, *BIOS Final Rpt.* 766 (1947); F. E. Hamerslag, *Technology and Chemistry of Alkaloids* (New York, 1950) chapter XI; Stoll, Jucker in *Ullmann's Encyklopädie der Technischen Chemie* **3**, Aufl., Bd. III (München-Berlin, 1953) p 231; *Beilstein* 2nd suppl., vol. **23**, 449. Structure: Robinson, *Nature* **162**, 524 (1948); Janot, *Bull. Soc. Chim. France* **1949**, 185. Stereochemistry: Battersby, *Chem. & Ind. (London)* **1958**, 1324; Battersby, Garrett, *J. Chem. Soc.* **1959**, 3512; Van Tamelen *et al., J. Am. Chem. Soc.* **81**, 6214 (1959). Total synthesis: Evstigneeva *et al., Proc. Acad. Sci. USSR* **75**, 539 (1950); Van Tamelen *et al., J. Am. Chem. Soc.* **91**, 7359 (1969); T. Kametani *et al., J. Chem. Soc. Perkin Trans. I* **1979**, 1211. Alternate syntheses: Burgstahler, Bithos, *J. Am. Chem. Soc.* **82**, 5446 (1960); Openshaw, Whittaker, *J. Chem. Soc.* **1963**, 1461; S. Takano *et al., J. Org. Chem.* **43**, 4169 (1978); T. Fujii, S. Yoshifuji, *Tetrahedron* **36**, 1539 (1980). *See also* Clark *et al., U.S.* pat. **3,102,118** (1963 to Glaxo). Review of syntheses: M. Shamma, *The Isoquinoline Alkaloids* (Academic Press, New York, 1972) pp 426-457. Toxicity: Radomski *et al., J. Pharmacol. Exp. Ther.* **104**, 421 (1952); of dihydrochloride: Child *et al., J. Pharm. Pharmacol.* **16**, 65 (1964). *General review:* Grollman, Jarkovsky, in *Antibiotics* Vol. **3**, J. W. Corcoran, F. E. Hahn, Eds. (Springer-Verlag, New York, 1975) pp 420-435. Comprehensive description: L. V. Feyns, L. T. Grady in *Analytical Profiles of Drug Substances* vol. **10**, K. Florey, Ed. (Academic Press, New York, 1981) pp 289-335.

White, amorphous powder. mp 74°. Turns yellow on exposure to light and heat. $[\alpha]_D^{20} -50°$ (c = 2 in CHCl$_3$). Strong alkaline reaction. pK$_1$ 5.77; pK$_2$ 6.64. Freely sol in methanol, ethanol, acetone, ethyl acetate, ether, chloroform. Sparingly sol in water, petr ether. Moderately sol in dil ammonium hydroxide, but sparingly in solns of KOH and NaOH. Absorption spectrum: Bayard, *Chem. Zentr.* **1943**, II, 305. LD$_{50}$ i.p. in rats: 12.1 mg/kg (Radomski).

Dihydrochloride, C$_{29}$H$_{42}$Cl$_2$N$_2$O$_4$, *emetine hydrochloride, Hemometina.* Contains water of crystallization varying from 3 to 8 H$_2$O. Clusters of needles after drying at 105°, mp 235-255° (dec). $[\alpha]_D +11°$ (c = 1) to $[\alpha]_D +21°$ (c = 8), calculated for the anhydrous salt. One gram of the hydrated salt dissolves in about 7 ml water. pH of aq soln (1 g in 50 ml) 5.6. Sol in alcohol. Solid and solutions turn yellow on exposure to light or heat. LD$_{50}$ (calculated as base) in mice (mg/kg): 32 s.c.; 30 orally (Child).

Acetarsone salt, *Arsemétine.*

THERAP CAT: Antiamebic.

THERAP CAT (VET): Hydrochloride has been used as an anti-amebic and in lung worm infection.

3518. Emodin. *1,3,8-Trihydroxy-6-methyl-9,10-anthracenedione; 1,3,8-trihydroxy-6-methylanthraquinone;* 4,5,7-trihydroxy-2-methylanthraquinone; frangula emodin; rheum emodin; archin; frangulic acid. C$_{15}$H$_{10}$O$_5$; mol wt 270.23. C 66.67%, H 3.73%, O 29.60%. Occurs mostly as the rhamnoside (*see* Frangulin) in rhubarb root, in alder buckthorn (*Rhamnus frangula L.*), in *Cascara sagrada* (*Rhamnus purshiana* DC., *Rhamnaceae*), also in *Rumex* and in other *Polygonaceae.* Isoln from rhubarb root: Tutin, Clewer, *J. Chem. Soc.* **99**, 946 (1911); Carelli, Giuliano, *Farmaco Ed. Prat.* **12**, 184 (1957); from bark of alder buckthorn: Bridel, Charaux, *Bull. Soc. Chim. Biol.* **15**, 648 (1933). Identity with archin: Chaudhry *et al., J. Sci. Ind. Res. (India)* **9B**, No. 6, 142 (1950), *C.A.* **44**, 9396h (1950). Synthesis from 3,5-dinitrophthalic anhydride and *m*-cresol: Elder, Widmer, *Helv. Chim. Acta* **6**, 966 (1923); from 2-methylanthraquinone: Ayyangar *et al., J. Sci. Ind. Res. (India)* **20B**, 493 (1961), *C.A.* **57**, 8514b (1962).

Orange needles from alc or by sublimation at 12 mm. mp 256-257°. Absorption max (ethanol): 222, 252, 265, 289, 437 nm (log ε 4.55, 4.26, 4.27, 4.34, 4.10). Practically insol in water; sol in alc, aq alkali hydroxide solns (cherry-red color), Na$_2$CO$_3$ and NH$_3$ solns. Soly at 25° (g/100 ml of satd soln): ether 0.140; chloroform 0.071; carbon tetrachloride 0.010; carbon bisulfide 0.009; benzene 0.041.

3-Methyl ether, C$_{16}$H$_{12}$O$_5$, *1,8-dihydroxy-3-methoxy-6-methylanthraquinone, rheochrysidin.* Brick-red, monoclinic needles, mp 207°. Occurs naturally as *physcione* or *parietin.* Trimethyl ether, C$_{18}$H$_{16}$O$_5$, pale yellow needles, mp 225°.

Note: See also Aloe-emodin.

THERAP CAT: Cathartic.

3519. Emorfazone. *4-Ethoxy-2-methyl-5-(4-morpholinyl)-3(2H)-pyridazinone;* M-73101; Nandron; Pentoyl. C$_{11}$H$_{17}$N$_3$O$_3$; mol wt 239.28. C 55.22%, H 7.16%, N 17.56%, O 20.06%. Prepn: K. Satoda *et al., Japan. pat.* 72 24,030

(1972 to Morishita), *C.A.* **77**, 164732f (1972); M. Takaya *et al., J. Med. Chem.* **22**, 53 (1979). Metabolism studies: T. Hayashi *et al., Chem. Pharm. Bull.* **26**, 3124 (1978); **27**, 317 (1979). General pharmacological study: M. Sato *et al., Nippon Yakurigaku Zasshi* **75**, 291 (1979), *C.A.* **91**, 117292 (1979). Mechanism of action studies: M. Sato, A. Yamaguchi, *Arzneimittel-Forsch.* **32**, 379 (1982). Antigenicity study: M. Sato *et al., Oyo Yakuri* **18**, 65 (1979), *C.A.* **92**, 104391 (1980). Toxicity studies: *eidem, ibid.* **16**, 1011 (1978), *C.A.* **90**, 197741 (1979); C. Onodera *et al., J. Toxicol. Sci.* **4**, 229 (1979); K. Shimpo *et al., ibid.* 255.

Cryst from methanol/isopropyl ether, mp 89-91°. LD$_{50}$ i.p. in mice: 700 mg/kg (Takaya).

THERAP CAT: Anti-inflammatory; analgesic.

3520. Emylcamate. *3-Methyl-3-pentanol carbamate; 1-ethyl-1-methylpropyl carbamate;* diethyl methyl carbinol urethan; 3-methyl-3-pentyl carbamate; methyl diethyl carbinol urethan; *tert*-hexanol carbamate; KABI-925; JD 91; Nuncital; Restetal; Statran; Striatran. C$_7$H$_{15}$NO$_2$; mol wt 145.20. C 57.90%, H 10.41%, N 9.65%, O 22.04%. Prepn: Ger. pat. 245,491 (1912 to Chinifabr. Zimmer); Ger. pat. 254,472 (1912 to E. Merck); Melander, Hanshoff, U.S. pat. 2,972,564 (1961 to Kabi).

Needles from 30% ethanol, mp 56-58.5°. bp$_{1.0}$ 35°; bp$_{0.7}$ 24°. Slight odor of camphor. Soly in water: 4.0 mg/ml. Freely sol in alcohol, ether, benzene, glycol ethers.

THERAP CAT: Anxiolytic.

3521. Enalapril. *(S)-1-[N-[1-(Ethoxycarbonyl)-3-phenylpropyl]-L-alanyl]-L-proline;* 1-[N-[(S)-1-carboxy-3-phenylpropyl]-L-alanyl]-L-proline 1'-ethyl ester. C$_{20}$H$_{28}$N$_2$O$_5$; mol wt 376.45. C 63.81%, H 7.50%, N 7.44%, O 21.25%. Angiotensin-converting enzyme inhibitor. Prepn: A. A. Patchett *et al., Nature* **288**, 280 (1980); *eidem,* **Eur. pat. Appl. 12,401**; E. E. Harris *et al.,* U.S. pat. **4,374,829** (1980, 1983 both to Merck & Co.). Biochemical and pharmacological profile: D. M. Gross *et al., J. Pharmacol. Exp. Ther.* **216**, 552 (1981). Antihypertensive activity in animals: C. S. Sweet *et al., ibid.* 558. Physiological disposition and metabolism in animals: D. J. Tocco *et al., Drug Metab. Dispos.* **10**, 15 (1982); in humans: E. H. Ulm, *Drug Metab. Rev.* **14**, 99 (1983). Pharmacodynamics: J. A. Millar *et al., Brit. J. Clin. Pharmacol.* **14**, 347 (1982). Pilot study in essential hypertension: R. K. Ferguson *et al., J. Clin. Pharmacol.* **22**, 281 (1982). Double-blind comparative trial with captopril, *q.v.,* in hypertension: P. H. Vlasses *et al., J. Am. Coll. Cardiol.* **9**, 651 (1986). Clinical trials in congestive heart failure: D. N. Sharpe *et al., Circulation* **70**, 271 (1984); K. Swedberg *et al., N. Engl. J. Med.* **316**, 1429 (1987). Comprehensive description: D. P. Ip, G. S. Brenner in *Analytical Profiles of Drug Substances,* Vol. 16, K. Florey, Ed. (Academic Press, New York, 1987) pp 207-243.

Maleate, C$_{24}$H$_{32}$N$_2$O$_9$, MK-421, Enap, Enapren, Innovace, Lotrial, Olivin, Pres, Renitec, Reniten, Renivace, Vasotec, Xanef. White to off-white crystalline powder, mp 143-144.5°. Soly (g/ml): water 0.025; alcohol 0.08; methanol

0.20. $[\alpha]_D^{25}$ $-42.2°$ (c = 1 in methanol). pH (1% water) 2.6. pKa_1 3.0; pKa_2 (25°) 5.4. Mixture with hydrochlorothiazide, *Co-Renitec, Vaseretic.*

THERAP CAT: Antihypertensive.

3522. Enalaprilat. *1-(N-1-Carboxy-3-phenylpropyl)-L-alanyl-L-proline dihydrate;* enalaprilic acid; MK-422; Vasotec IV. $C_{18}H_{24}N_2O_5.2H_2O$; mol wt 384.43. C 56.24%, H 7.34%, N 7.29%, O 29.13%. Nonsulfhydryl dipeptide angiotensin converting enzyme inhibitor. Prepn: A. A. Patchett *et al., Nature* **288**, 280 (1980); *eidem,* Eur. pat. **Appl. 12,401;** E. E. Harris *et al.,* U.S. pat. **4,374,829** (1980, 1983 both to Merck & Co.). Active metabolite of enalapril, *q.v.*: D. J. Tocco *et al., Drug Metab. Dispos.* **10**, 15 (1982). Pharmacology: D. M. Gross *et al., J. Pharmacol. Exp. Ther.* **216**, 552 (1981); M. A. Nelson *et al., Clin. Exp. Pharmacol. Physiol.* **Suppl. 7**, 87 (1982); T. A. Unger *et al., Biochem. Pharmacol.* **31**, 3063 (1982). Bioavailability: M. L. Cohen *et al., J. Pharmacol. Exp. Ther.* **226**, 192 (1983). Kinetics of enzyme inhibition: C. H. Reynolds, *Biochem. Pharmacol.* **33**, 1273 (1984). Pharmacokinetics: K. S. Pang *et al., Drug Metab. Dispos.* **12**, 309 (1984). Acute hemodynamic effects in essential hypertension: A. C. Simon *et al., Clin. Pharm. Ther.* **43**, 49 (1988). Clinical evaluation in severe and malignant hypertension: D. J. DiPette *et al., ibid.* **38**, 199 (1985).

Needles from H_2O, mp 148-151°. $[\alpha]_D$ $-67.0°$ (0.1M HCl).

THERAP CAT: Antihypertensive.

3523. Enallylpropymal. *1-Methyl-5-(1-methylethyl)-5-(2-propenyl)-2,4,6(1H,3H,5H)-pyrimidinetrione; 5-allyl-5-isopropyl-1-methylbarbituric acid;* N-methyl-5-allyl-5-isopropylbarbituric acid. $C_{11}H_{16}N_2O_3$; mol wt 224.25. C 58.91%, H 7.19%, N 12.49%, O 21.40%. Prepn: Schnider, U.S. pat. **2,072,829** and Brit. pat. **454,779** (1937, 1936, both to Hoffmann-La Roche).

Crystals, mp 56-57°. bp_{12} 176-178°. Soluble in organic solvents.
Sodium salt, $C_{11}H_{15}N_2NaO_3$, *Narconumal.* Sol in water.
Caution: May be habit forming. This is a controlled substance (depressant) listed in the U.S. Code of Federal Regulations, Title 21 Parts 329.1 and 1308.13 (1987).

THERAP CAT: Sedative, hypnotic.

3524. Enanthotoxin. *2,8,10-Heptadecatriene-4,6-diyne-1,14-diol;* oenanthotoxin. $C_{17}H_{22}O_2$; mol wt 258.35. C 79.03%, H 8.58%, O 12.39%. The poisonous principle of *Oenanthe crocata* L., *Umbelliferae,* the hemlock water dropwort, a toxic plant known since 1746 and believed to be the most poisonous plant in England. Isoln: Clarke *et al., J. Pharm. Pharmacol.* **1**, 377 (1949). Structure studies: Anet *et al., Chem. & Ind. (London)* **1952**, 757; *J. Chem. Soc.* **1953**, 309. Synthesis of DL-form: Bohlmann, Viehe, *Ber.* **88**, 1245 (1955). Fluorimetric determn: B. Del Castillo *et al., Ital. J. Biochem.* **29**, 233 (1980). Blocking of sodium current and intramembrane charge movement: J. M. Dubois, M. F. Schneider, *Nature* **289**, 685 (1981); *eidem, Toxicon* **20**, 49 (1982).

Large prisms (natural) or star-shaped crystals (synthetic). Unstable, dec by light and air to brown insol resin. mp 87° (natural), 68° (synthetic). $[\alpha]_D^{15}$ $+30.5°$ (c = 2.0 in methanol). uv max: 213, 252, 267, 281, 296, 315.5, 337.5 (ϵ × 10^{-3} 17.5, 33, 29, 17.5, 30.5, 40, 29). Practically insol in water, petr ether, alkalies, dil mineral acids. Readily sol in chloroform, ethanol, methanol, ether, benzene. Average LD i.p. in mice: 0.83 mg/kg, Clarke *et al., loc. cit.* LD_{50} reported as 2.94 mg/kg i.p. in rats, M. P. Martinez-Honduvilla *et al., Arch. Farmacol. Toxicol.* **7**, 197 (1981).
Caution: Very toxic. May cause convulsions, death.

3525. Encainide. *(±)-4-Methoxy-N-[2-[2-(1-methyl-2-piperidinyl)ethyl]phenyl]benzamide; (±)-2'-[2-(1-methyl-2-piperidyl)ethyl]-p-anisanilide; 4-methoxy-2'-[2-(1-methyl-2-piperidyl)ethyl]benzanilide.* $C_{22}H_{28}N_2O_2$; mol wt 352.48. C 74.96%, H 8.01%, N 7.95%, O 9.08%. Anti-arrhythmic benzanilide derivative. Prepn: S. J. Dykstra, J. L. Minielli, **Ger.** pat. **2,210,154** (1972 to Bristol-Myers), *C.A.* **78**, 4138j (1973); *eidem,* U.S. pat. **3,931,195** (1976 to Mead Johnson); S. J. Dykstra *et al., J. Med. Chem.* **16**, 1015 (1973). Anti-arrhythmic pharmacology in animals: J. E. Byrne *et al., J. Pharmacol. Exp. Ther.* **200**, 147 (1977). Clinical pharmacology and efficacy in chronic ventricular arrhythmia: D. M. Roden *et al., N. Engl. J. Med.* **302**, 877 (1980). Electrophysiology and effects on cardiac conduction: M. Sami *et al., Am. J. Cardiol.* **44**, 526 (1979). Hemodynamic effects: M. Sami *et al., ibid.* **52**, 507 (1983). Adverse effects: R. A. Winkle *et al., Am. Heart J.* **102**, 857 (1981). Comparison with other Class I anti-arrhythmic agents: A. Pottage, *Am. J. Cardiol.* **52**, 24C (1983). Series of articles on pharmacology, pharmacokinetics, metabolism, clinical safety and efficacy: *ibid.* **58**(5), 1C-116C (1986). *Review:* L. B. Mitchell, R. A. Winkle in *New Drugs Annual: Cardiovascular Drugs,* **vol. 1**, A. Scriabine, Ed. (Raven Press, New York, 1983) pp 93-107.

Hydrochloride, $C_{22}H_{29}ClN_2O_2$, *MJ 9067, Enkade, Enkaid.* Crystals, mp 131.5-132.5°. Freely sol in water; slightly sol in ethanol. Insol in heptane. LD_{50} in mice, dogs (mg/kg): 86, 43 orally; 16, 17 i.v. (Mitchell, Winkle).

THERAP CAT: Antiarrhythmic.

3526. Endiandric Acids. A group of novel polycyclic compounds isolated from leaves of the 'Dorrigo Plum', *Endiandra introrsa* C. T. White, *Lauraceae,* a large tree occurring in rain forests of Australia. Although they have a number of asymmetric centers, endiandric acids occur in nature in racemic rather than enantiomeric forms. This fact led to a proposed hypothesis for the "biogenesis" of these compounds in nature from achiral precursors by a series of non-enzymatic electrocyclizations called the "endiandric acid cascade". Isoln from *E. introrsa* leaves (obtained in 1958) and structure of endiandric acid A: W. M. Bandaranayake *et al., Chem. Commun.* **1980**, 162. Postulated electrocyclic reactions, prediction of endiandric acid D as a natural product: *eidem, ibid.* 902. X-ray crystallographic structural elucidation of A and detailed report of isoln: *eidem, Aust. J. Chem.* **34**, 1655 (1981). Isoln and structure of endiandric acid B: *eidem, ibid.* **35**, 557 (1982); of C: *eidem, ibid.* 567. Stereocontrolled total synthesis of A and B, description of "endiandric acid cascade": K. C. Nicolaou *et al., J. Am. Chem. Soc.* **104**, 5555 (1982). Stereocontrolled total synthesis of endiandric acids C-G: *eidem, ibid.* 5557. Synthesis of precursors and thermal studies: *eidem, ibid.* 5558, 5560.

endiandric acid A
R - CH₂COOH
endiandric acid B
R - CH₂CH=CHCOOH (E)

endiandric acid C

Endiandric Acid A, $C_{21}H_{22}O_2$, *1a,2,2a,5,5a,7a,7b,7c-octahydro-5-phenyl-1H-cyclobut[bc]acenaphthylene-1-acetic acid, 2-(6'-phenyltetracyclo[5,4,2,0³,¹³,0¹⁰,¹²]trideca-4',8'-dien-11'-yl)acetic acid.* Rods from aq ethanol, mp 147-149°. uv max (95% ethanol): 242, 255, 261, 268, 286 nm (log ε 2.19, 2.36, 2.45, 2.32, 1.45). pKa 5.1, 5.0. Gives a yellow color with tetranitromethane.

Endiandric Acid B, $C_{23}H_{24}O_2$, *4-(1a,2,2a,5,5a,7a,7b,7c-octahydro-5-phenyl-1H-cyclobut[bc]acenaphthylen-1-yl)-2-butenoic acid, (E)-4-(6'-phenyltetracyclo[5,4,2,0³,¹³,0¹⁰,¹²]-trideca-4',8'-dien-11'-yl)but-2-enoic acid.* Rosettes from aq ethanol and chloroform/petr ether, mp 163-165°. uv max (95% ethanol): 252, 258, 262, 265, 269 nm (log ε 3.11, 3.08, 3.07, 3.01).

Endiandric Acid C, $C_{23}H_{24}O_2$, *1,1a,2,3,3a,6,6a,6b-octahydro-1-(5-phenyl-2,4-pentadienyl)-3,6-methanocyclobut[cd]indene-7-carboxylic acid, 4-[(E,E)-5'-phenylpenta-2',4'-dien-1'-yl]tetracyclo[5,4,0,0²,⁵,0³,⁹]undec-10-ene-8-carboxylic acid.* Cryst from ethanol and methanol, mp 125-132° (variable and with dec). uv max (95% ethanol): 222, 228, 236, 280, 288 nm (log ε 4.10, 4.05, 3.89, 4.51, 4.53).

3527. Endobenzyline Bromide. *2-[(Bicyclo[2.2.1]hept-5-en-2-ylhydroxyphenylacetyl)oxy]-N,N,N-trimethylethanaminium bromide; choline bromide α-phenyl-5-norbornene-2-glycolate; N,N-dimethylaminoethyl α-(bicyclo-[2.2.1]-5-heptenyl)mandelate methyl bromide; α-(2,5-endo-methylene-Δ³-cyclohexenyl)mandelic acid β-dimethylaminoethyl ester methyl bromide; α-phenyl-5-norbornene-2-glycolic acid 2-dimethylaminoethyl ester methyl bromide; endobenziline bromide; PC-1238; Ulcyn.* $C_{20}H_{28}BrNO_3$; mol wt 410.37. C 58.54%, H 6.88%, Br 19.48%, N 3.41%, O 11.70%. Prepn: Kottler, Ohnacker, U.S. pat. **2,884,426** (1959 to Thomae).

Crystals from isopropanol, mp 170°.
THERAP CAT: Anticholinergic.

3528. Endorphins. Generic name derived from the term "endogenous morphine" and applied to a group of neuropeptides that are endogenous ligands of the opiate receptors. They are found in brain, pituitary gland and peripheral tissues of all vertebrates; the effects of endorphins on cells resemble those of opiates such as morphine. Their existence was postulated as a result of the discovery of stereospecific binding sites for narcotic analgesics in animal brain, *cf.* A. Goldstein *et al., Proc. Nat. Acad. Sci. USA* **68**, 1742 (1971); C. B. Pert, S. H. Snyder, *Science* **179**, 1011 (1973); E. J. Simon *et al., Proc. Nat. Acad. Sci. USA* **70**, 1947 (1973). Three types of stereochemically related opiate receptors have been postulated, *mu, delta, kappa:* W. R. Martin *et al., J. Pharm. Exp. Ther.* **197**, 517 (1976); P. E. Gilbert, W. R. Martin, *ibid.* **198**, 66 (1976); J. A. H. Lord *et al., Nature* **267**, 495 (1977). Of the endorphin group of peptides, several important members are known: α-, β- and γ-endorphins (β-endorphin being the most potent), α-neo-endorphin, β-neo-endorphin, dynorphin, *q.v.*, and met-enkephalin and leu-enkephalin, two naturally occurring pentapeptides belonging to the endorphin class. The amino acid sequences of

α-, β- and γ-endorphins are contained within the sequence 61-91 of β-lipotropin, *q.v.*, and, in each, the amino terminal pentapeptide sequence is that of met-enkephalin. Dynorphin and α- and β-neo-endorphins have common terminal leu-enkephalin sequences. Initial isoln of active oligopeptide: L. Terenius, A. Wahlstrom, *Acta Pharmacol. Toxicol.* **35**, Suppl. 1, 55 (1974); J. Hughes, *Brain Res.* **88**, 295 (1975). Isoln, characterization and synthesis of two active pentapeptides (*enkephalins*): J. Hughes *et al., Life Sci.* **16**, 1753 (1975); G. W. Pasternak *et al., ibid.* 1765; J. Hughes *et al., Nature* **258**, 577 (1975); J. D. Bower *et al., J. Chem. Soc. Perkin Trans. I* **1976**, 2488. Isoln and properties of β-endorphin from bovine pituitary: H. Teschemacher *et al., Life Sci.* **16**, 1771 (1975); B. M. Cox *et al., ibid.* 1777. Identity of β-endorphin with carboxy terminal amino acid sequence (61-91) of β-lipotropin: C. H. Li, D. Chang, *Proc. Nat. Acad. Sci. USA* **73**, 1145 (1976); B. M. Cox *et al., ibid.* 1821. Production of biologically active β-endorphin via expression of cloned gene sequences by *E. coli*: J. Shine *et al., Nature* **285**, 456 (1980). For a review of isoln, amino acid sequences, and synthesis of β-endorphins from various species, *see Hormonal Proteins and Peptides* vol. **X**, entitled "β-Endorphin", C. H. Li, Ed. (Academic Press, New York, 1981) pp 2-30.

Isoln and structure of α-*endorphin* or *LPH (61-76):* R. Guillemin *et al., Compt. Rend. Ser. D* **282**, 783 (1976); isoln, structure, synthesis of α- and γ-*endorphin* or *LPH (61-77):* N. Ling *et al., Proc. Nat. Acad. Sci. USA* **73**, 3942 (1976). High yield synthesis of met⁵-enkephalin: B. J. Dhotre *et al., J. Ind. Chem. Soc.* **55**, 1128 (1978). Rapid synthesis of leu⁵-enkephalin: E. Vilkas *et al., Int. J. Peptide Protein Res.* **15**, 29 (1980). Isoln of α-*neo-endorphin:* K. Kagawa *et al., Biochem. Biophys. Res. Commun.* **86**, 153 (1979). Amino acid sequence: *eidem, ibid.* **99**, 871 (1981). Isoln, purification, amino acid sequence of β-*neo-endorphin:* N. Minamino *et al., ibid.* 864. Metabolism and physiological effects of endorphins: H. W. Kosterlitz, J. Hughes, *Life Sci.* **17**, 91 (1975); L.-F. Tseng *et al., Proc. Nat. Acad. Sci. USA* **73**, 4187 (1976). Although the physiological role of the endorphins has not been completely elucidated, they are known to block inhibitory pathways in the vertebrate CNS and their possible effects on pain perception, addictive states, and psychiatric disorders have been investigated: D. T. Krieger, A. S. Liotta, *Science* **205**, 366 (1979); R. A. Nicoll *et al., Nature* **287**, 22 (1980). Studies have also suggested that endorphins may be involved in gluco-regulation: M. Feldman *et al., N. Engl. J. Med.* **308**, 350 (1983). *Reviews:* A. Goldstein, *Science* **193**, 1081-1086 (1976); *idem, Harvey Lectures* **79**, 291-314 (1979); R. A. North, *Life Sci.* **24**, 1527-1546 (1979); C. R. Beddell *et al., Progr. Med. Chem.* **17**, 1-39 (1980); several authors in *Advan. Biochem. Pharmacol.* **22**, 145-642 (1980); M. S. Gold *et al., Med. Res. Rev.* **2**, 211-246 (1982); S. Zakarian, D. G. Smyth, *Biochem. J.* **202**, 561-571 (1982). Books: *Opiates and Endogenous Opioid Peptides*, H. W. Kosterlitz, Ed. (Elsevier, New York, 1976) 466 pp; *Endorphins: Proceedings*, L. Graf *et al.*, Eds. (Elsevier, New York, 1979) 471 pp; *Endorphins and Opiate Antagonists in Psychiatric Research*, N. Shah, A. G. Donald, Eds. (Plenum, New York, 1982) 425 pp.

```
         61              65                  70                       76
H-Tyr-Gly-Gly-Phe-Met-Thr-Ser-Glu-Lys-Ser-Gln-Thr-Pro-Leu-Val-Thr-

         77               80                  85                       91
Leu-Phe-Lys-Asn-Ala-Ile-Ile-Lys-Asn-Ala-Tyr-Lys-Lys-Gly-Glu-OH

          Human β-Endorphin [βₕ-LPH (61-91)]
```

β-Endorphin, β-lipotropin C-fragment, β-lipotropin (61-91), LPH (61-91).

Met⁵-Enkephalin, $C_{27}H_{35}N_5O_7S$, *methionine⁵-enkephalin, met⁵-E,* L-tyrosylglycylglycyl-L-phenylalanyl-L-methionine, *LPH (61-55).* Needles from hot methanol, mp 196-198°. [α]$_{589}^{22}$ −21.9° (c = 1 in DMF); +14° (c = 1 in N HCl).

Leu⁵-Enkephalin, $C_{28}H_{37}N_5O_7$, *leucine⁵-enkephalin, leu⁵-E,* L-tyrosylglycylglycyl-L-phenylalanyl-L-leucine. White crystalline solid, mp 206° (dec). [α]$_{589}^{22}$ −23.4° (c = 1 in DMF); +18.3° (c = 1 in N HCl).

3529. Endosulfan. *6,7,8,9,10,10-Hexachloro-1,5,5a,6,9,-9a-hexahydro-6,9-methano-2,4,3-benzodioxathiepin 3-oxide; 1,4,5,6,7,7-hexachloro-5-norbornene-2,3-dimethanol cyclic sulfite;* 1,2,3,4,7,7-hexachlorobicyclo[2.2.1]-2-heptene-5,6-bisoxymethylene sulfite; chlorthiepin; Malix; Thiodan; Thionex. $C_9H_6Cl_6O_3S$; mol wt 406.95. C 26.56%, H 1.49%, Cl 52.28%, O 11.80%, S 7.88%. Prepd by reaction of hexachlorocyclopentadiene with *cis*-butene-1,4-diol to form the bicyclic dialcohol, followed by esterification and cyclization with $SOCl_2$: Geering, Nelson, U.S. pat. 2,983,732 (1961 to Hooker). Configuration studies: Reimschneider, Wuscherpfenning, *Z. Naturforsch.* **17b**, 585 (1962); Forman *et al., J. Org. Chem.* **30**, 169 (1965). Toxicity: T. B. Gaines, *Toxicol. Appl. Pharmacol.* **14**, 515 (1969).

Commercial product brown crystals. mp 70-100° (pure mp 106°). Practically insol in water. Sol in most organic solvents. Stable toward dil mineral acids; hydrolyzed rapidly by alkalies. Commercial product is a mixture of α-isomer, mp 108-110°, and β-isomer, mp 208-210°. LD_{50} orally in male, female rats: 43, 18 mg/kg (Gaines).

USE: Insecticide.

3530. Endothall. *7-Oxabicyclo[2.2.1]heptane-2,3-dicarboxylic acid;* 3,6-endoxohexahydrophthalic acid; endothal. $C_8H_{10}O_5$; mol wt 186.16. C 51.61%, H 5.41%, O 42.97%. Prepn: Olin, U.S. pat. 2,550,494 (1951 to Sharples Chemicals). Of the three possible isomers, the *exo-cis* form has the greatest biological activity. Fate in the environment and toxicity: Simsiman *et al., Residue Rev.* **62**, 131-174 (1976); T. B. Gaines, R. E. Linder, *Fundam. Appl. Toxicol.* **7**, 299 (1986).

Solubility of the acid in g/100 g at 20°: water 10; acetone 7; benzene 0.01; methanol 28. Converted to the anhydride at 90°. LD_{50} in adult male, female rats (mg/kg): 57, 46 orally (Gaines, Linder).
Disodium salt, $C_8H_8Na_2O_5$, *Aquathol.*
Dipotassium salt, $C_8H_8K_2O_5$, *Aquathol K.*
Caution: May be very irritating to skin, eyes, mucous membranes. Ingestion may cause severe G.I. inflammation with erosion: *Clinical Toxicology of Commercial Products,* R. E. Gosselin *et al.,* Eds. (Williams & Wilkins, Baltimore, 5th ed., 1984) Section II, p 342.

USE: Herbicide; defoliant.

3531. Endotoxins. Heat-stable lipopolysaccharide-protein complexes contained in cell walls of gram-negative bacteria including non-infectious gram-negatives. Endotoxins are pyrogenic and increase capillary permeability, exhibiting the same activity regardless of the species of bacteria from which they are derived. *Reviews:* A. I. Braude, *Sci. Am.* **210**, 36-45 (March 1964); C. Galanos *et al., Int. Rev. Biochem.* **14**, 239-335 (1977); *eidem, Prog. Clin. Biol. Res.* **29**, 321-332 (1979). Books: *Microbial Toxins,* vols. 4 and 5, G. Weinbaum *et al.,* Eds. (Academic Press, New York, 1971); *Bacterial Lipopolysaccharides,* E. H. Kass, S. M. Wolff, Eds. (Univ. of Chicago Press, Chicago, 1973) 312 pp; *Bacterial Endotoxins & Host Response,* M. K. Agarwal, Ed. (Elsevier/North-Holland, New York, 1980); J. A. Majde, R. J. Person, *Prog. Clin. Biol. Res.* **62** entitled "Pathophysical Effects of Endotoxins at the Cellular Level" (1981) 204 pp.

3532. Endralazine. *6-Benzoyl-5,6,7,8-tetrahydropyrido-[4,3-c]pyridazin-3(2H)-one hydrazone;* 6-benzoyl-3-hydrazino-5,6,7,8-tetrahydropyrido[4,3-c]pyridazine. $C_{14}H_{15}N_5O$; mol wt 269.31. C 62.44%, H 5.61%, N 26.01%, O 5.94%.

Prepn: E. Schenker, Ger. pat. 2,221,808 corresp to U.S. pat. 3,838,125 (1972, 1974 both to Sandoz); E. Schenker, R. Salzmann, *Arzneimittel-Forsch.* **29**, 1835 (1979). Pharmacology: R. Salzmann *et al., ibid.* 1843. Hemodynamic study: H. U. Lehmann *et al., Z. Cardiol.* **66**, 203 (1977). HPLC determn in human plasma: P. A. Reece *et al., J. Chromatog.* **225**, 151 (1981). Clinical pharmacology: F. C. Reubi, *Eur. J. Clin. Pharmacol.* **13**, 185 (1978). Clinical study in hypertension: H. U. Lehmann *et al., Med. Welt* **29**, 1007 (1978).

Crystals from acetonitrile/water, mp 220-223° (dec).
Methanesulfonate, $C_{15}H_{19}N_5O_4S$, *BQ 22-708, Miretilan.* Cryst, mp 185-188° (dec).
THERAP CAT: Antihypertensive.

3533. Endrin. *3,4,5,6,9,9-Hexachloro-1a,2,2a,3,6,6a,7,-7a-octahydro-2,7:3,6-dimethanonaphth[2,3-b]oxirene; 1,2,3,-4,10,10-hexachloro-6,7-epoxy-1,4,4a,5,6,7,8,8a-octahydro-endo,endo-1,4:5,8-dimethanonaphthalene;* mendrin; nendrin; hexadrin; compound 269; experimental insecticide no. 269; ENT 17251. $C_{12}H_8Cl_6O$; mol wt 380.93. C 37.84%, H 2.12%, Cl 55.85%, O 4.20%. Prepn from isodrin: H. Bluestone, U.S. pat. 2,676,132 (1954 to Shell); Marks, U.S. pat. 2,899,446 (1959 to Velsicol). Metabolism: M. K. Baldwin *et al., Pestic. Sci.* **7**, 575 (1976). Carcinogenicity studies: M. D. Rueber, *Sci. Total Environ.* **12**, 101 (1979). A stereoisomer of dieldrin, *q.v.*

Crystals, dec 245°. Vapor press at 25°: 2×10^{-7} mm Hg. Soly in g/100 ml solvent at 25°: acetone 17, benzene 13.8, carbon tetrachloride 3.3, hexane 7.1, xylene 18.3. LD_{50} orally in female, male rats: 7.5, 18 mg/kg, T. B. Gaines, *Toxicol. Appl. Pharmacol.* **14**, 515 (1969). Toxic to fish.

USE: Formerly as insecticide; manuf and use has been discontinued in the U.S. *Caution:* Symptoms similar to dieldrin and aldrin, *qq.v.*

3534. Enduracidin. Enramycin. Cyclodepsipeptide antibiotic produced by *Streptomyces fungicidicus,* strain no. B5477, from a soil sample collected in Nishinomiya, Hyogo Prefecture, Japan: Higashide *et al., J. Antibiot.* **21**, 126 (1968). Originally thought to be a single compound. Isoln and characterization: Asai *et al., ibid.* 138; Fr. pat. 1,514,-139 corresp to Brit. pat. 1,163,270 (1968, 1969 to Takeda). Manufacturing process: Japan. Kokai 82 79896 (1982 to Takeda), *C.A.* **97**, 108525s (1982). Separation into two components, enduracidins A and B: Hori *et al., Chem. Pharm. Bull.* **21**, 1171 (1973). The two components are each composed of seventeen amino acids, sixteen of which form a macrocyclic peptide lactone, and only differ by one methylene group in their fatty acid moieties. Structural studies: Hori *et al., ibid.* 1175. Final structure: Iwasaki *et al., ibid.* 1184. Antibacterial activity: Goto *et al., J. Antibiot.* **21**, 119 (1968); Tsuchiya *et al., ibid.* 147; M. Kawakami *et al., ibid.* **24**, 583 (1971). Toxicology: H. Yokotani *et al., Takeda Kenkyusho Nempo* **28**, 76 (1969), *C.A.* **72**, 77300s (1970).

Monohydrochloride, *Enradin.* Yellowish powder, dec 234-238°. Slightly sol in water, methanol. Practically insol in acetone, ethanol, chloroform, benzene. LD_{50} in mice and rats (mg/kg): >10000 orally; >5000 i.m.; >5000 s.c. (Yokotani).

Enduracidin A hydrochloride, $C_{107}H_{139}Cl_3N_{26}O_{31}$, mp 240-245°. $[\alpha]_D^{23}$ +92° (c = 0.5 in DMF). uv max (0.1*N* HCl): 231, 272 nm.

Enduracidin B hydrochloride, $C_{108}H_{141}Cl_3N_{26}O_{31}$, mp

238-241°. $[\alpha]_D^{23}$ +92° (c = 0.5 in DMF). uv max (0.1N HCl): 231, 272 nm.

THERAP CAT: Antibacterial.

3535. Enfenamic Acid. *2-[(2-Phenylethyl)amino]benzoic acid;* N-phenethyl anthranilic acid; RH 8; Tromaril. $C_{15}H_{15}NO_2$; mol wt 241.29. C 74.66%, H 6.27%, N 5.81%, O 13.26%. Nonsteroidal anti-inflammatory, antipyretic, analgesic. Prepn: S. R. Hashim *et al.*, **Indian** pats. **103,066** and **114,805** (both 1974 to Council Sci. Ind. Res.). Chemistry and pharmacology: P. Sisoda *et al.*, *Proc. Symp. on CNS Drugs* (Council of Scientific and Industrial Research, New Delhi, 1966) pp 238-248. Antithrombotic activity in rheumatoid arthritis and myocardial infarction: S. R. Manikeri *et al.*, *Ind. J. Med. Res.* **71**, 438 (1980). Natriuretic activity: N. Kshirsagar *et al.*, *Brit. J. Clin. Pharmacol.* **9**, 530 (1980). Clinical trial in nephrotic syndrome: *eidem*, *Ind. J. Med. Res.* **70**, 801 (1979); as post-partum analgesic: S. Tandon *et al.*, *Ind. Med. Gaz.* **116**, 182 (1982).

Crystals, mp 116-117°. Insol in water. LD_{50} in mice (mg/kg): 575 i.p., > 2000 orally (Sisoda).

THERAP CAT: Anti-inflammatory; analgesic.

3536. Enflurane. *2-Chloro-1-(difluoromethoxy)-1,1,2-trifluoroethane;* 2-chloro-1,1,2-trifluoroethyl difluoromethyl ether; methylflurether; compound 347; NSC-115944; Alyrane; Efrane; Ethrane. $C_3H_2ClF_5O$; mol wt 184.50. C 19.53%, H 1.09%, Cl 19.21%, F 51.49%, O 8.67%. CHF_2O-CF_2CHClF. Prepn by fluorination of the corresp dichloromethyl ether: Terrell, **Brit.** pat. **1,138,406** and **U.S.** pats. **3,469,011; 3,527,813** (1969, 1969, 1970 to Air Reduction). Synthesis and anesthetic properties: Terrell *et al.*, *J. Med. Chem.* **14**, 517 (1971); **15**, 604 (1972).

Stable, volatile, non-flammable liq. bp 56.5°. n_D^{20} 1.3025. d_{25}^{25} 1.5167. Does not degrade in the presence of alkali or light. Miscible with other organic liquids incl. fats and oils.

THERAP CAT: Anesthetic (inhalation).

3537. Enilconazole. *1-[2-(2,4-Dichlorophenyl)-2-(2-propenyloxy)ethyl]-1H-imidazole;* (±)-1-[β-(allyloxy)-2,4-dichlorophenethyl]imidazole; imazalil; R-23,979; Imaverol. $C_{14}H_{14}Cl_2N_2O$; mol wt 297.17. C 56.58%, H 4.75%, Cl 23.86%, N 9.43%, O 5.38%. Broad spectrum antimycotic with agricultural and veterinary applications. Prepn: E. F. Godefroi *et al.*, **Ger.** pat. **2,063,857** corresp to U.S. pat. **3,658,813** (1971, 1972 both to Janssen). Efficacy vs citrus decay: P. R. Harding, *Plant Dis. Rep.* **60**, 643 (1976); S. Ben-Yehoshua *et al.*, *Pestic. Sci.* **12**, 485 (1981). Biological and toxicological properties: D. Thienpont *et al.*, *Arzneimittel-Forsch.* **31**, 309 (1981). Residue analysis in treated grapefruit: E. R. Stein *et al.*, *J. Environ. Sci. Health [B]* **16**, 427 (1981).

Solidified oil. Slightly sol in organic solvents; poorly sol in water.

USE: As a disinfectant for stable and kennel equipment; exptly as an agricultural fungicide.

THERAP CAT: Antifungal.

THERAP CAT (VET): Antifungal.

3538. Enniatins. Ionophore antibiotics produced by the fungus *Fusarium orthoceras* var. *enniatum* and other *Fusaria*: Gaumann *et al.*, *Experientia* **3**, 202 (1947); Plattner, Nager, *ibid.* 325; Plattner *et al.*, *Helv. Chim. Acta* **31**, 594 (1948). They are cyclic depsihexapeptides related structurally to valinomycin, *q.v.* Enniatins A, B, and C are known; A and B are obtained from natural sources and C is a synthetic homolog. Structure and synthesis of enniatin A: Quitt *et al.*, *ibid.* **46**, 1715 (1963); **47**, 166 (1964); of enniatin B: Plattner *et al.*, *ibid.* **46**, 927 (1963); Shemyakin *et al.*, *Tetrahedron Letters* **1963**, 885. Conformation of enniatin B: Ovchinnikov *et al.*, *Biochem. Biophys. Res. Commun.* **37**, 668 (1969). Synthesis of C: *eidem*, *Izvest. Akad. Nauk SSSR, Ser. Khim.* **1964**, 1912, *C.A.* **62**, 28246b (1965). Enniatins are able to form "sandwich" complexes with cations, a factor used to explain their ability to induce permeability in biological membranes to the complexed ion, *cf.* V. T. Ivanov, *Ann. N.Y. Acad. Sci.* **264**, 221 (1975). Review: Y. A. Ovchinnikov, V. T. Ivanov, "The Cyclic Peptides: Structure, Conformation, and Function" in *The Proteins* vol. **V**, H. Neurath, R. L. Hill, Eds. (Academic Press, New York, 3rd ed., 1982) pp 365-373, 516-529.

enniatin A, R = $CH(CH_3)CH_2CH_3$
enniatin B, R = $CH(CH_3)_2$
enniatin C, R = $CH_2CH(CH_3)_2$

Enniatin A, $C_{36}H_{63}N_3O_9$. Long needles from ethanol + water, mp 122-122.5°. Very slowly sublimes at 10^{-4} mm and 127-128° (oil bath temp). $[\alpha]_D^{18}$ -91.9° (c = 0.926 in chloroform). Sol in ether, benzene, ethyl acetate; sparingly sol in water. Solns are stable to heat, inactivated by alkali.

Enniatin B, $C_{33}H_{57}N_3O_9$. Crystals from petr ether, mp 175-175.5°. $[\alpha]_D^{20}$ -107.9° (c = 0.63 in chloroform). Very sol in organic solvents; slightly sol in petr ether. Practically insol in water.

Enniatin C, $C_{34}H_{59}N_3O_9$, crystals, mp 160-161°. $[\alpha]_D^{23}$ -24°.

3539. Enocitabine. *N-(1-β-D-Arabinofuranosyl-1,2-dihydro-2-oxo-4-pyrimidinyl)docosanamide;* N(4)-behenoyl-1-β-D-arabinofuranosylcytosine; behenoylcytosine arabinoside; BH-AC; NSC-239336; Sunrabin. $C_{31}H_{55}N_3O_6$; mol wt 565.80. C 65.81%, H 9.80%, N 7.42%, O 16.97%. Deriv of cytarabine, *q.v.* Prepn: T. Ishida *et al.*, **Ger.** pat. **2,426,304** (1975 to Asahi), *C.A.* **82**, 171370 (1975); M. Akiyama *et al.*,

Chem. Pharm. Bull. **26**, 981 (1978). Anti-tumor activity: M. Aoshima *et al., Cancer Res.* **36**, 2726 (1976); *eidem, ibid.* **37**, 2481 (1977). Distribution, excretion: M. Fukama *et al., Gan to Kagaku Ryoho* **7**, 2109 (1980), *C.A.* **95**, 35186 (1981). Effect and mode of action: T. Kataoka, Y. Sakurai, *Recent Results Cancer Res.* **70**, 147 (1980). Pharmacological and clinical studies: K. Yamada *et al., ibid.* 219.

Cryst from DMSO, mp 141-142°. $[\alpha]_D$ +70° (c = 1 in THF, 22°). uv max (isopropyl alcohol): 216, 248, 303 nm (ϵ 16400, 15200, 8200).

THERAP CAT: Antineoplastic.

3540. Enoxacin. *1-Ethyl-6-fluoro-1,4-dihydro-4-oxo-7-(1-piperazinyl)-1,8-naphthyridine-3-carboxylic acid;* AT 2266; CI 919; PD 107779; Bactidan; Comprecin; Enoram; Flumark. $C_{15}H_{17}FN_4O_3$; mol wt 320.32. C 56.24%, H 5.35%, F 5.93%, N 17.49%, O 14.98%. Fluorinated quinolone antibacterial; nalidixic acid analog. Prepn and antibacterial activity: J. Matsumoto *et al.,* **Eur. pat. Appl. 9425;** *eidem,* **U.S. pat. 4,352,803;** *eidem,* **U.S. pat. 4,359,578** (1980, 1982, 1982 all to Dainippon and Roger Bellon); J. Matsumoto *et al., J. Med. Chem.* **27**, 292 (1984). HPLC determn in plasma, urine: T. B. Vree *et al., J. Chromatog.* **343**, 449 (1985). *In vitro* comparison with other quinolones and azaquinolones: S. Selwyn, M. Bakhtiar, *Drugs Exptl. Clin. Res.* **10**, 653 (1984). Pharmacokinetics, therapeutic efficacy: K. G. Naber *et al., Infection* **13**, 219 (1985). Efficacy in treatment of urinary tract infections: R. R. Bailey, B. A. Peddie, *N.Z. Med. J.* **98**, 286 (1985). Toxicology studies: H. Senda *et al., Chemotherapy (Tokyo)* **32**, Suppl. 3, 192 (1984). Series of articles on activity, pharmacology, metabolism, early clinical studies: *ibid.* 1-1093; *J. Antimicrob. Chemother.* **14**, Suppl. C, 1-96 (1984); on activity, pharmacokinetics, clinical safety and efficacy: *Infection* **14**, Suppl. 3, S183-S220 (1986).

Crystals from ethanol/methylene chloride, mp 220-224°. LD_{50} in male, female mice, male, female rats (mg/kg): 327, 391, 236, 294 i.v.; 1237, 1320, >2000, >2000 s.c.; all >5000 orally (Senda).

Sesquihydrate, $C_{15}H_{17}FN_4O_3.1\frac{1}{2}H_2O$, Enoxen, Gyramid, *Lentor.*

THERAP CAT: Antibacterial.

3541. Enoxaparin. PK 10169; Lovenox. Low molecular weight fragment of heparin, *q.v.,* having a 4-eno pyranosuronate sodium structure at the non-reducing end of the chain. Prepd by depolymerization of the benzylic ester of porcine mucosal heparin. Mean mol wt 4000-6000 daltons. Prepn: J. Mardiguian, **Eur. pat. Appl. 40,144** (1981 to Pharmindustrie), *C.A.* **96**, 218191s (1982). Effect on lipoprotein lipase activity in humans: J. Etienne *et al., Brit. J. Clin. Pharmacol.* **16**, 712 (1983). Exhibits antithrombotic activity comparable to standard heparin but with diminished hemorrhagic effects: M. Aiach *et al., Thromb. Res.* **31**, 611 (1983); H. Vinazzer, M. Woler, *ibid.* **40**, 135 (1985); M. Mestre *et al., ibid.* **38**, 389 (1985). Effects on human platelet aggregation: L. D. Brace, J. Fareed, *ibid.* **42**, 769 (1986). Clinical evaluation in postoperative thromboprophylaxis: A. G. G. Turpie *et al., N. Engl. J. Med.* **315**, 925 (1986). Determn in human plasma: L. Bara *et al., Haemostasis* **17**, 127 (1987). Symposium on pharmacology and clinical efficacy: *ibid.* **16**, 69-188 (1986).

THERAP CAT: Antithrombotic.

3542. Enoximone. *1,3-Dihydro-4-methyl-5-[4-(methylthio)benzoyl]-2H-imidazol-2-one;* 4-methyl-5-[p-(methylthio)benzoyl]-4-imidazolin-2-one; fenoximone; MDL 17043; RMI 17043; Perfan. $C_{12}H_{12}N_2O_2S$; mol wt 248.30. C 58.05%, H 4.87%, N 11.28%, O 12.89%, S 12.91%. Selective phosphodiesterase inhibitor with vasodilating and positive

inotropic activity. Prepn: **Belg. pat. 883,856** (1980 to Richardson-Merrell); R. A. Schnettler *et al.,* **U.S. pat. 4,405,635** (1983 to Merrell-Dow); *eidem, J. Med. Chem.* **25**, 1477 (1982). Cardiovascular pharmacology in animals: R. C. Dage *et al., J. Cardiovasc. Pharmacol.* **4**, 500 (1982); L. E. Roebel *et al., ibid.* 721. *In vitro* mechanism of action studies: T. Kariya *et al., ibid.* 509; H. S. Ahn *et al., Biochem. Pharmacol.* **35**, 1113 (1986). Acute hemodynamic effects in congestive heart failure: B. F. Uretsky *et al., Circulation* **67**, 823 (1983). HPLC determn in plasma: K. Y. Chan *et al., J. Chromatog.* **272**, 396 (1983). Pharmacokinetics in humans: R. G. Alken *et al., Clin. Pharmacol. Ther.* **36**, 209 (1984). Symposium on pharmacology and clinical efficacy in congestive heart failure: *Am. J. Cardiol.* **60**, 1C-90C (1987).

Crystals from isopropanol + water, mp 255-258° (dec). THERAP CAT: Cardiotonic.

3543. Enoxolone. *3-Hydroxy-11-oxoolean-12-en-29-oic acid; 3β-hydroxy-11-oxoolean-12-en-30-oic acid;* glycyrrhetic acid; 18β-glycyrrhetinic acid; uralenic acid; Arthrodont; Biosone; P.O. 12. $C_{30}H_{46}O_4$; mol wt 470.67. C 76.55%, H 9.85%, O 13.60%. From glycyrrhizic acid, *(q.v.).* Structure: Ruzicka *et al., Helv. Chim. Acta* **26**, 2143, 2278 (1943). Stereochemistry: Beaton, Spring, *J. Chem. Soc.* **1955**, 3126. Prepn from shredded licorice root: Mer, *Am. Perfum. Aromat.* **74**(6), 39 (1959). Manuf: **Brit. pat. 833,184** (1960 to Carlo Erba). Identity with uralenic acid: Belous *et al., Zh. Obshch. Khim.* **35**, 401 (1965). Metabolism: Parke *et al., J. Pharm. Pharmacol.* **15**, 500 (1963). Mechanism of action: Helbing, Berntsen, *Pharm. Weekbl.* **100**, 1438 (1965).

Needles from alcohol + petr ether, mp 296°. $[\alpha]_D^{21}$ +86° (alc); $[\alpha]_D^{20}$ +145.5° (dioxane); $[\alpha]_D^{20}$ +163° (chloroform). Freely sol in chloroform, dioxane. Soluble in alcohol, pyridine, acetic acid. Practically insol in petr ether.

18α-Hydrogen form, platelets from dil alcohol, mp 335°. $[\alpha]_D^{20}$ +140° (alcohol); $[\alpha]_D$ +98° (c = 0.1 in chloroform). Sol in alcohol, dioxane, chloroform.

THERAP CAT: Anti-inflammatory (topical).

3544. Enprofylline. *3,7-Dihydro-3-propyl-1H-purine-2,6-dione;* 3-propylxanthine; D-4018; Oxeze; Nilyph. $C_8H_{10}N_4O_2$; mol wt 194.19. C 49.48%, H 5.19%, N 28.85%, O 16.48%. Xanthine type antiasthmatic. Prepn: Y. Ohtsuka, *Bull. Chem. Soc. Japan* **46**, 506 (1973); *idem,* **Japan. pat. 74 04,469** (1974 to Sagami Chem. Res. Cent.), *C.A.* **81**, 152277f (1974). Prepn and use as bronchodilator: **Japan. Kokai 80 57,517;** P. G. Kjellin, C. G. A. Persson, **U.S. pat. 4,325,956** (1980, 1982 both to AB Draco). Comparative pharmacology and toxicology with theophylline, *q.v.:* C. G. A. Persson, G. Kjellin, *Acta Pharmacol. Toxicol.* **49**, 313 (1981). Effects on adenosine receptors: D. Ukena *et al., Eur. J. Pharmacol.* **117**, 25 (1985). HPLC determn in plasma: N. Grgurinovich, *J. Chromatog.* **380**, 431 (1986). Pharmacokinetics and bronchodilating activity in humans: T. B. R. Kluge *et al., Eur. J. Clin. Pharmacol.* **30**, 21 (1986). Clinical efficacy in acute asthma: J. Boe *et al., Ann. Allergy* **59**, 155 (1987); R. Ruffin *et al., Chest* **93**, 510 (1988). Cardiopulmonary effects in chronic lung disease: H. Vik-Mo *et al., ibid.* **94**, 354 (1988).

mp 287-289°. pKa 8.4. Soly in water (20°): 1.5 mg/ml. LD$_{50}$ in mice (mg/kg): ~157.4 i.v.; ~501.1 orally (Kjellin, Persson).

THERAP CAT: Bronchodilator.

3545. Enprostil. *7-[3-Hydroxy-2-(3-hydroxy-4-phenoxy-1-butenyl)-5-oxocyclopentyl]-4,5-heptadienoic acid methyl ester; (dl)-9-keto-11α,15α-dihydroxy-16-phenoxy-17,18,19,20-tetranorprosta-4,5,13-trans-trienoic acid methyl ester*; RS-84135; Fundyl; Gardrin; Syngard. C$_{23}$H$_{28}$O$_6$; mol wt 400.47. C 68.98%, H 7.05%, O 23.97%. Prostaglandin E$_2$ derivative with anti-ulcer activity. Prepn: A. R. Van Horn *et al.,* U.S. pat. **4,178,457** (1979 to Syntex). Pharmacokinetics: D. R. Stanski *et al., Clin. Pharmacol. Ther.* **31**, 273 (1982). Cytoprotective effect against aspirin induced gastromucosal injury in humans: M. M. Cohen *et al., Gastroenterology* **88**, 382 (1985). Clinical comparison with cimetidine of antisecretory and antigastrin activity in duodenal ulcer: V. Mahachai *et al., ibid.* **89**, 555 (1985); of efficacy in ulcer healing: L. Carling *et al., Scand. J. Gastroenterol.* **22**, 325 (1987). Series of articles on pharmacology, clinical efficacy and safety: *Am. J. Med.* **81**, Suppl. 2A, 1-88 (1986).

White to off-white waxy solid. Softens at 30°, liquid at 46°. Very slightly sol in water. Sol in alcohol, propylene glycol, propylene carbonate. uv max (methanol): 220, 265, 271, 277 nm (log ε 4.01, 3.14, 3.24, 3.16).

THERAP CAT: Anti-ulcerative. Antisecretory agent.

3546. Enrofloxacin. *1-Cyclopropyl-7-(4-ethyl-1-piperazinyl)-6-fluoro-1,4-dihydro-4-oxo-3-quinolinecarboxylic acid*; CFPQ; Bay Vp 2674; Baytril. C$_{19}$H$_{22}$FN$_3$O$_3$; mol wt 359.40. C 63.50%, H 6.17%, F 5.29%, N 11.69%, O 13.35%. Fluorinated quinolone antibacterial. Prepn: K. Grohe *et al.,* Ger. pat. **3,142,854**; *eidem,* U.S. pat. **4,670,444** (1983, 1987 both to Bayer AG); K. Grohe, H. Heitzer, *Ann.* **1987**, 29. Use as plant fungicide: K. Grohe *et al.,* U.S. pat. **4,563,459** (1986 to Bayer AG). Pharmacokinetics in calves: J. N. Davidson *et al., Proc. West. Pharmacol. Soc.* **29**, 129 (1986); in chickens: G. M. Conzelman *et al., ibid.* **30**, 393 (1987). Spectrofluorometric determn of residues in poultry tissues: T. B. Waggoner, M. C. Bowman, *J. Assoc. Offic. Anal. Chem.* **70**, 813 (1987). Toxicology and physical properties: P. Altreuther, *Vet. Med. Rev.* **2**, 87 (1987). Series of articles on pharmacology, *in vitro* antibacterial activity and field trials: *ibid.* 90-140.

Pale yellow crystals, mp 219-221°. Slightly sol in water at pH 7. LD$_{50}$ in male, female mice (mg/kg): > 5000, 4336 orally; ~200, ~200 i.v.; in male rats, male rabbits (mg/kg): > 5000, 500-800 orally (Altreuther).

THERAP CAT (VET): Antibacterial.

3547. Enterobactin. *N,N',N''-(2,6,10-Trioxo-1,5,9-trioxacyclododecane-3,7,11-triyl)tris[2,3-dihydroxy]benzamide; N,N',N''-(2,6,10-trioxo-1,5,9-trioxacyclododecane-3,7,11-triyl)tris-o-pyrocatechuamide*; enterochelin. C$_{30}$H$_{27}$N$_3$O$_{15}$; mol wt 669.57. C 53.81%, H 4.07%, N 6.28%, O 35.84%. Physiologically active macrocyclic iron sequestering agent of the phenolate type, involved in microbial transport and metabolism of iron; it is overproduced by *E. coli* and related enteric bacteria under low-iron stress. Isoln from culture medium of *Salmonella typhimurium* LT2 and structure: J. R. Pollack, J. B. Neilands, *Biochem. Biophys. Res. Commun.* **38**, 989 (1970); also produced by *Escherichia coli* and *Aerobacter aerogenes:* I. G. O'Brien, F. Gibson, *Biochim. Biophys. Acta* **215**, 393 (1970). Conformation: M. Leinás *et al., Biochemistry* **12**, 3836 (1973). Total synthesis: E. J. Corey, S. Bhattacharyya, *Tetrahedron Letters* **1977**, 3919. Synthesis of enterobactin and enantioenterobactin: W. H. Rastetter *et al., J. Org. Chem.* **45**, 5011 (1980); **46**, 3579 (1981). Biosynthetic study: J. R. Pollack *et al., J. Bacteriol.* **104**, 635 (1970). Stability constants and electrochemical behavior of ferric enterobactin: K. N. Raymond, *J. Am. Chem. Soc.* **101**, 6097 (1979). Kinetics and mechanism of iron removal from transferrin: C. J. Carrano *et al., ibid.* **103**, 5401. Review: J. B. Neilands in *Inorganic Biochemistry* **1**, G. Eichorn, Ed. (Elsevier, New York, 1973) pp 167-202; *idem, Bioinorganic Chemistry* **II**, K. N. Raymond, Ed. (A.C.S., Washington, 1977) pp 3-33.

Crystals from ethanol/water, mp 202-203°. [α]$_D^{25}$ +7.40° (ethanol). uv max (ethyl acetate): 316 nm (ε 9390). Sol in acetone, dioxane, dimethyl sulfoxide, methanol. Practically insol in water, soly increases at pH 7 to 8. Neutral on paper electrophoresis at pH 5. Displays bright blue fluorescence under uv light.

Ferric enterobactin, [C$_{30}$H$_{21}$FeN$_3$O$_{15}$]$^{3-}$. Anionic complex of enterobactin. uv max: 495 nm (ε 5600).

USE: As a growth-promoting agent in various organisms.

3548. Enterogastrone. Anthelone E; enteroanthelone; Duosan; Ileogastrone. An inhibitor of gastric secretion, apparently different from urogastrone. Obtained from extracts of the intestinal mucosa of mammals: Gray, Ivy, *Cold Spring Harbor Sympos. Quant. Biol.* **5**, 405 (1937); Ivy, Greengard, U.S. pat. **2,477,541** (1949 to Research Corp.). Comparison with urogastrone: Friedman, *Vitam. Horm. (New York)* **9**, 313 (1951); Gregory, *J. Physiol.* **129**, 528 (1955).

Amorphous buff-colored powder. Freely sol in water.

THERAP CAT: Gastric secretion inhibitor.

3549. Enterolactone. *trans-Dihydro-3,4-bis[(3-hydroxyphenyl)methyl]-2(3H)-furanone; trans-(±)-2,3-bis(3'-hydroxybenzyl)-γ-butyrolactone*; compound 180/442; HPMF; HBBL. C$_{18}$H$_{18}$O$_4$; mol wt 298.34. C 72.47%, H 6.08%, O 21.45%. Phenolic compound having 2,3-dibenzylbutane skeleton. First lignan, *q.v.,* found in animal species. Determn in female urine from humans, vervet monkeys and characterization: S. R. Stitch *et al., Nature* **287**, 738 (1980); K. D. R. Setchell *et al., ibid.* 740. Structure determn by GC-MS: *eidem, Biochem. J.* **197**, 447 (1981); *eidem, Biomed. Mass Spectrom.* **10**, 227 (1983). ^{1}H-NMR, X-ray crystal structure: G. Cooley *et al., J. Chem. Soc. Perkin Trans. I* **1984**, 489. Synthesis: M. B. Groen, J. Leemhuis, *Tetrahedron Letters* **21**, 5043 (1980); G. Cooley *et al., ibid.* **22**, 349 (1981); A. Pelter *et al., ibid.* 1549; P. A. Ganeshpure,

R. Stevenson, *Chem. & Ind. (London)* **1981**, 778; A. Pelter *et al., J. Chem. Soc. Perkin Trans. I* **1983**, 643. Lignan excretion peaks during luteal phase of menstrual cycle: K. D. R. Setchell *et al., J. Steroid Biochem.* **12**, 375 (1980). Excretion as glucuronide conjugate: M. Axelson, K. D. R. Setchell, *FEBS Letters* **122**, 49 (1980). Biosynthesis by intestinal bacteria: *eidem, ibid.* **123**, 337 (1981). Effect of diet on lignan excretion: H. Aldercreutz *et al., Med. Biol.* **59**, 337 (1981).

Gum, mp 141-143°. uv max (ethanol): 227, 261 nm (log ε 4.66, 4.64).

3550. Enteromycin. *3-[[(Methyl-aci-nitro)acetyl]amino]-2-propenoic acid; 3-[2-(methyl-aci-nitro)acetamido]acrylic acid; N-(O-methyl-aci-nitroacetyl)-3-aminoacrylic acid; seligocidin.* $C_6H_8N_2O_5$; mol wt 188.13. C 38.30%, H 4.28%, N 14.89%, O 42.52%. Antibiotic substance isolated from *Streptomyces albireticuli:* Nakazawa, *C.A.* **51**, 10647i (1957); from *S. achromogenes:* Herr *et al., Antimicrob. Ag. Chemother.* **1964**, 100. Structure: Mizuno, *Bull. Chem. Soc. Japan* **34**, 1419, 1425, 1631, 1633 (1961). Process for production: Shibata *et al.,* **Japan.** pat. **5747('58)** (to Takeda), *C.A.* **53**, 16481f (1959). Identification of carboxamide: S. E. DeVoe *et al., Antimicrob. Ag. Chemother.* **1964**, 105; L. A. Mitscher *et al., Tetrahedron* **21**, 267 (1965).

Two interchangeable crystal forms from methanol: needles or fine prisms. Indefinite mp. pKa 4.3. Unstable to heat, acid and alkali. Sparingly sol in water and other solvents. uv max (methanol): 230, 275, 298 nm (ε 9000, 13,000, 16,000).

Carboxamide, $C_6H_9N_3O_4$, rosettes from glacial acetic acid, dec 155°. uv max (methanol): 230, 300 nm (E$_{1cm}^{1\%}$ 650, 820). Sol in glacial acetic acid, dimethyl sulfoxide, dimethylformamide; slightly sol in methanol. Practically insol in the usual organic solvents.

Methyl ester, $C_7H_{10}N_2O_5$, crystals from ethanol, mp 141°.

3551. Entprol. *1,1',1'',1'''-(1,2-Ethanediyldinitrilo)-tetrakis[2-propanol]; 1,1',1'',1'''-(ethylenedinitrilo)tetra-2-propanol; N,N,N',N'-tetrakis(2-hydroxypropyl)ethylenediamine; Quadrol.* $C_{14}H_{32}N_2O_4$; mol wt 292.43. C 57.50%, H 11.03%, N 9.58%, O 21.89%. [CH$_3$CH(OH)CH$_2$]$_2$NCH$_2$-CH$_2$N[CH$_2$CH(OH)CH$_3$]$_2$. Prepd by the action of propylene oxide on ethylenediamine: Lundsted, Schulz, **U.S.** pat. **2,697,118** (1954 to Wyandotte).

Viscous liq. bp$_{1.0}$ 190°. Miscible with water. Sol in methanol, ethanol, toluene, ethylene glycol, perchloroethylene.

USE: Cross-linking agent and catalyst in the manuf of urethan foams; in epoxy resin curing; as complexing agent, humectant, emulsifier, plasticizer.

3552. Enviomycin. *(R)-1-(threo-4-Hydroxy-L-3,6-diaminohexanoic acid)-6-[L-2-(2-amino-1,4,5,6-tetrahydro-4-pyrimidinyl)glycine]viomycin; tuberactinomycin N.* $C_{25}H_{43}$-$N_{13}O_{10}$; mol wt 685.71. C 43.79% H 6.32%, N 26.55%, O 23.33%. Polypeptide antibiotic produced by a mutant of *Streptomyces griseoverticillatus* var. *tuberacticus.* Isoln and characterization: T. Ando *et al., J. Antibiot.* **24**, 680 (1971); A. Nagata *et al.,* in *Advan. Antimicrob. Antineopl. Chemo-*

ther. **vol. 1/2,** M. Hejzlar *et al.,* Eds. (University Park Press, Baltimore, 1972) pp 1039-1041. Prepn: J. Abe *et al.,* **Ger.** pat. **2,133,181;** *eidem,* **U.S.** pat. **3,892,732** (1972, 1975 both to Toyo Brewing Co., Ltd.). Structural studies: H. Yoshioka *et al., Tetrahedron Letters* **1971**, 2043; T. Wakamiya, T. Shiba, *J. Antibiot.* **28**, 292 (1975). Pharmacology: H. Hamakawa *et al., Oyo Yakuri* **8**, 817 (1974), *C.A.* **82**, 51530y (1975). Metabolism: T. Shimizu *et al., Japan. J. Antibiot.* **27**, 279 (1974). Total synthesis: T. Shiba *et al., J. Antibiot.* **32**, 1078 (1979). For structure *see* Tuberactinomycin, R$_1$ = OH, R = H.

Hydrochloride, $C_{25}H_{46}Cl_3N_{13}O_{10}$, white crystalline powder, mp >245° (dec). [α]$_D^{23}$ −19.1°. uv max (water, 0.1N HCl): 268 nm (E$_{1cm}^{1\%}$ 342); uv max (0.1N NaOH): 288 nm (E$_{1cm}^{1\%}$ 215). Very sol in water, slightly sol in methanol, ethanol. Insol in common organic solvents.

Sulfate, $C_{50}H_{92}N_{26}O_{32}S_3$, white crystalline powder. Soluble in water, slightly sol in usual organic solvents. LD$_{50}$ in mice, rats (mg/kg): 485, 680 i.v. (Hamakawa).

THERAP CAT: Antibacterial (tuberculostatic).

3553. Enviroxime. *6-[(Hydroxyimino)phenylmethyl]-1-[(1-methylethyl)sulfonyl]-1H-benzimidazol-2-amine; (E)-2-amino-6-benzoyl-1-(isopropylsulfonyl)benzimidazole oxime; anti-2-amino-1-(isopropylsulfonyl)-6-(α-hydroxyiminobenzyl)benzimidazole; LY 122772.* $C_{17}H_{18}N_4O_3S$; mol wt 358.42. C 56.97%, H 5.06%, N 15.63%, O 13.39%, S 8.94%. Benzimidazole deriv that inhibits rhinovirus multiplication. Prepn: C. J. Paget *et al.,* **Ger.** pat. **2,638,551** corresp to **U.S.** pat. **4,118,742** (1977, 1978 both to Lilly). Synthesis and sepn of *syn* and *anti* isomers: J. H. Wikel *et al., J. Med. Chem.* **23**, 368 (1980). Inhibition of rhinovirus replication in organ culture: D. C. De Long, S. E. Reed, *J. Infect. Dis.* **141**, 87 (1980). Metabolic studies: J. F. Quay *et al., Fed. Proc.* **39**, 3079 (1980); C. J. Parli *et al., ibid.* 214. Activity vs rhinovirus infection in man: R. J. Phillpotts *et al., Lancet* **1**, 1342 (1981). Prophylactic activity: F. G. Haden, J. M. Gwaltney, *Antimicrob. Ag. Chemother.* **21**, 892 (1982).

Cryst from acetonitrile, mp 198-199°. uv max (methanol): 218, 290 nm (ε 45600, 27100).

(Z)-isomer, zinviroxime. Cryst from methanol, mp 182-183°. uv max (methanol): 254, 285 nm (ε 20800, 13200).

3554. Eosine I Bluish. *4',5'-Dibromo-3',6'-dihydroxy-2',7'-dinitrospiro[isobenzofuran-1(3H),9'-[9H]xanthen]-3-one disodium salt; 4',5'-dibromo-2',7'-dinitrofluorescein disodium salt;* hydroxydibromodinitro-*o*-carboxyphenylfluorone sodium; C.I. Acid Red 91; C.I. 45400. $C_{20}H_6Br_2N_2$-Na$_2O_9$; mol wt 624.09. C 38.49%, H 0.97%, Br 25.62%, N 4.49%, Na 7.37%, O 23.07%. Prepd by nitration of 4',5'-dibromofluorescein: *Colour Index* **vol. 4** (3rd ed., 1971) p 4427.

Red powder. Freely sol in water with green fluorescence; sol in alc.

USE: Dyeing wool, cotton, and paper. In histology as a stain for epithelia, muscular fibers, nuclei, etc.

3555. Eosine Yellowish—(YS). *2',4',5',7'-Tetrabromo-3',6'-dihydroxyspiro[isobenzofuran-1(3H),9'-[9H]xanthen]-*

3-one disodium salt; 2',4',5',7'-tetrabromofluorescein; bromoeosine; tetrabromofluorescein sol; bromofluoresceic acid; D & C Red No. 22; C.I. Acid Red 87; Eosine; C.I. 45380. $C_{20}H_6Br_4Na_2O_5$; mol wt 691.91. C 34.72%, H 0.87%, Br 46.20%, Na 6.65%, O 11.56%. Prepd by bromination of fluorescein: *Colour Index* vol. 4 (3rd ed., 1971) p 4426.

Red crystals with bluish tinge, or brownish-red powder. Freely sol in water, less in alcohol; insol in ether. The concd aq soln is deep brownish-red, the dilute (1:500) soln is yellowish-red with greenish fluorescence; the alcoholic soln exhibits a strong green fluorescence.

USE: Lipstick and nail-polish coloring; dyeing wool, silk and paper; in red inks. Approved by FDA for use in drugs and cosmetics except for use in eye area: *Fed. Regist.* **47**, 53843 (1982).

3556. Epalrestat. *(E,E)-5-(2-Methyl-3-phenyl-2-propenylidene)-4-oxo-2-thioxo-3-thiazolidineacetic acid;* 5-[(E,E)-β-methylcinnamylidene]-4-oxo-2-thioxo-3-thiazolidineacetic acid; 3-carboxymethyl-5-(2-methylcinnamylidene)rhodanine; Ono-2235; Kinedak; Sorbistat. $C_{15}H_{13}NO_3S_2$; mol wt 319.39. C 56.41%, H 4.10%, N 4.38%, O 15.03%, S 20.08%. Aldose reductase inhibitor. Prepn: T. Tadao *et al.,* **Eur. pat. Appl. 47,109;** *eidem,* **U.S. pat. 4,464,382** (1982, 1984 both to Ono Pharmaceutical). Inhibitory effect on aldose reductase *in vitro:* H. Terashima *et al., J. Pharmacol. Exp. Ther.* **229**, 226 (1983). Effect on motor nerve conduction and sorbitol levels in diabetic rats: R. Kikkawa *et al., Diabetologia* **24**, 290 (1983). Effect in rats on peripheral nerve dysfunction: *eidem, Metabolism* **33**, 212 (1984); on retinal microangiopathy: K. Kojima *et al., Japan. J. Ophthalmol.* **29**, 99 (1985).

Crystals from ethanol-water, mp 210-217°.
N-Methyl-D-glucamine salt, $C_{22}H_{30}N_2O_8S_2$, crystals from methanol, mp 163-165°.

THERAP CAT: Treatment of diabetic neuropathy.

3557. Epanolol. *N-[2-[[3-(2-Cyanophenoxy)-2-hydroxypropyl]amino]ethyl]-4-hydroxybenzeneacetamide;* 1-(2-cyanophenoxy)-3-β-(4-hydroxyphenylacetamido)ethylamino-2-propanol; ICI 141292; Visacor. $C_{20}H_{23}N_3O_4$; mol wt 369.42. C 65.03%, H 6.28%, N 11.37%, O 17.32%. Cardioselective β_1-adrenergic blocker with intrinsic sympathomimetic activity. Prepn: L. H. Smith, **Ger. pat. 2,362,568;** *idem,* **U.S. pat. 4,167,581** (1974, 1979 both to ICI). Pharmacology in dogs: H. J. Smith *et al., J. Pharm. Exp. Ther.* **226**, 211 (1983); in humans: T. H. Pringle *et al., Brit. J. Clin. Pharmacol.* **21**, 249 (1986). Pharmacokinetics and bioavailability in dogs: J. McAinsh *et al., Eur. J. Drug Metabol. Pharmacokinet.* **9**, 129 (1984). Preliminary clinical trial in hypertension: B. Dahlof *et al., Brit. J. Clin. Pharmacol.* **18**, 831 (1984). Comparison with atenolol, *q.v.,* of bronchoconstrictor effect in asthmatics: S. Groth *et al., Eur. J. Clin. Pharmacol.* **30**, 653 (1986).

Crystals from acetonitrile, mp 118-120°.
THERAP CAT: Antihypertensive; anti-anginal.

3558. Eperisone. *1-(4-Ethylphenyl)-2-methyl-3-(1-piperidinyl)-1-propanone;* 4'-ethyl-2-methyl-3-piperidinopropiophenone. $C_{17}H_{25}NO$; mol wt 259.40. C 78.72%, H 9.71%, N 5.40%, O 6.17%. Spasmolytic agent related structurally to tolperisone, *q.v.* Prepn: E. Morita *et al.,* **Ger. pat. 2,458,638** corresp to **U.S. pat. 3,995,047** (1975, 1976 both to Eisai). Pharmacological study: K. Tanaka *et al., Nippon Yakurigaku Zasshi* **77**, 511 (1981), *C.A.* **95**, 35471t (1981). Absorption, distribution, excretion in rats and guinea pigs: T. Fujita *et al., Oyo Yakuri* **21**, 835 (1981), *C.A.* **96**, 28193w (1982). Toxicity study: H. Miyagawa *et al., ibid.* 939, *C.A.* **96**, 79722a (1982).

Hydrochloride, $C_{17}H_{26}ClNO$, *E-646, EMPP, Mional, Myonal.* Needles from isopropanol, mp 170-172°. LD_{50} in male S.D. rats, Wistar rats, mice (mg/kg): 1300, 1850, 1024 orally (Miyagawa).

THERAP CAT: Muscle relaxant (skeletal).

3559. Ephedra Equisetina. Ma Huang. Stems and leaves of *Ephedra equisetina* Bunge, *E. sinica* Stapf. and other species of *Ephedra, Gnetaceae.* Indigenous to China and India. It contains 0.75 to over 1% ephedrine; it is the source of natural ephedrine. Contains also variable quantities of pseudoephedrine.

3560. Ephedra Nevadensis. Cay note; canutillo; whorehouse tea; tapopote; teamsters' tea. Leaves and branches of *Ephedra nevadensis* S. Wats. *(E. antisyphilitica* C. A. Mey.), *Gnetaceae.* Habit. U.S. (Calif., Nevada). Contains little or no ephedrine.

3561. Ephedrine. α-[1-(Methylamino)ethyl]benzenemethanol; α-[1-(methylamino)ethyl]benzyl alcohol; 2-methylamino-1-phenyl-1-propanol; 1-phenyl-1-hydroxy-2-methylaminopropane; 1-phenyl-2-methylaminopropanol; α-hydroxy-β-methylaminopropylbenzene. $C_{10}H_{15}NO$; mol wt 165.23. C 72.69%, H 9.15%, N 8.48%, O 9.68%. α + β-Adrenergic agonist. Occurs in Ma Huang *(Ephedra vulgaris, E. sinica* Stapf., *E. equisetina* Bunge, *Gnetaceae)* and in several other Ephedra spp. Extraction procedure: Rymill, McDonald, *Quart. J. Pharm. Pharmacol.* **10**, 463 (1937). Isomeric forms include *d-* and *l*-ephedrine as well as *d-* and *l*-pseudoephedrine with *l*-ephedrine and *d*-pseudoephedrine as the naturally occurring isomers. Syntheses: Späth, Göhring, *Monatsh.* **41**, 319 (1920); Manske, Johnson, *J. Am. Chem. Soc.* **51**, 580, 1906 (1929). Stereochemistry: Freudenberg *et al., ibid.* **54**, 234 (1932); Freudenberg, Nikolai, *Ann.* **510**, 223 (1934); Witkop, Foltz, *J. Am. Chem. Soc.* **79**, 197 (1957); Pfanz, Kirchner, *Ann.* **614**, 149 (1958). Sepn of isomers: Paris *et al., Ann. Pharm. Franc.* **25**, 177 (1967). Mechanism of action of *l*-ephedrine: Drudi-Baracco *et al., Compt. Rend. Soc. Biol.* **158**, 259 (1964). Toxicity: M. D. Fairchild, G. A. Alles, *J. Pharmacol. Exp. Ther.* **158**, 135 (1967). Review of new methods and developments in stereochemistry: Fodor, *Recent Develop. Chem. Nat. Carbon Compounds* **1**, 15-160 (1965). Comprehensive description: S. A. Benezra, J. W. McRae in *Analytical Profiles of Drug Substances* vol. **8**, K. Florey, Ed. (Academic Press, New York, 1979) pp 489-507.

dl-Form, *racemic ephedrine, racephedrine.* Crystals, mp 79°. Sol in water, alcohol, ether, chloroform, oils.
dl-Form hydrochloride, $C_{10}H_{16}ClNO$, *Ephetonin, Race-*

phedrine Hydrochloride. Crystals, mp 187-188°. One gram dissolves in 4 ml water, in about 40 ml of 95% alc at 20°. Practically insol in ether. pH about 6.

dl-Form sulfate, $C_{20}H_{32}N_2O_6S$, *Racephedrine Sulfate.* Crystals, mp 247°. Sol in water and alcohol. pH about 6.

l-Form, L-*erythro-2-(methylamino)-1-phenylpropan-1-ol.* Waxy solid, crystals or granules. Soapy feel. Gradually dec on exposure to light. May contain up to ½ mole H_2O (5.2%). Anhydr material is hygroscopic. mp 34°. Absorption of water raises the mp to 40°. bp 255°. pH of aq solution (1 in 200) 10.8. One gram dissolves in about 20 ml water, 0.2 ml alcohol; sol in chloroform, ether, oils. *Keep well closed in a cool place.* Symposium on sympathomimetic agents: *Ind. Eng. Chem.* **37**, 116-148 (1945).

l-Form hydrochloride, *Biophedrin, Ephedral, Ephedrosst, Sanedrine.* Orthorhombic needles, affected by light, mp 216-220°. $[\alpha]_D^{25}$ −33° to −35.5° (c = 5). pH of aq soln (1 in 200) 5.9. One gram dissolves in 3 ml water, 14 ml alcohol. Practically insol in ether, chloroform.

l-Form sulfate, orthorhombic needles, affected by light, mp 245° (dec). $[\alpha]_D^{25}$ −29.5 to −32.0° (c = 5). One gram dissolves in 1.2 ml water, 95 ml alcohol. Freely sol in hot alcohol. pH about 6.

dl-Pseudoephedrine, DL-*threo-2-(methylamino)-1-phenylpropan-1-ol.* Crystals, mp 118°.

d-Pseudoephedrine, *d-ψ-ephedrine, d-isoephedrine.* Rhombic tablets from ether, mp 119°. $[\alpha]_D^{20}$ +51° (c = 0.6 in alc). pH of aq soln (1 in 200) 10.8. Sparingly sol in water (differs from *l*-ephedrine). Freely sol in alc or ether.

d-Pseudoephedrine hydrochloride, *Galpseud, Novafed, Rhinalair, Sinufed, Sudafed, Symptom 2.* Needles, mp 181-182°. $[\alpha]_D^{20}$ +62° (c = 0.8). uv max (ethanol): 208, 251, 257, 264 nm (ε 8300, 161, 201, 161). pKa 9.22. pH of aq soln (1 in 200) 5.9. Sol in water, alc, chloroform. LD$_{50}$ i.p. in mice: 1.0 mmole/kg (Fairchild, Alles).

d-Pseudoephedrine sulfate, *Afrinol.*

THERAP CAT: *l*-Form as bronchodilator. *d*-Pseudoephedrine as decongestant.

THERAP CAT (VET): Sympathomimetic, has been used to counteract hypotension associated with anesthesia; as a mydriatic; in allergic reactions and as a CNS stimulant.

3562. Epiandrosterone. *3β-Hydroxy-5α-androstan-17-one; 3β-hydroxy-17-androstanone;* isoandrosterone; *3β-androstanol-17-one; 3β-hydroxyetioallocholan-17-one.* C_{19}-$H_{30}O_2$; mol wt 290.43. C 78.57%, H 10.41%, O 11.02%. Present in normal human urine as a minor constituent, it is a less active androgen than androsterone. Synthesis: Cardwell *et al., J. Chem. Soc.* **1953**, 361; Johnson *et al., J. Am. Chem. Soc.* **75**, 2275 (1953); Johnson *et al., ibid.* **78**, 6331 (1956). *Review:* R. I. Dorfman, R. A. Shipley, *Androgens* (Wiley, New York, 1956).

dl-Form, crystals, mp 161-162°. Gives off a musk-like odor when hot.

d-Form, crystals from ethyl acetate + petr ether, mp 174.5°. $[\alpha]_D^{20}$ +88° (in methanol). Precipitated by digitonin. Practically insol in water. Soluble in organic solvents.

Acetate, $C_{21}H_{32}O_3$, stout prisms, mp 103-104°. $[\alpha]_D^{18}$ +68.5° (chloroform).

Benzoate, $C_{26}H_{34}O_3$, crystals, mp 210-212°.

3563. Epichlorohydrin. *Chloromethyloxirane; dl-α-epichlorohydrin;* 1-chloro-2,3-epoxypropane; γ-chloropropylene oxide. C_3H_5ClO; mol wt 92.53. C 38.94%, H 5.45%, Cl 38.32%, O 17.29%. Prepn: H. T. Clarke, W. W. Hartman, *Org. Syn.* coll. vol. I, 233 (2nd ed., 1941); G. Braun, *ibid.* coll. vol. II, 256 (1943). Manuf: Faith, Keyes & Clark's *Industrial Chemicals,* F. A. Lowenheim, M. K. Moran, Eds. (Wiley-Interscience, New York, 4th ed., 1975) pp 335-338.

Toxicity data: H. F. Smyth, C. P. Carpenter, *J. Ind. Hyg. Toxicol.* **30**, 63 (1948).

$$CH_2\text{---}CHCH_2Cl$$

Liquid. d_4^{20} 1.1812; d_4^{25} 1.1750; d_4^{50} 1.1436; d_4^{75} 1.1101. mp −25.6°. bp$_{760}$ 117.9°; bp$_{400}$ 98.0°; bp$_{200}$ 79.3°; bp$_{100}$ 62.0°; bp$_{40}$ 42.0°; bp$_{10}$ 16.6°; bp$_{1.0}$ −16.5°. $n_D^{11.6}$ 1.44195; n_D^{16} 1.43969; n_D^{25} 1.43585. Flash pt, open cup: 105°F (40°C). Insol in water. Misc with alcohol, ether, chloroform, trichloroethylene, carbon tetrachloride; immiscible with petr hydrocarbons. LD$_{50}$ orally in rats: 0.09 g/kg (Smyth, Carpenter).

Caution: Strong skin irritant, sensitizer. Chronic exposure can cause kidney injury: Patty's *Industrial Hygiene and Toxicology* Vol. 2A, G. D. Clayton, F. E. Clayton, Eds. (Wiley-Interscience, New York, 3rd ed., 1981) pp 2242-2247. This substance may reasonably be anticipated to be a carcinogen: *Fourth Annual Report on Carcinogens* (NTP 85-002, 1985) p 99.

USE: Solvent for natural and synthetic resins, gums, cellulose esters and ethers, paints, varnishes, nail enamels and lacquers, cement for Celluloid.

3564. Epicholestanol. *5α-Cholestan-3α-ol;* 3α-hydroxy-cholestane; ε-cholestanol. $C_{27}H_{48}O$; mol wt 388.65. C 83.43%, H 12.45%, O 4.12%. The 3α-hydroxy epimer of cholestanol. Prepn from cholestanone: Ruzicka, *Helv. Chim. Acta* **17**, 1407 (1934); *cf.* Marker, *et al., J. Am. Chem. Soc.* **57**, 2359 (1935); Barnett *et al., J. Chem. Soc.* **1940**, 1390.

Needles from alcohol, mp 185-186°. $[\alpha]_D^{20}$ +34° (c = 1.7 in chloroform). Less sol than cholestanol. Not precipitated by digitonin.

Acetate, $C_{29}H_{50}O_2$, mp 95.5-96°. Crystals from methanol.

Benzoate, $C_{34}H_{52}O_2$, mp 102-103°. $[\alpha]_{546}$ +27.2° in chloroform.

3565. Epicholesterol. *Cholest-5-en-3α-ol.* $C_{27}H_{46}O$; mol wt 386.64. C 83.87%, H 11.99%, O 4.14%. The 3α-hydroxy epimer of cholesterol. Prepd from 8-oxocholesteryl chloride or from cholest-5-en-3-one or by passing O_2 into a soln of cholesteryl Mg chloride: Ruzicka, Goldberg, *Helv. Chim. Acta* **19**, 1407 (1936); Marker, *et al., J. Am. Chem. Soc.* **58**, 481 (1936); Marker, U.S. pat. **2,117,355** (1938); Barnett *et al., J. Chem. Soc.* **1940**, 1390.

Crystals from alcohol, mp 141.5°. $[\alpha]_D^{30}$ −35° (c = 1 in alcohol). Not precipitated by digitonin.

Acetate, $C_{29}H_{48}O_2$, crystals from methanol, mp 85°.

3566. Epicillin. *6-[(Amino-1,4-cyclohexadien-1-ylacetyl)amino]-3,3-dimethyl-7-oxo-4-thia-1-azabicyclo[3.2.0]heptane-2-carboxylic acid;* 6-[D-α-amino-2-(1,4-cyclohexadien-1-yl)acetamido]penicillanic acid; D-α-amino-(1,4-cyclohexadien-1-yl)methylpenicillin; SQ 11302; Dexacilina; Dexacillin; Omnisan; Spectacillin. $C_{16}H_{21}N_3O_4S$; mol wt 351.43. C 54.69%, H 6.02%, N 11.96%, O 18.21%, S 9.12%. Semi-synthetic antibiotic related to penicillin. Prepn:

Weisenborn *et al.*, U.S. pat. **3,485,819** (1969 to Squibb); Dolfini *et al.*, *J. Med. Chem.* **14**, 117 (1971). Activity studies: Basch *et al.*, *Infec. Immunity* **4**, 44 (1971); Gadebusch *et al.*, *ibid.* 50. Clinical trials: Woodruff *et al.*, *N.Y. State J. Med.* **71**, 1087 (1971); Beck *et al.*, *Curr. Ther. Res.* **13**, 530 (1971); Reyes, *ibid.* 602; Landa, *ibid.* 654.

Crystals, dec 202° (hemihydrate).
THERAP CAT: Antibacterial.

3567. 16-Epiestriol. *Estra-1,3,5(10)-triene-3,16β,17β-triol;* $\Delta^{1,3,5}$-estratriene-3,16β,17β-triol; 16-epioestriol; Actriol. $C_{18}H_{24}O_3$; mol wt 288.37. C 74.97%, H 8.39%, O 16.64%. Isoln from urine of pregnant women: Marrian, Bauld, *Biochem. J.* **59**, 136 (1955); Watson, Marrian, *ibid.* **63**, 64 (1956). From human placenta: Diczfalusy, Halla, *Acta Endocrin.* **27**, 303 (1958). Prepn by reduction of 3,16β-diacetoxy-1,3,5(10)-estratrien-17-one: Biggerstaff, Gallagher, *J. Org. Chem.* **22**, 1220 (1957).

Crystals from methanol + benzene, mp 289-291°. $[\alpha]_D^{15}$ +76° (c = 0.297 in ethanol).
Epiestriol-3-allyl ether, crystals, mp 156-156.5°. Prepn: Huffman, U.S. pat. **3,002,009** (1961 to Lasdon Found.).

3568. Epimestrol. *3-Methoxyestra-1,3,5(10)-triene-16,17-diol;* 3-methoxy-17-epiestriol; ORG 817; NSC-55975; Stimovul. $C_{19}H_{26}O_3$; mol wt 302.42. C 75.46%, H 8.67%, O 15.87%. Estriol deriv, which stimulates ovulation. Prepn: J. de Visser, **Dutch** pat. **95,275** (1960 to Organon), *C.A.* **55**, 22378d (1961); L. Caglioti, M. Magi, *Tetrahedron Letters* **1962**, 1261; *eidem, Tetrahedron* **19**, 1127 (1963). ¹³C-NMR study: T. A. Wittstruck, K. I. H. Williams, *J. Org. Chem.* **38**, 1542 (1973). Proton NMR and configurational study: B. Schoenecker *et al.*, *J. Prakt. Chem.* **319**, 419 (1977). Clinical chemistry: G. Falkay *et al.*, *Clin. Chim. Acta* **44**, 209 (1973). Comparative study in rats: D. Bentue-Ferrer *et al.*, *Ann. Endocrinol.* **41**, 203 (1980). Effects in normal men: M. Luisi *et al.*, *Reproduccion* **4**, 77 (1980), *C.A.* **93**, 89260 (1980).

Cryst from acetone/heptane, mp 158-160°. $[\alpha]_D^{20}$ + 48° (chloroform).
THERAP CAT: Anterior pituitary activator.

3569. Epinephrine. *4-[1-Hydroxy-2-(methylamino)ethyl]-1,2-benzenediol; 3,4-dihydroxy-α-[(methylamino)methyl]benzyl alcohol;* 1-(3,4-dihydroxyphenyl)-2-(methylamino)ethanol; 3,4-dihydroxy-1-[1-hydroxy-2-(methylamino)ethyl]benzene; methylaminoethanolcatechol; Adrenalin; Epifrin; Glaucon; Simplene. $C_9H_{13}NO_3$; mol wt 183.20. C 59.00%, H 7.15%, N 7.65%, O 26.20%. The principal sympathomimetic hormone produced by the adrenal medulla in most species; occurs as the *l*-form in animals and man. Physiologic review: Malmejac, *Physiol. Rev.* **44**, 186 (1964). Isoln from animal adrenal glands: Takamine, *J. Soc. Chem.*

Ind. **20**, 746 (1901); Aldrich, *Am. J. Physiol.* **5**, 457 (1901). Synthesis of *dl*-form: Stolz, *Ber.* **37**, 4149 (1904); Payne, *Ind. Chem.* **37**, 523 (1961). Historic review of syntheses: Loewe, *Arzneimittel-Forsch.* **4**, 583 (1954). Resolution of *dl*-form: Flächer, *Z. Physiol. Chem.* **58**, 189 (1908); Fabian, **Brit.** pat. **816,857** (1959 to Carnegies of Welwyn). Configuration: Pratesi *et al.*, *J. Chem. Soc.* **1958**, 2069. Acute toxicity of *l*-form: A. M. Lands *et al.*, *J. Pharmacol. Exp. Ther.* **90**, 110 (1947). Comprehensive description: D. H. Szulczewski, W.-H. Hong, in *Analytical Profiles of Drug Substances* vol. 7, K. Florey, Ed. (Academic Press, New York, 1978) pp 193-229.

dl-Form, sparingly sol in water, alcohol.
dl-Form hydrochloride, $C_9H_{14}ClNO_3$, *Drenamist, Epiglaufrin.* Crystals from alcohol, mp 157°. Readily sol in water; sparingly sol in abs alc.
d-Form, crystals, dec 211-221°.
l-Form, *Adnephrine, Adrenal, Adrenamine, Adrenine, Adrin, Chelafrin, Epinephran, Epirenan, Hemisine, Hemostasin, Hemostatin, Hypernephrin, Levorenine, Nephridine, Nieraline, Paranephrin, Renaglandin, Renaleptine, Renalina, Renoform, Renostypticin, Renostyptin, Scurenaline, Styptirenal, Supracapsulin, Supranephrane, Suprarenaline, Suprarenin, Surrenine, Takamina, Vasoconstrictine, Vasotonin.* Minute crystals, gradually browning on exposure to light and air. mp 211-212°. mp about 215° (dec) when rapidly heated. $[\alpha]_D^{25}$ (1 gram/20 ml of 0.5N HCl, 200 mm tube) not < −50° and not > −53.5° (U.S.P.). Sparingly sol in water. Readily sol in aq solns of mineral acids and of NaOH and KOH. Insol in aq solns of ammonia and of the alkali carbonates. Insol in alcohol, chloroform, ether, acetone, oils. Aq solns are slightly alkaline to litmus. Combined with acids, forming water-sol salts. Solns undergo oxidation in the presence of oxygen esp in neutral or alkaline solns; the red color formed is adrenochrome. LD_{50} i.p. in mice: 4 mg/kg (Lands).
l-Form 3,4-cyclic borate, $C_9H_{12}BNO_4$, *epinephryl borate, Epinal, Eppy.*
l-Form *d*-bitartrate, $C_{13}H_{19}NO_9$, crystals, mp 147-154° (some dec). Darken slowly on exposure to air and light. One gram dissolves in about 3 ml water. Slightly sol in alc. Principal ingredient of *Epitrate* ophthalmic soln.
THERAP CAT: *l*-Form as adrenergic; bronchodilator; mydriatic; antiglaucoma agent.
THERAP CAT (VET): *l*-Form: Sympathomimetic, vasoconstrictor, cardiac stimulant, bronchodilator. To counter allergic reactions, anesthesia, cardiac arrest.

3570. Epiquinidine. *(9R)-6′-Methoxycinchonan-9-ol;* α-(6-methoxy-4-quinolyl)-5-vinyl-2-quinuclidinemethanol; 6-methoxy-α-(5-vinyl-2-quinuclidinyl)-4-quinolinemethanol. $C_{20}H_{24}N_2O_2$; mol wt 324.41. C 74.04%, H 7.46%, N 8.64%, O 9.86%. Occurs in cinchona bark. Isoln from quinoidine: Dirscherl, Thron, *Ann.* **521**, 48 (1938). By epimerization of quinine or quinidine: Rabe, Höter, *J. Prakt. Chem.* **154**, 66 (1940). Stereochemistry: Prelog, Zalán, *Helv. Chim. Acta* **27**, 535, 545 (1944); Prelog, Häfliger, *ibid.* **33**, 2021 (1950); Roth, *Pharmazie* **16**, 257 (1961). Synthesis: Grethe *et al.*, *Helv. Chim. Acta* **55**, 1044 (1972); Gutzwiller, Uskokovic, *ibid.* **56**, 1494 (1973).

Lustrous leaflets from ether, mp 111-113°. $[\alpha]_D^{25}$ +107.8°

(c = 1.02 in ethanol). Freely sol in alcohol; moderately sol in ether. Shows more blue fluorescence in H_2SO_4 than quinidine or quinine.

Dihydrochloride, $C_{20}H_{24}N_2O_2.2HCl$, crystals from alc, dec 195-196°. $[\alpha]_D^{20}$ +46° (c = 0.8 in 99% alc).

Neutral dibenzoyl-*d*-tartrate, $(C_{20}H_{24}N_2O_2)_2.C_{18}H_{14}O_8$, crystals from alcohol or acetone, dec 166-167°. $[\alpha]_D^{21}$ +3.7° (4:1 alcohol + chloroform).

Forms a double sulfate with epiquinine.

3571. Epiquinine. *(9S)-6'-Methoxycinchonan-9-ol.* For empirical formula, names, isoln and stereochemistry refs *see* Epiquinidine.

Viscous oil, $[\alpha]_D^{22}$ +43° (c = 0.95 in 99% alc). Freely sol in organic solvents. Shows more blue fluorescence in H_2SO_4 than quinine.

Dihydrochloride, $C_{20}H_{24}N_2O_2.2HCl$, crystals from acetone, dec 196°. $[\alpha]_D^{21}$ +33° (c = 0.8 in 99% alc).

Neutral dibenzoyl-*d*-tartrate, $(C_{20}H_{24}N_2O_2)_2.C_{18}H_{14}O_8$, crystals from acetone, dec 159°, $[\alpha]_D^{25}$ − 24.3 (c = 0.93 in ethanol).

Forms a double sulfate with epiquinidine.

3572. Epirizole. *4-Methoxy-2-(5-methoxy-3-methyl-1H-pyrazol-1-yl)-6-methylpyrimidine;* 2-(3-methyl-5-methoxy-1-pyrazolyl)-4-methoxy-6-methylpyrimidine; 2-(3-methoxy-5-methylpyrazol-2-yl)-4-methoxy-6-methylpyrimidine; 1-(4-methoxy-6-methyl-2-pyrimidinyl)-3-methyl-5-methoxypyrazole; mepirizole; DA-398; Mebron. $C_{11}H_{14}-N_4O_2$; mol wt 234.26. C 56.40%, H 6.02%, N 23.92%, O 13.66%. Prepn: Naito *et al.*, **S. Afr. pat. 67 04,936** (1968 to Daiichi Seiyaku), *C.A.* **70**, 57876q (1969); Naito *et al.*, *Chem. Pharm. Bull.* **17**, 1467 (1969). Pharmacology and metabolism: Oshima *et al.*, *ibid.* 1492; Takabatake *et al.*, *ibid.* **18**, 1900 (1970). Toxicity data: H. Ogura *et al.*, *J. Med. Chem.* **15**, 923 (1972).

Minute, white or cream-colored crystals from isopropyl ether, mp 90-92°. Characteristic odor, bitter taste. Sparingly sol in water. Sol in dil acids; freely sol in ethanol, benzene, dichloroethane. Also sol in ether, acetone. LD_{50} orally in mice: 820 mg/kg (Ogura).

THERAP CAT: Analgesic, antipyretic, anti-inflammatory.

3573. Epirubicin. *(8S-cis)-10-[(3-Amino-2,3,6-trideoxy-α-L-arabino-hexopyranosyl)oxy]-7,8,9,10-tetrahydro-6,8,-11-trihydroxy-8-(hydroxyacetyl)-1-methoxy-5,12-naphthacenedione;* 3-glycoloyl-1,2,3,4,6,11-hexahydro-3,5,12-trihydroxy-10-methoxy-6,11-dioxo-1-naphthacenyl-3-amino-2,3,6-tridioxy-α-L-arabino-hexopyranoside; 4'-epidoxorubicin; 4'-epiadriamycin; pidorubicin; 4'-epi-DX; IMI 28. $C_{27}H_{29}NO_{11}$; mol wt 543.53. C 59.66%, H 5.38%, N 2.58%, O 32.38%. Analog of the anthracycline antibiotic doxorubicin (adriamycin), *q.v.*, differing only in the position of the C-4 hydroxy group of the sugar moiety. Prepn: F. Arcamone *et al.*, **Ger. pat. 2,510,866;** *eidem,* **U.S. pat. 4,058,519** (1975, 1977 both to Soc. Farmaceut. Italia); *eidem, J. Med. Chem.* **18**, 703 (1975); S. Penco, *Process Biochem.*, **15**(5), 12 (1980). Purification: E. Oppici *et al.*, **Belg. pat. 898,506;**

eidem, **Brit.** pat. **Appl. 2,133,005** (both 1984 to Farmitalia Carlo Erba). Comparison with doxorubicin of *in vitro* activity: A. Di Marco *et al.*, *Cancer Res.* **36**, 1962 (1976). Tissue distribution: C. Italia *et al.*, *Brit. J. Cancer* **47**, 545 (1983); F. Arcamone *et al.*, *Cancer Chemother. Pharmacol.* **12**, 157 (1984). HPLC determn in biological fluids: P. E. Deesen, B. Leyland-Jones, *Drug. Metab. Dispos.* **12**, 9 (1984); G. Cassinelli *et al.*, *ibid.* 506. Clinical trials in solid tumors: K. Kolaric *et al.*, *J. Cancer Res. Clin. Oncol.* **106**, 148 (1983). Review of activity in experimental tumors: A. Goldin *et al.*, *Invest. New Drugs* **3**, 3-21 (1985). Review of pharmacology and clinical efficacy: F. Ganzina, *Cancer Treat. Rev.* **10**, 1-22 (1983); R. J. Cersosimo, W. K. Hong, *J. Clin. Oncol.* **4**, 425-439 (1986); P. Hurteloup, F. Ganzina, *Drugs Exp. Clin. Res.* **13**, 233-246 (1986). Review of epirubicin and other anthracycline antineoplastics: R. B. Weiss *et al.*, *Cancer Chemother. Pharmacol.* **18**, 185-97 (1986).

Hydrochloride, $C_{27}H_{30}ClNO_{11}$, **Farmorubicin, Pharmorubicin.** Red-orange crystals, mp 185° (dec). $[\alpha]_D^{20}$ +274° (c = 0.01 in methanol). Solution should be protected from sunlight.

THERAP CAT: Antineoplastic.

3574. Epithiazide. *6-Chloro-3,4-dihydro-3-[[(2,2,2-trifluoroethyl)thio]methyl]-2H-1,2,4-benzothiadiazine-7-sulfonamide 1,1-dioxide;* 3-[(2,2,2-trifluoroethylthio)methyl]-6-chloro-3,4-dihydro-2H-1,2,4-benzothiadiazine-7-sulfonamide 1,1-dioxide; 6-chloro-3,4-dihydro-7-sulfamoyl-3-(2,2,2-trifluoroethylthiomethyl)benzo-1,2,4-thiadiazine 1,1-dioxide; Thiaver. $C_{10}H_{11}ClF_3N_3O_4S_3$; mol wt 425.88. C 28.20%, H 2.60%, Cl 8.33%, F 13.38%, N 9.87%, O 15.03%, S 22.59%. Prepn: McManus, **U.S. pat. 3,009,911** (1961 to Pfizer).

Crystals from acetone + benzene, mp 206-207°.
THERAP CAT: Antihypertensive, diuretic.

3575. Epitiostanol. *2,3-Epithioandrostan-17-ol;* 10275-S; Thiodrol; $C_{19}H_{30}OS$; mol wt 306.51. C 74.45%, H 9.87%, O 5.22%, S 10.46%. Episulfide deriv of androstane, *q.v.* Prepn: **Brit. pat. 977,599** corresp to T. Komeno, **U.S. pat. 3,230,215** (1964, 1966 both to Shionogi); K. Takeda *et al., Tetrahedron* **1965**, 329; P. D. Klimstra *et al., J. Med. Chem.* **9**, 693 (1966). Antitumor effect in mice: A. Matsuzawa, T. Yamamoto, *Cancer Res.* **37**, 4408 (1977). Teratogenicity study: T. Minesita *et al., Oyo Yakuri* **7**, 723 (1973), *C.A.* **80**, 116474p (1974). Toxicity study: *eidem, ibid.* 805, *C.A.* **80**, 66865u (1974). Use in treatment of breast cancer: M. Fujimoro *et al., Cancer* **31**, 789 (1973).

Cryst from acetone, mp 127-128°. $[\alpha]_D^{27.5}$ +24.4° (c = 1.054 in chloroform). uv max (alcohol): 262 nm. LD_{50} in mice, rats: 1, 5 mg/kg i.p., T. Minesita et al., Oyo Yakuri **7**, 805 (1973).

THERAP CAT: Antineoplastic.

3576. EPN. *Phenylphosphonothioic acid O-ethyl O-(4-nitrophenyl) ester; O-ethyl O-p-nitrophenyl phenylphosphonothioate; ethyl p-nitrophenyl benzenethiophosphonate.* $C_{14}H_{14}NO_4PS$; mol wt 323.31. C 52.01%, H 4.36%, N 4.33%, O 19.80%, P 9.58%, S 9.92%. Prepn: A. G. Jelinek, U.S. pat. **2,503,390** (1950 to du Pont). Manufacturing process: N. Shindo et al., U.S. pat. **3,327,026** (1967 to Nissan). Insecticidal activity: R. L. McGarr, A. J. Chapman, J. Econ. Entomol. **59**, 1529 (1966); D. A. Wolfenbarger et al., ibid. **63**, 1568 (1970). Determn by HPLC: J. M. Lasker et al., Anal. Biochem. **109**, 369 (1980); X.-D. Ding, I. S. Krull, J. Agr. Food Chem. **32**, 622 (1984). Inhibition of acetylcholinesterase: O. Awad et al., Arzneimittel-Forsch. **22**, 1035 (1972); O. M. E. Awad, Enzyme **32**, 193 (1984). Metabolism in animals: R. L. Chrzanowski, A. G. Jelinek, J. Agr. Food Chem. **29**, 580 (1981). Toxicological study: H. C. Hodge et al., J. Pharmacol. Exp. Ther. **112**, 29 (1954). Teratogenic evaluation in mice: K. D. Courtney et al., J. Environ. Sci. Health **20B**, 373 (1985). Study of delayed neurotoxicity: B. M. Francis, Bull. Environ. Contam. Toxicol. **38**, 283 (1987). Toxicity data: T. B. Gaines, Toxicol. Appl. Pharmacol. **2**, 88 (1960).

Light yellow oil, aromatic odor. d^{25} 1.268. n_D^{25} 1.6021. Practically insol in water. Miscible with benzene, toluene, xylene, acetone, isopropanol, methanol. LD_{50} in female, male rats (mg/kg): 7.7, 36 orally; 25, 230 dermally (Gaines). *Caution:* Cholinesterase inhibitor. Skin absorption may be dangerous: Clinical Toxicology of Commercial Products, R. E. Gosselin et al., Eds. (Williams & Wilkins, Baltimore, 5th ed., 1984) Section II, p 294.

USE: Insecticide; acaricide.

3577. Eprazinone. *3-[4-(2-Ethoxy-2-phenylethyl)-1-piperazinyl]-2-methyl-1-phenyl-1-propanone; 3-[4-(β-ethoxyphenethyl)-1-piperazinyl]-2-methylpropiophenone.* $C_{24}H_{32}N_2O_2$; mol wt 380.53. C 75.75%, H 8.48%, N 7.36%, O 8.41%. Prepn of the dihydrochloride: R. Y. Mauvernay, **Neth. pat. Appl. 6,602,581** corresp to U.S. pat. **3,448,192** (1966, 1969 both to Riom). Pharmacodynamics: J. Vacher et al., Arch. Int. Pharmacodyn. Ther. **165**, 1 (1967). Pharmacological study: Y. Kase et al., Nippon Yakurigaku Zasshi **73**, 605 (1977), C.A. **88**, 15657 (1978). Efficacy: W. Spitzer, Fortschr. Med. **98**, 871 (1980). Metabolism: P. Toffel-Nadolny, W. Gielsdorf, Arzneimittel-Forsch. **31**, 719 (1981).

Dihydrochloride, $C_{24}H_{34}Cl_2N_2O_2$, *746 CE, Eftapan, Mucitux.* White, bitter-tasting crystalline powder, mp 201°, P. Toffel-Nadolny, W. Gielsdorf, loc. cit., also reported as crystals from methanol, mp 160°, R. Mauvernay, loc. cit. LD_{50} in mice (mg/kg): 729 orally, 38 i.v., J. Vacher et al., loc. cit.

THERAP CAT: Antitussive.

3578. Eprozinol. *4-(2-Methoxy-2-phenylethyl)-α-phenyl-1-piperazinepropanol; 1-(2-methoxy-2-phenylethyl)-4-(3-hydroxy-3-phenylpropyl)piperazine.* $C_{22}H_{30}N_2O_2$; mol wt 354.49. C 74.54%, H 8.53%, N 7.90%, O 9.03%. Prepn: Saunders, **Brit. pat. 1,188,505** (1970); to R. Y. Mauvernay, N. Busch, U.S. pat. **3,705,244** (1972). Pharmacology: Duchene-Marullaz et al., Therapie **26**, 155 (1971). Clinical studies: Sors, Dutarte, ibid. 163.

Dihydrochloride, $C_{22}H_{32}Cl_2N_2O_2$, *Alecor, Brovel, Eupnéron.* White crystalline powder, mp 164°. Sol in water and alcohol. LD_{50} orally in mice: 500 mg/kg (Mauvernay, Busch).

THERAP CAT: Bronchodilator.

3579. Eptazocine. *(1S)-2,3,4,5,6,7-Hexahydro-1,4-dimethyl-1,6-methano-1H-4-benzazonin-10-ol; (1S)-1,4-dimethyl-10-hydroxy-2,3,4,5,6,7-hexahydro-1,6-methano-1H-4-benzazonine; ST 2121; Sedapain.* $C_{15}H_{21}NO$; mol wt 231.35. C 77.88%, H 9.15%, N 6.05%, O 6.92%. Opioid agonist-antagonist analgesic, related to pentazocine, q.v. Prepn: M. Ikeda et al., **Belg. pat. 814,733** corresp to U.S. pat. **4,082,744** (1974, 1978 both to Nihon Iyakuhin Kogyo). Chromatographic study: I. Hayashi et al., Iyakuhin Kenkyu **12**, 442 (1981), C.A. **95**, 220196v (1981). Series of articles on pharmacological activities: Nippon Yakurigaku Zasshi **78**, 599-645 (1981), C.A. **96**, 62879, 79732, 79733 (1982). Opioid receptor binding study: T. Nabeshima et al., Res. Commun. Chem. Pathol. Pharmacol. **48**, 173 (1985).

Hydrobromide, $C_{15}H_{22}BrNO$, crystals, mp 207-210°.
THERAP CAT: Narcotic analgesic.

3580. EPTC. *Dipropylcarbamothioic acid S-ethyl ester; dipropylthiocarbamic acid S-ethyl ester; S-ethyl dipropylthiocarbamate; R-1608; FDA 1541; Eptam; Eradicane.* $C_9H_{19}NOS$; mol wt 189.31. C 57.10%, H 10.12%, N 7.40%, O 8.45%, S 16.93%. Selective pre-planting herbicide. Prepn: **Brit. pat. 808,753** corresp to H. Tilles, J. Antognini, U.S. pat. **2,913,327** (both 1959 to Stauffer). Activity: J. Antognini et al., Proc. Northeast. States Weed Control Conf. **1957**, 3. Metabolism: V. Y. Ong, S. C. Fang, Toxicol. Appl. Pharmacol. **17**, 418 (1970); J. P. Hubbell, J. E. Casida, J. Agr. Food Chem. **25**, 404 (1977).

Liquid, bp_{20} 127°. d^{30} 0.9546; n_D^{30} 1.4750. Vapor pressure at 35°: 3.4 × 10^{-2} mm Hg. Soly in water at 20°: 365 mg/l. Miscible with benzene, alc, toluene, xylene. LD_{50} orally in rats: 163l mg/kg (Antognini et al.).

USE: Herbicide.

3581. Equilenin. *3-Hydroxyestra-1,3,5,7,9-pentaen-17-one; 11,12,13,14,15,16-hexahydro-3-hydroxy-13-methyl-17H-cyclopenta[a]phenanthren-17-one; 1,3,5:10,6,8-estrapentaen-3-ol-17-one.* $C_{18}H_{18}O_2$; mol wt 266.32. C 81.17%, H 6.81%, O 12.01%. Estrogenic steroidal hormone isolated from urine of pregnant mares: Girard et al., Compt. Rend. **195**, 981 (1932). Occurs naturally in the d-form. Not found in human urine. Component of Conjugated Estrogenic Hormones, q.v. Isoln by chromatography: Duschinsky, Lederer, Bull. Soc. Chim. Biol. **17**, 1534 (1935). Total synthesis: Bachmann *et al., J. Am. Chem. Soc.* **61**, 974 (1939); **62**, 824 (1940); Johnson *et al., ibid.* **67**, 2274 (1945); **69**, 2942 (1947); **72**, 505 (1950); Hughes, Smith, *Chem. & Ind. (London)* **1960**, 1022; Bailey *et al., Chem. Commun.* **1967**, 1253; Stein *et al., Tetrahedron* **26**, 1917 (1970). Synthesis from estrone: Corbellini *et al., Farmaco Ed. Sci.* **19**, 913 (1964); O. N. Minailova *et al., Zh. Obschch. Khim.* **49**, 2633 (1979); A. R. Daniewski, T. Kowalczyk-Przewloka, *Tetrahedron Letters* **1982**, 2411. Review of synthetic studies: Taub, "Naturally Occurring Aromatic Steroids" in *The Total Synthesis of Natural Products* vol. 2, J. ApSimon, Ed. (John Wiley & Sons, New York, 1973) pp 642-663.

Needles from dil alc, mp 258-259°. Sublimes at 170-180° at 0.01 mm Hg. $[\alpha]_D^{16}$ +87° (12.8 mg made up to 1.8 ml in dioxane). uv max (ethanol): 231, 270, 282, 292, 325, 340 nm. Soly/100 ml alcohol 0.63 (18°); 2.5 g (boiling).

Acetate, $C_{20}H_{20}O_3$, mp 156-157°.

Benzoate, $C_{25}H_{22}O_3$, mp 222-223° (vac).

Methyl ether, $C_{19}H_{20}O_2$, needles from alc, mp 197-198°; mp 193-194° (vac).

THERAP CAT: Estrogen.

3582. Equilin. *3-Hydroxyestra-1,2,5(10),7-tetraen-17-one;* 1,3,5,7-estratetraen-3-ol-17-one. $C_{18}H_{20}O_2$; mol wt 268.34. C 80.56%, H 7.51%, O 11.92%. Steroidal hormone isolated from urine of pregnant mares: Girard *et al., Compt. Rend.* **195**, 981 (1932); Cartland, Meyer, *J. Biol. Chem.* **112**, 9 (1935/6). Structure: Serini, Logemann, *Ber.* **71**, 186 (1938); Pearlman, Wintersteiner, *J. Biol. Chem.* **130**, 35 (1939). Separation from estrone: Serini, Logemann, U.S. pat. **2,221,340** (1941 to Schering). Synthesis: Zderic *et al., J. Am. Chem. Soc.* **80**, 2596 (1958); Bagli *et al., Tetrahedron Letters* **1964**, 387; Stein *et al., ibid.* **1966**, 5015; eidem, *Tetrahedron* **26**, 1917 (1970). Prepn from testosterone by bacterial fermentation: Bowers *et al.*, U.S. pat. **3,067,212** (1962 to Syntex). Prepn by conversion of equilenin: Bailey *et al., Chem. Commun.* **1967**, 1253; Marshall, Deghenghi, *Can. J. Chem.* **47**, 3127 (1969). Component of Conjugated Estrogenic Hormones, *q.v.* Review of synthetic studies: Taub, "Naturally Occurring Aromatic Steroids" in *The Total Synthesis of Natural Products* vol. 2, J. ApSimon, Ed. (John Wiley & Sons, New York, 1973) pp 664-670.

Orthorhombic sphenoidal plates from ethyl acetate, mp 238-240°. $[\alpha]_D^{25}$ +308° (c = 2 in dioxane); $[\alpha]_D^{25}$ +325° (c = 2 in alc). uv max: 283-285 nm. Soluble in alcohol, dioxane, acetone, ethyl acetate, in other organic solvents; sparingly sol in water.

Benzoate, $C_{25}H_{24}O_3$, mp 196-197°.

Methyl ether, $C_{19}H_{22}O_2$, needles from alc, mp 161-162°.

THERAP CAT: Estrogen.

3583. Equol. *3,4-Dihydro-3-(4-hydroxyphenyl)-2H-1-benzopyran-7-ol;* 4',7-isoflavandiol; 4',7-dihydroxyisoflavan; 7-hydroxy-3-(4'-hydroxyphenyl)chroman. $C_{15}H_{14}O_3$; mol wt 242.28. C 74.36%, H 5.82%, O 19.81%. Isoln from mare's urine: G. F. Marrian, G. A. D. Haslewood, *Biochem. J.* **26**, 1226 (1932); from human urine: M. Axelson *et al., ibid.* **201**, 353 (1982). Physical description: *eidem, ibid.* **29**, 1586 (1935); F. Wessely *et al., Monatsh.* **71**, 215 (1938). Structure: F. Wessely, F. Prillinger, *Ber.* **72B**, 629 (1939); E. L. Anderson, G. F. Marrian, *J. Biol. Chem.* **127**,

649 (1939). Abs config: H. Suginome, *Bull. Chem. Soc. Japan* **39**, 409 (1966). Prepn of (±)-form: J. A. Lamberton *et al., Aust. J. Chem.* **31**, 455 (1978).

Crystals from aq ethanol, mp 189-190.5°. $[\alpha]_D$ −21.5°.

3584. Erabutoxins. Neurotoxic principles isolated from venom of the sea-snake, *Laticauda semifasciata* (in Japanese, *erabu-umihebi*). Their action on post-synaptic membrane blocks neuromuscular transmission. Separation and crystn of erabutoxins A and B: Tamiya, Arai, *Biochem. J.* **99**, 624 (1966). Structures are single polypeptide chains containing 62 amino acid residues with four S-S bridges, *erabutoxin A* differing from *erabutoxin B* only in the amino acid residue at position 26: Sato, Tamiya, *ibid.* **122**, 453 (1971); Endo *et al., ibid.* 463. Isoln and structure of the minor component, *erabutoxin C*: Tamiya, Abe, *ibid.* **130**, 547 (1972). Corrected amino acid sequences: N. Maeda, N. Tamiya, *Biochem. J.* **167**, 289 (1977). Three-dimensional structure of B, description of active site: B. W. Low, *Advan. Cytopharmacol.* **3**, 141 (1979). Molecular conformation: M. R. Kimball *et al., Biochem. Biophys. Res. Commun.* **88**, 950 (1979); F. Inagaki *et al., Eur. J. Biochem.* **109**, 129 (1980). Raman spectra: T. Takamatsu *et al., Biochim. Biophys. Acta* **622**, 189 (1980). Immunological studies: A. Tatsuya *et al., Toxicon* **17**, 571 (1979), *C.A.* **92**, 141459t (1980). Effect on transmission in autonomic ganglia: V. A. Chiappinelli *et al., Brain Res.* **211**, 107 (1981).

LD_{50} in mice, rats: 0.15, 0.07 µg/g i.m., Tamiya, Arai, *loc. cit.*

3585. Erbium. Er; at. wt 167.26; at. no. 68; valence 3. Rare earth metal of the yttrium group. Six naturally occurring isotopes: 162 (0.136%); 164 (1.56%); 166 (33.41%); 167 (22.94%); 168 (27.07%); 170 (14.88%): Hayden *et al., Phys. Rev.* **77**, 299 (1950). Artificial radioactive isotopes: 152-154; 157-161; 163; 165; 169; 170-171. Abundance in earth's crust: 2.5-6.5 ppm. Occurs in small quantities in all the rare earth minerals; main sources: xenotime, fergusonite, gadolinite, euxenite, polycrase, blomstrandine. Discovered by Mosander, *Skand. Naturför. Förh.* **3**, 387 (1842); *Phil. Mag.* [3] **23**, 241 (1843). Separation of the oxide by fractional crystn and precipitation: by the bromate process, James, *J. Am. Chem. Soc.* **32**, 517 (1910); **33**, 1332 (1911); **34**, 757 (1912); by the ethyl sulfate process, Urbain, *Compt. Rend.* **149**, 37 (1909); Hofmann, Burger, *Ber.* **41**, 308 (1908); Prandtl, *Z. Anorg. Chem.* **198**, 157 (1931). Sepn by ion-exchange: Spedding *et al., J. Am. Chem. Soc.* **69**, 2812 (1947); **76**, 2557 (1954). Prepn of the metal by heating anhydr erbium trichloride with potassium, cesium, or rubidium: Klemm, Bommer, *Z. Anorg. Chem.* **213**, 138 (1937). Reviews of prepn, properties and compounds of erbium and other lanthanides: *The Rare Earths*, F. H. Spedding, A. H. Daane, Eds. (Krieger, Huntington, N.Y., 1971, reprint of 1961 ed.) 641 pp; Hulet, Bode, "Separation Chemistry of the Lanthanides and Transplutonium Actinides" in *MTP Int. Rev. Sci.: Inorg. Chem., Ser. One* vol. 7, K. W. Bagnall, Ed. (University Park Press, Baltimore, 1972) pp 1-45; Moeller, "The Lanthanides" in *Comprehensive Inorganic Chemistry* vol. 4, J. C. Bailar Jr. *et al.*, Eds. (Pergamon Press, Oxford, 1973) pp 1-101.

Dark-gray metallic powder; hexagonal close-packed crystal lattice. d 9.062. mp 1497°. E°(aq) Er^{3+}/Er −230 V (calc). Similar to the other rare earth metals, possesses two reduction potentials 1.770 and 1.875 volts (ref to the normal calomel electrode), Noddack, Brukl, *Angew. Chem.* **50**, 362 (1937). Radioactivity induced by neutron bombardment: Sugden, *Nature* **135**, 469 (1935); McLennan, Rann, *ibid.* **136**, 831 (1935). Spectrum: Eder, *Ber. Wien. Akad.* [2a] **124**, 790 (1915); deGramont, *Compt. Rend.* **171**, 1106 (1920); Mott, McDonald, *Trans. Roy. Soc. Canada* [3] **21**, 230 (1927).

Oxide, Er_2O_3, *erbia.* Pinkish powder changing into cubic crystals on heating at 1300°; d 8.64; sp heat 0.065; prepd by igniting the oxalate or basic nitrate. Soly in water: 1.28×10^{-5} g-mol/l at 29°. Readily sol in acids.

Hydroxide, $Er(OH)_3$, pale pink gelatinous precipitate. Prepd by action of alkali hydroxide on a soln of erbium nitrate.

Chloride, $ErCl_3$, hexahydrate, deliquesc crystals. Sol in water; slightly sol in alcohol. Prepd by careful evaporation of a soln of erbium oxide in concd HCl: Jantsch et al., Z. Anorg. Chem. **207**, 353 (1932); dehydrated by heating in a stream of hydrogen chloride. The anhyd form is a pinkish powder, d 4.1. LD_{50} on mice: 535 mg/kg i.p.; 6.2 g/kg orally, Haley, J. Pharm. Sci. **54**, 663 (1965).

Bromide, $ErBr_3$, nonahydrate, deliquesc rose crystals. Prepn: Jantsch et al., loc. cit.

Nitrate, $Er(NO_3)_3$, pentahydrate, reddish, deliquesc cryst solid. Loses 4 mols of water on heating to 130°. LD_{50} (hexahydrate) in rats (female): 230 mg/kg i.p.; 35.8 mg/kg i.v., Haley, loc. cit.

Sulfate, see separate entry.

3586. Erbium Sulfate. $Er_2O_{12}S_3$; mol wt 622.58. Er 53.71%, O 30.84%, S 15.45%. $Er_2(SO_4)_3$. Prepared by slow evaporation of a soln of erbium oxide in sulfuric acid at 20-25°.

Powder. d 3.678. Hygroscopic; dissociates on heating; dissolves in water with evolution of heat; forms a basic sulfate on heating to 850°.

Octahydrate, pink monoclinic crystals; isomorphic with the corresponding yttrium, praseodymium and neodymium salts. d 3.205. Soly in water (parts/100 parts): 16 (20°); 6.53 (40°). On heating loses its water of hydration.

3587. Erbon. 2,2-Dichloropropanoic acid 2-(2,4,5-tri-chlorophenoxy)ethyl ester; 2,2-dichloropropionic acid 2-(2,4,-5-trichlorophenoxy)ethyl ester; 2-(2,4,5-trichlorophenoxy)-ethyl 2,2-dichloropropionate; Baron; Novon. $C_{11}H_9Cl_5O_3$; mol wt 366.47. C 36.05%, H 2.47%, Cl 48.38%, O 13.10%. Prepn: Brust, Seinkbeil, U.S. pat. **2,754,324** (1956 to Dow).

Crystals, mp 49-50°. $bp_{0.5}$ 161-164°. d_4^{50} 1.55. Sol in acetone, alcohol, kerosene, xylene. Insol in water. LD_{50} orally in rats: 1120 mg/kg, G. W. Bailey, J. L. White, Residue Rev. **10**, 97 (1965).

USE: Herbicide. Caution: Irritation on contact with eyes, skin. Repeated contact with concentrated material may cause dermatitis.

3588. Erdin. 5,7-Dichloro-4-hydroxy-6'-methoxy-6-methyl-3,4'-dioxospiro[benzofuran-2(3H),1'-[2,5]cyclohexa-diene]-2'-carboxylic acid. $C_{16}H_{10}Cl_2O_7$; mol wt 385.17. C 49.87%, H 2.62%, Cl 18.42%, O 29.08%. Chlorine containing antibiotic substance isolated from Aspergillus terreus Thom, together with geodin: H. Raistrick, G. Smith, Biochem. J. **30**, 1315 (1936). Molecular constitution of erdin and geodin: P. W. Clutterbuck et al., ibid. **31**, 1089 (1937); C. T. Calam et al., ibid. **33**, 579 (1939); **41**, 458 (1947). Structure: D. H. Barton, A. I. Scott, J. Chem. Soc. **1958**, 1767. Biosynthesis of geodin: Rhodes et al., Chem. & Ind. (London) **1962**, 611.

(±)-Form, yellow needles, mp 210-212°.

(+)-Form, prisms from chloroform + light petr, mp 210-212°. $[\alpha]_D$ +149° (c = 0.41 in dioxane). uv max: 284 nm (ϵ 21,000). Reacts as dibasic acid, dec slowly in presence of alkali. Solubilities of erdin are closely similar to those of geodin except that it is only very slightly sol in chloroform.

(+)-Form methyl ester, $C_{17}H_{12}Cl_2O_7$, geodin. Prisms from chloroform + ether, mp 228-231°. $[\alpha]_D$ +140° (c = 0.8 in chloroform). uv max: 284 nm (ϵ 19,000). Practically insol in water, petr ether; very sparingly sol in benzene, ether; somewhat more sol in alc, ethyl acetate; readily sol in chloroform, acetone, dioxane.

3589. Erdmann's Salt. Ammonium diamminetetrakis-(nitrito-N)cobaltate(1−). $CoH_{10}N_7O_8$; mol wt 295.06. Co 19.97%, H 3.42%, N 33.23%, O 43.38%. $NH_4[Co(NH_3)_2-(NO_2)_4]$. Prepared by evaporating a neutral aq soln of NH_4-NO_2 and $CoCl_2$: Erdmann, J. Prakt. Chem. **97**, 410 (1866); Gibbs, Proc. Am. Acad. **10**, 13 (1875); improved technique: Jörgensen, Z. Anorg. Chem. **17**, 477 (1898). Prepn from $CoCO_3$ and NH_4NO_2: Harkins et al., J. Am. Chem. Soc. **38**, 2647 (1916).

Dark, reddish-brown, orthorhombic bipyramidal crystals. d_4^{20} 1.972. Sol in water.

USE: A 1% aq soln is precipitated by $AgNO_3$, Tl_2SO_4, CsCl, RbCl, cinchonine sulfate.

3590. Ergochrysins. $C_{31}H_{28}O_{14}$; mol wt 624.56. C 59.62%, H 4.52%, O 35.87%. A mixture of yellow pigments from ergot: Bergmann, Ber. **65**, 1489 (1932). Two isomers, ergochrysin A and ergochrysin B, are known and differ in stereochemistry at positions 5, 6, and 10a: ApSimon et al., J. Chem. Soc. **1965**, 4144; Franck, Baumann, Ber. **99**, 3863 (1966); Hooper et al., J. Chem. Soc. (C) **1971**, 3580. Review of ergochrysins and other ergochromes: Franck, Flasch in Fortschr. Chem. Org. Naturst. **30**, 151-260 (1973).

Yellow crystals, dec 285°. $[\alpha]_D$ −68° (pyridine). Sol in alkali, warm pyridine, hot nitrobenzene; practically insol in all organic solvents.

3591. Ergocornine. 12'-Hydroxy-2',5'α-bis(1-methyl-ethyl)ergotaman-3',6',18-trione. $C_{31}H_{39}N_5O_5$; mol wt 561.66. C 66.29%, H 7.00%, N 12.47%, O 14.24%. Natural ergot alkaloid derived from lysergic acid; a member of the ergotoxine group. Isoln from ergot: Stoll, Hofmann, Helv. Chim. Acta **26**, 1570 (1943). Structure: Stoll et al., ibid. **34**, 1544 (1951). Separation and purification: Stoll, Hofmann, U.S. pat. **2,447,214** (1948 to Sandoz). Synthesis: Stadler et al., Helv. Chim. Acta **52**, 1549 (1969).

Solvated, polyhedra from methanol, dec 181° (contains 1 mole methanol). $[\alpha]_D^{20}$ −110° (pyridine); −175° (chloroform). uv max (methanol): 311 nm (log ϵ 3.91). Soluble in acetone, chloroform, ethyl acetate; slightly sol in ethyl and methyl alcohol; nearly insol in water. LD i.v. in rabbits: 1.17 mg/kg.

Ergocornine phosphate, crystals, dec 190-195°.
Ergocornine ethanesulfonate, crystals, dec 209°.
THERAP CAT: See Ergot.

3592. Ergocorninine. 12'-Hydroxy-2',5'α-bis(1-meth-ylethyl)-8α-ergotaman-3',6',18-trione. $C_{31}H_{39}N_5O_5$; mol wt 561.66. C 66.29%, H 7.00%, N 12.47%, O 14.24%. An alkaloid isomeric with ergocornine, *q.v.* Isoln, structure, separation, purification and synthesis *see* ergocornine.

Crystallizes solvent-free, unlike ergocornine which tends αto retain the solvent of crystn. Prisms from alc, dec 228°, $[\alpha]_D^{20}$ +409° (chloroform). uv max (methanol): 240.5, 312.5 (log ε 4.31, 3.92). Soluble in 15 parts boiling ethanol, 25 parts boiling methanol, 30 parts boiling benzene, 30 parts boiling ethyl acetate; freely sol in acetone, chloroform. Practically insol in water. Does not seem to form salts.

THERAP CAT: *See* Ergot.

3593. Ergocristine. *12'-Hydroxy-2'-(1-methylethyl)-5'-(phenylmethyl)ergotaman-3',6',18-trione.* $C_{35}H_{39}N_5O_5$; mol wt 609.74. C 68.94%, H 6.45%, N 11.49%, O 13.12%. Natural ergot alkaloid derived from lysergic acid; a member of the ergotoxine group. Isoln from ergot: Stoll, Burckhardt, *Z. Physiol. Chem.* **250**, 1 (1937); Stoll, Hofmann, *Helv. Chim. Acta* **26**, 1570 (1943). Structure: Stoll *et al.,* *ibid.* **34**, 1544 (1951). Separation and purification: Stoll, Hofmann, U.S. pat. 2,447,214 (1948 to Sandoz). Synthesis: Stadler *et al., Helv. Chim. Acta* **52**, 1549 (1969).

Orthorhombic crystals with $2C_6H_6$ from benzene. mp 155-157° (dec) (solvent-free base). $[\alpha]_D^{20}$ −183° (chloroform). Very sol in ethyl and methyl alcohol, acetone, chloroform, ethyl acetate. Slightly sol in ether. Practically insol in water, petr ether.

Ergocristine phosphate, crystals, dec 195°.

Ergocristine ethanesulfonate, crystals, dec 207°.

3594. Ergocristinine. *12'-Hydroxy-2'-(1-methylethyl)-5'α-(phenylmethyl)-8α-ergotaman-3',6',18-trione.* $C_{35}H_{39}N_5O_5$; mol wt 609.74. C 68.94%, H 6.45%, N 11.49%, O 13.12%. An alkaloid isomeric with ergocristine, *q.v.* Isoln, structure, separation and purification: *see* ergocristine.

Crystallizes solvent-free, unlike ergocristine which tends to retain the solvent of crystn. Long prisms from abs alc, mp 226° (dec). $[\alpha]_D^{20}$ +366° (c = 0.68 in chloroform); +471° (c = 0.35 in pyridine). uv max (methanol/methylene chloride): 313 nm (log ε 3.96). Much less sol than ergocristine. Does not seem to form salts.

3595. Ergocryptine. Ergokryptine. $C_{32}H_{41}N_5O_5$; mol wt 575.69. C 66.76%, H 7.18%, N 12.17%, O 13.90%. Two closely related isomers of the ergotoxine group which differ in the peptide portion of the molecule; α-ergocryptine yielding L-leucine upon hydrolysis, β-ergocryptine yielding L-isoleucine. The ergocryptine discussed in the literature prior to 1967 is now referred to as α-ergocryptine. Isoln from ergot: Stoll, Hofmann, *Helv. Chim. Acta* **26**, 1570 (1943). Structure: Stoll *et al., ibid.* **34**, 1544 (1951). Separation and purification: Stoll, Hofmann, U.S. pat. 2,447,214 (1948 to Sandoz). Separation of β-ergocryptine from α-ergocryptine: Schlientz *et al., Experientia* **23**, 991 (1967); *see also eidem, Pharm. Acta Helv.* **43**, 497 (1968). Synthesis of α- and β-ergocryptine: Stadler *et al., Helv. Chim. Acta* **52**, 1549 (1969).

α-Ergocryptine, *12'-hydroxy-2'(1-methylethyl)-5'α-(2-methylpropyl)ergotaman-3',6',18-trione.* R = CH_2CH-$(CH_3)_2$. Solvated prisms from acetone, benzene, methanol. With MeOH of crystn, mp 212° (dec). $[\alpha]_D^{20}$ −120° (pyridine); −198° (chloroform). uv max (methanol): 241, 312.5 nm (log ε 4.31, 3.95), Stadler *et al., loc. cit.* Freely sol in alcohol, chlorofom; almost insol in water. LD i.v. in rabbits: 1.05 mg/kg.

β-Ergocryptine, R = $CH(CH_3)CH_2CH_3$, rectangular plates from benzene, mp 173° (dec) (Schlientz). $[\alpha]_D^{20}$ −98° (c = 0.5 in pyridine); −179° (c = 0.5 in chloroform). uv max (methanol): 312 (log ε 3.93).

THERAP CAT: *See* Ergot.

3596. Ergocryptinine. Ergokryptinine. $C_{32}H_{41}N_5O_5$; mol wt 575.69. C 66.76%, H 7.18%, N 12.17%, O 13.90%. Alkaloid pair isomeric with α- and β-ergocryptine, resp., but differing by an α-configuration at C-8. The literature prior to 1967 refers to α-ergocryptinine as ergocryptinine. Isolation, structure, separation and purification *see* ergocryptine. Production: Abe *et al.,* U.S. pat. 2,835,675 (1958). Preparation of β-ergocryptinine from β-ergocryptine: Schlientz *et al., Experientia* **23**, 991 (1967). Synthesis of α- and β-ergocryptinin: Stadler *et al., Helv. Chim. Acta* **52**, 1549 (1969).

α-Ergocryptinine, *12'-hydroxy-2'-(1-methylethyl)-5'α-(2-methylpropyl)-8α-ergotaman-3',6',18-trione.* Crystallizes solvent-free, unlike ergocryptine which tends to retain the solvent of crystn. Fine needles from methanol, dec 240-242°. $[\alpha]_D^{20}$ +408° (chloroform); +485° (c = 0.5 in pyridine). uv max (methanol): 241.5, 312.5 nm (log ε 4.30, 3.94). Sol in 20 parts boiling ethanol, 50 parts boiling methanol; freely sol in acetone, chloroform; almost insol in water. Does not seem to form salts.

β-Ergocryptinine, colorless needles from methylene chloride-methanol, mp 217-218° (dec). $[\alpha]_D^{20}$ +421° (chloroform); +497° (pyridine). uv max (methanol): 240.5, 312 nm (log ε 4.31, 3.94). Differs from the α-isomer by the 1-methylpropyl group at the 5'-position.

3597. Ergoflavin. *[1S-[1α,3β,4β,4aα,7(1'R*,3'R*,-4'R*,4'aR*,9'aS*),9aβ]]-1,1',3,3',4,4',9a,9'a-Octahydro-4,4',8,8',9a,9'a-hexahydroxy-3,3'-dimethyl-[7,7'-bi-1,4a-(epoxymethano)-4aH-xanthene]-9,9',11,11'(2H,2'H)-tetrone; ergochrome CC(2,2').* $C_{30}H_{26}O_{14}$; mol wt 610.54. C 59.02%, H 4.29%, O 36.69%. Principal pigment from ergot: Freeborn, *Pharm. J.* **88**, 568 (1912); Eglinton *et al., J. Chem. Soc.* **1958**, 1833. Structure: McPhail *et al., ibid.* **1966**, Sect. B, 18. Review of ergoflavin and other ergochromes: Franck, Flasch in *Fortschr. Chem. Org. Naturst.* **30**, 151-206 (1973).

Yellow needles from methanol, decomp 350°. $[\alpha]_D^{21}$ +37.5° (c = 1.236 in acetone). uv max: 240, 260, 381 nm ($E_{1cm}^{1\%}$ 350, 346, 130). Sol in acetone, pyridine; moderately sol in methanol, alcohol, ethyl acetate, dioxane; sparingly sol in ether, benzene. Practically insol in $2N$ aq NaHCO$_3$.

Hexaacetate, $C_{42}H_{38}O_{20}$, prisms from chloroform + petr ether, dec 248-249°. $[\alpha]_D^{20}$ +61.2° (c = 0.62 in dioxane). uv max: 340, 338 nm ($E_{1cm}^{1\%}$ 343, 61).

3598. Ergoloid Mesylates. Dihydroergotoxine monomethanesulfonate (salt); co-dergocrine mesylate; CCK 179; Circanol; Coristin; Dacoren; DCCK; Deapril-ST; Decril; D-Ergotox forte L.U.T.; Dulcion; Ergodesit; Ergohydrin; Ergoplus; Hydergine; Hydro-Toxin-ratiopharm; Novofluen; Orphol; Perenan; Progeril; Redergin; Sponsin; Trigot. Hydrogenated ergot alkaloids, specifically an equiproportional mixture of dihydroergocornine methanesulfonate, dihydroergocristine methanesulfonate and α- and β-dihydroergocryptine methanesulfonate in the ratio of 1.5-2.5:1. Neuropharmacological investigations: D. M. Loew *et al., Postgrad. Med. J.* **52**, Suppl. 1, 40 (1976).

dihydroergocornine	R =	$-CH(CH_3)_2$
dihydroergocristine	R =	$-CH_2-C_6H_5$
dihydro-α-ergocryptine	R =	$-CH_2CH(CH_3)_2$
dihydro-β-ergocryptine	R =	$-CH(CH_3)CH_2CH_3$

Dihydroergocristine methanesulfonate, $C_{36}H_{45}N_5O_8S$, *Decme, Defluina, Enirant, Insibrin, Nehydrin, Simactil, Unergol.*

THERAP CAT: α-Adrenergic blocker (treatment of impaired mental function in the elderly).

3599. Ergometrinine. *[8α(S)]-9,10-Didehydro-N-(2-hydroxy-1-methylethyl)-6-methylergolinecarboxamide;* D-lysergic acid D-propanolamide; ergonovinine. $C_{19}H_{23}N_3O_2$; mol wt 325.39. C 70.13%, H 7.12%, N 12.91%, O 9.83%. An alkaloid isomeric with ergonovine, q.v. Ergometrinine and ergonovine can be interconverted by simple chemical procedures: Smith, Timmis, *J. Chem. Soc.* **1936**, 1166. Chromatographic determn in cereal grains: G. H. Ware *et al., J. Assoc. Off. Anal. Chem.* **69**, 697 (1986).

Stout prisms from acetone, dec 196°. $[\alpha]_D^{20}$ +416° (c = 0.45 in chloroform). pK 7.3. Freely sol in pyridine, moderately in chloroform; slightly sol in alcohol, acetone. Nearly insol in water.

Hydrobromide monohydrate, $C_{19}H_{24}BrN_3O_2.H_2O$, needles from aq acetone + ether, dec 190°.

Hydrochloride monohydrate, $C_{19}H_{24}ClN_3O_2.H_2O$, needles, dec 175-180°.

Mononitrate, $C_{19}H_{24}N_4O_5$, stout prisms from aq MeOH + ether, dec 235°. $[\alpha]_D^{20}$ +282° (c = 1).

Perchlorate, $C_{19}H_{24}ClN_3O_6$, needles, dec 225° (brown at 210°).

3600. Ergonovine. *[8β(S)]-9,10-Didehydro-N-(2-hydroxy-1-methylethyl)-6-methylergoline-8-carboxamide; N-[α-(hydroxymethyl)ethyl]-D-lysergamide;* D-lysergic acid L-2-propanolamide; ergometrine; Ergobasine; Ergotocine; Ergostetrine; Ergotrate; Ergoklinine; Syntometrine. $C_{19}H_{23}N_3O_2$; mol wt 325.39. C 70.13%, H 7.12%, N 12.91%, O 9.83%. From some ergots: Stoll *et al.,* U.S. pat. **2,809,920** (1957 to Saul & Co.). Prepn from D-lysergic acid and L(+)-2-amino-1-propanol: Stoll, Hofmann, *Helv. Chim. Acta* **26**, 956 (1943); Pioch, U.S. pat. **2,736,728** (1956 to Lilly); Patelli, Bernardi, U.S. pat. **3,141,887** (1964 to Farmitalia). Structure: Stoll *et al., Helv. Chim. Acta* **34**, 1544 (1951). Total synthesis: Kornfeld *et al., J. Am. Chem. Soc.* **78**, 3087 (1956). Metabolism: Slaytor, Wright, *J. Med. Pharm. Chem.* **5**, 483 (1962). Comprehensive description of the maleate: V. D. Reif in *Analytical Profiles of Drug Substances* vol. **11**, K. Florey, Ed. (Academic Press, New York, 1982) pp 273-312. Toxicity data: R. P. Beliles, *Toxicol. Appl. Pharmacol.* **23**, 537 (1972). Clinical use in diagnosis of angina: L. A. DiCarlo, Jr. *et al., Am. J. Cardiol.* **54**, 744

(1984); R. Nordlander, R. Orinius, *Acta Med. Scand.* **221**, 47 (1987).

Tetrahedra from ethyl acetate, fine needles from benzene. Tends to form solvated crystals, mp 162°. $[\alpha]_D^{20}$ +90°; −16° (c = 1 in pyridine). pK 6.8. Freely sol in lower alcohols, ethyl acetate, acetone; more sol in water than the other principal alkaloids of ergot; slightly sol in chloroform.

Hydrochloride, $C_{19}H_{24}ClN_3O_2$, needles from ethyl alcohol, dec 246°. $[\alpha]_D^{25}$ +63° (c = 0.9). More sol in water than the hydrobromide.

Maleate, $C_{23}H_{27}N_3O_6$, *Cornocentin, Ergotrate Maleate, Ermetrine.* Crystals, dec 167°. $[\alpha]_D^{25}$ +48° to +57°. One gram dissolves in 36 ml water, 120ml alcohol. Nearly insol in ether and chloroform. LD_{50} i.v. in mice: 8.26 mg/kg (Beliles).

Tartrate hydrate, $C_{42}H_{52}N_6O_{10}$, *Basergin, Neofemergen.* Crystals, slightly sol in water.

Hydracrylate, $C_{41}H_{52}N_6O_7$, *Ergotrate-H.*

THERAP CAT: Oxytocic.

THERAP CAT (VET): Oxytocic.

3601. Ergosine. *12'-Hydroxy-2'-methyl-5'α-(2-methylpropyl)ergotaman-3',6',18-trione.* $C_{30}H_{37}N_5O_5$; mol wt 547.64. C 65.79%, H 6.81%, N 12.79%, O 14.61%. Isoln of ergosine and ergosinine from ergot: Stoll, Timmis, *J. Chem. Soc.* **1937**, 396. Structure: Stoll *et al., Helv. Chim. Acta* **34**, 1544 (1951). Stereochemistry: Stoll *et al., ibid.* **37**, 2039 (1954). Total synthesis of ergosine and ergosinine: Stadler *et al., ibid.* **47**, 1911 (1964).

Prisms from ethyl acetate, dec 228°. $[\alpha]_D^{20}$ −161° (chloroform). Sol in chloroform; fairly sol in methanol, acetone; sparingly sol in ethyl acetate, benzene.

8α-Isomer, $C_{30}H_{37}N_5O_5$, *ergosinine, ergoclavinine.* Prisms from 90% alcohol, aq acetone, benzene or ethyl acetate, dec 228°. Also reported as colorless needles, mp 190-191° (dec), Stadler *et al., loc. cit.* $[\alpha]_D^{20}$ +420° (chloroform); $[\alpha]_D^{20}$ +380° (acetone). Very readily sol in chloroform; readily sol in acetone; less sol in ethyl acetate; sparingly sol in benzene; very sparingly sol in methyl alcohol; almost insol in water.

Note: Ergoclavine is an equimolar mixture of ergosine and ergosinine.

3602. Ergostane. $C_{28}H_{50}$; mol wt 386.71. C 86.97%, H 13.03%. Prepn from allocholanic acid: Fernholz, *Ber.* **69**, 1792 (1938).

Scales, plates from ether + methanol, mp 85°. $[\alpha]_D^{20}$ +17° (c = 2 in chloroform).

3603. Ergostanol. *Ergostan-3-ol.* $C_{28}H_{50}O$; mol wt 402.68. C 83.51%, H 12.52%, O 3.97%. Prepn by hydrogenation of ergosta-14,22-dien-7-one: Chen, *Ber.* **70**, 1432 (1937).

Crystals, mp 144-145°. $[\alpha]_D^{20}$ +15.9° (c = 1.8 in chloroform). Is precipitated by digitonin.
Acetate, $C_{30}H_{52}O_2$, mp 145°. $[\alpha]_D^{20}$ +6.0° (c = 1.8 in chloroform).
Benzoate, $C_{35}H_{54}O_2$, mp 163-165°.

3604. α-Ergostenol. *Ergost-8(14)-en-3β-ol.* $C_{28}H_{48}O$;
mol wt 400.66. C 83.93%, H 12.08%, O 3.99%. Prepd by
hydrogenation of ergosterol acetate in acetic acid soln: Hart
et al., J. Am. Chem. Soc. **52,** 2016 (1930); from 22-dihydro-
ergosterol: Windaus, Langer, *Ann.* **508,** 105 (1934).

Plates from methanol, mp 130-131°. $[\alpha]_D^{16}$ +11° (c = 0.9
in methanol). $[\alpha]_{546}$ −10.2° (c = 1 in methanol). Precipi-
tated by digitonin.
Acetate, $C_{30}H_{50}O_2$, mp 110°.
Benzoate, $C_{35}H_{52}O_2$, mp 118-120°.
Methyl ether, $C_{29}H_{50}O$, mp 56°. $[\alpha]_D^{20}$ +6.3° in chloroform.

3605. β-Ergostenol. *Ergost-14-en-3β-ol.* $C_{28}H_{48}O$; mol
wt 400.66. C 83.93%, H 12.08%, O 3.99%. From α-ergostenol by treatment with HCl-gas in chloroform: Reindel *et
al., Ann.* **452,** 34 (1927).

Plates, tablets from alcohol, mp 141°. $[\alpha]_D^{20}$ +21.2 (c =
0.9 in chloroform).
Acetate, $C_{30}H_{50}O_2$, mp 114°. $[\alpha]_D^{20}$ +13° (c = 1.2 in chloroform).
Benzoate, $C_{35}H_{52}O_2$, mp 158-160°. $[\alpha]_D^{20}$ +18.3, $[\alpha]_{546}^{20}$
+22.5° (c = 1 in ethyl acetate).

3606. γ-Ergostenol. *Ergost-7-en-3β-ol.* $C_{28}H_{48}O$; mol
wt 400.66. C 83.93%, H 12.08%, O 3.99%. From 22,23-di-
hydroergosterol: Lauch, *Z. Physiol. Chem.* **237,** 236 (1935);
from ergosterol: Wieland, Benend, *Ann.* **554,** 1 (1943); Bladon *et al., J. Chem. Soc.* **1951,** 2402.

Crystals from isopropanol, mp 146°. Seems optically inactive.
Acetate, $C_{30}H_{50}O_2$, mp 157°. $[\alpha]_D^{14}$ −5.3° (c = 0.7 in chloroform).
Benzoate, $C_{35}H_{52}O_2$, mp 179°.

3607. Ergosterol. *(3β,22E)-Ergosta-5,7,22-trien-3-ol;*
ergosta-5:6,7:8,22:23-trien-3-ol; ergosterin. $C_{28}H_{44}O$; mol
wt 396.63. C 84.78%, H 11.18%, O 4.03%. Most important
of the provitamins D. Usually obtained from yeast which
synthesizes it from simple sugars such as glucose. Damp
yeast yields about 2.5% ergosterol, the variety of the yeast
being very important. Isoln procedure: Green *et al.,* U.S.
pat. **3,006,932** (1961 to Vitamins Ltd.). When irradiated
with uv light, ergosterol develops powerful vitamin D_2, *q.v.,*
activity. Askew *et al.,* in England and Windaus and collaborators in Germany isolated the antirachitic vitamin D_2.
The main irradiation products of ergsterol are lumisterol → tachysterol → vitamin D_2. Structure: Chuang, *Ann.* **500,**
270 (1933). Oxidation products: Fuerst, *Arch. Pharm.* **300,**
144 (1967).

Small hydrated plates from alcohol, in hydrated needles
from ether. The best crystallized form contains 1½ mol
H_2O, mp 168°: Bills, Honeywell, *J. Biol. Chem.* **80,** 15
(1928). Complete removal of water is almost impossible and
results in an amorphous mass, melting range 166-183°.
$bp_{0.01}$ 250°. $[\alpha]_D^{20}$ −135° (c = 1.2 in $CHCl_3$ calcd as anhydr).
$[\alpha]_{546}^{20}$ −171° ($CHCl_3$). uv max (ethanol): 262, 271, 282, 293
nm, Hogness *et al., ibid.* **120,** 239 (1937). Practically insol in
water. One gram dissolves in 660 ml alcohol, in 45 ml boiling alcohol, in 70 ml ether, in 39 ml boiling ether, in 31 ml
chloroform. Precipitated by digitonin. Affected by light
and air, turns yellow. Oxygen forms peroxides and hydrogen may form polyhydro compds.
Acetate, $C_{30}H_{46}O_2$, mp 179°, clear at 181°.
Benzoate, $C_{35}H_{48}O_2$, plates or needles, mp 169-171°. $[\alpha]_D^{23}$
−71° (c = 1.1 in $CHCl_3$), $[\alpha]_{546}$ −88° ($CHCl_3$).
22,23-Dihydro analog, $C_{28}H_{46}O$, *22,23-dihydroergosterol.*
Prepn: A. Windhaus, R. Langer, *Ann.* **508,** 105 (1934); D.
H. R. Barton *et al., J. Chem. Soc., Perkin Trans. I* **1976,** 821;
D. J. Curry *et al., ibid.* **1977,** 822. uv irradiation gives vitamin D_4, *q.v.* Solvated needles from ethyl acetate + methanol, mp 152-153° (dried). $[\alpha]_D^{19}$ −109° ($CHCl_3$) (Windhaus);
also reported as needles from chloroform-methanol, mp
128-130°. $[\alpha]_D^{19}$ −121° (c = 0.1). uv max: 262, 272, 282,
294 nm (ε 8000, 11200, 11800, 6800) (Barton).
22,23-Dihydro analog acetate, $C_{30}H_{48}O_2$, mp 157-158°.
$[\alpha]_D^{17}$ −75° (c = 2.1 in $CHCl_3$).
THERAP CAT: Antirachitic vitamin.

3608. Ergot. Secale cornutum; spurred rye. Dried sclerotia of the fungus *Claviceps purpurea* (Fries) Tul., *Hypocreaceae,* parasitic on rye plants. *Habit.* Europe; cultivated in
Spain, Germany, and France. Four main classes of ergot
alkaloids can be distinguished: clavine alkaloids, lysergic
acids, lysergic acid amides and ergot peptide alkaloids.

There are ten ergot peptide alkaloids which are ergotamine, ergosine, ergocristine, ergocryptine, ergocornine, ergotaminine, ergosinine, ergocristinine, ergocryptinine, and ergocorninine, the last five alkaloids being isomers of the first five. These allkaloids are typified by a structure consisting of lysergic acid, dimethylpyruvic acid, proline, and phenylalanine joined in amide linkages. In 1943 A. Stoll and A. Hofmann, *Helv. Chim. Acta* **26**, 1570 (1943) made it clear that the ergotoxine reported by G. Barger and F. H. Carr, *J. Chem. Soc.* **91**, 377 (1907) and by F. Kraft, *Arch. Pharm.* **244**, 336 (1906) was but a mixture of ergocristine, ergocryptine, and ergocornine, and that the ergotinine, first reported by C. Tanret, *Compt. Rend.* **81**, 891 (1875), was also a mixture consisting of ergocristinine, ergocryptinine, and ergocorninine. Other constituents of ergot are ergonovine, ergometrinine, ergoclavine, elymoclavine, trimethylamine, putrescine, cadaverine, agmatine, histamine, tyramine, histidine, tyrosine, valine, leucine, betaine, choline, acetylcholine, ergothioneine, 15-30% fatty oil, ergosterol, mannitol, lactic acid, and succinic acid. Production of ergocryptinine and elymoclavine by cultures of fungi parasitic on *Elymus mollis* Trin: Abe *et al.*, U.S. pat. **2,835,675**. Production of ergot alkaloids by saprophytic cultures: Adams, U.S. pat. **3,117,917** (1964 to Miles Labs.). Ergot alkaloid fermentation: W. J. Kelleher, *Adv. Appl. Microbiol.* vol. **11**, 211-244 (1969). Biosynthesis of ergot alkaloids: H. G. Floss, *Tetrahedron* **32**, 873 (1976). *Reviews:* Gröger in *Microbial Toxins*, vol. **VIII**, S. Kadis *et al.*, Eds. (Academic Press, New York, 1972) pp 321-373; Stadler, Stütz, "The Ergot Alkaloids" in *The Alkaloids* vol. **15**, R. H. F. Manske, Ed. (Academic Press, New York, 1975) pp 1-40. *Books:* F. J. Bove, *The Story of Ergot* (Karger, Basel, 1970) 297 pp; *Handb. Exp. Pharmacol.* **49** entitled "Ergot Alkaloids and Related Compounds", B. Berde, H. O. Schild, Eds. (Springer-Verlag, New York, 1978) 1003 pp.

Human Toxicity: Ergot Poisoning: Acute: vomiting, diarrhea, thirst, tachycardia, confusion, coma. *Chronic:* CNS symptoms, G.I. disturbances, peripheral circulatory changes, gangrene, cf. *Clinical Toxicology of Commercial Products*, R. E. Gosselin *et al.*, Eds. (Williams & Wilkins, Baltimore, 4th ed., 1976) Section II, p 146.

THERAP CAT: Vasoconstrictor (specific in migraine).

THERAP CAT (VET): Has been used as an oxytocic.

3609. Ergotamine. *12'-Hydroxy-2'-methyl-5'α-(phenylmethyl)ergotaman-3',6',18-trione.* $C_{33}H_{35}N_5O_5$; mol wt 581.65. C 68.14%, H 6.07%, N 12.04%, O 13.75%. Vasoconstrictor found in ergot of Central Europe. Extraction procedure: Stoll, *Helv. Chim. Acta* **28**, 1283 (1945). Structure: Stoll *et al.*, *ibid.* **34**, 1544 (1951). Stereochemistry: Hofmann *et al.*, *ibid.* **46**, 2306 (1963). Total synthesis: Hofmann *et al.*, *Experientia* **17**, 206 (1961). Comprehensive description of the tartrate: B. Kreilgard in *Analytical Profiles of Drug Substances* vol. **6**, K. Florey, Ed. (Academic Press, New York, 1977) pp 113-159.

Elongated prisms from benzene. Becomes solvent-free only after prolonged heating in a high vacuum. Very hygroscopic. Darkens and dec on exposure to air, heat and light. Dec 212-214°. $[\alpha]_D^{20}$ −160° (chloroform). Sol in about 70 parts methanol, 150 parts acetone, 300 parts alcohol; freely sol in chloroform, pyridine, glacial acetic acid; moderately sol in ethyl acetate; slightly in benzene. Almost insol in water, petr ether.

Hydrochloride, $C_{33}H_{36}ClN_5O_5$, rectangular plates from 90% alc, mp 212° (dec). Sol in water-alcohol mixtures; sparingly in water or alcohol.

Succinate, $C_{70}H_{76}N_{10}O_{14}$, *Ergoton-A.*

Tartrate, $C_{70}H_{76}N_{10}O_{16}$, *Ergate, Ergomar, Ergostat, Ergotartrate, Exmigra, Femergin, Gynergen, Lingraine, Lingran.* Solvated crystals, e.g. the dimethanolate, heavy rhombic plates from methanol, mp 203° (dec). $[\alpha]_D^{25}$ −125° to −155° (c = 0.4 in chloroform). One gram dissolves in 500 ml water or 500 ml alc. *Protect from light and heat.* LD_{50} in mice, rats: 62, 80 mg/kg i.v., *RTECS* **Vol. I**, R. J. Lewis, R. L. Tatken, Eds. (1979) p 618.

THERAP CAT: Antimigraine.

THERAP CAT (VET): Tartrate has been used as an oxytocic.

3610. Ergotaminine. *12'-Hydroxy-2'-methyl-5'α-(phenylmethyl)-8α-ergotaman-3',6',18-trione.* $C_{33}H_{35}N_5O_5$; mol wt 581.65. C 68.14%, H 6.07%, N 12.04%, O 13.75%. An alkaloid isomeric with ergotamine, q.v. Isoln from ergot: Stoll, *Helv. Chim. Acta* **28**, 1283 (1945); Stoll *et al.*, U.S. pat. **2,809,920** (1957 to Saul & Co.); Tabor, Vining, *Can. J. Microbiol.* **3**, 55 (1957).

Crystallizes solvent-free, unlike ergotamine which tends to retain the solvent of crystn. Thin rhombic plates from methanol, dec 241-243°. $[\alpha]_D^{20}$ +369° (c = 0.5 in chloroform). Much less sol than ergotamine; sol in about 1000 parts boiling alcohol, 1500 parts boiling methanol; fairly sol in chloroform, pyridine, glacial acetic acid. Does not seem to form salts.

3611. Ergothioneine. *(S)-α-Carboxy-2,3-dihydro-N,N,N-trimethyl-2-thioxo-1H-imidazole-4-ethanaminium hydroxide inner salt; [1-carboxy-2-[2-mercaptoimidazol-4(or 5)yl]ethyl]trimethylammonium hydroxide, inner salt;* L(+)ergothioneine; thioneine; thiolhistidine-betaine; thiasine; sympectothion. $C_9H_{15}N_3O_2S$; mol wt 229.29. C 47.14%, H 6.59%, N 18.33%, O 13.95%, S 13.98%. First discovered in the sclerotia of the ergot fungus, *Claviceps purpurea,* by Tanret, *J. Pharm. Chim.* **30**, 145 (1909). Has since been found to occur in blood, semen and various mammalian tissues, principally liver and kidneys. Biosynthesis from histidine by *Claviceps purpurea* cultures: Heath, Wildy, *Nature* **179**, 196 (1957); also produced by *Neurospora crassa*: Melville *et al.*, *Fed. Proc.* **15**, 314 (1956). Chemical synthesis: Heath *et al.*, *J. Chem. Soc.* **1951**, 2215. Occurrence in the *Linulus polyphemus* L. (king crab): Ackermann, List, *Z. Physiol. Chem.* **313**, 30 (1958). *Review:* Melville, *Vitamins & Hormones* **17**, 155 (1959).

Dihydrate, needles or leaflets from dil ethanol, dec 256-257°. $[\alpha]_D^{20}$ +116.5°; $[\alpha]_D^{27}$ +115° (H_2O). uv max (water): 258 nm (ε 16,000). One gram dissolves in 5 ml at 25°, much more sol in hot water. Slightly sol in hot methanol, hot ethanol, acetone. Practically insol in ether, chloroform, benzene.

3612. Ergotinine. A 1:1:1 mixture of ergocornine, ergocristinine, and ergocryptinine. Isomeric with the ergotoxine mixture, q.v. First reported by Tanret, *Compt. Rend.* **81**, 891 (1875). Resolution into three alkaloids: Stoll, Hofmann, *Helv. Chim. Acta* **26**, 1570 (1943).

Crystallizes solvent-free, unlike ergotoxine which tends to retain the solvent of crystn. Long prisms from acetone. mp 229° with decompn. $[\alpha]_D^{20}$ +365° (c = 0.35 in chloroform). Very sol in chloroform. Sol in 25 parts acetone, 420 parts alcohol, 1000 parts abs ether.

3613. Ergotoxine. Ecboline. A mixture of ergocornine, ergocristine, and ergocryptine, q.q.v., with oxytocic activity. *See also* the isomeric ergotinine mixture. Prepn: Barger, Carr, *J. Chem. Soc.* **91**, 337 (1907); Smith, Timmis, *ibid.* **1930**, 1390; Kofler, *Arch. Pharm.* **275**, 455 (1937); Stoll, Hofmann, *Helv. Chim. Acta* **26**, 1570 (1943); Pitra, Sapara, *Cesk. Farm.* **5**, 585 (1956), *C.A.* **51**, 8369f (1957). Toxicity data: H. Kreitmair, *Arch. Exp. Pathol. Pharmakol.* **176**, 171 (1934).

Orthorhombic crystals with $2C_6H_6$ from benzene. The

solvent of crystn is given up only after long drying and heating in high vacuum. mp ~190° (dec) (solvent-free base). $[\alpha]_D^{20}$ −197° (chloroform). Very sol in ethyl and methyl alcohol, acetone, chloroform, ethyl acetate; slightly sol in ether. Almost insol in water, petr ether. LD in mice (mg/kg): 0.107 s.c.; 0.032 i.v. (Kreitmair).

Ergotoxine phosphate, clusters of needles, dec 187°. Sol in 320 parts water, 15 parts boiling alcohol.

Ergotoxine ethanesufonate, acicular crystals, dec 209°. Sol in methanol; slightly in alcohol. Almost insol in water. Solid and solns are sensitive to light and air.

3614. Erigeron. Fleabane; horseweed. Leaves and tops of *Conyza canadensis* (L.) Cron. *(Erigeron canadensis* L.), Compositae. *Habit.* Northern and central U.S. *Constit.* Volatile oil, tannin, gallic acid.

3615. Eriochrome® Black T. *3-Hydroxy-4-[(1-hydroxy-2-naphthalenyl)azo]-7-nitro-1-naphthalenesulfonic acid monosodium salt; C.I. Mordant Black 11; C.I. 14645.* $C_{20}H_{12}N_3NaO_7S$; mol wt 461.38. C 52.06%, H 2.62%, N 9.11%, Na 4.98%, O 24.28%, S 6.95%. Prepn: Hagenbach, U.S. pat. **790,363** (1904 to Geigy); *Colour Index* vol. 4 (3rd ed., 1971) p 4067.

Brownish-black powder with a faint metalic sheen. Sol in hot water giving a reddish-brown soln when cold. Violet-brown precipitate with excess HCl. Deep blue, then red in aq soln of NaOH. Sol in concd H_2SO_4 giving a blackish-blue soln which yields a brown precipitate on dilution.

USE: To dye wool from an acid bath reddish-black, which can be converted to blue-black by afterchroming. As indicator in the determination of the total calcium and magnesium content of water.

3616. Eriodictyol. *2-(3,4-Dihydroxyphenyl)-2,3-dihydro-5,7-dihydroxy-4H-1-benzopyran-4-one; 3',4',5,7-tetrahydroxyflavanone.* $C_{15}H_{12}O_6$; mol wt 288.25. C 62.50%, H 4.20%, O 33.30%. Isoln from *Eriodictyon californicum* (H. & A.) Greene, Hydrophyllaceae: Geissman, *J. Am. Chem. Soc.* **62**, 3258 (1940); from lemon: Mager, *Z. Physiol. Chem.* **274**, 109 (1942); Horowitz, *J. Am. Chem. Soc.* **79**, 6561 (1957); U.S. pat. **2,857,318** (1958 to U.S. Dept. Agr.). Structure and synthesis: Reichel *et al., Ann.* **550**, 146 (1942); Zemplén *et al., Ber.* **76B**, 1112 (1943); Pew, *J. Org. Chem.* **27**, 2935 (1962). Synthesis: G. Wurm, U. Geres, *Arch. Pharm.* **315**, 183 (1982). *See also* Bioflavanoids.

Needles with 1½ H_2O from dil alc, dec 257° (rapid heating). After drying in vacuo at 100° for 6 hours, dec 267°. uv max (alc): 290, 326 nm (log ε 2.54, 2.16). Sparingly sol in boiling water, hot alcohol, ether, glacial acetic acid; sol in dil alkalies.

7-L-Rhamnoside, $C_{21}H_{22}O_{10}$, *eriodictin.* Isoln from citrin: Bruckner, Szent-Györgyi, *Nature* **138**, 1057 (1936). Structure: Mager, *loc. cit.* Crystals from ethyl acetate, dec 184-186°. $[\alpha]_D^{20}$ −51.5° (in pyridine). Sol in water, acetone, alcohol; practically insol in ether.

3617. Eriodictyon. Yerba santa; consumptive's weed; bear's weed; mountain balm; gum plant. Dried leaves of

Eriodictyon californicum (H. & A.) Greene, *Hydrophyllaceae*. *Habit.* U.S. (California). *Constit.* Volatile oil, eriodictyol, homoeriodictyol, chrysoeriodictyol, xanthoeriodictyol, eriodonol, eriodictyonic acid, ericolin, resin.

USE: Pharmaceutic aid (flavor).

3618. Eritadenine. *(R)-6-Amino-α,β-dihydroxy-9H-purine-9-butanoic acid;* 2(R),3(R)-dihydroxy-4-(9-adenyl)-butyric acid, (D-*erythro*-form); lentinacin; lentysine. $C_9H_{11}N_5O_4$; mol wt 253.22. C 42.68%, H 4.38%, N 27.66%, O 25.27%. Isoln from the mushroom *Lentinus edodes* Sing. (Shiitake) and synthesis: Chibata *et al., Experientia* **25**, 1237 (1969). Syntheses: Kamiya *et al., Tetrahedron Letters* **1969**, 4729; *eidem, Chem. Ind.* **1970**, 652; Okumura *et al., Chem. Commun.* **1970**, 1045; Kawazu *et al., ibid.* **1970**, 1047. Synthesis of D-*threo*- and L-*threo*-forms: Hashimoto *et al., Tetrahedron Letters* **1970**, 1359.

Needles from hot water, mp 279° (dec), $[\alpha]_D$ +50° (0.1N NaOH) and +16° (N HCl). uv max (water): 261 nm (ε 14,300).

Sodium salt, $C_9H_{10}N_5NaO_4$, leaflets from 70% ethanol, mp 266-268° (dec), $[\alpha]_D^{20}$ +45.5° (H_2O). IR, NMR: Chibata, *loc. cit.*

THERAP CAT: Anticholesteremic.

3619. Erucic Acid. *13-Docosenoic acid;* $\Delta^{13:14}$-docosenoic acid. $C_{22}H_{42}O_2$; mol wt 338.56. C 78.04%, H 12.51%, O 9.45%. A monoethenoid acid found in the seed fats of Cruciferae and Tropaeolaceae. It constitutes 40 to 50% of the total fatty acids of rapeseed, mustard- and wallflower seed, and it represents up to 80% of fatty acids of nasturtium seeds, *cf.* K. S. Markley, *Fatty Acids* Part I (Interscience, New York, 2nd ed., 1960) p 138-139. Prepn of a crude product by alkaline hydrolysis of rapeseed oil: Noller, Talbot, *Org. Syn.* **coll. vol. II**, 258 (1943). A purer product is obtained by the method of Dorée and Pepper, *J. Chem. Soc.* **1942**, 477, which involves a fractional precipitation and crystallization. Prepn of a pure product by acid soap crystallization: Chobanov *et al., Chem. & Ind. (London)* **1965**, 606. Synthesis: Bowman, *J. Chem. Soc.* **1950**, 177, 325; Bounds *et al., ibid.* **1953**, 2393. Treatment with nitric acid yields the *trans* isomer, brassidic acid, *q.v.*: Dorée, Pepper, *loc. cit.* Brief review: E. Lower, *Manuf. Chem.* **56**(6), 61-63 (1985).

Needles from alcohol, mp 33.8°. Iodine value 74.98; neutralization value 165.72. d_4^{55} 0.860. bp_{760} 381.5° (decompn); bp_{400} 358.8°; bp_{100} 314.4°; bp_{60} 300.2°; bp_{20} 270.6°; bp_5 239.7°; $bp_{1.0}$ 206.7°. n_D^{45} 1.4534; n_D^{65} 1.44794. Insol in water. About 175 g dissolve in 100 ml ethanol and about 160 g dissolve in 100 ml methanol. Very sol in ether.

3620. Erythritol. *1,2,3,4-Butanetetrol;* meso-erythritol; tetrahydroxybutane; erythrol; erythrite; antierythrite; erythroglucin; phycite. $C_4H_{10}O_4$; mol wt 122.12. C 39.34%, H 8.25%, O 52.41%. Isoln from algae, lichens, grasses: Bamberger, Landsiedl, *Monatsh.* **21**, 571 (1900); Hesse, *J. Prakt. Chem.* **92**, 425 (1915); Hofmann, *Ber.* **7**, 508 (1874). Prepn by *Aspergillus niger*: Yuill, *Nature* **162**, 652 (1948); by *Penicillium herquei*: Galarraga *et al., Biochem. J.* **61**, 456 (1955); from 2-butene-1,4-diol: Reppe, Schnabel, *Ger.* pat. **734,025** (1943 to I. G. Farbenind.); from periodate-oxidized starch: Jeanes, Hudson, *J. Org. Chem.* **20**, 1565 (1955). Structure: Shimada, *Acta Cryst.* **11**, 748 (1958).

$$H_2COH$$
$$|$$
$$HCOH$$
$$|$$
$$HCOH$$
$$|$$
$$H_2COH$$

Tetragonal prisms, mp 121.5°. About twice as sweet as sucrose. bp 329-331°. Very sol in water (satd soln contains about 61% w/w); sol in pyridine (satd soln contains 2.5% w/w); slightly sol in alcohol. Practically insol in ether. Ka at 18°: 1.25×10^{-14}. LD i.v. in dogs: 5.0 g/kg.

THERAP CAT: Coronary vasodilator.

3621. Erythritol Anhydride. *(R*,S*)-2,2'-Bioxirane; 1,2:3,4-diepoxybutane;* butadiene diepoxide; butadiene dioxide; butanedione; Bioxiran. $C_4H_6O_2$; mol wt 86.09. C 55.80%, H 7.03%, O 37.17%. Prepn from erythrityl chlorohydrin: Przybytek, *Ber.* **17**, 1091 (1884); by refluxing 1,4-dichloro-2,3-butanediol in tetrahydrofuran with NaOH: Reppe, *Ann.* **596**, 141 (1955); *meso*-form from 1,4-dihydroxy-2-butene or from 3,4-epoxy-1-butene and *dl*-form from 1,4-dibromo-2-butene: Beech, *J. Chem. Soc.* **1951**, 2483. Use to prevent microbial spoilage: H. D. Michener, J. C. Lewis, U.S. pat. **2,934,439** (1960 to U.S. Secy of Agr.). Toxicity data: H. F. Smyth *et al., Arch. Ind. Hyg. Occup. Med.* **10**, 61 (1954).

$$O\overset{CH_2}{\underset{CH}{\diagdown}}$$
$$CH$$
$$O\overset{CH}{\underset{CH_2}{\diagdown}}$$

Liquid. d_4^{18} 1.113. bp 138°; bp$_{23}$ 50-51°. *meso*-form, bp$_{761}$ 140-142°. mp −19°. Compd with piperidine, mp 106° (Beech). Misc with water which hydrolyzes it to erythritol. LD$_{50}$ orally in rats: 0.078 g/kg (Smyth).
Note: This substance may reasonably be anticipated to be a carcinogen: *Fourth Annual Report on Carcinogens* (NTP 85-002, 1985) p 82.

USE: Curing polymers; crosslinking textile fibers.

3622. Erythrityl Tetranitrate. *(R*,S*)-1,2,3,4-Butanetetroltetranitrate;* erythritol tetranitrate; erythrol tetranitrate; tetranitrol; tetranitrin; nitroerythrite; Cardilate; Cardiloid. $C_4H_6N_4O_{12}$; mol wt 302.12. C 15.90%, H 2.00%, N 18.55%, O 63.55%. Made by nitration of erythritol.

Leaflets from alcohol, mp 61°. Sol in alcohol, ether, glycerol. Insol in water. Reduces Fehling's soln. Explodes on percussion. Its use in pharmaceutical preparations is only in admixture with carbohydrate substances such as lactose. Sold in tablet form only. Such tablets are nonexplosive.

THERAP CAT: Coronary vasodilator.
THERAP CAT (VET): Vasodilator.

3623. Erythrocentaurin. *3,4-Dihydro-1-oxo-1H-2-benzopyran-5-carboxaldehyde; 5-formyl-3,4-dihydroisocoumarin;* 5-formyl-3,4-dihydro-1H-2-benzopyran-1-one. $C_{10}H_8O_3$; mol wt 176.16. C 68.18%, H 4.58%, O 27.25%. From *Centaurium umbellatum* Gilib., (*Erythraea centaurium* Pers.), *Gentianaceae* or *Swertia japonica* (Maxim.) Makino, *Gentianaceae*. By hydrolysis of swertiamarin and erytaurin with emulsin. Isoln: Kariyone, Matsushima, *J. Pharm. Soc. Japan* **47**, 25 (1927). Structure: Kubota, Tomita, *Chem. & Ind. (London)* **1958**, 230; Kubota *et al., Tetrahedron Letters* **1961**, 223. Synthesis: Wenkert *et al., J. Org. Chem.* **29**, 2534 (1964).

Long needles, mp 140-141°. Turns red on exposure to sunlight. uv max: 223, 290 nm (log ε 4.30, 3.13).

THERAP CAT: Bitter tonic.

3624. α-Erythroidine. *(3β,12β)-1,2,6,7-Tetradehydro-12,17-dihydro-3-methoxy-16(15H)-oxaerythrinan-15-one.* $C_{16}H_{19}NO_3$; mol wt 273.32. C 70.31%, H 7.00%, N 5.13%, O 17.56%. Isoln from *Erythrina* spp, *Leguminosae:* Folkers, Major, U.S. pat. **2,373,952** (1945 to Merck & Co.); Boekelheide, Grundon, *J. Am. Chem. Soc.* **75**, 2563 (1953). Structure: Godfrey *et al., ibid.* **77**, 3342 (1955). Absolute configuration: Hill, Shearer, *J. Org. Chem.* **27**, 921 (1962); Wenzinger, Boekelheide, *Proc. Chem. Soc.* **1963**, 53. Conversion of α- to β-erythroidine: Boekelheide, Morrison, *J. Am. Chem. Soc.* **80**, 3905 (1958). Biosynthesis studies: Leete, Ahmad, *ibid.* **88**, 4722 (1966).

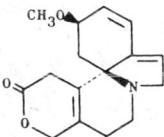

Needles from pentane, mp 58-60°. $[α]_D^{27}$ +136° (c = 0.5 in water). Unstable on exposure to air.
Hydrochloride, $C_{16}H_{20}ClNO_3$, prisms from ethanol, dec 226-228°. $[α]_D^{32}$ + 118° (c = 0.5 in water). uv max (ethanol): 224 nm (log ε 4.55).
Methiodide, $C_{17}H_{22}INO_3$, yellow prisms from ethanol, mp 219-220°.

3625. β-Erythroidine. *(3β)-1,2,6,7-Tetradehydro-14,17-dihydro-3-methoxy-16(15H)-oxaerythrinan-15-one; 12,13-didehydro-13,14-dihydro-α-erythroidine.* $C_{16}H_{19}NO_3$; mol wt 273.32. C 70.31%, H 7.00%, N 5.13%, O 17.56%. Skeletal muscle relaxant. Isolated from the seeds and other plant parts of *Erythrina* spp., *Leguminosae:* Folkers, Major, *J. Am. Chem. Soc.* **59**, 1580 (1937); *eidem,* U.S. pats. **2,373,952; 2,385,266; 2,407,713** (1945 and 1946 to Merck & Co.). Structural studies: Koniuszy, Folkers, *J. Am. Chem. Soc.* **72**, 5519 (1950); Boekelheide *et al., ibid.* **75**, 2550 (1953). Absolute configuration: Wenzinger, Boekelheide, *Proc. Chem. Soc.* **1963**, 53; *eidem, J. Org. Chem.* **29**, 1307 (1964). Toxicity data: F. M. Berger, R. P. Schwartz, *J. Pharmacol. Exp. Ther.* **93**, 362 (1948). Review: Boekelheide, *Record Chem. Progr.* **16**, 227-239 (1955).

Crystals from abs ethanol, mp 99.5-100°. $[α]_D^{25}$ +88.8°. Sol in water, benzene, chloroform, methanol, ethanol; moderately sol in diethyl ether. Reacts with NaOH to form sodium erythroidinate. LD$_{50}$ i.p. in mice: 24.0 mg/kg (Berger, Schwartz).
Hydrochloride, $C_{16}H_{20}ClNO_3$, small needles from abs ethanol, mp 232° (dec). $[α]_D^{25}$ +10°. Unstable in air and light. Bitter taste. Sol in water, benzene, chloroform, methanol, ethanol; moderately sol in diethyl ether. *Incompat.* with oxidizing agents, alkaline solns, and akaloidal reagents.
Hydrochloride hemihydrate, $C_{16}H_{20}ClNO_3$·½H$_2$O, small needles from abs ethanol, mp 229.5-230° (dec). $[α]_D$ +95°. Has curare-like action which is antagonized by prostigmine.
Hydrobromide, $C_{16}H_{20}BrNO_3$, crystals from abs ethanol, mp 222.5°. $[α]_D^{25}$ +111.2°.
Hydriodide, $C_{16}H_{20}INO_3$, crystals from abs ethanol, mp 206°. $[α]_d^{25}$ +108.1°.
Methiodide, $C_{17}H_{22}INO_3$, white prisms from alcohol, mp 211°.

3626. Erythromycin. Erythromycin A; Abomacetin; Ak-Mycin; Aknin; E-Base; EMU; E-Mycin; Eritrocina; Ery Derm; Erymax; Ery-Tab; Erythromast 36; Erythromid; ERYC; Erycen; Erycin; Erycinum; Ermysin; Ilotycin; Retcin; Staticin; Stiemycin; Torlamicina. $C_{37}H_{67}NO_{13}$; mol wt 733.92. C 60.55%, H 9.20%, N 1.91%, O 28.34%. Antibiotic substance produced by a strain of *Streptomyces erythreus* (Waksman) Waksman & Henrici, found in a soil sample from the Philippine Archipelago. Isoln: McGuire *et al.*, *Antibiot. & Chemother.* **2**, 281 (1952); Bunch, McGuire, U.S. pat. **2,653,899** (1953 to Lilly); Clark, Jr., U.S. pat. **2,823,203** (1958 to Abbott). Properties: Flynn *et al.*, *J. Am. Chem. Soc.* **76**, 3121 (1954). Solubility data: Weiss *et al.*, *Antibiot. & Chemother.* **7**, 374 (1957). Structure: Wiley *et al.*, *J. Am. Chem. Soc.* **79**, 6062 (1957). Configuration: Hofheinz, Grisebach, *Ber.* **96**, 2867 (1963); Harris *et al.*, *Tetrahedron Letters* **1965**, 679. There are three erythromycins produced during fermentation, designated A, B, and C; A is the major and most important component. Erythromycins A and B contain the same sugar moieties, desosamine, *q.v.*, and cladinose (3-O-methylmycarose). They differ in position 12 of the aglycone, erythronolide, A having an hydroxyl substituent. Component C contains desosamine and the same aglycone present in A but differs by the presence of mycarose, *q.v.*, instead of cladinose. Structure of B: P. F. Wiley *et al.*, *J. Am. Chem. Soc.* **79**, 6070 (1957); of C: *eidem, ibid.* 6074. Synthesis of the aglycone, erythronolide B: E. J. Corey *et al., ibid.* **100**, 4618, 4620 (1978); of erythronolide A: *eidem, ibid.* **101**, 7131 (1979). Asymmetric total synthesis of erythromycin A: R. B. Woodward *et al., ibid.* **103**, 3215 (1981). NMR spectrum of A: D. J. Ager, C. K. Sood, *Magn. Reson. Chem.* **25**, 948 (1987). Biosynthesis: Martin, Goldstein, *Progr. Antimicrob. Anticancer Chemother., Proc. 6th Int. Congr. Chemother.* **II**, 1112 (1970); Martin *et al., Tetrahedron*, **31**, 1985 (1975). Cloning and expression of clustered biosynthetic genes: R. Stanzak *et al., Biotechnology* **4**, 229 (1986). *Reviews:* T. J. Perun in *Drug Action and Drug Resistance in Bacteria* **1**, S. Mitsuhashi, Ed. (University Park Press, Baltimore, 1977) pp 123-152; Oleinick in *Antibiotics*, **vol. 3**, J. W. Corcoran, F. E. Hahn, Eds. (Springer-Verlag, New York, 1975) pp 396-419; *Infection* **10**, Suppl. 2, S61-S118 (1982). Comprehensive description: W. L. Koch in *Analytical Profiles of Drug Substances* **vol. 8**, K. Florey, Ed. (Academic Press, New York, 1979) pp 159-177.

erythromycin A

Hydrated crystals from water, mp 135-140°, resolidifies with second mp 190-193°. Melting point taken after drying at 56° and 8 mm. $[\alpha]_D^{25}$ −78° (c = 1.99 in ethanol). uv max (pH 6.3): 280 nm (ε 50). pKa$_1$ 8.8. Basic reaction. Readily forms salts with acids. Soly in water: ~2 mg/ml. Freely sol in alcohols, acetone, chloroform, acetonitrile, ethyl acetate. Moderately sol in ether, ethylene dichloride, amyl acetate.

Ethyl succinate, $C_{43}H_{75}NO_{16}$, *Anamycin, Arpimycin, Durapaediat, E.E.S., E-Mycin E, Eryliquid, Eryped, Erythro ES, Erythro-Holz, Erythroped, Esinol, Monomycin, Paediathrocin, Pediamycin, Refkas, Sigapedil, Wyamycin E.*

THERAP CAT: Antibacterial.

THERAP CAT (VET): Antibacterial.

3627. Erythromycin Acistrate. *Erythromycin 2'-acetate octadecanoate (salt); octadecanoic acid compd with erythromycin 2'-acetate;* 2'-acetylerythromycin stearate; Erasis. $C_{57}H_{105}NO_{16}$; mol wt 1060.46. C 64.56%, H 9.98%, N 1.32%, O 24.14%. Ester prodrug of erythromycin, *q.v.* Prepn: M. L. A. Marvola *et al.*, Belg. pat. **901,411**; E. J. Honkanen *et al.*, U.S. pat. **4,599,326** (1985, 1986 both to Orion-yhtymä Oy). Pharmacology, toxicity and clinical studies: *J. Antimicrob. Chemother.* **21**, Suppl D, 1-122 (1988).

R = COCH$_3$

THERAP CAT: Antibacterial.

3628. Erythromycin Estolate. *Erythromycin 2'-propanoate dodecyl sulfate (salt);* erythromycin propionate lauryl sulfate; lauryl sulfate salt of the propionic acid ester of erythromycin; propionylerythromycin lauryl sulfate; Eritroger; Eromycin; Ilosone; Lauromicina; Neo-Erycinum; PELS; Roxomicina; Stellamicina; Eriscel; Estomicina; Erytrarco; Marcoeritrex; Propiocine Enfant; Togiren. $C_{52}H_{97}NO_{18}S$; mol wt 1056.43. C 59.12%, H 9.26%, N 1.33%, O 27.26%, S 3.04%. $C_{40}H_{71}NO_{14}·C_{12}H_{26}SO_4$. Prepn from propionyl erythromycin: Stephens *et al., J. Am. Pharm. Assoc.* **48**, 620 (1959); Bray, Stephens, U.S. pat. **3,000,874** (1961 to Lilly). Toxicity: E. I. Goldenthal, *Toxicol. Appl. Pharmacol.* **18**, 185 (1971). Comprehensive description: J. M. Mann in *Analytical Profiles of Drug Substances* vol. 1, K. Florey, Ed. (Academic Press, New York, 1972) pp 101-117.

R = OCCH$_2$CH$_3$

Long needles, mp 135-140° (dec). pKa 6.9. Soly in water: 0.024 mg/ml; sol in alcohol, acetone, chloroform. LD$_{50}$ orally in rats: > 5000 mg/kg (Goldenthal).

THERAP CAT: Antibacterial.

3629. Erythromycin Glucoheptonate. D-*Glycero*-D-*gulo-heptonic acid compd with erythromycin (1:1);* erythromycin glucoheptonic acid salt; erythromycin gluceptate; Ilotycin Gluceptate. $C_{44}H_{81}NO_{21}$; mol wt 960.10. $C_{37}H_{67}NO_{13}$·$C_7H_{14}O_8$. Semi-synthetic macrolide antibiotic. Prepn: Shepler, U.S. pat. **2,852,429** (1958 to Lilly).

Crystals, mp 95-100°. Freely sol in water, alcohol, dioxane, acetone, propylene glycol. Practically insol in ether, carbon tetrachloride, toluene, benzene. 2% aq solns are levorotatory and have a pH of 6.0 to 7.5.

THERAP CAT: Antibacterial.

3630. Erythromycin Lactobionate. *4-O-β-*D-*Galactopyranosyl-*D-*gluconic acid compd with erythromycin (1:1);* Erythrocin Lactobionate. Semi-synthetic macrolide antibiotic prepd from erythromycin base and lactobiono-δ-lactone in water-acetone: Hoffhine, U.S. pat. **2,761,859** (1956 to Abbott); alternate prepn and antibacterial activity: S. K. Dutta, K. S. Basu, U.S. pat. **4,137,397** (1979 to Jadavpur Univ.). Pharmacokinetics: R. L. Parsons *et al., J. Int. Med. Res.* **8**, suppl. 2, 15 (1980).

White, amorphous powder, mp 145-150°. Soly in water about 200 mg/ml. Freely sol in alcohol; slightly sol in ether. pH of a 2% aq soln 6.0-7.5. A 5% stock soln can be prepd with water or glucose soln but not with NaCl isotonic soln, because NaCl at this concn precipitates erythromycin base. Lower concns of inorganic salts are compatible.

THERAP CAT: Antibacterial.
THERAP CAT (VET): Antibacterial.

3631. Erythromycin Propionate. *Erythromycin 2'-pro-panoate;* propionylerythromycin; monopropionylerythromy-cin; Propiocine. $C_{40}H_{71}NO_{14}$; mol wt 789.98. C 60.82%, H 9.06%, N 1.77%, O 28.36%. Semi-synthetic macrolide antibiotic. Prepn: V. C. Stephens, U.S. pat. **2,993,833** (1961 to Lilly); V. C. Stephens, J. W. Conine, *Antibiot. Ann.* **1958-59**, 346. Pharmacology and toxicology: C. Lee *et al., ibid.* 354. Clinical studies: R. S. Griffith, *ibid.* 364; D. M. Perry *et al., ibid.* 375. Pharmacokinetics in humans: M. Ducci *et al., Int. J. Clin. Pharmacol. Ther. Toxicol.* **19**, 494 (1981).

Monohydrate, crystals from acetone + water, mp 122-126°. $[\alpha]_D^{25}$ −81.6° (acetone). pKa = 6.9. Very slightly sol in water. Readily sol in methanol, ethanol, acetone, ethyl acetate, DMF. LD_{50} in mice, rats (g/kg): 2.87, > 5.0 orally; > 5.0, > 5.0 s.c. (Lee).

THERAP CAT: Antibacterial.

3632. Erythromycin Stearate. *Erythromycin octadeca-noate (salt);* Abboticine; Bristamycin; Dowmycin E; E-Mycin; Eratrex; Erypar; Eryprim; Erythro S; Erythrocin; Ethril; Ethryn; Gallimycin; Meberyt; Pantomicina; Pfizer-E; SK-Erythromycin; Wemid. $C_{55}H_{42}NO_{15}$; mol wt 836.83. C 64.59%, H 5.06%, N 1.67%, O 28.68%. Semi-synthetic macrolide antibiotic. Prepn: O. F. Walasek, U.S. pat. **2,881,163** (1959 to Abbott). Bioavailability: A.-S. Malmborg, *J. Antimicrob. Chemother.* **5**, 591 (1979). Pharmacokinetics and tolerance: D. C. Shanson *et al., ibid.* **14**, 157 (1984). Clinical efficacy in respiratory tract infections: J. P. Butler *et al., Chemotherapy (Basel)* **25**, 367 (1979); in acute otitis media: C. Rosen *et al., Acta Otolaryngol. (Stockholm)* Suppl. **407**, 23 (1984).

Note: The prepn of erythromycin stearate (ester) is described in Booth, Murray, U.S. pat. **2,862,921** (1958 to Upjohn).

Crystals. Slightly bitter taste. Slightly sol in alc, ether, chloroform. Practically insol in water.

THERAP CAT: Antibacterial.
THERAP CAT (VET): Antibacterial.

3633. Erythrophlamine. *[1R-(1α,2α,4aα,4bβ,7E,8β,-8aα,10aβ)]-7-[2-[2-(Dimethylamino)ethoxy]-2-oxoethylid-ene]tetradecahydro-2-hydroxy-1,4a,8-trimethyl-9-oxo-1-phenanthrenecarboxylic acid methyl ester.* $C_{25}H_{39}NO_6$; mol wt 449.57. C 66.79%, H 8.75%, N 3.12%, O 21.35%. Extracted from the bark of *Erythrophleum guineense* G. Don., and *E. couminga* Baillon, Leguminosae. Isoln: Ruzicka *et al., Experientia* **1**, 160 (1945); Engel, Tondeur, *Helv. Chim. Acta* **32**, 2364 (1949). Structure: Mathieson *et al., Experientia* **16**, 404 (1960); Arya, Engel, *Helv. Chim. Acta* **44**, 1650 (1961).

Crystals from ether + petr ether, mp 149-151°. $[\alpha]_D^{20}$ −62.5° (c = 0.911 in alc). uv max: 222 nm (log ε 4.2).

3634. Erythrophleine. *[1S-(1α,4aα,4bβ,7E,8β,8aα,9α,-10aB)]-Tetradecahydro-9-hydroxy-1,4a,8-trimethyl-7-[2-[2-(methylamino)ethoxy]-2-oxoethylidene]-1-phenanthrenecar-boxylic acid methyl ester;* norcassamidine. $C_{24}H_{39}NO_5$; mol wt 421.56. C 68.37%, H 9.33%, N 3.32%, O 18.98%. From the bark of *Erythrophleum guineense* G. Don., Leguminosae. Isoln: Harnack, *Arch. Pharm.* **234**, 561 (1896). Structural studies: Blount *et al., J. Chem. Soc.* **1940**, 286. Revised structure and identity with norcassamidine: Friedrich-Fiechtl, Spiteller, *Ber.* **104**, 3538 (1971). Activity: Flacke, *J. Pharmacol. Exp. Ther.* **125**, 49 (1959).

Solid from ethanol-ether, mp 115°. $[\alpha]_D^{20}$ −22.5° (c = 0.65 in ethanol). Sol in water and alcohol.

THERAP CAT: Cardiotonic.

3635. Erythropoietin. Erythropoiesis stimulating factor; hemopoietine; ESF; Ep; Epo; Epogen; Eprex; Erypo; Maro-gen. A circulating glycoprotein which stimulates red blood cell formation in higher organisms: P. Carnot, C. Deflandre, *Compt. Rend.* **143**, 384 (1906). Measurable amounts found in urine and plasma; thousand-fold concns found in animals in severely anemic or hypoxic states. Produced mainly in the kidneys. Contains hexoses, hexosamines, and sialic acid in varying amounts according to species. Isoln from anemic sheep plasma: White *et al., Recent Progr. Horm. Res.* **16**, 219 (1960); from urine of anemic human subjects: Espada, Gut-nisky, *Biochem. Med.* **3**, 475 (1970). Purification (human): T. Miyake *et al., J. Biol. Chem.* **252**, 5558 (1977). Extrarenal production: I. N. Rich *et al., Blood* **60**, 1007 (1982). Biogenesis: C. Peschle, M. Condorelli, *Science* **190**, 911 (1975); A. J. Erslev *et al., Exp. Hematol.* **8**, Suppl. 8, 1 (1980); F. Fyhrquist *et al., Nature* **308**, 649 (1984). Review of control of production: J. W. Fisher, *Proc. Soc. Exp. Biol. Med.* **173**, 289-305 (1983). Characterization of monoclonal antibody to human Ep: T. Weiss *et al., Proc. Natl. Acad. Sci. USA* **79**, 5465 (1982). Cloning and expression of human Ep cDNA: S. Lee-Huang, *ibid.* **81**, 2708 (1984). Reviews: Fisher, *Pharmacol. Rev.* **24**, 459 (1972); Gordon, *Vitam. Horm. (New York)* **31**, 105 (1973).

THERAP CAT: Hematinic.

3636. Erythropterin. *3-(2-Amino-4,5,6,8-tetrahydro-4,6-dioxo-7(3H)-pteridinylidene)-2-oxopropanoic acid; 2-amino-3,4,5,6-tetrahydro-4,6-dioxo-7-pteridinepyruvic acid.* $C_9H_7N_5O_5$; mol wt 265.19. C 40.76%, H 2.66%, N 26.41%, O 30.17%. Pigment responsible for the red and orange color spots on the wings of butterflies. Has been detected also in the integument of other insects and in *Mycobacterium lacticola.* Isoln: Schöpf, Becker, *Ann.* **524**, 49 (1936). Structure and synthesis: Tschesche, Ende, *Ber.* **91**, 2074 (1958). Revised structure: Pfleiderer, *ibid.* **95**, 2195 (1962); von Philipsborn *et al., Helv. Chim. Acta* **46**, 2592 (1963). Synthesis: Viscontini, Stierlin, *ibid.* **51**.

Monohydrate, red microcrystals from 0.5N HCl. R values: *n*-propanol/4% sodium citrate (1:2) 0.54, *n*-butanol/ethanol/water (2:2:1) 0.00, 4% ammonium acetate 0.17, 4% disodium phosphate 0.20, 4% sodium bicarbonate 0.27, 4% disodium phosphate satd with *n*-butanol 0.26. Abs max (pH 1.0): 450 nm (log ε 4.02).

3637. D-Erythrose. *(R)-2,3,4-Trihydroxybutanal.* C_4H_8-O_4; mol wt 120.11. C 40.00%, H 6.71%, O 53.29%. From calcium D-arabonate by oxidation with H_2O_2: Ruff, *Ber.* **32**, 3674 (1899); **33**, 1799 (1900). From D-glucose: Perlin, *Methods in Carbohydrate Chemistry* **1** (Academic Press, New York, 1962) p 64. Synthesis of DL-erythrose: Sonogashira, Nakagawa, *Bull. Chem. Soc. Japan* **45**, 2616 (1972).

CHO
|
HCOH
|
HCOH
|
CH₂OH

Syrup. Shows mutarotation. $[\alpha]_D^{20}$ +1° → −14.5° (3 days, c = 11). Sol in water. Slowly reduces cold Fehling's soln. Sodium amalgam reduces it to natural, inactive erythritol. No aldehyde reaction with benzenesulfhydroxamic acid. Not fermented by yeast.

Phenylosazone, $C_{16}H_{18}N_4O_2$, mp 164°.

3638. L-Erythrose. *[S-(R*,R*)]-2,3,4-Trihydroxybutanal.* $C_4H_8O_4$; mol wt 120.11. C 40.00%, H 6.71%, O 53.29%. From calcium L-arabonate by oxidation with H_2O_2: Ruff, Meusser, *Ber.* **34**, 1365 (1901). From L-arabinose oxime: Wohl, *Ber.* **32**, 3667 (1899). From L-arabonamide by treatment with alkaline NaOCl: Weerman, *Rec. Trav. Chim.* **37**, 35 (1918).

CHO
|
HOCH
|
HOCH
|
CH₂OH

Syrup. Sweet taste. Shows mutarotation. $[\alpha]_D^{24}$ +11.5° (8 min) → +15.2° (120 min) → +30.5° (final, c = 3): Felton, Freudenberg, *J. Am. Chem. Soc.* **57**, 1640 (1935). Soluble in water. Heating with HCl yields lactic acid. Oxidation with Br converts it to L-erythronic acid. Reduces Fehling's soln slowly in the cold. Not fermented by yeast.

Phenylosazone, $C_{16}H_{18}N_4O_2$, mp 164°.

3639. D-Erythrose 4-Phosphate. *[R-(R*,R*)]-2,3-Dihydroxy-4-(phosphonooxy)butanal; 4-D-erythrosephosphoric acid.* $C_4H_9O_7P$; mol wt 200.08. C 24.01%, H 4.53%, O 55.98%, P 15.48%. Occurs in minute amounts in muscle flesh of all animals. Important natural intermediate in the Embden-Meyerhof scheme of alcoholic fermentation and glycolysis. Prepn by chemical synthesis: Ballou *et al., J. Am. Chem. Soc.* **77**, 2658, 5967 (1955). Outline: *Chem. & Eng. News* **34**, 2506 (1956).

CHO
|
HCOH
|
HCOH
|
CH₂OPO₃H₂

Obtained in aq soln only. Shows no observable optical rotation. Is condensed with phosphoenolpyruvate by a cell-free extract from an *E. coli* mutant to give a 90% yield.

Note: D-Erythrose 4-phosphate is best stored and transported as the cyclohexylammonium salt of 4-phosphoryl-D-erythrose dimethyl acetal from which it can be prepared readily when needed.

3640. Erythrosine. *3′,6′-Dihydroxy-2′,4′,5′,7′-tetraiodospiro[isobenzofuran-1(3H),9′-[9H]xanthen]-3-one disodium salt; 2′,4′,5′,7′-tetraiodofluorescein disodium salt; erythrosine BS; erythrosine B; FD & C Red No. 3; C.I. Food Red 14; C.I. Acid Red 51; C.I. 45430.* $C_{20}H_6I_4Na_2O_5$; mol wt 879.92. C 27.30%, H 0.69%, I 57.70%, Na 5.23%, O 9.09%o. Prepn: Gilliard *et al., Ger. pat.* **108,838** (1899); *Frdl.* **5**, 215; Gomberg, Tabern, *J. Ind. Eng. Chem.* **14**, 1115 (1922); Dolinsky, Jones, *J. Assoc. Offic. Agr. Chem.* **34**, 114 (1951); *Colour Index* vol. **4** (3rd ed., 1971) p 4428. Structure: Holmes, Scanlan, *J. Am. Chem. Soc.* **49**, 1594 (1927).

Brown powder. Absorption max (water): 524 nm; in 95% alcohol: 531 nm. Soluble in water to cherry-red soln; sol in alcohol. HCl added to an aq soln produces a yellowish-brown ppt. Sodium hydroxide produces a red ppt sol in an excess of the reagent. LD_{50} orally in mice, rats: 2558, 2891 mg/kg, S. L. Yankell *et al., J. Periodontol.* **48**, 228 (1977).

USE: Biological stain; color additive. Certified by FDA for use in foods and drugs.

3641. L-Erythrulose. *(S)-1,3,4-Trihydroxy-2-butanone; L-glycero-tetrulose.* $C_4H_8O_4$; mol wt 120.11. C 40.00%, H 6.71%, O 53.29%. Prepn by bacterial oxidation of *meso*-erythritol: Müller *et al., Helv. Chim. Acta* **20**, 1468 (1937); Whistler, Underköfler, *J. Am. Chem. Soc.* **60**, 2507 (1937). Synthesis from D-fructose: Gorin *et al., J. Chem. Soc.* **1955**, 2699.

CH₂OH
|
CO
|
HOCH
|
CH₂OH

Syrup. $[\alpha]_D^{18}$ +11.4° (c = 2.4 in water). Very sensitive to alkali. Soluble in water, abs alc.

Phenylosazone, $C_{16}H_{18}N_4O_2$, crystals from aq ethanol, mp 162°. $[\alpha]_D$ +32° (10 min) → 0° (24 hr, constant) (c = 0.75 in pyridine + ethanol, 3:2 v/v).

3642. Esaprazole. *N-Cyclohexyl-1-piperazineacetamide; 1-piperazinyl-4-methylenecarbonylcyclohexylamine; N-(1-piperazinylacetyl)cyclohexylamine; N-[(N-cyclohexylcarbamoyl)methyl]piperazine; hexaprazol(e); exaprazole; C/63; Prazol.* $C_{12}H_{23}N_3O$; mol wt 225.33. C 63.96%, H 10.29%, N 18.65%, O 7.10%. Prepn: C. Corvi-Mora, **Ger. pat. 2,702,-537**; *idem, U.S. pat.* **4,123,530** (1977, 1978 both to Camillo Corvi). Metabolism in animals and man: P. Ventura *et al., Farmaco Ed. Prat.* **39**, 190 (1984). Cytoprotective activity: B. Lumachi, R. Scuri, *Drugs Exp. Clin. Res.* **11**, 693 (1985); and pharmacology: R. Scuri *et al., Boll. Chim. Farm.* **123**, 425 (1983). Pharmacology: G. Coruzzi *et al., Farmaco Ed. Prat.* **40**, 313 (1985); G. Coruzzi *et al., ibid.* 324. Antisecretory activity in humans: L. Erembourg *et al., Int. J. Clin. Pharm. Res.* **7**, 95 (1987). HPLC determn in biological fluids: M. A. Girometta *et al., J. Chromatog.* **383**, 85 (1986). Clinical trial as antiulcerative in comparison with cimetidine, *q.v.:* L. Capurso *et al., Clin. Trials J.* **23**, 293 (1986).

Residue, mp 111-112°. $bp_{0.5}$ 190°. LD_{50} in mice (mg/kg): 1974 orally; 271 i.v.; orally in rats: 3900 mg/kg (Corvi-Mora).

Hydrochloride, $C_{12}H_{24}ClN_3O$, *CO/1063.*

THERAP CAT: Antiulcerative.

3643. Escigenin. *16α,21α-Epoxyolean-12-ene-3β,22α,-23,28-tetrol; aescigenin.* $C_{30}H_{48}O_5$; mol wt 488.68. C 73.73%, H 9.90%, O 16.37%. Artifact previously thought to be the aglycon of escin: Ruzicka *et al., Helv. Chim. Acta* **25**, 1665 (1942); Wagner, Bosse, *Z. Physiol. Chem.* **320**, 27 (1960). Structure work: Ruzicka *et al., Helv. Chim. Acta* **32**, 2057, 2069 (1949). Structure: Cainelli *et al., ibid.* **40**, 2390 (1957); Nakano *et al., J. Org. Chem.* **34**, 3135 (1969). (See Escin.)

Fine needles from alcohol, mp 317-318°. $[\alpha]_D^{20}$ +46° (c = 1.52 in abs ethanol). uv max (abs ethanol): 275 nm (log ε 1.58).

Tetraacetate, $C_{38}H_{56}O_9$, flat needles, mp 207-208°. $[\alpha]_D^{20}$ +56.7° (c = 2.08). Sol in most organic solvents, except petr ether.

3644. Escin. Aescin; Aescusan; Reparil. A mixture of saponins occurring in the seed of the horse chestnut tree, *Aesculus hippocastanum* L., *Hippocastanaceae:* Winterstein, *Z. Physiol. Chem.* **199**, 25 (1931); Steiner, Holtzem in Paech-Tracey, *Moderne Methoden der Pflanzenanalyse III* (Springer-Verlag, 1955) p 117. Isoln by chromatography and purification: Fiedler, *Arzneimittel-Forsch.* **4**, 213 (1953); using ion-exchange resins: Erbring *et al.,* U.S. pat. **3,238,-190** (1966 to Madaus). Previously thought to be built up from the aglycon escigenin, *q.v.,* glucuronic acid, glucose and xylose: Jermstadt, Waaler, *Pharm. Acta Helv.* **28**, 265 (1953); Patt, Winkler, *Arzneimittel-Forsch.* **10**, 273 (1960); Tschesche *et al., Ann.* **669**, 171 (1963). Recent structural studies indicate that the two major glycosides in the mixture are built up from the aglycon, *protoescigenin,* which is acylated at C-22 by acetic acid, and from the sugar moiety, glucuronic acid and two D-glucose molecules. The two aglycons differ only at the C-21 position which is acylated by either angelic acid, *q.v.* or tiglic acid, *q.v.* Structure and stereochemistry: Wulff, Tschesche, *Tetrahedron* **25**, 415 (1969); Wagner *et al., Arzneimittel-Forsch.* **20**, 205 (1970); *eidem, Z. Physiol. Chem.* **351**, 1133 (1970). Early work identified two forms, α-escin and β-escin: Wagner, Basse, *ibid.* **320**, 27 (1960). Identity of *prosaponin B* with β-escin: Voigtlander, Rosenberg, *Arzneimittel-Forsch.* **13**, 385 (1963). β-Escin is the natural form and can be converted to α-escin: Wagner, Schlemmer, U.S. pat. **3,450,691** (1969 to Klinge); Wagner *et al., Arzneimittel-Forsch.,* loc. cit. Pharmacology: Hampel *et al., ibid.* **20**, 209 (1970); Lang, Mennicke, *ibid.* **22**, 1928 (1972). *Review:* Tschesche, Wulff in *Fortschr. Chem. Org. Naturst.* **30**, 461-606 (1973).

R = tiglic acid or angelic acid

Major glycosides of Escin

α-Escin, amorphous powder, mp 225-227°. $[\alpha]_D^{25}$ −13.5° (c = 5 in methanol). Very sol in water. Hemolytic

index: 1:20,000. LD_{50} in mice, rats, guinea pigs: 320, 720, 475 mg/kg orally; 3.2, 5.4, 15.2 mg/kg i.v.

Sodium salt, mp 251-252°.

β-Escin, *Flogencyl.* Leaflets from dil ethanol, mp 222-223°. $[\alpha]_D^{27}$ −23.7° (c = 5 in methanol). Practically insol in water. Hemolytic index: 1:40,000. LD_{50} in mice, rats, guinea pigs: 134, 400, 188 mg/kg orally; 1.4, 2.0, 7.2 mg/kg i.v.

THERAP CAT: In treatment of peripheral vascular disorders.

3645. Esculetin. *6,7-Dihydroxy-2H-1-benzopyran-2-one; 6,7-dihydroxycoumarin;* cichorigenin. $C_9H_6O_4$; mol wt 178.14. C 60.68%, H 3.40%, O 35.93%. The aglucon of esculin and of cichoriin. By hydrolysis of esculin or of cichoriin: Merz, *Arch. Pharm.* **270**, 486 (1932). By synthesis: Gattermann, Köbner, *Ber.* **32**, 288 (1899); Bert, *Compt. Rend.* **214**, 230 (1942). *Review:* Sethna, Shah, *Chem. Rev.* **36**, 1 (1945).

Prisms from glacial acetic acid; leaflets by vacuum sublimation. mp 268-270°. Soluble in dil alkalies with blue fluorescence; moderately sol in hot alcohol and in glacial acetic acid; almost insol in ether, in boiling water.

USE: In filters for absorption of ultraviolet light.

3646. Esculin. *6-(β-D-Glucopyranosyloxy)-7-hydroxy-2H-1-benzopyran-2-one;* 6,7-dihydroxycoumarin 6-glucoside; esculoside; bicolorin; enallachrome; polychrome; Escosyl. $C_{15}H_{16}O_9$; mol wt 340.28. C 52.94%, H 4.74%, O 42.32%. In leaves and bark of horse chestnut tree *Aesculus hippocastanum* L., *Hippocastanaceae.* Extraction procedure: Tumann, *Chem. Zentr.* **1916**, I, 1277. Synthesis: Merz, Hagemann, *Naturwiss.* **29**, 650 (1941); *eidem, Arch. Pharm.* **282**, 79 (1944); Amiard, Nominé, *Bull. Soc. Chim. France* **1948**, 476.

Sesquihydrate, needles from hot water, mp 204-206°. $[\alpha]_D^{18}$ −78.4° (c = 2.5 in 50% dioxane). One gram dissolves in 580 ml water, 13 ml boiling water. Sol in hot alcohol, methanol, pyridine, ethyl acetate, acetic acid. Aq solns show blue fluorescence above pH 5.8. Absorption spectrum: Merz, *Arch. Pharm.* **270**, 482 (1932); Goodwin, Pollock, *Arch. Biochim. Biophys.* **49**, 1 (1954). Has vitamin P activity.

Pentaacetate dihydrate, needles from methanol, mp 163-164°.

4-Methylesculin, $C_{16}H_{18}O_9$. Prepn: Velluz, Amiard, *Bull. Soc. Chim. France* **1948**, 1109.

THERAP CAT: Skin protectant.

3647. Eseridine. *2,3,4,4a,9,9a-Hexahydro-2,4aα,9α-trimethyl-1,2-oxazino[6,5-b]indol-6-ol methylcarbamate;* physostigmine aminoxide; eserine aminoxide; eserine oxide. $C_{15}H_{21}N_3O_3$; mol wt 291.34. C 61.83%, H 7.27%, N 14.42%, O 16.47%. From Calabar bean (*Physostigma venenosum* Balf., *Leguminosae*): Polonovski, Nitzberg, *Bull. Soc. Chim. France* [IV] **17**, 244 (1915); **21**, 191 (1917). Structure: Stedman, Barger, *J. Chem. Soc.* **127**, 247 (1925). Revised structure: Hootele, *Tetrahedron Letters* **1969**, 2713. Stereochemistry: Robinson, Moorcroft, *J. Chem. Soc. (C)* **1970**, 2077. Total synthesis of racemate: K. Shishido *et al., Chem. Commun.* **1986**, 904.

Rectangular prisms from ether, mp 129°. $[\alpha]_D^{15}$ —175° (alc). Weak base, just alkaline to litmus; does not form crystn salts with mineral acids. Almost insol in water. Sol in alc, chloroform, benzene, ether, petr ether, acetone, dil acids.

Salicylate, $C_{22}H_{27}N_3O_6$, Génésérine 3.

THERAP CAT: Cholinergic.

3648. Esmolol. *4-[2-Hydroxy-3-[(1-methylethyl)amino]propoxy]benzenepropanoic acid methyl ester;* (±)-methyl 3-[4-[2-hydroxy-3-(isopropylamino)propoxy]phenyl]propionate; methyl *p*-[2-hydroxy-3-(isopropylamino)propoxy]-hydrocinnamate. $C_{16}H_{25}NO_4$; mol wt 295.38. C 65.06%, H 8.53%, N 4.74%, O 21.67%. Cardioselective β-adrenergic blocker. Prepn: E. I. Carlsson *et al.*, Eur. pat. **Appl. 41,491** (1982 to A.B. Hässle), *C.A.* **96**, 122391f (1982); P. W. Erhardt *et al.*, *J. Med. Chem.* **25**, 1408 (1982). Use in treatment or prophylaxis of cardiac disorders: *eidem*, Eur. pat. **Appl. 53,435**; *eidem*, U.S. pat. **4,387,103** (1982, 1983 to Amer. Hosp. Supply). Pharmacology: R. J. Gorczynski *et al.*, *J. Cardiovasc. Pharmacol.* **5**, 668 (1983); *eidem, ibid.* **6**, 1548 (1984). Pharmacokinetics: A. Yacobi *et al.*, *J. Pharm. Sci.* **72**, 711 (1983). GC-MS determn in blood: C. Y. Sum, A. Yacobi, *ibid.* **73**, 1177 (1984). Clinical studies of effect in supraventricular tachycardia: G. Klein *et al.*, *Int. J. Clin. Pharmacol. Ther. Toxicol.* **22**, 112 (1984); R. J. Gray *et al.*, *J. Am. Coll. Cardiol.* **5**, 1451 (1985). Symposium on pharmacology and clinical efficacy: *Am. J. Cardiol.* **56**(11), 1F-62F (1985). Review of pharmacology and clinical efficacy: P. Benfield, E. M. Sorkin, *Drugs* **33**, 392-412 (1987).

Oil, gradually forming crystalline rosettes at room temp, mp 48-50°.

Hydrochloride, $C_{16}H_{26}ClNO_4$, ASL-8052, Brevibloc. Crystals from methanol-ether, mp 85-86°.

THERAP CAT: Anti-arrhythmic.

3649. Esparto Wax. Spanish grass wax; halfa wax. A wax derived from a tall, tough grass of the Mediterranean region (S. Europe and N. Africa and Libya). The grass is shipped to Scotland, where it is dewaxed and made into fine paper. The wax is a byproduct. Two species of grass are cultivated for their excellent cellulose content: *Stipa tenacissima* L. (*Macrochloa tenacissima* (L.) Kunth.), *Graminaceae* and *Lygeum spartum* L., *Graminaceae*. Constit. Esparto wax consists of 15-17% free wax acids, 20-22% alcohols and hydrocarbons, 63-65% esters. The principal hydrocarbon is hentriacontane ($C_{31}H_{64}$, mp 68°). The acids include cerotic, montanic, myricinic ($C_{30}H_{60}O_2$, mp 68°), lacceric ($C_{32}H_{64}O_2$, mp 70.5°) and hydroxy acids: A. H. Warth, *Chemistry and Technology of Waxes* (Reinhold, New York, 1947) p 139. Extraction of wax from grass: E. Ger. pat. **74,703** (1970). TLC identification: H. Schmidt, *Amer. Cosmet. Perfum.* **87**, 35 (1972).

Hard, tough wax. d_4^{25} 0.9887. mp 78.1°. Solidifies at 68.8°. Acid value 23.9. Saponification value 69.8. Soly in ethanol (25°): 0.244 g/100 ml; in ethylene chloride (37°): 1.48 g/100 ml.

USE: Substitute or extender for carnauba wax, *q.v.* Blends well with other waxes. Emulsifies easily and imparts smoothness to polishes. Preferred in the manufacture of carbon papers.

3650. Essential Fatty Acids. EFA. Essential dietary constituents belonging to the linoleic acid family. G. O. Burr and M. M. Burr described expts with albino rats which led them to conclude that in animals given a ration extreme-

ly poor in fat, a deficiency disease results which can be cured by linoleic or linolenic acids, *q.q.v.*, and arachidonic acid, *q.v.*, or fats contg these acids (such as lecithin from soybeans): *J. Biol. Chem.* **82**, 345 (1929); **86**, 587 (1930). Support for the conception of the essential nature of these two fatty acids was contained in two reports by Evans and Lepkovsky, *J. Biol. Chem.* **96**, 143, 157 (1932). It was stated that growth would not take place unless unsaturated fatty acids with two or more double bonds were present in the ration. In 1934 a series of three papers by Evans, Lepkovsky, and Murphy appeared, in which the name *vitamin F* was given to the above unsaturated fatty acids. There was a short period of commercial exploitation of vitamin F. Following investigations made by the American Medical Association in 1937, the term "vitamin F" became discredited. At the request of the Council on Foods and Nutrition of the American Medical Association, Hansen and Burr discussed the essentiality of fatty acids in human nutrition in their article "Essential Fatty Acids and Human Nutrition," *J. Am. Med. Assoc.* **132**, 855 (Dec. 7, 1946), in which they state: "In regard to clinical observations made thus far with human subjects, there is no evidence to indicate that a lack of the essential fatty acids produces a disturbance in growth, hematuria, kidney lesions, impaired reproduction or lactation, and sterility, which abnormalities, are attributed to the dietary lack of either linoleic or arachidonic acid in small experimental animals." *See also* the extensive reviews by Deuel and Greenberg in *Fortschr. Chem. Org. Naturst.* **6**, 1-86 (1950); Stangl, *Int. Rev. Vitamin Res.* **23**, 164-207 (1951); Rahm, Holman, *The Vitamins*, W. H. Sebrell, R. S. Harris, Eds., vol. 3 (Academic Press, New York, 1971) pp 303-339; W. H. Kunan, *Angew. Chem. Int. Ed.* **15**, 61 (1976); J. F. Mead, "Nutrients with Special Functions: Essential Fatty Acids" in *Human Nutrition, A Comprehensive Treatise*, vol. 3A, R. Alfin-Slater, D. Kritchevsky, Eds. (Plenum Press, New York, 1980) pp 213-238. Book: *Essential Fatty Acids and Prostaglandins*, R. J. Holman, Ed. (Pergamon Press, New York, 1982) 968 pp.

3651. Estazolam. *8-Chloro-6-phenyl-4H-[1,2,4]triazolo[4,3-a][1,4]benzodiazepine; 8-chloro-6-phenyl-4H-s-triazolo[4,3-a][1,4]benzodiazepine;* D-40TA; Esilgan; Eurodin; Julodin; Nuctalon; Somnatrol. $C_{16}H_{11}ClN_4$; mol wt 294.75. C 65.20%, H 3.76%, Cl 12.03%, N 19.01%. Prepn: J. B. Hester, Ger. pat. **2,012,190**; *idem*, U.S. pat **3,701,782** (1970, 1972 to Upjohn); Meguro, Kuwada, *Tetrahedron Letters* **1970**, 4039; *eidem*, Ger. pat. **1,955,349** (1971 to Toyama), *C.A.* **74**, 88078t (1971); Tawada *et al.*, Ger. pat. **2,114,441** (1971 to Takeda), *C.A.* **78**, 34320p (1972); Hester *et al.*, *J. Med. Chem.* **14**, 1078 (1971). Structure-activity studies: R. Nakajima *et al.*, *Japan. J. Pharmacol.* **21**, 489 (1971). Pharmacology: *eidem, ibid.* 497; *eidem, Takeda Kenkyusho Ho* **31**, 349 (1972), *C.A.* **78**, 24138m (1972). Toxicity: Yokotani *et al., ibid.* **32**, 152 (1973), *C.A.* **79**, 133006j (1973). Metabolism: Tanayama *et al., Xenobiotica* **4**, 33-64 (1974); *see also ibid.* 229, 441. Molecular structure studies: Kamiya *et al., Chem. Pharm. Bull.* **21**, 1520 (1973).

Crystals from ethyl acetate-Skellysolve B hexanes, mp 228-229°. LD_{50} orally in male mice, rats and rabbits: 740, 3200, 300 mg/kg (Yokotani).

Note: This is a controlled substance (depressant) listed in the U.S. Code of Federal Regulations, Title 21 Part 1308.14 (1987).

THERAP CAT: Sedative, hypnotic.

3652. Ester Gums. These are the glyceryl, methyl, and ethyl esters of rosin acids. They are made by heating rosin and the alcohol under pressure until condensation takes place. The gums are insol in water, but sol in amyl acetate, some oils, turpentine, carbon tetrachloride, etc.

USE: Instead of copal, damar, or kauri in making enamels,

paints, cellulose lacquers, and particularly with tung oil for waterproof varnishes.

3653. Estradiol. *(17β)-Estra-1,3,5(10)-triene-3,17-diol; β-estradiol;* α-estradiol (obsolete); *cis*-estradiol; 3,17-epidihydroxyestratriene; dihydrofollicular hormone; dihydrofolliculin; dihydroxyestrin; dihydrotheelin; Compudose 365; Dihydromenformon; Dimenformon; Diogyn; Estrace; Estraderm; Estrovite; Femestral; Gynergon; Gynoestryl; Lamdiol; Macrodiol; Oestrogel; Oestergon; Ovahormon; Ovasterol; Ovocyclin; Ovocylin; Perlatanol; Primofol; Profoliol; Progynon-DH. $C_{18}H_{24}O_2$; mol wt 272.37. C 79.37%, H 8.88%, O 11.75%. The most potent mammalian estrogenic hormone; formed by the ovary, placenta, testis and possibly by the adrenal cortex. Has been isolated from follicular liquor of sow ovaries; from pregnancy urine of mares. Isoln: MacCorquodale *et al., J. Biol. Chem.* **115**, 435 (1936). Numerous prepns from other steroids, *e.g.,* Butenandt, Georgens, *Z. Physiol. Chem.* **248**, 129 (1937); Hildebrandt, Schwenk; Logemann, Koester; Inhoffen; U.S. pats. **2,096,744; 2,225,-419; 2,361,847** (1938, 1941, 1944 all to Schering); Inhoffen, Zühlsdorff, *Ber.* **74**, 1914 (1941). Total syntheses: U. Eder *et al., Ber* **109**, 2948 (1976); W. Oppolzer, D. A. Roberts, *Helv. Chim. Acta* **63**, 1703 (1980). Pharmacology: R. W. Lievertz, *Am. J. Obstet. Gynecol.* **156**, 1289 (1987). Pharmacokinetics: B. Dusterberg, Y. Nishino, *Maturitas* **4**, 315 (1982). HPLC determn in plasma and urine: W. Slikker *et al., J. Chromatog.* **224**, 205 (1981). Comprehensive description of the valerate ester: K. Florey, Ed. in *Analytical Profiles of Drug Substances* vol. 4 (Academic Press, New York, 1975) pp 192-208; of estradiol: E. G. Salole, *ibid.* vol. 15 (1986) pp 283-318.

Prisms from 80% alc, stable in air, mp 173-179°. $[\alpha]_D^{25}$ +76 to +83° (dioxane). uv max: 225, 280 nm. Precipitated by digitonin. Almost insol in water. Freely sol in alcohol; sol in acetone, dioxane, other organic solvents; solns of fixed alkali hydroxides; sparingly sol in vegetable oils. 1 mg = 10,000 international units.

3-Benzoate, *see* Estradiol Benzoate.

17β-Cypionate, *see* Estradiol 17β-Cypionate.

17-Propionate, $C_{21}H_{28}O_3$, *Acrofollin, Akrofollin,* mp 199-200°. *See* Miescher, Scholz, *Helv. Chim. Acta* **20**, 263 (1937); U.S. pats. **2,160,555; 2,233,025** (1939, 1941 to Ciba).

Dipropionate, $C_{24}H_{32}O_4$, *Agofollin, Dimenformon Dipropionate, Diovocylin, Ovocyclin Dipropionate, Ovocyclin-P, Progynon-DP.* mp 104-105°. *See:* Miescher, Scholz, U.S. pats. **2,160,555; 2,205,627; 2,233,025** (1939, 1940, 1941 to Ciba).

Hemisuccinate, $C_{22}H_{28}O_5$, *Eutocol.*

17-Heptanoate, $C_{25}H_{36}O_3$, *estradiol enanthate, SQ 16150.* Crystals from diisopropyl ether, mp 94-96°. *See:* Gauthier *et al., Ann. Pharm. Franc.* **16**, 757 (1958).

17-Undecanoate, $C_{29}H_{44}O_3$, *estradiol undecylate, Delestrec.*

17-Undecenoate, $C_{29}H_{42}O_3$, small plates, mp 105-106°. $[\alpha]_D^{20}$ +42° (chloroform). uv max: 280-282 nm (log ϵ 3.30). Prepn: Ringold *et al.,* U.S. pat. **2,990,414** (1961 to Syntex).

17-Valerate, $C_{23}H_{32}O_3$, *Delestrogen, Progynova.* Crystals, mp 144-145°. *See:* Miescher, Scholz, U.S. pats. **2,205,-627; 2,233,025.**

Note: This substance may reasonably be anticipated to be a carcinogen: *Fourth Annual Report on Carcinogens* (NTP 85-002, 1985) p 100.

THERAP CAT: Estrogen.

THERAP CAT (VET): Estrogenic hormone therapy.

3654. α-Estradiol. *Estra-1,3,5(10)-triene-3,17α-diol;* 1,3,5-estratriene-3,17α-diol; 3,17-dihydroxyestratriene. $C_{18}H_{24}O_2$; mol wt 272.37. C 79.37%, H 8.88%, O 11.75%. Has been isolated from pregnancy urine of mares. Prepn from β-estradiol by inversion of the hydroxyl group at C-17

after tosylation: Allais, Hoffmann, U.S. pat. **2,835,681** (1958).

Needles with ½H_2O from 80% alcohol, mp 220-223°. $[\alpha]_D^{20}$ +53° to +56° (c = 0.9 in dioxane). Not precipitated by digitonin (in 80% alcoholic soln). Soluble in alcohol, acetone, aq alkalies. One gram dissolves in more than 100 ml of boiling benzene. Slightly sol in ether, chloroform. Insol in water, aq dil acids.

Diacetate, $C_{22}H_{28}O_4$, mp 140-142°.

3-Benzoate, $C_{25}H_{28}O_3$, mp 156-157°; also reported as polymorphous: I, mp 63°; II, mp 153°; III, mp 158°.

3655. Estradiol Benzoate. *(17β)-Estra-1,3,5(10)-triene-3,17-diol 3-benzoate;* β-estradiol 3-benzoate; oestradiol monobenzoate; Benovocylin; Benzhormovarine; Benzoestrofol; Benzofoline; Benzo-Gynoestryl; Benztrone; de Graafina; Diffolisterol; Difolliculine; Dimenformon benzoate; Diogyn B; Eston-B; Femestrone; Follicormon (ampuls); Follidrin (ampuls); Graafina; Gynécormone; Hidroestron; Hormogynon; Motovar; Oestroform; Ovasterol-B; Ovex B; Ovocyclin Benzoate; Ovocyclin M; Ovocyclin-MB; Primogyn B; Primogyn I; Progynon-B; Progynon Benzoate; Recthormone Oestradiol; Solestro; Unistradiol. $C_{25}H_{28}O_3$; mol wt 376.50. C 79.75%, H 7.50%, O 12.75%. Prepn: Schwenk, Hildebrandt, U.S. pat. **2,054,271** (1936 to Schering-Kahlbaum); Sandulesco, Brit. pat. **485,388** (1938 to Lab. Franc. Chimiother.). Soly of 17β-maltosides: Meystre, Miescher, *Helv. Chim. Acta* **27**, 235, 1154, 1157 (1944).

Crystals from alc, mp 191-196°. Stable in air. $[\alpha]_D^{25}$ +58 to +63° (c = 2 in dioxane). Sol in alc, acetone, dioxane; slightly sol in ether, vegetable oils. 1 mg = 10,000 international estradiol benzoate units. Precipitated by digitonin.

17β-Maltoside heptaacetate, $C_{51}H_{62}O_{20}$, dec 227-229°. $[\alpha]_D^{20}$ +56° (methanol). Sol in water; very sol in glucose solns.

17β-Maltoside hydrate, $C_{30}H_{44}O_{12} \cdot H_2O$, dec 272-282°. $[\alpha]_D^{20}$ +52° (c = 1.07 in methanol). Sol in water; very sol in glucose solns.

THERAP CAT: Estrogen.

THERAP CAT (VET): Estrogenic hormone therapy.

3656. Estradiol 17β-Cypionate. *(17β)-Estra-1,3,5(10)-triene-3,17-diol 17-cyclopentanepropanoate;* Estradiol 17β-cyclopentanepropionate; estradiol 17β-cyclopentylpropionate; ECP; Depoestradiol; Depofemin; Estradep. $C_{26}H_{36}O_3$; mol wt 396.55. C 78.74%, H 9.15%, O 12.10%. Prepd by treating estradiol 3,17β-dicyclopentanepropionate with potassium carbonate: Ott, U.S. pat. **2,611,773** (1952 to Upjohn).

Crystals from benzene + petr ether, mp 151-152°. $[\alpha]_D^{25}$ +45° (chloroform). Sol in ether, methanol, benzene, chloroform, peanut oil, cottonseed oil, corn oil, sesame oil. The limit of soly in the oils is about 400 mg/ml. Thixotropic gels may be prepd by adding aluminum monostearate to the oil solns.

THERAP CAT: Estrogen.

THERAP CAT (VET): Estrogenic hormone therapy.

3657. Estragole. *1-Methoxy-4-(2-propenyl)benzene; p-allylanisole;* chavicol methyl ether; esdragol. $C_{10}H_{12}O$; mol wt 148.20. C 81.04%, H 8.16%, O 10.80%. Main constituent of *tarragon oil (estragon oil),* the oil from *Artemisia dracunculus* L., *Compositae* (esdragon) where it occurs to an extent of 60-75%: Grimaux, *Bull. Soc. Chim.* [3] **11**, 34 (1894); Daufresne, *ibid.* [4] **3**, 333 (1908). Occurs also in pine oil and in American turpentine oil. Prepn: Tiffeneau, *Compt. Rend.* **139**, 482 (1904); Verley, **Ger.** pat. **154,654;** D. Wigfield, K. Taymaz, *Tetrahedron Letters* **1973**, 4841; P. Gramatica *et al., Gazz. Chim. Ital.* **104**, 629 (1974).

Liquid. d_4^{21} 0.9645. bp_{764} 216°; bp_{25} 108-114°; bp_{12} 95-96°. $n_D^{17.5}$ 1.5230. Sol in alcohol, chloroform. Forms azeotropic mixtures with water. LD_{50} orally in rats, mice: 1820, 1250 mg/kg, P. M. Jenner *et al., Food Cosmet. Toxicol.* **2**, 327 (1964).

USE: In perfumes and as flavor in foods and liqueurs.

3658. Estramustine. *Estra-1,3,5(10)-triene-3,17-diol 3-[bis(2-chloroethyl)carbamate]; estradiol 3-bis(2-chloroethyl)carbamate;* estra-1,3,5(10)triene-3,17β-diol 3-[N,N-bis(2-chloroethyl)carbamate]; Ro 21-8837. $C_{23}H_{31}Cl_2NO_3$; mol wt 440.41. C 62.72%, H 7.10%, Cl 16.10%, N 3.18%, O 10.90%. Estradiol to which nitrogen mustard is bound. Prepn: **Belg.** pat. **646,319** corresp to Fex *et al.,* **U.S.** pat. **3,299,104** (1963, 1967 to AB Leo); Niculescu-Duvaz *et al., J. Med. Chem.* **10**, 172 (1967). Clinical results: Anderes, *Praxis* **60**, 1375 (1971); Muntzing, Nilsson, *Z. Krebsforsch. Klin. Onkol.* **77**, 166 (1972).

Crystals from benzene-petr ether, mp 104-105°. $[\alpha]_D^{20}$ +50° (in dioxane). uv max (alcohol): 270.7, 276.5 nm. 17-Phosphate, $C_{23}H_{32}Cl_2NO_6P$, *Estracyt.* mp 155° (dec). $[\alpha]_D^{20}$ +30° (dioxane). Sol in aqueous and alkali solns. 17-(Dihydrogenphosphate), disodium salt, $C_{23}H_{30}Cl_2N-Na_2O_6P$, *Emcyt.*

THERAP CAT: Antineoplastic.

3659. Estriol. *Estra-1,3,5(10)-triene-3,16,17-triol;* 1,3,5-estratriene-3β,16α,17β-triol; 3,16α,17β-trihydroxy-Δ1,3,5-estratriene; 16α-hydroxyestradiol; follicular hormone hydrate; oestriol; trihydroxyestrin; Aacifemine; Colpogyn; Destriol; Gynäsan; Hormomed; Klimax E; Klimoral; Oekolp; Ortho-Gynest; Ovesterin; Ovestin; Ovo-Vinces; Theelol; Tridestrin; Triovex. $C_{18}H_{24}O_3$; mol wt 288.37. C 74.97%, H 8.39%, O 16.64%. A metabolite of, and considerably less potent than the hormone estradiol, *q.v.* It is usually the predominant estrogenic metabolite found in urine. During pregnancy the placenta produces relatively large amounts of estriol. Isoln from human pregnancy urine: Marrian, *Biochem. J.* **23**, 1090, 1233 (1929); probably occurs as a glycuronide: Cohen, Marrian, *ibid.* **29**, 1577 (1935). Isoln from human placenta: Collip, *Brit. Med. J.* **1930,** II,

1080; Collip *et al., Endocrinology* **18**, 71 (1934). Also obtained from plant sources. Isoln from pussywillows: Skarzynski, *Nature* **131**, 766 (1933). Structure: Huffman, Lott, *J. Am. Chem. Soc.* **69**, 1835 (1947). Crystal and molecular structure: Cooper *et al., Acta Crystallogr.* **25B**, 814 (1969). Soly studies: Ruchelman, Howe, *J. Chromatogr. Sci.* **7**, 340 (1969). Partial synthesis: Huffman, *J. Biol. Chem.* **169**, 167 (1947). Syntheses: Huffman, Lott, *J. Am. Chem. Soc.* **71**, 719 (1949); Leeds *et al., ibid.* **76**, 2943 (1954). Prepn of 16,17-bis(sodium hemisuccinate): **Brit.** pat. **879,014** (1960 to Organon).

Very small, monoclinic crystals from dil alc. d 1.27. During heating on the microscope heating stage, rearrangement of the crystal structure takes place at 270° and 275°. mp 282°. (Rate of heating, 4°/min, Kofler microscope heating stage). $[\alpha]_D^{25}$ +58° ±5° (0.04 g in 1 ml dioxane). uv max: 280 nm. Precipitated by digitonin. Practically insol in water. Sol in alcohol, dioxane, chloroform, ether, vegetable oils; freely sol in pyridine, in solns of fixed alkali hydroxides. Triacetate, $C_{24}H_{30}O_6$, crystals from 90% alc. mp 126°. 16,17-Bis(sodium hemisuccinate), $C_{26}H_{30}Na_2O_9$, *estriol succinate, Orgastyptin, Stiptanon, Synapause.*

THERAP CAT: Estrogen.

THERAP CAT (VET): Estrogenic hormone therapy.

3660. Estrone. *3-Hydroxyestra-1,3,5(10)-trien-17-one;* 1,3,5-estratrien-3-ol-17-one; oestrone; theelin; folliculin; follicular hormone; tokokin; thelykinin; ketohydroxyestrin; Hiestrone; Menformon; Glandubolin; Cristallovar; Destrone; Endofolliculina; Estrol; Femidyn; Folikrin; Ovex; Kolpon; Crinovaryl; Folisan; Disynformon; Hormovarine; Oestroperos; Wynestron; Thelestrin; Kestrone; Estrusol; Estrugenone; Femestrone Inj.; Folipex; Follestrine; Follidrin (tablets); Follicunodis; Hormofollin; Oestrin; Oestroform; Ovifollin; Perlatan; Ketodestrin; Unden. $C_{18}H_{22}O_2$; mol wt 270.36. C 79.96%, H 8.20%, O 11.84%. A metabolite of 17β-estradiol, *q.v.,* possessing considerably less biological activity. Occurs in pregnancy urine of women and mares, in follicular liquor of many animals, in human placenta, in urine of bulls and stallions, in palm-kernel oil. Isoln: Butenandt, *Naturwiss.* **17**, 879 (1929); Doisy *et al., Am. J. Physiol.* **90**, 329 (1929). Also isol from moghat roots and date palm pollen grains: Amin *et al., Phytochemistry* **8**, 295 (1969). Manuf: E. A. Doisy *et al.,* **U.S.** pats. **1,967,350, 1,967,351** (1934 both to St. Louis University); Joly *et al.,* **Fr.** pat. **1,305,992** (1962 to Roussel-Uclaf). Synthesis of a stereoisomer of estrone: W. E. Bachmann *et al., J. Am. Chem. Soc.* **64**, 974 (1942). Total synthesis of natural estrone: G. Anner, K. Miescher, *Helv. Chim. Acta* **31**, 2173 (1948); **32**, 1957 (1949); **33**, 1379 (1950); W. S. Johnson *et al., J. Am. Chem. Soc.* **72**, 1426 (1950); **74**, 2832 (1952); I. V. Torgov *et al., Doklady Akad. Nauk SSSR* **135**, 73 (1960), *C.A.* **55**, 11462f (1961). Stereochemistry of estrone isomers: W. S. Johnson *et al., J. Am. Chem. Soc.* **80**, 661 (1958). Review of estrone syntheses: L. F. Fieser, M. Fieser, *Steroids* (Reinhold, New York, 1959) pp 495-502; T. B. Windholz, M. Windholz, *Angew. Chem. Int. Ed.* **3**, 353 (1964); Taub, "Naturally Occurring Aromatic Steroids" in *The Total Synthesis of Natural Products* vol. 2, J. ApSimon, Ed. (John Wiley & Sons, New York, 1973) pp 670-725. Recent syntheses: P. A. Bartlett, W. S. Johnson, *J. Am. Chem. Soc.* **95**, 7501 (1973); S. Danishefsky, P. Cain, *ibid.* **98**, 4975 (1976); T. Kametani *et al., ibid.* **99**, 3461 (1977); P. A. Grieco *et al., J. Org. Chem.* **45**, 2247 (1980). Comprehensive description: D. Both in *Analytical Profiles of Drug Substances* vol. 12, K. Florey, Ed. (Academic Press, New York, 1983) pp 135-189.

dl-Form, crystals from acetone, mp 251-254°.

d-Form (natural), crystals from acetone, mp 254.5-256°. $[\alpha]_D^{22}$ +152° (c = 0.995 in CHCl₃). uv max (*p*-dioxane): 282, 296 nm (ε 2300, 2130); (conc H₂SO₄): 300, 450 nm; (0.1*M* NaOH): 239, 293 nm. Exists in three cryst phases, one monoclinic, the other two orthorhombic. Pptd by digitonin. Soly in water (25°): 0.003 g/100 ml. One gram of estrone dissolves in 250 ml of 96% alcohol at 15°, in 50 ml of boiling alcohol; in 50 ml acetone at 15°, in 110 ml chloroform at 15°, in 145 ml boiling benzene. Sol in dioxane, pyridine, fixed alkali hydroxide solns; slightly sol in ether, vegetable oils.

Methyl ether, C₁₉H₂₄O₂, *dl*-Form, crystals from acetone + methanol, mp 143.2-144.2°. *d*-Form, crystals from methanol, mp 164-165°. Synthesis: T. A. Bryson, C. J. Reichel, *Tetrahedron Letters* **21**, 2381 (1980).

Acetate, C₂₀H₂₄O₃, *Hogival*. Crystals from alcohol, mp 125-127°.

Propionate, C₂₁H₂₆O₃, mp 134-135°, U.S. pat. **2,156,599**.

Sulfate, *Conjugol*, Price, U.S. pat. **2,917,522** (1959 to Parke, Davis).

Sulfate piperazine salt, C₂₂H₃₂N₂O₅S, *estropipate, piperazine estrone sulfate, Harmogen, Ogen, Sulestrex Piperazine*. Cryst powder, melts at 190° to a liquid which resolidifies and then decomp at 245°. $[\alpha]_D^{25}$ +87.8° (c = 1 in 0.4% NaOH). uv max (0.04% NaOH): 275, 268 nm (ε 838, 851). Comprehensive description: Z. L. Chang in *Analytical Profiles of Drug Substances* vol. 5, K. Florey, Ed. (Academic Press, New York, 1976) pp 375-402.

Note: This substance may reasonably be anticipated to be a carcinogen: *Fourth Annual Report on Carcinogens* (NTP 85-002, 1985) p 102.

USE: In the prepn of commercial 19-nor steroids.

THERAP CAT: Estrogen.

THERAP CAT (VET): Estrogenic hormone therapy.

3661. Etafedrine. α-[1-(*Ethylmethylamino)ethyl]benzenemethanol; l-N-ethylephedrine; l-α-[1-(ethylmethylamino)ethyl]benzyl alcohol; 2-methylethylamino-1-phenyl-1-propanol; Menetryl; Novedrin.* C₁₂H₁₉NO; mol wt 193.28. C 74.57%, H 9.91%, N 7.25%, O 8.28%. β-Adrenergic agonist. Prepn: Skita, Keil, *Ber.* **63**, 34 (1930); Ueda *et al.*, *Pharm. Bull. (Tokyo)* **3**, 465 (1955).

Hydrochloride, C₁₂H₂₀ClNO, *Nethamine*. Crystals from acetone + alcohol, mp 183-184°. One gram dissolves in 1.5 ml water, 8.0 ml alcohol. Aq solns are very stable.

THERAP CAT: Bronchodilator.

3662. Etafenone. 1-[2-[2-(*Diethylamino)ethoxy]phenyl]-3-phenyl-1-propanone; 2'-[2-(diethylamino)ethoxy]-3-phenylpropiophenone; o-diethylaminoethoxy-β-phenylpropiophenone; 2'-(β-diethylaminoethoxy)-3-phenylpropiophenone; β-phenyl-o-(diethylaminoethoxy)propiophenone; LG 11457.* C₂₁H₂₇NO₂; mol wt 325.43. C 77.50%, H 8.36%, N 4.30%, O 9.83%. Prepn: G. DiPaco, S. C. Tauro, *Ann. Chim. (Rome)* **48**, 1215 (1958); *eidem*, Ger. pat. **1,265,758** (1968 to Guidotti), *C.A.* **69**, 58950a (1968). Series of articles on pharmacology, toxicology, metabolism: *Arzneimittel-Forsch.* **19**, 1664-1681 (1969).

Liquid, bp₃₀ 264-268°.

Hydrochloride, C₂₁H₂₈ClNO₂, *hetaphenone, Asamedol, Baxacor, Corodilan, Dialicor, Pagano-Cor, Relicor.* Crystals, mp 129-130°. LD₅₀ in rats (mg/kg): 716 orally; 20.8 i.v., H.-J. Hapke, W. Sterner, *Arzneimittel-Forsch.* **19**, 1664 (1969).

THERAP CAT: Vasodilator (coronary).

3663. Etamiphyllin. 7-[2-(*Diethylamino)ethyl]-3,7-dihydro-1,3-dimethyl-1H-purine-2,6-dione; 7-(2-diethylaminoethyl)theophylline; 7-(2-diethylaminoethyl)-1,3-dimethylxanthine; 1,3-dimethyl-7-(2-diethylaminoethyl)xanthine; dietamiphylline; millophylline.* C₁₃H₂₂ClN₅O₂; mol wt 279.34. C 55.89%, H 7.58%, N 25.07%, O 11.46%. Prepn: Quevauviller *et al.*, *Bull. Soc. Chim. Biol.* **31**, 532 (1946); Moussalli *et al.*, **Brit.** pat. **669,070** (1952); Klosa, *Arch. Pharm.* **288**, 301 (1955). Toxicity data on combinations: Baettig, *Arzneimittel-Forsch.* **21**, 354 (1971).

Waxy solid, mp 75°. Very sol in water, acetone; slightly sol in ethanol, ether.

Hydrochloride, C₁₃H₂₂ClN₅O₂, crystals, mp 239-241°. Very sol in water, forming stable neutral soln.

Camphorsulfonate, C₂₃H₃₇N₅O₆S, *Camphophyline*. Crystals, mp 174°.

Heparinate, *Milhéparine*.

Iodomethylate, *Iodaphyline, Iodafilina, Jodo-Metil-Fillina*.

Methesculetol salt, *see* metescufylline.

THERAP CAT: Bronchodilator.

3664. Etamycin. Etamycin A; neoviridogrisein IV; viridogrisein. C₄₄H₆₂N₈O₁₁; mol wt 879.04. C 60.12%, H 7.11%, N 12.75%, O 20.02%. Monocyclic peptide lactone antibiotic produced by *Streptomyces griseus* and allied *Streptomyces* spp., which shows activity vs gram positive bacteria, *Mycobacterium tuberculosis* and some fungi. Obtained together with griseoviridin, *q.v.*: B. Heinemann *et al.*, *Antibiot. Ann.* **1954-55**, 728; Q. R. Bartz *et al.*, *ibid.* 777. Chemical studies: T. H. Haskell *et al.*, *ibid.* 784. Biological studies: J. Ehrlich *et al.*, *ibid.* 790. Structure: Sheehan *et al.*, *J. Am. Chem. Soc.* **80**, 3349 (1958). Confirmation of structure: Arnold *et al.*, *J. Chem. Soc.* **1958**, 4466. Prodn with griseoviridin: Q. R. Bartz *et al.*, U.S. pat. **3,023,204** (1962 to Parke, Davis). Total synthesis: Sheehan, Ledis, *J. Am. Chem. Soc.* **95**, 875 (1973). Mode of action: Garcia-Mendoza, *Biochim. Biophys. Acta* **97**, 94 (1965). Mass spectra data: Rozynov *et al.*, *Zh. Obshch. Khim.* **39**, 891 (1969). Prodn of congeners by *S. griseoviridus*: C. Chopra *et al.*, *J. Antibiot.* **32**, 392 (1979). Biosynthetic studies of *neoviridogriseins*, homologs of etamycin: Y. Akumura *et al.*, *ibid.* 575, 1002, 1130.

Weakly basic polypeptide, dec 168-170°. $[\alpha]_D^{26}$ +62°, +31° (c = 5 in chloroform, alc). uv max (alc): 304.5 nm (log ϵ 3.91). pKa about 1 to 2. Soly in water about 1 mg/ml. Also sol in 1.0N HCl and 1.0N NaOH, in the lower alcohols and ketones, benzene, chloroform, carbon tetrachloride, carbon disulfide, ethyl acetate, peanut oil, ether. Practically insol in petr ether. LD_{50} in mice: > 2 g/kg s.c.; > 3 g/kg orally.

Hydrochloride, crystals, mp 163-170° (some dec). uv max: 304 nm ($E_{1cm}^{1\%}$ 86), after addition of alkali: 335 nm ($E_{1cm}^{1\%}$ 82). Soly in water about 4 mg/ml. Sol in methanol, ethanol, formamide. Slightly sol in acetone, ethyl acetate, or less polar solvents.

3665. Etaqualone. *3-(2-Ethylphenyl)-2-methyl-4(3H)-quinazolinone;* 2-methyl-3-(o-ethylphenyl)-4-quinazolone; ethinazone; Aolan. $C_{17}H_{16}N_2O$; mol wt 264.31. C 77.25%, H 6.10%, N 10.60%, O 6.05%. Prepd by condensation of N-acetylanthranilic acid and 2-ethylaniline with $POCl_3$: Belg. pat. 615,282 (1962 to P. Beiersdorf AG), *C.A.* **58**, 13971ef (1963). Structure-activity studies: Parmar *et al., J. Med. Chem.* **12**, 138 (1969).

mp 81°.
Hydrochloride, $C_{17}H_{16}N_2O \cdot HCl$, mp 247°.
THERAP CAT: Sedative, hypnotic.

3666. Eterobarb. *5-Ethyl-1,3-bis(methoxymethyl)-5-phenyl-2,4,6(1H,3H,5H)-pyrimidinetrione;* 5-ethyl-1,3-bis-(methoxymethyl)-5-phenylbarbituric acid; N,N'-dimethoxy-methylphenobarbital; eterobarbital; Ex 12-095; Antilon. $C_{16}H_{20}N_2O_5$; mol wt 320.35. C 59.99%, H 6.29%, N 8.74%, O 24.97%. Alkoxymethyl deriv of phenobarbital, q.v., reported to have little or no hypnotic activity. Prepn: C. M. Samour, J. A. Vita, Ger. pat. 1,939,987 corresp to U.S. pat. 3,595,862 (1970, 1971 both to Kendall); C. M. Samour *et al., J. Med. Chem.* **14**, 187 (1971); J. Gal, *J. Pharm. Sci.* **68**, 1562 (1979). Spectrophotographic and polarographic analysis: M. Romer *et al., Anal. Chim. Acta* **88**, 261 (1977). Pharmacologic study: J. A. Vida, M. L. Hasker, *J. Med. Chem.* **16**, 602 (1973). Metabolism: M. A. Goldberg *et al., Ann. Neurol.* **5**, 121 (1979). Efficacy in febrile convulsions: R. M. Julien, G. G. Fowler, *Neuropharmacology* **16**, 719 (1977).

Crystals from ethanol, mp 116-118°. LD_{50} in mice: 470 mg/kg orally (Vida, Hooker).

Note: This is a controlled substance (depressant) listed in the U.S. Code of Federal Regulations, Title 21 Part 1308.13 (1987).

THERAP CAT: Anticonvulsant.

3667. Etersalate. *2-(Acetyloxy)benzoic acid 2-[4-(acetyl-amino)phenoxy]ethyl ester;* 2-(4-acetamidophenoxy)ethyl 2-acetoxybenzoate; eterilate; etherylate; eterylate; Daital. $C_{19}H_{19}NO_6$; mol wt 357.37. C 63.86%, H 5.36%, N 3.92%, O 26.86%. Deriv of aspirin, q.v. Prepn: C. S. Letelier, F. C. Grafulia, Ger. pat. 2,538,206; *eidem,* U.S. pat. 4,014,921 (both 1977 to Alter); C. S. Letelier *et al., Arch. Farmacol. Toxicol.* **4**, 153 (1978). Effect on *in vitro* platelet function in human plasma: M. P. Ortega *et al., ibid.* **6**, 31 (1980). *In vivo* effects on primary immune response: I. Barasoain *et al., Immunopharmacology* **2**, 293 (1980). Comparative toxicity study and gastric tolerance: S. Alonso *et al., Arch. Farmacol. Toxicol.* **4**, 349 (1978).

Cryst from methanol, mp 139-141°. LD_{10} orally in rats: 7000 mg/kg (Leterier, Grafulia, 1977).

THERAP CAT: Analgesic; anti-inflammatory; antipyretic.

3668. Ethacridine. *7-Ethoxy-3,9-acridinediamine; 6,9-diamino-2-ethoxyacridine;* 2,5-diamino-7-ethoxyacridine; 2-ethoxy-6,9-diaminoacridine; etakridin. $C_{15}H_{15}N_3O$; mol wt 253.29. C 71.12%, H 5.97%, N 16.59%, O 6.32%. Prepn from 4'-ethoxy-5-nitrodiphenylamine-2-carboxylic acid: Ger. pats. 364,003; 364,037; 393,411; 395,683 (1922); *Frdl.* **14**, 804, 807-812; Albert, Gledhill, *J. Soc. Chem. Ind.* **61**, 159 (1942). Toxicity data: S. D. Rubbo, *Brit. J. Exp. Pathol.* **28**, 1 (1947).

Orange-yellow crystals or glistening scales from 50% alc. mp 226°. [The mp 126° found in the literature is due originally to a typographical error.]

Lactate monohydrate, $C_{18}H_{21}N_3O_4 \cdot H_2O$, acrinol, Rimaon, Rivanol, Vucine, Acrolactine, Ethodin, Metifex. Pale-yellow crystals from 90% alcohol + ether. Darkens at 200°, mp 235°. Slowly sol in 15 parts water; sol in 9 parts boiling water; in 110 parts alcohol at 22°, in 100 parts boiling alcohol. Solns are yellow, fluorescent and stable to boiling. LD_{50} s.c. in mice: 0.12 g/kg (Rubbo).

THERAP CAT: Antiseptic.

3669. Ethacrynic Acid. *[2,3-Dichloro-4-(2-methylene-1-oxobutyl)phenoxy]acetic acid; [2,3-dichloro-4-(2-methylene-butyryl)phenoxy]acetic acid;* [4-(methylenebutyryl)-2,3-dichlorophenoxy]acetic acid; MK-595; Crinuryl; Edecril; Edecrin; Endecril; Hydromedin; Reomax; Taladren; Uregit. $C_{13}H_{12}Cl_2O_4$; mol wt 303.15. C 51.51%, H 3.99%, Cl 23.39%, O 21.11%. Prepn: Schultz *et al., J. Med. Pharm. Chem.* **5**, 660 (1962); Schultz, Sprague, Belg. pat. 612,755, *C.A.* **59**, 12712b (1963) and U.S. pat. 3,255,241 (1962 and 1966, both to Merck & Co.). Pharmacology: Beyer *et al., J. Pharmacol. Exp. Ther.* **147**, 1 (1965). Crystal structure: J. Lamotte *et al., Acta Crystallogr.* **B34**, 2636 (1978). Review: Kim *et al., Am. J. Cardiol.* **27**, 407-415 (1971); H. E. Williamson, *J. Clin. Pharmacol.* **17**, 663-672 (1977).

Crystals, mp 121-122° (corr). pKa' 3.50. Sparingly sol in water, aq acids; sol in chloroform. LD$_{50}$ in mice (mg/kg): 176 i.v.; 627 orally, Peck *et al.*, *Fed. Proc.* **23**, 438 (1964).

Sodium salt, C$_{13}$H$_{11}$Cl$_2$NaO$_4$, *ethacrynate sodium, Lyovac Sodium Edecrin.* uv max (water): 225 nm (ϵ 15,287). Soly in water at 25°: up to 9%. Solns at pH 7 at 25° stable for short periods of time. Stability decreased with an increase in pH, temperature and time.

THERAP CAT: Diuretic.

THERAP CAT (VET): Diuretic.

3670. Ethadione. *3-Ethyl-5,5-dimethyl-2,4-oxazolidinedione;* 3-ethyl-5,5-dimethyl-2,4-diketooxazolidine; Dimedione; Petidiol; Didione; Petidion; Epinyl; Etydion; Petisan; Neo-Absentol. C$_7$H$_{11}$NO$_3$; mol wt 157.17. C 53.49%, H 7.05%, N 8.91%, O 30.54%. Prepn: Davis, Hook, **Brit.** pat. **626,971** (1949 to Brit. Schering Res. Labs.).

Flat, vitreous prisms from ether, mp 76-77°.

THERAP CAT: Anticonvulsant.

3671. Ethalfluralin. *N-Ethyl-N-(2-methyl-2-propenyl)-2,6-dinitro-4-(trifluoromethyl)benzenamine;* N-ethyl-N-α,α,α-trifluoro-N-(2-methylallyl)-2,6-dinitro-p-toluidine; N-ethyl-N-methallyl-4-trifluoromethyl-2,6-dinitroaniline; EL-161; Sonalan; Sonalen. C$_{13}$H$_{14}$F$_3$N$_3$O$_4$; mol wt 333.27. C 46.85%, H 4.23%, F 17.10%, N 12.61%, O 19.20%. Selective, pre-emergent herbicide. Prepn: H. D. Porter, **Ger.** pat. **2,511,897**; *idem*, **Brit.** pat. **1,505,249** (1975, 1978 both to Eli Lilly). Activity: G. Skylakakis *et al.*, *Proc. 12th Brit. Weed Contr. Conf.* **2**, 795 (1974). Review of chemistry and analytical methods: E. W. Day, "Ethalfluralin" in *Analytical Methods for Pesticides and Plant Growth Regulators* Vol. 10, G. Zweig, J. Sherma, Eds. (Academic Press, New York, 1978) pp 341-352. Persistence in soil: B. J. Hayden, A. E. Smith, *Bull. Environ. Contam. Toxicol.* **25**, 508 (1983).

Yellow crystals from petr ether, mp 54-57°. Readily sol in acetone, acetonitrile, benzene, chloroform, hexane, methanol, xylene. Soly in water (25°): 0.3 ppm. uv max (methanol): 374, 267 nm. Vapor pressure (25°): 8.2 × 10^{-5} torr. LD$_{50}$ in mice, rats (g/kg): > 10 orally; LC$_{50}$ in bluegill sunfish, rainbow trout, goldfish (ppb): 22, 37, 260 (Day).

USE: Pre-emergent herbicide.

3672. Ethambutol. *2,2'-(1,2-Ethanediyldiimino)bis-1-butanol; (+)-2,2'-(ethylenediimino)di-1-butanol; d-N,N'-*bis(1-hydroxymethylpropyl)ethylenediamine; Dadibutol; EMB; Sural; Tibutol. C$_{10}$H$_{24}$N$_2$O$_2$; mol wt 204.31. C 58.78%, H 11.84%, N 13.71%, O 15.66%. Prepn: Wilkinson *et al.*, *J. Am. Chem. Soc.* **83**, 2212 (1961); *eidem*, *J. Med. Chem.* **5**, 835 (1962). Pharmacology: Kulig *et al.*, *Diss. Pharm. Pharmacol.* **23**, 463 (1971); *C.A.* **76**, 81239d (1972). Comprehensive description: C. S. Lee, L. Z. Benet in *Analytical Profiles of Drug Substances* vol. 7, K. Florey, Ed. (Academic Press, N.Y., 1978) pp 231-249. Mechanism of

action: W. H. Beggs in *Antibiotics* **vol. 5**(pt. 1), F. E. Hahn, Ed. (Springer-Verlag, New York, 1979) pp 43-66.

mp 87.5-88.8°. [α]$_D^{25}$ +13.7° (c = 2 in water). Sol in chloroform, methylene chloride; less sol in benzene; sparingly sol in water.

Hydrochloride, *Ebutol, Etambol.*

Dihydrochloride, C$_{10}$H$_{26}$Cl$_2$N$_2$O$_2$, *Dexambutol, EMB-Fatol, Etibi, Etapiam, Myambutol, Mycobutol.* mp 198.5-200.3°; also reported as mp 201.8-202.6°. [α]$_D^{25}$ +7.6 (c = 2 in water). Sol in water, DMSO; sparingly sol in ethanol; difficultly sol in acetone, chloroform.

THERAP CAT: Antibacterial (tuberculostatic).

3673. Ethamivan. *N,N-Diethyl-4-hydroxy-3-methoxybenzamide; N,N-diethylvanillamide;* vanillic acid diethylamide; vanillic diethylamide; Vandid; Cardiovanil; Emivan. C$_{12}$H$_{17}$NO$_3$; mol wt 223.27. C 64.55%, H 7.67%, N 6.27%, O 21.50%. Prepd by dissolving vanillic acid *O*-acyl amide in NaOH and neutralizing with CO$_2$ or acid: Kratzl, Kvasnicka, *Monatsh.* **83**, 18 (1952); by refluxing a mixture of vanillic acid, P$_2$O$_5$ and powdered glass in xylene and treating with K$_2$CO$_3$: *eidem*, **U.S.** pat. **2,641,612** (1953 to Oesterr. Stickstoffwerke).

Needles from ligroin, mp 95-95.5°. LD$_{50}$ i.p. in rats: 28 mg/kg, Caujolle *et al.*, *Compt. Rend.* **243**, 609 (1956).

THERAP CAT: Central and respiratory stimulant.

3674. Ethamoxytriphetol. *α-[4-[2-(Diethylamino)ethoxy]phenyl]-4-methoxy-α-phenylbenzeneethanol;* 1-[p-(2-diethylaminoethoxy)phenyl]-2-(p-methoxyphenyl)-1-phenylethanol; etamsylate; 1-[p-(2-diethylaminoethoxy)phenyl]-1-phenyl-2-(p-anisyl)ethanol; 1-[p-(2-diethylaminoethoxy)phenyl]-1-phenyl-2-(p-methoxyphenyl)ethanol; MER-25. C$_{27}$H$_{33}$NO$_3$; mol wt 419.54. C 77.29%, H 7.93%, N 3.34%, O 11.44%. Prepn: Allen, Palopoli *et al.*, **U.S.** pat. **2,914,562** (1959 to Wm. S. Merrell).

Crystals, mp 104-106°. Soluble in alcohol, olive oil. Practically insol in water.

THERAP CAT: Estrogen antagonist.

3675. Ethamsylate. *2,5-Dihydroxybenzenesulfonic acid compd with N-ethylethanamine;* diethylammonium 2,5-dihydroxybenzenesulfonate; 1-hydroxy-4-oxo-2,5-cyclohexadiene-1-sulfonic acid compound with diethylamine; diethylammonium cyclohexadien-4-ol-1-one-4-sulfonate; cyclonamine; etamsylate; MD 141; E 141; Aglumin; Altodor; Biosinon; Dicynene; Dicynone. C$_{10}$H$_{17}$NO$_5$S; mol wt 263.33. C 45.61%, H 6.51%, N 5.32%, O 30.38%, S 12.18%. Prepn: **Brit.** pat. **895,709** (1962 to Labs. OM); **Span.** pat. **279,303** (1962 to Labs. Esteve), *C.A.* **60**, 1652g (1964). Pharmacology: Demars, *Med. Exp.* **4**, 173 (1961); Huguet *et al.*, *Therapie* **24**, 429 (1969). Toxicity: Esteve *et al.*, *Therapie* **15**, 110 (1960).

Crystals from ethanol, mp 125°. LD_{50} i.v. in mice, rats: 800, 1350 mg/kg (Esteve).

THERAP CAT: Hemostatic.

3676. Ethane. Bimethyl; dimethyl; methylmethane; ethyl hydride. C_2H_6; mol wt 30.07. C 79.89%, H 20.11%. CH_3-CH_3. Constituent of natural gas (about 9%). Can be recovered from the gases produced during the distillation of crude petroleum. *Review:* Smith, Hanson, *Oil and Gas J.* **44**, no. 10, 119 (1945).

Colorless, odorless gas. *Flammable asphyxiant.* Burns with a faintly luminous flame. d_4^0 1.0493 (air = 1) or 1.3562 g/l. d_4^0 liq 0.446. mp −172°. bp −88°. Crit temp 32°; crit press. 48.2 atm. Heat of combustion: 1727 Btu/cu ft at 25° (one pound of CH_3CH_3 yields 20,420 Btu [net] at 15.56°). Flammable limits in air: 3.2 to 12.5% by vol. Ignition temp (in air): 530°. Soly in water at 20°: 4.7 ml/100 ml H_2O; in alc at 4°: 46 ml/100 ml alc.

USE: In the manuf of chlorinated derivs; as refrigerant in some two-stage refrigeration systems where relatively low temps are produced; as fuel gas (so called "bottled gas" or "suburban propane" contains about 90% propane, 5% ethane, and 5% butane). *Caution:* Narcotic in high concns. A simple asphyxiant.

3677. Ethanearsonic Acid. *Ethylarsonic acid;* ethylarsinic acid. $C_2H_7AsO_3$; mol wt 153.39. C 15.60%, H 4.58%, As 48.65%, O 31.17%. $C_2H_5AsO(OH)_2$. Prepd from potassium arsenite and ethyl iodide: Dehn, *Am. Chem. J.* **33**, 131 (1905); from sodium arsenite and ethyl bromide: Quick, Adams, *J. Am. Chem. Soc.* **44**, 810 (1922).

Needles from alc, mp 99.5°. 70 g dissolve in 100 g water at 27°; 112 g dissolve in 100 g water at 40°; 39.4 g dissolve in 100 g of 95% alcohol at 25°.

Disodium salt, $C_2H_5AsNa_2O_3$, *Mon-Arsone.* Minute hygroscopic crystals, very sol in water.

THERAP CAT: Formerly as antisyphilitic.

3678. 1,2-Ethanedisulfonic Acid. 1,2-Ethylenedisulfonic acid. $C_2H_6O_6S_2$; mol wt 190.20. C 12.63%, H 3.18%, O 50.47%, S 33.72%. $HO_3SCH_2CH_2SO_3H$. Preparation of disodium salt from 1,2-dibromoethane and sodium sulfite: McElvain *et al., J. Am. Chem. Soc.* **67**, 1578 (1945); Ohba *et al.,* U.S. pat. 3,022,172 (1962 to Fuji Photo Film). Free acid obtained by conversion of disodium salt to barium salt followed by liberation with H_2SO_4: McElvain *et al., loc. cit.* Review of prepns: Suter, *Organic Chemistry of Sulfur* (Wiley, New York, 1944) p 166; Gilbert, *Sulfonation and Related Reactions* (Interscience, New York, 1965).

Dihydrate, crystals from acetic acid + acetic anhydride, mp 111-112°. Heating at 145° and 1 mm for 6 hours gives anhydr compd, mp 172-174°. Somewhat soluble in anhydr ether, very sol in dioxane.

Sodium salt dihydrate, $Na_2C_2H_4O_6S_2.2H_2O$, monoclinic prisms. Not very readily sol in water; practically insol in alcohol.

USE: Base in hypnotics, sedatives.

3679. 1,2-Ethanedithiol. Dithioethyleneglycol; ethylenedimercaptan. $C_2H_6S_2$; mol wt 94.20. C 25.50%, H 6.42%, S 68.08%. $HSCH_2CH_2SH$. Prepd by reacting ethanol, thiourea and ethylene dibromide and subsequent alkaline hydrolysis of the ethylenediisothiuronium bromide: Speciale, *Org. Syn.* **30**, 35 (1950), modified procedure: Grogan *et al., J. Org. Chem.* **20**, 50 (1955).

Liquid. $d^{23.5}$ 1.123. bp_{760} 146°; bp_{150} 76-81°; bp_{46} 63°; bp_{24} 51-52°. n_D^{20} 1.5589. Freely sol in alcohol and in alkalies. *Caution:* Vapors of ethanedithiol may cause severe headache and nausea.

3680. Ethanethiol. Ethyl mercaptan; mercaptoethane; ethyl sulfhydrate; thioethyl alcohol. C_2H_6S; mol wt 62.13. C 38.66%, H 9.74%, S 51.60%. CH_3CH_2SH. Found in urine of rabbits after ingestion of cabbage. Is formed in vinous fermentation. Occurs in illuminating gas, in "sour" natural gas of W. Texas; in petroleum distillates from which it may be separated by chemical or physical methods: Thompson *et al., Anal. Chem.* **27**, 175 (1955). Prepn from sodium ethylsulfate and KSH: Klason, *Ber.* **20**, 3407 (1887); catalytically from ethanol and hydrogen sulfide: Kramer, Reid, *J. Am. Chem. Soc.* **43**, 880 (1921). Review on occurrence, prepn, properties and reactions: E. Emmet Reid, *Organic Chemis-*

try of Bivalent Sulfur **vol. I**, 15-261 (Chemical Publishing Co., New York, 1958).

Liquid; penetrating leek-like odor: minimum detectable concn = 1 part in 50 billion parts of air. d_4^0 0.8617; d_4^{25} 0.83147. bp 34.7-35.04°; bp_{400} 17.7°; bp_{200} 1.5°; bp_{100} −13.0°; bp_{10} −50.2°; bp_1 −76.1°. mp −147.97 to −144.4°. n_D^{20} 1.431; n_D^{25} 1.420. Crit temp 225.5°. Crit pressure 54.2 atm. Explosive limits 2.8 and 18.2% by vol of vapor. Min ignition temp in air 299°, in oxygen 261°. Entropy at 25°: 70.77 cal/deg/mole. Heat capacity at 25°: 17.37 cal/deg/mole. Surface tension at 2°: 23.63 dynes/cm; at 16.7°: 21.62 dynes/cm. Viscosity at 20°: 0.003155 g/cm sec. Azeotrope with *n*-pentane (51% ethanethiol) bp 30.46°; with ether (40% ethanethiol) bp 31.50°. Soly in water at 20°: 6.76 g/l or 0.112 moles/l. Sol in alcohol, ether. *Keep tightly closed and in a cool place.*

Octadecahydrate, needles. Practically insol in water, ethanethiol.

Sodium salt, CH_3CH_2SNa, hydrolyzes instantly in water.

USE: Odorant for natural gas; intermediate and starting material in manuf of plastics, insecticides, antioxidants. *Caution:* May be narcotic in high concns. Irritating to mucous membranes.

3681. Ethanolamine. *2-Aminoethanol;* monoethanolamine; *β*-aminoethyl alcohol; 2-hydroxyethylamine; *β*-hydroxyethylamine; ethylolamine; colamine. C_2H_7NO; mol wt 61.08. C 39.33%, H 11.55%, N 22.93%, O 26.19%. $HOCH_2$-CH_2NH_2. Prepd on a large scale by ammonolysis of ethylene oxide: Knorr, *Ber.* **30**, 909 (1897); Fr. pat. 650,574 (1928 to I. G. Farben); Reid, Lewis, U.S. pat. 1,904,013 (1933 to Carbide & Carbon); Schwoegler, Olin, U.S. pat. 2,373,199 (1945 to Sharples). Also from nitromethane and formaldehyde: *Ullmann's Encyklopädie der technischen Chemie* **3**, 102 (3rd ed., 1953). Manuf: Faith, Keyes & Clark's *Industrial Chemicals,* F. A. Lowenheim, M. K. Moran, Eds. (Wiley-Interscience, New York, 4th ed., 1975) pp 339-344. Toxicity: H. F. Smyth *et al., J. Ind. Hyg. Toxicol.* **23**, 259 (1941).

Viscous, hygroscopic liq. Ammoniacal odor. Absorbs CO_2. d_4^{25} 1.0117; d_4^{40} 0.9998; d_4^{60} 0.9844. One gallon weighs 8.45 lbs in the U.S.A. Viscosity at 25°: 18.95 cps; at 60°: 5.03 cps. mp 10.3°. bp_{760} 170.8°; bp_{12} 70-72°. Strong base. pKa at 25°: 9.4. pH of 25% aq soln: 12.1; of 0.1N aq soln: 12.05. n_D^{20} 1.4539. Dipole moment 2.27. Flash pt 195°F. Misc with water, methanol, acetone. Soly at 25°: benzene 1.4%; ether 2.1%; carbon tetrachloride 0.2%; *n*-heptane <0.1%. LD_{50} orally in rats: 10.20 g/kg (Smyth).

Hydrochloride, C_2H_8ClNO, deliquesc crystals from alc, mp 75-77°.

Oleate, $C_{20}H_{41}NO_3$, *Antivariz, Esclerosina, Ethamolin.* Use as a sclerosing agent: S. E. Hedberg *et al., Am. J. Surgery* **143**, 426 (1982).

USE: To remove CO_2 and H_2S from natural gas and other gases; in the synthesis of surface active agents; in polishes, hair waving solns, in emulsifiers; as softening agent for hides; dispersing agent for agricultural chemicals. Is reacted with other substances to form an accelerator in the manuf of antibiotics. Pharmaceutic aid (surfactant).

THERAP CAT: Oleate as sclerosing agent.

3682. Ethaverine. *1-[(3,4-Diethoxyphenyl)methyl]-6,7-diethoxyisoquinoline;* 6,7-diethoxy-1-(3,4-diethoxybenzyl)-isoquinoline; ethylpapaverine; Dyscural. $C_{24}H_{29}NO_4$; mol wt 395.48. C 72.88%, H 7.39%, N 3.54%, O 16.18%. Tetraethyl homolog of papaverine. Synthesis: Wolf, U.S. pat. 1,962,-224 (1934).

     Consult the cross index before using this section.

Crystals from alcohol + ether, mp 99-101°. Insol in water; readily sol in hot alc; slightly in ether and chloroform.

Hydrochloride, $C_{24}H_{30}ClNO_4$, **Barbonin, Circubid, Diquinol, Ethabid, Isovex, Laverin, Papetherine, Perparin, Perperine**. Crystals from abs alc, mp 186-188° (dec). One gram dissolves in about 40 ml water. pH of 1% soln 3.6; of 0.1% soln 4.6. Prepn of stable solns: Hereld, U.S. pat. **2,971,888** (1961).

Acid oxalate, $C_{24}H_{29}NO_4$.$(COOH)_2$, crystals from alc, soluble in hot water.

Salt with 7-iodo-8-hydroxyquinoline-5-sulfonic acid, **Hung**. pat. **106,906** (1931); v. Issekutz *et al., Arch. Exp. Pathol. Pharmakol.* **164**, 158 (1932).

THERAP CAT: Antispasmodic.

3683. Ethchlorvynol. *1-Chloro-3-ethyl-1-penten-4-yl-3-ol;* 5-chloro-3-ethylpent-1-yn-4-en-3-ol; ethyl β-chlorovinyl ethynyl carbinol; ethclorvynol; Placidyl; Arvynol; Serenesil; Roeridorm; Normoson. C_7H_9ClO; mol wt 144.61. C 58.14%, H 6.27%, Cl 24.52%, O 11.07%. Prepn: W. M. McLamore *et al., J. Org. Chem.* **20**, 109 (1955); Bavley, McLamore, U.S. pat. **2,746,900** (1956 to Pfizer). Pharmacological studies: S. Y. P'an *et al., J. Pharmacol. Exp. Ther.* **114**, 326 (1955). Pharmacodynamics: Y. C. Martin, *Biochem. Pharmacol.* **16**, 2041 (1967). Evaluation as hypnotic: J. W. Middleton *et al., Curr. Ther. Res.* **8**, 391 (1966); J. H. Pattison *et al., J. Am. Geriatric Soc.* **20**, 398 (1972). Study of metabolites: J. P. Horwitz *et al., Drug Metab. Dispos.* **8**, 77 (1980).

$$CH_3CH_2-\underset{\underset{OH}{|}}{\overset{\overset{C\equiv CH}{|}}{C}}-CH=CHCl$$

Liquid, pungent aromatic odor. Slowly darkens on exposure to light and air. d_4^{25} 1.065-1.070 (also reported as d_4^{20} 1.070). bp_{760} 173-174° (also reported as bp_{760} 181°); $bp_{0.1}$ 28.5-30°. n_D^{20} 1.4675-1.4800. Immiscible with water; miscible with most organic solvents. LD_{50} in mice (mg/kg): 290 orally; 240 s.c. (P'an).

Caution: May be habit forming. This is a controlled substance (depressant) listed in the U.S. Code of Federal Regulations, Title 21 Part 1308.14 (1987).

THERAP CAT: Sedative, hypnotic.

3684. Ethebenecid. *4-[(Diethylamino)sulfonyl]benzoic acid;* p-(diethylsulfamoyl)benzoic acid; p-carboxy-N,N-diethylbenzenesulfonamide; p-carboxybenzenesulfo-di-N-ethylamide; Antidipsin; Longacid; Urelim. $C_{11}H_{15}NO_4S$; mol wt 257.32. C 51.34%, H 5.88%, N 5.44%, O 24.87%, S 12.46%. Prepn: Gilman, Arntzen, *J. Am. Chem. Soc.* **69**, 1537 (1947); Goldberg, Wragg, *J. Chem. Soc.* **1960**, 1408.

$$(C_2H_5)_2NSO_2-\text{⟨benzene ring⟩}-COOH$$

Crystals, mp 192-194° (with turbidity).

THERAP CAT: Uricosuric agent; inhibits penicillin excretion.

3685. Ethenzamide. *2-Ethoxybenzamide;* ethbenzamide; 2-ethoxybenzenecarbonamide; salicylamide o-ethyl ether; Lucamide; Protopyrin; Trancalgyl. $C_9H_{11}NO_2$; mol wt 165.19. C 65.44%, H 6.71%, N 8.48%, O 19.37%. Prepn: **Brit**. pat. **656,746** (1951 to H. Lundbeck); Shapiro *et al., J. Am. Chem. Soc.* **81**, 3728 (1959); Bryk, Osowski, **Pol**. pat. **47,822** (1961 to Polfa), *C.A.* **61**, 5573f (1964). Metabolism: Davison *et al., J. Pharmacol. Exp. Ther.* **136**, 226 (1962). Toxicity data: Starmer *et al., Toxicol. Appl. Pharmacol.* **19**, 20 (1971).

$$\text{⟨benzene ring with } CONH_2 \text{ and } OCH_2CH_3 \rangle$$

Crystals from ethyl acetate + hexane, mp 132-134°. LD_{50} orally in mice: 1160 mg/kg (Starmer).

THERAP CAT: Analgesic.

3686. Ethephon. *(2-Chloroethyl)phosphonic acid;* 2-chloroethanephosphonic acid; CEPA; camposan; CEPHA; Cerone; Florel; Ethrel. $C_2H_6ClO_3P$; mol wt 144.49. C 16.62%, H 4.19%, Cl 24.53%, O 33.22%, P 21.43%. $ClCH_2$-$CH_2PO(OH)_2$. Ethylene-releasing agent used to control growth and ripening of fruit crops. Prepn: Kabachnik, Rossiiskaya, *Izvest. Akad. Nauk S.S.S.R., Otdel Khim. Nauk* **1946**, 403-410, *C.A.* **42**, 7242e (1948); *Beilstein* EIII 4, 1780. Toxicity data: G. Hennighausen *et al., Pharmazie* **32**, 181 (1977). Review of use in viticulture: E. Szyjewicz *et al., Am. J. Enol. Viticult.* **35**, 117-123 (1984).

Very hygroscopic needles from benzene (must be dried over P_2O_5). mp 74-75°. Freely sol in water, methanol, acetone, ethylene glycol, propylene glycol. Slightly sol in benzene, toluene. Practically insol in petr ether. Aq solns are stable below pH 3.5. Above pH 3.5 hydrolysis begins with the release of free ethylene. LD_{50} orally in mice: 2850 mg/kg (Hennighausen).

USE: Plant growth regulator. *Caution:* Spray formulations are quite acidic, about pH 1.0. May be irritating to exposed skin and eyes, or if inhaled.

3687. Ethiazide. *6-Chloro-3-ethyl-3,4-dihydro-2H-1,2,-4-benzothiadiazine-7-sulfonamide 1,1-dioxide;* 6-chloro-3-ethyl-3,4-dihydro-7-sulfamoylbenzo-1,2,4-thiadiazine 1,1-dioxide; 3-ethyl-3,4-dihydro-6-chloro-7-sulfamyl-1,2,4-benzothiadiazine 1,1-dioxide; acthiazidum. $C_9H_{12}ClN_3O_4S_2$; mol wt 325.81. C 33.18%, H 3.71%, Cl 10.88%, N 12.90%, O 19.64%, S 19.68%. Prepn: Close *et al., J. Am. Chem. Soc.* **82**, 1132 (1960); Topliss *et al., J. Org. Chem.* **26**, 3842 (1961); **Brit**. pat. **861,367** (1961 to Ciba).

$$\text{⟨benzothiadiazine structure with } H_2NSO_2, Cl, \text{ and } C_2H_5 \rangle$$

Crystals from tetrahydrofuran + chloroform, mp 269-270°.

Component of *Hypertane*.

THERAP CAT: Diuretic.

3688. Ethinamate. *1-Ethynylcyclohexanol carbamate;* carbamic acid 1-ethynylcyclohexyl ester; ethynylcyclohexyl carbamate; Valamin; Valmid; Valmidate. $C_9H_{13}NO_2$; mol wt 167.20 C 64.65%, H 7.84%, N 8.38%, O 19.14%. Prepn: Junkmann, Pfeiffer, U.S. pat. **2,816,910** (1957 to Schering, AG).

$$\text{⟨cyclohexane ring with } HC\equiv C \text{ and } OCNH_2 \text{ (C=O)} \rangle$$

Needles from cyclohexane, mp 96-98°. bp_3 118-122°. Shows no optical activity. $n_D^{21.5}$ 1.4441 (20% in propylene glycol). Soly at 25° in water 0.25%, ethanol 35.0%, hexane 2.0%, sesame oil 0.7%, 1,2-propylene glycol 22%. Stable in $N\ H_2SO_4$ at 30°, but dec at 40° and in 0.02N acid at 60°. In 0.01N NaOH, CO_2 is split off at 30°. Lowers the surface tension in satd aq soln at 22° by 21%. The partition coefficient in oil/water is 1.48 (that of ethyl carbamate is 0.14).

Note: The trademark Valamin was used formerly to indicate a brand of *tert*-amyl isovalerate.

Caution: May be habit forming. This is a controlled substance (depressant) listed in the U.S. Code of Federal Regulations, Title 21 Part 1308.14 (1987).

THERAP CAT: Sedative, hypnotic.

3689. Ethinyl Estradiol. *(17α)-19-Norpregna-1,3,5(10)-trien-20-yne-3,17-diol;* 17α-ethynyl-1,3,5(10)-estratriene-3,17β-diol; 17-ethinylestradiol; ethynylestradiol; Diogyn E;

Dufaston; Dyloform; Esteed; Estigyn; Estinyl; Eston-E; Ethidol; Ethinyl-Oestradiol; Ethy 11; Eticyclol; Eticyclin; Etinestrol; Etinestryl; Etinoestryl; Etivex; Feminone; Ginestrene; Gynolett; Inestra; Kolpolyn; Linoral; Lynoral; Menolyn; Neo-Estrone; Norma-oestren; Novestrol; Oradiol; Orestralyn; Palonyl; Perovex; Primogyn; Primogyn C; Primogyn M; Progylut; Progynon C; Progynon M; Roldiol; Spanestrin; Ylestrol. $C_{20}H_{24}O_2$; mol wt 296.39. C 81.04%, H 8.16%, O 10.80%. Synthetic steroid with high oral estrogenic potency: Inhoffen, Hohlweg, *Naturwiss.* **26**, 96 (1938). Prepn from estrone: Inhoffen *et al.*, *Ber.* **71**, 1024 (1938). *See also* **Ger.** pat. **702,063**; **Brit.** pat. **516,444**; **U.S.** pats. **2,243,887**; **2,251,939**; **2,265,976**; **2,267,257**. Properties: Petit, Muller, *Bull. Soc. Chim. France* **1951**, 121; L. Ehmann, A. Wettstein, *Pharm. Acta Helv.* **25**, 297 (1950). NMR: Hampel, Kraemer, *Ber.* **98**, 3255 (1965). Toxicity: E. I. Goldenthal, *Toxicol. Appl. Pharmacol.* **18**, 185 (1971). Randomized double-blind clinical studies: S. Koetsawang *et al.*, *Contraception* **25**, 231 (1982); A. Sheth *et al.*, *ibid.* 243. Clinical evaluation in gonadal dysgenesis: L. Cuttler *et al.*, *J. Clin. Endocrinol. Metab.* **60**, 1087 (1985). Review of human metabolism and pharmacokinetics: K. Fotherby, *Methods Find. Exp. Clin. Pharmacol.* **4**, 133-141 (1982). Review of carcinogenicity studies: *IARC Monographs* **21**, 233-255 (1979); *ibid.* Suppl. 4, 186-188 (1982). General review: K. W. Thompson, *J. Clin. Pharmacol.* **8**, 1088-1098 (1948).

Hemihydrate, fine needles from methanol + water, mp 141-146°, $[\alpha]_D^{25}$ 0° ± 1° (dioxane). Dehydrates after melting and further heating, mp 182-184°. $[\alpha]_D^{24}$ 3.5° ± 0.5° (c = 2 in dioxane); −29.5° ± 1° (c = 2 in pyridine). uv max (ethanol): 281 nm (ε 2040 ± 60). Practically insol in water. Soly: 1 part in 6 of ethanol, 1 in 4 of ether, 1 in 5 of acetone, 1 in 4 of dioxane, and 1 in 20 of chloroform. Sol in vegetable oils, and in solns of fixed alkali hydroxides. LD_{50} in rats, mice (mg/kg): 2952, 1737 orally (Goldenthal).

3-Acetate, $C_{22}H_{26}O_3$, crystals, mp 152-153°. $[\alpha]_D^{20}$ +3° (chloroform).

3-Benzoate, $C_{27}H_{28}O_3$, needles from methanol, mp 200-202°.

Note: Also used in combination with chlormadinone acetate, desogestrel, ethynodiol, gestodene, lynestrenol, norethindrone or norgestrel, *q.q.v.* Has been used in combination with dimethisterone, medroxyprogesterone, or megestrol acetate, *q.q.v.* This substance may reasonably be anticipated to be a carcinogen: *Fourth Annual Report on Carcinogens* (NTP 85-002, 1985) p 103.

THERAP CAT: Estrogen. In combination with progestogen as oral contraceptive.

THERAP CAT (VET): Estrogen.

3690. Ethiodized Oil. Ethiodol; ethyl ester of iodinated fatty acid of poppyseed oil. Contains 37% organically bound iodine. Ref: Hom *et al.*, *J. Am. Pharm. Assoc. Sci. Ed.* **46**, 254 (1957). Use in lymphography: R. M. Paxton *et al.*, *Brit. Med. J.* **1**, 120 (1975).

Amber-colored oil, extremely fluid, sp gr 1.28. Soluble in acetone, chloroform, ether, hexane. Insol in water.

THERAP CAT: Diagnostic aid (radiopaque medium). Antineoplastic when part of the iodine is ^{131}I.

3691. Ethion. *Phosphorodithioic acid S,S'-methylene O,O,O',O'-tetraethyl ester; ethyl methylene phosphorodithioate; O,O,O',O'-tetraethyl S,S'-methylenediphosphorodithioate;* bis[S-(diethoxyphosphinothioyl)mercapto]methane; diethion; ENT 24105; FMC 1240; Niagara 1240; Nialate. $C_9H_{22}O_4P_2S_4$; mol wt 384.48. C 28.11%, H 5.77%, O 16.65%, P 16.11%, S 33.36%. Prepn: Willard, Henahan, **U.S.** pats. **2,873,228**; **3,014,058** (1959, 1961 to FMC).

Liquid, mp −12 to −13°. d_4^{20} 1.220. n_D^{20} 1.5490. Vapor press.: 1.5 × 10⁻⁶ mm Hg. Sol in xylene, methylated naphthalene, chloroform, acetone and kerosene + 1% methyl ethyl ketone or benzene; slightly sol in water. LD_{50} in female, male rats: 27, 65 mg/kg orally; 62, 245 mg/kg dermally, T. B. Gaines, *Toxicol. Appl. Pharmacol.* **14**, 515 (1969).

USE: Insecticide; acaricide. *Caution:* Cholinesterase inhibitor.

3692. Ethionamide. *2-Ethyl-4-pyridinecarbothioamide; 2-ethylthioisonicotinamide;* 3-ethylisothionicotinamide; 2-ethylisothionicotinamide; 2-ethyl-4-thiocarbamoylpyridine; α-ethylisonicotinoylthioamide; amidazine; ethioniamide; Bayer 5312; 1314 Th; Nisotin; Trecator; Trescatyl; Aetina; Ethimide; Iridocin; Tio-Mid. $C_8H_{10}N_2S$; mol wt 166.24. C 57.80%, H 6.06%, N 16.85%, S 19.29%. Obtained from the corresp nitrile by the action of H_2S in the presence of triethanolamine: Libermann *et al.*, *Compt. Rend.* **242**, 2409 (1956); *Bull. Soc. Chim. France* **1958**, 687; **Brit.** pat. **800,250** (1958 to Chimie et Atomistique).

Minute yellow crystals from ethanol, dec 164-166°. Very sparingly sol in water, ether. Sparingly sol in methanol, ethanol, propylene glycol. Sol in hot acetone, dichloroethane. Freely sol in pyridine.

THERAP CAT: Antibacterial (tuberculostatic).

3693. Ethionine. *S-Ethyl-L-homocysteine; 2-amino-4-(ethylthio)butyric acid;* α-amino-γ-(ethylmercapto)butyric acid; homocysteine S-ethyl ether. $C_6H_{13}NO_2S$; mol wt 163.23. C 44.15%, H 8.03%, N 8.58%, O 19.60%, S 19.64%. $CH_3CH_2SCH_2CH_2CH(NH_2)COOH$. Prepn from ethanethiol and acrolein, purification via Zn salt: Norton, **U.S.** pat. **2,840,587** (1958 to Dow). Prepn of DL-form and L-form: Armstrong, Lewis, *J. Org. Chem.* **16**, 749 (1951). Resolution of D- and L-isomers: Greenstein *et al.*, *J. Biol. Chem.* **204**, 307 (1953). *Review:* E. Farber, "Ethionine Carcinogenesis" in A. Haddow, S. Weinhouse, *Advan. Cancer Res.* vol. 7 (Academic Press, New York, 1963) pp 383-474.

DL-Form, crystals, dec 257-260°. Also reported as dec 272-284°, Armstrong, Lewis, *loc. cit.*

L-Form, crystals, dec 272-274°. $[\alpha]_D^{24}$ +25.1°. Also reported as $[\alpha]_D^{23}$ +20.1° (1N HCl), Armstrong, Lewis, *loc. cit.*

3694. Ethiozin. *4-Amino-6-(1,1-dimethylethyl)-3-(ethylthio)-1,2,4-triazin-5(4H)-one;* 3-ethylthio-4-amino-6-*tert*-butyl-1,2,4-triazine-5-one; ethyl metribuzin; SMY 1500; BAY SMY 1500; Tycor. $C_9H_{16}N_4OS$; mol wt 228.31. C 47.35%, H 7.06%, N 24.54%, O 7.01%, S 14.04%. Selective pre- and post-emergence herbicide; ethylthio analog of metribuzin, *q.v.* Prepn: **Belg.** pat. **697,083**; K. Westphal *et al.*, **U.S.** pat. **3,671,523** (1967, 1972 both to Bayer AG). Improved process: T. Schmidt, **Ger.** pat. **3,339,858**; *idem*, **U.S.** pat. **4,546,178** (both 1985 to Bayer AG). Physicochemical properties, toxicity and herbicidal activity: H. Hack *et al.*, *Proc. Brit. Crop Prot. Conf. - Weeds* **1985**, 35. Field trials in winter wheat: B. Bolton *et al.*, *ibid.* 915; R. Ratliff, T. F. Peeper, *Weed Technol.* **1**, 235 (1987).

Crystalline solid, mp 95°. Soly at 20°: 420 mg/l in water; 2500 mg/l in hexane; > 50% in dichloromethane; > 25% in

toluol; >15% in 2-propanol. Vapor pressure at 20°: 1 ×
10⁻³ Pa. LD_{50} in male, female rats (mg/kg): 2470, 1280
orally (Hack).
USE: Herbicide.

3695. Ethirimol. *5-Butyl-2-(ethylamino)-6-methyl-
4(1H)-pyrimidinone;* 2-ethylamino-4-methyl-5-*n*-butyl-6-
hydroxypyrimidine; PP 149; Milstem; Milgo; Milcurb Super.
$C_{11}H_{19}N_3O$; mol wt 209.29. C 63.13%, H 9.15%, N 20.08%,
O 7.64%. Prepn: B. K. Snell *et al., S. Afr.* pat. **67 01,373**
corresp to U.S. pat. **3,980,781** (1968, 1976 to ICI). Systemic
fungicidal activity: R. M. Bebbington, *Chem. & Ind. (London)* **1969**, 1512; K. J. Bent, *Ann. Appl. Biol.* **66**, 103 (1970).
Metabolism and mode of action: A. Calderbank, *Acta Phytopathol.* **6**, 355 (1971).

Cryst solid, mp 159-160°. Vapor press at 25°: 2 × 10⁻⁶
mm Hg. Soly in water at 25°: 200 mg/l. Sol in chloroform,
trichloroethylene, aq solns of strong acids and bases. Slightly sol in ethanol; sparingly sol in acetone. LD_{50} orally in
rats: 4000 mg/kg, R. M. Bebbington *et al., loc. cit.*
USE: Fungicide.

3696. Ethisterone. *17α-Hydroxypregn-4-en-20-yn-3-
one;* 17α-ethynyltestosterone; 17α-ethynyl-17β-hydroxy-4-
androsten-3-one; 17α-ethinyltestosterone; 17α-ethynyl-4-
androsten-17β-ol-3-one; anhydrohydroxyprogesterone;
pregneninolone; Progestoral; Pranone; Gestoral; Lutidon
Oral; Lucorteum Oral; Primolut C; Syngestrotabs; Trosinone; Proluton C; Lutocylol; Lutogyl (tabl); Prolidon (tabl);
Lutoral; Lutocyclin; Ora-Lutin. $C_{21}H_{28}O_2$; mol wt 312.44.
C 80.73%, H 9.03%, O 10.24%. Prepd by addition of acetylene to dehydroepiandrosterone followed by an Oppenauer
oxidation: Inhoffen *et al., Ber.* **71**, 1024 (1938).

Crystals from ethyl acetate, mp 269-275°. Sublimes in
high vacuum at 190-195°. $[\alpha]_D^{23}$ +23.8° (dioxane); $[\alpha]_D^{25}$
−32.0° (pyridine). uv max (methanol): 241 nm ($E_{1cm}^{1\%}$ 513).
Practically insol in water. Slightly sol in alcohol, acetone,
ether, chloroform, vegetable oils.
THERAP CAT: Progestogen.

3697. Ethofumesate. *2-Ethoxy-2,3-dihydro-3,3-dimethyl-5-benzofuranol methanesulfonate;* NC 8438; Nortron;
Tramat. $C_{13}H_{18}O_5S$; mol wt 286.34. C 54.53%, H 6.33%, O
27.94%, S 11.20%. Prepn: Gates *et al.,* U.S. pat. **3,689,507**
(1972 to Fisons).

White crystalline solid, mp 70-72°. Vapor press at 25°:
6.45 × 10⁻⁷ mm Hg. Soly in water at 25°: 110 mg/l. Soly
(g/kg): ethanol 100; acetone 400; hexane 4. Stable to hydrolysis in water at pH 7. LD_{50} orally in rats: 6400 mg/kg,
RTECS Vol. II, R. J. Lewis, R. L. Tatken, Eds. (1979) p 50.
USE: Herbicide.

3698. Ethoheptazine. *Hexahydro-1-methyl-4-phenyl-
1H-azepine-4-carboxylic acid ethyl ester;* 4-carbethoxy-1-
methyl-4-phenylhexamethylenimine; 4-carbethoxy-1-methyl-4-phenylazacycloheptane; ethyl heptazine; Wy-401; Zac-

tane. $C_{16}H_{23}NO_2$; mol wt 261.35. C 73.53%, H 8.87%, N
5.36%, O 12.24%. Prepd in several steps from 2-phenyl-4-
dimethylaminobutyronitrile as a starting material, the final
step being the hydrolysis and esterification of 1-methyl-4-
phenyl-4-cyanoazacycloheptane: Diamond *et al., J. Org.
Chem.* **22**, 399 (1957); Diamond, Bruce, U.S. pat. **2,666,050**
(1954 to Am. Home Prod.).

Liquid. d_D^{26} 1.038. bp_1 133-134°; $bp_{0.5}$ 127-129°; $bp_{0.3}$
128-130°. n_D^{26} 1.5210; n_D^{28} 1.5220.
Hydrochloride, $C_{16}H_{23}NO_2$.HCl, recrystallized from ethyl
acetate + ether, mp 151-153°.
Citrate, constituent of *Zactirin.*
Methobromide, $C_{16}H_{23}NO_2$.CH_3Br, crystals from ether,
mp 215-217°.
THERAP CAT: Analgesic.

3699. Ethohexadiol. 2-Ethyl-1,3-hexanediol; 2-ethyl-3-
propyl-1,3-propanediol; octylene glycol; Rutgers 612; 6-12.
$C_8H_{18}O_2$; mol wt 146.22. C 65.71%, H 12.41%, O 21.88%.
Prepn: Kulpinsky, Nord, *J. Org. Chem.* **8**, 256 (1943); B. G.
Wilkes, U.S. pat. **2,407,205** (1946 to Carbide & Carbon
Chemicals). Activity: P. Granett, H. L. Haynes, *J. Econ.
Entomol.* **38**, 671 (1945); W. V. King, *ibid.* **44**, 339 (1951); B.
V. Travis, C. N. Smith, *ibid.* 428.

Slightly oily liq. d_{20}^{20} 0.9422. d_4^{22} 0.9325. bp_{760} 244.2°; bp_{50}
163°; bp_{10} 129°; bp_3 102°; $bp_{0.5}$ 94-96°. n_D^{22} 1.4530. Vapor
pressure at 20° <0.01 mm. Abs viscosity: 271 cps. Soly in
water 0.6% w/w; soly of water in 2-ethyl-1,3-hexanediol
10.8% w/w; sol in ethanol, isopropanol, propylene glycol,
castor oil. LD_{50} orally in rats, mice: 1400, 1900 mg/kg,
RTECS Vol. I, R. J. Lewis, R. L. Tatken, Eds. (1979) p 742.
USE: Insect repellent. *Caution:* Moderately irritating to
eyes, mucous membranes, but not to skin. Ingestion causes
CNS depression.

3700. Ethomids®. A series of non-ionic surface active
agents. The Ethomids are *N*-substituted fatty acid amides,
the substituents being polyoxyethylene groups. In the Ethomid C series the fatty acids are from coconut oil, in the
Ethomid RT series from hydrogenated tallow, in the Ethomid RO series from red oil.

Example: Ethomid RO/60: d_{25}^{25} 1.143 (9.52 lbs/gal); surface tension of 0.1%, aq soln 47 dynes/cm, of 1.0% aq soln
46 dynes/cm. Sol in water (1% soln still clear at bp), benzene, ligroin, CCl_4, dioxane, acetone, isopropanol.
USE: Detergents, manuf feed supplements.

3701. Ethopabate. *4-Acetamido-2-ethoxybenzoic acid
methyl ester;* methyl 4-acetamido-2-ethoxybenzoate; 2-ethoxy-4-acetamidobenzoic acid methyl ester; ethyl pabate;
$C_{12}H_{15}NO_4$; mol wt 237.25. C 60.75%, H 6.37%, N 5.90%, O
26.98%. Prepn: Grimme, Schmitz, *Ber.* **87**, 179 (1954);
Rogers, Clark, *Belg.* pat. **613,166** and U.S. pat. **3,211,610**
(1962 and 1965 to Merck & Co.); Fr. pat. **1,407,055** corresp
to U.S. pat. **3,357,978**, and Thominet, *Brit.* pat. **1,019,781**
(1965, 1967 and 1966 to Soc. d'Etudes Sci. et Ind. de l'Ile
de France). Anticoccidial activity studies: Rogers *et al.,
Proc. Soc. Exp. Biol. Med.* **117**, 488 (1964). Metabolism in

chickens: Buhs *et al., J. Pharmacol. Exp. Ther.* **154**, 357 (1966).

White to pinkish-white, practically odorless crystals from methanol and water, mp 148-149°. uv max (methanol): 298, 267 nm ($A^{1\%}_{1cm}$ 805, 365). Sol in methanol, ethanol, acetone, acetonitrile. Sparingly sol in isopropanol, *p*-dioxane, ethyl acetate, methylene chloride; very slightly sol in water and isooctane.

Mixture with amprolium, *Amprol Plus*.

THERAP CAT (VET): Mixture as coccidiostat.

3702. Ethoprop. *Phosphorodithioic acid O-ethyl S,S-dipropyl ester; O-ethyl S,S-dipropylphosphorodithioate; ethoprophos;* VC9-104; Mocap. $C_8H_{19}O_2PS_2$; mol wt 242.32. C 39.65%, H 7.90%, O 13.21%, P 12.78%, S 26.46%. Prepn: L. E. Goyette, U.S. pat. **3,112,244** (1963 to Virginia-Carolina Chem. Corp.); J. H. Wilson, Jr., U.S. pat. **3,268,393** (1966 to Mobil). Activity: C. R. Harris, J. L. Hitchon, *J. Econ. Entomol.* **63**, 2 (1970). Metabolism: Z. M. Iqbal, R. E. Menzer, *Biochem. Pharmacol.* **21**, 1569 (1972). Soil degradation: J. H. Smelt *et al., Pestic. Sci.* **8**, 147 (1977).

Pale yellow oil, $bp_{0.2}$ 86-91°. d^{20}_4 1.094. Vapor press. at 26°: 3.5×10^{-4} mm Hg. Soly in water: 750 mg/l. Sol in most organic solvents.

USE: Insecticide; nematocide.

3703. Ethopropazine. *N,N-Diethyl-α-methyl-10H-phenothiazine-10-ethanamine; 10-(2-diethylaminopropyl)phenothiazine;* 10-(2-diethylamino-2-methylethyl)phenothiazine; 2-diethylamino-1-propyl-*N*-dibenzoparathiazine; phenopropazine; profenamine; RP 3356; W-483; Isothazine; Isothiazine; Parkin. $C_{19}H_{24}N_2S$; mol wt 312.46. C 73.03%, H 7.74%, N 8.97%, S 10.26%. Prepd from Grignard complexes of diethylaminopropyl halide and phenothiazine: Berg, Ashley, U.S. pat. **2,607,773** (1952 to Rhône-Poulenc).

Crystals, mp 53-55°. Usually obtained as an oil because of contamination with 10-(2-diethylamino-1-methylethyl)phenothiazine. LD_{50} orally in mice: 650 mg/kg.

Hydrochloride, $C_{19}H_{25}ClN_2S$, *Dibutil, Lysivane, Pardisol, Parphezein, Parphezin, Parsidol, Parsitan, Parsotil, Rodipal.* Crystals from ethylene dichloride, mp 223-225° (some decompn). Lower melting points reported are caused by admixture with 10-(2-diethylamino-1-methylethyl)phenothiazine-HCl which melts at 166-168°. One gram dissolves in 400 ml water at 20°, in 20 ml water at 40°. Sol in ethanol, chloroform. Soly in abs ethanol at 25° = 1.0 g/30 ml. Sparingly sol in acetone. Practically insol in ether, benzene. pH of a 5% aq soln is about 5.8.

THERAP CAT: Antiparkinsonian, anticholinergic.

3704. Ethosuximide. *3-Ethyl-3-methyl-2,5-pyrrolidinedione; 2-ethyl-2-methylsuccinimide; α-ethyl-α-methylsuccinimide; 3-methyl-3-ethylpyrrolidine-2,5-dione;* Atysmal; Capitus; Emeside; Epileo Petitmal; Ethymal; Mesentol;

Pemal; Peptinimid; Petinimid; Petnidan; Pyknolepsinum; Simatin; Succimal; Suxilep; Suximal; Suxinutin; Thilopemal; Zarontin. $C_7H_{11}NO_2$; mol wt 141.17. C 59.55%, H 7.85%, N 9.92%, O 22.67%. Prepn: Sircar, *J. Chem. Soc.* **1927**, 1252. Comparative clinical trial with valproic acid, *q.v.*, in epilepsy: S. Sato *et al., Neurology* **32**, 157 (1982). Acute toxicity data: H. Najer *et al., Bull. Soc. Chim. France* **3**, 1119 (1966). Brief review: S. J. Wallace, *Neurol. Clin.* **4**, 601 (1986).

Crystals from acetone + ether, mp 64-65°. Freely sol in water. LD_{50} in mice (g/kg): 1.65 i.p.; 1.75 orally (Najer).

THERAP CAT: Anticonvulsant.

3705. Ethotoin. *3-Ethyl-5-phenyl-2,4-imidazolidinedione; 3-ethyl-5-phenylhydantoin;* 1-ethyl-2,5-dioxo-4-phenylimidazolidine; Peganone. $C_{11}H_{12}N_2O_2$; mol wt 204.22. C 64.69%, H 5.92%, N 13.72%, O 15.67%. Prepd by heating the potassium salt of 5-phenylhydantoin with ethyl bromide in alc at 100° in a sealed tube: Pinner, *Ber.* **21**, 2320 (1888). Use as anticonvulsant: W. J. Close, U.S. pat. **2,793,157** (1957 to Abbott Labs.). Metabolism: Dudley *et al., J. Pharmacol. Exp. Ther.* **175**, 27 (1970).

Stout prisms from water, mp 94°. Sparingly sol in cold water, more sol in hot water. Freely sol in alcohol, ether, benzene, dil aq solns of alkali hydroxides.

THERAP CAT: Anticonvulsant.

3706. Ethoxazene. *4-[(4-Ethoxyphenyl)azo]-1,3-benzenediamine; 4-[(p-ethoxyphenyl)azo]-m-phenylenediamine;* 2,4-diamino-4'-ethoxyazobenzene; *p*-ethoxy-2,4-diaminoazobenzene; *p*-ethoxychrysoidine; 4-*p*-phenethylazo-*m*-phenylenediamine; SN-612; Carmurit; Cystural. $C_{14}H_{16}N_4O$; mol wt 256.30. C 65.60%, H 6.29%, N 21.86%, O 6.24%.

Hydrochloride, $C_{14}H_{17}ClN_4O$, *Serenium*. Reddish powder. Insol in water. Excreted in water-soluble form, giving a red color to urine. *Ref:* Lott and Christiansen, *J. Am. Pharm. Assoc.* **23**, 785 (1934).

USE: Acid-base indicator.

THERAP CAT: Analgesic.

3707. 2-Ethoxyethanol. Ethylene glycol monoethyl ether; Cellosolve; Oxitol. $C_4H_{10}O_2$; mol wt 90.12. C 53.31%, H 11.19%, O 35.51%. $HOCH_2CH_2OC_2H_5$. Manuf from ethylene oxide and ethanol: Faith, Keyes & Clark's *Industrial Chemicals*, F. A. Lowenheim, M. K. Moran, Eds. (Wiley-Interscience, New York, 4th ed., 1975) pp 403-407. Toxicity data: H. F. Smyth *et al., J. Ind. Hyg. Toxicol.* **23**, 259 (1941).

Colorless, practically odorless liq. d^{20}_{20} 0.931. bp 135°. mp −70°. Flash pt, closed cup: 112°F (44°C); open cup: 120°F (49°C); n^{25}_D 1.406. Misc with water, alc, ether, acetone, liq esters. It dissolves many oils, resins, waxes, etc. LD_{50} orally in rats: 3 g/kg (Smyth).

USE: Solvent for nitrocellulose, lacquers and dopes; in varnish removers; cleansing solns, dye baths; finishing leather with water pigments and dye solns; increasing stability of emulsions.

3708. 2-Ethoxyethyl Acetate. *2-Ethoxyethanol acetate;* cellosolve acetate; ethylene glycol monoethyl ether acetate. $C_6H_{12}O_3$; mol wt 132.16. C 54.53%, H 9.15%, O 36.32%. $C_2H_5OCH_2CH_2OOCCH_3$. Toxicity data: Smyth *et al., J. Ind. Hyg. Toxicol.* **23,** 259 (1941).

Colorless liq, pleasant odor. d_{20}^{20} 0.975. bp 156°. Flash pt, open cup: 134°F (56°C). Sol in about 6 parts water. LD_{50} orally in rats: 5.1 g/kg (Smyth).

USE: Automobile lacquers to retard evaporation and im-part high gloss.

3709. 2-Ethoxynaphthalene. Ethyl β-naphthyl ether; ethyl β-naphtholate; nerolin "new"; β-naphthyl ethyl ether; Bromelia. $C_{12}H_{12}O$; mol wt 172.22. C 83.69%, H 7.02%, O 9.29%. Prepd by heating β-naphthol with potassium ethyl sulfate, or with alcohol and H_2SO_4: Yokoyama *et al., Yaku-gaku Zasshi* **78,** 123 (1958); *C.A.* **52,** 10986c (1958); from β-naphthol + p-toluenesulfonyl chloride + ethyl alcohol: Drakowzal, Klamann, *Monatsh.* **82,** 588 (1951); from β-naphthol + H_2SO_4 + ethyl acetate: Patai, Bentov, *J. Am. Chem. Soc.* **74,** 6118 (1952); from β-naphthol + triethyl-phosphate + p-toluenesulfonic acid: Bell, U.S. pat. **2,683,-748** (1954 to Eastman Kodak).

Lustrous crystals, mp 37-38°. bp 28.2°. $n_D^{47.3}$ 1.5932. d_{20}^{20} 1.0640. Practically insol in water; sol in alcohol, chloro-form, ether, carbon disulfide, toluene, petr ether.

USE: In perfumery.

3710. Ethoxyquin. *6-Ethoxy-1,2-dihydro-2,2,4-trimeth-ylquinoline;* 1,2-dihydro-6-ethoxy-2,2,4-trimethylquinoline; EMQ; Santoflex; Santoquin. $C_{14}H_{19}NO$; mol wt 217.30. C 77.38%, H 8.81%, N 6.45%, O 7.36%. Prepn: Knoevenagel, *Ber.* **54,** 1722 (1921); Baird *et al., Brit.* pat. **505,113** (1939 to ICI); C. C. Tung, *Tetrahedron* **19,** 1685 (1963). Structure: Cliffe, *J. Chem. Soc.* **1933,** 1327. Metabolism: R. H. Wilson *et al., Agr. Food Chem.* **7,** 206 (1959); J. U. Skaare, E. Sol-heim, *Xenobiotica* **9,** 649 (1979). Toxicity data: V. I. Piul'-skaya *et al., C.A.* **90,** 4601c (1979).

Yellow liquid. bp_2 123-125°. n_D^{25} 1.569-1.672. d_{25}^{25} 1.029-1.031. LD_{50} orally in rats, mice: 1920, 1730 mg/kg (Piul'-skaya).

USE: Antioxidant in feed and food; antidegradation agent for rubber.

3711. Ethoxzolamide. *6-Ethoxy-2-benzothiazolesulfon-amide;* ethoxyzolamide; Cardrase; Ethamide; Glaucotensil; Redupresin. $C_9H_{10}N_2O_3S_2$; mol wt 258.33. C 41.85%, H 3.90%, N 10.85%, O 18.58%, S 24.82%. Prepn: **Brit.** pat. **795,174** (1958 to Upjohn); *C.A.* **52,** 20212 (1958).

Crystals from ethyl acetate + Skellysolve B, mp 188-190.5°.

THERAP CAT: Carbonic anhydrase inhibitor; diuretic.
THERAP CAT (VET): Diuretic.

3712. Ethybenztropine. *3-(Diphenylmethoxy)-8-ethyl-8-azabicyclo[3.2.1]octane; 3α-(diphenylmethoxy)-8-ethyl-1αH,5αH-nortropane;* N-ethylnortropine benzhydryl ether; tropethydrylin; N-ethyl-8-aza-3-bicyclo[3.2.1]octyl benzhy-dryl ether; N-ethylbenztropine; ethylbenzatropine. $C_{22}H_{27}$-NO; mol wt 321.47. C 82.20%, H 8.47%, N 4.36%, O 4.98%.

Prepn: **Brit.** pat. **804,837** (1958 to Sandoz); Boehringer *et al.,* **Brit.** pat. **824,875** (1959 to Boehringer, Ing.). Used as hydrochloride or hydrobromide salts.

Hydrochloride, $C_{22}H_{28}ClNO$, *Ponalid.* Crystals from acetone, mp 190-191°.
Hydrobromide, $C_{22}H_{28}BrNO$, *Panolid.* Crystals from methanol + ether, mp 226-228°.

THERAP CAT: Anticholinergic.

3713. Ethyl Acetate. *Acetic acid ethyl ester;* acetic ether; vinegar naphtha. $C_4H_8O_2$; mol wt 88.10. C 54.53%, H 9.15%, O 36.32%. $CH_3COOC_2H_5$. Obtained by the slow distillation of a mixture of acetic acid, ethyl alc, and sulfuric acid: Alheritiere, Mercier, U.S. pat. **2,787,636** (1957 to Usines de Melle); Faith, Keyes, & Clark's *Industrial Chem-icals,* F. A. Lowenheim, M. K. Moran, Eds. (Wiley-Intersci-ence, New York, 4th ed., 1975) pp 350-354. Toxicity: H. F. Smyth *et al., Am. Ind. Hyg. Assoc. J.* **23,** 95 (1962).

Clear, volatile, flammable liq; characteristic fruity odor; pleasant taste when diluted. Slowly dec by moisture, then acquires an acid reaction. Absorbs water (up to 3.3% w/w). d_4^{20} 0.902; d_{25}^{25} 0.898. bp 77°. mp −83°. Flash pt +7.2° (open cup). Ignition temp 800°F. Explosive limits (% vol in air): 2.2 to 11.5. n_D^{20} 1.3719. Vapor density 3.04 (air = 1). One ml dissolves in 10 ml water at 25°; more sol at lower and less sol at higher temps. Misc with alc, acetone, chloro-form, ether. Azeotropic mixture with water (6.1% w/w) bp 70.4°. Azeotropic mixture with water (7.8% w/w) and alc (9.0% w/w) bp 70.3°. *Keep tightly closed in a cool place and away from fire.* LD_{50} orally in rats: 11.3 ml/kg (Smyth).

USE: Pharmaceutic aid (flavor); artificial fruit essences; solvent for nitrocellulose, varnishes, lacquers, and aeroplane dopes; manuf smokeless powder, artificial leather, photogra-phic films and plates, artificial silk, perfumes; cleaning tex-tiles, etc.

3714. Ethyl Acetoacetate. *3-Oxobutanoic acid ethyl es-ter;* acetoacetic acid ethyl ester; acetoacetic ester; ethyl 3-oxobutanoate. $C_6H_{10}O_3$; mol wt 130.14. C 55.37%, H 7.75%, O 36.88%. $CH_3COCH_2COOC_2H_5$. Only the equilib-rium mixture of the keto and enol forms is described here. Prepd from ethyl acetate by the action of sodium, sodium ethoxide, sodamide, or calcium: Inglis, Roberts, *Org. Syn.* coll. vol. I, 235 (2nd ed., 1941); Hansley, Schott, U.S. pat. **2,843,623** (1958 to Natl. Distillers); Scheibler, *Ann.* **565,** 176 (1949); Gattermann-Wieland, *Praxis des Organischen Che-mikers* (de Gruyter, Berlin, 40th ed., 1961) p 218. Discus-sion of keto-enol tautomerism: Ward, *J. Chem. Ed.* **39,** 95 (1962).

Liq. Agreeable odor. d_4^{10} 1.0357; d_4^{17} 1.0288; d_4^{25} 1.0213; d_4^{54} 0.9924; d_4^{75} 0.9703. mp −45°. bp_{760} 180.8°; bp_{400} 158.2°; bp_{200} 138.0°; bp_{60} 106°; bp_{20} 81.1°; bp_5 54.0°; $bp_{1.0}$ 28.5°. n_D^{20} 1.41937. Absorption spectrum: Morton, Rosney, *J. Chem. Soc.* **1926,** 711. Flash pt, closed cup: 184°F. Sol in about 35 parts water; misc with the usual organic solvents. LD_{50} orally in rats: 3.98 g/kg, H. F. Smyth *et al., J. Ind. Hyg. Toxicol.* **31,** 60 (1949).

Caution: Moderately irritating to skin, mucous mem-branes.

3715. Ethyl Acrylate. *2-Propenoic acid ethyl ester;* acryl-ic acid ethyl ester. $C_5H_8O_2$; mol wt 100.11. C 59.98%, H 8.05%, O 31.96%. $CH_2=CHCOOCH_2CH_3$. Prepd from ethylene chlorohydrin or acrylonitrile, ethanol, and sulfuric acid; also by an oxo reaction from acetylene, carbon monox-ide, and ethanol in the presence of suitable catalysts. *See* the refs under Methyl Acrylate.

Monomer, liquid, acrid, penetrating odor, retained by clothing. *Lacrimator.* d_4^{20} 0.9405. fp below −72°. bp_{760} 99.4°; $bp_{39.2}$ 20° (polymerizes on distn). n_D^{20} 1.404. Specific heat at −60°: 0.442 cal/g/°C. Heat of vaporization 8.27

kcal/mol; heat of combustion 655.49 kcal/mol. Flash pt, open cup: 60°F (15°C). Vapor density 3.45 (air = 1). Soly in water at 20°: 2 g/100 ml. Soly of water in ethyl acrylate at 20°: 1.5 g/100 g. Sol in alcohol, ether. Azeotropes: 45.0% water = bp 81°; 56.8% ethanol = bp 76°. Easily polymerizes on standing; polymerization process speeded up by heat, light, and peroxides. If pure, the monomer can be stored below +10° without incurring polymerization.

Polymer, transparent, elastic substance. Practically no odor. Little adhesive power. Resists the usual solvents. USE: The monomer in the manuf of water emulsion paint vehicles, textile and paper coatings, leather finish resins and adhesives. Imparts flexibility to hard films. *Caution:* The monomer is highly irritating to eyes, skin, mucous membranes. Lethargy and convulsions may occur if vapors of the monomer are inhaled in high concns.

3716. Ethyl Alcohol. *Ethanol;* absolute alcohol; anhydrous alcohol; dehydrated alcohol; ethyl hydrate; ethyl hydroxide. C_2H_6O; mol wt 46.07. C 52.14%, H 13.13%, O 34.74%. C_2H_5OH. Manuf: by fermentation of starch, sugar, and other carbohydrates; from ethylene, acetylene, sulfite waste liquors, and synthesis gas (CO + H); by hydrolysis of ethyl sulfate, and oxidation of methane. Toxicity: G. S. Wiberg *et al., Toxicol. Appl. Pharmacol.* **16,** 718 (1970). Embryotoxicity in mammals: N. A. Brown *et al., Science* **206,** 573 (1979). Possible mechanism for actions of ethanol on the brain: G. Aston-Jones *et al., Nature* **296,** 857 (1982). Ethanol-induced chromosomal abnormalities in mice: M. H. Kaufman, *ibid.* **302,** 258 (1983). Disruption of reproductive function in female primates following alcohol self-administration: N. K. Mello *et al., Science* **221,** 677 (1983). Review of metabolism and toxicity: C. S. Lieber in *Reviews in Biochemical Toxicology* vol. 5, E. Hodgson *et al.,* Eds. (Elsevier, New York, 1983) pp 267-312. Review of pharmacology: L. Pohorecky, J. Brick, *Pharmacol. Ther.* **36,** 335-427 (1988). General reviews: P. Baud, "Ethyl Alcohol Industry" in Grignard, *Traité de Chimie Organique* vol. 5 (Masson, 1937) pp 841-975; Zabel, *Chem. Inds.* (now *Chem. Week*) **64,** 212 (1949); Faith, Keyes & Clark's *Industrial Chemicals,* F. A. Lowenheim, M. K. Moran, Eds. (Wiley-Interscience, New York, 4th ed., 1975) pp 355-364; P. D. Sherman, P. R. Kavasmaneck, "Ethanol" in Kirk-Othmer *Encyclopedia of Chemical Technology* vol. 9 (Interscience, New York, 3rd ed., 1980) pp 338-380.

Clear, colorless, very mobile, flammable liquid; pleasant odor; burning taste. Absorbs water rapidly from air. d_4^{20} 0.789. bp 78.5°. mp −114.1°. n_D^{20} 1.361. Flash pt, closed cup: 13°C. Miscible with water and with many organic liquids. *Keep tightly closed, cool, and away from flame!* LD_{50} in young, old rats (g/kg): 10.6, 7.06 orally (Wiberg).

The terms **95% alcohol** and **alcohol** (when used alone) refer to a binary azeotrope having a distillate composition of 95.57% ethyl alcohol (by wt) and bp 78.15°. *Alcohol, USP* is specified as containing not less than 92.3% and not more than 93.8% by weight, corresponding to not less than 94.9% and not more than 96.0% by vol of C_2H_5OH at 15.56°. d_{25}^{25} 0.810; d 0.816 at 15.56° (60°F). *Diluted alcohol,* prepd from equal vols 95% alcohol and water, contains about 41.5% by wt or about 48.9% by vol of C_2H_5OH. d_{25}^{25} 0.931; d 0.936 at 15.56° (60°F). *See U.S.P. XXI,* 22, 1530 (1985).

Human Toxicity: Nausea, vomiting, flushing, mental excitement or depression, drowsiness, impaired perception, incoordination, stupor, coma, death may occur, *cf. Clinical Toxicology of Commercial Products,* R. E. Gosselin *et al.,* Eds. (Williams & Wilkins, Baltimore, 5th ed., 1984) Section III, pp 166-171.

USE: Most ethyl alcohol is used in alcoholic beverages in suitable dilutions. Other uses are as solvent in laboratory and industry, in the manufacture of denatured alcohol, pharmaceuticals (rubbing compds, lotions, tonics, colognes), in perfumery, in organic synthesis. Octane booster in gasoline. Pharmaceutic aid (solvent).

THERAP CAT: Topical antiseptic.

THERAP CAT (VET): Antiseptic. To destroy nerve tissue. Solvent and dehydrating agent.

3717. Ethyl Alcohol, Denatured. Denatured alcohol. Ethyl alcohol to which has been added some substance or substances which, while allowing the use of the alcohol in

the most varied industries and arts, renders it entirely unfit for consumption as a beverage. The most commonly used denaturants, either alone or in combination, are the following: Methanol, camphor, Aldehol, amyl alcohol, gasoline, isopropanol, terpineol, benzene, castor oil, acetone, nicotine, aniline dyes, ether, cadmium iodide, pyridine bases, sulfuric acid, kerosene, diethyl phthalate. *Formula 1* is 5 gallons approved wood alcohol added to 100 gal of 95% ethanol. *Formula 2B* is 0.5 gal benzene added to 100 gal of 95% ethanol. Similarly *formula 3A* contains 5 gal commercial methanol, *formula 6B* contains 0.5 gal pyridine bases, *formula 12A* 5 gal benzene, *formula 13A* 10 gal ethyl ether, *formula 19* 4 gal methyl isobutyl ketone and 1 gal kerosene, *formula 20* 5 gal crude chloroform, *formula 23A* 10 gal acetone, *formula 28* 10 gal benzene, *formula 28A* 1 gal gasoline, *formula 30* 10 gal methanol, *formula 32* 5 gal ethyl ether, *formula 33* 30 lbs methyl violet, *formula 35A* 5 gal ethyl acetate, *formula 39C* 1 gal diethyl phthalate; *formula 44* contains 20 gal *n*-butanol. Additional permissible formulas are given in *Appendix to Regulations No. 3, Formulae for Completely and Specially Denatured Alcohol,* published by the U.S. Treasury Dept., Bureau of Industrial Alcohol. Reprinted in N. A. Lange, *Handbook of Chemistry.*

Caution: Denaturants, particularly methanol, may modify and increase toxic symptoms caused by ingestion and exposure to fumes.

3718. Ethylamine. *Ethanamine;* monoethylamine; aminoethane. C_2H_7N; mol wt 45.08. C 53.28%, H 15.65%, N 31.07%. $C_2H_5NH_2$. Prepn from ethyliodide + liq ammonia: Watt, Otto, *J. Am. Chem. Soc.* **69,** 836 (1947); from ethanol + ammonia: Davies *et al.,* U.S. pat. 2,609,394 (1952 to ICI); Lemon, Myerly, U.S. pat. 3,022,349 (1962 to Union Carbide).

Flammable, liq; ammonia odor; strong alkaline reaction. d_{15}^{15} 0.689. bp 16.6°. Solidif −80°. Miscible with water, alcohol, ether. *Keep tightly closed and in cold place.* LD_{50} orally in rats: 0.40 g/kg, H. F. Smyth *et al., Arch. Ind. Hyg. Occup. Med.* **10,** 61 (1954).

Hydrochloride, $C_2H_7N \cdot HCl$, crystals from ethanol + water, mp 110°. d 1.22. Soluble in 0.4 part water; freely sol in alcohol; slightly in chloroform or acetone. Practically insol in ether. *Keep well closed.*

Hydriodide, $C_2H_7N \cdot HI$, hygroscopic crystals, mp 188°. d 2.10. Freely sol in water or alcohol. Practically insol in chloroform, ether. *Keep well closed and protected from light.*

Oleate, *Etalate.* Commercial prepn is a 5% soln with 2% benzyl alcohol as anodyne.

USE: In resin chemistry; stabilizer for rubber latex; intermediate for dyestuffs, medicinals; in oil refining; in organic syntheses. *Caution:* Irritating to skin, mucous membranes, respiratory tract.

THERAP CAT: Oleate as a sclerosing agent.

3719. Ethyl Aminobenzoate. *4-Aminobenzoic acid ethyl ester;* Aethoform; Americaine; Anesthesin; Anesthone; Benzocaine; Orthesin; Parathesin. $C_9H_{11}NO_2$; mol wt 165.19. C 65.43%, H 6.71%, N 8.48%, O 19.37%. Prepd by the esterification of *p*-aminobenzoic acid: Salkowski, *Ber.* **28,** 1921 (1895); Vorländer, Meyer, *Ann.* **320,** 135 (1902); by the reduction of ethyl *p*-nitrobenzoate: Limpricht, *ibid.* **303,** 278 (1898); R. Adams, F. L. Cohen, *Org. Syn.* coll. vol. I, 240 (2nd ed., 1941). In industrial practice the reducing agent is usually iron and water in the presence of a little acid. Comprehensive description: S. L. Ali in *Analytical Profiles of Drug Substances* vol. 12, K. Florey, Ed. (Academic Press, New York, 1983) pp 73-104.

Rhombohedra from ether, mp 88-90°. Stable in air. One gram dissolves in about 2500 ml water, 5 ml alcohol, 2 ml chloroform, in about 4 ml ether, and in 30 to 50 ml of expressed almond oil or olive oil. Also sol in dil acids. pK_a 2.5.

THERAP CAT: Topical anesthetic.

THERAP CAT (VET): Local (usually surface) anesthetic.

3720. N-Ethylamphetamine. *N-Ethyl-α-methylbenzene-ethanamine; N-ethyl-α-methylphenethylamine; N-ethyl-ω-phenylisopropylamine; 2-ethylamino-1-phenylpropane; Adiparthrol; Apetinil.* $C_{11}H_{17}N$; mol wt 163.25. C 80.92%, H 10.50%, N 8.58%. Prepn: Keil, Dobke, **Ger. pat. 767,263** (1952 to Theodor H. Temmler); Leonard *et al., J. Am. Chem. Soc.* **80,** 4858 (1958). Separation of isomers: **Brit. pat. 814,339** (1959 to Sterling Drug).

bp$_{14}$ 104.5-106°. n_D^{25} 1.4986.
d-Form hydrochloride, $C_{11}H_{18}ClN$, mp 154-156°. $[\alpha]_D^{15}$ +17.2° (c = 2 in water).
l-Form hydrochloride, $C_{11}H_{18}ClN$, mp 155-156°. $[\alpha]_D^{25}$ −17.3° (c = 2 in water).
Note: This is a controlled substance (stimulant) listed in the U.S. Code of Federal Regulations, Title 21 Part 1308.11 (1985).

THERAP CAT: Anorexic.

3721. Ethyl Amyl Ketone. *5-Methyl-3-heptanone;* amyl ethyl ketone; EAK. $C_8H_{16}O$; mol wt 128.21. C 74.94%, H 12.58%, O 12.48%. *Review:* Buller, *Ind. & Eng. Chem.* **48,** 1323 (1956).

Liquid. Mild fruity odor. d_{20}^{20} 0.820-0.824. One gallon weighs 6.83 lbs at 20°. bp$_{760}$ 157-162°. Flash pt 59° (138°F). Evaporation rate 0.3 (*n*-butyl acetate = 1.0). n_D^{15} 1.4195. Slightly miscible with water. Compatible with alcohols, ketones, ethers, many other organic solvents.

USE: Solvent for nitrocellulose-alkyd, nitrocellulose-maleic, and vinyl resins. *Caution:* Narcotic in high concns.

3722. Ethylaniline. *N-Ethylbenzenamine;* ethylphenylamine. $C_8H_{11}N$; mol wt 121.18. C 79.29%, H 9.15%, N 11.56%. Produced by heating aniline hydrochloride and alcohol at 180°.

Very refractive liquid; rapidly becomes brown on exposure to light and air. Aniline-like odor. d_{25}^{25} 0.958. bp 204.5°. Solidifies below 80°. mp −63.5°. n_D^{20} 1.5559. Insol in water; miscible with alc, ether and many other organic solvents. *Keep well closed and protected from light.* LD$_{50}$ orally in rats: 1.1 g/kg.

3723. Ethylbenzene. C_8H_{10}; mol wt 106.16. C 90.50%, H 9.50%. $C_6H_5C_2H_5$. Prepn from acetophenone: Clemmensen, *Ber.* **46,** 1838 (1913); Gattermann-Wieland, *Praxis des organischen Chemikers* (de Gruyter, Berlin, 40th ed., 1961) p 332. By Huang-Minlon modification of Wolff-Kishner reduction: A. I. Vogel, *Practical Organic Chemistry* (Longmans, 3rd ed., 1959) p 516. Physical properties: L. C. Gibbons *et al., J. Am. Chem. Soc.* **68,** 1130 (1946). Manuf: Faith, Keyes & Clark's *Industrial Chemicals,* F. A. Lowenheim, M. K. Moran, Eds. (Wiley-Interscience, New York, 4th ed., 1975) pp 365-370. Toxicity: H. F. Smyth *et al., Am. Ind. Hyg. Assoc J.* **23,** 95 (1962).

Colorless liquid; flammable. d_{25}^{25} 0.866. bp 136.25°. mp −95.01°. n_D^{25} 1.4932. Flash pt, closed cup: 64°F (18°C). Practically insol in water; misc with the usual organic solvents. LD$_{50}$ orally in rats: 5.46 g/kg (Smyth).

Caution: Irritating to eyes, skin, mucous membranes, and, in high concns, narcotic.

USE: For conversion to styrene monomer; as resin solvent.

3724. Ethylbenzhydramine. *2-(Diphenylmethoxy)-N,N-diethylethanamine;* 2-(benzhydryloxy)triethylamine; β-diethylaminoethyl benzhydryl ether; 2-(benzhydryloxy)-N,N-diethylethylamine; benzhydryl 2-diethylaminoethyl ether; etanautine. $C_{19}H_{25}NO$; mol wt 283.40. C 80.52%, H 8.89%, N 4.94%, O 5.65%. Prepn: Martin *et al.,* **U.S. pat. 2,397,-799** (1946 to Geigy); Rieveschl, **U.S. pat. 2,421,714** (1947 to Parke, Davis).

bp$_{0.15}$ 140-142°.
Hydrochloride, $C_{19}H_{26}ClNO$, *Antiparkin, Rigidyl.* Prisms from alcohol + ethyl acetate, mp 140°. Bitter taste. Freely sol in water; sol in alcohol, acetone, chloroform; very slightly sol in benzene, ether. pH of 1% aq soln about 5.5.
Methyl iodide, $C_{20}H_{28}INO$, *Esyntin, Metropin.* A quaternary ammonium deriv of the base covered in the Geigy pat.

THERAP CAT: Hydrochloride as antiparkinsonian; methyl iodide as anticholinergic.

3725. Ethyl Benzoate. *Benzoic acid ethyl ester.* $C_9H_{10}O_2$; mol wt 150.17. C 71.98%, H 6.71%, O 21.31%. $C_6H_5COO-C_2H_5$.
Colorless, clear, refractive liq; aromatic odor; vapors cause cough. d_4^{25} 1.050. bp 211-213°. mp −34°. n_D^{20} 1.506. Almost insol in water; miscible with alcohol, chloroform, ether, petr ether. LD$_{50}$ orally in rats: 6.48 g/kg, Smyth *et al., Arch. Ind. Hyg. Occup. Med.* **10,** 61 (1954).

USE: In perfumery under the name *Essence de Niobe;* in manuf of *Peau d'Espagne;* artificial fruit essence.

3726. Ethyl Benzoylacetate. *β-Oxobenzenepropanoic acid ethyl ester.* $C_{11}H_{12}O_3$; mol wt 192.21. C 68.73%, H 6.29%, O 24.97%. $C_6H_5COCH_2COOC_2H_5$.
Liquid; pleasant odor; becomes yellow on exposure to air and light. d^{15} 1.122. bp 265-270° with decompn. Volatile with steam. n_D^{20} 1.5338. Insol in water; miscible with alcohol, ether. *Keep protected from air and light.*

3727. α-Ethylbenzyl Alcohol. *α-Ethylbenzenemethanol;* ω-ethylbenzyl alcohol; 1-phenyl-1-propanol; 1-phenylpropyl alcohol; ethyl phenyl carbinol; α-hydroxypropylbenzene; SH 261; Ejibil; Livonal; Phenycholon; Phenicol; Phenychol; Fenicol; Felicur; Felitrope. $C_9H_{12}O$; mol wt 136.19. C 79.37%, H 8.88%, O 11.75%. Prepd from benzaldehyde or from ethyl phenyl ketone: Norris, Cortese, *J. Am. Chem. Soc.* **49,** 2640 (1927).

Oily liquid. Weak, ester-like odor. Sweetish, slightly irritating taste. bp$_{760}$ 219°; bp$_{15}$ 107°; bp$_3$ 78°. d_4^{25} 0.9915. n_D^{23} 1.5169. uv max (methanol): 250, 260 nm (ε 173, 114). Misc with methanol, ethanol, ether, benzene, toluene, olive oil. LD$_{50}$ orally in rats: 1.6 ml/kg, O. Linét *et al., Arzneimittel-Forsch.* **12,** 347 (1962).

USE: As heat transfer medium; in perfumery.

THERAP CAT: Choleretic.

3728. Ethylbenzylaniline. *N-Ethyl-N-phenylbenzenemethanamine; N-ethyl-N-phenylbenzylamine.* $C_{15}H_{17}N$; mol wt 211.29. C 85.26%, H 8.11%, N 6.63%. Prepn: Martin, MacQueen, **U.S. pat. 1,887,772** (1933 to Dow); Burgstahler, *J. Am. Chem. Soc.* **73,** 3021 (1951).

$$C_6H_5CH_2 - \overset{\overset{\textstyle C_2H_5}{\textstyle |}}{N} - C_6H_5$$

Light yellow, oily liq. d_4^{19} 1.034. bp_{710} 287° with slight decompn; bp_{14} 170-180°. n_D^{23} 1.5938. Practically insol in water. One ml dissolves in 5.5 ml alcohol; sol in the usual organic solvents.

Picrate, crystals, mp 118-120°.

USE: Manuf dyes; in organic syntheses.

3729. Ethyl Biscoumacetate. *4-Hydroxy-α-(4-hydroxy-2-oxo-2H-1-benzopyran-3-yl)-2-oxo-2H-1-benzopyran-3-acetic acid ethyl ester; bis(4-hydroxy-2-oxo-2H-1-benzopyran-3-yl)acetic acid ethyl ester;* ethyl bis(4-hydroxycoumarinyl)acetate; ethyldicoumarol acetate; bis-3,3'-(4-hydroxycumarinyl)acetic acid ethyl ester; 3,3'-carboxymethylene bis(4-hydroxycoumarin)ethyl ester; ethyl 4,4'-dihydroxydicoumarinyl-3,3'-acetate; B.O.E.A.; Dicumacyl; Pelentan; Stabilene; Tromexan; Tromexan Ethyl Acetate. $C_{22}H_{16}O_8$; mol wt 408.35. C 64.70%, H 3.95%, O 31.34%. Prepn: Rosicky, *Cas. Lek. Cesk.* **83,** 1200 (1944); U.S. pats. **2,482,-510-12** (1949 to Spojené farm. zovody); Stahmann *et al., J. Am. Chem. Soc.* **65,** 2285 sqq (1943).

Crystals, dimorphous, mp 177-182° and mp 154-157°. Bitter, persistent taste. Practically insol in water. Sol in 20 parts acetone, also sol in benzene. Slightly sol in alcohol, ether. LD_{50} in mice, rats (g/kg): 0.88, 0.26 orally; 0.32, 1.1 i.p., C. M. Gruber *et al., Fed. Proc.* **10,** 303 (1951).

THERAP CAT: Anticoagulant.

3730. Ethyl Bromide. *Bromoethane;* monobromoethane; bromic ether; hydrobromic ether. C_2H_5Br; mol wt 108.98. C 22.04%, H 4.62%, Br 73.33%. CH_3CH_2Br. Made by distilling from a mixture of HBr, ethyl alcohol and H_2SO_4: Kamm, Marvel, *Org. Syn.* coll. vol. **I,** 29 (1941). By phosphorus and bromine method: Goshorn *et al., ibid.* 36. Absorption spectrum: Hantzsch, *Ber.* **58,** 619 (1925). Physical properties: Mumford, Philllips, *J. Chem. Soc.* **1950,** 75. Toxicity data: E. H. Vernot *et al., Toxicol. Appl. Pharmacol.* **42,** 417 (1977).

Colorless, flammable, volatile liq; ethereal odor; burning taste; becomes yellowish on exposure to air and light. Vapor harmful. d_4^{20} 1.4612; d_4^{25} 1.4515. bp 38.2°. mp $-119°$. n_D^{20} 1.4242. Soly in water (g/100 g) at 0°: 1.067; 10°: 0.965; 20°: 0.914; 30°: 0.896; miscible with alcohol, ether, chloroform and with other organic solvents. Explosive limits (% by vol in air), lower 6.75, upper 11.25. Autoignition temp 952°F (511°C). LC_{50} rats, mice (ppm): 27000, 16200 (Vernot).

Human Toxicity: Irritating to respiratory tract; narcotic in high concns.

USE: Ethylating agent in organic synthesis; as refrigerant. Formerly used as a topical and inhalation anesthetic.

3731. Ethyl α-Bromopropionate. *2-Bromopropanoic acid ethyl ester.* $C_5H_9BrO_2$; mol wt 181.04. C 33.17%, H 5.01%, Br 44.14%, O 17.68%. $CH_3CHBrCOOC_2H_5$.

Liquid; sharp, pungent odor; becomes yellow on exposure to light. d_{20}^{20} 1.447. bp 159-160°; also stated as 160-165°. n_D^{20} 1.4469. Insol in water; miscible with alcohol, ether. *Protect from light.*

3732. Ethyl tert-Butyl Ether. *2-Ethoxy-2-methylpropane;* ethyl *tert*-butyl oxide; 1,1-dimethylethyl ethyl ether; ethyl 1,1-dimethylethyl ether; ETBE. $C_6H_{14}O$; mol wt 102.18. C 70.53%, H 13.81%, O 15.66%. $(CH_3)_3C(OCH_2CH_3)$. Prepn J. U. Nef, *Ann.* **309,** 126 (1899). Synthesis: J. F. Norris, G. W. Rigby, *J. Am. Chem. Soc.* **54,** 2088 (1932). Physical properties: T. W. Evans, K. R. Edlund, *Ind. Eng. Chem.* **28,** 1186 (1936). Thermal decomposition: J. N. Daly, C. Wentrup, *Aust. J. Chem.* **21,** 1535 (1968). Brief review focusing on use as gasoline additive: M. Iborra *et al., Chemtech.* **18,** 120 (1988).

bp 69-71°. fp $-94.0°$. Also reported as bp 73.1° (Norris, Rigby). d_D^{25} 0.7364. n_D^{25} 1.3728. Also reported as bp 72.8° (Evans, Edlund). d_4^{15} 0.7456; d_4^{20} 0.7404; d_4^{25} 0.7353, d_4^{30} 0.7300. n_D^{20} 1.3760. Vapor pressure at 25°: 130 mm Hg. Heat vaporization: 74.3 cal/g. Specific heat (liquid) at 25°: 0.51 cal/g/°C. Surface tension at 24°: 19.8 dynes/cm. Soly in water (20°): 1.2 g/100 g soln. Soly of water in compound (20°): 0.5 g/100 g soln.

USE: Gasoline additive.

3733. Ethyl Butyrate. *Butanoic acid ethyl ester; butyric acid ethyl ester;* ethyl *n*-butyrate. $C_6H_{12}O_2$; mol wt 116.16. C 62.04%, H 10.41%, O 27.55%. $CH_3CH_2CH_2COOC_2H_5$. Toxicity data: P. M. Jenner *et al., Food Cosmet. Toxicol.* **2,** 327 (1964).

Colorless liq; pineapple odor. d_D^{20} 0.879. bp 120-121°. mp $-93°$. n_D^{20} 1.400. Flash pt, closed cup: 78°F (25°C); open cup: 85°F (29°C). Sol in about 150 parts water; misc with alcohol, ether. LD_{50} orally in rats: 13,050 mg/kg (Jenner).

USE: Manuf artificial rum; perfumery; the alcoholic soln constitutes the so-called "pineapple oil".

3734. Ethyl Caprate. *Decanoic acid ethyl ester;* ethyl decanoate. $C_{12}H_{24}O_2$; mol wt 200.31. C 71.95%, H 12.08%, O 15.97%. $CH_3(CH_2)_8COOC_2H_5$.

Colorless liq. d^{20} 0.862. bp 243-245°. Insoluble in water; miscible with alcohol, chloroform, ether.

USE: Manuf wine bouquets, cognac essence.

3735. Ethyl Caproate. *Hexanoic acid ethyl ester;* ethyl hexanoate. $C_8H_{16}O_2$; mol wt 144.21. C 66.63%, H 11.18%, O 22.19%. $CH_3(CH_2)_4COOC_2H_5$.

Colorless to yellowish liquid; pleasant odor. d^{20} 0.873. bp 166-167°. Insol in water; miscible with alcohol, ether.

USE: Manuf artificial fruit flavors.

3736. Ethyl Caprylate. *Octanoic acid ethyl ester;* ethyl octanoate; ethyl octylate. $C_{10}H_{20}O_2$; mol wt 172.26. C 69.72%, H 11.70%, O 18.58%. $CH_3(CH_2)_6COOC_2H_5$.

Colorless, clear, very mobile liquid; pleasant, pineapple odor. d^{17} 0.878. bp 207-209°. Insol in water; misc with alc, ether. LD_{50} orally in rats: 25,960 mg/kg, P. M. Jenner *et al., Food Cosmet. Toxicol.* **2,** 327 (1964).

USE: Manuf fruit ethers; constit of enanthic, cocoic, and cognac ethers.

3737. Ethyl β-Carboline-3-carboxylate. *9H-Pyrido-[3,4-b]indole-3-carboxylic acid ethyl ester;* ethyl norharmancarboxylate; β-CCE. $C_{14}H_{12}N_2O_2$; mol wt 240.26. C 69.99%, H 5.03%, N 11.66%, O 13.32%. Deriv of β-*carboline* that is a potent displacer of ^{3}H-diazepam from brain benzodiazepine receptors. Isoln from human urine and brain and binding site study: C. Braestrup *et al., Proc. Nat. Acad. Sci. USA* **77,** 2288 (1980). Initially thought to be an endogenous ligand for benzodiazepine receptors in mammalian CNS, it is now believed to be formed during isoln and extraction procedures, R. F. Squires in *GABA and Benzodiazepine Receptors,* E. Costa *et al.,* Eds. (Raven Press, New York, 1980) pp 129-138; M. Nielson *et al., J. Neurochem.* **36,** 276 (1981). Synthesis and psychotropic activity: **Japan. Kokai 81 43283** (to Schering AG), *C.A.* **95,** 115508a (1981); U. Eder *et al., Eur. Appl.* **30,254** (1981 to A/S Ferrosan; Schering AG). β-CCE has been shown to lower seizure threshold and to reverse the sedative effect of flurazepam, *q.v.:* P. J. Cowen *et al., Nature* **290,** 54 (1981). Neurochemical and pharmacological actions of β-CCE and other β-carbolines: M. Cain *et al., J. Med. Chem.* **25,** 1081 (1982). Anxiogenic and convulsant properties: L. Prado de Carvalho *et al., Nature* **301,** 64 (1983).

β-CCE

mp 229-233°. uv max (pH7): 215, 242, 279 nm.

3-Hydroxymethyl-β-carboline, $C_{12}H_{10}N_2O$, *9H-pyrido-[3,4-b]indole-3-methanol,* 3-HMC. Cryst, mp 225-228°.

Prepn: F. Hamaguchi, S. Ohki, *Heterocycles* **8**, 383 (1977); M. Cain *et al.*, *loc. cit.* Antagonism of anticonvulsant and anxiolytic actions of diazepam: P. Skolnick *et al.*, *Eur. J. Pharmacol.* **68**, 381 (1980).

USE: As tools for studying benzodiazepine receptors.

3738. Ethyl Carbonate. *Carbonic acid diethyl ester;* diethyl carbonate; Eufin. $C_5H_{10}O_3$; mol wt 118.30. C 50.83%, H 8.53%, O 40.63%. $(C_2H_5O)_2C=O$. Prepn: Palomaa *et al.*, *Ber.* **72**, 313 (1939). Manuf: Mador, Blackham, U.S. pat. **3,114,762** (1963 to Natl. Distillers & Chem.).

Liquid, bp 126°. Pleasant ethereal odor, mp −43°. Flash pt, closed cup: 77°F (25°C). d_4^{20} 0.9764. n_D^{20} 1.3843. Practically insol in water; miscible with alcohol, ether.

USE: Solvent for nitrocellulose; manuf radio tubes; fixing rare earths to cathode elements.

3739. Ethyl Cellulose. *Cellulose ethyl ether;* Ethocel. Prepd from wood pulp or chemical cotton by treatment with alkali and ethylation of the alkali cellulose with ethyl chloride. Review and bibliography: E. Ott, *Cellulose and Cellulose Derivatives* (New York, 2nd ed., 1955).

White granules. Soly is dependent upon the degree of substitution. Commercial ethyl cellulose has an ethoxy content of 43-50%. A 47% product softens at 140° and is sol in ethyl acetate, ethylene dichloride, benzene, toluene, xylene, butyl acetate, acetone, methanol, ethanol, butanol, carbon tetrachloride. To avoid brittleness, ethyl cellulose formulations usually include an antioxidant such as hydroquinone monobenzyl ether, 4-hexylpyrocatechol, or diphenylamine.

USE: In the manuf of plastics and lacquers. Pharmaceutic aid (tablet binder).

3740. Ethyl Chloride. *Chloroethane;* monochlorethane; chlorethyl; aethylis chloridum; ether chloratus; ether hydrochloric; ether muriatic; Kelene; Chelen; Anodynon; Chloryl Anesthetic; Narcotile. C_2H_5Cl; mol wt 64.52. C 37.23%, H 7.81%, Cl 54.96%. Prepd by the action of chlorine on ethylene in the presence of HCl and light: U.S. pat. **2,393,509** (1946); by the action of chlorine on ethylene in the presence of the chlorides of copper, iron, antimony, and calcium: Ger. pat. **298,931**; Bähr, Zieber, *Angew. Chem.* **1930**, 233, 286; by heating alcohol, HCl and $ZnCl_2$: U.S. pat. **2,396,-639** (1946). Review of mfg processes: Faith, Keyes & Clark's *Industrial Chemicals*, F. A. Lowenheim, M. K. Moran, Eds. (Wiley-Interscience, New York, 4th ed., 1975) pp 371-375.

Flammable gas at ordinary temp and pressure. Characteristic ethereal odor; burning taste. At low temps or under increased pressure, ethyl chloride is a mobile, very volatile liquid. d_4^0 0.9214. Vapor density: 2.22 (air = 1.00). mp −138.7°. bp_{760} 12.3°. When ethyl chloride is liberated at ordinary room temp from its sealed container (usually a tube with automatic closure) it vaporizes at once. $bp_{2\,atm}$ 32.5°; $bp_{10\,atm}$ 92.6°; $bp_{30\,atm}$ 149.5°; $bp_{50\,atm}$ 180.5°. Crit temp = 187.2°; crit press = 52.0 atm. Soly in water: 0.574 g/100 ml at 20°; in alcohol: 48.3 grams/100 ml. Miscible with ether. Ethyl chloride burns with a smoky, greenish flame, producing hydrogen chloride. Flash pt: −50° (−58°F) (closed cup): −43° (−45°F) (open cup). Explosive limits (% by vol in air): lower 3.6; upper 14.8. *Caution:* Keep away from heat, sparks, and open flame. Keep container closed, out of sun, and away from heat. Use with adequate ventilation. Carbon dioxide is recommended for extinguishing ethyl chloride fires.

Human Toxicity: Mildly irritating to mucous membranes. High concns of vapors cause narcosis, unconsciousness.

USE: Refrigerant, solvent, alkylating agent, starting point in the manuf of tetraethyl lead: U.S. pat. **1,907,701** (1933).

THERAP CAT: Topical anesthetic.

THERAP CAT (VET): Topical anesthetic.

3741. Ethyl Chloroacetate. *Chloroacetic acid ethyl ester.* $C_4H_7ClO_2$; mol wt 122.55. C 39.20%, H 5.76%, Cl 28.93%, O 26.11%. $CH_2ClCOOC_2H_5$. Prepn: Vogel, *J. Chem. Soc.* **1948**, 644.

Liquid; pungent odor. *Vapors irritate the eyes.* d_4^{20} 1.1498. mp −26°. bp 144-146°. Flash pt 54°. n_D^{20} 1.4227. Insol in water; miscible with alcohol, ether.

3742. Ethyl Chloroformate. *Carbonochloridic acid ethyl ester; chloroformic acid ethyl ester;* ethyl chlorocarbonate.

$C_3H_5ClO_2$; mol wt 108.53. C 33.20%, H 4.64%, Cl 32.69%, O 29.48%. $ClCOOC_2H_5$. Prepn from phosgene and alc: Dumas, *Ann.* **10**, 277 (1834).

Corrosive, flammable liquid. Flash pt slightly below 61°F. bp 95°. d_4^{20} 1.1403. n_D^{20} 1.3947. Practically insol and gradually decomp by water; miscible with alcohol, benzene, chloroform, ether.

Caution: Vapors strongly irritate eyes, mucous membranes, and skin.

3743. Ethyl α-Chloropropionate. *2-Chloropropanoic acid ethyl ester.* $C_5H_9ClO_2$; mol wt 136.58. C 43.97%, H 6.64%, Cl 25.96%, O 23.43%. $CH_3CHClCOOC_2H_5$. Liq; pleasant odor. d_4^{20} 1.087. bp 147-148°. n_D^{20} 1.4185. Insol in water; miscible with alcohol, ether.

3744. Ethyl Cyanoacetate. *Cyanoacetic acid ethyl ester;* cyanoacetic ester; ethyl cyanoethanoate; malonic acid ethyl ester nitrile. $C_5H_7NO_2$; mol wt 113.12. C 53.09%, H 6.24%, N 12.39%, O 28.29%. $CNCH_2COOC_2H_5$. Made by the interaction of sodium cyanide and chloroacetic acid and subsequent esterification of the cyanoacetic acid formed: Kohler, Allen, *Org. Syn.* **3**, 53 (1923); Inglis, *ibid.* **8**, 74 (1928).

Liquid. Slight, pleasant odor. d_4^{25} 1.0560; d_4^{50} 1.0306; d_4^{70} 1.0110. mp −22°. bp_{760} 206.0°; bp_{100} 152.8°; bp_{40} 133.8°; bp_{20} 119.8°; bp_{10} 106.0°; bp_5 93.5°; $bp_{1.0}$ 67.8°. $n_D^{20.5}$ 1.41793. Insol in water; miscible with alcohol, ether; sol in ammonia water, aq solns of alkalies.

3745. Ethyl Dibunate. *3,6-Bis(1,1-dimethylethyl)-1-naphthalenesulfonic acid ethyl ester; 3,6-di-tert-butyl-1-naphthalenesulfonic acid ethyl ester;* ethyl 3,6-di-tert-butyl-1-naphthalenesulfonate; 2,7-di-*tert*-butylnaphthalene-4-sulfonic acid ethyl ester; ethyl 2,7-di-*tert*-butylnaphthalene-4-sulfonate; dibunate ethyl; NDR 304; Neodyne. $C_{20}H_{28}O_3S$; mol wt 348.51. C 68.93%, H 8.10%, O 13.77%, S 9.20%. Prepd from 3,6-di-*tert*-butyl-1-naphthalenesulfonyl chloride and ethanol: Menard *et al.*, *Can. J. Chem.* **39**, 729 (1961). Pharmacology: Shemano, *Arch. Int. Pharmacodyn. Ther.* **165**, 410 (1967).

Crystals from ethanol, mp 138-139°.

THERAP CAT: Antitussive.

3746. Ethyl Diethylmalonate. *Diethylpropanedioic acid diethyl ester.* $C_{11}H_{20}O_4$; mol wt 216.27. C 61.09%, H 9.32%, O 29.59%. $C_2H_5COOC(C_2H_5)_2COOC_2H_5$. Produced by the action of sodium ethylate and ethylbromide on diethyl malonate.

Liquid. d 0.985-0.990. bp 228-230°. n_D^{20} 1.424. Insol in water; miscible with alcohol, ether.

USE: Manuf barbiturates.

3747. Ethyldimethyl-9-octadecenylammonium Bromide. *N-Ethyl-N,N-dimethyl-9-octadecen-1-aminium bromide;* Onyxide. $C_{22}H_{46}BrN$; mol wt 404.54. C 65.32%, H 11.46%, Br 19.76%; N 3.46%. Prepn: Du Bois, U.S. pat. **2,519,747** (1950 to Onyx Oil & Chem.).

USE: A cationic surface active agent. Used in the control of algae. Sold as a 75% concentrate in isopropanol or propylene glycol.

3748. Ethylene. *Ethene;* elayl; olefiant gas. C_2H_4; mol wt 28.05. C 85.63%, H 14.37%. $CH_2=CH_2$. Plant hormone; occurs in ripening fruit, in illuminating gas (up to 4%). Prepd by decompn of petr gases or by dehydration of alcohol. J. E. Faraday, *Encyclopedia of Hydrocarbon Com-*

pounds, lists 502 methods of making ethylene. Lab prepn from alc: Gattermann-Wieland, *Praxis des organischen Chemikers* (de Gruyter, Berlin, 40th ed, 1961) p 98; F. A. Wunder, E. I. Leupold, *Angew. Chem. Int. Ed.* **19**, 126 (1980). Biosynthesis from 1-aminocyclopropane-1-carboxylic acid: D. Adams, S. F. Yang, *Proc. Nat. Acad. Sci. USA* **76**, 170 (1979). Proposed mechanism of ethylene biosynthesis: M. C.Pirrung, *J. Am. Chem. Soc.* **105**, 7207 (1983). Comprehensive monograph: S. A. Miller, *Ethylene and its Industrial Derivatives* (Ernest Benn Ltd., London, 1969) 1321 pp. Review of ethylene in biological systems: Spencer, *Fortschr. Chem. Org. Naturst.* **27**, 31-80 (1969). Review of manuf: Faith, Keyes & Clark's *Industrial Chemicals,* F. A. Lowenheim, M. K. Moran, Eds. (Wiley-Interscience, New York, 4th ed., 1975) pp 376-384. *Review:* L. Kniel *et al.,* in Kirk-Othmer *Encyclopedia of Chemical Technology* vol. 9 (Wiley-Interscience, New York, 3rd ed., 1980) pp 393-431.

Colorless, flammable gas. Burns with a luminous flame. Solidif $-181°$ forming monoclinic prisms. mp $-169.4°$, $bp_{700} -102.4°$. Vapor density 0.978 (air = 1.00). One liter of ethylene gas at 760 mm and 0° weighs 1.260 g. $bp_{2\,atm} -90.8°$; $bp_{10\,atm} -52.8°$; $bp_{30\,atm} -14.2°$; $bp_{50\,atm} +8.9°$. $d_4^{6.5}$ (liquid): 0.30342. Critical temp $+9.6°$; crit press. 50.7 atm. One volume of ethylene gas dissolves in about 4 vols of water at 0°, in about 9 vols of water at 25°, in 0.5 vol alcohol at 25°, in about 0.05 vol ether at 15.5°. Sol in acetone, benzene. Explosive limits (% vol in air), lower: 3.02; upper: 34. Autoignition temp: 1009°F ($+543°C$). Lethal concn for mice in air: 950,000 ppm, Flury, *Arch. Exp. Pathol. Pharmacol.* **138**, 65 (1928).

Human Toxicity: Simple asphyxiant. High concns cause narcosis, unconsciousness.

USE: Oxyethylene welding and cutting metals; manuf alcohol, mustard gas, and many other organics. Manuf ethylene oxide (for plastics), "Polythene", polystyrene and other plastics. Plant growth regulator; used commercially to accelerate the ripening of various fruits.

THERAP CAT: Anesthetic (inhalation).

THERAP CAT (VET): Has been used as inhalation anesthetic.

3749. Ethylene Bromohydrin. *2-Bromoethanol; β-bromoethyl alcohol; glycol bromohydrin.* C_2H_5BrO; mol wt 124.98. C 19.22%, H 4.03%, Br 63.95%, O 12.80%. $BrCH_2-CH_2OH$. Prepd by the action of hydrobromic acid on ethylene oxide: Thayer *et al., Org. Syn.* **6**, 12 (1926).

Hygroscopic liq. $d^0 1.7902$; $d_4^{15} 1.7696$; $d_4^{20} 1.7629$; $d_4^{25} 1.7560$; $d_4^{30} 1.7494$. $bp_{750} 149-150°$ (decompn); $bp_{20} 56-57°$; $bp_{13} 48.5°$. $n_D^{20} 1.49361$. Miscible with water; sol in the usual organic solvents, except petr ether. Hydrolysis of aq solns is accelerated by heat, acids, alkalies. Aq solns have a sweet, burning taste.

Caution: Ethylene bromohydrin vapors are an irritant to the eyes and the mucosa.

3750. Ethylene Chlorohydrin. *2-Chloroethanol;* 2-chloroethyl alcohol; glycol chlorohydrin. C_2H_5ClO; mol wt 80.52. C 29.83%, H 6.26%, Cl 44.04%, O 19.87%. $ClCH_2-CH_2OH$. Made from ethylene by action of a hypochlorite. Toxicity data: H. F. Smyth *et al., J. Ind. Hyg. Toxicol.* **23**, 259 (1941). E. Browning, *Toxicity and Metabolism of Industrial Solvents* (Elsevier, New York, 1965) pp 397-401. *Review:* G. H. Riesser in Kirk-Othmer *Encyclopedia of Chemical Technology* vol. 5 (Wiley-Interscience, New York, 3rd ed., 1979) pp 848-864.

Colorless liq. *Poisonous!* $d_4^{20} 1.197$. bp 128-130°. mp $-67°$. $n_D^{20} 1.4419$. Flash pt, open cup: 105°F (40°C). Miscible with water, alcohol. LD_{50} orally in rats: 0.095 g/kg (Smyth). LC for rats in air: 32 ppm (Browning).

Caution: Inhalation of vapor has been fatal. May cause nausea, vomiting, pains in head and chest, stupefaction. More toxic by skin contact than orally. Irritates mucous membranes and causes kidney and liver degeneration, *cf.* Patty's *Industrial Hygiene and Toxicology* vol. 2C, G. D. Clayton, F. E. Clayton, Eds. (Wiley-Interscience, New York, 4th ed., 1982) pp 4675-4684.

USE: Solvent; manuf insecticides; treating sweet potatoes before planting.

3751. Ethylene Cyanohydrin. *3-Hydroxypropanenitrile;* hydracrylonitrile; glycol cyanohydrin; β-hydroxypropionitrile. C_3H_5NO; mol wt 71.08. C 50.69%, H 7.09%, N 19.71%, O 22.51%. $HOCH_2CH_2CN$. Prepd by the interaction of ethylene chlorohydrin and alkali cyanide: E. C. Kendall, B. McKenzie, *Org. Syn.* **coll. vol. I**, 256 (2nd ed., 1941).

Liquid. $d_4^{25} 1.0404$. mp $-46°$. $bp_{760} 228°$ (slight dec); $bp_{200} 178°$; $bp_{100} 157.7°$; $bp_{60} 144.7°$; $bp_{40} 134.1°$; $bp_{20} 117.9°$; $bp_{10} 102°$; $bp_5 87.8°$; $bp_{1.0} 58.7°$. $n_D^{25} 1.4241$. Miscible with water, acetone, methyl ethyl ketone, ethyl alcohol. Slightly sol in ether (2.3% w/w at 15°). Insol in benzene, petr ether, carbon disulfide, carbon tetrachloride. LD_{50} orally in rats: 10.0 g/kg, H. F. Smyth, C. P. Carpenter, *J. Ind. Hyg. Toxicol.* **26**, 269 (1944).

USE: Solvent for some cellulose esters and many inorganic salts.

3752. Ethylenediamine. *1,2-Ethanediamine;* 1,2-diaminoethane. $C_2H_8N_2$; mol wt 60.10. C 39.97%, H 13.41%, N 46.62%. $H_2NCH_2CH_2NH_2$. Manuf from ethylene dichloride and ammonia: Faith, Keyes & Clark's *Industrial Chemicals,* F. A. Lowenheim, M. K. Moran, Eds. (Wiley-Interscience, New York, 4th ed., 1975) pp 385-388. Toxicity: H. F. Smyth *et al., J. Ind. Hyg. Toxicol.* **23**, 259 (1941).

Colorless, clear, thick liq; ammonia odor. $d_4^{25} 0.898$. bp 116-117°. mp $+8.5°$. $n_D^{26} 1.4540$. Flash pt, closed cup: 110°F (43°C). Volatile with steam. Freely sol in water forming a hydrate, also in alcohol; slightly sol in ether; sol in benzene unless insufficiently methylated: S. G. Boas-Traube *et al., Nature* **162**, 960 (1948). *Note:* Strongly alkaline and may readily absorb CO_2 from air to form a nonvolatile carbonate. Protect against undue exposure to the atmosphere. Care must be taken in handling because of the caustic nature of ethylenediamine and the irritating properties of its vapor. LD_{50} orally in rats: 1.16 g/kg (Smyth).

Monohydrate, mp 10°; bp 118°.

Dihydrochloride, $C_2H_{10}Cl_2N_2$, crystals. Sublimes without melting. Freely sol in water. Practically insol in alcohol. Aq soln is practically neutral.

USE: Solvent for casein, albumin, shellac, and sulfur; emulsifier; stabilizing rubber latex; as inhibitor in antifreeze solns; in textile lubricants. Pharmaceutic aid (aminophylline injection stabilizer).

THERAP CAT (VET): Has been used as a urinary acidifier.

3753. Ethylene Dibromide. *1,2-Dibromoethane; sym*-dibromoethane; ethylene bromide; EDB; Dowfume W 85. $C_2H_4Br_2$; mol wt 187.88. C 12.78%, H 2.14%, Br 85.07%. $BrCH_2CH_2Br$. Made from ethylene and bromine; also from acetylene and HBr. Manuf: Faith, Keyes & Clark's *Industrial Chemicals,* F. A. Lowenheim, M. K. Moran, Eds. (Wiley-Interscience, New York, 4th ed., 1975) pp 389-391. Toxicity data: G. W. Fischer *et al., J. Prakt. Chem.* **320**, 133 (1978). Carcinogenicity studies: W. A. Olson *et al., J. Nat. Cancer Inst.* **51**, 1993 (1973); *Clin. Toxicol.* **14**, 473 (1979). History of controversial use as a fumigant: J. Walsh, *Science* **215**, 1592 (1982).

Heavy liquid; chloroform odor. $d_{25}^{25} 2.172$. bp 131-132°. mp $+9°$. $n_D^{20} 1.5379$. Vapor pressure at 25°: 11 mm Hg. Sol in about 250 parts water; misc with alcohol, ether. *Protect from light.* LD_{50} i.p. in mice: 220 mg/kg (Fischer).

Note: This substance may reasonably be anticipated to be a carcinogen: *Fourth Annual Report on Carcinogens* (NTP 85-002, 1985) p 77.

USE: Fumigant; in anti-knock gasolines. *Caution:* Severe skin irritant; may cause blistering. Inhalation causes delayed pulmonary lesions. Prolonged exposure may result also in liver and kidney injury.

3754. Ethylene Dichloride. *1,2-Dichloroethane; sym*-dichloroethane; ethylene chloride; EDC; Dutch liquid; Brocide. $C_2H_4Cl_2$; mol wt 98.96. C 24.27%, H 4.07%, Cl 71.66%. $ClCH_2CH_2Cl$. Made from ethylene and chlorine; also from acetylene and HCl: *Beilstein* **I**, 84 (1918). Manuf: Faith, Keyes & Clark's *Industrial Chemicals,* F. A. Lowenheim, M. K. Moran, Eds. (Wiley-Interscience, New York, 4th ed., 1975) pp 392-396. Toxicity data: Smyth *et al., Am. Ind. Hyg. Assoc. J.* **30**, 470 (1969). Review of carcinogenicity studies: *IARC Monographs* **20**, 429-444 (1979).

Heavy liq; burns with smoky flame; pleasant odor; sweet taste; vapors are irritating. $d_4^{20} 1.2569$. bp 83-84°. mp $\sim -40°$. Flash pt, closed cup: 56°F (13°C); open cup: 65°F (18°C). $n_D^{20} 1.4443$. Sol in about 120 parts water; misc with alcohol, chloroform, ether. LD_{50} orally in rats: 770 mg/kg (Smyth).

Caution: Vapors may produce irritation of respiratory tract and conjunctiva, corneal clouding, CNS depression, liver and kidney impairment: *Clinical Toxicology of Commercial Products,* R. E. Gosselin *et al.,* Eds. (Williams & Wilkins, Baltimore, 5th Ed., 1984) Section II, pp 163-164. This substance may reasonably be anticipated to be a carcinogen: *Fourth Annual Report on Carcinogens* (NTP 85-002, 1985) p 81.

USE: Solvent for fats, oils, waxes, gums, resins, and particularly for rubber; manuf acetyl cellulose, tobacco extract, etc. Also used as fumigant.

3755. Ethylene Glycol. *1,2-Ethanediol.* $C_2H_6O_2$; mol wt 62.07. C 38.70%, H 9.74%, O 51.56%. $HOCH_2CH_2OH$. Prepd on a large scale by the hydration of ethylene oxide. Flowsheets: *Ullmann's Encyklopädie der technischen Chemie,* **3** (3rd ed., 1953) p 137; M. Sittig, *Organic Chemical Process Encyclopedia* (Noyes Dev. Corp., Park Ridge, N.J., 1967) p 265. Manuf from ethylene and oxygen: Faith, Keyes & Clark's *Industrial Chemicals,* F. A. Lowenheim, M. K. Moran, Eds. (Wiley-Interscience, New York, 4th ed., 1975) pp 397-402. Monograph: G. O. Curme, F. Johnston, *Glycols* (Reinhold, New York, 1952).

Slightly viscous liq. Sweet taste *(Poisonous! Do not swallow!).* Considerably hygroscopic: Absorbs twice its weight of water at 100% relative humidity. d_4^0 1.1274. d_4^{10} 1.1204. d_4^{20} 1.1135. d_4^{30} 1.1065. One gallon weighs 9.3 lbs. Flash pt, open cup: 240°F (115°C). mp −13°. bp_{760} 197.6°; bp_{97} 140°; bp_{18} 100°; $bp_{3.0}$ 70°; $bp_{0.06}$ 20°. n_D^{15} 1.43312; n_D^{25} 1.43063. Viscosity in centipoises: 26 at 15°; 21 at 20°; 17.3 at 25°. Dielectric constant at 20° and 150 meters wavelength: 38.66 esu. Dipole moment 2.20. Spec heat at 20°: 0.561 cal/g/°C. Heat of formation −108.1 kcal/mol. Heat of fusion 44.7 cal/g. Heat of vaporization 191 cal/g. Heat of soln −6.5 cal/g of solute at 17° when 37 parts are mixed with 63 parts H_2O (w/w). Parachor 148.9 (theory 152.2). Surface tension at 20° = 48.4 dynes/cm. Miscible with water, lower aliphatic alcohols, glycerol, acetic acid, acetone and similar ketones, aldehydes, pyridine and similar coal tar bases. Slightly sol in ether (1:200). Practically insol in benzene and its homologs, chlorinated hydrocarbons, petr ether, oils. *Density and freezing point of ethylene glycol-water mixtures:* 10.15% ethylene glycol by wt (d_{20}^{20}, fp): 1.013, −3.5°; 20.44% ethylene glycol: 1.027, −8°; 29.88% ethylene glycol: 1.040, −15°; 40.23% ethylene glycol: 1.054, −24°; 50.18% ethylene glycol: 1.067, −36°; 58.37% ethylene glycol: 1.0770, −48°. LD_{50} orally in rats, guinea pigs: 8.54, 6.61 g/kg, Smyth *et al., J. Ind. Hyg. Toxicol.* **23,** 259 (1941); in mice: 13.79 ml/kg, Bornmann, *Arzneimittel-Forsch.* **4,** 643 (1954).

Human Toxicity: Constitutes a hazard when ingested; *e.g.,* drinking of antifreeze fluid. Transient stimulation of CNS followed by depression; vomiting, drowsiness, coma, respiratory failure, convulsions; renal damage, which may proceed to anuria, uremia, death. Lethal dose in humans about 1.4 ml/kg or 100 ml. *See* E. Browning, *Toxicity and Metabolism of Industrial Solvents* (Elsevier, New York, 1965) pp 594-600; 686-690; Patty's *Industrial Hygiene and Toxicology* vol. 2C, G. D. Clayton, F. E. Clayton, Eds. (Wiley-Interscience, New York, 3rd ed., 1982) pp 3817-3832.

USE: Antifreeze in cooling and heating systems. In hydraulic brake fluids. Industrial humectant. Ingredient of electrolytic condensers (where it serves as solvent for boric acid and borates). Solvent in the paint and plastics industries. In the formulation of printers' inks, stamp pad inks, and inks for ball-point pens. Softening agent for cellophane. Stabilizer for soybean foam used to extinguish oil and gasoline fires. In the synthesis of safety explosives, glyoxal, unsatd ester type alkyd resins, plasticizers, elastomers, synthetic fibers (Terylene, Dacron), and synthetic waxes.

3756. Ethylene Glycol Diacetate. *1,2-Ethanediol diacetate;* glycol diacetate; ethylene diacetate. $C_6H_{10}O_4$; mol wt 146.14. C 49.31%, H 6.90%, O 43.79%. $CH_3COOCH_2CH_2$-$OOCCH_3$. Prepn from ethylene bromide, glacial acetic acid and potassium acetate: Henry, *Bull. Soc. Chim.* [3] **17,** 207 (1897); Gattermann-Wieland, *Praxis des organischen Chemikers* (de Gruyter, Berlin, 40th ed., 1961) p 107; from synthe-

sis gas via homogeneous ruthenium catalysis: J. F. Knifton, *Chem. Commun.* **1981,** 188.

Liquid. d 1.104. bp 190-191°. mp −31°. n_D^{20} 1.415. Flash pt, open cup: 205°F (96°C). Sol in 7 parts water; miscible with alc, ether. LD_{50} orally in rats: 6.86 g/kg, Smyth *et al., J. Ind. Hyg. Toxicol.* **23,** 259 (1941).

USE: Solvent for oils, cellulose esters, explosives, etc.

3757. Ethylene Glycol Monoacetate. *1,2-Ethanediol monoacetate;* glycol-monoacetin. $C_4H_8O_3$; mol wt 104.10. C 46.15%, H 7.75%, O 46.11%. $CH_3COOCH_2CH_2OH$.

Liquid. d 1.108. bp 182°. Miscible with water, alcohol. LD_{50} orally in rats: 8.25 g/kg, Smyth *et al., J. Ind. Hyg. Toxicol.* **23,** 259 (1941).

3758. Ethylene Oxide. *Oxirane;* Anprolene. C_2H_4O; mol wt 44.05. C 54.53%, H 9.15%, O 36.32%. Prepd from ethylene chlorohydrin and KOH: A. Wurtz, *Ann.* **110,** 125 (1859). Manuf by catalytic oxidation of ethylene: T. E. Lefort, U.S. pat. **1,998,878** (1935 to Société Française de Catalyse Généralisée) reissued as U.S. pat. **20,370** (1937 to Carbide and Carbon). Physical props: W. H. Perkin, *J. Chem. Soc.* **63,** 488 (1893). Leukemia in workers exposed to ethylene oxide: C. Hogstedt *et al., J. Am. Med. Assoc.* **241,** 1132 (1979). *Review:* H. C. Schultze in *Glycols,* G. O. Curme, Jr., Ed. (Reinhold, New York, 1952) pp 74-113; J. N. Cawse *et al.,* in Kirk-Othmer *Encyclopedia of Chemical Technology* vol. **9** (Wiley-Interscience, New York, 3rd ed., 1980) pp 432-471.

Colorless, flammable gas at ordinary room temp and pressure; liquid below 12°. bp 10.7°. d_4^4 0.891, d_7^7 0.887, d_{10}^{10} 0.882. mp −111°. n_D^7 1.3597. Sol in water, alcohol, ether.

Caution: Explosive! Hazards and handling: L. G. Hess, V. V. Tilton, *Ind. Eng. Chem.* **42,** 1251 (1950). Highly irritating to eyes, mucous membranes. High concns can cause pulmonary edema. Exposure should not exceed 50 ppm on an 8-hour time-weighted average: *Prudent Practices for Handling Hazardous Chemicals in Laboratories* (National Academy Press, Washington, D.C., 1981) p 264. This substance may reasonably be anticipated to be a carcinogen: *Fourth Annual Report on Carcinogens* (NTP 85-002, 1985) p 106.

USE: Fumigant for foodstuffs and textiles; to sterilize surgical instruments; agricultural fungicide; (commercially as mixture with inert gases). In organic syntheses, esp in the production of ethylene glycol. Starting material for the manuf of acrylonitrile and nonionic surfactants.

3759. Ethylene Thiourea. *2-Imidazolidinethione; imidazoline-2-thiol;* 2-mercaptoimidazoline; ETU; Akrochem ETU-22; NA-22; Robac 22; Sanceller 22; Vulkacit NPV/C. $C_3H_6N_2S$; mol wt 102.17. C 35.26%, H 5.92%, N 27.42%, S 31.39%. Degradation product of ethylenebisdithiocarbamate fungicides such as mancozeb, maneb, zineb, *q.q.v.* Prepn: C. F. H. Allen *et al., Org. Syn.* coll. vol. **III,** 394 (1955); G. Matolcsy, *Ber.* **101,** 522 (1968). Identification as decomposition product: R. A. Ludwig *et al., Can. J. Bot.* **32,** 48 (1954); G. Czegledi-Janko, *J. Chromatog.* **31,** 89 (1967). Determn in various foods by GLC: W. H. Newsome, *J. Agr. Food Chem.* **20,** 967 (1972); by GLC, HPLC: J. H. Onley *et al., J. Assoc. Offic. Anal. Chem.* **60,** 1105 (1977); J. H. Onley, *ibid.* 1111. Review of determn methods: P. Bottomley *et al., Residue Rev.* **95,** 45-89 (1985). Metabolism in mice: K. Savolainen, H. Pyysalo, *J. Agr. Food Chem.* **27,** 1177 (1979); in cats and rats: F. Iverson, *Toxicol. Appl. Pharmacol.* **52,** 16 (1980). Aquatic toxicity data: C. J. VanLeeuwen, *Aquat. Toxicol.* **7,** 145 (1985). Goitrogenic activity: S. L. Graham, W. H. Hansen, *Bull. Environ. Contam. Toxicol.* **7,** 19 (1972); D. M. Smith, *Brit. J. Ind. Med.* **41,** 362 (1984). Review of toxicology: L. Fishbein, *J. Toxicol. Environ. Health* **1,** 713-756 (1976). Use as crosslinking agent in rubber industry: J. T. Oetzel, *Rubber World* **172,** 55 (1975); C. Hepburn, M. S. Mahdi, *Kaut. Gummi. Kunstst.* **39,** 629 (1986).

Needles, prisms from alc or amyl alc. mp 203-204°. Soly in 100 ml water: 2 g at 30°, 9 g at 60°, 44 g at 90°. Moderately sol in methanol, ethanol, ethylene glycol, and pyridine. Insol in acetone, ether, chloroform, benzene, ligroin. LD_{50} orally in rats: 1832 mg/kg (Graham, Hansen).

Note: This substance may reasonably be anticipated to be a carcinogen: *Fourth Annual Report on Carcinogens* (NTP 85-002, 1985) p 109.

USE: Accelerator in synthetic rubber production.

3760. Ethylenimine. *Aziridine;* azacyclopropane; dimethylenimine. C_2H_5N; mol wt 43.07. C 55.77%, H 11.70%, N 32.53%. Prepd by treating 2-chloroethylamine hydrochloride with NaOH: Wystrach *et al., J. Am. Chem. Soc.* **77,** 5915 (1955); **78,** 1263 (1956); from β-aminoethylsulfuric acid by reaction with NaOH: Allen *et al., Org. Syn.* **coll. vol. IV,** 433 (1963). Acute toxicity: H. F. Smyth *et al., J. Ind. Hyg. Toxicol.* **23,** 259 (1941). Toxicology: Weightman, Hoyle, *J. Am. Med. Assoc.* **189,** 543 (1964). Review of carcinogenicity studies: *IARC Monographs* **9,** 37 (1975).

Liquid. Intense odor of ammonia. *Poisonous! Handle in hood only!* Strongly alkaline. Polymerizes easily. d_4^{24} 0.8321. mp −73.96°. bp_{760} 56-57°. n_D^{25} 1.412. Vapor press. (20°) 160 mm Hg. Miscible with water. Sol in alc. Can be stored for some time over a few pellets of sodium hydroxide. LD_{50} orally in rats: 15 mg/kg (Smyth).

Caution: Strongly irritating to eyes, skin, mucous membranes. Can be a skin sensitizer. This substance has been listed as a carcinogen by OSHA: *Fed. Reg.* **39,** 3757 (1974).

USE: In the manuf of triethylenemelamine.

3761. Ethylestrenol. *19-Nor-17α-pregn-4-en-17-ol; 17α-ethylestr-4-en-17β-ol;* 17α-ethyl-17β-hydroxy-4-estrene; 17β-hydroxy-17α-ethyl-19-nor-4-androstene; Orabolin; Durabolin-O; Orgaboral; Maxibolin; Orgabolin (obsolete). $C_{20}H_{32}O$; mol wt 288.46. C 83.27%, H 11.18%, O 5.55%. Prepn: de Winter *et al., Chem. & Ind. (London)* **1959,** 905; Szpilfogel, de Winter, U.S. pat. **2,878,267** (1959 to Organon).

Crystals, mp 76-78°. $[\alpha]_D$ +31° (chloroform).
THERAP CAT: Anabolic.

3762. Ethyl Ether. *1,1'-Oxybisethane;* ethoxyethane; ether; diethyl ether; ethyl oxide; diethyl oxide; sulfuric ether; anesthetic ether. $C_4H_{10}O$; mol wt 74.12. C 64.81%, H 13.60%, O 21.59%. $C_2H_5OC_2H_5$. Produced on a large scale by dehydration of ethanol or by hydration of ethylene, both processes being carried out in the presence of sulfuric acid. Review of mfg processes: Himmler in *Ullmann's Encyklopädie der Technischen Chemie,* **vol. 5** (1954) pp 777-782; D. E. Keeley in Kirk-Othmer *Encyclopedia of Chemical Technology* **vol. 9** (Wiley-Interscience, New York, 3rd ed., 1980) pp 381-393.

Mobile, very volatile, highly flammable liq. Vapor heavier than air. Characteristic, sweetish, pungent odor, more agreeable than chloroform. Burning taste. Tends to form explosive peroxides under the influence of air and light, esp when evaporation to dryness is attempted. Peroxides may be removed from ether by shaking with 5% aq ferrous sulfate soln. Addition of naphthols, polyphenols, aromatic amines,

and aminophenols has been proposed for the stabilization of ethyl ether. d_4^0 0.7364; d_4^{10} 0.7249; d_4^{20} 0.7134; d_4^{30} 0.7019. Vapor density 2.55 (air = 1.0). mp −116.3° (stable crystals); mp −123.3° (metastable crystals). bp_{760} 34.6°; bp_{400} 17.9°; bp_{200} +2.2°; bp_{100} −11.5°; bp_{10} −48.1°; $bp_{1.0}$ −74.3°. Satd vapor press. at 0°: 184.9 mm; at 10°: 290.8 mm; at 20°: 439.8 mm; at 50°: 1276 mm; at 70°: 2304 mm. Critical temp 192.7°; crit press. 35.6 atm. Flash pt, closed cup: −49°F (−45°C). Air-ether mixtures containing more than 1.85 volume-% of ether vapor, are explosive hazards. Auto-ignition temp 180-190°. n_D^{15} 1.35555. Dielectric constant at 26.9° and 85.8 kilocycles = 4.197; good insulator. When shaken under absolutely dry conditions ether can generate enough static electricity to start a fire. Surface tension at 20°: 17.06 dynes/cm. Viscosity at 20°: 0.2448 cp. Heat of vaporization at 30°: 89.80 cal/g. Produces considerable coldness on quick evaporation. Heat of formation −907 cal/g; heat of combustion −8.807 kcal/g. Ether is slightly sol in water and water is slightly sol in ether. A satd water soln contains 8.43% (w/w) of ether at 15° and 6.05% (w/w) at 25°. Ether satd with water contains 1.2% H_2O at 20°. Soly in water increased by HCl. Sol in concd hydrochloric acid. May explode when brought in contact with anhydr nitric acid. Miscible with lower aliphatic alcohols, benzene, chloroform, petr ether, other fat solvents, many oils. Azeotrope with water (1.3%), bp 34.2°.

Human Toxicity: Mildly irritating to skin, mucous membranes. Inhalation of high concns causes narcosis, unconsciousness. Death may occur due to respiratory paralysis, cf. *Toxicity and Metabolism of Industrial Solvents,* E. Browning, Ed. (Elsevier, New York, 1965) pp 493-498.

USE: Solvent for waxes, fats, oils, perfumes, alkaloids, gums. Excellent solvent for nitrocellulose when mixed with alc. Important reagent in organic syntheses, esp in Grignard and Wurtz type reactions. Easily removable extractant of active principles (hormones, etc.) from plant and animal tissues. In the manuf of gun powder. As primer for gasoline engines. *Caution:* Explosion hazard.

THERAP CAT: Anesthetic (inhalation).

THERAP CAT (VET): Inhalation anesthetic. Has been used orally in colic, subcutaneously as a stimulant.

3763. Ethyl Formate. *Formic acid ethyl ester.* $C_3H_6O_2$; mol wt 74.08. C 48.64%, H 8.17%, O 43.20%. $HCOOC_2H_5$.

Mobile, flammable liq. d_4^{20} 0.917. bp 53-54°. mp −80°. Flash pt, closed cup: −4°F (−20°C). n_D^{20} 1.3597. Sol in about 10 parts water with gradual decompn into free acid and alcohol; miscible with alcohol, ether. *Keep tightly closed and preferably in contact with calcium chloride.* LD_{50} orally in rats: 4.29 g/kg, H. F. Smyth *et al., Arch. Ind. Hyg. Occup. Med.* **10,** 61 (1954).

USE: As flavor for lemonades and essences; for manuf artificial rum and arrac; also as a solvent for nitrocellulose; as fungicide and larvicide for tobacco, cereals, dried fruits, etc.; in organic synthesis. *Caution:* Irritating to skin, mucous membranes, and, in high concns, narcotic.

3764. 2-Ethyl-1-hexanol. 2-Ethylhexyl alcohol. $C_8H_{18}O$; mol wt 130.22. C 73.78%, H 13.93%, O 12.29%. Manuf from butyraldehyde: Faith, Keyes & Clark's *Industrial Chemicals,* F. A. Lowenheim, M. K. Moran, Eds. (Wiley-Interscience, New York, 4th ed., 1975) pp 413-417.

Colorless liq. d_{20}^{20} 0.8344. bp 184-185°. n_D^{20} 1.4300. Flash pt 81°. Soluble in about 720 parts water, in many organic solvents. It dissolves about 2.5% its weight of water at 25°. LD_{50} orally in rats: 12.46 ml/kg, H. F. Smyth *et al., Am. Ind. Hyg. Assoc. J.* **30,** 470 (1969).

USE: Mercerizing textiles; as a solvent for dyes, resins, oils; also claimed to possess antifoaming properties.

3765. Ethylhydrocupreine. *6'-Ethoxy-10,11-dihydro-8α-cinchonan-9R-ol; hydrocupreine ethyl ether;* Optoquine; Numoquin. $C_{21}H_{28}N_2O_2$; mol wt 340.45. C 74.08%, H 8.29%, N 8.23%, O 9.40%. Prepn: Giemsa, Halberkann, *Ber.* **51,** 1325 (1918); Thron, Freund, U.S. pat. **1,062,203**

(1913 to Ver. Chininfabr. Zimmer); Muira, *C.A.* **41**, 4582d (1947).

White bitter cryst powder. mp 123-128° when solvent-free. $[\alpha]_D^{25}$ −136.2° (alcohol). Practically insol in water; sol in alcohol, benzene, chloroform, ether, dil acids, oils, fats. LD s.c. in mice: 5.0 g/kg.

Hydrochloride, $C_{21}H_{29}ClN_2O_2$, *Neumolisina.* Rhombic crystals from acetone + ether, mp 252-254°. $[\alpha]_D^{21}$ −123.6° (water). One gram dissolves in 2 ml water, 5 ml alcohol, 2.5 ml chloroform; sparingly sol in dry acetone. Practically insol in ether, petr ether. Aq solns are sterilizable without decompn. *Protect from light.* LD s.c. in mice: 5.0 g/kg.

THERAP CAT: Antiseptic.

3766. Ethylidene Chloride. *1,1-Dichloroethane.* $C_2H_4Cl_2$; mol wt 98.97. C 24.27%, H 4.07%, Cl 71.65%. CH_3-$CHCl_2$. Prepd by the action of PCl_5 on acetaldehyde.

Oily liquid; odor and taste as of chloroform. d_4^{20} 1.1757; d_4^{25} 1.1680. bp 57.3°: Mumford, Phillips, *J. Chem. Soc.* **1950**, 75. mp ∼−98°. n_D^{20} 1.4167. Sol in about 200 parts water; miscible with alcohol.

Caution: Narcotic in high concns.

3767. Ethylidene Diacetate. *1,1-Ethanediol diacetate;* 1,1-diacetoxyethane. $C_6H_{10}O_4$; mol wt 146.14. C 49.31%, H 6.90%, O 43.79%. $(CH_3COO)_2CHCH_3$. Made by heating acetaldehyde and acetic anhydride. Control of fungal and bacterial growth in crops and animal feedstuffs: D. L. Kensler *et al.,* U.S. pats. **3,931,412** and **4,012,526** (1976 and 1977 to Chevron).

Liq; sharp, fruity odor. d_4^{12} 1.061. bp 167-169°. Slightly sol in water; miscible with alcohol. With NaOH it yields acetaldehyde.

USE: Agricultural fungicide.

3768. Ethylidene Dicoumarol. *3,3′-Ethylidenebis[4-hydroxy-2H-1-benzopyran-2-one];* 3,3′-ethylidenebis(4-hydroxycoumarin); 3,3-ethylidenebis(4-oxycoumarin); Pertrombon. $C_{20}H_{14}O_6$; mol wt 350.31. C 68.57%, H 4.03%, O 27.40%. Prepn: Jansen, Jensen, *Z. Phys. Chem.* **277**, 66 (1943); Sullivan *et al., J. Am. Chem. Soc.* **66**, 2288 (1943); Stahmann *et al., ibid.* **67**, 900 (1944); **Austrian pat. 196,874** (1958 to Spofa), *C.A.* **52**, 10212d (1958); Fucik, **Czech. pat. 85,918** (1957), *C.A.* **52**, 2930d (1958).

Crystals from ethanol + dioxane, mp 178°.
Dibenzoate, $C_{34}H_{22}O_8$, mp 209-210°.
THERAP CAT: Anticoagulant.

3769. Ethyl Iodide. *Iodoethane.* C_2H_5I; mol wt 155.98. C 15.40%, H 3.23%, I 81.37%. CH_3CH_2I. Made by the action of iodine on alcohol in the presence of red phosphorus. Ethyl iodide contains about 98% CH_3CH_2I and about 2% alcohol.

Heavy, clear, very refractive liquid; ethereal odor. When freshly prepared it is colorless, but soon becomes red (on exposure to light and air) due to liberation of iodine. Silver leaf prevents or greatly retards decompn. d_4^{20} 1.950. bp 72°. mp −108°. n_D^{15} 1.5168. Sol in 250 parts water with gradual decompn; miscible with alcohol and most organic solvents. *Keep dry and cool, tightly closed, protected from light.*

3770. Ethyl Isobutyrate. *2-Methylpropanoic acid ethyl*

ester. $C_6H_{12}O_2$; mol wt 116.16. C 62.04%, H 10.41%, O 27.55%. $(CH_3)_2CHCOOC_2H_5$.

Liquid; aromatic, fruity odor. d_{20}^{20} 0.870. bp 110-111°. mp −88°. n_D^{20} 1.3903. Slightly sol in water; miscible with alcohol, ether.

USE: Manuf flavoring compds and essences.

3771. Ethyl Isothiocyanate. *Isothiocyanatoethane;* ethyl mustard oil. C_3H_5NS; mol wt 87.15. C 41.34%, H 5.78%, N 16.08%, S 36.80%. C_2H_5NCS. Obtained by the action of mercuric chloride on the product of the reaction of carbon disulfide with ethylamine.

Colorless to yellowish liq; pungent odor. d_4^{18} 1.003. bp 130-132°. mp −6°. n_D^{18} 1.5142. Insol in water; miscible with alcohol, ether.

USE: Military poison gas. *Caution:* Vapors irritate the eyes; blister the skin.

3772. Ethyl Isovalerate. *3-Methylbutanoic acid ethyl ester.* $C_7H_{14}O_2$; mol wt 130.18. C 64.58%, H 10.84%, O 24.58%. $(CH_3)_2CHCH_2COOC_2H_5$.

Colorless, oily liq; apple odor. d_{20}^{20} 0.868. bp 135°. mp −99°. n_D^{20} 1.4009. Sol in about 350 parts water; miscible with alcohol, benzene, ether.

USE: In alc soln for flavoring confectionery and beverages.

3773. Ethyl Lactate. *2-Hydroxypropanoic acid ethyl ester;* ethyl α-hydroxypropionate. $C_5H_{10}O_3$; mol wt 118.13. C 50.83%, H 8.53%, O 40.63%. $CH_3CHOHCOOC_2H_5$.

Colorless liquid; characteristic odor. d_4^{14} 1.042. bp 154°. $[\alpha]_D^{14}$ −10°. Flash pt, closed cup: 117°F (47°C). Miscible with water (with partial decompn), alcohol, ether. *Keep well closed.*

USE: As solvent for nitrocellulose and cellulose acetate.

3774. Ethyl Laurate. *Dodecanoic acid ethyl ester.* $C_{14}H_{28}O_2$; mol wt 228.36. C 73.63%, H 12.36%, O 14.01%. $CH_3(CH_2)_{10}COOC_2H_5$.

Oil. d_{19}^{19} 0.867. mp −10°. bp 269°; bp_{25} 163°. n_D 1.4321. Insol in water; very sol in alcohol, ether.

3775. Ethyl Levulinate. *4-Oxopentanoic acid ethyl ester.* $C_7H_{12}O_3$; mol wt 144.17. C 58.31%, H 8.39%, O 33.29%. $CH_3COCH_2CH_2COOC_2H_5$.

Liquid. d_{20}^{20} 1.012. bp 205-206°. n_D^{20} 1.4229. Freely sol in water; miscible with alcohol.

3776. Ethyl Linoleate. *9,12-Octadecadienoic acid ethyl ester;* mandenol. $C_{20}H_{36}O_2$; mol wt 308.49. C 77.86%, H 11.76%, O 10.37%. Prepn from sunflower seed oil: McCutcheon, *Org. Syn.* coll. vol. III, 526 (1955); *see also* Parker *et al., Biochem. Prepn.* **4**, 88 (1955).

Colorless oil. More stable to air oxidation than linoleic acid. $d_4^{15.5}$ 0.8846; d^{20} 0.8919. bp_6 193°; $bp_{2.5}$ 175°; $bp_{0.001}$ 133°. n_D^{20} 1.46753; n_D^{48} 1.4489. Iodine value 162.5. Miscible with dimethylformamide, fat solvents, oils.

USE: In the vitamin industry.

3777. Ethyl Loflazepate. *7-Chloro-5-(2-fluorophenyl)-2,3-dihydro-2-oxo-1H-1,4-benzodiazepine-3-carboxylic acid ethyl ester;* ethyl 7-chloro-5-(o-fluorophenyl)-2,3-dihydro-2-oxo-1H-1,4-benzodiazepine-3-carboxylate; CM-6912; Meilax; Victan. $C_{18}H_{14}ClFN_2O_3$; mol wt 360.77. C 59.93%, H 3.91%, Cl 9.83%, F 5.27%, N 7.76%, O 13.30%. Benzodiazepine tranquilizer. Prepn: J. Hellerbach *et al.,* **Ger. pat. 2,001,276;** *eidem,* U.S. pat. **3,657,223** (1970, 1972 both to Hoffmann-La Roche). Analysis in plasma and urine: W. Cantreels, J. R. Jeanniot, *Biomed. Mass Spectrom.* **7**, 565 (1980). GC analysis of metabolites: J. P. Cano, *J. Chromatog.* **226**, 413 (1981). Toxicologic evaluation: G. Mozue *et al., Int. J. Clin. Pharmacol. Ther. Toxicol.* **19**, 453 (1981).

Cryst from ether, mp 193-194°.

Note: This is a controlled substance (depressant) listed in the U.S. Code of Federal Regulations, Title 21 Part 1308.14 (1985).

THERAP CAT: Anxiolytic.

3778. N-Ethylmaleimide. *1-Ethyl-1H-pyrrole-2,5-dione.* $C_6H_7NO_2$; mol wt 125.12. C 57.59%, H 5.64%, N 11.20%, O 25.58%. Prepd by heating N-ethylmaleamic acid in paraffin: Marrian, *J. Chem. Soc.* **1949**, 1515.

Crystals, mp 45°. Lacrimator when liquid.

USE: In cancer research (possible antimitotic activity.) *Caution:* Strong irritant.

3779. Ethyl Malonate. *Propanedioic acid diethyl ester;* diethyl malonate; malonic ester. $C_7H_{12}O_4$; mol wt 160.17. C 52.49%, H 7.55%, O 39.96%. $C_2H_5OOCCH_2COOC_2H_5$. Prepd from chloroacetic acid and sodium cyanide followed by esterification with ethanol and H_2SO_4: L. Gattermann, T. Wieland, *Praxis des Organischen Chemikers* (de Gruyter, Berlin, 40th ed., 1961) pp 220-221.

Liquid. Slightly aromatic, pleasant odor. d 1.055. bp 198-199°; bp_{20} 95°. mp $-50°$. n_D^{20} 1.4143. One gram dissolves in about 50 ml of water. Miscible with alcohol, ether.

USE: Manuf of barbiturates.

3780. Ethylmercuric Chloride. *Chloroethylmercury;* Granosan. C_2H_5ClHg; mol wt 265.13. C 9.06%, H 1.90%, Cl 13.37%, Hg 75.66%. CH_3CH_2HgCl. Prepd from ethylmercuric bromide by treating with methanolic KOH, filtering and neutralizing with HCl: Slotta, Jacobi, *J. Prakt. Chem.* [2] **120**, 249 (1929). Prepn from $HgCl_2$ and tetraethyl lead: Whelen, *Advances in Chemistry Series* **23**, entitled *Metal-Organic Compounds* (ACS, Washington, D.C., 1960) pp 82-86. For the prepn of the bromide from ethylmagnesium bromide and mercuric bromide *see* Slotta, Jacobi, *loc. cit.,* Marvel et al., *J. Am. Chem. Soc.* **47**, 3009 (1925). *Review:* Krause, von Grosse, *Die Chemie der Metall-organischen Verbindungen* (Berlin, 1937).

White, silvery leaflets from ethanol, mp 192°. Sublimes easily. Solubility in water at 18° = 1.4 × 10^{-4} g/100 ml, at 100° = 2.5 × 10^{-4} g/100 ml; in ethanol at 18° = 0.75 g/100 g; at 78° = 3.5 g/100 g; in chloroform at 18° = 2.6 g/100 g. Slightly sol in ether.

USE: Applied at 2% strength (soln or mixed with solids) as a fungicide for treating seeds. *Caution:* Highly toxic. Causes skin burns; is absorbed through the skin. Chronic exposure has caused permanent injury to brain. *Note:* A close analog, *chloro-(2-methoxyethyl)mercury, Ceresan Wet,* is also used as seed fungicide.

3781. N-(Ethylmercuri)-p-toluenesulfonanilide. *Ethyl-(4-methyl-N-phenylbenzenesulfonamidato-N)mercury; ethyl-(N-phenyl-p-toluenesulfonamido)mercury; Ceresan M.* $C_{15}H_{17}HgNO_2S$; mol wt 475.99. C 37.85%, H 3.60%, Hg 42.15%, N 2.94%, O 6.72%, S 6.74%.

Crystals. Pungent odor, somewhat reminiscent of garlic. Practically insol in water. LD_{50} orally in rats: 100 mg/kg, *Toxic Substances List,* H. E. Christensen, Ed. (1973) p 585.

USE: To control the various smuts which befall grain. Reduces bulb rot in gladiolus. *Caution:* Almost as toxic as mercuric chloride *(q.v.);* less irritating to the skin than ethyl mercury phosphate. Symptoms confined to CNS and consist of deafness, ataxia, dysarthria, progressive visual deterioration; dysphagia; sphincteric incontinence, mental confusion, stupor, death: Gleason et al., *Clinical Toxicology of Commercial Products* (Williams & Wilkins, Baltimore, 3rd ed., 1969) p 68, sect. II.

3782. Ethyl Methanesulfonate. *Methanesulfonic acid ethyl ester;* ethyl methanesulfonic acid; ethyl mesylate; EMS; NSC 26805. $C_3H_8O_3S$; mol wt 124.15. C 29.02%, H 6.49%, O 38.66%, S 25.82%. $CH_3SO_2OCH_2CH_3$. Prepn: O. C. Billeter, *Ber.* **38**, 2015 (1905). Mutagenicity studies: T. Alderson, *Nature* **207**, 164 (1965); J. B. Jenkins, *Mutat. Res.* **4**, 90 (1967); A. P. Schalet, *ibid.* **49**, 313 (1978). Review of carcinogenicity studies: *IARC Monographs* **7**, 245-252 (1974). Review of comparative mutagenicity of EMS and methyl methanesulfonate, *q.v.:* S. Kondo, *Environ. Sci. Res.* **24**, 743-785 (1981).

Liquid, bp_{761} 213-213.5°; bp_{10} 85-86°. d_4^{22} 1.1452.

USE: Exptlly as mutagen.

3783. Ethyl Methyl Ether. *Methoxyethane;* methyl ethyl ether. C_3H_8O; mol wt 60.09. C 59.96%, H 13.42%, O 26.62%. $C_2H_5OCH_3$.

Liquid. d_0^0 0.725. bp 10.8°. Sol in water; miscible with alcohol, ether.

3784. Ethylmethylthiambutene. *N-Ethyl-N-methyl-4,4-di-2-thienyl-3-buten-2-amine; N-ethyl-N,1-dimethyl-3,3-di-2-thienyl-2-propenamine; N-ethyl-N,1-dimethyl-3,3-di-2-thienylallylamine; 3-ethylmethylamino-1,1-di(2'-thienyl)but-1-ene; ethylmethiambutene;* 1C50; NIH-5145; Emethibutin. $C_{15}H_{19}NS_2$; mol wt 277.46. C 64.94%, H 6.91%, N 5.05%, S 23.11%. Prepn: Adamson, U.S. pat. **2,561,899** (1951 to Burroughs Wellcome). Toxicology: N. B. Eddy, D. Leimbach, *J. Pharmacol. Exp. Ther.* **107**, 385 (1953).

$bp_{0.01}$ 110-113°; bp_{18} 75-76°.

Hydrochloride, $C_{15}H_{19}NS_2$·HCl, crystals, mp 137-138°. LD_{50} in mice (mg/kg): 192 orally; 88 s.c. (Eddy, Leimbach). *Caution:* May be habit forming. This is a controlled substance (opiate) listed in the U.S. Code of Federal Regulations, Title 21 Part 1308.11 (1985).

THERAP CAT: Narcotic analgesic.

3785. Ethylmorphine. *7,8-Didehydro-4,5-epoxy-3-ethoxy-17-methylmorphinan-6-ol.* $C_{19}H_{23}NO_3$; mol wt 313.38. C 72.82%, H 7.40%, N 4.47%, O 15.32%. Prepd by ethylation of morphine: Baizer, Ellner, *J. Am. Pharm. Assoc.* **39**, 581 (1950); Gorecki, *Ann. Pharm. (Poznan)* **7**, 21 (1969). Toxicity data: M. Aurousseau, J. Navarro, *Ann. Pharm. Franc.* **15**, 640 (1957).

Crystals from ethanol, mp 199-201°.

Hydrochloride dihydrate, $C_{19}H_{24}ClNO_3$·$2H_2O$, *Codethyline, Dionin.* White to faintly yellow crystalline powder, mp about 123° (dec); anhydrous form, mp about 170° (dec). One gram dissolves in 10 ml water and in 25 ml alcohol; slightly sol in chloroform, ether. LD_{50} in mice: 264.6 mg/kg s.c.; 0.771 g/kg orally (Aurousseau, Navarro).

Methiodide, $C_{20}H_{26}INO_3$, *ethyl-N-methylmorphinium iodide, Trachyl.*

Caution: This is a controlled substance (opiate) listed in the U.S. Code of Federal Regulations, Title 21 Part 1308.12 (1987).

THERAP CAT: Narcotic analgesic; antitussive; chemotic.

THERAP CAT (VET): Narcotic analgesic; antitussive; mydriatic.

3786. Ethyl Nitrite. *Nitrous acid ethyl ester;* nitrous ether. $C_2H_5NO_2$; mol wt 75.07. C 32.00%, H 6.71%, N 18.66%, O 42.63%. C_2H_5ONO. Made by the action of sodium nitrite on a mixture of alcohol and H_2SO_4 in the cold. The article of commerce contains 90-95% ethyl nitrite, the remainder is chiefly alcohol as preservative. In the laboratory ethyl nitrite gas is conveniently generated by placing 20 liters of 90% ethanol in a suitable vessel, diluting with 200 liters of water, and while stirring, adding to the dil alc 18.3 kg of nitrosyl chloride at the rate of 2.25 kg per hour—see *also* Chase, U.S. pat. **2,615,896** (1952).

Colorless or yellowish, clear, flammable, exceedingly volatile liq. Characteristic odor; burning, sweetish taste. d_{15}^{15} 0.90. bp 17°. Slightly sol in water and dec by it; miscible with alcohol, ether. On keeping, it gradually dec, becoming acid and oxides of nitrogen form. Decompn is hastened by air, light, and moisture. *Keep tightly closed, in a cool place, protected from light.*

USE: For preparing spirit nitrous ether by mixing with 21 parts alcohol by weight. *Caution:* May cause methemoglobinemia and hypotension and, in high concns, narcosis.

3787. Ethyl Nitrobenzoate. $C_9H_9NO_4$; mol wt 195.17. C 55.38%, H 4.65%, N 7.18%, O 32.79%. $O_2NC_6H_4COOC_2H_5$.

Colorless crystals. Insol in water; freely sol in alc, ether.

o-Ester, mp 30°. bp_{18} 173°.

m-Ester, mp 47°. bp about 298°.

p-Ester, mp 57°.

3788. N-Ethyl-N-nitrosourea. ENU. $C_3H_7N_3O_2$; mol wt 117.10. C 30.77%, H 6.02%, N 35.88%, O 27.33%. $C_2H_5N(NO)CONH_2$. Potent mouse mutagen. Prepn: E. A. Werner, *J. Chem. Soc.* **115**, 1093 (1919); F. Arndt, H. Scholz, *Angew. Chem.* **46**, 47 (1933). Carcinogenicity studies: M. F. Rajewsky, R. Goth, *Mol. Base Malig., Sel. Pap. Int. Symp.* **1976**, 2. Mutagenicity studies: W. L. Russell *et al., Proc. Nat. Acad. Sci. USA* **16**, 5818 (1979).

Pale buff-yellow, hexagonal plates, mp 103-104°. LD_{50} i.v. in rats: 240 mg/kg, H. Druckrey *et al., Nature* **210**, 1378 (1966).

USE: Exptlly as mutagen; ethylating agent.

3789. Ethylnorepinephrine. *4-(2-Amino-1-hydroxybutyl)-1,2-benzenediol; α-(1-aminopropyl)-3,4-dihydroxybenzyl alcohol; α-(1-aminopropyl)protocatechuyl alcohol; 1-(3,4-dihydroxyphenyl)-2-aminobutanol; 1-(3,4-dihydroxyphenyl)-1-hydroxy-2-aminobutane; ethylnoradrenaline; ethylnorsuprarenin;* E.N.E.; E.N.S.; $C_{10}H_{15}NO_3$; mol wt 197.24. C 60.89%, H 7.67%, N 7.10%, O 24.34%. Prepd by catalytic hydrogenolysis of α-benzhydrylamino-3,4-dibenzyloxybutyrophenone-HCl: Suter, Ruddy, *J. Am. Chem. Soc.* **66**, 747 (1944); U.S. pat. **2,431,285** (1947 to Winthrop-Stearns).

$$\underset{\text{OH}}{\underset{|}{\text{HOCHCHCH}_2\text{CH}_3}}\overset{\text{NH}_2}{\underset{|}{}}$$

Hydrochloride, $C_{10}H_{16}ClNO_3$, *Bronkephrine.* Crystals, dec 199-200°. Soluble in water. Marketed as an 0.2% aq soln.

THERAP CAT: Bronchodilator.

3790. Ethyl Oenanthate. *Heptanoic acid ethyl ester;* ethyl heptanoate; ethyl *n*-heptoate; oenanthic ether; cognac oil, synthetic; oil of grapes; aether oenanthicus; oleum vitis viniferae. $C_9H_{18}O_2$; mol wt 158.23. C 68.31%, H 11.47%, O 20.22%. $CH_3(CH_2)_5COOC_2H_5$. Obtained commercially either by esterification of heptanoic acid, yielding a high-grade product; or by the direct esterification of coconut oil yielding a mixture of the ethyl esters of lauric, myristic, palmitic, and other higher fatty acids plus ethyl and isoamyl alcohol. For some flavoring applications the latter product is more desirable. Prepn: Rogers, *J. Am. Pharm. Assoc., Sci. Ed.* **12**, 503 (1923). Toxicity: P. M. Jenner *et al., Food Cosmet. Toxicol.* **2**, 327 (1964).

Pure ester, liquid. Fruity, wine-like odor and taste with burning aftertaste. d_4^{15} 0.8723; d_4^{25} 0.8630. bp_{760} 189°; bp_{35} 95°; bp_8 68°. Viscosity at 25°: 0.0111 g/cm/sec. Insol in water. Misc with alcohol, ether, chloroform. Forms an azeotrope with water, the azeotropic mixture boils at 98.5° and contains 72% w/w of ethyl oenanthate. LD_{50} orally in rats: > 34,640 mg/kg (Jenner).

Note: Artificial Ethyl Oenanthate, also known as *artificial cognac essence* or *oil of wine,* is a mixture of amyl and ethyl caprates with ethyl and isoamyl butyrates and caprylates.

USE: In the manuf of liqueurs. Plays an important part in the formulation of raspberry, gooseberry, grape, cherry, apricot, currant, bourbon, and other artificial essences.

3791. Ethyl Oxalacetate. *Oxobutanedioic acid diethyl ester;* diethyl oxalacetate; oxaloacetic ester. $C_8H_{12}O_5$; mol wt 188.18. C 51.06%, H 6.43%, O 42.51%. $C_2H_5OOCCH_2CO-COOC_2H_5$. Obtained by the action of sodium on a mixture of ethyl oxalate and ethyl acetate.

Oily liquid. d_4^{20} 1.131. bp_{24} 132°. $n_D^{16.6}$ 1.45614. Insol in water; miscible with alc, benzene, ether. Dec when boiled at ordinary pressure. With water it gradually dec into alc and acetic and oxalic acids.

Semicarbazone, crystals, mp 162°.

3792. Ethylparaben. *4-Hydroxybenzoic acid ethyl ester;* ethyl *p*-hydroxybenzoate; Nipagin A; Ethyl Parasept; Solbrol A. $C_9H_{10}O_3$; mol wt 166.17. C 65.05%, H 6.07%, O 28.89%. Prepd by esterification of *p*-hydroxybenzoic acid: Cavill, Vincent, *J. Soc. Chem. Ind. (London)* **66**, 175 (1947).

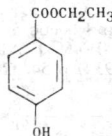

Crystals, mp 116°. bp 297-298° (decompn). Freely sol in alcohol, ether. Soly in water at 20°: 0.070% w/w; at 25°: 0.075% w/w.

USE: Preservative for pharmaceuticals.

3793. Ethyl Pelargonate. *Nonanoic acid ethyl ester;* ethyl nonanoate; wine ether. $C_{11}H_{22}O_2$; mol wt 186.29. C 70.92%, H 11.90%, O 17.18%. $CH_3(CH_2)_7COOC_2H_5$. Toxicity data: P. M. Jenner *et al., Food Cosmet. Toxicol.* **2**, 327 (1964).

Liquid. d_4^{18} 0.866. bp ~220°. mp −44°. Insol in water; misc with alc, ether. LD_{50} orally in rats: > 43,000 mg/kg (Jenner).

Note: The product of commerce is an alcoholic soln of various essences in form of a yellowish liq, d about 0.823. It is used in production of cognac-like beverages; 1 part ethyl pelargonate and 20 parts alcohol constitute one kind of artificial "Cognac Essence".

3794. Ethyl Phenylacetate. *Benzeneacetic acid ethyl ester; α-toluic acid ethyl ester.* $C_{10}H_{12}O_2$; mol wt 164.20. C 73.14%, H 7.37%, O 19.49%. $C_6H_5CH_2COOC_2H_5$. Prepd by heating benzyl cyanide with alcohol and H_2SO_4: Adams, Thal, *Org. Syn.* **2**, 27 (1922).

Liquid. Pleasant odor. d_4^{20} 1.0333. bp_{760} 226°; bp_{32} 135°; bp_{20} 121°. $n_D^{18.5}$ 1.49921. Absorption spectrum: Baly, Collie, *J. Chem. Soc.* **87**, 1344 (1905); Baly, Tryhorn, *ibid.* **107**, 1063 (1915).

USE: In perfumery.

3795. Ethyl Phosphorochloridite. *Phosphorochloridous*

acid diethyl ester; diethylchlorophosphinate; diethylchlorophosphonate; diethylphosphorous acid chloride; ethyl chlorophosphite. $C_4H_{10}ClO_2P$; mol wt 156.56. C 30.68%, H 6.44%, Cl 22.65%, O 20.44%, P 19.79%. $(C_2H_5O)_2PCl$. Prepd by passing chlorine into diethylphosphorous acid in ligroin contg Na wire: Arbusow, Arbusow, *Ber.* **65**, 195 (1932); by heating triethyl phosphite and phosphorus trichloride; or by adding phosphorus trichloride to a soln of ethanol and diethylaniline in ether: Cook *et al., J. Chem. Soc.* **1949**, 2021.

Liquid. Characteristic odor of acid chlorides. Fumes in moist air; reacts violently with water. d_4^{20} 1.0816; d_0^0 1.0962; d_4^{20} 1.0747. bp_{760} 153-155°; (also reported as bp_{760} 143-148°); bp_{30} 63-65°; bp_{15} 45-53°; bp_{10} 37-38°. n_D^{20} 1.4350.

USE: In peptide synthesis.

3796. 3-Ethyl-4-picoline. *3-Ethyl-4-methylpyridine; β-collidine; 3-ethyl-γ-picoline; β-ethyl-γ-methylpyridine.* $C_8H_{11}N$; mol wt 121.18. C 79.29%, H 9.15%, N 11.56%. Prepd from 2,6-dichloro-3-ethyl-4-methylpyridine by reduction with HI and red phosphorus: Ruzicka, Fornasir, *Helv. Chim. Acta* **2**, 338 (1919); from 3-acetyl-1,4-dimethyl-1,2,5,6-tetrahydropyridine: Prelog, Komzak, *Ber.* **74**, 1705 (1941); from γ-methylnicotinic acid: Rabe, Jantzen, *Ber.* **54**, 925 (1921); from β-ethylpyridine: Koenigs, Hoffmann, *Ber.* **58**, 194 (1925); from crotonaldehyde and ammonia: Tschitschibabin, Oparina, *Ber.* **60**, 1877 (1927).

Liquid, aromatic odor. d_4^{17} 0.9286. bp_{753} 195-196°; bp_{23} 88-90°; bp_{12} 76°. Volatile with steam. Sparingly sol in water. Soluble in alcohol, ether, chloroform, dil acids.

Hydrochloride, $C_8H_{11}N.HCl$, hygroscopic platelets, sol in water.

3797. 4-Ethyl-2-picoline. *4-Ethyl-2-methylpyridine; α-methyl-γ-ethylpyridine; 4-ethyl-α-picoline; α-collidine.* $C_8H_{11}N$; mol wt 121.18. C 79.29%, H 9.15%, N 11.56%. Prepd from 2-picoline and acetic anhydride in the presence of zinc dust: H. Maier-Bode, J. Altpeter, *Das Pyridin und Seine Derivate* (Halle, 1934) p 54; **Ger. pat. 390,333**.

Liquid. d_4^0 0.9291; d_4^{16} 0.9268. bp_{751} 179°. Volatile with steam. Sol in alcohol, ether, benzene, dil acids. Sparingly sol in water.

3798. 5-Ethyl-2-picoline. *5-Ethyl-2-methylpyridine;* aldehyde-collidine; aldehydine; 2-methyl-5-ethylpyridine; 5-ethyl-α-picoline; 3-ethyl-6-methylpyridine. $C_8H_{11}N$; mol wt 121.18. C 79.29%, H 9.15%, N 11.56%. Prepd by heating acetaldehyde ammonia in a double vol of abs alc: Ador, Bayer, *Ann.* **155**, 297 (1870); by heating ammonia water and paraldehyde in presence of ammonium acetate: Dürkopf, Schlaugk, *Ber.* **21**, 294 (1888); **Ger. pats. 347,820; 349,184; 349,267**; H. Maier-Bode, J. Altpeter, *Das Pyridin und Seine Derivate* (Halle, 1934) p 56; Frank *et al., J. Am. Chem. Soc.* **68**, 1368 (1946); Dunn, **U.S. pat. 2,717,897** (1955 to UCC); R. L. Frank *et al., Org. Syn.* **coll. vol. IV**, 451 (1963). Continuous process: Takeba *et al.,* **U.S. pat. 2,935,513** (1960 to Takeda).

Liquid. Aromatic odor. d_4^{23} 0.9184. bp_{747} 177.8°; bp_{20}

74-75°; bp_{17} 65-66°. n_D^{20} 1.4971. Volatile with steam. Practically insol in water. Sol in alcohol, ether, benzene, dil acids, concd H_2SO_4.

Hydrochloride, $C_8H_{11}N.HCl$, hygroscopic needles, freely sol in water.

3799. 1-Ethyl-3-piperidinol. 1-Ethyl-3-hydroxypiperidine. $C_7H_{15}NO$; mol wt 129.20. C 65.07%, H 11.70%, N 10.84%, O 12.38%. Prepn: Biel *et al., J. Am. Chem. Soc.* **74**, 1485 (1952); Biel, **U.S. pat. 2,802,007** (1957 to Lakeside).

Liquid. bp_{15} 93-95°. n_D^{14} 1.4777.

3800. 5-Ethyl-5-(1-piperidyl)barbituric Acid. *5-Ethyl-5-(1-piperidinyl)-2,4,6(1H,3H,5H)-pyrimidinetrione;* ethylpentamethyleneuramil; Eldoral. $C_{11}H_{17}N_3O_3$; mol wt 239.26. C 55.21%, H 7.16%, N 17.56%, O 20.06%. Prepd by the action of piperidine on 5-bromo-5-ethylbarbituric acid in presence of water and $CHCl_3$: **Belg. pat. 449,934** (1943 to Heyden); *cf.* **Fr. pat. 766,449** (1934); Räth, Gebauer, **Ger. pat. 589,146** (1933); Gebauer, **Ger. pat. 590,146** (1933). Synthesis of this type of compd: Busch, Pöhlmann, *Arch. Pharm.* **272**, 190 (1934); *cf.* Gros, *Münch. Med. Wochenschr.* **83**, 93 (1936).

Crystals, mp 215°. Sparingly sol in water (1:600); sol in alc, ether, alkaline liquids. Soly in 0.10N HCl: about 2.5%.

Caution: May be habit forming. This is a controlled substance (depressant) listed in the U.S. Code of Federal Regulations, Title 21 Parts 329.1 and 1308.13 (1987).

THERAP CAT: Sedative, hypnotic.

3801. Ethyl Propionate. *Propanoic acid ethyl ester.* $C_5H_{10}O_2$; mol wt 102.13. C 58.80%, H 9.87%, O 31.33%. $CH_3CH_2COOC_2H_5$.

Colorless liquid; fruity odor. d_4^{20} 0.891. bp 99°. mp −73°. n_D^{20} 1.3844. Flash pt, closed cup: 12°. Sol in about 60 parts water; miscible with alcohol, ether.

3802. 3-Ethylpyridine. "β-Lutidine". C_7H_9N; mol wt 107.15. C 78.46%, H 8.46%, N 13.08%.

Colorless to brownish liquid. d^{23} 0.940. bp 162-165°. n_D^{22} 1.5021. Slightly sol in water; freely sol in alcohol, or ether.

3803. 4-Ethylpyridine. γ-Ethylpyridine. C_7H_9N; mol wt 107.15. C 78.46%, H 8.47%, N 13.07%. Found in California petr: Hackmann *et al., Rec. Trav. Chim.* **62**, 229 (1943). Best prepd from pyridine by treatment with acetic anhydride and zinc: Dohrn, Horsters, **Ger. pat. 390,333** (1924 to Schering); Wibaut, Arens, *Rec. Trav. Chim.* **60**, 119 (1941); Frank, Smith, *Org. Syn.* **27**, 38 (1947). Using iron powder: Tenenbaum, Fand, **U.S. pat. 2,712,019** (1955 to Nepera Chem.).

Liquid. Obnoxious odor. Turns brown if not very pure. d_4^{22} 0.9404. bp_{750} 169.6-170.0°. n_D^{18} 1.5029. Sparingly sol in water; sol in alcohol, ether.

Picrate, mp 169-170°.

Caution: Symptoms similar to pyridine, *q.v.*

3804. Ethyl Salicylate. *2-Hydroxybenzoic acid ethyl ester;* Salicylic acid ethyl ester; salicylic ether; sal ethyl. $C_9H_{10}O_3$; mol wt 166.17. C 65.05%, H 6.07%, O 28.88%. $HOC_6H_4COOC_2H_5$.

Clear, very refractive liquid; pleasant odor resembling that of methyl salicylate. Becomes yellowish-brown on long exposure to air and light. d_4^{20} 1.131. bp 231-234°. mp +1°. n_D^{20} 1.5226. Slightly sol in water; miscible with alc, ether. *Protect from light.*

USE: Manuf artificial perfumes.

THERAP CAT (VET): Has been used as a counterirritant.

3805. Ethyl Silicate. *Silicic acid tetraethyl ester;* tetraethyl silicate. $C_8H_{20}O_4Si$; mol wt 208.30. C 46.13%, H 9.68%, O 30.72%, Si 13.47%. $Si(OC_2H_5)_4$. Prepn from abs alcohol and silicon tetrachloride: J. J. Ebelmen, *Ann.* **57**, 319 (1846); A. W. Dearing, E. E. Reid, *J. Am. Chem. Soc.* **50**, 3058 (1928); G. Sumrell, E. E. Ham, *ibid.* **78**, 5573 (1956). Toxicological studies: H. F. Smyth, J. Seaton, *J. Ind. Hyg. Toxicol.* **22**, 288 (1940); V. H. Rowe *et al., ibid.* **30**, 332 (1948). *Review:* H. D. Cogan, C. A. Setterstrom, *Ind. Eng. Chem.* **39**, 1364 (1947).

Colorless, flammable liquid. mp −77°. bp 165-166°. Flash point 125°F (52°C). d_4^{20} 0.933; n_D^{25} 1.3818. Viscosity 0.6 cps. Practically insol in water, and slowly dec by it. Miscible with alcohol.

USE: In weatherproofing and hardening stone, arresting decay and disintegration; manuf weatherproof and acidproof mortars and cements. In the "lost wax" process for casting of high-melting alloys.

Caution: Irritating to eyes, mucous membranes, and, in high concns, narcotic.

3806. Ethyl Silicone Resins. Polymeric ethyl silicon oxides, probably of a cross-linked siloxane structure. Prepd by hydrolyzing diethyl dichlorosilane or a diethyldialkyloxysilane: Ladenburg, *Ann.* **164**, 311 (1872); by oxidation of tetraethylsilane: Friedel, Crafts, *Ann. Chim. Phys.* [4] **9**, 5 (1866).

$$-(C_2H_5)_2Si-O-C_2H_5Si-O-(C_2H_5)_2Si-O-$$
$$| $$
$$O$$
$$| $$
$$-(C_2H_5)_2Si-O-C_2H_5Si-O-C_2H_5Si-O-$$
$$| $$
$$O$$
$$| $$

The preferred C_2H_5 to Si ratio for ethyl silicone resins is 0.5 to 1.5. Below 0.5, the products are brittle masses which shrink and crack as they condense; above 1.5 they are difficult to condense to the solid state; at a C_2H_5 to Si ratio of 2, the polymers remain liquid. Very stable to heat and most chemical reagents.

USE: In electrical insulation; in protective and decorative coatings.

3807. Ethylstibamine. 693 B; Neostibosan; Stibosamine; Astaril. A complex of *p*-aminobenzenestibonic acid (present largely as a tetramer), *p*-acetylaminobenzenestibonic acid (present largely as a dimer), antimonic acid and diethylamine in the approx molar ratio of 1:2:1:3. Contains 41 to 44% organic pentavalent antimony (the acetyl content is between 6 and 7%). The manuf requires special care since the compd is both a colloid and a molecular addition compd and slight variations affect its toxicity greatly: Schmidt, *Angew. Chem.* **43**, 963 (1930); U.S. pat. **1,988,632**, *C.A.* **29**, 1586 (1935).

Light-yellow to yellowish-brown powder. Freely sol in water forming neutral, colloidal solns. The pH of a 5% soln

is between 6.5 and 7.6. A 25% soln is approx isotonic. Solns for injection should be prepd immediately before use with sterile distilled water, as they are unstable on heating or standing.

THERAP CAT: Antiprotozoal (Leishmania).

3808. Ethyl Sulfate. *Sulfuric acid monoethyl ester;* ethyl hydrogen sulfate; ethylsulfuric acid; sulfethylic acid; sulfovinic acid. $C_2H_6O_4S$; mol wt 126.13. C 19.04%, H 4.79%, O 50.74%, S 25.42%. $C_2H_5OSO_2OH$. Has never been obtained 100% pure. Prepn: Hamid *et al., J. Chem. Soc.* **129**, 1098 (1926); Atwood, U.S. pat. **2,683,731** (1954 to Natl. Petro-Chemicals). *Review:* C. M. Suter, *The Organic Chemistry of Sulfur* (John Wiley, 1944) p 23-27. See also Diethyl Sulfate.

Barium salt, $C_4H_{10}BaO_8S_2$, *barium ethyl sulfate.* Dihydrate, lustrous leaflets. *Poisonous!* Freely soluble in water; slightly sol in alcohol. Not pptd by H_2SO_4 or sulfates.

Calcium salt, $C_4H_{10}CaO_8S_2$, *calcium ethyl sulfate, calcium sulfovinate.* Dihydrate, crystals. Soluble in water, alcohol.

USE: As intermediate in the manuf of ethanol from ethylene. *Caution:* Highly irritating to skin, mucous membranes.

3809. Ethyl Sulfide. *1,1'-Thiobisethane;* diethyl sulfide; thioethyl ether. $C_4H_{10}S$; mol wt 90.19. C 53.27%, H 11.18%, S 35.56%. $(C_2H_5)_2S$. Made by distillation of sodium ethyl sulfate with a soln of K_2S.

Liquid; ethereal odor. d_4^{20} 0.837. bp 92°. n_D^{20} 1.44233. Insol in water; miscible with alcohol, ether.

USE: Solvent for anhydr mineral salts; in plating baths for coating metals with gold or silver.

3810. 2-(Ethylsulfonyl)ethanol. Ethylsulfonylethyl alcohol. $C_4H_{10}O_3S$; mol wt 138.18. C 34.77%, H 7.29%, O 34.74%, S 23.20%. $CH_3CH_2SO_2CH_2CH_2OH$. Prepn from 2-(ethylthio)ethanol + H_2O_2: Lee, Muller, U.S. pat. **2,805,-964** (1957 to Pennsalt Chem.); Shostakovskii *et al., Zh. Obshch. Khim.* **30**, 1123 (1960).

Hygroscopic crystals from acetone, mp 46°. $bp_{2.5}$ 153°. d_4^{20} 1.25. Flash pt 188°C (371°F). Soluble in water and in many organic solvents. LD_{50} orally in mice: 18 g/kg.

Benzoate, $C_{11}H_{14}O_4S$, needles from alcohol, mp 118°.

USE: When liq, as solvent for polar org substances. Intermediate for pharmaceuticals, plasticizers, solvents. Humectant. Antistatic agent for synthetic fibers and fabrics.

3811. Ethyl Tartrate. *(R)-2,3-Dihydroxybutanedioic acid diethyl ester;* diethyl tartrate. $C_8H_{14}O_6$; mol wt 206.19. C 46.60%, H 6.84%, O 46.56%. $C_2H_5OOCCH(OH)CH-(OH)COOC_2H_5$.

Colorless, thick, oily liquid. d_4^{20} 1.204. bp 280°; bp_{11} 150°. mp 17°. $[\alpha]_D^{20}$ + 7.5°. n_D^{20} 1.4476. Slightly sol in water; miscible with alcohol or ether.

3812. Ethyl Tartrate, Acid. *(R)-2,3-Dihydroxybutanedioic acid monoethyl ester;* monoethyl tartrate. $C_6H_{10}O_6$; mol wt 178.14. C 40.45%, H 5.66%, O 53.89%. $HOOCCH-(OH)CH(OH)COOC_2H_5$.

Colorless, very hygroscopic crystals. Sweet taste. mp 90°. Sol in water, alcohol; insol in ether. *Keep well closed.*

USE: Printing with Indol Blue and Crystal Fast Blue on fustian, etc.

3813. 2-(Ethylthio)ethanol. *β-Hydroxydiethyl sulfide.* $C_4H_{10}OS$; mol wt 106.18. C 45.24%, H 9.49%, O 15.07%, S 30.19%. $CH_3CH_2SCH_2CH_2OH$. Prepn from 2-chloroethanol and ethyl mercaptan: Demuth, Meyer, *Ann.* **240**, 310 (1887); Steinkopf *et al., Ber.* **53**, 1010 (1920); from 2-thioethanol and diethyl sulfate: Bergmann, *J. Am. Chem. Soc.* **74**, 829 (1952).

Liquid. d_{20}^{20} 1.015-1.025. bp_{760} 184.5°; bp_{28} 99°. Soluble in ether, other organic solvents.

USE: Intermediate for pesticides, lubricating and cutting oil additives, flotation agents, plasticizers.

3814. Ethyl p-Toluenesulfonate. *4-Methylbenzenesulfonic acid ethyl ester.* $C_9H_{12}O_3S$; mol wt 200.25. C 53.98%, H 6.04%, O 23.97%, S 16.01%. $CH_3C_6H_4SO_3C_2H_5$.

Monoclinic crystals. d 1.17. mp 33°. bp_{15} 173°.

USE: For ethylation.

3815. Ethyl Vanillin. *3-Ethoxy-4-hydroxybenzaldehyde;* ethylprotocatechuic aldehyde; bourbonal; ethylprotal; vanillal; Ethavan; Ethovan. $C_9H_{10}O_3$; mol wt 166.17. C 65.05%, H 6.07%, O 28.88%.

Colorless flakes possessing a finer and more intense vanilla odor and taste than vanillin. mp 77-78°. Sparingly sol in water; sol in alc, ether, glycerol, propylene glycol. Soly in 95% alc about 1 g/2 ml. LD_{50} orally in rats: > 2000 mg/kg, P. M. Jenner et al., Food Cosmet. Toxicol. **2**, 327 (1964). USE: In flavoring and perfumery.

3816. Ethynodiol. (3β,17α)-19-Norpregn-4-en-20-yne-3,17-diol; 17α-ethynyl-19-norandrost-4-ene-3β,17β-diol; 17α-ethynyl-4-estrene-3β,17β-diol; ED. $C_{20}H_{28}O_2$; mol wt 300.42. C 79.95%, H 9.39%, O 10.65%. Prepn of the 3-acetate, 17-acetate, and diacetate: Klimstra, U.S. pat. **3,-176,013** (1965 to Searle); see also Sondheimer, Klibansky, Tetrahedron **5**, 15 (1959). Pharmacokinetics and metabolism: C. J. Lewis et al., Xenobiotica **10**, 705 (1980). Comparative clinical study in oral contraceptives: M. H. Briggs, J. Reprod. Med. **28**, Suppl. 1, 92 (1983). Review of carcinogenicity studies: IARC Monographs **21**, 387-398 (1979). Comprehensive description: E. P. K. Lau, J. L. Sutter, in Analytical Profiles of Drug Substances vol. 3, K. Florey, Ed. (Academic Press, New York, 1974) pp 253-279.

Diacetate, $C_{24}H_{32}O_4$, 3β,17β-diacetoxy-17α-ethynyl-4-estrene, SC 11,800, Femulen, Luteonorm, Luto-Metrodiol, Metrodiol. Crystals from methanol + water, mp about 126-127°. $[\alpha]_D$ −72.5° (chloroform). Mixture with mestranol, q.v., Luteolas, Metrulen, Ovaras, Ovulen. Mixture with ethinyl estradiol, q.v., Demulen, Miniluteolas.
THERAP CAT: Diacetate as a progestogen. In combination with estrogen as oral contraceptive.

3817. Ethynylbenzene. Phenylacetylene. C_8H_6; mol wt 102.13. C 94.08%, H 5.92%. $C_6H_5C\equiv CH$. Prepd by dropping β-bromostyrene on molten potassium hydroxide and distilling: Hessler, Org. Syn. **2**, 67 (1922).
Liquid. d_4^{20} 0.9300. mp −44.8°. bp_{760} 142.4°; bp_{90} 75°; bp_{15} 39°. n_D^{20} 1.5489. Insol in water; miscible with alcohol, ether, other organic solvents.

3818. Etidocaine. N-(2,6-Dimethylphenyl)-2-(ethylpropylamino)butanamide; 2-(ethylpropylamino)-2',6'-butyroxylidide; Duranest. $C_{17}H_{28}N_2O$; mol wt 276.42. C 73.87%, H 10.21%, N 10.14%, O 5.78%. Prepn: Adams et al., Ger. pat. **2,162,744** (1972 to Astra), C.A. **77**, 101244c (1972). Activity and toxicity studies: eidem, J. Pharm. Sci. **61**, 1829 (1972).

LD_{50} in female mice: 6.7 mg/kg i.v.; 99 mg/kg s.c. Hydrochloride, $C_{17}H_{29}ClN_2O$, W-19053.
THERAP CAT: Local anesthetic.

3819. Etidronic Acid. (1-Hydroxyethylidene)bisphosphonic acid; (1-hydroxyethylidene)diphosphonic acid; ethane-1-hydroxy-1,1-diphosphonic acid; EHDP. C_2H_8-

O_7P_2; mol wt 206.02. C 11.66%, H 3.91%, O 54.36%, P 30.07%. Bone resorption inhibitor structurally related to inorganic pyrophosphoric acid and to clodronic acid, q.q.v. Also a sequestering and chelating agent. Effect on hydroxy-apatite (durapatite, q.v.) crystal aggregation: N. M. Hansen et al., Biochim. Biophys. Acta **451**, 549 (1976); S. Bisaz et al., ibid. 560. Prepn by reaction of phosphorous trichloride, acetic acid, and aqueous alkali: H. von Bayer, K. A. Hoffmann, Ber. **30**, 1973 (1897); D. R. Peck, U.S. pat. **3,468,935** (1969 to Albright and Wilson Mfg.). Continuous prepn process: L. Rogovin et al., U.S. pat. **3,400,147** (1968 to Procter and Gamble). Prepn from acetic anhydride and dimethylphosphite: R. R. Irani, R. S. Mitchell, U.S. pat. **3,475,486** (1969 to Monsanto). Prepn and characterization of the acid and disodium salt: F. Kasparek, Monatsh. **99**, 2016 (1968); B. Blaser et al., Z. Anorg. Allgem. Chem. **381**, 247 (1971). Use as builder for detergents: F. L. Diehl, U.S. pat. **3,159,581** (1964 to Proctor & Gamble). Pharmacology: M. D. Francis et al., Science **165**, 1264 (1969); J. Jowsey et al., J. Lab. Clin. Med. **76**, 126 (1970); B. S. Strates et al., Biochim. Biophys. Acta **244**, 121 (1971). Toxicology: G. A. Nolen, E. V. Buehler, Toxicol. Appl. Pharmacol. **18**, 548 (1971). Metabolism: W. R. Michael et al., ibid. **21**, 503 (1972). Clinical study in postmenopausal osteoporosis: R. P. Heaney, P. D. Saville, Clin. Pharmacol. Ther. **20**, 593 (1976); in Paget's disease: R. D. Altman et al., N. Engl. J. Med. **289**, 1379 (1973); R. D. Altman, Am. J. Med. **79**, 583 (1985); C. J. Preston et al., Brit. Med.J. **292**, 79 (1986); in hypercalcemia of cancer: E. Ryzen et al., Arch. Intern. Med. **145**, 449 (1985). Symposium on clinical efficacy in malignancy-related hypercalcemia: Am. J. Med. **82**, Suppl 2A, 1-78 (1987).

Crystallizes from water as the monohydrate. pK_1 1.35 ±0.08; pK_2 2.87 ±0.01; pK_3 7.03 ±0.01; pK_4 11.3. Very sol in water (69% at 20°C). Insol in acetic acid.
Disodium salt, $C_2H_6Na_2O_7P_2$, disodium dihydrogen (1-hydroxyethylidene)bis[phosphonate], etidronate disodium, Calcimux, Didronel, Diphos, Etidron. Crystallizes from water as the di- or tetrahydrate.
THERAP CAT: Calcium regulator.

3820. Etifelmin. 2-Diphenylmethylenebutylamine; 2-ethyl-3,3-diphenyl-2-propenylamine; EDPA; Na III. $C_{17}H_{19}N$; mol wt 237.33. C 86.03%, H 8.07%, N 5.90%. Prepn: Heinrich, Heiger, Ger. pat. **1,122,514** (1962 to Giulini GmbH), C.A. **57**, 733c (1962).

Hydrochloride, $C_{17}H_{20}ClN$, Gilutensin (also as the nicotinate and gluconate), Tensinase D. Crystals, mp 232°. Freely sol in water.
THERAP CAT: Central stimulant, antihypotensive.

3821. Etifoxine. 6-Chloro-N-ethyl-4-methyl-4-phenyl-4H-3,1-benzoxazin-2-amine; 6-chloro-2-(ethylamino)-4-methyl-4-phenyl-4H-3,1-benzoxazine; HOE 36801; 36801. $C_{17}H_{17}ClN_2O$; mol wt 300.79. C 67.88%, H 5.70%, Cl 11.79%, N 9.31%, O 5.32%. Psychotropic agent with anxiolytic and anticonvulsant activity. Prepn: Fr. pat. **M7358** corresp to H. Kuch et al., U.S. pat. **3,725,404** (1969, 1973 both to Hoechst); I. Hoffmann et al., Arzneimittel-Forsch. **20**, 975 (1970). Exptl psychopharmacological study: J. R. Boissier et al., Therapie **27**, 325 (1972). Analysis of EEG effects: D. Bente *et al.*, *Arzneimittel-Forsch.* **25**, 944 (1975). Evaluation of psychotropic effect: R. Corsico *et al.*, *Psychopharmacologia* **45**, 301 (1976).

Colorless cryst from petr ether, mp 90-92°. uv max (ethanol): 273 nm (ε 21,200). LD$_{50}$ in mice: 12 g/kg orally, I. Hoffmann *et al.*, *loc. cit.*

Hydrochloride, $C_{17}H_{18}Cl_2N_2O$, *Stresam.* Cryst, mp 150-151°.

THERAP CAT: Anxiolytic.

3822. Etilefrin. α-[(Ethylamino)methyl]-3-hydroxyben-zenemethanol; α-[(ethylamino)methyl]-m-hydroxybenzyl alcohol; m-hydroxy-α-(ethylaminomethyl)benzyl alcohol; α-(m-hydroxyphenyl)-β-(ethylamino)ethanol; ethylphenyl-ephrine; etiladrianol. $C_{10}H_{15}NO_2$; mol wt 181.23. C 66.27%, H 8.34%, N 7.73%, O 17.66%. Prepn: Goto, *J. Pharm. Soc. Japan* **74**, 318 (1954), *C.A.* **49**, 3960 (1955); **Span.** pat. **273,-595** (1962 to Labs. Fher S.A.), *C.A.* **60**, 1649e (1964).

dl-Form, crystals, mp 147-148°.

dl-Form hydrochloride, $C_{10}H_{16}ClNO_2$, *Effontil, Effortil, Efortil, Ethyl Adrianol, Eti-Puren, Circupon, Apocretin, Pulsamin, Tonus-forte, Kertasin.* Crystals, mp 121°. Bitter taste. Freely sol in water; sol in alcohol. Practically insol in chloroform.

THERAP CAT: *dl*-Form hydrochloride as an antihypotensive.

3823. Etiocholane. 5β-Androstane; 5-epiandrostane. $C_{19}H_{32}$; mol wt 260.45. C 87.61%, H 12.39%. Parent compd of alkyl substituted etiocholanes, such as pregnane, cholane, coprostane. Prepd from etiocholane-17-one semicarbazone: Butenandt, Dannenbaum, *Z. Physiol. Chem.* **229**, 192 (1934).

Needles from acetone, mp 78-80°.

3824. Etiocholanic Acid. 5β-Androstane-17β-carboxylic acid; etianic acid; aetiocholanic acid; etiocholane-17β-carb-oxylic acid. $C_{20}H_{32}O_2$; mol wt 304.46. C 78.89%, H 10.59%, O 10.51%. Prepn: Wieland *et al.*, *Z. Physiol. Chem.* **161**, 80 (1926); Jacobs, Elderfield, *J. Biol. Chem.* **108**, 497 (1935); Tschesche, *Ber.* **68**, 9 (1935); Steiger, Reichstein, *Helv. Chim. Acta* **21**, 841 (1938).

Needles from glacial acetic acid; elongated leaflets from acetic acid. Sublimes at 0.002 mm press. and 160° bath temp, mp 228-229°. Insol in water; sol in pentane.

Methyl ester, $C_{21}H_{34}O_2$, needles from methanol, mp 99-101°.

Note: The stem name "etianic acid" was proposed by the Subcommittee on Steroid Nomenclature of the National Research Council as a replacement for "etiocholanic acid" in order to avoid the use of the same name for parent hydro-carbons of different carbon content: *J. Am. Chem. Soc.* **74**, 2817 (1952).

3825. Etiocobalamin. *Cobinamide dicyanide;* vitamin B$_{12p}$; Factor B. $C_{50}H_{72}CoN_{13}O_8$; mol wt 1042.17. C 57.62%, H 6.96%, Co 5.66%, N 17.47%, O 12.28%. Vitamin B$_{12}$ fac-tor obtained by removal of the nucleotide from cyanocobal-amin by acid hydrolysis. Isoln from calf feces: Ford, Porter, *Biochem. J.* **51**, V (1952). Prepn by acid hydrolysis: Gant *et al.*, *ibid.* **56**, XXXIV (1954); Friedrich, Bernhauer, *Angew. Chem.* **65**, 627 (1953). Prepn from factor V$_{1a}$ and D(—)-1-amino-2-propanol: Bernhauer *et al.*, *Helv. Chim. Acta* **43**, 696 (1960); Bernhauer, **U.S.** pat. **3,072,674** (1963 to Hoff-mann-La Roche).

Note: The term etiocobalamin is used by some authors to designate any cobalamin lacking the nucleotide group present in vitamin B$_{12}$.

3826. Etioporphyrin. 2,7,12,18-Tetraethyl-3,8,13,17-tet-ramethyl-21H,23H-porphine; 1,3,5,8-tetramethyl-2,4,6,7-tetraethylporphine; etioporphyrin III; mesoetioporphyrin. $C_{32}H_{38}N_4$; mol wt 478.65. C 80.29%, H 8.00%, N 11.71%. Occurs in Bavarian oil shale, crude petr, ozokerite, amber, cannel coal and other hard varieties of bituminous coal: Treibs, *Ann.* **510**, 60 (1934); **517**, 184 (1935); *Angew. Chem.* **49**, 682 (1936). Prepd by decarboxylation of mesoporphyrin IX: Fischer, Treibs, *Ann.* **466**, 191, 206 (1928). Synthesis: Fischer, Stangler, *Ann.* **462**, 265 (1928); Johnson *et al.*, *J. Chem. Soc.* **1959**, 3416. Structure: Abraham *et al.*, *ibid.* **1961**, 3468. *Review:* Rimington, Kennedy, in M. Florkin, H. S. Mason, *Comparative Biochemistry* (Academic Press, New York, 1962) pp 557-614.

Long prismatic needles or butterflies from pyridine or from chloroform-petr ether, mp 360-363°. pKa 18. Ab-sorption max: 246, 269, 396, 497, 532, 566, 620, 645 nm (log ε 3.90, 3.89, 5.22, 4.13, 3.99, 3.81, 3.65, 2.62).

Copper salt, $C_{32}H_{36}N_4Cu$, red needles from pyridine-acetic acid.

Magnesium salt, $C_{32}H_{36}N_4Mg$, crystals from methanol.

3827. Etiproston. *[1R-(1α(Z),2β(E),3α,5α)]-7-[3,5-Di-hydroxy-2-[2-[2-(phenoxymethyl)-1,3-dioxolan-2-yl]ethen-yl]cyclopentyl]-5-heptenoic acid; (5Z,13E)-(8R,9S,11R,12R)-9,11-dihydroxy-15,15-ethylenedioxy-16-phenoxy-17,18,-19,20-tetranorprostadienoic acid; 15-deoxy-15,15-ethylene-dioxy-16-phenoxy-17,18,19,20-tetranorprostaglandin $F_{2\alpha}$;* Prostavet. $C_{24}H_{32}O_7$; mol wt 432.51. C 66.65%, H 7.46%, O 25.89%. Prostaglandin $F_{2\alpha}$ analog with estrus cycle synchronizing activity. Prepn: W. Skuballa *et al.*, **Ger. pat. 2,434,-133;** *eidem,* **U.S. pat. 4,088,775** (1976, 1978 both to Schering AG); and biological activity: W. Skuballa *et al., J. Med. Chem.* **21,** 443 (1978).

Colorless oil.
THERAP CAT (VET): Luteolytic.

3828. Etiroxate. *O-(4-Hydroxy-3,5-diiodophenyl)-3,5-diiodo-α-methyltyrosine ethyl ester;* D,L*-α-methylthyroxine ethyl ester;* CG 635. $C_{18}H_{17}I_4NO_4$; mol wt 818.95. C 26.40%, H 2.09%, I 61.98%, N 1.71%, O 7.82%. Deriv of thyroxine, *q.v.* Prepn: **Neth. pat. Appl. 6,614,150** corresp to H. Kummer, R. Beckmann, **U.S. pat. 3,930,017** (1967, 1975 both to Grünenthal). Animal studies: R. Beckmann, *Arz-neimittel-Forsch.* **29,** 499 (1979). Effect on iodine metabolism in man: D. Emrich, *ibid.* **27,** 422 (1977). Use in hyper-lipoproteinemia: H. Banz, F. P. Gall, *Fortschr. Med.* **97,** 1942 (1979); *eidem, Med. Klin.* **75,** 51 (1980).

Cryst from ethanol, mp 156-157°.
Hydrochloride, $C_{18}H_{18}ClI_4NO_4$, *Skleronorm.*
THERAP CAT: Antihyperlipoproteinemic.

3829. Etisazol. *N-Ethyl-1,2-benzisothiazol-3-amine; 3-(ethylamino)-1,2-benzisothiazole;* Netrosylla. $C_9H_{10}N_2S$; mol wt 178.24. C 60.64%, H 5.66%, N 15.71%, S 17.99%. Prepd by the reaction of diphenyldisulfide-2,2'-dicarbonyl dichloride with ethylamine, followed by treatment with PCl_5 and ammonia: Boeshagen, *Ber.* **99,** 2566 (1966). Chemistry studies: Geiger *et al., ibid.* **102,** 1961 (1969); Boeshagen *et al., ibid.* **103,** 3166 (1970).

mp 78°.
Hydrochloride, $C_9H_{11}ClN_2S$, *BAY VA 5387, Ectimar.* Crystals from ethanol, mp 171°.
THERAP CAT (VET): Antifungal.

3830. Etizolam. *4-(2-Chlorophenyl)-2-ethyl-9-methyl-6H-thieno[3,2-f][1,2,4]triazolo[4,3-a][1,4]diazepine;* 1-meth-yl-6-o-chlorophenyl-8-ethyl-4H-s-triazolo[3,4-c]thieno-[2,3-e]-1,4-diazepine; Y-7131; Depas. $C_{17}H_{15}ClN_4S$; mol wt 342.85. C 59.56%, H 4.41%, Cl 10.34%, N 16.34%, S 9.35%. Prepn: M. Nakanishi *et al.,* **Ger. pat. 2,229,845;** *eidem,* **U.S. pat. 3,904,641** (1972, 1973 both to Yoshitomi). Pharmacology and toxicity studies: T. Tsumagari *et al., Arzneimittel-Forsch.* **28,** 1158 (1978). Effect on monoamine metabolism in brain: M. Setoguchi *et al., ibid.* 1165; on rage responses in cats: T. Fukuda, T. Tsumagari, *Japan. J. Pharmacol.* **33,** 885 (1983).

Crystals from toluene, mp 147-148°. LD_{50} in male, female rats, male, female mice (mg/kg): 3619, 3509, 4358, 4258 orally; 865, 825, 830, 783 i.p.; > 5000 s.c. (Tsumagari).
THERAP CAT: Anxiolytic.

3831. Etodolac. *1,8-Diethyl-1,3,4,9-tetrahydropyrano-[3,4-b]indole-1-acetic acid;* etodolic acid; AY-24236; Edo-lan; Lodine; Ramodar; Ultradol; Zedolac. $C_{17}H_{21}NO_3$; mol wt 287.37. C 71.05%, H 7.37%, N 4.88%, O 16.70%. Prepn: C. A. Demerson *et al.,* **Ger. pat. 2,226,340;** *eidem,* **U.S. pat. 3,843,681** (1973, 1974 both to Am. Home Products); *eidem, J. Med. Chem.* **19,** 391 (1976). Anti-inflammatory and anal-gesic properties: R. R. Martel, J. Klicius, *Can. J. Physiol. Pharmacol.* **54,** 245 (1976). Metabolic disposition in animals and man: M. N. Cayen *et al., Drug. Metab. Rev.* **12,** 339 (1981); E. S. Ferdinandi *et al., Xenobiotica* **16,** 153 (1986). Clinical comparison with sulindac in rheumatoid arthritis: G. Jacob *et al., Curr. Ther. Res.* **37,** 1124 (1985).

Crystals from hexane/chloroform, mp 145-148°.
THERAP CAT: Anti-inflammatory; analgesic.

3832. Etodroxizine. *2-[2-[2-[4-[(4-Chlorophenyl)-phenylmethyl]-1-piperazinyl]ethoxy]ethoxy]ethanol;* 1-(*p*-chlorobenzhydryl)-4-[2-[2-(2-hydroxyethoxy)ethoxy]-ethyl]piperazine; 1-(*p*-chloro-α-phenylbenzyl)-4-[2-[2-(2-hydroxyethoxy)ethoxy]ethyl]diethylenediamine; hydro-chlorbenzethylamine. $C_{23}H_{31}ClN_2O_3$; mol wt 418.98. C 65.93%, H 7.46%, Cl 8.46%, N 6.69%, O 11.46%. Prepn: Morren, **Brit. pat. 817,231** (1959). GC determn in plasma: R. Pentz, A. Schutt, *Arch. Toxicol.* **39,** 225 (1978). Clinical evaluations in insomnia: R. Loire, A. Perrin, *Lyon Med.* **219,** 1795 (1968); S. Fedeli, *Bruxelle Med. Belg.* **48,** 517 (1968). Toxicology: M. Giurgea, J. Puigdevall, *Proc. Eur. Soc. Study Drug Toxicity* **9,** 134 (1968).

Liquid, $bp_{0.01}$ 250°.
Dimaleate, $C_{31}H_{39}ClN_2O_{11}$, *Indunox, Drimyl,* LD_{50} orally in rats: 920 mg/kg (Giurgea, Puigdevall).
THERAP CAT: Hypnotic.

3833. Etofenamate. *2-[[3-(Trifluoromethyl)phenyl]ami-no]benzoic acid 2-(2-hydroxyethoxy)ethyl ester; N-(α,α,α-trifluoro-m-tolyl)anthranilic acid 2-(2-hydroxyethoxy)ethyl ester;* B 577; TV 485; Bayrogel; Rheumon gel; Traumon Gel. $C_{18}H_{18}F_3NO_4$; mol wt 369.35. C 58.54%, H 4.91%, F 15.43%, N 3.79%, O 17.33%. Percutaneously active anti-phlogistic agent. Prepn: K. H. Boltze *et al.,* **Ger. pat. 1,-939,112** corresp to **U.S. pat. 3,692,818** (1971, 1972 both to Troponwerke). Series of articles on chemistry, analysis, biochemistry, pharmacology, toxicology, and clinical studies: *Arzneimittel-Forsch.* **27**, 1300-1363 (1977). Toxicity: H. Jacobi *et al., ibid.* 1333. Metabolism and GC determn: *ibid.* **31**, 9-21 (1981).

Pale yellow viscous oil, thermolabile at 180°. $bp_{.001}$ 130-135°. uv max (methanol): 286 ($E_{1cm}^{1\%}$ 423). n_D^{25} 1.564. Sol in lower alcohols, ethyl acetate, acetone, chloroform, ether, benzene. Soly in water at 22°: 0.16 mg/100 ml. LD_{50} in male, female rats (mg/kg): 292, 470 orally; 140, 226 i.v.; 373, 397 i.p.; 643, 568 s.c. (Jacobi).

THERAP CAT: Anti-inflammatory.

3834. Etofibrate. *3-Pyridinecarboxylic acid 2-[2-(4-chlorophenoxy)-2-methyl-1-oxopropoxy]ethyl ester; nicotinic acid 2-hydroxyethyl ester 2-(p-chlorophenoxy)-2-methylpropionate (ester);* ethylene glycol 1-[2-(p-chlorophenoxy)-2-methylpropionate]-2-nicotinate; α-[(p-chlorophenoxy)isobutyroyl]-β-(nicotinoyl)glycol; ethofibrate; Lipo-Merz. $C_{18}H_{18}ClNO_5$; mol wt 363.80. C 59.43%, H 4.99%, Cl 9.74%, N 3.85%, O 21.99%. Glycol diester of nicotinic and clofibric acids, *q.q.v.* Prepn: A. Scherm, D. Peteri, Ger. pat. **1,941,-217**; *eidem,* U.S. pat. **3,723,446** (1971, 1973 both to Merz). Lipid lowering effects in rats: W. Sterner, A. Schultz, *Arzneimittel-Forsch.* **24**, 1990 (1974). Metabolism: H. Oelschläger *et al., ibid.* **30**, 984 (1980). *In vitro* effect on human platelet function: M. P. Ortega *et al., Thromb. Res.* **19**, 409 (1980). HPLC determn in plasma: E. R. Garrett, M. R. Gardner, *J. Pharm. Sci.* **71**, 14 (1982). Bioavailability and pharmacokinetics in humans: K. I. Johnson *et al., Arzneimittel-Forsch.* **34**, 1785 (1984). Review of pharmacology and comparison with other hypolipidemic agents: R. Paoletti *et al., Am. J. Cardiol.* **52**, Suppl B, 21B-27B (1983).

Crystalline powder, mp 100°.

THERAP CAT: Antihyperlipoproteinemic.

3835. Etofylline. *3,7-Dihydro-7-(2-hydroxyethyl)-1,3-dimethyl-1H-purine-2,6-dione; 7-(2-hydroxyethyl)theophylline;* oxyethyltheophylline; Oxytheonyl; Oxphylline; Phyllocormin N. $C_9H_{12}N_4O_3$; mol wt 224.22. C 48.21%, H 5.39%, N 24.99%, O 21.41%. Obtained by the action of glycol monochlorohydrin upon theophylline in an alkaline medium: Lespagnol *et al., Bull. Soc. Pharm. Lille* **1948**, no. 2, p 18; Roth, *Arch. Pharm.* **292**, 234 (1959); Fabbrini, Cencioni, *Farmaco Ed. Sci.* **17**, 660 (1962).

Crystals from abs ethanol, mp 158°. Freely sol in water. Moderately sol in alcohol. pH of a 5% aq soln 6.5-7.0. Solns may be sterilized by heating. Pharmacological action as of theophylline.

Combination with theophylline, *Cordalin.*

THERAP CAT: Bronchodilator.

3836. Etofylline Nicotinate. *3-Pyridinecarboxylic acid 2-(1,2,3,6-tetrahydro-1,3-dimethyl-2,6-dioxo-7H-purin-7-yl)ethyl ester; nicotinic acid ester with 7-(2-hydroxyethyl)-theophylline;* 7-(2-hydroxyethyl)theophylline nicotinate; Hesotanil; Hesotin. $C_{15}H_{15}N_5O_4$; mol wt 329.31. C 54.71%, H 4.59%, N 21.27%, O 19.43%. Prepn: Pongratz, Zirm, *Monatsh.* **88**, 330 (1957). Pharmacology: Fischbach, Haas, *Arzneimittel-Forsch.* **17**, 313 (1967).

Fine dendritic needles from abs alcohol, mp 151-152°. Hydrated crystals from water. One gram dissolves in 50 ml boiling water; moderately sol in alcohol.

THERAP CAT: Vasodilator.

3837. Etoglucid. *2,2'-(2,5,8,11-Tetraoxadodecane-1,12-diyl)bisoxirane; 1,2:15,16-diepoxy-4,7,10,13-tetraoxahexadecane;* 1,2-bis[2-(2,3-epoxypropoxy)ethoxy]ethane; triethylene glycol diglycidyl ether; TDE; ethoglucid; ICI 32865; Epodyl. $C_{12}H_{22}O_6$; mol wt 262.30. C 54.95%, H 8.45%, O 36.60%. Prepd from epichlorohydrin and triethylene glycol in the presence of $BF_3.Et_2O$ or from triethylene glycol diallyl ether and NaOCl: Greenshields *et al.,* **Brit.** pat. **901,876** (1962 to ICI); Blyakhman, *Zh. Org. Khim.* **3**, 1423 (1967), *C.A.* **68**, 59384k (1968). Pharmacology and metabolism: Duncan, Snow, *Biochem. J.* **82**, 8P (1962); James, Solheim, *Xenobiotica* **1**, 43 (1971). Clinical evaluation in bladder tumors: M. R. G. Robinson *et al., J. Urol.* **118**, 972 (1977); J. Flamm, F. Grof, *Urologe* **B27**, 26 (1987).

Liquid, mp —15° to —11°. $bp_{0.1}$ 133-149°; bp_2 195-197°. d^{20} 1.1312. n_D^{20} 1.4622.

THERAP CAT: Antineoplastic.

3838. Etomidate. *1-(1-Phenylethyl)-1H-imidazole-5-carboxylic acid ethyl ester; (R)-(+)-1-(α-methylbenzyl)imidazole-5-carboxylic acid ethyl ester;* R 16659; Amidate; Hypnomidate. $C_{14}H_{16}N_2O_2$; mol wt 244.29. C 68.83%, H 6.60%, N 11.47%, O 13.10%. Prepn of the (±)-form: Belg. pat. **662,474** corresp to E. F. Godefroi *et al.,* U.S. pat. **3,354,173** (1965, 1967 both to Janssen); *eidem, J. Med. Chem.* **8**, 220 (1965). Prepn of the (R)-(+)-form: L. F. C. Roevens *et al.,* Ger. pat. **2,609,573** corresp to U.S. pat. **3,991,072** (both 1976 to Janssen). Metabolism: E. Goetz, *Anaesthesist* **23**, 331 (1974). Pharmacokinetics: M. J. Van Hamme *et al., Anesthesiology* **49**, 274 (1978). Pharmacology: P. A. J. Janssen *et al., Arch. Int. Pharmacodyn. Ther.* **214**, 92 (1975). Clinical study: R. J. Fragen, N. Caldwell, *Anesthesiology,* **50**, 242 (1979). Comprehensive description: Z. L. Chang, J. B. Martin, in *Analytical Profiles of Drug Substances* vol. 12, K. Florey, Ed. (Academic Press, New York, 1983) pp 191-214.

Crystals from diisopropyl ether, mp 67°. $[\alpha]_D^{20}$ +66° (c = 1 in ethanol). uv max (isopropanol): 240 nm (ε 12200). Soly in water at 25°: 0.0045 mg/100 ml. Sol in chloroform, methanol, ethanol, propylene glycol, acetone. LD_{50} in mice, rats (mg/kg): 29.5, 14.8-24.3 i.v. (Janssen).

THERAP CAT: Hypnotic.

3839. Etomidoline. *2-Ethyl-2,3-dihydro-3-[[4-[2-(1-piperidinyl)ethoxy]phenyl]amino]-1H-isoindol-1-one; 2-ethyl-3-(β-piperidino-p-phenetidino)phthalimidine;* K 2680; Smedolin. $C_{23}H_{29}N_3O_2$; mol wt 379.50. C 72.79%, H 7.70%, N 11.07%, O 8.43%. Anticholinergic antispasmodic. Prepn: N. Giraldi, V. Mariotti, **S. Afr. pat. 6,705,704** corresp to **U.S. pat. 3,624,206** (1968, 1971 both to Erba). Quantitative evaluation in humans: G. Sacchetti *et al., Farmaco Ed. Prat.* **27,** 708 (1972). Spasmolytic properties: M. Takeda, *Oyo Yakuri* **7,** 903 (1973), *C.A.* **80,** 66678 (1974). Pharmacology: M. Takeda *et al., ibid.* 1073, *C.A.* **81,** 33285 (1974); T. Mukai *et al., Japan. J. Pharmacol.* **31,** 147 (1981), *C.A.* **95,** 18068 (1981).

Crystals from ligroin, mp 106-107°. LD_{50} in mice (mg/kg): 176 orally; 44.8 i.v. (**U.S. pat. 3,624,206**).
THERAP CAT: Anticholinergic.

3840. Etonitazene. *2-[(4-Ethoxyphenyl)methyl]-N,N-diethyl-5-nitro-1H-benzimidazole-1-ethanamine; 1-[(2-diethylamino)ethyl]-2-(p-ethoxybenzyl)-5-nitrobenzimidazole; 2-p-ethoxybenzyl-1-(2-diethylaminoethyl)-5-nitrobenzimidazole.* $C_{22}H_{28}N_4O_3$; mol wt 396.48. C 66.64%, H 7.12%, N 14.13%, O 12.11%. Prepn: Hunger *et al., Experientia* **13,** 400 (1957); Hoffmann *et al.,* **U.S. pat. 2,935,514** (1960 to Ciba). Crystal structure: C. Humblet *et al., Acta Crystallogr.* **B34,** 3828 (1978).

Hydrochloride, $C_{22}H_{29}ClN_4O_3$, mp 162-164°.
Caution: May be habit forming. This is a controlled substance (opiate) listed in the U.S. Code of Federal Regulations, Title 21 Part 1308.11 (1987).
THERAP CAT: Narcotic analgesic.

3841. Etoperidone. *2-[3-[4-(3-Chlorophenyl)-1-piperazinyl]propyl]-4,5-diethyl-2,4-dihydro-3H-1,2,4-triazol-3-one.* $C_{19}H_{28}ClN_5O$; mol wt 377.92. C 60.39%, H 7.47%, Cl 9.38%, N 18.53%, O 4.23%. Psychotropic drug related structurally to trazodone, *q.v.* Prepn: G. Palazzo, **Ger. pat. 2,351,739;** *idem,* **U.S. pat. 3,857,845** (both 1974 to Angelini). Pharmacology and acute toxicity: R. Lisciani *et al., Arzneimittel-Forsch.* **28,** 417 (1978); M. T. Ramacci *et al., ibid.* **29,** 294 (1979). Clinical evaluation: P. Bertoletti, *G. Clin. Med.* **58,** 393 (1977). Teratology study: S. Barcellona *et al., Toxicology* **8,** 87 (1977). Acute cardiovascular toxicity study: R. Lisciani *et al., ibid.* **10,** 151 (1978).

Liq, $bp_{0.5}$ 230°.
Hydrochloride, $C_{19}H_{29}Cl_2N_5O$, *AF-1191, ST-1191, Axiomin, Depracer, Etoran, Staff, Tropene.* Crystals from isopropanol, mp 197-198°. pH (2% aq soln): 4.1 ± 0.2. Sol in water, alc; slightly sol in chloroform. Practically insol in acetone, benzene, ether. LD_{50} in mice, rats (mg/kg): 580, 720 orally; 72, 62 i.v.; 135, 120 i.p. (Lisciani).

THERAP CAT: Antidepressant.

3842. Etoposide. *9-[(4,6-O-Ethylidene-β-D-glucopyranosyl)oxy]-5,8,8a,9-tetrahydro-5-(4-hydroxy-3,5-dimethoxyphenyl)furo[3',4':6,7]naphtho[2,3-d]-1,3-dioxol-6(5aH)-one; 4'-demethylepipodophyllotoxin 9-[4,6-O-ethylidene-β-D-glucopyranoside];* EPEG; NSC-141540; VP-16-213; Lastet; Vepesid. $C_{29}H_{32}O_{13}$; mol wt 588.58. C 59.18%, H 5.48%, O 35.34%. Semi-synthetic deriv of podophyllotoxin, *q.v.,* related structurally to teniposide, *q.v.* Prepn: M. Kuhn *et al.,* **Swiss pat. 514,578** (1971 to Sandoz), *C.A.* **76,** 99998k (1971); C. Keller-Juslén *et al., J. Med. Chem.* **14,** 936 (1971). Activity in exptl tumors: H. Stähelin, *Eur. J. Cancer* **9,** 215 (1973). Mechanism of action: S. B. Horwitz, J. D. Loike, *Lloydia* **40,** 82 (1977). Pharmacokinetics: F. R. Pelsor *et al., J. Pharm. Sci.* **67,** 1106 (1978). Clinical studies: R. L. Chard *et al., Cancer Treat. Rep.* **63,** 1755 (1979); R. E. Slayton *et al., ibid.* 2089. Teratogenicity and cytogenetic study: S. M. Sieber *et al., Teratology* **18,** 31 (1978). *Review:* B. F. Issell, S. T. Crooke, *Cancer Treat. Rev.* **6,** 107-124 (1979). Review of pharmacology and clinical experience: P. J. O'Dwyer *et al., N. Engl. J. Med.* **312,** 692-700 (1985). Review of clinical pharmacology, pharmacokinetics and assay methods: P. I. Clark, M. L. Slevin, *Clin. Pharmacokinet.* **12,** 223-252 (1987).

Crystals from methanol, mp 236-251°. $[\alpha]_D^{20}$ −110.5° (c = 0.6 in chloroform) (Keller-Juslén).
THERAP CAT: Antineoplastic.

3843. Etorphine. *[5α,7α(R)]-4,5-Epoxy-3-hydroxy-6-methoxy-α,17-dimethyl-α-propyl-6,14-ethenomorphinan-7-methanol; tetrahydro-7α-(1-hydroxy-1-methylbutyl)-6,14-endo-ethenooripavine; 7,8-dihydro-7α-[1(R)-hydroxy-1-methylbutyl]-O⁶-methyl-6,14-endo-ethenomorphine; 7α-[1(R)-hydroxy-1-methylbutyl]-6,14-endo-ethenotetrahydrooripavine; 19-propylorvinol; tetrahydro-7α-(2-hydroxy-2-pentyl)-6,14-endo-ethenooripavine.* $C_{25}H_{33}NO_4$; mol wt 411.52. C 72.96%, H 8.08%, N 3.40%, O 15.55%. Narcotic analgesic. Prepn: **Belg. pat. 618,392;** K. W. Bentley, **Brit. pat. 937,214** (1962, 1963 both to J. F. MacFarlan); and structure: K. W. Bentley, D. G. Hardy, *Proc. Chem. Soc.* **1963,** 220. Improved prepn: W. R. Hydro, **U.S. pat. 3,763,167** (1973 to U.S. Army). Determn by HPLC: I. Jane, A. McKinnon, *J. Chromatog.* **323,** 19 (1985). Analgesic activity: G. F. Blane, *J. Pharm. Pharmacol.* **19,** 367 (1967). Pharmacology: Lister, *ibid.* **16,** 364 (1964). Use as tranquilizer in dogs: J. L. Crooks *et al., Vet. Rec.* **87,** 498 (1970); M.

Grange *et al., Rev. Med. Vet.* **124**, 899 (1973); in large animals: A. M. Harthoorn, *J. Am. Vet. Med. Assoc.* **149**, 875 (1966); B. T. Alford *et al., ibid.* **164**, 702 (1974).

White crystals from aqueous ethoxyethanol, mp 214-217°. (7β-isomer melts at 280°).

Hydrochloride, $C_{25}H_{34}ClNO_4$, M 99. mp 266-267°. (7β-isomer hydrochloride, dec at 290°.) In combination with acepromazine, *q.v., Immobilon.*

3-Acetate, $C_{27}H_{35}NO_5$, *acetorphine, O^3-acetyl-7,8-dihydro-7α-[1(R)-hydroxy-1-methylbutyl]-O^6-methyl-6,14-endoethenomorphine, 3-O-acetyl-7α-[1(R)-hydroxy-1-methylbutyl]-6,14-endo-ethenotetrahydrooripavine, 3-O-acetyl-17-propylorvinol.* Crystals from methanol, mp 195°.

3-Acetate hydrochloride, $C_{27}H_{36}ClNO_5$, M 183. mp 206°.

Caution: May be habit forming. Acetorphine, etorphine and their salts are controlled substances (opium derivatives) listed in the U.S. Code of Federal Regulations, Title 21 Parts 1308.11, 1308.12 (1987). Use extreme care. Very small amounts will bring about respiratory paralysis and death: D. W. Upson, *Upson's Handbook of Clinical Veterinary Pharmacology* (VM Publishing, Bonner Springs, 1980) pp 380-381.

THERAP CAT (VET): Used to immobilize large animals.

3844. Etoxadrol. *(+)-2-(2-Ethyl-2-phenyl-1,3-dioxolan-4-yl)piperidine;* 2-ethyl-2-phenyl-4-(2-piperidyl)-1,3-dioxolane. $C_{16}H_{23}NO_2$; mol wt 261.37. C 73.53%, H 8.87%, N 5.36%, O 12.24%. Prepn of *dl*-form: Hardie *et al., J. Med. Chem.* **9**, 127 (1966); Hardie, Halverstadt, U.S. pat. **3,262,938** (1966 to Cutter Labs). Resolution of the α-racemates: Allen *et al.,* **Ger. Offen.** pat. **2,001,616** (1970 to Cutter Labs), *C.A.* **74**, 13129b (1971). Pharmacological studies: Traber *et al., J. Pharmacol. Exp. Ther.* **175**, 395 (1970); Hidalgo *et al., Anesth. Analg. (Cleveland)* **50**, 231 (1971); Kelly *et al., ibid.* **262.**

(+)-Hydrochloride, $C_{16}H_{24}ClNO_2$, *CL-1848C, Thoxan.* Crystals from isopropanol, mp 221.5-222°. $[\alpha]_D^{25}$ +16.63°.

THERAP CAT: Anesthetic (intravenous).

3845. Etozolin. *[3-Methyl-4-oxo-5-(1-piperidinyl)-2-thiazolidinylidene]acetic acid ethyl ester; 3-methyl-4-oxo-5-piperidino-$\Delta^{2,\alpha}$-thiazolidineacetic acid ethyl ester;* 2-carbethoxymethylene-3-methyl-5-piperidino-4-thiazolidone; Gö 687; W2900A; Elkapin. $C_{13}H_{20}N_2O_3S$; mol wt 284.39. C 54.90%, H 7.09%, N 9.85%, O 16.88%, S 11.28%. A member of a new class of heterocyclic compounds, the 2-methylene-thiazolidones; a diuretic with choleretic properties. Prepn: Satzinger, *Ann.* **665**, 150 (1963); U.S. pat. **3,072,653** (1963 to Warner-Lambert). Pharmacology: O. Heidenreich *et al., Arzneimittel-Forsch.* **14**, 1242 (1964). Series of articles on activity, metabolism and toxicology: G. Satzinger *et al., ibid.* **27**, 1742-1817 (1977).

Crystals from methanol, mp 140°. uv max (methanol): 283, 243 nm (log ε 4.32, 4.0). LD_{50} i.p. in mice, rats: 1.210, 1.575 g/kg.

Hydrochloride, $C_{13}H_{20}N_2O_3S$·HCl, crystals, mp 158-159°.

Free acid, $C_{11}H_{16}N_2O_3S$, *Gödecke 3282, ozolinone.*

THERAP CAT: Diuretic.

3846. Etretinate. *(all-E)-9-(4-Methoxy-2,3,6-trimethylphenyl)-3,7-dimethyl-2,4,6,8-nonatetraenoic acid ethyl ester;* Ro 10-9359; Tegison; Tigason. $C_{23}H_{30}O_3$; mol wt 354.49. C 77.93%, H 8.53%, O 13.54%. Aromatic analog of retinoic acid, *q.v.* Prepn: W. Bollag *et al.,* **Ger.** pat. **2,414,619**; *eidem,* **U.S.** pat. **4,105,681** (1974, 1978 both to Hoffmann-La

Roche). Effects on skin metabolism: W. P. Raab, B. M. Gmeiner, *Arch. Dermatol. Res.* **256**, 247 (1976). Toxicity study in organ culture: D. R. Bard, I. Lasnitzki, *Brit. J. Cancer* **35**, 110 (1977). Pharmacokinetics: R. Haenni, *Dermatologica* **157**, Suppl, 5 (1978). Mutation study: H. J. Juhl *et al., Mutat. Res.* **58**, 317 (1978). HPLC determn in plasma: R. Haenni *et al., J. Chromatog.* **162**, 615 (1979). Toxicology, carcinogenicity and teratogenicity study: J. J. Kamm, *J. Am. Acad. Dermatol.* **6**, 652 (1982). Clinical evaluation in psoriatic arthritis: M. L. Ciompi *et al., Int. J. Tissue React.* **10**, 25 (1988). *Review:* H. Mayer *et al., Experientia* **34**, 1105-1119 (1978). Review of pharmacology and therapeutic efficacy: A. Ward *et al., Drugs* **26**, 9-43 (1983); of clinical pharmacology: A. Vahlquist, O. Rollman, *Dermatologica* **175**, Suppl. 1, 20-27 (1987). Book: *The Retinoids,* **Vol. 1-2**, M. B. Sporn *et al.,* Eds. (Academic Press, New York, 1984).

Crystals, mp 104-105°. LD_{50} in mice (1 day): > 4000 mg/kg i.p. (Bollag). LD_{50} (20 day) in mice, rats (mg/kg): 1176, > 2000 i.p.; > 2000, > 4000 orally (Kamm).

Free acid, *see* acitretin.

THERAP CAT: Antipsoriatic.

3847. Etrimfos. *O-(6-Ethoxy-2-ethyl-4-pyrimidinyl)-phosphorothioic acid O,O-dimethyl ester; O,O-*dimethyl *O-(2-ethyl-4-ethoxypyrimidinyl)-6-thionophosphate;* SAN 197; Ekamet; Satisfar. $C_{10}H_{17}N_2O_4PS$; mol wt 292.29. C 41.09%, H 5.86%, N 9.58%, O 21.90%, P 10.60%, S 10.97%. Organophosphorous insecticide with stomach and contact activity against pests of fruit and vegetable crops. Prepn: K. Milzner, F. Reisser, **Ger.** pat. **2,209,554**; *eidem,* **U.S.** pat. **3,862,188** (1972, 1975 both to Sandoz). Physical properties and activity: H. J. Knutti, F. Reisser, *Proc. 8th Brit. Insect. Fungicide Conf.* **2**, 675 (1975). Residue analysis and toxicity study: M. C. Bowman *et al., J. Agr. Food Chem.* **26**, 35 (1978). Metabolism in plants: M. Akram *et al., ibid.* 925. Review of analytical methods: J. C. Karapally *et al., Anal. Methods Pestic. Plant Growth Regul.* **11**, 125-138 (1980).

Colorless oil, slight characteristic odor. mp −3.35°. d 1.195; n_D^{20} 1.5068. Vapor pressure at 20°: 6.5 × 10^{-5} mm Hg. Soly in water at 23-24°: 40 ppm. Miscible with ethanol, dimethyl sulfoxide, ethyl acetate, acetone, ether, chloroform, xylene, hexane. LD_{50} in male rats, mice (mg/kg): 1800, 437 orally; LC_{50} in carp (mg/l): 13.6 (48 hr); 13.3 (96 hr) (Knutti).

USE: Agricultural insecticide.

3848. Etryptamine. *α-Ethyl-1H-indole-3-ethanamine; 3-(2-aminobutyl)indole;* α-ethyltryptamine. $C_{12}H_{16}N_2$; mol wt 188.26. C 76.55%, H 8.57%, N 14.88%. Prepn: Heinzelman *et al., J. Org. Chem.* **23**, 1548 (1960); **Brit.** pat. **933,786** (1963 to ICI).

Crystals from ethyl acetate + petr ether, mp 97-99°. uv max: 220.5, 281, 289.5 nm (ε 20,500, 6000, 5200).

Acetate, $C_{14}H_{20}N_2O_2$, *Monase.* Crystals from ethyl acetate + methanol, mp 165-166°.

Hydrochloride, $C_{12}H_{16}N_2.HCl$, mp 215.5-218°.

THERAP CAT: Central stimulant.

3849. Etymemazine. *2-Ethyl-N,N,β-trimethyl-10H-phenothiazine-10-propanamine; 10-[3-(dimethylamino)-2-methylpropyl]-2-ethylphenothiazine; 3-ethyl-10-(3-dimethylamino-2-methylpropyl)phenothiazine; 3-(3-ethyl-10-phenothiazinyl)-2-methyl-1-dimethylaminopropane; ethotrimeprazine; ethylisobutrazine; RP 6484; Diquel.* $C_{20}H_{26}N_2S$; mol wt 326.51. C 73.57%, H 8.03%, N 8.58%, S 9.82%. Prepn: Jacob, Robert, U.S. pat. **2,837,518** (1958 to Rhône-Poulenc). Activity studies in pigs: Lambrerth, *Aust. Vet. J.* **44**, 333 (1968). Biotransformation studies: Robinson, *J. Pharm. Pharmacol.* **18**, 19 (1966).

Hydrochloride, $C_{20}H_{27}ClN_2S$, *Nuital, Sergetyl.* mp 160-163°. Soluble in water, ethanol, methanol, acetone, chloroform. Practically insol in ether.

THERAP CAT: Antihistaminic.

THERAP CAT (VET): Tranquilizer.

3850. β-Eucaine. *2,2,6-Trimethyl-4-piperidinol benzoate (ester); α-4-benzoyloxy-2,2,6-trimethylpiperidine; α-vinyldiacetonalkamine benzoate; benzamine; betacaine; eucaine B.* $C_{15}H_{21}NO_2$; mol wt 247.33. C 72.84%, H 8.56%, N 5.66%, O 12.94%. Prepn from lower melting form (α-form) of 2,2,6-trimethyl-4-piperidinol and benzoyl chloride: Harries, *Ann.* **417**, 107, *see* p 175 (1918). Stereochemistry: King, *J. Chem. Soc.* **125**, 41 (1924); Stenlake, *J. Pharm. Pharmacol.* **6**, 164 (1954). Configuration: Perks, Russell, *ibid.* **19**, 318 (1966).

Crystals from petr ether, mp 70-71° (corr).

Hydrochloride, *B.P.C. 1954.* Crystals from water, mp 277-279° (corr). pKa 9.4. One gram dissolves in 30 ml water, 35 ml alcohol, 6 ml chloroform; more sol in boiling water or boiling alcohol. Aq soln is neutral to litmus. MLD i.v. in rats: 15-25 mg/kg. *Incompat:* Salicylates.

d-Form, crystals from petr ether, mp 57-58° (corr).

d-Form hydrochloride, $[\alpha]_D$ +11.5° (water).

l-Form, crystals from petr ether, mp 57-58° (corr).

l-Form hydrochloride, crystals, mp 244-245° (corr). $[\alpha]_D$ −11.3° (water).

THERAP CAT: Local anesthetic.

THERAP CAT (VET): Has been used as a local anesthetic.

3851. Eucalyptol. *1,3,3-Trimethyl-2-oxabicyclo[2.2.2]octane; 1,8-epoxy-p-menthane; cineole; cajeputol.* $C_{10}H_{18}O$; mol wt 154.24. C 77.87%, H 11.76%, O 10.37%. The chief constituent of oil of eucalyptus: Cloez, *Ann.* **154**, 372 (1870); Berry, *Australasian J. Pharm.* **1929**, 203; from wormwood oil: Wallach, Brass, *Ann.* **225**, 291 (1884). Identity with cajeputol: Wallach, *ibid.* 314. Identity with cineole: Jahns, *Ber.* **17**, 2941 (1884). Structure: Wallach, *Ann.* **291**, 342 (1896). Biosynthesis: Birch *et al., Tetrahedron Letters,* no. 3, 1 (1959). Activity: Jori *et al., Eur. J. Pharmacol.* **9**, 362 (1970). Cockroach repellent activity: T. H. Maugh, *Science* **218**, 278 (1982).

Colorless liquid; camphor-like odor; spicy, cooling taste. d_{25}^{25} 0.921-0.923. bp 176-177°. mp +1.5°. n_D^{20} 1.455-1.460. Flash pt, closed cup: 118°F (48°C). Practically insol in water; miscible with alcohol, chloroform, ether, glacial acetic acid, oils.

USE: Pharmaceutic aid (flavor).

3852. Eucalyptus. Gum wood; Australian fever tree; blue-gum tree. Dried leaves of *Eucalyptus globulus* Labill., *Myrtaceae;* this is a dwarf species, known as Malee in Australia, and richest in oil of Eucalyptus. *Habit.* Australia, cultivated in subtropics, Europe, N. Africa and Southern U.S. *Constit.* 1-3% volatile oil, tannin, resins, eucalyptic acid, rutin.

USE: Pharmaceutic aid (flavor).

3853. Eucalyptus Gum. Eucalyptus kino; red gum. Dried gummy exudation from *Eucalyptus longirostris* F. Muell. (*E. rostrata* Schlecht.) and other species of *Eucalyptus, Myrtaceae. Habit.* Australia. *Constit.* About 45% kinotannic acid, kino red, glucoside, catechol, pyrocatechol.

THERAP CAT: Astringent, antidiarrheal.

3854. Eucatropine. *α-Hydroxybenzeneacetic acid 1,2,2,6-tetramethyl-4-piperidinyl ester;* mandelic acid *β-1,2,2,6-tetramethyl-4-piperidyl ester; β-4-hydroxy-1,2,2,6-tetramethylpiperidine mandelic acid ester; N-methyl-β-vinyldiacetonalkamine mandelate; β-4-mandeloyloxy-1,2,2,6-tetramethylpiperidine; β-1,2,2,6-tetramethyl-4-piperidinol mandelate; Euphthalmine.* $C_{17}H_{25}NO_3$; mol wt 291.38. C 70.07%, H 8.65%, N 4.81%, O 16.47%. Prepn from higher melting form (β-form) of 1,2,2,6-tetramethyl-4-piperidinol and mandelic acid: Harries, *Ann.* **296**, 328 (1897); Kipping, *J. Chem. Soc.* **123**, 3115 (1923). Description: Harries, *Ber.* **31**, 665 (1898).

dl-Form, prisms from petr ether sinters at 108°, mp 113°. Practically insol in cold water, more sol in hot water; readily sol in organic solvents other than petr ether.

dl-Form hydrochloride, crystals from alcohol + ether, sinters at 181°, mp 183-184°. Very sol in water; sol in alcohol (1 g dissolves in 2 ml boiling alcohol), in chloroform. Practically insol in ether. Aq solns are neutral to litmus.

THERAP CAT: *dl*-Form hydrochloride as anticholinergic (ophthalmic).

3855. Eugenol. *2-Methoxy-4-(2-propenyl)phenol; 4-allyl-2-methoxyphenol;* allylguaiacol; eugenic acid; caryophyllic acid. $C_{10}H_{12}O_2$; mol wt 164.20. C 73.14%, H 7.37%, O 19.49%. Obtained from many natural sources: *Beilstein* vol. **6**, 961. Prepn: Claisen, *Ann.* **418**, 113 (1919); from oil of cloves: Waterman, Priester, *Rec. Trav. Chim.* **48**, 1272 (1929).

Colorless or pale yellow liquid, bp 255°. Darkens and thickens on exposure to air. Odor of cloves; spicy, pungent taste. mp −9.2° to −9.1°. d_4^{20} 1.0664. n_D^{20} 1.5410. Practically insol in water. Misc with alcohol, chloroform, ether, oils. One ml dissolves in 2 ml 70% alcohol; sol in glacial acetic acid, in aq fixed alkali hydroxide solns. Ferric chloride, potassium permanganate. LD_{50} orally in rats, mice: 2680, 3000 mg/kg, E. C. Hagan *et al.*, *Toxicol. Appl. Pharmacol.* **7**, 18 (1965).

Benzoate, $C_{17}H_{16}O_3$, *O-benzoyleugenol.* Crystals, mp 69-70°. bp 360°. Practically insol in water; freely sol in benzene, chloroform, ether, hot alcohol.

USE: In perfumery instead of oil of cloves; manuf vanillin. As insect attractant.

THERAP CAT: Analgesic (dental).

3856. Euonymus. Wahoo; Indian arrow wood; bitter ash; burning bush; strawberry tree; spindle tree. Dried root bark of *Evonymus atropurpureus* Jacq., *Celastraceae. Habit.* U.S., Ontario to Florida east of Mississippi. *Constit.* Euonymol—an alcohol; euonysterol, atropurpurin, asparagine, citrullol, dulcite; malic, citric and tartaric acids.

THERAP CAT: Cathartic, diuretic.

3857. Euparin. *1-[6-Hydroxy-2-(1-methylethenyl)-5-benzofuranyl]ethanone; 6-hydroxy-2-isopropenyl-5-benzofuranyl methyl ketone;* 5-acetyl-6-hydroxy-2-isopropenylbenzofuran. $C_{13}H_{12}O_3$; mol wt 216.23. C 72.21%, H 5.59%, O 22.20%. From roots of *Eupatorium purpureum* L., *E. cannabinum* L., *Compositae:* Kamthong, Robertson, *J. Chem. Soc.* **1939**, 925; Jermanowska, *C.A.* **48**, 5848h (1954). Structure: Kamthong, Robertson, *J. Chem. Soc.* **1939**, 933. Synthesis: Ramachandran *et al.*, *J. Org. Chem.* **28**, 2744 (1963). *Review:* Gizycki, *Pharmazie* **6**, 686 (1951).

Yellow needles from alcohol, mp 121-122°. uv max (ethanol): 263, 358 nm (ε 34,400, 5900). Sol in alcohol, benzene, chloroform, ether. Practically insol in water, alkali.

Acetate, $C_{15}H_{14}O_4$, prisms from petr ether, mp 80°.

O-Methyl deriv, $C_{14}H_{14}O_3$, *O-methyleuparin*, *2-isopropenyl-6-methoxy-5-benzofuranyl methyl ketone.* Needles from dil alcohol, mp 76-77°. Sol in the usual organic solvents.

3858. Eupatorin. *5-Hydroxy-2-(3-hydroxy-4-methoxyphenyl)-6,7-dimethoxy-4H-1-benzopyran-4-one; 3',5-dihydroxy-4',6,7-trimethoxyflavone.* $C_{18}H_{16}O_7$; mol wt 344.31. C 62.79%, H 4.68%, O 32.53%. Isoln from *Eupatorium semiserratum* DC., *Compositae* and structure: Kupchan *et al.*, *J. Pharm. Sci.* **54**, 929 (1965).

Crystals from dioxane + water, mp 196-198°. uv max (alcohol): 243, 254, 274, 342 nm (ε 17,400; 19,300; 19,800; 27,700).

3859. Eupatorium. Boneset; thoroughwort. Flowering tops and dried leaves of *Eupatorium perfoliatum* L., *Compositae. Habit.* Canada to Florida and west to Texas and Nebraska. *Constit.* Eupatorin, volatile oil, resin, tannin, sugar, inulin, wax.

3860. Euphorbia. Pill-bearing spurge; snake-weed; cat's hair; Queensland asthma weed; flowery-headed spurge. Dried herb of *Euphorbia hirta* L. or *E. pilulifera* L., *Euphorbiaceae. Habit.* Queensland (Australia), India, widely distributed in tropical countries. *Constit.* Several resins, an unstable glucoside.

3861. Euprocin. *10,11-Dihydro-6'-(3-methylbutoxy)cin-*

chonan-9-ol; hydrocupreine isopentyl ether; isoamylhydrocupreine; isopentylhydrocupreine; eucupine; Eucupin. $C_{24}H_{34}N_2O_2$; mol wt 382.53. C 75.35%, H 8.96%, N 7.32%, O 8.36%. Prepd from hydrocupreine and *p*-toluenesulfonic acid isopentyl ester: Slotta, Behnisch, *Ber.* **66**, 360 (1933).

Almost tasteless powder, mp 152°. Sol in alc, ether, chloroform; practically insol in water. LD s.c. in mice: 300 mg/kg; LD i.v. in rabbits: 13 mg/kg, *Handbook of Toxicology* vol. 1, W. S. Spector, Ed. (Saunders, Philadelphia, 1956) pp 140-141.

Dihydrochloride monohydrate, $C_{24}H_{34}N_2O_2 \cdot 2HCl \cdot H_2O$, fine bitter needles. Sol in 15 parts water; freely sol in alc.

THERAP CAT: Anesthetic (topical), antiseptic.

3862. Europium. Eu; at. wt 151.96; at. no. 63; valences 2, 3. A lanthanide; belongs to cerium group of rare earths. Two naturally occurring isotopes: [151]Eu (47.77%); [153]Eu (52.23%); artificial, radioactive isotopes: 143-150; 152; 154-160. Abundance in earth's crust: 0.14-1.1 ppm. Occurs in monazite sand, gadolinite. Has been detected spectroscopically in the sun and in certain stars. Discovered and prepd in form of the oxide by Demarçay, *Compt. Rend.* **122**, 728 (1896); **130**, 1019, 1469 (1900); **132**, 1484 (1901). Prepn of a mixt of the metal with an alkali chloride: Klemm, Bommer, *Z. Anorg. Chem.* **231**, 138 (1937). Separation: Spedding *et al.*, *J. Am. Chem. Soc.* **76**, 2557 (1954); Henricks, U.S. pat. **2,747,973** (1956); by paper chromatography: Lederer, *Nature* **176**, 462 (1953). Reviews of prepn, properties, and compds of europium and other rare earths: *The Rare Earths*, F. H. Spedding, A. H. Daane, Eds. (Krieger, Huntington, N.Y., 1971, reprint of 1961 ed.) 641 pp; S. P. Sinha, *Europium* (Springer, New York, 1967) 164 pp; Hulet, Bode, "Separation Chemistry of the Lanthanides and Transplutonium Actinides" in *MTP Int. Rev. Sci.: Inorg. Chem.*, *Ser. One*, vol. 7, K. W. Bagnall, Ed. (University Park Press, Baltimore, 1972) pp 1-45; Moeller, "The Lanthanides" in *Comprehensive Inorganic Chemistry*, vol. 4, J. C. Bailar, Jr. *et al.*, Eds. (Pergamon Press, Oxford, 1973) pp 1-101.

Body-centered cubic crystal lattice; d 5.244; mp 826°. Sol in liq ammonia. Shows two reduction potentials −0.710 and −2.510 v. (referred to a normal calomel electrode): Noddack, Brukl, *Angew. Chem.* **50**, 362 (1937); gives two definite series of salts: one in which the metal is bivalent, and another in which it is trivalent.

Sesquioxide, Eu_2O_3, *europia.* Pink powder, d 7.42, prepd by heating the hydroxide, nitrate, oxalate or sulfate at 1600°. The oxide of the divalent metal is prepd by reduction of the sesquioxide at elevated temp.

Hydroxide, $Eu(OH)_3$, prepd by adding ammonia or an alkali hydroxide to a soln of an europic salt.

Europic chloride, $EuCl_3$, greenish-yellow needles; mp 623° in nitrogen (in a closed tube), d^{35} 4.471, prepd by passing sulfur chloride over the heated oxide at 200-500°. Reduction with hydrogen at 600° yields europous chloride, $EuCl_2$, white amorphous powder, sol in water. LD_{50} (trichloride) in mice: 550 mg/kg i.p.; 5 g/kg orally, Haley, *J. Pharm. Sci.* **54**, 663 (1965).

Europic sulfate, $Eu_2(SO_4)_3$, octahydrate, a pinkish cryst solid, prepd by dissolving the oxide in sulfuric acid. Soly in water: 2.56 parts per 100 parts at 20°, 1.93 parts per 100 parts at 40°. On heating at 375° yields the anhyd sulfate.

Europic nitrate, $Eu(NO_3)_3$, hexahydrate, mp 85° in its water of crystallization (sealed tube). LD_{50} in rats: 210 mg/kg i.p.; > 5000 mg/kg orally, Haley, *loc. cit.*

Europous sulfate, $EuSO_4$, colorless crystals. Insol in water and in dil acids, prepd by electrolytic reduction of europic salts.

USE: The salts in cathode ray tube coatings for color television receivers. Europium has a very high cross-section for the capture of thermal neutrons which is of value in the construction of electric atomic power stations. Organic derivs of europium are used as shift reagents in NMR spectroscopy: C. C. Hinckley, *J. Am. Chem. Soc.* **91**, 5160 (1969); R. E. Sievers, *Nuclear Magnetic Resonance Shift Reagents* (Academic Press, New York, 1973).

3863. Evan's Blue. *6,6′-[(3,3′-Dimethyl[1,1′-biphenyl]-4,4′-diyl)bis(azo)]bis[4-amino-5-hydroxy-1,3-naphthalenedisulfonic acid] tetrasodium salt;* C.I. Direct Blue 53; 4,4′-bis[7-(1-amino-8-hydroxy-2,4-disulfo)naphthylazo]-3,3′-bitolyl tetrasodium salt; C.I. 23860; T-1824; Azovan Blue. $C_{34}H_{24}N_6Na_4O_{14}S_4$; mol wt 960.83. C 42.50%, H 2.52%, N 8.75%, Na 9.57%, O 23.31%, S 13.35%. Prepd by coupling 1 mol of diazotized *o*-tolidine with 2 mols of Chicago acid (1-amino-8-naphthol-2,4-disulfonic acid): **Ger.** pats. **35,341; 38,802;** *Frdl.* **1,** 469, 488 (1877-1887); **3,949; 57,327; 75,469;** *Frdl.* **3,** 685, 687, 690 (1890-1894); Hartwell, Fieser, *Org. Syn.* **coll. vol. II,** 145 (1943). Diagnostic use: M. H. Nielsen, N. C. Nielsen, *Scand. J. Clin. Lab. Invest.* **14,** 605 (1962); O. Linderkamp *et al., Eur. J. Pediat.* **125,** 135 (1977).

Blue crystals with bronze to green luster. Sol in water, alcohol, acids, alkalies. Indicator changing color near pH 10. Destroyed by strong oxidizing and reducing agents and precipitated from soln by strong concns of neutral salts. Rather stable in aq soln, and may be autoclaved at 15 lbs pressure for 30 min. Dye made up in physiological saline should not be autoclaved.

THERAP CAT: Diagnostic aid (blood volume determination).

3864. Evening Primrose Oil. EPO; Efamol; Epogam. Seed oil of the evening primrose, *Oenothera biennis* L., *Onagraceae,* which contains approx 72% linoleic acid and approx 9% γ-linolenic acid, *q.q.v.* Unique among vegetable oils because of its high content of γ-linolenic acid. Effect on prostaglandin biosynthesis in rats: B. A. Schölkens *et al., Prostaglandins, Leukotienes Med.* **8,** 273 (1982). Clinical studies in atopic eczema: C. R. Lovell *et al., Lancet* **1,** 278 (1981); S. Wright, J. L. Burton, *ibid.* **2,** 1120 (1982); P. L. Biagi *et al., Drugs Exp. Clin. Res.* **14,** 285 (1988). Ingredient in cosmetics for aging skin: J. P. Marty, **Ger.** pat. **3,447,618** (1985 to Roussel-UCLAF), *C.A.* **103,** 146984r (1985). Brief review including discussion of uses: A. J. Barber, *Pharm. J.* **240,** 723-725 (1988).

Clear, golden yellow oil. d_{15} 0.9283. n_D^{25} 1.4782. Sapon. no. 287.8. Iodine no. 154.8.

USE: Dietary supplement.

THERAP CAT: In treatment of atopic eczema.

3865. Evodiamine. *8,13,13b,14-Tetrahydro-14-methylindolo[2′,3′:3,4]pyrido[2,1-b]quinazolin-5(7H)-one.* $C_{19}H_{17}N_3O$; mol wt 303.35. C 75.22%, H 5.65%, N 13.85%, O 5.27%. From *Evodia rutaecarpa* Hook. & Thoms and bark of *Zanthoxylum rhetsa* DC., *Rutaceae:* Y. Asahina, K. Kashiwaki, *J. Pharm. Soc. Japan* **1915,** 1293, *C.A.* **10,** 607 (1916); Gopinath *et al., Tetrahedron* **8,** 293 (1960). Structure: Y. Asahina *J. Pharm. Soc. Japan* **1924,** 1; Ohta, *J. Pharm. Soc. Japan* **65,** 15 (1945), *C.A.* **45,** 5697 (1951). Synthesis: Asahina, Ohta, *Ber.* **61B,** 319 (1928); T. Kametani *et al., J. Am. Chem. Soc.* **98,** 6186 (1976); *eidem, Heterocycles* **4,** 23 (1976). Biosynthesis: M. Yamazaki *et al., Tetrahedron Letters* **1966,** 3221; **1967,** 3317. Mass spec.: J. Tamas *et al., Acta Chim. Acad. Sci. Hung.* **89,** 85 (1976).

Yellow plates from alc, mp 278°. $[\alpha]_D^{15}$ +352° (acetone); $[\alpha]_D$ +440° (chloroform). uv max (acetonitrile): 272, 280, 291, 335 nm (log ε 4.06, 4.02, 3.90, 3.30). Sol in acetone; slightly sol in alcohol, ether, chloroform. Practically insol in water, petr ether, benzene. Does not seem to form salts.

3866. Exalamide. *2-(Hexyloxy)benzamide;* HBA; Hyperan. $C_{13}H_{19}NO_2$; mol wt 221.30. C 70.56%, H 8.65%, N 6.33%, O 14.46%. Deriv of salicylamide, *q.v.* Prepn: E. M. Bevin *et al., J. Pharm. Pharmacol.* **4,** 872 (1952); F. J. MacRae, D. E. Seymour, **Brit.** pat. **726,786** (1955 to Herts Pharm.), *C.A.* **50,** 5751i (1956); L. V. Coates *et al., J. Pharm. Pharmacol.* **9,** 855 (1957). Metabolism: Y. Kanaiwa *et al., Yakugaku Zasshi* **95,** 411 (1975), *C.A.* **83,** 71428v (1975). General pharmacologic effects: R. Inoki *et al., Oyo Yakuri* **10,** 165 (1975), *C.A.* **88,** 115369 (1976). Comparative study of antifungal efficacy vs clinical isolates: T. Kusunoki, S. Harada, *J. Dermatol.* **11,** 227 (1984).

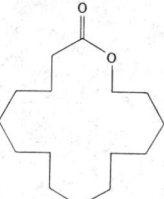

Cryst from ethanol, mp 71°. Sol in methanol, acetone, chloroform, benzene. Slightly sol in ether. Practically insol in water.

THERAP CAT: Topical antifungal.

3867. Exaltolide®. *Oxacyclohexadecan-2-one;* 15-hydroxypentadecanoic acid ε-lactone. $C_{15}H_{28}O_2$; mol wt 240.37. C 74.95%, H 11.74%, O 13.31%. Musk compd responsible for the pleasant odor of angelica root oil. Synthesis: Ruzicka, Stoll, *Helv. Chim. Acta* **11,** 1159 (1928); Carnduff *et al., Chem. & Ind. (London)* **1960,** 559; Dhenkne *et al., J. Chem. Soc.* **1962,** 2348; Mathur, Bhattacharya, *ibid.* **1963,** 3505; Becker, Ohloff, *Helv. Chim. Acta* **54,** 2889 (1971). Alternate syntheses: T. Mukaiyama *et al., Chem. Letters* **1977,** 441; K. Narasaka *et al., ibid.* 763; W. H. Kruizinga, R. M. Kellogg, *Chem. Commun.* **1979,** 286.

Thick oil. Odor of amber and musk. $bp_{0.25}$ 110°; bp_{15} 176°. d_4^{20} 0.9549. n_D^{20} 1.4708. When sublimed in vacuum, needles, mp 32°. Sol in alc.

USE: As a fixative in perfumery.

3868. Exifone. *(2,3,4-Trihydroxyphenyl)(3,4,5-trihydroxyphenyl)methanone;* 2,3,3′,4,4′,5′-hexahydroxybenzophenone; 4-galloylpyrogallol; Adlone. $C_{13}H_{10}O_7$; mol wt 278.22. C 56.12%, H 3.62%, O 40.25%. Cognition enhancing agent. Prepn: H. Bleuler, A. G. Perkin, *J. Chem. Soc.* **109,** 529 (1916); *see also* **Ger.** pat. **49,149** (1889 to BASF). Prepd but not claimed: J. M. Gazave *et al.,* **Ger.** pat. **2,501,443;** *eidem,* **U.S.** pat. **4,015,017** (1975, 1977 both to Pharmascience). Pharmacology: R. D. Porsolt *et al., Arzneimittel-Forsch.* **37,** 388 (1987). Animal studies in memory improvement: R. D. Porsolt *et al., Pharmacol. Biochem. Behav.* **27,**

253 (1987); in antagonism of benzodiazepine-induced amnesia: R. D. Porsolt *et al.*, *Psychopharmacol.* **95**, 291 (1988). Clinical evaluation in senile dementia: F. Piette *et al.*, *Rev. Geriatr.* **11**, 375 (1986); in Parkinson's disease: H. Allain *et al.*, *Fundam. Clin. Pharmacol.* **2**, 1 (1988).

Crystals from aqueous solution, mp 270° (Gazave); also reported as faintly yellow needles from water, mp 272-273° (Bleuler, Perkin). LD₅₀ in rats (g/kg): 0.355 i.p.; 1.425 orally (Gazave).

THERAP CAT: Nootropic.

3869. Exiproben. *2-[3-(Hexyloxy)-2-hydroxypropoxy]-benzoic acid.* $C_{16}H_{24}O_5$; mol wt 296.37. C 64.84%, H 8.16%, O 26.99%. Prepn: **Brit. pat.** **933,980** (1963 to Thomae), *C.A.* **60**, 2866a (1964); G. Ohnacher, **U.S. pat.** **3,198,827** (1965 to Boehringer, Ing.). Metabolism study: A. Pinelli, *Pharmacol. Res. Commun.* **5**, 17 (1973). Choleretic activity:

R. Bossa *et al.*, *Fegato* **17**, 21 (1971), *C.A.* **76**, 135808j (1972). Clinical study: N. D'Imperio *et al.*, *Clin. Ter.* **78**, 153 (1976).

Sodium salt, $C_{16}H_{23}NaO_5$, *DCH-21*, *Etopalin.* Solid, mp 147°.

THERAP CAT: Choleretic.

3870. Exotoxins. Disease-producing proteinaceous substances secreted by bacteria into surrounding medium. Generally produced by gram-positive organisms although at least two exotoxins, *Shigella dysenteriae* and *Vibrio cholerae*, are produced by gram-negative organisms. Each exotoxin has its own particular pharmacology. Less stable to heat and considerably more toxic than the endotoxins; detoxified by agents which do not affect endotoxins. *Reviews:* Braude, *Sci. Am.* **210**, 36-45 (March 1964); van Heyningen in *Microbial Toxins* **vol. 1**, S. J. Ajl *et al.*, Eds. (Academic Press, New York, 1970) pp 1-28; J. P. Arbuthnott, *J. Appl. Bacteriol.* **44**, 329-345 (1978).

F

3871. Factor V. *Blood-coagulation factor V;* proaccelerin; labile factor; accelerator globulin; Ac-globulin. Mol wt above 300,000. Participates in the later stages of blood coagulation. Activated by factor X, *q.v.*, and along with Ca^{2+} and phospholipid forms a complex which converts prothrombin, *q.v.*, to thrombin, *q.v.* Accelerates the interaction of prothrombin, thromboplastin and Ca^{2+} ions: Ware, Seegers, *Fed. Proc.* **7**, 131 (1948). Thrombin produces serum Ac-globulin, *accelerin,* a far more potent accelerator, from plasma Ac-globulin, apparently by catalytic action: Ware *et al., Science* **106**, 618 (1947); Rapaport *et al., Blood* **21**, 221 (1963). Purification: Ware, Seegers, *J. Biol. Chem.* **172**, 699 (1948); Blombäck, Blombäck, *Nature* **198**, 886 (1963). Role in the scheme of blood coagulation: Pavlovsky in *Fibrinogen and Fibrin Turnover of Clotting Factors,* R. B. Hunter *et al.,* Eds. (Schattauer-Verlag, Stuttgart, 1963) pp 443-454. Proaccelerin deficiency is characterized by prolonged Quick's prothrombin time, and cephalin time, abnormal thromboplastin generation test, abnormal prothrombin consumption, and prolonged whole blood clotting time: Owren in *Thrombolytic Activity and Related Phenomena,* I. S. Wright *et al.,* Eds. (Schattauer-Verlag, Stuttgart, 1961) pp 387-391. *Review:* R. W. Colman, *Prog. Hemostasis Thromb.* **3**, 109-143 (1976). Purification, stability, amino acid composition, physical properties of bovine and human factor V: R. W. Colman, R. M. Weinberg, *Methods Enzymol.* **45B**, 107-122 (1976).

3872. Factor VII. *Blood-coagulation factor VII;* proconvertin; co-thromboplastin; serum prothrombin conversion accelerator; SPCA. Enzyme which, in conjunction with tissue factor and Ca^{2+} ions initiates the activation of factor X, *q.v.* Isoln from human plasma or serum: Fantt, Osborn, *Thromb. Diath. Haemorrh.* **8**, 286 (1962); from bovine plasma: J. Jesty, Y. Nemerson, *J. Biol. Chem.* **249**, 509 (1974). Separation from factor X by DEAE cellulose: Hougie, Bunting, in *New Blood Clotting Factors,* I. S. Wright *et al.,* Eds. (Schattauer-Verlag, Stuttgart, 1960) pp 40-42. Purification and structure of bovine factor VII: J. Jesty, Y. Nemerson, *J. Biol. Chem.* **249**, 509 (1974); R. Radcliffe, Y. Nemerson, *ibid.* **250**, 388 (1975); *eidem, Methods Enzymol.* **45B**, 49 (1976). Its deficiency results in retarded prothrombin conversion and an elevated one-stage prothrombin time: Alexander in *Thrombolytic Activity and Related Phenomena,* I. S. Wright *et al.,* Eds. (Schattauer-Verlag, Stuttgart, 1961) pp 392-402. Factor VII deficiency is associated with severe liver disease, vitamin K deficiency and broad spectrum antibiotic therapy, or may follow anticoagulant coumarin therapy.

3873. Factor VIII. *Blood-coagulation factor VIII;* antihemophilic globulin; antihemophilic factor A; AHG; AHF; Factorate; Hemofil; Humafac; Koate; Monoclate; Profilate. Plasma protein which is involved in the intrinsic formation of thromboplastin by the activation of factor X, *q.v.*: Pavlovsky in *Fibrinogen and Fibrin Turnover of Clotting Factors,* R. B. Hunter *et al.,* Eds. (Schattauer-Verlag, Stuttgart, 1963), pp 443-454. Present at extremely low concn in plasma. Isoln and characterization of human and bovine factor VIII: Hershgold *et al., J. Lab. Clin. Med.* **77**, 185 (1971); Schmer *et al., J. Biol. Chem.* **247**, 2512 (1972); M. E. Legaz, E. W. Davie, *Methods Enzymol.* **45B**, 83 (1976). Deficiency of factor VIII results in classical hemophilia or hemophilia A, a sex-linked bleeding diathesis in which blood clots slowly, prothrombin is converted to thrombin at an abnormally slow rate and a rebleeding phenomenon is frequently seen: Wagner, Brinkhous in *Thrombolytic Activity and Related Phenomena,* I. S. Wright *et al.,* Eds. (Schattauer-Verlag, Stuttgart, 1961) pp 403-407. Molecular immunology of factor VIII: Zimmerman, Edgington, *Ann. Rev. Med.* **25**, 303 (1974). *Reviews:* Ingram, *Advan. Clin. Chem.* **8**, 189 (1965); Barrow, Graham, *Physiol. Rev.* **54**, 23 (1974); E. J. Hershgold, *Prog. Hemostasis Thromb.* **2**, 99-139 (1974); D. L. Aronson, *Semin. Thromb. Hemostasis* **6**, 12-27 (1979). Relatively stable in liquid, frozen, or lyophilized plasma when blood is properly drawn. Heat-labile in plasma.

About half the activity is lost from canine citrated plasma in 10 mins at 56°. Poorly adsorbed by $BaSO_4$ from citrated or oxalated plasma.
THERAP CAT: Antihemophilic factor (human).

3874. Factor IX. *Blood-coagulation factor IX;* Christmas factor; PTC; plasma thromboplastin component; antihemophilic factor B; autoprothrombin II. A plasma and serum glycoprotein that participates in the middle phases of blood coagulation. Activated by factor XI and Ca^{2+}. Interacts then with factor VIII, Ca^{2+}, and phospholipid to form the complex which converts factor X into factor X_a. Isoln and characterization of bovine factor IX: K. Fujikawa *et al., Biochemistry* **12**, 4938 (1973); K. Fujikawa, E. W. Davie, *Methods Enzymol.* **45B**, 74 (1976). Deficiency results in a congenital bleeding disorder known as Christmas disease or hemophilia B. *Review:* Macfarlane in *Thrombolytic Activity and Related Phenomena,* I. S. Wright *et al.,* Eds. (Schattauer-Verlag, Stuttgart, 1961) pp 408-415; B. N. Bouma, J. A. Van Mourik, *Haemostasis: Biochemistry, Physiology & Pathology,* D. Ogston, B. Bennett, Eds. (Wiley-Interscience, New York, 1977) pp 56-77.
Factor IX complex (human), Konyne, Proplex, Hemoplex *(obsolete).* A mixture of human factors II, VII, IX, and X.
THERAP CAT: Hemostatic.

3875. Factor X. *Blood-coagulation factor X;* Stuart-Prower factor; Stuart factor; Prower factor. Enzyme which is activated by a complex of factors IX, VIII, calcium ions and phospholipid, or by factor VII and tissue factor in the intrinsic and extrinsic pathways of blood coagulation resp. Non-physiological activators are trypsin and a specific protein from Russel's viper venom: *see* Fujikawa *et al., Biochemistry* **11**, 4892 (1972). Active factor X apparently acts upon factor V, *q.v.* in the presence of calcium ions and lipid to form a complex which activates the conversion of prothrombin to thrombin. Present in both plasma and serum because neither increases nor decreases during clotting. Presumably synthesized in the liver because level is reduced with coumarin drugs and in liver disease. Isoln and characterization of bovine factor X: *eidem., ibid.* 4882. Factor X_a is a serine protease which, in the presence of factor V, Ca^{2+} ions and phospholipids, activates prothrombin. Severe deficiency causes hemorrhagic state resembling hemophilia. Mild deficiency may result in easy bruising. In severe cases deep muscular and joint hemorrhages occur. Improved purification and bibliography of earlier isoln procedures: Bajaj, Mann, *J. Biol. Chem.* **248**, 7729 (1973). Comparison of the molecular forms generated by the different methods of activation: Radcliffe, Barton, *ibid.* 6788. *Review:* Graham, Hougie in *Thrombolytic Activity and Related Phenomena,* I. S. Wright *et al.,* Eds. (Schattauer-Verlag, Stuttgart, 1961) pp 416-420; C. M. Jackson, *Thromb. Diath. Haemorrh., Suppl.* **57**, 197-216 (1974). Reviews on prepn, physical properties and function in the coagulation process: K. Fujikawa, E. W. Davie, *Methods Enzymol.* **45B**, 89 (1976); J. Jesty, Y. Nemerson, *ibid.* p 95.
Stable at room temperature for several days. Adsorbed by $BaSO_4$, $Al(OH)_3$, $Ca_3(PO_4)_2$, Seitz filters, bentonite and DEAE cellulose. At 56° is destroyed rapidly, in serum, less rapidly as citrate eluate. At 65° is destroyed rapidly in any form. Stable between pH 6.1-9.0. Soluble in water. Precipitates from water between pH 8-4. Isoelec pt pH 4.3.

3876. Factor XI. *Blood-coagulation factor XI;* PTA; plasma thromboplastin antecedent. Participates in the initial phases of blood coagulation. In normal plasma, factor XI is present in a precursor form and is converted to an active form, factor XI_a, which in turn activates factor IX to factor IX_a in the presence of Ca^{2+} ions: Ratnoff, Davie, *Biochemistry* **1**, 677 (1962); Schiffman *et al., Blood* **22**, 733 (1963); K. Fujikawa *et al., Biochemistry* **13**, 4508 (1974). Deficiency is characterized by the moderate bleeding symptoms of hemophilia C; by pathological clotting and recalcification times, impaired prothrombin consumption, and delayed thromboplastin formation: Duckert, Soulier in *New Blood Clotting Factors,* I. S. Wright *et al.,* Eds. (Schattauer-Verlag, Stuttgart, 1960) pp 145-147, 123-131. Review on bovine factor XI: T. Koide, *Methods Enzymol.* **45B**, 65-73 (1976).

3877. Factor XII. *Blood-coagulation factor XII;* Hageman factor; HF. Mol wt about 82,000. An enzyme which circulates in zymogen form in blood and when activated initiates the first of a series of events in coagulation of blood plasma. Enzymatic nature of blood coagulation is described by the "cascade" theory in which clotting factors interact with one another in a stepwise manner, one acting as an enzyme, the other as substrate, in a sequence of reactions leading to the formation of thrombin, *q.v.:* Davie, Ratnoff, *Science* **145**, 1310 (1964); MacFarlane, *Nature* **202**, 498 (1964). Factor XII initiates clotting when blood or plasma comes into contact with glass, collagen, or surface-active agents. Once activated, it interacts with factor XI, *q.v.*, to convert it to an enzyme. Factor XII is a sialoglycoprotein with esterase activity. Its activation by glass may be viewed as a dynamic alteration of the tertiary structure of the protein moiety: Schoenmakers *et al.*, *Biochim. Biophys. Acta* **101**, 166 (1965). Isoln of human and rabbit factor XII in zymogen form: C. G. Cochrane, K. D. Wuepper, *J. Exp. Med.* **134**, 986 (1971); H. Saito *et al.*, *Circ. Res.* **34**, 641 (1974). Highly purified Hageman factor undergoes a change in physical properties during activation: Donaldson, Ratnoff, *Science* **150**, 754 (1965). Factor XII deficiency is termed the Hageman trait. Unlike deficiencies in all other clotting factors, persons with Hageman trait do not show significant bleeding tendencies. *Reviews:* Ratnoff in *New Blood Clotting Factors*, I. S. Wright *et al.*, Eds. (Schattauer-Verlag, Stuttgart, 1960) pp 116-122; Ratnoff *et al.*, in *Thrombolytic Activity and Related Phenomena*, I. S. Wright *et al.*, Eds. (Schattauer-Verlag, Stuttgart, 1961) pp 364-378; several authors in *Recent Advances in Blood Coagulation*, L. Poller, Ed. (J. & A. Churchill, London, 1969); J. Spragg, K. F. Austen, *Compr. Immunol.* **3**, 125-143 (1977); J. H. Griffin, C. G. Cochrane, *Methods Enzymol.* **45B**, 56-65 (1976).

Sedimentation coefficient, $s^{\circ}_{20,w} = 7.08$. Diffusion constant, $D^{\circ}_{20,w} = 7.14 \times 10^{-7}$ cm²/sec. uv max at pH 7: 280 nm ($E^{1\%}_{1cm}$ 12.0). Isoelectric pt pH 8.0.

3878. Factor XIII. *Blood-coagulation factor XIII;* fibrin-stabilizing factor; FSF; fibrinase; Laki-Lorand factor; LLF; Fibrogamin. Plasma enzyme precursor which, when activated by thrombin in the presence of Ca^{2+}, converts soluble fibrin, *q.v.* gel to a tough insoluble clot: Laki, Lorand, *Science* **108**, 280 (1948); Lorand, *Nature* **166**, 694 (1950). During the terminal stage of blood-clotting, factor XIII catalyzes the formation of cross-links in the fibrin gel by eliminating NH_3 in the transamidation reaction in which the ε-amino group of lysine from a fibrin molecule forms a peptide linkage with a glutaminyl residue from a nearby fibrin monomer: Lorand *et al.*, *Biochem. Biophys. Res. Commun.* **25**, 629 (1966); Fuller, Doolittle, *ibid.* 694. Trypsin, papain, and reptilase also activate factor XIII: Lorand *et al.*, *ibid.* **31**, 222 (1968). Structure contains two each of two types of soluble proteinaceous subunits called the α and β chains; mol wt of subunits each about 80,000: Schwartz *et al.*, *J. Biol. Chem.* **248**, 1395 (1973). Amino acid sequence studies: Holbrook *et al.*, *Biochem. J.* **135**, 901 (1973). Clinical studies: Cucuianu *et al.*, *Thromb. Diath. Haemorrh.* **30**, 480 (1973). *Review:* L. Lorand, *Ann. N.Y. Acad. Sci.* **202**, 6-30 (1972); A. G. Loewy, *ibid.* 41-58; C. G. Curtis, L. Lorand, *Haemostasis: Biochemistry, Physiology & Pathology*, D. Ogston, B. Bennett, Eds. (Wiley-Interscience, New York, 1977) pp 186-201; *eidem, Methods Enzymol.* **45B**, 177-191 (1976).

uv max: 280 nm ($E^{1\%}$ 13.8). Retains activity in plasma kept at 0°; inactivated in a few days at room temperature. Activity destroyed if serum heated at 60° for 10 min. Does not dialyse out from plasma. Inactivated by *Echis coloratus* venom. Pretreatment with compounds containing an SH-group, such as cysteine, reduced glutathione, DL-penicillamine, or mercapto-1-propanol prevents inactivation as well as restores activity after venom treatment.

THERAP CAT: Antihemorrhagic.

3879. Fagarine. *4,8-Dimethoxyfuro[2,3-b]quinoline;* γ-fagarine; 8-methoxydictamnine. $C_{13}H_{11}NO_3$; mol wt 229.23. C 68.11%, H 4.84%, N 6.11%, O 20.94%. Antiarrhythmic principle in *Fagara coco* (Gill.) Engl., *Rutaceae.* Isoln: Deulofeu *et al.*, *J. Am. Chem. Soc.* **64**, 2326 (1942); Briggs, Cambi, *Tetrahedron* **2**, 256 (1958). Structure: Berin-

zaghi *et al.*, *J. Org. Chem.* **10**, 181 (1945). Synthesis: Grundon, McCorkindale, *Chem. & Ind. (London)* **1956**, 1091.

Prisms from alcohol, mp 142°. uv max: 238, 332, 370 nm (log ε 4.76, 3.88, 3.89). Sol in chloroform, benzene, ether; slightly sol in water, petr ether.

Hydrochloride, $C_{13}H_{12}ClNO_3$, needles from chloroform + ether, mp 158-159°.

3880. Falone®. *2-(2,4-Dichlorophenoxy)ethanol phosphite (3:1);* tris(2,4-dichlorophenoxyethyl) phosphite; 3Y9. $C_{24}H_{21}Cl_6O_6P$; mol wt 649.14. C 44.41%, H 3.26%, Cl 32.77%, O 14.79%, P 4.77%. Prepn: Harris, Feldman, U.S. pat. **2,828,198** (1958 to U.S. Rubber).

Viscous liquid, n^{20}_D 1.5875. Practically insol in water; miscible with benzene, xylene, other aromatic hydrocarbons.

USE: Pre-emergence herbicide.

3881. Famotidine. *3-[[[2-[(Aminoiminomethyl)amino]-4-thiazolyl]methyl]thio]-N-(aminosulfonyl)propanimidamide;* [1-amino-3-[[[2-[(diaminomethylene)amino]-4-thiazolyl]methyl]thio]propylidene]sulfamide; N-sulfamoyl-3-[(2-guanidinothiazol-4-yl)methylthio]propionamide; YM-11170; MK-208; Dispronil; Famodil; Famosan; Famoxal; Fanosin; Fibonel; Ganor; Gaster; Gastridan; Gastridin; Gastropen; Lecedil; Motiax; Muclox; Nulcerin; Pepcid; Pepcidina; Pepcidine; Pepdine; Pepdul; Peptan; Ulceprax; Ulfamid; Ulfinol. $C_8H_{15}N_7O_2S_3$; mol wt 337.43. C 28.48%, H 4.48%, N 29.06%, O 9.48%, S 28.50%. Histamine H_2 receptor antagonist. Prepn, NMR and mass spectral data: H. Yasufumi *et al.*, Belg. pat. **882,071**; *eidem*, U.S. pat. **4,283,408**; Japan. Kokai 81 55383, *C.A.* **95**, 203930n (1980, 1981, 1981 all to Yamanouchi). Inhibition of gastric acid and pepsin secretion in rats: M. Takeda *et al.*, *Arzneimittel-Forsch.* **32**, 734 (1982); in man: M. Miwa *et al.*, *J. Clin. Pharmacol. Ther. Toxicol.* **22**, 214 (1984). Effect on disposition of antipyrine in liver: Ch. Staiger *et al.*, *Arzneimittel-Forsch.* **34**, 1041 (1984). Chromatographic determn in plasma and urine: W. C. Vincek *et al.*, *J. Chromatog.* **338**, 438 (1985). Pharmacokinetics: T. Takabatke *et al.*, *Eur. J. Clin. Pharmacol.* **28**, 327 (1985). Clinical trial in Zollinger-Ellison syndrome: J. M. Howard *et al.*, *Gastroenterology* **88**, 1026 (1985). Symposia on pharmacology and clinical efficacy: *Am. J. Med.* **81**, Suppl. 4B, 1-64 (1986); *Scand. J. Gastroenterol.* **22**, Suppl. 134, 1-62 (1987).

mp 163-164°. Soly at 20° (%, w/v): 80 in DMF; 50 in acetic acid; 0.3 in methanol; 0.1 in water; < 0.01 in ethanol, ethyl acetate, chloroform. LD_{50} i.v. in mice: 244.4 mg/kg (Yasufumi).

THERAP CAT: Anti-ulcerative.

3882. Famphur. *Phosphorothioic acid O-[4-[(dimethylamino)sulfonyl]phenyl] O,O-dimethyl ester; phosphorothioic acid O,O-dimethyl ester, O-ester with p-hydroxy-N,N-dimethylbenzenesulfonamide; O,O-dimethyl-O,p-(dimethylsulfam-*

oyl)phenyl phosphorothioate; *p*-hydroxy-*N*,*N*-dimethylbenzenesulfonamide ester with phosphorothioic acid *O*,*O*-dimethyl ester; famophos; American Cyanamid 38023; ENT 25644;Warbex. $C_{10}H_{16}NO_5PS_2$; mol wt 325.36. C 36.92%, H 4.96%, N 4.31%, O 24.59%, P 9.52%, S 19.71%. Prepn: Berkelhammer, U.S. pat. 3,005,004 (1961 to Am. Cyanamid). Metabolism: P. E. Gatterdam *et al.*, *J. Agr. Food Chem.* **15**, 845 (1967). Toxicology: W. D. Black *et al.*, *Toxicol. Appl. Pharmacol.* **50**, 167 (1979).

Crystals, mp 52.5-53.5°. LD_{50} orally in rats: 35 mg/kg, E. W. Schafer, *Toxicol. Appl. Pharmacol.* **21**, 315 (1972). *Human Toxicity:* Cholinesterase inhibitor. *See* Parathion.
USE: Insecticide.
THERAP CAT (VET): Insecticide (grubicide).

3883. α-Farnesene. *3,7,11-Trimethyl-1,3,6,10-dodecatetraene;* 2,6,10-trimethyl-2,6,9,11-dodecatetraene; farnesene. $C_{15}H_{24}$; mol wt 204.34. C 88.16%, H 11.84%. Isoln of (*E,E*)-form from natural coating of apples: Huelin, Murray, *Nature* **210**, 1260 (1966); Murray, *Aust. J. Chem.* **22**, 197 (1969); from Dufour's gland in ants: Cavill *et al.*, *Tetrahedron Letters* **1967**, 2201. Isoln of (*Z,E*)-form from oil of perilla: T. Sakai, Y. Hirose, *Bull. Chem. Soc. Japan* **42**, 3615 (1969). Oxidation products of farnesene are believed to cause scald, a serious storage disorder of apples. Four possible geometric isomers. Synthetic studies: Ruzicka, *Helv. Chim. Acta* **6**, 490, 501 (1923); Ruzicka, Capato, *ibid.* **8**, 267 (1925). Configuration of (*E,E*)-form confirmed by synthesis from *trans*-β-farnesene: Brieger *et al.*, *J. Org. Chem.* **34**, 3789 (1969). Stereospecific synthesis of (*E,Z*)- and (*Z,Z*)-isomers: Anet, *Aust. J. Chem.* **23**, 2101 (1970). Synthesis of (*E,E*)-form: Tanaka *et al.*, *J. Am. Chem. Soc.* **97**, 3252 (1975). (*E,E*)- and (*Z,E*)-forms are components of aphid alarm pheromones: J. A. Pickett, D. C. Griffiths, *J. Chem. Ecol.* **6**, 349 (1980); of the trail pheromone of red imported fire ants: R. K. Vandermeer *et al.*, *Tetrahedron Letters* **22**, 1651 (1981).

(*E̱,E̱*)-α-farnesene

Thin oil. bp_{12} about 125°. d_4^{20} 0.8410. n_D^{20} 1.4836. Practically insol in water; misc with hydrocarbon solvents. uv max of (*E,E*)-form (alc): 233 nm (ε 27,000), of (*Z,E*)-form (alc): 238 nm (ε 11300).

3884. β-Farnesene. *7,11-Dimethyl-3-methylene-1,6,10-dodecatriene.* $C_{15}H_{24}$; mol wt 204.34. C 88.16%, H 11.84%. The naturally occurring *trans* form or (*E*)-isomer is a constituent of various essential oils, *see* F. Sorm *et al.*, *Coll. Czech. Chem. Commun.* **14**, 699 (1949); **15**, 626 (1951); also an alarm pheromone of several aphid species: Bowers *et al.*, *Science* **177**, 1121 (1972); Edwards *et al.*, *Nature* **241**, 126 (1973); W. S. Bowers *et al.*, *J. Insect. Physiol.* **23**, 697 (1977). Isoln from leaves of wild potato, *Solanum berthaultii* Hawkes, *Solanaceae:* R. W. Gibson, J. A. Pickett, *Nature* **302**, 608 (1983). Prepd by dehydration of farnesol: Bhati, *Perfum. Essent. Oil Rec.* **54**, 376 (1963); Brieger, *J. Org. Chem.* **32**, 3720 (1967); alternate syntheses: Tanaka *et al.*, *J. Am. Chem. Soc.* **97**, 3252 (1975); S. Akutagawa *et al.*, *Chem. Letters* **1976**, 485; O. P. Vig *et al.*, *Indian J. Chem. Sect. B* **18**, 33 (1979). Synthesis of (*Z*)-isomer: Anet, *Aust. J. Chem.* **23**, 2101 (1970); O. P. Vig *et al.*, *J. Indian Chem. Soc.* **47**, 999 (1970); *eidem, Indian J. Chem.* **13**, 1244 (1975).

(*E*)-β-farnesene

(*E*)-Form. Oil, bp 124°. d_4^{20} 0.8310. n_D^{20} 1.4870. uv max (hexane): 224 nm (ε 14000).
(*Z*)-Form. Oil, bp_{3-4} 95-107°. n_D^{32} 1.4780. uv max (hexane): 224 nm (ε 17300).

3885. Farnesol. *3,7,11-Trimethyl-2,6,10-dodecatrien-1-ol.* $C_{15}H_{26}O$; mol wt 222.36. C 81.02%, H 11.79%, O 7.20%. Found in oils of citronella, neroli, cyclamen, lemon grass, tuberose, rose, musk, balsam Peru, and tolu. Isoln: Elge, *Chem. Ztg.* **34**, 857 (1910); **37**, 1422 (1913); Kerschbaum, *Ber.* **46**, 1732 (1913); Naves, *Helv. Chim. Acta* **32**, 1798, 2181 (1949); LaFace, *ibid.* **33**, 249 (1950). Synthesis: Ruzicka, *ibid.* **6**, 492 (1923); Ruzicka, Firmenich, *ibid.* **22**, 392 (1939); Nazarov *et al.*, *Zh. Obshch. Khim.* **28**, 1444 (1958); Shvarts, Petrov, *ibid.* **30**, 3598 (1960); Popjak *et al.*, *J. Biol. Chem.* **237**, 56 (1962). Four possible stereoisomers. Stereochemistry: Bates *et al.*, *Chem. & Ind. (London)* **1961**, 1907; *J. Org. Chem.* **28**, 1086 (1963). *trans-trans*-Farnesol is the only stereoisomer present in many essential oils but occurs mixed with *cis-trans*-farnesol in petitgrain oil and several other oils: Naves, *Compt. Rend.* **251**, 900 (1960). Stereospecific synthesis of *trans-trans*-farnesol: Corey *et al.*, *J. Am. Chem. Soc.* **92**, 6637 (1970).

trans-trans-Farnesol. Liquid. $bp_{0.35}$ 111°. n_D^{25} 1.4872. uv max: 192-196 nm (ε 28,500).
Commercial farnesol. $bp_{0.2}$ 110-113°. d_4^{20} 0.8871. n_D^{20} 1.4870.
USE: In perfumery, to emphasize the odor of sweet floral perfumes, such as lilac and cyclamen.

3886. Fast Green FCF. *N-Ethyl-N-[4-[[4-[ethyl[(3-sulfophenyl)methyl]amino]phenyl](4-hydroxy-2-sulfophenyl)-methylene]-2,5-cyclohexadien-1-ylidene]-3-sulfobenzenemethanaminium hydroxide inner salt, disodium salt;* FD & C Green No. 3; C.I. Food Green 3; C.I. 42053. $C_{37}H_{34}N_2Na_2-O_{10}S_3$; mol wt 808.85. C 54.94%, H 4.24%, N 3.46%, Na 5.68%, O 19.78%, S 11.89%. Prepn: H. Johnson, P. Staub, *Ind. Eng. Chem.* **19**, 497 (1927). Chronic toxicity study: W. H. Hansen *et al.*, *Food Cosmet. Toxicol.* **4**, 389 (1966). Review of carcinogenicity studies: *IARC Monographs* **16**, 187-197 (1978). *See also: Colour Index* vol. **4** (3rd ed., 1971) p 4383.

Dark green powder or granules with a metallic lustre. Absorption max: 628 nm. Very sol in water; sol in ethanol. Dull orange soln in conc H_2SO_4, changing to dull green on dilution. Orange soln in conc HCl or conc HNO_3. Bright blue soln in 10% aq NaOH. LD_{50} orally in rats: > 2 g/kg, F. C. Lu, A. Lavalle, *Can. Pharm. J.* **97**, 30 (1964).
USE: Biological stain. Approved by FDA for use in food, drugs and cosmetics excluding use in eye area: *Fed. Regist.* **47**, 52140 (1982).

3887. Fazadinium Bromide. *1,1'-Azobis[3-methyl-2-phenylimidazo[1,2-a]pyridinium] dibromide;* AH 8165;

Consult the cross index before using this section.

Fazadon. $C_{28}H_{24}Br_2N_6$; mol wt 604.36. C 55.65%, H 4.00%, Br 26.44%, N 13.91%. A short-acting curarimimetic with rapid onset. Prepn: D. Jack, E. E. Glover, **Ger.** pat. **2,127,-355** corresp to U.S. pats. **3,773,746** and **3,849,557** (1971, 1973 and 1974, all to Allen & Hanburys); E. E. Glover, M. Yorke, *J. Chem. Soc. (C)* **1971**, 3280. Pharmacology: L. Bolger *et al.*, *Nature* **238**, 354 (1972); R. T. Brittain, M. B. Tyers, *Brit. J. Anaesth.* **45**, 837 (1973); E. L. Post *et al.*, *Anesthesiology* **42**, 240 (1975); P. T. Hiser *et al.*, *ibid.* 245.

Dihydrate, yellow crysts from water, mp 215-219° (softens at 196°). Also reported as yellow solid from methanol/ethyl acetate, mp 218-220°. uv max (H_2O): 285, 297 nm (log ε 4.04, 4.34). LD_{50} in mice: 0.31 mg/kg i.v.

THERAP CAT: Skeletal muscle relaxant.

3888. Febantel. *[[2-[(Methoxyacetyl)amino]-4-(phenylthio)phenyl]carbonimidoyl]biscarbamic acid dimethyl ester;* dimethyl [[2-(2-methoxyacetamido)-4-(phenylthio)phenyl]-imidocarbonyl]dicarbamate; Bay Vh 5757; Bay h 5757; Rintal. $C_{20}H_{22}N_4O_6S$; mol wt 446.49. C 53.80%, H 4.97%, N 12.55%, O 21.50%, S 7.18%. Prepn: H. Koelling *et al.*, **Ger.** pat. **2,423,679** corresp to U.S. pat. **3,993,682** (1975, 1976 both to Bayer). Anthelmintic efficacy: H. Wollweber *et al.*, *Arzneimittel-Forsch.* **28**, 2193 (1977); H. Thomas, *Res. Vet. Sci.* **25**, 290 (1978). Safety evaluation in horses: J. A. Shmidl, *Vet. Med. Small Anim. Clin.* **73**, 775 (1978).

Crystals, mp 129-130°.

THERAP CAT (VET): Anthelmintic

3889. Febarbamate. *1-[2-[(Aminocarbonyl)oxy]-3-butoxypropyl]-5-ethyl-5-phenyl-2,4,6(1H,3H,5H)-pyrimidinetrione; 1-(3-butoxy-2-hydroxypropyl)-5-ethyl-5-phenylbarbituric acid carbamate;* 1-(3-butoxy-2-carbamoyloxypropyl)-5-phenyl-5-ethylbarbituric acid; Go-560; G-Tril; Tymium. $C_{20}H_{27}N_3O_6$; mol wt 405.44. C 59.24%, H 6.71%, N 10.37%, O 23.68%. Prepn: P. Gold-Aubert, E. Gysin, *Helv. Chim. Acta* **44**, 105 (1961); P. Gold-Aubert, U.S. pat. **3,075,983** (1963 to Sapos). Pharmacology: E. Frommel *et al.*, *Helv. Physiol. Acta* **19**, 241 (1961). Metabolism in rats: J. Vachta, P. Gold-Aubert, *Eur. J. Drug Metab. Pharmacokinet.* **7**, 147 (1982); in man: J. Vachta *et al.*, *ibid.* **8**, 297 (1983).

Syrup; could not be crystallized, dec ∼196°. Sol in alcohol, dioxane. Insol in water. LD_{50} orally in mice: 1065 mg/kg (Frommel).

THERAP CAT: Antidepressant; thymoanaleptic.

3890. Febrifugine. *3-[3-(3-Hydroxy-2-piperidinyl)-2-oxopropyl]-4(3H)-quinazolinone; 3-[3-(3-hydroxy-2-piperidyl)acetonyl]-4(3H)-quinazolinone;* 3-[β-keto-γ-(3-hydroxy-2-piperidyl)propyl]-4-quinazolone; β-dichroine. $C_{16}H_{19}$-

N_3O_3; mol wt 301.34. C 63.77%, H 6.36%, N 13.95%, O 15.93%. Isolated from the *Dichroa febrifuga* Louv., *Saxifragaceae* and from the common hydrangea: Koepfli *et al.*, *J. Am. Chem. Soc.* **71**, 1048 (1949); Ablondi *et al.*, *J. Org. Chem.* **17**, 14 (1952). Identity with β-dichroine: Fu, Yang, *C.A.* **43**, 1530a (1949). Structure: Koepfli *et al.*, *J. Am. Chem. Soc.* **72**, 3323 (1950). Synthesis: Baker *et al.*, *J. Org. Chem.* **17**, 132 (1952); **18**, 178 (1953); Koepfli *et al.*, U.S. pat. **2,504,847** (1950 to U.S. Gov't); Baker, Querry, U.S. pat. **2,651,632** (1953 to Am. Cyanamid). Config: Hill, Edwards, *Chem. & Ind. (London)* **1962**, 858. Revised stereochemistry: Barringer *et al.*, *J. Org. Chem.* **38**, 1937 (1973).

Dimorphic crystals: needles from ethanol, mp 139-140°; from chloroform, mp 154-156°. $[\alpha]_D^{25}$ +6° (c = 0.5 in chloroform); $[\alpha]_D^{25}$ +28° (c = 0.5 in ethanol). Freely sol in methanol + chloroform, water + ethanol; slightly sol in water, ethanol, acetone, chloroform. Practically insol in ether, benzene, petr ether. LD_{50} orally in white mice: 2.5-3.0 mg/kg, Koepfli *et al.*, *loc. cit.* (1949).

Dihydrochloride monohydrate, crystals from 90% ethanol, mp 220-222°. $[\alpha]_D^{25}$ +12.8° (c = 0.8).

THERAP CAT (VET): Coccidiostat.

3891. Febuprol. *1-Butoxy-3-phenoxy-2-propanol;* H-33; K-10033; Valbil. $C_{13}H_{20}O_3$; mol wt 224.30. C 69.61%, H 8.99%, O 21.40%. Prepn: W. F. Minor *et al.*, *J. Am. Chem. Soc.* **76**, 2993 (1954). Prepn and use as choleretic: H. Hoffmann *et al.*, **Ger.**pat. **2,120,396** corresp to U.S. pat. **3,839,-587** (1972, 1974 both to Klinge). Pharmacology: G. Hofrichter *et al.*, *Arzneimittel-Forsch.* **24**, 111 (1974). Pharmacokinetics, metabolism: K. Rehm *et al.*, *ibid.* **26**, 813 (1976). Spasmolytic action: R. Eisenburger, *ibid.* **27**, 1429 (1977). Clinical studies: C. Wolf *et al.*, *Münch. Med. Wochenschr.* **118**, 1285 (1976); W. E. Scholing, W. Weinert, *ibid.* **120**, 143 (1978); U. Ritter, H. J. Kyrein, *Arzneimittel-Forsch.* **33**, 891 (1983).

Colorless oil, bp_{11} 165°; $bp_{1.0}$ 125-132°. n_D^{20} 1.5004. d_4^{20} 1.027. LD_{50} in mice, rats (mg/kg): 3050, 2370 orally; 436, 400 i.p. (Hofrichter).

THERAP CAT: Choleretic.

3892. Feclemine. *2-(Cyclohexylphenylmethyl)-N,N,-N',N'-tetraethyl-1,3-propanediamine;* 1,3-bis(diethylamino)-2-(α-phenyl-α-cyclohexylmethyl)propane; 1-phenyl-1-cyclohexyl-2,2-bis(diethylaminomethyl)ethane; phenecyclamine; phenetamine; UCB 1545; Licaran; Spasmexan. $C_{24}H_{42}N_2$; mol wt 358.59. C 80.38%, H 11.81%, N 7.81%. Prepn: Morren *et al.*, *Ind. Chim. Belg.* **20**, 733 (1955); Morren, U.S. pat. **2,781,398** (1957 to U.C.B.).

Dihydrochloride, $C_{24}H_{44}Cl_2N_2$, *Licabile.* Hygroscopic crystals, mp below 70°. $bp_{0.05}$ 143°. Also used as pamoate.

THERAP CAT: Antispasmodic.

3893. Felbinac. *(l,1'-Biphenyl)-4-acetic acid; 4-biphenylacetic acid;* 4-carboxymethylbiphenyl; BPAA; LY 61017; L141; LJC 10141; Napageln; Traxam. $C_{14}H_{12}O_2$; mol wt 212.25. C 79.23%, H 5.70%, O 15.07%. One of the metabolites of fenbufen, *q.v.* Prepn: E. Schwenk, D. Papa, *J. Org.*

Chem. **11**, 798 (1946). Improved prepn: G. R. Malone, A. I. Meyers, *ibid.* **39**, 618 (1974). Pharmacology: A. E. Sloboda, A. C. Osterberg, *Inflammation* **1**, 415 (1976). Mode of action study in ocular inflammation: E. L. Tolman *et al., Invest. Ophthalmol.* **15**, 1005 (1976). HPLC determn in serum: J. S. Fleitman *et al., J. Chromatog.* **228**, 372 (1982).

White needles from ethyl ether, mp 164-165°. LD$_{50}$ orally in rats: 164 mg/kg (Sloboda, Osterberg).
THERAP CAT: Anti-inflammatory, analgesic.

3894. Felinine. *S-(3-Hydroxy-1,1-dimethylpropyl)-L-cysteine; L-3-[(3-hydroxy-1,1-dimethylpropyl)thio]alanine.* C$_8$H$_{17}$NO$_3$S; mol wt 207.31. C 46.34%, H 8.27%, N 6.76%, O 23.15%, S 15.47%. HOOCCH(NH$_2$)CH$_2$SC(CH$_3$)$_2$CH$_2$-CH$_2$OH. Isoln from cat urine: Westall, *Biochem. J.* **55**, 244 (1953). Synthesis of (±)-felinine: Trippett, *J. Chem. Soc.* **1957**, 1929. Synthesis of natural (−)-felinine: Eggerer, *Ann.* **657**, 212 (1962); Schöberl *et al., Ber.* **101**, 373 (1968).

(±)-Felinine, crystals from aq ethanol, dec 181°.
N-2,4-Dinitrophenylfelinine, C$_{14}$H$_{19}$N$_3$O$_7$S, crystals from aq ethanol, mp 120-122°.
(−)-Felinine, needles from water + acetone, dec 177°. [α]$_D^{25}$ −11.5° (c = 2.81 in water). Sol in water, ethanol; practically insol in ethanol, ethyl acetate, ether.
(−)-Felinine phosphate, C$_8$H$_{18}$NO$_6$PS, crystals from acetone, dec 120°. Hygroscopic. [α]$_D^{25}$ +5.5° (c = 2.54 in 1N NaOH). Sol in water, ethanol; practically insol in acetone, ethyl acetate, ether.

3895. Felodipine. *4-(2,3-Dichlorophenyl)-1,4-dihydro-2,6-dimethyl-3,5-pyridinedicarboxylic acid ethyl methyl ester;* H154/82; Agon; Hydac; Munobal; Plendil; Splendil. C$_{18}$H$_{19}$Cl$_2$NO$_4$; mol wt 384.26. C 56.26%, H 4.98%, Cl 18.45%, N 3.65%, O 16.66%. Dihydropyridine calcium channel blocker. Prepn: P. B. Berntsson *et al., Eur. pat. Appl.* **7293**; *eidem,* U.S. pat. **4,264,611** (1980, 1981 both to AB Hassle). Conformation: P. B. Berntsson, R. E. Carter, *Acta Pharm. Suec.* **18**, 221 (1981). Interaction with calmodulin, *q.v.* : S. L. Bostroem *et al., Nature* **292**, 777 (1981). Systemic and hemodynamic effects: A. C. Tweddel *et al., Circulation* **64**, 309 (1981). Diuretic-natriuretic properties: G. F. Dibona *et al., Clin. Res.* **30**, 571A (1982). Hemodynamic effects in coronary patients: P. Decoster *et al., Eur. J. Clin. Invest.* **12**, 43 (1982). Pharmacokinetics and pharmacodynamics: E. Boo *et al., Biopharm. Drug Dispos.* **8**, 235 (1987). Pharmacokinetics, vascular selectivity, effects on hypertension, hemodynamics and renal function: *Drugs* **29**, Suppl. 2, 1-212 (1985).

Crystals from isopropyl ether, mp 145°.
THERAP CAT: Antihypertensive; antianginal.

3896. Felypressin. *Vasopressin 2-(L-phenylalanine)-8-L-lysine; 2-(phenylalanine)-8-lysine vasopressin; Phe²-Lys⁸-vasopressin; Phe²-Phe²-Lys⁸-oxytocin; octapressin; PLV-2.* C$_{46}$H$_{65}$N$_{13}$O$_{11}$S$_2$; mol wt 1040.26. C 53.11%, H 6.30%, N 17.51%, O 16.92%, S 6.16%. Prepn: Boissonnas, Guttmann, *Helv. Chim. Acta* **43**, 190 (1960); Meienhofer, du Vigneaud, *J. Am. Chem. Soc.* **82**, 6336 (1960); **Brit. pat. 928,607,** corresp to U.S. pat. **3,232,923** (1963, 1966, both to Sandoz).

Cys-Phe-Phe-Gln-Asn-Cys-Pro-Lys-GlyNH$_2$

THERAP CAT: Vasoconstrictor.

3897. Femoxetine. *(3R-trans)-3-[(4-Methoxyphenoxy)-methyl]-1-methyl-4-phenylpiperidine; (+)-trans-1-methyl-3-(p-(methoxy)phenoxymethyl)-4-phenylpiperidine.* C$_{20}$H$_{25}$NO$_2$; mol wt 311.43. C 77.13%, H 8.09%, N 4.50%, O 10.28%. Serotonin uptake inhibitor. Prepn: J. A. Christensen, R. F. Squires, **Belg. pat. 810,310** corresp to U.S. pat. **3,912,743** (1974, 1975 to A/S Ferrosan). Neuropharmacology: J. B. Lassen *et al., Psychopharmacologia* **42**, 21 (1975). Pharmacology and toxicity: *eidem, Eur. J. Pharmacol.* **32**, 108 (1975). GC determn in body fluids: E. Bechgaard, J. Lund, *J. Chromatog.* **133**, 147 (1977). Metabolism: H. Larsson, J. Lund, *Acta Pharmacol. Toxicol.* **48**, 424 (1981). Bioavailability and pharmacokinetics: H. Mengel *et al., Arzneimittel-Forsch.* **33**, 462 (1983). Double-blind clinical trials in treatment of migraine: P. G. Andersson, E. N. Petersen, *Acta Neurol. Scand.* **64**, 280 (1981); in treatment of endogenous depression: L.-E. Dahl *et al., Acta Psychiatr. Scand.* **66**, 9 (1982); P. N. Reebye *et al., Pharmacopsychiatria* **15**, 164 (1982).

Hydrochloride, C$_{20}$H$_{26}$ClNO$_2$, FG 4963, Malexil. LD$_{50}$ in female, male mice (mg/kg): 48, 45 i.v.; 941, 723 s.c.; 1408, 1687 orally (Lassen).
THERAP CAT: Antidepressant.

3898. Fenadiazole. *2-(1,3,4-Oxadiazol-2-yl)phenol; 2-(o-hydroxyphenyl)-1,3,4-oxadiazole; Hypnazol.* C$_8$H$_6$N$_2$O$_2$; mol wt 162.14. C 59.26%, H 3.73%, N 17.27%, O 19.74%. Prepn: Maillard *et al., Bull. Soc. Chim. France* **1961**, 529; Vincent *et al., ibid.* **1962**, 1580; **Brit. pat. 902,388** and Maillard *et al., Fr. pat.* **M379,** *C.A.* **57**, 15251g (1962) (both 1962 to Logeais Labs.).

Crystals from methanol, mp 111-112°. bp$_{0.1}$ 180°. LD$_{50}$ i.p. in mice: 940 mg/kg.
THERAP CAT: Hypnotic.

3899. Fenalamide. *α-[[[2-(Diethylamino)ethyl]amino]-carbonyl]-α-ethylbenzeneacetic acid ethyl ester; N-[2-(diethylamino)ethyl]-2-ethyl-2-phenylmalonamic acid ethyl ester; ethyl N-[2-(diethylamino)ethyl]-2-ethyl-2-phenylmalon-amate; phenylethylmalonic acid (diethylamino)ethylamide ethyl ester; Spasmamide.* C$_{19}$H$_{30}$N$_2$O$_3$; mol wt 334.45. C 68.23%, H 9.04%, N 8.38%, O 14.35%. Smooth muscle relaxant. Prepn: Galimberti *et al.,* U.S. pat. **3,025,317** (1962 to Soc. Ital. Prod. Schering).

bp$_3$ 182-188°. Sol in acetone, methanol, ethanol, ethyl acetate, benzene, chloroform, ether, mineral acids. Practically insol in water, alkalies.
Hydrobromide, crystals, dec 74-80°.
Hydrochloride, crystals, dec 71-74°.
THERAP CAT: Antispasmodic.

3900. Fenalcomine. *α-Ethyl-4-[2-[(1-methyl-2-phenyl-ethyl)amino]ethoxy]benzenemethanol; α-ethyl-p-[2-[(α-methylphenethyl)amino]ethoxy]benzyl alcohol; 2-[N-[β-[4-(α-hydroxypropyl)phenoxy]ethyl]amino]-1-phenylpropane.* $C_{20}H_{27}NO_2$; mol wt 313.45. C 76.64%, H 8.68%, N 4.47%, O 10.21%. Prepn: Pinhas, **Fr. pat. M7255** (1969 to Laroche-Navarron), *C.A.* **76,** 14107w (1972). Pharmacology: Pham-Huu-Chanh *et al., Pharmacology* **6,** 137 (1971); *eidem, Arch. Int. Pharmacodyn. Ther.* **194,** 270 (1971).

CH₃CH₂CH— [benzene ring] —OCH₂CH₂NHCHCH₂— [benzene ring]
OH ... CH₃

Hydrochloride, $C_{20}H_{28}ClNO_2$, *Cordoxene.*
THERAP CAT: Cardiac stimulant; local anesthetic.

3901. Fenamiphos. *(1-Methylethyl)phosphoramidic acid ethyl 3-methyl-4-(methylthio)phenyl ester; isopropylphosphoramidic acid ethyl 4-(methylthio)-m-tolyl ester; ethyl 4-(methylthio)-m-tolyl isopropylphosphoramidate; ethyl 3-methyl-4-(methylthio)phenyl (1-methylethyl)phosphoramidate; phenamiphos;* Bay 68138; SRA 3886; B 68138; Bayer 68138; Nemacur. $C_{13}H_{22}NO_3PS$; mol wt 303.36. C 51.47%, H 7.31%, N 4.62%, O 15.82%, P 10.21%, S 10.57%. Systemic broad spectrum nematocide with anticholinesterase activity. Prepn: H. Kayser, G. Schrader, **U.S. pat. 2,978,-479** (1961 to Bayer). Nematocidal activity: W. M. Zeck, *Pflanzenschutz-Nach.* **24,** 114 (1971). Mode of action: B. Homeyer, K. Wagner, *Nematologica* **27,** 215 (1981). Effect on biological and chemical activity in soil: S. P. Mathur *et al., J. Environ. Sci. Health* **B15,** 61 (1980); D. J. Ross, T. W. Speir, *Soil. Biol. Biochem.* **17,** 123 (1985). GC determn: M. J. Brown, *J. Agr. Food Chem.* **29,** 1129 (1981). Residual toxicity: D. C. Read, *J. Econ. Entomol.* **69,** 429 (1976). Acute toxicity: E. F. Hill, M. B. Camardese, *Ecotoxicol. Environ. Safety* **8,** 551 (1984); E. W. Schafer, Jr., W. A. Bowles, Jr., *Arch. Environ. Contam. Toxicol.* **14,** 111 (1985). Solubility data: B. T. Bowman, W. W. Sans, *J. Environ. Sci. Health* **B18,** 221 (1983). *Review:* B. Homeyer, *Pflanzenschutz-Nach.* **24,** 48-67 (1971). Review of metabolism and biochemical properties, efficacy: T. B. Waggoner, A. M. Khasawinah, *Residue Review* **53,** 79-97 (1974).

CH₃
CH₃S— [benzene ring] —O—PNHCH(CH₃)₂ O ... OCH₂CH₃

Crystals, mp 49°. Soly in water: 329 μg/ml. LD_{50} in rats, birds (mg/kg): 10-25 (Waggoner, Khasawinah), 2.4 (Hill, Camardese) orally.
USE: Nematocide.

3902. Fenapanil. *α-Butyl-α-phenyl-1H-imidazole-1-propanenitrile;* phenapronil; RH 2161; Sisthane. $C_{16}H_{19}N_3$; mol wt 253.35. C 75.85%, H 7.56%, N 16.59%. Prepn: G. A. Miller *et al.,* **Ger. pat. 2,604,047** corresp to **U.S. pat. 4,073,921;** *eidem,* **U.S. pat. 4,225,723** (1976, 1978, 1980 all to Rohm & Haas).

CH₂CH₂CH₂CH₃
[benzene ring]—C—CH₂—N [imidazole]
CN

Hydrochloride, mp 160-162°.
USE: Fungicide.

3903. Fenarimol. *α-(2-Chlorophenyl)-β-(4-chlorophenyl)-5-pyrimidinemethanol;* 2,4'-dichloro-α-(pyrimidin-5-yl)benzhydryl alcohol; EL 222; Bloc; Rimidin(e); Rubigan. $C_{17}H_{12}Cl_2N_2O$; mol wt 331.20. C 61.65%, H 3.65%, Cl 21.41%, N 8.46%, O 4.83%. Prophylactic and curative fungicide; inhibits ergosterol biosynthesis. Prepn: **Neth. pat. Appl. 6,806,106;** H. M. Taylor *et al.,* **U.S. pat. 3,818,009**

(1968, 1974 both to Lilly). Improved prepn: *eidem,* **U.S. pat. 3,869,456;** use as fungicide: *eidem,* **U.S. pat. 3,887,708** (both 1975 to Lilly). Mode of action, efficacy against various fungi: H. Buchenauer, *Proc. Brit. Crop Prot. Conf.-Pests Dis.* **1977,** 699. Absorption by leaf tissue, local systemic activity: I. F. Brown, H. R. Hall, *ibid.* **1981**(2), 573. In control of apple scab: J. M. Olivier, J. Guillaumes, *Monogr. Brit. Crop Prot. Conf. - Fungic. Crop Prot.* **31,** 485 (1985). Post-infection efficacy: A. L. O'Leary, T. B. Sutton, *Phytopathol.* **76,** 119 (1986). Brief review: J.-M. Beraud *et al., Def. Veg.* **34,** 17 (1980).

[chemical structure]

White odorless crystals, mp 117-119°. Practically insol in water (13.7 ppm at pH 7). Sol in most organic solvents. Vapor pressure at 25° $< 10^{-7}$ millibar. LD_{50} in mice, rats (mg/kg): 4500, 2500 orally (Beraud).
USE: Plant fungicide.

3904. Fenbendazole. *[5-(Phenylthio)-1H-benzimidazol-2-yl]carbamic acid methyl ester;* 5-(phenylthio)-2-benzimidazolecarbamic acid methyl ester; methyl 5-(phenylthio)-2-benzimidazolecarbamate; HOE 881v; Panacur. $C_{15}H_{13}N_3$-O_2S; mol wt 299.35. C 60.19%, H 4.37%, N 14.04%, O 10.69%, S 10.71%. Prepn: **Belg. pat. 793,358** (1973 to Hoechst); Baeder *et al., Experientia* **30,** 753 (1974). Efficacy vs gastrointestinal nematodes in swine: Enigk *et al., Deut. Tierärztl. Wochenschr.* **81,** 177 (1974).

[chemical structure] H / NHCOOCH₃ ... S ... N

Light brownish-gray, odorless, tasteless crystalline powder, mp 233° (dec). Insol in water; insol or only very slightly sol in the usual solvents; freely sol in DMSO.
THERAP CAT (VET): Anthelmintic (swine).

3905. Fenbenicillin. *3,3-Dimethyl-7-oxo-6-[(phenoxyphenylacetyl)amino]-4-thia-1-azabicyclo[3.2.0]heptane-2-carboxylic acid;* 3,3-dimethyl-7-oxo-6-(2-phenoxy-2-phenylacetamido)-4-thia-1-azabicyclo[3.2.0]heptane-2-carboxylate; α-phenoxybenzylpenicillin; 6-(α-phenoxyphenylacetamido)penicillanic acid; 6-(α-phenoxy-α-phenylacetamido)penicillanate; phenbenicillin. $C_{22}H_{22}N_2O_5S$; mol wt 426.49. C 61.95%, H 5.20%, N 6.57%, O 18.76%, S 7.52%. Semisynthetic antibiotic related to penicillin. Prepn: **Brit. pat. 877,120** (1961 to Beecham). Bacteriological and pharmacological properties: Rollo, Burley, *Brit. Med. J.* **I,** 76 (1962).

[chemical structure] H H ... S CH₃ ... OCHCONH ... CH₃ ... N ... O ... COOH

Potassium salt, $C_{22}H_{21}KN_2O_5S$, *Penspek (formerly).* Powder, mp 88-95°, dec 120-125°. Readily sol in water. LD_{50} in mice (mg/kg): 225 i.v.; 520 i.p.; 3000 orally (Rollo, Burley).
THERAP CAT: Antibacterial.

3906. Fenbufen. *γ-Oxo-[1,1'-biphenyl]-4-butanoic acid;* 3-(4-biphenylylcarbonyl)propionic acid; β-p-phenylbenzoylpropionic acid; diphenyl-4-γ-oxo-γ-butyric acid; 4-(4-biphenylyl)-4-oxobutyric acid; CL 82204; Bufemid; Cinopal; Cinopol; Lederfen. $C_{16}H_{14}O_3$; mol wt 254.29. C

75.57%, H 5.55%, O 18.88%. Prepn: **Fr. pat. 798,941** (1936 to I. G. Farbenind.), *C.A.* **30**, 7729⁴ (1936); D. H. Hey, R. Wilkinson, *J. Chem. Soc.* **1940**, 1030; *cf.* M. Weizmann *et al.*, *Chem. & Ind.* **1940**, 402; W. Reppe *et al.*, *Ann.* **596**, 223 (1955); A. S. Tomcufcik *et al.*, **Ger. pat. 2,147,111** corresp to **U.S. pat. 3,784,701** (1972, 1974 to Am. Cyanamid). Series of articles on chemistry, pharmacology, reproductive toxicology and clinical experience: *Arzneimittel-Forsch.* **30**, 695-720, 725-746 (1980). Toxicity: H. F. Bolte *et al., ibid.* 721. Review of pharmacology and therapeutic use: R. N. Brogden *et al., Drugs* **21**, 1-22 (1981).

Crystals from ethanol, mp 185-187°. LD₅₀ in various strains of mice, rats (mg/kg): 795-1673, 200-720 orally; 506-811, 265-575 i.p. (Bolte).

THERAP CAT: Anti-inflammatory.

3907. Fenbutatin Oxide.

Hexakis(2-methyl-2-phenylpropyl)distannoxane; di[tri-(2-methyl-2-phenylpropyl)tin]-oxide; hexakis(β,β-dimethylphenethyl)distannoxane; SD 14114; Torque; Vendex. $C_{60}H_{78}OSn_2$; mol wt 1052.66. C 68.46%, H 7.47%, O 1.52%, Sn 22.55%. Selective organotin miticide. Prepn: **Ger. pat. 2,225,666;** C. A. Horne, **U.S. pat. 3,657,451** (1971, 1972 to Shell Oil Co.). Miticidal activity: L. R. Jeppson *et al., J. Econ. Entomol.* **68**, 707 (1975).

White cryst power, mp 138-139°. Soly in g/l at 23°: acetone 6; benzene 140; methylene chloride 380. Insol in water. Converts to the hydroxide in the presence of water. Nontoxic to bees, toxic to fish. LD₅₀ in rats: 2630 mg/kg orally, *RTECS* Vol. I, R. J. Lewis, R. L. Tatken, Eds. (1980) p 709.

Caution: Corrosive, skin and eye irritatant.

USE: Acaricide.

3908. Fenbutrazate.

α*-Ethylbenzeneacetic acid 2-(3-methyl-2-phenyl-4-morpholinyl)ethyl ester; 2-phenylbutyric acid 2-(3-methyl-2-phenylmorpholino)ethyl ester;* 2-(3-methyl-2-phenylmorpholino)ethyl 2-phenylbutyrate; phenbutrazate. $C_{23}H_{29}NO_3$; mol wt 367.47. C 75.17%, H 7.95%, N 3.81%, O 13.06%. Prepn: Siemer, Hengen, **U.S. pat. 3,018,222** (1962 to Ravensberg). Physical properties and pharmacology: Hengen, Siemer, *Arzneimittel-Forsch.* **5**, 526 (1955).

Honey colored, viscous oil, bp₀.₀₅ 235-240°. Sol in methanol with neutral reaction.

Hydrochloride, $C_{23}H_{30}ClNO_3$, *R 381.* Crystals, mp 154°. Slightly sol in water; sol in ethanol, acetone; practically insol in ether, benzene. LD₅₀ orally in mice, 3200 mg/kg.

Hydrochloride mixture with phenmetrazine theoclate, *Cafilon, Filon.*

THERAP CAT: Anorexic.

3909. Fencamfamine.

N-Ethyl-3-phenylbicyclo[2.2.1]heptan-2-amine; N-ethyl-3-phenyl-2-norbornanamine; 2-ethylamino-3-phenylnorcamphane; 2-phenyl-3-ethylaminonorbornane; 2-ethylamino-3-phenylnorbornane; 2-ethylamino-3-phenylbicyclo[2.2.1]heptane; Euvitol. $C_{15}H_{21}N$; mol wt 215.33. C 83.66%, H 9.83%, N 6.51%. Prepd from 3-amino-2-phenylnorbornane and acetaldehyde followed by hydrogenation in the presence of PtO_2: Thesing *et al.,* **Ger. pat. 1,110,159** (1961 to E. Merck), *C.A.* **56**, 2352g (1962).

bp₀.₁ 128-131°.

Hydrochloride, $C_{15}H_{22}ClN$, *H 610, Norcamphane.* Crystals from acetone, mp 192°. Freely sol in water, ethanol, methanol, chloroform; slightly sol in ether. Practically insol in ether. LD₅₀ orally in mice: 135 mg/kg.

Note: The hydrochloride is the principal ingredient of *Reactivan* which also contains vitamins B_1, B_6, B_{12} and C.

THERAP CAT: Central stimulant.

3910. Fencamine.

3,7-Dihydro-1,3,7-trimethyl-8-[[2-[methyl(1-methyl-2-phenylethyl)amino]ethyl]amino]-1H-purine-2,6-dione; 8-[[2-[methyl(α-methylphenethyl)amino]ethyl]amino]caffeine; 8-[2-(N,α-dimethyl-β-phenyl)ethylamino]-1,3,7-trimethyl-2,6-dioxopurine; N^1-(1,3,7-trimethyl-2,6-dioxopurin-8-yl)-N^2-(1-methyl)phenethyl-N^2-methylethylenediamine; N-8-caffeyl-N'-methyl-N'-(α-methylphenethyl)ethylenediamine; phencamine; ST-374. $C_{20}H_{28}N_6O_2$; mol wt 384.48. C 62.48%, H 7.34%, N 21.85%, O 8.32%. Prepn: **Span. pats. 347,509; 352,077** (both 1969 to Labs. Miquel); Pitarch *et al., Quim. Ind. (Madrid)* **17**, 71 (1971). Pharmacology: *eidem, ibid.* 76. Colorimetric and TLC determn in urine: J. Mallol *et al., Arzneimittel-Forsch.* **24**, 1301 (1974). Toxicity data: E. Usdin, D. H. Efron, *Psychotropic Drugs and Related Compounds* (National Institute of Mental Health, Rockville, MD, 2nd ed., 1972) p 229. Clinical trial in depression: P. G. Quiros e Isla, *Rev. Clin. Espan.* **119**, 437 (1970).

Crystals from methanol, mp 150-152°.

Hydrochloride, $C_{20}H_{29}ClN_6O_2$, *Altimina, Sicoclor.* Slightly hygroscopic, bitter crystals, decomp 278-279°. uv max (water): 296 nm (E₁%₁cm 398). Sol in water; slightly sol in many organic solvents. LD₅₀ in rats, mice (mg/kg): 93, 82 i.p.; 508, 418 orally (Usdin, Efron).

THERAP CAT: Analeptic.

3911. d-Fenchone.

1,3,3-Trimethylbicyclo[2.2.1]heptan-2-one; d-1,3,3-trimethyl-2-norbornanone; d-1,3,3-trimethyl-2-norcamphanone. $C_{10}H_{16}O$; mol wt 152.23. C 78.89%, H 10.60%, O 10.51%. Occurs in fennel oil and in the essential oil of *Lavandula stoechas* L., *Labiatae.* Isoln: Wallach, *Ann.* **263**, 129 (1891); **353**, 209 (1907); **369**, 63 (1909); Shavrygin, *J. Appl. Chem. USSR* **12**, 1201 (1939). Total synthesis: Boyle *et al., Chem. Commun.* **1971**, 395; G. Buchbauer, H. C. Rohner, *Ann.* **1981**, 2093. *Review:* Simonsen, *The Terpenes,* vol. II (Cambridge, 2nd ed., 1949) pp 560-580; D. L. J. Opdyke, *Food Cosmet. Toxicol.* **14**, Suppl., 769-771 (1976).

Oily liq. Camphor-like odor. d⁴₁₈ 0.948. mp 6.1°. bp₇₆₀ 193.5°; bp₁₀₀ 122°; bp₂₀ 82°; bp₁₅ 66°. [α]²₀D +66.9°. n¹⁸ 1.4636. Practically insol in water (pH of satd soln 6.82). Very sol in abs alcohol, ether. LD₅₀ orally in rats: 6.16 g/kg, P. M. Jenner *et al., Food Cosmet. Toxicol.* **2**, 327 (1964).

USE: As flavor in foods; in perfumes.

THERAP CAT: Counterirritant.

3912. Fencibutirol.

α*-Ethyl-1-hydroxy-4-phenylcyclohexaneacetic acid;* α-(1-hydroxy-4-phenylcyclohexyl)butyric acid; MG 4833; Biligen; Hepasil; Verecolene. $C_{16}H_{22}O_3$; mol wt 262.34. C 73.25%, H 8.45%, O 18.30%. Prepn: Carissimi, Ravenna, U.S. pat. 3,027,302 (1962 to Maggioni).

Crystals from ligroin, mp 157°.
THERAP CAT: Choleretic.

3913. Fenclofenac. *2-(2,4-Dichlorophenoxy)benzeneacetic acid; [o-(2,4-dichlorophenoxy)phenyl]acetic acid;* Rx 67408; R 67408; Flenac. $C_{14}H_{10}Cl_2O_3$; mol wt 297.13. C 56.59%, H 3.39%, Cl 23.86%, O 16.15%. Prepn: K. E. Godfrey, Ger. pat. 2,117,826 corresp to U.S. pats. 3,766,263 and 3,845,215 (1971, 1973 and 1974 all to Reckitt and Colman). Pharmacology: D. C. Atkinson *et al., J. Pharm. Pharmacol.* **26,** 357 (1974); D. C. Atkinson, E. C. Leach, *Agents Actions* **8,** 263 (1978). Metabolism: B. J. Jordan, M. J. Rance, *J. Pharm. Pharmacol.* **26,** 359 (1974); A. Garner, *Toxicol. Appl. Pharmacol.* **42,** 477 (1977). Clinical study: M. Thompson, M. S. Akyol, *J. Int. Med. Res.* **5** (suppl 2), 96 (1977).

Solid from carbon tetrachloride, mp 134-136°. LD_{50} in rats: 2280 mg/kg orally.
THERAP CAT: Anti-inflammatory.

3914. Fenclorac. *α,3-Dichloro-4-cyclohexylbenzeneacetic acid; α,m-dichloro-p-cyclohexylphenylacetic acid; chloro(3-chloro-4-cyclohexylphenyl)acetic acid;* WHR-539. $C_{14}H_{16}Cl_2O_2$; mol wt 278.18. C 58.55%, H 5.62%, Cl 24.69%, O 11.14%. Anti-inflammatory phenylacetic acid deriv with antipyretic and analgesic properties. Prepn: J. Diamond, N. J. Santora, Ger. pat. 2,122,273 corresp to U.S. pat. 3,825,553 (1972, 1974 both to Rorer). Pharmacology: G. W. Nuss *et al., Agents Actions* **6,** 735 (1976). Physiological disposition of ^{14}C-fenclorac in rats: P. P. Mathur, R. D. Smyth, *Res. Commun. Chem. Pathol. Pharmacol.* **25,** 23 (1979). GLC assay in human plasma: A. F. De Long *et al., J. Pharm. Sci.* **67,** 1171 (1978). Kinetics of hydrolysis: C. M. Won *et al., ibid.* **66,** 73 (1977). Inhibition of prostaglandin synthetase *in vitro:* R. Procaccini *et al., Biochem. Pharmacol.* **26,** 1051 (1977); *in vivo:* P. P. Mathur *et al., Agents Actions* **7,** 283 (1977).

Diethylammonium salt, $C_{18}H_{27}Cl_2NO_2$, *Fenbrac.* White crystals from *n*-hexane, mp 109-115°. LD_{50} in mice, rats: 430, 285 mg/kg orally, Nuss *et al., loc. cit.*
THERAP CAT: Anti-inflammatory.

3915. Fenclozic Acid. *2-(4-Chlorophenyl)-4-thiazoleacetic acid; 2-(p-chlorophenyl)thiazol-4-ylacetic acid;* acidum fenclozicum; ICI 54450; Myalex. $C_{11}H_8ClNO_2S$; mol wt 253.70. C 52.08%, H 3.18%, Cl 13.97%, N 5.52%, O 12.61%, S 12.64%. Prepn: Hepworth, Stacey, **Neth. Appl. 6,614,130** corresp to U.S. pat. 3,538,107 (1967, 1970 to ICI); Aries, **Fr. pat. 1,561,433** (1969), *C.A.* **72,** 43654y (1970). Pharmacology: Hepworth *et al., Nature* **221,** 582 (1969);

Newbould, *Brit. J. Pharmacol.* **35,** 487 (1969). Metabolic studies: Foulkes, *J. Pharmacol. Exp. Ther.* **172,** 115, 449 (1970). Clinical evaluation: Chalmers *et al., Ann. Rheum. Dis.* **28,** 590, 595 (1969).

Colorless crystalline solid from ethyl acetate, mp 155-156°. Soluble in most organic solvents; sparingly sol in water. LD_{50} in rats, mice: 850, 1000 mg/kg orally; 300, 250 mg/kg i.v, Hepworth *et al., loc. cit.* (1969).
THERAP CAT: Anti-inflammatory.

3916. Fendiline. *γ-Phenyl-N-(1-phenylethyl)benzenepropanamine; N-(3,3-diphenylpropyl)-α-methylbenzylamine; N-(1-phenylethyl)-3,3-diphenylpropylamine.* $C_{23}H_{25}N$; mol wt 315.46. C 87.57%, H 7.99%, N 4.44%. Calcium blocking agent. Prepn: Belg. pat. 621,300; Harsányi *et al.,* U.S. pat. 3,262,977 (1962, 1966 both to Chinoin); *eidem, J. Med. Chem.* **7,** 623 (1964); Klosa, *J. Prakt. Chem.* **34,** 312 (1966). Prepn and studies of the labelled compound: Volford, Harsányi, *J. Label. Compounds* **9,** 219 (1973). Structure-activity studies: Leszkovszky *et al., Acta Physiol. Acad. Sci. Hung.* **29,** 283 (1966). Pharmacologic properties: W. R. Kukovetz *et al., Arzneimittel-Forsch.* **26,** 1321 (1976); A. Fleckenstein *et al., ibid.* **27,** 562 (1977). Pharmacokinetics and tolerance: R. Weyhenmeyer *et al., ibid.* **37,** 58 (1987). Use in ischemic heart disease: Z. Antaloczy, I. Preda, *Ther. Hung.* **27,** 71 (1979). Assessment of Ca^{2+}-antagonist effects: M. Spedding, *Arch. Pharmacol.* **318,** 234 (1982).

$bp_{0.3}$ 206-210°.
Hydrochloride, $C_{23}H_{26}ClN$, *HK 137, Cordan, Difmecor, Fendilar, Sensit.* Almost white or slightly pink powder, mp 204-205°. Very slightly sol in water; easily sol in methanol, ethanol, chloroform. LD_{50} in mice (mg/kg): 14.5 i.v.; 950 orally (Harsányi).
THERAP CAT: Coronary vasodilator.

3917. Fendosal. *5-(4,5-Dihydro-2-phenyl-3H-benz[e]indol-3-yl)-2-hydroxybenzoic acid; 5-(4,5-dihydro-2-phenyl-3H-benz[e]indol-3-yl)salicylic acid; 3-(3-carboxy-4-hydroxyphenyl)-2-phenyl-4,5-dihydro-3H-benz[e]indole;* HP 129; P 71-0129; Alnovin. $C_{25}H_{19}NO_3$; mol wt 381.44. C 78.72%, H 5.02%, N 3.67%, O 12.58%. Prepn: R. C. Allen, B. Anderson, Ger. pat. 2,407,671 corresp to U.S. pat. 3,-878,225 (1974, 1975, both to Hoechst); V. B. Anderson *et al., J. Med. Chem.* **19,** 318 (1976). Pharmacology: H. B. Lassman *et al., Agents Actions* **8,** 209 (1978). Clinical study: S. S. Bloomfield *et al., Clin. Pharmacol. Ther.* **23,** 390 (1978).

Crystals from acetic acid, mp 223-225° (dec). LD_{50} in mice, rats: 740, 450 mg/kg orally.
THERAP CAT: Anti-inflammatory.

3918. Fenethazine. *N,N-Dimethyl-10H-phenothiazine-10-ethanamine; 10-(2-dimethylaminoethyl)phenothiazine; 10-(β-dimethylaminoethyl)phenothiazine;* RP 3015; SC 1627; Anergan; Anergen; Ethysine; Etisine; Lisergan; Lysergan; Phenethazinum; Rutergan. $C_{16}H_{18}N_2S$; mol wt 270.38. C 71.07%, H 6.71%, N 10.36%, S 11.86%. Prepd by the condensation of phenothiazine with dimethylaminoethyl chloride: Charpentier, U.S. pat. **2,519,886** (1950 to Rhône-Poulenc).

$$CH_2CH_2N(CH_3)_2$$

bp$_1$ 183-187°.
Hydrochloride, $C_{16}H_{18}N_2S.HCl$, mp 201-201.5°. LD$_{50}$ i.p. in mice: 115-120 mg/kg.
THERAP CAT: Antihistaminic.

3919. Fenethylline. *3,7-Dihydro-1,3-dimethyl-7-[2-[(1-methyl-2-phenylethyl)amino]ethyl]-1H-purine-2,6-dione; 7-[2-[(α-methylphenethyl)amino]ethyl]theophylline;* 7-[β-(α-methyl-β-phenylethylamino)ethyl]theophylline; 7-(phenylisopropylaminoethyl)theophylline; 7-(3-phenyl-2-propylaminoethyl)theophylline; theophyllineethylamphetamine. $C_{18}H_{23}N_5O_2$; mol wt 341.40. C 63.32%, H 6.79%, N 20.52%, O 9.37%. Prepn from 7-(β-chloroethyl)theophylline and amphetamine: Kholstaedt, Klinger, Ger. pat. **1,123,329** (1962 to Degussa), *C.A.* **57**, 5933c (1962). Metabolism in man: T. Ellison *et al., Eur. J. Pharmacol.* **13**, 123 (1970).

Hydrochloride, $C_{18}H_{24}ClN_5O_2$, *Captagon, Homburg 814.* Crystals, two different modifications, mp 227-229° and 237-239°.
d-Form hydrochloride, crystals, mp 246-247°.
l-Form hydrochloride, crystals, mp 246-247°.
Note: This is a controlled substance (stimulant) listed in the U.S. Code of Federal Regulations, Title 21 Part 1308.11 (1987).
THERAP CAT: CNS stimulant.

3920. Fenfluramine. *N-Ethyl-α-methyl-3-(trifluoromethyl)benzeneethanamine; N-ethyl-α-methyl-m-(trifluoromethyl)phenethylamine;* 2-ethylamino-1-(3-trifluoromethyl-phenyl)propane; S 768. $C_{12}H_{16}F_3N$; mol wt 231.27. C 62.32%, H 6.97%, F 24.65%, N 6.06%. Prepn: L. G. Beregi *et al.,* Fr. pat. **M1658**; *eidem,* U.S. pat. **3,198,833** (1963, 1965 both to Sci. Union et Cie Soc. Franc. Recherche Méd.). Prepn of optical isomers: *eidem,* U.S. pat. **3,198,834** (1965 to Sci. Union et Cie Soc. Franc. Recherche Med.). Pharmacology: *Presse Med.* **71**, 181 (1963). Pharmacology and toxicity of isomers and racemate: J. C. Le Douarec *et al., Arch. Int. Pharmacodyn. Ther.* **161**, 206 (1966). Pharmacokinetics: S. Caccia *et al., Eur. J. Clin. Pharmacol.* **29**, 221 (1985). Clinical trial of dextrofenfluramine in refractory obesity: N. Finer *et al., Curr. Ther. Res.* **38**, 847 (1985). Comprehensive review: Pinder *et al., Drugs* **10**, 241-323 (1975).

$$CH_2CHNHC_2H_5$$
$$|$$
$$CH_3$$
$$CF_3$$

dl-Form, bp$_{12}$ 108-112°. LD$_{50}$ orally in rats: 237.8 mg/kg (Le Douarec).
dl-Form hydrochloride, $C_{12}H_{17}ClF_3N$, *Acino, Adipomin, Ganal, Obedrex, Pesos, Ponderal, Ponderax, Ponderex, Pondimin, Rotondin.* Crystals from ethanol + ether, mp 166°.
d-Form, *dexfenfluramine, dextrofenfluramine, Adifax.* $[α]_D^{25}$ +9.5° (c = 8 in ethanol). LD$_{50}$ orally in rats: 114.6 mg/kg (Le Douarec).

d-Form hydrochloride, *Isomeride.* Crystals from ethyl acetate, mp 160-161°.
l-Form, $[α]_D^{25}$ −9.6° (c = 8 in ethanol). LD$_{50}$ orally in rats: 195 mg/kg (Le Douarec).
l-Form hydrochloride, crystals from ethyl acetate, mp 160-161°.
Note: This is a controlled substance listed in the U.S. Code of Federal Regulations, Title 21 Part 1308.14 (1987).
THERAP CAT: Anorexic.

3921. Fenipentol. *α-Butylbenzenemethanol; α-butylbenzyl alcohol;* phenylbutylcarbinol; 1-phenyl-1-hydroxypentane; phenylpentanol; PC 1; Ph BC; Pancoral. $C_{11}H_{16}O$; mol wt 164.25. C 80.44%, H 9.82%, O 9.74%. Synthesis: Fourneau, Puyal, *An. Soc. Espan. Fis. Quim.* **18**, 323 (1920); Adams, Vander Werf, *J. Am. Chem. Soc.* **72**, 4368 (1950); Protiva *et al., Chem. Listy* **46**, 37 (1952). Prepn of the drug: Scheffler, Engelhorn, Brit. pat. **915,815** corresp to U.S. pat. **3,084,100** (both 1963 to Thomae). Pharmacology: Engelhorn, *Arzneimittel-Forsch.* **10**, 255 (1960); Beck, Bierwisch, *ibid.* **20**, 693 (1970). Metabolism: Koss *et al., ibid.* **14**, 195 (1964).

$$OH$$
$$|$$
$$CH(CH_2)_3CH_3$$

Colorless or slightly yellow liquid. bp$_{12}$ 123-124°. n_D^{20} 1.5112. Miscible with organic liquids. Practically insol in water. LD$_{50}$ in mice: 1.03 g/kg i.p.; 3.10 g/kg orally.
THERAP CAT: Choleretic. Also used in treatment of mild chronic pancreatitis.

3922. Fenitrothion. *Phosphorothioic acid O,O-dimethyl O-(3-methyl-4-nitrophenyl) ester; O,O-*dimethyl *O*-4-nitro-*m*-tolyl phosphorothioate; *O,O*-dimethyl *O*-(3-methyl-4-nitrophenyl) phosphorothioate; *O,O*-dimethyl *O*-4-nitro-*m*-tolyl thiophosphate; MEP; metathion; Bayer 41831; Bayer S 5660; ENT 25715; OMS 45; AC 47300; Accothion; Cyten; Cyfen; Folithion; Sumithion. $C_9H_{12}NO_5PS$; mol wt 277.25. C 38.99%, H 4.36%, N 5.05%, O 28.85%, P 11.17%, S 11.57%. Prepn: Belg. pat. **594,669** (1960 to Sumitomo); Belg. pat. **596,091** (1960 to Bayer). Activity: Y. Nishizawa *et al., Agr. Biol. Chem.* **25**, 605 (1961).

$$CH_3O \quad S$$
$$\backslash \quad ||$$
$$P-O- \quad -NO_2$$
$$/$$
$$CH_3O \quad CH_3$$

Yellow oil. bp$_{0.05}$ 118°. n_D^{25} 1.5528. d_4^{25} 1.3227. Vapor press at 20°: 6 × 10^{-6} mm Hg. uv max: 269.5 nm (ε 6756). Practically insol in water; low soly in aliphatic hydrocarbons; sol in most organic solvents. LD$_{50}$ in rats: 250 mg/kg, G. Schrader, *Angew. Chem.* **73**, 331 (1961).
USE: Insecticide. *Caution:* Cholinesterase inhibitor.

3923. Fennel. Large fennel; sweet fennel. Dried, ripe fruit of cultivated varieties of *Foeniculum vulgare* Mill., *Umbelliferae. Habit.* Southern Europe, Western Asia, widely cultivated. Contains 3-4% volatile oil.

3924. Fenofibrate. *2-[4-(4-Chlorobenzoyl)phenoxy]-2-methylpropanoic acid 1-methylethyl ester;* isopropyl [4'-(*p*-chlorobenzoyl)-2-phenoxy-2-methyl]propionate; procetofen; procetofene; LF-178; Ankebin; Elasterin; Fenobrate; Lipanthyl; Lipantil; Lipidax; Lipidil; Lipoclar; Lipofene; Liposit; Lipsin; Nolipax; Procetoken; Protolipan; Secalip. $C_{20}H_{21}ClO_4$; mol wt 360.84. C 66.57%, H 5.87%, Cl 9.82%, O 17.74%. Prepn: A. Mieville, Ger. pat. **2,250,327** (1973 to Orchimed), *C.A.* **79**, 53029 (1973), addn to Ger. pat. **2,003,-430** corresp to U.S. pat. **3,907,792** (1970, 1975 both to Orchimed). Series of articles on synthesis, metabolism, pharmacology and clinical trials: *Arzneimittel-Forsch.* **26**, 885-909 (1976). Toxicity data: R. Sornay *et al., ibid.* 885. GC-MS of major metabolite in humans: L. F. Elsom *et al., J. Chromatog.* **123**, 463 (1976). Efficacy in hyperlipidemias: H. B. Stähelin *et al., Praxis* **68**, 24 (1979). Effect on human biliary lipids: E. M. Grandjean *et al., Schweiz. Med. Woch-*

enschr. **109**, 601 (1979). Mechanism of action: W. Wülfert *et al., Artery* **9**, 120 (1981).

Crystals from isopropanol, mp 80-81°. Practically insol in water. Slightly sol in methanol, ethanol. Sol in acetone, ether, benzene, chloroform. LD_{50} in mice: 1600 mg/kg orally (Sornay).

THERAP CAT: Antihyperlipoproteinemic.

3925. Fenoldopam. *6-Chloro-2,3,4,5-tetrahydro-1-(4-hydroxyphenyl)-1H-3-benzazepine-7,8-diol;* 6-chloro-7,8-dihydroxy-1-p-hydroxyphenyl-2,3,4,5-tetrahydro-1H-3-benzazepine; SKF 82526. $C_{16}H_{16}ClNO_3$; mol wt 305.76. C 62.85%, H 5.27%, Cl 11.60%, N 4.58%, O 15.70%. Dopamine D_1-receptor agonist. Prepn: J. Weinstock, Ger. pat. **2,751,258;** idem, U.S. pat. **4,197,297** (1978, 1980 both to SmithKline); J. Weinstock *et al., J. Med. Chem.* **23**, 973 (1980). HPLC determn in urine and plasma: V. K. Boppana *et al., J. Chromatog.* **317**, 463 (1984). Clinical pharmacology: R. M. Stote *et al., Clin. Pharmacol. Ther.* **34**, 309 (1983); R. M. Carey *et al., J. Clin. Invest.* **74**, 2198 (1984). Hemodynamic effects in hypertension: H. O. Ventura *et al., Circulation* **69**, 1142 (1984); M. P. Caruana *et al., Brit. J. Clin. Pharmacol.* **24**, 721 (1987). Clinical evaluation in congestive heart failure: G. S. Francis *et al., Am. Heart J.* **116**, 473 (1988).

Hydrobromide, $C_{16}H_{17}BrClNO_3$, mp 277° (dec).
Monomethanesulfonate, $C_{17}H_{20}ClNO_6S$, *SKF 82526-J, Corlopam.* mp 274° (dec).
THERAP CAT: Antihypertensive.

3926. Fenoprofen. *α-Methyl-3-phenoxybenzeneacetic acid;* (±)-m-phenoxyhydratropic acid; α-dl-2-(3-phenoxy-phenyl)propionic acid; Lilly 53838. $C_{15}H_{14}O_3$; mol wt 242.28. C 74.36%, H 5.82%, O 19.81%. Prepn: Marshall, Fr. pat. **2,015,718** corresp to U.S. pat. **3,600,437** (1970, 1971 to Lilly). Pharmacology: Rubin *et al., J. Pharm. Sci.* **60**, 1797 (1971); **61**, 800 (1972); Herrmann, *Proc. Soc. Exp. Biol. Med.* **139**, 548 (1972). Metabolism: Rubin *et al., J. Pharmacol. Exp. Ther.* **183**, 449 (1972). Toxicology: J. L. Emmerson *et al., Toxicol. Appl. Pharmacol.* **25**, 444 (1973). Comprehensive description: C. K. Ward, R. E. Schirmer in *Analytical Profiles of Drug Substances* vol. 6, K. Florey, Ed. (Academic Press, New York, 1977) pp 161-182. Review: R. N. Brogden *et al., Drugs* **13**, 241-265 (1977); R. Nickander *et al.,* in *Pharmacological and Biochemical Properties of Drug Substances* vol. 1, M. E. Goldberg, Ed. (Am. Pharm. Assoc., Washington, DC, 1977) pp 183-213.

Viscous oil, $bp_{0.11}$ 168-171°. n_D^{25} 1.5742. pKa 7.3.
Calcium salt dihydrate, $C_{30}H_{26}CaO_6 \cdot 2H_2O$, *Lilly 69323, Fenopron, Fepron, Feprona, Nalfon, Nalgesic, Progesic.* White crystalline powder. Soly in mg/ml at 37°: n-hexanol 11; methanol 8; water 2.5; chloroform 0.01. pKa 4.5. Aq

solns sensitive to intense uv light. LD_{50} orally in mice: 800 mg/kg (Emmerson).
THERAP CAT: Anti-inflammatory; analgesic.

3927. Fenoterol. *5-[1-Hydroxy-2-[[2-(4-hydroxyphen-yl)-1-methylethyl]amino]ethyl]-1,3-benzenediol; 3,5-dihydr-oxy-α-[[(p-hydroxy-α-methylphenethyl)amino]methyl]benzyl alcohol;* 1-(3,5-dihydroxyphenyl)-1-hydroxy-2-[(4-hydr-oxyphenyl)isopropylamino]ethane; 1-(p-hydroxyphenyl)-2-[[β-hydroxy-β-(3′,5′-dihydroxyphenyl)]ethyl]aminopro-pane; Th 1165. $C_{17}H_{21}NO_4$; mol wt 303.37. C 67.31%, H 6.98%, N 4.62%, O 21.09%. $β_2$-Adrenergic agonist. Prepn and sepn of stereoisomers: Belg. pat. **640,433;** Zelle *et al.,* U.S. pat. **3,341,593** (1962, 1967 to Boehringer, Ing.). Pharmacology: Schuster, Baum, *Arzneimittel-Forsch.* **19**, 1905 (1969); O'Donnell, *Eur. J. Pharmacol.* **12**, 35 (1970). Toxicity: E. I. Goldenthal, *Toxicol. Appl. Pharmacol.* **18**, 185 (1971). Metabolism in mice: S. Kojima *et al., Arzneimittel-Forsch.* **30**, 959 (1980). Clinical evaluation: Tweel, *Ann. Allergy* **29**, 142 (1971); Rebuck, Saunders, *Med. J. Aust.* **1**, 225 (1972). Comparison with ritodrine, *q.v.,* in preterm labor: J. Gerris *et al., Eur. J. Clin. Pharmacol.* **18**, 443 (1980). Review of pharmacology and therapeutic efficacy: R. C. Heel *et al., Drugs* **15**, 3-32 (1978); N. Svedmyr, *Pharmacother.* **5**, 109-126 (1985).

Hydrobromide, $C_{17}H_{22}BrNO_4$, *Th 1165a, Airum, Berotec, Dosberotec, Partusisten.* Crystals from methanol-ether, mp 222-223°. LD_{50} in mice (mg/kg): 1100 s.c.; 1990 orally (Goldenthal).
Hydrochloride, $C_{17}H_{21}NO_4 \cdot HCl$, mp 183° (acetonitrile-ether).
THERAP CAT: Bronchodilator; tocolytic.

3928. Fenoverine. *10-[[4-(1,3-Benzodioxol-5-ylmethyl)-1-piperazinyl]acetyl]-10H-phenothiazine;* 10-[4-(3,4-dioxy-methylenebenzyl)-1-piperazinylacetyl]phenothiazine; 10-[(4-piperonyl-1-piperazinyl)acetyl]phenothiazine; Spasmo-priv. $C_{26}H_{25}N_3O_3S$; mol wt 459.56. C 67.95%, H 5.48%, N 9.14%, O 10.44%, S 6.98%. Piperonylpiperazine derivative with spasmolytic activity. Prepn: A. Buzas, R. Pierre, Fr. pat. **2,092,639** (1972), *C.A.* **77**, 140143p (1972). Pharmacology, clinical studies in gastrointestinal disorders, dysmenorrhea: R. Pierre, R. Roustan, *Med. Interne* **15**, 49 (1980). In treatment of colon dysfunction: Claudon, *Rev. Fr. Gastro-Enterol.* **198**, 31 (1984).

Crystals from isopropyl ether, mp 141-142°. LD_{50} in mice (g/kg): ~1.50 orally, ~2.50 i.p. (Buzas, Pierre).
THERAP CAT: Antispasmodic.

3929. Fenoxaprop-ethyl. (±)-2-[4-[(6-Chloro-2-benz-oxazolyl)oxy]phenoxy]propanoic acid ethyl ester; HOE 33171; Acclaim; Furore; Puma; Whip. $C_{18}H_{16}ClNO_5$; mol wt 291.38. C 74.19%, H 5.53%, Cl 12.16%, N 4.83%, O 3.29%. Selective postemergent herbicide to control grassy weeds in broadleaved crops and established turfgrass. Prepn: Belg. pat. **858,618;** R. Handte *et al.,* U.S. pat. **4,130,413** (both 1978 to Hoechst). Metabolism in soybeans: O. Wink *et al., J. Agr. Food Chem.* **32**, 187 (1984). Persistence in soil: A. E. Smith, *ibid.* **33**, 483 (1985). Field trials in agricultural crops: G. W. Kerse *et al., Proc. 36th N.Z. Weed Pest Control Conf.,* 265 (1983); in turfgrass: P. H. Dernoeden, J. D. Fry, *Proc. Ann. Meet. Northeast Weed Sci. Soc.* **39**, 282 (1985). Brief description: J. Bieringer *et al., Proc. Brit. Crop Prot. Conf.-Weeds* **1982**, 11-17.

mp 84-85°. bp 200° at 100 Pa. Vapor pressure at 20°: 0.19×10^{-5} Pa. Soly at 20° (%): > 0.5 in hexane; > 1 in cyclohexane, ethanol, 1-octanol; > 20 in ethyl acetate; > 30 in toluene; > 50 in acetone. Soly in water at 25°: 0.9 mg/l. LD_{50} in male, female rats (mg/kg): 2357, 2500 orally; 739, 864 i.p. (Bieringer).

USE: Postemergent herbicide.

3930. Fenoxazoline. *4,5-Dihydro-2-[[2-(1-methylethyl)-phenoxy]methyl]-1H-imidazole; 2-[(o-cumenyloxy)methyl]-2-imidazoline;* 2-(o-isopropylphenoxymethyl)-2-imidazo-line; phenoxazoline. $C_{13}H_{18}N_2O$; mol wt 218.29. C 71.53%, H 8.31%, N 12.83%, O 7.33%. Preparation: Fr. pat. **1,365,-971** (1964 to Lab. Dausse and Soc. BMC). Use as sympa-thomimetic agent: Giudicelli, U.S. pat. **3,198,703** (1965 to Lab. Dausse).

Hydrochloride, $C_{13}H_{19}ClN_2O$, *Aturgyl, Snup.* Crystals, mp 174°. Sol in water, ethanol.

THERAP CAT: Sympathomimetic.

3931. Fenoxedil. *2-(4-Butoxyphenoxy)-N-(2,5-diethoxy-phenyl)-N-[2-(diethylamino)ethyl]acetamide; 2-(p-butoxy-phenyl)-N-[2-(diethylamino)ethyl]-2',5'-diethoxyacetani-lide.* $C_{28}H_{42}N_2O_5$; mol wt 486.66. C 69.11%, H 8.70%, N 5.75%, O 16.44%. Prepn: Ger. pat. **1,964,712** corresp to Thuillier, Geffroy, U.S. pat. **3,818,021** (1970, 1974, both to CERPHA).

Hydrochloride, $C_{28}H_{43}ClN_2O_5$, *Suplexedil.* mp 140°. LD_{50} in mice: 750 mg/kg orally, 17 mg/kg i.v., Thuillier, Geffroy, *loc. cit.*

THERAP CAT: Vasodilator.

3932. Fenozolone. *2-(Ethylamino)-5-phenyl-4(5H)-oxazolone;* 2-ethylamino-4-oxo-5-phenyl-Δ^2-oxazoline; 5-phenyl-2-(ethylimino)-4-oxazolidone; 5-phenyl-2-ethyl-amino-4-oxazolinone; LD 3394; Ordinator. $C_{11}H_{12}N_2O_2$; mol wt 204.23. C 64.69%, H 5.92%, N 13.71%, O 15.67%. Prepn: Najer, Giudicelli, *Bull. Soc. Chim. France* **1961**, 1231; Belg. pat. **613,985** and Ger. pat. **1,297,108** (1962 and 1969 to Dausse), *C.A.* **57**, 13761a (1962); **71**, 70586k (1969). Exists predominantly in the tautomeric ethylaminooxazolin-one form: Najer *et al., Compt. Rend.* **254**, 2173 (1962). Activity studies: Giudicelli *et al., ibid.* 2862.

Crystals from benzene, mp 148°. Stable in alkaline, un-stable in acid soln. uv max (alcohol): 221 nm (log ε 4.42). LD_{50} in white mice (g/kg): 0.425 orally; 0.175 i.p. (Belg. pat.).

THERAP CAT: Central stimulant.

3933. Fenpentadiol. *2-(4-Chlorophenyl)-4-methyl-2,4-pentanediol;* 2-methyl-4-(p-chlorophenyl)-2,4-pentanediol; RD 292; Tredum. $C_{12}H_{17}ClO_2$; mol wt 228.72. C 63.01%, H 7.49%, Cl 15.50%, O 13.99%. Prepn: Valette, Fr. pat. **M1984** (1963 to Labs Albert Rolland), *C.A.* **60**, 2828h (1964). Psychopharmacology: J. Ginet *et al., Arzneimittel-Forsch.* **21**, 1 (1971). Pharmacokinetics: Y. Dormand, J. C. Levron, *Eur. J. Toxicol.* **5**, 43 (1972). Clinical evaluations: P. Mouren *et al., Marseille Med.* **106**, 555 (1969); P. J. Par-quet, *Lille Med.* **18**, 836 (1973). Toxicity: H. Kriegel *et al., Arzneimittel-Forsch.* **21**, 9 (1971).

Crystals from petr ether, mp 76.5°. uv max (0.1% in abs ethanol): 228 nm. LD_{50} in male, female mice, rats (mg/kg): 940, 995, 1200, 1250 orally (Kriegel).

THERAP CAT: Antidepressant.

3934. Fenpiprane. *1-(3,3-Diphenylpropyl)piperidine;* 1,1-diphenyl-3-piperidinopropane; 1-(3,3-biphenylpropyl)pi-peridine; 1,1-diphenyl-3-pentamethyleneiminopropane. $C_{20}H_{25}N$; mol wt 279.41. C 85.97%, H 9.02%, N 5.01%. Prepn: Bockmühl, Ehrhart, *Ann.* **561**, 52 (1948); U.S. pat. **2,446,522** (1948 to Winthrop-Stearns); Ruddy, Buckley, *J. Am. Chem. Soc.* **72**, 718 (1950); Ruddy, Becker, U.S. pat. **2,662,886** (1953 to Winthrop-Stearns).

Crystals, mp 41-42.5°. bp_8 210-220°.
Hydrochloride, $C_{20}H_{25}N.HCl$, crystals, mp 216-217°.
Note: Active ingredient of *Aspasan.*
THERAP CAT: Antiallergic, antispasmodic.

3935. Fenpiverinium Bromide. *1-(4-Amino-4-oxo-3,3-diphenylbutyl)-1-methylpiperidinium bromide; 1-(3-carbam-oyl-3,3-diphenylpropyl)-1-methylpiperidinium bromide;* α,α-diphenyl-1-piperidinebutyramide methobromide; α,α-di-phenyl-γ-piperidylbutyramide methobromide; 2,2-diphenyl-2-(2-piperidinoethyl)acetamide methyl bromide; (diphenyl-piperidinoethyl)acetamide methyl bromide; fenpipramide methobromide; 12494 Hoechst; Resantin. $C_{22}H_{29}BrN_2O$; mol wt 417.41. C 63.30%, H 7.00%, Br 19.15%, N 6.71%, O 3.83%. Prepn: Brit. pat. **708,859** (1954 to Hoechst); Mof-fett, Aspergren, *J. Am. Chem. Soc.* **79**, 4451 (1957).

Crystals, mp 177.5-178.5° (dimorphic crystals from iso-propanol + ethyl acetate, mp 216-216.5°). Freely sol in water (water soln is neutral).

THERAP CAT: Antispasmodic.

3936. Fenpropathrin. *2,2,3,3-Tetramethylcyclopropane carboxylic acid cyano(3-phenoxyphenyl)methyl ester;* α-cy-ano-3-phenoxybenzyl 2,2,3,3-tetramethylcyclopropanecarb-oxylate; fenpropanate; S 3206; SD 41706; WL 41706; Dani-tol; Meothrin; Rody. $C_{22}H_{23}NO_3$; mol wt 349.43. C 75.62%,

H 6.63%, N 4.01%, O 13.74%. Synthetic pyrethroid insecticide with repellant and contact activity. Prepn: T. Matsuo *et al.*, **Ger. pat. 2,231,312**; *eidem*, **U.S. pat. 3,835,176** (1973, 1974 to Sumitomo). Commercial product is mixture of stereoisomers, the *(S)*-isomer primarily responsible for the bioactivity. Separation of enantiomers by chiral phase HPLC: R. A. Chapman, *J. Chromatog.* **258**, 175 (1983). Pesticidal activity: M. H. Breese, *Pestic. Sci.* **8**, 264 (1977). Metabolism: M. J. Crawford, D. H. Hutson, *ibid.* 579. Degradn in soil: T. R. Roberts, M. E. Standen, *ibid.* 600; R. A. Chapman, *Bull. Environ. Contam. Toxicol.* **26**, 513 (1981). Fish toxicity: J. R. Coats, *ibid.* **23**, 250 (1979). Mammalian toxicity study: R. D. Verschoyle, W. N. Aldridge, *Arch. Toxicol.* **45**, 325 (1980).

Pale yellow oil, n_D^{26} 1.5283. LC_{50} (24 hr) in rainbow trout: 76.7 ppb (Coats). LD_{50} in rats (mg/kg): 2.5 i.v. (Verschoyle); in male, female rats (mg/kg): 24-36, 18-24 orally (Crawford).

USE: Insecticide, acaricide.

3937. Fenpropidin. *1-[3-[4-(1,1-Dimethylethyl)phenyl]-2-methylpropyl]piperidine*; 1-[3-[4-(1,1-dimethylethyl)phenyl]-2-methylpropyl]piperidine; Patrol. $C_{19}H_{31}N$; mol wt 273.46. C 83.45%, H 11.43%, N 5.12%. Piperidine fungicide active against powdery mildews of grain. Prepn: **Belg. pat. 861,-002**; A. Pfiffner, **U.S. pat. 4,241,058** (1978, 1980 both to Hoffmann-La Roche). Improved process: N. Goetz, L. Hupfer, **Eur. pat. Appl. 17,893**; *eidem*, **U.S. pat. 4,283,534** (1980, 1981 both to BASF). Structure-activity relationships: W. Himmele, E.-H. Pommer, *Angew. Chem. Int. Ed.* **19**, 184 (1980). Inhibition of ergosterol biosynthesis: R. I. Baloch *et al.*, *Phytochemistry* **23**, 2219 (1984).

Oil, $bp_{0.2}$ 117°; $bp_{0.045}$ 125°; also reported as $bp_{0.032}$ 104°.
USE: Agricultural fungicide.

3938. Fenpropimorph. *4-[3-[4-(1,1-Dimethylethyl)-phenyl]-2-methylpropyl]-2,6-dimethylmorpholine*; cis-4-[3-(4-*tert*-butylphenyl)-2-methylpropyl]-2,6-dimethylmorpholine; BAS 42100F; Ro 14-3169/000; Corbel; Mistral. C_{20}-$H_{33}NO$; mol wt 303.49. C 79.15%, H 10.96%, N 4.62%, O 5.27%. Systemic fungicide for control of powdery mildew, rust in cereal crops. Prepn, fungicidal activity: W. Himmele *et al.*, **Ger. pat. 2,656,747** (1978 to BASF), *C.A.* **89**, 109522k (1978). Field trials in control of cereal diseases: J. C. Atkin *et al.*, *Proc. Brit. Crop Prot. Conf. - Pests Dis.* **1981**, 307. Inhibition of ergosterol biosynthesis: R. I. Baloch *et al.*, *Phytochemistry* **23**, 2219 (1984). Structure-activity relationships of 3-phenylpropylamines: W. Himmele, E.-H. Pommer, *Angew. Chem. Int. Ed.* **19**, 184 (1980); of substituted morpholines: E.-H. Pommer, *Pestic. Sci.* **15**, 285 (1984). Brief review: K. Bohnen *et al.*, *Proc. Brit. Crop Prot. Conf. - Pests Dis.* **1979**, 541-548.

Liquid, $bp_{0.05}$ 120°. Soly in water: 1 gm/l; sol in most organic solvents. Vapor pressure at 20°: 2.5×10^{-5} mm Hg. LD_{50} in male, female rats (mg/kg): 3650, 3420 orally; 4200,

4380 dermally; in male, female mice (mg/kg): 1180, 1270 i.p. (Bohnen).
USE: Systemic fungicide.

3939. Fenproporex. *3-[(1-Methyl-2-phenylethyl)amino]propanenitrile; 3-[(α-methylphenethyl)amino]propionitrile*; (±)-N-2-cyanoethylamphetamine. $C_{12}H_{16}N_2$; mol wt 188.27. C 76.55%, H 8.57%, N 14.88%. Deriv of amphetamine, *q.v.* Prepn: **Fr. pat. M4364** corresp to P. Pohrbach, J. Blum, **U.S. pat. 3,485,924** (1966, 1969 both to Bottu). Pharmacological studies: B. M. Beecham *et al.*, *J. Pharm. Pharmacol.* **23**, 140 (1971); A. H. Beckett *et al.*, *ibid.* **24**, 194 (1972). Peripheral effects in human and rat adipose tissue: M. Dubost *et al.*, *Brit. J. Pharmacol.* **58**, 436P (1976). Chromatographic identification of amphetamine in urine of patients treated with fenproporex: R. B. Sznelvar, *Eur. J. Toxicol. Environ. Hyg.* **8**, 5 (1975). Clinical trial: G. Hertel, W. Fallot-Burghardt, *Fortschr. Med.* **96**, 2380 (1978).

Liquid, bp_2 126-127°.
Hydrochloride, $C_{12}H_{17}ClN_2$, Gacilin, Solvolip. White, cryst, odorless powder from abs ethanol, mp 146°. Bitter taste. Sol in water, 95% ethanol.
Diphenyl acetate, $C_{26}H_{28}N_2O_2$, Fenproporex Retard Bottu.
THERAP CAT: Anorexic.

3940. Fenprostalene. *7-[3,5-Dihydroxy-2-(3-hydroxy-4-phenoxy-1-butenyl)cyclopentyl]-4,5-heptadienoic acid methyl ester*; (±)-9α,11α,15α-trihydroxy-16-phenoxy-17,-18,19,20-tetranorprosta-4,5,13-*trans*-trienoic acid methyl ester; RS-84043; Bovilene; Synchrocept B. $C_{23}H_{30}O_6$; mol wt 402.49. C 68.64%, H 7.51%, O 23.85%. Synthetic analog of prostaglandin $F_{2\alpha}$, related structurally to prostalene, *q.v.* Prepn: J. M. Muchowski, J. H. Fried, **U.S. pat. 3,985,791**; A. R. Van Horn *et al.*, **U.S. pat. 4,178,457** (1976, 1979 both to Syntex). Effect on pregnancy in beagles: B. Vickery, G. Mc Rae, *Biol. Reprod.* **22**, 438 (1980). Duration of action study: B. H. Vickery *et al.*, *Prostaglandins Med.* **5**, 93 (1980).

uv max (methanol): 220, 265, 271, 278 nm (log ε 3.99, 3.11, 3.23, 3.16).
THERAP CAT (VET): Luteolysin.

3941. Fenquizone. *7-Chloro-1,2,3,4-tetrahydro-4-oxo-2-phenyl-6-quinazolinesulfonamide; 7-chloro-2-phenyl-6-sulfamyl-1,2,3,4-tetrahydro-4-quinazolinone*; MG 13054. $C_{14}H_{12}ClN_3O_3S$; mol wt 337.78. C 49.78%, H 3.58%, Cl 10.50%, N 12.44%, O 14.21%, S 9.49%. Thiazide-like diuretic with saluretic and hypocalciuric activity. Prepn: M. G. Biressi *et al.*, *Farmaco Ed. Sci.* **24**, 199 (1969). Prepn of potassium salt: M. Carissimi, F. Ravenna, **Belg. pat. 799,-087** (1973 to Maggioni); *eidem*, **U.S. pat. 3,870,720** (1975). Gas-liquid chromatography: A. Marzo *et al.*, *J. Chromatog.* **272**, 95 (1983). Metabolism in dogs: D. Ceriani *et al.*, *Farmaco Ed. Sci.* **31**, 379 (1976). Human pharmacokinetics: G. C. Maggi *et al.*, *Arzneimittel-Forsch.* **35**, 994 (1985). Site of action: C. Ferrando *et al.*, *J. Pharm. Pharmacol.* **33**, 219 (1981). Clinical trials: O. Biadi *et al.*, *Drugs Exp. Clin. Res.* **7**, 763 (1981); N. Glorioso *et al.*, *Curr. Ther. Res.* **35**, 483 (1984); A. Cupisti *et al.*, *ibid.* **38**, 293 (1985).

mp > 310°. Insol in water.
Monopotassium salt, $C_{14}H_{12}ClKN_3O_3S$, *Idrolone*. White crystalline powder. Sol in water. pH ~9.9 (1% aqueous soln).

THERAP CAT: Diuretic.

3942. Fenspiride. *8-(2-Phenylethyl)-1-oxa-3,8-diaza-spiro[4.5]decan-2-one;* decaspiride; DESP. $C_{15}H_{20}N_2O_2$; mol wt 260.33. C 69.20%, H 7.74%, N 10.76%, O 12.29%. Prepn: **Neth. pat. Appl. 6,504,602** corresp to Regnier *et al.*, **U.S. pat. 3,399,192** (1965 and 1968, both to Sci. Union et Cie-Soc. Franc. Recherche Méd.). Pharmacology: LeDouarec *et al.*, *Arzneimittel-Forsch.* **19**, 1263 (1969); Duhault *et al.*, *ibid.* **22**, 1947 (1972).

Hydrochloride, $C_{15}H_{21}ClN_2O_2$, *NAT-333, NDR-5998A, Decaspir, Espiran, Fluiden, Pneumorel, Respiride, Tegencia, Viarespan*. Crystals decomp 232-233°. Soluble in water. LD_{50} i.v. in mice: 106 mg/kg; orally in rats: 437 mg/kg (LeDouarec).

THERAP CAT: Bronchodilator, α-adrenergic blocker.

3943. Fensulfothion. *Phosphorothioic acid O,O-diethyl O-[4-(methylsulfinyl)phenyl] ester; O,O-diethyl O-[p-(methylsulfinyl)phenyl] phosphorothioate;* BAY 25/141; Dasanit; Terracur P. $C_{11}H_{17}O_4PS_2$; mol wt 308.35. C 42.85%, H 5.56%, O 20.75%, P 10.04%, S 20.79%. Prepn: Homeyer, Schrader, **Belg. pat. 666,012** (1965 to Bayer), *C.A.* **64**, 20555f (1966).

Liquid. $b_{0.01}$ 138-141°. LD_{50} orally in rats: 5 mg/kg. *Note:* For application 1 part of fensulfothion may be mixed with 0.5 part xylene, 0.6 part emulsifier added and the mixture diluted with water. An effective concn of 10 to 40 ppm is claimed.

USE: Nematocide, insecticide; esp for the control of nematodes in golf courses and cemeteries. A 10% granular formulation is available for the control of onion maggots.

3944. Fentanyl. *N-Phenyl-N-[1-(2-phenylethyl)-4-piperidinyl]propanamide; N-(1-phenethyl-4-piperidyl)propionanilide; N-(1-phenethyl-4-piperidinyl)-N-phenylpropionamide;* R 4263; Leptanal. $C_{22}H_{28}N_2O$; mol wt 336.46. C 78.53%, H 8.39%, N 8.33%, O 4.76%. Prepn: Janssen, Gardocki, **U.S. pat. 3,141,823** (1964 to Janssen). Pharmacology: Gardocki, Yelnosky, *Toxicol. Appl. Pharmacol.* **6**, 48 (1964); Hess *et al.*, *J. Pharmacol. Exp. Ther.* **179**, 474 (1971). Effects on cerebral circulation and metabolism in rats: C. Carlsson *et al.*, *Anesthesiology* **57**, 375 (1982). Clinical studies: E. A. Welchew, J. A. Thornton, *Anaesthesia* **37**, 309 (1982); M. J. Stephens *et al.*, *Med. J. Aust.* **1**, 419 (1982).

Crystals, mp 83-84°.
Citrate, $C_{28}H_{36}N_2O_8$, *phentanyl, Fentanest, Pentanyl, Sublimaze*. Crystalline powder, mp 149-151°. Bitter taste. One gram dissolves in about 40 ml water. Sol in methanol, sparingly sol in chloroform. LD_{50} in mice (mg/kg): 11.2 i.v.; 62 s.c. (Gardocki, Yelnosky).
Caution: May be habit forming. This is a controlled substance (opiate) listed in the U.S. Code of Federal Regulations, Title 21 Part 1308.12 (1987).

THERAP CAT: Narcotic analgesic.
THERAP CAT (VET): The citrate as analgesic, tranquilizer.

3945. Fenthion. *Phosphorothioic acid O,O-dimethyl O-[3-methyl-4-(methylthio)phenyl] ester; O,O-*dimethyl *O-*(4-methylmercapto-3-methylphenyl) thionophosphate; *O,O-*dimethyl *O-*(3-methyl-4-methylthiophenyl) thiophosphate; *O,O-*dimethyl *O-*(4-methylthio-3-methylphenyl) thiophosphate; *O,O-*dimethyl *O-*[4-(methylthio)-*m*-tolyl] phosphorothioate; 4-methylmercapto-3-methylphenyl dimethyl thiophosphate; Bayer 29493; ENT 25540; S 1752; Baycid; Baytex; Entex; Lebaycid; Mercaptophos; Queletox; Spotton; Talodex; Tiguvon. $C_{10}H_{15}O_3PS_2$; mol wt 278.34. C 43.15%, H 5.43%, O 17.25%, P 11.13%, S 23.04%. Prepn: Schrader, *Die Entwicklung neuer insektizider Phosphorsäure Ester* (Verlag Chemie, 3rd ed., 1963) pp 298-313; Schlegk, Schrader, **Ger. pat. 1,116,656**, *C.A.* **56**, 14170g (1962); and **U.S. pat. 3,042,703** (1961 and 1962, both to Bayer); Nishizawa, Mizutani, **Japan. pat. 15,130('64)** (to Sumitomo), *C.A.* **61**, 16016h (1964). Properties: Jung, *Bull. World Health Org.* **21**, 215 (1959); Schrader, *loc. cit.*

Liquid, slight odor of garlic. $bp_{0.01}$ 87° (commercial product, $bp_{0.01}$ 105°). d_4^{20} 1.250. n_D^{20} 1.5698. Vapor press at 20°: 3×10^{-5} mm Hg. Thermally stable up to 210°; resistant to alkalies up to pH 9. Readily sol in methanol, ethanol, ether, acetone and many other organic solvents, esp chlorinated hydrocarbons. Practically insol in water (55 mg/l). LD_{50} orally in male, female rats: 215, 245 mg/kg, Gaines, *Toxicol. Appl. Pharmacol.* **2**, 88 (1960).

USE: Insecticide; acaricide. *Human Toxicity:* A cholinesterase inhibitor. *See* Parathion.
THERAP CAT (VET): Ectoparasiticide.

3946. Fentiazac. *4-(4-Chlorophenyl)-2-phenyl-5-thiazoleacetic acid;* BR 700; CH 800; Donorest; Flogene; Norvedan. $C_{17}H_{12}ClNO_2S$; mol wt 329.81. C 61.91%, H 3.67%, Cl 10.75%, N 4.25%, O 9.70%, S 9.72%. Prepn: K. Brown, **Brit. pat. 1,145,884** corresp to **U.S. pat. 3,476,766** (both 1969 to Wyeth). Anti-inflammatory activity: K. Brown *et al.*, *Nature* **219**, 164 (1968). Crystallographic study: R. Destro, *Acta Crystallogr.* **B 34**, 959 (1978). Absorption: G. Zanolo *et al.*, *Boll. Chim. Farm.* **119**, 209 (1980). Metabolism: S. Fumero *et al.*, *Arzneimittel-Forsch.* **30**, 1253 (1980). Pharmacokinetics in humans: M. Quattrini *et al.*, *ibid.* **31**, 1046 (1981). Toxicity study: D. A. Shriver *et al.*, *Toxicol. Appl. Pharmacol.* **42**, 75 (1977). Review of exptl and clinical pharmacology: E. Marmo, *Curr. Med. Res. Opin.* **6**, 53-63 (1979).

Colorless needles from benzene, mp 161-162°. LD_{50} in rats, mice (mg/kg): 661, 692 orally, E. Marmo, *loc. cit.*
THERAP CAT: Anti-inflammatory.

3947. Fenticlor. *2,2'-Thiobis[4-chlorophenol];* bis[2-hydroxy-5-chlorophenyl]sulfide; 2,2'-dihydroxy-5,5'-dichlorodiphenyl sulfide; S7; D 25-Antimykotikum; Novex. $C_{12}H_8Cl_2O_2S$; mol wt 287.18. C 50.19%, H 2.81%, Cl 24.69%, O 11.14%, S 11.17%. Prepn by chlorination of bis-[2-hydroxyphenyl]sulfide: Muth, **Ger. pat. 568,944** (1931 to I. G. Farben); Dunning *et al.*, *J. Am. Chem. Soc.* **53**, 3466 (1931).

Fine needles from toluene, mp 175°. Sol in aq solns of NaOH, in alcohol, in hot benzene.

Note: Novex is also a trade name for a thermoplastic resin.

USE: Fungicide; esp used against *Monosporium apiospermum:* Reifferscheid, Seeliger, *Deut. Med. Wochenschr.* **80,** 1841, 1850 (1955).

THERAP CAT: Anti-infective.

3948. Fenticonazole. *1-[2-(2,4-Dichlorophenyl)-2-[[4-(phenylthio)phenyl]methoxy]ethyl]-1H-imidazole;* 1-[2,4-dichloro-β-[[p-(phenylthio)benzyl]oxy]phenethyl]imidazole; 1-(2,4-dichlorophenyl)-2-(N-imidazolyl)ethyl-4-phenylthiobenzyl ether; 2,4-dichloro-4'-phenylthio-(N-imidazolylmethyl)dibenzyl ether. $C_{24}H_{20}Cl_2N_2OS$; mol wt 455.40. C 63.30%, H 4.43%, Cl 15.57%, N 6.15%, O 3.51%, S 7.04%. Broad spectrum antimycotic; also active as antibacterial. Prepn: D. Nardi *et al.,* Ger. pat. **2,917,244;** *eidem,* U.S. pat. **4,221,803** (1979, 1980 both to Recordati); *eidem, Arzneimittel-Forsch.* **31,** 2123 (1981). Series of articles on physico-chemical properties, *in vitro* antibacterial and antifungal activity, pharmacology and toxicology: *ibid.* 2127-2154. Toxicity data: G. Graziani *et al., ibid.* 2145. Preliminary clinical comparison with miconazole in vaginal candidiasis: A. Gastaldi, *Curr. Ther. Res.* **38,** 489 (1985). Comparative clinical trials in dermatomycoses: A. Finzi *et al., Mykosen* **29,** 41 (1986); E. M. Kokoschka *et al., ibid.* 45.

Mononitrate, $C_{24}H_{21}Cl_2N_3O_4S$, *Rec 15/1476, Falvin, Lomexin.* Odorless, white crystalline powder, mp 136°. uv max (methanol): 252 nm (ε = 13894). Soly at 20° (mg/ml): water < 0.1; ethyl ether < 0.1; ethanol 30; methanol 100; chloroform 300; DMF 600. pKa 6.54. LD_{50} in mice, male rats, female rats (mg/kg): 1191, 440, 309 i.p.; all > 3000 orally (Graziani).

THERAP CAT: Topical antifungal.

3949. Fentonium Bromide. *[3(S)-endo]-8-(2-[1,1'-Biphenyl]-4-yl-2-oxoethyl)-3-(3-hydroxy-1-oxo-2-phenylpropoxy)-8-methyl-8-azoniabicyclo[3.2.1]octane bromide; 3α-hydroxy-8-(p-phenylphenacyl)-1αH,5αH-tropanium bromide* (−)-*tropate;* N-(4-phenylphenacyl)-1-hyoscyaminium bromide; phentonium bromide; Fa 402; Z 326; Ketoscilium; Ulcesium; $C_{31}H_{34}BrNO_4$; mol wt 564.53. C 65.96%, H 6.07%, Br 14.15%, N 2.48%, O 11.34%. Prepn: U. Teotino, D. Della Bella, **Brit.** pat. **1,026,640;** *eidem,* U.S. pat. **3,356,-682** (1966, 1967 both to Whitefin Holding); Teotino *et al., Chim. Ther.* **3,** 453 (1968). Human pharmacology: Azzollini *et al., Curr. Ther. Res.* **12,** 734 (1970). Clinical studies: Alberto *et al., Minerva Med.* **62,** 852 (1971).

Crystals, mp 203-205° (dec). $[\alpha]_D^{23}$ −5.68° (c = 5 in DMF). Also reported as mp 193-194°. $[\alpha]_D^{25}$ −4.7° (c = 5 in DMF). LD_{50} in mice (mg/kg): 12.1 i.v.; > 400 s.c. and orally (Teotino, Della Bella).

THERAP CAT: Anticholinergic; antispasmodic.

3950. Fenugreek. Greek Hay. *Trigonella foenumgraecum* L., *Leguminosae,* an annual herb, cultivated in Southern Europe, Northern Africa and India for its seeds which are used in making curry. The seeds contain the alkaloid trigonelline *(q.v.)* and the steroidal sapogenin diosgenin *(q.v.):* Fazli, Hardman, *Trop. Sci.* **10,** 66-78 (1968).

USE: Has been used in making imitation maple syrup and for culinary spices other than curry.

THERAP CAT (VET): Emollient, flavoring agent.

3951. Fenuron. *N,N-Dimethyl-N-phenylurea;* N-phenyl-N',N'-dimethylurea; Dybar. $C_9H_{12}N_2O$; mol wt 164.20. C 65.83%, H 7.37%, N 17.06%, O 9.74%. $C_6H_5NHCON-(CH_3)_2$. Prepn: Crosby, Niemann, *J. Am. Chem. Soc.* **76,** 4458 (1954); Gilbert, Sorma, U.S. pat. **2,729,677** (1956 to Allied Chem.); Jones, U.S. pat. **2,768,971** (1956 to ICI); Applegath *et al.,* U.S. pat. **2,857,430** (1958 to Monsanto). Prepn of the trichloroacetate: Gilbert *et al.,* U.S. pat. **2,-782,112** (1957 to Allied Chem.). Toxicity data: G. W. Bailey, J. L. White, *Residue Rev.* **10,** 97 (1965).

Crystals, mp 131-133°. Sparingly sol in water (0.29% at 24°); in hydrocarbons. LD_{50} orally in rats: 7500 mg/kg (Bailey, White).

Trichloroacetate, $C_{10}H_{14}Cl_3N_2O_3$, *Urob, fenuron TCA.* mp 65-68°. Esp effective in woody plant control.

Caution: If hydrolyzed to aniline can cause methemoglobinemia: *Clinical Toxicology of Commercial Products,* R. E. Gosselin *et al.,* Eds. (Williams & Wilkins, Baltimore, 5th ed., 1984) Section II, p 331.

USE: Herbicide.

3952. Fenvalerate. *4-Chloro-α-(1-methylethyl)benzene-acetic acid cyano(3-phenoxyphenyl)methyl ester;* α-cyano-3-phenoxybenzyl α-(4-chlorophenyl)isovalerate; cyano(3-phenoxyphenyl)methyl 4-chloro-α-(1-methylethyl)benzeneacetate; α-cyano-3-phenoxybenzyl-2-(4-chlorophenyl)-3-methylbutyrate; phenvalerate; S 5602; SD 43775; WL 43775; Belmark; Pydrin; Pyridin; Sumicidin; Tirade. $C_{25}H_{22}ClNO_3$; mol wt 419.92. C 71.51%, H 5.28%, Cl 8.44%, N 3.34%, O 11.43%. A synthetic pyrethroid insecticide without the usual cyclopropane ring. Prepn: **Ger.** pat. **2,335,347;** K. Fujimoto *et al.,* U.S. pat. **3,996,244** (1974, 1976, both to Sumitomo); D. A. Wood, **Ger.** pat. **2,651,341;** *idem,* U.S. pat. **4,061,664** (1977, both to Shell). Has two chiral centers giving four possible optical isomers; prepn and comparative insecticidal activity of isomers: M. Hirano *et al.,* **Ger.** pat. **2,737,297;** *eidem,* U.S. pat. **4,503,071** (1978, 1985 both to Sumitomo). Absolute configuration of most active isomer, esfenvalerate: K. Aketa *et al., Agr. Biol. Chem.* **42,** 895 (1978). GC separation of isomers: G. R. Cayley, B. W. Simpson, *J. Chromatog.* **356,** 123 (1986); determn of esfenvalerate in technical prepn: S. Sakaue *et al., Agr. Biol. Chem.* **51,** 1671 (1987). Comprehensive description of activity and physical properties of esfenvalerate: H. Oo'uchi, *Japan. Pest. Info.* **1985,** 21-24. Comparative soil metabolism of isomers: P.W. Lee *et al., J. Agr. Food Chem.* **35,** 384 (1987).

Clear yellow viscous liquid at 23°. d^{23} 1.17. n_D^{20} 1.5533. Vapor press at 25° = 1.1 × 10⁻⁸ mm Hg. Soly at 20° (g/l): acetone, > 450; chloroform, > 450; methanol, > 450; hexane, 77. Insol in water. More stable in acidic soln than in alk soln. Dec gradually between 150-300°. No significant breakdown after 100 hrs at 75°. LD_{50} orally in rats: 451 mg/kg (DMSO); > 3200 mg/kg (aqueous suspension), Shell Technical Data Bulletin. Highly toxic to fish and bees.

(S,S)-Isomer, *esfenvalerate, Sumi-alpha,* Aα, OMS 3023, S-1844, S-5602α. White crystalline solid, mp 59-60.2°. $[\alpha]_D^{25}$ −15.0° (c = 2.0 in CH_3OH). d_{23}^{23} 1.163. Vapor pressure at 20°: 2.63 × 10⁻⁷ mm Hg; at 25°: 5.00 × 10⁻⁷ mm Hg. Soly (1%): acetonitrile > 60, chloroform > 60, DMF > 60, DMSO > 60, ethyl acetate > 60, acetone > 60, ethyl cellosolve 40-50, n-hexane 1-5, kerosene < 1, methanol 7-10, α-methylnaphthalene 50-60, xylene > 60.

Caution: Eye, skin irritant.

USE: Insecticide.

THERAP CAT (VET): Ectoparasiticide.

Feprazone

3953. Feprazone. *4-(3-Methyl-2-butenyl)-1,2-diphenyl-3,5-pyrazolidinedione;* 4-prenyl-1,2-diphenyl-3,5-pyrazolidinedione; 4-prenyl-1,2-diphenyl-3,5-dioxopyrazolidine; 4-(β-isoamylenyl)-1,2-diphenyl-3,5-pyrazolidinedione; 4-(2-isopentenyl)-1,2-diphenyl-3,5-pyrazolidinedione; phenylprenazone; prenazone; DA 2370; Analud; Methrazone; Zepelin. $C_{20}H_{20}N_2O_2$; mol wt 320.39. C 74.98%, H 6.29%, N 8.74%, O 9.99%. Non-steroidal anti-inflammatory drug. Prepn: Casadio, Pala, **Ger. pat.** 2,031,238 corresp to U.S. pat. 3,703,528 (1971, 1972 both to Ist. De Angeli); Casadio *et al.*, *Arzneimittel-Forsch.* **22**, 171 (1972). Series of articles on physical-chemical data, pharmacology, toxicology and clinical studies: *ibid.* 174-281. Toxicity data: C. Bianchi, G. Bonardi, *ibid.* 196.

Fine, white odorless and tasteless crystalline powder, mp 156.5° (ethanol). Very sol in acetone, chloroform, DMF; sparingly sol in acetonitrile, benzene, 10% NaOH. Slightly sol in ether, methanol, ethanol, cyclohexane. Practically insol in 10% HCl, 10% acetic acid, water. uv max (ethanol): 246 nm (log ϵ 4.19); (pH 9-12 buffer): 264 nm (log ϵ 4.322-4.326). pKa 5.09 $\pm$ 0.07. LD_{50} in male mice, rats (mg/kg): 408.8, 386.4 i.p.; 1067, >2000 orally (Bianchi, Bonardi).
THERAP CAT: Anti-inflammatory.

3954. Ferbam. *Tris(dimethylcarbamodithioato-S,S')-iron; tris(dimethyldithiocarbamato)iron;* dimethyldithiocarbamic acid, iron salt; ferric dimethyldithiocarbamate; Carbamate; Ferbeck; Fermate; Ferradow; Karbam Black. $C_9H_{18}FeN_3S_6$; mol wt 416.50. C 25.95%, H 4.36%, Fe 13.41%, N 10.09%, S 46.19%. $[(CH_3)_2NCS_2]_3Fe$. Prepd from Na dimethyldithiocarbamate and a soln of a ferric salt: Tisdale, Williams, **U.S. pat.** 1,972,961 (1934 to du Pont). Toxicity data: H. C. Hodge *et al.*, *J. Am. Pharm. Assoc.* **41**, 662 (1952).
Black solid. Melts with decompn above 180°. Solubility in water 120 ppm, pH of soln 5.0; sol in acetone, chloroform, pyridine, acetonitrile. Tends to decompose on prolonged storage or exposure to heat and moisture. LD_{50} in male mice, rats (mg/kg): 3000, 2700 i.p. (Hodge).
Caution: May cause irritation of skin and mucous membranes: *Clinical Toxicology of Commercial Products,* R. E. Gosselin *et al.,* Eds. (Williams & Wilkins, Baltimore, 5th ed., 1984) Section II, p 311.
USE: Fungicide.

3955. Fermium. Fm; at. wt (most stable known isotope) 257; at. no. 100; valence 3, also 2. Man-made, radioactive element. ^{255}Fm ($T_{1/2}$ 20.1 hrs, α-emitter) originally discovered in debris from a thermonuclear explosion in Nov. 1952. Reviews of discovery: Ghiorso *et al.*, *Phys. Rev.* **99**, 1048-1049 (1955); Ghiorso, Seaborg, *Sci. Am.* **195**(6), 67-80 (1956). Other known isotopes: 244-254; 256-258. ^{257}Fm ($T_{1/2}$ 80 days, α-emitter), produced by neutron bombardment of plutonium (or other elements with higher at. no.). Reviews: Seaborg, *J. Chem. Ed.* **36**, 38-44 (1959); C. Keller, *The Chemistry of the Transuranium Elements* (Verlag Chemie, Weinheim, English Ed., 1971) pp 591-594; Silva, "Trans-Curium Elements" in *MTP Int. Rev. Sci.: Inorg. Chem., Ser. One* vol. **8**, A. G. Maddock, Ed. (University Park Press, Baltimore, 1972) pp 71-105; *Comprehensive Inorganic Chemistry* vol. **5**, J. C. Bailar, Jr. *et al.*, Eds. (Pergamon Press, Oxford, 1973) passim; several authors, *Handb. Exp. Pharmakol.* **36**, 689-738 (1973).

3956. Ferredoxins. A group of electron transfer factors found in plants and bacteria, which are non-heme iron-sulfur proteins and which play an important role in photosynthesis, nitrogen and carbon dioxide fixation, and respiration. They are generally classified by the presence of either 2 or 4 iron atom clusters and an equivalent amount of inorganic or "acid-labile" sulfide bonded to the peptide chain through 4 cysteine sulfhydryl groups. The two-iron ferredoxins are found primarly in plants and blue-green algae and are sometimes referred to as chloroplast or "plant type" ferredoxins; the four-iron ferredoxins are predominant in bacteria. Mol wts of chloroplast ferredoxins are about 12,000; those of bacterial ferredoxins range from about 6,000 to 24,000. Isoln from *Clostridium pasteurianum:* L. E. Mortenson *et al.*, *Biochem. Biophys. Res. Commun.* **7**, 448 (1962). Prepn of bacterial ferredoxins: L. E. Mortenson, **U.S. pat.** 3,344,130 (1960 to duPont). The amino acid sequence of several bacterial and chloroplast ferredoxins has been elucidated. Amino acid sequence of ferredoxin from *Clostridium pasteurianum:* Tanaka *et al.*, *Biochem. Biophys. Res. Commun.* **16**, 422 (1964); *eidem*, *Biochemistry* **5**, 1666 (1966); from spinach: Matsubara, Sasaki, *J. Biol. Chem.* **243**, 1732 (1968). Synthesis of peptide chain of *C. pasteurianum* ferredoxins: Bayer *et al.*, *Tetrahedron* **24**, 4853 (1968); Trakatellis, Schwartz, *Proc. Nat. Acad. Sci. USA* **63**, 436 (1969). *Reviews:* Buchanan, *Struct. Bonding* **1**, 109 (1966); Malkin, Rabinowitz, *Ann. Rev. Biochem.* **36**, 113 (1967); Arnon, *Naturwiss.* **56**, 295 (1969); Buchanan, Arnon, *Advan. Enzymol.* **33**, 119 (1970); W. Lovenberg, "Ferredoxin and Rubredoxin" in *Microbial Iron Metabolism,* J. B. Neilands, Ed. (Academic Press, New York, 1974) pp 161-182; D. C. Yoch, R. P. Carithers, *Microbiol. Rev.* **43**, 384-421 (1979).
Absorption max of bacterial ferredoxins: 280, 385-400 nm; of plant ferredoxins: 280, 325, 420, 463 nm. All show negative oxidation-reduction potentials near that of the hydrogen electrode: −0.390 V *(C. pasteurianum);* −0.420 V (spinach): Tagawa, Arnon, *Biochim. Biophys. Acta* **153**, 602 (1968). Autoxidizable. Acidification, as well as treatment with iron-chelating agents or mercurials, results in evolution of H_2S and loss of the visible absorption.

3957. Ferric Acetate, Basic. $C_4H_7FeO_5$; mol wt 190.95. C 25.16%, H 3.70%, Fe 29.25%, O 41.90%. $Fe(OH)(CH_3COO)_2$. Prepd by heating ferric acetate soln: Casey, Doyle in Kirk-Othmer *Encyclopedia of Chemical Technology,* vol. **12** (Interscience, New York, 2nd ed., 1967) p 24.
Brownish-red scales or amorphous powder; faint acetic odor. Practically insol in water; sol in alcohol, acids. *Protect from light.*
USE: In the textile industry as a mordant in dyeing and printing, and for the weighting of silk and felt; as wood preservative; in leather dyes; as medicine.

3958. Ferric Albuminate. Albumized iron. Combinations of egg albumin and iron contg 17-19% Fe. The exact form in which iron exists in these compds is not known and probably varies with the method of manuf. Structure and composition of compds of Fe and albumin: Heymann, Oppenheimer, *Biochem. Z.* **199**, 468 (1928).
Reddish-brown, lustrous granules, or brown powder. Odorless. Freely sol in water; almost insol in alcohol.
THERAP CAT: Hematinic.

3959. Ferric and Ammonium Acetate Solution. Basham's mixture. Contains 0.16-0.20% Fe and about 3.5% ammonium acetate w/v. Prepn: *U.S.D.* 25th ed, p 722.
Clear, reddish-brown liq, having an aromatic odor. Acid to litmus.
THERAP CAT: Hematinic.
THERAP CAT (VET): In iron deficiency.

3960. Ferric Bromide. Br_3Fe; mol wt 295.60. Br 81.11%, Fe 18.89%. $FeBr_3$. Prepn from Fe and Br_2: Gregory, Thackrey, *J. Am. Chem. Soc.* **72**, 3176 (1950); Gregory, *ibid.* **73**, 472 (1951); from $Fe_2(SO_4)_3$ and LiBr: Lieser, Elias, *Z. Anorg. Allgem. Chem.* **316**, 208 (1962).
Dark-red or black hexagonal, rhombic plates. Very hygroscopic. Loses some Br_2 on exposure to air and light. Sol in water, alcohol, ether, acetic acid; slightly sol in liq NH_3; aq soln dec to $FeBr_2 + Br_2$ on boiling. *Keep cool, well-closed and protected from light.*
USE: Catalyst for organic reactions, particularly in bromination of aromatic compds. *Caution:* Irritant.

3961. Ferric Chloride. Flores martis. Cl_3Fe; mol wt 162.22. Cl 65.57%, Fe 34.43%. $FeCl_3$. Occurs in nature as the mineral *molysite.* Prepn: Tarr, *Inorg. Syn.* **3**, 191 (1950); Pray, *ibid.* **5**, 153 (1957); Epperson *et al.*, *ibid.* **7**, 163 (1963);

Consult the cross index before using this section.

Lieser, Elias, *Z. Anorg. Allgem. Chem.* **316,** 208 (1962); Attwood, Shelton, *J. Inorg. Nucl. Chem.* **26,** 1758 (1964); Bardawil *et al., Inorg. Chem.* **3,** 149 (1964). Acute toxicity: C. S. Hosking, *Aust. Paediatr. J.* **6,** 92 (1970). Use as clinical reagent: L. Cassidei *et al., Clin. Chim. Acta* **90,** 121 (1978).

Hexagonal, dark leaflets or plates. Red by transmitted light, green by reflected light; sometimes appears brownish-black. Very hygroscopic. Melts and volatilizes about 300°; bp about 316°; d^{25} 2.90. Vapor density measurements show that it is dimeric at about 400° but monomeric above 750°. Dissociates at high temps to $FeCl_2$ and Cl_2. Readily absorbs water in air to form the hexahydrate. Readily sol in water, alcohol, ether, acetone; slightly sol in CS_2. Practically insol in ethyl acetate. *Keep well closed.*

Hexahydrate, brownish-yellow or orange monoclinic crystals. d 1.82. Structure: *trans*-$[FeCl_2(H_2O)_4]Cl.2H_2O$: Lind, *J. Chem. Phys.* **47,** 990 (1967). Usually slight odor of HCl; very hygroscopic. mp about 37°. Readily sol in water, alcohol, acetone, ether; pH of 0.1 molar aq soln 2.0. *Keep well closed.* LD_{50} i.v. in mice: 0.049 mg Fe/g (Hosking). *Human Toxicity:* Anhydr form is irritant.

USE: In photoengraving; photography; manuf of other Fe salts, pigments, ink; as catalyst in organic reactions; purifying factory effluents and deodorizing sewage; chlorination of Ag and Cu ores; as mordant in dyeing and printing textiles; oxidizing agent in dye manuf. Clinical reagent (amino acids in urine, esp in phenylketonuria).

THERAP CAT: Hexahydrate as astringent, styptic.

THERAP CAT (VET): Styptic, astringent.

3962. Ferric Chromate(VI). *Chromic acid iron(3+) salt (3:2);* C.I. Pigment Yellow 45; C.I. 77505; Siderin yellow. $Cr_3Fe_2O_{12}$; mol wt 459.73. Cr 33.94%, Fe 24.30%, O 41.76%. $Fe_2(CrO_4)_3$. Preparation: Boericke, Bangert, *U.S. Bur. Mines, Rept. Investigations* **3813,** 19 pp (1945), *C.A.* **39,** 5164[7] (1945).

Yellow powder. Practically insol in water; sol in HCl.

USE: Pigment for ceramics, glass and enamels.

3963. Ferric Citrate. A combination of iron and citric acid of indefinite composition. Prepn: Belloni, *Gazz. Chim. Ital.* **50,** II, 159 (1920).

Garnet-red, transparent scales or pale-brown powder. Odorless; slight ferruginous taste. Slowly but completely sol in cold water, readily sol in hot water, but diminishing in soly with age. Practically insol in alc.

THERAP CAT: Hematinic.

THERAP CAT (VET): In iron deficiency.

3964. Ferric Ferrocyanide. *Hexakis(cyano-C)ferrate(4-) iron(3+) (3:4);* Ferric hexacyanoferrate(II); C.I. Pigment Blue 27; C.I. 77510; Berlin blue; Chinese blue; Hamburg blue; mineral blue; Paris blue; Prussian blue. $C_{18}Fe_7N_{18}$; mol wt 859.29. C 25.16%, Fe 45.50%, N 29.34%. $Fe_4[Fe(CN)_6]_3$. Usually contains some water and some alkali ferrocyanide. Prepn and list of refs: *Colour Index* vol. 4 (3rd ed, 1971) p 4673.

Dark-blue powder or lumps. Loses all of its water only at 250° with partial decompn. d 1.80. Practically insol in water, dil acids, most organic solvents; when freshly prepd is sol in aq oxalic acid.

USE: As pigment in printing inks, paints, alkyd resin enamels, linoleum, leathercloth, carbon papers, typewriter ribbons, rubbers, plastics, artists' colors; in removal of H_2S from gases.

3965. Ferric Fluoride. F_3Fe; mol wt 112.85. F 50.51%, Fe 49.49%. FeF_3. Prepn from $FeCl_3$ and HF: Kwasnik in *Handbook of Preparative Inorganic Chemistry* vol. 1, G. Brauer, Ed. (Academic Press, New York, 2nd ed., 1963) p 266; Shinn *et al., Inorg. Chem.* **5,** 1927 (1966); from Fe_2O_3 or Fe_2S_3 and SF_4: Oppegard *et al., J. Am. Chem. Soc.* **82,** 3835 (1960); from $FeCl_3$ and CH_3CHF_2: Natta *et al.,* **Brit. pat. 995,186** (1965 to Montecatini).

Green hexagonal crystals. Sublimes > 1000°. d 3.87. Very slightly sol in water; freely sol in dil HF; practically insol in alcohol, ether, benzene. Soly in liq HF: 0.008 g/100 g HF, Jache, Cady, *J. Phys. Chem.* **56,** 1106 (1952).

USE: As catalyst in organic reactions.

3966. Ferric Formate. $C_3H_3FeO_6$; mol wt 190.90. C 18.87%, H 1.58%, Fe 29.26%, O 50.29%. $Fe(HCOO)_3$.

Prepn: Weinland, Reihlin, *Ber.* **46,** 3144 (1913); Starke, *J. Inorg. Nucl. Chem.* **25,** 823 (1963). May contain 1 or 2 H_2O.

Red or yellow microcryst powder. Readily hydrolyzed to basic formates. Sol in water; very slightly sol in alcohol. *Protect from light.*

USE: Preservation of silage.

3967. Ferric Fructose. D-*Fructose iron(3+)-contg complex, potassium salt (2:1);* CB 302; Ferritose. Tentative formula where $n = 2$ to 200: $(C_6H_{10}FeO_7)_n K_{n/2}$. Prepn: Stitt *et al., Proc. Soc. Exp. Biol. Med.* **110,** 70 (1962); Saltman, Charley, **U.S. pats. 3,074,927** and **3,275,514** (1966).

THERAP CAT: Hematinic.

3968. Ferrichromes. Growth-promoting iron chelates. Ferrichrome and ferrichrome A first isolated from the rust fungus, *Ustilago sphaerogena:* Neilands, *J. Am. Chem. Soc.* **74,** 4846 (1952); Garibaldi, Neilands, *ibid.* **77,** 2429 (1955). Ferrichrome and four related compounds, *ferrichrysin, ferricrocin, ferrirubin* **and** *ferrirhodin* have also been isolated from *Aspergillaceae* and from *Penicillium resticulosum:* Zähner *et al., Arch. Mikrobiol.* **45,** 119 (1963); Keller-Schierlein, Deer, *Helv. Chim. Acta* **46,** 1907 (1963); Keller-Schierlein, *ibid.,* 1920. Ferrichromes are cyclic hexapeptides containing three small, neutral amino acids and three derivatives of δ-N-hydroxy-L-ornithine [$HONH(CH_2)_3CH-(NH_2)COOH$]. Structure: Rogers, Neilands, *Biochemistry* **3,** 1850 (1964). Crystal and molecular structure of ferrichrome A: Zalkin *et al., J. Am. Chem. Soc.* **88,** 1810 (1966). Synthesis of ferrichrome: Keller-Schierlein, Maurer, *Helv. Chim. Acta* **52,** 603 (1969); Isowa *et al., Bull. Chem. Soc. Japan* **47,** 215 (1974). Reviews of ferrichromes and other hydroxamic acids: Neilands, *Struct. Bonding* **1,** 59-108 (1966); Emery, *Advan. Enzymol. Relat. Areas Mol. Biol.* **35,** 135-185 (1971); J. B. Neilands, *Bioinorganic Chemistry* **II,** K. N. Raymond, Ed. (A.C.S., Washington, 1977), pp 3-32.

Ferrichrome, $C_{27}H_{42}FeN_9O_{12}$, R = CH_3, R' = R'' = H. Long, yellow needles from anhydr methanol, shrink and blacken at 240-242° without melting. $[\alpha]_D$ +300° (c = 0.04). Absorption max (methanol): 425 nm ($E^{1\%}_{1cm}$ 39.4). Sol in water and hot methanol; sparingly sol in ethanol, acetone, ether, chloroform.

Ferrichrome A, $C_{41}H_{58}FeN_9O_{20}$, R = *trans*-$HOOCCH_2$-$(CH_3)C{=}CH$, R' = H, R'' = $HOCH_2$. Crystals from water. Absorption max (0.1M phosphate buffer; pH 7): 440 nm ($E^{1\%}_{1cm}$ 33.8). Very sol in methanol, ethanol, propanol;

slightly sol in hot water. Practically insol in acetone, petr ether, ether, chloroform.

3969. Ferric Hydroxide. *Ferric hydroxide oxide;* hydrated ferric oxide. $FeHO_2$; mol wt 88.86. Fe 62.85%, H 1.13%, O 36.01%. FeO(OH). Occurs in nature as the minerals *goethite* [α-FeO(OH)], *lepidocrocite* [γ-FeO(OH)], and *limonite* [FeO(OH).nH_2O]. Other known allomorphic forms: β-FeO(OH); δ-FeO(OH). The hydroxide $Fe(OH)_3$ is not known. Prepn: Lux in *Handbook of Preparative Inorganic Chemistry*, vol 2, G. Brauer, Ed. (Academic Press, New York, 2nd ed., 1965) p 1499. Crystal structure of α-FeO(OH): Sampson, *Acta Cryst.* **25B**, 1683 (1969). Review: Bernal *et al.*, *Clay Miner. Bull.* **4**, 15-30 (1959).

Red to brown powder or crystals. Loses H_2O to form Fe_2O_3. d 3.4-3.9. Practically insol in water, alcohol; sol in mineral acids.

USE: In purifying water; as absorbent in chemical processing; as pigment; as catalyst.

3970. Ferric Hypophosphite. $FeH_6O_6P_3$; mol wt 250.83. Fe 22.27%, H 2.41%, O 38.27%, P 37.05%. $Fe(H_2PO_2)_3$. Prepn: *U.S.D.* 25th ed, p 573.

White or grayish-white powder. Odorless, tasteless. Sol in 2300 parts cold water, 1200 parts boiling water; more sol in water in presence of H_3PO_2; sol in warm concd solns of alkali citrates. *Protect from light.* Should not be heated or triturated with chlorates, nitrates, or other oxidizing agents.

USE: Formerly as dietary supplement for phosphorus.

3971. Ferriclate Calcium Sodium. *Pentaaqua*[D-*gluconato(4−)-O²,O⁴,O⁵]tetra-μ-hydroxydioxotriferrate(3−) calcium sodium (2:1:4);* monocalcium tetrasodium bis[pentaaqua-tetra-μ-hydroxy[D-gluconato(4−)]dioxotriferrate(3−)]; Kelfer. $C_{12}H_{44}CaFe_6Na_4O_{36}$; mol wt 1231.62. C 11.70%, H 3.60%, Ca 3.25%, Fe 27.21%, Na 7.47%, O 46.77%.

THERAP CAT: Hematinic.

3972. Ferric Nitrate. FeN_3O_9; mol wt 241.87. Fe 23.09%, N 17.37%, O 59.54%. $Fe(NO_3)_3$. Prepn: *Gmelin's, Iron* (8th ed.) **59**, part B, 161-172 (1932).

Nonahydrate, pale-violet to grayish-white, somewhat deliquesc crystals. mp 47°. Dec below 100°. d^{21} 1.68. Freely sol in water, alcohol, acetone; slightly sol in cold concd HNO_3. LD_{50} orally in rats: 3.25 g/kg, H. F. Smyth *et al.*, *Am. Ind. Hyg. Assoc. J.* **30**, 470 (1969).

USE: As mordant in dyeing, weighting silks, tanning; as reagent in analytical chemistry; as corrosion inhibitor.

3973. Ferric Oxide. Ferric sesquioxide; jeweler's rouge. Fe_2O_3; mol wt 159.70. Fe 69.94%, O 30.06%. α-Form occurs in nature as the mineral *hematite*. γ-Form occurs in nature as the mineral *maghemite*; prepd by dehydration of α-FeO(OH): Giovanoli, Brütsch, *Chimia* **28**, 188 (1974). Prepn of a third allomorphic form, ϵ-Fe_2O_3: Schrader, Büttner, *Z. Anorg. Allgem. Chem.* **320**, 220 (1963); Trautmann, Forestier, *Compt. Rend.* **261**, 4423 (1965). Color and appearance of Fe_2O_3 are dependent upon the size and shape of the particles and the amount of combined water. Preparation and properties: *Gmelin's, Iron* (8th ed.) **59**, part B, 63-94 (1932); Baudisch, Hartung, *Inorg. Syn.* **1**, 185 (1939); *Ullmann's Encyklopädie der Technischen Chemie* vol. **6**, 421-423 (1955); Bernal *et al.*, *Clay Miner. Bull.* **4**, 15-30 (1959).

Note: The composition of the substance called δ-Fe_2O_3 is actually FeO(OH): Bernal *et al.*, *loc. cit.*

Caution: Hematite dust causes benign pneumoconiosis: see L. T. Fairbanks, *Industrial Toxicology* (Hafner, New York, 2nd ed., 1969) pp 64-66.

USE: As pigment for rubber, paints, paper, linoleum, ceramics, glass; in paint for ironwork, ship hulls; as polishing agent for glass, precious metals, diamonds; in electrical resistors and semiconductors; in magnets, magnetic tapes; as catalyst; colloidal solns as stain for polysaccharides.

3974. Ferric Oxide, Saccharated. Saccharated iron; iron sugar; Colliron I.V.; Feojectin; Ferrivenin; Ferum Hausmann; Iviron; Neo-Ferrum; Proferrin; Sucrofer. Contains 2.8-3.2% Fe. Prepn: *U.S.D.* 26th ed., p 627 (1967). Prepn of soln contg 2% Fe suitable for i.v. injection: Slack, Wilkinson, *Lancet* **256**, 11 (1949).

Brown powder. Sol in water; practically insol in alcohol. Solns are unstable in the presence of electrolytes: *Do not mix with physiological saline!*

THERAP CAT: Hematinic.

3975. Ferric Phosphate. FeO_4P; mol wt 150.83. Fe 37.03%, O 42.43%, P 20.54%. $FePO_4$. Occurs in nature as the minerals *beraunite, cacoxenite, dufrenite, koninckite, phosphosiderite, strengite.* Prepn from $Fe(H_2PO_4)_3$: Remy, Boulle, *Compt. Rend.* **253**, 2699 (1961); from $Fe(CO)_5$ and H_3PO_4: Cate *et al.*, *Soil Sci.* **88**(3), 130 (1959); from phosphate rock: Vickery, U.S. pat. **2,914,380** (1959 to Horizons Inc.); from mill scale and H_3PO_4: Alexander, Mathes, U.S. pat. **3,070,423** (1962 to Chemetron).

Dihydrate, white, grayish-white, or light pink, orthorhombic or monoclinic crystals or amorphous powder. Loses water above 140°. d 2.87. Practically insol in water; slowly sol in HNO_3; readily sol in HCl.

USE: As food and feed supplement, particularly in bread enrichment; as fertilizer.

3976. Ferric Pyrophosphate. $Fe_4O_{21}P_6$; mol wt 745.25. Fe 29.98%, O 45.09%, P 24.94%. $Fe_4(P_2O_7)_3$. Prepn: *Gmelin's, Iron* (8th ed.) **59**, part B, 777 (1932); Knight, Kelly, U.S. pat. **3,014,784** (1962 to American Oil).

Nonahydrate, yellowish-white powder. Practically insol in water or acetic acid; sol in mineral acids.

USE: As catalyst; in fireproofing of synthetic fibers; in corrosion-preventing pigments.

THERAP CAT: Hematinic.

3977. Ferric Sodium Edetate. *[[N,N′-Ethanediylbis[N-(carboxymethyl)glycinato]](4−)]-N,N′,O,O′,O^N,O^N-ferrate-(1−) sodium;* sodium [(ethylenedinitrilo)tetraacetato]ferrate-(1−); (ethylenedinitrilo)tetraacetic acid sodium salt iron complex; ferric monosodium ethylenediaminetetraacetate; edetic acid sodium iron salt; sodium iron edetate; sodium feredetate; Ferrostrane; Ferrostrene; Sybron. $C_{10}H_{12}FeN_2$-NaO_8; mol wt 367.07. C 32.72%, H 3.30%, Fe 15.22%, N 7.63%, Na 6.26%, O 34.87%. Prepd from disodium ethylenediaminetetraacetic acid and ferric nitrate: Sawyer, McKinnie, *J. Am. Chem. Soc.* **82**, 4191 (1960).

Crystals from water + ethanol.

THERAP CAT: Iron source.

3978. Ferric Sodium Pyrophosphate. Sodium ferric pyrophosphate. $Fe_4Na_8O_{35}P_{10}$. Hydrate; $Na_8Fe_4(P_2O_7)_5.xH_2O$. The commercial product contains 15.5-16.5% Fe and 50.5-52.5% P_2O_5.

White powder. Bulk density 1.4-1.6. Sol in hydrochloric acid. Insol in water.

USE: Food enrichment. Less prone to induce rancidity than orthophosphates.

3979. Ferric Subsulfate Solution. Basic ferric sulfate soln; Monsel's soln. Approx: $Fe_4(OH)_2(SO_4)_5$. Prepn from $FeSO_4$ and HNO_3: *U.S.D.* 25th ed., p 574.

Reddish-brown liquid. Almost odorless; sour, strongly astringent taste. Acid to litmus. Affected by light. d^{25} 1.548. Miscible with water, alcohol. May crystallize or solidify at low temps. *Keep well closed, protected from light, in a warm place.*

USE: As mordant in dyeing textiles.

THERAP CAT: Styptic.

THERAP CAT (VET): Styptic, astringent.

3980. Ferric Sulfate. Ferric persulfate; ferric sesquisulfate; ferric tersulfate. $Fe_2O_{12}S_3$; mol wt 399.88. Fe 27.93%, O 48.01%, S 24.06%. $Fe_2(SO_4)_3$. Prepn: *Gmelin's, Iron* (8th ed.) **59**, part B, 439-462 (1932).

Grayish-white powder, or rhombic or rhombohedral crystals. Very hygroscopic. Commercial product usually contains about 20% water and is yellowish in color. d^{18} 3.097. Slowly sol in water, rapidly sol in the presence of a trace of $FeSO_4$; sparingly sol in alcohol; practically insol in acetone, ethyl acetate. Hydrolyzed slowly in aq soln. *Keep well closed and protected from light.*

USE: In preparation of iron alums, other iron salts and pigments; as coagulant in water purification and sewage treatment; in etching aluminum; in pickling stainless steel and copper; as mordant in textile dyeing and calico printing; in soil conditioners; as polymerization catalyst.

3981. Ferric Tannate. Ferric gallotannate. Variable composition. Contains 8-10% Fe, 70-80% tannin.

Bluish-black powder. Insol in water; sol in dil mineral acids.

USE: In inks.

3982. Ferric Thiocyanate. Ferric sulfocyanate; ferric sulfocyanide. $C_3FeN_3S_3$; mol wt 230.08. C 15.66%, Fe 24.27%, N 18.26%, S 41.81%. $Fe(SCN)_3$. Prepn: *Gmelin's Handb. Anorg. Chem., Iron* (8th ed.) part B, 747-761 (1932).

Sesquihydrate, red, deliquesc crystals. Dec on heating. Sol in water, alcohol, ether, acetone, pyridine, ethyl acetate; practically insol in $CHCl_3$, CCl_4, CS_2, toluene. *Keep well closed.*

USE: Analytical reagent.

3983. Ferrite. Ferrospinel. A crystalline, usually man-made material, having a spinel structure and consisting essentially of ferric oxide and at least one other metallic oxide which is usually, although not always, divalent in nature. When molded into compressed bodies, the material is characterized by high magnetic permeability. Typified composition: Fe_2O_3 67-70%; ZnO 10-10.5%; MnO_2 20-22.5%; CuO 0.1-10%; Co_3O_4 0.1%. Ferrites are prepd by ceramic techniques. The oxides or carbonates are milled in steel ball mills, and the mixture of very fine particles is dried and pre-fired in order to obtain a homogeneous end product: Hilpert, *Ber.* **42**, 2248 (1909). Examples of modern techniques: Simpkiss; Harvey, U.S. pats **2,723,238-9** (both 1955 to RCA). Prepn of single crystals: Rooymans, *Colloq. Int. Cent. Nat. Rech. Sci.* No. **205**, 151 (1972). Books: Snoek, *New Developments in Ferromagnetic Materials* (Elsevier, New York, 1947); Smit, Wijn, *Ferrites* (John Wiley, New York, 1959); Soohov, *Theory and Applications of Ferrites* (Prentice Hall, 1960); Standley, *Oxide Magnetic Materials* (Clarendon Press, Oxford, 1962); *Ferrites, Proc. Int. Conf.*, Y. Hoshino et al., Eds. (University Park Press, Baltimore, 1971) 670 pp; E. E. Riches, *Ferrites, A Review of Materials and Applications* (Mills and Boon, London, 1972) 88 pp. Reviews with bibliographies: Gorter, *Proc. I.R.E.* **43**, 1945-1973 (1955); Fresh, "Methods of Preparation and Crystal Chemistry of Ferrites," *ibid.* **44**, 1303-1311 (1956); Brailsford, *Magnetic Materials* (John Wiley, New York, 3rd ed., 1960) pp 160-181; Hogen, *Sci. Am.* **202**, 92-104 (1960); Economos in Kirk-Othmer's *Encyclopedia of Chemical Technology* vol. **8** (Interscience, New York, 2nd ed., 1965) pp 881-901; Gray, "Oxide Spinels" in *High Temperature Oxides*, Part IV, A. M. Alper, Ed. (Academic Press, New York, 1971) pp 77-107.

Note: The term "ferrites" has been expanded to mean any oxidic magnetic material.

USE: Magnetic cores for inductors and transformers; microwave devices; information storage; electromechanical transducers: E. E. Riches, *loc. cit.*; Brockman, *Ceram. Ind.* **99**, 24 (1972).

3984. Ferritin. Epadora; Ferrofolin; Ferrol; Ferrosprint; Sanifer; Sideros; Unifer. Major iron storage protein; found in spleen, liver and intestinal mucosa of vertebrates; widely distributed in the plant and animal kingdoms. Isoln and crystallization of horse spleen ferritin: Laufberger, *Bull. Soc. Chim. Biol.* **19**, 1575 (1937); Granick, *J. Biol. Chem.* **14**, 451 (1942); Crichton *et al., Biochem. J.* **131**, 51 (1973). Consists of a protein shell surrounding a crystalline, hydrated iron oxide/phosphate core. The core may contain up to 4500 Fe^{3+} ions; unfractionated, horse-spleen ferritin contains approx 20% iron on a dry weight basis. The protein shell, *apoferritin*, has a mol wt of ~445,000. Several proposed structures of protein moiety: Harrison, *J. Mol. Biol.* **6**, 404 (1963); Crichton, *Biochem. J.* **126**, 761 (1972); Niitsu *et al., Biochem. Biophys. Res. Commun.* **55**, 1134 (1973). Absorption spectra: Granick, *Chem. Rev.* **38**, 379 (1946). Use of serum ferritin and isoferritins in clinical medicine: J. W. Halliday, L. W. Powell, *Prog. Hematol.* **11**, 229 (1979). Reviews: Harrison, Hoy, "Ferritin" in *Inorganic Biochemistry* vol. 1, G. L. Eichhorn, Ed. (Elsevier, New York, 1973) pp 253-277; Crichton, *Angew. Chem. Int. Ed.* **12**, 57-65 (1973); P. Aisen, I. Listowsky, *Ann. Rev. Biochem.* **49**, 357 (1980).

Red-brown, water-soluble protein. Forms cubic and orthorhombic crystals.

THERAP CAT: Hematinic.

3985. Ferrocene. Dicyclopentadienyliron; biscyclopentadienyliron. $C_{10}H_{10}Fe$; mol wt 186.03. C 64.56%, H 5.42%, Fe 30.02%. Prepn: Kealy, Pauson, *Nature* **168**, 1039 (1951); Pauson, U.S. pat. **2,680,756** (1954 to du Pont); Miller *et al., J. Chem. Soc.* **1952**, 632; Anzilotti, Weinmayr, U.S. pat. **2,791,597** (1957 to du Pont). Other prepn: Pruett, Morehouse, *Advances in Chemistry Series* **23**, 368-371 (1959); Wilkinson, *Org. Syn.* coll. vol. IV, 473 (1963); Cordes, Fr. pat. **1,341,880** (1963 to BASF), *C.A.* **60**, 6873a (1964). Structure studies: Wilkinson *et al., J. Am. Chem. Soc.* **74**, 2125 (1952); Seibold, Sutton, *J. Chem. Phys.* **23**, 1967 (1955). Synthesis of a helical ferrocene: T. J. Katz, J. Pesti, *J. Am. Chem. Soc.* **104**, 346 (1982). *Reviews:* Rausch *et al., J. Chem. Ed.* **34**, 268-272 (1957); M. Rosenblum, *Chemistry of the Iron Group Metallocenes* (John Wiley, New York, 1965) 241 pp; Bruce, *Organometal. Chem. Rev.* (Sect. B) **10**, 75-122 (1972); B. W. Rockett, G. Marr, *J. Organometal. Chem.* **211**, 215-278 (1981).

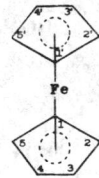

Orange needles from methanol or ethanol; odor of camphor. mp 173-174°. Sublimes above 100°. Volatile in steam. Practically insol in water, 10% NaOH, and concd boiling HCl. Sol in alcohol, ether, benzene. Dissolves in dil nitric and concd sulfuric acids forming a deep red soln with blue fluorescence. The molecule is diamagnetic and the dipole moment is effectively zero.

USE: Antiknock additive for gasoline; catalyst.

3986. Ferrocholinate. *2-Hydroxy-N,N,N-trimethylethanaminium (OC-6-44)-triaqua[2-hydroxy-1,2,3-propanetricarboxylato(4−)]ferrate(1−);* [hydrogen citrato(3−)]triaquoiron, choline salt; iron choline citrate; Chelafer; Chel-Iron; Ferrolip. $C_{11}H_{24}FeNO_{11}$; mol wt 402.17. C 32.85%, H 6.01%, Fe 13.89%, N 3.48%, O 43.76%. Chelate prepd by interaction of equimolar quantities of choline dihydrogen citrate and freshly prepd $Fe(OH)_3$ or $FeCO_3$: Bandelin, U.S. pat. **2,575,611** (1951 to Flint Eaton); by treatment of a freshly prepd solution of ferric citrate with an equimolar amount of choline: Chakrabarti, Sen, *Chem. & Ind. (London)* **1961**, 1407. Pharmaceutical prepn, characteristics and applications: G. N. Bookwalter *et al., J. Food Sci.* **38**, 618 (1973). Bioavailability in pigs: E. R. Miller *et al., J. Animal Sci.* **52**, 783 (1981).

Greenish-brown, reddish-brown or brown amorphous solid with glistening surface upon fracture. Freely sol in water, yielding stable solns; sol in acids, alkalies. One gram of pharmaceutical grade is equivalent to 120 mg of elemental iron and 360 mg of choline base.

Note: Other combinations of iron, choline and citric acid have been prepd for pharmaceutical use. A 1:2:2 chelate, $C_{22}H_{36}FeN_2O_{16}$, was reported by Chakrabarti and Sen, *loc. cit.*, and a 2:3:3 chelate, $C_{33}H_{57}Fe_2N_3O_{24}$, by Rosenfelder, **U.S. pat. 2,865,938** (1958 to H. Rosenstein).

THERAP CAT: Hematinic.

3987. Ferroglycine Sulfate. Ferroglycine sulfate complex; ferrous sulfate glycine complex; ferrous aminoaceto-sulfate; iron sulfate-glycine complex; glycine-ferrous sulfate complex; Plesmet; Kelferon; Ferronord; Ferro sanol; Glyferro; Pleniron. Prepn from glycine and ferrous sulfate: Rummell, **U.S. pats. 2,877,253** and **2,957,806** (1959, 1960 both to Schwarz Arzneimittelfabrik).

THERAP CAT: Hematinic.

3988. Ferrosoferric Oxide. Ferric ferrous oxide; triiron tetraoxide; black iron oxide; magnetic iron oxide; Ethiops iron. Fe_3O_4; mol wt 231.55. Fe 72.36%, O 27.64%. Occurs in nature as the mineral *magnetite* (red-black lumps). Prepn: *Gmelin's, Iron* (8th ed) part B, 36-62 (1932); *Ullmanns Encyklopädie der Technischen Chemie* vol. **6,** 420 (1955). Review: Robl, *Angew. Chem.* **70,** 367 (1958).

Black cubes or amorphous powder. mp 1538°; d 5.2. Oxidized to Fe_2O_3 on heating in air. Practically insol in water; sol in acids.

USE: Pigment in paints, linoleum, ceramic glazes; in coloring glass; as a polishing compd; in the textile industry; in cathodes; as catalyst.

3989. Ferrous Bromide. Br_2Fe; mol wt 215.68. Br 74.11%, Fe 25.89%. $FeBr_2$. Prepn: Baxter, *Z. Anorg. Chem.* **38,** 236 (1904); Baxter *et al., ibid.* **70,** 333 (1911); Schimmel, *Ber.* **62,** 963 (1929); Kühnl, Ernst, *Z. Anorg. Allgem. Chem.* **317,** 84 (1962).

Light yellow to dark brown hygroscopic crystals. mp 684°. d_4^{25} 4.63. Very sol in water and alcohol. *Keep tightly closed.*

Hexahydrate, pale green to bluish-green rhombic prisms. Forms tetrahydrate at 49°; dihydrate at 83°. Rapidly oxidized in moist air. *Keep tightly closed.*

USE: Polymerization catalyst.

3990. Ferrous Carbonate Mass. Blaud's mass; Vallet's mass; Fecarb. Contains 36-41% $FeCO_3$, the remainder consisting of honey and sugar. Prepn: *U.S.D.* 25th ed., p 575.

Dark greenish-gray to brown, moderately soft mass. Practically insol in water; appreciably sol in water satd with CO_2; sol in dil acids.

THERAP CAT: Hematinic.
THERAP CAT (VET): Has been used in iron deficiency.

3991. Ferrous Carbonate Saccharated. Freshly pptd $FeCO_3$ protected from oxidation by admixture with sugar. Contains not less than 15% $FeCO_3$. Prepn: *U.S.D.* 25th ed., p 575.

Olive-gray to greenish-brown powder. Odorless. Partially sol in water, completely sol in dil mineral acids.

THERAP CAT: Hematinic.
THERAP CAT (VET): In iron deficiency.

3992. Ferrous Chloride. Cl_2Fe; mol wt 126.76. Cl 55.94%, Fe 44.06%. $FeCl_2$. Occurs in nature as the mineral *lawrencite.* Preparation: Kovacic, Brace, *Inorg. Syn.* **6,** 172 (1960); Kühnl, Ernst, *Z. Anorg. Allgem. Chem.* **317,** 84 (1962). Prepn of dihydrate: Gayer, Woontner, *Inorg. Syn.* **5,** 179 (1957).

White rhombohedral crystals; may sometimes have a green tint. Very hygroscopic. mp 674°; bp 1023°; d^{25} 3.16. Can be sublimed in a stream of HCl at about 700°. Forms $FeCl_3$ and Fe_2O_3 on heating in air. Freely sol in water, alc, acetone; slightly sol in benzene; practically insol in ether.

Dihydrate, white monoclinic crystals with pale green tint. Loses $1H_2O$ at 120°: Gayer, Woontner, *loc. cit.;* also reported to lose $1H_2O$ at 150-160°: Schafer, *Z. Anorg. Allgem. Chem.* **258,** 69 (1949). Sol in water.

Tetrahydrate, pale green to blue-green, monoclinic crystals or cryst powder. Loses $2H_2O$ at about 105-115°: Schafer, *loc. cit.* d 1.93. Sol in water, alcohol. The technical product may not be completely sol without the addn of acid. Aq solns are readily oxidized.

USE: In metallurgy; as reducing agent; in pharmaceutical prepns; as mordant in dyeing. *Caution:* Mild irritant.

3993. Ferrous Citrate. Several forms of this salt are known. Prepn from citric acid and Fe powder: Oroshnik, Haffcke, **U.S. pat. 2,904,573** (1959 to Ortho); and ferrous salts: Carlson, **U.S. pat. 3,091,626** (1963 to Scherer).

Monohydrate, *monoferrous acid citrate monohydrate.* Powder. Practically insol in water, alcohol, acetone.

Decahydrate, *triferrous dicitrate decahydrate.* Very slightly colored powder or white crystals. Very stable to air oxidation. If H_2O is removed by vacuum desiccation, the dehydrated ferrous salt rapidly oxidizes to a ferric salt. Practically insol in water, acetone.

THERAP CAT: Hematinic.

3994. Ferrous Fluoride. F_2Fe; mol wt 93.85. F 40.49%, Fe 59.51%. FeF_2. Prepn from $FeCl_2$ and HF gas: Kwasnik in *Handbook of Preparative Inorganic Chemistry,* G. Brauer, Ed. (Academic Press, New York, 2nd ed., 1963) p 266; from Fe powder and liq HF: Muetterties, Castle, *J. Inorg. Nucl. Chem.* **18,** 148 (1961).

Tetragonal crystals (rutile type) or powder. Sublimes at about 1100°. d 4.09. Sparingly sol in water; more sol in dil HF; practically insol in alcohol, ether, benzene.

USE: As catalyst in organic reactions.

3995. Ferrous Fumarate. Cpiron; Erco-Fer; Feostat; Feroton; Ferrofume; Ferronat; Ferrone; Ferrotemp; Ferrum; Fersamal; Firon; Fumafer; Fumar F; Fumiron; Galfer; Heferol; Ircon; Meterfer; One-Iron; Palafer; Toleron; Tolferain; Tolifer. $C_4H_2FeO_4$; mol wt 169.91. C 28.27%, H 1.19%, Fe 32.87%, O 37.67%. Prepd by mixing hot aq solns of ferrous sulfate and sodium fumarate and separating the resulting slurry by filtration: Bertsch, Lemp, **U.S. pat. 2,848,366** (1958 to Mallinckrodt); Prabhakaran, Patel, *Indian J. Chem.* **7,** 266 (1969). The hot soln of sodium fumarate is preferably added to the ferrous sulfate soln. The commercial material contains a minimum of 31.3% total Fe and not less than 2.0% ferric iron. Toxicity: Weaver *et al., Am. J. Med. Sci.* **241,** 296 (1961).

Reddish-orange to reddish-brown, granular powder. d^{25} 2.435. Odorless; almost tasteless. Not melted at 280°. Soly at 25° in water: 0.14 g/100 ml; in alcohol < 0.01 g/100 ml. Soly in acid is limited by liberation of fumaric acid: up to 0.45 g can be dissolved in 100 ml of $1.0N$ HCl; up to 0.6 g in $0.1N$ HCl. LD_{50} i.p. in mice: 480 mg/kg (Weaver).

THERAP CAT: Hematinic.

3996. Ferrous Gluconate. Fergon; Ferlucon; Ferronicum; Gluco-Ferrum; Iromon; Irox; Nionate. $C_{12}H_{22}FeO_{14}$; mol wt 446.16. C 32.30%, H 4.97%, Fe 12.52%, O 50.21%. $Fe[HOCH_2(CHOH)_4CO_2]_2$. Prepd from Ba gluconate and $FeSO_4$. Prepn of isotonic solns: Hammarlund, *Pharm. Acta Helv.* **35,** 593 (1960).

Dihydrate, yellowish-gray or pale greenish-yellow powder. Slight odor of caramel. Acid to litmus. Sol in water; practically insol in alcohol. Aq solns are stabilized by the addition of glucose. Extensive stability studies on aq solns: Johnson, Thomas, *J. Pharm. Pharmacol.* **6,** 1037 (1954). LD_{50} in mice: 114 mg/kg i.v.; 3.7 g/kg orally, Hoppe *et al., Am. J. Med. Sci.* **230,** 491 (1955).

USE: As coloring and flavoring in foods.

THERAP CAT: Hematinic.

3997. Ferrous Hydroxide. FeH_2O_2; mol wt 89.97. Fe 62.15%, H 2.24%, O 35.61%. $Fe(OH)_2$. Prepn: Rihl, Fricke, *Z. Anorg. Allgem. Chem.* **251,** 406 (1943); Leussing, Kolthoff, *J. Am. Chem. Soc.* **75,** 2476 (1953). Oxidation products: Feitknecht, Keller, *Z. Anorg. Chem.* **262,** 61 (1950);

Mayne, *J. Chem. Soc.* **1953**, 129; Shipko, Douglas, *J. Phys. Chem.* **60**, 1519 (1956).

White, amorphous powder or white to pale green hexagonal crystals. Oxidized on exposure to air; may ignite spontaneously on exposure to air if finely divided. Practically insol in water; more sol in solns of NH₄ salts; sol in concd NaOH soln.

3998. Ferrous Iodide. FeI₂; mol wt 309.67. Fe 18.04%, O 81.96%. Prepn: Chaigneau, *Bull. Soc. Chim. France* **1957**, 886; Lieser, Elias, *Z. Anorg. Allgem. Chem.* **316**, 208 (1962); Lux in *Handbook of Preparative Inorganic Chemistry* vol. 2, G. Brauer, Ed. (Academic Press, New York, 2nd ed., 1965) p 1495.

Large, thin, red-violet crystals or black leaflets. Very hygroscopic. Sol in water, alcohol, ether; aq soln is readily oxidized by air. *Keep tightly closed and protected from light.*

USE: As catalyst for organic reactions.

THERAP CAT: Antitubercular.

THERAP CAT (VET): Source of iron and iodine.

3999. Ferrous Lactate. C₆H₁₀FeO₆; mol wt 233.99. C 30.80%, H 4.31%, Fe 23.87%, O 41.03%. Fe(C₃H₅O₃)₂. Prepn: *Gmelin's, Iron* (8th ed.) **59**, part B, p 532 (1932). Prepn of isotonic solns: Hammarlund, *Pharm. Acta Helv.* **35**, 593 (1960).

Trihydrate, greenish-white powder or cryst masses. Slight characteristic odor; mild, sweet and ferruginous taste. Sol in water; freely sol in alkali citrates forming a green soln; almost insol in alc. On exposure to air it becomes darker and incompletely sol. *Keep well closed and protected from light.* LD in rabbits: 578 mg/kg s.c.; 287 mg/kg i.v., *Handbook of Toxicology* vol. 1, W. S. Spector, Ed. (Saunders, Philadelphia, 1956) pp 142-143.

THERAP CAT: Hematinic.

4000. Ferrous Oxalate. Ferrox. C₂FeO₄; mol wt 143.87. C 16.70%, Fe 38.82%, O 44.48%. FeC₂O₄. Prepn of dihydrate: *Gmelin's, Iron* (8th ed.) **59**, part B, 532-533 (1932); prepn of anhydr salt: Waterman, Vivian, *J. Org. Chem.* **14**, 289 (1949).

Dihydrate, pale yellow, odorless, cryst powder. d 2.28. Dec at 150-160° on heating in air. Slightly sol in water; sol in dil mineral acids.

USE: As a photographic developer for silver bromide-gelatin plates; to impart a greenish-brown tint to optical glass (sunglasses, windshields, railroad car windows), for decorative glassware; pigment for plastics, paints, lacquers.

4001. Ferrous Oxide. FeO; mol wt 71.85. Fe 77.73%, O 22.27%. Prepn: Lux in *Handbook of Preparative Inorganic Chemistry* vol. 2, G. Brauer, Ed. (Academic Press, New York, 2nd ed., 1965) p 1497.

Jet-black powder. mp 1360°; d 5.7. Easily oxidized by air. It is a strong base and readily absorbs CO₂. Practically insol in water and alkalies; readily sol in acids.

USE: In manuf of green, heat-absorbing glass; in steel manuf; in enamels; as catalyst.

4002. Ferrous Phosphate. Fe₃O₈P₂; mol wt 357.50. Fe 46.87%, O 35.80%, P 17.33%. Fe₃(PO₄)₂. Prepn of octahydrate: *Gmelin's, Iron* (8th ed.) **59**, part B, 771 (1932). Due to unavoidable oxidation, the article of commerce contains basic ferric phosphate.

Octahydrate, grayish-blue powder or monoclinic crystals. d 2.58. Practically insol in water; sol in mineral acids. *Keep well closed and protected from light.*

USE: In ceramics, as catalyst.

4003. Ferrous Phosphide. Iron hemiphosphide; diiron phosphide. Fe₂P; mol wt 142.68. Fe 78.29%, P 21.71%. Prepn from Fe and red P: Rundqvist, Jellinek, *Acta Chem. Scand.* **13**, 425 (1959); Cadeville, Meyer, *Compt. Rend.* **252**, 1124 (1961).

Gray, hexagonal needles or blue-gray powder. d 6.85. Ferromagnetic. Insol in water, dil acid, dil alkali. Reacts with hot mineral acids.

4004. Ferrous Selenide. FeSe; mol wt 134.81. Fe 41.43%, Se 58.57%. Prepn from the elements: Gronvold, Westrum, *Acta Chem. Scand.* **13**, 241 (1959).

Black mass with metallic luster. Stable in air. Dec when heated in O₂. d 6.78. Practically insol in water; sol in HCl with the evolution of H₂Se.

USE: In electrical semiconductors.

4005. Ferrous Succinate. Cerevon; Ferromyn. C₄H₄FeO₄; mol wt 171.92. C 27.94%, H 2.34%, Fe 32.49%, O 37.23%. Prepn: Franke, *Ann.* **491**, 30 (1931).

$$\begin{matrix} CH_2COO \\ | \\ CH_2COO \end{matrix}\!\!\!\!Fe$$

Tetrahydrate, sparingly sol in water. Also obtained as a dihydrate.

THERAP CAT: Hematinic.

4006. Ferrous Sulfate. FeO₄S; mol wt 151.91. Fe 36.77%, O 42.13%, S 21.10%. FeSO₄. Hydrates occur in nature as the minerals: *melanterite, siderotil, szomolnikite, tauriscite.* Heptahydrate prepd commercially by the action of H₂SO₄ on Fe: Faith, Keyes & Clark's *Industrial Chemicals,* F. A. Lowenheim, M. K. Moran, Eds. (Wiley-Interscience, New York, 4th ed., 1975) pp 418-421. Crystal structure of heptahydrate: Baur, *Acta Cryst.* **17**, 1167 (1964). Acute toxicity: Hoppe *et al., Am. J. Med. Sci.* **230**, 491 (1955).

Monohydrate, *dried ferrous sulfate, exsiccated ferrous sulfate, Feromax, Feroritard, Ferro-Gradumet, Fer-Sul, Fespan, Tetucur.* White to yellow cryst powder. Loses H₂O at about 300°. Dec at higher temps. Sol in water.

Heptahydrate, *copperas, green vitriol, iron vitriol, Feosol, Feospan, Fesofor, Fesotyme, Fero-Gradumet, Fer-in-Sol, Haemofort, Ironate, Irosul, Mol-Iron, Presfersul, Sulferrous.* Blue-green, monoclinic, odorless crystals or granules. Efflorescent in dry air; oxidizes in moist air forming a brown coating of basic ferric sulfate. Forms tetrahydrate at 56.6° and monohydrate at 65°. d 1.897. Sol in water. Practically insol in alcohol. Aq solns are oxidized slowly by air when cold, rapidly when hot; rate of oxidation increased by addn of alkali or exposure to light. *Incompat.* Alkalies, sol carbonates, Au and Ag salts, Pb acetate, lime water, KI, K and Na tartrate, Na borate, tannin, vegetable astringent infusions and decoctions. LD₅₀ in mice: 65 mg/kg i.v.; 1.52 g/kg orally (Hoppe).

Human Toxicity: G.I. disturbances (e.g., gastric distress, colic, constipation, diarrhea) may occur. In children, ingestion of large quantities may cause vomiting, hematemesis, hepatic damage, tachycardia, peripheral vascular collapse: *Clinical Toxicology of Commercial Products,* R. E. Gosselin *et al.,* Eds. (Williams & Wilkins, Baltimore, 4th ed., 1976) Sect. III, pp 153-158.

USE: In manufacture of Fe, Fe compds, other sulfates; in Fe electroplating baths; in fertilizer; as food and feed supplement; in radiation dosimeters; as reducing agent in chemical processes; as wood preservative; as weed-killer; in prevention of chlorosis in plants; in other pesticides; in writing ink; in process engraving and lithography; as dye for leather; in etching aluminum; in water treatment; in qualitative analysis ("brown ring" test for nitrates); as polymerization catalyst.

THERAP CAT: Hematinic.

THERAP CAT (VET): In iron deficiency. Astringent.

4007. Ferrous Sulfide. Iron sulfuret. FeS; mol wt 87.92. Fe 63.52%, S 36.48%. Occurs in nature as the minerals: *magnetkies, pyrrhotine, troillite.* Usually prepd from the elements: Lux in *Handbook of Preparative Inorganic Chemistry* vol 2, G. Brauer, Ed. (Academic Press, New York, 2nd ed., 1965) p 1502. The commercial product is ~ 75-80% FeS.

Colorless hexagonal crystals when pure; usually gray to brownish-black lumps, rods or granular powder. mp 1194°; d 4.84. Trimorphic with transition points at 135° and 325°. Oxidized by moist air to S and Fe₃O₄. Practically insol in water; soluble in acids with the evolution of H₂S.

USE: As a laboratory source of H₂S; in the ceramic industry; as a paint pigment; in anodes; in lubricant coatings.

4008. Ferrous Thiocyanate. Ferrous sulfocyanate; ferrous sulfocyanide. C₂FeN₂S₂; mol wt 172.01. C 13.96%, Fe 32.47%, N 16.29%, S 37.28%. Fe(SCN)₂. Prepn: *Gmelin's, Iron* (8th ed) **59**, part B, 743-747 (1932); O'Brien, *Chem. & Eng. News* **33**, 2008 (1955).

Trihydrate, pale green monoclinic prisms. Rapidly oxidized on exposure to air; dec by heat. Freely sol in water, alcohol, ether. *Keep tightly closed and protected from light.*
USE: Indicator for peroxides in organic solns: O'Brien, *loc. cit.*

4009. Fertilysin. *N,N'-1,8-Octanediylbis[2,2-dichloro-acetamide]; N,N'-octamethylenebis(2,2-dichloroacetamide); N,N'*-bis(dichloroacetyl)-1,8-octamethylenediamine; WIN 18446. $C_{12}H_{20}Cl_4N_2O_2$; mol wt 366.10. C 39.37%, H 5.51%, Cl 38.73%, N 7.65%, O 8.74%. $Cl_2CHCONH(CH_2)_8NH$-$OCCHCl_2$. Inhibits spermatogenesis without affecting production of gonadotropins in male laboratory animals. Prepn and biological activity: Surrey, Mayer, *J. Med. Pharm. Chem.* **3**, 419 (1961); Surrey, U.S. pat. **3,143,566** (1964 to Sterling Drug). Assessment of antifertility activity in rats: S. Nag, J. J. Ghosh, *J. Steroid Biochem.* **11**(1B), 681 (1979). Teratological study: P. Taleporos *et al., Teratology* **18**, 5 (1978).
Crystals, mp 122.4-123.6°. LD_{50} orally in mice: > 16,000 mg/kg.

4010. Fertirelin. *9-(N-Ethyl-L-prolinamide)-10-degly-cinamide-luteinizing hormone-releasing factor(pig); Des-*Gly^{10}*-NH$_2$-LH-RH-ethylamide; TAP 031.* $C_{55}H_{76}N_{16}O_{12}$; mol wt 1153.31. C 57.28%, H 6.64%, N 19.43%, O 16.65%. Synthetic nonapeptide analog of LH-RH, *q.v.* Prepn: M. Fujino *et al.,* **Ger. pat. 2,321,174**; *eidem,* U.S. pat. **3,853,837** (1973, 1974 both to Takeda); S. Shinagawa, M. Fujino, *Chem. Pharm. Bull.* **23**, 229 (1975); *see also:* M. Fujino *et al., Biochem. Biophys. Res. Commun.* **49**, 863 (1972). Enzyme immunoassay in bovine plasma: J. Okada, S. Kondo, *ibid.* **33**, 4464 (1985). Field trial in bovine cystic ovarian disease: T. Nakao *et al., Japan. J. Vet. Sci.* **45**, 269 (1983); in induction of ovulation: T. Nakao *et al., Theriogenology* **20**, 111 (1983); D. A. Coleman *et al., ibid.* **30**, 149 (1988).

5-oxoPro-His-Trp-Ser-Tyr-Gly-Leu-Arg-Pro-NHCH$_2$CH$_3$

Monoacetate, $C_{57}H_{80}N_{16}O_{14}$, *U-69689E, Conceral, Ovalyse.* Pentahydrate, white, fluffy powder, $[\alpha]_D^{25}$ −53.6° (c = 0.5 in 5% acetic acid).
THERAP CAT (VET): Gonad stimulating principle.

4011. Ferulic Acid. *3-(4-Hydroxy-3-methoxyphenyl)-2-propenoic acid; 4-hydroxy-3-methoxycinnamic acid; 3-meth-oxy-4-hydroxycinnamic acid; caffeic acid 3-methyl ether.* $C_{10}H_{10}O_4$; mol wt 194.18. C 61.85%, H 5.19%, O 32.96%. Widely distributed in small amounts in plants. Isoln from *Ferula foetida* Reg. *Umbelliferae:* H. Hlasiwetz, L. Barth, *Ann.* **138**, 61 (1866); from *Pinus laricio* Poir. *Abietineae:* M. Bamberger, *Monatsh.* **12**, 441 (1891); *see also* Beilstein **10**, 436 (1927) and supplements. Prepd by the interaction of vanillin, malonic acid and piperidine in pyridine for 3 weeks, then precipitating with HCl: Vorsatz, *J. Prakt. Chem.* **145**, 265 (1936); Pearl, Beyer, *J. Org. Chem.* **16**, 216 (1951). Sepn of *cis* and *trans* isomers: Comte *et al., Compt. Rend.* **245**, 1144 (1957). ^{13}C NMR study: C. J. Kelley *et al., J. Org. Chem.* **41**, 449 (1976). Discovery as a component of cell walls in wheat and barley: M. G. Smart, T. P. O'Brien, *Aust. J. Plant Physiol.* **6**, 485 (1979). Use as food preservative: T. Tsuchiya, M. Takasawa, **Japan. Kokai 75 18621** (1975 to Kyokuto Shibosan), *C.A.* **83**, 7602v (1975).

CH=CHCOOH

OCH$_3$

OH

cis-Form, yellow oil. uv max (alcohol): 316 nm.
trans-Form, orthorhombic needles from water, mp 174°. uv max (alcohol): 236, 322 nm. Sol in hot water, alcohol, ethyl aetate. Moderately sol in ether. Sparingly sol in petr ether, benzene. Forms a sodium salt.
USE: Food preservative.

4012. Fervenulin. *6,8-Dimethylpyrimido[5,4-e]-1,2,4-triazine-5,7-(6H,8H)-dione;* 6,8-dimethyl-5,7-dioxo-5,6,7,8-

tetrahydropyrimido[5,4-e]-*as*-triazine; 1,3-dimethylazalumazine; planomycin. $C_7H_7N_5O_2$; mol wt 193.17. C 43.52%, H 3.65%, N 36.26%, O 16.57%. Antibiotic from culture filtrates of *Streptomyces fervens:* Eble *et al., Antibiot. Ann.* **1959-1960**, 227. Structure: Daves *et al., J. Org. Chem.* **26**, 5256 (1961). Synthesis: Pfleiderer, Schündehütte, *Ann.* **615**, 42 (1958); Daves *et al., J. Am. Chem. Soc.* **84**, 1724 (1962); Yoneda, Nagamatsu, *Bull. Chem. Soc. Japan* **48**, 2884 (1975); Taylor, Sowinski, *J. Org. Chem.* **40**, 2321 (1975); S. Senda *et al., J. Am. Chem. Soc.* **99**, 7358 (1977).

Yellow orthorhombic crystals, mp 178-179°. uv max (ethanol): 238, 275, 340 nm (ε 18,500, 1600, 4200). Sol in practically all of the common organic solvents; sol in cold water to about 2 mg/ml; in hot water to about 40 mg/ml. Practically insol in hydrocarbons. Labile to alkali; stable to acid.
4-Oxide, $C_7H_7N_5O_3$. Synthesis: M. Ichiba *et al., J. Heterocycl. Chem.* **14**, 175 (1977); K. Senga *et al., Heterocycles* **6**, 273 (1977); synthesis and conversion to fervenulin: M. Ichiba *et al., J. Org. Chem.* **43**, 469 (1978). Crystals from alc, mp 179-180°. uv max (alc): 240, 304 nm (log ε 4.10, 3.21).

4013. Feverfew. Featherfew; featherfoil; midsummer daisy. *Tanacetum parthenium* (L.) Sch. Bip., (formerly *Chrysanthemum parthenium* (L.) Bernh.) *Compositae;* a perennial, strongly aromatic herb found in Britain and the Balkan peninsula. Used medicinally since the Middle Ages as a febrifuge. Constituents include sesquiterpene lactones such as parthenolide, *q.v.*: P. J. Hylands, D. M. Hylands in *Development of Drugs and Modern Medicines,* J. W. Gorrod *et al.,* Eds. (Ellis Horwood, Chichester, 1986) pp 100-104. Inhibition of prostaglandin biosynthesis by feverfew extract: H. O. J. Collier *et al., Lancet* **2**, 922 (1980). Effect on human platelet phospholipase: A. N. Makheja, J. M. Bailey, *ibid.* 1054 (1981); *eidem, Prostaglandins, Leukotrienes Med.* **8**, 653 (1982); J. K. Thakkar *et al., Biochim. Biophys. Acta* **750**, 134 (1983). Inhibition of platelet secretory activity: S. Heptinstall *et al., Lancet* **1**, 1071 (1985); S. Heptinstall *et al., J. Pharm. Pharmacol.* **39**, 459 (1987); structure/anti-secretory activity study: W. A. Groenewegen *et al., ibid.* **38**, 709 (1986). Clinical trials in migraine using freeze dried feverfew leaves: E. S. Johnson *et al., Brit. Med. J.* **291**, 569 (1985). Use of oil extract in migraine: E. S. Johnson *et al.,* U.S. pat. **4,758,433** (1988 to R. P. Scherer). *Review:* M. I. Berry, *Pharm. J.* **232**, 611-614 (1984). Brief review of activity and possible side effects: C. A. Baldwin *et al., ibid.* **239**, 237-238 (1987).

4014. Fibrin. Fibrin monomer is fibrinogen from which one or two peptides have been removed by means of thrombin: Laki, Chandrasekhar, *Nature* **197**, 1267 (1963). The term fibrin is usually applied to polymerized fibrin monomer. Terminal clotting takes place in four steps: (1) fibrinogen hydrolyzes under the influence of thrombin into fibrin and fibrinopeptide fragments; (2) fibrin forms soft clots which can be readily dispersed; (3) thrombin activates fibrin-stabilizing factor, *q.v.,* an enzyme precursor, present in blood plasma; (4) fibrin in the networks cross-links under the influence of the activated FSF to give the final hard clots: *Chem. & Eng. News* **43**, no. 32, 38 (1965); O. D. Ratnoff, B. Bennett, *Science* **179**, 1291 (1973). Fibrin occurs in two principal forms, *fibrin-i,* "insoluble" fibrin, differing from *fibrin-s,* "soluble" fibrin, by urea solubility as well as other characteristics. Fibrin-i is formed through the reaction of a fibrinogen-like plasma protein, FSF, which in the presence of Ca^{2+} converts what would otherwise be a "soluble" weakly bonded gel into a covalently bonded, insol clot: Rosenberg, Carman, *Nature* **204**, 994 (1964). Chemical studies of crosslinking segments: Chen, Doolittle, *Proc. Nat. Acad. Sci. USA* **66**, 472 (1970); *eidem, Biochemistry* **10**, 4486 (1971); Doolittle *et al., Biochem. Biophys. Res. Commun.* **44**, 94 (1971). *Reviews:* W. H. Seegers, *Prothrombin* (Harvard

University Press, 1962) 728 pp; Laki, Gladner, *Physiol. Rev.* **44,** 127 (1964); Lorand, *Fed. Proc.* **24,** no. 4, part 1, 784 (1965); A. L. Copley, *Thromb. Res.* **14,** 249 (1979). Review of chemistry: several authors, *Thromb. Diath. Haemorrh., Suppl.* **39** (1970).

4015. Fibrinogen. Factor I; Parenogen. A plasma glycoprotein belonging structurally to the keratin-myosin group. Synthesized and secreted by hepatic parenchymal cells. Present to the extent of 0.3-0.4 g/100 ml in human plasma. Essential to the clotting of blood. Its synthesis is greatly increased during acute inflammatory challenge. The fibrinogen molecule consists of three peptide chains, α (A), β (B), and γ (C), crosslinked by several disulfide bonds. The mol wt of about 400,000 represents a dimeric form of the molecule. Thrombin releases fibrinopeptides A and B from the *N*-terminal ends of the α and β chains of fibrinogen in the formation of fibrin during coagulation. Because fibrinogen is less sol than other plasma proteins it is readily separated by precipitation with sodium chloride: Florkin, *J. Biol. Chem.* **87,** 629 (1930); or with ammonium sulfate: Nanninga, *Arch. Neer. Physiol.* **24,** 241 (1946). Prepn from human plasma: Edsall *et al., J. Am. Chem. Soc.* **69,** 2731 (1947). Purification: Cama *et al., Naturwiss.* **48,** 574 (1961). End group determination: Lorand, Middlebrook, *Science* **118,** 515 (1953). Structure studies: Schauenstein, Hochenegger, *Z. Naturforsch.* **8b,** 473 (1953); Edsall, *J. Polymer Sci.* **12,** 253 (1954). *Reviews:* Seegers, *Physiol. Rev.* **34,** 711 (1954); Lorand, *Fed. Proc.* **24,** no. 4, part 1, 784 (1965); several authors, *Thromb. Diath. Haemorrh., Suppl.* **39** (1970); A. L. Copley, *Thromb. Res.* **14,** 249 (1979). Fibrinogen has been shown to be the receptor for the endogenous lectin (agglutinin) secreted by activated platelets: T. K. Gartner *et al., Nature* **289,** 688 (1981). Review of biosynthesis: G. M. Fuller, D. G. Ritchie, *Ann. N.Y. Acad. Sci.* **389,** 308-322 (1982).

Sparingly sol in water. Aq solns are viscous. Isoelectric point 5.5. Readily denatured by heating to 56° or higher, and by chemical agents such as salicylaldehyde, naphthoquinone sulfonates, ninhydrin, and alloxan. Small amounts of papain will clot fibrinogen, but larger amounts will digest the clot.

THERAP CAT: Coagulant (clotting factor).

4016. Fibroblast Growth Factor. FGF. Growth stimulatory factor originally isolated from bovine brain and pituitary and found to stimulate DNA synthesis in cultured fibroblast cells. Isoln: D. Gospodarowicz, *Nature* **249,** 123 (1974). Mitogenic effect on cultured cell lines and induction of amphibian limb regeneration *in vivo:* D. Gospodarowicz *et al., Advan. Metab. Disord.* **8,** 301 (1975). Two closely related forms have been identified, known as basic (*bFGF*) and acidic (*aFGF*) fibroblast growth factors, having a total amino acid sequence homology of 55%. Both induce the proliferation and differentiation of a wide variety of cell types, including corneal and vascular endothelial cells, myoblasts, chondrocytes, osteoblasts and glial cells. FGF has neurotrophic and angiogenic activity and may play an important role in the wound healing process. Purification of bFGF from pituitary: D. Gospodarowicz, *J. Biol. Chem.* **250,** 2515 (1975); from brain: D. Gospodarowicz *et al., ibid.* **253,** 3736 (1978). Identification of aFGF from bovine brain: K. A. Thomas *et al., ibid.* **255,** 5517 (1980). Comparison of fibroblast growth factors: S. K. Lemmon *et al., J. Cell Biol.* **95,** 162 (1982). Purification and characterization of aFGF: K. A. Thomas *et al., Proc. Nat. Acad. Sci. USA* **81,** 357 (1984). Identity of bFGF from brain and pituitary: D. Gospodarowicz *et al., ibid.* 6963. Amino acid sequence of bFGF: F. Esch *et al., ibid.* **82,** 6507 (1985); of aFGF: G. Gimenez-Gallego *et al., Science* **230,** 1385 (1985); F. Esch *et al., Biochem. Biophys. Res. Commun.* **133,** 554 (1985). Possible identity of aFGF with *endothelial cell growth factor* (*ECGF*) and *eye-derived growth factor-II* (*EDGF-II*): A. B. Schreiber *et al., J. Cell Biol.* **101,** 1623 (1985); of bFGF with *macrophage-derived growth factor* (*MDGF*): A. Baird *et al., Biochem. Biophys. Res. Commun.* **126,** 358 (1985); of FGFs with *retina-derived endothelial cell growth factors*: A. Baird *et al., Biochemistry* **24,** 7855 (1985). Cloning of cDNA for bovine bFGF: J. A. Abraham *et al., Science* **233,** 545 (1986); for human bFGF: T. Kurokawa *et al., FEBS Letters* **213,**

189 (1987); M. Iwane *et al., Biochem. Biophys. Res. Commun.* **146,** 470 (1987). Expression of a chemically synthesized gene for bioactive bovine aFGF: D. L. Linemeyer *et al., Biotechnology* **5,** 960 (1987). Receptor binding study: G. Neufeld, D. Gospodarowicz: *J. Biol. Chem.* **261,** 5631 (1986).

FGF-like factors have been isolated from several human tumor cell lines: R. R. Lobb *et al., Biochem. Biophys. Res. Commun.* **139,** 861 (1986); M. Klagsbrun *et al., Proc. Nat. Acad. Sci. USA* **83,** 2448 (1986); D. Moscatelli *et al., J. Cell. Physiol.* **129,** 273 (1986). FGF has also been shown to be structurally homologous to the protein products of several oncogenes: C. Dickson, G. Peters, *Nature* **326,** 833 (1987); M. Taira *et al., Proc. Nat. Acad. Sci. USA* **84,** 2980 (1987); P. Delli Bovi *et al., Cell* **50,** 729 (1987). Review of tissue distribution and bioactivity of bFGF: A. Baird *et al., Rec. Prog. Horm. Res.* **42,** 143-205 (1986). Review of structural characterization and biological functions: D. Gospodarowicz *et al., Endocrine Rev.* **8,** 95-114 (1987). Potential role in the control of pituitary and gonad development: D. Gospodarowicz, N. Ferrara, *J. Steroid Biochem.* **32,** 183-191 (1989). *Reviews:* K. A. Thomas, G. Gimenez-Gallego, *Trends Biochem. Sci.* **11,** 81-84 (1986); D. Gospodarowicz *et al., Mol. Cell. Endocrinol.* **46,** 187-204 (1986); K. A. Thomas, *FASEB J.* **1,** 434-440 (1987).

Acidic fibroblast growth factor, pI 5-7. Exists in 2 microheterogeneous forms: *aFGF-1,* a 140 amino acid peptide, mol wt 15,900 daltons, and *aFGF-2,* an amino truncated form lacking 6 amino terminal residues, mol wt 15,200 daltons.

Basic fibroblast growth factor, pI 9.6. 146 amino acid peptide, mol wt ~16,000 daltons. Also exists as an amino truncated form, *des 1-15 bFGF,* lacking the first 15 amino acid residues.

4017. Fibroin. Protein filaments produced by members of the phylum *Arthropoda,* particularly by certain species belonging to the classes *Insecta* (insects) and *Arachnida* (spiders, etc.). Fibroin is the main protein of silk and is secreted by the insect in its silk glands together with *sericin,* the second silk protein, in aq soln and converted into silk by a process called "spinning." Organization of amino acid sequences in silk fibroin of *Bombyx mori* and general review of silk proteins: F. Lucas, K. M. Rudall in *Comprehensive Biochemistry,* M. Florkin, E. M. Stotz, Eds., **vol. 26B** (Elsevier, New York, 1968) pp 475-558. *See also* P. M. Lizardi, *Cell* **18,** 581 (1979); L. P. Gage, R. F. Manning, *J. Biol. Chem.* **255,** 9444 (1980). Chemical and crystalline structures: B. Lotz, F. Colonna Cesari, *Biochimie* **61,** 205 (1979).

Pale yellow mass resembling silk. Insol in water, alcohol, ether, dil alkalies. Sol in concd alkalies, concd mineral acids and in ammoniacal nickel oxide soln.

4018. Fibronectins. α_2-SB glycoproteins; α_2-opsonins; CIG; CSP; CAF; GAP A; LETS; Zeta protein. High mol wt multifunctional glycoproteins, found on cell surfaces, in body fluids (especially plasma), in soft connective tissue matrices, and in most basement membranes. Although fibronectins apparently function as adhesive ligand-like molecules, the full range of their biological activities and relationships are still being elucidated. Their importance in cell adhesion, oncogenic transformation, reticuloendothelial system function, embryonic differentiation, phagocytosis, hemostasis, and chemotaxis is being studied. Discovered as a result of isoln of a partially purified fraction of human plasma and initially termed "cold-insoluble globulin" or CIG: P. R. Morrison *et al., J. Am. Chem. Soc.* **70,** 3103 (1948). Subsequent studies described various proteins or factors, named according to sources or biological activities, that are now designated as fibronectins. At least two types are known to exist, termed plasma and cellular fibronectin, respectively. Both forms contain subunits of mol wt > 200,000, joined by disulfide bonds. They are similar in amino acid compositions, carbohydrate structures and secondary and tertiary structures; they cannot be distinguished in biological activity in assays of cell interactions with substrates or in opsonic activity for macrophages. They differ in their effects on cell morphology, on alignment of transformed cells and on hemagglutination; they also have differences in solubility and in the number of subunits linked by disulfide bonds. Monoclonal antibody studies have indicated that the two forms are distinct: B. T. Atherton, R. O. Hynes, *Cell* **25**, 133 (1981); K. D. Noonan *et al., J. Supramol. Struct. Cell Biochem.* **5**, Suppl, 302 (1981). Regulation of fibronectin biosynthesis: D. R. Senger *et al., Am. J. Physiol.* **245**, 144 (1983). Structure-function relationships: T. Vartio, A. Vaheri, *Trends Biochem. Sci.* **8**, 442 (1983). Role in cellular adhesion, spreading and cytoskeletal organization: I. Virtanen *et al., Nature* **298**, 660 (1982); in phagocytosis: L. Van de Water *et al., Science* **220**, 201 (1983); in wound healing: G. R. Martin *et al.,* "Regulation of Tissue Structure and Repair by Collagen and Fibronectin" in *The Surgical Wound,* P. Dineen, C. Hildrick-Smith, Eds. (Lea & Febiger, Philadelphia, 1981) pp 110-122. Use in treatment of corneal trophic ulcer therapy: T. Nishida *et al., Arch. Ophthalmol.* **101**, 1046 (1983). Review of role in cellular adhesion: S. K. Akiyama *et al., J. Supramol. Struct. Cell Biochem.* **16**, 345-358 (1981); role in inflammation: D. F. Mosher *et al., Adv. Inflamm. Res.* **2**, 187-207 (1981); C. Bianco, *Ann. N.Y. Acad. Sci.* **408**, 602-609 (1983); actvity in various disease states: S. K. Akiyama, K. M. Yamada, "Fibronectin in Disease" in *Monographs in Pathology* No. 24, N. Kaufman, Ed., entitled "Connective Tissue Diseases", B. M. Wagner *et al.,* Eds. (Williams & Wilkins, Baltimore, 1983). *General Reviews:* E. Pearlstein *et al., Mol. Cell. Biochem.* **29**, 103-128 (1980); M. W. Mosesson, D. L. Amrani, *Blood* **56**, 145-158 (1980); E. Ruoslahti, *J. Oral Pathol.* **10**, 3-13 (1981); R. O. Hynes, K. M. Yamada, *J. Cell Biol.* **95**, 369-377 (1982); R. O. Hynes, *Scientific American* **254**, 42-51 (1986).

4019. Fichtelite. 1,2,3,4,4a,4bα,5,6,7,8,8aβ,9,10,10aα-Tetradecahydro-7α-isopropyl-1β,4aβ-dimethylphenanthrene. $C_{19}H_{34}$; mol wt 262.46. C 86.94%, H 13.06%. From decayed wood of conifers: Bromeis, *Ann. Pharm.* **37**, 304 (1841). Structure: L. Ruzicka, E. Waldmann, *Helv. Chim. Acta* **18**, 611 (1935); Crowfoot, *J. Chem. Soc.* **1938**, 1241. Stereochemistry: Burgstahler, Marx, *Tetrahedron Letters* **1964**, 3333. Synthesis from abietic acid: Jensen, Johnson, *J. Org. Chem.* **32**, 2045 (1967); Burgstahler, Marx, ibid. **34**, 1562 (1969). Synthesis of *dl*-form: Johnson *et al., J. Am. Chem. Soc.* **88**, 3859 (1966); **90**, 5872 (1968); D. F. Taber, S. A. Saleh, ibid. **102**, 5085 (1980).

Crystals from methanol, mp 45-46°. bp₄₃ 235-236°. d_4^{22} 0.9380. n_D^{20} 1.5052. $[\alpha]_D$ +19°.

4020. Ficin. Ficus proteinase; Ficus protease; Debricin; Higueroxyl Delabarre. A proteolytic enzyme of est. mol wt 23,800-25,500 which requires a free sulfhydryl group for activity and as such is a member of a group which includes papain, *q.v.*, and bromelain, *q.v.* Occurs in the latex of tropical trees of the genus *Ficus* subgenus *Pharmacosyce,* Moraceae (Oje trees). The commercial product is a concentrate prepd by filtering and drying the latex of *Ficus glabrata* H. B. & K., Moraceae. First crystallized from fresh fig latex: Walti, *J. Am. Chem. Soc.* **60**, 493 (1938). Characterization: Cohen, *Nature* **182**, 659 (1958); Englund *et al., Biochemistry* **7**, 163 (1968). Purification: Gibian, Bratfisch, U.S. pat. 2,950,227 (1960 to Schering AG). Amino acid composition: Wong, Liener, *Biochem. Biophys. Res. Commun.* **17**, 470 (1964); Metrione *et al., Arch. Biochem. Biophys.* **122**, 137 (1967); Husain, Lowe, *Biochem. J.* **117**, 333 (1970). Pharmacodynamics: H. Heisto, M. K. Fagerhol, *Transfusion* **19**, 545 (1979); A. Perkash *et al., ibid.* **20**, 301 (1980). *Reviews:* Liener, Friedenson, *Methods Enzymol.* **19**, 260-273 (1970); Glazer, Smith, *The Enzymes* vol. III, P. D. Boyer, Ed. (Academic Press, New York, 3rd ed., 1971) pp 538-542.

Buff to cream-colored hygroscopic powder. Acrid odor, growing stronger with age. Bulky, approx 3 ml/g, not free-flowing. Appears dry, even with 15% H_2O present. Not completely sol in water, 2-10% insol material. The sol portion of 1 g dissolves in 3 ml H_2O. pH (2% soln): 4.1. Insol in usual organic solvents. Loses about 10-20% activity when stored 1-3 yrs at ordinary temp and atm conditions. Aq solns inactivated at 100°, solid partially inactivated within a few hours. Solns relatively stable between pH 4-8.5; incompatible with iron, copper, aluminum. Gelatin, coagulated egg white, casein, meat and most protein-like material hydrolyzed in aq ficin solns. *Caution:* Handle ficin carefully because of its tissue-dissolving properties. Ficin is 10-20 times as active as papain in regard to milk clotting; 4-10 times as active, in general. LD_{50} orally in rats and mice: approx 10 g/kg; in rabbits and guinea pigs: approx 5 g/kg.

Note: The name ficin is currently used to describe both the crude dried latex from different species of the genus *Ficus,* as well as the enzyme itself.

USE: Protein digestant. In the brewing industry as a chill-proofing agent in beer. In the cheese industry as a substitute for rennet in the coagulation of milk. In the meat industry as a meat tenderizer and as an agent for removing casings from formed sausage. In the leather industry for the bating of leather. In the textile industry for shrinkproofing wool, for removing gelatin from sized thread, and mixed with amylases and maltases as a spot remover. In the prepn of peptones. For solubilizing protein material in spent grains. Also used in the determination of the Rh factor. Speeds 10 times the agglutination of human blood cells by the Rh factor when in contact with the anti-Rh serum. *Human Toxicity:* Can cause irritation to skin, eyes, mucous membranes. Large doses by mouth cause purging.

THERAP CAT (VET): Has been used as a trichuricide.

4021. Filicinic Acid. 3,5-Dihydroxy-4,4-dimethyl-2,5-cyclohexadien-1-one; 1,1-dimethylcyclohexane-2,4,6-trione; gem-dimethylphloroglucinol; Filicinsäure (German). C_8H_{10}-O_3; mol wt 154.16. C 62.32%, H 6.54%, O 31.13%. Isolated from male fern and by hydrolytic decompn of filixic acid or aspidin: Boehm, *Ann.* **302**, 171 (1898); **329**, 321 (1903). Synthesis: Robertson, Sandrock, *J. Chem. Soc.* **1933**, 1617; Angus *et al., Chem. & Ind. (London)* **1954**, 546; Inagaki *et al., J. Pharm. Soc. Japan* **76**, 1258 (1956); Hoefer, Riedl, *Angew. Chem.* **74**, 501 (1962); *eidem, Ann.* **656**, 127 (1962).

Prisms from water or benzene, mp 214-215° (dec). pK 5.8. Practically insol in water, petr ether; sol in 70 parts boiling water, 10 parts boiling alcohol; sparingly sol in ether, benzene, glacial acetic acid. Reduces Tollen's reagent.

3,5-Diacetylfilicinic acid, $C_{12}H_{14}O_5$, crystals from methanol, mp 65-66°. pK 5.0. Sol in cold methanol, benzene, ether; slightly sol in hot water.

4022. Filipin. Filimarisin; U 5956; NSC-3364. Polyene antibiotic complex containing at least eight pentaene compounds, which has been resolved into three pure components, filipin II, III (major), and IV, which differ in the number of hydroxyl groups present (8, 9, 9, resp.) and filipin I, a mixture of at least 5 components. Described as a single entity in the earlier literature. Initial isolation from *Streptomyces filipenensis* in Philippine soil: Whitfield *et al., J. Am. Chem. Soc.* **77**, 4799 (1955). Early structural work: Dhar *et al., Proc. Chem. Soc.* **1960**, 310; Djerassi *et al., Tetrahedron Letters* **1961**, 383; Golding, Rickards, *ibid.* **1964**, 2615; Dhar *et al., J. Chem. Soc.* **1964**, 842; Ceder, Ryhage, *Acta Chem. Scand.* **18**, 588 (1964). Separation of filipin complex into components: Bergy, Eble, *J. Antibiot.* **23**, 414 (1970); *see also* Rickards *et al., ibid.* **603**. Mechanism of action: R. W. Holz in *Antibiotics* vol. 5, pt. 2, F. E. Hahn, Ed. (Springer-Verlag, New York, 1979) pp 313-340. Filipin also interacts specifically with 3β-hydroxysterols (e.g. cholesterol): P. M. Elias *et al., J. Histochem. Cytochem.* **27**, 1247 (1979); *see also* N. J. Severs, H. J. Simons, *Nature* **303**, 637 (1983).

filipin III

Yellow, feathery needles from chloroform, mp 195-205°. Sensitive to air. $[\alpha]_D^{22}$ −148.3° (c = 0.89 in methanol). uv max (methanol): 322, 338, 355 nm ($E_{1cm}^{1\%}$ 910, 1360, 1330). Freely sol in DMF, pyridine. Also sol in 95% ethanol, methanol, butanol, isopropanol, glacial acetic acid, ether. Practically insol in water, chloroform.

Filipin III. $C_{35}H_{58}O_{11}$, *4,6,8,10,12,14,16,27-octahydroxy-3-(1-hydroxyhexyl)-17,28-dimethyloxacyclooctacosa-17,19,21,-23,25-pentaen-2-one; 15-deoxylagosin.* Isomeric with filipin IV. Crystals from propanol, mp 163-180° (Bergy, Eble). $[\alpha]_D^{25}$ −245° (c = 0.8 in DMF). uv max (methanol): 243, 308, 321, 337, 354 nm ($E_{1cm}^{1\%}$ 62, 413, 851, 1368, 1343).

USE: Produces arthritis rapidly in rabbits: Pras, Weissman, *Drug Trade News*, July 4, 1966, p 40; as a sterol probe in freeze-fracture cytochemistry.

THERAP CAT: Antifungal.

4023. Filixic Acids. Filicin; filicic acid; Filixsäure (German). Natural filixic acid from male fern, *Dryopteris filixmas*, is a mixture of six homologues, the 3 main components (BBB, PBB, PBP) of which are obtained by recrystallization from ethyl acetate. When raw filixic acid is purified by treatments with methanol but without recrystns from ethyl acetate, a product is obtained which seems to contain 3 additional homologues with an analogous composition but with acetyl groups in the side chain. Isoln: Luck, *Ann.* **54**, 119 (1845); Boehm, *ibid.* **318**, 253 (1901). Structure studies: Riedl, *Ann.* **585**, 32 (1954); Chan, Hassal, *Experientia* **13**, 349 (1957). Structure of the three main filixic acids: Penttilä, Sundman, *Acta Chem. Scand.* **17**, 191 (1963).

Natural filixic acid, pale-yellow plates from ethyl acetate, mp 184-185°. uv max: 228, 288 nm (ϵ 41,000, 29,000). Practically insol in water, methanol, acetone; sol in chloroform, benzene, warm ethyl acetate, acetic acid; slightly sol in ether.

BBB, $C_{36}H_{44}O_{12}$, ($R_1 = R_2 = C_3H_7$); crystals from ethyl acetate, mp 172-174°.
PBB, $C_{35}H_{42}O_{12}$, ($R_1 = C_2H_5$; $R_2 = C_3H_7$); crystals from ethyl acetate, mp 184-186°.
PBP, $C_{34}H_{40}O_{12}$, ($R_1 = R_2 = C_2H_5$); crystals from ethyl acetate, mp 192-194°.

THERAP CAT (VET): Has been used as an anthelmintic.

4024. Fipexide. *1-(1,3-Benzodioxol-5-ylmethyl)-4-[(4-chlorophenoxy)acetyl]piperazine; 1-[(p-chlorophenoxy)acetyl]-4-piperonylpiperazine; 1-[2-(4-chlorophenoxy)acetyl]-4-(3,4-methylenedioxybenzyl)piperazine.* $C_{20}H_{21}ClN_2O_4$; mol wt 388.85. C 61.78%, H 5.44%, Cl 9.12%, N 7.20%, O 16.46%. Prepn of hydrochloride: A. Buzas, R. Pierre, *Fr. pat.* M7524 (1970 to Lab Bouchard), *C.A.* **75**, 112865r (1971). Manufacturing process: G. P. Gardini *et al.*, *U.S. pat.* 4,225,714 (1980 to Farmaceutici Geymon.). Improved prepn: G. P. Gardini *et al.*, *Syn. Commun.* **12**, 887 (1982). Determn by TLC: G. Musumarra *et al.*, *J. Chromatog.* **350**, 151 (1985). Toxicology and pharmacology: G. David *et al.*, *Acta Therap.* **11**, 387 (1985). Clinical trial in the elderly: B. Bompani, G. Scali, *Curr. Med. Res. Opin.* **10**, 99 (1986).

Hydrochloride, $C_{20}H_{22}Cl_2N_2O_4$, *BP 662, Attentil, Vigilor.* Crystals from ethanol, mp 230-232°. LD_{50} in Swiss mice, Sprague-Dawley rats, Wistar rats (mg/kg): 4150, 4482, 7000 orally; 499, 537, 450 i.p. (David).

THERAP CAT: Nootropic.

4025. Firefly Luciferin. *4,5-Dihydro-2-(6-hydroxy-2-benzothiazolyl)-4-thiazolecarboxylic acid;* 2-(6-hydroxybenzothiazol-2-yl)-2-thiazoline-4-carboxylic acid; D-(−)-luciferin. $C_{11}H_8N_2O_3S_2$; mol wt 280.33. C 47.13%, H 2.86%, N 10.00%, O 17.12%, S 22.88%. Light emission in the American firefly, *Photinus pyralis*, has been shown to involve the interaction of magnesium ion, oxygen, ATP, the enzyme luciferase, and the oxidizable substrate luciferin. Isoln from fireflies (yield from 15,000 active fireflies about 9 mg): Bitler, McElroy, *Arch. Biochem. Biophys.* **72**, 358 (1957); from Japanese firefly (*Luciola cruciata*): Kishi *et al.*, *Tetrahedron Letters* **1968**, 2847. Structure and synthesis: E. H. White *et al.*, *J. Am. Chem. Soc.* **83**, 2402 (1961); **85**, 337 (1963); *J. Org. Chem.* **30**, 2344 (1965); S. Seto *et al.*, *Bull. Chem. Soc. Japan* **36**, 331 (1963). Configuration: G. E. Blank *et al.*, *Biochem. Biophys. Res. Commun.* **42**, 583 (1971). *Review:* L. J. Bowie, *Methods Enzymol.* **57**, 15 (1978).

Pale yellow needles from methanol, dec 189.5-190°. Recrystallizes with difficulty and sublimes with decomposition and decarboxylation. $[\alpha]_D^{22}$ −36° (c = 1.2 in DMF). uv max (H₂O): 268, 327 nm (log ϵ 3.88, 4.27). Slightly sol in water at pH 6.5. Sol in alkaline aq solns, methanol, acetone, ethyl acetate, DMF. Aqueous solns are sensitive to extremes in pH, esp. in presence of light and oxygen. Racemization occurs rapidly in some solvents. Should be stored for extended periods dry, under nitrogen atmosphere in light-tight containers.

USE: In the assay of ATP. Solutions down to 10^{-11} molar can be analyzed.

4026. Fisetin. *2-(3,4-Dihydroxyphenyl)-3,7-dihydroxy-4H-1-benzopyran-4-one; 3,3',4',7-tetrahydroxyflavone;* 5-desoxyquercetin; fisidenolon 1521; C.I. 75620; C.I. Natural Brown 1. $C_{15}H_{10}O_6$; mol wt 286.23. C 62.94%, H 3.52%, O 33.54%. Coloring matter isolated from *Rhus cotinus* L. Anacardiaceae: Chevreul, *Lecons Chim. Appl. a la Teint.* **2**, 169 (1833) [cf. J. Schmid, *Ber.* **19**, 1734 (1886)]; from *R. rhodanthema* Müll: Perkin, *J. Chem. Soc.* **71**, 1194 (1897); Oyamada, *Ann.* **538**, 44 (1939); from heartwood of *Acacia* spp, Leguminosae: Roux, Paulus, *Biochem. J.* **77**, 315 (1960); Roux *et al.*, *ibid.* **78**, 834 (1961). Structure: Seshadri, *Ann. Rev. Biochem.* **20**, 492 (1951). Synthesis: Kostanecki *et al.*, *Ber.* **37**, 784 (1904); Allan, Robinson, *J. Chem. Soc.* **129**, 2334 (1926). Inhibition of aflatoxin cytotoxicity: A. G. Schwartz, W. R.Rate, *J. Environ. Pathol. Toxicol.* **2**, 1021 (1979). Mutagenicity studies: J. P. Brown *et al.*, *Biochem. Soc. Trans.* **5**, 1489 (1977); J. P. Brown, P. S. Dietrich, *Mutat. Res.* **66**, 223 (1979).

Yellow needles from dil alc, dec 330°. uv max (ethanol): 252, 320, 360 nm (log $E_{1cm}^{1\%}$ 2.62, 2.51, 2.73). Sol in alcohol, acetone, acetic acid, solns of fixed alkali hydroxides; practically insol in water, ether, benzene, chloroform and petr ether. Its soln in dil alcohol NaOH shows a dark green fluorescence. LD$_{50}$ in mice: 180 mg/kg i.v., *RTECS* Vol. I, R. J. Lewis, R. L. Tatken, Eds. (1980) p 686.

4027. Flavaspidic Acid. *3,5-Dihydroxy-4,4-dimethyl-2-(1-oxobutyl)-6-[[2,4,6-trihydroxy-3-methyl-5-(1-oxobutyl)-phenyl]methyl]-2,5-cyclohexadien-1-one;* 3′-[(5-butyryl-2,4-dihydroxy-3,3-dimethyl-6-oxo-1,4-cyclohexadien-1-yl)methyl]-5′-methylphlorobutyrophenone; polystichocitrin; Toxifren. C$_{24}$H$_{30}$O$_8$; mol wt 446.48. C 64.56%, H 6.77%, O 28.67%. Isolated from the rhizomes of male fern: Boehm, *Ann.* **318**, 253 (1901); **329**, 310 (1903). Structure and synthesis: McGookin *et al.*, *J. Chem. Soc.* **1953**, 1828; Riedl, *Ann.* **585**, 32 (1954); Aebi, *Helv. Chim. Acta* **39**, 153 (1956). Crystal structure of α- and β-forms: Erämetsä, Penttilä, *Acta Chem. Scand.* **24**, 3335 (1970).

α-Form, orthorhombic crystals from methanol or ethanol, mp 92°; solidifies again at 110° and melts again at 156°. LD$_{50}$ orally in mice: 690 mg/kg, Airaksinen *et al.*, *Acta Pharmacol. Toxicol.* **25**, 33 (1967).

β-Form, monoclinic crystals from benzene, xylene or acetic acid, mp 156°.

4028. Flavine-Adenine Dinucleotide. *Riboflavine 5′-(trihydrogen diphosphate) 5′ → 5′-ester with adenosine; adenosine 5′-(trihydrogen diphosphate) 5′ → 5′-ester with riboflavine;* FAD; riboflavine 5′-adenosine diphosphate; isoalloxazine-adenine dinucleotide; Fademin; Flamitajin; Flanin F; Flavitan. C$_{27}$H$_{33}$N$_9$O$_{15}$P$_2$; mol wt 785.56. C 41.28%, H 4.23%, N 16.05%, O 30.55%, P 7.89%. The prosthetic group of certain flavoproteins including D-amino acid oxidase, glucose oxidase, glycine oxidase, fumaric hydrogenase, histaminase, and xanthine oxidase. Isoln from yeast: Warburg *et al.*, *Biochem. Z.* **297**, 417 (1938). Structure and isoln from liver, kidneys, hearts, muscles: Warburg, Christian, ibid. **298**, 150 (1938); isoln from the mycelium of *Eremothecium ashbyii:* Yagi, Tada, *Biochem. Prepn.* **7**, 51 (1959); Masuda *et al.*, U.S. pat. 2,973,305 (1961 to Takeda). Synthesis: Christie *et al.*, *J. Chem. Soc.* **1954**, 46; Huennekens, Kilgour, *J. Am. Chem. Soc.* **77**, 6716 (1955); DeLuca, Kaplan, *J. Biol. Chem.* **223**, 569 (1956); Moffatt, Khorana, *J. Am. Chem. Soc.* **80**, 3756 (1958). Review: A. Holmgren, *Experientia* **36** (Suppl), 149-180 (1980). Review of FAD and other flavin coenzymes: Beinert, *The Enzymes* vol. 2, P. D. Boyer *et al.*, Eds. (Academic Press, New York, 2nd ed., 1960) pp 339-

416; *see also* vol. **XIII Part B** (Academic Press, New York, 3rd ed., 1975), several authors.

Isolated as the barium salt, C$_{27}$H$_{31}$BaN$_9$O$_{15}$P$_2$, small yellow spheres clustered like grapes. Absorption max: 366, 445 nm. The absorption curve is practically identical with that of riboflavine. There is some stronger absorption between 450 and 510 nm resulting in aq solns which are more reddish and less green than those of riboflavine. The appearance of a strong greenish fluorescence indicates decomposition and loss of catalytic activity.

4029. Flavipucine. *cis-(−)-6-Methyl-2-(3-methyl-1-oxobutyl)-1-oxa-5-azaspiro[2.5]oct-6-ene-4,8-dione;* 3′-isovaleryl-6-methylpyridine-3-spiro-2′-oxirane-2(1H),4(3H)-dione; glutamicine; glutamycin. C$_{12}$H$_{15}$NO$_4$; mol wt 237.26. C 60.75%, H 6.37%, N 5.90%, O 26.97%. Antibiotic substance produced by a strain of *Aspergillus flavipes;* (−)-form is naturally occurring. Isoln: C. G. Casinovi *et al.*, *Tetrahedron Letters* **1968**, 3175; C. G. Casinovi *et al.*, *Ann. Ist. Super. Sanita* **5**, 523 (1969). Revised structure: J. A. Findlay, L. Radics, *J. Chem. Soc. Perkin Trans. I* **1972**, 2071; J. A. Findlay, L. Kwan, ibid. 2962. X-ray crystal structure: P. S. White *et al.*, *Can. J. Chem.* **56**, 1904 (1978). Total synthesis of (±)-flavipucine: Girotra *et al.*, *Chem. Commun.* **1976**, 566; J. A. Findlay *et al.*, *Syn. Commun.* **7**, 149 (1977). Study on biosynthetic pathway: G. Grandolini *et al.*, *J. Antibiot.* **40**, 1339 (1987).

White needles from benzene, mp 130-131°. uv max (neutral ethanol): 330 nm (ε 5400). [α]$_D^{21}$ −71.8° (c = 1 in 95% alcohol) (Casinovi, 1968); [α]$_D^{21}$ −88° (c = 1 in 95% alcohol) (Findlay, Radics).

(±)-Form, plates from benzene, mp 154-155°.

4030. Flavone. *2-Phenyl-4H-1-benzopyran-4-one;* 2-phenylchromone; 2-phenyl-γ-benzopyrone; 2-phenyl-1,4-benzopyrone. C$_{15}$H$_{10}$O$_2$; mol wt 222.23. C 81.06%, H 4.54%, O 14.40%. Prepn: Feuerstein, Kostanecki, *Ber.* **31**, 1757 (1898); Kostanecki, Tambor, *Ber.* **33**, 330 (1900); Kostanecki, Szabranski, *Ber.* **37**, 2634 (1904); Ruhemann, *Ber.* **46**, 2192 (1913); Bogert, Marcus, *J. Am. Chem. Soc.* **41**, 89 (1919). Isoln from *Primula malacoides*, Franch., *Primulaceae* and biological properties: Weller *et al.*, *Antibiot. & Chemother.* **3**, 603 (1953). Syntheses: Wheeler, *Org. Syn.* coll. vol. **IV**, 478 (1963); Y. Ashihara *et al.*, *Bull. Chem. Soc. Japan* **50**, 3298 (1977); A. Banerji, N. C. Goomer, *Synthesis* **1980**, 874.

Crystals from petr ether, mp 99-100°. Practically insol in water; sol in most organic solvents. Absorption max: 350, 405 nm.

4031. Flavopereirine. *3-Ethyl-12H-indolo[2,3-a]quinolizin-5-ium;* melinonine G. C$_{17}$H$_{14}$N$_2$; mol wt 246.30. C 82.90%, H 5.73%, N 11.37%. From *Strychnos melinoniana* Baillon., *Loganiaceae:* Bächli *et al.*, *Helv. Chim. Acta* **40**, 1167 (1957); from *Geissospermum vellosii* Baillon., *Apocynaceae:* Rapoport *et al.*, *J. Am. Chem. Soc.* **80**, 1601 (1958). Bertho *et al.*, *Ber.* **91**, 2581 (1958). Structure: Bejar *et al.*, *Compt. Rend.* **244**, 2066 (1957). Synthesis: Le Hir *et al.*, *Bull. Soc. Chim. France* **1958**, 551; Prasad, Swan, *J. Chem. Soc.* **1958**, 2024; Thesing, Festag, *Experientia* **15**, 127 (1959); Ban, Seo, *Tetrahedron* **16**, 5 (1961); Wenkert *et al.*, *J.*

Am. Chem. Soc. **84**, 3732, 4914 (1962); J. Ninomiyo *et al., Heterocycles* **9**, 1527 (1978).

Orange crystals from acetone, mp 233-235°. uv max (ethanol): 230, 238, 248, 294, 351, 390 nm (log ε 4.40, 4.43, 4.39, 4.14, 4.25, 4.14).

Perchlorate, $C_{17}H_{15}ClN_2O_4$, crystals from methanol, mp 307-308°. uv max (0.015N HCl-ethanol): 238, 294, 350, 389 nm (log ε 4.57, 4.22, 4.31, 4.21).

4032. Flavoxanthin. *5,8-Epoxy-5,8-dihydro-β,ε-carotene-3,3'-diol.* $C_{40}H_{56}O_3$; mol wt 584.85. C 82.14%, H 9.65%, O 8.21%. Carotenoid pigment. Often found in plants, but in minute amounts only and never as a principal pigment. Has same structural formula, mp, rotation, and absorption as chrysanthemaxanthin, *q.v.*, but differs sterically. Isoln from *Ranunculus acris* L., *Ranunculaceae:* Kuhn, Brockmann, *Z. Physiol. Chem.* **213**, 192 (1932). Structure: Karrer, Rutschmann, *Helv. Chim. Acta* **25**, 1144 (1942). Partial synthesis: Karrer, Jucker, *ibid.* **28**, 300 (1945). Absolute configuration: H. Cadosch *et al., ibid.* **61**, 783 (1978).

Golden-yellow aggregates of prisms, mp 184°. Absorption max (chloroform): 430, 459 nm. $[\alpha]_{Cd}^{20}$ +190° (c = 0.04 in benzene). Freely sol in chloroform, benzene, acetone; less sol in methanol, ethanol. Almost insol in petr ether.

4033. Flavoxate. *3-Methyl-4-oxo-2-phenyl-4H-1-benzopyran-8-carboxylic acid 2-(1-piperidinyl)ethyl ester;* 3-methylflavone-8-carboxylic acid β-piperidinoethyl ester; 2-piperidinoethyl 3-methyl-4-oxo-2-phenyl-4H-1-benzopyran-8-carboxylate; 2-piperidinoethyl 3-methylflavone-8-carboxylate. $C_{24}H_{25}NO_4$; mol wt 391.45. C 73.63%, H 6.44%, N 3.58%, O 16.35%. Smooth muscle relaxant. Prepn of the hydrochloride: P. Da Re *et al., J. Med. Pharm. Chem.* **2**, 263 (1960); P. Da Re, U.S. pats. **2,921,070** and **3,350,411** (1960 to Recordati and 1967 to Seceph). Pharmacological studies: I. Setnikar *et al., J. Pharmacol. Exp. Ther.* **130**, 356 (1960). Mechanism of action: P. Cazzulani *et al., Arch. Int. Pharmacodyn.* **274**, 189 (1985). Pharmacokinetics in rats: I. Setnikar *et al., Arzneimittel-Forsch.* **25**, 1916 (1975); in man: M. Bertoli *et al., Pharmacol. Res. Commun.* **8**, 417 (1976). *In vivo* activity: C. Pietra, P. Cazzulani, *Farmaco Ed. Prat.* **41**, 267 (1986). Determn in tissues: A. Cova, I. Setnikar, *Arzneimittel-Forsch.* **25**, 1707 (1975). Clinical evaluations: D. V. Bradley, R. J. Cazort, *J. Clin. Pharmacol.* **10**, 65 (1970); A. Zanollo, F. Catanzaro, *Urol. Int.* **35**, 176 (1980). Clinical comparison with phenazopyridine, *q.v.*: S. Gould, *Urology* **5**, 612 (1975). Short review: R. Ruffmann, A. Sartini, *Drug Exp. Clin. Res.* **13**, 57 (1987).

Crystals, pK 7.3. Soly in water at 37°: 0.001% (w/v). Sol

in ethanol and chloroform. LD_{50} in rats (mg/kg): 1110 orally; 20.8 i.v. (Setnikar, 1975).

Hydrochloride, $C_{24}H_{26}ClNO_4$, *DW-61, Rec 7-0040, Bladderon, Genurin, Spasuret, Urispas.* Crystals from ethanol + ether, mp 232-234°. LD_{50} i.v. in rats: 27.4 mg/kg (Cazzulani).

Succinate, $C_{28}H_{32}NO_8$, soly in water at 37°: 33.7% (w/v).

THERAP CAT: Antispasmodic.

4034. Flecainide. *N-(2-Piperidinylmethyl)-2,5-bis(2,2,-2-trifluoroethoxy)benzamide.* $C_{17}H_{20}F_6N_2O_3$; mol wt 414.36. C 49.28%, H 4.87%, F 27.51%, N 6.76%, O 11.58%. Prepn: E. H. Banitt, W. R. Brown, U.S. pat. **3,900,481** (1975 to Riker); of the acetate: *eidem*, U.S. pat. **4,005,209** (1977 to Riker); E. H. Banitt *et al., J. Med. Chem.* **20**, 821 (1977). Preliminary pharmacological study: J. R. Schmid *et al., Fed. Proc.* **34**, 775 (1975). *In vitro* electrophysiological study: A. B. Hodess *et al., J. Cardiovasc. Pharmacol.* **1**, 427 (1979). Antiarrhythmic effects: P. Somani, *Clin. Pharmacol. Ther.* **27**, 464 (1980). Use in acute exptl myocardial infarction: H. Gülker *et al., Z. Cardiol.* **70**, 124 (1981). Clinical study in ventricular arrhythmias: J. L. Anderson *et al., N. Engl. J. Med.* **305**, 473 (1981). Measurement of the acetate in human plasma by spectrophotofluorometry: S. F. Chang *et al., Arzneimittel-Forsch.* **33**, 251 (1983). Review of pharmacology and clinical efficacy: D. M. Roden, R. L. Woosley, *N. Engl. J. Med.* **315**, 36-41 (1986). Symposium on clinical experience: *Am. J. Cardiol.* **62**, Suppl., 1D-67D (1988).

Monoacetate, $C_{19}H_{24}F_6N_2O_5$, *R-818, Almarytm, Apocard, Modicard, Tambocor.* White granular solid from isopropyl alcohol/isopropyl ether, mp 145-147°.

THERAP CAT: Cardiac depressant (antiarrhythmic).

4035. Fleroxacin. *6,8-Difluoro-1-(2-fluoroethyl)-1,4-dihydro-7-(4-methyl-1-piperazinyl)-4-oxo-3-quinolinecarboxylic acid;* AM-833; Ro 23-6240. $C_{17}H_{18}F_3N_3O_3$; mol wt 369.34. C 55.28%, H 4.91%, F 15.43%, N 11.38%, O 13.00%. Fluorinated quinolone antibacterial. Prepn: **Belg.** pat. **887,574**; T. Irikura *et al., U.S.* pat. **4,398,029** (1981, 1983 both to Kyorin). Antibacterial spectrum *in vitro* and *in vivo:* K. Hirai *et al., Antimicrob. Ag. Chemother.* **29**, 1059 (1986). *In vitro* activity vs anaerobic bacteria: J. Wüst, U. Hardegger, *Eur. J. Clin. Microbiol.* **6**, 688 (1987). HPLC determn in biological fluids: H. Kusajima *et al., J. Chromatog.* **381**, 137 (1986). Pharmacokinetics and bioavailability: E. Weidekamm *et al., Antimicrob. Ag. Chemother.* **31**, 1909 (1987). Preliminary clinical evaluation in gonorrhea: J. B. J. Boerema *et al., J. Antimicrob. Chemother.* **21**, 140 (1988).

Hydrochloride, $C_{17}H_{19}ClF_3N_3O_3$, crystals from water, mp 269-271° (dec).

THERAP CAT: Antibacterial.

4036. Flindersine. *2,6-Dihydro-2,2-dimethyl-5H-pyrano[3,2-c]quinolin-5-one;* 2,2'-dimethyl-α-pyrano(5',6',-3,4)-2(1H)-quinolone. $C_{14}H_{13}NO_2$; mol wt 227.25. C 73.99%, H 5.77%, N 6.16%, O 14.08%. From wood of *Flindersia australis* R. Br., *Rutaceae:* Matthes, Schreiber, *Ber. Deut. Pharm. Ges.* **24**, 385 (1914). Structure: Brown *et al., Aust. J. Chem.* **7**, 348 (1954). Synthesis: *eidem, ibid.* **9**, 277 (1956); Piozzi *et al., Gazz. Chim. Ital.* **99**, 711 (1969); Bowman *et al.*, *Chem. Commun.* **1970**, 666; Huffman, Hsu, *Tetrahedron Letters* **1972**, 141.

Crystals from methanol, dec 185-186°. uv max (methanol): 235, 333, 350, 365 nm (log ε 4.42, 4.00, 4.10, 3.93). Sol in alc, benzene, chloroform, glacial acetic acid, paraffin, fatty oils, alkali hydroxides, slightly sol in petr ether. Practically insol in water.

N-Methylflindersine, $C_{15}H_{15}NO_2$, crystals from petr ether, mp 84°.

Dihydroflindersine, $C_{14}H_{15}NO_2$, hexagons from ethyl acetate, mp 229°. uv max (methanol): 225, 272, 283, 312 nm (log ε 4.44, 3.94, 3.96, 3.93).

4037. Floctafenine. *2-[[8-(Trifluoromethyl)-4-quinolinyl]amino]benzoic acid 2,3-dihydroxypropyl ester; N-[8-(trifluoromethyl)-4-quinolyl]anthranilic acid 2,3-dihydroxypropyl ester;* 4-[o-(2',3'-dihydroxypropyloxycarbonyl)phenyl]-amino-8-trifluoromethylquinoline; 8-trifluoromethyl-7-deschloroglafenine; R 4318; RU 15750; Diralgan; Idalon; Idarac; Novodolan. $C_{20}H_{17}F_3N_2O_4$; mol wt 406.37. C 59.11%, H 4.22%, F 14.03%, N 6.89%, O 15.75%. Prepn: Allais *et al.*, Ger. pat. **1,815,467** corresp to U.S. pat. **3,644,368** (1969, 1972, both to Roussel-UCLAF); G. Mouzin *et al.*, *Synthesis* **1980**, 54. Analgesic activity: Allais *et al.*, *Chim. Ther.* **8**, 154 (1973); M. Peterfalvi *et al.*, *Arch. Int. Pharmacodyn. Ther.* **216**, 97 (1975). Clinical investigation: Stenport, *Curr. Ther. Res.* **18**, 303 (1975). Toxicology: R. Glomot *et al.*, *Toxicol. Appl. Pharmacol.* **36**, 173 (1976). Metabolism: R. K. Lynn *et al.*, *J. Clin. Pharmacol.* **19**, 20 (1979).

Crystals from methanol, mp 179-180°. Sol in alcohol, acetone; very slightly sol in ether, chloroform, methylene chloride. Insol in water. LD_{50} in male mice, rats (mg/kg): 3400, 960 orally; 180, 160 i.v. (Glomot).

THERAP CAT: Analgesic.

4038. Flomoxef. *(6R-cis)-7-[[[(Difluoromethyl)thio]-acetyl]amino]-3-[[[1-(2-hydroxyethyl)-1H-tetrazol-5-yl]-thio]methyl]-7-methoxy-8-oxo-5-oxa-1-azabicyclo[4.2.0]oct-2-ene-2-carboxylic acid;* 7-β-difluoromethylthioacetamido-7α-methoxy-3-[[1-(2-hydroxyethyl)-1H-tetrazol-5-yl]thio-methyl]-1-oxa-3-cephem-4-carboxylic acid. $C_{15}H_{18}F_2N_6O_7S_2$; mol wt 496.46. C 36.29%, H 3.65%, F 7.65%, N 16.93%, O 22.56%, S 12.92%. Semi-synthetic oxacephalosporin antibiotic. Prepn: T. Tsuji *et al.*, Belg. pat. **898,541**; *eidem*, U.S. pat. **4,532,233** (1984, 1985 both to Shionogi); *eidem*, *J. Antibiot.* **38**, 466 (1985). *In vitro* activity and β-lactamase stability: H. C. Neu, N.-X. Chin, *Antimicrob. Ag. Chemother.* **30**, 638 (1986).

Crystals from acetone + methylene chloride, mp 82.5°-87.5°.

Sodium salt, $C_{15}H_{17}F_2N_6NaO_7S_2$, *6315-S*, *Flumarin, Furumarin.*

THERAP CAT: Antibacterial.

4039. Flopropione. *1-(2,4,6-Trihydroxyphenyl)-1-pro-panone; 2',4',6,-trihydroxypropiophenone;* phloropropiophenone; RP 13907; Argobyl; Cospanon; Flopion; Gallepronin; Gasstenon; Labroda; Labrodax; Pasmus; Profenon; Spamorin; Spasmoril; Supanate; Supazlun. $C_9H_{10}O_4$; mol wt 182.17. C 59.33%, H 5.53%, O 35.13%. Prepn: Shinoda, *J. Pharm. Soc. Japan* **1927**, 111; Canter *et al.*, *J. Chem. Soc.* **1931**, 1245; Howells, Little, *J. Am. Chem. Soc.* **54**, 2451 (1932).

Monohydrate, needles from water. Anhydr compd, mp 175-176°. Sol in ethanol, ether, ethyl acetate, hot water; very slightly sol in cold water.

THERAP CAT: Antispasmodic.

4040. Florantyrone. *γ-Oxo-8-fluoranthenebutanoic acid; β-(8-fluoranthoyl)propionic acid;* Zanchol. $C_{20}H_{14}O_3$; mol wt 302.31. C 79.45%, H 4.67%, O 15.88%. Prepn from fluoranthene and succinic anhydride in the presence of aluminum chloride in nitrobenzene solution: Fancher, U.S. pat. **2,560,425** (1951 to Miles Labs).

Fine platelets from dioxane + ethanol, mp 195°. Soluble in methanol, ethanol, aq solns of sodium carbonate.

THERAP CAT: Choleretic.

4041. Floredil. *4-[2-(3,5-Diethoxyphenoxy)ethyl]mor-pholine.* $C_{16}H_{25}NO_4$; mol wt 295.39. C 65.06%, H 8.53%, N 4.74%, O 21.67%. Prepn by refluxing 3,5-diethoxyphenol with 4-(2-chloroethyl)morpholine in ethanol: Lefon, Ger. pat. **2,020,464** (1970 to Orsymonde), *C.A.* **74**, 31650u (1971).

Hydrochloride, $C_{16}H_{26}ClNO_4$, *Carfonal.*

THERAP CAT: Coronary vasodilator.

4042. Florfenicol. *[R-(R*,S*)]-2,2-Dichloro-N-[1-(flu-oromethyl)-2-hydroxy-2-[4-(methylsulfonyl)phenyl]ethyl]-acetamide;* D-(threo)-1-p-methylsulfonylphenyl-2-amino-3-fluoro-1-propanol; fluorothiamphenicol; Sch 25298. $C_{12}H_{14}Cl_2FNO_4S$; mol wt 358.21. C 40.24%, H 3.94%, Cl 19.79%, F 5.30%, N 3.91%, O 17.87%, S 8.95%. Fluorinated derivative of thiamphenicol, *q.v.* Prepn: T. Nagabhushan, Eur. pat. Appl. **14,437**; U.S. pat. **4,235,892** (both 1980 to Schering). *In vitro* antibacterial activity: H. C. Neu, K. P. Fu, *Antimicrob. Ag. Chemother.* **18**, 311 (1980); V. P. Syriopoulou *et al.*, *ibid.* **19**, 294 (1981). Pharmacokinetics in veal calves: K. J. Varma *et al.*, *J. Vet. Pharmacol. Therap.* **9**, 412 (1986).

SO$_2$CH$_3$

HOCH

HCNHCOCHCl$_2$

CH$_2$F

mp 153-154°. Sol in water.

THERAP CAT (VET): Antibacterial.

4043. Flosequinan. *7-Fluoro-1-methyl-3-(methylsulfinyl)-4(1H)-quinolinone;* flosequinon; BTS 49465; Manoplax. C$_{11}$H$_{10}$FNO$_2$S; mol wt 239.26. C 55.22%, H 4.21%, F 7.94%, N 5.85%, O 13.37%, S 13.40%. Mixed arterial and venous vasodilator. Prepn: R. V. Davies *et al.,* **Ger.** pat. **3,011,994;** *eidem,* **U.S.** pat. **4,302,460** (1980, 1981 both to Boots). Pharmacology: J. G. Smith, G. T. Kinasewitz, *J. Cardiovasc. Pharmacol.* **8,** 878 (1986); M. F. Sim *et al., Brit. J. Pharmacol.* **94,** 371 (1988). HPLC determn in plasma, serum and urine: M. B. Slegowski *et al., J. Chromatog.* **425,** 227 (1988). Pharmacokinetics and hemodynamics in humans: A. J. Cowley *et al., J. Hypertens.* **2,** Suppl. 3, 547 (1984); R. D. Wynne *et al., Eur. J. Clin. Pharmacol.* **28,** 659 (1985). Preliminary evaluation in hypertension: A. J. Cowley *et al., ibid.* **33,** 203 (1987); in congestive heart failure: P. D. Kessler, M. Packer, *Am. Heart J.* **113,** 137 (1987); P. D. Kessler *et al., J. Cardiovasc. Pharmacol.* **12,** 6 (1988). Clinical study in chronic heart failure: A. J. Cowley *et al., Brit. Med. J.* **297,** 169 (1988).

CH$_3$

F

N

SCH$_3$

O

Crystals, mp 226-228°.

THERAP CAT: Antihypertensive.

4044. Floxacillin. *6-[[[3-(2-Chloro-6-fluorophenyl)-5-methyl-4-isoxazolyl]carbonyl]amino]-3,3-dimethyl-7-oxo-4-thia-1-azabicyclo[3.2.0]heptane-2-carboxylic acid;* 3-(2-chloro-6-fluorophenyl)-5-methyl-4-isoxazolylpenicillin; 6-[3-(2-chloro-6-fluorophenyl)-5-methyl-4-isoxazolecarboxamido]penicillanic acid; flucloxacillin; BRL 2039; Abboflox; Flupen; Penplus; Staphlipen. C$_{19}$H$_{17}$ClFN$_3$O$_5$S; mol wt 453.88. C 50.28%, H 3.78%, Cl 7.81%, F 4.18%, N 9.26%, O 17.62%, S 7.06%. Semi-synthetic antibiotic active against penicillin-resistant staphylococci. Halogen-substituted derivative of oxacillin, *q.v.* Prepn: **S. Afr.** pat. **63 4,323** (1964 to Beecham); Nayler, **Brit.** pat. **978,299;** *idem,* **U.S.** pat. **3,239,507** (1964, 1966 both to Beecham). Pharmacology and toxicity: R. Sutherland *et al., Brit. Med. J.* **4,** 460 (1970). Clinical studies: Harding *et al., Clin. Trials J.* **7,** 368 (1970); Qureshi *et al., ibid.* 375.

F
H H
S
CH$_3$
CONH
CH$_3$
Cl
N
O
COOH
O
CH$_3$

Sodium monohydrate, C$_{19}$H$_{16}$ClFN$_3$NaO$_5$S.H$_2$O, *Culpen, Floxapen, Ladropen, Stafoxil, Staphcil, Staphylex.* Freely sol in water. LD$_{50}$ in mice (g/kg): 2.2 s.c., 3.8 orally (Sutherland).

THERAP CAT: Antibacterial.

4045. Floxuridine. *2'-Deoxy-5-fluorouridine;* 1-(2-deoxy-β-D-ribofuranosyl)-5-fluorouracil; 5-fluoro-2'-deoxy-

β-uridine; NSC-27640; FUDR. C$_9$H$_{11}$FN$_2$O$_5$; mol wt 246.21. C 43.91%, H 4.50%, F 7.72%, N 11.38%, O 32.49%. Prepn: Hoffer *et al., J. Am. Chem. Soc.* **81,** 4112 (1959); Heidelberger, Duschinsky, **U.S.** pat. **2,885,396** (1959); Duschinsky *et al.,* **U.S.** pat. **2,970,139** (1961); Hoffer, **U.S.** pats. **2,949,451; 3,041,335** (1960, 1962 both to Hoffmann-La Roche). Structure: Lemieux, Hoffer, *Can. J. Chem.* **39,** 110 (1961). Crystal and molecular structure: Harris, *Diss. Abstr.* **24,** 4425 (1964); Harris, McIntyre, *Biophys. J.* **4,** 203 (1964). Conformation of furanose ring in molecule: Sundaralingam, *J. Am. Chem. Soc.* **87,** 599 (1965).

O
F
NH
N
O
HOCH$_2$ O
H H
H H
OH H

Crystals from butyl acetate, mp 150-151° (Hoffer *et al., loc. cit.*) and 145° (**U.S.** pat. **3,041,335** above). [α]$_D$ +37° (water), +48.6° (DMF). uv max (pH 7.2): 268 nm (ε 7570); at pH 14: 270 nm (ε 6480).

α-Anomer, *1-(2-deoxy-α-D-erythro-pentofuranosyl)-5-fluorouracil.* Prepn: Hoffer *et al.,* and Hoffer, **U.S.** pats., *loc. cit.* Structure: Lemieux, Hoffer, *loc. cit.* Crystals from butyl acetate, mp 150-151°. [α]$_D^{25}$ −21° (c = 2 in water).

THERAP CAT: Antiviral; antineoplastic.

4046. Fluacizine. *10-[3-(Diethylamino)-1-oxopropyl]-2-(trifluoromethyl)-10H-phenothiazine; 10-(N,N-diethyl-β-alanyl)-2-(trifluoromethyl)phenothiazine;* fluoracisine; fluoracizine; ftoracizine; Phtorazisin. C$_{20}$H$_{21}$F$_3$N$_2$OS; mol wt 394.45. C 60.90%, H 5.37%, F 14.45%, N 7.10%, O 4.05%, S 8.13%. Phenothiazine derivative with psychotropic activity. Prepn: Y. I. Vikhlyaev *et al.,* **USSR** pat. **360,342** (1972), *C.A.* **78,** 97683w (1973); of hydrochloride: S. V. Zhuravlev *et al.,* **Brit.** pat. **1,191,800** (1970), *C.A.* **73,** 25493h (1970); Y. I. Vikhlyaev *et al.,* **Fr.** pat. **2,035,748** (1971), *C.A.* **75,** 140872j (1971); *eidem,* **Ger.** pat. **1,805,659** (1979), *C.A.* **75,** 49101w (1971). Use as antidepressant: *eidem,* **USSR** pat. **356,992** (1972), *C.A.* **78,** 75888q (1973) (all to Inst. Pharmacol. Chemother., Acad. Med. Sci., USSR). Mass spectrum: S. Morosawa *et al., Org. Mass Spec.* **17,** 309 (1982). Effect on drug metabolism: V. Avakumov *et al., Biochem. Pharmacol.* **27,** 2177 (1978). Pharmacokinetics in animals, man: V. P. Zherdev *et al., Farmakol. Toksikol. (USSR)* **45,** 83 (1982).

COCH$_2$CH$_2$N(C$_2$H$_5$)$_2$
N
CF$_3$
S

Hydrochloride, C$_{20}$H$_{22}$ClF$_3$N$_2$OS, *fluoracyzine, toracizin.* White cryst powder, mp 163-165°. Sol in water, warm alcohols.

THERAP CAT: Antidepressant.

4047. Fluanisone. *1-(4-Fluorophenyl)-4-[4-(2-methoxyphenyl)-1-piperazinyl]-1-butanone; 4'-fluoro-4-[4-(o-methoxyphenyl)-1-piperazinyl]butyrophenone;* haloanisone; R 2028; Sedalande. C$_{21}$H$_{25}$FN$_2$O$_2$; mol wt 356.45. C 70.76%, H 7.07%, F 5.33%, N 7.86%, O 8.98%. Prepn: Janssen, **U.S.** pat. **2,997,472** (1961). Determn in human plasma: M. P. Quaglio, A. M. Bellini, *Farmaco Ed. Prat.* **36,** 204 (1981). Comparison between biochemical and behavioral effects: G. B. Fregnan, R. Porta, *Arzneimittel-Forsch.* **31,** 70 (1981).

Crystals, mp 67.5-68.5°. Sol in chloroform; sparingly sol in methanol; slightly sol in ether. Practically insol in water. LD_{50} in mice: 200 mg/kg i.p., C. Cascio *et al., Farmaco Ed. Sci.* **35**, 605 (1980).

Hydrochloride, mp 205-205.5°.

THERAP CAT: Neuroleptic.

4048. Fluazacort. *(11β,16β)-21-(Acetyloxy)-9-fluoro-11-hydroxy-2'-methyl-5'H-pregna-1,4-dieno[17,16-d]oxazole-3,20-dione; 9-fluoro-11β,21-dihydroxy-2'-methyl-5'βH-pregna-1,4-dieno[17,16-d]oxazole-3,20-dione 21-acetate;* L 6400; Azacortid. $C_{25}H_{30}FNO_6$; mol wt 459.53. C 65.35%, H 6.58%, F 4.13%, N 3.05%, O 20.89%. Prepn of the 21-hydroxy compound: G. Nathansohn *et al., J. Med. Chem.* **10**, 799 (1967); of the 21-hydroxy and fluazacort: *eidem,* **Brit.** pat. **1,119,082;** *eidem,* **U.S.** pat. **3,461,119** (1968, 1969 to Lepetit); *eidem, Steroids* **13**, 383 (1969). Pharmacological and physico-chemical properties: *eidem, ibid.* 365. Percutaneous absorption: J. D. Lewis *et al., Arzneimittel-Forsch.* **25**, 1646 (1975).

Crystals from ethanol, mp 252-255°. $[\alpha]_D$ +54.8° (c = 0.5 in $CHCl_3$). uv max (methanol): 238-240 nm ($E_{1cm}^{1\%}$ 330).
THERAP CAT: Anti-inflammatory.

4049. Fluazifop-butyl. *2-[4-[[5-(trifluoromethyl)-2-pyridinyl]oxy]phenoxy]propanoic acid butyl ester;* butyl 2-[4-(5-trifluoromethyl-2-pyridyloxy)phenoxy]propionate; PP 009; TF 1169; Fusilade. $C_{19}H_{20}F_3NO_4$; mol wt 383.38. C 59.53%, H 5.26%, F 14.87%, N 3.65%, O 16.69%. Selective post-emergence herbicide. Prepn: R. Nishiyama *et al.,* **Ger.** pat. **2,812,571** (1979 to Ishihara Sangyo Kaisha), *C.A.* **90**, 152017g (1979). Activity: R. E. Plowman *et al., Proc. Brit. Crop Prot. Conf. — Weeds* **1980**, 29; J. R. Finney, P. B. Sutton, *ibid.* 429.

$bp_{0.05}$ 167°.
USE: Herbicide.

4050. Flubendazole. *[5-(4-Fluorobenzoyl)-1H-benzimidazol-2-yl]carbamic acid methyl ester; 5-(p-fluorobenzoyl)-2-benzimidazolecarbamic acid methyl ester;* R 17889; Flubenol; Flumoxal; Flumoxane; Fluvermal. $C_{16}H_{12}FN_3O_3$; mol wt 313.30. C 61.34%, H 3.86%, F 6.06%, N 13.41%, O 15.32%. Fluoro analog of mebendazole, *q.v.* Prepn: J. L. VanGelder *et al.,* **Ger.** pat. **2,029,637** corresp to **U.S.** pat. **3,657,267** (1971, 1972 both to Janssen); A. H. M. Raeymaekers *et al., Arzneimittel-Forsch.* **28**, 586 (1978). Pharmacological, biological properties: D. Thienpont *et al., ibid.* 605. Anthelmintic activity in rats: O. Vanparijs *et al., Vet. Parasitol.* **5**, 237 (1979); in pigs: E. Telléz-Giron *et al., Am. J. Trop. Med. Hyg.* **30**, 135 (1981).

Crystals, mp 260°. LD_{50} in mice, rats, guinea pigs (mg/kg): >2560 orally (Thienpont).
THERAP CAT (VET): Anthelmintic.

4051. Flubenzimine. *N-[3-Phenyl-4,5-bis[(trifluoromethyl)imino]-2-thiazolidinylidene]benzenamine; N^2,3-diphenyl-N^4,N^5-bis(trifluoromethyl)thiazolidine-2,4,5-triylidene triamine;* SLJ-312; Cropotex. $C_{17}H_{10}F_6N_4S$; mol wt 416.36. C 49.04%, H 2.42%, F 27.38%, N 13.46%, S 7.70%. Prepn: H.-J. Scholl *et al.,* **Ger.** pat. **2,062,348** corresp to **U.S.** pat. **3,895,020** (1972, 1975 to Bayer). Mode of action: G. Zoebelein *et al., Pflanzenschutz-Nachr.* **33**, 169 (1980); D. J. Bluett, A. Wainwright, *Proc. Brit. Crop Prot. Conf.-Pests Dis.* **1981**, 75.

Yellow crystals from methanol, mp 118-119°. Vapor pressure at 20°: 1×10^{-5} mm Hg. Soly in water at 20°: 30 mg/l. LD_{50} orally in rats: 3750 mg/kg (Bluett, Wainwright).
USE: Acaricide.

4052. Fluchloralin. *N-(2-Chloroethyl)-2,6-dinitro-N-propyl-4-(trifluoromethyl)benzenamine; N-(2-chloroethyl)-α,α,α-trifluoro-2,6-dinitro-N-propyl-p-toluidine;* BAS 392 H; Basalin. $C_{12}H_{13}ClF_3N_3O_4$; mol wt 355.70. C 40.52%, H 3.68%, Cl 9.97%, F 16.02%, N 11.81%, O 17.99%. Soil incorporated, pre-planting herbicide. Prepn: K. H. Karl, K. Kiehs, **Ger.** pat. **2,161,879** (1973 to BASF), *C.A.* **79**, 65990y (1973). Photochemistry: G. P. Nilles, M. J. Zabik, *J. Agr. Food Chem.* **22**, 684 (1974). Degradn and metabolism in soil: S. Otto, *Environ. Qual. Saf., Suppl.* **3**, 277 (1975).

Orange-yellow cryst solid, mp 42-43°. Soly in water: 10 ppm.
USE: Herbicide.

4053. Flucloronide. *9,11-Dichloro-6-fluoro-21-hydroxy-16α,17-[(1-methylethylidene)bis(oxy)]-pregna-1,4-diene-3,20-dione; 9,11β-dichloro-6α-fluoro-16α,17,21-trihydroxy-pregna-1,4-dien-3,20-dione cyclic 16,17-acetal with acetone; 9α,11β-dichloro-6α-fluoro-21-hydroxy-16α,17α-(isopropylidenedioxy)pregna-1,4-diene-3,20-dione; 6α-fluoro-9α,-11β-dichloro-1,4-pregnadiene-16α,17,21-triol-3,20-dione 16,17-acetonide;* fluclorolone acetonide; RS-2252; Topilar. $C_{24}H_{29}Cl_2FO_5$; mol wt 487.39. C 59.14%, H 6.00%, Cl 14.55%, F 3.90%, O 16.41%. Prepn: Bowers, **U.S.** pat. **3,201,391** (1965 to Syntex). Activity studies: Dorfman *et al., Acta Endocrinol.* **49**, 262 (1965); several authors in *Brit. J. Dermatol.* **82**, suppl. 6, 53-98 (1970). Clinical studies: several authors in *Clin. Trials J.* **8**, 35-63 (1971).

CH2OH

THERAP CAT: Glucocorticoid.

4054. Fluconazole. α-(2,4-Difluorophenyl)-α-(1H-1,2,4-triazol-1-ylmethyl)-1H-1,2,4-triazole-1-ethanol; 2,4-difluoro-α,α-bis(1H-1,2,4-triazol-1-ylmethyl)benzyl alcohol; 2-(2,4-difluorophenyl)-1,3-bis(1H-1,2,4-triazol-1-yl)-propan-2-ol; UK 49858; Diflucan; Triflucan. $C_{13}H_{12}F_2N_6O$; mol wt 306.27. C 50.98%, H 3.95%, F 12.41%, N 27.44%, O 5.22%. Orally active bistriazole antifungal agent. Prepn: K. Richardson, **Brit.** pat. **2,099,818**; idem, **U.S.** pat. **4,404,216** (1982, 1983 both to Pfizer). Antifungal activity in vivo: K. Richardson et al., Antimicrob. Ag. Chemother. **27**, 832 (1985); and in vitro: T. E. Rogers, J. N. Galgiani, ibid. **30**, 418 (1986). Pharmacokinetics: M. J. Humphrey et al., ibid. **28**, 648 (1985). GC determn in human plasma and urine: P. R. Wood, M. H. Tarbit, J. Chromatog. **383**, 179 (1986). Symposium on mode of action, toxicology, animal models and clinical studies: International Telesymposium on Recent Trends in the Discovery, Development and Evaluation of Antifungal Agents, **Section 2**, R. A. Fromtling, Ed. (J. R. Prous Science Publ., Barcelona, 1987) pp 77-174.

Crystals from ethyl acetate/hexane, mp 138-140°.
THERAP CAT: Antifungal.

4055. Flucythrinate. 4-(Difluoromethoxy)-α-(1-methylethyl)benzeneacetic acid cyano(3-phenoxyphenyl)methyl ester; (±)-cyano-(3-phenoxyphenyl)methyl (+)-4-(difluoromethoxy)-α-(1-methylethyl)benzeneacetate; AC-222705; Cybolt; Guardian; Pay-Off; Stock Guard. $C_{26}H_{23}F_2NO_4$; mol wt 451.48. C 69.17%, H 5.13%, F 8.42%, N 3.10%, O 14.18%. Synthetic pyrethroid insecticide with contact and stomach poison activity. Prepn: G. Berkelhammer, V. Kameswaran, **Ger.** pat. **2,757,066**; eidem, **U.S.** pat. **4,199,595** (1977, 1980 both to Am. Cyanamid). Activity: A. F. Saad et al., Proc. Brit. Crop Prot. Conf.-Pests Dis. **1981**, 381; K. Wettstein, ibid. 563. Efficacy in cattle fly control: S. M. Taylor et al., Vet. Rec. **116**, 566 (1985).

Viscous liquid. Sol in acetone, xylene, 2-propanol.
USE: Insecticide.
THERAP CAT (VET): Ectoparasiticide.

4056. Flucytosine. 4-Amino-5-fluoro-2(1H)-pyrimidinone; 5-fluorocytosine; 5-FC; Ro 2-9915; Alcobon; Ancobon; Ancotil. $C_4H_4FN_3O$; mol wt 129.09. C 37.21%, H 3.12%, F 14.72%, N 32.55%, O 12.39%. Prepn: Duschinsky et al., J. Am. Chem. Soc. **79**, 4559 (1957); Heidelberger, Duschinsky, U.S. pat. **2,802,005** (1957); Duschinsky, Heidelberger, U.S. pat. **2,945,038** and Duschinsky, U.S. pat.

3,040,026 (1960, 1962 both to Hoffmann-La Roche); Undheim, Gacek, Acta Chem. Scand. **23**, 294 (1969). Patents as a fungicide: Berger, Duschinsky, **Belg.** pat. **628,615** and **U.S.** pat. **3,368,938** (1963, 1968 both to Hoffmann-La Roche). Activity studies: Grunberg et al., Antimicrob. Ag. Chemother. **1963**, 566; Shadomy et al., ibid. **1968**, 452; Grunberg et al., in Vth Int. Congr. Chemother., Proc., **vol. IV**, K. Spitzy, Ed. (Verlag Wiener Med. Akad., 1967, Austria) p 69. Metabolic studies: Koechlin et al., Biochem. Pharmacol. **15**, 435 (1966). Clinical results: Utz et al., Antimicrob. Ag. Chemother. **1968**, 344; Warner et al., ibid. **1970**, 473. Comprehensive description: E. H. Waysek, J. H. Johnson in Analytical Profiles of Drug Substances **vol. 5**, K. Florey, Ed. (Academic Press, New York, 1976) pp 115-138.

Odorless, white crystalline solid, mp 295-297° (dec). uv max (0.1N HCl): 285 nm (ε 8900). Soly in water: 1.5 g/100 ml at 25°C. pKa_1: 3.26. LD_{50} in mice (mg/kg): > 2000 orally and s.c.; 1190 i.p.; 500 i.v. (Grunberg, 1963).
THERAP CAT: Antifungal.

4057. Fludarabine. 9-β-D-Arabinofuranosyl-2-fluoro-9H-purin-6-amine; 9-β-D-arabinofuranosyl-2-fluoroadenine; 2-fluorovidarabine; 2-fluoro-9-β-D-arabinofuranosyladenine; 2-F-araA; NSC-118218; NSC-118218-H. $C_{10}H_{12}FN_5O_4$; mol wt 285.23. C 42.11%, H 4.24%, F 6.66%, N 24.55%, O 22.44%. Adenosine deaminase-resistant purine nucleoside antimetabolite. Prepn and in vitro cytotoxicity: J. A. Montgomery, K. Hewson, J. Med. Chem. **12**, 498 (1969). Improved prepn: J. A. Montgomery et al., J. Heterocycl. Chem. **16**, 157 (1979); J. A. Montgomery, U.S. pat. **4,210,745** (1980 to U.S. Dept. Health, Education and Welfare). Inhibition of DNA synthesis and in vivo antileukemic activity: R. W. Brockman et al., Biochem. Pharmacol. **26**, 2193 (1977). Metabolized to 5'-monophosphate: R. W. Brockman et al., Cancer Res. **40**, 3610 (1980). HPLC determn in human leukemia cells: V. Gandhi et al., J. Chromatog. **413**, 293 (1987). Prepn of 5'-monophosphate: J. A. Montgomery, A. T. Shortnacy, U.S. pat. **4,357,324** (1982 to U.S. Dept. of Health and Human Services). Pharmacokinetics in humans: M. R. Hersh et al., Cancer Chemother. Pharmacol. **17**, 277 (1986). Evaluation of therapeutic efficacy and CNS toxicity in acute refractory leukemia: R. P. Warrell, Jr., E. Berman, J. Clin. Oncol. **4**, 74 (1986); H. G. Chun et al., Cancer Treat. Rep. **70**, 1225 (1986).

Crystals from ethanol + water, mp 260°. $[\alpha]_D^{25}$ +17 ±2.5° (c = 0.1 in ethanol). uv max (pH 1, pH 7, pH 13): 262, 261, 262 nm (ε × 10^{-3} 13.2, 14.8, 15.0). Sparingly sol in water, organic solvents.
5'-Monophosphate, $C_{10}H_{13}FN_5O_7P$, NSC-328002, NSC-312887, 2-F-ara-AMP. Sol in water.
THERAP CAT: Phosphate as antineoplastic.

4058. Fludiazepam. 7-Chloro-5-(2-fluorophenyl)-1,3-dihydro-1-methyl-2H-1,4-benzodiazepin-2-one; ID 540; Ro 5-3438; Erispan. $C_{16}H_{12}ClFN_2O$; mol wt 302.74. C 63.48%, H 4.00%, Cl 11.71%, F 6.28%, N 9.25%, O 5.28%. Fluorinated analog of diazepam, *q.v.* Prepn: E. Reeder *et al.*, **Ger. pat. 1,136,709** corresp to **U.S. pat. 3,051,701** (both 1962 to Hoffmann-LaRoche); L. H. Sternbach *et al.*, *J. Org. Chem.* **27**, 3788 (1962). Synthesis and pharmacology: Y. Asami *et al.*, *Arzneimittel-Forsch.* **24**, 1563 (1974). Pharmacodynamics: M. Nakamura, H. Fukushima, *J. Pharm. Pharmacol.* **30**, 56, 254 (1978); T. Tsuchiya, H. Fukushima, *Eur. J. Pharmacol.* **48**, 421 (1978). Anxiolytic vs. sedative properties: M. Babbini *et al.*, *Life Sci.* **25**, 15 (1979).

Colorless prisms from *n*-hexane/isopropanol, mp 88-92°, (Asami); also reported as mp 69-72° (Sternbach). LD_{50} in mice (mg/kg): 910 orally; 360 i.p.; 1150 s.c. (Asami).

Note: This is a controlled substance (depressant) listed in the U.S. Code of Federal Regulations, Title 21 Part 1308.14 (1987).

THERAP CAT: Anxiolytic.

4059. Fludrocortisone. *(11β)-9-Fluoro-11,17,21-tri-hydroxypregn-4-ene-3,20-dione;* 9α-fluorohydrocortisone; 9α-fluoro-17-hydroxycorticosterone; 9α-fluorocortisol; fluodrocortisone; fluohydrisone; fluohydrocortisone; Alflorone; Astonin-H; F-Cortef; Florinef; Fludrocortone. $C_{21}H_{29}FO_5$; mol wt 380.46. C 66.30%, H 7.68%, F 4.99%, O 21.03%. Prepd from 4-pregnen-9β,11β-oxido-17α,21-diol-3,20-dione 21-acetate: Fried, Sabo, *J. Am. Chem. Soc.* **76**, 1455 (1954). Series of articles on pharmacology and metabolism: *Arzneimittel-Forsch.* **21**, 1103-1158 (1971). Comprehensive description of the acetate: K. Florey, Ed. in *Analytical Profiles of Drug Substances* vol. 3 (Academic Press, New York, 1974) pp 281-306.

Crystals, dec 260-262°. $[\alpha]_D^{23}$ +139° (c = 0.55 in 95% ethanol). uv max (ethanol): 239 nm (ϵ 17600). Soly in water: 0.14 mg/ml.

21-Acetate, $C_{23}H_{31}FO_6$, *Scherofluron* (suspension or salve). Crystals, polymorphic, mp 233-234° (occasionally mp 205-208°, resolidifying on further heating, then mp 226-228°). Crystallizing procedure: Graber, Snoddy, U.S. pat. **2,957,013** (1960 to Merck & Co.). $[\alpha]_D^{23}$ +123° (c = 0.64 in chloroform). uv max (ethanol): 238 nm (ϵ 16800). Soly (mg/ml): water 0.04; acetone 56; alc 20; chloroform 20; ether 4.

21-*tert*-Butylacetate, $C_{27}H_{39}FO_6$, crystals, dec 225-227°. uv max ($E_{1cm}^{1\%}$ 377). Sol in boiling alcohol, chloroform. *Compare* Hydrocortisone Tebutate.

Water-sol 21-phosphates: Sarett, U.S. pat. **2,779,775** (1957 to Merck & Co.).

THERAP CAT: Mineralcorticoid.
THERAP CAT (VET): Mineralcorticoid.

4060. Flufenamic Acid. *2-[[3-(Trifluoromethyl)phenyl]-amino]benzoic acid; N-(α,α,α-trifluoro-m-tolyl)anthranilic acid;* 3'-trifluoromethyldiphenylamine-2-carboxylic acid; CI-440; INF 1837; Achless; Ansatin; Arlef; Fullsafe; Meralen; Paraflu; Parlef; Ristogen; Sastridex; Surika; Tecramine. $C_{14}H_{10}F_3NO_2$; mol wt 281.24. C 59.79%, H 3.58%, F 20.27%, N 4.98%, O 11.38%. Prepn from *m*-aminobenzotri-

fluoride and *o*-iodobenzoic acid: Wilkinson, Finar, *J. Chem. Soc.* **1948**, 32. Pharmacology: Winder *et al.*, *Arthritis Rheum.* **6**, 36 (1963); **12**, 472 (1969). Antiviral activity: Inglot, *J. Gen. Virol.* **4**, 203 (1969). Toxicity data: Zoni *et al.*, *Farmaco Ed. Sci.* **26**, 191 (1971). Comprehensive description: E. Abignente, P. deCaprariis, in *Analytical Profiles of Drug Substances* vol. **11**, K. Florey, Ed. (Academic Press, New York, 1982) pp 313-346.

Pale yellow needles from 50% alc, mp 125°. LD_{50} in mice: 715 mg/kg orally (Zoni).

Aluminum derivative, $C_{42}H_{27}AlF_9N_3O_6$, *aluminum flufenamate, Alfenamin, Opyrin.*

Butyl ester, $C_{18}H_{18}F_3NO_2$, *Fenazol, Combec.*

THERAP CAT: Anti-inflammatory. Analgesic.

4061. Fluindione. *2-(p-Fluorophenyl)-1,3-indandione;* LM 123; Previscan. $C_{15}H_9FO_2$; mol wt 240.24. C 75.00%, H 3.77%, F 7.91%, O 13.32%. Prepn: Geiger *et al.*, **Ger. pat. 1,130,439** (1962 to USV), *C.A.* **57**, 12403a (1962); Molho, Boschetti, **Fr. pats. 1,369,396** and **M 6913** (1964, 1969 to LIPHA), *C.A.* **62**, 3988g (1965); **74**, 141378u (1971). Alternate syntheses: Shapiro *et al.*, *J. Org. Chem.* **26**, 3580 (1961); Hrnciar, Kovalcik, *Chem. Zvesti* **16**, 200 (1962). Activity and toxicity data: Fontaine *et al.*, *Med. Pharmacol. Exp.* **17**, 497 (1967).

Crystals from acetic acid, mp 120°. LD_{50} orally in mice: 240 mg/kg (Fontaine).

THERAP CAT: Anticoagulant.

4062. Flumazenil. *8-Fluoro-5,6-dihydro-5-methyl-6-oxo-4H-imidazo[1,5-a][1,4]benzodiazepine-3-carboxylic acid ethyl ester;* ethyl-8-fluoro-5,6-dihydro-5-methyl-6-oxo-4H-imidazo[1,5-a][1,4]benzodiazepine-3-carboxylate; flumazepil; Ro 15-1788; Anexate; Lanexat. $C_{15}H_{14}FN_3O_3$; mol wt 303.29. C 59.40%, H 4.65%, F 6.26%, N 13.85%, O 15.83%. Imidazodiazepine which selectively blocks the central effects of classic benzodiazepines. Prepn: W. Haefely *et al.*, **Eur. pat. Appl. 27,214**; M. Gerecke *et al.*, **U.S. pat. 4,316,839** (1981, 1982 both to Hoffmann-La Roche). Specific inhibition of benzodiazepine-receptor binding: W. Hunkeler *et al.*, *Nature* **290**, 514 (1981). Electrophysiological study in animals: P. Pole *et al.*, *Arch. Pharmacol.* **316**, 317 (1981). Antagonist, agonist and inverse agonist properties: S. E. File *et al.*, *Neuropharmacol.* **89**, 113 (1986). HPLC determn in human plasma: U. Timm, M. Zell, *Arzneimittel-Forsch.* **33**, 358 (1983). Clinical reversal of benzodiazepine-induced sedation: A. Darragh *et al.*, *Lancet* **2**, 8 (1981); *eidem, ibid.* 1042; B. Ricou *et al.*, *Brit. J. Anaesth.* **58**, 1005 (1986). Clinical evaluation in drug overdose: G. F. O'Sullivan *et al.*, *Clin. Pharmacol. Ther.* **42**, 254 (1987).

Crystals from alcohol, mp 201-203°. LD_{50} in mice, rats (mg/kg): 4000, 1360 i.p.; 4300, 6000 orally (Hunkeler).

THERAP CAT: Benzodiazepine antagonist.

4063. Flumecinol. α-*Ethyl*-α-*phenyl*-3-*(trifluoromethyl)benzenemethanol;* α-ethyl-3-(trifluoromethyl)benzhydrol; RGH 3332; Zixoryn; Zyxorin. $C_{16}H_{15}F_3O$; mol wt 280.30. C 68.56%, H 5.39%, F 20.33%, O 5.71%. Hepatic microsomal enzyme inducer. Prepn: E. Toth *et al.*, **Ger.** pat. **2,438,399** corresp to **U.S.** pat. **4,039,589** (1975, 1977 both to Gedeon-Richter). Series of articles on metabolism, CNS effects, enzyme induction, pharmacological properties: *Arzneimittel-Forsch.* **28**, 663-679 (1978). Gas chromatographic determn in biological fluids: I. Klebovich, L. Vereczkey, *J. Chromatog.* **221**, 403 (1980).

Oil, bp$_{0.03}$ 106-108°. uv max (ethanol): 259, 265, 271 nm. LD$_{50}$ in adult rats: 2235 mg/kg orally, M. Ledniczky *et al.*, *Arzneimittel-Forsch.* **28**, 669 (1978).

THERAP CAT: Enzyme inducer (hepatic).

4064. Flumedroxone Acetate. 17-*(Acetyloxy)*-6α-*(trifluoromethyl)pregn-4-ene-3,20-dione;* 17-*hydroxy*-6α-*(trifluoromethyl)pregn-4-ene-3,20-dione acetate;* 17-acetoxy-6α-(trifluoromethyl)progesterone; 17-hydroxy-6α-(trifluoromethyl)progesterone acetate; 6α-(trifluoromethyl)-17-hydroxyprogesterone acetate; WG 537; Demigran. $C_{24}H_{31}F_3O_4$; mol wt 440.51. C 65.44%, H 7.09%, F 12.94%, O 14.53%. Prepn: Godtfredsen, Vangedal, *Acta Chem. Scand.* **15**, 1786 (1961); Godtfredsen, **Brit.** pat. **905,694** (1962 to Leo Pharm.). Solubilization study: T. Lovgren *et al.*, *Acta Pharm. Suec.* **15**, 233 (1978).

Solid, mp 206-207°. $[\alpha]_D^{20}$ +30°. uv max (ethanol): 234 nm (ε 15,600).

THERAP CAT: Treatment of migraine.

4065. Flumequine. 9-*Fluoro-6,7-dihydro-5-methyl-1-oxo-1H,5H-benzo[ij]quinolizine-2-carboxylic acid;* R-802; Apurone. $C_{14}H_{12}FNO_3$; mol wt 261.26. C 64.36%, H 4.63%, F 7.27%, N 5.36%, O 18.37%. Quinolone antibacterial. Prepn: J. F. Gerster, **Ger.** pat. **2,264,163;** *idem*, **U.S.** pat. **3,896,131** (1973, 1975 both to Riker). *In vitro* study: G. Stilwell *et al.*, *Antimicrob. Ag. Chemother.* **7**, 483 (1975). Bioevaluation: S. R. Rohlfing *et al.*, *J. Antimicrob. Chemother.* **3**, 615 (1977). Activity vs. *E. coli:* D. Greenwood, *Antimicrob. Ag. Chemother.* **13**, 479 (1978).

White microcrystalline powder, mp 253-255°. Sol in alkaline solns and alcohol. Insol in water.

THERAP CAT: Antibacterial.

4066. Flumethasone. 6,9-*Difluoro-11,17,21-trihydroxy-16-methylpregna-1,4-diene-3,20-dione;* 6α,9α-difluoro-16α-methylprednisolone; 6α-fluorodexamethasone; Aniprime; Cortexilar; Flucort; Methagon. $C_{22}H_{28}F_2O_5$; mol wt 410.46. C 64.38%, H 6.88%, F 9.26%, O 19.49%. Prepn of the 21-

acetate: Edwards *et al.*, *J. Am. Chem. Soc.* **81**, 3156 (1959); **82**, 2318 (1960).

21-Acetate, $C_{24}H_{30}F_2O_6$, crystals from acetone + hexane, mp 260-264°. $[\alpha]_D$ +91° uv max (ethanol): 237 nm (log ε 4.16).

21-Pivalate, $C_{27}H_{36}F_2O_6$, *Locacorten (formerly), Locorten, Lorinden, Losalen.*

THERAP CAT: Glucocorticoid; anti-inflammatory.

THERAP CAT (VET): Adrenocortical steroid.

4067. Flumethiazide. 6-*(Trifluoromethyl)-2H-1,2,4-benzothiadiazine-7-sulfonamide 1,1-dioxide;* 6-*trifluoromethyl-7-sulfamoyl-4H-1,2,4-benzothiadiazine 1,1-dioxide;* 6-trifluoromethyl-7-sulfamyl-1,2,4-benzothiadiazine 1,1-dioxide; trifluoromethylthiazide; Ademol; Fludemil. $C_8H_6F_3N_3O_4S_2$; mol wt 329.28. C 29.18%, H 1.84%, F 17.31%, N 12.76%, O 19.44%, S 19.48%. Synthesis: Holdrege *et al.*, *J. Am. Chem. Soc.* **81**, 4807 (1959); Yale *et al.*, *ibid.* **82**, 2042 (1960); **U.S.** pat. **3,040,042** (1962 to Olin Mathieson); **Brit.** pat. **861,809** (1961 to SK&F).

Crystals, dec 305.4-307.8°. uv max: 278 nm (E$_{1cm}^{1\%}$ 335) (50% diglyme + 50% 0.1N HCl). Sparingly sol in water (50 mg/ml in boiling water with decompn); sol in methanol, ethanol, DMF. Practically insol in ethyl acetate, methyl ethyl ketone, benzene, toluene. Unstable in alkaline soln (conversion to precursor α,α,α-trifluoro-3-amino-4,6-disulfamoyltoluene).

THERAP CAT: Carbonic anhydrase inhibitor.

4068. Flumethrin. 3-*[2-Chloro-2-(4-chlorophenyl)ethenyl]-2,2-dimethylcyclopropanecarboxylic acid cyano(4-fluoro-3-phenoxyphenyl)methyl ester;* 3′-phenoxy-4′-fluoro-α-cyanobenzyl 2,2-dimethyl-3-[2-(4-chlorophenyl)-2-chlorovinyl]cyclopropane carboxylate; Bayticol. $C_{28}H_{22}Cl_2FNO_3$; mol wt 510.40. C 65.89%, H 4.35%, Cl 13.89%, F 3.72%, N 2.74%, O 9.40%. Synthetic pyrethroid insecticide. Prepn: R. Fuchs *et al.*, **Ger.** pat. **2,730,515** corresp to **U.S.** pat. **4,276,306** (1979, 1981 to Bayer). Series of articles on laboratory evaluation, efficacy and toxicology: *Vet. Med. Rev.* **2**, 115-139, 158-177 (1982).

Oil, n_D^{25} 1.5831.

USE: Insecticide; acaricide.

THERAP CAT (VET): Ectoparasiticide.

4069. Flumetramide. 6-*[4-(Trifluoromethyl)phenyl]-3-morpholinone;* 6-(α,α,α-trifluoro-*p*-tolyl)-3-morpholinone; McN-1546; Duraflex. $C_{11}H_{10}F_3NO_2$; mol wt 245.21. C 53.88%, H 4.11%, F 23.25%, N 5.71%, O 13.05%. Prepn: Gannon, Poos, **U.S.** pat. **3,308,121** (1967 to McNeil).

mp 115.5-116.5°.
THERAP CAT: Skeletal muscle relaxant.

4070. Flunarizine. *(E)-1-[Bis(4-fluorophenyl)methyl]-4-(3-phenyl-2-propenyl)piperazine; (E)-1-[bis(p-fluorophenyl)-methyl]-4-cinnamylpiperazine;* 1-cinnamyl-4-(di-*p*-fluorobenzhydryl)piperazine. $C_{26}H_{26}F_2N_2$; mol wt 404.51. C 77.20%, H 6.48%, F 9.39%, N 6.92%. Calcium channel blocker; fluorinated deriv of cinnarizine, *q.v.* Prepn: P. A. J. Janssen, **Ger. pat. 1,929,330** and **Fr. pat. 2,014,487;** *idem*, **U.S. pat. 3,773,939** (1970, 1970, 1973, all to Janssen). Pharmacology: L. K. Desmedt *et al., Arzneimittel-Forsch.* **25,** 1408 (1975); T. Godfraind, *Eur. J. Pharmacol.* **53,** 273 (1979). Clinical studies: J. Schetz *et al., Curr. Ther. Res. Clin. Exp.* **23,** 131 (1978); G. Rudofsky *et al., Angiology* **30,** 479 (1979). Rheological effects in humans: J. D. Cree *et al., ibid.* 505. Mechanism of calcium blocking activity: T. Godfraind, *Fed. Proc.* **40,** 2866 (1981). Clinical study in retinal vasculopathy: P. Nihard, *Angiology* **33,** 37 (1982). Clinical trial in chronic cerebrovascular disorders: A. Agnoli *et al., Int. J. Clin. Pharm. Res.* **8,** 189 (1988). Series of articles on absorption, distribution, excretion and metabolism: *Arzneimittel-Forsch.* **33,** 1135-1151 (1983). Review of pharmacology, toxicology and clinical studies: B. Holmes *et al., Drugs* **27,** 6 (1984).

Dihydrochloride, $C_{26}H_{28}Cl_2F_2N_2$, *Dinaplex, Flugeral, Flunagen, Flunarl, Fluxarten, Gradient, Issium, Mondus, Sibelium.* Crystals from a mixture of 2-propanol/ethanol, mp 251.5°.

THERAP CAT: Cerebral and peripheral vasodilator.

4071. Flunisolide. *6-Fluoro-11,21-dihydroxy-16,17-[(1-methylethylidene)bis(oxy)]pregna-1,4-diene-3,20-dione; 6α-fluoro-11β,16α,17,21-tetrahydroxypregna-1,4-diene-3,20-dione cyclic 16,17-acetal with acetone;* 6α-fluoro-11β,21-dihydroxy-16α,17-isopropylidenedioxy-Δ^1,4-pregnadiene-3,20-dione; RS-3999; Lunis; Nasalide; Rhinalar; Syntaris; Yafrol. $C_{24}H_{31}FO_6$; mol wt 434.51. C 66.34%, H 7.19%1, F 4.37%, O 22.09%. Synthetic fluorinated corticosteroid related to prednisolone, *q.v.* Prepn using *S. roseochromogenes:* **Brit. pat. 933,867** (1963 to Am. Cyanamid), *C.A.* **60,** 3070f (1964). Prepn using *Cunninghamella blakesleeana* 8688b: H. J. G. Rosenkranz, **U.S. pat. 3,124,571** (1964 to Syntex). Metabolism: N. I. Chu *et al., Drug Metab. Dispos.* **7,** 81 (1979). Use in rhinitis: J. N. Sahay *et al., Clin. Allergy* **9,** 17 (1979); J. K. Sarsfield, G. E. Thomson, *Brit. Med. J.* **2,** 95 (1979). Use in bronchial asthma: D. R. Webb *et al., Ann. Allergy* **42,** 80 (1979). Clinical study in chronic asthma: R. G. Slavin *et al., J. Allergy Clin. Immunol.* **66,** 379 (1980). Review of pharmacology and therapeutic efficacy: G. E. Pakes *et al., Drugs* **19,** 397-411 (1980).

21-Acetate, $C_{26}H_{33}FO_7$, *RS-1320.*
THERAP CAT: Glucocorticoid; antiasthmatic.

4072. Flunitrazepam. *5-(2-Fluorophenyl)-1,3-dihydro-1-methyl-7-nitro-2H-1,4-benzodiazepin-2-one;* 1-methyl-7-nitro-5-(2-fluorophenyl)-3H-1,4-benzodiazepin-2(1H)-one; Ro 5-4200; Narcozep; Primun; Rohypnol; Roipnol. $C_{16}H_{12}FN_3O_3$; mol wt 313.30. C 61.34%, H 3.86%, F 6.06%, N 13.41%, O 15.32%. Prepn: L. H. Sternbach *et al., J. Med. Chem.* **6,** 261 (1963); J. Kariss, H. L. Newmark, **U.S. pats. 3,116,203** and **3,123,529;** O. Keller *et al.,* **U.S. pat. 3,203,990** (1963, 1964, 1965 all to Hoffmann-La Roche). Industrial prepn and purifn: J. M. Autin, **Fr. pat. 2,529,203** (1983 to Fabre), *C.A.* **100,** 191913r (1984). Pharmacology and clinical studies: E. Eidelberg *et al., Neurology* **5,** 223 (1965); Schuler *et al., Psychopharmacologia* **27,** 123 (1972); J. M. Monti, H. Altier, *ibid.* **32,** 343 (1973); Schallek *et al., Arch. Int. Pharmacodyn. Ther.* **206,** 161 (1973); Kaplan *et al., J. Pharm. Sci.* **63,** 527 (1974). Review of pharmacology and therapeutic use: M. A. K. Mattila, H. M. Larni, *Drugs* **20,** 353-374 (1980).

Pale yellow needles from methylene chloride-hexane, mp 166-167°. Also reported as crystals from acetonitrile and methanol, mp 170-172°.
Note: This is a controlled substance (depressant) listed in the U.S. Code of Federal Regulations, Title 21 Part 1308.14 (1985).

THERAP CAT: Hypnotic.

4073. Flunixin. *2-[[2-Methyl-3-(trifluoromethyl)phenyl]amino]-3-pyridinecarboxylic acid;* 2-(2-methyl-3-trifluoromethylanilino)nicotinic acid; 2-(α³,α³,α³-trifluoro-2,3-xylidino)nicotinic acid; Sch 14714. $C_{14}H_{11}F_3N_2O_2$; mol wt 296.25. C 56.76%, H 3.74%, F 19.24%, N 9.45%, O 10.80%. Prepn: M. H. Sherlock, N. Sperber, **Belg. pat. 679,271** corresp to **U.S. pat. 3,337,570** (1966, 1967 both to Schering); *eidem,* **U.S. pat. 3,689,653** (1972 to Schering). Prepn of the acid and meglumine salt: M. H. Sherlock, **Belg. pat. 812,-772** corresp to **U.S. pat. 3,839,344** (both 1974 to Schering). Pharmacology: V. B. Ciofalo *et al., J. Pharmacol. Exp. Ther.* **200,** 501 (1977); R. D. Sofia *et al., Pharmacol. Res. Commun.* **11,** 179 (1979). Use as anti-inflammatory analgesic in horses: J. W. Houdeshell, P. W. Hennessey, *J. Equine Med. Surg.* **1,** 57 (1977). Pharmacology review: T. Tobin, *ibid.* **3,** 298 (1979).

Cryst from acetone/hexane, mp 226-228°.

Meglumine salt, $C_{21}H_{28}F_3N_3O_7$, *Banamine, Finadyne.* Colorless cryst from acetonitrile or ethanol/ether, mp 135-137°.

THERAP CAT (VET): Anti-inflammatory, analgesic.

4074. Flunoxaprofen. (+)-2-(4-Fluorophenyl)-α-methyl-5-benzoxazoleacetic acid; RV-12424; Priaxim. $C_{16}H_{12}$-FNO$_3$; mol wt 285.28. C 67.36%, H 4.24%, F 6.66%, N 4.91%, O 16.83%. Nonsteroidal anti-inflammatory. Prepn of racemate: D. Evans *et al.*, Ger. pat. **2,324,443**; *eidem,* U.S. pat. **3,912,748** (1973, 1975 both to Lilly). Prepn of racemate and anti-inflammatory activity: D. W. Dunwell *et al., J. Med. Chem.* **18**, 53-58 (1975). Improved process for prepn of racemate and enantiomers: F. Mauri, R. Signorini, Belg. pat. **877,887**; *eidem,* U.S. pat. **4,304,918** (1979, 1981 both to Ravizza); R. Signorini, A. Verga, Ger. pat. **3,325,672** (1984 to Ravizza). HPLC determn of enantiomers in body fluids: S. Pedrazzini *et al., J. Chromatog.* **415**, 214 (1987). Prophylactic and therapeutic treatment of asthma and other immediate hypersensitivity diseases: W. Dawson, U.S. pat. **4,416,892** (1983 to Lilly).

Crystals from acetone-water or acetic acid, mp 162-164°. $[\alpha]_D^{20}$ +50° (c = 2% in DMF). LD$_{50}$ orally in mice: approx 1200 mg/kg (Dunwell).

THERAP CAT: Anti-inflammatory.

4075. Fluoboric Acid. *Hydrogen tetrafluoroborate;* borofluoric acid; hydrofluoboric acid. BF$_4$H; mol wt 87.83. B 12.32%, F 86.53%, H 1.15%. HBF$_4$. Prepd from H_3BO_3 + HF: Fichter, Thiele, *Z. Anorg. Allgem. Chem.* **67**, 302 (1910); Mathers *et al., J. Am. Chem. Soc.* **37**, 1516 (1915); Funk, Binder, *Z. Anorg. Allgem. Chem.* **155**, 327; **159**, 121 (1926); Sheintsis, *J. Appl. Chem. USSR* **13**, 1101 (1940); Wamser, Christian, *J. Am. Chem. Soc.* **70**, 1209 (1948); Kwasnik in *Handbook of Preparative Inorganic Chemistry,* vol. **1**, G. Brauer, Ed. (Academic Press, New York, 2nd ed., 1963) pp 221-222. For other methods of prepn *see* H. S. Booth, D. R. Martin, *Boron Trifluoride and Its Derivatives* (New York, 1949). *Review:* Sharp, *Advan. Fluorine Chem.* **1**, 68-128 (1960).

Liquid. *Poisonous!* bp 130° (decompn). Miscible with water, alcohol. Strong acid. n_D^{20} of a 20% aq soln 1.3284. Heat of formn 388.5 kcal. Undergoes limited hydrolysis in water to form hydroxyfluoborate ions; major product is BF$_3$OH$^-$. Pure HBF$_4$ may be stored in glass vessels at room temps. Forms cryst salts with metals. The heavy metal salts and LiBF$_4$ and NaBF$_4$ are very sol in water.

Note: See Potassium Fluoborate about color phenomena appearing in concd solns of fluoboric acid.

USE: As catalyst for preparing acetals, esterifying cellulose; to clean metal surfaces before welding; to brighten aluminum; as a solute in electrolytes for plating metals such as chromium, iron, nickel, copper, silver, zinc, cadmium, indium, tin, and lead (has a high throwing power). Reagent for sodium in the presence of magnesium and potassium ions; for making stabilized diazo salts (diazonium and tetrazonium fluoborates). An 0.1 to 0.5% soln retards fermentation: Homeyer, *Pharm. Ztg.* **34**, 761 (1889). *Caution:* Strong caustic action on skin, mucous membranes. Irritating to eyes, respiratory tract. Plating solns contg fluoborates are considered toxic.

4076. Fluocinolone Acetonide. 6,9-Difluoro-11,21-dihydroxy-16,17-[(1-methylethylidene)bis(oxy)]-pregna-1,4-diene-3,20-dione; 6α,9α-difluoro-11β,16α,17,21-tetrahydroxypregna-1,4-diene-3,20-dione cyclic 16,17-acetal with acetone; 6α,9α-difluoro-16α-hydroxyprednisolone 16,17-acetonide; 6α,9α-difluoro-16α,17α-isopropylidenedioxy-1,4-pregnadiene-3,20-dione; Coriphate; Cortiplastol; Dermalar;

Fluonid; Fluovitef; Fluocinil; Fluvean; Fluzon; Jellin; Localyn; Synalar; Synamol; Synandone; Synemol; Synotic; Synsac. $C_{24}H_{30}F_2O_6$; mol wt 452.50. C 63.70%, H 6.68%, F 8.40%, O 21.22%. Prepn: Mills *et al., J. Am. Chem. Soc.* **82**, 3399 (1960); Mills, Bowers, U.S. pat. **3,014,938** (1961 to Syntex). IR spectrum: Sammul *et al., J. Assoc. Offic. Agr. Chem.* **47**, 952 (1964).

Crystals from acetone + hexane, mp 265-266°. $[\alpha]_D$ +95° (chloroform). uv max: 238 nm (log ε 4.21).

THERAP CAT: Glucocorticoid; anti-inflammatory.

THERAP CAT (VET): Adrenocortical steroid (topical use).

4077. Fluocinonide. 21-(Acetyloxy)-6α,9-difluoro-11β-hydroxy-16,17-[(1-methylethylidene)bis(oxy)]pregna-1,4-diene-3,20-dione; 6α,9-difluoro-11β,16α,17,21-tetrahydroxypregna-1,4-diene-3,20-dione, cyclic 16,17-acetal with acetone, 21-acetate; 21-acetoxy-6α,9α-difluoro-11β-hydroxy-16α,17α-isopropylidenedioxy-1,4-pregnadiene-3,20-dione; 16α,17α-isopropylidene-6α-fluorotriamcinolone 21-acetate; fluocinolide (obsolete); fluocinolide acetate (obsolete); fluocinolone acetonide acetate; Biscosal; Dermaplus; Lidex; Metosyn; Synalate (rescinded); Straderm; Topsym; Topsymin; Topsyne; Topsyn. $C_{26}H_{32}F_2O_7$; mol wt 494.55. C 63.14%, H 6.52%, F 7.68%, O 22.65%. Prepn: **Brit. pat. 916,996** (1963 to Olin Mathieson), *C.A.* **59**, 1716b (1963); Ringold, Rosenkranz, U.S. pat. **3,124,571** (1964 to Syntex); Fried, U.S. pat. **3,197,469** (1965 to Pharm. Res. Prod.). NMR: Cross, Landis, *J. Am. Chem. Soc.* **86**, 4005 (1964).

Crystals from methanol. mp 308-311°. $[\alpha]_D$ +83° (chloroform). uv max: 237 nm (log ε 4.18).

THERAP CAT: Anti-inflammatory; glucocorticoid.

4078. Fluocortin Butyl. (6α,11β,16α)-6-Fluoro-11-hydroxy-16-methyl-3,20-dioxopregna-1,4-dien-21-oic acid butyl ester; butyl 6α-fluoro-11β-hydroxy-16α-methyl-3,20-dioxopregna-1,4-dien-21-oate; SH K 203; Varlane; Vaspit. $C_{26}H_{35}FO_5$; mol wt 446.57. C 69.93%, H 7.90%, F 4.25%, O 17.91%. The butyl ester deriv of fluocortolone-21-acid, a metabolite of fluocortolone, *q.v.,* in humans. Prepn: H. Laurent *et al.*, Ger. pats. **2,150,268, 2,150,270**; *eidem,* U.S. pat. **3,824,260** (1973, 1973, 1974, all to Schering AG); H. Laurent *et al., Angew. Chem.* **87**, 70 (1975). Description of this new corticoid structure type, the pregnan-21-oic esters, their activity and methods for synthesis: H. Laurent *et al., J. Steroid Biochem.* **6**, 185 (1975). Toxicity data: P. Günzel *et al., Arzneimittel-Forsch.* **27**, 2217 (1977). Series of articles on synthesis, pharmacology and clinical trials: *ibid.* 2185-2246.

Crystals from acetone/hexane, mp 195.1°. $[\alpha]_D^{25}$ +136° (c = 0.5 in chloroform). uv max (methanol): 242 nm (ϵ 16,800). Soluble in chloroform, ethanol; poorly sol in ethyl ether. Insol in water. LD_{50} in mice, rats (g/kg): >4 orally and s.c. (Günzel).

THERAP CAT: Anti-inflammatory.

4079. Fluocortolone. *6-Fluoro-11,21-dihydroxy-16-methylpregna-1,4-diene-3,20-dione;* 6α-fluoro-16α-methyl-1-dehydrocorticosterone; 6α-fluoro-16α-methyl-$\Delta^{1,4}$-pregnadiene-11β,21-diol-3,20-dione; Ultralan oral. $C_{22}H_{29}FO_4$; mol wt 376.47. C 70.19%, H 7.76%, F 5.05%, O 17.00%. Prepn: **Belg. pat. 614,196** corresp to Kieslich *et al.*, **U.S. pat. 3,232,839** (1962, 1966 to Schering AG); Doménico *et al.*, *Arzneimittel-Forsch.* **15**, 46 (1965); Kieslich *et al.*, *Ann.* **726**, 168 (1969). Pharmacology: Doménico *et al.*, *loc. cit*; Gillich *et al.*, *Int. Z. Klin. Pharmakol. Ther. Toxicol.* **1**, 197 (1968); Von Schoening *et al.*, *Klin. Wochenschr.* **48**, 1448 (1970). Metabolism: Gerhards *et al.*, *Acta Endocrinol. (Copenhagen)* **68**, 98 (1971). Clinical studies: Breuer *et al.*, *Arzneimittel-Forsch.* **15**, 50 (1965).

Crystals, mp 188-190.5°. Soly (mg/l): 295 in water (37°); 120 in ethanol (20°); 440 in toluene (20°). $[\alpha]_D^{20}$ +100° (dioxane). uv max (methanol): 242 nm (ϵ 16,300).

21-Acetate, $C_{24}H_{31}FO_5$, crystals, mp 237-239°. uv max (methanol): 240 nm (ϵ 15,860).

21-Hexanoate, $C_{28}H_{39}FO_5$, *fluocortolone 21-caproate, SH 770, Ficoid, Ultralanum.* Crystals, mp 242-245°. $[\alpha]_D^{20}$ +98.5° (dioxane). uv max (methanol): 242 nm (ϵ 16,200). Soly (mg/l): 7.8 in water (37°); 450 in ethanol (20°); 440 in toluene (20°).

21-Pivalate, $C_{22}H_{37}FO_5$, *fluocortolone trimethylacetate.* Crystals, mp 187°. Almost insol in water. Sol in chloroform, methanol. Slightly sol in ether.

THERAP CAT: Glucocorticoid.

4080. Fluometuron. *N,N-Dimethyl-N'-[3-(trifluoromethyl)phenyl]urea; 1,1-dimethyl-3-(α,α,α-trifluoro-m-tolyl)urea;* N-(3-trifluoromethylphenyl)-N',N'-dimethylurea; Ciba 2059; Cotoran; Cottonex; Lanex. $C_{10}H_{11}F_3N_2O$; mol wt 232.21. C 51.72%, H 4.77%, F 24.55%, N 12.06%, O 6.89%. Made by the reaction of dimethylamine on 3-trifluoromethylphenyl isocyanate: Abel, *Chem. & Ind. (London)* **1957**, 1106. Herbicidal preparations: Martin, Aebi, **U.S. pat. 3,134,665** (1964 to Ciba).

Crystals, mp 163-164.5°. Soly in water at 25°: 80 ppm; sol in acetone, ethanol, isopropanol, DMF and other organic

solvents. LD_{50} orally in male rats: 89 mg/kg, *RTECS* **Vol. II**, R. J. Lewis, R. L. Tatken, Eds. (1979) p 680.

USE: Herbicide.

4081. 9H-Fluorene. *o-Biphenylenemethane;* diphenylenemethane; 2,2'-methylenebiphenyl. $C_{13}H_{10}$; mol wt 166.21. C 93.94%, H 6.06%. Occurs in coal tar (about 1.6%). Isoln: Kruber, *Ber.* **70**, 1556 (1937). Also found in coke-oven tar: Weiss, Downs, *Ind. Eng. Chem.* **15**, 1022 (1923). From acetylene and hydrogen in red-hot tube: Meyer, *Ber.* **45**, 1609 (1912); Meyer, Taeger, *Ber.* **53**, 1261 (1921). From charcoal by boiling with fuming HNO_3: Dimroth, Kerkovins, *Ann.* **399**, 120 (1913). From 2,2'-dibromodiphenylmethane on boiling with hydrazine hydrate in presence of Pd: Busch, Weber, *J. Prakt. Chem.* [2] **146**, 47 (1936). Reactions: Rieveschl Jr., Ray, *Chem. Rev.* **23**, 287 (1938).

Dazzling white leaflets or flakes from alc. d 1.202. Sublimes easily in high vacuum. mp 116-117°. bp 295°. Absorption spectrum: Mayneord, Roe, *Proc. Roy. Soc. London* **A158**, 634 (1937). Freely sol in glacial acetic acid; sol in carbon disulfide, ether, benzene, hot alcohol. Soly data: Mortimer, *J. Am. Chem. Soc.* **45**, 633 (1923).

4082. 9H-Fluorene-2,7-diamine. 2,7-Diaminofluorene. $C_{13}H_{12}N_2$; mol wt 196.24. C 79.56%, H 6.16%, N 14.28%. Prepd by nitrating fluorene and reducing the 2,7-dinitrofluorene formed with tin and hydrochloric acid: Schmidt, Hinderer, *Ber.* **64**, 1793 (1931).

Needles from water, mp 165°. Slightly sol in cold water, more sol in hot water. Readily sol in alc.

Hydrochloride, $C_{13}H_{12}N_2$·HCl, crystals, readily sol in hot water.

USE: Detection of bromide, chloride, nitrate, persulfate, cadmium, copper, cobalt, zinc.

4083. N-2-Fluorenylacetamide. *N-9H-Fluoren-2-ylacetamide;* 2-acetylaminofluorene; AAF; 2-FAA. $C_{15}H_{13}$-NO; mol wt 223.26. C 80.69%, H 5.87%, N 6.28%, O 7.17%. Synthesis: Hayashi, Nakayama, *J. Soc. Chem. Ind. Japan* [Suppl] **36**, 127B (1933). Carcinogenicity studies: R. H. Wilson *et al.*, *Cancer Res.* **1**, 595 (1941). Toxicity studies: Haley *et al.*, *Proc. Soc. Exp. Biol. Med.* **143**, 1117 (1973); **146**, 648 (1974).

Crystals from alcohol + water, mp 194°. uv max: 285 nm. Insol in water. Sol in alcohols, glycols, fat solvents.

Note: This substance may reasonably be anticipated to be a carcinogen: *Fourth Annual Report on Carcinogens* (NTP 85-002, 1985) p 15.

4084. Fluorescamine. *4-Phenylspiro[furan-2(3H),1'-(3'H)-isobenzofuran]-3,3'-dione;* 4-phenylspiro[furan-2(3H),1'-phthalan]-3,3'-dione; Ro 20-7234; Fluram. $C_{17}H_{10}O_4$; mol wt 278.27. C 73.38%, H 3.62%, O 23.00%. Non-fluorescent reagent that reacts readily with primary amines to form highly fluorescent compds: S. Udenfriend *et al.*, *Science* **178**, 871 (1972). Prepn: M. Weigele *et al.*, *J. Am. Chem. Soc.* **94**, 5927 (1972); *eidem, J. Org. Chem.* **41**, 388 (1976). Use as fluorometric reagent: W. Leimgruber, M. Weigele, **Ger. pat. 2,350,179** corresp to **U.S. pat. 3,830,-629** (both 1974 to Hoffmann-La Roche). Review of analytical uses: C. Y. Lai, *Methods Enzymol.* **47**, 236-243 (1977); S. Stein, *Peptides in Neurobiology*, H. Gainer, Ed. (Plenum,

New York, 1977) pp 9-37; S. Udenfriend, *Pharmacology* **19**, 223-227 (1979).

mp 154-155°. uv max (ether): 235, 276, 284, 306 nm (ϵ 25900, 3950, 4100, 3800).

USE: Analytical reagent.

4085. Fluorescein. *3′,6′-Dihydroxyspiro[isobenzofuran-1(3H),9′-[9H]xanthen]-3-one;* 9-(o-carboxyphenyl)-6-hydroxy-3H-xanthen-3-one; 3′,6′-dihydroxyfluoran; 3′,6′-fluorandiol; 9-(o-carboxyphenyl)-6-hydroxy-3-isoxanthenone; resorcinolphthalein; C.I. Solvent Yellow 94; C.I. 45350:1; D & C Yellow no. 7. $C_{20}H_{12}O_5$; mol wt 332.30. C 72.28%, H 3.64%, O 24.07%. Prepd by heating phthalic anhydride with resorcinol: Fischer, Bollmann, *J. Prakt. Chem.* **104**, 123 (1922); McKenna, Sowa, *J. Am. Chem. Soc.* **60**, 124 (1938). Structure: Ramart-Lucas, *Compt. Rend.* **205**, 864 (1937); Nagase *et al., J. Pharm. Soc. Japan* **73**, 1033, 1039 (1953). Review of synthesis, properties and histological use: R. F. Steiner, H. Edelhoch, *Chem. Rev.* **62**, 457 (1962). Use as label in immunoassays: E. F. Ullman *et al., J. Biol. Chem.* **251**, 4172 (1976); Y. Suzuki *et al., Japan. J. Exp. Med.* **49**, 179 (1979). Toxicity studies in fish: L. L. Marking, *Progr. Fish Cult.* **31**, 139 (1969). Toxicity data: S. L. Yankell, J. J. Loux, *J. Periodontol.* **48**, 228 (1977). *See also: Colour Index* **vol. 4** (3rd ed., 1971) p 4424; *H. J. Conn's Biological Stains,* R. D. Lillie, Ed. (Williams & Wilkins, Baltimore, 9th ed., 1977) p 337.

Yellowish-red to red powder. mp 314-316° in sealed tube, with decompn. Insol in water, benzene, chloroform, ether. Sol in hot alcohol or glacial acetic acid; also sol in alkali hydroxides or carbonates with a bright green fluorescence appearing red by transmitted light. Absorption max: 493.5, 460 nm.

Disodium salt, $C_{20}H_{10}Na_2O_5$, *soluble fluorescein, resorcinol phthalein sodium, uranin(e), uranine yellow, D & C yellow No. 8, C.I. Acid Yellow 73, C.I. 45350, Ak-fluor, Fluorescite, Fluorets, Fluor-i-strip, Ful-glo, Funduscein, I-rescein.* Hygroscopic orange-red powder. Freely sol in water with yellowish-red color and intense yellowish-green-fluorescence perceptible down to a dil of 0.02 ppm under uv light. The fluorescence disappears when the soln is made acid, and reappears when the soln is again made neutral or alkaline. Absorption max (water): 493.5 nm. Slightly sol in alc. LD_{50} in mice, rats (mg/kg): 4738, 6721 orally (Yankel, Loux).

USE: In examining subterranean waters. Serves to ascertain source of springs, connections between streams and sea, determining approx vol of water delivered by a spring, detecting source of contamination of drinking water, infiltration of soil with waste waters of factories. Approved by FDA for use in externally applied drugs and cosmetics. Analytical reagent (protein label). Clinical reagent (immuno-histological stain, immuno-fluorescent label).

THERAP CAT: Diagnostic aid (corneal trauma indicator), ophthalmic angiography, contact lens fitting.

THERAP CAT (VET): Diagnostic aid (corneal lesions, intra-ocular inflammation).

4086. Fluorescein Paper. Zellner's paper. Paper charged with a black, substantive, neutral dye, then impregnated with a fluorescein soln and dried. Prepn and application: Zellner, **Ger.** pat. 124,922 (1901); *Chem. Zentralbl.* **1901**, II, 1032; *Pharm. Zentralh.* **1901**, 521; **1902**, 297; *E. Merck's Jahresber.* **1901**, 161-162.

USE: Exceedingly sensitive to alkalies (1:3,000,000) and particularly to ammonia (1:5,000,000) in spring or well waters; usable with dark or strongly colored liqs.

4087. Fluorescin. *2-(3,6-Dihydroxy-9H-xanthen-9-yl)-benzoic acid;* resorcinolphthalin. $C_{20}H_{14}O_5$; mol wt 334.31. C 71.85%, H 4.22%, O 23.93%. Obtained by heating fluorescein with NaOH and zinc dust. Formation by *Pseudomonas aeruginosa:* King *et al., Can. J. Res.* **26C**, 514 (1948); Totter, Moseley, *J. Bacteriol.* **65**, 45 (1953).

Bright yellow powder, mp 125-127°. Readily oxidizes to fluorescein. Practically insol in water. Sol in alkali hydroxides or carbonates, alcohol, ether. *Keep well closed.*

USE: Reagent for oxidases, peroxides.

4088. Fluoresone. *1-(Ethylsulfonyl)-4-fluorobenzene; ethyl p-fluorophenyl sulfone;* p-fluorophenyl ethyl sulfone; Bripadon; Caducid. $C_8H_9FO_2S$; mol wt 188.24. C 51.05%, H 4.82%, F 10.09%, O 17.00%, S 17.04%. Prepn: G. Thuillier *et al., Compt. Rend.* **248**, 2492 (1959); P. Rumpf, G. Thuillier, **Fr.** pat. **M399** corresp to **U.S.** pat. **3,084,101** (1962, 1963 both to Centre Nat. Recher. Scient.); A. A. Mignot, P. Rumpf, *Bull. Soc. Chim. France* **1968**, 435. Pharmacology: J. Thuillier *et al., Proc. Meeting Coll. Int. Neuropsychopharmacol., 3rd, Munich* **1962**, 317-326 (Publ. 1964). Clinical evaluation: H. Akimoto, S. Taen, *ibid.* 326. Gas chromatography: E. Marozzi *et al., Farmaco Ed. Prat.* **31**, 180 (1976).

Crystals, mp 41°. LD_{50} orally in mice: 2.5 g/kg (G. Thuillier); also reported as 850 mg/kg (J. Thuillier); 542 mg/kg (Akimoto, Taen).

THERAP CAT: Anticonvulsant; analgesic; anxiolytic.

4089. Fluoridamid. *N-[4-Methyl-3-[[(trifluoromethyl)-sulfonyl]amino]phenyl]acetamide;* 5-acetamido-2-methyltrifluoromethanesulfonanilide; 3-trifluorosulfonamido-p-acetotoluidide; Sustar. $C_{10}H_{11}F_3N_2O_3S$; mol wt 296.27. C 40.54%, H 3.74%, F 19.24%, N 9.45%, O 16.20%, S 10.82%. Prepn: Harrington *et al.,* **U.S.** pat. **3,639,474** (1972 to Minnesota Mining & Mfg).

mp 175-176.5°.

USE: Plant growth retardant.

4090. Fluorine. F; at. wt 18.998403; at. no. 9; valence 1; elemental state F_2. A halogen. Occurrence in earth's crust 0.065% by wt. Natural abundance of isotopes: ^{19}F 100%; ^{18}F ($T_{1/2}$ 109.7 min) is prepared in nuclear reactors. Does not occur in elemental state in nature. Most important sources are fluorite, cryolite and florapatite, q.v.: Finger, "Fluorine Resources and Fluorine Utilization" in *Advances in Fluorine*

Chemistry, vol. 2, M. Stacey *et al.*, Eds. (Butterworths, London, 1961) pp 35-54. Discovered in 1771 by Scheele. Isolated in 1886 by electrolyzing a soln of potassium fluoride in anhydr hydrogen fluoride at −23°, using platinum-iridium electrodes: Moissan, *Compt. Rend.* **102**, 1543 (1886); **103**, 202, 256. Subsequent methods of prepn: Ruff, *Ber.* **69**, 181 (1936); Henne, *J. Am. Chem. Soc.* **60**, 96 (1938). [18]F has been used to study fluorine exchange reactions: Dave, Sowerby, "Isotopic Halogen Exchange Reactions" in *Halogen Chemistry*, vol. **1**, V. Gutmann, Ed. (Academic Press, New York, 1967) pp 41-132. *Reviews:* A. J. Rudge, *The Manufacture and Use of Fluorine and its Compounds* (Oxford University Press, 1962); O'Donnell, "Fluorine" in *Comprehensive Inorganic Chemistry*, vol. 2, J. C. Bailar, Jr. *et al.*, Eds. (Pergamon Press, Oxford, 1973) pp 1009-1106; A. J. Woytek in Kirk-Othmer *Encyclopedia of Chemical Technology* vol. **10** (Wiley-Interscience, New York, 3rd ed., 1980) pp 630-654.

Pale yellow, diatomic gas. mp −219.61° (53.54°K); bp −188.13° (85.02°K); d (liq, −188.13°) 1.5127; vapor pressure data: Hu *et al.*, *J. Am. Chem. Soc.* **75**, 5642 (1953); White *et al.*, *ibid.* **76**, 2584 (1954). Crit temp: −129°; crit pressure: 55 atm. Most reactive nonmetal; higher oxidation potential than ozone; most electronegative element; E° (calc) ½F/F⁻ 2.9 V. F-F bond weaker than Cl-Cl and Br-Br bonds; enthalpy of dissociation: 37.7 kcal. Reacts vigorously with most oxidizable substances at room temp, frequently with ignition. Combines directly or indirectly, to form fluorides with all the elements except helium, neon and argon. Dec water, giving hydrofluoric acid, HF, oxygen fluoride, OF_2, hydrogen peroxide, oxygen and ozone. Reacts with nitric acid, forming the explosive gas, fluorine nitrate, NO_3F; with sulfuric acid, giving fluorosulfuric acid, $HFSO_3$. Yields the metal fluorides, water, oxygen and oxygen fluoride when made to react with metal hydroxides in the cold. Reacts violently with organic compds, usually with disintegration of the molecule. Under controlled conditions, however, hydrocarbon vapors may be fluorinated with elemental fluorine. Solid fluorine explodes when brought in contact with liquid hydrogen. Under ordinary conditions it does not react directly with oxygen, nor does it react with oxides of sodium, potassium or calcium. LC_{50} (1 hr) inhalation by rats, mice, guinea pigs: 185, 150, 170 ppm (by vol). Keplinger, Suissa, *Am. Ind. Hyg. Assoc. J.* **29**, 10 (1968).

Caution: Dangerous to inhale. Strong caustic action on eyes, skin, mucous membranes. Chronic absorption can cause mottled enamel of teeth, osteosclerosis and calcification of ligaments.

4091. Fluorine Dioxide. Dioxygen difluoride. F_2O_2; mol wt 70.00. F 54.29%, O 45.71%. Prepd from O_2 and F_2: Ruff, Menzel, *Z. Anorg. Allgem. Chem.* **211**, 204 (1933); **217**, 85 (1934); Goetschel *et al.*, *J. Am. Chem. Soc.* **91**, 4702 (1969). Chemical behavior: Streng, *ibid.* **85**, 130 (1963). *Review:* Kemmitt, Sharp, *Advan. Fluorine Chem.* **4**, 214-215 (1965).

Thermally unstable gas at room temp; begins to dec into fluorine and oxygen at −100°. Pale yellow solid or yellow liq at low temp. mp −154°C (119°K). Early prepns, described as brown gas, red liq and orange solid (mp −163.5°), probably contained other fluorine-oxygen compounds as impurities: Goetschel *et al.*, *loc. cit.* Strong oxidizing and fluorinating agent.

4092. Fluorine Monoxide. *Oxygen fluoride;* fluorine oxide. F_2O; mol wt 54.00. F 70.37%, O 29.63%. Prepd by passing fluorine slowly through an aq NaOH soln: Yost, *Inorg. Syn.* **1**, 109 (1939); Schnizlein *et al.*, *J. Phys. Chem.* **56**, 233 (1952). *Review:* Kemmitt, Sharp, *Advan. Fluorine Chem.* **4**, 213-314 (1965).

Colorless gas. Yellowish-brown when liq. Peculiar smell. *Attacks lungs. More poisonous than fluorine; delayed appearance of symptoms.* Does not attack glass in the cold. Corrodes mercury. Reacts very slowly with water. The gas may be kept over water unchanged for a month. d (liq; −224°) 1.90. mp −223.8°. bp −145.3°. Trouton constant 20.65. Soly in water (0°) 6.8 ml gas/100 ml H_2O. LC_{50} (1 hr) inhalation by rats, mice: 2.6, 1.5 ppm; Darmer *et al.*, *Am. Ind. Hyg. Assoc. J.* **33**, 661 (1972).

4093. Fluorine Nitrate. *Nitroxy fluoride;* nitrogen tri-

oxyfluoride; nitryl hypofluorite. FNO_3; mol wt 81.01. F 23.45%, N 17.29%, O 59.25%. $FONO_2$. Prepd by the action of fluorine on nitric acid: Cady, *J. Am. Chem. Soc.* **56**, 2635 (1934); Ruff, Kwasnik, *Angew. Chem.* **48**, 238 (1935); Kwasnik in *Handbook of Preparative Inorganic Chemistry*, vol. **1**, G. Brauer, Ed. (Academic Press, New York, 2nd ed., 1963) pp 187-189. *Reviews:* Kemmitt, Sharp, *Advan. Fluorine Chem.* **4**, 216-218 (1965); Woolf, *ibid.* **5**, 1-30 (1965); Schmutzler, *Angew. Chem. Int. Ed.* **7**, 440-455 (1968).

Colorless gas. Moldy, acrid odor. mp −175°. bp −45.9°. d (liq at bp) 1.507. d⁻¹⁹³·² (solid) 1.951. Trouton const 20.8. The liq explodes on slight percussion. Hydrolyzed by water to OF_2, O_2, HF and HNO_3. Sol in acetone. Conflagrates on contact with alcohol, ether, aniline. May be stored in vacuum-sealed glass ampuls cooled by liq oxygen. Powerful oxidizing agent.

USE: Oxidizing agent in rocket propellants.

4094. Fluorine Perchlorate. Chlorine tetroxyfluoride. $ClFO_4$; mol wt 118.46. Cl 29.93%, F 16.04%, O 54.03%. $FOClO_3$. Prepd by passing fluorine over cold 72% aq perchloric acid in platinum apparatus: Rohrback, Cady, *J. Am. Chem. Soc.* **69**, 677 (1947).

Colorless gas. Pungent, acrid odor. *Attacks lungs even in traces.* Explodes on the slightest provocation, *i.e.*, on contact with rough surfaces, dust, grease, rubber, on melting, distilling, etc. mp −167.3°. bp_{755} −15.9°.

4095. Fluoroacetamide. Fluoroacetic acid amide; monofluoroacetamide; 1081; Fluorakil 100; Fussol. C_2H_4FNO; mol wt 77.06. C 31.17%, H 5.23%, F 24.65%, N 18.18%, O 20.76%. CH_2FCONH_2. Numerous syntheses, e.g. from fluoroacetyl chloride and NH_3: Truce, *J. Am. Chem. Soc.* **70**, 2828 (1948). Mode of action study and toxicity data: F. Matsumura, R. D. O'Brien, *Biochem. Pharmacol.* **12**, 1201 (1963).

Crystals. Sublimes on heating. Freely sol in water; sol in acetone; sparingly sol in chloroform. LD_{50} i.p. in mice: 85 mg/kg (Matsumura, O'Brien).

USE: Rodenticide. Insecticide proposed mainly for use on fruits to combat scale insects, aphids, and mites.

4096. Fluoroacetic Acid. Fluoroethanoic acid; gifblaar poison. $C_2H_3FO_2$; mol wt 78.04. C 30.78%, H 3.87%, F 24.35%, O 41.00%. CH_2FCOOH. Occurs in "Gifblaar," *Dichapetalum cymosum* (Hook.) Engl. (*Chailletia cymosa* Hook.), *Dichapetalaceae*, one of the most poisonous plants, habitat South Africa. Prepd from methyl iodoacetate by heating with silver fluoride or mercurous fluoride or from methyl chloroacetate by heating with potassium fluoride, followed by saponification of the methyl ester with baryta: Swarts, *Bull. Soc. Chim.* [3] **15**, 1134 (1896); Gryszkiewicz-Trochimowski *et al.*, *Rec. Trav. Chim.* **66**, 419 (1947). Review and biochemical aspects: Peters, *Endeavour* **13**, 147 (1954); Peters *et al.*, *Biochem. J.* **77**, 17 (1960). Reviews of toxicity: *Clinical Toxicology of Commercial Products*, R. E. Gosselin *et al.*, Eds. (Williams & Wilkins, Baltimore, 4th ed., 1976) Section III, pp 163-166; Gribble, *J. Chem. Ed.* **50**, 460-462 (1973).

Crystals. Burns with a green flame.
Sodium salt, $C_2H_2FNaO_2$, *Compound 1080, 1080, Fratol*. A fine white powder, sol in water.

Methyl ester, $CH_2FCOOCH_3$, liq, odor of ethyl acetate. d_4^{15} 1.1613. bp 104.5°. Sol in water; slightly sol in petr ether.

Ethyl ester, $CH_2FCOOC_2H_5$, liq, odor of ethyl acetate. $d^{20.5}$ 1.0926. bp_{758} 121.6°. $n_D^{20.5}$ 1.3767. Sol in water.

USE: The sodium salt is used as a water-soluble rodenticide. *Caution:* Extremely toxic. Causes convulsions and ventricular fibrillation. Oral lethal dose of sodium salt estimated to be 2-5 mg/kg body wt. Should never be used where inhalation or food contamination may occur. Locally, it is an irritant.

4097. 3-Fluoro-D-alanine. *S*-2-Amino-3-fluoropropanoic acid; FA. $C_3H_6FNO_2$; mol wt 107.09. C 33.64%, H 5.65%, F 17.74%, N 13.08%, O 29.88%. Synthesis by C-photofluorination of D-alanine or by resolution of the DL-form: J. Kollonitsch, F. M. Kahan; J. Kollonitsch, **Ger.** pats. **2,136,067; 2,229,245,** *C.A.* **76**, 100053p (1972); **78**, 72586d (1973); *idem*, **U.S. pat. 3,839,170** (1972, 1972, 1974 all to Merck & Co.). The 2-deuterated analog has also been prepared. Antibacterial activity: J. Kollonitsch *et al.*, *Nature* **243**, 346 (1973).

$$H_2N - \underset{\underset{COOH}{|}}{\overset{\overset{CH_2F}{|}}{C}} - H$$

Crystals from water, dec 168°. $[\alpha]_D^{20}$ −10° (c = 3 in 1*N* HCl). Stable at acid pH; less stable at pH > 8.

4098. *p*-Fluoroaniline. *4-Fluorobenzenamine.* C_6H_6FN; mol wt 111.12. C 64.85%, H 5.44%, F 17.10%, N 12.61%. Prepd by reduction of 1-fluoro-4-nitrobenzene by Raney nickel: Benington *et al.*, *J. Org. Chem.* **18**, 1508 (1953); Finger *et al.*, *J. Am. Chem. Soc.* **81**, 98 (1959); by sulfurated sodium borohydride: Lalancette, Brindle, *Can. J. Chem.* **49**, 2990 (1971).

Liquid. d_4^{20} 1.1725; d_4^{25} 1.1690. mp −1.9°. bp 188°, bp_{20} 86°. n_D^{20} 1.51954. Heat of combustion 780.4 kcal/mol. Very slightly sol in water.

Picrate, $C_6H_6FN \cdot C_6H_3N_3O_7$, yellow crystals from alc, dec 214°.

USE: Intermediate in the manuf of herbicides and plant growth regulators.

4099. Fluorobenzene. C_6H_5F; mol wt 96.10. C 74.98%, H 5.24%, F 19.77%. Obtained by warming benzenediazonium chloride with concd HF. Pressure, vol, temp constants: Douslin *et al.*, *J. Am. Chem. Soc.* **80**, 2031 (1958).

Liquid; benzene odor. d_4^{20} 1.024. bp_{760} 84.73°; $bp_{13\ atm}$ 200°; $bp_{38\ atm}$ 275°. mp −40°. n_D^{20} 1.4677. Insol in water; miscible with alcohol, ether.

4100. *p*-Fluorobenzoic Acid. $C_7H_5FO_2$; mol wt 140.11. C 60.00%, H 3.60%, F 13.56%, O 22.84%. Prepd by treating *p*-carbethoxybenzenediazonium chloride with fluoboric acid, followed by thermal decompn of the resulting *p*-carbethoxybenzenediazonium fluoborate and by hydrolysis of the ensuing ethyl ester of *p*-fluorobenzoic acid: Schiemann, Winkelmüller, *Org. Syn.* **13**, 52 (1933).

Monoclinic prisms from water. Peculiar sweet taste. mp 182.6°. Sparingly sol in cold water; freely sol in hot water. Soly in water at 32°: 1.1 g/l. Water solns evaporate without leaving a residue. Sol in alc, ether. K at 25°: 1.4 × 10⁻⁴.

Ethyl ester, $C_9H_9FO_2$, crystals, mp 26°. bp 210°.

4101. 1-Fluoro-2,4-dinitrobenzene. 2,4-Dinitro-1-fluorobenzene; DNFB; DNB; Sanger's reagent. $C_6H_3FN_2O_4$; mol wt 186.11. C 38.72%, H 1.62%, F 10.21%, N 15.06%, O 34.39%. Prepn: A. F. Holleman, J. W. Beekman, *Rec. Trav. Chim.* **23**, 225 (1904); Cook, Saunders, *Biochem. J.* **41**, 558 (1947). Use in peptide analysis: Sanger, *Biochem. J.* **39**, 507 (1945); **40**, 261 (1946); **45**, 563 (1949); Porter, Sanger, *ibid.* **42**, 287 (1948). Tumor promoting activity: F. G. Bock *et al.*, *Cancer Res.* **29**, 179 (1969). Mutagenicity study: D. R. Jagannath *et al.*, *Mutat. Res.* **78**, 91 (1980). Review of uses: *Reagents for Organic Synthesis*, L. F. Fieser, M. Fieser, Eds. (Wiley, New York, 1967) pp 321-322.

Pale yellow crystals from ether, mp 26°. $bp_{2.0}$ 137°. Sol in benzene, ether, propylene glycol.

USE: Reagent for labeling a terminal amino acid group; in modified Wohl degradations of aldoses. As hapten. *Caution:* Vesicant. For proper handling see: J. S. Thompson, O. P. Edmunds, *Ann. Occup. Hyg.* **23**, 27 (1980).

4102. Fluoroform. *Trifluoromethane.* CHF_3; mol wt 70.02. C 17.15%, H 1.44%, F 81.41%. Prepd from $CHCl_3$ and HF: Meslans, *Compt. Rend.* **110**, 717 (1890); Valentiner, U.S. pat. 643,835 (1900); Ruff, *Ber.* **69**, 299 (1936); Whallay, *J. Soc. Chem. Ind. (London)* **66**, 429 (1947); Kwasnik in *Handbook of Preparative Inorganic Chemistry* vol. 1, G. Brauer, Ed. (Academic Press, New York, 2nd ed., 1963) pp 204-205. Early industrial prepn: Pearlson in *Fluorine Chemistry* vol. I, J. H. Simons, Ed. (Academic Press, New York, 1950) p 467.

Colorless, odorless gas. Stable up to 1150°. Chemically very inert. d (solid) 1.935. d (liq; −100°) 1.52. mp 160°. bp −84.4°. Critical temp 33°; critical pressure 47 atm, critical density 0.516. May be stored over water. Practically non-toxic: N. V. Sidgwick, *The Chemical Elements and Their Compounds*, vol. II (Oxford, 1950) p 1130. *Caution:* May be slightly irritating to respiratory tract, and, in high concns, narcotic.

USE: Refrigerant for low temps.

4103. Fluoromethane. Methyl fluoride. CH_3F; mol wt 34.03. C 35.29%, H 8.89%, F 55.83%. Prepn in 82% yield by heating fluorosulfonic acid methyl ester with KF: Zappel, Jonas, *Ger.* **1,131,197** (1962 to Bayer).

Gas. Agreeable ether-like odor. Burns with evolution of HF, the flame being about as colorless as that of alc: Dumas, Peligot, *Ann.* **15**, 59 (1835). d (liq; −78°) 0.8774. d (gas) 1.1951 (air = 1); d (gas) 1.0813 (oxygen = 1). Dipole moment 1.81. Molecular volume: 22.03. Van der Waals constants: 0.00923; 0.002350. Critical temp 44.9°; crit press. 62.0 atm. Dielectric constant (for wavelengths below 10⁴ cm) = 1.00948. mp −141.8°. bp_{872} −75.7°; bp_{760} −78.2°. bp_{143} −103.7°. One hundred vols of water dissolve 166 vols of the gas at 15°. Freely sol in alcohol, ether. *Caution:* Narcotic in high concns.

4104. Fluorometholone. *(6α,11β)-9-Fluoro-11,17-dihydroxy-6-methylpregna-1,4-diene-3,20-dione;* 21-desoxy-9α-fluoro-6α-methylprednisolone; 21-desoxy-6α-methyl-9α-fluoroprednisolone; fluormetholon; Cortilet; Delmeson; Efflumidex; Fluaton; Flumetholon; FML; Loticort; Oxylone; Ursnon. $C_{22}H_{29}FO_4$; mol wt 376.47. C 70.19%, H 7.76%, F 5.05%, O 17.00%. Prepn: Lincoln *et al.*, U.S. pat. **2,867,638** (1959 to Upjohn). Prepn of acetate and other esters: B. J. Magerlein *et al.*, U.S. pat. **3,038,914** (1962 to Upjohn). Pharmacology of acetate: A. Kupferman *et al.*, *Arch. Ophthalmol.* **100**, 640 (1982). HPLC determn: H. Tokunaga *et al.*, *Chem. Pharm. Bull.* **32**, 4012 (1984). Clinical evaluation in ocular inflammation: H. M. Leibowitz *et al.*, *Ann. Ophthalmol.* **16**, 1110 (1984).

Crystals from acetone, mp 292-303°.

17-Acetate, $C_{24}H_{31}FO_5$, crystals from ethyl acetate + hexanes, mp 230-232°. $[\alpha]_D$ +28° (chloroform).

THERAP CAT: Glucocorticoid; anti-inflammatory.

4105. *p*-Fluorophenylacetic Acid. *4-Fluorobenzeneacetic acid.* $C_8H_7FO_2$; mol wt 154.15. C 62.34%, H 4.58%, F 12.33%, O 20.76%. Prepd by treating *p*-fluorobenzyl cyanide with sulfuric acid: Olah *et al., J. Org. Chem.* **22**, 879 (1957).

CH₂COOH

Crystals from chloroform, mp 86°. bp₂ 164°.
β-Dimethylaminoethyl ester HCl, crystals, mp 103°.
USE: Intermediate in the manuf of fluorinated anesthetics.

4106. Fluorosalan. *3,5-Dibromo-2-hydroxy-N-[3-(trifluoromethyl)phenyl]benzamide; 3,5-dibromo-α,α,α-trifluoro-m-salicylotoluidide;* 3,5-dibromo-3'-trifluoromethylsalicylanilide; Fluorophene. $C_{14}H_8Br_2F_3NO_2$; mol wt 439.05. C 38.30%, H 1.84%, Br 36.40%, F 12.98%, N 3.19%, O 7.29%. Prepn: Stecker, U.S. pat. **3,041,236** (1962).

THERAP CAT: Disinfectant.

4107. Fluorosulfonic Acid. *Fluorosulfuric acid;* fluosulfonic acid. FHO_3S; mol wt 100.07. F 18.98%, H 1.01%, O 47.96%, S 32.04%. HSO_3F. Prepn: Thorpe, Kirman, *J. Chem. Soc.* **61**, 921 (1892); Meyer, Schramm, *Z. Anorg. Allgem. Chem.* **206**, 25 (1932); Kwasnik in *Handbook of Preparative Inorganic Chemistry,* vol. **1**, G. Brauer, Ed. (Academic Press, New York, 2nd ed., 1963) pp 177-178. Reviews of properties and chemistry: Gillespie, *Accounts Chem. Res.* **1**, 202-209 (1968); Thompson, "Fluorosulfuric Acid" in *Inorganic Sulfur Chemistry,* G. Nickless, Ed. (Elsevier, New York, 1968) pp 587-606; Jache in *Advan. Inorg. Chem. Radiochem.* **16**, 177-200 (1974).

Colorless liquid; fumes in moist air. d₄¹⁸ 1.740; d₄²⁵ 1.726. mp −89°. bp₇₆₀ 163°; bp₁₂₀ 110.0°; bp₁₉ 77.0°. Stable to 900°. Considerably more acidic than 100% H_2SO_4. Does not attack glass when anhydr and pure. Violent reaction with water although it is incompletely and reversibly hydrolyzed. Reddish-brown color with acetone. Forms stable salts which are little hydrolyzed by water and which may be recrystallized from water.
Methyl ester, *see* Methyl Fluorosulfonate.
Caution: May be highly irritating to skin, mucous membranes.
USE: Fluorinating agent. Catalyst in alkylation, acylation, polymerization and condensation reactions; in hydrofluorination of olefins; in production of substituted pyridines. In production of petroleum products. See Thompson, *loc. cit. Magic Acid,* a 1:1 HSO_3F-SbF_5 soln, is used in the study of stable solns of alkyl- and arylcarbonium ions: Olah, *Science* **168**, 1298 (1970).

4108. Fluorotoluene. *Fluoromethylbenzene.* C_7H_7F; mol wt 110.13. C 76.34%, H 6.41%, F 17.25%. Prepn of *o-, m-, p*-isomers: A. F. Holleman, J. W. Beekman, *Rec. Trav. Chim.* **23**, 225 (1904); A. F. Holleman, *ibid.* **25**, 330 (1906); E. D. Bergmann, S. Berkovic, *J. Org. Chem.* **26**, 919 (1961).

CH₃

Colorless liquids. Insol in water. Miscible with alc, ether. *o*-Form, d¹³ 1.004. mp about −80°. bp 114°. n_D²⁰ 1.4704. *m*-Form, d¹³ 0.997. mp −111°. bp 116°. n_D²⁰ 1.4691. *p*-Form, odor of oil of bitter almonds. d¹⁶ 1.001. bp 116°. n_D²⁰ 1.470.

4109. Fluorouracil. *5-Fluoro-2,4(1H,3H)-pyrimidinedione;* 2,4-dioxo-5-fluoropyrimidine; 5-FU; Ro 2-9757; NSC 19893; Adrucil; Arumel; Cinco F U; Efudex; Efudix; Fluril; Fluracil; Fluoroplex; Fluroblastin; Fluro Uracil; Timazin. $C_4H_3FN_2O_2$; mol wt 130.08. C 36.93%, H 2.32%, F 14.61%, N 21.54%, O 24.60%. Prepn: Duschinsky *et al., J. Am. Chem. Soc.* **79**, 4559 (1957); Heidelberger, Duschinsky, U.S. pats. **2,802,005** (1957) and **2,885,396** (1959); Barton *et al., J. Org. Chem.* **37**, 329 (1972); O. Miyashita *et al., Chem. Pharm. Bull.* **29**, 3181 (1981). Pharmacokinetics: W. Sadee, C. G. Wong, *Clin. Pharmacokinet.* **2**, 437 (1977). Site of action: F. Maley, G. F. Maley, *FEBS Symp.* **57**, 21 (1979). Comprehensive description: B. C. Rudy, B. Z. Senkowski, in *Analytical Profiles of Drug Substances* vol. **2**, K. Florey, Ed. (Academic Press, New York, 1973) pp 221-244.

Crystals from water or methanol-ether, dec 282-283°. Sublimes (0.1 mm) 190-200°. uv max (0.1N HCl): 265-266 nm (ε 7070).
Sodium salt, *Effluderm*.
THERAP CAT: Antineoplastic.

4110. Fluosilicic Acid. *Hydrogen hexafluorosilicate;* hexafluosilicic acid; hydrosilicofluoric acid; hydrofluosilicic acid; silicofluoric acid; fluorosilicic acid. F_6H_2Si; mol wt 144.08. F 79.12%, H 1.40%, Si 19.47%. H_2SiF_6. Prepd from HF + SiO_2; also prepd by the action of water on SiF_4; by the action of H_2SO_4 on $BaSiF_6$: Hempel, *Ber.* **18**, 1438 (1885); Baur, Glaessner, *Ber.* **36**, 4215 (1903); Söll, *FIAT-Review 23,* 257 (1946); Kwasnik in *Handbook of Preparative Inorganic Chemistry* vol. **1**, G. Brauer, Ed. (Academic Press, New York, 2nd ed., 1963) p 214-215; Lange in *Fluorine Chemistry,* vol. **I**, J. H. Simon, Ed. (New York, 1950) p 129. Commercial grades of fluosilicic acid soln are obtained as a by-product in the superphosphate industry. *Review:* Colton, *J. Chem. Ed.* **35**, 562-563 (1958).

Liquid, when anhydr dissociates almost instantly into SiF_4 and HF. Marketed as aq soln only. A 60-70% soln solidifies around 19°, forming a cryst dihydrate. May be distilled without decompn only as a 13.3% aq soln. Fairly strong acid. Sour, pungent odor. d₁₇.₅¹⁷·⁵ 5% soln: 1.0407; 10%: 1.0834; 15%: 1.1281; 20%: 1.1748; 25%: 1.2235; 30%: 1.2742; 34%: 1.3162. The more concd solns (but not the anhydr liq) can be stored in glass, although some etching will take place around the surface. Usually stored in iron containers.
USE: A 1-2% soln is used widely for sterilizing equipment in brewing and bottling establishments. Other concns are used in the electrolytic refining of lead, in electroplating, for hardening cement, crumbling lime or brick work, for the removal of lime from hides during the tanning process, to remove molds, as preservative for timber. *Caution:* Severe corrosive effect on skin, mucous membranes.

4111. Fluosol DA. Biologically inert fluorocarbon emulsion of perfluorodecalin and perfluorotripropylamine that performs the oxygen-carrying function of red blood cells. Has a particle size of 0.05 to 0.25μ and is stable at room temperature for years. Can be administered regardless of blood type; does not carry hepatitis or other infectious diseases. Prepn: K. Yokoyama *et al.*, **Ger. pat. 2,630,586;**

eidem, U.S. pat. **4,252,827** (1977, 1981 to Green Cross). Hemodynamic and oxygen transport effects: K. K. Tremper *et al.*, *Crit. Care Med.* **8**, 738 (1980). Clinical studies: H. Ohyanagi *et al.*, *Clin. Ther.* **2**, 306 (1979); T. Suyama *et al.*, *Prog. Clin. Biol. Res.* **55**, 609 (1981); T. Mitsuno *et al.*, *Ann. Surg.* **195**, 60 (1982). *Review: Int. Congr. Ser.-Excerpta Med.* **486**, 1-471 (1979). Review of use as red cell substitute: K. C. Lowe, *Comp. Biochem. Physiol.* **87A**, 825-838 (1987).

20% Emulsion, *FDA-20*. Viscosity at 37°: 2.3 cp. pH 7.4.

Perfluorodecalin, $C_{10}F_{18}$, *octadecafluorodecahydronaphthalene, perflunafene, FDC*. Prepn: E. T. McBee, L. D. Bechtol, *Ind. Eng. Chem.* **39**, 380 (1947). mp −7 to −10°. bp 140°. d_4^{20} 1.9456; n_D^{20} 1.3118. Vapor pressure at 37°: 12.7 mm Hg.

Perfluorotripropylamine, $C_9F_{21}N$, *1,1,2,2,3,3,3-heptafluoro-N,N-bis(heptafluoropropyl)-1-propanamine, heneicosafluorotripropylamine, perfluamine, FTPA*. Prepn: R. N. Haszeldine, *J. Chem. Soc.* **1951**, 102. bp 129.5-130.5°; d_4^{25} 1.822; n_D^{25} 1.279. Vapor press. at 37°: 20.0 mm Hg.

USE: Blood substitute.

4112. Fluoxetine. (±)-*N-Methyl-γ-[4-(trifluoromethyl)phenoxy]benzenepropanamine;* (±)-*N-methyl-3-phenyl-3-[(α,α,α-trifluoro-p-tolyl)oxy]propylamine; N-methyl-3-(p-trifluoromethylphenoxy)-3-phenylpropylamine*. $C_{17}H_{18}F_3NO$; mol wt 309.33. C 66.01%, H 5.86%, F 18.43%, N 4.53%, O 5.17%. Serotonin uptake inhibitor. Prepn: B. B. Malloy, K. K. Schmiegel, **Ger. pat. 2,500,110**; *eidem*, **U.S. pat. 4,314,081** (1975, 1982 both to Lilly). Pharmacology: D. T. Wong *et al.*, *Life Sci.* **15**, 471 (1974); *eidem*, *J. Pharmacol. Exp. Ther.* **193**, 804 (1975). GC determn in plasma: J. F. Nash *et al.*, *Clin. Chem.* **28**, 2100 (1982). Interaction with ethanol: L. Lemberger *et al.*, *Clin. Pharmacol. Ther.* **6**, 658 (1985). Clinical trial in depression: L. F. Fabre, L. Crismon, *Curr. Ther. Res.* **37**, 115 (1985). Pharmacology and toxicity data: P. Stark *et al.*, *J. Clin. Psychiatry* **46**, Suppl. 3(2), 7 (1985). Series of articles on clinical pharmacology and therapeutic efficacy: *ibid.* 14-67. Review of pharmacology, therapeutic efficacy: P. Benfield *et al.*, *Drugs* **32**, 481-508 (1986).

Hydrochloride, $C_{17}H_{19}ClF_3NO$, *LY-110140, Fontex, Prozac*. LD_{50} in mice, rats (mg/kg): 248, 452 orally (Stark). Oxalate, $C_{19}H_{18}F_3NO_5$, crystals from ethyl acetate-methanol, mp 179-182° (dec).

THERAP CAT: Antidepressant.

4113. Fluoxymesterone. *9-Fluoro-11,17-dihydroxy-17-methylandrost-4-en-3-one;* 11β,17β-dihydroxy-9α-fluoro-17α-methyl-4-androsten-3-one; 9α-fluoro-11β-hydroxy-17α-methyltestosterone; *Androsterolo; Androfluorene; Androfluorone; Fluotestin; Halotestin; Oratestin; Ora-Testryl; Testoral; Ultandren;* $C_{20}H_{29}FO_3$; mol wt 336.45. C 71.40%, H 8.69%, F 5.65%, O 14.27%. Prepn: Herr *et al.*, *J. Am. Chem. Soc.* **78**, 501 (1956). *cf.* Herr, U.S. pat. **2,813,881** (1957 to Upjohn). Comprehensive description: J. Kirschbaum in *Analytical Profiles of Drug Substances* vol. 7, K. Florey, Ed. (Academic Press, N.Y., 1978) pp 251-275.

Crystals, dec 270°. $[\alpha]_D$ +109° (ethanol). uv max (ethanol): 240 nm (ε 16,700). Sol in pyridine; slightly sol in acetone, chloroform; sparingly sol in methanol; practically insol in water, ether, benzene, hexanes.

THERAP CAT: Androgen.

4114. Flupentixol. *4-[3-[2-(Trifluoromethyl)-9H-thioxanthen-9-ylidene]propyl]-1-piperazineethanol;* 2-trifluoromethyl-9-[3-[4-(β-hydroxyethyl)-1-piperazinyl]propylidene]thioxanthene; flupenthixol; N-7009; LC 44. $C_{23}H_{25}F_3N_2OS$; mol wt 434.54. C 63.57%, H 5.80%, F 13.12%, N 6.45%, O 3.68%, S 7.38%. Neuroleptic agent related structurally to thiothixene, *q.v.* Prepn of *cis*- and *trans*-forms: Brit. pat. **925,538** (1963 to SKF). Metabolism of decanoate: Jorgensen *et al.*, *Acta Pharmacol. Toxicol.* **29**, 339 (1971). α-Flupentixol, the *cis*-isomer, shows greater pharmacological activity than β-flupentixol, the *trans*-isomer: I. Moller Nielsen *et al.*, *Acta Pharmacol. Toxicol.* **32**, 353 (1973). X-ray crystallography of isomers: M. L. Post *et al.*, *Nature* **256**, 342 (1975).

Dihydrochloride, $C_{23}H_{27}Cl_2F_3N_2O$, *Emergil, Fluanxol, Siplarol, Metamin*.

Decanoate, *Lu 5-110, Depixol, Fluanxol Dépot, α-FPTdec., Viscoleo*.

THERAP CAT: Antipsychotic.

4115. Fluperolone Acetate. *17-[2-(Acetyloxy)-1-oxopropyl]-9-fluoro-11,17-dihydroxyandrosta-1,4-dien-3-one; 9-fluoro-11β,17α-dihydroxy-17(S)-lactoylandrosta-1,4-dien-3-one 17β-acetate;* 9α-fluoro-11β,17α,21-trihydroxy-21-methylpregna-1,4-diene-3,20-dione 21-acetate; 17β-[2-acetoxypropionyl]-9α-fluoro-11β,17-dihydroxyandrosta-1,4-dien-3-one; 21-methyl-9α-fluoroprednisolone acetate; P-1742; ALAcortril; Methral. $C_{24}H_{31}FO_6$; mol wt 434.51. C 66.34%, H 7.19%, F 4.37%, O 22.09%. Prepn: Angello *et al.*, *Experientia* **16**, 357 (1960); *eidem*, *J. Org. Chem.* **28**, 1531 (1963). Activity studies: Child *et al.*, *Arch. Dermatol.* **97**, 407 (1968).

Crystals, mp 251-253°. $[\alpha]_D$ +87°. uv max: 239 nm (ε 15,350).

THERAP CAT: Glucocorticoid, anti-inflammatory.

4116. Fluphenazine. *4-[3-[2-(Trifluoromethyl)-10H-phenothiazin-10-yl]propyl]-1-piperazineethanol;* 1-(2-hydroxyethyl)-4-[3-(trifluoromethyl-10-phenothiazinyl)propyl]-piperazine; 10-[3'-[4''-(β-hydroxyethyl)-1''-piperazinyl]propyl]-3-trifluoromethylphenothiazine; 2-(trifluoromethyl)-10-[3-[1-(β-hydroxyethyl)-4-piperazinyl]propyl]phenothiazine; S94; SQ 4918; Elinol; Pacinol; Siqualine; Siqualon; Tensofin; Valamina; Vespazine. $C_{22}H_{26}F_3N_3OS$; mol wt 437.52. C 60.39%, H 5.99%, F 13.03%, N 9.60%, O 3.66%, S 7.33%. Prepn: Yale, Sowinski, *J. Am. Chem. Soc.* **82**, 2039 (1960); **Brit. pat. 829,246** (1960 to SKF), *C.A.* **54**, 17428h (1960); **Brit. pat. 833,473** (1960 to Scherico), *C.A.* **54**, 21143e (1960); Anderson *et al.*, *Arzneimittel-Forsch.* **12**, 937 (1962); Yale, Merrill, U.S. pat. **3,194,733** (1965 to Olin Mathieson). Metabolism: J. Dreyfuss, A. J. Cohen, *J. Pharm. Sci.* **60**, 826 (1971). Comprehensive description of the enanthate ester: K. Florey, Ed. in *Analytical Profiles of Drug Substances* vol. 2 (Academic Press, New York, 1973) pp 245-262; of the dihydrochloride: *idem, ibid.* pp 263-294; of the decanoate ester: G. Clarke, *ibid.* vol. 9 (1980) pp 275-294.

Dark brown viscous oil, $bp_{0.5}$ 268-274°; $bp_{0.3}$ 250-252°. Dihydrochloride, $C_{22}H_{28}Cl_2F_3N_3OS$, *Anatensol, Dapotum, Dapotum D, Lyogen, Moditen (tabl. or elixir), Omca, Permitil, Prolixin, Sevinol, Trancin*. Crystals from abs ethanol, mp 235-237°. Also reported as mp 224.5-226°.

Decanoate, $C_{32}H_{44}F_3N_3O_2S$, *SQ 10733, QD 10733, Modecate, Prolixin Decanoate*. Pale yellow-orange, viscous liquid. Slowly crystallizes at room temp. mp 30-32°. Very sol in chloroform, ether, cyclohexane, methanol, ethanol. Insol in water.

Enanthate, $C_{29}H_{38}F_3N_3O_2S$, *SQ 16144, Moditen Enanthate, Moditen-Retard, Prolixin Enanthate*. Pale yellow to yellow-orange viscous liquid or oily solid.

THERAP CAT: Antipsychotic.

4117. Flupirtine. *[2-Amino-6-[[(4-fluorophenyl)methyl]amino]-3-pyridinyl]carbamic acid ethyl ester; 2-amino-6-[(p-fluorobenzyl)amino]-3-pyridinecarbamic acid ethyl ester;* D 9998. $C_{15}H_{17}FN_4O_2$; mol wt 304.33. C 59.20%, H 5.63%, F 6.24%, N 18.41%, O 10.51%. Substituted pyridine with central analgesic properties. Prepn: K. Thiele, W. von Bebenburg, **S. Afr. pat. 69 02,364** (1970 to Degussa); W. von Bebenburg *et al., Chem. Ztg.* **103**, 387 (1979); *eidem, ibid.* **105**, 217 (1981). Prepn of maleate: W. von Bebenburg, S. Pauluhn, **Belg. pat. 890,331**; *eidem,* **U.S. pat. 4,481,205** (1980, 1984 both to Degussa). Comparison of pharmacology with other analgesics: V. Jakovlev *et al., Arzneimittel-forsch.* **35**, 30 (1985). Pharmacokinetic studies: K. Obermeier *et al., ibid.* 60. Effect on driving ability: B. Biehl, *ibid.* 77. Clinical trials in treatment of cancer pain: W. Scheef, D. Wolf-Gruber, *ibid.* 75. Efficacy in treatment of pain after hysterectomy: R. A. Moore *et al., Brit. J. Anaesth.* **55**, 429 (1983). Symposium on pharmacology and clinical efficacy: *Postgrad. Med. J.* **63**, Suppl. 3, 1-113 (1987).

Crystals from isopropanol, mp 115-116°. 5% ethanol soln is colorless, turns green on exposure to air for 20 hours. LD_{50} orally in mice, rats: 617, 1660 mg/kg (Jakovlev).

Hydrochloride, $C_{15}H_{18}ClFN_4O_2$, crystals from water, mp 214-215°. When prepd industrially contains intensely blue by-product.

Maleate, $C_{19}H_{21}FN_2O_6$, *Katadolon*. Colorless crystals from isopropanol, mp 175.5-176°. Formed as mixture of two crystalline forms A and B; mixtures containing 60-90% A are preferred.

THERAP CAT: Analgesic.

4118. Fluprednidene Acetate. *21-(Acetyloxy)-9-fluoro-11β,17-dihydroxy-16-methylenepregna-1,4-diene-3,20-dione; 9α-fluoro-11β,17,21-trihydroxy-16-methylenepregna-1,4-diene-3,20-dione 21-acetate;* 9α-fluoro-16-methylene-Δ^{1,4}-pregnadiene-11β,17,21-triol-3,20-dione 21-acetate; 16-methylene-9α-fluoroprednisolone 21-acetate; 9α-fluoro-16-methyleneprednisolone 21-acetate; fluprednylidene 21-acetate; StL 1106; Etacortin; Decoderm sine Gentamycin. $C_{24}H_{29}FO_6$; mol wt 432.49. C 66.65%, H 6.75%, F 4.39%, O 22.20%. Prepn: Taub *et al., J. Org. Chem.* **25**, 2258 (1960); **29**, 3486 (1964); v. Werder *et al., Arzneimittel-Forsch.* **18**, 7 (1968); Wendler, Taub; Taub, Wendler; Taub *et al.*; Wendler *et al.*, **U.S. pats. 3,065,239; 3,068,224; 3,068,226 and 3,136,-760** (1960 and 1964 to Merck & Co.).

Crystals, mp 231-234°. $[\alpha]_D$ +43° (CHCl₃); $[\alpha]_D$ +32° (dioxane) (v. Werder *et al.*). uv max (methanol): 238 nm (ϵ 15,700).

THERAP CAT: Anti-inflammatory (topical).

4119. Fluprednisolone. *6-Fluoro-11,17,21-trihydroxy-pregna-1,4-diene-3,20-dione;* 6α-fluoroprednisolone; 6α-fluoro-1,4-pregnadiene-11β,17α,21-triol-3,20-dione; 6α-fluoro-1-dehydrohydrocortisone; U-7800; NSC-47439; Alphadrol; Etadrol. $C_{21}H_{27}FO_5$; mol wt 378.45. C 66.65%, H 7.19%, F 5.02%, O 21.14%. Prepn: Hogg, Spero, **U.S. pat. 2,841,600** (1958 to Upjohn); Batres *et al.*, **Ger. pat. 1,079,042** (1960 to Syntex); Lettré, Hotz, **Ger. pat. 1,088,953** (1960 to Bayer).

Crystals, mp 208-213°. $[\alpha]_D$ +92°.
21-Acetate, $C_{23}H_{29}FO_6$, crystals, mp 235-238°.
THERAP CAT: Glucocorticoid; anti-inflammatory.
THERAP CAT (VET): Anti-inflammatory.

4120. Fluproquazone. *4-(4-Fluorophenyl)-7-methyl-1-(1-methylethyl)-2(1H)-quinazolinone;* 1-isopropyl-7-methyl-4-(p-fluorophenyl)-2(1H)-quinazolinone; RF 46-790; Arthrisin; Tormosyl. $C_{18}H_{17}FN_2O$; mol wt 296.35. C 72.95%, H 5.78%, F 6.41%, N 9.45%, O 5.40%. Fluoro analog of proquazone, *q.v.* Prepn: G. E. Hardtmann *et al.*, **Ger. pat. 2,230,393** (1973 to Sandoz); *C.A.* **78**, 111363a (1973); G. E. Hardtmann, **U.S. pat. 3,937,705** (1976 to Sandoz). Analgesic effect: U. Herrmann *et al., Clin. Pharmacol. Ther.* **27**, 379 (1980); B. von Graffenried, E. Nuesch, *Brit. J. Clin. Pharmacol.* **10**, Suppl. 2, 225 (1980). Series of articles on pharmacology, metabolism, analysis, clinical studies, toxicology: *Arzneimittel-Forsch.* **31**, 871-940 (1981). Toxicity: G. Rüttimann *et al., Arzneimittel-Forsch.* **31**, 882 (1981).

Crystals, mp 172-174°. LD_{50} orally in mice, rats, rabbits (mg/kg): > 5000, > 3000, 742 (Rüttimann).
THERAP CAT: Analgesic.

4121. Fluprostenol. *7-[3,5-Dihydroxy-2-[3-hydroxy-4-[3-(trifluoromethyl)phenoxy]-1-butenyl]cyclopentyl]-5-heptenoic acid;* ICI 81008; Equimate. $C_{23}H_{29}F_3O_6$; mol wt 458.48. C 60.25%, H 6.38%, F 12.43%, O 20.94%. Potent luteolytic agent related to prostaglandin $F_{2\alpha}$, *q.v.*: Dukes, Walpole, *Nature* **250**, 330 (1974). Prepn: Binder *et al., Prostaglandins*

6, 87 (1974). Induction of estrus and luteolysis in mares: Cooper, Farr, *Vet. Rec.* **94,** 161 (1974); Berwyn-Jones, Irvine, *N.Z. Vet. J.* **22,** 107 (1974).

Sodium salt, $C_{23}H_{28}F_3NaO_6$, *ICI 80008.*
THERAP CAT (VET): Treatment of infertility in mares.

4122. Flurandrenolide. *6α-Fluoro-11β,21-dihydroxy-16α,17-[(1-methylethylidene)bis(oxy)]pregn-4-ene-3,20-dione; 6α-fluoro-11β,16α,17,21-tetrahydroxypregn-4-ene-3,20-dione cyclic 16,17-acetal with acetone;* 6α-fluoro-16α,-17α-isopropylidenedioxy-4-pregnene-11β,21-diol-3,20-dione; 6α-fluoro-11β,16α,17,21-tetrahydroxyprogesterone cyclic 16,17-acetal with acetone; fludroxycortide; fluorandrenolone; flurandrenolone; flurandrenolone acetonide (obsolete); Cordran; Drenison; Drocort; Haelan; Sermaka. $C_{24}H_{33}FO_6$; mol wt 436.52. C 66.03%, H 7.62%, F 4.35%, O 21.99%. Prepn: Ringold *et al.,* **Ger.** pat. **1,131,213** corresp to **U.S.** pat. **3,126,375** (1962, 1964 both to Syntex).

Crystals from acetone + hexane, mp 247-255°. $[α]_D$ +140-150° ($CHCl_3$). uv max: 236 nm (log ε 4.17).
THERAP CAT: Glucocorticoid; anti-inflammatory.

4123. Flurazepam. *7-Chloro-1-[2-(diethylamino)ethyl]-5-(2-fluorophenyl)-1,3-dihydro-2H-1,4-benzodiazepin-2-one;* Felmane; Noctosom; Stauroderm. $C_{21}H_{23}ClFN_3O$; mol wt 387.89. C 65.03%, H 5.98%, Cl 9.14%, F 4.89%, N 10.83%, O 4.12%. Prepn: G. A. Archer *et al.,* **Belg.** pat. **629,005,** *C.A.* **60,** 15896f (1964); **Neth.** pat. **Appl. 6,401,335,** *C.A.* **62,** 5289a (1965); G. A. Archer *et al.,* **U.S.** pat. **3,299,-053** (1963, 1964, 1967 all to Hoffmann-La Roche). Manuf process: R. Fryer, L. H. Sternbach, **U.S.** pat. **3,567,710** (1971 to Hoffmann-La Roche). Synthesis: S. Inaba *et al., Chem. Pharm. Bull.* **19,** 263 (1971). Structure-activity studies: L. H. Sternbach *et al., J. Med. Chem.* **8,** 815 (1965); Armagnac *et al., Therapie* **26,** 439 (1971). Metabolism: Schwartz *et al., J. Med. Chem.* **11,** 770 (1968); *eidem, J. Pharm. Sci.* **59,** 1800 (1970). Pharmacology: L. O. Randall *et al., Arch. Int. Pharmacodyn. Ther.* **178,** 216 (1969). Toxicology: H. C. Rosenberg, *Pharmacol. Biochem. Behav.* **13,** 415 (1980). GC determn in plasma: Z. Salama *et al., Arzneimittel-Forsch.* **38,** 400 (1988). Comprehensive description: B. C. Rudy, B. Z. Senkowski in *Analytical Profiles of Drug Substances* vol. 3, K. Florey, Ed. (Academic Press, New York, 1974) pp 307-331.

White rods from ether-petr ether, mp 77-82°.
Dihydrochloride, $C_{21}H_{25}Cl_3FN_3O$, *flurazepam hydrochloride, Ro 5-6901, Benozil, Dalmadorm, Dalmane, Dalmate, Dormodor, Felison, Insumin, Lunipax, Somlan.* Pale yellow crystals from methanol + ether, mp 190-220°. LD_{50} in mice (mg/kg): 290 i.p.; 870 orally; 84 i.v. (Randall).
Caution: May be habit forming. This is a controlled substance (depressant) listed in the U.S. Code of Federal Regulations, Title 21 Part 1308.14 (1987).
THERAP CAT: Sedative, hypnotic.

4124. Flurbiprofen. *2-Fluoro-α-methyl[1,1'-biphenyl]-4-acetic acid; 2-fluoro-α-methyl-4-biphenylacetic acid;* 2-(2-fluoro-4-biphenylyl)propionic acid; 3-fluoro-4-phenylhydratropic acid; BTS 18322; U-27182; Adofeed; Ansaid; Cebutid; Froben; Flurofen; Ocufen; Stayban; Zepolas. $C_{15}H_{13}FO_2$; mol wt 244.27. C 73.76%, H 5.36%, F 7.78%, O 13.10%. Prepn: **Fr.** pat. **M5737;** Adams *et al.,* **U.S.** pat. **3,755,427** (1968 and 1973 to Boots Co., Ltd.). Pharmacology: Chalmers *et al., Ann. Rheum. Dis.* **31,** 319 (1972); *ibid.* **32,** 58 (1973); Glenn *et al., Agents Actions* **3,** 210 (1973); Nishizawa *et al., Thromb. Res.* **3,** 577 (1973). Symposium on pharmacokinetics and clinical efficacy in pain management: *Am. J. Med.* **80,** Suppl. 3A, 1-157 (1986).

Crystals from petr ether, mp 110-111°.
THERAP CAT: Anti-inflammatory. Analgesic.

4125. Flurogestone Acetate. *17-(Acetyloxy)-9-fluoro-11-hydroxypregn-4-ene-3,20-dione; 9-fluoro-11β,17-dihydroxypregn-4-ene-3,20-dione-17-acetate;* 17α-acetoxy-9α-fluoro-11β-hydroxyprogesterone; 9-fluoro-11β,17-dihydroxyprogesterone 17-acetate; flugestone acetate; SC-9880; Chronogest; Cronolone; Synchronate. $C_{23}H_{31}FO_5$; mol wt 406.50. C 67.96%, H 7.69%, F 4.67%, O 19.68%. Prepn: Bergstrom *et al., J. Am. Chem. Soc.* **81,** 4432 (1959); Bergstrom, Dodson; Bergstrom, Nicholson, U.S. pats. **2,892,851; 2,963,498** (1959, 1960 to Searle).

Crystals from benzene + petr ether, or ethyl acetate + petr ether, mp 266-269°. $[α]_D$ +77.6° (in chloroform). uv max (methanol): 238 nm (ε 17,500).
Component of *Syncro-Mate.*
THERAP CAT: Progestogen.
THERAP CAT (VET): Progestogen. Estrus regulation.

4126. Flurothyl. *1,1'-Oxybis[2,2,2-trifluoroethane]; bis-(2,2,2-trifluoroethyl) ether;* hexafluorodiethyl ether; bis(trifluoroethyl) ether; Indoklon. $C_4H_4F_6O$; mol wt 182.07. C 26.39%, H 2.21%, F 62.61%, O 8.79%. $CF_3CH_2OCH_2CF_3$. Prepn: **Brit.** pat. **814,493** (1959 to Pennsalt Chem.); **Brit.** pat. **889,282** (1962 to Air Reduction). Toxicology in animals: A. Cherkin, *Psychopharmacologia* **15,** 404 (1969). Review of pharmacology: L. Arce, *Psychosomatics* **11,** 358-360 (1970); of clinical trials in psychiatric convulsant therapy: J. G. Small, I. F. Small, *Sem. Psychiatry* **4,** 13-26 (1972).
Colorless, mobile liq, mild ethereal, pleasant odor. d_4^{20} 1.41. bp 63.9°. Practically insol in water. Sol in alcohol.
THERAP CAT: CNS stimulant.

4127. Fluroxene. *(2,2,2-Trifluoroethoxy)ethene; 2,2,2-trifluoroethyl vinyl ether;* Fluoromar. $C_4H_5F_3O$; mol wt 126.08. C 38.10%, H 4.00%, F 45.21%, O 12.69%. CF_3CH_2-

OCH=CH$_2$. Prepn: Shukys, **U.S.** pat. **2,830,007** (1958) and Townsend, **U.S.** pat. **2,870,218** (1959 to Air Reduction). Liquid. bp$_{751}$ 42.5°. n$_D^{20}$ 1.3192. d^{20} 1.135.
Caution: Potentially explosive.
THERAP CAT: Anesthetic (inhalation).

4128. Fluroxypyr. *[(4-Amino-3,5-dichloro-6-fluoro-2-pyridinyl)oxy]acetic acid;* Starane. C$_7$H$_5$Cl$_2$FN$_2$O$_3$; mol wt 255.03. C 32.97%, H 1.98%, Cl 27.80%, F 7.45%, N 10.98%, O 18.82%. Prepn and preliminary herbicidal activity: S. D. McGregor, **U.S.** pat. **3,761,486** (1973 to Dow). Activity of the 1-methylheptyl ester: W. G. Richardson *et al., Tech. Rept. Agr. Res. Counc. Weed Res. Org.* **63,** 45 (1981); A. R. Thompson, *Aspects Appl. Biol.* **9,** 221 (1985); specifically vs broad leaved weeds: *idem, Proc. Brit. Crop Prot. Conf. - Weeds* **1987,** 735. Mode of action: G. E. Sanders, K. E. Pallet, *Aspects Appl. Biol.* **9,** 179 (1985); *eidem, Ann. Appl. Biol.* **111,** 385 (1987). Uptake and metabolism: *eidem, Weed Res.* **27,** 159 (1987). Bioassay for interaction with other xenobiotics: H. F. Taylor *et al., Ann. Appl. Biol.* **103,** 311 (1983). Physical and toxicity data: J. A. Paul *et al., Proc. Brit. Crop Prot. Conf. - Weeds* **1985,** 939.

White crystalline solid, mp 232-233°. Vapor pressure at 25°: 9.42 × 10^{-7} mm Hg. Soly at 20° (g/l): water 0.091; acetone 41.6. LD$_{50}$ orally in rats: 2405 mg/kg; i.p. in male, female rats: 458, 519 mg/kg; percutaneous in rabbits > 5000 mg/kg (Paul).
1-Methylheptyl ester, C$_{15}$H$_{21}$Cl$_2$FN$_2$O$_3$, *Dowco 433.*
USE: Herbicide.

4129. Flurprimidol. α-*(1-Methylethyl)-α-[4-(trifluoromethoxy)phenyl]-5-pyrimidinemethanol;* α*-isopropyl-α-[p-(trifluoromethoxy)phenyl]-5-pyrimidinemethanol;* Compound 72500; EL 500; Cutless. C$_{15}$H$_{15}$F$_3$N$_2$O$_2$; mol wt 312.29. C 57.69%, H 4.84%, F 18.25%, N 8.97%, O 10.25%. Growth regulator for use on established turfgrass. Prepn: R. L. Benefiel, **Belg.** pat. **815,245** (1974 to Lilly), *C.A.* **83,** 97354t (1975); R. L. Benefiel, E. V. Krumkalns, **U.S.** pats. **3,967,949, 4,002,628** (1976, 1977 both to Lilly). Growth retardant properties, effect on turfgrass: L. G. Thompson, *Proc. West. Soc. Weed Sci.* **35,** 113 (1982). Effect on Kentucky bluegrass, comparison with other growth retardants: T. L. Watschke, *Proc. Ann. Meet. Northeast. Weed Sci. Soc.* **35,** 322 (1981); N. E. Christians, *J. Am. Soc. Hort. Sci.* **110,** 765 (1985). Long term study of effect on bluegrass-fescue turf: P. H. Dernoeden, *Agron. J.* **76,** 807 (1984). Efficacy in dwarfing southern pine seedlings: R. C. Hare, *Can. J. For. Res.* **14,** 123 (1984).

Non-volatile white crystals, mp 94-96°. Sol in acetone, ethanol, methanol, DMSO, diethyl ether. LD$_0$ orally in rats: > 500 mg/kg; LD$_{50}$ dermally in rabbits: > 2000 mg/kg (Thompson).
USE: Growth retardant for grasses.

4130. Flusilazole. *1-[[Bis(4-fluorophenyl)methylsilyl]methyl]-1H-1,2,4-triazole;* bis(4-fluorophenyl)methyl(1H-1,2,4-triazol-1-ylmethyl)silane; fluzilazol; DPX-H6573; Nustar; Olymp; Punch. C$_{16}$H$_{15}$F$_2$N$_3$Si; mol wt 315.40. C 60.93%, H 4.79%, F 12.05%, N 13.32%, Si 8.90%. Sterol-inhibiting, broad spectrum foliar fungicide. Prepn: W. K. Moberg, **Eur.** pat. **Appl.** **68,813;** *idem,* **U.S.** pat **4,510,136** (1983, 1985 both to Du Pont); W. K. Moberg *et al., Pest. Sci. Biotech., Proc. 6th Int. Congr. Pest. Chem.,* 57 (1987). Field performance, physical properties and toxicity: T. M. Fort, W. K. Moberg, *Proc. Brit. Crop Prot. Conf.-Pest Dis.*

1984, 413. Field trials against apple scab: A. R. Biggs, J. Warner, *Can. J. Plant Pathol.* **9,** 41 (1987); and mildew: A. A. J. Swait *et al., Tests Agrochem. Cultiv.* **8,** 42 (1987).

Crystalline solid, mp 55°. Soly: > 2 g/ml in many organic solvents. LD$_{50}$ male, female rats, rabbits (mg/kg): 1110, 674 orally; > 2000 dermally (Fort, Moberg).
USE: Agricultural fungicide.

4131. Fluspirilene. *8-[4,4-Bis(4-fluorophenyl)butyl]-1-phenyl-1,3,8-triazaspiro[4.5]decan-4-one;* R6128; Imap; Redeptin. C$_{29}$H$_{31}$F$_2$N$_3$O; mol wt 475.59. C 73.24%, H 6.57%, F 7.99%, N 8.83%, O 3.36%. Prepn: **Belg.** pat. **633,-914;** P. A. J. Janssen, **U.S.** pat. **3,238,216** (1963, 1966 both to Janssen). HPLC determn: A. H. Hikal, H. I. Al-Shoura, *J. Liq. Chromatog.* **5,** 2205 (1982). Metabolism in rat: J. P. P. Heykants, *Life Sci.* **8,** Part I, 1029 (1969). Pharmacology: P. A. J. Janssen *et al., Arzneimittel-Forsch.* **20,** 1689 (1970). Calcium channel blocking activity: R. J. Gould *et al., Proc. Nat. Acad. Sci. USA* **80,** 5122 (1983); J. P. Galizzi *et al., ibid.* **83,** 7513 (1986). Clinical trials in schizophrenia: H. Immich *et al., Arzneimittel-Forsch.* **20,** 1699 (1970); G. Chouinard *et al., J. Clin. Psychopharmacol.* **6,** 21 (1986).

White to yellowish amorphous or crystalline solid, mp 187.5-190°. Soly in water: 0.015-0.020 mg/ml. LD$_{50}$ i.m. in rats: 146±14 mg/kg (Janssen).
THERAP CAT: Neuroleptic.

4132. Flutamide. *2-Methyl-N-[4-nitro-3-(trifluoromethyl)phenyl]propanamide;* α,α,α*-trifluoro-2-methyl-4'-nitro-m-propionotoluidide;* 4'-nitro-3'-trifluoromethylisobutyranilide; niftolid; SCH-13521; Drogenil; Eulexin; Euflex; Flucinom; Flugeril; Fugerel; Sebatrol. C$_{11}$H$_{11}$F$_3$N$_2$O$_3$; mol wt 276.22. C 47.83%, H 4.01%, F 20.64%, N 10.14%, O 17.38%. Prepn: Baker *et al., J. Med. Chem.* **10,** 93 (1967); Neri, Topliss, **Ger.** pat. **2,130,450** (1972 to Sherico), *C.A.* **78,** 58091g (1973); Gold, **Ger.** pat. **2,261,293;** *idem,* **U.S.** pat. **3,847,988** (1973, 1974 both to Schering). Activity studies: Neri *et al., Endocrinology* **91,** 427 (1972); Peets *et al., ibid.* **94,** 532 (1974); B. A. Gladue, L. G. Clemens, *Endocrinology* **106,** 1917 (1980). Effect in prostatic cancer: P. C. Sogani, W. F. Whitmore, *J. Urol.* **122,** 640 (1979).

Crystals from benzene, mp 111.5-112.5°.
THERAP CAT: Antiandrogen; antineoplastic (hormonal).
THERAP CAT (VET): Anti-androgen.

4133. Flutazolam. *10-Chloro-11b-(2-fluorophenyl)-2,-3,7,11b-tetrahydro-7-(2-hydroxyethyl)oxazolo[3,2-d][1,4]-benzodiazepin-6(5H)-one;* MS 4101; Ro 07-6102; Coreminal. C$_{19}$H$_{18}$ClFN$_2$O$_3$; mol wt 376.81. C 60.56%, H 4.81%, Cl 9.41%, F 5.04%, N 7.43%, O 12.74%. Tricyclic benzodiazepine deriv. Prepn: **Belg.** pat. **765,253;** M. E. Derieg *et al.,* **U.S.** pat. **3,905,956** (1971, 1975 both to Hoffmann-La Roche). Physicochemical properties: A. Ito *et al., Iyakuhin Kenkyu* **9,** 787 (1978), *C.A.* **89,** 169013e (1979). Psychopharmacological profile: T. Mitsushima, S. Ueki, *Folia Pharmacol. Japon.* **74,** 959 (1978), *C.A.* **90,** 180083u (1979). Toxicological studies: K. Ishimura *et al., Oyo Yakuri* **15,** 1081 (1978); *eidem, ibid.* **16,** 693 (1978), *C.A.* **90,** 16379z, 145909p (1979).

White prisms from toluene, mp 142-147°. pKa 5.40. Freely sol in chloroform, ethanol. Moderately sol in acetone, benzene, methanol. Practically insol in water.

THERAP CAT: Anxiolytic.

4134. Flutoprazepam. *7-Chloro-1-(cyclopropylmethyl)-5-(2-fluorophenyl)-1,3-dihydro-2H-1,4-benzodiazepin-2-one;* KB-509; ID-1937; Restar; Restas. $C_{19}H_{16}ClFN_2O$; mol wt 342.80. C 66.57%, H 4.70%, Cl 10.34%, F 5.54%, N 8.17%, O 4.67%. Benzodiazepine derivative. Prepn: H. Yamamoto *et al.,* **Brit.** pat. **1,253,368;** *eidem,* **U.S.** pat. **3,925,364** (1971, 1975 both to Sumitomo). New process: T. Okamoto *et al.,* **Ger.** pat. **2,151,540;** *eidem,* **U.S.** pat. **3,832,-344** (1972, 1974 both to Sumitomo). Stability and degradation mechanism in aqueous solns: K. Akimoto *et al., Int. J. Pharmacol.* **3,** 109 (1979). Pharmacokinetics and metabolism: A. Kozaki *et al., Yakugaku Zasshi* **102,** 1177 (1982), *C.A.* **98,** 119080q (1983). Pharmacology of flutoprazepam and metabolites: K. Oki *et al., Arch. Int. J. Pharmacodyn.* **269,** 180 (1984). Acute toxicity study: H. Takebe *et al., Oyo Yakuri* **20,** 1055 (1980), *C.A.* **95,** 73630b (1981).

Crystals, mp 86-88°. LD$_{50}$ in male, female mice, rats (mg/kg): 2640, 2430, 13760, 10060 orally; 2400, 2110, 2460, 2230 i.p.; > 5000 s.c. (Takebe).

THERAP CAT: Anxiolytic.

4135. Flutriafol. *α-(2-Fluorophenyl)-α-(4-fluorophenyl)-1H-1,2,4-triazole-1-ethanol; (RS)-2,4'-difluoro-α-(1H-1,2,4-triazol-1-ylmethyl)benzhydryl alcohol;* flutriafen; R 152450; PP 450; Impact. $C_{16}H_{13}F_2N_3O$; mol wt 301.30. C 63.78%, H 4.35%, F 12.61%, N 13.95%, O 5.31%. Prepn: K. P. Parry *et al.,* **Eur.** pat. **Appl. 15,756** (1980 to ICI). Activity, physical properties, and toxicology: A. M. Skidmore *et al., Proc. 10th Int. Congr. Plant Prot.* **1,** 368 (1983). Mode of action: B. C. Baldwin *et al., Med. Fac. Landbouww. Rijksuniv. Gent* **49,** 303 (1984). Field trials: P. J. Northwood *et al., Proc. 10th Int. Congr. Plant Prot.* **3,** 930 (1983); J. S. Brown, *Crop Prot.* **6,** 157 (1987).

White crystalline solid, mp 130°. Vapor pressure at 20°: 3.00×10^{-9} mm Hg. Soly at 20° (g/l): acetone 190; dichloromethane 150; hexane 0.30; methanol 69; xylene 12; water (pH 7.0) 0.13. LD$_{50}$ male, female rats (mg/kg): 1140, 1480 orally; rats, rabbits: > 1000, > 2000 percutaneously (Skidmore).

USE: Agricultural fungicide.

4136. Flutropium Bromide. *(endo,syn)-8-(2-Fluoroethyl)-3-[(hydroxydiphenylacetyl)oxy]-8-methyl-8-azoniabicyclo[3.2.1]octane bromide;* (8r)-3α-benziloyloxy-8-(2-fluoroethyl)-8-methyl-8-azoniabicyclo[3.2.1]octane bromide; (8r)-8-(2-fluoroethyl)-3α-hydroxy-1αH,5αH-tropanium bromide benzilate; N-β-fluoroethylnortropine benzilate methobromide; benzilic acid N-β-fluoroethylnortropine ester methobromide; Ba 598Br; Flubron. $C_{24}H_{29}BrFNO_3$; mol wt 478.40. C 60.26%, H 6.11%, Br 16.70%, F 3.97%, N 2.93%, O 10.03%. Anticholinergic agent. Prepn: R. Banholzer *et al.,* **Ger.** pat. **2,540,633;** *eidem,* **U.S.** pat. **4,042,700** (both 1977 to Boehringer, Ing.). Synthesis: R. Banholzer *et al., Arzneimittel-Forsch.* **36,** 1161 (1986). Molecular and crystal structure: G. Kiel, *ibid.* 1166. Pharmacology in animals: S. Yanaura *et al., Japan. J. Pharmacol.* **33,** 971 (1983); R. Bauer, A. Fügner, *Arzneimittel-Forsch.* **36,** 1348 (1986). Pharmacodynamics: K. Yoshimura *et al., Iyakuhin Kenkyu* **18,** 240 (1987), *C.A.* **107,** 17197v (1987).

White crystals from acetonitrile, mp 192-193° (dec); from ethanol-acetone, mp 198-199° (dec). LD$_{50}$ in male, female mice, male, female rats (mg/kg): 12.5, 11.0, 16.4, 18.4 i.v.; 760, 810, 830, 740 orally (Bauer, Fügner).

THERAP CAT: Bronchodilator.

4137. Fluvalinate. *N-[2-Chloro-4-(trifluoromethyl)-phenyl]-DL-valine cyano(3-phenoxyphenyl)methyl ester;* ZR-3210; Mavrik. $C_{26}H_{22}ClF_3N_2O_3$; mol wt 502.93. C 62.09%, H 4.41%, Cl 7.05%, F 11.33%, N 5.57%, O 9.54%. Synthetic pyrethroid insecticide without the usual cyclopropane ring. Prepn: C. A. Henrick, B. A. Garcia, **Ger.** pat. **2,812,169** (1978 to Zoecon), *C.A.* **90,** 122072d (1970). Activity: C. A. Henrick *et al., Pestic. Sci.* **11,** 224 (1980).

Yellow-amber liquid. Vapor press. at 25°: 1×10^{-7} mm Hg. Soly in water: 2.0 ppb. Sol in organic solvents.

USE: Insecticide.

4138. Fluvoxamine. *5-Methoxy-1-[4-(trifluoromethyl)-phenyl]-1-pentanone O-(2-aminoethyl)oxime; 5-methoxy-4'-(trifluoromethyl)valerophenone (E)-O-(2-aminoethyl)-oxime.* $C_{15}H_{21}F_3N_2O_2$; mol wt 318.35. C 56.59%, H 6.65%,

F 17.90%, N 8.80%, O 10.05%. Serotonin uptake inhibitor. Prepn: **Neth.** pat. **Appl. 7,503,310**; H. B. A. Welle, V. Claassen, **U.S.** pat. **4,085,225** (1975, 1978 both to Philips-Duphar). Inhibition of 5-HT uptake: V. Claassen *et al.,* *Brit. J. Pharmacol.* **60**, 505 (1977). HPLC determn in plasma: G. J. De Jong, *J. Chromatog.* **183**, 203 (1980). Quantitative EEG, psychometric and pharmacokinetic studies in man: B. Saletu *et al., J. Neurol. Transm.* **49**, 63 (1980). Use in endogenous depression: J. E. De Wilde, D. P. Doogan, *J. Affective Disord.* **4**, 249 (1982). Effects on clonidine-induced depression in mice: J. Maj *et al., J. Neurol. Transm.* **55**, 19 (1982). *Review:* Brit. J. Clin. Pharmacol. **15**, Suppl. 3, 347S-450S (1983). Review of pharmacology, clinical efficacy: P. Benfield, A. Ward, *Drugs* **32**, 313 (1986).

F₃C─[benzene ring]─C─(CH₂)₄OCH₃ / N─O─CH₂CH₂NH₂

Maleate, $C_{19}H_{25}F_3N_2O_6$, *DU 23000, MK 264, Avoxin, Faverin, Fevarin, Floxyfral.* Crystals from acetonitrile, mp 120-121.5°.

THERAP CAT: Antidepressant.

4139. Folescutol. *6,7-Dihydroxy-4-(4-morpholinylmethyl)-2H-1-benzopyran-2-one; 6,7-dihydroxy-4-(morpholinomethyl)coumarin;* 4-morpholinomethylescutol; pholescutol. $C_{14}H_{15}NO_5$; mol wt 277.28. C 60.64%, H 5.45%, N 5.05%, O 28.85%. Prepn by the reaction of 4-chloromethyl-6,7-dihydroxycoumarin with morpholine: **Fr.** pat. **M2035** (1963 to Labs. Dausse), *C.A.* **60**, 4113 (1964). Pharmacological study: J. P. Tarayre *et al., Ann. Pharm. Franc.* **33**, 467 (1975).

Crystals from 50% ethanol, mp 232°.
Hydrochloride, $C_{14}H_{16}ClNO_5$, *LD 2988, Covalan.* Crystals from 75% ethanol, mp 259-261°.

THERAP CAT: Capillary protectant.

4140. Folic Acid. *N-[4-[[(2-Amino-1,4-dihydro-4-oxo-6-pteridinyl)methyl]amino]benzoyl]-L-glutamic acid; N-[p-[[(2-amino-4-hydroxy-6-pteridinyl)methyl]amino]benzoyl]-glutamic acid;* pteroylglutamic acid; N-(p-[(2-amino-4-hydroxypyrimido[4,5-b]pyrazin-6-yl)methylamino]benzoyl)-glutamic acid; PGA; liver *Lactobacillus casei* factor; vitamin Bc; vitamin M; folsäure; Cytofol; Folacin; Foldine; Folaemin; Folicet; Folipac; Folsan; Folvite; Incafolic; Millafol. $C_{19}H_{19}N_7O_6$; mol wt 441.40. C 51.70%, H 4.34%, N 22.22%, O 21.75%. Hematopoietic vitamin present, free or combined with one or more additional molecules of L(+)-glutamic acid, in liver, kidney, mushrooms, spinach, yeast, green leaves, grasses: Mitchell *et al., J. Am. Chem. Soc.* **63**, 2284 (1941). Isoln: Pfiffner *et al., ibid.* **69**, 1476 (1947); Stokstad *et al., ibid.* **70**, 3 (1948). Structure: Mowat *et al., ibid.* 14. Crystal structure: D. Mastropaolo *et al., Science* **210**, 334 (1980). History of the different folic acid factors: *Ann. N. Y. Acad. Sci.* **48**, 255-350 (1946). *See also* reviews by Subbarow *et al.* in *Vitam. Horm. (New York)* **3**, 237-296 (1945); Pfiffner, Hogan, *ibid.* **4**, 1-13 (1946). Several syntheses, *see* reviews by Gates, *Chem. Rev.* **41**, 63-95 (1947); Hutchings, Mowat, *Vitam. Horm. (New York)* **6**, 1-25 (1948); Sletzinger *et al., J. Am. Chem. Soc.* **77**, 6365 (1955); **U.S.** pats. **2,786,056; 2,816,109; 2,821,527-8** (all 1958 to Merck & Co.); Sadao Kawanishi, **U.S.** pat. **2,956,-057** (1960 to Kongo Kagaku Kabushiki Kaisha). Alternate syntheses: Bieri, Viscontini, *Helv. Chim. Acta* **56**, 2905 (1973); E. Khalifa *et al., ibid.* **59**, 242 (1976). Comprehensive reviews: Jaenicki, Kutzbach, *Fortschr. Chem. Org. Naturst.* **21**, 183-274 (1963); Marchetti, *Acta Vitaminol. Enzymol.* **25**, 41-64 (1971).

Yellowish-orange crystals. Extremely thin platelets (elongated at two ends) from hot water, no mp. Darkens and chars from about 250°. $[\alpha]_D^{25}$ +23° (c = 0.5 in 0.1N NaOH). uv max (pH 13): 256, 283, 368 nm (log ϵ 4.43, 4.40, 3.96). Very slightly sol in cold water (0.0016 mg/ml at 25°), sol to about 1% in boiling water. Slightly sol in methanol, appreciably less sol in ethanol and butanol. Insol in acetone, chloroform, ether, benzene. Relatively sol in acetic acid, phenol, pyridine, solns of alkali hydroxides and carbonates. Injectable solns are prepd by dissolving folic acid in normal sodium bicarbonate soln (which should be sterilized by filtration) or by preparing solns of the sodium or methylglucamine salt. A suspension of 1 g folic acid in 10 ml water has a pH of 4.0-4.8. Aq solns prepd with sodium bicarbonate have a pH between 6.5 and 6.8.

THERAP CAT: Hematopoietic vitamin.

THERAP CAT (VET): Nutritional factor (dietary requirement in poultry).

4141. Folinic Acid. *N-[4-[[(2-Amino-5-formyl-1,4,5,6,-7,8-hexahydro-4-oxo-6-pteridinyl)methyl]amino]benzoyl]-L-glutamic acid; N-[p-[[(2-amino-5-formyl-5,6,7,8-tetrahydro-4-hydroxy-6-pteridinyl)methyl]amino]benzoyl]glutamic acid;* 5-formyl-5,6,7,8-tetrahydropteroyl-L-glutamic acid; 5-formyl-5,6,7,8-tetrahydrofolic acid; CF; citrovorum factor; leucovorin. $C_{20}H_{23}N_7O_7$; mol wt 473.44. C 50.74%, H 4.90%, N 20.71%, O 23.66%. Intermediate product of the metabolism of folic acid; the active form into which that acid is converted in the body, ascorbic acid being a necessary factor in the conversion process. First reported as the *Leuconostoc citrovorum* 8081 growth factor: H. E. Sauberlich, C. A. Baumann, *J. Biol. Chem.* **176**, 165 (1948). Isoln from houseflies: Miller, Perry, *Life Sci.* **4**, 1573 (1965). Prepn from folic acid: Brockman, Jr., *et al., J. Am. Chem. Soc.* **72**, 4325 (1950); [*cf.* Keresztesy, Silverman, *ibid.* **73**, 1897 (1951)]; Flynn *et al., ibid.* 1979; Roth *et al., ibid.* **74**, 3247 (1952). Isomers: Cosulick *et al., ibid.* 4215. Structure: May *et al., ibid.* **73**, 3067 (1951); Pohland *et al., ibid.* 3247. Manuf: Shive, **U.S.** pat. **2,741,608** (1956 to Res. Corp.). Improved synthesis: E. Khalifa *et al., Helv. Chim. Acta* **63**, 2554 (1980). Used as an antidote to folic acid antagonists such as methotrexate, *q.v.,* which block the conversion of folic acid into folinic acid. Review of clinical combination therapy with methotrexate: J. R. Bertino *et al., Ann. N.Y. Acad. Sci.* **186**, 486-495 (1971). Pharmacokinetics, pharmacodynamics: P. F. Nixon, *Clin. Exp. Pharmacol. Physiol.* **Suppl. 5**, 35 (1979). Comprehensive description: L. O. Pont *et al.* in *Analytical Profiles of Drug Substances* **vol. 8**, K. Florey, Ed. (Academic Press, New York, 1979) pp 315-350. Review of clinical synergy with fluorouracil in cancer: R. J. DeLap, *Yale J. Biol. Med.* **61**, 23-34 (1988).

dl-L-Form, crystals, dec 240-250°. uv max (0.1N NaOH): 282 nm (% T = 27.0 for 10 mg/l). $[\alpha]_D^{20}$ +14.26° (c = 3.42 as anhydr Ca salt). pKa (3 groups): 3.1, 4.8, and 10.4. Sparingly sol in water. pH of satd aq soln 2.8-3.0 at which pH partial decompn takes place. More stable at neutral or mildly alkaline pH.
l-L-Form, $[\alpha]_D^{20}$ −15.1° (c = 1.82 as anhydr Ca salt).
d-L-Form, $[\alpha]_D^{20}$ +28.3° (c = 3.53 as anhydr Ca salt).
Calcium salt pentahydrate, $C_{20}H_{21}CaN_7O_7.5H_2O$, *calcium folinate, NSC-3590, Leucovorin, Leucosar, Rescufolin, Rescuvolin, Wellcovorin.* Off-white to light beige amorphous,

Consult the cross index before using this section.

odorless powder, freely sol in water. Practically insol in alc. $[\alpha]_D^{21}$ +14.9° (c = 1 in water).

THERAP CAT: Antidote to folic acid antagonists; antianemic (folate deficiency).

4142. Folpet. *2-[(Trichloromethyl)thio]-1H-isoindole-1,3(2H)-dione; N-(trichloromethylthio)phthalimide; N-(trichloromethylmercapto)phthalimide; Phaltan.* $C_9H_4Cl_3N$-O_2S; mol wt 296.58. C 36.45%, H 1.36%, Cl 35.87%, N 4.72%, O 10.79%, S 10.81%. *Cf:* Captan. Prepn: Kittleson, U.S. pat. 2,553,770 (1951 to Standard Oil). Toxicity: T. B. Gaines, R. E. Linder, *Fundam. Appl. Toxicol.* 7, 299 (1986).

Crystals from benzene, mp 177°. LD_{50} orally in adult male, female rats: > 5000 mg/kg (Gaines, Linder).

Caution: May irritate mucosal surfaces: *Clinical Toxicology of Commercial Products,* R. E. Gosselin *et al.,* Eds. (Williams & Wilkins, Baltimore, 5th ed., 1984) Section II, p 317.

USE: Agricultural fungicide.

4143. Fomecins. Antibacterial substances produced by several strains of the basidiomycete *Fomes juniperinus* Schrenk. Isoln of fomecin A (major) and fomecin B (minor) from cultures grown in corn steep liquor: M. Anchel *et al., Proc. Nat. Acad. Sci. USA* 38, 655 (1952). Structures: T. C. McMorris, M. Anchel, *Can. J. Chem.* 42, 1595 (1964). Synthesis of B: S. M. Al-Mousawi *et al., Bull. Soc. Chim. Belg.* 88, 883 (1979); of A and B: K. Hayashi *et al., Chem. Pharm. Bull.* 28, 1971 (1980).

Fomecin A, $C_8H_8O_5$, *2,3,4-trihydroxy-6-(hydroxymethyl)-benzaldehyde.* R = CH_2OH. Cream-colored to orange crystals from ethanol-water, ethanol-benzene, acetone-benzene, or ethyl acetate. Dec above 160° without melting. Optically inactive in ethanol. uv max (ethanol): 241, 304 nm (ϵ 10,800, 15,300). Weakly acidic. Sparingly sol in water (1 mg/ml); slightly more sol in ethanol, acetone, ethyl acetate. Less sol in chloroform, benzene. Aq solns at neutral or acid pH are stable. The activity is lost around pH 8. Intravenous injection in doses up to 1 mg per mouse (*ca.* 50 m/kg) had no apparent ill effect.

Fomecin B, $C_8H_6O_5$, *3,4,5-trihydroxy-1,2-benzenedicarboxaldehyde.* R = CHO. Yellow needles from ethyl acetate. Darkens on heating; mp ~230°. uv max (ethanol): 263, 336 nm (ϵ 26400, 9200).

4144. Fominoben. *N-[3-Chloro-2-[[methyl-[2-(4-morpholinyl)-2-oxoethyl]amino]methyl]phenyl]benzamide; 3'-chloro-α-[methyl[(morpholinocarbonyl)methyl]amino]-o-benzotoluidide; 3'-chloro-β-[N-methyl-N-[(morpholinocarbonyl)methyl]aminomethyl]benzanilide.* $C_{21}H_{24}ClN_3O_3$; mol wt 401.89. C 62.76%, H 6.02%, Cl 8.82%, N 10.46%, O 11.94%. Prepn: Fr. pats. 1,482,547 and Addn. 92,488 (1967, 1968, both to Thomae), *C.A.* 68, 95494e (1968); 72, 21938p (1970); Kruger *et al.,* U.S. pat. 3,661,903 (1972 to Boehringer, Ing.); *eidem, Arzneimittel-Forsch.* 23, 290 (1973). Series of articles on pharmacology and clinical findings: *ibid.* 296-375. Metabolism: Zimmer, *ibid.* 1798. Toxicity data: S. Püschmann, R. Engelhorn, *ibid.* 296.

Base, mp 122.5-123°.

Hydrochloride, $C_{21}H_{25}Cl_2N_3O_3$, *PB 89, Finaten, Noleptan, Oleptan, Terion, Tussirama.* Crystals, mp 206-208° (dec). Soly in water: 0.1 mg/100 ml; in 0.05N aq tartaric acid: 5 g/100 ml. LD_{50} in mice, rats (mg/kg): 630, 1201 i.p.; 2200, 1250 orally (Püschmann, Engelhorn).

THERAP CAT: Antitussive; respiratory stimulant.

4145. Fomocaine. *4-[3-[4-(Phenoxymethyl)phenyl]propyl]morpholine; 4-[3-(α-phenoxy-p-tolyl)propyl]morpholine; P 652; Erbocain.* $C_{20}H_{25}NO_2$; mol wt 311.43. C 77.13%, H 8.09%, N 4.50%, O 10.28%. Prepn: **Brit.** pat. 786,128 (1957 to Promonta), *C.A.* 52, 9225h (1958); H. Oelschläger, *Arzneimittel-Forsch.* 9, 313 (1959). Improved synthesis: *eidem, ibid.* 27, 1625 (1977). Pharmacology: O. Nieschulz *et al., ibid.* 8, 539 (1958). Metabolism: H. Blume, H. Oelschläger, *ibid.* 28, 956 (1978). Bioavailability and local anesthetic effect: H. Oelschläger, D. Rothley, *ibid.* 29, 693 (1979).

Colorless cryst from petr ether, mp 52-53°. $bp_{1.1}$ 238-240°. uv max (ethanol): 220, 269 nm (ϵ 15820, 1373). LD_{50} in mice: 175 mg/kg i.v., O. Nieschulz *et al., loc. cit.*

THERAP CAT: Anesthetic (local).

4146. Fonazine. *10-[2-(Dimethylamino)propyl]-N,N-dimethyl-10H-phenothiazine-2-sulfonamide; 10-(2-dimethylaminopropyl)-3-dimethylsulfamidophenothiazine; 3-dimethylsulfamido-10-(2-dimethylaminopropyl)phenothiazine; dimethothiazine; dimethiotazine; dimetiotazine; RP 8599.* $C_{19}H_{25}N_3O_2S_2$; mol wt 391.56. C 58.28%, H 6.44%, N 10.73%, O 8.17%, S 16.38%. Prepn: **Brit.** pat. 814,512 (1959 to Rhône-Poulenc). Pharmacokinetics and distribution in dog and rat: O. R. W. Lewellen, R. Templeton, *Adv. Biochem. Psychopharmacol.* 9, 213 (1974). Clinical evaluation in spasticity: W. B. Matthews *et al., Acta Neurol. Scand.* 48, 635 (1972). HPLC determn in plasma: J. Holt *et al., Brit. J. Clin. Pharmacol.* 13, 282 (1982).

Hydrochloride, $C_{19}H_{26}ClN_3O_2S_2$, crystals, dec 214°.

Methanesulfonate, $C_{20}H_{29}N_3O_5S_3$, *dimethothiazine mesylate, fonazine mesylate, Alius, Banistyl, Bonpac, Calsekin, Migristene, Neomestine, Promaquid, Yoristen.*

THERAP CAT: Serotonin inhibitor.

4147. Fonofos. *Ethylphosphonodithioic acid O-ethyl S-phenyl ester; O-ethyl S-phenyl ethylphosphonothiolothionate; N-2790; Dyfonate.* $C_{10}H_{15}OPS_2$; mol wt 246.32. C 48.76%, H 6.14%, O 6.50%, P 12.57%, S 26.03%. Prepn: Szabo *et al.,* U.S. pat. 2,988,474 (1961 to Stauffer); Pitt, Simone, **Ger.** pat. 2,002,629, *C.A.* 73, 77392u (1970), corresp to U.S. pat. 3,642,960 (1970, 1972, both to Stauffer).

Light yellow liquid, $bp_{0.1}$ 130°. d_{25}^{25} 1.16. n_D^{30} 1.5883. Practically insol in water. Miscible with organic solvents. LD_{50}

orally in rats: 3 mg/kg, *Toxic Substances List,* H. E. Christensen, Ed. (1974) p 602.

USE: Soil insecticide. *Caution:* Cholinesterase inhibitor.

4148. Formaldehyde, Gas. Methanal; oxomethane; oxymethylene; methylene oxide; formic aldehyde; methyl aldehyde. CH_2O; mol wt 30.03. C 39.99%, H 6.73%, O 53.29%. HCHO. Formed by incomplete combustion of many organic substances. Present in coal and wood smoke, esp in smoke as produced for smoking ham and fish. Found in the atm, esp over large cities. Prepd commercially by catalytic vapor phase oxidation of methanol using air as the oxidizing agent and heated silver, copper, alumina, or coke as catalysts. Process using molybdenum iron oxide catalyst: Allyn *et al.,* U.S. pat. 2,812,309 (1957) and U.S. pat. 2,849,492 (1958 to Reichhold Chem.). Prepn of stable formaldehyde by heating low molecular polyoxymethylenes with P_2O_5: Ger. pat. 1,070,611 (1959 to BASF). Mfg processes: Faith, Keyes & Clark's *Industrial Chemicals,* F. A. Lowenheim, M. K. Moran, Eds. (Wiley-Interscience, New York, 4th ed., 1975) pp 422-429. Carcinogenicity study: J. A. Swenberg *et al., Cancer Res.* 40, 3398 (1980). *Review:* H. R. Gerberich *et al.,* in Kirk-Othmer *Encyclopedia of Chemical Technology* vol. 11 (Wiley-Interscience, New York, 3rd ed., 1980) pp 231-250.

Flammable colorless gas at ordinary temp. Pungent suffocating odor. d 1.067 (air = 1.000). d_4^{-20} 0.815. mp $-92°$. bp_{760} $-19.5°$; bp_{400} $-33.0°$; bp_{200} $-46.0°$; bp_{100} $-57.3°$; bp_{60} $-65.0°$; bp_{40} $-70.6°$; bp_{20} $-79.6°$; bp_{10} $-88.0°$. Ignition temp about 300° (572°F). Very sol in water, up to 55%; sol in alcohol, ether. Very reactive, combines readily with many substances, and polymerizes easily. *See also* Paraformaldehyde and Formaldehyde Solution.

Caution: Intensely irritating to mucous membranes, high concns are intolerable. This substance may reasonably be anticipated to be a carcinogen: *Fourth Annual Report on Carcinogens* (NTP 85-002, 1985) p 110. For a discussion of the assessment, regulation and evaluation of carcinogenicity *see:* F. Perera, C. Petito, *Science* 216, 1285 (1982).

USE: In the prodn of phenolic, urea, melamine and acetal resins. In textiles, embalming fluids, fungicides, air fresheners, cosmetics.

4149. Formaldehyde Sodium Bisulfite. *Hydroxymethanesulfonic acid monosodium salt;* sodium formaldehydebisulfite; methylolsulfonic acid sodium salt. CH_3NaO_4S; mol wt 134.09. C 8.96%, H 2.25%, Na 17.15%, O 47.73%, S 23.92%. $HOCH_2SO_3Na$. Prepn: Skrabal, Skrabal, *Monatsh.* 69, 17 (1936); Chhabria *et al.,* Ger. pat. 1,173,059 (1964 to Schill & Seilacher). Structure: Lauer, Langkammerer, *J. Am. Chem. Soc.* 57, 2360 (1935); Caughlan, Tartar, *ibid.* 63, 1266 (1941).

Monohydrate, needles from water. K at 25°: 1.2×10^{-7}.

USE: Fixing agent for keratin-containing fibers. Flotation of lead-zinc ores. Protecting color pictures exposed to high humidities against fading and staining.

4150. Formaldehyde Solution. Formalin; Formol; Morbicid; Veracur. A soln of about 37% by wt of formaldehyde gas in water, usually with 10-15% methanol added to prevent polymerization. This soln is the full strength and also known as Formalin 100% or Formalin 40 which signifies that it contains 40 grams of formaldehyde within 100 ml of the soln. Toxicity data: H. F. Smyth *et al., J. Ind. Hyg. Toxicol.* 23, 259 (1941). Prepn of semicarbazone: M. Pomerantz *et al., J. Org. Chem.* 47, 2217 (1982).

Colorless liq; pungent odor. On standing, esp in the cold, may become cloudy and on exposure to very low temp a ppt of trioxymethylene is formed. When evaporated, some formaldehyde escapes, but most of it is changed to trioxymethylene. *See* Paraformaldehyde. It is a powerful reducing agent especially in presence of alkali. In the air it slowly oxidizes to formic acid. d_{25}^{25} 1.081-1.085. One gallon weighs 9.1 lbs. bp_{760} 96°. n_D^{20} 1.3746. Flash pt 60° (140°F). pH 2.8-4.0. Misc with water, alcohol, acetone. *Keep well closed in a moderately warm place.* LD_{50} orally in rats: 0.80 g/kg (Smyth).

Formaldehyde semicarbazone, $C_2H_5N_3O$, crystals from ethanol, mp 120-121°. uv max (CH_3CN): 227 nm (log ϵ 3.73).

Human Toxicity: Vapors intensely irritating to mucous membranes. Topical application may produce an irritant dermatitis. Ingestion may cause severe abdominal pain, hematemesis, hematuria, proteinuria, anuria, acidosis, vertigo, coma, death, *cf. Clinical Toxicology of Commercial Products,* R. E. Gosselin *et al.,* Eds. (Williams & Wilkins, New York, 4th ed., 1976) Section III, pp 166-168.

USE: Disinfecting dwellings, ships, storage houses, utensils, clothes, etc. As a germicide and fungicide for plants and vegetables; destroying flies and other insects. Manuf phenolic resins, artificial silk and cellulose esters, dyes, organic chemicals, glass mirrors, explosives; improving fastness of dyes on fabrics; tanning and preserving hides; mordanting and waterproofing fabrics; preserving and coagulating rubber latex; in embalming fluids. In photography for hardening gelatin plates and papers, toning gelatin-chloride papers, chrome printing and developing. To prevent mildew and spelt in wheat and rot in oats; to render casein, albumin and gelatin insol; also in chemical analysis. *Incompat:* Ammonia, alkalies, tannin, iron preparations, gelatin, bisulfides; salts of Cu, Fe and Ag; H_2O_2, iodine, potassium permanganate. Combines directly with albumin, casein, gelatin, agar and starch to form insol compds.

THERAP CAT: Disinfectant.

THERAP CAT (VET): Antiseptic. Fumigant. Has been used in tympany, diarrhea, mastitis, pneumonia, internal bleeding.

4151. Formamide. Methanamide; carbamaldehyde. CH_3NO; mol wt 45.04. C 26.66%, H 6.71%, N 31.10%, O 35.52%. $HCONH_2$. Prepd on a large scale from carbon monoxide and ammonia at high press. and temp: Meyer, Orthner, *Ber.* 54, 1705 (1921); 55, 857 (1922); Meyer, Ger. pat. 390,798 (1924); 392,409 (1924); 414,257 (1925 to BASF); Wietzel, Herbst, U.S. pat. 1,843,434 (1932 to I. G. Farben). *Review:* C. L. Eberling in Kirk-Othmer *Encyclopedia of Chemical Technology* vol. 11 (Wiley-Interscience, New York, 3rd ed., 1980) pp 258-263.

Slightly viscous, odorless, colorless liq. Industrial grades may have faint odor of ammonia. mp +2.55°. bp_{760} 210.5° (partial decomp into CO and NH_3 at atm press. beginning at 180°); bp_{400} 193.5°; bp_{200} 175.5°; bp_{60} 147.0°; bp_{20} 122.5°; bp_{10} 109.5°; $bp_{1.0}$ 70.5°. d_4^{15} 1.13756; d_4^{20} 1.13340; d_4^{30} 1.12483. Dielectric constant ϵ = 84. Viscosity $\eta \times 10^5$ at 15° = 4320, at 30° = 2926. Surface tension γ at 20° = 58.35. n_D^{15}1.44911; n_D^{20} 1.44754; n_D^{110} 1.4170; n_D^{130} 1.4095. Flash pt, open cup: 310°F (154°C). pH of 0.5 molar aq soln 7.1. Misc with water, methanol, ethanol, acetone, acetic acid, dioxane, ethylene glycol, U.S.P. glycerol, phenol. Very slightly sol in ether, benzene. Dissolves casein, glucose, zein, tannins, starch, lignin, polyvinyl alcohol, cellulose acetate, nylon, the chlorides of copper, lead, zinc, tin, cobalt, iron, aluminum, nickel, the acetates of the alkali metals, some inorganic sulfates and nitrates. LD_{50} i.p. in mice, rats: 4.6, 5.7 g/kg, Pham-Huu-Chanh *et al., Toxicol. Appl. Pharmacol.* 26, 596 (1973).

USE: As ionizing solvent, manuf formic esters, hydrocyanic acid by catalytic dehydration, as softener for paper, animal glues, water-sol gums. *Caution:* Moderately irritating to skin, mucous membranes.

4152. Formanilide. *N-Phenylformamide;* formylaniline. C_7H_7NO; mol wt 121.13. C 69.40%, H 5.83%, N 11.57%, O 13.21%. C_6H_5NHCHO. Review on methods of prepn: de Jonge *et al., Rec. Trav. Chim.* 75, 5 (1956).

Crystals from ether + petr ether, mp 46.6-47.5°. bp 271°; bp_{120} 216°; bp_{14} 166°. uv max (96% ethanol): 242-243 nm (ϵ 13,700); in water: 239-240 nm (ϵ 11,200); in cyclohexane: 240 nm (ϵ 12,600). Soly in water at 20°: 25.4 g/l; at 25°: 28.6 g/l.

4153. Formic Acid. Ameisensäure (German). CH_2O_2; mol wt 46.02. C 26.10%, H 4.38%, O 69.52%. HCOOH. Observed by S. Fisher in 1670 in the products resulting from the distillation of ants. Made by heating carbon monoxide and NaOH under pressure and dec the resulting sodium formate with H_2SO_4. Acute and chronic toxicity: G. Malorny, *Z. Ernaehrungswiss.* 9, 332 (1969). *Review:* F. S. Wagner in Kirk-Othmer *Encyclopedia of Chemical Technology* vol. 11 (Wiley-Interscience, New York, 3rd ed., 1980) pp 251-258.

Colorless liq; pungent odor. It is a strong reducing agent. Physicochemical properties for anhydr acid: d_4^{20} 1.220. bp

100.5°. mp 8.4°. Solidif +7°. n_D^{20} 1.3714. pKa 3.739±0.001 at 25°. It is miscible with water, alcohol, ether, glycerol. *Keep from contact with skin!* LD$_{50}$ in mice (mg/kg): 1100 orally; 145 i.v. (Malorny).

Human Toxicity: Dangerously caustic to skin! Chronic absorption has been reported to cause albuminuria, hematuria.

USE: Decalcifier, reducer in dyeing wool fast colors; dehairing and plumping hides, tanning, in sizes, electroplating, coagulating rubber latex, aid in regenerating old rubber; also in chemical analysis.

THERAP CAT: Counterirritant, astringent.

4154. Formicin. *N-(Hydroxymethyl)acetamide;* formaldehyde acetamide; methylal acetamide. $C_3H_7NO_2$; mol wt 89.10. C 40.44%, H 7.92%, N 15.72%, O 35.91%. $CH_3CO-NHCH_2OH$. Obtained by warming acetamide with formaldehyde and a small amount of potash: Fuchs, *Pharm. Ztg.* **1905**, 803; *Ger. pat.* **164,610** (1905 to Kalle).

Colorless, very hygroscopic mass. Marketed only as a slightly yellowish, syrupy liq. d 1.14-1.18. Very sol in water, alcohol; sol in chloroform, glycerol; insol in ether.

USE: Antiseptic, disinfectant esp for surgical instruments.

4155. Forminitrazole. *N-(5-Nitro-2-thiazolyl)formamide;* 2-formamido-5-nitrothiazole; 291 C 51; Aroxine. $C_4H_3N_3O_3S$; mol wt 173.15. C 27.74%, H 1.74%, N 24.27%, O 27.72%, S 18.52%. Prepn: Bushby, Copp, *J. Pharm. Pharmacol.* **7**, 112 (1955); Copp, *Brit. pat.* **723,948** (1955 to Wellcome Found.); Huffman, *J. Org. Chem.* **23**, 727 (1958).

Pale, straw-colored needles from ethanol or ethyl acetate, mp 192-194°.

THERAP CAT: Antiprotozoal (Trichomonas).

4156. Formocortal. *21-(Acetyloxy)-3-(2-chloroethoxy)-9-fluoro-11-hydroxy-16,17-[(1-methylethylidene)bis(oxy)]-20-oxopregna-3,5-diene-6-carboxaldehyde; 3-(2-chloroethoxy)-9-fluoro-11β,16α,17,21-tetrahydroxy-20-oxopregna-3,5-diene-6-carboxaldehyde, cyclic 16,17-acetal with acetone, 21-acetate;* 3-(2-chloroethoxy)-6-formyl-9α-fluoropregna-3,5-diene-11β,16α,17,21-tetrol-20-one 21-acetate 16α,17α-acetonide; 3-(2-chloroethoxy)-9α-fluoro-6-formyl-11β,21-dihydroxy-16α,17α-isopropylidenedioxypregna-3,5-dien-20-one; FI 6341; fluoroformylon; Cortocin-F; Cutisterol; Deflamene; Fluderma. $C_{29}H_{38}ClFO_8$; mol wt 569.07. C 61.21%, H 6.73%, Cl 6.23%, F 3.34%, O 22.49%. Prepn: Camerino *et al., Fr. pat.* **1,396,602**, *C.A.* **63**, 5713c (1965); *Neth. pat. Appl.* **6,508,458** corresp to *U.S. pat.* **3,314,945** (1965, 1966, 1967, all to Farmitalia); Baldratti *et al., Experientia* **22**, 468 (1966). Pharmacology: Suchowsky, Baldratti, *Proc. Int. Congr. Horm. Steroids, 2nd, Milan* **1966**, L. Martini, F. Fraschini, M. Motta, Eds. (Excerpta Medica Found., Amsterdam, 1967) pp 536-540; Tajima, Otsu, *Japan J. Pharmacol.* **62**, 115 (1966), *C.A.* **67**, 72329v (1967); Baldratti *et al., Arzneimittel-Forsch.* **21**, 1533 (1971). Toxicity studies: Bertazzoli *et al., J. Eur. Toxicol.* **4**, 88 (1971).

White crystalline powder from ether-petroleum ether, mp 180-182°. $[\alpha]_D^{20}$ +26° (chloroform). uv max (ethanol): 216, 324 nm (ε 12,100, 17,100). LD$_{50}$ in mice: >2000 mg/kg orally; 490 mg/kg s.c.; 537 mg/kg i.p.

THERAP CAT: Glucocorticoid.

4157. Formononetin. *7-Hydroxy-3-(4-methoxyphenyl)-4H-1-benzopyran-4-one; 7-hydroxy-4'-methoxyisoflavone;* biochanin B; formononetol; neochanin. $C_{16}H_{12}O_4$; mol wt 268.26. C 71.63%, H 4.51%, O 23.86%. Isoln from soybean meal *(Soja hispida):* Walz, *Ann.* **489**, 118 (1931); from clover species *Trifolium subterraneum* L. and *T. pratense* L., *Leguminosae* in which it is the major estrogenic factor: Bradbury, White, *J. Chem. Soc.* **1951**, 3447; Bate-Smith, Swain, *Chem. & Ind. (London)* **1953**, 1127. Identity with biochanin B: Bose, *J. Sci. Ind. Res.* **15B**, 325 (1956). Structure: Baker *et al., J. Chem. Soc.* **1933**, 274. Synthesis: Wessely *et al., Ber.* **66**, 685 (1933); Kagal *et al., Tetrahedron Letters* **1962**, 593. Biological half-life in *Cicer arietinum* L., *Papilionatae:* N. Amrhein, E. Diederich, *Naturwiss.* **67**, 40 (1980). HPLC analysis: R. E. Carlson, *J. Chromatog.* **198**, 193 (1980). ^{13}C-NMR study: H. C. Jha *et al., Can. J. Chem.* **58**, 1211 (1980). Mutagenicity study: R. M. Bartholomew *et al., Mutat. Res.* **78**, 317 (1980).

Needles from alcohol, mp 258°. uv max (ethanol): 250, 300 nm (ε 27,440, 11,240).

7-Glucoside, $C_{22}H_{22}O_9$, ononin, 4'-methyldaidzin. From *Ononis spinosa* L., *Leguminosae:* Hlasiwetz, *J. Prakt. Chem.* **65**, 415 (1855). Synthesis: Farkas, Varady, *Ber.* **92**, 819 (1959). Needles from water, $C_{22}H_{22}O_9.H_2O$, mp 210-214°. When anhydr, dec 245°. $[\alpha]_D^{25}$ −24.2° (pyridine). Sol in alcohol; slightly sol in water, ether.

4158. Formosulfathiazole. *4-Amino-N-2-thiazolylbenzenesulfonamide reaction products with formaldehyde; N^1-(2-thiazolyl)sulfanilamide condensation product with formaldehyde;* formaldehyde-sulfathiazole; Formo-Cibazol; Intraformazol. $(C_{10}H_9N_3O_2S_2)_x$. Contains about 11% formaldehyde. Prepn and constitution: Druey, Becker, *Helv. Chim. Acta* **31**, 2184 (1948).

Amorphous powder. Practically insol in water.

THERAP CAT: Antibacterial.

4159. Formoterol. *(R*,R*)-(±)-N-[2-Hydroxy-5-[1-hydroxy-2-[[2-(4-methoxyphenyl)-1-methylethyl]amino]-ethyl]phenyl]formamide;* 3-formylamino-4-hydroxy-α-[N-[1-methyl-2-(p-methoxyphenyl)ethyl]aminomethyl]benzyl alcohol; (±)-2'-hydroxy-5'-[(RS)-1-hydroxy-2-[[(RS)-p-methoxy-α-methylphenethyl]amino]ethyl]formanilide. $C_{19}H_{24}N_2O_4$; mol wt 344.41. C 66.26%, H 7.02%, N 8.13%, O 18.58%. β-Adrenoreceptor stimulating catecholamine analog with selective bronchodilator activity. Prepn: M. Murakami *et al., Ger. pat.* **2,305,092**; *eidem, U.S. pat.* **3,994,974**; K. Murase *et al., Japan. Kokai* **75 12,040** (1973, 1976, 1975 all to Yamanouchi), *C.A.* **83**, 113941q (1975); and separation of racemates (there are two pairs of enantiomers): *eidem, Chem. Pharm. Bull.* **25**, 1368 (1977). The following studies have been done on the "A" racemate fumarate dihydrate. GC-MS determn in urine: H. Kamimura *et al., J. Chromatog.* **229**, 337 (1982). Radioimmunoassay in plasma, urine: K. Yokoi *et al., Life Sci.* **33**, 1665 (1983). Pharmacology: H. Ida, *Arzneimittel-Forsch.* **26**, 839, 1337 (1976). Metabolism in rats, dogs: H. Sasaki *et al., Xenobiotica* **12**, 803 (1982). Anti-allergic activity in mice and guinea pigs: K. Tomioka *et al., Arch. Int. Pharmacodyn.* **250**, 279 (1981). Toxicity: T. Yoshida *et al., Pharmacometrics* **26**, 811 (1983).

Fumarate dihydrate, "A"-racemate; $C_{41}H_{52}N_4O_{12}.2H_2O$, **BD 40 A, Atock.** Crystals from 95% isopropyl alcohol, mp 138-140°. LD_{50} in male, female, rats, mice (mg/kg): 3130, 5580, 6700, 8310 orally; 98, 100, 72, 71 i.v.; 1000, 1100, 640, 670 s.c.; 170, 210, 240, 210 i.p. (Yoshida).

Fumarate dihydrate, "B"-racemate; mp 154-155°.

THERAP CAT: Anti-asthmatic.

4160. Formothion. *Phosphorodithioic acid S-[2-(formylmethylamino)-2-oxoethyl] O,O-dimethyl ester; phosphorodithioic acid O,O-dimethyl ester S-ester with N-formyl-2-mercapto-N-methylacetamide;* S-(N-formyl-N-methylcarbamoylmethyl) O,O-dimethyl phosphorodithioate; J-38; OMS-968; Aflix; Anthio. $C_6H_{12}NO_4PS_2$; mol wt 257.29. C 28.01%, H 4.70%, N 5.45%, O 24.87%, P 12.04%, S 24.93%. Prepn: Lutz, Schuler, **Brit.** pat. **900,557** (1962 to Sandoz). Toxicity: K. Liesche, *Deut. Apoth.* **108**, 604 (1968), *C.A.* **69**, 26287y (1968).

Yellowish liq, mp ~25°. Practically insol in water; miscible with alcohols, ether, chloroform, benzene. LD_{50} in rats, rabbits (mg/kg): 350, 410 orally (Liesche).

USE: Acaricide, systemic insecticide. *Human Toxicity:* Cholinesterase inhibitor. *See* Parathion.

4161. Formyldienolone. *11α,17β-Dihydroxy-17-methyl-3-oxoandrosta-1,4-diene-2-carboxaldehyde;* 2-formyl-17α-methylandrosta-1,4-diene-11α,17β-diol-3-one; 2-formyl-11α-hydroxy-Δ¹-methyltestosterone; Esiclene. $C_{21}H_{28}O_4$; mol wt 344.45. C 73.23%, H 8.19%, O 18.58%. Prepn: Canonica *et al., Gazz. Chim. Ital.* **95**, 138 (1968); Gomarasca, **Brit.** pat. **1,168,931;** Fr. pat. **1,584,960** (1969, 1970 both to Lab. Prod. Biol. Braglia); *C.A.* **72**, 90742g (1970); **73**, 131212a (1970). Pharmacology: *idem, Minerva Med.* **62**, 842 (1971); Vaccari, Livini, *ibid.* 846; De Marchi *et al., Arzneimittel-Forsch.* **23**, 1583 (1973). Toxicological studies: Travella, *Gazz. Int. Med. Chir.* **76**, 400 (1971), *C.A.* **75**, 30792a (1971).

Crystals from ethyl acetate, mp 209-212°. $[\alpha]_D^{25}$ −105° (CHCl₃). Water soluble. Approx LD_{50} in rats, mice (mg/kg): 104, 187 i.p.; 270, 293 s.c.; orally in rats: >1000 (Travella).

11-β-Epimer, mp 95-103° from ethyl ether-petr ether.

THERAP CAT: Anabolic.

4162. Formyl Fluoride. CHFO; mol wt 48.02. C 25.02%, H 2.10%, F 39.57%, O 33.31%. HCOF. Prepn by the interaction of anhydr formic acid, potassium fluoride and benzoyl chloride (16% yield): Nyesmejanov, Kahn, *Ber.* **67**, 203 (1946); from benzoyl fluoride and formic acid (36% yield): Masentshev, *J. Gen. Chem. USSR* **16**, 203 (1946); from benzoyl chloride, formic acid, and KHF₂ (35% yield): Olah *et al., Ber.* **89**, 862 (1956); Olah, Kuhn, *J. Am. Chem. Soc.* **82**, 2381 (1960); from acetic formic anhydride, CH₃CO-OOCH, and anhydr HF (67% yield): Olah, Kuhn, *ibid.* 2380.

Gas at ordinary temps and pressure: bp_{760} −29°. mp −142°. Trouton constant 20.8. Heat of vaporization at −24°: 5.175 kcal.

USE: Acylating agent in organic synthesis. *Caution:* Strong irritant.

4163. 4'-Formylsuccinanilic Acid Thiosemicarbazone. *4-[[4-[[(Aminothioxomethyl)hydrazono]methyl]phenyl]amino]-4-oxobutanoic acid;* p-succinylaminobenzaldehyde thiosemicarbazone; p-[(3-carboxypropionyl)amino]benzaldehyde thiosemicarbazone; MG 500; Idrotiobicina; Hydrothiobicin; TBS. $C_{12}H_{14}N_4O_3S$; mol wt 294.34. C 48.97%, H 4.79%, N 19.04%, O 16.31%, S 10.90%. Prepn: Cavallini, *Farm. Sci. e Tec.* (Pavia) **5**, 50 (1950); Bernstein *et al., J. Am. Chem. Soc.* **73**, 906 (1951); Behnisch *et al.,* **Ger.** pat. **852,086** (1952 to Bayer); Kuroda, Usui, **Japan.** pat. **4165**('52) (Takeda).

Crystals from isopropyl alc, mp 220-221°.

Sodium salt, $C_{12}H_{13}N_4NaO_3S$. Soluble in water.

THERAP CAT: Antitubercular.

4164. N²-Formylsulfisomidine. *N-(2,6-Dimethyl-4-pyrimidinyl)-4-(formylamino)benzenesulfonamide;* 4'-[(2,6-dimethyl-4-pyrimidinyl)sulfamoyl]formanilide; N-[4-[[(2,6-dimethyl-4-pyrimidinyl)amino]sulfonyl]phenyl]formamide; formylsulfamethine; FAK III; Wometin. $C_{13}H_{14}N_4O_3S$; mol wt 306.35. C 50.97%, H 4.61%, N 18.29%, O 15.67%, S 10.47%. Prepd by formylation of sulfisomidine: Brand, Rieckhoff, **Ger.** pats. **1,122,511; 1,126,857** (both 1962 to VEB Farbenfabrik Wolfen).

Crystals, mp 248.5-250.5°.

THERAP CAT: Antibacterial.

4165. Fortimicin(s). Aminoglycoside antibiotic complex produced by *Micromonospora olivoasterospora.* The main components, fortimicins A and B, both exhibit broad-spectrum antibacterial activity. Isoln of A: T. Nara *et al.,* **Japan. Kokai** 75 29,789 (to Kyowa); U.S. pat. **3,976,768** (1976 to Abbott); of B: *eidem,* **Japan. Kokai** 75 145,588 (to Kyowa), *C.A.* **84**, 162885d (1976). Manufacture of B: *eidem,* **Ger.** pat. **2,418,349** (1974 to Kyowa), *C.A.* **82**, 137735f (1975). Isoln and biological properties of A and B: *eidem, J. Antibiot.* **30**, 533 (1977); physical and chromatographic properties: R. Okachi *et al., ibid.* 541. Structures of A and B: R. S. Egan *et al., ibid.* 552. In vitro activity: R. Girolami, J. S. Stamm, *ibid.* 564. Isoln and properties of minor components, fortimicins C, D, KE: M. Sugimoto *et al., ibid.* **32**, 868 (1979). Synthesis of B: T. Suami, Y. Honda, *Chem. Letters* **1980**, 669; Y. Honda, T. Suami, *Bull. Chem. Soc. Japan* **55**, 1156 (1982).

Fortimicin A, $C_{17}H_{35}N_5O_6$, *astromicin, 4-amino-1-[(aminoacetyl)methylamino]-1,4-dideoxy-3-O-(2,6-diamino-2,3,-4,6,7-pentadeoxy-β-L-lyxo-heptopyranosyl)-6-O-methyl-L-chiro-inositol, Abbott-44747.* White amorphous powder, mp >200° (dec). $[\alpha]_D^{25}$ +87.5° (c = 0.1 in water). Sol in water and lower alcohols, insol in organic solvents. LD_{50} (of the sulfate salt) in mice: 380 mg/kg i.v.; 400 mg/kg s.c.

Fortimicin B, $C_{15}H_{32}N_4O_5$, *4-amino-1,4-dideoxy-3-O-(2,6-diamino-2,3,4,6,7-pentadeoxy-β-L-lyxo-heptopyranosyl)-6-O-methyl-1-(methylamino)-L-chiro-inositol.* White amorphous powder, mp 101-103°. $[\alpha]_D^{25}$ +22.2° (c = 0.1 in water). Sol in water and lower alcohols, insol in organic solvents.

THERAP CAT: Antibacterial.

4166. Foscarnet Sodium. *Dihydroxyphosphinecarboxylic acid oxide trisodium salt;* trisodium phosphonoformate; trisodium carboxyphosphate; A 29622; Foscavir. CNa_3O_5P; mol wt 191.95. C 6.26%, Na 35.93%, O 41.68%, P 16.13%. Inhibits viral DNA polymerase and reverse transcriptase. Prepn: P. Nylén, *Ber.* **57B**, 1023 (1924). Crystal structure: R. R. Naqvi *et al., J. Chem. Soc. (A)* **1971**, 2751; S. C. Abrahams, *Acta Crystallogr.* **B28**, 2886 (1972). Use as antiviral agent: B. F. H. Eriksson *et al.,* **Ger. pat. 2,728,685** corresp to **U.S. pat. 4,215,113** (1978, 1980 to Astra). Activity against herpes simplex virus: E. Helgstrand *et al., Science* **201**, 819 (1978); J. M. Reno *et al., Antimicrob. Ag. Chemother.* **13**, 188 (1978); S. Alenius *et al., Arch. Virol.* **60**, 197 (1979). Mechanism of inhibition of Epstein-Barr virus replication: A. K. Datta, R. E. Hood, *Virology* **114**, 52 (1981). Preclinical evaluation: B. Oeberg *et al., Int. Congr. Ser.-Excerpta Med.* **571**, 175 (1982). Inhibitory effect on HIV-1 (HTLV-III/LAV) virus replication *in vitro:* P. S. Sarin *et al., Biochem. Pharmacol.* **34**, 4075 (1985).

Hexahydrate, mp >250°. LD_{50} i.p. in mice: 2000-4000 μmol/kg (Eriksson).

THERAP CAT: Antiviral.

4167. Fosetyl Al. *Phosphonic acid monoethyl ester aluminum salt;* aluminum tris(ethyl phosphite); aluminum tris-(O-ethylphosphonate); efosite Al; phosethyl Al; LS 74-783; Aliette. $C_6H_{18}AlO_9P_3$; mol wt 354.12. C 20.35%, H 5.12%, Al 7.62%, O 40.66%, P 26.24%. Systemic fungicide translocated both upwards and downwards in plants. Prepn: V. V. Orlovskii *et al., J. Gen. Chem. U.S.S.R.* **42**, 1924 (1972). Alternate prepn, fungicidal properties: J. Ducret *et al.,* **Ger. pat. 2,456,627;** eidem, **U.S. pat. 4,139,616** (1975, 1979 both to Rhone-Poulenc); J. Abblard *et al.,* **U.S. pat. 4,143,059** (1979 to Philagro). Biological properties and field trials against vine mildew and other *Phycomycetes* pathogens: A. Bertrand *et al., Phytiatr.-Phytopharm.* **26**, 3 (1977). Review of properties and fungicidal action on tropical, temperate crops: D. J. Williams *et al., Proc. Brit. Crop Prot. Conf.-Pests Dis.* **1977**, 565.

White odorless crystals, mp >300°. Soly in water at 20°C: 120 g/l. Practically insol in acetonitrile, propylene glycol (<80 mg/l). LD_{50} in rats, mice, Japanese quail, rabbits (mg/kg): 5800, 3700, 4997, 2680 orally. LD_{50} i.p. in rats: 550 mg/kg (Williams).

USE: Systemic fungicide.

4168. Fosfestrol. *(E)-4,4'-(1,2-Diethyl-1,1-ethenediyl)-bisphenol bis(dihydrogen phosphate); α,α'-diethyl-4,4'-stilbenediol diphosphoric acid ester;* diethylstilbestrol 4,4'-diphosphoric ester; diethylstilbestrol diphosphate; diethylstilbestryl diphosphate; diethyldihydroxystilbene diphosphate; stilbestrol diphosphate; ST52-Asta; Honvol; Stilbostatin;

Stilphostrol. $C_{18}H_{22}O_8P_2$; mol wt 428.32. C 50.48%, H 5.18%, O 29.88%, P 14.46%. Prepd by treating stilbestrol with phosphorus oxychloride in pyridine: Miescher, Heer, **U.S. pat. 2,234,311** (1941); Arnold, **U.S. pat. 2,802,854** (1957 to Asta-Werke). Prepn of stable solns at pH 10: Fonner, Collins, **U.S. pat. 2,828,244** (1958 to Miles Labs.). Prepn of sodium salts: Dawson, **U.S. pat. 2,971,975** (1961 to Miles Labs.).

Voluminous white cryst powder from dil HCl. Dec 204-206°. Sparingly sol in water.

Disodium salt, $C_{18}H_{20}Na_2O_8P_2$, crystals. Softens at 190°, uniform melt at 230°. Sol in water. pH of aq soln 6.5. pH also reported as 4.85 to 5.00.

Tetrasodium salt, $C_{18}H_{20}Na_4O_8P_2$, *Cytonal, Honvan, Stilbostatin.*

THERAP CAT: Estrogen. Treatment of prostatic carcinoma.

4169. Fosfomycin. *(2R-cis)-(3-Methyloxiranyl)phosphonic acid; (−)-(1R,2S)-(1,2-epoxypropyl)phosphonic acid;* fosfonomycin; phosphonomycin; MK-955; Fosfocina. $C_3H_7O_4P$; mol wt 138.06. C 26.10%, H 5.11%, O 46.35%, P 22.44%. Antibiotic produced by *Streptomyces* strains: D. Hendlin *et al., Science* **166**, 122 (1969); B. G. Christensen *et al., ibid.* 123; also produced by *Pseudomonas syringae:* J. Shogi *et al., J. Antibiot.* **39**, 1011 (1986). Isoln: D. Hendlin *et al.,* **Belg. pat. 718,507;** eidem, **U.S. pat. 3,914,231** (1969, 1975 both to Merck & Co.). Synthesis and resolution: B. G. Christensen *et al.,* **Belg. pats. 723,072; 723,073** (both 1969 to Merck & Co.). Alternate synthesis: Girotra, Wendler, *Tetrahedron Letters* **1969**, 4647; Glamkowski *et al., J. Org. Chem.* **35**, 3510 (1970). Series of articles on characterization, activity and clinical testing: *Antimicrob. Ag. Chemother.* **1969**, 284-351; H. B. Woodruff *et al., Chemotherapy* **23**, Suppl. 1, 1-22 (1977). Mechanism of action: Kahan *et al., Ann. N. Y. Acad. Sci.* **235**, 354 (1974). Pharmacokinetics in humans: M. Goto *et al., Antimicrob. Ag. Chemother.* **20**, 393 (1981). Pharmacology of *cofosfolactamines,* fixed dose combinations of fosfomycin with β-lactam antibiotics: P. Periti, *Drugs Exp. Clin. Res.* **6**, 305 (1980). Clinical trial of cofosfolactamines: T. Barreca *et al., ibid.* **10**, 55 (1984).

Crystals, mp ~94°. Soluble in water.

Benzylammonium salt, $C_{10}H_{16}NO_4P$, mp 170-174°. $[\alpha]_{405}$ −9.1° (c = 5).

Calcium salt monohydrate, *Afos, Biocin, Biofos, Endociclina, Faremicin, Fonofos, Fosfobiotic, Fosfocin, Fosfogram, Fosfral, Fosfotricina, Fosmicin, Foximin, Francital, Gram-Micina, Ipamicina, Lancetina, Lofoxin, Neofocin, Palmofen, Priomicina, Selemicina, Ultramicina, Valemicina.*

Tromethanol, $C_7H_{18}NO_7P$, *(2R-cis)-(3-methyloxiranyl)-phosphonic acid compd with 2-amino-2-(hydroxymethyl)-1,3-propanediol (1:1); fosfomycin trometamol; Monuril.*

THERAP CAT: Antibacterial.

4170. Fosfosal. *2-(Phosphonooxy)benzoic acid; salicylic acid dihydrogen phosphate;* o-carboxyphenyl phosphate; salicyl phosphate; UR 1521; Disdolen. $C_7H_7O_6P$; mol wt 218.10. C 38.55%, H 3.23%, O 44.01%, P 14.20%. Salicylic acid deriv. Prepn: A. C. Marin Moga, E. Francia Barra, **Ger. pat. 2,641,526** (1978 to Uriach), *C.A.* **88**, 136317h (1978). Pharmacology and comparison with aspirin: J. Garcia Rafanell *et al., Arzneimittel-Forsch.* **30**, 1091 (1980). Toxicity studies: M. S. Sanchez *et al., ibid.* 1098. Pharmacokinetics: V. Rimbau *et al., Arch. Farmacol. Toxicol.* **7**, 61 (1981). Double-blind trials in musculoskeletal/articular pain: C. Diaz *et al.*, *Clin. Ther.* **4**, 121 (1981); L. Madera Cat, J. Garcia Rafanell, *Med. Clin.* **76**, 18 (1981).

White solid, mp 168-170°. Sol in water, ethanol, acetone. Insol in non-polar organic solvents. Hydrolyzes in aq soln. LD_{50} in male, female mice, male, female rats at pH 1.0, aq soln (mg/kg): 94, 105, 153, 257 i.v.; 352, 253, 338, 360 i.p.; 1455, 1539, 1104, 1213 orally; at pH 3.5 (mg/kg): 117, 118, 207, 215 i.v.; 1592, 1483, 1085, 1128 i.p.; 1702, 2007, 1685, 2225 orally (Sanchez).

THERAP CAT: Analgesic.

4171. Fosinopril. *(2α,4β)-4-Cyclohexyl-1-[[[2-methyl-1-(1-oxopropoxy)propoxy](4-phenylbutyl)phosphinyl]acetyl]-*L-*proline;* *(4S)*-4-cyclohexyl-1-[[[*(RS)*-1-hydroxy-2-methylpropoxy](4-phenylbutyl)phosphinyl]acetyl]-L-proline propionate (ester); fosenopril. $C_{30}H_{46}NO_7P$; mol wt 563.67. C 63.93%, H 8.23%, N 2.48%, O 19.87%, P 5.49%. Phosphinic acid containing angiotensin converting enzyme inhibitor. Prepn: E. W. Petrillo, Jr., U.S. pat. **4,337,201** (1982 to Squibb); of active diacid form: J. Krapcho *et al.*, *J. Med. Chem.* **31**, 1148 (1988). Metabolism and pharmacokinetics: S. M. Singhvi *et al.*, *Brit. J. Clin. Pharmacol.* **25**, 9 (1988). GC determn of diacid: M. Jemal *et al.*, *J. Chromatog.* **345**, 299 (1985). Clinical trial in hypertension: P. A. Sullivan *et al.*, *Am. J. Hypertension* **1**, 280S (1988). Brief description: E. W. Petrillo, Jr., *et al.*, *Clin. Exp. Theory Prac.* **A9**, 235 (1987).

Sodium salt, $C_{30}H_{45}NNaO_7P$, *SQ 28555, Monopril.* Diacid, $C_{23}H_{34}NO_5P$, *SQ 27519.* mp 149-153°. $[\alpha]_D$ —24° (c = 1 in methanol).

THERAP CAT: Antihypertensive.

4172. Fospirate. *Dimethylphosphoric acid 3,5,6-trichloro-2-pyridinyl ester; O,O*-dimethyl-*O*-3,5,6-trichloro-2-pyridyl phosphate; Dowco 217; Torelle. $C_7H_7Cl_3NO_4P$; mol wt 306.46. C 27.43%, H 2.30%, Cl 34.70%, N 4.57%, O 20.88%, P 10.11%. Prepn by reaction of 3,5,6-trihalo-2-pyridinol with phosphoryl chloride: Rigterink, Kenaga, *J. Agr. Food Chem.* **14**, 304 (1966).

Crystals from petr ether, mp 86.5-88°.
THERAP CAT (VET): Anthelmintic.

4173. Fosthietan. *1,3-Dithietan-2-ylidenephosphoramidic acid diethyl ester; phosphonodithioimidocarbonic acid cyclic methylene P,P-diethyl ester;* 2-(diethoxyphosphinylimino)-1,3-dithietane; AC 64475; CL 64475; Nem-A-Tak. $C_6H_{12}NO_3PS_2$; mol wt 241.26. C 29.87%, H 5.01%, N 5.81%, O 19.89%, P 12.84%, S 26.58%. Prepn: R. W. Addor, **S. Afr.** pat. **68 01,064** corresp to U.S. pat. **3,470,207** (1968, 1969 to Am. Cyanamid); *idem, J. Heterocycl. Chem.* **7**, 381 (1970). Activity: W. K. Whitney, J. L. Aston, *Proc. Brit. Insectic. Fungic. Conf.* **2**, 625 (1975).

Pale yellow oil, mercaptan-like odor. n_D^{25} 1.5348; d_{25} 1.3. Vapor press at 25°: 6.5×10^{-6} mm Hg. Soly in water at 25°: 50 g/kg. Sol in acetone, chloroform, methanol, toluene. LD_{50} orally in rats, mice: 5, 18 mg/kg, W. K. Whitney, J. L. Aston, *loc. cit.*

USE: Nematocide; insecticide.

4174. Fotemustine. *[1-[[[(2-Chloroethyl)nitrosoamino]-carbonyl]amino]ethyl]phosphonic acid diethyl ester;* (±)-diethyl [1-[3-(2-chloroethyl)-3-nitrosoureido]ethyl]phosphonate; 1-[*N*-(2-chloroethyl)-*N*-nitrosoureido]ethylphosphonic acid diethyl ester; S-10036. $C_9H_{19}ClN_3O_5P$; mol wt 315.69. C 34.24%, H 6.07%, Cl 11.23%, N 13.31%, O 25.34%, P 9.81%. Amino acid-linked nitrosourea alkylating agent. Prepn: G. Lavielle, C. Cudennec, **Fr.** pat. **2,536,075;** *eidem,* U.S. pat. **4,567,169** (1984, 1986 both to ADIR). Clinical evaluation in advanced cancers: D. Khayat *et al.*, *Cancer Res.* **47**, 6782 (1987); in disseminated malignant melanoma: D. Khayat *et al.*, *J. Nat. Cancer Inst.* **80**, 1407 (1988).

mp 85°.
THERAP CAT: Antineoplastic.

4175. Francium. Eka-cesium. Fr; at. no. 87. ^{223}Fr, *Actinium K,* is the most stable isotope ($T_{1/2}$ 21 min; β-emitter) formed by α-decay of actinium (^{227}Ac). Found in uranium minerals. First obtained in 1939 from an actinium prepn by Perey, *J. Phys. Radium* **10**, 439 (1939); *J. Chim. Phys.* **43**, 155, 262 (1946); Hyde, Ghiorso, *Phys. Rev.* **90**, 267 (1953); Hyde, *J. Am. Chem. Soc.* **74**, 4181 (1952). Isoln by paper chromatography: Perey, Adloff, *Compt. Rend.* **236**, 1163 (1953). Also obtainable by proton bombardment of thorium. Mass numbers of other known isotopes: 204-213; 217-222; 224. Most electropositive element. Chemical behavior similar to that of other alkali metals. *Reviews:* Hyde, *J. Chem. Ed.* **36**, 15-21 (1959); Whaley, "Sodium, Potassium, Rubidium, Cesium and Francium" in *Comprehensive Inorganic Chemistry* vol. 1, J. C. Bailar Jr., *et al.*, Eds. (Pergamon Press, Oxford, 1973) pp 369-529.

4176. Frangula. Buckthorn bark; alder buckthorn; black dogwood; berry alder; arrow wood; Persian berries. Dried bark of *Rhamnus frangula* L., Rhamnaceae. *Habit.* Europe, Russian Asia, Mediterranean coast of Africa. *Constit.* Frangulin, emodin, chrysophanic acid.

4177. Frangulin. Franguloside; avornin; Cascarin. In berries, bark, and rootbark of *Rhamnus* spp., especially in alder buckthorn (*Rhamnus frangula* L.), *Rhamnus cathartica* L., and *Rhamnus purshiana* DC. (*Cascara sagrada*), Rhamnaceae. Prepn from bark of alder buckthorn: Bridel, Charaux, *Bull. Soc. Chim. Biol.* **15**, 642 (1933); M. Kubiak, *Herba Pol.* **23**, 217 (1977); *C.A.* **88**, 166260b (1978). Consists of the two glucosides, frangulins A and B, which were originally thought to be isomeric. Structure and synthesis of frangulin A: Hörhammer, Wagner, *Z. Naturforsch.* **27B**, 959 (1972). Structure of frangulin B: Wagner, Demuth, *Tetrahedron Letters* **1972**, 5013.

frangulin A, R =

frangulin B, R =

Frangulin A, $C_{21}H_{20}O_9$. *1,3,8-Trihydroxy-6-methylan-thraquinone-l-rhamnoside, emodin-l-rhamnoside, rhamno-xanthin.* Crystals, mp 228°. Absorption max: 225, 264, 282, 300, 430 nm (log ϵ 4.52, 4.28, 4.15, 3.97, 4.05).

Frangulin B, $C_{20}H_{18}O_9$. *6-O-(D-Apiofuranosyl)-1,6,8-tri-hydroxy-3-methylanthraquinone.* mp 196°.

THERAP CAT: Cathartic.

4178. Fraxetin. *7,8-Dihydroxy-6-methoxy-2H-1-benzo-pyran-2-one; 7,8-dihydroxy-6-methoxycoumarin.* $C_{10}H_8O_5$; mol wt 208.16. C 57.70%, H 3.87%, O 38.43%. The aglucon of fraxin. Obtained by heating fraxin with dil H_2SO_4. Structure: Wessely, Demmer, *Ber.* **61**, 1279 (1928). Synthesis: Aghoramurthy, Seshadri, *J. Chem. Soc.* **1954**, 3065.

Plates from alc, mp 228°. Turns yellow at 150° and brown at mp. Sol in 10 liters cold water, in 300 ml boiling water; somewhat more sol in alcohol; hardly sol in ether.

Dimethyl ether, $C_{12}H_{12}O_5$, *6,7,8-trimethoxycoumarin.* Orthorhombic bipyramidal crystals, mp 104°. bp$_{0.2}$ 90-100°.

4179. Fraxin. *8-(β-D-Glucopyranosyloxy)-7-hydroxy-6-methoxy-2H-1-benzopyran-2-one; 7,8-dihydroxy-6-meth-oxycoumarin-8-β-D-glucoside; 8-(glucosyloxy)-6-methoxy-umbelliferone; fraxetin-8-glucoside; fraxoside; paviin.* $C_{16}H_{18}O_{10}$; mol wt 370.30. C 51.89%, H 4.90%, O 43.21%. From bark of the common European ash (*Fraxinus excelsior* L.): Salm-Horstmar, *Pogg. Ann.* **100**, 607 (1857); from *F. oxyphylla* Mar., *Oleaceae:* Paris, Stambouli, *Compt. Rend.* **253**, 313 (1961). Identity with paviin: Stokes, *J. Chem. Soc.* **12**, 126 (1859). From the horse chestnut tree (*Aesculus hip-pocastanum* L., *Hippocastanaceae):* Reppel, *Planta Med.* **9**, 199 (1956); from *Diervilla* spp., *Caprifoliaceae:* Charaux, *J. Pharm. Chim.* [7] **4**, 248 (1911). Structure: Wessely, Demmer, *Ber.* **62**, 120 (1929).

Yellow hydrated needles from water or dil alc. Slightly bitter, astringent taste. The water of crystn (about 3 mols) is removed at 130° and 0.2 mm Hg. The anhyd substance, mp 205° (rapid heating). Sparingly sol in cold water; freely sol in hot water, hot alcohol; practically insol in ether. Alkaline solns are sulfur-yellow and on dilution show characterisitic blue-green fluorescence.

4180. Fraxiparine®. CY-216. Low molecular weight fraction of heparin, *q.v.* Isolated from porcine mucosal heparin by selective ethanol precipitation; used in the form of the calcium salt. Meal mol wt 4500 daltons. Prepn: J.-C. Lormeau *et al.*, **Ger. pat.** **2,944,792**; *eidem,* **U.S. pats.**

4,486,420; 4,692,435 (1980, 1984, 1987 all to Choay); J. Choay *et al., Thromb. Res.* **18**, 573 (1980). Comparison of endothelial binding ability with unfractionated heparin: G. F. Gensini *et al., Haemostasis* **14**, 466 (1984). Exhibits anti-thrombotic activity comparable to standard heparin but with diminished hemorrhagic effects: M. D. Vandenbroek, B. Bordes, *J. Pharm. Clin.* **5**, 35 (1986); C. Doutremepuich *et al., Thromb. Res.* **43**, 691 (1986). Pharmacokinetics and bioavailability in humans: J. Harenberg *et al., ibid.* **44**, 549 (1986). Clinical trials in prophylaxis of postoperative thromboembolism: V. V. Kakkar, *Nouv. Rev. Fr. Hematol.* **26**, 277 (1984); V. V. Kakkar, W. J. G. Murray, *Brit. J. Surg.* **72**, 786 (1985).

THERAP CAT: Antithrombotic.

4181. Fredericamycin A. *(E,E)-6',7'-Dihydro-4,9,9'-trihydroxy-6-methoxy-3'-(1,3-pentadienyl)spiro[2H-benz(f)-indene-2,8'-(8H)cyclopent(g)isoquinoline]-1,1',3,5,8(2'H)-pentone;* FCRC-A48; NSC-305263. $C_{30}H_{21}NO_9$; mol wt 539.50. C 66.79%, H 3.92%, N 2.60%, O 26.69%. Antitu-mor antibiotic produced by *Streptomyces griseus* (FCRC-48); represents new structural class of antibiotics containing a spiro[4,4]nonane ring system. Isoln together with two bio-logically inactive components, fredericamycins B and C: R. C. Pandey *et al., J. Antibiot.* **34**, 1389 (1981). Antimicrobial and cytotoxic activity *in vitro* and antitumor activity *in vivo:* D. J. Warnick-Pickle *et al., ibid.* 1402. Prepn and biological activity of water soluble salts: R. Misra, *ibid.* **51**, 976 (1988). Structure determn by x-ray crystallography: R. Misra *et al., J. Am. Chem. Soc.* **104**, 4478 (1982). Spectro-scopic and mass spectral characterization: *eidem, J. Antibiot.* **40**, 786 (1987). Synthetic studies: A. V. R. Rao *et al., Chem. Commun.* **1984**, 1119; K. A. Parker *et al., Tetrahe-dron Letters* **26**, 2181 (1985). Synthesis of (±)-form: T. R. Kelly *et al., J. Am. Chem. Soc.* **108**, 7100 (1986). Biosynthe-tic study: K. M. Byrne *et al., Biochemistry* **24**, 478 (1985). Mechanism of action study: B. D. Hilton *et al., ibid.* **25**, 5533 (1986).

Thin, platelet-like crystals from acetonitrile + water, mp > 350° (dec). pKa (DMF) 6.80; 8.88. Sol in acetic acid, DMSO, DMF, pyridine; partially sol in acidic methanol, acidic chloroform, acidic ethyl acetate. Insol in water, petr ether, ether. Acts as an indicator: Red in acidic soln, green to blue in basic soln.

Potassium salt, soly (mg/ml): water (1.0); DMSO (8/1.5); 1:1 DMSO-H_2O (3.0); also readily sol in DMF, dimethyl-acetamide, pyridine. Sparingly sol in ethyl acetate, acetoni-trile, methanol, chloroform. Insol in hexanes, benzene, ace-tone, ether.

4182. Frenolicin. *3,4,5,10-Tetrahydro-9-hydroxy-5,10-dioxo-1-propyl-4a,10a-epoxy-1H-naphtho[2,3-c]pyran-3-acetic acid.* $C_{18}H_{18}O_7$; mol wt 346.32. C 62.42%, H 5.24%, O 32.34%. Antibiotic produced by *Streptomyces fradiae:* Van Meter *et al., Antimicrob. Ag. Ann.* **1960**, 77. Structure: Ellestad *et al., J. Am. Chem. Soc.* **88**, 4109 (1966); **90**, 1325 (1968). Total synthesis: A. Ichihara *et al., Tetrahedron Letters* **21**, 4469 (1980).

Pale yellow needles from benzene, mp 160-161°. $[\alpha]_D^{25}$ −37.7° (c = 1.5 in methanol). uv max (methanol): 234, 362 nm (ε 18,300; 5200). Sol in methanol, ethanol, ethyl acetate, acetone, glacial acetic acid, carbon tetrachloride; slightly sol in benzene; practically insol in water, cyclohexane, petr ether.

4183. Frequentin. *6-(1,3-Heptadienyl)-3,4-dihydroxy-2-oxocyclohexanecarboxaldehyde.* $C_{14}H_{20}O_4$; mol wt 252.30. C 66.64%, H 7.99%, O 25.37%. Mold metabolite with antineoplastic and antibiotic activity; isolated from *Penicillium frequentans* Westling and *P. palitans* Westling: Curtis *et al.*, *Nature* **167**, 557 (1951); Birkinshaw, *Biochem. J.* **51**, 271 (1952). Structure: Sigg, *Helv. Chim. Acta* **46**, 1061 (1963). Effect on Ehrlich ascites carcinoma cells: J. Fuska *et al.*, *Int. Congr. Chemother., Proc., 7th, Prague, 1971* vol. **2** (Univ. Park Press, Baltimore, 1972) pp 97-98. Synthetic study: C. G. Unson, *Diss. Abstr. B* **39**, 1296 (1978).

Needles from boiling water, mp 134.5°. When dissolved, exists as the 1-hydroxymethylene tautomer. $[\alpha]_D^{22}$ +65° (c = 0.95 in CHCl₃); $[\alpha]_D^{22}$ +61° (c = 0.96 in methanol). uv max (dioxane): 290, 232 nm (log ε 3.74, 4.52). Sol in acetone, dioxane; less sol in chloroform, ethanol, benzene; slightly sol in carbon tetrachloride, water.

4184. Friedelin. *D:A-Friedooleanan-3-one; friedelan-3-one.* $C_{30}H_{50}O$; mol wt 426.70. C 84.44%, H 11.81%, O 3.75%. Major triterpene constituent of cork, obtained by extraction of ground cork with alc. Isoln: Istrati, Ostrogovich, *Compt. Rend.* **128**, 1581 (1899). Isolated also from *Ceratopetalum apetalum* D. Don, Cunoniaceae: Jefferies, *J. Chem. Soc.* **1954**, 473. Structure, sterochemistry: Brownlie *et al.*, *Chem. & Ind. (London)* **1955**, 1156; Kane, Stevenson, *Tetrahedron* **15**, 223 (1961); Stevenson, *J. Org. Chem.* **28**, 188 (1963). Total synthesis of (±)-form: R. E. Ireland, D. M. Walba, *Tetrahedron Letters* **1976**, 1071.

Needles from ethyl acetate or alcohol, mp 263-263.5°. $[\alpha]_D$ −27.8° (chloroform). One gram dissolves in 8.6 ml chloroform, 264 ml 99% alcohol.

4185. Fructose. *D-Fructose; β-D-fructose; levulose; fruit sugar; Fructosteril; Laevoral; Levugen; Laevosan.* $C_6H_{12}O_6$; mol wt 180.16. C 40.00%, H 6.72%, O 53.29%. Occurs in a large number of fruits, honey, and as the sole sugar in bull and human semen: Auerbach, Bodlander, *Angew. Chem.* **36**, 602 (1923); Mann, *Nature* **157**, 79 (1946); Pryde, *ibid.* 660. Prepd by adding abs alcohol to the syrup obtained from the acid hydrolysis of inulin: Bates *et al.*, *Natl. Bur. Std. (U.S.),*

Circ. **C440**, 399 (1942). Prepn from dextrose: Cantor, Hobbs, U.S. pat. **2,354,664** (1944 to Corn Prod. Refining). From sucrose by enzymatic conversion: Koepsell *et al.*, U.S. pat. **2,729,587** (1956 to U.S.A.). Crystal and molecular structure: J. A. Kanters *et al., Acta Crystallogr.* **B33**, 665 (1977). *Review:* Barry, Honeyman, *Advan. Carbohyd. Chem.* **7**, 53-98 (1952); M. Chen, R. L. Whistler, *ibid.* **34**, 285-343 (1977). Occurs in both the furanose and pyranose forms. An aq soln at 20° contains about 20% of the furanose form.

Orthorhombic, bisphenoidal prisms from alc, dec 103-105°. Sweetest of the sugars. Shows mutarotation. $[\alpha]_D^{20}$ −132° → −92° (c = 2). Rapid and anomalous mutarotation involves pyranose-furanose interconversion. The final value is obtained instantly in the presence of hydroxyl ions. Ka at 18°: 8.8 × 10⁻¹³. Freely sol in water. One gram dissolves in 15 ml alc, in 14 ml methanol. Slightly sol in cold, freely in hot acetone; sol in pyridine, ethylamine, methylamine.

USE: To prevent sandiness in ice cream.

THERAP CAT: Fluid and nutrient replenisher.

THERAP CAT (VET): For bovine ketosis.

4186. DL-Fructose. α-Acrose; methose. $C_6H_{12}O_6$; mol wt 180.16. C 40.00%, H 6.72%, O 53.29%. Component of formose (polymerization product of formaldehyde): Vogel, *Helv. Chim. Acta* **11**, 370 (1928).

Needles from methanol. d_4^{16} 1.665. mp 129-130° (slow heating). Reduces Fehling's soln.

Phenylosazone, $C_{18}H_{22}N_4O_4$, mp 216-217°.

4187. Fructose-1,6-diphosphate. *D-Fructose 1,6-bis(dihydrogen phosphate); 1,6-D-fructosediphosphoric acid;* hexose diphosphate; Harden-Young ester; Esafosfina. $C_6H_{14}O_{12}P_2$; mol wt 340.13. C 21.19%, H 4.15%, O 56.45%, P 18.22%. Formed from fructose-6-phosphate in the presence of ATP, Mg^{2+} and the enzyme phosphohexokinase. Prepn from glucose, mannose, fructose, sucrose by the action of yeasts: A. Harden, *Alcoholic Fermentation* (Longmans, Green & Co., New York, 4th ed., 1932); v. Lebedev, *Biochem. Z.* **36**, 254 (1911); *cf* Ger. pats. **292,817; 293,864; 301,590.** Fructose-1,6-diphosphate is reversibly split in the presence of aldolase forming 1-phosphodihydroxyacetone and 3-phosphoglyceraldehyde: Meyerhof *et al., Biochem. Z.* **286**, 301 (1936). Metabolism regulation study: M. E. Kirtley, M. McKay, *Mol. Cell. Biochem.* **18**, 141 (1977).

Calcium salt monohydrate, $C_6H_{10}Ca_2O_{12}P_2 \cdot H_2O$, *Candiolin, Glucofos.* White powder.

Sodium salt, $C_6H_{12}Na_2O_{12}P_2$, in *Fructergyl.*

Trisodium salt, $C_6H_{11}Na_3O_{12}P_2$, *Hexaphosphin.*

THERAP CAT: Roborant; tonic.

4188. Fructose-6-phosphate. *D-Fructose 6-(dihydrogen phosphate);* D-fructose-6-phosphoric acid; fructose monophosphate; hexose phosphate; hexose monophosphate; Neuberg ester. $C_6H_{13}O_9P$; mol wt 260.14. C 27.70%, H 5.04%, P 11.91%, O 55.35%. Present in animal tissues as an equilibrium mixture with glucose-6-phosphate. The glucose-6-phosphate may be reversibly transformed into fructose-6-phosphate by the enzyme phosphohexose isomerase. Prepn by hydrolysis of 1,6-fructose diphosphate with dil acid: Neuberg, *Biochem. Z.* **88**, 432 (1917); Ger. pat. **334,250** (Bayer); *Chem. Zentr.* **1921**, II, 961; *Frdl.* **13**, 948. Role in metabolic regulation and heat generation: E. A. Newsholme, *Biochem. Soc. Trans.* **4**, 978 (1976).

Very sol in water. $[\alpha]_D^{21}$ +2.5° (c = 3), Meyerhof, Lohmann, *Biochem. Z.* **185**, 117 (1927). $[\alpha]_D$ +1.2° (c = 0.9), Lohmann, *ibid.* **262**, 145 (1933).

The magnesium and zinc salts also are sol in water. The equilibrium mixture of 75-80% glucose-6-phosphate and 20-25% fructose-6-phosphate is called *lactacidogen* or *Embden ester.*

4189. FSH. *Follicle-stimulating hormone;* urofollitrophin; Fertinorm; Follitropin; Luteoantine; Metrodin; Thylakentrin. Mol wt about 36,000. Glycoprotein gonadotropic hormone found in pituitary tissue of mammals that directly regulates the metabolic activity of granulosa cells of the ovary and Sertoli cells of the testis. In the female, FSH induces the maturation of the Graafian follicles of the ovary. The follicular cells which surround the growing ovum develop the capacity to produce estrogens which induce proliferative changes in the walls of the uterus and vagina. The rising titer of estrogen will bring on ovulation and the changeover to progestational activity. In the male, promotes the development of the germinal cells of the testes. Isoln procedures: Fevold *et al., Endocrinology* **26**, 999 (1940); Fraenkel-Conrat *et al., Proc. Soc. Exp. Biol. Med.* **45**, 627 (1940); McShan, Meyer, *J. Biol. Chem.* **135**, 473 (1940); Greep *et al., ibid.* **133**, 289 (1940); Li *et al., Science* **109**, 445 (1949). Prepn of human FSH from pituitaries and from postmenopausal urine: Roos, Gemzell in *Ciba Foundation Study Group* no. 22 (Little, Brown and Co., Boston, 1965). Isoln and a study of some chemical, physical, immunological and biological properties of FSH; evidence of subunit structure: Roos, *Acta Endocrinol. (Copenhagen)* **59**, Suppl. **131** (1968); Reichart; Rathnam, Saxena, in *Gonadotropins,* B. B. Saxena *et al.,* Eds. (Wiley-Interscience, New York, 1972) pp 107-131. Amino acid sequence of human FSH α-subunit: P. Rathnam, B. B. Saxena, *J. Biol. Chem.* **250**, 6735 (1975); of human β-subunit: B. B. Saxena, P. Rathnam, *ibid.* **251**, 993 (1976); of equine α: P. Rathnam *et al., ibid.* **253**, 5355 (1978); of equine β: Y. Fujiki *et al., ibid.* 5363. Physical and hydrodynamic properties: R. J. Ryan *et al., Recent Progr. Horm. Res.* **26**, 105 (1970). Effect on gonadal functions: J. H. Dorrington, D. T. Armstrong, ibid. **35**, 301 (1979). *Reviews:* A. R. Means *et al., Ann. Rev. Physiol.* **42**, 59-70 (1980); J. G. Pierce, T. F. Parsons, *Ann. Rev. Biochem.* **50**, 465-495 (1981). *See also* HCG, LH, TSH.

Solid. Isoelectric point 4.5. Soluble in water, physiological saline soln, 50% alc. Soluble in a pH 4.4 acetate buffer contg 20.5% sodium sulfate (different from LH which is insol in this solvent). A 1 mg-% soln in saline may be kept at 60° for 30 min. In solns of pH 7-8 the activity is retained at 75° for 30 min. In 50% alcoholic soln the activity is destroyed at 60° within 15 min.

Note: Menotropins is a term applied to an extract of postmenopausal urine containing FSH and LH activity. Marketed under the tradenames *Pergonal, Humegon,* and *Pregova.* Used to treat female sterility. For veterinary use, purified FSH is obtained from the pituitary gland of domesticated animals.

THERAP CAT: Gonad-stimulating principle.

THERAP CAT (VET): In certain reproductive disorders.

4190. Ftaxilide. *2-[[(2,6-Dimethylphenyl)amino]carbonyl]benzoic acid;* 2',6'-dimethylphthalanilic acid; phthalic 2,6-dimethylanilide; MP-12; Histanorm. $C_{16}H_{15}NO_3$; mol wt 269.31. C 71.36%, H 5.61%, N 5.20%, O 17.82%. Deriv of phthalic acid, *q.v.,* initially studied because of anti-inflammatory properties. Prepn: A. H. Sommers, *J. Am. Chem. Soc.* **78**, 2439 (1956); P. Grammaticakis, *Compt. Rend.* **251**, 179 (1960); G. Pagani *et al., Farmaco Ed. Sci.* **23**, 448 (1968). Synthesis and anti-ulcerogenic properties: C. G. Wermuth *et al.,* Ger. pat. **2,040,578** corresp to U.S. pat. **3,793,458** (1971, 1974 both to Socibre).

White solid, mp 178 ±1°, C. G. Wermuth *et al., loc. cit.,* also reported as mp 199°, P. Grammaticakis, *loc. cit.* and mp 188-189°, A. H. Sommers, *loc. cit.* Sol with heating in ethanol, isopropyl alcohol. Insol in chloroform. LD_{50} in male mice: 4500 mg/kg orally, C. G. Wermuth *et al., loc. cit.*

THERAP CAT: Anti-ulcerative.

4191. Fucosamine. *2-Amino-2,6-dideoxygalactose.* $C_6H_{13}NO_4$; mol wt 163.17. C 44.16%, H 8.03%, N 8.58%, O 39.22%. Isoln of D-form from lipopolysaccharide of *Chromobacterium violaceum:* Crumpton, Davies, *Biochem. J.* **70**, 729 (1958); from *Bacillus licheniformis:* Sharon *et al., ibid.* **93**, 210 (1964). Isoln of L-form from type V *Pneumococcus* capsular polysaccharide: Barker *et al., Nature* **189**, 303 (1961). Synthesis of L-form: Kuhn *et al., Ann.* **628**, 186 (1959); J. Lehmann *et al., Ber.* **112**, 1470 (1979); of D-form: Zehavi, Sharon, *J. Org. Chem.* **29**, 3654 (1964).

D-fucosamine

D-Form hydrochloride, $C_6H_{13}NO_4.HCl$, crystals from aq acetone, dec 170-175°. $[\alpha]_D^{20}$ +91° (water). Absorption max: 400 nm.

N-Acetyl-D-fucosamine, $C_8H_{15}NO_5$, crystals from ethanol, dec 196-197°. $[\alpha]_D^{22}$ +92° (c = 2 in water).

L-Form hydrochloride, $C_6H_{13}NO_4.HCl$, rods from methanol + isopropanol, mp 192-193°. $[\alpha]_D^{27}$ −92° (c = 0.89 in water).

N-Acetyl-L-fucosamine, $C_8H_{15}NO_5$, crystals from ethanol, mp 197-198°. $[\alpha]_D^{26}$ −82° (c = 1.46 in water).

4192. D-Fucose. *6-Deoxy-D-galactose;* D-galactomethylose; rhodeose. $C_6H_{12}O_5$; mol wt 164.16. C 43.90%, H 7.37%, O 48.73%. Obtained from glucosides found in various species of *Convolvulaceae, e.g.,* convolvulin, jalapin, β-turpethein. Isoln: Votocek, *Z. Zuckerind. Böhmen* **24**, 249; *Chem. Zentr.* **1901,** I, 1042; **1902,** II, 1361; **1905,** II, 1528. Isoln from jalap resin: Votocek, Bulir, *ibid.* **1906,** I, 1818. Prepn by acid hydrolysis of chartreusin: Sternbach *et al., J. Am. Chem. Soc.* **80**, 1639 (1958).

α-Form, needles from alc. Sweet taste. mp 144°. Shows mutarotation, $[\alpha]_D^{19}$ +127.0° (7 min) → +89.4° (31 min) → +77.2° (71 min) → +76.0° (final value 146 min, c = 10). Soluble in water; moderately sol in alcohol.

Pentaacetate, $C_{16}H_{22}O_{10}$, mp 115.5°.

Oxime, $C_6H_{13}NO_5$, mp 188.5°.

Phenylhydrazone, $C_{12}H_{18}N_2O_4$, mp 172°.

Phenylosazone, $C_{18}H_{22}N_4O_3$, mp 176.5°.

4193. L-Fucose. *6-Deoxy-L-galactose;* L-galactomethylose. $C_6H_{12}O_5$; mol wt 164.16. C 43.90%, H 7.37%, O 48.73%. Occurs in seaweed: *Ascophyllum nodosum* (L.) Ledol. *(Fucus nodosus* L.), *Fucus vesiculosus* L., *F. serratus* L., *F. virsoides* (Don) J. Ag., *Fucaceae,* and in gum tragacanth. Isoln from seaweed: Clark, *J. Biol. Chem.* **54**, 65

(1922); Hockett *et al., J. Am. Chem. Soc.* **61**, 1658 (1939). Manuf from fucoidan: Schweiger, U.S. pat. **3,240,775** (1966 to Kelco). Synthesis from D-galactose: Dejterjuszynski, Flowers, *Carbohydr. Res.* **28**, 144 (1973); from D-glucose: T. Chiba, S. Tejima, *Chem. Pharm. Bull.* **27**, 2838 (1979); from D-mannose: J. Defaye *et al., Carbohydr. Res.* **94**, 131 (1981). Review of chemistry and biochemistry: H. M. Flowers, *Adv. Carbohydr. Chem. Biochem.* **39**, 279-345 (1981).

α-Form, minute needles from abs alcohol, mp 140°. Shows mutarotation. $[\alpha]_D^{20}$ −124.1° (10 min) → −108.0° (20 min) → −91.5° (36 min) → −78.6° (70 min) → −75.6° (final value, 24 hrs, c = 9). Soluble in water and alcohol.

4194. Fucosterol. *Stigmasta-5,24(28)-dien-3-ol;* 24-ethylidenecholest-5-en-3β-ol. $C_{29}H_{48}O$; mol wt 412.70. C 84.40%, H 11.72%, O 3.88%. Isoln from *Fucus vesiculosus* L., *Fucaceae:* Heilbron *et al., J. Chem. Soc.* **1934**, 1572; from marine brown algae *(Phaeophyceae):* Tsuda, *Chem. Pharm. Bull.* **6**, 724 (1958). Structure: MacPhillamy, *J. Am. Chem. Soc.* **64**, 1732 (1942). Stereochemistry: C. Brooks *et al., Steroids* **20**, 487 (1972). Synthesis: Hayazu, *Pharm. Bull.* **5**, 452 (1957).

Crystals from methanol, mp 124°. $[\alpha]_D^{20}$ −38.4° (chloroform). Soluble in most organic solvents.

Acetate, $C_{31}H_{50}O_2$, crystals, mp 120-121°. $[\alpha]_D^{20}$ −45° (chloroform).

4195. Fucoxanthin. *3'-(Acetyloxy)-6',7'-didehydro-5,6-epoxy-5,5',6,6',7,8-hexahydro-3,5'-dihydroxy-8-oxo-β,β-carotene; 6',7'-didehydro-5,6-epoxy-4',5,5',6,7,8-hexahydro-3,3',5'-trihydroxy-8-oxo-α-carotene 3'-acetate.* $C_{42}H_{58}O_6$; mol wt 658.88. C 76.56%, H 8.87%, O 14.57%. Carotenoid pigment found in fresh brown algae: *Fucus virsoides* (Don) J. Ag., *Fucaceae.* Also found in *Zygnema pectinatum* (Vauch.) Ag., *Zygnemaceae* and in *Polysiphonia nigrescens* (Dillw.) Grev., *Rhodomelaceae.* Isoln: Karrer *et al., Helv. Chim. Acta* **14**, 628 (1931); Willstätter, Page, *Ann.* **404**, 237 (1914). Structure: Bonnet *et al., Chem. Commun.* **1966**, 515; *J. Chem. Soc. (C)* **1969**, 429. Abs config: DeVille *et al., Chem. Commun.* **1969**, 1311; K. Bernhard *et al., Tetrahedron Letters* **1976**, 115.

Needles from ether + petr ether, mp 160°. $[\alpha]_D^{18}$ +72.5° ±9° (chloroform). Abs max (chloroform): 492, 457 nm. Abs max (ethanol): 450 nm ($E_{1cm}^{1\%}$ 1140), Antia, *Can. J. Chem.* **43**, 302 (1965). Freely sol in ethanol; less sol in carbon disulfide; sparingly sol in ether. Practically insol in petr ether. 1.66 g dissolves in 100 g boiling methanol.

4196. Fucus. Bladder-wrack; sea-wrack; bladder fucus; kelpware; black-tang; cut-weed; sea-oak. Dried thallus of *Fucus vesiculosus* L., *F. serratus* L., or *F. siliquosus,* L. *Fucaceae. Habit.* Atlantic and Pacific Oceans. *Constit.* Algin, about 0.01% iodine and some bromine mannite.

4197. Fuller's Earth. Floridin. A nonplastic variety of kaolin containing an aluminum magnesium silicate. The name is derived from an ancient process of cleaning or fulling wool, to remove the oil and dirt particles, with a water slurry of earth or clay. At the present time, the term fuller's earth is applied to any clay that has adequate decolorizing and purifying capacity to be used commercially in oil refining without chemical treatment. It is sometimes considered to be synonymous with montmorillonite, *q.v.,* kaolinite $(Al_2-O_3.2SiO_2.2H_2O)$ and Halloysite $(Al_2O_3.2SiO_2.4H_2O)$. A long list of minerals likely to be found in fuller's earth is given by Porter, *U.S. Geol. Survey, Bull.* **315**, 268 (1907), *C.A.* **1**, 1684 (1907), and the opinion is expressed that fuller's earth results from the decompn of hornblendes and augites rather than from feldspars. Fuller's earth has for its base a series of amorphous, hydrous aluminum silicates that have a rather persistent colloidal (used in its widest sense) structure. It is to this colloidal structure, which is not lost at 130° and possibly higher, that the bleaching power is due. The bleaching efficiency of fuller's earth is usually increased by treatment with dilute acids.

USE: Decolorizer for oils and other liquids; filtering medium; filler for rubber; in agricultural formulations; also instead of absorbent charcoal.

4198. Fulvoplumierin. *(E,E)-7-(2-Butenylidene)-1,7-dihydro-1-oxocyclopenta[c]pyran-4-carboxylic acid methyl ester;* 3-(2-butenylidene)-2-carboxy-α-(hydroxymethylene)-1,4-cyclopentadiene-1-acetic acid δ-lactone methyl ester; methyl 7-crotonylidenecyclopenta[c]pyran-1-(7H)-one-4-carboxylate. $C_{14}H_{12}O_4$; mol wt 244.24. C 68.84%, H 4.95%, O 26.20%. Occurs together with plumieride and plumericin. Isoln from *Plumeria acutifolia* Poir., *Apocynaceae,* also from roots of *P. rubra* var *alba:* Grumbach *et al., Experientia* **8**, 224 (1952). Structure: Schmid, Bencze, *Helv. Chim. Acta* **36**, 206, 1468 (1953). Stereochemistry: Albers-Schönberg *et al., ibid.* **45**, 1406 (1962). Synthesis: Büchi, Carlson, *J. Am. Chem. Soc.* **90**, 5336 (1968); **91**, 6470 (1969).

Orange needles from chloroform + petr ether, ethyl acetate, or alcohol, dec 151-152°. Sublimes in high vacuum. uv max (ethanol): 272, 365 nm (ε 7,000; 33,700). Sol in chloroform, hot ethyl acetate, benzene, alcohol; less sol in pyridine, acetone. Practically insol in water, petr ether.

4199. Fumagillin. *2,4,6,8-Decatetraenedioic acid mono-[5-methoxy-4-[2-methyl-3-(3-methyl-2-butenyl)oxiranyl]-1-oxaspiro[2.5]oct-6-yl] ester; 2,4,6,8-decatetraenedioic acid mono[4-(1,2-epoxy-1,5-dimethyl-4-hexenyl)-5-methoxy-1-oxaspiro[2.5]oct-6-yl] ester;* Amebacilin; Fugillin; Fumadil B; Fumidil. $C_{26}H_{34}O_7$; mol wt 458.53. C 68.10%, H 7.47%, O 24.42%. Antibiotic substance produced by *Aspergillus fumigatus:* Eble, Hanson, *Antibiot. & Chemother.* **1**, 55 (1951); Peterson *et al.,* U.S. pat. **2,803,586** (1957 to Abbott). Purification: Tarbell *et al., J. Am. Chem. Soc.* **77**, 5613 (1955). Structure: *eidem, ibid.* **82**, 1005 (1960); **83**, 3096 (1961). Stereochemistry: McCorkindale, Sime, *Proc. Chem. Soc.* **1961**, 331; Turner, Tarbell, *Proc. Nat. Acad. Sci. USA* **48**, 733 (1962). Biosynthesis: Birch, Hussain, *J. Chem. Soc. (C)* **1969**, 1473. Total synthesis: Corey, Snider, *J. Am. Chem. Soc.* **94**, 2549 (1972). Anti-amebic activity: M. C. McCowen *et al., Science* **113**, 202 (1951). Antineoplastic activity: J. A. DiPaolo *et al.,* "Studies on the Carcinolytic Activity of Fumagillin and Some of Its Derivatives" in *Anti-*

biotics Annual **1958-1959**, H. Welch, F. Marti-Ibaney, Eds. (Medical Encyclopedia Inc., New York, 1959) pp 541-546. Review: Girolami, "Fumagillin" in Kavanagh's Anal. Microbiol. (Academic Press, New York, 1963) pp 295-301; Wilson, "Miscellaneous Aspergillus Toxins" in Microbial Toxins, vol. VI, A. Ciegler et al., Eds. (Academic Press, New York, 1971) pp 277-281.

Yellow needles from methanol, mp 194-195°. $[\alpha]_D^{25}$ −26.6° (c = 1 in 95% ethanol). Absorptivity: 156.0 at 335 nm and 146.5 at 351 nm (soln of 100 mg in 10 ml chloroform diluted with alcohol to 0.0004% fumagillin and 0.04% chloroform). Practically insol in water, dil acids, satd hydrocarbons. Sol in most other organic solvents, in aq solns of bicarbonates and alkali hydroxides. Best stored in dark, evacuated ampuls at low temps. It loses its activity against *E. histolytica* and *Staph. phage* as well as its uv absorption when precautions are not taken. Stability data: Eble, Garrett, J. Am. Pharm. Assoc. **43**, 536 (1954); Garrett, ibid. 539. LD_{50} in mice: ~800 mg/kg s.c. (DiPaolo).

Methyl ester, $C_{27}H_{36}O_7$, crystals from dil methanol, mp 147-150°.

THERAP CAT: Formerly used as antiamebic.

THERAP CAT (VET): Antiprotozoal. Control of *Nosema apis* in honey bees.

4200. Fumaric Acid. *(E)-2-Butenedioic acid; trans-*1,2-ethylenedicarboxylic acid; allomaleic acid; boletic acid. $C_4H_4O_4$; mol wt 116.07. C 41.39%, H 3.47%, O 55.14%. Occurs in many plants, e.g., in *Fumaria officinalis* L., *Fumariaceae*, in *Boletus scaber* Bull., *Boletaceae*, and in *Fomes igniarius* (Fries) Kickx., *Polyporaceae*. Essential to vegetable and animal tissue respiration. Prepd industrially from glucose by the action of fungi such as *Rhizopus nigricans*: Foster, Waksman, J. Am. Chem. Soc. **61**, 127 (1939). Laboratory prepn by the oxidation of furfural with sodium chlorate in the presence of vanadium pentoxide: Milas, Org. Syn. coll. vol. II, 302 (1943). Molecular structure: J. L. Derissen, J. Mol. Struct. **38**, 177 (1977). Review: W. D. Robinson, R. A. Mount in Kirk-Othmer Encyclopedia of Chemical Technology vol. **14** (Wiley-Interscience, New York, 3rd ed., 1981) pp 770-793.

Monoclinic, prismatic needles or leaflets from water. d 1.625. Sublimes at 200°. Sublimes at 165° at 1.7 mm pressure. Partial carbonization and formation of maleic anhydride occur at 230° (open vessel). mp 287° (closed capillary, rapid heating). K_1 at 25°: 9.3 × 10⁻⁴; K_2: 2.9 × 10⁻⁵. Absorption spectrum: Macbeth, Stewart, J. Chem. Soc. **111**, 830 (1917). Soly in 100 g water at 25°: 0.63 g; at 40°: 1.07 g; at 60°: 2.4 g; at 100°: 9.8 g; in 100 g 95% alcohol at 30°: 5.76 g; in 100 g acetone at 30°: 1.72 g; in 100 g ether at 25°: 0.72 g. Almost insol in olive oil, chloroform, carbon tetrachloride, benzene, xylene, molten camphor, liq ammonia.

Monomethyl ester, $C_5H_6O_4$, prisms from alc, mp 144.5°. Dimethyl ester, $C_6H_8O_4$, crystals, mp 102°. Subl, bp 142°.

USE: Substitute for tartaric acid in beverages and baking powders; as a replacement or partial replacement for citric acid in fruit drinks. As an antioxidant. Manuf polyhydric alcohols, synthetic resins. As mordant in dyeing.

4201. Fumigatin. *3-Hydroxy-2-methoxy-5-methyl-2,5-cyclohexadiene-1,4-dione; 3-hydroxy-2-methoxy-5-methyl-p-benzoquinone;* 6-hydroxy-5-methoxy-p-toluquinone; 3-hydroxy-4-methoxy-2,5-toluquinone. $C_8H_8O_4$; mol wt 168.14. C 57.14%, H 4.79%, O 38.06%. Fungal toxin with antibiotic properties isolated from metabolism soln of *Asper-*

gillus fumigatus Fres: Anslow, Raistrick, Biochem. J. **32**, 687 (1938); Waksman, Geiger, J. Bact. **47**, 391 (1944). Synthesis: Baker, Raistrick, J. Chem. Soc. **1941**, 670; Posternak, Ruelius, Helv. Chim. Acta **26**, 2045 (1943); Seshadri, Venkatasubramanian, J. Chem. Soc. **1959**, 1660. Biosynthesis: Pettersson, Acta Chem. Scand. **17**, 1323 (1963); Simonart, Verachtert, Bull. Soc. Chim. Biol. **49**, 543 (1967). Formation and polarographic assay: J. Lafond-Grellety et al., Ann. Microbiol. **129B**, 3 (1978). Review: Wilson, "Miscellaneous Aspergillus Toxins," in Microbial Toxins vol. VI, A. Ciegler et al., Eds. (Academic Press, New York, 1971) p 281.

Maroon-colored needles or hexagonal plates from petr ether, mp 116°. Sublimes in vacuo. Volatile with steam. Sparingly sol in water, petr ether; freely sol in acetone, ether, chloroform, benzene, ethyl acetate, alcohol.

4202. Fungichromin. Antibiotic A 246; cogomycin; lagosin; pentamycin; Cantricin. $C_{35}H_{58}O_{12}$; mol wt 670.85. C 62.66%, H 8.71%, O 28.62%. Antifungal polyene macrolide antibiotic, related structurally to filipin, q.v. Isoln from *Streptomyces cellulosae*: A. A. Tytell et al., Antibiot. Ann. **1954-1955**, 716. Isoln (as pentamycin) from S. penticus: S. Umezawa, Y. Tanaka, J. Antibiot. **11A**, 26 (1958). Isoln (as lagosin) from S. roseoluteus: C. J. Bessel et al., U.S. pat. **3,013,947** (1961 to Glaxo). Isoln from S. griseus (FCRC-21): R. C. Pandey et al., Biomed. Mass. Spectrom. **7**, 93 (1980). Structures: A. C. Cope, H. E. Johnson, J. Am. Chem. Soc. **80**, 1504 (1958); A. C. Cope et al., ibid. **84**, 2170 (1962); M. L. Dhar et al., J. Chem. Soc. **1964**, 842; M. P. Berry, M. C. Whiting, ibid. 862; V. Pozgay et al., J. Antibiot. **29**, 472 (1976). Identity of fungichromin with lagosin and cogomycin and comparison of reported physico-chemical constants: R. C. Pandey et al., J. Antibiot. **35**, 988 (1982). Biosynthesis and NMR assignment: H. Noguchi et al., J. Am. Chem. Soc. **110**, 2938 (1988).

Light yellow cryst. mp 157-162° (dec). $[\alpha]_D^{20}$ −227.7° (c = 0.53 in DMF). uv max (methanol): 357, 338, 322 nm ($E_{1cm}^{1\%}$ 1231, 1250, 786). LD_{50} in mice (mg/kg): 1624 orally; 33.3 i.p. (Umezawa, Tanaka).

THERAP CAT: Topical antifungal.

4203. Fungisterol. *Ergosta-6,8,22-trien-3β-ol.* $C_{28}H_{44}O$; mol wt 396.63. C 84.78%, H 11.18%, O 4.03%. Formed from ergosterol by fungi. Isoln from *Penicillium chrysogenum*: Saito, J. Ferment. Technol. **29**, 457 (1951), C.A. **47**, 12507f (1953). Structure: idem, ibid. **31**, 328 (1953); **32**, 138, 140 (1954), C.A. **48**, 5276d (1954); **49**, 9009i (1955).

Crystals from alcohol + ether + ethyl acetate, mp 147.5°. $[\alpha]_D^{15}$ −21.9° (chloroform).
Acetate, $C_{30}H_{46}O_2$, crystals, mp 158.5°. $[\alpha]_D^{15}$ −15.7° (chloroform).

4204. Funtumine. *3α-Amino-5α-pregnan-20-one;* 3α-amino-20-oxo-5α-pregnane. $C_{21}H_{35}NO$; mol wt 317.50. C 79.44%, H 11.11%, N 4.41%, O 5.04%. Steroidal alkaloid isolated from *Funtumia latifolia* Stapf., *Apocynaceae:* Janot *et al., Compt. Rend.* **246**, 3076 (1958); from leaves of *Holorrhena febrifuga* Stapf., *Apocynaceae:* H. Dodoun *et al., Phytochemistry* **12**, 923 (1973). Structure: Janot *et al., Bull. Soc. Chim. France* **1960**, 1640, 1669. Prepn: H. Kapnang *et al., Tetrahedron Letters* **1977**, 3469. *Review:* R. Goutarel, *Bull. Soc. Chim. France* **1960**, 769. Effect on liver carcinogenesis in rats: A. Lacassagne *et al., Compt. Rend. Ser. D* **274**, 2830 (1972).

Prisms from ethyl acetate, mp 126°. $[\alpha]_D$ +95° (c = 1.7 in chloroform). pK 9.18. Sol in the usual organic solvents. LD_{50} i.v. in mice: 30 mg/kg (Lacassagne).

4205. Furaltadone. *5-(4-Morpholinylmethyl)-3-[[(5-nitro-2-furanyl)methylene]amino]-2-oxazolidinone; 5-morpholinomethyl-3-(5-nitrofurfurylideneamino)-2-oxazolidinone;* 3-(5-nitro-2-furfurylideneamino)-5-(4-morpholinomethyl)-2-oxazolidone; furmethonol; nitrofurmethone; NF 260; Altafur; Altabactina; Furazolin; Ibifur; Medifuran; Nitraldone; Otifuril; Sepsinol; Ultrafur; Unifur; Valsyn. $C_{13}H_{16}N_4O_6$; mol wt 324.29. C 48.15%, H 4.97%, N 17.28%, O 29.60%. Prepn: Gever, U.S. pat. **2,802,002** (1957 to Norwich).

Yellow crystals from 95% ethanol, dec 206°. Sparingly sol in water: about 75 mg/100 ml at 25°.
THERAP CAT: Antibacterial.
THERAP CAT (VET): Antibacterial.

4206. Furan. Furfuran; oxole; tetrole; divinylene oxide. C_4H_4O; mol wt 68.07. C 70.57%, H 5.92%, O 23.50%. Occurs in oils obtained by the distillation of rosin contg pine wood. Prepd by decarboxylation of 2-furancarboxylic acid: Wilson, *Org. Syn.* **coll. vol. I** (2nd ed., 1941), p 274. Has been prepd directly from furfural over hot soda-lime or by dropping furfural on a fused mixt of sodium and potassium hydroxides: Hurd *et al., J. Am. Chem. Soc.* **54**, 2532 (1932). Toxicity data: Henderson, *J. Pharmacol. Exp. Ther.* **57**, 394 (1936). Thermodynamic properties: G. B. Guthrie, Jr. *et al., J. Am. Chem. Soc.* **74**, 4662 (1952).

Liquid. $d_4^{19.4}$ 0.9371. bp_{760} 31.36°; bp_{758} 32°. n_D^{20} 1.4216. Flash pt, closed cup: −32°F (−35°C). Absorption spectrum: Purvis, *J. Chem. Soc.* **97**, 1648, 1655 (1910). Insol in water. Freely sol in alcohol and ether. Stable to alkalies; resinifies on evaporation or when in contact with mineral acids. Lethal concn for rats in air: 30400 ppm (Henderson). *Caution:* The vapors are anesthetic. Can be absorbed

through skin. *See: Toxicity of Industrial Metals,* E. Browning (Appleton-Century-Crofts, New York, 2nd ed., 1969) p 698.

4207. 2-Furanacrylic Acid. *3-(2-Furanyl)-2-propenoic acid;* β-2-furylacrylic acid; furacrylic acid; 2-furalacetic acid; furfurylidene acetic acid. $C_7H_6O_3$; mol wt 138.12. C 60.87%, H 4.38%, O 34.75%. Prepd by the condensation of furfural with malonic acid in the presence of pyridine. Laboratory procedure: Rajagopalan, Raman, *Org. Syn.* **25**, 51 (1945). Large scale procedure: Johnson, *ibid.* **20**, 55 (1940).

Needles from water, mp 141°. Sublimes in high vacuum at 112°. bp_{760} 286° (also given as 226°); bp_8 117°. Volatile in steam. One gram dissolves in 500 ml water at 15°. 100 ml of a satd benzene soln contains 1.14 g at 19°. Sol in alcohol, ether, glacial acetic acid.
Labile form, mp 104°. Can be converted to the stable form by exposure to sunlight in benzene soln contg some iodine.

4208. 2-Furanacrylonitrile. *3-(2-Furanyl)-2-propenenitrile;* 3-(2-furyl)acrylonitrile. C_7H_5NO; mol wt 119.12. C 70.58%, H 4.23%, N 11.76%, O 13.43%. Prepd from furfural and cyanoacetic acid in the presence of ammonium acetate and pyridine: Patterson, *Org. Syn.* **40**, 46 (1960).

Liquid. bp_{760} 95-97°. n_D^{25} 1.5824. Miscible with toluene, dimethylformamide.

4209. Furazabol. *17α-Methyl-5α-androstano[2,3-c]-[1,2,5]oxadiazol-17β-ol;* 17β-hydroxy-17α-methyl-5α-androstano[2,3-c]furazan; androfurazanol; furazalon; DH 245; Frazalon; Miotolon; Myotolon. $C_{20}H_{30}N_2O_2$; mol wt 330.47. C 72.69%, H 9.15%, N 8.48%, O 9.68%. Prepn: Ohta *et al.,* **Belg.** pat. **645,743** corresp to U.S. pat. **3,245,988** (1964, 1966 both to Daiichi Seiyaku); Shimizu *et al., Chem. Pharm. Bull.* **13**, 895 (1965); Ohta *et al., ibid.* 1445. Pharmacological studies: Kasahara *et al., ibid.* **13**, 1460 (1965); **16**, 1456, 1460 (1968). Toxicity: *eidem, ibid.* **14**, 285 (1966). Metabolic studies: Takegoshi *et al., ibid.* **20**, 1243 (1972).

Needles from methanol, mp 152-153°. $[\alpha]_D$ +39.4° (c = 1.42 in CHCl₃). uv max (ethanol): 217 nm (ε 4,300). LD_{50} in mice: 2.330 g/kg orally; >4 g/kg s.c.; 0.494 g/kg i.p.
THERAP CAT: Anticholesteremic.

4210. Furazolidone. *3-[[(5-Nitro-2-furanyl)methylene]-amino]-2-oxazolidinone; 3-(5-nitrofurfurylideneamino)-2-oxazolidinone;* N-(5-nitro-2-furfurylidene)-3-amino-2-oxazolidone; NF 180; Furovag; Furoxane; Furoxone; Giarlam; Giardil; Medaron; Neftin; Nicolen; Nifulidone; Ortazol; Roptazol; Tikofuran; Topazone. $C_8H_7N_3O_5$; mol wt 225.16. C 42.67%, H 3.13%, N 18.66%, O 35.53%. Prepn: Drake, Hayes, U.S. pat. **2,759,931**; Gever, O'Keefe, U.S. pat. **2,927,110** (1956, 1960 to Norwich). Antimicrobial spectrum and toxicity: Yurchenco *et al., Antibiot. & Chemother.* **3**, 1035 (1953); Rogers *et al., ibid.* **6**, 231 (1956). Pharmacology and toxicology: B. H. Ali, *Vet. Res. Commun.* **6**, 1 (1983).

Yellow crystals, dec 275°. Darkens under strong light. Soly in water (pH 6): approx 40 mg/l. Dec by alkali.

THERAP CAT: Topical anti-infective; topical antiprotozoal (Trichmonas).

THERAP CAT (VET): Antimicrobial.

4211. Furazolium Chloride. *6,7-Dihydro-3-(5-nitro-2-furanyl)-5H-imidazo[2,1-b]thiazol-4-ium chloride;* NF-963; Dermafur. $C_9H_8ClN_3O_3S$; mol wt 273.71. C 39.49%, H 2.94%, Cl 12.96%, N 15.35%, O 17.54%, S 11.72%. Prepn: **Neth. pat. Appl. 6,400,380** corresp to Snyder, Jr., **U.S. pat. 3,169,970** (1964, 1965, both to Norwich Pharmacal); Snyder, Jr., Benjamin, *J. Med. Chem.* **9,** 402 (1966).

Crystals from methanol, dec > 250°.
Free base, $C_9H_7N_3O_3S$, crystals from benzene + hexane, dec 171-172°.

THERAP CAT: Antibacterial.

4212. Furcellaran. Furcellaria gum; Danish agar; Burtonite 44. A gum obtained from a seaweed of the *Rhodophyceae,* the red alga *Furcellaria fastigiata,* fam. *Furcellariaceae,* order *Gigartinales.* The weed is found primarily in Northern European waters, especially in the Kattegat (between Sweden and Denmark). The gum is the potassium salt of the sulfuric acid ester of a high molecular weight polysaccharide. Consists mainly of D-galactose, 3,6-anhydro-D-galactose, and the half-ester sulfates of these sugars; one sulfate group occurs for each three or four monomeric units, which are arranged in an alternating sequence of $(1 \rightarrow 3)$ and $(1 \rightarrow 4)$-linked units. *Review:* Bjerre-Petersen *et al.,* in *Industrial Gums,* R. L. Whistler, Ed. (Academic Press, New York, 2nd ed., 1973) pp 123-136.

The processed gum is a white, odorless powder. Sol in hot or warm water. Easily dispersed in cold water to a homogeneous suspension without lumps; the furcellaran particles hydrate, swell and become almost invisible but do not dissolve unless heated. Forms agar-like gels at low concns. The strength of the gel can be increased by adding salts, esp potassium salts. Highly viscous. Solns in neutral medium are not adversely affected by prolonged exposure to high heat. However, exposure to heat in acidic media results in rapid hydrolysis and loss of gelling power.

USE: Natural colloid, gelling agent, viscosity control agent used primarily in food products but also in pharmaceuticals. Also in products for diabetics, proprietaries for reducing excess body wt, toothpastes. As carrier for food preservatives, bactericides. In bacteriological culture media.

4213. Furethidine. *4-Phenyl-1-[2-[(tetrahydro-2-furanyl)methoxy]ethyl]-4-piperidinecarboxylic acid ethyl ester; 4-phenyl-1-[2-(tetrahydrofurfuryloxy)ethyl]isonipecotic acid ethyl ester;* ethyl 1-(tetrahydrofurfuryloxyethyl)-4-phenylpiperidine-4-carboxylate; 1-(2'-tetrahydrofurfuryloxyethyl)-norpethidine. $C_{21}H_{31}NO_4$; mol wt 361.47. C 69.77%, H 8.65%, N 3.88%, O 17.71%. Prepn: Frearson, Stern, **Brit. pat. 797,448** (1958 to J. F. Macfarlan & Co.); Frearson *et al., J. Chem. Soc.* **1960,** 2103.

$bp_{0.5}$ 210°; $bp_{0.3}$ 175-183°. mp about 28°. n_D^{20} 1.5219. pKa 7.48.

Methiodide, $C_{22}H_{34}INO_4$, crystals from ethyl acetate, mp 174°.

Caution: May be habit forming. This is a controlled substance (opiate) listed in the U.S. Code of Federal Regulations, Title 21 Part 1308.11 (1985).

4214. Furfural. *2-Furancarboxaldehyde; 2-furaldehyde;* pyromucic aldehyde; artificial oil of ants; "furfurol". $C_5H_4O_2$; mol wt 96.08. C 62.50%, H 4.19%, O 33.30%. Occurs in some essential oils. Prepd industrially from pentosans which are contained in cereal straws and brans. Laboratory prepn from corncobs: R. Adams, V. Voorhees, *Org. Syn.* **coll. vol. I,** 280 (2nd ed., 1941). May also be prepd from pyridine.

Colorless oily liq. Peculiar odor, somewhat resembling the odor of benzaldehyde. Turns yellow to brown on exposure to air and light and resinifies (the polymerization is greatly accelerated by hot alkali). d_4^{25} 1.1563. bp_{760} 161.8°; bp_{100} 103°; bp_{20} 67.8°; $bp_{1.0}$ 18.5° (mp −36.5°). Volatile in steam. n_D^{20} 1.5261. Absorption spectrum: Purvis, *J. Chem. Soc.* **97,** 1655 (1910). Sol in 11 parts water; very sol in alcohol, ether. Flash pt, closed cup, 140°F (60°C); open cup, 155°F (68°C). Lower explosive limit: 2.1% by vol in air. Autoignition temp 797°F (392°C). *Keep in airtight container and protect from light.* LD_{50} orally in rats: 127 mg/kg, P. M. Jenner *et al., Food Cosmet. Toxicol.* **2,** 327 (1964).

USE: In the manufacture of furfural-phenol plastics such as Durite; in solvent refining of petroleum oils; in the prepn of pyromucic acid. As a solvent for nitrated cotton, cellulose acetate, and gums; in the manuf of varnishes; for accelerating vulcanization; as insecticide, fungicide, germicide; as reagent in analytical chemistry. In the synthesis of furan derivatives. *Caution:* Irritates mucous membranes and acts on CNS. Causes lacrimation, inflammation of eyes, irritation of throat, headache.

4215. Furfuryl Alcohol. *2-Furanmethanol;* 2-furylcarbinol; 2-furancarbinol; α-furylcarbinol; furfuralcohol; 2-hydroxymethylfuran. $C_5H_6O_2$; mol wt 98.10. C 61.21%, H 6.17%, O 32.62%. Usually prepd from furfural which is obtained from the processing of corncobs. The oil obtained by steam distillation of roasted coffee bean meal consists of 50% furfuryl alcohol after all organic acids have been removed. Laboratory prepn from furfural by the Cannizzaro reaction: Wilson, *Org. Syn. coll. vol. I* (2nd ed., 1941) p 276; *cf.* **U.S. pat. 2,041,184** (to Quaker Oats), *C.A.* **30,** 4515 (1936). Has been obtained by yeast reduction of furfural. Prepd industrially by the catalytic reduction of furfural using nickel and Cu-CrO catalysts: Peters, **U.S. pat. 1,906,873** (1933 to Quaker Oats); Wojcik, *Ind. Eng. Chem.* **40,** 210 (1948).

Liquid. Faint burning odor. Bitter taste. *Poisonous!* d_4^{23} 1.1282. bp_{760} 170°; bp_{400} 151.8°; bp_{200} 133.1°; bp_{100} 115.9°; bp_{60} 104.0°; bp_{40} 95.7°; bp_{20} 81.0°; bp_{10} 68.0°; bp_5 56.0°; $bp_{1.0}$ 31.8°. n_D^{23} 1.48515. Miscible with water, but unstable in water. Very sol in alcohol and ether. Easily resinified by acids. Spontaneous ignition temp +490° (915°F). Flash pt +75° (167°F). LD_{50} orally in rats: 275 mg/kg (fed as 2% aq soln), *Handbook of Toxicology* vol. I, W. S. Spector, Ed. (Saunders, Philadelphia, 1956) pp 146-147.

USE: Solvent; manuf wetting agents, resins.

4216. 5-Furfuryl-5-isopropylbarbituric Acid. *5-(2-Furanylmethyl)-5-(1-methylethyl)-2,4,6(1H,3H,5H)-pyrimidinetrione;* 5-isopropyl-5-furfurylbarbituric acid; 5-(2-furylmethyl)-5-isopropylbarbituric acid; Dormovit. $C_{12}H_{14}N_2O_4$; mol wt 250.25. C 57.59%, H 5.64%, N 11.20%, O 25.58%. Prepn: Heilner, **U.S. pat. 2,035,317** (1935 to Wiernik AG).

Crystals, mp 168-170°. Practically insol in water. Sol in hot water, alcohol, ether, chloroform. Aq soln of the sodium salt has a pH of 9.47.

Note: This is a controlled substance listed in the U.S. Code of Federal Regulations Title 21, Part 1308.13 (1987).

THERAP CAT: Sedative, hypnotic.

4217. Furfurylmethylamphetamine. *N-Methyl-N-(1-methyl-2-phenylethyl)-2-furanmethanamine; N-methyl-N-(α-methylphenethyl)furfurylamine; N-methyl-N-(1-phenyl-2-propyl)furfurylamine; Furfenorex.* $C_{15}H_{19}NO$; mol wt 229.31. C 78.56%, H 8.35%, N 6.11%, O 6.98%. Prepn of *d*- and *dl*-forms: Boissier, Ratouis, *Fr.* pats. **M3332** and **1,407,075** (both 1965 to S.I.F.A.), *C.A.* **63**, 13214g, 13215c (1965).

dl-Form, bp$_{0.1}$ 86-88°.

dl-Form cyclohexylsulfamate, $C_{21}H_{32}N_2O_4S$, *Frugalan.* Crystals from acetone + ether, mp 113-114°.

d-Form, bp$_{0.1}$ 103-106°. n_D^{21} 1.5295.

d-Form cyclohexylsulfamate. Crystals from acetone, mp 90-92°. $[\alpha]_D^{20}$ +12.7° (c = 1 in water).

THERAP CAT: Anorexic.

4218. α-Furildioxime. *Di-2-furanylethanedione dioxime.* $C_{10}H_8N_2O_4$; mol wt 220.18. C 54.55%, H 3.66%, N 12.72%, O 29.07%.

Monohydrate, needle-like crystals, mp 166-168°. Very sol in alcohol, ether, slightly sol in benzene, petr ether.

USE: As a reagent, in alcohol soln, for nickel with which it gives an orange-red compd.

4219. 2-Furoic Acid. *2-Furancarboxylic acid;* α-furoic acid; pyromucic acid; Brenzschleimsäure (German). $C_5H_4O_3$; mol wt 112.08. C 53.58%, H 3.60%, O 42.82%. Prepd from furfural by a Cannizzaro reaction: Wilson, *Org. Syn.* **coll. vol. I** (2nd ed., 1941) p 276; *cf.* U.S. pat. **2,041,184** (to Quaker Oats), *C.A.* **30**, 4515 (1936); Wojcik, *Ind. Eng. Chem.* **40**, 210 (1948); Harrisson, Moyle, *Org. Syn.* **coll. vol. IV**, 493 (1963). Formed in man and several animal species as a metabolite of furfural and related compds.

Elongated monoclinic prisms from water or by sublimation at 130-140° and 50-60 mm pressure. K at 25°: 7.6×10^{-4}. mp 133-134°. bp$_{760}$ 230-232°; bp$_{20}$ 141-144°. Absorption spectrum: Hartley, Dobbie, *J. Chem. Soc.* **73**, 600 (1898). One gram dissolves in 26 ml water at 15°; in 4 ml boiling water. More sol in alcohol; freely sol in ether.

Ethyl ester, $C_7H_8O_3$, *ethyl furoate, ethyl pyromucate.* Colorless crystals. d_4^0 1.117. mp 34-36°. bp$_{706}$ 195°. Insol in water; sol in alcohol.

Methyl ester, $C_6H_6O_3$, *methyl furoate, methyl pyromucate.*

Liquid. Agreeable odor. bp 181.3°; d_4^{21} 1.1786; n_D^{21} 1.4871. Slightly sol in water; sol in alcohol.

4220. Furonazide. *4-Pyridinecarboxylic acid [1-(2-furanyl)ethylidene]hydrazide; isonicotinic acid α-methylfurfurylidenehydrazide;* 2-furyl methyl ketone isonicotinoylhydrazone; α-methylfurfurylidenehydrazide of isonicotinic acid; INF; Furilazone; Clitizina; Menazone. $C_{12}H_{11}N_3O_2$; mol wt 229.23. C 62.87%, H 4.84%, N 18.33%, O 13.96%. Prepn: Miyatake *et al.*, *J. Pharm. Soc. Japan* **75**, 1066 (1955).

Crystals, mp 199-201.5°.

THERAP CAT: Antibacterial (tuberculostatic).

4221. Furosemide. *5-(Aminosulfonyl)-4-chloro-2-[(2-furanylmethyl)amino]benzoic acid; 4-chloro-N-furfuryl-5-sulfamoylanthranilic acid;* 4-chloro-N-(2-furylmethyl)-5-sulfamoylanthranilic acid; frusemide; fursemide; LB-502; Aisemide; Aluzine; Beronald; Desdemin; Discoid; Diural; Dryptal; Durafurid; Endural; Errolon; Eutensin; Frusemin; Frusetic; Frusid; Fulsix; Fuluvamide; Furesis; Furo-Puren; Furosemide "Mita"; Furosedon; Hydroled; Hydro-rapid; Katlex; Lasilix; Lasix; Lowpstron; Macasirool; Mirfat; Nicorol; Odemase; Oedemex; Profemin; Rosemide; Rusyde; Transit; Trofurit; Urosemide; Urex. $C_{12}H_{11}ClN_2O_5S$; mol wt 330.77. C 43.57%, H 3.35%, Cl 10.72%, N 8.47%, O 24.19%, S 9.70%. Prepn: Sturm *et al.*, U.S. pat. **3,058,882** and Ger. pat. **1,122,541** (1962 to Hoechst). Molecular and crystal structure: J. Lamotte *et al.*, *Acta Crystallogr.* **B34**, 1657 (1978). Review of pharmacokinetics: L. Z. Benet, *J. Pharmacokinet. Biopharm.* **7**, 1-27 (1979); R. E. Cutler, A. D. Blair, *Clin. Pharmacokinet.* **4**, 279-296 (1979). Pharmacodynamics and pharmacokinetics of a slow-release formulation in man: A. Ebihara *et al.*, *Arzneimittel-Forsch.* **33**, 163 (1983). Toxicity: E. I. Goldenthal, *Toxicol. Appl. Pharmacol.* **18**, 185 (1971).

Crystals from aq ethanol, mp 206°. Slightly sol in water, chloroform. Sol in acetone, methanol, DMF, aq solns above pH 8.0. Less sol in ethanol. *Pharmaceut. Incompat:* Calcium gluconate, ascorbic acid, tetracyclines, urea, epinephrine. LD$_{50}$ orally in female, male rats: 2600, 2820 mg/kg (Goldenthal).

THERAP CAT: Diuretic. Antihypertensive.

THERAP CAT (VET): Diuretic.

4222. Furoyl Chloride. *Furancarbonyl chloride.* $C_5H_3ClO_2$; mol wt 130.53. C 46.00%, H 2.32%, Cl 27.16%, O 24.52%. C_4H_3OCOCl.

Colorless liquid; mp −2°. bp 170°. Dec by water or alcohol. Sol in ether. *Caution:* Vapors irritate the eyes.

4223. Fursultiamine. *N-[(4-Amino-2-methyl-5-pyrimidinyl)methyl]-N-[4-hydroxy-1-methyl-2-[[(tetrahydro-2-furanyl)methyl]dithio]-1-butenyl]formamide;* thiamine tetrahydrofurfuryl disulfide; Alinamin F; Diteftin; Judolor; TTFD. $C_{17}H_{26}N_4O_3S_2$; mol wt 398.56. C 51.23%, H 6.58%, N 14.06%, O 12.04%, S 16.09%. Prepn: S. Yurugi, T. Fushimi, U.S. pat. **3,016,380** (1962 to Takeda). Metabolism: S. Kikuchi *et al.*, *Eur. J. Pharmacol.* **9**, 367 (1970); C. Mitoma, *Drug Metab. Dispos.* **1**, 698 (1973). Series of papers on function of tetrahydrofurfuryl mercaptan moiety: *J. Biochem.* **74**, 717 (1973). Crystal structure: W. Shin, Y. C. Kim, *Bull. Kor. Chem. Soc.* **7**, 331 (1986).

Colorless prisms from ethyl acetate, mp 132° (dec). d 1.29. Sparingly sol in water; sol in organic solvents, dil mineral acids. LD_{50} in rats (mg/kg): 2200 orally; 540 i.p. (Yurugi, Fushimi).

THERAP CAT: Enzyme co-factor vitamin.

4224. Furterene. *6-(2-Furanyl)-2,4,7-pteridinetriamine; 2,4,7-triamino-6-(2-furyl)pteridine;* furamterene; F 6113. $C_{10}H_9N_7O$; mol wt 243.24. C 49.38%, H 3.73%, N 40.31%, O 6.58%. Prepn of the formate: **Fr.** pat. M2206 (1964 to Labs. Lumière), *C.A.* **61**, 670e (1964), corresp to **Brit.** pat. 988,481.

Formate, *Furedeme.* Crystals from aq formic acid, mp 283-284°.

THERAP CAT: Diuretic.

4225. Furtrethonium. *N,N,N-Trimethyl-2-furanmethanaminium; furfuryltrimethylammonium;* trimethylfurfuryl-ammonium; furtrimethonium; Furmethide. $[C_8H_{14}NO]^+$; mol wt 140.20. Prepn of salts: Nabenhauer, **U.S.** pat. 2,185,220 (1938 to SK & F); Weilmuenster, Jordan, *J. Am. Chem. Soc.* **67**, 415 (1945); Khromov-Borisov *et al., J. Gen. Chem. USSR* **24**, 2021 (1954).

Iodide, $C_8H_{14}INO$, *Furamon, Furanol.* Crystals from ethanol + ethyl acetate, mp 116-117° (Weilmuenster, Jordan, *loc. cit.*), mp 118-120° (Khromov-Borisov *et al., loc. cit.*). Soluble in water, alcohol. Practically insol in benzene. pH of 1% aq soln, 5.3-6.0.

Benzenesulfonate, $C_{14}H_{19}NO_4S$, *Benzamon.* Crystals from ethanol + butyl acetate, mp 134-135° (Kromov-Borisov *et al., loc. cit.*). Soluble in water, alcohol.

p-Toluenesulfonate, $C_{15}H_{21}NO_4S$, *Fisostina.*

THERAP CAT: Cholinergic. Iodide as parasympathomimetic.

4226. Fusafungine. S 314; Biofusal; Fusaloyos; Fusarine; Locabiotal. Antibiotic used as an aerosol for local application. Isoln from *Fusarium* species: Couchoud, **Fr.** pat. 1,164,181 (1958), *C.A.* **54**, 20074b (1960); Servier, **Belg.** pat. 612,474 (1962), *C.A.* **57**, 11320b (1962); *idem,* **Brit.** pat. 1,018,626 (1966 to Biofarma), *C.A.* **64**, 16585b (1966). Bacteriological activity: D. Haler, *J. Int. Med. Res.* **5**, 61 (1977); *idem, Vie Med.* **61**, 507 (1980).

Solid, mp 125-129°. Stable up to 180°. Practically insol in water; sol in glycols and fats.

THERAP CAT: Antibacterial.

4227. Fusaric Acid. *5-Butyl-2-pyridinecarboxylic acid; 5-butylpicolinic acid.* $C_{10}H_{13}NO_2$; mol wt 179.22. C 67.02%, H 7.31%, N 7.82%, O 17.85%. Antibiotic (wilting agent) first isolated from the fungus *Fusarium heterosporium*, Nees: Yabuta *et al., J. Agr. Chem. Soc. Japan* **10**, 1059 (1934). Isoln from other *Fusarium* species and from *Gibberella fujikuroi* and synthesis: Plattner *et al., Helv. Chim. Acta* **37**, 1379 (1954). Prepn: Hardegger, Nikles, *ibid.* **39**, 505 (1956); **40**, 2428 (1957); Schreiber, Adam, *Ber.* **93**, 1848 (1960); Umezawa, Nagatsu, **Ger.** pat. 2,005,255 (1970 to Microbiochem. Res. Found.); R. Tschesche, W. Führer, *Ber.* **111**, 3502 (1978). Dopamine β-hydroxylase inhibitor and hypotensive activity: Suda *et al., Chem. Pharm. Bull.* **17**,

2377 (1969); Nagatsu *et al., Biochem. Pharmacol.* **19**, 35 (1970).

Colorless crystals, mp 96-98°. LD_{50} orally in mice: 230 mg/kg, Ishii *et al., Arzneimittel-Forsch.* **25**, 55 (1975). Copper salt, bluish-violet cryst from water, mp 258-259°.

4228. Fusarubin. *3,4-Dihydro-3,6,9-trihydroxy-7-methoxy-3-methyl-1H-naphtho[2,3-c]pyran-5,10-dione; 5,8-dihydroxy-2-(hydroxymethyl)-6-methoxy-3-(2-oxopropyl)-1,4-naphthalenedione; 3-acetonyl-5,8-dihydroxy-2-(hydroxymethyl)-6-methoxy-1,4-naphthoquinone;* oxyjavanicin. $C_{15}H_{14}O_7$; mol wt 306.26. C 58.82%, H 4.61%, O 36.57%. Isoln from *Fusarium solani:* Ruelius, Gauhe, *Ann.* **569**, 38 (1950). Structure: Hardegger *et al., Helv. Chim. Acta* **47**, 2027 (1964). Efficient total synthesis: Y. Tanoue *et al., Bull. Chem. Soc.* **60**, 2927 (1987).

Red prisms from benzene, dec 218°. Absorption max (ether): 535, 499 nm. Sol in glacial acetic acid, tetrahydrofuran, acetone, dioxane, pyridine; slightly sol in cold chloroform, cold alcohol, ether. Practically insol in carbon disulfide, cyclohexane, cold benzene. Dissolves in dil NaOH with a violet color. Practically insol in bicarbonate soln.

4229. Fuscin. *9,10-Dihydro-5-hydroxy-4,8,8-trimethyl-2H,4H-benzo[1,2-b:4,3-c']dipyran-2,6(8H)-dione.* $C_{15}H_{16}O_5$; mol wt 276.28. C 65.21%, H 5.84%, O 28.95%. Antibacterial pigment produced by the fungus *Oidiodendron fuscom* Robak: Michael, *Biochem. J.* **42**, XL; **43**, 528 (1948). Derivatives and degradation products: Birkinshaw *et al., ibid.* **48**, 66 (1951). Structure and synthesis: Birch, *Chem. & Ind. (London)* **1955**, 682; Barton, Hendrickson, *J. Chem. Soc.* **1956**, 1028. Alternate synthesis: Pyuskyulev *et al., Tetrahedron* **29**, 2849 (1973). Biosynthesis: Birch, *Ciba Foundation Symposium, Quinones Electron Transport* **1960**, 233; Birch *et al., J. Chem. Soc.* **1965**, 1231.

Orange-colored, diamond-shaped plates from alc, mp 230°. uv max: 355, 283 nm (ε 27,600; 1000). Practically insol in water, petr ether. Sol in chloroform, acetone, ethyl acetate, acetic acid, ether, benzene. Slightly sol in cold ethanol, more sol in hot ethanol. Dissolves in aq solns of alkali hydroxides and pyridine with an intense purple color. Easily reduced by hydrosulfite or other reducing agents to the colorless dihydrofuscin, $C_{15}H_{18}O_5$, mp 206°, which is also found in the metabolic soln from the fungus grown on Czapek-Dox medium.

4230. Fusel Oil. A by-product of carbohydrate fermentations to produce ethyl alc. The material varies widely in composition, depending on the fermentation raw material used, but contains chiefly isopentyl alcohol and 2-methyl-1-butanol as well as isobutyl alcohol (20%), n-propyl alcohol

(3-5%), and small amounts of other alcohols, esters and aldehydes. Described as an oily liq with a disagreeable odor; 60% boils at 122-138°. *Amyl alcohol (commercial)* obtained by chemical treatment and refining of fusel oil contains about 85% isopentyl alcohol and 15% 2-methyl-1-butanol. *Ref: Industrial Chemicals,* W. L. Faith *et al.,* Eds. (John Wiley, New York, 2nd ed., 1957) pp 107-114.

Caution: Commercial amyl alcohol is considerably more toxic than ethyl alcohol. *See also* 1-Pentanol and Isopentyl Alcohol.

4231. Fusidic Acid. *(3α,4α,8α,9β,11α,13α,14β,16β,17Z)-16-(Acetyloxy)-3,11-dihydroxy-29-nordammara-17(20),24-dien-21-oic acid; 3α,11α,16β-trihydroxy-29-nor-8α,9β,13α,-14β-dammara-17(20),24-dien-21-oic acid 16-acetate; 3α,-11α,16β-trihydroxy-4α,8,14-trimethyl-18-nor-5α,8α,9β,-13α,14β-cholesta-17(20),24-dien-21-oic acid 16-acetate; 3,11,16-trihydroxy-4,8,10,14-tetramethyl-17-(1'-carboxy-isohept-4'-enylidene)cyclopentanoperhydrophenanthrene 16-acetate; ramycin; Fucithalmic.* $C_{31}H_{48}O_6$; mol wt 516.69. C 72.06%, H 9.36%, O 18.58%. Antibiotic substance isolated from the fermentation broth of *Fusidium coccineum.* Isoln and structure: W. O. Godtfredsen *et al., Nature* **193,** 987 (1962); *Lancet* **I,** 928 (1962); W. O. Godtfredsen, S. Vangedal, *Tetrahedron* **18,** 1029 (1962). Identity with ramycin: H. Vanderhaeghe *et al., Nature* **205,** 710 (1965). Structure: D. Arigoni *et al., Experientia* **19,** 521 (1963). Stereochemistry: W. O. Godtfredsen *et al., Tetrahedron* **21,** 3505 (1965). Synthetic studies: W. G. Dauben *et al., J. Am. Chem. Soc.* **94,** 8593 (1972); R. E. Ireland, U. Hengartner, *ibid.* 3652; M. Tanabe *et al., Tetrahedron Letters* **1977,** 1481. Total synthesis: W. G. Dauben *et al., J. Am. Chem. Soc.* **104,** 303 (1982). *Review:* Tanaka in *Antibiotics* vol. 3, J. W. Corcoran, F. E. Hahn, Eds. (Springer-Verlag, New York, 1975) pp 436-447; W. von Daehne *et al., Advan. Appl. Microbiol.* **25,** 95-146 (1979).

Crystals from ether or benzene, mp 192-193°. $[\alpha]_D^{20}$ −9° (chloroform). uv max: 204 nm (ε 9900). pK: 5.35 in water. Sol in alc, acetone, chloroform, pyridine, dioxane; sparingly sol in water, ether, hexane. LD_{50} in mice (g/kg): 1.2 s.c.; 1.5 orally (Godtfredsen).

Sodium salt, $C_{31}H_{47}NaO_6$, *sodium fusidate, ZN 6, Fucidin, Fucidina, Fucidine, Fucidin Intertulle.* Crystals, sol in water. LD_{50} in mice (g/kg): 0.2 i.v. (Godtfredsen, *Nature* 1962).

THERAP CAT: Antibacterial.

4232. Fustin. *(trans)-2-(3,4-Dihydroxyphenyl)-2,3-dihydro-3,7-dihydroxy-4H-1-benzopyran-4-one; 3,3',4',7-tetrahydroxyflavanone;* dihydrofisetin. $C_{15}H_{12}O_6$; mol wt 288.25. C 62.50%, H 4.20%, O 33.30%. From wood of *Rhus cotinus* L. (Venice sumac) and *R. succedanea* L., *Anacardiaceae:* Schmid, *Ber.* **19,** 1734 (1886); from *Gleditsia triacanthos* L., *Leguminosae:* Chadenson *et al., Compt. Rend.* **249,** 1362 (1955). Structure: Oyamada, *Ann.* **538,** 44 (1939). Stereochemistry: Weinges, *Ann.* **627,** 229 (1959); Roux, Paulus, *Biochem. J.* **77,** 315 (1960); Gaffield, *Tetrahedron* **26,** 4093 (1970).

(±)-Form, crystals, mp 226-228°. $[\alpha]_D^{21}$ −2.4°.
Tetraacetate, $C_{23}H_{20}O_{10}$, crystals from methanol, mp 147-148°.
(−)-Form, crystals, mp 216-218°. $[\alpha]_D^{25}$ −26° (c = 2 in 1:1 acetone, water).
Tetraacetate, crystals, mp 117-118°. $[\alpha]_D^{23}$ −25.2° (c = 0.8 in tetrachloroethane).
(+)-Form, needles from water, mp 228-229°. $[\alpha]_D^{23}$ +28.3° (c = 0.9 in 1:1 acetone, water).
Tetraacetate, crystals from ethanol, mp 116-119°. $[\alpha]_D^{25}$ +24.4° (c = 1.3 in tetrachloroethane).

4233. Fyrol FR-2®. *1,3-Dichloro-2-propanol phosphate (3:1);* phosphoric acid tris(1,3-dichloro-2-propyl)ester; tris(1,3-dichloroisopropyl)phosphate; tris[2-chloro-1-(chloromethyl)ethyl]phosphate; PF 38; TCPP; Emulsion 212. $C_9H_{15}Cl_6O_4P$; mol wt 430.88. C 25.09%, H 3.51%, Cl 49.36%, O 14.85%, P 7.19%. Flame retardant formerly used in children's sleepwear; once considered as a potential replacement for Tris-BP, *q.v.* Prepn: W. J. Jones *et al., J. Chem. Soc.* **1946,** 824. Use in flameproofing: R. J. Polacek, U.S. pat. **3,041,293** (1962 to Celanese). Thermal studies: K. Paciorek *et al., Am. Ind. Hyg. Assoc. J.* **39,** 633 (1978); N. Inagaki *et al., J. Appl. Polym. Sci.* **21,** 217 (1977); **24,** 1 (1979). Mutagenicity studies: M. D. Gold *et al., Science* **200,** 785 (1978); A. Nakamura *et al., Mutat. Res.* **66,** 373 (1979); D. Brusick *et al., J. Environ. Pathol. Toxicol.* **3,** 207 (1979).

Viscous liquid, bp_5 236-237°. n_D^{20} 1.5022. Soly in water ~100 ppm. LD_{50} orally in rats: 1.85 g/kg, J. K. Piotrowski *et al., C.A.* **85,** 104825u (1976).

USE: Flame retardant.

G

4234. Gabexate. *4-[[6-[(Aminoiminomethyl)amino]-1-oxohexyl]oxy]benzoic acid ethyl ester; p-hydroxybenzoic acid ethyl ester 6-guanidinohexanoate;* p-carbethoxyphenyl ε-guanidinocaproate. $C_{16}H_{23}N_3O_4$; mol wt 321.38. C 59.80%, H 7.21%, N 13.07%, O 19.91%. Non-peptide proteolytic enzyme inhibitor which also inhibits the hydrolytic effects of thrombin, plasmin, and kallikrein, but not chymotrypsin; *cf.* aprotinin. Prepn (as the p-toluenesulfonate salt): S. Fujii, T. Watanabe, **Ger. pat.** 2,050,484; *eidem*, **U.S. pat.** 3,751,447 (1971, 1973 both to Ono). Inhibitory effects on trypsin, plasmin, plasma kallikrein, thrombin: M. Muramatu, S. Fujii, *Biochim. Biophys. Acta* **268**, 221 (1972); S. Tamura *et al., ibid.* **484**, 417 (1977). Pharmacology: T. Okegada *et al., Nippon Yakurigaku Zasshi* **71**, 71 (1975), *C.A.* **84**, 218m (1976). Metabolism: M. Sugiyama *et al., Oyo Yakuri* **9**, 733 (1975), *C.A.* **83**, 188145s (1975). Metabolism of inhibitory effect on platelet aggregation: G. Kosaki *et al., Thromb. Res.* **20**, 587 (1980). Beneficial action in traumatic shock: A. M. Lefer *et al., IRCS Med. Sci.: Libr. Compend.* **8**, 278 (1980). Effect on plasma proteases in acute pancreatitis: S. Takasugi, N. Toki, *Hiroshima J. Med. Sci.* **29**, 189 (1980), *C.A.* **94**, 185445 (1981). Teratology and toxicity study: T. Fujita *et al., Oyo Yakuri* **9**, 743 (1975), *C.A.* **83**, 188322x (1975).

COOC₂H₅ / OOC—(CH₂)₅—NH—C=NH / NH₂

Monomethanesulfonate, $C_{17}H_{27}N_3O_7S$, *gabexate mesylate, Foy.* White crystals. Sol in water, ethanol, chloroform. Slightly sol in acetone. Practically insol in ether. pH of soln (1:100): 4.0-5.0. LD_{50} in mice (mg/kg): 8000 orally; 4700 s.c.; 25 i.v. (Fujita).

THERAP CAT: Enzyme inhibitor (proteinase).

4235. Gadolinium. Gd; at. wt 157.25; at. no. 64; valence 3. A lanthanide; belongs to yttrium group of rare earths. Seven naturally occurring isotopes: 152 (0.20%); 154 (2.15%); 155 (14.7%); 156 (20.47%); 157 (15.68%); 158 (24.9%); 160 (21.9%); ¹⁵²Gd is radioactive, $T_{1/2}$ 1.1 × 10¹⁴ years, α-emission. Artificial, radioactive isotopes: 145-151; 153; 159; 161; 162. Abundance in earth's crust: 4.5-6.4 ppm. Sources: samarskite, gadolinite (ytterbite), xenotime, and other rare earth minerals. Prepn of the metal by electrolysis of a fused mixture of gadolinium, potassium and lithium chlorides at 625-675°: Trombe, *Compt. Rend.* **200**, 459 (1935); Trombe, *Bull. Soc. Chim. France* **2**, 660 (1935); by heating the chloride with alkali metal: Klemm, Bommer, *Z. Anorg. Chem.* **231**, 138 (1937). Sepn from other rare earths: Spedding *et al., J. Am. Chem. Soc.* **69**, 2812 (1947); **76**, 2557 (1954). Spectrum: Albertson, *Phys. Rev.* **47**, 370 (1935); Spedding *et al., J. Chem. Phys.* **5**, 33 (1937). Reviews of prepn, properties and compds of gadolinium and other rare earths: *The Rare Earths,* F. H. Spedding, A. H. Daane (Krieger, Huntington, N.Y., 1971, reprint of 1961 ed.) 641 pp; Hulet, Bode, "Separation Chemistry of the Lanthanides and Transplutonium Actinides" in *MTP Int. Rev. Sci: Inorg. Chem., Ser. One,* vol. 7, K. W. Bagnall, Ed. (University Park Press, Baltimore, 1971) pp 1-45; Moeller, "The Lanthanides" in *Comprehensive Inorganic Chemistry,* vol. 4, J. C. Bailar, Jr. *et al.* (Pergamon Press, Oxford, 1973) pp 1-101.

Colorless or faintly yellowish metal; tarnishes in moist air. Hexagonal close-packed structure at room temp; d 7.886. mp 1312°. E°(aq) Gd³⁺/Gd −2.4 V (calc) Experimental reduction potentials (referred to a normal calomel electrode): −1.810, −1.955 V: Noddack, Brukl, *Angew. Chem.* **50**, 362 (1937).

Oxide, Gd_2O_3, *gadolinia.* Colorless, hygroscopic powder; prepd by igniting the hydroxide, nitrate, carbonate or oxalate, d¹⁵ 7.407, absorbs CO_2 from the air.

Hydroxide, $Gd(OH)_3$, gelatinous precipitate, prepd by the action of alkali or ammonium hydroxide on a soln of a gadolinium salt. Absorbs CO_2 from the air.

Chloride, $GdCl_3$, white monoclinic crystals, prepd by heating the oxide with excess of ammonium chloride above 200°. d⁰ 4.52. mp about 609°. Sol in water; forms double salts with platinic and auric chlorides. A hexahydrate, $GdCl_3.6H_2O$, deliquesc crystals, d 2.424, is obtained from the aq soln. LD_{50} in mice: 550 mg/kg i.p.; > 2000 mg/kg orally, Haley, *J. Pharm. Sci.* **54**, 663 (1965).

Sulfate, $Gd_2(SO_4)_3$, octahydrate, colorless monoclinic crystals. Soly in water decreases with rise in temp. On heating at 400° yields the anhydr sulfate, d 4.139; begins to dec at 500°.

Nitrate, $Gd(NO_3)_3$, hexahydrate, deliquesc triclinic crystals. d 2.332. mp 91°. Sol in water, in alcohol. A pentahydrate, prismatic crystals, mp 92°, d 2.406, very insol, has been prepd. LD_{50} (hexahydrate) in rats: 230 mg/kg i.p.; > 5000 mg/kg orally, Haley, *loc. cit.*

USE: Oxide in control rods of some nuclear power reactors.

4236. Gadopentetic Acid. *[N,N-Bis[2-[bis(carboxymethyl)amino]ethyl]glycinato(5−)]gadolinate(2−) dihydrogen; N,N-bis[2-[bis(carboxymethyl)amino]ethyl]glycine gadolinium complex; gadolinium diethylenetriaminepentaacetic acid;* Gd-DTPA. $C_{14}H_{20}GdN_3O_{10}$; mol wt 547.58. C 30.71%, H 3.68%, Gd 28.72%, N 7.67%, O 29.22%. Evaluation as water-soluble paramagnetic relaxation reagent (PARR) for NMR spectrometry: J. J. Dechter, G. C. Levy, *J. Magnet. Reson.* **39**, 207 (1980); and prepn: T. J. Wenzel *et al., Anal. Chem.* **54**, 615 (1982). Use as diagnostic medium for NMR imaging: H. Gries *et al.,* **Ger. pat.** 3,129,906; *eidem,* **U.S. pat.** 4,647,447 (1983, 1987 both to Schering AG). Physicochemical properties: H. Gries, H. Miklautz, *Physiol. Chem. Phys. Med. NMR* **16**, 105 (1984); E. Roux, L. De Broe, *J. Belg. Radiol.* **71**, 31 (1988). Evaluation as NMR contrast-enhancing agent in animals: R. C. Brasch *et al., Am. J. Roentgenol.* **142**, 625 (1984). Pharmacokinetics and toxicity: H. J. Weinmann *et al., ibid.* 619; in humans: H. J. Weinmann *et al., Physiol. Chem. Phys. Med. NMR* **16**, 167 (1984). HPLC determn: M. M. Vora *et al., J. Chromatog.* **369**, 187 (1986). Exptl use in imaging reperfused myocardium in dogs: S. Schaefer *et al., J. Am. Coll. Cardiol.* **12**, 1064 (1988). Clinical studies of diagnostic use in brain imaging: R. Felix *et al., Radiology* **156**, 681 (1985); J. P. Stack *et al., Neuroradiology* **30**, 145 (1988). Reviews of diagnostic use in intracranial lesions: J. R. Hesselink, G. A. Press, *Radiol. Clin. North Am.* **26**, 873-887 (1988); in spinal disease: G. Sze, *ibid.* 1009-1024.

⁻OOCCH₂ / Gd³⁺ / CH₂COO⁻ / NCH₂CH₂NCH₂CH₂N / HOOCCH₂ / CH₂COO⁻ / CH₂COOH

Dimeglumine salt, $C_{28}H_{54}GdN_5O_{20}$, *SHL 451 A, ZK 93035, Magnevist, Resovist.* Feely sol in water. LD_{50} i.v. in rats: 10 mmol/kg (Weinmann).

THERAP CAT: Diagnostic aid (NMR contrast medium).

4237. Galactaric Acid. Mucic acid; galactosaccharic acid; tetrahydroxyadipic acid; saccharolactic acid; Schleimsäure (German). $C_6H_{10}O_8$; mol wt 210.14. C 34.29%, H 4.80%, O 60.91%. Prepd by oxidation of lactose and of galactose: Kent, Tollens, *Ann.* **227**, 221 (1885); Maurer, Drefahl, *Ber.* **75B**, 1489 (1942). Manuf from wood sawdust: Acree, **Brit. pat.** 160,777 (1921). Review: B. A. Lewis *et al.,* "Galactaric Acid and Its Derivatives" in Whistler, Wolfrom, *Methods in Carbohydrate Chemistry,* vol. II (Academic Press, New York, 1963) pp 38-46.

$$HOOC-\overset{H}{\underset{OH}{C}}-\overset{OH}{\underset{H}{C}}-\overset{OH}{\underset{H}{C}}-\overset{H}{\underset{OH}{C}}-COOH$$

Cryst powder, dec about 255° when rapidly heated, also reported as 225°. Soluble in 300 parts cold water, 60 parts boiling water, alkalies; practically insol in alcohol, ether.

Ammonium salt, $(NH_4)_2C_6H_8O_8$, acicular crystals. Soluble in water.

USE: Has been proposed to replace potassium bitartrate in baking powder and for manuf of granular effervescing salts.

4238. Galactitol. Dulcitol; dulcite; dulcose; euonymit; melampyrite; melampyrum; melampyrin. $C_6H_{14}O_6$; mol wt 182.17. C 39.56%, H 7.75%, O 52.70%. Found in dulcite or Madagascar manna (*Melampyrum nemorosum* L.) and in other species of *Melampyrum, Scrophulariaceae*, and *Evonymus atropurpureus* Jacq., *Celastraceae*. Isoln: Hünefeld, *Ann.* **24**, 241 (1837); Bouchardat, *Ann. Chim. Phys.* **27**, 68 (1872); Fischer, Hertz, *Ber.* **25**, 1261 (1892); Rogerson, *J. Chem. Soc.* **101**, 1040 (1912). Prepn by catalytic isomerization of D-glucitol: Wright, Hartmann, *J. Org. Chem.* **26**, 1588 (1961). Synthesis: Lespieau, *Bull. Soc. Chim. France* [5] **1**, 1374 (1934); Delepine, Horeau, *ibid.* **4**, 1524 (1937); Wiemann, Gordon, *ibid.* **1958**, 433. Structure: R. L. Lohmar "The Polyols" in W. Pigman, *The Carbohydrates* (Academic Press, New York, 1957) p 247.

$$HOH_2C-\overset{H}{\underset{OH}{C}}-\overset{OH}{\underset{H}{C}}-\overset{OH}{\underset{H}{C}}-\overset{H}{\underset{OH}{C}}-CH_2OH$$

Crystals from methanol + water, mp 188-189°. Slightly sweet taste. d^{20} 1.47. bp_1 275-280°. One gram dissolves in 30 ml water, in 2 ml boiling water. Slightly sol in alc. Ka at 18° = 3.5 × 10⁻¹⁴.

Hexa-O-acetylgalactitol, $C_{18}H_{26}O_{12}$, crystals from ethanol, mp 168-169°.

Hexanitrate, *nitrodulcitol*, mp 94-95°. Has explosive properties: Taylor, Rinkenbach, *J. Franklin Inst.* **204**, 374 (1927).

4239. Galactoflavin. *1-Deoxy-1-(3,4-dihydro-7,8-dimethyl-2,4-dioxobenzo[g]pteridin-10(2H)-yl)-D-galactitol; 7,8-dimethyl-10-(D-galacto-2,3,4,5,6-pentahydroxyhexyl)benzo[g]pteridine-2,4(3H,10H)-dione; 7,8-dimethyl-10-(D-galacto-2,3,4,5,6-pentahydroxyhexyl)isoalloxazine; 7,8-dimethyl-10-(d-1'-dulcityl)isoalloxazine; 6,7-dimethyl-9-(d-1'-dulcityl)isoalloxazine; 6,7-dimethyl-9-(1-deoxy-D-galactitol-1-yl)isoalloxazine.* $C_{18}H_{22}N_4O_7$; mol wt 406.39. C 53.20%, H 5.46%, N 13.79%, O 27.56%. Prepd from 1-deoxy-1-(3,4-dimethyl-6-phenylazo)anilino-D-galactitol and barbituric acid: Berezovskii, Eremenko, *Zh. Obshch. Khim.* **32**, 4056 (1962), *C.A.* **59**, 736b (1963). Structure: Emerson *et al., J. Biol. Chem.* **160**, 165 (1945). Pharmacology: Lane, Brindley, *Proc. Soc. Exp. Biol. Med.* **116**, 57 (1964). Produces congenital malformations in animals: Nelson *et al., J. Nutr.* **58**, 125 (1956); Miller *et al., J. Biol. Chem.* **237**, 968 (1962); Mackler, *Pediatrics* **43**, 915 (1969).

Yellow crystals, dec 260°. Absorption max: 223, 267,

370, 445 nm (ε 2730, 28100, 9100, 10800). Compd has yellow-green fluorescence in water.

USE: Riboflavin antagonist.

4240. D-Galactosamine. *2-Amino-2-deoxy-D-galactose; chondrosamine;* GalN. $C_6H_{13}NO_5$; mol wt 179.17. C 40.22%, H 7.31%, N 7.82%, O 44.65%. Amino sugar isolated from chondroitin sulfate, *q.v.*: P. A. Levene, F. B. La Forge, *J. Biol. Chem.* **18**, 123 (1914). Sepn of α- and β-anomers: P. A. Levene, *ibid.* **57**, 337 (1923). Synthesis: S. P. James *et al., Nature* **156**, 308 (1945) *eidem, J. Chem. Soc.* **1946**, 625; R. Kuhn, W. Kirschenlohr, *Ann.* **600**, 126 (1956); P. A. Gent *et al., J. Chem. Soc. Perkin Trans. 1* **1972**, 277. Chemistry: D. Horton in *The Amino Sugars* Vol. 1A, R. W. Jeanloz, Ed. (Acdemic, New York, 1969) pp 133-145. Inducer of exptl hepatitis: D. Keppler *et al., Exp. Mol. Pathol.* **9**, 279 (1968); K. Decker, D. Keppler in *Progress in Liver Diseases* Vol. IV, H. Popper, F. Schaffner, Eds. (Grune & Stratton, New York, 1972) p 183. Powerful inhibitor of hepatic RNA synthesis: D. Keppler *et al., J. Biol. Chem.* **249**, 211 (1974); T. Anukarahanonta *et al., Eur. J. Cancer* **16**, 1171 (1980).

α-form

Hydrochloride, $C_6H_{14}ClNO_5$, crystals, mp 180° (dec). Shows mutarotation. α-Form: $[\alpha]_D^{23}$ +124° → +93° (water). β-Form: $[\alpha]_D^{23}$ +47° → 93° (water).

4241. Galactose. Cerebrose; brain sugar. $C_6H_{12}O_6$; mol wt 180.16. C 40.00%, H 6.72%, O 53.29%. Constituent of many oligo- and polysaccharides occurring in pectins, gums, and mucilages. Prepn: Kent, Tollens, *Ann.* **227**, 224 (1885); E. P. Clark, *J. Biol. Chem.* **47**, 2 (1921). Mutarotation and purification of β-form: C. S. Hudson, E. Yanosky, *J. Am. Chem. Soc.* **39**, 1021 (1917). Structural configuration: J. Pryde, *J. Chem. Soc.* **123**, 1809 (1923); W. Charlton *et al., ibid.* **1926**, 94; W. N. Haworth *et al., ibid.* **1927**, 2428; E. L. Jackson, C. S. Hudson, *J. Am. Chem. Soc.* **59**, 994 (1937); R. M. Hann *et al., ibid.* **66**, 1912 (1944). Isoln in the processing of the red alga, *Porphyra umbilicalis*: S. Peat *et al., J. Chem. Soc.* **1961**, 1590. Review: W. Pigman, *The Carbohydrates* (Academic Press, New York, 1957) pp 88-90. Review of diagnostic use: W. J. Schirmer *et al., J. Surg. Res.* **41**, 543 (1986).

α-D-galactose

α-Form, prisms from water or ethanol, mp 167°. $[\alpha]_D$ +150.7° → +80.2° (water). Soluble in about 0.5 parts water; freely sol in hot water; final soly in water at 25° = 68%; sol in pyridine; slightly sol in alcohol.

β-Form, crystals, mp 167°. $[\alpha]_D$ +52.8° → +80.2° (water). Sol in 1.7 parts water at 17°.

Monohydrate, prisms from water, mp 118-120°.

THERAP CAT: Diagnostic aid (hepatic function).

4242. D-Galacturonic Acid. $C_6H_{10}O_7$; mol wt 194.14. C 37.12%, H 5.19%, O 57.69%. Obtained by hydrolysis of pectin where it is present as polygalacturonic acid: Ehrlich, *Chem. Ztg.* **41**, 197 (1917); Ehrlich, Guttmann, *Biochem. Z.* **259**, 100 (1933); *Ber.* **66**, 220 (1933); Niemann, Link, *J. Biol. Chem.* **95**, 203 (1932); **104**, 743 (1934); Morell, Link, *ibid.* **100**, 385 (1933); Anderson, King, *J. Chem. Soc.* **1961**, 5333.

Isoln from mustard seeds: Goering, U.S. pat. **2,987,448** (1961 to Oil Seed Prod.).

α-Form, monohydrate, needles, mp 159°. $[\alpha]_D^{20}$ +98.0° → 50.9° (water). Soluble in water; slightly sol in hot alcohol. Practically insol in ether.

β-Form, mp 166°. $[\alpha]_D$ +27° → +55.6° (water). Phenylhydrazone, mp 141°.

4243. Galanga. Galangal; colic root; East India root; Chinese ginger. Dried rhizome of *Alpinia officinarum* Hance, *Zingiberaceae*. *Habit.* China. *Constit.* Volatile oil, resin, kaempferid, galangin, dioxyflavanol, galangol.

4244. Galangin. *3,5,7-Trihydroxy-2-phenyl-4H-1-benzopyran-4-one; 3,5,7-trihydroxyflavone;* norizalpinin. $C_{15}H_{10}O_5$; mol wt 270.25. C 66.67%, H 3.73%, O 29.60%. Isoln from galanga root, *Alpininia officinarum*, Hance and characterization: E. Jahns, *Ber.* **14**, 2807 (1881). Prepn: T. Heap, R. Robinson, *J. Chem. Soc.* **129**, 2336 (1926); J. J. Chavan, R. Robinson, *ibid.* **1933**, 368. Mutagenicity studies: J. T. MacGregor, L. Jurd, *Mutat. Res.* **54**, 297 (1978); J. P. Brown, P. S. Dietrich, *ibid.* **66**, 223 (1979).

Yellowish needles from ethanol, mp 214-215°. Moderately sol in ethanol, ether; insol in water. Very sol in chloroform, benzene.

4245. Galanthamine. *4a,5,9,10,11,12-Hexahydro-3-methoxy-11-methyl-6H-benzofuro[3a,3,2-ef][2]benzazepin-6-ol;* galantamine; lycoremine; Jilkon. $C_{17}H_{21}NO_3$; mol wt 287.35. C 71.05%, H 7.37%, N 4.87%, O 16.70%. From Caucasian snowdrops, *Galanthus woronowii* Vel., *Amaryllidaceae*: Proskurnina, Yakovleva, *J. Gen. Chem.* **22**, 1899 (1952); from *Narcissus* spp: Boit *et al.*, *Ber.* **90**, 725, 2197 (1957). Structure work: Kobayashi *et al.*, *Chem. & Ind. (London)* **1956**, 177. Synthesis and stereochemistry: Barton, Kirby, *Proc. Chem. Soc.* **1960**, 392; *J. Chem. Soc.* **1962**, 806; Williams, Rogers, *Proc. Chem. Soc.* **1964**, 357. Alternate total synthesis: Kametani *et al.*, *J. Chem. Soc. (C)* **1971**, 1043. Asymmetric synthesis of (+)- and (−)-forms from L-tyrosine: K. Shimizu *et al.*, *Heterocycles* **8**, 277 (1977). Biosynthesis studies: D. H. R. Barton *et al.*, *J. Chem. Soc.* **1963**, 4545; W. Döbke, *Heterocycles* **6**, 551 (1977).

Crystals from benzene, mp 126-127°. $[\alpha]_D^{20}$ −118.8° (c = 1.378 in ethanol). Monoacidic base. Fairly sol in hot water; freely sol in alcohol, acetone, chloroform. Less sol in benzene, ether.

Hydrochloride, $C_{17}H_{21}NO_3 \cdot$HCl, crystals from water, dec 256-257°. Sparingly sol in cold, more sol in hot water. Very sparingly sol in alcohol, acetone.

Hydrobromide, $C_{17}H_{22}BrNO_3$, *Nivalin.* Crystals from water, dec 246-247°. $[\alpha]_D^{20}$ −93.1° (c = 0.1015 in 15 ml H_2O). LD_{50} i.v. orally in mice: 8.0, 18.7 mg/kg, Umarova *et al.*, *C.A.* **66**, 53993v (1967).

THERAP CAT: Cholinesterase inhibitor.

4246. Galegine. *(3-Methyl-2-butenyl)guanidine; N-3,3-dimethylallylguanidine;* isoamyleneguanidine. $C_6H_{13}N_3$; mol wt 127.18. C 56.66%, H 10.30%, N 33.04%. Isoprenoid guanidine deriv from seeds of *Galega officinalis* L., *Leguminosae*: Tanret, *Compt. Rend.* **158**, 1182, 1426 (1914); **159**, 108 (1914); Markovic, Dittertová, *Chem. Zvesti* **9**, 576 (1955), *C.A.* **50**, 8137d (1956). Structure: Barger, White, *Biochem. J.* **17**, 827 (1923). Synthesis: Späth, Spitzy, *Ber.* **58**, 2273 (1925); Babor, Jezo, *Chem. Zvesti* **8**, 18 (1954), *C.A.* **49**, 7495f (1955). Metabolic effects: G. Weitzel *et al.*, *Z. Physiol. Chem.* **353**, 535 (1972). Effects on mitochondria: B. Lotina *et al.*, *Arch. Biochem. Biophys.* **159**, 520 (1973). Biosynthetic study: J. Steiniger, G. Reuter, *Biochem. Physiol. Pflanz.* **166**, 275 (1974). *Review:* Braun, *J. Chem. Ed.* **8**, 2175 (1931).

Hygroscopic, bitter crystals. mp 60-65°. Freely sol in water or alcohol, slightly in ether. *Keep well closed.*

4247. Galipine. *2-[2-(3,4-Dimethoxyphenyl)ethyl]-4-methoxyquinoline.* $C_{20}H_{21}NO_3$; mol wt 323.38. C 74.28%, H 6.55%, N 4.33%, O 14.84%. From Angostura bark (*Cusparia trifoliata* Engl., *Rutaceae*): Körner, Böhringer, *Gazz. Chim. Ital.* **13**, 363 (1883); Tröger, Kroseberg, *Arch. Pharm.* **250**, 494 (1912). Synthesis: Späth, Eberstaller, *Ber.* **57**, 1687 (1924); Späth, Pikl, *Ber.* **62**, 2244 (1929); Schläger, Leeb, *Monatsh.* **81**, 714 (1950).

Prismatic needles from alc, mp 116°. Soluble in alcohol, benzene, chloroform, ether; slightly sol in water, petr ether. The salts are more sol than those of cusparine.

Hydrochloride tetrahydrate, $C_{20}H_{21}NO_3 \cdot$HCl$\cdot 4H_2O$, plates, become anhydr at 100°, mp 165°.

Methiodide, $C_{20}H_{21}NO_3 \cdot CH_3I$, yellow needles, mp 146°.

4248. Gallacetophenone. *1-(2,3,4-Trihydroxyphenyl)ethanone; 2',3',4'-trihydroxyacetophenone;* Alizarine yellow C; C.I. 57000. $C_8H_8O_4$; mol wt 168.14. C 57.14%, H 4.79%, O 38.06%. Prepn: Hart, Woodruff, *J. Am. Chem. Soc.* **58**, 1957 (1936); Campbell, Coppinger, U.S. pat. **2,686,123** (1954 to U.S. Secy. Agr.); Knowles, U.S. pat. **2,763,691** (1956 to Kodak); Price, Israelstam *J. Org. Chem.* **29**, 2800 (1964).

White to brownish-gray, cryst powder, mp 173°. uv max (methanol): 237, 296 nm (ε 8560, 12,500). Sol in 600 parts cold water, more in hot water; sol in alcohol, ether, soln of sodium acetate.

USE: Antiseptic.

4249. Gallamine Triethiodide. *2,2',2''-[1,2,3-Benzenetriyltris(oxy)]tris[N,N,N-triethylethanaminium] triiodide; [v-phenenyltris(oxyethylene)]tris[triethylammonium triiodide];* 1,2,3-tris(2-triethylammonium ethoxy)benzene triiodide; 1,2,3-tris(2-diethylaminoethoxy)benzene tris(ethyl iodide); tri(β-diethylaminoethoxy)-1,2,3-benzene triiodoethylate; pyrogallol 1,2,3-(diethylaminoethyl ether) tris-(ethyl iodide); benzcurine iodide; RP 3697; F 2559; Tricuran; Retensin; Relaxan; Flaxedil. $C_{30}H_{60}I_3N_3O_3$; mol wt 891.56. C 40.41%, H 6.78%, I 42.71%, N 4.71%, O 5.38%. Curarizing properties: D. Bovet et al., Compt. Rend. **225**, 74 (1947); F. Depierre, ibid. 956. Prepn: E. Fourneau, U.S. pat. **2,-544,076** (1951 to Rhone-Poulenc). Comparatve clinical pharmacokinetics: W. Buzello, S. Agoston, Anaesthesist **27**, 313 (1978). Mode of action: D. Colquhoun, R. E. Sheridan, Brit. J. Pharmacol. **66**, 78 (1979); eidem, Proc. Roy. Soc. London, Ser. B **211**, 181 (1981). Effects in mammalian and amphibian nerve fibers: K. J. Smith, C. L. Schauf, Science **212**, 1170 (1981).

$$\left[\begin{array}{c} OCH_2CH_2\overset{+}{N}(C_2H_5)_3 \\ OCH_2CH_2\overset{+}{N}(C_2H_5)_3 \\ OCH_2CH_2\overset{+}{N}(C_2H_5)_3 \end{array} \right] 3I^-$$

White cryst from acetone/water, mp 152-153° (indefinite). Freely sol in water, alc, dil acetone; sparingly sol in anhydr acetone, ether, benzene, chloroform. *Pharmaceutical Incompat:* Meperidine hydrochloride (solns must not be mixed).

THERAP CAT: Skeletal muscle relaxant.

THERAP CAT (VET): Skeletal muscle relaxant.

4250. Gallein. *3',4',5',6'-Tetrahydroxyspiro[isobenzofuran-1(3H),9'-[9H]xanthen]-3-one; 3',4',5',6'-tetrahydroxyfluoran; 3',4',5',6'-tetrahydroxyspiro[phthalan-1,9'-xanthen]-3-one;* pyrogallolphthalein; C.I. 45445; mordant violet 25. $C_{20}H_{12}O_7$; mol wt 364.30. C 65.94%, H 3.32%, O 30.74%. Obtained by heating 1 part phthalic anhydride with 2 parts of pyrogallol or gallic acid: Baeyer, Ber. **4**, 457 (1871); Buchka, Ann. **209**, 261 (1881). Use as a biological stain: R. D. Lillie et al., Stain Technol. **49**, 339 (1974); R. Welsh, ibid. **52**, 261 (1977). See also H. J. Conn's Biol. Stains, R. D. Lillie, Ed. (Williams & Wilkins, Baltimore, 9th ed., 1977) p 351.

Brownish-red powder or crystals with 1½ H_2O, or red crystals with greenish-yellow color when anhydr. Loses the water of crystn at about 180° and blackens above this temp. Does not melt even at 300°. pH 3.8 brownish-yellow; pH 6.6 rose-red. Almost insol in water, benzene, chloroform. Slightly sol in ether; sol in alc, acetone, alkalies.

Disodium salt, $C_{20}H_{10}Na_2O_7$, *alizarin violet.* pH 10.6 rose; pH 13.0 violet.

USE: Clinical reagent (phosphates in urine). Monophosphates give a yellow, dibasic a red, tribasic a violet color. Used in soln of 0.5 g in 100 ml 50% alc; 2-3 drops for 100 ml liq. As sensitive indicator for acids, alkali hydroxides, NH_3, but not for carbonates. Biological stain.

4251. Gallic Acid. *3,4,5-Trihydroxybenzoic acid.* $C_7H_6O_5$; mol wt 170.12. C 49.42%, H 3.55%, O 47.02%. Obtained by alkaline or acid hydrolysis of the tannins from nutgalls; also by enzymatic hydrolysis using spent broths from Penicillium glaucum or Aspergillus niger which contain tannase: A. G. Perkin, O. Gunnell, J. Chem. Soc. **69**, 1303 (1896); Hsias, Huang Hai No. 7, 51 (1946), C.A. **42**, 3901i (1948); Cochrane, Econ. Bot. **2**, 145 (1948); Toth, Henster, Acta Chim. Acad. Sci. Hung. **2**, 209 (1952). Preparation from tannin containing materials: Krueger et al., U.S. pat. **2,723,992** (1955 to Mallinckrodt). Synthesis from aliphatic materials: Shipchandler et al., J. Chem. Soc. Perkin Trans. I **1975**, 1400. Biosynthesis: Haslam et al., J. Chem. Soc. **1961**,

1854. Study of polymorphic forms: E. Lindpainter, Mikrochemie **27**, 21 (1939). Toxicity studies: J. W. Dollahite et al., Am. J. Vet. Res. **23**, 1264 (1962).

Needles from abs methanol or chloroform, formerly reported as dec 235-240° (Perkin, Gunnell). Sublimes at 210° giving a stable form with mp 258-265° (dec) and an unstable form mp 225-230° (Lindpainter). One gram dissolves in 87 ml water, 3 ml boiling water, 6 ml alcohol, 100 ml ether, 10 ml glycerol, 5 ml acetone; practically insol in benzene, chloroform, petr ether. *Protect from light.* LD_{50} in rabbits (g/kg): 5.0 orally (Dollahite).

Methyl ester, $C_8H_8O_5$, *methyl gallate, gallicin.* Monoclinic prisms from methanol, often hydrated or solvated. When dry, mp 202°. Sol in hot water, alcohol, methanol, ether.

Propyl ester see Propyl Gallate.

USE: Manuf gallic acid esters, pyrogallol, inks; as photographic developer; in tanning; in dyeing; in testing for free mineral acids, dihydroxyacetone and alkaloids. Esters as antioxidants.

THERAP CAT: Formerly as astringent, styptic.

THERAP CAT (VET): Has been used as intestinal astringent.

4252. Gallium. Ga; at. wt 69.72; at. no. 31; valences 3, 2, 1. Natural isotopes: 69 (60.2%); 71 (39.8%); artificial radioactive isotopes: 63-68; 70; 72-76. Best source is the mineral germanite, a copper sulfide ore; occurs in very small quantities in zinc blendes, in aluminum clays, found in ores of iron, chromium, manganese; constitutes 5×10^{-4}% of the crust of the earth. Discovered by L. Boisbaudran, Compt. Rend. **81**, 493, 1100 (1875); **82**, 163, 1036 (1876); isolated pure by L. Boisbaudran and E. Jungfleisch, Bull. Soc. Chim. [2] **31**, 50 (1879). Isoln from rhenium-rich copper schist: Feit, Angew. Chem. **46**, 216 (1933). From bauxite: Chem. Eng. News **34**, 4300 (1956). Purification by zone melting: Chem. Ztg. **80**, 787 (1956). Alternate methods of purification: Gebauhr, U.S. pat. **2,928,731**; Merkel, U.S. pat. **2,927,853** (both 1960 to Siemens-Schuckert). Spectra: L. Boisbaudran et al., cited in Mellor's vol. **5**, 378 (1929). Reviews: Wagner, Gitzen, J. Chem. Ed. **29**, 162 (1952); Greenwood, Inorg. Chem. Radiochem. **5**, 91-134 (1963); Wade, Banister, Comprehensive Inorganic Chemistry vol. **1**, J. C. Bailar, Jr. et al., Eds. (Pergamon Press, Oxford, 1973) pp 997-1000, 1069-1117; P. de la Bretèque in Kirk-Othmer Encyclopedia of Chemical Technology vol. **11** (Wiley-Interscience, New York, 3rd ed., 1980) pp 604-620. Review of diagnostic use of ^{67}Ga citrate in tumor detection: C. Bekerman et al., Semin. Nucl. Med. **14**, 296-323 (1984); of pharmacology and clinical antitumor activity of nitrate: B. J. Foster et al., Cancer Treat. Rep. **70** 1311-1319 (1986).

Grayish metal; possesses a greenish-blue reflection; tin- or silver-like when molten; has a crystalline orthorhombic texture. mp 29.78°. bp approx 2400°: Cochran, Foster, J. Electrochem. Soc. **109**, 144 (1962). Shows a tendency to remain in supercooled state. Contracts on melting; $d^{29.65}$ (solid) 5.9037; $d^{29.8}$ (liq) 6.0947: Richards, Boyer, J. Am. Chem. Soc. **43**, 274 (1921). Heat capacity: 0.09 cal/g/°C (0-24°, solid). Latent heat of fusion 19.16 cal/g. Stable in dry air; tarnishes in moist air or oxygen. Reacts with alkalies with evolution of hydrogen; attacked by cold concd hydrochloric acid; rendered passive by hot nitric acid; readily attacked by halogens.

Sesquioxide, Ga_2O_3; obtained by thermal decompn of the salts; exists in many crystalline modifications; on heating with magnesium is violently reduced to the metallic state.

Suboxide, Ga_2O; brown powder; obtained by heating the sesquioxide and the metal at 700°; stable in dry air; dec above 800°; converted to the trivalent state by nitric acid or bromine.

Hydroxide, $Ga(OH)_3$, a gelatinous precipitate; obtained by the action of ammonia or alkali hydroxide on a soln of a gallic salt.

Chloride, GaCl$_3$, colorless needles, mp 78°; prepd by the action of chlorine or hydrogen chloride on the metal.

Dichloride, Ga$^+$[GaCl$_4$]$^-$, white crystals; mp 172.4°; prepd by heating the trichloride with the metal.

Sulfate octadecahydrate, Ga$_2$(SO$_4$)$_3$.18H$_2$O, octahedral crystals; formed by dissolving the sesquioxide or the hydroxide in sulfuric acid and precipitating with ether or alcohol; sol in water or 60% alcohol. Prepn: Reinmann, Tanner, *Z. Naturforsch.* **20b,** 71 (1965).

Citrate, C$_6$H$_5$^{67}GaO$_7$, *Neoscan.*

Nitrate, GaN$_3$O$_9$, *NSC-15200.* White crystalline powder. Sol in warm and cold aqueous solutions, absolute alcohol, ether. Clinical study in treatment of cancer related hypercalcemia: R. P. Warrell, Jr. *et al., Ann. Int. Med.* **108,** 669 (1988). LD$_{50}$ in mice, rats, rabbits (mg/kg): 55, 46, 43 i.v. (Foster).

Lactate, C$_9$H$_{15}$GaO$_9$, white amorphous powder, mp 270° (dec). Slightly hygroscopic. Soly (g/100 ml): water (20°) 8.9, water (100°) 21.5, 95% ethyl alcohol (20°) 1.5, ether, anhydrous (20°) 0.015, acetone, anhydrous (20°) 0.004. pH 2.7. LD$_{50}$ in rats, rabbits (mg/kg): 47, 43 i.v.; 121, 97 s.c. (Dudley, Levine). Synthesis and physical properties: H. C. Dudley, R. F. Garzoli, *J. Am. Chem. Soc.* **70,** 3942 (1948). Toxicity study: H. C. Dudley, M. D. Levine, *J. Pharmacol. Exp. Ther* **95,** 487 (1949).

Nitrate nonahydrate, Ga(NO$_3$)$_3$.9H$_2$O, obtained by dissolving the metal or the oxide in concd nitric acid, Reinmann, Tanner, *loc. cit.*

Human Toxicity: Administration to humans has caused metallic taste, skin rashes, bone marrow depression: H. E. Stokinger in Patty's *Industrial Hygiene and Toxicology* vol. **2A,** G. D. Clayton, F. E. Clayton, Eds. (Wiley-Interscience, New York, 3rd ed., 1981) pp 1630-1637.

THERAP CAT: ^{67}Ga citrate as diagnostic aid (neoplasm). Nitrate as antineoplastic, antihypercalcemic.

4253. Gallium Arsenide. AsGa; mol wt 144.64. Ga 48.20%, As 51.80%. GaAs. Prepd by passing a mixture of hydrogen and arsenic vapor over gallium(III) oxide at 600°: Goldschmidt, *Skr. Akad. Oslo* **1926,** no. 8, pp 34, 100; *Ber.* **60,** 1289 (1927). Simplified procedure: Juza, Schulz, *Z. Anorg. Allgem. Chem.* **275,** 65 (1954); Minden, *Sylvania Technologist* **11,** no. 1, p 19 (Jan. 1958). Vapor phase crystal growth: McAleer *et al., J. Electrochem. Soc.* **108,** 1168 (1961). Use in high-speed microcircuits: A. L. Robinson, *Science* **219,** 275 (1983).

Cubic crystals. mp 1238°. Dark gray with metallic sheen. Hardness 4.5. d$_4^{25}$ 5.31. Thermal expansion coefficient: 5.9 × 10^{-6}. Thermal conductivity 0.52 watt units. Specific heat 0.086 cal/g/°C. Molten gallium arsenide attacks quartz, therefore graphite boats or quartz boats, which are carbon coated by pyrolytic decompn of methane, should be used in zone refining. GaAs single crystals have been grown by the Czochralski technique and by the floating zone method. Extensive twinning occurs. Intrinsic electron concn 10^7. Energy gap at room temp 1.38 electron volts. Electron mobility 8800 cm^2/volt sec. Hole mobility 450 cm^2/volt sec. Effective mass for electrons 0.06 m_0. Lattice constant 5.6-54Å. Dielectric constant 11.1. Intrinsic resistivity at 300°K = 3.7 × 10^8 ohm-cm. Electron lattice mobility at 300°K = 10,000 cm^2/volt-sec. Intrinsic charge density at 300°K = 1.4 × 10^6/cm^3. Electron diffusion constant at 300°K = 310 cm^2/sec. Hole diffusion constant = 11.5 cm^2/sec.

USE: In semiconductor applications (transistors, solar cells, lasers).

4254. Gallium Phosphide. GaP; mol wt 100.69. Ga 69.24%, P 30.76%. Prepn and description: Folberth, Oswald, *Z. Naturforsch.* **9a,** 1050 (1954); Wolff *et al., Phys. Rev.* **94,** 753 (1954); Antell, Effer, *J. Electrochem. Soc.* **106,** 509 (1959); **107,** 252 (1960); Frosch, Derick, *ibid.* **108,** 251 (1961); Addamiano, *J. Am. Chem. Soc.* **82,** 1537 (1960); Gershenzon, Mikulyak, *ibid.* **108,** 548 (1961); Pizzarello, *ibid.* **109,** 226 (1962). Seguin, Gans, U.S. pat. **2,862,787** (1958); Chang, U.S. pat. **2,921,905** (1960 to Westinghouse).

Translucent, amber-colored crystals of the zinc blende type. Greenish-yellow, opaque cryst mass, when unreacted

gallium is present. mp 1465°. Dielectric constant 8.4. Energy gap 2.25 ev. Hole mobility 70. Electron mobility 1200. Current density at 4 volts = 25 amps/cm^2 at 25° (in rectifier circuitry).

USE: In semiconductor electronics.

4255. Gallium Trifluoride. F$_3$Ga; mol wt 126.72. F 44.98%, Ga 55.02%. GaF$_3$. Prepd by thermal decompn of ammonium hexafluogallate (NH$_4$)$_3$(GaF$_6$): Hannebohn, Klemm, *Z. Anorg. Allgem. Chem.* **229,** 342 (1936); Kwasnik in *Handbook of Preparative Inorganic Chemistry* vol. **1,** G. Brauer, Ed. (Academic Press, New York, 2nd ed., 1963) pp 227-228; from Ga and HF: Brewer *et al., J. Inorg. Nucl. Chem.* **9,** 56 (1959); Muetterties, Castle, *ibid.* **18,** 148 (1961).

White powder. d 4.47 (after heating to 630° in a current of F$_2$). mp >1000°. bp ~950°. Can be sublimed in N$_2$ current at about 800° without decomp. Soly in water (25°) 0.0024 g/100 ml. Soly in hot dil HCl 0.0028 g/100 ml.

Trihydrate, white substance, mp >140°. More sol in water than anhydr form.

4256. Gallocyanine. *1-Carboxy-7-(dimethylamino)-3,4-dihydroxyphenoxazin-5-ium chloride; C.I. Mordant Blue 10; C.I.* 51030. C$_{15}$H$_{13}$ClN$_2$O$_5$; mol wt 336.73. C 53.50%, H 3.89%, Cl 10.53%, N 8.32%, O 23.76%. Made by introducing nitrosodimethylaniline hydrochloride into a suspension of gallic acid in boiling methanol: Koechlin, *Ger. pat.* **19,-580** (1881), *Frdl.* **1,** 269; Nietzki, Otto, *Ber.* **21,** 1740 (1888); *Colour Index* vol. **4** (3rd ed., 1971) p 4460.

Green crystals. Practically insol in cold water, slightly sol in hot water; sol in alcohol, glacial acetic acid, in alkali carbonates with reddish color, in concd HCl with a blue color which becomes red upon diluting with water.

USE: As a dye; in alkali carbonate soln as a reagent for lead with which it forms a deep violet color.

4257. Gallopamil. *α-[3-[[2-(3,4-Dimethoxyphenyl)eth-yl]methylamino]propyl]-3,4,5-trimethoxy-α-(1-methylethyl)-benzeneacetonitrile; 5-[(3,4-dimethoxyphenethyl)methylamino]-2-isopropyl-2-(3,4,5-trimethoxyphenyl)valeronitrile; α-isopropyl-α-[(N-methyl-N-homoveratryl)-γ-aminopropyl]-3,4,5-trimethoxyphenylacetonitrile; methoxyverapamil; D 600.* C$_{28}$H$_{40}$N$_2$O$_5$; mol wt 484.64. C 69.39%, H 8.32%, N 5.78%, O 16.51%. Calcium channel blocking agent; methoxy derivative of verapamil, *q.v.* Prepn: F. Dengel, **Belg.** pat. **615,861;** *idem,* **U.S.** pat. **3,261,859** (1962, 1966 to Knoll); *idem,* **U.S.** pat. **4,115,432** (1978 to Knoll). Alternate process: G. Kastner *et al.,* **U.S.** pat. **4,350,636** (1982 to BASF). Synthesis of enantiomeric forms and of the hydrochloride salts from optically active materials: (*d*-form), **Belg.** pat. **776,219; Brit.** pat. **1,377,209** (1972, 1974 to Knoll); (*l*-form), **Belg.** pat. **776,218; Brit.** pat. **1,367,677** (1972, 1974 to Knoll). Prepn of enantiomeric forms from the racemate: S. Herrling, **Eur.** pat. **Appl.** 29175; *idem,* **U.S.** pat. **4,305,887** (1981, 1982 both to BASF). Pharmacology: H. Haas, E. Busch, *Arzneimittel-Forsch.* **17,** 257 (1967); H. Haas, *ibid.* **18,** 89 (1968); H. Haas, E. Busch, *ibid.* 407; A. Fleckenstein, *ibid.* **20,** 1317 (1970). Double-blind clinical studies in stable angina pectoris: G. Rettig *et al., Z. Kardiol.* **72,** 746 (1983); N. S. Khurmi *et al., Am. J. Cardiol.* **53,** 684 (1984).

Pale yellow viscous oil. n_D^{25} 1.5402.

Consult the cross index before using this section.

d-Form, $[\alpha]_D^{20}$ +5.5° (c = 15 mg/ml in ethanol).
l-Form, $[\alpha]_D^{20}$ −5.5° (c = 16.8 mg/ml in ethanol).
Hydrochloride, $C_{28}H_{41}ClN_2O_5$, *Algocor, Corgal, Procorum, Wingom.* mp 145-148°.
d-Form hydrochloride, crystals from isopropanol, mp 159-161°. $[\alpha]_D^{22}$ +13.3° (c = 1.2 in ethanol).
l-Form hydrochloride, crystals from isopropanol, mp 158-161°. $[\alpha]_D^{20}$ −13.4° (c = 1.06 in ethanol).
THERAP CAT: Anti-anginal; anti-arrhythmic.

4258. Gamabufotalin. *3β,11α,14-Trihydroxy-5β-bufa-20,22-dienolide;* gamabufogenin; gamabufagin. $C_{24}H_{34}O_5$; mol wt 402.51. C 71.61%, H 8.51%, O 19.87%. One of the genins found in the venom of a Japanese toad (gama; *Bufo vulgaris formosus*). Isoln and structure: Kotake, *Ann.* **465,** 11 (1928); Wieland, Vocke, *Ann.* **481,** 215 (1930); Chen *et al., J. Pharmacol. Exp. Ther.* **49,** 26 (1933); Kondo, Ohno, *J. Pharm. Soc. Japan* **59,** 186 (1939); Jensen, *J. Am. Chem. Soc.* **59,** 767 (1937); Kuno Meyer, *Helv. Chim. Acta* **32,** 1599 (1949). Also found in Ch'an-Su, A Chinese drug prepared from Chinese toads (*Bufo asiaticus = Bufo gargarizans* Cantor): Ruckstuhl, Meyer, *Helv. Chim. Acta* **40,** 1270 (1957).

Bitter, solvated prisms from alcohol + ether, mp 254°; when dry dec 262-263°. $[\alpha]_D^{18}$ +1.26° (c = 0.793 in methanol). Produces numbness of the tongue. uv max about 300 nm. Very sparingly sol in chloroform, acetone, water. Somewhat more sol in methanol.

4259. Gambir. Catechu; pale catechu; gambir catechu; terra japonica. Dried aqueous extract of leaves and twigs of *Uncaria gambier* (Hunter) Roxb. (*Ourouparia gambier* (Hunter) Baill.), *Rubiaceae.* *Habit.* Southern Asia. *Constit.* 7-30% catechol, 20-50% catechutannic acid; quercetin, catechu-red, gambir-fluorescein, fixed oil, wax. Contains not less than 70% water-sol substance and not less than 60% is sol in alcohol.
Incompat: Iron compds, gelatin, limewater, mercuric chloride, zinc sulfate, alkalies.
USE: Tanning, dyeing fabrics brown or black.
THERAP CAT: Astringent.
THERAP CAT (VET): Intestinal astringent.

4260. Gamboge. Cambogia. Gum-resin from *Garcinia hanburyi* Hook. f., *Guttiferae.* *Habit.* East Indies. *Constit.* 70-80% resin, cambogic acid, 15-25% gum.
THERAP CAT: Cathartic.
THERAP CAT (VET): Drastic purgative.

4261. Gambogic Acid. *[1R-[1α,1(Z),3aβ,5α,11β,14aS*]]-2-Methyl-4-[3a,4,5,7-tetrahydro-8-hydroxy-3,3,11-trimethyl-13-(3-methyl-2-butenyl)-11-(4-methyl-3-pentenyl)-7,15-dioxo-1,5-methano-1H,3H,11H-furo[3,4-g]pyrano[3,2-b]xanthen-1-yl]-2-butenoic acid.* $C_{38}H_{44}O_8$; mol wt 628.73. C 72.59%, H 7.05%, O 20.36%. Principal acidic component of gamboge from the latex of *Garcinia hanburyi* Hook. f., *Guttiferae:* Amorosa, Lipparini, *Ann. Chim. (Rome)* **45,** 977 (1955). Structure: Ollis *et al., Tetrahedron* **21,** 1453 (1965); Ahmed *et al., J. Chem. Soc. (C)* **1966,** 772.

Golden, friable mass. $[\alpha]_D^{20}$ −685° (methanol). uv max (ethanol): 217, 280, 291, 362 nm (ε 26000, 16700, 17000, 14900).
Methyl ester monomethyl ether, $C_{40}H_{48}O_8$, yellow prisms from methanol, mp 130-131°. $[\alpha]_D^{22}$ −560° (c = 0.7 in chloroform). uv max (ethanol): 224, 299 nm (ε 36000, 13600).
Pyridine salt, $C_{43}H_{49}NO_8$, orange needles from ether + petr ether, mp 147-149°. $[\alpha]_D$ −550° (chloroform). uv max (ethanol): 291.5, 359.5 nm (ε 22300, 18100).

4262. Ganciclovir. *2-Amino-1,9-[[2-hydroxy-1-(hydroxymethyl)ethoxy]methyl]-6H-purin-6-one;* 9-[(1,3-dihydroxy-2-propoxy)methyl]guanine; 2'-nor-2'-deoxyguanosine; DHPG; 2'NDG; BIOLF-62; BWB759U; BW-759; BW-759U; RS-21592; Cymevan; Cymevene; Cytovene. $C_9H_{13}N_5O_4$; mol wt 255.23. C 42.35%, H 5.13%, N 27.44%, O 25.07%. Nucleoside analog structurally related to acyclovir, *q.v.* Prepn: J. P. Verheyden, J. C. Martin, U.S. pat. **4,355,032** (1982 to Syntex); K. K. Ogilvie *et al., Can. J. Chem.* **60,** 3005 (1982). Prepn and antiherpetic activity: W. T. Ashton *et al., Biochem. Biophys. Res. Commun.* **108,** 1716 (1982); J. C. Martin *et al., J. Med. Chem.* **26,** 759 (1983). Antiviral spectrum *in vitro:* K. O. Smith *et al., Antimicrob. Ag. Chemother.* **22,** 55 (1982); Y.-C. Cheng *et al., Proc. Nat. Acad. Sci. USA* **80,** 2767 (1983). Inhibitory effect on human cytomegalovirus *in vitro:* E.-C. Mar *et al., Antimicrob. Ag. Chemother.* **24,** 518 (1983); M. J. Tocci *et al., ibid.* **25,** 247 (1984). Comparison with acyclovir in herpes-infected mice: P. Collins, N. M. Oliver, *Antiviral Res.* **5,** 145 (1985). Clinical treatment of cytomegalovirus infection in immunodeficient patients: S. H. Koretz *et al., N. Engl. J. Med.* **314,** 801 (1986); O. L. Raskin *et al., J. Infect. Dis.* **155,** 323 (1987). Symposium on pharmacology and clinical efficacy vs cytomegalovirus: *Rev. Infect. Dis.* **10,** Suppl. 3, S457-S572 (1988).

Crystals from methanol, mp 250° (dec) (Verheyden, Martin); also reported as crystalline monohydrate from water, mp 248-249° (dec) (Ashton); crystals from water, mp > 200° (Martin). uv max (methanol): 254 nm (ε 12880). Soly in water (25°): 4.3 mg/ml at pH 7. LD_{50} i.p. in mice: 1-2 g/kg (Martin).
THERAP CAT: Antiviral.

4263. Ganglefene. *4-(2-Methylpropoxy)benzoic acid 3-(diethylamino)-1,2-dimethylpropyl ester; p-isobutoxybenzoic acid 3-(diethylamino)-1,2-dimethylpropyl ester; 1,2-dimethyl-3-diethylaminopropyl p-isobutoxybenzoate; p-isobutoxybenzoic acid α,β-dimethyl-γ-diethylaminopropyl ester.* $C_{20}H_{33}NO_3$; mol wt 335.47. C 71.60%, H 9.92%, N 4.18%, O 14.31%. Prepn: A. L. Mndzhoyan *et al., U.S.S.R.* pat. **115,905** (1958), *C.A.* **53,** 17439a (1959). Pharmacologic studies: M. G. Udel'nov *et al., Farmakol. Toksikol.* **36,** 299 (1973), *C.A.* **79,** 61442y (1973); V. A. Gusel, *Pharmacol., Biochem. Behav.* **2,** 1 (1974), *C.A.* **81,** 114675b (1974). Stability study: V. G. Belikov *et al., Khromatogr. Metody Farm.* **1977,** 42, *C.A.* **88,** 141588 (1978).

LD$_{50}$ in mice: 530 mg/kg s.c., *RTECS* Vol. I, R. J. Lewis, R.C. Tatken, Eds. (1980) p 257.

Hydrochloride, C$_{20}$H$_{34}$ClNO$_3$, *Gangleron.*

THERAP CAT: Vasodilator (coronary).

4264. Gangliosides. Animal glycosphingolipids occurring in highest concentration in the central nervous system but widely distributed in other tissues. Primarily located in the outer surface of mammalian cell plasma membranes, they are also found in the synaptic membrane of the central nervous system. More than 60 gangliosides have been characterized. Each consists of a fatty acid (often stearic acid) and an oligosaccharide moiety, both attached to sphingosine. Sialic acid, *q.v.*, is the identifying sugar. The sialic acid residues of gangliosides and those of glycoprotein are the main cause of cell surface negative charge. Gangliosides are also thought to be involved in many different cell functions, e.g. metabolism, growth, malignant transformation. Ganglioside nomenclature: L. Svennerholm, "Ganglioside Metabolism" in *Comprehensive Biochemistry* vol. 18, M. Florkin, E. H. Stotz, Eds. (Elsevier, Amsterdam, 1970) pp 201-204. Isoln from normal brain and from the brain of patients with Tay-Sachs disease: Klenk, *Z. Physiol. Chem.* **268**, 50 (1941); **273**, 76 (1942); Trams, Lauter, *Biochim. Biophys. Acta* **60**, 350 (1962); Svennerholm, *Acta Chem. Scand.* **17**, 239 (1963). Occurrence of different gangliosides: Kuhn, Egge, *Angew. Chem.* **72**, 805 (1960). Chromatographic studies: Kuhn *et al., ibid.* **73**, 580 (1961); Klenk, Gielen, *Z. Physiol. Chem.* **326**, 144 (1961). Structure of ganglioside G$_I$ (or G$_{MI}$): Kuhn, Wiegandt, *Ber.* **96**, 866 (1963). Structure of gangliosides G$_I$ and G$_{II}$ (or G$_{D1a}$): Kuhn, Egge, *ibid.* **96**, 3338 (1963). Structure of gangliosides G$_{II}$, G$_{III}$ (or G$_{D1b}$) and G$_{IV}$ (or G$_{T1b}$): Kuhn, Wiegandt, *Z. Naturforsch.* **18b**, 541 (1963). Molecular diversity, biological implications of brain and thymus gangliosides: Y. Nogai, M. Iwamori, *Mol. Cell. Biochem.* **29**, 81 (1980). Biosynthesis in tissues: S. Basu *et al., Adv. Exp. Med. Biol.* **125**, 213 (1980). Action as biotransducers of membrane-mediated information: R. O. Brady, P. H. Fishman, *Adv. Enzymol.* **50**, 303 (1979). Reviews: Wiegandt, *Angew. Chem. Int. Ed.* **7**, 87-96 (1968). Collection of articles on structure and function: *Adv. Exp. Med. Biol.* **125**, 1-555 (1980); on structure analysis, function and properties, regeneration and recovery in the nervous system, clinical trials in amyotropic lateral sclerosis and diabetic neuropathy: *ibid.* **174**, 1-649 (1984). Book: *Gangliosides in Neurological and Neuromuscular Function, Development, and Repair,* M. M. Rapport, A. Gorio, Eds. (Raven Press, New York, 1981) 267 pp.

Gangliosides are colorless crystallizable substances which melt with decompn. Insol in non-polar solvents. Soly in polar solvents increases with the size of the sugar residue and the sialic acid content. Forms micelles in aq soln, having mol wt of about 200,000-250,000. Forms molecular soins in DMF or tetrahydrofuran, having mol wt of 1000-3000.

Ganglioside G$_{MI}$, C$_{73}$H$_{131}$N$_3$O$_{31}$, *siagoside, Sygen.* In combination with gangliosides G$_{D1a}$, G$_{D1b}$, and G$_{T1b}$, *Cronassial.*

4265. Gardenins. A group of flavones isolated from the resinous exudate of leaf buds of *Gardenia lucida* Roxb. *Rubiaceae*, a small spiny tree occurring throughout India.

Isoln of the main flavonoid pigment, gardenin A (originally thought to be the only component and referred to in early literature as "gardenin"): J. Stenhouse; C. G. Groves, *J. Chem. Soc.* **1877**, 552; *Ann.* **200**, 311 (1880); K. J. Balakrishna, T. R. Seshadri, *Proc. Indian Acad. Sci. Sect. A* **27**, 91 (1948). Isoln and structure of gardenins B, C, D, E: A. V. R. Rao *et al., Indian J. Chem.* **8**, 398 (1970). Isoln of B from *Citrus jambhiri* lush. *Rutaceae* and synthesis: B. P. Chaliha *et al., Tetrahedron* **21**, 1441 (1965). Structure of A: P. K. Bose, *J. Indian Chem. Soc.* **22**, 233 (1945). Revised structure: A. V. R. Rao, K. Venkataraman, *Indian J. Chem.* **6**, 677 (1968). Synthesis of A: M. Krishnamurti *et al., ibid.* **8**, 575 (1970); M. Kamalam, A. V. R. Rao, *ibid.* 573; of C: A. J. Kalra *et al., ibid.* **11**, 96 (1973); of E: *eidem, ibid.* 1092. Antineoplastic activity and cytotoxicity: J. Edwards *et al., J. Nat. Prod.* **42**, 85 (1979). ^{13}C-NMR study: M. Iinuma *et al., Chem. Pharm. Bull.* **28**, 708 (1980).

Gardenin A, C$_{21}$H$_{22}$O$_9$, *5-hydroxy-6,7,8-trimethoxy-2-(3,4,5-trimethoxyphenyl)-4H-1-benzopyran-4-one.* R = R' = OCH$_3$. Golden yellow needles from ethanol, mp 162-163°. Sol in alcohol, chloroform. All the gardenins give a green color with alcoholic ferric chloride.

Gardenin B, C$_{19}$H$_{18}$O$_7$, *5-O-desmethyltangeretin.* R = R' = H. Fine yellow needles from benzene/petr ether, mp 176-177°. uv max (ethanol): 292, 330 nm (log e 4.40, 4.35); (1% AlCl$_3$): 310, 350 nm (log ε 4.48, 4.49).

Gardenin C, C$_{20}$H$_{20}$O$_9$. R = OH, R' = OCH$_3$. Yellow flakes from ethyl acetate/petr ether, mp 179-180°. uv max (methanol) 304, 323 nm (log ε 4.18, 4.20).

Gardenin D, C$_{19}$H$_{18}$O$_8$. R = OH, R' = H. Cryst, mp 190-192°. uv max (ethanol): 256, 280, 343 nm.

Gardenin E, C$_{19}$H$_{18}$O$_9$. R = R' = OH. Golden yellow needles from ethyl acetate/petr ether, mp 234-235°. uv max (methanol): 280, 325 nm (log ε 4.37, 4.36).

4266. Gardinol Type Detergents. A mixture of the sodium salts of sulfated fatty alcohols made by reducing the mixed fatty acids of coconut oil or of cottonseed oil, and of fish oils. Sometimes natural waxes such as spermaceti, wool fat, and beeswax are sulfated directly. The mixture of the sulfated alcohols which goes by the commercial name "lauryl alcohol" consists of about 15% mixed C$_8$ and C$_{10}$ (octyl and decyl) alcohols, 40% C$_{12}$ (lauryl or dodecyl) alcohol, 30% C$_{14}$ (myristyl or tetradecyl) alcohol, and 15% mixed C$_{16}$ and C$_{18}$ (cetyl, stearyl, and oleyl) alcohols.

Some of the tradenames designating this type of detergent are: *Gardinol; Duponol; Modinal; Aurinol; Maprofix; Tergavon; Sadopan; Cyclopon; Cyclanon; Sapidan; Lissapol; Teepol. See also* Sodium Lauryl Sulfate.

USE: Detergents, wetting, emulsifying, dispersing agents for cosmetics, as dry cleaning aids, in fungicidal sprays, in metal cleaning, leather processing, in textile manuf. *Caution:* May cause local sensitivity reactions.

4267. Gardol®. *N-Methyl-N-(1-oxododecyl)glycine sodium salt; N-lauroylsarcosine sodium salt;* sodium *N-*lauroyl sarcosinate; Medialan LL-99. C$_{15}$H$_{28}$NNaO$_3$; mol wt 293.39. C 61.41%, H 9.62%, N 4.77%, Na 7.84%, O 16.36%. Prepn: Jungermann *et al., J. Am. Chem. Soc.* **78**, 172 (1956).

Aq soln, *Medialan LL-33.*

USE: Detergent, foaming agent, antienzyme for dentifrices: King, U.S. pat. **2,689,170** (1954 to Colgate-Palmolive).

4268. Garlic. Allium; Carisano. Fresh bulb of *Allium sativum* L., *Liliaceae*. *Habit.* Central Asia, Southern Europe,

U.S. *Constit.* Alliin, allicin, volatile and fatty oils, mucilage, albumin. Effect of garlic oil in hypercholesterolemic rabbits: R C. Jain, D. B. Konar, *Atherosclerosis* **29**, 125 (1978); in humans: A. Bordia, H. C. Bansal, *Lancet* **2**, 1491 (1973).

USE: As a spice and seasoning in foods.

4269. Garryine. $C_{22}H_{33}NO_2$; mol wt 343.49. C 76.92%, H 9.68%, N 4.08%, O 9.32%. From bark of *Garrya veatchii* Kellog, *Garryaceae*, where it occurs together with veatchine and other alkaloids: Oneto, *J. Am. Pharm. Assoc.* **35**, 204 (1946); Wiesner *et al.*, *Can. J. Chem.* **30**, 608 (1952). Structure: Wiesner *et al.: J. Am. Chem. Soc.* **76**, 6068 (1954); Djerassi *et al.*, *ibid.* **77**, 4801 (1955). Stereochemistry: Solo, Pelletier, *Chem. & Ind. (London)* **1960**, 1108. Racemic syntheses and resolution: Masamune, *J. Am. Chem. Soc.* **86**, 290 (1964); Nagata *et al.*, *ibid.* 929, **89**, 1499 (1967); Guthrie *et al.*, *Coll. Czech. Chem. Commun.* **31**, 602 (1966).

Monohydrate, crystals from dil acetone. Bitter taste, mp 74-82°. pK 8.70. $[\alpha]_D^{27.5}$ — 84.23° (c = 1.44 in ethanol). Sol in water, alcohol.

Hydrochloride, $C_{22}H_{33}NO_2 \cdot HCl$, crystals from abs ethanol + ether, dec 263-268°. Soluble in water.

4270. Gasoline. Petrol (British); Benzin (German). A mixture of C_4 to C_{12} hydrocarbons. Natural gasoline, obtained by fractional distillation of petroleum contains mostly saturated hydrocarbons; but the ordinary commercial grades of motor gasoline contain paraffins, olefins, naphthenes, and aromatics, all in substantial concns. Motor gasolines are made chiefly by cracking processes, in which heavier petr fractions are converted into more volatile fractions by thermal or catalytic decompn. (Where petr is scarce, as in Germany, gasoline also has been made commercially by catalytic high-pressure hydrogenation of soft coal and by catalytic synthesis of hydrocarbons from carbon monoxide and hydrogen.) Some gasolines sold in the U.S.A. contain a minor proportion of tetraethyllead, which is added in concns not exceeding 3 ml per gallon of motor gasoline to prevent "knock" in engines in which the gasoline is used as fuel. Knock is the audible manifestation of an excessive rate of pressure rise when the gasoline vapor is ignited under compression in an engine. (The relative knocking tendencies of gasolines are measured in terms of "Octane Number," which is defined as the percentage of iso-octane, having "100 Octane No.," to be blended with *n*-heptane, having "0 Octane No." by definition, in order to obtain the same degree of knock as is obtained with the gasoline being rated, under standard conditions in a standardized test engine.)

Commercial grades of tetraethyllead or Ethyl fluid typically contain about 63% tetraethyllead and about 35% ethylene dichloride or dibromide, which aids in evacuating the products of combustion of the lead compd from engines. In addition, the fluid contains a red or a blue dye. All leaded gasolines are dyed for recognition and should be used only as motor fuel. Other materials occasionally blended in gasoline to decrease knock, particularly in Europe, are benzene and ethanol. Comprehensive review: J. C. Lane, "Gasoline and Other Motor Fuels" in Kirk-Othmer *Encyclopedia of Chemical Technology* vol. 11 (Wiley-Interscience, New York, 3rd ed., 1980) pp 652-695.

Gasoline is a highly flammable, mobile liq with characteristic odor. Evaporates quickly: Flash pt ~—50°F (— 4-5°C). Explosive limits, vol % in air: lower 1.3, upper 6.0; sp gr 60/60°F: 0.72 to 0.76. Initial bp 39°C; after 10% distilled bp 60°; after 50% bp 110°; after 90% bp 170°; final bp 204°. Insol in water; freely sol in abs alcohol, ether, chloroform, benzene. Dissolves fats, oils, natural resins.

USE: As fuel in internal combustion engines of the spark-ignited, reciprocating type. *Caution:* Ingestion causes inebriation, vomiting, vertigo, fever, drowsiness, confusion,

cyanosis; aspiration causes bronchitis or pneumonia. Inhalation causes intense burning in throat and lungs; possibly bronchopneumonia.

4271. Gastrins. Gastrointestinal hormones isolated from the mucosal lining of the gastric antrum of various mammalian species. Highly potent gastric secretion stimulants, first discovered by J. S. Edkins: *Proc. Roy. Soc. London* **76B**, 376 (1905). Several gastrins have been identified; they are referred to as "little gastrin", "big gastrin" and "minigastrin". "Little gastrin" or G-17 exists in two forms, 18-34-gastrin I or G-17-I and 18-34-gastrin II or G-17-II, which are heptadecapeptides that are identical in amino acid sequence, with the latter having sulfated tyrosine residues. There are relatively small species differences in amino acid sequences of G-17, although there are differences in the ratio of the non-sulfated to the sulfated forms. This ratio is about 3:4 in hog compared to about 2:1 in man. "Big gastrin" or 1-34-gastrin, also found in two forms, consists of the heptadecapeptides of "little gastrins" extended from their *N*-termini by additional heptadecapeptides with amino acid compositions different from G-17. For each species, the G-17-I and G-17-II portions of the G-34 gastrin are identical. Pairs of shorter gastrins, or "minigastrins" (22-34-gastrin) are C-terminal tetradecapeptides.

```
5-oxo-Pro-Gly-Pro-Trp-Leu-Glu-Glu-Glu-Glu-Glu-

Ala-Tyr-Gly-Trp-Met-Asp-Phe-NH2

        18-34-gastrin I (human)
```

Structure of porcine "little gastrins": Gregory *et al.*, *Nature* **204**, 391 (1964); human: Bentley *et al.*, *ibid.* **209**, 583 (1966). Synthesis of porcine "little gastrins": Anderson *et al.*, *ibid.* **204**, 933 (1964); human: Beecham *et al.*, *ibid.* **209**, 585 (1966); human and canine: Agarwal, Kenner, *J. Chem. Soc. (C)* **1969**, 2213; ovine: Agarwal *et al.*, *ibid.* 954; feline: *eidem, Experientia* **25**, 346 (1969). General synthetic method: E. Brown *et al.*, *Chem. Commun.* **1980**, 1093. Isoln of "big gastrins" from Zollinger-Ellison tumor tissue: R. A. Gregory, H. J. Tracy, *Lancet* **2**, 797 (1972). Isoln of "minigastrins" from Zollinger-Ellison tumor tissue: *eidem, Gut* **15**, 683 (1974). Amino acid sequences of porcine and human "big gastrins": *eidem*, in *Gastrointestinal Hormones*, J. C. Thompson, Ed. (Univ. Texas Press, Austin, 1975) pp 13-14. Revised sequence of porcine: G. J. Dockray *et al.*, *Bioorg. Chem.* **8**, 465 (1979); of human: A. M. Choudhury *et al.*, *Z. Physiol. Chem.* **361**, 1719 (1980). Synthesis of human: G. Wendlberger *et al.*, *Monatsh.* **112**, 1297 (1981). Structure and synthesis of "minigastrins": R. A. Gregory *et al.*, *Z. Physiol. Chem.* **360**, 73 (1979). Review of chemical studies: Kenner, Sheppard, *Proc. Roy. Soc. London* **170B**, 89 (1968). Review of physiological advances: Gregory, *ibid.* 81. General reviews: Sanders, Schimmel, *Am. J. Med.* **49**, 380 (1970); McGuigan, *Vitam. Horm.* **32**, 47-88 (1974); V. Mutt, *ibid.* **39**, 231-426 (1982).

4272. Gaultherin. 2-[(6-O-β-D-*Xylopyranosyl*-β-D-*glu-copyranosyl*)oxy]benzoic acid methyl ester; methyl salicylate-2-glucoxyloside; methyl salicylate-2-primeveroside; monotropitin; monotropitoside. $C_{19}H_{26}O_{12}$; mol wt 446.40. C 51.12%, H 5.87%, O 43.01%. In the wintergreen plant, *Gaultheria procumbens* L., in *Monotropa hypopitys* L., *Ericaceae* in *Betula lenta* L., *Betulaceae*, in *Spiraea ulmaria* L. and *S. filipendula* L., *Rosaceae:* Bridel, *Compt. Rend.* **177**, 642 (1923); **179**, 991 (1924); **180**, 1421, 1864 (1925); Bridel, Grillon, *ibid.* **187**, 609 (1928). Synthesis: Robertson, Waters, *J. Chem. Soc.* **1931**, 1881. On hydrolysis with 3% H_2SO_4 gaultherin forms 1 mol methyl salicylate, 1 mol D-glucose and 1 mol D-xylose. Enzymatic hydrolysis gives methyl salicylate and primeverose (glucoxylose).

Needles in star formation from 99% acetone. mp 180°. $[\alpha]_D^{20}$ −58° (c = 2). Sol in water and alcohol, slightly sol in ethyl acetate, acetone. Insol in ether.

Note: The older literature regards gaultherin as methyl-salicylate-D-glucoside.

Gaultherioside, $C_{13}H_{24}O_{10}$, *ethyl primeveroside.* mp 185°. $[\alpha]_D$ −58°. Gaultherioside forms glucose, xylose, and EtOH on hydrolysis with 3% H_2SO_4.

4273. Gefarnate. *5,9,13-Trimethyl-4,8,12-tetradecatri-enoic acid 3,7-dimethyl-2,6-octadienyl ester; geranyl farnes-ylacetate; DA-688; Alsanate; Arsanyl; Dixnalate; Gefanil; Gefarnil; Gefarnyl; Gefulcer; Osteol; Salanil; Terpanil; Zackal.* $C_{27}H_{44}O_2$; mol wt 400.62. C 80.94%, H 11.07%, O 7.99%. Prepn: Cardani et al., J. Med. Chem. **6,** 457 (1963); **Belg. pat.** 617,994 (1962 to Ist. de Angeli). Metabolism: G. Coppi et al., Arzneimittel-Forsch. **19,** 1519 (1969). Clinical evaluation in duodenal ulcer: C. R. Newman, D. A. Montgomery, Brit. J. Clin. Pract. **27,** 85 (1973).

Slightly yellowish liq with weak terpenic smell. $bp_{0.05}$ 165-168°. n_D^{20} 1.4900. uv max: 204 nm ($E_{1cm}^{1\%}$ 486). Sol in alc, ether, dimethylformamide, acetone, fatty oils. Practically insol in water, formamide, ethylene glycol, propylene glycol, glycerine.

THERAP CAT: Antiulcerative.

4274. Geissoschizoline. *Curan-17-ol;* pereirine. $C_{19}H_{26}N_2O$; mol wt 298.41. C 76.47%, H 8.78%, N 9.39%, O 5.36%. From bark of *Geissospermum vellosii* Allem., Apocynaceae: Hesse, Ann. **202,** 141 (1880); Bertho, Moog, ibid. **509,** 241 (1934). Identity with pereirine: Bertho, Koll, Naturwiss. **48,** 49 (1961). Structure: Janot et al., Compt. Rend. **250,** 4383 (1960); Bertho, Koll Ber. **94,** 2737 (1961). Synthesis: Hymon, Schmid, Helv. Chim. Acta **49,** 2067 (1966); of dl-form: Dadson, Harley-Mason, Chem. Commun. **1969,** 665; Harley-Mason, Taylor, ibid. **1970,** 812.

Coarse crystals from abs methanol, mp 142.5-143°. $[\alpha]_D^{21}$ +32° (ethanol). uv max in (ethanol): 245, 300 nm (log ε 3.93, 3.47). Sol in alcohol, chloroform, ether; practically insol in water.

Methyl chloride, $C_{19}H_{26}N_2O \cdot CH_3Cl$, rods from abs methanol, mp 297°.

Methiodide, $C_{19}H_{26}N_2O \cdot CH_3I$, rods from abs methanol, mp 254°.

4275. Geissospermine. *(16R,19E)-19,20-Didehydro-16-[(10β,13β,21β)-23-deoxy-21,22-dihydro-11-oxa-12,14-seco-strychnidin-10-yl]corynan-17-oic acid methyl ester; 19,20-didehydro-16-(15-ethyl-3a,5,5a,7,8,13a-hexahydro-4H-4,6-ethano-1H,3H-[1,3]oxazino[3,4,5-lm]pyrrolo[2,3-d]carbazol-1-yl)-corynan-17-oic acid methyl ester; β,17-epoxy-α-(3-ethylidene-1,2,3,4,6,7,12,12b-octahydroindole[2,3-a]quinol-izin-2-yl)-curan-1-propanoic acid methyl ester.* $C_{40}H_{48}N_4O_3$; mol wt 632.82. C 75.91%, H 7.65%, N 8.85%, O 7.59%. From bark of *Geissospermum vellosii* Allem., Apocynaceae: Hesse, Ann. **202,** 141 (1880). Structure: Puisieux, LeHir, Compt. Rend. **252,** 902 (1961); Janot, Tetrahedron **14,** 113 (1961). Crystal structure: A. Chiaroni, C. Riche, Acta Crystallogr. **35B,** 1820 (1979). Pharmacology: M. M. Aurousseau, Ann. Pharm. Franc. **19,** 175 (1961). On hydrolysis with HCl splits into geissoschizine and geissoschizoline, q.v.

Anhydr form, crystals from abs acetone, mp 213-214° (dec). $[\alpha]_D^{20}$ −101° (ethanol). uv max (methanol): 251, 285, 293 nm (log ε 4.10, 3.91, 3.90). Slightly sol in water, ether; sol in alcohol.

Dihydrate, mp 207° (sinters at 145°).

4276. Gelatin. Gelfoam; Puragel. A heterogeneous mixture of water-soluble proteins of high average mol wt. Gelatin is not found in nature but derived from collagen, q.v. by hydrolytic action. Obtained by boiling skin, tendons, ligaments, bones, etc., with water. Approx amino acid content: glycine 25.5%, alanine 8.7%, valine 2.5%, leucine 3.2%, isoleucine 1.4%, cystine and cysteine 0.1%, methionine 1.0%, phenylalanine 2.2%, proline 18.0%, hydroxyproline 14.1%, serine 0.4%, threonine 1.9%, tyrosine 0.5%, aspartic acid 6.6%, glutamic acid 11.4%, arginine 8.1%, lysine 4.1%, histidine 0.8%. The total is over 100% because water is incorporated into the molecules of the individual amino acids. Nutritionally, gelatin is an incomplete protein lacking tryptophan and contg but small amounts of other important amino acids. Review of the chemistry and structure of collagen with emphasis on its transformation to gelatin: A. Veis, *The Macromolecular Chemistry of Gelatin* (Academic Press, New York, 1964) 433 pp.

Colorless or slightly yellow, transparent, brittle, practically odorless, tasteless sheets, flakes, or coarse powder. Swells up and absorbs 5-10 times its weight of water to form a gel in solutions below 35-40°. Sol in hot water, glycerol, acetic acid. Insol in organic solvents. Amphoteric.

USE: As stabilizer, thickener and texturizer in food; manuf rubber substitutes, adhesives, cements, lithographic and printing inks, plastic compds, artificial silk, photographic plates and films, matches, light filters for mercury lamps; clarifying agent; in hectographic masters; sizing paper and textiles; for inhibiting crystn in bacteriology, for preparing cultures. Pharmaceutic aid (suspending agent; encapsulating agent; tablet binder; tablet and coating agent). *Incompat:* Tannin, formaldehyde.

THERAP CAT (VET): Plasma expander; hemostasis (sponge).

4277. Gelsemine. $C_{20}H_{22}N_2O_2$; mol wt 322.40. C 74.50%, H 6.88%, N 8.69%, O 9.93%. CNS stimulant from roots and rhizome of *Gelsemium sempervirens* (L.) Ait., Loganiaceae. Isoln: Gerrard, Pharm. J. **13,** 641 (1883); Moore, J. Chem. Soc. **97,** 2223 (1910); **99,** 1231 (1911); Sayre, Watson, J. Am. Pharm. Assoc. **8,** 708 (1919); Chou, Chinese J. Physiol. **5,** 131 (1931), C.A. **25,** 4085[6] (1931); Schwarz, Marion, Can. J. Chem. **31,** 958 (1953). Structure: Conroy, Chakrabarti, Tetrahedron Letters **1959**(4), 6; Lovell et al., ibid. 1; Roe, Gates, Tetrahedron **11,** 148 (1960). NMR spectroscopic study: Y. Schun, G. A. Cordell, J. Nat. Prod. **48,** 969 (1985). Partial syntheses: W. E. Earley et al., Tetrahedron Letters **29,** 3781, 3785 (1988).

Crystals from acetone, mp 178°. *Poisonous!* $[\alpha]_D^{20}$ +13°

(c = 1.2 in chloroform). pKa 7.75 in 80% methylcellosolve. uv max (methanol): 210, 252, 280 nm (log ϵ 4.50, 3.87, 3.15). Slightly sol in water; sol in alcohol, benzene, chloroform, ether, acetone, dilute acids.

Hydrochloride, $C_{20}H_{23}ClN_2O_2$, prisms from methanol + ether, mp 326°. $[\alpha]_D$ +5° (c = 1.072 in water). Sol in water, slightly sol in alcohol.

4278. Gelsemium. Yellow jasmine; yellow jessamine; wild woodbine. CNS stimulant from dried rhizome and roots of *Gelsemium sempervirens* (L.) Ait., *Loganiaceae*. *Habit.* Southern U.S. *Constit.* Gelsemine, gelsemoidine, scopoletin, gelsemic acid, volatile oil, resin. *Poisonous!*

4279. Gemeprost. *11,15-Dihydroxy-16-16-dimethyl-9-oxoprosta-2,13-dien-1-oic acid methyl ester;* 16,16-dimethyl-*trans*-Δ^2-PGE$_1$ methyl ester; Ono 802; Cergem; Cervagem(e); Preglandin. $C_{23}H_{38}O_5$; mol wt 384.47. C 71.85%, H 7.34%, O 20.81%. Analog of prostaglandin E$_1$, *q.v.* Prepn: M. Hayashi *et al.*, **Ger. pat.** 2,700,021 corresp to U.S. pat. 4,052,512 (both 1977 to Ono); H. Suga *et al.*, *Prostaglandins* **15**, 907 (1978). Effects on uterine contractility and steroid hormone plasma levels: K. Oshima *et al.*, *J. Reprod. Fertil.* **55**, 353 (1979). Effects on reproductive function: K. Matsumoto *et al.*, *Nippon Yakurigaku Zasshi* **79**, 15 (1982), *C.A.* **96**, 98392 (1982). Use in termination of first trimester pregnancy: O. Reiertsen *et al.*, *Prostaglandins Leukotrienes Med.* **8**, 31 (1982).

THERAP CAT: Abortifacient; oxytocic.

4280. Gemfibrozil. *5-(2,5-Dimethylphenoxy)-2,2-dimethylpentanoic acid; 2,2-dimethyl-5-(2,5-xylyloxy)valeric acid;* CI-719; Decrelip; Gevilon; Lipozid; Lipur; Lopid. $C_{15}H_{22}O_3$; mol wt 250.35. C 71.97%, H 8.86%, O 19.17%. Serum lipid regulating agent. Prepn: P. L. Creger, **Ger. pat.** 1,925,423; *eidem*, U.S. pat. 3,674,836 (1969, 1972, both to Parke, Davis). Production: O. P. Goel, U.S. pat. 4,126,637 (1978 to Warner-Lambert). Pharmacology: A. H. Kissebach *et al.*, *Atherosclerosis* **24**, 199 (1976); M. T. Kahonen *et al.*, *ibid.* **32**, 47 (1979). Series of articles on metabolism, clinical pharmacology, kinetics and toxicology: *Proc. Roy. Soc. Med.* **69**, Suppl 2, 1-120 (1976). Toxicity data: S. M. Kurtz *et al.*, *ibid.* **15**. Controlled clinical trial in hyperlipidemia: J. E. Lewis *et al.*, *Pract. Cardiol.* **9**, 99 (1983). Multicenter clinical trial of long-term effect on coronary heart disease: V. Manninen *et al.*, *J. Am. Med. Assoc.* **260**, 641 (1988).

Crystals from hexane, mp 61-63°. bp$_{0.02}$ 158-159°. LD$_{50}$ in mice, rats (mg/kg): 3162, 4786 orally (Kurtz).

THERAP CAT: Antihyperlipoproteinemic.

4281. Genistein. *5,7-Dihydroxy-3-(4-hydroxyphenyl)-4H-1-benzopyran-4-one; 4',5,7-trihydroxyisoflavone;* prunetol; genisteol. $C_{15}H_{10}O_5$; mol wt 270.23. C 66.67%, H 3.73%, O 29.60%. The aglucon of genistin and of sophoricoside, *q.v.* Prepn from the glucoside by hydrolysis with emulsin: Charaux, Rabaté, *J. Pharm. Chim.* [9] **1**, 404 (1941); by hydrolysis with HCl in methanol: Walter, *J. Am. Chem. Soc.* **63**, 3273 (1941). Isoln from *prunus* spp., *Rosaceae*: Hasegawa *ibid.* **79**, 1738 (1957); from *Podocarpus spicata* R.Br., *Podocarpaceae*: Briggs, Cebalo, *Tetrahedron* **6**, 145 (1959). Structure: Baker, Robinson, *J. Chem. Soc.* **1925**, 1981; **1926**, 2713; Walz, *Ann.* **489**, 118 (1931). Synthesis: Baker, Robinson, *J. Chem. Soc.* **1928**, 3115; Narasimhachari

et al., *J. Sci. Ind. Res.* **12**, 287 (1953); Yoder *et al.*, *Proc. Iowa Acad. Sci.* **61**, 271 (1954); Zemplén *et al.*, *Acta Chim. Acad. Sci. Hung.* **19**, 277 (1959).

Rectangular or six-sided rods from 60% alcohol. Dendritic needles from ether. mp 297-298° (slight decompn). Sol in the usual organic solvents (Walter, *loc. cit.*); practically insol in water; sol in dil alkalies with yellow color. uv max: 262.5 nm (ϵ 138). Alkaline hydrolysis yields phloroglucinol and *p*-hydroxyphenylacetic acid.

Triacetate, $C_{15}H_7O_5(CH_3CO)_3$, clusters of needles from alcohol, mp 200-202°. Sol in organic solvents; practically insol in alkalies.

7-D-Glucoside,$C_{21}H_{20}O_{10}$, genistin. For isoln and structure *see* Walter, Hasegawa, Walz, *loc. cit.* Synthesis: Zemplén, Farkas, *Ber.* **76B**, 1110 (1943). Pale yellow plates from 80% ethanol, mp 256°. $[\alpha]_D^{21}$ −28° (c = 0.6 in 0.02N NaOH); $[\alpha]_D^{26}$ −21.4° (pyridine). uv max (85% ethanol): 262.5 nm (a 90.5). Practically insol in cold water; slightly sol in hot water, hot ethanol, hot methanol; sol in hot 80% ethanol, hot 80% methanol, hot acetone, pyridine.

4'-Methyl ether, $C_{16}H_{12}O_5$, *5,7-dihydroxy-4'-methoxyisoflavone; Biochanin A; olmelin.* Isoln from red clover: Pope *et al.*, *Chem. & Ind. (London)* **1953**, 1092; Wong, *J. Sci. Food Agr.* **13**, 304 (1962); from *Andira inermis* (Swartz) H.B.K., *Leguminosae*: Crocker *et al.*, *J. Chem. Soc.* **1962**, 4906. Identity with olmelin: Gakhokidze, *J. Appl. Chem. USSR* **23**, 789 (1950), *C.A.* **46**, 9098i (1952). Structure: Bose, Siddiqui, *J. Sci. Ind. Res.* **9B**, no. 1, 25 (1950). Synthesis: Baker *et al.*, *Nature* **169**, 706 (1952). Yellow needles from methanol, mp 212-216°.

4282. Genisteine-Alkaloid. *[7S-(7α,7aα,14α,14aα)]-Dodecahydro-7,14-methano-2H,6H-dipyrido[1,2-a:1',2'-e]-[1,5]diazocine;* 11-isosparteine; l-α-isosparteine. $C_{15}H_{26}N_2$; mol wt 234.37. C 76.86%, H 11.18%, N 11.95%. A stereoisomer of sparteine, *q.v.*, isolated from *Cytisus scoparius* (L.) Link, (*Spartium scoparium* L.), *Leguminosae:* Valeur, *Compt. Rend.* **167**, 23, 163 (1918); from *Lupinus caudatus* Kellog, *Leguminosae:* Marion *et al.*, *Can. J. Chem.* **29**, 22 (1951). Identity with l-α-isosparteine: Marion, Leonard, *ibid.* 297. Structure: Leonard *et al.*, *J. Am. Chem. Soc.* **77**, 1552 (1955). Absolute configuration: Okuda, Tsuda, *Chem. & Ind. (London)* **1961**, 1115.

Monohydrate, needles from boiling acetone, mp 108-110°. $[\alpha]_D^{22}$ −51.6° (c = 0.7 in abs ethanol).

4283. Genite®. *2,4-Dichlorophenol benzenesulfonate;* benzenesulfonic acid 2,4-dichlorophenyl ester; 2,4-dichlorophenyl benzenesulfonate; EM-923; Genitol 923. $C_{12}H_8Cl_2O_3S$; mol wt 303.17. C 47.54%, H 2.66%, Cl 23.39%, O 15.83%, S 10.58%. Prepn: Gilbert, U.S. pat. 2,618,583 (1952 to Allied Chemical).

Waxy solid. Practically insol in water; sol in most organic solvents.

USE: Acaricide, Gilbert, *loc. cit.*

4284. Gentamicin. *Gentamycin.* Antibiotic complex

produced by fermentation of *Micromonospora purpurea* or *M. echinospora* and variants thereof: M. J. Weinstein *et al.*, *Antimicrob. Ag. Chemother.* **1963**, 1. Isoln, purification and characterization: J. P. Rosselet *et al.*, *ibid.* 14. Industrial pats.: G. M. Luedemann, M. J. Weinstein; Charney, U.S. pats. **3,091,572; 3,136,704** (1963, 1964, both to Schering). Consists of three closely related components, gentamicins C_1, C_2, C_{1a}, and also gentamicin A which differs from the other members of the complex but is similar to kanamycin C, *q.v.* Separation of gentamicin C components: H. Maehr, C. P. Schaffner, *J. Chromatog.* **30**, 572 (1967); Wagman *et al.*, *ibid.* **34**, 210 (1968). Structures contain 2-deoxystreptamine, *q.v.*, linked to two saccharide units, these being *garosamine* and a *purpurosamine* in the C series gentamicins. Structure studies: D. J. Cooper *et al.*, *J. Chem. Soc. (C)* **1971**, 960, 2876, 3126. Structure of gentamicin A: H. Maehr, C. P. Schaffner, *J. Am. Chem. Soc.* **89**, 6787 (1967); **92** 1697 (1969). Separation and structures of gentamicins A_1, A_2, A_3, and A_4: Nagabhushan *et al.*, *J. Org. Chem.* **40**, 2830, 2835 (1975). Synthetic studies: W. Meyer zu Reckendorf, Bischof, *Ber.* **105**, 2546 (1972); M. Chmielewski *et al.*, *Carbohyd. Res.* **70**, 275 (1979). Pharmacology in man: L. J. Riff, G. G. Jackson, *J. Infec. Dis.* **124**, Suppl., S98 (1971); in animals and man: J. Black *et al.*, *Antimicrob. Ag. Chemother.* **1963**, 138; L. Hepding, H. Wahling, *Arzneimittel-Forsch.* **16**, 1 (1966). Comprehensive description: B. E. Rosenkrantz *et al.*, in *Analytical Profiles of Drug Substances* Vol. **9**, K. Florey, Ed. (Academic Press, New York, 1980) pp 295-340.

gentamicin C_1 $R_1 = R_2 = CH_3$
C_2 $R_1 = CH_3$; $R_2 = H$
C_{1a} $R_1 = R_2 = H$

White amorphous powder, mp 102-108°. $[\alpha]_D^{25}$ +146°. Freely sol in water; sol in pyridine, DMF, in acidic media with salt formation; moderately sol in methanol, ethanol, acetone. Practically insol in benzene, halogenated hydrocarbons.
Gentamicin C_1, $C_{21}H_{43}N_5O_7$, mp 94-100°. $[\alpha]_D^{25}$ +158°.
Gentamicin C_2, $C_{20}H_{41}N_5O_7$, mp 107-124°. $[\alpha]_D^{25}$ +160°.
Gentamicin C_{1a}, $C_{19}H_{39}N_5O_7$, *O-3-deoxy-4-C-methyl-3-(methylamino)-β-L-arabinopyranosyl-(1 → 6)-O-[2,6-diamino-2,3,4,6-tetradeoxy-α-D-erythro-hexopyranosyl-(1 → 4)]-2-deoxy-D-streptamine.* Also known as *gentamicin D.*
Gentamicin A, $C_{18}H_{36}N_4O_{10}$, *O-2-amino-2-deoxy-α-D-glucopyranosyl-(1 → 4)-O-[3-deoxy-3-(methylamino)-α-D-xylopyranosyl-(1 → 6)]-2-deoxy-D-streptamine.* Hydrochloride, mp 194-209°. $[\alpha]_D^{25}$ +113°. Freely sol in water, methanol; slightly in ether. Practically insol in other organic solvents.
C complex sulfate, *Alcomicin, Bristagen, Cidomycin, Duragentam, Garamycin, Garasol, Genoptic, Gentacin, Gentak, Gentalline, Gentalyn, Gentibioptal, Genticin, Gentocin, Gentogram, Gent-Ophtal, Lugacin, Nichogencin, Ophtagram, Pangram, Refobacin, Septopal, Sulmycin, U-gencin.* White, hygroscopic powder, mp 218-237°. $[\alpha]_D^{25}$ +102°. Sol in ethylene glycol, formamide. LD_{50} i.p. in mice: 430 mg/kg (Black); also reported as 490 mg/kg (Hepding, Wahling).
THERAP CAT: Antibacterial.
THERAP CAT (VET): Antibacterial.

4285. Gentian. Yellow gentian; pale gentian; bitter root. Dried rhizome and roots of *Gentiana lutea* L., Gentianaceae. *Habit.* Central and Southern Europe. *Constit.* Gentiin, gentiamarin, gentisin, gentisic acid, gentiopicrin, gentianose (a sugar), pectin.
THERAP CAT: Bitter tonic.
THERAP CAT (VET): Bitter tonic.

4286. Gentianine. *5-Ethenyl-3,4-dihydro-1H-pyrano-*

[3,4-c]pyridin-1-one; 4-(2-hydroxyethyl)-5-vinylnicotinic acid δ-lactone; erythricine. $C_{10}H_9NO_2$; mol wt 175.18. C 68.56%, H 5.18%, N 8.00%, O 18.27%. From *Gentiana kirilowi*, Gentianaceae: Proskurnina, *J. Gen. Chem. USSR* **14**, 1148 (1944), *C.A.* **40**, 7213 (1946); from *Anthocleista procera* Afz. and *Fagraea fragrans* Roxb., Loganiaceae: Lavie, Taylor-Smith, *Chem. & Ind. (London)* **1963**, 781; Wan, Chow, *J. Pharm. Pharmacol.* **16**, 484 (1964). Isoln from *Slevogtia orientalis* Gris., Gentianaceae, structure and synthesis: Govindachari *et al.*, *J. Chem. Soc.* **1957**, 551, 2725.

Needles from ether or petr ether, mp 82-83°. uv max: 220 nm (log ε 4.38). Sol in alkali.
Hydrochloride, $C_{10}H_9NO_2$.HCl, needles from alcohol + ether, dec 169-170°.
Picrate, $C_{10}H_9NO_2$.$C_6H_3N_3O_7$, yellow needles from water, mp 123-124°.
Dihydrogentianine, $C_{10}H_{11}NO_2$, crystals from ether + petr ether, mp 74-76°. uv max: 270 nm (log ε 3.4).

4287. Gentian Violet. *N-[4-[Bis[4-(dimethylamino)-phenyl]methylene]-2,5-cyclohexadien-1-ylidene]-N-methyl-methanaminium chloride; C.I. Basic Violet 3;* hexamethyl-pararosaniline chloride; hexamethyl-*p*-rosaniline chloride; aniline violet; crystal violet; methylrosaniline chloride; C.I. 42555; Adergon; Axuris; Badil; Gentiaverm; Meroxylan; Meroxyl; Pyoktanin; Vianin; Viocid. $C_{25}H_{30}ClN_3$; mol wt 408.00. C 73.60%, H 7.41%, Cl 8.69%, N 10.30%. Prepn and properties: *Colour Index* vol. **4** (3rd ed., 1971) p 4391. Toxicity studies: H. C. Hodge *et al.*, *Toxicol. Appl. Pharmacol.* **22**, 1 (1972).

Dark green powder or greenish, glistening pieces with metallic luster. Sol in water, chloroform. Practically insol in ether. One gram dissolves in about 10 ml alcohol, in about 15 ml glycerin. LD_{50} orally in mice, rats: 1.2, 1.0 g/kg (Hodge). Commercial product, which is usually admixed with *pentamethylpararosaniline chloride* and *tetramethylpararosaniline chloride*, contains not less than 96% gentian violet.
USE: As dye for wood, silk, paper; in inks; as biological stain.
THERAP CAT: Topical anti-infective. Has been used as anthelmintic (Nematodes).
THERAP CAT (VET): Topical antimicrobial.

4288. Gentiobiose. *6-O-β-D-Glucopyranosyl-D-glucose;* 6-(β-D-glucosido)-D-glucose; amygdalose. $C_{12}H_{22}O_{11}$; mol wt 342.30. C 42.10%, H 6.48%, O 51.42%. From gentianose by partial hydrolysis with 0.2% H_2SO_4 or with invertin. From D-glucose by enzymatic synthesis with emulsin: Helferich, Lette, *Org. Syn.* **22**, 53 (1942). Prepn and structure: Haworth, Wylam, *J. Chem. Soc.* **123**, 3120 (1923); Hudson, *J. Am. Chem. Soc.* **51**, 1708 (1930). Structure: Hassid, Ballou in W. Pigman, *The Carbohydrates* (Academic Press, New York, 1957) p 492. Synthesis: Helferich, Klein, *Ann.* **450**, 219 (1926); Reynolds, Evans, *J. Am. Chem. Soc.* **60**, 2559 (1938). Hydrolysis with almond emulsin gives 2 mols D-glucose.

α-Form, lentil-shaped crystals with 2CH$_3$OH from methanol. Bitter taste. Very hygroscopic. mp 86°. Shows mutarotation. $[\alpha]_D^{22}$ +16° (3 min) → +8.3° (3½ hrs c = 4). Sol in water, hot methanol, hot 90% alcohol.

β-Form, anhydr crystals from alc, mp 190-195°. Shows mutarotation. $[\alpha]_D^{22}$ − 5.9° (6 min) → +9.6° (6 hrs, c = 3). Sol in water, hot methanol, hot 90% alcohol.

4289. Gentiopicrin. *5-Ethenyl-6-(β-D-glucopyranosyl-oxy)-5,6-dihydro-1H,3H-pyrano[3,4-c]pyran-1-one;* gentiopicroside. C$_{16}$H$_{20}$O$_9$; mol wt 356.32. C 53.93%, H 5.66%, O 40.41%. The principal bitter glucoside of common gentians, which was isolated in 1862 from *Gentiana lutea*, L., *Gentianaceae:* Canonica *et al., Tetrahedron* **16**, 192 (1961). Isoln from roots of *G. lutea:* Korte, *Ber.* **87**, 512 (1954); Korte *et al., ibid.* **91**, 759 (1958). Structure: Canonica *et al., Tetrahedron Letters* **1960**(24), 7; *Tetrahedron* **16**, 192 (1961). Revised structure: Inouye *et al., Tetrahedron Letters* **1968**, 4429. Absolute configuration: Manitto, Pagnoni, *Gazz. Chim. Ital.* **94**, 229 (1964).

Crystals from anhydr ethyl acetate or abs alcohol, mp 191°. $[\alpha]_D^{20}$ − 199° (ethanol). uv max (c = 0.0285 g/l methanol): 270 nm (log ε 3.96).

Hemihydrate, crystals from 50% ethanol, mp 121°.

Tetraacetate, C$_{24}$H$_{28}$O$_{13}$, crystals from methanol. mp 138.5-139.5°. uv max (c = 0.0214 g/l methanol): 272 nm (log ε 3.84).

THERAP CAT: Antimalarial.

4290. Gentisic Acid. *2,5-Dihydroxybenzoic acid;* 5-hydroxysalicylic acid. C$_7$H$_6$O$_4$; mol wt 154.12. C 54.55%, H 3.92%, O 41.52%. Occurs in gentian: Redgrove, *Pharm. J.* **122**, 324 (1929). Found in urine of dogs after ingestion of salicylates: Neuberg, *Berl. Klin. Wochenschr.* **48**, 799 (1911). Prepd from hydroquinone: Senhofer, Sarlay, *Monatsh.* **2**, 448 (1881); Juch, *ibid.* **26**, 839 (1905); Brunner, *Ann.* **351**, 321 (1907); Zeltner, Landau, **Ger. pat. 258,887;** *Frdl.* **11**, 210; Dyson, *Manual of Organic Chemistry* vol. **I** (London, 1950) p 614; by oxidation of salicylic acid with potassium persulfate: **Ger. pat. 81,297** (to Schering AG); *Frdl.* **4**, 127; from *p*-hydroxyphenol: Meyer, **U.S. pat. 2,588,336** (1952 to Monsanto); from 5-bromo-2-hydroxybenzoic acid: Lowenthal, Pepper, *J. Am. Chem. Soc.* **72**, 3292 (1950); by Kolbe synthesis: Clemens, **U.S. pat. 2,816,137** (1957 to Eastman Kodak). Metabolic product of *Penicillium patulum:* Tanenbaum, Bassett, *Biochim. Biophys. Acta* **28**, 21 (1958); of *Polyporus tumulosus* Cooke: Crowden, Ralph, *Aust. J. Chem.* **14**, 475 (1961). Biosynthesis: Gatenback, Linnroth, *Acta Chem. Scand.* **16**, 2298 (1962).

Needles, monoclinic prisms from water, mp 199-200°. Crystals are dimorphic and undergo phase inversion upon heating. The stable phase begins to sublime at 200° and melts at 205°. K at 25° = 1.18 × 10^{-3}. Sol in water (about 1 part in 200 parts H$_2$O at 5°, much more sol in hot water), alcohol, ether. Practically insol in carbon disulfide, chloroform, benzene.

Sodium salt, C$_7$H$_5$NaO$_4$, *sodium gentisate, Gentinatre, Gentisod, Legential, Gentisine U.C.B.* Crystals with 5½ H$_2$O from water. Rapidly loses 3H$_2$O on exposure to air, but holds ½ H$_2$O tenaciously even at 100°. Sol in water.

THERAP CAT: Analgesic; anti-inflammatory.

4291. Gentisin. *1,7-Dihydroxy-3-methoxy-9H-xanthen-9-one;* 4,7-dihydroxy-2-methoxyxanthone; gentianic acid; gentianin; gentiin. C$_{14}$H$_{10}$O$_5$; mol wt 258.22. C 65.12%, H 3.90%, O 30.98%. From root of *Gentiana lutea* L., *Gentianaceae:* Henry, Caventou, *J. Pharm. Chim.* **1821**, 178; Canonica, Pelizzoni, *Gazz. Chim. Ital.* **85**, 1007 (1955). Structure: Korte, *Ber.* **87**, 1357 (1954). Synthesis: Anand, Venkataraman, *Proc. Indian Acad. Sci.* **25A**, 438 (1947); Rao, Seshadri, *ibid.* **37A**, 710 (1953).

Yellow needles from alc, mp 266-267°. Absorption max (methanol): 260, 275, 315, 410 nm (log ε 4.35, 4.30, 4.10, 3.70). Very slightly sol in water or organic solvents.

Diacetate, C$_{18}$H$_{14}$O$_7$, crystals from alc, mp 196-197°. uv max (methanol): 240, 270, 300 nm (log ε 4.58, 4.05, 4.10).

4292. Gentisyl Alcohol. *2-(Hydroxymethyl)-1,4-benzenediol;* 2,5-dihydroxybenzyl alcohol. C$_7$H$_8$O$_3$; mol wt 140.13. C 60.00%, H 5.75%, O 34.25%. A metabolic product of *Penicillium patulum:* Birkinshaw *et al., Biochem. J.* **37**, 726 (1943); Brack, *Helv. Chim. Acta* **30**, 1 (1947); Tanenbaum, Bassett, *Biochim. Biophys. Acta* **28**, 21 (1958). Prepn from gentisaldehyde: Birkinshaw, *loc. cit.*

Clusters of needles from chloroform. mp 100°. Sublimes in high vacuum at 70-80° (bath temp). Freely sol in water, alcohol, ether; slightly sol in benzene, chloroform. Insol in petr ether. Upon heating with 2N H$_2$SO$_4$, a white amorphous precipitate is obtained. With aq NaOH: yellow color deepening to reddish brown.

4293. Gentrogenin. *(25R)-3β-Hydroxyspirost-5-en-12-one;* 20α,22a,25D-spirost-5-en-3β-ol-12-one; botogenin. C$_{27}$H$_{40}$O$_4$; mol wt 428.59. C 75.66%, H 9.41%, O 14.93%. Isoln from acid hydrolysis mixture of sapogenins of *Dioscorea mexicana:* Marker, Lopez, *J. Am. Chem. Soc.* **69**, 2397 (1947); from crude sapogenin mixture extracted from tubers of *D. spiculiflora, Dioscoreaceae:* Walens *et al., J. Org. Chem.* **22**, 182 (1957); Townley, **U.S. pat. 2,954,374** (1960 to Schering); from *Heloniopsis orientalis* (Thunb.) C. Tanaka, *Liliaceae:* Okanishi *et al., Chem. Pharm. Bull.* **10**, 1195 (1962). Structure: Walens *et al.,* **U.S. pat. 2,830,986** (1958 to USDA).

Rectangular plates from methanol, mp 215-216°. $[\alpha]_D^{28}$ −56° (c = 1.02 in CHCl$_3$).

Acetate, C$_{29}$H$_{42}$O$_5$, crystals from ethyl acetate, mp 227°. $[\alpha]_D^{25}$ −58.1° (c = 1.02 in CHCl$_3$).

USE: Precursor for synthesis of pharmacologically active steroids.

4294. Geosmin. *Octahydro-4,8a-dimethyl-4a(2H)-naphthalenol; octahydro-4α,8aβ-dimethyl-4aα(2H)-naphthol; 1,10-trans-dimethyl-trans-(9)-decalol.* $C_{12}H_{22}O$; mol wt 182.31. C 79.06%, H 12.16%, O 8.78%. Major volatile component of beet essence, also determined to be the potent earthy odor contaminant of fish, beans, water: A. C. Thaysen, *Ann. Appl. Biol.* **23,** 99 (1936); J. Silvey, A. W. Roach, *J. Am. Water Works Assoc.* **56,** 60 (1964); A. A. Rosen *et al., Water Treat. Exam.* **19**(Pt. 2), 106 (1970); R. G. Buttery *et al., J. Agr. Food Chem.* **24,** 419, 1246 (1976). Isoln from *Actinomycetes, algae:* N. N. Gerber, H. A. Lechevalier, *Appl. Microbiol.* **13,** 935 (1965); A. A. Rosen *et al., ibid.* **16,** 178 (1968); R. S. Safferman *et al., Environ. Sci. Technol.* **1,** 429 (1967). Isoln from beets: K. E. Murray, P. A. Bannister, *Chem. & Ind. (London)* **1975,** 973; L. D. Tyler *et al., J. Agr. Food Chem.* **26,** 1466 (1978). Structure: N. N. Gerber, *Biotechnol. Bioeng.* **9,** 321 (1967); *eidem, Tetrahedron Letters* **1968,** 2971. Biological degradation: L. V. Narayan, W. J. Nunez, *J. Am. Water Works Assoc.* **66**(Pt. 1), 532 (1974).

Colorless neutral oil, bp 270°. $[\alpha]_D^{25}$ −16.5°. Threshold odor conc: 0.1 ppb. Decomp in acid to odorless compds.

4295. Gepefrine. *(S)-3-(2-Aminopropyl)phenol;* α-methyl-*m*-tyramine. $C_9H_{13}NO$; mol wt 151.20. C 71.49%, H 8.66%, N 9.27%, O 10.58%. Sympathomimetic isomer of hydroxyamphetamine, q.v. Prepn: W. S. Saari *et al., J. Med. Chem.* **11,** 1115 (1968); G. Petrik *et al., Ger. pat.* **2,712,860** (1978 to Helopharm), *C.A.* **88,** 136296a (1978). Molecular configuration and biological actions: M. L. Torchiana *et al., Arch. Int. Pharmacodyn. Ther.* **174,** 118 (1968). *In vitro* study of tritium-labeled gepefrine: R. L. Dorris, *Eur. J. Pharmacol.* **35,** 225 (1976). Biochemical-pharmacological study: R. R. Ruffolo *et al., Biochem. Pharmacol.* **25,** 399 (1976).

Crystals from ethanol/hexane, mp 155-158°. $[\alpha]_D^{25}$ +31.8° (c = 2 in methanol). Sublimes at 100° (0.3 mm). Tartrate, $C_{13}H_{19}NO_7$, **Pressionorm, Wintonin.**
THERAP CAT: Antihypotensive.

4296. Gephyrotoxin. *[1R-(1α,3aβ,5aα,6α(Z),9aα)]-Dodecahydro-6-(2-penten-4-ynyl)pyrrolo[1,2-a]quinoline-1-ethanol;* HTX D; histrionicotoxin D. $C_{19}H_{29}NO$; mol wt 287.45. C 79.39%, H 10.17%, N 4.87%, O 5.57%. Parent member of a class of tricyclic perhydrobenzoindolizine neurotoxin alkaloids isolated from the skin secretions of the Colombian poison-dart frogs *Dendrobates histrionicus* and *Dendrobates occultator,* family *Dendrobatidae.* The name gephyrotoxin was coined from the Greek word "gephyra" meaning "bridge", and refers to the bridge presumably formed by addition of the nitrogen function to one side of a 2,6-disubstituted piperidine precursor to form the bicyclic indolizine. Isolation of naturally occurring *l*-form: T. Tokuyama *et al., Helv. Chim. Acta* **57,** 2597 (1974). Structure and absolute configuration: J. W. Daly *et al., ibid.* **60,** 1128 (1977). Revised configuration: R. Fujimoto, Y. Kishi, *Tetrahedron Letters* **1981,** 4197. Total synthesis of (±)-form: R. Fujimoto *et al., J. Am. Chem. Soc.* **102,** 7154 (1980); D. J. Hart, *J. Org. Chem.* **46,** 3576 (1981); D. J. Hart, K. Kanai, *J. Am. Chem. Soc.* **105,** 1255 (1983); L. E. Overman *et al., ibid.* **108,** 5373. Complete proton and ^{13}C NMR assignment: M. W. Edwards, A. Bax, *ibid.* **108,** 918 (1986). Effect on neuromuscular transmission: C. Souccar *et al., Mol. Pharmacol.* **25,** 384, 395 (1984). Receptor binding

study: R. S. Aronstam *et al., Neurochem. Res.* **11,** 1227 (1986). Review including discussion of gephyrotoxin congeners: J. W. Daly, *Fortschr. Chem. Org. Naturst.* **41,** 283-300 (*see also* refs pp 326-340) (1982).

R = C ≡ CH₃

mp 231-232° (dec). uv max (ethanol): 225 nm (ε 8400). $[\alpha]_D^{25}$ −51.5° (c = 1 in ethanol).

4297. Gepirone. *4,4-Dimethyl-1-[4-[4-(2-pyrimidinyl)-1-piperazinyl]butyl]-2,6-piperidinedione;* 3,3-dimethyl-1-[4-[4-(2-pyrimidinyl)-1-piperazinyl]butyl]glutarimide; BMY 13805; MJ 13805. $C_{19}H_{29}N_5O_2$; mol wt 359.47. C 63.48%, H 8.13%, N 19.48%, O 8.90%. Nonbenzodiazepine tranquilizer structurally related to buspirone, q.v. Prepn: D. L. Temple, Jr., **Ger. pat.** **3,248,160** (1983 to Bristol-Myers); *idem,* **U.S. pat.** **4,423,049** (1983 to Mead Johnson); J. S. New *et al., J. Med. Chem.* **29,** 1476 (1986). Characterization of polymorphs: R. J. Behme *et al., J. Pharm. Sci.* **74,** 1041 (1985). Comparison of neuropharmacology with buspirone: B. A. McMillen, L. A. Mattiace, *J. Neural Transm.* **57,** 255 (1983). Metabolism and tissue disposition in rats: S. Caccia *et al., Xenobiotica* **15,** 835 (1985). Serotonergic effects in rats: A. S. Eison *et al., Pharmacol. Biochem. Behav.* **24,** 701 (1986). Preliminary clinical trials in anxiety: I. Csanalosi *et al., J. Clin. Psychopharmacol.* **7,** 31 (1987); N. E. Harto *et al., Psychopharmacol. Bull.* **24,** 154 (1988).

Crystals from acetonitrile, mp 97-99°.
Hydrochloride, $C_{19}H_{30}ClN_5O_2$, **BMY 13805-1.** Crystals from ethanolic HCl, mp 203-205° (Temple). Three polymorphic forms have been reported, mp 180°, 212°, 200°. pKa 7.42. Sol in water.
THERAP CAT: Anxiolytic.

4298. Geraniol. *3,7-Dimethyl-2,6-octadien-1-ol;* 2,6-dimethyl-2,6-octadien-8-ol; lemonol. $C_{10}H_{18}O$; mol wt 154.24. C 77.86%, H 11.76%, O 10.37%. An olefinic terpene alcohol constituting the chief part of oil of rose and oil of palmarosa; also found in many other essential oils such as citronella, lemon grass, etc. Isomeric with linalool. Isoln: Jacobsen, *Ann.* **157,** 234 (1871). Structure: Verley, *Bull. Soc. Chim. France* **25,** 68 (1919); J. L. Simonsen, *The Terpenes* vol. I (University Press, Cambridge, 2nd ed., 1947) pp 40-52. Stereochemistry: Burrell *et al., Proc. Chem. Soc.* **1959,** 263; Bates *et al., J. Org. Chem.* **28,** 1086 (1963). Synthesis: Burrell *et al., J. Chem. Soc. (C)* **1966,** 2144; K. Takabe *et al., Chem. Letters* **1977,** 1025; K. K. Mathew *et al., Indian J. Chem.* **B20,** 340 (1981).

Oily liq. Sweet rose odor. bp_{757} 229-230°; bp_{12} 114-115°. d_4^{20} 0.8894. n_D^{20} 1.4766. uv max: 190-195 nm (ε 18,000). Practically insol in water. Miscible with alcohol, ether.
Acetate, $C_{12}H_{20}O_2$, sweet, fragrant liq. bp about 242° with

decompn. d_{15}^{15} 0.9174. n_D^{15} 1.4628. Almost insol in water; very sol in alcohol; miscible with ether.

Butyrate, $C_{14}H_{24}O_2$, liq. Characteristic fragrant odor. bp_{18} 152°. d_4^{17} 0.901. Almost insol in water; sol in alc, ether.

Formate, $C_{11}H_{18}O_2$, liq. Odor of roses and of green rose leaves. bp_{15} 113-114°. d_4^{20} 0.927. Almost insol in alc, ether.

USE: In perfumery. As insect attractant. Butyrate for compounding artificial attar of rose. Formate as constituent of artificial neroli oil and of artificial orange blossom oil.

4299. Geranium. Cranesbill; storksbill; alum root. Dried rhizome of *Geranium maculatum* L., Geraniaceae. *Habit.* Canada and Eastern U.S., south to Georgia. *Constit.* 10-28% tannin, gallic, acid, resin, sugar, pectin.

THERAP CAT: Astringent.

THERAP CAT (VET): Intestinal astringent.

4300. Geranylhydroquinone. 2-(3,7-Dimethyl-2,6-octa-dienyl)-1,4-benzenediol; trans-(3,7-dimethyl-2,6-octadienyl)-hydroquinone; trans-1,4-dihydroxy-2-(3,7-dimethyl-2,6-octadienyl)benzene; geranyl-1,4-benzenediol; geroquinol; Béradia. $C_{16}H_{22}O_2$; mol wt 246.34. C 78.01%, H 9.00%, O 12.99%. Prepn and use: Baranger, *Fr.* pat. **M2694** (1964), *C.A.* **61**, 15940e (1964). Radioprotective activity: G. Rudali, *C. R. Soc. Biol.* **160**, 1365 (1966). Effects on cancers in mice: G. Rudali, L. Menetrier, *Therapie* **22**, 895 (1967). Discovery as a potent contact allergen from trichomes of *Phacelia crenulata* var. *funerea* J. Voss, *Hydrophyllaceae:* G. Reynolds, E. Rodriguez, *Phytochemistry* **18**, 1567 (1979).

Colorless needles from *n*-hexane/ethyl acetate, mp 61-62°, H. Inouye *et al.*, *Ber.* **101**, 4057 (1968).

USE: Exptly as a radioprotective agent.

4301. Germane. Germanium hydride. GeH_4; mol wt 76.63. Ge 94.74%, H 5.26%. Prepd by the action of lithium aluminum hydride on a germanium halide in ether soln: Finholt *et al.*, *J. Am. Chem. Soc.* **69**, 2692 (1947); by reduction of GeO_2 by sodium hydoborate: Griffiths, *Inorg. Chem.* **2**, 375 (1963).

Colorless gas. mp -165°, bp -90°, d_4^{-142} 1.523. Slightly sol in hot hydrochloric acid, dec in nitric acid.

Caution: Hemolytic gas: E. Browning, *Toxicity of Industrial Metals* (Appleton-Century-Crofts, New York, 2nd ed., 1969) pp 162-163.

4302. Germanium. Ge; at. wt 72.59; at. no. 32; valences 4, 2. Five naturally occurring isotopes: 70 (20.55%); 72 (27.37%); 73 (7.67%); 74 (36.74%); 76 (7.67%); artificial, radioactive isotopes: 65-69; 71; 75; 77; 78. Extent of occurrence in the earth's crust about 0.0007%. Predicted and called ekasilicon by Mendeléeff. Discovered in 1886 by Clemens Winkler: *J. Prakt. Chem.* **34**, 177 (1886). Obtained industrially from the flue dusts of smelters processing zinc-bearing ores: Jaffee *et al.*, *Trans. Electrochem. Soc.* **89**, 277 (1946). Purification by zone refining: Pfann, *J. Metals* **4**, 747 (1952). Physical properties: Hassion *et al.*, *J. Phys. Chem.* **59**, 1076 (1955). Review and description of modern isolation techniques: Pirest in L. P. Hunter, *Handbook of Semiconductor Electronics* (McGraw-Hill, New York, 1956), section 6. Comprehensive monograph: V. I. Davydov, *Germanium* (Gordon & Breach, New York, 1966) 417 pp. *Reviews:* Rochow in *Comprehensive Inorganic Chemistry* vol. **2**, J. C. Bailar, Jr. *et al.*, Eds. (Pergamon Press, Oxford, 1973) pp 1-41; J. H. Adams in Kirk-Othmer *Encyclopedia of Chemical Technology* vol. **11** (Wiley-Interscience, New York, 3rd ed., 1980) pp 791-802.

Grayish-white, lustrous, brittle metalloid. Diamond-cubic structure when cryst. Poor conductor of electricity. d_4^{25} 5.323. Reported melting points range from 925-975°; best value 937.2° (Hassion). Vol smaller by a few % when molten. bp 2700°. Thermal expansion coefficient (at ~25°): 6.1×10^{-6}/°C. Thermal conductivity (at 25°): 0.14 cal/sec

cm/°C. Specific heat (0-100°): 0.074 cal/g/°C. Lattice constant at 25°: 5.657 $\times$ 10^{-8} cm. Atoms/cc = 4.42 $\times$ 10^{22}. Volume compressibility: 1.3 $\times$ 10^{-12} cm^2/dyn. Dielectric constant: 16. Covalent bond ionization energy at 0°K = 1.2 ev. Band gap: 0.67 ev. Impurity atom ionization energy: ~0.01 ev. Intrinsic resistivity at 300°K = 47 ohm-cm. Electron mobility at 300°K = 3900 cm^2/v sec. Hole mobility at 300°K = 1900 cm/v sec. Magnetic (mass) susceptibility ($X \times 10^6$) = -0.12. Intrinsic charge density at 300°K = 2.4 $\times$ 10^{13}. Electron diffusion constant at 300°K = 100. Hole diffusion constant at 300°K = 49. Insol in water, hydrochloric acid, dil alkali hydroxides. Attacked by aqua regia, concd nitric or sulfuric acids, fused alkalies, alkali peroxides, nitrates, or carbonates. Relatively stable, unaffected by air, becomes oxidized above 600°; is slowly oxidized by hydrogen peroxide at room temp, fairly rapidly at 90°; is attacked by hydrogen above 1000°. When finely divided, burns in chlorine or bromine.

USE: In electronics: Manuf rectifying devices (germanium diodes), transistors, in red-fluorescing phosphors; in dental alloys; in the production of glass capable of transmitting infrared radiation. Review of uses: Aldington, Cumming, *Endeavour* **14**, 200-204 (1955); *New Uses for Germanium*, F. I. Metz, Ed. (Midwest Research Institute, 1974) 120 pp.

THERAP CAT (VET): Intestinal astringent.

4303. Germanium Dichloride. Cl_2Ge; mol wt 143.51. Ge 50.59%, Cl 49.41%. $GeCl_2$. Obtained as residue upon low temp distillation of $GeHCl_3$ (prepd by the reaction of hydrogen chloride with germanium monosulfide): Moulton, Miller, *J. Am. Chem. Soc.* **78**, 2702 (1956).

Unstable substance, dec into polymeric subchlorides even at low temps. Sol in ether, benzene.

4304. Germanium Dioxide. Germanic acid. GeO_2; mol wt 104.60. Ge 69.41%, O 30.59%.

White powder. Sol in about 250 parts cold water, 100 parts boiling water; sol in acids or in solns of the fixed alkalies. LD_{50} i.p. in rats: 750 mg/kg, Rosenfeld, Wallace, *Arch. Ind. Hyg. Occup. Med.* **8**, 466 (1953).

4305. Germanium Tetrachloride. Cl_4Ge; mol wt 214.43. Ge 33.86%, Cl 66.14%. $GeCl_4$. Prepd from Ge and Cl_2 or GeO_2 and HCl: Bauer, Burschkies, *Ber.* **66**, 277 (1933); Foster *et al.*, *Inorg. Syn.* **2**, 109 (1946).

Mobile liq. Fumes in air. Peculiar, acidic odor, but can be distinguished from that of concd HCl. Appreciably volatile at room temp. d_{20}^{20} 1.879. bp_{760} 83.1°. mp -49.5°. Hydrolyzed by water with a crackling noise. Is stable, but not very sol in 6N HCl. If more concd HCl is used, the soly of $GeCl_4$ is reduced. Sol in benzene, ether, other organic solvents.

Caution: Fumes irritating to eyes, mucous membranes of respiratory tract.

4306. Germanium Tetrafluoride. F_4Ge; mol wt 148.60. Ge 48.86%, F 51.14%. GeF_4. Prepd by heating $BaGeF_6$ to around 700°: Dennis, Laubengayer, *Z. Physik. Chem.* **130**, 520 (1927); Dennis, *Z. Anorg. Allgem. Chem.* **174**, 119 (1928); Biltz, *ibid.* **207**, 65 (1932); Hoffman, Gutowsky, *Inorg. Syn.* **4**, 147 (1953).

Colorless gas. Odor of garlic. *Very irritating to mucous membranes.* Thermally stable up to about 1000°. d^0 (liq) 2.162; d (solid; -195°) 3.148. mp -15° (under 3032 mm pressure). Sublimes at -36.5°. On contact with water, hydrolyzes to GeO_2 and H_2GeF_6. Does not attack glass when absolutely dry. Corrodes mercury, grease.

Trihydrate, obtained by slow evaporation of a soln of germanium dioxide in 20% HF. This is said to be water-soluble.

Caution: Highly irritating to eyes, skin, lungs, mucous membranes.

4307. Germine. 4α,9-Epoxycevane-3β,4,7α,14,15α,16β,-20-heptol. $C_{27}H_{43}NO_8$; mol wt 509.62. C 63.63%, H 8.51%, N 2.75%, O 25.12%. Alkamine present in many polyester alkaloids which occur in *Veratrum* and *Zygadenus* species. Isoln: Poethke, *Pharm. Monatsh.* **18**, 77 (1937); idem, *Arch. Pharm.* **275**, 571 (1937); Seiferle *et al.*, *J. Econ. Entomol.* **35**, 35 (1942); Klohs *et al.*, *J. Am. Chem. Soc.* **75**, 4925 (1953); Kupchan, Deliwala, *ibid.* **76**, 5545 (1954); Myers *et al.*, *ibid.*

77, 3348 (1955); **78**, 1621 (1956). Structure and configuration: Kupchan, Narayanan, *ibid.* **81**, 1913 (1959).

Crystals from methanol, mp 221.5-223°. $[\alpha]_D^{25}$ +4.5° (95% ethanol); $[\alpha]_D^{16}$ +23.1° (c = 1.13 in 10% acetic acid). Sol in chloroform, methanol, ethanol, acetone, water; slightly sol in ether.

3-Acetate, $C_{29}H_{45}NO_9$, needles from ether, mp 219-221°. $[\alpha]_D^{23}$ +10° (c = 1.05 in pyridine).

16-Acetate, $C_{29}H_{45}NO_9$, crystals from chloroform, mp 225-227°. $[\alpha]_D^{23}$ −19° (c = 0.93 in pyridine).

3,4,7,15,16-Pentaacetate, $C_{37}H_{53}NO_{13}$, prisms from acetone + petr ether, dec 285-287°. $[\alpha]_D^{23}$ −65° (c = 0.65 in pyridine).

4308. Gestodene. *(17α)-13-Ethyl-17-hydroxy-18,19-dinorpregna-4,15-dien-20-yn-3-one;* 17α-ethynyl-17β-hydroxy-18-methyl-4,15-estradien-3-one; 17α-ethynyl-13-ethyl-17β-hydroxy-4,15-gonadien-3-one; SH B 331. $C_{21}H_{26}O_2$; mol wt 310.44. C 81.25%, H 8.44%, O 10.31%. Orally active gestogen with progesterone-like profile of activity. Prepn: **Belg. pat. 847,090;** H. Hofmeister *et al.,* **U.S. pat. 4,081,537** (1977, 1978 both to Schering AG). Pharmacokinetics: B. Duesterberg *et al., Contraception* **24**, 673 (1981); B. Duesterberg, *Steroids* **43**, 43 (1984). Comparative study of anti-aldosterone activity: W. Losert *et al., Arzneimittel-Forsch.* **35**, 459 (1985). Clinical trial as oral contraceptive: G. Hoppe, *Adv. Contracept.* **3**, 159 (1987).

Crystals from acetone-hexane, mp 197.9°. Mixture with ethinyl estradiol, *Femodene, Femovan, Ginoden, Gynera, Minulet, Monodie, Phaeva.*
THERAP CAT: Progestogen. In combination with estrogen as oral contraceptive.

4309. Gestonorone Caproate. *17-[(1-Oxohexyl)oxy]-19-norpregn-4-ene-3,20-dione;* *17-Hydroxy-19-norpregn-4-ene-3,20-dione hexanoate;* 17α-hydroxy-19-norprogesterone caproate; 17β-acetyl-17-hydroxyestr-4-ene-3-one hexanoate; gestronol caproate; SH 582; Depostat. $C_{26}H_{38}O_4$; mol wt 414.59. C 75.32%, H 9.24%, O 15.43%. Prepn: Popper *et al., Arzneimittel-Forsch.* **19**, 352 (1969); Popper *et al.,* **Ger. pat. 1,074,582** (1960 to Schering). Pharmacological studies: Junkmann, *Anglo-Ger. Med. Rev.* **1**, 385 (1962). Metabolism: Breuer, Lisboa, *Acta Endocrinol.* **51**, 114 (1966).

Crystals, mp 123-124°. $[\alpha]_D$ +13° (chloroform). uv max: 239 nm (ε 17540).
THERAP CAT: Progestogen. In treatment of prostate hypertrophy.

4310. Gestrinone. *(17α)-13-Ethyl-17-hydroxy-18,19-dinorpregna-4,9,11-trien-20-yn-3-one;* 13β-ethyl-17α-ethynyl-17β-hydroxy-4,9,11-gonatrien-3-one; 13β-ethyl-17α-ethynyl-Δ⁴,⁹,¹¹-gonatriene-17β-ol-3-one; ethylnorgestrienone; A 46745; R 2323; RU 2323; Dimetrose; Nemestran; Trimodose. $C_{21}H_{24}O_2$; mol wt 308.42. C 81.78%, H 7.84%, O 10.38%. Steroidal antiestrogen, antiprogestogen, analog of norgestrienone, *q.v.* Prepn: G. Nomine *et al.,* **U.S. pat. 3,257,278;** **Neth.** pat. **Appl. 6,607,609,** *C.A.* **67**, 44029d (1967); D. Bertin, A. Pierdet, **U.S.** pat. **3,478,067** (1966, 1966, 1969 all to Roussel-UCLAF). Radioimmunoassay in human plasma: J. Frick *et al., Urol. Res.* **5**, 55 (1977). Clinical evaluation of contraceptive efficacy: G. Azadian-Boulanger *et al., Am. J. Obstet. Gynecol.* **125**, 1049 (1976); of use in fibrocystic breast disease: E. M. Coutinho, G. Azadian-Boulanger, *Int. J. Gynaecol. Obstet.* **22**, 363 (1984); in endometriosis: E. J. Thomas, I. D. Cooke, *Brit. Med. J.* **294**, 272 (1987); E. M. Coutinho, G. Azadian-Boulanger, *Fertil. Steril.* **49**, 418 (1988).

Crystals from ethyl acetate and benzene-cyclohexane (1:1), mp 154°. $[\alpha]_D^{20}$ +84.6° (c = 0.41 in methanol).
THERAP CAT: Antigonadotropin.

4311. Ghatti Gum. Gum Ghatti; Indian gum. The gummy exudate from stems of *Anogeissus latifolia* Wall., *Combretaceae,* abundant in India and Ceylon, *cf.* C. L. Mantell, *The Water-Soluble Gums* (New York, 1947). Name derived from the word *ghats,* meaning passes, and given to the gum because of its ancient mountain transportation routes. Structure is a complex water-soluble polysaccharide occurring as a calcium-magnesium salt; composed of L-arabinose, D-galactose, D-mannose, D-xylose, D-glucuronic acid, in a molar ratio of 10:6:2:1:2, and traces of 6-deoxyhexose: Aspinall *et al., J. Chem. Soc.* **1955**, 1160. Early investigation of chemistry and mol wt: Shaw *et al., Proc. S. Dakota Acad. Sci.* **15**, 46 (1935); **16**, 34 (1936); **17**, 27 (1937); **19**, 130 (1939); **21**, 78 (1941). *Review:* Meer *et al.,* in *Industrial Gums,* R. L. Whistler, Ed. (Academic Press, New York, 2nd ed., 1973) pp 265-271.

Ghatti gum sold in the U.S. usually has been autoclaved in order to make all of the gum water-sol. The *U.S. Dispensatory* (24th ed.) states that gum Ghatti suitable as clinical laboratory reagent "is entirely sol in 5 parts of cold water." Forms a very viscous mucilage, more viscous but less adhesive than acacia. Insol in 90% alcohol. $[\alpha]_D^{25}$ +42° (dil H_2SO_4). Gum ghatti solns may be colored slightly due to traces of pigment remaining in the gum. Does not form a true gel.
USE: As substitute for acacia. As emulsifying agent in pharmaceuticals, oils, waxes.

4312. Ghi. Ghee; samli; clarified butter. Prepd from cream or butter by melting and heating to 122° for about 25 mins. The water evaporates, and lactose, salt, and albuminous substances sink to the bottom. This treatment kills existing microorganisms by heat and deprives incoming microorganisms of moisture and necessary nutrients.

Properly clarified butter is of finely grained consistency, yellow color, mp 30° and retains most of the vitamin A content of the original butter. Keeps much longer than ordinary butter and contains less than 0.7% water. Used in India instead of butter.

4313. Gibberellic Acid. *2,4a,7-Trihydroxy-1-methyl-8-methylenegibb-3-ene-1,10-dicarboxylic acid 1,4a-lactone;* gibberellin X; gibberellin A_3. $C_{19}H_{22}O_6$; mol wt 346.37. C 65.88%, H 6.40%, O 27.72%. Plant hormone; most outstanding of the plant-growth promoting metabolites of *Gibberella fujikuroi.* Isoln: P. J. Curtis, B. E. Cross, *Chem. & Ind. (London)* **1954,** 1066; B. E. Cross, *J. Chem. Soc.* **1954,** 4670; P. W. Brian *et al., U.S.* pat. **2,842,051;** C. T. Calam, P. J. Curtis, U.S. pat. **2,950,288;** A. J. Birch *et al.,* U.S. pat. **2,977,285** (1958, 1960, 1961, all to ICI). Stereochemistry and structure: G. Stork, H. Newman, *J. Am. Chem. Soc.* **81,** 5518 (1959); B. E. Cross *et al., Proc. Chem. Soc.* **1959,** 302; F. McCapra *et al., ibid.* **1962,** 185; D. C. Aldridge *et al., J. Chem. Soc.* **1963,** 143; P. M. Bourn *et al., ibid.* **1963,** 154. Partial synthesis: E. J. Corey *et al., J. Am. Chem. Soc.* **93,** 7316 (1971). Stereospecific total synthesis: *eidem, ibid.* **100,** 8034 (1978). Promotes growth of seedlings: M. J. Bukovac, S. H. Wittwer, *Quart. Bull. Mich. Agr. Expt. Sta.* **39,** 307 (1956); J. M. Merritt, "Gibberellins for Agriculture", *J. Agr. Food Chem.* **6,** 184 (1958). *Reviews:* B. E. Cross *et al.* in *Gibberellins, Advances in Chemistry Series,* **vol. 28,** 13 (A.C.S., Washington, 1961); series of articles in *Plant Growth Subst., Proc. 7th Int. Conf.,* D. J. Carr, Ed. (Springer-Verlag, Berlin, 1972).

Crystals from ethyl acetate, mp 233-235° (effervescence). $[\alpha]_D^{19}$ +86° (c = 2.12). pK 4.0. Slightly sol in water, ether. Freely sol in methanol, ethanol, acetone. Moderately sol in ethyl acetate. Sol in aq solns of sodium bicarbonate and sodium acetate.

Methyl ester, $C_{20}H_{24}O_6$, needles from benzene + methanol, mp 209-210°. $[\alpha]_D^{20}$ +75° (c = 0.5).

Potassium gibberellate, $C_{19}H_{21}KO_6$, *Gibrel.* White, bulky powder. $[\alpha]_D^{25}$ +43° to +60° (c = 5 in water). pH of 5% soln 5.5-6.5. One gram dissolves in 20 ml water.

USE: Plant growth regulator. Promoting growth of plants, esp the growth of seedlings. Food additive in the malting of barley: *Fed. Reg.* **25,** 2162 (1960).

4314. Gibberellins. GAs. A class of plant growth hormones first isolated in 1938 from cultures of *Gibberella fujikuroi* (Sawada) Wollenweber *(Fusarium moniliforme* Sheldon) the fungus causing Bakanae disease in rice: Yabuta, Sumiki, *J. Agr. Chem. Soc. Japan* **14,** 1526 (1938). Isolated also from higher plants; for source references see review by Lang, *Ann. Rev. Plant Physiol.* **21,** 537 (1970). More than 60 gibberellins are known of which gibberellin A_3 (gibberellic acid, *q.v.*) is the most important. GA_3 and mixtures of GA_4 and GA_7 are available commercially. All gibberellins are diterpenoid acids based on the *gibberellane* skeleton containing the *gibbane* nucleus. Major structural differences lie in the substituents at positions 4a, 7, 8 (gibbane numbering) and the presence or absence of a γ-lactone ring. Total synthesis of racemic gibberellins A_2, A_4, A_9, A_{10}: Mori *et al., Tetrahedron* **25,** 1293 (1969); of gibberellin A_4: A. L. Cossey *et al., Tetrahedron Letters* **1980,** 4383; of gibberellin A_{15}: Nagata *et al., J. Am. Chem. Soc.* **93,** 5740 (1971); of gibberellins A_{15} and A_{37}: E. Fujita *et al., J. Chem. Soc., Perkin Trans. I* **1977,** 611. Stereochemistry: Meguro, Fuzimura, *Tetrahedron Letters* **1968,** 6305. Biosynthesis: Cross *et al., J. Chem. Soc. (C)* **1968,** 1054; Shechter, West, *J. Biol. Chem.* **244,** 3200 (1969). Nomenclature: MacMillan, Takahasni, *Nature* **217,** 170 (1968). Early reviews: Brian *et al., Fortschr. Chem. Org. Naturst.* **18,** 350 (1960); Paleg, *Ann. Rev. Plant*

Physiol. **16,** 291 (1965). Recent reviews: Cleland in *The Physiology of Plant Growth and Development,* M. B. Wilkins, Ed. (McGraw-Hill, New York, 1969) pp 49-81; L. Rappaport, "Applications of Gibberellins in Agriculture", in *Plant Growth Subst., Proc. Int. Conf. 10th,* F. K. Skoog, Ed. (Springer-Verlag, Berlin, 1980) pp 377-391; I. D. Railton, *Cell. Biol. Int. Rep.* **6,** 319-337 (1982). Comprehensive synthetic review: E. Fujita, M. Node, *Heterocycles* **7,** 709 (1977).

gibbane
CA numbering

gibberellane
alternate numbering

USE: Plant growth hormone. For specific agricultural uses *see* review by Turner, *Outlook Agr.* **7,** 14 (1972).

4315. Gibbs Reagent. *2,6-Dichloro-4-(chloroimino)-2,5-cyclohexadien-1-one;* 2,6-dichloro-*p*-benzoquinone-4-chloroimine; *N,*2,6-trichloro-*p*-quinoneimine; 2,6-dichloroquinone chloroimide; *N,*2,6-trichloroquinoneimine; *N,*2,6-trichlorobenzoquinone imine. $C_6H_2Cl_3NO$; mol wt 210.44. C 34.24%, H 0.96%, Cl 50.54%, N 6.66%, O 7.60%. Reagent used to determine the presence of phenols. The reaction with phenols unsubstituted in the *para* position is called the Gibbs Reaction: H. D. Gibbs, *Chem. Rev.* **3,** 291 (1927). Prepn of the reagent: *idem, J. Biol. Chem.* **72,** 649 (1927); G. I. Mikhailov, *Trans. Inst. Pure Chem. Reagents* **16,** 83 (1939). Analysis of the Gibbs color reaction: J. C. Dacre, *Anal. Chem.* **43,** 589 (1971).

Yellow needles from alcohol, mp 65-67°.
USE: In determination of phenols.

4316. Gigantine. *1,2,3,4-Tetrahydro-6,7-dimethoxy-1,2-dimethyl-5-isoquinolinol.* $C_{13}H_{19}NO_3$; mol wt 237.30. C 65.80%, H 8.07%, N 5.90%, O 20.23%. Hallucinogen from the cactus *Carnegia gigantea* Engelm. *Cactaceae:* Hodgkins *et al., Tetrahedron Letters* **1967,** 1321. Structure: Kapadia *et al., Chem. Commun.* **1970,** 856. Synthesis: *eidem, Chem. & Ind. (London)* **1970,** 1593; Choudhury *ibid.* **1971,** 578.

mp 151-152°. $[\alpha]_D^{25}$ +27.1° (c = 2 in chloroform).

4317. Ginger. Dried rhizome of *Zingiber officinale* Roscoe, *Zingiberaceae.* *Habit.* Southern Asia, West Indies, Africa, cultivated in all tropical countries. *Constit.* Volatile oil (1-3%), acrid resin, gingerol, *q.v.* Antiemetic effects in motion sickness: D. B. Mowrey, D. E. Clayson, *Lancet* **1,** 655 (1982). Study of mutagenic and antimutagenic principles of ginger juice: H. Nakamura, T. Yamamoto, *Mutat. Res.* **103,** 119 (1982).

USE: In foods and beverages. In oriental medicine.
THERAP CAT: Anti-emetic; carminative.
THERAP CAT (VET): Carminative, aromatic stimulant.

4318. [6]-Gingerol. *(S)-5-Hydroxy-1-(4-hydroxy-3-methoxyphenyl)-3-decanone.* $C_{17}H_{26}O_4$; mol wt 294.38. C

69.36%, H 8.90%, O 21.74%. The major phenol and most important of the pungent principles of ginger oil, isolated from rhizome of *Zingiber officinalis* Roscoe, *Zingiberaceae:* Thresh, *Pharm. J.* [3] **10,** 171 (1879); **12,** 721 (1881); **14,** 798 (1883); **15,** 208 (1884); Garnett, Grier, *ibid.* [4] **25,** 118 (1907); Nelson, *J. Am. Chem. Soc.* **39,** 1466 (1917). Structure studies: Lapworth *et al., J. Chem. Soc.* **111,** 777 (1917). Review of early literature: Redgrove, *Pharm. J.* **125,** 54 (1930); Jacobs, *Am. Perfumer* **48,** no. 7, 60, 62 (1946). Demonstration of the presence of an homologous series of phenolic ketones in the pungent constituents of ginger, [6]-gingerol being the major member: D. W. Connell, M. D. Sutherland, *Austral. J. Chem.* **22,** 1033 (1969). Synthesis of (±)-form: Hirao *et al., Chem. Pharm. Bull.* **20,** 2287 (1972); K. Banno, T. Mukaiyama, *Bull. Chem. Soc. Japan* **49,** 1453 (1976); P. Denniff, D. A. Whiting, *Chem. Commun.* **1976,** 712; P. Denniff *et al., J. Chem. Soc. Perkin Trans. I* **1981,** 82. Stereoselective synthesis of *S*(+)-form: D. Enders *et al., Ber.* **112,** 3703 (1979). Biosynthesis of (±)-form: P. Denniff, D. A. Whiting, *Chem. Commun.* **1976,** 711; I. Macleod, D. A. Whiting, *ibid.* **1979,** 1152; P. Denniff *et al., J. Chem. Soc. Perkin Trans. I* **1980,** 2637. Physical properties and cardiotonic effects of gingerols: N. Shoji *et al., J. Pharm. Sci.* **71,** 1174 (1982). Mutagenicity studies: H. Nakamura, T. Yamamoto, *Mutat. Res.* **103,** 119 (1982); *eidem, ibid.* **122,** 87 (1983).

$$CH_2CH_2COCH_2CH(CH_2)_4CH_3$$

Normally obtained as pungent, yellow oil. n_D^{25} 1.5224. uv max (ethanol) 282 nm (ε 2560) (Connell, Sutherland). Crystalline form, mp 30-32°. $[\alpha]_D$ +27.8° (c = 1 in CHCl$_3$). uv max (ethanol): 284 nm (ε 2700). Sol in 50% alcohol, ether, chloroform, benzene; moderately sol in hot petr ether.

4319. Ginkgo. Gingko; maidenhair tree; kew tree. *Ginkgo biloba* Linn, *Ginkgoacea.* The only living member of the Ginkgoaceae, valued as a street and park tree and in the Far East for its edible seeds. The ripe fruit gives off a foul odor. *Habit.* China, Japan, cultivated in Eastern U.S. and Canada. *Constit.* A wide variety of compounds are extracted from the fruits, leaves, and bark: see T. Karuyone *et al., Yakugaku Zasshi* **78,** 1152 (1958); **82,** 214, 888 (1962), *C.A.* **53,** 5221e (1950); **58,** 5728h (1963); **59,** 2864g (1963). Comprehensive list of components found in leaves: K. Weinges *et al., Arzneimittel-Forsch.* **18,** 537 (1968). Series of articles on the isoln, characterization, and structures of ginkgolides, the bitter substituents from the root bark: M. Maruyama *et al., Tetrahedron Letters* **1967,** 299-326.

4320. Ginkgo Biloba Extract. GBE; Ginkogink; rökan; Sophium; Tanakan; Tebonin. A defined extract obtained from the leaves of *Ginkgo biloba* L., also known as *Salisburia adiantifolia* Smith. Prepn: W. Schwabe, P. Kloss, *Fr. pat.* **2,007,352;** *see also: eidem, Ger. pat.* **1,767,098** (1970, 1972 both to Wilhelm Schwabe, GmbH), *C.A.* **78,** 133641s (1972). Isoln and characterization of the biflavonyl constituents of ginkgo leaves: W. Baker *et al., J. Chem. Soc.* **1963,** 1477; of the phenolic constituents: K. Weinges *et al., Arzneimittel-Forsch.* **18,** 539 (1968). HPLC analysis: A. Guth *et al., Planta Med.* **42,** 129 (1981). Pharmacological study: H. Peter *et al., Arzneimittel-Forsch.* **16,** 719 (1966). Mechanism of action study: M. Auguet *et al., Gen. Pharmacol.* **13,** 169 (1982). Cerebral effect: R. Hemmer, O. Tzavellas, *Arzneimittel-Forsch.* **17,** 491 (1967). Clinical efficacy in peripheral vascular diseases: G. Mussgnug, J. Alemany, *ibid.* **18,** 543 (1968); H. Salz, *Ther. Gegew.* **119,** 1345 (1980); U. Bauer, *Arzneimittel-Forsch.* **34,** 716 (1984).
THERAP CAT: In cerebral and peripheral circulatory disturbances.

4321. Ginseng. Panax; Energofit. Root of *Panax ginseng* C.A. Mey., *Araliaceae,* a perennial herb indigenous to Eastern Asia, *P. quinquefolium* L., found in Eastern U.S. and Canada, and *P. pseudoginseng* Wall, found in India,

China and Japan. The biologically active constituents are considered to be a series of saponin glycosides known as *ginsenosides, panaxosides* or *panaquilins.* Review of constituents: J. P. Hov, *Comp. Med. East West* **5,** 123-145 (1977). Isoln and identification of ginseng saponin glycosides: S. Shibata *et al., Tetrahedron Letters* **1962** 419; G. B. Elyakov *et al., ibid.* **1964,** 3591; R. Kasai *et al., Chem. Pharm. Bull.* **31,** 2120 (1983). Chemico-pharmacological study: T. Kaku, *Arzneimittel-Forsch.* **25,** 539 (1975). Effect on brain biogenic amines: V. Petkov, *ibid.* **28,** 388 (1978). TLC analysis of saponin content of commercial ginseng products: L. E. Liberti, A. D. Marderosian, *J. Pharm. Sci.* **67,** 1487 (1978). Use in oriental medicine as tonic: K. Chimin Wong, Wu Lien-teh, *History of Chinese Medicine* (Shanghai, 2nd ed., 1936) 906 pp. Comprehensive review of morphology, cultivation and uses: Baranov, *Econ. Bot.* **20,** 403-406 (1966). Review of chemical constituents: J. P. Hou, *Comp. Med. East West* **5,** 123-145 (1977). Brief reviews: W. E. Court, *Pharm. J.* **214,** 180-181 (1975); B. J. Spalding, *Chem. Week* **139,** 19-21 (1986).
Sweet, slightly aromatic taste.
THERAP CAT: Tonic.

4322. Giractide. *1-Glycine-18-*L-*argininamide-α*$^{(1\text{-}18)}$-*corticotropin.* $C_{100}H_{156}N_{34}O_{22}S$; mol wt 2218.75. C 54.13%, H 7.09%, N 21.47%, O 15.86%, S 1.45%. Polypeptide corresp to the first 18 amino acid residues in corticotropin, in which the 1-serine is replaced by glycine. Prepn: H. Otsuka *et al., Bull. Chem. Soc. Japan* **43,** 196 (1970); of the hexaacetate: *eidem, Japan.* **70 19,061** (1970 to Shionogi), *C.A.* **73,** 110138r (1970). Adrenocorticotropic potency in man: *J. Clin. Endocrinol. Metab.* **33,** 355 (1971). Fluorometric study: T. Muraki *et al., Experientia* **32,** 1605 (1976). Absorption: M. Hirata *et al., Chem. Pharm. Bull.* **26,** 1061 (1978). Chromatographic studies: S. Terabe *et al., J. Chromatog.* **172,** 163 (1979); M. Schoeneshoefer, A. Fenner, *ibid.* **224,** 472 (1981).

```
Gly-tyr-ser-met-glu-his-phe-arg
                                |
NH2-arg-arg-lys-lys-gly-val-pro-lys-gly-trp
```

$[\alpha]_D^{23.5}$ −51.4 ±1.9° (c = 0.472 in 0.1N acetic acid). uv max (0.1N NaOH): 281, 288 nm (ε 6750, 6490).
Hexaacetate, $C_{112}H_{180}N_{34}O_{34}S$, *S-50022, Acthormon.*
THERAP CAT: Adrenocorticotropic hormone.

4323. Girard Reagents. Quaternary ammonium acetylhydrazine chlorides which form water-soluble hydrazones with carbonyl compounds. The hydrazones formed can subsequently be hydrolyzed in order to regenerate the original carbonyl compounds. Prepn: A. Girard, G. Sandulesco, **Brit.** pat. **Appl. 6640** (1934); *eidem, Helv. Chim. Acta* **19,** 1095 (1936); A. Girard, *Org. Syn. coll. vol.* **II,** 85 (1943). *Review:* Wheeler, *Chem. Rev.* **62,** 205 (1962). Use of Girard reagents in separation of carbonyl compounds: Schubert, Wehrberger, *Endokrinologie* **48,** 70 (1965); R. E. J. Mitchel, H. C. Birnboim, *Anal. Biochem.* **81,** 47 (1977); W. Holstein, D. Severin, *Erdoel Kohle, Erdgas, Petrochem.* **32,** 487 (1979); *C.A.* **92,** 44207b (1980).

Girard reagent T   *Girard reagent P*

Girard reagent T, $C_5H_{14}ClN_3O$, *2-hydrazino-N,N,N-trimethyl-2-oxoethanaminium chloride, (carboxymethyl)trimethylammonium chloride hydrazide, betaine hydrazide hydrochloride.* Highly hygroscopic needles. May be stored in well-stoppered containers. Material that has developed an odor should be recrystallized from abs ethanol. mp 192° (slight decomp). Very freely sol in water; sol in about 150 parts of absolute ethanol; more sol in methanol. Also very sol in acetic acid, glycerol, ethylene glycol. Practically insol in organic solvents devoid of hydroxyl groups.

Girard reagent P, $C_7H_{10}ClN_3O$, *1-(2-hydrazino-2-oxoethyl)pyridinium chloride.* Non-hygroscopic crystals from methanol, dec 200°. Less soluble in polar solvents than Girard reagent T.

USE: In the isoln of 17-ketosteroids and other carbonyl compounds.

4324. Gitalin. Gitaligin. An extract of *Digitalis purpurea* L., *Scrophulariaceae*, once believed to be a single amorphous compd, is a stable, relatively constant mixture of digitoxin, gitoxin, and **gitaloxin (16-formylgitoxin, Cristaloxine)** with smaller amounts of several other glycosides and genins, together with minute amounts of other substances. Review of prepn and composition: Mitchell, *Am. J. Pharm.* **136**, 71 (1964).

Amorphous powder. Readily sol in alcohol, slightly (but slowly) sol in water. A satd aq soln has pH of about 7.0 and an intensely bitter taste. *Ref: N.N.D.* **1960**, 301.

THERAP CAT: Cardiotonic.

THERAP CAT (VET): Cardiotonic

4325. Gitogenin. *Spirostan-2,3-diol;* digin. $C_{27}H_{44}O_4$; mol wt 432.62. C 74.95%, H 10.25%, O 14.79%. Obtained from gitonin by heating with dil HCl. Structure: Tschesche, *Ber.* **68**, 1090 (1935); Jacobs, Simpson, *J. Biol. Chem.* **110**, 429 (1935); Marker, Rohrmann, *J. Am. Chem. Soc.* **61**, 2724 (1939); **62**, 647 (1940); Noller, Lieberman, *ibid.* **63**, 2131 (1941); Klass *et al*, *ibid.* **77**, 3829 (1955). Pharmacological study: H. K. Iwamoto *et al.*, *J. Pharmacol. Exp. Ther.* **91**, 130 (1947).

Leaflets from benzene, dec 271.5-275°. $[\alpha]_D^{20}$ −75° (c = 1.02 in CHCl$_3$). Sol in chloroform, hot alc; sparingly sol in cold ethyl acetate, in ether; practically insol in water. Not precipitated by digitonin.

Diacetate, $C_{31}H_{48}O_6$, long needles from ether + methanol, mp 251-254°. $[\alpha]_D^{20}$ −96° (c = 1.92 in CHCl$_3$).

4326. F-Gitonin. Gitogenin β-lycotetraoside. $C_{50}H_{82}$-O_{23}; mol wt 1051.21. C 57.13%, H 7.86%, O 35.01%. Leaf saponin from *Digitalis purpurea* L., *Scrophulariaceae:* Kawasaki, Nishioka, *Chem. Pharm. Bull.* **12**, 1311 (1964). Structure: Kawasaki *et al.*, *Tetrahedron* **21**, 299 (1965). A gitogenin tetraglycoside of which the sugar composition is 2 moles D-glucose, 1 mole D-galactose, and 1 mole D-xylose; differs from *gitonin* (the gitogenin tetraglycoside in *D. purpurea* seeds), in which the sugar composition is 2 galactose, 1 glucose, and 1 xylose: Tschesche, Wulff, *Ber.* **94**, 2019 (1961).

Dihydrate, needles from butanol + water, dec 252-255°. $[\alpha]_D^{25}$ −58.5° (c = 0.53 in pyridine).

4327. Gitoxigenin. *3,14,16-Trihydroxycard-20(22)-enolide;* $\Delta^{20,22}$-3,14,16,21-tetrahydroxynorcholenic acid lactone. $C_{23}H_{34}O_5$; mol wt 390.50. C 70.74%, H 8.78%,O 20.49%. The aglycon of gitoxin. By refluxing gitoxin in a mixture of water + alcohol + HCl: Smith, *J. Chem. Soc.* **1931**, 23. Structure: Jacobs, Elderfield, *J. Biol. Chem.* **100**, 671 (1933); Elderfield, *Chem. Rev.* **17**, 217 (1935); Henderson, Chen, *J. Med. Pharm. Chem.* **5**, 988 (1962). Configuration: Moore, *Helv. Chim. Acta* **37**, 659 (1954); Repke, Klesczewski, *Arch. Exp. Pathol. Pharmakol.* **239**, 131 (1960). *Cf.* ref under Digoxigenin.

Sesquihydrate, plates from dil alc. After drying at 100° *in vacuo* mp 234°. $[\alpha]_{545}^{20}$ +38.5° (c = 0.68 in methanol). Absorption max (96% H$_2$SO$_4$): 310, 485, 520 nm. Slightly sol in alcohol, acetone, ethyl acetate. Treatment with alcoholic HCl yields digitaligenin with loss of 2H$_2$O.

3,16-Diacetylgitoxigenin, mp 249-250°.

3,16-Dibenzoylgitoxigenin, mp 262°.

4328. Gitoxin. *3-[(O-2,6-Dideoxy-β-D-ribo-hexopyranosyl-(1 → 4)-O-2,6-dideoxy-β-D-ribo-hexopyranosyl-(1 → 4)-2,6-dideoxy-β-D-ribo-hexopyranosyl)oxy]-14,16-dihydroxycard-20(22)-enolide;* anhydrogitalin; bigitalin; pseudodigitoxin. $C_{41}H_{64}O_{14}$; mol wt 780.92. C 63.06%, H 8.26%, O 28.68%. R = H. Secondary glycoside mainly from *Digitalis purpurea* L., also from *D. lanata* Ehrh., *Scrophulariaceae.* Byproduct of digitoxin manuf. Isoln: Kraft, *Arch. Pharm.* **250**, 118 (1912); Cloetta, *Arch. Exp. Pathol. Pharmakol.* **112**, 261 (1926); Smith, *J. Chem. Soc.* **1931**, 23. Purification: McChesney *et al.*, *J. Am. Pharm. Assoc.* **37**, 364 (1948). Acid hydrolysis yields 1 mol gitoxigenin + 3 mols digitoxose, *q.q.v.*: Windaus, Schwarte, *Ber.* **58**, 1515 (1925). *See also* Satoh, Aoyama, *Chem. Pharm. Bull.* **18**, 94 (1970). Pharmacokinetics in man: M. Lesne, *Int. J. Clin. Pharmacol. Biopharm.* **16**, 456 (1978).

Stout prisms from chloroform + methanol, dec 285° (rapid heating). $[\alpha]_{546}^{20}$ +3.5° (c = 1.02 in pyridine). Absorption max (98% H$_2$SO$_4$): 315, 415, 495, 530 nm (E$_{1cm}^{1\%}$ 275, 185, 430, 505). Almost insol in chloroform, ethyl acetate and acetone, but dissolves in a mixture of chloroform and alcohol or in pyridine or in dil alcohol. Less sol in hot 80% alcohol than digoxin.

α- and β-Acetylgitoxin, $C_{43}H_{66}O_{15}$, obtained by enzymatic hydrolysis of digilanide B. β-Form: Long, thin, hair-like prisms from dil methanol, dec 220-225°. $[\alpha]_D^{20}$ 15.7° (c = 1.28 in pyridine). Sol in 80-100 parts methanol; sparingly sol in water and ether.

Pentaacetate, $C_{51}H_{74}O_{19}$, *pengitoxin, penta-O-acetylgitoxin, Carnacid-Cor, Cordoval, Pentagit.* R = —OCCH$_3$. Prepn: **Belg.** pat. **668,116** and **Neth.** pat. **Appl. 6,506,250** (1965, 1966, both to VEB Arzneimittelwerk Dresden), *C.A.* **65**, 23366 (1966); **67**, 91092d (1967). *Review:* R. Megges *et al.*, *Pharmazie* **32**, 665-667 (1977). Mass spectra studies: Blessington, Morton, *Org. Mass Spectrom.* **3**, 95 (1970). Rhomboid crystals, mp 151-155°. $[\alpha]_D^{20}$ 14.0 ±1.5° (c = 1.6 in pyridine). LD$_{50}$ i.p. in mice: 6.4 mg/kg; i.v. in rats: 21.0 mg/kg, Foerster *et al.*, *Arch. Int. Pharmacodyn.* **155**, 165 (1965).

THERAP CAT: Cardiotonic.

4329. Gladiolic Acid. *2,3-Diformyl-6-methoxy-5-methylbenzoic acid;* 4-methoxy-5-methyl-*o*-phthalaldehyde-3-carboxylic acid. $C_{11}H_{10}O_5$; mol wt 222.19. C 59.46%, H 4.54%, O 36.00%. Fungistatic antibiotic produced by *Penicillium gladioli* Machacek: Brian *et al.*, *Nature* **157**, 697 (1946). Active only in the form shown; the lactol form, 4-formyl-3-hydroxy-7-methoxy-6-methylphthalide, is inactive. Structure: Grove, *Biochem. J.* **50**, 648 (1952); **54**, 664 (1953); *J. Chem. Soc.* **1952**, 3345; Raistrick, Ross, *Biochem. J.* **50**, 635 (1952). Synthesis: Brown, Newbold, *J. Chem. Soc.* **1954**, 1076.

Silky needles from water, mp 160°. pK 4.4. uv max: 214, 271, 304 nm (ε 18,500, 6900, 3200). Sol in hot water.

4330. Glafenine. *2-[(7-Chloro-4-quinolinyl)amino]benzoic acid 2,3-dihydroxypropyl ester; N-(7-chloro-4-quinolyl)anthranilic acid 2,3-dihydroxypropyl ester;* 4-[(2-carboxyphenyl)amino]-7-chloroquinoline α-monoglyceride; 2-[(7-chloro-4-quinolyl)amino]benzoic acid α-glyceride; 2,3-dihydroxypropyl *N*-(7-chloro-4-quinolyl)anthranilate; glycerylaminophenaquine; glaphenine; R 1707; Glifan; Glifanan; Privadol. $C_{19}H_{17}ClN_2O_4$; mol wt 372.83. C 61.21%, H 4.60%, Cl 9.51%, N 7.52%, O 17.17%. Prepn: **Fr.** pat. **M2413** corresp to U.S. pat. **3,232,944** (1964, 1966 both to Roussel-UCLAF); G. Mouzin *et al.*, *Synthesis* **1980**, 54. Biotransformation in the rat and man: J. Pottier *et al.*, *Eur. J. Drug Metab. Pharmacokinet.* **4**, 109 (1979). Bioavailability, absorption: F. Moolenaar *et al.*, *Int. J. Pharm.* **4**, 195 (1980). Alkalimetric determn: M. S. Tawakkol *et al.*, *Pharmazie* **36**, 163 (1981).

Pale yellow prisms, mp 169-170°, G. Mouzin *et al.*, *loc. cit.*; also reported as mp 165°, U.S. pat. **3,232,944**. Practically insol in water. Slightly sol in abs alcohol, acetone, ether, benzene, chloroform; sol in dil aq alkalies and acids. LD$_{50}$ in mice: > 2 g/kg orally, U.S. pat. **3,232,944**.

THERAP CAT: Analgesic.

4331. Glaucarubin. *[1β,2α,11β,12α,15β(S)]-11,20-Epoxy-1,2,11,12-tetrahydroxy-15-(2-hydroxy-2-methyl-1-oxobutoxy)picras-3-en-16-one; 2-hydroxy-2-methylbutyric acid 4-ester with 1,2,3,3aβ,4,6aβ,7,7aα,10,11,11a,11bα-dodecahydro-1α,2α,4β,10α,11β-pentahydroxy-3α,8,11aβ-trimethyl-5H-1,11cβ-(epoxymethano)phenanthro[10,1-bc]pyran-5-one;* α-kirondrin; Glaumeba. $C_{25}H_{36}O_{10}$; mol wt 496.54. C 60.47%, H 7.31%, O 32.22%. From aceituno meal, the press cake from seeds of *Simaruba glauca* DC, *Simarubaceae:* Brink, *13th Int. Congr. Pure Appl. Chem.*

(Stockholm and Uppsala, 1953); Ham *et al.*, *J. Am. Chem. Soc.* **76**, 6066 (1954); Cuckler *et al.*, *Arch. Int. Pharmacodyn.* **114**, 307 (1958); Shafer, **U.S.** pat. **2,864,745** (1958 to Merck & Co.). Structure: Polonsky *et al.*, *Bull. Soc. Chim. France* **1964**, 1827. Identity of α- and β-kirondrin with glaucarubinone and glaucarubin, resp: Bourguignon, Polonsky, *Bull. Soc. Chim. Biol.* **46**, 1145 (1964). Crystal structure: Kartha *et al.*, *Nature* **202**, 389 (1964). Configuration: Nyburg *et al.*, *Chem. Commun.* **1965**, 203.

Bitter crystals from methanol, dec 250-255°. $[\alpha]_D^{25}$ +45° (c = 1.7 in pyridine); $[\alpha]_D^{25}$ +69° (c = 0.6 in methanol). Sol in 0.1*N* NaOH soln. Practically insol in aq sodium bicarbonate solns. Only slightly sol in water (less than 1.8 mg/ml). Prompt acidification of an alkaline soln regenerates glaucarubin, but standing in alkaline soln leads to decompn.

Tetramethoxyglaucarubin, $C_{29}H_{44}O_{10}$, crystals from ether, mp 202-205°. $[\alpha]_D$ +17.7° (c = 0.705 in pyridine); $[\alpha]_D$ +65.5° (c = 0.926 in methanol).

Pentaacetate, $C_{35}H_{46}O_{15}$, crystals from benzene + petr ether, mp 158-160°. $[\alpha]_D$ +49° (c = 0.534 in pyridine).

THERAP CAT: Antiamebic.

4332. Glaucine. *5,6,6a,7-Tetrahydro-1,2,9,10-tetramethoxy-6-methyl-4H-dibenzo[de,g]quinoline; 1,2,9,10-tetramethoxyaporphine;* boldine dimethyl ether; Bromcholitin; Glauvent. $C_{21}H_{25}NO_4$; mol wt 355.42. C 70.96%, H 7.09%, N 3.94%, O 18.01%. *d*-Form prevalent in nature. Found in *Glaucium flavum* Crantz (*G. luteum* Scop.), *Papaveraceae* and in *Dicentra* and *Corydalis* species, *Fumariaceae.* Isoln: Fischer, *Arch. Pharm.* **239**, 426 (1901). Structure and prepn: Gadamer, *ibid.* **249**, 680 (1911); Späth, Tharrer, *Ber.* **66**, 904 (1933). Configuration: Faltis, Adler, *Arch. Pharm.* **284**, 281 (1951). Synthesis of *dl*-form: Chan, Maitland, *J. Chem. Soc. (C)* **1966**, 753; Jackson, Martin, *ibid.* 2061; Cava *et al.*, *J. Org. Chem.* **35**, 175 (1970). Pharmacology and toxicity of *dl*-glaucine phosphate: Y. Kasé *et al.*, *Arzneimittel-Forsch.* **33**, 936 (1983). Sites of antitussive action: *eidem, ibid.* 947.

d-glaucine

d-Form, orthorhombic plates, prisms from ethyl acetate or ether, mp 120°. $[\alpha]_D^{20}$ +115° (c = 3 in alc). Sol in acetone, alcohol, chloroform, ethyl acetate. Moderately sol in ether, petr ether. Practically insol in water and benzene.

Hydrochloride trihydrate, $C_{21}H_{25}NO_4 \cdot HCl \cdot 3H_2O$, needles. mp 232° (anhydrous). Sol in water, alcohol, chloroform.

Hydrobromide, $C_{21}H_{25}NO_4 \cdot HBr$, crystals, mp 235°. Less sol than the hydrochloride.

dl-Form phosphate, $C_{42}H_{59}N_2O_{20}P_3$, DL-832. Cryst powder. LD$_{50}$ in mice (mg/kg): 98 i.v.; 401 orally (Kasé).

4333. Gliadin. A simple protein, one of the prolamins, derived from the gluten of wheat, rye, etc. May contain up to 43% glutamine (*C.A.* **50**, 15792a (1956)). Studies on the physical nature of gliadin: Holme, Briggs, *Cereal Chem.* **36**, 321 (1959). Use of deamidized gliadin in food products: McDonald, **U.S.** pat. **3,030,211** (1962 to USDA).

Practically insol in water, abs alcohol, and other neutral solvents. Sol in 70-80% alcohol, dil acid, dil alkali.

4334. Glibornuride. *N-[[(3-Hydroxy-4,7,7-trimethylbicyclo[2.2.1]hept-2-yl)amino]carbonyl]-4-methylbenzenesulfonamide; endo,endo-1-(2-hydroxy-3-bornyl)-3-(p-tolylsulfonyl)urea;* D-3-*endo*-p-tosylureidoborneol; 1-(p-tolylsulfonyl)-3-(2-*endo*-hydroxy-3-*endo*-D-bornyl)urea; Ro 6-4563; Gluborid; Glutril. $C_{18}H_{26}N_2O_4S$; mol wt 366.48. C 58.99%, H 7.15%, N 7.64%, O 17.46%, S 8.75%. Prepn: Bretschneider *et al., Monatsh.* **100**, 2133 (1969); *eidem,* **S. Afr.** pat. **67 06,161,** *C.A.* **71**, 61601j (1969); **S. Afr.** pat. **69 02,395** corresp to U.S. pat. **3,654,357** (1968, 1969, 1972, all to Hoffmann-La Roche). Comprehensive review of pharmacology: *Arzneimittel-Forsch.* **22**, 2153-2222 (1972).

Crystals, mp 192-195° (ethanol-water); also reported as 195-198°. [α]D +63.8° (ethanol).

THERAP CAT: Antidiabetic.

4335. Gliclazide. *N-[[(Hexahydrocylopenta[c]pyrrol-2(1H)-yl)amino]carbonyl]-4-methylbenzenesulfonamide; 1-(hexahydrocyclopenta[c]pyrrol-2(1H)-yl)-3-(p-tolylsulfonyl)urea; N-(4-methylbenzenesulfonyl)-N'-(3-azabicyclo-[3.3.0]oct-3-yl)urea;* 1-(3-azabicyclo[3.3.0]oct-3-yl)-3-(p-tolylsulfonyl)urea; S 1702; Diamicron; Glimicron; Nordialex. $C_{15}H_{21}N_3O_3S$; mol wt 323.42. C 55.71%, H 6.55%, N 12.99%, O 14.84%, S 9.91%. Prepn: Beregi *et al.,* **Fr.** pat. **1,510,714** and U.S. pat. **3,501,495** (1968 and 1970, both to Sci. Union & Co.-Soc. Franc. Rech. Med.). Series of articles on pharmacology: *Arzneimittel-Forsch.* **22**, 1686-1695 (1972). Toxicity data: J. Duhault *et al., ibid.* 1682. Review of pharmacology, efficacy: B. Holmes *et al., Drugs* **27**, 301 (1984).

Crystals from anhydrous ethanol, mp 180-182°. LD50 orally in mice: > 3 g/kg (Duhault).

THERAP CAT: Antidiabetic.

4336. Gliotoxin. *2,3,5a,6-Tetrahydro-6-hydroxy-3-(hydroxymethyl)-2-methyl-10H-3,10a-epidithiopyrazino[1,2-a]-indole-1,4-dione.* $C_{13}H_{14}N_2O_4S_2$; mol wt 326.39. C 47.84%, H 4.32%, N 8.58%, O 19.61%, S 19.65%. Antibiotic substance produced by various spp of *Trichoderma, Gladiocladium fimbriatum, Aspergillus fumigatus,* and *Penicillium* spp: Weindling, Emerson, *Phytopathology* **26**, 1068 (1936); **27**, 1175 (1937); Johnson *et al., J. Am. Chem. Soc.* **65**, 2005 (1943); Menzel *et al., J. Biol. Chem.* **152**, 419 (1944). Structure: Bell *et al., J. Am. Chem. Soc.* **80**, 1001 (1958); Beecham *et al., Tetrahedron Letters* **1966**, 3131. Crystallographic data: McCrone, *Anal. Chem.* **26**, 1662 (1954). Biosynthesis: Suhadolnik, Chenoweth, *J. Am. Chem. Soc.* **80**, 4391 (1958); Winstead, Suhadolnik, *ibid.* **82**, 1644 (1960); J. D. M. Herscheid *et al., J. Org. Chem.* **45**, 1885 (1980). Synthetic studies: Poisel, Schmidt, *Ber.* **104**, 1714 (1971); *ibid.* **105**, 625 (1972); Oehler *et al., ibid.* 625. Total synthesis of *dl*-form: T. Fukuyama, Y. Kishi, *J. Am. Chem. Soc.* **98**, 6723 (1976); T. Fukuyama *et al., Tetrahedron* **37**, 2045 (1981).

Monoclinic needles from methanol or benzene, dec 221°. $[\alpha]_D^{25}$ −290° (c = 0.08 in ethanol). uv max: 270 nm (ε 4500). Soly in mg/ml at 7°: acetic acid 12; acetone 9.0; acetonitrile 10.2; benzene 5.5; carbon tetrachloride 0.8; chloroform 20; dioxane 73 (decompn); dimethylformamide 17; ethyl acetate 8.5; ethanol 4.7; methanol 1.4; pyridine 77; water (30°): 0.07. Sensitive to oxidation and heat; inactivated by heating for 10 min at 100°.

Monoacetate, $C_{15}H_{16}N_2O_5S_2$, isolated from cultures of *Penicillium terlikowski* Zaleski: Johnson *et al., J. Am. Chem. Soc.* **75**, 2110(1953). Orthorhombic crystals from benzene, mp 162-163°.

4337. Glipizide. *N-[2-[4-[[[(Cyclohexylamino)carbonyl]amino]sulfonyl]phenyl]ethyl]-5-methylpyrazinecarboxamide; 1-cyclohexyl-3-[[p-[2-(5-methylpyrazinecarboxamido)ethyl]phenyl]sulfonyl]urea;* glydiazinamide; K 4024; Glibenese; Glucotrol; Mindiab; Minidiab. $C_{21}H_{27}N_5O_4S$; mol wt 445.55. C 56.61%, H 6.11%, N 15.72%, O 14.36%, S 7.20%. Second generation sulfonylurea with hypoglycemic activity. Prepn: Ambrogi, Logemann, **Ger.** pat. **2,012,138** corresp to U.S. pat. **3,669,966** (1970, 1972 both to Carlo Erba); Ambrogi *et al., Arzneimittel-Forsch.* **21**, 200 (1971). Pharmacology: *eidem, ibid.* 208; Marigo *et al., ibid.* 215. Metabolism: Goldaniga *et al., ibid.* **23**, 242 (1973); Fuccella *et al., J. Clin. Pharmacol.* **13**, 68 (1973). Toxicity: Ambrogi *et al., Arzneimittel-Forsch.* **21**, 208 (1971). Review of pharmacology and therapeutic efficacy: R. N. Brogden *et al., Drugs* **18**, 329-353 (1979); H. E. Lebovitz, *Pharmacother.* **5**, 63-77 (1985).

Crystals from ethanol, mp 208-209°. Also reported as mp 200-203°. LD50 in mice, rats (g/kg): > 3, 1.2 i.p. (Ambrogi).

THERAP CAT: Antidiabetic.

4338. Gliquidone. *N-[(Cyclohexylamino)carbonyl]-4-[2-(3,4-dihydro-7-methoxy-4,4-dimethyl-1,3-dioxo-2(1H)-isoquinolinyl)ethyl]benzenesulfonamide; 1-cyclohexyl-3-[[p-[2-(3,4-dihydro-7-methoxy-4,4-dimethyl-1,3-dioxo-2(1H)-iso-quinolyl)ethyl]phenyl]sulfonyl]urea;* 1,2,3,4-tetrahydro-2-[p-(N'-cyclohexylureido-N-sulfonyl)phenethyl]-4,4-dimethyl-7-methoxyisoquinoline-1,3-dione; AR-DF 26; Glurenorm. $C_{27}H_{33}N_3O_6S$; mol wt 527.64. C 61.46%, H 6.30%, N 7.96%, O 18.19%, S 6.08%. Prepn: Kutter *et al.,* **Ger.** pat. **2,000,339** (1971 to Thomae) corresp to U.S. pat. **3,708,486** (1973 to Boehringer, Ing.); *eidem,* **Ger.** pat. **2,011,126** (1971 to Thomae), *C.A.* **76**, 14359e (1972). Pharmacokinetics and metabolism: Kopitar, *Arzneimittel-Forsch.* **25**, 1455 (1975); Kopitar *et al., ibid.* 1933.

Crystals from boiling methanol, mp 180-182°. Sodium salt, sinters at 160°. LD50 in mice: > 2 g/kg orally; 234 mg/kg i.v., U.S. pat. **3,708,486.**

THERAP CAT: Antidiabetic.

4339. Glisoxepid. *N-[2-[4-[[[[(Hexahydro-1H-azepin-1-yl)amino]carbonyl]amino]sulfonyl]phenyl]ethyl]-5-methyl-3-isoxazolecarboxamide; 1-(hexahydro-1H-azepin-1-yl)-3-[[p-*

[2-(5-methyl-3-isoxazolecarboxamido)ethyl]phenyl]sulfonyl]-
urea; 4-[4-[β-(5-methylisoxazole-3-carboxamido)ethyl]-
phenylsulfonyl]-1,1-hexamethylenesemicarbazide; BS 4231;
RP 22410; Pro-Diaban. $C_{20}H_{27}N_5O_5S$; mol wt 449.54. C
53.43%, H 6.06%, N 15.58%, O 17.80%, S 7.13%. Prepn:
Plümpe, Puls, **S. Afr.** pat. **68 06,886** and U.S. pat. **3,668,215**
(1969, 1972 to Bayer). Pharmacology: Loubatieres *et al.,*
Compt. Rend. Ser. D **271**, 1446 (1970); *J. Pharmacol.* **3**, 171,
229 (1972). Series of articles: *Arzneimittel-Forsch.* **24**, 363-
452 (1974).

Colorless crystals from ethanol, mp 189°. LD_{50} orally in
mice, rats, cats, dogs: > 10.0, > 10.0, > 4.0, > 2.0 g/kg; i.v.
in mice, rats: 283, 196 mg/kg, Tettenborn, *Arzneimittel-*
Forsch. **24**, 409 (1974).
THERAP CAT: Antidiabetic.

4340. Globin. The colorless, basic protein of hemoglobin. Formed from tissue protein in the body, the globin part
of catabolized hemoglobin is re-used (unlike heme, the prosthetic group of hemoglobin). Prepn from ox hemoglobin:
Anson, Mirsky, *J. Gen. Physiol.* **13**, 469 (1930). Globin from
normal adult human hemoglobin consists of four polypeptide chains: two α-chains and two β-chains. The α-chain
contains 141, the β-chain 146 amino acids. Thus globin
contains 574 amino acids and has an approx mol wt of
62,000. Abnormal globins may contain γ- and δ-chains.
Structure: Braunitzer *et al., Z. Physiol. Chem.* **325**, 283
(1961); **331**, 1 (1963); Konigsberg *et al., J. Biol. Chem.* **237**,
1549, 2547 (1962); **238**, 2016, 2028 (1963). Review of prepn
and properties: Rossi Fanelli *et al., Advan. Protein Chem.*
19, 124 (1964). *Review:* Braunitzer *et al., ibid.* 1.
Denatures rapidly above 17°. At pH values near neutrality, combines with ferroprotoporphyrin to yield hemoglobin,
or with ferriprotoporphyrin to yield methemoglobin.

4341. Glucagon. Glukagon; hyperglycemic-glycogeno-
lytic factor; HG-factor; HGF. Mol wt approx 3550. A
polypeptide hormone produced in the alpha cells of the islets
of Langerhans in the pancreas. Existence of this hyperglycemic pancreatic factor was originally postulated to explain
the initial hyperglycemia following insulin administration.
Refs: Murlin *et al., J. Biol. Chem.* **56**, 252 (1923); Kimball,
Murlin, *ibid.* **58**, 337 (1923); Bürger, Kramer, *Arch. Exp.*
Pathol. Pharmakol. **156**, 1 (1930); Bürger, Brandt, *Z. Ges.*
Exp. Med. **96**, 375 (1935); Staub *et al., Science* **117**, 628
(1953); Ingle *et al., Proc. Soc. Exp. Biol. Med.* **84**, 232 (1953);
Geschwind, Staub, *ibid.* 244; Staub *et al., J. Biol. Chem.* **214**,
619 (1955); **Brit.** pat. **762,885** (1956 to Lilly). Amino acid
sequence of porcine, bovine, and human glucagon: Bromer
et al., J. Am. Chem. Soc. **79**, 2807 (1957); Bromer *et al., J.*
Biol. Chem. **246**, 2822 (1971); Thomsen *et al., FEBS Letters*
21, 315 (1972). Primary structure of glucagon from all species studied thus far is the same, with the exception of the
turkey which differs by one amino acid residue. Total synthesis: E. Wünsch, *Z. Naturforsch.* **22B**, 1269 (1967). *See*
also E. Wünsch, G. Wendlberger, *Ber.* **101**, 3659 (1968) and
preceding articles in the series which describe in detail the
prepn of the fragments of glucagon and the assembly of the
total sequence. Usually present in increased concns in diabetes. *Reviews:* Behrens, Bromer, *Vitamins and Hormones*
16, 263-301 (1958); A. C. Beynen, M. J. Geelen, *Vet. Res.*
Commun. **5**, 223-236 (1982). Books: P. J. Lefèbvre, R. H.
Unger, *Glucagon* (Pergamon Press, New York, 1972); *Gluca-*
gon **Pts. I and II**, P. Lefèbvre, Ed. (Springer, New York,
1983) 535 and 700 pp, resp.

1   15
His-Ser-Gln-Gly-Thr-Phe-Thr-Ser-Asp-Tyr-Ser-Lys-Tyr-Leu-Asp-

16   29
Ser-Arg-Arg-Ala-Gln-Asp-Phe-Val-Gln-Trp-Leu-Met-Asn-Thr

Rhombic dodecahedra. Stable. Practically insol in water.
Sol in acidic, basic media, *i.e.* below pH 3 and above pH 9.5.
THERAP CAT: Antidiabetic.

4342. Glucametacin. *2-[[[1-(4-Chlorobenzoyl)-5-meth-*
oxy-2-methyl-1H-indol-3-yl]acetyl]amino]-2-deoxy-D-glu-
cose; 2-[2-[1-(p-chlorobenzoyl)-5-methoxy-2-methylindol-
3-yl]acetamido]-2-deoxy-D-glucose; glucametacine; glucamethacin; indomethacin glucosamide. $C_{25}H_{27}ClN_2O_8$; mol
wt 518.96. C 57.86%, H 5.25%, Cl 6.83%, N 5.40%, O
24.66%. Deriv of indomethacin, *q.v.* Prepn: A. Demetrio *et*
al., Ger. pat. **2,223,051** (1973 to SIR Lab. Chem. Biol. SpA),
C.A. **80**, 83529e (1974). Pharmacological study: E. Paroli *et*
al., Arzneimittel-Forsch. **28**, 819 (1978). Clinical studies: P.
Petera *et al., Int. J. Clin. Pharmacol.* **15**, 581 (1977); L. Ca-
pelli *et al., Curr. Med. Res. Opin.* **7**, 227 (1981).

Monohydrate, $C_{25}H_{27}ClN_2O_8 \cdot H_2O$, *Euminex, Teorema,*
Teoremac.
THERAP CAT: Anti-inflammatory.

4343. Glucamine. *1-Amino-1-deoxy-D-glucitol;* 1-ami-
no-1-deoxysorbitol; glycamine; D-glucamine. $C_6H_{15}NO_5$;
mol wt 181.19. C 39.77%, H 8.34%, N 7.73%, O 44.15%.
Prepn from D-glucose: Holly *et al., J. Am. Chem. Soc.* **72**,
5416 (1950); from *N*-benzylglycamine: Kagan *et al., ibid.*
79, 3541 (1957); by catalytic reduction of glucose in the
presence of hydrazine: Lemieux, **U.S.** pat. **2,830,983** (1958
to Natl. Res. Council, Ottawa). Commercial prepn: Flint,
Salzberg, **U.S.** pat. **2,016,962** (1935 to du Pont); Groggins,
Stirton, *Ind. Eng. Chem.* **29**, 1358 (1937).

Crystals from methanol, mp 127. Sharp, slightly sweet
taste. $[\alpha]_D^{15}$ −7.95° (c = 10 in water). Very sol in water;
slightly sol in alcohol. Practically insol in ether.

4344. D-Glucaric Acid. Saccharic acid; D-glucosaccharic
acid; D-tetrahydroxyadipic acid. $C_6H_{10}O_8$; mol wt 210.14. C
34.29%, H 4.80%, O 60.91%. Best prepd by nitric acid oxidation of starch; yields as high as 65% are obtained in contrast to much lower yields from glucose or sucrose: Kiliani,
Ber. **58**, 2344 (1925); Schmidt *et al., ibid.* **70**, 2402 (1937).
Prepn from D-glucose: Mehltretter *et al.,* U.S. pat. **2,472,-**
168 (1949 to U.S. Secy. Agr.); Truchan, U.S. pat. **2,809,989**
(1957 to Cowles Chem.); Phillips *et al., J. Chem. Soc.* **1958**,
3522; **Belg.** pat. **615,023** (1962 to Ciba).

Needles from 95% ethanol, mp 125-126°. Shows mutarotation. $[\alpha]_D^{19}$ +6.86° → +20.60° (H_2O). $Ka_1 = 1.0 \times 10^{-5}$ at 25°. Soluble in water, ethanol. Sparingly sol in ether.

1,4-Lactone, *saccharolactone.* Monohydrate, crystals, mp 90°. Strong inhibitor or β-glucuronidase: Levvy, *Biochem. J.* **52**, 464 (1952); Boyland, Williams, *ibid.* **64**, 578 (1956).

4345. D-Glucoascorbic Acid. D-*arabino-Hept-2-enonic acid* γ-*lactone;* 3-keto-D-glucoheptonofuranolactone. C_7-$H_{10}O_7$; mol wt 206.15. C 40.78%, H 4.89%, O 54.33%. A physiologically inactive homolog of ascorbic acid. Prepn: Ault *et al., J. Chem. Soc.* **1933**, 741; Baird *et al., ibid.* **1934**, 62; Reichstein *et al., Helv. Chim. Acta* **17**, 510 (1934); Stacey, Turton, *J. Chem. Soc.* **1946**, 661; Stedehouder, *Rec. Trav. Chim.* **71**, 831 (1952). Review on analogs of ascorbic acid: Smith, *Advances in Carbohydrate Chemistry* **2**, 79 (1946).

Monohydrate, clusters of rod-like crystals with pointed ends from acetone + methanol + petr ether, mp 101-105° (Reichstein); mp 138° (Baird). Becomes anhyd at 70° under high vacuum, mp 191° (Ault). $[\alpha]_D^{20}$ −14° (c = 1 as hydrate); $[\alpha]_D^{20}$ −22° (c = 1 as hydrate in methanol); $[\alpha]_D^{14.5}$ −37.8° (c = 2.41 in 0.01N HCl); $[\alpha]_D^{20}$ −80° (c = 0.75 neutralized with NaOH). $pK_1 = 4.26$; $pK_2 = 11.58$. Soluble in water; moderately sol in alcohol.

4346. Glucofrangulin. Erbalax-N; Irgalax; Solco-Lax. $C_{27}H_{30}O_{14}$; mol wt 578.54. C 56.05%, H 5.23%, O 38.72%. Anthraquinone glycoside from bark of *Rhamnus frangula* L., *Rhamnaceae* (alder buckthorn): Casparis, Maeder, *Bull. Soc. Chim. Biol.* **9**, 324 (1927); Cucu, Jarpo, *Pharmazie* **14**, 316 (1959); Knap *et al., Czech.* pat. 110,024 (1964), *C.A.* **61**, 4159e (1964). Consists of two isomers, glucofrangulin A and glucofrangulin B, which differ by the linkage to the sugar moiety at the 3 position of the aglycone. Structure: Schindler, *Helv. Chim. Acta* **29**, 411 (1946); Hörhammer *et al., Naturwiss.* **51**, 310 (1964); Longo *et al., Arch. Pharm.* **297**, 248 (1964); Wagner, Hörhammer, *Naturwiss.* **53**, 585 (1966). Proof of structure of glucofrangulin A and partial synthesis: *eidem, Z. Naturforsch. B* **24**, 1408 (1969).

glucofrangulin A

Glucofrangulin A, 3-[(6-deoxy-α-L-mannopyranosyl)oxy]-1-(β-D-glucopyranosyloxy)-8-hydroxy-6-methyl-9,10-anthracenedione.

Glucofrangulin A octaacetate, $C_{43}H_{46}O_{22}$, needles from methanol, mp 228-230°. $[\alpha]_D^{20}$ −124° (c = 1.16 in acetone). uv max: 212, 264, 360 nm (log ε 4.57, 4.56, 4.26).

THERAP CAT: Cathartic.

4347. α-Glucogallin. α-D-*Glucopyranose 1-(3,4,5-trihydroxybenzoate);* α-D-*glucopyranosyl-1-gallate;* 1-galloyl-α-D-glucose. $C_{13}H_{16}O_{10}$; mol wt 332.26. C 46.99%, H 4.85%, O 48.15%. Synthesis: Schmidt, Herok, *Ann.* **587**, 63 (1954); Schmidt, Schmadel, *ibid.* **649**, 149 (1961).

Crystals, dec 179-181°. $[\alpha]_D^{20}$ +83° (c = 3 in methanol). Dihydrate, prisms from water, mp 171-173°. $[\alpha]_D^{25}$ +79.1° (water). Freely sol in water, methanol, ethanol, dioxane, acetic acid. Sparingly sol in ethyl acetate, ether and acetone.

4348. β-Glucogallin. β-D-*Glucopyranose 1-(3,4,5-trihydroxybenzoate);* β-D-*glucopyranosyl-1-gallate;* glucogallic acid; 1-galloyl-β-D-glucose. $C_{13}H_{16}O_{10}$; mol wt 332.26. C 46.99%, H 4.85%, O 48.15%. Glucoside or glucotannoid from chinese rhubarb, *Rheum officinale,* Baill., *Polygonaceae:* Gilson, *Compt. Rend.* **136**, 385 (1903). Structure and synthesis: Fischer, Bergmann, *Ber.* **51**, 1760 (1918); Schmidt, Schmadel, *Ann.* **649**, 149 (1961).

Bitter microscopic prisms from water, methanol or 80% ethanol, mp 207°. $[\alpha]_D^{20}$ −24.5° (c = 1.75 in water). Freely sol in hot water. Sparingly sol in cold water, methanol, ethanol, acetone, ethyl acetate. Practically insol in ether, benzene, chloroform, petr ether.

4349. Glucoheptonic Acid. D-*glycero-D-gulo-Heptonic acid;* α-glucoheptonic acid; glucosemonocarboxylic acid; glucomonocarbonic acid. $C_7H_{14}O_8$; mol wt 226.18. C 37.17%, H 6.24%, O 56.59%. Obtained by treating glucose with HCN yielding a cyanohydrin which is saponified to glucoheptonic acid: Kiliani, *Ber.* **19**, 769 (1886); Fischer, *Ann.* **270**, 71 (1892); Armestar, *C.A.* **45**, 2865 (1951). Process starting with calcium cyanide and glucose: Clevenot, U.S. pat. 2,735,866 (1956 to Lab. Clevenot). Diagnostic use of ^{99m}Tc complexes in renal scintigraphy: R. E. Boyd *et al., Brit. J. Radiol.* **46**, 604 (1973); in brain scanning: J. Léveillé *et al., J. Nucl. Med.* **18**, 957 (1977); T. W. Ryerson *et al., Radiology* **127**, 429 (1978). Subacute toxicity study: L. Belbeck *et al., Can. J. Comp. Med.* **45**, 299 (1981).

Lactonizes upon evapn. The lactone forms large sweetish crystals, mp 145-148°. $[\alpha]_D^{20}$ −56.0° (shows mutarotation). Sol in water.

Sodium salt, $C_7H_{13}NaO_8$, *gluceptate sodium, sodium glucoheptonate.* Prepn from corn syrup: Behnke, U.S. pat. 3,022,343 (1962 to Pfanstiehl Labs). Crystals (α-form), dec 161°. $[\alpha]_D^{20}$ +6.06° (c = 10 in H_2O). Freely sol in water.

Calcium salt, $C_{14}H_{26}CaO_{16}$, *gluceptate calcium, calcium glucoheptonate, calcium glucosemonocarbonate, calcium glucomonocarbonate, Calciforte, Calheptose.* Prepn from Na salt: Holstein, U.S. pat. 3,033,900 (1962 to Pfanstiehl Labs.). Hygroscopic crystals, somewhat acrid taste, dec 200°. Sol in water.

Magnesium salt, $C_{14}H_{26}MgO_{16}$, *magnesium glucoheptonate, magnesium glucosemonocarbonate, magnesium gluco-*

monocarbonate, Navolin. Prepn: Cipelli, **U.S. pat. 3,063,896** (1962 to Merck & Co.). Water-sol crystals, pleasant taste.

Complex with ^{99m}Tc, ^{99m}*Tc gluceptate*, ^{99m}*Tc gluheptonate*, *Glucoscan, TechneScan gluceptate*.

USE: Pharmaceutic aid.

THERAP CAT: ^{99m}Tc complex as diagnostic aid (radioactive imaging agent).

4350. Gluconic Acid. D-Gluconic acid; dextronic acid; maltonic acid; glyconic acid; glycogenic acid; pentahydroxycaproic acid. $C_6H_{12}O_7$; mol wt 196.16. C 36.74%, H 6.17%, O 57.10%. May be prepd from glucose by electrolytic oxidation in alkaline medium: Helwig, **U.S. pat. 1,937,273** (1933); by chemical oxidation with hypobromites: Stoll, Kussmaul, **U.S. pat. 1,648,368** (1927). At present produced in commercial quantities by the fermentative oxidation of the aldehyde group in glucose from corn using *Aspergillus niger, A. fumaricus, Acetobacter aceti, Penicillium chrysogenum*, and other *Penicillia*: Bernhauer, Schulof, **U.S. pat. 1,849,053** (1932 to Pfizer); Williams, *Mfg. Chemist* **16**, 239 (1945); Moyer *et al., Ind. Eng. Chem.* **32**, 1379 (1940); Prescott *et al., ibid.* **45**, 338 (1953); van Gelder, **U.S. pat. 2,916,515** (1959 to Noury & van der Lande, Holland); Roehr, *Naturwiss.* **48**, 478 (1961). Prepn by hydrolysis of α-D-glucose with bromine + H_2SO_4: Foster, Vardheim, *J. Chem. Soc.* **1957**, 989; by γ-irradiation of D-glucose: Phillips *et al., ibid.* **1958**, 3522; Grant, Ward, *ibid.* **1959**, 2871. Prepn of solid: Isbell, **U.S. pat. 1,985,255** (1934 to U.S. Secy of Commerce).

COOH
|
HCOH
|
HOCH
|
HCOH
|
HCOH
|
CH₂OH

Crystals, mp 131°. Mild acid taste. $[\alpha]_D^{20}$ −6.7° (c = 1). K at 25° = 2.5 × 10^{-4}. Freely sol in water, slightly sol in alcohol. Insol in ether and most other organic solvents. In aq solns the acid is partially transformed into an equilibrium mixt with gamma and delta gluconolactones. Because of the difficulties of preparing a solid cryst product, gluconic acid of commerce is a 50% aq soln, light amber color, faint odor of vinegar, d_4^{25} 1.24. May be stored in stainless steel drums.

Ammonium salt, $NH_4C_6H_{11}O_7$, needles, deliquesc in moist air and become yellow on exposure to light. Dec 154°. Neutral reaction. Soly in water: 31.6 g/100 ml at 25°. Slightly sol in alc. Practically insol in most other organic solvents. The salt dec on steaming with the formation of ammonia and gluconic acid.

Magnesium salt dihydrate, $C_{12}H_{22}MgO_{14}\cdot2H_2O$, *magnesium gluconate, Almora, Glucomag, Ultra-Mg*.

Zinc complex, $C_{12}H_{22}O_{14}Zn$, *Rubozinc*.

USE: *See* Gluconolactone. Ammonium salt used as latent acid catalyst in textile printing.

THERAP CAT: Magnesium salt as antispasmodic; magnesium replenisher.

4351. Gluconolactone. D-*Gluconic acid* δ-*lactone;* glucono delta lactone; delta gluconolactone. $C_6H_{10}O_6$; mol wt 178.14. C 40.45%, H 5.66%, O 53.89%. Prepn by oxidation of glucose with bromine water: Isbell, Pigman, *J. Res. Nat. Bur. Stand.* **10**, 337 (1933); by oxidation of glucose in *Acetobacter suboxydans:* King, Cheldelin, *Biochem. J.* **68**, 31P (1958). Structure: J. Staněk *et al., The Monosaccharides* (Academic Press, New York, 1963) p 271.

CH₂OH
|
H   O
H
OH H   O
HO
|
H   OH

Crystals, dec 153°. Sweet taste (different from gluconic

acid). $[\alpha]_D^{20}$ +61.7° (c = 1). Soly in water 59 g/100 ml; in alc about 1 g/100 g. Insol in ether. Hydrolyzed to gluconic acid by water. A freshly prepd 1% aq soln has a pH of 3.6 changing to pH 2.5 within 2 hrs.

USE: Component of many cleaning cmpds because of the sequestering ability of the gluconate radical which remains active in alk solns; in the dairy industry to prevent milkstone; in breweries to prevent beerstone; as latent acid catalyst for acid colloid resins, particularly in textile printing.

4352. Glucosamine. 2-*Amino-2-deoxyglucose;* chitosamine. $C_6H_{13}NO_5$; mol wt 179.17. C 40.22%, H 7.31%, N 7.82%, O 44.65%. Found in chitin, in mucoproteins, and in mucopolysaccharides. Isoln from chitin: Ledderhose, *Z. Physiol. Chem.* **2**, 213 (1878); Hackman, *Aust. J. Biol. Sci.* **7**, 168 (1954). Synthesis: Fischer, Leuchs, *Ber.* **35**, 3787 (1902); **36**, 24 (1903). Separation of α- and β-forms: Westphal, Holzmann, *ibid.* **75B**, 1274 (1942). Structure: Haworth *et al., J. Chem. Soc.* **1939**, 271; Cutler, Peat, *ibid.* 782; Cox, Jeffrey, *Nature* **143**, 894 (1939). Pharmacokinetics in dog and man: I. Setnikar *et al., Arzneimittel-Forsch.* **36**, 729 (1986). Clinical trials in arthrosis: Y. Vajarudal, *Clin. Ther.* **3**, 336 (1981); M. J. Tapadinhas *et al., Pharmatherapeutica* **3**, 157 (1982). *Review:* Foster, Stacey, "The Chemistry of the 2-Amino Sugars" in C. S. Hudson *et al., Advan. Carbohyd. Chem.* **vol. 7** (Academic Press, New York, 1952) pp 247-288.

CH₂OH
O
H   OH
H   H
OH H
HO
H   NH₂

α-Form, crystals, mp 88°. $[\alpha]_D^{20}$ +100° changing to +47.5° after 30 min (water).

β-Form, needles from methanol, dec 110°. $[\alpha]_D^{20}$ +28° changing to +47.5° after 30 min (water). Very sol in water, sol in about 38 parts boiling methanol; sparingly sol in cold methanol or ethanol. Practically insol in ether, chloroform.

N-Acetylglucosamine, $C_8H_{15}NO_6$, needles from methanol + ether, mp 205°. $[\alpha]_D^{18}$ +64° changing to +40.9° (in water).

Sulfate salt, $C_6H_{13}NO_5\cdot xH_2O_4S$, *Dona*.

USE: Pharmaceutic aid.

THERAP CAT: Antiarthritic.

4353. Glucose. D-Glucose; dextrose; blood sugar; grape sugar; corn sugar; Dextropur; Dextrosol; Glucolin. $C_6H_{12}O_6$; mol wt 180.16. C 40.00%, H 6.72%, O 53.29%. A main source of energy for living organisms. Occurs naturally and in the free state in fruits and other parts of plants. Combined in glucosides, in di- and oligosaccharides, in the polysaccharides cellulose and starch, and in glycogen. Normal human blood contains 0.08-0.1%. Manuf on a large scale from starch: Dean, Gottfried, *Advan. Carbohyd. Chem.* **5**, 127 (1950). Below 50°, α-D-glucose hydrate is the stable cryst form, above 50° the anhydr form is obtained and at still higher temps β-D-glucose is formed: W. Pigman, *The Carbohydrates* (Academic Press, New York, 1957) p 92. Structure: Kjaer, Lindberg, *Acta Chem. Scand.* **13**, 1713 (1959). Conformation: E. Percival, *Structural Carbohydrate Chemistry* (J. Garnet Miller, London, 1962) pp 51-57. Comprehensive monograph: H. Bartelheimer *et al., D-Glucose und verwandte Verbindungen in Medizin und Biologie* (Enke, Stuttgart, 1966) 1126 pp.

CH₂OH
O
H   H
H
OH H
HO   OH
H   OH

α-D-glucose

α-Form monohydrate, crystals from water, mp 83°. $[\alpha]_D$ +102.0° → +47.9° (water). 0.74 times as sweet as sucrose.

One gram dissolves in about 1 ml water and in about 60 ml alcohol. LD i.v. in rabbits: 35 g/kg.

α-Form anhydr, crystals from hot ethanol or water, mp 146°. $[\alpha]_D +112.2° \rightarrow +52.7°$ (c = 10 in water). The final value is obtained instantly in the presence of hydroxyl ions. Formula for varying concns: $[\alpha]_D^{20} +52.5° + 0.0188p$ (p = g/100 ml). pH of 0.5 molar aq soln 5.9. $d_{17.5}^{17.5}$ of water solns w/v: 5% = 1.019; 10% = 1.038; 20% = 1.076; 30% = 1.113; 40% = 1.149. n_D^{20} 10% soln 1.3479. One gram dissolves in 1.1 ml water at 25°; in 0.8 ml at 30°; in 0.41 ml at 50°; in 0.28 ml at 70°; in 0.18 ml at 90°; in 120 ml methanol at 20°. Very sparingly sol in abs alcohol, ether, acetone; sol in hot glacial acetic acid, pyridine, aniline. LD i.v. in rabbits: 35 g/kg.

β-Form, crystals from hot water + ethanol, from dil acetic acid, or from pyridine, mp 148-155°. $[\alpha]_D +18.7° \rightarrow +52.7°$ (c = 10 in water).

THERAP CAT: Fluid and nutrient replenisher.

THERAP CAT (VET): Nutrition (usually parenterally), hypoglycemia, ketosis, to counteract hepatotoxins.

4354. Glucose Oxidase. β-D-Glucopyranose aerodehydrogenase; P-FAD; corylophyline; microcide; mikrotsid; notatin. An enzyme obtained from mycelia of fungi, such as *Aspergilli* and *Penicillia*; a typical aerobic dehydrogenase which catalyzes the oxidation of glucose to gluconic acid (molecular oxygen is reduced to hydrogen peroxide). It is a flavoprotein, the prosthetic group being flavine-adenine dinucleotide (FAD). Commercial prepns frequently contain appreciable amounts of another enzyme, catalase, which is desirable for certain uses since it removes hydrogen peroxide aerobically generated by glucose oxidase. Names of some commercial prepns are: *DeeO, Fermcozyme, OxyBan, Ovazyme.* Isoln from *Penicillia* cultures: Coulthard *et al., Biochem. J.* **39,** 24 (1945). Commercial production from *Aspergilli* and *Penicillia:* Goldsmith *et al.,* U.S. pat. **2,926,122** (1960); from *Aspergillus niger:* Faucett *et al.,* U.S. pat. **3,102,081** (1963 to Miles Labs.). Removal of proteolytic enzymes from glucose oxidase (contg catalase) obtained from *Aspergilli* or *Penicillia* cultures: Ohlmeyer, U.S. pat. **2,940,904** (1960 to Ben L. Sarett). Separation from catalase: Pazur *et al., Biochem. Biophys. Acta* **65,** 369 (1962). Properties: Muller, *Enzymologia* **10,** 40 (1941); Keilin, Hartree, *Biochem. J.* **42,** 221 (1948), **50,** 331 (1952). *Reviews:* L. A. Underkofler "Glucose Oxidase: Production, Properties, Present and Potential Applications" in *Soc. Chem. Ind. (London) Monograph* **no. 11,** 72-86 (1961); R. Bentley, "Glucose Oxidase" in *The Enzymes* **vol. 7,** P. D. Boyer *et al.,* Eds. (Academic Press, New York, 1963) pp 567-586.

Amorphous powder or crystals. Abs max between 270-280, 375-380, and 450-460 nm (aq soln). Freely sol in water giving yellowish-green solns. Most active at pH 5.5-6.0 and 30-35°. Stable between pH 4.5 and 7.0. Stable to pepsin and trypsin. A glucose oxidase unit is defined as that quantity of enzyme which will cause the uptake of 10 mm³ oxygen per min in a Warburg manometer at 30° in the presence of excess air and excess catalase with a substrate contg 3.3% glucose monohydrate and 0.1M phosphate buffer, pH 5.9 with 0.4% sodium dehydroacetate: Scott, *J. Agr. Food Chem.* **1,** 727 (1953).

USE: Mainly in the protection of foods: for the removal of glucose from egg albumin and whole eggs prior to drying. To remove oxygen from canned foods, soft drinks, beer, and stored food. In the manuf of test papers for diabetes control and fertility tests. To stabilize ascorbic acid and vitamin B_{12} prepns. In combination with catalase, for treatment of food wrappers to prevent oxidative deterioration of food: Sarett, Scott, U.S. pat. **2,765,233** (1956 to Ben L. Sarett).

4355. Glucose-6-phosphate. D-*Glucose 6-(dihydrogen phosphate);* glucose-6-phosphoric acid; Robison ester. $C_6H_{13}O_9P$; mol wt 260.14. C 27.70%, H 5.04%, O 55.35%, P 11.91%. A normal constituent of resting muscle, probably always existing in equilibrium with fructose-6-phosphate. For the enzymatic conversion from the 1-phosphate *see* α-Glucose-1-phosphate. Isoln from a crude mixture of hexose phosphates, obtained by yeast fermentation: Robison, King, *Biochem. J.* **25,** 323 (1931). Prepn by the action of phosphoglucomutase on α-glucose-1-phosphate: Colowick, Sutherland, *J. Biol. Chem.* **144,** 423 (1942); from acetone

glucose: Levene, Raymond, *ibid.* **92,** 757 (1931); by phosphorylation of 1,2,3,4-tetraacetylglucose followed by deacetylation: Fischer, Lardy, *ibid.* **164,** 513 (1946); *Biochem. Prepn.* **2,** 39 (1952). Prepn from starch: de Chatelperron *et al., Fr.* pat. **1,379,068** (1964), *C.A.* **62,** 9394b (1965).

Barium salt, $C_6H_{11}O_5PO_4Ba$, nonhygroscopic, stable powder. $[\alpha]_D^{24} +17.9°$. Easily sol in water.

Dipotassium salt, $C_6H_{11}O_5PO_4K_2$, precipitate from methanol. $[\alpha]_D^{24} +21.2°$ (c = 1.3). Freely sol in water.

4356. α-Glucose-1-phosphate. α-Glucose-1-phosphoric acid; α-D-glucopyranose-1-phosphate; Cori ester. $C_6H_{13}O_9$-P; mol wt 260.14. C 27.70%, H 5.04%, O 55.35%, P 11.91%. Found widely in both plants and animals. In plants it is the immediate precursor of starch, and in animals of glycogen, being also the first product in the breakdown and utilization of these substances. Isoln from muscle and synthesis using trisilver phosphate: Cori *et al., J. Biol. Chem.* **121,** 465 (1937); Krahl, Cori, *Biochem. Prepn.* **1,** 33 (1949). Prepn from α-acetobromglucose + silver diphenyl phosphate: Posternak, *J. Am. Chem. Soc.* **72,** 4824 (1950); by phosphorolysis of starch using phosphorylase and orthophosphate: McCready, Hassid, *Biochem. Prepn.* **4,** 63 (1955). Structure: Wolfrom, Pletcher, *J. Am. Chem. Soc.* **63,** 1050 (1941). Configuration: Wolfrom *et al., ibid.* **64,** 23 (1942); Harmon, *Diss. Abstr.* **24,** 4400 (1964); Beevers, Maconochie, *Acta Cryst.* **18,** 232 (1965).

Free acid, $[\alpha]_D^{25} +120°$. $pK_1 = 1.11$; $pK_2 = 6.13$. Stronger acid than H_3PO_4. Extremely sol in water.

Barium salt trihydrate, nonhygroscopic, stable powder. $[\alpha]_D^{25} +75°$ (c = 1.26). Easily sol in water.

Calcium salt, *Actigam.*

Dipotassium salt dihydrate, crystals from ethanol. $[\alpha]_D^{20} +78°$ (c = 4); $[\alpha]_{549}^{20} +90°$ (c = 4). Freely sol in water.

THERAP CAT: Calcium salt as roborant.

4357. Glucosulfone Sodium. *1,1'-[Sulfonylbis(4,1-phenyleneimino)]bis[1-deoxy-1-sulfo-D-glucitol] disodium salt; p,p'-sulfonyldianiline N,N'-diglucoside disodium disulfonate; p,p'-*sulfonyldianiline-*N,N'-*di-D-glucose sodium bisulfite compd; *p,p'-*diaminodiphenylsulfone-*N,N'-*di(dextrose sodium sulfonate); disodium *p,p'-*diaminodiphenylsulfone-*N,N'-*diglucose sulfonate; 501 P; Protomin; Promin; Promanide; Angeli's sulfone. $C_{24}H_{34}N_2Na_2O_{18}S_3$; mol wt 780.70. C 36.92%, H 4.39%, N 3.59%, Na 5.89%, O 36.89%, S 12.32%. Prepd by refluxing a mixture of 4,4'-diaminodiphenylsulfone, glucose, sodium bisulfite, and 80% ethanol: Jain *et al., C.A.* **40,** 4687⁷ (1946); **Swiss** pat. **234,108** (1944 to B. Siegfried), *C.A.* **43,** 4297a (1949).

White, amorphous powder. Soluble in water; slightly sol in alc. Insol in ether, benzene, methanol, ethyl acetate, pyridine. Aq solns may be sterilized by autoclaving. Marketed as an injectable aq soln of a mixture of 88.5% glucosulfone sodium + 11.5% glucose. LD_{50} orally in mice: 3.93 g/kg.

THERAP CAT: Antibacterial (leprostatic).

4358. N^4-β-D-Glucosylsulfanilamide. *4-(β-D-Glucopyranosylamino)benzenesulfonamide;* N^4-β-d-glucosidosulfanilamide; sulfanilamide-d-glucoside; *N-p-sulfamylphenyl-*D-glucosylamine; Prontoglucal; Prontoglukal. $C_{12}H_{18}N_2O_7S$; mol wt 334.36. C 43.11%, H 5.43%, N 8.38%, O 33.50%, S 9.59%. Prepn: Kuhn, Birkofer, *Ber.* **71**, 621 (1938); Bognár, Nánási, *J. Chem. Soc.* **1953**, 1703; Bognár *et al., Magy. Kém. Foly.* **62**, 271 (1956), *C.A.* **54**, 4399g (1960). Structure: Braun *et al., J. Org. Chem.* **7**, 19 (1942).

$C_6H_{11}O_5$—NH—⟨ ⟩—SO_2NH_2

Fine needles from aq ethanol, mp 204°. $[\alpha]_D^{22}$ −117° (c = 0.9 in pyridine); $[\alpha]_D^{22}$ −128° (c = 0.9 in water).
2,3,4,6-Tetraacetate, $C_{20}H_{26}N_2O_{11}S$, crystals from ethanol, mp 204°. $[\alpha]_D^{22}$ −81° (pyridine).
$N,N',2,3,4,6$-Hexaacetate, $C_{24}H_{30}N_2O_{13}S$, crystals from ethanol, mp 115°.

THERAP CAT: Antibacterial.

4359. Glucovanillin. *4-(β-D-Glucopyranosyloxy)-3-methoxybenzaldehyde;* vanillin-D-glucoside; avenein; vanilloside. $C_{14}H_{18}O_8$; mol wt 314.28. C 53.50%, H 5.77%, O 40.73%. From green fruit of vanilla: Goris, *Compt. Rend.* **179**, 70 (1924). From coniferin by oxidation with CrO_3: Tremann, *Ber.* **18**, 1595 (1885). Structure and synthesis: Fischer, Raske, *Ber.* **42**, 1475 (1909); Thorpe, Williams, *J. Chem. Soc.* **1937**, 494.

Needles from methanol, mp 189-190°. Bitter taste. $[\alpha]_D^{20}$ −89.9° (water). Soluble in hot water, alcohol. Almost insol in ether.
Tetraacetate, $C_{22}H_{26}O_{12}$, crystals from dil alcohol, mp 142-143°. $[\alpha]_D$ −48.3° (chloroform).

USE: Pharmaceutic aid (flavor).

4360. D-Glucuronic Acid. $C_6H_{10}O_7$; mol wt 194.14. C 37.12%, H 5.19%, O 57.69%. Widely distributed in the plant and animal kingdoms. Usually occurs in "paired" form, *i.e.* as a glycosidic combination with phenols, alcohols, etc. Such glucuronides form in the liver to detoxify poisonous hydroxyl-containing substances. The glucuronides present in normal urine are those of phenol, cresol, and indoxyl. After the ingestion of poisons such as morphine, chloral hydrate, camphor, or turpentine, glucuronides formed with the poison or its hydroxylated derivatives appear in the urine. Review and bibliography: Stacey, *Advan. Carbohyd. Chem.* **2**, 161 (1946); Jones, Smith, *ibid.* **4**, 243 (1949). Structure: Pryde, Williams, *Nature* **128**, 187 (1931); Levene, Meyer, *J. Biol. Chem.* **92**, 257 (1931); Levene, Kreider, *ibid.* **120**, 597 (1937). Review of syntheses: Mehltretter, *Advan. Carbohyd. Chem.* **8**, 231 (1953). Prepn by irradiation of D-glucose in dil aq soln: Phillips *et al., J. Chem. Soc.* **1958**, 3522; by γ-irradiation of aq sucrose soln: Phillips, Moody, *ibid.* **1960**, 762. Electrophoretic sepn of D-glucuronic acid and its C-5 epimer, L-*iduronic acid:* I. Miyamoto, S. Nagase, *Anal. Biochem.* **115**, 308 (1981). Monographs: N. E. Artz, E. M. Osman, *Biochemistry of Glucuronic Acid* (Academic Press, New York, 1950); G. J. Dutton, Ed., *Glucuronic Acid, Free and Combined* (Academic Press, New York, 1966) 629 pp.

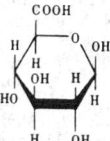

β-D-glucuronic acid

β-Form, needles from alcohol or ethyl acetate. mp 165°. Shows mutarotation: $[\alpha]_D^{24}$ +11.7° → +36.3° (2 hrs, c = 6). Soluble in water, alcohol. Reduces Fehling's soln.

4361. β-Glucuronidase. Glusulase. Glucuronide-splitting enzyme found in liver, spleen, and certain tissues of the endocrine and reproductive systems. Rats seem to have a higher concn than other mammals. Also found in fish liver, snails, mollusks, and some insects. Isoln from rat livers, kidneys, and spleens: Fishman, Talalay, *Science* **105**, 131 (1947). *Review and bibliography:* Fishman, *Advan. Enzymol.* **16**, 361-409 (1955).

White powder. Soluble in water.

USE: In the determination of urinary steroids and of steroid conjugates in blood.

4362. D-Glucuronolactone. *D-Glucuronic acid γ-lactone;* D-glucofuranurono-6,3-lactone; glucurolactone; glucurone; Dicurone; Glucoxy; Guronsan. $C_6H_8O_6$; mol wt 176.12. C 40.91%, H 4.58%, O 54.51%. Found in many plant gums in polymeric combination with other carbohydrates. Important structural constituent of practically all fibrous and connective tissues in the animal organism, *cf.* D-glucuronic acid. Prepd synthetically from many polysaccharides or suitable glucosides where the hydroxyl at carbon 6 may be oxidized while the other sensitive groups are protected. Prepn by oxidation of 1,2,3,4-tetraacetylglucose: Stacey, *J. Chem. Soc.* **1939**, 1529; by oxidation of β-methylglucoside with dinitrogen tetroxide: Hardegger, Spitz, *Helv. Chim. Acta* **33**, 337 (1950); with oxygen and platinum catalyst: Marsh, *Proc. Biochem. Soc.* [*Biochem. J.*], **50**, XI (1951); from 1,2-isopropylidene-D-glucose: Mehltretter *et al., J. Am. Chem. Soc.* **73**, 2424 (1951); by oxidizing suitable glucose derivatives: Benjamin, Kapranos, U.S. pat. **2,627,520** (1953 to Corn Prod.); by irradiation of sucrose: Phillips, Moody, *J. Chem. Soc.* **1960**, 762; from 1,2-cyclohexlidene-D-glucose: Mehltretter, U.S. pat. **3,012,041** (1961 to USDA). Structure: J. Stanek *et al., The Monosaccharides* (Academic Press, New York, 1963) p 259. For isoln procedures see the ref under glucuronic acid. For conversion of D-glucuronic acid to the lactone, see Mehltretter *et al., loc. cit.*

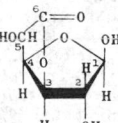

Crystals from ethanol, mp 176-178°. (Commercial grades, mp 172°.) d_4^{30} 1.76. $[\alpha]_D^{25}$ +19.8° (c = 5.19). Soluble in water (26.9 g/100 ml of soln); slightly sol in methanol (2.8

g/100 ml). Very slightly sol in abs ethanol (0.7 g/100 ml), in glacial acetic acid (0.3 g/100 ml). The free acid is more sol than the lactone. At room temp an aq soln of glucuronolactone reaches an equilibrium of about 20% lactone and 80% acid within 2 months. At 100° an equilibrium of 60% lactone and 40% free acid is reached within 2 hrs. Initial pH of 10% aq soln 3.5, after 1 week the pH is about 2.5.

Ferrous salt, *ferrous glucuronate, Guronsan Fe.*

THERAP CAT: Detoxicant.

4363. Glutamic Acid.
Glu (IUPAC abbrev); glutaminic acid; 2-aminopentanedioic acid; α-aminoglutaric acid; 1-aminopropane-1,3-dicarboxylic acid; Glutacid. C_5H_9-NO_4; mol wt 147.13. C 40.81%, H 6.16%, N 9.52%, O 43.50%. An amino acid classified as nonessential with respect to its growth effect in rats. Usually manufactured by the fermentation of a carbohydrate soln by a suitable microorganism such as *Micrococcus glutamicus;* or by either the acid hydrolysis of vegetable proteins, such as gluten, or the hydrolysis of casein, soybean cake, beet molasses. Also isolated from waste water from beet sugar manuf. Lab prepn from gluten flour: King, *Org. Syn.* **coll. vol. I,** 286 (2nd ed., 1941). Synthesis involving condensing the sodium deriv of ethyl benzamidomalonate with ethyl α-bromopropionate: Dunn *et al., J. Biol. Chem.* **94,** 599 (1931); Redemann, Dunn, *ibid.* **130,** 341 (1939). Other syntheses from methyl acrylate and phthalimidomalonate. From 2-cyclopentenylamine: Norman, U.S. pat. **2,900,391** (1959 to Int'l Minerals). Resolution of DL-form: Purvis, U.S. pat. **2,987,543** (1961 to Int'l Minerals). Identification as excitatory neurotransmitter: D. R. Curtis *et al., J. Physiol.* **150,** 656 (1960). Review of receptor binding study: P. A. Briley *et al., Mol. Cell. Biochem.* **39,** 347 (1981). Review of role as neurotransmitter: B. Engelsen, *Acta Neurol. Scand.* **74,** 337-355 (1986). *Reviews:* J. Greenstein, M. Winitz, *Chemistry of the Amino Acids* vol. 3 (Wiley, New York, 1961) pp 1929-1954; Huffman, Skelly, *Chem. Rev.* **63,** 625 (1963). *Books:* R. Powell, *Monosodium Glutamate and Glutamic Acid* (Noyes Development Corp., Park Ridge, N.J., 1968); *Glutamic Acid: Advances in Biochemistry & Physiology,* L. J. Filer, Jr. *et al.,* Eds. (Raven, New York, 1979).

L-glutamic acid

d-glutamic acid

D-glutamic acid

l-glutamic acid

L-Glutamic acid, *Glutaminol, Glutaton.* The naturally occurring dextrorotatory form with the L-configuration. Microbial conversion of α-ketoglutaric acid using *Aeromonas* spp: Good, U.S. pat. **2,933,434** (1960 to Int'l Minerals). Alternate method: Borel, U.S. pat. **3,022,224** (1962 to Hercules Powder). Using *B. megatherium-cereus:* Foster, U.S. pat. **3,032,474** (1962 to Merck & Co.). From fumaric acid using *B. pumilus:* Ogawa *et al.,* U.S. pat. **2,971,890** (1961 to Ajinomoto). From starch: Ogawa *et al.,* U.S. pat. **3,042,585** (1962 to Ajinomoto and Sanraku Dist.). From L-proline: S. Yoshifuji *et al., Tetrahedron Letters* **21,** 2963 (1980). Orthorhombic, bisphenoidal crystals from aq alc. d_4^{20} vac 1.538. Dec 247-249°. Sublimes at 200°. $[\alpha]_D^{22.4}$ +31.4° (6N HCl). uv spectrum: Marchlewski, Nowotnowna, *Bull. Soc. Chim.* [4] **39,** 162 (1926); *Chem. Zentr.* **1926, I,** 588. pK_1 2.19; pK_2 4.25; pK_3 9.67. Soly in water (g/l); 8.64 (25°); 21.86 (50°); 55.32 (75°); 140.00 (100°). Practically insol in methanol, ethanol, ether, acetone, cold glacial acetic acid.

Hydrochloride, $C_5H_{10}ClNO_4$, *Acidulin, Acidoride, Hypochylin, Antalka, Aciglumin, Pepsdol, Glutamidin, Acidogen, Aclor, Gastuloric, Glutan-HCl, Glutasin, Hydrionic, Muriamic, Recacid.* Orthorhombic bisphenoidal plates, dec 214°. $[\alpha]_D^{22}$ +24.4° (c = 6).

Sodium salt, *see* Monosodium Glutamate.

Magnesium salt hydrobromide, $C_{10}H_{17}BrMgN_2O_8$, *magnesium glutamate hydrobromide, magnesium bromoglutamate, Bromolate, Psicosoma, Psico-Soma, Psycho-Soma;* monohy-

drate, *Psychoverlan.* Prepn: Fischer *et al.,* **Ger.** pat. **2,228,**-**101** (1974 to Verla-Pharm), *C.A.* **80,** 96355v (1974).

DL-Glutamic acid, the synthetic racemic product. Orthorhombic crystals from water, 225-227° (dec). d_D^{20} 1.4601. Soly in water (g/l): 20.54 (25°); 49.34 (50°); 118.6 (75°); 284.9 (100°). Sparingly sol in alcohol, ether, petr ether.

D-Glutamic acid, the tasteless levorotatory D-isomer. Shiny leaflets from water. $[\alpha]_D^{20}$ −30.5° (c = 1.00 in 6N HCl).

USE: Only the L-form has the food flavor enhancement qualities. Chiefly used as the sodium salt to impart a meat flavor to foods. The hydrochloride has been used to improve the taste of beer.

THERAP CAT: Exptl antiepileptic; hydrochloride as gastric acidifier; magnesium salt hydrobromide as anxiolytic.

4364. L-Glutamic Acid 5-Ethyl Ester.
L-Glutamic acid γ-ethyl ester; γ-ethyl L-glutamate; Glutestere. $C_7H_{13}NO_4$; mol wt 175.18. C 47.99%, H 7.48%, N 8.00%, O 36.53%. $HOOCCH(NH_2)CH_2CH_2COOC_2H_5$. Prepn from glutamic acid, ethanol, and hydrogen chloride: Hegedus, *Helv. Chim. Acta* **31,** 737 (1948); Pravda, *Coll. Czech. Chem. Commun.* **24,** 2083 (1959); **Czech.** pat. **88,344** (1959), *C.A.* **54,** 8660h (1960).

Hydrochloride, $C_7H_{13}NO_4 \cdot HCl$, crystals from ethanol + ether, mp 134-136°.

4365. Glutamine.
Gln (IUPAC abbrev.); 2-aminoglutaramic acid; glutamic acid 5-amide; Cebrogen; Glumin; Levoglutamina; Stimulina. $C_5H_{10}N_2O_3$; mol wt 146.15. C 41.09%, H 6.90%, N 19.17%, O 32.84%. $HOOCCH(NH_2)$-$CH_2CH_2CONH_2$. Isoln from sugarbeet juice: Schulze, Bosshard, *Ber.* **16,** 312 (1879); Vickery *et al., J. Biol. Chem.* **109,** 39 (1935). Can be detected in most plants and animals (including bacteria). *Comprehensive review:* Archibald, *Chem. Rev.* **37,** 161-208 (1945). Synthesis: Bergmann *et al., Ber.* **66,** 1290 (1933); Nienburg, **Ger.** pat. **624,230** (1936 to I. G. Farben), *cf.* **Brit.** pat. **437,873;** Fruton, *J. Biol. Chem.* **165,** 333 (1946); King, Kidd, *J. Chem. Soc.* **1949,** 3315; Kline, Cox, *J. Org. Chem.* **26,** 1854 (1961); Norman, Joyce, U.S. pat. **2,790,000** (1957 to Int'l Minerals). *See also* Lepp, Dunn, *Biochem. Prepn.* **5,** 79 (1957).

Fine opaque needles from water or dil ethanol, dec 185-186°. $[\alpha]_D^{23}$ +6.1° (c = 3.6). pK_1 2.17; pK_2 9.13. One gram dissolves in 20.8 ml water at 30°, in 38.5 ml at 18°, in 56.7 ml at 0°. Practically insol in methanol (3.5 mg/100 ml at 25°), ethanol (0.46 mg/100 ml at 23°), ether, benzene, acetone, ethyl acetate, chloroform.

DL-Form, prisms from dil acetone, mp 185-186° (King); mp 173-174.5° (Kline). One gram dissolves in 38.5 ml water at 18°.

4366. Glutaraldehyde.
Pentanedial; glutaral; glutaric dialdehyde; 1,3-diformylpropane; Cidex; Glutarol; Verucasep. $C_5H_8O_2$; mol wt 100.11. C 59.98%, H 8.05%, O 31.97%. $OHCCH_2CH_2CH_2CHO$. Prepd by treating 2-ethoxy-3,4-dihydro-2H-pyran with aq HCl: A. C. Cope *et al., Org. Syn.* **coll. vol. IV,** 816 (1963). *See also* Beilstein **1,** 776 (1918) and supplements.

Oil, bp_{760} 187-189°; bp_{50} 106-108°; bp_{10} 71-72°. n_D^{25} 1.4338. Sol in water, volatile in steam. Polymerizes in water to a glassy form which regenerates the dialdehyde on vacuum distillation. LD_{50} of 25% soln orally in rats: 2.38 ml/kg; by skin penetration in rabbits: 2.56 ml/kg, H. F. Smyth *et al., Am. Ind. Hyg. Assoc. J.* **23,** 95 (1962).

Dioxime, $C_5H_{10}N_2O_2$, crystals from water or pyridine, mp 178°. Sublimes. Treatment with hot mineral acids gives pyridine.

USE: In sterilization of endoscopic instruments, thermometers, rubber or plastic equipment which cannot be heat sterilized; as a tanning agent for leather.

Caution: Causes severe eye irritation in rabbits.

THERAP CAT: Disinfectant.

4367. Glutaric Acid.
Pentanedioic acid; 1,3-propanedicarboxylic acid. $C_5H_8O_4$; mol wt 132.11. C 45.45%, H 6.10%, O 48.44%. $COOH(CH_2)_3COOH$. Occurs in green sugar beets; is found in water extracts of crude wool. Manuf from cyclopentanone by oxidative ring fission with hot 50% nitric acid in the presence of vanadium pentoxide. Lab prepn by acid hydrolysis of trimethylene cyanide: Marvel,

Tuley, *Org. Syn.* **5**, 69 (1925), or of methylenedimalonic ester: Otterbacher, *ibid.* **10**, 58 (1930). Several new methods: Paris *et al., Org. Syn.* **coll. vol. IV**, 496 (1963); English, Dayan, *ibid.* 499.

Large monoclinic prisms, mp 97.5-98°. d_4^{15} 1.429. bp_{760} 302-304° (very slight decompn); bp_{20} 200°; bp_{10} 195-198°, K_1 at 25°: 4.60 × 10^{-5}; K_2: 6.0 × 10^{-6}. n_D^{106} 1.41878. Absorption spectrum: *Compt. Rend.* **189**, 915 (1929). Soly in water (g/l): at 0°: 429; at 20°: 639; at 50°: 957; at 65°: 1118. Freely sol in abs alcohol, ether; sol in benzene, chloroform; slightly in petr ether.

Dimethyl ester, $C_7H_{12}O_4$, liquid, faint agreeable odor. d_4^{15} 1.0934. bp_{752} 213.5-214°; bp_{13} 93.5-94.5°. n_D^{20} 1.4246. Very sol in alcohol and ether.

4368. Glutaronitrile. *Pentanedinitrile;* glutaric acid dinitrile; trimethylene dicyanide; trimethylene cyanide. $C_5H_6N_2$; mol wt 94.12. C 63.80%, H 6.43%, N 29.77%. $CN(CH_2)_3$-CN. Prepd by the action of potassium cyanide on trimethylene bromide: Marvel, McColm, *Org. Syn.* **5**, 103 (1925). Viscous liquid. Bitter-sweet taste. d^{23} 0.9888. mp −29°. bp_{760} 286°; bp_{100} 206°; bp_{60} 190°; bp_{40} 176°; bp_{20} 157°; bp_{10} 140°; bp_5 124°; $bp_{1.0}$ 91.3°. n_D^{23} 1.4365. Soluble in water, alcohol, chloroform. Insol in ether, carbon disulfide.

4369. Glutathione. *N-(N-L-γ-Glutamyl-L-cysteinyl)glycine;* L-glutathione; glutathione-SH; Agifutol S; Copren; Deltathione; GSH; Glutathin; Glutathiol; Glutathion; Glutinal; Isethion; Neuthion; Tathiclon; Tathion; Triptide. C_{10}-$H_{17}N_3O_6S$; mol wt 307.33. C 39.08%, H 5.58%, N 13.67%, O 31.24%, S 10.43%. The major low mol wt thiol compound of the living plant or animal cell. Isoln from yeast: Hopkins, *J. Biol. Chem.* **84**, 269 (1929). Synthesis: du Vigneaud, Miller, *Biochem. Prepn.* **2**, 87 (1952); Goldschmidt *et al., Ber.* **97**, 2434 (1964); Y. Ozawa *et al., Bull. Chem. Soc. Japan* **53**, 2592 (1980). Review of early syntheses: Jeschkeit *et al., Pharmazie* **18**, 658 (1963). Review of metabolism: A. Meister, M. E. Anderson, *Ann. Rev. Biochem.* **52**, 711-760 (1983); of metabolic role in antineoplastic chemotherapy: B.A. Arrick, C. F. Nathan, *Cancer Res.* **44**, 4224-4232 (1984). *Monographs:* S. Colowick *et al., Glutathione* (Academic Press, New York, 1954); *Glutathione,* E. M. Crook, Ed., Biochem. Soc. Symposium No. 17, London, 1958 (Cambridge University Press, 1959); *Glutathione,* L. Flohe *et al.,* Eds. (Academic Press, New York, 1974); *Glutathione: Metabolism & Function,* I. M. Arias, W. B. Jackoby, Eds. (Raven, New York, 1976) .

$$\underset{\underset{COOH}{|}}{H_2NCHCH_2CH_2}CONHCH\underset{\underset{CH_2SH}{|}}{CONHCH_2}COOH$$

Crystals from 50% ethanol, mp 195°. $[\alpha]_D^{25}$ −18.9° (c = 4.653). $[\alpha]_D^{27}$ −21° (c = 2.74). pK_1' 2.12; pK_2' 3.53; pK_3' 8.66; pK_4' 9.12. Freely sol in water, dil alcohol, liq ammonia, dimethylformamide.

Disulfide, $C_{20}H_{32}N_6O_{12}S_2$, *N,N'-[dithiobis[1-[(carboxymethyl)carbamoyl]ethylene]]diglutamine, GSSG, oxidized glutathione.* Crystals, mp 123°. $[\alpha]_D^{20}$ −108° (c = 2 in water).

4370. Gluten. Protein substance of wheat which is intermixed with the starchy endosperm of the grain. Causes the carbon dioxide produced during dough fermentation to be retained by the dough in a manner which provides the porous and spongy structure of bread. Prepn from wheat: Rist, *Sugar J.* **11**, no. 9, 26 (1949), *C.A.* **43**, 9505c (1949); Christensen, **U.S.** pat. **2,583,684** (1952 to Gateway Chemurgic). Amino acid composition: Pence *et al., Cereal Chem.* **27**, 335 (1950). *Reviews:* M. J. Blish "Wheat Gluten" in M. L. Anson, J. T. Edsall, *Advan. Protein Chem.* **vol. II** (Academic Press, New York, 1945) pp 337-359; Meredith, *Cereal Sci. Today* **9**, 33, 54 (1964).

Yellowish-gray powder. Practically insol in water; partly sol in alcohol, dil acids; sol in alkalies.

USE: As adhesive and as substitute for flour.

4371. Glutethimide. *3-Ethyl-3-phenyl-2,6-piperidinedione; 2-ethyl-2-phenylglutarimide;* α-ethyl-α-phenylglutarimide; 3-ethyl-3-phenyl-2,6-dioxopiperidine; 3-ethyl-3-phenyl-2,6-diketopiperidine; Elrodorm; Doriden-Sed; Dori-

den. $C_{13}H_{15}NO_2$; mol wt 217.26. C 71.86%, H 6.96%, N 6.45%, O 14.73%. Synthesis: Tagmann *et al., Helv. Chim. Acta* **35**, 1541 (1952); Salmon-Legagneur, Neveu, *Compt. Rend.* **234**, 1060 (1952); *Bull. Soc. Chim. France* **1953**, 70; Hoffmann, Tagmann, **U.S.** pat. **2,673,205** (1954 to Ciba). Resolution: Kukalja *et al., Croat. Chem. Acta* **33**, 41 (1961), *C.A.* **55**, 27193g (1961). Abs config of antipodes: Finch *et al., Experientia* **31**, 1002 (1975). Comprehensive description: H. Y. Aboul-Enein in *Analytical Profiles of Drug Substances* **vol. 5**, K. Florey, Ed. (Academic Press, New York 1976) pp 139-187.

dl-Form, crystals from ether or from ethyl acetate + petr ether, mp 84°. uv max (methanol): 251, 257, 263 nm. Freely sol in ethyl acetate, acetone, ether, chloroform; sol in ethanol, methanol. Practically insol in water.

d-Form, crystals, mp 102.5-103°. $[\alpha]_D^{20}$ +176° (methanol).

l-Form, crystals, mp 102-103°. $[\alpha]_D^{20}$ −181° (methanol).

Caution: May be habit forming. This is a controlled substance (depressant) listed in the U.S. Code of Federal Regulations, Title 21 Part 1308.13 (1985).

THERAP CAT: Sedative. Hypnotic.

4372. Glyburide. *5-Chloro-N-[2-[4-[[[(cyclohexylamino)carbonyl]amino]sulfonyl]phenyl]ethyl]-2-methoxybenzamide; 1-[[p-[2-(5-chloro-o-anisamido)ethyl]phenyl]sulfonyl]-3-cyclohexylurea; N-[4-(β-(2-methoxy-5-chlorobenzamido)ethyl)benzosulfonyl]-N'-cyclohexylurea; N¹-[4-[β-(2-methoxy-5-chlorobenzoylamino)ethyl]benzenesulfonyl]-N²-cyclohexylurea;* glybenzcyclamide; glibenclamide; HB 419; U-26452; Adiab; Azuglucon; Bastiverit; Dia-basan; Diabeta; Daonil; Duraglucon; Euclamin; Euglucon; Gilemal; Gliben-Puren N; Glidiabet; Glimidstada; Glucoremed; Gluco-Tablinen; Glycolande; Hemi-Daonil; Lederglib; Libanil; Lisaglucon; Malix; Maninil; Micronase; Praeciglucon; Semi-Daonil; Semi-Euglucon; Semi-Gliben-Puren N. $C_{23}H_{28}ClN_3O_5S$; mol wt 494.00. C 55.92%, H 5.71%, Cl 7.17%, N 8.51%, O 16.19%, S 6.49%. Second generation sulfonylurea with hypoglycemic activity. Prepn: Aumuller *et al., Arzneimittel-Forsch.* **16**, 1640 (1966); **Neth.** pat. **Appl. 6,603,398** (1966 to Boehringer, Mann.), *C.A.* **66**, 65289h (1967); **Neth.** pat. **Appl. 6,610,580** corresp to H. Weber *et al.,* **U.S.** pat. **3,454,635** (1967, 1969 both to Hoechst). Pharmacology: Loubatières, Mariani, *C. R. Acad. Sci. Ser. D* **265**, 643 (1967). Toxicity: Mizukami *et al., Arzneimittel-Forsch.* **19**, 1413 (1969). Series of articles on synthesis, pharmacology, toxicology and clinical studies: *ibid.* 1323-1494. Effect on release of insulin, glucagon and somatostatin: S. Efendic *et al., Proc. Natl. Acad. Sci. USA* **76**, 5901 (1979). Symposium on pharmacology, mechanism of action and clinical trials: *Ann. Clin. Res.* **15**, Suppl. 37, 1-35 (1983). Comprehensive description: P. G. Takla in *Analytical Profiles of Drug Substances* **vol. 10**, K. Florey, Ed. (Academic, New York, 1981) pp 337-355. Review of pharmacology and clinical efficacy: J. M. Feldman, *Pharmacother.* **5**, 43-62 (1985).

Crystals from methanol, mp 169-170° (Weber); also reported as mp 172-174° (Aumüller). pKa 5.3. Sparingly sol in water, sol in the usual organic solvents. LD_{50} in rats and mice (g/kg): > 20 orally; > 12.5 i.p.; > 20 s.c. (Mizukami).

THERAP CAT: Antidiabetic.

4373. Glybuthiazol(e). *4-Amino-N-[5-(1,1-dimethylethyl)-1,3,4-thiadiazol-2-yl]benzenesulfonamide; N¹-(5-tert-butyl-1,3,4-thiadiazol-2-yl)sulfanilamide;* sulfatertiobutylthiadiazole; 2-(p-aminobenzenesulfamido)-5-tertiobutyl-1,3,4-thiadiazole; glybuthizol; RP 2259; Glipasol. $C_{12}H_{16}$-$N_4O_2S_2$; mol wt 312.40. C 46.13%, H 5.16%, N 17.93%, O 10.24%, S 20.53%. Prepn: **Brit. pat. 828,963** (1960 to Rhône-Poulenc).

H_2N—⟨ ⟩—SO_2NH—⟨S⟩—C(CH_3)_3
                        (N—N)

Needles from ethanol, mp 221-223°. Sol in ethanol (1.0 g/65 ml); in acetone (1.0 g/15 ml); in DMF (1.0 g/3 ml). Insol in water, ether, benzene. Forms a water-sol Na salt.
THERAP CAT: Antidiabetic.

4374. Glybuzole. *N-[5-(1,1-Dimethylethyl)-1,3,4-thia-diazol-2-yl]benzenesulfonamide; N-(5-tert-butyl-1,3,4-thia-diazol-2-yl)benzenesulfonamide;* 2-benzenesulfonamido-5-*tert*-butyl-1,3,4-thiadiazole; desaglybuzole; TH-1395; RP 7891; AN 1324; Gludiase. $C_{12}H_{15}N_3O_2S_2$; mol wt 297.39. C 48.46%, H 5.08%, N 14.13%, O 10.76%, S 21.56%. Prepn: Macrae, Drain, **Brit. pat. 822,947** (1959 to Smith & Nephew), *C.A.* **54**, 4622h (1960); **Fr. pat. M3389** (1965 to Rhône-Poulenc), *C.A.* **64**, 3553a (1966). Pharmacodynamics: Bargeton *et al., Arch. Int. Pharmacodyn.* **153**, 379 (1965).

(CH_3)_3C—⟨S⟩—NHSO_2C_6H_5
          (N—N)

Needles, mp 163°. LD_{50} orally in mice: 1315 mg/kg.
THERAP CAT: Antidiabetic.

4375. Glycarsamide. *[4-[(Hydroxyacetyl)amino]phenyl]-arsonic acid; N-glycoloylarsanilic acid;* N-glycolylarsanilic acid; *p*-glycolylaminobenzenearsonic acid; *p*-glycolylamino-phenylarsinic acid; Astryl. $C_8H_{10}AsNO_5$; mol wt 275.08. C 34.93%, H 3.66%, As 27.23%, N 5.09%, O 29.08%. Prepd by treating arsanilic acid with chloroacetyl chloride and NaOH: Streitwolf *et al.*, **Ger. pat. 513,210; Brit. pat. 344,532** (1929 to I. G. Farben).

O=As(OH)_2—⟨ ⟩—NHCOOCH_2OH

Crystals from dil acid or water, mp 202-203°. Freely sol in hot water, methanol. Sparingly sol in cold water, acetic acid. Practically insol in acetone, benzene.
Sodium salt, $C_8H_9AsNNaO_5$, *Quarcyl, SNGA, Allegan.* Crystals. Noticeable acid odor. Freely sol in water. Toxicity and physiological disposition: McChesney *et al., Toxicol. Appl. Pharmacol.* **4**, 14 (1962).
Nitroacridine complex, *see* Nitroakridin.
THERAP CAT (VET): Anthelmintic (Strongyloidosis).

4376. Glyceraldehyde. *2,3-Dihydroxypropanal;* glyceric aldehyde; α,β-dihydroxypropionaldehyde. $C_3H_6O_3$; mol wt 90.08. C 40.00%, H 6.71% O 53.28%. DL-glyceraldehyde together with its isomer dihydroxyacetone is obtained from glycerol by mild oxidation with hydrogen peroxide and ferrous salts as catalysts: Witzemann, *J. Am. Chem. Soc.* **36**, 2227 (1914). *See also Org. Syn.* **coll. vol. II**, 305 (1943). The equilibrium mixture of glyceraldehyde and dihydroxyacetone is called *glycerose.* The two isomers are convertible into another through a common enediol resulting from the migration of hydrogen atoms (Lobry de Bruyn-van Eckenstein rearrangement). The equilibrium mixture plays an important role in the fermentation of sugars and in the biogenesis of constituents of the animal organism, *cf.* L.F. Fieser, M. Fieser, *Advanced Organic Chemistry* (Reinhold, New York, 1961) pp 78, 284, 405. Glyceraldehyde is the simplest aldose and its D- and L-forms are taken as the configurational reference standard for carbohydrates. The two forms have been obtained through the action of nitrous acid on the corresponding form of 3-amino-2-hydroxypropanal: Wohl, Momber, *Ber.* **47**, 3346 (1914); Pictet, Barbier, *Helv. Chim.*

Acta **4**, 924 (1921); *cf.* Baer, Fischer, *Science* **88**, 108 (1938). Prepn of L-glyceraldehyde from L-sorbose and of D-glyceraldehyde from D-fructose: Perlin, *Methods in Carbohydrate Chemistry* **1**, 61 (1962).

CHO   CHO
HO—C—H   H—C—OH
CH_2OH   CH_2OH

L-form   D-form
==   ==

DL-Form, tasteless crystals from alcohol + ether, d_{18}^{18} 1.455. mp 145°. Distills at 140-150° (bath temp) and 0.8 mm pressure. 100 ml water dissolve 3 g at 18°. Insol in benzene, petr ether, pentane. Osazone, mp 132°.
L-Form, $[\alpha]_D^{25}$ —8.7° (c = 2 in H_2O). Its dimethylacetal, $HOCH_2CH(OH)CH(OCH_3)_2$, bp_{17-20} 126-129°. $[\alpha]_D^{26}$ —20.9° (p = 9.22).
D-Form, $[\alpha]_D^{25}$ +8.7° (c = 2 in H_2O). Its dimethylacetal, bp_{17} 127-129°, bp_{10} 123-126°. $[\alpha]_D^{15}$ +21.2° (c = 18).

4377. Glyceraldehyde 3-Phosphate. *2-Hydroxy-3-(phosphonooxy)propanal;* 3-phosphoglyceraldehyde. C_3H_7-O_6P; mol wt 170.07. C 21.19%, H 4.15%, O 56.45%, P 18.22%. An intermediate product of carbohydrate metabolism. Prepn of DL-form by enzymatic route: Meyerhof, Junowicz-Kocholaty, *J. Biol. Chem.* **149**, 71 (1943); by reductive cleavage of glyceraldehyde 1-benzyl ether 3-phosphoric acid: Fischer, Baer, *Ber.* **65**, 337, 1040 (1932); by hydrolysis of dimeric gylceraldehyde 1,3-diphosphoric acid: Baer, Fischer, *J. Biol. Chem.* **150**, 213 (1943); by hydrolysis of dimeric glyceraldehyde 1-bromide 3-phosphoric acid: *ibid.* 223. *See also* Baer in *Biochem. Prepn.* **1**, 50 (1949). Prepn of D-form by enzymatic route: Meyerhof, Junowicz-Kocholaty, *loc. cit.;* by synthetic route: Ballou, Fischer, *J. Am. Chem. Soc.* **77**, 3329 (1955).

CHO
CHOH   O   OH
CH_2O—P
           OH

DL-Form calcium salt dihydrate, $C_3H_5CaO_6P.2H_2O$, crystals, sol in water. Aq solns are not stable, particularly when alkaline. The dioxane addition compd of DL-glyceraldehyde 1-bromide 3-phosphoric acid described by Baer is relatively stable and may be stored in the refrigerator. An aq soln of the Na or K salt of glyceraldehyde-3-phosphoric acid (contg dioxane and bromide ion) is readily obtained by dissolving the dioxane compd in cold water and carefully neutralizing to pH 7.
D-Form calcium salt dihydrate, amorphous powder. $[\alpha]_D^{25}$ +14.5° (c = 1.2 in 0.1N HCl calcd as the free acid). Prepn of calcium-free aq soln: Ballou, Fischer, *loc. cit.*

4378. Glyceric Acid. *2,3-Dihydroxypropanoic acid;* α,β-dihydroxypropionic acid. $C_3H_6O_4$; mol wt 106.08. C 33.96%, H 5.70%, O 60.33%. $CH_2OHCH(OH)COOH$. Prepn from α,β-dibromopropionic acid by treatment with silver oxide: Karrer, Klarer, *Helv. Chim. Acta* **7**, 931 (1924); from isoserine and nitrous acid: Fischer, Jacobs, *Ber.* **40**, 1069 (1907); from glycerol and nitrous acid: Mulder, *Ber.* **9**, 1902 (1876); Beilstein, *Ann.* **120**, 229 (1861). Dextrorotatory glyceric acid is obtained by the action of *Penicillia* or *Aspergilli* on the DL-form: McKenzie, Harden, *J. Chem. Soc.* **83**, 431 (1903). Levorotatory glyceric acid has been obtained by the oxidation of D(+)-glyceric aldehyde: Wohl, Schellenberg, *Ber.* **55**, 1408 (1922).
DL-Form, syrup, dec on distn. K at 25°: 2.8×10^{-4}. Miscible with water, alcohol, acetone. Nearly insol in ether.
L(+)-Form, syrup. Its esters and salts are levorotatory and its salts are much more sol in water than those of the DL-form.
L-Form calcium salt, $Ca(C_3H_5O_4)_2.2H_2O$, monoclinic sphenoidal crystals, mp 137°. One gram dissolves in 10 ml water. $[\alpha]_D^{20}$ —14.6° (c = 5).
D(—)-Form, syrup. Its salts are dextrorotatory.

D-Form calcium salt, $Ca(C_3H_5O_4)_2.2H_2O$, prisms, mp 138°. $[\alpha]_D^{20}$ +14.5° (c = 5).

4379. Glycerol. *1,2,3-Propanetriol; glycerin;* glycerine; trihydroxypropane; incorporation factor; IFP; Ophthalgan. $C_3H_8O_3$; mol wt 92.09. C 39.12%, H 8.75%, O 52.12%. $CH_2OHCHOHCH_2OH$. Obtained from oils and fats as by-product in the manuf of soaps and fatty acids. During World War I supplementary quantities were produced by the "Protol" fermentation process from sugar, a process based upon the fixation of acetaldehyde by sodium sulfite. Just prior to World War II the synthesis of glycerol from propylene was announced. Production from sugars by fermentation: Onishi, U.S. pat. **3,012,945** (1961 to Noda). Identity with incorporation factor: Kuehl *et al., J. Am. Chem. Soc.* **82**, 2079 (1960). In nucleic acid the incorporation factor may exist as a bound form of glycerol. Acute toxicity: W. Bartsch *et al., Arzneimittel-Forsch.* **26**, 1581 (1976). Reviews and bibliographies: J. W. Lawrie, *Glycerol and the Glycols* (New York, 1928); G. Leffingwell, M. Lesser, *Glycerin* (Brooklyn, 1945); C. S. Miner, N. N. Dalton, *Glycerol* (New York, 1953); C. Lüttgen, *Glyzerin und glyzerinähnliche Stoffe* (Heidelberg, 2nd ed., 1955); J. C. Kern in Kirk-Othmer *Encyclopedia of Chemical Technology* vol. **11** (Wiley-Interscience, New York, 3rd ed., 1980) pp 921-932. Syrupy liquid. Sweet warm taste. About 0.6 times as sweet as cane sugar. Absorbs moisture from air; also absorbs H_2S, HCN, SO_2. *Caution: Contact with strong oxidizing agents such as chromium trioxide, potassium chlorate, or potassium permanganate may produce an explosion.* Neutral to litmus. Solidifies after prolonged cooling at 0° forming shiny orthorhombic crystals, mp 17.8°. bp_{760} 290.0° (dec); bp_{400} 263.0°; bp_{200} 240.0°; bp_{100} 220.1°; bp_{60} 208.0°; bp_{20} 182.2°; bp_{10} 167.2°; bp_5 153.8°; $bp_{1.0}$ 125.5°. n_D^{15} 1.4758; n_D^{20} 1.4746; n_D^{25} 1.4730. d_{15}^{15} 1.26557; d_{20}^{20} 1.26362; d_{25}^{25} 1.26201. Flash pt, open cup: 350°F (176°C). Specific gravities of 95% aq soln w/w (U.S.P. grade): d_{15}^{15} 1.25270; d_{20}^{20} 1.25075; d_{25}^{25} 1.24910; 90% aq soln w/w: d_{15}^{15} 1.23950; d_{20}^{20} 1.23755; d_{25}^{25} 1.23585; 80% d_{15}^{15} 1.213; 70% d_{15}^{15} 1.185; 60% d_{15}^{15} 1.157; 50% d_{15}^{15} 1.129; 20% d_{15}^{15} 1.049; 5% d_{15}^{15} 1.0122. Viscosity (cps. at 20°): 5% soln 1.143; 10% 1.311; 25% 2.095; 50% 6.050; 60% 10.96; 70% 22.94; 83% 111. Freezing points of aq solns w/w: 10% —1.6°; 30% —9.5°; 50% —23.0°; 66.7% —46.5°; 80% —20.3°; 90% —1.6°. Miscible with water, alcohol. One part dissolves in 11 parts ethyl acetate, in about 500 parts ethyl ether. Insol in benzene, chloroform, carbon tetrachloride, carbon disulfide, petr ether, oils. LD_{50} in rats (ml/kg): >20 orally; 4.4 i.v. (Bartsch).

USE: As solvent, humectant, plasticizer, emollient, sweetener, in the manuf of nitroglycerol (dynamite), cosmetics, liq soaps, liqueurs, confectioneries, blacking, printing and copying inks, lubricants, elastic glues, lead oxide cements; to keep fabrics pliable; to preserve printing on cotton; for printing rollers, hectographs; to keep frost from windshields; as antifreeze in automobiles, gas meters and hydraulic jacks, in shock absorber fluids. In fermentation nutrients in the production of antibiotics. Pharmaceutic aid (humectant; solvent). Leffingwell and Lesser (*op. cit.*) give 1583 different uses.

THERAP CAT: Diagnostic aid (ophthalmic).

THERAP CAT (VET): Emollient, emulcent, vehicle. As a source of glucose in bovine ketosis.

4380. Glycerol Formal. Methylidinoglycerol; Glicerinformal; Sericosol N. $C_4H_8O_3$; mol wt 104.10. C 46.15%, H 7.74%, O 46.11%. Mixture of isomeric α,α'- and α,β-forms prepd from glycerin and formaldehyde: M. Schultz, B. Tollens, *Ann.* **289**, 20 (1895). Prepn and separation of isomers: H. Hibbert, N. M. Carter, *J. Am. Chem. Soc.* **50**, 3120 (1928); J. D. van Roon, *Rec. Trav. Chim.* **48**, 173 (1929). Prepn and pharmacology: P. Gimeno, **Span.** pat. **475,962** (1979 to Calipe), *C.A.* **93**, 26445u (1980). Toxicological

α,α'-form   α,β-form

evaluation for use as pharmaceutical solvent: D. M. Sanderson, *J. Pharm. Pharmacol.* **11**, 150, 446 (1959). Teratogenicity studies: V. Aliverti *et al., Arch. Sci. Biol.* **61**, 89 (1977); E. Giavini, M. Prati, *Acta Anat.* **106**, 203 (1980).

Liquid, bp_{760} 191-195°, bp_{20} 95-97°. d_4^{20} 1.215; n_D^{20} 1.451. Sol in water, alc, chloroform. pH of 10% soln: 4-6.5. LD_{50} in rats (ml/kg): 8.6 orally, 3.5 i.v. (Gimeno).

α,α'-Form, *1,3-dioxan-5-ol, α,α'-methylene glycerin, α,α'-formaldehyde glycerol, 5-m-dioxanol.* bp 193.8°. d_4^{25} 1.2200. n_D^{25} 1.4527.

α,β-Form, *1,3-dioxolane-4-methanol, α,β-methylene glycerin, α,β-formaldehyde glycerol, 4-(hydroxymethyl)-1,3-dioxolane.* bp 192.5°. d_4^{25} 1.2008. n_D^{25} 1.4469.

USE: Pharmaceutic aid (solvent).

4381. Glycerophosphoric Acid. Phosphoric acid glycerol esters. $C_3H_9O_6P$; mol wt 172.08. C 20.94%, H 5.27%, O 55.79%, P 18.00%. Three isomers exist: β-glycerophosphoric acid, $(HOCH_2)_2CHOPO(OH)_2$ and the D(+)- and L(—)-forms of α-glycerophosphoric acid, $HOCH_2CH(OH)CH_2O-PO(OH)_2$. The L-$\alpha$-acid is the naturally occurring form; the β-acid, present in hydrolyzates of lecithins from natural sources, arises from migration of the phosphoryl group from the α-carbon atom. See review: Dawson, *Ann. Rept. Progr. Chem. (Chem. Soc. London)* **55**, 365 (1958). Prepn by phosphorylation of glycerol results in a mixture of the α- and β-acids: Cherbuliez, Weniger, *Helv. Chim. Acta* **29**, 2006 (1946). Prepn and configuration of L-α-acid: Baer, Fischer, *J. Biol. Chem.* **128**, 491 (1939). Prepn of D-α-acid: *eidem, ibid.* **135**, 321 (1940). Separation of α-acid from β- and polyglycerophosphoric acids: Carrara, **Ital.** pat. **460,219** (1950), *C.A.* **46**, 5077a (1952).

Absolute acid (commercial mixture of α- and β-acids); Clear syrupy liq, mp —25°. d_4^{14} 1.59. Tends to dec during concn; hence, usually marketed as a 25-50% soln. Soluble in water, alcohol.

α-Acid L-form, syrup. Readily sol in water, methanol, ethanol; practically insol in ether. $[\alpha]_D$ —1.45° (barium salt, c = 10.3 in 2N HCl).

Note: Phosphatidic acids are fatty acid diesters of glycerophosphoric acid.

USE: Absolute acid used to manuf certain glycerophosphates or to impart taste to solns of glycerophosphates which are generally used medicinally. *See also:* Calcium Glycerophosphate.

4382. Glyceryl p-Aminobenzoate. *1,2,3-Propanetriol 1-(4-aminobenzoate); p-aminobenzoic acid monoglyceryl ester;* monoglycerol p-aminobenzoate; Escalol 106. $C_{10}H_{13}NO_4$; mol wt 211.21. C 56.86%, H 6.20%, N 6.63%, O 30.30%. Prepd by controlled esterification of p-aminobenzoic acid with glycerol.

Semisolid, waxy mass or syrup. Faint aromatic odor. Liquefies and congeals very slowly. Soluble in methanol, ethanol, isopropanol, glycerol, propylene glycol. Insol in water, oils, fats.

USE: In cosmetic sunscreen prepns (up to 1%).

4383. Glyceryl Iodide. *3-Iodo-1,2-propanediol;* γ-iodopropyleneglycol; 3-iodo-1,2-dihydroxypropane. $C_3H_7IO_2$; mol wt 202.01. C 17.84%, H 3.49%, I 62.83%, O 15.84%. $ICH_2CH(OH)CH_2OH$. Prepd from 2,3-epoxy-1-propanol: Kratzl *et al., Monatsh.* **93**, 149 (1962).

Crystals, mp 48-49°. Freely sol in water, alcohol.

4384. Glyceryl Monostearate. *Octadecanoic acid monoester with 1,2,3-propanetriol;* Monostearin. The commercial product is a mixture of variable proportions of glyceryl monostearate and glyceryl monopalmitate.

White, wax-like solid or wax-like beads, or flakes, mp 56-58°. Saponification value 164-170. Iodine value not more than 6. Soluble in hot organic solvents such as alcohol, benzene, ether, acetone, mineral or fixed oils. Insol in

water, but may be dispersed in hot water with the aid of a small amount of soap or other suitable surface active agent.
USE: In pharmaceutical dispensing, see Green, *J. Am. Pharm. Assoc., Pract. Pharm. Ed.* **7**, 299 (1946).

4385. Glycidol. *Oxiranemethanol; 2,3-epoxy-1-propanol;* 3-hydroxypropylene oxide. $C_3H_6O_2$; mol wt 74.08. C 48.64%, H 8.16%, O 43.20%. May be prepd by the action of perbenzoic acid on allyl alc: Prileshajew, *Ber.* **42**, 4813 (1909); or from glycerol-1-monochlorohydrin by the action of KOH in alc, or by the action of sodium in ether: Rider, Hill, *J. Am. Chem. Soc.* **52**, 1521 (1930). L-Glycidol has been prepd from L-1-(*p*-toluenesulfonyl)glycerol: Sowden, Fischer, *ibid.* **64**, 1291 (1942).

DL-Form, slightly viscous liq. d_4^{25} 1.1143. bp_{760} 167° (decompn); $bp_{2.5}$ 66°; $bp_{0.9}$ 25°. Miscible with water.
L-Form, $[\alpha]_D^{20}$ +15° (neat). LD_{50} orally in rats: 0.85 g/kg. *Caution:* Moderately irritating to skin, mucous membranes. Can cause stimulation of CNS followed by depression. Absorbed through skin.

4386. Glycine. Gly (IUPAC abbrev.); aminoacetic acid; aminoethanoic acid; glycocoll; Gyn-Hydralin; Glycosthène. $C_2H_5NO_2$; mol wt 75.07. C 32.00%, H 6.71%, N 18.66%, O 42.63%. NH_2CH_2COOH. An amino acid classified as nonessential with respect to its growth effect in rats. Gelatin and silk fibroin are the best sources of this amino acid: D. M. Greenberg, *Amino Acids and Proteins* (Charles C. Thomas, Springfield, Illinois, 1951) p 232. Isoln from hydrolysis products of gelatin: Kingston, Schryver, *Biochem. J.* **18**, 1071 (1924); by hydrolysis of cattle ligaments: U.S. pat. 2,098,923 (1938 to Armour). Isoln from silk fibroin: Moore, Stein, *J. Biol. Chem.* **150**, 113 (1943). Prepn from hydrobromic acid + methyleneaminoacetonitrile: Clarke, Taylor, *Org. Syn.* **4**, 31 (1925); from chloroacetic acid + ammonia: Orten, Hill, *Org. Syn. coll. vol.* I, 2nd ed., 300 (1941); from ammonium bicarbonate + sodium cyanide: White, Wysong, U.S. pat. 2,663,713 (1953 to Dow); by catalytic cleavage of serine: Metzler *et al., J. Am. Chem. Soc.* **76**, 639 (1954). Exists in three polymorphic forms, α-, β-, and γ-: Iitaka, *Nature* **183**, 390 (1959).
Sweet, monoclinic prisms from alc, starts to dec at 233°, completely sintered at 290°. d 1.1607. pK_1' 2.34; pK_2' 9.60. pH of 0.2 molar soln in H_2O = 4.0. Adsorption on various chromatographic agents: Grettie, Williams, *J. Am. Chem. Soc.* **50**, 671 (1928). Soly in 100 ml water at 25°: 25.0 g; at 50°: 39.1 g; at 75°: 54.4 g; at 100°: 67.2 g. 100 g of abs alc dissolve about 0.06 g. Sol in 164 parts pyridine. Almost insol in ether.
Hydrochloride, $C_2H_6ClNO_2$, hygroscopic prisms from HCl, mp 182°.
Hemihydrochloride, $C_4H_{11}ClN_2O_4$, mp 189°.
Also forms a sodium salt, *sodium glycinate.*
THERAP CAT: Nutrient. Antipruritic.

4387. Glycine Sulfate. Triglycine sulfate. $C_6H_{17}N_3O_{10}S$; mol wt 323.29. C 22.29%, H 5.30%, N 13.00%, O 49.49%, S 9.92%. $(NH_2CH_2COOH)_3 \cdot H_2SO_4$. Prepn: Horsford, *Ann.* **60**, 1 (1846); *see also* Matthias, Miller, Remeika, *Phys. Rev.* **104**, 849 (1956). Crystal growing during prepn: Konstantinova *et al., Kristallografiya* **4**, 69-73 and 125-129 (1959).
Orthorhombic crystals. Very freely sol in water. Has ferroelectric properties: Curie point 47°. Spontaneous polarization at room temp: 2.2×10^{-6} coul/cm^2. Coercive field 220 v/cm.
USE: In electronics research.

4388. Glycinin. Chief protein constituent of soybeans: Osborne, Campbell, *J. Am. Chem. Soc.* **20**, 419 (1898); Smith, Circle, *Ind. Eng. Chem.* **31**, 1282 (1939). *Review:* H. Neurath, K. Bailey, Eds., *The Proteins* vol. I, Part A (Academic Press, New York, 1953) pp 208-209, 223; *eidem, ibid.* vol. II, part A, (1954) pp 503, 506.
Soluble in water in the pH range 1-4, as well as above 7.

4389. Glycobiarsol. *[[4-[(Hydroxyacetyl)amino]phenyl]-* *arsonato(1—)]oxobismuth; oxo(hydrogen N-glycoloylarsanilato)bismuth;* bismuth glycoloylarsanilate; bismuth *p*-glycolylaminophenylarsonate; bismuth glycolyl arsanilate; N-glycoloylarsanilic acid bismuth deriv; Broxolin; Dysentulin; Milibis; Viasept; Wintodon. $C_8H_9AsBiNO_6$; mol wt 499.07. C 19.25%, H 1.82%, As 15.01%, Bi 41.88%, N 2.81%, O 19.24%. Prepd from bismuth nitrate and sodium N-glycolylarsanilate: *Hagers Handb. Pharm. Praxis* **Band I** (Suppl. 2), 759 (Berlin, 1958).

Yellowish to pink powder. Dec on heating. Very slightly sol in water, alcohol. Practically insol in ether, chloroform, benzene. pH of satd aq soln 2.8-3.5.
THERAP CAT: Anti-amebic.
THERAP CAT (VET): For trichomoniasis.

4390. Glycocholic Acid. *N-[(3α,5β,7α,12α)-3,7,12-trihydroxy-24-oxocholan-24-yl]glycine;* N-cholylglycine. $C_{26}H_{43}NO_6$; mol wt 465.61. C 67.06%, H 9.31%, N 3.01%, O 20.62%. The product of conjugation of cholic acid with glycine; chief ingredient of the bile of herbivorous animals. In the weakly alkaline bile fluid glycocholic acid exists as the sodium salt. Prepn from bile: Hammarsten in *Abderhalden's Handbuch der Biol. Arbeitsmethoden*, Abt. I, Teil 6, p 211 (1925). Prepn from cholic acid: Cortese, *J. Am. Chem. Soc.* **59**, 2532 (1937). Synthesis: Cortese, Bauman, *ibid.* **57**, 1393 (1935); Bergstrom, Norman, *Acta Chem. Scand.* **7**, 1126 (1953). Separation: Antonides, Brit. pat. 928,635 (1963 to Armour). Metabolism: Norman, *Scand. J. Gastroenterol.* **5**, 231 (1970).

Sesquihydrate, crystals from 5% alc, mp about 130°. $[\alpha]_D^{23}$ +30.8° (c = 7.5 in 95% ethanol). Anhydr form, mp 165-168°. pK 4.4. Soly in water at 15°: 0.33 g/l; in boiling water: 8.3 g/l. Is hydrolyzed to cholic acid and glycine by acids and alkalies. Forms addition compds with nitrobenzene, aniline, benzyl alcohol, benzaldehyde, triolein.
Sodium salt, $C_{26}H_{42}NNaO_6$, crystals from 95% alcohol + ether, mp 230-240°. $[\alpha]_D^{24}$ +32° (water). Soly at 15° in water > 274 g/l; in alcohol > 340 g/l.

4391. Glycocyamine. *N-(Aminoiminomethyl)glycine;* N-amidinoglycine; guanidineacetic acid; guanidoacetic acid. $C_3H_7N_3O_2$; mol wt 117.11. C 30.77%, H 6.03%, N 35.88%, O 27.33%. Prepd from S-ethylthiourea hydrobromide, sodium hydroxide, and glycine: Brand, Brand, *Org. Syn.* **coll. vol. III**, 440 (1955). Crystal and molecular structure: Guha, *Acta Crystallogr.* **B29**, 2163 (1973); J. Berthou *et al., ibid.* **B32**, 1529 (1976).

Crystals, dec 280-284°. Appreciably sol in water.
THERAP CAT: In combination with betaine as cardiotonic.

4392. Glycogen. Animal starch; liver starch. $(C_6H_{10}O_5)_n$; mol wt from about 2.7×10^5 to 3.5×10^6. Reserve carbohydrate of the animal organism. High molecular wt polymer having branched-chain structure composed of D-glucopyranose residues. Distributed through the cell proto-plasm. Found esp in the liver and in rested muscle. Occurs also in insects and lower plants including fungi and yeasts. Isoln by alkaline destruction of the other cell constituents: Claude Bernard, *Lecons sur le diabete* (Paris, 1877) p 553; by destruction with trichloroacetic acid: Bell, Young, *Biochem. J.* **28**, 882 (1934); by centrifugation: Meyer, Jeanloz, *Advan. Enzymol.* **3**, 112 (1943); by hydraulic pressure: Stockhausen, Silbereisen, *Biochem. Z.* **287**, 276 (1936). For biological syn-thesis and lysis from the Cori ester (glucose-1-phosphate) *see* the review and bibliography by Meyer, *Advan. Enzymol.* **3**, 109 (1943); *see also* Nord, *Chem. Rev.* **26**, 423 (1940); Kalckar, *ibid.* **28**, 71 (1941). Isoln from the causal agent of cotton root rot, *Phymatotrichum omnivorum* (Shear) Dug-gar: Ergle, *J. Am. Chem. Soc.* **69**, 2061 (1947). Studies on linkages: Bahl, Smith, *J. Org. Chem.* **31**, 2915 (1966).

White powder. $[\alpha]_D^{25}$ +196° to +197°. Sol in water with opalescence. Insol in alc. Does not reduce Fehling's soln. With iodine, brown to violet colors are produced.

4393. Glycol Dilaurate. *Dodecanoic acid 1,2-ethanediyl ester;* ethylene dilaurate. $C_{26}H_{50}O_4$; mol wt 426.66. C 73.19%, H 11.81%, O 15.00%. $C_{11}H_{23}COOCH_2CH_2OOC$-$C_{11}H_{23}$.

Colorless, amorphous mass, mp 50-52°. bp_{20} 188°. Insol in alcohol, ether.

USE: In lacquers and varnishes as a plasticizer.

4394. Glycolic Acid. *Hydroxyacetic acid;* hydroxyetha-noic acid. $C_2H_4O_3$; mol wt 76.05. C 31.58%, H 5.30%, O 63.11%. $HOCH_2COOH$. Constituent of sugar cane juice. Made by the action of NaOH on monochloroacetic acid; also by electrolytic reduction of oxalic acid. *Review:* Sales brochure on hydroxyacetic acid from E. I. du Pont.

Odorless, somewhat hygroscopic crystals, mp 80°. K at 25°: 1.48×10^{-4}. Soluble in water, methanol, alcohol, ace-tone, acetic acid, ether. pH of aq solns: 2.5 (0.5%); 2.33 (1.0%); 2.16 (2.0%); 1.91 (5.0%); 1.73 (10.0%). LD_{50} orally in rats: 1.95 g/kg, H. F. Smyth *et al., J. Ind. Hyg. Toxicol.* **23**, 259 (1941).

USE: In the processing of textiles, leather, and metals; in pH control, and wherever a cheap organic acid is needed, e.g. in the manuf of adhesives, in copper brightening, decon-tamination cleaning, dyeing, electroplating, in pickling, cleaning and chemical milling of metals. *Caution:* Mild irri-tant to skin, mucous membranes.

4395. Glycol Salicylate. *2-Hydroxybenzoic acid 2-hydr-oxyethyl ester;* monoglycol salicylate; ethylene glycol mono-salicylate; 2-hydroxyethyl salicylate; Norges-ic; Phlogont (salve); Spirosal. $C_9H_{10}O_4$; mol wt 182.17. C 59.33%, H 5.53%, O 35.13%. $C_6H_4(OH)COOCH_2CH_2OH$.

Almost colorless, odorless liq. bp_{12} 169-172°. Soluble in about 110 parts water, 8 parts olive oil; very sol in alcohol, benzene, chloroform, ether.

THERAP CAT: Counterirritant, anti-inflammatory (topical).

4396. Glyconiazide. D-*Glucuronic acid* γ-*lactone 1-[(4-pyridinylcarboxyl)hydrazone];* D-glucuronolactone isonico-tinoylhydrazone; isonicotinylhydrazone of D-glucuronic acid lactone; isonicotinic acid hydrazide hydrazone with glucuronic acid lactone; Galatone; Gatalone; Glucazide; Gluconiazide; Gluronazide; Guidazide; Hydronsan; INH-G; Mycobactyl. $C_{12}H_{13}N_3O_6$; mol wt 295.25. C 48.81%, H 4.44%, N 14.23%, O 32.51%. Prepd by heating isonicotinic acid hydrazide with D-glucuronolactone in methanol: Sah, *J. Am. Chem. Soc.* **75**, 2512 (1953); Sah, U.S. pat. **2,940,899** (1960 to U. of Calif.).

Plates and rods from methanol, needles from abs ethanol. Dec 150-160°. Freely sol in water. Practically insol in cold alc; 1.2 g dissolve in 100 ml methanol at 66°.

THERAP CAT: Antibacterial (tuberculostatic).

4397. Glycopyrrolate. *3-[(Cyclopentylhydroxyphenylace-tyl)oxy]-1,1-dimethylpyrrolidinium bromide; 3-hydroxy-1,1-dimethylpyrrolidinium bromide* α-*cyclopentylmandelate;* α-*cyclopentylmandelic acid ester with 3-hydroxy-1,1-dimeth-ylpyrrolidinium bromide;* 1-methyl-3-pyrrolidyl α-cyclo-pentylmandelate methobromide; 1-methyl-3-pyrrolidyl α-phenyl-α-cyclopentylglycolate methobromide; 3-(2-phen-yl-2-cyclopentylglycoloyloxy)-1,1-dimethylpyrrolidinium bromide; glycopyrronium bromide; AHR 504; Nodapton; Robanul; Robinul; Tarodyl; Tarodyn. $C_{19}H_{28}BrNO_3$; mol wt 398.36. C 57.29%, H 7.08%, Br 20.06%, N 3.52%, O 12.05%. Prepn: Franko, Lunsford, *J. Med. Pharm. Chem.* **2**, 523 (1960); Lunsford, U.S. pat. **2,956,062** (1960 to A. H. Robins). Pharmacodynamics: E. Kaltiala *et al., J. Pharm. Pharmacol.* **26**, 352 (1974). Toxicology: B. V. Franko *et al., Toxicol. Appl. Pharmacol.* **17**, 361 (1970). Clinical compari-son with atropine, *q.v.,* in anaesthetic practice: F. Kongsrud, S. Sponheim, *Acta Anaesth. Scand.* **26**, 620 (1982); A. I. Webb, R. M. McMurphy, *Am. J. Vet. Res.* **48**, 1733 (1987); B. V. G. Malling *et al., Brit. J. Anaesth.* **60**, 426 (1988). Brief review of pharmacology and clinical use: R. K. Mirakhur, J. W. Dundee, *Anaesthesia* **38**, 1195-1204 (1983).

White crystals from butanone, mp 193.2-194.5°. Sol in water. LD_{50} (72 hr.) in female mice, female rats (mg/kg): 107, 196 i.p.; in male rats (mg/kg): 1150 orally (Franko).

THERAP CAT: Anticholinergic.

THERAP CAT (VET): Anticholinergic.

4398. Glycosine. *1-Methyl-2-(phenylmethyl)-4(1H)-quinazolinone;* 2-benzyl-1-methylquinazol-4-one; arborine. $C_{16}H_{14}N_2O$; mol wt 250.29. C 76.78%, H 5.64%, N 11.19%, O 6.39%. Found in the toothbrush plant, *Glycosmis penta-phylla* (Retz.) Corr., and *G. arborea* Corr., Rutaceae. Isoln from dried, powdered leaves: Chatterjee, Majumdar, *J. Am. Chem. Soc.* **76**, 2459 (1954). Identity of arborine and glyco-sine, structure: Chakravarti *et al., Tetrahedron* **16**, 224 (1961). Synthesis: Pakrashi *et al., Indian J. Chem.* **6**, 472 (1968); Ziegler *et al., Monatsh.* **100**, 948 (1969); T. Kametani *et al., Heterocycles* **9**, 1585 (1978).

Rhombohedral prisms from chloroform + ethyl acetate, mp 155-156°. uv max (ethanol): 231, 268, 277, 306 nm. Freely sol in chloroform, ethyl acetate, benzene, ethanol. Sparingly sol in ether.

Hydrochloride, $C_{16}H_{14}N_2O.HCl$, leaflets from 90% etha-nol, dec 209-210°.

Picrate, $C_{16}H_{14}N_2O.C_6H_3N_3O_7$, fine yellow needles from 90% ethanol, dec 171-172°.

4399. *N*-Glycylglycine. $C_4H_8N_2O_3$; mol wt 132.12. C 36.36%, H 6.10%, N 21.20%, O 36.33%. NH_2CH_2CONH-CH_2COOH. The simplest of all peptides. Prepn from 2,5-diketopiperazine: Schott *et al., J. Org. Chem.* **12**, 490 (1947); Greenstein, Winitz, *Chemistry of the Amino Acids* vol. **2**, (New York, 1961) p 803. From tritylglycylglycine: Zervas *et al., J. Am. Chem. Soc.* **78**, 1359 (1956). From phthalylgly-cylglycine: Sheehan, Frank, *ibid.* **71**, 1856 (1949). From the dicyclohexylamine salt of trifluoroacetylglycylglycine: Weygand, Reiher, *Ber.* **88**, 26 (1955).

Crystals from dil alc. Crystal shape described as small tetrahedral leaves with a lustrous ball in center. Dec 262-264°. pK$_1'$ 3.12; pK$_2'$ 8.17. Heat of combustion: 472.4 kcal/mole. Soluble in hot water; slightly sol in ethanol. Practically insol in ether.

Hydrochloride, $C_4H_8N_2O_3 \cdot HCl \cdot H_2O$, crystals from water + ethanol.

Ethyl ester hydrochloride, crystals from abs ethanol, dec 182°.

USE: In the synthesis of more complicated peptides.

4400. Glycyrrhiza. Licorice; liquorice; sweet root. Dried rhizome and roots of *Glycyrrhiza glabra* L., var. *typica* Regel & Herder (Spanish licorice), or of *G. glabra* L., var. *glandulifera* (Waldst. & Kit.) Regel & Herder (Russian licorice), or of other varieties of *G. glabra* yielding a yellow and sweet wood, *Leguminosae*. Habit. Southern Europe to Central Asia. Constit. 6-14% glycyrrhizin (the glucoside of glycyrrhetic acid), asparagine, sugars, resin. Used chiefly in the form of glycyrrhiza syrup. Incompat. Acids, metallic salts.

USE: Extract and syrup as pharmaceutic aids (flavor and flavored vehicles).

4401. Glycyrrhizic Acid. *20β-Carboxy-11-oxo-30-norolean-12-en-3β-yl-2-O-β-D-glucopyranuronosyl-α-D-glucopyranosiduronic acid;* glycyrrhizin; glycyrrhizinic acid; glycyrrhetinic acid glycoside. $C_{42}H_{62}O_{16}$; mol wt 822.92. C 61.30%, H 7.59%, O 31.11%. Extraction from *Glycyrrhiza glabra* L., *Leguminosae:* Karrer, Chao, *Helv. Chim. Acta* **4**, 100 (1921); Ruzicka, Louenberger, *ibid.* **19**, 1402 (1936). From commercial glycyrrhizinum ammoniacale: Tschirch, Cederberg, *Arch. Pharm.* **245**, 97 (1907); Voss *et al., Ber.* **70**, 122 (1937). Revised method of isoln: Conn, Conn, *J. Lab. Clin. Med.* **47**, 20 (1956). Structure: Lythgoe, Trippett, *J. Chem. Soc.* **1950**, 1983. Alternate view: Marsh, Levvy, *Biochem. J.* **63**, 9 (1956). Review: Nieman, *Chem. Weekbl.* **48**, 213 (1952). Synthesis of derivatives: Brieskorn, Sax, *Arch. Pharm.* **303**, 905 (1970).

glycyrrhetinic acid

Crystals from glacial acetic acid. Intensely sweet taste. $[\alpha]_D^{17}$ +46.2° (c = 1.5 in alc). Freely sol in hot water, alcohol; practically insol in ether.

Ammonium glycyrrhizinate pentahydrate, $C_{42}H_{65}NO_{16} \cdot 5H_2O$, needles from 75% aqueous ethanol, decomp 212-217°. $[\alpha]_D^{20}$ +46.9° (c = 1.5 in 40% ethanol). uv max: 248 nm (ϵ 11,400). Sol in ammonia water, glacial acetic acid.

Dipotassium salt, $C_{42}H_{60}K_2O_{16}$, *Rizinsan K2 A2.*

4402. Glyhexamide. *N-[(Cyclohexylamino)carbonyl]-2,3-dihydro-1H-indene-5-sulfonamide; 1-cyclohexyl-3-(5-indanylsulfonyl)urea;* 1-cyclohexyl-3-(5-hydrindenylsulfonyl)urea; SQ 15860; Subose. $C_{16}H_{22}N_2O_3S$; mol wt 322.45. C 59.60%, H 6.88%, N 8.69%, O 14.89%, S 9.95%. Prepd from hydrindene-5-sulfonamide and cyclohexyl isocyanate: Hoehn, Breuer, U.S. pat. **3,097,242** (1963 to Olin Mathieson). Clinical pharmacology: Grinnell *et al., Am. J. Med. Sci.* **253**, 312 (1967).

Crystals from 70% acetone, mp 153-155°.
THERAP CAT: Antidiabetic.

4403. Glymidine. *N-[5-(2-Methoxyethoxy)-2-pyrimidinyl]benzenesulfonamide;* 2-benzenesulfonamido-5-(β-meth-

oxyethoxy)pyrimidine; glycodiazine. $C_{13}H_{15}N_3O_4S$; mol wt 309.35. C 50.47%, H 4.89%, N 13.58%, O 20.69%, S 10.37%. Prepn: Belg. pat. **609,270** corresp to H. Priewe *et al.,* U.S. pat. **3,275,635** (1962, 1966 to Schering, AG); Gutsche *et al., Arzneimittel-Forsch.* **14**, 373 (1964). Series of articles on pharmacology: *ibid.* 377-412. Activity: Losert *et al., ibid.* **23**, 1251 (1973). Metabolism: Soyfer *et al., Chim. Ther.* **5**, 441 (1970).

Crystals, mp 152-154°. Soly in ethanol: 0.91%; in toluene: 0.67%.

Sodium salt, $C_{13}H_{14}N_3NaO_4S$, *SH 717, Glyconormal, Gondafon, Lycanol, Redul.* Crystals, mp 221-226°. Sparingly sol in alc. Soly in water at 37°: 70.5%. LD$_{50}$ in mice, rats (g/kg): 1.48, 2.00 i.v.; 5.30, 2.85 orally, Kramer *et al., Arzneimittel-Forsch.* **14**, 377 (1964).

THERAP CAT: Antidiabetic.

4404. Glyodin. *2-Heptadecyl-4,5-dihydro-1H-imidazole monoacetate;* 2-heptadecylglyoxalidine acetate; Crag Fruit Fungicide 341. $C_{22}H_{44}N_2O_2$; mol wt 368.59. C 71.68%, H 12.03%, N 7.60%, O 8.68%. Prepn from stearic acid and ethylenediamine: Kiff, U.S. pat. **2,540,171** (1951 to Union Carbide and Carbon).

Light orange crystals, mp 62-68°. d^{20} 1.035. Insol in water, acetone, toluene; sol in isopropanol. The base is a soft greasy wax, mp 94°.

USE: Fungicide.

4405. Glyoxal. *Ethanedial;* biformyl; diformyl; oxalaldehyde. $C_2H_2O_2$; mol wt 58.04. C 41.39%, H 3.48%, O 55.14%. OHCCHO. Prepd by the oxidation of acetaldehyde by nitric or selenious acid: Lubawin, *Ber.* **8**, 768 (1875); Wyss, *Ber.* **10**, 1366 (1877); Kölln, *Ann.* **416**, 230 (1918); Riley *et al., J. Chem. Soc.* **1932**, 1881; Ronzio, Waugh, *Org. Syn. coll. vol. III*, 438 (1955); by hydrolysis of dichlorodioxane: Butler, Cretcher, *J. Am. Chem. Soc.* **54**, 2988 (1932). Review of commercial development: J. F. Bohmfalk *et al., Ind. Eng. Chem.* **43**, 786 (1951). Review: A. B. Boese *et al.,* in *Glycols,* G. O. Curme, F. Johnston, Eds. (Reinhold, New York, 1952) pp 125-128.

Yellow prisms or irregular pieces turning white on cooling. d^{20} 1.14. Opaque at 10°, mp 15°. bp$_{776}$ 51°. The vapors are green and burn with a purple flame. *Caution:* Mixtures with air may explode! $n_D^{20.5}$ 1.3826. Sol in anhydr solvents. pH of a 40% aq soln: 2.1-2.7; d$_4^{20}$ 1.27. Polymerizes quickly on standing, on contact with water (violent reaction), or when dissolved in solvents contg water. The anhydr polymer changes to the monomer on heating. Solns of the monomer are obtained on heating the polymer with anethole, phenetole, safrole, methyl nonyl ketone, or benzaldehyde. LD$_{50}$ orally in rats, guinea pigs: 2020, 760 mg/kg, H. F. Smyth *et al., J. Ind. Hyg. Toxicol.* **23**, 259 (1941).

Dihydrate, $(OHCCHO)_2 \cdot 2H_2O$, cryst powder, nonhygroscopic. More sol in hot water than in cold water. Commercially available in anhydr form as cryst dihydrate, or as a 40% aq soln which may contain polymerization inhibitors.

Caution: Moderately irritating to skin, mucous membranes.

USE: In textiles, organic synthesis, glues, biocides.

4406. Glyoxal-Sodium Bisulfite. *1,2-Dihydroxy-1,2-ethanedisulfonic acid disodium salt; glyoxal compound with sodium bisulfite.* $C_2H_4Na_2O_8S_2$; mol wt 266.16. C 9.02%, H 1.51%, Na 17.28%, O 48.09%, S 24.09%. Prepn: Ronzio, Waugh, *Org. Syn. coll. vol. III*, 438 (1955).

```
        SO₃Na
         |
  HOCHCHOH
         |
        SO₃Na
```

Monohydrate, hard crystals. Faint SO₂ odor. Freely sol in water; practically insol in alcohol.

4407. Glyoxylic Acid. *Oxoacetic acid; formylformic acid; glyoxalic acid; oxoethanoic acid.* $C_2H_2O_3$; mol wt 74.04. C 32.44%, H 2.73%, O 64.83%. OHCCOOH. Occurs in unripe fruit and in young green leaves; has also been found in very young sugar beets. Prepd by heating dibromoacetic acid with some water: Grimaux, *Bull. Soc. Chim.* [2] **26**, 483; Cramer, *Ber.* **25**, 714 (1892); by electrolytic reduction of oxalic acid: Meyer, *Ber.* **37**, 3592 (1904); by the action of *Aspergillus niger* on calcium acetate, malonic or citric acid: Challenger *et al., J. Chem. Soc.* **1927**, 205, 207. Prepn of glyoxylic acid soln for analytical use: *Beilstein* vol. **III**, 594.

Hemihydrate, crystals from water, mp 70-75°. Also obtained in anhydr form as monoclinic crystals from water, mp 98°. Obnoxious odor. Strong, corrosive acid. K = 4.6 × 10⁻⁴. Deliquesces rapidly and forms a syrup on short exposure to air. Sparingly sol in alc, ether, benzene. Freely sol in water; aq solns tend to acquire a yellowish tint. Attacks most base metals except certain stainless steel alloys.

Monohydrate, crystals, mp ∼50°. Highly hygroscopic. *Caution:* Irritant, corrosive.

4408. Glyphosate. *N-(Phosphonomethyl)glycine;* MON-0573. $C_3H_8NO_5P$; mol wt 169.07. C 21.31%, H 4.77%, N 8.28%, O 47.32%, P 18.32%. Broad-spectrum post-emergence, translocated herbicide. Prepn: J. E. Franz, Ger. pat. **2,152,826** corresp to U.S. pats. **3,799,758** and **3,853,530** (1972, 1974, 1974 to Monsanto). Metabolism and degradation in soil and water: M. L. Rueppel *et al., J. Agr. Food Chem.* **25**, 517 (1977).

```
                    O
                    ||
 HOOCCH₂NHCH₂—P—OH
                    |
                   OH
```

White solid, mp 230° (dec). Soly in water at 25°: 12 g/l. Insol in most organic solvents. LD_{50} orally in rats, mice: 4873, 1568 mg/kg, E. A. Bababunmi *et al., Toxicol. Appl. Pharmacol.* **45**, 319 (1978).

Mono(isopropylamine) salt, $C_6H_{17}N_2O_5P$, *MON-2139, Round-up.* Very sol in water.
USE: Herbicide.

4409. Glyphosine. *N,N-Bis(phosphonomethyl)glycine; N-carboxymethyl-N,N-bis(methylenephosphonic acid)amine;* CP-41845; Polaris. $C_4H_{11}NO_8P_2$; mol wt 263.08. C 18.26%, H 4.22%, N 5.33%, O 48.65%, P 23.54%. Method of prepn: Irani, Moedritzer, U.S. pat. **3,288,846** (1966 to Monsanto). As ripening agent for sugar cane: Hamm, U.S. pat. **3,556,-762** (1971 to Monsanto). Brief review of ripening effect and toxicity: Porter, Ahlrichs, *Hawaii Sugar Technol. Rep.* **30**, 71 (1971) (publ. 1972).

```
                   O
                   ||
          CH₂—P—OH
         /         |
        /         OH
 HOOCCH₂N
        \          O
         \         ||
          CH₂—P—OH
                   |
                  OH
```

White crystals from aq ethanol. Soly in water at 20°: 248 g/l. Slightly corrosive to metals. LD_{50} orally in mammals: 3925 mg/kg, *RTECS* vol. **1**, R. J. Lewis, Sr., R. L. Tatken, Eds. (1979) p 719.
USE: Chemical ripener.

4410. Glypinamide. *4-Chloro-N-[[(hexahydro-1H-azepin-1-yl)amino]carbonyl]benzenesulfonamide;* 1-(p-chlorophenylsulfonyl)-3-(hexahydro-1H-azepin-1-yl)urea; azepin-

amide; Parinase. $C_{13}H_{18}ClN_3O_3S$; mol wt 331.83. C 47.05%, H 5.47%, Cl 10.69%, N 12.66%, O 14.47%, S 9.66%. Prepn: Wright, **Brit.** pat. **887,886** (1962 to Upjohn); Wright, Willette, *J. Med. Pharm. Chem.* **5**, 815 (1962).

```
 Cl—⬡—SO₂NHCONH—N⬡
```

Crystals from methyl ethyl ketone, mp 197-198.5°.
THERAP CAT: Oral hypoglycemic.

4411. Goitrin. *(S)-5-Ethenyl-2-oxazolidinethione;* 5-vinyl-2-thiooxazolidone; (−)-5-vinyl-2-oxazolidinethione. C_5H_7NOS; mol wt 129.19. C 46.49%, H 5.46%, N 10.84%, O 12.39%, S 24.82%. An antithyroid compd isolated from seeds of different species of *Brassica, Cruciferae:* Astwood *et al., J. Biol. Chem.* **181**, 121 (1949); Greer, *J. Am. Chem. Soc.* **78**, 1260 (1956). Stereochemistry: Kjaer *et al., Acta Chem. Scand.* **13**, 144 (1959). Activity: Langer *et al., Endokrinologie* **57**, 225 (1971).

```
        O    CH=CH₂
      /    \   /
    S        \/
      \   N  /
        |   H  H
        H
```

Large prisms from ether, mp 50°. $[\alpha]_D^{31}$ −70.5° (c = 2 in methanol). Behaves as a weak acid; pKa 10.5. Stable in alkali, but not in acid.

4412. Gold. Au; at. wt 196.9665; at. no. 79; valences 1, 3. Occurrence in the earth's crust: 0.005 ppm. One natural isotope: 197; artificial isotopes (mass numbers): 177-179, 181, 183, 185-196, 198-203. Probably the first pure metal known to man. Occurs in nature in its native form and in minute quantities in almost all rocks and in seawater. Gold ores include *calavarite,* (AuTe₂), *sylvanite,* [(Ag,Au)Te₂], *petzite,* [(Ag,Au)₂Te]. Methods of mining, extracting and refining: Hull, Stent, in *Modern Chemical Processes,* Vol. **5** (Reinhold, New York, 1958) pp 60-71. Lab prepn of gold powder from gold pieces: Block, *Inorg. Syn.* **4**, 15 (1953). Chemistry of gold drugs in the treatment of rheumatoid arthritis: D. H. Brown, W. E. Smith, *Chem. Soc. Rev.* **9**, 217 (1980). Use as catalyst in oxidation of organic compds by NO₂: R. E. Sievers, S. A. Nyarady, *J. Am. Chem. Soc.* **107**, 3726 (1985). *Reviews:* Gmelin's *Handb. Anorg. Chem., Gold* (8th ed.) **62**, parts 2, 3 (1954); Johnson, Davis, "Gold" in *Comprehensive Inorganic Chemistry,* vol. **3**, J. C. Bailar Jr. *et al.,* Eds. (Pergamon Press, Oxford, 1973) pp 129-186; J. G. Cohn, E. W. Stern in Kirk-Othmer *Encyclopedia of Chemical Technology* vol. **11** (Wiley-Interscience, New York, 3rd ed., 1980) pp 972-995.

Yellow, soft metal; face-centered cubic structure; when prepared by volatilization or precipitation methods, deep violet, purple, or ruby powder, mp 1,064.76°; bp 2700°. d 19.3. Hardness (Mohs') 2.5-3.0; (Brinell's) 18.5. Extremely inactive; not attacked by acids, air or oxygen. Superficially attacked by aq halogens at room temp. Reacts with aqua regia, with mixtures contg chlorides, bromides, or iodides if they can generate nascent halogens, with many oxidizing mixtures especially those contg halogens. Also with alkali cyanides, solns of thiocyanates and double cyanides.

USE: In manuf jewelry; in gold plating other metals; as a standard of currency; most frequently alloyed with silver and copper. For use in medicine, *see* Gold, Radioactive, Colloidal.

4413. Gold, Explosive. "Fulminating gold". An auric compd of nitrogen. Obtained by the action of ammonia on auric chloride or ammonium chloride on auric oxide: Raschig, *Ann.* **235**, 355 (1886); Weitz, *Ann.* **410**, 117 (1915).

Dark brown powder, which explodes on heating or rubbing to give gold, nitrogen, and ammonia. Exact compn of the compd is unknown since it is too explosive to be dried. Therefore, the only elements that can be determined are gold, nitrogen, and chlorine.

4414. Gold Monochloride. Aurous chloride. AuCl; mol

wt 232.46. Au 84.76%, Cl 15.24%. Prepd by thermal decompn of gold trichloride: Biltz, Wein, *Z. Anorg. Allgem. Chem.* **148**, 192 (1925); Capella, Schwab, *C.R. Acad. Sci.* **260**, 4337 (1965).

Yellowish powder. d 7.57. Dec at about 289° into gold and Cl_2. Practically insol in water, but slowly dec by it, more rapidly on heating, with formation of gold trichloride and separation of metallic gold; sol in alkali cyanides. With solns of alkali bromides, metallic gold and potassium auribromide are formed.

4415. Gold Monocyanide. Aurous cyanide; gold cyanide. CAuN; mol wt 223.02. Au 88.34%, C 5.39%, N 6.28%. AuCN. May be prepd by decompn of $Na[Au(CN)_2]$ with HCl: Wogrinz, *Metalloberfläche* **8**, B162 (1954).

Yellow, odorless powder; iridescent in sunlight; slowly dec in presence of moisture. When warmed with HCl, HCN is evolved. When ignited, dec into metal Au and CN. d_4^{20} 7.14. Hexagonal when cryst. Practically insol in water, alcohol, ether, dil acids; sol in ammonia, soln of NaCN; dissolved by aqua regia. On warming with concd H_2SO_4 half of the gold separates as metal.

Caution: May liberate HCN.

4416. Gold Monoiodide. Aurous iodide. AuI; mol wt 323.91. Au 60.84%, I 39.16%. Prepn: Weiss, Weiss, *Z. Naturforsch.* **11b**, 604 (1956); Grange, *Bull. Soc. Chim. France* **1964**, 2418.

Yellowish to greenish-yellow powder; dec slowly at ordinary temp, more rapidly at higher. d 8.25. Insol in water; sol in alkali iodide or cyanide solns; dec by warm acids.

4417. Gold Monosulfide. Aurous sulfide. Au_2S; mol wt 426.07. Au 92.48%, S 7.52%. May be prepd from potassium gold cyanide, H_2S and HCl: Hoffmann, Krüss, *Ber.* **20**, 2361 (1887); Glemser, Sauer in *Handbook of Preparative Inorganic Chemistry*, Vol. 2, G. Brauer, Ed. (Academic Press, New York, 2nd ed., 1965) pp 1061-1062.

Brownish-black powder. When freshly pptd it forms a colloidal soln in water. Practically insol in water or single acids; sol in aqua regia, alkali cyanide solns.

4418. Gold Monoxide. Aurous oxide. Existence doubtful; probably a mixture of gold and auric oxide: Pollard, *J. Chem. Soc.* **129**, 1347 (1926). "Au_2O" has been prepared by treating $K(AuBr_2)$ with alkali.

Pale gray-violet solid. May be dried at 200°, but gives off oxygen a few degrees above this, and rapidly at 250°. Soly in water (25°): 0.04 mg/l. When freshly precipitated, it dissolves in alkalies, but the soln deposits gold rapidly.

4419. Gold, Radioactive, Colloidal. Radioactive colloidal gold; colloidal gold—^{198}Au; radio-gold (^{198}Au) colloid; gold colloid ^{198}Au; Aurcoloid; Aurcoscan-198; Aureotope. Colloidal dispersion of radioactive gold (^{198}Au) for parenteral administration. Particle diameter range, 3-7 mμ. Has half-life of 2.7 days and emits beta and gamma radiation. Stable to heat except autoclaving under pressure. Compatible with saline solns, radiopaque media and other agents; flocculated by polyvalent metal ion.

THERAP CAT: Antineoplastic.

4420. Gold Selenate. Auric selenate. $Au_2O_{12}Se_3$; mol wt 822.88. Au 47.91%, O 23.33%, Se 28.77%. $Au_2(SeO_4)_3$. Prepd from gold and selenic acid: Lenher, *J. Am. Chem. Soc.* **24**, 354 (1902); Caldwell, Eddy, *ibid.* **71**, 2247 (1949).

Small, yellow crystals. Dec in light. Soluble in sulfuric and nitric acids; insol in water.

4421. Gold Selenide. Auric selenide. Au_2Se_3; mol wt 630.88. Au 62.48%, Se 37.52%. Prepd by adding an alcoholic soln of gold chloride to a soln of hydrogen selenide: Moser, Atynsky, *Monatsh.* **45**, 235 (1925); by passing hydrogen selenide into an aq soln of gold chloride: Uelsmann, *Ann.* **116**, 127 (1860); by combination of elements in solns or melts: Kulifay, U.S. pat. 3,026,175 (1962 to Monsanto).

Black amorphous solid. d^{22} 4.65. Dec by heat. Forms mercuric selenide when heated with mercury. Sol in aqua regia, soln of alkali sulfides.

4422. Gold Sodium Thiomalate. *Mercaptobutanedioic acid monogold(1+) sodium salt;* sodium aurothiomalate; Kidon; Myochrysine; Myocrisin; Shiosol; Taure(o)don. A

variable mixture of the mono- ($C_4H_4AuNaO_4S$) and disodium ($C_4H_3AuNa_2O_4S$) salts of gold thiomalic acid. Previously described as the monohydrate of the disodium salt. Prepd by the interaction of sodium thiomalate and a gold halide: U.S. pat. 1,994,213 (1935). Mechanism of action: P. E. Lipsky, M. Ziff, *Adv. Inflammation Res.* **3**, 219 (1982). Pharmacokinetics: E. S. Waller *et al., Res. Commun. Chem. Pathol. Pharmacol.* **37**, 33 (1982). Detection in blood by atomic absorption spectroscopy: A. I. A. Rodgers *et al., Anal. Proc.* **19**, 87 (1982).

$$Au{-}S{-}\underset{\displaystyle CHCOO^-}{\overset{\displaystyle CH_2COO^-}{|}} \cdot \ xNa^+ \cdot \ (2{-}x)H^+$$

White to yellowish-white powder, metallic taste. Very sol in water. Practically insol in alcohol, ether. Aq solns are colorless to pale yellow. The pH of a 5% aq soln: 5.8-6.5.

THERAP CAT: Antirheumatic.

4423. Gold Sodium Thiosulfate. *Bis[monothiosulfato-(2—)-O,S]aurate(3—) trisodium; bis(monothiosulfato)aurate(3—) trisodium;* aurothiosulfate natrium; aurothiosulfate sodium; double thiosulfate of gold and sodium; hyposulfite of gold and sodium; sodium aurothiosulfate; sel de Fordos et Gelis; Auricidine; Aurocidin; Aurolin; Auropex; Auropin; Aurosan; Aurothion; Crisalbine; Crytion; Novacrysin; Sanochrysine; Solfocrisol; Thiochrysine. $AuNa_3O_6S_4$; mol wt 490.21. Au 40.19%, Na 14.07%, O 19.58%, S 26.16%. Na_3-$Au(S_2O_3)_2$. Prepd as dihydrate by adding small amounts of gold chloride to a concd aq soln of sodium thiosulfate; after each addition of gold chloride a red color appears which must disappear before another addition can be made; the complex is finally precipitated by the addition of alc to the colorless aq soln. All the gold is reduced from the auric to the aurous state: Fordos, Gélis, *Ann. Chim. Phys.* **13**, 394 (1845); Brown, *J. Am. Chem. Soc.* **49**, 958 (1927); *Roger's Inorganic Pharmaceutical Chemistry*, T. O. Soine, C. O. Wilson, Eds. (Lea & Febiger, Philadelphia, 8th ed., 1967) pp 343-345. Crystal structure of dihydrate: Ruben *et al., Inorg. Chem.* **13**, 1836 (1974).

Dihydrate, white glistening crystals, needle-like or prismatic. d 3.09. Darkens slowly on exposure to light. The 2 moles of water of crystn are not given up until heated to 150 or 160°. One gram dissolves in 2 ml water. Insol in alc and most other organic solvents. A 1:20 aq soln is neutral or slightly alkaline to litmus. Aq solns dec on standing and turn yellow. LD i.v. in rabbits: 100 mg/kg.

Human Toxicity: Dermatitis, stomatitis, hepatitis, nephritis, G.I. disturbances, hematologic reaction may occur. *Antidote:* Dimercaprol (BAL).

THERAP CAT: Antirheumatic.

4424. Gold Stannate. Aurous stannate; C.I. Pigment Red 109; C.I. 77482; gold-tin precipitate; gold-tin purple; purple of Cassius. Contains gold, tin and oxygen; composition of commercial products varies and in some cases may be a complex mixture of gold and stannic acid. Prepn: *Colour Index*, vol. 4 (3rd ed., 1971) p 4669.

Brown powder; practically insol in water; sol in ammonia.

USE: Manuf ruby glass, colored enamels, and painting porcelain.

4425. Gold Tribromide. Auric bromide. $AuBr_3$; mol wt 436.75. Au 45.13%, Br 54.87%. Prepn: Burawoy, Gibson, *J. Chem. Soc.* **1935**, 217.

Brownish-black powder. mp about 160° with decompn. Sol in water with reddish-brown color; also sol in alcohol, glycerol, but gradually reduced by them. *Keep well closed and protected from light.*

4426. Gold Tribromide, Acid. *Hydrogen tetrabromoaurate(1—);* bromoauric acid. $AuBr_4H$; mol wt 517.65. Au 38.06%, Br 61.75%, H 0.19%. $HAuBr_4$. Prepd from $AuBr_3$ and HBr.

Pentahydrate, dark reddish-brown, odorless, deliquesc crystals or granular masses. mp about 27°. Very sol in water; sol in alcohol. *Keep tightly closed in a cool place and protected from light.*

4427. Gold Trichloride. Auric chloride. $AuCl_3$; mol wt 303.35. Au 64.94%, Cl 35.06%. Conveniently prepd in the laboratory from metallic gold and iodine monochloride: Gutmann, Z. Anorg. Allgem. Chem. **264**, 169 (1951).

Dihydrate, dark orange-red crystals, deliquesc in moist air. d_4^{20} 3.9. Sublimes at 180° (760 mm). bp 229°. Decomp 254°. Soluble in water, alcohol, ether. Keep well closed and protected from light. LD s.c. in mice: 1.5 g/kg.

4428. Gold Trichloride, Acid. Hydrogen tetrachloroaurate(1−); hydrochloroauric acid; aurochlorohydric acid; chloroauric acid. $AuCl_4H$; mol wt 339.81. Au 57.97%, Cl 41.73%, H 0.30%. $HAuCl_4$. Prepd according to the equation $2Au + 3Cl_2 + 2HCl \rightarrow 2HAuCl_4$: Thomsen, Ber. **16**, 1585 (1883); Biltz, Wein, Z. Anorg. Allgem. Chem. **148**, 192 (1925).

Tetrahydrate, golden-yellow to reddish-yellow, very hygroscopic and deliquesc monoclinic crystals; readily affected by sunlight. (Also available as brown crystals or cryst masses, contg 50-51% gold, and having the same properties as the yellow crystals.) Has caustic action on the skin (blisters) and then on exposure to light leaves violet-brown spots. Dec on strong heating to Cl_2, HCl, and metallic gold. d about 3.9. Very sol in water, alcohol; sol in ether. Keep tightly closed and protected from light.

USE: Photography, gold-plating, gilding glass and porcelain, manuf ruby glass; as a reagent for alkaloids.

4429. Gold Tricyanide. Auric cyanide. C_3AuN_3; mol wt 275.05. C 13.10%, Au 71.62%, N 15.28%. $Au(CN)_3$. Prepn of trihydrate: Chadwick, Sharpe, Adv. Inorg. Chem. Radiochem. **8**, 157-158 (1966).

Trihydrate, white, deliquesc crystals; dec at 50°. Poison! Very soluble in water, slightly sol in alcohol. Keep tightly closed and protected from light.

4430. Gold Trihydroxide. Auric hydroxide. AuH_3O_3; mol wt 248.02. Au 79.44%, H 1.22%, O 19.35%. $Au(OH)_3$. Usually contains about $3H_2O$ and hence about 65% gold. May be prepd in lab according to the equation $2KAuCl_4 + 3Na_2CO_3 + 3H_2O \rightarrow 2Au(OH)_3 + 6NaCl + 2KCl + 3CO_2$: Lydén, Z. Anorg. Allgem. Chem. **240**, 157 (1939).

Brown powder; dec by sunlight to metallic gold; also slowly dec with age or at 100°, and completely at 250°. Practically insol in water; sol in soln of NaCN, in HCl or concd HNO_3. With NH_3 yields gold fulminate which explodes easily in dry form. Protect from light.

USE: In gold-plating solns; for decorating porcelains.

4431. Gold Trioxide. Auric oxide; gold sesquioxide; gold oxide. Au_2O_3; mol wt 442.00. Au 89.15%, O 10.86%. Prepd from the hydroxide: Roseveare, Buehner, J. Am. Chem. Soc. **49**, 1221 (1927).

Brown powder; begins to evolve oxygen at 110°; at 250° it is entirely dec to metallic gold; also slowly dec by sunlight. Practically insol in water; sol in HCl, concd HNO_3 and in NaCN soln. Keep protected from light.

4432. Gold Trisulfide. Auric sulfide. Au_2S_3; mol wt 490.20. Au 80.39%, S 19.61%. Prepd by treating dry lithium aurichloride $Li(AuCl_4)$ with hydrogen sulfide at −10°. The reaction product consists of HCl, lithium chloride, and auric sulfide. The lithium chloride can be removed by extraction with alcohol. Ref: Antony, Lucchesi, Gazz. Chim. Ital. **19**, 552 (1889). Prepd from $AuCl_3$ or $HAuCl_4$ and H_2S in abs ether at low temp: Guthier, Dürrwachter, Z. Anorg. Allgem. Chem. **121**, 266 (1922).

Black powder. Heating to 200° dec it into its elements.

USE: In photography.

4433. Goserelin. 6-[O-(1,1-Dimethylethyl)-D-serine]-10-deglycinamideluteinizing hormone-releasing factor (pig) 2-(aminocarbonyl)hydrazide; D-Ser(But)6Azgly10-gonadorelin; D-Ser(But)6Azgly10-luliberin; ICI 118630. $C_{59}H_{84}$-$N_{18}O_{14}$; mol wt 1269.43. C 55.82%, H 6.67%, N 19.86%, O 17.65%. Synthetic peptide agonist analog of LH-RH, q.v. Prepn: A. S. Dutta et al., Ger. pat. **2,720,245**; eidem, U.S. pat. **4,100,274** (1977, 1978 both to I.C.I.); eidem, J. Med. Chem. **21**, 1018 (1978). Effect on ovulation in cattle: P. S. Jackson, B. J. A. Furr, Res. Vet. Sci. **34**, 182 (1983). Radioimmunoassay in human serum: R. N. Clayton et al., Clin. Endocrinol. **22**, 453 (1985). Pharmacokinetics, endocrinology in humans: T. J. Perren et al., Cancer Chemother. Pharmacol. **18**, 39 (1986). Preliminary clinical trials in prostatic

carcinoma: J. M. Allen et al., Brit. Med. J. **286**, 1607 (1983); K. J. Walker et al., Lancet **2**, 413 (1983); S. R. Ahmed et al., ibid. 415; in breast cancer: P. N. Plowman et al., Brit. J. Cancer **54**, 903 (1986). Suppression of human ovarian activity: C. P. West, D. T. Baird, Clin. Endocrinol. **26**, 213 (1987).

5-oxoPro-His-Trp-Ser-Tyr-D-Ser(t-Bu)-Leu-Arg-Pro-NHNHCONH₂

Acetate, $C_{61}H_{88}N_{18}O_{16}$, Zoladex.
THERAP CAT: Treatment of prostatic carcinoma.

4434. Gossyplure. 7,11-Hexadecadien-1-ol acetate. C_{18}-$H_{32}O_2$; mol wt 280.46. C 77.09%, H 11.50%, O 11.41%. Sex pheromone of pink bollworm, Pectinophora gossypiella (Saunders): Hummel et al., Science **181**, 873 (1973). Isoln and prepn of 1:1 mixture of (Z,Z) and (Z,E) isomers: B. A. Bierl et al., J. Econ. Entomol. **67**, 211 (1974). Improved prepn: R. J. Anderson, C. A. Hendrick, U.S. pat. **3,919,329** (1975 to Zoecon); eidem, J. Am. Chem. Soc. **97**, 4327 (1975). Stereoselective synthesis of isomers: K. Mori et al., Agr. Biol. Chem. **38**, 1551 (1974); eidem, Tetrahedron **31**, 1846 (1975); H. Su, P. G. Mahany, J. Econ. Entomol. **67**, 319 (1974); H. J. Bestmann et al., Tetrahedron Letters **1976**, 353; J. M. Muchowski, C. Venuti, J. Org. Chem. **46**, 459 (1981). Activity of isomers: H. M. Flint et al., Environ. Entomol. **6**, 274 (1977). Degradn: R. D. Henson, ibid. 821.

$$CH_3(CH_2)_3CH=CH(CH_2)_2CH=CH(CH_2)_5CH_2O\overset{\displaystyle O}{\overset{\|}{C}}CH_3$$

Yellow liquid. Sol in most org solvents. Extremely flammable.
(Z,Z)-Form, bp 130-132°. n_D^{21} 1.4592.
(Z,E)-Form, bp 132-134°. n_D^{21} 1.4591.
USE: Sex attractant for pink bollworm.

4435. Gossypol. 1,1',6,6',7,7'-Hexahydroxy-3,3'-dimethyl-5,5'-bis(1-methylethyl)[2,2'-binaphthalene]-8,8'-dicarboxaldehyde; 1,1',6,6',7,7'-hexahydroxy-5,5'-diisopropyl-3,3'-dimethyl[2,2'-binaphthalene]-8,8'-dicarboxaldehyde; 2,2'-bis[1,6,7-trihydroxy-3-methyl-5-isopropyl-8-aldehydonaphthalene]; 2,2'-bis[8-formyl-1,6,7-trihydroxy-5-isopropyl-3-methylnaphthalene]. $C_{30}H_{30}O_8$; mol wt 518.54. C 69.48%, H 5.83%, O 24.68%. Poisonous pigment found in cottonseed. Name derived from the botanical name of the cotton plant, Gossypium L., Malvaceae. Isoln (0.5% yield): Campbell et al., J. Am. Chem. Soc. **59**, 1723, 1726 (1937). Process for removal from cottonseed flour: Vix et al., J. Am.Oil Chem. Soc. **48**, 611 (1971). Isoln of (+)-form from Thespesia populnea, Malvaceae: King, de Silva, Tetrahedron Letters **1968**, 261; Datta et al., Indian J. Chem. **10**, 263 (1972). Structure: Adams, J. Am. Chem. Soc. **60**, 2193 (1938); Shirley, Dean, ibid. **79**, 1205 (1957). Probably exists in three tautomeric forms. Conformation: Wood et al., Chem. & Ind. (London) **1969**, 1738. Synthesis: Edwards, ibid. **80**, 3798 (1958); idem, J. Am. Oil Chem. Soc. **47**, 441 (1970). Toxic to nonruminant animals by reducing the oxygen-carrying capacity of blood. Metabolism in rats: Abou-Donia et al., Lipids **5**, 938 (1970). Potential use as a male contraceptive agent: H. Pösö et al., Lancet **1**, 885 (1980); T. H. Maugh, Science **212**, 314 (1981); in the treatment of genital herpes: V. Wichmann et al., Am. J. Obstet. Gynecol. **142**, 593 (1982). Biochemical mechanism of action study: C.-Y. G. Lee et al., Mol. Cell. Biochem. **47**, 65 (1982). Reviews: Adams et al., Chem. Rev. **60**, 555-631 (1960); L. C. Berardi, L. A. Goldblatt, "Gossypol" in Toxic Constituents of Plant Foodstuffs, I. E. Liener, Ed. (Academic Press, New York, 2nd ed., 1980) pp 183-237.

At least 3 cryst modifications: mp 184° from ether, mp

199° from chloroform, mp 214° from ligroin. Very slightly sol in petr ether. Sol in methanol, ethanol, ether, chloroform, DMF. Freely sol (with slow decompn) in dil aq solns of ammonia and sodium carbonate. Insol in water. LD_{50} orally in rats: 2.57 g/kg, El-Nockrashy *et al., J. Am. Oil Chem. Soc.* **40**, 14 (1963).

Toxicity: May be irritating to G.I. tract. In experimental animals large doses cause edema of lungs, shortness of breath, paralysis.

USE: Proposed as rubber antioxidant, stabilizer for vinyl polymers. Potential insecticide.

4436. Gougerotin. *(R)-1-(4-Amino-2-oxo-1(2H)-pyrimidinyl)-1,4-dideoxy-4-[[3-hydroxy-2-[[(methylamino)-acetyl]amino]-1-oxopropyl]amino]-β-D-glucopyranuronamide; 1-(4-amino-2-oxo-1(2H)-pyrimidinyl)-1,4-dideoxy-4-[D-2-[2-(methylamino)acetamido]hydracrylamido]glucopyranuronamide; 1-[4-deoxy-4-(sarcosyl-D-seryl)amino-β-D-glucopyranuronamide]cytosine; aspiculamycin; asteromycin.* $C_{16}H_{25}N_7O_8$; mol wt 443.43. C 43.34%, H 5.68%, N 22.11%, O 28.87%. Antibiotic substance with antibacterial and antineoplastic activity. Isoln from *Streptomyces gougerotii:* Kanzaki *et al., J. Antibiot.* **15A**, 93 (1962). Identity with asteromycin: Ikeuchi *et al., ibid.* **25**, 548 (1972). Structure: Iwasaki, *Yakugaku Zasshi* **82**, 1358 (1962). Revised structure: Fox *et al., Tetrahedron Letters* **1968**, 6029; Watanabe *et al., Chem. Pharm. Bull.* **17**, 416 (1969). Total synthesis: *eidem, J. Am. Chem. Soc.* **94**, 3272 (1972); Lichtenthaler *et al., Tetrahedron Letters* **1975**, 3527. Identity with aspiculamycin: Lichtenthaler *et al., ibid.* 665. Mechanism of action study: J. C. Lacal *et al., J. Antibiot.* **33**, 441 (1980). *Reviews:* Clark in *Antibiotics*, vol. 1, D. Gottlieb, P. D. Shaw, Eds. (Springer-Verlag, New York, 1967) pp 278-282; Yukioka, *ibid.* vol. 3, J. W. Corcoran, F. E. Hahn, Eds. (1975) pp 448-458.

Needles, mp 211-217° (dec). $[\alpha]_D^{27}$ +53° (c = 0.8). uv max (water): 267, 235 nm (ϵ 9400, 9300); in 0.1N HCl: 275 nm (ϵ 13600); in 0.1N NaOH: 267 nm (ϵ 9800). LD_{50} in mice (mg/kg): 57 i.v. (Kanzaki).

4437. G-Proteins. GTP binding proteins. Distinct class of membrane associated *guanine nucleotide binding proteins* characterized by their function as couplers between a wide variety of receptors and their effector molecules in transmembrane signalling pathways. An example is the retinal G-protein, *transducin*, which links the photon receptor, rhodopsin, *q.v.*, to cGMP phosphodiesterase. G-Proteins are heterotrimeric, with apparent mol wt of 100 kDa and composed of α, β, γ subunits. The α subunit contains the guanine nucleotide binding site, possesses GTPase activity, and is specific for each G-protein. β and γ subunits form a noncovalent, membrane attached complex. *Reviews:* A. M. Spiegel, *Mol. Cell. Endocrinol.* **49**, 1-16 (1987); *idem, Ann. Rep. Med. Chem.* **23**, 235-242 (1988); P. J. Casey, A. G. Gilman, *J. Biol. Chem.* **263**, 2577-2580 (1988); H. R. Bourne, *Nature* **337**, 504-505 (1989). Review of role in disease: A. C. Dolphin, *Trends Neurosci.* **10**, 53-57 (1987).

4438. Gramicidin(s). Gramicidin D (Dubos); linear gramicidins; Gramoderm. Polypeptide antibiotic complex first isolated from the mixture tyrothricin *(q.v.)* along with tyrocidine *(q.v.)* from cultures of *Bacillus brevis:* Dubos, Hotchkiss, *J. Exp. Med.* **73**, 629 (1941); *eidem, J. Biol. Chem.* **141**, 155 (1941). Commercial extraction: Baron, U.S. pat. **2,534,541** (1950 to Penick). Commercial prepara-

tion is a mixture of the four components, gramicidin A, B, C, and D, comprising about 87.5, 7.1, 5.1, 0.3 percent resp: Gross, Witkop, *Biochemistry* **4**, 2495 (1965). Each of the components A, B, and C consist of 2 chains, one with valine in position 1, comprising 80-95% of the component, and the other with isoleucine in position 1. Structure, characterization, and synthesis of valine- and isoleucine-gramicidin A: Sarges, Witkop, *J. Am. Chem. Soc.* **86**, 1862 (1964); **87**, 2011, 2020 (1965); Bauer *et al., Biochemistry* **11**, 3266 (1972). Structure of gramicidin B: Sarges, Witkop, *J. Am. Chem. Soc.* **87**, 2027 (1965); of gramicidin C: *eidem, Biochemistry* **4**, 2491 (1965). Synthesis of valine-gramicidin B and C: K. Noda, E. Gross in *Chemistry and Biology of Peptides, Proc. 3rd Am. Peptide Symp.*, J. Meienhofer Ed. (Ann Arbor Science Publishers, Michigan, 1972) pp 241-250. *Review:* Hunter, Schwartz, "Gramicidins" in *Antibiotics* I, S. Gottlieb, P. Shaw, Eds. (Springer-Verlag, New York, 1967) pp 642-648. Comprehensive description: G. A. Brewer in *Analytical Profiles of Drug Substances* vol. 8, K. Florey, Ed. (Academic Press, New York, 1979) pp 179-218.

valine-gramicidin A

Spear-shaped or lenticular platelets, mp 229-230°. Almost insol in water (0.6 mg/100 ml). Soluble in the lower alcohols, acetic acid, pyridine. Moderately sol in dry acetone and dioxane. Practically insol in ether, hydrocarbons. Tends to form colloidal suspensions in water.

THERAP CAT: Antibacterial.

THERAP CAT (VET): Antimicrobial.

4439. Gramicidin S. Gramicidin S (Soviet); gramicidin C (Soviet). $C_{60}H_{92}N_{12}O_{10}$; mol wt 1141.49. C 63.13%, H 8.12%, N 14.72%, O 14.02%. Cyclic decapeptide antibiotic produced by a strain of *Bacillus brevis*. Isoln: Gause *et al., Compt. Rend. Acad. Sci. USSR* **43**, 217 (1944), *C.A.* **39**, 1195 (1945); Gause, Brazhnikova, *Lancet* **247**, 715 (1944). More closely related to tyrocidines, *q.v.*, in biological and chemical properties than to true gramicidins, *q.v.* Structure: Synge, *Biochem. J.* **39**, 363 (1945); Consden *et al., ibid.* **40**, xliii (1946); **41**, 596 (1947); Battersby, Craig, *J. Am. Chem. Soc.* **73**, 1887 (1951); Erlanger, Goode, *Nature,* **174**, 840 (1954). Synthesis and absorption spectrum: Schwyzer, Sieber, *Helv. Chim. Acta* **40**, 624 (1957); Waki, Izuniya, *Bull. Chem. Soc. Japan* **40**, 1687 (1967). Solid phase synthesis: Losse, Neubert, *Tetrahedron Letters* **1970**, 1267; M. Ohno *et al., J. Am. Chem. Soc.* **93**, 5251 (1971). Improved synthesis via a linear pentapeptide: Y. Minematsu *et al., Tetrahedron Letters* **1980**, 2179; via a linear decapeptide: T. Mukaiyama *et al., Chem. Letters* **1981**, 1367. Industrial procedure: Brit. pat. **836,725** (1960 to Ciba). *Review:* Y. A. Ovchinnikov, V. T. Ivanov, "The Cyclic Peptides: Structure, Conformation, and Function" in *The Proteins* vol. V, H. Neurath, R. L. Hill, Eds. (Academic Press, New York, 3rd ed., 1982) pp 547-555.

Hydrochloride, $C_{60}H_{92}N_{12}O_{10} \cdot 2HCl$, prisms from ethanol + aq HCl, dec 277-278°. $[\alpha]_D^{24}$ −289° (c = 0.43 in 70% ethanol). Freely sol in alcohol; slightly sol in acetone. Practically insol in water, acids, alkalies. LD_{50} i.p. in rats: 17 mg/kg (Gause, Brazhnikova).

THERAP CAT: Topical antibacterial.

4440. Gramine. *N,N-Dimethyl-1H-indole-3-methanamine; 3-(dimethylaminomethyl)indole;* Donaxine. $C_{11}H_{14}$-N_2; mol wt 174.24. C 75.82%, H 8.10%, N 16.08%. In chlorophyll-deficient mutants of barley: Euler *et al., Z. Physiol. Chem.* **217,** 23 (1933). In the Asiatic reed *Arundo donax* L., *Gramineae:* Orechoff, Norkina, *Ber.* **68,** 436 (1935). From *Acer saccharinum* L. (the Silver Maple) and *A. rubrum* L., *Aceraceae:* Pachter *et al., J. Org. Chem.* **24,** 1285 (1959); Pachter, *J. Am. Pharm. Assoc., Sci. Ed.* **48,** 670 (1959). Synthesis: Kühn, Stein, *Ber.* **70,** 567 (1937). Biosynthesis from tryptophan in barley: Bowden, Marion, *Can. J. Chem.* **29,** 1037 (1951); O'Donovan, Leete, *J. Am. Chem. Soc.* **85,** 461 (1963); Gower, Leete, *ibid.* 3683; *see also* Gross *et al., Tetrahedron Letters* **1971,** 4047.

Shiny, flat needles or plates from acetone, mp 138-139°. Absorption spectrum: Kanakoa *et al., Chem. Pharm. Bull.* **8,** 294 (1960). Sol in alcohol, ether, chloroform; slightly sol in cold acetone; practically insol in petr ether, water.

Hydrochloride, $C_{11}H_{14}N_2 \cdot HCl$, crystals from ethanol + ether, dec 191°. Sol in water.

Methiodide, $C_{11}H_{14}N_2 \cdot CH_3I$, crystals from methanol + benzene, mp 168-169°. Sol in water: Geissman, Armen, *J. Am. Chem. Soc.* **74,** 3916 (1952).

4441. Granaticin. *[3aS-(3aα,5α,8α,9α,11β,13bβ,15S*)]-3,3a,5,8,11,13b-Hexahydro-7,8,12,15-tetrahydroxy-5,9-dimethyl-8,11-ethanofuro[2,3-e]naphtho[2,3-c:6,7-c']dipyran-2,6,13(9H)-trione;* antibiotic WR 141; litmomycin. $C_{22}H_{20}$-O_{10}; mol wt 444.40. C 59.46%, H 4.54%, O 36.00%. Antibiotic substance produced by *Streptomyces olivaceus* from soil of Portuguese West Africa. Isoln and antibacterial activity: R. Corbaz *et al., Helv. Chim. Acta* **40,** 1262 (1957). Determn by microbiological diffusion assay: A. Ricicova, M. Podojil, *Folia Microbiol.* **10,** 299 (1965). Isoln of *granaticin B*, the α-L-rhodinoside of granaticin: S. Barcza *et al., Helv. Chim. Acta* **49,** 1736 (1966). Fr. **pat. 1,525,993;** W. Keller, H. Zaehner, U.S. pat. **3,836,642** (1968, 1974 both to Ciba-Geigy). Structure of granaticin and granaticin B: W. Keller-Schierlein *et al., Helv. Chim. Acta* **51,** 1257 (1968); M. Brufani, M. Dobler, *ibid.* 1269; *Naturally Occurring Quinones,* R. H. Thomson, Ed. (Academic Press, New York, 2nd ed., 1971) pp 298-302. Identity of granaticin with antibiotic litmomycin: C.-J. Chang *et al., J. Antibiot.* **28,** 156 (1975). Biosynthesis: C. E. Snipes *et al., J. Nat. Prod.* **42,** 627 (1979); *eidem, J. Am. Chem. Soc.* **101,** 701 (1979). Total synthesis of (±)-form: K. Nomura *et al., ibid.* **109,** 3402 (1987); of the natural (−)-form: K. Okazaki *et al., Chem. Commun.* **1989,** 354. Cytotoxic action on carcinoma cells: E. Sturdik, L. Drobnica, *Neoplasma* **30,** 3 (1983). Inhibition of RNA synthesis: A. Ogilvie *et al., Biochem. J.* **152,** 517 (1975); P. Heinstein, *J. Pharm. Sci.* **71,** 197 (1982).

Deep red, garnet-like crystals from acetone, dec 204-206°. Also reported as mp 211-213° (dec). Acts as an indicator: red in acids, blue in alkalies. Absorption max (abs ethanol): 223, 286, 532, 576 nm (log ε 4.58, 3.76, 3.87, 3.75).

Tetraacetylgranaticin, $C_{30}H_{28}O_{14}$, yellow crystals from alc, mp 242-243°. $[\alpha]_D^{20}$ −100° (c = 0.818 in chloroform).

Granaticin B, $C_{28}H_{30}O_{12}$, red crystalline solid from methanolic-HCl, mp 117-119°. $[\alpha]_D^{22}$ 17.2° (c = 0.83 in pyridine). Absorption max (methanol): 223, 285, 527, 566 nm (log ε 4.42, 3.68, 3.76, 3.57).

4442. Grandisol. *(1R-cis)-1-Methyl-2-(1-methylethenyl)cyclobutaneethanol; cis-(+)-2-isopropenyl-1-methylcyclobutaneethanol;* (+)-(1R,2S)-1-(2'-hydroxyethyl)-1-methyl-2-isopropenylcyclobutane. $C_{10}H_{18}O$; mol wt 154.25. C 77.87%, H 11.76%, O 10.37%. Major component of *grandlure,* the sex attractant of the boll weevil (*Anthonomus grandis,* Boheman). Isoln and synthesis: J. H. Tumlinson *et al., Science* **166,** 1010 (1969); *eidem, J. Org. Chem.* **36,** 2616 (1971). Synthesis of optically active grandisol: P. D. Hobbs, P. D. Magnus, *Chem. Commun.* **1974,** 856; *eidem, J. Am. Chem. Soc.* **98,** 4594 (1976); K. Mori, *Tetrahedron* **34,** 915 (1978); of enantiomerically pure grandisol: J. B. Jones *et al., Can. J. Chem.* **60,** 2007 (1982). Synthesis of racemate: B. M. Trost *et al., J. Am. Chem. Soc.* **99,** 3088 (1977). Short stereoselective synthesis of (±)-grandisol: I. Aljancic-Solaja *et al., Helv. Chim. Acta* **70,** 1302 (1987). Review of syntheses: J. A. Katzenellenbogen, *Science* **194,** 139-148 (1976); J. M. Brand *et al., Fortschr. Chem. Org. Naturst.* **37,**18-29 (1979), *see also* refs pp 157-190; K. Mori, "The Synthesis of Insect Pheromones" in *The Total Synthesis of Natural Products* vol. **4,** J. ApSimon, Ed. (Wiley-Interscience, New York, 1981) pp 80-85.

Liquid, $bp_{1.0}$ 50-60°. $[\alpha]_D^{21.5}$ +18.5° (c = 1 in hexane). n_D^{20} 1.4748.

4443. Granisetron. *endo-1-Methyl-N-(9-methyl-9-azabicyclo[3.3.1]non-3-yl)-1H-indazole-3-carboxamide.* C_{18}-$H_{24}N_4O$; mol wt 312.41. C 69.20%, H 7.74%, N 17.93%, O 5.12%. Specific serotonin (5HT$_3$) receptor antagonist. Prepn: F. D. King, *Eur. pat. Appl.* **200,444** (1986 to Beecham). 5HT$_3$-receptor binding study: G. J. Kilpatrick *et al., Nature* **330,** 746 (1987). Clinical evaluations as antiemetic for cancer chemotherapy patients: R. A. Joss *et al., J. Nat. Cancer Inst.* **80,** 1340 (1988); J. Carmichael *et al., Brit. Med. J.* **297,** 110 (1988).

Hydrochloride, $C_{18}H_{25}ClN_4O$, *BRL 43694A.* mp 290-292°.

THERAP CAT: Antiemetic.

4444. Graphite. Plumbago; black lead; mineral carbon. Obtained by mining, especially in Canada and Ceylon. Monograph: A. R. Ubbelohde, F. A. Lewis, *Graphite and Its Crystal Compounds* (Oxford, 1960). *Review:* Holliday *et al.* in *Comprehensive Inorganic Chemistry* vol. **1,** J. C. Bailar, Jr. *et al.,* Eds. (Pergamon Press, Oxford, 1973) pp 1250-1294.

Crystallized carbon with traces of Fe, SiO_2, etc. Usually soft, black scales, crystals rare. d 2.09-2.23. Mohs' hardness = 1.0. Commercial varieties usually withstand temps up to 2820°. Sol in molten iron.

USE: For "lead" pencils, refractory crucibles, stove polish; as pigment, lubricant, graphite cement; for matches and explosives, commutator brushes, anodes, arc-lamp carbons, electroplating; polishing compds, rust and needle-paper; coating for cathode ray tubes; moderator in nuclear piles. *Caution:* The dust is mildly irritating to lungs.

4445. Graphitic Acid. Graphite oxide; graphitic oxide. This material, obtained by oxidation of graphite, was first prepd by Brodie in 1859; Hummers, Offeman, *J. Am. Chem. Soc.* **80**, 1339 (1958). Its composition is not well defined but usually given as $C_4O(OH)$: Aragon de la Cruz, Cowley, *Nature* **196**, 468 (1962), *Acta Cryst.* **16**, 531 (1963). Prepn and manuf: Hummers, U.S. pat. **2,798,878** (1957 to National Lead); Hummers, Offeman, *loc. cit.*; Ruskin, U.S. pats. **2,933,381** and **2,944,881** (both 1960 to Union Carbide). Crystal structure: Aragon de la Cruz, Cowley, *loc. cit.*

Very light to dark brown, or yellowish-brown solid.

USE: In rocket propellant mixtures, Ruskin, *loc. cit.*

4446. Gratiogenin. *20,24-Epoxy-3,25-dihydroxy-9-methyl-19-nor-9β-lanost-5-en-11-one.* $C_{30}H_{48}O_4$; mol wt 472.68. C 76.22%, H 10.24%, O 13.54%. Tetracyclic triterpene closely related to the cucurbitacins, *q.v.* Prepd by hydrolysis of gratioside: Retzlaff, *Arch. Pharm.* **240**, 561 (1902). Structural studies: Tschesche, Heesch, *Ber.* **85**, 1067 (1952). Revised structure: Tschesche *et al.*, *Ann.* **674**, 196 (1964).

Clusters of needles from methanol or ethanol, mp 191-196°. $[\alpha]_D^{20}$ +168° (c = 1.193 in chloroform); $[\alpha]_D^{20}$ +151° (c = 0.863 in methanol). Also reported as mp 202-203°. $[\alpha]_D^{25}$ +175 ± 3° (chloroform), Tschesche *et al.*, *loc. cit.* Freely sol in methanol, ethanol, chloroform, acetone, pyridine. Practically insol in water, ether, petr ether.

4447. Gratioside. $C_{42}H_{68}O_{14}$; mol wt 797.00. C 63.29%, H 8.60%, O 28.11%. Triterpene glycoside monohydrate from tubers of *Gratiola officinalis* L., *Scrophulariaceae:* Retzlaff, *Arch. Pharm.* **240**, 561 (1902); Tschesche *et al.*, *Ann.* **674**, 196 (1964). Dec on hydrolysis into 1 mol gratiogenin and 2 mol glucose: Tschesche, Heesch, *Ber.* **85**, 1067 (1952).

Monohydrate, clusters of crystals from ethanol, mp 268-274°; $[\alpha]_D^{20}$ +75° (c = 0.825 in alc). Also reported mp 286-289°; $[\alpha]_D^{21}$ 81° (c = 0.83 in ethanol), Tschesche *et al.*, *loc. cit.* uv max (methanol): 295 nm (log ε 1.84). Slightly sol in water, freely in alcohol; practically insol in ether.

4448. Gravitol(e). *2-(2-Allyl-6-methoxyphenoxy)triethylamine;* 2-methoxy-6-allylphenol diethylaminoethyl ether; Clavitol; Uterol. $C_{16}H_{25}NO_2$; mol wt 263.39. C 72.96%, H 9.57%, N 5.32%, O 12.15%. Prepn: Slotta, *Grundriss der modernen Arzneistoff-Synthese* (Stuttgart, 1931) p 118; H. Hahl, Ger. pat. **433,182** (1926 to I.G. Farbenind.), *Frdl.* **15**, 1500; *Chem. Zentr.* **1926,** II, 2223; H. Pal'gi, *Zh. Obshch. Khim. Eng. Ed.* **28**, 2275 (1958). Pharmacology: E. Käer, G. Barkan, *Arch. Exp. Pathol. Pharmakol.* **170**, 111 (1933).

Oily liq. bp_{10} 160-161°; bp_{2-3} 141-144°. $n_D^{24.5}$ 1.5075. Slightly sol in water.

Hydrochloride, $C_{16}H_{26}ClNO_2$, crystals, freely sol in water.

4449. Grayanotoxins. Toxic diterpenoids present in leaves of the various species of *Rhododendron, Kalmia,* and *Leucothoe, Ericaceae;* also found in honey from rhododen-

dron flowers. Eighteen grayanotoxins have been isolated, the first three being the most important. Isoln of grayanotoxins I, II, III: Kakisawa *et al.*, *Tetrahedron* **21**, 3091 (1965); of IV and V: Okuno *et al.*, *ibid.* **26**, 4765 (1970); of V, VI, and VII: Hikino *et al.*, *Chem. Pharm. Bull.* **18**, 2357 (1970); of VIII, IX, X, and XI: Hikino *et al.*, *ibid.* **19**, 1289 (1971); of XII and XIII: Hikino *et al.*, *ibid.* **20**, 422 (1972). Approaches to synthesis of the grayanotoxin skeleton: T. Shiozaki *et al.*, *Tetrahedron Letters* **1972**, 657; T. Kametani *et al.*, *Chem. Pharm. Bull.* **27**, 152 (1979); *eidem*, *Tetrahedron Letters* **22**, 2379 (1981); *eidem*, *Tetrahedron* **37**, 3813 (1981).

Grayanotoxin I, $C_{22}H_{36}O_7$, *grayanotoxane-3,5,6,10,14,16-hexol 14-acetate, G-I,* acetylandromedol, andromedotoxin, rhodotoxin, asebotoxin. R_1 = OH; R_2 = CH_3; R_3 = COCH₃. From *Leucothoe grayana* Max., *Ericaceae:* Miyajimi, Takei, *J. Agr. Chem. Soc. Japan* **10**, 1093 (1934); from *Rhododendron maximum* L., *Ericaceae:* Wood *et al.*, *J. Am. Chem. Soc.* **76**, 5689 (1954). Identity with acetylandromedol, andromedotoxin and rhodotoxin: Tallent *et al.*, *ibid.* **79**, 4548 (1957). Stereochemistry: Iwasa, Nakamura, *Tetrahedron Letters* **1969**, 3973; Narayanan *et al.*, *ibid.* **1970**, 3943; Hikino *et al.*, *Chem. Pharm. Bull.* **18**, 1071 (1970). Approach to synthesis: Okuno, Matsumoto, *Tetrahedron Letters* **1969**, 4077. Crystals from ethyl acetate, mp 258-260° to 267-270°, depending on rate of heating. $[\alpha]_D^{25}$ −8.8° (c = 2.3 in ethanol). Sol in hot water, alcohol, acetic acid, hot chloroform; very slightly sol in benzene, ether, petr ether. Has hypotensive action: Moran *et al.*, *J. Pharmacol. Exp. Ther.* **110**, 415 (1954). LD_{50} i.p. in mice: 1.31 mg/kg, H. Hikino *et al.*, *Toxicol. Appl. Pharmacol.* **35**, 303 (1976).

Grayanotoxin II, $C_{20}H_{32}O_5$, *grayanotox-10(20)-ene-3,5,6,-14,16-pentol, G-II, deacetylanhydroandromedotoxin.* R_1R_2 = CH_2=; R_3 = H. From *L. grayana* Max., *Ericaceae:* Miyajimi, Takei, *loc. cit.* Identity with deacetylanhydromedotoxin: Meguri, *Yakagaku Zasshi* **79**, 1060 (1959); *C.A.* **54**, 5599g (1960). Stereochemistry: Iwasa, Nakamura, *loc. cit.*; Kumazawa, Iriye, *Tetrahedron Letters* **1970**, 927; Yasue *et al.*, *Chem. Pharm. Bull.* **18**, 2586 (1970). Synthesis: S. Gasa *et al.*, *Tetrahedron Letters* **1976**, 553. Columns, mp 199-200°. $[\alpha]_D^{28}$ −41.88°. LD_{50} i.p. in mice: 26.1 mg/kg, H. Hikino *et al.*, *loc. cit.*

Grayanotoxin III, $C_{20}H_{34}O_6$, *grayanotoxane-3,5,6,10,14,16-hexol, G-III, deacetylandromedotoxin.* R_1 = OH; R_2 = CH_3; R_3 = H. From *L. grayana* Max., *Ericaceae:* Miyajimi, Takei, *J. Agr. Chem. Soc. Japan* **12**, 947 (1936), *C.A.* **30**, 6747⁹ (1936). Stereochemistry: Hikino *et al.*, *Chem. Pharm. Bull.* **18**, 1071 (1970). LD_{50} i.p. in mice: 0.84 mg/kg, H. Hikino *et al.*, *loc. cit.*

4450. Grindelia. Gum-plant (of California). Dried leaves and flowering tops of *Grindelia camporum* Greene or of *G. humilis* H. & A. (*G. cuneifolia* Auth.), *Compositae.* *Habit.* North America (California). *Constit.* Volatile oil, over 20% resin, grindelol, saponin, tannin, robustic acid.

THERAP CAT: Expectorant.

4451. Grindelic Acid. *4,4′a,5,5′,6′,7′,8′,8′a-Octahydro-2′,5,5′,5′,8′a-pentamethylspiro[furan-2(3H),1′(4′H)-naphthalene]-5-acetic acid; 9,13-epoxylabd-7-en-15-oic acid.* $C_{20}H_{32}O_3$; mol wt 320.46. C 74.96%, H 10.06%, O 14.98%. From the resin of *Grindelia robusta* Nutt., *Compositae.* Isoln and structure: Panizzi *et al.*, *Gazz. Chim. Ital.* **92**, 522 (1962). Configuration: Mangoni, Belardini, *ibid.* **92**, 1379 (1962); **93**, 455, 465 (1963). Synthesis: M. Adinolfi, *et al.*, *ibid.* **106**, 625 (1976).

Crystals from acetic acid, mp 100-101°. $[\alpha]_D$ −102.2°. Methyl ester, $C_{21}H_{34}O_3$, crystals from methanol, mp 70-70.5°. $[\alpha]_D$ −134.1° (c = 1.46 in methanol).

4452. Grisein. Antibiotic substance produced by strains of *Streptomyces griseus.* Isoln: Reynolds *et al., Proc. Soc. Exp. Biol. Med.* **64,** 50 (1947); Reynolds, Waksman, *J. Bacteriol.* **55,** 739 (1948). Improved method of isoln: F. A. Kuehl, L. Chaiet, U.S. pat. **2,505,053** (1950 to Merck); F. A. Kuehl *et al., J. Am. Chem. Soc.* **73,** 1770 (1951). Analysis of composition: $C_{40}H_{61}FeN_{10}O_{20}S$. Degradation of grisein by acid hydrolysis yielded 3-methyluracil and at least two amino acids. One of the acids appears to be glutamic acid: F. A. Kuehl *et al., loc. cit.* Probably is a mixture of components; similar or identical to albomycin (*q.v.*): Stapley, Ormond, *Science* **125,** 587 (1957); Turková *et al., Coll. Czech. Chem. Commun.* **31,** 2444 (1966). Toxicity study: V. I. Aksenov, *Veterinariya (Moscow)* **12,** 93 (1974), *C.A.* **83,** 54127d (1975). Biosynthesis: V. V. Kuklin *et al., Antibiotiki* **25,** 403 (1980), *C.A.* **93,** 146173a (1980).

Amorphous red powder. Sol in water; slightly sol in 95% alcohol; insol in abs alcohol, ether, acetone, chloroform, benzene. The activity remains unchanged when an aq soln is heated to 100° for 10 min. LD_{50} in mice (mg/kg): 600 orally; 34 s.c., V. I. Aksenov, *loc. cit.*

4453. Griseofulvin. *7-Chloro-2',4,6-trimethoxy-6'-methylspiro[benzofuran-2(3H),1'-[2]cyclohexene]-3,4'-dione; 7-chloro-4,6-dimethoxycoumaran-3-one-2-spiro-1'-(2'-methoxy-6'-methylcyclohex-2'-en-4'-one);* amudane; Curling factor; Fulcin; Fulvicin; Grifulvin; Grisactin; Griséfuline; Grisovin; Gris-PEG; Grysio; Lamoryl; Likuden; Neo-Fulcin; Polygris; Poncyl; Spirofulvin; Sporostatin. C_{17}-$H_{17}ClO_6$; mol wt 352.77. C 57.88%, H 4.86%, Cl 10.05%, O 27.21%. Antibiotic substance produced by *Penicillium griseofulvum* Dierckx and by *P. janczewskii* Zal. [= *P. nigricans* (Banier)Thom]. Isoln: Oxford *et al., Biochem. J.* **33,** 240 (1939); Brian *et al., Trans. Brit. Mycol. Soc.* **29,** 173 (1946); Hockenhull, Dorey *et al.,* U.S. pats. **3,069,328, 3,069,329** (both 1962 to Glaxo). Structure: Grove *et al., Chem. & Ind. (London)* **1951,** 219; *J. Chem. Soc.* **1952,** 3977. Stereochemistry: MacMillan, *ibid.* **1959,** 1823; Brown, Sim, *ibid.* **1963,** 1050. Total synthesis: Brossi *et al., Helv. Chim. Acta* **43,** 1444, 2071 (1960); Taub *et al., Tetrahedron* **19,** 1 (1963); Stork, Tomasz, *J. Am. Chem. Soc.* **86,** 471 (1964); S. Danishefsky, F. J. Walker, *ibid.* **101,** 7018 (1979). Conformation: Levine, Hicks, *Tetrahedron Letters* **1971,** 311. Crystal structure: G. Malmros *et al., Cryst. Struct. Commun.* **6,** 463 (1977). Review and evaluation of studies of carcinogenic action in laboratory animals: *IARC Monographs* **10,** 153-161 (1976). *Review:* Grove, *Quart. Rev.* **17,** 1 (1963); Huber in *Antibiotics* vol. **3,** J. W. Corcoran, F. E. Hahn, Eds. (Springer-Verlag, New York, 1975) pp 606-613. Comprehensive description: E. R. Townley in *Analytical Profiles of Drug Substances* vol. **8,** K. Florey, Ed. (Academic Press, New York, 1979) pp 219-249.

Stout octahedra or rhombs from benzene, mp 220°. $[\alpha]_D^{17}$ +370° (satd $CHCl_3$ soln). uv max: 286, 325 nm. Soly in DMF at 25°: 12 to 14 g/100 ml. Slightly sol in ethanol, methanol, acetone, benzene, $CHCl_3$, ethyl acetate, acetic acid. Practically insol in water, petr ether.

THERAP CAT: Antifungal.

THERAP CAT (VET): Antifungal antibiotic.

4454. Griseoviridin. GV. $C_{22}H_{27}N_3O_7S$; mol wt 477.55. C 55.33%, H 5.70%, N 8.80%, O 23.45%, S 6.72%. Antibiotic substance obtained together with Etamycin (viridogrisein) from *Streptomyces griseus:* Bartz *et al., Antibiot. Ann.* **1954-1955,** 777. Characterization: Ames *et al., J. Chem. Soc.* **1955,** 4260; Ames, Bowman, *ibid.* **1955,** 4264. Structure studies: Fallona *et al., J. Am. Chem. Soc.* **84,** 4162 (1962); *eidem, Can. J. Chem.* **42,** 371, 394 (1964). Revised structure: G. I. Birnbaum, S. R. Hall, *J. Am. Chem. Soc.* **98,** 1926 (1976). Absolute configuration: B. W. Bycroft, T. J. King, *J. Chem. Soc. Perkin Trans. I* **1976,** 1996. Production of griseoviridin and viridogrisein: Bartz *et al.,* U.S. pat. **3,023,-204** (1962 to Parke, Davis). Synthetic study: A. I. Meyers, R. A. Amos, *J. Am. Chem. Soc.* **102,** 870 (1980).

Polymorphic crystals, dec 158-166°, or 194-200°, or 230-240° depending on the crystal modification. $[\alpha]_D^{27}$ −237° (c = 0.5 in methanol). uv max (methanol): 221 nm ($E_{1cm}^{1\%}$ 870). Sol in pyridine, moderately sol in lower alcohols, sparingly sol in water and nonpolar solvents.

Diacetate, $C_{26}H_{31}N_3O_9S$, needles from methanol + ether + petr ether, dec 137-140°, $[\alpha]_D^{27}$ −230° (c = 0.44 in methanol). uv max (ethanol): 218 nm (ϵ 41,800).

4455. Guaiac. Gum guaiac; resin guaiac; guaiacum. Resin from wood of *Guajacum officinale* L. or *G. sanctum* L., *Zygophyllaceae. Constit.* About 70% α- and β-guaiaconic acids, about 11% guaiacic acid, related compds and guaiaretic acid, 15% vanillin, guaiac yellow, guaiac saponin (guaiacin). Use as clinical reagent for occult blood: R. H. Wilkinson, W. A. F. Penfold, *Lancet* **2,** 847 (1969). Acute toxicity: P. M. Jenner *et al., Food Cosmet. Toxicol.* **2,** 327 (1964).

Brown or greenish-brown, irregular lumps. mp 85-90°. Insol in water. Freely sol in alcohol, chloroform, ether, creosote, solns of chloral hydrate, alkalies; slightly sol in benzene, carbon disulfide. *Incompat:* Of liquid preparations: Mineral acids, acacia, ferric chloride, gold chloride, permanganates, spirit nitrous ether, water. LD_{50} orally in rats: > 5000 mg/kg (Jenner).

USE: Clinical reagent (blood or hemoglobin).

4456. Guaiac-Copper Sulfate Paper. Schönbein-Pagenstecher's paper. White filter paper impregnated first with an alc soln of guaiac resin, then after drying, with an aq soln of cupric sulfate. Used for detecting HCN, a trace of which colors the paper blue.

4457. Guaiacol. *2-Methoxyphenol;* methylcatechol; o-hydroxyanisole; 1-hydroxy-2-methoxybenzene; Anastil. $C_7H_8O_2$; mol wt 124.13. C 67.73%, H 6.49%, O 25.78%. Isolated from guaiac resin: Sobrero, *Ann.* **48,** 19 (1843); from hardwood tar: McGinness *et al., Tappi* **43,** 1027 (1960). Prepd by mercuric oxide oxidation of lignin: Lewis, Pearl, U.S. pat. **2,433,227** (1947 to Sulphite Prod.); by oxidation of anisole with trifluoroperoxyacetic acid: McClure, Williams, *J. Org. Chem.* **27,** 627 (1962); from acetovanillone + $ZnCl_2$: Read, U.S. pat. **3,057,927** (1962 to Ontario Res. Found.); from the diazonium salt of o-anisidine: Herbst, Ger. pat. **1,148,236** (1963 to Hoechst). Toxicity data: Taylor *et al., Toxicol. Appl. Pharmacol.* **6,** 378 (1964).

White or slightly yellow cryst mass or colorless to yellowish, very refractive liquid; characteristic odor. Darkens on exposure to air and light. d (crystals) 1.129; d (liq) ~1.112. Solidif 28°, but may remain liq for a long time even at a much lower temp. bp 204-206°; bp_4 53-55°. One gram dissolves in 60-70 ml water, 1 ml glycerol; miscible with alcohol, chloroform, ether, oils, glacial acetic acid. Slightly sol in petr ether; sol in NaOH soln; with moderately concd KOH it forms a sparingly sol compd. *Protect from light.* LD_{50} orally in rats: 725 mg/kg (Taylor).

Phenylacetate, $C_{15}H_{14}O_3$, *Gujaphenyl, Gunyl.*

THERAP CAT: Expectorant.

THERAP CAT (VET): Expectorant.

4458. Guaiacol Benzoate. *2-Methoxyphenol benzoate;* benzoylguaiacol; Benzosol. $C_{14}H_{12}O_3$; mol wt 228.24. C 73.67%, H 5.30%, O 21.03%. $CH_3OC_6H_4OOCC_6H_5$. Prepn: Lynch, Moore, *Can. J. Chem.* **40**, 1461 (1962).

Odorless, almost tasteless, cryst powder. mp 57-58°. Slightly sol in water; sol in hot alcohol, ether, chloroform.

THERAP CAT: Expectorant.

4459. Guaiacol Carbonate. Carbonic acid bis(2-methoxyphenyl) ester; guaiacol carbonic acid neutral ester; carbonic acid guaiacol ether; Duotal. $C_{15}H_{14}O_5$; mol wt 274.26. C 65.69%, H 5.14%, O 29.17%. Prepd by the action of phosgene on a concd soln of guaiacol in aq NaOH: **Ger. pat. 58,129** (to Bayer); S. P. Schotz, *Synthetic Organic Compounds* (London, 1925) p 137; J. Schwyzer, *Die Fabrikation Pharmazeutischer und Chemischtechnischer Produkte* (Berlin, 1931) p 212; H. P. Kaufmann, *Arzneimittel-Synthese* (Springer-Verlag, 1953) p 607.

Odorless needles from ethanol, mp 88.1°. Faint taste of guaiacol. Practically insol in water. At room temp 1 g dissolves in 60 ml ethanol, 1 ml chloroform, 18 ml ether. Much more sol in hot alcohol, hot benzene, hot chloroform. Slightly sol in liq fatty acids. At 20°, 100 g of glycerol (d 1.2612) will dissolve 0.043 g of guaiacol carbonate. Difficult to hydrolyze. Upon ingestion much is excreted unchanged.

THERAP CAT: Expectorant.

THERAP CAT (VET): Expectorant.

4460. Guaiacol Phosphate. $C_{21}H_{21}O_7P$; mol wt 416.36. C 60.57%, H 5.08%, O 26.90%, P 7.44%. $(CH_3OC_6H_4O)_3PO$. Obtained by the action of $POCl_3$ upon guaiacol sodium: Kucherov, *Zh. Obshch. Khim.* **19**, 126 (1949). Prepn of sodium salt: Auger, Dupuis, *Compt. Rend.* **146**, 1151 (1908).

Crystals, mp 91°. bp_3 275-280°. Practically insol in water, ether. Sol in hot alc, freely in chloroform, toluene, acetone.

Sodium salt, *Novocol.*

THERAP CAT: Expectorant.

4461. Guaiacol Valerate. $C_{12}H_{16}O_3$; mol wt 208.25. C 69.21%, H 7.75%, O 23.05%. $CH_3OC_6H_4OOCC_4H_9$. Has antibacterial properties: Kellner, Kober, *Arzneimittel-Forsch.* **5**, 224 (1955).

Yellowish liq. d 1.05. bp 265°. Practically insol in water. Freely sol in alcohol, benzene, chloroform, ether.

USE: Detection of copper: Shapiro, *Zh. Analit. Khim.* **17**, 248 (1962).

4462. Guaiactamine. *N,N-Diethyl-2-(2-methoxyphenoxy)ethanamine; 2-(o-methoxyphenoxy)triethylamine;* 2-(2-methoxyphenoxy)ethyldiethylamine. $C_{13}H_{21}NO_2$; mol wt 223.31. C 69.92%, H 9.48%, N 6.27%, O 14.33%. Prepn: Wright, Moore, *J. Am. Chem. Soc.* **73**, 2281 (1951).

bp_2 126-127°.

Hydrochloride, $C_{13}H_{22}ClNO_2$, crystals, mp 114-115°.

Citrate, combination with secobarbital as *Surnox.*

THERAP CAT: Spasmolytic.

4463. Guaiapate. *1-[2-[2-[2-(2-Methoxyphenoxy)ethoxy]ethyl]piperidine;* M.G. 5454; Klamar. $C_{18}H_{29}NO_4$; mol wt 323.44. C 66.84%, H 9.04%, N 4.33%, O 19.79%. Basic ether deriv of guaiacol, *q.v.* Prepn: M. Carissimi, F. Ravenna, **Fr. pat. 1,386,633** corresp to **U.S. pat. 3,320,254** (1965, 1967 both to Maggioni); M. Carissimi *et al., J. Med. Chem.* **8**, 542 (1965). Pharmacological and clinical studies: G. Barbi, *Gazz. Med. Ital.* **127**, 13 (1968). Analytical study: S. Cavicchi *et al., Boll. Chim. Farm.* **108**, 682 (1969).

Liq, $bp_{0.5}$ 190-193°. LD_{50} in mice: 254 μ moles/kg i.p., M. Carissimi *et al., loc. cit.*

THERAP CAT: Antitussive.

4464. Guaiazulene. *1,4-Dimethyl-7-(1-methylethyl)azulene; 7-isopropyl-1,4-dimethylazulene;* AZ 8; AZ 8 Beris; S-guaiazulene; Eucazulen; Kessazulen; Vaumigan. $C_{15}H_{18}$; mol wt 198.29. C 90.85%, H 9.15%. Isoln from chamomile oil: Sorm *et al., Coll. Czech. Chem. Commun.* **16**, 626 (1951); from guaiac wood oil: Joos, **Swiss pat. 314,487** (1956), *C.A.* **52**, 443b (1958). Total synthesis: Plattner *et al., Helv. Chim. Acta* **32**, 2452 (1949); Sorm *et al., Coll. Czech. Chem. Commun.* **16**, 168 (1951); Jacob *et al., Tetrahedron* **20**, 2821 (1964); J. Mukherjee *et al., J. Am. Chem. Soc.* **101**, 251 (1979). Pharmacokinetics of guaiazulene soluble in animals: H. Mukai *et al., J. Pharmacobio-Dyn.* **8**, 329, 337 (1985). Effect on gastric and duodenal ulcers in rats: S. Okabe *et al., Nippon Yakurigaku Zasshi* **88**, 467 (1986), *C.A.* **106**, 43769 (1987).

Blue oil. bp_{10} 165-170°.

3-Sulfonate sodium salt, $C_{15}H_{17}NaO_3S$, *5-isopropyl-3,8-dimethyl-1-azulenesulfonic acid sodium salt, sodium gualenate, guaiazulene soluble, Azulon.*

Trinitrobenzene deriv, $C_{21}H_{21}N_3O_6$, violet to black needles from ethanol, mp 151°.

THERAP CAT: Anti-inflammatory. Anti-ulcerative.

4465. Guaifenesin. *3-(2-Methoxyphenoxy)-1,2-propanediol;* glycerol mono(2-methoxyphenyl) ether; glycerol α-(2-methoxyphenyl) ether; guaiacyl glyceryl ether; glyceryl guaiacyl ether; glycerol guaiacolate; α-glyceryl guaiacol ether; o-methoxyphenyl glyceryl ether; 1,2-dihydroxy-3-(2-methoxyphenoxy)propane; guaiacol glyceryl ether; guaiphenesin; guaiacuran; My 301; XL-90; Actifed-C; Calmipan; Colrex Expectorant; Equicol; Gecolate; Glycodex; Guaiamar; Guaiatuss; Guayanesin; Miocaina; Myocaine; Myoscain; Neuro-ton; Oresol; Oreson; Relaxil G; Reorganin; Respenyl; Resyl; Robitussin; Sirotol; Tenntuss; Tulyn. $C_{10}H_{14}O_4$; mol wt 198.21. C 60.59%, H 7.12%, O 32.29%. Centrally acting muscle relaxant with expectorant properties. Prepn: Marle, *J. Chem. Soc.* **101**, 305 (1912); Yale *et al., J. Am. Chem. Soc.* **72**, 3710 (1950); Roviralta, Astoul, **Span. pat. 212,920** (1954), *C.A.* **49**, 8332b (1955). Prepn from 2-methoxyphenol and glycidol: W. Merk *et al.,* **Ger. pat. 3,106,995** corresp to **U.S. pat. 4,390,732** (1982, 1983 to Degussa AG). GLC determn in blood: W. R. Maynard, R. B. Bruce, *J. Pharm. Sci.* **59**, 1346 (1970). Clinical use in chronic respiratory disease: D. G. Workman *et al., Curr. Ther. Res.* **7**, 665 (1965). Clinical efficacy as antitussive: J. J. Kuhn *et al., Chest* **82**, 713 (1982). Pharmacokinetics and cardiopulmonary effects in horses: J. A. E. Hubbell *et al., Am. J. Vet. Res.* **41**, 1751 (1980). Use in equine anesthesia: J. L. Grandy, W. N. McDonell, *J. Am. Vet. Med. Assoc.* **176**, 619 (1980); G. J. Brouwer, *Eq. Vet. J.* **17**, 133 (1985).

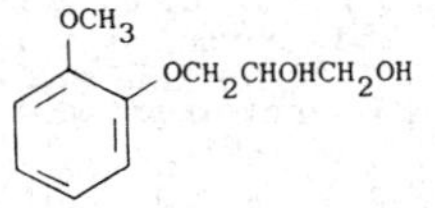

Minute rhombic prisms from ether, mp 78.5-79°. bp$_{19}$ 215°. Slightly bitter aromatic taste. One gram dissolves in 20 ml water at 25°. Much more sol in hot water; freely sol in ethanol; sol in chloroform, glycerol, propylene glycol, DMF; moderately sol in benzene. Practically insol in petr ether.
THERAP CAT: Expectorant.
THERAP CAT (VET): Expectorant; muscle relaxant.

4466. Guaiol. *[3S]-1,2,3,4,5α,6,7,8-Octahydro-α,α,3α,-8α-tetramethyl-5-azulenemethanol;* 3,8-dimethyl-5-(α-hydroxyisopropyl)-Δ⁹-octahydroazulene; champaca camphor; champacol; guaiac alcohol; Guajol. C$_{15}$H$_{26}$O; mol wt 222.36. C 81.02%, H 11.79%, O 7.20%. A sesquiterpene alc from guaiac wood: *Michelia champaca* L., *Magnoliaceae*: also from oil of wood of *Bulnesia sarmienti* Lorentz, *Zygophyllaceae*. Isoln: Plattner, Lemay, *Helv. Chim. Acta* **23**, 897 (1940). Structure: Plattner, Magyar, *ibid.* **25**, 581 (1942). Stereochemistry: Takeda, Minato, *Tetrahedron Letters* **1960**(22), 33; Minato, *Chem. Pharm. Bull.* **9**, 625 (1961); *idem, Tetrahedron* **18**, 365 (1962). Total synthesis of *dl*-form: Buchanan, Young, *Chem. Commun.* **1971**, 643; *eidem, J. Chem. Soc. Perkin Trans. I* **1973**, 2404; Marshall *et al., Tetrahedron Letters* **1971**, 885; Marshall, Greene, *J. Org. Chem.* **37**, 982 (1972); Andersen, Uh, *Tetrahedron Letters* **1973**, 2079.

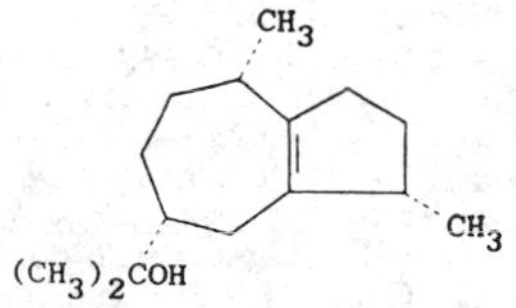

Trigonal pyramidal crystals from alc, mp 91°. d$_{20}^{100}$ 0.9074. bp$_{760}$ 288° (slight decompn); bp$_{17}$ 165°; bp$_{10}$ 148°. [α]$_D^{20}$ −30° (c = 4 in alc). n$_D^{20}$ 1.4716. Sol in alc, ether. Insol in water. Methyl ether, C$_{16}$H$_{28}$O, liq, d$_4^{25}$ 0.9332. bp$_9$ 142°. [α]$_D^{20}$ −31.8°. n$_D^{18.5}$ 1.4896.
Note: The name "guaiol" is also applied to 1,2-dimethyl-acrolein, isolated from guaiacum resin.

4467. Guaithylline. *3,7-Dihydro-1,3-dimethyl-1H-purine-2,6-dione compd with 3-(2-methoxyphenoxy)-1,2-propanediol (1:1);* 3-(o-methoxyphenoxy)-1,2-propanediol compd with theophylline; theophylline compd with 3-(o-methoxyphenoxy)-1,2-propanediol; theophylline compd with glyceryl guaiacolate; Eclabron. C$_{17}$H$_{22}$N$_4$O$_6$; mol wt 378.36. C 53.96%, H 5.86%, N 14.81%, O 25.37%. Solubilization of theophylline with glyceryl guaiacolate to provide an improved acid stable oral form of theophylline: Brit. pat. 932,874 (1963 to Mead Johnson).

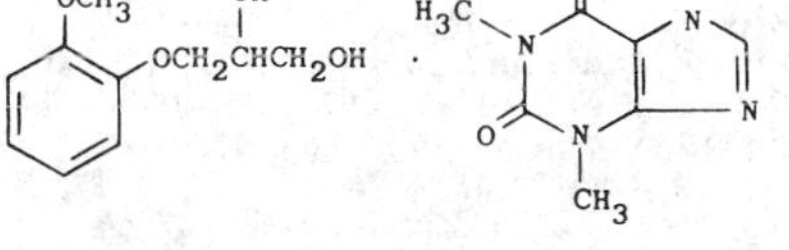

THERAP CAT: Bronchodilator; expectorant.

4468. Guamecycline. *N-[[4-[[(Aminoiminomethyl)amino]iminomethyl]-1-piperazinyl]methyl]-4-(dimethylamino)-1,4,4a,5,5a,6,11,12a-octahydro-3,6,10,12,12a-pentahydroxy-6-methyl-1,11-dioxo-2-naphthacenecarboxamide;* tetrabiguanide; "xanthomycin"; xantomicina. C$_{29}$H$_{38}$N$_8$O$_8$; mol wt 626.68. C 55.58%, H 6.11%, N 17.88%, O 20.43%. Prepn:

Brit. pat. 1,042,207 (1966 to Societa Prodotti Antibiotici), *C.A.* **65**, 16922g (1966). Clinical studies: O. Restivo, G. Sindoni, *Gazz. Med. Ital.* **127**, 17 (1968).

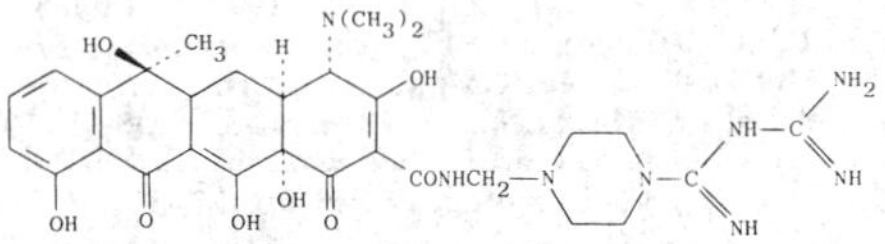

Dihydrochloride, C$_{29}$H$_{40}$Cl$_2$N$_8$O$_8$, Terratrex, Xantociclina.
THERAP CAT: Antibacterial.

4469. Guanabenz. *2-[(2,6-Dichlorophenyl)methylene]-hydrazinecarboximidamide; [(2,6-dichlorobenzylidene)amino]guanidine; N-(2,6-dichlorobenzylidene)-N'-amidinohydrazine;* NSC-68982. C$_8$H$_8$Cl$_2$N$_4$; mol wt 231.07. C 41.58%, H 3.49%, Cl 30.68%, N 24.25%. α$_2$-Adrenergic agonist. Prepn: J. Yates, E. Haddock, Brit. pat. 1,019,120 (1966 to Shell), *C.A.* **64**, 11132h (1966). Use as antihypertensive: W. J. Houlihan *et al.,* Ger. pat. 1,804,634 (1969 to Sandoz), *C.A.* **71**, 89976j (1969). Pharmacology: T. Baum *et al., J. Pharmacol. Exp. Ther.* **171**, 276 (1970); E. Lampa *et al., Experientia* **36**, 228 (1980). Disposition of ¹⁴C-guanabenz in humans: R. H. Meacham *et al., Clin. Pharmacol. Ther.* **27**, 44 (1980). Mechanism of action: G. F. DiBona, *J. Cardiovasc. Pharmacol.* **6**, Suppl. 3, S543 (1984). Radioimmunoassay determn in plasma: H. Tatsumi, *Arzneimittel-Forsch.* **34**, 1704 (1984). Clinical studies: A. Reppelli *et al., Boll. Soc. Ital. Cardiol.* **23**, 177 (1978); C. V. Ram *et al., J. Clin. Pharmacol.* **19**, 148 (1979). Clinical studies in opiate withdrawal: F. S. Tennant, R. A. Rawson, *Nat. Inst. Drug Abuse Res. Monogr. Ser.* **49**, 338 (1984); J. T. Murphy, *Drug Intell. Clin. Pharm.* **19**, 32 (1985). Review of pharmacodynamic properties and therapeutic efficacy: B. Holmes *et al., Drugs* **26**, 212-229 (1983). Comprehensive description: C. M. Shearer in *Analytical Profiles of Drug Substances* vol. **15**, K. Florey, Ed. (Academic Press, New York, 1986) pp 319-336.

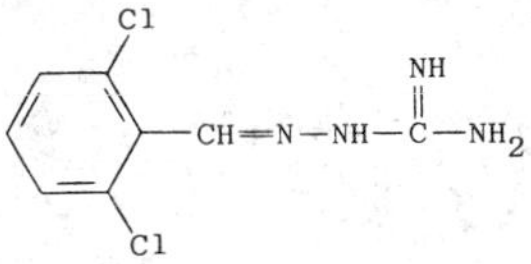

White solid from acetonitrile, mp 227-229° (dec).
Monoacetate, C$_{10}$H$_{12}$Cl$_2$N$_4$O$_2$, Wy-8678, Rexitene, Tenelid, Wytensin. Solid, mp 192.5° (dec). Soly at 25° (mg/ml): water 11; alcohol 50; propylene glycol 100; chloroform 0.6; ethyl acetate 1.
THERAP CAT: Antihypertensive.

4470. Guanacline. *[2-(3,6-Dihydro-4-methyl-1(2H)-pyridinyl)ethyl]guanidine; N-(2-guanidinoethyl)-4-methyl-Δ³-piperidine; N-(2-guanidinoethyl)-4-methyl-1,2,3,6-tetrahydropyridine; 1-(2-guanidinoethyl)-1,2,3,6-tetrahydro-4-picoline; cyclazenin;* FBA 1464. C$_9$H$_{18}$N$_4$; mol wt 182.27. C 59.30%, H 9.95%, N 30.74%. Prepn: Fr. pat. M3016 (1965 to Bayer), *C.A.* **62**, 16206b (1965).

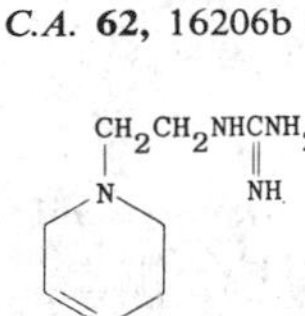

Monosulfate, C$_9$H$_{20}$N$_4$O$_4$S, Leron. Crystals from ethanol + ethyl acetate, dec 185-186°. Unstable in air.
Neutral sulfate, crystals, mp 232.5°. Stable in air.
THERAP CAT: Antihypertensive.

4471. Guanadrel. *(1,4-Dioxaspiro[4.5]dec-2-ylmethyl)-guanidine.* C$_{10}$H$_{19}$N$_3$O$_2$; mol wt 213.28. C 56.32%, H 8.98%, N 19.70%, O 15.00%. Orally active postganglionic synpathetic inhibitor. Prepn: W. R. Hardie, J. E. Aaron, S.

Afr. pat. **67 06,328** (1968 to Cutter), *C.A.* **70**, 57808u (1969); *eidem*, U.S. pat. **3,547,951** (1970 to Cutter). Antihypertensive activity: L. Hansson *et al.*, *Clin. Pharmacol. Ther.* **14**, 204 (1973). *In vitro* adrenergic neuron blocking activity: L. Roller, *Aust. J. Pharm. Sci.* **5**, 35 (1976). Pharmacologic study: E. M. Johnson, F. E. Hunter, *Biochem. Pharmacol.* **28**, 1525 (1979). Effect on patients with thyrotoxicosis: S. Rubenfeld *et al.*, *Arch. Intern. Med.* **138**, 1106 (1978). Review of pharmacology and efficacy in hypertension: F. A. Finnerty Jr., R. N. Brogden, *Drugs* **30**, 22-31 (1985).

Sulfate, $C_{20}H_{40}N_6O_8S$, *CL-1388R*, *U-28,288D*, *Anarel*, *Hylorel*. Cryst from methanol/ethanol, mp 213.5-215°.
THERAP CAT: Antihypertensive.

4472. Guanazodine. *[(Octahydro-2-azocinyl)methyl]-guanidine;* α-guanidinomethylheptamethylenimine; 1-azacyclooct-2-ylmethylguanidine. $C_9H_{20}N_4$; mol wt 184.28. C 58.66%, H 10.94%, N 30.40%. Hypotensive agent, related structurally to guanethidine, *q.v.* Prepn of the sulfate salt: J. Rakoczi *et al.*, *Hung. pat.* **155,990** corresp to U.S. pat. **3,856,778** (1969, 1974 both to EGYT). Pharmacokinetics in man: T. Past *et al.*, *Int. J. Clin. Pharmacol. Ther. Toxicol.* **8**, 111 (1973). General pharmacology: H. Iwata *et al.*, *Oyo Yakuri* **14**, 235 (1977), *C.A.* **88**, 16070 (1977). Analytical study: E. Zollner-Ivan, *Acta Pharm. Hung.* **48**, 76 (1978). Clinical studies in hypertension: S. Dobi *et al.*, *Ther. Hung.* **28**, 60 (1980), Z. Herpai *et al.*, *ibid.* 181.

Sulfate monohydrate, $C_9H_{22}N_4O_4S.H_2O$, *EGYT-739*, *Calnegyt*, *Sanegyt*. White cryst from water, mp 239-241°. LD_{50} in mice (mg/kg): 2450 orally, 700 s.c., 165 i.v., J. Rakoczi *et al.*, *loc. cit.*
THERAP CAT: Antihypertensive.

4473. Guanethidine. *[2-(Hexahydro-1(2H)-azocinyl)eth-yl]guanidine;* [2-(octahydro-1-azocinyl)ethyl]guanidine; 1-(2-guanidinoethyl)octahydroazocine; 2-(1'-azacyclooct-yl)ethylguanidine; *N*-(2-perhydroazocin-1-ylethyl)guani-dine; 2-(1-*N,N*-heptamethylenimino)ethylguanidine; okta-din; oktatenzin; Su 5864; Eutensol; Dopom; Octatensine; Oktatensin; Sanotensin; Abapresin. $C_{10}H_{22}N_4$; mol wt 198.31. C 60.56%, H 11.18%, N 28.26%. Prepn: Maxwell *et al.*, *Experientia* **15**, 267 (1959); Mull, U.S. pat. **2,928,829** (1960 to Ciba). Improved process: Mull, U.S. pats. **3,006,-913; 3,055,882** (1961, 1962 to Ciba).

Sulfate, $C_{20}H_{46}N_8O_4S$, *Guethine*, *Iporal*, *Isobarin*, *Ismelin*. Crystals from dil ethanol, mp 276-281° (dec).
Monosulfate, $C_{10}H_{24}N_4O_4S$, component of *Esimil*.
THERAP CAT: Antihypertensive.

4474. Guanfacine. *N-(Aminoiminomethyl)-2,6-dichlo-robenzeneacetamide; N-amidino-2-(2,6-dichlorophenyl)acet-amide;* [(2,6-dichlorophenyl)acetyl]guanidine. $C_9H_9Cl_2N_3O$; mol wt 246.08. C 43.92%, H 3.69%, Cl 28.81%, N 17.08%, O 6.50%. Centrally acting α$_2$-adrenoceptor agonist. Prepn: J. B. Bream, C. W. Picard, **Fr.** pat. **1,584,670**; *eidem*, U.S. pat. **3,632,645** (1969, 1972, both to Wander). Pharmacology: D. M. Coward *et al.*, *Arzneimittel-Forsch.* **27**, 2326 (1977); H. F. Oates *et al.*, *Arch. Int. Pharmacodyn. Ther.* **231**, 148 (1978); *eidem*, *Clin. Exp. Pharmacol. Physiol.* **6**, 61

(1979). Determn in biological fluids: M. Guerrat *et al.*, *J. Pharm. Sci.* **68**, 219 (1979). Kinetics: Y. A. Weiss *et al.*, *Clin. Pharmacol. Ther.* **25**, 283 (1979). Clinical studies: U. C. Dubach *et al.*, *Arzneimittel-Forsch.* **27**, 674 (1977); P. MacCarthy *et al.*, *Clin. Exp. Pharmacol. Physiol.* **5**, 187 (1978). Symposium on pharmacology, pharmacokinetics, clinical efficacy and comparison with clonidine: *Am. J. Cardiol.* **57**, 1E-61E (1986). *Review: Brit. J. Clin. Pharma-col.* **10**, Suppl. 1, 1S-208S (1980).

White grains from methanol/ether, mp 225-227°.
Hydrochloride, $C_9H_{10}Cl_3N_3O$, *BS 100-141*, *LON 798*, *Estulic*, *Tenex*. White needles, mp 213-216°. LD_{50} in mice: 165 mg/kg orally (Coward).
THERAP CAT: Antihypertensive.

4475. Guanidine. Aminomethanamidine; carbamami-dine; carbamidine; aminoformamidine; iminourea. CH_5N_3; mol wt 59.07. C 20.33%, H 8.53%, N 71.14%. $(NH_2)_2$-C=NH. Strong organic base existing primarily as guani-dinium ion at physiological pH. Found in turnip juice, mushrooms, corn germ, rice hulls, mussels, earthworms. Has antiviral, antifungal, antipyretic and muscle stimulatory activity. Prepn of guanidine nitrate from dicyanodiamide and ammonium nitrate: Smith *et al.*, *Ind. Eng. Chem.* **23**, 1124 (1931); Davis, *Org. Syn.* coll. vol. I (2nd ed., 1941) p 302; from SO_2, CO_2, and NH_3: Boivin, U.S. pat. **2,762,843** (1956); from urea: Mackay, U.S. pat. **2,590,257** (1952 to Am. Cyanamid); Craig, Minor, U.S. pat. **3,009,949** (1961 to Deere); Shaver, U.S. pat. **3,108,999** (1963 to Monsanto); from ammonium thiocyanate or thiourea + ammonia: Watt, Makosky, *Ind. Eng. Chem.* **46**, 2599 (1954). Review of mode of action: D. R. Tershak *et al.*, "Guanidine" in *Handbook of Experimental Pharmacology* Vol. **61**, G. V. R. Born *et al.*, Eds. entitled "Chemotherapy of Viral Infections" P. E. Came, L. A. Caliguiri, Eds. (Springer-Verlag, New York, 1982) pp 343-375. Review of experimental and clin-ical uses: F. Davidoff, *N. Engl. J. Med.* **289**, 141-146 (1973). General reviews: M. Schenck, *Pharmazie* **3**, 5 (1948); G. Schaefer, "Guanidines and Biguanidines" in *International Encyclopedia of Pharmacology and Therapeutics* **107**, M. Erecinska, Ed. (Pergamon, Oxford 1981) pp 165-185.

Deliquescent, cryst mass, mp about 50°. pKa ~12.5. Absorbs CO_2 from air. Very sol in water, alcohol. On heat-ing to 160° it is converted to melamine and NH_3. *Keep well closed.* LD orally in rabbits: 500 mg/kg, *Handbook of Tox-icology* vol. I, W. S. Spector, Ed. (Saunders, Philadelphia, 1956) p 152.

Hydrochloride, $CH_5N_3.HCl$, cryst powder. Freely sol in water, alcohol. Aq soln is neutral.
Nitrate, $CH_5N_3.HNO_3$, cryst powder, mp 214°. Soluble in 10 parts water; in alcohol. Aq soln is neutral.

4476. Guanidinium Aluminum Sulfate Hexahydrate. *Guanidine, compd with aluminum sulfate;* GASH. CH_{18}-$AlN_3O_{14}S_2$; mol wt 387.29. C 3.10%, H 4.68%, Al 6.97%, N 10.85%, O 57.84%, S 16.56%. $CH_5N_3.AlH(SO_4)_2.6H_2O$. Prepd from an aq soln of an equimolecular mixture of guani-dine sulfate and aluminum sulfate: Ferraboschi, *Proc. Cambridge Phil. Soc.* **14**, 473 (1908); Holden *et al.*, *Phys. Rev.* **101**, 962 (1956); Wieder, *Proc. IRE* **45**, 1094 (1957). Crystal structure: Schein *et al.*, *J. Chem. Phys.* **47**, 5183 (1967). Crystal growth: T. A. Zarembovskaya, *Zh. Fiz. Khim.* **45**, 2504 (1971). Optical properties: P. M. Nikolic *et al.*, *Fizika* (Zagreb) **12**, Suppl. 1, 165 (1980).

Large hexagonal plates belonging to the trigonal system with perfect basal cleavage. Has ferroelectric properties.

4477. Guanine. *2-Amino-1,7-dihydro-6H-purin-6-one; 2-aminohypoxanthine.* $C_5H_5N_5O$; mol wt 151.13. C 39.73%, H 3.33%, N 46.34%, O 10.59%. Constituent of nucleic acids; widespread in animal and plant kingdom. First isolated from guano. Syntheses: Fisher, *Ber.* **30**, 2226 (1897); Traube, *ibid.* **33**, 1371 (1900); **Ger.** pats. **134,984** (1902), **158,591** (1903), **162,336** (1904). Prepn of ^{15}N-isotopic guanine following Traube's synthesis: Plentl, Schoenheimer, *J. Biol. Chem.* **153**, 203 (1944). Several desmotropic forms. Crystal structure of hydrochloride monohydrate: Broomhead, *Acta Cryst.* **4**, 92 (1951). *Reviews:* Shapiro, *Progr. Nucleic Acid Res. Mol. Biol.* **8**, 73-112 (1968); Ts'o, "Bases, Nucleosides and Nucleotides" in *Basic Principles in Nucleic Acid Chemistry* vol. 1, P. O. P. Ts'o, Ed. (Academic Press, New York, 1974) pp 453-584. *See also* Nucleic Acids.

Usually amorphous. Small rhombic crystals by slow evaporation of aq soln contg large excess of NH_3. Dec above 360° with partial sublimation. uv max (pH 6.2): 246, 275 nm ($\epsilon \times 10^{-3}$ 10.7, 8.1). Freely sol in ammonia water, aq KOH solns, dil acids; sparingly sol in alcohol, ether. Almost insol in water. Kb 6.09×10^{-4}, Ka 1.19×10^{-10} detd at 40°. Many compds with acids, bases and metals have been prepared.

Hydrochloride monohydrate, $C_5H_5N_5O \cdot HCl \cdot H_2O$, cryst powder. Loses H_2O at 100°, HCl at 200°. Practically insol in water, alcohol, ether; sol in acidulated water.

4478. Guano. Bird manure. The dried excrements of sea birds (cormorants) and bats from coastal islands of Peru, Chile, West Indies, and Africa. Usually mixed with feathers and bones. Contains about 9% nitrogen, 6% phosphorus, 2% potassium, and 15-20% moisture. Used as fertilizer.

4479. Guanochlor. *2-[2-(2,6-Dichlorophenoxy)ethyl]hydrazinecarboximidamide; [[2-(2,6-dichlorophenoxy)ethyl]amino]guanidine.* $C_9H_{12}Cl_2N_4O$; mol wt 263.14. C 41.08%, H 4.60%, Cl 26.95%, N 21.30%, O 6.08%. Prepn of free base and sulfate: **Belg.** pat. **629,613** (1963 to Pfizer), *C.A.* **60**, 14437d (1964).

Sulfate, $C_{18}H_{26}Cl_4N_8O_6S$, *Vatensol.* Crystals from water, mp 214°.
THERAP CAT: Sulfate as antihypertensive.

4480. Guanosine. *2-Amino-9-β-D-ribofuranosyl-9H-purine-6(1H)-one; guanine riboside; vernine.* $C_{10}H_{13}N_5O_5$; mol wt 283.24. C 42.40%, H 4.63%, N 24.73%, O 28.24%. Constituent of nucleic acids. Prepn from yeast nucleic acid: P. A. Levene, L. W. Bass, *Nucleic Acids* (New York, 1931) p 163. Prepn from plants: H. Stendel, E. Peiser in G. Klein, *Handbuch der Pflanzenanalyse* IV (Vienna, 1933) p 448. Structure: Levene, Tipson, *J. Biol. Chem.* **97**, 491 (1932); Gulland *et al., J. Chem. Soc.* **1934**, 1639; Tsuboi *et al., Biochim. Biophys. Acta* **55**, 1 (1962). Synthesis: Davoll, *J. Chem. Soc.* **1958**, 1593. Tautomerism in aq soln: Miles *et al., Science* **142**, 1458 (1963). Crystal structure and conformation: Bugg *et al., Biochem. Biophys. Res. Commun.* **3**, 436 (1968). *Review: Basic Principles in Nucleic Acid Chemistry* vol. 1, P. O. P. Ts'o, Ed. (Academic Press, New York, 1974) *passim.* *See also* Nucleic Acids.

Dihydrate, needles from water. Anhydr at 110°. Dec 240° in sealed tube (rapid heating). $[\alpha]_D^{20}$ −60.5° (p = 3 in 0.1N NaOH). uv max (pH 5.5): 188.3, 252.5 nm ($\epsilon \times 10^{-3}$ 26.8, 13.7), Voet *et al., Biopolymers* **1**, 193 (1963). One gram dissolves in 1320 ml water at 18°, in 33 ml boiling water. Soluble in dil mineral acids, in hot acetic acid, and in dil bases. Insol in alcohol, ether, chloroform, benzene.

4481. Guanoxabenz. *2-[(2,6-Dichlorophenyl)methylene]-N-hydroxyhydrazinecarboximidamide; 1-[(2,6-dichlorobenzylidene)amino]-3-hydroxyguanidine;* Compound 43-663. $C_8H_8Cl_2N_4O$; mol wt 211.62. C 45.40%, H 3.81%, Cl 16.75%, N 26.48%, O 7.56%. Prepn: W. J. Houlihan, R. E. Manning, **Ger.** pat. **1,902,449** corresp to **U.S.** pat. **3,591,-636** (1969, 1971 both to Sandoz). Pharmacological study: J. C. Doxey, A. S. Hersom, *Brit. J. Pharmacol.* **70**, 171 (1980).

Hydrochloride, $C_8H_9Cl_3N_4O$, *Benzerial.* Crystals from ethanol/ether, mp 173-175°.
THERAP CAT: Antihypertensive.

4482. Guanoxan. *[(2,3-Dihydro-1,4-benzodioxin-2-yl)methyl]guanidine; (1,4-benzodioxan-2-ylmethyl)guanidine; 2-guanidinomethyl-1,4-benzodioxan.* $C_{10}H_{13}N_3O_2$; mol wt 207.23. C 57.96%, H 6.32%, N 20.28%, O 15.44%. Prepn: **Belg.** pat. **632,701**; J. Augstein, S. M. Green, **U.S.** pat. **3,247,221** (1963, 1966, both to Pfizer); Monro, *Chem. & Ind.* (London) **1964**, 1806; Gardner, **Brit.** pat. **996,708** (1965 to SKF), *C.A.* **63**, 14877d (1965); Yu, Shen, *C.A.* **66**, 94969k (1967). Prepn and properties of (+) and (−)-forms: Stenlake *et al., J. Pharm. Pharmacol.* **20** (Suppl), 82 (1968). Pharmacology: Augstein *et al., J. Med. Chem.* **8**, 446 (1965); Cession-Fossion, *Arch. Int. Pharmacodyn. Ther.* **164**, 419 (1966); Vidal-Beretewide *et al., Arzneimittel-Forsch.* **19**, 947 (1969). Metabolic studies: Jack *et al., J. Pharm. Pharmacol.* **23**, 2225 (1971); *Xenobiotica* **2**, 35 (1972).

Crystals, mp 164-165°.
Sulfate, $C_{20}H_{28}N_6O_8S$, *Envacar.*
THERAP CAT: Antihypertensive.

4483. 3'-Guanylic Acid. Guanosine 3'-monophosphate; guanine riboside-3-phosphoric acid; guanylic acid b. $C_{10}H_{14}N_5O_8P$; mol wt 363.23. C 33.06%, H 3.88%, N 19.28%, O 35.24%, P 8.53%. From yeast or pancreas. Prepn: P. A. Levene, L. W. Bass, *Nucleic Acids* (New York, 1931) pp 224-227. Structure: Levene, Jorpes, *J. Biol. Chem.* **81**, 579 (1929); Levene, Harris, *ibid.* **95**, 755 (1932); **98**, 9 (1932). Early work probably done on a mixture of 2'- and 3'-guanylic acids; see physical data below. Separation of two isomers: Cohn, *J. Am. Chem. Soc.* **72**, 1471 (1950); Khym, Cohn, *ibid.* **76**, 1818 (1954); *eidem, Biol. Prepn.* **5**, 40 (1957).

Absorption spectrum: Voet *et al.*, *Biopolymers* **1**, 193 (1963). *Reviews:* see Guanine, Nucleic Acids.

Dihydrate, long prisms from water. The water of crystn is given up at 118° and is taken up again at room temp. When anhydrous, dec 180° (closed tube). $[\alpha]_D^{25}$ − 8° (c = 2); −65° (c = 2 in 5% NaOH). Acid to litmus. Soluble in cold water, freely sol in hot water. Boiling with dil mineral acids yields guanine, H_3PO_4, and D-ribose.

Neutral sodium salt, $Na_2C_{10}H_{12}N_5O_8P$, flakes from water, contains 21.1% H_2O. Sol in cold, freely sol in hot water.

Brucine salt heptahydrate, $C_{10}H_{14}N_5O_8P.(C_{23}H_{26}N_2O_4)_2$·7$H_2O$, rectangular leaflets from alc. When anhydr, dec 233-240°. $[\alpha]_D^{20}$ − 26° (35% alc). One gram dissolves in 100 ml water.

4484. 5'-Guanylic Acid. Guanosine 5'-monophosphate; GMP; guanosine 5'-phosphate; guanine riboside-5-phosphoric acid. $C_{10}H_{14}N_5O_8P$; mol wt 363.23. C 33.06%, H 3.88%, N 19.28%, O 35.24%, P 8.53%. Nucleotide widely distributed in nature; found in hydrolyzates of RNA. Isolated together with inosinic acid from sardines or yeast extract: Kuninaka *et al.*, *New Food Ind. (Tokyo)* **3**, no. 1, 21-47 (1961). Also by direct biosynthesis using microorganisms or enzymes: Abrams, Bentley, *Arch. Biochem. Biophys.* **79**, 91 (1959); Magasanik, Karibian, *J. Biol. Chem.* **235**, 2672 (1960); Okumura *et al.*, U.S. pat. **3,249,511** (1966). Chemical synthesis: Michelson, Todd, *J. Chem. Soc.* **1949**, 2483; Chambers *et al.*, *J. Am. Chem. Soc.* **79**, 3747 (1957); Gilham, Tener, *Chem. & Ind. (London)* **1959**, 542; Tener, *J. Am. Chem. Soc.* **83**, 159 (1961); Koransky *et al.*, *Z. Naturforsch.* **17B**, 291 (1962). Prepn of Na salt: Ishibashi, Ito, U.S. pat. **3,190,877** (1965 to Takeda). Monograph on synthesis of nucleotides: G. R. Pettit, *Synthetic Nucleotides* vol. 1 (Van Nostrand Reinhold, New York, 1972) 252 pp. *Reviews:* See Guanidine; Nucleic Acids.

Microcrystals, dec 190-200°. Sparingly sol in cold water.

Barium salt octahydrate, $C_{10}H_{12}N_5O_8PBa.8H_2O$, white powder. uv max (pH 2): 256 nm (ϵ 12,400); (pH 12): 260 nm (ϵ 12,100).

Disodium salt monohydrate, $C_{10}H_{12}N_5O_8PNa_2.H_2O$, hygroscopic crystals, decomp at about 250°. Characteristic meaty taste. a_M (molar absorbancy): 13.7 × 10^3 at 252.5 nm (pH 7). Soly in water at 25° about 25 g/100 ml. Practically insol in alcohol, acetone, ether.

USE: The disodium salt as flavor intensifier, like sodium inosinate and sodium glutamate. Said to be more effective than either.

4485. Guaran. Principal polysaccharide from endosperm of guar seeds, *Cyamopsis tetragonaloba* (L.) Taub., *Leguminosae:* Heyne, Whistler, *J. Am. Chem. Soc.* **70**, 2249 (1948). Structure: Whistler, Durso, *ibid.* **74**, 5140 (1952). Configuration: Koleske, Kurath, *J. Polymer Sci.* Pt. A, **2**, 4123 (1964). *Review:* Deuel *et al.*, *Chimia* **8**, 64 (1954).

$[\alpha]_D^{25}$ +53° (1N NaOH). Sol in cold water.

Triacetate, fibrous material, mp 226-227°. Can be formed into strong films which can be elongated 550%. Becomes birefringent and does not develop crystallinity.

USE: In textile and paper industry.

4486. Guar Gum. Guar flour; Decorpa; Jaguar; Gum cyamopsis; Cyamopsis gum; Burtonite V-7-E; Guarina; Glucotard; Guarem. Mol wt about 220,000. The ground endosperms of *Cyamopsis tetragonolobus* (L.) Taub., *Leguminosae* which is cultivated in India as livestock feed. The water soluble fraction (85%) of guar flour is called guaran which consists of linear chains of (1→4)-β-D-mannopyranosyl units with α-D-galactopyranosyl units attached by (1→6) linkages. Ratio of D-galactose to D-mannose is 1:2. Effect on lipid metabolism: D. J. A. Jenkins *et al.*, *Brit. Med. J.* **2**, 1555 (1979); on glucose and lipid levels in diabetic and healthy volunteers: U. Smith, G. Holm, *Atherosclerosis* (Shannon, Ire.) **45**, 1 (1982); on renal tumors in diabetic rats: B. C. Chin *et al.*, *Biomed. Res.* **5**, 273 (1984). As source of fiber in patients with non-insulin dependent diabetes: M. E. McIvor *et al.*, *Am. J. Clin. Nutr.* **41**, 891 (1985). Toxicology studies: S. L. Graham *et al.*, *Food Cosmet. Toxicol.* **19**, 287 (1981). Comprehensive monograph: F. Smith, R. Montgomery, *The Chemistry of Plant Gums and Mucilages* (Reinhold, New York, 1959) 627 pp. *Review:* Goldstein *et al.* in *Industrial Gums*, R. L. Whistler, Ed. (Academic Press, New York, 2nd ed., 1973) p 303-321.

Free flowing powder. Completely sol in cold and hot water; practically insol in oils, greases, hydrocarbons, ketones, esters. Water solns are tasteless, odorless, nontoxic, of a pale, translucent gray color, and neutral. Stable to heat. Has five to eight times the thickening power of starch. Water solns may be converted to a gel by small amounts of borax. Aq dispersions are neutral. *Cf.* "A Comparative Study of Commercially Available Guar Gums" by I. A. Schlakman, A. J. Bartilucci, *Drug Standards* **25**, 149-154 (1957). LD_{50} in male, female rats (g/kg): 7.35, 6.77 orally (Graham).

USE: In paper sizing; as a protective colloid, stabilizer, thickening and film forming agent for cheese, salad dressings, ice cream, soups; as a binding and disintegrating agent in tablet formulations; in pharmaceutical jelly formulations; in suspensions, emulsions, lotions, creams, toothpastes; in the mining industry as a flocculant, as a filtering agent; in water treatment as a coagulant aid.

THERAP CAT: Adjunct to diet, insulin or oral hypoglycemics in control of diabetes.

4487. Guinea Green B. *N-Ethyl-N-[4-[[4-[ethyl[(3-sulfophenyl)methyl]amino]phenyl]phenylmethylene]-2,5-cyclohexadien-1-ylidene]-3-sulfobenzenemethanaminium hydroxide inner salt, sodium salt; C.I. Acid Green 3; C.I. 42085; C.I.*

Food Green 1; FD & C Green 1. $C_{37}H_{35}N_2NaO_6S_2$; mol wt 690.80. C 64.33%, H 5.11%, N 4.06%, Na 3.33%, O 13.90%, S 9.28%. Prepn: Jones *et al.*, *J. Assoc. Offic. Agr. Chem.* **38**, 977 (1955). Chronic toxicity study: W. H. Hansen *et al.*, *Food Cosmet. Toxicol.* **4**, 389 (1966). *See also: Colour Index* vol. **4** (3rd ed., 1971) p 4385.

A dull, dark green powder, or a bright, cryst solid. Sol in water to a green soln which becomes brownish-yellow on addn of HCl and blackish-green with NaOH. An excess of NaOH decolorizes the soln. Sparingly sol in alcohol; it dissolves in concd H_2SO_4 to a yellow soln which, when diluted with water, turns first yellowish-red, then green. LD_{50} orally in rats: > 2 g/kg, F. C. Lu, A. Lavalle, *Can. Pharm. J.* **97**, 30 (1964).

USE: Limited use as a dye for silk and wool fabrics; as biological stain. Delisted by FDA in 1966 for use in foods, drugs, and cosmetics.

4488. D-Gulonic Acid. $C_6H_{12}O_7$; mol wt 196.16. C 36.74%, H 6.17%, O 57.10%. Prepd as the sodium salt by reduction of sodium glucuronate with sodium amalgam in alkaline medium: Fischer, Piloty, *Ber.* **24**, 525 (1891); from D-gulonic acid γ-lactone: Rehorst, Naumann, *ibid.* **77**, 24 (1944).

$[\alpha]_D^{20}$ −6° (10 min) → −38.6° (15 days). The free acid forms the lactone spontaneously. K at 25° = 2.1 × 10⁻⁴.
Sodium salt, $C_6H_{11}NaO_7$, crystals. $[\alpha]_D^{20}$ +11.5°. Sol in water.
Calcium salt, $Ca(C_6H_{11}O_7)_2$. $[\alpha]_D^{21}$ −14.45° (c = 1.73). Precipitated from aq soln by alc.

4489. L-Gulonic Acid. Xylosecarboxylic acid. $C_6H_{12}O_7$; mol wt 196.16. C 36.74%, H 6.17%, O 57.10%. Prepd from L-xylose and HCN followed by hydrolysis of the nitrile: Fischer, Stahel, *Ber.* **24**, 529 (1891). Prepn from D-glucuronic acid: **Ger. pat. 618,907** (1935 to Hoffmann-La Roche); from L-gulonolactone: Ishidate *et al.*, *Chem. Pharm. Bull.* **13**, 173 (1965).

Crystallizes as the lactone on evapn of an aq soln.
Sodium salt, $[\alpha]_D^{20}$ +12.7° (c = 9). Freely sol in water.

4490. D-Gulose. $C_6H_{12}O_6$; mol wt 180.16. C 40.00%, H 6.72%, O 53.29%. Prepd by sodium amalgam reduction of an acid soln of the γ-lactone of D-gulonic acid: Fischer,

Stahel, *Ber.* **24**, 532 (1891); van Ekenstein, Blanksma, *Rec. Trav. Chim.* **27**, 3 (1908). Alternate synthesis: Meyer zu Reckendorf, *Angew. Chem. Int. Ed.* **6**, 177 (1967); *idem*, *Methods Carbohyd. Chem.* **6**, 129 (1972); R. Köster *et al.*, *Angew. Chem. Int. Ed.* **19**, 547 (1980).

Syrup. Sweet taste. $[\alpha]_D^{20}$ −20.4°. Sol in water, slightly sol in alcohol. Not fermentable by yeast.

4491. L-Gulose. $C_6H_{12}O_6$; mol wt 180.16. C 40.00%, H 6.72%, O 53.29%. Prepd by sodium amalgam reduction of an acid soln of the γ-lactone of L-gulonic acid: Fischer, Piloty, *Ber.* **24**, 526 (1891). *See also* van Ekenstein, Blanksma, *Rec. Trav. Chim.* **27**, 3 (1908); Levene, LaForge, *J. Biol. Chem.* **20**, 430 (1915); Talen, *Rec. Trav. Chim.* **44**, 891 (1925); Isbell, *J. Am. Chem. Soc.* **55**, 2167 (1933). Synthesis from D-mannose: Evans, Parrish, *Carbohyd. Res.* **28**, 359 (1973); from D-glucose: D. K. Minster, S. M. Hecht, *J. Org. Chem.* **43**, 3987 (1978).

Syrup. $[\alpha]_D^{20}$ +61.6°. $[\alpha]_D$ +21.3° (c = 4.58) (Evans, Parrish, *loc. cit.*). Freely sol in water; slightly sol in alcohol. Not fermentable by yeast.

4492. Gum Benzoin. Resin benzoin; resin benjamin; gum benjamin. Balsamic resin from *Styrax benzoin* Dryand., known as Sumatra benzoin, or from *S. tonkinensis* (Pierre) Craib, *Styracaceae*, or other species of *Styrax* known as Siam benzoin. *Habit.* Thailand, Cambodia, S. Vietnam, Sumatra, Java, and Sunda Islands. *Constit.* Ethereal oil, free and combined benzoic and cinnamic acids up to 39%, vanillin, coniferyl benzoate, resin (a mixture of benzoresinol and benzoresinotannol) esterified with benzoic acid, styrol, styracin. Not less than 90% of Siam and not less than 75% of Sumatra benzoic is sol in alc (U.S.P.). *Ref:* Reinitzer, *Arch. Pharm.* **264**, 131 (1926); Brans, *Pharm. Weekbl.* **73**, 374 (1936); Freudenberg, Bittner, *Ber.* **83**, 600 (1950).

USE: Preserving ointments; preparing natural benzoic acid; for fumigating pastilles; in perfumery and cosmetics.

THERAP CAT: Topical protectant.

THERAP CAT (VET): Tincture is used topically as an antiseptic and to promote healing; as an inhalant for bronchitis, and orally as an expectorant.

4493. Gum Tragacanth. Tragacanth. Mol wt about 840,000. The dried gummy exudation from *Astragalus gummifer* Labill. (white gavan) or other Asiatic species of *Astragalus, Leguminosae*, found largely in Iran, also in Asia Minor and in Syria. When mixed with water gives a soluble fraction, as a hydrosol, called *tragacanthin* which is a complex mixture of polysaccharides containing D-galacturonic acid, other sugars, and traces of starch and cellulose. The insoluble fraction swells to a gel and consists of 60-70% bassorin, *q.v.* Structural studies: Norman, *Biochem. J.* **25**, 200 (1931); James, Smith, *J. Chem. Soc.* **1945**, 739, 749; Aspinall, Baillie, *ibid.* **1963**, 1702, 1714. *Reviews:* Beach, in *Natural Plant Hydrocolloids*, Advances in Chemistry Series **11** (A.C.S., Washington, 1954) pp 38-44; Meer *et al.* in *Industrial Gums*, R. L. Whistler, Ed. (Academic Press, New York, 2nd ed., 1973) pp 289-299. *Book:* F. Smith, R. Montgomery, *The Chemistry of Plant Gums and Mucilages* (Reinhold, New York, 1959) 627 pp.

Odorless. Insipid, mucilaginous taste. Acid reaction. One gram requires 0.9 ml 0.1N NaOH for neutralization to phenolphthalein: Gabel, *J. Am. Pharm. Assoc.* **23**, 341 (1934). Viscosity of tragacanth mucilages is reduced by adding acid, alkali, and NaCl particularly if the mucilage is heated: Mantell, *The Water-Soluble Gums* (New York, 1947). Maximum initial viscosity of solns at pH 8; maximum stable viscosity near pH 5. Forms a deep yellow stringy precipitate when a soln is boiled with a few drops of 10% aqueous ferric chloride soln. A stringy precipitate formed also on heating a soln with Schweitzer reagent. Tragacanth is entirely insol in alcohol.

USE: In pharmaceutical compounding and dispensing, *e.g.*, to suspend heavy insol powders, as an excipient for tablets and to impart consistence to troches; also in making emulsions and emulsifying agents; as stabilizer, thickener, texturizer in food; in adhesives (mucilages, pastes); in textile sizing, textile printing and general printing inks, and in dyeing with insol color lakes.

4494. Gutta-Percha. The purified, coagulated, milky exudate of various trees of the genus *Palaquium, Sapotaceae. Habit.* Malayan Archipelago. Extensive review: Williams, *Econ. Bot.* **18**, 5-26 (1964). Defined as a *trans* isomer of rubber. Rubber has a repeat period of 8.2 Å, whereas α-gutta-percha has 8.7 Å and β-gutta-percha has 4.8 Å. The short period of β-gutta-percha identifies it almost uniquely as an all-*trans* polyisoprene.

Becomes pliable at 25-30°, plastic at 60°. mp 100° (partial dec). On exposure to air and sunlight, it absorbs oxygen and becomes brittle. Insol in water; partially sol in hot alcohol; 90% or more dissolves in chloroform, carbon disulfide, petr ether, oil of turpentine. *Keep under water and protected from light.*

USE: Insulator in electrotechnics, as dental cement; in orthopedics for fracture splints; manuf surgical instruments; covering golf balls.

4495. Guvacine. *1,2,5,6-Tetrahydro-3-pyridinecarboxylic acid; 1,2,5,6-tetrahydronicotinic acid.* $C_6H_9NO_2$; mol wt 127.14. C 56.68%, H 7.13%, N 11.02%, O 25.17%. From betel nuts, the seeds of *Areca catechu* L., *Palmae*. Extraction: Jahns, *Ber.* **24**, 2615 (1891). Synthesis: Freudenberg, *ibid.* **51**, 976, 1669 (1918); Hess, Leibbrandt, *ibid.* **51**, 806; **52**, 206 (1919).

Prisms from water, dec 295°. Neutral to litmus. Sol in water. Almost insol in abs alc, ether, chloroform, benzene.

Hydrochloride, $C_6H_{10}ClNO_2$, needles from water, dec 318°. Sol in water.

Hydrobromide, $C_6H_{10}BrNO_2$, needles from abs alcohol, dec 280°. Almost insol in acetone.

USE: Has been proposed as growth factor for *Staphylococcus aureus* and *Proteus vulgaris* instead of nicotinic acid.

4496. Gymnemic Acid. Gymnemin. Antisaccharin principle occurring as the potassium salt (gymnemin) in the leaves of *Gymnema sylvestre* R.Br. and allied *Asclepiadaceae:* Hooper, *Chem. News* **59**, 159 (1889); Mhaskar, Caius, *Indian J. Med. Res.* No. **16**, 1 (1930); Warren, Pfaffman, *J. Appl. Physiol.* **14**, 40 (1959); Stöcklin *et al., Helv. Chim. Acta* **50**, 474 (1967). A complex mixture of at least nine closely related acidic glycosides, the major active component being gymnemic acid A_1: Stöcklin, *J. Agr. Food Chem.* **17**, 704 (1969); Dateo, Long, *ibid.* **21**, 899 (1973). Separation of major components: Sinsheimer *et al., J. Pharm. Sci.* **59**, 622, 629 (1970). Completely obtunds taste for several hours for bitter or sweet, *e.g.*, quinine or sugar, but not for sour, astringent or pungent substances.

The acid is a yellow to brown amorphous, bitter powder. Almost insol in water; sol in alcohol. The potassium salt is a reddish-brown cryst mass, sol in water or alcohol.

4497. Gypsogenin. *3β-Hydroxy-23-oxoolean-12-en-28-oic acid;* githagenin; albasapogenin; gypsophilasapogenin. $C_{30}H_{46}O_4$; mol wt 470.67. C 76.55%, H 9.85%, O 13.60%. From *Agrostemma githago* (L.) Scop., *Caryophyllaceae:* Wedekind, Krecke, *Z. Physiol. Chem.* **155**, 122 (1925); from *Gypsophila oldhamiana, Caryophyllaceae:* Kutani, Karr, *J. Pharm. Soc. Japan* **64**, 18 (1944), *C.A.* **45**, 2961d (1951). Structure: Ruzicka, Giacomello, *Helv. Chim. Acta* **20**, 299 (1937). Identity with githagenin: Kon, Soper, *J. Chem. Soc.* **1940**, 617.

Needles or leaflets from methanol, mp 274-276°. $[\alpha]_D$ +91° (alc).

Acetate, $C_{32}H_{48}O_5$, square tablets from methanol, mp 176-177°. $[\alpha]_D^{20}$ +78° $(CHCl_3)$.

Methyl ester, $C_{31}H_{48}O_4$, crystals, mp 192°.

H

4498. H3®. Gerovital; Gerovital H3; Gerontex H3; GH 3; Sex-Ex. A factor claimed to exist in commercial procaine hydrochloride prepns according to the Romanian M.D. Anna Aslan [Institute of Geriatrics, Bucharest]. H3 is Dr. Aslan's name for the factor in procaine HCl which she found to be effective in achieving an apparent reversal of phenomena previously considered irreversible, such as those encountered in cerebral arteriosclerosis. *Ref: Publishers' Weekly,* Dec. 7, 1959, p 36; Aslan, *C.A.* **53,** 1465b (1959). Historical review of controversy: P. M. McGrady, *The Youth Doctors* (Coward-McCann, New York, 1968) pp 181-192; A. Hecht, *FDA Consumer* (March 1980) p 17. Clinical study: M. R. Hall *et al., Age Ageing* **12,** 302 (1983); and a response: P. H. Millard, *Br. Med. J.* **289,** 1094 (1984).
Note: Oral version is marketed as *KH3.* Each KH3 capsule contains: Procaine hydrochloride 0.05 g, hematoporphyrin 0.0002 g, magnesium carbonate 0.08 g.

4499. Hachimycin. *Trichomycin;* Cabimicina; Trichonat. Heptaene macrolide antibiotic substance produced by *Streptomyces hachijoensis* from soil of the Pacific Island Hachijo Jima: S. Hosoya *et al., J. Antibiot.* **5,** 564 (1952); T. Yamaguchi, *ibid.* **7A,** 10 (1954). Purification and prepn of insol derivs: S. Hosoya *et al., ibid.* **8A,** 5 (1955). At one time, trichomycin was believed to be identical to candicidin and hamycin, *q.q.v.:* H. Burrows, D. Calam, *J. Chromatog.* **53,** 566 (1970). Subsequent HPLC studies have shown the three polyene antibiotics to be different entities: P. Helboe *et al., ibid.* **189,** 249 (1980). Active against *Trichomonas vaginalis, Treponema pallidum, Trichophyton, Candida,* and yeast, weak activity against *Aspergillus* and *Penicillium:* T. Yamaguchi, *J. Antibiot.* **7A,** 10 (1954). Toxicity: S. Hosoya *et al., J. Antibiot.* **6A,** 98 (1953). *Review:* S. A. Waksman, H. A. Lechevalier, *The Actinomycetes* vol. III (Williams & Wilkins, Baltimore, 1962) pp 397-399.
Yellow crystals. Acid reaction. Forms a water-sol sodium salt. Can be pptd as a water-insol salt by acridine derivs such as acriflavine, by sulfa drugs such as homosulfanilamide, by basic antibiotics such as streptomycin base, by enzymes such as papain and lysozyme, by dyes such as methylene blue, and by metallic salts such as CaCl$_2$. LD$_{50}$ i.p. in mice: 0.05 mg/10-12 g mouse (Hosoya).
THERAP CAT: Antifungal. Antiprotozoal (Trichomonas).

4500. Hadacidin. *N-Formyl-N-hydroxyglycine; N-*formyl-*N*-hydroxyaminoacetic acid; *N*-hydroxyformamidoacetic acid. C$_3$H$_5$NO$_4$; mol wt 119.08. C 30.26%, H 4.23%, N 11.76%, O 53.75%. Antitumor antibiotic originally isolated from cultures of *Penicillium frequentans* Westling. Synthesis: Kaczka *et al., Biochemistry* **1,** 340 (1962); Kinnel, Schoenewaldt, U.S. pat. **3,154,578** (1964 to Merck & Co.). Biosynthesis: Stevens, Emery, *Biochemistry* **5,** 74 (1966). *Review:* Shigeura, "Hadacidin" in *Antibiotics* I, D. Gottlieb, P. Shaw, Eds. (Springer-Verlag, New York, 1967) pp 451-456.

OH
|
OHC—NCH$_2$COOH

Unstable crystals, mp 119-120°. Turns brown and liquefies on standing. The decompn products are formic acid and *N*-hydroxyglycine. Dibasic acid, potentiometric titration shows pH peak at 3.5 and 9.1. Soluble in water, methanol, ethanol, acetone, ether.
Monosodium salt, C$_3$H$_4$NNaO$_4$, crystals. Easily forms a hydrate, very freely sol in water.

4501. Hafnium. Hf; at. wt 178.49; at. no. 72; valences 4; also 2, 3. Six naturally occurring isotopes: 180 (35.22%); 178 (27.1%); 177 (18.56%); 179 (13.75%); 176 (5.21%); 174 (0.163%; α-emitter, T$_{1/2}$ 2 × 10^{15} years); artificial isotopes: 157; 158; 168-173; 175; 181-183. Abundance in earth's crust: 5 ppm. Found in all zirconium-contg minerals. Discovered in 1923 by Coster and Hevesy: *Nature* **111,** 79 (1923). Extraction from the mineral cyrtolite: Larsen *et al.,*

Inorg. Syn. **3,** 67 (1950). Prepd by thermal decompn of its iodide; by reduction of the tetrachloride or of the hydrofluohafniate with metallic sodium; by reduction of the oxide with a mixture of calcium and sodium: van Arkel, de Boer, Z. *Anorg. Chem.* **148,** 345 (1925); de Boer, Fast, *ibid.* **187,** 193 (1930); U.S. pat. **2,741,628.** Spectra: Coster *et al.,* cited by Mellor, *A Comprehensive Treatise on Inorganic and Theoretical Chemistry* **7,** 170 (1930). Reviews of hafnium and its compds: Larsen, "Zirconium and Hafnium Chemistry" in *Advan. Inorg. Chem. Radiochem.* **13,** 1-133 (1970); Bradley, Thornton, "Zirconium and Hafnium" in *Comprehensive Inorganic Chemistry,* vol. 3, J. C. Bailar, Jr. *et al.,* Eds. (Pergamon Press, Oxford, 1973) pp 419-490; R. H. Nielson in Kirk-Othmer *Encyclopedia of Chemical Technology* vol. **12** (Wiley-Interscience, New York, 3rd ed., 1980) pp 67-80.
Highly lustrous, ductile metal of hexagonal cryst structure. mp 2227°; d 13.3. Resembles zirconium and thorium.
Dioxide, HfO$_2$, obtained by igniting the hydroxide, the oxalate, or the sulfate. d^{20} 9.68. mp 2774°.
Tetrachloride, HfCl$_4$, white cryst mass, obtained by heating the oxide in the presence of chlorine and a reducing agent. Hydrolyzed to hafnyl chloride, HfOCl$_2$, by water. Volatilizes appreciably at 250°.
Sulfate, Hf(SO$_4$)$_2$, prepd by the action of fuming sulfuric acid on the tetrachloride, dec above 500°.
A dibromide and tribromide have been prepd: Schumb, Morehouse, *J. Am. Chem. Soc.* **69,** 2696 (1947).

4502. Halazepam. *7-Chloro-1,3-dihydro-5-phenyl-1-(2,2,2-trifluoroethyl)-2H-1,4-benzodiazepin-2-one;* Sch 12041; Paxipam. C$_{17}$H$_{12}$ClF$_3$N$_2$O; mol wt 352.75. C 57.88%, H 3.43%, Cl 10.05%, F 16.16%, N 7.94%, O 4.54%. Polyfluoroalkyl analog of diazepam, *q.v.* Prepn: J. G. Topliss, U.S. pat. **3,429,874** (1969 to Schering); M. Steinman *et al., J. Med. Chem.* **16,** 1354 (1973). Pharmacology: J. B. Petel *et al., Psychopharmacology* **61,** 25 (1979). Use in schizophrenia: K. Y. Ota *et al., Curr. Ther. Res.* **15,** 327 (1973); in anxiety: K. Rickels *et al., Int. Pharmacopsychiatry* **13,** 118 (1978).

Crystals from acetone-hexane, mp 164-166°. LD$_{50}$ in mice: >4000 mg/kg orally (Steinman).
Caution: May be habit forming. This is a controlled substance (depressant) listed in the U.S. Code of Federal Regulations, Title 21 Part 1308.14 (1987).
THERAP CAT: Anxiolytic.

4503. Halazone. *4-[(Dichloroamino)sulfonyl]benzoic acid; p-(dichlorosulfamoyl)benzoic acid; p-sulfondichloramidobenzoic acid; p-carboxybenzenesulfondichloroamide;* Pantocid. C$_7$H$_5$Cl$_2$NO$_4$S; mol wt 270.10. C 31.13%, H 1.87%, Cl 26.26%, N 5.19%, O 23.69%, S 11.87%. Prepd by chlorination of *p*-sulfamylbenzoic acid in alkaline medium. The commercially available product is a mixture of mono- and dichloroamides with the dichloro compd predominating, *see* Proschko, **Ger.** pat. **492,249;** *Chem. Zentr.* **101,** I, 2630 (1930); U.S. pat. **1,697,139,** *C.A.* **23,** 1138 (1929); Claas, **Ger.** pat. **318,899;** *Chem. Zentr.* **91,** IV, 14 (1920). Review of use in water disinfection: O'Connor, Kapoor, *J. Am. Water Works Assoc.* **62,** 80 (1970).

Crystals or white powder, dec about 195°. Odor of chlorine. Slightly sol in water, chloroform (differing from di-chloramine-T which is very sol in chloroform). Sol in glacial acetic acid and in solns of alkali hydroxides and of alkali carbonates with the formation of a salt. MLD orally in rats: 3.5 g/kg.

Sodium salt, $C_7H_4Cl_2NNaO_4S$, *Aseptamide, Gynamide.*

USE: Has been proposed for use in the determination of iodine numbers of fats and oils. Used as water disinfectant.

4504. Halcinonide. *21-Chloro-9-fluoro-11-hydroxy-16,17-[(1-methylethylidene)bis(oxy)]pregn-4-ene-3,20-dione; 21-chloro-9-fluoro-11β,16α,17-trihydroxypregn-4-ene-3,20-dione cyclic 16,17-acetal with acetone;* 21-chloro-9α-fluoro-11β-hydroxy-16α,17α-isopropylidenedioxy-4-pregnene-3,20-dione; 9α-fluoro-21-chloro-11β,16α,17α-trihydroxy-pregn-4-ene-3,20-dione 16,17-acetonide; SQ-18566; Halcidern; Halcimat; Halcort; Halog. $C_{24}H_{32}ClFO_5$; mol wt 454.97. C 63.36%, H 7.09%, Cl 7.79%, F 4.17%, O 17.58%. Prepn: Bernstein *et al., J. Org. Chem.* **27**, 690 (1962); Difazio, Augustine, **Ger. pat. 2,355,710** corresp to **U.S. pat. 3,892,857** (1972, 1975 both to Squibb). Pharmacological evaluation: Bagatell, Augustine, *Curr. Ther. Res.* **16**, 748 (1974); R. C. Millonig, E. Yiakas, in *Pharmacological and Biochemical Properties of Drug Substances* **vol. 1**, M. E. Goldberg, Ed. (Am. Pharmaceut. Assoc., Washington, DC, 1977) pp 215-231. Comprehensive description: J. Kirshbaum in *Analytical Profiles of Drug Substances* **vol. 8**, K. Florey, Ed. (Academic Press, New York, 1979) pp 251-281.

Crystals from acetone-petr ether, mp 264-265° (dec). uv max (methanol): 238 nm (ε 16,400). $[\alpha]_D^{25}$ +155° (CHCl$_3$). Sol in acetone, chloroform, DMSO. Slightly sol in benzene, ethanol, ethyl ether, methanol. Insol in water, 0.1M HCl, 0.1M NaOH, hexanes. LD$_{50}$ i.p. in mice: 150 mg/kg, R. C. Millonig, E. Yiakas, loc. cit.

THERAP CAT: Anti-inflammatory (topical).

4505. Halethazole. *5-Chloro-2-[p-(diethylaminoethoxy)-phenyl]benzothiazole;* 2-[p-(diethylaminoethoxy)phenyl]-5-chlorobenzothiazole; haletazole; Episol. $C_{19}H_{21}ClN_2OS$; mol wt 360.92. C 63.23%, H 5.87%, Cl 9.82%, N 7.76%, O 4.43%, S 8.89%. Prepn: Stephens, **U.S. pat. 2,996,512** (1961 to Crookes Labs.).

Crystals from ethanol, mp 93-94°.
Citrate, $C_{19}H_{21}ClN_2OS \cdot C_6H_8O_7$, crystals from water, mp 167°.

THERAP CAT: Antiseptic, antifungal.

4506. Halibut Liver Oil. From the liver of the halibut, *Hippoglossus hippoglossus* L. It usually contains not less than 60,000 U.S.P. vitamin A units and not less than 500 U.S.P. vitamin D units per gram.

Pale yellow liq; slight fishy odor and taste. d 0.922-0.925. n_D^{40} 1.470-1.478. Sapon no. 170-180. Iodine no. 120-136. Unsaponifiable matter 8-13%.

THERAP CAT: Source of vitamins A and D.
THERAP CAT (VET): Source of vitamins A and D.

4507. Halimide®. *Dodecylbenzyltrimethylammonium chloride.* $C_{22}H_{40}ClN$; mol wt 354.03. C 74.64%, H 11.39%, Cl 10.02%, N 3.96%. Prepn: Darragh, Stayner, *Ind. Eng.*

Chem. **46**, 254 (1954); Stayner, **U.S. pat. 2,692,286** (1954 to Calif. Res. Corp.).

Pale yellow, glassy solid. Phenol coefficient *(S. aureus):* 408; *(S. typhosa):* 367.

USE: Disinfectant; cationic surface active agent.

4508. Halofantrine. *1,3-Dichloro-α-[2-(dibutylamino)-ethyl]-6-(trifluoromethyl)-9-phenanthrenemethanol;* 1-(1,3-dichloro-6-trifluoromethyl-9-phenanthryl)-3-di(n-butyl)-aminopropanol; γ-(dibutylamino)-1,3-dichloro-6-(trifluoro-methyl)-9-phenanthrenepropanol. $C_{26}H_{30}Cl_2F_3NO$; mol wt 500.43. C 62.40%, H 6.04%, Cl 14.17%, F 11.39%, N 2.80%, O 3.20%. Prepn of the hydrochloride: W. T. Colwell *et al., J. Med. Chem.* **15**, 771 (1972); of the β-glycerophosphate salt: J. F. Rossignol, **Eur. pat. Appl. 138,374;** idem, **U.S. pat. 4,507,288** (both 1985 to Smithkline Beckman). Antimalarial activity *in vivo:* L. H. Schmidt *et al., Antimicrob. Ag. Chemother.* **14**, 292 (1978). HPLC determn in human blood: J. W. Hines *et al., J. Pharm. Sci.* **74**, 433 (1985). Antimalarial activity in clinically-induced infection in humans: J. Rinehart *et al., Am. J. Trop. Med. Hyg.* **25**, 769 (1976); T. M. Cosgriff *et al., ibid.* **31**, 1075 (1982); in naturally acquired disease: J. P. Coulaud *et al., Trans. Roy. Soc. Trop. Med. Hyg.* **80**, 615 (1986).

Hydrochloride, $C_{26}H_{31}Cl_3F_3NO$, *SKF 102886, WR-171669, Halfan.* Two crystalline forms have been reported: mp 93-96°, mp 203-204°.
β-Glycerophosphate, $C_{29}H_{39}Cl_2F_3NO_7P$, white crystals, mp 60-65°.

THERAP CAT: Antimalarial.

4509. Halofuginone. *7-Bromo-6-chloro-3-[3-(3-hydr-oxy-2-piperidinyl)-2-oxopropyl]-4(3H)-quinazolinone; (±)-trans-7-bromo-6-chloro-3-[3-(3-hydroxy-2-piperidyl)aceto-nyl]-4(3H)-quinazolinone;* 7-bromo-6-chlorofebrifugine. $C_{16}H_{17}BrClN_3O_3$; mol wt 414.70. C 46.34%, H 4.13%, Br 19.27%, Cl 8.55%, N 10.13%, O 11.57%. Halogenated deriv of febrifugine, *q.v.* Prepn with 6-bromoisatin and sulfuryl chloride as starting materials: E. Waletzky *et al.,* **U.S. pat. 3,320,124** (1967 to Am. Cyanamid). Analysis in chicken feed: A. Anderson *et al., J. Chromatog.* **168**, 471 (1979). *In vivo* and *in vitro* activity: J. G. Ryley *et al., Parasitology* **70**, 203 (1975). Toxicity to fresh water organisms: *Bull. Environ. Contam. Toxicol.* **15**, 720 (1976).

Hydrobromide, $C_{16}H_{18}Br_2ClN_3O_3$, *RU-19110, Stenorol.* Crystals, mp 247° (dec).

THERAP CAT (VET): Antiprotozoal (coccidiostat).

4510. Halometasone. *(6α,11β,16α)-2-Chloro-6,9-di-fluoro-11,17,21-trihydroxy-16-methylpregna-1,4-diene-3,20-dione;* 2-chloroflumethasone; C-48401-Ba. $C_{22}H_{27}ClF_2O_5$;

mol wt 444.90. C 59.39%, H 6.12%, Cl 7.97%, F 8.54%, O 17.98%. Synthetic corticosteroid. Prepn: G. G. Anner *et al.*, **Neth. pat. Appl. 540,244** corresp to **U.S. pat. 3,652,554**; *eidem*, **Swiss pat. 551,399**, *C.A.* **81**, 120864d (1974) (1968, 1972, 1974 all to Ciba-Geigy). Series of clinical trials in eczema, psoriasis, dermatomycoses, bacterial infections: *J. Int. Med. Res.* **11**, Suppl. 1 (1983).

mp 220-222° (dec).
Monohydrate, $C_{22}H_{27}ClF_2O_5 \cdot H_2O$, *Sicorten*.
THERAP CAT: Topical anti-inflammatory, antipruritic.

4511. Haloperidol. *4-[4-(4-Chlorophenyl)-4-hydroxy-1-piperidinyl]-1-(4-fluorophenyl)-1-butanone; 4-[4-(p-chlorophenyl)-4-hydroxypiperidino]-4'-fluorobutyrophenone; 4'-fluoro-4-(4-hydroxy-4-p-chlorophenylpiperidino)butyrophenone; 1-(3-p-fluorobenzoylpropyl)-4-p-chlorophenyl-4-hydroxypiperidine;* R 1625; *Aloperidine; Bioperidolo; Brotopon; Dozic; Einalon S; Eukystol; Fortunan; Haldol; Halosten; Keselan; Linton; Peluces; Serenace; Serenase.* $C_{21}H_{23}ClFNO_2$; mol wt 375.88. C 67.10%, H 6.17%, Cl 9.43%, F 5.05%, N 3.73%, O 8.51%. Prepn: P. A. J. Janssen *et al.*, *J. Med. Pharm. Chem.* **1**, 281 (1959); P. A. J. Janssen, **Belg. pat. 577,977** (1959), *C.A.* **54**, 4630c (1960); *idem*, **Brit. pat. 895,309**; *idem*, **U.S. pat. 3,438,991** (1962, 1969 both to Janssen). Toxicity: E. I. Goldenthal, *Toxicol. Appl. Pharmacol.* **18**, 185 (1971); A. J. Collins, M. Horlington, *Brit. J. Pharmacol.* **37**, 140 (1969). Metabolism in man: A. Forsman *et al.*, *Curr. Ther. Res., Clin. Exp.* **21**, 606 (1977). Comprehensive description: C. A. Janicki, C. Y. Ko in *Analytical Profiles of Drug Substances* vol. 9, K. Florey, Ed. (Academic Press, New York, 1980) pp 341-369. Review of pharmacology and therapeutic efficacy of decanoate in psychosis: R. Beresford, A. Ward, *Drugs* **33**, 31-49 (1987).

Crystals, mp 148.0-149.4°. uv max (9:1 0.1M HCl:methanol): 247, 221 nm (ε 13300, 15000). pKa 8.3. Soly in water: 1.4 mg/100 ml. Freely sol in chloroform, methanol, acetone, benzene, dil acids. LD_{50} orally in rats: 165 mg/kg (Goldenthal); i.p. in mice: 60 mg/kg (Collins, Horlington).
Hydrochloride, $C_{21}H_{24}Cl_2FNO_2$, crystals, mp 226-227.5°. Soly in water: 300 mg/100 ml.
Decanoate, $C_{31}H_{41}ClFNO_3$, *KD-136, Haldol Decanoate, Halomonth, Neoperidole.*
THERAP CAT: Antidyskinetic (in Gilles de la Tourette's disease); antipsychotic.

4512. Halopredone Acetate. *17,21-Bis(acetyloxy)-2-bromo-6,9-difluoro-11-hydroxypregna-1,4-diene-3,20-dione; 2-bromo-6β,9-difluoro-11β,17,21-trihydroxypregna-1,4-diene-3,20-dione 17,21-diacetate; Haloart; Topicon.* $C_{25}H_{29}BrF_2O_7$; mol wt 559.42. C 53.68%, H 5.23%, Br 14.28%, F 6.79%, O 20.02%. Prepn: M. Riva, L. Toscano, **Ger. pat. 2,508,136** (1975 to Pierrel), *C.A.* **84**, 31311r (1975); L. Toscano *et al.*, *J. Med. Chem.* **20**, 213 (1977). Physico-chemical and analytical studies: *eidem*, *Arzneimittel-Forsch.* **27**, 1636 (1977). Pharmacology: A. Bianchetti *et al.*, *ibid.* 2096. Clinical study: B. Palmerio, P. Magnani, *ibid.* 2404. Toxicological study: L. Casilli *et al.*, *ibid.* 2102.

Cryst from benzene, mp 290-292° (dec). $[\alpha]_D^{24}$ −36° (c = 1 in chloroform). uv max (methanol): 246 nm (ε 12500).
THERAP CAT: Topical anti-inflammatory.

4513. Haloprogesterone. *17-Bromo-6-fluoropregn-4-ene-3,20-dione; 17α-bromo-6α-fluoroprogesterone; 6α-fluoro-17α-bromoprogesterone; Prohalone.* $C_{21}H_{28}BrFO_2$; mol wt 411.37. C 61.31%, H 6.86%, Br 19.43%, F 4.62%, O 7.78%. Prepn: Marshall, **U.S. pat. 2,924,610** (1960 to Am. Home Prod.); Mills *et al.*, *J. Org. Chem.* **25**, 1056 (1960); Engel, Deghenghi, *Can. J. Chem.* **38**, 452 (1960).

Cryst from dil acetone, mp 180-181° (dec). $[\alpha]_D^{24}$ +12.1° (c = 2.46 in CCl_4). uv max (ethanol): 236 nm (log ε 4.20).
THERAP CAT: Progestogen.

4514. Haloprogin. *1,2,4-Trichloro-5-[(3-iodo-2-propynyl)oxy]benzene; 3-iodo-2-propynyl 2,4,5-trichlorophenyl ether; 2,4,5-trichlorophenyl γ-iodopropargyl ether;* M 1028; *Halotex; Mycanden; Mycilan; Polik.* $C_9H_4Cl_3IO$; mol wt 361.41. C 29.91%, H 1.12%, Cl 29.43%, I 35.12%, O 4.43%. Prepn: Seki *et al.*, *Agr. Biol. Chem. (Tokyo)* **27**, 150 (1963), *C.A.* **58**, 14635g (1963). Pharmacology: Seki *et al.*, *Antimicrob. Ag. Chemother.* **1963**, 569; Harrison *et al.*, *Appl. Microbiol.* **19**, 746 (1970).

White or pale yellow crystals, mp 113-114°; dec 190°. uv max (anhydr ethanol): 288.5, 298.5 nm. Easily soluble in methanol, ethanol; very slightly sol in water. LD_{50} in mice: >3 g/kg orally and s.c.; 510 mg/kg i.p.
THERAP CAT: Antibacterial.

4515. Halopropane. *3-Bromo-1,1,2,2-tetrafluoropropane; 1-bromo-2,2,3,3-tetrafluoropropane; 1,1,2,2-tetrafluoro-3-bromopropane; FHD-3; Tebron.* $C_3H_3BrF_4$; mol wt 194.97. C 18.48%, H 1.55%, Br 40.99%, F 38.98%. $CHF_2CF_2CH_2Br$. Prepn: Cohen, **U.S. pat. 3,080,430** (1963 to du Pont). Use in the study of cardiac arrhythmias: Katz, *J. Pharmacol. Exp. Ther.* **152**, 88 (1966).
Liquid, bp 74°; d_{20} 1.81; n_D^{25} 1.3558.
THERAP CAT: Anesthetic (inhalation).

4516. Halostachine. *α-[(Methylamino)methyl]benzenemethanol; l-α-(methylaminomethyl)benzyl alcohol; l-1-hydroxy-1-phenyl-2-methylaminoethane; l-phenyl(methylaminomethyl)carbinol; 2-methylamino-1-phenylethanol.* $C_9H_{13}NO$; mol wt 151.20. C 71.49%, H 8.66%, N 9.27%, O 10.58%. Isoln from *Halostachis caspica (Halostachis caspia)*: Menshikov, Rubinshtein, *J. Gen. Chem. USSR* **13**, 801

(1943), *C.A.* **39**, 1172 (1945). Synthesis: Menshikov, Borodina, *J. Gen. Chem. USSR* **17**, 1569 (1947), *C.A.* **42**, 2245 (1948). Configuration: Lukes *et al., Coll. Czech. Chem. Commun.* **26**, 466 (1961). As a possible cause of ryegrass staggers: Aasen *et al., Aust. J. Agr. Res.* **20**, 71 (1969).

$$OH$$
$$C_6H_5CHCH_2NHCH_3$$

Crystals, mp 43-45°. $[\alpha]_D^{20}$ $-47.03°$. Soluble in water, alc, ether.
Hydrochloride, crystals from alcohol, mp 113-114°. $[\alpha]_D^{20}$ $-52.46°$. Freely sol in water.

4517. Halothane. *2-Bromo-2-chloro-1,1,1-trifluoroethane;* bromochlorotrifluoroethane; 1,1,1-trifluoro-2,2-chlorobromoethane; Fluothane; Rhodialothan. $C_2HBrClF_3$; mol wt 197.39. C 12.17%, H 0.51%, Br 40.48%, Cl 17.96%, F 28.87%. $CF_3CHClBr$. Prepn from a mixt of F_3CCH_2Cl and F_3CCBr_2Cl: Suckling, Raventos, U.S. pat. **2,921,098** (1960 to I.C.I.); by rearrangement of $F_2BrCCHFCl$: Scherer, Kühn, U.S. pat. **2,959,624** (1960 to Hoechst); from BrCl-CHCBrCl₂: Chapman, McGinty, Brit. pat. **805,764** (1958 to I.C.I.); from Br_2ClCCF_3: McGinty, U.S. pat. **3,082,263** (1963 to I.C.I.); Madai, Muller, *J. Prakt. Chem.* **19**, 83 (1963). Comprehensive description: R. D. Daley in *Analytical Profiles of Drug Substances* vol. 1, K. Florey, Ed. (Academic Press, New York, 1972) pp 119-147.
Non-flammable, highly volatile liquid. Characteristic, sweetish, not unpleasant odor. d_4^{20} 1.871. bp 50.2°; bp_{243} 20°. n_D^0 1.3697. Sensitive to light, may be stabilized with 0.01% thymol. Soly in water 0.345%. Miscible with petr ether, other fat solvents.
THERAP CAT: Anesthetic (inhalation).
THERAP CAT (VET): Inhalation anesthetic.

4518. Haloxazolam. *10-Bromo-11b-(2-fluorophenyl)-2,3,7,11b-tetrahydrooxazolo[3,2-d][1,4]benzodiazepin-6(5H)-one;* CS-430; Somelin. $C_{17}H_{14}BrFN_2O_2$; mol wt 377.22. C 54.13%, H 3.74%, Br 21.18%, F 5.04%, N 7.43%, O 8.48%. Sleep-inducing agent, related structurally to oxazolam and cloxazolam, *q.v.* Prepn of analogous compounds: T. Miyadera *et al., J. Med. Chem.* **14**, 520 (1971). Structure-activity study: M. Yoshimoto *et al., Chem. Pharm. Bull.* **25**, 1378 (1977). Pharmacological studies: T. Kamioka *et al., Arzneimittel-Forsch.* **28**, 838 (1978). Metabolism in animals: R. Hayashi *et al., Oyo Yakuri* **17**, 617 (1979), *C.A.* **91**, 204161s (1979).

Colorless cryst, mp 185°. Sparingly sol in water. LD_{50} in mice: 1850 mg/kg orally (Kamioka).
Note: This is a controlled substance (depressant) listed in the U.S. Code of Federal Regulations, Title 21 Part 1308.14 (1985).
THERAP CAT: Sedative. Hypnotic.

4519. Haloxon. *Phosphoric acid bis(2-chloroethyl) 3-chloro-4-methyl-2-oxo-2H-1-benzopyran-7-yl ester;* 7-[[bis-(2-chloroethoxy)phosphinyl]oxy]-3-chloro-4-methyl-2H-1-benzopyran-2-one; 3-chloro-7-hydroxy-4-methylcoumarin bis(2-chloroethyl) phosphate; O,O-bis(β-chloroethyl) O-(3-chloro-4-methyl-7-coumarinyl) phosphate; 3-chloro-7-hydroxy-4-methyl-2H-1-benzopyran-2-one bis(chloroethyl) phosphate; 3-chloro-4-methylumbelliferone di(2-chloroethyl) phosphate; galoxone; helmirone; #96H60; Galloxon; Loxon; Luxon. $C_{14}H_{14}Cl_3O_6P$; mol wt 415.61. C 40.46%, H 3.40%, Cl 25.59%, O 23.10%, P 7.45%. Prepn: Belg. pat. **610,896** (1962 to Cooper, McDougall & Robertson), *C.A.* **57**, 13729g (1962). Toxicity data: Brown *et al., Nature* **194**, 379 (1962).

Crystals from ethanol, mp 91°. LD_{50} orally in rats: 900 mg/kg (Brown).
THERAP CAT (VET): Anthelmintic.

4520. Halquinol. *5,7-Dichloro-8-quinolinol mixt with 5-chloro-8-quinolinol and 7-chloro-8-quinolinol;* chlorquinol; SQ 16401; CHQ; Halquivet; Quinolor; Quixalin; Quixalud; Tarquinor. Proportions resulting naturally from chlorination of 8-quinolinol. Prepn of products contg 57-74 wt-% 5,7-dichloro-8-quinolinol, 23-40 wt-% 5-chloro-8-quinolinol, and up to 3 wt-% 7-chloro-8-quinolinol: Fr. pat. **1,372,414** (1964 to Olin Mathieson), *C.A.* **62**, 13130c (1965).
THERAP CAT: Topical anti-infective.
THERAP CAT (VET): Antimicrobial (intestinal infections).

4521. Hamamelis. Witch hazel; winter bloom; snapping hazel; striped alder; spotted alder; tobacco wood. Dried leaves of *Hamamelis virginiana* L., *Hamamelidaceae,* collected in autumn. History of hamamelis extract and distillate: Lloyd, Lloyd, *J. Am. Pharm. Assoc.* **24**, 220 (1935). Habit. North America (New England to Minnesota, southward to Louisiana). *Constit.* Hamamelitannin, *q.v.*, gallic acid, volatile oil, bitter principle.
THERAP CAT: Astringent.
THERAP CAT (VET): Has been used as astringent, hemostatic, sedative.

4522. Hamamelitannin. *2-C-[[(3,4,5-Trihydroxybenzoyl)oxy]methyl]-D-ribofuranose 5-(3,4,5-trihydroxybenzoate);* 2-C-(hydroxymethyl)-D-ribofuranose 2',5-digallate. $C_{20}H_{20}O_{14}$; mol wt 484.36. C 49.59%, H 4.16%, O 46.25%. From bark of *Hamamelis virginiana* L., *Hamamelidaceae:* Grüttner, *Arch. Pharm.* **236**, 278 (1898); from bark of *Castanea dentata* (Marsh.) Borkh., *Fagaceae:* Mayer, Kunz, *Naturwiss.* **46**, 206 (1959). Structure: Mayer *et al., Ann.* **688**, 232 (1965). Synthesis: Ezekiel *et al., Carbohyd. Res.* **11**, 233 (1969).

Prisms from water, mp 145-147°. $[\alpha]_D^{19}$ +32.6° (c = 1.5). Hexamethyl ether, $C_{26}H_{32}O_{14}$, needles from dioxane + water, mp 73-76°. $[\alpha]_{578}^{25}$ +10.0° (c = 2.0 in dioxane). Sol in acetone, dioxane, methanol. Practically insol in water.

4523. Hamamelose. *2-C-(Hydroxymethyl)-D-ribose.* $C_6H_{12}O_6$; mol wt 180.16. C 40.00%, H 6.71%, O 53.29%. From tannin of witch hazel (*Hamamelis virginiana* L., *Hamamelidaceae*): Fischer, Freudenberg, *Ber.* **45**, 2709 (1912); Freudenberg, Peters, *ibid.* **53**, 953 (1920); Anderson, *U.S.A.E.C.* UCRL-8870, 114 (1959). Structure: Schmidt, *Ann.* **476**, 250 (1929). Configuration: Schmidt, Heintz, *ibid.* **515**, 77 (1934). Synthesis: Novák, Sorm, *Coll. Czech. Chem. Commun.* **30**, 3303 (1965); Paulsen *et al., Ber.* **105**, 1978 (1972). Synthesis of L-hamamelose: Burton *et al., Proc. Chem. Soc.* **1962**, 181. Synthesis and ¹H, ¹³C NMR study of aqueous equilibrium: W. A. Szarek *et al., Can. J. Chem.* **61**, 461 (1983).

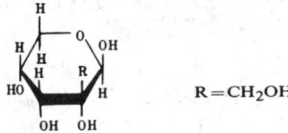

R=CH₂OH

D-Form, crystals from abs ethanol, mp 111°. $[\alpha]_D^{21}$ −7.4° (equilib in water).
L-Form, crystals from ethanol + ethyl acetate, mp 110-111°. $[\alpha]_D^{22}$ +1.3° (3 min) → +7.3° (equilib after 17 min).

4524. Hamycin. Primamycin. Polyene antibiotic complex produced by *Streptomyces pimprina*: M. J. Thirumalachar *et al.*, *Hindustan Antibiot. Bull.* **3**, 136 (1961), *C.A.* **55**, 27515i (1961); M. J. Thirumalachar, U.S. pat. **3,261,751** (1966 to Hindustan Antibiot.). At one time, hamycin was believed to be identical to hachimycin and candicidin, *q.q.v.*: H. Burrows, D. Calam, *J. Chromatog.* **53**, 566 (1970). Subsequent HPLC studies have shown the three polyene antibiotics to be different entities: P. Helboe *et al.*, *ibid.* **189**, 249 (1980). Structural study showing hamycin to be a mixture of components A, B, C, D: R. C. Pandey, K. L. Rinehart, *16th Interscience Conference on Antimicrob. Ag. Chemother.*, Chicago, 1976, *Abstracts of Papers*, no. 41. Absorption, excretion, tissue concentration: M. G. Phatak *et al.*, *Indian J. Exp. Biol.* **12**, 284 (1974), *C.A.* **82**, 118703j (1975). Toxicity and antifungal activity study: A. C. Parekh, C. V. Dave, *Life Sci.* **19**, 1737 (1976). HPLC study: W. Mechlinski, C. P. Schaffner, *J. Antibiot.* **33**, 591 (1980). Review of use in mycoses in man: M. J. Thirumalachar, *J. Sci. Ind. Res.* **31**, 542-544 (1972), *C.A.* **79**, 13308 (1973).
Yellow amorphous powder, no definite mp, decomp 160°. $[\alpha]_D^{25}$+216°. uv max (80% methanol): 383 nm ($E_{1cm}^{1\%}$ 916). An amphoteric compd. Almost insol in water, benzene, chloroform, dry lower aliphatic alcohols, ether; sol in basic solvents such as pyridine, collidine, and in aq lower alcohols. In concd H_2SO_4 gives stable blue color; no coloration with ferric chloride or with HCl. LD_{50} i. v. in mice (using sodium carboxymethyl cellulose as a vehicle): 6.16 mg/kg, Williams *et al.*, *Antimicrob. Ag. Chemother.* **1964**, 737-741. Also reported as LD_{50} i.v. in Swiss mice (using a colloidal suspension): 1.20 mg/kg, A. C. Parekh, C. V. Dave, *loc. cit.*
Hamycin A, mp >300° (dec), $[\alpha]_D^{24}$ +181.1° (c = 0.6 in DMF). uv max: 380 nm ($E_{1cm}^{1\%}$ 989).
THERAP CAT: Antifungal.

4525. Haplophytine. *[15(3aS,7R)]-3,4-Didehydro-19-hydroxy-16,17-dimethoxy-1-methyl-15-(2,3,5,6-tetrahydro-11-hydroxy-4-methyl-1,13-dioxo-1H-3a,7-methanopyrrolo-[1,2-a][1,3]benzodiazocin-7(4H)-yl)aspidospermidin-21-oic acid γ-lactone.* $C_{37}H_{40}N_4O_7$; mol wt 652.76. C 68.08%, H 6.18%, N 8.58%, O 17.16%. Insecticidal alkaloid from the Mexican plant *Haplophyton cimicidum* A. DC., *Apocynaceae*: Rogers *et al.*, *J. Am. Chem. Soc.* **74**, 1987 (1952); **76**, 2819 (1954). Partial structure: Snyder *et al.*, *ibid.* **80**, 3708 (1958). Structure: Rae *et al.*, *ibid.* **89**, 3061 (1967). Crystal structure and absolute config.: Zacharias, *Acta Crystallogr.* **26B**, 1455 (1970). Synthetic approaches: P. Yates, D. A. Schwartz, *Can. J. Chem.* **61**, 509 (1983). Review: Yates *et al.*, *J. Am. Chem. Soc.* **95**, 7842 (1973).

Crystals from ethanol + chloroform, mp 290-293° (rapid heating, starting at 250°). Also reported as mp 300-302° (Yates *et al.*, *loc. cit.*). $[\alpha]_D^{25}$ +109.0° (chloroform). uv max

(ethanol): 220, 265, 305 nm (ε 48500, 14300, 4500). Very sol in chloroform, benzene, dioxane, ethyl acetate. Moderately sol in acetone, methanol, somewhat less in ethanol. Practically insol in water, ether, petr ether. Readily sol in dil acids or alkalies.
Dihydrochloride, $C_{27}H_{31}N_3O_5$·2HCl, dec 208-218° (darkens at 200°).
O-Methylhaplophytine, $C_{28}H_{33}N_3O_5$, crystals from ether + ethanol, dec 288-291°. $[\alpha]_D^{24}$ +12° (c = 4.37 in chloroform).

4526. Haptens. Small molecules, generally of low mol wt, that form antigens with proteins, polypeptides or other carrier substances to produce antibodies. Although they do not induce antibody formation by themselves, haptens, as a portion of the complete antigen, enter into the combining sites of the antibodies, hence, they are also referred to as antigenic determinants. Pioneering work of K. Landsteiner to prepare "artificial conjugated antigens" is described in *The Specificity of Serological Reactions* (Harvard University Press, Cambridge, 2nd ed., 1945). Chemical nature of antigens and antibodies: D. H. Campbell, N. Bulman in *Fortschr. Chem. Org. Naturst.* **9**, 443-484 (1952). Structural features for antibody-hapten interaction: *The Structural Basis of Antibody Specificity*, Pressman, Grossberg, Eds. (W. A. Benjamin, New York, 1968) pp 30-147. Immune response to haptens: S. Leskowitz in *Immunogenicity*, F. Borek, Ed. (Elsevier, New York, 1972) pp 131-151. Prepn of antigenic hapten-carrier conjugates: B. F. Erlanger, *Methods Enzymol.* **70**, 85-104 (1980).

4527. Haptoglobins. α_2-Globulins which combine with hemoglobin to form a weak peroxidase and which constitute about one-quarter of the α_2-globulin fraction of human plasma: Jayle, Conas, *Bull. Soc. Chim. Biol.* **34**, 65 (1952). Prepn: Herman-Boussier *et al.*, *ibid.* **48**, 817, 837 (1960). For each individual there are three genetic types of haptoglobins, each reacting identically with antibodies to each of the others. Structure of this serum glycoprotein is a tetramer composed of α- and β-polypeptide chains, the α-chain varying in the three phenotypes, the β-chains being identical. Amino acid sequence of α-chains: Black, Dixon, *Nature* **218**, 736 (1968); Malchy, Dixon, *Can. J. Biochem.* **51**, 321 (1973). C- and N-terminal sequences of the β-chain: Barnett *et al.*, *Biochemistry* **11**, 1189 (1972); A. Kurosky *et al.*, *Comp. Biochem. Physiol.* **55B**, 453 (1976); *eidem*, *Biochemistry* **15**, 5326 (1976). Lowered levels of haptoglobins found in individuals with acute hepatitis and pernicious anemia. Review: Laurell, Grönvall, *Advan. Clin. Chem.* **5**, 135-172 (1962). Monograph: Kirk, *The Haptoglobin Groups in Man* (S. Karger, New York, 1968) 77 pp. Review on genetics, biochemistry and physiology of human haptoglobins: J. Javid, *Curr. Top. Hematol.* **1**, 151-192 (1978).

4528. Harmaline. *4,9-Dihydro-7-methoxy-1-methyl-3H-pyrido[3,4-b]indole;* 1-methyl-7-methoxy-3,4-dihydro-β-carboline; 3,4-dihydroharmine; harmidine; harmalol methyl ether; *O*-methylharmalol. $C_{13}H_{14}N_2O$; mol wt 214.26. C 72.87%, H 6.59%, N 13.08%, O 7.47%. CNS stimulant from seeds of *Peganum harmala* L., *Zygophyllaceae*: Goebel, *Ann.* **38**, 363 (1841); from *Banisteria caapi* Spruce, *Malpighiaceae*: Hochstein, Paradies, *J. Am. Chem. Soc.* **79**, 5735 (1957). Structure: Manske *et al.*, *J. Chem. Soc.* **1927**, 1. Synthesis: Späth, Lederer, *Ber.* **63**, 120, 2102 (1930); Spenser, *Can. J. Chem.* **37**, 1851 (1959). Identity with harmidine: Robinson, *Chem. & Ind. (London)* **1965**, 605. Pharmacology: Fuentes, Longo, *Neuropharmacology* **10**, 15 (1971). Metabolism: Ho *et al.*, *Biochem. Pharmacol.* **20**, 1313 (1971). Review of structure and synthesis work: Hofmann, *Sv. Farm. Tidskr.* **75**, 933 (1971), *C.A.* **76**, 149609g (1972).

Orthorhombic bipyramidal prisms, tablets from methanol, rhombic octahedra from ethanol, mp 229-231°. Solns fluoresce blue. pK 4.2. uv max (methanol): 218, 260, 376 nm (log ε 4.27, 3.90, 4.02). Slightly sol in water, alcohol, ether; quite sol in hot alcohol, dil acids.

Hydrochloride dihydrate, slender, yellow needles, moderately sol in water, alcohol.

N-Acetylharmaline, needles, mp 204-205°.

4529. Harmalol. 4,9-Dihydro-1-methyl-3H-pyrido[3,4-b]indol-7-ol; 3,4-dihydro-1-methyl-9H-pyrido[3,4-b]indol-7-ol. $C_{12}H_{12}N_2O$; mol wt 200.24. C 71.97%, H 6.04%, N 13.99%, O 7.99%. From seeds of *Peganum harmala* L., *Zygophyllaceae:* Göbel, *Ann.* **38**, 363 (1841); Fischer, *Ber.* **18**, 400 (1885); by demethylation of harmaline: Coulthard *et al., Biochem. J.* **27**, 727 (1933).

Trihydrate, red needles from water. Dec 212° (anhydr). Readily sol in hot water, acetone, chloroform, alkali hydroxides, but not carbonates. Aq solns are yellow with green fluorescence.

Lactate monohydrate, bright yellow leaflets, mp 116-120°; when anhydr mp 174-176°. Sol in water and alcohol.

O-Ethylharmalol, brown needles from alc, mp 237-239°.

O-n-Propylharmalol, brown needles from alc, mp 195-197°.

O-n-Butylharmalol, colorless needles from alc, mp 173°.

O-Methylharmalol, see harmaline.

4530. Harman. 1-Methyl-9H-pyrido[3,4-b]indole; 3-methyl-4-carboline; 2-methyl-β-carboline; aribine; loturine; passiflorin. $C_{12}H_{10}N_2$; mol wt 182.22. C 79.09%, H 5.53%, N 15.38%. From bark of *Sickingia rubra* (Mart.) K. Schum. (*Arariba rubra* Mart.), *Rubiaceae; Symplocus racemosa* Roxb., *Symplocaceae;* and *Passiflora incarnata* L., *Passifloraceae:* Rieth, Wohler, *Ann.* **120**, 247 (1861); Späth, *Monatsh.* **40**, 351; **41**, 401 (1920); Neu, *Arzneimittel-Forsch.* **4**, 601 (1954). Structure: Neu, *ibid.* **6**, 94 (1956). Synthesis: Harvey, Robson, *J. Chem. Soc.* **1938**, 97; Snyder *et al., J. Am. Chem. Soc.* **70**, 222 (1948); Clemo, Holt, *J. Chem. Soc.* **1953**, 1313; Kametani *et al., ibid.* (C) **1968**, 1006. Isoln from cigarette smoke: Poindexter, Carpenter, *Chem. & Ind. (London)* **1962**, 176.

Bitter orthorhombic crystals from heptane + cyclohexane, mp 237-238°. Exhibits bright blue fluorescence in uv light. pKa's 7.37, 144.6. uv max (methanol): 234, 287, 347 nm (log ε 4.57, 4.21, 3.66). Absorption and fluorescence spectra: O. S. Wolfbeis *et al., Monatsh. Chem.* **113**, 509 (1982). Practically insol in water; sol in dil acids. LD_{50} i.p. in mice: 50 mg/kg, E. B. Sigg *et al., Arch. Int. Pharmacodyn.* **149**, 164 (1964).

Hydrochloride, $C_{12}H_{10}N_2$·HCl, rosettes of needles from ethanol + 20% HCl in water, sublimes at 120-130°.

4531. Harmine. 7-Methoxy-1-methyl-9H-pyrido[3,4-b]indole; banisterine; yageine; telepathine; leucoharmine. $C_{13}H_{12}N_2O$; mol wt 212.25. C 73.56%, H 5.70%, N 13.20%, O 7.54%. CNS stimulant isolated from seeds of *Peganum harmala* L., *Zygophyllaceae:* Göbel, *Ann.* **38**, 363 (1841); Reinhard *et al., Phytochemistry* **7**, 503 (1968); from *Banisteria caapi* Spruce, *Malpighiaceae:* Hochstein, Paradies, *J. Am. Chem. Soc.* **79**, 5735 (1957); from *Banisteriopsis inebrians* Morton, *Malpighiaceae:* O'Connell, Lynn, *J. Am. Pharm. Assoc.* **42**, 753 (1953). Structure: Manske *et al., J. Chem. Soc.* **1927**, 1, 240. Synthesis: Späth, Lederer, *Ber.* **63B**, 120 (1930); Akaboro, Saito, *ibid.* **2245**; Hahn *et al., Ber.* **67B**, 2031 (1934), *Ann.* **520**, 107, 123 (1935), *Ber.* **71B**, 2163, 2175 (1938); Harvey, Robson, *J. Chem. Soc.* **1938**, 97.

Metabolism: Slotkin, DiStefano, *J. Pharmacol. Exp. Ther.* **174**, 456 (1970).

Slender, orthorhombic prisms from methanol, mp 261° (dec). Sublimes. pKa 7.70. uv max (methanol): 241, 301, 336 nm (log ε 4.61, 4.21, 3.69). Absorption and fluorescence spectra: O. S. Wolfbeis, E. Fürlinger, *Z. Physikal. Chem.* **129**, 171 (1982). Slightly sol in water, alcohol, chloroform, ether.

Hydrochloride dihydrate, crystals, mp 262° (dec), mp 321° when anhydr. Sol in 40 parts water, freely sol in hot water. Aq solns have blue fluorescence. LD_{50} i.v. in mice: 38 mg/kg, K. K. Chen *et al., J. Pharmacol. Exp. Ther.* **79**, 127 (1943).

4532. Hashish. Hasach; kif. Purified alcoholic extract of *Cannabis sativa* L. (*Cannabis sativa* var. *indica* Auth.), *Moraceae* deprived of its volatile oil; resin secreted by the flowering tops of the female plant. Principal active constituents are tetrahydrocannabinols, *q.v.* Review and recent advances in the chemistry of hashish: Mechoulam, Gaoni, *Fortschr. Chem. Org. Naturst.* **25**, 175 (1967). *See also* Cannabis.

4533. Hasubanonine. 7,8-Didehydro-3,4,7,8-tetramethoxy-17-methylhasubanan-6-one. $C_{21}H_{27}NO_5$; mol wt 373.43. C 67.54%, H 7.29%, N 3.75%, O 21.42%. Alkaloid having a modified morphinan skeleton with a five membered heterocyclic ring, hitherto unknown in natural sources. (Alkaloids with this unique skeleton are called *hasubanan* alkaloids; *see* S. Shiotani, T. Kometani, *Tetrahedron Letters* **1976**, 767 for synthesis of hasubanan skeleton). Isolated from *Stephania japonica* Miers, *Menispermaceae:* Kondo *et al., Ann. Rept. ITSUU Lab.* **2**, 35 (1951); Satomi, *ibid.* **3**, 37 (1953); Kondo, Satomi, *ibid.* **8**, 41 (1957); *C.A.* **47**, 5951f (1953); **48**, 2728c (1954); **51**, 17956i (1957). Structure: Tomita *et al., Tetrahedron Letters* **1964**, 2937; *eidem, Chem. Pharm. Bull.* **13**, 538 (1965). Total synthesis: Ibuka *et al., Tetrahedron Letters* **1970**, 4811; *Chem. Pharm. Bull.* **22**, 782 (1974). Biosynthesis: A. R. Battersby *et al., Chem. Commun.* **1974**, 773; *eidem, J. Chem. Soc. Perkin Trans. I* **1981**, 2010, 2016, 2030. *Review:* Y. Inubushi, T. Ibuka in *The Alkaloids,* vol. **XVI**, R. H. F. Manske, Ed. (Academic Press, New York, 1977) pp 393-428.

Prisms from methanol, mp 116°. $[\alpha]_D$ −219° (ethanol).

4534. HCG. Human Chorionic Gonadotrop(h)in; chorionic gonadotropin; pregnancy urine extract; Choriogonin; Follutein; Coriantin; Endocorion; Antuitrin S; Choragon; Riogon; Antophysin; Prolan; A.P. L.; Predalon; Pregnesin; Profasi; Ferti-Cept; Gravimun; Glanduantin-Ch; Gonadotraphon L.H.; Gonan; Lutormone; Physostab; Pregnyl; Primogonyl; Ambinon; Gonadyl-Chorionic; Antèparsine; Physex; Randonos; Antelobine; Apoidina; Korotrin; Glukor Injection; Gestasol Dry; Libigen; Profasi HP. A hormone synthesized by chorionic tissue of the placenta and found in urine during pregnancy. Found also in body fluids of persons having trophoblastic disease or embryonic testicular or ovarian tumors. HCG, together with LH, FSH, and TSH, *q.q.v.*, comprise the glycoprotein hormone family. All are dimeric and contain α and β subunits. Within a species, the peptide portions of the α-subunit are essentially identical between the four hormones, while the β-subunits, although exhibiting varying degrees of homology, are unique and believed to be responsible for the biological specificity of the individual hormones. Isoln procedures: Zondek, Aschheim, *Klin. Wochenschr.* **7**, 831 (1928); Gurin *et al., J. Biol. Chem.* **128**, 525 (1939); Katzman *et al., ibid.* **148**, 501 (1943); Claesson *et al., Acta Endocrinol.* **1**, 1 (1948). Isoln of α and β subunits: N. Swaminathan, O. P. Bahl, *Biochem. Biophys. Res. Commun.* **40**, 422 (1970); F. J. Morgan, R. E. Canfield, *Endocrinology* **88**, 1045 (1971). Amino acid sequence of α-subunit: R. Bellesario *et al., J. Biol. Chem.* **248**, 6796 (1973); of the β-subunit: R. B. Carlsen *et al., ibid.* 6810; F. J. Morgan *et al., ibid.* **250**, 5247 (1975). Supplements the hypophysis in maintaining growth of the corpus luteum during pregnancy; disappears from maternal circulation after delivery of the placenta. Stimulates Leydig tissue in the male which results in the secretion of androgens. Stability: Bischoff, *J. Biol. Chem.* **165**, 406 (1946); C. H. Li, *Vitam. Horm. (New York)* **7**, 223 (1949). Comprehensive monograph: H. H. Cole, Ed., *Gonadotropins, Their Chemical and Biological Properties and Secretory Control* (W. H. Freeman, San Francisco, 1964) 252 pp. *Reviews:* Brody in *Foetus and Placenta,* A. Klopper, E. Diczfalusy, Eds. (Blackwell Scientific Pubs., Oxford, 1969) pp 301-373; Canfield *et al., Recent Progr. Horm. Res.* **27**, 121-164 (1971); Hellema, *J. Endocrinol.* **49**, 393 (1971); Saxena, *Vitam. Horm. (New York)* **29**, 97-115 (1971); J. G. Pierce, T. F. Parsons, *Ann. Rev. Biochem.* **50**, 465-495 (1981).

Long thin rods or needles from 60% alc. Stable in dry form. Freely sol in water; sol in aq glycerol and glycols. Insol in the usual anhydr organic solvents. Isoelec pt pH 3.2-3.3. Precipitated by 70% alcohol or by acetone from neutral or slightly acid solns. Also precipitated by phosphomolybdic and phosphotungstic acids, but not by trichloroacetic, sulfosalicylic, flavianic, picric, or picrolonic acids. Dil aq solns lose their activity rapidly on storage even at 0°, or by heating, or by excess acid or base. Protective colloids (gelatin, serum protein) help to prevent decompn of aq solns. HCG is stable in glycerol soln at 100° for one hour and seems compatible with sodium lauryl sulfate.

The medicinal product of commerce is standardized in international units. One I.U. equals 0.1 mg of a standardized powder, *see* Council Report *J. Am. Med. Assoc.* **113**, 2418 (1939).

THERAP CAT: Gonad-stimulating principle.

THERAP CAT (VET): In certain reproductive disorders.

4535. HCS. *Lactogen (human placental); lactogenic hormone (placental human);* human chorionic somatomammotropin; chorionic growth hormone-prolactin; CGP; human placental lactogen; HPL. Mol wt about 21,000. Growth hormone from human placenta. Exhibits lactogenic as well as growth-promoting activity: Josimovich, MacLearen, *Endocrinology* **71**, 209 (1962); Kaplan, Grumbach, *J. Clin. Endocrinol. Metab.* **24**, 80 (1964). Structure is a single polypeptide chain containing 190 amino acid residues: Li *et al., Science* **173**, 56 (1971); Sherwood *et al., Nature New Biol.* **233**, 59 (1971). Structural revision to 191 amino acid residues and comparison with human growth hormone and human prolactin: Li *et al., Arch. Biochem. Biophys.* **155**, 95 (1973); Niall *et al., Recent Progr. Horm. Res.* **29**, 387 (1973). Purification: Parcells, Dahlgren, U.S. pat. **3,687,833** (1972 to Upjohn). Terminology: Li *et al., Experientia* **24**, 1288 (1968). *Review:* M. Chatterjee, H. N. Munro in *Vitamins and Hormones* vol. **35**, P. L. Munson *et al.,* Eds. (Academic Press, New York, 1977) pp 149-208.

4536. Heart Muscle Extract. Usually prepd from the hearts of calf embryos. *Ref:* Strauss, *Deut. Med. Wochenschr.* **1952**, 1284; Umhau, *Landarzt* **1952**, 100; Kristen, *Deut. Med. J.* **1952**, 250; Willems, *Die Medizinische* **1953**, no. 13; Jasinski, *Cardiologia* **23**, 49 (1953); Brenner, *Klin. Wochenschr.* **1954**, 295; Witzleb *et al., ibid.* 297. Commercial preparations: *Myocardone; Recosen; Rocosenin; Corhormon; Lysomiol; Hormocardiol; Herzolan; Cordiomon* (inj.).

THERAP CAT: Coronary vasodilator.

4537. Hecogenin. (25R)-3β-Hydroxy-5α-spirostan-12-one. $C_{27}H_{42}O_4$; mol wt 430.61. C 75.30%, H 9.83%, O 14.86%. A steroidal sapogenin which has been isolated from plants, particularly from numerous *Agave* species. Isoln: Marker *et al., J. Am. Chem. Soc.* **69**, 2167 (1947); Rubin,

U.S. pat. **3,303,186** (1967). Synthesis and configuration: Mazur *et al., J. Am. Chem. Soc.* **82**, 5889 (1960). Separation of an optically inactive product from sapogenin mixtures: Cardenas, U.S. pat. **3,013,010** (1961 to Searle). *Review:* L. F. Fieser, M. Fieser, *Steroids* (Reinhold, New York, 1959) pp 667-671 sqq.

Crystals from acetone, mp 264-266°. $[\alpha]_D$ +8° ($CHCl_3$) (Mazur *et al.*); mp 245°, 253°, 268° (Marker *et al.*); mp 240-245°, 245-250°. $[\alpha]_D$ ±0° ($CHCl_3$) (Cardenas).

USE: In prepn of steroidal hormones.

4538. Hectorite. Strese & Hofmann's hectorite. Swelling and gelling clay of the montmorillonite group. Approx formula: $Na_{0.67}(Mg,Li)_6Si_8O_{20}(OH,F)_4$. Some analyses show neither Li nor F, whether by oversight or by variation in the mineral is uncertain: M. H. Hey, *Mineral Species and Varieties* (Brit. Museum, London, 2nd ed., 1962) p 205.

USE: In the chill-proofing of beer: Shaler *et al.,* U.S. pat. **3,100,707** (1963 to American Tansul). *Caution:* Dust can be irritating to respiratory tract.

4539. Hedaquinium Chloride. 2,2'-(1,16-Hexadecanediyl)bisisoquinolinium dichloride; 2,2'-hexadecamethylenediisoquinolinium dichloride; hexadecamethylene-1,16-bis-(isoquinolinium chloride); BIQ 16; Teoquil. $C_{34}H_{46}Cl_2N_2$; mol wt 553.67. C 73.76%, H 8.38%, Cl 12.81%, N 5.06%. Prepn of similar salts: Austin *et al., J. Chem. Soc.* **1958**, 1489; Brit. pat. **839,505** (1960 to Allen & Hanburys).

THERAP CAT: Antiseptic.

4540. Hederagenin. 3,23-Dihydroxyolean-12-en-28-oic acid; caulosapogenin; melanthigenin. $C_{30}H_{48}O_4$; mol wt 472.68. C 76.22%, H 10.24%, O 13.54%. Occurs as glycoside in many saponins, *see* α-Hederin. Isoln and structure: van der Haar, *Arch. Pharm.* **250**, 424 (1912); Ruzicka *et al., Helv. Chim. Acta* **28**, 380 (1945); Haynes *et al., J. Chem. Soc.* **1963**, 744. Identity with melanthigenin: Mustafa, Soliman, *ibid.* **1943**, 70. Identity with caulosapogenin: McShefferty, Stenlake, *ibid.* **1956**, 2314. Molecular and crystal structure: R. Roques *et al., Acta Crystallogr.* **B34**, 1634 (1978). *See also* α- and β-Amyrin and Oleanolic Acid.

Crystals from alc, mp 332-334°. $[\alpha]_D^{20}$ +81° (c = 0.7 in pyridine). Freely sol in pyridine; sol in chloroform-alcohol mixtures; slowly sol in alcohol; practically insol in water; sol in dil alcoholic NaOH, but not in water solns of alkalies.

Diacetate, mp 172-174°. $[\alpha]_D^{20}$ +64° (CHCl$_3$).
Methyl ester, mp 240°. $[\alpha]_D^{23}$ +76° (c = 0.8 in CHCl$_3$).

4541. α-Hederin. Helixin (the saponin). $C_{41}H_{66}O_{12}$; mol wt 750.98. C 65.57%, H 8.86%, O 25.57%. Isoln from ivy leaves (*Hedera helix* L., *Araliaceae*): van der Haar, *Arch. Pharm.* **250**, 424 (1912); **251**, 632, 650; idem, *Biochem. Z.* **76**, 335 (1916); idem, *Ber.* **54**, 3142 (1921). Isoln and structure: Tschesche et al., *Z. Naturforsch.* **20b**, 708 (1965).

R = rhamnose (1 → 2), arabinose (1 →)

Precipitated from ethanol by addition of ether, mp 256-259°. $[\alpha]_D^{20}$ +14.5° (c = 0.92 in methanol).

4542. Helenalin. 3,3a,4,4a,7a,8,9,9a-Octahydro-4-hydroxy-4a,8-dimethyl-3-methyleneazuleno[6,5-b]furan-2,5-dione; 6α,8β-dihydroxy-4-oxoambrosa-2,11(13)-dien-12-oic acid 12,8-lactone. $C_{15}H_{18}O_4$; mol wt 262.29. C 68.68%, H 6.92%, O 24.40%. Pseudoguaianolide sesquiterpenoid lactone from *Helenium autumnale* L., *H. amarum* (Raf.) H. Roch, *H. microcephalum* DC., *Compositae*. Isoln: Clark, *J. Am. Chem. Soc.* **58**, 1982 (1936); Adams, Herz, ibid. **71**, 2546 (1949). Structure: Büchi, Rosenthal, ibid. **78**, 3860 (1956); Barton, de Mayo, *Quart. Rev. (London)* **11**, 189 (1957); Herz, *J. Org. Chem.* **27**, 4043 (1962). Abs config studies: Herz, Kagan, ibid. **32**, 216 (1967). Synthesis of racemic form: Y. Ohfune et al., *J. Am. Chem. Soc.* **100**, 5946 (1978); M. R. Roberts, R. H. Schlessinger, ibid. **101**, 7626 (1979); C. H. Heathcock et al., ibid. **104**, 1907 (1982). Toxicity study: D. A. Witzel et al., *Am. J. Vet. Res.* **37**, 859 (1976).

Bitter, sternutative crystals from benzene, mp 167-168°. $[\alpha]_D^{25}$ −102.8° (c = 3.64 in 95% ethanol). uv max: 223 nm (ε 11,900). Absorption spectrum: Adams, Herz, loc. cit. Slightly sol in water; sol in alcohol, chloroform, hot benzene. LD$_{50}$ in mice: 150 mg/kg orally.
Acetylhelenalin, $C_{17}H_{20}O_5$, crystals from aq methanol, mp 184°. uv max: 221, 316 nm (ε 12,600, 61).
Human Toxicity: Intensely poisonous, capable of causing paralysis of voluntary and cardiac musculature and fatal gastroenteritis.

4543. Helenynolic Acid. 9-Hydroxy-10-octadecen-12-ynoic acid. $C_{18}H_{30}O_3$; mol wt 294.42. C 73.43%, H 10.27%, O 16.30%. From seed oil of *Helichrysum bracteatum* Andr., *Compositae*: R. G. Powell et al., *J. Am. Oil. Chem. Soc.* **42**, 165 (1965). Structure: eidem, *J. Org. Chem.* **30**, 610 (1965). Absolute configuration: J. C. Craig et al., ibid. 4342. Partial synthesis of racemic helenynolic acid: H. R. S. Conacher, Gunstone, *Lipids* **5**, 137 (1970). Synthesis of (±)-form: T. B. Patrick, G. F. Melm, *J. Org. Chem.* **44**, 645 (1979).

$$CH_3(CH_2)_4C\equiv CCH=CHCH(CH_2)_7COOH$$
$$|$$
$$OH$$

Methyl ester, $C_{19}H_{32}O_3$. uv max (isooctane): 228, 238 nm (ε 17,400, 14,300). $[\alpha]_{600}^{26}$ −7° (c = 3.6 in ethanol).

4544. Helicin. 2-(β-D-Glucopyranosyloxy)benzaldehyde; salicylaldehyde β-D-glucoside. $C_{13}H_{16}O_7$; mol wt 284.26. C 54.93%, H 5.67%, O 39.40%. Prepn by the oxidation of salicin with dil nitric acid: Schiff, *Ann.* **154**, 19 (1870); from salicylaldehyde + O-tetraacetyl-α-glucosidyl bromide: Robertson, Waters, *J. Chem. Soc.* **1930**, 2729. ORD and stereochemical studies: Tsuzuki et al., *Bull. Chem. Soc. Japan* **44**, 526 (1971).

Needles with 0.75 mol H$_2$O from H$_2$O; mp 175-176° when dried at 100°. $[\alpha]_D^{20}$ −60° (c = 1.4). One gram dissolves in 55 ml water; freely sol in hot water, alcohol. Forms compds with urea, thiourea, certain amino acids.
Tetraacetate, $C_{21}H_{24}O_{11}$, needles from alc, mp 142°. $[\alpha]_D^{20}$ −37° (acetone).
Note: The same formula is ascribed to spirein, found in *Spiraea camtschatica* Pall. and in *S. ulmaria* L., *Rosaceae*. Emulsin hydrolyzes helicin and spirein, yielding D-glucose and salicylaldehyde.

4545. Heliosupine. 2-Methyl-2-butenoic acid 7-[[2,3-dihydroxy-2-(1-hydroxyethyl)-3-methyl-1-oxobutoxy]methyl]-2,3,5,7a-tetrahydro-1H-pyrrolizin-1-yl ester; cynoglossophine. $C_{20}H_{31}NO_7$; mol wt 397.46. C 60.44%, H 7.86%, N 3.52%, O 28.18%. Hepatotoxic pyrrolizidine alkaloid isolated from *Heliotropium supinum* L., *Boraginaceae*: Denisova et al., *Doklady Akad. Nauk SSSR* **93**, 59 (1953); from *Cynoglossum officinale* L., *Boraginaceae*: Man'ko, Borisyuk, *Ukrain. Khim. Zhur.* **23**, 362 (1957), C.A. **52**, 2188a (1958); Man'ko, Marchenko, *Khim. Prir. Soedin.* **7**, 537 (1971), C.A. **75**, 126598t (1971). Identity with cynoglossophine: Man'ko, *Ukrain. Khim. Zhur.* **25**, 627 (1959), C.A. **54**, 12494d (1960). Structure: Crowley, Culvenor, *Aust. J. Chem.* **12**, 694 (1959). Biosynthesis studies: Crout, *J. Chem. Soc. (C)* **1967**, 1233. Reviews: see Lasiocarpine.

Colorless gum. $[\alpha]_D^{20}$ −4.3° (c = 5.1 in ethanol).
Methyl ether, see Lasiocarpine.

4546. Helium. He; at. wt 4.0026; at. no. 2. The natural gas is ^{4}He with a trace of ^{3}He. Three short-lived artificial isotopes are known: 5; 6; 8. Noble gas discovered in the sun by Lockyer and Frankland in 1868. Obtained by Hillebrand in 1890 by heating uranium minerals and identified by Ramsay in 1895. Found in natural gas from which it is extracted on a commercial scale. Produced in the decay of radioactive elements: 1 kg of uranium in its conversion into 865 g of lead forms 756 l of helium; also produced by the bombardment of beryllium, lithium, and other light elements with cosmic rays, x-rays and high-speed protons and deuterons. Such bombardments take place in nature and give a constant supply. Monograph: G. A. Cook, *Argon, Helium and the Rare Gases* (Interscience, New York, 1961). Review: Cockett, Smith, "The Monatomic Gases" in *Comprehensive Inorganic Chemistry* vol. 1, J. C. Bailar, Jr. et al., Eds. (Pergamon Press, Oxford, 1973) pp 139-211; E. Cook, *Science* **206**, 1141-1146 (1979).
Colorless, odorless, nonflammable, inert gas. d^0 (gas) 0.17847 g/l. bp −268.9° (4.215°K). Crit temp −267.9°; crit press. 2.26 atm. Heat of vaporization 19.6 cal/g-atom. Trouton's const 4.64. Two liq forms exist: He I above ∼2.2°K; He II below ∼2.2°K. d (liq; at bp): 0.1249 g/cc. He II is a superconducting liq; has very low viscosity; superfluid. Helium cannot be frozen by lowering the temp at ordinary press.; no triple point. mp at 25.05 atm 1.00°K.

Heat of fusion 4.35 cal/g-atom. Very slightly sol in water: at 0° = 0.97 ml/100 ml; at 50° = 1.08 ml/100 ml.

USE: Liquid helium (the most volatile liq known) is used for the production of low temps. The inert, nonflammable gas is used for balloons and airships, lifting power is 0.93 if hydrogen is taken as 1.00. In nucleonics and rocket research; in lasers. Diluent for gases. *Q-Gas*, a mixture of 98.7% helium and 1.3% butane, has been used as a filling for gas-flow Geiger counters.

4547. Hellebrin. *3-[(6-Deoxy-4-O-β-D-glucopyranosyl-α-L-mannopyranosyl)oxy]-5,14-dihydroxy-19-oxobufa-20,-22-dienolide;* hellebrigenin glucorhamnoside. $C_{36}H_{52}O_{15}$; mol wt 724.82. C 59.66%, H 7.23%, O 33.11%. Cardiac glycoside isolated from rhizome of *Helleborus niger* L., *Ranunculaceae:* Karrer, *Festschrift E. C. Barell* **1936**, 238, *C.A.* **31**, 2348[8] (1937). Structure: Karrer, *Helv. Chim. Acta* **26**, 1353 (1943); Wissner, Kating, *Planta Med.* **20**, 344 (1971).

rhamnose-glucose

Crystals from hot methanol, mp 283-284°. $[\alpha]_D^{20}$ −23.4° (50% methanol). Sol in dil alcohol; less sol in methanol, ethanol; slightly sol in water. Practically insol in ether. MLD in guinea pigs: 0.85 μMol/kg, Klup, *Arch. Exp. Pathol. Pharmakol.* **252**, 314 (1966).

4548. Helminthosporal. *1,7-Dimethyl-4-(1-methyleth-yl)bicyclo[3.2.1]oct-6-ene-6,8-dicarboxaldehyde.* $C_{15}H_{22}O_2$; mol wt 234.33. C 76.88%, H 9.46%, O 13.66%. Toxin produced by the fungus *Bipolaris sarokiniana* Shoemaker *(Helminthosporium sativum).* Isoln: deMayo *et al., Can. J. Chem.* **39**, 1608 (1961). Structure: deMayo *et al., J. Am. Chem. Soc.* **84**, 494 (1962). Stereochemistry: deMayo *et al., Can. J. Chem.* **41**, 2996 (1963). Total synthesis and absolute configuration of the (−)-form: Corey, Nozoe, *J. Am. Chem. Soc.* **87**, 5728 (1965). Synthesis of the (+)-form: E. Piers, H. P. Isenring, *Can. J. Chem.* **55**, 1039 (1977). Synthesis of racemic mixture: M. Yanagiya *et al., Tetrahedron Letters* **1979**, 1761.

Crystals from light petr, mp 56-59°. $[\alpha]_D$ −49° (c = 1.18 in CHCl₃). bp$_{0.015}$ 115-120°. uv max (ethanol): 266 nm (ε 11,000).

Bis-oxime, $C_{15}H_{24}N_2O_2$, crystals from alc, mp 177-182°.

4549. Helminthosporol. *8-(Hydroxymethyl)-1,7-dimeth-yl-4-(1-methylethyl)bicyclo[3.2.1]oct-6-ene-6-carboxalde-hyde; 8-(hydroxymethyl)-4-isopropyl-1,7-dimethylbicyclo-[3.2.1]oct-6-ene-6-carboxaldehyde.* $C_{15}H_{24}O_2$; mol wt 236.34. C 76.22%. H 10.24%, O 13.54%. A natural plant-growth regulator isolated from *Helminthosporium sativum.* Prepn: Tamura *et al., Agr. Biol. Chem. (Tokyo)* **29**, 216 (1965); Tamura, Sakurai, *ibid.* **28**, 337 (1964); Tamura *et al., ibid.* **27**, 738 (1963). Synthesis of (+)-form: E. Piers, H. P. Isenring, *Can. J. Chem.* **55**, 1039 (1977). Structure: Tamura *et al., loc. cit.; see also* Helminthosporal.

Crystals from n-hexane, mp 98°. $[\alpha]_D^{23}$ −28.7° (c = 1.93 in CHCl₃). uv max (ethanol): 267 nm.

4550. Helonias. False unicorn; blazing star; starwort. Rhizome and roots of *Chamaelirium luteum* (L.) A. Gray, *Liliaceae. Habit.* Ontario and Eastern U.S.

4551. Helveticoside. *3β-[(2,6-Dideoxy-β-D-ribo-hexopy-ranosyl)oxy]-5,14-dihydroxy-19-oxo-5β-card-20(22)-enolide;* erisimin; erysimin; alleoside A. $C_{29}H_{42}O_9$; mol wt 534.63. C 65.15%, H 7.92%, O 26.93%. A β-glycoside consisting of one mole strophanthidin and one mole D-digitoxose. Isoln from *Erysimum helveticum* (Jacq.) DC. and *Erysimum crepidifolium* Reichenb., *Cruciferae,* and structure: Nagata *et al., Helv. Chim. Acta* **40**, 41 (1957); *eidem, Festschrift Arthur Stoll* **1957**, 715; Gmelin, Ger. pat. **1,221,764** (1966). From *Erysimum* spp.: Kowalewski, *Helv. Chim. Acta* **43**, 1314 (1960).

Dihydrate, needles from dil methanol, mp 153-157°. $[\alpha]_D^{25}$ +30.7° (c = 1.5 in methanol); $[\alpha]_D^{24}$ +26.0° (c = 1.1 in chloroform). uv max (ethanol): 217, 304 nm (log ε 4.27, 1.55). *Violent poison!* LD$_{100}$ i.v. in cats: 0.104 mg/kg, Graebner, Giesel, *Arzneimittel-Forsch.* **22**, 1854 (1972).

3',4'-Diacetate, $C_{33}H_{46}O_{11}$, prisms from methanol + ether, mp 237-244°. $[\alpha]_D^{25}$ +41.3° (chloroform).

4552. Helvolic Acid. *6,16-Bis(acetyloxy)-3,7-dioxo-29-nordammara-1,17(20),24-trien-21-oic acid; (Z)-6β,16β-di-hydroxy-3,7-dioxo-29-nor-8α,9β,13α,14β-dammara-1,17-(20),24-trien-21-oic acid diacetate;* fumigacin. $C_{33}H_{44}O_8$; mol wt 568.68. C 69.69%, H 7.80%, O 22.51%. Antibiotic substance of the fusidane class. Produced by *Aspergillus fumigatus:* Waksman *et al., J. Bact.* **45**, 233 (1943); by *A. fumigatus* mut. *helvola:* Chain *et al., Brit. J. Exp. Pathol.* **24**, 108 (1943). Structural studies: Okuda *et al., Chem. Pharm. Bull.* **12**, 121 (1964); Oxley, *Chem. Commun.* **1966**, 729; Okuda *et al., Tetrahedron Letters* **1967**, 2295. Revised structure and stereochemistry: Iwaki *et al., Chem. Commun.* **1970**, 1119. Chemical and biological data: Reshetova, *Antibiotiki* **14**, 554 (1969). Review of literature until 1960: Wilson, "Miscellaneous *Aspergillus* Toxins" in *Microbial Toxins,* vol. VI, A. Ciegler *et al.,* Eds. (Academic Press, New York, 1971) p 265.

Needles from methanol, mp 215°. $[\alpha]_D^{25}$ −121° (chloroform). uv max (ethanol): 231 nm (log ε 4.24). Very slightly sol in water; slightly sol in petr ether, methanol, ethanol. More sol in hot methanol, ethanol; sol in chloroform, acetone, ethyl acetate, benzene, pyridine, glacial acetic acid, dioxane, ether.

Methyl ester, $C_{34}H_{46}O_8$, *methyl helvolate.* Crystals from methanol, mp 257°. $[\alpha]_D^{25}$ −140° (chloroform).

4553. Hematein. *6a,7-Dihydro-3,4,6a,10-tetrahydroxy-benz[b]indeno[1,2-d]pyran-9(6H)-one;* hydroxybrasilein; hydroxybrazilein. $C_{16}H_{12}O_6$; mol wt 300.26. C 64.00%, H 4.03%, O 31.97%. Not to be confused with hematin. From hematoxylin or logwood extract and NH₃ by treatment with air: Engels *et al., J. Chem. Soc.* **93**, 1115 (1908); Rolland,

Teintex **3**, 261, 322, 460 (1938); Justin-Mueller, *Melliand Textilber.* **30**, 26, 63 (1949), *C.A.* **46**, 8375h (1952).

Reddish-brown crystals with yellowish-green metallic luster. mp above 200°; also stated as 250° with decompn. Sol in about 1700 parts water; slightly sol in alcohol, ether; insol in benzene, chloroform; freely sol in ammonia with brownish-violet color and in dil NaOH with bright red color. Forms salts with heavy metals.

USE: As an indicator like hematoxylin; for staining animal tissue, particularly cell nuclei.

4554. Hematin. *[7,12-Diethenyl-3,8,13,17-tetramethyl-21H,23H-porphine-2,18-dipropanoato(4—)-N²¹,N²²,N²³,N²⁴]-hydroxyferrate(2—) dihydrogen; [dihydrogen 3,7,12,17-tetramethyl-8,13-divinyl-2,18-porphinedipropionato(2—)]hydroxyiron;* hydroxy[dihydrogen protoporphyrin IX-ato(2—)]-iron; ferriheme hydroxide; ferriporphyrin hydroxide; ferriprotoporphyrin basic; hydroxyhemin; phenodin. $C_{34}H_{33}$-FeN_4O_5; mol wt 633.49. C 64.46%, H 5.25%, Fe 8.82%, N 8.84%, O 12.63%. Found in the body in pathological conditions, *e.g.*, after phosgene poisoning, in pernicious anemia. Prepn: Fischer-Orth, *Die Chemie des Pyrrols* II, 1, 386 (Leipzig, 1937). Brief review including structure: J. E. Falk, *Porphyrins and Metalloporphyrins* (Elsevier, New York, 1964) pp 17, 23, 46. Toxicology: D. L. Lips *et al.*, *Toxicol. Letters* **2**, 329 (1978). Use in treatment of hepatic porphyria: C. J. Watson *et al.*, *Ann. Intern. Med.* **79**, 80 (1973); G. J. Dhar *et al.*, *ibid.* **83**, 20 (1975); J. M. Lamon *et al.*, *Medicine* **58**, 252 (1979).

Solvated crystals from pyridine, dry at 40° *in vacuo*. Freely sol in dil solns of alkali hydroxides, slightly sol in hot pyridine. Unstable. Absorption max (10% aq NaOH): 580 nm (E_{mM} 10.5). LD_{50} i.v. in rats: 4.32 mg/100 g (Lips).

4555. Hematoporphyrin. *7,12-Bis(1-hydroxyethyl)-3,8,-13,17-tetramethyl-21H,23H-porphine-2,18-dipropanoic acid; 7,12-bis(1-hydroxyethyl)-3,8,13,17-tetramethyl-2,18-porphinedipropionic acid;* 1,3,5,8-tetramethyl-2,4-bis(α-hydroxyethyl)porphine-6,7-dipropionic acid; hematoporphyrin IX; Photodyn. $C_{34}H_{38}N_4O_6$; mol wt 598.68. C 68.21%, H 6.40%, N 9.36%, O 16.03%. Prepd from hemin by the action of hydrobromic acid in glacial acetic acid: Fischer-Orth, *Die Chemie des Pyrrols* II, 1, 421 (Leipzig, 1937). Review and prepn: J. E. Falk, *Porphyrins and Metalloporphyrins* (Elsevier, New York, 1964) pp 175-177.

Deep red crystals. Absorption max (0.1N KOH): 615.5, 565, 534.4, 499.5 nm. Insol in water. Sol in alc. Sparingly sol in ether, chloroform.

Hydrochloride monohydrate, $C_{34}H_{39}ClN_4O_6.H_2O$, *Sensibion*. Dark red crystals. Prepn of a stable product: Woods, Steigman, U.S. pat. **2,858,320** (1958 to Baxter Labs.).

THERAP CAT: Antidepressant.

4556. Hematoxylin. *7,11b-Dihydrobenz[b]indeno[1,2-d]pyran-3,4,6a,9,10(6H)-pentol;* hematoxiline; hydroxybrazilin; hydroxybrasilin. $C_{16}H_{14}O_6$; mol wt 302.29. C 63.57%, H 4.67%, O 31.76%. From the heart-wood of logwood (*Haematoxylon campechianum* Linn., *Leguminosae*): Chevreul, *Ann. Chim. Phys.* **82**, 54, 126 (1810). Structure: Perkin, Robinson, *J. Chem. Soc.* **93**, 489 (1908). Synthesis: Dann, Hofmann, *Angew. Chem.* **75**, 1125 (1963); Morsingh, Robinson, *Tetrahedron* **26**, 281 (1970); Kirkiacharian, Billet, *Bull. Soc. Chim. France* **1972**, 3292. *Review:* Robinson, *ibid.* **1958**, 125. Stereochemistry: Craig *et al.*, *J. Org. Chem.* **30**, 1573 (1965).

Trihydrate, white to yellowish crystals; redden on exposure to light, mp 100-120°; also stated as 140°. Slightly sol in cold water, ether; sol in hot water, hot alc, also in alkali hydroxides, borax, glycerol. Its solns darken on standing.

USE: Chiefly as a stain in microscopy; also in manufacture of ink.

4557. Hematoxylon. Logwood. Heart-wood of *Haematoxylon campechianum* L., *Leguminosae*. *Habit.* Central America; grown in the West Indies. *Constit.* About 10% hematoxylin; tannin, resin.

USE: Largely as dye.

4558. Heme. *[7,12-Diethenyl-3,8,13,17-tetramethyl-21H,23H-porphine-2,18-dipropanoato(4—)-N²¹,N²²,N²³,N²⁴]-ferrate(2—) dihydrogen; [dihydrogen 3,7,12,17-tetramethyl-8,13-divinyl-2,18-porphinedipropionato(2—)]iron;* 1,3,5,8-tetramethyl-2,4-divinylporphine-6,7-dipropionic acid ferrous complex; ferroheme; hem; protoheme; protoheme IX; reduced hematin; ferroprotoporphyrin. $C_{34}H_{32}FeN_4O_4$; mol wt 616.48. C 66.24%, H 5.23%, Fe 9.06%, N 9.09%, O 10.38%. Heme occurs free in tissues in the presence of certain pathological conditions, and in normal tissues; it occurs as the prosthetic group of a number of hemoproteins. It has been identified as the prosthetic group of hemoglobins, erythrocruorins (the hemoglobin analog of many invertebrates), myoglobins, some peroxidases, catalases, and cytochromes b. It is the color-furnishing portion of hemoglobin. Obtained when a soln of hematin in alkali is reduced in the absence of nitrogenous substances: Bertin-Sans, de Moitessier, *Compt. Rend.* **114**, 923 (1892); Dhéré *et al.*, *ibid.* **165**, 515 (1917). Synthesis: Fischer-Orth, *Die Chemie des Pyrrols* II, 1, 384 (Leipzig, 1937). Biosynthesis: Shemin, *Naturwiss.* **57**, 185 (1970). *Review:* J. E. Falk, *Porphyrins and Metallo-*

porphyrins (Elsevier, NewYork, 1964) pp 8, 94, 183. Comprehensive monograph: Chance *et al., Hemes and Hemoproteins* (Academic Press, New York, 1966) 624 pp; *Handbook of Experimental Pharmacology* vol. **44,** entitled "Heme and Hemoproteins", F. DeMatteis, W. N. Aldridge, Eds. (Springer-Verlag, New York, 1978) 449 pp.

Fine brown needles with a dark violet sheen. Absorption max in phosphate buffer at pH 7: ~550, 575 nm ($E_{mM}^{572} = 5.5$). Sparingly sol in glacial acetic acid; freely sol in the presence of oxygen. Very unstable.

4559. Hementin. Anticoagulant enzyme from the salivary glands of the South American giant leech *Haementeria ghiliarii.* Extraction: R. T. Sawyer *et al.,* U.S. pat. **4,390,-630** (1983 to Univ. California, Berkeley). Anticoagulant, fibrinolytic properties: A. Z. Budzynski *et al., Proc. Soc. Exp. Biol. Med.* **168,** 266 (1981). Purification and identification as single polypeptide chain with mol wt approx 120,000: S. M. Malinconico *et al., J. Lab. Clin. Med.* **103,** 44 (1984). Proteolytic degradation of fibrinogen by hementin: *eidem, ibid.* **104,** 842 (1984). Protease inhibition: E. H. Murer *et al., Thromb. Haemostasis* **51,** 24 (1984).

4560. Hemerythrin. Mol wt 108,000. A non-heme, oxygen-carrying protein found in members of four invertebrate phyla: sipunculids, polychaetes, priapulids and brachiopods. Oxygenated form is called *oxyhemerythrin.* Isoln from the sipunculid, *Golfingia goldii* (or *Phascolosoma goldii):* Klotz *et al., Arch. Biochem. Biophys.* **68,** 284 (1957). Consists of eight subunits, each containing two ferrous ions which form a complex with one molecule of oxygen: Klotz, Keresztes-Nagy, *Biochemistry* **2,** 445, 923 (1963). Each subunit is built up from 113 amino acids. Amino acid sequence: Klippenstein *et al., ibid.* **7,** 3868 (1968); Klippenstein, *ibid.* **11,** 372 (1972). Structure of trimeric hemerythrin: J. L. Smith *et al., Nature* **303,** 86 (1983). *Reviews:* Okamura, Klotz, "Hemerythrin" in *Inorganic Biochemistry,* **vol. 1,** G. L. Eichhorn, Ed. (Elsevier, New York, 1973) pp 320-343; Klotz *et al., Science* **192,** 335-344 (1976); J. S. Loehr, T. M. Loehr in *Advances in Inorganic Biochemistry* vol. 1, G. L. Eichhorn, L. G. Marzilli, Eds. (Elsevier, New York, 1979) pp 235-252.

Hemerythrin is colorless. Oxyhemerythrin forms violet-pink crystals. Spectral data: Garbett *et al., Arch. Biochem. Biophys.* **135,** 419 (1969).

4561. Hemicelluloses. Large group of polysaccharides found, in association with lignin, in the primary and secondary cell walls of all plants and of some seaweeds. Whether lignin and hemicellulose are chemically bonded or lignin mechanically entraps hemicellulose molecules is still unknown. There are variations in the hemicellulosic composition of plants and even between the organs of the same plant. Hemicellulosic composition of plants changes with growth and maturation and is influenced also by environmental factors during growth. Principal sugar residues present in hemicellulose are: D-xylose, D-glucose, D-galactose, D-mannose, L-arabinose, D-glucuronic acid, D-galacturonic acid, 4-O-methyl-D-glucuronic acid, L-rhamnose and L-fucose. The most ubiquitous and abundant hemicelluloses are the *xylans* which are composed of linear and/or branched chains of β-(1→4) linked D-xylopyranosyl units. Hemicelluloses from woods: T. E. Timell, *Adv. Carbohyd. Chem.* **19,** 247 (1964); *idem,* **20,** 409 (1965). Hemicelluloses

from grasses and cereals, K. C. B. Wilke, *ibid.* **36,** 215 (1979).

4562. Hemi-Dewar Biphenyl. *2-Phenylbicyclo[2.2.0]-hexa-2,5-diene.* $C_{12}H_{10}$; mol wt 154.20. C 93.46%, H 6.54%. Prepn by decompn of cyclobutadieneiron tricarbonyl with ceric ammonium nitrate in the presence of phenylacetylene: Burt, Pettit, *Chem. Commun.* **1965,** 517.

Oil. n_D^{26} 1.5834. Upon heating to 90° for several minutes is converted quantitatively into biphenyl.

4563. Hemin. *Chloro[7,12-diethenyl-3,8,13,17-tetramethyl-21H,23H-porphine-2,18-dipropanoato(4−)-N21,N22,-N23,N24]ferrate(2−) dihydrogen; chloro[dihydrogen 3,7,12,17-tetramethyl-8,13-divinyl-2,18-porphinedipropionato(2−)]-iron;* chlorohemin; 1,3,5,8-tetramethyl-2,4-divinylporphine-6,7-dipropionic acid ferrichloride; Teichmann's crystals; ferriheme chloride; ferriprotoporphyrin chloride; ferriporphyrin chloride. $C_{34}H_{32}ClFeN_4O_4$; mol wt 651.96. C 62.64%, H 4.95%, Cl 5.44%, Fe 8.57%, N 8.59%, O 9.81%. Prepd from hemoglobin soln by heating with acetic acid and sodium chloride. Practical procedure for its prepn from ox blood: Schalfejeff, *J. Russ. Physiol. Chem. Soc.* **1885,** 30; *Ber.* **18,** 232 (1885); Gattermann-Wieland, *Praxis des Organischen Chemikers* 23rd ed., p 407; Fischer-Orth, *Die Chemie des Pyrrols* II, 1, 377 (Leipzig, 1937); H. Fischer, *Org. Syn.* coll. vol. III, 442 (1955); Labbe, Nishida, *Biochim. Biophys. Acta* **26,** 437 (1957). Biosynthesis: Karlzeile, *Angew. Chem.* **66,** 729 (1954). *Review:* Stoll, *Experientia* **4,** 6 (1948); H. H. Inhoffen, *Naturwiss.* **55,** 457 (1968); W. S. Caughey in *Inorganic Biochemistry* vol. 2, G. L. Eichhorn, Ed. (Elsevier, New York, 1973) pp 797-831.

Long, thin blades from glacial acetic acid or from chloroform-pyridine-acetic acid, appearing brown in transmitted light and steel-blue in reflected light. Sinters at 240° but is not melted at 300°. Absorption spectrum: Fischer-Orth, *loc. cit.* Freely sol in dil ammonia water, also in solns of sodium hydroxide with hematin formation, *i.e.,* the chlorine is displaced by an OH group. Practically insol in carbonate solns, dil acid solns. Sol in strong organic bases such as trimethylamine, *p*-toluidine, dimethylaniline. Sol in concd H_2SO_4 with loss of Fe. Sparingly sol in 70-80% alc. Practically insol but stable in water.

Dimethyl ester, $C_{36}H_{36}ClFeN_4O_4$, needles from benzene, not melted at 300°. Freely sol in acetic acid, benzene, chloroform, acetone.

Diethyl ester, $C_{38}H_{40}ClFeN_4O_4$, crystals, freely sol in chloroform.

4564. Hemipyocyanine. *1-Phenazinol;* α-hydroxyphenazine; 1-hydroxy-5,10-diazoanthracene; pyoxanthose. $C_{12}H_8N_2O$; mol wt 196.20. C 73.46%, H 4.11%, N 14.28%, O 8.15%. Found in old cultures of *Pseudomonas pyocyanea.* Isoln: Fordos, *Compt. Rend.* **56,** 1128 (1863). Prepd from pyocyanine by the action of 2% NaOH: Wrede, Strack, *Z. Physiol. Chem.* **140,** 12 (1924); by the reaction of pyrogallol monomethyl ether and *o*-phenylenediamine followed by demethylation of the formed α-methoxyphenazine: Surrey, *Org. Syn. coll. vol. III*, 753 (1955). Alternate syntheses: Hegedüs, *Helv. Chim. Acta* **33**, 766 (1950); Vivian, *Nature* **178**, 753 (1956).

Yellow needles from benzene, mp 159-160°. Sublimes easily *in vacuo*. Slightly sol in hot water; freely sol in the usual organic solvents except petr ether. Sol in aq alkaline solns with purplish-red color which turns yellow on neutralization. Forms red salts with mineral acids.

4565. Hemisulfur Mustard. *2-(2-Chloroethylthio)ethanol;* 2-chloro-2'-hydroxydiethyl sulfide; 2-chloroethyl 2-hydroxyethyl sulfide; β-chloroethyl β-hydroxyethyl thioether; mustard chlorohydrin; semi-mustard gas; semisulfur mustard. C_4H_9ClOS; mol wt 140.64. C 34.16%, H 6.45%, Cl 25.21%, O 11.38%, S 22.80%. $ClCH_2CH_2SCH_2CH_2OH$. Prepn: Ogston *et al., Trans. Faraday Soc.* **44**, 45 (1948); Grant, Kinsey, *J. Am. Chem. Soc.* **68**, 2075 (1946); Fuson, Ziegler, *J. Org. Chem.* **11**, 510 (1946); Rueggeberg *et al., ibid.* **13**, 110 (1948); Tsou *et al., ibid.* **26**, 4987 (1961).

Oily, somewhat hygroscopic liquid. $bp_{0.6}$ 100° (Fuson); $bp_{0.5-0.75}$ 87°; bp_6 44.5° (Rueggeberg). n_D^{20} 1.5188; $n_D^{24.5}$ 1.5205. Sol in water; miscible with bis(2-chloroethyl) sulfide, thiodiglycol. Unstable, forming sulfonium salts at room temp or in solns with ethanol or chloroform.

4566. Hemocyanins. A non-heme, oxygen carrying copper protein found in arthropods and mollusca of which *keyhole-limpet hemocyanin (KLH)* is an example. Dissolved in the hemolymph; not found in blood cells. Mol wt ranges from 4.5×10^5 to 1.3×10^7. One molecule of oxygen is bound by two copper atoms. The oxygenated form, *oxyhemocyanin* is blue, while *deoxyhemocyanin* is colorless. The copper can be removed reversibly to form *apohemocyanin*. Structure of snail, *Helix pomatia*, hemocyanin: J. E. Mellema, A. Klug, *Nature* **239**, 146 (1972). The role of copper in hemocyanin: R. Lontie, L. Vanquickenborne, *Met. Ions Biol. Syst.* **3**, 183 (1974). *Reviews:* K. E. van Holde, E. F. J. van Bruggen in *Subunits in Biological Systems vol. 5*, pt. A, S. N. Timasheff, G. D. Fasman, Eds. (Dekker, New York, 1971) pp 1-53; R. Lontie, R. Witters in *Inorganic Biochemistry, vol. I*, G. L. Eichhorn, Ed. (Elsevier, New York, 1973) pp 344-358; E. F. J. van Bruggen, *Trends Biochem. Sci.* **5**, 185-8 (1980). Review of respiratory function: J. Bonaventura, C. Bonaventura, *Am. Zool.* **20**, 7-17 (1980); C. P. Magnum, *ibid.* 19-38.

USE: KLH as an experimental antigen in animals.

4567. Hemoglobin. Hb; ferrohemoglobin. The major component of red blood cells which transports oxygen from the lungs to body tissues and facilitates the return transport of carbon dioxide. Mammalian hemoglobins have mol wts of about 64,500. Composed of four peptide chains called globins, *q.v.*, each of which is bound to a heme, *q.v.* Normal human hemoglobin is composed of a pair of two identical chains. Iron is coordinated to four pyrrole nitrogens of protoporphyrin IX, and to an imidazole nitrogen of a histidine residue from the globin side of the porphyrin. The sixth coordination position is available for binding with oxygen and other small molecules. Called *oxyhemoglobin*, HbO_2, in the oxygenated form and *carboxyhemoglobin*, HbCO, when oxygen is displaced by carbon monoxide. Binds reversibly with oxygen while the heme iron remains in the ferrous state. Autoxidation is prevented by the cover of hydrophobic groups of the globin. When the iron in hemoglobin is oxidized from the ferrous to the ferric state the compd is called methemoglobin, *q.v.* and is accompanied by a loss of oxygen-binding capacity. Hemoglobin is usually prepd by separating the red blood corpuscles from the lighter plasma by centrifuging; the plasma is siphoned off, and on adding ether to the blood corpuscle paste, the cells burst. After another centrifugation to remove the ruptured cell envelopes, a clear red soln of the protein is obtained [Fieser, Fieser, *Org.*

Chem. (New York, 3rd ed., 1956) p 455]. Prepn of cryst HbO_2 from washed horse or dog erythrocytes: Heidelberger, *J. Biol. Chem.* **53**, 31 (1922); *see also* Ferry, Green, *ibid.* **81**, 175 (1929); from human blood: Drabkin, *ibid.* **164**, 703 (1946). Structure studies: Muirhead, Perutz, *Nature* **199**, 633 (1963); Perutz *et al., ibid.* **219**, 131 (1968); Perutz, *Proc. Roy. Soc. London* **173B**, 113 (1969). Respiratory properties of hemoglobin and its function as a carrier of oxygen and carbon dioxide: Peters, Van Slyke, *Quantitative Clinical Chemistry vol. I* (Baltimore, 1932). Mechanism of action: Arnone in *Ann. Rev. Med.* **25**, 123-130 (1974). *Reviews:* Lemberg, Legge, *Hematin Compounds and Bile Pigments* (New York, 1949); F. W. Sunderman, *Hemoglobin—its Precursors and Metabolites* (Lippincott, Philadelphia, 1964); H. Lehmann, R. G. Huntsman, *Man's Hemoglobins (ibid.* 1966); Huisman, Schroeder, *New Aspects of the Structure, Function, and Synthesis of Hemoglobins* (Butterworth, London, 1971); M. F. Perutz, *Ann. Rev. Biochem.* **48**, 327-386 (1979); G. Fermi, M. F. Perutz, *Haemoglobin & Myoglobin* (Oxford Univ. Press, New York, 1982) 104 pp; R. E. Dickerson, I. Geis, *Hemoglobin: Structure, Function, Evolution, and Pathology* (Benjamin-Cummings, Menlo Park, Calif., 1983) 176 pp.

Crystal form, solubility, affinity for oxygen, absorption spectra differ quantitatively in hemoglobins of different species, due to the variation in amino acid sequence of the protein moiety since the same heme group is present in all vertebrate and many invertebrate hemoglobins. Human HbO_2 = tetragonal crystals; horse HbO_2 = orthorhombic crystals from citrated blood, monoclinic crystals from oxalated blood. Absorption spectra, *see* Lemberg, Legge, *loc. cit.,* 228. Oxyhemoglobin is an article of commerce where it is called hemoglobin. Brownish-red powder or scales. Soluble in about 7 parts water, slowly sol in glycerol.

4568. Hempa. *Hexamethylphosphoric triamide;* hexamethylphosphoramide; hexametapol; HMPA; HMPT; ENT 50882. $C_6H_{18}N_3OP$; mol wt 179.20. C 40.22%, H 10.12%, N 23.45%, O 8.93%, P 17.28%. $[(CH_3)_2N]_3PO$. Prepn: Saul, Godfrey, U.S. pat. **2,752,392** (1956 to Monsanto); Godfrey, U.S. pat. **2,852,550** (1958 to Monsanto); Miller, Lomonte, U.S. pat. **3,084,190** (1963 to Dow); Vetter, Noeth, *Ber.* **96**, 1308 (1963). Solvation effects: J. E. Dubois, A. Bienvenue, *Tetrahedron Letters* **1966**, 1809. Toxicity studies: T. B. Gaines, *Toxicol. Appl. Pharmacol.* **14**, 515 (1969). Carcinogenicity studies: J. A. Zapp, *Science* **190**, 422 (1975); *IARC Monographs* **15**, 211 (1977). *Review:* H. Normant, *Angew. Chem. Int. Ed.* **6**, 1047 (1967).

Colorless liquid, completely miscible with water. mp 7.20°. bp_{760} 235°. bp_{11} 105-107°; bp_6 97-99°; $bp_{2.5}$ 78°. n_D^{21} 1.4572. d_{20} 1.03. Dipole moment at 25°: 5.37D. LD_{50} in male, female rats (mg/kg): 2650, 3360 orally (Gaines).

Note: This substance may reasonably be anticipated to be a carcinogen: *Fourth Annual Report on Carcinogens* (NTP 85-002, 1985) p 114.

USE: As aprotic solvent in organic synthesis. As de-icing additive for jet fuels. Chemosterilant for a number of insect pests; also as a chemical mutagen.

4569. Henna. Dried powdered leaves of *Lawsonia alba* Lam., *L. inermis* L., and *L. spinosa* L., Lythraceae. Obtained from North Africa or India. Contains about 1% of lawsone, *q.v.* Ref: Cox, *Analyst* **63**, 397 (1938); Talaat, *Brit. Med. J.* **II**, 944 (1960).

USE: For dyeing hair and nails auburn to red, in the Orient together with "reng," the dried, powdered leaves of the indigo plant, in order to produce darker and even bluish-black shades. For relatively permanent dyeing the pH must be about 5.5; this is achieved by the addition of citric, boric, or adipic acid. Ingredient of many commercial hair rinses.

4570. Heparamine. *N*-Desulfoheparin. The $—NHSO_3H$ groups in heparin are replaced by $—NH_2$ groups. Prepn of sodium salt by methanolysis of sodium heparinate and saponification: Velluz *et al., Compt. Rend.* **247**, 1521 (1958); from sodium heparinate heated with HCl, neutralized with $NaHCO_3$ and dialyzed against running water: Foster *et al., J. Chem. Soc.* **1961**, 1204; from sodium heparinate demineralized with Dowex 50 resin: **Brit. pat. 863,235** (1961 to UCLAF).

Sodium salt, solid, sol in water. $[\alpha]_D^{20}$ +67 ± 2°.

N-Acyl derivatives, *heparides*. Similar to heparin in clearing alimentary lipemia but are practically devoid of its anticoagulant activity.

4571. Heparin. Heparinic acid. Glycosaminoglycan with anticoagulant activity. Heterogenous mixture of variably sulfated polysaccharide chains composed of repeatng units of D-glucosamine and either L-iduronic or D-glucuronic acids. Mol wt ranges from 6000-30000 Da. Biosynthesized and stored in mast cells of various animal tissues, particularly liver, lung or gut. Commercial heparin is isolated from beef lung or pork intestinal mucosa. Isoln from mammalian tissue: Howell, *Am. J. Physiol.* **63**, 434 (1922-23); **71**, 553 (1924-25); Korn, *J. Biol. Chem.* **234**, 1325 (1959); L. B. Jaques, *Can. J. Biochem. Physiol.* **37**, 1183 (1959); J. A. Bush *et al.*, U.S. pat. **2,884,358** (1959 to So. Calif. Gland). Purification: G. Nominé *et al.*, U.S. pat. **2,989,438** (1961 to UCLAF); Toccacele, U.S. pat. **3,016,331** (1962 to Ormonoterapia Richter); L. Roden *et al.*, *Methods Enzymol.* **26**, 73 (1972). Structural studies: M. L. Wolfrom, *J. Am. Chem. Soc.* **72**, 5796 (1950); Velluz *et al.*, *Compt. Rend.* **247**, 1521 (1958); M. L. Wolfrom *et al.*, *J. Org. Chem.* **29**, 540 (1964). Configuration of glycosidic linkages: M. L. Wolfrom *et al.*, *ibid.* **31**, 1173 (1966); A. S. Perlin *et al.*, *Can. J. Chem.* **48**, 2260 (1970); T. Helting, U. Lindahl, *J. Biol. Chem.* **246**, 5442 (1971). Identification of L-iduronic acid residues: A. S. Perlin *et al.*, *Carbohyd. Res.* **7**, 369 (1968). Antithrombotic activity results from the binding and activation of *antithrombin III*, a plasma protein which inhibits several enzymes in the coagulation cascade: R. D. Rosenberg, *Fed. Proc.* **36**, 10 (1977). Anticoagulant activity is related to the mol wt of the polysaccharide fragments; low molecular weight components exhibit decreased hemorrhagic effects while retaining antithrombin binding ability: L.-O. Andersson *et al.*, *Thromb. Res.* **15**, 531 (1979); T. W. Barrowcliffe *et al.*, *Brit. J. Haematol.* **41**, 573 (1979); J. Hirsch *et al.*, *Sem. Thromb. Hemostas.* **11**, 13 (1985). Characterization of the antithrombin binding site: U. Lindahl *et al.*, *Proc. Nat. Acad. Sci. USA* **76**, 3198 (1979); J. Choay *et al.*, *Thromb. Res.* **18**, 573 (1980). Synthesis of the pentasaccharide corresponding to the binding site sequence: *eidem, Biochem. Biophys. Res. Commun.* **116**, 492 (1983).
Symposium on structure, activity and clinical applications: *Fed. Proc.* **36**, 9-116 (1977). Review of mechanism of action: I. Björk, U. Lindahl, *Mol. Cell. Biochem.* **48**, 161-182 (1982); of structure-activity relationships and prepn of low mol wt fractions: B. Casu, *Advan. Carbohyd. Chem. Biochem.* **43**, 51-134 (1985); of biosynthesis: U. Lindahl *et al.*, *Trends Biochem. Sci.* **11**, 221-225 (1986). Comprehensive description: F. Nachtmann *et al.*, in *Analytical Profiles of Drug Substances* vol. **12**, K. Florey, Ed. (Academic Press, New York, 1983) pp 215-276. Overview of clinical results in pulmonary embolism and venous thrombosis: R. Collins *et al.*, *N. Engl. J. Med.* **318**, 1162-1170 (1988); of clinical studies with low mol wt heparinoids: H. ten Cate *et al.*, *Am. J. Hematol.* **27**, 146-153 (1988).

R = H or SO₃⁻   R' = SO₃⁻ or COCH₃

Antithrombin Binding Site of Heparin

Heparin has a rotation of $[\alpha]_D^{20}$ +55°.
Calcium salt, *Calciparine, Ecasolv*.
Magnesium salt, *magnesium heparinate, Cutheparine*. Sol in water. Insol in organic solvents.
Potassium salt, *Clarin* (formerly).
Sodium salt, *heparin sodium, Heprinar, Hepsal, Lipo-Hepin, Liquémin, Lipo-Hepinette, Longheparin, Monoparin, Panheprin, Pularin, Liquaemin Sodium, Minihep, Thrombo-Hepin, Thromboliquine, Thrombophob, Unihep*. White to grayish-brown amorphous powder. Odorless, hygroscopic. $[\alpha]_D^{25}$ +47° (c = 1.5 in water). One gram dissolves in 20 ml

water. Sol in saline soln. Practically insol in alcohol, acetone, benzene, chloroform, ether. pH of 1% aq soln = 6.0 to 7.5. Absorption spectrum: Burson *et al.*, *J. Am. Chem. Soc.* **78**, 5874 (1956). Ampuled solns may be stored at room temp for at least 12 months. Commercially available ampuled, sterile solns contain 0.5% phenol or chlorobutanol as preservative.
THERAP CAT: Anticoagulant.
THERAP CAT (VET): Anticoagulant.

4572. Hepaxanthin. *5,6-Epoxy-5,6-dihydroretinol;* 5,6-monoepoxyvitamin A; vitamin A epoxide; 574-chromogen. $C_{20}H_{30}O_2$; mol wt 302.44. C 79.42%, H 10.00%, O 10.58%. Isoln from cod liver oil and formation from vitamin A: Karrer, Jucker, *Helv. Chim. Acta* **28**, 717 (1945); **30**, 559 (1947). Prepn and properties: Jungalwala, Cama, *Biochem. J.* **95**, 17 (1965). *See also* Lakshmanan *et al.*, *ibid.* **27**; Tsukida *et al.*, *J. Vitaminol.* **14**, 95 (1968).

Viscous yellow oil. uv max (ethanol): 272 nm. Freely sol in chloroform, ether, benzene. Moderately sol in alcohol. Slightly sol in petr ether.

4573. HEPES. *4-(2-Hydroxyethyl)-1-piperazineethanesulfonic acid;* N-(2-hydroxyethyl)piperazine-N'-2-ethanesulfonic acid. $C_8H_{18}N_2O_4S$; mol wt 238.30. C 40.32%, H 7.61%, N 11.76%, O 26.86%, S 13.45%. One of several zwitterionic N-substituted aminosulfonic acids known as "*Good*" *buffers*, active in the pH range of 6-8.5. Prepn: N. E. Good *et al.*, *Biochemistry* **5**, 467 (1966). Buffering characteristics: W. J. Ferguson *et al.*, *Anal. Biochem.* **104**, 300 (1980). Temperature effects on pKa and pH: R. N. Roy *et al.*, *Cryo-Letters* **6**, 285 (1985). Interaction with hydroxyl radicals: M. Hicks, J. M. Gebicki, *FEBS Letters* **199**, 92 (1986). Interference with Lowry protein determination: H. M. Himmel, W. Heller, *J. Clin. Chem. Clin. Biochem.* **25**, 909 (1987). Use as biological buffer: H. Eagle, *Science* **174**, 500 (1971); K. V. Rao, *Ind. J. Exp. Biol.* **15**, 552 (1977); E. D. Lalague *et al.*, *J. Microsc.* **127**, 307 (1982); G. I. McFadden, M. Melkonian, *Phycologia* **25**, 551 (1986).

Crystals from alcohol + water, mp 234°. pKa₁ ≃3, pKa₂ (20°) 7.55. ΔpKa/°C −0.014. Saturated aqueous soln is 2.25M at 0°.
USE: Biological buffer.

4574. Hepronicate. *3-Pyridinecarboxylic acid 2-hexyl-2-[[(3-pyridinylcarbonyl)oxy]methyl]-1,3-propanediyl ester;* nicotinic acid triester with 2-hexyl-2-(hydroxymethyl)-1,3-propanediol; 2-hexyl-2-(hydroxymethyl)-1,3-propanediol trinicotinate; 2,2-dihydroxymethyl-n-octanol trinicotinate; 1,1,1-trimethylolheptane trinicotinate; 1,1,1-(trihydroxymethyl)heptane trinicotinate; Megrin. $C_{28}H_{31}N_3O_6$; mol wt 505.58. C 66.52%, H 6.18%, N 8.31%, O 18.99%. Prepn: M. Nakanishi *et al.*, Neth. pat. Appl. **6,514,807**; *eidem*, U.S. pat. **3,384,642**; *eidem*, Japan. pat. 67 **10988** (1966, 1968 and 1967, all to Yoshitomi).

Crystals from ethanol, mp 94-96°.
THERAP CAT: Peripheral vasodilator.

4575. Heptabarbital. *5-(1-Cyclohepten-1-yl)-5-ethyl-*

2,4,6(1H,3H,5H)-pyrimidinetrione; 5-(1-cyclohepten-1-yl)-5-ethylbarbituric acid; 5-ethyl-5-cycloheptenylbarbituric acid; heptabarb; Heptadorm; Medomin. $C_{13}H_{18}N_2O_3$; mol wt 250.29. C 62.38%, H 7.25%, N 11.19%, O 19.18%. Prepn: **Fr. pat. 870,714** (1942 to Geigy); Taub, **U.S. pat. 2,501,551** (1950).

Crystals, mp 174°. Slightly bitter taste. uv max (0.2N NaOH): 218.5, 254 nm. Very sparingly sol in water, more sol in alcohol. At 25° 100 ml of soln contains: 4.0 g in alcohol; 5.7 g in acetone; 1.4 g. in chloroform. Soluble in alkaline solns. Forms water-soluble sodium, magnesium, and calcium salts.

Caution: May be habit forming. This is a controlled substance (depressant) listed in the U.S. Code of Federal Regulations, Title 21 Parts 329.1 and 1308.13 (1987).

THERAP CAT: Sedative, hypnotic.

4576. Heptachlor. *1H-1,4,5,6,7,8,8-Heptachloro-3a,4,-7,7a-tetrahydro-4,7-methanoindene;* E 3314; Velsicol 104; Drinox; Heptamul; Heptamul. $C_{10}H_5Cl_7$; mol wt 373.35. C 32.17%, H 1.35%, Cl 66.48%. Prepn: Bluestone *et al.,* **U.S. pat. 2,576,-666** (1951 to Julius Hyman); McKenna *et al.,* **U.S. pat. 2,661,377-8** (1953 to Shell); Kleiman, Tapas, **U.S. pat. 2,904,599** (1959 to Velsicol). Toxicity data: T. B. Gaines, *Toxicol. Appl. Pharmacol.* **14,** 515 (1969).

Crystals, mp 95-96°. Vapor pressure at 25° = 3 × 10⁻⁴ mm Hg. Soly in g/100 ml solvent at 27°: acetone 75, benzene 106, carbon tetrachloride 112, cyclohexanone 119, alcohol 4.5, xylene 102. LD_{50} in male, female rats (mg/kg): 100, 162 orally (Gaines).

Caution: Poisoning may occur by ingestion, inhalation, skin contamination. May cause blood dyscrasias and liver necrosis. *See: Clinical Toxicology of Commercial Products,* R. E. Gosselin *et al.,* Eds. (Williams & Wilkins, Baltimore, 5th ed., 1984) Section II, pp 285-286.

Note: The EPA has cancelled registration of pesticides containing this compound with the exception of its use through subsurface ground insertion for termite control and the dipping of roots or tops of non-food plants, *Fed. Reg.* **vol. 40,** p 28850 (July 9, 1975).

USE: Insecticide for control of cotton boll weevil.

4577. Heptaminol. *6-Amino-2-methyl-2-heptanol;* 2-methyl-6-amino-2-heptanol; 6-methyl-2-amino-6-heptanol. $C_8H_{19}NO$; mol wt 145.24. C 66.15%, H 13.19%, N 9.64%, O 11.02%. The hydrochloride is prepd from 2-formamido-6-methyl-5-heptene by boiling with HCl: Doeuvre, Poizat, *Compt. Rend.* **224,** 286 (1947).

Hydrochloride, $C_8H_{20}ClNO$, *RP 2831, Cortensor, Eoden, Hept-a-myl, Heptylon.* Crystals, mp 178-180° (also reported as mp 150°). Freely sol in water. Sol in alcohol. Practically insol in acetone, benzene, ether. pH of 2% aq soln: 4.5-5.5. 5'-Adenylate, *Ampecyclal.*

THERAP CAT: The hydrochloride as cardiotonic.

4578. Heptanal. Heptaldehyde; Aldehyde C-7; heptylaldehyde; oenanthal; enanthal; oenanthol; oenanthaldehyde; enanthaldehyde. $C_7H_{14}O$; mol wt 114.18. C 73.63%, H 12.36%, O 14.01%. $CH_3(CH_2)_5CHO$. Obtained by distilling castor oil under reduced pressure: Rogers, *J. Am. Pharm. Assoc., Sci. Ed.* **12,** 503 (1923); Dominguez *et al., J. Chem. Ed.* **29,** 446 (1952). Catalytic dehydration of ricinoleic acid methyl ester yields heptanal as a cleavage product in almost quantitative yield: Panjutin, *Chem. Zentr.* **1928,** II, 747.

Liquid. Penetrating fruity odor. d_4^0 0.83423; d_4^{15} 0.82162; d_4^{30} 0.80902. mp −43.3°. bp_{760} 152.8°; bp_{30} 59.6°; bp_{10} 42.5°. n_D^{20} 1.42571. Viscosity at 15°: 0.977 cp; at 30°: 0. 791 cp. Surface tension (γ) at 30° = 25.68. Surface tension against water at 30°: 14.41. Heat of combustion (liq) −1062.4 kcal/mol. Slightly sol in water. Misc with alc, ether. Sol in 3 vols of 60% alc.

USE: Manufacture of 1-heptanol; ethyl oenanthate.

4579. Heptanal Sodium Bisulfite. *1-Hydroxy-1-heptanesulfonic acid monosodium salt;* heptaldehyde sodium bisulfite; Hepbisul. $C_7H_{15}NaO_4S$; mol wt 218.26. C 38.52%, H 6.93%, Na 10.53%, O 29.32%, S 14.69%. $CH_3(CH_2)_5CHOH-SO_3Na$. Prepd by shaking a 30% ether soln of heptanal with an excess of 40% aq sodium bisulfite. Soln: Garay, U.S. pat. **3,019,161** (1962 to Baker Castor Oil Co., compd and method not claimed). General method of prepn: Lauer, Langkammerer, *J. Am. Chem. Soc.* **57,** 2360 (1935).

Moderately sol in water. Somewhat sol in methanol. Practically insol in ethanol, ether.

USE: In antifungal compositions.

4580. n-Heptane. C_7H_{16}; mol wt 100.20. C 83.90%, H 16.10%. $CH_3(CH_2)_5CH_3$. A hydrocarbon from petroleum. Volatile, flammable liquid. d_4^{20} 0.684. bp 98.4°. mp −90.7°. Flash pt, open cup: 30°F (−1°C); closed cup: 25°F (−4°C). n_D^{25} 1.3855. Insol in water; sol in alcohol, chloroform, ether. Lethal concn for mice in air: 15,900 ppm, *Handbook of Toxicology* **vol 1,** W. S. Spector, Ed. (Saunders, Philadelphia, 1956) pp 338-339.

USE: As standard in testing knock of gasoline engines.

Caution: May be irritating to respiratory tract, and, in high concns, narcotic.

4581. Heptanoic Acid. Enanthic acid; oenanthic acid; oenanthylic acid; n-heptoic acid; n-heptylic acid. $C_7H_{14}O_2$; mol wt 130.18. C 64.58%, H 10.84%, O 24.58%. $CH_3-(CH_2)_5COOH$. Found in the various fusel oils in appreciable amounts. Has been observed in rancid oils. Prepd by the oxidation of heptaldehyde with potassium permanganate in dil sulfuric acid: Ruhoff, *Org. Syn. coll. vol. II,* 315 (1943).

Oily liquid. Disagreeable, rancid odor. Faint, tallow-like odor when spectroscopically pure. d_4^0 0.9345; d_4^{15} 0.9222; d_2^{20} 0.9181; d_4^{30} 0.9099. mp −7.5°. bp_{760} 223.01°, also reported as 221.9°; bp_{256} 187.5°; bp_{64} 150.8°. Specific heat 0.54 cal/g. Heat of combustion −986.1 cal/g (20°). n_D^{20} 1.42162. Viscosity 3.40 cp at 30°; 0.82 cp at 120°. Interfacial tension against water: 7.0 dynes/cm. K at 25° = 1.42 × 10⁻⁵. Soly in water (15°): 0.2419 g/100 ml H_2O. Sol in ethanol, ether, DMF, DMSO. LD_{50} i.v. in mice: 1200±56 mg/kg, L. Orö, A. Wretlind, *Acta Pharmacol. Toxicol.* **18,** 141 (1961).

Methyl ester, $C_8H_{16}O_2$, liquid, d_2^{20} 0.8815; mp −55.8°; bp 173.8°. n_D^{20} 1.41152.

Ethyl ester, *see* separate entry as Ethyl Oenanthate.

4582. 1-Heptanol. n-Heptyl alcohol; enanthic alcohol; 1-hydroxyheptane. $C_7H_{16}O$; mol wt 116.20. C 72.35%, H 13.88%, O 13.77%. $CH_3(CH_2)_5CH_2OH$. Prepd from heptaldehyde by reduction with iron filings in dil acetic acid: Clarke, Dreger, *Org. Syn.* **6,** 52 (1926); *cf.* Noller, Bannerot, *ibid.* **14,** 91 (1934); coll. vol. **I** (2nd ed., 1941) p 304. Other methods include the reaction between pentane and ethylene oxide in the presence of anhydr aluminum bromide: I. G. Farbenind., **Fr. pat. 716,604,** *C.A.* **26,** 2198 (1932); and the action of amyl magnesium bromide on ethylene oxide: Vaughn *et al., J. Am. Chem. Soc.* **55,** 4207 (1933).

Liquid. d_4^{25} 0.8187. mp −34.6°. bp_{760} 175.8°; bp_{400} 155.6°; bp_{200} 136.6°; bp_{100} 119.5°; bp_{40} 9.98°; bp_{20} 85.8°; bp_{10} 74.7°; bp_5 64.3°; $bp_{1.0}$ 42.4°. n_D^{25} 1.4224. Water solns are colloidal: Traube, Klein, *Koll. Z.* **29,** 236 (1921), *Chem. Zentr.* **1922, I,** 233. One liter of water dissolves 1.0 g at 18°; 2.85 g at 100°; 5.15 g at 130°. Miscible with alcohol, ether.

4583. 2-Heptanol. Amylmethylcarbinol; 2-hydroxyheptane. $C_7H_{16}O$; mol wt 116.20. C 72.35%, H 13.88%, O

13.77%. $CH_3CH(OH)(CH_2)_4CH_3$. Occurs in oil of cloves: Masson, *Compt. Rend.* **149**, 630 (1907). Prepd by the action of amylmagnesium bromide on acetaldehyde: Henry, de Wael, *Rec. Trav. Chim.* **28**, 446 (1909); by the reduction of methyl amyl ketone with sodium in alcoholic soln: Thoms, Mannich, *Ber.* **36**, 2544 (1903); Pickard, Kenyon, *J. Chem. Soc.* **99**, 58 (1911); **105**, 849 (1914); Whitmore, Otterbacher, *Org. Syn.* **10**, 60 (1930); by the action of *Penicillium palitans* on coconut oil: Stokoe, *Biochem. J.* **22**, 82, 84 (1928). Toxicity data: Smyth *et al.*, *Arch. Ind. Hyg. Occup. Med.* **10**, 61 (1954).

dl-Form, liquid. d^0 0.8344; d^{20} 0.8193. bp_{760} 158-160°, n_D 1.42131. Almost insol in water: 3.5 g/l. Sol in alc, ether, benzene. LD_{50} orally in rats: 2.58 g/kg (Smyth).

d-Form, liquid. d_4^{20} 0.8190; d_4^{35} 0.8050; d_4^{51} 0.7920; d_4^{64} 0.7815; d_4^{110} 0.7417. bp_{20} 73.5°. $[\alpha]_D^{20}$ +11.45° (1.039 g in 20 ml abs ethanol); $[\alpha]_4^{20}$ +13.71° (0.992 g in 20 ml benzene).

l-Form, liquid. d_4^{20} 0.8184. bp_{23} 74.5°. $[\alpha]_D^{17}$ −10.48°.

4584. 2-Heptanone. Methyl amyl ketone. $C_7H_{14}O$; mol wt 114.18. C 73.63%, H 12.36%, O 14.01%. $CH_3(CH_2)_4$-$COCH_3$. Found in oil of cloves and in cinnamon-bark oil. Responsible for the "peppery" odor in cheeses of the Roquefort type: Hammer, Bryant, *Iowa State Coll. J. Sci.* **11**, 281 (1937). Prepd by the ketone decompn of ethyl butylacetoacetate: Drake, Riemenschneider, *J. Am. Chem. Soc.* **52**, 5005 (1930); Dehn, Jackson, *ibid.* **55**, 4285 (1933); Johnson, Hager, *Org. Syn.* **coll. vol. I** (2nd ed., 1941) p 351; by the hydration of 1-heptyne and 2-heptyne: Thomas *et al.*, *J. Am. Chem. Soc.* **60**, 719 (1938).

Liquid. Penetrating fruity odor. d_4^0 0.8324; d_4^{15} 0.8197; d_4^{30} 0.8068. bp_{760} 151.5°; bp_{21} 111°. n_D^{15} 1.41156; n_D^{25} 1.40729. Very slightly sol in water. Sol in alc, ether. LD_{50} orally in rats: 1.67 g/kg, Smyth *et al.*, *Am. Ind. Hyg. Assoc. J.* **23**, 95 (1962).

USE: In perfumery as constituent of artificial carnation oils; as industrial solvent.

4585. Heptenophos. *Phosphoric acid 7-chlorobicyclo-[3.2.0]hepta-2,6-dien-6-yl dimethyl ester; 7-chlorobicyclo-[3.2.0]hepta-2,6-dien-6-yl dimethylphosphate;* Hoe 2982; Ragadan; Hostaquick. $C_9H_{12}ClO_4P$; mol wt 250.61. C 43.13%, H 4.82%, Cl 14.15%, O 25.54%, P 12.36%. Systemic and contact insecticide of short persistence. Prepn: B. Böhner, K. Rüfenacht, **S. Afr. pat. 67 06,947** corresp to U.S. pat. **3,600,474** (1968, 1971 to Geigy). Activity: R. T. Hewson, *Proc. Brit. Insectic. Fungic. Conf., 8th* **2**, 697 (1975); W. Bonin, *ibid.* 705.

Pale amber liquid, $bp_{0.001}$ 94-95°. d_4^{20} 1.294. Vapor press. at 20°: 7.5×10^{-4} mm Hg. Sol in xylene, acetone, methanol. LD_{50} orally in rats: 96-117 mg/kg, R. T. Hewson, *loc. cit.*

USE: Insecticide.

THERAP CAT (VET): Ectoparasiticide.

4586. Heptoxime. *1,2-Cycloheptanedione dioxime.* $C_7H_{12}N_2O_2$; mol wt 156.18. C 53.83%, H 7.74%, N 17.94%, O 20.49%. Prepn from cycloheptanone: Vander *et al.*, *J. Org. Chem.* **14**, 836 (1949); Belcher *et al.*, *J. Chem. Soc.* **1958**, 2743. *Review:* Banks, Nicholas, *USAEC* ISC-737 (1956).

Crystals from benzene, mp 182°. pK_1: 10.65 ± 0.2; pK_2: 12.21 ±0.2. From water crystallizes in form of the monohydrate, mp 179-180°.

USE: As reagent in quantitative determination of nickel.

4587. D-*manno*-Heptulose. D-*manno*-Ketoheptose. C_7-$H_{14}O_7$; mol wt 210.18. C 40.00%, H 6.71%, O 53.29%. Isoln from the wet pulp of the fruit of the avocado tree, *Persea gratissima* Gaertn. f. (*P. americana* Mill.), *Lauraceae:* La-Forge, *J. Biol. Chem.* **28**, 511 (1917); Montgomery, Hudson, *J. Am. Chem. Soc.* **61**, 1654 (1939); Richtmyer in *Methods in Carbohydrate Chemistry*, R. L. Whistler *et al.*, Eds. **vol. 1**, 173 (1962).

Large, transparent prisms from methanol, mp 151-152°. $[\alpha]_D^{20}$ +29° (c = 2 in H_2O). Sol in water. Usually accompanied by perseitol. Although perseitol is much more insol in water and methanol than D-*manno*-heptulose, clean separation of the two substances is not always easy and may require patient fractional crystn.

4588. Hercynine. *α-Carboxy-N,N,N-trimethyl-1H-imidazole-4-ethanamium hydroxide inner salt; (1-carboxy-2-imidazol-4-ylethyl)trimethyl ammonium hydroxide inner salt;* histidine-betaine; histidine trimethylbetaine. $C_9H_{15}N_3O_2$; mol wt 197.23. C 54.80%, H 7.67%, N 21.31%, O 16.22%. Present in many fungi, especially in *Amanita muscaria* Fr. and *Agaricus compestris* L., *Agaricaceae.* Isoln: Kutscher, *Zentr. Physiol.* **24**, 775 (1910); **26**, 569 (1912); Barger, Ewins, *J. Chem. Soc.* **99**, 2340 (1911); Küng, *Z. Physiol. Chem.* **91**, 249 (1914). Occurrence in the *Limulus polyphemus* L. (king crab), and prepn: Ackermann, List, *ibid.* **313**, 30 (1958). Synthesis: Reinhold *et al.*, *J. Med. Chem.* **11**, 258 (1968). Biosynthesis by fungi and *Actinomycetales:* Genghof, *J. Bacteriol.* **103**, 475 (1969).

White crystals from methanol and ether, mp 237-238° (dec). $[\alpha]_D^{22}$ +44.5° (5N HCl). Sol in water, alc.

4589. Herqueinone. *8,9-Dihydro-4,6,7a-trihydroxy-5-methoxy-1,8,8,9-tetramethyl-3H-phenaleno[1,2-b]furan-3,7-(7aH)-dione.* $C_{20}H_{20}O_7$; mol wt 372.36. C 64.51%, H 5.41%, O 30.08%. Red fungal pigment isolated from *Penicillium herquei:* Stodola *et al.*, *Nature* **167**, 773 (1951); Galarraga *et al.*, *Biochem. J.* **61**, 456 (1955); Harman *et al.*, *J. Org. Chem.* **20**, 1260 (1955). Structure: Cason *et al.*, *Tetrahedron* **18**, 839 (1962); Paul, Sim, *Proc. Chem. Soc.* **1962**, 352; Cason *et al.*, *J. Org. Chem.* **35**, 179 (1970); J. S. Brooks, G. A. Morrison, *Tetrahedron Letters* **1970**, 963; *eidem, J. Chem. Soc. Perkin Trans. I* **1972**, 421; **1974**, 2114. Crystal structure and abs config: A. Quick *et al.*, *Chem. Commun.* **1980**, 1051. ^{13}C-NMR study: T. Suga *et al.*, *Chem. Letters* **1981** 1063.

Red needles from alc, dec 226°. Sublimes in high vacuum

at 175-190°. Absorption max (ethanol): 220, 250, 314, 416 nm (log ε 4.29, 4.09, 4.47, 3.66). Sol in acetone, dimethylformamide, 2N NaOH, concd HCl, cold concd H_2SO_4; fairly sol in methanol, ethanol, ether, ethyl acetate, chloroform; slightly sol in benzene, carbon disulfide. Practically insol in petr ether, carbon tetrachloride, water, aq Na_2CO_3.

4590. Hesperetin. (S)-2,3-Dihydro-5,7-dihydroxy-2-(3-hydroxy-4-methoxyphenyl)-4H-1-benzopyran-4-one; 3',5,7-trihydroxy-4'-methoxyflavanone; cyanidanon 4'-methyl ether 1626. $C_{16}H_{14}O_6$; mol wt 302.27. C 63.57%, H 4.67%, O 31.76%. The aglucon of hesperidin. Prepd by hydrolysis of hesperidin or by synthesis: Shinoda, Kawagoye, C.A. 23, 2957 (1929); Seka, Prosche, Monatsh. 69, 284 (1936). Sepn of isomers: Arthur et al., J. Chem. Soc. 1956, 632. Structure and configuration: Arakawa, Nakazaki, Ann. 636, 111 (1960). See also Bioflavonoids.

(±)-Form, prisms from alc, mp 226-228°. Freely sol in alc, moderately in ether. Slightly sol in water, chloroform, benzene; sol in dil alkalies. Precipitated by carbonates.
Triacetate, $C_{22}H_{20}O_9$, crystals, mp 139-141°.
(−)-Form, plates from alc, mp 216-218°. $[\alpha]_D^{27}$ −37.6° (c = 1.80 in alc).
Triacetate, crystals, mp 130-132°. $[\alpha]_D^{26}$ +21.1° (c = 1.28 in chloroform).

4591. Hesperidin. (S)-7-[[6-O-(6-Deoxy-α-L-mannopyranosyl)-β-D-glucopyranosyl]oxy]-2,3-dihydro-5-hydroxy-2-(3-hydroxy-4-methoxyphenyl)-4H-1-benzopyran-4-one; hesperetin 7-rhamnoglucoside; cirantin; hesperetin-7-rutinoside. $C_{28}H_{34}O_{15}$; mol wt 610.55. C 55.08%, H 5.61%, O 39.31%. Predominant flavonoid in lemons and sweet oranges (Citrus sinensis). Extraction procedures: Higby, J. Am. Pharm. Assoc., Sci. Ed. 30, 629 (1941); U.S. pat. 2,421,-061 (1947); Baier, U.S. pat. 2,442,110 (1948 to Calif. Fruit Growers Exchange). Structure: King, Robertson, J. Chem. Soc. 1931, 1704; Arthur et al., ibid. 1956, 632; Horowitz, Gentili, Tetrahedron 19, 773 (1963). Synthesis: Zemplen, Bognar, Ber. 75, 1043 (1943); 76, 773 (1943). Identity with cirantin: Manwaring et al., Phytochemistry 7, 1881 (1968). See also Bioflavonoids, Rutinose.

Fine, dendritic needles by precipitation at pH 6-7, mp 258-262° (softens at 250°). $[\alpha]_D^{20}$ −76° (c = 2 in pyridine). One gram dissolves in 50 l water. Sol in formamide, dimethylformamide at 60°. Slightly sol in methanol, hot glacial acetic acid. Almost insol in acetone, benzene, chloroform. Freely sol in dil alkalies, pyridine.
Octaacetate, $C_{44}H_{50}O_{23}$, needles from methanol, mp 176-178°. $[\alpha]_D^{20}$ −47.3° (pyridine).

4592. Hetacillin. 6-(2,2-Dimethyl-5-oxo-4-phenyl-1-imidazolidinyl)-3,3-dimethyl-7-oxo-4-thia-1-azabicyclo-[3.2.0]heptane-2-carboxylic acid; 6-(2,2-dimethyl-5-oxo-4-phenyl-1-imidazolidinyl)penicillanic acid; phenazacillin; BRL 804; Penplenum; Versapen; Versatrex. $C_{19}H_{23}N_3O_4S$; mol wt 389.48. C 58.59%, H 5.95%, N 10.79%, O 16.43%, S 8.23%. Semi-synthetic antibiotic related to penicillin. Prepn

and structure: Hardcastle et al., J. Org. Chem. 31, 897 (1966); Johnson, Panetta, U.S. pat. 3,198,804 (1965 to Bristol-Myers). Pharmacology: Kirby, Kind, Ann. N.Y. Acad. Sci. 145, 291 (1967); Ueda et al., J. Antibiot. 20B, 206 (1967), C.A. 69, 95016w (1968). Stability studies: Saccani, Pansera, Boll. Chim. Farm. 107, 640 (1968). Epimerization at C-6 to epihetacillin: Johnson et al., Tetrahedron Letters 1968, 1903.

Rectangular plates from water + methyl isobutyl ketone, dec 182.8-183.9°; also reported as mp 189.2-191.0°. $[\alpha]_D^{25}$ +366° (pyridine). Practically insol in most organic solvents and water; sol in dil aq NaOH soln (pH 7-8), DMF, DMSO, pyridine, methanol (with dec).
Potassium salt, $C_{19}H_{22}KN_3O_4S$, Uropen, Versapen K, Hetacin K, Natacillin.
Methyl ester, $C_{20}H_{25}N_3O_4S$, crystals from carbon tetrachloride, mp 101.5-102°. Sol in most organic solvents.
Epihetacillin, crystals, mp 164-165°. $[\alpha]_D^{23}$ +232° (pyridine).

THERAP CAT: Antibacterial.
THERAP CAT (VET): Antibacterial.

4593. Hetastarch. Starch 2-hydroxyethyl ether; hydroxyethyl starch; HES; 6-H.E.S.; Hespan; Hespander; Hestar; Hestat; Hestsol; Plasmasteril; Volex. A starch derivative of undetermined exact composition. Comprised of more than 90% amylopectin and etherified to the extent that an average of 7 to 8 of the OH groups present in every ten D-glucopyranose units of the polymer have been converted to OCH_2-CH_2OH groups. Preparation by treating starch with pyridine and ethylene chlorohydrin: Mima, Yokoyama, Japan. pat. 06,556('70) (to Green Cross), C.A. 73, 36822r (1970). Structure evaluation studies: Banks et al., Brit. J. Pharmacol. 47, 172 (1973). Physicochemical and biological properties: Tamada et al., Chem. Pharm. Bull. 19, 286 (1971); Banks et al., Staerke 24, 181 (1972). Pharmacology: Irikura et al., Oyo Yakuri 6, 985 sqq (1972). Metabolism: Bogan et al., Toxicol. Appl. Pharmacol. 15, 206 (1969); Ryan et al., Xenobiotica 2, 141 (1972). Toxicity studies: Irikura et al., Oyo Yakuri 6, 1023 (1972). Introduction as a plasma expander: Wiedersheim, Arch. Int. Pharmacodyn. 111, 353 (1957). Review of production and uses: Hjermstad, Starch: Chem. Technol. 2, 423-432 (1967).

Amylose derivative. R or R' = H or CH_2CH_2OH.
Amylopectin derivative. Differs from the amylose deriv in that the sequence is frequently interrupted by a unit in which R' is the residue of an additional O-hydroxyethylated α-D-glucopyranosyl moiety that constitutes the first unit in a branch or sub-branch of the polymer.
USE: Cryoprotective agent for erythrocytes.
THERAP CAT: Plasma volume extender.

4594. 5-HETE. 5-Hydroxy-6,8,11,14-eicosatetraenoic acid; (S)-5-hydroxy-6-trans-8,11,14-cis-eicosatetraenoic acid. $C_{20}H_{32}O_3$; mol wt 320.48. C 74.96%, H 10.06%, O 14.98%. Important intermediate in a series of biosynthetic processes leading from arachidonic acid, q.v., to a number of biologically active compounds. First discovered during the transformation of arachidonic acid by rabbit peritoneal polymorphonuclear leukocytes: P. Borgeat et al., J. Biol. Chem. 251, 7816 (1976). Revised description: eidem, ibid.

252, 8772 (1977). It has also been found during metabolism of arachidonic acid by human peripheral blood polymorpho-nuclear leukocytes: P. Borgeat, B. Samuelsson, *Proc. Nat. Acad. Sci. USA* **76**, 2148 (1979), and has been shown to be strongly chemotactic for human eosinophils and neutrophils: E. J. Goetzl *et al.*, *Immunology* **39**, 491 (1980); E. J. Goetzl, *N. Engl. J. Med.* **303**, 822 (1980). 5-HETE is formed via *5-HPETE (5-hydroperoxy-6,8,11,14-eicosatetraenoic acid)*, a precursor of the leukotrienes, *q.v.* Chemical and enzymic syntheses of 5-HETE and 5-HPETE: E. J. Corey *et al.*, *J. Am. Chem. Soc.* **102**, 1435 (1980). Synthesis via phenyl-selenylation of arachidonic acid: J. E. Baldwin *et al.*, *Tetrahedron* **37**, Suppl. 1, 263 (1981). Large-scale synthesis of 5-HETE: E. J. Corey, S. Hashimoto, *Tetrahedron Letters* **22**, 299 (1981). Biosynthetic study: R. C. Mebane, *Diss. Abstr. B* **42**, 2839 (1982). Selected synthesis of octadeuter-ated (±)-5-HETE: W. C. Hubbard *et al.*, *Prostaglandins* **23**, 61 (1982). Stereospecific syntheses of 5-*(S)*- and 5-*(R)*-HETE and transformation to (±)-5-HPETE: R. Zamboni, J. Rokach, *Tetrahedron Letters* **24**, 999 (1983). Review of syntheses: J. G. Atkinson, J. Rokach, in *Handbook of Prostaglandins and Related Lipids: The Eicosanoids*, A. L. Willis *et al.*, Eds. (CRC Press, Boca Raton, 1987). *See* arachi-donic acid and leukotrienes for additional refs.

Methyl ester, $C_{21}H_{34}O_3$, colorless oil. $[\alpha]_D^{23}$ +14.0°; $[\alpha]_{436}^{23}$ +35.7° (c = 2.02 in benzene).

4595. Heteronium Bromide. *3-[(Hydroxyphenyl-2-thi-enylacetyl)oxy]-1,1-dimethylpyrrolidinium bromide; 3-hydr-oxy-1,1-dimethylpyrrolidinium bromide α-phenyl-2-thio-pheneglycolate*; 3-hydroxy-1,1-dimethylpyrrolidinium bro-mide α-phenyl-α-(2-thienyl)glycolate; 1-methyl-3-pyrrolid-yl phenyl-2-thienylglycolate methobromide; Hetrum bro-mide. $C_{18}H_{22}BrNO_3S$; mol wt 412.38. C 52.43%, H 5.38%, Br 19.38%, N 3.40%, O 11.64%, S 7.78%. Prepn: Ryan, Ainsworth, *J. Org. Chem.* **27**, 2901 (1962); of α- and β-dia-stereoisomers, Ainsworth, Ryan, U.S. pat. **3,138,614** (1964 to Lilly). Toxicity data: E. I. Goldenthal, *Toxicol. Appl. Pharmacol.* **18**, 185 (1971).

Crystals from abs ethanol; α- and β-diastereoisomers, separated by fractional crystn, have different crystal forms and different infrared absorption spectra. LD_{50} in female, male rats(mg/kg): 3399±624, 3576±954 (Goldenthal).

α-Diastereoisomer, crystals from methanol + ethyl ace-tate, mp 210-211°.

β-Diastereoisomer, crystals from methanol + ethyl ace-tate, mp 182-184°.

THERAP CAT: Anticholinergic.

4596. Hetolin®. *1-Methyl-4-[3,3,3-tris(4-chlorophen-yl)-1-oxopropyl]piperazine; 1-methyl-4-[3,3,3-tris(p-chloro-phenyl)propionyl]piperazine;* 1-[β,β,β-tris(p-chlorophenyl)-propionyl]-4-methylpiperazine; LZ 544. $C_{26}H_{25}Cl_3N_2O$; mol wt 487.85. C 64.01%, H 5.17%, Cl 21.80%, N 5.74%, O 3.28%. Prepn: **Belg**. pat. **634,833** (1964 to Hoechst), *C.A.* **61**, 16076h (1964). Pharmacology: Schorr *et al.*, *Arzneimit-tel-Forsch.* **14**, 1151 (1964).

Crystals from ethyl acetate, mp 213-215°.

Hydrochloride, $C_{26}H_{26}Cl_4N_2O$, *Dicroden*. Crystals from ethanol, mp 267-269°. Freely sol in hot water and alcohol. LD_{50} orally in mice: 610 mg/kg.

THERAP CAT (VET): Anthelmintic (flukicide).

4597. Hexaaminecobalt Trichloride. Luteocobaltic chlo-ride; hexammino-cobalt chloride. $Cl_3CoH_{18}N_6$; mol wt 267.50. Cl 39.76%, Co 22.03%, H 6.78%, N 31.42%. [Co-$(NH_3)_6]Cl_3$. Prepd from $CoCl_2$, NH_4Cl, NH_3 and O_2: Bjer-rum, McReynolds, *Inorg. Syn.* **2**, 217 (1946).

Wine-red or brownish-orange-red, monoclinic crystals. Sol in water; on long boiling with water, NH_3 is evolved and $Co(OH)_3$ pptd.

USE: As reagent for pyrophosphoric acid, for the estima-tion of phosphate.

4598. Hexaborane(10). Hexaboron decahydride; boro-hexane. B_6H_{10}; mol wt 75.00. B 86.56%, H 13.44%. Prepd by the reaction of magnesium boride with hydrochloric or phosphoric acid: Stock, Kuss, *Ber.* **56B**, 789 (1923).

Liquid. mp −62.3°; bp 108°; vapor pressure (0°): 7.5 mm: Burg, Kratzer, *Inorg. Chem.* **1**, 725 (1962). d^0 0.69. Slowly dec at room temp. Hydrolyzes in water after long heating.

4599. Hexacarbacholine Bromide. *N,N,N,N′,N′,N′-Hexamethyl-4,13-dioxo-3,14-dioxa-5,12-diazahexadecane-1,16-diaminium dibromide; 2,2′-[1,6-hexanediylbis(imino-carbonyloxy)]bis[N,N,N-trimethylethanaminium] dibromide;* choline bromide hexamethylenedicarbamate; hexamethyl-enedicarbamic acid choline bromide diester; hexamethylene-1,6-bis(carbamoylcholine bromide); *N,N′-hexamethylene-bis[(2-carbamoyloxyethyl)trimethylammonium bromide];* BC 16; Imbretil. $C_{18}H_{40}Br_2N_4O_4$; mol wt 536.38. C 40.31%, H 7.52%, Br 29.80%, N 10.45%, O 11.93%. Prepn: Schmied *et al.*, **Austrian** pat. **185,371** (1956); **Ger**. pat. **1,021,842** (1958 to Oesterreichische Stickstoffwerke).

Crystals from ethanol, mp 174-176°.

THERAP CAT: Muscle relaxant (skeletal).

4600. Hexachlorobenzene. Perchlorobenzene; Anticarie; Bunt-cure; Bunt-no-more; Julin's carbon chloride. C_6Cl_6; mol wt 284.80. C 25.30%, Cl 74.70%. Not to be confused with benzene hexachloride, *see* Lindane. Prepn: Becke, Sperber, U.S. pat. **2,792,434** (1957 to BASF). Teratogenici-ty study: K. D. Courtney *et al.*, *Toxicol. Appl. Pharmacol.* **35**, 239 (1976). Carcinogenicity studies: J. R. P. Cabral *et al.*, *Nature* **269**, 510 (1977); D. L. Arnold *et al.*, *Food Chem. Toxicol.* **23**, 779 (1985) (corr. *ibid.* **26**, 169 (1988)). uv spec-trum: H. Conrad-Billroth, *Z. Phys. Chem.* **19**, 76 (1932); O. Schnepp, R. Kopelman, *J. Chem. Phys.* **30**, 868 (1959). Toxicological evaluation: G. Vettorazzi, *Residue Rev.* **56**, 107 (1975).

Needles. d^{23} 2.044. mp 231°. bp 323-326°. Vapor pres-sure at 20°: 1.09 × 10^{-5} mm Hg. Sublimable. Insol in water. Sparingly sol in cold alcohol; sol in benzene, chloroform, ether. Maximum acceptable daily intake: 0.0006 mg/kg body weight/day (Vettorazzi).

Caution: Cutaneous porphyria may result from prolonged periods of ingestion, R. Ockner, R. Schmid, *Nature* **189**, 499 (1961). This substance may reasonably be anticipated to be a carcinogen: *Fourth Annual Report on Carcinogens* (NTP 85-002, 1985) p 113.

USE: In organic syntheses. Fungicide.

4601. Hexachloroethane. Carbon hexachloride; perchloroethane. C_2Cl_6; mol wt 236.74. C 10.15%, Cl 89.85%. CCl_3CCl_3. Prepn: *Beilstein* **1**, 87 (1918) and suppls.

Crystals; camphoraceous odor. d 2.09. Readily sublimes without melting. bp 186.8° (triple point). Heat of sublimation 12.2 kcal/mol. Sol in alcohol, benzene, chloroform, ether, oils. Insol in water. MLD i.v. in dogs: 325 mg/kg, Barsoum, Saad, *Quart. J. Pharm. Pharmacol.* **7**, 205 (1934).

USE: Solvent; in explosives; as camphor substitute in celluloid; rubber vulcanizing accelerator. *Caution:* May be moderately irritating to skin, mucous membranes.

THERAP CAT (VET): Anthelmintic (flukicide).

4602. Hexachlorophene. *2,2'-Methylenebis[3,4,6-trichlorophenol];* 2,2'-dihydroxy-3,3',5,5',6,6'-hexachlorodiphenylmethane; bis(3,5,6-trichloro-2-hydroxyphenyl)methane; G-11; AT-7; Bilevon; Dermadex; Exofene; Gamophen; Hexosan; pHisohex; Surgi-Cen; Surofene. $C_{13}H_6Cl_6O_2$; mol wt 406.92. C 38.37%, H 1.49%, Cl 52.28%, O 7.86%. Prepd by the condensation of 2 mols of 2,4,5-trichlorophenol with 1 mol formaldehyde in the presence of concd sulfuric acid: Gump, U.S. pat. **2,250,480** (1941 to Burton T. Bush). Improved procedures: U.S. pat. **2,435,593** (1948) and **2,812,-365** (1957 to Givaudan). Acute toxicity: T. B. Gaines, R. E. Linder, *Fundam. Appl. Toxicol.* **7**, 299 (1986).

Crystals from benzene, mp 164-165°. Practically insol in water. Sol in alcohol, acetone, ether, chloroform, propylene glycol; polyethylene glycols; olive oil; cottonseed oil; dil aq solns of the alkalies. Forms salts with alkalies and alkaline earths. Phenol coefficient ~125 (monopotassium salt). Incompatible with Tweens from bacteriological point of view. LD$_{50}$ in adult male, female rats (mg/kg): 66, 57 orally (Gaines, Linder).

Monophosphate, *Hepadist.*

Toxicity: Excessive dosage to animals results in symptoms of neurotoxicity. Reversible vacuolar changes mainly affecting the myelin of the brain and spinal cord have been reported. Because of potential neurotoxicity in humans, the FDA has regulated use. *See* Lockhart, *Pediatrics* **50**, 229 (1972).

USE: Chiefly in the manuf of germicidal soaps.

THERAP CAT: Topical anti-infective.

THERAP CAT (VET): Anthelmintic (flukicide).

4603. Hexacyclonate Sodium. *1-(Hydroxymethyl)cyclohexaneacetic acid sodium salt;* sodium 3,3-pentamethylene-4-hydroxybutyrate; sodium β,β-pentamethylene-γ-hydroxybutyrate; β,β-pentamethylene-γ-hydroxybutyric acid sodium salt; Neuryl. $C_9H_{15}NaO_3$; mol wt 194.21. C 55.66%, H 7.78%, Na 11.84%, O 24.71%. Prepn: Van Wessem, Sakal, Shavel *et al.*, U.S. pats. **2,960,441; 3,007,940** (1960; 1961 to Warner-Lambert).

Monohydrate, platelets from *n*-butanol + benzene, mp 106-108°. The anhydr salt is hygroscopic. Readily sol in water, methanol, ethanol; sparingly sol in ether, acetone.

THERAP CAT: Central stimulant.

4604. Hexadecyl 3-Hydroxy-2-naphthoate. *3-Hydroxy-2-naphthalenecarboxylic acid hexadecyl ester.* $C_{27}H_{40}O_3$; mol wt 412.59. C 78.59%, H 9.77%, O 11.63%. Prepd by the action of 3-hydroxy-2-naphthoyl chloride on cetyl alc: Oshima, Hayashi, *J. Soc. Chem. Ind. Japan* **44**, 821 (1941).

Greenish-white, flaky crystals, mp 72-73°. Soluble in benzene, glacial acetic acid, petr ether. Sparingly sol in cold alcohol. Insol in water.

USE: As waterproofing agent for rayon.

4605. Hexadimethrine Bromide. *N,N,N',N'-Tetramethyl-1,6-hexanediamine polymer with 1,3-dibromopropane;* polymer of *N,N,N',N'*-tetramethylhexamethylenediamine and trimethylene bromide; poly(*N,N,N',N'*-tetramethyl-*N*-trimethylenehexamethylenediammonium dibromide); Polybrene. $(C_{13}H_{30}Br_2N_2)_x$.

White, hygroscopic, amorphous polymer. Soluble in water up to 10%. pH of 1% saline soln 5-9. Stable in soln and when autoclaved. Polymers with mol wt of 5000-10,000 have LD$_{50}$ i.v. in mice of 25-40 mg/kg. Ref: Kimura *et al.*, *Toxicol. Appl. Pharmacol.* **1**, 185 (1959).

THERAP CAT: Heparin antagonist.

4606. Hexafluorenium Bromide. *N,N'-Di-9H-fluoren-9-yl-N,N,N',N'-tetramethyl-1,6-hexanediaminium dibromide;* hexamethylenebis[9-fluorenyldimethylammonium bromide]; hexamethylenebis(dimethyl-9-fluorenylammonium bromide); Mylaxen. $C_{36}H_{42}Br_2N_2$; mol wt 662.59. C 65.26%, H 6.39%, Br 24.12%, N 4.23%. Prepn: Cavallito *et al.*, *J. Am. Chem. Soc.* **76**, 1862 (1954); Cavallito, Gray, U.S. pat. **2,783,237** (1957 to Irwin, Neisler).

Crystals from *n*-propanol, mp 188-189°.

THERAP CAT: Skeletal muscle relaxant; succinylcholine synergist.

4607. Hexahydroequilenin. *Estra-5,7,9-triene-3β,17β-diol.* $C_{18}H_{24}O_2$; mol wt 272.37. C 79.37%, H 8.88%, O 11.75%. Prepn from material isolated from equine pregnancy urine: Heard, Hoffman, *J. Biol. Chem.* **138**, 662 (1941). May also be prepd from equilenin by hydrogenation in acid ethanol: Ruzicka *et al.*, *Helv. Chim. Acta* **21**, 1399 (1938).

Crystals from ethyl acetate. Sublimes as plates at 150-

160° and 0.01 mm Hg, mp 168-168.5°. $[\alpha]_D^{23}$ —16° (ethanol). Not precipitated by digitonin.

Diacetate, $C_{18}H_{22}(OCOCH_3)_2$, platelets from dil alcohol, sublimes, mp 115-116°.

4608. Hexalure. *7-Hexadecen-1-ol acetate; cis-7-hexadecenyl acetate; cis-1-acetoxy-7-hexadecene; hexalene.* $C_{18}H_{34}O_2$; mol wt 282.47. C 76.54%, H 12.13%, O 11.33%. Synthetic sex attractant for pink bollworm moths, *Pectinophora gossypiella* (Saunders). Discovery and prepn: N. Green *et al.*, *Experientia* **25**, 682 (1969); N. Green, J. C. Keller, **Ger. pat.** 1,960,155 corresp to **U.S. pat.** 3,586,712 (1970, 1971 to U.S. Sec. Agric.). Field trials: J. C. Keller *et al.*, *J. Econ. Entomol.* **62**, 1520 (1969). *See also* Gossyplure, Propylure.

Clear oily liquid, bp$_{0.001}$ 100-104°. n_D^{25} 1.4484. Insol in water. Sol in hexane, ether, acetone, benzene. LD$_{50}$ orally in rats: >34,600 mg/kg, M. Beroza *et al.*, *Toxicol. Appl. Pharmacol.* **31**, 421 (1975).

USE: Insect sex attractant.

4609. Hexamethonium. *N,N,N,N',N',N'-Hexamethyl-1,6-hexanediaminium; hexamethylenebis(trimethylammonium);* α,ω-bis(trimethylammonium)hexane; hexathonide; hexamethone. $[C_{12}H_{30}N_2]^{2+}$; mol wt 202.38. $[(CH_3)_3N-(CH_2)_6N(CH_3)_3]^{2+}$. Prepn: H. J. Barber, **U.S. pat.** 2,641,-610 (1953 to May & Baker); H. J. Barber, K. Gaimster, *J. Pharm. Pharmacol.* **3**, 663 (1951). Ganglion blocking agent: R. Wien, D. F. J. Mason, *Brit. J. Pharmacol.* **6**, 611 (1951).

Bromide, $C_{12}H_{30}Br_2N_2$, *C-6, Bistrium bromide, Esametina, Gangliostat, Hexameton bromide, Hexanium bromide, Simpatoblock, Vegolysen, Vegolysin.* Hygroscopic crystals, mp 274-276°. Sol in water, alc. Insol in acetone, chloroform, ether. pH of 1% soln: 6.2-7.0.

Chloride, $C_{12}H_{30}Cl_2N_2$, *Bistrium chloride, Chloor-hexaviet, Depressin, Esomid chloride, Hestrium chloride, Hexameton chloride, Hexone chloride, Hiohex chloride, Methium chloride, Meton.* Hygroscopic crystals, dec 289-292°. Sol in water, alc. Practically insol in chloroform, ether. pH of 10% soln: 5.5-6.5.

Iodide, $C_{12}H_{30}I_2N_2$, *Hexathide.* Hygroscopic cryst powder. Sol in water; practically insol in alc.

Tartrate, $C_{20}H_{40}N_2O_{12}$, *Vegolysen-T.* Hygroscopic powder, dec 186°. Sol in water; practically insol in alc. pH of 10% soln: 3.8.

THERAP CAT: Antihypertensive.

4610. Hexamethylene Glycol. *1,6-Hexanediol;* 1,6-dihydroxyhexane. $C_6H_{14}O_2$; mol wt 118.17. C 60.98%, H 11.94%, O 27.08%. HOCH$_2$(CH$_2$)$_4$CH$_2$OH. Prepared by reduction of ethyl adipate with copper chromite: Lazier *et al.*, *Org. Syn.* coll. vol. II, 325 (1943); from 2,5-tetrahydrofurandimethanol: Utne *et al.*, **U.S. pat.** 3,070,633 (1962 to Merck & Co.).

Crystals, mp 42.8°. bp$_{760}$ 208°; bp$_{10}$ 134°. n_D^{25} 1.4579. Dipole moment: 2.48 (dioxane). Sol in water, alcohol; sparingly sol in hot ether. LD$_{50}$ orally in rats: 3.73 g/kg, *Toxic Substances List*, H. E. Christensen *et al.*, Ed. (1974) p 399.

Bis(3,5-dinitrobenzoate), $C_{20}H_{18}N_4O_{12}$, crystals from dioxane + ethanol, mp 169-171°.

USE: Intermediate in the production of nylon; to make hexamethylenediamine, polyesters, polyurethans; in gasoline refining; as plasticizer.

4611. Hexamethylolmelamine. *(1,3,5-Triazine-2,4,6-triyltrinitrilo)hexakismethanol; (s-triazine-2,4,6-triyltrinitrilo)hexamethanol;* hexakis(hydroxymethyl)melamine; Resloom M 75. $C_9H_{18}N_6O_6$; mol wt 306.28. C 35.29%, H 5.92%, N 27.44%, O 31.34%. Prepn: Gams *et al.*, *Helv. Chim. Acta* **24**, 302E (1941); Widmer, Fisch, **U.S. pat.** 2,387,547 (1945 to Ciba).

Amorphous to cryst, white mass, mp 135-139°. Sol in water. Easily forms insol polymerization products. For the prepn of aq solns double distilled water and hard glass vessels are recommended. A 1.4-2% aq soln is thus possible.

USE: In fireproofing and creaseproofing of cottons, rayons.

4612. Hexamidine. *4,4'-[1,6-Hexanediylbis(oxy)]bisbenzenecarboximidamide;* 4,4'-(hexamethylenedioxy)dibenzamidine; 4,4'-diamidino-α,ω-diphenoxyhexane. $C_{20}H_{26}N_4O_2$; mol wt 354.45. C 67.77%, H 7.39%, N 15.81%, O 9.03%. Prepn: A. J. Ewins *et al.*, **Brit. pat.** 507,565 (1939 to May & Baker); and trypanocidal activity: J. N. Ashley *et al.*, *J. Chem. Soc.* **1942**, 103. Activity in fibrinolytic systems: J. D. Geratz, *Thromb. Diath. Haemorrh.* **29**, 154 (1973). Antibacterial activity: G. Michel *et al.*, *J. Int. Med. Res.* **14**, 205 (1986). Antifungal activity: M. C. Reynaud, C. Chauve, *Bull. Soc. Fr. Mycol. Med.* **15**, 269 (1986). HPLC determn in pharmaceutics: P. Taylor *et al.*, *J. Pharm. Sci.* **72**, 1477 (1983); in cosmetics: B. Wyhowski de Bukanski, M. O. Masse, *Int. J. Cosmet. Sci.* **6**, 283 (1984). Clinical use as a topical antiseptic: M. J. Fénelon, *Bordeaux Med.* **3**, 867 (1970). Use in treatment of acne: P. Taylor, A. A. Levy, **Eur. pat. Appl.** 93,186 (1983 to Richardson-Vicks).

Isethionate, $C_{24}H_{38}N_4O_{10}S_2$, *RF 2535, Desomedine, Esomedine, Hexomedine, Ophtamedine.* Prisms from HCl, mp 246-247° (dec).

USE: Preservative in cosmetics.

THERAP CAT: Topical antiseptic.

4613. *n*-Hexane. C_6H_{14}; mol wt 86.17. C 83.62%, H 16.38%. CH$_3$(CH$_2$)$_4$CH$_3$. Chief constituent of petr ether or ligroin. Toxicity data: *Handbook of Toxicology* vol. 1, W. S. Spector, Ed. (Saunders, Philadelphia, 1956) pp 338-339; E. T. Kimura *et al.*, *Toxicol. Appl. Pharmacol.* **19**, 699 (1971).

Colorless, very volatile liquid; faint, peculiar odor. d$_4^{20}$ 0.660. bp 69°. mp —95° to —100°. n_D^{20} 1.375. Insol in water; miscible with alcohol, chloroform, ether. LC for mice in air ~40,000 ppm (Spector). LD$_{50}$ orally in young adult rats: 49.0 ml/kg (Kimura).

USE: Determining refractive index of minerals; filling for thermometers instead of mercury, usually with a blue or red dye. *Caution:* May be irritating to respiratory tract and, in high concns, narcotic.

4614. 1,6-Hexanediamine. Hexamethylenediamine; 1,6-diaminohexane. $C_6H_{16}N_2$; mol wt 116.20. C 62.01%, H 13.88%, N 24.11%. NH$_2$(CH$_2$)$_6$NH$_2$. Prepd by reducing adiponitrile with sodium and alcohol: Slotta, Tschesche, *Ber.* **62**, 1404 (1929). Manuf: Faith, Keyes & Clark's *Industrial Chemicals*, F. A. Lowenheim, M. K. Moran, Eds. (Wiley-Interscience, New York, 4th ed., 1975) pp 442-444.

Platelets, leaflets. Sublimes as long needles. Odor of piperidine. Absorbs water and carbon dioxide from air. mp 42°. bp 205°; bp$_{20}$ 100°. Freely sol in water; slightly sol in alcohol, benzene.

Dihydrochloride, $C_6H_{16}N_2 \cdot 2HCl$, needles from water or alcohol, mp 248°. Freely sol in water.

USE: Intermediate in the manuf of nylon.

4615. 1-Hexanol. *n*-Hexyl alcohol; amylcarbinol; pentylcarbinol; 1-hydroxyhexane. $C_6H_{14}O$; mol wt 102.17. C 70.53%, H 13.81%, O 15.66%. CH$_3$(CH$_2$)$_4$CH$_2$OH. Occurs as the acetate in seeds and fruits of *Heracleum sphondylium* L. and *H. giganteum* Fisch., *Umbelliferae*. Lab prepn by the action of butylmagnesium bromide on ethylene oxide: Dreger, *Org. Syn.* coll. vol. I, 306 (2nd ed., 1941). Reduction of 1,3-hexadienal with iron wire in the presence of nickel acetate: Zeisel, Neuwirth, *Ann.* **433**, 127 (1923); *cf.* Baumgarten, Glatzel, *Ber.* **59**, 2659 (1926); Kuhn, Hoffer, *Ber.* **63**, 2165 (1930). Industrial prepn by reducing ethyl caproate with sodium in abs alcohol: Bouveault, Blanc, **Ger. pat. 164,294** (1903).

Liquid. d_4^{25} 0.8153; d_4^{35} 0.8082. mp $-51.6°$. bp_{760} 157°; bp_{400} 138°; bp_{200} 119.6°; bp_{100} 102.8°; bp_{60} 92°; bp_{40} 83.7°; bp_{20} 70.3°; bp_{10} 58.2°; bp_5 47.2°; $bp_{1.0}$ 24.4°. n_D^{25} 1.4162. Absorption spectrum: Massol, Faucon, *Bull. Soc. Chim.* [4] **11**, 932. Flash pt, closed cup: 145°F (63°C). Slightly sol in water; miscible with alcohol, ether. LD_{50} orally in rats: 4.59 g/kg, H. F. Smyth *et al., Arch. Ind. Hyg. Occup. Med.* **4**, 119 (1951).

USE: Manuf antiseptics, hypnotics.

4616. Hexapropymate. *1-(2-Propynyl)cyclohexanol carbamate; carbamic acid 1-(2-propynyl)cyclohexyl ester;* 1-(2-propynyl)cyclohexyl carbamate; 1-carbamoyloxy-1-(2-propynyl)cyclohexane; hexopropynate; L 2103; Modirax; Lunamin; Merinax. $C_{10}H_{15}NO_2$; mol wt 181.23. C 66.27%, H 8.34%, N 7.73%, O 17.66%. Prepn: Lauger *et al., Helv. Chim. Acta* **42**, 2379 (1959); Prost, **U.S. pat. 2,931,828** (1960 to Soc. Belge des Labs. Labaz).

Crystals from petr ether or dioxane, mp 99°. Slight characteristic odor. Sol in ethanol, glycerol, propylene glycol.

THERAP CAT: Sedative, hypnotic.

4617. Hexazinone. *3-Cyclohexyl-6-(dimethylamino)-1-methyl-1,3,5-triazine-2,4(1H,3H)-dione;* DPX 3674; Velpar. $C_{12}H_{20}N_4O_2$; mol wt 252.32. C 57.12%, H 7.99%, N 22.20%, O 12.68%. Broad spectrum, pre- and post-emergence herbicide effective against woody and herbaceous weeds. Prepn: Neth. pat. Appl. **7,307,218**; J. J. Fuchs, J. B. Wommack, **U.S. pat. 3,850,924**; K. Lin, **U.S. pat. 3,983,116** (1973, 1974, 1976 all to Du Pont). Herbicidal activity: D. A. Allison, T. D. Joyce, *Proc. Brit. Weed Control Conf.,* **1974**, 279; J. D. Riggleman, *Proc. South Weed Sci. Soc.* **31**, 141 (1978). Field trials: S. J. B. Hay, R. G. Jones, *Proc. Symp. Methods Weed Control* **1977**, 139; D. E. Yarborough *et al., Weed Sci.* **34**, 723 (1986). HPLC determn in soil and water: D. C. Bouchard, T. L. Lavy, *J. Chromatog.* **270**, 396 (1983). Residues in ground water: D. G. Neary, *South. J. Appl. Forestry* **7**, 217 (1983). Retention and degradation in soil: K. I. N. Jensen, E. R. Kimball, *Bull. Environ. Contam. Toxicol.* **38**, 232 (1987).

White crystalline solid. mp 97-100.5°. Very soluble in water, 330 g/l at 25°. LD_{50} orally in rats: 1690 mg/kg (Neary).

USE: Herbicide.

4618. Hexazole. *4-Cyclohexyl-3-ethyl-4H-1,2,4-triazole;* 3-ethyl-4-cyclohexyl-4H-1,2,4-triazole; Azoman; Triazol Ingelheim; T 156. $C_{10}H_{17}N_3$; mol wt 179.27. C 67.00%, H 9.56%, N 23.44%. Description: Behrens *et al., Klin. Wochenschr.* **16**, 944 (1937).

Stout prisms from ether, mp 89-90°. bp_{10} 227°. Can be distilled in vacuum. Slightly basic reaction. Freely sol in water; also sol in benzene, more sol in chloroform.

4619. Hexedine. *2,6-Bis(2-ethylhexyl)hexahydro-7a-methyl-1H-imidazo[1,5-c]imidazole;* 3,7-bis(2-ethylhexyl)-5-methyl-1,3,7-triazabicyclo[3.3.0]octane; Sterisol (Domestic). $C_{22}H_{45}N_3$; mol wt 351.60. C 75.15%, H 12.90%, N 11.95%. Prepn: McMillan, **Ger. pat. 1,209,249** corresp to **U.S. pat. 3,357,886** (1966, 1967, both to Warner-Lambert). Compound claimed but not described: Senkus, **U.S. pat. 2,393,826** (1946 to CSC).

Liquid, $bp_{0.025}$ 131°. n_D^{22} 1.4660.

THERAP CAT: Antibacterial.

4620. 3-Hexen-1-ol. Leaf alcohol; Blätteralkohol. C_6-$H_{12}O$; mol wt 100.16. C 71.95%, H 12.08%, O 15.97%. $CH_3CH_2CH=CHCH_2CH_2OH$. Occurs in leaves of odoriferous plants (including shrubs and trees). Isoln from Japanese oil of peppermint: Walbaum *J. Prakt. Chem.* [2] **96**, 245 (1917); Walbaum, Rosenthal, *Ber. Schimmel* 1929 Jubiläums-Ausg., p 205; from raspberry juice: Bohnsack, *Ber.* **75**, 72 (1942). The natural product has the *cis* configuration: Crombie, Harper, *J. Chem. Soc.* **1950**, 873; Harper, Smith, *ibid.* **1955**, 1512.

Liquid. Strong odor resembling that of isoamyl alc, approaching the odor of green leaves when highly dil. d_{15}^{22} 0.846. bp 156-157°; bp_9 55-56°. n_D^{20} 1.4389.

4621. Hexestrol. *4,4'-(1,2-Diethyl-1,2-ethanediyl)bisphenol; 4,4'-(1,2-diethylethylene)diphenol;* meso-3,4-bis(p-hydroxyphenyl)-n-hexane; 4,4'-dihydroxy-γ,δ-diphenylhexane; 4,4'-dihydroxy-α,β-diethyldiphenylethane; dihydrodiethylstilbestrol; hexoestrol; Synthovo; Cycloestrol; Hexanoestrol; Hormoestrol; Syntrogène. $C_{18}H_{22}O_2$; mol wt 270.36. C 79.96%, H 8.20%, O 11.84%. From anethole HBr: Campbell *et al., Proc. Roy. Soc.* **B128**, 253 (1940); Bernstein, Wallis, *J. Am. Chem. Soc.* **62**, 2871 (1940); **U.S. pat. 2,357,985** (1944); Buu-Hoi, Hoán, *J. Org. Chem.* **14**, 1023 (1949). By the reduction of Grignard compds with cobaltous chloride: Kharasch *et al., J. Am. Chem. Soc.* **65**, 491 (1943). Proof of meso configuration: Wessely, Welleba, *Ber.* **74**, 777 (1941). Comprehensive description: H. Y. Aboul-Enein *et al.,* in *Analytical Profiles of Drug Substances* vol. 11, K. Florey, Ed. (Academic Press, New York, 1982) pp 347-374.

Needles from benzene, thin plates from dil alc, mp 185-188°. Freely sol in ether; sol in acetone, alcohol, methanol; slightly sol in benzene, chloroform. Sol in vegetable oils upon slight warming, also in dil solns of alkali hydroxides. Practically insol in water and in dil mineral acids.

Diacetate, *Retalon-Lingual*, crystals, mp 137-139°.

Dipropionate, *Retalon Oleosum*, crystals from petr ether, mp 127-128°.

Diphosphate, $C_{18}H_{24}O_8P_2$, *hexestrol 4,4'-diphosphoric ester, Cytostatin.* Prepn: **Brit. pat 593,480** (1947 to Roche); Atherton *et al.,* **U.S. pat. 2,490,573** (1949 to Hoffmann-La Roche).

THERAP CAT: Estrogen. Antineoplastic (hormonal).

THERAP CAT (VET): Estrogen.

4622. Hexestrol Bis(β-diethylaminoethyl ether). *2,2'-[(1,2-Diethyl-1,2-ethanediyl)bis(4,1-phenyleneoxy)]bis[N,N-diethylethanamine]; 2,2'''-[(1,2-diethylethylene)bis(p-phenyleneoxy)]bis(triethylamine);* 4,4'-bis(β-diethylaminoethoxy)-α,β-diethyldiphenylethane; 3,4-bis[p-(β-diethylaminoethoxy)phenyl]hexane; α,α'-diethyl-4,4'-bis(β-diethylaminoethoxy)bibenzyl. $C_{30}H_{48}N_2O_2$; mol wt 468.70. C 76.87%,

H 10.32%, N 5.98%, O 6.83%. Prepn: Lowe *et al.*, *J. Chem. Soc.* **1951**, 3286.

$(C_2H_5)_2NCH_2CH_2O$ $OCH_2CH_2N(C_2H_5)_2$

Dihydrochloride, $C_{30}H_{50}Cl_2N_2O_2$, *Coralgil, Coralgina.* Needles from alc + ethyl acetate, mp 226-227°. Freely sol in water, methanol, chloroform, hot alc.

THERAP CAT: Coronary vasodilator.

4623. Hexethal Sodium. *5-Ethyl-5-hexyl-2,4,6(1H,3H,5H)-pyrimidinetrione monosodium salt; sodium 5-ethyl-5-hexylbarbiturate;* Ortal Sodium; Hebaral. $C_{12}H_{19}N_2NaO_3$; mol wt 262.29. C 54.95%, H 7.30%, N 10.68%, Na 8.77%, O 18.30%. Prepn: A. W. Dox, U.S. pat. **1,624,546.** Anesthetic properties and toxicity data: C. M. Gruber, J. T. Brundage, *J. Pharmacol. Exp. Ther.* **60**, 439 (1937).

White or slightly yellowish powder, mp 126°. Bitter taste. Very sol in water; sol in alcohol. Insol in ether, benzene. Aq solns are alkaline to litmus. Aq solns are unstable and dec on standing. Boiling evolves ammonia and causes precipitation of a solid. MLD i.p. in rats: 240-250 mg/kg (Gruber, Brundage).

Caution: May be habit forming. This is a controlled substance (depressant) listed in the U.S. Code of Federal Regulations, Title 21 Parts 329.1 and 1308.13 (1987).

THERAP CAT: Sedative, hypnotic.

THERAP CAT (VET): Sedative, hypnotic.

4624. Hexetidine. *1,3-Bis(2-ethylhexyl)hexahydro-5-methyl-5-pyrimidinamine; 5-amino-1,3-bis(2-ethylhexyl)-hexahydro-5-methylpyrimidine;* 1,3-bis(β-ethylhexyl)-5-methyl-5-aminohexahydropyrimidine; 5-amino-1,3-di(β-ethylhexyl)hexahydro-5-methylpyrimidine; Glypesin; Hexigel; Hexocil; Hexoral; Hextril; Oraldene; Sterisil; Sterilate; Sterisol; Triocil. $C_{21}H_{45}N_3$; mol wt 339.59. C 74.27%, H 13.36%, N 12.37%. Prepn: Senkus, *J. Am. Chem. Soc.* **68**, 1611 (1946). U.S. pat. **2,415,047** and Bell, Necker, U.S. pat. **3,054,797** (1947, 1962 both to Comm. Solvents Corp.). Comprehensive description: G. Satzinger *et al.*, in *Analytical Profiles of Drug Substances* **vol. 7**, K. Florey, Ed. (Academic Press, New York, 1978) pp 277-295.

Liquid. d_{20}^{20} 0.8889. $bp_{0.4}$ 160°. n_D^{20} 1.4668. Sol in petr ether, methanol, benzene, acetone, ethanol, *n*-hexane, chloroform. Practically insol in water. pKa 8.3. Has good thermal stability.

THERAP CAT: Antifungal.

THERAP CAT (VET): Antifungal.

4625. Hexobarbital. *5-(1-Cyclohexen-1-yl)-1,5-dimethyl-2,4,6(1H,3H,5H)-pyrimidinetrione; 5-(1-cyclohexen-1-yl)-1,5-dimethylbarbituric acid;* 5-cyclohexenyl-3,5-dimethylbarbituric acid; methylhexabital; methexenyl; enhexymal; hexobarbitone; Somnalen; Sombucaps; Sombulex; Evipan; Evipal; Citodon; Noctivane; Hexenal; Citopan; Cyclonal; Dorico; Hexanastab Oral. $C_{12}H_{16}N_2O_3$; mol wt 236.26. C

61.00%, H 6.83%, N 11.86%, O 20.32%. Prepd by condensation of monomethylurea with methyl cyclohexenylmethylcyanoacetate in abs alcohol in the presence of sodium: U.S. pat. **1,947,944;** *cf.* Ger. pat. **595,175;** Fr. pat. **753,178.** Comparative clinical trial in anesthesia: D. W. Barron *et al.*, *Brit. J. Anaesth.* **38**, 802 (1966). GC determn in human plasma: D. D. Breimer, J. M. Van Rossum, *J. Chromatog.* **88**, 235 (1974). Pharmacokinetics: N. P. E. Vermeulen *et al.*, *Brit. J. Clin. Pharmacol.* **15**, 459 (1983); M. Van der Graaff *et al.*, *Biopharm. Drug Dispos.* **7**, 265 (1986).

Prismatic crystals, practically tasteless, mp 145-147°. Practically insol in water: One gram dissolves in about 3 liters of water. Sol in methanol, hot ethanol, ether, chloroform, acetone, benzene, aq solns of alkali hydroxides, but not carbonates.

Sodium deriv, $C_{12}H_{15}N_2NaO_3$, *hexobarbital soluble, Cyclonal Sodium, Dorico Soluble, Evipal Sodium, Evipan Sodium, Narcosan Soluble, Privenal, Hexanastab, Hexanal, Methexenyl Sodium, Noctivane Sodium.* White, bitter, very hygroscopic powder. Very sol in water, freely sol in alcohol, methanol, acetone. Practically insol in chloroform, ether, benzene. A soln in water absorbs carbon dioxide from the air, causing precipitation of the insol free acid. Aq and alcoholic solns are alkaline to litmus and phenolphthalein. The pH of a 10% w/v soln of hexobarbital soluble is about 11.5.

Caution: May be habit forming. This is a controlled substance listed in the U.S. Code of Federal Regulations, Title 21 Parts 329.1 and 1308.13 (1987).

USE: Indicator of hepatic drug metabolism.

THERAP CAT: Sedative, hypnotic. Anesthetic (intravenous).

THERAP CAT (VET): Sedative, hypnotic.

4626. Hexobendine. *3,4,5-Trimethoxybenzoic acid 1,2-ethanediylbis[(methylimino)-3,1-propanediyl] ester; 3,4,5-trimethoxybenzoic acid diester with 3,3'-[ethylenebis(methylimino)]di-1-propanol; N,N'-dimethyl-N,N'-bis[3-(3',4',-5'-trimethoxybenzoxy)propyl]ethylenediamine;* hexabendin. $C_{30}H_{44}N_2O_{10}$; mol wt 592.57. C 60.79%, H 7.48%, N 4.73%, O 27.00%. Prepn: O. Kraupp, K. Schlögl, Austrian pat. **231,432** corresp to U.S. pat. **3,267,103** (1964, 1966 to OSSW). Pharmacology: Rudolph *et al.*, *Arzneimittel-Forsch.* **20**, 637 (1970); Kolassa, Pfleger, *Biochem. Pharmacol.* **20**, 490 (1971); H. Rameis *et al.*, *Arzneimittel-Forsch.* **30**, 671 (1980).

mp 75-77°.

Dihydrochloride, $C_{30}H_{46}Cl_2N_2O_{10}$, *Reoxyl, ST 7090, Ustimon, Andiamine.* Crystals, mp 170-174°. uv max: 267 nm. Freely sol in water, less sol in alc, practically insol in ether.

THERAP CAT: Vasodilator.

4627. Hexocyclium Methyl Sulfate. *4-(2-Cyclohexyl-2-hydroxy-2-phenylethyl)-1,1-dimethylpiperazinium methyl sulfate (salt); N-(β-cyclohexyl-β-hydroxy-β-phenylethyl)-N¹-methylpiperazine;* N^1-methylpiperazine methosulfate; *N-(β-cyclohexyl-β-hydroxy-β-phenylethyl)-N¹-methylpiperazine dimethylsulfate; 4-(β-cyclohexyl-β-hydroxy-β-phenethyl)-1,1-dimethylpiperazinium methyl sulfate;* Tral; Tralin. $C_{21}H_{36}N_2O_5S$; mol wt 428.59. C 58.85%, H 8.47%, N 6.53%, O 18.67%, S 7.48%. Description: Helgren *et al.*, *J. Am. Pharm. Assoc., Sci. Ed.* **46**, 639 (1957). Prepn: Weston, U.S. pat. **2,907,765** (1959 to Abbott).

Consult the cross index before using this section.

Crystals, mp 200-210°. uv max (0.1N H_2SO_4): 252, 257, 263 nm. Soly in water about 50% w/v. Slightly sol in chloroform. Insol in ether.

THERAP CAT: Anticholinergic.

4628. Hexoprenaline. *4,4'-[1,6-Hexanediylbis[imino(1-hydroxy-2,1-ethanediyl)]]bis-1,2-benzenediol; α,α'-[hexamethylenebis(iminomethylene)]bis[3,4-dihydroxybenzyl alcohol]; N,N'-bis[2-(3,4-dihydroxyphenyl)-2-hydroxyethyl]-hexamethylenediamine;* BYK 1512; Broncholysin. $C_{22}H_{32}N_2O_6$; mol wt 420.51. C 62.84%, H 7.67%, N 6.66%, O 22.83%. β$_2$-Adrenergic agonist. Prepn: Schmid *et al.,* Austrian pat. **241,436** corresp to U.S. pat. **3,329,709** (1965, 1967, both to OSSW); Schmid *et al.,* Ger. pat. **1,215,729** (1966 to Lentia GmbH). Pharmacology: Thiede *et al., Arzneimittel-Forsch.* **21,** 416 (1971). Clinical studies: Schindl, *ibid.* **20,** 1755 (1970); several authors in *Wien. Klin. Wochenschr.* **83,** 75, 80, 101, 114, 117, 130 (1971). Clinical evaluation for interruption of labor: J. Lipshitz *et al., J. Reprod. Med.* **31,** 1023 (1986).

Crystals, mp 162-165° (hemihydrate).
Dihydrochloride, $C_{22}H_{34}Cl_2N_2O_6$, *ST 1512, Ipradol.* Crystals from methanol-ether, mp 197.5-198°.
Sulfate, $C_{22}H_{34}N_2O_{10}S$, *Bronalin, Etoscol, Gynipral, Ipradol, Leanol.* Crystals from water-alcohol, mp 222-228°.

THERAP CAT: Bronchodilator. Tocolytic.

4629. Hexylcaine Hydrochloride. *1-(Cyclohexylamino)-2-propanol benzoate (ester) hydrochloride;* D109; Cyclaine. $C_{16}H_{24}ClNO_2$; mol wt 297.83. C 64.53%, H 8.12%, Cl 11.90%, N 4.70%, O 10.74%. Prepn: Cope, Hancock, *J. Am. Chem. Soc.* **66,** 1453 (1944); Cope, U.S. pat. **2,486,374** (1949 to Sharp & Dohme). Pharmacokinetics: R. M. Rodgers *et al., Res. Commun. Chem. Pathol. Pharmacol.* **29,** 99 (1980).

Crystals from abs alc, mp 177-178.5°. Soluble in water to the extent of about 12% w/w. A 1% soln is stable to boiling and autoclaving for sterilization purposes.

THERAP CAT: Local anesthetic.
THERAP CAT (VET): Local and topical anesthetic.

4630. 2-Hexyldecanoic Acid. $C_{16}H_{32}O_2$; mol wt 256.42. C 74.94%, H 12.58%, O 12.48%. Prepn: Lederer *et al., Bull. Soc. Chim. France* **1952,** 413.

Viscous oil, bp$_{0.02}$ 140-150°. n_D^{24} 1.4432.
Sodium salt, *Devaricin.* White powder. Sol in water.
THERAP CAT: Sclerosing agent.

4631. Hexylene Glycol. *2-Methyl-2,4-pentanediol; α,α,-α'-trimethyltrimethyleneglycol;* pinakon. $C_6H_{14}O_2$; mol wt 118.17. C 60.98%, H 11.94%, O 27.08%. Prepn: Franke, *Monatsh.* **22,** 1067 (1901); Leopold, Ger. pat. **486,767** (1925 to I. G. Farben), *Frdl.* **16,** 679; Adkins, Cramer, *J. Am. Chem. Soc.* **52,** 4349 (1930); Arundale, Mikeska, U.S. pat. **2,367,324** (1945 to Standard Oil).

Liquid. Mild sweetish odor. d$_{15}^{15}$ 0.924. bp$_{760}$ 198°; bp$_{10}$ 97°. Flash pt: ca. 93° (200°F). n_D^{20} 1.4276. Dipole moment: 2.8. Viscosity at 20° = 34 cp. LD$_{50}$ orally in rats: 4.70 g/kg, Smyth, Carpenter, *J. Ind. Hyg. Toxicol.* **30,** 63 (1948). Diacetate, $C_{10}H_{18}O_4$. Liquid. bp$_{12}$ 95°.

USE: Cosmetics, hydraulic brake fluids (as coupling agent to castor oil).

4632. Hexyl Methyl Ketone. *2-Octanone;* methyl hexyl ketone. $C_8H_{16}O$; mol wt 128.21. C 74.94%, H 12.58%, O 12.48%. $CH_3(CH_2)_5COCH_3$.

Liquid; apple odor; camphor taste. d$_4^{20}$ 0.820. mp −16°. bp 172-173°. n_D^{20} 1.41512. Insol in water; miscible with alc, ether.

4633. 4-Hexylresorcinol. *4-Hexyl-1,3-benzenediol;* 4-hexyl-1,3-dihydroxybenzene; ST 37; Ascaryl; Caprokol; Crystoids; Gelovermin; Sucrets; Worm-agen. $C_{12}H_{18}O_2$; mol wt 194.26. C 74.19%, H 9.34%, O 16.47%. Prepd by reduction of hexanoylresorcinol with zinc amalgam + dil HCl: Dohme *et al., J. Am. Chem. Soc.* **48,** 1688 (1926); Twiss, *ibid.* 2206; Ger. pats. **488,419** and **489,117** (1929 to Sharp & Dohme). Toxicity: Lamson *et al., J. Pharmacol. Exp. Ther.* **53,** 198 (1935).

Pale yellow, heavy liq becoming solid on standing at room temp. Needles from benzene or petr ether, mp 67.5-69°. bp$_{6.7}$ 178-180°; bp$_{13-14}$ 198-200°; bp$_{760}$ 333-335°. Pungent odor; sharp astringent taste. Sol in ether, chloroform, acetone, alcohol, vegetable oils; slightly sol in petr ether; sol in about 2000 parts water. LD$_{50}$ orally in rats: 550 mg (Lamson).

Human Toxicity: Concd solns can cause irritation of skin, mucous membranes: *Clinical Toxicology of Commercial Products,* R. E. Gosselin *et al.,* Eds. (Williams & Wilkins, Baltimore, 5th ed., 1984) Section II, p 190.

THERAP CAT: Anthelmintic (Nematodes). Topical antiseptic.

THERAP CAT (VET): Anthelmintic.

4634. Hexythiazox. *trans-5-(4-Chlorophenyl)-N-cyclohexyl-4-methyl-2-oxo-3-thiazolidinecarboxamide; trans-4-methyl-5-(4-chlorophenyl)-3-cyclohexylcarbamoyl-2-thiazolidone;* HTZ; NA 73; DPX Y5893-9; Nissorun; Acarflor; C'esar; Savey; Zeldox. $C_{17}H_{21}ClN_2O_2S$; mol wt 352.88. C 57.86%, H 6.00%, Cl 10.05%, N 7.94%, O 9.07%, S 9.08%. Prepn: I. Iwataki *et al.,* Ger. pat. **3,037,105;** eidem, U.S. pat. **4,442,116** (1981, 1984 to Nippon Soda). Miticidal activity: M. A. Hoy, Y.-L. Ouyang, *J. Econ. Entomol.* **79,** 1377 (1986); C. Welty *et al., ibid.* **81,** 586 (1988). Residue determn in crops: M. Tokieda *et al., J. Pestic. Sci.* **12,** 711 (1987), *C.A.* **108,** 162770b (1988). Brief description of physical properties, toxicity and efficacy: Nippon Soda Co., Ltd., *Japan. Pest. Info.* **44,** 21 (1984).

White odorless crystals, mp 105.5°. Vapor pressure at 20°: 2.54 × 10^{-8} mm Hg. Soly at 20° (g/100 ml): acetone 16,

Consult the cross index before using this section.

chloroform 137.9, methanol 2.06, *n*-hexane 0.39, xylene 36.2, acetonitrile 2.86; water 0.5 ppm. LD_{50} in male, female mice, male, female rats (mg/kg): all > 5000 orally; all > 5000 dermally (Nippon Soda Co.).

USE: Acaricide.

4635. Hinderin. *4-[3,5-Bis(1,1-dimethylethyl)-4-hydroxyphenoxy]-3,5-diiodobenzenepropanoic acid; β-[3,5-diiodo-4-(4'-hydroxy-3',5'-di-tert-butylphenoxy)phenyl]propionic acid; 3,5-diiodo-4-(3',5'-di-tert-butyl-4'-hydroxyphenoxy)hydrocinnamic acid.* $C_{23}H_{28}I_2O_4$; mol wt 622.27. C 44.39%, H 4.54%, I 40.79%, O 10.28%. Prepn: Kharasch *et al., Science* **127**, 756 (1958).

Needles from dil alc, mp 197-198°. Also used as the ethyl ester.

THERAP CAT: Thyroid inhibitor.

4636. Hippuric Acid. *N-Benzoylglycine; benzoylaminoacetic acid; benzamidoacetic acid.* $C_9H_9NO_3$; mol wt 179.17. C 60.33%, H 5.06%, N 7.82%, O 26.79%. $C_6H_5CONHCH_2$-COOH. Present in the urine of herbivorous animals; also in smaller amounts in human urine. Prepd from benzoyl chloride and glycine in NaOH soln: Ingersoll, Babcock, *Org. Syn.* coll. vol. **II**, 328 (1943).
Crystals. mp 187-188°. One gram dissolves in about 250 ml cold water, 1000 ml chloroform, 400 ml ether, 60 ml amyl alcohol; slightly sol in cold, freely in hot alcohol or hot water; also sol in aq soln of sodium phosphate; practically insol in benzene, carbon disulfide, petr ether.
Ammonium salt, $NH_4C_9H_8NO_3$, crystals. Freely sol in water; sol in alcohol.
Potassium salt monohydrate, $KC_9H_{10}NO_4 \cdot H_2O$, cryst powder. Very sol in water; sol in alcohol.

4637. Hirsutic Acid C. *Decahydro-2-hydroxy-3a,5-dimethyl-3-methylenecyclopenta[4,5]pentaleno[1,6a-b]oxirene-5-carboxylic acid; 6β,6aβ-epoxy-2,3,3aα,3b,4,5,6,6a,7,7aα-decahydro-5β-hydroxy-2β,3bβ-dimethyl-4-methylene-1H-cyclopenta[a]pentalene-2-carboxylic acid.* $C_{15}H_{20}O_4$; mol wt 264.31. C 68.16%, H 7.63%, O 24.21%. Antibiotic metabolite isolated from the fungus *Stereum hirsutum*: Heatley *et al., Brit. J. Exp. Pathol.* **28**, 35 (1947). Structure and stereochemistry: Comer *et al., Chem. Commun.* **1965**, 310; Comer, Trotter, *J. Chem. Soc. (B)* **1966**, 11; Comer *et al., Tetrahedron* **23**, 4761 (1967). Synthetic studies: Lansbury *et al., Tetrahedron Letters* **1971**, 1829; Sakar, *J. Chem. Soc. Perkin Trans. I,* **1973**, 2454. Total synthesis of *dl*-hirsutic acid: Hashimoto *et al., Tetrahedron Letters* **1974**, 3745; M. Yamazaki *et al., Chem. Letters* **1981**, 1245. Stereocontrolled total synthesis of (±)-form: B. M. Trost *et al., J. Am. Chem. Soc.* **101**, 1284 (1979); of (+)-form: M. Shibasaki *et al., Tetrahedron Letters* **23**, 5311 (1982).

Prisms from ethanol, mp 179-182°. $[\alpha]_D^{23}$ +116° (c = 1.05).
Methyl ester, $C_{16}H_{22}O_4$, colorless prisms, mp 161-162°. $[\alpha]_D^{20}$ +119° (c = 2.25).

4638. Hirudin. *Exhirud; Exhirudine; Hirudex.* The dried and somewhat refined extract from leeches; shows anticoagulant activity. The medicinal leech of Central Europe, *Hirudo medicinalis* Linn. (*Sanguisuga officinalis* Savigny) yields about 3 mg per head. American leeches contain much less hirudin. Isoln: Haycraft, *Arch. Exp. Pathol. Pharmakol.* **18**, 209 (1884); Bock, *ibid.* **41**, 160 (1898); Franz, *ibid.* **49**,

342 (1903); **Ger.** pats. **136,103** (1902); **150,805** (1903); Bodong, *Arch. Exp. Pathol. Pharmakol.* **52**, 242 (1905); Abel *et al., J. Pharmacol.* **5**, 302 (1913/14); Marshall, *ibid.* **7**, 157 (1915); Yanagisawa, Yokoi, *Proc. Imp. Acad. Japan* **14**, 69 (1938); Kirsanov, Bystritskaya, *Biokhimiya* **5**, 596 (1940), *C.A.* **36**, 6308; Markwardt, *Naturwiss.* **42**, 537 (1955). Revised extraction procedure: Schremmer *et al., Monatsh.* **87**, 87 (1956). Purification: Jutisz *et al., Bull. Soc. Chim. Biol.* **45**, 55 (1963), *C.A.* **58**, 10426c (1963). Hirudin is a polypeptide with mol wt about 10,800, based on amino acid composition. C-terminal sequence: Llosa *et al., Biochim. Biophys. Acta* **93**, 40 (1964); N-terminal sequence: Graf *et al., ibid.* **310**, 416 (1973). Hirudin is characterized by a high proportion of dicarboxylic acids, which explains its acid character, and by the absence of tryptophan, methionine and arginine: Llosa *et al., Bull. Soc. Chim. Biol.* **45**, 69 (1963). A specific inhibitor of thrombin, *q.v.,* hirudin inhibits coagulation in the initial stages of clotting and does not require presence of other coagulation factors or plasma constituents: Markwardt, *Methods Enzymol.* **19**, 924 (1970). Pharmacokinetics in humans: F. Markwardt *et al., Thromb. Haemostasis* **52**, 160 (1984). Chromogenic assay in plasma: U. Griessbach *et al., Thromb. Res.* **37**, 347 (1985). Review of pharmacology: F. Markwardt, *Biomed. Biochim. Acta* **44**, 1007 (1985).
Gray or white flakes or powder. Isoelec. pt. pH 3.9. Sol in water, physiol saline soln, in pyridine. Practically insol in alcohol, ether, acetone, benzene. Deteriorates on storage in sealed ampuls, on exposure to heat and when in soln with dil acids.

THERAP CAT: Anticoagulant.

4639. Histaminase. *Diamine oxidase; benzylamine oxidase; histamine deaminase; histamine oxidase; E.C. 1.4.3.6.* A copper contg enzyme present in tissues, esp in kidneys and in the intestinal mucosa. Attacks diamines such as histamine in the body by oxidative deamination: Best, *J. Physiol.* **67**, 256 (1929); Zeller, *Helv. Chim. Acta* **21**, 880 (1938); *Advan. Enzymol.* **2**, 93 (1942). Extraction from hog kidneys: Swedin, *Acta Med. Scand.* **114**, 21 (1943). Identity with diamine oxidase: Zeller, *Fed. Proc.* **24**, 764 (1965). Appears to have the general properties of a flavoprotein. *Reviews:* E. A. Zeller "Diamine Oxidases" in *The Enzymes,* vol. **8**, P. D. Boyer *et al.,* Eds. (Academic Press, New York, 2nd ed., 1963) pp 313-335; Buffoni, *Pharmacol. Rev.* **18**, 1163-1199 (1966); Hansson, *Scand. J. Clin. Lab. Invest.* vol. **31**, suppl. 129, 7 (1973).
Has been sold in enteric-coated tablets as *Torantil* (Torantyl) and *Metoryl* (Metoral).

4640. Histamine. *1H-Imidazole-4-ethanamine; 2-(4-imidazolyl)ethylamine; 4-imidazoleethylamine; 5-imidazoleethylamine; β-aminoethylimidazole; β-aminoethylglyoxaline.* $C_5H_9N_3$; mol wt 111.15. C 54.03%, H 8.16%, N 37.81%. A potent vasodilator found in normal tissues and blood. Occurs widely in nature as a result of putrefactive processes. Stimulates the secretion of pepsin and acid by the stomach; eating and vagal stimulation cause the release of histamine from gastric mucosa. The flavoprotein diamine oxidase converts histamine to the corresponding aldehyde and ammonia. Some undegraded histamine, in the form of the N-acetyl and N'-methyl derivs, is excreted in the urine. Produced from histidine by *pneumococci* or *B. coli:* Lévy-Brühl, Ungar, *Ann. Inst. Pasteur* **61**, 828 (1938). Synthesis from imidazolylpropionic acid: Pyman, *J. Chem Soc.* **99**, 668 (1911); Koessler, Hanke, *J. Am. Chem. Soc.* **40**, 1716 (1918); from α-aminobutyrolactone: Garforth, Pyman, *J. Chem. Soc.* **1935**, 489. Diagnostic use as gastric stimulant: O. M. Laudano, *Gastroenterol.* **50**, 653 (1966); in pheochromocytoma: T. Nakai, R. Yamada, *J. Clin. Endocrinol. Met.* **57**, 19 (1983). Role of histamine in induction of drinking by food intake: F. S. Kraly, *Nature* **302**, 65 (1983). *Review:* H. Wetterqvist, *Handb. Exp. Pharmakol.* **18**(pt. 2), 131-150

(1978). Toxicity: K. Nagai *et al.*, *Arzneimittel-Forsch.* **17**, 1575 (1967).

Deliquescent needles from chloroform, mp 83-84°. bp_{18} 209-210°. Freely sol in water, alcohol, hot chloroform. Sparingly sol in ether. LD_{50} i.p. in mice: 2020 mg/kg (Nagai).

Dihydrochloride, $C_5H_{11}Cl_2N_3$, *Amin-Glaukosan*, *Imido*, *Imadyl* (obsolete), *Ergamine*, *Peremine*. Prisms from ethanol, mp 244-246°. Freely sol in water, methanol; sol in ethanol.

Phosphate, $C_5H_{15}N_3O_8P_2$, *Histapon*. Prismatic crystals, mp 140°. Readily sol in water.

THERAP CAT: Diagnostic aid (gastric secretion, pheochromocytoma). Hyposensitization therapy.

4641. Histapyrrodine. *N-Phenyl-N-(phenylmethyl)-1-pyrrolidineethanamine; 1-[2-(N-benzylanilino)ethyl]pyrrolidine; N-pyrrolidylethyl-N-phenylbenzylamine; N-benzyl-N-phenylpyrrolidinoethylamine; N-benzyl-N-pyrrolidinoethylaniline.* $C_{19}H_{24}N_2$; mol wt 280.40. C 80.80%, H 9.28%, N 9.92%. Prepn: Hopff *et al.*, U.S. pat. **2,623,880** (1952). Chemical properties: Auterhoff, *Arch. Pharm.* **284**, 123 (1951).

bp_1 198-205°.

Hydrochloride, $C_{19}H_{25}ClN_2$, *Calcistin*, *Domistan*, *Luvistin*. Crystals from alc, mp 196-197°. Soly in water (18°): 2 g/100 ml.

THERAP CAT: Antihistaminic.

4642. Histidine. His (IUPAC abbrev.); *(S)-α-amino-1H-imidazole-4-propanoic acid; α-amino-4(or 5)-imidazolepropionic acid; glyoxaline-5-alanine; L-histidine.* C_6H_9-N_3O_2; mol wt 155.16. C 46.44%, H 5.85%, N 27.08%, O 20.62%. An amino acid classified as essential with respect to its growth effect in rats. The L-form is naturally occurring. Isolated from blood corpuscles by an electrical transport method: Cox *et al.*, *J. Biol. Chem.* **81**, 755 (1929); by precipitation with mercuric chloride: Gilson, *ibid.* **124**, 281 (1938); Vickery, Leavenworth, *ibid.* **72**, 403 (1927); **78**, 627 (1928); **83**, 523 (1929); Gebauer-Fülnegg, Kendall, *Ber.* **64**, 1067 (1931). Convenient laboratory procedures: Foster, Shemin, *Org. Syn.* coll. vol. II, 330 (1943). Synthesis: Pyman, *J. Chem. Soc.* **99**, 668, 1386 (1911); **109**, 186 (1916). Metabolism: F. B. Stifel, R. H. Herman, *Am. J. Clin. Nutr.* **24**, 207 (1971). Crystal structure: H. Fuess *et al.*, *Acta Crystallogr.* **B33**, 654 (1977).

Sweet needles or plates, dec 287° (softens at 277°). $[\alpha]_D^{20}$ −39.74° (c = 1.13). pK_1' 1.78; pK_2' 5.97; pK_3' 8.97. Soly in water at 25°: 41.9 g/l. Very slightly sol in alcohol. Insol in ether.

Monohydrochloride, $C_6H_{10}ClN_3O_2$, *Ecristidine*, *Laristine*, *Larostidin*, *Plexamine*. Rhombic crystals, dec 251-252° (also forms a monohydrate, mp 80°, anhydrous, mp at 140°). $[\alpha]_D^{26}$ +8.0° (c = 2 in 3 mols HCl). Fairly sol in water. Insol in alcohol, ether.

Dihydrochloride, $C_6H_{11}Cl_2N_3O_2$, rhombic crystals, isomorphous with the mono-HCl, dec 245° (also reported as 196°). $[\alpha]_D^{20}$ +47.6° (c = 2). Sol in water.

DL-Form, quadrilateral plates or tetragonal prisms, dec 285°. Sol in water.

USE: Dietary supplement.

4643. Histones. Small chromosomal proteins (mol wt 12,000-20,000) possessing an open, unfolded structure, attached to DNA of cell nuclei by ionic linkages. First isolated from bird erythrocytes: Kossel, *Z. Physiol. Chem.* **8**, 511 (1884). Prepd also from cell nuclei of calf thymus, spleen, and liver, liver and spleen of leukemic rats, bovine spermatozoa: Hnilica, *Biochem. J.* **82**, 123 (1962); Berry, Mayer, *Exp. Cell Res.* **20**, 116 (1960). Classification is based on relative amounts of lysine and arginine: Histone I is very rich in lysine and has several subtypes; histone II, moderately rich in lysine with two subtypes; histone III, moderately rich in arginine, contains cysteine; histone IV, very rich in arginine and in glycine. Histones of the same type obtained from various plant and animal sources are very similar in amino acid sequence. Amino acid composition of animal histones: Hnilica *et al.*, *Biochim. Biophys. Acta* **124**, 109 (1966); of plant histones and similarity of plant and animal histones: Fambrough, Bonner, *Biochemistry* **5**, 2563 (1966). Complete amino acid sequence of calf thymus histone IV: DeLange *et al.*, *J. Biol. Chem.* **244**, 319 (1969); Ogawa *et al.*, *ibid.* 4387; of histone III: DeLange *et al.*, *Proc. Nat. Acad. Sci. USA* **69**, 882 (1972); *eidem*, *J. Biol. Chem.* **248**, 3261 (1973); Hooper *et al.*, *ibid.* 3275; of the two subtypes of histone II: Iwai *et al.*, *Nature* **226**, 1056 (1970); Yeoman *et al.*, *J. Biol. Chem.* **247**, 6018 (1972). Review of structural studies: DeLange, *Accounts Chem. Res.* **5**, 368 (1972). Review of histone in the chromosome structure: Bradbury *et al.*, *Ann. N.Y. Acad. Sci.* **222**, 266 (1973). Review of biological functions: Binner, Garrard, *Life Sci.* **14**, 209 (1974). General review: Butler *et al.*, *Progr. Biophys. Mol. Biol.* **18**, 211 (1968); R. J. DeLange, E. L. Smith in *Proteins* vol. 4, H. Neurath, R. L. Hill, Eds. (Academic, New York, 3rd ed., 1979) pp 134-243.

Histones are susceptible to enzymatic cleavage; sol in an Hg_2SO_4-H_2SO_4 medium. Infrared spectra: de Lozé, *Compt. Rend.* **246**, 417 (1958). LD_{50} i.v. in rats of lysine-rich, slightly lysine-rich, and arginine-rich histones: 90, 60-70, 60 mg/kg, Starbuck *et al.*, *Arch. Int. Pharmacodyn. Ther.* **165**, 374 (1967).

4644. HN1. *2-Chloro-N-(2-chloroethyl)-N-ethylethanamine; 2,2′-dichlorotriethylamine; bis(2-chloroethyl)ethylamine; ethylbis(2-chloroethyl)amine.* $C_6H_{13}Cl_2N$; mol wt 170.08. C 42.37%, H 7.70%, Cl 41.69%, N 8.24%. C_2H_5N-$(CH_2CH_2Cl)_2$. A nitrogen mustard. Prepn: Hanby, Rydon, *J. Chem. Soc.* **1947**, 513. *Review:* Sartori, *Chem. Rev.* **48**, 225 (1951).

Liquid; faint fishy, amine odor. *Deadly vesicant!* bp_3 66°, bp_{12} 85.5°. mp −34°. d_4^{23} 1.0861. n_D^{25} 1.4653. Volatility at 25°: 2.29 mg/l. Practically insol in water; miscible with many organic solvents.

Hydrochloride, $C_6H_{13}Cl_2N.HCl$, crystals, from acetone, mp 141°.

Caution: Highly irritating to skin, eyes, and mucous membranes.

4645. Holarrhenine. *3β-(Dimethylamino)con-5-enin-12β-ol; 12β-hydroxyconessine.* $C_{24}H_{40}N_2O$; mol wt 372.58. C 77.36%, H 10.82%, N 7.52%, O 4.29%. From leaves and bark of *Holarrhena congolensis* Stapf, *Apocynaceae.* Isoln: Pyman, *J. Chem. Soc.* **115**, 163 (1919). Structure: Uffer, *Helv. Chim. Acta* **39**, 1834 (1956); van Hove, *Tetrahedron* **7**, 104 (1959).

Needles from acetic acid or ethyl acetate, mp 197-198°. $[\alpha]_D^{24}$ −7° (chloroform); $[\alpha]_D^{24}$ +9° (alcohol). Practically insol in water; sol in alcohol, chloroform, slightly in ether, acetone, ethyl acetate.

O-Acetylholarrhenine, $C_{26}H_{42}N_2O_2$, crystals from pentane or acetone, mp 177-178°.

4646. Holmium. Ho; at. wt 164.9304; at. no. 67; valence 3. A rare earth metal of the yttrium group. One naturally occurring isotope: ^{165}Ho; artificial radioactive isotopes: 150-164; 166-170. Estimated abundance in earth's crust 0.7-1.2 ppm; occurs in rare earth minerals. Discovered by Soret in 1878 and independently by Cleve in 1879: Cleve, *Compt. Rend.* **89**, 478, 708 (1879). Purification of holmium salts: Holmberg, *Z. Anorg. Chem.* **71**, 226 (1911). Separation from erbium by crystn of the bromates: Driggs, Hopkins, *J. Am. Chem. Soc.* **47**, 363 (1925). Sepn from other rare earths by ion exchange: Spedding *et al.*, *ibid.* **69**, 2812 (1947); **76**, 2557 (1954). Absorption spectrum: Severin, *Z. Physik* **125**, 455 (1949). Analysis by means of emission spectra: Smith, Wiggins, *Analyst* **74**, 95 (1949). Reviews of prepn, properties and compounds of holmium and other lanthanides: *The Rare Earths*, F. H. Spedding, A. H. Daane, Eds. (Krieger, Huntington, N.Y., 1971, reprint of 1961 ed.) 641 pp; Hulet, Bode, "Separation Chemistry of the Lanthanides and Transplutonium Actinides" in *MTP Int. Rev. Sci.: Inorg. Chem., Ser. One*, vol. 7, K. W. Bagnall, Ed. (University Park Press, Baltimore, 1972) pp 1-45; Moeller, "The Lanthanides" in *Comprehensive Inorganic Chemistry*, vol. **4**, J. C. Bailar, Jr. *et al.*, Eds. (Pergamon Press, Oxford, 1973) pp 1-101.

Metal; hexagonal close-packed structure. d 8.799. mp 1461°. Forms yellow-green salts.

Oxide, Ho_2O_3, *holmia*. Yellow solid. Obtained by igniting the hydroxide, nitrate, sulfate, oxalate. Dissolves in acids with formation of a yellow salt.

Chloride, $HoCl_3$, bright yellow, cryst solid, mp 718°. Formed by heating the hydrated salt in a current of hydrogen chloride at 350°. LD_{50} in mice: 560 mg/kg i.p.; 7.2 g/kg orally, Haley *et al.*, *Toxicol. Appl. Pharmacol.* **8**, 37 (1966).

Bromide, $HoBr_3$, mp 914°.

Iodide, HoI_3, light-yellow solid. mp 1010° ± 10°. Obtained by passing hydrogen iodide over the anhydr chloride at 600°: Jantsch *et al.*, *Z. Anorg. Chem.* **207**, 353 (1932).

4647. Holomycin. *N-(4,5-Dihydro-5-oxo-1,2-dithiolo-[4,3-b]pyrrol-6-yl)acetamide; N-(4,5-dihydro-5-oxo-1,2-dithiolo[4,3-b]pyrrol-6-yl)-N-methylformamide;* 6-acetamido-1,2-dithiolo[4,3-*b*]pyrrol-5(4*H*)-one; *N*-demethylthiolutin. $C_7H_6N_2O_2S_2$; mol wt 214.26. C 39.23%, H 2.82%, N 13.08%, O 14.94%, S 29.93%. Antibiotic substance produced by a strain of *Streptomyces griseus* (Krainski) Waksman et Henrici. Isoln and structure: Ettlinger *et al.*, *Helv. Chim. Acta* **42**, 563 (1959). Crystal structure: Jensen, *J. Antibiot.* **22**, 231 (1969). Activity: Von Daehne *et al.*, *ibid.* 233. Manuf process: Gaumann *et al.*, U.S. pat. **3,014,922** (1961 to Ciba). Total synthesis: Schmidt, Geiger, *Ann.* **664**, 168 (1963); Büchi, Lukas, *J. Am. Chem. Soc.* **86**, 5654 (1964); Hagio, Yoneda, *Bull. Chem. Soc. Japan* **47**, 1484 (1974); J. E. Ellis *et al.*, *J. Org. Chem.* **42**, 2891 (1977).

Orange-yellow flakes from methanol + ethyl acetate, dec 268-270. uv max: 245, 302, 290 nm (log ε 3.78, 3.51, 4.05). LD_{50} i.v. in mice: 5-10 mg/kg, Von Daehne *et al., loc. cit.*

4648. Homarine. *2-Carboxy-1-methylpyridinium hydroxide, inner salt;* 1-methyl-2-pyridinium carboxylate; picolinic acid *N*-methylbetaine; *N*-methylpicolinic acid. C_7H_7-NO_2; mol wt 137.13. C 61.31%, H 5.15%, N 10.21%, O 23.33%. Quaternary ammonium base occurring in tissues of marine animals such as sea urchin, *Arabacia pustulosa*, jellyfish, *Velella spirans*, lugworm, *Arenicola marina*. Isoln: Hoppe-Seyler, *Z. Physiol. Chem.* **222**, 105 (1933). Prepn: Kosower, Patton, *J. Org. Chem.* **26**, 1318 (1961); Quast, Schmitt, *Ann.* **732**, 64 (1970). Review of distribution and function in animals: Brodzicki, *Kosmos (Warsaw)* **16A**, 431-438 (1967), *C.A.* **68**, 18617v (1968).

Crystals from methanol. Does not have a melting point, slowly carbonizes when heated. Absorption spectrum: Gasteiger *et al., Ann. N.Y. Acad. Sci.* **90**, 624 (1962).

Hydrochloride, $C_7H_7NO_2$.HCl, fine needles, dec 170-175°. Freely sol in water, less sol in methanol and ethanol.

4649. Homatropine. *endo-*(±)-*α-Hydroxybenzeneacetic acid 8-methyl-8-azabicyclo[3.2.1]oct-3-yl ester; 1αH,5αH-tropan-3α-ol mandelate;* mandelyltropeine; tropine mandelate. $C_{16}H_{21}NO_3$; mol wt 275.33. C 69.79%, H 7.69%, N 5.09%, O 17.43%. Prepd from mandelic acid and tropine: Ladenburg, *Ann.* **217**, 82 (1883); Chemnitius, *J. Prakt. Chem.* **117**, 142 (1927). Prepn of D(−)- and L(+)-forms: Werner, Miltenberger, *Ann.* **631**, 163 (1960). Comprehensive description: F. J. Muhtadi *et al.* in *Analytical Profiles of Drug Substances* vol. **16**, K. Florey, Ed. (Academic Press, New York, 1987) pp 245-290. Toxicity: R. L. Cahen, K. Tvede, *J. Pharmacol. Exp. Ther.* **105**, 166 (1952).

DL-Form, prisms from ether, mp 99-100°. *Poisonous!* Hygroscopic, but only slightly sol in water; sol in alcohol, benzene, chloroform, ether, acetone, dil acids.

DL-Hydrobromide, $C_{16}H_{22}BrNO_3$, *Homatrisol, Bufopto Homatrocel.* Orthorhombic, bipyramidal prisms from water, mp about 212° (partial decompn). *Poisonous!* One gram dissolves in 6 ml water, 40 ml alcohol (12 ml at 60°), 420 ml chloroform. Insol in ether. pH (1% aq soln): 5.4.

DL-Hydrochloride, $C_{16}H_{22}ClNO_3$, prisms from water, mp 217-220° (dec). *Poisonous!* Freely sol in water, alcohol. pH (1% aq soln): 5.4.

DL-Methylbromide, $C_{17}H_{24}BrNO_3$, *Arkitropin, Homapin, Malcotran, Mesopin, Novatrin, Novatropine, Sethyl.* Minute needles from alc + ether, mp 191-192° (slight dec). *Poisonous!* Freely sol in water, dil alc; slightly sol in abs alc. Insol in ether. pH (1% aq soln) 5.9; pH (10% aq soln) 4.5. LD_{50} in mice (mg/kg): 1400 orally, 60 i.p. (Cahen, Tvede).

THERAP CAT: Anticholinergic (ophthalmic).

THERAP CAT (VET): Anticholinergic (ophthalmic).

4650. Homidium. *3,8-Diamino-5-ethyl-6-phenylphenanthridinium;* 2,7-diamino-9-phenyl-10-ethylphenanthridinium; 2,7-diamino-10-ethyl-9-phenylphenanthridinium; ethidium; RD 1572; Novidium; Babidium. $[C_{21}H_{20}N_3]^+$; mol wt 314.41. C 80.22%, H 6.41%, N 13.36%. Prepn: T. I. Watkins, *J. Chem. Soc.* **1952**, 3059; Short *et al.*, U.S. pat. **2,662,082** (1953 to Boots Pure Drug). uv spectrum: B. Hudson, R. Jacobs, *Biopolymers* **14**, 1309 (1975). Inhibition of DNA synthesis: B. A. Newton, *J. Gen. Microbiol.* **17**, 718 (1957); of DNA polymerase: W. H. Elliott, *Biochem.* **86**, 562 (1963). Intercalation of double-stranded DNA: M. J. Waring, *J. Mol. Biol.* **13**, 269 (1965); J.-B. LePecq, C. Paoletti, *ibid.* **27**, 87 (1967). Mutagenicity studies: J. T. MacGregor, I. J. Johnson, *Mutat. Res.* **48**, 103 (1977); G. S. Probst *et al.*, *Environ. Mutagen.* **3**, 11 (1981). Review of interactions with nucleic acids: J.-B. LePecq in *Methods of Biochemical Analysis* vol. **20**, D. Glick, Ed. (Wiley-Interscience, New York, 1971) pp 41-86; M. Waring in *Antibiotics* vol. **3**, J. W. Corcoran, F. E. Hahn, Eds. (Springer-Verlag, New York, 1975) pp 141-165.

Bromide, $C_{21}H_{20}BrN_3$, *Dromilac.* Bitter tasting dark red crystals from alc, mp 238-240°. uv max in water: 210, 285, 316, 343 nm (ε 200-500, 5000-10000, 50000, 40000). Sol in 20 parts water and 750 parts chloroform at 20°.

Chloride, $C_{21}H_{20}ClN_3$, dark red cryst powder. Crystallizes with 1 mol of ethanol. Sol in 5 parts of water at room temp.

USE: In sepn and determn of nucleic acids.

THERAP CAT (VET): Antiprotozoal (Trypanosoma).

4651. Homocamfin. *3-Methyl-5-(1-methylethyl)-2-cyclohexen-1-one; m-menth-6-en-5-one;* methylisopropylcyclohexenone; 5-isopropyl-3-methyl-Δ²-cyclohexenone; Cyclosal; Hexetone. $C_{10}H_{16}O$; mol wt 152.24. C 78.90%, H 10.59%, O 10.51%. Prepn: Horning *et al., J. Org. Chem.* **9,** 547 (1944); Whitmore, Roberts, *ibid.* **13,** 31 (1948).

Pale yellow, bitter, oily liq. bp_{18} 127-128°; bp_{16} 124-126°; bp_{15} 120-122°; $bp_{11.5}$ 113-116°; bp_9 108-110°. Slightly sol in water; sol in alc, benzene, ether. Marketed in aq soln each 100 ml containing 10 g of the oil and 25 g sodium salicylate.

THERAP CAT: Central stimulant.

THERAP CAT (VET): Has been used as a CNS stimulant.

4652. Homochelidonine. *4b,5,6,11b,12,13-Hexahydro-1,2-dimethoxy-12-methyl[1,3]benzodioxolo[5,6-c]phenanthridin-5-ol;* α-homochelidonine. $C_{21}H_{23}NO_5$; mol wt 369.42. C 68.28%, H 6.28%, N 3.79%, O 21.65%. From root of *Chelidonium majus* L., *Papaveraceae:* Schmidt, Selle, *Arch. Pharm.* **228,** 441 (1890). Structure: Späth, Kuffner, *Ber.* **64,** 1123 (1931); H. W. Bersch, *Arch. Pharm.* **291,** 491 (1958). Total synthesis: I. Ninomiya *et al., Heterocycles* **7,** 137 (1977).

Orthorhombic prisms from ethyl acetate, mp 182°. $[\alpha]_D$ +118° (alc). Very sol in chloroform. Sol in alcohol, acetic acid, ethyl acetate, dil mineral acids; sparingly sol in ether.

4653. Homochlorcyclizine. *1-[(4-Chlorophenyl)phenylmethyl]hexahydro-4-methyl-1H-1,4-diazepine;* N-(p-chlorobenzhydryl)-N'-methylhomopiperazine; 1-(4-chlorodiphenylmethyl)-4-methyl-2,3,4,5,6,7-hexahydro-1,4-diazepine; 1-(p-chloro-α-phenylbenzyl)-4-methylhomopiperazine; SA 97; Curosajin; Homorestar. $C_{19}H_{23}ClN_2$; mol wt 314.86. C 72.48%, H 7.36%, Cl 11.26%, N 8.90%. Prepn: Weston, Sommers, U.S. pat. **2,655,498** (1953 to Abbott); eidem, *J. Am. Chem. Soc.* **76,** 5805 (1954).

Hydrochloride, $C_{19}H_{24}Cl_2N_2$, *Homoginin.*

Dihydrochloride, $C_{19}H_{25}Cl_3N_2$, *Homoclomin.* Crystals from ethanol, mp 227-228°. Base $bp_{0.8}$ 177°; n_D^{25} 1.5804.

THERAP CAT: Serotonin antagonist.

4654. Homocysteine. *2-Amino-4-mercaptobutyric acid.* $C_4H_9NO_2S$; mol wt 135.19. C 35.53%, H 6.71%, N 10.36%, O 23.67%, S 23.72%. $HSCH_2CH_2CH(NH_2)COOH$. A sulfur containing amino acid, produced by the demethylation of methionine and an intermediate in the biosynthesis of cysteine from methionine. Originally obtained from the liver of mammals. Has also been obtained from cystathionine: Binkley, *Methods Enzymol.* **2,** 314 (1955). D-Homocysteine may be prepd from S-benzyl-D-homocysteine, while L-homocysteine is best obtained from L-homocystine: du Vigneaud, Brown, *Biochem. Prepn.* **5,** 93 (1957), see also Patterson, du Vigneaud, *J. Biol. Chem.* **111,** 393 (1935); Riegel, du Vigneaud, *ibid.* **112,** 149 (1935). Origin and biochemical conversions in review by Shapiro, Schlenk, *Advan. Enzymol.* **22,** 264-268 (1960). Use of L-form sodium salt as flavor enhancer similar to monosodium glutamate: Kaneko *et al.*, U.S. pat. **3,259,505** (1966 to Ajinomoto Kabushiki Kaisha).

DL-Form, platelets from dil ethanol, mp 232-233°. pK_1 2.22; pK_2 8.87; pK_3 10.86.

Thiolactone hydrochloride, C_4H_8ClNOS, crystals from abs ethanol. L-form: $[\alpha]_D^{26}$ +21.5° (c = 1); D-form: $[\alpha]_D^{26}$ −21.5° (c = 1).

4655. Homocystine. *4,4'-Dithiobis[2-aminobutanoic acid];* 4,4'-dithiobis[2-aminobutyric acid]. $C_8H_{16}N_2O_4S_2$; mol wt 268.36. C 35.80%, H 6.01%, N 10.44%, O 23.85%, S 23.90%. Synthesis by the "malonate" method: Patterson, du Vigneaud, *J. Biol. Chem.* **111,** 393 (1935); Shiroishi *et al., Rept. Food Res. Inst. (Tokyo)* **5,** 75 (1951); du Vigneaud, Brown, *Biochem. Prepns.* **5,** 93 (1957); from 3,6-bis(β-chloroethyl)-2,5-dioxopiperazine: Snyder, Cannon, *J. Am. Chem Soc.* **66,** 511 (1944).

DL-Form, platelets from water, dec 263-265°. pK_1 1.59; pK_2 2.54; pK_3 8.52; pK_4 9.44.

L-Form, crystals, dec 281-284°. $[\alpha]_D^{21}$ −16° (H_2O); $[\alpha]_D^{26}$ +79° (1.0N HCl).

D-Form, crystals, dec 281-284°. $[\alpha]_D^{21}$ +16° (H_2O); $[\alpha]_D^{26}$ −79° (1.0N HCl).

4656. Homoeriodictyol. *2,3-Dihydro-5,7-dihydroxy-2-(4-hydroxy-3-methoxyphenyl)-4H-1-benzopyran-4-one; 4',5,7-trihydroxy-3'-methoxyflavanone;* eriodictyonone; cyanidanon-3-methyl ether 1625. $C_{16}H_{14}O_6$; mol wt 302.27. C 63.57%, H 4.67%, O 31.76%. From *Eriodictyon californicum* (H. & A.) Torr., and *E. angustifolium* Nutt., *Hydrophyllaceae:* Geissman, *J. Am. Chem. Soc.* **62,** 3258 (1940); Hadley, Gisvold, *J. Am. Pharm. Assoc.* **33,** 275 (1944). Purification: Seka, Prosche, *Monatsh.* **69,** 284 (1936). Structure: Shinoda, Sato, *J. Pharm. Soc. Japan* **49,** 64 (1929), *C.A.* **23,** 4210 (1929). Synthesis: Farooq *et al., Naturwiss.* **46,** 76 (1959).

Crystals from 70% acetic acid. Needles by sublimation in high vacuum (0.003-0.005 mm Hg) at 190-195°. Plates from dil alc, dec 225° (after drying *in vacuo* at 110°). $[\alpha]_D^{20}$ −28° (alc). uv max (alc): 290, 328 nm (log ε 2.26, 2.33). Moderately sol in alcohol, acetic acid. Nearly insol in water, ethyl acetate; practically insol in benzene, chloroform.

Triacetate, mp 115-116°.

4657. Homofenazine. *Hexahydro-4-[3-[2-(trifluoro-methyl)phenothiazin-10-yl]propyl]-1H-1,4-diazepine-1-ethanol;* 1-(β-hydroxyethyl)-4-[3-(2-trifluoromethylphenothiazin-10-yl)propyl]hexahydro-1,4-diazepine; 1-(2-hydroxyethyl)-4-[3-(3-trifluoromethylphenothiazin-10-yl)propyl]homopiperazine; 3-trifluoromethyl-10-[3-[4-(2-hydroxyethyl)homopiperazino]propyl]phenothiazine. $C_{23}H_{28}F_3N_3OS$; mol wt 451.56. C 61.18%, H 6.25%, F 12.62%, N 9.31%, O 3.54%, S 7.10%. Prepn: Schuler *et al.,* U.S. pat. **3,040,043** (1962 to Degussa).

bp$_1$ 230-240°.
Dihydrochloride, $C_{23}H_{30}Cl_2F_3N_3OS$, *Pasaden.*
Difumarate, $C_{23}H_{28}F_3N_3OS \cdot 2C_4H_4O_4$, crystals, mp 148°.
THERAP CAT: Sedative.

4658. Homogentisic Acid. *2,5-Dihydroxybenzeneacetic acid; 2,5-dihydroxyphenylacetic acid;* 2,5-dihydroxy-α-toluic acid. $C_8H_8O_4$; mol wt 168.14. C 57.14%, H 4.80%, O 38.06%. An important intermediate in metabolism of tyrosine *(q.v.)* and phenylalanine *(q.v.).* Occurs in plants and in the urine of alkaptonurics: Garrod, *Inborn Errors of Metabolism* (Oxford Medical Publications, London, 1923, and later). Isoln from alkaptonuric urine: Mörner, *Z. Physiol. Chem.* **117,** 85 (1921). Synthesis from 2,5-dihydroxyman-delic acid or from 2,5-dihydroxyphenylglyoxylic acid by boiling with fuming HI: Neubauer, Flatow, *ibid.* **52,** 395 (1907). Alternate syntheses: L. DeForrest Abbott, J. D. Smith, *J. Biol. Chem.* **179,** 365 (1949); S. B. Bostock, A. H. Renfrew, *Synthesis* **1978,** 66; J. L. Bloomer, K. M. Damodaran, *ibid.* **1980,** 111. Biosynthesis from tyrosine: Davies *et al., J. Chem. Soc.* **1964,** 3126. Metabolic studies: W. E. Knox, M. LeMay-Knox, *Biochem. J.* **49,** 686 (1951); B. N. LaDu, V. G. Zannoni, *J. Biol. Chem.* **217,** 777 (1955).

Monohydrate, prisms from water. Anhydrous leaflets from hot alcohol + chloroform, mp 152°. Freely sol in water, alcohol, ether; insol in chloroform, benzene. Easily dehydrated to the lactone. Aq solns are stable.
Dimethyl ether, $C_{10}H_{12}O_4$, mp 124.5°.
Methyl ester dimethyl ether, $C_{11}H_{14}O_4$, mp 45°.

4659. Homonicotinic Acid. *3-Pyridineacetic acid; liox-*

one; Minedil; Piristerol. $C_7H_7NO_2$; mol wt 137.13. C 61.31%, H 5.15%, N 10.21%, O 23.33%. Prepn: Hartmann *et al.,* U.S. pat. **2,408,020** (1946 to Ciba); Carboni, *Gazz. Chim. Ital.* **85,** 1194 (1955).

Crystals from ethyl acetate or alcohol, mp 144°.
Hydrochloride, $C_7H_7NO_2 \cdot HCl$, crystals from methanol, mp 153-154°.
Methyl ester, $C_8H_9NO_2$, bp$_{11}$ 112°.
THERAP CAT: Vasodilator.

4660. Homosalate. *2-Hydroxybenzoic acid 3,3,5-tri-methylcyclohexyl ester; salicylic acid 3,3,5-trimethylcyclohexyl ester;* 3,3,5-trimethylcyclohexyl salicylate; homomenthyl salicylate; Heliophan. $C_{16}H_{22}O_3$; mol wt 262.36. C 73.25%, H 8.45%, O 18.30%. Prepn: Stockelbach, U.S. pat. **2,369,-084** (1945 to Fries Bros.).

bp$_4$ 161-165°. n^{20} 1.516 to 1.518. d$_{25}^{25}$ 1.045.
A component of *Coppertone* products.
USE: Ultraviolet screen.

4661. Homoserine. *2-Amino-4-hydroxybutanoic acid;* 2-amino-4-hydroxybutyric acid; α-amino-γ-hydroxy-*n*-butyric acid. $C_4H_9NO_3$; mol wt 119.12. C 40.33%, H 7.62%, N 11.76%, O 40.29%. Principal free amino acid occurring in pea plants: A. I. Virtanen, *Acta Chem. Scand.* **7,** 1423 (1953); J.A. Bakhuis, *Nature* **180,** 713 (1957). Prepn: Fischer, Blumenthal, *Ber.* **40,** 106 (1907); Armstrong *J. Am. Chem. Soc.* **70,** 1756 (1948); Birnbaum, Greenstein, *Arch. Biochem. Biophys.* **42,** 212 (1953); M. Frankel, Y. Knobler, *J. Am. Chem. Soc.* **80,** 3147 (1958). Review of homoserine production by fermentation: T. Nara in *Microbial Prod. Amino Acids,* K. Yamada, Ed. (Wiley, New York, 1972) pp 417-434.

L-Homoserine, flat prisms from 90% alc. Dec 203°. $[M]_D$ +21.8° (5N HCl), $[M]_D$ +14.3° (glacial acetic acid). $[\alpha]_D^{26}$ −8.8° (c = 5 in H_2O); $[\alpha]_D^{26}$ +18.3° (c = 2 in 2N HCl). On standing for 8 hrs at 26° the $[\alpha]_D$ of the HCl soln decreases to nearly zero as the corresponding levorotatory-γ-butyro-lactone is formed.

L-Homoserine γ-lactone monohydrochloride (prepd by refluxing L-homoserine with 2N HCl for 2 hrs), crystals, $[\alpha]_D^{26}$ −27.0° (c = 5).

D-Homoserine, crystals, dec 203°. $[\alpha]_D^{26}$ +8.8° (c = 5).
DL-Homoserine, crystals from dil ethanol, dec 186-187°.

4662. Homovanillic Acid. *4-Hydroxy-3-methoxybenz-eneacetic acid; (4-hydroxy-3-methoxyphenyl)acetic acid;* 4-hydroxy-3-methoxy-α-toluic acid. $C_9H_{10}O_4$; mol wt 182.17. C 59.33%, H 5.53%, O 35.13%. Metabolite found in human urine. Prepn: F. Tiemann, N. Nagai, *Ber.* **10,** 201 (1877); F. Mauthner, *Ann.* **370,** 368 (1909); H. E. Fisher, H. Hibbert, *J. Am. Chem. Soc.* **69,** 1208 (1947). End product of dopamine, *q.v.*: K. Shaw *et al., J. Biol. Chem.* **226,** 255 (1957); M. Goodall, H. Alton, *Biochem. Pharmacol.* **25,** 2635 (1968).

White crystals, mp 143°. Sol in water, benzene. Slightly sol in alcohol, ether. Insol in cyclohexane.

4663. HON. *2-Amino-5-hydroxy-4-oxopentanoic acid; 2-amino-5-hydroxylevulinic acid;* δ-hydroxy-γ-oxo-L-nor-valine. $C_5H_9NO_4$; mol wt 147.13. C 40.81%, H 6.17%, N 9.52%, O 43.50%. $HOCH_2COCH_2CH(NH_2)COOH$. Antitubercular antibiotic substance isolated from *Streptomyces akiyoshiensis nova* sp.: Tatsuoka *et al.*, *J. Antibiot.* **14A**, 39 (1961). Synthesis and structure: Miyake, *Chem. Pharm. Bull.* **8**, 1071, 1074, 1079 (1960).

Needles from water + acetone, no definite mp. $[\alpha]_D^{20}$ −6° (water); $[\alpha]_D^{17}$ −8.2° (c = 3.4 in water). pKa 2.91. uv max (water): 271 nm (ε 24). Stable in pure, dry state. Less stable in basic than in acidic solns. Colors red when heated in caustic soln, losing antibiotic activity. The DL-form is half as active as L-HON. LD_{50} in mice: 5200 mg/kg i.v.; 8000 mg/kg s.c.; 7600 mg/kg orally, S. Tatsuoka *et al.*, *loc. cit.*

4664. Hopantenic Acid. *4-[(2,4-Dihydroxy-3,3-dimethyl-1-oxobutyl)amino]butanoic acid; 4-(2,4-dihydroxy-3,3-dimethylbutyramido)butyric acid;* D-homopantothenic acid. $C_{10}H_{19}NO_5$; mol wt 233.27. C 51.49%, H 8.21%, N 6.01%, O 34.29%. Homolog of pantothenic acid, *q.v.* Prepn: R. Fuerst, L. L. Li, *Biochem. Biophys. Acta* **86**, 26 (1964); C. M. DeSha, R. Fuerst, *ibid.* 33. Prepn of the calcium salt: Y. Nishizawa *et al.*, *Japan. pat.* 732('66) (to Tanabe), *C.A.* **64**, 12555c (1965). Series of articles on analysis, physicochemical properties, electrophysiology, pharmacology, toxicity studies, teratology: *Bitamin* **33**, 603-632 (1966), *C.A.* **65**, 5949a-g (1966). Colorimetric determn, properties: T. Kodama *et al.*, *J. Vitaminol.* **13**, 298 (1967), *C.A.* **68**, 62755n (1968). Peri- and postnatal studies in rats: Y. Asano *et al.*, *Oyo Yakuri* **19**, 1011 (1980), *C.A.* **94**, 58160 (1981). Mechanism of action: V. M. Arakumov, M. A. Kovler, *Farmakol. Toksikol.* **44**, 30 (1981), *C.A.* **94**, 114577 (1981). Determn in plasma: Y. Umeno *et al.*, *J. Chromatog.* **226**, 333 (1981). Acute toxicity study and neuroleptic activity: B. F. Dorofeev, *C.A.* **87**, 33526a (1977).

Sol in water; stable at pH 5-6. $[\alpha]_D^{20}$ +23.8°. pKa 4.52 at 25°.

Calcium salt hemihydrate, $C_{20}H_{36}CaNO_5 \cdot \frac{1}{2}H_2O$, *calcium homopantothenate, hopantenate calcium, pantogam, Hopate.* White powder, mp 155-165°. $[\alpha]_D^{20}$ +24.19° (c = 2 in water). Bitter taste. Sol in water, slightly sol in methaol, practically insol in organic solvents. Stable at pH 5-6. LD_{50} in mice: 1.80 g/kg i.p., B. F. Dorofeev, *loc. cit.*

THERAP CAT: Cerebral activator.

4665. Hops. Carefully dried strobiles of *Humulus lupulus* L., *Moraceae,* bearing their glandular trichomes. *Habit.* Europe, Asia, North America, cultivated widely. *Constit.* Volatile oil (0.3-1%), lupulone, humulone, xanthohumol, lactaric acid, ceryl alcohol, cerotic acid, soft and hard hopresins. Review of chemistry and constituents: Stevens, *Chem. Rev.* **67**, 19 (1967).

USE: Beer brewing.

THERAP CAT: Aromatic bitter.

4666. Hordenine. *4-[2-(Dimethylamino)ethyl]phenol; N,N-dimethyltyramine; p-hydroxy-N,N-dimethylphenethylamine; anhaline; eremursine; peyocactine.* $C_{10}H_{15}NO$; mol wt 165.23. C 72.68%, H 9.15%, N 8.48%, O 9.68%. Isoln from barley germs: Leger, *Compt. Rend.* **142**, 108 (1906); **143**, 234, 916 (1906); **144**, 488 (1907); Erspamer, Falconieri, *Naturwiss.* **39**, 431 (1952). Structure: Leger, *Bull. Soc. Chim. France* [3] **35**, 868 (1906); [4] **1**, 148 (1907); Gaebel, *Arch. Pharm.* **244**, 441 (1906). Synthesis from phenethyl

alcohol: Barger, *J. Chem. Soc.* **95**, 2193 (1909); from tyrosine: Raoul, *Compt. Rend.* **204**, 74 (1937); from *p*-(β-hydroxyethyl)anisole: Cheng *et al.*, *J. Am. Chem. Soc.* **73**, 4081 (1951).

Orthorhombic prisms from alcohol or from benzene + petr ether, needles from water, mp 117-118°. bp$_{11}$ 173°. Sublimes 140-150°. Very sol in alcohol, chloroform, ether. 7 grams dissolve in 1000 ml water. Sparingly sol in benzene, toluene, xylene. Practically insol in petr ether.

Hydrochloride, $C_{10}H_{15}NO \cdot HCl$, needles from alcohol, mp 177°. Very sol in water. LD_{50} in mice: 113.5 mg/kg.

4667. Horehound. Hoarhound. *Marrubium vulgare* (Tourn.) L., *Labiatae. Habit.* Europe, Asia, naturalized in the U.S. *Constit.* Bitter principle marrubiin, volatile oil.

THERAP CAT: Expectorant.

4668. Horse-radish. Raphanus rusticanus; Kren; Maliner Kren. The root of *Radicula armoracia* (L.) Robinson [*Cochlearia armoracia* L.; *Armoracia lopathifolia* Gilib.; *Roripa armoracia* Hitch.; *Nasturtium armoracia* Fries], *Cruciferae. Habit.* Europe, naturalized in North America. Contains ascorbic acid and sinigrin which yields allyl isothiocyanate on hydrolysis with peroxidase or myrosinase, an enzyme from black mustard: Stoll, Seebeck, *Helv. Chim. Acta* **31**, 1432 (1948).

Active ingredient of *Rasapen,* a urinary antiseptic.

USE: Condiment.

4669. HPA-23. *Ammonium antimony tungsten oxide;* 5-tungsto-2-antimonate; 21-tungsto-9-antimonate. $(NH_4)_{17}Na(NaSb_9W_{21}O_{86}) \cdot 14H_2O$. Heteropolyanion (mineral condensed ion) with significant *in vitro* and *in vivo* antiviral activity. Synthesis: M. Michelon, G. Hervé, *Compt. Rend. Ser. C* **274**, 209 (1972); M. Michelon *et al.*, *J. Inorg. Nucl. Chem.* **42**, 1583 (1980). Crystal structure: J. Fischer *et al.*, *J. Am. Chem. Soc.* **98**, 3050 (1976). Laser Raman studies of intracellular localization: C. Cibert, C. Jasmin, *Biochim. Biophys. Acta* **108**, 1424 (1982). Effect on encephalomyocarditis, vesicular stomatitis virus infections in mice, and toxicity studies: G. H. Werner *et al.*, *J. Gen. Virol.* **31**, 59 (1976). *In vivo* effect on murine leukemia, sarcoma viruses: C. Jasmin *et al.*, *J. Natl. Cancer Inst.* **53**, 469 (1976). *In vitro* inhibition of rabies virus: H. Tsiang *et al.*, *J. Gen. Virol.* **40**, 665 (1978). Modulates Epstein-Barr virus: M. Souyri-Caporale *et al.*, *ibid.* **65**, 831 (1984). Effect against scrapie in sheep: R. H. Kimberlin, C. A. Walker, *Lancet* **2**, 591 (1979); *eidem, Arch. Virol.* **78**, 9 (1983). Clinical evaluation in AIDS: W. Rozenbaum *et al.*, *Lancet* **1**, 450 (1985); B. L. Moskovitz, *Antimicrob. Ag. Chemother.* **32**, 1300 (1988).

Crystals from water. LD_{50} i.p. in mice: 750 mg/kg.

4670. HQNO. *2-Heptyl-4-quinolinol 1-oxide; 2-heptyl-4-hydroxyquinoline N-oxide.* $C_{16}H_{21}NO_2$; mol wt 259.36. C 74.10%, H 8.16%, N 5.40%, O 12.34%. Inhibitor of electron transport through the cytochrome bc_1 segment of the respiratory chain. Isoln of the naturally occurring antagonist to dihydrostreptomycin from *Pseudomonas pyocyanea*: Hays *et al.*, *J. Biol. Chem.* **159**, 725 (1948); J. Lightbown, *J. Gen. Microbiol.* **11**, iv (1954). Properties: J. W. Cornforth, A. T. James, *Biochem. J.* **58**, xlviii (1954). Synthesis: *eidem, ibid.* **63**, 124 (1956); D. E. Ames *et al.*, *J. Chem. Soc.* **1956**, 3079. Inhibition of electron transport: J. W. Lightbown, F. L. Jackson, *Biochem. J.* **63**, 130 (1956); M. Avron, *ibid.* **78**, 735 (1961); N. J. Jacobs, M. J. Wolin, *Biochim. Biophys. Acta* **69**, 29 (1963). Effect on proton permeability of the mitochondrial membrane: K. Krab, M. Wikström, *Biochem. J.* **186**, 637 (1980). Sites of inhibition: G. Izzo *et al.*, *FEBS Letters* **93**, 320 (1978); M. Droppa *et al.*, *Z. Naturforsch.* **36C**, 109 (1981).

Crystals from methylethyl ketone, mp 156-157°.

4671. Humic Acids. Allomelanins found in soils, coals, and peat, resulting from the decompn of organic matter, particularly dead plants. Consists of a mixture of complex macromolecules having polymeric phenolic structures with the ability to chelate with metals, esp iron. *Review:* R. A. Nicolaus, *Melanins* (Hermann, Paris, 1968) pp 147-153; W. Flaig *et al.* in *Soil Components* vol. 1, J. E. Gieseking, Ed. (Springer, New York, 1975) pp 1-211.

Chocolate-brown, dust-like powder. Slightly sol in water, usually with much swelling; sol in alkali hydroxides and carbonates; also sol in hot concd HNO_3 with dark-red color.

USE: In mud baths, drilling muds, pigments for printing inks, fertilizers, growth hormones for plants, transporters of trace minerals in soil: Steelnick, *J. Chem. Ed.* **40**, 379 (1963).

4672. Humulene. *2,6,6,9-Tetramethyl-1,4,8-cycloundeca-triene;* α-humulene; α-caryophyllene. $C_{15}H_{24}$; mol wt 204.36. C 88.16%, H 11.84%. Sesquiterpenoid isomer of caryophyllene, *q.v.* occurring in many essential oils, especially oil of hops *(Humulus lupulus* L. *Moraceae)* and leaves of *Lindera strychnifolia* (F.) Will *Lauraceae.* Occurs in nature as a mixture with β-humulene. Isolation of mixture: A. C. Chapman, *J. Chem. Soc.* **67**, 54, 780 (1895). Identity with α-caryophyllene: F. Sorm *et al., Coll. Czech. Chem. Commun.* **14**, 693, 699, 716 (1949). Structure: F. Sorm *et al., ibid.* **19**, 570 (1954). Stereochemistry: A. T. McPhail, G. A. Sim, *J. Chem. Soc. (B)* **1966**, 112. Synthesis: E. J. Corey, E. Hanamaka, *J. Am. Chem. Soc.* **89**, 2758 (1967). Stereoselective synthesis: Y. Kitagawa *et al., ibid.* **99**, 3864 (1977). Conversion to β-humulene by chromatography on alkaline alumina: V. Benesova *et al., Coll. Czech. Chem. Commun.* **26**, 1832 (1961). *Reviews:* F. Sorm in *Fortschr. Chem. Org. Naturst.* **19**, 1-32 (1961); *Rodd's Chemistry of Carbon Compounds,* S. Coffey, Ed., vol. IIC (Elsevier, New York, 2nd ed., 1969) pp 282-283.

Liquid. bp_5 106-107°. n_D^{30} 1.5004, N. P. Damodaran, S. Dev, *Tetrahedron* **24**, 4113 (1968). Also reported as bp_{10} 123°. n_D^{25} 1.5015. d_D^{25} 0.8865, R. P. Hildebrand *et al., Chem. Ind. (London)* **1959**, 489. NMR spectrum: S. Dev *et al., J. Am. Chem. Soc.* **90**, 1246 (1968).

Silver nitrate complex, $C_{15}H_{24} \cdot 2AgNO_3$, crystals from aq ethanol, mp 175°.

β-Humulene, (E,E)-1,4,4-trimethyl-8-methylene-1,5-cycloundecadiene. Liquid. n_D^{20} 1.5014. d_4^{20} 0.8905.

4673. Humulon. *3,5,6-Trihydroxy-4,6-bis(3-methyl-2-butenyl)-2-(3-methyl-1-oxobutyl)-2,4-cyclohexadien-1-one;* α-bitter acid; α-lupulic acid. $C_{21}H_{30}O_5$; mol wt 362.45. C 69.58%, H 8.34%, O 22.07%. Antibiotic constituent of hops *(Humulus lupulus* L., *Moraceae). See also* Lupulon. Isoln from commercial hops: Bungener, *Bull. Soc. Chim.* [2] **45**, 487 (1886); Barth, Lintner, *Ber.* **31**, 2022 (1898); Wollmer, *Ber.* **49**, 780 (1916); Lewis *et al., J. Clin. Invest.* **28**, 916 (1949). Structure: Riedl, *Ber.* **85**, 692 (1952); Carson, *J. Am. Chem. Soc.* **73**, 4652 (1951). Absolute configuration and structure of preferred isomer: DeKeukeleire, Verzele, *Tetrahedron* **26**, 385 (1970).

Crystals from ether, mp 65-66.5°. Bitter taste, esp in alcoholic soln. More stable to air than lupulon. Monobasic acid. $[\alpha]_D^{20}$ −212° (1.0 g in 15.5 g 96% alc). uv max (ethanol): 237, 282 nm (ε 13,760; 8330). Soluble in the usual organic solvents. Slightly sol in boiling water from which it separates as a milky precipitate on cooling. Forms a sodium salt which is readily sol in water. Suffers no loss of bacteriostatic potency against *Staphylococcus aureus* upon autoclaving 40 ppm in phosphate buffer at pH 6.5 or 8.5. The presence of ascorbic acid in low concns extends the duration of bacteriostatic action.

4674. Hyalobiuronic Acid. *2-Amino-2-deoxy-3-O-β-glucopyranuronosyl-*D-*glucose; 3-O-(β-*D-glucopyranosyl-uronic acid)-2-amino-2-deoxy-*D-glucose. $C_{12}H_{21}NO_{11}$; mol wt 355.31. C 40.56%, H 5.96%, N 3.94%, O 49.53%. Disaccharide unit of hyaluronic acid. Isoln from hyaluronic acid: Rapport *et al., Nature* **168**, 996 (1951). Structure: Weissman, Meyer, *J. Am. Chem. Soc.* **76**, 1753 (1954). Synthesis: Takanashi *et al., ibid.* **84**, 3029 (1962).

Rectangular prisms from water, darken at 190° with no characteristic melting or dec point. pK_1' = 2.6, pK_2' = 7.1. Shows mutarotation: $[\alpha]_D^{20}$ +34° → +30° (c = 1.08 in 0.1N HCl). Sparingly sol in hot water, dilute HCl, dil $NaHCO_3$; practically insol in water, glacial acetic acid, ethanol, methanol and pyridine.

N-Acetylhyalobiuronic acid, $C_{14}H_{23}NO_{12}$, amorphous. pK' = 3.3. $[\alpha]_D^{24}$ −32° (c = 2.0 in water).

4675. Hyaluronic Acid. Mol wt is within the range of 50,000 to 8 × 10^6 depending on source, methods of prepn, and determination. A natural high viscosity mucopolysaccharide with alternating β (1-3) glucuronidic and β (1-4) glucosaminidic bonds. Found in the umbilical cord, in vitreous humor, in synovial fluid, in pathologic joints, in group A and C hemolytic streptococci and in Wharton's jelly. Isoln and characterization: Meyer, Palmer, *J. Biol. Chem.* **107**, 629 (1934); **114**, 689 (1936); Balazs, *Fed. Proc.* **17**, 1086 (1958); Laurent *et al., Biochim. Biophys. Acta* **42**, 476 (1960). Structure: Weissman, Meyer, *J. Am. Chem. Soc.* **76**, 1753 (1954); Meyer, *Fed. Proc.* **17**, 1075 (1958). Crystal structure of hyaluronate films: Dea *et al., Science* **179**, 560 (1973); Atkins, Sheehan, *ibid.* 562. Possible role in determining blood vessel location in the embryo: R. N. Feinberg, D. C. Beebe, *Science* **220**, 1177 (1983). *Reviews:* Tauber, *Chemistry and Technology of Enzymes* (New York, 1946); Meyer, Rapport in *Advan. Enzymol.* **13**, 199 (1952); Whistler, Olson in *Advan. Carbohyd. Chem.* **12**, 299 (1957). Review of role in various developmental processes: B. P. Toole, *Cell Biology of Extracellular Matrix,* E. D. Hay, Ed. (Plenum Press, New York, 1981) pp 259-288.

Sodium salt, *ARTZ, Connettivina, Equron, Healon, Healonid, Hyalgan, Hyalovet, Ial, Opegan, Synacid.* $[\alpha]_D^{25}$ $-74°$ (c = 0.25 in water): Rapport *et al., J. Am. Chem. Soc.* **73**, 2416 (1951). Most viscosity determinations of hyaluronic acid vary from 1-8: Jensen, *Acta Chem. Scand.* **7**, 603 (1953). Infrared absorption spectra: Orr, *Biochim. Biophys. Acta* **14**, 173 (1954).

USE: Surgical aid (ophthalmological).

THERAP CAT (VET): Adjunct in treatment of noninfectious synovitis.

4676. Hyaluronidases. Spreading factor; diffusing factor; invasin; Alidase; Apertase; Diffusin; Enzodase; Harodase; Hyalase; Hyalozima; Hyalidase; Hyasmonta; Hyason; Hyazyme; Infiltrase; Jalovis; Kinaden; Kinetin-Schering; Luronase; Permease; Rondase; Ronidase; Thiomucase; Unidasa; Wydase. Enzymes which have in common the cleavage of glycosidic bonds of hyaluronic acid, *q.v.*, and, to a variable degree, of some other acid mucopolysaccharides of connective tissue. The skin is probably the largest store of hyaluronidase in the body; the enzyme although generally present in an inactive form, may be supposed to regulate the velocity of water and metabolite exchange by decreasing the viscosity of the intercellular matrix. Also has a physiological role in fertilization: The sperm is rich in the enzyme and can thus advance better in the cervical canal and reach the ovum. Found in the type II pneumococci, in group A and C hemolytic streptococci, *Staphylococcus aureus* and *Clostridium welchii*: Linker *et al., J. Biol. Chem.* **219**, 13 (1956); in heads of leeches: Linker *et al., Nature* **180**, 810 (1957); in snake venoms: Favilli, *ibid.* **145**, 866 (1940); in testes: Hahn, *Biochem. Z.* **315**, 83 (1943); Högberg, *Acta Chem. Scand.* **8**, 1098 (1954). Biochemical properties: D. Platt, *Arzneimittel-Forsch.* **20**, 1836 (1970). *Review:* Meyer, Rapport in *Advan. Enzymol.* **13**, 199-236 (1952); Meyer *et al., The Enzymes* vol. **4**, P. D. Boyer *et al.,* Eds. (Academic Press, New York, 2nd ed., 1960) pp 447-460; Meyer, *ibid.* vol. **5** (3rd ed., 1971) pp 307-320. Reviews of clinical trials in myocardial infarction: G. S. May *et al., Progr. Cardiovasc. Dis.* **25**, 335-359 (1983); A. B. Saunders, *Emerg. Med. Clin. North Am.* **6**, 361-372 (1988). Hyaluronidase manufacturers define their product in terms of turbidity-reducing (TR) units or in viscosity units. Prepd solns for injection usually contain 150 turbidity-reducing units or 500 viscosity units dissolved in 1 ml of isotonic NaCl soln.

USE: Pharmaceutical aid (diffusing agent—s.c. injections).

THERAP CAT: Spreading agent.

THERAP CAT (VET): To promote diffusion, absorption, resorption.

4677. Hycanthone. *1-[[2-(Diethylamino)ethyl]amino]-4-(hydroxymethyl)-9H-thioxanthen-9-one.* $C_{20}H_{24}N_2O_2S$; mol wt 356.48. C 67.39%, H 6.78%, N 7.86%, O 8.98%, S 8.99%. Metabolite of lucanthone, *q.v.*: Rosi *et al., Nature (London)* **208**, 1005 (1965). Prepn by oxidative fermentation of lucanthone and schistosomicidal activity: Rosi *et al., J. Med. Chem.* **10**, 867 (1967); **Neth. pat. Appl. 6,410,359,** and Rosi, Peruzzotti, **U.S. pats. 3,294,803; 3,312,598** (1965, 1966, 1967 all to Sterling Drug). Alternate synthesis: Laidlaw *et al., J. Org. Chem.* **38**, 1743 (1973).

Crystals, mp 100.6-102.8°. Absorption max (ethanol): 233, 258, 329, 438 nm (ϵ 19,400, 37,000, 9,700, 6,600). Extremely sensitive to acid.

Hydrochloride, mp 173-176° (dec).

Mesylate, *Etrenol.*

THERAP CAT: Anthelmintic (Schistosoma).

4678. Hydantoin. *2,4-Imidazolidinedione; 2,4-(3H,5H)-imidazoledione; glycolylurea.* $C_3H_4N_2O_2$; mol wt 100.08. C 36.00%, H 4.03%, N 27.99%, O 31.97%. Prepn: Baeyer, *Ann.* **130**, 129 (1864). Manuf: Gresham, Schweitzer, U.S. pat. **2,402,134** (1946 to du Pont); White, Wysong, U.S. pat. **2,663,713** (1953 to Dow). *Review:* J. H. Bateman in Kirk-Othmer *Encyclopedia of Chemical Technology* vol. **12** (Wiley-Interscience, New York, 3rd ed., 1980) pp 692-711.

Needles from methanol, mp 220°. Slightly sol in water or ether; sol in alcohol, in solns of fixed alkali hydroxides.

4679. Hydnocarpic Acid. *2-Cyclopentene-1-undecanoic acid;* 11-(2-cyclopenten-1-yl)undecanoic acid. $C_{16}H_{28}O_2$; mol wt 252.38. C 76.14%, H 11.18%, O 12.68%. Component of chaulmoogra oil. Isoln from seeds of *Hydnocarpus wightiana* Blume or *H. anthelmintica* Pierre, *Flacourtiaceae*, or from the seeds of *Taraktogenos kurzii* King, *Bixaceae*: Power, Barrowcliff, *J. Chem. Soc.* **87**, 884 (1905); **91**, 557 (1907). Structure: Shriner, Adams, *J. Am. Chem. Soc.* **47**, 2727 (1925). Synthesis of *dl*-hydnocarpic acid: Diaper, Smith, *Biochem. J.* **42**, 581 (1948). Toxicity: *Handbook of Toxicology*, vol. 1, W. S. Spector, Ed. (Saunders, Philadelphia, 1956) pp 274-275.

dl-Form, pearly plates from alcohol, ethyl actate or petr ether + ethyl acetate, mp 59-59.5°.

d-Form, leaflets from petr ether + ethyl acetate, mp 59.5-60°. $[\alpha]_D^{20}$ +70° (chloroform).

Sodium salt, *sodium hydnocarpate, hydnocarpate sodium, sodium gynocardate, Alepol.* Yellowish powder. Sol in water, alc. The aq soln is alkaline. MLD i.v. in rats: 100-125 mg/kg (Spector).

THERAP CAT: Antibacterial (leprostatic).

4680. Hydracarbazine. *6-Hydrazino-3-pyridazinecarboxamide;* 3-hydrazino-6-carbamoylpyridazine; 3-hydrazinopyridazine-6-carboxamide. $C_5H_7N_5O$; mol wt 153.15. C 39.21%, H 4.61%, N 45.73%, O 10.45%. Prepn: Libermann, Rouaix, *Bull. Soc. Chim. France* **1959**, 1793; **Brit. pat. 856,-409** (1960 to Chimie et Atomistique).

Crystals, dec 249-250°.

Note: A component of *Normatensyl.*

THERAP CAT: Antihypertensive; diuretic.

4681. Hydracrylic Acid. *3-Hydroxypropanoic acid; β-hydroxypropionic acid; ethylene lactic acid.* $C_3H_6O_3$; mol wt 90.08. C 40.00%, H 6.72%, O 53.28%. CH_2OHCH_2COOH. Prepd by alkaline hydrolysis of the nitrile: R. R. Read, *Org. Syn. coll. vol. I*, 321 (2nd ed., 1941).

Viscous liq. Strong acid, K at 25° = 3.11 × 10⁻⁵. On distn or boiling with 50% H_2SO_4 dec into water and acrylic acid. Very sol in water, sol in alcohol, miscible with ether.

Sodium salt, $NaC_3H_5O_3$, deliquescent crystals, mp 143°.

Calcium salt dihydrate, $Ca(C_3H_5O_3)_2 \cdot 2H_2O$, prisms, mp 140-145°; freely sol in cold water.

4682. Hydralazine. *1(2H)-Phthalazinone hydrazone;* 1-hydrazinophthalazine; Ciba 5968; Präparat 5968; C-5968; Hipoftalin; Hypophthalin; Apresoline. $C_8H_8N_4$; mol wt 160.18. C 59.98%, H 5.03%, N 34.98%. Prepd by the action of hydrazine hydrate on 1-chloro- or 1-phenoxyphthalazine: Hartmann, Druey, U.S. pat. **2,484,029** (1949 to Ciba); Druey, Ringier, *Helv. Chim. Acta* **34**, 204 (1951). Metabolism: Z. H. Israile, P. G. Dayton, *Drug Metab. Rev.* **6**, 283 (1977); K. Schmid *et al., Arzneimittel-Forsch.* **31**, 1143 (1981). Pharmacology: J. L. Cangiano *et al., J. Lab. Clin. Med.* **92**, 516 (1978). Clinical paper: R. F. Albrecht *et al., Int. Anesthesiol. Clin.* **16**, 299 (1978). Acute toxicity: L. Dorigotti *et al., Pharmacol. Res. Commun.* **8**, 295 (1976). Comprehensive description: C. E. Orzech *et al.,* in *Analytical Profiles of Drug Substances* vol. 8, K. Florey, Ed. (Academic Press, New York, 1979) pp 283-314.

Yellow needles from methanol, mp 172-173° (rapid heating). One gram dissolves in 3 ml 2N acetic acid, in 12 ml warm methanol. Forms a red compd (phthalazinylhydrazone) with acetone at 60° in presence of 2N acetic acid. LD_{50} in mice, rats (mg/kg): 122, 90 orally; 101, 40 i.p. (Dorigotti).

Hydrochloride, $C_8H_9ClN_4$, *Lopress.* Yellow crystals, dec 273°. Soly in water (g/100 ml) at 15°: 3.01; at 25°: 4.42; in 95% ethanol: 0.2 g/100 ml. Very slightly sol in ether. pH of a 2% aq soln 3.5 to 4.5. uv max (0.001% aq soln): 211, 240, 260, 304, 315 nm. Aq solns containing 20 mg/ml may be preserved with 0.5% chlorobutanol.

Sulfate, *Depressan.*
THERAP CAT: Antihypertensive.

4683. Hydrallostane. *11β,17α,21-Trihydroxy-5α-pregnane-3,20-dione;* 11β,17α,21-trihydroxyallopregnane-3,20-dione; allodihydrohydrocortisone; allopregnane-11β,17α,21-triol-3,20-dione; 4,5α-dihydrocortisol; allodihydro F. $C_{21}H_{32}O_5$; mol wt 364.47. C 69.20%, H 8.85%, O 21.95%. Isoln from beef and hog adrenals: Neher, Wettstein, *Helv. Chim. Acta* **39**, 2062 (1956). Prepn: Pataki *et al., J. Biol. Chem.* **195**, 753 (1952); Fukushima, Daum, *J. Org. Chem.* **26**, 520 (1961); Gould, Oliveto, U.S. pats. **2,783,254; 2,897,216** (1957; 1959 to Schering); Gould, Herzog, U.S. pat. **2,783,-226** (1957 to Schering); Brit. pat. **742,888** (1956 to Syntex).

Crystals, mp 234-240°. $[\alpha]_D^{25}$ +83° (acetone). Practically insol in water. Sol in methanol, acetone, chloroform.

21-Acetate, $C_{23}H_{34}O_6$, crystals from hexane + ethyl acetate, mp 211-213°. $[\alpha]_D^{20}$ +69° (chloroform).

21-tert-Butylacetate, $C_{27}H_{42}O_6$, crystals from ethanol, mp 256-264° (also crystallizes as a hydrate). Sol in chloroform, oils, fats.

4684. Hydramethylnon. *Tetrahydro-5,5-dimethyl-2(1H)-pyrimidinone[3-[4-(trifluoromethyl)phenyl]-1-[2-[4-(trifluoromethyl)phenyl]ethenyl]-2-propenylidene]hydrazone;* 1,5-bis(α,α,α-trifluoro-p-tolyl)-1,4-pentadiene-3-one (1,4,-5,6-tetrahydro-5,5-dimethyl-2-pyrimidinyl)hydrazone; AC 217300; Amdro; Combat; Maxforce. $C_{25}H_{24}F_6N_4$; mol wt 494.50. C 60.72%, H 4.89%, F 23.05%, N 11.33%. Slow activating stomach poison insecticide. Prepn: J. B. Lovell, U.S. pat. **4,087,525** (1975 to Am. Cyanamid). Activity: idem, *Proc. Brit. Crop Prot. Conf.-Pests Dis.* **1979**, 575. Field

trials for control of red imported fire ant: D. P. Harlan *et al., Southwest. Entomol.* **6**, 150 (1981); W. A. Banks *et al.,* ibid. 158; for control of cockroaches: J. F. Milio *et al., J. Econ. Entomol.* **79**, 1280 (1986).

Crystals from isopropanol, mp 189-191°. Insol in water; sol in alc, acetone.
USE: Insecticide.

4685. Hydramitrazine. *4,6-Bis(diethylamino)-1,3,5-triazin-2(1H)-one hydrazone; 2,4-bis(diethylamino)-6-hydrazino-s-triazine;* 2-hydrazino-4,6-bis(diethylamino)-1,3,5-triazine; Meladrazine. $C_{11}H_{23}N_7$; mol wt 253.35. C 52.15%, H 9.15%, N 38.70%. Prepn: Hüni, Staehelin, U.S. pat. **2,824,103** (1958 to Ciba).

d-Tartrate, $C_{15}H_{29}N_7O_6$, *Lisidonil.*
THERAP CAT: Antispasmodic.

4686. Hydrangea. Seven barks. Dried rhizome and roots of *Hydrangea arborescens* L., *Saxifragaceae.* Habit. Eastern U.S. *Constit.* The glucoside hydrangin, saponin, resins, fixed and volatile oils, starch.

4687. Hydrargaphen. [μ-[[3,3'-Methylenebis[2-naphthalenesulfonato]](2−)]]diphenyldimercury; phenylmercuric methylenebis(2-naphthyl-3-sulfonic acid); phenylmercury 2,2'-dinaphthylmethane-3,3'-disulphonate; phenylmercuric dinaphthylmethanedisulfonate; methylenedinaphthylenesulfonic acid bisphenylmercuri salt; phenylmercury methylenedinaphthalenesulfonate; bis(phenylmercuri)methylenedinaphthalenesulfonate; phenyl mercuric Fixtan; Conotrane; P.M.F.; Versotrane; Hydraphen; Penotrane; Fibrotan; Septotan. $C_{33}H_{24}Hg_2O_6S_2$; mol wt 981.87. C 40.36%, H 2.46%, Hg 40.86%, O 9.78%, S 6.53%. Prepn: Bywater, Pritchard, Brit. pats. **584,196** and **2,555,114** (1949 and 1951 to Ward, Blenkinsop). Review and properties: Goldberg, *Mfg. Chemist* **22**, 182 (1951); Hopf, ibid. **24**, 444 (1953).

Amorphous powder. Practically insol in water but readily goes into colloidal soln in alkali metal dinaphthylmethane disulfonates. The colloidal solns have strong tendency to absorb at interfaces and to form charged hydrated aggregates. LD_{100} orally in mice: 80 mg/kg.
USE: Bactericide and fungicide for the treatment of wool, hides, leather, textiles, timber and wood-pulp, paints, adhesives.
THERAP CAT: Topical anti-infective.

4688. Hydrastine. *6,7-Dimethoxy-3-(5,6,7,8-tetrahydro-6-methyl-1,3-dioxolo[4,5-g]isoquinolin-5-yl)-1(3H)-isobenzofuranone; l-β-hydrastine.* $C_{21}H_{21}NO_6$; mol wt 383.39. C 65.78%, H 5.52%, N 3.65%, O 25.04%. Isoln of naturally occurring l-β-form from *Hydrastis canadensis* L., *Ranunculaceae* together with berberine and canadine. Synthesis of diastereoisomeric mixtures of hydrastines: Hope *et al., J. Chem. Soc.* **1931**, 236; Marshall *et al.,* ibid. **1934**, 1315; M. Hanaoka *et al., Chem. Pharm. Bull.* **27**, 1947 (1979); J. R. Falck, S. Manna, *Tetrahedron Letters* **22**, 619 (1981). Resolution to the *l*-β-form: Haworth, Pinder, *J. Chem. Soc.* **1950**, 1776; Haworth *et al.*, *Nature* **165**, 529 (1950). Structure: Knabe, *Arch. Pharm.* **293**, 121 (1960). Abs config: Ohta *et al.*, *Tetrahedron Letters* **1963**, 859; Blaha *et al.*, *Coll. Czech. Chem. Commun.* **29**, 2328 (1964); Snatzke *et al.*, *Tetrahedron* **25**, 5059 (1969). Biosynthesis: Gear, Spenser, *Can. J. Chem.* **41**, 783 (1963).

Orthorhombic prisms from alc, mp 132°. $[\alpha]_D^{20} - 50°$ (c = 0.3 in abs alc). uv max (ethanol): 202, 218, 238, 298, 316 nm (log ε 4.79, 4.53, 4.15, 3.86, 3.63). pK 7.8. Freely sol in acetone and benzene; insol in water. The salts hydrolyze easily and do not crystallize well.

Hydrochloride, $C_{21}H_{22}ClNO_6$. Prepn: R. Paech, M. V. Tracy, *Modern Methods of Plant Analysis* vol. **IV** (Springer-Verlag, Berlin, 1955) p 383. Powder, mp 116°. $[\alpha]_D^{17} +127°$ (c = 4 in dil HCl). Very sol in water, alcohol; slightly sol in CHCl₃; very slightly in ether. pH (0.5% aq soln): 4.2. Hygroscopic. *Keep well closed.*

THERAP CAT: Hydrochloride formerly as uterine hemostatic, antiseptic.

4689. Hydrastinine. *5,6,7,8-Tetrahydro-6-methyl-1,3-dioxolo[4,5-g]isoquinolin-5-ol;* 1-hydroxy-6,7-methylenedioxy-2-methyl-1,2,3,4-tetrahydroisoquinoline. $C_{11}H_{13}$-NO_3; mol wt 207.22. C 63.75%, H 6.32%, N 6.76%, O 23.16%. By oxidation of hydrastine: Freund, Will, *Ber.* **20**, 88 (1887). From berberine: Freund, *Ger. pat.* **241,136** (1910), *C.A.* **6**, 2145 (1912). From cotarnine: Pyman, Remfry, *J. Chem. Soc.* **101**, 1595 (1912); Topchiev, *J. Appl. Chem. USSR* **6**, 529 (1933). From formylhomopiperonylamine: Decker *et al.*, *Ann.* **395**, 299, 321, 328 (1913); Rosenmund, *Ber. Deut. Pharm. Ges.* **29**, 200 (1919). From safrole: Kindler, Peschke, *Arch. Pharm.* **270**, 353 (1932). Structure study: Schneider, Müller, *Ann.* **615**, 34 (1958).

Needles from petr ether, mp 117°. Freely sol in alcohol, chloroform, ether, dil acids; practically insol in cold, moderately sol in hot water. Solns with water or alcohol are yellow and fluorescent, those with nonpolar organic solvents are colorless. pK 2.62. K 2.4 × 10⁻³. Absorption spectra: Dobbie, Tinkler, *J. Chem. Soc.* **85**, 1005 (1904).

THERAP CAT: The hydrochloride as cardiotonic; uterine hemostatic.

4690. Hydrastis. Golden seal; orange root; yellow root; yellow puccoon; Indian turmeric. Dried rhizome and roots of *Hydrastis canadensis* L., *Ranunculaceae*, contg not less than 2.5% ether-soluble alkaloids. *Habit.* North America. *Constit.* 2-4% hydrastine, 2-3% berberine; canadine, volatile oil, resin.

4691. Hydrazine. Hydrazine anhydrous. H_4N_2; mol wt 32.05. H 12.58%, N 87.41%. H_2NNH_2. Prepn from hydrazine hydrate: Sisler *et al.*, *J. Am. Chem. Soc.* **76**, 3914 (1954); Schenk in *Handbook of Preparative Inorganic Chemistry* vol. **1**, G. Brauer, Ed. (Academic Press, New York, 1963) pp 469-472. Toxicity data: Witkin, *Arch. Ind. Health* **13**, 34 (1956). Toxicology study: Back, Thomas, *Ann. Rev. Pharmacol.* **10**, 395 (1970). Review of carcinogenicity: *IARC Monographs* **4**, 127-136 (1974). Books: L. F. Audri-

eth, B. A. Ogg, *The Chemistry of Hydrazine* (Wiley, New York, 1951); C. C. Clark, *Hydrazine* (Mathieson Chem., Baltimore, 1953). *Reviews:* Troyan, *Ind. Eng. Chem.* **45**, 2608-2612 (1953); Zimmer, *Chem. Ztg.* **79**, 599-605 (1955); Hudson *et al.*, "Hydrazine" in *Mellor's* vol. **VIII**, suppl. II, *Nitrogen* (Part 2), 69-114 (1967); Jones in *Comprehensive Inorganic Chemistry* vol. **2**, J. C. Bailar, Jr. *et al.*, Eds. (Pergamon Press, Oxford, 1973) p 250-265; H. W. Schiessl in Kirk-Othmer *Encyclopedia of Chemical Technology* vol. **12** (Wiley-Interscience, New York, 3rd ed., 1980) pp 734-771.

Colorless oily liq, fuming in air. Penetrating odor resembling that of ammonia. Burns with violet flame. Explodes during distn if traces of air are present, also affected by uv and metal ion catalysts. Flash and fire pt 126°F (52°C). Can be stored for years, if sealed in glass and kept in a cool, dark place. Contracts on freezing. d_4^{-5} 1.146; d_4^0 1.0253; d_4^2 1.024; d_4^{15} 1.011; d_4^{25} 1.0036; d_4^{35} 0.9955. One gallon of the commercial product weighs 8.38 lbs. mp 2.0°. bp_{760} 113.5°; bp_{71} 56°; $bp_{5\,atm}$ 170°; $bp_{10\,atm}$ 200°; $bp_{20\,atm}$ 236°. $n_D^{22.3}$ 1.46979; n_D^{35} 1.46444. Dipole moment 1.83-1.90. Dielectric constant (25°): 51.7. Latent heat of fusion (mp): 3.025 kcal/mole; latent heat of vaporization (bp): 9760 kcal/mole (calc). Crit temp 380°; crit pressure 14 atm. Diacidic base. K_1 (25°): about 9×10^{-7}. Forms salts with inorganic acids. Highly polar solvent. Powerful reducing agent. Dissolves many inorganic substances. Misc with water, methyl, ethyl, propyl, isobutyl alcohols. Forms an azeotropic mixture with water, bp_{760} 120.3°, which contains 55 mole-% (68.5 weight-%) N_2H_4. LD_{50} in mice (mg/kg): 57 i.v.; 59 orally (Witkin).

Dihydrochloride, $Cl_2H_6N_2$, white crystalline powder, mp 198°. d 1.42. Freely sol in water, slightly in alcohol.

Caution: Direct liquid contact with skin or eyes may produce severe burns. Vapors are highly irritating to eyes, nose, throat. May cause injury to lungs, liver, kidneys: Patty's *Industrial Hygiene and Toxicology*, vol. **2A**, G. D. Clayton, F. E. Clayton, Eds. (Wiley-Interscience, New York, 3rd ed., 1981) pp 2798-2800. This substance may reasonably be anticipated to be a carcinogen: *Fourth Annual Report on Carcinogens* (NTP 85-002, 1985) p 115.

USE: Reducing agent; organic hydrazine derivs; rocket fuel. Dihydrochloride as chlorine scavenger for HCl gas streams.

4692. Hydrazine Hydrate. H_6N_2O; mol wt 50.06. H 12.09%, N 55.95%, O 31.96%. $H_2NNH_2.H_2O$. Prepd from hydrazine sulfate by the action of NaOH, followed by distn under nitrogen.

Fuming refractive liquid, faint characteristic odor. *Violent poison! Causes delayed eye irritation.* d^{21} 1.03. mp −51.7° or below −65° (two eutectics). bp_{740} 118-119°; bp_{26} 47°. n_D^{20} 1.42842. Strong base, very corrosive, attacks glass, rubber, cork, but not stainless V_2A steel or Allegheny stainless 304 and 347. Molybdenum steels such as Allegheny stainless 316 should not be used. Very powerful reducing agent. Miscible with water and alcohol. Insol in chloroform and ether.

Mixture with methanol, *C-Stuff.*

USE: Reducing agent, solvent for inorganic materials. Manuf "Helman" catalyst, consisting of 80% hydrazine hydrate, 19.5% ethanol, 0.5 to 0.05% copper, used to dec hydrogen peroxide in V-2 type rockets. Mixture with methanol as propellant for rocket engines.

4693. Hydrazine Sulfate. Hydrazinium sulfate; hydrazonium sulfate. $H_6N_2O_4S$; mol wt 130.12. H 4.65%, N 21.53%, O 49.18%, S 24.64%. $H_2NNH_2.H_2SO_4$. Prepd by Raschig synthesis: $2NH_3.aq + [Ca(OCl)_2/Na_2CO_3/colloid]$ and treatment with H_2SO_4. Starch, glue, or gelatin are used as colloids, and sodium hypochlorite may be used instead of bleaching powder: Adams, Brown, *Org. Syn.* **2**, 37 (1922); Audrieth, Nickles, *Inorg. Syn.* **1**, 90 (1939). Industrial prepn by the action of sodium hypochlorite on urea in the presence of NaOH: *B.I.O.S. Final Report 369*; Moncrieff, *Manuf. Chem.* **18**, 177 (1947). Revised lab procedures: Pfeiffer, Simons, *Ber.* **80**, 127 (1947); Adams, Brown, *Org. Syn.* **coll. vol. I**, 2nd ed. (1941), p 309. Crystal structure: Nitta *et al.*, *Acta Cryst.* **4**, 289 (1951); Jönsson, Hamilton, *ibid.* **26B**, 536 (1970). Review of activity and clinical studies in cancer cachexia: J. Gold, *Nutr. Cancer* **9**, 59-66 (1987).

Orthorhombic crystals. Glass-like plates or prisms. d 1.378: Curtis, Jay, *J. Prakt. Chem.* **39**, 39 (1889); d[7] 2.016: Nitta *et al., loc. cit.* mp 254°. Sol in about 33 parts water; freely sol in hot water. Insol in alcohol. pH of 0.2 molar aq soln 1.3.

Note: This substance may reasonably be anticipated to be a carcinogen: *Fourth Annual Report on Carcinogens* (NTP 85-002, 1985) p 115.

USE: In the gravimetric estimation of nickel, cobalt and cadmium; in the refining of rare metals; as antioxidant in soldering flux for light metals; as reducing agent in the analysis of minerals and slags; in separating polonium from tellurium; in tests for blood; for destroying fungi and molds; in the prepn of hydrazine hydrate.

4694. Hydrazine Tartrate. Hydrazine acid tartrate; hydrazine hydrogen tartrate; hydrazine bitartrate. $C_4H_{10}N_2O_6$; mol wt 182.14. C 26.36%, H 5.53%, N 15.39%, O 52.71%. $H_2NNH_2 \cdot C_4H_6O_6$.

Crystals, mp 182-183°. $[\alpha]_D^{20}$ +22.5°. Soly in water at 0° about 6 g/100 ml. pH of a satd aq soln 3.6.

USE: In chemical deposition of metals (silvering mirrors, etc.): Owen, U.S. pat. **2,801,935** (1957 to Merck & Co.).

4695. 4-Hydrazinobenzenesulfonic Acid. *p*-Sulfophenylhydrazine; phenylhydrazine-*p*-sulfonic acid. $C_6H_8N_2O_3S$; mol wt 188.20. C 38.29%, H 4.28%, N 14.88%, O 25.50%, S 17.04%. Prepn by sulfonation of phenylhydrazine: L. Claisen, P. Roosen, *Ann.* **278**, 296 (1894); by the reduction of *p*-diazobenzenesulfonic acid: Th. Zincke, A. Kuchenbecker, *Ann.* **330**, 1 (1903); L. V. Lazeeva *et al.*, USSR pat. **1,057,493** (1983 to Tambov Pigment), *C.A.* **100**, 138755q (1984). Used in resoln of 2-pyrazoline cmpds: M. Mukai *et al., Can. J. Chem.* **57**, 360 (1979); in isoln of volatile ketones: W. Treibs, H. Röhnert, *Ber.* **84**, 433 (1951); in analysis of trace amounts of selenium: T. Kawashima *et al., Anal. Chim. Acta* **49**, 443 (1970); *eidem, ibid.* **89**, 65 (1977).

$$HSO_3-\!\!\bigcirc\!\!-NHNH_2$$

Needles from water, mp 286°. Slightly sol in water, alcohol.

4696. 2-Hydrazinoethanol. 2-Hydroxyethylhydrazine; β-hydroxyethylhydrazine; Omaflora. $C_2H_8N_2O$; mol wt 76.10. C 31.56%, H 10.60%, N 36.81%, O 21.02%. $HOCH_2CH_2NHNH_2$. Prepn from hydrazine monohydrate and 2-chloroethanol: Gansser, Rumf, *Helv. Chim. Acta* **36**, 1423 (1953); from hydrazine monohydrate and ethylene oxide: Gever, O'Keefe, U.S. pat. **2,660,607** (1953 to Eaton Labs.); from hydrazine and ethylene oxide: **Brit.** pat. **776,113** (1957 to Olin Mathieson).

Colorless, slightly viscous liquid. d 1.11. One gallon weighs 9.26 lbs. mp −70°. bp$_{17.5}$ 110-130°; bp$_{25}$ 145-153°. Flash pt 224°F (106°C). Misc with water. Sol in the lower alcohols. Slightly sol in ether.

USE: Plant growth regulant.

4697. Hydrazoic Acid. Hydrogen azide; hydronitric acid; triazoic acid; stickstoffwasserstoffsäure (German). HN_3; mol wt 43.03. H 2.34%, N 97.66%. Produced by the action of sulfuric acid on sodium azide: L. F. Audrieth, C. F. Gibbs, *Inorg. Syn.* **1**, 77 (1939); using stearic acid: Günther, Meyer, *Z. Elektrochem.* **41**, 541 (1935). Prepn of water and ether solns of hydrazoic acid: W. S. Frost *et al., J. Am. Chem. Soc.* **55**, 3516 (1933); L. F. Audrieth, C. F. Gibbs, *loc. cit.*; P. W. Schenk in *Handbook of Preparative Inorganic Chemistry* vol. 1, G. Brauer, Ed. (Academic Press, New York, 2nd ed., 1963) pp 472-474. GC determn: J. M. Zehner, R.A. Simonaitis, *J. Chrom. Sci.* **14**, 493 (1976). Toxicity study: Graham *et al., J. Ind. Hyg. Toxicol.* **30**, 98 (1948). Review of toxicology: C. F. Reinhardt, M. R. Brittelli, in Patty's *Industrial Hygiene & Toxicology* Vol. 2A, G. D. Clayton, F. E. Clayton, Eds. (Wiley-Interscience, New York, 1981) pp 2779-2784. *Reviews:* Mason in *Mellor's Comprehensive Treatise on Inorganic and Theoretical Chemistry* vol. VIII, suppl. II, *Nitrogen* (part II), 1-15 (1967); Jones in *Comprehensive Inorganic Chemistry* vol. 2, J. C. Bailar Jr. *et al.*, Eds. (Pergamon Press, Oxford, 1973) pp 276-293.

Mobile liquid. Intolerable pungent odor. *Extremely explosive!* mp −80°. bp 37°. pKa 4.72. LD$_{50}$ in mice (mg/kg): 21.5 i.p. (Graham).

Human Toxicity: Acute: eye irritation, cough, headache, fall in blood pressure, weakness, collapse. *Chronic:* hypotension, weakness, palpitation, ataxia (Graham).

USE: Industrially in prepn of heavy metal azides for shell detonators.

4698. Hydrindantin. *2,2'-Dihydroxy-[2,2'-bi-1H-indene]-1,1',3,3'-(2H,2'H)-tetrone; 2,2'-dihydroxy-[2,2'-biindan]-1,1',3,3'-tetrone;* reduced ninhydrin. $C_{18}H_{10}O_6$; mol wt 322.26. C 67.08%, H 3.13%, O 29.79%. Formed by the action of potassium cyanide on ninhydrin: Bruice, Richards, *J. Org. Chem.* **23**, 145 (1958). Convenient prepn by reduction of ninhydrin with ascorbic acid: Moore, Stein, *J. Biol. Chem.* **211**, 907 (1954).

Dihydrate, prisms from acetone, anhydr at 100°, turns reddish-brown at 200°, dec 249-254°. Very sparingly sol in hot water; sol in Methyl Cellosolve. Sol with decompn in aq Na_2CO_3 solns (deep red color), and in NaOH solns (deep blue color). Can be precipitated from carbonate solns by the addn of acid. Deep purple color with ammonia and blue color with amino acids.

USE: Reagent for the photometric determination of amino acids and similar compds.

4699. Hydriodic Acid. A soln of hydrogen iodide in water. Marketed in various concns, *e.g.*, 57% HI, d 1.7; 47%, d 1.5; 10%, d 1.1. Prepd by absorption of hydrogen iodide gas in water or by the action of iodine on hydrogen sulfide according to the eq $H_2S + I_2 \rightarrow 2HI + S$: Heisig, Frykholm, *Inorg. Syn.* **1**, 157 (1939). *See also* Hydrogen Iodide.

Colorless when freshly made, but rapidly turns yellowish or brown on exposure to light and air. This discoloration can be prevented by the addition of about 1.5% hypophosphorous acid (H_3PO_2). Concd solns that have been stored for some time are usually opaque from oxidation, they may be regenerated with hypophosphorous acid: Foster, Nahas, Jr., *Inorg. Syn.* **2**, 210 (1946). *Keep protected from light and air, preferably not above 30°.* Miscible with water or alcohol. Dissolves iodine. The azeotrope (constant-boiling acid) bp$_{760}$ 127°, d 1.70, contains 56.9% HI. Strong, corrosive acid, attacks natural rubber. 0.1 molar soln, pH 1.0.

USE: Reducing agent, manuf of inorganic iodides, pharmaceuticals, disinfectants. The 57% acid is also used for analytical purposes, such as methoxyl determinations. *Caution:* Strong irritant.

THERAP CAT: Expectorant.

4700. Hydrobenzoin. *1,2-Diphenyl-1,2-ethanediol;* diphenylethyleneglycol. $C_{14}H_{14}O_2$; mol wt 214.27. C 78.48%, H 6.59%, O 14.93%. Prepn of *d-, l-, dl-* and *meso*-forms: Forst, Zincke, *Ann.* **182**, 262, 275 (1876); Irvine, Weir, *J. Chem. Soc.* **91**, 1390 (1907); Buck, Jenkins, *J. Am. Chem. Soc.* **51**, 2163 (1929); L. F. Fieser, *Organic Experiments* (D. C. Heath, Boston, 1964) pp 210, 214, 216-217, 229-231. Improved method for prepn of *dl*-form from the *meso* isomer: Collet, *Synthesis* **1973**, 664.

$$\begin{matrix} HO & OH \\ | & | \\ C_6H_5CHCHC_6H_5 \end{matrix}$$

dl-Form, crystals from ether + petr ether, mp 120°. Resolved into *d*- and *l*-forms by slow, repeated crystallizations from ether.

d-Form, $[\alpha]_D^{20}$ +97.6° (chloroform).
l-Form, $[\alpha]_D^{20}$ −97.0° (chloroform).
meso-Form, monoclinic leaflets from alcohol or water, mp 139°. Soly in water: 0.25% at 20°, 1.25% at 100°. Freely sol in hot alcohol, chloroform. Infrared absorption (chloroform): 2.82, 2.96μ. Dipole moment 2.67.

4701. Hydrobromic Acid. A soln of hydrogen bromide gas in water. Marketed in various concns, e.g., 50% HBr, d 1.517; 40% HBr, d 1.38; 34%, d 1.31; 10%, d 1.08. Lab prepn according to the eq $H_2SO_4 + KBr \rightarrow KHSO_4 + HBr$: Heisig, Amdur, *Inorg. Syn.* **1**, 155 (1939). *See also* Hydrogen Bromide.

Colorless or faintly yellow; slowly darkens on exposure to air and light. Miscible with water, alcohol. *Keep protected from light.* When dil hydrobromic acid is distilled, a weaker acid comes over first and when a very concd acid is boiled, HBr gas chiefly distills over first; in both cases a "constant boiling" acid contg about 47.5% HBr remains which distills unchanged at 126°. The bp and compn of the azeotrope varies with pressure: at 100 mm, bp 74.12°, compn 49.80% HBr; at 400 mm, bp 107.00°, compn 48.47%; at 700 mm, bp 122°, compn 47.74%; at 800 mm, bp 125.79°, compn 47.56%: Bonner *et al.*, *J. Am Chem. Soc.* **55**, 1406 (1933). Aq solns are strongly acid. The satd aq soln contains 68.85% HBr at 0° and 66% at 25°.

USE: The concd acid is used principally in analytical chemistry and organic prepns. *Caution:* Strong irritant.

THERAP CAT: Sedative.

THERAP CAT (VET): Has been used as a sedative.

4702. Hydrocarbostyril. *3,4-Dihydro-2(1H)-quinolinone;* 3,4-dihydrocarbostyril; 3,4-dihydro-2-quinolinol; *o*-aminohydrocinnamic acid lactam; 2-oxo-1,2,3,4-tetrahydroquinoline; dihydro-α-quinolone. C_9H_9NO; mol wt 147.17. C 73.45%, H 6.16%, N 9.52%, O 10.87%. Prepd by the catalytic reduction of *o*-nitrocinnamic acid: Blout, Silverman, *J. Am. Chem. Soc.* **66**, 1442 (1944).

Prisms from methanol + water, mp 165-166.5°. Freely sol in alcohol, ether, dimethylformamide. Practically insol in water. Sol in hot aq NaOH solns.

4703. Hydrochloric Acid. Muriatic acid. A soln of hydrogen chloride gas (HCl) in water. Prepn and reviews: *see* Hydrogen Chloride.

Fumes in air. May be colored yellow by traces of iron, chlorine, and organic matter. Reagent grade concd hydrochloric acid contains close to 38.0% HCl. 83 ml of concd HCl poured into sufficient water to make 1 liter yields approx 1.0N HCl. The pH of 1.0N HCl is 0.10; of 0.1N = 1.10; of 0.01N = 2.02; of 0.001N = 3.02; of 0.0001N = 4.01. n_D^{18} (1.0N soln) 1.34168. d_4^{15} 1.05 (10.17% w/w soln); 1.10 (20%); 1.15 (29.57%); 1.20 (39.11%). Freezing pt: −17.14° (10.81% soln); −62.25° (20.69%); −46.2° (31.24%); −25.4° (39.17%), *Gmelin's, Chlorine* (8th ed.) **6**, 136-137 (1927). Constant boiling azeotrope with water bp$_{760}$ 108.58° contg 20.22% HCl, d_4^{15} 1.096. Boiling weaker or stronger aq solns results in loss of either component until the constant boiling acid is obtained.

Human Toxicity: External—concd solns cause severe burns; permanent visual damage may occur. Dermatitis and photosensitization may result from industrial contact. *Inhalation*—cough, choking; inflammation and ulceration of respiratory tract may occur. *Ingestion*—corrosion of mucous membranes, esophagus, stomach; dysphagia, nausea, vomiting, intense thirst, diarrhea. Circulatory collapse and death may occur: *Clinical Toxicology of Commercial Compounds*, R. E. Gosselin *et al.*, Eds. (Williams & Wilkins, Baltimore, 5th ed., 1984) Section III, pp 8-11.

USE: In the production of chlorides; refining ore in the production of tin and tantalum; for the neutralization of basic systems; as laboratory reagent; hydrolyzing of starch and proteins in the prepn of various food products; pickling and cleaning of metal products; as catalyst and solvent in organic syntheses. Also used for oil- and gas-well treament and in removing scale from boilers and heat-exchange equipment. Pharmaceutic aid (acidifier).

THERAP CAT (VET): Has been used as gastric acidifier.

4704. Hydrochlorothiazide. *6-Chloro-3,4-dihydro-2H-*

1,2,4-benzothiadiazine-7-sulfonamide 1,1-dioxide; 6-chloro-3,4-dihydro-7-sulfamoyl-2H-1,2,4-benzothiadiazine 1,1-dioxide; 6-chloro-7-sulfamyl-3,4-dihydro-1,2,4-benzothiadiazine 1,1-dioxide; 3,4-dihydrochlorothiazide; chlorsulfonamidodihydrobenzothiadiazine dioxide; chlorosulthiadil; Aquarius; Bremil; Chlorzide; Cidrex; Dichlorosal; Dichlotride; Diclotride; Direma; Diu-melusin; Disalunil; Esidrex; Esidrix; Fluvin; Hidroronol; Hydril; Hydro-Aquil; Hydro-Diuril; Hydrosaluric; Hydrothide; Hypothiazide; Ivaugan; Jen-Diril; Maschitt; Nefrix; Neo-Codema; Neoflumen; Oretic; Panurin; Ro-Hydrazide; Thiaretic; Thiuretic; Urodiazin; Vetidrex. $C_7H_8ClN_3O_4S_2$; mol wt 297.72. C 28.24%, H 2.71%, Cl 11.91%, N 14.11%, O 21.49%, S 21.53%. Prepn: de Stevens *et al.*, *Experientia* **14**, 463 (1958); Werner *et al.*, *J. Am. Chem. Soc.* **82**, 1161 (1960); Jones, Novello, U.S. pat. **3,025,292** (1962 to Merck & Co.); de Stevens, Werner, U.S. pat. **3,163,645** (1964 to Ciba); Klosa, Ger. pat. **1,163,332** (1964); Irons *et al.*, U.S. pat. **3,164,588** (1965 to Merck & Co.). Purification: Downing, U.S. pat. **3,043,840** (1962 to Merck & Co.). Toxicity data: J. J. Piala *et al.*, *J. Pharmacol. Exp. Ther.* **134**, 273 (1961). Comprehensive description: H. P. Deppeler in *Analytical Profiles of Drug Substances* vol. 10, K. Florey, Ed. (Academic Press, New York, 1981) pp 405-441.

Crystals, mp 273-275°. uv max (methanol + trace HCl): 317, 271, 226 nm ($A_{1cm}^{1\%}$ 130, 654, 1280). pKa 7.9, 9.2. Sol in dil ammonia, or sodium hydroxide; also sol in methanol, ethanol, acetone. Practically insol in water. LD_{50} in mice (mg/kg): 590 i.v.; > 8000 orally (Piala).

THERAP CAT: Diuretic.

THERAP CAT (VET): Diuretic.

4705. Hydrocinchonidine. *(8α,9R)-10,11-Dihydrocinchonan-9-ol;* (−)-dihydrocinchonidine; cinchamidine. $C_{19}H_{24}N_2O$; mol wt 296.40. C 76.99%, H 8.16%, N 9.45%, O 5.40%. Found in cinchona barks: Forst, Böhringer, *Ber.* **14**, 1270 (1881); Hesse, *ibid.* 1683; *idem, Ann.* **214**, 1 (1882); Skita, Nord, *Ber.* **45**, 3312 (1912); Heidelberger, Jacobs, *J. Am. Chem. Soc.* **41**, 817 (1919). Configuration: Ochiai *et al.*, *J. Pharm. Soc. Japan* **67**, 211 (1947). Stereospecific synthesis from secologanin: R. T. Brown, D. Curless, *Tetrahedron Letters* **27**, 6005 (1986). Total synthesis: M. Ihara *et al.*, *J. Chem. Soc. Perkin Trans. I* **1988**, 1277. Stereoisomer of hydrocinchonine, *q.v.*

Needles or leaflets, mp 230°. $[\alpha]_D^{15}$ −98° (alcohol). Practically insol in water. Sol in alcohol, chloroform; slightly sol in ether.

4706. Hydrocinchonine. *(9S)-10,11-Dihydrocinchonan-9-ol;* cinchotine; cinconifine; (+)-dihydrocinchonine; pseudocinchonine. $C_{19}H_{24}N_2O$; mol wt 296.40. C 76.99%, H 8.16%, N 9.45%, O 5.40%. From cinchona barks: Caventou, Willm, *Compt. Rend.* **69**, 284 (1869). Prepn from cinchonine: Hesse, *Ann.* **300**, 46 (1898); Pum, *Monatsh.* **16**, 68 (1895); Arlt, *ibid.* **20**, 426, 439 (1899); Heidelberger, Jacobs, *J. Am. Chem. Soc.* **41**, 817 (1919). Structure: Rabe, *Ber.* **55**, 522 (1922). Conversion of hydroquinidine to hydrocinchonine: King, *J. Chem. Soc.* **1946**, 523. HPLC determn: C.-T. A. Chung, E. J. Staba, *J. Chromatog.* **295**, 276 (1984). Stereospecific synthesis from secologanin: R. T. Brown, D. Curless, *Tetrahedron Letters* **27**, 6005 (1986). Total synthesis: M. Ihara *et al., J. Chem. Soc. Perkin Trans. I* **1988**, 1277. Prisms, mp 268-269°. $[\alpha]_D^{14}$ +204° (c = 0.6 in alc). Almost insol in water, ether. Sol in alcohol.

Hydrochloride, $C_{19}H_{25}ClN_2O$, crystals, mp 220-221°. $[\alpha]_D^{23}$ +155° (c = 0.8 in water).

4707. Hydrocinnamic Acid. *Benzenepropanoic acid;* 3-phenylpropionic acid; β-phenylpropionic acid; benzylacetic acid. $C_9H_{10}O_2$; mol wt 150.18. C 71.98%, H 6.71%, O 21.31%. $C_6H_5CH_2CH_2COOH$. Prepn by reduction of cinnamic acid: Ingersoll, *Org. Syn.* **9**, 42 (1929); from propiophenone: Schwenk, Papa, *J. Org. Chem.* **11**, 798 (1946); by Mauer oxidation of 1-phenyl-3-propanol: Langenbeck, Richter, *Ber.* **89**, 202 (1956); by chromate oxidation of propylbenzene: Reitsema, Allphin, *J. Org. Chem.* **27**, 27 (1962).

White, cryst powder, mp 47-48°. bp 280°; bp$_{75}$ 194-197°; bp$_{18}$ 145-147°; bp$_6$ 125-129°. Sol in 170 parts cold, more sol in hot water, in alcohol, benzene, chloroform, ether, glacial acetic acid, petr ether, carbon disulfide.

4708. Hydrocodone. *4,5-Epoxy-3-methoxy-17-methylmorphinan-6-one;* dihydrocodeinone; Bekadid; Dicodid. $C_{18}H_{21}NO_3$; mol wt 299.36. C 72.21%, H 7.07%, N 4.68%, O 16.03%. Prepn by hydrogenation of codeinone: Mannich, Löwenheim, *Arch. Pharm.* **258**, 295 (1920); by oxidation of dihydrocodeine, Ger. pat. **415,097** (1925 to E. Merck), *Frdl.* **15**, 1518 (1925-1927); by catalytic rearrangement of codeine: Ger. pat. **623,821**. Industrial prepn from dihydrocodeine: Pfister, Tishler, U.S. pat. **2,715,626** (1955 to Merck & Co.). Toxicity data: Eddy, Reid, *J. Pharmacol. Exp. Ther.* **52**, 468 (1934). *Review:* Small, Lutz, "Chemistry of the Opium Alkaloids," Suppl. No. 103, Public Health Reports, Washington (1932).

Prisms from alcohol, mp 198°. Sol in alcohol, dil acids. Insol in water. uv max: 280 nm (ϵ 1310). LD$_{50}$ s.c. in mice: 8.57 mg/kg (Eddy, Reid).

Hydrochloride monohydrate, $C_{18}H_{21}NO_3 \cdot HCl \cdot H_2O$, crystals, mp 185-186° dec. $[\alpha]_D^{27}$ -130° (c = 2.877). Very sol in water.

Bitartrate hemipentahydrate, $C_{22}H_{27}NO_9 \cdot 2\frac{1}{2}H_2O$, *Calmodid, Codinovo, Duodin, Kolikodal, Orthoxycol, Mercodinone, Synkonin, Norgan, Dicodrine, Hydrokon.* Needles, mp 146-148° (dry). (This melting range is reported, but actual tests do not confirm it.) One gram dissolves in 16 ml water, in 150 g 95% ethanol. Almost insol in ether, chloroform. pH of a 2% aq soln about 3.6.

Hydriodide, $C_{18}H_{21}NO_3 \cdot HI$, mp 219-220°.

Methiodide, $C_{18}H_{21}NO_3 \cdot CH_3I$, mp 250-255°.

Caution: May be habit forming. This is a controlled substance (opiate) listed in the U.S. Code of Federal Regulations, Title 21 Parts 329.1 and 1308.12 (1985).

THERAP CAT: Narcotic analgesic. Antitussive.

4709. Hydrocortamate. *N,N-Diethylglycine (11β)-11,17-dihydroxy-3,20-dioxopregn-4-en-21-yl ester; cortisol 21-ester with N,N-diethylglycine;* cortisol 21-(N,N-diethyl)-glycinate; hydrocortisone 21-diethylaminoacetate; Ulcort. $C_{27}H_{41}NO_6$; mol wt 475.61. C 68.18%, H 8.69%, N 2.95%, O 20.18%. Prepn: Pinson, Laubach, Ger. pat. **1,016,708** (1957 to Pfizer), *C.A.* **54**, 22737a (1960); Richter, Schenck, Ger. pat. **1,037,451** (1958 to Schering AG), *C.A.* **54**, 22730h (1960).

Crystals from ethyl acetate, mp 162-163°.

Hydrochloride, $C_{27}H_{42}ClNO_6$, *ethamicort, Magnacort.* Crystals from ethyl acetate, dec 222°.

THERAP CAT: Glucocorticoid.

4710. Hydrocortisone. *11,17,21-Trihydroxypregn-4-ene-3,20-dione; cortisol;* 4-pregnene-11β,17α,21-triol-3,20-dione; 17-hydroxycorticosterone; hydrocortisone free alcohol; anti-inflammatory hormone; Ala-Cort; Anflam; Cleiton; Cremesone; Cobadex; Cort-Dome; Cortef; Cortifoam; Cortril; Dermacort; Dermolate; Dermolen; Dioderm; Dome-Cort; Efcorbin; Efcorlin; EF-Cortelan; Efcortelin; Evacort; Eye-Cort; Epicort; Ficortril; Genacort (Lotion); Hidro-Colisona; HVB; Hydro-Adreson; Hydrocort; Hydrocortisyl; Hydrocortone; Incortin-H; Kendall's compound F; Lubricort; Medicort; Meusicort; Maintasone; Mildison; Reichstein's substance M; Scheroson F; Sigmacort; Texacort; Timocort; Proctocort; Rectoid; Zenoxone. $C_{21}H_{30}O_5$; mol wt 362.47. C 69.59%, H 8.34%, O 22.07%. Isoln from adrenal glands: Reichstein, *Helv. Chim. Acta* **20**, 953 (1937); Mason *et al., J. Biol. Chem.* **124**, 459 (1938); from urine of a case of Cushing's syndrome: Mason, Sprague, *ibid.* **175**, 451 (1948); from blood obtained from adrenal veins of dogs: Reich *et al., ibid.* **187**, 411 (1950). Configuration: von Euw, Reichstein, *Helv. Chim. Acta* **25**, 988 (1942); **30**, 205 (1947). Synthesis: Wendler *et al., J. Am. Chem. Soc.* **72**, 5793 (1950). Biosynthesis: *Chem. & Eng. News* **29**, 4000 (1951); cf. Zaffaroni *et al., J. Am. Chem. Soc.* **73**, 1390 (1951); Colingsworth *et al., J. Biol. Chem.* **203**, 807 (1953); Murray, Peterson, U.S. pats. **2,602,769** (1952 to Upjohn); **2,649,400** (1953); **2,649,402** (1953). Comprehensive description: K. Florey, Ed. in *Analytical Profiles of Drug Substances* vol. **12** (Academic Press, New York, 1983) pp 277-324.

Bitter-tasting crystalline, striated blocks from abs ethanol or isopropanol, mp 217-220° with some decompn. Commercial samples, mp 212-213°. $[\alpha]_D^{22}$ +167° (abs ethanol). Commercial samples: $[\alpha]_D^{22}$ +163° (c = 0.5 in methanol). uv max: 242 nm ($E_{1cm}^{1\%}$ 445). Soly (mg/ml) at 25°: water 0.28; ethanol 15.0; methanol 6.2; acetone 9.3; chloroform 1.6; propylene glycol 12.7; ether about 0.35. Sol in concd sulfuric acid with intense green fluorescence. Yields adrenosterone upon treatment with CrO₃.

21-Acetate *see* Hydrocortisone Acetate.

21-Bendazac, $C_{37}H_{42}N_2O_7$, *AF-2071, bendacort, cortazac, Versacort.*

17-Butyrate, $C_{25}H_{36}O_6$, *Alfason, Locoid, Plancol.*

21-Sodium succinate, $C_{25}H_{33}NaO_8$, *Corlan, Efcortelan.*

17-Valerate, $C_{26}H_{38}O_6$, *hydrocortisone valerate, cortisone 17-valerate, Westcort Cream.*

THERAP CAT: Glucocorticoid; anti-inflammatory.

THERAP CAT (VET): Adrenocortical steroid; glucocorticoid; topical anti-inflammatory.

4711. Hydrocortisone Acetate. *Cortisol acetate;* 17-hydroxycorticosterone 21-acetate; Biocortar; Carmol HC; Colifoam; Colofoam; Cortaid; Cortes; Epifoam; Hc45; Hydrocortisat; Hydrocortistab; Hydrocortone Acetate; Lanacort; Lenirit; Litraderm; Pabracort; Sintotrat; Velopural. $C_{23}H_{32}O_6$; mol wt 404.51. C 68.29%, H 7.97%, O 23.73%. ROCCH$_3$ where R is the 21-hydrocortisone radical.

Monoclinic, sphenoidal, tabular crystals from dil acetone. Tasteless. Somewhat hygroscopic. d$_4^{20}$ 1.289; dec 223°. $[\alpha]_D^{25}$ +166° (c = 0.4 in dioxane); $[\alpha]_D^{25}$ +150.7° (c = 0.5 in acetone). uv max (methanol): 242 nm (E$_{1cm}^{1\%}$ 390). Soly in water: 1 mg/100 ml; in ethanol: 0.45 g/100 ml; in methanol: 3.9 mg/ml; in acetone: 1.1 mg/g; in ether: 0.15 mg/ml. One gram dissolves in about 200 ml chloroform (U.S.P.). Very sol in DMF. Also sol in dioxane. pH of aq suspension in water 5.5-7.5. Sol in concd sulfuric acid with yellowish-green fluorescence.

THERAP CAT: Glucocorticoid; anti-inflammatory.

THERAP CAT (VET): Adrenocortical steroid; local anti-inflammatory and anti-allergic agent.

4712. Hydrocortisone Phosphate. *11β,17-Dihydroxy-21-(phosphonooxy)pregn-4-ene-3,20-dione; 21-hydrocortisonephosphoric acid;* hydrocortisone 21-(dihydrogen phosphate); *Cortiphate Injectable.* $C_{21}H_{31}O_8P$; mol wt 442.45. C 57.01%, H 7.06%, O 28.93%, P 7.00%. RPO(OH)$_2$ where R is the 21-hydrocortisone radical. Prepn: Conbere, Pfister; Christensen *et al.;* Conbere *et al.,* U.S. pats. **2,870,177; 2,932,657; 3,068,223** (1959, 1960, 1962 all to Merck & Co.); Elks, Phillipps, U.S. pat. **2,936,313** (1960 to Glaxo).

Disodium salt, $C_{21}H_{29}Na_2O_8P$, *Actocortin, Efcortesol, Hydrocortone Sodium Phosphate.* White powder. uv max (methanol): 242 nm (A$_{1cm}^{1\%}$ 298-341). $[\alpha]_D^{25}$ +120° (H$_2$O). Soly in water at 25°: >500 mg/ml. pH of a 1% aq soln 7.5-8.5. Aq solns may be sterilized by aseptic filtration.

THERAP CAT: Glucocorticoid.

4713. Hydrocortisone 21-Sodium Succinate. Sodium 17-hydroxycorticosterone 21-succinate; hydrocortisone hemisuccinate sodium salt; A-hydro Cort; Solu-Cortef; Corlan; EF-Cortelan Soluble; Hycorace; Buccalsone; Solu-Glyc; Intracort; Nordicort; Saxizon; Suxizon 300. $C_{25}H_{33}$-NaO$_8$; mol wt 484.50. C 61.97%, H 6.86%, Na 4.75%, O 26.42%. ROCCH$_2$CH$_2$COONa where R is the 21-hydrocortisone radical. 133.7 mg of hydrocortisone 21-sodium succinate equals 100 mg of hydrocortisone (free alcohol): Orr *et al., J. Clin. Endocrinol. Metab.* **15,** 763 (1955).

Amorphous, hygroscopic, white powder, mp 169.0-171.2°. Soly in water or physiological saline about 500 mg/ml. Similarly sol in methanol and 95% ethanol. Sparingly sol in chloroform. The ampuled dry material may be stored for an indefinite length of time. Solns may be kept under refrigeration for 3 to 5 days.

THERAP CAT: Glucocorticoid.

4714. Hydrocortisone Tebutate. Hydrocortisone 21-(3,3-dimethylbutanoic acid ester); hydrocortisone 21-(3,3-dimethylbutyrate); hydrocortisone 21-β,β-dimethylbutyrate; hydrocortisone *tert*-butylacetate; hydrocortisone TBA; Hydrocortone-TBA. $C_{27}H_{40}O_6$; mol wt 460.62. C 70.40%, H 8.75%, O 20.84%. ROCCH$_2$C(CH$_3$)$_3$ where R is the 21-hydrocortisone radical. Prepn: Rogers, Conbere, U.S. pat. **2,736,733** (1956 to Merck & Co.); Brit. pat. **765,505** (1964); Fr. pat. **1,349,103** (1964).

Dimorphic crystals from ethanol, mp 168-169° and 229-230°. $[\alpha]_D^{25}$ +152° (chloroform). uv max (methanol): 242 nm (E$_{1cm}^{1\%}$ 372). Soly in boiling alc about 8%.

THERAP CAT: Glucocorticoid.

THERAP CAT (VET): Adrenocortical steroid, local anti-inflammatory agent.

4715. Hydrocotarnine. *5,6,7,8-Tetrahydro-4-methoxy-6-methyl-1,3-dioxolo[4,5-g]isoquinoline;* 8-methoxy-5,6-methylenedioxy-2-methyl-1,2,3,4-tetrahydroisoquinoline. $C_{12}H_{15}NO_3$; mol wt 221.25. C 65.14%, H 6.83%, N 6.33%, O 21.69%. Found in mother liquors from morphine extraction. It is not certain whether it is formed from narcotine during the extraction or whether it exists in the poppy plant. May also be prepd by reduction of cotarnine: Topchiev, *J. Appl. Chem. USSR* **6,** 529 (1933), *C.A.* **28,** 2718 (1934); Schneider, Müller, *Ann.* **615,** 34 (1958); Knabe, *Arch. Pharm.* **292,** 652 (1959). Reduction of hydrocotarnine with sodium in alcohol leads to replacement of the methoxyl group by hydrogen, with formation of hydrohydrastinine. Review and bibliogra-

phy: Small, Lutz, "Chemistry of the Opium Alkaloids," Suppl. No. 103, *Public Health Reports,* Washington (1932).

Hemihydrate, plates from petr ether, mp 56°. Loses water of crystn at 60°. May be distilled with little decompn at 100°: Hesse, *Ber.* **4,** 693 (1871). Absorption spectrum: Hantzsch, *Ber.* **44,** 1816 (1911); Steiner, *Compt. Rend.* **176,** 244, 1379 (1923); Csokán, *Z. Anal. Chem.* **124,** 344 (1942). Almost insol in water, alkaline solns; sol in alcohol, acetone, chloroform, benzene, ether.

Hydrochloride monohydrate, prisms; sol in water.
Hydrobromide, crystals, mp 237°; sparingly sol in water.
Hydriodide, needles from methanol, mp 196°; sol in hot water.
Methiodide, needles from water, plates from alc, mp 206°.
Methobromide, needles from chloroform, dec 221°.

4716. Hydroflumethiazide. *3,4-Dihydro-6-(trifluoromethyl)-2H-1,2,4-benzothiadiazine-7-sulfonamide 1,1-dioxide;* 6-trifluoromethyl-3,4-dihydro-7-sulfamoyl-2H-1,2,4-benzothiadiazine 1,1-dioxide; 3,4-dihydro-7-sulfamyl-6-trifluoromethyl-1,2,4-benzothiadiazine 1,1-dioxide; trifluoromethylhydrothiazide; dihydroflumethiazide; methforylthiazidine; metflorylthiazidine; Bristab; Bristurin; Di-Ademil; Diucardin; Elodrine; Finuret; Hydol; Hydrenox; Leodrine; NaClex; Olmagran; Rodiuran; Rontyl; Saluron; Sisuril; Vergonil. $C_8H_8F_3N_3O_4S_2$; mol wt 331.29. C 29.00%, H 2.43%, F 17.21%, N 12.68%, O 19.32%, S 19.36%. Synthesis: Holdrege *et al., J. Am. Chem. Soc.* **81,** 4807 (1959); Close *et al.,* ibid. **82,** 1132 (1960); Yale *et al., ibid.* 2042; Novello *et al., J. Org. Chem.* **25,** 970 (1960). Numerous patents, *e.g.,* Lund *et al.,* U.S. pat. **3,254,076** (1966 to Leo Pharm.). Pharmacology: J. J. Piala *et al., J. Pharmacol. Exp. Ther.* **134,** 273 (1961). Comprehensive description: C. E. Orzech *et al.,* in *Analytical Profiles of Drug Substances* Vol. 7, K. Florey, Ed. (Academic Press, New York, 1978) pp 297-317.

Crystals, mp 272-273°. uv max (methanol): 272.5 nm (log ε 4.286). Soly in mg/ml at 25°: acetone > 100; methanol 58; acetonitrile 43; water 0.3; ether 0.2; benzene < 0.1. pK$_1$ 8.9; pK$_2$ 10.7. Forms water-sol salts with bases. LD$_{50}$ in mice (mg/kg): > 8000 orally, 750 i.v., 6280 i.p. (Piala).

THERAP CAT: Antihypertensive. Diuretic.

4717. Hydrofluoric Acid. Fluohydric acid. A soln of hydrogen fluoride gas in water. Obtained by distilling calcium fluoride with H_2SO_4. Usually marketed in concns of about 47% and 53%. d 1.15-1.18. Refs: See Hydrogen Fluoride.

Colorless or almost colorless, fuming liquid. *Poisonous! Handle with care as it causes painful sores on the skin usually noticed on the next day only; avoid inhaling the vapors.* Miscible with water. Vapor pressure data: Brosheer *et al., Ind. Eng. Chem.* **39,** 423 (1947). Compn of liq and vapor: Munter *et al., ibid.* 427. A weak acid: Ka 6.46 × 10^{-4} mole/l. The 38.2% (w/w HF) soln is a binary azeotrope; bp 112.2°. It attacks glass or stoneware, dissolving the silica. *Keep in plastic, lead, wax, or paraffin paper bottles.*

USE: Cleaning cast iron, copper, brass; removing efflorescence from brick and stone, or sand particles from metallic castings; working over too heavily weighted silks; frosting, etching glass and enamel; polishing crystal glass; decomposing cellulose; enameling and galvanizing iron; increasing porosity of ceramics. Its salts are used as insecticides and to arrest undesirable fermentation in brewing. Also used in analytical work to determine SiO_2, etc. *Caution: Acute ef-*

fects: External contact—liquid or vapor causes severe irritation of eyes and eyelids which may result in prolonged or permanent visual defects or total destruction of eyes. Skin contact may result in severe burns. Inhalation—extreme irritation of respiratory tract, pulmonary inflammation, congestion. Ingestion—necrosis of esophagus and stomach with nausea, vomiting, diarrhea, circulatory collapse, death. *Chronic effects:* (from inhalation or ingestion), may cause fluorosis. Symptoms—wt loss, malaise, anemia, leukopenia, discoloration of teeth, osteosclerosis.

4718. Hydrofuramide. *1-(2-Furanyl)-N,N'-bis(2-furanylmethylene)methanediamine; N,N'-difurfurylidene-2-furanmethanediamine;* furfuramide; $C_{15}H_{12}N_2O_3$; mol wt 268.26. C 67.15%, H 4.51%, N 10.44%, O 17.89%. Prepn: Hartley, Dobbie, *J. Chem. Soc.* **73**, 598 (1898); Taniyama, *J. Chem. Soc. Japan,* Ind. Chem. Sect. **51**, 33 (1948); Kapur *et al., J. Sci. Ind. Res. (India)* **19B**, 509 (1960); Kamal *et al., Tetrahedron* **19**, 869 (1963). Structure: Soundararajan, Anantakrishnan, *Proc. Indian Acad. Sci.* **38A**, 176 (1953).

Brownish crystals from abs alcohol, mp 117°. bp about 250° with decompn. uv max: 259, 215 nm (log ε 4.18, 4.16). Practically insol in water; freely sol in alc, ether; readily dec by acids.

USE: Vulcanization accelerator.

4719. Hydrogen. Protium. H; at. wt 1.0079; at. no. 1; valence 1; elemental state: H_2. Isotopes: 1 (protium 99.9844%); 2 (deuterium 0.0156%); 3 (tritium, traces only). The most abundant element in the known universe. Occurrence in the earth's atmosphere 0.00005% H_2. First recognized as an element by Cavendish in 1766; named by Lavoisier. Obtained by passing H_2O vapors over heated iron; by electrolysis of water or by action of HCl or H_2SO_4 on Fe or Zn; by hydrolysis of metal hydrides. Produced industrially by electrolysis; from methane or coke and steam. *Reviews: Nouveau Traité de Chimie Minérale* vol. 1, P. Pascal, Ed. (Masson, Paris, 1956) pp 565-675; Mackay in *Comprehensive Inorganic Chemistry* vol. 1, J. C. Bailar, Jr. *et al.,* Eds. (Pergamon Press, Oxford, 1973) pp 1-76; B. G. Mandelik, D. S. Newsome in Kirk-Othmer *Encyclopedia of Chemical Technology* vol. 12 (Wiley-Interscience, New York, 3rd ed., 1980) pp 938-982. *See also* Deuterium and Tritium.

Colorless, odorless, tasteless gas; flammable or explosive when mixed with air, oxygen, chlorine, etc. mp −259.2° (13.96°K) at 54 mm (triple point). bp −252.77° (20.39°K). d_{gas} 0.069 (air = 1); d_{liq} 0.0700 (at bp); d_{sol} 0.0763 (13°K). A liter of the gas at 0° weighs 0.08987 g. Crit. temp −239.9°; crit press. 12.8 atm. Sol in about 50 vols of water at 0°. *Caution:* No specific toxic action. In high concns can act as a simple asphyxiant.

USE: In oxy-hydrogen blowpipe (welding) and limelight; autogenous welding of steel and other metals; manuf ammonia, synthetic methanol, HCl; hydrogenation of oils, fats, naphthalene, phenol; in balloons and airships; manuf tungsten. In thermonuclear reactions: the hydrogen atom has a single peripheral electron (1s). It ionizes to form protons, deuterons (D) or tritons (T). The ionization potential of the H atom is 13.59 eV. Accelerated protons bring about extremely varied nuclear reactions. Liq hydrogen used in bubble chambers to study subatomic particles; as a coolant.

4720. Hydrogen Bromide. *Anhydrous hydrobromic acid.* BrH; mol wt 80.92. H 1.25%, Br 98.75%. HBr. Prepd commercially by direct combination of the elements at 375° preferably over a catalyst such as platinized silica gel or platinized asbestos: Richards, Hönigschmid, *J. Am. Chem. Soc.* **32**, 1581 (1910); Smyth, Hitchcock, *ibid.* **55**, 1830 (1933); Schneider, Johnson, *Inorg. Syn.* **1**, 152 (1939). Lab procedure from tetrahydronaphthalene and bromine: Müller, *Monatsh.* **49**, 29 (1928); Duncan, *Inorg. Syn.* **1**, 151 (1939); Schmeisser in *Handbook of Preparative Inorganic Chemistry*

vol. 1, G. Brauer, Ed. (Academic Press, New York, 2nd ed., 1963) pp 282-286. Detailed description of laboratory methods of prepn: Houben-Weyl, *Methoden der organischen Chemie,* vol **5/4** (Thieme, Stuttgart, 4th ed., 1960) p 16-20. Review of prepn and properties of HBr and other hydrogen halides: Woolf in *Mellor's Comprehensive Treatise on Inorganic and Theoretical Chemistry* vol. **II**, Suppl I (originally published as Suppl II, part 1) 724-741 (1956); John in *Bromine and its Compounds,* Z. E. Jolles, Ed. (Ernest Benn, London, 1966) pp 81-105; Downs, Adams in *Comprehensive Inorganic Chemistry* vol. 2, J. C. Bailar Jr., *et al.,* Eds. (Pergamon Press, Oxford, 1973) pp 1280-1329.

Colorless, corrosive, nonflammable gas. Acrid odor. Fumes in moist air forming clouds which have a sour taste. d 2.71 (air = 1.00). mp −86.9°. bp_{760} −66.8°; $bp_{11.0\,atm}$ −4.8°; $bp_{17.1\,atm}$ +12°; $bp_{30.0\,atm}$ 36°; $bp_{59.2\,atm}$ 70°. Crit temp 89.8°; crit press. 84.5 atm. Sp heat (cal/g/°C): solid (−91°) 0.152; liq 0.176; gas (27°) 0.085. Heat of fusion at mp: 7.44 cal/g. Heat of vaporization at bp 51.3 cal/g. Freely sol in water: One vol H_2O dissolves 600 vols HBr gas at 0°. Also sol in alc. Soly in organic solvents: Fernandes, *J. Chem. Eng. Data* **17**, 377 (1972); Gerrard, *Chem. & Ind. (London)* **1969**, 295; Ahmed *et al., J. Appl. Chem.* **20**, 109 (1970). Aq solns are strongly acid. The satd aq soln contains 68.85% HBr at 0° and 66% at 25°. The boiling point of a constant-boiling mixture is 122.5° at 740 mm and 126° at 760 mm. The composition of the constant-boiling mixture is 47.38% HBr at 752 mm. For complete tables *see* Bonner *et al., J. Am. Chem. Soc.* **55**, 1406 (1943). *See also* Hydrobromic Acid. Anhydr HBr is marketed in steel cylinders in the form of a gas over liquid. LC_{50} in mice, rats: 814, 2858 ppm by inhalation, K. C. Back *et al., Reclassification of Materials Listed as Transportation Health Hazards* (TSA-20-72-3, PB 214-270).

USE: Mfg organic and inorganic bromides, hydrobromic acid, as reducing agent and as catalyst in controlled oxidations, in the alkylation of aromatic compds, in the isomerization of conjugated diolefins. *Caution:* May be highly irritating to eyes, skin, mucous membranes, respiratory tract.

4721. Hydrogen Chloride. *Anhydrous hydrochloric acid.* ClH; mol wt 36.47. Cl 97.23%, H 2.76%. HCl. *See also* Hydrochloric Acid. Produced industrially by the interaction of NaCl and H_2SO_4; from NaCl, SO_2, air and water vapor; by controlled combination of the elements; or as a by-product of the synthesis of chlorinated hydrocarbons: A. C. Cumming, *Hydrochloric Acid and Salt Cake* (Gurney and Jackson, London, 1923); N. A. Laury, *Hydrochloric Acid and Sodium Sulfate* (Chem. Catalog Co., New York, 1927); Maude, *Chem. Eng. Progress* **44**, 179 (1948); Faith, Keyes & Clark's *Industrial Chemicals,* F. A. Lowenheim, M. K. Moran, Eds. (Wiley-Interscience, New York, 4th ed., 1975) pp 454-461. Prepn of pure HCl for research purposes: Hönigschmid *et al., Z. Anorg. Allgem. Chem.* **163**, 315 (1927); Kemp, *J. Chem. Ed.* **37**, 142 (1960); Schmeisser in *Handbook of Preparative Inorganic Chemistry* vol. 1, G. Brauer, Ed. (Academic Press, New York, 2nd ed., 1963) pp 280-282. Toxicity: K. I. Darmer *et al., Am. Ind. Hyg. Assoc. J.* **35**, 623 (1974). Reviews of prepn and properties: Addison, Lewis in *Mellor's Comprehensive Treatise on Inorganic and Theoretical Chemistry* vol. **II**, suppl. I (originally published as suppl. II, part I) 402-475 (1956); Downs, Adams in *Comprehensive Inorganic Chemistry* vol. 2, J. C. Bailar, Jr. *et al.,* Eds. (Pergamon Press, Oxford, 1973) pp 1280-1329; D. S. Rosenberg in Kirk-Othmer *Encyclopedia of Chemical Technology* vol. 12 (Wiley-Interscience, New York, 3rd ed., 1980) pp 983-1015.

Colorless, corrosive, nonflammable gas. Characteristic pungent odor. Fumes in air. d 1.268 (air = 1.000) or 1.639 g/l. mp −114.22°. bp_{760} −85.05°; bp_{100} −114.61°; bp_{10} −137.77°; $bp_{1.0}$ −154.37°. Critical temp 51.4°; critical pressure 81.6 atm; critical density 0.42 g/ml. n_D^{20} (liquid under pressure) 1.256. Heat capacity at constant volume (15°): 0.1939 cal/g/°C. Heat capacity at constant pressure (15°): 0.1375 cal/g/°C. Heat of vaporization at −85°: 3860 cal/-mole; heat of soln (infinite dilution) −17.88 kcal/mole; heat of formation of gas at 25° −22.063 kcal/mole. Dielectric constant (gas at 0°) 1.0046; dipole moment 1.07. Soly in water (g/100 g H_2O): 82.3 (0°); 67.3 (30°); 63.3 (40°); 59.6 (50°); 56.1 (60°). Forms a const boiling mixture: 20.22

g/100 g soln; *see* Hydrochloric Acid. Soly in methanol (g/100 g soln): 54.6 (−10°); 51.3 (0°); 47.0 (20°); 43.0 (30°); in ethanol: 45.4 (0°); 42.7 (10°); 41.0 (20°); 38.1 (30°); in ether: 37.52 (−10°); 35.6 (0°); 24.9 (20°); 19.47 (30°). LC$_{50}$ (30 min) in mice, rats: 2142, 5666 ppm (Darmer).

Human Toxicity: See Hydrochloric Acid.

USE: In the manuf of pharmaceutical hydrochlorides, vinyl chloride from acetylene, alkyl chlorides from olefins, and arsenious chloride from arsenious oxide. In the chlorination of rubber, as a gaseous flux for babbitting operations. In organic reactions involving isomerization, polymerization, and alkylation. For making chlorine where economical.

4722. Hydrogen Cyanide. *Hydrocyanic acid;* Blausäure (German); prussic acid. CHN; mol wt 27.03. C 44.44%, H 3.73%, N 51.83%. HCN. Prepd on a large scale by the catalytic oxidation of ammonia-methane mixtures (Andrussow Process): *see* Andrussow, *Angew. Chem.* **48**, 593 (1935); Maffezzoni, *Chim. Ind. (Milan)* **34**, 460 (1952); Faith, Keyes & Clark's *Industrial Chemicals,* F. A. Lowenheim, M. K. Moran, Ed. (Wiley-Interscience, New York, 4th ed., 1975) pp 482-486. May also be prepd by the catalytic decompn of formamide. Conveniently prepd in the laboratory by acidifying NaCN or K$_4$[Fe(CN)$_6$]: Glemser in *Handbook of Preparative Inorganic Chemistry* vol. **1**, G. Brauer, Ed. (Academic Press, New York, 2nd ed., 1963) pp 658-660.

Colorless gas or liquid; characteristic odor; very weakly acid (does not redden litmus); burns in air with a blue flame; *intensely poisonous* even when mixed with air. d(gas) 0.941 (air = 1); d(liq) 0.687. mp −13.4°. bp 25.6°. Miscible with water, alc; slightly sol in ether. LC$_{50}$ in rats, mice, dogs: 544 ppm (5 min), 169 ppm (30 min), 300 ppm (3 min), K. C. Back *et al., Reclassification of Materials Listed as Transportation Health Hazards* (TSA-20-72-3; PB214-270).

Human Toxicity: High concn produces tachypnea (causing increased intake of cyanide); then dyspnea, paralysis, unconsciousness, convulsions and respiratory arrest. Headache, vertigo, nausea and vomiting may occur with lesser concns. Chronic exposure over long periods may cause fatigue, weakness. Exposure to 150 ppm for ½ to 1 hr may endanger life. Death may result from a few min exposure to 300 ppm. Average fatal dose: 50 to 60 mg. *Antidote:* Sodium nitrite and sodium thiosulfate, *cf.* Patty's *Industrial Hygiene and Toxicology* vol. **2C**, G. D. Clayton, F. E. Clayton, Eds. (Wiley-Interscience, NewYork, 3rd ed., 1982) pp 4850-4853.

USE: The compressed gas is used for exterminating rodents and insects in ships and for killing insects on trees, etc. *Must be handled by specially trained experts.*

4723. Hydrogen Fluoride. *Hydrofluoric acid gas;* fluohydric acid gas; anhydr hydrofluoric acid. FH; mol wt 20.01. F 94.96%, H 5.04%. HF. Obtained by the action of sulfuric acid on fluorspar (calcium fluoride): Faith, Keyes & Clark's *Industrial Chemicals,* F. A. Lowenheim, M. K. Moran, Eds. (Wiley-Interscience, New York, 4th ed., 1975) pp 462-467; prepn of pure HF: Simons, *Inorg. Syn.* **1**, 134 (1939); Shamir, Netzer, *J. Sci. Instrum.* (Ser. 2) **1**, 770 (1968). Exists as hydrogen-bonded polymers: Simons, Hildebrand, *J. Am. Chem. Soc.* **46**, 2183 (1924); Jarry, Davis, *J. Phys. Chem.* **57**, 600 (1953); Atoji, Lipscomb, *Acta Cryst.* **7**, 173 (1954). Reviews of prepn, properties and chemistry: Simons in *Fluorine Chemistry* vol. **1**, J. H. Simons, Ed. (Academic Press, New York, 1950); Hyman, Katz, "Liquid Hydrogen Fluoride" in *Non-Aqueous Solvent Systems,* T. C. Waddington, Ed. (Academic Press, New York, 1965) pp 47-81; O'Donnell in *Comprehensive Inorganic Chemistry* vol. **2**, C. Bailar, Jr. *et al.,* Eds. (Pergamon Press, Oxford, 1973) pp 1038-1054; J. F. Gall in Kirk-Othmer *Encyclopedia of Chemical Technolgy* vol. **10** (Wiley-Interscience, New York, 3rd ed., 1980) pp 733-753.

Colorless gas. Fumes in air. *Highly irritating, corrosive and poisonous!* d^{34} 1.27 (air = 1); d$_4^0$ 1.002. mp −83.55°: Gillespie, Humphreys, *J. Chem. Soc. (A)* **1970**, 2311. bp 19.51°; bp$_{400}$ +2.5°; bp$_{200}$ −13.2°; bp$_{100}$ −28.2°; bp$_{40}$ −45.0°; bp$_{20}$ −56.0°; bp$_5$ −74.7°. Very sol in water, alcohol. Slightly sol in ether. Sol in many organic solvents; soly (wt % at 5°): in benzene 2.54; toluene 1.80; *m*-xylene 1.28; tetralin 0.27. Many compds are sol in HF. See Simons, *loc. cit.;* Hyman, Katz, *loc. cit.* Anhyd HF is one of the most

acidic substances known; Hammett acidity function (H$_0$) −10.98: Hyman, Katz, *loc. cit.* In aqueous soln, it is a weak acid: Ka 6.46 × 10$^{−4}$ mole/l, Vanderborgh, *Talanta* **15**, 1009 (1968). Forms a constant boiling mixture with water, *see* Hydrofluoric Acid. Dissolves silica, silicic acid, glass. Should be stored in steel cylinders. LC$_{50}$ (1 hr) in rats, mice, monkeys: 1278, 500, 1780 ppm by inhalation, K. C. Back *et al., Reclassification of Materials Listed as Transportation Health Hazards* (TSA-20-72-3, PB 214-270).

USE: Catalyst, especially in the petroleum industry (paraffin alkylation); in fluorination processes, especially in the aluminum industry; in the manuf of fluorides; for separating uranium isotopes; in making fluorine contg plastics; in dye chemistry. *Caution:* Extremely corrosive to skin and eyes. Causes severe burns which may not be painful or visible for several hours. *See also* Hydrofluoric Acid.

4724. Hydrogen Iodide. *Anhydrous hydriodic acid.* HI; mol wt 127.93. I 99.21%, H 0.79%. Prepd by catalytic union of the elements: Caley, Burford, *Inorg. Syn.* **1**, 159 (1939); Powell, Campbell, *J. Am. Chem. Soc.* **69**, 1227 (1947). May also be prepd by treating concd hydriodic acid solns with P$_2$O$_5$: Schmeisser in *Handbook of Preparative Inorganic Chemistry* vol. **1**, G. Brauer, Ed. (Academic Press, New York, 2nd ed., 1963) pp 286-289. Lab prepn: Hoffman, *Inorg. Syn.* **7**, 180 (1963). Purification: A. Klemenc, *Die Behandlung und Reindarstellung von Gasen* (Vienna, 2nd ed., 1948) p 239; Irving, Wilson, *Chem. & Ind. (London)* **1964**, 653. Reviews of prepn and properties of HI and other hydrogen halides: Hills in *Mellor's Comprehensive Treatise on Inorganic and Theoretical Chemistry* vol. **II**, suppl. 1 (originally published as suppl. II, part 1) 857-869 (1956); Downs, Adams, in *Comprehensive Inorganic Chemistry* vol. **2**, J. C. Bailar, Jr. *et al.,* Eds. (Pergamon Press, Oxford, 1973) pp 1280-1329.

Colorless, acrid, non-flammable gas. Fumes in moist air. Decomposed by light. mp −50.8°. bp$_{760}$ −35.1°; bp$_{2\,atm}$ −18.9°; bp$_{5\,atm}$ +7.3°; bp$_{10\,atm}$ 32.0°; bp$_{60\,atm}$ 127.5°. d^0 5.66; d^{25} 5.23 g/l. Crit temp 151.0°, crit press. 82.0 atm. Sp heat (25°) 0.0545 cal/g/°C. Extremely sol in water. Soly (g/100 g H$_2$O): 234 (10°); 900 (0°). Soly in organic solvents: Gerrard, *Chem. & Ind. (London)* **1969**, 295; Ahmed *et al., J. Appl. Chem.* **20**, 109 (1970). Forms an azeotrope with water, *see* Hydriodic Acid. Reacts with the lower aliphatic alcohols forming the corresponding iodo compds. Forms a colorless liquid at atm pressure when cooled with dry ice and ether or similar cooling mixture. Attacks natural rubber.

USE: Manuf of hydriodic acid, organic iodo compds, to remove iodine from iodo compds. *Caution:* Strong irritant.

4725. Hydrogen Peroxide. Hydrogen dioxide; hydroperoxide; Albone; Hioxyl. H$_2$O$_2$; mol wt 34.02. H 5.94%, O 94.06%. First reported by Thenard in 1818; prepd by treating barium peroxide with acid. Manuf of aqueous solns: Faith, Keyes & Clark's *Industrial Chemicals,* F. A. Lowenheim, M. K. Moran, Eds. (Wiley-Interscience, New York, 4th ed., 1975) pp 487-495; R. Powell, *Hydrogen Peroxide Manufacture* (Noyes Dev. Corp., Park Ridge, N.J., 1968) 221 pp. Production of anhydr. hydrogen peroxide by continuous fractional crystn: Crewson, Ryan, U.S. pat. **2,724,-640** (1955 to Becco). Reviews: W. C. Schumb *et al., Hydrogen Peroxide* A.C.S. Monograph Series no. **128** (Reinhold, New York, 1955) 759 pp; Ebsworth *et al.,* in *Comprehensive Inorganic Chemistry* vol. **2**, J. C. Bailar, Jr. *et al.,* Eds. (Pergamon Press, Oxford, 1973) pp 771-778; J. R. Kirchner in Kirk-Othmer *Encyclopedia of Chemical Technology* vol. **13** (Wiley-Interscience, New York, 3rd ed., 1981) pp 12-38.

Colorless, rather unstable liquid; bitter taste; caustic to the skin. Distillable in high vacuum. May dec violently if traces of impurities are present. d^0 1.463. mp −0.43°. bp 152°. Misc with water; sol in ether; insol in petr ether; decomposed by many organic solvents.

Marketed as a soln in water in concns of 3-90% by wt. Solns of hydrogen peroxide gradually deteriorate and are usually stabilized by the addition of acetanilide or similar organic materials. Agitation or contact with rough surfaces, metals or many other substances accelerates decomposition. Rapidly dec by alkalies, finely divided metals; the presence of mineral acid renders it more stable.

USE: A 90% soln is used in rocket propulsion. As dough

conditioner, maturing and bleaching agent in food. *Caution:* Strong oxidizer. Undiluted form can cause burns of skin, mucous membranes.

THERAP CAT: Anti-infective.

THERAP CAT (VET): Topical antiseptic and cleansing agent (as a dilute soln).

4726. Hydrogen Peroxide Solution 3%. Hydrogen dioxide soln; oxydol. Contains 2.5-3.5% by wt of H_2O_2 = 8-12 vols oxygen.

Colorless, slightly acid liq. d about 1.00. Foams in the mouth. *Keep protected from light and in a cool place. Incompat:* Alkalies, ammonia and their carbonates, albumin, balsam Peru, phenol, charcoal, chlorides, alkali citrates; ferrous, mercurous or gold salts; hypophosphites, iodides, lime water, permanganates, sulfites, tinctures, and organic matter in general.

USE: In the plastics industry; white discharge printing on indigo-dyed wool; bleaching feathers, hair, silk, straw, ivory, flour, bone, gelatin, and textile fabrics; renovating old paintings, engravings; as oxidizer in manuf dyes; disinfecting water and hides; artificially aging wines, liquors, etc.; refining oils and fats; as antichlor; with paraphenylenediamine as a dye for furs, dead hair, etc.; in photography as hypo eliminant; with NaOH for cleaning metal surfaces, for gilding, silvering, etc. In pharmaceutical prepns, mouthwashes, dentifrices, sanitary lotions.

THERAP CAT: Topical anti-infective.

THERAP CAT (VET): Topical antiseptic and cleansing agent.

4727. Hydrogen Peroxide Solution 30%. Superoxol. Contains 30% by wt of H_2O_2 = 100 vols of oxygen.

Clear, colorless liquid. d about 1.11. Miscible with water. Now replacing the 3% soln for industrial uses; diluted to the required strength immediately before use. It also is used for making the 3% soln.

Caution! Strong oxidizing agent. Avoid contact with skin and eyes—wear rubber gloves and goggles. Avoid contact with combustible materials. Drying of concd product on clothing or other combustible materials may cause fire. In case of contact, immediately flush with plenty of water for at least 15 min; for eyes, get medical attention. Avoid contamination from any source, including metals, dust, etc. Such contamination may cause rapid decompn, generation of large quantities of oxygen gas and high pressures.

Store in original closed container. Be sure that the container vent is working satisfactorily. Do not add any other product to container. When empty, rinse thoroughly with clean water.

4728. Hydrogen Selenide. Selenium hydride. H_2Se; mol wt 80.98. H 2.49%, Se 97.51%. Prepd by heating selenium and hydrogen in a sealed tube at 440°: Hautefeuille, *Bull. Soc. Chim.* [2] **7**, 198 (1867); by passing a mixture of hydrogen and selenium vapor over pumice stone at 440°: Corenwinder, *Ann. Chim. Phys.* [3] **34**, 77 (1852); by warming potassium or ferrous selenide with hydrochloric acid: Berzelius, *Acad. Handl. Stockholm* **39**, 13 (1818); by the action of water on aluminum selenide: Fonzes-Diacon, *Traité de Chimie Minérale, Paris* **1**, 469 (1904); Waitkins, Shutt, *Inorg. Syn.* **2**, 183 (1946).

Gas. Disagreeable odor. d_4^{-42} 2.12. bp -41.3°. Liquefies at 0° under a pressure of 6.6 atm; at 18°, 8.6 atm; at 52°, 21.5 atm; at 100°, 47.1 atm; at the crit temp 137°, 91.0 atm. mp -65.73°. v.p. at -30°, 1.75 atm; v.p. at 0.2°, 4.5 atm; v.p. at 30.8°, 12 atm. K_1 at 25° = 1.30 × 10^{-4}; K_2 at 25° = 1 × 10^{-11}. Soly in water (ml/100 ml): 377 (4°); 270 (22.5°). Sol in carbonyl chloride and carbon disulfide. Unites directly with most metals to form metal selenides. Approx LC_{50} (30 min) in guinea pigs: 6 ppm, *Handbook of Toxicology vol.* **1**, W. S. Spector, Ed. (Saunders, Philadelphia, 1956) pp 340-341.

Caution: Irritating to eyes, mucous membranes. Causes garlic odor of breath, dizziness, nausea.

4729. Hydrogen Sulfide. Sulfureted hydrogen; "hydrosulfuric acid". H_2S; mol wt 34.08. H 5.92%, S 94.09%. In coal pits, gas wells, sulfur springs, from decaying organic matter contg sulfur. Produced by reacting dil sulfuric acid with iron sulfide, by reacting hydrogen and sulfur in the vapor phase, by heating sulfur with paraffin. Lab prepn

from CaS and $MgCl_2$ in water: Bickford, Wilkinson, *Inorg. Syn.* **1**, 111 (1939). Purification: Ward *et al., ibid.* **3**, 14 (1950).

Flammable, poisonous gas with characteristic odor of rotten eggs, perceptible in air in a dilution of 0.002 mg/l, sweetish taste. Burns in air with pale blue flame. Ignition temp 260°C. Explosive limits when mixed with air: lower limit 4.3% by vol, upper limit 46% by vol. mp -85.49°; bp -6 0.33°: Giauque, Blue, *J. Am. Chem. Soc.* **58**, 831 (1936). Heavier than air; 1.5392 g/l (0°; 760 mm). d^{gas} 1.19 (air = 1.00). One gram H_2S dissolves in 187 ml water at 10°, in 242 ml water at 20°, in 314 ml water at 30°; in 94.3 ml abs alcohol at 20°; in 48.5 ml ether at 20°. Also sol in glycerol. Water solns of H_2S are not stable, absorbed oxygen causes the formation of elemental sulfur, and the solns become turbid rapidly. In a 50:50 v/v mixture of glycerol and water the precipitation of sulfur is retarded considerably. pH of freshly prepd satd water soln 4.5. LC_{50} (1 hr) in mice, rats: 673, 713 ppm by inhalation, K. C. Back *et al., Reclassification of Materials Listed as Transportation Health Hazards* (TSA-20- 72-3; PB 214-270).

Human Toxicity: Extremely hazardous. Collapse, coma, and death from respiratory failure may come within a few seconds after one or two inspirations. Insidious poison, since sense of smell may be fatigued and fail to give warning of high concns. Low concns produce irritation of conjunctiva and mucous membranes. Headache, dizziness, nausea, lassitude may appear after exposure, *Clinical Toxicology of Commercial Products*, R. E. Gosselin *et al.*, Eds. (Williams & Wilkins, Baltimore, 4th ed., 1976) Section III, pp 169-173.

USE: In the manuf of chemicals, in metallurgy; as analytical reagent.

4730. Hydrogen Telluride. H_2Te; mol wt 129.63. H 1.56%, Te 98.44%. Prepd by the action of H_2O or HCl on aluminum telluride; by electrolysis of a 50% soln of sulfuric or phosphoric acid with a Te cathode: Dennis, Anderson, *J. Am. Chem. Soc.* **36**, 882 (1914); Fehér in *Handbook of Preparative Inorganic Chemistry* vol. **1**, G. Brauer, Ed. (Academic Press, New York, 2nd ed., 1963) pp 438-441.

Colorless gas. Offensive, garlic-like odor. *Highly poisonous!* mp -49°. bp -2°. d_{liq}^{-12} 2.68. Wt of one liter of the gas: 6.234 g. Liquid H_2Te is dec immediately by light. The dry gas is stable to light, but dec in the presence of dust, traces of moisture, rubber, cork, etc. Sol in water with fairly quick decompn. A satd aq soln is about 0.1N.

Caution: Imparts offensive odor to breath. Symptoms similar to hydrogen selenide, *q.v.*

4731. Hydrogen Tetracarbonylferrate(II). Iron hydrocarbonyl; iron tetracarbonyl dihydride. $C_4H_2FeO_4$; mol wt 169.91. C 28.27%, H 1.19%, Fe 32.87%, O 37.67%. $H_2Fe(CO)_4$. Prepn from $Fe(CO)_5$: Blanchard, Coleman, *Inorg. Syn.* **2**, 243 (1946); Sternberg *et al., J. Am. Chem. Soc.* **79**, 6116 (1957); Bishop *et al., J. Chem. Soc.* **1959**, 2484.

Colorless crystals, mp -70°; colorless liq, very unstable, dec below -10° on exposure to air. Slight traces of iron carbonyls impart a red or yellow color. Gas has extremely nauseating odor. Sol in alkalies; neutral aq solns of salts unstable at room temp; strongly alkaline solns of salts, e.g. $K_2Fe(CO)_4$, are stable at room temp when all traces of air are excluded but dec gradually at 100° to CO and H_2.

4732. Hydrohydrastinine. 5,6,7,8-*Tetrahydro-6-methyl-1,3,dioxolo[4,5-g]isoquinoline.* $C_{11}H_{13}NO_2$; mol wt 191.22. C 69.09%, H 6.85%, N 7.33%, O 16.73%. See ref under Hydrastinine, Cotarnine, Hydrocotarnine. Prepn from cotarnine: Topchiev, *J. Appl. Chem. USSR* **6**, 529 (1933), *C.A.* **28**, 2718 (1934); Clayson, *J. Chem. Soc.* **1949**, 2016; Schneider, Müller, *Ann.* **615**, 34 (1958); Knabe, *Arch. Pharm.* **292**, 652 (1959).

Crystals from petr ether, mp 66°. bp$_{752}$ 303°. Absorption spectrum: Dobbie, Tinkler, *J. Chem. Soc.* **85**, 1007 (1904). Sol in alc, ether, acetone, benzene, CS_2, ethyl acetate.

Hydrochloride, $C_{11}H_{13}NO_2 \cdot HCl$, crystals from water or alcohol, dec 278°. Sol in water.

Hydrobromide, $C_{11}H_{13}NO_2 \cdot HBr$, needles from water, dec 272°.

Hydriodide, $C_{11}H_{13}NO_2 \cdot HI$, crystals from water, dec 242°. Sol in water or alc.

Platinichloride, $(C_{11}H_{13}NO_2)_2 \cdot H_2PtCl_6$, yellow tablets, dec 222°. Sparingly sol in alc.

Picrate, $C_{11}H_{13}NO_2 \cdot C_6H_3N_3O_7$, yellow needles, dec 176°. Fairly sol in alc.

4733. Hydromorphone. *4,5-Epoxy-3-hydroxy-17-methylmorphinan-6-one;* dihydromorphinone; Dimorphone; Novolaudon. $C_{17}H_{19}NO_3$; mol wt 285.33. C 71.56%, H 6.71%, N 4.91%, O 16.82%. Prepn by electrolytic reduction of morphine: Takagi, Ueda, *J. Pharm. Soc. Japan* **56**, 44 (1936); Nakamura, *ibid.* **62**, 347 (1942); by oxidation of dihydromorphine: Rapoport *et al., J. Org. Chem.* **15**, 1103 (1950); Homeyer, De la Mater, U.S. pats. **2,628,962** and **2,654,756** (both 1953 to Mallinckrodt); Rapoport, U.S. pat. **2,649,454** (1953 to Univ. of California). Crystal structure: Steinmetz, *Z. Krist.* **67**, 434 (1928). Toxicity data: M. E. Buchwald, G. S. Eadie, *J. Pharm. Exp. Ther.* **71**, 197 (1941). *Review:* King *et al., U.S. Pub. Health Repts.* Suppl. no. 113, 38 pp (1935).

Crystals from ethanol, mp 266-267°. $[\alpha]_D^{25}$ −194° (c = 0.98 in dioxane).

Hydrochloride, $C_{17}H_{20}ClNO_3$, *Dilaudid, Laudicon, Hymorphan.* Crystals, dec 305-315° (evacuated tube). $[\alpha]_D^{25}$ −133°. Sol in 3 parts water; sparingly sol in alc. LD_{50} in mice (mg/kg): 61-96 i.v. (Buchwald, Eadie).

Caution: May be habit forming. This is a controlled substance (opiate) listed in the U.S. Code of Federal Regulations, Title 21 Parts 329.1 and 1308.12 (1985).

THERAP CAT: Narcotic analgesic.

4734. Hydrone®. An alloy of 35% Na and 65% Pb. Small, irregular lumps which on contact with water yield pure hydrogen gas (2.6 cu ft per lb); leaves a residue of lead-sponge and NaOH soln.

USE: As a source of hydrogen for laboratory work, portable blowpipes, and burners.

4735. Hydroorotic Acid. *Hexahydro-2,6-dioxo-4-pyrimidinecarboxylic acid;* 4,5-dihydroorotic acid. $C_5H_6N_2O_4$; mol wt 158.11. C 37.98%, H 3.83%, N 17.72%, O 40.48%. Prepn from carbethoxyasparagine: Miller *et al., J. Am. Chem. Soc.* **75**, 6086 (1953); U.S. pat. **2,773,872** (1956 to Merck & Co.).

L-Form, crystals, dec 266°. $[\alpha]_D^{25.3}$ +33.23° (c = 1.992 in 1% $NaHCO_3$).

D-Form, crystals from water, dec 266°. $[\alpha]_D^{25.3}$ −31.54° (c = 2.01 in 1% $NaHCO_3$).

DL-Form, crystals, dec 259°. Forms a water-sol salt.

USE: The L- and DL-dihydroorotic acids are precursors to the biological pyrimidines, such as thymine, uracil, cytosine and hydroxymethylcytosine. When employed in large amounts D-dihydroorotic acid is an antimetabolite of L-dihydroorotic acid in pyrimidine utilization. It has also been found to inhibit the orotic acid utilization of those organisms which have a requirement for it in order to grow. In general it may be said to inhibit bacterial growth.

4736. Hydroquinidine. *(9S)-10,11-Dihydro-6'-methoxycinchonan-9-ol;* dihydroquinidine; hydroconchinine. $C_{20}H_{26}N_2O_2$; mol wt 326.40. C 73.59%, H 8.03%, N 8.58%, O 9.80%. An alkaloid of cinchona, stereoisomeric with hydroquinine. Usually prepd by hydrogenation of quinidine: Heidelberger, Jacobs, *J. Am. Chem. Soc.* **41**, 826 (1919). Conversion to dihydrocinchonine by removal of the methoxy group: King, *J. Chem. Soc.* **1946**, 523. Manuf pat.: Gutzwiller, Uskokovic, **Ger.** pat. **1,933,599** (1970 to Hoffmann-La Roche), *C.A.* **72**, 90696v (1970). Pharmacology: Cosnier *et al., Therapie* **26**, 97 (1971).

Plates from ether, needles from alcohol, mp 169°. $[\alpha]_D^{20}$ +231° (c = 2.02 in alc); +299° (c = 0.82 in 0.1N H_2SO_4). Readily sol in hot alcohol; slightly sol in water and ether.

Hydrochloride, $C_{20}H_{27}ClN_2O_2$, *Serecor.* Rhombic plates, mp 273-274°. $[\alpha]_D^{26}$ +184° (c = 1.3). Freely sol in methanol, chloroform; less readily in water or abs alcohol; difficultly sol in dry acetone.

THERAP CAT: Antiarrhythmic.

4737. Hydroquinine. *(8α,9R)-10,11-Dihydro-6'-methoxycinchonan-9-ol;* dihydroquinine. $C_{20}H_{26}N_2O_2$; mol wt 326.40. C 73.60%, H 8.03%, N 8.58%, O 9.80%. An alkaloid of cinchona, found in quinine sulfate mother liquors. Stereoisomeric with hydroquinidine, *q.v.* Usually prepd by careful hydrogenation of quinine: Heidelberger, Jacobs, *J. Am. Chem. Soc.* **41**, 819 (1919). Total synthesis: Rabe *et al., Ber.* **64B**, 2487 (1931). Synthesis of isomers: Rubtsov, *J. Gen. Chem. USSR* **9**, 1493 (1939), *C.A.* **34** 2850 (1940); *ibid.* **13**, 593, 702 (1943), *C.A.* **39**, 705 (1945). LC determn in quinine beverages: L. P. Valenti, *J. Assoc. Offic. Anal. Chem.* **68**, 782 (1985).

Needles from ether or benzene, mp 172°. $[\alpha]_D^{18}$ −142° (alc); $[\alpha]_D^{20}$ −236° (c = 0.82 in 0.1N H_2SO_4). pK_1 = 5.33; K_1 = 4.7 × 10^{-6}. Freely sol in acetone, alcohol, chloroform, ether, petr ether; fairly sol in ammonia water; almost insol in water (290 mg/l).

Hydrochloride hemihydrate, $C_{20}H_{27}ClN_2O_2 \cdot \frac{1}{2}H_2O$, prisms from water, mp 208° (Heidelberger). $[\alpha]_D^{21}$ −124° (c = 1.1). pH of 0.005 molar soln 5.85. Freely sol in water, alcohol, methanol, acetone. Almost insol in ether. Crystallizes also with 2 mols H_2O. Infrared spectrum: Suszko, Dega-Szafran, *C.A.* **62**, 7819b (1965).

THERAP CAT: Depigmentor.

4738. Hydroquinone. *1,4-Benzenediol;* p-dihydroxybenzene; hydroquinol; quinol; Aida; Black and White Bleaching Cream; Eldoquin; Eldopaque; Quinnone; Tecquinol. $C_6H_6O_2$; mol wt 110.11. C 65.44%, H 5.49%, O 29.06%. Prepd by the oxidation of aniline: L. Gattermann, T. Wieland, *Die Praxis des Organischen Chemikers* (de Gruyter, Berlin, 40th ed., 1961) p 266; by reduction of quinone: Kitchen, U.S. pat. **1,322,580** (1920); Seyewetz, Miodon, *Bull. Soc. Chim. France* **33**, 449 (1923); by Elbs persulfate oxidation of phenol: Baker, Brown, *J. Chem. Soc.* **1948**, 2303; Forrest, Petrow, *ibid.* **1950**, 2340; from acetylene + CO: Howk, Sauer, U.S. pat. **3,055,949** (1962 to du Pont). Toxicity data: Woodard *et al.*, *Fed. Proc.* **8**, 348 (1949). *Review:* J. Varagnat in Kirk-Othmer *Encyclopedia of Chemical Technology* vol. **13** (Wiley-Interscience, New York, 3rd ed., 1981) pp 39-69.

Crystals, mp 170-171°. d_{15} 1.332. bp 285-287°. Sol in 14 parts water; freely sol in alcohol, ether, slightly in benzene. Its soln becomes brown in the air due to oxidation; the oxidation is very rapid in presence of alkali. *Keep well closed and protected from light.* LD_{50} orally in rats: 320 mg/kg (Woodard).

Human Toxicity: Relatively safe in very low concns. Under conditions of ordinary use, hydroquinone does not present a serious hazard. Ingestion of 1 g has caused tinnitus, nausea, vomiting, sense of suffocation, shortness of breath, cyanosis, convulsions, delirium, collapse. Death has followed ingestion of 5 g. Irritation of intestinal tract occurs with oral ingestion. Dermatitis can result from skin contact. Staining and opacification of cornea occur in workers exposed for prolonged periods to concns of vapor not high enough for production of systemic effects, W. B. Deichmann, M. L. Keplinger in Patty's *Industrial Hygiene and Toxicology* vol. **2A**, G. D. Clayton, F. E. Clayton, Eds. (Wiley-Interscience, New York, 3rd ed., 1981) pp 2589-2592.

USE: As photographic reducer and developer; as reagent in the determination of small quantities of phosphate; as antioxidant.

THERAP CAT: Depigmentor.

4739. Hydroxocobalamin. *Cobinamide dihydroxide dihydrogen phosphate (ester), mono(inner salt), 3'-ester with 5,6-dimethyl-1-α-D-ribofuranosyl-1H-benzimidazole;* α-(5,6-dimethylbenzimidazolyl)hydroxocobamide; vitamin B_{12a}; hydroxocobemine; OHB_{12}; Alpha Cobione; AlphaRuvite; Axion; Axlon; Ciplamin H; Cobalin-H; Cobalex; Codroxomin; Depogamma; Docelan; Docevita; Droxomin; Ducobee-Hy; Duradoce; Duralta-12; Hydrogrisevit; Hydrovit; Hyxobamine; Idrogriseovit; Lyovit-H; Neo-Betalin 12; Neo-Cytamen; Neo-Macrabin; Neo-Rojamin; OH-Duphar; Oxobemin; Primabalt RP; Oxolamine; alphaRedisol; Redisol H; Sytobex-H; Vitadurin. $C_{62}H_{89}CoN_{13}O_{15}P$; mol wt 1346.41. C 55.31%, H 6.66%, Co 4.38%, N 13.52%, O 17.83%, P 2.30%. An analog of vitamin B_{12} where the CN group is replaced with OH. Prepn: E. A. Kaczka *et al.*, U.S. pat. **2,738,301** (1956 to Merck & Co.). Prepn of stable solns of the acetate: Marcus *et al.*, Fr. pat. **1,336,671** (1963 to Merck & Co.), *C.A.* **60**, 380a (1964). Apparently hydroxocobalamin exists in aq soln as an equilibrium mixture of the hydroxy isomer and the ionic aquo isomer (*see* Aquocobalamin); $[CoX]^+$ equals vitamin B_{12} without the cyanide group. Acids replace the hydroxo group with the anion of the acid: E. A. Kaczka *et al.*, *Science* **112**, 354 (1950).

$$[\text{HO-CoX}] \xrightleftharpoons{\text{H}_2\text{O}} [\text{H}_2\text{O} \cdot \text{CoX}]^+ \text{OH}^-$$

Dark red, orthorhombic needles or platelets from water + acetone. Darkens at 200°, but not melted at 300°. Absorption max (H₂O): 270-277, 351-354, and 520-530 nm. Moderately sol in water and in the lower aliphatic alcohols. Practically insol in acetone, ether, petr ether, halogenated hydrocarbons, benzene.

Acetate, $C_{64}H_{91}CoN_{13}O_{16}P$, *cobinamide acetate phosphate*

3'-ester with 5,6-dimethyl-1-α-D-ribofuranosylbenzimidazole inner salt, acetatocobalamin, Idoxo-B_{12}, Novidroxin.

THERAP CAT: Hematopoietic vitamin.

THERAP CAT (VET): In cobalt or vitamin B_{12} deficiency.

4740. Hydroxyamphetamine. (±)-4-(2-Aminopropyl)-phenol; *dl*-*p*-hydroxy-α-methylphenethylamine; *dl*-1-*p*-hydroxyphenyl-2-propylamine; *p*-hydroxyphenylisopropylamine; α-methyltyramine; Paredrine; Paredrinex; Pulsoton. $C_9H_{13}NO$; mol wt 151.20. C 71.49%, H 8.67%, N 9.26%, O 10.58%. Prepn from oxime of *p*-methoxyphenyl acetone: Mannich, Jacobsohn, *Ber.* **43**, 189 (1910), **Ger. pat. 243,546.** From *p*-nitrobenzyl chloride and a salt of nitroethane: Hoover, Hass, *J. Org. Chem.* **12**, 501 (1947).

Crystals (rosettes) from benzene, mp 125-126°. Sol in water, alcohol, chloroform, ethyl acetate.

Iodide, $C_9H_{14}INO$, stout prisms, mp 155°. Freely sol in water, alcohol, acetone.

Hydrochloride, $C_9H_{14}ClNO$, crystals from HCl, mp 171-172°. Sol in water, alcohol. Practically insol in ether.

Hydrobromide, $C_9H_{14}BrNO$, crystals. Freely sol in water, alcohol, acetone. Sold as Paredrine Hydrobromide Aqueous—a 1% aq soln made isotonic with sodium chloride and preserved with sodium ethylmercuri thiosalicylate. Also sold as Paredrine Hydrobromide Ophthalmic 1%, with boric acid—a 1% aq soln made tear-isotonic with 2% boric acid and preserved with sodium ethylmercuri thiosalicylate.

THERAP CAT: Adrenergic (ophthalmic); mydriatic.

4741. *p*-Hydroxybenzaldehyde. 4-Formylphenol. $C_7H_6O_2$; mol wt 122.12. C 68.84%, H 4.95%, O 26.20%. Widely distributed in plants in very small amounts. Obtained as a byproduct from phenol: L. F. Fieser, M. Fieser, *Organic Chemistry* (Reinhold, New York, 3rd ed., 1956) p 681; *cf.* Reimer, Tiemann, *Ber.* **9**, 824 (1876); Herzfeld, Tiemann, *Ber.* **10**, 64 (1877).

Needles from water. Slight, agreeable, aromatic odor, mp 116°. Sublimes at atmospheric pressure without decomposition. Reacts like a weak monobasic acid: Ka at 25° = 2.2 × 10⁻⁸. Dipole moment: 4.19. Sparingly sol in cold water: 1.38 g/100 ml H₂O at 30.5°; more sol in hot water. Freely sol in alc, ether. Slightly sol in benzene: 3.68 g/100 ml C_6H_6 at 65°.

4742. *p*-Hydroxybenzoic Acid. $C_7H_6O_3$; mol wt 138.13. C 60.86%, H 4.38%, O 34.75%. Prepn from *p*-bromophenol: Gilinan, Arntzen, *J. Am. Chem. Soc.* **69**, 1537 (1947); from *p*-hydroxybenzaldehyde: Pearl, *J. Org. Chem.* **12**, 85 (1947); from phenol + potassium ethyl carbonate: Jones, *Chem. & Ind. (London)* **1958**, 228; from potassium phenolate + CO₂: **Brit. pat. 942,418** (1963 to Inventa). A metabolic product of *Penicillium patulum:* Tanenbaum, Bassett, *Biochim. Biophys. Acta* **28**, 21 (1958).

Crystals, mp 213-214°. d 1.46. Soluble in about 125 parts water, freely in alcohol, slightly in chloroform; sol in ether, acetone. Practically insol in carbon disulfide. Ferric chloride does not color its aq soln.

USE: In org. syntheses; intermediate for dyes, fungicides.

4743. *p*-Hydroxybenzylpenicillin Sodium. *6-[[(Hydroxyphenyl)acetyl]amino]-3,3-dimethyl-7-oxo-4-thia-1-aza-bicyclo[3.2.0]heptane-2-carboxylic acid sodium salt;* sodium *p*-hydroxybenzylpenicillinate; sodium penicillin X; sodium penicillin III. $C_{16}H_{17}N_2NaO_5S$; mol wt 372.40. C 51.61%, H 4.60%, N 7.52%, Na 6.17%, O 21.48%, S 8.61%. Produced by a mutant of *Penicillium chrysogenum:* MacCorquodale *et al.,* in *The Chemistry of Penicillin* (Princeton Univ. Press, Princeton, 1949) pp 95-98; Cartland *et al.,* U.S. pat. 2,487,-018 (1949 to Upjohn).

Crystals (compact rosettes) from water + *n*-butanol. $[\alpha]_D^{25}$ +267° (c = 2). uv max: 278 nm. pK at 23° = 2.62. pH 5.5-6.5. Very sol in water, in isotonic sodium chloride soln and in glucose solns; sol in alcohol (fairly stable in methanol, a property not shared by the other penicillins). Stability of the solid and of its aq solns is similar to that of benzylpenicillin sodium, *cf.* Benedict *et al., J. Bacteriol.* **51,** 291 (1946).

4744. α-Hydroxybenzylphosphinic Acid. $C_7H_9O_3P$; mol wt 172.13. C 48.85%, H 5.27%, O 27.89%, P 18.00%. Prepd from benzaldehyde (3 moles) and H_3PO_2 (1 mole): Ville, *Ann. Chim. (Paris)* [6] **23,** 305 (1891); *Beilstein* **7,** 232; Viout, *J. Rech. Cent. Nat. Rech. Sci.* no. **28,** 15-31 (1954), *C.A.* **50,** 7078d (1956).

Leaflets from butanol, mp 110°. Freely sol in water, alcohol; also sol in benzene. Practically insol in anhydr ether.

Sodium salt, *Phos, Phoselit, Phosilite,* very freely sol in water. The pH is on the acid side, but aq solns are suitable for injection.

THERAP CAT: Sodium salt as nutrient.

4745. 4′-Hydroxybutyranilide. *N-(4-Hydroxyphenyl)-butanamide;* p-hydroxybutyranilide; *N*-butyroyl-*p*-aminophenol; Suconox-4. $C_{10}H_{13}NO_2$; mol wt 179.21. C 67.02%, H 7.31%, N 7.82%, O 17.86%. Prepn: Kuhn *et al., Z. Physiol. Chem.* **247,** 197 (1937); Fierz-David, Kuster, *Helv. Chim. Acta* **22,** 82 (1939); Rohmann, Friedrich, *Arch. Pharm.* **278,** 456 (1940).

Needles from water, mp 139-140°. Slightly sol in cold water, more sol in hot water, sol in alcohol.

USE: As antioxidant: Young, Cottle, U.S. pat. 2,654,722 (1953 to Standard Oil).

4746. β-Hydroxybutyric Acid. *3-Hydroxybutanoic acid.* $C_4H_8O_3$; mol wt 104.10. C 46.15%, H 7.75%, O 46.11%. $CH_3CHOHCH_2COOH$.

d-Form, prepd by the action of *Aspergillus griseus* on the *dl*-form: McKenzie, Harden, *J. Chem. Soc.* **83,** 430 (1903). Crystals. $[\alpha]_D^{10}$ +24.3° (c = 2.226). Sol in water, alcohol, ether.

l-Form, found in the urine of diabetics (as much as 30 g per day). Isoln: Fischer, Scheibler, *Ber.* **42,** 1221 (1909); Shaffer, Marriott, *J. Biol. Chem.* **16,** 268 (1913). Hygro-

scopic, monoclinic crystals, mp 45.5-48°. $[\alpha]_D^{25}$ −24.5° (c = 5). K at 22° = 3.86 × 10^{-5}. Freely sol in water, alcohol, ether. Sparingly sol in benzene. On distn it dec into crotonic acid and water.

dl-Form, prepd from acetoacetic ester by the action of sodium amalgam: Wislicenus, *Ann.* **149,** 207 (1869); Marian, *Biochem. Z.* **150,** 283 (1924); by the oxidation of aldol: Wurtz, *Compt. Rend.* **76,** 1167 (1873); from crotonic acid by heating with dil acid: Wacker, **Ger.** pat. **441,003;** *Frdl.* **15,** 135; by heating crotonitrile with KOH soln: Bruylants, *Bull. Soc. Chim. Belg.* **31,** 182 (1922). Hygroscopic syrup. Volatile with steam. Sol in water, alcohol, ether. On distn it dec into crotonic acid and water.

4747. 3-Hydroxycamphor. *3-Hydroxy-1,7,7-trimethylbi-cyclo[2.2.1]heptan-2-one;* oxycamphor; Oxaphor. $C_{10}H_{16}O_2$; mol wt 168.24. C 71.39%, H 9.59%, O 19.02%. Prepn: Lucius, Brüning, **Ger.** pat. **91,718** (1897); Manasse, *Ber.* **30,** 659 (1897); Lapworth, Chapman, *J. Chem. Soc.* **79,** 377 (1901).

Needles from benzene + petr ether, mp 205-206°. Sol in 50 parts water, more in hot water; very sol in alcohol, chloroform, ether.

THERAP CAT: Topical antipruritic.

4748. Hydroxychloroquine. *2-[[4-[(7-Chloro-4-quino-linyl)amino]pentyl]ethylamino]ethanol;* 7-chloro-4-[4-[ethyl-(2-hydroxyethyl)amino]-1-methylbutylamino]quinoline; 7-chloro-4-[4-(*N*-ethyl-*N*-β-hydroxyethylamino)-1-methyl-butylamino]quinoline; 7-chloro-4-[5-(*N*-ethyl-*N*-2-hydr-oxyethylamino)-2-pentyl]aminoquinoline; oxychloroquine; oxichlorochine. $C_{18}H_{26}ClN_3O$; mol wt 335.87. C 64.36%, H 7.80%, Cl 10.56%, N 12.51%, O 4.76%. Prepd by reacting a mixture of 4,7-dichloroquinoline, phenol and *N′*-ethyl-*N′*-β-hydroxyethyl-1,4-pentadiamine at 125-130°: Surrey, Hammer, *J. Am. Chem. Soc.* **72,** 1814 (1950); Surrey, U.S. pat. **2,546,658** (1951 to Sterling Drug). Use in combination with cyclophosphamide and azathioprine, *q.q.v.* in the treatment of rheumatoid arthritis: D. J. McCarty, G. F. Carrera, *J. Am. Med. Assoc.* **248,** 1718 (1982). Reassessment in the treatment of rheumatoid arthritis: *Am. J. Med.* **75,** no. 1A, 1-56 (1983). Series of articles on clinical use: *ibid.* **85,** Suppl. 4A, 1-71 (1988).

Crystals from ethylene dichloride and Skellysolve B; mp 89-91°.

Diphosphate, $C_{18}H_{32}ClN_3O_9P_2$. Recrystallized from ethanol, mp 168-170° (dec).

Sulfate, $C_{18}H_{28}ClN_3O_5S$, *Ercoquin, Plaquenil Sulfate, Quensyl.* White crystalline powder; odorless but has a bitter taste. pH of aq solns about 4.5. Exists in two forms, the usual form mp ∼240°, the other mp ∼198°. Freely sol in water. Practically insol in alcohol, chloroform, ether.

THERAP CAT: Antimalarial; antirheumatic; lupus erythematosus suppressant.

4749. 1α-Hydroxycholecalciferol. *(1α,3β,5Z,7E)-9,10-Secocholesta-5,7,10(19)-triene-1,3-diol;* 1α-hydroxyvitamin D₃; 1α-OH-CC; alfacalcidol; Alfarol; Alpha D₃; EinsAlpha; Etalpha; One-Alpha. $C_{27}H_{44}O_2$; mol wt 400.65. C 80.94%, H 11.07%, O 7.99%. Synthetic analog of calcitriol, *q.v.,* the hormonal form of vitamin D₃, which shows identical potency with respect to stimulation of intestinal calcium absorption and bone mineral mobilization. This common activity is thought to be due either to the presence in both of a 1α hydroxy group or more likely, to the conversion of the 1α-OH-CC *in vivo* to 1α,25-dihydroxycholecalciferol: Haussler *et al., Proc. Nat. Acad. Sci. USA* **70**, 2248 (1973). Synthesis from cholesterol: Holick *et al., Science* **180**, 190 (1973); Barton *et al., J. Am. Chem. Soc.* **95**, 2748 (1973); Fürst *et al., Helv. Chim. Acta* **56**, 1708 (1973); M. Morisaki *et al., Chem. Pharm. Bull.* **23**, 3272 (1975); T. Sato *et al., ibid.* **26**, 2933 (1979). Total synthesis: R. G. Harrison *et al., Tetrahedron Letters* **1973**, 3649; *eidem, J. Chem. Soc. Perkin Trans. I* **1974**, 2654; P. J. Kocienski, B. Lythgoe, *ibid.* **1980**, 1400. Synthesis of the 1β-epimer from the 1α-form and biological properties: H. E. Paaren *et al., Chem. Commun.* **1977**, 890.

mp 134-136° (Harrison); also reported as mp 138-139.5° (Fürst). [α]$_D^{25}$ +28° (ether). uv max (ether): 264 nm [ε 18000 (Harrison); ε 20200 (Barton)].

1β-Epimer, *1β-hydroxycholecalciferol, 1β-hydroxyvitamin D$_3$*.

THERAP CAT: Vitamin D source.

4750. 24-Hydroxycholesterol. *Cholest-5-ene-3β,24-diol.* C$_{27}$H$_{46}$O$_2$; mol wt 402.64. C 80.54%, H 11.52%, O 7.95%. Two stereoisomers exist: *cholest-5-ene-3β,24β-diol* and *cholest-5-ene-3β,24α-diol.* Isoln of the naturally occurring 24β-epimer, from horse brain: Ercoli *et al., Boll. Soc. Ital. Biol. Sper.* **29**, 494 (1953), *C.A.* **49**, 4744i (1955); from beef spinal cord, Fieser *et al., J. Org. Chem.* **22**, 1380 (1957). Prepn and separation of isomers: Ercoli, Ruggieri, *Gazz. Chim. Ital.* **83**, 720 (1953); *eidem, J. Am. Chem. Soc.* **75**, 3284 (1953). Configurations at C-24: Klyne, Stokes, *J. Chem. Soc.* **1954**, 1979.

24β-Epimer, *cerebrostenediol, cerebrosterol.* Needles from acetone, mp 175-176°. [α]$_D^{20}$ −48.2° (c = 1.06 in CHCl$_3$).
24β-Epimer dibenzoate, C$_{41}$H$_{54}$O$_4$, crystals from ether + acetone, mp 182-183°. α$_D$ −19° (c = 1.20 in CHCl$_3$). Absorption max in chloroform: 5.85, 6.2, 6.86, 7.8 μ.
24α-Epimer, crystals, mp 182-183°. [α]$_D^{20}$ −26.8° (c = 0.672 in CHCl$_3$).
24α-Epimer dibenzoate, crystals, mp 141-142°. [α]$_D^{26}$ −11.8° (c = 1 in CHCl$_3$).

4751. 25-Hydroxycholesterol. *Cholest-5-ene-3,25-diol;* Δ^5-cholestene-3β,25-diol. C$_{27}$H$_{46}$O$_2$; mol wt 402.67. C 80.54%, H 11.51%, O 7.95%. An important intermediate in the synthesis of 25-hydroxycholecalciferol, *q.v.* Synthesis: A. Ryer *et al., J. Am. Chem. Soc.* **72**, 4247 (1950); W. G. Dauben, H. L. Bradlow, *ibid.* 4248. Isoln from autoxidation products of cholesterol, *q.v.:* L. F. Fieser *et al., J. Org. Chem.* **22**, 1380 (1957). Prepn from pregnenolone, *q.v.:* T. A. Narwid, M. R. Uskokovic, U.S. pat. **3,856,780** (1974). Alternate syntheses: M. Morisaki *et al., Chem. Pharm. Bull.*

21, 457 (1973); A. Rotman, Y. Mazur, *Chem. Commun.* **1974**, 15; J. J. Partridge *et al., Helv. Chim. Acta* **57**, 764 (1974); T. A. Narwid *et al., ibid.* 771; W. G. Salmond *et al., Tetrahedron Letters* **1977**, 987, 1237, 1695; K. Ochi *et al., Chem. Pharm. Bull.* **27**, 252 (1979); M. Riediker, J. Schwartz, *Tetrahedron Letters* **22**, 4655 (1981).

Colorless needles from methanol, mp 178-180°. [α]$_D^{25}$ −39.0° (c = 1.05 in chloroform).
20(S)-isomer, crystals from chloroform, mp 189.5-190.5°. [α]$_D^{25}$ −41.50°(c = 0.9278 in chloroform).

4752. Hydroxycodeinone. *5α-7,8-Didehydro-4,5α-epoxy-14-hydroxy-3-methoxy-17-methylmorphinan-6-one;* 14-hydroxycodeinone. C$_{18}$H$_{19}$NO$_4$; mol wt 313.34. C 68.99%, H 6.11%, N 4.47%, O 20.42%. Prepn from thebaine: Freund, Speyer, *J. Prakt. Chem.* **94**, 135 (1916). From codeine: Merck, Ger. pat. **411,530**; *Frdl.* **15**, 1516; K. W. Bentley, *The Chemistry of the Morphine Alkaloids* (Oxford, 1954). Improved synthesis: F. M. Hauser *et al., J. Med. Chem.* **17**, 1117 (1974). Isoln from *Papaver bracteatum* Lindl.: H. G. Theuns *et al., Phytochemistry* **16**, 753 (1977).

Plates from 96% alcohol + few drops chloroform, dec 275°. Freely sol in CHCl$_3$, methyl Cellosolve, petr ether, ethyl acetate; slightly sol in alcohol. Practically insol in water, ether; also insol in aq alkaline solns.
Hydrochloride monohydrate, C$_{18}$H$_{20}$ClNO$_4$.H$_2$O, rods from water, dec 285-286°. [α]$_D^{20}$ −150° (c = 2.5).
Note: This is a controlled substance (opiate) listed in the U.S. Code of Federal Regulations, Title 21 Part 1308.12 (1987).

4753. Hydroxydione Sodium. *21-(3-Carboxy-1-oxopropoxy)-5β-pregnane-3,20-dione sodium salt;* 21-hydroxypregnane-3,20-dione sodium hemisuccinate; hydroxydione succinate; Presuren; Viadril. C$_{25}$H$_{35}$NaO$_6$; mol wt 454.52. C 66.06%, H 7.76%, Na 5.06%, O 21.12%. Prepd by palladium reduction of deoxycorticosterone followed by treatment with succinic anhydride and formation of the sodium salt: Laubach, U.S. pat. **2,708,651** (1955 to Pfizer); Laubach *et al., Science* **122**, 78 (1955).

Lyophilized, fluffy white powder. Dec 193-203°. uv max: 280 nm (ε 93.2). Sol in water, in mildly alkaline buffer solns, acetone, chloroform. pH of 2% aq soln 8.5-9.8.
Free acid, mp 195-197°. [α]$_D^{20}$ +95° in chloroform.
THERAP CAT: Anesthetic (intravenous).

4754. *p*-Hydroxyephedrine. *4-Hydroxy-α-[1-(methyl-*

amino)ethyl]benzenemethanol; p-hydroxy-α-[1-(methylami-no)ethyl]benzyl alcohol; 1-(4-hydroxyphenyl)-2-methylaminopropanol; α-(1-methylaminoethyl)-*p*-hydroxybenzyl alcohol; oxyephedrin; Carnigen; Edornat; Methylsympatol; Suprifen. $C_{10}H_{15}NO_2$; mol wt 181.23. C 66.27%, H 8.34%, N 7.73%, O 17.66%. Prepn: Stolz, U.S. pat. 1,878,021 (1932 to Winthrop). Prepn of hydrochlorides of *d*- and *l*-forms: Takamatsu, Minaki, *J. Pharm. Soc. Japan* 76, 1230 (1956), *C.A.* 51, 4305e (1957).

Cryst powder, mp 152-154°. Sparingly sol in water, alcohol, ether; readily sol in NaOH soln and dil acids.

Hydrochloride, $C_{10}H_{15}NO_2 \cdot HCl$, crystals, mp 209-211°. Sol in 3 parts water, 5 parts glycerol, 10 parts 90% alcohol. Sparingly sol in abs alcohol, acetone.

d-Form hydrochloride, crystals, dec 223-224°. $[\alpha]_D^{22}$ +33.8°.

l-Form hydrochloride, crystals, dec 222-223°. $[\alpha]_D^{22}$ −35.5°.

THERAP CAT: Adrenergic.

4755. N-Hydroxyethylpromethazine Chloride. *N*-(2-*Hydroxyethyl*)-*N,N,α-trimethyl-10H-phenothiazine-10-ethanaminium chloride; (2-hydroxyethyl)dimethyl(1-methyl-2-phenothiazin-10-ylethyl)ammonium chloride; N,N*-dimethyl-*N*-[(α-methyl-β-phenothiazin-10-yl)ethyl]-*N*-(β-hydroxyethyl)ammonium chloride; [1-(10-phenothiazinylmethyl)-ethyl]-2-hydroxyethyldimethylammonium chloride; Aprobit. $C_{19}H_{25}ClN_2OS$; mol wt 364.96. C 62.53%, H 6.90%, Cl 9.72%, N 7.68%, O 4.38%, S 8.79%. Prepd from *N,N*-dimethyl-*N*-[(α-methyl-β-phenothiazin-10-yl)ethyl]amine and glycol chlorohydrin in methyl ethyl ketone: Carlsson, Karlson, **Brit.** pat. 881,379 (1961 to Aktiebolaget Recip).

Crystals from isopropanol, mp 233°.
THERAP CAT: Antihistaminic.

4756. Hydroxyglutamic Acid. β-Hydroxyglutamic acid; α-amino-β-hydroxyglutaric acid. $C_5H_9NO_5$; mol wt 163.14. C 36.82%, H 5.56%, N 8.59%, O 49.04%. An amino acid classified as nonessential with respect to its growth effect in rats. "Not among the naturally occurring hydrolytic products of proteins"—Schmidt, *Chemistry of the Amino Acids and Proteins* (Springfield, Ill., 2nd ed., 1944) p 1090. Synthesis from ethyl α-isonitrosoacetone dicarboxylate: Harington, Randall, *Biochem. J.* 25, 1917 (1931); Levene, Schormüller, *J. Biol. Chem.* 106, 595 (1934).

HOCHCH₂COOH
H₂NCHCOOH

L-Form, prisms from water. Dec 135° (if completely anhydrous; hydrated crystals soften at 100-105°). $[\alpha]_D^{20}$ +1.2° (c = 2); $[\alpha]_D^{20}$ +17.6° (c = 2 in 6N HCl). pK_1' 2.09; pK_2' 4.18; pK_3' 9.20. Freely sol in water, acetic acid. Insol in alc, ether.

DL-Form, orthorhombic prisms, needles from water. Dec 198°. Very sol in hot water.

4757. 8-Hydroxy-7-iodo-5-quinolinesulfonic Acid. 7-Iodo-8-hydroxyquinoline-5-sulfonic acid; *m*-iodo-*o*-hydroxyquinolineanasulfonic acid; Anayodin; Ferron; Loretin. $C_9H_6INO_4S$; mol wt 351.13. C 30.78%, H 1.72%, I 36.15%, N 3.99%, O 18.23%, S 9.13%. Obtained from the potassium salt of 8-hydroxy-5-quinolinesulfonic acid by the action of KI, bleaching powder, and HCl: Claus, *Arch. Pharm.* 231, 704 (1893); **Ger.** pat. 72,942 (1893); by the action of KI and Ca(OCl)₂ on 8-hydroxy-5-quinolinesulfonic acid: Botton, *C.A.* 52, 18407e (1958).

Sulfur yellow, almost odorless and tasteless, crystalline powder. mp 260-270° (dec). One gram dissolves in 500 ml cold, 170 ml boiling water; slightly sol in alcohol. Practically insol in ether or oils.

Compd with chloroquine (2:1), *chloquinate, chlochinate, cloquinate, Resotren.* Prepn: Koenig, Andersag, U.S. pat. 2,650,224 (1953 to Schenley Industries).

Mixture of 80% of 8-hydroxy-7-iodoquinoline-5-sulfonic acid and 20% sodium bicarbonate: *chiniofon; quiniofon; Yochinol; Quinoxyl; Sefona; Yatren.*

Note: The name Anayodin is also used to designate sodium iodide.

USE: As colorimetric reagent for ferric ion.
THERAP CAT: Antiamebic; topical antiseptic.

4758. 4-Hydroxyisophthalic Acid. *4-Hydroxy-1,3-benzenedicarboxylic acid.* $C_8H_6O_5$; mol wt 182.13. C 52.75%, H 3.32%, O 43.92%. Prepd by high pressure carbonation of potassium phenate (Kolbe-Schmitt reaction): Baine *et al., J. Org. Chem.* 19, 510 (1954) or by boiling salicylic acid with carbon tetrachloride in alkaline medium in the presence of Cu powder: **Ger.** pat. 258,887 (1913 to Zeltner & Landau); *Chem. Zentr.* 1913, I, 1641; *Frdl.* 11, 210. Major constituent of the brown dust residue from sublimation of salicylic acid: Hunt *et al., Chem. & Ind. (London)* 1955, 417.

Branched needles from water, platelets from dil alc. Dec 314-315°. One gram dissolves in 3 liters of water at 24°, in 160 ml at 100°. Freely sol in alcohol, ether.

Dimethyl ester, crystals, mp 97.5°.
Diethyl ester, crystals, mp 54-55°.

Note: 4-Hydroxyisophthalic acid has been proposed as an improvement over aspirin: Chesher *et al., Nature* 175, 206 (1955).

4759. Hydroxylamine. H_3NO; mol wt 33.03. H 9.15%, N 42.41%, O 48.44%. NH₂OH. Prepn as the hydrochloride: W. L. Semon, *Org. Syn. coll. vol.* I, 318 (1932). Prepn: Hurd, *Inorg. Syn.* 1, 87 (1939); Benson *et al., J. Am. Chem. Soc.* 78, 4202 (1956). Crystal structure: Meyers, Lipscomb, *Acta Cryst.* 5, 583 (1955). Toxicity data: Riemann, *Acta Pharmacol. Toxicol.* 6, 285 (1950); R. P. Smith, W. R. Layne, *J. Pharmacol. Exp. Ther.* 165, 30 (1969). Mutagenic action: Phillips, Brown in *Progr. Nucl. Acid Res. Mol. Biol.* 7, 349-368 (1967). Reviews: Mason, "Hydroxylamine" in *Mellor's* vol. VIII, supplement 2, *Nitrogen* (part 2), 115-157 (1967); Jones in *Comprehensive Inorganic Chemistry* vol. 2, J. C. Bailar, Jr. *et al.,* Eds. (Pergamon Press, Oxford, 1973) pp 265-276.

Unstable, large white flakes or needles, mp 33°, bp_{22} 58°. d_4^0 1.2255; d_4^{40} 1.204. K at 20° = 1.07×10^{-8}. Very sol in water. Very sol in liq ammonia and methanol. The soly in the higher alcohols decreases with increasing mol wt. Sparingly sol in ether, benzene, carbon disulfide, chloroform. Very hygroscopic. Dec by hot water. Undergoes rapid decompn at room temps esp in the presence of atm moisture and CO_2. Detonates in test tube heated with flame. LD_{50} i.p. in mice: 1.83 mmol/kg (Smith, Layne).

Hydrochloride, ClH_4NO, *oxammonium hydrochloride.* Monoclinic columnar crystals; sowly dec when moist. d_{17} 1.67. mp about 151°. One gram dissolves in about 1 ml water (83 g in 100 ml water at 17°); 19 ml alcohol; 8 ml methanol. Sol in glycerol, propylene glycol. Insol in cold ether. pH of 0.2 molar aq soln 3.2. Keep well closed. LD_{50} orally in mice: 408 mg/kg (Riemann).

Sulfate, $H_8N_2O_6S$, *oxammonium sulfate.* Crystals, mp about 170°. Freely sol in water.

Caution: Skin irritant. May cause methemoglobinemia, sulfhemoglobinemia, cyanosis, convulsions, hypotension and coma: *Clinical Toxicology of Commercial Products,* R. E. Gosselin *et al.,* Eds. (Williams & Wilkins, Baltimore, 5th ed., 1984) Section II, p 117.

USE: As reducing agent in photography; in synthetic and analytical chemistry; to purify aldehydes and ketones. As antioxidant for fatty acids and soaps. As dehairing agent for hides.

4760. Hydroxylupanine. *Dodecahydro-9-hydroxy-7,14-methano-4H,6H-dipyrido[1,2-a:1',2'-e][1,5]diazocin-4-one;* octalupine; oxylupanine. $C_{15}H_{24}N_2O_2$; mol wt 264.36. C 68.15%, H 9.15%, N 10.60%, O 12.10%. From seed of *Lupinus perennis* L., *L. angustifolius* L. and *L. polyphyllus* Lindl., *Leguminosae:* Rink, Schäfer, *Arch. Pharm.* **287,** 290 (1954); Winterfeld, Pies, *ibid.* **290,** 537 (1957); Bohlmann, Winterfeldt, *Ber.* **93,** 1956 (1960). Structure: Galinovsky *et al., Monatsh.* **81,** 77 (1950). Biosynthesis: Schuette *et al., Arch. Pharm.* **296,** 438 (1963).

Crystals from dry acetone, mp 169-170°. $[\alpha]_D^{20}$ +45.6° (c = 1.49 in ethanol); $[\alpha]_D^{20}$ +65.1° (c = 1.09 in water). Sol in water, alc, chloroform; slightly sol in benzene, ether.

Methiodide, $C_{15}H_{24}N_2O_2 \cdot CH_3I$, needles from methanol + ether, mp 254-255°.

4761. Hydroxymercurichlorophenols. Prepared from *o*-chlorophenol and mercuric oxide or mercuric salts: F. C. Whitmore, *Organic Compounds of Mercury* (Chemical Catalog Co., New York, 1921) p 269; **D.R.P.** 234,851 (1910 to Bayer) in *Frdl.* **10,** 1272.

p-Form (n = 1), $C_6H_5ClHgO_2$, *(3-chloro-4-hydroxyphenyl)hydroxymercury.* Major ingredient of *Semesan.*

Sulfate, *Uspulun.* See *Beilstein,* 2nd suppl., vol. **16,** 683.

USE: Seed disinfectant.

4762. 1-(Hydroxymethyl)-5,5-dimethylhydantoin. *1-(Hydroxymethyl)-5,5-dimethyl-2,4-imidazolidinedione;* monomethyloldimethylhydantoin; MDMH; methylol dimethylhydantoin. $C_6H_{10}N_2O_3$; mol wt 158.16. C 45.56%, H 6.37%, N 17.71%, O 30.35%. Prepd by treating 5,5-dimethylhydantoin with formaldehyde: Mackey, **U.S. pat.** 2,762,-708 (1956 to GAF).

Crystals, mp 100°. Freely sol in water, methanol, ethanol, acetone; slightly sol in ethyl acetate; practically insol in ether, trichloroethylene, carbon tetrachloride.

USE: As preservative in cosmetic prepns. The compound liberates formaldehyde steadily at a very slow rate in the presence of water at pH 6. The available formaldehyde is about 19% (w/w), and 0.1-10% is usuallly mixed with the cosmetic product. When heated in the dry state MDMH forms a water-soluble *dimethylhydantoin formaldehyde resin,* compatible with gelatin, polyvinyl acetate, ethyl cellulose. The resin is used in hair lacquers.

4763. 17-Hydroxy-16-methylene-Δ^6-progesterone. *17-Hydroxy-16-methylenepregna-4,6-diene-3,20-dione;* 6-dehydro-17α-hydroxy-16-methyleneprogesterone. $C_{22}H_{28}O_3$; mol wt 340.44. C 77.61%, H 8.29%, O 14.10%. Prepn: Syhora, Mazac, *Coll. Czech. Chem. Commun.* **29,** 2351 (1964). Prepn of the 17-acetate: **Brit. pat.** 901,293 (1962 to Upjohn); Syhora, Mazac, *loc. cit.;* **Brit.** pat. 963,427 (1964 to E. Merck).

Crystals from methanol, mp 196.5-197°. $[\alpha]_D^{20}$ −72.5°. uv max: 283 nm (log ϵ 4.51).

17-Acetate, $C_{24}H_{30}O_4$, *17-acetoxy-16-methylenepregna-4,6-diene-3,20-dione, 17α-acetoxy-6-dehydro-16-methyleneprogesterone, Superlutin.* Crystals from ethyl acetate, mp 233-234°. $[\alpha]_D^{20}$ −132°. uv max: 283 nm (log ϵ 4.50). THERAP CAT: Oral progestogen.

4764. 5-(Hydroxymethyl)-2-furaldehyde. *5-(Hydroxymethyl)-2-furancarboxaldehyde;* 5-(hydroxymethyl)-2-furancarbonal; 5-(hydroxymethyl)-2-furfural; HMF; 5-hydroxymethyl-2-formylfuran. $C_6H_6O_3$; mol wt 126.11. C 57.14%, H 4.80%, O 38.06%. Prepn from the fructose portion of the sugar molecule in 57% yield: Haworth, Jones, *J. Chem. Soc.* **1944,** 667; Haworth, Wiggins, **Brit.** pats. 591,858 (1947) and 600,871 (1948). Improved process: Garber, Jones, **U.S. pat.** 2,929,823 (1960 to Merck & Co.). Purification: Jones, Lange, **U.S. pat.** 2,994,645 (1961 to Merck & Co.). Outline of process using cornstarch, glucose and sucrose as raw materials: Medwick, *Chem. Eng. News* (Sept. 11, 1961) 75. Prepn from molasses: Jones, Lange, **U.S.** pat. 3,066,150 (1962 to Merck & Co.); Hales *et al.,* **U.S.** pat. 3,071,599 (1963 to Atlas Chem.).

Needles from ether + petr ether, mp 31.5°. Avoid contact with eyes. Produces harmless yellow stains on skin. Odor of chamomile flowers. Very slightly volatile with steam (as compared with furfural). d_4^0 1.2062. $bp_{0.02}$ 110°. n_D^{18} 1.5627. uv max: 283 nm. Heat of combustion: 665 kcal/mol. Freely sol in water, methanol, ethanol, acetone, ethyl acetate, dimethylformamide. Sol in ether, benzene, chloroform. Less sol in carbon tetrachloride. Sparingly sol in petr ether. *Keep protected from light and air.*

USE: In the synthesis of dialdehydes, glycols, ethers, aminoalcohols, acetals. Aq acid catalyzes ring opening.

4765. *N*-(Hydroxymethyl)nicotinamide. *N*-(*Hydroxymethyl)-3-pyridinecarboxamide;* 3-pyridinecarboxylic acid hydroxymethylamide; pyridine-3-carboxylic acid *N*-methylolamide; Bilamid(e); Felosan; Nikoform; Choligen. $C_7H_8N_2O_2$; mol wt 152.15. C 55.25%, H 5.30%, N 18.41%, O 21.03%. Prepn: Graf, *J. Prakt. Chem.* **138,** 292 (1933); Chechelska, Urbanski, *Rocz. Chem.* **27,** 396 (1953), *C.A.* **49,** 1033f (1955).

Crystals, mp 141-142°. Freely sol in hot alcohol and water; sparingly sol in cold alcohol and water.

Picrate, $C_7H_8N_2O_2.C_6H_3N_3O_7$, crystals, mp 140-142°.

Hydrochloride, $C_7H_8N_2O_2.HCl$, crystals, mp above 120° (dec).

THERAP CAT: Cholagogue.

4766. 1-Hydroxy-2-naphthoic Acid. 1-Naphthol-2-carboxylic acid. $C_{11}H_8O_3$; mol wt 188.17. C 70.21%, H 4.29%, O 25.51%. Obtained by the action of CO_2 on the sodium salt of α-naphthol under pressure at 120-140°.

White to reddish crystals. mp 191-192°. Almost insol in cold water; freely sol in alcohol, benzene, ether, alkalies.

4767. 3-Hydroxy-2-naphthoic Acid. *3-Hydroxy-2-naphthalenecarboxylic acid;* 3-naphthol-2-carboxylic acid. $C_{11}H_8O_3$; mol wt 188.17. C 70.21%, H 4.29%, O 25.51%. Obtained by the action of CO_2 on the sodium salt of β-naphthol under pressure at 280-290°: Schmitt, Burkard, *Ber.* **20**, 2702 (1887); **Ger.** pat. **50,341**; *Frdl.* **2**, 133; **Ger.** pat. **436,034**; *Chem. Zentr.* **1926**, I, 1717; *Frdl.* **15**, 295; **Ger.** pat. **436,524**; *Chem. Zentr.* **1927**, I, 182; *Frdl.* **15**, 298; Schwenk, *Chem. Ztg.* **53**, 335 (1929).

Very pale yellow crystals, mp 222-223°. Practically insol in cold water. Slightly sol in hot water. Sol in benzene, chloroform. Freely sol in alc, ether. Sol in alkaline solns.

Magnesium salt, *Regacholyl.*

USE: The water-sol sodium salt has been used to solubilize riboflavin: Arnold *et al., Poultry Sci.* **31**, 350 (1952).

4768. 4-Hydroxy-19-nortestosterone. *4,17β-Dihydroxyestr-4-en-3-one;* 19-norandrost-4-ene-4,17β-diol-3-one. $C_{18}H_{26}O_3$; mol wt 290.39. C 74.44%, H 9.03%, O 16.53%. Prepn: Camerino *et al.*, U.S. pats. **2,999,870** and **3,145,201** (1961, 1964, both to Farmitalia).

Solid, mp 188-190°. uv max: 278 nm (11,600).

17-Cyclopentanepropionate, $C_{26}H_{38}O_4$, 17β-(2-cyclopentylpropionyloxy)-4-hydroxyestr-4-en-3-one, *Steranabol longacting, Steranabol ritardo.* Prepn: Camerino, Patelli, **Brit.** pat. **879,622** and **U.S.** pat. **3,020,295** (1961, 1962, both to Farmitalia). Crystalline powder, mp 158-160°. Practically insol in water, hexane. Slightly sol in methanol; sol in chloroform, dioxane, benzene. $[\alpha]_D^{20}$ +30° (c = 1 in chloroform). uv max (alc): 276 nm ($E_{1cm}^{1\%}$ 315).

THERAP CAT: The 17-cyclopentanepropionate as anabolic.

4769. Hydroxypethidine. *4-(3-Hydroxyphenyl)-1-methyl-4-piperidinecarboxylic acid ethyl ester;* 4-(m-hydroxyphenyl)-1-methylisonipecotic acid ethyl ester; ethyl 4-(m-hydroxyphenyl)-1-methylisonipecotate; demidone; oxipethidine; oxypetidin; Hoechst 10446; Win 771; Bemidone; Biphenal.

$C_{15}H_{21}NO_3$; mol wt 263.33. C 68.41%, H 8.04%, N 5.32%, O 18.23%. Prepn: Morrison, Rinderknecht, *J. Chem. Soc.* **1950**, 1467; Eisleb, **Ger.** pat. **752,755** (1952 to I. G. Farben), *C.A.* **52**, 7361e (1958).

Crystals from ethanol, mp 110°.

Hydrochloride, $C_{15}H_{21}NO_3.HCl$, crystals, mp 173-174°. Soluble in water; slightly sol in alc.

Caution: May be habit forming. This is a controlled substance (opiate) listed in the U.S. Code of Federal Regulations, Title 21, Part 1308.11 (1985).

THERAP CAT: Narcotic analgesic.

4770. Hydroxyphenamate. *2-Phenyl-1,2-butanediol 1-carbamate;* carbamic acid β-ethyl-β-hydroxyphenethyl ester; β-ethyl-β-hydroxyphenethyl carbamic acid ester; β-ethyl-β-hydroxyphenethyl carbamate; 2-hydroxy-2-phenylbutyl carbamate; Al 0361; Listica. $C_{11}H_{15}NO_3$; mol wt 209.24. C 63.14%, H 7.23%, N 6.69%, O 22.94%. Prepd from β-ethyl-β-hydroxyphenethyl alcohol and ethyl chloroformate followed by reaction with ammonia: Sifferd, Braitberg, U.S. pat. **3,066,164** (1962 to Armour-Pharm.). Pharmacology and toxicology: Bastian, Clements, *Dis. Nerv. Sys.* **22**, 9 (1961).

Crystals, mp 55-56.5°. Soly in water at 25°: 2.5% w/v. LD_{50} orally in mice: 830 mg/kg.

THERAP CAT: Anxiolytic.

4771. N-(4-Hydroxyphenyl)glycine. *p*-Hydroxyphenylaminoacetic acid; *p*-hydroxyanilinoacetic acid; photoglycine; Glycin; Iconyl; Monazol. $C_8H_9NO_3$; mol wt 167.16. C 57.48%, H 5.43%, N 8.38%, O 28.71%. Prepd from *p*-aminophenol and chloracetic acid: Vater, *J. Prakt. Chem.* **29**, 291 (1884); Meldola *et al., J. Chem. Soc.* **111**, 552 (1917); Galatis, *Helv. Chim. Acta* **4**, 576 (1921).

Shiny leaflets from water, browns at 200°, begins to melt at 220°, completely melted at 245-247° (decompn). Sparingly sol in water, alcohol, acetone, ether, chloroform, ethyl acetate, benzene, glacial acetic acid. Sol in alkalies and mineral acids. Freely sol in warm 20% hydrochloric acid.

USE: Photographic developer. In determination of iron; detection and determination of phosphorus and silicon. Acid indicator in bacteriology.

4772. Hydroxyprocaine. *Diethylaminoethyl p-aminosalicylate;* Oxycaine; Oxyprocain. $C_{13}H_{20}N_2O_3$; mol wt 252.31. C 61.88%, H 7.99%, N 11.10%, O 19.02%. Prepd by the addition of diethylaminoethanol to an H_2SO_4 suspension of *p*-aminosalicylic acid: Grimme, Schmitz, *Ber.* **84**, 734 (1951); Keil, Rademacher, *Arzneimittel-Forsch.* **1**, 154, 218 (1951); Grimme *et al., ibid.* 326; *cf.* **Swiss** pat. **270,986** (1948); *Chem. Zentr.* **1951**, II, 102.

Oily liquid. Soluble in chloroform.

Hydrochloride, $C_{13}H_{20}N_2O_3 \cdot HCl$, prisms from ethanol, mp 154°. Soluble in water.

Penicillin-G salt, $C_{13}H_{20}N_2O_3 \cdot C_{16}H_{18}N_2O_4S$, dec 112-113°. Sparingly sol in water (7.5 g/l).

THERAP CAT: Local anesthetic.

4773. 17α-Hydroxyprogesterone. *17-Hydroxypregn-4-ene-3,20-dione;* 4-pregnen-17α-ol-3,20-dione; Gestageno; Prodox. $C_{21}H_{30}O_3$; mol wt 330.45. C 76.32%, H 9.15%, O 14.52%. Isoln from adrenal glands: Pfiffner, North, *J. Biol. Chem.* **132**, 459 (1940); **139**, 855 (1941); von Euw, Reichstein, *Helv. Chim. Acta* **24**, 879 (1941). Prepn: Julian *et al.*, U.S. pat. 2,648,662 (1953 to Glidden); Ringold *et al.*; Stork *et al.*, U.S. pats. 2,802,839 and 2,805,203 (both 1957 to Syntex); Chemerda *et al.*; Cutler, Chemerda; Dulaney, McAleer; U.S. pats. 2,777,843; 2,786,856/7; 2,813,060 (all 1957 to Merck & Co.); Cutler *et al.*, *J. Org. Chem.* **24**, 1629 (1959); Pederson, U.S. pat. 3,000,883 (1961 to Upjohn).

Rhombic or hexagonal leaflets from acetone or alcohol, mp 222-223° (rapid heating). With slow heating the substance undergoes molecular rearrangement accompanied by partial resolidification and becomes completely molten only at 276°. $[\alpha]_D^{17}$ +105.6° (c = 1.0417 in chloroform).

Acetate, $C_{23}H_{32}O_4$, *17α-acetoxyprogesterone.* Crystals from chloroform + methanol, mp 239-240°. uv max: 240 nm (log ε 4.33). *Ref:* Stork *et al., loc. cit.*

THERAP CAT: Progestogen.

THERAP CAT (VET): Estrus regulator.

4774. 17α-Hydroxyprogesterone Caproate. *17-[(1-Oxohexyl)oxy]pregn-4-ene-3,20-dione; 17-hydroxypregn-4-ene-3,20-dione hexanoate;* 17α-hydroxyprogesterone hexanoate; Delalutin; Hyproval P.A.; Lentogest; Pharlon; Proge; Proluton Depot; Teralutil. $C_{27}H_{40}O_4$; mol wt 428.59. C 75.66%, H 9.41%, O 14.93%. Prepn: Kaspar *et al.*, U.S. pat. 2,753,360 (1956 to Schering AG). Comprehensive description: K. Florey, Ed. in *Analytical Profiles of Drug Substances* vol. 4 (Academic Press, New York, 1975) pp 209-224.

Dense needles from isopropyl ether or methanol, mp 119-121°. $[\alpha]_D^{20}$ +61° (c = 1 in chloroform). Soly (mg/ml): sesame oil 25-29; levulinic acid butyl ester 350-400.

THERAP CAT: Progestogen.

4775. 4-Hydroxy-L-proline. Hyp; L_s-hydroxyproline; *trans*-hydroxyproline; 4-hydroxy-2-pyrrolidinecarboxylic acid. $C_5H_9NO_3$; mol wt 131.13. C 45.79%, H 6.92%, N 10.68%, O 36.60%. An amino acid classified as nonessential with respect to its growth effect in rats. Constituent of collagen, *q.v.* Isoln from gelatin hydrolyzates: E. Fischer, *Ber.* **35**, 2660 (1902); Klabunde, *J. Biol. Chem.* **90**, 293 (1931).

Synthesis: Leuchs, *Ber.* **38**, 1937 (1905); R. Gaudry, C. Godin, *J. Am. Chem. Soc.* **76**, 139 (1954); C. Eguchi, A. Kakuta, *Bull. Chem. Soc. Japan* **47**, 1704 (1974); S. G. Ramaswamy, E. Adams, *J. Org. Chem.* **42**, 3440 (1977). Flow sheets of four different syntheses: *Chem. & Eng. News* **40**, 40 (Nov. 12, 1962). Structure based on crystallographic data: Zussman, *Acta Cryst.* **4**, 72 (1951); Donohue, Trueblood, *ibid.* **5**, 414 (1952). Stereochemistry: Hudson, Neuberger, *J. Org. Chem.* **15**, 24 (1950). In plant glycoproteins: D. Ashford, A. Neuberger, *Trends in Biochem. Sci.* **5**, 245 (1980). Isoln of *cis*-form from *Santalum album* L.: A. N. Radhakrishnan, K. V. Giri, *Biochem. J.* **58**, 57 (1954). Detection of *cis*- and *trans* isomers in collagen hydrolysates: G. Bellon *et al.*, *Anal. Biochem.* **137**, 151 (1984). Review of metabolism: E. Adams, L. Frank, *Ann. Rev. Biochem.* **49**, 1005-1061 (1980).

Rhombs or needles from water, mp 274°. $[\alpha]_D$ −76.5° (c = 2.5 in water). pK_1' 1.82; pK_2' 9.65. Soly in water at 0°: 288.6 g/l; at 25°: 361.1 g/l; at 50°: 451.8 g/l; at 65°: 516.7 g/l. Very slightly sol in alcohol; insol in ether.

cis-Form, *allohydroxyproline.* mp 238-241°. $[\alpha]_D^{18}$ −58.1° (c = 5.2 in water).

4776. Hydroxypropyl Cellulose. *Cellulose 2-hydroxypropyl ether;* oxypropylated cellulose; Klucel; Lacrisert. Nonionic water soluble ether of cellulose, *q.v.* that produces solns having a wide range of viscosity (200-2500 cp). Prepn: **Neth. pat. Appl. 6,401,036;** E. D. Klug, U.S. pats. **3,278,-520, 3,278,521** (1964, 1966, 1966 all to Hercules). Use in the treatment of dry eye syndrome: T. P. Werblin *et al.*, *Ophthalmology* **88**, 78 (1981); P. Huguet *et al.*, *Bull. Soc. Ophthalmol. Fr.* **81**, 1173 (1981). Review of chemistry, physical properties and uses: E. D. Klug in *Encyclopedia of Polymer Science and Technology* vol. 15 (Interscience, New York, 1971) pp 307-314; A. J. Desmarais, *Industrial Gums,* R. L. Whistler, Ed. (Academic Press, New York, 2nd ed., 1973) pp 649-672.

Off-white powder, softens at 130°. Sol in many polar organic solvents. Ppts from water at 40-45°. Thermoplastic.

USE: As emulsifier, stabilizer, whipping aid, protective colloid, film former or thickener in foods; as binder in ceramics and glazes; in hair and cosmetic prepns; in vacuum-formed containers and blow-molded bottles; as suspending agent in PVC polymerization. Pharmaceutic aid (tablet coating agent).

THERAP CAT: Protectant (topical).

4777. Hydroxypropyl Methylcellulose. *Cellulose 2-hydroxypropyl methyl ether;* hypromellose; Gonak; Goniosol; Lacril; Tearisol; Methocel HG; Ultra Tears. Non-ionic water soluble ether of methylcellulose, *q.v.* that produces solns having a wide range of viscosity (400-15,000 cp). Prepn: A. B. Savage, U.S. pat. **2,949,252** (1960 to Dow). Review of chemistry, physical properties and use: *idem, Encyclopedia of Polymer Science and Technology* vol. 3 (Interscience, New York, 1965); pp 496-511; G. K. Greminger, A. B. Savage, *Industrial Gums,* R. L. Whistler, Ed. (Academic Press, New York, 1973) pp 619-647.

Powder. Dissolves slowly in cold water. Insol in hot water. Sol in most polar organics. Has thermogelling properties. Has higher salt tolerance and is more sol than methylcellulose.

USE: As emulsifier, film former, protective colloid, stabilizer, suspending agent, or thickener in foods. Pharmaceutic aid (suspending agent; tablet excipient; demulcent; viscosity increasing agent); ophthalmic lubricant. In adhesives, asphalt emulsions, caulking compounds, tile mortars, plastic mixes, cements, paints. As sticker for agricultural sprays and dusts.

4778. 8-Hydroxyquinoline. *8-Quinolinol;* oxyquinoline; hydroxybenzopyridine; oxybenzopyridine; phenopyridine; oxychinolin; oxine; Bioquin; Quinophenol. C_9H_7NO; mol wt 145.15. C 74.47%, H 4.86%, N 9.65%, O 11.02%. Prepn from o-aminophenol, glycerol and H_2SO_4: Z. H. Skraup, *Monatsh.* **1**, 316 (1880); **3**, 536 (1882); R. H. F. Manske *et al., Can. J. Res.* **27F**, 359 (1949). *Review:* J. P. Phillips, *Chem. Rev.* **56**, 271-297 (1956). Book: R. G. W. Hollingshead, *Oxine and Its Derivatives,* **I-IV** (Butterworth, London, 1954/56).

White crystals or cryst powder. mp 76°. bp ~267°. Almost insol in water, ether; freely sol in alc, acetone, chloroform, benzene, aq mineral acids. LD_{50} i.p. in mice: 48 mg/kg, Bernstein *et al., Toxicol. Appl. Pharmacol.* **5**, 599 (1963).

USE: As fungistat; chelating agent in determn of trace metal ions.

THERAP CAT: Disinfectant.

4779. 8-Hydroxyquinoline Sulfate. *8-Quinolinol sulfate;* oxyquinoline sulfate; oxine sulfate; 8-hydroxyquinoline sulfuric acid salt; Quinosol; Chinosol. $C_{18}H_{16}N_2O_6S$; mol wt 388.40. C 55.66%, H 4.15%, N 7.21%, O 24.72%, S 8.25%. $(C_9H_7NO)_2 \cdot H_2SO_4$.

Pale yellow, cryst powder; slight saffron odor; burning taste. mp 175-178°. Freely sol in water; sol in about 100 parts glycerol, slightly in alcohol; insol in ether.

Aluminum salt, $C_{27}H_{24}AlN_3O_{15}S_3$, *Nyxolan, Aloxyn.*

USE: Antiseptic, antiperspirant, deodorant.

THERAP CAT: Topical antiseptic, disinfectant.

4780. 8-Hydroxy-5-quinolinesulfonic Acid. $C_9H_7NO_4S$; mol wt 225.22. C 47.99%, H 3.13%, N 6.22%, O 28.42%, S 14.24%. Prepn: K. Matsumura, *J. Am. Chem. Soc.* **49**, 810 (1927); N. K. Chawla, M. M. Jones, *Inorg. Chem.* **3**, 1549 (1964).

Pale yellow, needle-like crystals or cryst powder; odorless. mp 322-324°. Freely sol in water, slightly in organic solvents.

USE: In determn of trace metal ions.

4781. Hydroxystilbamidine. *4-[2-[4-(Aminoiminomethyl)phenyl]ethenyl]-3-hydroxybenzenecarboximidamide;* 2-hydroxy-4,4'-stilbenedicarboxamidine; 2-hydroxy-4,4'-diamidinostilbene; 2-hydroxy-4,4'-diguanylstilbene; 2-hydroxystilbamide. $C_{16}H_{16}N_4O$; mol wt 280.33. C 68.55%, H 5.75%, N 19.99%, O 5.71%. Prepn: J. N. Ashley, J. O. Harris, *J. Chem. Soc.* **1946**, 567; A. J. Ewins *et al.,* **Brit.** pat. **574,486**; A. J. Ewins, U.S. pat. **2,510,047** (1946, 1950 both to May & Baker). Organ and tissue distribution in animals: I. Snapper *et al., Cancer* **4**, 1246 (1951). Pharmacology and antiprotozoal activity: I. Snapper *et al., Trans. N.Y. Acad. Sci.* **14**, 269 (1952). Probe for studying nucleic acid conformation: B. Festy, *C.R. Acad. Sci. Ser. D* **266**, 1433 (1968); B. Festy, M. Daune, *Biochemstry* **12**, 4827 (1973); B. Festy *et al., Biochim. Biophys. Acta* **407**, 24 (1975). Crystal structure: C. Courseille *et al., C.R. Acad. Sci. Ser. C* **274**, 1921 (1972). Use as a fluorochrome for selective staining of nuclei: L. B. Murgatroyd, *Histochemstry* **74**, 107 (1982). *Review:* B. Festy in *Antibiotics* vol. 5, pt. 2, F. E. Hahn, Ed. (Springer-Verlag, New York, 1979) pp 223-235.

Yellow microcrystals from nitrobenzene, mp 235°. LD_{50} in mice (mg/g); 0.027 i.v.; 0.14 s.c. (Ewins, 1950).

Isethionate, $C_{20}H_{28}N_4O_9S_2$, yellow crystals, discolored by light, mp 286° (dec). Freely sol in water. Soly in alcohol ~1.0 g/100 ml. Practically insol in ether. pH of a 1% aq soln 3.3 to 5.3. Solns show strong yellow fluorescence under uv light. Solns should be freshly prepared. Although the hydroxy compd is more stable in soln, such solns should not be stored and must not be used if cloudy.

THERAP CAT: Antiprotozoal (Leishmania).

4782. Hydroxystreptomycin. Reticulin (the antibiotic). $C_{21}H_{39}N_7O_{13}$; mol wt 597.60. C 42.21%, H 6.58%, N 16.41%, O 34.81%. Antibiotic substance produced by *Streptomyces reticuli:* Hosoya *et al., Japan. J. Exp. Med.* **20**, 327 (1949), *C.A.* **45**, 3459i (1951); by *S. griseocarneus* (from a Japanese soil): Stodola *et al., J. Am. Chem. Soc.* **73**, 2290 (1951); Benedict, Stodola, U.S. pat. **2,617,755** (1952 to U.S. Secy. Agr.); by *Streptomyces* NA 232-M1: Grundy *et al., Antibiot. & Chemother.* **1**, 309 (1951); by *S. subrutilus:* Arai *et al., J. Antibiot.* **17A**, 23 (1964). Identity of reticulin (the antibiotic) and hydroxystreptomycin: Hosoya *et al., Japan. J. Exp. Med.* **22**, 303 (1952), CA. **48**, 3477a (1954). Toxicity: Ambrose, *Proc. Soc. Exp. Biol. Med.* **76**, 466 (1951).

Trihydrochloride, $C_{21}H_{42}Cl_3N_7O_{13}$. The physical characteristics approx those of streptomycin. The specific rotation in water is 91° under conditions which give 86.1° for streptomycin trihydrochloride. Hydroxystreptomycin trihydrochloride, when assayed against *Bacillus subtilis,* was found to be equiv to 784 μg of streptomycin base/mg. The corresponding value of streptomycin is 842 μg/mg. LD_{50} s.c. in mice: 865 mg/kg (Ambrose).

4783. Hydroxytetracaine. *4-Butylamino-2-hydroxybenzoic acid 2-dimethylaminoethyl ester; p-butylaminosalicylic acid 2-dimethylaminoethyl ester;* 2-dimethylaminoethyl p-butylaminosalicylate; hydroxamethocaine; Rhenocain; Salicain. $C_{15}H_{24}N_2O_3$; mol wt 280.36. C 64.26%, H 8.63%, N 9.99%, O 17.12%. Prepn: **Brit.** pats. **736,960** (1955) and **760,003** (1956 to Rheinpreussen AG); Grimme, Schmitz, *Ber.* **84**, 734 (1951).

Hydrochloride, $C_{15}H_{24}N_2O_3 \cdot HCl$, crystals from water, mp 157°. Soly in water at 20°: about 4%.

Hemihydrate, prisms from ligroin, mp 48°.

THERAP CAT: Topical anesthetic.

4784. 5-Hydroxytryptophan. 5-HTP. $C_{11}H_{12}N_2O_3$; mol wt 220.22. C 59.99%, H 5.49%, N 12.72%, O 21.80%. Precursor of serotonin. Synthesis from 5-benzyloxyindole: Ek, Witkop, *J. Am. Chem. Soc.* **76,** 5579 (1954); Shaw, Morris, *Biochem. Prepns.* **9,** 92 (1962); from 5-benzyloxytryptophan: Frangatos, Chubb, *Can. J. Chem.* **37,** 1374 (1959); Frangatos, **Can.** pat. **619,472** (1961 to Frank W. Horner); Ash, **Brit.** pat. **845,034** (1960 to May & Baker); from tryptophan: Renson *et al., Biochem. Biophys. Res. Commun.* **6,** 20 (1961). Prepn of 5-hydroxy-L and D-tryptophan: A. J. Morris, M. D. Armstrong, *J. Org. Chem.* **22,** 306 (1957). Crystal and molecular structure of DL-form: Wakahara *et al., Tetrahedron Letters* **1970,** 3003. Use of L-5HTP in treatment of myoclonus, a neuromuscular disease: M. H. Van Woert, D. Rosenbaum, *Adv. Neurol.* **26,** 107 (1979); L. J. Thal *et al., Ann. Neurol.* **7,** 570 (1980). Orphan drug under development by Bolar. *Review:* M. H. Van Woert, *Orphan Drugs,* F. E. Karch, Ed. (Marcel Dekker, New York, 1982) pp 13-31.

DL-Form, **Prétonine.** Minute rods or needles from ethanol, dec 298-300°. uv max (H_2O at pH 6.0): 278 nm. Soly in water at 5°: 1.0 g/100 ml; at 100°: 5.5 g/100 ml. Soly in 50% boiling alc: 2.5 g/100 ml. Aq solns are stable at low pH.

L-Form, **oxitriptan, L-5HTP, Levothym, Quietim, Tript-Oh.** Crystals, $[\alpha]_D^{20}$ −32.5° (H_2O); $[\alpha]_D^{20}$ +16.0° (4N HCl).

D-Form, crystals, $[\alpha]_D^{20}$ +32.2° (H_2O).

THERAP CAT: L-Form as antidepressant; antiepileptic.

4785. Hydroxyurea. Hydroxycarbamide; Hydrea; Litalir. $CH_4N_2O_2$; mol wt 76.06. C 15.79%, H 5.30%, N 36.84%, O 42.07%. $H_2NCONHOH$. Prepn from hydroxylamine HCl and KCN: Hantzsch, *Ann.* **299,** 99 (1898). Alternate route: Graham, **U.S.** pat. **2,705,727** (1955 to du Pont).

Needles from alc, mp 133-136°. Freely sol in water, hot alcohol.

THERAP CAT: Antineoplastic.

4786. Hydroxyzine. *2-[2-[4-[(4-Chlorophenyl)phenylmethyl]-1-piperazinyl]ethoxy]ethanol; 1-(p-chloro-α-phenylbenzyl)-4-(2-hydroxyethoxyethyl)piperazine;* 1-(p-chlorodiphenylmethyl)-4-[2-(2-hydroxyethoxy)ethyl]piperazine; *N*-(4-chlorobenzhydryl)-*N*'-(hydroxyethyloxyethyl)piperazine; 1-(p-chlorobenzhydryl)-4-[2-(2-hydroxyethoxy)ethyl]-diethylenediamine; UCB 4492; Tran-Q; Tranquizine. $C_{21}H_{27}ClN_2O_2$; mol wt 374.92. C 67.28%, H 7.26%, Cl 9.46%, N 7.47%, O 8.54%. H_1 receptor antagonist. Outline of commercial prepn: *Chem. Week* **79**(5), 70 (Aug. 4, 1956); Morren, **U.S.** pat. **2,899,436** (1959 to UCB). Pharmacology and metabolism: Cannizaro, *Boll. Chim. Farm.* **104,** 39 (1965); Close *et al., Ind. Chim. Belge* **33,** 94 (1968); *eidem, Proc. Eur. Soc. Study Drug Toxicity* **9,** 144 (1968); S. F. Pong, C. L. Huang, *J. Pharm. Sci.* **63,** 1527 (1974). Pharmacokinetics and antihistaminic activity: F. E. R. Simons *et al., J. Allergy Clin. Immunol.* **73,** 69 (1984); S. Ting *et al., ibid.* **75,** 63 (1985). Clinical trials of efficacy in allergic rhinitis: L. Wong *et al., ibid.* **67,** 223 (1981); in urticaria, R. P. Harvey *et al., ibid.* **68,** 262 (1981); as anti-emetic: R. McKenzie *et al., Anesth. Analg.* **60,** 783 (1981); as pre-surgical sedative: G. Wallace, L. J. Mindlin, *ibid.* **63,** 571 (1984). Toxicity data: E. I. Goldenthal, *Toxicol. Appl. Pharmacol.* **18,** 185 (1971). Comprehensive description: J. Tsau, N. DeAngelis in *Analytical Profiles of Drug Substances* Vol. 7, K. Florey, Ed. (Academic Press, New York, 1978) pp 319-341.

Dihydrochloride, $C_{21}H_{29}Cl_3N_2O_2$, *Alamon, Atarax, Aterax, Durrax, Orgatrax, Quiess, Vistaril Parenteral.* Crystals, mp 193°. Bitter taste. Soly in mg/ml: water < 700; chloroform 60; acetone 2; ether < 0.1. Solns are unstable to intense uv light. LD_{50} in rats (mg/kg): 126 i.p.; 950 orally (Goldenthal).

Pamoate, $C_{44}H_{43}ClN_2O_8$, *Equipose, Masmoran, Paxistil, Vistaril Pamoate.* Crystals. Practically insol in water.

THERAP CAT: Anxiolytic. Antihistaminic.

THERAP CAT (VET): Has been used as a tranquilizer.

4787. Hyenanchin. *Hexahydro-1b,6,8-trihydroxy-6a-methyl-8-(1-methylethenyl)spiro[2,5-methano-7H-oxireno-[3,4]cyclopent[1,2-d]oxepin-7,2'-oxiran]-3(2H)-one;* hyaenanchin; hyenancin; mellitoxin. $C_{15}H_{18}O_7$; mol wt 310.29. C 58.06%, H 5.85%, O 36.09%. Isolated from fruit of *Hyaenanche globosa* Lamb., *Euphorbiaceae:* Henry, *J. Chem. Soc.* **117,** 1619 (1920). Structure and identity with mellitoxin: Jommi *et al., Chim. Ind. (Milan)* **46,** 549 (1964), *C.A.* **61,** 5697 (1964).

Crystals, mp 225-235°. Soly in water at 15°: 1.18%; more sol in hot water; sparingly sol in alcohol, acetone, ethyl acetate. $[\alpha]_D^{15}$ +14.7° (water).

4788. Hygrine. *(R)-1-(1-Methyl-2-pyrrolidinyl)-2-propanone;* 2-acetonyl-1-methylpyrrolidine; *N*-methyl-2-acetonylpyrrolidine. $C_8H_{15}NO$; mol wt 141.21. C 68.04%, H 10.71%, N 9.92%, O 11.33%. Occurs in leaves of *Erythroxylon coca* Lam., *Erythroxylaceae* of diverse origin: Liebermann, *Ber.* **22,** 677 (1889). Synthesis: Galinovsky *et al., Monatsh.* **82,** 551 (1951); Lukes *et al., Coll. Czech. Chem. Commun.* **24,** 2433 (1959); Leonard, Cook, *J. Am. Chem. Soc.* **81,** 5627 (1959). Enzymatic synthesis: Tuppy, Faltaous, *Monatsh.* **91,** 167 (1960). Stereochemistry: Galinovsky *et al., ibid.* **84,** 798 (1953). Absolute configuration: Lukes *et al., Coll. Czech. Chem. Commun.* **25,** 483 (1960).

Liquid. bp_{11} 76.5°; bp_{14} 81°. n_D^{20} 1.4555. Sol in alcohol, chloroform, dil acids; slightly sol in water.

Picrate, $C_{14}H_{18}N_4O_8$, crystals from alc, mp 149-151°.

Oxime, $C_8H_{16}N_2O$, crystals from ether, mp 123-124°.

Styphnate, $C_{14}H_{18}N_4O_9$, crystals from ethanol, mp 137°.

Reineckate, $C_{12}H_{21}CrN_7OS_4$, needles from methanol, mp 249-251°.

4789. Hygromycin. *5-Deoxy-5-[[3-[4-[(6-deoxy-β-D-arabino-hexofuranos-5-ulos-1-yl)oxy]-3-hydroxyphenyl]-2-methyl-1-oxo-2-propenyl]amino]-1,2-O-methylene-D-neo-inositol;* homomycin; hygromycin A; 1703-18B; St-4331. $C_{23}H_{29}NO_{12}$; mol wt 511.47. C 54.01%, H 5.71%, N 2.74%, O 37.54%. Antibiotic substance produced by *Streptomyces hygroscopicus* (Jensen) Waksman & Henrici, from forest soil near Indianapolis, Ind: R. L. Mann *et al., Antibiot. &*

Chemother. **3**, 1279 (1953). Produced also by *Streptomyces noboritoensis* from soil collected in Kanawaga Prefecture. Identity of homomycin and hygromycin: K. Isono *et al., J. Antibiot.* **10A**, 21 (1957). Structure: *eidem, ibid.* 160; R. L. Mann, D. O. Woolf, *J. Am. Chem. Soc.* **79**, 120 (1957). Abs config: K. Kakinuma, Y. Sakagami, *Agr. Biol. Chem.* **42**, 279 (1978). Synthesis of the sugar component: M. Nakajima, S. Takahashi, *ibid.* **31**, 1082 (1967). Total synthesis: N. Chida *et al., Chem. Commun.* **1989**, 436. Peptide synthesis inhibitor: M. Guerrero, J. Modolell, *Eur. J. Biochem.* **107**, 409 (1980).

Amorphous white powder. mp 105-109°, dec above 160°. Also reported as mp 110-112° (dec) (Chida). Weakly acidic. pKa 8.9. $[\alpha]_D^{25}$ −126° (c = 1 in water). Freely sol in water, alc. Practically insol in the less polar solvents. uv max (dil HCl): 272, 214 nm ($E_{1cm}^{1\%}$ 306, 416).

4790. Hygromycin B. *O-6-Amino-6-deoxy-L-glycero-D-galacto-heptopyranosylidene-(1 → 2-3)-O-β-D-talopyranosyl-(1 → 5)-2-deoxy-N³-methyl-D-streptamine;* Hygromix. C_{20}-$H_{37}N_3O_{13}$; mol wt 527.54. C 45.54%, H 7.07%, N 7.96%, O 39.43%. Antibiotic substance produced by *Streptomyces hygroscopicus* (Jensen) Waksman & Henrici, from soil samples from Nebraska and Indiana: Mann, Bromer, *J. Am. Chem. Soc.* **80**, 2714 (1958). Prodn: McGuire, Mann, U.S. pat. **3,018,220** (1962 to Lilly). Structure: Neuss *et al., Helv. Chim. Acta* **53**, 2314 (1970). Studies on mode of action: A. Gonzalez *et al., Biochim. Biophys. Acta* **521**, 459 (1978).

Amorphous powder, dec 160-180°. Weakly basic. pKa 7.1, 8.8. $[\alpha]_D^{20}$ +20.2° (c = 1 in water). Freely sol in water, methanol and ethanol. Practically insol in less polar solvents. Absorption spectrum: Mann, Bromer, *loc. cit.*

THERAP CAT (VET): Anthelmintic (swine and chicken).

4791. Hygrophylline. *1,2-Dihydro-12,14α-dihydroxysenecionan-11,16-dione.* $C_{18}H_{27}NO_6$; mol wt 353.40. C 61.17%, H 7.70%, N 3.96%, O 27.16%. Pyrrolizidine alkaloid isolated from *Senecio hygrophylus* Dyer and Sm., *Compositae:* Richardson, Warren, *J. Chem. Soc.* **1943**, 452. Structure and stereochemistry: Schlosser, Warren, *ibid.* **1965**, 5707. NMR spectrum: S. E. Drewes *et al., J. Chem. Soc. Perkin Trans. I* **1981**, 287.

Prisms from acetone, mp 173-174°. $[\alpha]_D^{20}$ −67.3° (c = 2.9 in ethanol).

4792. Hymecromone. *7-Hydroxy-4-methyl-2H-1-benzopyran-2-one; 7-hydroxy-4-methylcoumarin;* 7-hydroxy-4-methyl-2-oxo-3-chromene; 4-methylumbelliferone; β-methylumbelliferone; imecromone; 4-MU; Bilcolic; Bilicante; Cantabilin; Cantabiline; Cholonerton; Cholspasmin; Cumarote C; Eurogale; Himecol; Medilla; Mendiaxon. $C_{10}H_8O_3$; mol wt 176.16. C 68.18%, H 4.58%, O 27.25%. Prepd from resorcinol and ethyl acetoacetate: Pechmann, Duisberg, *Ber.* **16**, 2119 (1883); Russell, Frye, *Org. Syn.* **21**, 23 (1941); Woods, Sapp, *J. Org. Chem.* **27**, 3703 (1962). Use in determn of nitric acid: F. J. Welcher, *Organic Analytical Reagents* vol. 1 (Van Nostrand, New York, 1947) pp 214-215.

Crystals from alc, mp 194-195° (Woods, Sapp), formerly reported as 185-186° (Pechmann, Duisberg). uv max (methanol): 221, 251, 322.5 nm. Blue fluorescence in alcohol + water. Sol in methanol, glacial acetic acid; slightly sol in ether, chloroform. Practically insol in cold water.

USE: Analytical reagent.

THERAP CAT: Choleretic; antispasmodic.

4793. Hymecromone O,O-Diethyl Phosphorothioate. *7-Hydroxy-4-methylcoumarin O-ester with O,O-diethyl phosphorothioate; O,O-diethyl O-(4-methyl-7-coumarinyl) phosphorothioate; O,O-diethyl O-(4-methyl-7-coumarinyl) phosphothioate; O,O-diethyl O-(4-methylumbelliferone) phosphorothioate;* 4-methylumbelliferone *O,O-diethyl phosphorothioate;* E 838; Potasan. $C_{14}H_{17}O_5PS$; mol wt 328.32. C 51.21%, H 5.22%, O 24.37%, P 9.43%, S 9.77%. Prepn: Schrader, Kükenthal, U.S. pat. **2,583,744** (1952 to Bayer). *Review:* G. Schrader, *Die Entwicklung neuer insektizider Phosphorsäure Ester* (Verlag Chemie, 1963) p 187.

Long needles from petr ether. Weak aromatic odor. mp 38°. bp$_{1.0}$ 210° (dec). d$_4^{38}$ 1.260 (liq); n$_D^{37}$ 1.5685 (liq). Very sparingly sol in water. Slightly sol in petr ether. Sol in most other organic solvents. Aq solns of emulsions adjusted to pH 7 to 8 show a blue fluorescence. LD$_{50}$ i.p. in rats: 15 mg/kg.

USE: Systemic insecticide, especially effective against the Colorado beetle. *Caution:* Acts as cholinesterase inhibitor *in vivo;* symptoms similar to Parathion.

4794. Hyodeoxycholic Acid. *(3α,5β,6α)-3,6-Dihydroxycholan-24-oic acid;* 3α,6α-dihydroxy-5β-cholanic acid; α-hyodeoxycholic acid; hyodesoxycholic acid. $C_{24}H_{40}O_4$; mol wt 392.56. C 73.43%, H 10.27%, O 16.30%. Isoln from pig bile: Wieland, Gumlich, *Z. Physiol. Chem.* **215**, 18 (1933); Trickey, U.S. pat. **2,547,726** (1951 to U.S. Rubber); Fogle, U.S. pat. **2,745,849** (1956 to Armour); Buckley, Ziegler, U.S. pat. **2,758,120** (1956 to Canada Packers); Liebig, U.S. pat. **3,006,927** (1961 to Riedel-de Haen). Configuration: Moffett, Hoehn, *J. Am. Chem. Soc.* **69**, 1995 (1947).

Crystals from ethyl acetate, mp 196-197°. $[\alpha]_D^{20}$ +8° (alc). Moderately sol in ethanol, glacial acetic acid. Less sol in ether, acetone, ethyl acetate, benzene.

4795. Hyoscyamine. *α-(Hydroxymethyl)benzeneacetic acid 8-methyl-8-azabicyclo[3.2.1]oct-3-yl ester; 1αH,5αH-tropan-3α-ol(−)-tropate; 3α-tropanyl S-(−)-tropate; l-tropic acid ester with tropine; l-tropine tropate; daturine; duboisine; l-hyoscyamine; Cystospaz; Levsin.* $C_{17}H_{23}NO_3$; mol wt 289.36. C 70.56%, H 8.01%, N 4.84%, O 16.59%. From *Hyoscyamus niger* L., *Atropa belladonna* L., *Datura stramonium* L., and other *Solanaceae:* Ladenburg, *Ann.* **206**, 274 (1881). Identity with duboisine and daturine: Beckurts, *Apoth. Ztg.* **27**, 683 (1912). Obtained by resolution of atropine: Werner, Miltenberger, *Ann.* **631**, 163 (1960). Prepd from (−)-acetyltropoyl chloride and atropine hydrochloride: Fodor *et al., Acta Chim. Acad. Sci. Hung.* **28**(4), 409 (1961), *C.A.* **61**, 1903g (1964). Configuration: Fodor, Csepreghy, *Tetrahedron Letters* no. 7, 16 (1959).

Silky, tetragonal needles from evaporating alc. *Keep well closed and protect from light and heat;* easily racemized. mp 108.5°. $[\alpha]_D^{20}$ −21.0° (alc). K at 19° is 1.9 × 10⁻¹². One gram dissolves in 281 ml water (pH 9.5), 69 ml ether, 150 ml benzene, 1 ml chloroform. Freely sol in alcohol, dil acids. Absorption spectra: Dobbie, Fox, *J. Chem. Soc.* **103**, 1194, 1195 (1913).

Hydrobromide, $C_{17}H_{23}NO_3$·HBr, deliquescent crystals, mp 152°. pH 5.4 (1 in 100). Levorotatory. Very sol in water. One gram dissolves in 3 ml alcohol, 1.2 ml chloroform, 2260 ml ether.

Hydrochloride, $C_{17}H_{23}NO_3$·HCl. Crystals. *Poisonous!* mp 149-151°. Freely sol in water, alcohol.

Methyl bromide, $C_{18}H_{26}BrNO_3$, *N-methylhyoscyaminium bromide.* Crystals, mp 210-212°. Freely sol in water, dil alc; slightly sol in abs alc.

Sulfate dihydrate, $C_{34}H_{48}N_2O_{10}S$·2H$_2$O, *Egacene, Egazil Duretter, Peptard.* Needles from alcohol, mp 206° when dry. $[\alpha]_D^{15}$ −29° (c = 2). pH 5.3 (1 in 100). One gram dissolves in 0.5 ml water, about 5.0 ml alcohol. Very slightly sol in chloroform, ether.

THERAP CAT: Anticholinergic.

4796. Hyoscyamus. Henbane; hog's bean; insane root; poison tobacco; black henbane. Dried leaves, with or without the tops, of *Hyoscyamus niger* L., *Solanaceae,* yielding 0.040% hyoscyamine alkaloids (chiefly hyoscyamine and scopolamine). The leaves and seeds have been used as smooth muscle relaxants. The alkaloid content of the seeds is less than that of the leaves. Egyptian henbane *(Hyoscyamus muticus, Solanaceae)* contains about 0.5% alkaloids. *Habit.* Europe, Africa, Asia, naturalized in U.S.; cultivated in England. *Constit.* Leaves: scopolamine, hyoscyamine, hyoscipicrin, choline.

4797. Hypalon®. Elastomer made by substituting chlorine and sulfonyl chloride groups into polyethylene. Prepd from polyethylene of average mol wt of about 20,000; contains approx one chlorine atom for each seven carbon atoms and one sulfonyl chloride group for every ninety carbon

atoms. Three types of Hypalon are available: 20, 30, and 40 with sp gr of 1.12, 1.28, and 1.18, respectively. Prepn: Reed, Horn, U.S. pat. **2,046,090** (1933). *Reviews:* Ullmann's *Encyclopädie der Technischen Chemie* **vol. 9**, 342 (1957); Keeley, "Hypalon Synthetic Rubber" in *Introduction to Rubber Technology,* Morton, Ed. (Reinhold, New York, 1959) p 349.

4798. Hypaphorine. *α-Carboxy-N,N,N-trimethyl-1H-indole-3-ethanaminium hydroxide inner salt.* $C_{14}H_{18}N_2O_2$; mol wt 246.30. C 68.27%, H 7.37%, N 11.37%, O 12.99%. The betaine of tryptophane. From seeds of *Erythrina americana* Mill., *E. sandwicensis* Degener, *E. crista-galli* L., and *Abrus precatorius* Linn., *Leguminosae:* Folkers, Koniuszy, *J. Am. Chem. Soc.* **61**, 1232 (1939); **62**, 1677 (1940); Deulofeu *et al., J. Chem. Soc.* **1939**, 1841; Tung, Lao, *C.A.* **55**, 17770i (1961). Structure and synthesis: von Romburgh, Barger, *J. Am. Chem. Soc.* **99**, 2068 (1911).

Crystals from dil alc, dec 237°; after purification through the nitrate, dec 255°. $[\alpha]_D^{27}$ +113° (c = 0.52). Sol in 12 parts water, slightly in alc; almost insol in other usual solvents. A dihydrate has been reported, but Folkers, Koniuszy, *loc. cit.* (1939), found no evidence of hydrate formation.

Hydrochloride, $C_{14}H_{18}N_2O_2$·HCl, crystals from water, dec 231-232°, $[\alpha]_D^{32}$ +90° (c = 0.5). Moderately sol in water.

Nitrate, $C_{14}H_{18}N_2O_2$·HNO$_3$, dec 224°, $[\alpha]_D^{20}$ +95°. Sol in 170 parts water.

Caution: A convulsive poison.

4799. Hypericin. *1,3,4,6,8,13-Hexahydroxy-10,11-dimethylphenanthro[1,10,9,8-opqra]perylene-7,14-dione; 4,5,-7,4′,5′,7′-hexahydroxy-2,2′-dimethylnaphthodianthrone;* hypericum red; Cyclo-Werrol; Cyclosan. $C_{30}H_{16}O_8$; mol wt 504.43. C 71.43%, H 3.20%, O 25.38%. Occurs in *Hypericum* spp. (St. John's wort). Isoln from *Hypericum perforatum* L., *Hypericaceae:* Brockmann *et al., Ann.* **553**, 1 (1942). Synthesis from bromoemodin trimethylether: Brockmann, Muxfeld, *Naturwiss.* **40**, 411 (1953). Also obtained by irradiation of oxypenicilliopsin produced by the fungus *Penicilliopsis clavariaeformis* Solms-Laubach: Brockmann, Neef, *ibid.* **38**, 47 (1951). Structure: Brockmann *et al., ibid.* **37**, 540 (1950); Brockmann, Sanne, *ibid.* **40**, 509 (1953). Total synthesis: Brockmann, Kluge, U.S. pat. **2,707,704** (1955 to Schenley); Brockmann *et al., Ber.* **90**, 2302 (1957); Brockmann, Eggers, *ibid.* **91**, 547 (1958).

Solvated blue-black needles from pyridine + methanolic HCl, dec 320°. Freely sol in pyridine and other organic bases yielding cherry-red solns with red fluorescence. Almost insol in most other organic solvents. Sol in alkaline aq solns; below pH 11.5 solns are red, above pH 11.5 they are green with red fluorescence. Absorption and fluorescence spectra: Scheibe, Schöntag, *Ber.* **75**, 2019 (1942); Brockmann, Neef, *loc. cit.* Produces photosensitivity upon ingestion: Oxford, Raistrick, *Biochem. J.* **34**, 790 (1940). Very small quantities appear to have a tonic and tranquilizing action on the human organism.

THERAP CAT: Antidepressant.

4800. Hypochlorous Acid. ClHO; mol wt 52.47. Cl 67.58%, H 1.92%, O 30.50%. HClO. Known in aq soln only. Formed by the action of water on chlorine: Cady, *Inorg. Syn.* **5**, 160 (1957). A 25% soln of hypochlorous acid is obtained when chlorine hydrate and mercuric oxide are distilled at low pressure according to the eq $2Cl_2.6H_2O + HgO \rightarrow 2HClO + HgCl_2 + 11H_2O$: Goldschmidt, *Ber.* **52**, 753 (1919). *Review:* J. A. Wojtowicz, "Chlorine Monoxide, Hypochlorous Acid, and Hypochlorites", in Kirk-Othmer *Encyclopedia of Chemical Technology* vol. 5 (Wiley-Interscience, New York, 3rd ed., 1979) pp 580-611.

Aq soln (25%), greenish yellow liquid. May be stored for a few days if kept at $-20°$. Dec slowly to Cl_2, O_2 and $HClO_4$. Very weak acid. K at $25° = 3.2 \times 10^{-8}$. Strong oxidizing agent. When a 25% soln is evapd in a vacuum, chlorine monoxide, Cl_2O, is given off. Cl_2O in soln with HClO may be removed by shaking with carbon tetrachloride, but will form again. *Protect from light.*

Note: An aq soln of the potassium salt is known as *Javelle water* or *Eau de Javelle* (also used for sodium hypochlorite, *q.v.*).

USE: Disinfectant.

4801. Hypoglycine A. *α-Amino-2-methylenecyclopropanepropanoic acid; 2-methylenecyclopropanealanine;* 2-amino-4,5-methylenehex-5-enoic acid; α-amino-β-(2-methylenecyclopropyl)propionic acid; hypoglycin; hypoglycin A. $C_7H_{11}NO_2$; mol wt 141.17. C 59.55%, H 7.85%, N 9.92%, O 22.67%. Hypoglycemic principle from the akee plant, *Blighia sapida* Kon., *Sapindaceae:* Hassall *et al., Nature* **173**, 356 (1954); Hassall, Reyle, *Biochem. J.* **60**, 334 (1955). Structure: de Ropp *et al., J. Am. Chem. Soc.* **80**, 1004 (1958); Renner *et al., Helv. Chim. Acta* **41**, 589 (1958); Ellington *et al., J. Chem. Soc.* **1959**, 80. Synthesis: Carbon *et al., J. Am. Chem. Soc.* **80**, 1002 (1958); Black, Landor, *Tetrahedron Letters* **1963**, 1065.

Yellow plates from methanol + water, mp 280-284°. $[α]_D^{32}$ +9.2°.

Methyl ester hydrochloride, $C_8H_{13}NO_2$.HCl, needles from methanol + ether, mp 151-152°. $[α]_D^{22}$ +36° (c = 2.0).

4802. Hypoglycine B. *N-L-γ-Glutamyl-3-(2-methylenecyclopropyl)alanine;* γ-L-glutamyl-α-amino-β-(2-methylenecyclopropyl)propionic acid dipeptide; γ-L-glutamylhypoglycine. $C_{12}H_{18}N_2O_5$; mol wt 270.28. C 53.32%, H 6.71%, N 10.37%, O 29.60%. Isoln *see* Hypoglycine A. Structure: Jöhl, Stoll, *Helv. Chim. Acta* **42**, 156 (1959); Hassall, John, *J. Chem. Soc.* **1960**, 4112. Synthesis: Jöhl, Stoll, *Helv. Chim. Acta* **42**, 716 (1959).

Yellow needles from acetone + water, mp 194-195°, 200-206°. $[α]_D^{32}$ +9.6 (c = 1.12). Neutralization equivalent 175.

N-2,4-Dinitrophenylhydrazone, $C_{18}H_{20}N_4O_9$, needles from methanol + water, mp 170-172°. $[α]_D^{19}$ −84° (c = 0.6 in 4% aq $NaHCO_3$). uv max: 359 nm (ε 16,600).

Phenylhydantoin derivative, $C_{19}H_{21}N_3O_4S$, needles from alc + water, mp 186-188°. uv max: 268-269 nm (ε 16,000).

4803. Hypophosphoric Acid. $H_4O_6P_2$; mol wt 161.99. H 2.49%, O 59.27%, P 38.25%. Prepn from its salts: Salzer,

Ann. **187**, 322 (1877); **211**, 1 (1882). *Review:* Ohashi, "Lower Oxo Acids of Phosphorus and Their Salts" in *Topics in Phosphorus Chemistry,* Vol. 1, M. Grayson, E. J. Griffith, Eds. (Interscience, New York, 1964) pp 113-187.

Orthorhombic plates, mp 70° (easily forms a dihydrate, mp 55°), usually available only in aq soln. The aq acid is colorless and odorless; dec on concn at atm pressure. Readily forms crystallizable normal and acid sodium salts.

USE: Acid sodium salts in baking powders.

4804. Hypophosphorous Acid. H_3O_2P; mol wt 66.00. H 4.58%, O 48.49%, P 46.94%. Conveniently prepd by treating NaH_2PO_2 with an ion-exchange resin: Klement, *Z. Anorg. Allgem. Chem.* **260**, 267 (1949). *Review:* Ohashi, "Lower Oxo Acids of Phosphorus and Their Salts" in *Topics in Phosphorus Chemistry,* Vol. 1, M. Grayson, E. J. Griffith, Eds. (Interscience, New York, 1964) pp 113-187.

The water-free acid forms deliquesc crystals; supercools to a colorless, odorless, oily liquid. d 1.493. mp 26.5°. Dec by heat into H_3PO_4 and spontaneously flammable PH_3. Miscible with water, alcohol, ether. Oxidized by hot H_2SO_4; SO_2 and S formed. It is a powerful reducing agent. Marketed in aq solns of various concns, *e.g.*, 50%, d 1.274; 30-32%, d 1.13; 10%, d 1.04. $K_1 = 8 \times 10^{-2}$.

4805. Hypoxanthine. *1,7-Dihydro-6H-purin-6-one; purin-6(1H)-one;* sarcine; sarkin. $C_5H_4N_4O$; mol wt 136.11. C 44.12%, H 2.96%, N 41.17%, O 11.75%. Desmotropic forms: purin-6-ol; 1*H*-purin-6-ol; 3*H*-purin-6-ol; 9*H*-purin-6-ol; purin-6(3*H*)-one; 9*H*-purin-6(1*H*)-one. Formed in the animal body during the breakdown of nucleic acids; formation from adenosine continues after death. Also widely distributed in the vegetable kingdom. Numerous syntheses, *e.g.*, from 2,6,8-trichloropurine: Fischer, *Ber.* **30**, 2226 (1897); by oxidation of adenine: Krüger, *Z. Physiol. Chem.* **18**, 445 (1894); by reduction of uric acid: Sundwik, *ibid.* **76**, 486 (1912); by condensing ethyl cyanoacetate and thiourea in the presence of Na-ethoxide: Traube, *Ann.* **331**, 64 (1904); from mercapto-4-hydroxy-6-aminopyrimidine: Taylor, Cheng. *J. Org. Chem.* **25**, 148 (1960). *Reviews:* Levene, Bass, *Nucleic Acids,* A.C.S. Monograph Series no. 56 (New York 1931); Chargaff, Davidson, *Nucleic Acids,* **vols. I & II** (Academic Press, New York, 1955).

Small octahedra, needles from water, dec 150° without melting. Sol in 1400 parts water, 70 parts boiling water; sol in dil acids, alkalies. Combines with one equivalent of acid or with two equivalents of base. pK_b (25°): 8.7. Absorption spectrum: Dhere, *Compt. Rend.* **141**, 720.

I

4806. Ibogaine. *12-Methoxyibogamine.* $C_{20}H_{26}N_2O$; mol wt 310.42. C 77.38%, H 8.44%, N 9.03%, O 5.15%. Indole alkaloid of the *iboga* group. Isoln from root (1.27%), root-bark (2 to 6%), stems (1.95%) and leaves (0.35%) of the shrub *Tabernanthe iboga* Baill., *Apocynaceae*, found in Africa: Dybowski, Landrin, *Compt. Rend.* **133**, 748 (1901); Haller, Heckel, *ibid.* 850, 1236; from other *Apocynaceae:* H. Achenbach, B. Raffelsberger, *Z. Naturforsch.* **35B**, 219, 885 (1980); N. Ghorbel *et al., J. Nat. Prod.* **44**, 717 (1981); T. Mulamba *et al., ibid.* 184; B. Richard *et al., ibid.* **46**, 283 (1983). Purification: Schlittler *et al., Helv. Chim. Acta* **36**, 1341 (1953). Revised extraction procedure: Dickel *et al., J. Am. Chem. Soc.* **80**, 123 (1958). Review of early isolation work: Lebeau, Janot, *Traité de Pharmacie Chimique* vol. 4 (Masson et Cie., Paris, 1956) pp 2982-2988. Structure: Bartlett *et al., J. Am. Chem. Soc.* **80**, 126 (1958). Mass spectrum: Biemann, Friedmann-Spiteller, *ibid.* **83**, 4805 (1961). Synthesis: Büchi *et al., ibid.* **88**, 3099 (1966); Rosenmund *et al., Ber.* **108**, 1871 (1975). Derivs: Taylor, U.S. pat. **2,877,-229** (1959 to Ciba). Absolute configuration: K. Blàha *et al., Tetrahedron Letters* **1972**, 2763. Interatomic distances similar to those of serotonin: J. M. Kelley, R. H. Adamson, *Pharmacology* **10**, 28 (1973). NMR spectrum: E. Wenkert *et al., Helv. Chim. Acta* **59**, 2437 (1976). Determn in biological fluids: E. Bertol *et al., J. Chromatog.* **117**, 239 (1976). Iboga extracts said to be used by African natives while stalking game, to enable them to remain motionless for as long as 2 days while retaining mental alertness. Neuropharmacological studies: Schneider, Sigg, *Ann. N.Y. Acad. Sci.* **66**, 765 (1957); S. Gershon, W. J. Lang, *Arch. Int. Pharmacodyn. Ther.* **135**, 31 (1962). Cardiovascular effects: J. A. Schneider, R. K. Rinehart, *ibid.* **110**, 92 (1957). Serotonergic properties: R. S. Sloviter *et al., J. Pharmacol. Exp. Ther.* **214**, 231 (1980). Experimental use in treatment of heroin addiction: H. S. Lotsof, U.S. pat. **4,499,096** (1985). *Reviews:* W. I. Taylor, "The Iboga and Voacanga Alkaloids" in *The Alkaloids, Chemistry and Physiology* Vol. 8, R. H. F. Manske, Ed. (Academic Press, New York, 1965) p 203-235, *idem, ibid.* **Vol. 11** (1968), pp 79-98.

Prismatic needles from abs ethanol, mp 152-153°. Sublimes$_{0.01}$ 150°. $[\alpha]_D^{20}$ − 53° (in 95% ethanol). pKa 8.1 in 80% methylcellosolve. uv max (methanol): 226, 298 nm (log ϵ 4.39, 3.93). Sol in ethanol, ether, chloroform, acetone, benzene. Practically insol in water.
Hydrochloride, $C_{20}H_{27}ClN_2O$, crystals. Dec 299-300°. $[\alpha]_D^{25}$ − 63° (ethanol); $[\alpha]_D^{25}$ − 49° (H_2O). Soluble in water, methanol, ethanol. Slightly sol in acetone, chloroform. Practically insol in ether.
Caution: This is a controlled substance (hallucinogen) listed in the U.S. Code of Federal Regulations, Title 21 Part 1308.11 (1985).

4807. Ibopamine. *2-Methylpropanoic acid 4-[2-(methylamino)ethyl]-1,2-phenylene ester;* 4-[2-(methylamino)ethyl]-*o*-phenylene diisobutyrate; *N*-methyldopamine diisobutyric ester; 3,4-di-*o*-isobutyryl epinine. $C_{17}H_{25}NO_4$; mol wt 307.39. C 66.42%, H 8.20%, N 4.56%, O 20.82%. Inotropic agent with dopaminergic and adrenergic agonist activities. Prepn: C. Casagrande, G. Ferrari, **Ger.** pat. **2,734,678;** *eidem,* **U.S.** pat. **4,218,470** (1978, 1980 both to Simes). Pharmacology: G. F. Melloni *et al., Curr. Ther. Res.* **25**, 406 (1979); *eidem, ibid.* **26**, 466 (1979). Diuretic effect in chronic renal failure: S. Stefoni *et al., Brit. J. Clin. Pharmacol.* **11**, 69 (1981); in ascitic liver disease: G. F. Melloni *et al., ibid.* **12**, 813 (1981). Acute hemodynamic effects in congestive heart failure: P. Ghirardi *et al., ibid.* **19**, 613 (1985). α- and β-adrenergic activity: A. J. Nichols, R. R. Ruffolo, Jr., *J. Pharmacol. Exp. Ther.* **242**, 455 (1987). Series of articles on

synthesis, pharmacology, clinical efficacy: *Arzneimittel-Forsch.* **36**, 285-408 (1986). Post-marketing surveillance study: D. Sher, V. Ferrari, *ibid.* **37**, 869 (1987). Review of pharmacodynamics, pharmacokinetics and therapeutic efficacy: J. M. Henwood, P. A. Todd, *Drugs* **36**, 11-31 (1988).

Hydrochloride, $C_{17}H_{26}ClNO_4$, *SB 7505, Inopamil, Scandine.* Crystals from ethyl acetate, mp 132°.
THERAP CAT: Cardiotonic.

4808. Ibotenic Acid. *α-Amino-2,3-dihydro-3-oxo-5-isoxazoleacetic acid; α-amino-3-hydroxy-5-isoxazoleacetic acid;* amino-(3-hydroxy-5-isoxazolyl)acetic acid. C_5H_6-N_2O_4; mol wt 158.11. C 37.98%, H 3.83%, N 17.71%, O 40.48%. Fly-killing and narcosis-potentiating amino acid structurally similar to kainic acid, *q.v.*, extracted from poisonous mushroom species. Isoln from *Amanita pantherina* (DC.) Fr., and *A. muscaria* (L.) Fr., *Agaricaceae:* Takemoto *et al., J. Pharm. Soc. Japan* **84**, 1233 (1964); Eugster *et al., Tetrahedron Letters* **1965**, 1813. Structure: Takemoto *et al., J. Pharm. Soc. Japan* **84**, 1186, 1232 (1964). Syntheses: Gagneux *et al., Tetrahedron Letters* **1965**, 2081; Sirakawa *et al., Chem. Pharm. Bull.* **14**, 89 (1966); Kishida *et al., ibid.* **14**, 92 (1966); **15**, 1025 (1967). Improved synthesis: Nakamura, *ibid.* **19**, 46 (1971). Industrial pats: **Belg.** pat. **665,249**, *C.A.* **65**, 2266e (1966); Gagneux *et al.,* **U.S.** pat. **3,459,862** (1965, 1969, both to Geigy); Kishida *et al.,* **Japan.** pats. **15,975**-('68) and **25,780**('69) (both to Sankyo), *C.A.* **70**, 77944p (1969); **72**, 13054g (1970). Pharmacology: Theobald *et al., Arzneimittel-Forsch.* **18**, 311 (1968); Johnston *et al., Biochem. Pharmacol.* **17**, 2488 (1968). Exhibits potent neuroexcitatory activity: *eidem, Nature* **248**, 804 (1974). Chemistry review: Eugster, *Fortschr. Chem. Org. Naturst.* **27**, 261-321 (1969); Catalfomo, Eugster, *Bull. Narcotics* **22**, 33-41 (1970). Excitatory and possible sedative actions on spinal neurons: D. R. Curtis *et al., J. Physiol.* **291**, 19 (1979); in cerebral cortex: E. Puil, *Can. J. Physiol. Pharmacol.* **59**, 1025 (1981). Use as experimental neurotoxic agent: A. Contestabile *et al., Experientia* **40**, 524 (1984).

Crystals from water or methanol, mp 151-152° (anhydrous); mp 144-146° (monohydrate). LD_{50} in mice, rats (mg/kg): 15, 42 i.v.; 38, 129 orally (Theobald).
USE: Neurobiological tool.

4809. Ibrotamide. *2-Bromo-2-ethyl-3-methylbutanamide;* α-bromo-α-isopropylbutyramide; α-ethyl-α-isopropyl-α-bromoacetamide; 2-bromo-2-ethylisovaleramide; Vagoprol. $C_7H_{14}BrNO$; mol wt 208.12. C 40.40%, H 6.78%, Br 38.40%, N 6.73%, O 7.69%. Prepn: Hildebrandt *et al.,* U.S. pat. **1,780,131** (1931 to Knoll); Safir *et al., J. Am. Chem. Soc.* **77**, 4840 (1955).

Crystals, mp 51°. Soluble in the usual organic solvents and in oil.
THERAP CAT: Sedative, hypnotic.

4810. Ibudilast. *2-Methyl-1-[2-(1-methylethyl)pyrazolo-*

[1,5-a]pyridin-3-yl]-1-propanone; 3-isobutyryl-2-isopropyl-pyrazolo[1,5-*a*]pyridine; KC-404; Ketas. $C_{14}H_{18}N_2O$; mol wt 230.31. C 73.01%, H 7.88%, N 12.16%, O 6.95%. Leukotriene D_4 antagonist. Prepn: T. Irikura *et al.,* **Ger.** pat. **2,315,801;** *eidem,* **U.S.** pat. **3,850,941** (1973, 1974 both to Kyorin). Pharmacology and antiallergic activity: K. Nishino *et al., Japan. J. Pharmacol.* **33,** 267 (1983); H. Nagai *et al., ibid.* 1215. *In vitro* cerebral vasodilating activity: M. Ohashi *et al., Arch. Int. Pharmacodyn.* **280,** 216 (1986); *in vivo* activity: W. M. Armstead *et al., J. Pharmacol. Exp. Ther.* **244,** 138 (1988). Bronchodilating activity in animals: S. Mue *et al., Arch. Int. Pharmacodyn.* **283,** 153 (1986). Antiplatelet activity in animals: M. Ohashi *et al., ibid.* 321; M. Ohashi *et al., Gen. Pharmacol.* **17,** 385 (1986).

Crystals from hexane, mp 53.5-54°. Slightly sol in water, freely sol in organic solvents. LD_{50} i.v. in mice: 260 mg/kg (Irikura, 1973).

THERAP CAT: Antiallergic; antiasthmatic; cerebral vasodilator.

4811. Ibufenac. *4-(2-Methylpropyl)benzeneacetic acid; (p-isobutylphenyl)acetic acid; p-isobutyl-α-toluic acid;* Dytransin; Ibunac. $C_{12}H_{16}O_2$; mol wt 192.25. C 74.97%, H 8.39%, O 16.65%. Prepn: J. S. Nicholson, S. S. Adams, **Brit.** pat. **971,700** corresp to U.S. pat. **3,228,831** (1964, 1966 to Boots Pure Drug). Pharmacology: S. S. Adams *et al., J. Pharm. Pharmacol.* **20,** 305 (1968).

Crystals, mp 85-87°. Very slightly sol in water. Freely sol in many organic solvents. LD_{50} orally in mice: 1.8 g/kg, S. S. Adams *et al., Nature* **200,** 271 (1962).

THERAP CAT: Analgesic, anti-inflammatory.

4812. Ibuprofen. *α-Methyl-4-(2-methylpropyl)benzeneacetic acid; p-isobutylhydratropic acid;* 2-(4-isobutylphenyl)-propionic acid; RD 13621; Adran; Advil; Anco; Amibufen; Anflagen; Apsifen; Artril 300; Bluton; Brufanic; Brufen; Brufort; Buburone; Butylenin; Dolgin; Dolgirid; Dolgit; Dolocyl; Dolo-Dolgit; Ebufac; Emodin; Epobron; Femadon; Fenbid; Haltran; Ibu-Attritin; Ibumetin; Ibuprocin; Ibu-slo; Ibutid; Ibutop; Inabrin; Inoven; Lamidon; Lebrufen; Liptan; Lobufen; Medipren; Mono-Attritin; Motrin; Mynosedin; Napacetin; Nobfelon; Nobfen; Nobgen; Novogent N; Nuprin; Nurofen; Opturem; Paxofen; Prontalgin; Rebugen; Recidol; Roidenin; Seclodin; Suspren; Tabalon; Trendar; Urem. $C_{13}H_{18}O_2$; mol wt 206.27. C 75.69%, H 8.80%, O 15.51%. Prepn: Nicholson, Adams, **Brit.** pat. **971,700;** *eidem,* **U.S.** pats. **3,228,831** and **3,385,886** (1964, 1966, 1968 all to Boots Pure Drug); T. Shiori, N. Kawai, *J. Org. Chem.* **43,** 2936 (1978); J. T. Pinhey, B. A. Rowe, *Tetrahedron Letters* **21,** 965 (1980). Pharmacology: Adams *et al., Arch. Pharmacodyn. Ther.* **178,** 115 (1969). Metabolism and toxicity: Adams *et al., Toxicol. Appl. Pharmacol.* **15,** 310 (1969). Toxicity: G. Orzalesi *et al., Arzneimittel-Forsch.* **27,** 1006 (1977). Series of articles on pharmacology, pharmacokinetics, adverse effects, and clinical efficacy: *Am. J. Med.* **77**(1A), 1-125 (1984). *Review:* L. Cavallini, G. Lucchetti, *Gazz. Med. Ital.* **134,** 7 (1975).

Colorless, crystalline stable solid, mp 75-77°. Relatively insol in water. Readily sol in most organic solvents. LD_{50} in male mice, rats (mg/kg): 495, 626 i.p.; 1255, 1050 orally (Orzalesi).

Lysine salt, $C_{19}H_{32}N_2O_4$, *Solprofen.*

Methylglucamine salt, $C_{20}H_{35}NO_7$, *Artrene.*

THERAP CAT: Anti-inflammatory.

4813. Ibuproxam. *N-Hydroxy-α-methyl-4-(2-methylpropyl)benzeneacetamide; dl*-2-(4-isobutylphenyl)propiono-hydroxamic acid; G 277; Ibudros. $C_{13}H_{19}NO_2$; mol wt 221.30. C 70.56%, H 8.65%, N 6.33%, O14.46%. Hydroxylamine deriv of ibuprofen, *q.v.,* to which it is converted *in vivo.* Prepn: G. Orzalesi, R. Selleri, **Ger.** pat. **2,400,531** corresp to U.S. pat. **4,082,707** (1974, 1978 both to Manetti & Roberts). Metabolism in rats: G. Orzalesi *et al., Arzneimittel-Forsch.* **27,** 1012 (1977); in humans: *eidem, ibid.* **30,** 1607 (1980). Pharmacological study: *eidem, ibid.* **27,** 1006 (1977). Thermal decomposition: S. Chimichi *et al., J. Pharm. Sci.* **69,** 521 (1980). Physico-chemical properties: M. Mannelli *et al., Boll. Chim. Farm.* **119,** 203 (1980).

Cryst from acetone/petr ethr, mp 119-121°. Sol in methanol, ethanol, acetone, ethyl ether. Practically insol in water and petr ether. LD_{50} in mice, rats: > 2 g/kg, > 3 g/kg orally, G. Orzalesi, R.Selleri, *loc. cit.*

THERAP CAT: Anti-inflammatory.

4814. Iceland Moss. *Cetraria islandica* (L.) Ach., *Parmeliaceae,* a lichen growing in all northern countries. Exported from Iceland, Norway, and Sweden. The gum from the powdered plant appears to be a hemicellulose contg uronic acid, galactose, mannose, and glucose; *cf.* Mantell, *The Water-Soluble Gums,* New York, 1947.

About 60% of dried Iceland moss dissolves when boiled with water contg a little sodium bicarbonate. The soln forms a jelly when cold.

USE: Manuf sea biscuits which are somewhat more resistant to weevil infestation than when wheat flour alone is used. In foods for convalescents. Manuf sizing agents for rayon; hair-setting lotions, other cosmetics.

4815. Ichthammol. Ammonium bituminosulfonate; ammonium ichthosulfonate; ammonium sulfobituminate; ammonium sulfoichthyolate; bitumol; bituminol; ichthammonium; ammonium bithiolicum; ichthosulfol; Ichthyol; Hirathiol; Ichden; Ichtammon; Ichthadone; Ichthymall; Ichthysalle; Ichthalum; Ichthium; Ichtopur; Ichthosan; Ichthynat; Ichthyopon; Lithol; Petrosulpho; Perichthol; Piscarol; Pisciol; Saurol; Subitol; Sulfogenol; Thilaven; Thiolin; Thiozin; Trasulphane; Tumenol; Leukochthol; Ichthosauran; Amsubit; Bitulan. Obtained by sulfation and ammoniation of a distillate from mineral deposits (bituminous schists) originally found near Seefeld, Tyrol. Contains satd and unsatd hydrocarbons, nitrogenous bases, acids, and several thiophene derivs. Analysis shows at least 2.5% NH_3 and at least 10% S. Also contains traces of some 20 minerals and "*zoomelanoidic*" acids. Method of prepn: Schröter, **Ger.** pat. **35,216** (1885); Helmers, **Ger.** pat. **76,128** (1892). Similar deposits occur in Asia east of Lake Baikal where the oil is known as *stone oil, barakshin, Asil;* sold in India for remedial purposes as *saladjidi:* Gerbrein, *Photo-Journal* (Montreal, 1969, July 2-9) p 19. *Review:* Wernicke, *Chem. Ztg.* **60,** 85-87 (1936).

Pale yellow or (usually) brownish-black, thick, viscous liquid. Bituminous odor. Miscible with water, glycerol, propylene glycol, fats, oils, carbowaxes, lanolin. Partially sol in alcohol, ether.

An injectable form is marketed as *Adnexol.*

THERAP CAT: Topical anti-infective.

THERAP CAT (VET): Demulcent, emollient, antiseptic.

4816. Ichthyopterin. *2-Amino-6-(1,2-dihydroxypropyl)-4,7(1H,8H)-pteridinedione;* 6-(1,2-dihydroxypropyl)isoxanthopterine; fluorescyanine. $C_9H_{11}N_5O_4$; mol wt 253.22. C 42.69%, H 4.38%, N 27.67%, O 25.27%. Found as a chromoprotein in skin and scales of fish. Isoln from minnows: Huttel, Springling, *Ann.* **554,** 69 (1943); from carp: Polonovski *et al., Helv. Chim. Acta* **29,** 1328 (1946). Structure: Kauffmann, *Ann.* **625,** 133 (1959).

Rosettes of thin needles. Thermostable. uv max (0.1N NaOH): 343, 276, 258, 220 nm; (0.1N HCl): 343, 289, 212 nm. Maximum fluorescence (blue) between 265 and 390 nm at a pH between 5 and 8. Sol in water (0.3 mg/ml), hot ethanol, methanol, and pyridine. Practically insol in organic solvents not miscible with water.

4817. Idebenone. *2-(10-Hydroxydecyl)-5,6-dimethoxy-3-methyl-2,5-cyclohexadiene-1,4-dione;* 6-(10-hydroxydecyl)-2,3-dimethoxy-5-methyl-1,4-benzoquinone; 2,3-dimethoxy-5-methyl-6-(10'-hydroxydecyl)-1,4-benzoquinone; 6-(10-hydroxydecyl)ubiquinone; CV-2619; Avan. $C_{19}H_{30}O_5$; mol wt 338.44. C 67.43%, H 8.93%, O 23.64%. Ubiquinone derivative with protective effects against cerebral ischemia. Prepn: H. Morimoto *et al.,* **Ger.** pat. **2,519,-730;** *eidem,* **U.S.** pat. **4,271,083** (1975, 1981 both to Takeda); K. Okamoto *et al., Chem. Pharm. Bull.* **30,** 2797 (1982); C.-A. Yu, L. Yu, *Biochemistry* **21,** 4096 (1982). *In vitro* inhibition of mitochondrial lipid peroxidation: M. Suno, A. Nagaoka, *Biochem. Biophys. Res. Commun.* **125,** 1046 (1984). Inhibitory effect on cerebrovascular lesions in hypertensive rats: A. Nagaoka *et al., Japan. J. Pharmacol.* **36,** 291 (1984). Effect on ischemia-induced amnesia: N. Yamazaki *et al., ibid.* 349. Metabolism in animals: T. Kobayashi *et al., J. Pharmacobio.-Dyn.* **8,** 448 (1985). Disposition: H. Torii *et al., ibid.* 457. Pharmacokinetics and tolerance in humans: M. F. Barkworth *et al., Arzneimittel-Forsch.* **35,** 1704 (1985).

Orange needles from ligroin, mp 46-50° (Morimoto); also reported as crystals from hexane + ethyl acetate, mp 52-53° (Okamoto).

THERAP CAT: Nootropic.

4818. Idose. $C_6H_{12}O_6$; mol wt 180.16. C 40.00%, H 6.71%, O 53.29%. Prepn of D-idose by reduction of D-idonolactone: Fischer, Fay, *Ber.* **28,** 1975 (1895); from D-galactose: Sorking, Reichstein, *Helv. Chim. Acta* **28,** 1 (1945); from tri-*O*-acetyl-1,6-anhydro-β-D-idopyranose: *eidem, ibid.* 662. Prepn of L-idose by reduction of L-idonolactone: Fischer, Fay, *loc. cit.;* by hydrolysis of 1,2-*O*-isopropylidene-L-idofuranose: Meyer, Reichstein, *Helv. Chim. Acta* **29,** 152 (1946); von Vargha, *Ber.* **87,** 1351 (1954). Improved synthesis: M. Blanc-Muesser, J. Defaye, *Synthesis* **1977,** 568. Structure: S. F. Dyke, *The Carbohydrates* (Interscience, New York, 1960) p 45. Conformation: Reeves, *J. Am. Chem. Soc.* **72,** 1499 (1950). *Review:* R. L. Whistler, M. L. Wolfrom, *Methods in Carbohydrate Chemistry* (Academic Press, New York, 1962) pp 140-145.

β-D-idose

D-Form: Syrup. $[\alpha]_D^{13}$ +15.8° (c = 2.3).
Phenylosazone, $C_{18}H_{22}N_4O_4$, yellow needles from alc, mp 168-169°.
L-Form: Syrup. $[\alpha]_D^{20}$ −17.4° (c = 3.6).

1,2-*O*-Isopropylidene-L-idofuranose, plates from ethyl acetate, mp 113-114°. $[\alpha]_D^{26}$ −20° (c = 2.7 in methanol).

4819. Idoxuridine. *2'-Deoxy-5-iodouridine;* 1-(2-deoxy-β-D-ribofuranosyl)-5-iodouracil; 5-iodo-2'-deoxyuridine; IDU; IDUR; IUDR; Dendrid; Emanil; Herpes-Gel; Herplex; Idexur; Idoxene; Idulea; Idu Oculos; Iduridin; Kerecid; Ophthalmadine; Stoxil; Virudox. $C_9H_{11}IN_2O_5$; mol wt 354.12. C 30.53%, H 3.13%, I 35.84%, N 7.91%, O 22.59%. Prepn: Prusoff, *Biochim. Biophys. Acta* **32,** 295 (1959); Cheong *et al., J. Biol. Chem.* **235,** 1441 (1960); Chang, Welch, *J. Med. Chem.* **6,** 428 (1963); Amiard, Torelli, **Fr.** pat. **1,336,-866** (1963 to Roussel-UCLAF), *C.A.* **60,** 3082g (1964), corresponds to **Brit.** pat. **1,024,156;** Prystas, Sorm, *Coll. Czech. Chem. Commun.* **29,** 121 (1964). Crystal and molecular structure: Camerman, Trotter, *Acta Cryst.* **18,** 203 (1965). *Review:* W. H. Prusoff *et al.* in *Antibiotics* vol. 5(pt. 2), F. E. Hahn, Ed. (Springer-Verlag, New York, 1979) pp 236-261.

Crystals from water, triclinic, dec 160° (Prusoff; Chang, Welch), 190-195° (Cheong *et al.*), 240° (Amiard, Torelli), over 175° (Prystas, Sorm). uv max (water): 288 nm (log ϵ 3.87). $[\alpha]_D^{25}$ +7.4° (c = 0.108 in water); $[\alpha]_D^{20}$ +29° (N soda). Following properties, *see* Ravin, Gulesich, *J. Am. Pharm. Assoc.* [NS] **4,** 122 (1964). pKa 8.25. pH of 0.1% aq soln, about 6. Soly at 25° in mg/ml: 2.0 in water; 2.0 in 0.2N HCl; 74.0 in 0.2N NaOH; 4.4 in methanol; 2.6 in alc; 0.014 in ether; 0.003 in chloroform; 1.6 in acetone; 1.8 in ethyl acetate; 5.7 in dioxane. LD_{50} i.p. in mice: 1800 mg/kg.
Note: α-Anomer, *1-(2-deoxy-α-D-erythro-pentofuranosyl)-5-iodouracil,* 1-(2-deoxy-α-D-ribofuranosyl)-5-iodouracil, α-2'-deoxy-5-iodouridine. Prepn: Prystas, Sorm, *loc. cit.* Crystals from water, dec 170°. $[\alpha]_D^{25}$ +21.8° (c = 0.170). uv max (water): 288 nm (log ϵ 3.88).

THERAP CAT: Antiviral.

4820. Idrocilamide. *N-(2-Hydroxyethyl)-3-phenyl-2-propenamide;* N-(2-hydroxyethyl)cinnamamide; LCB 29; Brolitène; Srilane. $C_{11}H_{13}NO_2$; mol wt 191.23. C 69.09%, H 6.85%, N 7.33%, O 16.73%. Prepn: M. Bayssat *et al.,* **Ger.** pat. **2,015,447** corresp to **U.S.** pat. **3,659,014** (1970, 1972 to LIPHA); *Chim. Ther.* **8,** 202 (1973). Pharmacology: Grand *et al., Eur. J. Med. Chem.* **9,** 205 (1974). Metabolism: Belleville *et al., Therapie* **29,** 829 (1974).

White crystals from ethyl acetate or acetone, mp 100-102°. Soluble in alcohol; slightly sol in water. LD_{50} orally in mice, rats: > 2950, > 3000 mg/kg (Grand).

THERAP CAT: Muscle relaxant (skeletal).

4821. Ifenprodil. *α-(4-Hydroxyphenyl)-β-methyl-4-(phenylmethyl)-1-piperidineethanol;* 4-benzyl-α-(p-hydroxyphenyl)-β-methyl-1-piperidineethanol; 2-(4-benzylpiperidino)-1-(4-hydroxyphenyl)-1-propanol; 1-methyl-2-hydroxy-2-(4-hydroxyphenyl)ethyl-1-(4-benzylpiperidine); RC 61-91. $C_{21}H_{27}NO_2$; mol wt 325.46. C 77.50%, H 8.36%, N 4.30%, O 9.83%. Prepn and pharmacology: Carron *et al.,* **Fr.** pat. **M5733;** **U.S.** pat. **3,509,164** (1968, 1970 to Robert et Carrière); *eidem, Arzneimittel-Forsch.* **21,** 1992 (1971).

mp 114°.

Neutral tartrate, $C_{46}H_{60}N_2O_{10}$, *Cerocral, Dilvax, Vadilex, Validex.* Crystals from methanol, mp 178-180°. Sol in alc, water. Very slightly sol in acetone, chloroform. Practically insol in ether. LD_{50} in male Swiss mice (mg/kg): 17 i.v.; 120 i.p.; 275 orally (Carron, 1970).

THERAP CAT: Cerebral and peripheral vasodilator.

4822. Ifosfamide. *N,3-Bis(2-chloroethyl)tetrahydro-2H-1,3,2-oxazaphosphorin-2-amine 2-oxide; 3-(2-chloroethyl)-2-[(2-chloroethyl)amino]tetrahydro-2H-1,3,2-oxazaphosphorin-2-oxide;* iphosphamid(e); isoendoxan; isophosphamide; A 4942; Asta Z 4942; MJF 9325; NSC-109724; Z 4942; Cyfos; Holoxan; Ifex; Mitoxana; Naxamide. $C_7H_{15}Cl_2N_2O_2P$; mol wt 261.07. C 32.20%, H 5.79%, Cl 27.16%, N 10.73%, O 12.26%, P 11.86%. Cytostatic agent, related structurally to cyclophosphamide, *q.v.* Prepn: **Fr. pat.** **1,530,962** (1968 to Asta), *C.A.* **71**, 49998m (1969); H. Arnold *et al.*, U.S. pat. **3,732,340** (1973 to Asta). Chemical properties: H. Arnold, *Proc. 5th Int. Congr. Chemother. Vienna* (Verhandlungen, Vienna, 1967) **2**, pp 751-754. Pharmacology: N. Brock, *ibid.* pp 155-161. Molecular structure and conformation: H. A. Brassfield *et al.*, *J. Am. Chem. Soc.* **97**, 4143 (1975). Mass spectrometry: M. Przybylski *et al.*, *Biomed. Mass Spectrom.* **4**, 209 (1977). Mechanism of action: S. Tomita *et al.*, *Chemother. (Tokyo)* **25**, 3014 (1977). Metabolism: A. Takamizawa *et al.*, *Chem. Pharm. Bull.* **25**, 2900 (1977); *eidem, J. Med. Chem.* **17**, 1237 (1974). Toxicity studies: R. Marcy *et al.*, *IRCS Med. Sci. Libr. Compend.* **5**, 427, 478 (1977). Mutagenicity studies: D. Wald, *J. Mutat. Res.* **56**, 319 (1978); G. R. Mohn, *J. Ellenberger, ibid.* **32**, 331 (1976). Clinical studies: P. J. Creaven *et al., Cancer Treat. Rep.* **60**, 445, 451 (1976); J. Schnitker *et al., Arzneimittel-Forsch.* **26**, 1793 (1976). Symposium on clinical efficacy and comparison with cyclophosphamide: *Cancer Chemother. Pharmacol.* **18**, Suppl. 2, S1-S58 (1986). Review of pharmacology, toxic effects and clinical activity: M. Zalupski, L. H. Baker, *J. Nat. Cancer Inst.* **80**, 556-566 (1988).

Crystals from anhyd ether, mp 39-41°. LD_{50} in rats (mg/kg): 160 i.p. (Arnold, 1973); also reported as 150 i.p. (Brock).

THERAP CAT: Antineoplastic.

4823. Ignatia. Ignatius bean; St. Ignatius' bean. Dried, ripe seed of *Strychnos ignatii* Berg., *Loganiaceae.* Habit. Philippine Islands, naturalized in Vietnam, Cambodia. *Constit.* 2-3% strychnine, about 1% brucine, igasuric acid, loganin.

THERAP CAT: Bitter tonic.

4824. Illudins. Anti-tumor antibiotic substances produced by the poisonous basidiomycetes *Clitocybe illudens:* Anchel *et al., Proc. Nat. Acad. Sci. USA* **36**, 300 (1950); **38**, 927 (1952); and *Lampteromyces japonicus:* Nakanishi *et al., Nature* **197**, 292 (1963); Endo *et al., Chem. Commun.* **1970**, 309. Structure: McMorris, Anchel, *J. Am. Chem. Soc.* **87**, 1594 (1965); Matsumoto *et al., Tetrahedron* **21**, 2671 (1965); Tada *et al., Chem. Pharm. Bull.* **12**, 853 (1964). Stereochemistry: Nakanishi *et al., ibid.* 856; *Tetrahedron* **21**, 1231 (1965); Matsumoto *et al., Bull. Chem. Soc. Japan* **37**, 1716 (1964). Abs config of illudin S: Harada, Nakanishi, *Chem. Commun.* **1970**, 310. Total synthesis of illudin M: Matsumoto *et al., J. Am. Chem. Soc.* **90**, 3280 (1968); *Tetrahedron Letters* **1970**, 1171; of illudin S: *eidem, ibid.* **1971**, 2049.

Illudin M. $C_{15}H_{20}O_3$. *2',3'-Dihydro-3'β,6'α-dihydroxy-2',2',4',6'-tetramethylspiro[cyclopropane-1,5'-[5H]inden]-7'(6'H)-one.* R = H. Rods, mp 120-122.5° (Matsumoto); from ethanol-water, mp 128-130° (McMorris). uv max (ethanol): 228, 318 nm (ε 13,900, 3600).

Monoacetate, $C_{17}H_{22}O_4$, mp 75-76° from petr ether.

Illudin S. $C_{15}H_{20}O_4$. *2',3'-Dihydro-3'β,6'-dihydroxy-2'-(hydroxymethyl)-2',4',6'-trimethylspiro[cyclopropane-1,5'-[5H]inden]-7'(6'H)-one, lampterol, lunamycin* (obsolete). R = OH. Needles from acetone, mp 124-126°. uv max (ethanol): 233, 319 nm (ε 13,200, 3600). Sol in polar organic solvents.

Diacetate, $C_{19}H_{24}O_5$, crystals from petr ether, mp 99-100°. uv max (ethanol): 227, 313 nm (ε 12,900, 3400).

4825. Imazamethabenz. *2-[4,5-Dihydro-4-methyl-4-(1-methylethyl)-5-oxo-1H-imidazol-2-yl]-4(and 5)-methylbenzoic acid methyl ester;* imazamethabenz methyl; imazethabenz; AC 222293; AC 293; CL 222293; Assert; Dagger. $C_{16}H_{20}N_2O_3$; mol wt 288.35. C 66.65%, H 7.00%, N 9.71%, O 16.64%. Selective, post-emergence imidazolinone herbicide; mixture of *methyl 2-(4-isopropyl-4-methyl-5-oxo-2-imidazolin-2-yl)-p-toluate* and *methyl 6-(4-isopropyl-4-methyl-5-oxo-2-imidazolin-2-yl)-m-toluate* (approx 3:2). Prepn of isomeric mixture: M. Los, U.S. pat. **4,188,487** (1980 to American Cyanamid). Activity, physical properties and toxicity: D. L. Shaner *et al., Proc. Brit. Crop Prot. Conf.-Weeds* **1982**, 25; K. Hedlund, L. Andersson, *Weeds Weed Control* **28**, 1 (1987). Activity of component isomers: D. L. Shaner *et al., Proc. Brit. Crop Prot. Conf.-Weeds* **1982**, 333. Mechanism of action: J. B. Pillmoor, J. C. Caseley, *Pestic. Biochem. Physiol.* **27**, 340 (1987). Persistence and mobility in soil: R. Allen, J. C. Caseley, *Proc. Brit. Crop Prot. Conf.-Weeds* **1987**, 569. Field studies: K. Kirkland, N. E. Shafer, *ibid.* **1982**, 33; A. A. Hudson, S. C. E. Townsend, *ibid.* **1985**, 923; in food crops: S. D. Miller, H. P. Alley, *Weed Technol.* **1**, 29 (1987).

m-form   p-form

Off-white fine powder with a tendency to form easily friable aggregates; slight musty odor. Softening begins at 108-117°, melting starts at 113-122° and is completed at 144-153°. Soly (g/100 ml) at 25°: acetone 18.2, DMSO 23.8, distilled water 0.13 (p-isomer), 0.22 (m-isomer), n-heptane 0.04, isopropyl alcohol 14.4, methanol 24.4, methylene chloride 30.0, toluene 3.9. Soly (g/100 g) at 25°: xylene < 5, DMF 30. Partition coefficient in n-octanol/water: 35 (p-isomer), 66 (m-isomer). LD_{50} in rats, rabbits (mg/kg): > 5000, 4500 orally, > 2000, > 2000 dermally (Hedlund, Anderson).

USE: Herbicide.

4826. Imazaquin. *2-[4,5-Dihydro-4-methyl-4-(1-methylethyl)-5-oxo-1H-imidazol-2-yl]-3-quinolinecarboxylic acid;* 2-(5-isopropyl-5-methyl-4-oxo-2-imidazolin-2-yl)-3-quinolinecarboxylic acid; AC 252214. $C_{17}H_{17}N_3O_3$; mol wt 311.34. C 65.58%, H 5.50%, N 13.50%, O 15.42%. Pre- and post-emergence imidazolinone herbicide especially for use in soybean crops. Prepn: M. Los, **Eur. pat. Appl. 41,623** (1981 to Am. Cyanamid), *C.A.* **96**, 199687q (1982). Alternate prepn: D. R. Maulding, R. F. Doehner, Jr., U.S. pat. **4,459,408** (1984 to Am. Cyanamid). Inhibition of branched-chain amino acid biosynthesis: D. L. Shaner *et al., Plant*

Physiol. **76**, 545 (1984). Metabolism by plants: D. L. Shaner, P. A. Robson, *Weed. Sci.* **33**, 469 (1985). Persistence in soil: G. Basham *et al., ibid.* **35**, 576 (1987). Herbicidal activity: M. W. Beale *et al., Proc. Ann. Meet. Northeast. Weed Sci. Soc.* **38**, 36 (1984); W. F. Congleton *et al., Weed Technol.* **1**, 186 (1987).

Crystals from hexane + ethyl acetate, mp 219-222° (dec). Slightly sol in some organic solvents. Soly in water at 25°: 60-120 ppm. LD_{50} orally in rats: 5000 mg/kg; LC_{50} (96 hr) in rainbow trout: 100 mg/l (Congleton).

Ammonium salt, $C_{17}H_{20}N_4O_3$, *Scepter.* Sol in water.

USE: Herbicide.

4827. Imiclopazine. *1-[2-[4-[3-(2-Chloro-10H-phenothiazin-10-yl)propyl]-1-piperazinyl]ethyl]-3-methyl-2-imidazolidinone;* 3-chloro-10-[γ-[N'-[β-(1-methyl-2-oxoimidazolidin-3-yl)ethyl]-N-piperazinyl]propyl]phenothiazine; chlorimpiphenine. $C_{25}H_{32}ClN_5OS$; mol wt 486.11. C 61.77%, H 6.64%, Cl 7.29%, N 14.41%, O 3.29%, S 6.60%. Prepn: **Belg. pat. 668,972** (1965 to Asta), *C.A.* **65**, 8934f (1966).

Free base, $bp_{0.01}$ 260°.
Dihydrochloride, $C_{25}H_{34}Cl_3N_5OS$, *P4241, Ponsital.* Crystals from dioxane or aq acetone.

THERAP CAT: Antipsychotic.

4828. Imidazole. Glyoxaline; 1,3-diazole; iminazole; miazole; pyrro[b]monazole; 1,3-diaza-2,4-cyclopentadiene. $C_3H_4N_2$; mol wt 68.08. C 52.92%, H 5.92%, N 41.15%. Prepd by the action of ammonia on glyoxal: Debus, *Ann.* **107**, 204 (1858); from glyoxal, ammonia, and formaldehyde: Radziszewski, *Ber.* **15**, 1493 (1882); Behrend, Schmitz, *Ann.* **277**, 338 (1893); vapor phase synthesis from formamide and ethylenediamine in presence of a dehydrogenation catalyst: Green, **U.S. pat. 3,255,200** (1966 to Air Products and Chemicals). Crystal structure: B. M. Craven *et al., Acta Crystallogr.* **33B**, 2585 (1977). Acute toxicity: Nishie *et al., Toxicol. Appl. Pharmacol.* **14**, 301 (1969). *Review:* Pyman, *J. Soc. Dyers Colourists* **36**, 107 (1920). *Monograph:* K. Hofmann, *Imidazole and Its Derivatives* (Interscience, New York, 1953). Review of imidazole chemistry: Grimmett, *Advan. Heterocyclic Chem.* **12**, 103-183 (1970).

Stout prisms from benzene. mp 90-91°. bp_{760} 257°; bp_{20} 165-168°; bp_{12} 138.2°. Weak base. K at 25° = 1.2×10^{-7}. Absorption spectrum: Rosanov, *J. Russian Physiol. Chem. Soc.* **48**, 1241 (1916); *Chem. Zentr.* **1923**, III, 1080. Freely sol in water, alcohol, ether, chloroform, pyridine; slightly sol in benzene; very sparingly sol in petr ether. LD_{50} in mice (mg/kg): 610 i.p.; 1880 orally (Nishie).

4829. Imidazole Salicylate. *2-Hydroxybenzoic acid compd with 1H-imidazole (1:1); mono(2-hydroxybenzoate)-1H-imidazole;* salizolo; ITF-182; Selezen. $C_{10}H_{10}N_2O_3$; mol wt 206.20. C 58.25%, H 4.89%, N 13.58%, O 23.28%. Prepn from equimolar amounts of salicylic acid with imidazole: M. Brissemoret, *Bull. Soc. Chim. France* Ser. 3, **35**, 316 (1906).

Use as anti-inflammatory agent: **Belg. pat. 889,704** corresp to G. Sportoletti, **U.S. pat. 4,329,340** (1981, 1982 both to Italfarmaco). Pharmacology: P. G. Pagella *et al., Arzneimittel-Forsch.* **33**, 716 (1983). Penetration of inflamed sites: eidem, *ibid.* **34**, 208 (1984). Pharmacokinetics: H. P. Kuemmerle *et al., Int. J. Clin. Pharmacol. Ther. Toxicol.* **22**, 521 (1984). Series of clinical studies: *Boll. Chim. Farm.* **122**, 37S-63S (1983). *Review:* R. Fantozzi, *Drugs Exptl. Clin. Res.* **10**, 853-856 (1984).

Crystals from methanol-ether, mp 123-124°. uv max: 300 nm ($E_{1cm}^{1\%}$ 182.5). Soly in water > 100 mg/cc. LD_{50} in male, female rats, mice (mg/kg): 763, 724, 595, 685 s.c.; 422, 434, 462, 435 i.v.; 1211, 1430, 1034, 1091 orally (Pagella).

USE: In heat sensitive copying materials: **Belg. pat. 766,987** corresp to R. C. Desjariais, **U.S. pat. 3,694,247** (1971, 1972 both to Scott Graphics).

THERAP CAT: Anti-inflammatory. Antipyretic. Analgesic.

4830. 2-Imidazolidinone. 2-Imidazolidone; ethylene urea. $C_3H_6N_2O$; mol wt 86.10. C 41.85%, H 7.03%, N 32.54%, O 18.58%. Prepd from ethylenediamine and carbon dioxide under the influence of heat and pressure: Mulvaney, Evans, *Ind. Eng. Chem.* **40**, 393 (1948); from ethylenediamine and urea: Schweitzer, *J. Org. Chem.* **15**, 471 (1950). Systematic survey and bibliography: Klaus Hofmann, *Imidazole*, Part I (Interscience, New York, 1953).

Needles from chloroform, mp 131°. Very sol in water and in hot alc; difficultly sol in ether.

USE: Manuf high polymers, finishing agents for textiles and leather. In the formulation of plasticizers, lacquers, and adhesives. Insecticide: Simkover, **U.S. pat. 3,242,044** (1966 to Shell).

4831. Imidocarb. *N,N'-Bis[3-(4,5-dihydro-1H-imidazol-2-yl)phenyl]urea; 3,3'-di-2-imidazolin-2-ylcarbanilide.* $C_{19}H_{20}N_6O$; mol wt 348.41. C 65.50%, H 5.78%, N 24.13%, O 4.59%. Prepn: **Brit. pat. 1,007,334** corresp to R. Fischer, R. Hirt, **U.S. pat. 3,338,917** (1965, 1967 both to Wander). Babesicidal effect in mice and rats: G. Schmidt *et al., Res. Vet. Sci.* **10**, 530 (1969); E. Beveridge, *ibid.* 534. Effect on exptl anaplasmosis in calves: T. O. Roby, *ibid.* **13**, 519 (1972). Comparison of the dipropionate and tetracycline, *q.v.,* in canine ehrlichiosis: J. E. Price, T. T. Dolan, *Vet. Rec.* **107**, 275 (1980). Efficacy in *Babesia felis* infection: F. T. Potgieter, *J. S. Afr. Vet. Assoc.* **52**, 289 (1981). Effect on *B. ovis* infection in sheep: S. A. Michael, A. H. El Refaii, *Trop. Animal Health Prod.* **14**, 1 (1982), *C.A.* **96**, 192991 (1982).

Dihydrochloride, $C_{19}H_{22}Cl_2N_6O$, *4A65, imizocarb.* Solid, mp 350° (dec). LD_{50} in mice, rats (mg/kg): 107, 150 s.c., E. Beveridge, *loc. cit.*

Dipropionate, $C_{25}H_{32}N_6O_7$, *imizol, Imizad Equine Injection.*

THERAP CAT (VET): Antiprotozoal (*Babesia*).

4832. Iminodiacetic Acid. *N-(Carboxymethyl)glycine;* iminodiethanoic acid; diglycine; IDA. $C_4H_7NO_4$; mol wt 133.10. C 36.10%, H 5.30%, N 10.52%, O 48.08%. HOOC-CH_2NHCH_2COOH. Obtained from nitrilotriacetic acid,

N(CH$_2$COOH)$_3$, by HCl-hydrolysis in a bomb tube: Schwarzenbach *et al.*, *Helv. Chim. Acta* **28**, 1133 (1945); by oxygenation in presence of palladium/ carbon catalyst: Tetenbaum, Stone, *Chem. Commun.* **1970**, 1699. Iminodiacetic acid and nitrilotriacetic acid are formed upon boiling chloroacetic acid with concd aq ammonia: Heintz, *Ann.* **149**, 88 (1869). *See also* Martell, Bersworth, *J. Org. Chem.* **15**, 46 (1950).

Orthorhombic crystals, dec 247.5° (commercial grade, mp 220-250°). pKa$_1$ 2.98; pKa$_2$ 9.89. Forms salts with acids and bases. Soly in water at 5°: 2.43 g/100 ml. Practically insol in acetone, methanol, ether, benzene, carbon tetrachloride, heptane.

Hydrochloride, C$_4$H$_8$ClNO$_4$, crystals, dec 238-239°. Concd aq solns yield the free acid when adjusted to pH 2 with NaOH.

Sodium salt monohydrate, C$_4$H$_6$NNaO$_4$.H$_2$O, freely sol in water.

Forms complexes with Mg, Ca, Ba.

USE: Has been suggested as intermediate in the manuf of surface active agents, complex salts, chelating agents.

4833. 4,4′-Iminodicyclohexanecarboxylic Acid. 4,4′-Dicarboxydicyclohexylamine. C$_{14}$H$_{23}$NO$_4$; mol wt 269.33. C 62.43%, H 8.61%, N 5.20%, O 23.76%. Prepd by hydrogenation of *p*-aminobenzoic acid: Ferber, Brückner, *Ber.* **76**, 1019 (1943); Chinoporos, *Anal. Chem.* **34**, 437 (1962).

HOOC—⟨cyclohexyl⟩—NH—⟨cyclohexyl⟩—COOH

Minute needles from dil ethanol, dec 188-190°. Sol in abs alcohol.

USE: Detection of ferric ions: Chinoporos, *loc. cit.*

4834. Imipenem. *[5R-[5α,6α(R*)]]-6-(1-Hydroxyethyl)-3-[[2-[(iminomethyl)amino]ethyl]thio]-7-oxo-1-azabicyclo-[3.2.0]hept-2-ene-2-carboxylic acid monohydrate;* N-formimidoylthienamycin monohydrate; imipemide; MK-787. C$_{12}$H$_{17}$N$_3$O$_4$S.H$_2$O; mol wt 317.36. C 45.42%, H 6.03%, N 13.24%, O 25.21%, S 10.10%. Extremely broad-spectrum semi-synthetic antibiotic; first stable derivative of thienamycin, *q.v.* Prepn: W. J. Leanza *et al.*, *J. Med. Chem.* **22**, 1435 (1979); T. W. Miller, *Eur. pat. Appl.* **6639** (1980 to Merck & Co.), *C.A.* **93**, 155845y (1980); B. G. Christensen *et al.*, *U.S. pat.* **4,194,047** (1980 to Merck & Co.). Totally synthetic prepn without formation of thienamycin: I. Shinkai *et al.*, *Tetrahedron Letters* **23**, 4903 (1982). HPLC determn in serum: C. M. Myers, J. L. Blumer, *Antimicrob. Ag. Chemother.* **26**, 78 (1984). Series of articles on *in vitro* activity, pharmacokinetics, clinical efficacy of combination with cilastatin sodium, *q.v.*, a renal dehydropeptidase I inhibitor: *J. Antimicrob. Chemother.* **12**, Suppl. D, 1-155 (1983); *Rev. Infect. Dis.* **7**, Suppl. 3, S389-S536 (1985); *Am. J. Med.* **78**, Suppl. 6A, 1-167 (1985); *Infection* **14**, Suppl. 2, S111-S180 (1986). Comprehensive description: E. R. Oberholtzer in *Analytical Profiles of Drug Substances* vol. 17, K. Florey, Ed. (Academic Press, New York, 1988) pp 73-114.

Crystals from water-ethanol. [α]$_D^{25}$ +86.8° (c = 0.05 in 0.1*M* phosphate, pH 7). pKa$_1$ ~3.2, pKa$_2$ ~9.9. uv max (water): 299 nm (ε 9670, 98% NH$_2$OH ext). Soly (mg/ml): water 10, methanol 5, ethanol 0.2, acetone <0.1, dimethylformamide <0.1, dimethylsulfoxide 0.3.

Combination with cilastatin sodium, *q.v.*, Primaxin, Tenacid, Tienam, Zienam.

THERAP CAT: Antibacterial.

4835. Imipramine. *10,11-Dihydro-N,N-dimethyl-5H-dibenz[b,f]azepine-5-propanamine; 5-(3-dimethylaminopropyl)-10,11-dihydro-5H-dibenz[b,f]azepine;* N-(γ-dimethyl-

aminopropyl)iminodibenzyl; imizin; G 22355; Antideprin. C$_{19}$H$_{24}$N$_2$; mol wt 280.40. C 81.38%, H 8.63%, N 9.99%. Tricyclic antidepressant. Prepn: Haefliger, Schindler, *U.S. pat.* **2,554,736** (1951 to Geigy); *eidem, Helv. Chim. Acta* **37**, 472 (1954). Reviews of pharmacology: Crismon, *Psychopharmacol. Bull.* **4**, 151 pp (Oct. 1967); Glassman, Perel, *Arch. Gen. Psychiat.* **28**, 649 (1973). Toxicity data: A. Tobe *et al.*, *Arzneimittel-Forsch.* **31**, 1278 (1981). Comparative clinical trials in depression: R. S. Lipman *et al.*, *Arch. Gen. Psychiatry* **43**, 68 (1986); in anxiety: R.J. Kahn *et al.*, *ibid.* 79. Comprehensive description: D. N. Kender, R. E. Schiesswohl in *Analytical Profiles of Drug Substances* vol. 14, K. Florey, Ed. (Academic Press, New York, 1985) pp 37-75.

Free base, bp$_{0.1}$ 160°.

Hydrochloride, C$_{19}$H$_{25}$ClN$_2$, *Berkomine, Deprinol, Dyna-Zina, Efuranol, Imavate, Tofranil, Chrytemin, Imiprin, Iramil, Imidol, Melipramin, Dimipressin, Pryleugan, Presamine, Feinalmin, Imilanyle, Iprogen, Janimine.* Crystals from acetone, mp 174-175°. Acquires a yellow to reddish discoloration under the influence of light. Freely sol in water, less sol in alcohol, sparingly sol in acetone. LD$_{50}$ in mice, rats (mg/kg): 400, 490 orally; 110, 90 i.p. (Tobe).

Pamoate, *Tofranil-PM.*

THERAP CAT: Antidepressant.

4836. Imipramine N-Oxide. *10,11-Dihydro-N,N-dimethyl-5H-dibenz[b,f]azepine-5-propanamine 5-oxide; 5-[3-(dimethylamino)propyl]-10,11-dihydro-5H-dibenz[b,f]azepine 5-oxide;* N-(γ-dimethylaminopropyl)iminobenzyl N-oxide. C$_{19}$H$_{24}$N$_2$O; mol wt 296.41. C 76.99%, H 8.16%, N 9.45%, O 5.40%. Identification as a metabolite of imipramine, *q.v.*: Fishman, Goldenberg, *Proc. Soc. Exp. Biol. Med.* **110**, 187 (1962). Prepn: **Fr. pat. M2508**; H. Dyrsting, J. B. Pederson, *U.S. pat.* **3,574,852** (1964, 1971, both to Aktieselskabet Dumex); Dietrich, **Swiss** pat. **408,018** (1966 to Geigy), *C.A.* **66**, 10857n (1967). Pharmacology: Buech *et al.*, *Med. Pharmacol. Exp.* **15**, 187 (1966); Bickel, Weder, *J. Pharm. Pharmacol.* **21**, 160 (1969). Metabolism: Bickel, Baggiolini, *Biochem. Pharmacol.* **15**, 1155 (1966); Minder *et al.*, *Arch. Pharmakol. Exp. Pathol.* **268**, 334 (1971).

White needle-shaped crystals, mp 120-123° (dec). (Monohydrate, mp 75-79°.) Sol in methanol, ether, acetone and benzene. Strongly hygroscopic.

Hydrochloride, C$_{19}$H$_{25}$ClN$_2$O, *Imiprex.* White crystals, mp 153.1-155° (dec). LD$_{50}$ in rats, mice (mg/kg): 90, 150 i.p. (Dyrsting, Pederson).

THERAP CAT: Antidepressant.

4837. Immunoglobulins. *Immune globulins;* gamma globulins; γ-globulins; Ig's; Allergam; Cytotect; Endobulin; Gamastan; Gamimune; Gammagard; Gammagee; Gammabulin; Gammar; Immuglobin; Intraglobin; Kabiglobulin; Sandoglobulin. A heterogeneous group of proteins synthesized by lymphocytes and plasma cells and found in human and animal blood serum, urine, spinal fluid, lymph nodes, spleen, and other body fluids and tissues. They function as antibodies by recognizing determinants on the surfaces of foreign substances (antigens) and binding to them. Immunoglobulins are divided into classes arising from differences in structure and in biological properties; major classes are designated as *IgG, IgM, IgA, IgD, IgE.* IgG, IgM, and IgA constitute about 80%, 10% and <10%, resp, of the total γ-globulins. Normal serum concentrations of IgD and IgE are low, but IgE levels are elevated in allergic persons. All immunoglobulins have a general structure consisting of two heavy chains and two light chains of varying lengths held together by disulfide bonds. Heavy chains have mol weights of approx 55,000 and are differentiated into the γ, α, δ, ϵ, or μ types; light chains have mol weights of approx 20,000 and are of the K (k) or L (λ) type. IgM and IgA contain an additional J chain, cf. S. P. Hauptman, T. B. Tomas, *J. Biol. Chem.* **250**, 3891 (1975). System of nomenclature: Rowe, *Nature* **228**, 509 (1970). Variations in amino acid sequence occur in the N-terminal half of the chains, while the C-terminal half is constant: Hieschmann, Craig, *Proc. Nat. Acad. Sci. USA* **53**, 1403 (1965). These variations, due to genetic factors and antigen specificity, are infinite in normal Ig's, making structural determinations by usual methods of protein chemistry difficult. First complete amino acid sequence of an IgG molecule: G. E. Edelman *et al., ibid.* **63**, 78 (1969). Review of structural studies: Cohen, Milstein, *Nature* **214**, 449 (1967); Porter in *MTP Int. Rev. Sci.: Biochem. Ser. One* vol. 10, R. R. Porter, Ed. (Univ. Park Press, Baltimore, 1973) pp 159-197; R. S. Nezlin, Y. K. Sykulev, *Mol. Immunol.* **19**, 347-356 (1982). Review of therapeutic uses: E. R. Stiehm, *Pediatrics* **63**, 301-319 (1979). Comprehensive review: Finger, Seeliger, in *Research in Immunochemistry and Immunobiology*, vol. 1, J. B. G. Kwapinski, Ed. (University Park Press, Baltimore, 1972) pp 3-70; J. N. Goldman, M. B. Goldman, *J. Am. Med. Assoc.* **251**, 774-787 (1984). Books: R. Grubb, *The Genetic Markers of Human Immunoglobulins* (Springer-Verlag, New York, 1970) 152 pp; L. E. Glynn, M. W. Steward, *Structure and Functions of Antibodies* (Wiley-Interscience, New York, 1981) 231 pp; *Pathology of Immunoglobulins: Diagnostic and Clinical Aspects* vol. 2, S. E. Ritzmann, Ed. (G. R. Liss, New York, 1982) 396 pp.

Monoclonal antibodies. Identical antibodies (immunoglobulins) produced against a single antigenic determinant by a single cell's progeny, or clone. When cells from myelomas (malignant tumors of the immune system) are fused with lymphocytes obtained from animals that have been immunized with a particular antigen, the resulting hybrid cells, termed *hybridomas*, possess the "immortality" of the myeloma cells and the lymphocytes' ability to produce specific antibodies, cf. G. Kohler, C. Milstein, *Nature* **256**, 495 (1975). Monoclonal antibodies produced by this method are now widely used in research. Exptl use in detection of metastatic tumor: P. J. Moldofsky *et al., N. Engl. J. Med.* **311**, 106 (1984). *Reviews:* A. S. Tung, *Ann. Rep. Med. Chem.* **16**, 243-256 (1981); *Monoclonal Hybridoma Antibodies: Techniques and Applications*, J. G. R. Hurrell, Ed. (CRC Press, Boca Raton, 1982) 240 pp.

USE: Monoclonal antibodies as tools in biological research and clinical medicine.

THERAP CAT: Passive immunizing agent.

4838. Imolamine. *4-[2-(Diethylamino)ethyl]-5-imino-3-phenyl-Δ^2-1,2,4-oxadiazoline;* 3-phenyl-4-diethylaminoethyl-5-imino-1,2,4-oxadiazoline. $C_{14}H_{20}N_4O$; mol wt 260.33. C 64.59%, H 7.74%, N 21.52%, O 6.15%. Prepn: Aron-Samuel, Sterne, *Fr. pat.* M2023 (1963), *C.A.* **60**, 2952f (1964). Pharmacology: Molino, Belluardo, *Gazz. Int. Med. Chir.* **71**, 2100 (1966), *C.A.* **66**, 74830e (1967).

$bp_{0.2}$ 165°.
Hydrochloride, $C_{14}H_{21}ClN_4O$, *LA 1211, Angolon, Irrigor*. Crystals, mp 154-155°.

THERAP CAT: Antianginal.

4839. Imperatorin. *9-[(3-Methyl-2-butenyl)oxy]-7H-furo[3,2-g][1]benzopyran-7-one;* 6-hydroxy-7-(3-methyl-2-butenyloxy)-5-benzofuranacrylic acid δ-lactone; 8-isoamylenoxypsoralen; marmelosin; ammidin; pentosalen. $C_{16}H_{14}$-O_4; mol wt 270.27. C 71.10%, H 5.22%, O 23.68%. From roots of *Imperatoria osthruthium* L., *Umbelliferae:* Späth,

Holzen, *Ber.* **66**, 1137 (1933); from seeds of *Angelica archangelica* L.: Späth, Vierhapper, *ibid.* **71**, 1667 (1938); from fruit of *Pastinaca sativa* L., *Umbelliferae:* Soine *et al., J. Am. Pharm. Assoc.* **45**, 426 (1956). Identity with marmelosin: Späth *et al., Ber.* **70**, 1021 (1937). Identity with ammidin: Fahmy, Abu-Shady, *Quart. J. Pharm. Pharmacol.* **21**, 499 (1948). Partial synthesis: Späth, Holzen, *Ber.* **68**, 1123 (1935). Synthesis: Späth, Vierhapper, *ibid.* **70**, 248 (1937).

Prisms from ether, long fine needles from hot water, mp 102°. Practically insol in cold water. Very sparingly sol in boiling water; freely sol in chloroform; sol in benzene, alcohol, ether, petr ether, alkali hydroxides. uv max: 302, 265, 250 nm (log ϵ 3.95, 4.00, 4.24).

4840. Imperialine. *3,20-Dihydroxycevan-6-one;* raddeamine; sipeimine. $C_{27}H_{43}NO_3$; mol wt 429.62. C 75.48%, H 10.09%, N 3.26%, O 11.17%. Isolation from bulb of *Fritillaria imperialis* L., *Liliaceae:* Fragner, *Ber.* **21**, 3284 (1888); Boit, *ibid.* **87**, 472 (1954); from *Petilium eduardi:* Shakirov *et al., C.A.* **60**, 13280h (1964). Identity with sipeimine and raddeamine: Chu *et al., C.A.* **53**, 7503e (1959); Nurridinov, Yunusov, *C.A.* **57**, 15165h (1962). Structure: Nuriddinov *et al., Khim. Prir. Soedin* **3**, 316 (1967); *C.A.* **68**, 69168g (1968). Prepn of the β-D-glucopyranoside: Shakirov *et al., C.A.* **63**, 1858e (1965). Isolation and structure of the β-D-glucopyranoside: *eidem, C.A.* **63**, 3007f (1965). Absolute configuration: S. Itô *et al., Tetrahedron Letters* **1976**, 3161.

Prisms from methanol, mp 267°. $[\alpha]_D^{22}$ $-38.5°$ (c = 1.5 in chloroform).

β-D-Glucopyranoside, $C_{33}H_{53}NO_8$, *edpetiline*. Crystals from methanol, mp 272-276°. $[\alpha]_D$ $-57.89°$ (c = 0.449 in methanol).

4841. Improsulfan. *3,3'-Iminobis-1-propanol dimethanesulfonate (ester);* N,N-bis(3-methylsulfonyloxypropyl)-amine; bis(3-mesyloxypropyl)amine. $C_8H_{19}NO_6S_2$; mol wt 409.48. C 52.80%, H 4.68%, N 3.42%, O 23.44%, S 15.66%. Alkylating agent related to busulfan, *q.v.* Prepn and antitumor activity of the hydrochloride: M. M. El-Merzabani, Y. Sakurai, *Gann.* **56**, 589 (1965), *C.A.* **64**, 8793b (1965); Y. Sakurai, **Japan. pat.** 67 26,616 (1967), *C.A.* **68**, 51596f (1968). Prepn of the tosylate: M. Nakanishi *et al., Ger. pat.* **2,059,377** (1971 to Yoshitomi), *C.A.* **75**, 63127r (1972). Absorption, distribution, excretion, metabolism: Y. Kato *et al., Yakugaku Zasshi* **94**, 1107 (1974), *C.A.* **82**, 38441a (1975). Pharmacology: N. Brock, J. Kuhlmann, *Arzneimittel-Forsch.* **24**, 1139 (1974); H. Imamura *et al., Oyo Yakuri* **9**, 861 (1975), *C.A.* **84**, 12492j (1976). Antitumor effect: T. Okumoto *et al., Chem. Pharm. Bull.* **25**, 3003 (1977); T. Iwaguchi, H. Kitagawa, *Gann.* **69**, 123 (1968), *C.A.* **89**, 157282 (1978). Clinical study: S. J. Altman *et al., Cancer* **35**, 1145 (1975). Toxicity study: N. Rakieten *et al., C.A.* **69**, *U.S. NTIS Rep.* **PB-196468** (1970) 17 pp, *C.A.* **75**, 47359n (1971). Mutagenicity study: W. F. Benedict *et al., Cancer Res.* **37**, 2209 (1977).

Hydrochloride, $C_8H_{20}ClNO_6S_2$, *Yoshi 864, NSC-102627*. Crystals, mp 94-95°. LD_{50} i.v. in rats: 75 mg/kg (Brock, Kuhlmann).

Tosylate, $C_{15}H_{27}NO_9S_3$, *864T, NSC-140117, Protecton*. White crystals, mp 113-118°. Faint, characteristic odor and bitter taste. Sol in water, methanol, acetonitrile. Slightly sol in chloroform. LD_{50} in male Wistar rats (mg/kg): 137 orally; 73 i.v.; 131 i.p.; 147 s.c., *Japan. Med. Gaz.* **17**(3), 5 (1980).

THERAP CAT: Antineoplastic.

4842. Indaconitine. *16-Ethyl-1,6,19-trimethoxy-4-(methoxymethyl)aconitane-3,8,10,11-tetrol 8-acetate 10-benzoate;* acetylbenzoylpseudaconine. $C_{34}H_{47}NO_{10}$; mol wt 629.72. C 64.84%, H 7.52%, N 2.22%, O 25.41%. From root of *Aconitum chasmanthum* Stapf. *Ranunculaceae:* Dunstan, Andrews, *J. Chem. Soc.* **87**, 1620 (1905). Structure: Gilman, Marion, *Can. J. Chem.* **42**, 2700 (1964).

Needles from ether, mp 202-203°. $[\alpha]_D^{21}$ +18.28° (c = 2.18 in alc). Indaconitine and its salts are highly toxic: Cash, Dunstan, *Proc. Roy. Soc. (London)* **B176**, 468 (1905). Sol in acetone, chloroform, alcohol, ether; practically insol in water, petr ether.

Hydrobromide, prisms from water, mp 183-187°; mp 217-218° from alcohol + ether. $[\alpha]_D$ −17.27° (c = 2.991).

Hydrochloride trihydrate, needles from alcohol + ether. Anhyd salt mp 166-171°. Very hygroscopic. $[\alpha]_D^{20}$ −15.83° (c = 1.921).

4843. Indalpine. *3-[2-(4-Piperidinyl)ethyl]-1H-indole;* 4-[2-(3-indolyl)ethyl]piperidine; LM-5008; Upstene. $C_{15}H_{20}N_2$; mol wt 228.34. C 78.90%, H 8.83%, N 12.27%. Selective serotonin uptake inhibitor. Prepn: A. A. Champseix *et al.*, *Ger.* pat. 2,618,152; *eidem*, *U.S.* pat. 4,064,255 (both 1977 to Mar-Pha Soc. Etud. Exploit. Marques). HPLC determn: C. Jozefczak *et al.*, *J. Chromatog.* **230**, 87 (1982). Spectrofluorimetric determn and clinical pharmacology: A. Iliopoulou *et al.*, *Pharmatherapeutica* **2**, 613 (1981). Pharmacology: G. LeFur *et al.*, *Life Sci.* **23**, 1959 (1978); C. Gueremy *et al.*, *J. Med. Chem.* **23**, 1306 (1980); R. Ashkenazi *et al.*, *Brit. J. Pharmacol.* **79**, 765, 915 (1983). Clinical study: A. Wauters, *Acta Psychiat. Belg.* **83**, 69 (1983).

Monohydrochloride, $C_{15}H_{21}ClN_2$, mp 167°.

THERAP CAT: Antidepressant.

4844. Indan. *2,3-Dihydro-1H-indene;* hydrindene. C_9H_{10}; mol wt 118.17. C 91.47%, H 8.53%. Occurs in coal tar: Krämer, Spilker, *Ber.* **29**, 552 (1896). Prepn: *eidem, ibid.* **33**, 2257 (1900); Grosse *et al.*, *Ind. Eng. Chem.* **38**, 1041 (1946); Carpenter, Easter, *J. Org. Chem.* **19**, 87 (1954); Juday, *ibid.* **22**, 532 (1957); Hunter, Aldridge, *U.S.* pat. 3,082,267 (1963 to Esso); Bestmann *et al.*, *Ann.* **718**, 33 (1968).

Liquid. d_4^{20} 0.9639; d_4^{50} 0.9378. mp −51.4°. bp_{762} 176.5°; bp_{12} 61-64°. n_D^{20} 1.5383. uv max (isooctane): 272, 265, 258 nm (log ϵ 3.18, 3.08, 2.90). Insol in water. Sol in organic solvents.

1-Methylindan, $C_{10}H_{12}$. Liq, d_4^{20} 0.9402. bp_{760} 186-190°; bp_{10} 60°. n_D^{20} 1.5222.

4845. Indanazoline. *N-(2,3-Dihydro-1H-inden-4-yl)-4,5-dihydro-1H-imidazol-2-amine;* N-(2-imidazolin-2-yl)-N-(4-indanyl)amine; E-VA-16. $C_{12}H_{15}N_3$; mol wt 201.27. C 71.61%, H 7.51%, N 20.88%. Prepn: H. J. May, A. Berg, *Ger.* pat. 2,136,325 corresp to *U.S.* pat. 3,882,229 (1973, 1975 both to Nordmark); H. J. May, *Arzneimittel-Forsch.* **30**, 1733 (1980). Series of articles on galenical development, pharmacology, toxicity, pharmacokinetics, decongestant effect: *ibid.* 1738-1787.

Cryst from petr ether, mp 109-113°.

Hydrochloride, $C_{12}H_{16}ClN_3$, *Farial*. Cryst from isopropanol, mp 182-184°. LD_{50} in male, female mice (mg/kg): 179, 233 orally; 22.3, 26.9 i.v.; in male, female rats: 481, 542 orally; 16.3, 17.6 i.v.; W. Worstmann *et al.*, *Arzneimittel-Forsch.* **30**, 1760 (1980).

THERAP CAT: Nasal decongestant; vasoconstrictor.

4846. Indanthrene®. *6,15-Dihydro-5,9,14,18-anthrazinetetrone; N,N'*-dihydro-1,2,1',2'-anthraquinonazine. $C_{28}H_{14}N_2O_4$; mol wt 442.41. C 76.01%, H 3.19%, N 6.33%, O 14.47%. Vat dye discovered by René Bohn (1901). Prepn: Fierz-David, Blangey, *Grundlegende Operationen der Farbenchemie* (Vienna, 5th ed., 1943) pp 304-305; Thielert, Bauman, *U.S.* pat. 2,693,469 (1954 to Bayer); Sutter, Fioroni, *U.S.* pat. 2,831,860 (1958 to Ciba); Kastner, *U.S.* pat. 3,138,612 (1964 to Allied Chem.). Structure: Weinstein, Merrit, *J. Am. Chem. Soc.* **81**, 3759 (1959).

Blue powder, dec 470-500°. uv max on cellophane film: 278 nm. Practically insol in organic solvents. Sol in concd H_2SO_4, in dil alkali solns.

USE: Mainly to dye cotton. Indanthrene Blue R, the usual commercial grade, is extremely stable to light and heat, but sensitive to chlorine. A purer grade, which is not as sensitive to chlorine, is sold as Indanthrene Brilliant Blue FF.

4847. Indapamide. *3-(Aminosulfonyl)-4-chloro-N-(2,3-dihydro-2-methyl-1H-indol-1-yl)benzamide; 4-chloro-N-(2-methyl-1-indolinyl)-3-sulfamoylbenzamide;* N-(3-sulfamyl-4-chlorobenzamido)-2-methylindoline; S 1520; SE 1520; Bajaten; Damide; Fludex; Indaflex; Indamol; Ipamix; Lozol; Natrilix; Noranat; Tandix; Veroxil. $C_{16}H_{16}ClN_3O_3S$; mol wt 365.84. C 52.53%, H 4.41%, Cl 9.69%, N 11.49%, O 13.12%, S 8.76%. Prepn: Beregi *et al.*, *Fr.* pat. 2,003,311 corresp to *U.S.* pat. 3,565,911 (1969, 1971, both to Sci. Union et Cie, Soc. Franc. Recherche Med.). Pharmacology:

Leary et al., Curr. Ther. Res. Clin. Exp. **15**, 571 (1973); D. B. Campbell, R. A. Moore, Postgrad. Med. J., Suppl. **57**, 7 (1981). Acute toxicity data: J. Kyncl et al., Arzneimittel-Forsch. **25**, 1491 (1975). Symposium on pharmacology and clinical efficacy: Am. J. Med. **84**, Suppl. 1B, 1-111 (1988).

Crystals from isopropanol/water, mp 160-162°. LD_{50} in rats, mice, guinea pigs (mg/kg): 393-421, 410-564, 347-416 i.p.; 394-440, 577-635, 272-358 i.v.; > 3000 all species orally (Kyncl).

Hemihydrate, $C_{16}H_{16}ClN_3O_3S$.½H_2O, *Pressural*.

THERAP CAT: Diuretic. Antihypertensive.

4848. 1H-Indazole. Isoindazole; benzopyrazole. C_7H_6-N_2; mol wt 118.13. C 71.16%, H 5.12%, N 23.72%. Prepn: Stephenson, Org. Syn., coll. vol. III, 475 (1955); Ainsworth, Org. Syn. **39**, 27 (1959); Huisgen, Bast, ibid. **42**, 69 (1962).

Needles from hot water, mp 146.5°. bp_{743} 267-270°. Sol in hot water, alcohol, ether.

Picrate, yellow needles or orange prisms from ether, mp 136°.

4849. Indecainide. *9-[3-[(1-Methylethyl)amino]propyl]-9H-fluorene-9-carboxamide;* 9-carbamoyl-9-(3-isopropyl-aminopropyl)fluorene; 9-[3-(isopropylamino)propyl]-9-(aminocarboyl)fluorene; ricainide. $C_{20}H_{24}N_2O$; mol wt 308.42. C 77.89%, H 7.84%, N 9.08%, O 5.19%. Prepn: W. B. Lacefield, R. L. Simon, U.S. pats. **4,197,313** and **4,452,-745** (1980, 1984 both to Lilly). Pharmacokinetics in animals: T. L. Lindstrom, G. W. Whitaker, Drug Metab. Dispos. **12**, 683 (1984). Metabolism: eidem, ibid. 691. LC determn in biological fluids: K. Z. Farid et al., J. Chromatog. **337**, 329 (1985). Toxicological studies: G. E. Sandusky, Jr., D. B. Meyers, Fundam. Appl. Toxicol. **5**, 175 (1985). Clinical evaluation in cardiac arrhythmias: P. F. Nestico et al., Am. J. Cardiol. **59**, 1332 (1987); P. J. Podrid et al., ibid. **61**, 764 (1988).

Crystals from Skelly B, mp 94-95°.

Hydrochloride, $C_{20}H_{25}ClN_2O$, *LY 135837*, *Decabid.* Crystals from chloroform, mp 216.5-217°; also reported as crystals from fresh ethanol and diethyl ether, mp 203-204°. LD_{50} orally in male, female mice, rats: 100, 96, 103, 82 mg/kg (Sandusky, Meyers).

THERAP CAT: Cardiac depressant (antiarrhythmic).

4850. Indeloxazine Hydrochloride. *2-[(1H-Inden-7-yl-oxy)methyl]morpholine hydrochloride;* (R,S)-2-[(7-indenyl-oxy)methyl]morpholine hydrochloride; YM-08054-1; Elen; Noin. $C_{14}H_{18}ClNO_2$; mol wt 267.76. C 62.80%, H 6.78%, Cl 13.24%, N 5.23%, O 11.95%. Serotonin uptake inhibitor. Prepn: M. Masuo et al., Ger. pat 2,601,703; eidem, U.S. pat. **4,109,088** (1976, 1978 both to Yamanouchi). Synthesis and resolution of isomers: T. Kojima et al., Chem. Pharm. Bull. **33**, 3766 (1985). Pharmacology and toxicity: S. Tachikawa et al., Arch. Int. Pharmacodyn. **238**, 81 (1979). Inhibition of synaptosomal uptake of serotonin and noradrenaline: M. Harada, H. Maeno, Biochem. Pharmacol. **28**, 2645 (1979).

GLC determn in human plasma: A. G. Hayes, T. Chang, J. Chromatog. **272**, 176 (1983).

(±)-Form: Polymorphic; pale yellow needles from methanol, mp 169-170°; colorless, acicular crystals from acetone, mp 155-156°. LD_{50} in mice (mg/kg): 47 i.v. (Tachikawa).

(+)-Form, crystals from ethanol, mp 112-113°. $[\alpha]_D^{21}$ +4.9° (c = 5 in methanol).

(−)-Form, crystals from isopropanol, mp 142-142.5°. $[\alpha]_D^{20}$ −4.9° (c = 5 in methanol).

THERAP CAT: Antidepressant; nootropic.

4851. Indene. Indonaphthene. C_9H_8; mol wt 116.15. C 93.06%, H 6.94%. Found in the tars from coal, lignite, and crude petr. Isoln from coal tar: Weissgerber, Ber. **42**, 569 (1909); Brennstoff-Chemie **5**, 208 (1924); Weissgerber, Seidler, Ber. **60**, 2088 (1927). Prepn from acetylene over activated charcoal at 625°: Zelinsky, Ber. **57**, 264 (1924). Manuf from tetrahydronaphthalene by passing over SiO_2-Al_2O_3 catalyst at 670°: Brit. pat. **578,083** (1946). Prepn from o-$BrCH_2C_6H_4CH_2CH_2Br$ and triphenylphosphine: Bestmann et al., Ann. **718**, 33 (1968).

Liquid. d_4^4 1.0081; d_4^{20} 0.9968; d_4^{50} 0.9692. mp −1.8°. bp_{760} 181.6°; bp_{400} 157.8°; bp_{200} 135.6°; bp_{100} 114.7°; bp_{60} 100.8°; bp_{40} 90.7°; bp_{20} 73.9°; bp_{10} 58.5°; bp_5 44.3°; $bp_{1.0}$ 16.4°. $n_D^{18.5}$ 1.5773. Absorption spectrum: Morton, de Gouveia, J. Chem. Soc. **1934**, 911. Insol in water. Miscible with most organic solvents. Polymerizes and oxidizes on standing. Concd H_2SO_4 forms metaindene, $(C_9H_8)_{16-22}$.

4852. Indenolol. *1-[1H-Inden-4(or 7)-yloxy]-3-[(1-methylethyl)amino]-2-propanol;* (±)-1-[inden-4(or 7)-yl-oxy]-3-(isopropylamino)-2-propanol; YB-2; Sch 28316Z. $C_{15}H_{21}NO_2$; mol wt 247.35. C 72.84%, H 8.56%, N 5.66%, O 12.94%. Non-selective β-adrenergic blocker. Prepn: M. Murakami et al., Ger. pat. **1,955,229**; eidem, U.S. pat. **4,045,482** (1970, 1977 both to Yamanouchi); K. Murase et al., Yakugaku Zasshi **92**, 1358 (1972), C.A. **78**, 71723 (1973). Indenolol is a tautomeric mixture of the 7- and 4-indenyl-oxy isomers in a 2:1 ratio, respectively. Unambiguous synthesis of the two isomers: eidem, Chem. Pharm. Bull. **24**, 552 (1976). β-Blocking and cardiovascular properties: T. Takenaka, S. Tachikawa, Arzneimittel-Forsch. **22**, 1864 (1972). General pharmacology: M. Takeda et al., Oyo Yakuri **7**, 469 (1973), C.A. **80**, 44015 (1974). Comparative pharmacological study with other β-blockers: W. Bartsch et al., Arzneimittel-Forsch. **27**, 1022 (1977). Dose-response studies: F. E. Okupa et al., Clin. Pharmacol. Ther. **29**, 434 (1981). Pharmacokinetics: R. Sega et al., J. Clin. Pharmacol. **25**, 337 (1985). Clinical evaluation in hypertension: B. Trimarco et al., ibid. 328; L. Poggesi et al., Clin. Pharmacol. Ther. **41**, 344 (1987).

Crystals from n-hexane/ether, mp 88-89°.

Hydrochloride, $C_{15}H_{22}ClNO_2$, *Pulsan, Securpres.* Crystals from ethanol/ether, mp 147-148°. LD_{50} in mice: 26 mg/kg i.v. (Bartsch).

THERAP CAT: Antihypertensive, antiarrhythmic, antianginal.

4853. Indican (Metabolic Indican). *1H-Indol-3-ol hydrogen sulfate ester;* indol-3-yl sulfate; 3-indoxylsulfuric acid. $C_8H_7NO_4S$; mol wt 213.23. C 45.06%, H 3.31%, N 6.57%, O 30.01%, S 15.04%. Unstable as the free acid. Occurs in urine of mammals, also in blood plasma. Prepn of potassium salt from potassium indoxyl and potassium bisulfate: Baeyer, *Ber.* **14**, 1745 (1881); from chlorosulfonic acid and *N*-acetylindoxyl in pyridine: Schwenk, Jolles, *Biochem. Z.* **69**, 467 (1915); by persulfate oxidation of indole: Boyland *et al.*, *Biochem. J.* **62**, 546 (1956).

Potassium salt, $C_8H_6KNO_4S$, *indol-3-yl potassium sulfate, urinary indican, potassium indoxyl sulfate.* Light brown plates from aq alc, dec 179-180° with sublimation. Very sol in water; practically insol in cold alcohol.

4854. Indican (Plant Indican). *1H-Indol-3-yl-β-D-glucopyranoside;* 3-(β-glucosido)indole; indoxyl-β-D-glucoside. $C_{14}H_{17}NO_6$; mol wt 295.28. C 56.94%, H 5.80%, N 4.74%, O 32.51%. Isolated from *Indigofera* spp., *Leguminosae* and from *Polygonium tinctorium* Ait., *Polygonaceae.* Extraction procedure for the glucoside: Hoogewerff, Meulen, *Rec. Trav. Chim.* **19**, 166 (1900); Perkin, Bloxam, *J. Chem. Soc.* **91**, 1715 (1907); Perkin, Thomas, *ibid.* **95**, 793 (1909). About 3% indican is found in 4-day old air-dried leaves of *Indigofera tinctoria.* Synthesis: Robertson, *J. Chem. Soc.* **1927**, 1937; Robertson, Waters, *ibid.* **1933**, 30; Freudenberg *et al.*, *Ber.* **85**, 641 (1952).

Trihydrate, orthorhombic spear-shaped needles, mp 57-58°. $[\alpha]_{546}^{19}$ —66°. After heating *in vacuo* (oil pump) over H_2SO_4 for 48 hrs the hemihydrate is obtained, mp 101°. Robertson obtained the anhydrous compound by heating in the steam oven for 3 hrs, at 110° for 1 hr, and at 160° for 5 minutes. Perkin and Bloxam obtained anhydr indican by treating the trihydrate with boiling alcohol and boiling benzene. Anhydr indican decomposes at 178-180°. $[\alpha]_D^{15}$ —66° (c = 2); $[\alpha]_{546}^{19}$ —78° (c = 0.6). Sol in water, methanol, alcohol, acetone; slightly sol in ether, benzene, chloroform, ethyl acetate, carbon disulfide. Anhydr indican is sparingly sol in anhydr alcohol.

4855. Indigo. *2-(1,3-Dihydro-3-oxo-2H-indol-2-ylidene)-1,2-dihydro-3H-indol-3-one; [Δ²,²'-biindoline]-3,3'-dione;* Δ²,²'-bipseudoindoxyl; indigotin; indigo blue; D & C Blue No. 6; C.I. Pigment Blue 66; C.I. Vat Blue 1; C.I. 73000. $C_{16}H_{10}N_2O_2$; mol wt 262.26. C 73.27%, H 3.84%, N 10.68%, O 12.20%. Probably the oldest known coloring matter. Originally obtained from various species of *Indigofera, Leguminosae,* indigenous to Bengal, Java, Guatemala, in which it occurs as a glucoside. First synthesis: Baeyer, *Ber.* **11**, 1296 (1878); **12**, 456 (1879). First commercial process: Heumann, *Ber.* **23**, 3043, 3431 (1890). Newer prepns: Harley-Mason, *J. Chem. Soc.* **1950**, 2907; Ziegler, Kappe, *Monatsh.* **96**, 889 (1965); J. Gosteli, *Helv. Chim. Acta* **60**, 1980 (1977). Indigo occurs in *cis* and *trans* forms, in solid state it is in the *trans* form: Posner, *Ber.* **56**, 31 (1923); **57**, 1313 (1924); **59**, 1799 (1926). Reviews on indigo and indigoid dyes: Lubs, *The Chemistry of Synthetic Dyes*, ACS Monograph Series no. **127** (Reinhold, New York, 1955) pp 551-576; Schweizer, *Künstliche Organische Farbstoffe und ihre Zwischenprodukte* (Springer Verlag, 1964) pp 320-334; *Colour Index* vol. 4 (3rd ed., 1971) p 4595.

Dark-blue powder with coppery luster. Sublimes at about 300°; dec 390°. Practically insol in water, alcohol, ether, and dil acids; dissolves in nonpolar solvents with red and in polar solvents with blue color. With fuming H_2SO_4 it forms a sol sulfonic acid.

USE: As textile dye. Approved by FDA for use in sutures.

4856. Indigo Carmine. *2-(1,3-Dihydro-3-oxo-5-sulfo-2H-indol-2-ylidene)-2,3-dihydro-3-oxo-1H-indole-5-sulfonic acid disodium salt; 3,3'-dioxo-[Δ²,²'-biindoline]-5,5'-disulfonic acid disodium salt;* disodium 5,5'-indigotin disulfonate; sodium indigotin disulfonate; soluble indigo blue; indigotine; Acid Blue 74; C.I. Acid Blue 74; C.I. Food Blue 1; FD & C Blue No. 2; C.I. 73015. $C_{16}H_8N_2Na_2O_8S_2$; mol wt 466.37. C 41.20%, H 1.73%, N 6.01%, Na 9.86%, S 13.75%, O 27.45%. Synthesis and structure determination: Vorlander, Schubart, *Ber.* **34**, 1860 (1901). Prepn and properties: Matthews, *Color Trade J.* **6**, 96 (1920). See also *Colour Index* vol. 4 (3rd ed., 1971) p 4597.

Dark-blue powder with coppery luster. Sensitive to light. Its solns have a blue or bluish-purple color. One gram dissolves in about 100 ml water at 25°. Slightly sol in alcohol; practically insol in most other organic solvents. It is also marketed as a paste with water, the dye contents varying according to specification or requirements of the user. It almost always contains sodium chloride or sulfate used for "salting" it out. Indigo carmine is very sensitive to oxidizing agents. The color is readily discharged by nitric acid, chlorates, etc. The color of the aq soln fades on standing.

USE: As dye; in a functional kidney test; in coloring nylon surgical sutures. As a reagent for the detection of nitrate, chlorate, and in testing milk. Approved by the FDA for use in food and ingested drugs: *Fed. Regist.* **48**, 5252 (1982).

4857. Indium. In; at. wt 114.82; at. no. 49; valence 3, 2, 1. Natural isotopes: 115 (95.77%); 113 (4.23%); [115]In is a β-emitter, $T_{1/2}$ 6 × 10^{14} years. Artificial radioactive isotopes: 107-112; 114; 116-124. Occurrence in the earth's crust: 1 × 10^{-5}%. Discovered in sphalerite ore by Reich and Richter in 1863. Generally found in zinc blendes. Monograph: M. T. Ludwick, *Indium* (Indium Corp. of America, Utica, N.Y., 1950). *Review:* Wade, Banister in *Comprehensive Inorganic Chemistry* vol. 1, J. C. Bailar, Jr. *et al.*, Eds. (Pergamon Press, Oxford, 1973) pp 997-1000, 1065-1117; E. F. Milner, C. E. T. White in Kirk-Othmer *Encyclopedia of Chemical Technology* vol. 13 (Wiley-Interscience, New York, 3rd ed., 1981) pp 207-212 .

Soft, white metal with bluish tinge. Emits a "tin cry" on bending. Ductile, malleable, softer than lead, leaves a mark on paper. Quite stable in air. Crystallizes and is diamagnetic. d^{20} 7.3. mp 155°. bp 2000°. Sp heat: 0.0568 cal/g/°C. Hardness (Mohs') = 1.2. Unaffected by water; attacked by mineral acids. Very resistant to alkalies.

Toxicity: Indium salts are relatively nontoxic when administered orally; highly toxic when administered subcutaneously or intravenously. Experimental animal poisoning has produced injury to blood, heart, liver, kidneys: E. Browning, *Toxicity of Industrial Metals* (Appleton-Century-Crofts, New York, 2nd ed., 1969) pp 164-168.

USE: In bearing alloys; as a thin film on moving surfaces made from other metals. In dental alloys. In semiconductor research. In nuclear reactor control rods (in the form of an Ag-In-Cd alloy).

4858. Indium Antimonide. InSb; mol wt 236.58. In 48.53%, Sb 51.47%. Prepd by melting together stoichiometric amounts of indium and antimony in evacuated ampuls or in zone-refining apparatus: Kleppa, *J. Am. Chem. Soc.* **77**, 897 (1955); Harmon, *J. Electrochem. Soc.* **103**, 128 (1956). *Reviews:* Welker, Weiss in *Solid State Physics*, vol. 3 (Academic Press, New York, 1956); Minden, *Sylvania Technologist* **11**, no. 1, 13-25 (Jan. 1958); Hulme, Mullin, *Solid State Electronics* **5** (Pergamon Press, 1962) 211-247.

Crystals (zinc blende structure). mp 535°. d at mp 5.74 (solid); 6.48 (liq). Dielectric constant = 15.9. Energy gap at 25° = 0.18 ev. Electron mobility approx 80,000 cm^2/volt-sec. Hole mobility 1250 cm^2/volt-sec.

USE: In semiconductor electronics. Grown p-n junctions have been made by doping a melt with an acceptor impurity such as zinc or cadmium, and dipping in an n-type crystal. Rate-grown junctions have also been made. Broad-area surface junctions have been produced by out-diffusing antimony in vacuum from the surface of an n-type crystal, producing a p-n junction just inside the surface. Also has photoconductive, photoelectromagnetic, and magnetoresistive properties. Useful as an infrared detector and filter, and in Hall effect devices.

4859. Indium Arsenide. AsIn; mol wt 189.73. As 39.48%, In 60.52%. InAs. Prepd by fusion of the elements in an evacuated, sealed tube: Gans *et al.*, *Compt. Rend.* **237**, 310 (1953); Talley, Enright, *Phys. Rev.* **94**, 1931 (1954); Harmon *et al.*, *ibid.* **104**, 1562 (1956); Minden, *Sylvania Technologist* **11**, no. 1, 17 (Jan. 1958).

Metallic appearance. Small single crystals have been grown by a modification of the Czochralski technique: Gremmelmaier, *Z. Naturforsch.* **11A**, 463 (1956). mp 943°. Hardly attacked by mineral acids. Energy gap: 0.35 ev. Electron mobility: approx 33,000 cm^2/volt-sec. Hole mobility: 460 cm^2/volt-sec. Dielectric constant 11.7.

USE: In semiconductor electronics.

4860. Indium Oxide. Indium sesquioxide. In_2O_3; mol wt 277.64. In 82.71%, O 17.29%.

White to pale-yellow powder. d 7.18. Volatilizes at 850°. Insol in water; sol in hot mineral acids.

USE: In glass manufacture.

4861. Indium Phosphide. InP; mol wt 145.80. In 78.75%, P 21.25%. Prepd from white phosphorus and indium iodide at 400°: Thiel, Koelsch, *Z. Anorg. Chem.* **66**, 319 (1910); from phosphorus vapor and heated indium metal: Jandelli, *Gazz. Chim. Ital.* **71**, 58 (1941). Synthesis in zone melting furnace at 1010° from a non-stoichiometric melt: Minden, *Sylvania Technologist* **11**, no. 1, 18 (Jan 1958).

Brittle mass with metallic appearance, not easily attacked by mineral acids. mp 1070°. Dielectric constant: 10.8. Energy gap: 1.3 ev at 25°. Electron mobility: approx 4600 cm^2/volt-sec. Hole mobility: approx 150 cm^2/volt-sec. Solid solns of InP can cover the energy gap continuously from 0.3 to 1.3 ev. Rectification has been observed in InP although it is more characteristic of a Schottky type barrier than the minority carrier injection phenomenon observed in germanium.

USE: In electronics for research on semiconductors.

4862. Indium Selenide. InSe; mol wt 193.78. In 59.25%, Se 40.75%. Prepd from the elements: Klemm *et al.*, *Z. Anorg. Allgem. Chem.* **219**, 45 (1934); Schubert *et al.*, *Naturwiss.* **41**, 448 (1954).

Black crystals, rhombohedric structure, mp 660°.

USE: In semiconductor research.

4863. Indium Sulfate. $In_2O_{12}S_3$; mol wt 517.82. In 44.35%, O 37.08%, S 18.57%. $In_2(SO_4)_3$.

White, hygroscopic powder. d 3.44. Sol in water. *Keep well closed.* MLD in rabbits: 1.8 g/kg orally; 0.67 mg/kg i.v., McCord *et al.*, *J. Ind. Hyg. Toxicol.* **24**, 243 (1942).

4864. Indium Telluride. In_2Te_3; mol wt 612.47. In 37.49%, Te 62.51%. Prepn of black brittle crystals by double furnace technique: Inuzuka, Sugaike, *Proc. Japan. Acad.* **30**, 383 (1954), *C.A.* **49**, 2922e (1955).

Pycnometric d 5.78; x-ray d 5.798. mp 667°: Klemm, Ulrich, *Z. Anorg. Allgem. Chem.* **219**, 45 (1934), *C.A.* **29**,

1730 (1935). Activation energies: 0.94 ev at 300-600°K, 2.4 ev at 600-900°K.

USE: In semiconductor research.

4865. Indium Trichloride. Indium chloride. Cl_3In; mol wt 221.13. Cl 48.10%, In 51.90%. $InCl_3$.

Yellowish, deliquesc crystals. d 4.0. mp 586°; sublimes at 500°. Freely sol in water. *Keep tightly closed.* MLD s.c. in rats: 10.2 mg/kg; MLD i.v. in rabbits: 0.64 mg/kg, McCord *et al.*, *J. Ind. Hyg. Toxicol.* **24**, 243 (1942).

USE: In electroplating using a soln of the salt with dextrose and NaCN. This soln is stable, though it turns dark on standing and deposits a mud which, however, contains no indium.

4866. Indium Trifluoride. Indium fluoride; indic fluoride. F_3In; mol wt 171.82. F 33.19%, In 66.81%. InF_3. Prepd by heating In_2O_3 or $(NH_4)_3InF_6$ in a current of fluorine: Hannebohn, Klemm, *Z. Anorg. Allgem. Chem.* **229**, 342 (1936); Ensslin, Dreyer, *ibid.* **249**, 119 (1942); Kwasnik in *Handbook of Preparative Inorganic Chemistry*, vol. 1, G. Brauer, Ed. (Academic Press, New York, 1963) pp 228-229; from In and HF: Brewer *et al.*, *J. Inorg. Nucl. Chem.* **9**, 56 (1959); Muetterties, Castle, *ibid.* **18**, 148 (1961).

Colorless substance, d 4.39. mp 1170°. bp > 1200°. Soly in water (25°) 0.040 g/100 ml. Freely sol in dil acids. Stable in hot and cold water.

Trihydrate, crystals. Soly in water (22°) 8.49 g/100 ml, indicating a complex.

4867. Indobufen. (±)-4-(1,3-Dihydro-1-oxo-2H-isoindol-2-yl)-α-ethylbenzeneacetic acid; (±)-2-[p-(1-oxo-2-isoindolinyl)phenyl]butyric acid; 1-oxo-2-[p-[(α-ethyl)carboxymethyl]phenyl]isoindoline; 2-[4-(1-carboxypropyl)phenyl]-1-isoindolinone; K-3920; Ibustrin. $C_{18}H_{17}NO_3$; mol wt 295.34. C 73.20%, H 5.80%, N 4.74%, O 16.25%. Platelet aggregation inhibitor with anti-inflammatory and analgesic activity. Prepn: R. W. J. Carney, G. de Stevens, **Ger. pat. 2,034,240** (1971 to Ciba), *C.A.* **74**, 12547p (1971); P. N. Giraldi *et al.*, **Ger. pat. 2,154,525** (1972 to Carlo Erba), *C.A.* **77**, 88292v (1972); G. Nannini *et al.*, *Arzneimittel-Forsch.* **23**, 1090 (1973). New process: K. Noda *et al.*, **Eur. pat. Appl. 47,674** corresp to **U.S. pat. 4,400,520** (1982, 1983 to Hisamitsu). Use as anti-inflammatory agent: R. W. J. Carney, G. de Stevens, **U.S. pat. 3,997,669** (1976 to Ciba-Geigy). Use as platelet aggregation inhibitor: P. N. Giraldi *et al.*, **Belg. pat. 818,033** corresp to **U.S. pat. 4,010,274** (1974, 1977 both to Carlo Erba). Pharmacokinetics and bioavailability: L. M. Fuccella *et al.*, *Eur. J. Clin. Pharmacol.* **15**, 323 (1979); V. Tamassia *et al.*, *ibid.* **329.** Pharmacology: M. Bergameaschi *et al.*, *Pharmacol. Res. Commun.* **16**, 979 (1984). HPLC determn in plasma: E. Wahlin-Boll *et al.*, *Eur. J. Clin. Pharmacol.* **20**, 375 (1981). Clinical trial in cerebrovascular disease: P. Carrieri *et al.*, *Pharmatherapeutica* **3**, 410 (1983). Clinical comparison with aspirin: G. Orefice *et al.*, *Acta Neurol.* (Napoli) **6**, 97 (1984).

Crystals from ethanol, mp 182-184°.

THERAP CAT: Antithrombotic.

4868. Indocyanine Green. 2-[7-[1,3-Dihydro-1,1-dimethyl-3-(4-sulfobutyl)-2H-benz[e]indol-2-ylidene]-1,3,5-heptatrienyl]-1,1-dimethyl-3-(4-sulfobutyl)-1H-benz[e]indolium hydroxide inner salt sodium salt; anhydro-3,3,3',3'-tetramethyl-1,1'-bis(4-sulfobutyl)-4,5,4',5'-dibenzoindotricarbocyanine hydroxide inner salt sodium salt; Cardio-Green; Fox green. $C_{43}H_{47}N_2NaO_6S_2$; mol wt 774.99. C 66.64%, H 6.11%, N 3.62%, Na 2.97%, O 12.39%, S 8.27%. Prepn: Heseltine, Brooker, **U.S. pat. 2,895,955** (1959 to Kodak).

A tricarbocyanine type of dye with infrared absorbing properties; peak absorption at about 800 nm. Has little or no absorption in the visible. Commercial product contains moisture and about 5% sodium iodide as contaminant.

USE: In infrared photography; in prepn of Wratten filters.

THERAP CAT: Diagnostic aid (blood volume determination, cardiac output, hepatic function).

4869. Indole. 2,3-Benzopyrrole. C_8H_7N; mol wt 117.14. C 82.02%, H 6.02%, N 11.96%. Obtained from the 240-260° fraction from coal tar: Weissgerber, *Ber.* **43**, 3520 (1910); from feces: Bergeim, *J. Biol. Chem.* **32**, 17 (1917). Prepn from *o*-formotoluide: Tyson, *Org. Syn.* **coll. vol. III**, 479 (1955); by dehydrocyclizing ortho alkyl anilines: Erner *et al.*, U.S. pat. **2,953,575** (1960 to Houdry Process); from *N*-(2-tolyl)-*N'*-methyl-*N'*-phenylformamidine: Lorenz *et al.*, *J. Org. Chem.* **30**, 2531 (1965). Toxicity data: Smyth *et al.*, *Am. Ind. Hyg. Assoc. J.* **23**, 95 (1962). Comprehensive review: Van Order, Lindwall, *Chem. Rev.* **30**, 69 (1942); D. W. Bannister in Kirk-Othmer *Encyclopedia of Chemical Technology* **vol. 13** (Wiley-Interscience, New York, 3rd ed., 1981) pp 213-222.

Leaflets, mp 52°. bp$_{762}$ 253°; bp$_{28}$ 128-133°. Intense fecal odor. Volatile with steam. Soluble in hot water, hot alcohol, ether, benzene. LD$_{50}$ orally in rats: 1 g/kg (Smyth).

USE: In highly dil solns the odor is pleasant, hence indole has been used in perfumery.

4870. Indoleacetic Acid. *1H-Indole-3-acetic acid;* heteroauxin; IAA. $C_{10}H_9NO_2$; mol wt 175.18. C 68.56%, H 5.18%, N 8.00%, O 18.27%. Plant hormone; recognized as the principal auxin of higher plants. Prepd by the reaction of indole with potassium glycolate at 250°: Johnson, Crosby, *J. Org. Chem.* **28**, 1246 (1963). From indole and chloroacetic acid: Shagalov *et al.*, U.S. pat. **3,320,281** (1967). *Reviews:* Leopold in *The Hormones*, **vol. IV**, G. Pincus *et al.*, Eds. (Academic Press, New York, 1964) pp 1-66; Thimann in *The Physiology of Plant Growth and Development*, M. B. Wilkins, Ed. (McGraw-Hill, New York, 1969) pp 1-45.

Leaflets or cryst powder from water, mp 168-170°. pK 4.75. Sparingly sol in water or chloroform; freely sol in alc; sol in acetone, ether.

USE: Plant growth regulator.

4871. Indolebutyric Acid. *1H-Indole-3-butanoic acid;* indole-3-butyric acid; 4-(3-indolyl)butyric acid. $C_{12}H_{13}NO_2$; mol wt 203.23. C 70.91%, H 6.45%, N 6.89%, O 15.74%. Preparation: Jackson, Manske, *J. Am. Chem. Soc.* **52**, 5029 (1930); by heating indole, γ-butyrolactone, and sodium hydroxide, followed by acidification of the product: Fritz, U.S. pat. **3,051,723** (1962 to Union Carbide); by decarboxylation of 2-carboxyindole-3-butyric acid: Bowman, Islip, *Chem. & Ind. (London)* **1971**, 154. Toxicity: causes tumors in rats, Pesonen, *Acta Endocrinol.* **5**, 409 (1950); hypoglycemic effect in rats, Mirsky *et al.*, *Endocrinology* **59**, 715 (1956).

White or slightly yellow crystals. Slight characteristic odor. mp 123-125°. Practically insol in water, chloroform; sol in alc, ether, acetone. LD$_{50}$ i.p. in mice: 100 mg/kg: Anderson *et al.*, *Proc. Soc. Exp. Biol. Med.* **34**, 138 (1936).

USE: Indolebutyric acid suitably diluted (*Hormodin, Seradix*) is used for promoting and accelerating root formation of plant clippings.

4872. Indolmycin. *5-[1-(1H-Indol-3-yl)ethyl]-2-(methylamino)-4(5H)-oxazolone;* 5-(1-indol-3-ylethyl)-2-(methylamino)-2-oxazolin-4-one; 2-methylamino-5α-(β-indolyl)-ethyl-2-oxazolin-4-one; Pa-155A. $C_{14}H_{15}N_3O_2$; mol wt 257.28. C 65.35%, H 5.88%, N 16.33%, O 12.44%. Antibiotic substance produced by *Streptomyces albus*: Rao, *Antibiot. & Chemother.* **10**, 312 (1960); Marsh *et al.*, *ibid.* 316; **Brit.** pat. **862,685** (1961 to Pfizer). Structure: M. Schach von Wittenau, H. Els, *J. Am. Chem. Soc.* **83**, 4678 (1961). Total synthesis: *eidem*, *ibid.* **85**, 3425 (1963); Preobrazhenskaya *et al.*, *Tetrahedron* **24**, 6131 (1968); T. Takeda, T. Mukaiyama, *Chem. Letters* **1980**, 163. Abs config: T. H. Chan, R. K. Hill, *J. Org. Chem.* **35**, 3519 (1970). Biosynthetic studies: Hornemann *et al.*, *Chem. Commun.* **1969**, 245; *eidem*, *J. Am. Chem. Soc.* **93**, 3028 (1971).

Long rectangular prisms from methanol or ethyl acetate, mp 209-210°. $[\alpha]_D^{25}$ −214° (c = 2 in methanol). uv max: 218 nm (E$_{1cm}^{1\%}$ 1960). Weakly basic, stable to heat. Slightly sol in water, benzene, ether; moderately in lower alcohols, acetone.

4873. 3-Indolylacetone. *Indol-3-yl-2-propanone.* $C_{11}H_{11}NO$; mol wt 173.21. C 76.27%, H 6.40%, N 8.09%, O 9.24%. Prepn: Brown *et al.*, *J. Chem. Soc.* **1952**, 3172; Williamson, *ibid.* **1962**, 2834; Morris *et al.*, **Brit.** pat. **974,895** (1964 to Parke, Davis).

Brownish rhombs from benzene or needles from aq methanol. mp 115-117.5°. uv max (ethanol): 221, 280, 289 nm (ε 35100, 6400, 5300).

4874. Indomethacin. *1-(4-Chlorobenzoyl)-5-methoxy-2-methyl-1H-indole-3-acetic acid;* 1-(p-chlorobenzoyl)-5-methoxy-2-methyl-3-indolylacetic acid; Amuno; Argun; Artracin; Artrinovo; Artrivia; Catolep; Chibro-Amuno; Chrono-Indocid; Confortid; Dolcidium; Durametacin; Elmetacin; Idomethine; Imbrilon; Inacid; Indacin; Indocid; Indocin; Indocollyre; Indomed; Indomee; Indomethine; Indomod; Indo-Phlogont; Indoptic; Indorektal; Indo-Tablinen; Indoxen; Inflazon; Infrocin; Inteban SP; Lausit; Mezolin; Mikametan; Mobilan; Rheumacin LA; Tannex; Vonum. $C_{19}H_{16}ClNO_4$; mol wt 357.81. C 63.78%, H 4.51%, Cl 9.91%, N 3.91%, O 17.89%. Prepn: T. Y. Shen *et al.*, *J. Am. Chem. Soc.* **85**, 488 (1963); T. Y. Shen, U.S. pat. **3,161,654** (1964 to Merck & Co.). Alternate process: **Belg.** pat. **679,678** (1966 to Sumitomo). Pharmacology: Winter *et al.*, *J. Pharmacol. Exp. Ther.* **141**, 369 (1963). Metabolic studies: Yesair *et al.*, *Biochem. Pharmacol.* **19**, 1579 (1970). Indomethacin blocks prostaglandin biosynthesis: see *Prostaglandin Synthetase Inhibitors—Their Effects on Physiological Func-*

tions and Pathological States, H. J. Robinson, J. R. Vane, Eds. (Raven Press, New York, 1974) 395 pp. Toxicity: C. D. Klaassen, *Toxicol. Appl. Pharmacol.* **38,** 127 (1976). *Review:* T. Y. Shen, C. A. Winter in *Advances in Drug Research* vol. 12, A. B. Simmons, Ed. (Academic Press, New York, 1977) pp 89-245; *Semin. Arthritis Rheum.* **12,** Suppl. 1, 77-151 (1982).

Crystals exhibiting polymorphism, mp for one form ~155°, for the other ~162°. uv max (ethanol): 230, 260, 319 nm (ε 20800, 16200, 6290). pKa 4.5. Sol in ethanol, ether, acetone, castor oil. Practically insol in water. Stable in neutral or slightly acidic media; dec by strong alkali. LD_{50} i.p. in rats: 13 mg/kg (Klaassen).
Sodium salt trihydrate, $C_{19}H_{15}ClNNaO_4 \cdot 3H_2O$, *Osmosin.* Pale yellow crystalline powder. pH of 1% soln: 8.4. Very sol in methanol; sol in water, ethanol. Very slightly sol in chloroform, acetone.
Meglumine salt, *Liometacen.*
THERAP CAT: Anti-inflammatory, antipyretic, analgesic.

4875. Indoprofen. *4-(1,3-Dihydro-1-oxo-2H-isoindol-2-yl)-α-methylbenzeneacetic acid; p-(1-oxo-2-isoindolinyl)hydratropic acid;* 2-[4-(1-carboxyethyl)phenyl]-1-isoindolinone; 1-oxo-2-[p-[(α-methyl)carboxymethyl]phenyl]isoindoline; 2-[4-(1-oxo-2-isoindolinyl)phenyl]propionic acid; IPP; K 4277; Bor-Ind; Flosin; Flosint; Isindone; Praxis; Reumofene. $C_{17}H_{15}NO_3$; mol wt 281.32. C 72.58%, H 5.37%, N 4.98%, O 17.06%. Nonsteroidal anti-inflammatory agent. Prepn: R. W. J. Carney, G. de Stevens, Ger. pat. **2,034,240** (1971 to Ciba), *C.A.* **74,** 125471p (1971); P. N. Giraldi *et al.,* Ger. pat. **2,154,525** (1972 to Carlo Erba), *C.A.* **77,** 88292v (1972); G. Nannini *et al., Arzneimittel-Forsch.* **23,** 1090 (1973). Synthesis and optical resolution of enantiomers: F. Buzzetti *et al.,* Ger. pat. **2,258,088** (1974 to Carlo Erba), *C.A.* **81,** 13382y (1974); T. Kametani *et al., J. Heterocycl. Chem.* **15,** 369 (1978). Abs config: S. De Munari *et al., Tetrahedron Letters* **21,** 2273 (1980). Pharmacology: Buttinoni *et al., Arzneimittel-Forsch.* **23,** 1100 (1973). Absorption, excretion and metabolism: Goldaniga *et al., ibid.* **24,** 1603 (1974); Chasseaud *et al., ibid.* 1606. GLC determn of enantiomers in plasma: G. P. Tosolini *et al., J. Pharm. Sci.* **63,** 1072 (1974). Biological activity and toxicity data: A. Buttinoni *et al., J. Pharm. Pharmacol.* **35,** 603 (1983).

Colorless scales from ethanol, mp 213-214°. LD_{50} in rats (mg/kg): 58.66 i.v.; 60.83 orally (Buttinoni).
(+)-Form, *dexindoprofen, Nedius.* Colorless scales from isopropanol, mp 205-207°. $[\alpha]_D^{20} +48°$ (c = 0.05 in DMSO). LD_{50} in rats (mg/kg): 31.98 i.v.; 33.75 orally (Buttinoni).
(−)-Form, colorless scales from isopropanol, mp 205-207°. $[\alpha]_D^{20} -48°$ (c = 0.05 in DMSO). LD_{50} in rats (mg/kg): 555.39 i.v.; 538.02 orally (Buttinoni).
THERAP CAT: Analgesic; anti-inflammatory.

4876. Indoramin. *N-[1-[2-(1H-Indol-3-yl)ethyl]-4-piperidinyl]benzamide;* 3-[2-(4-benzamidopiperidino)ethyl]indole; Wy 21901. $C_{22}H_{25}N_3O$; mol wt 347.46. C 76.05%, H 7.25%, N 12.09%, O 4.60%. $α_1$-Adrenergic blocking agent with antihypertensive and bronchodilating activity. Prepn: J. L. Archibald, J. L. Jackson, S. Afr. pat. **68 03204;** *eidem,* U.S. pat. **3,527,761** (1969, 1970 both to Wyeth); J. L. Archibald *et al., J. Med. Chem.* **14,** 1054 (1971). Pharmacological study: R. B. Royds *et al., Clin. Pharmacol. Ther.* **13,** 380

(1972). Pharmacokinetics: G. H. Draffan *et al., Brit. J. Clin. Pharmacol.* **3,** 489 (1976). Antihypertensive activity: G. S. Stokes *et al., Clin. Pharmacol. Ther.* **25,** 783 (1979). Cardiovascular effects in man: A. J. Coleman *et al., J. Int. Med. Res.* **7,** 511 (1979). Effects on respiration in guinea pigs: C. Hamer, D. M. Temple, *Agents Actions* **10,** 399 (1980). Determn of therapeutic concentrations: A. J. Swaisland, *Analyst* **106,** 717 (1981). Symposium on pharmacology and clinical studies: *Brit. J. Clin. Pharmacol.* **12,** Suppl. 1, 1S-140S (1981). Review of pharmacology, therapeutic efficacy: B. Holmes, E. M. Sorkin, *Drugs* **31,** 467-499 (1986).

Crystals from ethanol, mp 208-210°.
Hydrochloride, $C_{22}H_{26}ClN_3O$, *Baratol, Doralese, Vidora, Wydora, Wypres, Wypresin.* Crystals, mp 230-232°. Recryst from isopropanol gives different cryst modification, mp 258-260°.
THERAP CAT: Antihypertensive.

4877. Indospicine. *(S)-2,7-Diamino-7-iminoheptanoic acid;* L-*6-amidinonorleucine;* L-2-amino-6-amidinohexanoic acid; L-α-amino-ε-amidinocaproic acid. $C_7H_{15}N_3O_2$; mol wt 173.21. C 48.53%, H 8.73%, N 24.26%, O 18.47%. H_2N-$C=NH(CH_2)_4CH(NH_2)COOH.$ Teratogenic and hepatotoxic factor found in extracts of *Indigofera spicata* Forsk (*Indogofera endecaphylla* Jacq.), Leguminosae. Isoln: Hegarty, Pound, *Nature* **217,** 354 (1968). Isoln, structure and biological studies: *eidem, Aust. J. Biol. Sci.* **23,** 831 (1970). Total synthesis: Culvenor *et al., Aust. J. Chem.* **24,** 371 (1971). *Review:* Hegarty: *Australas. J. Dermatol.* **14,** 35-38 (1973).
Monohydrochloride monohydrate, $C_7H_{15}N_3O_2 \cdot HCl \cdot H_2O$, needle-like crystals from aq ethanol, mp 131-134°. $[\alpha]_D^{22}$ +18° (c = 1.1 in 5N HCl).

4878. Infusorial Earth. Siliceous earth; diatomaceous earth; fossil flour; kieselguhr; Celite; Super-Cel. Siliceous frustules and fragments of various species of diatoms. *See also* Silicon Dioxide.
White to light gray to pale buff powder. Insol in water, acids, or dil alkalies. Capable of taking up and holding about four times its wt of water.
USE: Clarifying agent. Largely used as an absorbent for liquids and for dispensing fluid extracts in powder form; also in cataplasms and as constituent of and excipient for pill masses. Clarifying oils, varnishes; filtering liquids; manuf heat insulators, fire brick, and fire- and acid-proof packing materials; filler for paper, paints; adsorbent dynamite; in metal polishes, dentifrices, nail polishes; in chromatography.

4879. Inhibin. Gonadal polypeptide hormone produced by Sertoli cells in males and granulosa cells in females, which selectively inhibits the secretion of FSH, *q.v.* Dimer, mol wt 32,000 Da, isolated in two isoforms, inhibin A and inhibin B, which have the same α subunit linked to similar but distinct β subunits. Dimers of the β subunits have been isolated which stimulate the secretion of FSH, *see* activins. Name "inhibin" coined upon identification as a water-soluble, testicular factor with pituitary inhibitory activity: D. R. McCullagh, *Science* **76,** 19 (1932). Specific inhibition of FSH: H. H. F. Klinefelter *et al., J. Clin. Endocrinol.* **2,** 615 (1942); B. P. Setchell, F. Jacks, *J. Endocrinol.* **62,** 675 (1974). Identification in ovarian follicular fluid: F. H. de Jong, R. M. Sharpe, *Nature* **263,** 71 (1976); N. B. Schwartz, C. P. Channing, *Proc. Nat. Acad. Sci. USA* **74,** 5721 (1977). Localization of production to Sertoli cells: A. Steinberger, E. Steinberger, *Endocrinology* **99,** 918 (1976); to granulosa cells: G. F. Erickson, A. J. W. Hsueh, *ibid.* **103,** 1960 (1978). Immunohistochemical confirmation of production sites: P. Cuevas *et al., Biochem. Biophys. Res. Commun.* **142,** 23 (1987). Identification of inhibin-like proteins from human seminal plasma: N. G. Seidah *et al., FEBS Letters* **175,** 349 (1984); C. H. Li *et al., Proc. Nat. Acad. Sci. USA* **82,** 4041 (1985). Isolation of inhibins and higher mol wt precursors: K. Miyamoto *et al., Biochem. Biophys. Res. Commun.* **129**, 396 (1985). Purification and characterization of inhibins A and B: J. Rivier *et al., ibid.* **133**, 120 (1985); N. Ling *et al., Proc. Nat. Acad. Sci. USA* **82**, 7217 (1985). Each subunit is synthesized as a larger precursor molecule, then processed to the mature form. Complete amino acid sequence of mature form α, β_A and β_B subunits: A. J. Mason *et al., Nature* **318**, 659 (1985); A. J. Mason *et al., Biochem. Biophys. Res. Commun.* **135**, 957 (1986). Demonstration of negative feedback mechanism between secretion of FSH and inhibin: S.-Y. Ying *et al., Proc. Nat. Acad. Sci. USA* **84**, 4631 (1987). Proposed nomenclature: H. G. Burger, *J. Endocrinol.* **117**, 159 (1988). Approach to nonradiometric assays: R. Schwall *et al., Prog. Clin. Biol. Res.* **285**, 205 (1988). Review of early bioassays: B. Hudson *et al., J. Reprod. Fert., Suppl.* **26**, 17-29 (1979). Reviews of purification and physiology: C. P. Channing *et al., Proc. Soc. Exp. Biol. Med.* **178**, 339-361 (1985); F. H. de Jong, D. M. Robertson, *Mol. Cell. Endocrinol.* **42**, 95-103 (1985). Review of possible use as contraceptive: A. R. Sheth, S. B. Moodbidri, *Adv. Contracept.* **2**, 131-139 (1986); P. Franchimont, *Male Contraception: Advances and Future Prospects*, G. I. Zatuchni *et al.*, Eds. (Harper & Row, Philadelphia, 1986) pp 408-418. *Reviews:* F. H. de Jong, *Oxford Rev. Reprod. Biol.* **9**, 1-53 (1987); N. Ling *et al., Vitam. Horm.* (New York) **44**, 1-46 (1988).

4880. Inosine. Hypoxanthine riboside; 9-β-D-ribofuranosylhypoxanthine; hypoxanthosine; Aminosin; Inosie; Oxiamine; Ribonosine; Trophicardyl. $C_{10}H_{12}N_4O_5$; mol wt 268.23. C 44.78%, H 4.51%, N 20.89%, O 29.82%. In meat and meat extracts, in sugar beets. Prepd from adenosine by incubation with purified adenosine deaminase from intestine: Kalckar, *J. Biol. Chem.* **167**, 445 (1947); also by the action of sodium nitrite and acetic acid on adenosine: Levene, Jacobs, *Ber.* **43**, 3161 (1910); by the use of barium nitrite and H_2SO_4: Reiff *et al.*, U.S. pat. **3,049,536** (1962 to Zellstoff-Fabrik Waldhof). Fermentation method: Motozaki *et al.*, U.S. pat. **3,111,459** (1963 to Ajinomoto). Structure: Bredereck, *Ber.* **66**, 198 (1933); Z. *Physiol. Chem.* **223**, 61 (1934); Gulland, Holiday, *J. Chem. Soc.* **1936**, 765.

Dihydrate, long rectangular plates from water, mp 90°. Anhydrous needles from 80% alc, dec 218° (rapid heating). $[\alpha]_D^{18}$ −49.2° (c = 0.9 in H_2O). $[\alpha]_{white}^{20}$ −73° (0.5 g + 2 ml N NaOH + 1 ml H_2O). 100 ml of the satd water soln at 20° contain 1.6 g inosine. Absorption spectrum: Kalckar, *loc. cit.* uv max (pH 6.0): 248.5 nm (ϵ 12,200). Boiling with $0.1N$ H_2SO_4 yields hypoxanthin and D-ribose.

THERAP CAT: Activates cellular functions.

4881. Inosine Pranobex. *Inosine, mono[4-(acetylamino)benzoate] (salt), compd with 1-(dimethylamino)-2-propanol (1:3);* p-acetamidobenzoic acid salt of dimethylaminoisopropanol:inosinate complex 3:1 molar ratio; inosine:dimethylaminoisopropanol acetamidobenzoate (1:3); inosiplex; methisoprinol; NP 113; NPT 10381; Aviral; Delimmun; Imunovir; Imunoviral; Isoprinosin; Isoprinosina; Isoprinosine; Isoviral; Modimmunal; Pranosina; Pranosine; Viruxan. $C_{52}H_{78}N_{10}O_{17}$; mol wt 1115.26. C 56.00%, H 7.05%, N 12.56%, O 24.39%. Immunopotentiator. Prepn: P. Gordon, **Ger. pat. 1,965,431;** idem, U.S. pat. **3,646,007** (1971, 1972 both to Newport Pharm.). Metabolic studies in monkeys: D. G. Streeter, E. H. Pfadenhauer: *Drug Metab. Dispos.* **12**, 199 (1984). Antiviral activity: E. R. Brown, P. Gordon, *Can. J. Microbiol.* **18**, 1463 (1972); R. L. Muldoon *et al., Antimicrob. Ag. Chemother.* **2**, 224 (1972). Mode of antiviral action: H. Ohnishi *et al., Infection Immunity* **38**, 243 (1982). Effects in immunosuppressed rats: L. Binderup, *Int. J.*

Immunopharmacol. **7**, 93 (1985). *In vitro* studies of effect on interleukin II production, *q.v.:* K. Y. Tsang *et al., N. Engl. J. Med.* **310**, 987 (1984); M. Wiranowska-Stewart, J. W. Hadden: *Int. J. Immunopharmacol.* **8**, 63 (1986); on lymphocyte response to Epstein-Barr virus: S. K. Sundar, *ibid.* **7**, 187 (1985); on HIV-1 (HTLV-III/LAV) virus infected lymphocytes: A. Pompidou *et al., Lancet* **2**, 1423 (1985). Clinical trials in subacute sclerosing panencephalitis (SSPE): C. E. Jones *et al., ibid.* **1**, 1034 (1982); in genital herpes: O. Salo, A. Lassus, *Eur. J. Sexually Trans. Dis.* **1**, 101 (1983); in lymphadenopathy: J. I. Wallace, J. G. Bekesi, *Clin. Immun. Immunopath.* **39**, 179 (1986). Review of immunomodulation in various pathologies: A. J. Glasky, J. Gordon: *Meth. Find. Exp. Clin. Pharmacol.* **8**, 35-40 (1986); of pharmacodynamics and efficacy: D. M. Campoli-Richards, *Drugs* **32**, 383 (1986).

Neutral water-soluble solid. LD_{50} in mice and rats (mg/kg): >4000 orally and i.p. (Gordon).

THERAP CAT: Immunomodulator. Antiviral.

4882. Inosinic Acid. *5′-Inosinic acid;* 5-inosinic acid; muscle inosinic acid; t-inosinic acid; hypoxanthine riboside-5-phosphoric acid; IMP. $C_{10}H_{13}N_4O_8P$; mol wt 348.22. C 34.49%, H 3.76%, N 16.09%, O 36.76%, P 8.90%. Prepn from meat extract: Levene, Bass, *Nucleic Acids* (New York, 1931) p 229; from dried sardines: Yoshida, Kageyama, **Japan. pat. 732('56)** (to Ajinomoto), *C.A.* **51**, 3870b (1957). Structure: Levene, Bass, *op. cit.*, pp 187-192; Bredereck, *Ber.* **66**, 198 (1933); Levene, Tipson, *J. Biol. Chem.* **111**, 313 (1935). Also prepd from muscle by enzymatic deamination of muscle adenylic acid: Ostern, *Biochem. Z.* **254**, 65 (1932); by hydrolysis of inosine triphosphate: Kleinzeller, *Biochem. J.* **36**, 729 (1942). Studies on the enzymatic synthesis: Greenberg, *J. Biol. Chem.* **190**, 611 (1951); Korn *et al., ibid.* **217**, 875 (1955). Microbial fermentation method using mutant strains of *Micrococcus glutamicus:* Kinoshita *et al.*, U.S. pat. **3,232,844** (1966 to Kyowa).

Syrup, solidifies to a glass when dried over H_2SO_4. Agreeable sour taste. $[\alpha]_D^{20}$ −18.5° (0.3 g of anhydr Ba salt in 10 ml of 2.5% HCl). pK_1 = 2.4; pK_2 = 6.4. Absorption spectrum: Kalckar, *J. Biol. Chem.* **167**, 445 (1947). Freely sol in water, in formic acid; very sparingly sol in alcohol, ether. On boiling with acid hydrolyzes to 1 mol H_3PO_4, 1 mol hypoxanthine, 1 mol D-ribose.

Disodium salt dihydrate, $C_{10}H_{11}N_4O_8PNa_2 \cdot 2H_2O$, barely sol in alcohol, ether, acetone; soly in water at 20° about 13 g/100 ml. Kawasaki, *New Food Ind.* (Tokyo) **3**, no. 1, 17 (1961).

Barium salt, $C_{10}H_{11}BaN_4O_8P$. Hemipentadecahydrate, lustrous leaflets. Becomes anhydr at 100° *in vacuo*.

USE: Its salts as flavor intensifiers, like sodium glutamate. Examples of mixtures of sodium inosinate and sodium glutamate or other salts: Toi *et al.*, U.S. pat. **3,109,741** (1963 to Ajinomoto).

4883. Inositol. *myo-Inositol; meso-*inositol; i-inositol;

hexahydroxycyclohexane; cyclohexanehexol; cyclohexitol; meat sugar; inosite; mesoinosite; phaseomannite; dambose; nucite; bios I; rat antispectacled eye factor; mouse antialope-cia factor. $C_6H_{12}O_6$; mol wt 180.16. C 40.00%, H 6.71%, O 53.29%. Widely distributed in plants and animals. Growth factor for animals and microorganisms. Isoln from heart muscle: Scherer, *Ann.* **73**, 322 (1850); from liver: Woolley, *J. Biol. Chem.* **139**, 29 (1941). Synthesis: Wieland, Wishart, *Ber.* **47**, 2082 (1914); Anderson, Wallis, *J. Am. Chem. Soc.* **70**, 2931 (1948). Obtained commercially from corn steep liquor, since inositol is present as phytic acid in corn: Bartow, Walker, *Ind. Eng. Chem.* **30**, 300 (1938); U.S. pat. **2,112,553** (1938); Hoglan, Bartow, *J. Am. Chem. Soc.* **62**, 2397 (1940); Elkin, Meadows, U.S. pat. **2,414,365** (1947); **Brit.** pat. **601,273** (1948 to Corn Prod. Refining). Nine possible stereoisomers: Seven are optically inactive or *meso*. Two optically active forms, the racemic form, and several *cis,trans*-isomers occur naturally. The prevalent natural form is *cis*-1,2,3,5-*trans*-4,6-cyclohexanehexol which is described here. *Reviews:* R. Beckmann, *m-Inosit* (Editio Cantor, Aulendorf, 1953); several authors in *The Vitamins,* **vol. 2**, W. H. Sebrell, Jr., R. S. Harris, Eds. (Academic Press, New York, 1954) pp 321-386; *ibid.* **vol. 3** (2nd ed., 1971) pp 340-415.

Anhyd, non-hygroscopic crystals from water or acetic acid above 80°. Sweet taste. d 1.752. mp 225-227°. Optically inactive. Soly in water at 25°: 14 g/100 ml soln; at 60°: 28 g/100 ml soln. Slightly sol in alc. Practically insol in ether and other common organic solvents. Aq solns are neutral to litmus.

Dihydrate, efflorescent crystals from water below 50°. d 1.524. mp 218°. Becomes anhyd at 100°.

THERAP CAT: Vitamin B complex; lipotropic.

4884. Inositol Monophosphate. *myo-Inositol 1-(dihydro-gen phosphate).* $C_6H_{13}O_9P$; mol wt 260.14. C 27.70%, H 5.04%, O 55.35%, P 11.91%. Prepd by enzymatic hydrolysis of sodium phytate using the phytase of wheat bran: Posternak, Posternak, *Helv. Chim. Acta* **12**, 1165 (1929); McCormick, Carter, *Biochem. Prepn.* **2**, 65 (1952).

Crystals from water + alcohol, dec 195-197°. Titrates as a dibasic acid. Freely soluble in water (1 g dissolves in 3 ml H_2O). Practically insol in abs ethanol, ether. Remarkably resistant to hydrolysis by boiling with strong alkali. May be hydrolyzed by boiling with 6N HCl for 14 hrs.

4885. Inositol Niacinate. *myo-Inositol hexa-3-pyridine-carboxylate;* hexanicotinoyl inositol; hexanicotinyl *cis*-1,2,-3,5-*trans*-4,6-cyclohexane; inositol hexanicotinate; *meso*-inositol hexanicotinate; Dilcit; Dilexpal; Mesotal; Esantene; Hämovannid; Hexanicit; Hexanicotol; Hexopal; Linodil; Mesonex; Palohex. $C_{42}H_{30}N_6O_{12}$; mol wt 810.71. C 62.22%, H 3.73%, N 10.37%, O 23.68%. Prepn: Badgett, Woodward, *J. Am. Chem. Soc.* **69**, 2907 (1947).

Crystals, mp 254.3-254.9°. Practically insol in water; sol in dil acids.

THERAP CAT: Vasodilator (peripheral).

4886. Insularine. $C_{38}H_{40}N_2O_6$; mol wt 620.72. C 73.52%, H 6.50%, N 4.51%, O 15.47%. From root of *Cissampelos insularis* Makino and *C. ochiaiana* Yamamoto, *Menispermaceae.* Isoln: Kondo, Yano, *J. Pharm. Soc. Japan* **47**, 815 (1927); Kondo, Tomita, *Arch. Pharm.* **274**, 76 (1936). Structure: Tomita, Kikuchi, *J. Pharm. Soc. Japan* **77**, 997 (1957).

Amorphous powder, $[\alpha]_D^7$ +28°, mp 160°. Readily sol in ether. Absorption spectrum: Ochai, *J. Pharm. Soc. Japan* **49**, 425 (1929).

4887. Insulin. Inutral; Optisulin long. Polypeptide hormone produced in the beta cells of the islets of Langerhans situated in the pancreas of all vertebrates. Synthesized *in vivo* via a single chain polypeptide precursor, proinsulin, *q.v.*, which is converted by proteolysis to insulin after the native conformation has been established. Secreted directly into the bloodstream where it regulates carbohydrate metabolism, influences the synthesis of protein and of RNA, and the formation and storage of neutral lipids. The first protein for which chemical structure and precise mol wt were determined: Sanger, Tuppy, *Biochem. J.* **49**, 463, 481 (1951); Sanger, Thompson, *ibid.* **53**, 353, 366 (1953). Minimum mol wt 6000. All cryst insulin prepns contain small amounts of heavy metals, such as zinc, nickel, cobalt, or cadmium as components of the molecule. Isoln: Banting, Best, *J. Lab. Clin. Med.* **7**, 251 (1921-22); Abel, *Proc. Nat. Acad. Sci. USA* **12**, 132 (1926); Scott, *Biochem. J.* **28**, 1592 (1934); Lens, *Biochim. Biophys. Acta* **2**, 76 (1948); Maxwell, Hinkel, U.S. pat. **2,695,861** (1954 to Armour). Structure of cattle insulin: Ryle *et al., Biochem. J.* **60**, 541 (1955); of pig and sheep insulins: Brown *et al., ibid.* **556**; of horse and whale insulins: Harris *et al., Arch. Biochem. Biophys.* **65**, 427 (1956). Structure of human insulin: Nichol, Smith, *Nature* **187**, 483 (1960). The amino acid composition and physical structure of mammalian insulin varies only slightly from one species to the next, except for the rat and the guinea pig which have large structural differences: Smith, *Am. J. Med.* **40**, 662 (1966). Chemistry: Sanger, *Science* **129**, 1340 (1959). Crystal structure: Hodgkin, *Verh. Schweiz. Naturforsch. Ges.* **150**, 93 (1970). Partial synthesis: Shields, Carpenter, *J. Am. Chem. Soc.* **83**, 3066 (1961). Synthesis of sheep insulin: Meinhofer *et al., Z. Naturforsch.* **18b**, 1120 (1963); of bovine insulin: Kung *et al., Scientia Sinica* **14**, 1710 (1965). Synthesis of human insulin: Katsoyannis *et al., J. Am. Chem.*

Soc. **88,** 164, 166 (1966); Ruttenberg, *Science* **177,** 623 (1972) (from porcine insulin). Synthesis of sheep and human insulins: Katsoyannis, *Am. J. Med.* **40,** 652 (1966). Improved synthesis of human insulin: Sieber *et al., Helv. Chim. Acta* **57,** 2617 (1974). Prior to the discovery of recombinant DNA technology, the major source of insulin for human consumption was the pancreases of slaughtered animals. Human insulin was the first commercial health care product produced by recombinant DNA technology. For a review of the research, development, and production of human insulin by this technology *see* I. S. Johnson, *Science* **219,** 632-637 (1983). Synthesis of human insulin gene: H. M. Hsiung *et al., Nucleic Acid Res.* **6,** 137 (1979); **7,** 2199 (1979); **8,** 5753 (1980); S. A. Narang *et al., Nucleic Acids Symp. Ser.* **7,** 377 (1980).

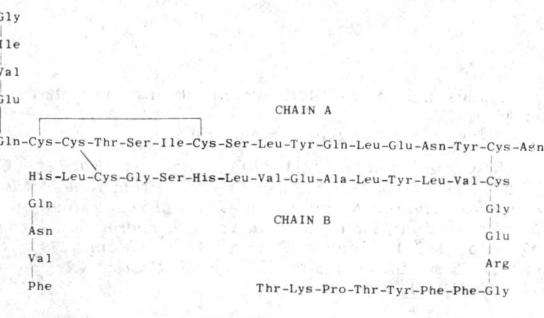

HUMAN INSULIN

Molecular basis of insulin action: M. P. Czech, *Ann. Rev. Biochem.* **46,** 359 (1977). Popularized history: G. A. Wrenshall, *The Story of Insulin* (Indiana Univ. Press, 1963) 232 pp; *Brit. Med. Bull.* **vol. 16,** no. 3 (Sept. 1960) entire issue. Review of biosynthesis: D. F. Steiner *et al., Recent Progr. Horm. Res* **25,** 207-282 (1969). Review of bioactivity, pharmacokinetics and therapeutic efficacy of human insulin: R. N. Brogden, R. C. Heel, *Drugs* **34,** 350-371 (1987). *General reviews:* Klostermeyer, Humbel, *Angew. Chem. Int. Ed.* **5,** 807-822 (1966) [332 refs.]; Trakatellis, Shwartz, *Fortschr. Chem. Org. Naturst.* **26,** 120-160 (1969); Grodsky, *Vitam. Horm. (New York)* **28,** 37-101 (1970). Books: *Insulins, Growth Hormone, and Recombinant DNA Technology,* J. L. Gueriguian, Ed. (Raven Press, New York, 1981) 227 pp; M. Bliss, *The Discovery of Insulin* (Univ. Chicago Press, Chicago, 1982) 304 pp.

Crystals, hexagonal system, usually obtained as flat rhombohedra and contg 0.4% Zn. Readily sol in dil acids and alkalies. Exists as the dimer in mild acid and as the monomer in 30% acetic acid. Practically insol from pH 4.5 to 7.0. Isoelectric point 5.30 to 5.35.

Crystals contg 2% Zn from acetate buffer contg excess Zn ions at pH 7 are much less sol and have delayed action.

Human insulin, *biosynthetic human insulin, insulin (prb), Huminsulin, Humulin,* human insulin prepd by recombinant DNA technology. *Semi-synthetic human insulin, insulin (emp), Biohulin, Orgasuline,* human insulin prepd by enzymatic modification of porcine insulin.

Insulin [131]*I, Radio-Iodinated Insulin.* Prepn: Burrows *et al., J. Clin. Invest.* **36,** 393 (1957); Grodsky *et al., Arch. Biochem. Biophys.* **81,** 264 (1959).

USE: Insulin ^{131}I used in the study of insulin binding factors from insulin resistant sera.

THERAP CAT: Antidiabetic.

4888. Insulinase. An enzyme that hydrolyzes insulin and is prepd from hog pancreas: Brink, Lewis, U.S. pat. **2,957,-809** (1960 to Merck & Co.). May be obtained from commercial pancreatin or trypsin. Even the purified crystals contain large amounts of elastase. *Review:* Thomas, *Postgrad. Med. J. Suppl.* **49,** 940 (1973).

4889. Insulin Injection. Actrapid; Endopancrine; Human Actrapid; Humulin S; Iletin; Insular; Insulyl; Iszilin; Regular Iletin; Velosulin. Sterile, acidified or neutral soln of insulin. Has a potency of not less than 95.0% and not more than 105% of the potency on the label, expressed in USP

Insulin Units. The potency is 40, 80, 100, or 500 USP Insulin Units in each ml.

Insulin injection containing not more than 100 U.S.P. units/ml is colorless or almost colorless; that containing 500 units may be straw-colored. Substantially free from turbidity and from insol matter. pH 2.5-3.5. Contains 0.1 to 0.25% (w/v) of either phenol or cresol and 1.4 to 1.8% (w/v) of glycerin. Onset of action occurs within 1 hr after s.c. injection, reaching its max in 2 to 3 hrs. Duration of effect, 6 to 8 hrs.

THERAP CAT: Antidiabetic.

4890. Insulin Zinc Suspension. Lente Insulin; Lente Iletin; Insulin Novo Lente; Human Monotard; Insulin Ultracard; Lentard; Monotard; Semitard; Ultratard. Sterile suspension, in a buffered water medium, of insulin modified by the addition of zinc chloride in a manner such that the solid phase of the suspension consists of a mixture of crystals and amorphous material in a ratio of approx 7:3: Hallas-Moller, *Lancet* **267,** 1029 (1954). Contains 40 or 80 U.S.P. insulin units/ml. Prepn of amorphous zinc insulin suspension: Petersen *et al.,* U.S. pat. **2,882,203** (1959 to Novo Terapeutisk Lab. A/S).

Almost colorless suspension of a mixture of characteristic crystals predominantly 10 to 40μ in max dimension and many particles which have no uniform shape and do not exceed 2μ in max dimension. pH 7.1-7.5. Contains 0.15 to 0.17% (w/v) of sodium acetate, 0.65 to 0.75% (w.v) of sodium chloride and 0.09 to 0.11% (w/v) methylparaben. Also contains, for each 100 U.S.P. insulin units, 0.20 to 0.25 mg zinc of which 40 to 65% is in the supernatant liquid.

Note: Insulin Zinc Suspension Prompt; Insulin Zinc Suspension (Amorphous); Semilente Iletin and *Insulin Zinc Suspension Extended; Insulin Zinc Suspension (Crystalline); Ultralente Iletin* are suspensions of minute particles of zinc insulin and differ only in the particle size which determines the duration of their action. Contain 40 or 80 U.S.P. insulin units/ml.

THERAP CAT: Antidiabetic.

4891. Interferon. IFN. A family of species-specific vertebrate proteins that confer non-specific resistance to a broad range of viral infections, affect cell proliferation and modulate immune responses. Discovered by A. Isaacs and J. Lindenmann, *Proc. Roy. Soc.* **B147,** 258 (1957) while studying viral interference. Originally produced by the interaction of inactivated influenza virus with chick chorioallantoic membranes; subsequently found to be inducible by viable virus in a variety of cells: D. C. Burke, A. Isaacs, *Brit. J. Exp. Pathol.* **39,** 452 (1958); in human leukocytes: I. Gresser, *Proc. Soc. Exp. Biol. Med.* **108,** 799 (1961). Host cell specificity and physical properties: D. A. J. Tyrrell, *Nature* **184,** 452 (1959); T. C. Merigan, *Science* **145,** 811 (1964). Production of acid-labile interferon by mitogen-stimulated human leukocytes: E. F. Wheelock, *ibid.* **149,** 310 (1966). Three major interferons, alpha, beta and gamma, *q.q.v.,* have been identified based on antigenic and physicochemical properties, the nature of the inducer, and the cellular source from which they are derived, *cf. Nature* **286,** 110 (1980). Known collectively as type I interferon, IFNs-α and -β are structurally related, are stable at pH 2, and compete for the same cell surface receptor. IFN-γ, also known as type II interferon, is structurally unrelated to type I IFNs, is acid labile, and has a different receptor. Receptor binding study: A. A. Branca, C. Baglioni, *Nature* **294,** 768 (1981).

Reviews: S. Baron, F. Dianzoni, Eds., *Tex. Rep. Biol. Med.* **35,** 1-573 (1977); *Ann. N.Y. Acad. Sci.* **350,** entitled "Regulatory Function of Interferons", J. Vilcek *et al.,* Eds. (1980) 643 pp ; S. Baron *et al.,* Eds., *Tex. Rep. Biol. Med.* **41,** 1-715 (1982). Review of immunobiology and clinical significance: E. R. Stiem *et al., Ann. Intern. Med.* **96,** 80-93 (1982); of pharmacology and toxicology: G. J. Mannering, L. B. Deloria, *Ann. Rev. Pharmacol. Toxicol.* **26,** 455-515 (1986); of IFN-gene family: C. Weissmann, H. Weber, *Prog. Nucl. Acid Res. Mol. Biol.* **33,** 251-300 (1986). Symposium on antiviral activity of natural and recombinant IFNs: *Antiviral Res.* **5,** Suppl. 1, 131-257 (1985). Review of structure, function and nomenclature: K. C. Zoon, *Interferon* **9,** 1-12 (1987). Books: *Interferons and Interferon Inducers,* N. Finter, Ed. (North Holland Publ. Co., Amsterdam, 1973) 598 pp; W. E. Stewart, *The Interferon System* (Springer, New York, 1979) 421 pp; *Interferons*, T. C. Merigan, R. M. Friedman, Eds. (Academic Press, New York, 1982) 481 pp; *Interferon*, K. Munk, H. Kirchner, Eds. (S. Karger, New York, 1982) 233 pp; H. Strander, *Interferon Treatment of Human Neoplasia* (Academic Press, New York, 1986) 265 pp.

4892. Interferon-α. Alfa-interferon; alpha-interferon; IFN-α; LeIF; leukocyte interferon; lymphoblastoid interferon. Family of highly homologous species-specific proteins that inhibits viral replication and cellular proliferation and modulates the immune response. One of the type I interferons. Produced by peripheral blood leukocytes or lymphoblastoid cells upon exposure to live or inactivated virus, double-stranded RNA, or bacterial products. Multiple subtypes have been identified which contain 165-166 amino acids; mol wt approximately 18,000-20,000 daltons. Production by virus stimulated human leukocytes: I. Gresser, *Proc. Soc. Exp. Biol. Med.* **108**, 799 (1961). Purification: C. B. Anfinsen *et al.*, *Proc. Nat. Acad. Sci. USA* **71**, 3139 (1974); K. Berg *et al.*, *J. Immunol.* **114**, 640 (1975). Comparison with interferon-β, *q.v.*: R. L. Cavalieri *et al.*, *Proc. Nat. Acad. Sci. USA* **74**, 3287 (1977). Amino acid analysis: M. Rubenstein *et al.*, *ibid.* **76**, 640 (1979). Partial sequence of human leukocyte IFN: W. P. Levy *et al.*, *ibid.* **77**, 5102 (1980); of human lymphoblastoid IFN: K. C. Zoon *et al.*, *Science* **207**, 527 (1980). Production by recombinant DNA technology: S. Nagata *et al.*, *Nature* **284**, 316 (1980); D. V. Goeddel *et al.*, *ibid.* **287**, 411 (1980). Comparison of structures and biological activities of HuIFN-α subtypes: M. Streuli *et al.*, *Science* **209**, 1343 (1980); D. V. Goeddel *et al.*, *Nature* **290**, 20 (1981). Receptor binding study: A. A. Branca, C. Baglioni, *ibid.* **294**, 768 (1981). Symposia on clinical antineoplastic activity: *Sem. Oncol.* **12**, Suppl. 5, 1-34 (1985); *ibid.* **13**, Suppl. 2, 1-101 (1986); *Invest. New Drugs* **5**, Suppl, S1-S77 (1987). Review of clinical efficacy in viral infections: M. Ho, *Ann. Rev. Med.* **38**, 51-59 (1987); in Kaposi's sarcoma: S. E. Krown, *Sem. Oncol.* **14**, Suppl. 3, 27-33 (1987); in hairy cell leukemia: B. D. Cheson, A. Martin, *Ann. Int. Med.* **106**, 871-878 (1987); in multiple sclerosis: R. L. Knobler, *Neurology* **38**, Suppl. 2, 58-61 (1988).

Interferon alfa-2a, $C_{860}H_{1353}N_{227}O_{255}S_9$, *IFN-αA, Ro 22-8181, Canferon, Roferon-A.*

Interferon alfa-2b, $C_{860}H_{1353}N_{229}O_{255}S_9$, *IFN-α₂, SCH-30500, YM-14090, Cibian, Introna, Intron A.*

Interferon alfa-2c, *Berofor Alpha 2.*

Interferon alfa-n1, *NSC-339140, Sumiferon, Wellferon.* Mixture of natural α-interferons.

Interferon alfa-n3, *Alferon.* Mixture of natural α-interferons.

THERAP CAT: Antiviral; antineoplastic.

4893. Interferon-β. Beta-interferon; fibroblast interferon; FIF; IFN-β; IFN-β₁; Feron; Fiblaferon; Frone; Naferon. Cytokine with antiviral, antiproliferative and immunomodulatory activity produced by fibroblasts in response to stimulation by live or inactivated virus or by certain synthetic polynucleotides. One of the type I interferons. Glycoprotein containing 166 amino acids; mol wt approx 20,000 daltons. High yield production by human fibroblast cell cultures: E. A. Havell, J. Vilcek, *Antimicrob. Ag. Chemother.* **2**, 476 (1972). Partial purification: C. B. Anfinsen *et al.*, *Proc. Nat. Acad. Sci USA* **71**, 3139 (1974); K. Berg *et al.*, *J. Immunol.* **114**, 640 (1975). Purification and initial characterization: E. Knight, Jr., *Proc. Nat. Acad. Sci. USA* **73**, 520 (1976). Comparison with interferon-α, *q.v.*: R. L. Cavalieri *et al.*, *ibid.* **74**, 3287 (1977). Review of production, purification and potential applications: W. A. Carter, J. S. Horoszewicz, *Pharmacol. Ther.* **8**, 359-377 (1980). Amino acid analysis, partial sequence: E. Knight, Jr., *et al.*, *Science* **207**, 525 (1980); S. Stein *et al.*, *Proc. Nat. Acad. Sci. USA* **77**, 5716 (1980). Amino acid sequence: T. Taniguchi *et al.*, *Gene* **10**, 11 (1980). Production by recombinant DNA technology: T. Taniguchi *et al.*, *Proc. Nat. Acad. Sci. USA* **77**, 5230 (1980); R. Derynck *et al.*, *Nature* **285**, 542 (1980); D. V. Goeddel *et al.*, *Nucleic Acids Res.* **8**, 4057 (1980). Comparative antiproliferative activity of natural IFNs-α and -β: E. C. Borden *et al.*, *Cancer Res.* **42**, 4948 (1982). Clinical evaluation of recombinant HuIFN-β in colorectal cancer: P. K.

Lillis *et al.*, *Cancer Treat. Rep.* **71**, 965 (1987); P. L. Triozzi *et al.*, *ibid.* 983; of natural form in herpes simplex infections: M. Glezerman *et al.*, *Lancet* **1**, 150 (1988). *Note:* A second factor produced by fibroblasts under conditions for IFN-β production has been identified: P. B. Sehgal, A. D. Sagar, *Nature* **288**, 95 (1980). Originally designated IFN-β₂, subsequently shown to be identical to the human *β-cell differentiation factor* or BSF-2 and to the *hybridoma growth factor* or HGF: P. B. Sehgal *et al.*, *Science* **235**, 731 (1987). Referred to in the literature as *interleukin-6*: J. Van Damme *et al.*, *Eur. J. Biochem.* **168**, 543 (1987); P. Poupart *et al.*, *EMBO J.* **6**, 1219 (1987).

IFN-beta_ser, *Betaseron.* Synthetic mutein having a serine substituted for the cysteine residue at position 17 of the native molecule. Prepn: D. F. Mark *et al.*, *Proc. Nat. Acad. Sci. USA* **81**, 5662 (1984).

THERAP CAT: Antineoplastic; antiviral.

4894. Interferon-γ. Gamma-interferon; IFN-γ; immune IFN; ImIFN; type II interferon. Lymphokine produced by antigen or mitogen stimulated T-cells with antiviral, antiproliferative, and immunomodulatory activity. Natural IFN-γ is composed of 143 amino acids. Mol wt ranges from ~16,000 to 25,000 daltons depending on the degree of glycosylation. Forms non-covalent dimers of approx mol wt 40,000 to 50,000 daltons. IFN-γ is structurally unrelated to interferons-α and -β, *q.q.v.*, is acid-labile and has a different cell surface receptor. Discovery of interferon produced by mitogen stimulated human leukocytes: E. F. Wheelock, *Science* **149**, 310 (1965). Characterization of the immune interferon response: J. A. Green *et al.*, *ibid.* **164**, 1415 (1969). Large-scale production by cell culture: M. P. Langford *et al.*, *Infect. Immun.* **26**, 36 (1979). Purification and characterization: Y. K. Yip *et al.*, *Proc. Nat. Acad. Sci. USA* **79**, 1820 (1982). Cloning and expression of a 146 amino acid recombinant HuIFN-γ: P. W. Gray *et al.*, *Nature* **295**, 503 (1982). Structure of HuIFN-γ gene: P. W. Gray, D. V. Goeddel, *ibid.* **298**, 859 (1982). Complete amino acid sequence of natural IFN-γ and comparison with recombinant form: E. Rinderknecht *et al.*, *J. Biol. Chem.* **259**, 6790 (1984). Comparison with type I interferons of antiviral and antiproliferative effects: B. Y. Rubin, S. L. Gupta, *Proc. Nat. Acad. Sci. USA* **77**, 5928 (1980). Receptor binding study: A. A. Branca, C. Baglioni, *Nature* **294**, 768 (1981). Enhancement of macrophage activation and proposed identity with *macrophage activating factor* (MAF): L. F. Nathan *et al.*, *J. Exp. Med.* **158**, 670 (1983); J. Le *et al.*, *J. Immunol.* **131**, 2821 (1983). Pharmacokinetics and toxicology in cancer patients: R. Kurzrock *et al.*, *Oncology* **42**, Suppl. 1, 41 (1985). Clinical evaluation in cancer: S. Sarhan-Raj *et al.*, *Cancer Treat. Rep.* **70**, 609 (1986); in rheumatoid arthritis: E. M. Lemmel *et al.*, *Rheumatol. Int.* **8**, 87 (1988). Review of biological activities: G. R. Adolf, *Oncology* **42**, Suppl. 1, 33-40 (1985); E. M. Bonnem, R. K. Oldham, *J. Biol. Resp. Mod.* **6**, 275-301 (1987). Review of IFN-γ induction and interrelationship with other cytokines: H. M. Johnson *et al.*, *Ann. N.Y. Acad. Sci.* **524**, 208-217 (1988).

Interferon Gamma-1b, $C_{734}H_{1166}N_{204}O_{216}S_5$, *N²-L-methionyl-1-139-interferon γ (human lymphocyte protein moiety reduced).*

Cys-Tyr-Cys-interferon γ, *N²-[N-(N-L-cysteinyl-L-tyrosyl)-L-cysteinyl]-interferon γ (human lymphocyte protein moiety reduced), S-6810, Immuneron, Polyferon.*

THERAP CAT: Antiviral, antineoplastic, immunomodulator.

4895. Interleukin-1. IL-1; lymphocyte activating factor; LAF; B-cell activating factor; BAF; T-cell replacing factor; TRF; endogenous pyrogen; leukocytic endogenous mediator; mononuclear cell factor. Mol wt 12,000-18,000. Immunoenhancing, pyrogenic, polypeptide factor produced in blood and in a variety of tissues by mononuclear phagocytes responding to antigens or inflammatory agents. Responsible for a wide variety of bioactivities. Elicits a non-antigen specific amplification of cellular and humoral immune responses. Induces production of interleukin-2 (IL-2), collagenase and prostaglandins, *q.q.v.* Initial identification of factor inducing T-cell proliferation: I. Gery *et al.*, *J. Exp. Med.* **136**, 128, 143 (1972); of factor enhancing antibody production: D. D. Wood, S. L. Gaul, *J. Immunol.*

113, 925 (1974). Isoln and preliminary chemical characterization: I. Gery, R. E. Handschumacher, *Cell. Immunol.* **11,** 162 (1974); D. D. Wood, *J. Immunol.* **123,** 2395 (1979). Definition and nomenclature: L. A. Aarden *et al., ibid.* 2928. Identity with endogenous pyrogen: L. J. Rosenwasser, C. A. Dinarello, *Cell. Immunol.* **63,** 134 (1981). Purification of murine IL-1: S. B. Mizel, D. Mizel, *J. Immunol.* **126,** 834 (1981); of human IL-1: J. A. Schmidt, *J. Exp. Med.* **160,** 772 (1984). Exists in several biochemically distinct forms exhibiting charge heterogeneity and differing in amino acid sequence. Produced as ~30,000 mol wt precursor which is subsequently converted to low mol wt form. Cloning and expression of cDNA for murine precursor IL-1: P. T. Lomedico *et al., Nature* **312,** 458 (1984); for human precursor IL-1: P. E. Auron *et al., Proc. Natl. Acad. Sci. USA* **81,** 7907 (1984); for 2 distinct forms of human IL-1 (IL-1α and IL-1β): C. J. March *et al., Nature* **315,** 641 (1985). Amino acid sequence of dominant species corresponding to IL-1β: P. Cameron *et al., J. Exp. Med.* **162,** 790 (1985). Crystallization of recombinant human IL-1β: D. B. Carter *et al., Proteins: Struct. Funct. Genet.* **3,** 121 (1988). Characterization of membrane-associated IL-1: E. A. Kurt-Jones *et al., Proc. Natl. Acad. Sci. USA* **82,** 1204 (1985). Induction of IL-2 production by IL-1: K. A. Smith *et al., J. Exp. Med.* **151,** 1551 (1980); S. Gillis, S. B. Mizel, *Proc. Natl. Acad. Sci. USA* **78,** 1122 (1981). Effect on B-cells in antibody production: D. D. Wood, *J. Immunol.* **123,** 2400 (1979); P. Lipsky, *Contemp. Topics Mol. Immunol.* **10,** 195 (1985). Effect in inflammatory response: S. B. Mizel *et al., Proc. Natl. Acad. Sci. USA* **78,** 2474 (1981). Role in the pathogenesis of the acute-phase response: C. A. Dinarello, *N. Engl. J. Med.* **311,** 1413 (1984). Comparison of bioactivity of 4 forms of human IL-1: D. D. Wood *et al., J. Immunol.* **134,** 895 (1985). Review of inter-relationship with lymphokines: N. M. Kouttab *et al., Clin. Chem.* **30,** 1539-1545 (1984). *Reviews:* J. J. Oppenheim *et al., Fed. Proc.* **41,** 257-262 (1982); S. B. Mizel, *Immunol. Rev.* **63,** 51-72 (1982); K. Bendtzen, *Allergy* **38,** 219-226 (1983); C. A. Dinarello, *Rev. Infect. Dis.* **6,** 51-95 (1984).
  Isoelectric point (human): 6.8-7.3 (dominant species); 5.3-5.8 (minor species). Isoelectric point (murine).: 4.5-5.5. Sensitive to trypsin, chymotrypsin, sodium dodecyl sulfate. Insensitive to 2-mercaptoethanol, neuraminidase, sodium periodate, iodacetamide. Stable at pH 3-10; stable at 56° for 60 min; unstable after 2 min at 100°.

4896. Interleukin-2. T-Cell Growth Factor; TCGF; Thymocyte Stimulating Factor; Costimulator; Lymphocyte Mitogenic Factor; IL-2; Proleukin. Mol wt ~15,000 (human); ~30,000 (murine). Soluble immunoenhancing glycoprotein (lymphokine) produced by T-lymphocytes following activation by antigens or mitogens in the presence of IL-1, *q.v.* Induces T-cell growth and proliferation; enhances natural killer cell activity. Potentiates the release of γ-interferon, B-cell growth factor, and B-cell differentiation factor. Restores T-cell function in immunodeficient disease states. Identification of soluble mitogenic factor produced by T-cells in response to antigen: R. S. Geha, E. Merler, *Cell. Immunol.* **10,** 86 (1974). Identification of T-cell growth promoting factor produced by mitogen stimulated lymphocytes: D. A. Morgan *et al., Science* **193,** 1007 (1976). Definition and nomenclature: L. A. Aarden *et al., J. Immunol.* **123,** 2928 (1979). Induction of IL-2 production by IL-1: K. A. Smith *et al., J. Exp. Med.* **151,** 1551 (1980). Review of early studies: J. Watson, D. Mochizuki, *Immunol. Rev.* **51,** 257 (1980). Purification and partial amino acid sequence: R. J. Robb *et al., Proc. Natl. Acad. Sci. USA* **80,** 5990 (1983). Production of human IL-2 by cultured malignant T-cells: S. Gillis, **Japan. Kokai 82 175,120** corresp to U.S. pat. **4,401,-756** (1982, 1983 to Immunex); by T-cell hybridoma: *idem,* **Japan. Kokai 82 181,698** corresp to U.S. pat. **4,473,642** (1982, 1984 to Immunex). Expression in monkey cells of cloned cDNA for human IL-2 and predicted amino acid sequence: T. Taniguchi *et al., Nature* **302,** 305 (1983); microbial expression of human IL-2: *eidem,* **Eur. pat. Appl. 142,268** (1985 to Ajinomoto; Japanese Foundation for Cancer Research), *C.A.* **103,** 49160m (1985). Isolation of human IL-2 gene: S. Mita *et al., Biochem. Biophys. Res. Commun.* **117,** 114 (1983). Synthesis of IL-2 peptides: A. Altman *et al., Proc. Natl. Acad. Sci. USA* **81,** 2176 (1984).

Biological activity of recombinant IL-2: S. A. Rosenberg *et al., Science* **223,** 1412 (1984). Structure of murine IL-2 and comparison with human IL-2: N. Kashima *et al., Nature* **313,** 402 (1985). Purification and cloning of the T-cell surface receptor for IL-2: W. J. Leonard *et al., Nature* **311,** 626 (1984). Studies on the mechanism of IL-2 signaling: M. A. Tigges *et al., Science* **243,** 781 (1989). Potential therapeutic use: A. W. Boylston, B. M. Vose, *Clin. Immunol. Allergy* **3,** 229 (1983). Clinical trial in AIDS: J. D. Lifson *et al., Lancet* **1,** 698 (1984); in metastatic cancer: S. A. Rosenberg *et al., N. Engl. J. Med.* **313,** 1485 (1985); R. L. Kradin *et al., Lancet* **1,** 577 (1989). Review of clinical studies: D. R. Parkinson, *Semin. Oncol.* **15,** Suppl. 6, 10-26 (1988). *Reviews:* K. Bendtzen, *Allergy* **38,** 219-226 (1983); J. J. Ferrar *et al., Ann. Rep. Med. Chem.* **19,** 191-200 (1984); N. M. Kouttab *et al., Clin. Chem.* **30,** 1539-1545 (1984). *Book: Contemporary Topics in Molecular Immunology,* F. P. Inman *et al.,* Ed., Vol. **10,** "The Interleukins", S. Gillis, F. P. Inman, Eds. (Plenum Press, New York, 1985).
  Isoelectric point: 6.5-6.8 (human); 3.9-5.0 (mouse). Sensitive to trypsin, chymotrypsin, sodium dodecyl sulfate. Stable in 2-mercaptoethanol, urea (2-4 mol/liter), neuraminidase. Stable at pH 3-10; at 56°C for 60 min. Unstable after 30 min. at 70°C.
  THERAP CAT: Antineoplastic; immunomodulator.

4897. Inula. Elecampane; scabwort; elfwort; horseheal. Dried rhizome and roots of *Inula helenium* L., *Compositae. Habit.* Central Asia, Europe; naturalized in U.S. *Constit.* Inulin, volatile oil, alantol, helenin, alantic acid, acrid resin.

4898. Inulin. Dahlin; alantin; alant starch. Mol wt approx 5000. Polysaccharide of *Compositae* which partially or completely replaces starch as a reserve food. Isoln from dahlia tubers: McDonald, "Polyfructosans and Difructose Anhydrides" in *Advan. Carbohyd. Chem.* vol. **2,** 254 (1946); from Jerusalem artichoke tubers: Bacon, Edelman, *Biochem. J.* **48,** 114 (1951). Structure: E. G. V. Percival, *Structural Carbohydrate Chemistry* (J. Garnet Miller, London, 2nd ed., 1962) p 274.

n = approx. 35

Spherical crystals from water. Hygroscopic in moist air. $[\alpha]_D^{20}$ −40° (c = 2) for the anhydr. Sol in hot water; slightly sol in cold water and organic solvents. Yields D-fructose and D-glucose upon acid hydrolysis.
  Acetate, fine powder from methanol. $[\alpha]_D^{20}$ −34° (c = 1.5 in chloroform); $[\alpha]_D^{20}$ −43° (c = 1.8 in acetic acid).
  Trimethylinulin, powder from hot water + acetone, mp 140°. $[\alpha]_D^{20}$ −55° (c = 1.03 in chloroform); $[\alpha]_D^{20}$ −54° (c = 1.09 in benzene).
  THERAP CAT: Diagnostic aid (renal function).

4899. Invertase. Invertin; saccharase; sucrase. Enzymes, obtained primarily from yeast as well as other sources, which catalyze the hydrolysis of sucrose into fructose and glucose. Since sucrose is both a β-fructofuranoside and an α-glucoside, it is split by two different types of enzymes, *β-fructofuranosidases* or *β-h-fructosidases* and certain α-

glucosidases or *α-n-glucosidoinvertases*, which attack the sucrose molecule from the fructose and glucose end, respectively. *β*-Fructofuranosidase, generally obtained from yeast, is characterized by its ability to hydrolyze raffinose, while *α*-glucosidase is inactive toward raffinose. *Reviews:* K. Myrbäck, "Invertases" in *The Enzymes*, vol. 4, P. D. Boyer *et al.*, Eds. (Academic Press, New York, 1960) pp 379-396; Lampen, "Yeast and Neurospora Invertases", *ibid.*, vol. **5** (3rd ed., 1971) pp 291-305; Cochrane, *Soc. Chem. Ind. (London) Monograph* no. **11**, 25-31 (1959) (Pub. 1961); Meister, *Wallerstein Lab. Commun.* **28**, 7 (1965).

USE: For preparation of invert sugar from sucrose; as analytical reagent for sucrose.

4900. Invert Sugar. Nulomoline; Calorose; Invesol; Invertogen; Insubeta; Travert. A mixture of about 50% glucose (dextrose) and 50% fructose (levulose) obtained by hydrolysis of sucrose. Hydrolysis of the sucrose may be carried out with acids or enzymes. Invert sugar is slightly levorotatory, reduces Fehling's soln and can be fermented. Honey is mostly invert sugar. Due to the levulose, it is somewhat sweeter than sucrose. The commercial product is obtained by inversion of a 96% cane sugar soln. The inversion is carried out at pH 3 to 4 by means of invertase and dil HCl. The acid is usually neutralized with sodium carbonate to pH 6.5. At this point, the dextrose crystallizes and the entire mass is beaten into a creamy, plastic product.

USE: In food products, in confectionery. As a humectant, like glycerol, to hold moisture and to prevent drying out. In brewing.

THERAP CAT: Parenteral nutrient.

4901. Iobenzamic Acid. *N-(3-Amino-2,4,6-triiodobenzoyl)-N-phenyl-β-alanine;* *β*-[N-(3-amino-2,4,6-triiodobenzoyl)phenylamino]propionic acid; isobenzamic acid; Bilibyk; Osbil. $C_{16}H_{13}I_3N_2O_3$; mol wt 662.03. C 29.03%, H 1.98%, I 57.51%, N 4.23%, O 7.25%. Prepn: **Brit. pat. 870,321** (1961 to Oesterreichische Stickstoffwerke). Toxicity: Lindner *et al., Arzneimittel-Forsch.* **11**, 384 (1961).

Crystals, mp 133-134.5°. Sol in acetone, dioxane, DMF; sparingly sol in cold ethanol, ether, benzene, toluene. Practically insol in water. LD_{50} orally in mice: 2.87 g/kg (Lindner).

Methyl ester, $C_{17}H_{15}I_3N_2O_3$, crystals, mp 156-157°.

THERAP CAT: Diagnostic aid (radiopaque medium—cholecystographic).

4902. Iocarmic Acid. *3,3'-[(1,6-Dioxo-1,6-hexanediyl)diimino]bis[2,4,6-triiodo-5-[(methylamino)carbonyl]benzoic acid];* *5,5'-(adipoyldiimino)bis[2,4,6-triiodo-N-methylisophthalamic acid];* Myelotrast. $C_{24}H_{20}I_6N_4O_8$; mol wt 1253.85. C 22.99%, H 1.61%, I 60.72%, N 4.47%, O 10.21%. Prepn: **Brit. pat. 1,033,695; U.S. pat. 3,290,366** (both 1966 to Mallinckrodt); G. B. Hoey *et al., J. Med. Chem.* **9**, 964 (1966). Clinical studies: Kunze, Schiefer, *Deut. Med. Wochenschr.* **97**, 245 (1972). Diagnostic use in myelography: P. Ahlgren, *Acta Radiol. Diag.* **13**, 753 (1972); in hysterosalpingography: H. E. Schutte, *Diag. Imaging* **51**, 277 (1982). Spinal toxicity in rats: I. O. Skalpe, A. Torvik, *Invest. Radiol.* **10**, 154 (1975).

Crystals from dimethylformamide, mp 302° (dec). Di-*N*-methylglucamine salt, $C_{38}H_{54}I_6N_6O_{18}$, *dimeglumine*

iocarmate, LM 280, Dimeray, Dimer-X, Dirax. Soly in water (25°): 65 g/100 ml.

THERAP CAT: Diagnostic aid (radiopaque medium).

4903. Iocetamic Acid. *3-[Acetyl-(3-amino-2,4,6-triiodophenyl)amino]-2-methylpropanoic acid; N-acetyl-N-(3-amino-2,4,6-triiodophenyl)-2-methyl-β-alanine; N-acetyl-N-(2,4,6-triiodo-3-aminophenyl)-β-aminoisobutyric acid;* DRC 1201; MP 620; Cholebrine; Cholimil. $C_{12}H_{13}I_3N_2O_3$; mol wt 613.94. C 23.48%, H 2.13%, I 62.01%, N 4.56%, O 7.82%. Prepn: **Neth. pat. Appls. 6,515,305; 6,607,275** (both 1967 to Dagra), *C.A.* **67**, 108422m (1967); **69**, 10263b (1968); Korver, *Rec. Trav. Chim.* **87**, 308 (1968). Pharmacology and toxicology: J. M. Janbroers *et al., Toxicol. Appl. Pharmacol.* **14**, 232, 246 (1969). Metabolic studies: Neleman, *Pharm. Weekblad.* **102**, 1039 (1967). Clinical studies: Hekster, *Radiol. Clin. Biol.* **37**, 338 (1968). Use in cholecystography: B. Goldberg, *Radiol. Clin.* **46**, 42 (1977).

White to light cream-colored powder, mp 224-225° (Neth. pat. **Appl. 6,515,305**). Also reported as mp range 191-212° (Korver). Practically insol in water. Very slightly sol in ether, ethanol, benzene; slightly sol in acetone, chloroform (Janbroers). LD_{50} in rats (g/kg): 7.1 orally; 0.70 i.v. (Janbroers). Consists of 2 isomers having mps 232° and 200-201° and pKa's of 4.25 and 4.0 respectively. The lower melting compd is approx two times as soluble.

THERAP CAT: Diagnostic aid (radiopaque medium).

4904. Iodamide. *3-(Acetylamino)-5-[(acetylamino)methyl]-2,4,6-triiodobenzoic acid; α,5-diacetamido-2,4,6-triiodo-m-toluic acid;* 3-acetamido-5-(acetamidomethyl)-2,4,6-triiodobenzoic acid; ametriodinic acid; SH 926. $C_{12}H_{11}I_3N_2O_4$; mol wt 627.93. C 22.95%, H 1.77%, I 60.63%, N 4.46%, O 10.19%. Prepn: Felder, Pitre, **Fr. pat. 1,382,277** and **U.S. pat. 3,360,436** (1964 and 1967 to Eprova); Felder *et al., Helv. Chim. Acta* **48**, 259 (1965). Pharmacology and toxicology: F. Bonati *et al., Arzneimittel-Forsch.* **15**, 222 (1965); Z. B. Zsebök, L. Szlavy, *ibid.* **17**, 1380 (1965); eidem, *Int. Z. Klin. Pharmakol. Ther. Toxikol.* **3**, 157 (1970). Clinical trials: Grothuesmann, *Arzneimittel-Forsch.* **15**, 233 (1965). Comprehensive description: D. Pitré in *Analytical Profiles of Drug Substances* vol. **15**, K. Florey, Ed. (Academic Press, New York, 1986) pp 337-365.

Crystals from acetic acid, mp 255-257°. Soly in water: 0.3 g/100 ml (22°).

N-Methyl-D-glucamine salt, $C_{19}H_{28}I_3N_3O_9$, *iodamide meglumine, Isteropac E.R., Jodomiron, Opacist E.R., Renovue-65, Renovue-DIP, Uromiro.* Sparingly sol in methanol, slightly sol in water, ethanol. Practically insol in ether, chloroform. LD_{50} in mice, rats, rabbits (g/kg): 9.0, 11.4, 13.2 i.v.; in rats, guinea pigs (g/kg): 17.9, 15.0 i.p. (Bonati).

USE: Diagnostic aid (radiopaque medium).

4905. Iodic Acid. HIO_3; mol wt 175.93. H 0.57%, I 72.14%, O 27.29%. Prepd by the oxidation of iodine with nitric acid, perchloric acid, or hydrogen peroxide: Baxter, Tilley, *Z. Anorg. Allgem. Chem.* **61**, 295 (1909); Lamb *et al., J. Am. Chem. Soc.* **42**, 1643 (1920); Bray, Caulkins, *ibid.* **53**, 44 (1931). Electrolytic method: Willard, Ralston, *Trans. Am. Electrochem. Soc.* **62**, 239 (1932).

Orthorhombic crystals. Darkens upon exposure to light. d_4^0 4.629. mp 110° (dec). Decompn to $HIO_3.I_2O_5$ starts at

70° and decompn to I_2O_5 is complete at 220°. Not hygroscopic, yet very freely sol in water: 269 g/100 ml H_2O at 20°; 295 g/100 ml H_2O at 40°. Density of aq solns (w/w) at 18°: 1% = 1.0071; 10% = 1.0900; 20% = 1.1969; 40% = 1.4640; 75% = 2.4710. Strong electrolyte, especially when dil. Sol in nitric acid, dil alcohol. Insol in abs alc, ether, chloroform. *Keep well closed and protected from light.*

THERAP CAT: Astringent, disinfectant.

4906. Iodinated Glycerol. Iodopropylidene glycerol; Organidin. $C_6H_{11}IO_3$; mol wt 258.07. C 27.93%, H 4.30%, I 49.18%, O 18.60%. Isomeric mixture of 67-75% of *2-(1-iodoethyl)-1,3-dioxolane-4-methanol* or *2,3-(2-iodopropylidenedioxy)propanol* and 33-25% of *2-(2-iodoethyl)-1,3-dioxolane-4-methanol* or *2,3-(3-iodopropylidenedioxy)propanol.* Prepn: Manchey *et al.,* U.S. pat. **2,872,378** (1959 to The Denver Chem. Manuf.).

Pale yellow liquid. Pungent, bitter aftertaste. d 1.797. n_D 1.547. Sol in ether, chloroform, isobutyl alc, methyl acetate, ethyl acetate, methyl formate, tetrahydrofuran.

THERAP CAT: Expectorant.

4907. Iodine. I; at. wt 126.9045; at. no. 53; valences 1 to 7 (usually monovalent); elemental state: I_2. A halogen. Abundance in igneous rocks: 3×10^{-5}% by wt; in seawater 5×10^{-8}% by wt. Natural isotope: 127 (100%); isotopes range in mass number from 117 to 139; radioactive tracer elements: 124, 125, 128, 131, 132. Discovered in 1811 by Courtois. Classed among the rarer elements. Extracted from Chilean nitrate-bearing earth (caliche) and from seaweed. Prepn of ultra-pure iodine for research purposes according to the equation $2KI + CuSO_4.5H_2O \rightarrow CuI + K_2SO_4 + \frac{1}{2}I_2 + 5H_2O$: Schmeisser in *Handbook of Preparative Inorganic Chemistry,* Vol. **1**, G. Brauer, Ed. (Academic Press, New York, 2nd ed., 1963) p 275. Reviews: *MTP Int. Rev. Sci.: Inorg. Chem., Ser. One,* Vol. **3**, V. Gutmann, Ed. (Butterworths, London, 1972); Downs, Adams, "Chlorine, Bromine, Iodine and Astatine" in *Comprehensive Inorganic Chemistry* vol. **2**, J. C. Bailar, Jr. *et al.,* Eds. (Pergamon Press, Oxford, 1973) p 1107-1594; C. J. Mazac in Kirk-Othmer *Encyclopedia of Chemical Technology* vol. **13** (Wiley-Interscience, New York, 3rd ed., 1981) pp 649-677.

Bluish-black scales or plates; diatomic; metallic luster; characteristic odor; sharp, acrid taste; violet corrosive vapor. mp 113.60°. bp 185.24°. d (solid, 25°) 4.93. Vapor pressure (solid): 0.030 mm (0°); 0.305 mm (25°); 2.154 mm (50°); 26.78 mm (90°). Heat capacity at constant pressure (25°) 13.011 cal/mole/°C: Shirley, Giauque, *J. Am. Chem. Soc.* **81**, 4778 (1959). Total soly in water (25°) 0.0013 moles/l with negligible formation of HOI (6.4×10^{-6} moles/l); freely sol in aq solns of HI or iodides. Soly in organic solvents (g I_2/100 g soln, 25°): benzene 14.09; CS_2 16.47; ethanol 21.43; ethyl ether 25.20; cyclohexane 2.719; CCl_4 (35°) 2.603: Hildebrand, Jenks, *J. Am. Chem Soc.* **42**, 2180 (1920); Hildebrand *et al., ibid.* **72**, 1017 (1950); sol in chloroform, glacial acetic acid, glycerol oils. Solutions of iodine in aq solns of inorganic iodides are brown or deep brown, depending on the concn of the iodine. Solvents contg nitrogen atoms, such as pyridine, quinoline, or amines, dissolve iodine to form brown solns. Chloroform, carbon tetrachloride, carbon disulfide, and especially phosphorus trichloride give violet solns. The violet color is also given by fluorinated -amines. The soly in water is also increased by alkali bromides, but decreased by sulfates and nitrates. Less reactive than bromine; E° (aq) $\frac{1}{2}I_2/I^-$ 0.535 V dissociation energy (25°): 36.115 kcal. For reactions, complexes, and formation of monatomic iodine: *see* Chlorine. Iodine stains may be removed with sodium thiosulfate soln or ammoniated alc. *Incompat:* alkaloids, starch, tannins.

Note: Tamed Iodine, *see* Iodophor.

Human Toxicity: Ingestion of large quantities causes abdominal pain, nausea, vomiting, diarrhea. In severe cases, purging, excessive thirst and circulatory failure may develop. Amounts of 2 to 4 g have been fatal. The solid element is intensely irritating to eyes, skin, mucous membranes. *See Clinical Toxicology of Commercial Products,* R. E. Gosselin *et al.,* Eds. (Williams & Wilkins, Baltimore, 4th ed., 1976) Sect. III, pp 181-182.

USE: Manuf iodine compds, germicides, antiseptics. To reduce friction of hard surfaces, including stainless steel and glass. Catalyst in the alkylation and condensation of aromatic amines; in sulfations and sulfonations. Artificial isotopes of iodine are used in biological, biochemical and chemical structure research. Important reagent in analytical chemistry.

THERAP CAT: Antihyperthyroid; topical anti-infective.

THERAP CAT (VET): Internally for goiter, hypothyroidism, in iodine deficiency. Topically as antiseptic, disinfectant, counterirritant and to promote absorption.

4908. Iodine Colloidal. Elemental iodine in a colloidal state, stabilized with a suitable "protecting" colloid. Prepn: Chandler, Miller, *J. Phys. Chem.* **31**, 1091 (1927). This iodine is of a brick-red color and was supplied as a powder contg 25% iodine or in the form of an aq suspension contg 16% elemental iodine.

THERAP CAT (VET): Has been used as an anthelmintic in poultry.

4909. Iodine Heptafluoride. F_7I; mol wt 259.91. F 51.17%, I 48.83%. IF_7. Prepd by passing fluorine through liq IF_5 at 90°, then heating the vapors to 270°: Ruff, Keim, *Z. Anorg. Allgem. Chem.* **193**, 176 (1930); Kwasnik in *Handbook of Preparative Inorganic Chemistry,* vol. **1**, G. Brauer, Ed. (Academic Press, New York, 2nd ed., 1963) pp 160-161. May also be prepd from fluorine and dried PdI_2 or KI to minimize formation of the impurity, IOF_5: Bartlett, Levchuk, *Proc. Chem. Soc.* **1963**, 342; Selig *et al., J. Phys. Chem.* **71**, 2739 (1967). *Reviews:* Kemmitt, Sharp, *Advan. Fluorine Chem.* **4**, 246-247 (1965); Stein, "Physical and Chemical Properties of Halogen Fluorides" in *Halogen Chemistry,* Vol. **1**, V. Gutmann, Ed. (Academic Press, New York, 1967) pp 133-224; Meinert, *Z. Chem.* **7**, 41-57 (1967).

Colorless gas. Crystals when solid. Moldy, acrid odor. *Attacks glass and quartz.* d (liq; 6°) 2.8. mp 6.45°. Sublimes at 4.77°. Easily forms a supercooled liquid which boils at +4.5°. Trouton constant 26.4. Water dissolves the gas without violence, some decompn. Readily absorbed by NaOH soln.

USE: Fluorinating agent. *Caution:* Highly irritating to skin, mucous membranes. *See also* Fluorine.

4910. Iodine Monobromide. BrI; mol wt 206.84. I 61.36%, Br 38.64%. IBr. Prepd from the elements, excess Br is removed by heating to 50° in a current of CO_2: Bornemann, *Ann.* **189**, 202 (1877); Terwogt, *Z. Anorg. Allgem. Chem.* **47**, 203 (1905); *Gmelin's, Iodine* (8th ed.) **8**, p 631 (1933); Schmeisser in *Handbook of Preparative Inorganic Chemistry* vol. **1**, G. Brauer, Ed. (Academic Press, New York, 2nd ed., 1963) p 291-292. Crystal structure: Swink, Carpenter, *Acta Cryst.* **24B**, 429 (1968).

Brownish-black crystals or very hard solid. d 4.416. mp 40°. bp 116° (decompn). Sol in water, alcohol, ether, carbon disulfide, glacial acetic acid. May be stored in brown glass-stoppered bottles. *Keep in a cool place.*

Caution: Vapors are corrosive to the eyes and mucous membranes.

4911. Iodine Monochloride. Wijs' chloride. ClI; mol wt 162.38. Cl 21.84%, I 78.16%. ICl. Prepd from the elements: Cornog, Karges, *Inorg. Syn.* **1**, 165 (1939); Buckles, Bader, *ibid.* **9**, 130 (1967).

Black crystals or reddish-brown liquid. The crystals occur in two modifications: α-form (stable), black needles (ruby red by transmitted light), mp 27.2°. β-Form (labile), black platelets (brownish red by transmitted light), mp 13.9°. bp 97° (decompn). d_4^{29} 3.10. Sol in water, alc, ether, CS_2, acetic acid. May be stored in brown glass-stoppered bottles. Not hygroscopic, but forms I_2O_5 in the presence of air.

USE: In Wijs' soln (iodine monochloride in glacial acetic acid), used to determine iodine values of fats and oils. *Caution:* Attacks the skin, forming dark, painful patches (effective countermeasure: immediate washing with 20% HCl).

THERAP CAT: Topical anti-infective.

4912. Iodine Pentafluoride. F_5I; mol wt 221.91. F 42.81%, I 57.19%. IF_5. Prepd by passing fluorine over iodine with cooling: Gore, *Phil. Mag.* [4] **41**, 309 (1871); Moissan, *Compt. Rend.* **135**, 563 (1902); *Bull. Soc. Chim. France* [3], **29**, 6 (1903); Ruff, Keim, *Z. Anorg. Allgem. Chem.* **201**, 245 (1931); Ruff, Braida, *ibid.* **220**, 43 (1934); Kwasnik in *Handbook of Preparative Inorganic Chemistry*, vol. 1, G. Brauer, Ed. (Academic Press, New York, 2nd ed., 1963) pp 159-160. *Reviews:* Kemmitt, Sharp, *Advan. Fluorine Chem.* **4**, 247-248 (1965); Stein, "Physical and Chemical Properties of Halogen Fluorides" in *Halogen Chemistry*, Vol. 1, V. Gutmann, Ed. (Academic Press, New York, 1967) pp 133-224; Meinert, *Z. Chem.* **7**, 41-57 (1967).

Liquid. mp 9.43°. bp 100.5°. d^{25} 3.19. Thermodynamic data: Osborne *et al., J. Chem. Phys.* **54**, 3790 (1971). *Fumes in air!* Attacks glass, especially when hot. Violent reaction with water. Organic compds carbonize on contact, sometimes with conflagration. Instant reaction with S, P, Si, Bi, W, As, usually with incandescence.

USE: Mild fluorinating agent.

4913. Iodine Pentoxide. Iodic anhydride. I_2O_5; mol wt 333.84. I 76.04%, O 23.97%. Prepd by thermal dehydration of iodic acid at 195°: Lamb *et al., J. Am. Chem. Soc.* **42**, 1644 (1920). Crystal structure: Selte, Kjekshus, *Acta Chem. Scand.* **24**, 1912 (1970).

Needle shaped, hygroscopic crystals (an impure product may have a pink cast). d^{25} 5.08. Decompn into I_2 and O_2 begins at 275° and proceeds rapidly at 350°. Very sol in water (187.4 g/100 ml at 13°) forming HIO_3. Insol in abs alc, ether, chloroform, carbon disulfide. Dissolves in nitric acid from which it crystallizes as I_2O_5 if the HNO_3 concn is > 50%. Strong oxidizing agent, may detonate when triturated with oxidizable matter. *Keep well closed and protected from light.*

USE: Oxidizes carbon monoxide to carbon dioxide with the formn of iodine (Ditte's reaction, which proceeds rapidly at 65° or above, but slowly at room temp). This reaction is used in gas analysis and for removing carbon monoxide from the air (in respirators preferably in presence of some H_2SO_4).

4914. Iodine Trichloride. Cl_3I; mol wt 233.39. Cl 45.60%, I 54.40%. ICl_3. Prepd by adding finely powdered iodine to an excess of liq chlorine, which is then boiled away: Thomas, Depuis, *Compt. Rend.* **143**, 282 (1906); Booth, Morris, *Inorg. Syn.* **1**, 167 (1939). Alternate process: Birk, *Angew. Chem.* **41**, 751 (1928); *Z. Anorg. Allgem. Chem.* **172**, 399 (1928); Wilke-Dörfurt, Wolff, *ibid.* **185**, 333 (1930).

Long yellow or brownish needles or fluffy powder. Pungent, irritating odor. *Corrosive to human skin. Must be stored in well closed amber glass bottles, preferably in refrigerator.* d^{-4} 3.203. mp about 33°. Volatile at room temp. mp 101° at 16 atm (pressure of its satd vapor). A pressure of 1 atm is reached at 64°. Complete decompn into ICl and Cl_2 at 77°. Generally used as a 20 to 35% soln in concd HCl.

USE: Chlorinating and oxidizing agent. *Caution:* Concd solns are strongly irritating.

THERAP CAT: Topical anti-infective.

4915. Iodinin. *1,6-Phenazinediol 5,10-dioxide;* 1,5-dihydroxyphenazine *N,N'*-dioxide. $C_{12}H_8N_2O_4$; mol wt 244.20. C 59.02%, H 3.30%, N 11.47%, O 26.21%. Antibiotic pigment from *Chromobacterium iodinum:* Clemo, McIlwain, *J. Chem. Soc.* **1938**, 479; Hegedüs in *Emil Barell Jubilee Volume* (1946), p 388; from *Waksmania aerata* and *Pseudomonas iodinum:* Gerber, Lechevalier, *Biochemistry* **3**, 598 (1964). Structure: Clemo, Daglish, *J. Chem. Soc.* **1950**, 1481. Synthesis: Matsumura, Takeda, *Nippon Kagaku Zasshi* **81**, 515 (1960), *C.A.* **56**, 470a (1962). Biosynthetic studies on incorporation of shikimic acid, *q.v.,* into iodinin:

U. Hollstein *et al., Tetrahedron Letters* **1978**, 2987; R. B. Herbert *et al., J. Chem. Soc. Perkin Trans. I* **1979**, 2411; T. Etherington *et al., ibid.* 2416.

Purple crystals with coppery sheen from chloroform, dec 236°. pK 12.5. Stable in acid, unstable in alkali. Sol in benzene, toluene, xylene, carbon disulfide, chloroform, ethyl acetate. Slightly sol in hot alc. Practically insol in cold alc, ether, acetic acid, petr ether, amyl alcohol. Insol in water. Sol in concd sulfuric acid and in glacial acetic acid with red color. Sol in NaOH solns giving brilliantly blue solns which deposit green crystals of the unstable sodium deriv.

4916. Iodipamide. *3,3'-[(1,6-Dioxo-1,6-hexanediyl)diimino]bis[2,4,6-triiodobenzoic acid]; 3,3'-(adipoyldiimino)-bis[2,4,6-triiodobenzoic acid]; N,N'*-adipylbis(3-amino-2,-4,6-triiodobenzoic acid); adipic acid di(3-carboxy-2,4,6-triiodoanilide); adipiodone; Cholografin; Cholospect. $C_{20}H_{14}I_6N_2O_6$; mol wt 1139.81. C 21.08%, H 1.24%, I 66.81%, N 2.46%, O 8.42%. Prepn: Priewe, Rutkowski, U.S. pat. **2,776,241** (1957 to Schering AG); Kotler-Brajtburg *et al., Roczniki Chem.* **36**, 763 (1962), *C.A.* **58**, 5568a (1963). Purification: Cassebaum, Drux, **East Ger.** pat. **33,738** (1964), *C.A.* **63**, 11441b (1965). Pharmacology: Fischer, Varga, *Acta Physiol.* **38**, 135 (1970). Metabolic studies: Kiyono *et al., Radioisotopes* **20**, 78 (1971). Pharmacology and toxicity study: F. J. Rosenberg *et al., Invest. Radiol.* **15**, S142 (1980). Comprehensive description: H. H. Lerner in *Analytical Profiles of Drug Substances* vol. 3, K. Florey, Ed. (Academic Press, New York, 1974) pp 333-363.

Crystals, dec 306-308°. $n_D^{21.5}$ 1.3294 (c = 0.445 in methanol). Soly at 20°: methanol 0.8%; ethanol 0.3%; acetone 0.2%; ether 0.1%. Practically insol in water, benzene. LD_{50} in mice, rats (mg/kg): 2380 ±290, 4430 ±310 i.v. (Rosenberg).

Disodium salt, $C_{20}H_{12}I_6N_2Na_2O_6$. Prepd by dissolving the free acid in a dil aq soln of NaOH and buffering to pH 6.5-7.7. Used as 20% soln.

Bis[N-methylglucamine] salt, $C_{34}H_{48}I_6N_4O_{16}$, Biligrafin, Cholegrafin, Endocistobil, Endografin, Intrabilix, Transbilix. More sol in water than the disodium salt. Used as 40% soln.

THERAP CAT: Diagnostic aid (radiopaque medium—cholecystographic).

THERAP CAT (VET): Diagnostic aid (radiopaque medium).

4917. Iodized Oil. Lipiodol. An iodine addition product of vegetable oils, contg 38-42% organically combined iodine.

Thick, viscous, oily liq; alliaceous odor; oleaginous taste; dec on exposure to air and light, becoming brown. *Keep in small, nearly full containers, protected from light.*

THERAP CAT: Diagnostic aid (radiopaque medium).

THERAP CAT (VET): X-ray contrast medium.

4918. Iodoacetic Acid. $C_2H_3IO_2$; mol wt 185.96. C 12.92%, H 1.63%, I 68.25%, O 17.21%. CH_2ICOOH. Prepd by treating chloroacetic acid in acetone soln with NaI.

Colorless or white crystals. mp 82-83°. Sol in water, alc; very slightly sol in ether.

4919. Iodoalphionic Acid. *4-Hydroxy-3,5-diiodo-α-phenylbenzenepropanoic acid; β-(4-hydroxy-3,5-diiodophenyl)-α-phenylpropionic acid;* 3,5-diiodo-α-phenylphloretic acid; Coletrast; Priodax; Pheniodol; Dikol; Iodobil; Jodobil; Biliognost; Tenicid; Biliselectan. $C_{15}H_{12}I_2O_3$; mol wt 494.08. C 36.46%, H 2.45%, I 51.38%, O 9.72%. Prepd by iodinating β-(4-hydroxyphenyl)-α-phenylpropionic acid with iodine in an aq soln of alkali iodide in the presence of NH_3 or with iodine in acetic anhydride as the iodinating agent, *cf.* **Brit.** pat. **559,024** (1944), *C.A.* **40**, 1883 (1946). Prepn of optically active forms: Tullar, Hoppe, U.S. pat. **2,552,696** (1951 to Sterling Drug). Toxicity data: J. O. Hoppe, S. Archer, *Am. J. Roentgenol. Rad. Ther.* **69**, 630 (1953).

dl-Form, faintly yellowish powder. dec 157-162°. Crystals from acetic acid, mp 163.5-163.8°. Insol in water. Sol in alcohol, ether, alkali carbonate and in alkali hydroxide solns; slightly sol in benzene and chloroform. Forms a water-sol disodium salt. LD_{50} in mice (g/kg): 3.8 orally (Hoppe, Archer).

l-Form, hydrated crystals, mp 80-85°, $[\alpha]_D^{25} - 59°$ (c = 1.5 in 95% alcohol). Disodium salt, mp >230°, $[\alpha]_D^{26} -56.6°$ (c = 1.5 in water).

d-Form, hydrated crystals, mp 80-85°, $[\alpha]_D^{25} +59°$ (c = 1.5 in 95% alcohol). Disodium salt, mp >230°, $[\alpha]_D^{26}$ +56.6° (c = 1.5 in H_2O).

THERAP CAT: Diagnostic aid (radiopaque medium—cholecystographic).

4920. *p*-Iodoaniline. C_6H_6NI; mol wt 219.04. C 32.90%, H 2.76%, I 57.94%, N 6.40%.

White crystals. mp 67-68°. Slightly sol in water, freely in alcohol, chloroform, ether.

4921. *o*-Iodoanisole. C_7H_7IO; mol wt 234.04. C 35.92%, H 3.01%, I 54.23%, O 6.84%. Prepd by diazotizing *o*-anisidine and decomposing the diazonium salt with KI.

Yellow liquid; darkens on exposure to air. d^{20} 1.80. bp_{730} 238-240°. Insol in water; miscible with alcohol, chloroform, ether. *Protect from light.*

4922. Iodobenzene. C_6H_5I; mol wt 204.02. C 35.32%, H 2.47%, I 62.23%. Obtained by diazotizing aniline and then treating with an aq soln of KI, or by the action of HNO_3 on a mixture of C_6H_6 and iodine.

Colorless liquid; rapidly becomes yellow; characteristic odor. d_4^{15} 1.8384. bp 188-189°. mp −30°. n_D^{18} 1.621. Insol in water; miscible with alcohol, chloroform, ether. *Protect from light.*

4923. *o*-Iodobenzoic Acid. $C_7H_5IO_2$; mol wt 248.03. C 33.90%, H 2.03%, I 51.17%, O 12.90%. C_6H_4ICOOH. Prepd by diazotizing *o*-aminobenzoic acid and treating the resulting diazonium compound with aq KI.

White needles. d 2.25. mp 162°. Sparingly sol in water; sol in alcohol, ether.

4924. Iodochlorhydroxyquin. *5-Chloro-7-iodo-8-quinolinol;* 5-chloro-8-hydroxy-7-iodoquinoline; chloroiodoquin; iodochlorohydroxyquinoline; iodochloroxyquinoline; quinoform; Amebil; Alchloquin; Amoenol; Bactol; Barquinol; Budoform; Chinoform; Clioquinol; Cliquinol; Eczecidin; Enteroquinol; Entero-Septol; Entero-Vioform; Enterozol; Entrokin; Hi-Enterol; Iodoenterol; Nioform; Quinambicide; Rometin; Vioform. C_9H_5ClINO; mol wt 305.52. C 35.38%, H 1.65%, Cl 11.61%, I 41.54%, N 4.59% O 5.24%. Prepn: Ger. pat. 117,767 (1899 to Ciba); U.S. pat. 641,491 (1900); A. Das, S. L. Mukherji, *J. Org. Chem.* 22, 1111 (1957).

Brownish-yellow, bulky powder. Practically odorless. Dec about 178-179°. One part dissolves in 43 parts of boil-

ing alc, 128 parts chloroform, 17 parts of boiling ethyl acetate, 170 parts of cold acetic acid, 13 parts of boiling acetic acid. Almost insol in water, cold alc, ether. Darkens upon exposure to light. LD_{50} orally in cats: 400 mg/kg, Davis *et al., Am. J. Trop. Med.* 24, 29 (1944).

Human Toxicity: Has been linked with the occurrence of subacute myelo-optic neuropathy (S.M.O.N. syndrome) in Japan, T. Tsubaki *et al., Lancet* 1, 696 (1971); *see also ibid.* 1, 534 (1977); J. Tateishi in *Drug-Induced Sufferings,* T. Soda, Ed. (Excerpta Medica, New York, 1980) pp 464-472.

THERAP CAT: Topical anti-infective; anti-amebic.

THERAP CAT (VET): Has been used as an intestinal anti-infective.

4925. Iodofenphos. *Phosphorothioic acid O-(2,5-dichloro-4-iodophenyl) O,O-dimethyl ester; O-(2,5-dichloro-4-iodophenyl) O,O-dimethyl phosphorothioate;* C 9491; Nuvanol N. $C_8H_8Cl_2IO_3PS$; mol wt 412.99. C 23.27%, H 1.95%, Cl 17.17%, I 30.73%, O 11.62%, P 7.50%, S 7.76%. Organophosphorous insecticide with stomach and contact activity. Prepn: E. Beriger *et al.,* **S. Afr. pat.** 65 06,107; *eidem,* **U.S. pat.** 3,468,984 (1965, 1969 both to Ciba). Metabolism and anti-cholinesterase activity: F. R. Johannsen, C. O. Knowles, *J. Econ. Entomol.* 63, 693 (1970). GLC determn in cattle urine and tissues: M. C. Ivey, D. D. Oehler, *J. Agr. Food Chem.* 24, 1049 (1976). Crystal and molecular structure: R. G. Baughman, P.-J. Yu, *ibid.* 30, 293 (1982). Use to control flies and ticks in cattle: R. O. Drummond *et al., J. Econ. Entomol.* 65, 1354 (1972). Review: B. C. Haddow, T. G. Marks, *Proc. 5th Brit. Insect. Fungic. Conf., 1969* 2, 531 (1970).

Pale yellow oil, solidifies to crystals, mp 72-73° (Beriger). Also reported as white crystals, mp 75° (Haddow). Sol in acetone, xylene. Slightly sol in alcohol. Soly in water: <2 ppm. LD_{50} in rats, female mice (mg/kg): 2100, >10000, orally (Haddow, Marks).

USE: Insecticide.

THERAP CAT (VET): Ectoparasiticide.

4926. Iodoform. *Triiodomethane.* CHI_3; mol wt 393.78. C 3.05%, H 0.26%, I 96.69%. Prepn from acetone, sodium hypochlorite, potassium iodide, and sodium hydroxide: Glass, *Quart. J. Pharm. Pharmacol.* 8, 351 (1935); from chloroform + methyl iodide: Soroos, Hinkamp, *J. Am. Chem. Soc.* 67, 1642 (1945); by electrolysis: Glasstone, *Ind. Chem.* 7, 315 (1931). Description of the iodoform reaction: Seelye, Turney, *J. Chem. Ed.* 36, 572 (1959).

Yellow powder or crystals; mp about 120°; dec at high temp with evolution of iodine. Unctuous touch; characteristic, disagreeable odor. Volatile with steam. d 4.1. Very slightly sol in water. One gram dissolves in 60 ml cold alcohol, 16 ml boiling alcohol, 10 ml chloroform, 7.5 ml ether, 80 ml glycerol, 3 ml carbon disulfide, 34 ml olive oil; freely sol in benzene, acetone, slightly sol in petr ether. LD_{50} s.c. in mice: 1.6 mmoles/kg, Kutob, Plaa, *Toxicol. Appl. Pharmacol.* 4, 354 (1962).

Incompat: Mercuric oxide, calomel, silver nitrate, tannin, balsam Peru directly mixed.

THERAP CAT: Topical anti-infective.

THERAP CAT (VET): Antiseptic, disinfectant for superficial lesions and in the female reproductive tract.

4927. *o*-Iodohippurate Sodium. *N-(2-Iodobenzoyl)glycine monosodium salt;* sodium *o*-iodohippurate; Hippodin; Jodairol. $C_9H_7INNaO_3$; mol wt 327.05. C 33.05%, H 2.16%, I 38.80%, N 4.28% O 7.03%, Na 14.68%. "Contains not less than 38.5% nor more than 39% iodine, when calcd to the dried substance." Prepd by the condensation of *o*-iodobenzoic acid with glycine, and treatment with NaOH soln: A. P. Sachs, U.S. pat. 2,135,474 (1938 to Zonite Prods.). Diagnostic use in posttransplant renography: H.

Huland, K. Bischoff, *J. Urology* **129**, 925 (1983); J.-I. D. Jorgensen, J. Ladefoged, *Dan. Med. Bull.* **34**, 50 (1987).

(1970). Metabolism in rabbits: D. Pitre, L. Fumagalli, *Farmaco Ed. Sci.* **32**, 76 (1977).

Dihydrate, crystals. Freely sol in water; sol in alcohol and in dil solns of alkalies. An aq soln is neutral or faintly alkaline to litmus.

[131]I labelled form, *Hippuran I 131, Hipputope.*
[123]I labelled form, *Nephroflow.*

USE: Diagnostic aid (radiopaque medium); labelled form as diagnostic aid (radioactive imaging agent).

4928. o-Iodophenol. C_6H_5IO; mol wt 220.02. C 32.75%, H 2.29%, I 57.69%, O 7.27%. Prepd by the action of iodine on o-chloromercuriphenol in chloroform: Whitmore, Hanson, *Org. Syn.* **4**, 37 (1925).

Plates from ligroin, mp 43°. d^{80} 1.8757. bp_{160} 186-187°. Sol in alcohol, ether, chloroform, carbon disulfide, benzene; fairly sol in hot water.

4929. p-Iodophenol. C_6H_5IO; mol wt 220.02. C 32.75%, H 2.29%, I 57.69%, O 7.27%. Obtained by action of iodine in alkaline KI upon phenol; or by diazotization of p-aminophenol and subsequent replacement of the diazonium group by iodine.

White or reddish crystals; characteristic odor. mp 93-94°. Slightly sol in water, freely in alc, ether. *Protect from light.*

Note: The designation "phenol iodide", or more properly "iodized phenol", is applied to a soln of 1 part iodine in 4 parts phenol.

4930. Iodophors. Tamed Iodine. The term iodophor may be applied to any product in which surface active agents (such as nonoxynol, *q.v.,* or Monosan-IOD) act as carriers and solubilizing agents for iodine, *cf.* povidone-iodine. Other examples: *Biopal CVL-10; Dermevan; Idonyx; Iobac; Ioprep; Iosan; Kleenodyne; Rhudane; Showersan; Wescodyne; Westamine X.* An iodophor usually enhances the bactericidal activity of iodine, reduces vapor pressure and odor. Staining is almost nonexistent and wide dilution with water is possible. Examples of prepn: Shelanski, Winicov, U.S. pat. **2,710,277** (1955 to West Labs); Scheib *et al.,* U.S. pat. **2,977,315** (1961 to Lazarus Labs). Additional literature: G. F. Reddish, *Antiseptics, Disinfectants, Fungicides* (Philadelphia, 1954) p 195; General Aniline & Film Corp.: *Brochure on Biopal CVL-10.* Practical formulations for making solid and liq iodophor-containing sanitizers and detergents: *Chem. Process.* **19**, no. 5, 56-57 (May 1956).

THERAP CAT: Topical anti-infective.
THERAP CAT (VET): Skin disinfectant.

4931. Iodophthalein Sodium. Soluble iodophthalein; tetraiodophenolphthalein sodium; T.I.P.P.S.; tetraiodophthalein sodium; tetiothalein sodium; Iodeikon; Cholepulvis; Keraphen; Shadocol; Bilitrast; Iodognost; Stipolac; Tetraiode; Foriod; Iodtetragnost; Antinosin; Cholumbrin; Iodorayoral; Opacin; Photobiline; Piliophen; Videophel; Radiotetrane; Nosophene sodium; Iodophene sodium. $C_{20}H_8I_4Na_2O_4$; mol wt 865.86. C 27.74%, H 0.93%, I 58.63%, Na 5.31%, O 7.39%. Isomeric with phentetiothalein sodium, *q.v.* Prepd by the action of iodine (in KI soln) on phenolphthalein in aq alk soln: Classen, Löb, *Ber.* **28**, 1603 (1895); **Ger.** pats. **85,930** (1894); **88,390** (1895). Also obtained by treating an alk soln of phenolphthalein with iodine monochloride: Kalle, **Ger.** pat. **143,596** (1900). Prepn of free acid: Orndorff, Mahood, *J. Am. Chem. Soc.* **40**, 937 (1918). Toxicity data: J. O. Hoppe *et al., J. Med. Chem.* **13**, 997

Trihydrate, pale-blue to violet crystals. Saline and astringent taste. Somewhat hygroscopic. Gradually decomp on exposure to air due to absorption of CO_2 and becomes incompletely sol. One g dissolves in about 7 ml water, forming a clear, deep-blue dichroic soln. Slightly sol in alc. *Keep well closed.* LD_{50} in mice (mg/kg): 360 i.v., 3800 orally (Hoppe).

Free acid, $C_{20}H_{10}I_4O_4$, *iodophthalein, tetraiodophthalein, Iodophene, Nosophen.* Light-yellow, odorless, tasteless powder; dec at about 200°. Practically insol in water. Slightly sol in alcohol; sol in alkalies, chloroform, ether.

THERAP CAT: Diagnostic aid (radiopaque medium—cholecystographic).

4932. Iodopsin. One of 3 visual pigments found in retinal cone cells. Absorption maximum approx 560 nm. Biological activity is similar to that of rhodopsin, *q.v.* Composed of the chromophore, 11-*cis* retinal, *q.v.,* bound to photopsin, the specific protein component of cone pigments. (See Opsins.) Isoln from chicken retinas: G. Wald, *Nature* **140**, 545 (1937). Prepn from 11-*cis* retinal and opsin: G. Wald *et al., J. Gen. Physiol.* **38**, 623 (1955). Methods for purification, prepn, and assay: R. Hubbard *et al.,* "Methodology of Vitamin A and Visual Pigments", in *Methods in Enzymology* Vol. 18, D. B. McCormick, L. D. Wright, Eds. (Academic Press, New York, 1971) pp 615-653. Exposure to light initiates the conversion of iodopsin through a series of distinct intermediates to yield opsin and *trans*-retinal. Photochemistry: R. Hubbard, A. Kropf, *Nature* **183**, 448 (1959); T. Yoshizawa, G. Wald, *ibid.* **214**, 566 (1967). Studies on the iodopsin binding site: H. Matsumoto *et al., Biochim. Biophys. Acta* **404**, 300 (1975). Three distinct pigments are responsible for color vision: W. B. Marks *et al., Science* **143**, 1181 (1964); P. K. Brown, G. Wald, *ibid.* **144**, 45 (1964). The variation in the absorption maximum of the 3 pigments is regulated by differences in the apoprotein portion of the molecule: R. Hubbard, L. Sperling, *Exp. Eye Res.* **17**, 581 (1973); B. Honig *et al., J. Am. Chem. Soc.* **101**, 7084 (1979). *Reviews:* G. Wald, *Science* **162**, 230-239 (1968); H. Matsumoto, T. Yoshizawa, *Methods in Enzymology* Vol. 81, L. Packer, Ed. (Academic Press, New York, 1982) pp 154-160. General review of the visual process: P. S. Zurer, *Chem. & Eng. News* **61**, 24-35 (Nov. 28, 1983).

4933. Iodopyracet. *3,5-Diiodo-4-oxo-1(4H)-pyridineacetic acid compound with 2,2'-iminobis[ethanol] (1:1);* bis-(hydroxyethyl)ammonium 3,5-diiodo-4-pyridone-N-acetate; diethanolamine 3,5-diiodo-4-pyridone-N-acetate; 3,5-diiodo-4-pyridone-N-acetic acid diethanolamine salt; diodone; RP 3203; Diatrast; Diodrast; Iopyracil; Nosylan; Neo-Methiodal; Neo-Skiodan; Neo-Tenebryl; Nosydrast; Oparenol; Per-Abrodil; Per-Radiographol; Pyelosil; Pylumbrin; Savac; Umbradil; Uriodone; Vasiodone; Xumbradil. $C_{11}H_{16}I_2N_2O_5$; mol wt 510.09. C 25.90%, H 3.16%, I 49.76%, N 5.49%, O 15.68%. Prepn: **Fr.** pat. **728,634** (1931 to I. G. Farbenhind AG); J. Reitmann, U.S. pat. **1,993,039** (1935 to Winthrop). Pharmacology, toxicology and radiographic studies: R. St. A. Heathcote, R. A. Gardner, *Brit. J. Radiol.* **6**, 304 (1933). HPLC determn in biological fluids: P. Hekman, C. A. M. Van Ginneken, *J. Chromatog.* **182**, 492 (1980).

Odorless powder, dec 155-157°. Soly: about 36% in water; about 12% in methanol, much less sol in ice-cold methanol. Practically insol in acetone, ether, chloroform. pH of aq solns: 5 to 8, depending on concn. 3,5-Diiodo-4-pyridone-N-acetic acid separated by pptn with dil HCl and dried at 100°, mp 245-249°, and contains not less than 61.5% and not more than 63.5% iodine: *N.F. XI.*

Iodopyracet Injection is a 35% (w/v) soln of the diethanolamine salt of 3,5-diiodo-4-pyridone-N-acetic acid. Contains about 22% I (w/v). d_4^{25} 1.185. Neutral to litmus.

Iodopyracet Concentrated Solution; Diodrast Concentrated Solution is a 70% soln (w/v) of the diethanolamine salt of 3,5-diiodo-4-pyridone-N-acetic acid. Contains about 44% I (w/v). Neutral to litmus.

Iodopyracet Compound Solution; Diodrast Compound Solution is a soln contg approx 40.5% (w/v) of the diethanolamine salt and approx 9.5% (w/v) of the diethylamine salt of 3,5-diiodo-4-pyridone-N-acetic acid. Contains about 25% I. Neutral to litmus. d_4^{25} 1.270.

THERAP CAT: Diagnostic aid (radiopaque medium—urographic).

THERAP CAT (VET): Diagnostic aid (radiopaque medium).

4934. Iodopyrrole. *2,3,4,5-Tetraiodo-1H-pyrrole;* Iodol. C_4HI_4N; mol wt 570.74. C 8.42%, H 0.18%, I 88.95%, N 2.45%. Prepd by iodination of pyrrole: Potts, *J. Chem. Soc.* **1953**, 3711; Treibs, Kolm, *Ann.* **614**, 176 (1958). Isoln from shale-oil naphtha: Janssen *et al., J. Am. Chem. Soc.* **73**, 4040 (1951).

Needles from ethanol, dec 162-164°, also reported as dec 150-160° and 140-150°. One gram dissolves in 4900 ml water, 9 ml alcohol, 1.5 ml ether, 105 ml chloroform, 155 ml glycerol; sol in fixed oils.

THERAP CAT: Topical antiseptic.

THERAP CAT (VET): Antiseptic, disinfectant for superficial lesions and in the female reproductive tract.

4935. Iodoquinol. *5,7-Diiodo-8-quinolinol;* diiodohydroxyquin; diiodo-oxyquinoline; 5,7-diiodo-8-hydroxyquinoline; SS 578; Diodoquin; Di-Quinol; Disoquin; Floraquin; Dyodin; Dinoleine; Searlequin; Diodoxylin; Moebiquin; Rafamebin; Ioquin; Direxiode; Stanquinate; Quinadome; Yodoxin; Zoaquin; Enterosept; Embequin. $C_9H_5I_2NO$; mol wt 396.98. C 27.23%, H 1.27%, I 63.94%, N 3.53%, O 4.03%. Prepd by the action of iodine monochloride on 8-hydroxyquinoline: Papesch, Burtner, *J. Am. Chem. Soc.* **58**, 1314 (1936); by the action of KIO_3 on 8-hydroxyquinoline: Zeifman, *C.A.* **34**, 3745. Electrolytic prepn: Brown, Berkowitz, *Trans. Electrochem. Soc.* **75**, 385 (1939). *See also* Claus, **Ger. pat. 78,880;** Passek, **Ger. pat. 411,050;** Matsumura, *C.A.* **21**, 1461 (1927); Pirrone, Cherubino, *C.A.* **28**, 3073 (1934).

Crystals from xylene. The medicinal grade is a yellowish-brown powder. mp 200-215° (extensive decompn). Almost insol in water; sparingly sol in alcohol, ether, and acetone; sol in hot pyridine and in hot dioxane.

THERAP CAT: Anti-amebic.

4936. N-Iodosuccinimide. Succiniodimide; NIS. $C_4H_4INO_2$; mol wt 224.99. C 21.35%, H 1.79%, I 56.40%, N 6.23%, O 14.22%. Prepd by iodination of silver succinimide: Djerassi, Lenk, *J. Am. Chem. Soc.* **75**, 3493 (1953).

Needles from dioxane + carbon tetrachloride. mp 200-201°. Sol in acetone, methanol; practically insol in CCl_4, ether. Dec in water. *Protect from light and avoid contact with skin and mucous membranes.*

USE: Iodination of ketones and aldehydes.

4937. 3-Iodotyrosine. Monoiodotyrosine. $C_9H_{10}INO_3$; mol wt 307.10. C 35.20%, H 3.28%, I 41.33%, N 4.56%, O 15.63%. Prepn of L-form: Hillman, Hillman-Elies, *Z. Physiol. Chem.* **305**, 177 (1956); Pitt-Rivers, *Chem. & Ind. (London)* **1956**, 21. Prepn of L- and DL-form: Harington, Rivers, *Biochem. J.* **38**, 320 (1944).

L-Form, crystals from water, dec 204-206°. $[\alpha]_D^{20}$ −4.4° (c = 5 in 1N HCl). Dissolves in 15 parts boiling water.

DL-Form monohydrate, crystals from water, dec 200-201°.

4938. Iofetamine ^{123}I. (±)-4-(Iodo-^{123}I)-α-methyl-N-(1-methylethyl)benzeneethanamine; (±)-p-iodo-^{123}I-N-isopropyl-α-methylphenethylamine; [^{123}I](±)-N-isopropyl-p-iodoamphetamine. $C_{12}H_{18}^{123}$IN. Lipid soluble radioactive brain imaging agent. Prepn: R. M. Baldwin *et al., Eur. pat. Appl. 11,858* (1980 to Hoffmann-La Roche), *C.A.* **94**, 83749r (1981). Detailed synthesis: L. Carlsen, K. Andresen, *Eur. J. Nucl. Med.* **7**, 280 (1982). Modified synthesis for improved yields: A. Najafi, *J. Labelled Compd. Radiopharm.* **24**, 1167 (1987). Preparative HPLC purification: J. Mertens *et al., Nucl. Med. Commun.* **5**, 705 (1984). Brain uptake and localization: H. S. Winchell *et al., J. Nucl. Med.* **21**, 940 (1980); H. S. Winchell *et al.,* ibid. 947. Distribution kinetics in animals: P. Som *et al., Int. J. Nucl. Med. Biol.* **12**, 185 (1985); in humans: B. L. Holman *et al., J. Nucl. Med.* **25**, 25 (1984). Diagnostic use for cerebral function in Alzheimer's disease: K. A. Johnson *et al., Arch. Neurol.* **45**, 392 (1988). Reviews of radiopharmaceutical production and diagnostic use with single photon emission computed tomography (SPECT) imaging: M. B. Cohen *et al., Appl. Radiat. Isot.* **37**, 749-763 (1986); of diagnostic applications in stroke: C. H. Park *et al., RadioGraphics* **8**, 305-326 (1988); of metabolism and kinetics: R. M. Baldwin, J.-L. Wu, *J. Nucl. Med.* **29**, 122-124 (1988).

$bp_{0.05}$ 91-92°. Also reported as $bp_{0.6}$ 98-99° (Najafi). Hydrochloride, $C_{12}H_{19}Cl^{123}IN$, *IMP,* ^{123}*I labeled IMP,* ^{123}*I-M123, Perfusamine, Spectamine.* Crystals from ether, mp 156-158°.

THERAP CAT: Diagnostic aid (radioactive imaging agent).

4939. Ioglycamic Acid. *3,3'-[Oxybis[(1-oxo-2,1-ethanediyl)imino]]bis[2,4,6-triiodobenzoic acid]; 3,3'-[oxybis(methylenecarbonylimino)]bis[2,4,6-triiodobenzoic acid]; 3,3'-(diglycoloyldiimino)bis(2,4,6-triiodo-3-carboxanilide); N,N'-(oxydiacetyl)-bis[3-amino-2,4,6-triiodobenzoic acid].* $C_{18}H_{10}I_6N_2O_7$; mol wt 1127.69. C 19.17%, H 0.89%, I 67.52%, N 2.48%, O 9.93%. Prepn: Priewe *et al., Ber.* **87**, 651 (1954); Priewe,

Rutkowski, U.S. pats. **2,776,241; 2,853,424** (1957, 1958 both to Schering AG).

Occurs in three cryst modifications: Neudert, Röpke, *Helv. Chim. Acta* **41**, 855 (1958). mp with baking at 222°, sintering at 227°, splitting of iodine at 245°, dec 281°. Meglumine salt, $C_{25}H_{27}I_6N_3O_{12}$, **Biligram.** Water soluble.

THERAP CAT: Diagnostic aid (radiopaque medium—cholecystographic).

4940. Iohexol. *5-[Acetyl(2,3-dihydroxypropyl)amino]-N,N'-bis(2,3-dihydroxypropyl)-2,4,6-triiodo-1,3-benzenedicarboxamide; N,N'-bis(2,3-dihydroxypropyl)-5-[N-(2,3-dihydroxypropyl)acetamido]-2,4,6-triiodoisophthalamide;* WIN 39424; compound 545; Omnipaque; Exypaque. $C_{19}H_{26}I_3N_3O_9$; mol wt 821.14. C 27.79%, H 3.19%, N 46.36%, N 5.12%, O 17.54%. Nonionic radio-contrast medium. Prepn: V. Nordal, H. Holtermann, **Ger. pat. 2,726,196** corresp to U.S. pat. **4,250,113** (1977, 1981 both to Nyegaard). Pharmacology and toxicology: *Acta Radiol. (Stockholm)* **Suppl. 362**, 1-134 (1980). Acute toxicity: S. Salvesen, *ibid.* 73. Fibrillatory potential in dogs: G. L. Wolf *et al.*, *Invest. Radiol.* **16**, 320 (1981). Comparative clinical studies in coronary angiography: G. B. J. Mancini *et al.*, *Am. J. Cardiol.* **51**, 1218 (1983); I. D. Sullivan *et al.*, *Br. Heart J.* **51**, 643 (1984); M. A. Bettmann *et al.*, *Radiology* **153**, 583 (1984). *Review:* T. Almén, *Acta Radiol. (Stockholm)* **Suppl. 366**, 9-19 (1983).

Crystals from butanol, mp 174-180°. Sol in water. Stable in aqueous solutions. Viscosity (cP): 6.2 at 37°; 12.6 at 20° (c = 200 mg Iodine/ml). LD_{50} in male, female rats, mice (g Iodine/kg): 15.0, 12.3, 24.3, 25.1 i.v. (Salvesen).

THERAP CAT: Diagnostic aid (radiopaque medium).

4941. Iomeglamic Acid. *5-[(3-Amino-2,4,6-triiodophenyl)methylamino]-5-oxopentanoic acid; 3'-amino-2',4',6'-triiodo-N-methylglutaranilic acid; N-methyl-N-(3-amino-2,4,6-triiodophenyl)glutaramic acid; N-methyl-N-(2,4,6-triiodo-3-aminophenyl)glutaramidic acid;* RG 270; Falignost. $C_{12}H_{13}I_3N_2O_3$; mol wt 613.94. C 23.48%, H 2.13%, I 62.01%, N 4.56%, O 7.82%. Prepn: Cassebaum, Dierbach, **East Ger. pat. 67,209** (1969), *C.A.* **72**, 66654j (1970). Biotransformation: Pfeifer *et al.*, *Pharmazie* **27**, 403 (1972). Pharmacology: H. Bekker *et al.*, *ibid.* 411. Clinical evaluation: Barke, *Zbl. Pharm. Pharmakother. Laboratoriumsdiagn.* **110**, 1117 (1971).

Yellow to brown crystals from acetic acid, mp 169°. Free-

ly sol in DMF; slightly sol in ethanol, chloroform. Practically insol in water. LD_{50} i.v. in mice: 500 mg/kg (Bekker).

THERAP CAT: Diagnostic aid (radiopaque medium—cholecystographic).

4942. Ionone. Irisone. $C_{13}H_{20}O$; mol wt 192.29. C 81.20%, H 10.48%, O 8.32%. Mixture in which the major constituents are α-ionone, *4-(2,6,6-trimethyl-2-cyclohexen-1-yl)-3-buten-2-one*, and β-ionone, *4-(2,6,6-trimethyl-1-cyclohexen-1-yl)-3-buten-2-one*. The latter is a key intermediate in the synthesis of vitamin A. Isoln from the volatile oil of *Boronia megastigma* Nees., *Rutaceae:* Naves, Parry, *Helv. Chim. Acta* **30**, 419 (1946). Prepn by condensing citral with acetone: Tremann, Schmidt, *Ber.* **33**, 3703 (1900); Hibbert, Cannon, *J. Am. Chem. Soc.* **46**, 119 (1924); Krishna, Joshi, *J. Org. Chem.* **22**, 224 (1957). Prepn from isobutylene and vinyl methyl ketone: Pasedach, Seefelder, **Ger. pat. 1,000,374** (1957 to BASF), *C.A.* **54**, 1595g (1960). Prepn by cyclization of pseudoionone with H_2SO_4: Kimel *et al.*, *J. Org. Chem.* **22**, 1611 (1957); Kaiser, Kimel, U.S. pat. **2,877,271** (1959 to Hoffmann-La Roche). *Review:* Naves, *J. Soc. Cosmet. Chem.* **22**, 439 (1971).

Liquid. Odor reminiscent of cedar wood; in very dilute alcoholic soln it resembles the odor of violets. d_{25}^{25} 0.933-0.937. n_D^{20} 1.503-1.508. bp_{12} 126-128°. uv max: α-ionone 228.5 nm (ε 14,300); β-ionone 293.5 nm (ε 8,700). Miscible with abs alc. Soluble in 2 to 3 parts of 70% alcohol, ether, chloroform, benzene. Very slightly sol in water.

USE: In perfumery. *Caution:* May cause allergic reactions, Shultheiss, *Dermatol. Wochenschr.* **135**, 629 (1967).

4943. Iopamidol. *(S)-N,N'-Bis[2-hydroxy-1-(hydroxymethyl)ethyl]-5-[(2-hydroxy-1-oxopropyl)amino]-2,4,6-triiodo-1,3-benzenedicarboxamide; (S)-N,N'-bis[2-hydroxy-1-(hydroxymethyl)ethyl]-2,4,6-triiodo-5-lactamidoisophthalamide; 5-(α-hydroxypropionylamino)-2,4,6-triiodoisophthalic acid di(1,3-dihydroxyisopropylamide); iomapidol;* B-15,000; SQ 13,396; Iopamiro; Iopamiron; Isovue; Jopamiro; Niopam; Solutrast. $C_{17}H_{22}I_3N_3O_8$; mol wt 777.09. C 26.28%, H 2.85%, I 48.99%, N 5.41%, O 16.47%. Non-ionic radiocontrast medium. Prepn: E. Felder, D. Pitre, **Ger. pat. 2,547,789;** *eidem,* U.S. pat. **4,001,323** (1976, 1977 both to Savac). Physicochemical properties, preclinical studies: E. Felder *et al.*, *Farmaco Ed. Sci.* **32**, 835 (1977). Determn of optical purity: *eidem, Farmaco Ed. Prat.* **32**, 3 (1982). *In vitro* effects: B. Schulze, H. K. Beyer, *Arzneimittel-Forsch.* **31**, 1067 (1981). Development, initial testing: M. Sovak *et al.*, *Radiology* **142**, 115 (1982). Clinical studies in angiography: A. J. Molyneux, P. W. Sheldon, *Brit. J. Radiol.* **55**, 117 (1982); G. H. Whitehouse, S. L. Snowden, *Clin. Radiol.* **33**, 231 (1982); in myelography: B. Drayer *et al.*, *Am. J. Neuroradiol.* **3**, 59 (1982). Comprehensive description: E. Felder *et al.*, in *Analytical Profiles of Drug Substances* vol. 17, K. Florey, Ed. (Academic Press, New York, 1988) pp 115-154.

White, odorless cryst powder. Dec at about 300° without melting. $[α]_D^{20}$ −2.01° (c = 10 in water). pKa (25°) 10.70. Very sol in water, methanol. Practically insol in chloroform. Miscible with boiling ethanol. LD_{50} in mice, rats, rabbits, dogs (g/kg): 44.5, 28.2, 19.6, 34.7 i.v. (Felder). Also reported as: LD_{50} in mice (mg iodine/kg body wt): 21,800 i.v.; 20,000 i.p.; 1500 intracerebral (Felder, Pitre).

THERAP CAT: Diagnostic aid (radiopaque medium).

4944. Iopanoic Acid. *3-Amino-α-ethyl-2,4,6-triiodobenzenepropanoic acid; 3-amino-α-ethyl-2,4,6-triiodohydrocinnamic acid;* iodopanoic acid; 3-(3-amino-2,4,6-triiodophenyl)-2-ethylpropanoic acid; 3-(3-amino-2,4,6-triiodophenyl)-2-ethylpropionic acid; Cistobil; Colepax; Telepaque; Teletrast. $C_{11}H_{12}I_3NO_2$; mol wt 570.93. C 23.14%, H 2.12%, I 66.69%, N 2.45%, O 5.60%. Oral radiocontrast medium. Prepn: S. Archer, U.S. pat. **2,705,726** (1955 to Sterling Drug). Optical resolution: Pitré, Boveri, *J. Med. Chem.* **11**, 406 (1968). Toxicity data: P. Tirone, G. Rosati, *Farmaco Ed. Prat.* **31**, 397 (1976). Review of pharmacology and clinical efficacy in oral cholecystography: R. N. Berk *et al., N. Engl. J. Med.* **290**, 204 (1974). Comprehensive description: D. Pitre in *Analytical Profiles of Drug Substances* vol. **14**, K. Florey, Ed. (Academic Press, New York, 1985) pp 181-206.

dl-Form, cream-colored solid, mp 155.2-157°. pKa 4.8. Insol in water. Sol in dil alkali, in 95% alc, in other organic solvents. LD_{50} in mice, rats (mg/kg): 285, 320 i.v.; 1540, 2870 orally (Tirone).

l-Form, crystals, mp 162-163°. $[\alpha]_D^{20}$ $-5.2° \pm 0.1°$ (c = 2 in ethanol).

d-Form, crystals, mp 162°. $[\alpha]_D^{20}$ $+5.1° \pm 0.1°$ (c = 2 in ethanol).

Sodium salt, *Bilijodon-Natrium*.

THERAP CAT: Diagnostic aid (radiopaque medium—cholecystographic).

4945. Iopentol. *5-[Acetyl(2-hydroxy-3-methoxypropyl)-amino]-N,N′-bis(2,3-dihydroxypropyl)-2,4,6-triiodo-1,3-benzenedicarboxamide; N,N′-bis(2,3-dihydroxypropyl)-5-[N-(2-hydroxy-3-methoxypropyl)acetamido]-2,4,6-triiodoisophthalamide.* $C_{20}H_{28}I_3N_3O_9$; mol wt 835.17. C 28.76%, H 3.38%, I 45.59%, N 5.03%, O 17.24%. Nonionic, water soluble vascular contrast medium. Prepn: K. Wille, Eur. pat. **Appl. 105,752** (1984 to Nyegaard A/S), *C.A.* **101**, 72456e (1984). Synthesis and preparative LC purification: W. Skjold, A. Berg, *J. Chromatog.* **366**, 299 (1986). Effects on blood-brain barrier in animals: A. A. Michelet, *Acta Radiol.* **28**, 329 (1987).

Osmolality (300 mg I/ml): 0.64 mol/kg H_2O. Viscosity at 20° (300 mg I/ml): 13.2 mPa. Partition coefficient in 1-octanol/water: 0.007.

THERAP CAT: Diagnostic aid (radiopaque medium).

4946. Iophendylate. *4-Iodo-iota-methylbenzenedecanoic acid ethyl ester;* ethyl 10-(p-iodophenyl)undecylate; ethyl 10-(p-iodophenyl)hendecanoate. $C_{19}H_{29}IO_2$; mol wt 416.34. C 54.81%, H 7.02%, I 30.48%, O 7.69%. Chief component of *Ethiodan, Mulsopaque, Myodil, Neurotrast, Pantopaque;* may also contain other isomers including 11-(p-iodophenyl)-

undecylate. Prepn: W. H. Strain *et al., J. Am. Chem. Soc.* **64**, 1436 (1942); *eidem,* U.S. pat. **2,348,231** (1944 to Noned Corp. and Kodak); W. Baker *et al., J. Soc. Chem. Ind.* **63**, 223 (1944). Pharmacology and toxicity: T. B. Steinhausen *et al., Radiology* **43**, 230 (1944). Clinical evaluation of diagnostic use in myelography: G. H. Ramsey *et al., ibid.* 236. Clinical comparison with metrizamide, *q.v.,* in diagnostic radiology: S. A. Keiffer *et al., ibid.* **129**, 695 (1978). Evaluation of use in MR imaging of the spine: I. F. Braun *et al., Am. J. Neuroradiol.* **7**, 997 (1986); A. K. Anand *et al., Comput. Radiol.* **11**, 165 (1987).

Viscous liquid. Darkens slowly in air. d_{20}^{20} 1.240-1.263. n_D^{25} 1.5230-1.5280. Saponif equiv 395-420. Very slightly sol in water. Freely sol in alcohol, benzene, chloroform, ether. LD_{50} in mice, rats (g/kg): 4.6, 19 i.p. (Steinhausen).

THERAP CAT: Diagnostic aid (radiopaque medium).

THERAP CAT (VET): Diagnostic aid (radiopaque medium).

4947. Iophenoxic Acid. *α-Ethyl-3-hydroxy-2,4,6-triiodobenzenepropanoic acid; α-ethyl-3-hydroxy-2,4,6-triiodohydrocinnamic acid;* α-(2,4,6-triiodo-3-hydroxybenzyl)butyric acid; α-ethyl-β-(3-hydroxy-2,4,6-triiodophenyl)propionic acid; α-ethyl-β-(2,4,6-triiodo-3-hydroxyphenyl)propionic acid; triiodoethionic acid; Teridax. $C_{11}H_{11}I_3O_3$; mol wt 571.94. C 23.10%, H 1.94%, I 66.57%, O 8.39%. Prepn: Papa *et al., J. Am. Chem. Soc.* **75**, 1107 (1953); **Brit. pat. 726,987** (1955 to Sterling Drug). Toxicity data: J. O. Hoppe *et al., J. Med. Chem.* **13**, 997 (1970).

Crystals from benzene + petr ether, mp 143-144°. LD_{50} i.v. in mice: 374 mg/kg (Hoppe).

THERAP CAT: Diagnostic aid (radiopaque medium—cholecystographic).

4948. Iopromide. *N,N′-Bis(2,3-dihydroxypropyl)-2,4,6-triiodo-5-[(methoxyacetyl)amino]-N-methyl-1,3-benzenedicarboxamide; N,N′-bis(2,3-dihydroxypropyl)-2,4,6-triiodo-5-(2-methoxyacetamido)-N-methylisophthalamide;* 5-methoxyacetylamino-2,4,6-triiodoisophthalic acid [(2,3-dihydroxy-N-methylpropyl)-(2,3-dihydroxypropyl)]diamide; Ultravist. $C_{18}H_{24}I_3N_3O_8$; mol wt 791.12. C 27.33%, H 3.06%, I 48.12%, N 5.31%, O 16.18%. Nonionic, injectable radio-contrast medium. Prepn: U. Speck *et al.,* Ger. pat. **2,909,439**; *eidem,* U.S. pat. **4,364,921** (1980, 1982 both to Schering AG). Physicochemical properties and pharmacological profile: W. Muetzel, U. Speck, *Am. J. Neuroradiol.* **4**, 350 (1983); P. Dawson *et al., Acta Radiol. Diagn.* **25**, 253 (1984). Clinical experience: K.-J. Wolf *et al., ibid.* **24**, 55 (1983).

Colorless solid. Stable in aqueous solutions. Viscosity

(cP): 4.8 at 37°; 10.2 at 20° (c = 300 mg iodine/ml). LD$_{50}$ in mice, rats (g iodine/kg body weight): 16.5, 11.4 i.v. (Muetzel).

THERAP CAT: Diagnostic aid (radiopaque medium).

4949. Iopronic Acid. *2-[[2-[3-(Acetylamino)-2,4,6-triiodophenoxy]ethoxy]methyl]butanoic acid;* (±)-2-[[2-(3-acetamido-2,4,6-triiodophenoxy)ethoxy]methyl]butyric acid; 3-[2-(3-acetylamino-2,4,6-triiodophenoxy)ethoxy]-2-ethyl-propionic acid; B-11420; SQ-21983; Bilimiro; Bilimiron; Oravue; Videobil. C$_{15}$H$_{18}$I$_3$NO$_5$; mol wt 673.02. C 26.77%, H 2.69%, I 56.57%, N 2.08%, O 11.89%. Oral cholecysto-graphic agent. Prepn: E. Felder, D. Pitre, **Ger.** pat. **2,128,-902;** *eidem*, **U.S.** pat. **3,842,124** (1972, 1974 both to Bracco); E. Felder *et al., Farmaco Ed. Sci.* **31,** 349 (1976). Pharmacology, toxicology: P. Tirone, G. Rosati, *Farmaco Ed. Prat.* **31,** 397, 437 (1976). Series of articles on metabolism: *ibid.* 516-546; 755. Intestinal absorption: J. R. Amberg *et al., Invest. Radiol.* **15,** S136 (1980). Effect on bile flow and composition: J. L. Barnhart *et al., ibid.* S124. Toxicity: P. Tirone, G. Rosati, *Farmaco Ed. Prat.* **31,** 397 (1976).

Crystals from 50% ethanol, mp 130°. LD$_{50}$ in mice, rats, dogs (mg/kg): 1950, 5650, > 3000 orally; 1090, 1000, 835 i.v. (Tirone).

THERAP CAT: Diagnostic aid (radiopaque medium—cholecystographic).

4950. Iopydol. *1-(2,3-Dihydroxypropyl)-3,5-diiodo-4(1H)-pyridinone.* C$_8$H$_9$I$_2$NO$_3$; mol wt 420.99. C 22.82%, H 2.15%, I 60.29% N 3.33%, O 11.40%. Prepn: Reitmann, **Ger.** pat. **579,224** (1933 to I.G. Farbenind.), *C.A.* **27,** 4880 (1933).

Solid, mp 161°.

THERAP CAT: Diagnostic aid (radiopaque medium—bronchographic).

4951. Iopydone. *3,5-Diiodo-4(1H)-pyridinone.* C$_5$H$_3$-I$_2$NO; mol wt 346.91. C 17.31%, H 0.87%, I 73.17%, N 4.04%, O 4.61%. Prepn: Dohrn, Diedrich, *Ann.* **494,** 284 (1932).

Crystals, dec 321°. Practically insol in ordinary solvents. Sol in caustic alkalies.

THERAP CAT: Diagnostic aid (radiopaque medium—bronchographic).

4952. Iothalamic Acid. *3-(Acetylamino)-2,4,6-triiodo-5-[(methylamino)carbonyl]benzoic acid;* 5-acetamido-2,4,6-triiodo-N-methylisophthalamic acid. C$_{11}$H$_9$I$_3$N$_2$O$_4$; mol wt 613.94. C 21.52%, H 1.48%, I 62.01%, N 4.56%, O 10.42%. Synthesis: Hoey *et al., J. Med. Chem.* **6,** 24 (1963).

Crystals, dec about 285°.

N-Methylglucamine salt, C$_{18}$H$_{26}$I$_3$N$_3$O$_9$, *meglumine iothalamate, Conray, Contrix "28", Cysto-Conray.*

Sodium salt, C$_{11}$H$_8$I$_3$N$_2$NaO$_4$, *sodium iothalamate, Angio-Conray, Angio-Contrix "48", Conray-400, Medio-Contrix "38", Glofil-131* (labeled with ^{131}I).

THERAP CAT: Diagnostic aid (radiopaque medium).

4953. Iothion. *1,3-Diiodo-2-propanol;* iopropane; 1,3-diiodo-2-hydroxypropane; α-diiodohydrin; 1,3-diiodoisopropyl alcohol. C$_3$H$_6$I$_2$O; mol wt 311.89. C 11.55%, H 1.94%, I 81.38%, O 5.13%. CH$_2$ICH(OH)CH$_2$I. Prepn from dichlorohydrin and KI: Claus, *Ann.* **168,** 24 (1873); from glycerol, HCl and NaI: Lusignani, *Boll. Chim. Farm.* **78,** 557 (1939), *C.A.* **34,** 2322 (1940).

Yellowish, heavy, oily liquid. d 2.4-2.5. One gram dissolves in 80 ml water, 20 ml glycerol; freely sol in alc, benzene, CHCl$_3$, ether, carbon disulfide, oils; dec by alkalies.

4954. Iothiouracil. *2,3-Dihydro-5-iodo-2-thioxo-4(1H)-pyrimidinone;* 5-iodo-2-thiouracil; 4-hydroxy-5-iodo-2-mercaptopyrimidine; iodothiouracil; Itrumil. C$_4$H$_3$-IN$_2$OS; mol wt 254.06. C 18.91%, H 1.19%, I 49.95%, N 11.03%, O 6.30%, S 12.62%. Prepn: H. W. Barrett *et al., J. Am. Chem. Soc.* **70,** 1753 (1948); H. W. Barrett, **U.S.** pat. **2,585,615** (1952 to Chemical Found.); A. E. Lanzilotti *et al.,* **U.S.** pat. **2,666,764** (1954 to Ciba); *eidem, J. Am. Chem. Soc.* **76,** 3666 (1954).

Crystals, mp 228-230° (dec). Light-sensitive.

Sodium salt monohydrate, crystals, dec 235°.

THERAP CAT: Thyroid inhibitor.

4955. Iotrolan. *5,5'-[(1,3-Dioxo-1,3-propanediyl)bis(methylimino)]bis[N,N'-bis[2,3-dihydroxy-1-(hydroxymethyl)propyl]-2,4,6-triiodo-1,3-benzenedicarboxamide];* 5,5'-[malonylbis(methylimino)]bis[N,N'-bis[2,3-dihydroxy-1-(hydroxymethyl)propyl]-2,4,6-triiodoisophthalamide]; iotrol; DL-3117; SH 437; ZK 39482; Isovist; Iotrovist. C$_{37}$-H$_{48}$I$_6$N$_6$O$_{18}$; mol wt 1626.24. C 27.33%, H 2.97%, I 46.82%, N 5.17%, O 17.71%. Nonionic, isotonic contrast medium. Prepn: M. Sovak, R. Ranganathan, **Eur.** pat. **Appl. 33,426;** *eidem,* **U.S.** pat. **4,341,756** (1981, 1982 both to Univ. Cal., Berkeley). Neurotoxicity study: M. Sovak *et al., Radiology* **142,** 115 (1982). Neuropharmacology: M. Sovak *et al., Am. J. Neuroradiol.* **4,** 319 (1983); and cerebral distribution: J.-C. Castel *et al., Neuroradiol.* **29,** 206 (1987). Arthrographic evaluation in animals: J. Guerra *et al., Invest. Radiol.* **19,** 228 (1983). Cardiovascular effects in animals: P. Lanzer *et al., ibid.* **20,** 746 (1985). Clinical study of patient tolerance and diagnostic adequacy for intrathecal use: B. Hammer, E. Deisenhammer, *Neuroradiol.* **27,** 337 (1985); B. Hoffmann *et al., ibid.* **29,** 380 (1987); M. D. Malnor *et al., Radiology* **158,** 845 (1986).

Colorless glassy solid. Soly at 20° (mgI/ml): water > 400. LD_{50} in mice, rats (g I/kg): > 26, 12.7 i.v. (Sovak, 1982). THERAP CAT: Diagnostic aid (radiopaque medium).

4956. Ioxaglic Acid. 3-*[[[[3-(Acetylmethylamino)-2,4,6-triiodo-5-[(methylamino)carbonyl]benzoyl]amino]acetyl]-amino]-5-[[(2-hydroxyethyl)amino]carbonyl]-2,4,6-triiodobenzoic acid;* N-*(2-hydroxyethyl)-2,4,6-triiodo-5-[2-[2,4,6-triiodo-3-(N-methylacetamido)-5-(methylcarbamoyl)benzamido]acetamido]isophthalamic acid;* P-286. $C_{24}H_{21}I_6N_5O_8$; mol wt 1268.90. C 22.72%, H 1.67%, I 60.00%, N 5.52%, O 10.09%. Ionic contrast medium of low osmolality. Prepn: G. Tilly *et al.,* Ger. pat. **2,523,567;** *eidem,* U.S. pat. **4,014,-986** (1975, 1977 both to Guerbet). Effect on peripheral arterial blood flow in dogs: R. M. Steiner *et al., Clin. Radiol.* **31,** 621 (1980). Effect on thyroid function: R. G. Grainger, G. W. Pennington, *Brit. J. Radiol.* **54,** 768 (1981). Clinical study in femoral angiography: S. Suzuki *et al., Acta Radiol.* **23,** 87 (1982); in cardiac radiology: R. L. Feldman, *Am. J. Cardiol.* **61,** 1334 (1988). Toxicity study: M. Sovak *et al., Invest. Radiol.* **16,** 438 (1981). Mutation study: W. Hadnagy *et al., Mutat. Res.* **104,** 249 (1982).

Mixture of *ioxaglate meglumine* and *ioxaglate sodium,* MP 302, Hexabrix.
THERAP CAT: Diagnostic aid (radiopaque medium).

4957. Ipecac. Ipecacuanha; hippo. Dried rhizome and roots of *Uragoga ipecacuanha* (Brot.) Baill. [*Cephaelis ipecacuanha* (Brot.) A. Rich.] known as Rio or Brazilian Ipecac; or of *Uragoga acuminata* (Benth.) Kuntze, known as Cartagena, Nicaragua or Panama Ipecac, *Rubiaceae. Habit.* Brazil to Bolivia; cultivated in India. Cartagena ipecac from Colombia. *Constit.* Emetine, cephaeline, emetamine, psychotrine, methyl psychotrine, protoemetine and resin. Cartagena or Nicaragua ipecac generally contains 2.6-3% total alkaloids, cephaeline being present to a larger extent than in Rio ipecac. The Rio variety usually contains only a little over 2% alkaloids of which about 60-75% is emetine; the Cartagena variety contains 40-50% emetine. *Ref:* "Cephaelis" in *Hagers Handbuch der pharmazeutischen Praxis,* **vol. 3,** P. H. List, L. Hörhammer, Eds. (Springer-Verlag, Berlin, 4th ed., 1971) pp 795-809.
THERAP CAT: Emetic.
THERAP CAT (VET): Emetic, expectorant. Has been used as a ruminatoric and as an antihistomonad.

4958. Ipodate. 3-*[[(Dimethylamino)methylene]amino]-2,4,6-triiodobenzenepropanoic acid;* 3-*[(dimethylaminomethylene)amino]-2,4,6-triiodohydrocinnamic acid;* 2,4,6-triiodo-3-[(dimethylaminomethylene)amino]hydrocinnamic acid; β-(3-dimethylaminomethyleneamino-2,4,6-triiodophenyl)-propionic acid. $C_{12}H_{13}I_3N_2O_2$; mol wt 597.97. C 24.10%, H

2.19%, I 63.67%, N 4.69%, O 5.35%. Prepn: Priewe, Poljak, *Ber.* **93,** 2347 (1960). Toxicity data: J. O. Hoppe *et al., J. Med. Chem.* **13,** 997 (1970).

Crystals, mp 168-169° (dec 225°). Practically insol in water; very sol in methanol, ethanol, chloroform, acetone, dil sulfuric acid. Estimated pK 5-5.5.
Sodium salt, $C_{12}H_{12}I_3N_2NaO_2$, *sodium ipodate, Biloptin, Oragrafin-Sodium.* Bitter leaflets from water + acetone, mp 303-304° (decompn with evolution of iodine). Very freely sol in water; freely sol in methanol, ethanol. Practically insol in acetone, ether. Soly in DMF and DMSO about 33 g/100 ml; in dimethylacetamide about 66 g/100 ml. LD_{50} in mice (mg/kg): 290 i.v.; 2570 orally (Hoppe).
Calcium salt, $C_{24}H_{24}CaI_6N_4O_4$, *calcium ipodate, Solu-Biloptin, Oragrafin-Calcium.* Crystals, mp 298-302°. Soly in water at 20°: 0.1%. Sol in chloroform, dimethylformamide, hot propylene glycol.
Ethyl ester, $C_{14}H_{17}I_3N_2O_2$, *SH 617L, Myelographin.*
THERAP CAT: Diagnostic aid (radiopaque medium—cholecystographic).

4959. Ipomea. Mexican scammony (root); Orizaba jalap root. Dried root of *Ipomoea orizabensis* Ledenois, *Convolvulaceae.* Active constituent is the resin. Yields not less than 15% total ipomea resins. Different from *Ipomoea violacea* var. *Pearly Gates* Hort., *Convolvulaceae* and *Ipomoea rubrocoerulea* var. *praecox, morning-glory, ololiuqui,* which contain ergot alkaloids. Occurrence of lysergic acid derivatives and of ergolines in Ipomea: A. Hofmann, H. Tscherter, *Experientia* **16,** 414 (1960); D. Stauffacher *et al., Helv. Chim. Acta* **48,** 1379 (1965).
THERAP CAT: Cathartic.

4960. Ipratropium Bromide. (*endo,syn*)-(±)-3-(3-*Hydroxy-1-oxo-2-phenylpropoxy)-8-methyl-8-(1-methylethyl)-8-azoniabicyclo[3.2.1]octane bromide;* (8r)-3α-*hydroxy-8-isopropyl-1αH,5αH-tropanium bromide* (±)-*tropate;* 8-isopropylnoratropine methobromide; N-isopropylnoratropinium bromomethylate; Sch 1000; Atem; Atrovent; Bitrop; Itrop; Rinatec. $C_{20}H_{30}BrNO_3$; mol wt 412.38. C 58.25%, H 7.33%, Br 19.38%, N 3.40%, O 11.64%. Anticholinergic. Prepn: K. Zeile *et al.,* S. Afr. pat. **67 07,766;** *eidem,* U.S. pat. **3,505,337** (1968, 1970, both to Boehringer, Ing.); W. Schulz *et al., Arzneimittel-Forsch.* **26,** 960 (1976). Chemistry and pharmacokinetics: W. Deckers, *Postgrad. Med. J.* **51,** Suppl. 7, 76 (1975). Pharmacology and toxicology: A. Engelhardt, H. Klupp, *ibid.* 82. Series of articles on pharmacology, toxicology, pharmacokinetics and clinical studies: *Arzneimittel-Forsch.* **26,** 974-985, 989-1020 (1976). Toxicity data: L. Sarafana *et al., ibid.* 985. *Review:* R. Bauer *et al.,* in *Pharmacological and Biochemical Properties of Drug Substances* vol. 2, M. E. Goldberg, Ed. (Am. Pharm. Assoc., Washington, DC, 1979) pp 489-515. Symposium on pharmacology, toxicology and clinical efficacy: *Am. J. Med.* **81**(5A), 1-102 (1986).

White crystals from *n*-propanol, mp 230-232°. Freely sol in water and lower alcohols. Insol in ether, chloroform and fluorohydrocarbons. Fairly stable in neutral and acid solns; rapidly hydrolyzed in alkaline soln. LD_{50} in male, female mice (mg/kg): 1001, 1083 orally; 12.29, 14.97 i.v.; 300, 340 s.c. LD_{50} in male, female rats (mg/kg): 1663, 1779 orally; 15.89, 15.70 i.v. (Sarafana).

THERAP CAT: Bronchodilator. Antiarrhythmic.

4961. Ipriflavone. *7-(1-Methylethoxy)-3-phenyl-4H-1-benzopyran-4-one;* 7-isopropoxy-3-phenyl-4H-1-benzopyran-4-one; 7-isopropoxy-3-phenylchromone; 7-isopropoxy-isoflavone; FL-113; TC-80; Osten; Yambolap. $C_{18}H_{16}O_3$; mol wt 280.33. C 77.12%, H 5.75%, O 17.12%. Isoflavone derivative with anti-anginal and anti-osteopenic activity. Prepn: L. Feuer *et al.*, Ger. pat. 2,125,245 (1971 to Chinoin), *C.A.* **76**, 72407e (1972). Use as anabolic in animals: *eidem,* U.S. pat. 3,833,730; in humans; *eidem,* U.S. pat. 3,-949,085 (1974, 1976 both to Chinoin). Cardiovascular properties in stable angina: V. Grubich *et al.*, *Lancet* **1**, 211 (1979). Cardiological effects in animals: L. Feuer *et al.*, *Arzneimittel-Forsch.* **31**, 953 (1981). Metabolism and disposition in rats: K. Yoshida *et al.*, *Radioisotopes* **34**, 612, 618 (1985). Effect in rats on estrogen-stimulated calcitonin secretion: I. Yamazaki, *Life Sci.* **38**, 757 (1986); I. Yamazaki, M. Kinoshita, *ibid.* 1535; on glucocorticoid-induced osteoporosis: I. Yamazaki *et al.*, *ibid.* 951.

Crystals from acetone, mp 115-117°.
THERAP CAT: Calcium regulator.

4962. Iprindole. *6,7,8,9,10,11-Hexahydro-N,N-dimethyl-5H-cyclooct[b]indole-5-propanamine;* 5-[3-(dimethylamino)-propyl]-6,7,8,9,10,11-hexahydro-5H-cyclooct[b]indole; 1-(3-dimethylaminopropyl)-2,3-hexamethyleneindole; pramindole (obsolete); Tertran. $C_{19}H_{28}N_2$; mol wt 284.43. C 80.23%, H 9.92%, N 9.85%. Prepn: Rice, Freed, **Belg. pat. 623,933** corresp to U.S. pat. 3,282,942 (1963, 1966, both to Am. Home Prods.); Rice *et al.*, *J. Med. Chem.* **7**, 313 (1964). Pharmacology: Gluckman, Baum, *Psychopharmacologia* **15**, 169 (1969); Baum *et al.*, *Eur. J. Pharmacol.* **13**, 287 (1971). Mechanism of action: Miller *et al.*, *Experientia* **26**, 863 (1970); Lemberger *et al.*, *Biochem. Pharmacol.* **19**, 3021 (1970). Freeman, Sulser, *J. Pharmacol. Exp. Ther.* **183**, 307 (1972).

Hydrochloride, $C_{19}H_{29}ClN_2$, WY-3263, Prondol, Galatur. Crystals from methanol + acetone, mp 146-147°.
THERAP CAT: Antidepressant.

4963. Iproclozide. *4-(Chlorophenoxy)acetic acid 2-(1-methylethyl)hydrazide;* Sursum. $C_{11}H_{15}ClN_2O_2$; mol wt 242.72. C 54.43%, H 6.23%, Cl 14.61%, N 11.54%, O 13.18%. Monoamine oxidase inhibitor. Prepn: Libermann, Denis, *Bull. Soc. Chim. France* **1961**, 1952. GC determn in urine: R. M. DeSagher *et al.*, *Anal. Chem.* **47**, 1144 (1976).

Crystals, mp 93-94°.
THERAP CAT: Antidepressant.

4964. Iprodione. *3-(3,5-Dichlorophenyl)-N-(1-methyl-ethyl)-2,4-dioxo-1-imidazolidinecarboxamide;* glycophene;

promidione; RP 26019; ROP 500F; NRC 910; LFA 2043; FA 2071; Rovral; CHIPCO-26019. $C_{13}H_{13}Cl_2N_3O_3$; mol wt 330.16. C 47.29%, H 3.97%, Cl 21.47%, N 12.73%, O 14.54%. Prepn: M. Sauli, Ger. pat. 2,149,923 corresp to U.S. pat. 3,755,350 (1972, 1973 to Rhône-Poulenc). Activity: L. Lacroix *et al.*, *Phytiatr.-Phytopharm.* **23**, 165 (1974). Structural rearrangement in ethanolic soln: B. K. Cooke *et al.*, *Pestic. Sci.* **10**, 393 (1979). Thermal degradn: J. Gomez *et al.*, *J. Agr. Food Chem.* **30**, 180 (1982). *Review:* L. Lacroix *et al.*, *Anal. Methods Plant Growth Regul.* **11**, 247-261 (1980).

Colorless crystals, mp ~136°. Vapor press. at 20°: < 10^{-6} mm Hg. Soly in water at 20°: 13 mg/l. Soly at 20° (g/l): ethanol 25; methanol 25; acetone 300; dichloromethane 500; DMF 500. LD_{50} orally in mice, rats: 3500, 4000 mg/kg, *RTECS* Vol. I, R. J. Lewis, R. L. Tatken, Eds. (1980) p 929.
USE: Fungicide.

4965. Iproniazid. *4-Pyridinecarboxylic acid 2-(1-methyl-ethyl)hydrazide;* isonicotinic acid 2-isopropylhydrazide; 1-isonicotinoyl-2-isopropylhydrazine; 1-isonicotinyl-2-iso-propylhydrazine; Euphozid; Marsilid. $C_9H_{13}N_3O$; mol wt 179.22. C 60.31%, H 7.31%, N 23.45%, O 8.93%. Prepd by treating isoniazid with isopropyl bromide in abs alcohol in the presence of sodium: McMillan *et al.*, *J. Am. Pharm. Assoc., Sci. Ed.* **42**, 457 (1953). Alternate synthesis: Fox, Gibas, *J. Org. Chem.* **18**, 994 (1953).

Needles from benzene + ligroin, mp 112.5-113.5°. Freely sol in water, alcohol; pH of aq solns about 6.7.
Dihydrochloride, $C_9H_{13}N_3O.2HCl$, rhomboid crystals from isopropanol, dec at 227-228°. Freely sol in water; aq solns are very acidic.
THERAP CAT: Monoamine oxidase inhibitor; antidepressant.

4966. Ipronidazole. *1-Methyl-2-(1-methylethyl)-5-ni-tro-1H-imidazole;* 2-isopropyl-1-methyl-5-nitroimidazole; Ro 7-1554; Ipropran. $C_7H_{11}N_3O_2$; mol wt 169.18. C 49.69%, H 6.55%, N 24.84%, O 18.91%. Prepn: Hoffer, Mitrovic, **Brit. pat. 1,119,636** and U.S. pat. 3,502,776 (1968, 1970 to Hoffmann-La Roche); Butler, **S. Afr. pat. 66 07,466** (1968 to Pfizer), *C.A.* **71**, 3384e (1969). Structure-activity relationship: Butler *et al.*, *J. Med. Chem.* **10**, 891 (1967). Activity studies as an antihistomonal agent: Mitrovic *et al.*, *Antimicrob. Ag. Chemother.* **1968**, 445 and *Poultry Sci.* **49**, 86 (1970); as a growth promotant: Marusich *et al.*, *ibid.* 98. Toxicity studies: *eidem, ibid.* 92.

White plates, mp 60°. LD_{50} orally in poults: 640 ± 25 mg/kg, Marusich *et al.*, *loc. cit.* 92.
Hydrochloride, mp 177-182°. Water-soluble.
THERAP CAT: Antiprotozoal (Histomonas).
THERAP CAT (VET): Antimicrobial.

4967. Ipsapirone. *2-[4-[4-(2-Pyrimidinyl)-1-piperazin-yl]butyl]-1,2-benzisothiazolin-3(2H)-one 1,1-dioxide;* isapirone. $C_{19}H_{23}N_5O_3S$; mol wt 401.48. C 56.84%, H 5.77%, N 17.44%, O 11.96%, S 7.99%. Nonbenzodiazepine anxiolytic; 5-hydroxytryptamine (5-HT_1) receptor agonist. Prepn: W. U. Dompert et al., Ger. pat. **3,321,969** (1984 to Tropon-werke), C.A. **102**, 220896m (1985). Pharmacology and sero-tonin (5-HT_1) receptor binding study: J. Traber et al., Brain Res. Bull. **12**, 741 (1984). 5-HT_1 specific binding activity: W. U. Dompert et al., Arch. Pharmacol. **328**, 467 (1985); T. Glaser, J. Traber, ibid. **329**, 211 (1985). Behavioral effects in mice: R. J. DeSouza et al., Brit. J. Pharmacol. **89**, 377 (1986). Neurophysiological effects in comparison with bu-spirone, q.v.: M. J. Rowan, R. Anwyl, ibid. **132**, 93 (1987). HPLC determn in biological samples: G. Bianchi, S. Caccia, J. Chromatog. **431**, 477 (1988).

Crystals from isopropanol, mp 137-138°.
Hydrochloride, $C_{19}H_{24}ClN_5O_3S$, Bay q 7821, TVX Q 7821. Crystals from isopropanol, mp 221-222°.
THERAP CAT: Anxiolytic.

4968. Iridium. Ir; at. wt 192.22; at. no. 77; valences 1, 3, 4; also 2, 5, 6. Two naturally occurring isotopes: 191 (38.5%); 193 (61.5%); artificial, radioactive isotopes: 182-190; 191; 192; 194-198. Occurrence in earth's crust about 0.001 ppm. Discovered in 1804 by Tennant. Occurs in nature in the metallic state, usually as a natural alloy with osmium (osmiridium); found in small quantities alloyed with native platinum (platinum mineral) or with native gold. Recovery and purification from osmiridium: Deville, De-bray, Ann. Chim. Phys. **61**, 84 (1861); from the platinum mineral: Wichers, J. Res. Nat. Bur. Stand. **10**, 819 (1933). Reviews of prepn, properties and chemistry of iridium and other platinum metals: Gilchrist, Chem. Rev. **32**, 277-372 (1943); W. P. Griffith, The Chemistry of the Rare Platinum Metals (John Wiley, New York, 1967) pp 1-41, 227-312; Livingstone in Comprehensive Inorganic Chemistry vol. **3**, J. C. Bailar Jr. et al., Eds. (Pergamon Press, Oxford, 1973) pp 1163-1189, 1254-1274.

Silver-white, very hard metal; face-centered cubic lattice. mp 2450°; bp about 4500°. d_4^{20} 22.65; highest sp gr of all elements. Specific heat 0.0307 cal/g/°C. Mohs' hardness 6.5. Pure iridium is not attacked by any acids including aqua regia; only slightly by fused (non-oxidizing) alkalies. It is superficially oxidized on heating in the air. Is attacked by fluorine and chlorine at a red heat; attacked by potassium sulfate or by a mixture of potassium hydroxide and nitrate on fusion; attacked by lead, zinc or tin. The powdered metal is oxidized by air or oxygen at a red heat to the dioxide, IrO_2, but on further heating the dioxide dissociates into its constituents.
USE: In manufacturing crucibles; in hardening platinum; in making nibs for fountain-pen points.

4969. Iridium Hexafluoride. F_6Ir; mol wt 306.20. Ir 62.77%, F 37.23%. IrF_6. Prepd by direct fluorination of iridium: Ruff, Fischer, Z. Anorg. Allgem. Chem. **179**, 161 (1929); Robinson, Westland, J. Chem. Soc. **1956**, 4481. Crystal structure: Siegel, Northrop, Inorg. Chem. **5**, 2187 (1966).
Golden yellow, cubic crystals. d (calc) 4.82. Very hy-groscopic. mp 44.4°. bp 53°. Volatilizes on slow heating. Reduced to IrF_4 by halogens. Decomposes in aqueous soln. Does not react with dry glass below 150°. May be stored in evacuated quartz ampuls.

4970. Iridium Sesquioxide. Ir_2O_3; mol wt 432.40. Ir 88.90%, O 11.10%.
Blue-black powder. Decomposes at about 1000° into metal and oxygen. Insol in water; slowly dissolved by boil-ing HCl. Oxidized to IrO_2 by HNO_3.

4971. Iridium Trichloride. Cl_3Ir; mol wt 298.57. Cl 35.63%, Ir 64.37%. $IrCl_3$. Prepn: Dillamore, Edwards, J. Inorg. Nucl. Chem. **31**, 2427 (1969). Several known modifi-

cations. Crystal structure: Brodersen et al., Naturwiss. **52**, 205 (1966); Babel, Deigner, Z. Anorg. Allgem. Chem. **339**, 57 (1965).
α-Form; brown, monoclinic crystals. β-Form; red, ortho-rhombic crystals. Olive-green modification also prepd: Dillamore, Edwards, loc. cit. Decomposes at 763°. Insol in water, acids, alkalies.
USE: Catalyst for chlorination and polymerization reac-tions.

4972. Iridomyrmecin. Hexahydro-4,7-dimethylcyclopen-ta[c]pyran-3(1H)-one; 2-(hydroxymethyl)-α,3-dimethylcyclo-pentaneacetic acid δ-lactone; Iridomyrmexin; Iridomirme-cina. $C_{10}H_{16}O_2$; mol wt 168.23. C 71.39%, H 9.59%, O 19.02%. Isoln from the Argentine ant Iridomyrmex humilis Mayr., Dolichoderinae: Pavan, Ricerca Sci. **19**, 1011 (1949); Chim. Ind. (Milan) **37**, 625 (1955); Fusco et al., ibid. **251**, 958; Cavill et al., Austr. J. Chem. **9**, 288 (1956); **10**, 352 (1957). Synthesis: Korte et al., Tetrahedron **6**, 201 (1959); Sisido et al., J. Org. Chem. **29**, 3361 (1964); R. S. Matthews, J. K. Whitesell, ibid. **40**, 3312 (1975); Y. Yamada et al., Chem. Letters **1978**, 1405; P. Callant et al., Tetrahedron **37**, 2085 (1981). Stereochemistry: Korte et al., Ber. **94**, 1952 (1961); McConnell et al., Tetrahedron Letters **1962**, 445; Büchi, Manning, Tetrahedron **18**, 1049 (1962). Activity as bactericide and insecticide: M. Pavan, Z. Hyg. Infektion-Krankh. **134**, 136 (1952), C.A. **46**, 9796f (1952).

Prisms from petr ether. Aromatic odor, resembling cat-nip. Salty taste. mp 60-61°. Sublimes$_{0.01}$ 50-55°. bp$_{1.5}$ 104-108°. $[\alpha]_D^{20}$ +210° (c = 4 in ethanol); $[\alpha]_D^{17}$ +205° (c = 0.223 in carbon tetrachloride). n_D^{65} 1.4607. Sparingly sol in water: 0.2 g/100 ml H_2O at 25°. Sol in fats and fat solvents. Freely sol in ether.
USE: Insecticide.
THERAP CAT: Antibacterial.

4973. Irigenin. 5,7-Dihydroxy-3-(3-hydroxy-4,5-di-methoxyphenyl)-6-methoxy-4H-1-benzopyran-4-one; 3',5,7-trihydroxy-4',5',6-trimethoxyisoflavone. $C_{18}H_{16}O_8$; mol wt 360.31. C 60.00%, H 4.48%, O 35.53%. From rhizome of Iris florentina L., Iridaceae: de Laire, Tiemann, Ber. **26**, 2010 (1893). Structure: Baker, J. Chem. Soc. **1928**, 1022. Synthesis: Baker et al., Tetrahedron Letters no. **5**, 6 (1960); Farkas, Várady, Ber. **93**, 2685 (1960); Baker et al., J. Chem. Soc. (C) **1970**, 1219.

Yellow plates or needles from dil alc, mp 185°. uv max (abs ethanol): 267 nm. Sol in warm alcohol, benzene, chlo-roform; practically insol in water, ether, petr ether.
Triacetate, $C_{24}H_{22}O_{11}$, prisms from dil acetic acid, mp 127-128°.
7-Glucoside, $C_{24}H_{26}O_{13}$, iridin, 7-(glucosyloxy)-3',5-di-hydroxy-4',5',6-trimethoxyisoflavone. Monohydrate, nee-dles from dil alc, mp 208°. When anhydrous, mp 217°. One gram of anhydrous iridin dissolves in 500 ml water, 43 ml acetone. Sol in hot alcohol; practically insol in ether, ethyl acetate, benzene, chloroform.
Note: There also is an oleoresin called iridin (Extractum Iridis) from Iris versicolor (blue flag) by extraction with 60% alc. Another substance named iridine is a protamine from

the sperm of rainbow trout: Felix, Mager, *Z. Physiol. Chem.* **249**, 124 (1937).

4974. Irisolone. *7-(4-Hydroxyphenyl)-9-methoxy-8H-1,3-dioxolo[4,5-g][1]benzopyran-8-one;* "Sosan"; *4′-hydroxy-5-methoxy-6,7-methylenedioxyisoflavone.* $C_{17}H_{12}O_6$, mol wt 312.27. C 65.38%, H 3.87%, O 30.74%. From rhizomes of *Iris nepalensis* D. Don, *Iridaceae.* Isoln and structure: Gopinath *et al., Tetrahedron* **16**, 201 (1961). Synthesis: Fukui, Matsumato, *Bull. Chem. Soc. Japan* **38**, 887 (1965).

Pale yellow crystals from ethyl acetate, mp 269-270°. uv max: 270 nm (log ε 4.62), inflection 330 nm (log ε 3.97). Methyl ether, $C_{18}H_{14}O_6$, plates, mp 184-185°.

4975. Iron. Fe; at. no. 26; at. wt 55.847; valences 2, 3; seldom 1, 4, 6. Four naturally occurring isotopes: 54 (5.82%), 56 (91.66%), 57 (2.19%), 58 (0.33%); artificial, radioactive isotopes: 52; 53; 55; 59-61. Second most abundant metal in earth's crust after aluminum: about 5%. The earth's core is believed to consist mainly of iron. Important ores include hematite (Fe_2O_3), magnetite (Fe_3O_4), limonite [$FeO(OH).nH_2O$] and *siderite* ($FeCO_3$). Study of iron and its compds by Mössbauer spectroscopy: Danon, "57Fe: Metal, Alloys and Inorganic Compounds" in *Chemical Applications of Mössbauer Spectroscopy,* V. I. Goldanskii, R. H. Herber, Eds. (Academic Press, New York, 1968) p 159-313. Ions involved in oxygen transport, electron transport, nitrogen fixation and a number of other biological processes: Nielands, "Evolution of Biological Iron Binding Centers" in *Struct. Bonding* **11**, 145-170 (1972). Review of biology, pharmacology and toxicity of iron compounds: several authors, *Clin. Toxicol.* **4**, 525-642 (1971). Comprehensive reviews: Feldmann, Schenck in *Ullmanns Encyklopädie der Technischen Chemie* vol. 6 (München-Berlin, 1955) pp 261-407; Nicholls in *Comprehensive Inorganic Chemistry* vol. 3, J. C. Bailar, Jr. *et al.,* Eds. (Pergamon Press, Oxford, 1973) pp 979-1051; W. A. Knepper in Kirk-Othmer *Encyclopedia of Chemical Technology* vol. 13 (Wiley-Interscience, New York, 3rd ed., 1981) pp 735-753.

Silvery-white or gray, soft, ductile, malleable, somewhat magnetic metal. Holds magnetism only after hardening (as alloy steel, *e.g.,* Alnico). Supplied as ingots, powder, wire, sheets, etc. Takes a bright polish; can be rolled, hammered, bent, particularly when red hot. Stable in dry air but readily oxidizes in moist air, forming "rust" (chiefly oxide, hydrated). In powder form it is black to gray. Commercial iron usually contains some C, P, Si, S and Mn. d pure 7.86; cast 7.76; wrought 7.25-7.78; steel 7.6-7.78. mp pure 1535°; cast 1000-1300°; wrought 1500°; steel 1300°. bp 3000°. Electrical resistivity (20°): 9.71 microhm-cm. Readily attacked by dil mineral acids and attacked or dissolved by organic acids; not appreciably attacked by cold concd H_2SO_4 or HNO_3, but is attacked by the hot acids.

USE: Alloyed with C, Mn, Cr, Ni, and other elements to form steels. 55Fe and 59Fe used in tracer studies; the former in biological studies.

4976. Irone. 6-Methylionone. $C_{14}H_{22}O$; mol wt 206.32. C 81.50%, H 10.75%, O 7.76%. The fragrant principle of violets, best isolated from the rhizomes of iris or from orris oil: F. Tiemann, P. Kruger, *Ber.* **26**, 2675 (1893); Ruzicka *et al., Helv. Chim. Acta* **16**, 1143 (1933); Naves, Mazuyer, *Les Parfums Naturels* (Paris, 1939). Irone, as isolated, is an isomeric mixture of α-, β-, and γ-irones, *q.v.,* with each having the same molecular formula. Synthesis of isomeric mixture: Eschinazi, U.S. pat. **3,019,265** (1962 to Givaudan).

Light yellow, highly viscous liquid. Very stable, does not discolor. In dil alcoholic soln the characteristic odor of violets is best appreciated.

USE: In perfumery for orris and violet compositions.

4977. α-Irone. *4-(2,5,6,6-Tetramethyl-2-cyclohexen-1-yl)-3-butene-2-one.* Main perfume ingredient of violets. Both the *d*-(1,5)-*cis*- and *d*-(1,5)-*trans*-forms occur in nature. Isoln from irone mixture and synthesis: Naves, *Helv. Chim. Acta* **31**, 893, 1103 (1948); Ruzicka *et al., ibid.* **31**, 257 (1948). Synthesis: Barton, Mousseron-Canet, *J. Chem. Soc.* **1960**, 271. Industrial preparation: Kaiser, Kimel, U.S. pat. **2,877,271** (1959 to Hoffmann-La Roche); Eschenmoser *et al.,* U.S. pats. **3,413,351; 3,470,241** (1968, 1969 to Firmenich & Cie). Nomenclature in early papers: Naves, *Helv. Chim. Acta* **32**, 969 (1949). Stereochemistry: Tribolet, Schinz, *ibid.* **37**, 2184 (1954); Naves, Ardizio, *Bull. Soc. Chim. France* **1955**, 1479; V. Rautenstrauch, G. Ohloff, *Helv. Chim. Acta* **54**, 1776 (1971); V. Rautenstrauch *et al., ibid.* **67**, 325 (1984). Stereoselective synthesis of *cis*- and *trans*-α-irones: S. Torii *et al., J. Org. Chem.* **45**, 16 (1980); of (±)-*cis*-α-irone: C. Nussbaumer, G. Fráter, *ibid.* **52**, 2096 (1987).

(+)-cis-α-irone   (+)-trans-α-irone

All forms are viscous liquids with characteristic odors. *dl-cis-α*-Irone, d_4^{20} 0.9360; n_D^{20} 1.50098. uv max (ethanol): 227 nm (ε 15,400), Naves, Bachmann, *Helv. Chim. Acta* **32**, 394 (1949).

d-cis-α-Irone, $[\alpha]_D^{20}$ +109° (CH_2Cl_2).

dl-trans-α-Irone, d_4^{20} 0.9347; n_D^{20} 1.50119. uv max (ethanol): 229 (ε 15,450).

d-trans-α-Irone, $[\alpha]_D^{20}$ +420° (CH_2Cl_2).

4978. β-Irone. *4-(2,5,6,6-Tetramethyl-1-cyclohexen-1-yl)-3-buten-2-one.* Frequent byproduct in synthesis of α-irone. Stereoselective synthesis of (±)-form: S. Torii *et al., J. Org. Chem.* **45**, 16 (1980). For general refs *see* α-irone.

(+)-β-irone

Oil with odor similar to that of β-ionone. $bp_{0.1}$ 85-90°; $bp_{0.7}$ 99-104°; bp_{11} 125°. $[\alpha]_D^{20}$ +59° (CH_2Cl_2). d_4^{21} 0.9434; n_D^{21} 1.5178; n_D^{25} 1.5162. uv max: 295 nm (log ε 4.05).

4979. γ-Irone. *4-(2,2,3-Trimethyl-6-methylenecyclohexyl)-3-buten-2-one.* Principal fragrant constituent of natural iris oil. Isoln from irone mixture: Ruzicka *et al., Helv. Chim. Acta* **30**, 1807 (1947); Ruzicka *et al., ibid.* **31**, 257 (1948). Synthesis of *dl-γ*-irone: Favre, Schinz, *ibid.* **41**, 1368 (1958). Synthesis of (±)-*cis*- and *trans*-γ-irone: F. Leyendecker, M.-T. Comte, *Tetrahedron* **43**, 85 (1987); T. Kawanobe *et al., Agr. Biol. Chem.* **51**, 791 (1987). Stereoselective synthesis of (±)-*cis*-γ-irone: C. Nussbaumer, G. Fráter, *Helv. Chim. Acta* **71**, 619 (1988). Stereochemistry: *See* α-irone.

(+)-cis-γ-irone

Oil. $bp_{0.06}$ 85-88°. bp_2 114-116°. d_4^{15} 0.939; n_D^{15} 1.505. $[\alpha]_D^{20}$ +2° (methylene chloride). uv max: 230 nm (log ε 4.2).

USE: In perfumery.

4980. Iron Pentacarbonyl. Iron carbonyl; pentacarbonyliron. C_5FeO_5; mol wt 195.90. C 30.65%, Fe 28.51%, O 40.84%. $Fe(CO)_5$. Prepn from CO and Fe ore: Wallis, Townshend, U.S. pat. **2,378,053** (1945 to International Nickel); from CO and Fe or $FeSO_4.7H_2O$: Reppe *et al., Ann.* **582**, 116 (1953); from CO and Fe amalgams: Ettmayer, Jangg, *Monatsh.* **92**, 834 (1961); from CO and steel turnings: Shipman, **Brit.** pat. **897,204** (1962 to ICI). Convenient lab prepn of small quantities of $Fe(CO)_5$ from $Fe(CO)_4I_2$: Hieber, Lagally, *Z. Anorg. Allgem. Chem* **245**, 295 (1940). Thermodynamic data: Cotton *et al., J. Am. Chem. Soc.* **81**, 800 (1959); Leadbetter, Spice, *Can. J. Chem.* **37**, 1923 (1959). Toxicity: Sunderman *et al., Arch. Ind. Health* **19**, 11 (1959). *Reviews:* Cable, Sheline, *Chem. Rev.* **56**, 1 (1956); Wender *et al., The Chemistry and Catalytic Properties of Cobalt and Iron Carbonyls* (U.S. Govt. Printing Office, Washington, 1962) 83 pp; H. Alper, "Organic Syntheses with Iron Pentacarbonyl" in *Organic Syntheses via Metal Carbonyls* vol. 2, I. Wender, P. Pino, Eds. (John Wiley, New York, 1977) pp 545-593. Review of *iron tetracarbonyl*, the photochemically produced intermediate: M. Poliakoff, *Chem. Soc. Rev.* **7**, 527-540 (1978).

Colorless to yellow, oily liquid. *Pyrophoric in air;* burns to Fe_2O_3. Dec by light to $Fe_2(CO)_9$ and CO. mp −20°. bp 103°. d_4^{20} 1.46-1.52; n_D^{22} 1.453. Critical temp 285-288°; critical pressure 29.6 atm. Flash pt −15°. Heat capacity at constant pressure (14°) 56.9 cal/mole/°C. Latent heat of fusion 3161 cal/mol; latent heat of vaporization 9.6 kcal/mole. Heat of combustion −386.9 kcal/mole; heat of formation $[Fe(CO)_5(liq)]$ −182.6 kcal/mole. Practically insol in water; readily sol in most organic solvents including ether, benzene, petr ether, acetone, ethyl acetate, carbon tetrachloride, carbon disulfide; slightly sol in alcohol; practically insol in liquid ammonia. *Protect from light and air.* LD_{50} in mice, rats: 2.19, 0.91 mg/l inhalation, exposure for 30 min. (Sunderman).

USE: To make finely divided iron, so-called carbonyl iron, which is used in the manuf of powdered iron cores for high frequency coils used in the radio and television industry; as antiknock agent in motor fuels; as catalyst and reagent in organic reactions. *Caution:* Lung irritant. Affects central nervous system; causes liver and kidney damage: Gage, *Brit. J. Ind. Med.* **27**, 1 (1970).

4981. Iron Sorbitex. *Glucitol iron complex, compound with citric acid;* iron sorbitol; iron sorbitol citrate; Jectofer. Sterile soln (pH 7.2-7.9) of a complex of iron, sorbitol, and citric acid, stabilized with dextrin in a dil soln of sorbitol. Contains 50 ± 2 mg/ml of elemental iron. Average mol wt <5000. Prepn: Lindvall, Andersson, *Brit. J. Pharmacol. Chemother.* **17**, 358 (1961).

The prepn is hypertonic, stable in serum and rapidly absorbed from muscle, does not produce hemolysis, affects coagulation only at very high concns.

THERAP CAT: Hematinic.

4982. Irsogladine. *6-(2,5-Dichlorophenyl-1,3,5-triazine-2,4-diamine;* 2,4-diamino-6-(2,5-dichlorophenyl-s-triazine; dicloguamine. $C_9H_7Cl_2N_5$; mol wt 256.09. C 42.21%, H 2.75%, Cl 27.69%, N 27.35%. Gastric cytoprotectant with benzoguanamine skeleton. Prepn: H. Murai *et al.,* **Ger.** pat. **2,506,814;** *eidem,* U.S. pat. **3,966,728** (1975, 1976 both to Nippon Shinyaku). Series of articles on pharmacology: F. Ueda *et al., Arzneimittel-Forsch.* **34**, 474-491 (1984). Tissue distribution studies: T. Ando *et al., ibid.* **36**, 1221 (1984). Metabolism: M. Sugiyama *et al., ibid.* 1229. Phase I study: M. Nakashima *et al., ibid.* **34**, 492 (1984). Mechanism of antiulcer action study: F. Ueda *et al., Oyo Yakuri* **33**, 157 (1987), *C.A.* **107**, 17604a (1987).

Colorless crystals from dioxane, mp 268-269°.

Maleate, $C_{13}H_{11}Cl_2N_5O_4$, *MN-1695, Gaslon.* Crystals from dioxane, mp 205° (dec). LD_{50} in male, female mice, rats (mg/kg): 6035, 5697, 3898, 2917 orally; 2841, 3216, 1600, 1524 s.c.; 775, 1006, 558, 545 i.p. (Ueda, 1984 p 474).

THERAP CAT: Antiulcerative.

4983. Isanic Acid. *17-Octadecene-9,11-diynoic acid;* bolecic acid; erythrogenic acid. $C_{18}H_{26}O_2$; mol wt 274.39. C 78.79%, H 9.55%, O 11.66%. $CH_2=CH(CH_2)_4C\equiv CC\equiv C-(CH_2)_7COOH$. Found in vegetable oils obtained from tropical and equatorial regions. Isoln from boleko oil: De Vries, *Oléagineux* **12**, 427 (1957); U.S. pat. **2,789,993** (1957 to UCB). Structure: Steger, Van Loon, *Rec. Trav. Chim.* **59**, 1156 (1940); Gunstone, Sealy, *J. Chem. Soc.* **1963**, 5772; Morris, *ibid.* 5779. Synthesis: Black, Weedon, *ibid.* **1953**, 1785.

Crysts from petr ether, mp 39.5°. d_4^{45} 0.9309; d_4^{60} 0.91966; d_4^{78} 0.9095; n_D^{50} 1.49148; n_D^{78} 1.4860. Turns pink on exposure to air and light. Tendency to polymerize and explode, especially when heated above 250°. Sol in acetone, ethanol, isopropanol. Moderately sol in petr ether.

4984. Isatide. *1,1',3,3'-Tetrahydro-3,3'-dihydroxy-[3,3'-bi-2H-indole]-2,2'-dione; 3,3'-dihydroxy-[3,3'-biindoline]-2,2'-dione; 3,3'-dihydroxy-3,3'-dihydroisoindigo;* isatyde. $C_{16}H_{12}N_2O_4$; mol wt 296.27. C 64.86%, H 4.08%, N 9.46%, O 21.60%. Prepn by dimerization reduction of isatin with ammonium hydrosulfide: Laurent, *Ann. Chim.* **3**, 372 (1841); by piperidine-catalyzed condensation of isatin and dioxindole: Heller, *Ber.* **37**, 943 (1904); Wahl, Hansen, *Compt. Rend.* **178**, 393 (1924); from isatin and acenaphthene in sunlight: Oliveri-Mandala, Deleo, *Gazz. Chim. Ital.* **79**, 337 (1949). Structure: Lefèvre, *Bull. Soc. Chim. France* **19**, 113 (1916); Sumpter, *J. Am. Chem. Soc.* **54**, 2917 (1932); Bergmann, *ibid.* **77**, 1549 (1955).

Prisms from alcohol or ether, dec 245°. Difficultly sol in alcohol and ether; practically insol in water.

4985. Isatin. *Indole-2,3-dione;* 2,3-indolinedione; 2,3-diketoindoline. $C_8H_5NO_2$; mol wt 147.14. C 65.30%, H 3.43%, N 9.52%, O 21.75%. May be obtained by oxidation of indigo or of oxygenated indoles such as indoxyl, oxindole, or dioxindole: Erdmann, *J. Prakt. Chem.* [1] **24**, 1 (1841); Laurent, *ibid.* **25**, 430 (1842); **Ger.** pat. **229,815** in *Frdl.* **10**, 353 (1910); **Japan.** pat. **152,932** (1942 to ICI); *C.A.* **44**, 1544d (1950). Synthesis: Sandmeyer, *Helv. Chim. Acta* **2**, 234 (1919); **Brit.** pat. **128,122** in *C.A.* **13**, 2375 (1919); Marvel, Hiers, *Org. Syn.* **coll. vol. I** (2nd ed., 1941) p 327; Wibaut, Gerling, *Rec. Trav. Chim.* **50**, 41 (1931); Neunhoeffer, Lehmann, *Ber.* **94**, 2960 (1961); Ziegler *et al., Monatsh.* **94**, 453 (1963). May be isolated from the urine of rabbits that are fed *o*-nitrophenylglyoxylic acid: Bohm, *Z. Physiol. Chem.* **265**, 210 (1940). Pharmacology: Singh, *Indian Vet. J.* **48**, 672 (1971). *Reviews:* Heller, *Ueber Isatin, Isatyd, Dioxindol und Indophenin* (F. Enke, Stuttgart, 1931); Sumpter, *Chem. Rev.* **34**, 393 (1944). Discussion of chemistry of isatin: Morton, *Chemistry of Heterocyclic Compounds* (New York, 1946) pp 126-132.

Orange-colored monoclinic prisms. mp 203.5° (partial sublimation). Absorption spectrum: Hartley, Dobbie, *J. Chem. Soc.* **75**, 647, 656. Freely sol in boiling alcohol; sol in ether and in boiling water with reddish-brown color; sol in alkali hydroxide solns with a violet color becoming yellow on standing. The alc soln imparts a persistent, disagreeable odor to the human skin. Extremely weak base, forms a crystalline perchlorate, $C_8H_5NO_2 \cdot HClO_4 \cdot 2H_2O$.

USE: Manuf vat dyes. In analytical chemistry as a reagent for cuprous ions, mercaptans, thiophene, and indican.

4986. Isatropic Acid. *1,2,3,4-Tetrahydro-1-phenyl-1,4-naphthalenedicarboxylic acid;* 1-phenyltetrahydro-1,4-naphthalenedicarboxylic acid; 1-phenyltetralin-1,4-dicarboxylic acid; diatropic acid; α-isatropic acid; *trans*-isatropic acid. $C_{18}H_{16}O_4$; mol wt 296.31. C 72.96%, H 5.44%, O 21.60%. Prepn: Staudinger, Ruzicka, *Ann.* **380**, 296 (1911); Smith, Forsberg, *C.A.* **42**, 7283 (1948). Structure: Voigtlander, Engelbert, *Ann.* **625**, 196 (1959).

dl-Form, crystals from chloroform + petr ether, mp 238.5-239°. Sparingly sol in hot water; moderately sol in alcohol, acetic acid; practically insol in ether, benzene, ligroin, carbon disulfide.
d-Form, crystals, sinters at 234°, dec at 230°. $[\alpha]_D^{20}$ +9.44° (1 g/15 ml abs alc). K at 18° = 1.20×10^{-4}. Soly in water 0.0047%.

4987. Isaxonine. *N-(1-Methylethyl)-2-pyrimidinamine;* 2-(isopropylamino)pyrimidine. $C_7H_{11}N_3$; mol wt 137.18. C 61.28%, H 8.08%, N 30.63%. Neurotropic agent that promotes neurite outgrowth. Prepn: D. J. Brown, J. S. Harper, *J. Chem. Soc.* **1965**, 5542. Prepn of pharmacologically active salts: A. Esanu, **Belg. pat. 788,648** corresp to **U.S. pat. 3,984,414** and **Ger. pat. 2,707,095** corresp to **U.S. pat. 4,073,895** (1973, 1976, 1977, 1978 all to Soc. d'Etudes Prod. Chim.). Nerve growth promoting action: A. Hugelin *et al.*, *Compt. Rend. Ser. D* **285**, 1339 (1977); *eidem, Experientia* **35**, 626 (1979). Enhancement of muscle re-innervation in man: A. Sebille, A. Hugelin, *Brit. J. Pharmacol.* **9**, 275 (1980). Series of articles on *in vitro* effects, pharmacology, clinical studies: *Nouv. Presse Med.* **11**, 1189-1280 (1982).

mp 27-28°. bp$_{12}$ 92-93°.
Phosphate, $C_7H_{14}N_3O_4P$, *Nerfactor.* White cryst powder, mp 125°. LD$_{50}$ in mice (g/kg): 2.83 orally; 1.53 i.p., A. Esanu, **U.S. pat. 4,073,895.**
THERAP CAT: In treatment of peripheral neuropathies.

4988. Isazofos. *Phosphorothioic acid O-[5-chloro-1-(1-methylethyl)-1H-1,2,4-triazol-3-yl] O,O-diethyl ester;* O-(5-chloro-1-isopropyl-1H-1,2,4-triazol-3-yl) O,O-diethyl phosphorothioate; isazophos; CGA 12223; Brace; Miral; Triumph. $C_9H_{17}ClN_3O_3PS$; mol wt 313.74. C 34.46%, H 5.46%, Cl 11.30%, N 13.39%, O 15.30%, P 9.87%, S 10.22%. Organophosphate pesticide. Prepn: D. Dawes, B. Böhner, **Ger. pat. 2,260,015;** *eidem,* **U.S. pat. 3,867,396** (1973, 1975 both to Ciba-Geigy); B. Böhner *et al., Kem.-Kemi* **1**, 585 (1974). Physical properties and nematocidal activity: D. Dawes *et al., Meded. Fac. Landbouwwet., Rijksuniv. Gent* **39**, 727 (1974). Persistence in turfgrass thatch and soil: H. D. Niemczyk, H. R. Krueger, *J. Econ. Entomol.* **80**, 950 (1987).

Amber liquid, bp$_{0.001}$ 100°; bp$_{760}$ 170°. n_D^{20} 1.4867. d^{20} 1.22. Vapor pressure at 20°: 1.3×10^{-4} mm Hg. Soly (20°): 150 ppm in water. Miscible with methanol, chloroform, benzene, hexane. LD$_{50}$ in rats (mg/kg): 60 orally; 250-700 dermally (Dawes). *Caution:* Cholinesterase inhibitor.
USE: Nematocide; turf insecticide.

4989. Isepamicin. *(S)-O-6-Amino-6-deoxy-α-D-glucopyranosyl-(1→4)-O-[3-deoxy-4-C-methyl-3-(methylamino)-β-L-arabinopyranosyl-(1→6)]-N^1-(3-amino-2-hydroxy-1-oxopropyl)-2-deoxy-D-streptamine;* 1-N-[(S)-3-amino-2-hydroxypropionyl]gentamicin B; HAPA-B; Sch-21420. $C_{22}H_{43}N_5O_{12}$; mol wt 569.61. C 46.39%, H 7.61%, N 12.29%, O 33.71%. Aminoglycoside antibiotic; semisynthetic derivative of gentamicin B. Prepn: M. J. Weinstein *et al.*, **Belg. pat. 818,431;** J. J. Wright, P. J. L. Daniels, **Belg. pat. 824,657** (both 1975 to Sherico); J. J. Wright *et al.*, **U.S. pat. 4,002,742** (1977 to Schering Corp.); T. L. Nagabhushan *et al., J. Antibiot.* **31**, 681 (1978). Antibacterial spectrum *in vitro* and *in vivo*: G. H. Miller *et al., ibid.* 688. Comparison with amikacin of nephrotoxic potential: L. I. Rankin *et al., Antimicrob. Ag. Chemother.* **16**, 491 (1979). Series of articles on antibacterial activity, toxicology and clinical efficacy: *Chemotherapy (Tokyo)* **35**, Suppl 5, 1-610 (1985); on toxicology: *Japan. J. Antibiot.* **39**, 3164-3328 (1986), *C.A.* **107**, 32666r-32673r (1987). Acute toxicity: T. Morino *et al., ibid.* 3164, *C.A.* **107**, 32667s (1987).

Sulfate, *Exacin, Isepacin.* $[\alpha]_D^{26}$ 110.9° (c = 1 in water as the disulfate hydrate). LD$_{50}$ in male, female mice, rats (mg/kg): 234, 236, 489, 476 i.v.; 3312, 3320, 3451, 3392 s.c.; > 5000 orally both species (Morino).
THERAP CAT: Antibacterial.

4990. Isethionic Acid. *2-Hydroxyethanesulfonic acid;* hydroxyethylsulfonic acid. $C_2H_6O_4S$; mol wt 126.13. C 19.04%, H 4.79%, O 50.74%, S 25.42%. $HOCH_2CH_2SO_3H$. Obtained by heating ethylene, chlorosulfonic acid, and water: Klason, *J. Prakt. Chem.* [2] **19**, 234 (1879); from ether and SO$_3$: Hubner, *Ann.* **223**, 198 (1884); from ethylenesulfite + NaHCO$_3$: Smith, **U.S. pat. 2,899,461** (1959 to Dow).
Colorless, syrupy, strongly acid liquid. Kept over H_2SO_4, it solidifies to a very hygroscopic, cryst mass. Miscible with water, alc. Forms readily cryst salts with organic bases. *Caution:* Irritating to skin, mucous membranes.

4991. Isinglass. Ichthyocolla; fish glue. The inner membrane of the swimming bladder of *Acipenser huso* and other species of sturgeon and of hake. *Constit.* Chiefly glutin.
Thin, white or yellowish, semi-transparent, pearly iridescent, horny sheets. Sol in hot water, hot dil alcohol. In cold water, softens and becomes adhesive.
USE: As nutrient and food instead of gelatin; as adhesive, in "court plaster"; for clarifying (fining) wine and beer; as protective colloid in manuf various chemicals; in glass and porcelain cements; for lustering and stiffening silks and other fabrics; with pyroxylin for waterproofing fabrics.

4992. Isoaminile. *α-[2-(Dimethylamino)propyl]-α-(1-methylethyl)benzeneacetonitrile; 4-(dimethylamino)-2-isopropyl-2-phenylvaleronitrile;* 3-cyano-5-dimethylamino-3-phenyl-2-methylhexane; α-(β-dimethylaminopropyl)-α-isopropylphenylacetonitrile; Aprecon; Dimyril; Nullatuss. $C_{16}H_{24}N_2$; mol wt 244.37. C 78.63%, H 9.90%, N 11.46%. Prepn: **Brit. pat. 822,695;** Stühmer, Funke, **U.S. pat. 2,934,557** (1959, 1960, both to Kali-Chemie).

Consult the cross index before using this section.

$$CH_3CHCH_2-\overset{\overset{\displaystyle CH(CH_3)_2}{|}}{\underset{\underset{\displaystyle N(CH_3)_2}{|}}{C}}-CN$$
$$C_6H_5$$

Liquid, bp$_3$ 138-146°.
Citrate, $C_{22}H_{32}N_2O_7$, *Perocan.* Crystals from alcohol + ether, mp 63-64°.
Cyclamate, $C_{22}H_{37}N_3O_3S$, *Mucalan, Peracon.* Prepn: Dickinson, U.S. pat. 3,074,996 (1963 to Abbott).
Tartrate, $C_{16}H_{24}N_2 \cdot C_4H_6O_6$, crystals, mp 64°.
THERAP CAT: Antitussive.

4993. Isoamyl Acetate. Amylacetic ester. $C_7H_{14}O_2$; mol wt 130.18. C 64.58%, H 10.84%, O 24.58%. CH_3COOCH_2-$CH_2CH(CH_3)_2$. The technical product is also known as *pear oil* or *banana oil.*
Colorless, neutral liq; pear-like odor and taste. d$_4^{15}$ 0.876. Pure isoamyl acetate bp 142°. n$_D^{21}$ 1.400; the ordinary grade of commerce boils between 120-145°. Flash pt, closed cup: 92° F (33° C); open cup: 100° F (38° C). Sol in 400 parts water; miscible with alcohol, ether, ethyl acetate, amyl alcohol. Soly of water in isoamyl acetate (25°) 1.6% by volume.
USE: In alcohol solution as a pear flavor in mineral waters and syrups; as solvent for old oil colors, for tannins, nitrocellulose, lacquers, celluloid, and camphor; swelling bath sponges; covering unpleasant odors, perfuming shoe polish; manuf artificial silk, leather or pearls, photographic films, celluloid cements, waterproof varnishes, bronzing liquids, and metallic paints; dyeing and finishing textiles. A special grade of the amyl acetate has been used for burning in the Hefner lamp serving as a photometric standard.

4994. Isoamylamine. *3-Methyl-1-butanamine; isopentylamine;* isobutylcarbylamine; 3-methylbutylamine. $C_5H_{13}N$; mol wt 87.16. C 68.90%, H 15.03%, N 16.07%. $(CH_3)_2$-$CHCH_2CH_2NH_2$. Prepn from acetamide + isoamylbromide: Erickson, *Ber.* **59B,** 2665 (1926); from α-methylhydroxylamine + isoamylmagnesium chloride: Sheverdina, Kocheshkov, *J. Gen. Chem. USSR* **8,** 1825 (1938). Isoln from *Clostridium sordelli:* Prevot, Thouvenot, *Ann. Inst. Pasteur* **103,** 925 (1962).
Colorless liq; strong ammonia odor. d^{18} 0.751. bp 95°. n$_D^{18}$ 1.4096. Miscible with water, alcohol, chloroform, ether.
Hydrochloride, $C_5H_{13}N \cdot HCl$, crystals from acetone, mp 215°. One gram dissolves in 0.5 ml water, 20 ml chloroform; sol in alcohol.
Caution: Irritating to skin, mucous membranes.

4995. Isoamyl Benzoate. $C_{12}H_{16}O_2$; mol wt 192.25. C 74.96%, H 8.39%, O 16.64%. $C_6H_5COOCH_2CH_2CH(CH_3)_2$.
Colorless liquid, d$_4^{15}$ 0.993. bp 260-262°. Insol in water; miscible with alcohol.
USE: In perfumery and cosmetics.

4996. Isoamyl Bromide. *1-Bromo-3-methylbutane.* C_5-$H_{11}Br$; mol wt 151.05. C 39.75%, H 7.34%, Br 52.92%. $(CH_3)_2CHCH_2CH_2Br$. Prepd from isoamyl alcohol, bromine and phosphorus or by refluxing isoamyl alcohol with HBr in presence of H_2SO_4: Kamm, Marvel, *Org. Syn.* **vol. 1,** p 4 (1921).
Colorless liquid. d$_4^{15}$ 1.210. bp 120-121° (mp -112°). n$_D^{15}$ 1.4433. Slightly sol in water (0.02 g/100 ml at 16.5°); miscible with alcohol, ether.
USE: In organic synthesis.

4997. Isoamyl Butyrate. $C_9H_{18}O_2$; mol wt 158.23. C 68.31%, H 11.47%, O 20.22%. $CH_3(CH_2)_2COOCH_2CH_2CH$-$(CH_3)_2$.
Colorless liq; aromatic pear-like odor. d$_{15}^{19}$ 0.866. bp 179°. Slightly sol in water; miscible with alcohol, ether.
USE: Manuf artficial rum and fruit essences.

4998. Isoamyl Chloride. *1-Chloro-3-methylbutane.* C_5-$H_{11}Cl$; mol wt 106.60. C 56.33%, H 10.40%, Cl 33.26%. $(CH_3)_2CHCH_2CH_2Cl$. Prepd by refluxing isoamyl alcohol and HCl in presence of $ZnCl_2$; also by direct chlorination of isopentane.
Liquid. d^0 0.893. bp about 100°. n$_D^{20}$ 1.4103. Slightly sol in water; miscible with alcohol, ether.

4999. Isoamyl Cyanide. *4-Methylpentanenitrile;* isocapronitrile. $C_6H_{11}N$; mol wt 97.16. C 74.17%, H 11.41%, N 14.42%. $(CH_3)_2CHCH_2CH_2CN$. Prepd from KCN and isoamyl chloride in boiling alcohol.
Liquid, very disagreeable odor. d$_4^{20}$ 0.806. bp 155-156°. mp -51°. n$_D^{20}$ 1.406. Insol in water; miscible with alc, ether.

5000. Isoamyl Ether. *1,1'-Oxybis[3-methylbutane];* isoamyl oxide; diisoamyl ether. $C_{10}H_{22}O$; mol wt 158.28. C 75.88%, H 14.01%, O 10.11%. $[(CH_3)_2CHCH_2CH_2]_2O$.
Colorless liquid; pleasant, fruity odor. d$_4^{12}$ 0.783. bp 172°. n$_D^{20}$ 1.408. Insol in water; miscible with alc, chloroform, ether.
USE: Solvent in Grignard reaction; also as solvent of odorous principles; manuf lacquers; regenerating rubber.

5001. Isoamyl Formate. $C_6H_{12}O_2$; mol wt 116.09. C 62.07%, H 9.55%, O 28.38%. $HCOOCH_2CH_2CH(CH_3)_2$.
Colorless liquid; fruity odor. d^{20} 0.877. bp 123-124°. n$_D^{20}$ 1.391. Sol in 300 parts water; miscible with alc, ether. LD$_{50}$ orally in rats: 9840 mg/kg, P. M. Jenner *et al., Food Cosmet. Toxicol.* **2,** 327 (1964).
USE: Artificial fruit syrups.

5002. Isoamyl Iodide. *1-Iodo-3-methylbutane.* $C_5H_{11}I$; mol wt 198.06. C 30.32%, H 5.60%, I 64.08%. $(CH_3)_2$-$CHCH_2CH_2I$. Obtained from isoamyl alcohol, iodine, and phosphorus, or by distillation of the alcohol with HI.
Liquid; quickly becomes brown on exposure to air and light. d$_{15}^{18}$ 1.515. bp 147°. Slightly sol in water; miscible with alc, ether. *Keep well closed and protected from light.*

5003. Isoamyl Isovalerate. Isoamyl valerianate; amyl valerate; amyl isovalerate; apple oil. $C_{10}H_{20}O_2$; mol wt 172.26. C 69.72%, H 11.70%, O 18.58%. $(CH_3)_2CHCH_2$-$COOCH_2CH_2CH(CH_3)_2$.
Colorless liquid; apple-like odor. d$_4^{19}$ 0.858. bp 191-194°. n$_D^{19}$ 1.413. Very slightly sol in water; miscible with alc, ether.
USE: As apple essence for flavoring liqueurs and candy.

5004. Isoamyl Nitrate. *3-Methyl-1-butanol nitrate.* C_5-$H_{11}NO_3$; mol wt 133.15. C 45.10%, H 8.33%, O 36.05%, N 10.52%. $(CH_3)_2CHCH_2CH_2NO_3$.
Colorless liquid. d$_4^{22}$ 0.996. bp 147-148°. n$_D^{22}$ 1.4122. Slightly sol in water; miscible with alcohol, ether.
Caution: Do not confuse with isoamyl *nitrite, q.v.*

5005. Isoamyl Nitrite. Isopentyl nitrite. $C_5H_{11}NO_2$; mol wt 117.15. C 51.26%, H 9.47%, N 11.96%, O 27.32%. $(CH_3)_2CHCH_2CH_2ONO$. Prepd by nitrosation of isopentyl alcohol: Bevillard, Choucroun, *Bull. Soc. Chim. France* **1957,** 337.
Yellowish, transparent, flammable liquid; penetrating fragrant, somewhat fruity odor. Unstable and dec on exposure to air and light. The pure nitrite has d$_{25}^{25}$ 0.875. bp 97-99°, but volatilizes readily at much lower temps. n$_D^{21}$ 1.3871. Very slightly sol in water; miscible with alc, chloroform, ether. *Caution: Forms an explosive mixture with air or oxygen. Keep in tightly closed containers protected from light and in cool place. Incompat.* Alcohol, antipyrine, caustic alkalies, alkaline carbonates, potassium iodide, bromides, ferrous salts.

5006. Isoamyl Phthalate. *1,2-Benzenedicarboxylic acid bis(3-methylbutyl) ester;* diisoamyl phthalate; amyl phthalate. $C_{18}H_{26}O_4$; mol wt 306.41. C 70.56%, H 8.55%, O 20.89%.
Colorless, practically odorless liquid. d 1.028. bp$_{40}$ 225°. Insol in water; sol in organic solvents.
USE: Plasticizer for nitrocellulose and resin lacquers; preventing foam in manuf of glue; in rubber cements.

5007. Isoamyl Salicylate. $C_{12}H_{16}O_3$; mol wt 208.25. C 69.21%, H 7.74%, O 23.05%. $C_6H_4(OH)COOC_5H_{11}$.
Colorless liquid; pleasant odor. d$_{15}^{19}$ 1.048. bp 274-278°. Flash pt 132°. n$_D^{20}$ 1.506. Almost insol in water, miscible with alcohol, chloroform, ether.
USE: In perfumery and soaps.

5008. Isoapo-β-erythroidine. $C_{15}H_{15}NO_2$; mol wt 241.28. C 74.66%, H 6.26%, N 5.81%, O 13.26%. Separation from apo-β-erythroidine: Sauvage *et al., Science* **109,** 627 (1949); Sauvage, Boekelheide, *J. Am. Chem. Soc.* **72,**

2062 (1950); Koniuszy, Folkers, *ibid.* **73**, 333 (1951). Synthesis and structure: Blake *et al.*, *ibid.* **87**, 1397 (1965).

Crystals, mp 146-148°. uv max (ethanol): 379, 288, 253 nm (ε 6500, 10,800, 16,800).

5009. Isoascorbic Acid. D-*erythro-Hex-2-enonic acid* γ-*lactone;* D-araboascorbic acid; erythorbic acid; isovitamin C; saccharosonic acid; glucosaccharonic acid; D-*erythro*-3-ketohexonic acid lactone; D-*erythro*-3-oxohexonic acid lactone; erycorbin; Mercate "5"; Neo-Cebicure. $C_6H_8O_6$; mol wt 176.12. C 40.91%, H 4.58%, O 54.51%. Epimer of L-ascorbic acid. Has one-twentieth of the vitamin C activity of L-ascorbic acid. Prepd by treating methyl 2-keto-D-gluconate with sodium methoxide: Maurer, Schiedt, *Ber.* **66**, 1054 (1933); **67**, 1239 (1934); Ohle *et al.*, *ibid.* **67**, 324 (1934); Baird *et al.*, *J. Chem. Soc.* **1934**, 63; Reichstein *et al.*, *Helv. Chim. Acta* **17**, 516 (1934). Synthesis from sucrose: Heimann, Reiff, *Pharm. Zentralh.* **93**, 97 (1954). Production by *Penicillium* spp: Takahashi *et al.*, *Nature* **188**, 411 (1960); U.S. pat. **3,052,609** (1962 to Sankyo); *Bull. Agr. Chem. Soc. Japan* **24**, 533 (1960), *C.A.* **55**, 1788 (1961). Conformation: Matsui *et al.*, *Agr. Biol. Chem.* **27**, 185 (1963).

Shiny granular crystals from water or dioxane. Dec 174°. $[\alpha]_D^{16.5}$ −17° (c = 1.8 in 0.01N HCl); $[\alpha]_D^{20}$ −16.6° (H_2O). Soluble in water, alc, pyridine. Moderately sol in acetone. Slightly sol in glycerol.

Sodium salt, *sodium erythorbate, Mercate "20", Neo-Cebitate.* Crystals, sol in water. pH of aq solns of the sodium salt between 5 and 6. A 10% soln, made from commercial grade, may have a pH of 7.2 to 7.9. The free acid is more sol in water (40 g/100 ml H_2O) than the sodium salt (16 g/100 ml H_2O).

USE: Antioxidant and antimicrobial agent for foods: Kadin, Osadca, *J. Agr. Food Chem.* **7**, 358 (1959).

5010. Isobenzan. *1,3,4,5,6,7,8,8-Octachloro-1,3,3a,4,7,-7a-hexahydro-4,7-methanoisobenzofuran;* 1,3,4,5,6,7,10,10-octachloro-4,7-endomethylene-4,7,8,9-tetrahydrophthalan; SD 4402; R 6700; Omtan; Telodrin. $C_9H_4Cl_8O$; mol wt 411.79. C 26.25%, H 0.98%, Cl 68.88%, O 3.89%. Prepn: Brit. pat. **772,212** (1957 to Ruhrchemie); Korte, Stiasni, *Ann.* **656**, 140 (1962); Feichtinger, Linden, *Chem. & Ind. (London)* **1965**, 1938.

Crystals from heptane, mp 120-122°. Sol in acetone, benzene, toluene, ether, xylene, heavy aromatic naphtha. LD_{50} i.v. in rats: 1.8 mg/kg, Tauberger, *Arch. Exp. Pathol. Pharmakol.* **232**, 227 (1957).

USE: Insecticide.

5011. Isoborneol. *exo-1,7,7-Trimethylbicyclo[2.2.1]heptan-2-ol;* exo-2-bornanol; exo-2-camphanol. $C_{10}H_{18}O$; mol wt 154.24. C 77.86%, H 11.76%, O 10.37%. Prepn of *dl*-form: Pickard, Littlebury, *J. Chem. Soc.* **91**, 1973 (1907); Truett, Moulton, *J. Am. Chem. Soc.* **73**, 5913 (1951); Ziegler,

Brit. pat. **803,178** (1958). Prepn of the *l*-form by reduction of *d*-camphor with lithium aluminum hydride: Trevoy, Brown, *J. Am. Chem. Soc.* **71**, 1675 (1949). Separation of isoborneol from its *endo*-isomer, borneol, via the *p*-nitrobenzoate deriv: Truett, Moulton, *loc. cit.* Resolution of the *dl*-form: Pickard, Littlebury, *loc. cit.*; Kenyon, Priston, *ibid.* **127**, 1472 (1925). Configuration (isoborneol = *exo*-form; borneol = *endo*-form): Toivonen *et al.*, *Acta Chem. Scand.* **3**, 991 (1949). *Review:* J. L. Simonsen, *The Terpenes* **vol. II** (University Press, Cambridge, 2nd ed., 1949) pp 365-367; A. R. Pinder, *The Chemistry of the Terpenes* (Chapman & Hall, London, 1960) pp 22-24, 101, 103, 105-107, 111.

l-isoborneol

dl-Form, crystals from petr ether. Sublimes on heating, mp 212° (in a sealed tube). Practically insol in water. Readily sol in alcohol, ether, chloroform.

d-Form, crystals from petr ether, mp 214°. Approx $[\alpha]_D$ +34.3° in alc soln: Picard, Littlebury, *loc. cit.*

l-Form, crystals from petr ether, mp 214°. Approx $[\alpha]_D$ −34.3° in alc soln.

5012. Isobornyl Thiocyanoacetate, Technical. *Thiocyanatoacetic acid 1,7,7-trimethylbicyclo[2.2.1]hept-2-yl ester;* thanite; terpinyl thiocyanoacetate. $C_{13}H_{19}NO_2S$; mol wt 253.36. C 61.62%, H 7.56%, N 5.53%, O 12.63%, S 12.66%. *See also* Borneol. The technical grade contains 82% or more of isobornyl thiocyanoacetate, with other terpenes. May be prepared by treating isoborneol with chloroacetyl chloride and KCNS: Borglin, U.S. pats. **2,217,611-15** (1940 to Hercules).

Yellow, oily liq. Terpene-like odor. $bp_{0.06}$ 95°. Flash pt 82° (180°F). d_4^{25} 1.1465. Acid no. 1.19. n_D^{25} 1.512. Very sol in alcohol, benzene, chloroform, ether; practically insol in water. LD_{50} orally in rats: 1 g/kg, *Toxic Substances List*, H. E. Christensen, Ed. (1973) p 917.

USE: Insecticide, especially in cattle sprays.

THERAP CAT: Pediculicide.

5013. Isobutol. *4-Pyridinecarboxylic acid 2-(sulfomethyl)hydrazide compd with [S-(R*,R*)]-2,2'-(1,2-ethanediyldiimino)bis(1-butanol) (2:1);* (+)-2,2'-(ethylenediimino)dibutanol diisoniazide methanesulfonate; ethambutol isoniazid methanesulfonate; isoetham; Isoetam. $C_{24}H_{42}N_8O_{10}S_2$; mol wt 666.77. C 43.23%, H 6.35%, N 16.81%, O 23.99%, S 9.62%. Tuberculostatic conjugate of isoniazid and ethambutol. Prepn: C. Ferrer-Salat *et al.*, Belg. pat. **753,862**; U.S. pat. **3,718,655** (1971, 1973 both to Lab. Ferrer). New process: S.-W. Park, Japan. Kokai **83 8065**; U.S. pat. **4,450,274** (1983, 1984 both to Korean Adv. Inst. Sci. Technol.). Pharmacokinetics: J. E. Beneyto *et al.*, *Actas 1st Congr. Nac. Biofarm. Farmacocinet.*, **1977**, 283, *C.A.* **89**, 99680u (1978). Clinical experience: T. A. Loenin, *Kekkaku* **56**, 63 (1981).

White powder, crystallizes as needles, mp 121-122° (dec). Sol in water, methanol, *N,N*-dimethylformamide. Insol in ethanol, ethyl ether, chloroform. uv max: 265 nm. $[\alpha]_D^{20}$

+3° (c = 5% in water). LD_{50} in mice (g/kg): 2.800 orally; 2.210 i.p. (Ferrer-Salat).

THERAP CAT: Antibacterial (tuberculostatic).

5014. Isobutyl Acetate. $C_6H_{12}O_2$; mol wt 116.16. C 62.04%, H 10.41%, O 27.55%. $CH_3COOCH_2CH(CH_3)_2$. Prepn from isobutyl alcohol + acetic acid: **Belg. pat. 505,-023** (1951 to Soc. Belge de l'Azote et des Prod. Chim. du Marly); from methyl isobutyl ketone: White, Emmons, *Tetrahedron* **17**, 31 (1962). Isoln from wood-rotting fungus, *Endoconidiophora coerulescens* Münch.: Birkinshaw, Morgan, *Biochem. J.* **47**, 55 (1950).

Colorless liquid. d_{4}^{20} 0.871. bp 118°. mp −99°. Flash pt, closed cup: 64° F (18° C). n_D^{19} 1.3907. Sol in 180 parts water, freely in alc.

USE: Flavoring, solvent.

5015. Isobutyl Alcohol. *2-Methyl-1-propanol;* isopropylcarbinol; 1-hydroxymethylpropane; fermentation butyl alcohol. $C_4H_{10}O$; mol wt 74.12. C 64.81%, H 13.60%, O 21.59%. $(CH_3)_2CHCH_2OH$. Present in fusel oil; also produced by fermentation of carbohydrates: Baraud, Genevois, *Compt. Rend.* **247**, 2479 (1958); Sukhodol, Chatskii, *Spirt. Prom.* **28**, 35 (1962), *C.A.* **57**, 5124e (1962). Prepn: Wender *et al., J. Am. Chem. Soc.* **71**, 4160 (1949); Schreyer, U.S. pat. **2,564,130** (1951 to du Pont); Pistor, U.S. pat. **2,753,366;** Harrer, Rühl, **Ger. pat. 1,011,865;** Himmler, Schiller, U.S. pat. **2,787,628** (1956, 1957, 1957 all to BASF).

Colorless, refractive liq; flammable; odor like that of amyl alcohol, but weaker. d^{15} 0.806. bp 108°. mp −108°. Flash pt, closed cup: 82°F (28° C). n_D^{15} 1.3976. Sol in about 20 parts water; misc with alcohol, ether. LD_{50} orally in rats: 2.46 g/kg, Smyth *et al., Arch. Ind. Hyg. Occup. Med.* **10**, 61 (1954).

USE: Manuf esters for fruit flavoring essences; solvent in paint, varnish removers. *Caution:* May be mildly irritating to skin, mucous membranes, and, in high concns, narcotic.

5016. Isobutylamine. *2-Methyl-1-propanamine;* 2-methylpropylamine; 1-amino-2-methylpropane. $C_4H_{11}N$; mol wt 73.14. C 65.68%, H 15.16%, N 19.17%. $(CH_3)_2CH-CH_2NH_2$. Prepn from isobutyl alcohol + NH_3: **Brit. pat. 847,799** (1960 to Cellulose-Polymères et Dérivés "Cepede"); Shirley, Speranza, U.S. pat. **3,128,311** (1964 to Jefferson Chem.). Isoln from fungi: von Kamienski, *Planta* **50**, 331 (1958). Acute toxicity: K. L. Cheever *et al., Toxicol. Appl. Pharmacol.* **62**, 150 (1982).

Liquid. d_4^{25} 0.724. mp −85°. bp 68-69°. n_D^{17} 1.3988. Miscible with water, alcohol, ether. LD_{50} (14 day) in male, female rats (mg/kg): 224.4, 231.8 orally (Cheever).

Caution: Skin contact can result in erythema, blistering. Inhalation causes headache, dryness of nose and throat.

5017. Isobutyl p-Aminobenzoate. *4-Aminobenzoic acid 2-methylpropyl ester;* Cycloform; Isobutyl Kelo-form; Isocaine. $C_{11}H_{15}NO_2$; mol wt 193.24. C 68.37%, H 7.82%, N 7.25%, O 16.56%. Prepn: Adams *et al., J. Am. Chem. Soc.* **48**, 1758 (1926).

Crystals from benzene, mp 65°. Slightly sol in water; sol in alcohol, benzene, ether, acetone, olive oil.

USE: Sunscreen.

THERAP CAT: Topical anesthetic.

5018. Isobutylbenzene. *(2-Methylpropyl)benzene;* 2-methyl-1-phenylpropane. $C_{10}H_{14}$; mol wt 134.21. C 89.49%, H 10.51%. $C_6H_5CH_2CH(CH_3)_2$. Prepd from bromobenzene, isobutyl iodide, and sodium preferably in the presence of benzene: Wreden, Znatovicz, *Ber.* **9**, 1606 (1876); Schramm, *Monatsh.* **9**, 616 (1888); by passing vapors of dimethylbenzyl carbinol and hydrogen over activated charcoal: Zelinsky, Gawerdowskaya, *Ber.* **61**, 1052 (1928).

Liquid. d_4^{20} 0.8673. bp_{760} 170.5°; bp_{400} 145.2°; bp_{200} 120.7°; bp_{100} 99.0°; bp_{60} 84.1°; bp_{40} 73.2°; bp_{20} 54.7°; bp_{10} 37.3°; bp_5 21.1°; $bp_{1.0}$ −9.8°. n_D^{20} 1.4928. Critical temp 377.1°; crit pressure 23,636 mm Hg.

5019. Isobutyl Bromide. *1-Bromo-2-methylpropane.* C_4H_9Br; mol wt 137.03. C 35.06%, H 6.62%, Br 58.32%. $(CH_3)_2CHCH_2Br$. Obtained from isobutyl alc and PBr_3: Noller, Dinsmore, *Org. Syn.* **coll. vol. II,** 358 (1943).

Colorless liq. d^{15} 1.272. bp 91.5°. mp −119°. n_D^{15} 1.4391. Slightly sol in water (0.6 g/l); miscible with alcohol, ether.

5020. Isobutyl n-Butyrate. $C_8H_{16}O_2$; mol wt 144.21. C 66.63%, H 11.18%, O 22.19%. $CH_3CH_2CH_2COOCH_2CH-(CH_3)_2$.

Liquid. d 0.866. bp 157°. n_D^{20} 1.4035. Slightly sol in water; miscible with alcohol, ether.

5021. Isobutyl Carbamate. $C_5H_{11}NO_2$; mol wt 117.15. C 51.26%, H 9.46%, N 11.96%, O 27.32%. $H_2NCOOCH_2-CH(CH_3)_2$. Prepd from isobutyl chloroformate and NH_3 in benzene soln.

Crystals, d 0.956. mp 67°; also stated at 61°. bp 206-207°. Insol in water; sol in alcohol, ether.

5022. Isobutyl Chloride. *1-Chloro-2-methylpropane.* C_4H_9Cl; mol wt 92.53. C 51.90%, H 9.80%, Cl 38.30%. $(CH_3)_2CHCH_2Cl$. Formed by the action of HCl or PCl_5 on isobutyl alcohol.

Liquid. d^{15} 0.883. bp 68-69°. mp −131°. n_D^{15} 1.40096. Insoluble in water; miscible with alcohol, ether.

5023. Isobutyl Chlorocarbonate. *Carbonochloridic acid 2-methylpropyl ester;* isobutyl chloroformate. $C_5H_9ClO_2$; mol wt 136.58. C 43.97%, H 6.64%, Cl 25.96%, O 23.43%. $ClCO_2CH_2CH(CH_3)_2$. Prepd from isobutyl alcohol and phosgene.

Clear liquid; vapors irritate eyes and mucous membranes. d 1.040. bp 130°. Gradually dec by water and alcohol. Miscible with benzene, chloroform, ether.

5024. Isobutylene. *2-Methylpropene;* isobutene. C_4H_8; mol wt 56.10. C 85.63%, H 14.37%. $CH_2=C(CH_3)_2$. Obtained from refinery streams by absorption on 65% H_2SO_4 at about 15°: Packie, Rupp, U.S. pats. **2,424,186, 2,509,885;** Draeger, U.S. pat. **2,456,260;** Steele, Epps Jr., U.S. pat. **2,497,191** (1947, 1950, 1948, 1950, all to Standard Oil); Peters, Gothman, Edwards, Wesselhoft, U.S. pats. **2,962,-537; 3,129,265** (1960, 1964, both to Esso). Separation from a mixed C_4 stream using 50% H_2SO_4: **Brit. pats. 824,573, 858,645** and **Fr. pat. 1,337,232** (1959, 1961 and 1963 to Compagnie Francaise de Raffinage); Valet *et al., Hydrocarbon Process. Petrol. Refiner* **41**, No. 5, 119 (1962); Martel, *Chem. Eng.* **72**, No. 7, 66 (1965). Prepn: Verdol, U.S. pat. **3,170,000** (1965 to Sinclair). Review: Kennedy, Kirshenbaum, "Isobutylene" in *Vinyl and Diene Monomers* (part 2), E. C. Leonard, Ed. (Wiley-Interscience, New York, 1971) pp 691-756.

Gas. bp_{760} −6.900°; bp_{100} −49.309°; bp_{30} −67.90°; bp_{10} −81.95°; bp_1 −105.06°. d_4^{20} 0.5942; d_4^{25} 0.5879; d_4^{40} 0.5815. Practically insol in water. Very sol in alc, ether, sulfuric acid.

Caution: Simple asphyxiant.

USE: Primarily used to produce diisobutylene, trimers, butyl rubber, and other polymers; also to produce antioxidants for foods, packaging, food supplements, and for plastics: Hatch, *Petrol. Refiner* **39**, No. 6, 207 (1960).

5025. Isobutyl Ether. *1,1'-Oxybis[2-methylpropane];* diisobutyl ether. $C_8H_{18}O$; mol wt 130.22. C 73.78%, H 13.93%, O 12.29%. $(CH_3)_2CHCH_2OCH_2CH(CH_3)_2$.

Colorless liquid, characteristic odor. d^{15} 0.761. bp 122-124°. Insol in water, miscible with alcohol, ether.

5026. Isobutyl Formate. Tetryl formate. $C_5H_{10}O_2$; mol wt 102.13. C 58.80%, H 9.87%, O 31.33%. $HCOOCH_2CH-(CH_3)_2$.

Liquid. d_4^{20} 0.885. bp 98°. mp −95°. n_D^{20} 1.3858. Sol in 100 parts water; miscible with alc, ether.

5027. Isobutyl Iodide. *1-Iodo-2-methylpropane.* C_4H_9I; mol wt 183.03. C 26.10%, H 4.93%, I 68.97%. $(CH_3)_2CH-CH_2I$. Obtained by distilling isobutyl alcohol and HI, or from isobutyl alcohol, iodine and phosphorus.

Liquid; becomes brown on exposure. d^{20} 1.605. bp 120°. mp −93°. n_D^{20} 1.4960. Insol in water; miscible with alcohol, ether. *Keep well closed and protected from light.*

5028. Isobutyl Isobutyrate. *2-Methylpropanoic acid 2-methylpropyl ester.* $C_8H_{16}O_2$; mol wt 144.21. C 66.63%, H 11.18%, O 22.19%. $(CH_3)_2CHCOOCH_2CH(CH_3)_2$.
Liquid. d_4^0 0.875. bp 147°. mp −81°. n_D^{20} 1.3999. Insol in water; miscible with alcohol.

5029. Isobutyl Isovalerate. Isobutyl valerate. $C_9H_{18}O_2$; mol wt 158.23. C 68.31%, H 11.47%, O 20.22%. $(CH_3)_2CH-CH_2COOCH_2CH(CH_3)_2$.
Liquid; ethereal odor. d^{20} 0.853. bp 170-172°. n_D^{20} 1.4064. Insol in water; miscible with alcohol, ether.
USE: Flavoring and manuf fruit essences.

5030. Isobutyl Mercaptan. *2-Methyl-1-propanethiol.* $C_4H_{10}S$; mol wt 90.19. C 53.26%, H 11.18%, S 35.56%. $(CH_3)_2CHCH_2SH$. May be prepd from isobutyl bromide and KHS in alcohol, see ref *under* other butyl mercaptans.
Mobile liq. Heavy skunk odor. mp −79°. bp 88°. Flammable. d_4^{20} 0.8357. n_D^{20} 1.43859. Slightly sol in water; very sol in alcohol, ether, liquid hydrogen sulfide.

5031. Isobutyl Nitrate. $C_4H_9NO_3$; mol wt 119.12. C 40.33%, H 7.62%, N 11.76%, O 40.29%. $(CH_3)_2CHCH_2O-NO_2$.
Liquid. d_4^2 1.015. bp 123-125°. n_D^{20} 1.4028. Insol in water; miscible with alcohol, ether.

5032. Isobutyl Nitrite. $C_4H_9NO_2$; mol wt 103.12. C 46.59%, H 8.80%, O 31.03%, N 13.58%. $(CH_3)_2CHCH_2NO_2$. Prepd from isobutyl alcohol, $NaNO_2$, and dil H_2SO_4.
Colorless liq. d_4^{22} 0.870. bp 67°. n_D^{22} 1.3715. Slightly sol and gradually dec by water; miscible with alcohol.

5033. Isobutyl Propionate. $C_7H_{14}O_2$; mol wt 130.18. C 64.58%, H 10.84%, O 24.58%. $CH_3CH_2COOCH_2CH(CH_3)_2$.
Liquid; agreeable, ethereal odor. d_4^0 0.888. bp 137°. mp −71°. n_D^{20} 1.3975. Insol in water; miscible with alc.
USE: Manuf fruit essences.

5034. Isobutyl Stearate. $C_{22}H_{44}O_2$; mol wt 340.57. C 77.58%, H 13.02%, O 9.40%. $CH_3(CH_2)_{16}COOCH_2CH-(CH_3)_2$.
Paraffin-like, crystal substance at low temp, mp about 20°.
USE: Waterproof coatings, polishes, face creams, rouges, ointments, soaps, rubber manuf, dye solns, inks, lubricants.

5035. Isobutyl Sulfide. *1,1'-Thiobis[2-methylpropane];* diisobutyl sulfide. $C_8H_{18}S$; mol wt 146.30. C 65.68%, H 12.40%, S 21.93%. $[(CH_3)_2CHCH_2]_2S$. Obtained by heating isobutyl chloride or potassium isobutyl sulfate with K_2S.
Liquid. d_4^{10} 0.836. bp 171-173°. Insol in water; miscible with alcohol, ether.

5036. Isobutyl Thiocyanate. Isobutyl sulfocyanate. C_5H_9NS; mol wt 115.19. C 52.13%, H 7.87%, N 12.16%, S 27.83%. $(CH_3)_2CHCH_2SCN$. Obtained by distilling potassium isobutyl sulfate with NaCSN.
Liquid, bp 174-176°. Almost insol in water; miscible with alcohol.

5037. Isobutyl Urethane. *(2-Methylpropyl)carbamic acid ethyl ester.* $C_7H_{15}NO_2$; mol wt 145.20. C 57.90%, H 10.41%, N 9.65%, O 22.04%. $(CH_3)_2CHCH_2NHCOOC_2H_5$. Obtained by shaking isobutylamine with ethyl chloroformate and aq KOH in the cold.
Liquid; apple-like odor. d_4^{20} 0.943. Solidif below −65°. bp_{17} 96°. n_D^{20} 1.4288. Insol in water; sol in alcohol.

5038. Isobutyraldehyde. *2-Methylpropanal;* isobutyl-aldehyde; isobutyric aldehyde. C_4H_8O; mol wt 72.10. C 66.63%, H 11.18%, O 22.19%. $(CH_3)_2CHCHO$. Manuf by the oxo process from propylene, carbon monoxide, and hydrogen at 130-160° and 1500-3000 psi in the presence of a cobalt catalyst: Roelen, U.S. pat. **2,327,066** (1943); also prepd by air oxidation of isobutyl alcohol: Fossek, *Monatsh.* **2**, 614 (1881); Lipp, *Ann.* **205**, 2 (1880). *Review:* P. D. Sherman in Kirk-Othmer *Encyclopedia of Chemical Technology* vol. 4 (Wiley-Interscience, New York, 3rd ed., 1978) pp 376-386.
Flammable liq. Pungent odor. d^{20} 0.7938. mp −65.9°. bp_{760} 64°. n_D^{20} 1.3730. Heat of combustion 599.9 kcal. Flash pt (open cup) <20°F. Soly in water at 20°: 11 g/100 ml H_2O. Miscible with ethanol, ether, carbon disulfide, ace-tone, benzene, toluene, chloroform. Forms an azeotrope with water contg 94% isobutyraldehyde. The azeotrope bp_{760} 59°. Weight per gallon of isobutyraldehyde: 6.55 lbs. Oxidizes slowly on exposure to air, forming isobutyric acid. LD_{50} orally in rats: 3.7 g/kg, Smyth *et al., Arch. Ind. Hyg. Occup. Med.* **10**, 61 (1954).
USE: In the synthesis of pantothenic acid, valine, leucine, cellulose esters, perfumes, flavors, plasticizers, resins, gasoline additives.

5039. Isobutyric Acid. *2-Methylpropanoic acid;* isopropylformic acid. $C_4H_8O_2$; mol wt 88.10. C 54.53%, H 9.15%, O 36.32%. $(CH_3)_2CHCOOH$. Prepn from 1-nitroisobutane: Lippincott, Hass, *Ind. Eng. Chem.* **31**, 118 (1939); from methallyl chloride: Towle, Hall, U.S. pat. **2,667,508** (1954 to Standard Oil Co. of Indiana); from propylene: Alderson, U.S. pat. **3,020,314** (1962 to du Pont); from 2-methylpropane + CO with HF catalyst: Friedman, Cotton, *J. Org. Chem.* **27**, 481 (1962).
Liquid; pungent odor like that of butyric acid, but not as unpleasant. bp_{760} 152-155°. mp −47°. d_4^{20} 0.950. n_D^{20} 1.3930. Flash pt, open cup: 170° F (77° C). Sol in 6 parts of water; misc with alcohol, chloroform, ether.
Caution: Mild irritant.

5040. Isocarboxazid. *5-Methyl-3-isoxazolecarboxylic acid 2-benzylhydrazide;* 3-(N-benzylhydrazinocarbonyl)-5-methylisoxazole; 1-benzyl-2-(5-methyl-3-isoxazolylcarbonyl)hydrazine; Ro 5-0831; Marplan. $C_{12}H_{13}N_3O_2$; mol wt 231.25. C 62.32%, H 5.67%, N 18.17%, O 13.84%. Monoamine oxidase inhibitor. Prepn: Gardner, Wenis, U.S. pat. **2,908,688** (1959 to Hoffmann-La Roche); *J. Med. Pharm. Chem.* **2**, 133 (1960). Comprehensive description: B. C. Rudy, B. Z. Senkowski in *Analytical Profiles of Drug Substances* vol. 2, K. Florey, Ed. (Academic Press, New York, 1973) pp 295-314.

Crystals from methanol, practically tasteless. mp 105-106°. Very sparingly sol in hot water (0.05%), somewhat more (1 to 2%) in 95% alc, in glycerol, in propylene glycol.
THERAP CAT: Antidepressant.

5041. Isochondrodendrine. $O^7,O^{7'}$-Didemethylcycleanine; isobebeerine. $C_{36}H_{38}N_2O_6$; mol wt 594.68. C 72.70%, H 6.44%, N 4.71%, O 16.14%. From *Chondodendron tomentosum* Ruiz and Pav., *Menispermaceae:* Faltis, *Monatsh.* **33**, 873 (1912); Faltis, Neumann, *ibid.* **42**, 311 (1921); Dutcher, *J. Am. Chem. Soc.* **68**, 419 (1946); from the drug *Radix pareirae bravae:* King, *J. Chem. Soc.* **1940**, 737. Structure: Faltis, Dietreich, *Ber.* **67**, 231 (1934); Jeffreys, *J. Chem. Soc.* **1956**, 4451.

Needles from methanol, dec 288°. $[\alpha]_D^{20}$ −29° (c = 1.3 in chloroform). Sol in alcohol, benzene, chloroform.

5042. Isocil. *5-Bromo-6-methyl-3-(1-methylethyl)-2,4-(1H,3H)-pyrimidinedione; 5-bromo-3-isopropyl-6-methyluracil;* Hyvar, General Weed Killer. $C_8H_{11}BrN_2O_2$; mol wt 247.12. C 38.88%, H 4.49%, Br 32.34%, N 11.34%, O 12.95%. Prepn: Bucha *et al., Science* **137**, 537 (1962). Superseded by bromacil, *q.v.*

Crystals from 20% aq ethanol, mp 158-159°. LD_{50} orally in rats: 3400 mg/kg, G. W. Bailey, J. L. White, *Residue Rev.* **10**, 97 (1965).

USE: Herbicide.

5043. Isocinchomeronic Acid. *2,5-Pyridinedicarboxylic acid.* $C_7H_5NO_4$; mol wt 167.12. C 50.31%, H 3.02%, N 8.38%, O 38.29%. Prepd by treating 5-ethyl-2-picoline (aldehyde-collidine) with suitable oxidizing agents such as HNO_3, $KMnO_4$, or H_2SeO_3: Meyer, Staffen, *Monatsh.* **34**, 517 (1913); Jordan, *Ind. Eng. Chem.* **44**, 332 (1952); Kato, *Bull. Chem. Soc. Japan* **34**, 636 (1961); by oxidation of 5-acetyl-2-methylpyridine: Binns, Swan, *J. Chem. Soc.* **1962**, 2831; from the lutidine fraction of coal tar lutidine: Lukes *et al., Coll. Czech. Chem. Commun.* **26**, 3044 (1961).

Triclinic leaflets or prisms from dilute HCl, mp 254°. (Monohydrate, crystals from water or alcohol, dec 238°.) Sublimes as nicotinic acid when heated above mp. Practically insol in cold water, alcohol, ether, benzene. Slightly sol in boiling water, boiling alcohol. Appreciably sol in hot, dil aq solns of mineral acids.

Dihydrazide, crystals, mp 268-269°.

USE: Intermediate in the manuf of nicotinic acid.

5044. Isoconazole. *1-[2-(2,4-Dichlorophenyl)-2-[(2,6-dichlorophenyl)methoxy]ethyl]-1H-imidazole; 1-[2,4-dichloro-β-[(2,6-dichlorobenzyl)oxy]phenethyl]imidazole.* $C_{18}H_{14}$-Cl_4N_2O; mol wt 416.12. C 51.96%, H 3.39%, Cl 34.08%, N 6.73%, O 3.84%. Prepn: E. F. Godefroi *et al., J. Med. Chem.* **12**, 784 (1969); *eidem, Ger.* pat. **1,940,388**; *eidem,* U. S. pats. **3,717,655** and **3,839,574** (1970, 1973, 1974 all to Janssen). *In vitro* activity: H. J. Kessler, *Arzneimittel-Forsch.* **29**, 1344 (1979). Animal study: H. J. Kessler *et al., ibid.* 1352. Antimicrobial activity in humans: H. Wendt, H. J. Kessler, *ibid.* 846. Bioavailability: U. Täuber, M. Rzadkiewicz, *Mykosen* **22**, 201 (1979).

Nitrate, $C_{18}H_{15}Cl_4N_3O_4$, *R 15454, Fazol, Gyno-Travogen, Travogen, Travogyn.* Solid, mp 182-183°.

THERAP CAT: Antibacterial; antifungal.

5045. Isocorybulbine. *5,8,13,13a-Tetrahydro-3,9,10-trimethoxy-13-methyl-6H-dibenzo[a,g]quinolizin-2-ol; 3,9,-10-trimethoxy-13-methyl-13a-berbin-2-ol; 2-hydroxy-13-methyl-3,9,10-trimethoxyberbine.* $C_{21}H_{25}NO_4$; mol wt 355.42. C 70.96%, H 7.09%, N 3.94%, O 18.01%. From tubers of *Corydalis cava* (L.) Schweigg. & Korte (*C. tuberosa* DC., *Fumariaceae):* Gadamer, *Arch. Pharm.* **240**, 19 (1902). Structure: Späth, Dobrowsky, *Ber.* **58**, 1274 (1925). Synthesis: Späth, Holter, *ibid.* **59**, 2800 (1926). Config: Kondo, *Yakugaku Zasshi* **83**, 1017 (1963), *C.A.* **60**, 8073e (1964).

Leaflets, mp 179-180°; also given as 187-188°. $[\alpha]_D^{15}$ +301° in chloroform. Sol in alcohol, dil acids.

Methiodide, $C_{21}H_{25}NO_4.CH_3I$, crystals, mp 218-221°.

5046. Isocorydine. *(S)-5,6,6a,7-Tetrahydro-1,2,10-trimethoxy-6-methyl-4H-dibenzo[de,g]quinolin-11-ol; 1,2,10-trimethoxy-6aα-aporphin-11-ol;* 11-hydroxy-1,2,10-trimethoxyaporphine; artabotrine; luteanine. $C_{20}H_{23}NO_4$; mol wt 341.39. C 70.36%, H 6.79%, N 4.10%, O 18.75%. From tubers of *Corydalis cava* (L.) Schweigg. & Korte (*C. tuberosa* DC., *Fumariaceae); Artabotrys suaveolens* Blume, *Anonaceae; Papaver oreophilum* Rupr.; *Phoebe clemensii* Allen (*Lauraceae*) and others. Isoln: Gadamer, *Arch. Pharm.* **249**, 669 (1911); Johns, Lamberton, *Aust. J. Chem.* **20**, 1277 (1967); Pfeifer, Mann, *Pharmazie* **23**, 82 (1968). Structure: Späth, Berger, *Ber.* **64**, 2038 (1931); Gulland *et al., J. Chem. Soc.* **1931**, 2885; Barger, Sargent, *ibid.* **1939**, 991. Identity with artabotrine: Schlittler, Huber, *Helv. Chim. Acta* **35**, 111 (1952); with luteanine: Manske, *Can. J. Res.* **21B**, 13 (1943). Total synthesis: Kametani *et al., Tetrahedron* **27**, 5367 (1971). Pharmacology: Berezhinskaya *et al., Farmakol. Toksikol. (Moscow)* **31**, 44 (1968), *C.A.* **68**, 94521z (1968).

Plates from ethanol or acetone, mp 185°. $[\alpha]_D^{20}$ +195° (chloroform). Sol in ether, chloroform, alcohol, acetone, alkali hydroxide. Practically insol in water, alkali carbonates.

5047. Isocorypalmine. *5,8,13,13a-Tetrahydro-3,9,10-trimethoxy-6H-dibenzo[a,g]quinolizin-2-ol; 3,9,10-trimethoxy-13aα-berbin-2-ol; d-tetrahydrocolumbamine.* $C_{20}H_{23}NO_4$; mol wt 341.39. C 70.36%, H 6.79%, N 4.10%, O 18.75%. From tubers of *Corydalis cava* (L.) Schweigg. & Korte (*C. tuberosa* DC.) and *C. nobilis* Pers., *Fumariaceae:* Gadamer *et al., Arch. Pharm.* **265**, 675 (1927); Manske, *Can. J. Res.* **18B**, 288 (1940). Structure: Späth, Burger, *Ber.* **59**, 1486 (1926). Synthesis of *dl*-form: Govindachari *et al., ibid.* **92**, 1654 (1959). Configuration: Corrodi, Hardegger, *Helv. Chim. Acta* **39**, 889 (1956).

Crystals, mp 239-241°. $[\alpha]_D^{20}$ 30.3° (c = 0.4 in chloroform). Sol in alcohol, ether.

dl-Form, crystals from methanol, mp 216°.

5048. Isocrotonic Acid. *cis-2-Butenoic acid.* $C_4H_6O_2$; mol wt 86.09. C 55.80%, H 7.03%, O 37.17%. CH_3-$CH=CHCOOH$. Prepd by reacting acetaldehyde with malonic acid in the presence of pyridine: Auwers *et al., Ann.* **432**, 46 (1923); from ethyl acetoacetate: Hatch, Nesbitt, *J.*

Am. Chem. Soc. **72**, 727 (1950); from but-2-ynoic acid: Allan *et al.*, *J. Chem. Soc.* **1955**, 1862; from $CH_3CH=CH$-$MgBr + CO_2$: Normant, Maitte, *Bull. Soc. Chim. France* **1956**, 1439. Structure: Auwers, Wissebach, *Ber.* **56**, 715 (1923); Plisov, Bogatskii, *Zh. Obshch. Khim.* **27**, 360 (1957), *C.A.* **51**, 15401c (1957).

Liquid. bp_5 55-56°; bp_{10} 62-64°; bp_{760} 168-169°. mp 13.5-14.5°; crystallized from pentane. n_D^{20} 1.4450. uv max (95% ethanol): 205.5 nm (ϵ 13,500).

Methyl ester, $C_5H_8O_2$, liquid. bp 118°. n_D^{20} 1.4175. uv max (95% ethanol): 205.5 nm (ϵ 14,000).

Benzyl ester, $C_{11}H_{12}O_2$, liquid. bp_{10} 121-122°. n_D^{20} 1.5110. *Caution:* Can cause severe irritation of skin, mucous membranes.

5049. Isocyanic Acid. Carbimide. CHNO; mol wt 43.03. C 27.91%, H 2.34%, N 32.56%, O 37.19%. HN=C=O. Prepn: J. Liebig, F. Wöhler, *Ann.* **39**, 29 (1846); Steyermark, *J. Org. Chem.* **28**, 586 (1963); R. Vorhoeve, L. E. Trimble, *Science* **202**, 525 (1978). Crystal structure: Von Dohlen, Carpenter, *Acta Cryst.* **8**, 646 (1955). Review of prepn and properties: D. J. Belson, A. N. Strachan, *Chem. Soc. Reviews* **11**, 41-56 (1982). *See also* Cyanic Acid.

Store in dil solns of carbon tetrachloride or ether at −30° to slow polymerization. The free acid in the vapor phase or in ether soln gives no indication of being a mixture, and all the evidence supports the iso structure, HNCO, whereas in aq soln cyanic acid HOCN is present: N. V. Sidgwick, *The Chemical Elements and Their Compounds* **vol. I** (Oxford, 1950) p 673. *Caution:* Strongly acidic, will blister skin.

5050. Isodurene. *1,2,3,5-Tetramethylbenzene.* $C_{10}H_{14}$; mol wt 134.21. C 89.49%, H 10.51%. Occurs in coal tar. Prepd by the action of dimethyl sulfate on 2,4,6-trimethylphenylmagnesium bromide: Smith, *Org. Syn.* **11**, 66 (1931).

Liquid. d_4^0 0.8961; d_4^{20} 0.8906. mp −24°. bp_{760} 197.9°; bp_{400} 173.7°; bp_{200} 149.9°; bp_{100} 128.3°; bp_{60} 115.4°; bp_{40} 105.8°; bp_{20} 91.0°; bp_{10} 77.8°; bp_5 65.8°; $bp_{1.0}$ 40.6°. n_D^{20} 1.5134; n_{He}^{20} 1.51126. Insol in water; sol in alcohol; very sol in ether.

5051. Isoestradiol. *8α-Estra-1,3,5(10)-triene-3,17β-diol;* $\Delta^{1,3,5}$-8α-epiestratriene-3,17β-diol; 8α-estradiol; 8-epiestradiol; 8-isoestradiol-17β. $C_{18}H_{24}O_2$; mol wt 272.37. C 79.37%, H 8.88%, O 11.75%. Prepn by high pressure hydrogenation of dihydroequilin: Serini, Logemann, *Ber.* **71**, 186 (1938); by hydrogenation of equilenin with Raney nickel at 2800 psi and 85°: Dauben, Ahramjian, *J. Am. Chem. Soc.* **78**, 633 (1956). Prepn of *dl*-form from *dl*-equilenin and stereochemistry: Johnson *et al.*, *ibid.* **80**, 661 (1958).

d-Form, crystals from dil methanol + chloroform, mp 181°. $[\alpha]_D^{20}$ +18° (16 mg in 2 ml dioxane).

3-Benzoate, $C_{25}H_{28}O_3$, crystals from ethyl acetate, mp 190°. $[\alpha]_D^{20}$ +9.5° (17 mg in 2 ml dioxane).

dl-Form, needles from dil methanol, mp 213.5-214°.

5052. 8-Isoestrone. *3-Hydroxyestra-1,3,5(10)-trien-17-one;* $\Delta^{1,3,5}$-8-epiestratrien-3-ol-17-one; 8α-estrone; 8-epiestrone. $C_{18}H_{22}O_2$; mol wt 270.36. C 79.96%, H 8.20%, O 11.84%. Prepn: Serini, Logemann, *Ber.* **71**, 186 (1938); Johnson *et al.*, *J. Am. Chem. Soc.* **80**, 661 (1958); Ananchenko, Torgov, *Tetrahedron Letters* **1963**, 1553.

dl-Form, prisms from methanol, mp 254-255°.

dl-Methyl ether, $C_{19}H_{24}O_2$, blades from methanol, mp 153-155°.

dl-Benzoate, $C_{25}H_{26}O_3$, prisms from methanol, mp 197-198°.

5053. Isoetharine. *4-[1-Hydroxy-2-[(1-methylethyl)-amino]butyl]-1,2-benzenediol; 3,4-dihydroxy-α-[1-(isopropylamino)propyl]benzyl alcohol;* α-(1-isopropylaminopropyl)-protocatechuyl alcohol; N-isopropylethylnorepinephrine; 1-(3,4-dihydroxyphenyl)-2-isopropylamino-1-butanol; etyprenaline; isoetarine; Win 3046; Dilabron; Neoisuprel. $C_{13}H_{21}NO_3$; mol wt 239.31. C 65.24%, H 8.85%, N 5.85%, O 20.06%. Prepn: Bockmühl *et al.*, **Ger. pat.** 638,650 (1936 to I. G. Farben.), *C.A.* **31**, 3209⁴ (1937). Pharmacology: Lands *et al.*, *J. Pharmacol. Exp. Ther.* **99**, 45 (1950); *eidem*, *J. Am. Pharm. Assoc.* **47**, 744 (1958).

Hydrochloride, $C_{13}H_{22}ClNO_3$, *Asthmalitan, Numotac.* Crystals from methanol + ether. mp 212-213° (dec).

THERAP CAT: Bronchodilator.

5054. Isoeugenol. *2-Methoxy-4-propenylphenol;* 4-hydroxy-3-methoxy-1-propenylbenzene; 4-propenylguaiacol. $C_{10}H_{12}O_2$; mol wt 164.20. C 73.14%, H 7.37%, O 19.49%. Occurs in ylang-ylang and other essential oils. Prepd from eugenol: West, *J. Soc. Chem. Ind.* **59**, 275 (1940); Pal'gi, *Zh. Obshch. Khim.* **28**, 2239 (1958). Stereochemistry: von Auwers, *Ber.* **68**, 1346 (1935); Puxeddu, Rattu, *Gazz. Chim. Ital.* **67**, 647 (1937).

Oily liquid; easily becomes somewhat yellow. d_4^{25} 1.080. bp 266°; bp_8 128-130°. mp −10°. n_D^{19} 1.5739. Slightly sol in water; misc with alcohol, ether. LD_{50} orally in rats: 1560 mg/kg, P. M. Jenner *et al.*, *Food Cosmet. Toxicol.* **2**, 327 (1964).

trans-Form, crystals, mp 33°, bp_{12} 140°. d_4^{20} 1.087. n_{He}^{20} 1.5778.

cis-Form, liquid, bp_{11} 133°. d_4^{20} 1.088. n_{He}^{20} 1.5724.

USE: Manuf vanillin.

5055. Isofenphos. *2-[[Ethoxy[(1-methylethyl)amino]-phosphinothioyl]oxy]benzoic acid 1-methylethyl ester; salicylic acid isopropyl ester O-ester with O-ethyl isopropylphosphoramidothioate;* 1-methylethyl 2-[[ethoxy[(1-methylethyl)amino]phosphinothioyl]oxy]benzoate; O-ethyl O-2-isopropoxy-carbonylphenyl isopropylphosphoramidothioate; isophenphos; Bay 92114; SRA 12869; Amaze; Oftanol. $C_{15}H_{24}NO_4PS$; mol wt 345.40. C 52.16%, H 7.00%, N 4.06%, O 18.53%, P 8.97%, S 9.28%. Selective soil insecticide. Prepn: G. Schrader *et al.*, S. Afr. pat. 68 7,378, *C.A.* **72**, 3239g (1970); Fr. pat. **1,600,932** corresp to *eidem*, U.S. pat. **3,621,082** (1969, 1970, 1971 to Bayer). Determn of residues: M. J. Brown, I. H. Williams, *Pestic. Sci.* **7**, 545 (1976).

Oil, $bp_{0.01}$ 120°. d_4^{20} 1.13. Vapor press at 20°: 4×10^{-6} mm Hg. Soly in water at 20°: 23.8 mg/kg. Sol in dichloromethane, cyclohexanone, acetone, alc, ether, benzene. LD_{50} orally in rats, mice: 28, 91.3 mg/kg, *RTECS* **Vol. I**, R. J. Lewis, R. L. Tatken, Eds. (1980) p 310.

USE: Insecticide.

5056. Isofezolac. *1,3,4-Triphenyl-1H-pyrazole-5-acetic acid;* LM 22102; Sofenac. $C_{23}H_{18}N_2O_2$; mol wt 354.40. C 77.95%, H 5.12%, N 7.90%, O 9.03%. Non-steroidal anti-inflammatory agent; prostaglandin synthetase inhibitor. Prepn: C. Gueremy, C. Renault, **Ger. pat. 2,312,256;** *eidem,* **U.S. pat. 3,984,431** (1973, 1976 to Société Générale de Recherches et d'Applications Scientifiques). Pharmacology: J. Mizoule *et al., Arch. Int. Pharmacodyn.* **238,** 305 (1979). HPLC determn in biological fluids: A. Bannier *et al., J. Chromatog.* **227,** 213 (1982).

Crystals from acetonitrile, mp 200°. LD_{50} in mice, rats (mg/kg): 215, 13 orally (Mizoule).

THERAP CAT: Anti-inflammatory; antipyretic; analgesic.

5057. Isoflavone. *3-Phenyl-4H-1-benzopyran-4-one;* 3-phenylchromone. $C_{15}H_{10}O_2$; mol wt 222.23. C 81.06%, H 4.54%, O 14.40%. Prepn from o-hydroxyphenyl benzyl ketone, sodium dust and ethyl formate: Joshi, Venkataraman, *J. Chem. Soc.* **1934,** 513. Structure: Warburton, *Quart. Rev. (London),* **8,** 67 (1954).

Leaflets and needles from petr ether, mp 148°. uv max: 245, 307 nm (log ϵ 4.41, 3.82).

5058. Isoflupredone. *9-Fluoro-11,17,21-trihydroxypregna-1,4-diene-3,20-dione;* 1-dehydro-9α-fluorohydrocortisone; 9-fluoroprednisolone. $C_{21}H_{27}FO_5$; mol wt 378.45. C 66.65%, H 7.19%, F 5.02%, O 21.14%. Prepn: J. Fried *et al., J. Am. Chem. Soc.* **77,** 4181 (1955). Microbiological prepn: A. Nobile *et al., J. Am. Chem. Soc.* **77,** 4184 (1955); E. Vischer *et al., Helv. Chim. Acta* **38,** 1502 (1955); T. H. Stoudt *et al., Arch. Biochem. Biophys.* **59,** 304 (1955). Prepn of the 21-acetate: R. F. Hirschmann *et al., ibid.* 3166; J. A. Hogg *et al., ibid.* 4438; Ch. Meystre *et al., Helv. Chim. Acta* **39,** 734 (1956); A. Wellstein *et al.,* **Ger. pat. 1,020,329** (1957 to Ciba), *C.A.* **54,** 25562h (1960); **Brit. pat. 843,214** (1960 to Schering), *C.A.* **55,** 5597b (1961). Absorption spectra: L. L. Smith, W. H. Muller, *J. Org. Chem.* **23,** 960 (1958). Mass spectrometry: P. Toft *et al., Can. J. Pharm. Sci.* **7,** 53 (1972). Urinary excretion in horses: D. I. Chapman *et al., Vet. Rec.* **100,** 447 (1977).

Cryst from acetone, mp 263-266° (dec). $[\alpha]_D^{23}$ +108° (c = 0.611 in ethanol). uv max (ethanol): 240 nm (ϵ 15,800), E. Vischer *et al., loc. cit.,* also reported as mp 274-275° (dec), $[\alpha]_D^{23}$ +94° (alcohol), J. Fried *et al., loc. cit.*

21-acetate, $C_{23}H_{29}FO_6$, *U-6013,* **Predef.** Cryst from acetone/isopropyl ether or methanol, mp 244-246° (dec). $[\alpha]_D^{23}$ +108° (c = 0.735 in dioxane). uv max (ethanol): 240 (ϵ 16,250).

THERAP CAT (VET): Anti-inflammatory.

5059. Isoflurane. *2-Chloro-2-(difluoromethoxy)-1,1,1-trifluoroethane; 1-chloro-2,2,2-trifluoroethyl difluoromethyl ether;* Compound 469; Aerrane; Forane; Forene. $C_3H_2Cl-F_5O$; mol wt 184.50. C 19.53%, H 1.09%, Cl 19.21%, F 51.49%, O 8.67%. $CF_3CHClOCHF_2$. Prepn: Croix, Terrell, **Ger. pat. 1,814,962** (1969) corresp to Terrell, **U.S. pats. 3,535,388; 3,535,425** (both 1970, to Air Reduction); Terrell *et al., J. Med. Chem.* **14,** 517 (1971). Series of articles on pharmacology: several authors in *Anesthesiology* **35,** 8-53 (1971); Byles *et al., Can. Anaesth. Soc. J.* **18,** 376-407 (1971).

Clear, colorless liquid having a slight odor. bp 48.5°. Vapor pressure at 25°: 330 mm. sp gr 1.45. n_D^{20} 1.3002. Nonflammable; soda lime stable. Easily miscible with organic liquids including fats and oils.

USE: Solvent and dispersant for fluorinated materials.

THERAP CAT: Anesthetic (inhalation).

5060. Isoflurophate. *Phosphorofluoridic acid bis(1-methylethyl) ester; phosphorofluoridic acid diisopropyl ester;* isopropyl fluophosphate; diisopropyl fluorophosphonate; diisopropyl fluorophosphate; diisopropylphosphorofluoridate; fluostigmine; isofluorphate; DFP; Diflupyl; Dyflos; Floropryl; Fluropryl. $C_6H_{14}FO_3P$; mol wt 184.15. C 39.13%, H 7.66%, F 10.32%, O 26.07%, P 16.82%. Prepd by the action of PCl_3 on isopropanol, chlorinating the resulting intermediate, and converting the diisopropyl chlorophosphate by means of sodium fluoride: Hardy, Kosolapoff, **U.S. pat. 2,409,039** (1946); *see also* Edgewood Arsenal, *Chemical Warfare Service TDMR* **832** (April, 1944); **Brit. pat. 601,210** (1948); Lange, von Krueger, *Ber.* **65,** 1598 (1932); Saunders, Stacey, *J. Chem. Soc.* **1948,** 695; B. C. Saunders, *Some Aspects of the Chemistry and Toxic Action of Organic Compounds Containing Phosphorus and Fluorine* (Cambridge, 1957) p 46. Toxicity data: R. G. Horton *et al., J. Pharmacol. Exp. Ther.* **87,** 414 (1946).

Liquid. Forms HF in presence of moisture. d 1.055. mp $-82°$. bp_5 46°; bp_9 62°; bp_{760} 183° (by extrapolation). Vapor pressure at 20°: 0.579 mm. n_D^{25} 1.3830. Soly in water at 25°: 1.54% w/w (dec; pH about 2.5). Sol in vegetable oils. Not very sol in mineral oils. The anhydr compd or oil solns are stable in glass containers at room temp. LD_{50} in mice (mg/kg): 3.71 s.c.; 36.8 orally (Horton).

Toxicity: Highly toxic; cholinesterase inactivator. Do not inhale vapors. Avoid contact with skin.

THERAP CAT: Cholinergic (ophthalmic).

THERAP CAT (VET): Has been used as a miotic.

5061. L-**Isoglutamine.** *4,5-Diamino-5-oxopentanoic acid;* 4-aminoglutaramic acid; glutamic acid 1-amide. $C_5H_{10}-N_2O_3$; mol wt 146.15. C 41.09%, H 6.90%, N 19.17%, O 32.84%. Prepn by the action of NH_3 on *N*-carbobenzyloxy-L-glutamic acid anhydride: Kozo Narita, *J. Chem. Soc. Japan, Pure Chem. Sect.* **74,** 832 (1953); from 1-tosylpyroglutamide by reduction with Na in NH_3: Swan, du Vigneaud, *J. Am. Chem. Soc.* **76,** 3110 (1954); from the γ-methyl ester of *N*-tritylglutamic acid: Amiard, Heymes, **U.S. pat. 2,927,118** (1960 to UCLAF).

Crystals from acetone, dec 181°. $[\alpha]_D^{21}$ +20.5° (c = 6.1 in H_2O). pK_1' 3.81; pK_2' 7.88. Sol in water. Sparingly sol in organic solvents.

5062. Isoladol. β-*Amino-4-methoxy-α-(4-methoxyphenyl)benzeneethanol; 2-amino-1,2-bis(p-methoxyphenyl)-ethanol;* isodianisylethanolamine; di-*p*-methoxyphenylhydroxyethylamine; Evadol. $C_{16}H_{19}NO_3$; mol wt 273.32. C 70.31%, H 7.01%, N 5.13%, O 17.56%. Prepd by the condensation of anisaldehyde and glycine: Read, Campbell, *J. Chem. Soc.* **1930**, 2676; Lespagnol, *Pharmacie Chimique* (Paris, 3rd ed., 1950) p 695.

dl-Form, glistening prisms from alcohol, mp 135.5°. Sol in ethyl acetate.

dl-Form hydrochloride, *LC 2, Evaclin.* Crystals, dec about 200°. Sol in water.

l-Form, needles from abs alcohol, mp 111-112°. $[\alpha]_D^{20}$ −150° (c = 0.952 in abs alcohol).

l-Form hydrochloride, crystals from ethyl acetate, dec 188-190°. $[\alpha]_D^{20}$ −99.0° (c = 0.5805). Sol in water.

THERAP CAT: Analgesic.

5063. Isolan®. *Dimethylcarbamic acid 3-methyl-1-(1-methylethyl)-1H-pyrazol-5-yl ester;* 1-isopropyl-3-methyl-5-pyrazolyl dimethylcarbamate; G-23611. $C_{10}H_{17}N_3O_2$; mol wt 211.27. C 56.85%, H 8.11%, N 19.89%, O 15.15%. Carbamate insecticide: Gerguson, Alexander, *J. Agr. Food Chem.* **1**, 288 (1953); Muller, Spindler, *Experientia* **10**, 91 (1954). Prepd by treating the K salt of the enol form of 1-isopropyl-3-methyl-5-pyrazolone with dimethylcarbamoyl chloride: **Swiss pats.** 279,553 and 282,655 and **Brit. pat.** 681,376 (all 1952 to Geigy); Kost, Sagitullin, *Zh. Obshch. Khim.* **33**, 867 (1963), *C.A.* **59**, 8724 (1963).

Liquid, $bp_{0.7}$ 103°, $bp_{2.5}$ 117.5-118°. d 1.07. Vapor press. at 20° = 0.001 mm Hg. Unlike most insecticides, Isolan is more toxic to rats by the dermal route than orally: Gaines *et al.*, *Nature* **209**, 88 (1966). LD_{50} in male, female rats: 23, 13 mg/kg orally; 5.6, 6.2 mg/kg dermally, T. B. Gaines, *Toxicol. Appl. Pharmacol.* **14**, 515 (1969).

USE: Insecticide.

5064. Isoleucine. Ile (IUPAC abbrev.); 2-amino-3-methylvaleric acid; α-amino-β-methylvaleric acid; 2-amino-3-methylpentanoic acid; iLeu; Ileu. $C_6H_{13}NO_2$; mol wt 131.17. C 54.94%, H 9.99%, N 10.68%, O 24.39%. CH_3-$CH_2CH(CH_3)CH(NH_2)COOH$. An amino acid classified as essential with respect to its growth effect in rats. Essential component in human nutrition, not synthesized by the human body. Isoln from beet-sugar mother liquors: Ehrlich, *Ber.* **37**, 1809 (1904); from casein, ovalbumin, blood fibrin: Weitzenböck, *Monatsh.* **27**, 831 (1906); from yeast proteins: Abderhalden, Bahn, *Z. Physiol. Chem.* **245**, 246 (1937). Synthesis: van Romburgh, *Rec. Trav. Chim.* **6**, 150 (1887); Bouveault, Locquin, *Bull. Soc. Chim. France* [3] **35**, 965 (1906); Ehrlich, *Ber.* **41**, 1453 (1908). *See also Org. Syn.* **21**, 60 (1941). Optical and enzymatic characterization: Ehrlich, *Ber.* **40**, 2538 (1907); Greenstein *et al.*, *J. Biol. Chem.* **204**, 307 (1953). Production from tiglaldehyde: Livak, Britton, U.S. pat. 2,553,055 (1951 to Dow); from hydantoin: White, U.S. pats. 2,557,920; 2,700,054 (1951, 1955 to Dow). Biosynthesis and metabolism: R. Cahill *et al.*, *Chem. Commun.* **1980**, 419.

L-Form, the natural form. Waxy, shiny, rhombic leaflets from alc. Bitter taste. Sublimes 168-170°. Dec 284°. $[\alpha]_D^{20}$ +11.29° (c = 3); +40.61° (c = 4.6 in 6.1N HCl); +11.09° (c = 3.3 in 0.33N NaOH). $[M]_D$ + 53.5° (5N HCl); +64.2° (glacial acetic acid). Soly in water (g/l): 37.9 at 0°; 41.2 at

25°; 48.2 at 50°; 60.8 at 75°; 82.6 at 100°. Sparingly sol in hot alc (0.13% w/w at 80°), hot acetic acid. Insol in ether.

DL-Form, (±)-*erythro-2-amino-3-methylpentanoic acid.* Glistening rhombic or monoclinic plates from dil alc. Dec 292°. pK_1' 2.32; pK_2' 9.76. Soly in water (g/l): 18.3 at 0°; 22.3 at 25°; 30.3 at 50°; 46.1 at 75°; 78.0 at 100°.

L-allo-Form, α-*Ile;* L(+)-*alloisoleucine,* (+)-*threo-2-amino-3-methylpentanoic acid.* The dextrorotatory diastereoisomer. Waxy leaflets. Sweet taste. Dec 280°. $[\alpha]_D^{20}$ +14.0° (c = 2); +38.1° (c = 2 in 6N HCl). $[M]_D$ +53.1° (5N HCl); +55.7° (glacial acetic acid). pK_1' 2.27; pK_2' 9.62. One part dissolves in 34.2 parts water at 20°; 0.82 part dissolves in 100 parts 80% alcohol at 20°; 1.97 parts dissolve in 100 parts 80% alcohol; 0.19 part dissolves in 100 parts abs alcohol at 90°.

THERAP CAT: Nutrient.

THERAP CAT (VET): An essential dietary amino acid.

5065. Isolysergic Acid. 9,10-*Didehydro-6-methylergoline-8α-carboxylic acid.* $C_{16}H_{16}N_2O_2$; mol wt 268.34. C 71.62%, H 6.01%, N 10.44%, O 11.93%. Parent compd of the ergotinine group of alkaloids from ergot. Prepn: Smith, Timmis, *J. Chem. Soc.* **1936**, 1440; Craig *et al.*, *J. Biol. Chem.* **125**, 289 (1938). Differs from lysergic acid by the α-configuration at C-8. Stereochemical studies: Stenlake, *J. Chem. Soc.* **1955**, 1626; Leemann, Fabbri, *Helv. Chim. Acta* **42**, 2696 (1959).

Dihydrate from water, mp 218° (dec). $[\alpha]_D^{20}$ +281° (pyridine). pKa 3.44, pKb 8.61. More sol in water and pyridine than lysergic acid.

Methyl ester, rods from benzene, mp 170-174°. $[\alpha]_D^{20}$ +179° (c = 0.5 in chloroform).

5066. Isomaltol. 1-(3-*Hydroxy-2-furanyl)ethanone;* 3-hydroxy-2-furyl methyl ketone. $C_6H_6O_3$; mol wt 126.11. C 57.14%, H 4.80%, O 38.06%. Trace constituent of bread. Obtained by the action of enzymes on starch: Backe, *Compt. Rend.* **151**, 78 (1910). Synthesis from α-lactose and a secondary amine: Hodge, Nelson, U.S. pat. 3,054,805 (1962 to USDA). Structure: Hodge, Nelson, *Cereal Chem.* **38**, 207 (1961); Fischer, Hodge, *J. Org. Chem.* **29**, 776 (1964).

Crystals from water or ether, mp 98-103°. uv max (abs methanol): 280 nm ($E_{1cm}^{1\%}$ 1270). Sol in alcohols, acetone, chloroform, benzene, ethyl acetate, hot water. Practically insol in cold water and petr ether.

Benzoate, $C_{13}H_{10}O_4$, crystals from benzene-toluene, mp 100-101°.

p-Nitrophenylhydrazone, $C_{12}H_{11}N_3O_4$, deep red crystals from nitrobenzene, mp 100-101°.

5067. Isometamidium Chloride. 3-*Amino-8-[3-[3-(aminoiminomethyl)phenyl]-1-triazenyl]-5-ethyl-6-phenylphenanthridinium chloride;* 8-[3-(*m-amidinophenyl)-2-triazeno]-3-amino-5-ethyl-6-phenylphenanthridinium chloride;* 7-*m*-amidinophenyldiazoamino-2-amino-10-ethyl-9-phenylphenanthridinium chloride; M & B 4180 A; Samorin. $C_{28}H_{26}ClN_7$; mol wt 496.04. C 67.80%, H 5.28%, Cl 7.15%, N 19.77%. Prepn: Wragg *et al.*, *Nature* **182**, 1005 (1958). Structure: Berg, *ibid.* **188**, 1106 (1960).

Red crystals from aq methanol, dec 244-245°.
THERAP CAT (VET): Antitrypanosomal agent.

5068. Isomethadol. β-[2-(Dimethylamino)-1-methyleth-yl]-α-ethyl-β-phenylbenzeneethanol; 6-(dimethylamino)-5-methyl-4,4-diphenyl-3-hexanol; 1-dimethylamino-2-methyl-3,3-diphenyl-4-hexanol; 3-hydroxy-4,4-diphenyl-4-methyl-5-dimethylaminohexane. $C_{21}H_{29}NO$; mol wt 311.47. C 80.98%, H 9.39%, N 4.50%, O 5.14%. Prepn of α-dl-form by lithium aluminum hydride reduction of dl-isomethadone: May, Mosettig, *J. Org. Chem.* **13**, 663 (1948). Prepn of α-isomers by lithium aluminum hydride reduction of the optical isomers and of β-isomers by sodium-propanol reduction: May, Eddy, *ibid.* **17**, 1210 (1952). Stereochemistry of isomers: Portoghese, Williams, *Tetrahedron Letters* **1966**, 6299.

$$CH_3CH_2CH(OH)\underset{\underset{C_6H_5}{|}}{\overset{\overset{C_6H_5}{|}}{C}}CH(CH_3)CH_2N(CH_3)_2$$

β-dl-Form, prisms from alcohol + water, mp 107-108.5°. Hydrochloride, crystals from alcohol + ether, mp 252-254°.

β-d-Form, slim rods from alcohol + water, mp 94-95°; $[\alpha]_D^{20}$ +13.3° (c = 1.65). Hydrochloride, thin prisms from alcohol + ether, mp 241-243°; $[\alpha]_D^{20}$ +13.5° (c = 1.78).

β-l-Form, mp 93.5-94.5°; $[\alpha]_D^{20}$ -13.8° (c = 1.09). Hydrochloride, 241-243°; $[\alpha]_D^{20}$ -13.5° (c = 1.56).

α-dl-Form, mp 100-102.5°. Hydrochloride, mp 231-233° (sinters at 200°).

α-d-Form, mp 124.5-125.5°; $[\alpha]_D^{20}$ +19.1° (c = 0.63). Hydrochloride hemihydrate, mp 202-204°; $[\alpha]_D^{20}$ +10.1° (c = 1.89).

α-l-Form, prisms from alcohol + water, mp 125-126°; $[\alpha]_D^{20}$ -19.7° (c = 1.43). Hydrochloride hemihydrate, rectangular plates, mp 202-204°; $[\alpha]_D^{20}$ -9.6° (c = 2.93).

5069. Isomethadone. 6-(Dimethylamino)-5-methyl-4,4-diphenyl-3-hexanone; 6-dimethylamino-4,4-diphenyl-5-methyl-3-hexanone; 1-dimethylamino-2-methyl-3,3-diphenyl-4-hexanone; isoamidone; Isoadanone. $C_{21}H_{27}NO$; mol wt 309.43. C 81.51%, H 8.80%, N 4.53%, O 5.17%. Prepn of dl-form: Easton *et al.*, *J. Am. Chem. Soc.* **70**, 76 (1948); Walton *et al.*, *J. Chem. Soc.* **1949**, 648; Larsen, Tullar, **U.S.** pat. 2,773,901 (1956 to Sterling Drug). Prepn of d- and l-forms and resolution of dl-form: Larsen *et al.*, *J. Am. Chem. Soc.* **70**, 4194 (1948); Howe, Sletzinger, *ibid.* **71**, 2935 (1949); Larsen, Tullar, *loc. cit.* Configuration of l-form, the isomer with greater analgesic activity: Beckett *et al.*, *Chem. & Ind. (London)* **1960**, 1418. Pharmacology: Eddy *et al.*, *J. Pharmacol. Exp. Ther.* **98**, 121 (1950); Winter, Flataker, *ibid.* 305.

$$CH_3CH_2CO\underset{\underset{C_6H_5}{|}}{\overset{\overset{C_6H_5}{|}}{C}}CH(CH_3)CH_2N(CH_3)_2$$

dl-Form, slightly yellow, very viscous liq, bp_{12} 215°.

dl-Form hydrobromide, $C_{21}H_{27}NO.HBr$, crystals from water, mp 149-150°.

d-Form, oil, $[\alpha]_D^{25}$ +20.8°.

d-Form hydrochloride, $C_{21}H_{28}ClNO$, mp 231-232°. $[\alpha]_D^{25}$ +70°.

d-Form hydrochloride monohydrate, $C_{21}H_{27}NO.HCl.H_2O$, mp 176-177°. $[\alpha]_D^{25}$ +66°.

l-Form, **Liden.** Oily liq, $bp_{0.6}$ 162-165°. $[\alpha]_D^{25}$ -20° (c = 1.5 in 95% ethanol).

l-Form hydrobromide, $C_{21}H_{28}BrNO$, crystals, mp 217-218°. $[\alpha]_D^{25}$ -59° (c = 1.5).

l-Form hydrochloride, mp 231-233°. $[\alpha]_D^{25}$ -70° (c = 1.5 in water), -90° (methanol). Sol in water, alcohol. pH of 1% aq soln: 5 to 6.5. Solutions and tablets are stable. LD_{50} s.c. in mice: 21 mg/kg (Winter, Flataker).

l-Form hydrochloride monohydrate, mp 173-174°.

Caution: May be habit forming. This is a controlled substance (opiate) listed in the U.S. Code of Federal Regulations, Title 21 Part 1308.12 (1985).

THERAP CAT: Narcotic analgesic.

5070. Isometheptene. N,6-Dimethyl-5-hepten-2-amine; N,1,5-trimethyl-4-hexenylamine; 6-methylamino-2-methyl-heptene; 2-methyl-6-methylamino-2-heptene; methylisooc-tenylamine; methyloctenylamine; Octin; Octon; Octanil. $C_9H_{19}N$; mol wt 141.25. C 76.52%, H 13.56%, N 9.92%. $(CH_3)_2C=CHCH_2CH_2CH(NHCH_3)CH_3$. Prepn: Klavehn, Wolf, **U.S.** pats. 2,230,753-4 (1941 to E. Bilhuber). Toxicity data: Walton *et al.*, *J. Pharmacol. Exp. Ther.* **92**, 214 (1948).

Colorless, oily liq; characteristic amine odor. Strong base. Volatile with steam. d 0.795. bp 176-178°; bp_7 58-59°. n_D^{15} 1.4472. Practically insol in water. Freely sol in alc, ether, acetone, chloroform.

Hydrochloride, $C_9H_{20}ClN$, crystals, very hygroscopic, mp 68-69°. Sol in water, alc. LD_{50} in mice (mg/kg): 17.5 i.v.; 171 s.c. (Walton).

Bitartrate (acid tartrate), $C_{13}H_{25}NO_6$, crystals, mp 78-80° (preliminary sintering). Freely sol in water, alcohol.

Mucate, $C_{24}H_{48}N_2O_8$, bitter crystals, mp 152° (rapid heating). Freely sol in water. Soly in alc about 5 g/100 ml. Almost insol in ether, chloroform. LD_{50} orally in dogs: 148 mg/kg (Walton).

THERAP CAT: Adrenergic.

THERAP CAT (VET): Sympathomimetic. Antispasmodic for gut and urinary tract.

5071. Isoniazid. 4-Pyridinecarboxylic acid hydrazide; isonicotinic acid hydrazide; isonicotinoylhydrazine; isonico-tinylhydrazine; INH; rimitsid; RP 5015; FSR-3; INH-Burg-thal; Cedin (Aerosol); Isocid; Neoxin; Hidrasonil; Ertuban; Antimicina; Hyzyd; Isonex; Unicozyde; Zonazide; Hycozid; Niconyl; Isonicazide; Isonicid; Isonicotan; Tubazid; Tibi-zide; Isobicina; Isozid; Isonilex; Isonindon; Isotebezid; Lani-azid; Nicetal; Nikozid; Nitadon; Nyscozid; Pelazid; Rau-manon; Retozide; RU-EF-Tb; Tebecid; Tisiodrazida; Tizide; Isozyd; Sauterazid; Niplen; TB-Vis; Tekazin; Isidrina; Hydrazid; Nevin; Cotinazin; Dinacrin; Ditubin; Mybasan; Neoteben; Niadrin; Nicozide; Nydrazid; Nidaton; Nicizina; Nicotibina; Pycazide; Pyricidin; Isolyn; Pyrizidin; Rimifon; Robisellin; Isonizide; Neumandin; Isocotin; Tubicon; Tyvid; Tisin; Tibinide; Tubilysin; Tubomel; Tubeco; Atcotibine; Vazadrine; Vederon; Isonidrin; Zinadon. $C_6H_7N_3O$; mol wt 137.15. C 52.54%, H 5.15%, N 30.64%, O 11.67%. Prepn: Meyer, Mally, *Monatsh.* **33**, 400 (1912); Lock, *Pharm. Ind.* **14**, 366 (1952); Urbanski *et al.*, *Rocz. Chem.* **27**, 161 (1953); Gasson, **U.S.** pat. 2,830,994 (1958 to Distillers). Compositions for combating tuberculosis: H. H. Fox, **U.S.** pat. 2,-596,069 (1952 to Hoffmann-La Roche). Pharmacokinetics: W. W. Weber, D. W. Hein, *Clin. Pharmacokinet.* **4**, 401 (1979). Toxicity data: E. H. Jenney, C. C. Pfeiffer, *J. Pharmacol. Exp. Ther.* **122**, 110 (1958). Review of carcinogenicity studies: *IARC Monographs* **4**, 159 (1974). Comprehensive description: G. A. Brewer in *Analytical Profiles of Drug Substances* vol. 6, K. Florey, Ed. (Academic Press, New York, 1977) pp 183-258. *Review:* Krishna Murti in *Antibiotics* vol. 3, J. W. Corcoran, F. E. Hahn, Eds. (Springer-Verlag, New York, 1975) pp 623-652; K. Takayama, L. A. Davidson, *ibid.* vol. 5(pt. 1), F. E. Hahn, Ed. (1979) pp 98-119.

Crystals from alc, mp 171.4°. uv max (water): 266 nm ($E_{1cm}^{1\%}$ 378); (0.01N HCl): 265 nm ($E_{1cm}^{1\%}$ ~420). Soly in water at 25°: about 14%; at 40°: about 26%; in alc at 25°: about 2%; in boiling alc: about 10%; in chloroform: about 0.1%. Practically insol in ether, benzene. pH of a 1% aq soln 5.5 to 6.5. Aqueous solns may be sterilized at 120° for 30 min. LD_{50} in mice (mg/kg): 151 i.p., 149 i.v. (Jenney, Pfeiffer).

THERAP CAT: Antibacterial (tuberculostatic).

THERAP CAT (VET): Antituberculous and anti-actinomycotic agent.

5072. Isoniazid Methanesulfonate. *4-Pyridinecarboxylic acid 2-(sulfomethyl)hydrazide; isonicotinic acid 2-sulfomethylhydrazide;* isoniazid mesylate; isonicotinic acid hydrazide methanesulfonic acid. $C_7H_9N_3O_4S$; mol wt 231.24. C 36.36%, H 3.92%, N 18.17%, O 27.68%, S 13.87%. Prepn: Logemann, U.S. pat. 2,759,944 (1956 to Carlo Erba).

CONHNHCH₂SO₃H

Crystals, dec 187-189°.

Sodium salt, $C_7H_8N_3NaO_4S$, *sodium isonicotinylhydrazide methanesulfonate, IHMS, Neo-Iscotin, Neo-Tizide (amp.).* Yellow crystals from water, dec 164-167°.

Calcium salt, $C_{14}H_{16}CaN_6O_8S_2$, *Neo-Tizide (tabl.).* Crystals, dec 215-220°.

THERAP CAT: Antibacterial (tuberculostatic).

5073. Isonicotinic Acid. *4-Pyridinecarboxylic acid;* γ-pyridinecarboxylic acid; γ-picolinic acid. $C_6H_5NO_2$; mol wt 123.11. C 58.53%, H 4.09%, N 11.38%, O 25.99%. Prepd by permanganate oxidation of γ-picoline: Gilman, Broadbent, *J. Am. Chem. Soc.* **70**, 2757 (1948); Fields, U.S. pat. 2,946,-801 (1960 to Standard Oil); from 4-ethylpyridine: Wibaut, Arens, *Rec. Trav. Chim.* **60**, 137 (1941); from citric acid to citrazinic acid to 2,6-dichloroisonicotinic acid: Behrmann, Hofmann, *Ber.* **17**, 2681 (1884); Wibaut, *Rec. Trav. Chim.* **63**, 141 (1944). Prepn of ethyl ester: Burrus, Powell, *J. Am. Chem. Soc.* **67**, 1469 (1945).

COOH

Platelets, mp 319°. Sublimes at 260° and 15 mm pressure. K at 25° = 1.1 × 10⁻⁵. pH of satd aq soln 3.6. Sparingly sol in cold water (0.52 g/100 ml at 20°), more sol in hot water. Practically insol in benzene, ether, boiling alcohol.

Methyl ester, $C_7H_7NO_2$, liq, slight odor, similar to mint or oil of wintergreen. mp 8.5°. bp_{21} 104°; bp_{760} 209° (slight decompn).

Ethyl ester, liq, ester-like odor. Needles, when cooled by salt-ice mixture, mp 23°. d_4^{15} 1.0091 (liq). bp_5 78.5°; bp_{15} 110°; bp_{760} 220°. Sol in alcohol, ether, chloroform, benzene. Practically insol in water.

5074. Isonicotinic Acid Diethylamide. *N,N-Diethylisonicotinamide;* pyridine-4-carboxylic acid diethylamide. $C_{10}H_{14}N_2O$; mol wt 178.23. C 67.39%, H 7.92%, N 15.72%, O 8.98%. Prepd by the action of phosphorus oxychloride upon a mixture of isonicotinic acid and diethylamine: Carrara, U.S. pat. 2,858,317 (1958 to Lepetit).

CON(C₂H₅)₂

Slightly viscous liq. $bp_{1.0}$ 119-120°. n_D^{20} 1.525. Miscible with water, ether, chloroform, acetone, alcohol.

5075. Isonipecotic Acid. *4-Piperidinecarboxylic acid;* hexahydroisonicotinic acid. $C_6H_{11}NO_2$; mol wt 129.16. C 55.79%, H 8.58%, N 10.85%, O 24.78%. Prepd by reduction of isonicotinic acid in glacial acetic acid in the presence of platinum oxide: Wibaut, *Rec. Trav. Chim.* **63**, 141 (1944); by reduction of isonicotinic acid in H_2O and concd HCl with PtO_2: Sperber *et al.,* U.S. pat. 2,739,968 (1956 to Schering); Freifelder, U.S. pat. 3,159,639 (1964 to Abbott).

COOH

Needles. Darkens at about 300°; mp 336°.

Hydrochloride, $C_6H_{11}NO_2.HCl$, mp 300° (Wibaut); mp 293° with decompn (Sperber). Sol in ethanol and methanol; freely sol in water.

5076. Isonitrosoacetone. *2-Oxopropanal 1-oxime;* pyruvaldoxime; propanone 1-oxime. $C_3H_5NO_2$; mol wt 87.08. C 41.38%, H 5.79%, N 16.09%, O 36.75%. CH_3CO-CH=NOH. Prepd from acetone by treatment with sodium nitrite in acetic acid at 0°: Küster, *Z. Physiol. Chem.* **155**, 174 (1926); by the action of methyl nitrite on acetone in ether in the presence of HCl: Slater, *J. Chem. Soc.* **117**, 589 (1920).

Leaflets from ether + petr ether or from carbon tetrachloride, mp 69°. Sublimes easily forming shiny needles. Volatile with steam. Freely sol in water, ether. Moderately sol in warm chloroform, benzene, carbon tetrachloride. Practically insol in petr ether. Sol in alkalies forming yellow solns. Faintly acid to litmus. K at 25° = 4.1 × 10⁻⁹.

5077. Isonitrosoacetophenone. α-Oxobenzeneacetaldehyde aldoxime; benzoylformaldoxime. $C_8H_7NO_2$; mol wt 149.14. C 64.42%, H 4.73%, N 9.39%, O 21.45%. C_6H_5-COCH=NOH.

Plates or prisms, mp 126-128°. Slightly sol in cold water; more sol in hot water; sol in alkalies and alkali solutions.

USE: As a reagent for the detection of ferrous ions with which it gives a blue color sol in chloroform.

5078. Isonixin. *N-(2,6-Dimethylphenyl)-1,2-dihydro-2-oxo-3-pyridinecarboxamide;* 2-pyridone-3-carboxylic acid 2,6-xylidide; 2-hydroxy-2',6'-nicotinoxylidide; Nixyn. $C_{14}H_{14}N_2O_2$; mol wt 274.27. C 61.31% H 5.14%, N 10.21%, O 23.33%. Non-steroidal anti-inflammatory agent. Prepn: J.-A. Canicio, **Belg.** pat. 820,578; *idem,* U.S. pat. 3,984,423 (1975, 1976 both to Hermes). Prepn, pharmacology and toxicity: R. Cadena *et al., Arzneimittel-Forsch.* **27**, 1457 (1977). Toxicological study: M. T. Mitjavila *et al., ibid.* 1460. Absorption, excretion studies: G. Carrera *et al., ibid.* **29**, 1401 (1979). Tolerance in humans: T. M. Serra *et al., Therapie* **35**, 173 (1980). HPLC determn in plasma: F. González Lopez *et al., Farmaco, Ed. Prat.* **38**, 273 (1983).

White crystalline powder from ethanol, mp 266-267°. Sol in chloroform. Practically insol in water. LD_{50} in male, female mice, male, female rats (mg/kg): 7000, 8000, >6000, >6000 orally; all >2000 i.p. (Cadena).

THERAP CAT: Analgesic, anti-inflammatory.

5079. Isooctane. *2,2,4-Trimethylpentane;* isobutyltrimethylmethane. C_8H_{18}; mol wt 114.22. C 84.12%, H 15.88%. $(CH_3)_3CCH_2CH(CH_3)_2$. Produced by the refining of petroleum.

Highly flammable, mobile liquid. Odor of gasoline. Antiknock octane no. 100. mp −107.45°. bp 99.3°. d_4^{20} 0.69194.

n_D^{20} 1.39157. Dipole moment: 0. Flash pt, closed cup: 10°F (−12°C). Practically insol in water; somewhat sol in abs alcohols; sol in benzene, toluene, xylene, chloroform, ether, carbon disulfide, carbon tetrachloride, DMF and oils, except castor oil.

Note: The name isooctane has also been applied to 2-methylheptane.

USE: In determining octane numbers of fuels; in spectrophotometric analysis; as solvent and thinner.

5080. Isooctyl Alcohol. $C_7H_{15}CH_2OH$. A mixture of closely related isomeric branched-chain primary alcohols: RCH_2OH where R represents a branched heptyl radical. The branching consists mostly of methyl groups located in the 3-, 4-, or 5-positions.

5081. Isopentyl Alcohol. *3-Methyl-1-butanol;* isoamyl alcohol; isobutyl carbinol; primary isoamyl alcohol; fermentation amyl alcohol. $C_5H_{12}O$; mol wt 88.15. C 68.13%, H 13.72%, O 18.15%. $(CH_3)_2CHCH_2CH_2OH$. A major component of commercial amyl alcohol, fusel oil *q.v.*, and *potato-spirit oil.* Prepn from butadiene, CO, H_2O, plus catalyst: Alderson, U.S. pat. 3,020,314 (1962 to du Pont); from methylbutenes: Brown, Zweifel, *J. Am. Chem. Soc.* **82**, 1504 (1960). Early review of metabolism and toxicity: H. W. Haggard *et al., J. Ind. Hyg. Toxicol.* **27**, 1 (1945). Review of manuf by fractionation of fusel oil and via chlorination of pentanes, and properties: W. L. Faith *et al.,* Eds., *Industrial Chemicals* (John Wiley, New York, 2nd ed., 1957) pp 107-114. Acute toxicity data: Smyth *et al., Am. Ind. Hyg. Assoc. J.* **30**, 470 (1969).

Liquid; characteristic, disagreeable odor; pungent, repulsive taste. mp −117.2°. d_4^{15} 0.813. n_D^{20} 1.4075. Flash pt, closed cup: 114°F (45°C); open cup: 132°F (55°C). Slightly sol in water (2 g/100 ml at 14°); misc with alcohol, ether, benzene, chloroform, petr ether, glacial acetic acid, oils. LD_{50} orally in rats: 7.07 ml/kg (Smyth).

USE: Solvent for fats, resins, alkaloids, etc.; manuf isoamyl (amyl) compds, isovaleric acid, mercury fulminate, pyroxylin, artificial silk, lacquers, smokeless powders; in microscopy; for dehydrating celloidin solns; for determining fat in milk. *Caution:* May be moderately irritating to mucous membranes. High concns may cause CNS depression, narcosis; lower concns, headache, dizziness: *Patty's Industrial Hygiene and Toxicology* **vol. 2C**, G. D. Clayton, F. E. Clayton, Eds. (Wiley-Interscience, New York, 3rd ed., 1982) pp 4588-4599.

5082. Isophane Insulin Suspension. Humulin I; Insulatard; Isophane Insulin; Isophane Insulin Injection; NPH Insulin; NPH Iletin. Sterile suspension, in a buffered water medium, of insulin made from zinc-insulin crystals modified by the addition of protamine sulfate in a manner such that the solid phase of the suspension consists of crystals composed of insulin, protamine, and zinc. The protamine sulfate is prepared from the sperm or from the mature testes of fish belonging to the genus *Oncorhynchus* Suckley, or *Salmo* Linné, *Salmonidae.* The designation NPH is derived as follows: N refers to the prepn being neutral; P stands for protamine, of which about 0.5 mg per 100 units is present (the designation NPH-50 has been used); H refers to Hagedorn, in whose laboratory the preparation was developed. Contains 40, 80 or 100 U.S.P. insulin units/ml.

White suspension of rod-shaped crystals approx 30μ in length, contg not more than traces of amorphous material. pH 7.1-7.4. Contains either *(1)* 1.4-1.8% (w/v) of glycerin, 0.15-0.17% (w/v) of metacresol, and 0.06-0.07% (w/v) of phenol or *(2)* 0.42-0.45% (w/v) of sodium chloride, 0.7-0.9% (w/v) of glycerol, and 0.18-0.22% (w/v) of metacresol. Contains 0.15-0.25% (w/v) of dibasic sodium phosphate. Also contains 0.016-0.04 mg of zinc and 0.3-0.6 mg of protamine per 100 U.S.P. insulin units. Onset of action occurs within 2 hrs after s.c. injection, reaching its maximum in 10-20 hrs. Duration of action is 20-32 hrs.

THERAP CAT: Antidiabetic.

5083. Isophthalic Acid. *1,3-Benzenedicarboxylic acid;* *m*-phthalic acid. $C_8H_6O_4$; mol wt 166.13. C 57.83%, H 3.64%, O 38.52%. Obtained by the oxidation of *m*-xylene: Smith, *J. Am. Chem. Soc.* **43**, 1920 (1921); Weisemann, Fragen, *World Petrol. Congr., Proc. 5th, N.Y. 1959* **4**, 197

(1960); Brill, *Ind. Eng. Chem.* **52**, 837 (1960); Bhattacharyya, Ganguly, *J. Indian Chem. Soc.* **38**, 463 (1961); Saffer, Barker, U.S. pat. 3,089,906 (1963 to Mid-Century Corp.); Hay, **Brit.** pat. 951,192 (1964 to General Electric).

Cryst powder, mp 345-348°; sublimes without decompn. Sol in 8000 parts cold water, 460 boiling water; freely sol in alcohol, glacial acetic acid; practically insol in benzene, petr ether.

5084. Isophytol. *3,7,11,15-Tetramethyl-1-hexadecen-3-ol;* 2,6,10,14-tetramethylhexadec-15-en-14-ol; 2,6,10-trimethyl-14-vinylpentadecan-14-ol. $C_{20}H_{40}O$; mol wt 296.52. C 81.01%, H 13.60%, O 5.40%. Decompn product of chlorophyll. Synthesis from linalool or citral: Fischer, Löwenberg, *Ann.* **475**, 183 (1929); Karrer *et al., Helv. Chim. Acta* **26**, 1741 (1943); Sarycheva *et al., Zh. Obshch. Khim.* **28**, 647 (1958); Maurit *et al., ibid.* **32**, 2483 (1962); from acetylene: Nazarov *et al., ibid.* **28**, 1444 (1958); Blaha, Weichet, **Czech.** pat. 88,887 (1959), *C.A.* **54**, 2167c (1960); from pseudoionone and propargyl alcohol: Sato *et al., J. Org. Chem.* **28**, 45 (1963).

Oily liquid. d_4^{20} 0.8519. n_D^{20} 1.4571. $bp_{0.01}$ 107-110°; $bp_{0.06}$ 125-128°. Practically insol in water; sol in the usual organic solvents.

USE: Prepn of vitamins E and K_1.

5085. Isopilosine. *Dihydro-3-(hydroxyphenylmethyl)-4-[(1-methyl-1H-imidazol-5-yl)methyl]-2(3H)-furanone;* carpiline; carpidine. $C_{16}H_{18}N_2O_3$; mol wt 286.32. C 67.11%, H 6.34%, N 9.78%, O 16.76%. This compound was originally called pilosine (the *cis*-isomer of isopilosine): H. W. Voigtländer, W. Rosenberg, *Arch. Pharm.* **292**, 579 (1959). Isoln from leaves of *Pilocarpus microphyllus* Stapf, *Rutaceae:* F. L. Pyman, *J. Chem. Soc.* **101**, 2260 (1912). Synthesis and abs config: Link, Bernauer, *Helv. Chim. Acta* **55**, 1053 (1972). Crystal structure: W. E. Oberhaensli, *Cryst. Struct. Commun.* **1**, 203 (1972).

Needles from alcohol, mp 182-182.5°. $[\alpha]_D^{20}$ +37.6° (alcohol).

Pilosine, needles from alcohol, mp 179°. $[\alpha]_D^{20}$ +83.9° (alcohol). uv max (alcohol): 210 nm (log ε 4.10).

5086. Isopimaric Acid. *7-Ethenyl-1,2,3,4,4a,4b,5,6,7,8,-10,10a-dodecahydro-1,4a,7-trimethyl-1-phenanthrenecarboxylic acid; 13β-methyl-13-vinylpodocarp-7-ene-15-oic acid;* miropinic acid. $C_{20}H_{30}O_2$; mol wt 302.44. C 79.42%, H 10.00%, O 10.58%. Isoln from *Dacrydium biforme* Pilg., *Coniferae:* Hosking, Brandt, *Ber.* **68**, 1311 (1935); from the bled resin of *Podocarpus ferrugineus, Taxaceae:* Brandt, Neubauer, *J. Chem. Soc.* **1940**, 683; from *Pimus palustris* Mill., *Pinaceae:* Harris, Sanderson, *J. Am. Chem. Soc.* **70**, 2079, 2081 (1948). Identity of isopimaric and miropinic acids: Brossi, Jeger, *Helv. Chim. Acta* **33**, 722 (1950). Structure: Antkowiak *et al., J. Org. Chem.* **27**, 1930 (1962). Stereochemistry: Ireland, Newbould, *ibid.* 1931; Antkowiak *et al., Can. J. Chem.* **43**, 1257 (1965); Bose *et al., Indian J. Chem.* **5**, 228 (1967).

Consult the cross index before using this section.

Needles from methanol or ethanol, mp 160°. $[\alpha]_D^{16}$ −3.6° (10.4% soln in 1:1 alcohol : chloroform). Freely sol in chloroform, benzene; moderately sol in ether, alcohol; slightly sol in light petroleum. Absorption spectrum: Brossi, Jeger, *loc. cit.*

5087. Isoprene. *2-Methyl-1,3-butadiene.* C_5H_8; mol wt 68.11. C 88.17%, H 11.83%. Isoln from the products of the pyrolysis of natural rubber: Williams, *Trans. Roy. Soc. London* **150**, 241 (1860); Boonstra, van Amerongen, *Ind. Eng. Chem.* **41**, 161 (1949). Structure: Euler, *Ber.* **30**, 1989 (1897). Prepn from turpentine: Tilden, *J. Chem. Soc.* **45**, 410 (1884); Bibb, U.S. pat. **2,386,537** (1945 to Newport Ind.); from terpenes: Davis *et al., Ind. Eng. Chem.* **38**, 53 (1946); Bourbon, U.S. pat. **2,547,684** (1951 to Manufacture de Caoutchouc Michelin); by condensation of isobutylene and formaldehyde in acetic acid: Blomquist, Verdol, *J. Am. Chem. Soc.* **77**, 78 (1955); Chaffe *et al., Fr.* pat. **1,294,716** (1962 to Inst. Francais du Petrole); by dehydrogenation of isoamylenes from cracked gasolines: **Brit.** pat. **875,346** (Appl. Apr. 21, 1960 to Shell Int. Res. Maatschappij); Voge *et al.,* U.S. pat. **3,110,746** (1963 to Shell); by dimerization of propylene followed by pyrolysis of the resulting 2-methyl-2-pentene: Gorin, Oblad, U.S. pat. **2,404,056** (1946 to Socony-Vacuum Oil); **Brit.** pat. **913,852** (1962 to Goodyear); by dehydrogenation of isopentane: Dempsey, U.S. pat. **2,914,-588** (1959 to Sun Oil); Owen, U.S. pat. **2,982,795** (1961 to Phillips Petroleum); Stevenson, U.S. pat. **3,088,986** (1963 to Air Products & Chem.); by reacting isoamylenes with methanol and pyrolizing and dehydrogenating the resulting tertiary ethers: Verdol, Walker, U.S. pat. **2,972,645** (1961 to Sinclair Refining). Reviews of isoprene and its polymers: Bean *et al.,* "Isoprene Polymers" in *Encyclopedia of Polymer Science and Technology* **vol. 7** (Interscience, New York, 1967) pp 782-855; Bailey, "Isoprene" in *Vinyl and Diene Monomers* (part 2), E. C. Leonard, Ed. (Wiley-Interscience, New York, 1971) pp 997-1148; W. M. Saltman in Kirk-Othmer *Encyclopedia of Chemical Technology* **vol. 13** (Wiley-Interscience, New York, 3rd ed., 1981) pp 818-837. Toxicity: Gostinskii, *C.A.* **62**, 15338a (1965).

$$CH_2{=}\underset{\underset{CH_3}{|}}{C}{-}CH{=}CH_2$$

Unstable, oxidizable liquid. bp_{760} 34.067. mp −145.95°. d_4^{20} 0.681; d_{20}^{20} 0.6805. n_D^{20} 1.42160. Practically insol in water; miscible with alcohol or ether. LD_{50} for mice: 144 mg isoprene vapors/l air (Gostinskii).

USE: Manuf "synthetic" natural rubber. Manuf butyl rubber. Copolymer in the production of synthetic elastomers. *Caution:* Irritating to skin, mucous membranes. Narcotic in high concns.

5088. Isopromethazine. *N,N,β-Trimethyl-10H-phenothiazine-10-ethanamine;* *10-(2-dimethylamino-1-methylethyl)-phenothiazine;* 10-(2-dimethylaminoisopropyl)phenothiazine; isomethazine; Lilly 01526; RP 4460; Isophenergan; Fen-Bridal. $C_{17}H_{20}N_2S$; mol wt 284.41. C 71.79%, H 7.09%, N 9.85%, S 11.27%. Prepn: Charpentier, *Compt. Rend.* **225**, 306 (1947); Charpentier, Ducrot, *ibid.* **232**, 415 (1951); Charpentier *et al., ibid.* 2232; Berg, Ashley, U.S. pat. **2,607,773** (1952 to Rhône-Poulenc); Edge, Wragg, *J. Pharm. Pharmacol.* **5**, 279 (1953).

Hydrochloride, $C_{17}H_{21}ClN_2S$, *RP 3389, diprazin.* Crystals, mp 193-194°. More sol than promethazine.

THERAP CAT: Antihistaminic.

5089. Isopropalin. *4-(1-Methylethyl)-2,6-dinitro-N,N-dipropylbenzenamine;* *2,6-dinitro-N,N-dipropylcumidine;* 4-isopropyl-2,6-dinitro-N,N-dipropylaniline; EL-179; Paarlan. $C_{15}H_{23}N_3O_4$; mol wt 309.37. C 58.24%, H 7.49%, N 13.58%, O 20.69%. Selective pre-plant herbicide. Prepn: Q. F. Soper, U.S. pat. **3,257,190** (1966 to Lilly). Persistence in soil: T. Golab, W. A. Althaus, *Weed Sci.* **23**, 165 (1975); L. L. Gingerich, R. L. Zimdahl, *ibid.* **24**, 431 (1976); R. R. Romanowski, A. W. Libik, *ibid.* **26**, 258 (1978).

Red-orange liquid. Soly in water at 25°: 0.1 mg/l. Readily sol in organic solvents. Decomp by uv light. LD_{50} orally in rats: 5000 mg/kg, *RTECS* **Vol. I**, R. J. Lewis, R. L. Tatken, Eds. (1980) p 163.

USE: Herbicide.

5090. Isopropamide. *γ-(Aminocarbonyl)-N-methyl-N,N-bis(1-methylethyl)-γ-phenylbenzenepropanaminium iodide;* (3-carbamoyl-3,3-diphenylpropyl)diisopropylmethylammonium iodide; 2,2-diphenyl-4-diisopropylaminobutyramide methiodide; R 79; Darbid; Priamide; Tyrimide. $C_{23}H_{33}IN_2O$; mol wt 480.42. C 57.50%, H 6.92%, I 26.42%, N 5.83%, O 3.33%. Prepn: Janssen *et al., Arch. Int. Pharmacodyn.* **103**, 82 (1955). Crystal and molecular structure: N. Datta *et al., J. Chem. Soc., Perkin Trans. II* **1977**, 781. Comprehensive description: R. S. Santoro *et al.,* in *Analytical Profiles of Drug Substances* **vol. 2**, K. Florey, Ed. (Academic Press, New York, 1973) pp 315-338; A. Post, R. S. Santoro, *ibid.* **vol. 12** (1983) pp 721-732.

$$\left[\;H_2NCOCCH_2CH_2\underset{\underset{C_6H_5}{|}}{\overset{\overset{C_6H_5}{|}}{N}}CH_3\;\overset{\overset{CH(CH_3)_2}{|}}{\underset{\underset{CH(CH_3)_2}{|}}{}}\;\right]I^-$$

Crystals or amorphous powder, mp 198-201° (dec). Sensitive to light. Free base, $C_{22}H_{30}N_2O$, mp 84-86°. The methiodide is freely sol in boiling water, in methanol, ethanol, chloroform. Practically insol in ether.

Note: Ingredient of *Stelabid* (also contg trifluoperazine).

THERAP CAT: Anticholinergic.

5091. Isopropenyl Acetate. *1-Propen-2-ol acetate;* 1-propen-2-yl acetate. $C_5H_8O_2$; mol wt 100.11. C 59.98%, H 8.05%, O 31.96%. Prepn from ketene + acetone: Hull, Agett, Hagemeyer, U.S. pats. **2,481,669**; **2,476,860** (both 1949 to Eastman Kodak); Young, U.S. pats. **2,461,016**; **2,511,423** (1949, 1950 both to Carbide & Carbon Chem.); Mawer, U.S. pat. **2,684,980** (1954 to I.C.I.); Buttner, Enk, U.S. pat. **2,867,653** (1959 to Wacker-Chemie).

$$CH_2{=}\underset{\underset{CH_3}{|}}{C}{-}\overset{\overset{OCOCH_3}{|}}{}{-}CH_3$$

Liquid. bp 97°; bp_{200} 58-60°. n_D^{20} 1.4001. LD_{50} orally in rats: 3.0 g/kg, Smyth *et al., J. Ind. Hyg. Toxicol.* **31**, 60 (1949).

USE: Reagent for acylation of potential enols: Hagemeyer, Hull, *Ind. Eng. Chem.* **41**, 2920 (1949); Hull, Agett, U.S. pat. **2,482,066** (1949 to Eastman Kodak); Jeffery, Satchell, *J. Chem. Soc.* **1962**, 1876, 1906. *Caution:* Mild irritant. Narcotic in high concns.

5092. 4-(5-Isopropenyl-2-methyl-1-cyclopenten-1-yl)-2-butanone. *4-[2-Methyl-5-(1-methylethenyl)-1-cyclopenten-1-yl]-2-butanone;* 4-(2-methyl-5-isopropenyl-1-cyclopenten-1-yl)-2-butanone; Pentione. $C_{13}H_{20}O$; mol wt 192.29. C 81.20%, H 10.48%, O 8.32%. Prepn: Kimel, Sax, U.S. pat. **2,799,706** (1957 to Hoffmann-La Roche).

Pale yellow liquid. Citrus-like odor with woody background. d_{25}^{25} 0.9218. $bp_{0.5}$ 72-75°. n_D^{25} 1.4800. Freely sol in diethyl phthalate. At least 10% soly in 70% ethanol, in mineral oil and in corn oil.

USE: In perfumery, i.e., in milled soaps, in modified eau de Cologne, in neroli and orange-blossom compositions. Blends well with ionones.

5093. Isopropyl Acetate. $C_5H_{10}O_2$; mol wt 102.13. C 58.80%, H 9.87%, O 31.33%. $CH_3COOCH(CH_3)_2$.
Colorless liquid. d_4^{20} 0.870. bp 89°. Flash pt, open cup: 40°F (4°C); closed cup: 36°F (2°C). Sol in 23 parts water at 27°; miscible with alc, ether. LD_{50} orally in rats: 6.75 g/kg, Smyth et al., Arch. Ind. Hyg. Occup. Med. **10,** 61 (1954).
USE: Solvent for cellulose derivatives, plastics, oils and fats; in perfumery.

5094. Isopropyl Acetoacetate. *3-Oxobutanoic acid 1-methylethyl ester.* $C_7H_{12}O_3$; mol wt 144.17. C 58.31%, H 8.39%, O 33.29%. $CH_3COCH_2COOC_3H_7$.
Liquid. d_{25}^{25} 0.957. bp about 205° with decompn. Slightly sol in water, freely in alcohol, ether.

5095. Isopropylacetone. *4-Methyl-2-pentanone;* methyl isobutyl ketone; hexone. $C_6H_{12}O$; mol wt 100.16. C 71.94%, H 12.08%, O 15.97%. $CH_3COCH_2CH(CH_3)_2$. Manuf: Faith, Keyes & Clark's *Industrial Chemicals,* F. A. Lowenheim, M. K. Moran, Eds. (Wiley-Interscience, New York, 4th ed., 1975) pp 543-546. Toxicity data: Smyth et al., Arch. Ind. Hyg. Occup. Med. **4,** 119 (1951).
Colorless liquid; faint, ketonic and camphor odor. d_4^{20} 0.801. bp 117-118°. n_D^{20} 1.396. Flash pt, closed cup: 73°F (23°C). mp −84.7°. Moderately sol in water (1.91%); misc with alcohol, benzene, ether. LD_{50} orally in rats: 2.08 g/kg (Smyth).
USE: Solvent for gums, resins, nitrocellulose, etc.

5096. Isopropyl Alcohol. *2-Propanol;* isopropanol; secondary propyl alcohol; dimethyl carbinol; petrohol. C_3H_8O; mol wt 60.09. C 59.96%, H 13.42%, O 26.62%. CH_3-CHOHCH$_3$. Manuf from propylene: Faith, Keyes & Clark's *Industrial Chemicals,* F. A. Lowenheim, M. K. Moran, Eds. (Wiley-Interscience, New York, 4th ed., 1975) pp 496-501. Monograph: L. F. Hatch, *Isopropyl Alcohol* (McGraw-Hill, New York, 1961) 184 pp. Toxicity: Smyth, Carpenter, J. Ind. Hyg. Toxicol. **30,** 63 (1948).
Flammable liq. Slight odor resembling that of a mixture of ethanol and acetone. Slightly bitter taste (*not potable!*). mp −88.5°; fp −89.5°. bp_{760} 82.5°; bp_{400} 67.8°; bp_{200} 53.0°; bp_{100} 39.5°; bp_{60} 30.5°; bp_{40} 23.8°, bp_{20} 12.7°; bp_{10} +2.4°; bp_5 −7.0°; $bp_{1.0}$ −26.1°. d_4^{20} 0.78505; d_4^{25} 0.78084; d_4^{83} 0.728. Flash pt, closed cup: +11.7° (53°F). Autoignition temp 455.6° (852°F). Lower explosive limit in air: 2.5% (v/v). n_D^8 1.3852; n_D^{15} 1.3802; n_D^{20} 1.37723; n_D^{25} 1.3749. Absorption spectrum: Brode, J. Phys. Chem. **30,** 61 (1926). Miscible with water, alcohol, ether, chloroform. Insol in salt solns. May be recovered from aq mixtures by salting out with sodium chloride, sodium sulfate, sodium hydroxide, etc. Forms an azeotrope with water (bp_{760} 80.37°, d_4^0 0.83361) isopropanol 87.7% (w/w). Freezing points of mixtures with water (v/v, per cent by volume): Isopropanol 15% = −3.3°; 25% = −6.7°; 30% = −8.3°; 35% = −11.1°; 40% = −13°; 45% = −17.8°; 50% = −23°; 60% = −32°. LD_{50} orally in rats: 5.8 g/kg (Smyth, Carpenter).
Human Toxicity: Ingestion or inhalation of large quantities of the vapor may cause flushing, headache, dizziness, mental depression, nausea, vomiting, narcosis, anesthesia, coma. 100 ml can be fatal. See: E. Browning, *Toxicity and Metabolism of Industrial Solvents* (Elsevier, New York, 1965) pp 335-341.
USE: In antifreeze compositions; as solvent for gums, shellac, essential oils; in the extraction of alkaloids; in quick-

drying oils; in quick-drying inks; in denaturing ethyl alcohol; in body rubs; hand lotions, after-shave lotions and similar cosmetics. Solvent for creosote, resins, gums; in manuf of acetone, glycerol, isopropyl acetate. Pharmaceutic aid (solvent).
THERAP CAT: Antiseptic.
THERAP CAT (VET): Antiseptic, rubefacient.

5097. Isopropylamine. *2-Propanamine;* 2-aminopropane. C_3H_9N; mol wt 59.08. C 60.95%, H 15.35%, N 23.70%. $(CH_3)_2CHNH_2$. Prepn from acetone + NH_3: Skita, Keil, Ber. **61,** 1682 (1928); Norton et al., J. Org. Chem. **19,** 1054 (1954); from acetone oxime: Winans, Adkins, J. Am. Chem. Soc. **55,** 2056 (1933).
Colorless, flammable liquid; ammonia odor; strong base. d_4^{15} 0.694. mp −101°. bp 33-34°. n_D^{15} 1.3770. Flash pt, open cup: −15°F (−26°C). Miscible with water, alcohol, ether. LD_{50} orally in rats: 820 mg/kg, Smyth et al., Arch. Ind. Hyg. Occup. Med. **4,** 119 (1951).

5098. Isopropyl Bromide. *2-Bromopropane.* C_3H_7Br; mol wt 123.00. C 29.29%, H 5.74%, Br 64.97%. CH_3CHBr-CH_3. Obtained by heating isopropyl alcohol with HBr.
Colorless liquid. d_4^{20} 1.31. mp −89°. bp 59-60°. n_D^{20} 1.4251. Slightly sol in water; miscible with alcohol, benzene, chloroform, ether.

5099. Isopropyl Chloride. *2-Chloropropane.* C_3H_7Cl; mol wt 78.54. C 45.87%, H 8.98%, Cl 45.15%. CH_3CHCl-CH_3. Prepd by refluxing isopropyl alcohol with concd HCl and $ZnCl_2$.
Liquid. d^{15} 0.868. mp −117°. bp 35-36°. Flash pt, closed cup: −26°F (−32°C). Slightly sol in water; miscible with alcohol, ether.

5100. Isopropyl Ether. *2,2'-Oxybis[propane]; 2-isopropoxypropane;* diisopropyl ether. $C_6H_{14}O$; mol wt 102.17. C 70.53%, H 13.81%, O 15.66%. $(CH_3)_2CHOCH(CH_3)_2$. Usually contains p-benzylaminophenol as stabilizer.
Liquid. d_4^{20} 0.7258. mp −60°. bp 68-69°. n_D^{23} 1.3678. Flash pt, open cup: 15°F (−9°C). Slightly sol in water (0.2% at 20°); misc with alcohol, ether. *Caution:* Unstabilized material readily forms peroxides and may explode on shaking. Test all isopropyl ether for peroxides before using and destroy peroxides with sodium sulfite soln.
Human Toxicity: Toxic symptoms similar to ethyl ether, q.v.

5101. Isopropylidene Glycerol. *2,2-Dimethyl-1,3-dioxolane-4-methanol;* 2,2-dimethyl-4-hydroxymethyl-1,3-dioxolane; glycerol dimethylketal; acetone glycerol; Solketal. $C_6H_{12}O_3$; mol wt 132.16. C 54.53%, H 9.15%, O 36.32%. Prepn: Fisher, Ber. **28,** 1167 (1895); Fisher, Pfähler, Ber. **53,** 1607 (1920); Hibbert, Morazain, Can. J. Res. **2,** 38 (1930); Smith, Lindberg, Ber. **64,** 510 (1931); M. M. Maglio, C. A. Burger, J. Am. Chem. Soc. **68,** 529 (1946); M. Renoll, M. S. Newman, Org. Syn. coll. **vol. III,** 502 (1955); Mikschik, **Austrian** pat. **180,926** (1955 to Chemomedica Chemikalien), C.A. **49,** 15951 (1955); Williams, **Brit.** pat. **802,022** (1958 to Peter Spence & Sons); **Brit.** pat. **819,835** (1959 to Bayer); Perez et al., **Span.** pat. **499,129** (1982 to Calipe), C.A. **97,** 55794v (1982). Synthesis of D(+)-form: E. Baer, H. Fischer, J. Biol. Chem. **128,** 463 (1939); of L(−)-form: idem, J. Am. Chem. Soc. **61,** 761 (1939). Conversion of D(+)- to L(−)-form: idem, ibid. **67,** 944 (1945).

Practically odorless liq of medium mobility. d_4^{20} 1.064. bp_{10} 82°. n_D^{20} 1.4383. Viscosity at 20° = 11 cp. Practically nonflammable at ordinary storage temps: Flash pt 90°C (194°F). Evaporation no. about 600 (ether = 1). Miscible with water, alcohols, acetals, esters, ethers, aromatic hydrocarbons, chlorinated hydrocarbons, gasolines, petr ether, turpentine, oils. LD_{50} in rats (g/kg): 7 orally; 3 i.p. (Perez).
D(+)-Form, α_D +13.6°; $[\alpha]_D$ +10.8° (c = 15.19 in benzene).

L(−)-Form, α_D −13.6°; $[\alpha]_D$ −10.8° (c = 15.19 in benzene).

USE: Versatile solvent and plasticizer. Pharmaceutic aid (solubilizing and suspending agent).

5102. Isopropyl Iodide. *2-Iodopropane.* C_3H_7I; mol wt 170.01. C 21.19%, H 4.15%, I 74.65%. CH_3CHICH_3. Prepared by distilling isopropyl alcohol with HI; or from glycerol, iodine, water, and phosphorus.

Colorless liquid but readily discolors in air and light. d_4^{20} 1.703. mp −90°. bp 89-90°. n_D^{20} 1.5026. Sol in 720 parts water; miscible with alcohol, benzene, chloroform, ether.

5103. Isopropyl Myristate. *Tetradecanoic acid 1-methylethyl ester.* $C_{17}H_{34}O_2$; mol wt 270.44. C 75.50%, H 12.67%, O 11.83%. $CH_3(CH_2)_{12}COOCH(CH_3)_2$. The commercial product usually appears as a mixture of myristate with small amounts of esters of palmitic and other satd fatty acids. Physical properties: Bonhorst *et al., Ind. Eng. Chem.* **40**, 2379 (1948).

Liquid of low viscosity, practically odorless. mp ~+3°. bp_2 140.2°; bp_{20} 192.6°; dec 208°. d^{20} 0.8532; d^{99} 0.7942. n_D^{25} 1.432-1.434. Withstands oxidation and does not readily become rancid. Sol in castor oil, cottonseed oil, acetone, chloroform, ethyl acetate, ethanol, toluene, and mineral oil. Practically insol in water, glycerol, propylene glycol. Dissolves many waxes, cholesterol, lanolin.

USE: In cosmetic and topical medicinal prepns where good absorption through the skin is desired. A jellied isopropyl myristate was marketed as *Estergel.*

5104. Isopropyl Nitrite. *Nitrous acid 1-methylethyl ester; 2-propanol nitrite; nitrous acid isopropyl ester.* $C_3H_7NO_2$; mol wt 89.09. C 40.44%, H 7.92%, N 15.72%, O 35.92%. $(CH_3)_2CHONO$. Prepd by treating isopropyl alcohol with nitrosyl chloride: Kornblum *et al., J. Am. Chem. Soc.* **77**, 5531 (1955); by adding concd sulfuric acid and isopropyl alcohol to sodium nitrite: Levin, Hartung, *Org. Syn.* **24**, 26 (1944).

Pale yellow oil. d_4^0 0.856; d_4^{25} 0.844. bp_{752} 39-39.5°; bp_{745} 39-40°. n_D^{20} 1.3520.

USE: Jet propellant. *Caution:* Can cause vasodilation with fall in blood pressure, tachycardia, headache. Large doses can cause methemoglobinuria with cyanosis. Severe poisoning results in shock which can end fatally.

5105. Isoproterenol. *4-[1-Hydroxy-2-[(1-methylethyl)-amino]ethyl]-1,2-benzenediol; 3,4-dihydroxy-α-[(isopropylamino)methyl]benzyl alcohol; α-(isopropylaminomethyl)protocatechuyl alcohol; isoprenaline; isopropylarterenol; 1-(3,4-dihydroxyphenyl)-2-isopropylaminoethanol; isopropylaminomethyl-(3,4-dihydroxyphenyl)carbinol; N-isopropyl-β-dihydroxyphenyl-β-hydroxyethylamine; dihydroxyphenyl-ethanolisopropylamine; N-isopropylnoradrenaline; epinephrine isopropyl homolog; A 21; Aludrine; Aleudrin; Isuprel; Norisodrine; Asiprenol; Asmalar; Neo-Epinine; Novodrin; Isupren; Neodrenal; Lomupren; Isopropydrin; Assiprenol; Respifral; Bellasthman; Saventrine; Proternol; Isorenin; Vapo-N-Iso; Isonorin.* $C_{11}H_{17}NO_3$; mol wt 211.24. C 62.54%, H 8.11%, N 6.63%, O 22.72%. β-Adrenergic agonist. Catecholamine derivative prepd by the reaction of 3,4-dihydroxy-α-haloacetophenone with an excess of isopropylamine: Scheuing, Thomä, *Ger. pat.* **723,278** (1942 to Boehringer, Ing.); *U.S. pat.* **2,308,232** (1943); from guaiacol and chloral hydrate: Beke *et al., Pharm. Zentralh.* **92**, 237 (1953). Resolution: Kerschbaum, Benedikt, *Monatsh.* **83**, 1090 (1952); Beccari *et al., Science* **118**, 249 (1953); Delmar *et al., U.S. pat.* **2,715,141** (1955 to Delmar Chem.). Configuration: Pratesi *et al., Farmaco Ed. Sci.* **15**, 3 (1960). Toxicity: E. I. Goldenthal, *Toxicol. Appl. Pharmacol.* **18**, 185 (1971). Effect on gastric acid secretion: M. J. Daly, *Scand. J. Gastroenterol.* **89**, 3 (1984). Use in treatment of primary pulmonary hypertension: D. A. Pietro *et al., N. Engl. J. Med.* **310**, 1032 (1984). Review of pharmacology and comparison with other β-adrenoceptor agonists: V. T. Popa, *J. Asthma* **21**, 183-207 (1984). Comprehensive description: M. Tariq, A. A. Al-Badr in *Analytical Profiles of Drug Substances* vol. 14, K. Florey, Ed. (Academic Press, New York, 1985) pp 391-422.

dl-Form, crystals from alc, mp 155.5°. pKa 8.64. LD_{50} in male, female rats (mg/kg): 3675, 4282 orally (Goldenthal).

dl-Form hydrochloride, $C_{11}H_{18}ClNO_3$, *Aerolone, Aerotrol, Euspiran, Isomenyl, Isovon, Mistarel, Suscardia.* Crystals from alc, mp 170-171°. Soly: one gram dissolves in 3 ml water, in 50 ml ethanol (95%). Less sol in abs ethanol. Practically insol in chloroform, ether, benzene. pH of 1% aq soln about 5. Aq solns turn brownish-pink upon prolonged exposure to air or upon addition of alkali. LD_{50} orally in rats: 2221 ±93 mg/kg (Goldenthal).

dl-Form sulfate dihydrate, $C_{22}H_{36}N_2O_{10}S.2H_2O$, *Aludrin, Isomist, Propal.* Crystals from acetone + methanol, mp 128° (some decomp). One gram dissolves in about 4 ml water. Slightly sol in alc. Practically insol in chloroform, ether, benzene. pH of a 1% aq soln about 5.

l-Form, crystals, mp 164-165°. $[\alpha]_D^{19}$ −45.0° (c = 2 in 2N HCl).

l-Form hydrochloride, crystals, dec 162-164°. $[\alpha]_D^{20}$ −50°.

l-Form *d*-bitartrate dihydrate, *Isolevin.* Crystals, mp 80-83° (sinters at 78°). $[\alpha]_D^{19}$ −14.9° (c = 2.31).

THERAP CAT: Bronchodilator.

THERAP CAT (VET): Sympathomimetic. Chiefly as bronchodilator.

5106. Isoproturon. *N,N-Dimethyl-N'-[4-(1-methylethyl)phenyl]urea; 3-p-cumenyl-1,1-dimethylurea; N-(4-isopropylphenyl)-N',N'-dimethylurea;* I.P.U.; Hoe 16410; CGA 18731; Arelon; Graminon; Tolkan. $C_{12}H_{18}N_2O$; mol wt 206.29. C 69.87%, H 8.79%, N 13.58%, O 7.76%. Pre- and post-emergence herbicide for control of annual grasses and broad-leaved weeds. Prepn and herbicidal properties: C. W. Todd, *U.S. pat.* **2,655,447** (1953 to du Pont); Belg. pat. **770,928** (1971 to Pepro); A. Thizy *et al., U.S. pat.* **4,295,877** (1981 to Philagro); Selective control of weeds: D. Dürr, *Ger. pat.* **2,107,774** (1971 to Ciba-Geigy); Belg. pat. **770,918** (1971 to Hoechst). Physical properties, herbicidal activity: A. Thizy *et al., Compt. Rend. Ser. D* **274**, 2053 (1972); J. Rognon *et al., Meded. Fac. Landbouwwetensch. Rijksuniv. Gent* **37**, 663 (1972), *C.A.* **79**, 1260c (1973). Microbial degradation of labelled isoproturon: J.-C. Fournier *et al., Chemosphere* **4**, 207 (1975). Efficacy in winter cereals: R. T. Hewson, *Proc. 12th Brit. Weed Contr. Conf.* 75 (1974); P. Gonzales *et al., Weed Res.* **23**, 39 (1983). HPLC determn in soil: G. Kulshrestha, R. Khazanchi, *J. Chromatog.* **318**, 144 (1985).

White crystals from benzene, mp 158°. Soly in water at 22°: 70 mg/l. LD_{50} in mice, rats (mg/kg): 3350, 3600 orally (Thizy, 1972).

USE: Herbicide.

5107. Isopyrocalciferol. *9β-Ergosta-5,7,22-trien-3β-ol; 9β-ergosterol.* $C_{28}H_{44}O$; mol wt 396.63. C 84.78%, H 11.18%, O 4.03%. Differs sterically from pyrocalciferol at C-10, from ergosterol at C-9. *See ref under* Pyrocalciferol. Stereochemistry: Castells *et al., J. Chem. Soc.* **1959**, 1159. Dehydrogenation with selenium gives Diel's hydrocarbon (3'-methyl-1,2-cylopentenophenanthrene).

Prisms from ether-methanol. mp 112-115°. uv max: 262, 280 nm. $[\alpha]_D^{20}$ +332°, $[\alpha]_{546}^{20}$ +415° (c = 1.5 in chloroform). Precipitated by digitonin.

Acetate, $C_{30}H_{46}O_2$, mp 113-115°, $[\alpha]_D^{20}$ +333°, $[\alpha]_{546}^{20}$ +423° (c = 2 in chloroform).

5108. Isoquassin. Picrasmin. $C_{22}H_{28}O_6$; mol wt 388.44. C 68.02%, H 7.27%, O 24.71%. Bitter principle from Jamaica quassia, *Picrasma excelsa* (Sw.) Planch., *Simaroubaceae.* Isoln: Clark, *J. Am. Chem. Soc.* **60**, 1146 (1938). Identity of isoquassin and picrasmin: Adams, Whaley, *ibid.* **72**, 375 (1950). Isomer of quassin (q.v.): Valenta *et al., Tetrahedron* **18**, 1433 (1962).

Crystals from methanol, mp 222-225°. $[\alpha]_D^{20}$ +45.4° (chloroform). uv max: 258 nm (ϵ 12,500).

5109. Isoquercitrin. *2-(3,4-Dihydroxyphenyl)-3-(β-D-glucofuranosyloxy)-5,7-dihydroxy-4H-1-benzopyran-4-one; 3,3',4',5,7-pentahydroxyflavone-3-glucoside;* quercetin-3-glucoside; isotrifoliin; trifoliin. $C_{21}H_{20}O_{12}$; mol wt 464.37. C 54.31%, H 4.34%, O 41.35%. From flowers of *Gossypium herbaceum* L., *Malvaceae:* Perkin, *J. Chem. Soc.* **95**, 2181 (1909); from flowers of *Aesculus hippocastanum* L., *Hippocastanaceae:* Hörhammer, Wagner, *Arch. Pharm.* **290**, 224 (1957); from *Tropaeolum majus* L., *Tropaeolaceae:* Delaveau, *Compt. Rend.* **252**, 1510 (1961); from *Arnica montana* L., *Compositae:* Friedrick, *Naturwiss.* **49**, 541 (1962). Structure: Attree, Perkin, *J. Chem. Soc.* **1927**, 234. Identity with trifoliin and isotrifoliin: Hattori *et al., J. Chem. Soc. Japan* **58**, 844 (1937), *C.A.* **32**, 219c (1938). Synthesis: Ice, Wender, *J. Am. Chem. Soc.* **74**, 4606 (1952); *eidem*, U.S. pat. **2,727,890** (1955 to the USAEC).

Yellow needles from water, mp 225-227°. uv max: 257, 369 nm. Practically insol in cold but sparingly sol in boiling water; sol in alkaline solns with a deep yellow tint.

5110. Isoquinoline. 2-Benzazine; benzo[c]pyridine. C_9H_7N; mol wt 129.15. C 83.69%, H 5.46%, N 10.85%. Occurs in coal tar: Hoogewerff, van Dorp, *Rec. Trav. Chim.* **4**, 125, 285 (1885); Weissgerber, *Ber.* **47**, 3175 (1914); Harris, Pope, *J. Chem. Soc.* **121**, 1029 (1922). Modern isoln processes: H. J. V. Winkler, *Der Steinkohlenteer und seine Aufarbeitung* (Essen, 1951). Isoln by a freeze-out method: Kjellman, U.S. pat. **2,483,420** (1946 to Koppers). Purification: Freiser, Glowacki, *J. Am. Chem. Soc.* **71**, 514 (1949). Toxicity data: Smyth *et al., Arch. Ind. Hyg. Occup. Med.* **4**, 119 (1951). Synthesis: D. L. Boger *et al., Tetrahedron* **37**, 3977 (1981). Toxicity data: Smyth *et al., Arch. Ind. Hyg. Occup. Med.* **4**, 119 (1951). Comprehensive survey of isoquinoline alkaloids: Gensler, *Heterocyclic Compounds* **vol. 4**, R. C. Elderfield, Ed. (Wiley, New York, 1952) pp 344-490; T. Kametani, *Total Synthesis of Natural Products* **vol. 3**, J. ApSimon, Ed. (Wiley, New York, 1977) pp 1-272.

Liquid. Pungent odor resembling that of a mixture of anise oil and benzaldehyde. Hygroscopic platelets when solid. d_4^{30} 1.09101; d_4^{80} 1.05143. mp 26.48°. n_D^{30} 1.62078. Viscosity at 30°: 3.2528 cp, at 100°: 1.0230 cp, at 200°: 0.4223 cp. bp_{743} 242.2°; bp_{760} 243.25°. Dipole moment: 2.49. More basic than quinoline, pKa (25°): 8.60. Almost insol in water. Miscible with many organic solvents. Sol in dil acids. LD_{50} orally in rats: 360 mg/kg (Smyth).

Sulfate, $C_9H_9NO_4S$, crystals, mp 209-209.5°. Sol in water.

USE: Synthesis of dyes, insecticides, antimalarials, rubber accelerators.

5111. Isorubijervine. *Solanid-5-ene-3β,18-diol.* $C_{27}H_{43}NO_2$; mol wt 413.62. C 78.40%, H 10.48%, N 3.39%, O 7.74%. From various species of *Veratrum, Liliaceae.* Isoln and structure: Jacobs, Craig, *J. Biol. Chem.* **148**, 41 (1943); **159**, 617 (1945); Pelletier, Jacobs, *J. Am. Chem. Soc.* **75**, 4442 (1953). Stereochemistry: Sato, Latham, *ibid.* **78**, 3146 (1956); Höhne *et al., Tetrahedron* **22**, 673 (1966). *Review:* Morgan, Barltrop, *Quart. Revs.* **12**, 34 (1958).

Crystals from alcohol, mp 235-237°. $[\alpha]_D^{25}$ +6.5° (c = 0.97 in abs ethanol). Forms a sparingly sol digitonide.

Hydrobromide, $C_{27}H_{44}BrN_2O$, fine needles from alcohol + ether, dec 290-295°.

5112. Isosafrole. *5-(1-Propenyl)-1,3-benzodioxole; 1,2-(methylenedioxy)-4-propenylbenzene.* $C_{10}H_{10}O_2$; mol wt 162.18. C 74.05%, H 6.22%, O 19.73%. Purification and separation from safrole: Balbiano, *Ber.* **42**, 1505 (1911); Hoering, Baum, *ibid.* 3082. Prepn: Bert, *Compt. Rend.* **213**, 873 (1941); Naves, Ardizio, *Bull. Soc. Chim. France* **1957**, 1053; Fengeas, *ibid.* **1964**, 1892; Cabiddu *et al., Ann. Chim. (Rome)* **52**, 1261 (1962). Review and evaluation of studies of carcinogenic action in laboratory animals: *IARC Monographs* **10**, 231-241 (1976).

trans-Form, liquid, odor of anise. bp_{760} 253°; bp_{100} 179.5°; bp_{20} 135.6°; $bp_{3.4}$ 85-86°. mp 8.2°. d_4^{20} 1.1206. n_D^{20} 1.5782. uv max (96% alc): 305, 267, 259.5 nm (ϵ 5340; 11,600; 12,160). Miscible with alc, ether, benzene. Sol in 8 parts of 90% alcohol.

cis-Form, liq. $bp_{3.5}$ 77-79°. mp −21.5°. d_4^{20} 1.1182. n_D^{20} 1.5691. uv max (96% alc): 296.5, 259 nm (ϵ 4450; 10,000).

USE: Manuf heliotropin; to modify oriental perfumes; to strengthen soap perfumes; in small quantities together with methyl salicylate in root beer and sarsaparilla flavors.

5113. Isosorbide. *1,4:3,6-Dianhydro-D-glucitol;* 1,4:3,6-dianhydrosorbitol; AT-101; NSC-40725; Hydronol; Ismotic; Isobide. $C_6H_{10}O_4$; mol wt 146.14. C 49.31%, H 6.90%, O 43.79%. Prepd by acid dehydration of D-glucitol: Haworth, Wiggins, **Brit. pat.** 600,870 (1948). Purification: Hartmann, U.S. pat. **3,160,641** (1964 to Atlas Chemical Industries). Review of prepn, structure, properties: Wiggins, *Advan. Carbohyd. Chem.* **5**, 191-228 (1950). Stereochemistry: Cope, Shen, *J. Am. Chem. Soc.* **78**, 3177 (1956). Conformation studies: Hopton, Thomas, *Can. J. Chem.* **47**, 2395 (1969). Pharmacodynamics in man: Nodine *et al., Clin. Pharmacol. Ther.* **14**, 196 (1973).

Crystals, mp 61-64°. $[\alpha]_D$ +44°.

THERAP CAT: Diuretic.

5114. Isosorbide Dinitrate. *1,4:3,6-Dianhydro-D-gluci-tol dinitrate;* dinitrosorbide; 1,4:3,6-dianhydrosorbitol 2,5-dinitrate; Astridine; Cardio 10; Cardis; Carvanil; Carvasin; Cedocard; Corovliss; Dignonitrat; Dilatrate; Disorlon; Duranitrat; EureCor; Frandol; Glentonin-retard; Harrical; IBD; Isdin; ISDN; Iso-Bid; Isoket; IsoMack; Iso-Puren; Isorbid; Isordil; Isostenase; Isotrate; Langoran; Laserdil; Monoclair; Maycor; Myorexon; Nitorol; Nitrol; Nitrosorbon; Nosim; Rifloc Retard; Rigedal; Risordan; Soni-Slo; Sorbangil; Sorbichew; Sorbidilat; Sorbid SA; Sorbislo; Sorbitrate; Sorquad; Vascardin; Vasorbate; Vasotrate. $C_6H_8N_2O_8$; mol wt 236.14. C 30.52%, H 3.41%, N 11.86%, O 54.20%. Coronary vasodilator. Prepn from sorbitol: Goldberg, *Acta Physiol. Scand.* **15**, 173 (1948), *C.A.* **42**, 5564 (1948); Kochergin, Titkova, *C.A.* **54**, 8647h (1960). Absorption and disposition in man: H. Laufen *et al., Arzneimittel-Forsch.* **33**, 980 (1983). Comprehensive description: L. A. Silvieri, N. J. DeAngelis in *Analytical Profiles of Drug Substances* **vol. 4**, K. Florey, Ed. (Academic Press, New York, 1975) pp 225-244.

Hard colorless crystals, mp 70°. $[\alpha]_D^{20}$ +135° (alc). Sparingly sol in water (1.089 mg/ml, also given as 1.0 g dissolves in 900 ml H_2O). Freely sol in organic solvents, such as acetone, alc, ether.

Isosorbide-5-mononitrate, $C_6H_9NO_6$, *Elantan, Imdur, ISMO, Monicor, Monit, Mono-Cedocard, Monoket, Mono Mack, Monosorb, Olicard, Pentacard.* Metabolite of isosorbide dinitrate.

THERAP CAT: Anti-anginal.

5115. β-Isosparteine. *[7S-(7α,7aβ,14α,14aβ)]-Dodeca-hydro-7,14-methano-2H,6H-dipyrido[1,2-a:1',2'-e][1,5]di-azocine;* pusilline. $C_{15}H_{26}N_2$; mol wt 234.40. C 76.86%, H 11.18%, N 11.95%. Isoln from *Lupinus pusillus*, Pursh., *Leguminosae:* Marion, Fenton, *J. Org. Chem.* **13**, 780 (1948). Identity with pusilline: Greenhalgh, Marion, *Can. J. Chem.* **34**, 456 (1956); identity of monohydrochloride with *nonalupine* and *spathulatine:* Marion, *ibid.* **29**, 959 (1951). Synthesis: Bohlmann, *Ber.* **90**, 653 (1957). Absolute configuration: Okuda, Tsuda, *Chem. & Ind. (London)* **1961**, 1115.

Oil, $bp_{0.2}$ 100-110°. $[\alpha]_D^{32}$ −15.3° (c = 2.3 in abs alc). Sol in methanol, ethanol. Practically insol in water.

Hydrochloride monohydrate, $C_{15}H_{26}N_2 \cdot HCl \cdot H_2O$, prismatic needles from alcohol or acetone. Melts at 88-90°, resolidifies, and remelts at 235.5-236.5°. $[\alpha]_D$ −1.88° (c = 5.90 in chloroform); −2.32° (c = 6.42 in alc); −2.44° (c = 6.42 in water). Sol in water, benzene, alcohol, chloroform, carbon tetrachloride, acetone, ethyl acetate; slightly sol in petr ether, ether. The aq soln is slightly alkaline to litmus.

Dihydrochloride, $C_{15}H_{26}N_2 \cdot 2HCl$, clusters of fine needles from ethyl acetate, mp 270-273° (sinters at 262°).

Hydriodide, $C_{15}H_{26}N_2 \cdot HI$, prismatic needles from acetone, mp 252.5° (sinters at 250.5°).

Methiodide, $C_{15}H_{26}N_2 \cdot CH_3I$, diamond-shaped crystals from ethyl acetate, mp 250-252°.

5116. Isothebaine. *5,6,6a,7-Tetrahydro-2,11-dimethoxy-6-methyl-4H-dibenzo[de,g]quinolin-1-ol; 2,11-dimethoxy-6aα-aporphin-1-ol;* 1-hydroxy-2,11-dimethoxyaporphine. $C_{19}H_{21}NO_3$; mol wt 311.37. C 73.29%, H 6.80%, N 4.50%, O

15.42%. From the root and whole plant of *Papaver orientale* L., *Papaveraceae*, collected in late autumn. Isoln: Gadamer, *Arch. Pharm.* **249**, 39 (1911). Structure: Bentley, Blues, *J. Chem. Soc.* **1956**, 1732; Bentley, Dyke, *J. Org. Chem.* **22**, 429 (1957). Synthesis of *dl*-form: Battersby, Brown, *Proc. Chem. Soc. (London)* **1964**, 85; H. Hara *et al., Chem. Pharm. Bull* **29**, 1083 (1981). Synthesis of natural and racemic forms: Battersby *et al., J. Chem. Soc.* **1965**, 4550. Biosynthesis: *eidem, Chem. Commun.* **1965**, 230.

Light-sensitive, rhombic crystals. mp 203-204°. $[\alpha]_D^{18}$ +285° in alc. Sol in alcohol, chloroform, slightly in ether.

5117. Isothipendyl. *N,N,α-Trimethyl-10H-pyrido[3,2-b][1,4]benzothiazine-10-ethanamine;* 10-(2-dimethylamino-2-methylethyl)-10H-pyrido[3,2-b][1,4]benzothiazine; 10-(2-dimethylaminopropyl)-9-thia-1,10-diazaanthracene; 10-(2-dimethylaminopropyl)-1-azaphenothiazine; *N*-dimethylaminoisopropylthiophenylpyridylamine; D 201. $C_{16}H_{19}N_3S$; mol wt 285.42. C 67.33%, H 6.71%, N 14.72%, S 11.24%. Prepn: Yale, Sowinski, *J. Am. Chem. Soc.* **80**, 1651 (1958). Pharmacology: A. von Schlichtegroll, *Arzneimittel-Forsch.* **8**, 489 (1958).

Hydrochloride, $C_{16}H_{20}ClN_3S$, *Andantol, Andanton, Nilergex, Theruhistin.* Crystals from acetonitrile, mp 222-223°. Sol in water. LD_{50} in mice (mg/kg): 62 i.p.; 222 orally (Schlichtegroll)

THERAP CAT: Antihistaminic.

5118. Isovaleraldehyde. *3-Methylbutanal;* isovaleral; isovaleric aldehyde. $C_5H_{10}O$; mol wt 86.13. C 69.72%, H 11.70%, O 18.58%. $(CH_3)_2CHCH_2CHO$. Occurs in orange, lemon, peppermint, eucalyptus and other oils. Made by oxidation of isoamyl alcohol with $Na_2Cr_2O_7$ and H_2SO_4.

Colorless liquid; pungent apple-like odor. d_{20}^{20} 0.785. mp −51°. bp 92-93°. n_D^{20} 1.3902. Sparingly sol in water; miscible with alcohol, ether.

USE: In artificial flavors and perfumes.

5119. Isovaleramide. $C_5H_{11}NO$; mol wt 101.15. C 59.37%, H 10.96%, N 13.85%, O 15.82%. $(CH_3)_2CHCH_2CONH_2$.

Crystals. d 0.965. mp 135-137°. bp 232°. Sol in water, alcohol.

5120. Isovaleric Acid. *3-Methylbutanoic acid;* isovalerianic acid; isopropylacetic acid; delphinic acid. $C_5H_{10}O_2$; mol wt 102.13. C 58.80%, H 9.88%, O 31.33%. $(CH_3)_2CHCH_2COOH$. Occurs in hop oil, tobacco and several other plants: Schmeltz *et al., J. Assoc. Offic. Agr. Chem.* **46**, 779 (1963). Prepn: Caldwell, U.S. pat. **2,484,486** (1949 to Kodak); Gustak, Sajko, *Arkiv. Kemi.* **24**, 11 (1952), *C.A.* **49**, 163e (1955); L'vov *et al., Zh. Prikl. Khim.* **35**, 700 (1962). "Valeric" acid of commerce is isovaleric acid. Toxicity data: L. Orö, A. Wretlind, *Acta Pharmacol. Toxicol.* **18**, 141 (1961).

Liquid. Acid taste; disagreeable, rancid-cheese odor. d_4^{20} 0.931. mp −37°. bp 175-177°. n_D^{20} 1.4043. Sol in 24 parts water; sol in alcohol, chloroform, ether. *Keep tightly closed.* LD_{50} i.v. in mice: 1120±30 mg/kg (Orö, Wretlind).

USE: In flavors, perfumes, manuf sedatives.

5121. Isovaleryl Chloride. C_5H_9ClO; mol wt 120.58. C 49.80%, H 7.52%, Cl 29.41%, O 13.27%. $(CH_3)_2CHCH_2$-COCl.

Liquid. d_4^{20} 0.985. bp 114-116°. n_D^{20} 1.4156. Dec by water, alcohol.

5122. Isovaleryl Diethylamide. *N,N-Diethyl-3-methylbutyramide; N,N-diethylisovaleramide; Valyl; Xalyl.* C_9H_{19}-NO; mol wt 157.25. C 68.74%, H 12.18%, N 8.91%, O 10.17%. Prepd from isovaleric acid and diethylamine: Liebrecht, **Ger. pat.** 129,967 (1902); *Frdl.* **6,** 1226; from ethyl isovalerate and lithium diethylamide: Hamell, Levine, *J. Org. Chem.* **15,** 162 (1950); by reduction of *N,N*-diethyl-β-methylcrotonamide with lithium aluminum hydride: Snyder, Putnam, *J. Am. Chem. Soc.* **76,** 1893 (1954).

$$CH_3CHCH_2CON(C_2H_5)_2$$
$$\overset{|}{CH_3}$$

Liquid, aromatic odor, peppery taste. bp 210-212°, bp_{14} 93-95°, bp_5 77-78°. n_D^{20} 1.4422. d_{20} 0.8764. One gram dissolves in about 25 ml water; sol in alcohol, ether.

THERAP CAT: Sedative.

5123. 2-Isovalerylindane-1,3-dione. *2-(3-Methyl-1-oxobutyl)-1H-indene-1,3(2H)-dione;* PMP; Valone. $C_{14}H_{14}$-O_3; mol wt 230.25. C 73.02%, H 6.13%, O 20.85%. Prepn: Shapiro *et al., J. Org. Chem.* **25,** 1860 (1960).

Yellow solid from methanol, mp 68-69°. Practically insol in water, sol in most organic solvents.

USE: Insecticide, rodenticide.

5124. Isovaline. *2-Amino-2-methylbutyric acid; α-amino-α-methylbutyric acid.* $C_5H_{11}NO_2$; mol wt 117.15. C 51.26%, H 9.46%, N 11.96%, O 27.32%. Prepn: Kurono, *Biochem. Z.* **134,** 427 (1922); Levene, Steiger, *J. Biol. Chem.* **76,** 299 (1928); Bucherer, Steiner, *J. Prakt. Chem.* **141,** 5 (1934); Upham, Dermer, *J. Org. Chem.* **22,** 799 (1957). Optical enantiomorphs: Ehrlich, Wendel, *Biochem. Z.* **8,** 438 (1908); Fischer, von Gravenitz, *Ann.* **406,** 5 (1914); Baker *et al., J. Am. Chem. Soc.* **74,** 4701 (1952); Greenstein *et al., J. Biol. Chem.* **204,** 307 (1953).

$$CH_3CH_2\overset{\overset{\displaystyle NH_2}{|}}{\underset{\underset{\displaystyle CH_3}{|}}{C}}COOH$$

L-Form, crystals. $[\alpha]_D^{25}$ +11.13° (c = 5 in H_2O). $[M]_D$ +9.7° (5N HCl); $[M]_D$ +26.3° (glacial acetic acid).

D-Form, long needles from water + acetone. $[\alpha]_D^{25}$ −11.28° (c = 5 in H_2O).

DL-Form, monoclinic prisms, mp 307-308° (closed tube). Sublimes at around 300°. Soly in cold water about 39 g/100 ml; in alc about 6.6 g/100 g alc at 75°. Slightly sol in ether.

5125. Isoxaben. *N-[3-(1-Ethyl-1-methylpropyl)-5-isoxazolyl]-2,6-dimethoxybenzamide;* benzamizole; EL-107; NA 8318; Flexidor. $C_{18}H_{24}N_2O_4$; mol wt 332.40. C 65.04%, H 7.28%, N 8.43%, O 19.25%. Selective, pre-emergent herbicide for use in cereal crops. Prepn: K. W. Burow, **Brit. pat. Appl.** 2,084,140; *see also: idem,* **U.S. pat.** 4,636,243 (1982, 1987 both to Lilly). Physical properties and herbicidal activity: F. Huggenberger *et al., Proc. Brit. Crop Prot. Conf. - Weeds* **1982,** 47. Mobility and degradation in soil: F. Huggenberger, P. J. Ryan, *ibid.* **1985,** 947. Effect on cellular growth and metabolism: A. Lefebvre *et al., Weed Res.* **27,** 125 (1987). Field trials: F. Huggenberger, F. Gueguen, *Crop Prot.* **6,** 75 (1987); F. O. Colbert, D. H. Ford, *Proc. West. Soc. Weed Sci.* **40,** 155 (1987).

White crystalline solid, mp 172-174°. Soly at 25° (mg/ml): methanol 50-100; ethyl acetate 50-100; acetonitrile 30-50; toluene 4-5; hexane 0.07-0.08; water 0.001-0.002. Vapor pressure at 30°: $< 3.9 \times 10^{-7}$ mm Hg. LD_{50} in mice, rats (mg/kg): $> 10,000$ orally both species; LC_{50} in rats by inhalation: > 1.99 mg/l (Huggenberger, 1982).

USE: Herbicide.

5126. Isoxepac. *6,11-Dihydro-11-oxodibenz[b,e]oxepin-2-acetic acid;* oxepinac; HP 549; P 720549; Artil. $C_{16}H_{12}O_4$; mol wt 268.27. C 71.64%, H 4.51%, O 23.85%. Non-steroidal anti-inflammatory with analgesic and antipyretic activity. Prepn: K. Ueno *et al.,* **Japan. Kokai** 74 124,086 (1974 to Daiichi), *C.A.* **82,** 170735d (1975); G. C. Helsley *et al.,* **Ger. pat.** 2,442,060 (1975 to Hoechst), *C.A.* **83,** 97063 (1975); K. Ueno *et al., J. Med. Chem.* **19,** 941 (1976); D. E. Aultz *et al., ibid.* **20,** 66 (1977). Pharmacology: H. B. Lassman *et al., Arch. Int. Pharmacodyn. Ther.* **227,** 142 (1977). Disposition and metabolism in animals and humans: H. P. A. Illing, J. M. Fromson, *Drug Metab. Dispos.* **6,** 510 (1978). HPLC determn in plasma: J. A. Slack, *J. Chromatogr.* **221,** 431 (1980); H. K. L. Hundt, L. W. Brown, *ibid.* **225,** 482 (1981). Clinical studies: S. Sasaki, *Arzneimittel-Forsch.* **28,** 462 (1978); L. B. Svendsen *et al., Scand. J. Rheumatol.* **10,** 186 (1981); J. Scott, E. C. Huskisson, *Rheumatol. Rehabil.* **21,** 48 (1982).

Crystals from ethyl acetate, mp 131-132.5°. LD_{50} orally in rats: 199 mg/kg (Ueno).

THERAP CAT: Anti-inflammatory.

5127. Isoxicam. *4-Hydroxy-2-methyl-N-(5-methyl-3-isoxazolyl)-2H-1,2-benzothiazine-3-carboxamide 1,1-dioxide;* 4-hydroxy-3-(5-methyl-3-isoxazolocarbamyl)-2-methyl-2H-1,2-benzothiazine 1,1-dioxide; W8495; Floxicam; Maxicam; Pacyl; Vectren. $C_{14}H_{13}N_3O_5S$; mol wt 335.33. C 50.14%, H 3.91%, N 12.53%, O 23.86%, S 9.56%. Nonsteroidal anti-inflammatory drug with antipyretic and analgesic properties. Prepn: H. Zinnes *et al.,* **Ger. pat.** 2,208,351; *eidem,* **U.S. pat.** 3,787,324 (1972, 1974 both to Warner-Lambert). Anti-inflammatory properties: G. DiPasquale *et al., Agents Actions* **5,** 256 (1975); *ibid.* **6,** 748 (1976). Pharmacological studies: G. DiPasquale, D. Mellace, *ibid.* **7,** 481 (1977); K. Rainsford, *ibid.* 573; G. DiPasquale *et al., Res. Commun. Chem. Pathol. Pharmacol.* **19,** 529 (1978). Pharmacokinetics in man: E. U. Kölle *et al., Arzneimittel-Forsch.* **33,** 582 (1983). Series of articles on pharmacology, safety and clinical efficacy: *Am. J. Med.* **79,** Suppl. 4B, 1-42 (1985); *Brit J. Clin. Pharmacol.* **22,** Suppl. 2, 107S-190S (1986). *Review: Semin. Arthritis Rheum.* **12,** Suppl. 2, 153-183 (1982).

Crystals from 1,4-dioxane, mp 265-271° (dec).

Sodium salt, $C_{14}H_{12}N_3NaO_5S$, crystals from ethanol. mp 270-272° (dec).

THERAP CAT: Anti-inflammatory.

5128. Isoxsuprine. *4-Hydroxy-α-[1-[(1-methyl-2-phen-*

oxyethyl)amino]ethyl]benzenemethanol; p-hydroxy-α-[1-[(1-methyl-2-phenoxyethyl)amino]ethyl]benzyl alcohol; p-hydr-oxy-N-(1-methyl-2-phenoxyethyl)norephedrine; 1-(p-hydr-oxyphenyl)-2-(1-methyl-2-phenoxyethylamino)-1-propa-nol; 2-(3-phenoxy-2-propylamino)-1-(p-hydroxyphenyl)-1-propanol. $C_{18}H_{23}NO_3$; mol wt 301.37. C 71.73%, H 7.69%, N 4.65%, O 15.93%. Prepn: Moed, van Dijk, *Rec. Trav. Chim.* **75**, 1215 (1956); **Brit.** pats. **832,286** and **832,287** (both 1960 to N. V. Philip Gloeilampenfabrieken); Moed, **U.S.** pat. **3,056,836** (1962 to N. Am. Philips). Configuration: Dirkx, *Rec. Trav. Chim.* **83**, 535 (1964). Toxicity data: Goldenthal, *Toxicol. Appl. Pharmacol.* **18**, 185 (1971).

Crystals, mp 102.5-103.5°.

Hydrochloride, $C_{18}H_{24}ClNO_3$, *Dilavase, Duvadilan, Iso-lait, Suprilent, Vadosilan, Vasodilan, Vasoplex, Vasotran.* Bitter crystals from water, mp 203-204°. Soly in water at 25° about 2.0% w/v; also sol in ethanol. LD_{50} in rats (mg/kg): 1750 orally; 164 i.p. (Goldenthal).

Resinate, *Defencin.*

THERAP CAT: Vasodilator.

5129. Isradipine. *4-(4-Benzofurazanyl)-1,4-dihydro-2,6-dimethyl-3,5-pyridinedicarboxylic acid methyl 1-methyl-ethyl ester;* isopropyl 4-(2,1,3-benzoxadiazol-4-yl)-1,4-di-hydro-5-methoxycarbonyl-2,6-dimethyl-3-pyridinecarbox-ylate; *4-(2,1,3-benzoxadiazol-4-yl)-2,6-dimethyl-1,4-di-hydro-3-isopropyloxycarbonylpyridine-5-carboxylic acid methyl ester; isrodipine; PN 200-110; DynaCirc; Dynacrine; Lomir; Prescal; Rebinden.* $C_{19}H_{21}N_3O_5$; mol wt 371.39. C 61.45%, H 5.70%, N 11.31%, O 21.54%. Dihydropyridine calcium channel blocker. Prepn: P. Neumann, **Ger.** pat. **2,949,491**; idem, **U.S.** pat. **4,466,972** (1980, 1984 both to Sandoz). Prepn of enantiomers: A. Vogel, **Ger.** pat. **3,320,-616** (1983 to Sandoz), *C.A.* **101**, 7162s (1984). Comparative study of *in vitro* effects on human and canine cerebral arter-ies: E. Müller-Schweinitzer, P. Neumann, *J. Cereb. Blood Flow Metab.* **3**, 354 (1983). Effect on α-adrenoceptor medi-ated vasoconstriction in rats: K. Jie *et al.*, *Arch. Int. Phar-macodyn.* **278**, 72 (1985). Pharmacokinetics: F. L. S. Tee, J. M. Jaffe, *Eur. J. Clin. Pharmacol.* **32**, 361 (1987). Clinical evaluation in angina and coronary artery disease: C. E. Handler, E. Sowton, *ibid.* **27**, 415 (1984); in hypertension: E. B. Nelson *et al.*, *Clin. Pharmacol. Ther.* **40**, 694 (1986). Comparison of hemodynamic effects of enantiomers: R. P. Hof *et al.*, *J. Cardiovasc. Pharmacol.* **8**, 221 (1986). Series of articles on pharmacology and clinical use: *Am. J. Med.* **86**, 1-146 (1989).

mp 168-170°.

S(+)-Form, *PN 205-033.* Crystals from ether + hexane, mp 142°. $[\alpha]_D^{20}$ +6.7° (c = 1.5 in ethanol).

R(−)-Form, *PN 205-034.* Crystals from ether + hexane, mp 140°. $[\alpha]_D^{20}$ −6.7° (c = 1.67 in ethanol).

THERAP CAT: Antihypertensive; antianginal.

5130. Itaconic Acid. *Methylenesuccinic acid;* propylene-dicarboxylic acid. $C_5H_6O_4$; mol wt 130.10. C 46.16%, H 4.65%, O 49.19%. Obtained by dry distillation of citric acid and subsequent treatment of the anhydride with water. Pro-duced on a large scale by submerged aerobic fermentation using *Aspergillus terreus* and low cost carbohydrates from beet or cane: Kane *et al.*, **U.S.** pat. **2,385,283** (1945 to Pfi-

zer). Synthesis from propargyl chloride, carbon monoxide, nickel carbonyl and water: Chiusoli, **U.S.** pat. **3,025,320** (1962 to Montecatini).

Hygroscopic crystals; characteristic odor. d 1.63. mp 162-164° with decompn. Also reported as mp 172° [Kino-shita, *Acta Phytochem. (Japan)* **5**, 273 (1931)]. One gram dissolves in 12 ml water, 5 ml alcohol; very slightly sol in benzene, chloroform, ether, carbon disulfide, petr ether. *Keep well closed.*

5131. Itraconazole. *4-[4-[4-[4-[[2-(2,4-Dichlorophen-yl)-2-(1H-1,2,4-triazol-1-ylmethyl)-1,3-dioxolan-4-yl]methoxy]phenyl]-1-piperazinyl]phenyl]-2,4-dihydro-2-(1-methylpropyl)-3H-1,2,4-triazol-3-one;* (±)-1-sec-butyl-4-[p-[4-[p-[[(2R*,4S*)-2-(2,4-dichlorophenyl)-2-(1H-1,2,4-triazol-1-ylmethyl)-1,3-dioxolan-4-yl]methoxy]phenyl]-1-piperazinyl]phenyl]-Δ²-1,2,4-triazolin-5-one; *oriconazole; R 51211; Sporanos; Sporanox.* $C_{35}H_{38}Cl_2N_8O_4$; mol wt 705.64. C 59.57%, H 5.43%, Cl 10.05%, N 15.88%, O 9.07%. Orally active antimycotic structurally related to ketocona-zole, *q.v.* Prepn: J. Heeres, L. J. J. Backx, **Eur.** pat. **Appl. 6,711;** idem, **U.S.** pat. **4,267,179** (1980, 1981 both to Jans-sen); J. Heeres *et al.*, *J. Med. Chem.* **27**, 894 (1984). *In vitro* activity: A. Espinel-Ingroff *et al.*, *Antimicrob. Ag. Chemo-ther.* **26**, 5 (1984). Effect in animals vs systemic aspergillosis: J. Van Cutsem *et al.*, *ibid.* 527; vs systemic candidosis: J. Van Cutsem *et al.*, *Chemotherapy (Basel)* **33**, 52 (1987). HPLC determn in human plasma and animal tissues: R. Woestenborghs *et al.*, *J. Chromatog.* **413**, 332 (1987). Sym-posium on pharmacology and clinical efficacy: *Rev. Infect. Dis.* **9**, Suppl 1, S1-S152 (1987). Pharmacokinetics and tox-icity: H. Van Cauteren *et al.*, *ibid.* S43.

Crystals from toluene, mp 166.2°. LD_{50} (14 day) in mice, rats, dogs (mg/kg): >320, >320, >200 orally (Van Cau-teren).

THERAP CAT: Antifungal.

5132. Itramin Tosylate. *2-Aminoethanol nitrate mono-(4-methylbenzenesulfonate); 2-aminoethanol nitrate mono-p-toluenesulfonate;* 2-nitratoethylaminotoluene-p-sulfonate; *Cardisan; Tostram; Nilatil.* $C_9H_{14}N_2O_6S$; mol wt 278.30. C 38.84%, H 5.07%, N 10.07%, O 34.50%, S 11.52%. Prepn: **Swed.** pat. **168,308** (1959 to Aktiebolaget Pharmacia), *C.A.* **54**, 24405d (1960).

Crystals from ethanol, mp 132-133°.

THERAP CAT: Vasodilator.

5133. Ivermectin. 22,23-Dihydroabamectin; 22,23-dihydroavermectin B_1; 22,23-dihydro C-076B_1; MK-933; Cardomec; Eqvalan; Heartgard 30; Ivomec; Mectizan; Zimecterin. Semi-synthetic deriv of abamectin, *q.v.*, one of the avermectins, *q.v.* Ivermectin contains at least 80% of 22,23-dihydroavermectin B_{1a} and not more than 20% of 22,23-dihydroavermectin B_{1b}. Prepn: **Japan. Kokai 79 61198;** J. C. Chabala, M. H. Fisher, **U.S.** pat. **4,199,569** (1979, 1980 both to Merck & Co.); J. C. Chabala *et al.*, *J. Med. Chem.* **23**, 1134 (1980). Approaches to synthesis: H. Mrozik *et al.*, *Tetrahedron Letters* **23**, 2377 (1982); M. E. Jung *et al.*, *ibid.* **28**, 5977 (1987); M. E. Jung, L. J. Street,

Heterocycles **27**, 45 (1988). Antiparasitic activity: J. R. Egerton *et al., Brit. Vet. J.* **136**, 88 (1980); E. T. Lyons *et al., Am. J. Vet. Res.* **41**, 2069 (1980); L. S. Blair, W. C. Campbell, *ibid.* 2108. Acaricidal efficacy: C. A. Wilkins *et al., Am. J. Vet. Res.* **41**, 2112 (1980); R. O. Drummond *et al., J. Econ. Entomol.* **74**, 432 (1981). Metabolism in animals: S.-H. L. Chiu *et al., Drug Metab. Dispos.* **14**, 590 (1986). Pharmacokinetics in horses and sheep: S. E. Marriner *et al., J. Vet. Pharmacol. Ther.* **10**, 175 (1987). Evaluation in human onchocerciasis (river blindness): M. A. Aziz *et al., Lancet* **2**, 171 (1982); P. T. Soboslay *et al., Trop. Med. Parasitol.* **38**, 8 (1987). Clinical trial in onchocerciasis: A. T. White *et al., J. Infec. Dis.* **156**, 463 (1987). HPLC determn in cattle and sheep tissues: P. C. Tway *et al., J. Agr. Food Chem.* **29**, 1059 (1981); in human plasma and milk: R.

Chiou *et al., J. Chromatog.* **416**, 196 (1987). Review of chemistry and biology: W. C. Campbell *et al., Science* **221**, 823-828 (1983). Review of use in treatment of heartworm in dogs: W. C. Campbell, *Sem. Vet. Med. Surg.* **2**, 48-55 (1987). Review of clinical use and pharmacology: W. C. Campbell, G. W. Benz, *J. Vet. Pharmacol. Ther.* **7**, 1-16 (1984); T. B. Barragry, *Can. Vet. J.* **28**, 512-517 (1987); of pharmacology: J. L. Bennett *et al., Parasitol. Today* **4**, 226-228 (1988).

Off-white powder. $[\alpha]_D$ +71.5 ± 3° (c = 0.755 in chloroform). uv max (methanol): 238, 245 nm (ϵ 27100, 30100). Soly in water: ~4 μg/ml. Virtually insol in satd hydrocarbons such as cyclohexane. Highly sol in methyl ethyl ketone, propylene glycol, polyethylene glycol.

Component B$_{1a}$, C$_{48}$H$_{74}$O$_{14}$, *5-O-demethyl-22,23-dihydroavermectin A$_{1a}$, 22,23-dihydroavermectin B$_{1a}$, 22,23-dihydro C-076B$_{1a}$.* Crystals from ethanol/water, mp 155-157°.

Component B$_{1b}$, C$_{47}$H$_{72}$O$_{14}$, *5-O-demethyl-25-de(1-methylpropyl)-22,23-dihydro-25-(1-methylethyl)avermectin A$_{1a}$, 22,23-dihydroavermectin B$_{1b}$.*

THERAP CAT: Anthelmintic (Onchocerca).

THERAP CAT (VET): Anthelmintic, insecticide, acaricide.

component B$_{1a}$, R=C$_2$H$_5$ component B$_{1b}$, R=CH$_3$

5134. Ixbut. A native name for *Euphorbia lancifolia* Schlecht, *Euphorbiaceae.* Used as a galactagogue: Serrano, *Time,* August 1st, **1949**, p 38; F. Rosengarten, Jr., *Botan. Museum Leaflets (Harvard Univ.)* **26**(9-10), 277-309 (Nov.-Dec. 1978).

5135. Ixodin. Hirudin-like anticoagulant from *Ixodes ricinus* L., *Acari* and other blood-sucking ticks: Sabbatani, *Arch. Ital. Biol.* **31**, 37 (1899). Thrombokinase inhibitor. Extraction procedure: F. Markwardt, *Blutgerinnungshemmende Wirkstoffe aus blutsaugenden Tieren* (Jena, 1963) pp 88-92.

White powder. Soluble in 80% ethanol, 90% methanol and 75% acetone. Practically insol in ether. Stable at pH 1.0-12.0. Adsorbed on Amberlite IRC 50 (H-form). Acts like a high-molecular weight polypeptide. Attacked by papain.

J

5136. Jalap. Dried tuberous root of *Exogonium purga* (Hayne) Lindl. *(E. jalapa* Baill., *Ipomoea purga* Hayne), *Convolvulaceae. Habit.* Mexico; culivated in India. *Constit.* 7-12% resin, gum, sugar.

THERAP CAT: Cathartic.

5137. Jambul. Jamboo; Java plum; jumbul. Bark, fruit and seeds of *Syzygium jambolanum* (Lam.) DC. *(Eugenia jambolana* Lam.), *Myrtaceae. Habit.* East Indies. *Constit.* Bark: resin, tannin. Fruit: volatile and fixed oils, resin, tannin. Seeds: resin, fat, gallic acid, albumin.

THERAP CAT: Antidiarrheal.

5138. Janus Green B. *3-(Diethylamino)-7-[[4-(dimeth-ylamino)phenyl]azo]-5-phenylphenazinium chloride;* C.I. 11050. $C_{30}H_{31}ClN_6$; mol wt 511.09. C 70.50%, H 6.11%, Cl 6.94%, N 16.45%. Prepd by diazotizing 3-amino-7-(diethyl-amino)-5-phenylphenazinium chloride *(N,N-diethylpheno-safranine)* and coupling resulting diazo compd with *N,N*-di-methylaniline: *Colour Index* **vol. 4** (3rd ed., 1971) p 4015.

USE: As dye for cotton, wool; as biological stain; in elec-trodeposition of copper, Brown, Fellows, U.S. pat. **2,882,209** (1959 to Udylite Res. Corp.).

5139. Japan Wax. Vegetable wax; sumach wax; Japan tallow. A fat expressed from mesocarp of the fruit of *Rhus succedanea* L., *Anacardiaceae. Habit.* Japan and China. *Constit.* 10-15% palmitin; stearin, olein; 1% japanic acid and homologs. Brief review: C. S. Letcher in Kirk-Othmer *Encyclopedia of Chemical Technology* vol. 24 (Wiley-Inter-science, New York, 3rd ed., 1984) p 470.

Pale yellow, flat cakes, disks or squares with a greasy feel; somewhat tallow-like, rancid odor and taste. d 0.97-0.98. mp 53.5-55°. Acid number 22-23. Saponification number 217-237. Iodine number 10-15. Insol in water or cold alco-hol; sol in benzene, carbon disulfide, petr ether, ether, hot alc, alkalies.

USE: As a substitute for beeswax in wax varnishes or can-dles, ingredients in plasters, ointments; floor waxes, furni-ture polish. As a plasticizer in dental impression com-pounds.

5140. Jasmolins. Active insecticidal constituents of py-rethrum flowers. Isoln and structure: Godin *et al., J. Econ. Entomol.* **58**, 548 (1965); Godin *et al., J. Chem. Soc. (C)* **1966**, 322. Stereochemistry: Begley *et al., Chem. Commun.* **1972**, 1276.

Jasmolin I, $C_{21}H_{30}O_3$, *2,2-dimethyl-3-(2-methyl-1-propen-yl)cyclopropanecarboxylic acid 2-methyl-4-oxo-3-(2-penten-yl)-2-cyclopenten-1-yl ester, 4',5'-dihydropyrethrin I.* R = CH₃. Liquid. uv max (Spectrosol hexane): 219 nm (ε 21,500).
Jasmolin II, $C_{22}H_{30}O_5$, *3-(3-methoxy-2-methyl-3-oxo-1-*

propenyl)-2,2-dimethylcyclopropanecarboxylic acid 2-methyl-4-oxo-3-(2-pentenyl)-2-cyclopenten-1-yl ester, 4',5'-dihy-dropyrethrin II. R = COOCH₃. Liquid. uv max (Spectro-sol hexane): 229 nm (ε 22,900).

5141. Jasmone. *3-Methyl-2-(2-pentenyl)-2-cyclopenten-1-one.* $C_{11}H_{16}O$; mol wt 164.24. C 80.44%, H 9.82%, O 9.74%. Found in the volatile portion of oil from jasmine flowers. Natural jasmone is the *cis*-ketone. Isoln and struc-ture: Ruzicka, Pfeiffer, *Helv. Chim. Acta* **16**, 1208 (1933). Stereochemistry: Crombie, Harper, *J. Chem. Soc.* **1952**, 869. Synthesis of *cis*-jasmone: H. Hunsdiecker, *Ber.* **75**, 447 (1942); Stork, Borch, *J. Am. Chem. Soc.* **86**, 936 (1964); Büchi, Wuest, *J. Org. Chem.* **31**, 977 (1966); Sakan *et al., Chem. Letters* **1973**, 713; P. Bakuzis, M. L. F. Bakuzis, *J. Org. Chem.* **42**, 2362 (1977). Synthesis of dihydrojasmone and *cis*-jasmone: C. S. Subramaniam *et al., J. Chem. Soc. Perkin Trans. I* **1979**, 2346; T. Kato *et al., Chem. Pharm. Bull.* **28**, 349 (1980). Synthesis of *trans*-jasmone: Sisido *et al., J. Org. Chem.* **29**, 2290 (1964). Comprehensive synthetic reviews: R. A. Ellison, *Synthesis* **1973**, 397; T. L. Ho, *Synth. Commun.* **4**, 265 (1974).

cis-Form, oil, odor of jasmine, bp_{27} 146°. n_D^{20} 1.4978. uv max: 235 nm (ε 12,000).
trans-Form, oil, odor of jasmine, bp_{23} 142°. n_D^{20} 1.4974. uv max: 234 nm (ε 12,300).

USE: In perfumery.

5142. Jatrorrhizine. *5,6-Dihydro-3-hydroxy-2,9,10-tri-methoxydibenzo[a,g]quinolizinium; 7,8,13,13a-tetradehydro-3-hydroxy-2,9,10-trimethoxyberbinium;* jateorrhizine; nepro-tin. $[C_{20}H_{20}NO_4]^+$; mol wt 338.39. From root of *Jateorhiza palmata* (DC.) Miers *(J. columba* Miers), *Menispermaceae:* Feist, *Arch. Pharm.* **245**, 586 (1907); from *Berberis asiatica* Roxb. ex DC. and *B. thunbergii* DC., *Berberidaceae:* Chat-terjee *et al., J. Indian Chem. Soc.* **31**, 83 (1954); Tomita, Kikuchi, *J. Pharm. Soc. Japan* **76**, 597 (1956); from *Coptis teeta* Wall., *Ranunculaceae:* Chatterjee *et al., J. Indian Chem. Soc.* **29**, 97 (1952); from *Mahonia acanthifolia* Wall., *M. borealis* Takeda and *M. simonsii* Takeda, *Berberidaceae:* Chatterjee, Guha, *J. Am. Pharm. Assoc.* **39**, 577 (1950); Chatterjee *et al., ibid.* **40**, 36 (1951). Structure: Späth, Duschinsky, *Ber.* **58**, 1939 (1925).

Iodide, $C_{20}H_{20}INO_4$, reddish-yellow needles, mp 208-210°. Sol in water and alcohol.

5143. Javanicin. *5,8-Dihydroxy-6-methoxy-2-methyl-3-(2-oxopropyl)-1,4-naphthalenedione; 3-acetonyl-5,8-dihydr-oxy-6-methoxy-2-methyl-1,4-naphthoquinone.* $C_{15}H_{14}O_6$; mol wt 290.26. C 62.06%, H 4.86%, O 33.07%. Antibiotic substance produced by *Fusarium javanicum:* Arnstein, Cook, *J. Chem. Soc.* **1947**, 1021. Prepd by reduction of fusarubin: Ruelius, Gauhe, *Ann.* **569**, 38 (1950). Structure: Birch, Donovan, *Chem. & Ind. (London)* **1954**, 1047; Whal-ley, *ibid.* **1958**, 131; Hardegger *et al., Helv. Chim. Acta* **47**, 2027 (1964).

Red crystals with a coppery luster from ethanol, decomp 207.5-208°. Absorption max (alc): 303, 305 nm (log ϵ 3.97, 3.90); in chloroform: 307, 510 nm (log ϵ 3.99, 3.86).

Diacetyljavanicin, $C_{19}H_{18}O_8$, needles from dil acetone, dec 207-208°. Absorption max (alc): 221, 290, 426 nm (log ϵ 4.57, 4.17, 3.79).

5144. Jerusalem Artichoke. Topinambur. The subterranean stem tuber of *Helianthus tuberosus* L., *Compositae*, a kind of sunflower native to North America. Used as food by American Indians. Contains up to 20% (wet basis) of polysaccharides usable as sweetening agents: Rubin, U.S. pat. 2,782,123 (1957).

5145. Jervine. *(3β,23β)-17,23-Epoxy-3-hydroxyveratraman-11-one.* $C_{27}H_{39}NO_3$; mol wt 425.59. C 76.19%, H 9.24%, N 3.29%, O 11.28%. Steroidal alkaloid found in *Veratrum grandiflorum* (Maxim.) Loes F., *V. album* L., *V. viride* Sol., *Liliaceae*: Saito, *Bull. Chem. Soc. Japan* **9**, 15 (1934), *C.A.* **28**, 2463 (1934); Poethke, *Arch. Pharm.* **275**, 357, 571 (1937); **276**, 170 (1938); **282**, 56 (1944); Seiferle *et al.*, *J. Econ. Entomol.* **35**, 35 (1942); Jacobs, Craig, *J. Biol. Chem.* **160**, 555 (1945). Structure: Fried *et al.*, *J. Am. Chem. Soc.* **73**, 2970 (1951). *Reviews:* Wintersteiner in Graff, *Essays in Biochemistry* (Wiley, New York, 1956) pp 308-321; *idem, Festschrift Arthur Stoll* (Basel, 1957) pp 166-176; L. F. Fieser, M. Fieser, *Steroids* (Reinhold, New York, 1959) pp 870-877. Stereochemistry: Sicher, Tichy, *Tetrahedron Letters* no. 12, 6 (1959); Augustine, *Chem. & Ind. (London)* **1961**, 1448; Mitsuhashi, Shimizu, *Tetrahedron* **19**, 1027 (1963); Bailey *et al.*, *Tetrahedron Letters* **1963**, 555; Masamune *et al., ibid.* **1965**, 489. Revised stereochemistry: Scott *et al., ibid.* **1967**, 2381; Kupchan, Suffness, *J. Am. Chem. Soc.* **90**, 2730 (1968); Sprague *et al.*, *Tetrahedron* **27**, 4857 (1971). Total synthesis: Kutney *et al.*, *Can. J. Chem.* **53**, 1796 (1975).

Needles from methanol + water, mp 243.5-244.5° (Saito). $[\alpha]_D^{20}$ −150° (ethanol) (Saito); $[\alpha]_D^{20}$ −167.6° (chloroform) (Poethke). uv max: 250, 360 nm (ϵ 15000, 60). LD_{50} in mice: 9.3 mg/kg i.v.

Diacetyljervine, $C_{31}H_{43}NO_5$, mp 173-175°. $[\alpha]_D$ −112°. uv max (ethanol): 250, 360 nm (ϵ 16400, 80).

Hydrochloride, mp 300-302°.

5146. Jesaconitine. *(1α,3α,6α,14α,15α,16β)-20-Ethyl-1,6,16-trimethoxy-4-(methoxymethyl)aconitane-3,8,13,14,15-pentol 8-acetate 14-(4-methoxybenzoate).* $C_{35}H_{49}NO_{12}$; mol wt 675.75. C 62.20%, H 7.31%, N 2.07%, O 28.41%. From *Aconitum fischeri* Reich. and *A. sachalinense* F. Schmidt, *Ranunculaceae*: Makoshi, *Arch. Pharm.* **247**, 243 (1909); Majima, Morio, *Ber.* **57**, 1472 (1924). Structure: Ochiai *et al., J. Pharm. Soc. Japan* **75**, 545 (1955); Keith, Pelletier, *Chem. Commun.* **1967**, 993 (corr. *ibid.* **1968**, 1739); *eidem, J. Org. Chem.* **33**, 2497 (1968).

Amorphous powder, dec 128-131°. Sol in ether, dil acids. Perchlorate, $C_{35}H_{49}NO_{12}.HClO_4$, prisms from alcohol, dec 230-232°. $[\alpha]_D$ −16.7° (methanol).

5147. Jojoba Oil. Oil of jojoba. A liquid wax ester mixture extracted from ground or crushed seeds from *Simmondsia chinensis* and *S. californica* Nutt. *Buxaceae*, desert shrubs native to Arizona, California, and northern Mexico: Greene, Foster, *Bot. Gaz.* **94**, 826 (1933); Green *et al., J. Chem. Soc.* **1936**, 1750; McKinney, Jamieson, *Oil Soap (Chicago)* **13**, 289 (1936). Similar to sperm whale oil, it is composed essentially of C_{20} and C_{22} straight chain monoethylene acids and alcohols in the form of esters: Molaison *et al., J. Am. Oil Chem. Soc.* **36**, 379 (1959); Miwa, *ibid.* **48**, 259 (1971). Solvent effects in extraction: Knoepfler *et al., ibid.* **36**, 644 (1959). Comparison of sulfurized jojoba and sperm whale oils as high pressure lubricants: T. K. Miwa *et al., ibid.* **56**, 765 (1979). Potential chemical utilization studies: Fore *et al., ibid.* **37**, 387 (1960); J. D. Johnson, C. W. Hinman, *Science* **208**, 460 (1980). Possible uses: J. H. Brown, *Mfg. Chem.* **50**(6), 47 (1979). *Reviews:* Knoepfler, Vix, *J. Agr. Food Chem.* **6**, 118 (1958); *Products from Jojoba: A Promising New Crop for Arid Lands*, Committee on Jojoba Utilization, Natl. Res. Council, 1975; *Jojoba: New Crop for Arid Lands, New Raw Material for Industry*, Natl. Res. Council, 1985.

Liquid wax. fp 10.6 to 7°; mp 6.8 to 7°, bp_{757} (under N_2) 398°. Fire point 338°. Pour point 10°. d^{25} 0.8642. n^{25} 1.4648. Iodine no. 81.7. Saponification value 92.2. Acid value 0.32. Avg mol wt of wax esters: 606. Highly stable, and resistant to bacterial degradation; can be stored for years without becoming rancid.

USE: Potentially as lubricant, fuel, chemical feedstock, substitute for sperm whale oil. For other potential uses *see* Fore *et al., loc. cit.*, Knoepfler, Vix, *loc. cit*, Johnson, Hinman, *loc. cit.*

5148. Josamycin. *Leucomycin V 3-acetate 4B-(3-methylbutanoate);* leucomycin A_3; EN-141; Iosalide; Jomybel; Josamina. $C_{42}H_{69}NO_{15}$; mol wt 828.02. C 60.92%, H 8.40%, N 1.69%, O 28.98%. Macrolide antibiotic produced by *Streptomyces narbonensis* var. *josamyceticus* nov. var. Isoln and characterization: T. Osono *et al., J. Antibiot.* **20A**, 174 (1967). Manuf process: H. Umezawa,T. Osono, **Japan.** pat. **66 21,759** (1966 to Microbiochemical Res. Found.), *C.A.* **66**, 54258w (1967). Prepn: Y. Oka *et al.*, **Japan. Kokai 76 41,-497** (to Yamanouchi), *C.A.* **85**, 121788b (1976). Identification with leucomycin A_3: S. Omura *et al., J. Antibiot.* **23**, 511 (1970). Structure: *eidem, ibid.* **27**, 366 (1974). Absolute configuration: A. Ducruix *et al., Chem. Commun.* **1976**, 947. Stereospecific total synthesis: K. Tatsuta *et al., Tetrahedron Letters* **1980**, 2837. Retrosynthetic studies: K. C. Nicolaou *et al., J. Am. Chem. Soc.* **103**, 1222 (1981). A 17-membered aglycone was proposed at one time, *see* T. Osono, H. Umezawa in *Drug Action and Drug Resistance in Bacteria*, **1**, S. Mitsuhashi, Ed. (University Park Press, Balti-

R = $COCH_2CH(CH_3)_2$

more, 1971) pp 41-120; T. Osono *et al.*, *J. Antibiot.* **27**, 366 (1974). Pharmacology: K. Kuriaki *et al.*, *Japan. J. Antibiot.* **22**, 232 (1969). *Review:* T. Osono, H. Umezawa, *loc. cit.*

Colorless needles from benzene, mp 130-133° (after drying under reduced pressure at 100° for 5 hrs). $[\alpha]_D^{25}$ —70° (c = 1 in ethanol). uv max (0.001N HCl): 232 nm ($E_{1cm}^{1\%}$ 325). pKa 7.1 (40% aq methanol). Very sol in methanol, ethanol, acetone, chloroform, ethyl acetate, dioxane, acidic water. Sol in butanol, ether, CCl$_4$, benzene, toluene. Practically insol in water, petr ether, ligroin, *n*-hexane.

Propionate, *Josacine, Josamy, Josaxin, Wilprafen.*

THERAP CAT: Antibacterial.

THERAP CAT (VET): Antibacterial.

5149. Juglans. Butternut; white walnut; lemon walnut; oil nut. Dried inner root bark of *Juglans cinerea* L., *Juglandaceae*, collected in the autumn. *Habit.* North America. *Constit.* Resinoid juglandin, juglone, juglandic acid, fixed and volatile oils, tannin.

5150. Juglone. *5-Hydroxy-1,4-naphthalenedione; 5-hydroxy-1,4-naphthoquinone;* 8-hydroxy-1,4-naphthoquinone; C.I. 75500; C.I. Natural Brown 7; nucin; regianin. $C_{10}H_6O_3$; mol wt 174.15. C 68.96%, H 3.47%, O 27.56%. Coloring matter occurring in various *Juglandaceae* spp: Brissemoret, Combes, *Compt. Rend.* **141**, 838 (1905). Isoln from walnut shells: Combes, *Bull. Soc. Chim.* **1**(4), 800 (1907). Isoln and description of sedative properties in fish, mammals: B. A. Westfall *et al.*, *Science* **134**, 1617 (1961). Isoln from pecans and identification as inhibitory agent of mycelial growth of *Fusicladium effusum:* P. A. Hedin *et al.*, *J. Agr. Food Chem.* **28**, 340 (1980). Synthesis: Bernthsen, Semper, *Ber.* **20**, 938 (1887); Willstätter, Wheeler, *Ber.* **47**, 2798 (1914); Teuber, Götz, *Ber.* **87**, 1236 (1954); C. Grundmann, *Synthesis* **1977**, 644. Tumor-promoting activity study: B. L. Van Duuren *et al.*, *J. Med. Chem.* **21**, 26 (1978). Use as pH indicator: S. S. Sawhney, B. M. L. Bhatia, *J. Indian Chem. Soc.* **57**, 438 (1980).

Yellow needles from benzene + petr ether, mp 155°. Sublimes. Absorption max (methanol): 420 nm (log ε 3.56). Volatile with steam. Slightly sol in hot water; freely sol in chloroform, benzene; sol in alcohol, ether; sol in aq solns of alkalies giving a purplish-red soln.

5151. Julocrotine. *N-[2,6-Dioxo-1-(2-phenylethyl)-3-piperidinyl]-2-methylbutanamide;* yulocrotine; β-phenylethylimide of *N*-α-methylbutyrylglutamic acid. $C_{18}H_{24}N_2O_3$; mol wt 316.39. C 68.33%, H 7.65%, N 8.85%, O 15.17%. Isoln from root of *Julocroton montevideensis* Klotzsch, *Euphorbiaceae:* Anastasi, *Anales Asoc. Quim. Argentina* **23**, 348 (1925), *C.A.* **20**, 2332 (1926). Structure: Nakano *et al.*, *J. Org. Chem.* **26**, 1184 (1961).

Crystals from ether + petr ether, mp 108-109°. $[\alpha]_D$ —9° (c = 1.24 in chloroform), —50.1° (c = 1.19 in methanol). uv max (ethanol): 252, 258, 264, 268 nm (log ε 2.43, 2.44, 2.30, 2.18). Sol in alc, benzene, chloroform, ether, slightly in petr ether.

5152. Juniper Berries. Dried ripe fruit of *Juniperus communis* L., *Cupressaceae*. The wood and tops are also employed. *Habit.* Northern Europe, Asia, North America. *Constit.* Volatile oil, juniperin, resin, proteins, formic, acetic and malic acids.

USE: In liquor manuf, perfumery and cosmetics: A. Pomini, *Minerva Farm.* **11**, 207 (1962).

THERAP CAT: Diuretic.

5153. Juniper Tar. Oil of cade; empyreumatic oil of juniper; oil of juniper tar; Haarlem oil; Harlem oil; Tilly drops; Holland balsam; silver drops; silver balsam; Kaparlem; Caparlem. Volatile oil from wood of *Juniperus oxycedrus* L., *Cupressaceae*. *Constit.* Chiefly cadinene. Toxicity data: P. M. Jenner *et al.*, *Food Cosmet. Toxicol.* **2**, 327 (1964). *Review:* D. L. J. Opdyke, *ibid.* **13**, 733-734 (1975).

Dark brown, more or less viscid liq; smoky odor; acrid, slightly aromatic taste. d_{25}^{25} 0.950-1.055. n_D^{20} 1.510- 1.530. Very slightly sol in water; sol in 3 vols ether, in chloroform, amyl alcohol, glacial acetic acid, oil turpentine; partly sol in alcohol or petr ether. LD$_{50}$ orally in rats: 8014 mg/kg (Jenner).

USE: As a gin-like flavor; in perfumery.

THERAP CAT: Topical anti-eczematic.

5154. Justicidins. Lignans from various species of *Justicia, Acanthaceae*. Isoln and structures for justicidins A, B, C, D, E and F have been elucidated. Isoln of A and B: Munakata *et al.*, *Tetrahedron Letters* **1965**, 4167; Ohta *et al.*, *Agr. Biol. Chem.* **33**, 610 (1969); Okigawa *et al.*, *Tetrahedron* **26**, 430 (1970). Structure of A: Govindachari *et al.*, *Tetrahedron Letters* **1967**, 3517; Horii *et al.*, *Chem. Pharm. Bull.* **19**, 535 (1971). Synthesis of B: Munakata *et al.*, *Tetrahedron Letters* **1967**, 2831; Munakata, Katsura, **Brit.** pat. **1,-178,341** (1970), *C.A.* **72**, 100365c; T. Momose *et al.*, *Chem. Pharm. Bull.* **26**, 3195 (1978). Isoln and structures of C and D: Ohta, Munakata, *Tetrahedron Letters* **1970**, 923. Synthesis of C: Horii *et al.*, *Chem. Pharm. Bull.* **17**, 1878 (1969). Structure of E: Wada, Munakata, *Tetrahedron Letters* **1970**, 2017. Synthesis of E: Holmes, Stevenson, *J. Org. Chem.* **36**, 3450 (1971). Synthesis of D, E, and F: Z. I. Horii *et al.*, *Chem. Pharm. Bull.* **25**, 1803 (1977).

Justicidin A, $C_{22}H_{18}O_7$, *9-(1,3-benzodioxol-5-yl)-4,6,7-trimethoxynaphtho[2,3-c]furan-1(3H)-one, diphyllin methyl ether.* R = OCH$_3$. Crystals, mp 263°. uv max (CHCl$_3$): 265, 295, 315, 335 nm (log ε 4.35, 4.13, 4.13, 3.33).

Justicidin B, $C_{21}H_{16}O_6$, *dehydrocollinusin.* R = H. Crystals, mp 240°. uv max (CHCl$_3$): 260, 295, 310, 350 nm (log ε 4.52, 4.13, 4.13, 3.41).

USE: Piscicidal agent.

5155. Juvenile Hormones. JH. Hormones controlling the larval metamorphosis of insects; so named since they induce the retention of insects' juvenile characteristics and prevent maturation. Produced by the small gland, corpora allata, of the male wild silk moth, *Hyalophora cecropia* L. The isolate consists of a 9:1 mixture of *methyl 12,14-dihomojuvenate* also called *C-18 JH* and *methyl 12-homojuvenate* or *C-17 JH*. Isoln and structure: Williams, *Nature* **178**, 212 (1956); Röller *et al.*, *Angew. Chem.* **79**, 190 (1965); *Intl. Ed.* **6**, 179 (1967); cf. *Chem. Eng. News* (April 10, 1967) 48-49. Abs config: Meyer *et al.*, *Proc. Nat. Acad. Sci. USA* **68**, 2312 (1971); Nakanishi *et al.*, *Chem. Commun.* **1971**, 1235. First syntheses of racemic C-18 JH: Dahm *et al.*, *J. Am. Chem. Soc.* **89**, 5292 (1967); Corey *et al.*, *ibid.* **90**, 5618 (1968); for bibliography of syntheses *see* Henrick *et al.*, *ibid.* **94**, 5374 (1972). Synthesis of optically active C-18 JH: Loew, Johnson, *ibid.* **93**, 3765 (1971); Faulkner, Peterson, *ibid.* 3767. Synthesis allowing for structural variations in the alkyl residues: Mori *et al.*, *Agr. Biol. Chem.* **35**, 1116 (1971); Cochrane, Hanson, *J. Chem. Soc. Perkin Trans. I* **1972**, 361.

Synthesis involving dihydrothiopyran intermediates: Kondo et al., *Chem. Commun.* **1972,** 1311; Stotter, Hornish, *J. Am. Chem. Soc.* **95,** 4444 (1973). Synthesis of C-17 JH racemate: Johnson et al., *Proc. Nat. Acad. Sci. USA* **62,** 1005 (1969); Anderson et al., *J. Am. Chem. Soc.* **94,** 5379 (1972); Mori et al., *Agr. Biol. Chem.* **36,** 1085 (1972). Biological activity, specificity, low mammalian toxicity, and short environmental persistence, indicate potential use as insecticides. Toxicity in mice: Siddall, Slade, *Nature New Biol.* **229,** 158 (1971). Structural requirements for JH activity: Wigglesworth, *Nature* **221,** 190 (1969); Slama, *Ann. Rev. Biochem.* **40,** 1079 (1971). Reviews: Röller, Dahm, *Recent Progr. Horm. Res.* **24,** 651 (1968); Trost, *Accounts Chem. Res.* **3,** 120 (1970); L. M. Riddiford, *Ann. Rev. Physiol.* **42,** 511 (1980). Book: *Insect Juvenile Hormones*, J. L. Menn, M. Beroza, Eds. (Academic Press, New York, 1972).

C-18 JH, $C_{18}H_{30}O_3$, *7-ethyl-9-(3-ethyl-3-methyloxiranyl)-3-methyl-2,6-nonadienoic acid methyl ester; cis-10,11-epoxy-7-ethyl-3,11-dimethyl-trans,trans-2,6-tridecadienoic acid; methyl cis-10,11-oxido-3,11-dimethyl-7-ethyltrideca-trans,-trans-2,6-dienoate.* R = C_2H_5.

C-17 JH, $C_{17}H_{28}O_3$, *cis-10,11-epoxy-3,7,11-trimethyl-trans,trans-2,6-tridecadienoic acid methyl ester.* R = CH_3.

K

Kainic Acid as a Tool in Neurobiology, E. G. McGeer *et al.,* Eds. (Raven Press, New York, 1978). *Review:* J. T. Coyle, *Ciba Found. Symp.* **126,** 186-203 (1987).

5156. Kaempferol. *3,5,7-Trihydroxy-2-(4-hydroxyphenyl)-4H-1-benzopyran-4-one; 3,4',5,7-tetrahydroxyflavone;* swartziol; trifolitin; nimbecetin; populnetin; robigenin; rhamnolutein; pelargidenolon 1497. $C_{15}H_{10}O_6$; mol wt 286.23. C 62.94%, H 3.52%, O 33.54%. Plant flavonoid isolated from *Delphinium consolida* L., *Ranunculaceae:* A. G. Perkin, E. J. Wilkinson, *J. Chem. Soc.* **81,** 585 (1902); from grapefruit (*Citrus paradisi* Macf., *Rutaceae):* Dunlap, Wender, *Anal. Biochem.* **4,** 110 (1962); from stems and seeds of *Cuscuta reflexa* Roxb., *Convolvulaceae:* K. W. Gopinath *et al., J. Sci. Ind. Res.* **21B,** 601 (1962), *C.A.* **58,** 7134e (1963). Identity with swartziol: M. R.-R. Paris, L. Bézanger-Beauquesne, *Compt. Rend.* **242,** 1761 (1956). Structure: S. von Kostanecki, A. Rozycki, *Ber.* **34,** 3721 (1901). Total synthesis: Y.-H. Lu *et al., Yao Hsueh Hsueh Pao* **15,** 477 (1980), *C.A.* **94,** 174808a (1981). Convenient synthesis: M. Ichikawa *et al., Org. Prep. Proc. Int.* **14,** 183 (1982). Mutagenicity studies: J. T. MacGregor, L. Jurd, *Mutat. Res.* **54,** 297 (1978); A. A. Hardigree, J. L. Epler, *ibid.* **58,** 231 (1978). HPLC determination in soybean: J. D. Gaynor *et al., Chromatographia* **25,** 1049 (1988).

Yellow needles, mp 276-278°; also reported as light yellow powder from ethanol-water, mp 278-280° (dec) (Ichikawa). uv max: 265, 365 nm. Slightly sol in water; sol in hot alcohol, ether or alkalies.

3,7-Dirhamnoside, $C_{27}H_{30}O_{14}$, *kaemferitrin, lespedin, Lespenephryl.* Isoln from the leaves of *Indigo arrecta* Benth., *Leguminosae:* Perkin, *J. Chem. Soc.* **91,** 435 (1907); from leaves of *Trichosanthus cucumeroides* Maxim., *Cucurbitaceae:* Nakaoki, Morita, *J. Pharm. Soc. Japan* **77,** 108 (1957). Identity with lespedin: Hatlori, *Nature* **168,** 788 (1952). Crystals from dil alc, mp 201-203°. Slightly sol in boiling water and cold alcohol.

3-Glucoside, $C_{21}H_{20}O_{11}$, *astragalin.* From *Podophyllum peltatum* L. and *P. emodi* Wall., *Berberidaceae.* Isoln and structure: von Wartburg, Kuhn, *Experientia* **21,** 67 (1965). Crystals from methanol, mp 175-178°. $[\alpha]_D^{20}$ −16.9° (c = 0.45 in methanol).

5157. Kainic Acid. *2-Carboxy-4-(1-methylethenyl)-3-pyrrolidineacetic acid;* 2-carboxy-3-carboxymethyl-4-isopropenylpyrrolidine; digenic acid; α-kainic acid; L_s-*xylo*-kainic acid; Digenin; Helminal. $C_{10}H_{15}NO_4$; mol wt 213.23. C 56.32%, H 7.09%, N 6.57%, O 30.01%. Anthelmintic principle from the dried red alga *Digenea simplex* (Wulf.) Ag., *Rhodomelaceae:* Murakami *et al., J. Pharm. Soc. Japan* **73,** 1026 (1953); **Japan.** pat. **4947('54),** *C.A.* **49,** 13604i (1955); Katsuya *et al.,* **Japan.** pat. **1942('64)** (to New-Japan Pharmaceutical Co). Structure: Watase, Nitta, *Bull. Chem. Soc. Japan* **30,** 889 (1957); Watase *et al., ibid.* **31,** 714 (1958). Eight theoretical stereoisomers, Nitta *et al., Nature* **181,** 761 (1958); **Brit.** pat. **795,750** (1958 to Takeda). Synthesis: Ueno; Tatsuoka *et al.,* **U.S.** pats. **2,902,492** and **2,954,384** (1959, 1960 both to Takeda); W. Oppolzer, H. Andres, *Helv. Chim. Acta* **62,** 2282 (1979). Excitotoxic amino acid used to identify a specific subset of EAA receptors. Consequently the receptors are known as kainate receptors. Neurotoxic activity: J. V. Nadler, *Life Sci.* **24,** 289 (1979). Mechanism of neurotoxicity: E. G. McGeer *et al., Adv. Neurol.* **23,** 577 (1979); J. T. Coyle *et al., ibid.* 593; J. W. Ferkany *et al., Nature* **298,** 757 (1982); J. Garthwaite, G. Garthwaite, *ibid.* **305,** 138 (1983). Induces epileptogenic lesions: J. V. Nadler, *Neurosci. Res. Program Bull.* **19,** 369 (1981). Autoradiographic characterization of binding sites: J. T. Greenamyre *et al., J. Pharmacol. Exp. Ther.* **233,** 254 (1985). *Book:*

Needles, dec 251°. $[\alpha]_D^{24}$ −14.8° (c = 1.01). Intense absorption at 6.05 and 11.2 μ. Sol in water. Insol in ethanol. Stable in boiling aq solns.

USE: Neurobiological tool.

THERAP CAT: Anthelmintic (Nematodes).

5158. Kallidin. *N²-L-Lysylbradykinin;* kallidin-10; kallidin II. $C_{56}H_{85}N_{17}O_{12}$; mol wt 1188.44. C 56.60%, H 7.21%, N 20.04%, O 16.16%. A hypotensive and smooth muscle-stimulating principle formed by the proteolysis of kininogen by glandular and other kallikreins (*q.v.*). The decapeptide structure is a homologue of bradykinin (*q.v.*): Werle *et al., Z. Physiol. Chem.* **326,** 174 (1961). Synthesis: Nicolaides *et al., Biochem. Biophys. Res. Commun.* **6,** 210 (1961); Pless *et al., Helv. Chim. Acta* **45,** 394 (1962).

Lys-Arg-Pro-Pro-Gly-Phe-Ser-Pro-Phe-Arg

Amorphous precipitate. $[\alpha]_D^{21}$ −57° (c = 1 in N acetic acid). R_f value in a butanol/acetic acid/water system (70:10:20) 0.15.

THERAP CAT: Vasodilator.

5159. Kallikrein. Kallidinogenase; Callicrein; Padutin; Padreatin; Padukrein; Glumorin; Depot-Glumorin; Diminal; Circuletin; Onokrein P; Prokrein; Promotin. Hypotensive enzyme which releases kinins from plasma proteins. Major sources in the body are blood plasma, glandular tissues, and urine, occurring abundantly in the pancreas, parotid and submaxillary glands, in intestinal wall, in feces, in duodenal juice, and to a lesser degree in kidney. Isoln from mammalian pancreas or urine: Abelous, Bardier, *C.R. Soc. Biol.* **64,** 848 (1908); Webster *et al., Proc. Soc. Exp. Biol. Med.* **93,** 181 (1956). Separation into two components, kallikreins A and B: E. Habermann, *Z. Physiol. Chem.* **328,** 15 (1962); F. Fiedler, E. Werle, *ibid.* **348,** 1087 (1967); C. Kutzbach, G. Schmidt-Kastner, *ibid.* **353,** 1099 (1972). Prepn of high purity material: Werle, Trautschold, **Ger.** pat. **1,102,973** (1960 to Bayer). Review on pig pancreatic kallikrein: F. Fiedler, *Methods Enzymol.* **45B,** 289-303 (1976); on human kallikrein and prekallikrein: R. W. Colman, A. Bagdasarian, *ibid.* 303-322. Plasma kallikrein differs from glandular or urinary kallikrein. The latter two liberate kallidin, *q.v.;* the former releases bradykinin, *q.v.,* both from the common precursor, kininogen. Pharmacology: Franz, Marquardt, *Arzneimittel-Forsch.* **10,** 779 (1960). *Reviews:* Schachter, *Physiol. Rev.* **49,** 509 (1969); Suzuki *et al., Advan. Exp. Med. Biol.* **8,** 15 (1970).

THERAP CAT: Vasodilator.

5160. Kamala. Kamila; kameela; spoonwood. Glands and hairs covering the fruits of *Mallotus philippinensis* Muell.-Arg., *Euphorbiaceae.* Habit. Philippine Islands, India, China, Australia. Constit. Rottlerin, isorathlerin, resins, wax: Khorana, Motiwala, *Indian J. Pharm.* **11,** 37 (1949).

THERAP CAT: Anthelmintic.

THERAP CAT (VET): Has been used as purgative, teniacide, flukicide.

5161. Kanamycin. Antibiotic complex produced by *Streptomyces kanamyceticus* Okami & Umezawa from Japanese soil: Umezawa *et al., J. Antibiot.* **10A,** 181 (1957); **U.S.** pat. **2,931,798** (1960). Comprised of three components, kanamycin A, the major component (usually designated as kanamycin) and kanamycins B and C, two minor congeners. Isoln and purification of kanamycins A and B and their salts: Johnson *et al.,* and Johnson, Hardcastle, **U.S.** pats.

2,936,307 and 2,967,177 (1960, 1961 both to Bristol-Myers). Separation process: Rothrock, Putter, U.S. pat. 3,032,547 (1962 to Merck & Co.). Prepn of kanamycin C: Murase *et al., J. Antibiot.* **14A**, 156 (1961). Studies on kanamycin B: Wakazawa *et al., ibid.* 180, 187. Structure of kanamycin A: Ogawa *et al., ibid.* **11A**, 169 (1958); Cron *et al., J. Am. Chem. Soc.* **80**, 4741 (1958). Structure of kanamycin B: Ito *et al., J. Antibiot.* **17A**, 189 (1964). Structure of kanamycin C: Murase, *ibid.* **14A**, 367 (1961). Abs config of kanamycin A: Hichens, Rinehart, *J. Am. Chem. Soc.* **85**, 1547 (1963); Umezawa *et al., Bull. Chem. Soc. Japan* **39**, 1244 (1966). Crystal structure of kanamycin A: Koyama *et al., Tetrahedron Letters* **1968**, 1875. Monograph: *Ann. N.Y. Acad. Sci.* vol. **76**, Art. 2, pp 17-408 (1958). Synthesis of kanamycin A: Umezawa *et al., J. Antbiot.* **21**, 367 (1968); Nakajima *et al., Tetrahedron Letters,* **1968**, 623; Umezawa *et al., Bull. Chem. Soc. Japan* **42**, 533 (1969). Synthesis of kanamycin B: eidem, *J. Antibiot.* **21**, 424 (1968); *Bull. Chem. Soc. Japan* **42**, 537 (1969). Chemical conversion of kanamycin B to kanamycin C: S. Toda *et al., J. Antibiot.* **30**, 1002 (1977). Synthesis of kanamycin C: Umezawa *et al., Bull. Chem. Soc. Japan* **41**, 533 (1968); *J. Antibiot.* **21**, 162 (1968). Effects on protein synthesis: Suzuki *et al., ibid.* **23**, 99 (1970). Toxicity data (kanamycin A sulfate): Zel'tser *et al., Antibiotiki* **19**, 552 (1974). Comprehensive description: P. J. Claes *et al.,* in *Analytical Profiles of Drug Substances* vol. **6**, K. Florey, Ed. (Academic Press, New York, 1977) pp 259-296.

Kanamycin A, $C_{18}H_{36}N_4O_{11}$, *O-3-amino-3-deoxy-α-D-glucopyranosyl-(1→6)-O-[6-amino-6-deoxy-α-D-glucopyranosyl-(1→4)]-2-deoxy-D-streptamine.* R = NH_2; R' = OH. Crystals from methanol + ethanol. $[\alpha]_D^{24}$ +146° (0.1N H_2SO_4). LD_{50} i.v. in mice: 583 mg/kg (Wakazawa).

Kanamycin A sulfate, *Cantrex, Cristalomicina, Enterokanacin, Kamycin, Kamynex, Kanabristol, Kanacedin, Kanamytrex, Kanasig, Kanatrol, Kanicin, Kannasyn, Kantrex, Kantrox, Klebcil, Otokalixin, Resistomycin (Bayer), Ophtalmokalixan, Kantrexil, Kano, Kanescin, Kanaqua.* (U.S.P. requires that kanamycin sulfate contains not less than 75% kanamycin A on an anhydr basis). Irregular prisms, dec over a wide range above 250°C. Freely sol in water. Practically insol in the common alcohols and nonpolar solvents. LD_{50} in mice: 20.7 g/kg orally; 1450 mg/kg i.p. (Zel'tser).

Kanamycin B, $C_{18}H_{37}N_5O_{10}$, *NK 1006, bekanamycin, aminodeoxykanamycin.* R = R' = NH_2. Crystals, mp 178-182° (dec). $[\alpha]_D^{18}$ +130° (c = 0.5 in water). $[\alpha]_D^{21}$ +114° (c = 0.98 in water). Sol in water, formamide; slightly sol in chloroform, isopropyl alcohol. Practically insol in the common alcohols and nonpolar solvents. LD_{50} i.v. in mice: 136 mg/kg (Wakazawa).

Kanamycin B sulfate, *Coltericin, Kanendomycin, Kanendos.* Pharmacokinetics: F. Di Nola *et al., Minerva Med.* **70**, 1803 (1979).

Kanamycin C, $C_{18}H_{36}N_4O_{11}$. R = OH; R' = NH_2. Crystals from methanol + ethanol, dec above 270°. $[\alpha]_D^{20}$ +126° (H_2O). Sol in water; slightly sol in formamide. Practically insol in the common alcohols and nonpolar solvents.

THERAP CAT: Antibacterial.

THERAP CAT (VET): Antibacterial.

5162. Kaolin. Bolus alba; China clay; porcelain clay; white bole; argilla. Essentially a hydrated aluminum silicate, approximately $H_2Al_2Si_2O_8 \cdot H_2O$. Prepared for pharmaceutical and medicinal purposes by levigating with water to remove sand, etc.

White or yellowish-white, earthy mass or white powder; unctuous when moist. Insol in water, cold acids or in alkali hydroxides.

USE: Manuf porcelain, pottery, bricks, Portland cement; ultramarine, color lakes, refractory mortar; plaster material, filler for paper; electric and heat insulators; clarifying liquids; drying and emollient agent.

THERAP CAT: Adsorbent.

THERAP CAT (VET): Topical and G.I. adsorbent. Poultice.

5163. Karanjin. *3-Methoxy-2-phenyl-4H-furo[2,3-h]-1-benzopyran-4-one.* $C_{18}H_{12}O_4$; mol wt 292.28. C 73.96%, H 4.14%, O 21.90%. From *Pongamia glabra* Vent., *Leguminosae:* Beal, Katti, *J. Am. Pharm. Assoc.* **14**, 1086 (1925); Rao, Rao, *J. Indian Chem. Soc.* **17**, 526 (1940); Bhat *et al., J. Am. Oil Chem. Soc.* **33**, 197 (1956). Structure: Limaye, *Rasayanam* **1**, 1 (1936), *C.A.* **31**, 2206[9] (1937); Manjunath *et al., Ber.* **72B**, 39 (1939). Synthesis: Seshadri, Venkateswarlu, *Proc. Indian Acad. Sci.* **13A**, 404 (1941); **17A**, 16 (1943); Kawase *et al., Bull. Chem. Soc. Japan* **28**, 273 (1955); Rao, Seshadri, *Proc. Indian Acad. Sci.* **33A**, 168 (1951); Aneja *et al., Tetrahedron* **2**, 203 (1958); Raizada *et al., J. Sci. Ind. Res.* **19B**, 76 (1960).

Needles from methanol, mp 157-158°. Sol in methanol, ethanol, chloroform, benzene, ether, concd H_2SO_4, HNO_3, HOAc, HCl; practically insol in petr ether, dil mineral acids.

5164. Karaya Gum. Gum karaya; kadaya; katilo; kullo; kuteera; sterculia; Indian tragacanth; mucara. The dried exudate of the tree *Sterculia urens* Roxb., *Sterculiaceae,* found in India, especially in the Gujerat region and in the central provinces: Toothaker, *The Soluble Gums* (Philadelphia, 1921); Mantell, *The Water-Soluble Gums* (New York, 1947). Constituents and structure: Hirst, Dunstan, *J. Chem. Soc.* **1953**, 2332. Structure is a partially acetylated polysaccharide containing about 8% acetyl groups and about 37% uronic acid residues. *Reviews:* F. Smith, R. Montgomery, *The Chemistry of Plant Gums and Mucilages* (Reinhold, New York, 1959); Goldstein, Alter, in *Industrial Gums,* R. L. Whistler, Ed. (Academic Press, New York, 2nd ed., 1973) pp 273-287.

Finely ground white powder, faint odor of acetic acid. Acid to litmus. Absorbs water rapidly to form viscous mucilages at low concs. Viscosity decreases on addn of acid or alkali. Color of the soln lightens in acidic media and darkens in alkaline soln due to the presence of tannins. Gum karaya loses viscosity forming ability when stored in the dry state, the loss being greater for a powdered material than for the crude gum. Cold storage inhibits this degradation.

Note: Karaya gum occurring in broken irregular pieces having a somewhat crystalline appearance has been referred to commercially as 'crystal' gum.

USE: As denture adhesive; as binder in paper manuf; as stabilizer, thickener, texturizer, emulsifier in foods; as thickening agent for dyes in textile industry. A substitute for gum tragacanth.

THERAP CAT: Cathartic.

5165. Karsil. *N-(3,4-Dichlorophenyl)-2-methylpentanamide; 3',4'-dichloro-2-methylvaleranilide;* Niagara 4562. $C_{12}H_{15}Cl_2NO$; mol wt 260.17. C 55.40%, H 5.81%, Cl 27.26%, N 5.38%, O 6.15%. Prepd from 3,4-dichloroaniline and 2-methylvaleryl chloride: Dorschner *et al.,* **Brit.** pat. **869,169** (1961 to FMC).

Crystals, mp 106-107°.
USE: Herbicide.

5166. Kava. Kava-kava; ava-ava; kawa. Dried rhizome and roots of *Piper methysticum* Forst., *Piperaceae*. *Habit*. Polynesia. Most important constituents are: kawain, dihydrokawain, methysticin, dihydromethysticin, and yangonin: Borsche, Lewinsohn, *Ber.* **66**, 1792 (1933) and references to preceding papers therein. Chemical and pharmacological investigation of the kava constituents: Klohs *et al.*, *J. Med. Pharm. Chem.* **1**, 95 (1959); Meyer, Kretzschmar, *Klin. Wochenschr.* **44**, 902 (1966). Review of chemistry, pharmacology and historical sketch: *U.S. Public Health Service Publ. No. 1645*, D. H. Efron, Ed., pp 103-181 (1967).
Note: Kava is also the popular name for the intoxicating drink prepared from the plant's roots.

5167. Kawain. *5,6-Dihydro-4-methoxy-6-(2-phenylethenyl)-2H-pyran-2-one;* 5-hydroxy-3-methoxy-7-phenyl-2,6-heptadienoic acid δ-lactone; 4-methoxy-6-(β-phenylvinyl)-5,6-dihydro-α-pyrone; 4-methoxy-6-styryl-5,6-dihydro-α-pyrone; kavain; gonosan. $C_{14}H_{14}O_3$; mol wt 230.25. C 73.02%, H 6.13%, O 20.85%. From the rhizome and roots of *Piper methysticum* Forst., *Piperaceae* (kava): Borsche, Peitzsch, *Ber.* **63**, 2414 (1930); Hänsel, Beiersdorff, *Naturwiss.* **45**, 573 (1958); *eidem*, *Arzneimittel-Forsch.* **9**, 581 (1959). Synthesis: Fowler, Henbest, *J. Chem. Soc.* **1950**, 3642; Kostermans, *Nature* **166**, 788 (1950); *idem*, *Rec. Trav. Chim.* **70**, 79 (1951); Z. H. Israili, E. E. Smissman, *J. Org. Chem.* **41**, 4070 (1976). Abs config: Snatzke, Hänsel, *Tetrahedron Letters* **1968**, 1797. Crystal and molecular structure: A. Yoshino, W. Nowacki, *Z. Kristallogr., Kristallgeom., Kristallphys., Kristallchem.* **136**, 66 (1972), *C.A.* **78**, 63591z (1973).

C_6H_5—CH=CH

OCH₃

(+)-Form, rods from methanol + ether, mp 105-106°. $bp_{0.1}$ 195-197°. $[\alpha]_D^{20}$ +105° (abs alc). uv max (methanol): 210, 245, 282 nm (log ε 4.38, 4.44, 2.81). Practically insol in water. Sol in acetone, ether, methanol; slightly sol in hexane.
(±)-Form, needles from methanol, mp 146-147°.
Dihydrokawain, $C_{14}H_{16}O_3$, *marindinin*. Crystals from ether, mp 58-60°. $[\alpha]_D^{24}$ +31° (methanol). uv max (methanol): 236 nm (log ε 4.14). Sol in alc, chloroform; moderately sol in ether. Practically insol in petr ether, water.

5168. Kebuzone. *4-(3-Oxobutyl)-1,2-diphenyl-3,5-pyrazolidinedione;* 1,2-diphenyl-4-(γ-ketobutyl)-3,5-pyrazolidinedione; 1,2-diphenyl-4-(3'-oxobutyl)-3,5-dioxopyrazolidine; ketophenylbutazone; KPB; Chebutan; Chepirol; Chetazolidin; Chetil; Copirene; Hichillos; Ketason; Ketazone; Ketobutane-Jade; Pecnon; Phloguron; Recheton. $C_{19}H_{18}N_2O_3$; mol wt 322.35. C 70.79%, H 5.63%, N 8.69%, O 14.89%. Prepn: Deuss *et al.*, U.S. pat. **2,910,481** (1959 to Geigy). Review of pharmacology: Horakova *et al.*, *Pharmacotherapeutica* **1950-1959**, 335-350 (1963), *C.A.* **60**, 6072g (1964). Metabolism: Nemecek *et al.*, *Arzneimittel-Forsch.* **16**, 1339 (1966); Queisnerova, Nemecek, *Cesk. Farm.* **20**, 55 (1971), *C.A.* **75**, 47007u (1971).

C_6H_5
N
C_6H_5
N
CH₃COCH₂CH₂
O

Crystals, mp 115.5-116.5° or 127.5-128.5° depending on cryst form.
THERAP CAT: Antirheumatic.

5169. Kefir Fungi. Kefir grains; kefir seeds. A conglomeration of various fungi, including *Dispora caucasica*, *Schizomycetes* and a species of *Saccharomyces*.
Grayish-yellow lumps, irregular in size; firm, toughly gelatinous consistency, becoming cartilaginous and brittle when dry.
USE: Preparing a nutritious beverage of fermented milk known as kefir.

5170. Keratin. A protein obtained from hair, wool, horn, nails, claws, beaks, scales, membranes of egg shells and nerve tissue. There are two types of keratins—hard keratin of hair, horn, nails, etc., and soft keratin (pseudokeratin) of the epidermis, and whalebone: Rudall, *Advan. Protein Chem.* **7**, 253-290 (1952). The keratins contain all of the common amino acids and differ from other fibrous structural proteins chiefly by their high cystine content. Sequence studies revealed no preferable grouping or periodicity. Amino acid compn of a few keratins: Tristram in H. Neurath, K. Bailey, *The Proteins* vol. **1A** (Academic Press, New York, 1953) p 220. Stability of the protein is due to frequent primary valence cross-links (disulfide bonds) and secondary valence cross-links (hydrogen bonds) between neighboring polypeptide chains. Prepn of different forms for different purposes: Grassmann, Ger. pats. **673,203** and **682,257** (both 1939). Molecular structure of α-keratin: Fraser *et al.*, *Nature* **203**, 1231 (1964); of β-keratin: M. L. Huggins, *Macromolec.* **13**, 465 (1980). *Reviews:* Ward, Lundgren, *Advan. Protein Chem.* **9**, 243-297 (1954); Crewther *et al.*, *ibid.* **20**, 191 (1965); Bradbury, *ibid.* **27**, 111 (1973).
Characteristic properties ascribed to keratins: *(a)* insolubility in water, including aq solns of salts, hydrotropic substances, and dil acids and bases at temps not much above room temp; *(b)* resistance to proteolytic enzymes; *(c)* resistance to hydrolysis; *(d)* solvolysis by mixtures of substances which break the —S—S— bonds and the hydrogen bonds.
USE: Coating "enteric" pills which are unaffected in the stomach, but dissolved by the alkaline intestinal secretions. Detailed coating instructions: Dale, *Pharm. J.* **129**, 494 (1932). Other uses are in formulations of foam type extinguishers and in the production of protein hydrolyzates.

5171. Keratinase. M-Zyme. A proteolytic enzyme, commercially produced by a strain of *Streptomyces fradiae:* Nickerson, Noval, U.S. pat. **2,988,487** (1961 to Rutgers Res. and Educ. Found.). Can convert about 50% of the dry wt of wool into a water-sol form, permitting hair separation from hides. The amount of enzyme needed to dehair hides is not enough to cause much digestion of the hair itself, which is recovered and used in felt making. The enzyme attacks at the base of the hair shaft, which is particularly sensitive to the action of keratinase.
Soluble in water. Non-dialyzable. Optimum efficiency at pH 8.5-9.5. Becomes inactive by heating to 100° for 5 min. The ability to digest keratin appears to be dependent on the presence of at least trace amounts of metal ions which form chelates.
USE: In the dehairing of hides and skins: Robison, Nickerson, U.S. pat. **2,988,488** (1961 to Mearl). In depilatory compositions for human use: Mattin, Greenstein, U.S. pat. **2,988,485** (1961 to Mearl).

5172. Kermesic Acid. *9,10-Dihydro-3,5,6,8-tetrahydroxy-1-methyl-9,10-dioxo-2-anthracenecarboxylic acid;* C.I. Natural Red 3; C.I. 75460. $C_{16}H_{10}O_8$; mol wt 330.26. C 58.19%, H 3.05%, O 38.76%. Principle constituent of kermes, one of the oldest known insect dyes. Isoln and characterization: O. Dimroth, *Ber.* **43**, 1387 (1910); O. Dimroth, W. Scheurer, *Ann.* **399**, 43 (1913). Structure: O. Dimroth, R. Fick, *Ann.* **411**, 315 (1916). Revised structure: D. D. Gadgil *et al.*, *Tetrahedron Letters* **1968**, 2223. Synthesis: D. W. Cameron *et al.*, *Chem. Commun.* **1978**, 688; *eidem*, *Aust. J. Chem.* **34**, 2401 (1981).

HO   O   CH₃
                COOH
HO
HO   O   OH

Dark red rosettes from acetic acid, mp > 320° (dec). Absorption max: 276, 312, 498 nm (log ε 4.52, 4.12, 3.96).

Slightly sol in cold water; sol in hot water (yellowish-red soln). Violet-red in conc H_2SO_4, turning blue on addn of boric acid. Violet in aq NaOH.

USE: As brilliant scarlet dye.

5173. Kerosene. Kerosine. A mixture of petroleum hydrocarbons, chiefly of the methane series having from 10 to 16 carbon atoms per molecule. It constitutes the fifth fraction in the distillation of petroleum (after the petr ethers and before the oils). A typical analysis of the kerosene fraction from Midcontinent crude includes *n*-dodecane, three alkyl derivatives of benzene, naphthalene, 1- and 2-methyl-5,6,7,8-tetrahydronaphthalene.

Pale yellow or water-white, mobile, oily liquid. Characteristic, not altogether disagreeable odor. d about 0.80. bp 175-325°. Flash pt 150-185°F. (65-85°C). Insol in water. Misc with other petroleum solvents. A water-white, deodorized form of kerosene is marketed under the trade name *Deobase*. Kerosene is deodorized and decolorized by washing with (fuming) sulfuric acid, followed by sodium plumbite soln and sulfur (Doctor sweetening). LD_{50} orally in rabbits: 28 ml/kg, W. B. Deichmann *et al., Ann. Int. Med.* **21**, 803 (1944).

USE: In kerosene lamps, flares, and stoves; as degreaser and cleaner; Deobase formerly used as a solvent in cosmetics and in fly spray. *Human Toxicity:* Defatting action on skin can lead to irritation, infection. Inhalation of high concns causes headache, drowsiness, coma. Swallowing causes G.I. irritation with vomiting, diarrhea. Aspiration of vomitus causes serious pneumonitis. Toxicological studies and recommended treatment of kerosene poisoning: H. W. Gerarde, *Toxicol. Appl. Pharmacol.* **1**, 462 (1959). *Caution:* Induction of vomiting following ingestion is contraindicated.

5174. Ketamine. (±)-2-(2-*Chlorophenyl*)-2-(*methyl-amino)cyclohexanone.* 2-(methylamino)-2-(2-chlorophenyl)-cyclohexanone. $C_{13}H_{16}ClNO$; mol wt 237.74. C 65.68%, H 6.78%, Cl 14.92%, N 5.89%, O 6.73%. Prepn: Stevens, *Belg.* pat. **634,208** (1963) corresp to U.S. pat. **3,254,124** (1966 to Parke, Davis). Prepn and characterization of T-labelled compd: Blackburn, Ober, *J. Label. Compounds* **3**, 38 (1967). Isoln of optical isomers: Hudyma *et al.,* **Ger.** pat. **2,062,620** (1971 to Bristol-Myers), *C.A.* **75**, 118119x (1971). Pharmacology: Domino *et al., Clin. Pharmacol. Ther.* **6**, 279 (1966); Chen *et al., J. Pharmacol. Exp. Ther.* **152**, 332 (1966). Toxicity: E. J. Goldenthal, *Toxicol. Appl. Pharmacol.* **18**, 185 (1971). Comprehensive description: W. C. Sass, S. A. Fusari in *Analytical Profiles of Drug Substances* vol. 6, K. Florey, Ed. (Academic Press, New York, 1977) pp 297-322.

Crystals from pentane-ether, mp 92-93°. uv max (0.01*N* NaOH in 95% methanol): 301, 276, 268, 261 nm ($A_{1cm}^{1\%}$ 5.0, 7.0, 9.8, 10.5). pKa 7.5.

Hydrochloride, $C_{13}H_{17}Cl_2NO$, *CI-581, Ketaject, Ketalar, Ketanest, Ketaset, Ketavet, Vetalar.* White crystals, mp 262-263°. Soly in water: 20 g/100 ml. LD_{50} in adult mice, rats (mg/kg): 224 ±4, 229 ±5 i.p. (Goldenthal).

THERAP CAT: Anesthetic (intravenous).

THERAP CAT (VET): Anesthetic.

5175. Ketanserin. 3-[2-[4-(4-*Fluorobenzoyl*)-1-piperi-dinyl]ethyl]-2,4[1H,3H]-quinazolinedione;* R 41468; Ket; Serefrex; Serepress; Sufrexal. $C_{22}H_{22}FN_3O_3$; mol wt 395.43. C 66.82%, H 5.61%, F 4.80%, N 10.63%, O 12.14%. Specific serotonin S_2-receptor antagonist with hypotensive properties. Prepn: J. Vandenbeck *et al., Eur.* pat. Appl. **13,612**; *eidem,* **U.S.** pat. **4,335,127** (1980, 1982 both to Janssen). X-ray structure: O. M. Peeters *et al., Cryst. Struct. Commun.* **11**, 375 (1982). Receptor binding profile, physical properties: J. E. Leysen *et al., Life Sci.* **28**, 1015 (1981). Antihypertensive properties: J. M. Van Neuten *et al., J. Pharmacol. Exp. Ther.* **218**, 217 (1981). Use in treatment of cardiac failure: J.-C. Demoulin *et al., Lancet* **1**, 1186 (1981). HPLC determn in plasma: A. T. Kacprowicz *et al., J. Chro-*

matog. **272**, 417 (1983); C. L. Davies, *ibid.* **275**, 232 (1983). Efficacy in treatment of intermittent claudication: J. De Cree *et al., Lancet* **2**, 775 (1984). Long term treatment of hypertensive patients: A. Amery *et al., J. Cardiovasc. Pharmacol.* **6**, 182 (1984). In treatment of Raynaud's phenomenon: E. Stranden *et al., Brit. Med. J.* **285**, 1069 (1982); J. R. Seibold, A. H. M. Jageneau, *Arthritis Rheum.* **27**, 139 (1984). Review of antihypertensive properties: P. M. Vanhoutte *et al., Fed. Proc.* **42**, 182-185 (1983); of pharmacology: J. M. Van Neuten, P. M. Vanhoutte, *Bibl. Cardiol.* **38**, 222-233 (1984). Series of articles on cardiovascular effects: *J. Cardiovasc. Pharmacol.* **7**, Suppl. 7, S1-S182 (1985); on pharmacokinetics and metabolism in animals and man: *Arzneimittel-Forsch.* **38**, 775-800 (1988).

Crystals from 4-methyl-2-pentanone, mp 227-235°. Soly (g/100 ml): 0.001 in water; 0.038 in ethanol; 2.34 in DMF. pKa 7.5.

THERAP CAT: Antihypertensive.

5176. Ketazolam. 11-Chloro-8,12b-dihydro-2,8-dimeth-yl-12b-phenyl-4H-[1,3]oxazino[3,2-d][1,4]benzodiazepine-4,7(6H)-dione;* U-28,774; Anseren; Ansieten; Anxon; Contamex; Loftran; Unakalm. $C_{20}H_{17}ClN_2O_3$; mol wt 368.82. C 65.13%, H 4.65%, Cl 9.61%, N 7.59%, O 13.01%. Prepn: J. Szmuszkovicz, **Ger.** pat. **1,947,226** corresp to U.S. pat. **3,573,282** (1970, 1971 both to Upjohn); U.S. pat. **3,575,965** (1971 to Upjohn). Synthesis and structure: J. Szmuszkovicz *et al., Tetrahedron Letters* **1971**, 3665. Analysis by HPLC: D. J. Weber, *J. Pharm. Sci.* **61**, 1797 (1972). Pharmacology: V. H. Sethy, *Arch. Exp. Pathol. Pharmakol.* **301**, 157 (1978). Clinical studies: L. F. Fabre, *J. Int. Med. Res.* **4**, 50 (1976); L. A. Gottschalk, J. B. Cohn, *Psychopharmacol. Bull.* **14**, 39 (1978).

Colorless prisms from chloroform/ether, mp 182-183.5° (sinters at 170°). uv max (ethanol): 202, 241 nm (ε 40600; 18400).

Note: This is a controlled substance (depressant) listed in the U.S. Code of Federal Regulations, Title 21 Part 1308.14 (1987).

THERAP CAT: Anxiolytic.

5177. Ketene. *Ethenone;* carbomethene. C_2H_2O; mol wt 42.04. C 57.14%, H 4.80%, O 38.06%. $CH_2=C=O$. Prepd by the thermal decompn of acetone, diketene or acetic anhydride: Hurd, *Org. Syn.* coll. vol. **I**, 330 (2nd ed., 1941); S. Andreades, H. D. Carlson, *ibid.* coll. vol. **V**, 679 (1973). Structure of ketene dimer as 3-buteno-β-lactone: Blomquist, Baldwin, *J. Am. Chem. Soc.* **70**, 29 (1948); Hurd, Blanchard, *ibid.* **72**, 1461 (1950); Katz, Lipscomb, *J. Org. Chem.* **17**, 515 (1953). Toxicology: J. F. Treon *et al., J. Ind. Hyg. Toxicol.* **31**, 209 (1949). *Review* on the prepn of stable ketenes: R. S. Ward in *The Chemistry of Ketenes, Allenes and Related Compounds,* **Part 1**, S. Patai, Ed. (Wiley, New York, 1980) pp 223-277; on synthetic uses of ketenes: W. T. Brady, *ibid.* pp 279-308.

Gas. Penetrating odor. mp −150°. bp −56°. Electron diffraction pictures: Beach, Stevenson, *J. Chem. Physics* **6**, 75 (1938). Fairly sol in acetone.

Human Toxicity: Poisonous gas of the same order of toxicity as phosgene, *q.v.* All operations with ketene should be carried out in an efficient hood, S. Andreades, H. D. Carlson, *loc. cit.*

USE: For the conversion of higher acids into their anhydrides; for acetylation in the manuf of cellulose acetate and aspirin.

5178. Kethoxal. *3-Ethoxy-1,1-dihydroxy-2-butanone; β-ethoxy-α-ketobutyraldehyde;* 3-ethoxy-2-oxobutyraldehyde hydrate; U 2032. $C_6H_{12}O_4$; mol wt 148.16. C 48.64%, H 8.17%, O 43.19%. Also depicted as $CH_3CH(OC_2H_5)$-COCHO monohydrate. Prepn: Rappen, *J. Prakt. Chem.* **157**, 177 (1941); Tiffany *et al., J. Am. Chem. Soc.* **79**, 1682 (1957). Antiviral activity: H. E. Rennis, *Ann. N.Y. Acad. Sci.* **173/1**, 527 (1970).

$$CH_3CHCOCHOH$$
$$CH_3CH_2O \quad OH$$

Yellow oil. bp_{11} 54-58°. bp_{760} 145°. Initial $n_D^{23.7}$ 1.4348. Miscible with alc; sol in benzene with yellow color; one part dissolves in ten parts of water.

Bis(thiosemicarbazone), $C_8H_{16}N_6OS_2$, *BW 356-C-61, NSC 82116, 2,2'-[1-(1-ethoxyethyl)-1,2-ethanediylidene]bishydrazinecarbothioamide, gloxazone, Contrapar.*

THERAP CAT: Antiviral.

THERAP CAT (VET): Bis(thiosemicarbazone) as anaplasmodastat.

5179. Ketipic Acid. *3,4-Dioxohexanedioic acid;* 2,3-diketoadipic acid; oxalodiacetic acid. $C_6H_6O_6$; mol wt 174.11. C 41.39%, H 3.47%, O 55.14%. HOOCCH₂COCOCH₂-COOH. Prepn: Fittig *et al., Ann.* **249**, 182 (1888); Stachel, *Arch. Pharm.* **295**, 735 (1932); **296**, 479 (1963). Structure studies; occurs in the solid form as the γ-lactol: Stachel, *Ann.* **689**, 118 (1965).

Dec at 150° forming diacetyl. Sparingly sol in ether. Sol in glacial acetic acid, concd hydrochloric acid. Practically insol in water, alc, chloroform, CS_2, benzene, petr ether.

Diethyl ester, $C_{10}H_{14}O_6$, prisms from alcohol, mp 83°. bp_{30} 230°. Soluble in ether, chloroform.

5180. Ketobemidone. *1-[4-(3-Hydroxyphenyl)-1-methyl-4-piperidyl]-1-propanone;* 4-(*m*-hydroxyphenyl)-1-methyl-4-piperidyl ethyl ketone. $C_{15}H_{21}NO_2$; mol wt 247.33. C 72.84%, H 8.56%, N 5.66%, O 12.94%. Prepn: Suter, *Abstracts of First National Medicinal Chemistry Symposium of the A.C.S.* (Ann Arbor, 1948) p 20; **Brit. pats. 591,992** and **609,763** (1947 and 1948 to Ciba); Eisleb, **U.S. pats. 2,167,-351; 2,242,575** and **2,248,018** (1939 and 1941 to Winthrop); **Ger. pat. 752,755** (1952 to I. G. Farben). Metabolism in man: U. Bondesson *et al., Drug Metab. Dispos.* **9**, 376 (1981).

Crystals, mp 156-157°.

Hydrochloride, $C_{15}H_{22}ClNO_2$, *Hoechst 10720, Win 1539, Cliradon, Cymidon, Ketogan, Ketogin.* Crystals, mp 201-202°. Sol in water; slightly sol in alcohol. Aq solns may be sterilized by boiling for short periods.

Caution: May be habit forming. This is a controlled substance (opiate) listed in the U.S. Code of Federal Regulations, Title 21 Part 1308.11 (1985).

THERAP CAT: Narcotic analgesic.

5181. Ketoconazole. *cis-1-Acetyl-4-[4-[[2-(2,4-dichlorophenyl)-2-(1H-imidazol-1-ylmethyl)-1,3-dioxolan-4-yl]methoxy]phenyl]piperazine;* R 41400; Fungarest; Fungoral; Ketoderm; Ketoisdin; Nizoral; Orifungal M; Panfungol. $C_{26}H_{28}Cl_2N_4O_4$; mol wt 531.44. C 58.76%, H 5.31%, Cl 13.34%, N 10.54%, O 12.04%. Orally active, broad-spec-

trum antimycotic. Prepn: J. Heeres *et al.,* **Ger. pat. 2,804,-096,** *eidem,* **U.S. pats. 4,144,346** and **4,223,036** (1978, 1979, 1980, all to Janssen); *eidem, J. Med. Chem.* **22**, 1003 (1979). Pharmacokinetics: E. W. Gascoigne *et al., Clin. Res. Rev.* **1**, 177 (1981); C. Brass *et al., Antimicrob. Ag. Chemother.* **21**, 151 (1982). HPLC determn in human serum: V. L. Pascucci *et al., J. Pharm. Sci.* **72**, 1467 (1983). Series of articles on animal and human studies: *Rev. Infect. Dis.* **2**, 519-692 (1980). Effect on hepatic enzymes *in vitro* and *in vivo:* K. N. Buchi *et al., Biochem. Pharmacol.* **35**, 2845 (1986); J. K. Ritter, M. R. Franklin, *Toxicol. Letters* **36**, 51 (1987). Case reports of hepatic toxicity: J. K. Heiberg, E. Svejgaard, *Brit. Med. J.* **283**, 825 (1981); R. Rollman, L. Loof, *Brit. J. Dermatol.* **108**, 376 (1983). Controlled clinical trials: E. A. Petersen *et al., Ann. Intern. Med.* **93**, 791 (1980); W. T. Hughes *et al., J. Infect. Dis.* **147**, 1060 (1983); H. W. Jolly *et al., Cutis* **31**, 208 (1983). Clinical evaluation as inhibitor of steroid synthesis: N. Sonino, *N. Engl. J. Med.* **317**, 812 (1987). Review of pharmacology and therapeutic efficacy: C. A. Sohn, *Clin. Pharm.* **1**, 217 (1982); R. C. Heel *et al., Drugs* **23**, 1-36 (1982). Series of articles on clinical efficacy and therapeutic experience: *Drugs Exptl. Clin. Res.* **12**, 397-427 (1986).

Crystals from 4-methyl-2-pentanone, mp 146°. LD_{50} in mice, rats, guinea pigs, dogs (mg/kg): 44, 86, 28, 49 i.v.; 702, 227, 202, 780 orally (Heel).

THERAP CAT: Antifungal.

5182. α-Ketoglutaric Acid. *2-Oxopentanedioic acid; 2-oxoglutaric acid;* 2-oxo-1,5-pentanedioic acid. $C_5H_6O_5$; mol wt 146.10. C 41.10%, H 4.14%, O 54.76%. HOOC-CH₂CH₂COCOOH. Plays an important role in amino acid metabolism (transamination) *see* Severo Ochoa, "Enzymic Mechanisms in the Citric Acid Cycle" in *Advances in Enzymology* **15**, 183-270 (1954). Prepn: Friedman, Kosower, *Org. Syn. coll. vol. III,* 510 (1955); Bottorff, Moore, *ibid.* **coll. vol. V,** 687 (1973). Microbial synthesis using a strain of *Pseudomonas:* Lockwood *et al.,* **U.S. pat. 2,443,919** (1948); Berger, Witt, **U.S. pat. 2,841,616** (1958).

Crystals from acetone-benzene, mp 113.5°. Freely sol in water, alcohol. Very sparingly sol in ether.

Diethyl ester, $C_9H_{14}O_5$, liq, bp_{23} 160°, bp_{13} 144°.

Compound with L(+)-ornithine, $C_{10}H_{18}N_2O_7$, *L(+)-ornithine α-ketoglutarate, Ornicetil.*

USE: Can be converted to L-glutamic acid by *Aeromonas* spp. *see* Good, **U.S. pat. 2,933,434** (1960 to International Minerals & Chem. Corp.).

5183. 2-Keto-L-gulonic Acid. L-*xylo-2-Hexulosonic acid;* 2-oxo-L-gulonic acid. $C_6H_{10}O_7$; mol wt 194.14. C 37.12%, H 5.19%, O 57.69%. Important intermediate in vitamin C manuf. From sorbitol, which is first oxidized to L-sorbose by bacteria. Then by condensation of L-sorbose with acetone in the presence of sulfuric acid followed by oxidation with permanganate and hydrolysis of the diisopropylidene derivative by boiling: Reichstein, Grüssner, *Helv. Chim. Acta* **17**, 311 (1934); Reichstein, **U.S. pat. 2,301,811** (1942). By careful oxidation of the L-sorbose with nitric acid: Haworth *et al.,* **Brit. pat. 443,901** (1936). By biochemical oxidation using *Pseudomonas:* Huang, **U.S. pat. 3,043,749** (1962 to Pfizer); using *Pseudomonas* and *Acetobacter:* Motizuki *et al.,* **U.S. pat. 3,234,105** (1966 to Takeda).

COOH
|
CO
|
HOCH
|
HCOH
|
HOCH
|
CH₂OH

Crystals from water (may be washed with acetone). mp 171° (slight decompn). $[\alpha]_D^{18}$ −48° (c = 1). Moderately sol in water. Strong acid (Reichstein). Reduces boiling Fehling's soln rapidly.

Methyl ester, crystals, mp 155-157°. $[\alpha]_D^{18}$ −25° (c = 1 in methanol).

Ethyl ester, crystals, $[\alpha]_D^{20}$ −14.5° (c = 0.63 in abs alc).

5184. Ketoprofen. *3-Benzoyl-α-methylbenzeneacetic acid; m-benzoylhydratropic acid;* 2-(3-benzoylphenyl)propionic acid; RP 19583; Alrheumat; Alrheumun; Bi-Profenid; Capisten; Dexal; Epatec; Fastum; Iso-K; Kefenid; Kephina; Ketopron; Kevadon; Lertus; Menamin; Meprofen; Miltax; Mohrus; Orudis; Orugesic; Oruvail; Oscorel; Profenid; Sectorgel. $C_{16}H_{14}O_3$; mol wt 254.29. C 75.57%, H 5.55%, O 18.88%. Prepn: **Fr. pat. M6444**, *C.A.* **75**, 5528m (1971); Farge *et al.,* **S. Afr.** pat. **68 00,524** corresp to **U.S.** pat. **3,641,127** (1968, 1968, 1972 all to Rhône-Poulenc); G. A. Pinna *et al., Farmaco Ed. Sci.* **35**, 684 (1980). Clinical data: Gyory *et al., Brit. Med. J.* **4**, 398 (1972); Cathcart *et al., Ann. Rheum. Dis.* **32**, 62 (1973). Pharmacokinetics in man: T. Ishizaki *et al., Eur. J. Clin. Pharmacol.* **18**, 407 (1980). Toxicity data: K. Ueno *et al., J. Med. Chem.* **19**, 941 (1976). Comprehensive description: G. G. Liversidge in *Analytical Profiles of Drug Substances* **vol. 10**, K. Florey, Ed. (Academic Press, New York, 1981) pp 443-471.

CH₃
|
CHCOOH
(benzophenone structure)

Crystals from 6:20 benzene-petr ether, mp 94°. uv max (methanol): 255 nm (log ε 4.33). Sol in ether, alc, acetone, chloroform, DMF, ethyl acetate. Slightly sol in water. LD_{50} orally in rats: 101 mg/kg (Ueno).

Lysine salt, $C_{22}H_{28}N_2O_5$, *Artrosilene.*

THERAP CAT: Anti-inflammatory; analgesic.

5185. 11-Ketoprogesterone. *Pregn-4-ene-3,11,20-trione;* 11-oxoprogesterone; U-1258; Ketogestin. $C_{21}H_{28}O_3$; mol wt 328.46. C 76.79%, H 8.59%, O 14.61%. Prepn: Reichstein, Fuchs, *Helv. Chim. Acta* **23**, 688 (1940); from 11α-hydroxyprogesterone: Whiting, Schneider, **U.S.** pat. **3,019,238** (1962 to Upjohn).

(steroid structure)

Crystals, mp 171.5-173°. $[\alpha]_D^{25}$ +270° (chloroform). Practically insol in water. Sol in acetone, chloroform.

THERAP CAT (VET): Has been used in ketosis.

5186. Ketorolac. (±)-*5-Benzoyl-2,3-dihydro-1H-pyrrolizine-1-carboxylic acid;* 5-benzoyl-1,2-dihydro-3H-pyrrolo[1,2-a]pyrrole-1-carboxylic acid; RS 37619. $C_{15}H_{13}NO_3$; mol wt 255.27. C 70.58%, H 5.13%, N 5.49%, O 18.80%. Prostaglandin biosynthesis inhibitor. Prepn and separation of isomers: **Belg. pat. 856,681**; J. M. Muchowski, A. F. Kluge, **U.S.** pat. **4,089,969** (both 1978 to Syntex).

Alternate processes: J. M. Muchowski, R. Greenhouse, **U.S.** pat. **4,347,186** (1982 to Syntex); F. Franco *et al., J. Org. Chem.* **47**, 1682 (1982); J. B. Doherty, **U.S.** pat. **4,496,741** (1985 to Merck & Co.). Absolute configuration: A. Guzman *et al., J. Med. Chem.* **29**, 589 (1986). Structure-activity relationships: J. M. Muchowski *et al., ibid.* **28**, 1037 (1985). Pharmacology and analgesic, anti-inflammatory profile of ketorolac and its tromethamine salt: W. H. Rooks *et al., Agents Actions* **12**, 684 (1982); *eidem, Drugs Exp. Clin. Res.* **11**, 479 (1985). Clinical comparison with acetaminophen in post-operative pain: H. J. McQuay *et al., Clin. Pharmacol. Ther.* **39**, 89 (1986).

(structure)

(±)-Form, crystals from ethyl acetate+ ether, mp 160-161°. uv max in methanol: 245, 312 nm (ε 7080, 17400). pKa 3.49 ±0.02. LD_{50} orally in mice: ∼200 mg/kg (Rooks).

(±)-Form tromethamine salt, $C_{19}H_{24}N_2O_6$, *Toradol.*

(+)-Form, crystals from hexane + ethyl acetate, mp 174° (Guzman); also reported as mp 154-156° (Muchowski, Kluge). $[\alpha]_D$ +173° (c = 1 in methanol).

(−)-Form, crystals from hexane + ethyl acetate, mp 169-170° (Guzman); also reported as mp 153-155° (Muchowski, Kluge). $[\alpha]_D$ −176° (c = 1 in methanol).

THERAP CAT: Analgesic; anti-inflammatory.

5187. Ketotifen. *4,9-Dihydro-4-(1-methyl-4-piperidinylidene)-10H-benzo[4,5]cyclohepta[1,2-b]thiophen-10-one;* HC 20-511. $C_{19}H_{19}NOS$; mol wt 309.43. C 73.75%, H 6.19%, N 4.53%, O 5.17%, S 10.36%. Prepn: J. P. Bourquin, **Ger.** pat. **2,111,071** corresp to **U.S.** pat. **3,682,930** (1971, 1972 both to Sandoz); E. Waldvogel *et al., Helv. Chim. Acta* **59**, 866 (1976). Pharmacology: U. Martin, D. Roemer, *Arzneimittel-Forsch.* **28**, 770 (1978). Clinical studies: B. Wüthrick, P. Radielovic, *Deut. Med. Wochenschr.* **103**, 1865 (1978); H. Gmür, M. Scherrer, *Schweiz. Med. Wochenschr.* **109**, 881 (1979). Review: U. Martin *et al.,* in *Pharmacological and Biochemical Properties of Drug Substances* **vol. 3**, M. E. Goldberg, Ed. (Am. Pharm. Assoc., Washington, DC, 1981) pp 424-460.

(structure)

Crystals from ethyl acetate, mp 152-153°.

Fumarate, $C_{23}H_{23}NO_5S$, *Allerkif, Totifen, Zaditen, Zasten.* Crystals, mp 192° (dec).

THERAP CAT: Anti-asthmatic.

5188. Khat. Chat; quat. Leaves of *Catha edulis* Forsk., *Celastraceae;* widely used in E. Africa and Yemen as an amphetamine-like stimulant. Most important constituent is the alkaloid cathinone, *q.v.;* also norpseudoephedrine, cathidine, cathedulin. Isoln and characterization of constituents of khat: O. Wolfes, *Arch. Pharm.* **268**, 81 (1930); H. Friebel, R. Brilla, *Naturwiss.* **50**, 354 (1963); M. Cais *et al., Tetrahedron* **31**, 2727 (1975); R. L. Baxter *et al., Chem. Commun.* **1976**, 463. Pharmacology of khat extracts: G. A. Alles *et al., J. Med. Pharm. Chem.* **3**, 323 (1961). *Reviews:* A. Getahun, A. D. Krikorian, *Econ. Bot.* **27**, 353, 378 (1973); P. Kalix, *Gen. Pharmacol.* **15**, 179 (1984).

5189. Khellin. *4,9-Dimethoxy-7-methyl-5H-furo[3,2-g][1]benzopyran-5-one;* 5,8-dimethoxy-2-methyl-4′,5′-

furo-6,7-chromone; 5,8-dimethoxy-2-methyl-6,7-furano-chromone; 4,9-dimethoxy-7-methyl-5-oxofuro[3,2-g]-1,2-chromene; 4,9-dimethoxy-7-methyl-5-oxofuro[3,2-g][1]-benzopyran; 4,9-dimethoxy-7-methyl-5-oxo-1,8-dioxa-benz[*f*]indene; visammin; Kellin; Kelamin; Kelicor; Gyno-khellan; Kelicorin; Keloid; Norkel; Simeskellina; Vasokel-lina; Visnagalin; Visnagen; Methafrone; Eskel; Amicardine; Viscardan; Corafurone; Cardio-Khellin; Benecardin; Ammi-visnagen; Khelfren; Lynamine; Coronin; Ammicardine; Ammipuran; Ammivin. $C_{14}H_{12}O_5$; mol wt 260.24. C 64.61%, H 4.65%, O 30.74%. Found in seeds of *Ammi visnaga* Lam., *Umbelliferae* (toothpick ammi; chellah; khella). Isoln and structure: Späth, Gruber, *Ber.* **71**, 106 (1938). Synthesis: Baxter *et al., J. Chem. Soc.* **1949**, S 30; Gardner *et al., J. Org. Chem.* **15**, 841 (1950); Schönberg, Sina, *J. Am. Chem. Soc.* **72**, 1611, 3396 (1950); M. W. Reed, H. W. Moore, *J. Org. Chem.* **53**, 4166 (1988). Crystal structure: J. P. Beale, *Cryst. Struct. Commun.* **2**, 125 (1973). Comprehensive description: M. A. Hassan, M. U. Zubair in *Analytical Profiles of Drug Substances* **vol. 9**, K. Florey, Ed. (Academic Press, New York, 1980) pp 371-396.

Crystals from methanol. Bitter taste. mp 154-155°. $bp_{0.05}$ 180-200°. uv max (alc): 250, 338 nm ($E_{1cm}^{1\%}$ 1600, 200). Soly in g/100 ml at 25°: water 0.025; acetone 3.0; methanol 2.6; isopropanol 1.25; ether 0.5; Skellysolve B 0.15. Much more sol in hot water and hot methanol. Stable when mixed with the usual tabletting excipients. LD_{50} orally in rats: 80 mg/kg.

THERAP CAT: Vasodilator.

5190. Khellol Glucoside. *7-[(β-D-Glucopyranosyloxy)-methyl]-4-methoxy-5H-furo[3,2-g][1]benzopyran-5-one;* 2-hydroxymethyl-5-methoxyfuranochrome glucoside; khellinin. $C_{19}H_{20}O_{10}$; mol wt 408.35. C 55.88%, H 4.94%, O 39.18%. From seeds of *Ammi visnaga* Lam., *Umbelliferae*: Späth, Gruber, *Ber.* **74**, 1549 (1941); Illing, *Arzneimittel-Forsch.* **7**, 497 (1957); from *Eranthis hyemalis* L., *Ranuncula-ceae*: Egger, *Z. Naturforsch.* **16B**, 697 (1962). Structure: Fabbrini, Franchi, *Ann. Chim. (Rome)* **49**, 894 (1959), *C.A.* **54**, 4558f (1960).

Crystals from ethanol, mp 179°. uv max (ethanol): 250, 325 nm. Soluble in acetic acid, hot ethanol; slightly sol in hot methanol. Practically insol in acetone, ethyl acetate, chloroform, ether, cold alkali.

Tetraacetate, flakes from ethanol, mp 153°. Sol in ethanol, acetone, ethyl acetate. Practically insol in petr ether.

5191. Kiku Oil. Obtained by distillation from leaves and flowers of *Chrysanthemum indicum* L., *Compositae*. Produced in Japan: Perrier, *Bull. Soc. Chim. France* [3] **23**, 216 (1900). *Constit.* *l*-Camphene, camphor carvone, xanthophyll, coumarin, angelic acid esters.

Colorless or greenish oil. Odor reminiscent of oil of Eucalyptus. d_4^{15} 0.932. n_D^{18} 1.4931. Acid to moist litmus paper. Ten grams dissolves in 100 grams abs alc at 95°. Almost insol in alc at 70°. Used as a folk remedy in Japan in a manner comparable with the use of chamomile and mint in Europe, but also against intestinal worms.

5192. Kinetin. *N-(2-Furanylmethyl)-1H-purin-6-amine;* N^6-*furfuryladenine*; 6-furfurylaminopurine. $C_{10}H_9N_5O$; mol wt 215.21. C 55.81%, H 4.22%, N 32.54%, O 7.43%. A cell

division factor found in various plant parts and in yeast. Isoln from autoclaved water slurries of deoxyribonucleic acid: Miller *et al., J. Am. Chem. Soc.* **77**, 1392 (1955). Structure and synthesis from furfurylamine and 6-methyl-mercaptopurine: Miller *et al., ibid.* 2662; **78**, 1375 (1956); U.S. pat. 2,903,455 (1959 to Wisc. Alumni Res. Found.). Physiologically active at very great dilutions, but only in presence of auxin, see Indoleacetic Acid. Crystal structure: M. Soriano-Garcia, R. Parthasarathy, *Acta Crystallogr.* **33B**, 2674 (1977). *Review:* Miller, *Ann. Rev. Plant Physiol.* **12**, 395 (1961).

Platelets from abs ethanol, mp 266-267° (sealed tube). Sublimes at 220°. pKa_1 2.7; pKa_2 9.9. uv max (ethanol): 268 nm (ε 18,650). Slightly soluble in cold water, methanol, ethanol. Freely soluble in dil aq HCl or NaOH. Can be extracted from neutral aq solns by shaking with ether.

Note: Kinetin is also a brand of hyaluronidase, *q.v.*

USE: Plant growth regulator. To augment growth of microbial cultures: **Belg. pat. 632,589** (1963 to Hoechst).

5193. Kino. Resin kino; gum kino. Dried juice from trunk of *Pterocarpus marsupium* Roxb., *Leguminosae*. *Habit.* Western Africa, East India, Ceylon. *Constit.* 70-80% kino-tannic acid; kino-red, pyrocatechol, kinoin, gum. Contains not less than 45% alcohol-soluble and not less than 80% water-soluble material.

THERAP CAT: Astringent.

THERAP CAT (VET): Has been used as astringent.

5194. Kinoprene. *3,7,11-Trimethyl-2,4-dodecadienoic acid 2-propynyl ester;* 2-propynyl (E,E)-3,7,11-trimethyl-2,4-dodecadienoate; ENT 70531; ZR 777; Enstar. $C_{18}H_{28}$-O_2; mol wt 276.42. C 78.21%, H 10.21%, O 11.58%. Juvenile hormone mimic. Prepn: C. A. Hendrick, **Ger. pat. 2,-246,924** (1973 to Zoecon), *C.A.* **79**, 41975w (1973). Prepd but not claimed: *idem*, **U.S. pat. 3,833,635** (1974 to Zoecon). Activity: R. A. Hamlen, *J. Econ. Entomol.* **68**, 223 (1975); A. DeLoof *et al., Entomol. Exp. Appl.* **26**, 301 (1979).

Amber liquid, $bp_{0.04}$ 115-116°. Vapor press at 20°: 7.19 × 10^{-6} mm Hg. Soly in water: 5.22 ppm. Sol in most organic solvents. LD_{50} orally in rats: 4900 mg/kg, *RTECS Vol. I*, R. J. Lewis, Sr., R. L. Tatken (1979) p 600.

USE: Insect growth regulator.

5195. Kitol. *3,6-Dimethyl-5-[2-methyl-4-(2,6,6-trimethyl-1-cyclohexen-1-yl)-1,3-butadienyl]-6-[4-methyl-6-(2,6,6-trimethyl-1-cyclohexen-1-yl)-1,3,5-hexatrienyl]-3-cyclohex-ene-1,2-dimethanol.* $C_{40}H_{60}O_2$; mol wt 572.88. C 83.86%, H 10.56%, O 5.59%. One of the provitamins A. Obtained from mammalian liver oil: Embree, Shantz, *J. Am. Chem. Soc.* **65**, 910 (1943); Clough *et al., Science* **105**, 436 (1947); Barua, Morton, *Biochem. J.* **45**, 309 (1949); Chatan, Fridenson, *Compt. Rend.* **234**, 1094 (1952). Purification: Tawara, Fukazawa, *C.A.* **47**, 12483e (1953). Alpha particle bombardment with radon effects the conversion of kitol to vitamin A: Embree, Shantz, *loc. cit.;* **U.S. pats. 2,414,458** (1947); **2,434,687** (1948); Libermann, Grundland, *Compt. Rend.* **224**, 1033 (1947). Structure: Burger *et al., Chem. Commun.* **1965**, 588; Giannotti *et al., ibid.* **1966**, 28; Giannotti *et al., Bull. Soc. Chim. France* **1966**, 3299; Burger, Garbers, *J. Chem. Soc. Perkin Trans. I* **1973**, 590.

Crystals from methanol, mp 88-90° (Embree); 98-100° (Chatan); 72° (Tawara). $[\alpha]_D$ −2.6° (c = 1.1 in chloroform). Labile in light, air, petr ether. Kitol has no vitamin A activity. uv max: 290 nm ($E_{1cm}^{1\%}$ 586).

Diphenylazobenzoate, exists in two forms, probably geometric isomers, one mp 125-126°, the other 153-155°.

5196. Kodel®. Polyester staple and filament fiber. Prepn from dimethyl terephthalate and 1,4-cyclohexanedimethanol: Kibler *et al.*, U.S. pat. **2,901,466** (1959 to Eastman Kodak). Review and structure: R. W. Moncrieff, *Man-Made Fibres* (John Wiley, New York, 4th ed., 1963) pp 389-393.

Solid, mp 290-295°. Burns slowly. Sp gr 1.22. Fiber has good resistance to acids and alkalies. Treatment of fabric with trichloroethylene and methylene chloride causes it to shrink; boiling water and perchloroethylene induce slight shrinkage. Fabric has good crease resistance.

5197. Kojic Acid. *5-Hydroxy-2-(hydroxymethyl)-4H-pyran-4-one*; 5-hydroxy-2-(hydroxymethyl)-4-pyrone; 2-hydroxymethyl-5-hydroxy-γ-pyrone. $C_6H_6O_4$; mol wt 142.11. C 50.71%, H 4.26%, O 45.03%. Antibiotic substance produced in an aerobic process by a variety of microorganisms from a wide range of carbon sources. Isoln from *Aspergillus oryzae*: Saito, *Bot. Mag. Tokyo* **21**, 249 (1907). Structure: Yabuta, *J. Chem. Soc.* **125**, 575 (1924); Heyns, Vogelsang, *Ber.* **87**, 13 (1954). Synthesis: Stacey, Turton, *J. Chem. Soc.* **1946**, 661; Lichtenthaler, Heidel, *Angew. Chem. Int. Ed.* **8**, 978 (1968). Industrial prepn: **Brit. pat. 826,244** (1959 to Pfizer), *C.A.* **54**, 11372e (1960). *Review:* Beélik, *Advan. Carbohyd. Chem.* **11**, 145-183 (1956); Wilson, "Miscellaneous *Aspergillus* Toxins" in *Microbial Toxins*, vol. VI, A. Ciegler *et al.*, Eds. (Academic Press, New York, 1971) pp 235-250.

Prismatic needles from acetone, ethanol + ether or methanol+ ethyl acetate, mp 153-154°. pKa 7.90, 8.03. Freely sol in water, ethanol, acetone; sparingly sol in ether, ethyl acetate, chloroform, pyridine. Absorption spectrum: Stacey, Turton, *loc. cit.*

USE: Converted to maltol and ethyl maltol, flavor-enhancing additives.

5198. Kola. Cola; Soudan coffee; Bissy nuts; gooroo nuts; guru nuts. Dried cotyledons of *Cola nitida* Schott and Endl. or of other species of *Cola, Sterculiaceae. Habit.* West Africa; naturalized in West Indies, India, Ceylon. *Constit.* About 1.5% caffeine; theobromine, kola-red, kolatin; the

glucoside kolanin; kolazyme—an enzyme; kola-tannin, glucose, gum.

THERAP CAT: Analeptic.

5199. Kopsine. *3-Hydroxy-22-oxokopsan-1-carboxylic acid methyl ester.* $C_{22}H_{24}N_2O_4$; mol wt 380.43. C 69.45%, H 6.36%, N 7.36%, O 16.82%. From *Kopsia fructicosa* A.D., *Apocynaceae:* Bhattacharya *et al., J. Am. Chem. Soc.* **71**, 3370 (1949). Structure: Spiteller, *Monatsh.* **93**, 1220 (1962); Govindachari *et al., Helv. Chim. Acta* **45**, 1146 (1962); **46**, 572 (1963); Guggisberg *et al., ibid.* **52**, 76 (1969).

Crystals from alc, dec 217-218°. $[\alpha]_D^{27}$ −14.3° ± 1° (c = 2 in chloroform). uv max (ethanol): 240, 278, 285-286 nm (log ϵ 4.08, 3.37, 3.35). Sol in chloroform; sparingly sol in methanol, ethanol, ethyl acetate, benzene, ether. Practically insol in petr ether, water.

Methiodide, $C_{23}H_{27}IN_2O_4$, crystals from methanol, dec 194-196°.

Oxalate, $C_{24}H_{26}N_2O_8$, prisms from alcohol + acetone, dec 154°. Sol in water.

5200. Kosin. From flowers of *Hagenia abyssinica* J. J. Gmel. (*Brayera anthelmintica* Kunth.), *Rosaceae:* Leichsenring, *Arch. Pharm.* **232**, 54 (1894). Occurs as a mixture of α- and β-kosin: Lobeck, *ibid.* **239**, 672 (1901); Hems, Todd, *J. Chem. Soc.* **1937**, 562. Structure of α-kosin: Riedl, *Ber.* **89**, 2600 (1956). Synthesis of α-kosin and probable structure of β-kosin: Orth, Riedl, *Ann.* **663**, 83 (1963).

α-kosin: R = CH_3; R' = H
β-kosin: R = H; R' = CH_3

α-Kosin, $C_{25}H_{32}O_8$, 5,5'-methylene-bis[4,6-dihydroxy-2-methoxy-3-methylisobutyrophenone]. Yellow needles from ethanol, mp 160-160.5°. uv max: 227, 290 nm (ϵ 30800, 24400). Soluble in alcohol, benzene, chloroform, ether, glacial acetic acid, alkalies.

Tetraacetate, $C_{33}H_{40}O_{12}$, needles from benzene + hexane, mp 124°.

β-Kosin. Yellow prisms from methanol, mp 120°. uv max: 228, 292 nm (ϵ 30300, 21260).

THERAP CAT: Has been used as anthelmintic (Cestodes).

5201. Krebiozen®. A white powder "chemically separated from horses' serum after stimulation of their cell network by the injection of *Actinomyces bovis.*" Claimed to be a lipopolysaccharide. Developed by Stevan Durovic. Used experimentally in the treatment of cancer: Szujewski, *J. Am. Med. Assoc.* **148**, 929 (1952). Monographs: H. Bailey, "*K*" *for Krebiozen* (Hermitage House, 1955); G. W. Stoddard, *Krebiozen* (Beacon Press, 1955); A. C. Ivy *et al., Observations on Krebiozen in the Management of Cancer* (Henry Regnery Co., 1956). Reports on current status: *Ca (Cancer J. for Clinicians)* **13**, no. 2 (March-April 1963), pp 76-78; *Science* **140**, 1291 sqq (1963). Popularized history: *Consumer Rep.*, Sept. 1963, pp 441-443. "Andrew C. Ivy, William F. P. Phillips, Stevan Durovic, and Marko Durovic were found not guilty of fraud in selling the drug as an agent for suppressing cancer": *N.Y. Times*, Feb. 1st, 1966. Quackery

Congress Symposium: James F. Holland, "The Krebiozen Story", *J. Am. Med. Assoc.* **200**, 213-218 (1967).

5202. Krypton. Kr; at. wt 83.30; at. no. 36. Six stable isotopes: 78 (0.354%); 80 (2.27%); 82 (11.56%); 83 (11.55%); 84 (56.90%); 86 (17.37%); radioactive, artificial isotopes: 74-77; 79; 81; 85; 87-95; 97. Noble gas; discovered in 1898 by Ramsay and Travers. Isoln from the final residues obtained after evaporation of liq air: Ramsay, Travers, *Proc. Roy. Soc.* **63 [A]**, 405 (1898). Occurs in the atmosphere to the extent of about 1×10^{-6} parts per unit volume of air. Prepn: Lepape, *Compt. Rend.* **187**, 231 (1928); Claude, *ibid.* 581. Monograph: *Argon, Helium and the Rare Gases,* vols. **1, 2,** G. A. Cook, Ed. (Interscience, New York, 1963) 818 pp. Review of chemistry: Bartlett, Sladky, "The Chemistry of Krypton, Xenon and Radon" in *Comprehensive Inorganic Chemistry,* vol. **1,** J. C. Bailar, Jr. *et al.,* Eds. (Pergamon Press, Oxford, 1973) pp 213-249. Comprehensive review: Cockett, Smith, "The Monatomic Gases" *ibid.* pp 139-211.
Gas; condenses to a colorless liquid. d^0 (gas) 3.7493g/l. bp $-153.35°$ (119.8°K). d (liq at bp) 2.413 g/cc. Crit temp $-63.8°$; crit press. 54.3 atm; crit density 0.9085. Triple pt $-157.20°$ (115.95°K); pressure 548 mm; d (solid at triple pt) 2.823. Excitation potentials: 9.91; 10.03; 10.56; 10.64 ev; ionization potentials: 14.00; 14.66 ev: Brocklehurst, *Quart. Rev.* **22**, 147 (1968). Slightly soluble in water with formation of a hydrate; ideal formula Kr.5.75H₂O. A hexadeuterate has been prepd. Soluble in liquid oxygen.
Krypton difluoride, F₂Kr. First prepd by Turner, Pimentel, *Science* **140**, 974 (1963). *Review:* Bartlett, Sladky, *loc. cit.* Colorless solid; decomposes rapidly at room temperature. d (calc) 3.24. Vapor pressure at 0°: 29 ± 2 mm Hg; at 15°: 73 ± 3 mm Hg.
USE: In certain types of electric bulbs (both incandescent and fluorescent).

5203. Kurchessine. *N,N,N′,N′-Tetramethylpregn-5-ene-3β,20α-diamine; 3β,20α-bis(dimethylamino)-5-pregnene.* C₂₅H₄₄N₂; mol wt 372.65. C 80.58%, H 11.90%, N 7.51%. A steroidal alkaloid from Kurchi bark. Isoln from *Holarrhena antidysenterica* Wall., *Apocynaceae:* Tschesche, Otto, *Ber.* **95**, 1144 (1962); Labler, Sorm, *Coll. Czech. Chem. Commun.* **28**, 2345 (1963). Synthesis by methylation of kurchamine: Tschesche, Wiensz, *Ber.* **91**, 1504 (1958). Structure: Labler, Sorm, *loc. cit.* Possible identity with *sarcodinine* from *Saracococa pruniformis* Lindl., *Euphorbiaceae:* Chatterjee *et al., Tetrahedron Letters* **1965,** 67. Related alkaloids, *kurchiline, kurchiphylline, kurchiphyllamine,* from Kurchi leaves: Janot *et al., Bull. Soc. Chim. France* **1964,** 2158.

Needles. mp 140-141° (from acetone). $[\alpha]_D^{20}$ $-37°$ (c = 1.9); $[\alpha]_D^{22}$ $-36°$ (c = 1.112 in CHCl₃); $[\alpha]_D^{20}$ $-17°$ (c = 1.050 in methanol).

5204. Kurcholessine. *3β-(Dimethylamino)-4α-methyl-5α-conanine-5,7β-diol.* C₂₅H₄₄N₂O₂; mol wt 404.65. C 74.23%, H 10.96%, N 6.93%, O 7.88%. A steroidal alkaloid

from Kurchi bark. Isoln from *Holarrhena antidysenterica* Wall., *Apocynaceae:* Tschesche, Otto, *Ber.* **95**, 1144 (1962). Structure: Tschesche *et al., Tetrahedron Letters* **1964,** 1659.
Slender needles from ethanol; microcryst powder from acetone. mp 218.5-221.5°. $[\alpha]_D^{20}$ $-4°$ $\pm 2°$ (CHCl₃), $+5°$ $\pm 2°$ (c = 0.01, abs ethanol). Very sol in chloroform, methanol, ethanol, pyridine; slightly sol in warm acetone; practically insol in ethyl acetate, benzene, petr ether, ether, water.

5205. Kyanmethin. *2,6-Dimethyl-4-pyrimidinamine; 4-amino-2,6-dimethylpyrimidine;* 6-amino-2,4-dimethylpyrimidine. C₆H₉N₃; mol wt 123.16. C 58.51%, H 7.37%, N 34.12%. Prepn from acetonitrile and potassium methoxide: Ronzio, Cook, *Org. Syn.* **coll. vol. III,** 71 (1955).

Needles from alcohol or scales from benzene, mp 182-183°. One gram dissolves in 0.64 ml water at 18° or in 5.25 parts alcohol at 18°.

5206. Kynurenic Acid. *4-Hydroxy-2-quinolinecarboxylic acid; 4-hydroxyquinaldic acid.* C₁₀H₇NO₃; mol wt 189.16. C 63.49%, H 3.73%, N 7.41%, O 25.37%. Found in the urine of some animals as a metabolic product of tryptophan; its excretion is stepped up in avitaminoses B₁, B₂ and B₆. Isoln and syntheses: Späth, *Monatsh.* **42**, 89 (1921); Besthorn, *Ber.* **54**, 1330 (1921). Alternate syntheses: Benassi, *Gazz. Chim. Ital.* **91**, 1097 (1961); Wald, Joullie, *J. Org. Chem.* **31**, 3369 (1966); Jordanides, *Ann.* **729**, 244 (1969).

Yellow needles, mp 282-283°. Soly in water: about 0.9% at 100°. Sol in hot alc. Insol in ether.
Methyl ester, C₁₁H₉NO₃, yellow crystals, mp 224°. *Cf.* xanthurenic acid.
USE: In nutrition studies, specifically in vitamin B deficiency diseases.

5207. Kynurenine. *α,2-Diamino-γ-oxobenzenebutanoic acid; 3-anthraniloylalanine.* C₁₀H₁₂N₂O₃; mol wt 208.21. C 57.68%, H 5.81%, N 13.46%, O 23.05%. An amino acid produced in the body from tryptophan. Isoln from urine of rabbits that had been fed tryptophan: Matsuoka, Yoshimatsu, *Z. Physiol. Chem.* **143**, 206 (1925); Butenandt *et al., ibid.* **279**, 27 (1943); Heidelberger *et al., J. Biol. Chem.* **179**, 143 (1949). Structure and synthesis: Butenandt *et al., loc. cit.* Laboratory prepn by oxidation of L-tryptophan with a *Pseudomonas* sp.: Hayaishi, Meister, *Biochem. Prepn.* **3**, 108 (1953). From acetyltryptophan: Warnell, Berg, *J. Am. Chem. Soc.* **76**, 1708 (1954); Auerbach, Knox, *Methods Enzymol.* **3**, 620 (1957).

L-Kynurenine hydrate, 3C₁₀H₁₂N₂O₃.H₂O, leaflets from water, dec 180-190°. $[\alpha]_D^{20}$ $-29°$ (c = 0.4). Slightly sol in water (more sol than the DL-form). Forms a molecular compound with sucrose, C₂₂H₃₄N₂O₁₄.H₂O, rosettes; dec 145-153°, $[\alpha]_D^{21}$ $+14.5°$ (c = 0.7). Soluble in water.
L-Kynurenine sulfate monohydrate, C₁₀H₁₂N₂O₃.H₂SO₄.-H₂O, needles from water + alcohol. Darkens at 165°, dec 195°. $[\alpha]_D^{20}$ $+7.3°$ (c = 1). uv max (pH 7.0): 230, 257, 360 nm (ε 18,900, 7500, 4500). Soluble in water, slightly in alcohol. Kynurenine sulfate requires 3 molecules alkali for neutralization in alcoholic soln.

L-Kynurenine diacetate, $C_{14}H_{16}N_2O_5$, obtained by acetylating L-kynurenine with ketene. Needles, mp 198°.

Anhydro-L-kynurenine monoacetate, $C_{12}H_{12}N_2O_3$, obtained by acetylating L-kynurenine with acetic anhydride in pyridine. Needles from alc, darkens at 215°, dec 237°.

L-Kynurenine picrate monohydrate, $(C_{10}H_{12}N_2O_3)_2.C_6H_3N_3O_7.H_2O$, red needles, dec 181°. $[\alpha]_D^{20}$ −19.6° (c = 0.5).

DL-Kynurenine sulfate, $C_{10}H_{12}N_2O_3.H_2SO_4$, crystals from water + alcohol. Darkens at 166°, dec 194°. Soluble in water. Slightly sol in alcohol.

DL-Kynurenine picrate, $(C_{10}H_{12}N_2O_3)_2.C_6H_3N_3O_7$. Yellow needles (rosettes) from water, dec 188.5-189.5°.

USE: In biochemical investigations.

L

5208. Labetalol. *2-Hydroxy-5-[1-hydroxy-2-[(1-methyl-3-phenylpropyl)amino]ethyl]benzamide; 5-[1-hydroxy-2-[(1-methyl-3-phenylpropyl)amino]ethyl]salicylamide;* ibidomide. $C_{19}H_{24}N_2O_3$; mol wt 328.41. C 69.49%, H 7.36%, N 8.53%, O 14.62%. A specific competitive antagonist at both α- and β-adrenergic receptor sites. Prepn: L. H. Lunts, D. T. Collin, *Ger.* pat. **2,032,642**; *eidem, U.S.* pat. **4,012,444** (1971, 1977 both to Allen & Hanburys); J. E. Clifton *et al., J. Med. Chem.* **25,** 670 (1982). Pharmacology: J. B. Farmer *et al., Brit. J. Pharmacol.* **45,** 660 (1972). Clinical studies: J. G. Collier *et al., ibid.* **44,** 286 (1972); D. A. Richards *et al., Brit. J. Clin. Pharmacol.* **1,** 505 (1974). Metabolism in animals and man: R. Hopkins *et al., Biochem. Soc. Trans.* **4,** 726 (1976). HPLC determn in serum: T. F. Woodman, B. Johnson, *Ther. Drug Monit.* **3,** 371 (1981). Synthesis of stereoisomers and comparison of cardiovascular properties: *J. Med. Chem.* **25,** 1363 (1982). Series of articles on pharmacology, metabolism and clinical studies: *Brit. J. Clin. Pharmacol.* **3,** Suppl. 3, 681S-824S (1976); *ibid.* **8,** Suppl. 2, 85S-244S (1979); *ibid.* **13,** Suppl. 1, 1S-141S (1982). Toxicity: K. Shimpo *et al., Hokkaido Igaku Zasshi* **53,** 15 (1978), *C.A.* **90,** 66465v (1974). *Review:* R. T. Brittain *et al.* in *Pharmacological and Biochemical Properties of Drug Substances* vol. 2, M. E. Goldberg, Ed. (Am. Pharm. Assoc., Washington, DC, 1979) pp 229-254.

Hydrochloride, $C_{19}H_{25}ClN_2O_3$, *AH 5158A, SCH 15719W, Amipress, Ipolab, Labelol, Labrocol, Normadate, Normodyne, Presdate, Pressalolo, Trandate.* White crystalline solid from ethanol-ethyl acetate, mp 187-189°. Sol in water, ethanol. Insol in ether, chloroform. LD_{50} in male, female mice, male, female rats (mg/kg): 114, 120, 113, 107 i.p.; 47, 54, 60, 53 i.v.; 1450, 1800, 4550, 4000 orally (Shimpo).

(*R,R*)-Isomer, *see* Dilevalol.

THERAP CAT: Antihypertensive.

5209. Laccaic Acid. *C.I. Natural Red 25;* lac dye; *C.I.* 75450. Pigment found in the lac resin produced by the insect *Coccus laccae* (*Laccifer lacca* Kerr) on certain trees in India. Isoln: R. E. Schmidt, *Ber.* **20,** 1285 (1887). Chemistry: O. Dimroth, S. Goldschmidt, *Ann.* **399,** 62 (1913). Originally thought to be one compound, laccaic acid has been separated into four components: A (major), B, C and D. Isoln of A: R. Burwood *et al., J. Chem. Soc.* (C) **1965,** 6067. Structure of A: *eidem, Tetrahedron Letters* **1966,** 3059; *eidem, J. Chem. Soc.* (C) **1967,** 842; E. D. Pandhare *et al., Indian J. Chem.* **7,** 977 (1969). Isoln and structure of B: *eidem, Tetrahedron Letters* **1967,** 2437; N. S. Bhide *et al., Indian J. Chem.* **7,** 987 (1969); of C: A. V. R. Rao *et al., ibid.* 188; of D: A. R. Mehandale *et al., Tetrahedron Letters* **1968,** 2231. Synthesis of D: D. W. Cameron *et al., Chem. Commun.* **1978,** 688; *eidem, Aust. J. Chem.* **34,** 2401 (1981).

laccaic acids A, B, C   laccaic acid D

Laccaic acid A, $C_{26}H_{19}NO_{12}$, *laccaic acid* A_1, R = CH_2-$CH_2NHCOCH_3$. Red platelets from methanol, chars at 230°. Absorption max (conc H_2SO_4): 302, 361, 518, 558 nm (log ϵ 4.33, 4.11, 4.32, 4.37).

Laccaic acid B, $C_{24}H_{16}O_{12}$, R = CH_2CH_2OH. Red needles from methanol.

Laccaic acid C, $C_{25}H_{17}NO_{13}$, R = $CH_2CH(NH_2)COOH$. Dark red needles from methanol, dec > 360°.

Laccaic acid D, $C_{16}H_{10}O_7$, *xanthokermesic acid.* Yellow needles from water, dec > 300°.

USE: As crimson dye.

5210. Lachesine. *N-Ethyl-2-[(hydroxydiphenylacetyl)oxy]-N,N-dimethylethanaminium chloride; ethyl(2-hydroxyethyl)dimethylammonium chloride benzilate;* E-3. $C_{20}H_{26}$-$ClNO_3$; mol wt 363.90. C 66.01%, H 7.20%, Cl 9.74%, N 3.85%, O 13.19%. Prepn: Ford-Moore, Ing, *J. Chem. Soc.* **1947,** 55. Muscarine agonist: F. Roberts, R. P. Stephenson, *Brit. J. Pharmacol.* **57,** 395 (1976); D. A. Brown, S. Fatherazi, *ibid.* **59,** 500P (1977); N. J. M. Birdsall *et al., Mol. Pharmacol.* **14,** 723 (1978).

Crystals from ethanol + acetone, mp 213°. LD_{50} in mice: 0.8 mg/20 g i.p.; 3.2 mg/20 g s.c.; 20.0 mg/20 g orally.

5211. Lacmoid. Resorcin blue; Iris Blue B; Fluorescent Blue; *C.I.* 51400. Prepn: Traub, Hock, *Ber.* **17,** 2615 (1884); Musso *et al., Angew. Chem.* **73,** 434 (1961). Discussion of structure controversy: H. J. Conn's *Biological Stains,* R. D. Lillie, Ed. (Williams & Wilkins, Baltimore, 8th ed., 1969) pp 283-284. *See also: Colour Index* vol. 4 (3rd ed., 1971) p 4468.

Dark-violet, lustrous scales or granules. Freely sol in methanol, ethanol, amyl alcohol, glacial acetic acid, acetone, phenol. Sparingly sol in water, ether. Practically insol in chloroform, benzene, petr ether.

USE: As acid-base indicator in 0.2% soln in alcohol. pH: 4.4 red; 6.4 blue. Satisfactory for titrating mineral acids, strong bases, many alkaloids; determining alkalinity and temporary hardness in water analysis. Biological stain; dye for wool, silk. Not adapted for carbonates, weak inorganic and organic acids, weak bases. Lacmoid is more sensitive than litmus, particularly in form of test paper.

5212. Lactaroviolin. *4-Methyl-7-(1-methylethenyl)-1-azulenecarboxaldehyde; 7-isopropenyl-4-methyl-1-azulene-carboxaldehyde; 1-formyl-4-methyl-7-isopropenylazulene.* $C_{15}H_{14}O$; mol wt 210.26. C 85.68%, H 6.71%, O 7.61%. Antibiotic pigment produced by the fungus *Lactarius deliciosus:* Willstaedt, *Ber.* **68,** 333 (1935); **69,** 997 (1936); Benesová *et al., Chem. Listy* **48,** 882 (1954). Structure: Plattner *et al., Chem. & Ind.* (London) **1954,** 1202; Sorm *et al., ibid.* 1511; Heilbron, Schmid, *Helv. Chem. Acta* **37,** 2018 (1954).

Purple crystals from petr ether, mp 58°. Sol in the usual organic solvents. Practically insol in water.

5213. Lactate Dehydrogenase. Lactic dehydrogenase; serum lactic dehydrogenase. Enzyme found in almost all animal tissues, in microorganisms, and in plants. Isoln from heart muscle, rat skeletal muscle, and Jensen sarcoma: Straub, *Biochem. J.* **34,** 483 (1940); Kubowitz, Ott, *Biochem. Z.* **314,** 94 (1943); Meister, *Biochem. Prepn.* **2,** 18 (1952). Catalyzes the equilibrium reaction of pyruvic acid to lactic acid. Plays an important role in the equilibrium of carbohydrate catabolism and anabolism. Used as fuel for aerobic tissues such as the heart. Structure is a tetramer of mol wt about 140,000. Consists of units of mol wt about

35,000. Two types of subunits are distinguishable: M (muscle) type and H (heart) type. Lactate dehydrogenases of heart and muscle are mainly H_4 and M_4; all other possible hybrids have been found in various tissues. Elevations of lactate dehydrogenase activity have been found in myocardial infarction, hepatocellular necrosis, metastatic carcinoma, diabetic ketosis, sickle cell anemia, malignant lymphoma, infectious mononucleosis, and cerebral infarction: Standjord et al., J. Am. Med. Assoc. **182**, 1099 (1962). Comprehensive reviews: Everse, Kaplan, Advan. Enzymol. Relat. Areas Mol. Biol. **37**, 61 (1973); Holbrook et al., in The Enzymes, vol. XI (part A), P. D. Boyer, Ed. (Academic Press, New York, 3rd ed., 1975) pp 191-292.

USE: In the determination of pyruvate (used in conjunction with reduced coenzyme). In the diagnosis of myocardial infarction and leukemia.

5214. D-Lactic Acid. (R)-2-Hydroxypropanoic acid; D(−)-lactic acid; levorotatory lactic acid; l-lactic acid; D-Milchsäure (German). $C_3H_6O_3$; mol wt 90.08. C 40.00%, H 6.71%, O 53.29%. Obtained by resolution of DL-lactic acid: Purdie, Walker, J. Chem. Soc. **61**, 754 (1892); Borsook et al., J. Biol. Chem. **102**, 449 (1933). Convenient laboratory prepn from glucose using Lactobacillus leichmannii: Brin, Biochem. Prepn. **3**, 61 (1953).

$$\begin{array}{c} COOH \\ | \\ H\!\!\leftarrow\!\!C\!\!\rightarrow\!\!OH \\ | \\ CH_3 \end{array}$$

Crystals from ether + isopropyl ether, mp 52.8°. $[\alpha]_{546}^{21.5}$ −2.6° (c = 8). pK = 3.83. Sol in water, alcohol, acetone, ether, glycerol. Practically insol in chloroform.

Forms salts with many metals. Most of these salts are dextrorotatory.

Zinc D(+)-lactate, $Zn(C_3H_5O_3)_2 \cdot 2H_2O$, crystals, $[\alpha]_D^{14}$ +8.18° (c = 2.5).

5215. DL-Lactic Acid. 2-Hydroxypropanoic acid; racemic lactic acid; ordinary lactic acid; α-hydroxypropionic acid; Milchsäure (German); Lactovagan; Tonsillosan (Lösung). $C_3H_6O_3$; mol wt 90.08. C 40.00%, H 6.71%, O 53.29%. Occurs in sour milk as a result of lactic acid bacteria; also found in molasses due to partial conversion of sugars, in apples and other fruits, tomato juice, beer, wines, opium, ergot, foxglove, and several higher plants, especially during germination. Lactic acid is prepd technically by "lactic acid fermentation" of carbohydrates such as glucose, sucrose, lactose with Bacillus acidi lacti or related organisms such as Lactobacillus delbrueckii, L. bulgaricus etc. The fermentation is carried out at relatively high temps. Produced commercially by fermentation of whey, cornstarch, potatoes, molasses. Review on the production of lactic acid by fermentation: S. C. Prescott, C. G. Dunn, Industrial Microbiology (McGraw-Hill, New York, 3rd ed., 1959) pp 304-331. Chem prepns from acetaldehyde and CO in dil H_2SO_4 at 130-200° and 900 atm: Loder, U.S. pat. **2,265,945** (1938 to du Pont); by hydrolysis of hexoses with NaOH: Lock, U.S. pat. **2,382,889** (1943). Prepn of crystalline lactic acid: Borsook et al., J. Biol. Chem. **102**, 449 (1933). Toxicity data: Smyth et al., J. Ind. Hyg. Toxicol. **23**, 259 (1941).

$$\begin{array}{c} COOH \\ | \\ CHOH \\ | \\ CH_3 \end{array}$$

Crystals, mp 16.8°. bp_{14-15} 122°; $bp_{0.5-1}$ 82-85°. K at 25° 1.38×10^{-4}. Heat of combustion at constant pressure 3615 cal/kg. Volatile with superheated steam. Sol in water, alc, furfurol; less sol in ether. Practically insol in chloroform, petr ether, carbon disulfide. Pharm. Incompat: Oxidizing agents, iodides, HNO_3, albumin. LD_{50} orally in rats: 3.73 g/kg (Smyth).

Barium salt, $C_6H_{10}BaO_6$, barium lactate. Powder. Poisonous! Sol in water, dil alcohol.

Copper salt dihydrate, $C_6H_{10}CuO_6 \cdot 2H_2O$, cupric lactate.

Green to blue crystals. Readily sol in water; practically insol in alcohol.

USE: In dyeing baths, as mordant in printing woolen goods, solvent for water-insoluble dyes (alcohol-soluble induline, nigrosine, spirit-blue); reducing chromates in mordanting wool; manuf cheese, confectionery; acidulant in beverages; for acidulating worts in brewing, for removing Clostridium butyricum in manuf of yeast; dehairing, plumping, and decalcifying hides; solvent for cellulose formate; flux for soft solder; manuf lactates which are used in food products, in medicine, and as solvents; plasticizer, catalyst in the casting of phenolaldehyde resins. Caution: Caustic in concd solns.

THERAP CAT: Acidulant.

THERAP CAT (VET): Has been used as a caustic, and in dilute solutions to irrigate tissues; as an intestinal antiseptic and antiferment.

5216. L-Lactic Acid. (S)-2-Hydroxypropanoic acid; L-(+)-lactic acid; dextrorotatory lactic acid; d-lactic acid; sarcolactic acid; paralactic acid; Fleishmilchsäure; L-Milchsäure. $C_3H_6O_3$; mol wt 90.08. C 40.00%, H 6.71%, O 53.29%. Occurs in small quantities in the blood and muscle fluid of man and animals. The lactic acid concn increases in muscle and blood after vigorous activity. L(+)-Lactic acid is also present in liver, kidney, thymus gland, human amniotic fluid, and other organs and body fluids. Obtained by resolution of DL-lactic acid: Purdie, Walker, J. Chem. Soc. **61**, 754 (1892); Borsook et al., J. Biol. Chem. **102**, 449 (1933). Convenient laboratory prepn from glucose by fermentation by Lactobacillus delbrueckii: Brin, Biochem. Prepn. **3**, 61 (1953). Prepn from hexoses using B. dextrolacticus: Andersen, Greaves, Ind. Eng. Chem **34**, 1522 (1942). Monograph: M. Brin, R. H. Dunlop, "Chemistry and Metabolism of L-and D-Lactic Acids", Ann. N.Y. Acad. Sci. vol. **119**, art. 3, 851-1165 (1965).

$$\begin{array}{c} COOH \\ | \\ HO\!\!\leftarrow\!\!C\!\!\rightarrow\!\!H \\ | \\ CH_3 \end{array}$$

Crystals from acetic acid or chloroform, mp 53°. $[\alpha]_{546.1}^{21.22}$ +2.6° (c = 2.5). pK at 25°, 3.79. Forms salts with many metals. The salts are more sol in water than the salts of the racemic acid. Most of the salts are levoratory.

Zinc L(−)-lactate dihydrate, $Zn(C_3H_5O_3)_2 \cdot 2H_2O$, prisms. $[\alpha]_D^{25}$ −8.2° (c = 2.5 in water).

5217. Lactic Acid Lactate. 2-Hydroxypropanoic acid 1-carboxyethyl ester; 2-(lactoyloxy)propanoic acid; 2-(2-hydroxypropanoyloxy)propanoic acid. $C_6H_{10}O_5$; mol wt 162.14. C 44.44%, H 6.22%, O 49.34%. Prepd by heating lactic acid at 120° for 10 hours: Dietzel, Krug, Ber. **58**, 1307 (1925).

$$\begin{array}{c} CH_3 \quad CH_3 \\ | \quad\quad | \\ CHOOCCHOH \\ | \\ COOH \end{array}$$

Pale yellow, clear, odorless oil. Sol in water and in the usual organic solvents.

Methyl ester, $C_7H_{12}O_5$. Prepn: Claborn, U.S. pat. **2,371,-281** (1945 to the people of the U.S.). $bp_{7.8}$ 107°; n_D^{20} 1.4313.

USE: The methyl ester as a solvent or plasticizer.

5218. Lactobacillic Acid. 2-Hexylcyclopropanedecanoic acid; 11,12-methyleneoctadecanoic acid; phytomonic acid. $C_{19}H_{36}O_2$; mol wt 296.48. C 76.97%, H 12.24%, O 10.79%. A lipid constituent of various microorganisms. Isoln from Lactobacillus arabinosus: K. Hofmann, R. A. Lucas, J. Am. Chem. Soc. **72**, 4328 (1950); from Agrobacterium tumefaciens and identity of phytomonic acid with lactobacillic acid: K. Hofmann, F. Tausig, J. Biol. Chem. **213**, 425 (1955). Structure: K. Hofmann et al., J. Am. Chem. Soc. **80**, 5717 (1958). Abs config: J. F. Tocanne, Tetrahedron **28**, 363 (1972).

$$CH_3(CH_2)_4CH_2 \overline{} (CH_2)_9COOH$$

Crystals from acetone, mp 27.8-28.8°. Soluble in acetone, petr ether.

Methyl ester, $C_{20}H_{38}O_2$, liq, bp$_3$ 187-187.5°. Soluble in many fat solvents.

Amide, $C_{19}H_{37}NO$, *lactobacillamide*. Crystals, mp 79.4-81.5°. Soluble in dimethylformamide.

5219. Lactobionic Acid. *4-O-β-D-Galactopyranosyl-D-gluconic acid;* 4-(β-D-galactosido)-D-gluconic acid. C_{12}-$H_{22}O_{12}$; mol wt 358.30. C 40.22%, H 6.19%, O 53.59%. Obtained by oxidation of lactose: Fischer, Meyer, *Ber.* **22**, 362 (1889); Ruff, Ollendorff, *ibid.* **33**, 1806 (1900); Isbell, *J. Res. NBS* **11**, 713 (1933); Margariello, U.S. pat. **2,746,916** (1956 to Nat. Dairy Res. Labs.); Eddy, *Nature* **181**, 904 (1958); Nishizuka *et al., J. Biol. Chem.* **235**, PC13 (1960). Manuf from lactose: Y. Sato *et al.,* Ger. pat. **2,038,230** (1971 to Hayashibara Co.), *C.A.* **74**, 142296c (1971). Crystal structure of calcium salt: W. J. Cook, C. E. Bugg, *Acta Crystallogr.* **B29**, 215 (1973). NMR studies: T. Taga *et al., Bull. Chem. Soc. Japan* **51**, 2278 (1978). For therapeutic use *see* Erythromycin Lactobionate.

H H OH H
HOH₂C - C - C - C̣ C - COOH
 | | |
 CH₂OH O H OH

HO O
H
OH H
H H
H OH

Syrup. Freely sol in water, slightly sol in methanol, ethanol, glacial acetic acid. Dehydration by distillation with dioxane yields *lactobionic δ-lactone,* $C_{12}H_{20}O_{11}$, non-deliquescent crystals, dec 195-196°. Shows mutarotation. $[\alpha]_D^{20}$ +53.0° initial (c = 8.8) → $[\alpha]_D^{20}$ +22.6° final (240 minutes).

Calcium salt, $C_{24}H_{42}CaO_{24}$, *calcium lactobionate.* Pentahydrate, hairlike needles in brushlike groups. When anhydr, slender needles from small amts of anhydr ethanol. $[\alpha]_D^{20}$ +23.7° (c = 6.28). n_D^{20} 1.4583 (concd syrup just before crystallization). Freely sol in water.

5220. p-Lactophenetide. *N-(4-Ethoxyphenyl)-2-hydroxypropanamide; p-lactophenetidide;* lactyl-*p*-phenetidin-*N*-(*p*-ethoxyphenyl)lactamide; Fenolactine; Lactophenin; Phenolactine. $C_{11}H_{15}NO_3$; mol wt 209.24. C 63.14%, H 7.23% N 6.69%, O 22.94%. Prepn: Shapiro *et al., J. Am. Chem. Soc.* **81**, 6322 (1959).

OH
|
C₂H₅O─⟨ ⟩─NHOCCHCH₃

Slightly bitter crystals from ethyl acetate + hexane, mp 117-118°. One gram dissolves in 330 ml cold, 55 ml boiling water, 8.5 ml alcohol; slightly sol in ether, petr ether.
THERAP CAT: Analgesic, antipyretic.

5221. Lactose. *4-O-β-D-Galactopyranosyl-D-glucose;* 4-(β-D-galactosido)-D-glucose; milk sugar. $C_{12}H_{22}O_{11}$; mol wt 342.30. C 42.10%, H 6.48%, O 51.42%. Present in milk of mammals: human 6.7%; cow's 4.5%. Milk at body temp contains lactose as an equilibrium mixture of 2 parts of α-lactose and 3 parts of β-lactose. By-product of the cheese industry, produced from whey: Davis, *Can. Dairy and Ice Cream J.* **19**, 52 (1940); *Milk Trade Gaz.* **12**, 4 (1941); F. Ullmann, *Encyklopädie der Technischen Chemie,* **VII**, 579 (2nd ed.), 1931. Structure and configuration: Zemplén, *Ber.* **59**, 2402 (1926); Levene, Sobotka, *J. Biol. Chem.* **71**, 471 (1926); Levene, Wintersteiner, *ibid.* **75**, 315 (1927); Haworth, Long, *J. Chem. Soc.* **1927**, 544; Hudson, *J. Am. Chem. Soc.* **52**, 1712 (1930); Hassid, Ballou in *The Carbohydrates,* W. Pigman, Ed. (Academic Press, New York, 1957) p 495. Synthesis: Haskins *et al., J. Am. Chem. Soc.* **64**, 1852 (1942).

Reviews: Whittier, *Chem. Rev.* **2**, 85-125 (1926); *J. Dairy Sci.* **27**, 505-537 (1944); Weisberg, *ibid.* **37**, 1106-1115 (1954); L. A. W. Thelwall, *Dev. Food Carbohyd.* **2**, 275-326 (1980).

CH₂OH CH₂OH
HO O H O OH
H H
OH H OH H
O
H H H H
H OH H OH

α-Lactose monohydrate, is the usual milk sugar and the lactose of pharmacy. Monoclinic sphenoidal crystals from water. Faintly sweet taste. Stable in air, but readily absorbs odors. d²⁰ 1.53. Becomes anhydrous at 120°. mp 201-202° (rapid heating). Shows mutarotation. $[\alpha]_D^{20}$ +92.6° → +83.5° (10 min.) → +69° (50 min) → +52.3° (22 hrs, c = 4.5). The final value is obtained instantly in the presence of a trace of NH_3. U.S.P. requires +52.2° to +52.5° (c = 10). One gram dissolves in 5 ml water, in 2.6 ml boiling water; very slightly sol in alcohol. Insol in chloroform, ether. Ka at 16.5° = 6.0 × 10⁻¹³. d²⁰ of aq solns calcd for the monohydrate: 5.2% = 1.018; 10.2% = 1.038; 20.0% = 1.078; 30.2% = 1.123; 50.9% = 1.226; 60.8% = 1.281; 69.1% = 1.330.

β-Lactose, $C_{12}H_{22}O_{11}$. Obtained by crystallizing concd solns of α-lactose above 93.5°. Somewhat sweeter than the α-form. $[\alpha]_D^{25}$ +34° (3 min) → +39° (6 min) → +46° (1 hr) → +52.3° (22 hrs). One gram dissolves in 2.2 ml water at 15°, in 1.1 ml boiling water. After a few days crystals of the less sol α-monohydrate appear from satd solns.

On hydrolysis with 2% H_2SO_4 or with emulsin lactose yields 1 mol D-glucose and 1 mol D-galactose. Reduces Fehling's soln.

USE: Both forms of lactose are employed, with the α-form predominating: as a nutrient in preparing modified milk and food for infants and convalescents (Whittier, "Lactose and Its Utilization," *loc. cit;* review with 327 ref). In baking mixtures. Pharmaceutic aid (tablet and capsule diluent). To produce lactic acid fermentation in ensilage and food products. As chromatographic adsorbent in analytical chemistry. In culture media. For many other uses see the comprehensive review by Weisberg "Recent Progress in the Manufacture and Use of Lactose," *loc. cit.*

THERAP CAT (VET): Added to cow's milk for feeding orphan foals.

5222. Lactucarium—"French". Thridace. Inspissated juice of *Lactuca sativa* L., var. *capitata* L., *Compositae.* Constit. Lactucin, hyoscyamine, mannite.

Brown pieces or powder; bitter taste; opium-like odor. Partly sol in water, alcohol, ether.

5223. Lactucarium—"German". "Lettuce opium". Dried milk-juice of *Lactuca virosa* L., *Compositae* (wild lettuce). Constit. About 0.2% lactucin; about 50% lactucerol; hyoscyamine, lactucic acid, caoutchouc, volatile oil, mannite.

Brown powder or irregular pieces; wax-like when cut; bitter taste. Partly sol in water, alcohol, ether. *Keep dry.*

THERAP CAT: Sedative.

5224. Lactucin. *3,3a,4,5,9a,9b-Hexahydro-4-hydroxy-9-(hydroxymethyl)-6-methyl-3-methyleneazuleno[4,5-b]furan-2,7-dione.* $C_{15}H_{16}O_5$; mol wt 276.30. C 65.21%, H 5.84%, O 28.95%. From various *Lactuca* spp and *Cichorium intybus* L., *Compositae.* Isoln: Schenck, Graf, *Arch. Pharm.* **274**, 537 (1936); **275**, 36 (1937); Schenck *et al., ibid.* **294**, 17 (1961). Purification: Späth *et al., Monatsh.* **82**, 114 (1951). Structure: Dolejs *et al., Coll. Czech. Chem. Commun.* **23**, 2195 (1958); Barton, Narayanan, *J. Chem. Soc.* **1958**, 963; Michl, Högenauer, *Monatsh.* **89**, 317 (1958). Revised stereochemistry: Bachelor, Itô, *Can. J. Chem.* **51**, 3626 (1973).

Crystals from methanol, sinters at 218°, mp 228-233°. $[\alpha]_D$ +49° (c = 0.90 in methanol), +77.9° (c = 3.44 in pyridine). uv max: 257 nm (ϵ 14,000). Soluble in water, ethanol, methanol, ethyl acetate, dioxane anisol.

p-Hydroxyphenylacetate hydrate, $C_{23}H_{22}O_7$, *intybin, lactucopicrin.* Identity with intybin: Schmitt, *Bot. Arch.* **40**, 516 (1940), *C.A.* **36**, 5616 (1942). Structure: Michl, Högenauer, *Monatsh.* **91**, 500 (1960). Crystals from water, dec 148-151°. $[\alpha]_D^{7.5}$ +67.3° (pyridine).

5225. Lactulose.

*4-O-β-*D*-Galactopyranosyl-*D*-fructose;* 4-D-galactopyranosyl-4-D-fructofuranose; 4-*O-β-*D*-galactosyl-*D*-fructose; 4-β-*D*-galactosido-*D*-fructose; Bifiteral; Cephulac; Duphalac; Generlac; Lactuflor; Laevilac; Normase. $C_{12}H_{22}O_{11}$; mol wt 342.30. C 42.10%, H 6.48%, O 51.42%. Prepd from lactose: Montgomery, Hudson, *J. Am. Chem. Soc.* **52**, 2101 (1930); Oosten, *Rec. Trav. Chim.* **86**, 673 (1967); K. B. Hicks, F. W. Parrish, *Carbohyd. Res.* **82**, 393 (1980).

Hexagonal clustered plates from methanol, mp 169.0° (1°/min). Sweeter than lactose, but not as sweet as sucrose. Shows mutarotation; constant value after 24 hrs: −51.4° (c = 4 in H_2O at pH 4.8 and 20°). Soly in water (w/w) at 30°: 76.4%; at 60°: 81%; at 90°: >86%. Acid hydrolysis yields galactose and fructose.

THERAP CAT: Laxative.

5226. Laminaran.

Laminarin. A polysaccharide found in brown seaweed and occurring principally in the *Laminaria* spp. Linear polymer composed of β-(1 → 3)-linked glucose residues; may contain small amounts of β-(1 → 6) linkages as interresidue linkages or as branch points and 2-3% D-mannitol as end groups. Two forms of laminaran are recognized; they are referred to as soluble and insoluble laminaran: Percival, Ross, *J. Chem. Soc.* **1951**, 720. Structure: Peat *et al.*, *ibid.* **1958**, 724, 729; **1960**, 175; Goldstein *et al.*, *Chem. & Ind. (London)* **1959**, 124; Annan *et al.*, *ibid.* **1962**, 984; Annan *et al.*, *J. Chem. Soc.* **1965**, 885; Maeda, Nisizawa, *Carbohyd. Res.* **7**, 97 (1968). Structure of soluble laminaran from *Eisenia bicyclis:* T. Usui *et al.*, *Agr. Biol. Chem.* **43**, 603 (1979). NMR studies of laminaran: D. Gagnaire, *Org. Magn. Res.* **11**, 344 (1978); H. Friebolin *et al.*, *ibid.* **12**, 216 (1979). Review: W. A. P. Black, E. T. Dewar in *Industrial Gums,* R. L. Whistler, Ed. (Academic Press, New York, 2nd ed., 1973) pp 137-145.

The water-insol laminaran, isolated from *L. cloustoni* Edmondst., *Laminariaceae*, is pptd spontaneously from the aq acid extract of the plant. Has lower degree of branching than the sol form. Typical analysis of the dry material: 92.5% polyglucose, 0.4% non-volatile matter; $[\alpha]_D^{16}$ −13.4° (c = 0.9). Amorphous triacetate, $(C_{12}H_{16}O_8)_n$, $[\alpha]_D^{16}$ −63.5° (c = 0.4 in chloroform).

The sol form, isolated from *L. digitata,* is separated from the acidified extract only after addition of a precipitant such as ethanol. Typical analysis (on dry basis): 91.2% polyglucose, 1.0% non-volatile matter; $[\alpha]_D^{18}$ −11.9° (c = 2.1).

Sulfate, *laminaran hydrogen sulfate.* Laminaran can be sulfated to varying degrees. Highly sulfated products have anticoagulant properties comparable to heparin, while laminarans with few sulfate groups are antilipemic only: Besterman, Evans, *Brit. Med. J.* **1957**, I, 310.

5227. Lamoparan.

Org-10172. Low molecular weight heparinoid derived from porcine intestinal mucosa; mixture of sulfated glycosaminoglycans with mean mol wt 6500 daltons. Prepn: A. L. M. Sanders *et al.*, **Eur. pat. Appl. 66,908**; *eidem,* U.S. pat. **4,438,108** (1982, 1984 both to Akzo). Exhibits antithrombotic activity comparable to standard heparin but with diminished hemorrhagic effects: D. G. Meuleman *et al.*, *Thromb. Res.* **27**, 353 (1982); H. ten Cate *et al.*, *ibid.* **38**, 211 (1985). Preliminary pharmacokinetics and tissue distribution in humans: I. D. Bradbrook *et al.*, *Brit. J. Clin. Pharmacol.* **23**, 667 (1987). Clinical evaluation during hemodialysis: C. P. Henny *et al.*, *Lancet* **1**, 890 (1983); in acute thrombotic stroke: A. G. G. Turpie *et al.*, *ibid.* **1**, 523 (1987).

White amorphous, slightly hygroscopic powder. $[\alpha]_D^{20}$ +30° to +70°.

THERAP CAT: Antithrombotic.

5228. Lamotrigine.

6-(2,3-Dichlorophenyl)-1,2,4-triazine-3,5-diamine; 3,5-diamino-6-(2,3-dichlorophenyl)-1,2,4-triazine; LTG; BW 430C; Lamictal. $C_9H_7Cl_2N_5$; mol wt 256.10. C 42.21%, H 2.76%, Cl 27.69%, N 27.34%. New structural class of antiepileptics. Prepn: M. G. Baxter *et al.*, **Eur. pat. Appl. 21,121** (1981 to Wellcome Foundation); D. A. Sawyer *et al.*, U.S. pat. **4,602,017** (1986). Anticonvulsant activity: A. A. Miller *et al.*, *Epilepsia* **27**, 483 (1986). Mechanism of action study: M. J. Leach *et al.*, *ibid.* 490. Evaluation of CNS effects: A. F. Cohen *et al.*, *Brit. J. Clin. Pharmacol.* **20**, 619 (1985). Pharmacokinetics: A. F. Cohen *et al.*, *Clin. Pharmacol. Ther.* **42**, 535 (1987). Clinical evaluations in epilepsy: C. D. Binnie *et al.*, *Epilepsia* **27**, 248 (1986); C. D. Binnie *et al.*, *Epilepsy Res.* **1**, 202 (1987).

Crystals from isopropanol, mp 216-218° (uncorr). LD_{50} in mice, rats (mg/kg): 250, >640 orally (Sawyer).

THERAP CAT: Anticonvulsant.

5229. Lanatosides.

Family of four glycosides, A, B, C, D, isolated from various species of *Digitalis* including *D. lanata* Ehrh., *Scrophulariaceae:* Stoll, Kreis, *Helv. Chim. Acta* **16**, 1049 (1933); Ligeti, *Pharmazie* **12**, 433 (1957); **14**, 162 (1959); Angliker *et al.*, *Ann.* **607**, 131 (1957); *D. Lutea* L., *Scrophulariaceae:* Cole, Gisvold, *J. Am. Pharm. Assoc., Sci. Ed.* **47**, 654 (1958). Structure: Tschesche *et al.*, *Ber.* **92**, 2258 (1959); Uskert, *Ann.* **638**, 199 (1960); Kuhn *et al.*, *Helv. Chim. Acta* **45**, 881 (1962). Absorption spectrum: Bell, *J. Am. Pharm. Assoc., Sci. Ed.* **49**, 277 (1960). Cardioactivity of lanatoside C: K.-O. Haustein, *Pharmacology* **11**, 117 (1974). Mitogenicity: L. L. G. Hammarström, C. I. E. Smith, *J. Immunol.* **120**, 694 (1978). HPLC determn: V. Y. Davydov *et al.*, *J. Chromatog.* **248**, 49 (1982); of lanatoside C: F. Orosz *et al.*, *Anal. Biochem.* **156**, 171 (1986).

Lanatoside A	R=digitoxigenin
Lanatoside B	R=gitoxigenin
Lanatoside C	R=digoxigenin
Lanatoside D	R=diginatigenin

Lanatoside A, $C_{49}H_{76}O_{19}$, *Digilanide A, Adigal.* Long, flat prisms from methanol, dec 245-248°. $[\alpha]_D^{20}$ +31.6° (0.48 g in 25 ml 95% alc); $[\alpha]_D^{20}$ +23.2° (0.95 g in 25 ml dioxane). Soluble in 20 parts methanol, 40 parts alcohol, 225 parts chloroform, 16,000 parts water.

Lanatoside B, $C_{49}H_{76}O_{20}$, *Digilanide B.* Long, flat prisms from alcohol, dec 245-248° after drying in high vacuum at 150° $[\alpha]_D^{20}$ +36.7° (0.47 g in 25 ml 95% alc); $[\alpha]_D^{20}$ +31.8° (0.25 g in 14.1 ml dioxane). Soluble in 20 parts methanol, 40 parts alcohol, 550 parts chloroform. Nearly insol in water.

Lanatoside C, $C_{49}H_{76}O_{20}$, *Allocor, Cedilanid, Ceglunat, Celadigal, Cetosanol, Digilanide C, Lanimerck (Suppositories).* Long, flat prisms from alcohol, dec 248-250°, after drying in vacuum at 150°. $[\alpha]_D^{20}$ +33.4° to +33.7° (200 mg dry weight in 10 ml alcohol). One gram dissolves in 20,000 ml methanol, in 2000 ml chloroform. Freely sol in pyridine, dioxane. Practically insol in ether, petr ether.

Lanatoside D, $C_{49}H_{76}O_{21}$, *Digilanide D.* Needles from methanol and water, dec 242-250°. $[\alpha]_D^{20}$ +40.5° (c = 5.95 in methanol). uv max: 220 nm (log ε 4.16).

THERAP CAT: Cardiotonic.

5230. Lankamycin. Kujimycin B. $C_{42}H_{72}O_{16}$; mol wt 833.04. C 60.56%, H 8.71%, O 30.73%. Macrolide antibiotic produced by *Streptomyces violaceoniger* from soil of Ceylon. Isolation: Gaumann *et al., Helv. Chim. Acta* **43**, 601 (1960). Yields on hydrolysis an aglucone, monoacetyllankolid, $C_{26}H_{48}O_{10}$, and two sugar-like substances, lankavose (D-chalcose, *q.v.*) and acetylarcanose: Keller-Schierlein, Roncari, *ibid.* **45**, 138 (1962). Structure: *eidem, ibid.* **47**, 78 (1964); Egan, Martin, *J. Am. Chem. Soc.* **92**, 4129 (1970). Stereochemistry: Muntwyler, Keller-Schierlein, *Helv. Chim. Acta* **55**, 460 (1972). Identity with kujimycin B: Omura *et al., J. Antibiot.* **22**, 629 (1969).

Crystals from ether + petr ether. Double mp, 147-150° and 181-182°. $[\alpha]_D^{20}$ −94° (c = 1.23 in alc.). uv max: 289 nm (log ε 1.50).

5231. Lanolin. Wool fat; oesipos; agnin; alapurin; Agnolin; Lanum; Lanain; Lanalin; Lanesin; Lanichol; Laniol. Lanolin is the "fat-like" secretion of the sebaceous glands of sheep which is deposited onto the wool fibers. Chemically it is a wax rather than a fat, being a complex mixture of esters and polyesters of 33 high-molecular-weight alcohols and 36 fatty acids. The alcohols are of three types: aliphatic alcohols, steroid alcohols, and triterpenoid alcohols; the acids are also of three types: saturated nonhydroxylated acids, unsaturated nonhydroxylated acids and hydroxylated acids. Liquid lanolin is rich in low molecular weight, branched aliphatic acids and alcohols while waxy lanolin is rich in high molecular weight, straight-chain acids and alcohols. Reviews on composition, derivs, modifications and uses: Barnett, *Drug & Cosmet. Ind.* **80**, 744 (1957); **83**, 292 (1958); Leideritz, *Chemiker Ztg.* **83**, 707 (1959); F. Fawaz *et al., Ann. Pharm. Franc.* **31**, 217-226 (1973). Rheological properties of lanolin: F. Puisieux, *Pharm. Acta Helv .* **51**, 289 (1976). Monograph: E. V. Truter, *Wool Wax, Chemistry and Technology* (Interscience, 1956).

Lanolin contains about 25-30% water. It is a yellowish-white unctuous mass; slight odor. Practically insol in water. Sol in chloroform or ether with the separation of the water.

Anhydr lanolin is a yellowish tenacious, semisolid fat; slight odor or practically odorless. mp 38-42°. Practically insol in, but mixes with about twice its weight of water, without separation. Sparingly sol in cold, more in hot alcohol; freely sol in benzene, chloroform, ether, carbon disulfide, acetone, petr ether.

Acetylated lanolin, *Acetadeps, Modulan.* Almost odorless, pale yellow, semi-solid, unctuous mass. mp 36°. Sol in mineral oils, in some vegetable oils. Readily dispersed in oil-in-water emulsions. Will not emulsify to form water-in-oil emulsions.

USE: Pharmaceutic aid (ointment base).

THERAP CAT (VET): Ointment base, hoof dressing.

5232. Lanosterol. *Lanosta-8,24-dien-3-ol;* kryptosterol. $C_{30}H_{50}O$; mol wt 426.70. C 84.44%, H 11.81%, O 3.75%. The core steroid from which all others are derived by biological modification. From wool fat of sheep: Windaus, Tschesche, *Z. Physiol. Chem.* **190**, 51 (1930). Identity with kryptosterol: Ruzicka *et al., Helv. Chim. Acta* **28**, 759 (1945). Structure: Voser *et al., ibid.* **35**, 2414 (1952); Barnes *et al., J. Chem. Soc.* **1953**, 571. Stereochemistry: *eidem, ibid.* **1953**, 576. Prepn from isocholesterol: Bloch, Urech, *Biochem. Prepn.* **6**, 32 (1958). Prepn by cyclization of squalene: Cornforth *et al., Ciba Foundation Symposium of Terpenes and Sterols* **1958**, 119; van Tamelen *et al., J. Am. Chem. Soc.* **88**, 4752 (1966). Mechanism of the squalene to lanosterol conversion: *eidem, ibid.* **104**, 6479, 6480 (1982).

Crystals, mp 138-140°. $[\alpha]_D^{20}$ +62.0° (chloroform). Infrared absorption max 6.124; 9.69; 12.22 microns (KBr disks of lanosterol).

Lanosteryl acetate, crystals, mp 131.5-133°, $[\alpha]_D^{20}$ +62.5° (c = 1.12 in chloroform).

5233. Lanthanum. La; at. wt 138.9055; at. no. 57; valence 3. A rare earth metal. Two naturally occurring isotopes: ^{139}La (99.911%); ^{138}La (0.089%); ^{138}La is radioactive, $T_{1/2}$ 1.12 × 10^{11} years; artificial radioactive isotopes: 125-137; 140-144. Estimated abundance in earth's crust: 5-18 ppm. Found in association with cerium and other light lanthanons. Minerals of commercial interest are monazite, bastnaesite and cerite. Discovery and isoln: Mosander, *Pogg. Ann.* **47**, 207 (1839). Separation by fractional crystallization and pptn: James, *J. Am. Chem. Soc.* **34**, 757 (1912); prepn of the metal by electrolysis of the fused chloride at 700-1000°: Mazzi, *Atti X. Congr. Internaz. Chim.* **3**, 604 (1938); prepn by reduction of the chloride by calcium: Spedding *et al., Ind. Eng. Chem.* **44**, 553 (1952). Reviews of prepn, properties and compounds of lanthanum and other rare earth metals: *The Rare Earths,* F. H. Spedding, A. H. Daane, Eds. (Krieger, Huntington, N.Y., 1971, reprint of 1961 ed.) 641 pp; Hulet, Bode, "Separation Chemistry of the Lanthanides and Transplutonium Actinides" in *MTP Int. Rev. Sci.: Inorg. Chem., Ser. One,* Vol. 7, K. W. Bagnall, Ed. (University Park Press, Baltimore, 1972) pp 1-45; Vickery, "Scandium, Yttrium, Lanthanum" in *Comprehensive Inorganic Chemistry* vol. 3, J. C. Bailar Jr. *et al.,* Eds. (Pergamon Press, Oxford, 1973) pp 329-353; Moeller, "The Lanthanides" *ibid.* **vol. 4**, 1-101.

White, malleable metal; tarnishes in air; has three crystalline forms: the α-modification, hexagonal crystals, stable at ordinary temp, d 6.17; the β-form, obtained on heating the α-form at 350°, face-centered cubic crystals, d 6.19; the high temp γ-form, exists above 868°, body-centered cubic crystals, d 5.98. The metal melts at 920°. E°(aq) La^{3+}/La −2.52 V (calc). Very active; dec water slowly in the cold, more readily on heating. Readily attacked by mineral acids; not attacked by cold concd H_2SO_4. Burns in air at about

450° producing a mixture of oxide and nitride; forms the hydride on heating in hydrogen. Forms alloys with several metals.

Oxide, La_2O_3, *lanthana, lanthanum sesquioxide, lanthanum trioxide*, almost white, amorphous powder. d 6.51. mp above 2000°. Insol in water; sol in dil mineral acids with formation of salts. Absorbs CO_2 from the air.

Hydroxide, $La(OH)_3$, white, amorphous precip, prepd by adding excess of caustic alkali to a lanthanum salt soln. Strongly basic, displaces ammonia from ammonium salts, absorbs CO_2 from air. On dehydration yields $La_2O_3.H_2O$.

Chloride, $LaCl_3$, heptahydrate, triclinic crystals, sol in water or alc. On heating the anhydr salt (mp 852°) is formed. LD_{50} in rats: 4.2 g/kg orally; 350 mg/kg i.p., Cochran *et al., Arch. Ind. Hyg. Occup. Med.* **1**, 637 (1950).

Carbonate octahydrate, $C_3La_2O_9$, *artificial lanthanite*. White, cryst powder, insol in water, freely sol in dil mineral acids.

Sulfate, $La_2(SO_4)_3$, nonahydrate, hexagonal prisms. Prepd by treating a lanthanum salt with a slight excess of sulfuric acid. Dec at white heat. Is the least sol of the rare earth sulfates; soly in water decreases with increase in temp. Insol in alc. Forms double salts with alkali or ammonium hydroxide. Anhydr salt prepd by heating hydrate. LD_{50} in rats: > 5000 g/kg orally; 275 mg/kg i.p. (Cochran).

Nitrate, $La(NO_3)_3$, hexahydrate, white deliquesc crystals, mp about 40°, at higher temp forms a basic salt. bp 126°. Very sol in water, alc. Forms double salts with bivalent ion nitrates and ammonium nitrates. *Keep well closed.* LD_{50} in rats: 4.5 g/kg orally; 450 mg/kg i.p. (Cochran).

USE: Oxide in glass to improve optical properties. La^{3+} used in experimental biology as a specific antagonist of Ca: Weiss, *Ann. Rev. Pharmacol.* **14**, 343 (1974).

5234. Lanthionine. *S-(2-Amino-2-carboxyethyl)cysteine; 3,3'-thiodialanine;* bis(2-amino-2-carboxyethyl)sulfide. $C_6H_{12}N_2O_4S$; mol wt 208.24. C 34.60%, H 5.81%, N 13.45%, O 30.73%, S 15.40%. An amino acid found in proteins. First isolated as an artifact from wool hydrolysates: Horn *et al., J. Biol. Chem.* **138**, 141 (1941). Isoln of naturally occurring L- and meso-lanthionine: N. H. Sloane, K. G. Utich, *Biochemistry* **5**, 2658 (1966). Synthesis: du Vigneaud, Brown, *ibid.* 151; J. T. Snow *et al., Int. J. Peptide Protein Res.* **8**, 57 (1976). Synthesis of L-lanthionine from cystine: D. N. Harpp, J. G. Gleason, *J. Org. Chem.* **36**, 73 (1971). Stereoisomerism: *eidem, J. Biol. Chem.* **140**, 767 (1941). Structure studies: I. W. Stapleton, O. A. Weber, *Int. J. Peptide Protein Res.* **3**, 243 (1971).

$$HOOCCHCH_2SCH_2CHCOOH$$
$$\quad | \qquad\qquad\qquad | $$
$$\quad NH_2 \qquad\qquad\quad NH_2$$

meso-Form, the form obtained from wool by the procedure of Horn *et al.:* Six-sided plates having a triangular appearance from dil NH_3. Softens at 207°, dec 304°. Stable to alkalies. Sol in dil acids and alkalies. Sparingly sol in water. Insol in alcohol, ether, chloroform, acetone.

DL-Form, elongated hexagonal plates. Chars at 240°, dec 286-292°.

L-Form, elongated hexagonal plates. mp 295-296° (dec). $[\alpha]_D^{25}$ +9.4° (c = 1.4 in 2.4N NaOH); $[\alpha]_D^{22}$ +8.6° (c = 5 in 2.4N NaOH); $[\alpha]_D^{22}$ +6.0° (c = 1 in 1N NaOH).

D-Form, elongated hexagonal plates. Darkens at 245°, dec 293-295°. $[\alpha]_D^{21}$ −8.0° (c = 5 in 2.4N NaOH).

5235. Lapachol. *2-Hydroxy-3-(3-methyl-2-butenyl)-1,4-naphthalenedione;* lapachic acid; taiguic acid; tecomin; greenhartin; NSC-11905. $C_{15}H_{14}O_3$; mol wt 242.26. C 74.36%, H 5.83%, O 19.81%. Yellow crystalline material derived from the heartwood of Asian and South American bignoniaceous plants, esp Surinam greenheart, Taigu wood, Lapacho heartwood and Bethabarra wood: Arnoudon, *Compt. Rend.* **41**, 1152 (1857); Stein, *J. Prakt. Chem.* **99**, 1 (1866); Paterno, *Gazz. Chim. Ital.* **9**, 506 (1879); Greene, Hooker, *Am. Chem. J.* **11**, 267 (1889). Structure: Paterno, *Gazz. Chim. Ital.* **12**, 337 (1882); Hooker *J. Chem. Soc.* **61**, 611 (1892); **69**, 1355 (1896); *J. Am. Chem. Soc.* **58**, 1168 (1936). Synthesis: Fieser, *ibid.* **49**, 857 (1927); Hooker, *ibid.* **58**, 1181 (1936); G. R. Pettit, L. E. Houghton, *J. Chem. Soc. (C)* **1971**, 509. Although it is related structurally to vitamin

K, *q.v.*, it does not possess antihemorrhagic activity: H. J. Almquist, A. A. Klose, *J. Am. Chem. Soc.* **61**, 1923 (1939); L. F. Fieser *et al., J. Biol. Chem.* **137**, 659 (1941). It is reported to be an inhibitor of respiratory processes: E. G. Ball *et al., ibid.* **168**, 257 (1947). Lapachol has also exhibited antitumor activity vs Walker 256 carcinoma: K. V. Rao *et al., Proc. Am. Assoc. Cancer Res.* **8**, 55 (1967). Mass spectrometry: T. A. Elwood *et al., Org. Mass Spectrom.* **3**, 841 (1970). Chromatographic detection: M. H. Simatupang *et al., J. Chromatog.* **52**, 180 (1970). Pharmacology: S. M. Sieber *et al., Cancer Treat. Rep.* **60**, 1127 (1976). Review of antitumor activity: K. V. Rao, *Cancer Chemother. Rep. (pt. 2)* **4**(4), 11-17 (1974). Toxicity study: R. K. Morrison *et al., Toxicol. Appl. Pharmacol.* **17**, 1 (1970).

Yellow prisms from alcohol or ether, mp 140°. uv max: 251.5, 278, 331 nm (log ϵ 4.38, 4.28, 3.43). Soluble in alcohol, chloroform, benzene, acetic acid, slightly sol in ether, hot water. Sol in aq NaOH solns forming a bright red sodium salt. LD_{50} in BALB/c mice (g/kg): 0.487 (males); 0.792 (females); 0.621 (combined), R. K. Morrison *et al., loc. cit.*

5236. Lappa. Burdock; clotbur; bardana. Dried first-year root of *Arctium lappa* L., or of *Arctium minus* Bernh., *Compositae. Habit.* Europe, Northern Asia; naturalized in N. America. *Constit.* Volatile oil, bitter principle, inulin, tannin.

THERAP CAT: Dermatologic.

5237. Lappaconitine. $(1\alpha,14\alpha,16\beta)$-20-Ethyl-1,14,16-trimethoxyaconitane-4,8,9-triol 4-[2-(acetylamino)benzoate]. $C_{32}H_{44}N_2O_8$; mol wt 584.69. C 65.73%, H 7.59%, N 4.79%, O 21.89%. From tubers and herb of *Aconitum septentrionale* Kölle, *A. orientale* Mill., *A. excelsum* Reichb., *A. ranunculaefolium, Ranunculaceae:* Schulze, Ulfert, *Arch. Pharm.* **260**, 230 (1922); Kuzovkov, Massagetov; Platonova *et al., Zh. Obshch. Khim.* **25**, 178 (1955); **28**, 258 (1958), *C.A.* **50**, 1852e (1956), *C.A.* **52**, 12883i (1958), resp.; Mollov *et al., Compt. Rend. Acad. Bulgare Sci.* **17**, 251 (1964), *C.A.* **61**, 12324g (1964). Structure: Khaimova *et al., Tetrahedron Letters* **1964**, 2711. Revised structure: V. A. Tel'nov *et al., Khim. Prir. Soedin.* **6**, 583 (1970), *C.A.* **74**, 42527k (1971). Analgesic activity studies in mice and rats: N. Ono, J. Satoh, *Arzneimittel-Forsch.* **38**, 892 (1988). Toxicology: F. Dybing *et al., Acta Pharmacol. Toxicol.* **7**, 337 (1951).

Bitter crystals, mp 217-218°. $[\alpha]_D^{18}$ +27° (chloroform). Sol in benzene; slightly sol in alcohol, ether. Practically insol in water. LD_{50} in mice (mg/kg): 6.9 i.v.; 9.1 i.p.; approx 20 orally (Dybing).

5238. Lapyrium Chloride. *1-[2-Oxo-2-[[2-[(1-oxodecyl)oxy]ethyl]amino]ethyl]pyridinium chloride; 1-[[(2-hydroxyethyl)carbamoyl]methyl]pyridinium chloride laurate (ester);* N-(lauroylcolaminoformylmethyl)pyridinium chloride; N-(acylcolaminoformylmethyl)pyridinium chloride; emulsept (obsolete); E 607; Emcol E-607. $C_{21}H_{35}ClN_2O_3$; mol wt 398.97. C 63.22%, H 8.84%, Cl 8.89%, N 7.02%, O 12.03%. Cationic surfactant. Prepn: A. K. Epstein, B. R.

Harris, U.S. pat. **2,290,173** (1942). Bactericidal potency: A. K. Epstein *et al., Proc. Soc. Exp. Biol. Med.* **53**, 238 (1943). Toxicity study: A. E. Vivino, T. Koppanyi, *J. Am. Pharm. Assoc., Sci. Ed.* **35**, 169 (1946).

Powder.

Stearoyl analog, $C_{27}H_{47}ClN_2O_3$, *Emcol E-607S, quaternium-7.*

USE: Cationic emulsifier; deodorant; detergent-germicide; antistatic agent. Pharmaceutic aid (surfactant).

5239. Lard. Adeps; axungia porci. Purified internal fat from abdomen of the hog.

Soft, white unctuous mass; slight characteristic odor, bland taste. d 0.917. mp 36-42°. Insoluble in water; very slightly sol in alcohol, freely in benzene, chloroform, ether, carbon disulfide, petr ether. Sapon. no. 195-203. Iodine no. 46-70. *Keep cool and in tight containers.*

USE: In manuf ointments, particularly when absorption is desired.

5240. Lard, Benzoinated. Prepd by heating lard with 1% benzoin powder at temps not above 60°. Less likely to become rancid than pure lard.

USE: As vehicle for medicinal agents and in manufacture of ointments.

5241. Lard Oil. The oil expressed from lard at a low temp. *Constit.* Olein, stearin.

Colorless or pale yellow liq. d 0.905-0.915. Solidif −2° to +4°. n_D^{20} 1.470-1.472. Sapon no. 195-197. Iodine no. 56-74. Acid no. 1.5-2.5. Insol in water or cold alc; freely sol in benzene.

USE: As lubricant, manuf soap, oiling wool, illuminant.

5242. Larkspur. Delphinium; knight's spur; lark's-heel; lark's-claw; staggerweed. Dried ripe seeds of *Delphinium ajacis L., Ranunculaceae. Habit.* Central Europe, cultivated in U.S. *Constit.* Calcatripine, volatile oil, gum, resin, fixed oil, gallic and aconitic acids.

Caution: Toxic. Poisoning from percutaneous absorption may occur.

THERAP CAT: Pediculicide.

5243. Lasalocid A. *6-[7R-[5S-Ethyl-5-(5R-ethyltetrahydro-5-hydroxy-6S-methyl-2H-pyran-2R-yl)tetrahydro-3S-methyl-2S-furanyl]-4S-hydroxy-3R,5S-dimethyl-6-oxononyl]-2-hydroxy-3-methylbenzoic acid;* 3-methyl-6-[7-ethyl-4-hydroxy-3,5-dimethyl-6-oxo-7-[5-ethyl-3-methyl-5-(5-ethyl-5-hydroxy-6-methyl-2-tetrahydropyranyl)-2-tetrahydrofuryl]heptyl]salicylic acid; antibiotic X-537A; ionophore X-4537A; Ro 2-2985; X-537A; Bovatec. $C_{34}H_{54}O_8$; mol wt 590.80. C 69.12%, H 9.21%, O 21.67%. Ionophorous (transport-inducing) antibiotic isolated from an unidentified *Streptomyces* from soil samples of Hyde Park, Mass.: Berger *et al., J. Am. Chem. Soc.* **73**, 5295 (1951). Prepn and activity of title compd and derivs: Stempel, Westley, **Ger.** pat. **2,040,998** corresp to **U.S.** pat. **3,715,372** (1971, 1973 to Hoffmann-La Roche); Westley *et al., J. Med. Chem.* **16**, 397 (1973). Structure: *eidem, Chem. Commun.* **1970**, 71. Crystal and molecular structure studies: Johnson *et al., ibid.* 72; *J. Am. Chem. Soc.* **92**, 4428 (1970). Total synthesis: T. Nakata *et al., ibid.* **100**, 2933 (1978); R. E. Ireland *et al., ibid.* **102**, 1155 (1980). Complete assignment of ^{13}C-NMR spectra of lasalocid A and its sodium salt: H. Seto *et al., J. Antibiot.* **31**, 289 (1978). Effect in coccidiosis: G. M. J. Horton, P. H. G. Stockdale, *Am. J. Vet. Res.* **42**, 433 (1981). Biosynthesis: Westley *et al., Chem. Commun.* **1970**, 1467; C. R. Hutchinson *et al., J. Am. Chem. Soc.* **103**, 5953, 5956 (1981). Mode of action studies: Lin, Kun, *Biochem. Biophys. Res. Commun.* **50**, 820 (1973). Isoln and structure of four homologs, lasalocids B, C, D, E from *Streptomyces lasaliensis:* J. W. Westley *et al., J. Antibiot.* **27**, 744 (1974).

Crystals, mp 110-114°; also reported as mp 100-109° (unsharp). $[\alpha]_D^{25}$ −7.55° (methanol). uv max (50% aq isopropanol): 248, 318 nm (ε 6750, 4200). Sol in organic solvents; insol in water. LD_{50} in mice (mg/kg): 40 i.p., Berger *et al., loc. cit.,* also reported as 146 orally, 64 i.p., J. W. Westley, "Antibiotics (Polyether)" in Kirk-Othmer *Encyclopedia of Chemical Technology,* **Vol.** 3 (Interscience, New York, 3rd ed., 1978) p 61.

Sodium salt, $C_{34}H_{53}NaO_8$, *Avatec.* Crystals from benzene-ligroin, mp (open capillary) 191-192° (dec), also reported as mp 168-171°. $[\alpha]_D^{25}$ −30° (c = 1 in methanol). uv max (50% aq isopropanol): 308 nm (ε 4100).

THERAP CAT (VET): Coccidiostat (for poultry).

5244. Laserpitin. *[1R-[1α,3aα,4β(Z),6β,8β(Z),8aβ]]-2-Methyl-2-butenoic acid decahydro-1,6-dihydroxy-3a,6-dimethyl-1-(1-methylethyl)-5-oxo-4,8-azulenediyl ester.* $C_{25}H_{38}O_7$; mol wt 450.55. C 66.64%, H 8.50%, O 24.86%. From root of *Laserpitium latifolium* L., *Umbelliferae:* Külz, *Arch. Pharm.* **221**, 161 (1883); Morgenstern, *Monatsh.* **33**, 709 (1912). Structure: M. Holub *et al., Tetrahedron Letters* **1965**, 1441, 2855. Absolute configuration: M. Holub *et al., Coll. Czech. Chem. Commun.* **35**, 3597 (1970).

Crystals, mp 118°. Practically insol in water. Sol in alcohol, chloroform, ether, petr ether, fatty oils.

USE: Used as seasoning and flavoring agent in antiquity (Greece and Rome).

5245. Lasiocarpine. *2-Methyl-2-butenoic acid 7-[[2,3-dihydroxy-2-(1-methoxyethyl)-3-methyl-1-oxobutoxy]methyl]-2,3,5,7a-tetrahydro-1H-pyrrolizin-1-yl ester.* $C_{21}H_{33}NO_7$; mol wt 411.50. C 61.30%, H 8.08%, N 3.40%, O 27.22%. Hepatotoxic pyrrolizidine alkaloid isolated from *Heliotropium lasiocarpum* Fish. et C. Mey. *Boraginaceae:* G. Menschikoff, *Ber.* **65**, 974 (1932), *C.A.* **26**, 4818[3] (1932), G. Menschikoff, J. Schdanowitsch, *Ber.* **69**, 1110 (1936), *C.A.* **30**, 5227[9] (1936). Structure: L. J. Drummond, *Nature* **167**, 41 (1951); R. Adams, B. L. VanDuvren, *J. Am. Chem. Soc.* **76**, 6379 (1954). Review and evaluation of studies of carcinogenicity and toxicity in laboratory animals: *IARC Monographs* **10**, 281-290, 333-342 (1976). Comprehensive reviews: L. Bull *et al., The Pyrrolizidine Alkaloids* (North-Holland, Amsterdam, 1965) 293 pp; F. L.Warren in *The Alkaloids* vol. **12**, R. H. F. Manske, Ed. (Academic Press, New York, 1970) pp 245-331; D. J. Robins, *Fortschr. Chem. Org. Naturst.* **41**, 115-203 (1982).

Colorless leaflets from petr ether. mp 94-95.5°. $[\alpha]_D$ −4° (10% alc). Sol on ether, alc, benzene; difficultly sol in water.

5246. Latrunculins. Highly potent toxins isolated from the Red Sea sponge *Latrunculia magnifica* Keller. When *L. magnifica* is squeezed into an aquarium, exudation of the toxins results, causing agitation of the fish, hemorrhage, loss of balance and death within 4-6 minutes. Isoln of the toxins and biological activities: I. Néeman *et al.*, *Marine Biol.* **30**, 293 (1975). The latrunculins are the first marine macrolides known to contain 16- and 14-membered rings, as well as the unusual 2-thiazolidine moiety. Structural elucidation of latrunculins A, B, and C (a stereoisomer of A): Y. Kashman *et al.*, *Tetrahedron Letters* **21**, 3629 (1980). Although their mode of action is unknown, they have been shown to disrupt microfilament organization in cultured cells at concentrations 1/10 to 1/100 of the cytochalasins, *q.v.*: I. Spector *et al.*, *Science* **219**, 493 (1983).

latrunculin A   latrunculin B

Latrunculin A, $C_{22}H_{31}NO_5S$, *4-(17-hydroxy-5,12-dimethyl-3-oxo-2,16-dioxabicyclo[13.3.1]nonadeca-4,8,10-trien-17-yl)-2-thiazolidinone*, *LAT-A*. Oil. $[\alpha]_D^{24}$ +152° (c = 1.2 in chloroform). uv max (methanol): 218 nm (ε 23500).

Latrunculin B, $C_{20}H_{29}NO_5S$, *4-(15-hydroxy-5,10-dimethyl-3-oxo-2,14-dioxabicyclo[11.3.1]heptadeca-4,8-dien-15-yl)-2-thiazolidinone*, *LAT-B*. Crystals. $[\alpha]_D^{24}$ +112° (c = 0.48 in chloroform). uv max (methanol): 212 nm (ε 17200).

USE: In elucidation of molecular mechanisms of motile processes.

5247. Laudanidine. *(R)-2-Methoxy-5-[(1,2,3,4-tetrahydro-6,7-dimethoxy-2-methyl-1-isoquinolinyl)methyl]phenol*; *l*-laudanine; tritopine. $C_{20}H_{25}NO_4$; mol wt 343.43. C 69.95%, H 7.34%, N 4.08%, O 18.64%. The *l*-form of laudanine, *q.v.* Traces in opium: Hesse, *Ann.* **282**, 208 (1894); Kauder, *Arch. Pharm.* **228**, 419 (1890). Structure and synthesis: Späth, Bernhauer, *Ber.* **58**, 200 (1925); Späth, Seka, *ibid.* 1272; Späth, Burger, *Monatsh.* **47**, 733 (1926); Frydman *et al.*, *Tetrahedron* **4**, 342 (1958). Configuration: Corrodi, Hardegger, *Helv. Chim. Acta* **39**, 889 (1956).

Prisms from alcohol, mp 185°. $[\alpha]_D^{15}$ −88° (p = 5 in chloroform). $[\alpha]_D^{22}$ −94.7 ± 1.3°. The hydrochloride is more sol than laudanine hydrochloride and makes possible a separation of the two alkaloids.

5248. Laudanine. *(±)-2-Methoxy-5-[(1,2,3,4-tetrahydro-6,7-dimethoxy-2-methyl-1-isoquinolinyl)methyl]phenol*; *dl*-laudanidine. $C_{20}H_{25}NO_4$; mol wt 343.43. C 69.95%, H 7.34%, N 4.08%, O 18.64%. In the alkaline mother liquors from morphine extraction, in amounts corresp to about 0.005% of the opium employed: L. F. Small, R. E. Lutz, *Chemistry of the Opium Alkaloids*, Supplement No. 103, Public Health Reports, Washington (1932) p 34. Structure: Späth, *Monatsh. Chem.* **41**, 297 (1920). Synthesis: Späth, Lang, *ibid.* **42**, 273 (1921); Frydman *et al.*, *Tetrahedron* **4**, 342 (1958).

Orthorhombic prisms from alcohol or chloroform, mp 167°. d_4^{20} 1.26. uv max: 284 nm (log ε 3.78). Sol in benzene, chloroform, hot alcohol; sparingly sol in ether, practically insol in water.

5249. Laudanosine. *1-[(3,4-Dimethoxyphenyl)methyl]-1,2,3,4-tetrahydro-6,7-dimethoxy-2-methylisoquinoline; 1,2,-3,4-tetrahydro-6,7-dimethoxy-2-methyl-1-veratrylisoquinoline*; *N-methyltetrahydropapaverine*. $C_{21}H_{27}NO_4$; mol wt 357.43. C 70.56%, H 7.61%, N 3.92%, O 17.90%. Occurs in opium (0.0008%). It is the last alkaloid to be separated from morphine extraction mother liquors; occurs as (+)-form. Synthesis: Pictet, Finkelstein, *Ber.* **42**, 1979 (1909); Frydman *et al.*, *Tetrahedron* **4**, 342 (1958); Elliott, *J. Heterocycl. Chem.* **9**, 853 (1972). Asymmetric synthesis of *(S)*-(+)-laudanosine: M. Konda *et al.*, *Chem. Pharm. Bull.* **23**, 1025 (1975); of *(R)*-(−)-laudanosine: M. Konda *et al.*, *ibid.* **25**, 69 (1977); R. E. Gawley, G. A. Smith, *Tetrahedron Letters* **29**, 301 (1988). Configuration: Leithe, *Ber.* **64**, 2827 (1931); Faltis, Adler, *Arch. Pharm.* **284**, 281 (1951); Corrodi, Hardegger, *Helv. Chim. Acta* **39**, 889 (1956).

(±)-Form, crystals from dil alc, mp 114-115.5°.

(+)-Form, crystals from light petr (30-60°), mp 89°. $[\alpha]_D^{16}$ +106° (c = 1.6 in 97% alc); $[\alpha]_D^{16}$ +130° (chloroform); $[\alpha]_D^{22}$ +52.2 ±1.3° (chloroform). $[\alpha]_D^{20}$ +82.5° (ethanol). Absorption spectrum: Dobbie, Lauder, *J. Chem. Soc.* **83**, 626 (1903). Practically insol in water. Freely sol in alcohol, chloroform, ether, hot petr ether.

(−)-Form, colorless needles from ethanol, mp 83-85°. $[\alpha]_D^{20}$ −84.8° (c = 0.466 in ethanol).

5250. Laudexium Methyl Sulfate. *2,2'-(1,10-Decanediyl)bis[1-[(3,4-dimethoxyphenyl)methyl]-1,2,3,4-tetrahydro-6,7-dimethoxy-2-methylisoquinolium] bis(methyl sulfate)*; *2,2'-decamethylenebis[1,2,3,4-tetrahydro-6,7-dimethoxy-2-methyl-1-veratrylisoquinolinium methyl sulfate]*; *decamethylenebis[1,2,3,4-tetrahydro-6,7-dimethoxy-1-(3,4-dimethoxybenzyl)-2-methylisoquinolinium methyl sulfate]*; *decamethylene-α-ω-bis[1-(3',4'-dimethoxybenzyl)-1,2,3,4-tetrahydro-6,7-dimethoxy-2-methylisoquinolinium methosulfate]*; *Compound 20; Laudissine; Laudolissin*. $C_{54}H_{80}N_2O_{16}S_2$; mol wt 1077.38. C 60.20%, H 7.48%, N 2.60%, O 23.76%, S 5.95%. Preparation: Taylor, *J. Chem. Soc.* **1952**, 142; Eastland *et al.*, *Brit. pat.* 695,298 (1953 to Allen & Hanbury).

Cream colored granules from alcohol + ether, mp 172-174° after darkening at 164-166°.

THERAP CAT: Skeletal muscle relaxant.

5251. Laureline. *6,7,7a,8-Tetrahydro-11-methoxy-6-methyl-5H-benzo[g]-1,3-benzodioxolo[6,5,4-de]quinoline; 10-methoxy-1,2-(methylenedioxy)aporphine*. $C_{19}H_{19}NO_3$; mol wt 309.35. C 73.76%, H 6.19%, N 4.53%, O 15.52%. From bark of *Laurelia novae-zelandiae* A. Cunn., *Lauraceae*: Aston, *J. Chem. Soc.* **97**, 1381 (1910). Synthesis: Schlittler,

Helv. Chim. Acta **15**, 394 (1932); Faltis *et al.*, *Ber.* **77B**, 686 (1945); Gibson *et al.*, *J. Chem. Soc. (C)* **1970**, 2234; Govindachari *et al.*, *Indian J. Chem.* **8**, 475 (1970).

dl-Form, rough needles from petr ether, mp 115-116°.
d-Form, cubes from petr ether, mp 114-115°. $[\alpha]_D^{20}$ +97.9° (c = 1.12 in abs alc); $[\alpha]_D^{20}$ +100° (c = 0.560 in CS_2).
l-Form, cubes from petr ether. $[\alpha]_D^{20}$ −99.2° (c = 0.736 in 50% alc). Absorption spectrum: Girardet, *J. Chem. Soc.* **1931**, 2636. Feebly basic. Readily oxidizes in air. Sol in alcohol, ether; practically insol in water.
Hydrochloride, $C_{19}H_{19}NO_3 \cdot HCl$, crystals, mp 280°. $[\alpha]_D^{20}$ −57°. Sparingly sol in water.
Nitrate, $C_{19}H_{19}NO_3 \cdot HNO_3$, crystals, mp 238-240°. Sparingly sol in water.
Hydriodide, $C_{19}H_{19}NO_3 \cdot HI$, crystals. Practically insol in water.
Oxalate, $(C_{19}H_{19}NO_3)_2 \cdot (COOH)_2$; moderately sol in water.

5252. Laurel Oil. Laurel berry oil. Fixed oil from fresh fruit of *Laurus nobilis* L., *Lauraceae*. *Constit.* Chiefly the lauryl alcohol esters of lauric, stearic, etc., acids and a volatile oil, the so-called laurel camphor.
Greenish, fatty solid; the green color due to presence of some chlorophyll. d about 0.88. mp about 40°. n_D^{25} 1.4783. Sapon no. 198-199. Iodine no. 68-80. Insol in water; sparingly sol in alcohol; sol in benzene, ether, carbon disulfide.

5253. Laurepukine. *(7R-cis)-6,7,7a,8-Tetrahydro-7-methyl-5H-benzo[g]-1,3-benzodioxolo[6,5,4-de]quinoline-12-ol 7-oxide.* $C_{18}H_{17}NO_4$; mol wt 311.32. C 69.44%, H 5.50%, N 4.50%, O 20.56%. From bark of *Laurelia novae-zelandiae* A. Cunn., *Lauraceae*. Isoln: Aston, *J. Chem. Soc.* **97**, 1381 (1910). Structure: Girardet, *Helv. Chim. Acta* **14**, 504 (1931); Govindachari *et al.*, *Ber.* **91**, 36 (1958). Revised structure: Weiss *et al.*, *Helv. Chim. Acta* **54**, 1342 (1971).

Needles from chloroform + ether, mp 230-231°. $[\alpha]_D$ −222° (c = 0.022 in chloroform). Practically insol in water; sol in benzene, chloroform.

5254. Lauric Acid. *Dodecanoic acid;* laurostearic acid; dodecoic acid. $C_{12}H_{24}O_2$; mol wt 200.31. C 71.95%, H 12.08%, O 15.97%. $CH_3(CH_2)_{10}COOH$. Isoln from coconut oil: Dale, Meara, *J. Sci. Food Agr.* **6**, 162 (1955); Naudet *et al.*, *Bull. Soc. Chim. France* **1959**, 718; from arecanut fat: Pathak, Mathur, *J. Sci. Food Agr.* **5**, 461 (1954); from *Holoptelea integrifolia* seed fat: Badami, *ibid.* **13**, 297 (1962). Synthesis: Ballard *et al.*, *U.S. pat.* **2,572,238** (1951 to Shell); Langenbeck, Richter, *Ber.* **89**, 202 (1956); Sprowls, *U.S. pat.* **2,782,214** (1957 to Baker Castor Oil); Zapesochnoya *et al.*, *Zh. Obshch. Khim.* **33**, 2552 (1963).
White, cryst powder; slight odor of bay oil, mp 44°; also stated at 48°. d_4^5 0.869. bp_{100} 225°; bp_{20} 160-165°. n_D^{82} 1.4183. Insol in water; 1 g dissolves in 1 ml alcohol, 2.5 ml propyl alcohol; freely sol in benzene, ether. LD_{50} i.v. in mice: 131 ±5.7 mg/kg, L. Orö, A. Wretlind, *Acta Pharmacol. Toxicol.* **18**, 141 (1961).

5255. Laurocapram. *1-Dodecylhexahydro-2H-azepin-2-one;* 1-dodecylazacycloheptan-2-one; *N*-dodecyl-ε-caprolactam; N 0252; Azone. $C_{18}H_{35}NO$; mol wt 281.49. C 76.81%, H 12.53%, N 4.98%, O 5.68%. Caprolactam derivative used to enhance percutaneous absorption of physiologically active agents. Also exhibits intrinsic anti-inflammatory activity. Prepn: A. P. Swain *et al.*, *J. Org. Chem.* **18**, 1087 (1953). Improved synthesis: V. J. Rajadhyaksha *et al.*, Eur. pat. Appl. **95,096**; *eidem*, U.S. pat. **4,422,970** (both 1983 to Nelson). Use as skin penetrant: V. J. Rajadhyaksha, U.S. pats. **3,989,816; 4,405,616** (1976, 1983 to Nelson). Anti-inflammatory activity: E. L. Nelson, U.S. pat. **4,310,-525** (1982 to Nelson). Pharmacology: R. B. Stoughton, *Arch. Dermatol.* **118**, 474 (1982). Comprehensive description: *idem, Drug Dev. Ind. Pharm.* **9**, 725 (1983). Topical antiviral activity *in vivo*: M. F. Leonard *et al.*, *Chemotherapy (Basel)* **33**, 151 (1987).

Clear, colorless liquid, mp −7°. $bp_{50\mu}$ 160°. d 0.91; n 1.4701. Insol in water. Freely sol in most organic solvents. LD_{50} rats, mice (g/kg): 8 i.v., i.p. (Stoughton).
USE: Pharmaceutic aid (excipient).

5256. Lauroguadine. *N,N'''-[4-(Dodecyloxy)-1,3-phenylene]bisguanidine; 1,1'-[4-(dodecyloxy)-m-phenylene]-diguanidine;* 2,4-diguanidinophenyl lauryl ether; 2,4-di-guanidino-1-dodecyloxybenzene; 2,4-diguanidino-1-lauryl-oxybenzene; 2,4-diguanidinophenyl dodecyl ether; dodecyl 2,4-diguanidinophenyl ether. $C_{20}H_{36}N_6O$; mol wt 376.54. C 63.79%, H 9.64%, N 22.32%, O 4.25%. Prepn of the dihydrochloride: Pasini, *Farmaco Ed. Sci.* **8**, 646 (1953); *idem, Rend. Sci. Farmitalia* **1**, 405 (1954); Brit. pat **730,394** (1955 to Farmitalia).

Dihydrochloride monohydrate, $C_{20}H_{38}Cl_2N_6O \cdot H_2O$, P7, Farmidril. Crystals from water + ethanol. Sinters from 135-210°, dec 250°. Soly (g/100 ml): 0.2 in cold water; ~10 in boiling water; 84.2 in methanol (25°). Practically insol in acetone, benzene. Aq solns are neutral to litmus.
THERAP CAT: Antiprotozoal (Trichomonas).

5257. Laurolinium Acetate. *4-Amino-1-dodecyl-2-methylquinolinium acetate; 4-amino-1-dodecyquinaldinium acetate;* 1-dodecyl-4-aminoquinaldinium acetate; Laurodin. $C_{24}H_{38}N_2O_2$; mol wt 386.56. C 74.57%, H 9.91%, N 7.25%, O 8.28%. Prepn: D. Caldwell *et al.*, *J. Pharm. Pharmacol.* **13**, 554 (1961); D. Caldwell, L. R. Rowe, U.S. pat. **2,997,-476** (1961 to Allen & Hanburys). Toxicity data: W. A. Cox, P. F. D'Arcy, *J. Pharm. Pharmacol.* **15**, 129 (1963).

Crystals from acetone, mp 170-171°. Freely sol in water. About 1 g dissolves in 2 ml of water at 20°. LD_{50} in mice (mg/kg): 132 orally; 30 s.c.; 6 i.v.; 2 s.c. (Cox, D'Arcy).
THERAP CAT: Antiseptic.

5258. Laurotetanine. 5,6,6a,7-Tetrahydro-1,2,10-tri-methoxy-4H-dibenzo[de,g]quinolin-9-ol; 1,2,10-trimethoxy-6aα-noraporphin-9-ol; Litsoeine. $C_{19}H_{21}NO_4$; mol wt 327.37. C 69.70%, H 6.47%, N 4.28%, O 19.55%. From the bark of Litsea citrata Blume (Tetranthera citrata (Blume) Nees), Lauraceae and allied plants. Isoln: Greshoff, Ber. **23**, 3537 (1890); Filippo, Arch. Pharm. **236**, 601 (1898). Structure: Barger et al., Ber. **66**, 450 (1933). Synthesis: Kikkawa, C.A. **53**, 17163i (1959).

Monohydrate, needles, mp 125°. $[\alpha]_D^{25}$ +98.5°. Practically insol in water; freely sol in alcohol, chloroform, ethyl acetate, slightly in ether.

5259. 3-O-Lauroylpyridoxol Diacetate. Dodecanoic acid 4,5-bis[(acetyloxy)methyl]-2-methyl-3-pyridinyl ester; lauric acid ester with pyridoxol diacetate (ester); 5-lauroyloxy-6-methyl-3,4-pyridinedimethanol diacetate; 3-lauroyloxy-2-picoline-4,5-dimethanol diacetate; 2-methyl-3-lauroyloxy-4,5-diacetoxymethylpyridine; Epixine; Rosamit. $C_{24}H_{37}$-NO_6; mol wt 435.54. C 66.18%, H 8.56%, N 3.22%, O 22.04%. Prepn: **Belg. pat. 640,827** (1964 to Soc. Belge Azote Prod. Chim. Marly), C.A. **63**, 587h (1965).

Crystals, mp 44°. Practically insol in water; sol in ether, chloroform, ethanol, ethylene dichloride.
THERAP CAT: Antiseborrheic.

5260. Lauryl Bromide. 1-Bromododecane; dodecyl bromide. $C_{12}H_{25}Br$; mol wt 249.24. C 57.82%, H 10.11%, Br 32.06%. $CH_3(CH_2)_{10}CH_2Br$. Prepd by the action of hydrobromic acid on primary n-lauryl alcohol in the presence of sulfuric acid: Kamm, Marvel, Org. Syn. **1**, 7 (1921).
Liquid. bp_{45} 175-180°. Insol in water. Sol in alc, ether.

5261. Lavender. Garden lavender; true lavender. Flowers of Lavandula officinalis Chaix (L. vera DC.), Labiatae. Habit. Mediterranean region. Constit. Volatile oil.
USE: For fumigating; in perfumery; to keep moths from clothes; manuf oil lavender. Pharmaceutic aid (perfume).

5262. Lawrencium. Lr; formerly Lw; at. wt (longest-lived known isotope, $T_{1/2} \sim 3$ minutes) 260; at. no. 103; valence 3. Known isotopes 255-260. Discovery of first isotope claimed by Ghiorso et al., Phys. Rev. Letters **6**, 473 (1961). Prepared by bombardment of californium with boron ions; originally assigned mass number 257; later changed to 258 ($T_{1/2}$ 4.2 seconds, α-emitter): Eskola et al., Phys. Rev. C **4**, 632 (1971). Prepn of ^{256}Lr ($T_{1/2} \sim 45$ seconds) by irradiating ^{243}Am with ^{18}O ions: Donets et al., At. Energ. (USSR) **19**, 109 (1965), C.A. **64**, 1542c (1966). Prepn of isotopes 255-260 by bombardment of transuranium elements with heavy ions: Eskola et al., loc. cit. Reviews of history, prepn and properties: C. Keller, The Chemistry of the Transuranium Elements (Verlag Chemie, Weinheim, English Ed., 1971) pp 609-612; Silva, "Trans-Curium Elements" in MFP Int. Rev. Sci.: Inorg. Chem., Ser. One vol. **8**, A. G. Maddock, Ed. (University Park Press, Baltimore, 1972) pp 71-105; Ghiorso, Handb. Exp. Pharmakol. **36**, 691-715 (1973); Taylor, ibid. 717-738.

5263. Lawsone. 2-Hydroxy-1,4-naphthalenedione; 2-hydroxy-1,4-naphthoquinone. $C_{10}H_6O_3$; mol wt 174.15. C 68.96%, H 3.47%, O 27.56%. From leaves of Lawsonia iner-

mis L. and L. alba Lam., Lythraceae: Latif, Indian J. Agr. Sci. **29**, No. 2-3, 147 (1959), C.A. **55**, 14828g (1961). Synthesis: Fieser, J. Am. Chem. Soc. **70**, 3165 (1948); Jain, Seshadri, Proc. Indian Acad. Sci. **35A**, 233 (1952); Eistert, Müller, Ber. **92**, 2071 (1959).

Yellow prisms from acetic acid, dec 195-196°.
THERAP CAT: Ultraviolet screen.

5264. Lazaroids. Novel class of nonglucocorticoid, 21-aminosteroid antioxidants which inhibit lipid peroxidation. A representative compound is known as U74006F. Prepn: J. M. McCall et al., PCT Int. pat. **Appl. 87 01,706** (1987 to Upjohn), C.A. **108**, 6287u (1987). Inhibition of iron-dependent lipid peroxidation in vitro: J. M. Braughler et al., J. Biol. Chem. **262**, 10438 (1987). Endocrinological profile in mice: J. M. Braughler et al., J. Pharmacol. Exp. Ther. **244**, 423 (1988). HPLC determn in plasma: J. W. Cox, R. H. Pullen, J. Chromatog. **424**, 293 (1988). In vivo attenuation of vasogenic brain edema: E. D. Hall, M. A. Travis, Brain Res. **451**, 350 (1988). Effects on experimental head injury in mice: E. D. Hall et al., J. Neurosurg. **68**, 456 (1988); in post-traumatic spinal cord ischemia in cats: E. D. Hall, ibid. 462. Review of development and potential clinical applications in trauma and stroke: J. M. McCall et al., Acta Anaesthesiol. Belg. **38**, 417-420 (1987).

U74006F

U74006F, $C_{39}H_{56}N_6O_5S$, 21-[4-(2,6-di-1-pyrrolidinyl-4-pyrimidinyl)-1-piperazinyl]-16α-methylpregna-1,4,9(11)-triene-3,20-dione monomethanesulfonate. Monohydrate, mp 181-185° (dec). uv max: 234, 285 nm (ε 52000, 17000).

5265. Lazurite. Lapis lazuli; lasurite. Composition: $(Na,Ca)_4(AlSiO_4)_3(SO_4,S,Cl)$, E. S. Dana, A System of Mineralogy (John Wiley, New York, 6th ed., 1901) pp 432-433; C. S. Hurlbut, Jr., Dana's Manual of Mineralogy (John Wiley, New York, 17th ed., 1959) p 503.
Blue, blue-violet or greenish-blue, translucent, cubic or dodecahedral crystals. d 2.4. Dec by HCl with pptn of SiO_2 and evolution of H_2S.
USE: In manuf of vases, ornamental furniture, mosaics; in paints, jewelry.

5266. LBF. Lactobacillus bulgaricus factor. Growth factor occurring in products derived from both animal and plant sources and in culture filtrate of certain microorganisms: Williams et al., J. Biol. Chem. **177**, 933 (1949); Vitucci et al., Arch. Biochem. Biophys. **34**, 409 (1951); Peters et al., J. Am. Chem. Soc. **75**, 1688 (1953). Contains pantetheine q.v. which is oxidized during purification to the disulfide, pantethine q.v. Natural occurrence of several different forms of LBF each being a mixed disulfide of pantetheine: Rasmussen et al., Proc. Soc. Exp. Biol. Med. **73**, 658 (1950); Brown, Snell, J. Biol. Chem. **198**, 375 (1952). Coenzyme A digested with intestinal phosphatase shows 2-4 LBF-active components: Long, Williams, J. Bacteriol. **61**, 195 (1951). Review:

Consult the cross index before using this section.

Snell, Brown, "Pantethine and Related Forms of the Lacto-bacillus Bulgaricus Factor (LBF)" in *Advan. Enzymol.* **14**, 49-71 (1953).

5267. Lead. Pb; at. wt 207.2; at. no. 82; valence 2, 4. Four naturally occurring isotopes: 204 (1.40%); 206 (25.2%); 207 (21.7%); 208 (51.7%); artificial, radioactive isotopes: 195-203; 205; 209-214. One of the metals known to the ancient world. Extent of occurrence in earth's crust about 15 g/ton, also expressed as 0.002% (depth of crust: 16 km). Occurs chiefly as sulfide in galena, other minerals include anglesite (PbSO$_4$), *cerussite* (PbCO$_3$), *mimetite* [PbCl$_2$.3Pb$_3$-(AsO$_4$)$_2$] and *pyromorphite* [PbCl$_2$.3Pb$_3$(PO$_4$)$_2$]. Recovery from ore and purification: Heuser, *Metall.* **9**, 675 (1955), *C.A.* **49**, 14609 (1955); Ziegfeld, *Eng. Mining J.* **153**, 82 (1952), *C.A.* **46**, 2975 (1952). Prepn of high purity lead: Piontelli, Fagnani, *Chim. Ind. (Milan)* **34**, 629 (1952), *C.A.* **47**, 12062 (1953); Giesen, *Technik (Berlin)* **2**, 393 (1947), *C.A.* **42**, 852 (1948); Hughes *J. Electrochem. Soc.* **101**, 267 (1954); Baralis, Marone, *Met. Ital.* **59**, 494 (1967), *C.A.* **67**, 119613a (1967). Reviews of lead, its alloys and compds: W. Hofmann, *Lead and Lead Alloys, Properties and Technology* (Springer, New York, Eng. Ed., 1970) 551 pp; Abel in *Comprehensive Inorganic Chemistry* vol. **2**, J. C. Bailar, Jr. *et al.*, Eds. (Pergamon Press, Oxford, 1973) pp 105-146; H. E. Howe in Kirk-Othmer *Encyclopedia of Chemical Technology* vol. **14** (Wiley-Interscience, New York, 3rd ed., 1981) pp 98-139. Review of carcinogenicity studies of lead and lead compds: *IARC Monographs* **23**, 325-415 (1980).

Bluish-white, silvery, gray metal. Highly lustrous when freshly cut, tarnishes upon exposure to air. Very soft and malleable, easily melted, cast, rolled, and extruded. Cubic crystal structure. mp 327.4°; bp 1740°. d$_4^{20}$ 11.34; d at mp 10.65: Schneider *et al., Naturwiss.* **41**, 326 (1954). Heat of vaporization (1740°) 206 cal/g. Heat capacity (20°): 0.031 cal/g/°C. Resistivity (μ-ohm-cm) at 20°: 20.65; at 100°: 27.02; at 320°: 54.76; at 330°: 96.74. Vapor pressure at 1000°: 1.77 mm Hg. E° (aq) Pb/Pb^{2+} +0.126 v. Coefficient of linear expansion (0-100°) 29 $\times$ 10^{-6}, (20-300°) 31.3 $\times$ 10^{-6}, (−183° to +14°) 27 $\times$ 10^{-6}; thermal conductivity varies from 0.083 at 50° to 0.077 at 225°: Francl, Kingery, *J. Am. Ceram. Soc.* **37**, 80 (1954); viscosity of molten lead (327.4°) 3.2 centipoises, (400°) 2.32 cp, (600°) 1.54 cp, (800°) 1.23 cp. Heat capacity and heat of fusion study: Douglas, Dever, *J. Am. Chem. Soc.* **76**, 4824 (1954); hardness 1 on Mohs' scale; Brinell hardness (high purity Pb) 4.0: McLellan, *Am. Mineralogist* **30**, 635 (1945). Reacts with hot concd nitric acid, with boiling concd hydrochloric or sulfuric acid. Attacked by pure water, weak organic acids in the presence of oxygen. Resistant to tap water, hydrofluoric acid, brine, solvents.

Human Toxicity: Acute: most common in young children with history of pica; anorexia, vomiting, malaise, convulsions due to increased intracranial pressure. May leave permanent brain damage. Blood lead increased above 0.05 mg %. *Chronic:* children show weight loss, weakness, anemia. Lead poisoning in adults is usually occupational due mainly to inhalaton of lead dust or fumes. Wristdrop and colic rarely occur. More often there are vague G.I. and CNS complaints. Pb content of blood >0.05 mg % and of urine >0.08 mg per liter support a diagnosis of Pb poisoning. Provocative excretion test using Edathamil may be helpful in confirming excess Pb absorption. Review of toxicity: *Clinical Toxicology of Commercial Products*, R. E. Gosselin *et al.*, Eds. (Williams & Wilkins, Baltimore, 4th ed., 1976) Section III, pp 194-202; *Lead Toxicity*, R. L. Singhal, J. A. Thomas, Eds. (Urban & Schwarzenberg, Baltimore, 1980) 514 pp.

USE: Construction material for tank linings, piping, and other equipment handling corrosive gases and liqs used in the manuf of sulfuric acid, petr refining, halogenation, sulfonation, extraction, condensation; for x-ray and atomic radiation protection; manuf of tetraethyllead, pigments for paints, and other organic and inorganic lead compds; bearing metal and alloys; storage batteries; in ceramics, plastics, and electronic devices; in building construction; in solder and other lead alloys; in the metallurgy of steel and other metals. *Review* of uses, corrosion metallurgy: Mullarkey, *Ind. Eng. Chem.* **49**, 1607 (1957).

5268. Lead Acetate. Neutral lead acetate; normal lead acetate; sugar of lead; salt of Saturn. C$_4$H$_6$O$_4$Pb; mol wt 325.28. C 14.77%, H 1.86%, Pb 63.70%, O 19.67%. Pb-(CH$_3$COO)$_2$. Toxicity data: Bradley, Frederick, *Ind. Med.* **10**, *Ind. Hyg. Sect.* **2**, 15 (1941).

Trihydrate, colorless crystals or white granules or powder; slight acetic odor; slowly effloresces. *Poisonous!* Takes up CO$_2$ from air and becomes incompletely sol. d 2.55. mp 75° when rapidly heated; at a little above 100° it begins to lose acetic acid; dec completely above 200°. One gram dissolves in 1.6 ml water, 0.5 ml boiling water, 30 ml alcohol; freely sol in glycerol. Aq solns of lead acetate dissolve lead monoxide. pH of 5% aq soln at 25° = 5.5-6.5. *Keep well closed. Incompat:* Acids, sol sulfates, citrates, tartrates, chlorides, carbonates, alkalies, tannin, phosphates, resorcinol, salicylic acid, phenol, chloral hydrate, sulfites, vegetable infusions, tinctures. LD$_{50}$ i.p. in rats: 200 mg/kg (Bradley, Frederick).

Caution: Avoid breathing dust. This substance may reasonably be anticipated to be a carcinogen: *Fourth Annual Report on Carcinogens* (NTP 85-002, 1985) p 121.

USE: Dyeing and printing cottons; weighting silks; manuf lead salts, chrome-yellow; also for various analytical procedures, e.g. detection of sulfide, determination of CrO$_3$, MoO$_3$.

THERAP CAT: Astringent.

THERAP CAT (VET): Astringent and sedative (usually in lotions) for bruises and superficial inflammation. Has been used internally in diarrheas.

5269. Lead Antimonate(V). Naples yellow. Approx Pb$_3$(SbO$_4$)$_2$.

Orange-yellow powder. Insol in water, dil acids.

USE: As pigment in oil painting, staining glass, crockery and porcelain.

5270. Lead Arsenate. Approx PbHAsO$_4$. Occurs in nature as the mineral *schultenite*.

White, heavy powder. *Poisonous!* d 5.79. At about 280° loses H$_2$O and is converted into pyroarsenate. Insol in water; sol in HNO$_3$, caustic alkalies. LD$_{50}$ in rats, rabbits: approx 825, 125 mg/kg orally, Voigt *et al., J. Am. Pharm. Assoc.* **37**, 122 (1948).

USE: As constituent of various insecticides for larvae of gypsy moth, boll weevil, etc.

THERAP CAT (VET): Has been used as a teniacide; insecticide.

5271. Lead Arsenite. Approx Pb(AsO$_2$)$_2$.

White powder. *Poisonous!* d 5.85. Insol in water; sol in dil HNO$_3$.

USE: As insecticide like the arsenate.

5272. Lead Azide. N$_6$Pb; mol wt 291.26. N 28.86%, Pb 71.14%. Pb(N$_3$)$_2$. Prepd from sodium azide and lead nitrate: Schenk in *Handbook of Preparative Inorganic Chemistry* vol. **1**, G. Brauer, Ed. (Academic Press, New York, 2nd ed., 1963) p 763. Most complete description: B. T. Fedoroff *et al., Encyclopedia of Explosives and Related Items* vol. **1** (Picatinny Arsenal, Dover, N.J., 1960) pp A545-A587.

Needles or white powder. Explodes at 350° or on percussion. Heat of formation (25°): +110.5 kcal/mol. Soly in water: 0.023% at 18°; 0.09% at 70°. Freely sol in acetic acid. Insol in NH$_4$OH.

USE: As primer in explosives. Generally used in the form of dextrinated lead azide.

5273. Lead Borate. Approx Pb(BO$_2$)$_2$.H$_2$O.

White powder. *Poisonous!* Insoluble in water; soluble in dil HNO$_3$.

USE: Drier for varnishes and paints; with other metals (e.g., Ag) in galvanoplasty for production of conducting coatings on glass, pottery, porcelain, and chinaware.

5274. Lead Bromate. Br$_2$O$_6$Pb; mol wt 463.01. Br 34.52%, O 20.73%, Pb 44.75%. Pb(BrO$_3$)$_2$.

Monohydrate, colorless crystals. *Poisonous!* d 5.53. Dec at 180°. Slightly sol in cold water, moderately in hot water.

Note: Pure lead bromate is not dangerous, but when made by pptng lead acetate with an alkali bromate, it may detonate or explode on heating, striking or rubbing because some acetate is occluded.

5275. Lead Bromide. Br$_2$Pb; mol wt 367.04. Br 43.55%, Pb 56.45%. PbBr$_2$.

White, cryst powder. *Poisonous!* d 6.66. mp 373°. On solidifying forms a horn-like mass. Sol in about 200 parts cold water, 20 parts boiling water; insol in alcohol.

5276. Lead Butyrate. *Butyric acid, lead salt.* C$_8$H$_{14}$-O$_4$Pb; mol wt 381.40. C 25.19%, H 3.70%, O 16.78%, Pb 54.33%. Pb(C$_4$H$_7$O$_2$)$_2$.

Colorless scales or viscid mass. *Poisonous!* mp about 90°. Insol in water; sol in dil HNO$_3$.

5277. Lead Chlorate. Cl$_2$O$_6$Pb; mol wt 374.12. Cl 18.95%, O 25.66%, Pb 55.39%. Pb(ClO$_3$)$_2$.

Colorless, deliquesc crystals. *Poisonous!* d 3.9. Dec at 230°. Sol in 0.7 part water, freely in alcohol.

5278. Lead Chloride. Cl$_2$Pb; mol wt 278.12. Cl 25.49%, Pb 74.50%. PbCl$_2$. Occurs in nature as the mineral *cotunnite*.

White, cryst powder. *Poisonous!* d 5.85. mp 501°. bp 950°. Sol in 93 parts cold water, 30 parts boiling water; readily sol in soln of NH$_4$Cl, NH$_4$NO$_3$, alkali hydroxides; slowly in glycerol. MLD orally in guinea pigs: 1.5-2.0 g/kg, *Handbook of Toxicology* vol. 1, W. S. Spector, Ed. (Saunders, Philadelphia, 1956) pp 176-177.

USE: Manuf Pattison's white lead, Verona Yellow, Turner's Patent Yellow, lead oxychloride; as solder and flux.

5279. Lead Chromate(VI). Chrome yellow; Cologne yellow; King's yellow; Leipzig yellow; Paris yellow; C.I. Pigment Yellow34; C.I. 77600. CrO$_4$Pb; mol wt 323.22. Cr 16.09%, O 19.80%, Pb 64.11%. PbCrO$_4$. Occurs in nature as the minerals *crocoite, phoenicochroite*. Ref: *Colour Index* vol. 4 (3rd ed., 1971) p 4677.

Yellow or orange-yellow powder. d 6.3. mp 844°. It is one of the most insol salts (0.2 mg/l H$_2$O). Insol in acetic acid; sol in solns of fixed alkali hydroxides, in dil HNO$_3$. LD$_{75}$ i.p. in guinea pigs: 156 mg/kg, *Handbook of Toxicology* vol. 1, W. S. Spector, Ed. (Saunders, Philadelphia, 1956) pp 176-177.

USE: Pigment in oil and water colors; printing fabrics; decorating china and porcelain; in chemical analysis of organic substances; in traffic paints.

Note: Basic lead chromates of various shades of color from brown-yellow to red are used as pigments.

5280. Lead Chromate(VI) Oxide. *Chromic acid lead-(2+) salt (1:2);* basic lead chromate; red lead chromate; chrome red; chromium lead oxide; Persian red; Austrian cinnabar. CrPb$_2$O$_5$; mol wt 546.40. C 9.52%, Pb 75.84%, O 14.64%. PBCrO$_4$.PbO. See: *Colour Index* vol. 4, (3rd ed., 1971) p 4677.

Red powder. Insol in water.

USE: As pigment.

5281. Lead Dioxide. Lead oxide brown; lead peroxide; lead superoxide. O$_2$Pb; mol wt 239.21. O 13.38%, Pb 86.62%. Occurs in nature as the mineral *plattnerite.* Lab prepn from lead acetate and calcium hypochlorite: Newell, Maxson, *Inorg. Syn.* **1**, 45 (1939); by hydrolysis of lead acetate: Kuhn, Hammer, *Ber.* **83**, 413 (1950).

Dark-brown powder; evolves oxygen when heated, first forming Pb$_3$O$_4$, at high temp PbO. d 9.38. Insol in water; sol in HCl with evolution of Cl; in dil HNO$_3$ in presence of H$_2$O$_2$, oxalic acid, or other reducers; sol in alkali iodide solns with liberation of iodine; soluble in hot caustic alkali solns. LD$_{50}$ i.p. in guinea pigs: 200 mg/kg, *Handbook of Toxicology* vol. 1, W. S. Spector, Ed. (Saunders, Philadelphia, 1956) pp 176-177.

USE: Electrodes in batteries; oxidizing agent in manuf dyes; as discharge in dyeing with indigo; manuf rubber substitutes; with amorphous phosphorus as ignition surface for matches; pyrotechny; manuf pigments; in anal. chemistry.

5282. Lead Fluoride. Lead difluoride; plumbous fluoride. F$_2$Pb; mol wt 245.21. F 15.50%, Pb 84.50%. PbF$_2$. Prepd by treating lead carbonate or hydroxide with hydrogen fluoride and evaporating the soln; by mixing solns of potassium fluoride and lead acetate; by precipitation from the soln of a lead salt by HF; by the action of fluorine on lead: Ruff, *Die Chemie des Fluors* (Springer, Berlin, 1920)

p 33; Eméleus in *Fluorine Chemistry* vol. 1, J. H. Simons, Ed. (Academic Press, New York, 1950) p 51; Kwasnik in *Handbook of Preparative Inorganic Chemistry* vol. 1, G. Brauer, Ed. (Academic Press, New York, 1963) pp 218-219. *Review:* Kemmitt, Sharp, *Advan. Fluorine Chem.* **4**, 188 (1965).

White to colorless crystals. *Poisonous!* Dimorphous: orthorhombic, converted to cubic above 316°. d (orthorhombic) 8.445; d (cubic) 7.750. mp 824°. bp 1293°. Soly in water (g/100 ml): 0.057 (0°); 0.065 (20°). The soly increases in the presence of HNO$_3$ or nitrates.

5283. Lead Formate. C$_2$H$_2$O$_4$Pb; mol wt 297.25. C 8.08%, H 0.68%, O 21.53%, Pb 69.71%. Pb(CHO$_2$)$_2$.

White, lustrous prisms or needles. *Poisonous!* d 4.63. Dec at 190°. Sol in 65 parts cold water, 6 parts boiling water; insol in alcohol.

5284. Lead Hexafluorosilicate. Lead fluosilicate; lead silicofluoride. F$_6$PbSi; mol wt 349.28. C 32.64%, Pb 59.32%, Si 8.04%. PbSiF$_6$.

Dihydrate, colorless crystals. Sol in water. *Poisonous!*

USE: In refining lead by electrolytic methods.

5285. Leadhillite. Pb(OH)$_2$.PbSO$_4$.2PbCO$_3$—basic lead carbonate sulfate.

5286. Lead Hydroxide. Lead oxide hydrate; basic lead hydroxide. H$_2$O$_4$Pb$_3$; mol wt 687.59. H 0.29%, O 9.30%, Pb 90.40%. 3PbO.H$_2$O. Prepn and structure: Oswald *et al., Helv. Chim. Acta* **51**, 1389 (1968).

White powder; absorbs CO$_2$ from the air. d 7.41. *Poisonous!* Insol in H$_2$O; sol in dil acids, fixed alkali hydroxides.

5287. Lead Hypophosphite. H$_4$O$_4$P$_2$Pb; mol wt 337.20. H 1.20%, O 18.98%, P 18.37%, Pb 61.45%. Pb(H$_2$PO$_2$)$_2$.

Hygroscopic, cryst powder; dec at elevated temps. *Poisonous!* Slightly sol in cold water; more in hot water; insol in alcohol. *Keep well closed and protected from light.*

5288. Lead Iodide. I$_2$Pb; mol wt 461.05. I 55.06%, Pb 44.94%. PbI$_2$.

Bright yellow, heavy odorless powder. *Poisonous!* d 6.16. mp 402°. One gram dissolves in 1350 ml cold, 230 ml boiling water; sol in concd solns of alkali iodides; freely sol in soln of sodium thiosulfate; sol in 200 parts cold 90 parts hot aniline; insol in alcohol or cold HCl. *Protect from light.*

USE: Bronzing, gold pencils, mosaic gold, printing, photography.

5289. Lead Lactate. C$_6$H$_{10}$O$_6$Pb; mol wt 385.35. C 18.70%, H 2.62%, O 24.91%, Pb 53.77%. Pb(C$_3$H$_5$O$_3$)$_2$.

White, heavy, cryst powder. *Poisonous!* Slowly absorbs CO$_2$ from the air, becoming partly insol. Sol in water, hot alcohol. *Keep well closed.*

5290. Lead Molybdate(VI). MoO$_4$Pb; mol wt 367.16. Mo 26.13%, O 17.43%, Pb 56.44%. PbMoO$_4$. Occurs as the mineral *wulfenite.*

Yellow powder. Insol in water; sol in HNO$_3$ or NaOH when freshly pptd.

USE: In pigments.

5291. Lead Monoxide. Lead oxide yellow; plumbous oxide; litharge; massicot; lead protoxide. OPb; mol wt 223.21. Pb 92.83%, O 7.17%. PbO. Prepn of high purity PbO: Kwestroo, Huizing, *J. Inorg. Nucl. Chem.* **27**, 1591 (1965); Kwestroo *et al., ibid.* **29**, 39 (1967). Mfg processes: Faith, Keyes & Clark's *Industrial Chemicals,* F. A. Lowenheim, M. K. Moran, Eds. (Wiley-Interscience, New York, 4th ed., 1975) pp 509-513.

Exists in two forms: red to reddish-yellow, tetragonal crystals, stable at ordinary temp; yellow, orthorhombic crystals, stable above 489°: Petersen, *J. Am. Chem. Soc.* **63**, 2617 (1941). *Poisonous!* At 300-450° in the air, it converts slowly into Pb$_3$O$_4$, but at higher temp reverts to PbO. d 9.53. mp 888°. Insol in water, alcohol; sol in acetic acid, dil HNO$_3$, in warm solns of fixed alkali hydroxides. LD$_{50}$ i.p. in rats: 430 mg/kg, Bradley, Fredrick, *Ind. Med.* **10**, *Ind. Hyg. Sect.* **2**, 15 (1941).

Caution: Avoid breathing dust. Wear dust mask approved by U.S. Bureau of Mines for this purpose. Wash thoroughly before eating or smoking. Keep away from feed or food products.

USE: In ointments, plasters; preparing soln of lead subace-tate. Glazing pottery; glass flux for painting on porcelain and glass; lead glass; varnishes; with glycerol as metal cement; producing iridescent colors on brass and bronze; coloring sulfur-containing substances, *e.g.*, hair, nails, wool, horn; manuf artificial tortoise shell and horn; pigment for rubber; manuf boiled linseed oil; in assay of gold and silver ores.

5292. Lead Nitrate. N_2O_6Pb; mol wt 331.23. N 8.46%, O 28.98%, Pb 62.55%. $Pb(NO_3)_2$.
White or colorless translucent crystals. d 4.53. *Poisonous!* One g dissolves in 2 ml cold, 0.75 ml boiling water, in 2500 ml abs alcohol, 75 ml abs methanol; insol in concd HNO_3. The aq soln is slightly acid. pH of 20% aq soln at 25° = 3.0-4.0.
USE: Manuf matches and special explosives; as mordant in dyeing and printing on textiles; mordant for staining horn, mother-of-pearl; oxidizer in dye industry; sensitizer in photography; process engraving.
THERAP CAT (VET): Has been used as a caustic in equine canker.

5293. Lead Oleate. *Oleic acid lead salt.* Approx $Pb(C_{18}H_{33}O_2)_2$.
Granular, wax-like mass. Insol in water; sol, when fresh, in alc, benzene, ether, oil turpentine. LD orally in guinea pigs: 8.0 g/kg, *Handbook of Toxicology* vol. 1, W. S. Spector, Ed. (Saunders, Philadelphia, 1956) pp 176-177.
USE: In varnishes; in extreme pressure lubricants.

5294. Lead Oxalate. C_2O_4Pb; mol wt 295.23. C 8.14%, O 21.68%, Pb 70.19%. PbC_2O_4.
White, heavy powder. Dec at 300°. d 5.28. *Poisonous!* Insol in water; sol in dil HNO_3, fixed alkali hydroxides; sparingly sol in acetic acid.

5295. Lead Phosphate. $O_8P_2Pb_3$; mol wt 811.54. O 15.77%, P 7.63%, Pb 76.59%. $Pb_3(PO_4)_2$.
White powder. d 6.9. mp 1014°. *Poisonous!* Insol in water, alcohol. Sol in HNO_3, fixed alkali hydroxides.
Note: This substance may reasonably be anticipated to be a carcinogen: *Fourth Annual Report on Carcinogens* (NTP 85-002, 1985) p 121.
USE: Stabilizer for plastics.

5296. Lead Selenate. O_4PbSe; mol wt 350.17. O 18.28%, Pb 59.17%, Se 22.55%. $PbSeO_4$. Prepd by adding a soln of lead nitrate to sodium selenate: Lenher, Kao, *J. Am. Chem. Soc.* 47, 1521 (1925).
Orthorhombic crystals. d_4^{20} 6.37. Dec by heat. Sol in concd acids; insol in water.

5297. Lead Selenite. O_3PbSe; mol wt 334.17. O 14.36%, Pb 62.01%, Se 23.63%. $PbSeO_3$. Prepd by adding selenious acid or an alkali selenite to a soln of lead chloride or nitrate: Berzelius, cited in *Mellor's* vol. X, 833 (1930).
Powder. mp ~500°. Melts to a yellow liquid. Dec at a bright-red heat giving off selenium oxide. Very sparingly sol in water; difficultly dec by boiling sulfuric acid.

5298. Lead Sesquioxide. Lead trioxide; plumbous plumbate. O_3Pb_2; mol wt 462.42. O 10.38%, Pb 89.62%. Pb_2O_3.
Reddish-yellow powder; converted at 370° in air to Pb_3O_4, dec at about 530° to PbO. Insol in water; dec by concd HCl or H_2SO_4 with the liberation of Cl or oxygen, respectively.

5299. Lead Sodium Thiosulfate. Lead sodium hyposulfite. $Na_4O_9PbS_6$; mol wt 635.59. Na 14.47%, O 22.66%, Pb 32.60%, S 30.27%. $Na_4Pb(S_2O_3)_3$.
White, small, heavy crystals. *Poisonous!* Sparingly sol in water, more in thiosulfate solns.
USE: Manuf matches.

5300. Lead Stearate. *Stearic acid lead salt.* Approx $Pb(C_{18}H_{35}O_2)_2$.
White powder, mp about 125°. *Poisonous!* Insol in water; sol in hot alcohol.
USE: In extreme pressure lubricants; as drier in varnishes.

5301. Lead Subacetate. Lead monosubacetate; monobasic lead acetate. $C_4H_{10}O_8Pb_3$; mol wt 807.75. C 5.95%, H 1.25%, O 15.85%, Pb 76.96%. $Pb(C_2H_3O_2)_2.2Pb(OH)_2$.
White, heavy powder. *Poisonous!* Sol in 16 parts cold, 4

parts boiling water with alkaline reaction. On exposure to air absorbs CO_2 and becomes incompletely sol. *Keep well closed.*
USE: In sugar analysis to remove coloring matters, etc., from solns before polarizing; for clarifying and decolorizing other solns of organic substances.

5302. Lead Sulfate. O_4PbS; mol wt 303.28. O 21.10%, Pb 68.32%, S 10.57%. $PbSO_4$. Occurs as the minerals: *anglesite, lanarkite.*
White, heavy, cryst powder. *Poisonous!* d 6.2. mp 1170°. Sol in about 2225 parts water; more soluble in dil HCl or HNO_3, less in dil H_2SO_4; sol in NaOH, ammonium acetate or tartrate soln; sol in concd hydriodic acid; insol in alcohol. LD_{75} i.p. in guinea pigs: 290 mg/kg, *Handbook of Toxicology* vol. 1, W. S. Spector, Ed. (Saunders, Philadelphia, 1956) pp 178-179.
USE: Instead of white lead as pigment; with zinc in galvanic batteries; manuf minium, in lithography; preparing rapidly drying oil varnishes; weighting fabrics.

5303. Lead Sulfide. PbS; mol wt 239.28. Pb 86.60%, S 13.40%. Occurs as the mineral *galena.*
Black powder. Insol in water; sol in HNO_3, hot, dil HCl. LD_{50} i.p. in rats: 1.8 g/kg, Bradley, Fredrick, *Ind. Med.* 10, *Ind. Hyg. Sect.* 2, 15 (1941).
USE: Glazing earthenware.

5304. Lead Telluride. PbTe; mol wt 334.82. Pb 61.89%, Te 38.11%. Found in nature as the mineral *altaite.* Prepd from lead nitrate, sodium carbonate and powdered tellurium: Montignie, *Bull. Soc. Chim. France* 1947, 750. Prepn of single crystals by heating stoichiometric quantities of the elements in a graphite cup or fused quartz tube: Brady, *J. Electrochem. Soc.* 101, 466 (1954).
Silver-gray cubic crystals. d_4^{20} 8.16. mp 905°. Most of the crystal is *p*-type, the *n*-type material being present in the surface layer. Energy gap 0.27 ev. Electron mobility 2240 cm^2/volt-sec. Hole mobility 860 cm^2/volt-sec. Resistivity 0.005 ohm-cm (*p*-type), 0.00090 ohm-cm (*n*-type). Not attacked by hydrochloric, hydrofluoric, perchloric and acetic acids or their mixtures; not attacked by solns of 30% potassium hydroxide or of alkali metal sulfides. Dil nitric acid turns the surface black, while concd nitric acid produces lighter gray surface and turns the black surface to gray. Hot concd sulfuric acid produces a reddish-violet surface.
USE: In photoconductor cells; in semiconductor research.

5305. Lead Tetraacetate. $C_8H_{12}O_8Pb$; mol wt 443.39. C 21.67%, H 2.73%, O 28.87%, Pb 46.73%. $Pb(CH_3COO)_4$. Prepd from Pb_3O_4 and glacial acetic acid preferably in the presence of some acetic anhydride: Dimroth, Schweizer, *Ber.* 56, 1375 (1923); Bailar, *Inorg. Syn.* 1, 47 (1939); in *Handbook of Preparative Inorganic Chemistry* vol. 1, G. Brauer, Ed. (Academic Press, New York, 2nd ed., 1963) p 767. Prepn by electrolysis: Fioshin, Gus'kov, *Dokl. Akad. Nauk SSSR* 112, 303 (1957), *C.A.* 51, 16146 (1957); Sataev *et al.*, *Khim. Prom.* (Moscow) 46, 892 (1970), *C.A.* 74, 49005x (1971). Reviews of prepn and use as oxidizing agent: Criegee "Oxidations with Lead Tetraacetate" in *Oxidation in Organic Chemistry*, Part A, K. B. Wiberg, Ed. (Academic Press, New York, 1965) pp 277-366; Zyka, *Pure Appl. Chem.* 13, 569-581 (1966).
Colorless monoclinic prisms from glacial acetic acid. Turns pink easily. Unstable in air. Hydrolyzed by water with the formation of brown lead dioxide and acetic acid. *Avoid contact with skin.* d_4^{17} 2.228. mp 175-180°. Sol in hot glacial acetic acid, benzene, chloroform, tetrachloroethane, nitrobenzene. Dissolves in concd halogen acids with the formation of haloplumbic acids, H_2PbX_6. The dry material can be stored in sealed, evacuated ampuls.
USE: Selective oxidizing agent in organic syntheses: Criegee, *Angew. Chem.* 53, 321 (1940); *Newer Methods of Preparative Organic Chemistry* (Interscience, N. Y., 1948) pp 1-17.

5306. Lead Tetrafluoride. Plumbic fluoride. F_4Pb; mol wt 283.21. F 26.84%, Pb 73.16%. PbF_4. Prepd by passing fluorine diluted with CO_2 or N_2 over PbF_2 at 300°: v. Wartenberg, *Z. Anorg. Allgem. Chem.* 244, 339 (1940). *Review:* Kemmitt, Sharp, *Advan. Fluorine Chem.* 4, 187 (1965).
White, tetragonal crystals, d 6.7. mp about 600°. Readily

hydrolyzes and turns brown (forms PbO_2) in the presence of moisture.

USE: Has been proposed as a fluorinating agent for hydrocarbons.

5307. Lead Tetroxide. Lead oxide red; red lead; minium; lead orthoplumbate; mineral orange; mineral red; Paris red; Saturn red; C.I. Pigment Red 105; C.I. 77578. O_4Pb_3; mol wt 685.63. O 9.33%, Pb 90.67%. Pb_3O_4. The article of commerce contains about 90% Pb_3O_4; the remainder being chiefly lead monoxide. Prepn: M. Baudler in *Handbook of Preparative Inorganic Chemistry* vol. 1, G. Brauer, Ed. (Academic Press, New York, 1963) pp 755-757. Structure: S. T. Gross, *J. Am. Chem. Soc.* **65**, 1107 (1943). *Review: Mellor's* vol. **7** (1930) pp 672-680.

Bright-red, heavy powder. *Poisonous!* Dec at about 500° with evolution of oxygen. d 9.1. Insol in water or alcohol; sol in excess glacial acetic acid, in hot HCl with evolution of Cl, in dil HNO_3 in presence of H_2O_2. LD_{50} i.p. in guinea pigs: 220 mg/kg, *Handbook of Toxicology* vol. 1, W. S. Spector, Ed. (Saunders, Philadelphia, 1956) pp 176-177.

USE: Plasters and ointments; manuf colorless glass; glaze for faience; flux for porcelain painting, protective paint for iron and steel; oil-color for ship paints, varnishes; coloring rubber; cement for glass, gas and steam pipes; storage batteries; pencils for writing on glass; manuf lead peroxide, matches.

5308. Lead Thiocyanate. Lead sulfocyanate. $C_2N_2PbS_2$; mol wt 323.38. C 7.43%, N 8.66%, Pb 64.08%, S 19.83%. $Pb(SCN)_2$. Prepn: Gardner, Weinberger, *Inorg. Syn.* **1**, 85 (1939).

White, odorless powder. d 3.82. *Poisonous!* Sol in about 200 parts cold, 50 parts boiling water; also sol in alkali hydroxide and thiocyanate solns.

USE: Reverse dyeing with aniline black; manufacture of safety matches and cartridges.

5309. Lead Tungstate(VI). O_4PbW; mol wt 455.07. O 14.06%, Pb 45.53%, W 40.41%. $PbWO_4$. Occurs as the minerals *raspite, scheelite, stolzite.*

White powder. Insol in water or cold HNO_3; sol in fixed alkali hydroxide solns.

5310. Lead Vanadate(V). Lead metavanadate. O_6PbV_2; mol wt 405.11. O 23.70%, Pb 51.15%, V 25.15%. $Pb(VO_3)_2$. Yellow powder, insol in water; dec by HNO_3.

USE: Manuf other vanadium compds; as pigment.

5311. Lecithin. Phosphatidylcholine; Lecithol; Vitellin; Kelecin; Granulestin. Phosphatide found in all living organisms (plants and animals). Significant constituent of nervous tissue and brain substance. A mixture of the diglycerides of stearic, palmitic, and oleic acids, linked to the choline ester of phosphoric acid. Commercial grades contain 2.2% P. Isoln from eggs: Sinclair, *Can. J. Res.* **26B**, 777 (1948). Product of commerce is predominantly soybean lecithin obtained as a by-product in the manuf of soybean oil: Stanley in K. S. Markley, *Soybeans* vol. II (Interscience, New York, 1951) pp 593-647. Soybean lecithin contains palmitic acid 11.7%, stearic 4.0%, palmitoleic 8.6%, oleic 9.8%, linoleic 55.0%, linolenic 4.0%, C_{20} to C_{22} acids (includes arachidonic) 5.5%. Synthesis of a mixed acid α-lecithin: de Haas, van Deenen, *Tetrahedron Letters* **1960** (no. 9), 1. Synthetic L-α-(distearoyl)lecithin is identical with hydrogenated egg yolk lecithin and L-α-(dipalmitoyl)lecithin is identical with a natural phosphatide of brain, lung, and spleen. (*See also* Phosphatidic Acid.) Commercial grades of natural lecithin are reported to contain a potent vasodepressor substance: McQuarrie, Andersen, U.S. pat. **2,931,818** (1960 to Cutter Labs.). Comprehensive monograph: G. B. Ansell, J. N. Hawthorne, *Phospholipids* (Elsevier, New York, 1964) 439 pp; J. Eichberg, "Lecithin" in Kirk-Othmer *Encyclopedia of Chemical Technology* vol. **14** (Wiley-Interscience, New York, 3rd ed., 1981) pp 250-269.

```
     CH₂OCOR
     |
     CHOCOR   O⁻
     |        |
     CH₂O — P — OCH₂CH₂N⁺(CH₃)₃
              ‖
              O
```

Waxy mass when the acid value is about 20. Pourable, thick fluid when the acid value is around 30. Color is nearly white when freshly made, but rapidly becomes yellow to brown in air. d_4^{24} 1.0305. Iodine value 95; saponification value 196. Insoluble but swells up in water and in NaCl soln forming a colloidal suspension. Soluble in about 12 parts cold, abs alcohol; sol in chloroform, ether, petr ether, in mineral oils and fatty acids; sparingly sol in benzene. Insol in acetone; practically insol in cold vegetable and animal oils.

USE: Edible and digestible surfactant and emulsifier of natural origin. Used in margarine, chocolate and in the food industry in general. In pharmaceuticals and cosmetics. Many other industrial uses, e.g. treating leather and textiles.

THERAP CAT: Lipotropic.

5312. Lectins. Agglutinins; affinitins; phytoagglutinins; phasins; protectins. A group of proteins, widely distributed in nature, that have the ability to agglutinate erythrocytes and many other types of cells. Although their existence has been known since 1899, when Stillmark isolated a hemagglutinin from castor beans, the term "lectin" (from the Latin *legere*, to choose) was first introduced by W. C. Boyd and E. Slapleigh in *Science* **119**, 419 (1954). It is now used to designate "a sugar-binding protein or glycoprotein of non-immune origin which agglutinates cells and/or precipitates glycoconjugates", I. J. Goldstein *et al.*, *Nature* **285**, 66 (1980). Lectins are found primarily in seeds of plants, but also occur in roots, leaves and bark. In addition, they are present in invertebrates such as clams, snails, and horseshoe crabs, and in several vertebrate species. The term *phytohemagglutinin* is used to refer to plant lectins. Important members of the lectin family include concanavalin A, abrin, ricin, *q.q.v.*, as well as *soybean agglutinin* or *SBA* and *wheat germ agglutinin* or *WGA*. Lectins vary considerably in chemical and physical properties; only a limited number have been purified. Mol wts of 17,000 to 400,000 have been reported and most lectins have been found to contain Mn^{2+} and Ca^{2+}. Nearly all lectins can be inhibited by free oligo- or monosaccharides of appropriate specificity. Although their physiological functions in plants or in other organisms are unknown, lectins exhibit a variety of unusual biological properties. Some are specific in their reactions with human blood groups; some induce mitosis in lymphocytes. WGA from wheat germ lipase has been shown to agglutinate mouse tumor cells more readily than cells from normal tissue: J. C. Aub *et al.*, *Proc. Nat. Acad. Sci. USA* **50**, 613 (1963); M. M. Burger, A. R. Goldberg, *ibid.* **57**, 359 (1967). Soybean agglutinin and concanavalin A have been shown to agglutinate cell lines transformed by viral or chemical carcinogens: M. Inbar, L. Sachs, *Nature* **223**, 710 (1969); *eidem, Proc. Nat. Acad. Sci. USA* **63**, 1418 (1969); B. A. Sela *et al., J. Membrane Biol.* **3**, 267 (1970). Some plant lectins mimic the direct effects of insulin on nuclear envelope phosphorylation: F. Purrello *et al., Science* **221**, 462 (1983). Soybean agglutinin has also been used in bone marrow transplants in patients with severe combined immunodeficiency: Y. Reisner *et al., Blood* **61**, 341 (1983). *Reviews:* N. Sharon, H. Lis, *Science* **177**, 949-955 (1972); *eidem, Ann. Rev. Biochem.* **42**, 541-574 (1973); L. Sequeira, *Ann. Rev. Phytopathol.* **16**, 453-481 (1978). Book: *Lectins: Biology, Biochemistry, Clinical Biochemistry* vol. 1, T. C. Bog-Hansen, Ed. (de Gruyter, New York, 1981) 414 pp.

USE: As tools for studying cell surface properties; in cancer research.

5313. Ledol. *Decahydro-1,1,4,7-tetramethyl-1H-cycloprop[e]azulen-4-ol;* "Ledum camphor". $C_{15}H_{26}O$; mol wt 222.36. C 81.02%, H 11.79%, O 7.20%. Occurs in the essential oil from leaves of *Ledum palustre* L.: Grassmann, *Re-*

pert. Pharm. **38**, 53 (1931); Hjelt, *Ber.* **28**, 3087 (1895); from *L. groenlandicum* Veder; *L. columbianum* Piper, *Ericaceae:* Cain, Lynn, *J. Am. Pharm. Assoc.* **23**, 666 (1934); Penfold, *J. Proc. Roy. Soc. N.S. Wales* **59**, 206 (1925). Structure: Büchi *et al., Tetrahedron Letters* **1959** (no. 6), 14; Graham *et al., Aust. J. Chem.* **13**, 372 (1960). Stereochemistry: Dolejs, Sorm, *Tetrahedron Letters* **1959** (no. 17), 1.

Needles from alc, mp 104-105°. Sublimes easily, even below the mp. bp₇₆₀ 292°. n_D^{110} 1.4667. $[\alpha]_D^{20}$ +28° (c = 10 in chloroform). Practically insol in water. Sol in alc (about 10% w/v). Soluble in other organic solvents.

Chromate, $C_{30}H_{50}O_4Cr$, ruby-red prisms, mp 92°. $[\alpha]_{671}^{20}$ +30° (c = 2 in chloroform).

5314. Lefetamine. *(R)-N,N-Dimethyl-α-phenylbenzene-ethanamine; (−)-N,N-dimethyl-1,2-diphenylethylamine; (-)-N,N*-dimethyl-α-phenylphenethylamine. $C_{16}H_{19}N$; mol wt 225.33. C 85.28%, H 8.50%, N 6.22%. Centrally acting analgesic with stereochemical resemblance to morphine, *q.v.* Prepn and activity: K. Ogiu *et al., J. Pharm. Soc. Japan* **80**, 283 (1960); *eidem,* **Japan.** pat. **23,087('61)** (to Res. Found. Practical Life). Absolute configuration: M. Nakazaki, *Chem. & Ind. (London)* **1962**, 1577. NMR data indicates 22% eclipsed conformation which appears to enhance stereoselectivity for morphine receptors: T. Sasaki *et al., J. Med. Chem.* **9**, 847 (1966). Pharmacology: H. Nakamura, M. Shimizu, *Arch. Ind. Pharmacodyn. Ther.* **221**, 105 (1976). Clinical study: J. P. Famaey, T. L. Peeters, *Brux. Med.* **56**, 21 (1976). Analgesic activity: M. Nozaki *et al., Life Sci.* **33**, Suppl. 1, 431 (1983). Opiate agonist activity: M. Graziella de Montis *et al., Pharmacol. Res. Commun.* **17**, 471 (1985).

bp₆ 142-147°. $[\alpha]_D^{20}$ −124.2° (ethanol).
Hydrochloride, $C_{16}H_{20}ClN$, *Santenol, SPA.* mp 218-219°. $[\alpha]_D^{20}$ −91.7° (water).
THERAP CAT: Analgesic.

5315. Leghemoglobin. *Legoglobin.* Hemoglobin-like red pigment present in the root nodules of leguminous plants. Isolation from soya beans: Keilin, Wang, *Nature* **155**, 227 (1945); Appleby, *Biochim. Biophys. Acta* **60**, 226 (1962). Mol wt is approx one-fourth that of hemoglobin: Ehrenberg, Ellfolk, *Acta Chem. Scand.* **17**, S343 (1963). Resolved into four components on DEAE-cellulose column: Ellfolk, *ibid.* **14**, 609 (1960). Suggested to act as an oxido-reduction catalyst in the symbiotic nitrogen fixation: *idem, ibid.* **15**, 975 (1961). Primary structure of soybean leghemoglobin: Ellfolk, Sievers, *ibid.* **25**, 3532 (1971).

5316. Leiopyrrole. *N,N-Diethyl-2-[2-(2-methyl-5-phenyl-1H-pyrrol-1-yl)phenoxy]ethanamine; 1-[o-(2-diethylaminoethoxy)phenyl]-2-methyl-5-phenylpyrrole;* 2-methyl-1-(2-β-diethylaminoethoxyphenyl)-5-phenylpyrrole; *DV 714.* $C_{23}H_{28}N_2O$; mol wt 348.47. C 79.27%, H 8.10%, N 8.04%, O 4.59%. Preparation: Buu-Hoi *et al., J. Med. Pharm. Chem.* **1**, 23 (1959).

Hydrochloride, $C_{23}H_{29}ClN_2O$, *Leioplegil.* Needles from carbon tetrachloride, mp 138°. (Base, bp₁₃ 232°; n_D^{20} 1.6025.) Readily sol in water; aq soln turns yellow on exposure to air or light.
THERAP CAT: Antispasmodic.

5317. Lemon Peel. Outer rind of fresh ripe fruit of *Citrus limonum* (L.) Risso *(C. medica* var. *limon* L.), *Rutaceae. Habit.* Northern India; cultivated in California, West Indies, Italy, Spain. *Constit.* Volatile oil, hesperidin, bitter extractive.

USE: As a flavor in medicines; also in beverages, confectionery, and cooking.

5318. Lenacil. *3-Cyclohexyl-6,7-dihydro-1H-cyclopenta-pyrimidine-2,4-(3H,5H)-dione;* 3-cyclohexyl-5,6-trimethyleneuracil; 3-cyclohexyl-1,5,6,7-tetrahydro-2H-cyclopenta-pyrimidine-2,4(3H)-dione; *du Pont 634; Venzar.* $C_{13}H_{18}N_2O_2$; mol wt 234.29. C 66.64%, H 7.74%, N 11.96%, O 13.66%. Prepn: Senda, Fujimura, **Japan.** pat. **4892('62).**

Solid, mp 290° (Senda, Fujimura, *loc. cit.*). Commercial product: mp 316-317°; sp gr 1.32; solubility in water 6 ppm; slightly sol in most organic solvents; sol in pyridine; approx lethal dose in rats: > 11,000 mg/kg.
USE: Herbicide.

5319. Lenampicillin. *[2S-[2α,5α,6β(S*)]]-6-[(Amino-phenylacetyl)amino]-3,3-dimethyl-7-oxo-4-thia-1-azabicyclo[3.2.0]heptane-2-carboxylic acid (5-methyl-2-oxo-1,3-dioxol-4-yl)methyl ester;* ampicillin (5-methyl-2-oxo-1,3-dioxolen-4-yl)methyl ester; 6-[D(−)-α-aminophenylacetamido]penicillanic acid (5-methyl-2-oxo-1,3-dioxol-4-yl)methyl ester. $C_{21}H_{23}N_3O_7S$; mol wt 461.49. C 54.66%, H 5.02%, N 9.10%, O 24.27%, S 6.95%. Orally active ampicillin prodrug. Prepn: F. Sakamoto *et al.,* **Eur.** pat. **Appl. 39,086;** *eidem,* **U.S.** pat. **4,342,693** (1981, 1982 both to Kanebo); F. Sakamoto *et al., Chem. Pharm. Bull.* **32**, 2241 (1984); S. Ikeda *et al., ibid.* 4316. Metabolism in man, dogs, rats: N. Awata *et al., Japan. J. Antibiot.* **38**, 1776, 1785, *C.A.* **104**, 161479u, 122550r (1985). Human pharmacokinetics: A. Saito, M. Nakashima, *Antimicrob. Ag. Chemother.* **29**, 948 (1986). Series of articles on antibacterial activity, mutagenicity, pharmacology, clinical trials: *Chemotherapy* (Tokyo) **32**, Suppl. 8, pp 1-772 (1984). Acute toxicity data: F. Ogino *et al., ibid.* 31.

Hydrochloride, $C_{21}H_{24}ClN_3O_7S$, *KB 1585, KBT 1585, Varacillin, Takacillin.* Crystals from isopropanol-ethyl acetate, mp 145° (dec). LD_{50} in male, female rats, male, female mice (mg/kg): approx 10000, approx 10000, 8294, 8492 orally; 4362, 4471, 3576, 4284 s.c.; 876, 838, 711, 775 i.v.; in dogs (mg/kg): > 300 orally (Ogino).
THERAP CAT: Antibacterial.

5320. Lenperone. *4-[4-(4-Fluorobenzoyl)-1-piperidinyl]-1-(4-fluorophenyl)-1-butanone; 4'-fluoro-4-[4-(p-fluorobenzoyl)piperidino]butyrophenone.* $C_{22}H_{23}F_2NO_2$; mol wt 371.43. C 71.14%, H 6.24%, F 10.23%, N 3.77%, O 8.62%. Prepn of the hydrochloride: R. L. Duncan, G. C. Grover, **Ger.** pat. **1,930,818** corresp to **U.S.** pat. **3,576,810** (1970, 1971 both to Robins). Synthesis and preliminary pharmacology of the HCl salt: R. L. Duncan *et al., J. Med. Chem.* **13**, 1 (1970). Pharmacodynamics: D. N. Johnson *et al., Arch. Int. Pharmacodyn. Ther.* **194**, 197 (1971). Study of lenperone and its metabolites by field ionization GC/mass spectrometry: D. E. Games *et al., Advan. Mass Spectrom.* **7B**, 1616 (1978).

Hydrochloride, $C_{22}H_{24}ClF_2NO_2$, *AHR-2277*, *Elanone*. Cryst from isopropyl alc/methanol, mp 255-257°. LD_{50} in mice: 135 mg/kg orally, 50 mg/kg i.v., R. L. Duncan *et al.*, *loc. cit.*

THERAP CAT (VET): Tranquilizer; anti-emetic.

5321. Lenthionine. *1,2,3,5,6-Pentathiepane.* $C_2H_4S_5$; mol wt 188.38. C 12.75%, H 2.14%, S 85.11%. Odorous principle from the edible mushroom *Shiitake Lentinus edodes* (Berk.) Sing. Isoln and synthesis: Morita, Kobayashi, *Tetrahedron Letters* **1966**, 573. Simple, efficient synthesis: I. Still, G. W. Kutney, *ibid.* **22**, 1939 (1981).

Crystals from methylene chloride, mp 60-61°.

5322. Lentinan. LC-33. $(C_6H_{10}O_5)_n$; mol wt 400,000-800,000. Neutral polysaccharide isolated from the edible mushroom, *Lentinus edodes* (Berk.) Sing. Primary structure is a β-1,3-D-glucan having 2 β-1,6-glucopyranoside branchings for every 5 β-1,3 linear linkages. Isoln and antitumor activity: G. Chihara *et al.*, *Nature* **222**, 637 (1969); *eidem*, *Cancer Res.* **30**, 2776 (1970). Immunostimulant activity: Y. Y. Maeda, G. Chihara, *Nature* **229**, 634 (1971). Prepn of stable aqueous solution: M. Fujii *et al.*, **Neth.** pat. **Appl. 7,601,114** corresp to U.S. pat. **4,207,312** (1976, 1980 both to Ajinomoto; Morishita). Structural study: T. Sasaki *et al.*, *Carbohydrate Res.* **47**, 99 (1976). Preclinical evaluation and toxicity studies: G. Chihara, *Adv. Exp. Med. Biol.* **166**, 189 (1983). Mechanism of action studies: H. Miyakoshi, T. Aoki, *Int. J. Immunopharmacol.* **6**, 365, 373 (1984). Clinical trial in gastric and colorectal cancer: T. Taguchi *et al.*, *Adv. Exp. Med. Biol.* **166**, 181 (1983); K. Okuyama *et al.*, *Cancer* **55**, 2498 (1985). *Review:* G. Chihara, *Int. J. Tissue Res.* **4**, 207-225 (1982).

White powder, dec 250°. Sol in aqueous alkali, formic acid. Slightly sol in hot water, DMSO. Insol in cold water, alcohol, ether, chloroform, pyridine, hexamethylphosphoramide. Stable against sulfuric and hydrochloric acids. $[\alpha]_D^{20}$ +13.5-14.5° (in 2% NaOH); +19.5-21.5° (in 10% NaOH).

THERAP CAT: Antineoplastic. Immunostimulant.

5323. Leonurine. *4-Hydroxy-3,5-dimethoxybenzoic acid 4-[(aminoiminomethyl)amino]butyl ester;* 4-hydroxy-3,5-dimethoxybenzoic acid δ-guanidinobutyl ester; syringic acid δ-guanidinobutyl ester; [4-(4-hydroxy-3,5-dimethoxybenzoyloxy)butyl]guanidine; 4-guanidino-1-butanol syringate. $C_{14}H_{21}N_3O_5$; mol wt 311.33. C 54.01%, H 6.80%, N 13.50%, O 25.70%. Isoln from leaves of *Leonurus sibiricus* L., *Labiatae:* Kubota, Nakajima, *C.A.* **25**, 771 (1931). Structure: Goto *et al.*, *Tetrahedron Letters* **1962**, 545. Revised structure: Kishi *et al.*, *ibid.* **1968**, 631. Structure and synthesis: Sugiura *et al.*, *Tetrahedron* **25**, 5155 (1969). Alternate synthesis: K. F. Cheng *et al.*, *Experientia* **35**, 571 (1979). Biological effect: H. W. Yeung *et al.*, *Planta Med.* **31**, 51 (1977).

Hydrochloride monohydrate, $C_{14}H_{21}N_3O_5 \cdot HCl \cdot H_2O$, mp 193-194°. pKa 7.9 in water.

5324. Lepidine. *4-Methylquinoline;* cincholepidine. $C_{10}H_9N$; mol wt 143.18. C 83.88%, H 6.34%, N 9.78%. Prepn from 2-chlorolepidine: Neumann *et al.*, *Org. Syn.* **26**, 45 (1946); from aniline hydrochloride: Campbell, U.S. pat. **2,451,611** (1948 to du Pont); Bach, Rast, *J. Prakt. Chem.* **17**, 63 (1962); from ethylacetanilide: Ardashev, Minkin, *Zh. Obshch. Khim.* **28**, 1578 (1958).

Colorless, oily liq; quinoline odor; turns reddish-brown in light. bp 261-263°; bp_{14-15} 126-127°; $bp_{6.7}$ 115-120°; $bp_{1.5.2}$ 90-95°. d_4^{20} 1.0826. n_D^{20} 1.6190. mp 0°. Slightly sol in water; miscible with alc, benzene, ether. *Protect from light.*

5325. Leptandra. Culver's root; black root. Dried rhizome and roots of *Veronicastrum virginicum* (L.) Farw. (*Leptandra virginica* (L.) Nutt.), *Scrophulariaceae. Habit.* North America. *Constit.* A sterol; esters of cinnamic, methoxycinnamic and fatty acids; resin saponin, tannin, sugars.

THERAP CAT: Cathartic.

5326. Leptodactyline. *3-Hydroxy-N,N,N-trimethylbenzeneethanaminium; (m-hydroxyphenethyl)trimethylammonium.* $[C_{11}H_{18}NO]^+$. Occurs in skin of the amphibian species of the genus *Leptodactylus:* its probable precursor is *m*-tyrosine. Isoln as picrate and chloride: Erspamer, *Arch. Biochem. Biophys.* **82**, 431 (1959). Prepd as chloride: Buck *et al.*, *J. Am. Chem. Soc.* **60**, 1789 (1938). Pharmacologic activity of the picrate: Erspamer, Glässer, *Brit. J. Pharmacol.* **15**, 14 (1960).

Chloride, $C_{11}H_{18}ClNO$, nodules of small spindles from alcohol + ether, mp 220°. Soluble in water, alcohol, HCl; practically insol in ether, ethyl acetate.

Picrate, $C_{17}H_{20}N_4O_8$, gold-yellow or orange-yellow needles from water, mp 198-200°. Less than 0.1% sol in cold water, more sol in hot water. Stable on storage; resistant to acids and alkalies. LD_{50} in mice: 3.3 mg/kg i.v.; about 325 mg/kg orally.

THERAP CAT: Neuromuscular blocker.

5327. Leptophos. *Phenylphosphonothioic acid O-(4-bromo-2,5-dichlorophenyl) O-methyl ester; O-(4-bromo-2,5-dichlorophenyl) O-methyl phenylphosphonothioate; O-methyl O-2,5-dichloro-4-bromophenyl phenylthiophosphonate;* VCS 506; Abar; Phosvel. $C_{13}H_{10}BrCl_2O_2PS$; mol wt 412.06. C 37.89%, H 2.45%, Br 19.39%, Cl 17.21%, O 7.76%, P 7.52%, S 7.78%. Prepn and use as an insecticide: Richter, Hanna; U.S. pats. **3,459,836, 3,551,563, 3,577,482** (1969, 1970, 1971, all to Velsicol). Metabolism in mice and on cotton plants: Holmstead *et al.*, *Arch. Environ. Contam. Toxicol.* **1**, 133 (1973). Toxicological studies in mammals: Kamel *et al.*, *J. Eur. Toxicol.* **6**, 70 (1973); Abou-Donia *et al.*, *Experientia* **30**, 63 (1974); T. B. Gaines, R. E. Linder, *Fundam. Appl. Toxicol.* **7**, 299 (1986).

Tan waxy solid. mp range 55-67°. Sp gr at 25° = 1.53. Solubility at 25° (gm/100 ml): benzene, 133; xylene, 73; acetone, 62; cyclohexane, 14; heptane, 7; isopropanol, 2.4; water, 0.03 ppm. Stable to acids; hydrolyzed slowly under strongly alkaline conditions. LD_{50} in adult male, female rats (mg/kg): 19, 20 orally (Gaines, Linder).

USE: Insecticide.

5328. Lethane® 60. *Dodecanoic acid 2-thiocyanatoethyl ester; lauric acid 2-thiocyanatoethyl ester;* lauric acid ester with 2-hydroxyethyl thiocyanate; 2-thiocyanoethyl dodecanoate; 2-thiocyanoethyl laurate. $C_{15}H_{27}NO_2S$; mol wt 285.47. C 63.11%, H 9.54%, N 4.91%, O 11.21%, S 11.23%. $CH_3(CH_2)_{10}COOCH_2CH_2SCN$. Prepn: Hester, **U.S.** pat. **2,220,521** (1941 to Rohm & Haas).

Liquid, $bp_{0.1}$ 160-190°. LD_{50} orally in rats: 500 mg/kg, *Bull. Entomol. Soc. Amer.* **12**, 178 (1966).

Human Toxicity: Moderately irritating to skin and mucous membranes, and, in high concns, narcotic.

USE: Insecticide.

5329. Lethane® 384. *Thiocyanic acid 2-(2-butoxyethoxy)ethyl ester;* 2-butoxy-2'-thiocyanodiethyl ether; 2-[2-(butoxy)ethoxy]ethyl ester of thiocyanic acid; butyl carbitol rhodanate; 1-butoxy-α-(2-thiocyanoethoxy)ethane. C_9H_{17}-NO_2S; mol wt 203.32. C 53.17%, H 8.43%, N 6.89%, O 15.74%, S 15.77%. $C_4H_9OCH_2CH_2OCH_2CH_2SCN$. Prepn: Bruson, **U.S.** pat. **2,372,809** (1945 to Rohm & Haas).

Liquid, $bp_{0.25}$ 120-125°. Commercial product, brownish oil, is a mixture with petroleum distillate which constitutes 47% by weight or 50% by volume. Practically insol in water; miscible with hydrocarbons and most organic solvents. LD_{50} orally in rats: 90 mg/kg, *RTECS* Vol. II, R. J. Lewis, R. L. Tatken, Eds. (1980) p 712.

Human Toxicity: Effects similar to Lethane 60 but may be more severe. Absorbed through intact skin.

USE: Insecticide.

5330. Letosteine. *2-[2-[(2-Ethoxy-2-oxoethyl)thio]ethyl]-4-thiazolidinecarboxylic acid;* 2-[2-[(carboxymethyl)-thio]ethyl]-4-thiazolidinecarboxylic acid 2-ethyl ester; Viscotiol. $C_{10}H_{17}NO_4S_2$; mol wt 279.37. C 42.99%, H 6.13%, N 5.01%, O 22.91%, S 22.95%. Cyclic cysteine derivative. Prepn: M. X. Chodkiewicz, **Ger.** pat. **2,410,307**; *eidem,* **U.S.** pat. **4,032,534** (1974, 1979 to Ferlux-Chemie). Effect on viscoelasticity of mucus: E. Puchelle *et al., Eur. J. Respir. Dis.* **61** (Suppl 110), 195 (1980). In treatment of bronchitis: V. Macquet *et al., Lille Med.* **24**, 735 (1979); P. Freour, J. Vergeret, *Bordeaux Med.* **12**, 1553 (1979); P. Freour, *ibid.* **13**, 1431 (1980); P. Meroni *et al., Clin. Ter.* **105**, 109 (1983). Comparison with sulfur-containing compds in bronchitis: D. Rodde, *C.R. Ther. Pharmacol. Clin.* **1**(5), 31 (1983).

Crystals from ethanol, mp 142°.
THERAP CAT: Mucolytic.

5331. Leucine. Leu (IUPAC abbrev.); 2-amino-4-methylvaleric acid; α-aminoisocaproic acid; 2-amino-4-methyl-pentanoic acid. $C_6H_{13}NO_2$; mol wt 131.17. C 54.94%, H 9.99%, N 10.67%, O 24.40%. $(CH_3)_2CHCH_2CH(NH_2)$-COOH. An amino acid classified as essential for maintaining growth in rats. Essential component in human nutrition, not synthesized by the human body. Isoln from gluten, casein, keratin: Barnett, *J. Biol. Chem.* **100**, 543 (1933); **U.S.** pat. **2,009,868;** by electrical transport from blood-corpuscle paste: Cox *et al., J. Biol. Chem.* **81**, 755 (1929). Purification and separation from methionine: Fox, *Science* **84**, 163

(1936); Bergmann, Stein, *J. Biol. Chem.* **129**, 609 (1939). Synthesis by the action of ammonia on α-bromoisocaproic acid: Fischer, *Ber.* **37**, 2486 (1904); Marvel, *Org. Syn.* coll. vol. III, 523 (1955). Prepn from hydantoin: White, **U.S.** pat. **2,557,920** (1951 to Dow).

L-Leucine, the natural form. Glistening hexagonal plates from aq alc. d_{18} 1.293. Sublimes at 145-148°. Dec 293-295° (rapid heating, sealed tube). $[M]_D$ +21.0° (5N HCl); $[M]_D$ +29.5° (glacial acetic acid). $[\alpha]_D^{25}$ −10.8° (c = 2.2); $[\alpha]_D^{26}$ +15.1° in 6N HCl (38 mols HCl per mol leucine); $[\alpha]_D^{26}$ +7.6° in 3N NaOH (30 mols NaOH per mol leucine). Ka at 25° = 2.5 × 10⁻¹⁰; Kb = 2.3 × 10⁻². R_f value 0.79. Soly in water (g/l): 22.7 (0°); 24.26 (25°); 28.87 (50°); 38.23 (75°); 56.38 (100°); in 99% alcohol: 0.72; in acetic acid: 10.9; insol in ether.

DL-Leucine, the synthetic form. Leaflets fom water. Sweet taste. Dec 332° (also reported as 290°). Sublimes. pK_1 2.36; pK_2 9.60. Soly in water (g/l): 7.97 (0°); 9.91 (25°); 14.06 (50°); 22.76 (75°); 42.06 (100°); in 90% alcohol: 1.3; insol in ether.

Hydrochloride, $C_6H_{13}NO_2$.HCl, crystals. Freely sol in water.

THERAP CAT: Nutrient.

5332. Leucinocaine Mesylate. *2-(Diethylamino)-4-methyl-1-pentanol 4-aminobenzoate (ester) monomethanesulfonate (salt);* 2-diethylamino-4-methylpentyl p-aminobenzoate methanesulfonate; p-aminobenzoic acid β-diethylamino-isohexyl ester methanesulfonate; p-aminobenzoic acid N,N-diethylleucinol ester methanesulfonate; p-aminobenzoyl-N-1-diethylamino-1-isobutylethanol methanesulfonate; 2-methyl-4-diethylaminopentan-5-ol p-aminobenzoate; Panthesin. $C_{18}H_{32}N_2O_5S$; mol wt 388.52. C 55.64%, H 8.30%, N 7.21%, O 20.59%, S 8.25%. Prepn: Lebeau, Janot, *Traité de Pharmacie Chimique* vol. III (Masson et Cie, Paris, 1955-1956) p 1296. Prepn of hydrochloride: Karrer, **U.S.** pat. **1,555,217** and *see also* **Ger.** pat. **464,484** (1925, 1928 to Chem. Fabrik Flora). Efficacy and toxicity in animals: R. N. Bieter *et al., J. Pharmacol. Exp. Ther.* **57**, 221 (1936).

Crystals, mp 171°. Sol in 3 parts water, alcohol. Practically insol in ether, petr ether, ethyl acetate, acetone, benzene. MLD in rabbits (mg/kg): 20 i.v.; 10.8 intraspinally (Bieter).

THERAP CAT: Local anesthetic.

5333. Leucite. Amphigene. $KAl(SiO_3)_2$—potassium aluminum silicate.

5334. Leucocyanidin. *2-(3,4-Dihydroxyphenyl)-3,4-di-hydro-2H-1-benzopyran-3,4,5,7-tetrol; 3,3',4,4',5,7-flavan-hexol;* 3,3',4,4',5,7-hexahydroxyflavane; leucocyanidol; Flavan; Hamaméliode P; Résivit. $C_{15}H_{14}O_7$; mol wt 306.26. C 58.82%, H 4.61%, O 36.57%. From petals of Asiatic cotton flowers (*Gossypium* spp): Stephens, *Arch. Biochem. Biophys.* **18**, 449 (1948); from *Butea frondosa* Koen. ex Roxb., *Leguminosae:* Ganguly, Seshadri, *Tetrahedron* **6**, 21 (1959). Prepn from quercetin: Bauer *et al., Chem. & Ind. (London)* **1954**, 433; from taxifolin: Freudenberg, Weinges, *Ann.* **613**, 61 (1958). Metabolism: Claveau, Masquelier, *Can. J. Pharm. Sci.* **1**, 74 (1966). *See also* Bioflavonoids.

Monohydrate, crystals from ethyl acetate + petr ether, mp above 355°. uv max (ethanol): 285 nm. Soluble in water, alcohol, acetone. Practically insol in ether, chloroform, petr ether.

Dihydrate, $C_{15}H_{14}O_7.2H_2O$, *Pygnoforton*. Crystals from hot water, mp above 300°.

Hexaacetate, $C_{27}H_{26}O_{13}$, crystals from methanol, mp 142-144°.

THERAP CAT: Capillary protectant.

5335. Leucodrin. *8-(1,2-Dihydroxyethyl)-9-hydroxy-4-(4-hydroxyphenyl)-1,7-dioxaspiro[4.4]nonane-2,6-dione.* $C_{15}H_{16}O_8$; mol wt 324.28. C 55.55%, H 4.97%, O 39.47%. From leaves of *Leucadendron concinnum* R. Br., *Proteaceae*: Hesse, *Ann.* **290**, 314 (1896); Rapson, *J. Chem. Soc.* **1938**, 282. Structure: Rapson, *ibid.* **1940**, 1271; Perold, Pachler, *Proc. Chem. Soc.* **1964**, 62; *eidem, J. Chem. Soc. (C)* **1966**, 1918.

Prisms from water, mp 212-212.5°. $[\alpha]_D^{17}$ − 19.2° (c = 3.5 in 40% alc). Slightly sol in water, alcohol.

5336. Leucoglycodrin. *8-(1,2-Dihydroxyethyl)-4-[4-(D-glucopyranosyloxy)phenyl]-9-hydroxy-1,7-dioxaspiro[4.4]nonane-2,6-dione.* $C_{21}H_{26}O_{13}$; mol wt 486.44. C 51.85%, H 5.39%, O 42.76%. From leaves of *Leucadendron concinnum* R. Br., *Proteaceae*: Merck, *Merck's Jahresber.* **1895**, 3. Structure: Murray, Bradshaw, *Tetrahedron Letters* **1966**, 3773.

Hemihydrate, $C_{21}H_{26}O_{13}.\frac{1}{2}H_2O$, amorphous solid, mp 220-222°. uv max: 201, 226, 276, 280-284 nm (ε 19880, 9240, 2100, 1960-1820).

5337. Leucomycins. Kitasamycin; C 637; Ayermicina; Sineptina; Stereomycine; Syneptine. Macrolide antibiotic complex similar to carbomycin, *q.v.* and erythromycin, *q.v.*, produced by *Streptomyces kitasatoensis* Hata: T. Hata *et al., J. Antibiot.* **6A**, 87 (1953); Sano, *ibid.* **7A**, 93 (1954). Early work performed on a mixture of at least six components, A_1, A_2, B_1-B_4, of which A_1 was the most biologically active: J. Abe *et al., J. Chem. Soc. Japan, Pure Chem. Sect.* **81**, 969 (1960); T. Watanabe *et al., Bull. Chem. Soc. Japan* **33**, 1100, 1104 (1960). Leucomycin isolated from an improved strain of *S. kitasatoensis* is a mixture of at least eight biologically active components, A_1 and new substances A_3-A_9: Hata *et al.,* U.S. pat. **3,535,309** (1970 to Kitasato Institute and Toyo Jozo). Leucomycins A_1, A_3-A_9, U and V are substituted variations of the structure represented below. Structure of A_1: T. Watanabe *et al., Angew. Chem.* **76**, 792 (1964); T. Hata *et al., Chem. Pharm. Bull. (Japan)* **15**, 358 (1967); of A_3: S. Omura *et al., Tetrahedron Letters* **1967**, 609, 1267; of A_4 through A_9: *eidem, J. Antibiot.* **20A**, 234 (1967); **21**, 272 (1968). Revised configuration at C-9: L. A. Freiberg *et al., J. Org. Chem.* **39**, 2474 (1974). Conformation: *eidem, Tetrahedron* **28**, 2839 (1972). Liquid chromatography of A_1, A_{3-9}, U, V: S. Omura *et al., J. Antibiot.* **26**, 795 (1973). Antibacterial and antimycoplasmal activity: Iwata, Akiba, *ibid.* **15A**, 258 (1962); Omura *et al., ibid.* **21**, 532 (1968); **25**, 105 (1972). *In vitro* and preliminary clinical studies: B. C. Stratford, S. Dixson, *Med. J. Australia* **1**, 1029 (1974).

Antimicrobial transformation of A_5 to V: K. Singh, S. Rakhit, *ibid.* **32**, 78 (1979). Biosynthesis: C. Kitao *et al., ibid.* **32**, 1055 (1979); S. Omura *et al., ibid.* **36**, 611 (1983). *Reviews*: Desvignes, *Ann. Pharm. Franc.* **21**, 569 (1963); Toju, Omura, "Chemical and Biological Studies on Leucomycins (Kitasamycins)" in *Drug Action and Drug Resistance in Bacteria*, **vol. I**, S. Mitsuhashi, Ed. (University Park Press, Baltimore, 1971) pp 267-291; Keller-Schierlein in *Fortschr. Chem. Org. Naturst.* **30**, 313-460 (1973); S. Omura, A. Nakagawa, *J. Antibiot.* **28**, 401-433 (1975).

Leucomycin A_1, A_2, B_1-B_4 complex (Sano, *loc. cit.*). Powder from acetone; stable at room temp; dec 125°. $[\alpha]_D^{20}$ − 67.1 (ethanol). uv max (ethanol): 232, 285 nm ($E_{1cm}^{1\%}$ 228, 8.6). Freely sol in most organic solvents. Insol in petr ether. Slightly sol in water.

Leucomycin A_3-A_9 complex (Hata *et al.,* U.S. pat. **3,535,-309**). White powder; mp 128-145°. uv max (methanol): 231 nm ($E_{1cm}^{1\%}$ 353). $[\alpha]_D^{25}$ − 53° ($CHCl_3$). Sol in methanol, ethanol, ethyl acetate, butyl acetate, acetone, benzene, and chloroform; difficultly sol in water. Insol in petr ether. LD_{50} in mice: > 650 mg/kg i.v.; > 1000 mg/kg orally.

Leucomycin A_1, $C_{40}H_{67}NO_{14}$. R = H; R' = $COCH_2CH-(CH_3)_2$. $[\alpha]_D^{25}$ − 66.0° ($CHCl_3$). uv max (methanol): 232 nm ($E_{1cm}^{1\%}$ 400). pKa' (50% ethanol): 6.69.

Triacetylleucomycin A_1, $C_{46}H_{73}NO_{17}$, crystals, mp 125-126°. $[\alpha]_D^{25}$ − 82.5° (c = 1.3 in $CHCl_3$).

Leucomycin A_3, see Josamycin.

THERAP CAT: Antibacterial.

5338. Leucopterin. *2-Amino-5,8-dihydro-4,6,7(1H)-pteridinetrione; 2-amino-4,6,7-pteridinetriol; 2-amino-4,6,7-trihydroxypteridine; 2-amino-4,6,7-trihydroxypyrimido-[4,5-b]pyrazine.* $C_6H_5N_5O_3$; mol wt 195.14. C 36.93%, H 2.58%, N 35.89%, O 24.60%. A colorless substance found in the wings of butterflies (especially of white-winged butterflies). May be obtained from xanthopterin by dehydrogenation: Wieland, Purrmann, *Ann.* **544**, 172 (1940). Synthesis by fusing 2,4,5-triamino-6-hydroxypyrimidine with an excess of oxalic acid: Purrmann, *ibid.* 188. *See also ref under* Xanthopterin and review on pterins by Purrmann, *Fortschr. Chem. Org. Naturst.* **4**, 64 (1945).

Fine colorless crystals forming yellow Na and Ag salts, and a sparingly sol NH_4 salt. Soluble in alkaline solns with blue fluorescence.

5339. Leukotrienes. LTs. A family of endogenous metabolites of arachidonic acid via the lipoxygenase pathway, chemically related to the prostaglandins and thromboxanes, *q.q.v.* The name leukotrienes was applied because of their origin in leukocytes and their conjugated triene structures. [For a description of the nomenclature of individual leukotrienes *see* B. Samuelsson, S. Hammarström, *Prostaglandins* **19**, 645 (1980)]. Members of the group are potent bronchoconstrictors that play an important pathophysiological role in immediate hypersensitivity reactions; they have been proposed as mediators of the inflammatory process and some are potent chemotactic agents. Initial studies of novel arachidonate metabolites in rabbit polymorphonuclear leukocytes: P. Borgeat *et al., J. Biol. Chem.* **251**, 7816 (1976); **252**, 8772 (1977). Subsequent studies and structure of a dihydroxyeicosatetraenoic acid (leukotriene B,

Consult the cross index before using this section.

LTB₄): P. Borgeat, B. Samuelsson, *ibid.* **254,** 2643 (1979). LTB₄ is formed from an unstable intermediate oxido-eicosatetraenoic acid, leukotriene A or LTA₄: *eidem, Proc. Nat. Acad. Sci. USA* **76,** 3213 (1979). Stereochemistry and enzymatic conversion of LTA₄ to LTB₄: O. Radmark *et al., Biochem. Biophys. Res. Commun.* **92,** 954 (1980). Synthesis of the four optical isomers of LTA₄: J. Rokach *et al., Tetrahedron Letters* **22,** 2759, 2763 (1981). Total synthesis of LTB₄: E. J. Corey *et al., J. Am. Chem. Soc.* **102,** 7984 (1980); Y. Guindon *et al., Tetrahedron Letters* **23,** 739 (1982). Formation of the 20-hydroxy and 20-carboxy metabolites of LTB₄: G. Hansson *et al., FEBS Letters* **130,** 107 (1981); total synthesis: R. Zamboni, J. Rokach, *Tetrahedron Letters* **23,** 4751 (1982). Earlier studies had described a "slow-reacting substance of anaphylaxis" (SRS-A or SRS) released from guinea pig and cat lung by cobra venom and in guinea pig lung after anaphylactic shock, *cf.* W. S. Feldberg, C. H. Kellaway, *J. Physiol.* **94,** 187 (1938); C. H. Kellaway, E. R. Trethewie, *Quart. J. Exp. Physiol.* **30,** 121 (1940). The relationship between SRS and members of the leukotriene family was established following publication of the uv spectrum of purified SRS-A, which showed the presence of the conjugated triene, *cf.* H. R. Morris *et al., FEBS Letters* **87,** 203 (1978). Purification and structure of LTC₄, an SRS from mouse mastocytoma cells: R. C. Murphy *et al., Proc. Nat. Acad. Sci. USA* **76,** 4275 (1979); S. Hammarström *et al., Biochem. Biophys. Res. Commun.* **91,** 1266 (1979). Total synthesis of LTC₄: E. J. Corey *et al., J. Am. Chem. Soc.* **102,** 1436, 3663 (1980); J. Rokach *et al., Tetrahedron Letters* **21,** 1485 (1980). Identity of synthetic LTC₄ with SRS from mouse mastocytoma cells: S. Hammarström *et al., Biochem. Biophys. Res. Commun.* **92,** 946 (1980). Structure of the SRS from rat basophil leukemia cells (RBL-1) and identification as a leukotriene (LTD₄): H. R. Morris *et al., Prostaglandins* **19,** 185 (1980). It was subsequently proposed that LTC₄ was an intermediate in the biosynthesis of LTD₄. Identity of SRS-A released in sensitized guinea pig lung perfusates and LTD₄: *eidem, Nature* **285,** 104 (1980). Assignment of stereochemistry: *eidem, Prostaglandins* **20,** 601 (1980). Detection of LTA₄ as an intermediate in the biosynthesis of LTC₄ and LTD₄: S. Hammarström, B. Samuelsson, *FEBS Letters* **122,** 83 (1980).

The sulfone of LTC₄, which has also been proposed as a natural product, *cf.* H. Ohnishi *et al., Prostaglandins* **20,** 655 (1980), has been found to be as potent as LTC₄: T. Jones *et al., ibid.* **24,** 279 (1982). Synthesis: Y. Girard *et al., Tetrahedron Letters* **23,** 1023 (1982). Discovery of LTF₄: M. E. Anderson *et al., Proc. Nat. Acad. Sci. USA* **79,** 1088 (1982). Synthesis: F. Ellis *et al., Tetrahedron Letters* **23,** 3735 (1982). It is now known that SRS-A is made up of varying amounts of cysteine-containing members of the leukotrienes, i.e. leukotrienes C₄, D₄, and E₄. These three LTs are generally found to be 100 to 1000 times more potent, on a molar basis, than histamine or prostaglandins in their effects on pulmonary airways. Review of chemistry and structure elucidation: D. A. Clark, A. Marfat, *Ann. Rep. Med. Chem.* **17,** 291-300 (1982). Comprehensive review of synthesis of leukotrienes and other lipoxygenase-derived products: J. G. Atkinson, J. Rokach, in *Handbook of Eicosanoids: Prostaglandins and Related Lipids IB,* A. L. Willis, Ed. (CRC Press, Boca Raton, 1987) pp 175-263. General reviews: P. Sirois, P. Borgeat, *Int. J. Immunopharmacol.* **2,** 281-293 (1980); P. Borgeat, P. Sirois, *J. Med. Chem.* **24,** 121-126 (1981); B. Samuelsson, *Int. Arch. Allergy Appl. Immunol.* **66,** Suppl. 1, 98-106 (1981); L. S. Wolfe, *J. Neurochem.* **38,** 1-14 (1982); J. L. Marx, *Science* **215,** 1380-1384 (1982); B. Samuelsson, S. Hammarström, *Vitam. Horm.* **39,** 1-30 (1982). Books: *SRS-A and Leukotrienes,* P. J. Piper, Ed. (Research Studies Press, London, 1981) 279 pp; *Advances in Prostaglandin, Thromboxane, and Leukotriene Research,* vol. 9, B. Samuelsson, R. Paoletti, Eds. (Raven Press, New York, 1982) 341 pp.

Leukotriene A₄, C₂₀H₃₀O₃, 3-(1,3,5,8-tetradecatetraenyl)-oxiranebutanoic acid, leukotriene A, LTA₄. Unstable. Characterized as its methyl ester: mp 28-32°. [α]²⁵_D −27° (hexane). uv max (methanol): 270, 278, 290 nm (ε 43900, 56700, 43100), *cf.* J. Rokach *et al., Tetrahedron Letters* **22,** 2759 (1981); I. Ernest *et al., ibid.* **23,** 167 (1982).

Leukotriene B₄, C₂₀H₃₂O₄, 5,12-dihydroxy-6,8,10,14-eico-

satetraenoic acid, leukotriene B, LTB₄. uv max (methanol): 260, 270.5, 281 nm (ε 38000, 50000, 39000).

Leukotriene C₄, C₃₀H₄₇N₃O₉S, *N-[S-[1-(4-carboxy-1-hydroxybutyl)-2,4,6,9-pentadecatetraenyl]-N-L-γ-glutamyl-L-cysteinyl]glycine, leukotriene C, leukotriene Cₚ, LTC₄.* uv max (methanol): 270, 280, 290 nm (ε 32000, 40000, 31000). Can be stored for several days without appreciable decomposition in frozen (−20°) pH 6.8 phosphate buffer under argon or as the tripotassium salt frozen in water. Biological activity destroyed after incubation with soybean lipoxygenase and uv max shifts to 308 nm.

Leukotriene D₄, C₂₅H₄₀N₂O₆S, *N-[S-[1-(4-carboxy-1-hydroxybutyl)-2,4,6,9-pentadecatetraenyl]-L-cysteinyl]glycine, leukotriene D, LTD₄.* uv max (methanol): 270, 280, 290 nm (ε 31000, 40000, 31000). Storage, destruction of biological activity, uv shift are the same as for LTC₄.

Leukotriene E₄, C₂₃H₃₇NO₅S, *6-[(2-amino-2-carboxyethyl)thio]-5-hydroxy-7,9,11,14-eicosatetraenoic acid, leukotriene E, LTE₄.* uv max (ethanol) of methyl ester: 269, 280, 291 nm (ε 28200, 35200, 28900).

5340. Leupeptins. Class of modified tripeptide protease inhibitors produced by various species of *Actinomyces*. Two major components, *leupeptin Ac-LL* and *leupeptin Pr-LL,* consisting of L-leucyl-L-leucyl-DL-argininal modified at the amino terminal by acetyl- or propionyl-, respectively, have been isolated. Various minor analogs, in which valine or isoleucine replaces either or both leucines, have also been found. Isolation from *Actinomyces:* T. Aoyagi *et al., J. Antibiot.* **22,** 283 (1969); S.-I. Kondo *et al., Chem. Pharm. Bull.* **17,** 1896 (1969). Structure and synthesis of leupeptins Ac-LL and Pr-LL: K. Kawamura *et al., ibid.* 1902; K. Maeda *et al., J. Antibiot.* **24,** 402 (1971). Improved purification of leupeptin Ac-LL: M. C. Y. Ning, R. J. Beynon, *Int. J. Biochem.* **18,** 813 (1986). Inhibition of proteases: T. Aoyagi *et al., J. Antibiot.* **22,** 558 (1969); and synthesis of analogs: G. Borin *et al., Z. Physiol. Chem.* **362,** 1435 (1981). Effects on protein degradation in normal and diseased muscle: P. Libby, A. L. Goldberg, *Science* **199,** 534 (1978); I. Nonaka *et al., Acta Neuropathol.* **58,** 279 (1982); R. P. Hummel, III *et al., J. Surg. Res.* **45,** 140 (1988). HPLC de-

termn of leupeptin Ac-LL in serum and muscle: M. Kai *et al., J. Chromatog.* **345**, 259 (1985).
Note: In some sources leupeptin refers only to leupeptin Ac-LL.
USE: Enzyme inhibitor in biological preparations.

5341. Leuprolide. *6-D-Leucine-9-(N-ethyl-L-prolinamide)-10-deglycinamideluteinizing hormone-releasing factor (pig);* leuprorelin; (D-Leu⁶)-des-Gly¹⁰-LH-RH-ethylamide. $C_{59}H_{84}N_{16}O_{12}$; mol wt 1209.42. C 58.59%, H 7.00%, N 18.53%, O 15.87%. Synthetic nonapeptide agonist analog of LH-RH, *q.v.* Prepn: M. Fujino *et al., Ger. pat.* **2,446,005** (1975 to Takeda), *C.A.* **83**, 10895y (1975); R. L. Gendrich *et al.,* U.S. pat. **4,005,063** (1977 to Abbott). Synthesis: J. A. Vilchez-Martinez *et al., Biochem. Biophys. Res. Commun.* **59**, 1226 (1974); M. Fujino *et al., ibid.* **60**, 406 (1974). Comparison of biological activity with natural LH-RH: D. H. Coy *et al., ibid.* **67**, 576 (1975). Use in inducing ovulation: R. L. Gendrich *et al.,* U.S. pat. **3,914,412** (1975 to Abbott). Use as antineoplastic: E. S. Johnson, J. H. Seely, U.S. pat. **4,002,738** (1977 to Abbott). Pharmacokinetics: H. Okada *et al., J. Pharm. Sci.* **73**, 298 (1984). Clinical efficacy in prostatic carcinoma: M. A. Vance, J. A. Smith, *Clin. Pharmacol. Ther.* **36**, 350 (1984); H. Yamanaka *et al., Prostate* **6**, 27 (1985). Review of clinical applications of leuprolide and other LH-RH analogs: J. Sandow, *Clin. Endocrinol.* **18**, 571-592 (1983).

5-oxoPro-His-Trp-Ser-Tyr-D-Leu-Leu-Arg-Pro-NHC₂H₅

Fluffy solid. $[\alpha]_D^{25}$ −31.7° (c = 1 in 1% acetic acid). Monoacetate (salt), $C_{61}H_{88}N_{16}O_{14}$, *Abbott-43818, A 43818, TAP-144, Carcinil, Lucrin, Lupron.*
THERAP CAT: Treatment of prostatic carcinoma.

5342. Levallorphan. *17-(2-Propenyl)morphinan-3-ol; l,N-allyl-3-hydroxymorphinan; l-3-hydroxy-N-allylmorphinan.* $C_{19}H_{25}NO$; mol wt 283.40. C 80.52%, H 8.89%, N 4.94%, O 5.65%. Prepn from 3-hydroxy-N-methylmorphinan: Schnider, Grüssner, *Helv. Chim. Acta* **34**, 2211 (1951). Alternate route: Hellerbach *et al., ibid.* **39**, 429 (1956). Comprehensive description: B. C. Rudy, B. Z. Senkowski in *Analytical Profiles of Drug Substances* vol. 2, K. Florey, Ed. (Academic Press, New York, 1973) pp 339-361.

Crystals from dilute ethanol, mp 180-182°. $[\alpha]_D^{20}$ −88.9° (c = 3 in methanol).
Tartrate, $C_{23}H_{31}NO_7$, *Lorfan.* Crystals from ethanol, mp 176-177°. $[\alpha]_D^{16}$ −39°. Soluble in water.
THERAP CAT: Narcotic antagonist.
THERAP CAT (VET): Narcotic antagonist.

5343. Levobunolol. *(S)-5-[3-[(1,1-Dimethylethyl)amino]-2-hydroxypropoxy]-3,4-dihydro-1(2H)-naphthalenone; (−)-5-[3-(tert-butylamino)-2-hydroxypropoxy]-3,4-dihydro-1(2H)-naphthalenone; l-bunolol; W-6421A.* $C_{17}H_{25}NO_3$; mol wt 291.39. C 70.07%, H 8.65%, N 4.81%, O 16.47%. Nonselective β-adrenoceptor antagonist. Prepn of racemate: C. F. Schwender *et al., J. Med. Chem.* **13**, 684 (1970); J. Shavel, Jr., S. Farber, *Ger. pat.* **1,948,144**; *eidem,* U.S. pat. **3,641,152** (1970, 1972 both to Warner-Lambert). Resolution of enantiomers: J. Shavel, Jr., C. F. Schwender, *Ger. pat.* **2,046,043**; *eidem,* U.S. pat. **3,649,691** (1971, 1972 both to Warner-Lambert); R. D. Dennis *et al.,* U.S. pat. **4,463,-176** (1984 to Bristol Myers, Mead Johnson). Cardiovascular pharmacology, anti-arrhythmic activity of racemate and isomers: R. D. Robson, H. R. Kaplan, *J. Pharmacol. Exp. Ther.* **175**, 157, 168 (1970). Adrenoceptor blocking activity: H. R. Kaplan *et al., Eur. J. Pharmacol.* **16**, 237 (1971). Metabolism in humans: F. J. DiCarlo *et al., Clin. Pharma-*

col. Ther. **22**, 858 (1977); F.-J. Leinweber *et al., Pharmacology* **16**, 70 (1978). Corneal permeability: R. D. Schoenwald, H.-S. Huang, *J. Pharm. Sci.* **72**, 1266, 1272 (1983). HPLC determn in biological fluids: H. Hengy, E. U. Kolle, *J. Chromatog.* **338**, 444 (1985); D. D. S. Tang-Liu *et al., J. Liq. Chromatog.* **9**, 2237 (1986). Clinical evaluation in hypertension: E. Arce-Gomez *et al., Curr. Ther. Res.* **19**, 386 (1976); comparison with timolol, *q.v.,* in glaucoma and ocular hypertension: A. Cinotti *et al., Am. J. Ophthalmol.* **99**, 11 (1985); D. Long *et al., ibid.* 18. Review: H. R. Kaplan, "Levobunolol" in *Pharmacology of Antihypertensive Drugs,* A. Scriabine, Ed. (Raven Press, New York, 1980) pp 317-323. Brief review of long-term treatment of glaucoma: G. D. Novack, *Gen. Pharmacol.* **17**, 373-377 (1986).

LD₅₀ in male rats, mice (mg/kg): 700 orally, 25 i.v.; 1530 orally, 78 i.v. (Kaplan review).
Hydrochloride, $C_{17}H_{26}ClNO_3$, *W 7000A, Betagan, Vistagan, Gotensin.* Crystals from methanol-ether, mp 209-211°. $[\alpha]_{589}^{24}$ −19.6±0.7° (c = 2.90 in methanol). uv max (NaOH): 221, 253, 310 mμ (ε 24700, 9000, 2400) (Shavel *et al.*).
THERAP CAT: Antiglaucoma agent.

5344. Levodopa. *3-Hydroxy-L-tyrosine; (−)-3-(3,4-dihydroxyphenyl)-L-alanine; L-dopa; β-(3,4-dihydroxyphenyl)-α-alanine; 2-amino-3-(3,4-dihydroxyphenyl)propanoic acid; Bendopa; Biodopa; Brocadopa; Cerepar; Deadopa; Dopaflex; Dopal; Dopaidan; Dopalina; Dopar; Doparkine; Doparl; Dopasol; Dopaston; Dopastral; Cidandopa; Doprin; Eldopal; Eldopar; Eldopatec; Eurodopa; Helfo-Dopa; Laradopa; Maipedopa; Larodopa; Ledopa; Parda; Levopa; Sobiodopa; Veldopa (formerly Weldopa).* $C_9H_{11}NO_4$; mol wt 197.19. C 54.82%, H 5.62%, N 7.10%, O 32.46%. Naturally occurring form of *dopa, q.v.;* the biological precursor of the catecholamines. Prepn from *l*-3-nitrotyrosin: Wasser, Lewandowski, *Helv. Chim. Acta* **4**, 657 (1921); from 3-(3,4-methylenedioxyphenyl)-L-alanine: Yamada *et al., Chem. Pharm. Bull.* **10**, 693 (1962); from L-tyrosine: Vorbrüggen, Krolikiewicz, *Ber.* **105**, 1168 (1972); Bretschneider *et al., Helv. Chim. Acta* **56**, 2857 (1973); from *Vicia faba* beans: Wysong, U.S. pat. **3,253,023** (1966 to Dow Chem.); by fermentation of L-tyrosine: Sih *et al., J. Am. Chem. Soc.* **91**, 6204 (1969); Florent, Renaut, *Ger. Offen.* **2,102,793** (1971 to Rhône-Poulenc), *C.A.* **75**, 108505f (1971). Sepn from racemate: Vogler, Baumgartner, *Helv. Chim. Acta* **35**, 1776 (1952); *Neth. pat. Appl.* **6,514,950** corresp to U.S. pat. **3,405,159** (1966 and 1968 to Merck & Co.). Molecular conformation: Becker *et al., Biochem. Biophys. Res. Commun.* **41**, 444 (1970). Metabolism studies: Shaw *et al., J. Biol. Chem.* **226**, 255 (1957); Calne *et al., Brit. J. Pharmacol.* **37**, 57 (1969). Acute toxicity: *Rx Bulletin* **1**, 16 (November, 1970). Hemodynamic effects in congestive heart failure: S. I. Rajfer *et al., N. Engl. J. Med.* **310**, 1357 (1984). Series of articles on clinical efficacy in Parkinson's disease: *Advan. Neurol.* **45**, 457-510 (1986). Reviews on L-dopa and parkinsonism: Barbeau, *Can. Med. Assoc. J.* **101**, 791 (1969); Pletscher *et al., Schweiz. Med. Wochenschr.* **100**, 797 (1970); Calne, Sandler, *Nature* **226**, 21 (1970); *L-Dopa and Parkinsonism,* A. Barbeau, Ed. (F. A. Davis, Philadelphia, 1970). Comprehensive description: R. Gomez *et al.,* in *Analytical Profiles of Drug Substances* vol. 5, K. Florey, Ed. (Academic Press, New York, 1976) pp 189-223.

Colorless to white, odorless and tasteless crystals or crystalline powder. Needles from water, mp 276-278° (dec) (Yamada); also reported as mp 284-286° (Wysong). $[\alpha]_D^{13}$ −13.1° (c = 5.12 in 1N HCl). uv max (0.001N HCl): 220.5, 280 nm (log ε 3.79, 3.42). Readily sol in dil HCl and formic acid. Soly in water: 66 mg/40 ml. Practically insol in ethanol, benzene, chloroform and ethyl acetate. In the presence of moisture, L-dopa is rapidly oxidized by atmospheric oxygen and darkens. LD_{50} in mice, rats, rabbits (mg/kg): 3650, 4000, 609 orally (Rx Bulletin).

THERAP CAT: Antiparkinsonian.

5345. Levomepate. α-(Hydroxymethyl)-α-methylbenzeneacetic acid 8-methyl-8-azabicyclo[3.2.1]oct-3-yl ester; 1αH,5αH-tropan-3α-ol (−)-2-methyl-2-phenylhydracrylate; (−)-2-methyl-2-phenylhydracrylic acid 3α-tropanyl ester; (−)-α-methylhyoscyamine; 3α-tropanyl (−)-2-methyl-2-phenylhydracrylate; tropine (−)-α-methyltropate; Dispan. $C_{18}H_{25}NO_3$; mol wt 303.39. C 71.25%, H 8.31%, N 4.62%, O 15.82%. Prepd from tropine and (−)-β-acetoxy-α-methyl-α-phenylpropionyl chloride: Vecchi et al., Brit. pat. **874,015** (1959 to Lepetit); Melone et al., J. Org. Chem. **25**, 859 (1960). Pharmacology: Bianchi et al., Toxicol. Appl. Pharmacol. **10**, 424 (1967).

Hydrochloride, $C_{18}H_{25}NO_3 \cdot HCl$, crystals from ethyl acetate, mp 210-212°. $[\alpha]_D^{20}$ −6.8°.

THERAP CAT: Anticholinergic.

5346. Levomethadyl Acetate. [S-(R*,R*)]-β-[2-(Dimethylamino)propyl]-α-ethyl-β-phenylbenzeneethanol acetate (ester); 6-(dimethylamino)-4,4-diphenyl-3-heptanol acetate (ester); α-l-acetylmethadol; levo-alpha-acetylmethadol; LAAM. $C_{23}H_{31}NO_2$; mol wt 353.51. C 78.15%, H 8.84%, N 3.96%, O 9.05%. Longest-acting enantiomer of methadyl acetate, q.v. Duration of action due to active metabolites. Prepn: A. Pohland et al., J. Am. Chem. Soc. **71**, 460 (1949). Synthesis of metabolites: F. I. Carroll et al., J. Org. Chem. **41**, 3521 (1976). Metabolism in rats: G. L. Henderson et al., Drug Metab. Dispos. **5**, 321 (1977); M. Man et al., ibid. **8**, 55 (1980); in man: B. S. Finkle et al., J. Anal. Toxicol. **6**, 100 (1982). Determn of LAAM and metabolites by HPLC: C.-H. Kiang et al., J. Chromatog. **222**, 81 (1981); by GLC: K. Verebey et al., ibid. **343**, 339 (1985). Analgesic activity in mice: N. B. Eddy et al., J. Org. Chem. **17**, 321 (1952); in man: A. S. Keats, H. K. Beecher, J. Pharmacol. Exp. Ther. **105**, 210 (1952). Pharmacology in monkeys: S. J. Mule, A. L. Misra, Ann. N.Y. Acad. Sci. **311**, 199 (1978). Comparison with methadone in the treatment of heroin addiction: T. J. Crowley et al., Psychopharmacology **86**, 458 (1985). Review of use in treatment of narcotic addiction: J. D. Blaine et al., Ann. N.Y. Acad. Sci. **362**, 101-115 (1981).

Hydrochloride, $C_{23}H_{32}ClNO_2$. Crystals from ethanol-ether, mp 215°. $[\alpha]_D^{25}$ −60° (c = 0.2). Sol in water. LD_{50} in mice (mg/kg): 110.0 s.c., 172.8 orally (Eddy).

Caution: May be habit forming. This is a controlled substance (opiate) listed in the U.S. Code of Federal Regulations, Title 21 Part 1308.11 (1987).

5347. Levophacetoperane. α-Phenyl-2-piperidinemethanol acetate; threo-1-acetoxy-1-phenyl-1-(2-piperidyl)methane; acetic acid α-phenyl-2-piperidylmethyl ester; 1-phenyl-1-(2′-piperidyl)-1-acetoxymethane; phacetoperane; RP 8228. $C_{14}H_{19}NO_2$; mol wt 233.30. C 72.07%, H 8.21%, N 6.00%, O 13.72%. Prepn: Jacob, Joseph, U.S. pat. **2,928,835** (1960 to Rhône-Poulenc).

Hydrochloride, $C_{14}H_{20}ClNO_2$, Lidepran. Crystals from acetone + ether, mp 229-230°. Levorotatory.

THERAP CAT: Antidepressant, anorexic.

5348. Levopimaric Acid. (1R)-1,2,3,4,4a,4bα,5,9,10,-10aα-Decahydro-1,4aβ-dimethyl-7-(1-methylethyl)-1α-phenanthrenecarboxylic acid; 13-isopropylpodocarpa-8(14),-12-dien-15-oic acid; $\Delta^{6,8(14)}$-abietadienoic acid; l-pimaric acid; β-pimaric acid; l-sapietic acid. $C_{20}H_{30}O_2$; mol wt 302.44. C 79.42%, H 10.00%, O 10.58%. Isolation from American pine oleoresin: Palkin, Harris, J. Am. Chem. Soc. **55**, 3677 (1933); from French galipot, from Pinus maritima: Ruzicka, Bacon, Helv. Chim. Acta **20**, 1542 (1937); from Pinus palustris: Harris, Sanderson, J. Am. Chem. Soc. **70**, 334, 3671 (1948). Structure: Ruzicka, Kaufmann, Helv. Chim. Acta **23**, 1346 (1940): cf. Arbuzov, Chem. Zentr. **1942**, II, 893. Stereochemistry: Schuller, Lawrence, J. Am. Chem. Soc. **83**, 2563 (1961); Burgstahler et al., ibid. **83**, 4660 (1961); Weiss et al., Chem. & Ind. (London) **1962**, 1286; Dauben, Coates, J. Org. Chem. **28**, 1698 (1963). Conformation: Burgstahler et al., Chem. Commun. **1971**, 121; Weiss et al., ibid. **1972**, 17.

Orthorhombic crystals, mp 150°; $[\alpha]_D^{20}$ −280.4° (c = 0.7 in alcohol); $[\alpha]_D^{14}$ −266.6° (c = 0.4 in chloroform). Absorption max 273 nm: Kraft, Ann. **520**, 133 (1935). Practically insol in water, sol in most organic solvents. Forms a crystalline ammonium salt.

Methyl ester, $C_{21}H_{32}O_2$, crystals from methanol, mp 64°. $[\alpha]_D^{20}$ −268° (c = 1 in ether).

Molecular compd with quinone, $C_{26}H_{34}O_4$, yellow prisms from methanol, mp 214°. $[\alpha]_D^{20}$ −148° (c = 0.7 in chloroform).

5349. Levopropoxyphene. α-[2-(Dimethylamino)-1-methylethyl]-α-phenylbenzeneethanol propanoate(ester); α-l-4-dimethylamino-3-methyl-1,2-diphenyl-2-butanol propionate; α-l-4-dimethylamino-1,2-diphenyl-3-methyl-2-butanol propionate; l-propoxyphene. $C_{22}H_{29}NO_2$; mol wt 339.48. C 77.83%, H 8.61%, N 4.13%, O 9.43%. Prepn: Pohland, Sullivan, J. Am. Chem. Soc. **77**, 3400 (1955). Stereoselective synthesis: Pohland et al., J. Org. Chem. **28**, 2483 (1963). Toxicity data: E. I. Goldenthal, Toxicol. Appl. Pharmacol. **18**, 185 (1971). See also Propoxyphene.

Crystals from petr ether, mp 75-76°. $[\alpha]_D^{25}$ −68.2° (c = 0.6 in chloroform).

Hydrochloride, $C_{22}H_{29}NO_2 \cdot HCl$, crystals from methanol + ethyl acetate, mp 163-164°. $[\alpha]_D^{25}$ −60.1° (c = 0.7).

2-Naphthalenesulfonate, $C_{32}H_{37}NO_5S$, *levopropoxyphene napsylate, Novrad, Letusin, Contratuss*. LD_{50} orally in female rats: 1455 ±77 mg/kg (Goldenthal).

THERAP CAT: Antitussive.

5350. Levorphanol. *17-Methylmorphinan-3-ol;* (−)-3-hydroxy-*N*-methylmorphinan; levorphan; lemoran; Ro 1-5431. $C_{17}H_{23}NO$; mol wt 257.38. C 79.33%, H 9.01%, N 5.44%, O 6.22%. Orally active synthetic morphine analog. Prepn of racemate from 2-methyl-1-benzyl-1,2,3,4,5,6,7,8-octahydroisoquinoline: Grewe, *Naturwiss.* **33**, 333 (1946); *Angew. Chem.* **A59**, 198 (1947); Grewe, Mondon, *Ber.* **81**, 279 (1948); **Swiss** pat. **280,674** (1952 to Hoffmann-La Roche), *C.A.* **47**, 7554 (1953). Prepn of isomers: Schnider, Grüssner, *Helv. Chim. Acta* **34**, 2211 (1951); Vogler, U.S. pat. **2,744,112** (1956 to Hoffmann-La Roche). Absolute configuration: Corrodi *et al., Helv. Chim. Acta* **42**, 212 (1959). Analgesic activity and toxicity data: L. O. Randall, G. Lehmann, *J. Pharmacol. Exp. Ther.* **99**, 163 (1950). HPLC determn in plasma: R. Lucek, R. Dixon, *J. Chromatog.* **341**, 239 (1985). Clinical pharmacokinetics: R. Dixon *et al., Res. Commun. Chem. Pathol. Pharmacol.* **41**, 3 (1983).

Crystals, mp 198-199°. $[\alpha]_D^{20}$ −56° (c = 3 in absolute alcohol).

Tartrate dihydrate, $C_{21}H_{29}NO_7 \cdot 2H_2O$, *Ro 1-5431/7, Dromoran, Levo-Dromoran*. Crystals, mp 113-115° (when anhydrous, mp 206-208°). $[\alpha]_D^{20}$ −14° (c = 3 in water). pH of a 0.2% aq soln 3.4 to 4.0. One gram dissolves in 45 ml water, in 110 g alcohol, in 50 g ether.

dl-Form, *racemorphan, methorphinan*. Crystals from anisole and dil alcohol, mp 251-253°.

dl-Form hydrobromide, $C_{17}H_{24}BrNO$, *NU-2206*. Crystals, mp 193-195°. Sol in water; sparingly sol in alcohol. Practically insol in ether. LD_{50} i.v. in mice: 41 mg/kg (Randall, Lehmann).

d-Form, *Ro 1-6794, dextrorphan*. Crystals, mp 198-199°. $[\alpha]_D^{20}$ +56.3° (c = 3 in abs alcohol).

d-Form tartrate monohydrate, $C_{21}H_{29}NO_7 \cdot H_2O$, crystals, mp 183-185°. $[\alpha]_D^{20}$ +34.6° (c = 3 in water). Sol in water.

Caution: May be habit forming. Levorphanol, racemorphan and their salts are controlled substances (opiates) listed in the U.S. Code of Federal Regulations, Title 21 Part 1308.12 (1987).

THERAP CAT: Narcotic analgesic.

5351. Levothyroxine Sodium. *O-(4-Hydroxy-3,5-diiodophenyl)-3,5-diiodo-*L*-tyrosine monosodium salt;* L-*thyroxine sodium salt;* sodium levothyroxine; L-3,3′,5,5′-tetraiodothyronine sodium salt; Eferox; Eltroxin; Euthyrox; Laevoxin; Letter; Levaxin; Levothroid; Levothyrox; Oroxine; Synthroid Sodium; Thyroxevan. $C_{15}H_{10}I_4NNaO_4$; mol wt 798.85. C 22.55%, H 1.26%, I 63.54%, N 1.75%, Na 2.88%, O 8.01%. The sodium salt of the amino acid L-thyroxine, *q.v.*, an active physiological principle obtained from the thyroid gland of domesticated animals used for food by man, or prepd synthetically. Prepn: Chalmers *et al., J. Chem. Soc.* **1949**, 3424. Solubility in phosphate buffer solns: Evert, *J. Phys. Chem.* **64**, 478 (1960). Clinical pharmacology in hypothyroidism: P. W. Ladenson *et al., Am. J. Med.* **73**, 467 (1982). Bioavailability and metabolism in hypothyroidism: L. H. Fish *et al., N. Engl. J. Med.* **316**, 764 (1987). Comprehensive description: A. Post, R. J. Warren in *Analytical Profiles of Drug Substances* vol. 5, K. Florey, Ed. (Academic Press, New York, 1976) pp 225-281.

Pentahydrate, triclinic crystals or cream-colored powder. Odorless and tasteless. Somewhat hygroscopic. d_4^{20} 2.381. $[\alpha]_D^{20}$ −4.4° (c = 3 in 70% ethanol). Soly at 25° in water: about 15 mg/100 ml. (Higher solys have been reported). Sol in mineral acids and in solns of alkali hydroxides and carbonates. More sol in alcohol. Very slightly sol in chloroform, ether. pH of a satd water soln: 8.35 to 9.35.

THERAP CAT: Treatment of hypothyroidism.

THERAP CAT (VET): In thyroid deficiency (dogs). Has been used in obesity, renal insufficiency, chronic skin conditions. Has also been used in lowered fertility in bulls; and to stimulate lactation.

5352. Levulinic Acid. *4-Oxopentanoic acid;* laevulinic acid; β-acetylpropionic acid. $C_5H_8O_3$; mol wt 116.11. C 51.72%, H 6.94%, O 41.34%. $CH_3COCH_2CH_2COOH$. Laboratory procedure from starch or cane sugar by boiling with HCl: McKenzie, *Org. Syn.* **coll. vol. I**, 335 (1941). Produced commercially from low grade cellulose. By-product of furfural manuf. Extensive review: Leonard, *Ind. Eng. Chem.* **48**, 1331 (1956).

Plates or leaflets (commercial product is yellow), mp 33-35°. bp 245-246°. d 1.1447. n_D^{16} 1.442. Freely sol in water, alcohol, ether; essentially insol in aliphatic hydrocarbons. *Protect from light.*

Phenylhydrazone, $C_{11}H_{14}N_2O_2$, *Antithermin*. Leaflets, mp 108°. Slightly sol in cold water, more sol in hot water; freely sol in alcohol, chloroform, ether, dil acids.

USE: In organic syntheses; in the manuf of nylon, synthetic rubbers, plastics, and medicinals.

5353. LH. *Luteinizing hormone;* ICSH; interstitial cell stimulating hormone. Mol wt about 30,000. A glycoprotein gonadotrop(h)ic hormone found in the anterior lobe of the pituitary gland. First isolated from sheep pituitaries: Li *et al., Endocrinology* **27**, 803 (1940); *Science* **92**, 355 (1940); J. Am. Chem. Soc. **64**, 367 (1942); from pig pituitaries: Shedlovsky *et al., Science* **92**, 178 (1940); Chow *et al., Endocrinology* **30**, 650 (1942). Stimulates the synthesis of progesterone in the ovaries. Together with FSH stimulates the release of estrogen from Graafian follicles. Also induces the process of ovulation in which the mature ovum is extruded from the follicle and, following this, the cells which hitherto surrounded it are converted under the influence of LH into lutein cells (corpus luteum). In the male, stimulates the interstitial cells of testes to secrete testosterone. The glycoprotein structure contains two dissimilar subunits designated as α and β, *cf.* HCG, FSH, TSH. Each of the dissociated subunits shows little of the biological activity associated with the native hormone. Complete amino acid sequence of ovine LH subunits: Papkoff *et al., J. Am. Chem. Soc.* **93**, 1531 (1971); Liu *et al., J. Biol. Chem.* **247**, 4351, 4365 (1972); of human LH subunits: H. T. Keutmann *et al., Endocrin. Res. Commun.* **5**, 57 (1978); eidem, *Biochem. Biophys. Res. Commun.* **90**, 842 (1979). Reviews: Ward *et al., Recent Progr. Horm. Res.* **29**, 533 (1973); Papkoff *et al.*, ibid. 563; J. G. Pierce, T. F. Parsons, *Ann. Rev. Biochem.* **50**, 465-495 (1981).

White powder. Isoelectric point 4.6 (sheep), 7.45 (pig). Soluble in water. Chymotrypsin, trypsin and pepsin destroy the gonadotrop(h)ic action of LH. Picrolonic, flavianic, picric, and trichloroacetic acids precipitate LH with retention of its physiological activity.

THERAP CAT: Gonadotropic hormone.

5354. LH-RH. *Luteinizing hormone-releasing factor;* LH-RF; luteinizing hormone-releasing hormone; LRF; LRH; gonadorelin; gonadotropin-releasing factor; gonadoliberin; luliberin; LH-RH/FSH-RH; Fertagyl; Fertiral; Kryptocur; Relefact LH-RH. $C_{55}H_{75}N_{17}O_{13}$; mol wt 1182.33. C 55.87%, H 6.39%, N 20.14%, O 17.59%. Neurohumoral hormone produced in the hypothalamus which stimulates the secretion of the pituitary hormones, LH (luteinizing hormone) and FSH (follicle-stimulating hormone) *q.q.v.*, which in turn produce changes resulting in the induction of ovulation. Isoln from porcine hypothalamic extracts: A. V. Schally *et al., Biochem. Biophys. Res. Commun.* **43**, 393 (1971). Structure: Matsuo *et al., ibid.* 1334; Baba *et al., ibid.* **44**, 459 (1971); Burgus *et al., C.R. Acad. Sci., Ser. D.* **273**, 1611 (1971). Solid phase synthesis: Monahan *et al., ibid.* 508; Monahan, Rivier, *Biochem. Biophys. Res. Commun.* **48**, 1100 (1972); Coy *et al., Methods Enzymol.* **37**, 416 (1975); D. H. Rich, S. K. Gurwara, *Tetrahedron Letters* **1975**, 301. Industrial prepn: Geiger *et al.*, Ger. pat. **2,213,737** (1973 to Hoechst). Confirmation of biological activity: A. V. Schally *et al., Science* **173**, 1036 (1971). Pharmacokinetics: T. W. Redding, *J. Clin. Endocrinol. Metab.* **37**, 626 (1973). Reviews of physiology and implications in fertility control: A. V. Schally *et al., Fert. Steril.* **22**, 703 (1971); R. Guillemin, *Contraception* **6**, 1 (1972). Review of LH-RH and other hypothalamic hormones: A. V. Schally *et al., Science* **179**, 341 (1973); A. V. Schally, *ibid.* **202**, 18-28 (1973); of LH-RH and somatostatin, *q.v.:* S. M. McCann, *Ann. Rev. Pharmacol. Toxicol.* **22**, 491-515 (1982). General reviews: J. Sandow, *Clin. Endocrinol.* **18**, 571-592 (1983); S. M. McCann, *J. Endocrinol. Invest.* **6**, 243-251 (1983).

5-oxoPro-His-Trp-Ser-Tyr-Gly-Leu-Arg-Pro-GlyNH$_2$

$[\alpha]_D^{25}$ —50° (1% acetic acid). Chymotrypsin, papain, subtilisin, thermolysin destroy the pituitary hormone releasing action of LH-RH.

Acetate (salt) hydrate, C$_{55}$H$_{75}$N$_{17}$O$_{13}$·xC$_2$H$_4$O$_2$·yH$_2$O, *gonadorelin acetate, Cystorelin, Hypocrine, Lutrelef, Ovarelin.* As the diacetate tetrahydrate or a mixture of monoacetate and diacetate hydrates.

Hydrochloride, *AY-24031, H.R.F., Factrel.*

THERAP CAT: Gonad-stimulating principle.

THERAP CAT (VET): Gonad-stimulating principle.

5355. Liatris. Deer's tongue; vanilla plant. Leaves of *Trilisa odoratissima* (Walt.) Cass. (*Liatris odoratissima* (Walt.) Willd.), *Compositae.* Habit. U.S., Virginia to Florida and Louisiana. Constit. Volatile oil, coumarin.

USE: In perfumery and for perfuming smoking, chewing and snuff tobacco.

5356. Licheniformins. Antibiotic substances produced by *Bacillus licheniformis.* Isoln: Callow *et al., Brit. J. Exp. Pathol.* **28**, 418 (1947). Separation of licheniformins A, B, and C: Callow, Work, *Biochem. J.* **51**, 558 (1952). Mol wt as measured by sedimentation: licheniformin A, 4400; licheniformin B, 3800; and licheniformin C, 4800: Ogston, *ibid.* **51**, 569 (1952).

All three constituents in the form of hydrochlorides are white, amorphous, slightly hygroscopic powders, melting with decompn at indefinite temperatures.

5357. Lichenin. Moss starch. C$_6$H$_{10}$O$_5$; mol wt 162.14. C 44.44%, H 6.22%, O 49.34%. Polyglucan from *Cetraria islandica* (L.) Ach., *Parmeliaceae* (Iceland Moss). Isoln: Peat *et al., J. Chem. Soc.* **1957**, 3916. Linear polysaccharide structure composed of regular sequences of two (1 → 4)- and one (1 → 3)-β-D-glucopyranosyl residues: Perlin, Suzuki, *Can. J. Chem.* **40**, 50 (1962); Fleming, Manners, *Biochem. J.* **100**, 4, 24 (1966). Exhibits antineoplastic activity: Shibata, *Japan.* pat. **17,147**('71), *C.A.* **75**, 67474z (1971); Tadahiro *et al., Chem. Pharm. Bull.* **20**, 2445 (1972). NMR spectrum: D. Gagnaire, M. Vincedon, *Bull. Soc. Chim. France* **1977**, 479.

White powder. Sol in boiling water, in HCl. $[\alpha]_D$ +18.4°.

5358. Lidamidine. *N-(2,6-Dimethylphenyl)-N'-[imino-(methylamino)methyl]urea;* 1-(2,6-dimethylphenyl)-3-methylamidinourea. C$_{11}$H$_{16}$N$_4$O; mol wt 220.27. C 59.98%, H 7.32%, N 25.44%, O 7.26%. Amidinourea with antisecretory, antimotility properties. Prepn as hydrochloride: Belg. pat. **844,832** (1977 to Rorer), *C.A.* **88**, 22432m (1977). *See also:* J. Diamond, G. H. Douglas, U.S. pat. **4,147,804** (1979 to Rorer). Prepn, structure activity relationship: G. H. Douglas *et al., Arzneimittel-Forsch.* **28**, 1435 (1978). Physical-chemical properties: J. J. Zalipsky *et al., ibid.* 1441. Effect on α$_2$-adrenergic receptors and electrolyte absorption: T. Durbin *et al., Gastroenterology* **82**, 1352 (1982). Series of

articles on pharmacology, metabolism, pharmacokinetics: *Arzneimittel-Forsch..* **28**, 1448-1480 (1978). Toxicity: B. J. Chou *et al., ibid.* 1471. Clinical comparison with loperamide, *q.v.*, in acute diarrhea: G. Gasbarrini *et al., ibid.* **36**, 1843 (1986). Brief review: G. Friedman, *Am. J. Gastroenterol.* **80**, 143 (1985).

Hydrochloride, C$_{11}$H$_{17}$ClN$_4$O, *WHR 1142A, Lidarral, Smodin.* White powder, mp 194-197°. uv max (H$_2$O): 262, 271 nm (ε 626, 524). Soly at 25° (mg/ml): water 153.55, methanol 297.94, ethanol 88.55, chloroform 4.62, hexane 0.01. LD$_{50}$ in male mice, male, female rats (mg/kg): 260, 267, 160 orally; in mice (mg/kg): 56 i.v. (Chou).

THERAP CAT: Antiperistaltic; antidiarrheal.

5359. Lidocaine. 2-(Diethylamino)-N-(2,6-dimethylphenyl)acetamide; 2-diethylamino-2',6'-acetoxylidide; ω-diethylamino-2,6-dimethylacetanilide; lignocaine; Xylocaine; Xylotox; Leostesin; Rucaina; Isicaine; Cuivasal; Duncaine; Xylestesin; Anestacon; Gravocain; Lidothesin; Xylocitin. C$_{14}$H$_{22}$N$_2$O; mol wt 234.33. C 71.75%, H 9.46%, N 11.96%, O 6.83%. Long-acting, membrane stabilizing agent against ventricular arrhythmia. Originally developed as a local anesthetic. Prepn: N. M. Löfgren, B. J. Lundqvist, U.S. pat. **2,441,498** (1948 to Astra); A. D. H. Self, A. P. T. Easson, Brit. pat. **706,409** (1954 to May & Baker); I. P. S. Hardie, E. S. Stern, Brit. pat. **758,224** (1956 to J. F. Macfarlane & Co.); Zhuravlev, Nikolaev, *Zh. Obshch. Khim.* **30**, 1155 (1960). Toxicity studies: E. R. Smith, B. R. Duce, *J. Pharmacol. Exp. Ther.* **179**, 580 (1971); G. H. Kronberg *et al., J. Med. Chem.* **16**, 739 (1973). Review of pharmacokinetics: N. L. Benowitz, W. Meister, *Clin. Pharmacokinet.* **3**, 177 (1978). Review of action as local anesthetic: Löfgren, *Studies on Local Anesthetics: Xylocaine, A New Synthetic Drug* (Hoeggstroms, Stockholm, 1948); Cooper, *Pharm. J.* **171**, 68 (1953). Reviews of anti-arrhythmic agents: J. L. Anderson *et al., Drugs* **15**, 271 (1978); L. H. Opie, *Lancet* **1**, 861 (1980); E. Carmeliet, *Ann. N.Y. Acad. Sci.* **427**, 1 (1984). Comprehensive description: K. Groningsson *et al.*, in *Analytical Profiles of Drug Substances* vol. **14**, K. Florey, Ed. (Academic Press, New York, 1985) pp 207-243; M. F. Powell, *ibid.* vol. **15**, (1986) pp 761-779.

Needles from benzene or alcohol, mp 68-69°. bp$_4$ 180-182°; bp$_2$ 159-160°. Insol in water. Sol in alcohol, ether, benzene, chloroform, oils.

Hydrochloride monohydrate, C$_{14}$H$_{23}$ClN$_2$O.H$_2$O, *Lignavet, Odontalg, Xylocard, Xyloneural.* Crystals, mp 77-78°; anhydrous, mp 127-129°. Very sol in water, alcohol; sol in chloroform. Insol in ether. pH of 0.5% aq soln: 4.0-5.5. LD$_{50}$ in mice (mg/kg): 292 orally (Smith, Duce); 105 i.p.; 19.5 i.v. (Kronberg).

THERAP CAT: Local anesthetic. Cardiac depressant (antiarrhythmic).

THERAP CAT (VET): Local anesthetic.

5360. Lidoflazine. 4-[4,4-Bis(4-fluorophenyl)butyl]-N-(2,6-dimethylphenyl)-1-piperazineacetamide; 4-[4,4-bis(p-fluorophenyl)butyl]-1-piperazineaceto-2',6'-xylide; 1-[4,4-di(4-fluorophenyl)butyl]-4-[(2,6-dimethylanilinocarbonyl)methyl]piperazine; McN-JR-7094; R 7904; Angex; Clinium; Klinium; Ordiflazine; Corflazine. C$_{30}$H$_{35}$F$_2$N$_3$O; mol wt 491.63. C 73.29%, H 7.18%, F 7.73%, N 8.55%, O 3.25%. Calcium blocking agent. Prepn: **Neth.** pat. **Appl. 6,507,312** corresp to H. K. F. Hermans, W. K. Schaper,

U.S. pat. **3,267,104** (1965, 1966, both to Janssen). Crystal structure: G. Germain *et al., Acta Crystallogr.* **33B,** 1971 (1977). Tissue specificity of calcium-blocking properties: J. M. Van Neuten, D. Wellens, *Arch. Int. Pharmacodyn. Ther.* **242,** 329 (1979). Cardioprotective effects: W. Daenen, W. Flameng, *Angiology* **32,** 543 (1981). Effects in angina: F. L. Gobel *et al., Circulation* **65,** 1 Pt. 2, 127 (1982); W. Shapiro *et al., ibid.* 143.

Crystals, mp 159-161°. Almost insol in water (<0.01%), very sol in chloroform (>50%), but much less sol in other common organic solvents: W. K. Schaper *et al., J. Pharmacol. Exp. Ther.* **152,** 265 (1966).

THERAP CAT: Vasodilator (coronary).

5361. Light Green SF Yellowish. *N-Ethyl-N-[4-[[4-[ethyl[(3-sulfophenyl)methyl]amino]phenyl](4-sulfophenyl)-methylene]-2,5-cyclohexadien-1-ylidene]-3-sulfobenzene-methanaminium hydroxide, inner salt, disodium salt; C.I. Acid Green 5;* ethyl[4-[p-[ethyl(m-sulfobenzyl)amino]-α-(p-sulfophenyl)benzylidene]-2,5-cyclohexadien-1-ylidene](m-sulfobenzyl)ammonium hydroxide inner salt disodium salt; C.I. 42095; FD & C Green No. 2; Lissamine Green SF. $C_{37}H_{34}N_2Na_2O_9S_3$; mol wt 792.85. C 56.05%, H 4.32%, N 3.53%, Na 5.80%, O 18.16%, S 12.13%. Prepn and refs: *Colour Index* vol. **4** (3rd ed., 1971) p 4385. Biological use: H. J. Conn's *Biological Stains,* R. D. Lillie, Ed. (Williams & Wilkins, Baltimore, 9th ed., 1977) pp 257-258, 582. Toxicity data: F. C. Lu, A. Lavalle, *Can. Pharm. J.* **97,** 30 (1964). Chronic toxicity study: W. H. Hansen *et al., Food Cosmet. Toxicol.* **4,** 389 (1966).

A reddish-brown powder. Sol in water to a green soln which turns yellowish-brown with HCl and then gradually fades. Addition of NaOH almost completely decolorizes the soln yielding a dull violet ppt. LD_{50} orally in rats : >2 g/kg (Lu, Lavalle).

USE: As dye, biological stain. *Caution:* Delisted by FDA in 1966 for use in foods, drugs and cosmetics.

5362. Lignans. Plant products of low molecular weight formed primarily by the oxidative coupling of p-hydroxyphenylpropene units in which the two units may be linked by an oxygen bridge. The monomeric precursor units are cinnamic acid, cinnamyl alcohol, propenylbenzene and allylbenzene. The term lignan or *Haworth lignan* is applied to compounds derived by coupling acid and/or alcohol while the compounds derived by coupling propenyl and/or allyl derivatives are called *neolignans:* O. R. Gottlieb, *Fortschr. Chem. Org. Naturst.* **35,** 1-72 (1978). Lignans occur widely and have been obtained from roots, heartwood, foliage, fruit and resinous exudates of plants. Lignans are optically active compounds. They represent the dimer stage intermediate between monomeric propylphenol units and lignin. Naturally occurring trimers and tetramers have not been reported. Occurrence of lignans, enterolactone, *q.v.,* and enterodiol, in man and animal species: S. R. Stitch *et al., Nature*

287, 238 (1980); K. D. R. Setchell, *ibid.* 740. Synthesis of first lignans found in man and animals: G. Cooley *et al., Tetrahedron Letters* **22,** 349 (1981).

5363. Lignin. The most abundant natural aromatic organic polymer found in all vascular plants. Lignin together with cellulose, *q.v.,* and hemicellulose, *q.v.,* are the major cell wall components of the fibers of all wood and grass species. Lignin is composed of coniferyl, p-coumaryl and sinapyl alcohols in varying ratios in different plant species. Monographs: F. E. Brauns, D. A. Brauns, *The Chemistry of Lignin,* Supplement Volume covering the literature 1949-1958 (Academic Press, New York, 1960) 804 pp; I. A. Pearl, *The Chemistry of Lignin* (Marcel Dekker, New York, 1967) 360 pp. Structural aspects and applications: H. Veeramani, G. A. Wani, *Chem. Ind. Dev.* **11,** 13-25 (1977). Chemistry and structure: C. A. Reddy, L. Forney, *Dev. Ind. Microbiol.* **19,** 27-34 (1978). *Reviews:* Nord, Shubert, *Sci. Am.* **199,** no. 4, 104-113 (Oct. 1958); I. A. Pearl, "Lignin as a Raw Material for the Production of Pure Chemicals," *J. Chem. Ed.* **35,** 502 (1958); D. W. Goheen, C. H. Hoyt in Kirk-Othmer *Encyclopedia of Chemical Technology* vol. **14,** (Wiley-Interscience, New York, 3rd ed., 1981) pp 294-312.

USE: Source of vanillin, syringic aldehyde, dimethyl sulfoxide. Extender for phenolic plastics, to strengthen rubber (esp for shoe soles), as oil mud additive, to stabilize asphalt emulsions, to precipitate proteins.

5364. Lignoceric Acid. *Tetracosanoic acid.* $C_{24}H_{48}O_2$; mol wt 368.62. C 78.19%, H 13.13%, O 8.68%. $C_{23}H_{47}$-COOH. Obtained from beechwood tar or by the distillation of rotten oak wood: Sullivan, *Ind. Eng. Chem.* **8,** 1027 (1916). Most natural fats contain small amounts (0.2-1%). The seed fat of the Indian tree *Adenanthera pavonina* is said to contain 25%. Synthesis: Fieser, Szmuszkovicz, *J. Am. Chem. Soc.* **70,** 3352 (1948).

Crystals, mp 84.15°. n_D^{100} 1.4287. Soly in 91.53% ethanol: 0.182 g/100 ml. Neutralization value 152.2.

Methyl ester, $C_{25}H_{50}O_2$, platelets, mp 58-59.8°.

5365. Lignum Vitae. The wood of *Guaiacum officinale* L., or *G. sanctum* L., Zygophyllaceae. *See also* Guaiac.

5366. Ligroin. V.M.&P. naphtha; varnish makers' and painters' naphtha; refined solvent naphtha; solvent naphtha; Benzoline; Canadol. Term that has been applied to petroleum fractions of the same nature as described for petroleum benzin, *q.v.,* but of higher density, higher boiling range and higher flash pt. Defined by ASTM prior to 1950 as synonymous with petroleum benzin and petroleum ether: ASTM standard specification D **288-49,** 865-867 (1949).

Refined solvent, mobile, flammable liquid. $d_{15.6}^{15.6}$ 0.850 to 0.870. Distillation range at 760 mm: percentage recovered at 130° = not more than 5; percentage recovered at 145° = not less than 90. End point (dry point) = not above 155°. Technical benzin (high boiling petr ether) usually has a $d_{15.6}^{15.6}$ 0.730-0.750 and bp_{760} 90-120°.

5367. Limaprost. *(2E,11α,13E,15S,17S)-11,15-Dihydroxy-17,20-dimethyl-9-oxoprosta-2,13-dien-1-oic acid; 17S,-20-dimethyl-trans-2,3-didehydro-PGE$_1$; 9-oxo-11α,15α-dihydroxy-17S,20-dimethylprosta-trans-2,trans-13-dienoic acid; 17S-methyl-ω-homo-trans-Δ²-PGE$_1$;* ONO-1206; OP-1206; Opalmon; Prorenal. $C_{22}H_{36}O_5$; mol wt 380.52. C 69.44%, H 9.54%, O 21.02%. Derivative of prostaglandin E$_1$, *q.v.* Prepn: M. Hayashi *et al.,* **Ger.** pat. **3,002,677** (1980 to Ono); *eidem,* U.S. pat. **4,294,849** (1981 to Warner-Lambert). Cardiovascular pharmacology in animals: T. Tsuboi *et al., Arch. Int. Pharmacodyn.* **247,** 89 (1980). Effect on smooth muscle *in vitro:* P. G. Adaikan, S. M. M. Karim, *Prostaglandins Med.* **6,** 449 (1981). Coronary vasodilating effects in primates: S. R. Kottegoda *et al., Prostaglandins Leukotrienes Med.* **8,** 343 (1982). Clinical hemodynamics in chronic lung disease: T. Ishizaki *et al., Chest* **85,** 382 (1984).

White crystals, mp 97-100°.
THERAP CAT: Anti-anginal.

5368. Limestone. Natural calcium carbonate; agricultural limestone; Agstone; lithographic stone; Solnhofen stone. A term originally applied only to minerals consisting largely of $CaCO_3$, such as Portland stone, dolomite, marble, and chalk, now used indiscriminately to designate technical and agricultural grades of calcium carbonate. *Review:* R. S. Boynton in Kirk-Othmer *Encyclopedia of Chemical Technology* vol. **14** (Wiley-Interscience, New York, 3rd ed., 1981) pp 343-382.

5369. Lime Sulfurated Solution. Vleminckx's soln; Vleminckx's lotion; calcium oxysulfide solution. Made by boiling 16.5 parts freshly slaked lime with 25 parts sublimed sulfur and water to make 100 ml; the resulting soln contg calcium polysulfide and thiosulfate.
Brown, clear liquid.
THERAP CAT: Topical antiseptic; scabicide.
THERAP CAT (VET): Insecticide, scabicide, has been used in boils and fistulas.

5370. Limettin. *5,7-Dimethoxy-2H-1-benzopyran-2-one; 5,7-dimethoxycoumarin;* citropten. $C_{11}H_{10}O_4$; mol wt 206.19. C 64.07%, H 4.89%, O 31.04%. From rind of fruit of *Citrus limetta* Lunan (*C. limetta* Auth.), Rutaceae (lime); Tilden, Beck, *J. Chem. Soc.* **57**, 323 (1890); from W. Indian lime oil: Caldwell, Jones, *ibid.* **1945**, 570; from citrus oils: Stanley, Vannier, U.S. pat. **2,889,337** (1959 to U.S.D.A.). Synthesis: Schmidt, *Arch. Pharm.* **242**, 288 (1904); Heyes, Robertson, *J. Chem. Soc.* **1936**, 1831.

Needles from methanol, mp 147-148°. uv max (alcohol): 222, 247, 250.5, 324 nm (log ε 4.03, 3.84, 3.84, 4.18). Almost insol in boiling water, ether, petr ether; freely sol in alcohol, chloroform, acetone.

5371. Limonene. *1-Methyl-4-(1-methylethenyl)cyclohexene; p-mentha-1,8-diene;* cinene; cajeputene; kautschin. $C_{10}H_{16}$; mol wt 136.23. C 88.16%, H 11.84%. Occurs in various ethereal oils, particularly in oils of lemon, orange, caraway, dill and bergamot. Isoln of *d*-limonene from mandarin peel oil (*Citrus reticulata* Blanco, Rutaceae): Kugler, Kováts, *Helv. Chim. Acta* **46**, 1480 (1963). *Review:* J. L. Simonsen, *The Terpenes* vol. I (University Press, Cambridge, 2nd ed., 1947) pp 143-165.

dl-Form, **inactive limonene, dipentene.** Liquid. Pleasant lemon-like odor. bp_{763} 175.5-176.5°. $d^{20.85}$ 0.8402. n_D 1.4744. Practically insol in water; miscible with alcohol. With dry HCl or HBr it forms monohalides, and with aq HCl or HBr, the dihalide.
d-Form, liquid. bp_{763} 175.5-176°. d_4^{21} 0.8402. n_D^{21} 1.4743. $[\alpha]_D^{19.5}$ +123.8°.
l-Form, liquid. bp_{763} 175.5-176.5°. $d_4^{20.5}$ 0.8407. n_D^{21} 1.474. $[\alpha]_D^{19.5}$ −101.3°.

USE: Solvent, manuf resins; wetting and dispersing agent. *Caution:* Skin irritant, sensitizer.

5372. Limonin. *Limonoic acid di-δ-lactone; limonoic acid 3,19:16,17-dilactone; 8-(3-furyl)decahydro-2,2,4a,8a-tetramethyl-11H,13H-oxireno[d]pyrano[4',3':3,3a]isobenzofuro[5,4-f][2]benzopyran-4,6,13(2H,5aH)-trione.* $C_{26}H_{30}O_8$; mol wt 470.50. C 66.37%, H 6.43%, O 27.21%. Bitter principle of lemon and other *Rutaceae.* Isoln: Bernays, *Ann.* **40**, 317 (1841). Structure and stereochemistry: Melera *et al., Helv. Chim. Acta* **40**, 1420 (1957); Arigoni *et al., Experientia* **16**, 41 (1960); Arnott *et al., ibid.* 49; Barton *et al., J. Chem. Soc.* **1961**, 255; Arnott *et al., ibid.* 4183. Synthetic studies: Schlatter *et al., Helv. Chim. Acta* **57**, 1044 (1975); Lüthy *et al., ibid.* 1060.

Bitter crystals from methylene chloride + isopropanol or acetic acid, mp 298°. $[\alpha]_D$ −128° (c = 1.21 in acetone). uv max: 207, 285 nm (ε 7000, 38). Slightly sol in water, ether; sol in alcohol, glacial acetic acid.

5373. Linalool. *3,7-Dimethyl-1,6-octadien-3-ol;* 2,6-dimethyl-2,7-octadien-6-ol; linalol. $C_{10}H_{18}O$; mol wt 154.24. C 77.87%, H 11.76%, O 10.37%. $(CH_3)_2C=CHCH_2CH_2-C(CH_3)(OH)CH=CH_2$. Chief constituent of linaloe oil; also occurs in oils of Ceylon cinnamon, sassafras, orange flower, bergamot, *Artemisia balchanorum,* ylang ylang, etc.: Tiemann, *Ber.* **31**, 808 (1898); Walbaum, Stephan, *ibid.* **33**, 2305 (1900); Hesse, Zeitschel, *J. Prakt. Chem.* **66**, 493 (1902); Rafanova *et al.,* U.S.S.R. pat. **103,725** (1956); *C.A.* **51**, 3656c (1957); Naves, *Helv. Chim. Acta* **42**, 1692 (1959). Presence in essential oils: *idem, Compt. Rend.* **251**, 900 (1960). Absolute configuration: Prelog, Watanabe, *Ann.* **603**, 1 (1957). Synthesis of *dl*-linalool: Ruzicka, Fornasir, *Helv. Chim. Acta* **2**, 182 (1919); Surmatis, U.S. pat. **2,848,502** (1958 to Hoffmann-La Roche); Nair, Pandit, *Tetrahedron Letters* **1966**, 5097. *Review:* J. L. Simonsen, *The Terpenes* vol. I (University Press, Cambridge, 2nd ed., 1947), pp 57-68.

l-Form, **licareol.** Colorless liq. bp_{760} 198°; bp_{25} 98-98.3°; bp_{14} 86-87°. d^{20} 0.8622. n_D^{22} 1.4604. $[\alpha]_D^{20}$ −20.1°. Practically insol in water; miscible with alcohol, ether.
d-Form, **coriandrol.** bp_{760} 198-200°; bp_{26} 114-114.5°; $bp_{15.5}$ 93-94°; bp_{12} 86°. d_4^{20} 0.8733. n_D^{20} 1.4673. $[\alpha]_D^{20}$ +19.3°. Soluble in 10 vol 50% alc, 4 vol 60% alc.
dl-Form, bp_{720} 194-197°; bp_{14} 89-91°. d^{15} 0.865.
USE: In perfumery instead of bergamot or French lavender oil since it has an odor similar to these oils.

5374. Linalyl Acetate. *3,7-Dimethyl-1,6-octadien-3-yl acetate;* bergamol. $C_{12}H_{20}O_2$; mol wt 196.28. C 73.43%, H 10.27%, O 16.30%. $CH_3COOC_{10}H_{17}$. Most valuable constituent of bergamot and lavender oils, also found in many other volatile oils.
Liquid; bergamot odor. d_4^{20} 0.895. bp 220°. n_D^{20} 1.4460. Insol in water; miscible with alcohol, ether.
USE: In perfumery.

5375. Linamarin. *2-(β-D-Glucopyranosyloxy)-2-methylpropanenitrile;* phaseolunatin. $C_{10}H_{17}NO_6$; mol wt 247.24. C 48.58%, H 6.93%, N 5.67%, O 38.83%. From the seed skins or embryos of flax: Jorissen, Hairs, *Bull. Acad. Roy. Sci. Belg.* [3] **21**, 529 (1891); André *et al., Compt. Rend.* **231**, 590 (1950); Lüdtke, *Biochem. Z.* **323**, 428 (1953). Synthesis: Fischer, Anger, *Ber.* **52**, 854 (1919). Biosynthesis in white clover: Butler, Butler, *Nature* **187**, 780 (1960).

Bitter needles, mp 142-143°. $[\alpha]_D^{18}$ $-29°$. Freely sol in water, cold alcohol, hot acetone; slightly in hot ethyl acetate, ether, benzene, chloroform; practically insol in petr ether. Evolves HCN with linseed meal but not with emulsin.

Tetraacetate, $C_{18}H_{25}NO_{10}$, needles from alcohol, mp 140-141°. $[\alpha]_D^{14}$ $-10.8°$ (acetone). Sol in acetone, ethyl acetate, chloroform, glacial acetic acid, benzene, warm methanol and ethanol; practically insol in petr ether.

5376. Linarin. 7-[[6-O-(6-Deoxy-α-L-mannopyranosyl)-β-D-glucopyranosyl]oxy]-5-hydroxy-2-(4-methoxyphenyl)-4H-benzopyran-4-one; acacetin-β-rutinoside; linarigenin-glucoside; 5,7-dihydroxy-4'-methoxyflavone-D-glucosido-L-rhamnoside; buddleoflavonoloside. $C_{28}H_{32}O_{14}$; mol wt 592.54. C 56.75%, H 5.44%, O 37.80%. From the flowers of *Linaria vulgaris* Mill., *Scrophulariaceae*: Merz, Wu, *Arch. Pharm.* **274,** 126 (1936); from *Cirsium oleraceum* Scop., *Compositae*: Wagner et al., ibid. **293,** 1053 (1960). Structure: Baker et al., *J. Chem. Soc.* **1951,** 691. Synthesis: Zemplén, Bognàr, *Ber.* **74,** 1818 (1941).

Monohydrate, needles from methanol, mp 268-270°. $[\alpha]_D^{16}$ $-100°$ (0.07 g in 10 ml glacial acetic acid); $[\alpha]_D^{24}$ $-87°$ (0.05 g in pyridine). Practically insol in water and the usual organic solvents. Sol in nitrobenzene, phenol, aniline, pyridine, concd acids and alkalies. The water of crystn cannot be removed at 100° *in vacuo* over P_2O_5 (Merz); may be removed at 138° in high vacuum (Zemplén). Hydrolysis gives 5,7-dihydroxy-4'-methoxyflavone, D-glucose, and L-rhamnose.

5377. Linatine. 1-[(4-Amino-4-carboxy-1-oxobutyl)-amino]-D-proline; N-(D-2-carboxy-1-pyrrolidinyl)-L-glut-amine; 1-[N-(γ-L-glutamyl)amino]-D-proline. $C_{10}H_{17}N_3O_5$; mol wt 259.26. C 46.32%, H 6.61%, N 16.21%, O 30.86%. A vitamin B_6 antagonist. The first reported naturally occurring hydrazino acid having antibacterial properties. Isoln from flaxseed (*Linum usitatissimum*), characterization, and synthesis: Klosterman et al., *Biochemistry* **6,** 170 (1967); Lamoureux, *Diss. Abstr. B* **28,** 4908 (1968). Bacterial inhibition: Parsons et al., *Antimicrob. Ag. Chemother.* **1967,** 415.

Amorphous powder. $[\alpha]_D^{25}$ $+46.4°$ (c = 2.8 in water). Attempts to obtain well-defined crystals have failed; melts with dec over a wide temperature range. Very sol in water. Practically insol in anhydr organic solvents.

L-Isomer, amorphous solid. $[\alpha]_D^{24}$ $-34.6°$ (c = 2 in water).

5378. Lincomycin. Methyl 6,8-dideoxy-6-[[(1-methyl-4-propyl-2-pyrrolidinyl)carbonyl]amino]-1-thio-D-erythro-α-D-galacto-octopyranoside; lincolnensin; U-10149; NSC-70731; Lincolcina. $C_{18}H_{34}N_2O_6S$; mol wt 406.56. C 53.18%, H 8.43%, N 6.89%, O 23.61%, S 7.89%. Antibiotic produced by *Streptomyces lincolnensis* var. *lincolnensis*. Isoln: Mason et al., *Antimicrob. Ag. Chemother.* **1962,** 555; Herr, Bergy,

ibid. 560; Bergy, Herr and Bergy et al., U.S. pats. **3,086,912** and **3,155,580** (1963, 1964, both to Upjohn). Structure: Hoeksema et al., *J. Am. Chem. Soc.* **86,** 4223 (1964). Prepn of derivs: Argoudelis et al., U.S. pat. **3,380,992** (1968 to Upjohn). Synthesis: Magerlein, *Tetrahedron Letters* **1970,** 33. Synthesis of carbohydrate moiety: I. Hoppe, U. Scholl-kopf, *Ann.* **1980,** 1474. Prepn of hydrochloride hemihydrate: **Neth. pat. Appl. 6,409,689** (1965 to Upjohn), *C.A.* **63,** 5458f (1965). Clinical study: F. Puleo et al., *Gazz. Med. Ital.* **138,** 401 (1979). Mechanism of action: F. N. Chang in *Antibiotics* vol. **5**(pt. 1), F. E. Hahn, Ed. (Springer-Verlag, New York, 1979) pp 127-134. Toxicity data: Gray et al., *Toxicol. Appl. Pharmacol.* **6,** 476 (1964).

Free base, pKa' 7.6. More stable in salt form. Sol in methanol, lower alcohols, ethyl acetate, acetone, chloroform. Slightly sol in water. (Bergy)

Hydrochloride hemihydrate, $C_{18}H_{35}ClN_2O_6S.\frac{1}{2}H_2O$, *Frademicina, Lincocin, Mycivin, Waynecomycin.* Formerly obtained as needle-like crystals of low sp gr from aq soln by rapid addition of acetone at low temps; now obtained as crystals of higher sp gr, with cubic crystal structure and greater soly in HCl, by slow addition of acetone (Upjohn pat. 1965). mp 145-147°. $[\alpha]_D^{25}$ $+137°$ (water). Freely sol in water, methanol, ethanol; sparingly sol in most organic solvents other than hydrocarbons. LD_{50} in mice, rats (g/kg): 1 i.p.; 4 orally (Gray).

Hydrochloride monohydrate, *Albiotic, Cillimycin, Lincomix.*

THERAP CAT: Antibacterial.

THERAP CAT (VET): Antibacterial.

5379. Lindane. 1α,2α,3β,4α,5α,6β-Hexachlorocyclo-hexane; γ-HCH; γ-benzene hexachloride; gamma benzene hexachloride; gamma hexachlor; ENT 7796; Aparasin; Aphtiria; Esoderm; γ-BHC; Gammalin; Gamene; Gamiso; Gammexane; Gexane; Jacutin; Kwell; Lindafor; Lindatox; Lorexane; Quellada; Streunex; Tri-6; Viton. $C_6H_6Cl_6$; mol wt 290.85. C 24.78%, H 2.08%, Cl 73.14%. Lindane, the gamma isomer, is the active isomer among the eight well-described stereoisomers of hexachlorocyclohexane. Prepns sold for pharmaceutical or medicinal purposes now contain at least 99% pure γ-isomer. Prepn: T. Hardie, U.S. pat. **2,218,148** (1940 to ICI); R. E. Slade, *Chem. & Ind. (London)* **1945,** 314; F. A. Gunther, ibid. **1946,** 399. Metabolism: R. Engst et al., *Residue Rev.* **68,** 59 (1977); **72,** 71 (1979). Acute toxicity: T. B. Gaines, *Toxicol. Appl. Pharmacol.* **14,** 515 (1969). Reviews of carcinogenicity studies: M. D. Reuber, *Environ. Res.* **19,** 460-481 (1979); S. D. Vesselino-vitch, F. W. Carlborg, *Toxicol. Pathol.* **11,** 12 (1983). Review of isomers: I. Hornstein, *Science* **121,** 206 (1955); J. G Colson in Kirk-Othmer *Encyclopedia of Chemical Technology* vol. 5 (Wiley-Interscience, New York, 3rd ed., 1979) pp 808-818.

Crystals, mp 112.5°. Slight musty odor. Vapor pressure at 20°: 9.4 × 10⁻⁶ mm Hg. n_D^{20} 1.644. Soly in g/100 g at

20°: acetone 43.5, benzene 28.9, $CHCl_3$ 24.0, ether 20.8, ethanol 6.4. Insol in water. LD_{50} in male, female rats (mg/kg): 88, 91 orally (Gaines).

Human Toxicity: Poisoning may occur by ingestion, inhalation, or percutaneous absorption. *Acute:* Dizziness, headache, nausea, vomiting, diarrhea, tremors, weakness, convulsions, dyspnea, cyanosis, circulatory collapse. *Chronic:* Hepatic damage has occurred in exptl animals. Topical use may cause local sensitivity reactions. Vapors may irritate eyes, nose, throat. *See: Clinical Toxicology of Commercial Products,* R. E. Gosselin *et al.,* Eds. (Williams & Wilkins, Baltimore, 5th ed., 1984) Section III, pp 239-241. Lindane and other hexachlorocyclohexane isomers may reasonably be anticipated to be carcinogens: *Fourth Annual Report on Carcinogens* (NTP 85-002, 1985) p 123.

USE: Insecticide.

THERAP CAT: Pediculicide; scabicide.

THERAP CAT (VET): Ectoparasiticide.

5380. Lindlar Catalyst. $Pd-Pb-CaCO_3$. Prepn: Lindlar, *Helv. Chim. Acta* **35,** 446 (1952); Lindlar, Dubuis cited by Fieser, Fieser, *Reagents for Organic Synthesis* (New York, 1967) p 566.

USE: In selective hydrogenation of triple bonds to *cis*-double bonds.

5381. Lineatin. *[1R-(1α,2β,5α,7β)]-3,3,7-Trimethyl-4,9-dioxatricyclo[3.3.1.0²ʼ⁷]nonane;* 3,3,7-trimethyl-2,9-di-oxatricyclo[3.3.1.0⁴ʼ⁷]nonane; 4,6,6-lineatin. $C_{10}H_{16}O_2$; mol wt 168.24. C 71.39%, H 9.59%, O 19.02%. Isoln of the unique tricyclic aggregation pheromone from ambrosia beetles, *Trypodendron lineatum* (Olivier): J. G. MacConnell *et al., J. Chem. Ecol.* **3,** 549 (1977). Synthesis of (±)-form: K. Mori, M. Sasaki, *Tetrahedron Letters* **1979,** 1329; K. N. Slessor *et al., J. Org. Chem.* **45,** 2290 (1980); K. Mori *et al., Tetrahedron Letters* **23,** 1921 (1982); L. Skattebol, Y. Stenstrom, *ibid.* **24,** 3021 (1983); B. D. Johnston *et al., J. Org. Chem.* **50,** 114 (1985). Synthesis of racemate and optical isomers: K. Mori, M. Sasaki, *Tetrahedron* **36,** 2197 (1980). Short stereoselective synthesis: I. Aljancic-Solaja *et al., Helv. Chim. Acta* **70,** 1302 (1987). Comparative activity of the isomers: J. H. Borden *et al., Can. Entomol.* **112,** 107 (1980).

Oil, bp_{10} 70°. $[α]_D^{24}$ +66.3° (c = 3.1 in $CHCl_3$).

USE: Insect sex attractant.

5382. Linoleic Acid. *(Z,Z)-9,12-Octadecadienoic acid;* 9,12-linoleic acid; linolic acid. $C_{18}H_{32}O_2$; mol wt 280.44. C 77.09%, H 11.50%, O 11.41%. An essential fatty acid, *q.v.* Major constituent of many vegetable oils, *e.g.,* cottonseed, soybean, peanut, corn, sunflower seed, safflower, poppy seed, linseed, and perilla oils, where it occurs as a glyceride. Characteristic ingredient of semi-drying oils. Isoln: Swern, Parker, *J. Am. Oil Chem. Soc.* **30,** 5 (1953); Parker *et al., Biochem. Prepn.* **4,** 86 (1955); McCutcheon, *Org. Syn.* coll. vol. III, 526 (1955). Summary of work on structure: T. P. Hilditch, *The Chemical Constitution of Natural Fats* (Chapman & Hall, London, 2nd ed. 1956). Synthesis: Raphael, Sondheimer, *J. Chem. Soc.* **1950,** 2102; Gensler, Thomas, *J. Am. Chem. Soc.* **73,** 4601 (1951); Walborsky *et al., ibid.* 2590; Nigam, Weedon, *J. Chem. Soc.* **1956,** 4052; Osbond, Wickens, *Chem. & Ind. (London)* **1959,** 1288. Review of physiological role in mammals: H. S. Hansen, *Trends Biochem. Sci.* **11,** 263 (1986).

Colorless oil. Easily oxidized by air, cannot be distilled without decompn. Storage in ester form is recommended.

d_4^{18} 0.9038; d_4^{22} 0.9007. mp −12°. $bp_{1.4}$ 202°; bp_{16} 230°. $n_D^{11.5}$ 1.4715; n_D^{20} 1.4699; $n_D^{21.5}$ 1.4683; n_D^{50} 1.4588. Iodine value: 181.1. Thiocyanogen value 96.7. Freely sol in ether. Sol in abs alc. One ml dissolves in 10 ml petr ether. Miscible with dimethylformamide, fat solvents, oils.

Aluminum salt, $Al(C_{18}H_{29}O_2)_3$. Yellow lumps or powder; linseed oil odor. Practically insol in water. Sol in oils, fixed alkali hydroxides.

Methyl Ester *see* Methyl Linoleate.

Ethyl Ester *see* Ethyl Linoleate.

Cyclohexylamide, $C_{24}H_{43}NO$, *linolexamide, N-cyclohexyl-linoleamide, Clinolamide.*

USE: Manuf paints, coatings, emulsifiers, vitamins. Aluminum salt used to manuf lacquers.

THERAP CAT: Nutrient (essential fatty acid).

5383. Linolenic Acid. *(Z,Z,Z)-9,12,15-Octadecatrienoic acid;* α-linolenic acid. $C_{18}H_{30}O_2$; mol wt 278.42. C 77.65%, H 10.86%, O 11.49%. An essential fatty acid, *q.v.* Occurs as the glyceride in most drying oils. Synthesis: Nigam, Weedon, *J. Chem. Soc.* **1956,** 4049; Osbond, Wickens, *Chem. & Ind. (London)* **1959,** 1288. Biosynthetic studies: C. G. Kannangara *et al., Biochem. Biophys. Res. Commun.* **52,** 648 (1973); B. S. Jacobson *et al., ibid.* 1190; C. J. Bedord *et al., Arch. Biochem. Biophys.* **185,** 15 (1978). Effects on lipid metabolism in rat tissue: M. L. Garg *et al., Lipids* **23,** 847 (1988). Review of dietary linolenic acid in mammals: J. Tinoco *et al., ibid.* **14,** 166-171 (1979); in man: N. Zöllner, *Prog. Lipid Res.* **25,** 177-180 (1986).

Colorless liquid. d_4^{18} 0.914. bp_1 230-232°. Insol in water. Sol in organic solvents.

THERAP CAT: Nutrient (essential fatty acid).

5384. γ-Linolenic Acid. *(Z,Z,Z)-6,9,12-Octadecatrienoic acid; cis-6,cis-9,cis-12-octadecatrienoic acid;* gamolenic acid; GLA. $C_{18}H_{30}O_2$; mol wt 278.44. C 77.65%, H 10.86%, O 11.49%. Polyunsaturated fatty acid produced in the body as the $Δ^6$-desaturase metabolite of linoleic acid, *q.v.* Converted to dihomo-γ-linolenic acid, a biosynthetic precursor of monoenoic prostaglandins such as PGE_1. Present to varying extents in the fatty acid fraction of evening primrose oil (7-10%), in borage oil (18-26%), in black currant oil (15-20%) and in oils from different fungal sources (6-24%). Isoln from evening primrose oil, *q.v.*: A. Heiduschka, K. Luft, *Arch. Pharm.* **257,** 33 (1919). Proposed structure: Eibner *et al., Chem. Umschau.* **34,** 312 (1927). Confirmation of structure: J. P. Riley, *J. Chem. Soc.* **1949,** 2728. Discussion of occurrence, esp. in fungi: R. Shaw, *Biochim. Biophys. Acta* **98,** 230 (1965). Synthesis: J. M. Osbond *et al., J. Chem. Soc.* **1961,** 2779; J. M. Osbond, *ibid.* 5270. Metabolism studies: J. F. Mead, D. R. Howton, *J. Biol. Chem.* **229,** 575 (1957); K. J. Stone *et al., Lipids* **14,** 174 (1979). Effect of source on essential fatty acid and prostanoid metabolite formation: D. K. Jenkins *et al., Med. Sci. Res.* **16,** 525 (1988).

Hexabromide deriv, $C_{18}H_{30}Br_6O_2$, crystals from ethyl methyl ketone, mp 201-202°.

USE: Nutrient.

THERAP CAT: In treatment of atopic eczema.

5385. Linseed. Flaxseed; linum. Dried ripe seeds of *Linum usitatissimum* L., Linaceae. Source of linseed oil. *Constit.* 30-40% oil, about 6% mucilage, about 25% proteins and linamarin.

USE: Emollient.

THERAP CAT (VET): Poultice (crushed seeds), demulcent (boiled in water).

5386. Linseed Oil. A drying oil obtained by expression of linseed. *Constit.* Glycerides of linolenic, linoleic, oleic, stearic, palmitic and myristic acids. *Ref:* T. P. Hilditch, *The Chemical Constitution of Natural Fats* (London, 3rd ed.,

1956) p 175 sqq; E. W. Eckey, *Vegetable Fats and Oils* (New York, 1954) pp 535-547.

Yellowish liquid, peculiar odor, bland taste. Exposed to air it gradually thickens, becomes darker, and acquires a more pronounced odor and taste. d 0.925-0.935. n_D^{40} 1.4725-1.4750. Does not congeal above $-20°$. Sapon no.: 187-195. Iodine no. not below 170. Unsaponifiable matter not over 1.5%. Slightly sol in alcohol, miscible with chloroform, ether, petr ether, carbon disulfide, oil turpentine.

USE: In varnishes, paints, putty, oilcloths, linoleum, printing inks, artificial rubber, tracing cloth, tanning and enameling leather; applied to paper and fabrics to render them waterproof and tough. Emollient.

THERAP CAT (VET): Laxative.

5387. Linuron. *N'-(3,4-Dichlorophenyl)-N-methoxy-N-methylurea;* methoxydiuron; du Pont Herbicide 326; Hoe 2810; Afalon; Linurex; Lorox. $C_9H_{10}Cl_2N_2O_2$; mol wt 249.11. C 43.39%, H 4.05%, Cl 28.47%, N 11.25%, O 12.85%. Selective pre- and post-emergence herbicide. Prepn: Sherer, Heller, U.S. pat. 2,960,534 (1960 to Hoechst). Degradn in soil: G. F. Kempson-Jones, R. J. Hance, *Pestic. Sci.* **10,** 449 (1979).

mp 93-94°. Vapor press. at 24°: 1.5×10^{-5} mm Hg. Soly in water: 75 ppm. Partially sol in acetone, alcohol, benzene, toluene, xylene. LD_{50} orally in rats: 1500 mg/kg, G. W. Bailey, J. L. White, *Residue Rev.* **10,** 97 (1965).

USE: Herbicide.

5388. Liothyronine. *O-(4-Hydroxy-3-iodophenyl)-3,5-diiodo-L-tyrosine;* L-3-[4-(4-hydroxy-3-iodophenoxy)-3,5-diiodophenyl]alanine; 4-(3-iodo-4-hydroxyphenoxy)-3,5-diiodophenylalanine; 3,5,3'-triiodothyronine; T-3; Triothyrone. $C_{15}H_{12}I_3NO_4$; mol wt 651.01. C 27.68%, H 1.86%, I 58.48%, N 2.15%, O 9.83%. Amino acid found in human plasma and thyroid gland, similar to thyroxine *q.v.* Gross, Pitt-Rivers, *Lancet* **1952,** I, 439. Formation from diiodothyronine: Roche *et al., Biochem. Biophys. Acta* **11,** 215 (1953). Isoln from thyroid gland and synthesis: Gross, Pitt-Rivers, *Biochem. J.* **53,** 645 (1953); Roche *et al., Bull. Soc. Chim. France* **4,** 462 (1957); Plati, Wenner, U.S. pat. 2,784,222 (1957 to Hoffmann-La Roche); Pitt-Rivers, Gross, U.S. pat. 2,823,164 (1958 to Nat. Res. Dev. Corp.); Razdan, Wetherill, U.S. pat. 2,993,928 (1961 to Glaxo). Elevated levels of T-3 in victims of sudden infant death syndrome: G. Kocsard-Varo, *Med. J. Aust.* **2,** 789 (1973); M. A. Chacon, J. T. Tildon, *J. Pediatr.* **99,** 758 (1981). Monograph: Pitt-Rivers, R. Tata, *The Thyroid Hormones* (Pergamon Press, 1959).

Crystals, dec 236-237°. $[\alpha]_D^{29.5} +21.5°$ (c = 4.75 in a mixture of 1 part *N* HCl + 2 parts ethanol). Possesses 5 times the activity of L-thyroxine (goiter prevention test in rats). Insol in water, alc, propylene glycol. Sol in dil alkalies with the formation of a brownish, water-soluble, sodium salt.

Sodium salt, $C_{15}H_{11}I_3NNaO_4$, *liothyronine sodium, sodium* L-*triiodothyronine, Cytobin, Cytomine, Cytomel, Cynomel, Tertroxin, Triothyrone.* The commonly used form.

Hydrochloride, $C_{15}H_{13}ClI_3NO_4$, *Thybon.* Long birefringent needles, decomp 202-203°. $[\alpha]_D^{29.5} +21.5°$ (c = 4.75 in a mixture of 1 vol *N* HCl and 2 vols ethanol).

DL-Triiodothyronine hydrochloride, *Trionine.*

THERAP CAT: Thyroid hormone.

THERAP CAT (VET): Thyroid hormone.

5389. Lipase. *Triacylglycerol lipase.* An enzyme (or more exactly a group of enzymes) belonging to the esterases.

Hydrolyzes fat (present in ester form, such as glycerides) yielding fatty acids and glycerol. Catalyzes digestion. Widely distributed in the plant world, also in molds, bacteria, milk and milk products, and in animal tissues, especially in the pancreas. Isoln from castor beans: H. Gibian in *Ullmann's Encyklopädie der technischen Chemie,* 3rd ed., vol. 7, 406-407 (1956). Purification of pancreatic lipase: Marchis-Mouren *et al., Arch. Biochem. Biophys.* **83,** 309 (1959). Review of milk lipases: Chandan, Shahani, *J. Dairy Sci.* **47,** 471 (1964). Comprehensive reviews: Wills, *Advan. Lipid Res.* **3,** 197-240 (1965); Desnuelle in *The Enzymes* vol. 7, P. D. Boyer, Ed. (Academic Press, New York, 3rd ed., 1972) pp 575-616.

The optimum temp for enzyme action is between 35° and 37° at pH 5-6. Lipase contains sulfhydryl groups and is inactivated by substances that inhibit such compds. It is activated by substances that keep SH groups in the reduced state, such as glutathione, cysteine, and ascorbic acid. The addition of acid activates lipase preparations. Castor-oil lipase is activated by sulfuric, oxalic, formic, acetic and butyric acids. Acetic, salicylic and hydrochloric acids increase the action of lipase derived from various organs of the pig. Caprylic and caproic acids increase the action of lipase derived from certain mold fungi. Almost all organic solvents decrease lipase activity, petr ether being an exception.

USE: To split fats without damaging sensitive constituents, such as vitamins or unsaturated fatty acids. In food processing for flavor improvement; in detergents for the improvement of cleaning action. For review of industrial applications of microbial lipases, *see* Seitz, *J. Am. Oil Chem. Soc.* **51,** 12 (1974).

THERAP CAT: Digestive enzyme.

5390. Lipoprotein Lipase. *Diacylglycerol lipase; clearing factor.* A specific lipase which preferentially hydrolyzes triglycerides in the presence of a lipoprotein complex, forming glycerol and unesterified fatty acids: Korn, *Methods Enzymol.* **5,** 542 (1962). The normal substrate is turbid lipemic or chylomicron-containing plasma. This substance is clarified by the enzyme, hence the term clearing factor. First detected in the plasma of animals injected intravenously with heparin: Hahn, *Science* **98,** 19 (1943); Robinson, French, *Phamacol. Rev.* **12,** 241 (1960). Prepn from post heparin human plasma: Baskys, *Diss. Abstr.* **20,** 1146 (1959). It has since been prepd from several tissues of normal, untreated animals. Prepn from chicken adipose tissue: Korn, Quigley, *J. Biol. Chem.* **226,** 833 (1957); from human heart: Schnatz *et al., Am. J. Physiol.* **205,** 401 (1963). Role in the metabolism of triglycerides by adipose tissues: Rodbell, Scow in K. Rodahl, *Fat as a Tissue* (McGraw-Hill, New York, 1964) pp 110-126. *Reviews:* Jensen, *Progr. Chem. Fats Other Lipids* **11,** 347 (1971); Desnuelle in *The Enzymes,* vol. 7, P. D. Boyer, Ed. (Academic Press, New York, 3rd ed., 1972) pp 606-609.

5391. Liposomes. Liposom Forte. Structures (smectic mesomorphs) formed spontaneously by polar lipids in aqueous media at temperatures allowing the fatty acyl chains to be fluid. Lipid molecules are arranged in concentric bilayers separated from each other by aqueous compartments. Within each bilayer, the molecules are arranged so that their hydrophobic tails point away from the aqueous compartment and their hydrophilic heads point toward it. Liposomes are usually spherical but shape can vary. The largest are fractions of a millimeter; unilamellar liposomes (vesicles) can be as small as several microns. Liposomes are able to entrap solutes and can interact with cellular structures. They can carry into cells biologically active materials which do not normally penetrate the plasma membrane. Encapsulation in liposomes protects these materials from contact with tissue and organ cultures, from inactivation within a living organism's circulatory system or, for pesticides and fertilizers, from removal by rain or irrigation. Liposomes may interact with cells by stable adsorption onto the cell surface, endocytosis (the intact liposome is delivered to the lysosomal apparatus), fusion with the plasma membrane (liposome content is released into the cytoplasm) and lipid exchange between liposome and cell surface without cell association of liposome contents. Liposomes are used as models in biological membrane research, as experimental targeted drug delivery systems and as carriers for other biologically active materials. *See also* Phospholipids. Geometry and thermodynamics: J. N. Israelachvili *et al., Biochim. Biophys. Acta* **470**, 185 (1977). Effect on brain metabolism: G. Toffano *et al.,* "Phospholipid Liposomes Effect on Brain Dopamine Metabolism and on the Secretion of GH" in *Neuroendocrinology: Biological and Clinical Aspects,* A. Polleri, R. M. Macleod, Eds. (Academic Press, New York, 1979) pp 345-349. Interaction with nucleic acids: B. Brosius *et al., J. Biomolec. Struct. Dyn.* **1**, 1535 (1984). Synthesis and use as carriers for biologically active materials: D. P. Papahadjopoulos, F. C. Szoka, Jr., *Belg.* pat. **874,408** corresp to U.S. pat. **4,235,871** (1979, 1980); for pesticides: *eidem,* U.S. pat **4,241,046** (1980). Inhibition of erythrocytic malaria by liposomes containing neutral glycolipids: C. R. Alving *et al., Science* **205**, 1142 (1979). In ophthalmology: H. E. Schaeffer, D. L. Krohn, *Invest. Ophthalmol. Vis. Sci.* **22**, 220 (1982). In delivery of spermatozoal antigen for fertility control: L. Mettler, A. B. Czuppon, *Ger.* pat. **3,301,951** (1984 to Nattermann).

In transfer of nucleic acids: I. J. Mettler, **Brit.** pat. **Appl.** **2,140,822** (1984 to Stauffer). In gene transfer (preproinsulin I): C. Nicolau, *Biochem. Soc. Trans.* **12**, 349 (1984). Rigid, non-phospholipid liposomes as drug delivery system in hepatoma ascites tumor in mice: K. R. Patel *et al., Biochim. Biophys. Acta* **797**, 20 (1984). Review of early studies: D. A. Tyrrell *et al., ibid.* **457**, 259-302 (1976); of pharmacology of phospholipid liposomes: G. Toffano, A. Bruni, *Pharm. Res. Comm.* **12**, 829-845 (1980). Books: *Liposomes in Biological Systems,* G. Gregoriadis, A. C. Allison, Eds. (Wiley, New York, 1980) 412 pp; *Liposomes: From Physical Structure to Therapeutic Applications,* C. G. Knight, Ed. (Elsevier, Amsterdam, 1981) pp 323-348; *Liposome Technology,* G. Gregoriadis, Ed. (CRC Press, Boca Raton, 1984) 3 vols.

Immunosome. Immunogenic structures consisting of liposomes attached to antigens. Liposomes as immunological adjuvant: A. C. Allison, G. Gregoriadis, *Nature* **252**, 252 (1974). Immunosomes in influenza virus vaccine: L. Thibodeau *et al.,* **Eur.** pat. **Appl.** **47,480** (1982 to Inst. Armand Frappier); rabies vaccine: P. Perrin *et al., Ann. Virol.* **135E**, 183 (1984); vaccine against polysaccharide encapsulated bacteria: H. Snippe *et al.,* **Eur.** pat. **Appl.** **97,407** (1984 to Rijksuniv. Utrecht). Antibodies to hepatitis B surface antigen elicited: A. R. Neurath *et al., J. Gen. Virol.* **65**, 1009 (1984).

5392. Lipotropic Hormone. *Lipotropin; pituitary lipotropic hormone; lipid-mobilizing hormone;* adipokinetic hormone; lipolytic hormone; lipotrophin; LPH. Hypophysial hormone which stimulates the release of fatty acids from adipose tissue. β-*LPH* is a single chain polypeptide containing 89-93 amino acid residues and has been proposed as the biosynthetic precursor for β-MSH and β-endorphin, *q.v., see* M. Chretien *et al., Can. J. Biochem.* **57**, 1111 (1979). γ-*LPH* consists of 58 amino acids and is identical to the sequence of the first 58 residues of β-LPH. Both contain sequences common to ACTH, *q.v.,* and β-melanotropin, *q.v.* Sequences differ slightly among mammalian species. Isoln and proposed structure of ovine β-LPH: C. H. Li, *Nature* **201**, 924 (1964); Y. Birk, C. H. Li, *J. Biol. Chem.* **239**, 1048 (1964); C. H. Li *et al., Nature* **208**, 1093 (1965); C. H. Li, *Arch. Biol. Med. Exp.* **5**, 55 (1968); revised structure: Chretien *et al., Int. J. Peptide Protein Res.* **4**, 263 (1972). Isoln and characterization of porcine β-LPH: Gilardeau, Chretien, *Can. J. Biochem.* **48**, 1017 (1970). Amino acid sequence of porcine γ-LPH: Graf *et al., Acta Biochim. Biophys.* **5**, 305 (1970). Isoln, characterization of bovine β-LPH: P. Lohmar, C. H. Li, *Biochem. Biophys. Res. Commun.* **77**, 1088 (1977); of rat β-LPH: M. Rubinstein *et al., Proc. Nat. Acad. Sci. USA* **74**, 3052 (1977); of whale β-LPH: H. Kawauchi *et al., Int. J. Peptide Protein Res.* **15**, 171 (1980); of turkey β-LPH: W. C. Chang *et al., ibid.* 561. Initial isoln, partial structure of human β-LPH: G. Cseh *et al., FEBS Letters* **2**, 42 (1968); *eidem, ibid.* **21**, 344(1972). Isoln, characterization, amino acid sequence of human β-LPH: C. H. Li, D. Chung, *Int. J. Peptide Protein Res.* **17**, 131 (1981). *Review:* D. T. Krieger *et al., Recent Progr. Horm. Res.* **36**, 277-344 (1980).

5393. Lisinopril. *(S)-1-[N²-(1-Carboxy-3-phenylprop-*

yl)-L-lysyl]-L-proline dihydrate; MK-521; Acerbon; Carace; Novatec; Prinivil; Renacor; Vivatec; Zestril. $C_{21}H_{31}N_3$-$O_5.2H_2O$; mol wt 441.52. C 57.13%, H 7.99%, N 9.52%, O 25.36%. Orally active angiotensin-converting enzyme inhibitor. Prepn: A. A. Patchett *et al.,* **Eur.** pat. **Appl. 12,401**; E. E. Harris *et al.,* U.S. pat. **4,374,829**; A. A. Patchett, M. T. Wu, U.S. pat. **4,555,502** (1980, 1983, 1985, all to Merck & Co.). Prepn and activity in rats, dogs: A. A. Patchett *et al., Nature* **288**, 280 (1980); M. T. Wu *et al., J. Pharm. Sci.* **74**, 352 (1985). Hemodynamic effects in heart failure in dogs: C. S. Sweet *et al., J. Cardiovasc. Pharmacol.* **8**, Suppl. 1, S15 (1986). Activity in normotensive man: J. Biollaz *et al., Clin. Pharmacol. Ther.* **29**, 665 (1981); *eidem, Brit. J. Clin. Pharmacol.* **14**, 363 (1982); G. P. Hodsman *et al., ibid.* **17**, 233 (1984). Human pharmacodynamics: J. A. Millar *et al., ibid.* **14**, 347 (1982). Metabolism in man: E. H. Ulm *et al., ibid.* 357. Radioimmunoassay: P. J. Worland, B. Jarrott, *J. Pharm. Sci.* **75**, 512 (1986). Preliminary clinical trials in hypertension: H. H. Rotmensch *et al., Am. J. Cardiol.* **53**, 116 (1984); in congestive heart failure: K. Dickstein *et al., Am. Heart J.* **112**, 11 (1986). Series of articles in hypertension and in congestive heart failure: *Am. J. Med.* **85**, Suppl. 3B, 1-59 (1988).

Soly (mg/ml): water 97; methanol 14; ethanol < 0.1; acetone < 0.1; acetonitrile < 0.1; chloroform < 0.1; *N,N*-dimethylformamide < 0.1. pKa_1 (25°) 2.5; pKa_2 4.0; pKa_3 6.7; pKa_4 10.1.

THERAP CAT: Antihypertensive.

5394. Lisuride. *N'-[(8α)-9,10-didehydro-6-methylergolin-8-yl]-N,N-diethylurea;* 9-(3,3-diethylureido)-4,6,6a,7,-8,9-hexahydro-7-methylindolo[4,3-f.g]quinoline; 1,1-diethyl-3-(D-6-methylisoergolen-8-yl)urea; *N-(D-6-methyl-8-isoergolenyl)-N,N-diethylurea;* methylergol carbamide; lysuride. $C_{20}H_{26}N_4O$; mol wt 338.46. C 70.97%, H 7.74%, N 16.55%, O 4.73%. Dopamine D_2-receptor agonist. Prepn: Zikan, Semonsky, *Coll. Czech. Chem. Commun.* **25**, 1922 (1960); *eidem, Pharmazie* **23**, 147 (1968). Pharmacology and toxicity: Z. Votava, I. Lamplova, *Physiol. Bohemoslov.* **12**, 37 (1963), *C.A.* **59**, 9221d (1963).

Crystals from benzene, mp 186°. $[\alpha]_D^{20}$ +313° (c = 0.60 in pyridine).

Acid maleate, $C_{24}H_{30}N_4O_5$, *Cuvalit, Dopergin, Eunal, Lysenyl.* Prisms from ethanol, mp 200° (dec). $[\alpha]_D^{20}$ +288° (c = 0.5 in methanol). uv max (methanol): 313 nm. LD_{50} i.v in mice: 14.4 mg/kg (Votava, Lamplova).

THERAP CAT: Antimigraine; prolactin inhibitor.

5395. Lithium. Li; at. wt 6.941; at. no. 3; valence 1. Alkali metal. Occurrence in earth's crust: 0.005% by wt. Natural isotopes: 7 (92.58%); 6 (7.42%); artificial radioactive isotopes: 5, 8, 9; all are unstable ($T_{1/2}$ < 1 sec). Discovered as salt in 1817 by Arfvedson: *Ann. Chim. Phys.* [2] **10**, 82 (1819); metal prepared independently by Davy and Brandé in 1818. Occurs in a number of minerals; *spodumene* (LiAl-S_2O_6), *lepidolite* [K(Li,Al)$_3$(Si,Al)$_4O_{10}$(F,OH)$_2$], *petalite*

(LiAlSi$_4$O$_{10}$), *amblygonite* (AlPO$_4$,LiF), and *triphylite* (LiFe-PO$_4$) contain 3-10% Li$_2$O. Also recovered from natural brines. Prepn of the metal by electrochemical processes: Guntz, *Compt. Rend.* **117**, 732 (1893); Ruff, Johannsen, *Z. Electrochem.* **12**, 186 (1906); by reduction of the oxide with magnesium or aluminum: Warren, *Chem. News* **74**, 6 (1896); Hanson, U.S. pat. 2,028,390 (1936). Reviews of biology, pharmacology and toxicity of lithium ion: Schou, "Lithium in Psychiatry—A Review" in *Psychopharmacology, A Review of Progress 1957-1967*, D. H. Efron, Ed. (Public Health Service Publication No. 1836, 1968) pp 701-718; Doig *et al.*, *J. Chem. Ed.* **50**, 343-345 (1973); Samuel, Gottesfeld, *Endeavour* **32**, 122-128 (1973); Saran, Gaind, *Clin. Toxicol.* **6**, 257-269 (1973); Schou, *Ann. Rev. Pharmacol. Toxicol.* **16**, 231-243 (1976). *Reviews:* Hart, Beumel, "Lithium and its Compounds" in *Comprehensive Inorganic Chemistry* vol. **1**, J. C. Bailar, Jr. *et al.*, Eds. (Pergamon Press, Oxford, 1973) pp 331-367; R. Bach, J. R. Wasson in Kirk-Othmer *Encyclopedia of Chemical Technology* vol. **14** (Wiley-Interscience, New York, 3rd ed., 1981) pp 448-476.

Silvery-white metal; body-centered cubic structure; becomes yellowish on exposure to moist air. mp 180.54°. bp 1336 ± 5°: Hartman, Schneider, *Z. Anorg. Chem.* **180**, 275 (1929). d 0.534. Heat capacity at constant pressure (25°): 5.892 cal/mole deg. *See:* Douglass *et al.*, *J. Am. Chem. Soc.* **77**, 2144 (1955). E° (aqueous) Li/Li$^+$ 3.045 V. Reacts with H$_2$O forming the hydroxide and H$_2$. Attacked rapidly by dil HCl and H$_2$SO$_4$; slowly by concd H$_2$SO$_4$. Reacts vigorously with HNO$_3$. Sol in liq ammonia forming a blue soln. Does not react with oxygen at room temp; forms Li$_2$O when heated to 100° or higher. Reacts with hydrogen at a red heat; combines directly with nitrogen, halogens and sulfur under proper conditions. Imparts a carmine-red color to a bunsen flame. *Keep under mineral oil or other liquid free from oxygen or water.*

USE: In manuf of alloys, especially lithium-hardened bearing metals; as a "getter" in vacuum tubes. In making catalysts for the polyolefin plastics industry, in fuels for aircraft and missiles. Lithium salts are used in porcelain enamels, in air-conditioning, and for making multi-purpose greases. *Human Toxicity:* Li ion may injure kidneys, especially if sodium intake is limited.

5396. Lithium Acetate. Quilonorm; Quilonum. C$_2$H$_3$-LiO$_2$; mol wt 65.98. C 36.40%, H 4.58%, Li 10.52%, O 48.50%. Prepn: *Gmelin's, Lithium* (8th Ed.) **20**, pp 230-232 (1927). Prepn and solubilities: N. V. Sidgwick, J. A. H. R. Gentle, *J. Chem. Soc.* **121**, 1837 (1922). Crystal structure of anhydrous: C. Saunderson, R. B. Ferguson, *Acta Crystallogr.* **14**, 321 (1961). Pharmacokinetics in psychiatric patients: J.-L. Evrard *et al.*, *Acta Psychiatr. Scand.* **58**, 67 (1978).

mp 286° (slight dec). Changes to dihydrate at 56.5°.
Dihydrate, rhombic crystals. Begins to melt at 49° (Gmelin). Congruent mp 57.8° (Sidgwick, Gentle). Freely sol in water or alcohol. Its aq soln is practically neutral.
THERAP CAT: Antimanic.

5397. Lithium Acetylsalicylate. Hydropyrin; Litmopyrine; Tyllithin. C$_9$H$_7$LiO$_4$; mol wt 186.08. C 58.09%, H 3.79%, Li 3.73%, O 34.39%. CH$_3$COOC$_6$H$_4$COOLi.
White, slightly hygroscopic powder; dec in moist air. Sol in 1 part water, 4 parts alcohol.

5398. Lithium Amide. Lithamide. H$_2$LiN; mol wt 22.96. H 8.78%, Li 30.23%, N 61.01%. LiNH$_2$. Prepd from Li + NH$_3$: Titherley, *J. Chem. Soc.* **65**, 517 (1894); Ruff, Geisel *Ber.* **39**, 840 (1906); **44**, 505 (1911); Campbell *et al.*, *Proc. Indiana Acad. Sci.* **50**, 123 (1940); Schenk in *Handbook of Preparative Inorganic Chemistry* vol. **1**, G. Brauer, Ed. (Academic Press, New York, 2nd ed., 1963) p 463. *See also Inorg. Syn.* **2**, 135. Crystal structure: Juza, Opp, *Z. Anorg. Allgem. Chem.* **266**, 313 (1951).
Tetragonal crystal structure, d 1.18. mp 380-400°. Sublimes in NH$_3$ current. When heated *in vacuo* to 450°, NH$_3$ is given off and Li$_2$NH forms, which dec about 750-800°. Heat of formation: 42 kcal/mole at 18° and 760 mm Hg. Stable in air-tight containers at room temp. Dec by water to form LiOH + NH$_3$, slowly when in lumps, faster as the particle size decreases. Insol in anhyd ether, benzene, toluene.

USE: In Claisen condensations, alkylation of nitriles and ketones, synthesis of ethynyl compds, acetylenic carbinols.

5399. Lithium Benzoate. C$_7$H$_5$LiO$_2$; mol wt 128.05. C 65.65%, H 3.94%, Li 5.42%, O 24.99%.
White, cryst powder. One gram dissolves in 3 ml water, 13 ml alcohol, 10 ml boiling alcohol. Soly in water is increased by sodium benzoate. The aq soln is slightly alkaline to litmus; pH about 8.
USE: Has been used as a lubricant for compressing tablets.

5400. Lithium Bitartrate. Tartarlithine. C$_4$H$_5$LiO$_6$; mol wt 156.02. C 30.79%, H 3.23%, Li 4.45%, O 61.53%.
Monohydrate, white cryst powder, sol in water.

5401. Lithium Borate. Lithium biborate; lithium tetraborate. B$_4$Li$_2$O$_7$; mol wt 169.10. B 25.57%, Li 8.20%, O 66.23%. Li$_2$B$_4$O$_7$.
Pentahydrate, white, cryst powder. Slightly sol in water; insol in alcohol.
USE: In enamels.

5402. Lithium Borohydride. *Lithium tetrahydroborate.* BH$_4$Li; mol wt 21.79. B 49.66%, H 18.50%, Li 31.85%. LiBH$_4$. Prepd by the action of diborane on ethyllithium; by the action of aluminum borohydride on ethyllithium: Schlesinger, Brown, *J. Am. Chem. Soc.* **62**, 3429-35 (1940). Review of lithium and other metal tetrahydroborates: James, Wallbridge, *Prog. Inorg. Chem.* **11**, 99-231 (1970).
Orthorhombic crystals, mp 268°; dec 380°. d 0.66. Stable under ordinary conditions, but dec in moist air. Reacts with hydrogen chloride to form hydrogen, diborane, and lithium chloride; forms lithium boromethoxide and hydrogen with methyl alcohol. Sol in water at pH above 7. Aq solns dec slowly if the pH is above 7, and rather vigorously when acidified. Sol in ether, tetrahydrofuran, aliphatic amines.
USE: Strong reducing agent. Used in the reduction of compds contg ketonic, aldehydic, or ester carbonyls and a nitrile group, where reduction of the carbonyl, but not of the nitrile group, is wanted. In the determination of free carboxyl groups in peptides and proteins; after esterification and acetylation, only the ester groups, and none of the peptide bonds are reduced.

5403. Lithium Bromide. BrLi; mol wt 86.84. Br 92.02%, Li 7.98%. LiBr. Usually contains water and 85-90% LiBr.
White, very deliquesc, slightly bitter, granular powder. mp 547° when anhydr. Sol in 0.6 part water, 0.4 part boiling water; freely sol in alcohol, glycol; sol in ether, amyl alcohol. The aq soln is neutral or only slightly alkaline to litmus. *Keep well closed.*
Human Toxicity: Large doses may cause CNS depression. Chronic absorption may cause skin eruptions and CNS disturbances due to bromide. May also cause disturbed blood electrolyte balance.
USE: As humectant, in air-conditioning systems.
THERAP CAT: Sedative, hypnotic.

5404. Lithium Carbonate. Camcolit; Candamide; Carbolith; Carbolithium; Ceglution; Eskalith; Hypnorex; Limas; Liskonum; Lithane; Lithobid; Lithonate; Lithotabs; Phasal; Plenur; Priadel; Quilonorm-retard; Quilonum retard. CLi$_2$-O$_3$; mol wt 73.89. C 16.25%, Li 18.78%, O 64.96%. Li$_2$CO$_3$. Purification: Caley, Elving, *Inorg. Syn.* **1**, 1 (1939). Initial report of effect in mental illness: J. F. J. Cade, *Med. J. Aust.* **2**, 349 (1949). Early review of biology and pharmacology of lithium: M. Schou, *Pharmacol. Rev.* **9**, 17-58 (1957); of toxicity: *idem, Acta Pharmacol. Toxicol.* **15**, 70-84 (1958). Acute toxicity data: H. F. Smyth *et al.*, *Am. Ind. Hyg. Assoc. J.* **30**, 470 (1969). Toxicological studies in mammals: Gralla, McIlhenny, *Toxicol. Appl. Pharmacol.* **21**, 428 (1972). Pharmacokinetics: J. S. Ku *et al.*, *Int. J. Clin. Pharmacol. Ther. Toxicol.* **25**, 648 (1987). Clinical studies: J. Rybakowski *et al.*, *Int. Pharmacopsychiatry* **15**, 86 (1980); B. F. Kjellman *et al.*, *Acta Psychiatr. Scand.* **62**, 32 (1980). Review of development of use: G. Chouinard, *Union Med. Can.* **109**, 221-224, 226, 304 (1980). Comprehensive description: H. C. Stober in *Analytical Profiles of Drug Substances* vol. **15**, K. Florey, Ed. (Academic Press, New York, 1986) pp 367-391.
White, light, alkaline powder. d 2.11. Melting points of 618-720° have been reported: A. Reisman, *J. Am. Chem. Soc.* **80**, 3558 (1958). One gram dissolves in 78 ml cold, 140

ml boiling water; dissolved by dil acids. Practically insol in alcohol. LD_{50} orally in rats: 0.71 g/kg (Smyth).
USE: In the production of glazes on ceramic and electrical porcelain.
THERAP CAT: Antimanic.

5405. Lithium Chloride. ClLi; mol wt 42.40. Cl 83.63%, Li 16.37%. LiCl. Prepn: Pray, *Inorg. Syn.* **5**, 153 (1957). Review of prepn and properties: Oliver in *Mellor's* vol. II, supplement II, *The Alkali Metals* (part 1) 179-215 (1961).
Deliquesc, cubic crystals, granules or cryst powder; sharp saline taste. mp 613° when anhydr; bp 1360°; d 2.07. One gram dissolves in 1.3 ml cold, 0.8 ml boiling water; sol in alcohol, acetone, amyl alcohol, pyridine. The aq soln is neutral or slightly alkaline. *Keep well closed.* LD_{50} i.p. in mice: 1.06 g/kg, *Handbook of Toxicology* vol. **1**, W. S. Spector, Ed. (Saunders, Philadelphia, 1956) pp 178-179.
Human Toxicity: Prolonged absorption may cause disturbed electrolyte balance, impaired renal function, CNS disturbances.
USE: Manuf mineral waters; in pyrotechnics; soldering aluminum; in refrigerating machines.
THERAP CAT: Antidepressant.

5406. Lithium Chromate(VI). $CrLi_2O_4$; mol wt 129.87. Cr 40.04%, Li 10.69%, O 49.27%. Li_2CrO_4.
Dihydrate, yellow, deliquesc, cryst powder. Becomes anhydr at 74.6°. Very sol in water. The anhydr salt is appreciably sol in methanol and ethanol. Eutectic with water below —60°. *Keep well closed.*
USE: Corrosion inhibitor for water-cooled atomic reactors. Soln as low temp heat transfer medium.

5407. Lithium Citrate. *2-Hydroxy-1,2,3-propanetricarboxylic acid trilithium salt; citric acid trilithium salt;* Litarex; Lithonate S. $C_6H_5Li_3O_7$; mol wt 209.92. C 34.33%, H 2.40%, Li 9.92%, O 53.35%.
Tetrahydrate, white granules or cryst powder; feebly alkaline taste; deliquesces on exposure to moist air; loses all of its water at 105°. Sol in 1.5 parts water, slightly in alc. pH about 8. *Keep well closed.*
USE: Formerly in soft drinks.
THERAP CAT: Antidepressant.

5408. Lithium Dichromate(VI). Lithium bichromate. $Cr_2Li_2O_7$; mol wt 229.87. Cr 45.24%, Li 6.04%, O 48.72%. $Li_2Cr_2O_7$.
Dihydrate, yellowish-red, hygroscopic, cryst powder. Sol in less than 1 part water. *Keep well closed.*

5409. Lithium Fluoride. FLi; mol wt 25.94. Li 26.75%, F 73.25%. LiF. Prepd from LiOH and HF or by dissolving Li_2CO_3 in excess HF, evaporating to dryness, and heating to red heat: v. Wartenberg, Schultz, *Z. Elektrochem.* **27**, 568 (1921); Kwasnik in *Handbook of Preparative Inorganic Chemistry* vol. **1**, G. Brauer, Ed. (Academic Press, New York, 2nd ed., 1963) p 235.
Cubic crystals (NaCl lattice) or white fluffy powder. d^{20} 2.640. mp 848°; bp 1681°. Volatilizes at 1100-1200°. Soly in water (25°): 0.13 g/100 ml. With hydrofluoric acid it forms lithium bifluoride, $LiHF_2$. With lithium hydroxide it forms a double salt LiF.LiOH, mp 462°. LD orally in guinea pigs: 200 mg/kg.
USE: As flux for soldering and welding aluminum, in the manuf of vitreous enamels and glazes. Lithium fluoride prisms are used in infrared spectrophotometers.

5410. Lithium Formate. $CHLiO_2$; mol wt 51.96. C 23.11%, H 1.94%, Li 13.36%, O 61.59%. HCOOLi.
Monohydrate, colorless or white crystals. d 1.46. Sol in 3 parts water. The aq soln is practically neutral.

5411. Lithium Hydride. HLi; mol wt 7.95. H 12.70%, Li 87.30%. LiH. Prepd by the direct combination of hydrogen and lithium: Gibb, *Trans. Electrochem. Soc.* **93**, 198-211 (1948). Review of prepn and properties: Truter in *Mellor's* vol. **II**, supplement II, *The Alkali Metals* (part 1) 131-145 (1961).
Cubic crystals, darkens rapidly on exposure to light, the commercial product is usually gray. mp 680°. d 0.76-0.77. No solvent known. Rapidly dec in water to form lithium hydroxide and hydrogen. Reacts with the lower alcohols,

carboxylic acids, chlorine, and ammonia at 400° to liberate hydrogen.
USE: Reducing agent; condensing agent with ketones and acid esters; desiccant; in hydrogen generators; 1 g in water liberates approx 2.8 liters of hydrogen at STP.

5412. Lithium Hydroxide. Lithium hydrate. HLiO; mol wt 23.95. H 4.21%, Li 28.98%, O 66.80%. LiOH. Prepn: Barnes, *J. Chem. Soc.* **1931**, 2605; Cohen, *Inorg. Syn.* **5**, 3 (1957); Bravo, *ibid.* **7**, 1 (1963). Review of prepn, properties and uses: Oliver in *Mellor's* vol. **II**, supplement II, *The Alkali Metals* (part 1) 159-171 (1961).
Granular, free-flowing powder; acrid, strongly alkaline. Readily absorbs CO_2 and water from air. d 2.54. mp 471°. Sol in water, slightly sol in alcohol. *Keep tightly closed.* Monohydrate, small monoclinic crystals. d^{20} 1.51. Heat of formation —188.9 kcal/mol at 25°. Heat of soln —0.87 kcal/mol at 25°. Soly in water (w/w) at 0°: 10.7%; at 20°: 10.9%; at 100°: 14.8%. Slightly sol in alcohol. pH of a 1.0N soln about 14.
USE: In photographic developers; in alkaline storage batteries; in prepn of other lithium salts where use of carbonate is not practical; as catalyst in the production of alkyd resins, in esterifications; in the production of lithium soaps, greases, sulfonates. *Caution:* Strongly alkaline and hence caustic. Very irritating to skin. Systemic toxicity similar to other lithium compds. *See* Lithium.

5413. Lithium Iodide. ILi; mol wt 133.83. I 94.82%, Li 5.18%. LiI.
Trihydrate, white, deliquesc granules or fused masses; becomes yellow on exposure to air, due to liberation of iodine. mp 73°; mp 446° when anhydr. Sol in about 0.5 part water or alcohol, freely in amyl alcohol or acetone. The aq soln is neutral or slightly alkaline. *Keep tightly closed and protected from light.*
USE: In photography.

5414. Lithium Nitrate. $LiNO_3$; mol wt 68.95. Li 10.07%, N 20.32%, O 69.62%.
Colorless, deliquesc granules. d 2.38. mp about 255°. Sol in about 2 parts water; sol in alcohol. The aq soln is neutral. *Keep well closed.*

5415. Lithium Oxalate. $C_2Li_2O_4$; mol wt 101.88. C 23.58%, Li 13.62%, O 62.81%. $Li_2C_2O_4$.
White crystals. d 2.12. Sol in 15 parts water.

5416. Lithium Oxide. Li_2O; mol wt 29.88. Li 46.45%, O 53.55%. Prepd from lithium peroxide: Cohen, *Inorg. Syn.* **5**, 5 (1957); from lithium hydroxide: Bravo, *ibid.* **7**, 3 (1963). Review of prepn and properties: Oliver in *Mellor's* vol. **II**, supplement II, *The Alkali Metals* (part 1) 146-158 (1961).
Finely divided powder or crusty material. d^{25} 2.013. mp 1570°: van Arkel *et al.*, *Can. J. Chem.* **31**, 1009 (1953); 1427° (1700°K): Brewer, Margrave, *J. Phys. Chem.* **59**, 421 (1955). Readily absorbs carbon dioxide and water from the atm. At elevated temp attacks glass, silica, many metals.

5417. Lithium Perchlorate. $ClLiO_4$; mol wt 106.40. Li 6.52%, Cl 33.33%, O 60.15%. $LiClO_4$.
Small crystals. d^{25} 2.43. mp 236°. Decompn starts at about 400° and becomes rapid at 430° yielding lithium chloride and oxygen. Heat of formation: —99.94 kcal/mol at 25°. Soly in water (w/w) at 0°: 29.9%; at 25°: 37.5%; at 100°: 71.5%. Appreciably sol in alcohol, acetone, ether, ethyl acetate.
USE: Oxidizing agent. *Caution:* May be irritating on contact with skin, mucous membranes.

5418. Lithium Selenate. LiO_4Se; mol wt 149.90. Li 4.63%, O 42.70%, Se 52.67%. Prepd by roasting lithium selenite in air or by roasting lithium carbonate with selenium or selenium oxide: Lenher, Wechter, *J. Am. Chem. Soc.* **47**, 1522 (1925).
Monohydrate, monoclinic crystals. d 2.565. *Poisonous!* Stable in air. Readily sol in water.

5419. Lithium Selenite. LiO_3Se; mol wt 133.90. Li 5.18%, O 35.85%, Se 58.97%. $LiSeO_3$. Prepd from a soln of selenious acid in lithium hydroxide at 60°: Nilson, *Bull. Soc. Chim.* [2] **21**, 253 (1874); *ibid.* [2] **23**, 262 (1875).

Monohydrate, acicular crystals. Hygroscopic. More sol in cold water than in hot water.

5420. Lithium Silicate. Lithium metasilicate. Li_2O_3Si; mol wt 89.97. Li 15.43%, O 53.35%, Si 31.22%. Li_2SiO_3. Prepd by fusing Li_2CO_3 with SiO_2: Schwarz, Sturm, *Ber.* **47**, 1737 (1914).

Orthorhombic needles. d_4^{25} 2.52. mp 1201°. Heat of formation (solid): -434.9 kcal/mol. Heat of formation (liq) -374.6 kcal/mol. Latent heat of fusion (1177°) = 7.24 kcal/mol, also reported as -80.2 cal/g. Insol in cold water, dec by boiling water, dilute hydrochloric acid.

USE: To calibrate thermoelements.

5421. Lithium Sulfate. Lithiophor; Lithium-Duriles. Li_2O_4S; mol wt 109.88. Li 12.63%, O 58.25%, S 29.12%. Li_2SO_4.

Monohydrate, colorless crystals; loses the water at 130°. d 2.06. Sol in 2.6 parts water; almost insol in alcohol. The aq soln is neutral.

THERAP CAT: Antidepressant.

5422. Lithium Tetracyanoplatinate(II). Platinous lithium cyanide; lithium platinocyanide. $C_4Li_2N_4Pt$; mol wt 313.04. C 15.35%, Li 4.43%, N 17.90%, Pt 62.32%. $Li_2Pt(CN)_4$.

Pentahydrate, greenish-yellow crystals. Slightly sol in water.

USE: In x-ray photography.

5423. Lithocholic Acid. *3α-Hydroxy-5β-cholan-24-oic acid; 3α-hydroxycholanic acid; 17β-(1-methyl-3-carboxy-propyl)etiocholan-3α-ol.* $C_{24}H_{40}O_3$; mol wt 376.56. C 76.55%, H 10.71%, O 12.75%. Found in ox bile, human bile, rabbit bile, and in ox and pig gallstones. Isoln: Fischer, *Z. Physiol. Chem.* **73**, 234 (1911). Characterization: Wieland, Weyland, *ibid.* **110**, 123 (1920). Prepn from cholic or from desoxycholic acid: Hoehn, Mason, *J. Am. Chem. Soc.* **62**, 569 (1940); Sarel, Yanuka, *J. Org. Chem.* **24**, 2018 (1959). Crystal and molecular structure by x-ray diffraction: S. K. Arora et al., *Acta Crystallogr.* **32B**, 415 (1976).

Hexagonal leaflets from alcohol, prisms from acetic acid, mp 184-186°. $[\alpha]_D^{20}$ +33.7° (c = 1.5 in abs ethanol); $[\alpha]_D^{19}$ +23.3° (Wieland); $[\alpha]_D^{20}$ +32.1° (Fischer). Freely sol in hot alc. More sol in ether than cholic or desoxycholic acid. Sol in about 10 times its weight of ethyl acetate. Slightly sol in glacial acetic acid (about 0.2 g in 3 ml). More sol in benzene than desoxycholic acid. Insol in petr ether, gasoline, ligroin, water.

Methyl ester, $C_{25}H_{42}O_3$, crystallizes with ½ mol methanol, mp 125-127°.

Ethyl ester, $C_{26}H_{44}O_3$, crystals, mp 92-93°.

Benzyl ester, $C_{31}H_{46}O_3$, crystals, mp 145-148°.

Acetyllithocholic acid, $C_{26}H_{42}O_4$, crystals, mp 169°.

Acetyllithocholic acid methyl ester, $C_{27}H_{44}O_4$, flat needles from pentane, mp 123-130°.

Acetyllithocholic acid ethyl ester, $C_{28}H_{46}O_4$, crystals, mp 90-91°.

5424. Lithopone. *C.I. Pigment White 5;* Griffith's zinc white. A white pigment consisting of a mixture of zinc sulfide, barium sulfate and some zinc oxide. Several grades of lithopone are commercially available. The zinc sulfide content varies from 26 to 60%. The commercial importance of lithopone has decreased since the introduction of titanium pigments.

USE: In water and oil paints to provide thixotropy, improve gloss and flow.

5425. Litmocidin. Antibiotic substance produced by *Proactinomyces cyaneusantibioticus.* Isoln: Gause, *J. Bacte-*

riol. **51**, 649 (1946); Brazhnikova, *ibid.* 655; Abou-Zeid, El-Gammal, *Z. Allg. Mikrobiol.* **11**, 5 (1971). Belongs to the class of pigments and shows same qualitative reactions as anthocyanidine: Brazhnikova, *C.A.* **41**, 5576h (1947). Approx mol wt of 398-418: Paskhina, *Biokhimiya* **21**, 448 (1956).

5426. Litmus. Lacmus; tournesol; turnsole; lacca musica; lacca coerulea. Mol wt about 3300. Blue coloring matter from various species of lichens, particularly *Variolaria, Lecanora,* and *Roccella.* *Habit.* Scandinavia, shores of Mediterranean, Azores, California, East India, Madagascar. *Constit.* Chiefly azolitmin and erythrolitmin combined with alkalies: lecanoric acid, orcein, erythrolein. Manuf almost exclusively in Holland. Structure studies: Beecken et al., *Angew. Chem.* **73**, 665 (1961). Contains in small amounts α,β,γ-amino and hydroxyorcein.

Blue powder, lumps or cubes. Partly soluble in water or alcohol.

USE: As acid-base indicator; pH: 4.5 red, 8.3 blue. For preparing litmus papers; in microscopy to color culture media for diagnostic purposes. Has been used for coloring beverages.

5427. Liver Extract. An extract made from the livers of mammals. Upon ingestion or injection in a suitable dosage form it increases the number of red blood corpuscles in the blood of persons afflicted with pernicious anemia. Contains folic acid and vitamin B_{12} activity.

Some commercial products are: *Intraheptol; Pernaemon; Desiver; Anahaemin; Campolon; Campovit; Cromaton; Curethyl; Examen; Ficalon; Hepalon; Hepatopron; Hormantoxone; Hoban; Neo-Hepatex; Pernaemyl; Pernexin; Perniciosan; Plexan; Prolex; Reticulogen; Ripason; Sykoton; Tenelon; Hepol.*

Note: The definition given here comprises all types of liver extracts and does not differentiate between oral, injectable crude and injectable refined forms, or between solids and solns.

THERAP CAT: Hematinic; folic acid and vitamin B_{12} source.

5428. Livetins. Major water-soluble proteins found in egg yolk: Shepard, Hottle, *J. Biol. Chem.* **179**, 349 (1949). Prepn and mol wts of α- and β-livetin: Martin et al., *Can. J. Biochem. Physiol.* **35**, 241 (1957). Prepn and mol wt of γ-livetin: Martin, Cook, *ibid.* **36**, 153 (1958). Separation of four fractions, α_1-, α_2-, β-, and γ-livetins, and preparation of β- and γ-livetins: Oberdorfer, *Z. Physiol. Chem.* **331**, 280 (1963). Protein constituents of the livetin fraction of egg yolk were identified as follows: α-livetin identified as serum albumin; β-livetin as an α_2-glycoprotein with a sedimentation coeff of 2.95; and γ-livetin as serum γ-globulin: Williams, *Biochem. J.* **83**, 346 (1962). Heterogeneity of livetin proteins in egg yolk: S. F. Hui, R. H. Common, *Can. J. Biochem.* **44**, 1357 (1966); W. M. McIndoe, J. Culbert, *Int. J. Biochem.* **10**, 659 (1979).

5429. Lobelanidine. *1-Methyl-α,α-diphenyl-2,6-piperidinediethanol; α,α'-[(1-methyl-2,6-piperidinediyl)dimethylene]dibenzyl alcohol.* $C_{22}H_{29}NO_2$; mol wt 339.46. C 77.84%, H 8.61%, N 4.13%, O 9.43%. From herb of *Lobelia inflata* L., *Lobeliaceae:* Wieland et al., *Ann.* **444**, 40 (1925). Synthesis: Wieland, Drishaus, *ibid.* **473**, 102 (1929). Comprehensive review of H. Wieland's work on Lobelia alkaloids with complete references is given by Schöpf, *Naturwiss.* **30**, 359 (1942).

Scales from alc, mp 150°. Distills unchanged *in vacuo.* Freely sol in benzene, chloroform, acetone; slighty sol in ether, petr ether; almost insol in water.

Hydrochloride, $C_{22}H_{29}NO_2$.HCl, needles from alcohol, mp 135-138°.

Hydrobromide, $C_{22}H_{29}NO_2$.HBr, crystals, mp 189°.

5430. Lobelanine. *2,2-(1-Methyl-2,6-piperidinediyl)bis [1-phenylethanone]; 2,2''-(1-methyl-2,6-piperidinediyl)di-*

acetophenone. $C_{22}H_{25}NO_2$; mol wt 335.43. C 78.77%, H 7.51%, N 4.18%, O 9.54%. After lobeline, the most abundant alkaloid of *Lobelia inflata* L., *Lobeliaceae.* Isoln: Wieland *et al., Ann.* **444**, 40 (1925). Synthesis: Schöpf, Lehmann, *ibid.* **518**, 1 (1935); Parker *et al., J. Chem. Soc.* **1959**, 2433.

CH_3 structure diagram

Rosettes of needles from petr ether or ether, mp 99°. Freely sol in alcohol, acetone, benzene, chloroform; slightly sol in water, ether.
Hydrochloride, $C_{22}H_{25}NO_2 \cdot HCl$, crystals from dil alcohol, dec 188°. Sol in chloroform; slightly sol in abs alcohol, cold water.
Hydrobromide, $C_{22}H_{25}NO_2 \cdot HBr$, dec 188°.
Nitrate, $C_{22}H_{25}NO_2 \cdot HNO_3$, mp 153-154°.

5431. Lobelia. Indian tobacco; wild tobacco; emetic herb; asthma weed; bladder pod; vomit wort. Dried leaves and tops of*Lobelia inflata* L., *Lobeliaceae.* (The seeds are also used.) *Habit.* Canada, U.S. *Constit.* Leaves and tops: chiefly lobeline, also lobelidine, lobelanine, lobelanidine, and other alkaloids. Seeds: lobeline, fixed oil.
THERAP CAT: Expectorant.
THERAP CAT (VET): Has been used as an expectorant.

5432. Lobeline. *2-[6-(2-Hydroxy-2-phenylethyl)-1-methyl-2-piperidinyl]-1-phenylethanone; 2-[6-(β-hydroxyphenethyl)-1-methyl-2-piperidyl]acetophenone.* $C_{22}H_{27}NO_2$; mol wt 337.47. C 78.30%, H 8.07%, N 4.15%, O 9.48%. From herb and seeds of *Lobelia inflata* L., *Lobeliaceae* (Indian tobacco): Wieland, Ber. **54**, 1784 (1921). Structure: Wieland, Dragendorff, *Ann.* **473**, 83 (1929); Wieland , Drishaus, *ibid.* 102; Wieland *et al., ibid.* 118; Schering, Winterhalder, *ibid.* 126; Thomä, *Ann.* **540**, 99 (1939). Absolute configuration: Schöpf, Müller, *Ann.* **687**, 241 (1965). Pharmacology: Cambar *et al., Arch. Int. Pharmacodyn. Ther.* **177**, 1 (1969); Korczyn *et al., ibid.* **182**, 370 (1969); Mansuri *et al., Arzneimittel-Forsch.* **23**, 1721 (1973).

CH_3 structure diagram

L-Form, *α-lobeline; inflatine.* Needles from alc, ether, benzene, mp 130-131°. $[\alpha]_D^{15}$ −43° (alc). Very slightly sol in water or in petr ether; sol in hot alcohol, or in chloroform, benzene, ether.
Hydrochloride, $C_{22}H_{28}ClNO_2$, *Zoolobelin, Lobron.* Rosettes of slender needles from alcohol, mp 178-180°. $[\alpha]_D^{20}$ −43° (c = 2). One gram dissolves in 40 ml water, 12 ml alcohol; very sol in chloroform. Water solns are slightly acid to litmus.
Sulfate, $C_{44}H_{56}N_2O_8S$, *Bantron, Unilobin, Lobeton, Lobidan, Toban o-t-c.* Hygroscopic. Crystals from alc. $[\alpha]_D^{20}$ −25° (c = 2). Soluble in about 30 parts water, slightly in alcohol.
DL-Form, *lobelidine.* Prisms, mp 110°. Has approx same solubilities as L-form.
THERAP CAT: Respiratory stimulant.
THERAP CAT (VET): Has been used as a respiratory stimulant, ruminatoric.

5433. Lobenzarit. *2-[(2-Carboxyphenyl)amino]-4-chlorobenzoic acid; N-(2-carboxyphenyl)-4-chloroanthranilic acid; 4-chloro-2,2′-iminodibenzoic acid; CCA.* $C_{14}H_{10}ClNO_4$; mol wt 291.69. C 57.65%, H 3.46%, Cl 12.15%, N 4.80%, O 21.94%. Anti-arthritic agent with an immunomodulating mechanism of action. Prepn: M. Tanemura *et al.,* Ger. pat. **2,526,092**; *eidem,* U.S. pat. **4,092,426** (1976, 1978 both to Chugai). CCA lacks anti-inflammatory activity in carrageenin paw edema assay or cotton pellet granuloma test, but suppresses secondary inflammation in adjuvant

arthritic rats: Y. Ohsugi *et al., J. Pharm. Pharmacol.* **29**, 636 (1977). Prevention of autoimmune kidney disease in mice: *eidem, ibid.* **30**, 126 (1978). Immunoregulating activity: *eidem, Int. J. Immunopharmacol.* **2**, 224 (1980). Multicenter clinical trial in rheumatoid arthritis: Y. Shiokawa *et al., J. Rheumatol.* **11**, 615 (1984).

COOH COOH structure diagram with Cl

Crystals, mp > 306°. LD_{50} in male, female rats (mg/kg): 2100, 2600 orally (Tanemura).
Disodium salt, $C_{14}H_8ClNNa_2O_4$, *Carfenil, Cerfenil.* Solid, mp 388° (dec).
THERAP CAT: Antirheumatic.

5434. Lochnericine. *2,3-Didehydro-6,7-epoxyaspidospermidine-3-carboxylic acid methyl ester.* $C_{21}H_{24}N_2O_3$; mol wt 352.42. C 71.57%, H 6.86%, N 7.95%, O 13.62%. R = H. From *Vinca rosea* Linn. (*Catharanthus roseus* G. Don.), *Apocynaceae:* Gorman *et al., J. Am. Pharm. Assoc.* **48**, 256 (1959); Svoboda *et al., ibid.* **48**, 659 (1959); Moza, Trojánek, *Coll. Czech. Chem. Commun.* **28**, 1419 (1963). Structure: Moza *et al., Tetrahedron Letters* **1964**, 2561.

structure diagram with C_2H_5, R, COOCH_3

Crystals from methanol, dec 190-193°. $[\alpha]_D^{27}$ −432° (chloroform). pKa in 66% DMF: 4.2. uv max (ethanol): 227, 299, 328 nm (log ε 4.10, 4.15, 4.32).
Methoxylochnericine, $C_{22}H_{26}N_2O_4$, *lochnerinine.* R = OCH_3. Isoln: Moza, Trojánek, *loc. cit.* Structure: Moza *et al., loc. cit.* Prisms from methanol + acetone, mp 168-169°. $[\alpha]_D^{16}$ −424° (chloroform). uv max (methanol): 247, 326 nm (log ε 4.20, 4.33).

5435. Lochneridine. *2,16-Didehydro-20-hydroxycuran-17-oic acid methyl ester.* $C_{20}H_{24}N_2O_3$; mol wt 340.41. C 70.56%, H 7.11%, N 8.23%, O 14.10%. From *Vinca rosea* Linn., *Apocynaceae:* Svoboda *et al., J. Pharm. Sci.* **50**, 409 (1961). Structure: Nakagawa *et al., Chem. & Ind. (London)* **1962**, 1986.

structure diagram with COOCH_3, H, OH, CH_2CH_3

Prisms from methanol, dec 211-214°. $[\alpha]_D^{26}$ +607.5° (chloroform). pKa 5.5 in 66% DMF. uv max (ethanol): 230, 293, 328 nm (log ε 4.04, 3.94, 4.07).

5436. Locust Bean Gum. Carob flour; Johannisbrotmehl; Arobon. Mol wt about 310,000. The ground kernel endosperms of tree pods of *Ceratonia siliqua* L., *Leguminosae* (St. John's bread). Consists of proteins such as albumins, globulins, prolamins, gluteline; carbohydrates such as reducing sugars, sucrose, dextrins, pentosans; ash; fat; crude fiber; moisture. *Refs:* Plaut *et al., Bull. Res. Council Israel* **3**, No. 11, 129 (1953), *C.A.* **48**, 5397 (1954); Griffiths, *Mfg. Chemist* **20**, 321 (1949); Coit, *Econ. Bot.* **5**, 82 (1951). Use as coffee, chocolate, cocoa substitute, extender: W. A. Meer, *Manuf. Confectioner* **59**, 41 (1979). Caffeine, theobromine content: W. J. Craig, T. T. Nguyen, *J. Food Sci.* **49**, 302 (1984). Re-

view: F. Rol in *Industrial Gums,* R. L. Whistler, Ed. (Academic Press, New York, 2nd ed., 1973) pp 323-337.

Yellow-green color, is odorless and tasteless, but acquires a leguminous taste when boiled in water.

USE: Stabilizer, thickener, and binder in foods and cosmetics. Coffee, chocolate, cocoa substitute. Sizing and finishing agent in textiles. As fiber bonding in paper manuf. Drilling mud additive.

THERAP CAT: Adsorbent-demulcent.

5437. Lofentanil. *3-Methyl-4-[(1-oxopropyl)phenylamino]-1-(2-phenylethyl)-4-piperidinecarboxylic acid methyl ester;* (−)-*cis*-3-methyl-1-phenethyl-4-(N-phenylpropionamido)isonipecotic acid methyl ester. $C_{25}H_{32}N_2O_3$; mol wt 408.55. C 73.50%, H 7.90%, N 6.85%, O 11.75%. Prepn: P. A. J. Janssen, G. H. P. Van Daele, Ger. pat. **2,610,228** corresp to U.S. pat. **3,998,834** (both 1976 to Janssen). Receptor affinity and pharmacological potency: K. D. Stahl *et al., Eur. J. Pharmacol.* **46**, 199 (1977). Structural study: C. De Ranter *et al., Arch. Int. Physiol. Biochim.* **87**, 1031 (1979). Antinociceptive effects in cats: A. S. Tung, T. L. Yaksh, *Pain* **12**, 343 (1982). Opiate receptor binding studies of ³H-lofentanil: W. Gommeren, J. E. Leysen, *Arch. Int. Pharmacodyn. Ther.* **258**, 171 (1982); B. Ilien *et al., ibid.* 313; P. M. Laduron, P. Janssen, *Life Sci.* **31**, 457 (1982).

Oxalate, $C_{27}H_{34}N_2O_7$, *R 34995.* Solid, mp 177°.
THERAP CAT: Narcotic analgesic.

5438. Lofepramine. *1-(4-Chlorophenyl)-2-[[3-(10,11-dihydro-5H-dibenz[b,f]azepin-5-yl)propyl]methylamino]-ethanone; 4'-chloro-2-[[3-(10,11-dihydro-5H-dibenz[b,f]azepin-5-yl)propyl]methylamino]acetophenone; N-methyl-N-(4-chlorobenzoylmethyl)-3-(10,11-dihydro-5H-dibenzo-[b,f]azepin-5-yl)propylamine;* lopramine. $C_{26}H_{27}ClN_2O$; mol wt 418.97. C 74.54%, H 6.50%, Cl 8.46%, N 6.68%, O 3.82%. Psychotropic drug related to imipramine, *q.v.* Prepn: E. Eriksoo *et al.,* Brit. pat. **1,177,525**; *eidem,* U.S. pat. **3,-637,660** (1970, 1972 both to AB Leo). Chemistry and pharmacology: E. Eriksoo, O. Rohte, *Arzneimittel-Forsch.* **20**, 1561 (1970). Absorption and metabolism: J. R. Tulic *et al., Acta Pharmacol. Toxicol.* **32**, 304 (1973). Distribution and excretion: G. Plym Forshell, *Xenobiotica* **5**, 73 (1975). Pharmacokinetics: G. Plym Forshell *et al., Eur. J. Clin. Pharmacol.* **9**, 291 (1976). Clinical study: S. Wright, L. Herrmann, *Arzneimittel-Forsch.* **26**, 1167 (1976).

Crystals from methanol or acetone, mp 104-106°. Easily oxidized by air and other oxidizing agents to desipramine and *p*-chlorobenzoic acid.

Hydrochloride, $C_{26}H_{28}Cl_2N_2O$, *Leo 640, Amplit, Gamanil, Gamonil, Timelit, Tymelyt.* Crystals from butanone, mp 152-154°. Sol in methanol, ethanol, chloroform. Practically insol in water. LD_{50} in mice, rats (mg/kg): > 2500, > 1000 orally; 920, > 1000 i.p.; > 1000, > 1000 s.c. (Eriksoo, Rohte).
THERAP CAT: Antidepressant.

5439. Lofexidine. *2-[1-(2,6-Dichlorophenoxy)ethyl]-4,5-dihydro-1H-imidazole; 2-[1-(2,6-dichlorophenoxy)eth-yl]-2-imidazoline.* $C_{11}H_{12}Cl_2N_2O$; mol wt 259.13. C 50.99%, H 4.67%, Cl 27.36%, N 10.81%, O 6.17%. Vasoactive agent related structurally to clonidine, *q.v.* Prepn of the HCl salt: H. Baganz, H. J. May, S. Afr. pat. **68 00,850** cor-

resp to U.S. pat. **3,966,757** (1968, 1976 both to Nordmark); of the free base: *eidem,* Ger. pat. **1,935,479** (1971 to Nordmark), *C.A.* **74**, 87979 (1971). Pharmacological studies: J. Velly, *J. Pharmacol.* **8**, 351 (1977); B. Jarrot *et al., Biochem. Pharmacol.* **28**, 141 (1979). NMR data and cardiovascular effects: P. B. M. Timmermans, P. A. Van Zwieten, *Eur. J. Med. Chem.* **15**, 323 (1980). Hypotensive and sedative properties: P. Birch *et al., Brit. J. Pharmacol.* **68**, 107 (1980). Effects in hypertension: N. D. Vlachakis *et al., Fed. Proc.* **39**, 4844 (1980). Series of articles on pharmacology, toxicity studies, clinical studies: *Arzneimittel-Forsch.* **32**, 915-993 (1982). Toxicity: T. H. Tsai *et al., ibid.* 955.

Cryst, mp 126-128°.

Hydrochloride, $C_{11}H_{13}Cl_3N_2O$, *BA 168, MDL-14042A, Lofetensin, Loxacor.* Cryst from ethanol/ether or 2-propanol, mp 221-223° (U.S. patent); also reported as mp 230-232° (Ger. patent). Very sol in water, ethanol. Slightly sol in 2-propanol. Practically insol in ether. LD_{50} in mice, rats, dogs (mg/kg): between 74-147 orally; between 8-18 i.v. (Tsai).
THERAP CAT: Antihypertensive.

5440. Loflucarban. *N-(3,5-Dichlorophenyl)-N'-(4-fluorophenyl)thiourea; 3,5-dichloro-4'-fluorothiocarbanil-ide;* Fluonilid. $C_{13}H_9Cl_2FN_2S$; mol wt 315.21. C 49.53%, H 2.88%, Cl 22.50%, F 6.03%, N 8.89%, S 10.17%. Prepd from *p*-fluorophenyl isothiocyanate and 3,5-dichloroaniline or from 3,5-dichlorophenyl isothiocyanate and *p*-fluoroaniline: Belg. pat. **613,154** (1962 to Madan), *C.A.* **58**, 474f (1963).

Crystals from ethanol, mp 148°. Soluble in ethyl oleate, isopropyl myristate.
THERAP CAT: Antifungal.

5441. Loganin. *1-(β-D-Glucopyranosyloxy)-1,4a,5,6,7,-7a-hexahydro-6-hydroxy-7-methylcyclopenta[c]pyran-4-carboxylic acid methyl ester;* 7-hydroxy-6-desoxyverbenalin. $C_{17}H_{26}O_{10}$; mol wt 390.40. C 52.30%, H 6.71%, O 40.99%. Key intermediate in the biosynthesis of indole alkaloids. First isolated from the seeds but chiefly from the pulp of the fruit of *Strychnos nux-vomica* L., *Loganiaceae:* Dunstan, Short, *Pharm. J.* **14**, 1025 (1883); Merz, Krebs, *Arch. Pharm.* **275**, 217 (1937); Merz, Lehmann, *ibid.* **290**, 543 (1957). Structure: Sheth *et al., Tetrahedron Letters* **1961**, 394; Büchi, Manning, *Tetrahedron* **18**, 1049 (1962). Crystal structure: Lentz, Rossmann, *Chem. Commun.* **1969**, 1269; P. G. Jones *et al., Acta Crystallogr.* **B36**, 481 (1980). Abs config: Inouye *et al., Tetrahedron* **26**, 3905 (1970). Total synthesis: Büchi *et al., J. Am. Chem. Soc.* **92**, 2165 (1970); Partridge *et al., ibid.* **95**, 532 (1973); Büchi *et al., ibid.* 540; B.-W. Au-Yeung, I. Fleming, *Chem. Commun.* **1977**, 81; I. Fleming, B.-W. Au-Yeung, *Tetrahedron* **37**, Suppl. 9, 13 (1981); K. Hiroi *et al., Chem. Letters* **1981**, 559. Biosynthetic studies: Battersby, "Biosynthesis II—Terpenoid Indole Alkaloids", in *The Alkaloids* vol. 1, The Chemical Society (Burlington House, London, 1971) pp 31-47.

Crystals, mp 222-223°. $[\alpha]_D^{20}$ —82.1° (water). Freely sol in water; less sol in 96% alcohol; sparingly in abs alcohol. Practically insol in ether, petr ether, ligroin, ethyl acetate, acetone, chloroform.

5442. Loline. *Hexahydro-N-methyl-2,4-methano-4H-furo[3,2-b]pyrrol-3-amine;* festucine; methyl-*N*-depropion-yldecorticasine. $C_8H_{14}N_2O$; mol wt 154.21. C 62.30%, H 9.15%, N 18.17%, O 10.38%. One of a group of pyrrolizidine alkaloids whose common feature is a unique oxygen bridge. Isoln from seeds of *Lolium cuneatum* Nevski, *Gramineae:* S. Y. Yunusov, S. T. Akramov, *J. Gen. Chem. USSR (Engl. Transl.)* **25,** 1765 (1955); from tall fescue hay, *Festuca arundinaceae* Schreb., *Gramineae:* Yates, Tookey, *Aust. J. Chem.* **18,** 53 (1965). Identity with festucine: Aasen, Culvenor, *ibid.* **22,** 2021 (1969). Identity with methyl-*N*-depropionyl-decorticasine: Ribas *et al., An. Quim.* **64,** 516 (1968), *C.A.* **69,** 109764c (1968). Abs config: R. B. Bates, S. R. Morehead, *Tetrahedron Letters* **1972,** 1629. Synthesis of bisquaternary salts: N. P. Abdullaev *et al., Khim. Prir. Soedin.* **3,** 371 (1977), *C.A.* **88,** 23216 (1978). Synthetic and structural studies: S. R. Wilson *et al., J. Org. Chem.* **46,** 3887 (1981).

Dihydrochloride, $C_8H_{14}N_2O.2HCl$, needles from abs ethanol, dec 237-242°. $[\alpha]_D^{25}$ +4.6° (c = 4.37 in water). pKa 8.25 and 2.5-3.0.

5443. Lomefloxacin. *1-Ethyl-6,8-difluoro-1,4-dihydro-7-(3-methyl-1-piperazinyl)-4-oxo-3-quinolinecarboxylic acid.* $C_{17}H_{19}F_2N_3O_3$; mol wt 351.35. C 58.11%, H 5.45%, F 10.81%, N 11.96%, O 13.66%. Fluorinated quinolone antibacterial. Prepn: Y. Itoh *et al.,* **Ger.** pat. 3,433,924; *eidem,* **U.S.** pat. 4,528,287 (both 1985 to Hokuriku Pharm.). *In vitro* and *in vivo* activity: T. Hirose *et al., Antimicrob. Ag. Chemother.* **31,** 854 (1987). Comparative antibacterial spectrum *in vitro:* R. Wise *et al., ibid.* **32,** 617 (1988). HPLC determn in urine and bile and preliminary pharmacokinetics in rats: A. Saito *et al., ibid.* 156. Supplement on antibacterial spectrum, pharmacokinetics and clinical efficacy: *Chemotherapy (Japan)* **36,** Suppl. 2, 1-1418 (1988).

Colorless needles from ethanol, mp 239-240.5°. LD_{50} in mice (mg/kg): 245.6 i.v.; >4000 orally (Itoh).

Monohydrochloride, $C_{17}H_{20}ClF_2N_3O_3$, *NY-198, SC-47111, Bareon.* Colorless needles from water, mp 290-300° (dec).

THERAP CAT: Antibacterial.

5444. Lomustine. *N-(2-Chloroethyl)-N'-cyclohexyl-N-nitrosourea;* 1-(2-chloroethyl)-3-cyclohexyl-1-nitrosourea; NSC 79037; RB 1509; Belustine; CCNU; Cecenu; CeeNU; CiNU. $C_9H_{16}ClN_3O_2$; mol wt 233.69. C 46.25%, H 6.90%, Cl 15.17%, N 17.98%, O 13.69%. Chloroethylnitrosourea derivative with antitumor activity. Similar to carmustine, chlorozotocin, nimustine, ranimustine, *q.q.v.* Synthesis: T. P. Johnston *et al., J. Med. Chem.* **9,** 892 (1966). Clinical review: S. K. Carter, J. W. Newman, *Cancer Chemother. Rep. Part 3* **1,** 136 (1968). Physiological disposition: V. T. Oliverio *et al., Cancer Res.* **30,** 1330 (1970). Toxicology and pharmacology: G. R. Thompson, R. E. Larson, *Toxicol. Appl. Pharmacol.* **21,** 405 (1972); V. T. Oliverio, *Cancer Chemother. Rep. Part 3* **4,** 13 (1973).

Yellow powder, mp 90°. Soly in water, 0.1*N* NaOH, 0.1*N* HCl or 10% ethanol: <0.05 mg/ml; in absolute ethanol: 70 mg/ml. The bulk drug should be stored in a deep freeze and protected from moisture. LD_{50} in male mice (mg/kg): 51 orally; 56 i.p.; 61 s.c. (Thompson, Larson).

Note: This substance may reasonably be anticipated to be a carcinogen: *Fourth Annual Report on Carcinogens* (NTP 85-002, 1985) p 55.

THERAP CAT: Antineoplastic.

5445. Lonazolac. *3-(4-Chlorophenyl)-1-phenyl-1H-pyrazole-4-acetic acid.* $C_{17}H_{13}ClN_2O_2$; mol wt 312.75. C 65.29%, H 4.19%, Cl 11.33%, N 8.96%, O 10.23%. Prepn: R. A. Newberry, **Brit.** pat. **1,373,212** (1974), *C.A.* **82,** 72987t (1975); G. Rainer, **U.S.** pat. **4,146,721** (1979 to Byk Gulden); G. Rainer *et al., Arzneimittel-Forsch.* **31,** 649 (1981). Pharmacology: R. Riedel, *ibid.* 655. Clinical studies: G. Lonauer *et al., Z. Rheumatol.* **40,** 161 (1981); W. Siegmeth, P. Placheta, *Wien Klin. Wochenschr.* **94,** 145 (1982).

Cryst from ethanol/water, mp 150-151°. uv max (methanol): 281 nm (ϵ 24,800); (0.1*N* NaOH): 281 nm (ϵ 23,700). pKa 4.3. LD_{50} in male mice, rats: 195, 165 mg/kg i.v., R. Riedel, *loc. cit.*

Calcium salt, $C_{34}H_{24}CaCl_2N_4O_4$, *Irritren.* Solid, melts between 270-290° (dec). uv max (0.1*N* NaOH): 280 nm (ϵ 46,400). LD_{50} orally in mice, rats (mg/kg): 670, 845 (males); 730, 1000 (females), R. Riedel, *loc. cit.*

THERAP CAT: Anti-inflammatory.

5446. Longifolene. *Decahydro-4,8,8-trimethyl-9-methylene-1,4-methanoazulene;* junipene; kuromatsuene. $C_{15}H_{24}$; mol wt 204.36. C 88.16%, H 11.84%. Tricyclic sesquiterpene present to the extent of 5-10% in Indian turpentine oil which is produced commercially from the Himalayan pine, *Pinus longifolia* Roxb., *Pinaceae:* J. L. Simonsen, *J. Chem. Soc.* **117,** 570 (1920). Structure: R. H. Moffett, D. Rogers, *Chem. & Ind. (London)* **1953,** 916; P. Naffa, G. Ourisson, *ibid.* 917. Abs config: G. Ourisson, *Bull. Soc. Chim. France* **1955,** 895. Identity of longifolene with junipene and kuromatsuene: S. Akiyoshi *et al., Tetrahedron* **9,** 237 (1960). Total synthesis of naturally occurring (+)-form: E. J. Corey *et al., J. Am. Chem. Soc.* **86,** 478 (1964); of the (±)-form: *eidem, ibid.* **83,** 1251 (1961); J. E. McMurray, S. J. Isser, *ibid.* **94,** 7132 (1972); R. A. Volkmann *et al., ibid.* **97,** 4777 (1975). Review of chemistry of longifolene and its derivs: S. Dev, *Accounts Chem. Res.* **14,** 82-88 (1981); *idem, Fortschr. Chem. Org. Naturst.* **40,** 50-104 (1981).

Viscous oil, bp_{706} 254-256°, bp_{15} 126-127°. $[\alpha]_D^{18}$ +42.73. d_4^{18} 0.9319; n_D^{20} 1.5040. Insol in water; sol in benzene.

Borane deriv, $C_{30}H_{51}B$, *dilongifolylborane.* Heavy, snow-white, shiny plates, mp 160-161° (sealed, evacuated capillary). Strongly dimeric. Sparingly sol in common organic solvents. Prepn and use in hydroboration: P. K. Jadhav, H. C. Brown, *J. Org. Chem.* **46,** 2988 (1981).

USE: Borane deriv as a chiral hydroborating agent.

5447. Lonidamine. *1-[(2,4-Dichlorophenyl)methyl]-1H-indazole-3-carboxylic acid; 1-(2,4-dichlorobenzyl)-1H-indazole-3-carboxylic acid; diclondazolic acid; DICA; AF 1890; Doridamina.* $C_{15}H_{10}Cl_2N_2O_2$; mol wt 321.16. C 56.10%, H 3.14%, Cl 22.08%, N 8.72%, O 9.96%. Prepn: G. Palazzo, B. Silvestrini, **Ger. pat. 2,310,031;** *eidem,* **U.S. pat. 3,895,026** (1973, 1975 both to Angelini Francesco); G. Corsi, G. Palazzo, *J. Med. Chem.* **19,** 778 (1976). Antispermatogenic activity and mechanism of action study: T. J. Lobl, *Arch. Androl.* **2,** 353 (1979); S. K. Singh, C. J. Dominic, *Exp. Clin. Endocrinol.* **83,** 291 (1984). *In vitro* and *in vivo* antitumor activity: B. Silvestrini *et al., Brit. J. Cancer* **47,** 221 (1983). Effects on protein synthesis in neoplastic cells: A. Floridi *et al., Exp. Mol. Pathol.* **42,** 293 (1985). Pharmacological activities: V. Cioli *et al., Arzneimittel-Forsch.* **34,** 455 (1984). Clinical evaluations in treatment of renal and breast cancers: B. H. Weinerman *et al., Cancer Treat. Rep.* **70,** 751 (1986); P. R. Band *et al., ibid.* 1305. HPLC determn in plasma and urine: R. Leclaire *et al., J. Chromatog.* **277,** 427 (1983). Toxicological study: R. Heywood *et al., Chemotherapy* **27,** Suppl. 2, 91 (1981). Series of articles on mechanism of action, pharmacokinetics and clinical applications in spermatogenesis and tumors: *ibid.* 5-120. Series of articles on pharmacology and clinical evaluation in cancers: *Oncology* **41,** Suppl. 1, 1-121 (1984). *Review:* B. Silvestrini *et al.,* "Lonidamine and Related Compounds" in *Prog. Med. Chem.* **21,** G. P. Ellis, G. B. West, Eds. (Elsevier Science Publishers, Amsterdam, 1984) pp 111-135.

Crystals from ethanol, mp 207°. Sol in methanol, acetic acid. LD_{50} in mice, rats (mg/kg): 900, 1700 orally; 435, 525 i.p. (Heywood).

THERAP CAT: Antineoplastic.

5448. Lonomycins. Polyether antibiotic complex with anticoccidial activity. Isoln from *Streptomyces ribosidificus* strain TM-481 as sodium salt: M. Shibata *et al.,* **Japan. Kokai 75 49,495** (to Taisho Pharm.), *C.A.* **83,** 191339p (1975); S. Omura, *J. Antibiot.* **29,** 15 (1976). The major component is lonomycin A (initially named lonomycin). Lonomycin B and C are minor congeners. Identity of A with *emericid, antibiotic DE-3936:* M. Oshima *et al., ibid.* 354. Structure of A: M. Otake, M. Koenuma, *Tetrahedron Letters* **1975,** 4147; C. Riche, C. Pascard-Billy, *Chem. Commun.* **1975,** 951; H. Seto *et al., J. Antibiot.* **33,** 979 (1980); of B, C: *eidem, ibid.* **31,** 929 (1978); T. Mizutani *et al., ibid.* **33,** 1224 (1980). Determn by reaction with vanillin: R. L. Brown, *J. Assoc. Off. Anal. Chem.* **62,** 266 (1979). Ionophoric activity: M. Mitani, N. Otake, *J. Antibiot.* **31,** 750 (1978). Cardiovascular effects in dogs: K. Tsuchida *et al., Japan. J. Pharmacol.* **38,** 109 (1985). Toxoplasmacidal activity: T. Miyagami *et al., J. Antibiot.* **34,** 218 (1981).

Lonomycin A

Lonomycin A sodium salt, $C_{44}H_{75}NaO_{14}$, colorless prisms, mp 188-189°. $[\alpha]_D^{25}$ +47° (c = 1 in methanol).
Lonomycin B sodium salt, $C_{44}H_{75}NaO_{14}$, mp 181-182°.
Lonomycin C sodium salt, $C_{43}H_{73}NaO_{14}$, mp 186-187°.

5449. Looplure. *7-Dodecen-1-ol acetate; cis-7-dodecenyl acetate; cis-1-acetoxy-7-dodecene;* ENT 33266. $C_{14}H_{26}O_2$; mol wt 226.36. C 74.28%, H 11.58%, O 14.14%. Sex pheromone of the female cabbage looper moth, *Trichoplusia ni* (Hübner). Isoln, identification and synthesis: R. S. Berger, *Ann. Entomol. Soc. Am.* **59,** 767 (1966). Synthesis and

activity of isomers: N. Green *et al., J. Med. Chem.* **10,** 533 (1967); H. H. Toba *et al., J. Econ. Entomol.* **63,** 1048 (1970). Alternate syntheses: W. Seidel *et al., Ber.* **110,** 3544 (1977); H. J. Bestmann *et al., Ber.* **112,** 1923 (1979); M. Horiike, T. Hirano, *Agr. Biol. Chem.* **44,** 2229 (1980); D. Basavaiah *et al., J. Org. Chem.* **47,** 1792 (1982).

Oil, $bp_{0.05}$ 98-100°. n_D^{25} 1.4420. LD_{50} orally in rats: > 13,430 mg/kg, M. Beroza *et al., Toxicol. Appl. Pharmacol.* **31,** 421 (1975).

USE: Insect sex attractant.

5450. Loperamide. *4-(4-Chlorophenyl)-4-hydroxy-N,N-dimethyl-α,α-diphenyl-1-piperidinebutanamide; 4-(p-chlorophenyl)-4-hydroxy-N,N-dimethyl-α,α-diphenyl-1-piperidinebutyramide.* $C_{29}H_{33}ClN_2O_2$; mol wt 477.04. C 73.02%, H 6.97%, Cl 7.43%, N 5.87%, O 6.71%. Prepn: Janssen *et al.,* **Fr. pat. 2,100,711** corresp to **U.S. pat. 3,714,159** (1972, 1973 to Janssen); Stokbroekx *et al., J. Med. Chem.* **16,** 782 (1973). Series of articles on pharmacology, toxicology, metabolism, and clinical studies: *Arzneimittel-Forsch.* **24,** 1640-1665 (1974). Toxicity studies: C. J. E. Niemegeers *et al., ibid.* 1633, 1636. *Review:* D. A. Shriver *et al.,* in *Pharmacological and Biochemical Properties of Drug Substances* **vol. 3,** M. E. Goldberg, Ed. (Am. Pharm. Assoc., Washington, DC, 1981) pp 461-476.

Hydrochloride, $C_{29}H_{34}Cl_2N_2O_2$, PJ 185, R 18553, *Arret, Blox, Brek, Dissenten, Fortasec, Imodium, Imosec, Lopemid, Lopemin, Loperyl, Suprasec, Tebloc.* Crystals from isopropanol, mp 222-223°. Practically insol in water at physiological pH (0.002%). Stable, can be stored for several years under normal conditions; not hygroscopic; not affected by light. LD_{50} in mice (mg/kg): 75 s.c.; 28 i.p.; 105 orally; in rats (mg/kg): 185 orally (Niemegeers).

THERAP CAT: Antidiarrheal.

5451. Lophophorine. *(S)-6,7,8,9-Tetrahydro-4-methoxy-8,9-dimethyl-1,3-dioxolo[4,5-h]isoquinoline; N-methylanhalonine.* $C_{13}H_{17}NO_3$; mol wt 235.27. C 66.36%, H 7.28%, N 5.95%, O 20.40%. In *Lophophora williamsii* (Lemaire) Coult., Cactaceae. Extraction procedure: Kauder, *Arch. Pharm.* **237,** 191 (1899). Structure: Späth, Gangl, *Monatsh.* **44,** 104, 106 (1923). Synthesis: Späth, Kesztler, *Ber.* **68,** 1667 (1935).

Oily liquid. $bp_{0.02}$ 140-145° (air bath temp). $[\alpha]_D^{25}$ −47° (c = 1 in chloroform). Sol in ether, chloroform.
Hydrochloride, $C_{13}H_{17}NO_3 \cdot HCl$, needles from alcohol. $[\alpha]_D^{17}$ −9.5° (c = 1). Sol in water, alcohol.
Picrate, mp 162-163°.

5452. Lophotoxin. *2-(Acetyloxy)-12-methyl-4-(1-methylethenyl)-17-oxo-11,16,18,19-tetraoxapentacyclo[12.2.2.1^{6,9}.0^{1,15}.0^{10,12}]nonadeca-6,8-diene-7-carboxaldehyde;* LTX. $C_{22}H_{24}O_8$; mol wt 416.43. C 63.45%, H 5.81%, O 30.74%. A novel neuromuscular toxin isolated from several spp of Pacific gorgonians (sea fans and whips) of the genus *Lophogorgia.* Although it belongs to the cembrene class of diterpenoids, the ability of LTX to cause irreversible postsynaptic blockade at the neuromuscular junction resembles the mode of action of the protein snake venom α-bungarotoxin (*cf.* bungarotoxins). Isoln and structure: W. Fenical *et al.*, *Science* **212**, 1512 (1981). Molecular mode of action: J. R. Smythies, *Med. Hypotheses* **7**, 1457 (1981), *C.A.* **96**, 137527 (1982). Toxicity study: P. Culver, R. S. Jacobs, *Toxicon* **19**, 825 (1981). For a description of the cytotoxic, ichthytoxic, and antibacterial activity of alcoholic extracts of gorgonians, *see* R. S. Jacobs *et al.*, *Fed. Proc.* **40**, 26 (1981); V. J. Paul, W. Fenical, *J. Org. Chem.* **45**, 3401 (1981).

White needles, mp 164-166°. $[\alpha]_D^{27}$ +14.2° (c = 1.7 in chloroform). LD_{50} in mice: 8.9 mg/kg s.c., P. Culver, R. S. Jacobs, *loc. cit.*

5453. Loprazolam. *6-(2-Chlorophenyl)-2,4-dihydro-2-[(4-methyl-1-piperazinyl)methylene]-8-nitro-1H-imidazo-[1,2-a][1,4]benzodiazepin-1-one.* $C_{23}H_{21}ClN_6O_3$; mol wt 464.91. C 59.42%, H 4.55%, Cl 7.63%, N 18.08%, O 10.32%. Annelated 1,4-benzodiazepine deriv. Prepn: J. B. Taylor, D. R. Harrison, **Ger. pat. 2,605,652;** *eidem,* **U.S. pat. 4,044,-142** (1976, 1977 both to Roussel-UCLAF). CNS activity: J. R. Agar *et al.*, *J. Med. Chem.* **20**, 1035 (1977). Pharmacological profile: T. G. Johns *et al.*, *Arch. Int. Pharmacodyn. Ther.* **240**, 53 (1979). Effects on sleep in normal volunteers: I. Hindmarch, C. A. Clyde, *Drugs Exp. Clin. Res.* **2**, 61 (1980). Review of pharmacology and efficacy in insomnia: B. G. Clark *et al.*, *Drugs* **31**, 500-516 (1986).

Crystals from chloroform/ether, mp 214-215°. LD_{50} in mice: > 1000 mg/kg orally (Agar).

Methanesulfonate, $C_{24}H_{25}ClN_6O_6S$, *HR-158, RU-31158, Avlane, Dormonoct, Somnovit, Sonin.* Crystals from methylene chloride, mp 205-210°.

Note: This is a controlled substance (depressant) listed in the U.S. Code of Federal Regulations, Title 21 Part 1308.14 (1985).

THERAP CAT: Hypnotic.

5454. Lorajmine. *(17R,21α)-Ajmalan-17,2l-diol 17-chloroacetate;* 17-monochloroacetylajmaline; 17-chloroacetylajmaline; MCAA. $C_{22}H_{27}ClN_2O_3$; mol wt 412.92. C 65.58%, H 6.75%, Cl 8.80%, N 6.95%, O 11.91%. Semisynthetic ajmaline derivative. Prepn from ajmaline: **Belg. pat. 622,395** (1962 to Inverni Della Beffa), *C.A.* **59**, 8816d (1963); A. Bonati, A. Bocchia, *Farmaco Ed. Sci.* **18**, 84 (1963). Improved prepn: A. Bonati, **Ger. pat. 2,023,949;** *eidem,* **U.S. pat. 3,741,972** (1970, 1973 both to Inverni Della Beffa). In vitro and in vivo anti-arrhythmic activity: C. Capra, *Farmaco Ed. Sci.* **19**, 865 (1964); cardiac action: Y. Katano *et al.*, *Arzneimittel-Forsch.* **23**, 483 (1973); cardiovascular effects and toxicity: C. Capra *et al.*, *Farmaco Ed. Prat.* **35**, 49 (1980). Determination in plasma by TLC fluorescence: L. J. Dombrowski *et al.*, *J. Pharm. Sci.* **64**, 643 (1975). GC-MS determination in blood: St. D. Clemans *et al.*, *Arzneimittel-Forsch.* **27**, 1128 (1977). Comparison with other ajmaline esters: M. Salmona *et al.*, *J. Pharm. Sci.* **64**,

1561 (1975); P. Jaillon *et al.*, *Arch. Int. Pharmacodyn.* **224**, 310 (1976). Clinical studies: M. Cafiero *et al.*, *Minerva Cardioangiol.* **32**, 59 (1984). Book: G. G. Gensini, Ed., *Concepts on the Mechanism and Treatment of Arrhythmias* (Futura Publishing Co., Mount Kisco, N.Y., 1974) 281 pp.

Crystals, mp 232-238°. $[\alpha]_D^{20}$ +27.5° (c = 1 in chloroform).

Hydrochloride, $C_{22}H_{28}Cl_2N_2O_3$, *WIN 11831, Nevergor, Ritmos Elle.* Crystals from methyl ethyl ketone, mp 243-246°. $[\alpha]_D^{20}$ +29° (c = 1 in ethanol). LD_{50} in mice, rats (mg/kg): 176, 139 i.p.; 370, 480 orally (Capra).

THERAP CAT: Anti-arrhythmic (cardiac depressant).

5455. Loratadine. *4-(8-Chloro-5,6-dihydro-11H-benzo-[5,6]cyclohepta[1,2-b]pyridin-11-ylidene)-1-piperidinecarboxylic acid ethyl ester;* 11-[N-(ethoxycarbonyl)-4-piperidylidene]-8-chloro-6,11-dihydro-5H-benzo[5,6]cyclohepta-[1,2-b]pyridine; Sch-29851; Clantin; Claritin; Clarityne; Lisino. $C_{22}H_{23}ClN_2O_2$; mol wt 382.89. C 69.01%, H 6.05%, Cl 9.26%, N 7.32%, O 8.36%. Nonsedating-type histamine H_1-receptor antagonist. Prepn: F. J. Villani, **U.S. pat. 4,282,233** (1981 to Schering Corp.); F. J. Villani *et al.*, *Arzneimittel-Forsch.* **36**, 1311 (1986). Pharmacology in animals: A. Barnett *et al.*, *Agents Actions* **14**, 590 (1984). Anti-allergic effects in humans: R. L. Batenhorst *et al.*, *Eur. J. Clin. Pharmacol.* **31**, 247 (1986); I. J. Roman *et al.*, *Ann. Allergy* **57**, 253 (1986). Evaluation of CNS effects in humans: C. M. Bradley, A. N. Nicholson, *Eur. J. Clin. Pharmacol.* **32**, 419 (1987). Pharmacokinetics in humans: E. Radwanski *et al.*, *J. Clin. Pharmacol.* **27**, 530 (1987). Clinical trial in allergic rhinitis: R. J. Dockhorn *et al.*, *Ann. Allergy* **58**, 407 (1987). Clinical comparison with terfenadine, *q.v.*: F. Horak *et al.*, *Arzneimittel-Forsch.* **38**, 124 (1988); G. Bruttmann *et al.*, *J. Allergy Clin. Immunol.* **83**, 411 (1989). Brief review of pharmacodynamics, pharmacokinetics and therapeutic efficacy: S. P. Clissold *et al.*, *Drugs* **37**, 42-57 (1989).

Crystals from acetonitrile, mp 134-136°.
THERAP CAT: Antihistaminic.

5456. Lorazepam. *7-Chloro-5-(2-chlorophenyl)-1,3-dihydro-3-hydroxy-2H-1,4-benzodiazepin-2-one;* Wy-4036; Almazine; Ativan; Bonatranquan; Emotival; Lorasolid; Lorax; Lorsilan; Pro Dorm; Psicopax; Punktyl; Securit; Sedatival; Somagerol; Tavor; Temesta; Wypax. $C_{15}H_{10}Cl_2$-N_2O_2; mol wt 321.16. C 56.10%, H 3.14%, Cl 22.08%, N 8.72%, O 9.96%. Prepn and activity studies: Childress, Gluckman, *J. Pharm. Sci.* **53**, 577 (1964); Bell, **Belg. pat. 621,819,** *C.A.* **60**, 2992h (1964); **Brit. pat. 1,057,492; U.S. pat. 3,296,249** (1963, 1967, 1967 all to Am. Home Prod.); *idem, J. Med. Chem.* **11**, 457 (1968). Determn by HPLC: I. Jane, A. McKinnon, *J. Chromatog.* **323**, 191 (1985). Psychopharmacological studies: Stein, Berger, *Science* **166**, 253 (1969); Elliot *et al.*, *Clin. Pharmacol. Ther.* **12**, 468 (1971). Series of articles on pharmacology, metabolism, clinical studies: *Arzneimittel-Forsch.* **21**, 1047-1065, 1073-1102 (1971). Toxicity study: G. Owen *et al.*, *ibid.* 1065. Comprehensive description: J. G. Rutgers, C. M. Shearer in *Analytical Profiles of Drug Substances* vol. 9, K. Florey, Ed. (Academic Press, New York, 1980) pp 397-426. Review of clinical pharmacology and therapeutic use: B. Ameer, D. J. Greenblatt, *Drugs* **21**, 161-200 (1981).

Crystals, mp 166-168°. uv max (methanol): 229 nm, (1*N* NaOH): 233 nm, (1*N* HCl): 237 nm. Soly (mg/ml): water 0.08, chloroform 3, alcohol 14, propylene glycol 16, ethyl acetate 30. pK_1 13; pK_2 11.5. LD$_{50}$ in mice, rats (mg/kg): 3178, > 5000 orally (Owen).

Note: This is a controlled substance (depressant) listed in the U.S. Code of Federal Regulations, Title 21 Part 1308.14 (1987).

THERAP CAT: Anxiolytic.

5457. Lorcainide. *N-(4-Chlorophenyl)-N-[1-(1-methylethyl)-4-piperidinyl]benzeneacetamide;* 4'-chloro-*N*-(1-isopropyl-4-piperidyl)-2-phenylacetanilide. $C_{22}H_{27}ClN_2O$; mol wt 370.92. C 71.24%, H 7.34%, Cl 9.56%, N 7.55%, O 4.31%. Prepn: S. Sanczuk, H. K. F. Hermans, **Ger. pat. 2,642,856** (1977 to Janssen), *C.A.* **87**, 53094 (1977); *eidem,* U.S. pat. **4,126,689** (1978 to Janssen). Determn of lorcainide and its metabolites by GLC: R. Woestenborghs *et al., J. Chromatog.* **164**, 169 (1979). Disposition and anti-arrhythmic effects: U. Klotz *et al., Int. J. Clin. Pharmacol. Biopharm.* **17**, 152 (1979). Pharmacokinetics and tissue distribution: U. Klotz, E. Golbs, *Arzneimittel-Forsch.* **30**, 619 (1980). Clinical study: P. Somani, S. di Giorgi, *Chest* **78**, 658 (1980). Toxicology: G. J. Barton, R. Marsboom, "A Brief Summary of Animal Toxicology Studies on Lorcainide", in *Prognosis and Pharmacotherapy of Life-Threatening Arrhythmias,* Jähnchen *et al.,* Eds. (Royal Soc. Med. London, 1981) pp 23-25. Review of pharmacology and efficacy: C. E. Eriksson, R. N. Brogden, *Drugs* **27**, 279-300 (1984). Symposium: *Am. J. Cardiol.* **54**, 1B-54B (1984).

LD$_{50}$ in male mice, male and female rats (mg/kg): 18.8, 19.3, 18.6 i.v.; 483, 395, 435 orally (Barton, Marsboom).

Hydrochloride, $C_{22}H_{28}Cl_2N_2O$, *R 15,889, Ro 13-1042, Lopantrol, Lorivox, Remivox.* Crystals from 2-propanone and 2-propanol, mp 263°.

THERAP CAT: Cardiac depressant (anti-arrhythmic).

5458. Lormetazepam. *7-Chloro-5-(2-chlorophenyl)-1,3-dihydro-3-hydroxy-1-methyl-2H-1,4-benzodiazepin-2-one;* *N*-methyllorazepam; Wy-4082; Ergocalm; Loramet; Noctamid. $C_{16}H_{12}Cl_2N_2O_2$; mol wt 335.19. C 57.33%, H 3.61%, Cl 21.15%, N 8.36%, O 9.55%. Analog of lorazepam, *q.v.* Prepn: S. C. Bell *et al.,* **Belg. pat. 621,819** (1963 to Am. Home Products), *C.A.* **60**, 2993b (1964); *eidem, J. Med. Chem.* **11**, 457 (1968); A. Nudelman *et al., J. Pharm. Sci.* **63**, 1886 (1974). Pharmacokinetics and biotransformation in humans: M. Hümpel *et al., Eur. J. Drug Metab. Pharmacokinet.* **4**, 237 (1979). Hypnotic effect: A. Doenicke *et al., Anaesthesist* **28**, 578 (1979). Comparative study: H. Ott *et al., ibid.* 29. Absorption, distribution, excretion of ^{14}C-lormetazepam: R. Girkin *et al., Xenobiotica* **10**, 401 (1980).

Radioimmunologic study: M. Hümpel *et al., Clin. Pharmacol. Ther.* **28**, 673 (1980).

Crystals from ethanol/THF, mp 205-207°.

Note: This is a controlled substance (depressant) listed in the U.S. Code of Federal Regulations, Title 21 Part 1308.14 (1985).

THERAP CAT: Sedative. Hypnotic.

5459. Lotrifen. *2-(4-Chlorophenyl)-(1,2,4)triazolo-[5,1-a]isoquinoline;* 2-(p-chlorophenyl)-s-triazolo[5,1-a]isoquinoline; L 12717; DL 717-IT; Canocenta; Privaprol. $C_{16}H_{10}ClN_3$; mol wt 279.73. C 68.70%, H 3.60%, Cl 12.67%, N 15.02%. Non-hormonal antifertility agent. Prepn: **Belg. pat. 815,498**; A. Omodei-Salé *et al.,* U.S. pat. **4,075,341** (1974, 1978 both to Lepetit). Pharmacokinetics: G. Galliani *et al., J. Pharmacobio.-Dyn.* **5**, 55 (1981). Pregnancy-terminating effect in dogs: G. Galliani, A. Omodei-Salé, *J. Small Anim. Pract.* **23**, 295 (1982). Effect on subsequent fertility: G. Galliani *et al., IRCS Med. Sci.* **12**, 433, 435 (1984). *Review:* A. Assandri *et al., Rev. Drug Metab. Drug Interact.* **4**, 237 (1982).

Crystals mp 238-240°.
THERAP CAT (VET): Abortifacient.

5460. Lovastatin. *[1S-[1α(R*)*,3α,7β,8β(2S*,4S*),8aβ]]-2-Methylbutanoic acid 1,2,3,7,8,8a-hexahydro-3,7-dimethyl-8-[2-(tetrahydro-4-hydroxy-6-oxo-2H-pyran-2-yl)ethyl]-1-naphthalenyl ester;* (1S,3R,7S,8S,8aR)-1,2,3,7,8,8a-hexahydro-3,7-dimethyl-8-[2-[(2R,4R)-tetrahydro-4-hydroxy-6-oxo-2H-pyran-2-yl]ethyl]-1-naphthalenyl (S)-2-methylbutyrate; 1,2,6,7,8,8a-hexahydro-β,δ-dihydroxy-2,6-dimethyl-8-(2-methyl-1-oxobutoxy)-1-naphthaleneheptanoic acid δ-lactone; 2β,6α-dimethyl-8α-(2-methyl-1-oxobutoxy)-mevinic acid lactone; mevinolin; 6α-methylcompactin; monacolin K; MK 803; Mevacor; Mevinacor; Mevlor. $C_{24}H_{36}O_5$; mol wt 404.55. C 71.25%, H 8.97%, O 19.78%. Fungal metabolite; potent inhibitor of HMG-CoA reductase, the rate controlling enzyme in cholesterol biosynthesis. Isoln from *Monascus ruber:* A. Endo, *J. Antibiot.* **32**, 852 (1979); from *Aspergillus terreus:* R. L. Monaghan *et al.,* U.S. pat. **4,231,938** (1980 to Merck & Co.). Structure and biochemical properties: A. W. Alberts *et al., Proc. Nat. Acad. Sci. USA* **77**, 3957 (1980). Synthesis: M. Hirama, M. Iwashita, *Tetrahedron Letters* **24**, 1811 (1983); D. L. J. Clive *et al., J. Am. Chem. Soc.* **110**, 6914 (1988). Biosynthesis: M. D. Greenspan, J. B. Yudkovitz, *J. Bacteriol.* **162**, 704 (1985); R. N. Moore *et al., J. Am. Chem. Soc.* **107**, 3694 (1985). HPLC determn in plasma and bile: R. J. Stubbs *et al., J. Chromatog.* **383**, 438 (1986). Stimulation of receptor-mediated clearance of low density lipoproteins: D. W. Bilheimer *et al., Proc. Nat. Acad. Sci. USA* **80**, 4124 (1983). Effects on lipoprotein metabolism: S. M. Grundy, G. L. Vega, *J. Lipid Res.* **26**, 1464 (1985). Multicenter clinical comparison with gemfibrozil, *q.v.:* M. J. Tikkanen *et al., Am. J. Cardiol.* **62**, 35J (1988). Review of syntheses: T. Rosen, C. H. Heathcock, *Tetrahedron* **42**, 4909-4951 (1986).

Review of clinical experience: J. A. Tobert, *Circulation* **76**, 534-538 (1987); *idem, Am. J. Cardiol.* **62**, 28J-34J (1988).

White crystals, mp (under N$_2$): 174.5°. [α]$_D^{25}$ +323° (c = 0.5 g in 100 ml acetonitrile). uv max: 231, 238, 247 nm (A$^{1\%}$ 532, 621, 418). Freely sol in chloroform; sol in acetone; sparingly sol in acetonitrile, methanol, ethanol. Insol in water. LD$_{50}$ orally in mice: > 1000 mg/kg (Endo).
THERAP CAT: Antihypercholesterolemic.

5461. Loxapine. *2-Chloro-11-(4-methyl-1-piperazinyl)-dibenz[b,f][1,4]oxazepine;* oxilapine; CL-62362; S-805; SUM-3170. C$_{18}$H$_{18}$ClN$_3$O; mol wt 327.81. C 65.95%, H 5.53%, Cl 10.81%, N 12.82%, O 4.88%. Prepn. Neth. pat. Appl. **6,406,089** corresp to Schmutz *et al.*, U.S. pat. **3,546,-226** (1964, 1970 both to Wander); *eidem Helv. Chim. Acta* **50**, 245 (1967); Coppola, U.S. pat. **3,412,193** (1968 to Am. Cyanamid). Crystal structure: D. B. Cosulich, F. M. Lovell, *Acta Crystallogr.* **33B**, 1147 (1977). Pharmacology: Schmutz *et al., Chim. Ther.* **2**, 424 (1967); Latimer, *J. Pharmacol. Exp. Ther.* **166**, 151 (1969). Toxicity data: Stille *et al., Arzneimittel-Forsch.* **15**, 841 (1965). Toxicity studies: Mineshita *et al., Oyo Yakuri* **4**, 293 (1970), *C.A.* **76**, 81145v (1972). Review of pharmacology and therapeutic efficacy: R. C. Heel *et al., Drugs* **15**, 198-217 (1978).

Pale yellowish crystals from petr ether, mp 109-110°. LD$_{50}$ orally in mice: 65 mg/kg (Stille).
Hydrochloride, C$_{18}$H$_{19}$Cl$_2$N$_3$O, Loxitane C.
Succinate, C$_{22}$H$_{24}$ClN$_3$O$_5$, CL-71563, Daxolin, Loxapac, *Loxitane.*
THERAP CAT: Anxiolytic.

5462. Loxoprofen. *α-Methyl-4-[(2-oxocyclopentyl)meth-yl]benzeneacetic acid;* (±)-p-[(2-oxocyclopentyl)methyl]-hydratropic acid. C$_{15}$H$_{18}$O$_3$; mol wt 246.31. C 73.14%, H 7.37%, O 19.49%. Non-steroidal anti-inflammatory prodrug; the active metabolite is the *trans*-cyclohydroxypentane. Prepn of racemate: A. Terada *et al.*, Ger. pat. **2,814,-556**; *eidem,* U.S. pat. **4,161,538** (1978, 1979 both to Sankyo); of enantiomers: A. Terada, E. Misaka, Eur. pat. Appl. **55,-588**; *eidem,* U.S. pat. **4,400,534** (1982, 1983 both to Sankyo). Use as topical anti-inflammatory: A. Terada *et al.*, Belg. pat. **889,149** (1981 to Sankyo), *C.A.* **96**, 199326w (1982). Metabolism: Y. Tanaka *et al., Chem. Pharm. Bull.* **31**, 3656 (1983); *eidem, ibid.* **32**, 1040 (1984). Structure of metabolites: S. Naruto *et al., ibid.* 258. Optical inversion of (2R)- to (2S)-isomer after oral administration to rats: H. Nagashima *et al., ibid.* 251. Inhibition of *in vivo* and *in vitro* prostaglandin synthesis: K. Matsuda *et al., Biochem. Pharmacol.* **33**, 2473 (1984).

Colorless oil. bp$_{0.3}$ 190-195°. mp 108.5-111°. Sodium salt, C$_{15}$H$_{17}$NaO$_3$, CS-600, Loxonin.
THERAP CAT: Anti-inflammatory. Analgesic.

5463. Lucanthone Hydrochloride. *1-[[2-(Diethylami-no)ethyl]amino]-4-methyl-9H-thioxanthen-9-one hydrochloride;* 1-(2-diethylaminoethylamino)-4-methylthiaxanthone hydrochloride; Ms 752; RP 3735; Miracil D; Nilodin; Miracol; Tixantone. C$_{20}$H$_{25}$ClN$_2$OS; mol wt 376.94. C 63.72%, H 6.69%, Cl 9.41%, N 7.43%, O 4.24%, S 8.51%. Prepd by the reaction of 1-chloro-4-methylthiaxanthone with *asym*-diethylethylenediamine: Mauss, *Naturwiss.* **33**, 253 (1946); *idem, Ber.* **81**, 19 (1948); Sharp, *J. Chem. Soc.* **1951**, 2961; Archer, Suter, *J. Am. Chem. Soc.* **74**, 4296 (1952). Review of mode of action: Weinstein, Hirschberg, *Progr. Mol. Subcell. Biol.* **2**, 232 (1971). *Review:* Hirschberg in *Antibiotics* vol. 3, J. W. Corcoran, F. E. Hahn, Eds. (Springer-Verlag, New York, 1975) pp 274-303.

Yellow crystals from alcohol, mp 195-196°. Freely sol in water. Aq soln (orange) is neutral. Slightly sol in alcohol. Free base, mp 64-65°. Sol in the usual organic solvents.
THERAP CAT: Anthelmintic (Schistosoma).

5464. Lucensomycin. FI 1163; Etruscomicina; Etrusco-mycin; Antibiotic FI 1163. C$_{36}$H$_{53}$NO$_{13}$; mol wt 707.83. C 61.09%, H 7.55%, N 1.98%, O 29.39%. Polyene antifungal antibiotic isolated from cultures of *Streptomyces lucensis:* Arcamone *et al., Giorn. Microbiol.* **4**, 119 (1957); Arcamone, Perego, *Ann. Chim. (Rome)* **49**, 345 (1959); Marini, Pennella, *Proc. Symp. Antibiotics Prague* (May 1959) p 148; Arcamone *et al.*, U.S. pat. **3,170,837** (1965 to Farmitalia). Structure: Guadiano *et al., Tetrahedron Letters* **1966**, 3559, 3567; *Gazz. Chim. Ital.* **96**, 1470 (1966); *Chim. Ind. (Milan)* **48**, 1327 (1966). Revised structure: R. C. Pandey, K. L. Rinehart, *J. Antibiot.* **29**, 1035 (1976).

Crystalline powder. [α]$_D^{20}$ +296° (pyridine), +50° (methanolic 0.1N HCl). uv max: 218, 278, 290, 303, 318 nm (E$_{1cm}^{1\%}$ 300, 370, 780, 1170, 1098). Practically insol in water, anhydr alcohol, non-polar solvents; sol in pyridine, dimethyl-formamide. Unstable beyond pH 6-8, and to heat, light, or air. LD$_{50}$ orally in mice: 1263 mg/kg.
THERAP CAT: Antifungal.

5465. Luciferin. A generic term referring to a substrate which, upon oxidation by the enzyme luciferase, produces bioluminescence. Luciferins isolated from different species may vary greatly in structure, although in many cases identical structures have been found in widely diverse animals. The most widely studied luciferins are those isolated from the sea pansey, *Renilla reniformis,* the ostracod, *Cypridina hilgendorfii,* the limpet, *Latia neritoides,* and the firefly, *Photinus pyralis (see* separate entry, Firefly Luciferin). Structure of *Renilla reniformis* luciferin: K. Hori *et al., Proc. Nat. Acad. Sci. USA* **74,** 4285 (1977). *Review:* M. J. Cormier *et al., Fortschr. Chem. Org. Naturst.* **30,** 1-60 (1973).

5466. Lucifer Yellow CH. *6-Amino-2-[(hydrazinocarbonyl)amino]-2,3-dihydro-1,3-dioxo-1H-benz[de]isoquinoline-5,8-disulfonic acid dilithium salt.* $C_{13}H_9Li_2N_5O_9S_2$; mol wt 457.25. C 34.15%, H 1.98%, Li 3.04%, N 15.32%, O 31.49%, S 14.02%. Highly fluorescent dye that reveals functional connection between cells by its movement from cell to cell, termed "dye-coupling": W. W. Stewart, *Cell* **14,** 741 (1978). Prepn: *idem, J. Am. Chem. Soc.* **103,** 7615 (1981). As tracer for electron microscopy: A. R. Maranto, *Science* **217,** 953 (1982). Selective uptake by retinal cells: P. V. Sarthy *et al., J. Comp. Neurol.* **206,** 371 (1982). *Review:* W. W. Stewart, *Nature* **292,** 17-21 (1981).

Fluffy orange hygroscopic powder. Absorption max (water): 280, 428 nm (ϵ 24200, 11900). Sol in water.
USE: As intracellular marker in biological systems.

5467. Lumazine. *2,4(1H,3H)-Pteridinedione.* $C_6H_4N_4O_2$; mol wt 164.12. C 43.91%, H 2.46%, N 34.14%, O 19.50%. Prepn from 4,5-diamino-2,6-dihydoxypyrimidine and glyoxal sodium bisulfate in the presence of ammonia: Weijlard *et al., J. Am. Chem. Soc.* **67,** 804 (1945); U.S. pat. **2,479,442** (1950 to Merck & Co.); from 5,6-diaminouracil hydrosulfite + ethylene glycol disulfonic acid disodium salt + HCl: Dallacker, Steiner, *Ann.* **660,** 98 (1962).

Yellow to orange needles from water, mp 348-349°. Soly in water at 25° = 0.125%. Sol in acetic acid. Aq solns have a bluish-green fluorescence when neutral, a green fluorescence when alkaline, and a blue fluorescence when acid.

5468. Lumichrome. *7,8-Dimethylbenzo[g]pteridine-2,4-(1H,3H)-dione; 7,8-dimethylalloxazine;* 6,7-dimethylalloxazine. $C_{12}H_{10}N_4O_2$; mol wt 242.23. C 59.50%, H 4.16%, N 23.13%, O 13.21%. Irradiation product of riboflavine. Isoln: Karrer *et al., Helv. Chim. Acta* **17,** 1010 (1934). Prepn: Tishler, Wellman, U.S. pat. **2,417,143** (1947 to Merck & Co.); Cresswell, Wood, *J. Chem. Soc.* **1960,** 4768; Bardos *et al.,* U.S. pat. **3,057,865** (1962 to Armour Pharm. Co.); Seng, Ley, *Angew. Chem.* **84,** 1061 (1972).

Pale yellow crystals from pyridine or pyridine + alcohol, mp >300°. Sparingly sol in methanol and 90% hot ethanol,

water and chloroform. The aqueous, alcoholic and chloroformic solns fluoresce blue. Absorption spectrum: Karrer *et al., loc. cit.*

5469. Lumiflavine. *7,8,10-Trimethylbenzo[g]pteridine-2,4(3H,10H)-dione;* 7,8,10-trimethylisoalloxazine. $C_{13}H_{12}N_4O_2$; mol wt 256.26. C 60.93%, H 4.72%, N 21.87%, O 12.49%. Irradiation product of riboflavine. Isoln: Warburg, Christian, *Biochem. Z.* **266,** 377 (1933). Structure: Kuhn, Rudy, *Ber.* **67,** 1298 (1934). Prepn: Tishler, Wellman, U.S. pat. **2,417,143** (1947 to Merck & Co.); Birch, Moye, *J. Chem. Soc.* **1958,** 2622; Yoneda *et al., Chem. Pharm. Bull.* **20,** 1832 (1972).

Orange crystals from 12% acetic acid, mp 320°. Purified by sublimation in vacuum, mp 330°. Freely sol in chloroform; very sparingly sol in water and most organic solvents. Aq and chloroformic solns fluoresce green. uv max: 269, 355, 445 nm (ϵ 38,800, 11,700, 11,800) in 0.1N NaOH and 264, 373, 440 nm (ϵ 34,700, 11,400, 10,400) in 0.1N HCl.

5470. Luminol. *5-Amino-2,3-dihydro-1,4-phthalazinedione;* o-aminophthalylhydrazide; 3-aminophthalic hydrazide; o-aminophthaloyl hydrazide. $C_8H_7N_3O_2$; mol wt 177.16. C 54.23%, H 3.98%, N 23.72%, O 18.06%. Prepn from 3-nitrophthalic acid: Huntress *et al., J. Am. Chem. Soc.* **56,** 241 (1934); Redemann, Redemann, *Org. Syn.* **29,** 78, 8 (1949); L. Fieser, *Organic Experiments* (Boston, 1964) pp 240-242.

Crystals, mp 319-320°. Chemiluminescent: Weber, *Ber.* **75,** 565 (1942); Weber *et al., ibid.* **76,** 366 (1943). Oxidation of luminol is accompanied by a striking emission of light.
Note: Not to be confused with Luminal.
USE: Detection of copper, iron, peroxides, cyanides.

5471. Lumisterol. *9β,10α-Ergosta-5,7,22-trien-3β-ol.* $C_{28}H_{44}O$; mol wt 396.63. C 84.78%, H 11.18%, O 4.03%. Differs from ergosterol by spatial arrangement involving the methyl group at C-10. Prepd by ultraviolet irradiation of a benzene-alcohol soln of ergosterol: Askew *et al., Proc. Roy. Soc. (London)* **B109,** 488 (1932). Stereochemistry: Castells *et al., J. Chem. Soc.* **1959,** 1159. Dehydrogenation with selenium gives Diel's hydrocarbon, 3'-methyl-1,2-cyclopentenophenanthrene.

Needles from acetone-methanol, mp 118°. $[\alpha]_D^{19}$ +191.5°, $[\alpha]_{546}^{19}$ +235.4° (c = 2 in acetone). uv max: 265, 280 nm: Windaus *et al., Ann.* **493,** 259 (1932); Heilbron *et al., J. Chem. Soc.* **1937,** 411. Practically insol in water, sol in organic solvents. Forms a monomolecular compd with calciferol, mp 122°.

Consult the cross index before using this section.

Acetate, $C_{30}H_{46}O_2$, mp 100°. $[\alpha]_D^{19}$ +130.5°, $[\alpha]_{546}^{19}$ +163° (c = 1.8 in acetone).

5472. Lunacridine. *(R)-3-(2-Hydroxy-3-methylbutyl)-4,8-dimethoxy-1-methyl-2(1H)-quinolinone.* $C_{17}H_{23}NO_4$; mol wt 305.36. C 66.86%, H 7.59%, N 4.59%, O 20.96%. From bark of *Lunasia costulata* Miq., and *L. quercifolia* K. Schum. et Lauterb., *Rutaceae.* (+)-Form is naturally occurring. Isoln: Boorsma, *Bull. Inst. Bot. Buitenzorg* **21**, 8 (1904). Structure: Goodwin, Horning, *J. Am. Chem. Soc.* **81**, 1908 (1959). Synthesis: Clarke, Grundon, *J. Chem. Soc.* **1964**, 438. Synthesis and stereochemistry: R. M. Bowman *et al., J. Chem. Soc. Perkin Trans. I* **1973**, 1051. Short synthesis of racemate: M. Ramesh, P. Shanmugam, *Ind. J. Chem.* **24B**, 767 (1985).

Crystals from methanol + water, mp 86-87°. $[\alpha]_{589}^{25}$ +28.1°; $[\alpha]_{436}^{25}$ +76.5° (c = 0.935 in ethanol). uv spectra: Goodwin, Horning, *loc. cit.* Very slightly sol in water; freely sol in alcohol, benzene, chloroform, ether, ethyl acetate, carbon disulfide, petr ether.

Hydroperchlorate, $C_{17}H_{24}ClNO_8$, shiny flakes from methanol + ether, mp 146-148°; remelts at 193-195°. $[\alpha]_{589}^{25}$ +22.3°; $[\alpha]_{436}^{25}$ +60.7° (c = 0.750 in ethanol).

(±)-Form, prisms from petr ether, mp 72-74°. uv max (methanol): 239, 258, 285, 292, 333 nm (ε 24000, 26300, 8300, 7800, 3100).

5473. Lunacrine. *3,9-Dihydro-8-methoxy-9-methyl-2-(1-methylethyl)furo[2,3-b]quinolin-4(2H)-one.* $C_{16}H_{19}NO_3$; mol wt 273.32. C 70.31%, H 7.01%, N 5.13%, O 17.56%. From bark of *Lunasia costulata* Miq., and *L. amara* Blanco, *Rutaceae.* Isoln: Boorsma, *Bull. Inst. Bot. Buitenzorg* **6**, 15 (1900). Structure: Goodwin, Horning, *J. Am. Chem. Soc.* **81**, 1908 (1959). Synthesis: Clarke, Grundon, *J. Chem. Soc.* **1964**, 438. Stereochemistry: Bowman *et al., J. Chem. Soc. Perkin Trans. I* **1973**, 1051.

dl-Form, prisms from ethyl acetate + pentane, mp 146-148°. $[\alpha]_{589}^{25}$ +2.3°; $[\alpha]_{436}^{25}$ +4.2° (c = 0.864 in ethanol).

(−)-Form, fine needles from ethyl acetate, mp 117-119°. $[\alpha]_{589}^{25}$ −50.4°; $[\alpha]_{436}^{25}$ −116° (c = 0.806 in ethanol). uv spectra: Goodwin, Horning, *loc, cit.* Practically insol in water. Freely sol in alcohol, benzene, chloroform, ether, carbon disulfide, ethyl acetate; slightly sol in petr ether.

5474. Lunasine. *2,3-Dihydro-4,8-dimethoxy-9-methyl-2-(1-methylethyl)furo[2,3-b]quinolinium.* $[C_{17}H_{22}NO_3]^+$. From bark of *Lunasia costulata* Miq., and *L. quercifolia* K. Schum. et Lauterb., *Rutaceae.* Isoln: Boorsma, *Bull. Inst. Bot. Buitenzorg* **6**, 14 (1900). Structure: Price, *Aust. J. Chem.* **12**, 458 (1959). Stereochemistry: Bowman *et al., J. Chem. Soc. Perkin Trans I* **1973**, 1051.

(−)-Lunasine picrate, $C_{17}H_{22}NO_3 \cdot C_6H_2N_3O_7$, yellow needles from methanol, mp 143.5-144°. $[\alpha]_D^{20}$ −20.3° (c = 0.98 in methanol).

(−)-Lunasine perchlorate, $C_{17}H_{22}NO_3 \cdot ClO_4$, microcrystalline powder from methanol + ether, mp 195-196°. $[\alpha]_D^{20}$ −29.3° (c = 0.46 in methanol).

5475. Lunine. *7,10-Dihydro-10-methyl-8-(1-methylethyl)-1,3-dioxolo[4,5-h]furo[2,3-b]quinolin-6(8H)-one.* $C_{16}H_{17}NO_4$; mol wt 287.30. C 66.88%, H 5.96%, N 4.88%, O 22.28%. From leaves of *Lunasia costulata* Miq., and *L. amara* Blanco, *Rutaceae:* Boorsma, *Bull. Inst. Bot. Buitenzorg* **21**, 8 (1904); Goodwin *et al., J. Am. Chem. Soc.* **81**, 6209 (1959). Structure: Goodwin *et al., ibid.* **81**, 3065 (1959).

Prisms from ethyl acetate, mp 227-230°; or mixture of prisms and needles from methanol + water, mp 224-226°. $[\alpha]_{589}^{25}$ −38.5°; $[\alpha]_{436}^{25}$ −87.4° (c = 0.926 in chloroform). uv max (ethanol): 222, 247, 314, 325 nm (log ε 4.31, 4.60, 4.02, 4.01). Practically insol in water; freely sol in alcohol, benzene, chloroform, slightly in ether, ethyl acetate.

5476. Lunularic Acid. *2-Hydroxy-6-[2-(4-hydroxyphenyl)ethyl]benzoic acid; 6-(p-hydroxyphenethyl)salicylic acid.* $C_{15}H_{14}O_4$; mol wt 258.27. C 69.76%, H 5.46%, O 24.78%. Dihydrostilbene growth inhibitor isolated from liverworts and algae. Isoln from *Lunularia cruciata* (L.) Dum., *Lunulariaceae,* and characterization: I. F. M. Valio *et al., Nature* **223**, 1178 (1969); I. F. M. Valio, W. W. Schwab, *J. Exp. Bot.* **21**, 138 (1970); from more than 70 species of liverwort: J. Gorham, *Phytochemistry* **16**, 249 (1977); from cultures of *Marchantia polymorpha:* S. Abe, Y. Ohta, *ibid.* **22**, 1917 (1983). Synthesis: Y. Arai *et al., Tetrahedron Letters* **1972**, 1615; S. Haneck, K. Schreiber, **East Ger.** pat. **126,866** (1977 to Akad. Wissenschaft. D.D.R.), *C.A.* **88**, 152249t (1978). Effect on liverwort growth: J. Gorham, *Phytochemistry* **17**, 99 (1978). Metabolism by *L. cruciata:* R. J. Pryce, *ibid.* **11**, 1355 (1972).

Pale yellow crystals from methanol/water, mp 192°. uv max in neutral ethanol: 280, 287, 308 nm (ε 3300, 3600, 4200); in weakly alkaline ethanol: 300 nm (ε 6600).

USE: Growth inhibitor for lower plants.

5477. Lupanine. *Dodecahydro-7,14-methano-4H,6H-dipyrido[1,2-a:1',2'-e][1,5]diazocin-4-one.* $C_{15}H_{24}N_2O$; mol wt 248.36. C 72.54%, H 9.74%, N 11.28%, O 6.44%. Racemic lupanine is found in white lupins, *d*-lupanine is found in blue lupins, *l*-lupanine has been prepd from the natural racemic form. Structure: Davis, *Arch. Pharm.* **235**, 199, 218, 229 (1897); Clemo, Leitch, *J. Chem. Soc.* **1928**, 1811; Clemo *et al., ibid.* **1931**, 429; **1933**, 644; Ing. *ibid.* **1933**, 504; Karrer *et al., Helv. Chim. Acta* **11**, 1062 (1928), **13**, 1292 (1930); Winterfeld, Holschneider, *Ber.* **64**, 137, 2415 (1931), **66**, 1751 (1933); *Ann.* **499**, 109 (1932); Winterfeld, Kneuer, *Ber.* **64**, 150 (1931); Winterfeld, Hoffman, *Arch. Pharm.* **275**, 5, 65 (1937). Abs config of *l*-form: Okuda *et al., Chem. & Ind.* (*London*) **1961**, 1116. Crystal structure of *dl*-form: H. Doucerain *et al., Acta Crystallogr.* **B32**, 3213 (1976). Biosynthetic study of *d*-form: W. M. Golebewski, I. D. Spenser, *J. Am. Chem. Soc.* **106**, 7925 (1984). Further ref. *under* Sparteine.

dl-Form, orthorhombic prisms from acetone, mp 98-99°. bp$_{1.0}$ 185-195°. Sol in water, alcohol, ether, chloroform. Insol in petr ether. *See:* Soldaini, *Arch. Pharm.* **231**, 321 (1893); Couch, *J. Am. Chem. Soc.* **56**, 1423 (1934).

Dihydrochloride, $C_{15}H_{26}Cl_2N_2O$, deliquescent prisms, mp 185° (dec).

Hydrochloride dihydrate, $C_{15}H_{25}ClN_2O.2H_2O$, crystals, mp 177-178°. mp 250-252° (dry).

d-Form, sirupy liq crystallizing with difficulty in hygroscopic needles, mp 40-44°. bp$_3$ 190-193°. n_D^{24} 1.5444. $[\alpha]_D^{25}$ +84° (c = 4.8 in alc). Freely sol in water, alcohol, chloroform, ether. Sol in petr ether.

Hydrochloride dihydrate, $C_{15}H_{25}ClN_2O.2H_2O$, rhombic crystals from water, mp 127° (dry).

l-Form, viscous oil, bp$_{1.0}$ 186-188°. $[\alpha]_D$ about −61° in acetone.

α-*Isolupanine.* cis-cis-Isomer of lupanine: Marion, Leonard, *Can. J. Chem.* **29**, 355 (1951). Isoln of *d*-form from *Lupinus sericeus* Pursh., *L. perennis* L. and *L. augustifolius*, Leguminosae: Marion et al., *ibid.* **31**, 181 (1953); Rink, Schäfer, *Arch. Pharm.* **287**, 290 (1954); Winterfeld, Pies, *ibid.* **290**, 537 (1957). Needles from petr ether, mp 75-76°. $[\alpha]_D^{26}$ +39° (c = 0.77). $[\alpha]_D^{26}$ +65.9° (c = 3.4 in abs ethanol).

5478. Lupeol. *Lup-20(29)-en-3β-ol; monogynol B; β-viscol; fagarasterol.* $C_{30}H_{50}O$; mol wt 426.70. C 84.44%, H 11.81%, O 3.75%. Abundant plant triterpene. Occurs in the skin of lupin seeds, in chicle, in the latex of fig trees and of rubber plants. Was detected in cocoons of *Bombyx mori.* Isoln: Cohen, *Rec. Trav. Chim.* **28**, 368 (1909); Ruzicka et al., *Helv. Chim. Acta* **20**, 1567 (1937). Identity with β-viscol: Meyer, Jeger, *ibid.* **31**, 1868 (1948). Identity with monogynol B: Chatterji et al., *J. Sci. Ind. Res.* **18B**, 262 (1959). Structure: Ruzicka et al., *Helv. Chim. Acta* **28**, 942 (1945); Ames et al., *J. Chem. Soc.* **1951**, 450. Structure of hydrochloride: Halsall et al., *ibid.* **1952**, 2862. Configuration: Barton, Holmes, *ibid.* **1952**, 78; Djerassi et al., *J. Am. Chem. Soc.* **77**, 5330 (1955). NMR studies: Buckley et al., *Chem. & Ind. (London)* **1971**, 298. Total synthesis: Stork et al., *J. Am. Chem. Soc.* **93**, 4945 (1971).

Needles from alcohol or acetone, mp 215°. $[\alpha]_D^{20}$ +27.2° (c = 4.8 in chloroform). Freely sol in ether, benzene, petr ether, warm alcohol. Practically insol in water, dil acid and alkalies.

Acetate, $C_{32}H_{52}O_2$, needles from acetone, mp 218°. $[\alpha]_D^{20}$ +47.3° (c = 2 in chloroform).

Benzoate, $C_{37}H_{54}O_2$, prisms from acetone, mp 273-274°. $[\alpha]_D^{20}$ +61° (0.78 g in 25 ml chloroform).

Hydrochloride, $C_{30}H_{50}O.HCl$, needles from ethanol, mp 211-212°. $[\alpha]$ −31° (c = 1.1 in chloroform).

5479. Lupinine. *[1R-trans]-Octahydro-2H-quinolizine-1-methanol. l-lupinine; (−)-lupinine.* $C_{10}H_{19}NO$; mol wt 169.27. C 70.96%, H 11.31%, N 8.27%, O 9.45%. Naturally occurring *l*-form isolated from seeds and herb of *Lupinus luteus* L. and other *L.* species, Leguminosae also found in *Anabasis aphylla* L., Chenopodiaceae. Extraction procedure: J. F. Couch, *J. Am. Chem. Soc.* **56**, 2434 (1934). Structure and synthesis: R. W. Willstätter, E. Fourneau, *Ber.* **35**, 1910 (1902); P. Karrer et al., *Helv. Chim. Acta* **11**, 1062 (1928); K.

Winterfeld, F. W. Holschneider, *Ber.* **64B**, 137, 692 (1931); K. Winterfeld, *ibid.* 692. Synthesis of racemic lupinine: F. W. Holschneider, K. Winterfeld, *Arch. Pharm.* **277**, 192 (1939); G. C. Gerrans et al., *Tetrahedron Letters* **1975**, 4171; T. Iwashita et al., *J. Org. Chem.* **47**, 230 (1982). Synthesis of racemic lupinine and *epi-lupinine:* G. R. Clemo et al., *J. Chem. Soc.* **1937**, 965; H. Takahata et al., *Chem. Pharm. Bull.* **34**, 4523 (1986); of *l-epi-lupinine:* M. L. Bremmer, S. M. Weinreb, *Tetrahedron Letters* **24**, 261 (1983). Absolute configuration of (−)-form: R. C. Cookson, *Chem. & Ind. (London)* **1953**, 339. Crystal structure: A. Koziol et al., *Acta Crystallogr.* **B34**, 3491 (1978). Biosynthesis: Soucek, Schutte, *Angew. Chem.* **74**, 901 (1962); W. M. Golebiewski, I. D. Spenser, *J. Am. Chem. Soc.* **106**, 1441 (1984).

Stout, orthorhombic prisms from acetone, mp 68.5-69.2°. bp$_4$ 160-164°; bp$_{755}$ 269-270°. $[\alpha]_D^{26}$ −25.9° (c = 3 in water); $[\alpha]_D^{28}$ −21° (c = 9.5 in alcohol). Sol in water, alcohol, chloroform, ether. It is a strong base.

l-Form hydrochloride, $C_{10}H_{20}ClNO$, orthorhombic prisms, mp 208-213°. $[\alpha]_D$ −14°.

dl-Form, crystals from acetone, mp 58.5-59.5°.

5480. Luprostiol. *7-[2-[[3-(3-Chlorophenoxy)-2-hydroxypropyl]thio]-3,5-dihydroxycyclopentyl]-5-heptenoic acid; 9α,11α,15-trihydroxy-16-m-chlorophenoxy-13-thia-17,18,-19,20-tetranor-5-prostenoic acid; prostianol.* EMD-34946; Pronilin; Prosolvin; Reprodin. $C_{21}H_{29}ClO_6S$; mol wt 444.97. C 56.68%, H 6.57%, Cl 7.97%, O 21.57%, S 7.20%. Analog of prostaglandin F$_{2\alpha}$, *q.v.* Prepn: J. Kraemer et al., *Ger. pat.* **2,513,371** (1976 to E. Merck), *C.A.* **86**, 55069k (1977); eidem, *U.S. pat.* **4,309,441** (1982 to E. Merck). Luteolytic study in cattle: H. De Vries, H. Feenstra, *J. Vet. Pharmacol. Ther.* **2**, 223 (1979). Estrus cycle regulation: D. Schams, H. Karg, *Theriogenology* **17**, 499 (1982), *C.A.* **97**, 50430p (1982).

THERAP CAT (VET): Luteolytic.

5481. Lupulin. Glandular trichomes separated from strobiles of *Humulus lupulus* L., Moraceae (hops). *Habit.* Europe, Asia, North America; widely cultivated. *Constit.* Lupamaric acid, humulol, about 3% volatile oil; resin, choline, wax (myricin), tannin, asparagin. Not less than 60% of lupulin is sol in ether.

THERAP CAT: Aromatic bitter, sedative.

THERAP CAT (VET): Formerly in nymphomania.

5482. Lupulon. *3,5-Dihydroxy-2,6,6-tris(3-methyl-2-butenyl)-4-(3-methyl-1-oxobutyl)-2,4-cyclohexadien-1-one; β-bitter acid; β-lupulic acid.* $C_{26}H_{38}O_4$; mol wt 414.56. C 75.32%, H 9.24%, O 15.44%. Antimicrobial constituent of hops. Isoln from commercial hops: Bungener, *Bull Soc. Chim. France* [2] **45**, 487 (1886); Barth, Lintner, *Ber.* **31**, 2022 (1898); Wöllmer, *Ber.* **49**, 780 (1916); Lewis, et al., *J. Clin. Invest.* **28**, 916 (1949). Structure: Wöllmer, *Ber.* **58**, 672 (1925); Wieland, *ibid.* **102**, 2012; Govaert, Verzele, *Bull. Soc. Chim. Belg.* **58**, 432 (1949); Riedl, *Ber.* **85**, 692 (1952).

Prisms from 90% methanol, mp 92-94°. Bitter taste esp in alc soln. Turns yellow and amorphous within a few days with development of an odor. Perfectly stable *in vacuo* even at 60°. Slightly acid reaction. Monobasic acid. Optically inactive. Soluble in methanol, ethanol, petr ether, hexane, isooctane. Slightly sol in neutral or acidic aq solns. Forms a sodium salt which is readily sol in water. The addition of 0.1% ascorbic acid exerts a marked protective action on the bacteriostatic activity of lupulon steamed or autoclaved at a concentration of 4 ppm in phosphate buffers at pH 6.5 and 8.5. LD_{50} orally in rats: 1.8 g/kg.

5483. Luteolin. *2-(3,4-Dihydroxyphenyl)-5,7-dihydroxy-4H-1-benzopyran-4-one; 3',4',5,7-tetrahydroxyflavone;* digitoflavone; cyanidenon 1470. $C_{15}H_{10}O_6$; mol wt 286.23. C 62.94%, H 3.52%, O 33.54%. Found in many plants in glycosidic combination, *e.g.*, as the arabinoside: Perkin, *J. Chem. Soc.* **69**, 800 (1896); Fleischer, *Ber.* **32**, 1186 (1899); Perkin, Horsfall, *J. Chem. Soc.* **77**, 1315 (1900); Hayashi, Inoue, *Acta Phytochim. (Japan)* **15**, 53 (1949); *C.A.* **43**, 8450 (1949). Identity with digitoflavone: Kiliani, Mayer, *Ber.* **34**, 3577 (1901). Synthesis: Hutchins, Wheeler, *J. Chem. Soc.* **1939**, 91. *See also* Bioflavonoids.

Monohydrate, yellow needles from alc, dec 328-330°. Sublimes in high vacuum. Sparingly sol in water. Sol in alkalies forming yellow solns.
5-Glucoside, $C_{21}H_{20}O_{11}$, *galuteolin.* From seeds of *Galega officinalis* L., *Leguminosae:* Barger, White, *Biochem. J.* **17**, 836 (1923). Structure: Nakamura, Hukuti, *J. Pharm. Soc. Japan* **60**, 449 (1940); *C.A.* **34**, 7910³ (1940); Nordström, Swain, *J. Chem. Soc.* **1953**, 2764. Yellow needles from hot dil alc, dec 280°. Practically insol in water; slightly sol in abs alcohol; sol in hot dil alcohol.
7-Glucoside, $C_{21}H_{20}O_{11}$, *cynaroside.* From *Achillea millefolium* L., *Compositae:* Hörhammer *et al., Acta Chim. Acad. Sci. Hung.* **40**, 463 (1964). Yellow needles from alc, mp 254-256°. uv max (CH_3OH): 350, 255 nm (log ε 4.30, 4.27).

5484. Lutetium. Lutecium. Lu; at. wt 174.967; at. no. 71; valence 3. Belongs to the yttrium group of rare earths. Two naturally occurring isotopes: ¹⁷⁵Lu (97.40%); ¹⁷⁶Lu (2.60%), the latter is radioactive, $T_{1/2}$ 2.2 × 10¹⁰ years, β⁻ emission. Artificial radioactive isotopes: 155; 156; 167-174; 177-180. Abundance in earth's crust: 0.8-1.7 ppm. Occurs in xenotime, gadolinite and other rare-earth minerals. Discovered in 1907 by Urbain and independently in 1908 by von Welsbach, who called it *cassiopeium:* Urbain, *Compt. Rend.* **145**, 759 (1907); **146**, 406 (1908); **152**, 141 (1911); von Welsbach, *Sitzungsber. Akad. Wiss. Wien* **1907**, 468; *idem, Monatsh.* **29**, 181 (1908); **30**, 695 (1909). Isoln in form of salt: Prandtl, *Z. Anorg. Chem.* **238**, 321 (1938); Marsh, *J. Chem. Soc.* **1952**, 4804. Sepn by ion exchange: Spedding *et al., J. Am. Chem. Soc.* **76**, 2557 (1954). Reviews of prepn, properties, and compds of lutetium and other rare earths: *The Rare Earths,* F. H. Spedding, A. H. Daane, Eds. (Krieger, Huntington, N.Y., 1971, reprint of 1961 ed.) 641 pp; Hulet, Bode, "Separation Chemistry of the Lanthanides and Transplutonium Actinides" in *MTP Int. Rev. Sci: Inorg. Chem., Ser. One* vol. 7, K. W. Bagnall, Ed. (University Park Press, Baltimore, 1972) pp 1-45; Moeller, "The Lanthanides" in *Comprehensive Inorganic Chemistry* vol. 4, J. C. Bailar, Jr. *et al.,* Eds. (Pergamon Press, Oxford, 1973) pp 1-101.

Silvery-white metal. Hexagonal, close-packed structure, d 9.842. mp 1652°.
Oxide, Lu_2O_3, cubic crystals.
Chloride, $LuCl_3$, colorless crystals. Sublimes above 750°. mp 892° ± 2°; sol in water. LD_{50} in mice: 315 mg/kg i.p.; 7.1 g/kg orally: Haley, *J. Pharm. Sci.* **54**, 663 (1965).

Sulfate, $Lu_2(SO_4)_3$, octahydrate, soly in water (g/100 g): 42.27 (20°); 16.93 (40°).

5485. 2,6-Lutidine. *2,6-Dimethylpyridine;* α,α'-lutidin. C_7H_9N; mol wt 107.15. C 78.46%, H 8.47%, N 13.07%. Isolated from the basic fraction of coal tar: Heap *et al., J. Am. Chem. Soc.* **43**, 1936 (1921); from a bone oil fraction: Ladenburg, Roth, *Ber.* **18**, 51 (1885). Synthesis from ethyl acetoacetate, formaldehyde, and ammonia: Singer, McElvain, *Org. Syn. coll. vol. II*, 214 (1943).

Oily liquid. Odor of pyridine + peppermint. d_4^{20} 0.9252. mp −5.8°. bp_{760} 144°; bp_{87} 79°. n_D^{20} 1.49797. Solubility in water at 45.3° = 27.2% (w/w); at 48.1° = 18.1%; at 57.5° = 12.1%; at 74.5° = 9.5%. Also sol in alcohol, ether. Miscible with dimethylformamide and tetrahydrofuran.

5486. Lututrin. Lutrexin. A uterus-relaxing factor obtained from the corpus luteum of sow ovaries and standardized for potency in terms of units of activity on the guinea pig uterus. *Ref: J. Am. Med. Assoc.* **156**, 1252 (1954).
THERAP CAT: Uterine relaxant.

5487. Lyapolate Sodium. *Ethenesulfonic acid homopolymer sodium salt;* sodium lyapolate; sodium apolate; sodium polyethylene sulfonate; polyethylene sodium sulfonate; PES; Peson. $(C_2H_3NaO_3S)_n$.

THERAP CAT: Anticoagulant.

5488. Lycoctonine. *20-Ethyl-4-(hydroxymethyl)-1,6,-14,16-tetramethoxyaconitane-7,8-diol;* royline. $C_{25}H_{41}NO_7$; mol wt 467.59. C 64.21%, H 8.84%, N 3.00%, O 23.95%. Isomeric with delsoline. Originally isolated from *Aconitum lycoctonum* L., *Ranunculaceae:* Hubschmann, *Schweiz. Wochschr. Pharm.* **3**, 269 (1865). Widely distributed in *Aconitum* and *Delphinium* spp., *Ranunculaceae.* Structure: Edwards *et al., Can. J. Chem.* **34**, 1315 (1956); Anet *et al., ibid.* **35**, 400 (1957). Identity with royline: Edwards, Rodger, *ibid.* **37**, 1187 (1959). Stereochemistry: Przybylska, Marion, *ibid.* 1843. Revised configuration: S. W. Pelletier *et al., J. Am. Chem. Soc.* **103**, 6536 (1981).

Crystals. Bitter taste. mp 143°. $[\alpha]_D^{20}$ +53° (ethanol). Slightly sol in water, ether, benzene; freely sol in alcohol, chloroform.
Hydrochloride, mp 152°.
N-Acetylanthranilic acid ester, *See* Ajacine.
N-Succinylanthranilic acid ester, $C_{36}H_{48}N_2O_{10}$, *lycaconitine.* White, amorphous powder, mp 111-114°. $[\alpha]_D$ +42°; also reported as +31.5%. Insol in water. Sol in alcohol, chloroform, ether. Isoln: Schulze, Bierling, *Arch. Pharm.* **251**, 8 (1913).

5489. Lycodine. 2,3,4,4a,5,6-Hexahydro-12-methyl-1H-5,10b-propano-1,7-phenanthroline. $C_{16}H_{22}N_2$; mol wt 242.37. C 79.29%, H 9.15%, N 11.56%. Isoln from *Lycopodium annotinum* L., *Lycopodiaceae:* F. A. L. Anet, C. R. Eves, *Can. J. Chem.* **36**, 902 (1958). Structure: W. A. Ayer,

G. G. Iverach, *ibid.* **38**, 1823 (1960); F. A. L. Anet, M. V. Rao, *Tetrahedron Letters* **1960**, 9. Synthesis of (±)-form: E. Kleinman, C. H. Heathcock, *ibid.* **1979**, 4125; C. H. Heathcock *et al.*, *J. Am. Chem. Soc.* **104**, 1054 (1982).

Cryst powder, mp 118-119°. $[\alpha]_D$ −10° (c = 1.01 in ethanol). pKa's 3.97, 8.08.

5490. Lycofawcine. *(5β,8R,15S)-15-Methyllycopodane-5,8,12-triol 5-acetate;* base L. $C_{18}H_{29}NO_4$; mol wt 323.42. C 66.84%, H 9.04%, N 4.33%, O 19.79%. From *Lycopodium fawcettii* Lloyd and Underwood, *Lycopodiaceae:* Burnell *et al.*, *Can. J. Chem.* **41**, 3091 (1963). Structure: Ayer *et al.*, *ibid.* **43**, 328 (1965).

Liquid. pKa 9.7.
Perchlorate, crystals, dec 290-294°.
Methiodide, $C_{19}H_{32}INO_4$, crystals from acetone, mp 281-282°.

5491. Lycomarasmine. *N-[2-[(2-Amino-2-oxoethyl)amino]-2-carboxyethyl]-L-aspartic acid;* Welkstoff. $C_9H_{15}N_3O_7$; mol wt 277.23. C 38.99%, H 5.45%, N 15.16%, O 40.40%. Antibiotic peptide (tomato-wilting agent) produced by the fungus *Fusarium lycopersici.* Isoln: Plattner, Clauson-Kaas, *Helv. Chim. Acta* **28**, 188 (1945). Structure: *eidem, Experientia* **1**, 195 (1945); Hardegger *et al.*, *Helv. Chim. Acta* **46**, 60 (1963).

$$HOOCCH_2CHNHCH_2CHNHCH_2CONH_2$$
$$\text{COOH} \qquad \text{COOH}$$

Crystals, dec 227-229°. $[\alpha]_D^{20}$ −42° to −48° (aq soln at pH 7). Acid reaction. Sparingly sol in water; freely sol in dil acids or alkalies.

5492. Lycopene. *ψ,ψ-Carotene.* $C_{40}H_{56}$; mol wt 536.85. C 89.48%, H 10.51%. Carotenoid occurring in ripe fruit, especially in tomatoes. One kg of fresh ripe tomatoes yields 0.02 g lycopene. Isoln procedure: Willstätter, Escher, *Z. Physiol. Chem.* **64**, 47 (1910). Chromatographic sepn from other carotenoids: Winterstein, *ibid.* **215**, 52 (1933); Winterstein, Stein, *ibid.* **220**, 250 (1933). Structure: Willstätter, Escher, *loc. cit.;* Karrer and collaborators: *Helv. Chim. Acta* **11**, 751, 1201 (1928); **12**, 285 (1929); **13**, 1084 (1930); **14**, 435 (1931); Kuhn, Grundmann, *Ber.* **65**, 898, 1880 (1932). Synthesis: Karrer *et al.*, *Helv. Chim. Acta* **33**, 1349 (1950); Isler *et al.*, *ibid.* **39**, 463 (1956). Commercial prepn: Kabbe *et al.*, **Ger. pat. 1,168,890** (1964 to Bayer).

Long, deep red needles from carbon disulfide + ethanol, from methylene chloride + methanol, mp 172-173°. Ab-

sorption max (*trans*-form): 446, 472, 505 nm ($E_{1cm}^{1\%}$ 2250, 3450, 3150). 15,15'-*cis*-Form: mp about 105° then solidifies again; absorption max: 361, 444, 470, 502 nm ($E_{1cm}^{1\%}$ 1110, 1280, 1660, 1280). One gram dissolves in 50 ml carbon disulfide, in 3 l boiling ether, in 12 l boiling petr ether, in 14 l hexane at 0°. Sol in chloroform, benzene. Almost insol in methanol, ethanol.

5493. Lycophyll. *ψ,ψ-Carotene-16,16'-diol; (all-trans)-lycopene-16,16'-diol.* $C_{40}H_{56}O_2$; mol wt 568.85. C 84.45%, H 9.92%, O 5.63%. Carotenoid pigment, isoln from *Solanum dulcamara* L. and *Lycopersicum esculentum* Mill., *Solanaceae* and structure: Zechmeister, Cholnoky, *Ber.* **69**, 422 (1936). Revised structure: Cholnoky, Szabolcs, *Tetrahedron Letters* **1968**, 1931. Stereochemistry: Kelly *et al.*, *Acta Chem. Scand.* **25**, 1607 (1971). Synthesis: Kjosen, Liaaen-jensen, *ibid.* 1500.

Purple leaflets from benzene + methanol, needles from benzene + petr ether, mp 179°. Absorption max (benzene): 521, 487, 456 nm. Freely sol in carbon disulfide, less in benzene, ethanol, very slightly sol in petr ether.
Dipalmitate, purple needles from benzene + methanol, mp 76°.

5494. Lycopodine. *15-Methyllycopodan-5-one.* $C_{16}H_{25}$-NO; mol wt 247.37. C 77.68%, H 10.19%, N 5.66%, O 6.47%. From herb of *Lycopodium complanatum* L., *Lycopodiaceae:* Bödeker, *Ann.* **208**, 363 (1881). Structure: Harrison *et al.*, *Can. J. Chem.* **39**, 2086 (1961). Stereochemistry: Anet, *Tetrahedron Letters* **1960**, no. 20, 13. Synthesis of (±)-form: G. Stork *et al.*, *J. Am. Chem. Soc.* **90**, 1647 (1968); W. A. Ayer *et al.*, *ibid.* 1648; C. H. Heathcock *et al.*, *ibid.* **100**, 8036 (1978); **104**, 1054 (1982); D. Schumann *et al.*, *Ann.* **1982**, 1700. Review of synthetic studies: Stork, *Pure Appl. Chem.* **17**, 383 (1968).

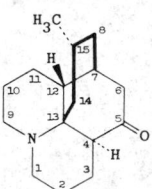

Bitter prisms, mp 114-115°; mp (racemate) 130-131°. $[\alpha]_D$ −24° (alc). Sol in water, alc, benzene, chloroform, ether.

5495. Lycopodium. Club-moss spores; Lycopodium seed (spores); vegetable sulfur. Spores of *Lycopodium clavatum* L., *Lycopodiaceae.* The spores of *Lycopodium annotinum* L., and of *L. anceps* Wall., *Lycopodiaceae* can also be used. The spores contain a substance called *selaginine.* Habit. North America, Europe, Asia; cultivated in Russia.
The spores are a fine yellowish powder which is highly flammable. Odorless and tasteless. Unctuous to the touch and easily sticking to the fingers. Lycopodium powder is very mobile and when poured on a flat surface should form an even layer, without visible lumps or dimples. When observed in chloral hydrate soln it is seen that the powder consists of unicellular lycopod spores, about 30 μ in diameter, in the shape of triangular pyramids with a convex base and rounded angles; a three-radical suture runs from the top of the pyramid along its facets. After warming and crushing the spores between glass slides, they burst along the suture and yield drops of oil, assuming a red color with alkalies. Adulteration usually consists of the admixture of flour (detected microscopically with iodine soln which stains the starch grains of the flour violet). Other admixtures may be

Consult the cross index before using this section.

pine pollen and sawdust. When mounted with chloral hydrate, pine pollen is larger than lycopodium, it is oval and has two lateral flying sacs, filled with air, and appearing black at the beginning of the observation. Sawdust is easily detected by the phloroglucinol test.

USE: As covering for pills or suppositories; in explosives, pyrotechnics, flashlight powders; as "dry parting compound" in foundry work for ornamental and nameplate castings. *Review:* Appel, *Scientific Monthly* **78**, 268 (1954).

THERAP CAT: Adsorbent.

5496. Lycopus. Bugleweed; sweet bugle; water bugle. Whole plant of *Lycopus virginicus* L., Labiatae. *Habit.* N. America. *Constit.* Volatile oil, resin, tannin, glucoside.

5497. Lycoramine. *1,2-Dihydrogalanthamine; 3,4-dihydrogalanthamine.* $C_{17}H_{23}NO_3$; mol wt 289.36. C 70.56%, H 8.01%, N 4.84%, O 16.59%. One of the minor alkaloids of *Lycoris radiata* Herb., Amaryllidaceae: Kondo, Ishiwata, *Ber.* **70B**, 2427 (1937). From *Narcissus* spp.: Boit et al., *ibid.* **90**, 725 2197 (1957). Tentative structure: Kobayashi, Uyeo, *Chem. & Ind. (London)* **1956**, 177. Structure elucidation: William, Rogers, *Proc. Chem. Soc.* **1964**, 357. Synthesis of the naturally occurring (−)-form: Barton, Kirby, *J. Chem. Soc.* **1962**, 806; of the (±)-form: N. Hazama et al., *J. Chem. Soc. (C)* **1968**, 2947; Y. Misaka et al., *ibid.* 2954; A. G. Schultz et al., *J. Am. Chem. Soc.* **99**, 8065 (1977); S. F. Martin, P. J. Garrison, *J. Org. Chem.* **46**, 3567 (1981); **47**, 1513 (1982).

Plates from acetone, mp 121°. $[\alpha]_D^{27}$ −98° (alcohol). Freely sol in water; sol in alc, acetone. Its salts hydrolyze easily.

5498. Lycorine. *3,12-Didehydro-9,10-[methylenebis-(oxy)]galanthan-1,2-diol; 3,3a-didehydrolycoran-1α,2β-diol; 2,4,5,7,12b,12c-hexahydro-1H-[1,3]dioxolo[4,5-j]pyrrolo-[3,2,1-de]phenanthridine-1,2-diol;* amarylline; belamarine; narcissine; galanthidine. $C_{16}H_{17}NO_4$; mol wt 287.30. C 66.88%, H 5.96%, N 4.88%, O 22.27%. Alkaloid isolated from the bulbs of *Lycoris radiata* L., *Narcissus pseudonarcissus* L., *N. tazetta* L., from *Buphane disticha* Herb., in *Crinum* spp., *Amaryllis belladonna* L., *Clivia miniata* Regel and other Amaryllidaceae. Extraction procedure: Cook, Loudon in Manske-Holmes, *Alkaloids* vol. II (Academic Press, 1952) p 336. Identity of lycorine and galanthidine: Proskurnina, *Doklady Akad. Nauk SSSR* **90**, 565 (1953), *C.A.* **49**, 12500c (1955). Structure: Takeda et al., *J. Am. Chem. Soc.* **80**, 2562 (1958). Stereochemistry: Nakagawa, Uyeo, *J. Chem. Soc.* **1959**, 3736; K. Kotera et al., *Tetrahedron Letters* **1966**, 2009. Crystal structure: R. Roques, *Acta Crystallogr.* **30B**, 296 (1974). Attempted synthesis: Dyke et al., *Tetrahedron* **29**, 213 (1973). Synthesis: Y. Tsuda et al., *Chem. Commun.* **1975**, 933; eidem, *J. Chem. Soc. Perkin Trans. I* **1979**, 1358; T. Sano et al., *Heterocycles* **14**, 1097 (1980); S. F. Martin, C. Tu, *J. Org. Chem.* **46**, 3763 (1981). Biosynthesis: Archer et al., *Proc. Chem. Soc.* **1963**, 168; Fugati, Mazza, *Chem. Commun.* **1972**, 936. *Review:* W. C. Wildman in *The Alkaloids,* R. H. F. Manske, Ed., **vol. XI** (Academic Press, New York, 1968) pp 307-400.

Stout prisms from alcohol, mp 275-280° (dec). $[\alpha]_D^{16}$ −129° (c = 0.16 in 98% alc). Alkaline reaction to litmus; salts hydrolyze easily. Sparingly sol in alcohol, chloroform, petr ether. Sol in dilute acids. Practically insol in water, alkalies.

Hydrochloride, $C_{16}H_{17}NO_4$.HCl, long needles from water, mp 217° (dec with slight preliminary sintering). $[\alpha]_D^{20}$ +43°. Also a hydrochloride monohydrate, elongated prisms from water, mp 206°. Soluble in 20 parts water.

Methiodide, $C_{16}H_{17}NO_4$.CH_3I, two forms exist which may be stereoisomeric about the nitrogen atom. α- Form: polyhedra from alc, dec 247°. $[\alpha]_D^{20}$ −46° (c = 1.52). Freely sol in water; sparingly sol in hot alcohol. β-Form: prisms contg $1H_2O$ from water, dec 198°, after recryst from alcohol, dec 281°. $[\alpha]_D^{20}$ +123° (c = 2.44). Freely sol in water and hot alcohol.

5499. Lycoxanthin. ψ,ψ-*Carotene-16-ol; (all-trans)-lycopen-16-ol.* $C_{40}H_{56}O$; mol wt 552.85. C 86.90%, H 10.21%, O 2.89%. Carotenoid isolated from *Solanum dulcamara* L., *Lycopersicum esculentum* Mill., Solanaceae; *Tamus communis* L., Dioscoreaceae. Isoln by chromatography and structure: Zechmeister, Cholnoky, *Ber.* **69**, 422 (1936). Structure: Karrer et al., *Helv. Chim. Acta* **13**, 268, 1084 (1930); **14**, 614, 843 (1931); Winterstein, *Angew. Chem.* **72**, 902 (1960). Revised structure: Cholnoky, Szabolcs, *Tetrahedron Letters* **1968**, 1931. Stereochemistry: Kelly et al., *Acta Chem. Scand.* **25**, 1607 (1971). Total synthesis: Kjosen, Liaaen-jensen, *ibid.* 1500.

Purple needles from carbon disulfide. Reddish-brown round or acicular cryst aggregates from benzene + petr ether, mp 168°. Absorption max (acetone): 448, 474 ($E_{1cm}^{1\%}$ 3080), 505 nm. Sol in carbon disulfide and benzene. Moderately sol in petr ether. Sparingly sol in alc.

Monoacetate, deep purple needles from benzene+ methanol, mp 137°. Freely sol in carbon disulfide; sparingly sol in alcohol, petr ether.

5500. Lymecycline. N^6-*[[[[4-(Dimethylamino)-1,4,4a,-5,5a,6,11,12a-octahydro-3,6,10,12,12a-pentahydroxy-6-methyl-1,11-dioxo-2-naphthacenyl]carbonyl]amino]methyl]-*L-*lysine;* tetracyclinemethylene lysine; *N*-lysinomethyltetracycline; Armyl; Ciclolysal; Mucomycin; Tetralisal; Tetramyl; Tetralysal. $C_{29}H_{38}N_4O_{10}$; mol wt 602.63. C 57.79%, H 6.36%, N 9.30%, O 26.55%. Semi-synthetic antibiotic related to tetracycline, q.v. Prepn: Blackwood et al., U.S. pat. 3,042,716 (1962 to Pfizer); Lauria, Logemann, Ger. pat. 1,134,071 (1962 to Carlo Erba), *C.A.* **58**, 492 (1963); Tubaro, Raffaldoni, *Boll. Chim. Farm.* **100**, 9 (1961).

Sodium salt, $C_{29}H_{37}N_4NaO_{10}$. uv max (CH_3OH): 376 nm.

THERAP CAT: Antibacterial.

5501. Lynestrenol. (17α)-*19-Norpregn-4-en-20-yn-17-ol;* 17α-ethinyl-17β-hydroxyestr-4-ene; 17α-ethynylestr-4-en-17β-ol; 3-desoxynorlutin; ethinylestrenol; Exluton(a); Exlutena; Orgametril; Orgametil. $C_{20}H_{28}O$; mol wt 284.42. C 84.45%, H 9.92%, O 5.63%. Prepn: de Winter et al., *Chem. & Ind. (London)* **1959**, 905. Metabolism: Kamyab et al., *Biochem. J.* **103**, 14P (1967); *J. Endocrinol.* **42**, 337 (1968). Pharmacokinetics and metabolism: H. Kuhl et al., *Contraception* **26**, 303 (1982). Clinical trial of use in normophasic oral contraceptives: N. Dombrowicz et al., *ibid.* **22**,

537 (1980). Review of carcinogenicity studies: *IARC Monographs* **21**, 407-415 (1979).

Solid, mp 158-160°. $[\alpha]_D$ -13° (chloroform). Mixture with ethinyl estradiol, *q.v.*, *Anacyclin, Fysionorm, Minilyn, Noracyclin, Ovanon, Ovoresta, Yermonil*. Mixture with mestranol, *q.v.*, *Lyndiol*.

THERAP CAT: Progestogen. In combination with estrogen as oral contraceptive.

5502. Lyovac® Antivenin. Black widow spider antivenin. Prepd from the serum of horses immunized against the venom of the New World spider *Latrodectus mactans* Fabr., *Araneae*.

One unit of antivenin neutralizes one average mouse lethal dose of black widow spider venom when the antivenin and the venom are injected simultaneously in mice under suitable conditions.

THERAP CAT: Antivenin.

5503. Lypressin. 8-L-*Lysinevasopressin;* Schweine-Vasopressin; Phe³-Lys⁸-oxytocin; Diapid; Postacton; Syntopressin. $C_{46}H_{65}N_{13}O_{12}S_2$; mol wt 1056.26. C 52.31%, H 6.20%, N 17.24%, O 18.18%, S 6.07%. Purification after isolation from hog pituitary glands: Ward, du Vigneaud, *J. Biol. Chem.* **222**, 951 (1956). Structure: du Vigneaud *et al., J. Am. Chem. Soc.* **75**, 4880 (1953); Bartlett *et al., ibid.* **78**, 2905 (1956). Synthesis: du Vigneaud *et al., loc. cit.;* Bartlett *et al., loc. cit.;* Bodanszky *et al., ibid.* **82**, 3195 (1960); Boissonnas, Huguenin, *Helv. Chim. Acta* **43**, 182 (1960); Zaoral, *Coll. Czech. Chem. Commun.* **30**, 1853 (1965); Meienhofer, Sano, *J. Am. Chem. Soc.* **90**, 2996 (1968); Meienhofer, Trzeciak, *Proc. Nat. Acad. Sci. USA* **68**, 1006 (1971). Configuration: Schally, Barret, *J. Am. Chem. Soc.* **87**, 2497 (1965). *Review:* E. Schröder, K. Lübke, *The Peptides* **vol. II** (Academic Press, New York, 1966) pp 336-350.

```
Cys-Tyr-Phe-Gln-Asn-Cys-Pro-Lys-GlyNH₂
```

THERAP CAT: Antidiuretic; vasopressor.

5504. Lysalbinic Acid. A colloidal substance prepd by the action of caustic alkalies on albumin, usually egg albumin, although it also may be obtained from serum albumin or from casein (caseolysalbinic acid): Paal, *Ber.* **35**, 2195 (1902); **Ger. pats. 129,031** (1902) and **132,322** (1902). Composition: C 52.9%, H 7.0%, N 14.9%, S 1.2%.

Usually isolated as the Na salt, a brittle yellowish mass, readily sol in water, forming a colloidal soln which foams. It is precipitated by alc, and it may be purified by dialysis.

USE: As a protective colloid for metal sols, especially gold, silver, and mercury sols. "Colloidal gold" is prepd by adding a soln of gold chloride to a soln of lysalbinic acid. The intensely red hydrosol is precipitated with alcohol and dried. The precipitate forms hard, bronze-colored granules which may be stored for a long time, yielding a red, colloidal gold soln upon addn of water. Analysis of dried granules shows about 30% Au. See Paal, *Ber.* **35**, 2236-2244 (1902), and **Ger. pats. 170,433** and **180,730** (1906 and 1907 to Kalle).

5505. Lysergamide. *9,10-Didehydro-6-methylergoline-8β-carboxamide;* lysergic acid amide; ergine. $C_{16}H_{17}N_3O$; mol wt 267.32. C 71.88%, H 6.41%, N 15.72%, O 5.99%. Isoln from *Rivea corymbosa* (L.) and from *Ipomoea tricolor* Cav., *Convolvulaceae:* Hofmann, Tscherter, *Experientia* **16**, 414 (1964). Prepn from lysergic acid hydrazide: Ainsworth, **U.S. pat. 2,756,235** (1956 to Lilly); from lysergic acid and phosgene-dimethylformamide complex: Patelli, Bernardi, **U.S. pat. 3,141,887** (1964 to Farmitalia). Microbiological production: Rutschmann, Kobel, **U.S. pat. 3,219,545** (1965 to Sandoz).

Prisms from methanol, dec 242°. $[\alpha]_{5461}^{20}$ +15° (c = 0.5 in pyridine).

Methanesulfonate, $C_7H_{21}N_3O_4S$, prisms from methanol + acetone, dec 232°.

Note: This is a controlled substance (depressant) listed in the U.S. Code of Federal Regulations, Title 21 Part 1308.13 (1987).

5506. Lysergic Acid. *9,10-Didehydro-6-methylergoline-8-carboxylic acid.* $C_{16}H_{16}N_2O_2$; mol wt 268.32. C 71.62%, H 6.01%, N 10.44%, O 11.93%. Lysergic acid and isolysergic acid are the main cleavage products formed on alkaline hydrolysis of the alkaloids which are characteristic of ergot. Jacobs, Craig *et al., J. Biol. Chem.* **104**, 547 (1934); **125**, 289 (1938); **130**, 399 (1939); **145**, 487 (1942); *J. Org. Chem.* **10**, 76 (1945). High-yield production by *Claviceps paspali:* Arcamone *et al., Proc. Roy. Soc. (London), Ser. B,* **155**, 26 (1961). Total synthesis: Kornfeld *et al., J. Am. Chem. Soc.* **76**, 5256 (1954); **78**, 3087 (1956); M. Julia *et al., Tetrahedron Letters* **1969**, 1569; V. W. Armstrong *et al., ibid.* **1976**, 4311; W. Oppolzer *et al., Helv. Chim. Acta* **64**, 478 (1981); R. Ramage *et al., Tetrahedron* **37**, Suppl. 9, 157 (1981); J. Rebek, D. F. Tai, *Tetrahedron Letters* **24**, 859 (1983). Stereochemistry: Stoll *et al., Helv. Chim. Acta* **37**, 2039 (1954); Stenlake, *J. Chem. Soc.* **1955**, 1626; Leeman, Fabbri, *Helv. Chim. Acta* **42**, 2696 (1959). Absolute configuration: Stadler, Hofmann, *ibid.* **45**, 2005 (1962).

Hexagonal scales, plates with one or two moles H_2O from water, mp 240° (dec). $[\alpha]_D^{20}$ +40° (c = 0.5 in pyridine). Behaves as an acid and base, pKa 3.44, pKb 7.68. Moderately sol in pyridine. Sparingly sol in water and in neutral organic solvents; sol in NaOH, NH_4OH, Na_2CO_3, and HCl solns. Slightly sol in dil H_2SO_4.

Methyl ester, thin leaflets from benzene, mp 168°.

Note: This is a controlled substance (depressant) listed in the U.S. Code of Federal Regulations, Title 21 Part 1308.13 (1987).

5507. Lysergide. *9,10-Didehydro-N,N-diethyl-6-methylergoline-8β-carboxamide; N,N-diethyl-D-lysergamide;* D-lysergic acid diethylamide; LSD; LSD-25; Lysergsäure Diethylamid. $C_{20}H_{25}N_3O$; mol wt 323.42. C 74.27%, H 7.79%, N 12.99%, O 4.95%. Microbial formation by *Claviceps paspali* over the hydroxyethylamide: Arcamone *et al., Proc. Roy. Soc. (London)* **155B**, 26 (1961). Partial synthesis: Stoll, Hofmann, *Helv. Chim. Acta* **26**, 944 (1943); **38**, 421 (1955). Industrial prepn: Pioch; Garbrecht, **U.S. pats. 2,736,728; 2,774,763** (both 1956 to Lilly); Patelli, Bernardi, **U.S. pat. 3,141,887** (1964 to Farmitalia). Isotope-labeled LSD: Stoll *et al., Helv. Chim. Acta* **37**, 820 (1954). Toxicity data: E. Rothlin, *Ann. N.Y. Acad. Sci.* **66**, 668 (1957). *Review:* Hoffer, *Clin. Pharmacol. Ther.* **6**, 183 (1965). Book: *The Use of LSD in Psychotherapy and Alcoholism,* H. A. Abramson, Ed. (Bobbs-Merrill, Indianapolis, 1967) 697 pp.

Pointed prisms from benzene, mp 80-85°. $[\alpha]_D^{20}$ +17° (c = 0.5 in pyridine). uv max (ethanol): 311 nm ($E_{1cm}^{1\%}$ 257). LD_{50} in mice, rats, rabbits (mg/kg): 46, 16.5, 0.3 i.v. (Rothlin).

D-Tartrate, $C_{46}H_{64}N_6O_{10}$, solvated, elongated prisms from methanol, mp 198-200°. $[\alpha]_D^{20}$ +30°. Soluble in water.

Caution: This is a controlled substance (hallucinogen) listed in the U.S. Code of Federal Regulations, Title 21 Part 1308.11 (1987).

USE: In biochemical research as antagonist to serotonin. Has been used experimentally as adjunct in study and treatment of mental disorders.

5508. Lysidine. *4,5-Dihydro-2-methyl-1H-imidazole; 2-methyl-2-imidazoline;* methylglyoxalidine. $C_4H_8N_2$; mol wt 84.12. C 57.11%, H 9.59%, N 33.30%. Prepn: Chitwood, Reid, *J. Am. Chem. Soc.* **57**, 2424 (1935); King, McMillan, *ibid.* **68**, 1774 (1946); Kyrides, U.S. pats. **2,392,-326** and **2,404,299** (both 1946 to Monsanto); Ahrens, U.S. pat. **2,813,862** (1957 to Organon).

Needles from benzene, mp 105°. bp 198-200°. Soluble in water, alc, chloroform; less sol in benzene, carbon tetrachloride, petr ether; practically insol in ether. *Keep well closed. Incompat:* Acids, metallic salts, alkaloids.

5509. Lysine. Lys (IUPAC abbrev.); 2,6-diaminohexanoic acid; α,ε-diaminocaproic acid. $C_6H_{14}N_2O_2$; mol wt 146.19. C 49.29%, H 9.65%, N 19.16%, O 21.89%. NH_2-$(CH_2)_4CH(NH_2)COOH$. An amino acid classified as essential with respect to its growth effect in rats. Essential in human nutrition, not synthesized by the human body. Isoln from acid-hydrolyzed proteins (casein, fibrin or blood corpuscle paste): Rice, *J. Biol. Chem.* **131**, 1 (1939); *Biochem. Prepns.* **I**, 63 (1949). Synthesis: Fischer, Weigert, *Ber.* **35**, 3772 (1902); von Braun, *Ber.* **42**, 839 (1909); Eck, Marvel, *J. Biol. Chem.* **106**, 387 (1934); *Org. Syn. coll. vol. II*, 374 (1943). Total synthesis: Tuites, U.S. pat. **2,934,541** (1960 to du Pont). Production: Broquist *et al.,* U.S. pat. **2,980,590** (1961 to Am. Cyanamid); Seto, U.S. pat. **3,056,729** (1962 to Pfizer). Resolution of DL-lysine: Emmick, Rogers, U.S. pats. **2,556,907** and **2,657,230** (1951 and 1953, both to du Pont). Recent synthesis: K. Warning *et al., Ann.* **1978**, 112.

L-Lysine. Needles from water, hexagonal plates from dil alcohol. Darkens at 210°; dec 224.5°. $[\alpha]_D^{20}$ +14.6° (c = 6.5); $[\alpha]_D^{23}$ +25.9° (c = 2 in 6.0N HCl). pK_1 2.20; pK_2 8.90; pK_3 10.28 at 38°. Very freely sol in water, very slightly sol in alcohol, practically insol in ether.

L-Lysine dihydrochloride, $C_6H_{14}N_2O_2.2HCl$, crystals from ethanol + ether, mp 193°. $[\alpha]_D^{20}$ +15.3° (c = 2).

L-Lysine monohydrochloride, $C_6H_{15}ClN_2O_2$, *Darvyl, Enisyl, Lyamine.* Prepd by treating an ethanol soln of the dihydrochloride with pyridine. Crystals from dil ethanol. mp 263-264° when anhydrous (L-lysine dihydrochloride dihydrate, large crystals from water, mp 257°). $[\alpha]_D^{25}$ +14.6° (c = 2 in 0.6N HCl).

L-Lysine calcium salt, $C_{12}H_{26}CaN_4O_4$, *calcium lysinate.* Prepn: Hause, U.S. pat. **2,833,821** (1958 to du Pont). White powder, freely sol in water. Solubilizes acetylsalicylic acid.

L-Lysine monoorotate, $C_{11}H_{18}N_4O_6$, *Lysortine.* Prepn: **Brit.** pat. **922,361** (1962 to A.E.C. Soc. Chim. Org. Biol.). Crystals, dec 315°.

L-Lysine succinate, crystals, decomp 250°, salty taste approaching that of NaCl.

DL-Lysine dihydrochloride, crystals, mp 187-189°.

DL-Lysine monohydrochloride, crystals, mp 260-263°.

USE: Food enrichment. Supplementation of wheat-based foods with lysine improves their protein quality and results in improved growth and tissue synthesis: Feldberg, Hetzel, *Food Technol.* **12**, 496 (1958). Method of incorporation: Hause, Todd, U.S. pat. **3,015,567** (1962 to du Pont).

THERAP CAT: Nutrient.

5510. Lysine Acetylsalicylate. DL-*Lysine mono[2-(acetyloxy)benzoate]; lysine monosalicylate acetate;* aspirin lysine salt; LAS; Aspegic; Aspidol; Aspisol; Delgesic; Flectadol; Venopirin. $C_{15}H_{22}N_2O_6$; mol wt 326.35. C 55.21%, H 6.79%, N 8.58%, O 29.42%. Water soluble, injectable aspirin derivative. Prepn: **Fr.** pat. **1,295,304** (1962 to Equilibre Biologique), *C.A.* **58**, 1536f (1963). Prepn of D,L isomers and racemate: **Fr.** pat. **2,115,060** (1972 to Metabio), *C.A.* **78**, 75877k (1972). Comparison with aspirin: S. Rampon *et al., Rhumatologie* **24**, 141 (1972). Pharmacokinetics: H. von Ross *et al., Klin. Wochenschr.* **56**, 1119 (1978); F. Gentit, *Gaz. Med. France* **86**, 4539 (1979). Veterinary use as analgesic: P. Richez *et al., J. Vet. Pharmacol. Therap.* **2**, 231 (1979); *eidem, ibid.* **3**, 121 (1980). Analgesic activity in postoperative pain: K. Korttila *et al., Br. J. Anaesth.* **52**, 613 (1980); in articular pain: C. Diaz *et al., Clin. Ther.* **4**, 121 (1981). Use in treatment of premature labor: F. Wolff *et al., Arch. Gynecol.* **233**, 15 (1982). Comparison with aspirin in production of gastromucosal damage: J.-F. Bretagne *et al., Gastroenterol. Clin. Biol.* **8**, 28 (1984).

Crystals from ethanol, mp 154-156°. Sol in water; slightly sol in ethanol. Insol in methanol, acetone, ether.

THERAP CAT: Analgesic; antipyretic; anti-inflammatory.

5511. L-Lysine L-Glutamate. L-Glutamic acid L-lysine salt. $C_{11}H_{23}N_3O_6$; mol wt 293.32. C 45.04%, H 7.90%, N 14.33%, O 32.73%. $H_2N(CH_2)_4CH(NH_2)COOH.HOOC-(CH_2)_2CH(NH_2)COOH$. Prepd from DL-lysine and L-glutamic acid as an intermediate in the resolution of DL-lysine: Emmick, Rogers, U.S. pats. **2,556,907**; **2,657,230** (1951, 1953, both to du Pont).

Monohydrate, crystals, $[\alpha]^{22}$ +3.73 (12% in water): Rogers, *loc. cit.*

USE: As a flavor and nutritive additive to food: **Brit.** pat. **882,163** (1961 to Stamicarbon N.V.).

5512. Lysosomes. A group of cytoplasmic, subcellular particles present, with few exceptions, in all animal cells and involved in intracellular digestive processes. Functionally different from mitochondria. Lysosomes contain protein-dissolving enzymes, viz. acid phosphates, ribonuclease, β-glucuronidase, and cathepsin. Lysosomes released from their container inside the cell by the activity of some germs, notably *Streptococci*, can cause the cell to be digested by its own lysosomes. Comprehensive reviews: de Duve, *Sci. Am.* **208**, 64-72 (May 1963); Weissmann, *Blood* **24**, 594 (1964). Reviews of isoln and purification procedures: Several authors, *Methods Enzymol.* **31A**, 330-356 (1974).

5513. Lysostaphin. Antibiotic protein complex produced by *Staphylococcus staphyloliticus* with highly specific lytic activity against other *Staphylococcus* species. Contains three enzymes: a hexosaminidase, an amidase, and the major component, an endopeptidase which cleaves the polyglycine cross-linkages in the staphylococcal cell wall. Isoln and antibacterial spectrum: Schindler, Schuhardt, *Proc. Nat. Acad. Sci. USA* **51**, 414 (1964). Prepn by fermentation: *eidem,* U.S. pat. **3,278,378** (1966); Zygmunt, Browder, U.S. pat. **3,398,056** (1968 to Mead Johnson). The endopeptidase is a single polypeptide chain of mol wt 25,000: H. R. Trayer, C. E. Buckley, *J. Biol. Chem.* **245**, 4842 (1970). Identification of active principle: H. P. Browder *et al., Biochem. Biophys. Res. Commun.* **19**, 383 (1965). Characterization of three enzyme components: T. Wadstrom, O. Vesterberg, *Acta Pathol. Microbiol. Scand.* **79**, 248 (1971); O.-J. Iversen, A. Grov, *Eur. J. Biochem.* **38**, 293 (1973). Clinical studies to eradicate *Staph. aureus* from nasal carriers: R. L. Harris *et al., Antimicrob. Ag. Chemother.* **1967**, 110; K. E. Quickel *et al., Appl. Microbiol.* **1971**, 446. Use as identification method for *Staphylococcus* sp.: K. H. Schleifer, W. E. Kloos, *J. Clin. Microbiol.* **1**, 337 (1975); P. J. Severance *et al., ibid.* **11**, 724 (1980); disk modification: B. Poutrel, J.-P. Caffin, *ibid.* **13**, 1023 (1982). *Review:* Zygmunt, Tavormina, *Progress in Drug Research* **Vol. 16**, E. Jucker, Ed. (Birkhauser Verlag, Basel, 1972) pp 309-333.

Isoelectric pt: pH 10.4-11.4. uv max: 278 nm. Sedimentation coeff 2.35 S. Destroyed by pepsin or trypsin. Inhibited by Hg^{2+}, Cu^{2+}, Zn^{2+}. LD_{50} (7-day) in mice, rats (mg/kg): 820, 530 i.v. (Zygmunt).

Note: The previous nonproprietary name, staphcidin, was found to be in conflict with an established trademark.

5514. Lysozyme. Muramidase; *N*-acetylmuramide glycanohydrolase; *N*-acetylmuramyl hydrolase; globulin G_1. Mol wt about 14,400 ± 100. Mucolytic enzyme with antibiotic properties, first discovered by A. Fleming: *Proc. Roy. Soc. London* **93B**, 306 (1922). Found in tears, nasal mucus, milk, saliva, blood serum, in a great number of tissues and secretions of different animals, vertebrates and invertebrates, in egg white, in some molds, and in the latex of different plants. Isoln from egg white: Alderton *et al., J. Biol. Chem.* **157**, 43 (1945); Alderton, Fevold, *ibid.* **164**, 1 (1946); *Biochem. Prepns* **1**, 67 (1949); Sophianopoulos *et al., J. Biol. Chem.* **237**, 1107 (1962); from *Ficus* latex: Meyer *et al., ibid.* **163**, 733 (1946). Structure consists of a single polypeptide chain of 129 amino acid subunits of 20 different kinds crosslinked by four disulfide bridges. Chromatographic studies: Goncalves *et al., Arch. Biochem. Biophys.* **60**, 171 (1956); King, Craig, *J. Am. Chem. Soc.* **80**, 3366 (1958). Complete amino acid sequence of egg white lysozyme: Jolles *et al., Biochem. Biophys. Acta* **78**, 668 (1963); Canfield, *J. Biol. Chem.* **238**, 2698 (1963). Synthetic studies of egg white lysozyme: L. E. Barston *et al.* in *Chemistry and Biology of Peptides, Proc. 3rd Am. Peptide Symp.,* J. Meienhofer, Ed. (Ann Arbor Science Publishers, Inc., Michigan, 1972) pp 231-233; J. J. Sharp *et al., J. Am. Chem. Soc.* **95**, 6097 (1973). Three dimensional structure: Blake *et al., Nature* **206**, 757 (1965); North, *Science J.* **2**(11), 55 (1966); Phillips, *Sci. Am.* **215**, 78 (Nov. 1966); Blake *et al., Proc. Roy. Soc. London* **167B**, 365 (1967). Primary chemical structure and tentative complete sequence of human milk lysozyme: Jolles, Jolles, *Helv. Chim. Acta* **52**, 2671 (1969). Complete primary structure of human milk lysozyme and comparison with lysozymes of various origins: *eidem, ibid.* **54**, 2668 (1971). Dissolves bacterial cell wall mucopolysaccharides by hydrolyzing the β-(1→4) linkages between *N*-acetyl-D-muramic acid and 2-acetylamino-2-deoxy-D-glucose residues. Also acts on chitin. *Reviews:* Salton, *Bacteriol. Rev.* **21**, 82 (1957); Acker, Hartsell, *Sci. Am.* **202**, 132 (June 1960); Jolles, "Lysozyme" in P. D. Boyer, H. Lardy, K. Myrback, *The Enzymes,* **vol. 4,** (Academic Press, New York, 2nd ed., 1960) pp 431-445; *idem, Angew. Chem.* **76**, 20 (1964); Raftery, Dahlquist, *Fortschr. Chem. Org. Naturst.* **27**, 340 (1969); Hamaguchi, Hayashi, "Lysozyme" in *Proteins, Structure and Function,* **vol. 1,** M. Funatsu *et al.,* Eds. (Kodansha, Tokyo, Wiley,

New York, 1972) pp 85-222. Book: *Lysozyme,* E. F. Osserman *et al.,* Eds. (Academic Press, New York, 1974).

Crystals. Isoelectric pt: pH 10.5-11.0. Fairly stable in acid soln. Not affected by heat up to 55°: Cotterill, Winter, *Poultry Sci.* **33**, 1185 (1954).

Hydrochloride, *Acdeam, Antalzyme, Lanzyme, Leftose, Likinozym, Lisozima, Murazyme, Neutase, Neuzyme, Toyolysom-DS.*

THERAP CAT: Mucolytic enzyme. Antiviral.

5515. Lyxoflavine. *5-Deoxy-5-(3,4-dihydro-7,8-dimethyl-2,4-dioxobenzo[g]pteridin-10(2H)-yl)-*L*-arabinitol; 6,7-dimethyl-9-(1-*L*-lyxityl)isoalloxazine;* L*-lyxoflavine.* C_{17}-$H_{20}N_4O_6$; mol wt 376.36. C 54.25%, H 5.36%, N 14.89%, O 25.51%. The L-lyxose analog of riboflavine. Isoln from human myocardium: Pallares, Garza, *Arch. Biochem. Biophys.* **22**, 63 (1949); Gardner, *ibid.* **34**, 98 (1951). Synthesis: Wenis, U.S. pat. **2,734,054** (1956 to Hoffmann-La Roche); Folkers, U.S. pat. **2,760,865** (1956 to Merck & Co.).

Yellow to orange needles from water, dec 283-284°. $[\alpha]_D^{23}$ −49° (c = 0.26 in 0.05N NaOH). Sparingly sol in water, even less sol in alcohol. Freely sol with decompn in alkaline aq solns.

USE: Growth-promoting agent in feedstuffs.

5516. D**-Lyxose.** α-D-Lyxose. $C_5H_{10}O_5$; mol wt 150.13. C 40.00%, H 6.71%, O 53.29%. Prepd by the oxidation of calcium D-galactonate: Clark, *J. Biol. Chem.* **31**, 605 (1917); Fletcher *et al., J. Am. Chem. Soc.* **72**, 4546 (1950); by γ-irradiation of lactose: Adachi, *J. Dairy Sci.* **45**, 1427 (1962). Structure of β-D-lyxose: Hordvik, *Acta Chem. Scand.* **15**, 1781 (1961).

α-D-lyxose

Hygroscopic monoclinic prisms from ethanol + ether, mp 106-107°. Sweet taste. d^{20} 1.545. Shows mutarotation: $[\alpha]_D^{20}$ +5.5° → −14.0° (c = 0.82 in water). Freely sol in water. One part dissolves in 38 parts abs alc at 17°; 100 ml of 90% alc satd at 20° contain 7.9 g.

M

5517. Mabuterol. *4-Amino-3-chloro-α-[[(1,1-dimethylethyl)amino]methyl]-5-(trifluoromethyl)benzenemethanol;* 4-amino-α-[(*tert*-butylamino)methyl]-3-chloro-5-(trifluoromethyl)benzyl alcohol; 1-(4'-amino-3'-chloro-5'-trifluoromethylphenyl)-2-*tert*-butylaminoethanol; ambuterol. $C_{13}H_{18}ClF_3N_2O$; mol wt 310.75. C 50.25%, H 5.84%, Cl 11.41%, F 18.34%, N 9.01%, O 5.15%. Orally active β_2-adrenergic agonist related to clenbuterol, *q.v.* Prepn and resolution of isomers: **Belg. pat. 808,743** (1974 to Thomae); G. Engelhardt *et al.*, **U.S. pat. 4,119,710** (1978 to Boehringer Ing.); *eidem, Arzneimittel-Forsch.* **34**, 1612 (1984). Series of articles on pharmacology, pharmacokinetics, toxicology and clinical studies: *ibid.* 1625-1700. Toxicity data: K. Amemiya *et al., ibid.*, 1680. Teratological study: A. M. Hoberman *et al., J. Am. Coll. Toxicol.* **4**, 91 (1985). Determn in human urine and plasma by enzyme immunoassay: I. Yamamoto *et al., J. Immunoassay* **6**, 261 (1985).

dl-Form hydrochloride, $C_{13}H_{19}Cl_2F_3N_2O$, *KF 868, PB 868Cl, Broncholin.* Crystals from ethyl acetate + ether, mp 205-206°. Fairly sol in water. LD_{50} in male, female mice, male, female rats (mg/kg): 41.5, 51.1, 26.4, 28.1 i.v.; 60.3, 60.0, 76.3, 78.3 i.p.; 113.0, 125.7, 117.2, 123.1 s.c.; 220.8, 199.9, 319.3, 305.6 orally (Amemiya).

d-Form hydrochloride, mp >194° (slow dec). $[\alpha]_{364}^{20}$ +154.9° (c = 1 in methanol).

l-Form hydrochloride, mp >194° (slow dec). $[\alpha]_{364}^{20}$ −154.8° (c = 1 in methanol).

THERAP CAT: Bronchodilator; anti-asthmatic.

5518. Macassar Oil. Kusum oil; Kon oil; Paka oil. From the nut kernels of *Schleichera trijuga* Willd., *Sapindaceae* (Ceylon oak). *Habit.* India, Burma, Ceylon, Java. *Constit.* Glycerides of oleic acid 60%, arachidic 20-25%, palmitic 5-8% stearic 2-6%; free acetic acid 1-2%. Unsaponifiable matter 1.5-3% (this may include cyanide-containing glycosides which are removed by careful refining). *Reviews:* Sen-Gupta, *J. Soc. Chem. Ind. (London)* **39**, 88T (1920); Dhingra *et al., ibid.* **48**, 281T (1929).

Yellowish-white oil. Pleasant odor. Melting range: initial 21°, complete transparency 31°. d_{15}^{99} 0.860. n_D^{21} 1.46757; n_D^{27} 1.46655; $n_D^{31.5}$ 1.4646; n_D^{45} 1.4636. Acid value 10-70; saponification value 220-230; iodine value 48-58; Reichert-Meissl value 16. Miscible with ether, chloroform, other vegetable oils, petrolatum, lanolin.

USE: In hair oil formulations.

5519. Maclurin. *(3,4-Dihydroxyphenyl)(2,4,6-trihydroxyphenyl)methanone; 2,3',4,4',6-pentahydroxybenzophenone;* morintannic acid; moritannic acid; laguncurin; kinoyellow; C.I. Natural Yellow 11; C.I. 75240. $C_{13}H_{10}O_6$; mol wt 262.21. C 59.54%, H 3.84%, O 36.61%. From wood of *Chlorophora tinctoria* (L.) Gaud. (*Morus tinctoria* L., *Maclura tinctoria* (L.) D. Don), *Moraceae* (old fustic): Haley, Bassin, *J. Am. Pharm. Assoc.* **40**, 111 (1951); Laidlow, Smith, *Chem. & Ind. (London)* **1959**, 1604; from *Morus alba* Linn., *Moraceae:* Spada *et al., Gazz. Chim. Ital.* **86**, 46 (1956). Identity with laguncurin and kino-yellow: Nierenstein, *Quart. J. Pharm. Pharmacol.* **16**, 11 (1943). *See also Colour Index* vol. **4** (3rd ed., 1971) p 4627.

Yellow needles from ethanol, mp 222-222.5°. Sol in 190 parts water; freely sol in alcohol or ether.

USE: Dyeing fabrics.

5520. Macromerine. *α-[(Dimethylamino)methyl]-3,4-dimethoxybenzenemethanol; α-[(dimethylamino)methyl]veratryl alcohol;* 3,4-dimethoxy-α-[(dimethylamino)methyl]benzyl alcohol; dimethylaminomethyl 3,4-dimethoxyphenyl carbinol. $C_{12}H_{19}NO_3$; mol wt 225.28. C 63.97%, H 8.50%, N 6.22%, O 21.31%. Prepn of *dl*-form: La Manna *et al., Farmaco Ed. Sci.* **15**, 9 (1960); Chapman *et al., Proc. Roy. Soc. London, Ser. B,* **163**, (990), 116 (1965); Hodgkins *et al., Tetrahedron Letters* **1967**, 1321; of *d*- and *l*-forms: La Manna *et al., loc. cit.* Isoln of *l*-form from cactus, *Coryphantha macromeris* (Engelm.) Lem., Britton and Rose, *Cactaceae:* Hodgkins *et al., loc. cit.*

dl-Form, crystals, mp 46-47°.

l-Form, crystals, mp 66-67.5°. $[\alpha]_D^{25}$ −147.01° (c = 0.0390 g/ml chloroform), −42.61° (c = 0.0200 g/ml abs alcohol).

d-Form, crystals, mp 60-62°.

dl-Form hydrochloride, $C_{12}H_{19}NO_3$.HCl, crystals from ethanol, mp 163-164°.

l-Form hydrochloride, crystals, mp 178-179°. $[\alpha]_D^{18}$ −41.3° (c = 2.04 in 50% ethanol).

d-Form hydrochloride, crystals from ethanol, mp 178-179°. $[\alpha]_D^{18}$ +51.5° (c = 2.2 in 50% ethanol).

Note: Both the *dl*- and *l*-forms are physiologically active, and caused hallucinogenic reactions when tested in animals (Hodgkins *et al., loc. cit.*).

5521. Macusines. Isoln of A and B from *Strychnos toxifera* Schomb., *Loganiaceae:* A. R. Battersby *et al., J. Chem. Soc.* **1960**, 1848. Structure of A: A. T. McPhail *et al., Proc. Chem. Soc.* **1961**, 223; of B: A. R. Battersby, D. A. Yeowell, *ibid.* 17. Config of A: M.-M. Janot *et al., Bull. Soc. Chim. France* **1962**, 1079. Crystal and molecular structure of A: A. T. McPhail *et al., J. Chem. Soc.* **1963**, 1832. Sepn of C and abs config of A, B and C: A. R. Battersby, D. A. Yeowell, *ibid.* **1964**, 4419.

Macusine A, $[C_{22}H_{27}N_2O_3]^+$, *17-hydroxy-16-(methoxycarbonyl)-4-methylsarpaganium,* R_1 = CH_2OH, R_2 = COOCH$_3$. Chloride, prisms from ethanol + ether, mp 252°. $[\alpha]_D^{25}$ −57.5° (c = 1.45).

Macusine B, $[C_{20}H_{25}N_2O]^+$, *17-hydroxy-4-methylsarpaganium,* R_1 = H, R_2 = CH_2OH. Chloride, prisms from ethanol + ether, dec 248-249°. $[\alpha]_D^{22}$ +15.6° (c = 1.21). uv max (water): 222, 273, 280, 291 nm (log ε 4.61, 3.84, 3.82, 3.74).

Macusine C, $[C_{22}H_{27}N_2O_3]^+$, the C-16 epimer of A, R_1 = COOCH$_3$, R_2 = CH_2OH. Chloride, crystals from ethanol + ether, dec 264-265°. $[\alpha]_D^{24}$ −60.8° (c = 2.13). uv max (water): 222, 272, 278, 289 nm (log ε 4.63, 3.88, 3.87, 3.76).

5522. Maddrell's Salt. Polymeric sodium metaphosphate. Upon heating NaH_2PO_4 or $NaNH_4HPO_4$ to 250° and then slowly to 350°, first $Na_2H_2P_2O_7$ and then the insol Maddrell's salt is obtained, which is stable to about 500°; at 505° it changes to sodium trimetaphosphate, and above 607° to sodium hexametaphosphate. According to Maddrell, *Ann.* **61**, 63 (1847) the salt may be prepd by heating two parts $NaNO_3$ with one part H_3PO_4. An improved procedure is described by v. Knorre, *Z. Anorg. Chem.* **24**, 397 (1900). Structure: Jost, *Acta Cryst.* **16**, 428 (1963). *Reviews: see* Sodium Metaphosphate.

White powder. Practically insol in water and in aq solns of pyrophosphates and hexametaphosphates. Sol in mineral acids. Does not form complexes with Fe(II), Fe(III), U(IV) salts.

USE: In dentifrices, polishing agents, abrasive detergents.

5523. MADU. 2'-*Deoxy*-5-(*methylamino*)*uridine;* 5-(methylamino)-2'-deoxyuridine. $C_{10}H_{15}N_3O_5$; mol wt 257.24. C 46.69%, H 5.88%, N 16.34%, O 31.10%. Prepn: Visser *et al., Biochim. Biophys. Acta* **76**, 463 (1963). Active *in vitro* against herpes infections: Nemes, Hilleman, *Proc. Soc. Exp. Biol. Med.* **119**, 515 (1965); Shen, U.S. pat. **3,322,-627** (1967 to Merck & Co.).

Crystals from ethanol + water, mp 180-182°.
THERAP CAT: Antiviral.

5524. Maduramicin. (3R,4S,5S,6R,7S,22S)-23,27-Dide-methoxy-2,6,22-tridemethyl-11-O-demethyl-22-[(2,6-dideoxy-3,4-di-O-methyl-β-L-arabino-hexopyranosyl)oxy]-6-methoxylonomycin A monoammonium salt; antibiotic X-14868A ammonium salt; CL 273,703; Cygro. $C_{47}H_{83}NO_{17}$; mol wt 934.17. C 60.43%, H 8.95%, N 1.50%, O 29.12%. Polyether antibiotic chemically related to the lonomycins, *q.v.* Isoln as sodium salt from *Nocardia* sp X-14868 and biological activity: C.-M. Liu *et al.,* U.S. pat. **4,278,663** (1981 to Hoffmann-La Roche); as free acid hydrate from *Actinomadura yumaense* sp nov.: D. P. Labeda *et al.,* U.S. pat. **4,407,946** (1983 to Am. Cyanamid). Fermentation and properties: C.-M. Liu *et al., J. Antibiot.* **36**, 343 (1983). ^{13}C-NMR spectrum: S. Rajan, *ibid.* **37**, 1495 (1984). Biosynthetic studies: H.-R. Tsou *et al., ibid.* 1651; H.-R. Tsou *et al., ibid.* **40**, 94 (1987). Antimalarial activity: L. Oronsky, U.S. pat. **4,496,549** (1985 to Am. Cyanamid). Nematocidal activity: I. B. Wood, U.S. pat. **4,510,134** (1985 to Am. Cyanamid).

Sodium salt, $C_{47}H_{79}NaO_{17}$, crystals from ethyl acetate + n-hexane, mp 193-195°. $[\alpha]_D$ +40.6° (chloroform). $[\alpha]_D$ +23.8° (methanol).

Note: Not to be confused with *maduramycin* isolated from *Actinomadura rubra,* $C_{28}H_{22}O_{10}$: W. F. Fleck *et al., 16th Interscience Conference on Antimicrob. Ag. Chemother.,* Chicago, 1976, *Abstracts of Papers,* no. 51.
THERAP CAT (VET): Coccidiostat.

5525. Mafenide. 4-(*Aminomethyl*)*benzenesulfonamide;* α-*amino*-p-*toluenesulfonamide;* p-(aminomethyl)benzenesulfonamide; 4-homosulfanilamide; maphenide; Marfanil; Mesudrin; Mesudin; Sulfamylon; Homosulfamine; Ambamide; Neofamid; Septicid; Emilene; Homonal; Paramenyl. $C_7H_{10}N_2O_2S$; mol wt 186.25. C 45.14%, H 5.41%, N 15.04%, O 17.18%, S 17.22%. Prepn: Miller *et al., J. Am. Chem. Soc.* **62**, 2099 (1940); Klarer, U.S. pat. **2,288,531**

(1942 to Winthrop); Bergeim, Braker, *J. Am. Chem. Soc.* **66**, 1459 (1944); Kusami *et al., J. Pharm. Soc. Japan* **64**, no. 9a, 51 (1944); Komokina *et al., J. Appl. Chem. (USSR)* **21**, 681 (1948); Angyal, Jenkin, *Aust. J. Sci. Res.* **3A**, 461 (1950). Toxicity study: T. W. Skulan, J. O. Hoppe, *Life Sci.* **5**, 2279 (1966).

Crystals from alc, mp 151-152°. Sol in dil alkali, acid. Hydrochloride, $C_7H_{11}ClN_2O_2S$, crystals from 95% alc, mp 256°. Neutral in soln. LD_{50} in rats, mice (mg/kg): 1170, 900 i.v. (Skulan, Hoppe).
Acetate, $C_9H_{14}N_2O_4S$, *Mafatate, Mefamide.* LD_{50} in rats, mice (mg/kg): 2040, 1580 i.v. (Skulan, Hoppe).
Propionate, $C_{10}H_{16}N_2O_4S$, *Sulfomyl.* Crystals, mp 158°. Readily sol in water.
THERAP CAT: Antibacterial.

5526. Magainins. A class of potent antimicrobial peptides isolated from the glanular gland of clawed toad *Xenopus laevis.* Name coined by Zasloff from the Hebrew word for "shield". Consists of magainins I and II, both containing 23 amino acid residues and differing only at the 10 (I: Gly, II: Lys) and 22 (I: Lys, II: Asn) positions. Isolation and characterization: M. Zasloff, *Proc. Nat. Acad. Sci. USA* **84**, 5449 (1987). Identification of post-secretory degradation product: T. W. Schwartz *et al., Nature* **329**, 494 (1987). Brief overview: J. Alper, *Chem. Week* (November 4, 1987) p 66; M. Cannon, *Nature* **328**, 478 (1987).

1   10
Gly-Ile-Gly-Lys-Phe-Leu-His-Ser-Ala-Gly-Lys-Phe-

13   22
Gly-Lys-Ala-Phe-Val-Gly-Glu-Ile-Met-Lys-Ser

Magainin I

5527. Magaldrate. *Aluminum magnesium hydroxide* ($AlMg(OH)_7$) *monohydrate;* magnesium aluminate hydrate; monalium hydrate; AY 5710; Dynese; Riopan; Ripon. Formerly treated as $[Mg(OH)]_4[(HO)_4Al(OH)(HO)Al(OH)_4]$.-$2H_2O$. Byk Gulden company information states that based on ^{27}Al broadline nuclear resonance spectra, radiographic investigation and chemical analytical finds, the chemical formula for Riopan powder is: $Mg_{1.5+x}Al(OH)_6(SO_4)_x$-$(H_2O)_{0.5-x}$ with 0.325 < x < 0.462. Prepd by mixing together and allowing to react at 0° to 50° an alkali aluminate soln (contg 3-5 mols of Na_2O or K_2O per mol Al_2O_3) with a magnesium salt soln in such proportions that 1 g of Al combines with 0.9 to 3.0 g of Mg, and washing the pptd product: Hallmann, U.S. pat. **2,923,660** (1960 to Byk-Gulden Lomberg).

White, odorless, crystalline powder which contains the equivs of not < 28.0% and not > 39.0% MgO and not < 17.0% and not > 25.0% Al_2O_3.
THERAP CAT: Antacid.

5528. Magenta I. 4-[(4-*Aminophenyl*)(4-*imino*-2,5-*cyclohexadien*-1-*ylidene*)*methyl*]-2-*methylbenzenamine* mono-hydrochloride; C.I. Basic Violet 14; α^4-(p-aminophenyl)-α^4-(4-imino-2,5-cyclohexadien-1-ylidene)-2,4-xylidine hydrochloride; 2-methyl-4,4'-[[(4-imino-2,5-cyclohexadien-1-ylidene)methylene]dianiline hydrochloride; fuchsine; magenta; rosaniline hydrochloride; C.I. 42510. $C_{20}H_{20}ClN_3$; mol wt 337.85. C 71.10%, H 5.97%, Cl 10.49%, N 12.44%. Prepn: Fischer, Fischer, *Ann.* **194**, 276, 290 (1878); *Ber.* **13**, 2204 (1880); J. T. Scalan, *J. Am. Chem. Soc.* **57**, 887 (1935). Evaluation of carcinogenicity studies: *IARC Monograph* **4**, 57 (1974). *See also: Colour Index* vol. 4 (3rd ed., 1971) p 4389.

Metallic green, lustrous crystals, dec above 200°. Absorption max (ethanol): 543 nm (ϵ 93,000). 2.65 parts dissolve in 1000 parts water. Practically insol in ether; sol in alcohol with a carmine red color.

USE: As a dye or in manuf of other dyes.

THERAP CAT: Antifungal.

5529. Magnesium. Mg; at. wt 24.305; at. no. 12; valence 2. An alkaline earth metal. One of the most common elements in the earth's crust: 2.1% by weight. Isotopes: 24 (78.70%); 25 (10.13%); 26 (11.17%). Found naturally only in the form of its compounds in magnesite, carnallite, *dolomite* [$CaMg(CO_3)_2$], epsomite, kieserite, and many other minerals; found in sea-water; in animal and vegetable kingdom. First obtained in metallic form by Davy in 1808 by electrolysis of a mixture of magnesia and mercuric oxide. Methods of prepn: Deville, Caron, etc, cited in *Gmelin's, Magnesium* (8th ed.) 27A, 121 (1937). Prepd industrially by reduction of MgO-contg materials. Examples of large-scale processes: Weiss, U.S. pat. **3,264,097** (1966 to Ver. Aluminium-Werke). Review of magnesium and its compounds: Goodenough, Stenger, "Magnesium, Calcium, Strontium, Barium and Radium" in *Comprehensive Inorganic Chemistry*, vol. **1**, J. C. Bailar, Jr. *et al.*, Eds. (Pergamon Press, Oxford, 1973) pp 591-664; L. F. Lockwood *et al.*, in Kirk-Othmer *Encyclopedia of Chemical Technology* vol. **14** (Wiley-Interscience, New York, 3rd ed., 1981) pp 570-615.

Silvery-white metal; hexagonal close-packed structure. Slowly oxidizes in moist air. Available as bars, ribbons, wire and powder. mp 651°. bp 1100°. d^{20} 1.738. Sp heat (20°) 0.245 cal/g. Heat of fusion 88 cal/g. Electrical resistivity 4.46 μohm-cm. E° (aq) Mg^{2+}/Mg -2.37 V. Reacts very slowly with water at ordinary temp, less slowly at 100°. Reacts readily with dil acids with liberation of hydrogen; reacts with aq solns of ammonium salts, forming a double salt. Reduces carbon monoxide, carbon dioxide, sulfur dioxide, nitric oxide, and nitrous oxide at a red heat. Burns in air; continues to burn in a current of steam. Combines directly with nitrogen, sulfur, the halogens, phosphorus, and arsenic. Reacts with methyl alcohol at 200° giving magnesium methylate.

Human Toxicity: Particles embedded in skin can produce gaseous blebs with a protracted course. Inhalation of the dust is irritating; fumes can cause metal fume fever.

USE: As a constituent of light alloys; for manuf precision instruments, optical mirrors; in pyrotechnics; in metallurgy as deoxidizing and desulfurizing agent; instead of zinc in dry batteries; for flash bulbs and flares, alumino-thermics, ignition of thermite mixture, intense signal lights; for Grignard reagents; in the recovery of titanium.

5530. Magnesium Acetate. Cromosan. $C_4H_6MgO_4$; mol wt 142.40. C 33.73%, H 4.25%, Mg 17.08%, O 44.94%. $Mg(C_2H_3O_2)_2$.

Tetrahydrate, colorless or white, deliquesc crystals. d 1.45. mp about 80°. Very sol in water or alcohol. The aq soln is neutral or slightly acid. *Keep well closed.* LD_{50} i.v. in mice: 18 mg/kg.

5531. Magnesium Acetylsalicylate. 2-*(Acetyloxy)benzoic acid magnesium salt;* magnesium aspirin; Apyron; Fyracyl; Magisal; Magnespirin; Novacetyl. $C_{18}H_{14}MgO_8$; mol wt 382.61. C 56.50%, H 3.69%, Mg 6.36%, O 33.45%. $(CH_3$-$COOC_6H_4COO)_2Mg$.

White, nonhygroscopic, almost tasteless and odorless powder. Freely soluble in water; less sol in alcohol.

THERAP CAT: Has been used as analgesic, antipyretic.

5532. Magnesium Amide. Magnesium diamide. H_4-MgN_2; mol wt 56.37. Mg 43.15%, N 49.70%, H 7.15%.

$Mg(NH_2)_2$. Prepd by the action of ammonia on an ether soln of magnesium diethyl or on magnesium activated with iodine (at 400°): Terentiew, *Z. Anorg. Chem.* **162**, 351 (1927); Schlenk, Jr., *Ber.* **64**, 738 (1931). Prepd from metal and liquid ammonia at high pressure: Juza, Jacobs, *Angew. Chem., Int. Ed.* **5**, 247 (1966); eidem, *Z. Anorg. Allgem. Chem.* **370**, 245 (1969).

White powder or crystals. d_4^{25} 1.39. Catches fire in air. Dec on heating. Reacts violently with water, evolving ammonia.

USE: Polymerization catalyst.

5533. Magnesium Benzoate. $C_{14}H_{10}MgO_4$; mol wt 320.60. C 63.08%, H 3.78%, Mg 9.12%, O 24.01%. Mg-$(C_7H_5O_2)_2$.

Trihydrate, white, odorless powder. mp about 200°. Sol in 20 parts water, freely in hot water, alcohol. The aq soln is neutral or slightly acid.

5534. Magnesium Bisulfate. Magnesium hydrogen sulfate; acid magnesium sulfate. $H_2MgO_8S_2$; mol wt 218.47. H 0.92%, Mg 11.13%, O 58.59%, S 29.36%. $Mg(HSO_4)_2$.

White powder. Sol in water.

5535. Magnesium Borate. "Antifungin". B_2MgO_4; mol wt 109.93. B 19.67%, Mg 22.11%, O 58.22%. $Mg(BO_2)_2$. Various magnesium borate minerals occur in nature. These include: *ascharite, camsellite, inderite, kotoite, kurnakovite, paternoite, pinnoite, szaibelyite*.

Octahydrate, white powder. Slightly sol in water.

USE: Antiseptic; fungicide.

5536. Magnesium Bromate. Br_2MgO_6; mol wt 280.13. Br 57.05%, Mg 8.68%, O 34.27%. $Mg(BrO_3)_2$.

Hexahydrate, colorless or white crystals; loses all water at about 200°; dec at higher temp. Sol in 1.5 parts water.

5537. Magnesium Bromide. Br_2Mg; mol wt 184.13. Br 86.80%, Mg 13.20%. $MgBr_2$.

Hexahydrate, colorless, very deliquesc crystals or white granules; bitter taste. mp about 165° with decompn. Sol in 0.3 part water; sol in alcohol. The aq soln is neutral. *Keep well closed.*

USE: In organic syntheses.

THERAP CAT: Sedative, anticonvulsant.

5538. Magnesium Carbonate Hydroxide. Marinco C. Usually a basic carbonate, approx $(MgCO_3)_4.Mg(OH)_2.$-$5H_2O$; approx mol wt 485; contains 40-42% MgO. Prepn of various hydrates: Pond, Heneghan, U.S. pat. **3,169,826** (1965 to Merck & Co.). Magnesium carbonate minerals include *magnesite* and *lansfordite*.

White, odorless, bulky powder or light friable masses; at about 700° is converted into MgO. Sol in about 3300 parts CO_2-free water; more sol in water contg CO_2; sol in dil acids with effervescence; insol in alcohol. Imparts slight alkaline reaction to water.

Hydrotalcite, Altacite, Hi-Ti, Nacid, Talcid, an aluminum magnesium hydroxide carbonate hydrate of the general formula $Al_2O_3.6MgO.CO_2.12H_2O$. Prepn: Kumuru *et al.,* U.S. pat. **3,539,306** (1970 to Kyowa).

USE: Fireproofing, heat insulating; preparing effervescent magnesium citrate; clarifying liqs by filtration; in tooth and face powders, in polishing compds; manuf mineral waters, pigments, paper; filler for rubber.

THERAP CAT: Magnesium carbonate hydroxide as antacid, cathartic; hydrotalcite as antacid.

THERAP CAT (VET): Laxative.

5539. Magnesium Chlorate. Cl_2MgO_6; mol wt 191.21. Cl 37.08%, Mg 12.71%, O 50.21%. $Mg(ClO_3)_2$.

Hexahydrate, white, very deliquesc crystals or cryst powder; bitter taste; d 1.80. mp about 35°. Sol in 0.9 part water; slightly sol in alcohol. *Keep well closed.* LD_{50} orally in rats: 5.25 g/kg, Ulrich, *J. Pharmacol. Exp. Ther.* **35**, 1 (1929).

5540. Magnesium Chloride. Magnogene. Cl_2Mg; mol wt 95.23. Cl 74.46%, Mg 25.54%. $MgCl_2$. Prepd from magnesium ammonium chloride hexahydrate in the presence of HCl: Bryce-Smith, *Inorg. Syn.* **6**, 9 (1960).

Soft, highly deliquescent leaflets. mp 712° (rapid heating). d 2.41, also reported as d 2.325. Slow heating releases chlorine at 300°. Can be distilled in a stream of hydrogen. Attacks fused silica when melted. Sol in water, with the evolution of much heat, giving a clear soln. Easily forms alcoholates and etherates.

Hexahydrate, deliquescent crystals. d 1.56. At 100° loses $2H_2O$ (17.7%); at 110° begins to lose some HCl. By strong ignition is converted into oxychloride. mp, when rapidly heated, at about 118° with decompn. One gram dissolves in 0.6 ml water, 0.3 ml boiling water, 2 ml alcohol. Its aq soln is neutral. *Keep well closed.* LD_{50} orally in rats: 8.1 g/kg, H. F. Smyth *et al., Am. Ind. Hyg. Assoc. J.* **30**, 470 (1969).

USE: Fireproofing wood; in disinfectants; various magnesia cements; fire extinguishers; dressing cotton fabrics; in floor-sweeping compds; carbonizing wool; manuf parchment paper or artificial leather; as addition to casein glue; as a reagent in analytical chemistry.

THERAP CAT: Cathartic.

THERAP CAT (VET): Has been used in bovine hypomagnesemia.

5541. Magnesium Citrate. *2-Hydroxy-1,2,3-propanetricarboxylic acid magnesium salt (2:3).* $C_{12}H_{10}Mg_3O_{14}$; mol wt 451.13. C 31.95%, H 2.23%, Mg 16.17%, O 49.65%. Mg_3-$(C_6H_5O_7)_2$.

Tetradecahydrate, white, odorless, cryst powder or granules. Slightly sol in water, the solv is increased by citric acid or alkali citrates; sol in dil acids.

THERAP CAT: Cathartic.

5542. Magnesium Citrate, Dibasic. *2-Hydroxy-1,2,3-propanetricarboxylic acid magnesium salt (1:1);* magnesium citrate sol; acid magnesium citrate; Citresia. $C_6H_6MgO_7$; mol wt 214.42. C 33.61%, H 2.82%, Mg 11.34%, O 52.23%. Prepn of stable, hydrated compd: Davenport, Costa, U.S. pat. **2,260,004** (1941 to Pfizer).

Pentahydrate, white or slightly yellow, odorless, granules or powder. Sol in about 5 parts water, becoming less sol with age or by loss of water; insol in alcohol.

5543. Magnesium Fluoride. Afluon; Sellaite. F_2Mg; mol wt 62.32. F 60.98%, Mg 39.02%. MgF_2. Prepd from $MgCO_3$ + HF: Klemm *et al., Z. Anorg. Allgem. Chem.* **176**, 13 (1928).

Colorless substance (rutile-type lattice). Slight violet fluorescence. d 3.148. Mohs' hardness: 6. mp 1248°. bp 2260°. Very sparingly sol in water 87 mg/l (18°). Slightly sol in dil acids, especially nitric acid. May be stored in glass bottles. LD orally in guinea pigs: 1.0 g/kg, *Handbook of Toxicology,* **vol. 1,** W. S. Spector, Ed. (Saunders, Philadelphia, 1956) pp 180-181.

USE: In the ceramics and glass industry.

5544. Magnesium Formate. $C_2H_2MgO_4$; mol wt 114.35. C 21.01%, H 1.76%, Mg 21.27%, O 55.96%. $Mg(HCOO)_2$.

Dihydrate, crystals or granules. Sol in water; insol in alcohol. Its aq soln is practically neutral.

5545. Magnesium Germanide. $GeMg_2$; mol wt 121.24. Mg 40.12%, Ge 59.88%. Mg_2Ge. Prepd by fusion of the components: Winkler, *Helv. Phys. Acta* **28**, 633 (1955).

Crystals, mp 1115°.

USE: In semiconductor research.

5546. Magnesium Hexafluorosilicate. Magnesium silicofluoride; magnesium fluosilicate. F_6MgSi; mol wt 166.40. F 68.51%, Mg 14.61%, Si 16.88%. $MgSiF_6$.

Hexahydrate, white, efflorescent, odorless crystals. d 1.788. At about 120° loses SiF_4. Soluble in water; insol in alcohol. pH of 1% aq soln: 3.1. LD orally in guinea pigs: 200 mg/kg, *Handbook of Toxicology,* **vol. 1,** W. S. Spector, Ed. (Saunders, Philadelphia, 1956) pp 180-181.

USE: For mothproofing of textile fabrics.

5547. Magnesium Hydride. H_2Mg; mol wt 26.34. H 7.65%, Mg 92.35%. MgH_2. Prepd by thermal decompn of diethylmagnesium at 200° in high vacuum: Wiberg, Bauer, *Ber.* **85**, 593 (1952); from the elements: Ellinger *et al., J. Am. Chem. Soc.* **77**, 2647 (1955).

White, nonvolatile mass or tetragonal crystals. d 1.45. Dec at 280° in high vacuum. Strong reducing agent. Ignites spontaneously on contact with air, forming MgO and H_2O. On contact with water a violent evolution of hydrogen takes place. On contact with methanol, magnesium alcoholate

and hydrogen are formed. Forms double hydrides with boron hydride and aluminum hydride.

5548. Magnesium Hydroxide. Magnesium hydrate; Marinco H. H_2MgO_2; mol wt 58.34. H 3.46%, Mg 41.69%, O 54.85%. $Mg(OH)_2$. Occurs in nature as the mineral *brucite.* Prepn from magnesium chloride or sulfate and NaOH: Perlard, Waldron, U.S. pat. **3,127,241** (1964 to Dow). Industrial method for producing magnesium hydroxide having variable particle sizes: Chisholm, U.S. pat. **3,232,708** (1966 to FMC).

Amorphous powder. Practically insol in water (1:80,000); sol in dil acids. Imparts slight alkaline reaction to water. pH of aq slurry: 9.5-10.5. Absorbs CO_2 in the presence of water. *Keep well closed.*

Suspension of 30% magnesium hydroxide in water: *Hydro-Magma.*

THERAP CAT: Antacid, cathartic.

THERAP CAT (VET): Laxative.

5549. Magnesium Iodide. I_2Mg; mol wt 278.11. I 91.26%, Mg 8.74%. MgI_2.

Octahydrate, white, deliquesc powder; readily discolors in air and light. Very sol in water; sol in alcohol. The aq soln is neutral or slightly alkaline. *Keep well closed and protected from light.*

5550. Magnesium Lactate. *2-Hydroxypropanoic acid magnesium salt.* $C_6H_{10}MgO_6$; mol wt 202.45. C 35.59%, H 4.98%, Mg 12.01%, O 47.42%. $(CH_3CHOHCOO)_2Mg$.

Trihydrate, white cryst, very bitter powder. One gram dissolves in 25 ml cold water, 3.5 ml boiling water; slightly sol in alcohol. The aq soln is slightly acid.

THERAP CAT: Cathartic.

5551. Magnesium Mandelate. Magdelate. $C_{16}H_{14}MgO_6$; mol wt 326.61. C 58.84%, H 4.32%, Mg 7.45%, O 29.39%. $Mg[C_6H_5CH(OH)COO]_2$.

White, odorless powder. Stable in air. Slightly sol in cold water; sol in about 250 parts boiling water; insol in alcohol. The aq soln is practically neutral. Dec by acids forming free mandelic acid.

THERAP CAT: Urinary antiseptic.

5552. Magnesium Nitrate. MgN_2O_6; mol wt 148.32. Mg 16.39%, N 18.88%, O 64.73%. $Mg(NO_3)_2$. Hydrated form occurs in nature as the mineral *nitromagnesite.*

Hexahydrate, colorless, clear, deliquesc crystals. d 1.464. mp about 95°. Sol in 0.8 part water; freely sol in alcohol. The aq soln is neutral. *Keep well closed.*

USE: In pyrotechnics; in the concentration of nitric acid.

5553. Magnesium Oleate. *9-Octadecenoic acid magnesium salt.* $C_{36}H_{66}MgO_4$; mol wt 587.21. C 73.63%, H 11.33%, Mg 4.14%, O 10.90%. $Mg(C_{18}H_{33}O_2)_2$. Commercial form usually contains some stearate and palmitate.

Yellowish powder or mass. Insol in water; partly sol in alcohol, ether, petr ether.

USE: As an addition to naphtha to prevent spontaneous ignition in cleaning establishments.

5554. Magnesium Oxalate. *Ethanedioic acid magnesium salt.* C_2MgO_4; mol wt 112.33. C 21.38%, Mg 21.65%, O 56.97%. MgC_2O_4.

Dihydrate, white powder. Sol in about 1500 parts water, in dil mineral acid; insol in alcohol.

5555. Magnesium Oxide. Magnesia; calcined magnesia; magnesia usta; Magcal; Maglite. MgO; mol wt 40.32. Mg 60.32%, O 39.68%. Occurs in nature as the mineral *periclase.* Commercial prepn from magnesite ores: Adams, U.S. pat. **3,320,029** (1967 to Northwest Magnesite).

White, very fine, odorless powder. Available in a very bulky form termed "Light" or in a dense form termed "Heavy." Takes up CO_2 and H_2O from the air, the light form more readily than the heavy. mp 2800°. Highly reflective in visible and near uv region. Combines with water to form magnesium hydroxide. Very slightly sol in pure water, solv increased by CO_2; sol in dil acids; insol in alcohol. Imparts slight alkaline reaction to water. pH of satd aq soln 10.3. *Keep well closed.*

USE: Manuf refractory crucibles, fire bricks, magnesia cements and boiler scale compounds, "powdered" oils, casein glue. Reflector in optical instruments; white color standard. Insulator at low temp.

THERAP CAT: Antacid.

THERAP CAT (VET): Antacid, laxative, in hypomagnesemia.

5556. Magnesium Perborate. B_2MgO_6; mol wt 141.93. B 15.23%, Mg 17.13%, O 67.64%. $Mg(BO_3)_2$.

Heptahydrate, white, bulky powder; gradually dec with loss of active oxygen. Sparingly sol in water with partial decompn; sol in dil acids.

USE: As of sodium perborate; with pptd chalk in tooth powders.

5557. Magnesium Perchlorate. Anhydrone; Dehydrite. Cl_2MgO_8; mol wt 223.23. Mg 10.89%, Cl 31.77%, O 57.34%. $Mg(ClO_4)_2$.

White, very hygroscopic, granular or flaky powder; dec above 250°. Dissolves in water with evolution of a considerable amount of heat. Crystallizes from water with $6H_2O$. *Keep tightly closed.*

USE: As a drying agent for gases. The article of commerce may contain an amount of water equivalent to a dihydrate, but even the trihydrate is said to be effective for drying gases. *Caution:* Can cause irritation of skin, mucous membranes.

5558. Magnesium Permanganate. $MgMn_2O_8$; mol wt 262.19. Mg 9.27%, Mn 41.91%, O 48.82%. $Mg(MnO_4)_2$.

Bluish-black, deliquesc crystals. Freely sol in water. *Keep well closed.*

USE: As polymerization catalyst: Pengilly et al., U.S. pat. 3,576,790 (1971 to Goodyear). In purification of benzene: Ingwalson et al., U.S. pat. 3,478,092 (1969 to Velsicol).

5559. Magnesium Peroxide. Magnesium dioxide; Magnesium Perhydrol; Magnesium Superoxol. MgO_2; mol wt 56.31. Mg 43.17%, O 56.83%. The article of commerce contains 15-25% MgO_2, the balance being magnesium hydroxide. For medicinal use the 25% article is preferred.

White, tasteless, odorless powder. Insol in water and gradually dec by it with liberation of oxygen; sol in dil acids, forming hydrogen peroxide; when strongly heated loses all the peroxide oxygen. *Keep well closed.*

THERAP CAT: Antacid, anti-infective.

5560. Magnesium Phosphate, Dibasic. Magnesium hydrogen phosphate; secondary magnesium phosphate. $HMgO_4P$; mol wt 120.29. H 0.84%, Mg 20.21%, O 53.20%, P 25.74%. $MgHPO_4$. Trihydrate occurs in nature as the minerals *newberyite, phosphor-roesslerite.*

Trihydrate, white, cryst powder. d 2.13. Slightly sol in water; sol in dil acids.

THERAP CAT: Cathartic.

5561. Magnesium Phosphate, Monobasic. Magnesium biphosphate; acid magnesium phosphate; primary magnesium phosphate. $H_4MgO_8P_2$; mol wt 218.28. H 1.84%, Mg 11.14%, O 58.64%, P 28.38%. $Mg(H_2PO_4)_2$.

Trihydrate, white powder. Soluble in water.

USE: In fireproofing wood.

5562. Magnesium Phosphate, Tribasic. "Neutral" magnesium phosphate; tertiary magnesium phosphate; trimagnesium phosphate. $Mg_3O_8P_2$; mol wt 262.86. Mg 27.74%, O 48.70%, P 23.56%. $Mg_3(PO_4)_2$. Octahydrate occurs in nature as the mineral *bobierrite.*

Pentahydrate, white, cryst powder. Loses last mol of water at about 400°. Insol in water; sol in dil mineral acids.

THERAP CAT: Antacid.

5563. Magnesium Potassium Selenate. Potassium magnesium selenate. $K_2MgO_8Se_2$; mol wt 388.43. K 20.13%, Mg 6.26%, Se 40.66%, O 32.95%. $K_2Mg(SeO_4)_2$. Prepd from a soln of aq magnesium and potassium selenates: Tutton, cited in *Mellor's* 10, 864 (1930).

Hexahydrate, monoclinic prisms. d_4^{20} 2.365. Freely sol in water.

5564. Magnesium Pyrophosphate. $Mg_2O_7P_2$; mol wt 222.55. Mg 21.84%, O 50.33%, P 27.83%. $Mg_2P_2O_7$.

Trihydrate, white powder; loses its water at 100°. d 2.56. Insol in water; sol in dil mineral acids.

5565. Magnesium Salicylate. *2-Hydroxybenzoic acid*

magnesium salt; Analate; Lorisal; Magan; Mobidin; Triact. $C_{14}H_{10}MgO_6$; mol wt 298.54. C 56.32%, H 3.38%, Mg 8.15%, O 32.15%. $Mg[C_6H_4(OH)COO]_2$.

Tetrahydrate, white, odorless, efflorescent, cryst powder. Sol in 13 parts water; sol in alc. The aq soln is slightly acid.

THERAP CAT: Intestinal anti-infective.

5566. Magnesium Selenate. MgO_4Se; mol wt 167.27. Mg 14.53%, O 38.26%, Se 47.21%. $MgSeO_4$. Prepn: Mitscherlich, *Pogg. Ann.* 9, 623 (1827); Berzelius, *ibid.* 32, 11 (1834); Huff, McCrosky, *J. Am. Chem. Soc.* 51, 1457 (1929).

Hexahydrate, monoclinic crystals. d 1.928. Stable in air. Sol in water.

5567. Magnesium Selenide. MgSe; mol wt 103.28. Mg 23.55%, Se 76.45%. Prepd by the action of hydrogen selenide on anhydr magnesium chloride at a red heat and by the action of selenium vapor carried in a current of nitrogen on powdered magnesium: Fonzes-Diacon, *Contribution a l'Etude des Séléniures Métalliques* (Montpellier, 1901); by dropping selenium into molten magnesium: Liddell, *Chem. Met. Eng.* 25, 102, 263, 453 (1921).

Light brown powder. d 4.21. Unstable in air. Dec in water.

5568. Magnesium Selenite. MgO_3Se; mol wt 151.27. Mg 16.07%, O 31.73%, Se 52.20%. $MgSeO_3$. Hexahydrate prepd by adding sodium selenite to a soln of magnesium chloride: Nilson, cited in *Mellor's* 10, 826 (1930); by treating a soln of magnesium chloride or sulfate with selenious acid and adding sodium carbonate to start precipitation: Boutzoureano, *Recherches sur les Sélénites* (Paris, 1889).

Orthorhombic crystals. Insol in water; sol in dil acids. Forms monoclinic prisms of the dihydrate when heated in a sealed tube to 150°. When heated to 100°, it loses 5 mols of water leaving the monohydrate.

5569. Magnesium Silicates. Several varieties of magnesium silicate are known. *See also* asbestos, talc.

Magnesium metasilicate, MgO_3Si. Occurs in nature as the minerals *clinoenstatite, enstatite, protoenstatite.* White monoclinic crystals, dec at 1557°. d_4^{25} 3.192. Practically insol in water; very slightly sol in HF.

Magnesium orthosilicate, Mg_2O_4Si. Occurs in nature as the mineral *forsterite.* White orthorhombic crystals. mp 1910°; d 3.21. Practically insol in water.

Magnesium trisilicate, $Mg_2O_8Si_3$, *magnesium mesotrisilicate,* Magnosil, Petimin, Trisomin. Occurs in nature as the minerals *meerschaum, parasepiolite, sepiolite.* Prepd from sodium silicate and magnesium sulfate: Glass, *Quart. J. Pharm. Pharmacol.* 9, 445 (1936); Uyeda, U.S. pat. 3,272,-594 (1966 to Merck & Co.). Odorless, tasteless, slightly hygroscopic powder. Usually contains some water of hydration. Practically insol in water.

Magnesium trisilicate pentahydrate, *Sellagen.*

Serpentine (mineral), $Mg_3O_7Si_2$. Occurs in nature as the dihydrate, $(OH)_4MgSi_2O_5$; exists in two forms: *antigorite,* a platy variety, and *chrysotile,* a fibrous variety. The latter is the most common form of asbestos, q.v.

USE: An activated magnesium silicate, *Florisil,* a hard, porous, granular substance, is used in vitamin analysis, chromatography, antibiotic processing: Simons, U.S. pat. 2,393,625 (1946 to Floridin).

THERAP CAT: Trisilicate as antacid.

THERAP CAT (VET): Trisilicate as antacid; gastric sedative.

5570. Magnesium Silicide. Mg_2Si; mol wt 76.73. Mg 63.39%, Si 36.61%. Prepd by heating finely powdered Mg and Si in a proportion of 20 to 6: Gire, *Compt. Rend.* 196, 1404 (1933), *C.A.* 27, 3678 (1933); Winkler, *Helv. Phys. Acta* 28, 633 (1955), *C.A.* 50, 6908c (1956). Prepn and conversion to silanes: Emeleus, Maddock, *J. Chem. Soc.* 1946, 1131.

Slate-blue, cubic crystals. d_4^{20} about 2.0. mp 1085°. Dec at 550° *in vacuo* yielding Mg_3Si_2. Heat of formation: 18.5 cal/mol. Dec by water, HCl.

USE: In semiconductor research. Has been used to build Mg-Si rectifiers.

5571. Magnesium Stannide. Mg_2Sn; mol wt 167.34. Mg 29.07%, Sn 70.93%. Prepn by fusion of a stoichiometric mixture of the elements: Winkler, *Helv. Phys. Acta* 28, 633

(1955); Korenblit, Kolesnikov, *Zh. Tekh. Fiz.* **26**, 941 (1956), *C.A.* **50**, 11475b (1956).

Bluish-white, metallic substance. Calcium fluoride crystal structure. mp 778°. Magnetic susceptibility and electrical conductivity measurements: Boltaks, *Doklady Akad. Nauk SSSR* **64**, 487, 653, (1949), *C.A.* **43**, 4528f, 4533f (1949). Resistivity (25°): 42,000 microhm-cm; thermoelectric power 90 microvolts per degree for 15-100°. Energy of activation: 22,100 cal/mole. Electrical conductivity, magnetoresistance: Frederikse *et al.*, *Phys. Rev.* **103**, 67 (1956). Soluble in water, dilute HCl.

USE: In semiconductor research.

5572. Magnesium Stearate. *Octadecanoic acid magnesium salt.* $C_{36}H_{70}MgO_4$; mol wt 591.27. C 73.13%, H 11.93%, Mg 4.11%, O 10.82%. $Mg(C_{18}H_{35}O_2)_2$. The commercial preparation also contains palmitate.

White powder. Insoluble in water; dec by dil acids.

USE: In baby dusting powders; as tablet lubricant.

5573. Magnesium Sulfate. MgO_4S; mol wt 120.38. Mg 20.20%, O 53.16%, S 26.63%. $MgSO_4$. Monohydrate occurs in nature as the mineral *kieserite*.

Trihydrate, odorless crystals. Prepn: Bennett, U.S. pat. **3,297,413** (1967 to Dow).

Heptahydrate, *bitter salts, epsom salts, Mg 5-Sulfat.* Occurs in nature as the mineral *epsomite*. Efflorescent crystals or powder; bitter, saline, cooling taste. d 1.67. Soly in water (g/100 ml): at 20 ° = 71; at 40° = 91. Slightly sol in alcohol. Its aq soln is neutral. pH 6-7. On exposure to dry air at ordinary temp it loses about $1H_2O$; at 70-80° loses $4H_2O$; at 100° loses $5H_2O$; at 120° loses $6H_2O$, rapidly reabsorbing water when exposed to moist air; loses the last mol of water at about 250°: *Gmelin's, Magnesium* (8th ed) **27**, 210-211, 223-226 (1939). *Dried magnesium sulfate* is prepd by heating the heptahydrate until approx 25% of its weight is lost. Used in *Morison's paste. Keep well closed.*

Human Toxicity: Parenteral use or use in presence of renal insufficiency may lead to magnesium intoxication.

USE: Weighting cotton and silk; increasing the bleaching action of chlorinated lime; manuf mother-of-pearl and frosted papers; fire-proofing fabrics; dyeing and printing calicos; in fertilizers; explosives; matches; mineral water; tanning leather.

THERAP CAT: Heptahydrate as anticonvulsant; cathartic.

THERAP CAT (VET): Purgative, in general anesthesia, in hypomagnesemia; externally (in strong solns) in local inflammations, infected wounds.

5574. Magnesium Sulfite. MgO_3S; mol wt 104.37. Mg 23.29%, O 45.99%, S 30.72%. $MgSO_3$. Occurs as tri- and hexahydrates.

Hexahydrate, colorless crystals or white, cryst powder; gradually oxidizes to sulfate on exposure to air; loses all its water at 200°; dec at higher temp. Sol in about 150 parts water, slightly more in boiling water.

USE: A soln of magnesium sulfite in sulfurous acid (magnesium bisulfite) is used in the manuf of paper pulp.

5575. Magnesium Thiocyanate. *Magnesium sulfocyanate.* $C_2MgN_2S_2$; mol wt 140.46. C 17.10%, Mg 17.31%, N 19.94%, S 45.65%. $Mg(SCN)_2$.

Tetrahydrate, colorless or white deliquesc crystals. Freely sol in water or alcohol. *Keep well closed.*

5576. Magnesium Thiosulfate. *Magnesium hyposulfite;* Magnosulf; Antichoc Hipmag. MgO_3S_2; mol wt 136.43. Mg 17.82%, O 35.18%, S 47.00%. MgS_2O_3.

Hexahydrate, colorless or white crystals; loses $3H_2O$ at 170°. d 1.82. Sol in 2 parts water; insol in alcohol. Its aq soln is neutral.

5577. Magnesium Tungstate(VI). MgO_4W; mol wt 272.18. Mg 8.93%, O 23.51%, W 67.56%. $MgWO_4$. Prepn of doped single crystals: Van Vitert, U.S. pat. **3,003,112** (1961 to Bell).

White, cryst powder. Insol in water.

USE: In solid state maser.

5578. Magneson. *4-[(4-Nitrophenyl)azo]-1,3-benzenediol; 4-(p-nitrophenylazo)resorcinol;* 2,4-dihydroxy-4'-nitroazobenzene. $C_{12}H_9N_3O_4$; mol wt 259.22. C 55.60%, H

3.50%, N 16.21%, O 24.69%. *Ref:* Suitzu, Okuma, *J. Soc. Chem. Ind. (Japan)* **29**, 132 (1926); Ruigh, *J. Am. Chem. Soc.* **51**, 1456 (1929); Engel, *ibid.* **52**, 1812 (1930).

Brownish-red powder. Practically insol in water; sol in dil aq NaOH.

USE: For the detection of magnesium with which it yields a bright blue color in alkaline soln. Also used to determine molybdenum with which it forms a red-violet complex: Nikitina, Andrianova, *C.A.* **75**, 136789v (1971).

5579. Magnoflorine. *5,6,6a,7-Tetrahydro-1,11-dihydroxy-2,10-dimethoxy-6,6-dimethyl-4H-dibenzo[de,g]quinolinium;* thalictrine. $[C_{20}H_{24}NO_4]^+$; mol wt 342.50. From *Magnolia grandiflora* L., *Magnoliaceae.* Isoln and structure: Nakano, *Chem. Pharm. Bull.* **2**, 329 (1954). From *Cocculus trilobus* D.C., *Menispermaceae:* Nakano, *ibid.* **4**, 69 (1956); from *Thalictrum thunbergii* D.C., *Ranunculaceae:* Fujita, Tomimatsu, *ibid.* **6**, 107 (1958); from *Aristolochia clematitis* L., *Aristolochiaceae:* Pailer, Pruckmayr, *Monatsh.* **90**, 145 (1959). Identity with thalictrine: Gopinath *et al., J. Sci. & Ind. Res. (India)* **18B** 444 (1959). Synthesis: Tomita, Kikkawa, *J. Pharm. Soc. Japan* **77**, 195 (1957).

Iodide, $C_{20}H_{24}INO_4$, crystals from methanol + acetone, dec 248-249°. $[\alpha]_D^{15}$ 220.1° (methanol). uv max: 270, 310 nm (log ε 3.75, 3.59).

Picrate, crystals from methanol, dec 206-207°.

5580. Magnoline. *1,2,3,4-Tetrahydro-1-[[4-[2-hydroxy-5-[(1,2,3,4-tetrahydro-7-hydroxy-6-methoxy-2-methyl-1-isoquinolinyl)methyl]phenoxy]phenyl]methyl]-6-methoxy-2-methyl-7-isoquinolinol.* $C_{36}H_{40}N_2O_6$; mol wt 596.70. C 72.46%, H 6.76%, N 4.70%, O 16.09%. From bark of *Magnolia macrophylla* Michx. and *M. fuscata* Andr., *Magnoliaceae:* Lloyd, *Am. J. Pharm.* **1891**, 438; Proskurnina, Orékhov, *Bull. Soc. Chim. France* **5**, 1357 (1938). Structure: *eidem, C.A.* **35**, 2520⁹ (1941).

Crystals, mp 178-179°. $[\alpha]_D$ −9.6° (pyridine). Slightly sol in alcohol, acetone, chloroform; practically insol in water, benzene, ether.

Trimethyl ether, $C_{39}H_{46}N_2O_6$, crystals, mp 109-110°.

Picrate, crystals from alcohol, dec 160-162°.

5581. Malachite Green. *N-[4-[[4-(Dimethylamino)-phenyl]phenylmethylene]-2,5-cyclohexadien-1-ylidene]-N-methylmethanaminium chloride; bis[p-(dimethylamino)-phenyl]phenylmethylium chloride;* C.I. Basic Green 4; C.I. 42000; Aniline Green; China Green; Victoria Green B; Victoria Green WB; New Victoria Green Extra O; New Victoria Green Extra I; New Victoria Green Extra II; Diamond Green B; Diamond Green Bx; Diamond Green P Extra; Solid Green O; Light Green N; Benzal Green; Benzaldehyde Green; Fast Green. $C_{23}H_{25}ClN_2$; mol wt 364.90. C 75.70%,

H 6.91%, Cl 9.72%, N 7.68%. Triphenylmethane dye with fungicidal and limited antiseptic activity. Prepn: *Colour Index* Vol. **4** (3rd ed., 1971) p 4380. For dyeing it is prepd as a double salt with $ZnCl_2$. Conformational changes in excited state: S. Saikan, J. Sei, *J. Chem. Phys.* **79**, 4154 (1983). Use in electron microscopy: R. G. Pourcho *et al.*, *Stain Technol.* **53**, 29 (1978). As stain in determn of prostaglandin E and F compounds: E. J. Singh *et al.*, *J. Chromatog.* **105**, 195 (1975). In determn of inorganic phosphate: W. Hohenwallner, E. Wimmer, *Clin. Chim. Acta* **45**, 169 (1973); S. G. Carter, D. W. Karl, *J. Biochem. Biophys. Methods* **7**, 7 (1982). Antiseptic activity: J. E. Madden *et al.*, *Surg. Forum* **22**, 63 (1971). Toxicology: N. Brock, A. Erhardt, *Arzneimittel-Forsch.* **1**, 5 (1951); T. D. Bills, L. L. Marking, *Investigations in Fish Control* **75**, 6 (1977); S. Clemmensen *et al.*, *Arch. Toxicol.* **56**, 43 (1984). Reviews of use as fungicide in fish culture: N. C. Nelson, *U.S. NTIS Reps.* PB-235450 (1974) 79 pp, *C.A.* **82**, 150006p (1975), PB-235451 (1974) 33 pp, *C.A.* **83**, 1547j (1975); of use as biological stain: R. D. Lillie, *H. J. Conn's Biological Stains* (Williams & Wilkins Co., Baltimore, 9th ed., 1977) 248-250, 579-580.

Green crystals with metallic luster. Very sol in water; sol in alcohol, methanol, amyl alcohol. Water solns are blue-green, absorption max 616.9 nm; yellow below pH 2. LD_{50} in mice (mg/kg): 80 orally; 4.2 i.p. (Brock, Erhardt). LC_{50} in various fish species (mg/l): 0.0305-0.383 (Bills, Marking).

Note: The term malachite green applies to the oxalate as well as the chloride.

USE: For directly dyeing silk, wool, jute and leather; dyeing cotton after mordanting. Biological stain. Clinical reagent (inorganic phosphate assay). As spot test reagent for detecting sulfurous acid and cerium. As acid-base indicator: pH: 0.0 yellow, 2.0 green; 11.6 blue-green, 14 colorless.

THERAP CAT (VET): Fungicide and parasiticide in fish.

5582. Malathion. *[(Dimethoxyphosphinothioyl)thio]butanedioic acid diethyl ester; mercaptosuccinic acid diethyl ester; S-ester with O,O-dimethyl phosphorothioate; S-(1,2-dicarbethoxyethyl) O,O-dimethyldithiophosphate; O,O-dimethyl dithiophosphate of diethyl mercaptosuccinate;* insecticide no. 4049; carbofos; malathon (obsolete); mercaptothion; phosphothion; ENT 17034; Cythion; Derbac-M; Malamar 50; Malaspray; Organoderm; Prioderm; Suleo-M. $C_{10}H_{19}O_6PS_2$; mol wt 330.36. C 36.35%, H 5.80%, O 29.06%, P 9.38%, S 19.41%. Prepn: Johnson *et al.*, *J. Econ. Entomol.* **45**, 279 (1952); Cassaday, U.S. pat. 2,578,652 (1951 to Am. Cyanamid). Purification: Usui, U.S. pat. 2,962,521 (1960 to Sumitomo). Treatment of *Pediculus humanus* (head lice) infestation: D. Taplin *et al.*, *J. Am. Med. Assoc.* **247**, 3103 (1982). Toxicity data: T. B. Gaines, *Toxicol. Appl. Pharmacol.* **14**, 515 (1969). Review of distribution, transport and fate in the environment: M. S. Mulla *et al.*, *Residue Rev.* **81**, 1-159 (1981).

Deep brown to yellow liq, mp 2.9°. $bp_{0.7}$ 156-157°. Characteristic odor. d_4^{25} 1.23. n_D^{25} 1.4985. Vapor pressure at 30°: 4×10^{-5} mm Hg. Slightly sol in water (145 ppm). Misc with many organic solvents including alcohols, esters, ketones, ethers, aromatic and alkylated aromatic hydrocarbons

and vegetable oils. Limited soly in certain paraffin hydrocarbons. Petroleum ether is sol to about 35% in malathion. Hydrolyzed at pH > 7.0 or < 5.0. Stable in an aq soln buffered to pH 5.26. LD_{50} in female, male rats (mg/kg): 1000, 1375 orally (Gaines).

USE: Insecticide. *Caution:* Toxic symptoms similar to parathion, *q.v.* Malathion is considered to be less toxic.

THERAP CAT: Pediculicide.

THERAP CAT (VET): Ectoparasiticide.

5583. Maleamic Acid. *(Z)-4-Amino-4-oxo-2-butenoic acid;* maleic acid monoamide. $C_4H_5NO_3$; mol wt 115.09. C 41.74%, H 4.38%, N 12.17%, O 41.71%. Prepd by passing ammonia into a mixture of maleic anhydride and an inert solvent such as xylene, dioxane, etc.: Robinson, Humburger, U.S. pat. 2,459,964 (1949 to Beck, Koller).

Crystals from alcohol, mp 172-173°. Also reported as mp 178-180° (dec). Very sol in water, hot alcohol. Practically insol in ether, chloroform, benzene.

5584. Maleanilic Acid. *(Z)-4-Oxo-4-(phenylamino)-2-butenoic acid;* N-phenylmaleamic acid. $C_{10}H_9NO_3$; mol wt 191.18. C 62.32%, H 4.75%, N 7.33%, O 25.11%. Use of esters as fungicides: Ligett *et al.*, U.S. pat. 2,885,319 (1959 to Pittsburgh Coke & Chem.).

Monoclinic yellow crystals, dec 192°. Apparent density 1.418 g/ml at 30°. Soly in g/100 ml soln at 27°: acetonitrile 0.2; acetone 0.3; benzene < 0.1; butyl Cellosolve 1.1; carbon tetrachloride < 0.1; chloroform < 0.1; dimethylformamide 12.9; dioxane 0.9; ethanol 0.2; ether < 0.1; methanol 0.3; toluene < 0.1; water < 0.1. Shows hydrolysis in aq soln. Dissociates to maleic anhydride and aniline when heated above 100° *in vacuo.*

USE: Esters as fungicides.

5585. Maleic Acid. *(Z)-Butenedioic acid;* toxilic acid; cis-1,2-ethylenedicarboxylic acid. $C_4H_4O_4$; mol wt 116.07. C 41.39%, H 3.47%, O 55.14%. Prepd by the catalytic oxidation of benzene over vanadium pentoxide: Bhattacharyya, Venkataraman, *J. Appl. Chem. (London)* **8**, 728 (1958); Saffer, Olenberg, Fr. pat. 1,321,416 (1963 to Scientific Design), *C.A.* **59**, 11265d (1963). Crystal structure: Shahat, *Acta Cryst.* **5**, 763 (1952). *Review:* W. D. Robinson, R. A. Mount in Kirk-Othmer *Encyclopedia of Chemical Technology* vol. **14** (Wiley-Interscience, New York, 3rd ed., 1981) pp 770-793.

White crystals from water, mp 138-139°; from alcohol and benzene, mp 130-131°. Faint, acidulous odor; characteristic repulsive, astringent taste. d 1.59. Is converted in part into the much higher-melting fumaric acid (mp 287°) when heated to a temp slightly above the melting point. Freely sol in water or alcohol. Sol in acetone, glacial acetic acid. Slightly sol in ether. Practically insol in benzene.

USE: Manuf artificial resins; to retard rancidity of fats and oils in 1:10,000 (these are said to keep 3 times longer than those without the acid); dyeing and finishing wool, cotton, and silk; preparing the maleate salts of antihistamines and similar drugs. *Caution:* Strong irritant.

5586. Maleic Anhydride. *2,5-Furandione;* cis-butenedioic anhydride; toxilic anhydride. $C_4H_2O_3$; mol wt 98.06. C 48.99%, H 2.06%, O 48.95%. May be prepd by sublimation of maleic acid and P_2O_5 under reduced pressure: Kempf, *J. Prakt. Chem.* [2] **78**, 239 (1908). Commercial production by catalytic vapor-phase oxidation of benzene or other suitable hydrocarbons: Weiss, Downs, *Ind. Eng. Chem.* **12**, 228 (1920); U.S. pat. **1,318,633** (1920 to Barrett Co.). Many other syntheses. Review of commercial methods of manufacture: Ashcroft, Clifford, *Chem. Prods.* **24**, 11 (1961), *C.A.* **55**, 9724d (1961); Faith, Keyes & Clark's *Industrial Chemicals*, F. A. Lowenheim, M. K. Moran, Eds. (Wiley-Interscience, New York, 4th ed., 1975) pp 514-518. *Review:* W. D. Robinson, R. A. Mount in Kirk-Othmer *Encyclopedia of Chemical Technology* vol. **14** (Wiley-Interscience, New York, 3rd ed., 1981) pp 770-793. Book: B. C. Trivedi, B. M. Culbertson, *Maleic Anhydride* (Plenum, New York, 1982) 872 pp.

Orthorhombic needles from chloroform; also readily by sublimation. Commercial grades are furnished in fused form, as briquettes. d 1.48. mp 52.8°. bp_{760} 202.0°; bp_{400} 179.5°; bp_{200} 155.9°; bp_{100} 135.8°; bp_{60} 122.0°; bp_{40} 111.8°; bp_{20} 95.0°; bp_{10} 78.7°; bp_5 63.4°. Specific heat: 0.285 (solid); 0.396 (liq). Sol in water, forming maleic acid. Soly at 25° (g/100 g): acetone 227; ethyl acetate 112; chloroform 52.5; benzene 50; toluene 23.4; o-xylene 19.4; carbon tetrachloride 0.60; ligroin 0.25. Sol in dioxane. Sol in alc with ester formation.
Caution: Powerful irritant. *Causes burns.* Avoid contact with skin, eyes, clothing. Inhalation can cause pulmonary edema. Avoid exposure to concd vapor. In case of contact, immediately flush skin or eyes with plenty of water for at least 15 min.
USE: In Diels-Alder syntheses (as a dienophile), in copolymerization reactions, manuf alkyd-type of resins, dye intermediates, pharmaceuticals, agricultural chemicals (maleic hydrazide, Malathion).

5587. Maleic Hydrazide. *1,2-Dihydro-3,6-pyridazinedione;* maleic acid hydrazide; MH; Fazor; Malazide; Regulox. $C_4H_4N_2O_2$; mol wt 112.09. C 42.86%, H 3.60%, N 24.99%, O 28.55%. Prepd by treating maleic anhydride with hydrazine hydrate in alcohol: Arndt *et al., C.A.* **43**, 579 (1949); *see also* Curtius, Foerstinger, *J. Prakt. Chem.* [2] **51**, 391 (1895). From maleic acid and a hydrazine salt of a strong inorganic acid: Harris, Schoene, U.S. pat. **2,575,954** (1951 to U.S. Rubber). Alternate prepn from hydrazine sulfate and maleic anhydride in aqueous NaOH: Amatsu, Karasawa, *C.A.* **51**, 18014c (1957); from hydrazine hydrate and maleic anhydride in glacial acetic acid: Feuer *et al., J. Am. Chem. Soc.* **80**, 3790 (1958). Has the ability to inhibit growth of plants without killing them: Schoene, Hoffmann, *Science* **109**, 588 (1949). Toxicity data: R. Ben-Dyke *et al., World Rev. Pest Contr.* **9**, 119 (1970). *Review:* Massey, *Mfg. Chem.* **26**, 197-200 (1955). Review and chromatographic studies: Fishbein, *Chromatography of Environmental Hazards* (Elsevier, New York, 1972) pp 161-166.

Crystals from water, dec 260°. mp over 300° (Feuer). Slightly sol in hot alcohol, more sol in hot water. LD_{50} orally in rats: 3800-6800 mg/kg; dermally in rabbits: > 4000 mg/kg (Ben-Dyke).
USE: Experimentally in horticulture and agriculture. To control suckering of tobacco. In the synthesis of pyridazine.

5588. Maleuric Acid. *(Z)-4-[(Aminocarbonyl)amino]-4-oxo-2-butenoic acid; N-carbamoylmaleamic acid;* maleylurea. $C_5H_6N_2O_4$; mol wt 158.11. C 37.98%, H 3.83%, N 17.72%, O 40.48%. Prepd from maleic anhydride and urea: Dunlap, Phelps, *Am. Chem. J.* **19**, 492 (1897); Batt *et al., J. Am. Chem. Soc.* **76**, 3663 (1954).

Crystals from hot water, dec 158.5°. uv max: 235 nm (ϵ 8720). Sol in hot acetic acid. Practically insol in cold water, cold acetic acid, acetone, ligroin, chloroform, alc, ether.
Methyl ester, crystals, mp 113-114°. Sol in hot water, methanol, ethanol, acetone, dioxane.
Butyl ester, crystals, mp 95-98°. Sol in ethanol, acetone, benzene, chloroform.

5589. Malic Acid. *Hydroxybutanedioic acid;* hydroxysuccinic acid. $C_4H_6O_5$; mol wt 134.09. C 35.83%, H 4.51%, O 59.66%. The naturally occurring isomer is the L-form which has been found in apples and many other fruits and plants. Prepn of D- and DL-forms, and resolution of racemic mixture: McKenzie *et al., J. Chem. Soc.* **123**, 2875 (1923). Solubilities: Descamps, *Bull. Soc. Chim. Belg.* **49**, 91 (1940). Microbial production of L-form: Kitahara; Abe *et al.,* U.S. pats. **2,972,566; 3,063,910** (1961, 1962, both to Kyowa). Configuration: J. A. Mills, W. Klyne in *Progress in Stereochemistry* vol. **1**, W. Klyne, Ed. (Academic Press, New York, 1954) pp 182-183; E. L. Eliel, *Stereochemistry of Carbon Compounds* (McGraw-Hill, New York, 1962) pp 97-98; Cymerman-Craig, Roy, *Tetrahedron* **21**, 1847 (1965).

DL-Form, crystals, mp 131-132°. Soly in g/100 g solvent at 20°: methanol 82.70, diethyl ether 0.84, ethanol 45.53, acetone 17.75, dioxane 22.70, water 55.8. Practically insol in benzene.
D-(+)-Form, crystals, mp 101°.
L-(−)-Form, *apple acid.* Crystals from acetone, or acetone + CHCl$_3$, mp 100°. Dec about 140°. $[\alpha]_D$ −2.3° (c = 8.5). Soly in g/100 g solvent at 20°: methanol 197.22, diethyl ether 2.70, ethanol 86.60, acetone 60.66, dioxane 74.35, water 36.35. Practically insol in benzene.
USE: *See: Mfg Chemist Aerosol News* **35**, 56 (1964) for review of uses.

5590. Mallow. Common mallow; high mallow; cheeseflower. Leaves of *Malva sylvestris* L., and *M. rotundifolia* L., Malvaceae. *Habit.* Europe, Asia, naturalized in U.S. *Constit.* Pectin, tannin, coloring matter.

5591. Malonic Acid. *Propanedioic acid;* methanedicarboxylic acid. $C_3H_4O_4$; mol wt 104.06. C 34.62%, H 3.87%, O 61.50%. $HOOCCH_2COOH$. Prepd from malic acid: Dessaignes, *Ann.* **107**, 251 (1858). Made by the interaction of monochloroacetic acid and NaCN followed by hydrolysis of the resulting cyanoacetic acid: Weiner, *Org. Syn. coll. vol.* **II**, 376 (1943). Prepn from diethyl malonate: Britton, Monroe, U.S. pat. **2,373,011** (1945 to Dow); from ligneous wastes: Grangaard, U.S. pat. **2,928,868** (1960 to Kimberly-Clark); from sodium acetate: Normant, Angelo, *Bull. Soc. Chim. France* **1962**, 810. *Review:* D. W. Hughes in Kirk-Othmer *Encyclopedia of Chemical Technology* vol. **14** (Wiley-Interscience, New York, 3rd ed., 1981) pp 794-810.

Small crystals, mp about 135° with decompn; sublimes *in vacuo.* d 1.63. One gram dissolves in 0.65 ml water, about 2 ml alcohol, 1.1 ml methanol, 3 ml propyl alcohol, 13 ml ether, 7 ml pyridine.
Diethyl Ester *see* Ethyl Malonate.
USE: In manuf of barbiturates. *Caution:* Strong irritant.

5592. Malononitrile. *Propanedinitrile;* methylene cyanide; dicyanomethane; cyanoacetonitrile. $C_3H_2N_2$; mol wt 66.06. C 54.54%, H 3.06%, N 42.40%. $CH_2(CN)_2$. Prepn: Henri, *Compt. Rend.* **102**, 1394 (1886); B. B. Corson *et al., Org. Syn. coll. vol.* **2**, 379 (1943); A. R. Surrey, *J. Am. Chem. Soc.* **65**, 2471 (1943); idem, U.S. pat. **2,389,217** (1945 to Winthrop); idem, *Org. Syn. coll. vol.* **3**, 535 (1955). *Reviews:* F. Freeman, *Chem. Rev.* **69**, 591-624 (1969); A. J. Fatiadi, *Synthesis* **1978**, 165-204, 241-282.
Colorless solid, mp 32°. bp_{760} 218-219°, bp_{20} 109°. d_4^{20} 1.1910; n_D^{34} 1.4146. Very sol in alc, ether; sol in water, ace-

tone, benzene. LD_{50} i.p. in mice: 12.9 mg/kg, G. R. N. Jones, M. S. Israel, *Nature* **228**, 1315 (1970).

USE: In organic synthesis.

5593. Malotilate. *1,3-Dithiol-2-ylidenepropanedioic acid bis(1-methylethyl) ester;* diisopropyl 1,3-dithiol-2-ylidenemalonate; NKK 105; Hepation; Kantec. $C_{12}H_{16}O_4S_2$; mol wt 288.38. C 49.98%, H 5.59%, O 22.19%, S 22.23%. Prepn from diisopropyl malonate: K. Taninaka *et al.*, **Belg.** pat. **834,631** corresp to U.S. pat. **4,035,387** (1976, 1977 both to Nihon Nohyaku); from the corresponding ketene mercaptide: H. Matsui *et al.*, U.S. pat. **4,327,223** (1982 to Nihon Nohyaku). Effect on CCl_4-induced liver injury in rats: Y. Imaizumi *et al.*, *Japan. J. Pharmacol.* **31**, 15 (1981). Enhancement of rat liver protein synthesis: *eidem, ibid.* **32**, 369 (1982). Pharmacokinetics and pharmacodynamics: M. Buhrer *et al.*, *Eur. J. Clin. Pharmacol.* **30**, 407 (1986). Clinical evaluation in liver cirrhosis: S. Takase *et al.*, *Gastroenterol. Japan* **23**, 639 (1988).

Pale yellow crystals, mp 60.5°. Sol in benzene, cyclohexane, *n*-hexane, ether.

THERAP CAT: Hepatoprotectant.

5594. Maltol. *3-Hydroxy-2-methyl-4H-pyran-4-one;* 3-hydroxy-2-methyl-4-pyrone; 3-hydroxy-2-methyl-γ-pyrone; larixinic acid; Palatone; Veltol. $C_6H_6O_3$; mol wt 126.11. C 57.14%, H 4.80%, O 38.06%. Found in the bark of young larch trees (*Larix decidua* Mill.), in pine needles (*Abies alba* Mill., *Pinaceae*), in chicory, in wood tars and oils, in roasted malt. Isoln from these sources and structure: Kiliani, Bazlen, *Ber.* **27**, 3115 (1894); Feuerstein, *Ber.* **34**, 1804 (1901); Erdmann, Schaefer, *Ber.* **43**, 2398 (1910); Reichstein, Beitter, *Ber.* **63**, 824 (1930), *cf.* Peratoner, Tamburello, *Chem. Zentr.* **76**, II 680 (1905). Also obtained by alkaline hydrolysis of streptomycin salts: Schenck, Spielman, *J. Am. Chem. Soc.* **67**, 2276 (1945). Synthesis: Spielman, Freifelder, *ibid.* **69**, 2908 (1947); Chawla, McGonigal, *J. Org. Chem.* **39**, 3281 (1974). Novel high-yield synthesis: T. M. Brennan *et al.*, *Tetrahedron Letters* **1978**, 331; P. D. Weeks *et al.*, *J. Org. Chem.* **45**, 1109 (1980). History and comparison with isomaltol: Hodge, Nelson, *Cereal Chem.* **38**, 207 (1961).

Monoclinic prisms from chloroform, orthorhombic bipyramidal crystals + monoclinic prisms from 50% alcohol, mp 161-162°. Fragrant, caramel-like odor. Begins to sublime at 93°. Volatile with steam. uv max (0.1N HCl): 274 nm (E_m 8400); (0.1N NaOH): 317 nm (E_m 7300). pH of 0.5% aq soln 5.3. One gram dissolves in 85 ml water. Freely sol in hot water, chloroform; sol in alcohol; sparingly sol in benzene, ether, petr ether; sol in aq alkali hydroxides giving yellow solns.

USE: Flavoring agent, to impart "freshly baked" odor and flavor to bread and cakes.

5595. Maltose. *4-O-α-D-Glucopyranosyl-D-glucose;* malt sugar; maltobiose; 4-(α-D-glucosido)-D-glucose; Maltos; Martos-10. $C_{12}H_{22}O_{11}$; mol wt 342.31. C 42.10%, H 6.48%, O 51.42%. Monohydrate obtained in about 80% yield by enzymatic (diastase) degradation of starch: Gore, U.S. pat. **1,657,079** (1928). Structure: Haworth, Peat, *J. Chem. Soc.* **1926**, 3094; Hassid, Ballou in W. Pigman, *The Carbohydrates* (Academic Press, New York, 1957) p 498. Crystal and molecular structure: F. Takusagawa, R. A. Jacobson, *Acta Crystallogr.* **B34**, 213 (1978). *Review:* E.

Tarelli, *Dev. Food Carbohydr.* **2**, 187-227 (1980); R. Khan, *Carbohydr. Chem. Biochem.* **39**, 213-278 (1981).

Monohydrate, crystals from water or dil alc, mp 102-103°. Does not lose its water of crystn by drying at room temp *in vacuo* over H_2SO_4 or P_2O_5. About one-third as sweet as sucrose. *Cf.* Isbell, Pigman, *J. Res. NBS* **18**, 141 (1937). Shows mutarotation. $[\alpha]_D^{20} +111.7° \rightarrow +130.4°$ (c = 4). Ka at 21° = 9.0×10^{-13}. Sol in water, slightly sol in alcohol. Practically insol in ether.

USE: Nutrient, sweetener, in culture media, in prepd bee food. Parenteral supplement of sugar for diabetics. Fermentable intermediate in brewing. Stabilizer for polysulfides. In pharmaceutical dispensing.

5596. Malvidin Chloride. *3,5,7-Trihydroxy-2-(4-hydroxy-3,5-dimethoxyphenyl)-1-benzopyrylium chloride; 3,4',5,7-tetrahydroxy-3',5'-dimethoxyflavylium chloride;* enidin; primulidin; syringidin; 3,4',5,7-tetrahydroxy-3',5'-dimethoxy-2-phenylbenzopyrylium chloride. $C_{17}H_{15}ClO_7$; mol wt 366.75. C 55.67%, H 4.12%, Cl 9.67%, O 30.54%. Found as the diglucoside (malvin) in wild malve (*Primula viscosa* All., *Primulaceae*) and as the monoglucoside in blue grapes: Willstätter, Mieg, *Ann.* **408**, 122 (1915); Karrer, Widmer, *Helv. Chim. Acta* **10**, 5 (1927). Structure: Anderson, Nabenhauer, *J. Am. Chem. Soc.* **48**, 2997 (1926). Synthesis: Bradley, Robinson, *J. Chem. Soc.* **1928**, 1541.

Prisms or rhombic tablets, appearing red by transmitted light with a green luster when dry or with a steel-blue luster when in contact with solvent. Usually obtained as the mono- or dihydrate. The anhyd salt is very hygroscopic. Not melted at 300°. Sol in abs alc, giving a violet-red soln. Sparingly sol in water. Also sol in amyl alc. In methanol the substance is first sol with purple color, then begins to separate as red crystals which are violet by transmitted light.

3-β-Glucoside tetrahydrate, $C_{23}H_{25}ClO_{12}.4H_2O$, 3-(glucosyloxy)-4',5,7-trihydroxy-3',5'-dimethoxyflavylium chloride, enin, cyclamin. Synthesis: Levy *et al.*, *J. Chem. Soc.* **1931**, 2701. Dark prisms with a green metallic shine, from alcohol + HCl. Sparingly sol in water, alcohol, glycerol; practically insol in benzene, chloroform, ether.

3,5-Diglucoside, $C_{29}H_{35}ClO_{17}$, 3,5-bis(glucosyloxy)-4',7-dihydroxy-3',5'-dimethoxyflavylium chloride, malvin, malvoside. Synthesis: Robinson, Todd, *ibid* **1932**, 2299. Reddish-brown prisms or needles with a green shine from dil HCl, dec 165°.

3-Galactoside hemihendecahydrate, $C_{23}H_{25}ClO_{12}.5\frac{1}{2}H_2O$, *primulin.* Bronze metallic needles or prisms from methanol + HCl. Bluish-violet by transmitted light. Very sol in methanol, ethanol, cold water, forming deep red solns; slightly sol in acetone; practically insol in ethyl acetate.

5597. Manaca. Manacan; camganiba; geratacaca; vegetable mercury. Dried root of *Brunfelsia hopeana* Benth. (*Franciscea uniflora* Pohl), *Solanaceae*. Habit. Brazil. *Constit.* The alkaloid manacine, a second principle probably identical with gelsemic acid, and a fluorescent substance probably identical with esculetin.

THERAP CAT: Formerly as antisyphilitic.

5598. Mancozeb. *[[1,2-Ethanediylbis(carbamodithio-ato)](2−)]manganese mixt. with [[1,2-ethanediylbis(carbamodithioato)](2−)]zinc; ethylenebis(dithiocarbamic acid) manganese zinc complex;* manzeb; manganese ethylenebis-(dithiocarbamate) (polymeric) complex with zinc salt; zinc manganese ethylenebisdithiocarbamate; Dithane M-45; Dithane 945; Dithane LF; Manzate 200; Manzin 80; Nemispor; Penncozeb; Vondozeb Plus; Policar MZ; Policar S. Polymeric salt of *ethylenebisdithiocarbamic acid* containing approx 20% manganese and 2.5% zinc; related to maneb and zineb, *q.q.v.* Displays activity against a wide range of foliage fungal diseases. Prepn: **New Zealand** pat. **131,543;** C. B. Lyon *et al.,* U.S. pat. **3,379,610** (1965, 1968 both to Rohm & Haas). Metabolic fate in plants, animals: W. R. Lyman, "Metabolic fate of Dithane M-45," in *Pestic. Terminal Residues,* Invited Pap. Int. Symp., A. S. Tahori, Ed. (Butterworth, New York, 1971) pp 243-256. Residue studies: *WHO Pestic. Residues Ser.* vol. **4,** (WHO, FAO, 1975) 451; W. H. Newsome, *J. Agric. Food Chem.* **27,** 1188 (1979). HPLC determn: K. Gustafsson *et al., ibid.* **29,** 729 (1981). Demonstrates low phytotoxicity: L. H. Cornford, *Proc. 36th N.Z. Weed and Pest Control Conf.,* 258 (1983). In potato blight: H. W. Platt, *Can. J. Plant Pathol.* **5,** 38 (1983). *In vitro* activity against *Alternaria alternata:* V. A. Bourbos *et al., Brit. Crop Prot. Council Monograph* **No. 31** (Fungic. Crop Prot., Vol. 2), 461 (1985). Aquatic toxicology: C. J. Van Leeuwen *et al., Aquat. Toxicol.* **7,** 145 (1985).

$$\left[-MnSCNHCH_2CH_2NHCS- \right]_x \quad \left[Zn^{2+} \right]_y$$

x:y 10:1

Insol in water.
USE: Fungicide.

5599. Mandelic Acid. *α-Hydroxybenzeneacetic acid; dl-*mandelic acid; racemic mandelic acid; *α-*hydroxy-*α-*toluic acid; *α-*hydroxyphenylacetic acid; phenylhydroxyacetic acid; phenylglycolic acid; amygdalic acid; amygdalinic acid; para-mandelic acid; Uromaline. $C_8H_8O_3$; mol wt 152.14. C 63.15%, H 5.30%, O 31.55%. $C_6H_5CH(OH)COOH$. Prepd by the action of warm, dil alkali upon dichloroacetophenone: *Org. Syn.* **23,** 48 (1943); by hydrolysis of mandelonitrile (prepd from benzaldehyde and hydrogen cyanide or from benzaldehyde, sodium bisulfite, and sodium cyanide): *Org. Syn.* coll. vol. **I,** 336 (1941); *ibid.* coll. vol. **III,** 538 (1955); L. F. Fieser, *Organic Experiments* (Boston, 1964) p 109. May be prepd by boiling amygdalin with HCl.

Orthorhombic plates from water, mp 119°. Darkens and dec on prolonged exposure to light. d 1.30. Can be distilled rapidly *in vacuo* at 2 mm without much decomp. Acid to litmus. K at 25° = 4.3 × 10⁻⁴. One gram dissolves in 6.3 ml water, 1 ml alc; freely sol in ether, isopropyl alcohol.

Calcium salt, $C_{16}H_{14}CaO_6$, *calcium mandelate, Camdelate.* Contains 88.88% mandelic acid. White powder. One gram dissolves in 80 ml boiling water. Slightly sol in cold water. Insol in alc. The aq soln is slightly acid.

Sodium salt, $C_8H_7NaO_3$, *sodium mandelate.* Cryst powder; slight aromatic odor. Very sol in water; sol in alcohol; the aq soln is neutral or slightly alkaline to litmus.

THERAP CAT: Urinary antiseptic.

5600. Mandelic Acid Isoamyl Ester. *α-Hydroxybenzene-acetic acid 3-methylbutyl ester; mandelic acid isopentyl ester;* isoamyl mandelate; Atractyl; Spasmol; Mandaverm; Vermiparin; Spasmostenyl. $C_{13}H_{18}O_3$; mol wt 222.29. C 70.24%, H 8.16%, O 21.59%. Prepd by esterification of DL-mandelic acid with an excess of isoamyl alcohol in the presence of concd H_2SO_4: Rona *et al., Biochem. Z.* **181,** 50 (1927). Improved procedure: Brock *et al., Arzneimittel-Forsch.* **2,** 166 (1952).

Oily liquid. bp₁₁ 172° (Rona); bp₁₂ 155° (Brock). Practically insol in water. Miscible with fat solvents. Easily hydrolyzed by esterases. *The pure, undiluted liquid is a skin irritant.*

THERAP CAT: Adjunct in the treatment of gastric and intestinal spasms.

THERAP CAT (VET): Has been used as an anthelmintic.

5601. Mandelonitrile. *α-Hydroxybenzeneacetonitrile;* mandelic acid nitrile; benzaldehyde cyanohydrin. C_8H_7NO; mol wt 133.14. C 72.16%, H 5.30%, N 10.52%, O 12.02%. $C_6H_5CH(OH)CN$. Isoln of *d*-form from peach flower buds: Jones, Enzie, *Science* **134,** 284 (1961). Prepn of *d*-form by hydrolysis of amygdalin: Auld, *J. Chem. Soc.* **95,** 927 (1909); Smith, *Ber.* **64,** 427 (1931). Resolution of *dl*-form: Feist, *Arch. Pharm.* **247,** 226 (1909). Synthesis of *l*-form using cotton fibers: Bredig, Gerstner, *Biochem. Z.* **250,** 414 (1932). Asymmetric synthesis: Krieble, Wieland, *J. Am. Chem. Soc.* **43,** 164 (1921); Prelog, Wilhelm, *Helv. Chim. Acta* **37,** 1634 (1954); Tsuboyama, *Bull. Chem. Soc. Japan* **35,** 1004 (1962). It has been suggested that mandelonitrile may be responsible for the alleged anticancer activity of Laetrile (*q.v.*): see Culliton, *Science* **182,** 1000 (1973).

dl-Form, yellow, oily liquid, mp −10°. Decomp at 170°. d 1.115-1.120. Almost insol in water; freely sol in alcohol, chloroform, ether. LD s.c. in rabbits: 6 mg/kg.

d-Form, [α]$_D^{25}$ +43.75° (c = 5.006 in benzene).

USE: Preparing bitter almond water, by mixing 11 g with 500 g alc and 1489 g water; the mixture contains 0.1% HCN.

5602. Mandelonitrile Glucoside. *α-(β-D-Glucopyranos-yloxy)benzeneacetonitrile.* $C_{14}H_{17}NO_6$; mol wt 295.28. C 56.94%, H 5.80%, N 4.74%, O 32.51%. Prepn of *d*-form by action of yeast on amygdalin: Fischer, *Ber.* **28,** 1508 (1895); Auld, *J. Chem. Soc.* **93,** 1276 (1908). Isoln of *d*-form from *Prunus serotina,* Ehrk, *P. macrophylla* Sieb et Zucc., *Rosaceae:* Power, Moore, *ibid.* **95,** 243 (1909); **97,** 1099 (1910); Kariyone, Matsushima, *J. Pharm. Soc. Japan* no. **514,** 1061 (1924); from *Eucalyptus corynocalyx* F.v.M., *Myrtaceae:* Finnemore *et al., J. Proc. Roy. Soc. N.S. Wales* **69,** 209 (1936); from *Pteridium aquilinum* (L.) Kühn, *Polypodiaceae:* Kofod, Eyjolfssen, *Tetrahedron Letters* **1966,** 1289. Isoln of *dl*-form from *Prunus laurocerasus* L. and *Cotoneaster microphylla* Wall., *Rosaceae:* Winkler, Simon, *Ann.* **31,** 263 (1839); Hérissey, *J. Pharm. Chim.* **24,** 537 (1906). Isoln of *l*-form from *Sambucus nigra* L., *Caprifoliaceae:* Guignart, *Compt. Rend.* **141,** 16 (1905); Bourquelot, Danjot, *ibid.* **598** (1905). Prulaurasin is *dl*-mandelonitrile glucoside and sambunigrin is *l*-mandelonitrile glucoside: Caldwell, Courtauld, *J. Chem. Soc.* **91,** 671 (1907). Biosynthesis of *d*-mandelonitrile glucoside: Mentzer, Favre-Bonvin, *Compt. Rend.* **253,** 1072 (1961). Absolute configuration of isomers: U. Schwarzmaier, *Ber.* **109,** 3250 (1976).

d-Form, *prunasin.* Needles from chloroform, mp 147-148°. [α]$_D^{14}$ −29.94°. Soluble in water, alcohol, acetone. Converted by alkalies to prulaurasin.

dl-Form, *prulaurasin.* Slightly bitter needles from ethyl acetate + ether, mp 123-125°. [α]$_D$ −54° (water). Soluble in water, alcohol; practically insol in ether. On hydrolysis yields *dl*-mandelic acid.

l-Form, *sambunigrin.* Bitter needles from hot ethyl acetate, mp 151-152°. [α]$_D^{18}$ −75.1°. Soluble in water, alcohol, ethyl acetate. On hydrolysis yields *l*-mandelic acid.

5603. Maneb. *[[1,2-Ethanediylbis[carbamodithioato]]-(2−)]manganese; [ethylenebis(dithiocarbamato)]manganese;* ethylenebis[dithiocarbamic acid] manganous salt; manganous ethylenebis[dithiocarbamate]; Manzate; Dithane M-22; Nespor. $C_4H_6MnN_2S_4$. Polymeric salt of *ethylenebisdithiocarbamic acid;* related to zineb and mancozeb, *q.q.v.* Prepd

Consult the cross index before using this section.

from aq nabam, *q.v.*, by neutralizing with acetic acid and adding MnCl$_2$ soln: A. L. Flenner, U.S. pat. **2,504,404** (1950 to du Pont), *see also* W. F. Hester, U.S. pat. **2,317,765** (1943 to Rohm & Haas). Fungicidal activity is due to degradation products, principally *ethylenethiuram monosulfide:* R. A. Ludwig *et al., Can. J. Bot.* **33,** 42 (1955). *In vitro* activity against *Alternaria alternata:* V. A. Bourbos *et al., Brit. Crop. Prot. Council Monograph No. 31* (Fungic. Crop Prot., Vol. 2), 461 (1985). Decomposition of maneb to ethylenebisdithiocarbamate and ethylene thiourea, *q.v.:* W. J. Trotter, J. Pardue, *J. Food Safety* **4,** 59 (1982). HPLC determn: K. Gustafsson *et al., J. Agr. Food Chem.* **29,** 729 (1981). Toxicological studies in rats: D. W. R. Bleyl, H. Seidler, *Nährung* **29,** 421 (1985); in aquatic organisms: C. J. Van Leeuwen *et al., Aquat. Toxicol.* **7,** 145 (1985).

$$\left[-MnSCNHCH_2CH_2NHCS- \right]_n$$

Yellow powder. Crystals from alcohol. Moderately sol in water. Sol in chloroform, pyridine.
USE: Agricultural fungicide.

5604. Manganese. Mn; at. wt 54.9380; at. no. 25; valences 2, 4, 7; 1, 3, 5, 6 rare. One stable isotope: 55; artificial radioactive isotopes 49-54; 56-58. Widely-distributed, abundant element; constitutes 0.085% of earth's crust. Occurs in the minerals pyrolusite, hausmannite, manganite, *manganosite* (MnO), *braunite* (3Mn$_2$O$_3$.MnSiO$_3$), and in several others; occurs in minute quantities in water, plants and animals. First isolated by Gahn in 1774. Prepn of the metal: John *et al.*, cited by Mellor, *A Comprehensive Treatise on Inorganic and Theoretical Chemistry,* **12,** 163 (1932); A. H. Sully, *Manganese* (Academic Press, New York, 1955) 305 pp. Review of physical properties: Meaden, *Met. Rev.* **13,** 97-114 (1968). Review of manganese and its compds: Kemmitt in *Comprehensive Inorganic Chemistry,* **vol. 3,** J. C. Bailar Jr. *et al.*, Eds. (Pergamon Press, Oxford, 1973) pp 771-876; L. R. Matricardi, J. H. Downing in Kirk-Othmer *Encyclopedia of Chemical Technology* **vol. 14,** (Wiley-Interscience, New York, 3rd ed., 1981) pp 824-843.

Steel gray, lustrous, hard, brittle metal. Exists in four allotropic forms: α-form, body-centered cubic, stable below 710° (approx), d^{20} 7.47; β-form, cubic, stable in the range 710-1079°, d^{20} 7.26; γ-form or electrolytic Mn, face-centered cubic, stable in the range 1079-1143°C, d^{1100} 6.37; δ-form, body-centered cubic, stable from 1143° to mp, d^{1143} 6.28. γ-Mn when stabilized at room temp is face-centered tetragonal, d^{20} 7.21. mp 1244°. bp 2095°. Sp heat 0.115 cal/g/°C; latent heat of fusion: 3.5 kcal/g-atom. Mohs' hardness 5.0. Superficially oxidized on exposure to air. Burns with an intense white light when heated in air. Dec water slowly in the cold, rapidly on heating; pure electrolytic manganese is not attacked by water at ordinary temp; slightly attacked by steam. Reacts with dil mineral acids with evolution of hydrogen and formation of divalent manganous salts. Reacts with aq solns of sodium or potassium bicarbonate. When heated in nitrogen above 2000° burns to form a nitride. Converted by fluorine into the di- and trifluoride, by chlorine into the dichloride. In form of powder reduces most metallic oxides on heating. On heating, reacts directly with carbon, phosphorus, antimony or arsenic.

Human Toxicity: Occurs by inhalation of the dust or fumes. Symptoms: languor, sleepiness, weakness, emotional disburbances, spastic gait, paralysis. Picture resembles parkinsonism: E. Browning, *Toxicity of Industrial Metals* (Appleton-Century-Crofts, New York, 2nd ed., 1969) pp 213-225.
USE: In manuf of steel; for rock crushers, railway points and crossings, wagon buffers; as a constituent of several alloys, *e.g.*, ferromanganese, copper manganese, Manganin.

5605. Manganese Acetate. C$_4$H$_6$MnO$_4$; mol wt 173.03. C 27.76%, H 3.50%, Mn 31.75%, O 36.99%. Mn(CH$_3$COO)$_2$.
Tetrahydrate, pale red, transparent monoclinic crystals.

d 1.59. Sol in water or alcohol. LD$_{50}$ orally in rats: 3.73 g/kg, Smyth *et al., Am. Ind. Hyg. Assoc. J.* **30,** 470 (1969).
USE: As mordant in dyeing; manuf bister; drier for paints and varnishes (1-1.5:1000).

5606. Manganese Borate. Approx MnB$_4$O$_7$.8H$_2$O.
Brownish-white powder. Insol in water or alcohol. Dec on long contact with water; sol in dil acids.
USE: In drying varnishes and oils; as drier for linseed oil, 1:500; also in leather industry.

5607. Manganese Bromide. Manganese dibromide. Br$_2$-Mn; mol wt 214.76. Br 74.42%, Mn 25.58%. MnBr$_2$.
Tetrahydrate, rose-red, slightly deliquesc crystals. mp 64° with some decompn. Sol in 0.5 parts water; sol in alcohol. The aq soln is slightly acid. *Keep well closed.*

5608. Manganese Carbonate. CMnO$_3$; mol wt 114.94. C 10.45%, Mn 47.79%, O 41.76%. MnCO$_3$. (Usually contains some H$_2$O). Occurs in nature as the mineral *rhodochrosite.*
Pink to almost white powder when freshly pptd, but gradually becomes light brown in the air. Rhombohedral, calcite structure. d 3.1. Insol in water or alcohol; sol in dil acid. *Keep well closed.*
USE: As pigment—"manganese white"; drier for varnishes; in feeds.

5609. Manganese Carbonyl. *Decacarbonyldimanganese.* C$_{10}$Mn$_2$O$_{10}$; mol wt 389.99. C 30.80%, Mn 28.17%, O 41.03%. Mn$_2$(CO)$_{10}$. First prepd by reduction of MnI$_2$ with a Grignard reagent under CO pressure: Hurd *et al., J. Am. Chem. Soc.* **71,** 1899 (1949). Improved prepns: Brimm *et al., ibid.* **76,** 3831 (1954); Closson, *ibid.* **80,** 6167 (1958); Podall *et al., ibid.* **82,** 1325 (1960); Calderazzo, *Inorg. Chem.* **4,** 293 (1965). Crystal structure: Dahl, Rundle, *Acta Cryst.* **16,** 419 (1963); molecular structure: Almenningen *et al., Acta Chem. Scand.* **23,** 685 (1969). *Reviews: Organic Syntheses via Metal Carbonyls* vol. 1, I. Wender, P. Pino, Eds. (Interscience, New York, 1968) *passim;* Anisimov *et al., Usp. Khim.* **37,** 380-408 (1968); *Russ. Chem. Rev.* **37,** 184-197 (1968) (English translation).
Golden yellow, monoclinic crystals. d^{25} 1.75. mp 154-155°. Stable under an atm of CO. In absence of CO, begins to dec at 110°. Sol in organic solvents, insol in water. Less stable to air, heat and light in soln.
USE: Catalyst; antiknock additive.

5610. Manganese Chloride. Manganous chloride; manganese dichloride. Cl$_2$Mn; mol wt 125.84. Cl 56.34%, Mn 43.66%. MnCl$_2$.
Tetrahydrate, reddish, slightly deliquesc, monoclinic crystals. d 2.01. mp 58°. Sol in 0.7 part water; sol in alcohol; insol in ether. pH of 0.2 molar aq soln 5.5. *Keep well closed.* LD s.c. in mice: 180-250 mg/kg; i.v. in dogs: 201.6 mg/kg, *Handbook of Toxicology,* Vol. 1, W. S. Spector, Ed. (Saunders, Philadelphia, 1956) pp 182-183.
USE: In dyeing (manganese bister); disinfecting; purifying natural gas; linseed oil drier; in electric batteries.

5611. Manganese Difluoride. Manganous fluoride; manganese fluoride. F$_2$Mn; mol wt 92.93. F 40.89%, Mn 59.11%. MnF$_2$. Prepd from manganese carbonate and hydrogen fluoride: Moissan, Venturi, *Compt. Rend.* **130,** 1158 (1900); Kwasnik in *Handbook of Preparative Inorganic Chemistry,* Vol. 1, G. Brauer, Ed. (Academic Press, New York, 2nd ed., 1963) pp 262-263.
Pink, quadratic prisms (tetragonal structure, rutile type) or reddish powder. *Poisonous!* d 3.98. mp 856°. Soly in water (g/100 ml): 0.66 (40°); 0.44 (60°); 0.48 (100°). Insol in alc. Sol in dil hydrofluoric acid, concd hydrochloric or nitric acid. Lowest published lethal dose in guinea pigs: 200 mg/kg orally; 700 mg/kg s.c., *Toxic Substances List,* H. E Christensen, Ed. (1973) p 574.
Tetrahydrate, obtained by dissolving manganese carbonate in hydrofluoric acid, evaporating, and drying *in vacuo.*

5612. Manganese Dioxide. Manganese binoxide; manganese peroxide; manganese superoxide; black manganese oxide. MnO$_2$; mol wt 86.94. Mn 63.19%, O 36.81%. Occurs in nature as the mineral *pyrolusite,* or made artificially (pptd). The native product is heavy, steel-gray when in lumps, black when powdered; the pptd product is a brownish-black, fine powder. Both usually contain some Mn_3O_4 and some water. When ignited evolves oxygen, leaving Mn_3O_4. Lab prepn: Moore et al., J. Am. Chem. Soc. **72**, 856 (1950); Covington et al., Trans. Faraday Soc. **58**, 1975 (1962). Review of use as reagent: J. S. Pizey, Synthetic Reagents vol. **2** (John Wiley, New York, 1974) pp 143-174.

Tetragonal crystals (rutile structure). Insol in water, nitric or cold sulfuric acid; slowly dissolves in cold HCl with evolution of Cl_2; in presence of hydrogen peroxide or oxalic acid it dissolves in dil H_2SO_4 or HNO_3. Caution: Manganese dioxide is a strong oxidizer, hence it should not be heated or rubbed with organic matter or other oxidizable substances, e.g., sulfur, sulfides, phosphides, hypophosphites, etc. LD i.v. in rabbits: 45 mg/kg, Handbook of Toxicology, Vol. 1, W. S. Spector, Ed. (Saunders, Philadelphia, 1956) pp 182-183.

USE: The mineral is the source of manganese and all its compds; largely used in manuf manganese steel; oxidizer; in alkaline batteries (dry cells); for making amethyst glass, decolorizing glass; painting on porcelain, faience and majolica. The ppt is used in electrotechnics, pigments, browning gun barrels, drier for paints and varnishes, printing and dyeing textiles.

5613. Manganese Hypophosphite. $H_4MnO_4P_2$; mol wt 184.91. H 2.18%, Mn 29.71%, O 34.61%, P 33.50%. $Mn(H_2PO_2)_2$.

Monohydrate, pink, odorless, almost tasteless crystals or powder. When heated evolves spontaneously flammable phosphine. One gram dissolves in 6.5 ml water, 6 ml boiling water. Insol in alcohol.

5614. Manganese Iodide. Manganese diiodide. I_2Mn; mol wt 308.74. I 82.21%, Mn 17.79%. MnI_2.

Tetrahydrate, rose-red crystals; rapidly becomes brown on exposure to air and light, due to liberation of iodine. Very sol in water with gradual decompn; sol in alcohol. The aq soln is slightly acid. Keep tightly closed and protected from light.

5615. Manganese Nitrate. MnN_2O_6; mol wt 178.95. Mn 30.70%, N 15.65%, O 53.65%. $Mn(NO_3)_2$. Prepn: Dehnicke, Strähle, Ber. **97**, 1502 (1964). Review: Gmelin's, Manganese (8th ed.) **56**, part C3, 267-281 (1975).

Colorless; hygroscopic. Sol in water, dioxane, tetrahydrofuran, acetonitrile.

Tetrahydrate, pink, deliquesc cryst masses at temp below 20°. d 2.129. mp 37.1°. Very sol in water, sol in alcohol. The aq soln is slightly acid.

Hexahydrate, rose-colored, deliquesc, monoclinic needles (colorless after several crystns). d 1.8. mp 25.0°. Freely sol in water, alcohol.

USE: Intermediate in manuf of reagent grade MnO_2; in prepn of porcelain colorants.

5616. Manganese Oleate. Approx $Mn(C_{18}H_{33}O_2)_2$.

Brown, granular mass. Insol in water; sol in ether, oleic acid; slightly sol in alcohol.

USE: As a drier in varnishes.

5617. Manganese Oxalate. C_2MnO_4; mol wt 142.96. C 16.81%, Mn 38.43%, O 44.76%. MnC_2O_4.

Dihydrate, white cryst powder; dec at 150°. Slightly sol in water; sol in acids.

5618. Manganese Oxide. Manganomanganic oxide. Mn_3O_4; mol wt 228.79. Mn 72.03%, O 27.98%. Occurs in nature as the mineral hausmannite. Prepn: Moore et al., J. Am. Chem. Soc. **72**, 856 (1950).

Brownish-black powder. d 4.7. Insol in water; sol in HCl with evolution of chlorine.

5619. Manganese Phosphate, Dibasic. $HMnO_4P$; mol wt 150.92. H 0.67%, Mn 36.40%, O 42.41%, P 20.52%. $MnHPO_4$. Trihydrate. Exists in two modifications: gray form prepd by decompn of $MnCO_3$ by phosphoric acid; pink form prepd by reaction of $Mn_3(PO_4)_2$ with phosphoric acid (P_2O_5 concn 0.6-28%): Goloshchapov et al., Russ. J. Inorg Chem. **11**, 504 (1966).

Pink form, usually contains some tribasic phosphate. Very slightly sol in water; sol in dil acids. Gray form, sol only in hot concd HCl.

5620. Manganese Pyrophosphate. MnO_7P_2; mol wt 228.88. Mn 24.00%, O 48.93%, P 27.06%. MnP_2O_7.

Trihydrate, white or nearly white powder. Insol in water; sol in excess of alkali pyrophosphate, in acids.

5621. Manganese Selenide. MnSe; mol wt 133.89. Mn 41.02%, Se 58.98%. Prepd by the action of manganese salts on alkali selenides: Berzelius et al., cited by Mellor, A Comprehensive Treatise on Inorganic and Theoretical Chemistry **10**, 798 (1930); by the action of a manganese salt on a soln of hydrogen selenide out of contact with air: Moser, Atynsky, Monatsh. **45**, 235 (1925).

Gray-black cubic crystals. d^{15} 5.59. Completely oxidized when heated to redness in oxygen. Insol in cold water. Dec in hot water and dil acids.

5622. Manganese Sesquioxide. Mn_2O_3; mol wt 157.86. Mn 69.59%, O 30.41%. Occurs in nature as the mineral manganite, $Mn_2O_3 \cdot H_2O$. Prepn: Moore et al., J. Am. Chem. Soc. **72**, 856 (1950).

Black, fine powder. d 4.50. Insol in water; sol in HCl with evolution of chlorine.

5623. Manganese Silicate. Approx $MnSiO_3$. Occurs in nature as the minerals rhodonite, manganjustite, tephroite.

Red crystals or yellowish-red powder. The pptd article is a yellowish-red powder. d of the mineral 3.48. Insol in water.

USE: As color for special glass; producing red glazes on pottery.

5624. Manganese Sulfate. MnO_4S; mol wt 151.00. Mn 36.38%, O 42.38%, S 21.23%. $MnSO_4$. Forms several hydrates. The article of commerce is usually a mixture of the tetra- and pentahydrates.

Monohydrate, pale red, slightly efflorescent crystals; loses all water at 400-450°. Sol in about 1 part cold, 0.6 part boiling water; insol in alcohol.

USE: In dyeing; for red glazes on porcelain; boiling oils for varnishes; in fertilizers for vines, tobacco; in feeds.

THERAP CAT (VET): Nutritional factor (essential trace element in all animals); prevention of perosis in poultry.

5625. Manganese Sulfide. Manganese monosulfide. MnS; mol wt 87.00. Mn 63.14%, S 36.86%. Occurs in nature as the mineral alabandite or manganblende. Prepn: Mehmed, Haraldsen, Z. Anorg. Chem. **235**, 194 (1938).

Pink, green or brown-green powder. Three cryst modifications: α-form, green cubic crystals; β-form, red cubic crystals; γ-form, red hexagonal crystals. Practically insol in water; sol in dil acids. In moist condition it readily oxidizes in air to the sulfate.

5626. Manganese Trifluoride. Manganic fluoride. F_3Mn; mol wt 111.93. F 50.93%, Mn 49.07%. MnF_3. Prepd by fluorination of manganese iodide according to the equation $2MnI_2 + 13F_2 \rightarrow 2MnF_3 + 4IF_5$: Moissan, Compt. Rend. **130**, 622 (1900); v. Wartenberg, Z. Anorg. Allgem. Chem. **244**, 346 (1940); Kwasnik in Handbook of Preparative Inorganic Chemistry, vol. **1**, G. Brauer, Ed. (Academic Press, New York, 2nd ed., 1963) pp 263-264.

Red mass; monoclinic crystals. d 3.54. Stable to 600°. Hydrolyzed by water. May be stored in glass ampuls.

USE: Fluorinating agent in organic chemistry.

5627. Mangostin. 1,3,6-Trihydroxy-7-methoxy-2,8-bis-(3-methyl-2-butenyl)-9H-xanthen-9-one; 1,3,6-trihydroxy-7-methoxy-2,8-di(3-methyl-2-butenyl)xanthone. $C_{24}H_{26}O_6$; mol wt 410.45. C 70.23%, H 6.39%, O 23.39%. From various parts of the mangosteen tree (Garcinia mangostana L., Guttiferae): Schmid, Ann. **93**, 83 (1855); Dragendorff, ibid. **482**, 280 (1930). Structure: Yates, Stout, J. Am. Chem. Soc. **80**, 1691 (1958); Scheinmann, Chem. Commun. **1967**, 1015; Stout et al., ibid. **1968**, 211.

Consult the cross index before using this section.

Yellow crystals from benzene, mp 181.6-182.6°. uv max (ethanol): 243, 259, 318, 351 nm (log ε 4.54, 4.44, 4.38, 3.86). Practically insol in water; sol in alcohol, ether, acetone, chloroform, ethyl acetate.

3,6-Dimethylmangostin, $C_{26}H_{30}O_6$, pale yellow needles from ethanol, mp 123.3-123.8°. uv max (ethanol): 245, 262, 314, 350 nm (log ε 4.50, 4.53, 4.36, 3.81).

5628. Manna. Dried exudation of *Fraxinus ornus* L., *Oleaceae. Habit.* Mediterranean Basin, Asia Minor, Spain. *Constit.* 40-60% mannitol; 10-16% mannotetrose; 6-16% mannotriose; glucose, mucilage, fraxin. One gram dissolves in 5 ml water, 150 ml 90% alcohol.

THERAP CAT: Cathartic.

5629. Mannitol. D-Mannitol; mannite; manna sugar; cordycepic acid; Diosmol; Manicol; Mannidex; Osmitrol; Osmosal; Resectisol. $C_6H_{14}O_6$; mol wt 182.17. C 39.56%, H 7.74%, O 52.70%. Widespread in plants and plant exudates; obtained from manna and seaweeds: *The Carbohydrates,* W. Pigman, Ed. (Academic Press, New York, 1957) pp 249-250. Forms a stable, equimolar compound with H_2O_2: S. Tanatar, *J. Russ. Phys. Chem. Soc.* **40,** 376, *C.A.* **3,** 883 (1909). Prepd by electrolytic reduction of glucose: Creighton, *Can. Chem. Process Inds.* **26,** 690 (1942); *C.A.* **37,** 1088⁵ (1943); Wolfrom *et al., J. Am. Chem. Soc.* **68,** 578 (1946). Prepn from seaweed: Sorensen, Kristensen, U.S. pat. **2,516,350** (1950); by hydrogenation of invert sugar, monosaccharides, and sucrose: Kasehagen and Kasehagen, Luskin, U.S. pats. **2,642,462, 2,749,371,** and **2,759,024** (1953, 1956, and 1956, all to Atlas Powder). Review of prepn: Pigman, *loc. cit.* Novel synthesis: M. Makkee *et al., Chem. Commun.* **1980,** 930.

$$CH_2OH$$
$$|$$
$$HOCH$$
$$|$$
$$HOCH$$
$$|$$
$$HCOH$$
$$|$$
$$HCOH$$
$$|$$
$$CH_2OH$$

Orthorhombic needles from alc, mp 166-168°. Sweetish taste. d^{20} 1.52. $bp_{3.5}$ 290-295°. Is inactive or very slightly levorotatory in distilled water. Forms sodium mannitoborate on addition of borax giving a greater rotation: $[\alpha]_D^{20}$ +23° to 24° after 1 hr in a soln of 10 g mannitol + 12.8 g borax + sufficient H_2O to make 100 ml. One gram dissolves in about 5.5 ml water, 83 ml alcohol; more sol in hot water. Insol in ether. Sol in pyridine, aniline, aq solns of alkalies. One gram dissolves in 18 ml glycerol (d 1.24). Soly tables: Creighton, Klauder, *J. Franklin Inst.* **195,** 687 (1923). pKa at 18° = 13.50.

USE: Used with boric acid in the manuf of dry electrolytic condensers for radio applications; in making artificial resins and plasticizers; in pharmacy as excipient and diluent for solids and liqs; in analytical chemistry for boron determinations; in the manuf of mannitol hexanitrate. Used in the food industry as anticaking and free-flow agent, flavoring agent, lubricant and release agent, stabilizer and thickener and nutritive sweetener.

THERAP CAT: Diuretic. Diagnostic aid (renal function).

5630. Mannitol Hexanitrate. Mannitol nitrate; nitromannite; nitromannitol; Maxitate; Medemanol; Dilangil; Moloid; Mannitrin; Nitranitol; Manexin. $C_6H_8N_6O_{18}$; mol wt 452.17. C 15.94%, H 1.78%, N 18.59%, O 63.69%. Made by nitration of mannitol: Fleury *et al., Mem. Poudres* **31,** 107 (1949).

Long needles in regular clusters from alc. mp 106-108°. Soluble in alc, in ether; insol in water. Explodes on percussion. Its stability at ordinary temps is such that it may be used commercially, but it is distinctly less stable than nitroglycerol at 75°. Its employ for pharmaceutical prepns is only in admixture with carbohydrate substances in dilutions corresponding to 1 part of mannitol hexanitrate to 9 or more parts of carbohydrate. In such dilutions mannitol hexanitrate is considered nonexplosive.

THERAP CAT: Vasodilator.

5631. Mannomustine. *1,6-Bis(2-chloroethylamino)-1,6-dideoxy-*D-*mannitol dihydrochloride;* 1,6-dideoxy-1,6-di(2-chloroethylamino)-D-mannitol dihydrochloride; 1,6-di(2-chloroethylamino)-1,6-dideoxy-D-mannitol dihydrochloride; mannitol nitrogen mustard; BCM; Degranol. $C_{10}H_{24}Cl_4N_2O_4$; mol wt 378.13. C 31.76%, H 6.40%, Cl 37.50%, N 7.41%, O 16.93%. Prepn: Vargha *et al., J. Chem. Soc.* **1957,** 805. Toxicity: Scherf *et al., Arzneimittel-Forsch.* **20,** 1467 (1970).

$$\left[\begin{array}{c} \overset{+}{C}H_2NH_2CH_2CH_2Cl \\ | \\ HOCH \\ | \\ HOCH \\ | \\ HCOH \\ | \\ HCOH \\ | \\ \overset{+}{C}H_2NH_2CH_2CH_2Cl \end{array} \right] \quad 2Cl^-$$

Crystals from 80% ethanol, dec 239-241°. $[\alpha]_D^{20}$ +18.46° (c = 1.81 in H_2O). [Free base, crystals, dec 278°.] Sol in water. Slightly sol in ethanol. Aq solns are mildly acid and stable at room temps. Claimed to be considerably less toxic than mechlorethamine. LD_{50} i.v. in rats: 56 mg/kg (Scherf).

THERAP CAT: Antineoplastic.

5632. D-**Mannose.** Seminose; carubinose. $C_6H_{12}O_6$; mol wt 180.16. C 40.00%, H 6.71%, O 53.29%. Prepn of α-form by treating ivory nut shavings with H_2SO_4: Isbell, *J. Res. Nat. Bur. Stand.* **26,** 47 (1941); Isbell, Frush in *Methods in Carbohydrate Chemistry,* R. L. Whistler, M. L. Wolfrom, Eds. (Academic Press, New York, 1962) pp 145-147. Prepn and stability of α- and β-forms: Reeves, *J. Am. Chem. Soc.* **72,** 1499 (1950); J. Sowden in *The Carbohydrates,* W. Pigman, Ed. (Academic Press, New York, 1957) pp 94-95.

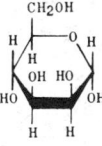

α-D-mannose

α-Form, crystals from methanol, mp 133°. $[\alpha]_D$ +29.3° → +14.2° (water).

β-Form, orthorhombic, bisphenoidal needles from alcohol or acetic acid, dec 132°. Sweet taste with bitter aftertaste. d^{20} 1.54. Shows mutarotation. $[\alpha]_D^{20}$ −17.0° → +14.2° (c = 4). One gram dissolves in 0.4 ml water, 120 ml methanol, 250 ml abs ethanol, 3.5 ml pyridine. Ka at 18° = 10.9 × 10^{-13}. Reduces Fehling's soln; is fermented by yeast.

Phenylhydrazone, $C_{12}H_{18}N_2O_5$, crystals from dil ethanol, mp 199-200°. $[\alpha]_D^{20}$ +26.3° → +33.8° (pyridine).

$CaCl_2$-addition cpd tetrahydrate, $C_6H_{12}O_6 \cdot CaCl_2 \cdot 4H_2O$, mp 101-102°. $[\alpha]_D^{20}$ −31.3° → +6.0° (c = 9).

5633. Maprotiline. *N-Methyl-9,10-ethanoanthracene-9(10H)-propanamine;* 9-(γ-methylaminopropyl)-9,10-dihydro-9,10-ethanoanthracene; 1-(3-methylaminopropyl)dibenzo[*b,e*]bicyclo[2.2.2]octadiene. $C_{20}H_{23}N$; mol wt 277.41.

C 86.59%, H 8.35%, N 5.05%. Prepn: Wilhelm *et al.*, **Swiss pats. 467,237** and **467,747** (both 1969 to Ciba), *C.A.* **71**, 70397z, 49652u (1969). Synthesis, NMR, mass spectra: Wilhelm, Schmidt, *Helv. Chim. Acta* **52**, 1385 (1969). Toxicity: R. Hess *et al.*, *Boll. Chim. Farm.* **112**, 782 (1973), *C.A.* **81**, 33479p (1974). Review of pharmacology and therapeutic efficacy: R. M. Pinder *et al.*, *Drugs* **13**, 321 (1977). Comprehensive description: S. K. Suh, J. B. Smith in *Analytical Profiles of Drug Substances* vol. 15, K. Florey, Ed. (Academic Press, New York, 1986) pp 393-426.

CH$_2$CH$_2$CH$_2$NHCH$_3$

mp 92-94°.

Hydrochloride, C$_{20}$H$_{24}$ClN, *Ba 34276*, *Ludiomil*. Crystals from isopropanol, mp 230-232°. Freely sol in methanol and chloroform, slightly sol in water. Practically insol in isooctane. LD$_{50}$ in mice, rats (mg/kg): ~750, ~900 orally (Hess).

THERAP CAT: Antidepressant.

5634. Margaric Acid. *Heptadecanoic acid.* C$_{17}$H$_{34}$O$_2$; mol wt 270.44. C 75.49%, H 12.67%, O 11.83%. CH$_3$-(CH$_2$)$_{15}$COOH. Prepn: Kaufmann, Stamm, *Ber.* **91**, 2121 (1958); Bhattacharyya *et al.*, *Chem. & Ind. (London)* **1959**, 1352; Hünig, Ledle, *Ber.* **93**, 913 (1960). Metabolism: Boyer, Scheig, *Lipids* **4**, 615 (1969).

Crystals from alcohol, mp 61°. d 0.853. bp$_{100}$ 227°. n_D^{60} 1.4342. Insol in water. Freely sol in ether; slightly sol in alcohol. LD$_{50}$ i.v. in mice: 36±0.3 mg/kg, L. Orö, A. Wretlind, *Acta Pharmacol. Toxicol.* **18**, 141 (1961).

5635. Marrubiin. *6-[2-(3-Furanyl)ethyl]decahydro-6-hydroxy-2a,5a,7-trimethyl-2H-naphtho[1,8-bc]furan-2-one; 15,16-epoxy-6β,9-dihydroxy-8βH-labda-13(16),14-dien-19-oic acid γ-lactone; 5-[2-(3-furyl)ethyl]decahydro-5,8-dihydroxy-1,4a,6-trimethyl-1-naphthoic acid γ-lactone.* C$_{20}$-H$_{28}$O$_4$; mol wt 332.42. C 72.26%, H 8.49%, O 19.25%. Diterpene lactone principle isolated from white horehound, *Marrubium vulgare* (Tourn.) L., *Labiatae*: Harms, *Arch. Pharm.* **83**, 144 (1842); Ludwig, Kromayer, *ibid.* **158**, 257 (1861); Nicholas, *J. Pharm. Sci.* **53**, 895 (1964). Alternate view that marrubiin is an artefact generated from *premarrubiin* during the isoln: Henderson, McCrindle, *J. Chem. Soc. (C)* **1969**, 2014. Structure and stereochemistry: Cocker *et al.*, *J. Chem. Soc.* **1953**, 2540; *Chem. & Ind. (London)* **1954**, 1561; **1955**, 1484; Fulke, McCrindle, *ibid.* **1965**, 647. Total stereochemistry: Wheeler *et al.*, *Tetrahedron* **23**, 3909 (1967); Appleton *et al.*, *J. Chem. Soc. (C)* **1967**, 1943. Synthesis: Mangoni *et al.*, *Tetrahedron* **28**, 611 (1972).

Crystals from alc, mp 160°. [α]$_D^{20}$ +35.8° (c = 3.1 in chloroform); [α]$_D^{24}$ +45° (acetone). uv max: 208, 212, 216 nm (log ε 3.75, 3.75, 3.70). One gram dissolves in 60 ml alc. Freely sol in chloroform, acetone, hot alc, pyridine; sparingly sol in ether, benzene. Practically insol in water.

5636. Mastic. Balsam tree; pistachia galls; mastiche; mastix; lentisk; Mastisol. Concrete resinous exudation from *Pistacia lentiscus* L., *Anacardiaceae*. *Habit.* Mediterranean Islands, especially Chios. *Constit.* Volatile oil (about 2%); masticinic, masticonic acids; masticoresene.

Pale yellow or greenish-yellow, globular, elongated or pear-shaped tears; slightly balsamic odor and terebene taste. Almost insol in water. Nearly completely sol in alcohol; 1 g

dissolves in 0.5 ml chloroform, 0.5 ml ether; partially sol in oil turpentine.

USE: In tooth cements, plasters, lacquers, chewing gums, and incense; also for retouching negatives.

5637. Maté. Paraguay tea; yerba maté; St. Bartholomew's tea; Jesuit's tea. Leaves of *Ilex paraguensis* St. Hil., *Aquifoliaceae*, grown in Brazil, Uruguay, Argentina, and Paraguay. *Review:* Cheney, *Econ. Botany* **1**, 243 (1947). For constituents *see:* Siesto, *Quaderni Nutr.* **18**, 35 (1959), *C.A.* **57**, 13133d (1962); Roberts, *Chem. & Ind. (London)* **1956**, 985; Paula, *Rev. Brasil. Quim.* (Sao Paulo) **42**, 202 (1956), *C.A.* **51**, 3167i (1957). Among its constituents are caffeine, other purines, tannins or tannoid substances, such as chlorogenic acid and derivatives.

5638. Matico. Dried leaves of *Piper elongatum* Vahl. (*P. angustifolium* R. & P.), *Piperaceae*. *Habit.* Peru, Bolivia, Brazil, Mexico, Cuba. *Constit.* 1-3.5% volatile oil, maticin (a bitter principle); artanthic acid, tannin, mucilage, resin.

THERAP CAT: Astringent.

5639. Matricaria. German chamomile; Hungarian chamomile; wild chamomile. Dried flower heads of *Matricaria chamomilla* L., *Compositae*. *Habit.* Europe, Western Asia, cultivated in U.S. *Constit.* Volatile oil, anthemic acid, anthemidine, tannin, matricarin. *Ref.* Herz, Ueda, *J. Am. Chem. Soc.* **83**, 1139 (1961).

THERAP CAT: Carminative; topical counterirritant.

5640. Matricarin. *4-(Acetyloxy)-3,3a,4,5,9a,9b-hexahydro-3,6,9-trimethylazuleno[4,5-b]furan-2,7-dione.* C$_{17}$H$_{20}$-O$_5$; mol wt 304.35. C 67.09%, H 6.62%, O 26.29%. Isoln from *Matricaria chamomilla* L. and *Artemisia tilesii* Ledeb, *Compositae*: Cekan *et al.*, *Coll. Czech. Chem. Commun.* **24**, 1554 (1959). Structure: Herz, Ueda, *J. Am. Chem. Soc.* **83**, 1139 (1961).

Plates from benzene+ petr ether and acetone + petr ether, mp 190-191°. [α]$_D^{23}$ +23.5° (c = 0.65 in chloroform). uv max: 255 nm (ε 15,100).

Desacetylmatricarin monohydrate, C$_{15}$H$_{18}$O$_4$.H$_2$O, crystals from benzene + acetone, mp 123-125°, resolidifies and mp 143-146°.

Tetrahydromatricarin, C$_{17}$H$_{24}$O$_5$, needles from benzene + petr ether, mp 175-178°. [α]$_D^{22}$ +32.3° (c = 1.92 in chloroform).

5641. Matrine. *Matridin-15-one*; sophocarpidine. C$_{15}$-H$_{24}$N$_2$O; mol wt 248.36. C 72.54%, H 9.74%, N 11.28%, O 6.44%. Occurs naturally as the (+)-form in the Chinese drug *Kuh Seng* or the Japanese *Shinkyogan*, the dried roots of *Sophora angustifolia* Sieb. & Zucc., *S. flavescens* Ait. and other *Sophora* spp., *Leguminosae*. Isomeric with lupanine, *q.v.* Isoln: Kondo, *Arch. Pharm.* **266**, 1 (1928); Winterfeld, Kneier, *Ber.* **64**, 150 (1931); Orechov, Proskurnina, *Ber.* **67**, 77 (1934); Briggs, Ricketts, *J. Chem. Soc.* **1937**, 1795. Identity with sophocarpidine: Orechov *et al.*, *Ber.* **68**, 429 (1935). Alternate method of isoln: F. Bohlmann *et al.*, *Ber.* **91**, 2189 (1958). Structure: Tsuda, Murakami, *Ber.* **69**, 429 (1936). Configuration: Bohlmann *et al.*, *Ber.* **91**, 2176 (1958). Absolute configuration: Okuda *et al.*, *Chem. Pharm. Bull.* **14**, 314 (1966); Cervinka, *Z. Chem.* **7**, 190 (1967). Synthetic studies: Tsuda *et al.*, *J. Org. Chem.* **21**, 1481 (1956); **23**, 1179 (1958). Synthesis of *matridine (deoxymatrine)*: Mandell, Singh, *J. Am. Chem. Soc.* **83**, 1766 (1961). Total synthesis of (±)-matrine: Mandell *et al.*, *ibid.* **87**, 5234 (1965); J. Chen *et al.*, *Chem. Commun.* **12**, 905 (1986). Synthesis of (+)-matrine: Okuda *et al.*, *Chem. Pharm. Bull.* **14**, 275 (1966). Biosynthesis: Schütte *et al.*, *Arch. Pharm.* **295**, 34 (1962).

Has been obtained in four forms: α-Form: needles or flat prisms, mp 76°. β-Form: orthorhombic prisms, mp 87°, $[\alpha]_D^{20}$ +38° (alc). γ-Form: liquid, bp_6 223°, d_4^{20} 1.088, n_D^{85} 1.5287. δ-Form: prisms, mp 84°. Sol in water, benzene, chloroform, ether, carbon disulfide; slightly sol in petr ether.

5642. Maytansine. Maitansine; NSC 153858. $C_{34}H_{46}$-ClN_3O_{10}; mol wt 692.21. C 59.00%, H 6.70%, Cl 5.12%, N 6.07%, O 23.11%. Anti-leukemic ansa macrolide, member of a class of compounds that includes the rifamycins, streptovaricins etc. First ansa compound isolated from a plant rather than produced by a microorganism. Isoln from *Maytenus ovatus* Loes., *Celastraceae* and structure elucidation: Kupchan *et al., J. Am. Chem. Soc.* **94,** 1354 (1972); Kupchan, *Ger. pat.* **2,241,418** (1974). Crystal structure and absolute configuration of the (3-bromopropyl) ether: Bryan *et al., J. Chem. Soc. Perkin Trans. II* **1973,** 897. Synthetic studies: Meyers *et al., Tetrahedron Letters* **1974,** 717; **1975,** 1745, 1749; Corey, Bock, *ibid.* **1975,** 2643; E. J. Corey *et al., ibid.* **1978,** 1051; A. I. Meyers, J. P. Hudspeth, *ibid.* **22,** 3925 (1981). Total synthesis of (±)-form: A. I. Meyers *et al., J. Am. Chem. Soc.* **102,** 6597 (1980); of naturally occurring (−)-form: E. J. Corey *et al., ibid.* 6613. Cytotoxic action: Wolpert-Defilippes, *Biochem. Pharmacol.* **24,** 751 (1975). Clinical studies in carcinoma, melanoma: J. A. Neidhart *et al., Cancer Treat. Rep.* **64,** 675 (1980); D. L. Ahmann *et al., ibid.* 721. Toxicity: G. M. Mugera, J. M. Ward, *ibid.* **61,** 1333 (1977).

mp 171-172°. $[\alpha]_D^{26}$ − 145° (c = 0.055 in chloroform). uv max (ethanol): 233, 254, 282, 290 nm (ε 29800, 27200, 5690, 5520). LD_{50} in rats (mg/kg): 0.48 s.c. (Mugera, Ward).

5643. Mazindol. 5-(4-Chlorophenyl)-2,5-dihydro-3H-*imidazo[2,1-a]isoindol-5-ol;* 5-(4-chlorophenyl)-2,3-dihydro-5-hydroxy-5H-imidazo[2,1-*a*]isoindole; SaH 42548; Dimagrir; Magrilon; Mazildene; Sanorex; Terenac; Teronac. $C_{16}H_{13}ClN_2O$; mol wt 284.74. C 67.49%, H 4.60%, Cl 12.45%, N 9.83%, O 5.62%. Prepn: W. J. Houlihan, **Ger. pat. 1,814,540,** *C.A.* **71,** 81368s (1969); W. J. Houlihan, M. K. Eberle, **Ger. pat. 1,930,488;** *eidem,* **U.S. pat. 3,597,445** (1969, 1970, 1970 all to Sandoz). Clinical trials as appetite depressant: C. Sirtori *et al., Am. J. Med. Sci.* **261,** 341 (1971); A. J. Hadler, *J. Clin. Pharmacol.* **12,** 453 (1972). Effect on human prolactin and growth hormone responses: D. A. Thompson *et al., Metabolism* **30,** 1015 (1981). Clinical evaluation in Duchenne muscular dystrophy: P. J. Collipp *et al., J. Med. Genet.* **21,** 254 (1984).

Crystals from acetone-hexane, mp 198-199°.
Note: This is a controlled substance (stimulant) listed in the U.S. Code of Federal Regulations, Title 21 Part 1308.14 (1987).
THERAP CAT: Anorexic. CNS stimulant.

5644. Mazipredone. *11β,17-Dihydroxy-21-(4-methyl-1-piperazinyl)pregna-1,4-diene-3,20-dione hydrochloride;* 11β,-17α-dihydroxy-3,20-dioxo-21-(4-methyl-1-piperazinyl)-pregna-1,4-diene hydrochloride; 11β,17-dihydroxy-21-(4-methyl-1-piperazinyl)-Δ¹-progesterone hydrochloride; Depersolone. $C_{26}H_{39}ClN_2O_4$; mol wt 479.08. C 65.19%, H 8.21%, Cl 7.40%, N 5.85%, O 13.36%. Prepn: Tuba *et al.,* **Hung. pat. 150,350** (1963 to Gedeon Richter), *C.A.* **60,** 3057h (1964).

Crystals, dec 246°. (Free base crystals from tetrahydrofuran + ligroin, dec 199°.) Sol in water.
THERAP CAT: Anti-inflammatory.

5645. MCPA. *(4-Chloro-2-methylphenoxy)acetic acid;* (4-chloro-*o*-toloxy)acetic acid; 2-methyl-4-chlorophenoxyacetic acid; MCP; Agritox; Agroxone; Cornox; Methoxone. $C_9H_9ClO_3$; mol wt 200.63. C 53.88%, H 4.52%, Cl 17.68%, O 23.92%. Prepn: Synerholm, Zimmerman, *Contrib. Boyce Thompson Inst.* **14,** 91 (1945); Templeman, Sexton, *Proc. Roy. Soc.* **133B,** 300 (1946); Foster, **Brit. pats. 573,479** and **573,510** (1945 to ICI); Skeeters, **U.S. pat. 2,740,810** (1956 to Diamond Alkali).

Plates from benzene, mp 120°. Practically insol in water. LD_{50} orally in rats: 700 mg/kg, Rowe, Hymas, *Am. J. Vet. Res.* **15,** 622 (1954).
Used as the sodium salt. Very sol in water.
USE: Powerful selective weed killer.

5646. MDMA. *N,α-Dimethyl-1,3-benzodioxole-5-ethanamine;* 3,4-methylenedioxymethamphetamine; *N*-methyl-3,4-methylenedioxyphenylisopropylamine; *N,α*-dimethylhomopiperonylamine. $C_{11}H_{15}NO_2$; mol wt 193.25. C 68.37%, H 7.82%, N 7.25%, O 16.56%. A centrally active phenethylamine derivative. Prepn: **Ger. pat. 274,350** (1914 to E. Merck), *C.A.* **8,** 3350 (1914); Y. Kasuya, *Yakugaku Zasshi* **78,** 509 (1958), *C.A.* **52,** 17196f (1958); S. Biniecki, E. Krajewski, *Acta Polon. Pharm.* **17,** 421 (1960), *C.A.* **55,** 14350e (1961). Relationship of structure of MDMA and other mescaline analogs to behavior and toxicity in laboratory animals: H. F. Hardman *et al., Toxicol. Appl. Pharmacol.* **25,** 299 (1973). Identification by TLC: M. A. Shaw, H. W. Peel, *J. Chromatog.* **104,** 201 (1975). Pharmacological activity of *N*-substituted amphetamine analogs in men and animals: U. Braun *et al., J. Pharm. Sci.* **69,** 192 (1980); *eidem, Arzneimittel-Forsch.* **30,** 825 (1980). Relationship between configuration and psychomimetic activity in rabbits: G. M. Anderson *et al., Nat. Inst. Drug Abuse Res. Monogr. Ser.* **22,** 8 (1978). Identification of *(S)*-(+)-MDMA as the more potent isomer in inducing serotonin release from rat brain synaptosomes: D. E. Nichols *et al., J. Med. Chem.* **25,** 530 (1982). Receptor binding affinities for various brain recognition sites: G. Battaglia *et al., Eur. J. Pharmacol.* **149,** 159 (1988).

Oil, $bp_{0.4}$ 100-110°. n_D^{19} 1.5311.

Hydrochloride, $C_{11}H_{16}ClNO_2$, *ecstasy*. Crystals, mp 148-149°. LD_{50} in mice, rats (mg/kg): 97, 49 i.p.; in dogs, monkeys: 14, 22 i.v. (Hardman).

Note: This is a controlled substance (hallucinogen) listed in the U.S. Code of Federal Regulations, Title 21, Part 1308.11 (1987).

5647. Mebendazole. *(5-Benzoyl-1H-benzimidazol-2-yl)carbamic acid methyl ester;* 5-benzoyl-2-benzimidazolecarbamic acid methyl ester; methyl 5-benzoyl-2-benzimidazolecarbamate; R 17635; Bantenol; Lomper; Mebenvet; Noverme; Ovitelmin; Pantelmin; Telmin; Vermicidin; Vermirax; Vermox. $C_{16}H_{13}N_3O_3$; mol wt 295.30. C 65.08%, H 4.44%, N 14.23%, O 16.25%. Prepn: J. L. H. Van Gelder *et al.*, **Ger. pat. 2,029,637;** *eidem,* U.S. pat. **3,657,267** (1971, 1972 to Janssen). Activity studies: Walker, Knight, *Vet. Rec.* **90,** 58 (1972). Clinical studies: Brugmans *et al., J. Am. Med. Assoc.* **217,** 313 (1971); Callear, Neave, *Brit. Vet. J.* **127,** xli (1971); *eidem, ibid.* **129,** 79 (1973). Comprehensive description: A. A. Al-Badr, M. Tariq, in *Analytical Profiles of Drug Substances* **vol. 16,** K. Florey, Ed. (Academic Press, New York, 1987) pp 291-326.

Crystals from acetic acid and methanol, mp 288.5°. Soluble in formic acid. Practically insol in water, ethanol, ether, chloroform. LD_{50} orally: > 80 mg/kg in sheep; > 40 mg/kg in mice, rats and chickens (Van Gelder).

THERAP CAT: Anthelmintic (Nematodes).

THERAP CAT (VET): Anthelmintic.

5648. Mebeverine. *3,4-Dimethoxybenzoic acid 4-[ethyl-[2-(4-methoxyphenyl)-1-methylethyl]amino]butyl ester;* veratric acid 4-[ethyl(p-methoxy-α-methylphenethyl)amino]butyl ester; 3,4-dimethoxybenzoic acid 4-[ethyl(p-methoxy-α-methylphenethyl)amino]butyl ester; 4-[ethyl(p-methoxy-α-methylphenethyl)amino]butyl 3,4-dimethoxybenzoate; 4-[N-[2-(p-methoxyphenyl)-1-methylethyl]-N-ethylamino]-butyl 3,4-dimethoxybenzoate. $C_{25}H_{35}NO_5$; mol wt 429.54. C 69.90%, H 8.21%, N 3.26%, O 18.62%. Smooth muscle relaxant. Prepn: **Belg. pat. 609,490,** *C.A.* **59,** 517b (1963) and T. Kralt *et al.,* **Ger. pat. 1,126,889** corresp to U.S. pat. **3,265,577** (1962, 1962, 1966 to N. V. Philips). Pharmacology: G. Bertaccini *et al., Farmaco Ed. Sci.* **30,** 823 (1975).

Hydrochloride, $C_{25}H_{36}ClNO_5$, *Colofac, Duspatalin, Duspatal*. Crystals from ethyl methyl ketone, mp 105-107° (**Ger.** patent); also reported as mp 129-131° (**Belg.** patent).

THERAP CAT: Antispasmodic.

5649. Mebhydroline. *2,3,4,5-Tetrahydro-2-methyl-5-(phenylmethyl)-1H-pyrido[4,3-b]indole;* 5-benzyl-2,3,4,5-tetrahydro-2-methyl-1H-pyrido[4,3-b]indole; 3-methyl-9-benzyl-1,2,3,4-tetrahydro-γ-carboline; 5-benzyl-1,2,3,4-tetrahydro-2-methyl-γ-carboline; N-methyl-9-benzyltetrahydro-γ-carboline; Incidal. $C_{19}H_{20}N_2$; mol wt 276.37. C 82.57%, H 7.29%, N 10.14%. Prepn: Hörlein, *Ber.* **87,** 463 (1954); **Brit. pat. 721,171** (1954 to Bayer); Hörlein, U.S. pat. **2,786,059** (1957 to Schenley). Pharmacology of Diazoline: Kharkevich, *Farmakol. Toksikol.* **20**(no. 6), 46 (1957).

Minute crystals, mp 95°. bp_1 207-215°. Practically insol in water. Freely sol in methanol, ethanol, acetone, chloroform. Slightly sol in ether.

1,5-Naphthalenedisulfonate salt, $C_{48}H_{48}N_4O_6S_2$, *Diazoline, Fabahistin, Omeril*. White powder, dec 280°. Practically insol in water. Sparingly sol in hot glacial acetic acid. Sol in hot formamide, giving a yellow soln. One of the active ingredients of *Refagan*.

THERAP CAT: Antihistaminic.

5650. Mebiquine. *8-[(Dihydroxybismutho)oxy]-6-methylquinoline; dihydroxy(6-methyl-8-quinolinolato)bismuth;* 6-methyl-8-bismuthohydroxyquinoline; DV1; Diarétyl. $C_{10}H_{10}BiNO_3$; mol wt 401.18. C 29.94%, H 2.51%, Bi 52.09%, N 3.49%, O 11.96%. Prepn by the addition of 6-methyl-8-hydroxyquinoline to a solution of bismuth nitrate: Riviere *et al.,* **Ger. pat. 1,811,581;** *eidem,* U.S. pat. **3,591,591** (1969, 1971 to Ugine Kuhlmann). Parasiticidal properties: R. Cavier *et al., Ann. Pharm. Franc.* **31,** 273 (1973).

Yellow powder, insol in water, alc, weak alkaline solns. Sol in strong acids. LD_{50} orally in mice: > 10,000 mg/kg (Cavier).

THERAP CAT: Antidiarrheal.

5651. MeBmt. *[2S-(2R*,3S*,4S*,6E)]-3-Hydroxy-4-methyl-2-(methylamino)-6-octenoic acid;* (4R)-4-[(E)-2-butenyl]-4,N-dimethyl-L-threonine. $C_{10}H_{19}NO_3$; mol wt 201.27. C 59.68%, H 9.51%, N 6.96%, O 23.85%. Characteristic C_9 amino acid of cyclosporins, *q.v.* Identification in cyclosporin A: A. Ruegger *et al., Helv. Chim. Acta* **59,** 1075 (1976). Syntheses: R. M. Wenger, *ibid.* **66,** 2308 (1983); D. A. Evans, A. E. Weber, *J. Am. Chem. Soc.* **108,** 6757 (1986); U. Schmidt, W. Siegel, *Tetrahedron Letters* **28,** 2849 (1987). Appears to be necessary for full biological activity of cyclosporin: R. M. Wenger, *Angew. Chem. Int. Eng. Ed.* **24,** 77 (1985); D. H. Rich *et al., J. Med. Chem.* **29,** 978 (1986).

Crystals from ethanol, mp 240-241° (Wenger). $[\alpha]_D^{20}$ +13.5° (c = 0.50 in water, pH 7). Also reported as mp 242-243° (Evans, Weber). $[\alpha]_D$ +17° (c = 0.51 in 0.4N aq HCl).

5652. Mebrofenin. *N-[2-[(3-Bromo-2,4,6-trimethylphenyl)amino]-2-oxoethyl]-N-(carboxymethyl)glycine;* [[[(3-bromomesityl)carbamoyl]methyl]imino]diacetic acid; N-(3-bromo-2,4,6-trimethylacetanilide)iminodiacetic acid; trimethylbromo-IDA; Choletec. $C_{15}H_{19}BrN_2O_5$; mol wt

387.23. C 46.53%, H 4.95%, Br 20.63%, N 7.23%, O 20.66%. Prepn: A. D. Nunn, M. D. Loberg, **Belg. pat. 891,534**; *eidem*, U.S. pat. **4,418,208** (1982, 1983 both to Squibb). Physicochemical properties of ^{99m}technetium complex used for hepatobiliary tract imaging: A. D. Nunn *et al., J. Nucl. Med.* **24**, 423 (1983). Pharmacokinetics in rats: A. R. Fritzberg *et al., J. Pharm. Sci.* **73**, 1861 (1984). Evaluation as cholescintigraphic imaging agent in animals: A. van Aswegen *et al., Nucl. Med. Biol.* **13**, 509 (1986); J. Kapuscinski *et al., Nucl. Med.* **25**, 188 (1986); in humans: W. C. Klingensmith *et al., Radiology* **146**, 181 (1983). Toxicity study: M. Jiang *et al., Zhonghua Heyixue Zazhi* **4**, 214 (1984), *C.A.* **102**, 42241j (1985).

H$_3$C — [structure: benzene ring with CH$_3$ top, CH$_3$ bottom, Br bottom left, and NHCOCH$_2$—N(CH$_2$COOH)$_2$ side chain]

Crystals from ethanol, mp 198-200°. LD$_{50}$ in mice, rats (mg/kg): 213.8, 226.4 i.v. (Jiang).

THERAP CAT: ^{99m}Tc complex as diagnostic aid (radioactive imaging agent).

5653. Mebutamate. *2-Methyl-2-(1-methylpropyl)-1,3-propanediol dicarbamate; 2-sec-butyl-2-methyl-1,3-propanediol dicarbamate;* carbamic acid 2-*sec*-butyl-2-methyltrimethylene ester; 2-*sec*-butyl-2-methyltrimethylenecarbamate; 2-methyl-2-*sec*-butyl-1,3-propanediol dicarbamate; 2,2-dicarbamoyloxymethyl-3-methylpentane; dicamoylmethtane; W-583; Capla; Butatensin; Carbuten; Dormate; Mebutina; Prean; Sigmafon; Vallene; Mega; No-Press; Axiten; Ipotensivo. C$_{10}$H$_{20}$N$_2$O$_4$; mol wt 232.28. C 51.70%, H 8.68%, N 12.06%, O 27.55%. Prepn: Berger, Ludwig, **U.S. pat. 2,878,280** (1959 to Carter Prod.). Mechanism of action: Kletzkin, *Arch. Int. Pharmacodyn. Ther.* **164**, 71 (1966). Metabolism: Edelson, Douglas, *ibid.* **173**, 182 (1968).

[structure: H$_2$NCOOCH$_2$CCH$_2$OOCNH$_2$ with CH$_3$ above and CH$_3$CHCH$_2$CH$_3$ below]

Crystals, mp 77-79°. Soly in water: ~0.1%. Sol in most organic solvents. On heating or boiling with acid or alkali, hydrolyzes to the alcohol, ammonia and carbon dioxide. *Note:* This is a controlled substance (depressant) listed in the U.S. Code of Federal Regulations, Title 21 Part 1308.14 (1985).

THERAP CAT: Antihypertensive.

5654. Mecamylamine. *N,2,3,3-Tetramethylbicyclo[2.2.1]-heptan-2-amine; N,2,3,3-tetramethyl-2-norbornanamine; N,2,3,3-tetramethyl-2-norcamphanamine;* 3-methylaminoisocamphane; 2-methylaminoisocamphane; 2-methylamino-2,3,3-trimethylnorbornane; N-methyl-2-isocamphanamine; 3β-methylamino-2,2,3-trimethylbicyclo[2.2.1]heptane; mecamine; Inversine; Mekamine; Mevasine; Plegangin; Revertina; Versamine. C$_{11}$H$_{21}$N; mol wt 167.29. C 78.97%, H 12.65%, N 8.37%. Prepn from 3-formamidoisocamphane: Stein *et al., J. Am. Chem. Soc.* **78**, 1514 (1956); Stein *et al., J. Med. Pharm. Chem.* **5**, 665 (1962); Pfister, Stein, **U.S. pat. 2,831,027** (1958 to Merck & Co.). From 3-aminoisocamphane: **Brit. pat. 856,862** (1961 to Lepetit).

[structure: bicyclic norbornane with NHCH$_3$, CH$_3$, CH$_3$, CH$_3$ substituents]

dl-Form, oily liquid. bp$_{4.0}$ 72°. n$_D^{25}$ 1.4881. Slightly sol in water.
Hydrochloride, C$_{11}$H$_{22}$ClN, *Iversine*. Crystals, dec 245.5-246.5°. Bittersweet. Soly in water (g/100 ml): 21.2; in alc: 8.2; in glycerol: 10.4; in isopropanol: 2.1. pH of 1% aq soln 6.0-7.5. Solns can be sterilized by autoclaving.

THERAP CAT: The hydrochloride as antihypertensive.

5655. Mechlorethamine. *2-Chloro-N-(2-chloroethyl)-N-methylethanamine; 2,2'-dichloro-N-methyldiethylamine;* N-methyl-2,2'-dichlorodiethylamine; di(chloroethyl)methylamine; methylbis(β-chloroethyl)amine; methyldi(2-chloroethyl)amine; chlormethine; nitrogen mustard; MBA; HN2; Stickstofflost [Ebewe]. C$_5$H$_{11}$Cl$_2$N; mol wt 156.07. C 38.48%, H 7.10%, Cl 45.44%, N 8.98%. CH$_3$N(CH$_2$CH$_2$Cl)$_2$. A nitrogen mustard prepd by action of thionyl chloride on 2,2'-(methylimino)diethanol in trichloroethylene: Prelog, Stepan, *Coll. Czech. Chem. Commun.* **7**, 93 (1935); Hanby, Rydon, *J. Chem. Soc.* **1947**, 513; Abrams *et al., J. Soc. Chem. Ind. (London)* **68**, 280 (1949); Witten in Kirk-Othmer, *Encyclopedia of Chemical Technology* vol. **7** (Interscience, New York, 1951) p 130. Toxicity data: W. P. Anslow *et al., J. Pharmacol. Exp. Ther.* **91**, 224 (1947).

Mobile liquid. Faint odor of herring. *Vesicant, necrotizing irritant. Never use without appropriate gas mask.* Volatility at 25° = 3.581 mg/l. d$_4^{25}$ 1.118. mp -60°. bp$_{18}$ 87°; bp$_{10}$ 75°; bp$_5$ 64°; bp$_2$ 59°. Very slightly sol in water. Miscible with DMF, CS$_2$, CCl$_4$, many other organic solvents and oils. The undiluted liq dec on standing and forms polymeric quaternary ammonium salts which are insol in the free base.

Hydrochloride, C$_5$H$_{12}$Cl$_3$N, *Caryolysine, Cloramin, Dichloren, Embichen, Embikhine, Erasol, Mustargen hydrochloride, Mustine hydrochloride, Nitrogranulogen.* Hygroscopic leaflets from acetone or chloroform, mp 109-111°. Very sol in water; sol in alcohol. Initial pH of a 2% aq soln: 3.0-4.0. Dry crystals are stable at temps up to 40°. LD$_{50}$ in rats (mg/kg): 1.1 i.v.; 1.9 s.c. (Anslow). *Caution:* Very irritating to skin, mucous membranes; esp. harmful to eyes: Patty's *Industrial Hygiene and Toxicology* vol. **2A**, G. D. Clayton, F. E. Clayton, Eds. (Wiley-Interscience, New York, 3rd ed., 1981) pp 2680-2684.

N-Oxide hydrochloride, *see* separate entry.

Note: This substance may reasonably be anticipated to be a carcinogen: *Fourth Annual Report on Carcinogens* (NTP 85-002, 1985) p 144.

USE: Base as a gas warfare agent.

THERAP CAT: Hydrochloride as an antineoplastic.

THERAP CAT (VET): Hydrochloride has been used as an antineoplastic.

5656. Mechlorethamine Oxide Hydrochloride. *2-Chloro-N-(2-chloroethyl)-N-methylethanamine N-oxide hydrochloride; 2,2'-dichloro-N-methyldiethylamine N-oxide hydrochloride;* N-methyl-2,2'-dichlorodiethylamine N-oxide hydrochloride; methylbis(β-chloroethyl)amine N-oxide hydrochloride; methyldi(2-chloroethyl)amine N-oxide hydrochloride; N-Oxyd-Lost·HCl; Nitromin; Mitomen; Mustron. C$_5$H$_{12}$Cl$_3$NO; mol wt 208.53. C 28.80%, H 5.80%, Cl 51.01%, N 6.72%, O 7.67%. Prepd by treating 2,2'-dichloro-N-methyldiethylamine with hydrogen peroxide and acetic anhydride in ether, followed by shaking with 10% hydrochloric acid: Aiko *et al., J. Pharm. Soc. Japan* **72**, 1297 (1952, English Text), *C.A.* **47**, 1289 (1953). *Cf.* Stahmann, Bergmann, *J. Org. Chem.* **11**, 586 (1946).

[structure: CH$_3$—N(CH$_2$CH$_2$Cl)$_2$ with O above N, ·HCl]

Prisms from acetone, mp 109-110°. Sol in water.
Picrate, mp 122-124°.

THERAP CAT: Antineoplastic.

5657. Meclizine. *1-[(4-Chlorophenyl)phenylmethyl]-4-[(3-methylphenyl)methyl]piperazine; 1-(p-chloro-α-phenylbenzyl)-4-(m-methylbenzyl)piperazine;* 1-(p-chlorobenzhydryl)-4-(m-methylbenzyl)piperazine; 1-(p-chlorobenzhydryl)-4-(m-methylbenzyl)diethylenediamine; meclozine; parachloramine. C$_{25}$H$_{27}$ClN$_2$; mol wt 390.96. C 76.80%, H 6.96%, Cl 9.07%, N 7.17%. Synthesis: Morren, **Belg. pat. 502,889** (1951); **Brit. pat. 705,979** (1954), **U.S. pat. 2,709,-169** (1955 to U.C.B.). Outline of reactions: Grivsky, *Ind. Chim. Belge* **17**, 735 (1952). Clinical evaluation in vertigo: B. Cohen, J. M. B. Vianney de Jong, *Arch. Neurol.* **27**, 129 (1972).

bp$_2$ 230°.

Dihydrochloride monohydrate, $C_{25}H_{29}Cl_3N_2 \cdot H_2O$, *UCB 5062, Ancolan, Antivert, Bonamine, Bonine, Calmonal, Diadril, Histametizine, Navicalm, Neo-Istafene, Peremesin, Postafen Supp., Sabari, Sea-Legs, Veritab.* Freely sol in chloroform, pyridine; slightly sol in dil acids, alc. Practically insol in water (0.1 gm/100 ml), ether.

THERAP CAT: Antiemetic.
THERAP CAT (VET): Antiemetic.

5658. Meclocycline. *7-Chloro-4-(dimethylamino)-1,4,-4a,5,5a,6,11,12a-octahydro-3,5,10,12,12a-pentahydroxy-6-methylene-1,11-dioxo-2-naphthacenecarboxamide;* 7-chloro-6-methylene-5-hydroxytetracycline; GS 2989; NSC-78502. $C_{22}H_{21}ClN_2O_8$; mol wt 476.87. C 55.41%, H 4.44%, Cl 7.43%, N 5.87%, O 26.84%. Semi-synthetic antibiotic derived from tetracycline, q.v. Prepn: R. K. Blackwood *et al.*, U. S. pat. **2,984,686** (1961 to Pfizer); *eidem, J. Am. Chem. Soc.* **83**, 2773 (1961); of the free base and 5-sulfosalicylate salt: *eidem, ibid.* **85**, 3943 (1963). *In vitro* study: L. Lucca, G. Vittadini, *G. Ital. Chemioter.* **26**, 203 (1979). ^{13}C-NMR study: E. Mazzola *et al., J. Pharm. Sci.* **69**, 229 (1980). Toxicity study: F. Bernardi *et al., G. Ital. Chemioter.* **17**, 276 (1970).

uv max (methanol, 0.01N HCl): 245, 347 nm (log ϵ 4.34, 410). LD$_{50}$ in mice (mg/kg): > 5000 orally, 425 i.p. (Bernardi).

5-Sulfosalicylate, $C_{29}H_{27}ClN_2O_{14}S$, *Meclan, Mecloderm, Meclosorb, Meclutin, Traumatociclina.* uv max (methanol, 0.01N HCl): 239, 268, 346 nm (log ϵ 4.46, 4.07, 4.11).

THERAP CAT: Antibacterial.

5659. Meclofenamic Acid. *2-[(2,6-Dichloro-3-methylphenyl)amino]benzoic acid; N-(2,6-dichloro-m-tolyl)anthranilic acid;* meclophenamic acid; CI-583; INF-4668; Arquel. $C_{14}H_{11}Cl_2NO_2$; mol wt 296.15. C 56.78%, H 3.74%, Cl 23.94%, N 4.73%, O 10.81%. Prepn: Scherrer, Short, **Ger. pat. 1,149,015,** *C.A.* **61**, 1801d (1964) and **U.S. pat. 3,313,-848** (1963, 1967 to Parke, Davis); Juby *et al., J. Med Chem.* **11**, 111 (1968). Pharmacology and toxicology: Winder *et al., J. Pharmacol. Exp. Ther.* **148**, 422 (1965); Winder, *Ann. Phys. Med.* **1966**, 7; Kaump, *ibid.* 16. Series of articles on structure-activity relationships, mechanism of action, clinical studies: *Arzneimittel-Forsch.* **33**, 619-680 (1983). Effect on prostaglandin E receptor binding: M. C. P. Rees *et al., Lancet* **2**, 541 (1988).

White crystals from acetone-water, mp 257-259°; also reported as mp 248-250°. Soly (mg/ml): water 0.03; 0.1N NaOH 28. pH of satd aq soln: ~6.9.

Sodium salt monohydrate, $C_{14}H_{10}Cl_2NNaO_2 \cdot H_2O$, *Meclomen, Movens.* mp 289-291°. Soly in water: 15 mg/ml (slightly turbid). pH 8.7.

THERAP CAT: Anti-inflammatory; antipyretic.
THERAP CAT (VET): Anti-inflammatory.

5660. Meclofenoxate. *(4-Chlorophenoxy)acetic acid 2-(dimethylamino)ethyl ester;* dimethylaminoethyl p-chlorophenoxyacetate; centrophenoxine; meclofenoxane; acephen;

ANP 235; Analux; Cerebon; Cetrexin; Proseryl. $C_{12}H_{16}$-$ClNO_3$; mol wt 257.73. C 55.92%, H 6.26%, Cl 13.76%, N 5.44%, O 18.62%. Prepn: Rumpf, Thuillier, **Fr. pat. M398** (1962 to Centre Nat'l. Recherche Sci.), *C.A.* **57**, 16768e (1962); Thuillier *et al., Compt. Rend.* **249**, 2081 (1959); Thuillier, Rumpf, *Bull. Soc. Chim. France* **1960**, 1786. Pharmacology: Petkov, *C.A.* **65**, 20717g (1966); Liberman, *Farmakol. Toksikol.* **30**, 409 (1967). Auxin activity: Conti, *Boll. Chim. Farm.* **107**, 325 (1968).

Hydrochloride, $C_{12}H_{17}Cl_2NO_3$, *Cellative, Clocete, Lucidril, Methoxynal, Proserout, Brenal, Marucotol, Helfergin.* Crystals from isopropyl alcohol or acetone, mp 135-139°. Sol in cold water. Sparingly sol in cold isopropyl alcohol, acetone. Practically insol in benzene, ether, chloroform. A 5% soln in water has pH 6. LD$_{50}$ in mice: 330 mg/kg i.v.; 1750 mg/kg orally; 845 mg/kg i.p., Rumpf, Thuillier, *loc. cit.*

USE: Plant growth regulator.
THERAP CAT: Cerebral stimulant.

5661. Mecloqualone. *3-(2-Chlorophenyl)-2-methyl-4(3H)-quinazolinone;* 2-methyl-3-(2-chlorophenyl)-4-quinazolone; Nubarene. $C_{15}H_{11}ClN_2O$; mol wt 270.74. C 66.55%, H 4.10%, Cl 13.10%, N 10.35%, O 5.90%. Prepd from acetylanthranilic acid and o-chloroaniline in the presence of POCl$_3$: Jackman *et al., J. Pharm. Pharmacol.* **12**, 529 (1960); Closa, *J. Prakt. Chem.* [4] **14**, 84 (1961). Metabolism: Dubnick, Towne, *Toxicol. Appl. Pharmacol.* **22**, 82 (1972); Daenens, Van Bovan, *Arzneimittel-Forsch* **24**, 195 (1974).

Crystals mp 126-128°.
Hydrochloride, $C_{15}H_{12}Cl_2N_2O$, crystals, mp 239-241°.
Note: This is a controlled substance (depressant) listed in the U.S. Code of Federal Regulations, Title 21 Part 1308.11 (1987).

THERAP CAT: Sedative, hypnotic.

5662. Mecloralurea. *N-Methyl-N'-(2,2,2-trichloro-1-hydroxyethyl)urea;* trichloroethylolmethylurea; Heraldium. $C_4H_7Cl_3N_2O_2$; mol wt 221.46. C 21.69%, H 3.19%, Cl 48.02%, N 12.65%, O 14.45%. Prepn from chloral hydrate and N-methylurea: Chattaway, James, *J. Chem. Soc* **1934**, 109; **Fr. pat. M5341; Brit. pat. 1,129,437** corresp to Castaigne, **U.S. pat. 3,510,557** (1967, 1968 and 1970 to Centre d'Etudes l'Ind. Pharm.).

White crystalline powder from boiling water, mp 135-140°. Partly sol in water, acetic acid, ether; sol in alcohol, pyridine. Practically insol in hydrocarbons and chloroform.

THERAP CAT: Anxiolytic.

5663. Mecloxamine. *2-[1-(4-Chlorophenyl)-1-phenylethoxy]-N,N-dimethyl-1-propanamine; 2-[(p-chloro-α-methyl-α-phenylbenzyl)oxy]-N,N-dimethylpropylamine;* β-dimethylaminoisopropyl 4-chloro-α-methylbenzhydryl ether. $C_{19}H_{24}ClNO$; mol wt 317.87. C 71.79%, H 7.61%, Cl 11.16%, N 4.41%, O 5.03%. Prepn: **Brit. pat. 875,060** (1960 to Astra Werke).

bp$_{0.6}$ 154-160°.

Citrate, $C_{25}H_{32}ClNO_8$, component of *Melidorm*. Crystals, mp 120-124°.

THERAP CAT: Anticholinergic; sedative; hypnotic.

5664. Meconic Acid. *3-Hydroxy-4-oxo-4H-pyran-2,6-dicarboxylic acid;* oxychelidonic acid. $C_7H_4O_7$; mol wt 200.10. C 42.01%, H 2.01%, O 55.97%. From opium which contains 4 to 6%. Prepn: Thoms, Pietrulla, *Ber. Pharm. Ges.* **31**, 4 (1921); Wibaut, Kleinpool, *Rec. Trav. Chim.* **66**, 24 (1947).

Monohydrate, prisms from concd aq solns. One gram dissolves in 4 ml hot water, 50 ml methanol, 50 ml ethyl acetate, 100 ml acetone; freely sol in ethanol, benzene.

Trihydrate, orthorhombic pyramidal prisms from dil aq solns. Becomes anhydr when heated to 100-102° for 20 min. Dec with evolution of CO_2 when heated to 120° or when boiled with water. The slight brown tint often characteristic of meconic acid prepns is due to traces of iron.

5665. Meconin. *6,7-Dimethoxy-1(3H)-isobenzofuranone; 6,7-dimethoxyphthalide;* opianyl; meconinic acid lactone. $C_{10}H_{10}O_4$; mol wt 194.18. C 61.85%, H 5.19%, O 32.96%. Isolated from opium by Dublanc in 1832. Occurs also in the root of *Hydrastis canadensis* L., *Ranunculaceae:* Freund, *Ber.* **22**, 456, 459 (1889). Is formed in oxidation of noscapine with HNO_3. Synthesis from *o*-veratric acid: Edwards *et al.*, *J. Chem. Soc.* **1925**, 195; Wilson *et al.*, *J. Org. Chem.* **16**, 792 (1951). From opianic acid: Brown, Newbold, *J. Chem. Soc.* **1952**, 4878.

White, optically inactive needles; sharp bitter taste. mp 102-103°. uv max: 213, 308 nm (ε 25,000, 3800). Sol in 700 parts cold, 22 parts boiling water; sol in alcohol, benzene, chloroform, ether, glacial acetic acid; slowly sol in alkalies with formation of alkali salt of meconinic acid, (CH_2OH)-$(CH_3O)_2C_6H_2COOH$,[2,5,6,1]. The acid itself is unstable, rapidly changing to lactone.

5666. Mecoprop. (±)-2-(*4-Chloro-2-methylphenoxy*)-*propanoic acid;* (±)-2-[(*4-chloro-o-tolyl*)*oxy*]*propionic acid;* mechlorprop; MCPP; CMPP; RD 4593; Astix CMPP; Iso-Cornox; Compitox; Compitox Plus; Proponex-Plus. $C_{10}H_{11}$-ClO_3; mol wt 214.65. C 55.96%, H 5.16%, Cl 16.52%, O 22.36%. Prepn: M. E. Synerholm, P. W. Zimmerman, *Contrib. Boyce Thompson Inst.* **14**, 91 (1945). Studies on plant growth regulation: C. H. Fawcett *et al.*, *Ann. Appl. Biol.* **40**, 231 (1953); and comparison of enantiomers: M. Matell, *Kungl. Lantbruks-Hogsk. Ann.* **20**, 207 (1953); B. Aberg, *ibid.* 241. GLC determn: H. G. Higson, D. Butler, *Analyst* **85**, 657 (1960). Crystal structure: G. Smith *et al.*, *Acta Crystallogr.* **B36**, 992 (1980). Herbicidal activity: G. B. Lush, *Proc. 3rd Brit. Weed Contr. Conf.* 625 (1956); E. L. Leafe, *ibid.* 633; B. Wallgren, *Weeds Weed Contr.* **24th** *Swedish Weed Conf.* 30 (1983); of (+)-enantiomer: J. Toll, *Weeds Weed Contr.* **28th** *Swedish Weed Conf.* 100 (1987). Degradation in soils: L. Lindholm *et al.*, *Acta Agr. Scand.* **32**, 429 (1982); A. E. Smith, *Bull. Environ. Contam. Toxicol.* **34**, 656 (1985). Toxicological studies: M. R. Gurd *et al.*, *Food Cosmet. Toxicol.* **3**, 883 (1965); H. G. Verschuuren *et al.*, *Toxicology* **3**, 349 (1975); R. Roll, G. Matthiaschk, *Arzneimittel-Forsch.* **33**, 1479 (1983). EC-GLC determn in tissues and biological fluids: J. De Beer *et al.*, *Vet. Hum. Toxicol.* **21**, Suppl., 172 (1979). HPLC resolution of enantiomers: B. Blessington *et al.*, *J. Chromatog.* **396**, 177 (1987).

Solid, mp 93-94°. LD$_{50}$ in rats (mg/kg): 1210 orally, 402 i.p. (Verschuuren).

(+)-Form, *Mecoprop-P, Duplosan KV.* Solid, mp 95-96°. $[\alpha]_D^{25}$ +19° (alcohol).

Sodium salt, $C_{10}H_{11}ClNaO_3$. LD$_{50}$ i.p. in rats, mice: 500, 600 mg/kg; orally in mice: 650 mg/kg (Gurd).

Diethylamine salt, $C_{14}H_{22}ClNO_3$, *Mecopar.* LD$_{50}$ in rats, mice (mg/kg): 1060 ±120, 600 ±35 orally; 350, 400 i.p. (Gurd).

Potassium salt, $C_{10}H_{11}ClKO_3$, *Mecomec, Hedonal.*

USE: Herbicide.

5667. Mecrylate. *2-Cyano-2-propenoic acid methyl ester; 2-cyanoacrylic acid methyl ester;* methyl 2-cyanoacrylate; methyl α-cyanoacrylate; AD/here; Coapt. $C_5H_5NO_2$; mol wt 111.10. C 54.05%, H 4.54%, N 12.61%, O 28.80%. CH_2=$C(C≡N)COOCH_3$. Prepn: McKeever, U.S. pat. **2,912,454** (1959 to Rohm & Haas); McKeever, Raterink, U.S. pat. **2,926,188** (1960 to Rohm & Haas).

Liquid, bp$_{1.8}$ 47-49°. n_D^{25} 1.443.

USE: Manuf of polymers and adhesives, see U.S. pats. **2,776,232** and **2,794,788** (1957 to Eastman Kodak). Surgical aid (tissue adhesive).

5668. Mecysteine Hydrochloride. L-*Cysteine methyl ester hydrochloride;* methyl cysteine hydrochloride; methyl β-mercaptoalanine hydrochloride; methyl α-amino-β-mercaptopropionate hydrochloride; LJ 48; Acdrile; Visclair. C_4H_{10}-$ClNO_2S$; mol wt 171.66. C 27.99%, H 5.87%, Cl 20.66%, N 8.16%, O 18.64%, S 18.68%. $HSCH_2CH(NH_2)COOCH_3$·HCl. Prepn: Bergmann, Michalis, *Ber.* **63**, 987 (1930); Zervas, Theodoropoulos, *J. Am. Chem. Soc.* **78**, 1359 (1956). Crystals from methanol, mp 140-141°. $[\alpha]_D^{20}$ −2.9° (methanol).

THERAP CAT: Mucolytic.

5669. Medazepam. *7-Chloro-2,3-dihydro-1-methyl-5-phenyl-1H-1,4-benzodiazepine;* Ansilan; Diepin; Elbrus; Esmail; Medazepol; Mezepan; Megasedan; Narsis; Nobrium; Pazital; Psiquium; Resmit; Rudotel; Serenium; Siman; Tranquilax. $C_{16}H_{15}ClN_2$; mol wt 270.76. C 70.98%, H 5.58%, Cl 13.09%, N 10.34%. Prepn: L. H. Sternbach *et al.*, *J. Org. Chem.* **28**, 2456 (1963); G. A. Archer *et al.*, *Belg. pat.* **620,-773,** *C.A.* **59**, 10095b (1963); E. Reeder, L. H. Sternbach, U.S. pats. **3,109,843** and **3,141,890** (1963, 1963, 1964 all to Hoffmann-La Roche); S. Inaba *et al.*, *Chem. Pharm. Bull.* **20**, 1628 (1972); M. Mihalic *et al.*, *J. Heterocycl. Chem.* **14**, 941 (1977). Pharmacology: L. O. Randall *et al.*, *Arch. Int. Pharmacodyn. Ther.* **185**, 135 (1970). Crystal structure: G. Gilli *et al.*, *Acta Crystallogr.* **B34**, 3793 (1978).

Colorless prismatic crystals from ether + petr ether, mp 95-97°. LD$_{50}$ in mice (mg/kg): 360 i.p., 1070 orally (Randall).

Hydrochloride, $C_{16}H_{16}Cl_2N_2$, orange-red crystalline powder. Freely sol in water, alcohol.

Note: This is a controlled substance (depressant) listed in the U.S. Code of Federal Regulations, Title 21 Part 1308.14 (1987).

THERAP CAT: Anxiolytic.

5670. Medetomidine. *4-[1-(2,3-Dimethylphenyl)ethyl]-1H-imidazole;* (±)-4-(α,2,3-trimethylbenzyl)imidazole; 4-[(α-methyl)-2,3-dimethylbenzyl]imidazole. $C_{13}H_{16}N_2$; mol wt 200.28. C 77.96%, H 8.05%, N 13.98%. α$_2$-Adrenergic agonist. Prepn: A. J. Karjalainen *et al.*, **Brit.** pat.

Appl. 2,101,114; A. J. Karjalainen, K. O. A. Kurkela, **U.S. pat. 4,544,664** (1983, 1985 both to Farmos). Receptor binding study: R. Virtanen *et al., Eur. J. Pharmacol.* **150,** 9 (1988). Sedative and cardiovascular effects in humans: M. Scheinin *et al., Brit. J. Clin. Pharmacol.* **24,** 443 (1987). Veterinary evaluation in cats: D. Stenberg *et al., J. Vet. Pharmacol. Ther.* **10,** 319 (1987).

Hydrochloride, $C_{13}H_{17}ClN_2$, *MPV-785, Domitor.*
THERAP CAT (VET): Sedative, analgesic.

5671. Medibazine. *1-(1,3-Benzodioxol-5-ylmethyl)-4-(diphenylmethyl)piperazine; 1-(diphenylmethyl)-4-piperonylpiperazine;* 1-benzhydryl-4-piperonylpiperazine. $C_{25}H_{26}N_2O_2$; mol wt 386.47. C 77.69%, H 6.78%, N 7.25%, O 8.28%. Prepn: **Belg. pat. 616,371;** Regnier *et al.,* **U.S. pat. 3,119,-826** (1962, 1964, both to Science Union). Pharmacology: Laubie *et al., Arch. Int. Pharmacodyn. Ther.* **151,** 313 (1964); Laubie, Schmitt, *ibid.* **155,** 1 (1965).

Dihydrochloride, $C_{25}H_{28}Cl_2N_2O_2$, *Vialibran.* Solid, mp 288°.
THERAP CAT: Coronary vasodilator; bronchodilator.

5672. Medicagol. *3-Hydroxy-6H-[1,3]dioxolo[5,6]benzofuro[3,2-c][1]benzopyran-6-one;* 7-hydroxy-11,12-(methylenedioxy)coumestan; 7-hydroxy-5′,6′-methylenedioxybenzofurano(3′,2′:3,4)coumarin. $C_{16}H_8O_6$; mol wt 296.22. C 64.87%, H 2.72%, O 32.41%. Occurs in alfalfa having viral leafspot infections. Synthesis: Livingston *et al., J. Org. Chem.* **30,** 2353 (1965); Jurd, *J. Pharm. Sci.* **54,** 1221 (1965); Fukui *et al., Experientia* **24,** 536 (1968).

mp 326-327°. uv max (ethanol): 245, 270, 297, 310, 348 nm (log ε 4.29, 3.91, 3.85, 4.03, 4.46).

5673. Medicarpin. *(6aR-cis)-6a,11a-Dihydro-9-methoxy-6H-benzofuro[3,2-c][1]benzopyran-3-ol;* (−)-3-hydroxy-9-methoxypterocarpan; demethylhomopterocarpin. $C_{16}H_{14}O_4$; mol wt 270.28. C 71.10%, H 5.22%, O 23.68%. Antifungal phytoalexin produced by leguminous species. Isoln from the heartwood of *Swartzia madagascariensis* Desv., *Caesalpinioidae:* S. H. Harper *et al., Chem. & Ind. (London)* **1965,** 562; from alfalfa, *Medicago sativa* L., *Leguminosae:* D. G. Smith *et al., Physiol. Plant Pathol.* **1,** 41, (1971); from red clover, *Trifolium pratense* L., *Leguminosae:* V. J. Higgins, D. G. Smith, *Phytopathology* **62,** 235 (1972). ^{13}C-NMR: A. A. Chalmers *et al., Tetrahedron* **33,** 1735 (1977). HPLC: J. Koster *et al., J. Chromatog.* **270,** 392 (1983). Synthesis of racemic form: W. Cocker *et al., J. Chem. Soc. (C)* **1965,** 1034. Biosynthetic studies: P. M. Dewick, *Chem. Commun.* **1975,** 656; S. W. Banks *et al., ibid.* **1982,** 157; H. A. M. Al-Ani, P. M. Dewick, *J. Chem. Soc. Perkin Trans. I* **1984,** 2831. Antifungal properties: L. J. Duczek, V. J. Higgins, *Can. J. Bot.* **54,** 2620 (1976); H. D. Van Etten, *Phytochemistry* **15,** 655 (1976); A. O. Latunde-Dada, J. A. Lucas, *Physiol. Plant Pathol.* **26,** 31 (1985).

Prisms from benzene, mp 127.5-128.5°. $[\alpha]_D^{22}$ −226° (chloroform). uv max: 207, 282, 287, 310 nm (log ε 4.86, 3.97, 4.01, 3.38).
(±)-Form, prisms from ethyl acetate/light petroleum, mp 194-195°.

5674. Medifoxamine. *N,N-Dimethyl-2,2-diphenoxyethanamine; (dimethylamino)acetaldehyde diphenyl acetal; N,N′-dimethyl-2,2-diphenoxyethylamine.* $C_{16}H_{19}NO_2$; mol wt 257.33. C 74.68%, H 7.44%, N 5.44%, O 12.43%. Prepn: **Fr. pat. M5498** (1967 to Gerda), *C.A.* **72,** 12358x (1970); M. A. Brunet *et al., Bull. Soc. Chim. France* **6,** 2000 (1967). Pharmacology and toxicity: A. Vagne *et al., Therapie* **26,** 553 (1971). Clinical pharmacology: M. A. Randhawa *et al., Human Psychopharmacol.* **3,** 195 (1988). Comparative clinical trial with clomipramine, *q.v.,* in depression: H. Scharbach *et al., Psychol. Med.* **18,** 1485 (1986).

Fumarate, $C_{20}H_{23}NO_6$, *LG 152, Clédial, Gerdaxyl.* Crystals from 95% ethanol, mp 128.5°. LD_{50} orally in rats: 750 mg/kg (Vagne).
THERAP CAT: Antidepressant.

5675. Medmain. *3-Ethyl-N,N,2-trimethyl-1H-indol-5-amine; 5-dimethylamino-3-ethyl-2-methylindole;* 3-ethyl-2-methyl-5-dimethylaminoindole; 2-methyl-3-ethyl-5-dimethylaminoindole; antiserotonin. $C_{13}H_{18}N_2$; mol wt 202.29. C 77.18%, H 8.97%, N 13.85%. Prepn: Shaw, *J. Am. Chem. Soc.* **76,** 1384 (1954).

Crystals, mp 100-102°.
THERAP CAT: Serotonin inhibitor.

5676. Medrogestone. *6,17-Dimethylpregna-4,6-diene-3,20-dione;* 6,17α-dimethyl-6-dehydroprogesterone; AY 62022; Colpro; Colprone; Prothil. $C_{23}H_{32}O_2$; mol wt 340.51. C 81.13%, H 9.47%, O 9.40%. Prepn: Deghenghi, Gaudry, *J. Am. Chem. Soc.* **83,** 4668 (1961); Deghenghi *et al., Tetrahedron* **19,** 289 (1963); *J. Med. Chem.* **6,** 301 (1963); Deghenghi, **U.S. pats. 3,133,913, 3,210,387;** Morand, Deghenghi, **U.S. pat. 3,170,936** (1964, 1965, and 1965, all to Am. Home Prod.). HPLC determn in serum: W. T. Robinson, L. Cosyns, *Arzneimittel-Forsch.* **29,** 882 (1979). Comparative clinical evaluation in benign prostatic hyperplasia: D. F. Paulson, R. D. Kane, *J. Urol.* **113,** 811 (1975); in endometrial carcinoma: K. C. Podratz *et al., Obstet. Gynecol.* **66,** 106 (1985).

Crystals from ether, mp 144-146°. $[\alpha]_D^{23}$ +79° (c = 1 in chloroform). uv max: 288 nm (ϵ 25000).
THERAP CAT: Progestogen.

5677. Medroxyprogesterone. *(6α)-17-Hydroxy-6-meth-ylpregn-4-ene-3,20-dione;* 17α-hydroxy-6α-methylproges-terone; 6α-methyl-17α-hydroxyprogesterone; 6α-methyl-4-pregnen-17α-ol-3,20-dione. $C_{22}H_{32}O_3$; mol wt 344.48. C 76.70%, H 9.36%, O 13.93%. Orally active progestogen; formerly used in combinations as oral contraceptive. Prepn: Babcock *et al., J. Am. Chem. Soc.* **80,** 2904 (1958); Mira-montes *et al.,* **U.S.** pat. **3,000,914** (1961 to Searle); **Brit.** pat. **868,303** (1961 to Syntex); Ruggieri, Ferrari, **U.S.** pat. **3,043,832** (1962 to Ormonoterapia Richter); Camerino *et al.,* **U.S.** pat. **3,061,616** (1962 to Farmitalia). Prepd as an inter-mediate: Patchett, Hoffman; Beyler, **U.S.** pats. **3,084,174** and **3,105,840** (both 1963 to Merck & Co.); Spero, **U.S.** pat. **3,147,290** (1964 to Upjohn). HPLC determn in plasma: J. Read, G. Mould, *J. Chromatog.* **341,** 437 (1985). Compara-tive clinical trial with norethisterone as injectable contracep-tive: H. K. Toppozada *et al., Contraception* **28,** 1 (1983). Clinical trial in advanced breast cancer: M. Izuo *et al., Cancer* **56,** 2576 (1985). Review of pharmacology and clini-cal uses: I. S. Fraser, E. Weisberg, *Med. J. Aust.* **1,** Suppl. 1, 3-19 (1981).

Crystals from chloroform, mp 220-223.5°. $[\alpha]_D^{25}$ +75° (in chloroform). uv max (ethanol): 241 nm (ϵ 16000).
17-Acetate, $C_{24}H_{34}O_4$, *17α-acetoxy-6α-methylprogesterone, 6α-methyl-17α-acetoxyprogesterone, MAP, Agestal, Amen, Clinovir, Curretab, Cycrin, Depo-Clinovir, Depo-Provera, Deporone, Farlutal, Gestapuran, Gestapuron, G-Farlutal, Hysron H, Lutoral, Nidaxin, Oragest, Perlutex, Prodasone, Provera, Sodelut "G", Veramix.* Crystals from methanol, mp 207-209°. $[\alpha]_D$ +61° (in chloroform). uv max (ethanol): 240 nm (ϵ 15900). Mixture with with ethinyl estradiol, *q.v., Provest.*
THERAP CAT: The 17-acetate as progestogen. Injectable contraceptive. Palliative treatment of breast and endometri-al carcinoma.
THERAP CAT (VET): Progestogen. Estrus regulator.

5678. Medrylamine. *2-[(4-methoxyphenyl)phenylmethox-y]-N,N-dimethylethanamine;* 2-(p-methoxy-α-phenylbenzyl-oxy)-N,N-dimethylethylamine; 4-methoxy-Benadryl; *p*-methoxybenzhydryl *β*-dimethylaminoethylether; *β*-(p-meth-oxybenzhydryloxy)ethyldimethylamine; phenyl-*p*-methoxy-phenylcarbinyl dimethylaminoethyl ether; Histaphen(e); Postafen Salve. $C_{18}H_{23}NO_2$; mol wt 285.37. C 75.75%, H 8.12%, N 4.91%, O 11.21%. Prepn: Morren, **U.S.** pat. **2,-668,856** (1954 to U.C.B.).

Hydrochloride, $C_{18}H_{24}ClNO_2$, crystals, mp 141°.
THERAP CAT: Antihistaminic.

5679. Medrysone. *11-Hydroxy-6-methylpregn-4-ene-3,20-dione;* 11β-hydroxy-6α-methylprogesterone; 6α-meth-yl-11β-hydroxyprogesterone; hydroxymesterone; U-8471; HMS; Medrocort; Ophtocortin; Spectamedryn. $C_{22}H_{32}O_3$; mol wt 344.48. C 76.70%, H 9.36%, O 13.93%. Prepn and use as an intermediate: Sebek *et al.,* Spero, Thompson, **U.S.** pats. **2,864,837** and **2,968,655** (1958 and 1961, both to

Upjohn). In treatment of ocular inflammation: Bedrossian, *Arch. Ophthalmol.* **81,** 184 (1969).

Crystals, mp 155-158°. $[\alpha]_D$ +189° (in chloroform).
THERAP CAT: Glucocorticoid.

5680. Mefenamic Acid. *2-[(2,3-Dimethylphenyl)amino]-benzoic acid; N-(2,3-xylyl)anthranilic acid;* CI-473; INF 3355; Bafhameritin-M; Bonabol; Coslan; Lysalgo; Nam-phen; Parkemed; Ponalar; Ponstan; Ponstel; Ponstil; Pon-styl; Pontal; Tanston; Vialidon. $C_{15}H_{15}NO_2$; mol wt 241.28. C 74.66%, H 6.27%, N 5.81%, O 13.26%. Prepn: **Belg.** pat. **605,302** (1961 to Parke, Davis). Pharmacology: C. V. Winder *et al., J. Pharmacol. Exp. Ther.* **138,** 405 (1962); Mokhort, Korkhova, *Farmakol. Toksikol. (Kiev)* **1968,** 85, *C.A.* **71,** 29080c (1969); C. V. Winder *et al., Arthritis Rheum.* **12,** 472 (1969). Crystal structure: J. F. McConnell, F. Z. Company, *Cryst. Struct. Commun.* **5,** 861 (1976).

Crystals, mp 230-231° (effervescence). pKa 4.2. uv max (0.1N NaOH): 285, 340 nm. Soly in H_2O, pH 7.1 (g/100 ml): 0.0041 (25°); 0.008 (37°). Soluble in solns of alkali hydroxides; sparingly sol in ether, chloroform; slightly sol in ethanol. LD_{50} orally in mice, rats: 630, 790 mg/kg, U. Jahn, R. W. Adrian, *Arzneimittel-Forsch.* **19,** 36 (1969).
Sodium salt, $C_{15}H_{14}NNaO_2$, white powder. Soly in H_2O: >5 g/100 ml. LD_{50} in mice: 600 mg/kg orally; 150 mg/kg i.p., Mokhort, Korkhova, *loc. cit.*
THERAP CAT: Anti-inflammatory.

5681. Mefenorex. *N-(3-Chloropropyl)-α-methylphen-ethylamine;* 1-phenyl-2-(3-chloropropylamino)propane. $C_{12}H_{18}ClN$; mol wt 211.74. C 68.07%, H 8.57%, Cl 16.75%, N 6.62%. $C_6H_5CH_2CH(CH_3)NH(CH_2)_3Cl$. Prepn of hydro-chloride salt: Beschke *et al.,* **Ger.** pat. **1,210,873** (1966 to Hoffmann-La Roche), *C.A.* **64,** 19486c (1966).
Hydrochloride, $C_{12}H_{19}Cl_2N$, *Doracil, Pondinil, Pondinol, Rondimen.* Solid, mp 128-130°.
THERAP CAT: Anorexic.

5682. Mefexamide. *N-[2-(Diethylamino)ethyl]-2-(4-methoxyphenoxy)acetamide;* (p-methoxyphenoxy)acetic acid *N-[2-(diethylamino)ethyl]amide;* mexephenamide; NP 297; Mefexadyne; Timodyne. $C_{15}H_{24}N_2O_3$; mol wt 280.36. C 64.26%, H 8.63%, N 9.99%, O 17.12%. Prepn: Thuillier, Rumpf, *Bull. Soc. Chim. France* **1960,** 1786. Pharmacology: Thuillier, *Arzneimittel-Forsch.* **14,** 556 (1964). Clinical study in psychiatric patients: V. Faust *et al., Schweitz. Rundsch. Med. Prax.* **61,** 1177 (1972).

Hydrochloride, $C_{15}H_{25}ClN_2O_3$, mp 112°.
THERAP CAT: CNS stimulant.

5683. Mefloquine Hydrochloride. *(R*,S*)-(±)-α-2-Piperidinyl-2,8-bis(trifluoromethyl)-4-quinolinemethanol monohydrochloride;* DL-*erythro-α-2-piperidyl-2,8-bis(tri-*

fluoromethyl)-4-quinolinemethanol monohydrochloride; WR 142490; Ro 21-5998; Lariam. $C_{17}H_{17}ClF_6N_2O$; mol wt 414.77. C 49.23%, H 4.13%, Cl 8.55%, F 27.48%, N 6.75%, O 3.86%. Quinoline derivative effective as single-dose treatment of drug-resistant malaria. Prepn: R. E. Lutz *et al., J. Med. Chem.* **14**, 926 (1971); G. Grethe, T. Mitt, **Ger.** pat. **2,806,909** (1979 to Hoffmann-La Roche), *C.A.* **90**, 22838q (1979); E. Hickmann *et al.,* **Ger.** pat. **2,940,443;** *eidem,* **U.S.** pat. **4,327,215** (1981, 1982 both to BASF). Stereochemistry, synthesis and antimalarial activity of all stereoisomers: F. I. Carroll, J. T. Blackwell, *J. Med. Chem.* **17**, 210 (1974). Does not intercalate with DNA: M. W. Davidson *et al., J. Med. Chem.* **20**, 1117 (1977). Antibacterial activity and mode of action: R. E. Brown *et al., Life Sci.* **25**, 1857 (1979). Photochemistry: G. A. Epling, U. C. Yoon, *Chem. Letters* **1982** (2), 211. Single dose kinetics in man: D. E. Schwartz *et al., Chemotherapy* **28**, 70 (1982). HPLC determn in blood and plasma: I. M. Kapetanovic *et al., J. Chromatog.* **277**, 209 (1983). Clinical studies: G. M. Trenholme *et al., Science* **190**, 792 (1975); F. Tin *et al., Bull. WHO* **60**, 913 (1982); J. M. Kofe Ekue *et al., ibid.* **61**, 713 (1983); J.-M. de Souza *ibid.* 809, 815. Comprehensive description: P. Lim in *Analytical Profiles of Drug Substances* **vol. 14**, K. Florey, Ed. (Academic Press, New York, 1985) pp 157-180.

mp 259-260° (dec).

Note: Fansimef is a combination of mefloquine with *Fansidar* (pyrimethamine and sulfadoxine, *q.q.v.*).

THERAP CAT: Antimalarial.

5684. Mefluidide. *N-[2,4-Dimethyl-5-[[(trifluoromethyl)sulfonyl]amino]phenyl]acetamide;* 5-acetamido-2,4-dimethyltrifluoromethanesulfonanilide; methafluoridamid; MBR 12325; VEL 3973; Vistar; Embark. $C_{11}H_{13}F_3N_2O_3S$; mol wt 310.29. C 42.58%, H 4.22%, F 18.37%, N 9.03%, O 15.47%, S 10.33%. Prepn: T. L. Fridlinger, **Ger.** pat. **2,406,-475** corresp to **U.S.** pat. **3,894,073** (1974, 1975 to 3M). Metabolism: G. W. Ivie, *J. Agr. Food Chem.* **28**, 1286 (1980).

Cryst solid, mp 183-185°. Vapor press at 25°: $<10^{-4}$ mm Hg. pKa 4.6. Soly at 25° (g/l): water 0.18; benzene 0.31; dichloromethane 2.1; 1-octanol 17; methanol 310; acetone 350. Aq solns decomp in uv light. Mildly corrosive to metals.

USE: Plant growth regulator; herbicide.

5685. Mefruside. *4-Chloro-N¹-methyl-N¹-[(tetrahydro-2-methyl-2-furanyl)methyl]-1,3-benzenedisulfonamide; 4-chloro-N¹-methyl-N¹-(tetrahydro-2-methylfurfuryl)-m-benzenedisulfonamide;* 2-[(4-chloro-N¹-methyl-3-sulfamoylbenzenesulfonamido)methyl]-2-methyltetrahydrofuran; N-(4-chloro-3-sulfamoylbenzenesulfonyl)-N-methyl-2-furfurylamine; B-1500; Baycaron. $C_{13}H_{19}ClN_2O_5S_2$; mol wt 382.90. C 40.78%, H 5.00%, Cl 9.26%, N 7.32%, O 20.89%, S 16.75%. Prepn: H. Horstmann *et al.,* **Brit.** pat. **1,031,916** corresp to **U.S.** pat. **3,356,692** (1966, 1967 to Bayer AG); of *dl-, d-,* and *l*-forms, *eidem, Arzneimittel-Forsch.* **17**, 653 (1967).

dl-Form, crystals, mp 149-150°.
d-Form, crystals, mp 146°. $[\alpha]_{578}^{20}$ +5.4° (c = 2.026 in methanol).
l-Form, crystals, mp 146°. $[\alpha]_{578}^{20}$ −5.5° (c = 2.100 in methanol). More active as a diuretic than the *d*-form, but the difference is of no practical importance: Horstmann *et al., loc. cit.* (1967).

THERAP CAT: Diuretic.

5686. Megacins. Antibiotic proteins produced by strains of *Bacillus megaterium,* which are highly specific in their antibacterial activity against strains of homologous species. Isoln of megacin A from *B. megaterium:* Ivanovics, Alföldi, *Nature* **174**, 465 (1954); Holland, *Biochem. J.* **78**, 641 (1961). Destroys the cytoplasmic membrane of sensitive bacteria: Nagy, *Acta Microbiol. Acad. Sci. Hung.* **6**, 337 (1959). Isoln of megacin C from *B. megaterium:* Holland, Roberts, *J. Gen. Microbiol.* **35**, 271 (1964). Has a mode of action quite different from megacin A and is more reminiscent of colicins, *q.v. Review:* Holland in *Antibiotics* **vol. 1,** D. Gottlieb, P. Shaw, Eds. (Springer-Verlag, New York, 1967) pp 688-695.

5687. Megestrol Acetate. *17-Hydroxy-6-methylpregna-4,6-diene-3,20-dione acetate;* 17α-acetoxy-6-methylpregna-4,6-diene-3,20-dione; 6-dehydro-6-methyl-17α-acetoxy-progesterone; 6-methyl-Δ⁴,⁶-pregnadien-17α-ol-3,20-dione acetate; Maygace; Megace; Megestat; Nia; Niagestin; Ovaban. $C_{24}H_{32}O_4$; mol wt 384.50. C 74.97%, H 8.39%, O 16.65%. Orally active progestogen; formerly used in combinations as oral contraceptive. Prepn: Ringold *et al., J. Am. Chem. Soc.* **81**, 3712 (1959); Dodson, Sollman, **U.S.** pat. **2,891,079** (1959 to Searle); Kirk *et al.,* **Brit.** pat. **870,286,** *C.A.* **56**, 10248i (1962), and **U.S.** pat. **3,356,573** (1967 to Brit. Drug Houses); Cross, **U.S.** pat. **3,400,137** (1968 to Syntex). Biological effectiveness studies: Chang, Kincl, *Steroids* **12**, 689 (1968). Metabolic studies: Cooper, Kellie, *ibid.* **11**, 133 (1968). Soly and diffusion studies: Sundaram, Kincl, *ibid.* **12**, 517 (1968). Comparative clinical trial with tamoxifen in metastatic breast cancer: J. C. Allergra *et al., Semin. Oncol.* **12**, Suppl. 6, 61 (1985). Review of carcinogenicity studies: *IARC Monographs* **21**, 431-439 (1979).

Crystals from methanol, mp 214-216°. Soly at 37°: in water (2 μg/ml); in plasma (24 μg/ml). $[\alpha]_D^{24}$ +5° (chloroform). uv max (ethanol): 287 nm (log ε 4.40). Mixture with ethinyl estradiol, *q.v., Co-Ervonum, Kombiquens, Noval, Nuvacon, Ovex, Planovin, Tri-Ervonum, Volidan, Weradys.* Mixture with mestranol, *q.v., Delpregnin.*

THERAP CAT: Progestogen. Palliative treatment of breast and endometrial carcinoma.

THERAP CAT (VET): Progestogen. Estrus regulator.

5688. Meglumine Acetrizoate. *1-Deoxy-1-(methylamino)-D-glucitol 3-(acetylamino)-2,4,6-triiodobenzoate (salt);* 3-acetamido-2,4,6-triiodobenzoic acid methylglucamine salt; 3-acetylamino-2,4,6-triiodobenzoate of methylglucamine; methylglucamine 3-acetylamino-2,4,6-triiodobenzoate; methylglucamine acetrizoate; Fortombrine M; Fortoshade M; Jodozoat Meglumin; Vasurix. $C_{16}H_{23}I_3N_2O_8$; mol wt 752.11. C 25.55%, H 3.08%, I 50.62%, N 3.73%, O 17.02%. $C_9H_6I_3NO_3 \cdot C_7H_{17}NO_5$.

Minute crystals. Freely sol in water. Aq solns have a pH of about 7.0. The iodine content of the dry powder is reported as 50.67%.

THERAP CAT: Diagnostic aid (radiopaque medium).

5689. Meglumine Diatrizoate. *3,5-Diacetamido-2,4,6-triiodobenzoic acid methylglucamine salt;* methylglucamine 3,5-diacetamido-2,4,6-triiodobenzoate; urografic acid methylglucamine salt; diatrizoate meglumine; diatrizoate methylglucamine; methylglucamine diatrizoate; meglumine amidotrizoate; Angiografin; Cardiografin; Cystografin; Hypaque Cysto; Hypaque Meglumine; Renografin; Reno M; Urovist. $C_{18}H_{26}I_3N_3O_9$; mol wt 809.13. C 26.72%, H 3.24%, I 47.05%, N 5.19%, O 17.80%. $C_{11}H_9I_3N_2O_4 \cdot C_7H_{17}NO_5$. Distribution and metabolism: H. Langecker *et al., Arch. Exp. Pathol. Pharmakol.* **222**, 584 (1954). Clinical evaluation of diagnostic use in contrast enhanced computer tomography: J. S. Morrow, *Curr. Ther. Res.* **27**, 229 (1980).

Rhombic needles, slightly sweet taste. mp 189-193° (dec). Soly in water at 20°: 89 g/100 ml.

Note: Gastrografin, MD-76, Renovist, Urografin, mixtures of sodium diatrizoate and meglumine diatrizoate.

THERAP CAT: Diagnostic aid (radiopaque medium).
THERAP CAT (VET): X-ray contrast medium.

5690. Meglutol. *3-Hydroxy-3-methylpentanedioic acid; 3-hydroxy-3-methylglutaric acid;* dicrotalic acid; medroglutaric acid; HMG; HMGA; CB-337; Lipoglutaren; Mevalon. $C_6H_{10}O_5$; mol wt 162.14. C 44.44%, H 6.22%, O 49.34%. Hypolipidemic agent that decreases the rate of cholesterol synthesis. Prepn via Reformatsky reaction between ethyl acetoacetate and ethyl bromoacetate: R. Adams, B. L. Van Duuren, *J. Am. Chem. Soc.* **75**, 2377 (1953). Synthesis via oxidation of diallylmethylcarbinol with ozone and hydrogen peroxide: H. J. Klosterman, F. Smith, *ibid.* **76**, 1229 (1954); J. L. Rabinowitz, *Biochem. Prepns.* **6**, 25 (1958). [*Caution:* attempts to increase batch size of the ozonolysis resulted in an explosion, *cf. Chem. & Eng. News* **51**(6), 29 (1973).] Improved synthesis: A. Yavrouian *et al., Synthesis* **1981**, 791. Hypocholesteremic properties: Z. H. Beg, M. Siddiqi, *Experientia* **23**, 380 (1967); M. Siddiqi, Z. H. Beg, U.S. pat. **3,629,449** (1971); Z. H. Beg, P. J. Lupien, *Biochim. Biophys. Acta* **260**, 439 (1972); A. N. K. Yusufi, M. Siddiqi, *Atherosclerosis* **20**, 517 (1974); C. D. Padova *et al., Life Sci.* **30**, 1907 (1982). Biosynthetic studies: L. W. White, H. Rudney, *Biochemistry* **9**, 2713 (1970); L. Hagenfeldt, K. Hellstrom, *Life Sci.* **9**, Pt. 2, 991 (1970). Organ distribution: L. L. Savoie, P. J. Lupien, *Can. J. Physiol. Pharmacol.* **53**, 638 (1975). Effectiveness in familial hypercholesteremia: P. J. Lupien *et al., J. Clin. Pharmacol.* **19**, 120 (1979). GC study in patients with organic acidurias: K. Tanaka, D. J. Hine, *J. Chromatog.* **239**, 301 (1982). Toxicological study: L. L. Savoie, P. J. Lupien, *Arzneimittel-Forsch.* **25**, 1284 (1975).

$$HOOCCH_2\overset{\overset{\displaystyle CH_3}{|}}{\underset{\underset{\displaystyle OH}{|}}{C}}CH_2COOH$$

Cryst from ether/petr ether, mp 108-109°. Sol in water. Stable when stored dry. LD$_{50}$ in mice (g/kg): 7.33 orally; 3.23 i.p. (Savoie, Lupien).

THERAP CAT: Antihyperlipoproteinemic.

5691. Melamine. *1,3,5-Triazine-2,4,6-triamine;* 2,4,6-triamino-s-triazine; cyanurotriamide. $C_3H_6N_6$; mol wt 126.13. C 28.57%, H 4.80%, N 66.64%. Usually prepd by heating dicyandiamide, $H_2NC(=NH)NHC\equiv N$, under pressure: Mackay, U.S. pat. **2,737,513** (1956 to Am. Cyanamid). Alternate methods starting with urea: Mackay, U.S. pat. **2,760,961** (1956 to Am. Cyanamid); Pomot *et al.,* U.S. pat. **3,111,519** (1963 to Office Natl. Ind. de l'Azote). X-ray and neutron crystal structure: J. N. Varghese *et al., Acta Crystallogr.* **33B**, 2102 (1977). Review of mfg processes: Faith, Keyes & Clark's *Industrial Chemicals,* F. A. Lowenheim, M. K. Moran, Eds. (Wiley-Interscience, New York, 4th ed., 1975) pp 519-523.

Monoclinic prisms. mp <250°. Sublimes. d^{250} 1.573. Slightly sol in water; very slightly sol in hot alc; insol in ether.

USE: Forms synthetic resins with formaldehyde.

5692. Melanins. Pigments responsible for the dark color of skin, hair, feathers, fur, insect cuticle, soil; found also in fungi, bacteria, and pathological human urine where it is an indication of melanotic tumors. Structures are highly irregular polymers produced in the form of granules which may be bound to protein material. *Allomelanins* are found in the plant kingdom and are produced from nitrogen-free precursors. *See:* aspergillin, humic acid. *Eumelanins* and *phaeomelanins* are found in the animal kingdom. Eumelanins are black or brown, insoluble, nitrogenous pigments produced by the oxidative polymerization of 5,6-dihydroxyindoles derived enzymatically from tyrosine via dopa. One of the best characterized is sepiomelanin, *q.v.* Phaeomelanins are sulfur-containing, alkali-soluble, yellow to reddish-brown pigments produced by oxidative polymerization of cysteinyl-dopas via 1,4-benzothiazine intermediates. Biosynthesis from tyrosine: H. S. Raper, *Physiol. Rev.* **8**, 245 (1928); H. S. Mason, *J. Biol. Chem.* **172**, 83 (1948). *In vitro* prepn from dopa: L. E. Arnow, *Science* **87**, 308 (1938). Series of articles on structure and biosynthesis: M. Piattelli, R. A. Nicolaus, *Tetrahedron* **15**, 66 (1961); M. Piattelli *et al., ibid.* **18**, 941 (1962); *eidem, ibid.* **19**, 2061 (1963); R. A. Nicolaus *et al., ibid.* **20**, 1163 (1964). NMR-study: G. A. Duff *et al., Biochemistry* **27**, 7112 (1988). Determn of eumelanin metabolites in human urine: S. Pavel, W. van der Slik, *J. Chromatog.* **375**, 392 (1986). Review of biosynthesis: J. M. Pawelek, A. M. Körner, *Am. Scientist* **70**, 136-145 (1982); of photochemistry and photobiology: M. R. Chedekel, *Photochem. Photobiol.* **35**, 881-885 (1982); M. R. Chedekel, L. Zeise, *Lipids* **23**, 587-591 (1988). *Reviews:* R. A. Nicolaus, *Melanins* (Hermann, Paris, 1968); G. Prota, *Med. Res. Rev.* **8**, 525-556 (1988).

5693. Melanostatin. *Melanocyte-stimulating hormone-release inhibiting factor;* MIF; MRIH; MSH-release inhibiting hormone; melanotropic inhibiting factor. An inhibiting factor which mediates hypothalamic control of MSH, *q.v.,* a pituitary hormone: A. J. Kastin, A. V. Schally, *Gen. Comp. Endocrinol.* **7**, 452 (1966); A. V. Schally, A. J. Kastin, *Endocrinology* **79**, 768 (1966). Although several substances have MIF activity, the exact nature of MSH release inhibition is not known, *cf.* A. J. Kastin *et al., Yale J. Biol. Med.* **46**, 617 (1973); *eidem, Fed. Proc.* **39**, 2931 (1980). The first factor with melanocyte-release inhibiting activity to be isolated was the side-chain tripeptide of oxytocin, *q.v.,* now referred to as MIF-I: M. E. Celis *et al., Proc. Nat. Acad. Sci. USA* **68**, 1428 (1971). Isoln of bovine MIF-I and identity with the tripeptide (Pro-Leu-GlyNH$_2$): R. M. G. Nair *et al., Biochem. Biophys. Res. Commun.* **43**, 1376 (1971). Synthesis: H. Irie *et al., Chem. Letters* **1980**, 705. Extrapituitary effects of MIF-I are known, particularly potentiation of DOPA-induced behavioral changes: N. P. Plotnikoff *et al., Life Sci.* **10**, 1279 (1971); *eidem, Neuroendocrinology* **14**, 271 (1974). Clinical studies in Parkinson's disease: A. J. Kastin, A. Barbeau, *Can. Med. Assoc. J.* **107**, 1079 (1972); A. Barbeau, *Lancet* **2**, 683 (1975); V. E. Schneider *et al., Arzneimittel-Forsch.* **28**, 1296 (1978). *Reviews:* A. V. Schally *et al., Recent Progr. Horm. Res.* **24**, 497-581 (1968); A. J. Kastin *et al., Life Sci.* **25**, 401-414 (1979); *eidem* in *Polypeptide Hormones,* R. F. Beers, E. G. Bassett, Eds. (Raven Press, New York, 1980) pp 223-224; *eidem, Fed. Proc.* **39**, 2931-2936 (1980).

MIF-I, $C_{13}H_{24}N_4O_3$, L-*prolyl*-L-*leucylglycinamide, melanostatin I (ox), MSH-release inhibiting factor.*

5694. Melarsoprol. *2-[4-[(4,6-Diamino-1,3,5-triazin-2-yl)amino]phenyl]-1,3,2-dithiarsolane-4-methanol;* p-[(4,6-diamino-s-triazin-2-yl)amino]dithiobenzenearsonous acid 3-hydroxypropylene ester; 2-p-(4,6-diamino-s-triazin-2-ylamino)phenyl-4-hydroxymethyl-1,3,2-dithiarsoline; 2-(4-melamin-2-ylphenyl)-4-hydroxymethyl-1,3-dithia-2-arsolane; Mel B; Arsobal. $C_{12}H_{15}AsN_6OS_2$; mol wt 398.34. C 36.18%, H 3.80%, As 18.81%, N 21.10%, O 4.02%, S 16.10%. Prepn: Friedheim, **U.S.** pats. **2,659,723** (1953) and **2,772,-303** (1956).

Practically insol in water, cold ethanol, methanol. Sol in propylene glycol.

THERAP CAT: Antiprotozoal (Trypanosoma).

5695. Melatonin. *N-[2-(5-Methoxy-1H-indol-3-yl)eth-yl]acetamide; N-acetyl-5-methoxytryptamine.* $C_{13}H_{16}N_2O_2$; mol wt 232.27. C 67.22%, H 6.94%, N 12.06%, O 13.78%. A hormone of the pineal gland, also produced by extra-pineal tissues, that lightens skin color in amphibians by reversing the darkening effect of MSH, *q.v.* Melatonin has been postulated as the mediator of photic-induced antigona-dotrophic activity in photoperiodic mammals and has also been shown to be involved in thermoregulation in some ectotherms and in affecting locomotor activity rhythms in sparrows. Isoln from the pineal glands of beef cattle: Lerner *et al., J. Am. Chem. Soc.* **80,** 2587 (1958); Wurtman *et al., Science* **141,** 277 (1963). Structure: Lerner *et al., J. Am. Chem. Soc.* **81,** 6084 (1959). Crystal and molecular structure: A. Wakahara, *Chem. Lett.* **1972,** 1139. Synthesis from 5-methoxyindole as starting material by two different routes: Szmuszkovicz *et al., J. Org. Chem.* **25,** 857 (1960). Biochemical role of melatonin: *Chem. & Eng. News* **45,** 40 (May 1, 1967). Pharmacological studies: Barchas *et al., Nature* **214,** 919 (1967). Identification of antigonadal action sites in mouse brain: J. D. Glass, G. R. Lynch, *Science* **214,** 821 (1981). Binding studies in human hypothalamus: S. M. Reppert *et al., Science* **242,** 78 (1988). *Reviews:* M. K. Vaughn, *Int. J. Rev. Physiol.* **24,** 41-95 (1981); D. C.Klein *et al., Life Sci.* **28,** 1975-1986 (1981). Book: *Advan. Biosci.* **vol. 29,** N. Birau, W. Schlott, Eds. (Pergamon Press, New York, 1981) 420 pp. Review of etiological role in clinical disease: A. Miles, D. Philbrick, *CRC Crit. Rev. Clin. Lab. Sci.* **25,** 231-253 (1987); in psychiatric disorders: *eidem, Biol. Psychiat.* **23,** 405-425 (1988).

Pale yellow leaflets from benzene, mp 116-118°. uv max: 223, 278 nm (ε 27550, 6300).

5696. Meldrum's Acid. *2,2-Dimethyl-1,3-dioxane-4,6-dione; malonic acid cyclic isopropylidene ester;* isopropylidene malonate; 2,2-dimethyl-4,6-diketo-1,3-dioxane. $C_6H_8O_4$; mol wt 144.12. C 50.00%, H 5.60%, O 44.41%. Prepd from malonic acid and acetone in the presence of acetic anhydride: Meldrum, *J. Chem. Soc.* **93,** 598 (1908). Prepn and structure: Davidson, Bernhard, *J. Am. Chem. Soc.* **70,** 3426 (1948); P. Schuster *et al., Monatsh.* **95,** 53 (1964); G. H. Bihlmayer *et al., ibid.* **98,** 564 (1967). *Review:* H. McNab, *Chem. Soc. Rev.* **7,** 345-358 (1978).

Crystals from acetone + water, dec 94-95°. pKa 5.1.

USE: In organic synthesis as a substitute for acyclic malonic esters and generally as a C_3O_2 synthon.

5697. Melengestrol. *17α-Hydroxy-6-methyl-16-methyl-enepregna-4,6-diene-3,20-dione;* 6-methyl-16-methylene-17α-hydroxy-Δ⁶-progesterone. $C_{23}H_{30}O_3$; mol wt 354.47. C 77.93%, H 8.53%, O 13.54%. Prepn of the acetate: Kirk *et al.;* Petrow, **Brit.** pat. **886,619** and U.S. pat. **3,117,966** (1962, 1964, both to Brit. Drug Houses).

Acetate, $C_{25}H_{32}O_4$, *17α-acetoxy-6-methyl-16-methylene-4,6-pregnadiene-3,20-dione, MGA.* Crystals, mp 224-226°. $[\alpha]_D^{23}$ -127° (c = 0.31 in chloroform). uv max (ethanol): 287 nm (log ε 4.35).

THERAP CAT: Antineoplastic; progestogen.

THERAP CAT (VET): Progestogen.

5698. Melezitose. *O-α-D-Glucopyranosyl-(1 → 3)-β-D-fructofuranosyl-α-D-glucopyranoside.* $C_{18}H_{32}O_{16}$; mol wt 504.44. C 42.86%, H 6.39%, O 50.75%. Trisaccharide built from 2 mols glucose and 1 mol fructose. Acid hydrolysis yields at first glucose and turanose. Occurs in a manna that forms upon the Douglas fir, jack pine *(Pinus virginiana* Mill., *Pinaceae)* and other trees. During periods of drought, bees collect it from the manna in sufficient quantity to change the character of the honey, making it unsuitable for maintaining the bees, but providing a source of melezitose: Hudson, Sherwood, *J. Am. Chem. Soc.* **42,** 116 (1920). Procedure: Bates, *NBS Circular* **C440,** p 472 (1942). Structure: Hehre, Carlson, *Arch. Biochem. Biophys.* **36,** 158 (1952); Hehre, *Advan. Carbohyd. Chem.* **8,** 277 (1953). *Review:* E. B. Rathbone, *Dev. Food Carbohyd.* **2,** 145-185 (1980).

Dihydrate, crystals from water. The water of crystn is given up at 110°. When anhydr, mp 153-154°. $[\alpha]_D^{20}$ +88° (c = 4). Sol in water, very sparingly sol in alc. Not fermented by baker's yeast; does not reduce Fehling's soln.

5699. Melibiose. *6-O-α-D-Galactopyranosyl-D-glucose;* 6-(α-D-galactosido)-D-glucose. $C_{12}H_{22}O_{11}$; mol wt 342.30. C 42.10%, H 6.49%, O 51.42%. Prepd from raffinose by fermentation with top yeast which removes the fructose: Hudson, Harding, *J. Am. Chem. Soc.* **37,** 2734 (1915); Fletcher, Diehl, *ibid.* **74,** 5774 (1952). Structure: Haworth, Leitch, *J. Chem. Soc.* **113,** 188 (1918); Charlton *et al., ibid.* **1926,** 99; Charlton *et al., ibid.* **1927,** 1527; Haworth *et al., ibid.* 3146; Levene, Jorpes, *J. Biol. Chem.* **86,** 403 (1930). Synthesis: Helferich, Bredereck, *Ann.* **465,** 166 (1928).

Dihydrate, monoclinic crystals from water or dil alcohol. mp 84-85°. Shows mutarotation. $[\alpha]_D^{20} +111.7° \rightarrow 129.5°$ (c = 4). One gram dissolves in 0.4 ml water, 8.5 ml methanol, 220 ml abs alcohol. Dilute acids hydrolyze melibiose to D-glucose and D-galactose. Also split by emulsin and by bottom yeast. Reduces Fehling's soln. 3.5 g melibiose dihydrate are about as sweet as 1.0 g sucrose.

5700. Melilot. Sweet clover; yellow Melilot; yellow sweet clover. Dried leaves and flowering tops of *Melilotus officinalis* (L.) Lam., *Leguminosae.* Habit. Europe, Asia, natural to some extent in U.S. Constit. Coumarin, resin, volatile oil.

An extract is marketed as *Esberiven.*

5701. Melilotoside. *3-[2-(β-D-Glucopyranosyloxy)phenyl]-2-propenoic acid.* $C_{15}H_{18}O_8$; mol wt 326.31. C 55.21%, H 5.56%, O 39.23%. From flowers of *Melilotus altissimus* Thuill. and *M. arvensis* Wallr., *Leguminosae:* Charaux, *Bull. Soc. Chim. Biol.* 7, 1056 (1925). Synthesis: Shinoda, Imaida, *J. Pharm. Soc. Japan* 54, 107 (1934), *C.A.* 31, 100⁵ (1937).

Monohydrate, yellowish needles; slightly bitter, acidulous, astringent taste. mp 240-241° with decompn. $[\alpha]_D -60.9°$ (50% alcohol). Sol in water, alcohol; slightly sol in acetone, ethyl acetate.

5702. Melinamide. *(Z,Z)-N-(1-Phenylethyl)-9,12-octadecadienamide; N-(α-methylbenzyl)linoleamide;* MBLA; AC 223; Artes. $C_{26}H_{41}NO$; mol wt 383.62. C 81.41%, H 10.77%, N 3.65%, O 4.17%. Linoleic acid derivative. Prepn and use in lowering blood cholesterol levels: **Fr. pat. 1,476,-596** (1967 to Sumitomo), *C.A.* **68**, 87027q (1967). Physical properties and effect on cholesterol levels in rabbits: K. Toki *et al., J. Atheroscler. Res.* 7, 708 (1967). Comparative activity of melinamide and its optical isomers: H. Fukushima, K. Nakatani, *ibid.* 9, 65 (1969); H. Fukushima *et al., ibid.* 10, 403 (1969). Mechanism of action studies: D. Kritchevsky, S. A. Tepper, *Arzneimittel-Forsch.* 21, 1024 (1971); D. Kritchevsky *et al., Lipids* 12, 16 (1977); K. Natori *et al., Japan. J. Pharmacol.* 42, 517 (1986). Metabolism: A. Hirohashi *et al., Xenobiotica* 6, 329 (1976).

Oil. mp <4°. bp$_{0.03}$ 200-215°; bp$_{0.07}$ 200-204°. n_D^{23} 1.5050; n_D^{30} 1.4863. Saponification value 0.9. Iodine value 127.0.

THERAP CAT: Antihyperlipoproteinemic.

5703. Melinonine A. *16,17-Didehydro-16-(methoxycarbonyl)-4,19α-dimethyloxayohimbanium.* $[C_{22}H_{27}N_2O_3]^+$. Calabash-curare alkaloid from *Strychnos melinoniana* Baillon, *Loganiaceae.* Isoln and structure: Schlittler, Hohl, *Helv. Chim. Acta* 35, 29 (1952); Bächli *et al., ibid.* 40, 1167 (1957).

Chloride, needles from alcohol + acetone, mp 260-261°. $[\alpha]_D^{20} -107°$ (c = 0.65 in water). uv max (50% ethanol): 225 nm (log ε 4.68).

Iodide, mass spectrum: Hesse *et al., Helv. Chim. Acta* 48, 674 (1965).

5704. Melitracen. *3-(10,10-Dimethyl-9(10H)-anthracenylidene)-N,N-dimethyl-1-propanamine; N,N,10,10-tetramethyl-Δ⁹⁽¹⁰ᴴ⁾,ᵞ-anthracenepropylamine;* 9,10-dihydro-10,10-dimethyl-9-(3-dimethylaminopropylidene)anthracene; 9-[3-(dimethylamino)propylidene]-10,10-dimethyl-9,10-dihydroanthracene; *N,N-dimethyl-3-(10,10-dimethyl-9(10H)-anthrylidene)propylamine.* $C_{21}H_{25}N$; mol wt 291.42. C 86.55%, H 8.65%, N 4.81%. Prepn of the hydrochloride: Holm, *Acta Chem. Scand.* 17, 2437 (1963); idem, **Brit. pat. 939,856** corresp to **U.S. pat. 3,177,209** (1963, 1965, both to Kefalas A/S). Crystal structure: J. Lopez de Lerma *et al., Acta Crystallogr.* **B35**, 1739 (1979). Toxicity data: P. V. Petersen *et al., Acta Pharmacol. Toxicol.* 24, 121 (1966).

Hydrochloride, $C_{21}H_{26}ClN$, *U-24973A, Melixeran, Trausabun, Dixeran.* Crystals from acetone, mp 245-248°. LD$_{50}$ i.v. in mice: 52 mg/kg (Petersen).

THERAP CAT: Antidepressant.

5705. Melittin. Melittin I; mellitin; Forapin. $C_{131}H_{229}$-$N_{39}O_{31}$; mol wt 2846.54. C 55.28%, H 8.11%, N 19.19%, O 17.42%. Strongly basic polypeptide, the principal component of the venom of the honey bee, *Apis mellifera),* comprising 40-50% of the dried venom. A "direct" hemolysin, one of the hemolytic principles present in the venom, the other being "indirect" acting *phospholipase A.* Sepn and isoln: Neumann *et al., Naturwiss.* 39, 286 (1952); Neumann, Haberman, *Arch. Exp. Pathol. Pharmakol.* 222, 367 (1954); Habermann, Reiz, *Biochem. Z.* 341, 451 (1965). Melittin is the first polypeptide whose biological effects can be understood on the basis of its primary structure. Elucidation of structure and correlation with activity: E. Habermann, J. Jentsch, *Z. Physiol. Chem.* 348, 37 (1967). Conformation studies: R. Bazzo *et al., Eur. J. Biochem.* 173, 139 (1988). About 10% of the melittin is thought to be formylated at the *N*-terminus: Kreil, Kreil-Kiss, *Biochem. Biophys. Res. Commun.* 27, 275 (1967). Isoln and structure of $N^α$*-formyl melittin:* Lübke *et al., Experientia* 27, 765 (1971). Synthesis of melittin and related peptides: Lübke, Schröder, *Peptides,* H. C. Beyerman, A. van der Linde, W. M. van den Brink, Eds. (North-Holland Publishing Company, Amsterdam, 1967) pp 271-279; Dorman, Markley, *J. Med. Chem.* 14, 5 (1970); Schröder *et al., Experientia* 27, 764 (1971). Solid phase synthesis and purification: M. T. Tosteson *et al., Biochemistry* 26, 6627 (1987). Review of biochemistry and pharmacology: Habermann, *Science* 177, 314 (1972).

Gly-Ile-Gly-Ala-Val-Leu-Lys-Val-Leu-Thr-Thr-Gly-Leu-Pro-
1   7   14

Ala-Leu-Ile-Ser-Trp-Ile-Lys-Arg-Lys-Arg-Gln-Gln-NH₂
15   26

Cream white, water soluble powder. $[\alpha]_D^{21} -89.52°$ (c = 0.409).

THERAP CAT: Antirheumatic.

5706. Mellitic Acid. *Benzenehexacarboxylic acid;* mellic acid. $C_{12}H_6O_{12}$; mol wt 342.17. C 42.12%, H 1.77%, O 56.11%. $C_6(COOH)_6$. Preparation from carbonaceous material: Kiebler, U.S. pat. **2,461,749** (1949 to Carnegie Inst. of Tech.); Germain *et al.*, *Bull. Soc. Chim. France* **1962**, 779; from tetrahalophthalic acid: Brusset, Uny, *ibid.* **1951**, 565; Juettner, U.S. pat. **3,067,246** (1962).

Crystals. mp 286-288° in sealed tube with decompn. Freely sol in water or alcohol; sol in boiling concd H_2SO_4 without decompn.

5707. Melperone. *1-(4-Fluorophenyl)-4-(4-methyl-1-piperidinyl)-1-butanone; 4'-fluoro-4-(4-methylpiperidino)-butyrophenone; γ-(4-methylpiperidino)-p-fluorobutyrophenone;* methylperone; flubuperone. $C_{16}H_{22}FNO$; mol wt 263.37. C 72.97%, H 8.42%, F 7.21%, N 5.32%, O 6.08%. Neuroleptic agent related structurally to haloperidol, *q.v.* Prepn: *Belg.* pat. **651,144** (1964 to Ferrosan), *C.A.* **63**, 13224c (1965). Distribution of ^{14}C melperone: N. Einer-Jensen, E. Hansson, *Acta Pharmacol. Toxicol.* **23**, 65 (1965). Pharmacological and toxicological studies: J. A. Christensen *et al.*, *ibid.* 109; R. Heywood, A. K. Palmer, *Farmaco Ed. Prat.* **29**, 586 (1974). Dopamine-receptor binding in relation to clinical effect: I. Creese *et al.*, *Science* **192**, 481 (1976). Sedative and sleep-inducing properties: R. Kretzschmer *et al.*, *Arzneimittel-Forsch.* **26**, 1073 (1976). Clinical studies in anxiety: W. J. Poeldinger, *Therapiewoche* **30**, 4862 (1980); L. F. Fabre, M. J. Napoliello, *Curr. Ther. Res.* **30**, 427 (1981). Melperone has also been shown to have anti-arrhythmic effects: E. S. Platou *et al.*, *Acta Pharmacol. Toxiol.* **50**, 108 (1982).

Liquid, bp$_{0.1}$ 120-125°.

Hydrochloride, $C_{16}H_{23}ClFNO$, *FG 5111, Buronil, Eunerpan.* Cryst, mp 209-211°. LD_{50} in rats, mice (mg/kg): 330, 230 orally; 40, 35 i.v., J. A. Christensen *et al.*, *loc. cit.*

THERAP CAT: Neuroleptic.

5708. Melphalan. *4-[Bis(2-chloroethyl)amino]-L-phenylalanine; p-di(2-chloroethyl)amino-L-phenylalanine; L-phenylalanine mustard; alanine nitrogen mustard; L-PAM;* melfalan; L-sarcolysine; NSC-8806; CB 3025; Alkeran; Sarcoclorin. $C_{13}H_{18}Cl_2N_2O_2$; mol wt 305.20. C 51.16%, H 5.94%, Cl 23.23%, N 9.18%, O 10.48%. Syntheses: Bergel, Stock, *J. Chem. Soc.* **1954**, 2409; **1955**, 1223; *eidem,* U.S. pats. **3,032,584; 3,032,585** (both 1962 to NRDC); Larionov, *Lancet* **2**, 169 (1955). Toxicity: W. C. J. Ross, *Biochem. Pharmacol.* **13**, 969 (1964). Neurotoxicity study: M. G. Donelli *et al.*, *J. Pharm. Pharmacol.* **18**, 760 (1966). Mutation study: J. McCann *et al.*, *Proc. Nat. Acad. Sci. USA* **72**, 5135 (1975). Biliary excretion in rats: K. H. Byington *et al.*, *Biochem. Pharmacol.* **29**, 2518 (1980). Review of carcinogenicity studies: *IARC Monographs* **9**, 167-180 (1975). Review: R. L. Furner, R. K. Brown, *Cancer Treat. Rep.* **64**, 559-574 (1980).

Needles from methanol (monosolvate), mp 182-183° (dec). $[\alpha]_D^{25}$ +7.5° (c = 1.33 in 1.0N HCl); $[\alpha]_D^{22}$ −31.5° (c = 0.67 in methanol). Soluble in ethanol, propylene glycol. Practically insol in water. LD_{50} i.p. in rats: 14.7 μmol/kg (Ross).

D-Form, *CB 3026, NSC-35051,* D-*sarcolysine, medphalan.* Needles from methanol (monosolvate), mp 181.5-182° (dec). $[\alpha]_D^{21}$ −7.5° (c = 1.26 in 1.0N HCl).

DL-Form, *merphalan, sarcolysine.* Tiny needles from methanol, mp 180-181°.

Human Toxicity: Bone marrow depression may occur. This substance has been listed as a known carcinogen: *Fourth Annual Report on Carcinogens* (NTP 85-002, 1985) p 125.

THERAP CAT: Antineoplastic.

5709. Memantine. *3,5-Dimethyltricyclo[3.3.1.1^{3,7}]decan-1-amine; 3,5-dimethyl-1-adamantanamine;* 1-amino-3,5-dimethyladamantane; DMAA; D 145. $C_{12}H_{21}N$; mol wt 179.31. C 80.38%, H 11.81%, N 7.81%. Deriv of adamantine, *q.v.,* with anti-parkinson activity. Prepn of the hydrochloride: K. Gerzon *et al.*, *J. Med. Chem.* **6**, 760 (1963); of the free base and hydrochloride: J. Mills, E. Krumkalns, U.S. pat. **3,391,142** (1968 to Lilly). GC and mass spec studies of memantine metabolites: W. Wesemann *et al.*, *Arzneimittel-Forsch.* **27**, 1471 (1977). Effects in parkinsonian patients: P.-A. Fischer *et al.*, *ibid.* 1487. Series of articles on distribution, effects on neurobiological processes, clinical studies in control of micturition and limb muscle mobility: *ibid.* **32**, 1236-1276 (1982). Clinical studies as antispasmodic agent: H. Rohde, *Fortschr. Med.* **100**, 2023 (1982). Pharmacodynamics and pharmacokinetics: W. Wesemann *et al.*, *Arzneimittel-Forsch.* **33**, 1122 (1983).

Oil, n_D^{25} 1.4941.

Hydrochloride, $C_{12}H_{22}ClN$, *Akatinol.* Cryst from alcohol/ether, mp 258° (Mills, Krumkalns); also reported as mp 290-295° (Gerzon).

THERAP CAT: Muscle relaxant (skeletal).

5710. Menadiol Diacetate. *2-Methyl-1,4-naphthalenediol diacetate;* acetomenaphthone; 2-methyl-1,4-naphthohydroquinone diacetate; 1,4-diacetoxy-2-methylnaphthalene; vitamin K$_4$; Kapilin; Kapilon; Prokayvit Oral; Vitavel K; Davitamon K; Kappaxan; Kayvite. $C_{15}H_{14}O_4$; mol wt 258.26. C 69.76%, H 5.46%, O 24.78%. Prepd from naphthalene: Sah *et al.*, *Ber.* **73**, 762 (1940); *Rec. Trav. Chim.* **59**, 461 (1940); by reductive acetylation of menadione: Horii *et al.*, *Pharm. Bull. (Tokyo)* **5**, 82 (1957).

Crystals, mp 112-114°. Almost insol in water; slightly sol in cold alc; sol in 3.3 parts boiling alc, in acetic acid.

THERAP CAT: Prothrombogenic vitamin.

5711. Menadiol Dibutyrate. *2-Methyl-1,4-naphthalenediol dibutyrate;* 2-methyl-1,4-naphthohydroquinone dibutyrate; Karanum. $C_{19}H_{22}O_4$; mol wt 314.37. C 72.59%, H 7.05%, O 20.36%. Prepn: von Werder, *Ger.* pat. **734,220** (1943 to E. Merck).

Crystals, mp about 53°. Practically insol in water; sol in alcohol, benzene, oils and fats.

THERAP CAT: Prothrombogenic vitamin.

5712. Menadiol Diphosphate (Tetrasodium Salt). *2-Methyl-1,4-naphthalenediol diphosphoric acid ester tetrasodium salt;* 2-methyl-1,4-naphthohydroquinone diphosphoric acid ester tetrasodium salt; tetrasodium 2-methyl-1,4-naphthohydroquinone diphosphoric acid ester; menadiol sodium diphosphate; menadiol tetrasodium diphosphate; menadione diphosphate tetrasodium salt; Kappadione; Kipca, Water Soluble; Procoagulo; Synka-Vit; Synkayvite. $C_{11}H_8Na_4O_8P_2$;

mol wt 422.09. C 31.30%, H 1.91%, Na 21.79%, O 30.32%, P 14.68%. Prepn from hydroquinone + phosphorus oxychloride: Fieser, Fry, *J. Am. Chem. Soc.* **62**, 228 (1940); from 2-methyl-1,4-naphthohydroquinone + phosphorus oxychloride: Kudryashov *et al.*, *Voprosy Med. Khim.* **5**, No. 4, 279 (1959), *C.A.* **55**, 9517g (1961).

Dihydrate, crystals from methanol + water. Very hygroscopic. Liquefies on exposure to moist air.

Hexahydrate, white to pinkish powder. Salty taste. Very sol in water. Practically insol in methanol, ethanol, ether, acetone. Aq solns are slightly alkaline, pH 7.0 to 9.0.

THERAP CAT: Prothrombogenic vitamin.

5713. Menadiol Disulfate. *2-Methyl-1,4-naphthalenediol bis(hydrogen sulfate).* $C_{11}H_{10}O_8S_2$; mol wt 334.33. C 39.52%, H 3.01%, O 38.29%, S 19.18%. Prepn of salts: Fieser, Fry, *J. Am. Chem. Soc.* **62**, 228 (1940); Fieser, *J. Biol. Chem.* **133**, 391 (1940); Oxley, Short, **Brit.** pat. **623,242** (1949 to Boots Pure Drug).

Dipotassium salt, $C_{11}H_8K_2O_8S_2$, *potassium menaphthosulfate, Vikastab.*

Disodium salt dihydrate, $C_{11}H_8Na_2O_8S_2 \cdot 2H_2O$, *sodium menaphthosulfate.* Solid. Sol in water, methanol. Practically insol in alcohol, ether.

THERAP CAT: Vitamin (prothrombogenic).

5714. Menadione. *2-Methyl-1,4-naphthalenedione; 2-methyl-1,4-naphthoquinone;* menaphthone; Vitamin K$_{2(0)}$; Vitamin K$_3$; Kayklot; Kativ-G; Kipca, Oil Soluble; Kappaxin; Aquakay; Aquinone; Kaergona; Kareon; Kayquinone; Kolklot; Panosine; Synkay; Thyloquinone; Kanone; Klottone; K-Thrombyl; Prokayvit; Koaxin; K-Vitan. $C_{11}H_8O_2$; mol wt 172.17. C 76.73%, H 4.68%, O 18.58%. A synthetic naphthoquinone derivative having physiologic properties of vitamin K. Prepd from β-methylnaphthalene by oxidation of chromic oxide under mild conditions: Fieser, *J. Biol. Chem.* **133**, 391 (1940).

Bright yellow crystals. Very faint acrid odor. Stable in air, but is dec by sunlight. mp 105-107°. Insol in water. One gram dissolves in about 60 ml alcohol, in 10 ml benzene, in 50 ml vegetable oils; moderately sol in chloroform and in carbon tetrachloride. The alcoholic soln is neutral to litmus. Solutions may be heated to 120° without dec. Destroyed by alkalies and reducing agents. *Keep protected from light.* LD$_{50}$ orally in mice: approx 0.5 g/kg, Molitor, Robinson, *Proc. Soc. Exp. Biol. Med.* **43**, 725 (1940).

Note: Mikhlin, *C. R. Acad. Sci. USSR* **37**, 191 (1941); *Biokhimiya* **8**, 158 (1943) reported that corn stigma contains an alcohol sol substance termed by him vitamin K$_3$; however, its differentiation from vitamin K$_1$ is not clear. In Anglo-

American literature the term K$_3$ designates menadione. This designation should be discontinued since the derivatives of 1,4-naphthoquinone without a sidechain in the 3-position cannot exert all the functions of the K vitamins: Isler, Wiss, *Vitam. Horm.* *(New York)* **17**, 55 (1959).

Human Toxicity: Irritating to mucous membranes, respiratory passages and skin.

THERAP CAT: Vitamin (prothrombogenic).

THERAP CAT (VET): In hypoprothrombinemia and bishydroxycoumarin poisoning, including sweet clover poisoning.

5715. Menadione Dimethylpyrimidinol Bisulfite. *1,2,-3,4-Tetrahydro-2-methyl-1,4-dioxo-2-naphthalenesulfonic acid compd with 4,6-dimethyl-2(1H)-pyrimidinone (1:1);* 2-methyl-1,4-naphthoquinone 2-hydroxy-4,6-dimethylpyrimidine bisulfite; MPB; Hetrazeen. $C_{17}H_{18}N_2O_6S$; mol wt 378.40. C 53.96%, H 4.79%, N 7.40%, O 25.37%, S 8.47%. Naphthoquinone deriv with high vitamin K activity. Prepn: J. B. Nanninga *et al.*, U.S. pat. **3,328,169** (1967 to Heterochemical). Vitamin K activity in poultry: P. Griminger, *Poultry Sci.* **44**, 210 (1965); P. N. Dua, E. J. Day, *ibid.* **45**, 94 (1966). Efficacy in swine diets: R. W. Seerley *et al.*, *J. Anim. Sci.* **42**, 599 (1976).

Cryst powder, mp 215-217°. Soly in water ~1 g/100 ml. Slightly sol in alcohol; insol in ether, benzene.

THERAP CAT (VET): Vitamin (prothrombogenic).

5716. Menadione Sodium Bisulfite. *1,2,3,4-Tetrahydro-2-methyl-1,4-dioxo-2-naphthalenesulfonic acid sodium salt;* 2-methyl-1,4-naphthoquinone sodium bisulfite; sodium 1,2,3,4-tetrahydro-2-methyl-1,4-dioxo-2-naphthalenesulfonate; Hemodal; Hykinone; Ido-K; Kavitan; Klotogen. $C_{11}H_9NaO_5S$; mol wt 276.24. C 47.83%, H 3.28%, Na 8.32%, O 28.96%, S 11.61%. Prepn: Moore, *J. Am. Chem. Soc.* **63**, 2049 (1941); *see also ibid.* **64**, 1096 (1942); **65**, 1209 (1943); Moore, Kirchmeyer, U.S. pat. **2,367,302** (1945 to Abbott). Structure: Carmack *et al.*, *J. Am. Chem. Soc.* **72**, 844 (1950).

Trihydrate, white, hygroscopic crystals. Discolors and may turn purple under the influence of light. One gram dissolves in about 2 ml water. Slightly sol in alcohol; almost insol in ether, benzene. Addition of NaOH soln to a soln of menadione sodium bisulfite produces a bright yellow precipitate, of menadione.

THERAP CAT: Vitamin (prothrombogenic).

5717. Menadoxime. *[[(3-Methyl-4-oxo-1(4H)-naphthalenylidene)amino]oxy]acetic acid ammonium salt; menadione carboxymethoxime ammonium salt;* menaphthone carboxymethoxime ammonium salt; carboxymethylmenadione monoxime ammonium salt; Kapilon injectable. $C_{13}H_{14}N_2O_4$; mol wt 262.26. C 59.53%, H 5.38%, N 10.68%, O 24.40%. Prepn: Holland, **Brit.** pat. **621,934** (1949 to Glaxo).

Crystals from alc. Sol in water. Aq solns are neutral.

May be sterilized by autoclaving and stored in ampuls if protected from light. Shows high antihemorrhagic activity. Free acid, $C_{13}H_{11}NO_4$, yellow platelets from alc, mp 162-163°. Forms water-sol salts.

THERAP CAT: Vitamin (prothrombogenic).

5718. Menazon. *S-[(4,6-Diamino-1,3,5-triazin-2-yl)-methyl]phosphorodithioic acid O,O-dimethyl ester; S-[(4,6-diamino-s-triazin-2-yl)methyl] O,O-dimethyl phosphorodithioate; O,O-dimethyl S-[(4,6-diamino-s-triazin-2-yl)methyl] phosphorodithioate; 2-dimethoxyphosphinothioylthiomethyl-4,6-diamino-s-triazine;* ENT 25760; PP 175; Saphicol; Saphizon-DP; Sayfos. $C_6H_{12}N_5O_2PS_2$; mol wt 281.32. C 25.62%, H 4.30%, N 24.90%, O 11.37%, P 11.01%, S 22.80%. Prepn: Calderbank *et al., Ger.* pat. **1,118,789** corresp to U.S. pat. **3,169,964** (1961, 1965 both to I.C.I.). Activity: *eidem, Chem. & Ind. (London)* **1961**, 630. Toxicity: T. B. Gaines, *Toxicol. Appl. Pharmacol.* **14**, 515 (1969).

Crystals, dec 160-162°, mp 164-166°. Vapor press. at 25°: 1×10^6 mm Hg. Low soly in water and organic solvents. LD_{50} orally in male, female rats: 1020, 1450 mg/kg (Gaines).

USE: Insecticide; acaricide. *Caution:* Cholinesterase inhibitor.

5719. Menbutone. *4-Methoxy-γ-oxo-1-naphthalenebutanoic acid; 3-(4-methoxy-1-naphthoyl)propionic acid; β-(1-methoxy-4-naphthoyl)propionic acid; γ-oxo-4-methoxy-1-naphthalenebutyric acid;* Ictéryl. $C_{15}H_{14}O_4$; mol wt 258.26. C 69.75%, H 5.46%, O 24.78%. Prepn: Ruzicka, Waldman, *Helv. Chim. Acta* **15**, 907 (1932); Fieser, Hershberg, *J. Am. Chem. Soc.* **58**, 2314 (1936); Burtner, U.S. pat. **2,623,065** (1952 to Searle).

Crystals, mp 172-173°.
Magnesium salt, $C_{30}H_{26}MgO_8$, *Hepalande.*
THERAP CAT: Choleretic.

5720. Mendelevium. Md; formerly Mv; at. wt (most stable known isotope) 258; at. no. 101; valence 3, also 2. Man-made radioactive element. ^{256}Md ($T_{1/2}$ 1.5 hrs), first produced in 1955 by bombarding a target of ^{253}Es with helium ions; decays by electron capture to ^{256}Fm: Ghiorso *et al., Phys. Rev.* **98**, 1518 (1955), *C.A.* **49**, 12149f (1955). Known isotopes: 248-258. ^{258}Md ($T_{1/2}$ 56 days) produced by irradiating ^{255}Es with ^{4}He ions: Fields *et al., Nucl. Phys. A* **154**, 407 (1970). The behavior of mendelevium is typical of the actinide family: It ionizes to the "tripositive" state (*i.e.*, it acquires a triple positive charge); also has the expected chemical kinship with thulium, its counterpart in the rare-earth family. Review on chemistry of mendelevium: Maly, Brandstetr, *Chem. Listy* **58**, 751 (1964). *Reviews:* G. T. Seaborg, *Man-Made Transuranium Elements* (Prentice-Hall, Englewood Cliffs, N. J., 1963) 120 pp; C. Keller, *The Chemistry of the Transuranium Elements* (Verlag Chemie, Weinheim, English Ed., 1971) pp 595-600; Silva, "Trans-Curium Elements" in *MTP Int. Rev. Sci.: Inorg. Chem. Ser. One* vol. **8**, A. G. Maddock, Ed. (University Park Press, Baltimore, 1972) pp 71-105; *Comprehensive Inorganic Chemistry* vol. **5**, J. C. Bailar, Jr. *et al.,* Eds. (Pergamon Press, Oxford, 1973) *passim;* several authors, *Handb. Exp. Pharmakol.* **36**, 689-738 (1973).

5721. Menhaden Oil. Pogy oil; mossbunker oil. Obtained along the East Coast of North America from the menhaden fish, *Brevoortia tyrannis,* somewhat larger than a herring. The oil contains (in the form of glycerides) about 6% myristic acid, 16% palmitic acid, 30% linoleic acid, 19% C_{20}- and 11% C_{22}-acids (highly unsaturated).

Reddish oil. Characteristic, distasteful fishy odor and taste. d 0.925-0.933. mp 38.5-47.2°. n_D^{20} 1.480. Sapon no. 191-200. Iodine no. 139-180. Acid no. 3.0-11.6. Sol in ether, benzene, petr ether, naphtha, kerosene, CS_2.

USE: Substitute for linseed oil. In leather dressing formulations. Hydrogenated menhaden oil can be used as a substitute for tallow in soap-making.

5722. Menichlopholan. *5,5'-Dichloro-3,3'-dinitro[1,1'-biphenyl]-2,2'-diol; 4,4'-dichloro-6,6'-dinitro-o,o'-biphenol; 5,5'-dichloro-2,2'-dihydroxy-3,3'-dinitrobiphenyl;* niclofolan; Bayer 9015; Me 3625; Bilevon-M. $C_{12}H_6Cl_2N_2O_6$; mol wt 345.09. C 41.77%, H 1.75%, Cl 20.55%, N 8.12%, O 27.82%. Prepn of analogous 4,4'-bromo compd: Yamashiro, *Bull. Chem. Soc. Japan* **17**, 10 (1942), *C.A.* **42**, 4574b (1948). Anthelmintic activity: Meiser, Federmann, U.S. pat. **3,082,151** (1963 to Bayer); Lane, Stewart, *Vet. Rec.* **80**, 702 (1967).

THERAP CAT (VET): Anthelmintic (fasciolicide).

5723. Menthol. *(1α,2β,5α)-5-Methyl-2-(1-methylethyl)-cyclohexanol; 3-p-menthanol; l-menthol; hexahydrothymol;* peppermint camphor. $C_{10}H_{20}O$; mol wt 156.26. C 76.86%, H 12.90%, O 10.24%. Obtained from peppermint oil or other mint oils, or prepd synthetically by hydrogenation of thymol. Chromatographic sepn from mint oils: Chang, U.S. pat. **2,760,993** (1956 to Iowa State College). Toxicity data: P. M. Jenner *et al., Food Cosmet. Toxicol.* **2**, 327 (1964).

Crystals or granules; peppermint taste and odor. d 0.890. mp 41-43°. bp 212°. n_D^{25} 1.458. $[\alpha]_D^{18}$ −50° (10% alc soln). Slightly sol in water; very sol in alcohol, chloroform, ether, petr ether; freely sol in glacial acetic acid, liq petrolatum. LD_{50} orally in rats: 3180 mg/kg (Jenner).

Incompat: Butylchloral hydrate, camphor, phenol, chloral hydrate, Exalgine, betanaphthol, resorcinol or thymol in triturations; potassium permanganate, chromium trioxide, pyrogallol.

USE: In liqueurs, confectionery, perfumery, cigarettes, cough drops, and nasal inhalers.

THERAP CAT: Topical antipruritic.

THERAP CAT (VET): Has been used as a mild local anesthetic, antiseptic; internally as a carminative and gastric sedative.

5724. l-Menthone. *(−)-5-Methyl-2-(1-methylethyl)cy-clohexanone; l-p-menthan-3-one;* 1-methyl-4-isopropylcy-clohexan-3-one. $C_{10}H_{18}O$; mol wt 154.24. C 77.87%, H 11.76%, O 10.37%. Of the four optically active isomers of methone, the one occuring most frequently in nature. Found in various volatile oils, such as pennyroyal, peppermint, geranium: Simonsen, *The Terpenes* vol. **I** (University Press, Cambridge, 2nd ed., 1947) pp 314-327. Prepd by chromic acid oxidation of *l*-menthol: Hussey, Baker, *J. Org. Chem.* **25**, 1434 (1960); Brown, Garg, *J. Am. Chem. Soc.* **83**, 2952 (1961).

Consult the cross index before using this section.

Bitter liq; slight peppermint odor. bp 207°. bp$_{41}$ 116-119°. mp −6°. d$_4^{20}$ 0.895. n$_D^{20}$ 1.4505; n$_D^{23}$ 1.4490. [α]$_D^{20}$ −24.8°; [α]$_D^{27}$ −28.9°. Slightly sol in water; sol in organic solvents.
Note: Apinol obtained by dry distln of wood of *Pinus palustris* Mill. (*P. australis* Michx.), *Pinaceae* is chiefly *l*-menthone: *J. Pharm. Chim.* **18**, 139, 177, 208 (1918), *C.A.* **13**, 56^9 (1919). Amber-colored oil, bp about 182.2°, d 0.946.
USE: In perfume and flavor compositions.

5725. Menthyl Acetate. *5-Methyl-2-(1-methylethyl)cyclohexanol acetate.* C$_{12}$H$_{22}$O$_2$; mol wt 198.30. C 72.68%, H 11.18%, O 16.14%. CH$_3$COOC$_{10}$H$_{19}$. Present in peppermint oil.
Colorless liquid, characteristic odor. d$_4^{20}$ 0.919. bp 227°. n$_D^{20}$ 1.4468. [α]$_D^{20}$ −79.42°. Slightly sol in water; miscible with alcohol, ether.
USE: In perfumery; emphasizes floral notes, especially that of rose, used in toilet waters having a lavender odor. Has been suggested for flavoring extracts having caraway or mint flavors.

5726. Menthyl Borate. *5-Methyl-2-(1-methylethyl)cyclohexanol monoester with boric acid.* Estoral. C$_{30}$H$_{57}$BO$_3$; mol wt 476.58. C 75.60%, H 12.06%, B 2.27%, O 10.07%. (C$_{10}$H$_{19}$)$_3$BO$_3$.
White, tasteless cryst powder; faint menthol odor. Insol in water or alcohol; freely sol in chloroform, ether. Dec into its constituents when in soln.

5727. Menthyl Salicylate. *2-Hydroxybenzoic acid 5-methyl-2-(1-methylethyl)cyclohexyl ester.* C$_{17}$H$_{24}$O$_3$; mol wt 276.36. C 73.88%, H 8.75%, O 17.37%. HOC$_6$H$_4$COO-C$_{10}$H$_{19}$.
Clear, yellowish, syrupy liquid; odorless or slight fruity odor. d$_{25}^{25}$ 1.045. Insol in water; miscible with most organic solvents. *Keep well closed and protected from light.*
USE: As a "sun screen" to filter out ultraviolet light in prepns for preventing sunburn.

5728. Menthyl Valerate. *3-Methylbenzoic acid 5-methyl-2-(1-methylethyl)cyclohexyl ester; isovaleric acid p-menth-3-yl ester.* C$_{15}$H$_{28}$O$_2$; mol wt 240.37. C 74.95%, H 11.74%, O 13.31%. Contains about 30% free menthol.
Liquid; menthol and valerian odor; cooling, faintly bitter taste. d 0.906-0.908. Insol in water; freely sol in alcohol, chloroform, ether, oils; dec by alkalies.
THERAP CAT: Sedative.

5729. Menyanthes. Buck bean; bog bean; marsh trefoil; water shamrock. Dried leaves or roots of *Menyanthes trifoliata* L., *Gentianaceae*. Habit. Europe, Asia, N. America. Constit. Menyanthin.

5730. Meobentine. *N-[(4-Methoxyphenyl)methyl]-N',-N''-dimethylguanidine; 1-(p-methoxybenzyl)-2,3-dimethylguanidine.* C$_{11}$H$_{17}$N$_3$O; mol wt 207.28. C 63.74%, H 8.27%, N 20.27%, O 7.72%. Antidysrhythmic, antifibrillatory deriv of guanidine, *q.v.* Prepn of the sulfate: R. A. Maxwell, E. Walton, **Ger. pat.** 2,030,693 corresp to **U.S. pat.** 3,949,089 (1971, 1976 both to Burroughs Wellcome). Antidysrhythmic effects and tissue concentration: K. B. Touw *et al., Pharmacologist* **19**, 268 (1977); K. B. Touw, *Diss. Abstr. B* **39**, 5340 (1979). Pharmacokinetics by radioimmunoassay: J. W. A. Findlay *et al., Pharmacologist* **21**, 337 (1979). Pharmacological study: W. B. Wastila *et al., J. Pharm. Pharmacol.* **33**, 594 (1981).

Sulfate, C$_{22}$H$_{36}$N$_6$O$_6$S, *Rythmatine.* Cryst, mp 273-274°.
THERAP CAT: Cardiac depressant (anti-arrhythmic).

5731. Meparfynol. *3-Methyl-1-pentyn-3-ol;* ethyl ethynyl methyl carbinol; 2-ethynyl-2-butanol; methylparafynol; methylpentynol; Allotropal; Anti-Stress; Apridol; Atemorin; Atempol; Dalgol; Dorison; Dormalest; Dormidin; Dormigen; Dormiphen; Dormison; Dormosan; Formison; Hesofen; Hexofen; Imnudorm; Oblivon; Pentadorm; Perlopal; Riposon; Seral; Somnesin. C$_6$H$_{10}$O; mol wt 98.14. C 73.43%, H 10.27%, O 16.30%. Prepd from methyl ethyl ketone and sodium acetylide in liquid ammonia or by a Grignard reaction: **Ger. pat.** 285,770 (1913 to Bayer); also **Ger. pats.** 289,800 and 291,185 see Frdl. **12**, 55, 56, 57; Sung Wouseng, *Ann. Chim.* [10] **1**, 343 (1924); Rupe, Vonaesch, *Ann.* **442**, 80 (1925); Carothers, Coffman, *J. Am. Chem. Soc.* **54**, 4071 (1932); Campbell *et al., ibid.* **60**, 2882 (1938); Campbell, Campbell, *Proc. Indiana Acad. Sci.* **50**, 123-127 (1940), *C.A.* **35**, 5471; Smith, **U.S. pat.** 2,385,547 (1945 to Commercial Solvents); A. W. Johnson, *The Chemistry of the Acetylenic Compounds* vol. I (Edward Arnold & Co., London, 1946) p 278; Hurd, McPhee, *J. Am. Chem. Soc.* **69**, 239 (1947); P. Piganiol, *Acetylene Homologs and Derivatives* (Brooklyn, 1950); Papa *et al., Arch. Biochem. Biophys.* **33**, 482 (1951); Thiele, Martinez, *Ciencia (Mex.)* **15**, 70 (1955). An optically active form has been described by Hickmann, Kenyon, *J. Chem. Soc.* **1955**, 2051. Toxicity data: Margolin *et al., Science* **114**, 384 (1951).

$$CH_3CH_2\underset{\underset{CH_3}{|}}{\overset{\overset{OH}{|}}{C}}C\equiv CH$$

Mobile liquid. Acrid odor, burning taste. d$_4^{20}$ 0.8688. d$_{20}^{20}$ 0.8721; 7.28 lbs/U.S. gal. bp$_{760}$ 121-122°; bp$_{37}$ 50°; bp$_{6.5}$ 20°. mp −30.6°. n$_D^{20}$ 1.4318. Flash pt 101.3°F. Surface tension at 25°: 23.8 dynes/cm; 5% aq soln: 34.1 dynes/cm. Soly in water at 25°: 12.8 g/100 ml. Sol in ether. Miscible with acetone, benzene, carbon tetrachloride, Cellosolve, cyclohexanone, diethylene glycol, ethyl acetate, kerosene, methyl ethyl ketone, mineral spirits, ethanolamine, neatsfoot oil, petr ether, soybean oil, Stoddard solvent. LD$_{50}$ orally in mice, rats, guinea pigs: 600-900 mg/kg (Margolin).
THERAP CAT: Hypnotic; sedative.

5732. Meparfynol Carbamate. *3-Methyl-1-pentyn-3-ol carbamate; carbamic acid 1-ethyl-1-methyl-2-propynyl ester;* 1-ethyl-1-methyl-2-propynyl carbamate; methylpentynol carbamate; N-Oblivon; Oblivon C; Trusono. C$_7$H$_{11}$NO$_2$; mol wt 141.17. C 59.55%, H 7.85%, N 9.92%, O 22.67%. Prepd by conversion of the *tert*-carbinol to the corresponding phenyl carbonate with phenyl chloroformate, followed by ammonolysis: McLamore *et al., J. Org. Chem.* **20**, 1379 (1955); from 3-methyl-1-pentyn-3-ol, trichloroacetic acid, KOCN: Marshall *et al.,* **U.S. pat.** 2,814,637 (1957 to British Schering); McCrea *et al.,* **Brit. pat.** 761,817 (1956 to British Schering). Acute toxicity: Soehring *et al., Arzneimittel-Forsch.* **5**, 161 (1955).

$$CH_3CH_2\underset{\underset{CH_3}{|}}{\overset{\overset{OOCNH_2}{|}}{C}}C\equiv CH$$

Crystals from ether + petr ether. mp 55.8-57° (McLamore); from cyclohexane, mp 53.5-55° (McCrea). bp$_{16}$ 120-121°; bp$_{0.01}$ 95°. Solubility in water: 1.6 g/100 ml. LD$_{50}$ (4 hr) s.c. in mice: 0.56 g/kg (Soehring).
THERAP CAT: Sedative, hypnotic.

5733. Mepartricin. Partricin methyl ester; methylpartricin; SN 654; SPA-S160; Ipertrofan; Orofungin; Tricandil; Tricangine. Methyl ester of the heptaene macrolide antibiotic complex, partricin, *q.v.* Prepn: T. Bruzzese *et al., Experientia* **28**, 1515 (1972); T. Bruzzese, R. Ferrari, **Ger. pat.** 2,154,436 (1972 to SPA); eidem, **U.S. pat.** 3,780,173 (1973). Alternate prepn: R. C. Pandey, *J. Antibiot.* **30**, 158 (1977); R. C. Tweit *et al., ibid.* **35**, 997 (1982). Structure of mepartricins A and B: J. Golik *et al., ibid.* **33**, 904 (1980).

Antimycotic activity: W. Ritzerfield, *Farmaco Ed. Sci.* **27**, 235 (1972); activity vs *Trichomonas vaginalis:* G. Pucci, S. Ripa, *Farmaco Ed. Prat.* **28**, 293 (1973). Hemolytic activity studies: B. Cybulska *et al., Biochem. Pharmacol.* **33**, 41 (1984). Stability data: S. Mizuba, K. Lee, *Dev. Ind. Microbiol.* **16**, 380 (1975). Use of mepartricin and its complexes in treatment of benign prostatic hypertrophy: T. Bruzzese, L. Ferrari, U.S. pat. **4,237,117** (1980 to SPA). Clinical studies: M. De Bernardi *et al., Curr. Ther. Res.* **43**, 1159 (1988); T. Lotti *et al., ibid.* **44**, 402 (1988).

mepartricin A: R' = CH₃
mepartricin B: R' = H

Deep yellow crystalline material. uv max (ethanol): 401, 378, 359, 340 nm. Slightly sol in water, aq alkali, ether, petr ether, benzene. Sol in acetone, alcohols, pyridine, DMF, DMSO. LD_{50} in mice: >2 g/kg orally; 200 mg/kg i.p. (Bruzzese, 1972).

Mepartricin A, $C_{60}H_{88}N_{20}O_{19}$, *40-demethyl-3,7-dideoxo-3,7-dihydroxy-N^{47}-methyl-5-oxocandicidin D methyl ester cyclic 15,19-hemiacetal, gedamycin methyl ester.* Lemon yellow powder from ether, mp 145-149° (dec). uv max (methanol): 400, 377, 357, 339, 287, 240, 234, 204 nm (ε 79326, 92454, 68094, 51685, 14199, 24612, 26505, 16092).

Mepartricin B, $C_{59}H_{86}N_{20}O_{19}$. Lemon yellow powder from ether, mp 154-158° (dec). uv max (methanol): 402, 379, 359, 340, 285, 233, 204 nm (ε 81101, 94729, 64171, 41558, 16196, 23835, 21696).

Complex with sodium lauryl sulfate, *SPA-S-222, Montricin.*

THERAP CAT: Antifungal; antiprotozoal (Trichomonas). In treatment of benign prostatic hypertrophy.

5734. Mepazine. *10-[(1-Methyl-3-piperidinyl)methyl]-10H-phenothiazine;* (N-methyl-3-piperidyl)methylphenothiazine; mepasin; pecazine; MPMP; P 391; III-2318; Paxital; Lacumin; Pacatal; Nothiazine; Pacatol. $C_{19}H_{22}N_2S$; mol wt 310.47. C 73.51%, H 7.14%, N 9.02%, S 10.33%. Prepn: Schuler, U.S. pat. **2,784,185** (1952 to Promonta).

bp₄ 230-235°.

Hydrochloride monohydrate, $C_{19}H_{23}ClN_2S.H_2O$, crystals, mp 180-181°. Photosensitive. Protect from light. Slightly bitter taste. Very slightly sol in water; freely sol in abs ethanol; sol in chloroform. Practically insol in ether, benzene.

THERAP CAT: Antipsychotic.

THERAP CAT (VET): Tranquilizer.

5735. Mepenzolate Bromide. *3-[(Hydroxydiphenylacetyl)oxy]-1,1-dimethylpiperidinium bromide;* N-methyl-3-piperidyl benzilate methyl bromide; N-methyl-3-piperidyl diphenylglycolate methobromide; Cantil; Cantril; Gastropidil; Trancolon. $C_{21}H_{26}BrNO_3$; mol wt 420.37. C 60.00%, H 6.23%, Br 19.01%, N 3.33%, O 11.42%. Prepn: Biel, U.S. pat. **2,918,408** (1959 to Lakeside Labs.). Prepn of base: Biel

et al., J. Am. Chem. Soc. **77**, 2250 (1955). Crystal structure: J. M. Leger *et al., Acta Crystallogr.* **B35**, 886 (1979). Toxicity data: E. I. Goldenthal, *Toxicol. Appl. Pharmacol.* **18**, 185 (1971).

Crystals, mp 228-229° (dec), (free base, bp₀.₀₃ 175-176°). LD_{50} orally in rats: 742±47 mg/kg (Goldenthal).

Note: N-Methyl-3-piperidyl benzilate is a controlled substance (hallucinogen) listed in the U.S. Code of Federal Regulations, Title 21, Part 1308.11 (1987).

THERAP CAT: Anticholinergic.

5736. Meperidine. *1-Methyl-4-phenyl-4-piperidinecarboxylic acid ethyl ester; 1-methyl-4-phenylisonipecotic acid ethyl ester;* N-methyl-4-phenyl-4-carbethoxypiperidine; ethyl 1-methyl-4-phenylpiperidine-4-carboxylate; isonipecaine; pethidine. $C_{15}H_{21}NO_2$; mol wt 247.35. C 72.84%, H 8.56%, N 5.66%, O 12.94%. Prepn: Eisleb, U.S. pat. **2,167,-351** (1939 to Winthrop); Smissman, Hite, *J. Am. Chem. Soc.* **81**, 1201 (1959). Pharmacokinetics: L. E. Mather, P. J. Meffin, *Clin. Pharmacokinet.* **3**, 352 (1978). Metabolism: S. Y. Yeh *et al., J. Pharm. Sci.* **70**, 867 (1981). Toxicity data: O. W. Barlow, J. R. Lewis, *J. Pharmacol. Exp. Ther.* **103**, 147 (1951). Comprehensive description: N. P. Fish, N. J. DeAngelis in *Analytical Profiles of Drug Substances* vol. 1, K. Florey, Ed. (Academic Press, New York, 1972) pp 175-205.

Hydrochloride, $C_{15}H_{22}ClNO_2$, *Algil, Alodan, Dolantin, Dolestine, Dolenal, Demerol hydrochloride, Dolosal, Endolat, Dispadol, Dolvanol, Dolcontral, Doloneurin(e), Centralgin, Dolopethin, Mefedina, Pantalgine, Sauteralgyl, Spasmedal, Spasmodolin, Mephedine, Lydol, Lidol, Synlaudine.* Minute crystals, mp 186-189°. Slightly bitter taste. Stable to air. Just acid to litmus (1:10 soln). Sol in water, acetone, ethyl acetate; slightly sol in alcohol, isopropanol. Insol in benzene, ether. Aq solns may be sterilized by boiling for short periods without dec. LD_{50} orally in rats: 170 mg/kg (Barlow, Lewis).

Caution: May be habit forming. This is a controlled substance (opiate) listed in the U.S. Code of Federal Regulations, Title 21 Part 1308.12 (1985).

THERAP CAT: Narcotic analgesic.

THERAP CAT (VET): Narcotic, sedative, analgesic, anesthetic.

5737. Mephenesin. *3-(2-Methylphenoxy)-1,2-propanediol; 3-(o-tolyloxy)-1,2-propanediol;* 1,2-dihydroxy-3-(2-methylphenoxy)propane; α-(o-tolyl)glyceryl ether; glyceryl o-tolyl ether; o-cresyl glycerol ether; cresoxypropanediol; cresoxydiol; BDH 312; Atensin; Avosyl; Avoxyl; Curythan; Daserol; Decontractyl; Dioloxol; Glyotol; Glykresin; Kinavosyl; Lissephen; Memphenesin; Mepherol; Mephesin; Mephson; Mervaldin; Myanesin; Myanol; Myodetensine; Myolysin; Myopan; Myoserol; Myoten; Myoxane; Oranixon; Prolax; Relaxar; Relaxil; Renarcol; Rhex "Hobein"; Sansdolor; Sinan; Spasmolyn; Stilalgin; Thoxidil; Tolansin; Tolax; Tolcil; Tolhart; Tolosate; Toloxyn; Tolserol; Tolseron; Tolulexin; Tolulox; Tolyspaz; Walconesin. $C_{10}H_{14}O_3$; mol wt 182.20. C 65.91%, H 7.74%, O 26.34%. Prepn: P. Morch, *Arch. Pharm. Chem.* **54**, 327 (1947), *C.A.* **42**, 2058 (1948); **Brit.** pat. **589,821** (1947 to Carroll and Boake Roberts). Prepn of the carbamate: Lott, Pribyl, U.S. pat. **2,609,386** (1952 to Squibb). Toxicity data: Dresel, Slater, *Proc. Soc.*

Exp. Biol. Med. **79,** 286 (1952); A. P. Roszkowski, *J. Pharmacol. Exp. Ther* **129,** 75 (1960).

Crystals. Bitter taste. Produces numbness of the tongue. mp 70-72°. uv max (0.005% aq soln): 270 nm (E 0.395). Freely sol in alcohol, propylene glycol, chloroform. At 20° one part dissolves in 85 parts water, in 11 parts ether. Urea and its derivatives, particularly urethan, increase the water soly. One part dissolves in 60 parts of 5% urethan soln, in 40 parts of 10%, in 4.5 parts of 25%. Aq solns are stable, can be sterilized by heating, and are compatible and freely miscible with solns of sodium chloride, glucose, and derivatives of barbituric and thiobarbituric acids. pH of satd aq soln about 6. LD_{50} in mice, rats (mg/kg): 471, 283 i.p.; 990, 945 orally (Roszkowski).

Carbamate, $C_{11}H_{15}NO_4$, *Tolseram.* Crystals from water, mp 93°. (Also a hemihydrate, mp 80-84°.) uv max (ethanol): 271, 277 nm ($A_{1cm}^{1\%}$ 72.7, 64.1). Has a lower water-solubility and a higher oil-solubility than mephenesin. Soly in water about 0.3%, in chloroform about 2.0%. Freely sol in alcohol. LD_{50} in mice (mg/kg): 7.67 orally; 2.77 i.p. (Dresel, Slater).

THERAP CAT: Muscle relaxant (skeletal).
THERAP CAT (VET): Muscle relaxant.

5738. Mephenhydramine. *2-(1,1-Diphenylethoxy)-N,N-dimethylethanamine;* α-methylbenzhydryl 2-dimethylaminoethyl ether; α-methyldiphenhydramine; Spofa 325; Alfadryl "Spofa"; Alphadryl "Spofa"; Alphadril "Spofa". $C_{18}H_{23}NO$; mol wt 269.37. C 80.25%, H 8.61%, N 5.20%, O 5.94%. Prepn: Protiva *et al., Chem. Listy* **43,** 257 (1949); Czech. pat. **86,516** (1957), *C.A.* **53,** 1261 (1959).

$$CH_3COCH_2CH_2N(CH_3)_2 \quad (C_6H_5)_2$$

$bp_{0.15}$ 129-136°.
Hydrochloride, $C_{18}H_{23}NO.HCl$, mp 168°.
THERAP CAT: Antihistaminic.

5739. Mephenoxalone. *5-[(2-Methoxyphenoxy)methyl]-2-oxazolidinone;* 5-(o-methoxyphenoxymethyl)-2-oxazolidinone; metoxadone; methoxydon(e); methoxadone; AHR 233; OM-518; Lenetran; Trepidone; Ekilan; Dorsiflex; Control-Om; Tranpoise; Dorsilon; Placidex; Riself; Xerene. $C_{11}H_{13}NO_4$; mol wt 223.22. C 59.18%, H 5.87%, N 6.28%, O 28.67%. Prepn: Lunsford *et al., J. Am. Chem. Soc.* **82,** 1166 (1960); Lunsford, U.S. pat. **2,895,960** (1959 to A. H. Robins). Toxicity data: E. I. Goldenthal, *Toxicol. Appl. Pharmacol.* **18,** 185 (1971).

Crystals from 95% ethanol, mp 143-145°. Water insol. LD_{50} orally in rats: 3820 ±17 mg/kg (Goldenthal).
THERAP CAT: Anxiolytic, muscle relaxant (skeletal).

5740. Mephentermine. *N,α,α-Trimethylbenzeneethanamine; N,α,α-trimethylphenethylamine;* 2-methylamino-2-methyl-1-phenylpropane; N-methyl-ω-phenyl-*tert*-butylamine; Wyamine; Vialin; Mephine. $C_{11}H_{17}N$; mol wt 163.25. C 80.92%, H 10.50%, N 8.58%. Prepd by hydrogenating 1-chloro-2-methylamino-2-methyl-1-phenylpropane hydrochloride in methanol in the presence of palladium barium carbonate: Bruce, Szabo, U.S. pat. **2,597,445** (1952 to Wyeth). Alternate process: Abel *et al.,* U.S. pat. **2,590,079** (1952 to Wyeth).

$$C_6H_5-CH_2\underset{CH_3}{\overset{CH_3}{C}}NHCH_3$$

Liquid. Fishy amine odor. Alkaline reaction. Freely sol in alc. Sol in ether. Practically insol in water. LD_{50} i.p. in mice: 100-110 mg/kg, *Handbook of Toxicology* **vol. 1,** W. S. Spector, Ed. (Saunders, Philadelphia, 1956) pp 184-185.

Sulfate dihydrate, $(C_{11}H_{17}N)_2.H_2SO_4.2H_2O$, crystals. One gram dissolves in 20 ml water, about 150 ml ethanol (95%). Practically insol in chloroform. pH of aq soln about 6. Picrate, yellow crystals, m 155-156°.
THERAP CAT: Adrenergic (vasopressor).

5741. Mephenytoin. *5-Ethyl-3-methyl-5-phenyl-2,4-imidazolidinedione;* 5-ethyl-3-methyl-5-phenylhydantoin; 3-methyl-5,5-phenylethylhydantoin; "methyl hydantoin"; phenylethylmethylhydantoin; 3-ethylnirvanol; methoin; Insulton; Mesontoin; Mesantoin; Phenantoin; Sedantoinal; Gerot-Epilan; Sacerno. $C_{12}H_{14}N_2O_2$; mol wt 218.25. C 66.03%, H 6.47%, N 12.84%, O 14.66%. Prepd from 5,5-ethylphenylhydantoin by the action of 1 mol of dimethylsulfate: **Fr. pat. 769,667; Swiss pat. 166,004** (both 1934 to Sandoz). By treating phenylmethylureidoacetonitrile with HCl: **Swiss pat. 179,692** (1935 to Sandoz).

Crystals, mp 136-137°. Insol in water. Forms a water-soluble sodium salt which has an alkaline reaction. LD_{100} i.p. in rats: 270 mg/kg.
THERAP CAT: Anticonvulsant.

5742. Mephobarbital. *5-Ethyl-1-methyl-5-phenyl-2,4,6(1H,3H,5H)-pyrimidinetrione; 5-ethyl-1-methyl-5-phenylbarbituric acid;* 5-phenyl-5-ethyl-3-methylbarbituric acid; N-methylethylphenylbarbituric acid; methylphenobarbital; Phemiton; Prominal; Mebaral; Isonal. $C_{13}H_{14}N_2O_3$; mol wt 246.26. C 63.40%, H 5.73%, N 11.38%, O 19.49%. Prepn: Taub, Kropp, **Ger. pat. 537,366** (1929 to I. G. Farben.).

White, tasteless crystals. mp 176°. Slightly sol in cold, freely in hot water or in alcohol.
Caution: May be habit forming. This is a controlled substance (depressant) listed in the U.S. Code of Federal Regulations, Title 21 Parts 329.1 and 1308.14 (1987).
THERAP CAT: Anticonvulsant; sedative, hypnotic.
THERAP CAT (VET): Anticonvulsant; sedative, hypnotic.

5743. Mephosfolan. *(4-Methyl-1,3-dithiolan-2-ylidene)phosphoramidic acid diethyl ester; phosphonodithioimidocarbonic acid cyclic propylene P,P-diethyl ester;* 2-diethoxyphosphinylimino-4-methyl-1,3-dithiolane; cyclic propylene (diethoxyphosphinyl)dithioimidocarbonate; EI 47470; ENT 25991; Cytrolane. $C_8H_{16}NO_3PS_2$; mol wt 269.31. C 35.68%, H 5.99%, N 5.20%, O 17.82%, P 11.50%, S 23.81%. Insecticidal contact and stomach poison with systemic activity: D. L. Bull *et al., J. Econ. Entomol.* **57,** 112 (1964). Prepn: J. B. Lovell, U.S. pat. **3,197,365** (1965 to Am. Cyanamid). Metabolism: J. Zulalian, R. C. Blinn, *J. Agr. Food Chem.* **25,** 1033 (1977). Photodegradation: C. C. Ku *et al., ibid.* **27,** 1046 (1979).

Yellow to amber liquid, $bp_{0.001}$ 120°. n_D^{26} 1.5354. Soly in water at 25°: 57 g/kg. Sol in acetone, ethanol, benzene, 1,2-dichloroethane. Stable at neutral pH; hydrolyzed by acid or alkali.
USE: Insecticide, acaricide.

5744. Mepindolol. *1-[(1-Methylethyl)amino]-3-[(2-methyl-1H-indol-4-yl)oxy]-2-propanol; 1-(isopropylamino)-3-[(2-methylindol-4-yl)oxy]-2-propanol.* $C_{15}H_{22}N_2O_2$; mol wt 262.36. C 68.67%, H 8.45%, N 10.68%, O 12.20%. β-Adrenergic blocker; the 2-methyl analog of pindolol, *q.v.* Prepn: F. Troxler, **Swiss** pats. **469,002** and **472,404** (both 1969 to Sandoz); F. Seeman *et al., Helv. Chim. Acta* **54**, 2411 (1971). Pharmacokinetics: R. Gugler *et al., Arzneimittel-Forsch.* **25**, 1067 (1975). Pharmacodynamics: J. Bonelli *et al., Eur. J. Clin. Pharmacol.* **15**, 1 (1979). Clinical studies: H. M. Beumer *et al., Int. J. Clin. Pharmacol. Biopharm.* **16**, 249 (1978); M. Sukerman *et al., Curr. Ther. Res.* **25**, 384 (1979).

Crystals from ethyl acetate, mp 100-102° (Seeman). Also reported as 95-97° (Troxler).
Sulfate salt, $C_{30}H_{46}N_4O_8S$, *SH-E-222, Betagon, Caridian, Corindolan.*
THERAP CAT: Antihypertensive, antianginal.

5745. Mepiprazole. *1-(3-Chlorophenyl)-4-[2-(5-methyl-1H-pyrazol-3-yl)ethyl]piperazine; 3-[2-(N'-m-chlorophenylpiperazino)ethyl]-5-methylpyrazole;* EMD-16923. $C_{16}H_{21}ClN_4$; mol wt 304.83. C 63.04%, H 6.94%, Cl 11.63%, N 18.38%. Psychotropic agent with CNS-depressant effects. Prepn: K. Volker *et al., Brit.* pat. **1,124,710** corresp to **U.S.** pat. **3,491,097** (1968, 1970 both to E. Merck); *eidem, Eur. J. Med. Chem.* **10**, 154, 162 (1975). Pharmacological study: M. A. Ruch-Monachon *et al., Arch. Int. Pharmacodyn. Ther.* **219**, 326 (1976). Effects on brain biogenic amine metabolism: C. Seyfried *et al., Arzneimittel-Forsch.* **26**, 1088 (1976). EEG effects in rabbits: S. Watanabe *et al., ibid.* **29**, 274 (1979).

Cryst, mp 106°.
Dihydrochloride, $C_{16}H_{23}Cl_3N_4$, *H-4007, Psigodal.* Cryst from ethanol, mp 234°.
THERAP CAT: Tranquilizer.

5746. Mepiquat Chloride. *1,1-Dimethylpiperidinium chloride;* BAS-083; BAS 85559X; Pix. $C_7H_{16}ClN$; mol wt 149.66. C 56.17%, H 10.78%, Cl 23.68%, N 9.36%. Prepn: B. Zeeh *et al.,* **Ger.** pat. **2,207,575** corresp to **U.S.** pat. **3,-905,798** (1973, 1975 to BASF).

mp >350°. LD_{50} orally in rats: 1420 mg/kg, *RTECS* Vol. II, R. J. Lewis, R. L. Tatken, Eds. (1980) p 390.
USE: Plant growth regulator.

5747. Mepitiostane. *2,3-Epithio-17-[(1-methoxycyclopentyl)oxy]androstane; cyclopentanone 2α,3α-epithio-5α-androstan-17β-yl methyl acetal;* 10364-S; Thiodelone; Thioderon. $C_{25}H_{40}O_2S$; mol wt 404.66. C 74.21%, H 9.96%, O 7.91%, S 7.92%. Orally active deriv of epitiostanol, *q.v.* Prepn: T. Komeno, **S. Afr.** pat. **68 00,565** corresp to **U.S.** pat. **3,567,713** (1968, 1971 both to Shionogi); T. Komeno *et al., Shionogi Kenkyusho Nempo* **19**, 3 (1969), *C.A.* **72**, 28554u (1970). Inhibitory effect on mammary tumors in rats: O. Takatani, S. Kumaoka, *Gann* **68**, 337 (1977), *C.A.* **87**, 112187u (1977). Clinical study in advanced breast cancer: K. Inoue *et al., Cancer Treat. Rep.* **62**, 743 (1978). Toxicity studies: *Oyo Yakuri* **16**, 739-812 (1978), *C.A.* **90**, 146152f, 146153g (1979).

Cryst, mp 98-101°. $[\alpha]_D^{20}$ +22.5 ±0.5° (c = 1 in chloroform).
THERAP CAT: Antineoplastic.

5748. Mepivacaine. *N-(2,6-Dimethylphenyl)-1-methyl-2-piperidinecarboxamide; 1-methyl-2',6'-pipecoloxylidide; dl-N-methylpipecolic acid 2,6-dimethylanilide; dl-N-methylhexahydropicolinic acid 2,6-dimethylanilide.* $C_{15}H_{22}N_2O$; mol wt 246.34. C 73.13%, H 9.00%, N 11.37%, O 6.50%. Prepn: Ekenstam *et al., Acta Chem. Scand.* **11**, 1183 (1957); **U.S.** pat. **2,799,679** (1957 to A. B. Bofors). Prepn of other salts: Rinderknecht, *Helv. Chim. Acta* **42**, 1324 (1959); **Brit.** pat. **826,668** (1960 to Crookes Labs.). Resolution of isomers: Tullar, *J. Med. Chem.* **14**, 891 (1971); Friberger, Aberg, *Acta Pharm. Suecica* **8**, 361 (1971). Pharmacology: Helmy *et al., J. Egypt. Med. Assoc.* **50**, 688 (1967). Metabolism: Reynolds, *Brit. J. Anaesth.* **43**, 33 (1971). Toxicity data: G. Aberg, *Acta Pharmacol. Toxicol.* **31**, 273 (1972).

Crystals from ether, mp 150-151°.
Hydrochloride, $C_{15}H_{23}ClN_2O$, *Carbocaina, Carbocaine hydrochloride, Chlorocain, Meaverin, Mepicaton, Mepivastesin, Scandicain.* mp 262-264°. Sol in water. LD_{50} in mice, rats (mg/kg): 280, 500 s.c. (Aberg).
(+)-Form, *dexivacaine.*
THERAP CAT: Local anesthetic.
THERAP CAT (VET): Local anesthetic.

5749. Mepixanox. *3-Methoxy-4-(1-piperidinylmethyl)-9H-xanthen-9-one; 3-methoxy-4-piperidinylmethyl-9-oxo-10-oxa-9,10-dihydroanthracene; mepixanthone; Pimexone.* $C_{20}H_{21}NO_3$; mol wt 323.39. C 74.28%, H 6.55%, N 4.33%, O 14.84%. Prepn: M. Sparaci, **S. Afr.** pat. **69 02,150** (1969 to

Mondi), *C.A.* **72**, 111,300d (1970). Alternate process: D. Milani, U.S. pat. **3,646,030** (1972 to Mondi). CNS activity: P. Da Re *et al., J. Med. Chem.* **13**, 527 (1970). Pharmacology: R. Guira *et al., Minerva Pneumolog.* **21**, 317 (1982). Pharmacokinetics in man: G. Grossi *et al., Curr. Ther. Res.* **38**, 141 (1985). HPLC determn in serum: G. Grossi *et al., J. Chromatog.* **309**, 214 (1984). Clinical comparison with doxapram, *q.v.*, in chronic bronchopulmonary disease: M. Parziale *et al., G. Ital. Mal. Torace* **38**, 323 (1984). Clinical evaluations in chronic obstructive lung disease: D. Olivieri *et al., Arch. Monaldi Tisiol. Mal. Appar. Respir.* **35**, 117 (1980); L. Bertoli *et al., Minerva Pneumolog.* **23**, 323 (1984).

White crystalline powder from ethyl acetate, mp 159-160°. LD_{50} i.p. in mice: 70.73 mg/kg (Sparaci).

Hydrochloride, $C_{20}H_{22}ClNO_3$, white crystalline solid from ethanol, mp > 200° (dec).

THERAP CAT: Respiratory stimulant.

5750. Meprednisone. *17,21-Dihydroxy-16β-methylpregna-1,4-diene-3,11,20-trione;* 16β-methylprednisone; Betapred; Deltacortene Beta; Betalone; Deltisona B. $C_{22}H_{28}O_5$; mol wt 372.46. C 70.94%, H 7.58%, O 21.48%. Prepn: Taub *et al., J. Am. Chem. Soc.* **80**, 4435 (1958); Oliveto *et al., ibid.* 4428; Taub *et al., ibid.* **82**, 4012 (1960); Nathanson *et al., Experientia* **17**, 448 (1961).

Crystals, mp 200-205°. $[\alpha]_D$ +200° (dioxane). uv max (methanol): 239 nm ($E_{1cm}^{1\%}$ 416).

21-Acetate, $C_{24}H_{30}O_6$, mp 232-235°. $[\alpha]_D$ +210° (dioxane). uv max (methanol): 238 nm ($E_{1cm}^{1\%}$ 358).

THERAP CAT: Glucocorticoid.

5751. Meprobamate. *2-Methyl-2-propyl-1,3-propanediol dicarbamate; carbamic acid 2-methyl-2-propyltrimethylene ester;* 2,2-di(carbamoyloxymethyl)pentane; 2-methyl-2-propyltrimethylene carbamate; procalmadiol; procalmidol; Amosene; Anastress; Andaxin; Aneural; Apascil; Arcoban; Artolon; Atraxin; Ayeramate; Bamo 400; Biobamate; Calmax; Calmiren; Cap-O-Tran; Cirpon; Crestanil; Cyrpon; Ecuanil; Equanil; Equinil; Fas-Cile 200; Gadexyl; Harmonin; Hartol; Holbamate; Kesso-Bamate; Klort; Larten; Lepetown; Libiolan; Mar-Bate; Mepantin; Mepavlon; Meposed; Meprin; Meprindon; Meproban; Meprocompren; Meprol; Meprosin; Meprospan; Meprotabs; Meprotan; Meprotil; Meptran; Mesmar; Miltaun; Miltown; Morbam; My-trans; Nervonus; Oasil; Panediol; Perequil; Perquietil; Pertranquil; Placidon; Placitate; Probamyl; Promate; Protran; Quaname; Quanil; Reostral; Restenil; Robamate; Sedazil; Seril; Setran; Sowell; Tamate; Trankvilan; Tranlisant; Tranquilan; Tranquilax; Tranquiline; Urbil; Urbilat; Viobamate. $C_9H_{18}N_2O_4$; mol wt 218.25. C 49.53%, H 8.31%, N 12.84%, O 29.32%. Prepn: Ludwig, Piech, *J. Am. Chem. Soc.* **73**, 5779 (1951); Berger, Ludwig, U.S. pat. **2,724,720** (1955 to Carter Prod.). Synthesis of 2-methyl-2-propyl-1,3-propanediol: Fries, Mönkemeyer, **Swiss** pat. **373,026** (1963 to Chemische Werke Hüls). Pharmacology: F. M. Berger, *J. Pharmacol. Exp. Ther.* **112**, 413 (1954). Comprehensive description: C. Shearer, P. Rulon in *Analytical Profiles of Drug Substances* vol. 1, K. Florey, Ed. (Academic

Press, New York, 1972) pp 207-232; C. Shearer, *ibid.* **vol. 11** (1982) pp 587-591.

Crystals from hot water, mp 104-106°. Characteristic bitter taste. Soly in water at 20°: 0.34% (w/w), at 37°: 0.79% (w/w). Easily forms supersatd solns with hot water. Freely sol in most organic solvents. Aq solns are neutral. The substance is stable in dilute acid and alkali and is not broken down in gastric or intestinal juices. LD_{50} i.p. in mice: 800 mg/kg (Berger).

Caution: May be habit forming. This is a controlled substance (depressant) listed in the U.S. Code of Federal Regulations, Title 21 Part 1308.14 (1987).

THERAP CAT: Anxiolytic.

5752. Meprylcaine Hydrochloride. *2-Methyl-2-propylamino-1-propanol benzoate hydrochloride;* β-(n-propylamino)-β,β-dimethylethyl benzoate hydrochloride; 2-methyl-2-propylaminopropyl benzoate hydrochloride; Oracaine hydrochloride. $C_{14}H_{22}ClNO_2$; mol wt 271.80. C 61.87%, H 8.16%, Cl 13.05%, N 5.15%, O 11.77%. Prepn: Reasenberg, U.S. pat. **2,767,207** (1956 to Mizzy).

Crystals, mp 150-151°. Sol in water, alcohol. pH of 1:50 aq soln: 5.7.

Base, $C_{14}H_{21}NO_2$, oil. Practically insol in water. Sol in alcohol, ether, acetone, oils.

THERAP CAT: Local anesthetic.

5753. Meptazinol. *3-(3-Ethylhexahydro-1-methyl-1H-azepin-3-yl)phenol;* 1-methyl-3-ethyl-3-(m-hydroxyphenyl)hexahydro-1H-azepine. $C_{15}H_{23}NO$; mol wt 233.36. C 77.21%, H 9.93%, N 6.00%, O 6.86%. Mixed opioid agonist-antagonist. Prepn of the hydrobromide: J. F. Cavalla, A. C. White, **Ger.** pat. **1,941,534** corresp to U.S. pat. **3,729,465** (1970, 1973 both to Wyeth); of the free base: *eidem,* U.S. pat. **4,197,241** (1980 to Wyeth). Properties: P. G. Goode, A. C. White, *Brit. J. Pharmacol.* **43**, 462 (1971). HPLC determn in plasma: T. Frost, *Analyst* **106**, 999 (1981). Pharmacokinetics in man: G. Davies, *Eur. J. Clin. Pharmacol.* **23**, 535 (1982). Human pharmacology and abuse potential: R. E. Johnson, D. R. Jasinski, *Clin. Pharmacol. Ther.* **41**, 426 (1987). Clinical studies: R. K. Price, A. N. Latham, *Curr. Med. Res. Opin.* **8**, 54 (1982); C. E. Parker, A. F. Langrick, *J. Int. Med. Res.* **10**, 408 (1982). Symposium on clinical studies: *Postgrad. Med. J.* **59**, Suppl. 1, 1-94 (1983). Review of pharmacology and clinical efficacy: B. Holmes, A. Ward, *Drugs* **30**, 285-312 (1985).

Cryst from acetonitrile, mp 127.5-133°.

Hydrochloride, $C_{15}H_{24}ClNO$, *WY 22811, Meptid.*

THERAP CAT: Narcotic analgesic.

5754. Mequitazine. *10-(1-Azabicyclo[2.2.2]oct-3-ylmethyl)-10H-phenothiazine; 10-(3-quinuclidinylmethyl)-phenothiazine;* LM 209; Metaplexan; Mircol; Primalan; Zesulan. $C_{20}H_{22}N_2S$; mol wt 322.47. C 74.49%, H 6.88%, N 8.69%, S 9.94%. Prepn: G. Gueremy *et al.,* **Ger.** pat. **2,009,-555** corresp to U.S. pat. **3,987,042** (1970, 1976 to Sogeras). Absorption, distribution, and excretion: A. Uzan *et al., Xenobiotica* **6**, 633 (1976). Biotransformation: *eidem, ibid.* 649. Pharmacologic study: *eidem, J. Pharm. Pharmacol.* **31**,

701 (1979). Clinical studies: J. Blamoutier, *Curr. Med. Res. Opin.* **5**, 366 (1978); P. Laugier, M. Orusco, *ibid.* 371.

Crystals from acetonitrile, mp 130-131°.
THERAP CAT: Antihistaminic.

5755. Meralein Sodium. (*3',6'-Dihydroxy-2',7'-diiodospiro[3H-2,1-benzoxanthiole-3,9'-[9H]xanthen]-4'-yl)-hydroxymercury S,S-dioxide monosodium salt; o-[6-hydroxy-5-(hydroxymercuri)-2,7-diiodo-3-oxo-3H-xanthen-9-yl]-benzenesulfonic acid sodium salt; 2,7-diiodo-4-hydroxymercuriresorcinsulfonphthalein monosodium salt; monohydroxymercuridiiodoresorcinsulfonphthalein sodium salt; sodium meralein; Merodicein.* $C_{19}H_9HgI_2NaO_7S$; mol wt 858.77. C 26.57%, H 1.06%, Hg 23.36%, I 29.56%, Na 2.68%, O 13.04%, S 3.73%. Prepn: Dunning, Farinholt, *J. Am. Chem. Soc.* **51**, 804 (1929). Pharmacology and toxicology: Macht, Cook, *J. Pharmacol. Exp. Ther.* **43**, 571 (1931).

Green scales, turn dark red on pulverizing. Sol in water; aq soln slightly fluorescent.
THERAP CAT: Topical anti-infective.

5756. Meralluride. *[3-[[[(3-Carboxy-1-oxopropyl)amino]carbonyl]amino]-2-methoxypropyl]hydroxymercury, mixture with 3,7-dihydro-1,3-dimethyl-1H-purine-2,6-dione; [3-[3-(3-carboxypropionyl)ureido]-2-methoxypropyl](theophyllinato)mercury; N-[[2-methoxy-3-[(1,2,3,6-tetrahydro-1,3-dimethyl-2,6-dioxopurin-7-yl)mercuri]propyl]carbamoyl]succinamic acid.* $C_{16}H_{22}HgN_6O_7$; mol wt 610.99. C 31.45%, H 3.63%, Hg 32.83%, N 13.75%, O 18.33%. Structure: Pearson, Sigal, *J. Org. Chem.* **15**, 1055 (1950).

White to slightly yellow powder, slowly dec on exposure to light. Saturated soln is acid to litmus. Slightly sol in water; sol in hot water, glacial acetic acid, and solns of alkali hydroxides. Almost insol in alc, chloroform, ether. LD_{50} s.c. in rats: 28±7 mg/kg, E. I. Goldenthal, *Toxicol. Appl. Pharmacol.* **18**, 185 (1971).
Sodium salt, $C_{16}H_{21}HgN_6NaO_7$, *Dilurgen, Mercardan, Mercuhydrin, Mercuretin.*
THERAP CAT: Diuretic.
THERAP CAT (VET): Diuretic.

5757. Merbromin. *(2'7'-Dibromo-3',6'-dihydroxy-3-oxospiro[isobenzofuran-1(3H),9'-[9H]xanthen]-4'-yl)hydroxymercury disodium salt; [2,7-dibromo-9-(o-carboxyphenyl)-6-hydroxy-3-oxo-3H-xanthen-4-yl]hydroxymercury disodium salt;* mercurochrome; dibromohydroxymercurifluo-

rescein disodium salt; no. 220 sol; Mercurochrome-220 Soluble; Chromargyre; Planochrome; Flavurol; D.O.M.F.; Mercurophage; Mercurocol; Gallochrome; Gynochrome; Mercurome; Asceptichrome; Mercuranine. $C_{20}H_8Br_2Hg-Na_2O_6$; mol wt 750.70. C 32.00%, H 1.07%, Br 21.29%, Hg 26.72%, Na 6.13%, O 12.79%. Prepd by treating dibromofluorescein with mercuric acetate and sodium hydroxide: White, *J. Am. Chem. Soc.* **42**, 2355 (1920); also prepd by the action of mercuric acetate on dibromofluorescein sodium: Rymill, Corran, *Quart. J. Pharm. Pharmacol.* **7**, 543 (1934), see also U.S. pat. **1,535,003** (1925).

Trihydrate, iridescent green scales or granules. Freely sol in water, giving a carmine-red soln. Very dil solns (1:2000) possess a yellow-green fluorescence. pH of 0.5% soln: 8.8. One gram dissolves in 50 grams of 94% alcohol, in 8.1 grams methanol: Denoel, *J. Pharm. Belg.* **22**, 423 (1940); **23**, 75 (1941). Practically insol in alcohol, acetone, chloroform, ether. *Incompat.* Acids, most alkaloidal salts and most local anesthetics. Colors the skin carmine-red. The stains may be removed by washing first with permanganate soln and then with oxalic acid soln.
THERAP CAT: Antibacterial.
THERAP CAT (VET): Antiseptic.

5758. Mercamphamide. *3-[[3-(Hydroxymercuri)-2-methoxypropyl]carbamoyl]-1,2,2-trimethylcyclopentanecarboxylic acid; N-[3-(hydroxymercuri)-2-methoxypropyl]camphoramic acid; N-(γ-hydroxymercuri-β-methoxy)propylcamphoramic acid; Mercurin.* $C_{14}H_{25}HgNO_5$; mol wt 487.97. C 34.46%, H 5.16%, Hg 41.11%, N 2.87%, O 16.39%. Prepn: Molnar, U.S. pat. **2,117,901** (1938); Werner, Scholz, *J. Am. Chem. Soc.* **76**, 2453 (1954).

Sodium salt, $C_{14}H_{24}HgNNaO_5$. Prepd as a mixture of 20% of the free acid plus 80% of the sodium salt. Mixture contains about 40% Hg and is described as a white, bitter powder; very slightly sol in water or ether; sol in alcohol and dil solns of sodium hydroxide. An aq suspension is practically neutral. pH about 8.
Mixture of sodium salt with theophylline, *mercurophyllin, Novurit.*
THERAP CAT: Diuretic.

5759. 2-Mercaptobenzothiazole. *2(3H)-Benzothiazolethione; 2-benzothiazolethiol;* MBT; Captax; Dermacid; Mertax; Thiotax. $C_7H_5NS_2$; mol wt 167.25. C 50.27%, H 3.01%, N 8.38%, S 38.34%. Prepd industrially by reacting aniline, carbon disulfide, and sulfur at elevated press. and temps: Kelly, U.S. pat. **1,631,871**. Purification by treatment with a per-acid salt in alkaline medium: Weyker, Ebel, U.S. pat. **2,730,528** (1956 to Am. Cyanamid). Improved process: Szlatinay, U.S. pat. **3,031,073** (1962 to Monsanto).

Pale yellow, monoclinic needles or leaflets. Disagreeable

odor. d 1.42. mp 180.2-181.7° (the technical product mp 170-175°). Practically insol in water. Soly at 25° (g/100 ml) in alcohol: 2.0; ether 1.0; acetone 10.0; benzene 1.0; carbon tetrachloride <0.2; naphtha <0.5. Moderately sol in glacial acetic acid. Sol in alkalies and alkali carbonate solns.

Zinc salt, *Bantex*. Light yellow powder, d_4^{25} 1.70.

Sodium salt, *Nuodex 84*.

USE: Rubber vulcanization accelerator. Salts used as fungicide.

5760. 2-Mercaptoethanol. β-Mercaptoethanol; 2-hydroxy-1-ethanethiol; 2-hydroxyethyl mercaptan; monothioethyleneglycol; thioglycol. C_2H_6OS; mol wt 78.13. C 30.74%, H 7.74%, O 20.48%, S 41.03%. $SHCH_2CH_2OH$. Prepn: Woodward, *J. Chem. Soc.* **1948**, 1892; Peppel, Signaigo; Jones, U.S. pats. **2,402,665; 3,394,192** (1946, 1968 both to du Pont). Properties: Bennett, *J. Chem. Soc.* **121**, 2139 (1922). IR: Thompson, *J. Am. Chem. Soc.* **61**, 1398 (1939). Toxicology: K. White *et al.*, *J. Pharm. Sci.* **62**, 237 (1973).

Liquid with very strong, disagreeable odor. bp_{742} 157-158° (dec). d_4^{20} 1.1143. n_D^{20} 1.4996. The pure liquid is miscible with water, alc, ether and benzene. LD_{50} in mice (mg/kg): 322.0 i.p.; 344.8 orally (White).

USE: In organic synthesis; as biochemical research tool.
Caution: Irritating to eyes, nose and skin

5761. Mercaptomerin Sodium. *[3-[[(3-Carboxylato-2,2,3-trimethylcyclopentyl)carbonyl]amino]-2-methoxypropyl][mercaptoacetato(2−)-O,S]mercurate(2−) disodium(T-4); N-(γ-carboxymethylmercaptomercuri-β-methoxy)propylcamphoramic acid disodium salt;* Diucardyn sodium; Thiomerin sodium. $C_{16}H_{25}HgNNa_2O_6S$; mol wt 606.04. C 31.71%, H 4.16%, Hg 33.10%, N 2.31%, Na 7.59%, O 15.84%, S 5.29%. Prepn: Lehman, U.S. pat. **2,576,349** (1951 to Wyeth); Wendt, Bruce, *J. Org. Chem.* **23**, 1448 (1958); Wendt, U.S. pat. **2,834,795** (1958 to Am. Home Prod.).

Hygroscopic white powder. Dec 150-155°. Freely soluble in water. Soluble in alcohol. Practically insoluble in ether, benzene, chloroform.

THERAP CAT: Diuretic.

THERAP CAT (VET): Diuretic.

5762. 6-Mercaptopurine. *6-Purinethiol;* 6MP; Leukerin; Mercaleukin; Purinethol; Puri-Nethol. $C_5H_4N_4S$; mol wt 152.19. C 39.46%, H 2.65%, N 36.82%, S 21.07%. Prepd from hypoxanthine and phosphorus pentasulfide: Elion *et al.*, *J. Am. Chem. Soc.* **74**, 411 (1952); U.S. pats. **2,721,866, 2,724,711** (1955 to Burroughs Wellcome). Improved procedure: Beaman, Robins, *J. Am. Chem. Soc.* **83**, 4042 (1961). Prepn from 7-aminothiazolo[5,4-*d*]pyrimidine: Hitchings, Elion, U.S. pat. **2,933,498** (1960 to Burroughs Wellcome). Metabolized in the body to 6-thiouric acid (6-mercapto-2,8-purinediol): Elion *et al.*, *J. Am. Chem. Soc.* **81**, 3042 (1959). Pharmacology: H. Froberg, M. S. Schencking, *Arch. Toxicol.* **32**, 1 (1974). Comprehensive description: S. A. Benezra, P. R. B. Foss in *Analytical Profiles of Drug Substances* vol. 7, K. Florey, Ed. (Academic Press, New York, 1978) pp 343-357.

Monohydrate, yellow prisms from water. Becomes anhydrous at 140°, decomp 313-314°. uv max (0.1N NaOH): 230, 312 nm (ε 14000, 19600); (0.1N HCl): 222, 327 nm (ε 9240, 21300); (methanol): 216, 329 nm (ε 8940, 19300). pKa_1 7.77, pKa_2 11.17. Insol in water, acetone, ether. Sol in hot ethanol, in alkaline solns with slow decompn. LD_{50} in

mice, hamsters (mg/kg): 157, 364 i.p. (Froberg, Schencking).

THERAP CAT: Antineoplastic.

THERAP CAT (VET): Antineoplastic.

5763. Mercufenol Chloride. *o-(Chloromercuri)phenol; o*-hydroxyphenylmercuric chloride. C_6H_5ClHgO; mol wt 329.18. C 21.89% H 1.53%, Cl 10.77%, Hg 60.94%, O 4.86%. Prepd from phenol and mercuric acetate: Whitmore, Hansen, *Org. Syn.* coll. vol. I, 161 (1941); Kaplan, Mellick, U.S. pat. **2,502,382** (1950 to Edwal Labs.).

Feathery crystals, mp 150-152°. Slightly sol in cold water, moderately sol in boiling water; freely sol in alcohol and hot benzene, sparingly sol in chloroform. *Poisonous!*

USE: Disinfectant.

5764. Mercumallylic Acid. *[3-(3-Carboxy-2-oxo-2H-1-benzopyran-8-yl)-2-methoxypropyl]hydroxymercurate(1−) hydrogen; [3-[3-(hydroxymercuri)-2-methoxypropyl]salicylidene]malonic acid δ-lactone;* 8-[3-(hydroxymercuri)-2-methoxypropyl]-2-oxo-2H-1-benzopyran-3-carboxylic acid; 8-(2-methoxy-3-hydroxymercuripropyl)coumarin-3-carboxylic acid; 8-[3-(hydroxymercuri)-2-methoxypropyl]-3-carboxycoumarin; 8-[3-(hydroxymercuri)-2-methoxypropyl]coumarin-3-carboxylic acid. $C_{14}H_{14}HgO_6$; mol wt 478.86. C 35.11%, H 2.95%, Hg 41.89%, O 20.05%. Prepn: Schlesinger *et al.*, U.S. pat. **2,667,442** (1954 to Endo); Werner, Scholz, *J. Am. Chem. Soc.* **76**, 2453 (1954).

Powder, bitter taste. mp 155-160° (Schlesinger); mp 197° (Werner, Scholz). One gram dissolves in about 4.2 parts of 1N NaOH. Slightly sol in acetic acid. Very slightly sol in water, alcohol, chloroform. Practically insol in ether.

USE: The water-soluble sodium salt is used in the prepn of diuretics.

THERAP CAT: Diuretic.

5765. Mercumatilin Sodium. *[3-(3-Carboxy-2-oxo-2H-1-benzopyran-8-yl)-2-methoxypropyl]hydroxymercury sodium salt compd with theophylline;* mercumallylic acid theophylline sodium; 8-(γ-hydroxymercuri-β-methoxypropyl)-3-coumarincarboxylic acid theophylline sodium salt; Cumertilin Sodium. $C_{21}H_{21}HgN_4NaO_8$; mol wt 681.01. C 37.04%, H 3.11%, Hg 29.45%, N 8.23%, Na 3.38%, O 18.79%. Prepd by adding just enough sodium hydroxide soln to mercumatilin to effect soln. The salt is not isolated. An excess over 1 mol of theophylline may be added. *See also* Mercumallylic Acid.

Used as an aq soln contg 0.132 g mercumatilin sodium (equivalent to 39 mg Hg) and 11 mg of excess theophylline in each ml. LD_{50} orally in rats: 238 mg/kg, Blumberg *et al.*, *J. Pharmacol. Exp. Ther.* **105**, 336 (1952).

THERAP CAT: Diuretic.

5766. Mercuric Acetate. $C_4H_6HgO_4$; mol wt 318.70. C

15.07%, H 1.90%, Hg 62.95%, O 20.08%. $Hg(CH_3COO)_2$. Prepd from HgO and acetic acid: Wagenknecht, Juza in *Handbook of Preparative Inorganic Chemistry*, vol. 2, G. Brauer, Ed. (Academic Press, New York, 2nd ed., 1965) pp 1120-1121.

Crystals or cryst powder; slight acetic odor; sensitive to light. *Poison!* d 3.28. mp 178-180° (overheating results in decompn). One g dissolves in 2.5 ml cold, 1 ml boiling water; sol in alcohol. *Keep well closed and protected from light.* Aq solns decomp on standing, yielding a yellow ppt.

USE: Chiefly for mercuration of organic compounds; for the absorption of ethylene.

5767. Mercuric Arsenate. $AsHHgO_4$; mol wt 340.53. As 22.00%, H 0.30%, Hg 58.91%, O 18.79%. $HgHAsO_4$.

Yellow powder. *Poison!* Insoluble in water; sol in HCl or HNO_3.

5768. Mercuric Benzoate. $C_{14}H_{10}HgO_4$; mol wt 442.82. C 37.97%, H 2.28%, Hg 45.30%, O 14.45%. $Hg(C_7H_5O_2)_2$.

Monohydrate, odorless, cryst powder; sensitive to light. *Poison!* Sol in 90 parts cold, 40 parts boiling water; freely sol in NaCl soln; slightly sol in alcohol. When boiled with water or alc it hydrolyzes to a basic salt and free benzoic acid. *Protect from light.*

THERAP CAT: Formerly as antisyphilitic.

5769. Mercuric Bromide. Br_2Hg; mol wt 360.44. Br 44.34%, Hg 55.66%. $HgBr_2$. Usually prepd from the elements: Jander, Brodersen, *Z. Anorg. Chem.* **261**, 261 (1950).

White crystals or cryst powder; sensitive to light. *Poison!* d 6.05. mp 237°; sublimes at higher temp. Sol in about 200 parts cold, 25 parts boiling water; freely sol in hot alcohol, in methanol, HCl, HBr, alkali bromide solns; slightly sol in chloroform. *Protect from light.*

5770. Mercuric Chloride. Mercury bichloride; corrosive sublimate; mercury perchloride; corrosive mercury chloride. Cl_2Hg; mol wt 271.52. Cl 26.12%, Hg 73.88%. $HgCl_2$.

Crystals or white granules or powder. *Violent poison! May be fatal if swallowed.* d 5.4. mp 277°; volatilizes unchanged at about 300°; also slightly volatile at ordinary temp; appreciably so at 100°. One gram dissolves in 13.5 ml water (soly is increased by HCl or alkali chlorides), in 2.1 ml boiling water, 3.8 ml alc, 1.6 ml boiling alc, 200 ml benzene, 22 ml ether, 12 ml glycerol, 40 ml acetic acid; also sol in methanol, acetone, ethyl acetate; slightly sol in carbon disulfide, pyridine. pH about 4.7. Also reported as 3.2 for 0.2 molar aq soln. Coagulates albumin; produces with NaOH a yellow ppt (difference from calomel, which turns black). *Incompat:* Formates, sulfites, hypophosphites, phosphates, sulfides, albumin, gelatin, alkalies, alkaloid salts, ammonia, lime water, antimony and arsenic, bromides, borax, carbonates; reduced iron; copper, iron, lead, silver salts; infusions of cinchona, columbo, oak bark or senna; tannic acid; vegetable astringents.

Human Toxicity: Highly toxic. Corrosive to mucous membranes. Ingestion may cause severe nausea, vomiting, hematemesis, abdominal pain, diarrhea, melena, renal damage, prostration. 1 or 2 g is frequently fatal. Poisoning and death also have occurred from intrauterine douches and application of alcoholic soln to large areas of skin. Do not breathe dust. Keep away from feed or food products. Wash hands before eating or smoking. *Antidote:* Dimercaprol (BAL). Review of clinical toxicology of mercury and its compounds: H. B. Gerstner, J. E. Huff, *J. Toxicol. Environ. Health* **2**, 491-526 (1977).

USE: Preserving (kyanizing) wood and anatomical specimens; also embalming; disinfecting; browning and etching steel and iron; intensifier in photography; white reserve in fabric printing; tanning leather; electroplating aluminum; depolarizer for dry batteries; freeing gold from lead; magic photograms; mordant for rabbit and beaver furs; staining wood and vegetable ivory pink; manuf of ink for mercurography; treating seed potatoes; manuf other mercury compds. As an important reagent in anal. chemistry.

THERAP CAT: Topical antiseptic, disinfectant.

THERAP CAT (VET): Caustic, antiseptic, disinfectant.

5771. Mercuric Chloride, Ammoniated. *Mercury amide chloride;* aminomercuric chloride; mercury ammonium chloride; ammoniated mercury; white precipitate; white mercuric

precipitate. ClH_2HgN; mol wt 252.09. Cl 14.07%, H 0.80%, Hg 79.58%, N 5.56%. $HgNH_2Cl$. Prepn: Sen, *Z. Anorg. Allgem. Chem.* **33**, 197 (1903); Wagenknecht, Juza in *Handbook of Preparative Inorganic Chemistry*, vol 2, G. Brauer, Ed. (Academic Press, New York, 2nd ed., 1965) p 1114.

Odorless powder; d 5.38; earthy, styptic, metallic taste. *Poison!* Volatile without melting at dull red heat. Insoluble in water or alcohol; sol in warm HCl, HNO_3, acetic acid, in cold soln ammonium carbonate or sodium thiosulfate.

Caution: French "White Precipitate" (Précipité Blanc) is calomel. Do not confuse the two when French prescriptions are filled.

Human Toxicity: May produce allergic dermatitis; prolonged use may cause local pigmentation of skin or eyelids. Absorption following vigorous dermal application has been reported with resulting ptyalism. Oral ingestion causes epigastric pain, nausea, purging. *Antidote:* Dimercaprol (BAL). *Caution:* Not to be used in conjunction with sulfur or iodine.

THERAP CAT: Topical anti-infective.

THERAP CAT (VET): Topical anti-infective.

5772. Mercuric Cyanide. Cianurina. C_2HgN_2; mol wt 252.65. C 9.51%, Hg 79.40%, N 11.09%. $Hg(CN)_2$. Obtained by evaporating a soln of HgO in aq HCN: Biltz, *Z. Anorg. Allgem. Chem.* **170**, 161 (1928).

Colorless, odorless, tetragonal crystals or white powder; darkens on exposure to light. *Violent poison!* d 3.996. Dec at 320°. One gram dissolves in 13 ml water, 3 ml boiling water, 13 ml alcohol, 4 ml methanol; slowly sol in glycerol; slightly sol in ether. *Protect from light.*

THERAP CAT: Topical antiseptic.

THERAP CAT (VET): Has been used as a topical antiseptic.

5773. Mercuric Dichromate(VI). Mercury bichromate. Cr_2HgO_7; mol wt 416.63. Cr 24.97%, Hg 48.15%, O 26.88%. $HgCr_2O_7$.

Red, heavy, cryst powder. Practically insol in water; sol in HCl or HNO_3. *Poison!*

5774. Mercuric Fluoride. Mercury difluoride. F_2Hg; mol wt 238.61. F 15.93%, Hg 84.07%. HgF_2. Prepd from $HgCl_2$ and F_2: Henne, Midgley, *J. Am. Chem. Soc.* **58**, 886 (1936).

White powder or cubic crystals. Very sensitive to moisture. d^{15} 8.95. mp 645°. bp above 650°. Turns yellow and hydrolyzes in the prolonged presence of water.

Dihydrate, obtained when mercuric oxide is dissolved in excess of 50% HF.

USE: In the fluorination of organic compds.

5775. Mercuric Iodate. HgI_2O_6; mol wt 550.45. Hg 36.44%, I 46.11%, O 17.44%. $Hg(IO_3)_2$.

White powder. *Poisonous!* On warming, dec at 175° to give mercuric iodide and oxygen. Soly in water (20°) 0.002 g/100 g; soly increased by NaCl or KI.

5776. Mercuric Iodide, Red. Mercury biniodide. HgI_2; mol wt 454.45. Hg 44.14%, I 55.86%.

Scarlet-red, heavy, odorless, almost tasteless powder; sensitive to light; at 130° becomes yellow, and then red on cooling. *Poison!* d 6.28. mp 259°. bp about 350° and sublimes. Soly in water (25°) 0.006 g/100 g. 1 g dissolves in 115 ml alcohol, 20 ml boiling alcohol, about 120 ml ether, about 60 ml acetone, 910 ml chloroform, 75 ml ethyl acetate, 260 ml carbon disulfide, readily in alkali iodides, $HgCl_2$, $Na_2S_2O_3$; sol in 230 ml olive oil, 50 ml castor oil. *Protect from light.*

USE: In anal. chemistry for preparation of Nessler's Reagent (alkaline mercuric potassium iodide solution).

THERAP CAT: Topical antiseptic.

THERAP CAT (VET): Has been used as a topical antiseptic, counterirritant, vesicant.

5777. Mercuric Nitrate. Mercury pernitrate. HgN_2O_6; mol wt 324.66. Hg 61.80%, N 8.63%, O 29.57%. $Hg(NO_3)_2$.

Hydrate, white or slightly yellow, deliquesc, cryst powder; odor of nitric acid. *Poison!* d 4.3. Sol in a small amount of water; with much water or on boiling with water, an insol basic salt is formed; sol in dil acids. *Keep well closed and protect from light.*

USE: Manufacture of felt; mercury fulminate; destroying phylloxera.

5778. Mercuric Oleate. *9-Octadecenoic acid mercury salt;* oleate of mercury. $C_{36}H_{66}HgO_4$; mol wt 763.52. C 56.63%, H 8.71%, Hg 26.27%, O 8.38%. $Hg(C_{17}H_{33}COO)_2$. Commercial product prepd by dissolving yellow mercuric oxide in oleic acid; contains 24-26% HgO and about 10% excess oleic acid.

Yellowish-brown, somewhat transparent, ointment-like mass; odor of oleic acid. *Poisonous!* On keeping, the mercury is partly reduced to mercurous state or metal. Practically insol in water. Slightly sol in alcohol or ether, freely in fixed oils. *Protect from light.*

THERAP CAT: Has been used as ectoparasiticide.

5779. Mercuric Oxide, Red. Red precipitate. HgO; mol wt 216.61. Hg 92.61%, O 7.39%. Contains 99-99.5% HgO.

Bright red or orange-red, heavy, odorless, crystalline powder or scales; orthorhombic structure; yellow when finely powdered. *Poison!* Dec on exposure to light into mercury and oxygen; at 400° becomes almost black, but red again on cooling; at 500° dec into mercury and oxygen. d 11.14. Practically insol in water; sol in dil HCl or HNO_3 or in solns of alkali cyanides or iodides, slowly in solns of alkali bromides; insol in alcohol. *Protect from light.*

USE: In marine bottom paints, diluting pigments for painting on porcelain, with graphite as depolarizer in dry batteries. In Kjeldahl nitrogen determination; and as reagent for citric acid, thiophene, glucose, aldehyde, urea, acetone. As reagent and catalyst in organic reactions: J. S. Pizey, *Synthetic Reagents,* **vol. 1** (John Wiley, New York, 1974) pp 295-319.

THERAP CAT: Topical antiseptic.

THERAP CAT (VET): Has been used as a topical antiseptic.

5780. Mercuric Oxide, Yellow. Yellow precipitate. HgO; mol wt 216.61. Hg 92.61%, O 7.39%. Contains 99-99.5% HgO.

Yellow or orange-yellow, heavy, odorless powder, orthorhombic structure. *Poison!* Becomes red on heating and yellow again on cooling; more finely divided and more reactive than red mercuric oxide. Other physical properties and solubilities: *see* Mercuric Oxide, Red. *Protect from light. Incompat:* Reducing agents.

USE: Similar to that of the red oxide; in the mfr of organic mercurials. In anal. chemistry for determining Zn or HCN; detecting acetic acid in formic acid, CO in gas mixtures.

THERAP CAT: Topical antiseptic (ophthalmic).

THERAP CAT (VET): Has been used as a topical antiseptic, fungicide, in chronic skin conditions, conjunctivitis, corneal ulcers.

5781. Mercuric Oxycyanide. *Mercury cyanide oxide.* $C_2Hg_2N_2O$; mol wt 469.26. C 5.12%, Hg 85.50%, N 5.97%, O 3.41%. $HgO.Hg(CN)_2$. For reasons explained below, the article of commerce contains about 33% mercuric oxycyanide and about 67% mercuric cyanide.

White, orthorhombic crystals or cryst powder d 4.44. One gram dissolves in 80 ml cold water, more sol in hot water. *Violent poison! It explodes when touched with a flame or by percussion;* hence for commerce it is made with an excess of mercuric cyanide which eliminates the danger of explosion.

THERAP CAT: Topical antiseptic.

5782. Mercuric Salicylate. Mercury subsalicylate. $C_7H_4HgO_3$; mol wt 336.69. C 24.97%, H 1.20%, Hg 59.58%, O 14.25%. Approx $C_7H_4O_3Hg$. Article of commerce contains 54-58% Hg. Prepn: C. H. Rogers *et al., Inorganic Pharmaceutical Chemistry* (Lea & Febriger, Philadelphia, 4th ed., 1948) pp 409-411.

White or slightly yellowish or pinkish, odorless powder. *Poison!* Insoluble in water or alcohol; sol in warm solns of alkali halides, fixed alkali hydroxides and carbonates. *Incompat:* Alkali iodides.

THERAP CAT: Topical antiseptic.

5783. Mercuric Sodium *p*-Phenolsulfonate. Mercury and sodium phenolsulfonate; mercuriphenoldisulfonate sodium; Hermophényl. $C_{12}H_8HgNa_2O_8S_2$; mol wt 590.92. C 24.39%, H 1.36%, Hg 33.95%, Na 7.78%, O 21.66%, S 10.85%. Prepd by the action of the sodium salt of *p*-phenolsulfonic acid on yellow mercuric oxide: Lumière, Chevrot-

tier, *Compt. Rend.* **132,** 145 (1901); Lumière, Perrin, *Pharmacie* **1,** 102 (1943). The commercial product is a mixture of the monosulfonate (80%) and the disulfonate.

White powder. One gram dissolves in 5 ml water. The aq soln does not coagulate albuminous matter.

Human Toxicity: Ingestion or percutaneous absorption may cause mercury poisoning.

USE: As germicide in soaps and lotions. Usual concn in soap about 1:100.

THERAP CAT: Local antiseptic.

5784. Mercuric Stearate. *Octadecanoic acid mercury salt.* $C_{36}H_{70}HgO_4$; mol wt 767.55. C 56.34%, H 9.19%, Hg 26.13%, O 8.34%. $Hg(C_{17}H_{35}COO)_2$.

Yellowish, granular powder. Toxic because of its solubility in lipids. Practically insol in water or alcohol; sparingly sol in fatty oils.

5785. Mercuric Subsulfate. *Mercury oxide sulfate;* mercury oxonium sulfate; Turpeth mineral; basic mercuric sulfate. Hg_3O_6S; mol wt 729.90. Hg 82.45%, O 13.15%, S 4.39%. $HgSO_4.2HgO$. Proved to be mercury oxonium sulfate in x-ray studies of its structure. The polymeric oxonium cations $(Hg_3O_2)_x$ form 2-dimensional infinite layers. The sulfate ions lie inside the loops of the wide-meshed cation lattice.

Lemon-yellow, heavy, odorless powder. *Poison!* Practically insol in water; sol in acids.

5786. Mercuric Succinimide. Mercuric imidosuccinate. $C_8H_8HgN_2O_4$; mol wt 396.77. C 24.22%, H 2.03%, Hg 50.56%, N 7.06%, O 16.13%.

White, small crystals or cryst powder; odorless and stable in air; affected by light. *Poison!* One gram dissolves in 20 ml water, 5 ml boiling water, 300 ml alcohol; practically insol in ether. *Protect from light.* On adding NaOH to its aq soln a yellowish-white ppt forms which is reduced to metallic Hg by heat; heated with zinc dust, pyrrole evolves which colors a pine splinter moistened with HCl red. The aq soln is not pptd by albumin.

THERAP CAT: Antibacterial.

5787. Mercuric Sulfate. Mercury bisulfate. HgO_4S; mol wt 296.68. Hg 67.62%, O 21.57%, S 10.81%. $HgSO_4$.

White, odorless granules or cryst powder. *Poison!* d 6.47. Dec by water into a yellow insol, basic sulfate and free H_2SO_4; sol in HCl, hot dil H_2SO_4, concd soln of NaCl. *Protect from light.*

USE: Electrolyte for primary batteries; with NaCl for extracting gold and silver from roasted pyrites; as a reagent for wine coloring, barbital, and cystine.

5788. Mercuric Sulfide, Black. Ethiops mineral. HgS; mol wt 232.68. Hg 86.22%, S 13.78%. Occurs as a mineral in California. Prepd by passing hydrogen sulfide through a soln of mercuric chloride in hydrochloric acid.

Black or grayish-black, heavy, odorless, tasteless, amorphous powder. Also occurs as black, cubic crystals (β-form). Transition temp (red to black) 386°. Black form can exist indefinitely in metastable state at room temp. Insol in water, alcohol, dil mineral acids.

USE: As pigment for horn, rubber, etc.

5789. Mercuric Sulfide, Red. Vermilion; Chinese red; C.I. Pigment Red 106; C.I. 77766. HgS; mol wt 232.68. Hg

86.22%, S 13.78%. Occurs in nature as the mineral *cinnabar*. Prepd from mercuric acetate, ammonium thiocyanate, glacial acetic acid and hydrogen sulfide: Newell *et al.*, *Inorg. Syn.* **1**, 19 (1939). *See also Colour Index* vol. **4** (3rd ed., 1971) p 4682.

Bright scarlet-red powder, lumps, hexagonal crystals (α-form); blackens on exposure to light, particularly in presence of H_2O or alkali hydroxides. At about 250° becomes brownish, at higher temp black, but red again on cooling. When ignited in air it dec into metal and sulfur, the latter burning to SO_2. Practically insol in water; not attacked by HNO_3 or cold HCl; dec by hot concd H_2SO_4; sol in aqua regia with separation of S, in warm hydriodic acid with evolution of H_2S. *Protect from light.*

USE: For coloring plastics, sealing wax, and with $FeSO_4$ for marking linen; manuf fancy colored papers; as pigment.

THERAP CAT: Antibacterial.

5790. Mercuric Thiocyanate. Mercuric sulfocyanate; mercuric sulfocyanide. $C_2HgN_2S_2$; mol wt 316.79. C 7.58%, Hg 63.33%, N 8.84%, S 20.25%. $Hg(SCN)_2$. Prepd from $Hg(NO_3)_2$ + KSCN: Peters, *Z. Anorg. Allgem. Chem.* **77**, 157 (1912).

Odorless powder. When crystalline, usually in radially arranged needles. *Poisonous!* When heated it swells up to many times its original vol, decomposing finally into mercury, nitrogen, etc., at about 165°. Slightly sol in cold water (0.069 g/100 ml H_2O at 25°), more sol in boiling water with decompn; sol in dil HCl, in solns of alkali cyanides, chlorides. *Protect from light.*

USE: For Pharaoh's serpents (fireworks); intensifier in photography.

5791. Mercurophen. Sodium 4-(hydroxymercuri)-2-nitrophenolate; sodium hydroxymercuri-*o*-nitrophenolate; 4-(hydroxymercuri)-2-nitrophenol sodium salt. C_6H_4HgN-NaO_4; mol wt 377.70. C 19.08%, H 1.07%, Hg 53.11%, N 3.71%, Na 6.09%, O 16.94%. Prepn from *o*-nitrophenol and $Hg(OCOCH_3)_2$: Schamberg *et al.*, U.S. pat. **1,390,972** (1922). Structure: Malcolm, *J. Bacteriol.* **22**, 403 (1931).

Brick-red, odorless powder. *Poison!* Sol in hot water. Solns are deep amber in color.

THERAP CAT: Local antiseptic; disinfectant.

THERAP CAT (VET): Antiseptic, disinfectant.

5792. Mercurous Acetate. $C_4H_6Hg_2O_4$; mol wt 519.31. C 9.25%, H 1.17%, Hg 77.26%, O 12.32%. $Hg_2(CH_3COO)_2$. Prepd from $Hg_2(NO_3)_2$ + CH_3COONa: Wagenknecht, Juza, in *Handbook of Preparative Inorganic Chemistry* vol. 2, G. Brauer, Ed. (Academic Press, New York, 2nd ed., 1965) p 1120.

Lustrous leaflets or cryst powder. Darkens on exposure to light. Sol in about 100 parts water; sol in dil acetic acid. Practically insol in alcohol, ether. Aq solns dec quickly under the influence of light and heat. *Keep well closed and protected from light.*

THERAP CAT: Antibacterial.

5793. Mercurous Bromide. Br_2Hg_2; mol wt 560.99. Br 28.49%, Hg 71.51%. Hg_2Br_2.

White, odorless powder; darkens on exposure to light. d 7.3. Sublimes at approx 390° (dec). Insol in water, alcohol, ether; dec by hot HCl or alkali bromides. *Protect from light.*

5794. Mercurous Chlorate. $Cl_2Hg_2O_6$; mol wt 568.08. Cl 12.48%, Hg 70.62%, O 16.90%. $Hg_2(ClO_3)_2$.

White crystals; dec at about 250° to form oxygen, mercuric oxide and mercuric chloride. d 6.41. Sparingly sol in water; hydrolyzed in hot water to form basic salts.

5795. Mercurous Chloride. Calomel; mild mercury chloride; mercury monochloride; mercury protochloride; mercu-

ry subchloride; precipité blanc; Calogreen; Cyclosan. Cl_2-Hg_2; mol wt 472.09. Cl 15.02%, Hg 84.98%. Hg_2Cl_2.

White, odorless, tasteless, heavy powder; slowly dec by sunlight into mercuric chloride and metallic mercury; sublimes at 400-500° without melting. d 7.15. Practically insol in water (0.00020g/100 ml H_2O at 25°); HCl or alkali and alkaline earth chlorides increase soly in water. Insol in alcohol, ether. Dec by solns of alkali iodides, bromides or cyanides into the mercuric salt and metallic mecury; solns of alkali chlorides act similarly but slowly. It is blackened by ammonia, caustic alkali and alkaline earth solns. *Protect from light. Incompat:* Bromides, iodides, alkali chlorides, sulfates, sulfites, carbonates, hydroxides, lime water, acacia, ammonia, golden antimony sulfide, cocaine, cyanides, copper salts, hydrogen peroxide, iodine, iodoform, lead salts, silver salts, soap, sulfides.

Human Toxicity: Excessive doses may cause mercury poisoning. *Antidote:* Dimercaprol (BAL). *Caution:* If laxation from oral mercurous chloride should not occur, saline laxative must be administered to prevent possibility of mercury poisoning, cf. *Clinical Toxicology of Commercial Products*, R. E. Gosselin *et al.*, Eds. (Williams & Wilkins, Baltimore, 4th ed., 1976) Section II, p 95.

USE: Dark green Bengal lights; calomel paper, mixed with gold in painting on porcelain; for calomel electrodes; as fungicide; in agriculture to control root maggots on cabbage and onions.

THERAP CAT: Cathartic, diuretic, antiseptic, anti-syphilitic.

THERAP CAT (VET): Has been used as a cathartic, and locally as an antiseptic and desiccant.

5796. Mercurous Fluoride. F_2Hg_2; mol wt 439.22. F 8.65%, Hg 91.35%. Hg_2F_2. Prepd from Hg_2CO_3 + HF: Henne, Renoll, *J. Am. Chem. Soc.* **60**, 1060 (1938); also by addition of sodium fluoride to mercurous nitrate soln. Probably formed in the reaction between fluorine and mercury at room temp: Emeléus in *Fluorine Chemistry* vol. 1, J. H. Simons, Ed. (New York, 1950) p 39. The prepn should be carried out in the dark.

Small, yellow cubic crystals, d_4^{15} 8.73. mp 570° (dec). Sublimes at around 240°. The crystals blacken in light or with ammonia fumes. Hydrolyzed in water to form mercury, mercuric oxide and hydrofluoric acid.

5797. Mercurous Iodide. Yellow mercury iodide; mercury protoiodide. Hg_2I_2; mol wt 654.99. Hg 61.25%, I 38.75%.

Bright-yellow, amorphous, heavy, odorless powder. Darkens or becomes greenish on exposure to light, HgI_2 and metallic mercury being formed. d 7.70. mp 290° when rapidly heated with partial decompn into Hg and HgI_2. Insol in water, alcohol or ether; sol in solns of mecurous or mercuric nitrates; cold ammonia, its solns or alkali iodide dec it into mercury and mercuric iodide. *Protect from light.* "Green" mercury iodide is made from metallic mercury and iodine, the green color being due to presence of some uncombined mercury. *Incompat:* Soluble iodides. *Caution:* Never prescribe mercury iodide yellow (or green) with a sol iodide, because a highly *poisonous* mercuric iodide is formed!

THERAP CAT: Antibacterial.

5798. Mercurous Nitrate. Mercury protonitrate. Hg_2-N_2O_6; mol wt 525.19. Hg 76.39%, N 5.33%, O 18.28%. $Hg_2(NO_3)_2$. Normally exists as dihydrate. Prepn of anhydr salt: Potts, Allred, *Inorg. Chem.* **5**, 1066 (1966).

Dihydrate, colorless crystals, usually with slight odor of HNO_3. mp at about 70° with decompn. *Poisonous!* d 4.78. Sol in 13 parts water contg 1% HNO_3; with water alone a basic salt is formed. Blackened by ammonia, caustic alkali, and alkaline earth solns. *Keep well closed and protected from light.*

USE: Fire gilding, blackening brass.

5799. Mercurous Oxide. Mercury oxide black. No evidence that Hg_2O has been isolated. X-ray data shows that solid formed by addition of sodium hydroxide to mercurous nitrate is an intimate mixture of mercury and mercuric oxide: Fricke, Ackemann, *Z. Anorg. Allgem. Chem.* **211**, 233 (1933).

Black or brownish-black powder. d 9.8. Insol in water; sol in HNO_3. HCl converts it into calomel. *Protect from light.*

5800. Mercurous Sulfate. Hg_2O_4S; mol wt 497.29. Hg 80.68%, O 12.87%, S 6.45%. Hg_2SO_4.

White to slightly yellow cryst powder. Becomes gray on exposure to light with production of mercury and mercuric sulfate. d 7.56. Slightly sol in water (0.06 g/100g at 25°); sol in dil HNO_3.

USE: For making electric batteries; with zinc sulfate in the standard Clark cell and with cadmium sulfate in the standard Weston cell.

5801. Mercury. Hydrargyrum; liquid silver; quicksilver. Hg; at. wt 200.59; at. no. 80; valences 1, 2. Group 2b element. Abundance in earth's crust 0.5 ppm. Natural isotopes: 202 (29.80%); 200 (23.13%); 199 (16.84%); 201 (13.22%); 198 (10.02%); 204 (6.85%); 196 (0.146%); known isotopes range in mass number from 189 to 206. Obtained by roasting cinnabar (mercuric sulfide). General reviews: Roberts, *Advan. Inorg. Chem. Radiochem.* **11** (Academic Press, New York, 1968) pp 309-339; Aylett, "Group IIB" in *Comprehensive Inorganic Chemistry* vol. 3 (Pergamon Press, Oxford, 1973) pp 187-328; H. J. Drake in Kirk-Othmer *Encyclopedia of Chemical Technology* vol. 15 (Wiley-Interscience, New York, 3rd ed., 1981) pp 143-156. Review of clinical toxicology: H. B. Gerstner, J. E. Huff, *J. Toxicol. Environ. Health* **2**, 491-526 (1977).

Silver-white, heavy, mobile, liquid metal; slightly volatile at ordinary temp; solid mercury is a tin-white, ductile, malleable mass which may be cut with a knife. mp −38.87°; bp 356.72°; d^{25} 13.534. Heat capacity at constant pressure (25°) 6.687 cal/mole deg. Vapor pressure (25°): 2×10^{-3} mm; heat of vaporization (25°): 14.652 kcal/mole: Busey, Giauque, *J. Am. Chem. Soc.* **75**, 806 (1953). Surface tension (25°): 484 dynes/cm; electrical resistivity (20°): 95.76 μohm cm. When pure does not tarnish on exposure to air at ordinary temp, but when heated to near the boiling point slowly oxidizes to HgO. Forms alloys with most metals except iron and combines with sulfur at ordinary temperatures. E^0 (aq) Hg/Hg^{2+} −0.854 V; E^0 (aq) $2 Hg/Hg_2^{2+}$ −0.789 V. Soly in water (25°): 0.28 μmoles/l; data on soly in organic solvents: Spencer, Voigt, *J. Phys. Chem.* **72**, 464 (1968). Reacts with HNO_3 and hot, concd H_2SO_4; does not react with dil HCl, cold H_2SO_4, or alkalies. Reacts with ammonia solns in air to form Hg_2NOH, *Millon's base.* Mercury salts when heated with Na_2CO_3 yield metallic Hg and are reduced to metal by H_2O_2 in the presence of alkali hydroxide. Cu, Fe, Zn and many other metals ppt metallic Hg from neutral or slightly acid solns of mercury salts. Soluble ionized *mercuric* salts give a yellow ppt of HgO with NaOH and a red ppt of HgI_2 with alkali iodide. *Mercurous* salts give a black ppt with alkali hydroxides and a white ppt of calomel with HCl or sol chlorides. They are slowly dec by sunlight. *Poisonous!*

Human Toxicity: Readily absorbed via respiratory tract (elemental mercury vapor, mercury compd dusts), intact skin, and G.I. tract, although occasional incidental swallowing of metallic mercury is without harm. Spilled and heated elemental mercury is particularly hazardous. *Acute:* sol salts have violent corrosive effects on skin and mucous membranes; severe nausea, vomiting, abdominal pain, bloody diarrhea; kidney damage; death usually within 10 days. *Chronic:* inflammation of mouth and gums, excessive salivation, loosening of teeth; kidney damage; muscle tremors, jerky gait, spasms of extremities; personality changes, depression, irritability, nervousness. Phenyl and alkyl mercurials can cause skin burns and be absorbed by the skin. Burning sensation is delayed several hours and thus gives no warning. Alkyls have affinity for brain tissue and may cause permanent damage. Phenyls are no more toxic than inorganic Hg. *Antidote:* Dimercaprol (BAL). *See* E. Browning, *Toxicity of Industrial Metals* (Appleton-Century Crofts, New York, 2nd ed., 1969) pp 226-242.

USE: In barometers, thermometers, hydrometers, pyrometers; in mercury arc lamps producing ultraviolet rays; in switches, fluorescent lamps; in mercury boilers; manuf all mercury salts, mirrors; as catalyst in oxidation of organic compds; extracting gold and silver from ores; making amalgams, electric rectifiers, mercury fulminate; also in dentistry; in determining N by Kjeldahl method, for Millon's reagent; as cathode in electrolysis, electroanalysis, and many other uses. Also in pharmaceuticals, agricultural chemicals, antifouling paints.

5802. Mercury Mass. Blue pill; blue mass. Contains 32-34% metallic Hg. Bluish-gray mass. The rest is honey, licorice, althea, glycerol, and some mercury oleate.

Caution: Can cause systemic mercury poisoning.

THERAP CAT: Cathartic.

THERAP CAT (VET): Has been used as a laxative.

5803. Merisoprol Hg 197. *Hydroxy(2-hydroxypropyl)-mercury-^{197}Hg; 1-(hydroxymercuri-^{197}Hg)-2-propanol;* Merprane. $C_3H_8HgO_2$.

$$\underset{\text{OH}}{\overset{\text{OH}}{\text{CH}_3\text{CHCH}_2}}-^{197}\text{Hg}-\text{OH}$$

THERAP CAT: Diagnostic aid (renal function).

5804. Merphyrin. *[7,12-Bis(1-hydroxyethyl)-3,8,13,17-tetramethyl-21H,23H-porphine-2,18-dipropanoato(4−)-N^{21},-N^{22},N^{23},N^{24}]mercurate(2−) disodium;* mercuri-hematoporphyrin disodium salt; M. H.; hematoporphyrinmercury disodium salt. $C_{34}H_{34}HgN_4Na_2O_6$; mol wt 841.26. C 48.54%, H 4.07%, Hg 23.85%, N 6.66%, Na 5.47%, O 11.41%. Prepd by action of mercuric acetate upon hematoporphyrin and treating the product with sodium methoxide: **Japan.** pat. **1273('59); Brit.** pat. **824,705** (1959 to Daiichi), *C.A.* **54**, 9972 (1960); Obika *et al.,* U.S. pat. **3,004,985** (1961 to Daiichi).

Dark brown powder. Sol in water (a 1:1000 soln is slightly alkaline). Aq solns are affected by light. Slightly sol in methanol. Practically insol in acetone, benzene, chloroform.

5805. Mersalyl. *[3-[[2-(Carboxylatomethoxy)benzoyl]-amino]-2-methoxypropyl]hydroxymercurate(1−) sodium; o-[[3-(hydroxymercuri)-2-methoxypropyl]carbamoyl]phenoxyacetic acid sodium salt;* sodium *o-[(3-hydroxymercuri-2-methoxypropyl)carbamoyl]phenoxyacetate; N-(γ-hydroxy-mercuri-β-methoxypropyl)salicylamide-O-acetic acid sodium salt;* mercuramide; Mercusal; Mersalin; Salurin; Salyrgan. $C_{13}H_{16}HgNNaO_6$; mol wt 505.87. C 30.86%, H 3.19%, Hg 39.66%, N 2.77%, Na 4.55%, O 18.98%. Prepn: Diels, Beccard, *Ber.* **39**, 4125 (1906); Bockmühl, Schwarz, **Ger.** pat. **423,031** (1925 to Hoechst); *Frdl.* **15**, 1609; B.I.O.S. *Final Report* no. **766**, p 132; Foye *et al., J. Am. Pharm. Assoc.* **41**, 273 (1952); Werner, Scholz, U.S. pat. **2,705,716** (1955 to Ciba). Toxicity data: E. B. Robbins, K. K. Chen, *J. Am. Pharm. Assoc.* **40**, 249 (1951).

Bitter crystals. *Poison!* Somewhat deliquescent. Gradually dec by light. One gram dissolves in about 1 ml water, in about 3 ml ethanol (95%), in 2 ml abs methanol. Practically insol in ether, chloroform. Aq solns are alkaline to litmus. Solns of mersalyl contg sodium chloride or other salts may become toxic unless some substance, such as theophylline which inhibits the decompn of the mercurial complex, is present. LD_{50} i.v. in mice, rats: 72.6, 17.7 mg/kg (Robbins, Chen).

Note: *Neptal* and *Diursal* also contain theophylline; *cf* U.S. pat. 2,213,457 (1940).

THERAP CAT: Diuretic.

THERAP CAT (VET): Diuretic.

5806. Mesaconic Acid. *(E)-2-Methyl-2-butenedioic acid;* methylfumaric acid; *trans*-1-propene-1,2-dicarboxylic acid. $C_5H_6O_4$; mol wt 130.10. C 46.16%, H 4.65%, O 49.19%. Prepd by heating citraconic anhydride with nitric acid: Shriner *et al., Org. Syn.* **11**, 74 (1931); from ethyl itaconate: Jones *et al., J. Chem. Soc.* **1954**, 1865; from *N,N*-diethyl-5-diethylamino-2,3-dihydro-3-furamide: Sauer *et al., J. Am. Chem. Soc.* **81**, 693 (1959). Obtained almost instantly by the action of sunlight on citraconic acid in ether + chloroform in the presence of a little bromine: Fittig, Langworthy, *Ber.* **26**, 46 (1893); *Ann.* **304**, 119, 149 (1899); by action of β-radiation on citraconic acid + bromine: Lavigne, Levine, U.S. pat. 2,979,445 (1961 to California Res. Corp.).

$$CH_3CCO_2H$$
$$\|$$
$$HO_2CCH$$

Orthorhombic needles from alcohol, monoclinic tablets from ethyl acetate, mp 204-205°. d 1.466. Sublimes. bp 250° (decompn). K_1 at 25° = 8.5 × 10^{-4}; K_2 = 15 × 10^{-6}. 100 g water dissolves 2.7 g at 18° and 117.9 g at the boiling point; 100 g 90% alc dissolves 30.6 g at 17° and 95.7 g at the boiling point. Sol in ether; sparingly sol in chloroform, carbon disulfide and ligroin.

Dimethyl ester, $C_7H_{10}O_4$, liq, bp 206°. d_4^{21} vac 1.12011. n_D^{20} 1.45575. Sol in 122 parts water at 15°.

Diethyl ester, $C_9H_{14}O_4$, liq, bp 229°. d_4^{20} vac 1.04675. n_4^{20} 1.44936.

5807. Mesalamine. *5-Amino-2-hydroxybenzoic acid;* 5-aminosalicylic acid; 5-amino-2-hydroxybenzene-1-carboxylic acid; *m*-aminosalicylic acid; fisalamine; mesalazine; 5-ASA; Asacol; Asacolitin; Claversal; Lixacol; Mesasal; Pentasa; Rowasa; Salofalk. $C_7H_7NO_3$; mol wt 153.13. C 54.90%, H 4.61%, N 9.15%, O 31.34%. Prepn by reduction of *m*-nitrobenzoic acid with Zn dust and HCl: H. Weil *et al., Ber.* **55B**, 2664 (1922); by electrolytic reduction: Le Guyader, Peltier, *Compt. Rend.* **253**, 2544 (1961). Active metabolite of sulfasalazine, *q.v.:* A. K. Azad Khan *et al., Lancet* **2**, 892 (1977); P. A. M. Van Hees *et al., Gut* **21**, 632 (1980). Determn by HPLC in serum: E. Brendel *et al., J. Chromatog.* **385**, 299 (1987). Bioavailability, plasma level and excretion: S. N. Rasmussen *et al., Gastroenterology* **83**, 1062 (1982). Clinical trials in ulcerative colitis: M. Campieri *et al., Lancet* **2**, 270 (1981); *eidem, Digestion* **29**, 204 (1984); L. S. Friedman *et al., Am. J. Gastroenterol.* **81**, 412 (1986); in comparison with olsalazine, *q.v.:* S. S. Rao *et al., Scand. J. Gastroenterol.* **22**, 332 (1987). Experimental treatment of Crohn's disease: S. N. Rasmussen *et al., Gastroenterology* **85**, 1350 (1983). Brief review: G. Friedman, *Am. J. Gastroenterol.* **81**, 141 (1986). Use in manufacture of dyes: Brit. pat. 751,386 (1956 to J. R. Geigy).

COOH
OH
H₂N

White to pinkish crystals, dec about 280°. Slightly sol in cold water, alc; more sol in hot water; sol in HCl.

USE: In manuf of light-sensitive paper, azo and sulfur dyes.

THERAP CAT: Anti-inflammatory (gastrointestinal). Treatment of ulcerative colitis.

5808. Mescaline. *3,4,5-Trimethoxybenzeneethanamine;*

3,4,5-*trimethoxyphenethylamine;* Mezcaline. $C_{11}H_{17}NO_3$; mol wt 211.25. C 62.54%, H 8.11%, N 6.63%, O 22.72%. Psychotomimetic alkaloid isolated from *peyote* (mescal buttons), the flowering heads of *Lophophora williamsii* (Lemaire) Coult., *Cactaceae.* Isoln: A. Heffter, *Ber.* **29**, 221 (1896). Structure and synthesis: E. Späth, *Monatsh.* **40**, 129 (1919); K. H. Slotta, H. Heller, *Ber.* **63**, 3029 (1930); E. Späth, F. Becke, *Monatsh.* **66**, 327 (1935); M. U. Tsao, *J. Am. Chem. Soc.* **73**, 5495 (1951); K. Banholzer *et al., Helv. Chim. Acta* **35**, 1577 (1952). Novel synthesis: M. N. Aboul-Enein, A. I. Eid, *Acta Pharm. Suec.* **16**, 267 (1979). MS determn: S. P. Jindal, T. Lutz, *Eur. J. Mass Spectrom. Biochem. Med. Environ. Res.* **2**, 117 (1982). Pharmacokinetics in rabbits: C. Van Peteghem *et al., Eur. J. Drug Metab. Pharmacokinet.* **7**, 1 (1982). Mode of action study: M. E. Trulson *et al., Eur. J. Pharmacol.* **96**, 151 (1983). Use in evaluating serotonin S_2 antagonists: C. J. E. Niemegeers *et al., Drug Dev. Res.* **3**, 123 (1983). Evaluation of use with chlorpromazine, *q.v.,* in various psychoses: H. C. B. Denber, S. Merlis: *J. Nerv. Ment. Dis.* **122**, 463 (1955). Toxicity data: L. B. Speck, *J. Pharmacol. Exp. Ther.* **119**, 78 (1957); H. F. Hardman *et al., Toxicol. Appl. Pharmacol.* **25**, 299 (1973). *Reviews:* A. R. Patel, *Progress in Drug Research* vol. **11**, E. Jucker, Ed. (Birkhäuser Verlag, Basel, 1968) pp 11-47; G. J. Kapadia, M. B. E. Fayez, *J. Pharm. Sci.* **59**, 1699-1727 (1970).

CH₂CH₂NH₂
CH₃O   OCH₃
OCH₃

Crystals, mp 35-36°. bp₁₂ 180°. Moderately sol in water; sol in alcohol, chloroform, benzene. Practically insol in ether, petr ether. Takes up CO_2 from the air and forms a crystalline carbonate. LD_{50} i.p. in rats: 370 mg/kg (Speck).

Hydrochloride, $C_{11}H_{18}ClNO_3$, needles, mp 181°. Sol in water, alcohol. LD_{50} in mice, rats, guinea pigs (mg/kg): 212, 132, 328 i.p. (Hardman).

Sulfate dihydrate, $C_{22}H_{36}N_2O_{10}S.2H_2O$, prisms, mp 183-186°. Sol in hot water, methanol; sparingly sol in cold water, ethanol.

Acid sulfate, $C_{11}H_{19}NO_7S$, crystals, mp 158°.

N-Benzoylmescaline, needles from aq alc, mp 121°. Very sol in alcohol, ether.

N-Methylmescaline, bp 130-140° (picrate mp 178°) and *N*-acetylmescaline, mp 94°, occur naturally.

Caution: This is a controlled substance (hallucinogen) listed in the U.S. Code of Federal Regulations, Title 21 Part 1308.11 (1987).

5809. Mesembrine. *3a-(3,4-Dimethoxyphenyl)octahydro-1-methyl-6H-indol-6-one;* 3a-(3,4-dimethoxyphenyl)tetrahydro-1-methyl-6(3a*H*)-indolinone. $C_{17}H_{23}NO_3$; mol wt 289.36. C 70.56%, H 8.01%, N 4.84%, O 16.59%. Alkaloid used in preparing *Channa*, a drug of Southwest Africa. Occurs naturally as the (−)-form. From *Sceletium expansum* L., *S. tortuosum* L., *L. bolus* (formerly called *Mesembryanthemum expansum* L., *M. tortuosum* L.) *Ficodaceae* or *Aizoaceae:* Hartwick, Zwicky, *Apoth. Ztg.* **29**, 925 (1914); Rimington *et al., J. Vet. Sci. Animal Ind.* **9**, 187 (1938), *C.A.* **32**, 4279⁹ (1938). Structure: Popelak *et al., Naturwiss.* **47**, 156 (1960). Configuration: P. W. Jeffs *et al., J. Am. Chem. Soc.* **91**, 3831 (1969). Synthesis of (±) form: Shamma, Rodriguez, *Tetrahedron Letters* **1965**, 4847; O. Hoshino *et al., Heterocycl.* **10**, 61 (1978); of (±)-form and *trans* isomer: Oh-Ishi, Kugita, *Chem. Pharm. Bull.* **18**, 299 (1970). Synthesis of (+)-form: Yamada, Otani, *Tetrahedron Letters* **1971**, 1133; *eidem, Chem. Pharm. Bull.* **21**, 2130 (1973). Stereoselective synthesis of (±)-form: Wijnberg, Speckamp, *Tetrahedron Letters* **1975**, 3963; *eidem, Tetrahedron* **34**, 2579 (1978); S. F. Martin *et al., J. Org. Chem.* **44**, 3391 (1979); S. Takano *et al., Chem Letters* **1981**, 1385. Enantioselective synthesis of natural mesembrine: *eidem, Tetrahedron Letters* **22**, 4479 (1981). Biosynthesis: Jeffs *et al., J. Am. Chem. Soc.* **93**, 3752 (1971); *eidem, Chem. Commun.* **1977**, 60. Review of mesembrine alkaloids: A. Popelak, G. Lettenbauer in *The*

Alkaloids, R. H. F. Manske, Ed., **vol. IX** (Academic Press, New York, 1967) pp 467-481; R. V. Stevens in *The Total Synthesis of Natural Products* vol. 3, J. ApSimon, Ed. (Wiley, New York, 1977) pp 443-453.

(-)-mesembrine

Pale yellow oil. $bp_{0.3}$ 186-190°. $[\alpha]_D^{20}$ −55.4° (CH_3OH). Freely sol in alcohol, chloroform, acetone; slightly sol in ether. Practically insol in benzene, petr ether, alkalies.

Hydrochloride, $C_{17}H_{23}NO_3 \cdot HCl$, mp 205-206°. $[\alpha]_D^{20}$ −8.4° (CH_3OH).

(+)-Form. (Partially optically active). Pale yellow oil. $[\alpha]_D^{20}$ +16.1° (c = 1.32 in CH_3OH).

(+)-Form hydrochloride, $C_{17}H_{23}NO_3 \cdot HCl$, crystals from 2-propanol, mp 206.5-207.5°. $[\alpha]_D^{20}$ +7.3° (c = 0.465 in CH_3OH).

(±)-Form. Colorless oil. $bp_{0.07}$ 178°.

(±)-Form hydrochloride, $C_{17}H_{23}NO_3 \cdot HCl$, mp 179-181°.

5810. Mesitylene. *1,3,5-Trimethylbenzene; sym*-trimethylbenzene. C_9H_{12}; mol wt 120.19. C 89.93%, H 10.06%. Occurs in coal tar and in petroleum crudes; prepd by dehydrating acetone with H_2SO_4: Adams, Hufferd, *Org. Syn.* **2**, 41 (1922).

Liquid; peculiar odor. d^{20} 0.8637. mp −44.8°. bp_{760} 164.7°; bp_{100} 98.9°; bp_{20} 61°; bp_{10} 47.4°; $bp_{1.0}$ 9.6°. n_D^{18} 1.49541. Practically insol in water (100 g H_2O dissolve 0.002 g). Miscible with alcohol, ether, benzene.

5811. Mesityl Oxide. *4-Methyl-3-penten-2-one;* isopropylideneacetone. $C_6H_{10}O$; mol wt 98.14. C 73.43%, H 10.27%, O 16.30%. $(CH_3)_2C=CHCOCH_3$. Made by distilling diacetone alcohol with a small amount of iodine: Conant, Tuttle, *Org. Syn.* **1**, 53 (1921). Condensation of acetone to mesityl oxide using sulfonated polystyrene-divinylbenzene resin as ion exchange catalyst: Klein, Banchero, *Ind. Eng. Chem.* **48**, 1278 (1956). Believed to be a mixture of two isomers. Toxicity data: *Handbook of Toxicology* vol. **1**, W. S. Spector, Ed. (Saunders, Philadelphia, 1956) pp 342-343.

Colorless, oily liq; honey-like odor. d^{15}_4 0.8592. bp_4 130°; bp_{100} 72.1°; bp_{20} 26°; $bp_{1.0}$ −8.7°. mp −41.5°; also reported as mp −59°. Can be made to crystallize at low temp in petr ether. n_D^{22} 1.4425. Absorption spectrum: Morton, *J. Chem. Soc.* **1926**, 719. Sol in about 30 parts water; miscible with most organic liqs. Flash pt: 87°F (30.6°C). LC for rats in air: 2500 ppm (Spector).

USE: Solvent for nitrocellulose, many gums and resins, particularly vinyl resins. In lacquers, varnishes and enamels. In making methyl isobutyl ketone.

5812. Mesna. *2-Mercaptoethanesulfonic acid sodium salt;* sodium mercaptoethanesulfonate; UCB 3983; Mesnex; Mistabron; Mistabronco; Mucofluid; Uromitexan. C_2H_5-NaO_3S_2; mol wt 164.17. C 14.63%, H 3.07%, Na 14.00%, O 29.24%, S 39.06%. $[HSCH_2CH_2SO_3]^-Na^+$. Sulfhydryl donor used to reduce the urotoxic effects of antineoplastic alkylating agents; also has mucolytic activity. Prepn: I. M. Lipovich, *J. Appl. Chem. USSR* **18**, 718 (1945); C. H.

Schramm *et al., J. Am. Chem. Soc.* **77**, 6231 (1955). Synthesis and properties: V. E. Petrun'kin, *C.A.* **51**, 5693a (1957); *ibid.* **54**, 24379c (1960). Use as mucolytic: **Neth.** pat. **Appl. 6,605,816;** H. Morren, **U.S.** pat. **3,567,835** (1966, 1971 both to U.C.B.); as uroprotective agent: N. Brock, **Ger.** pat. **2,756,018;** *idem,* **U.S.** pat. **4,220,660** (1979, 1980 both to Asta-Werke AG). Pharmacology, toxicity, and uroprotective effects in animals: N. Brock *et al., Eur. J. Cancer Clin. Oncol.* **18**, 1377 (1982). HPLC determn in plasma and urine: C. A. James in H. J. Rogers, *J. Chromatog.* **382**, 394 (1986). Pharmacokinetics in humans: C. A. James *et al., Brit. J. Clin. Pharmacol.* **23**, 561 (1987). Clinical study of mucolytic effects: M. Tekeres *et al., Clin. Ther.* **4**, 56 (1981). Symposium on pharmacology, toxicity and clinical uroprotective efficacy: *Cancer Treat. Rev.* **10**, Suppl. A, 1-192 (1983). *Review:* I. C. Shaw, M. I. Graham, *ibid.* **14** 67-86 (1987).

LD_{50} in male, female mice, male, female rats (mg/kg): 1887, 2048, 2098, 1683 i.v.; 2005, 2098, 1529, 1251 i.p.; 6102, >7200, 4440, 4679 orally (Brock).

THERAP CAT: Mucolytic; uroprotective (cancer chemotherapy).

5813. Mesoridazine. *10-[2-(1-Methyl-2-piperidinyl)ethyl]-2-(methylsulfinyl)-10H-phenothiazine;* thioridazine-2-sulfoxide; TPS-23. $C_{21}H_{26}N_2OS_2$; mol wt 386.59. C 65.24%, H 6.78%, N 7.25%, O 4.14%, S 16.59%. Dopamine receptor blocking agent; analog of thioridazine. Prepn: Renz *et al.,* **U.S.** pat. **3,084,161** (1963 to Sandoz). Pharmacology and toxicology: Loew *et al., Boll. Chim. Farm.* **106**, 332-371 (1967). Effects on dopaminergic function in comparison with sulforidazine and thioridazine, *q.q.v.:* D. M. Niedzwiecki *et al., J. Pharmacol. Exp. Ther.* **228**, 636 (1984); C. D. Kilts *et al., ibid.* **231**, 334 (1984). GLC determn in plasma: E. C. Dinovo *et al., J. Pharm. Sci.* **65**, 667 (1976). Clinical evaluation in sleep disorders: K. Adam *et al., Brit. J. Clin. Pharmacol.* **3**, 157 (1976); as antipsychotic: R. Axelsson, *Curr. Ther. Res.* **21**, 587 (1977). Toxicity studies: S. Maruyama *et al., Niigata Igakkai Zasshi* **81**, 611 (1967), *C.A.* **68**, 76856h (1968). Brief description: *J. Am. Med. Assoc.* **216**, 313 (1971).

Oily product.

Benzenesulfonate, $C_{27}H_{32}N_2O_4S_3$, *mesoridazine besylate,* NC-123, *Lidanar, Lidanil, Serentil.* LD_{50} in mice (mg/kg): 33 i.v.; 611 s.c.; 346 orally (Maruyama).

Tartrate, $C_{25}H_{32}N_2O_7S_2$, crystals from ethyl acetate, mp 115-120°.

THERAP CAT: Antipsychotic.

5814. Mesoxalic Acid. *Oxopropanedioic acid;* ketomalonic acid; oxomalonic acid. $C_3H_2O_5$; mol wt 118.05. C 30.52%, H 1.71%, O 67.77%. HOOCCOCOOH. Occurs in *Medicago sativa* L., *Leguminosae;* has been found in beet molasses. Prepd by boiling a soln of alloxan and lead acetate: Deichsel, *J. Prakt. Chem.* [1] **93**, 194 (1864). By electrolysis of *d*-tartaric acid in alkaline soln: *Chem. Zentr.* **1922, III,** 871. Laboratory prepn from dibromomalonic acid: Conrad, Reinbach, *Ber.* **35**, 1819 (1902); from malonic ester and N_2O_3: Curtiss, *Am. Chem. J.* **35**, 477 (1906).

Monohydrate, $C_3H_2O_5 \cdot H_2O$, *dihydroxymalonic acid.* Begins to melt at 113-114° and is clear at 121°. Very sol in water; sol in alc, ether.

Diethyl ester, $C_7H_{10}O_5$, *ethyl oxomalonate.* Liquid. bp_{19} 105-107°. $d_4^{15.6}$ 1.1419. $n_D^{15.6}$ 1.419. Prepn: A. W. Dox, *Org. Syn. coll.* **vol. I**, 266 (2nd ed., 1941). Diethyl ester of dihydroxymalonic acid ($C_7H_{12}O_6$) is obtained from the crude ester mixture by fractional distn. Crystals, mp 57°. Sol in water, alcohol.

5815. Mesquite Gum. Sonora; Prosopis gum. Gathered

from small thorny trees abundant in the arid regions of the Western United States and as far south as Chile: *Prosopis juliflora* (Swartz) DC., *P. dulcis* Kunth., *P. horrida* Kunth., *P. inermis* H.B.K., *P. glandulosa* Torr., *P. pubescens* Benth., *P. spicigera* L., and other species of *Prosopis, Leguminosae*. Mesquite gum resembles acacia (gum arabic) in its physical and chemical characteristics. Review of structure work: F. Smith, R. Montgomery, *The Chemistry of Plant Gums and Mucilages* (Reinhold, New York, 1959) pp 175, 288-291.

USE: Substitute for acacia. Potential source of L-arabinose and D-glucuronic acid, *cf.* C. L. Mantell, *The Water-Soluble Gums* (Reinhold, New York, 1947) pp 72-73.

5816. Mestanolone. *17β-Hydroxy-17-methyl-5α-andro-stan-3-one;* 17β-hydroxy-17α-methyl-3-androstanone; 17α-methylandrostan-17β-ol-3-one; 17α-methylandrostan-3-on-17β-ol; Androstalone. $C_{20}H_{32}O_2$; mol wt 304.46. C 78.89%, H 10.59%, O 10.51%. Prepd by the oxidation of 17-methyl-3,17-androstanediol: Ruzicka *et al., Helv. Chim. Acta* **18**, 1487 (1935); **Swiss. pat. 208,080** (1940 to Ciba).

Crystals from ethyl acetate, mp 192-193°. Insol in water. Sol in acetone, alcohol, ether, ethyl acetate.

THERAP CAT: Androgen.

5817. Mesterolone. *17-Hydroxy-1-methylandrostan-3-one;* 1α-methyl-5α-androstan-17β-ol-3-one; 1α-methyl-5α-dihydrotestosterone; Androviron; Proviron; Mestoranum. $C_{20}H_{32}O_2$; mol wt 304.46. C 78.89%, H 10.59%, O 10.51%. Prepn of acetate: R. Wiechert, **Ger. pat. 1,122,944** corresp to **U.S. pat. 3,361,773** (1962, 1968 to Schering, AG).

Crystals from ethyl acetate, mp 203.5-205.0°. $[\alpha]_D^{20}$ +17.6° (c = 0.875 in CHCl₃).

Acetate, $C_{22}H_{34}O_3$, *17β-acetoxy-1α-methyl-5α-androstan-3-one.* Crystals, mp 169-170°. $[\alpha]_D^{25}$ +16.5° (c = 0.88 in CHCl₃).

THERAP CAT: Androgen.

5818. Mestilbol. *4-[1-Ethyl-2-(4-methoxyphenyl)-1-butenyl]phenol;* α,α'-diethyl-4'-methoxy-4-stilbenol; diethylstilbestrol monomethyl ether; 3-p-hydroxyphenyl-4-p-methoxyphenyl-3-hexene; monomestrol. $C_{19}H_{22}O_2$; mol wt 282.37. C 80.81%, H 7.85%, O 11.33%. Prepn: Reid, Wilson, *J. Am. Chem. Soc.* **64**, 1625 (1942); **U.S. pat. 2,385,468** (1945). The monoether is sepd from the diether by its greater soly in 0.4N alcoholic KOH. Other syntheses: **Ger. pat. 708,202** (1941); Wiles, Biggerstaff, *J. Am. Chem. Soc.* **67**, 789 (1945). *See also* Dimestrol.

Needles from benzene + petr ether, mp 116-117.5°; leaflets from 70% alc, mp 114°; v. Pallos, *Arch. Gynäkol.* **170**, 355, 385 (1940), reports mp 120-121°. Distills at 185-195° at 0.3 mm Hg. Is generally more sol than the dimethyl ether of diethylstilbestrol. Practically insol in water. Sol in alcohol, dil aq or alcoholic solns of alkali hydroxides, and in vegetable oils; freely sol in acetone, ether.

5819. Mestranol. *(17α)-3-Methoxy-19-norpregna-1,3,5(10)-trien-20-yn-17-ol;* 17α-ethynyl-3-methoxy-1,3,5(10)-estratrien-17β-ol; 17α-ethynylestradiol 3-methyl ether; Menophase; Norquen; Ovastol. $C_{21}H_{26}O_2$; mol wt 310.42. C 81.25%, H 8.44%, O 10.31%. Orally active estrogenic steroid. Prepn: Colton, **U.S. pat. 2,666,769** (1954 to Searle); *J. Am. Chem. Soc.* **79**, 1123 (1957). Comprehensive description: H. A. El-Obeid, A. A. Al-Badr, in *Analytical Profiles of Drug Substances* vol. 11, K. Florey, Ed. (Academic Press, New York, 1982) pp 375-406. Clinical pharmacokinetics: J. W. Goldzieher *et al., Contraception* **21**, 17 (1980). Effect on carbohydrate metabolism: W. N. Spellacy *et al., Metabolism* **31**, 106 (1982). Randomized, double-blind clinical trials: S. Koetsawang *et al., Contraception* **25**, 231 (1982); A. Sheth *et al., ibid.* 243. Evaluation of carcinogenic risk: *IARC Monographs* **21**, 257 (1979).

Crystals from methanol or acetone, mp 150-151°. uv max (methanol): 279, 287.5 nm ($E_{1cm}^{1\%}$ 82, 14.4). Sol in ethanol, ether, chloroform, dioxane, acetone. Slightly sol in methanol. Practically insol in water.

Note: Also used in combination with chlormadinone acetate, ethynodiol, lynestranol, norethindrone or norethynodrel, *q.q.v.* Has been used in combination with megestrol acetate, *q.v.* This substance may reasonably be anticipated to be a carcinogen: *Fourth Annual Report on Carcinogens* (NTP 85-002, 1985) p 105.

THERAP CAT: Estrogen; in combination with progestogen as oral contraceptive.

5820. Mesulergine. *N'-[(8α)-1,6-Dimethylergolin-8-yl]-N,N-dimethylsulfamide;* 1,6-dimethyl-8α-(N,N-dimethylsulfamoylamino)ergoline. $C_{18}H_{26}N_4O_2S$; mol wt 362.49. C 59.64%, H 7.23%, N 15.46%, O 8.83%, S 8.84%. Ergoline deriv with antiparkinson activity. Prepn: P. L. Stütz *et al.,* **Ger. pat. 2,656,344** corresp to **U.S. pat. 4,348,391** (1978, 1982 both to Sandoz). Simplifed synthesis and central dopaminergic activity: P. L. Stütz *et al., Eur. J. Med. Chem.* **17**, 537 (1982). Inhibits prolactin release: E. Flückige *et al., Experientia* **35**, 1677 (1979). Dopamine agonist potency *in vivo:* A. Enz, *Life Sci.* **29**, 2227 (1981); R. Markstein, *Eur. J. Pharmacol.* **95**, 101 (1983); A. Enz *et al., J. Neural Transm.* **60**, 225 (1984). Central dopamine agonist activity: K. Fuxe *et al., Brain Res.* **328**, 325 (1985). Comparison of hypotensive action with bromocriptine: F. J. Morales-Olivas *et al., Arch. Int. Pharmacodyn. Ther.* **372**, 71 (1984). Clinical studies in treatment of Parkinson's disease: K. Jellinger, *J. Neurol.* **227**, 75 (1982); E. Schneider *et al., Neurology* **33**, 468 (1983).

Yellowish resin.

Hydrochloride, $C_{18}H_{27}ClN_4O_2S$, *CU 32-085.* Crystals from ethanol, mp 226-228°. $[\alpha]_D^{20}$ −23° (c = 0.3 in pyridine).

5821. Mesulphen. *2,7-Dimethylthianthrene;* 2,6-dimethylthianthrene; 2,6-dimethyldiphenylene disulfide; Mitigal; Odylen; Sudermo; Peligal; Neosulfine. $C_{14}H_{12}S_2$; mol wt 244.38. C 68.81%, H 4.95%, S 26.24%. Prepn: Cohen, Skirrow, *J. Chem. Soc.* **75**, 890 (1899); Barber, Smiles, *ibid.* **1928**, 1149; Rumpf, *Bull. Soc. Chim. France* **7**, 632 (1940).

Needles from acetic acid, ethyl acetate, or alcohol, mp 123°. bp$_3$ 184°; bp$_{14}$ 228-231°. Freely sol in acetone, chloroform, ether, petr ether; moderately sol in abs alcohol, ethyl acetate; practically insol in water.

THERAP CAT: Scabicide, antipruritic.

THERAP CAT (VET): Scabicide, antipruritic.

5822. Metabutoxycaine Hydrochloride. *3-Amino-2-butoxybenzoic acid 2-diethylaminoethyl ester hydrochloride;* 2-butoxy-3-aminobenzoic acid β-diethylaminoethyl ester hydrochloride; β-diethylaminoethyl 2-butoxy-3-aminobenzoate hydrochloride; Primacaine hydrochloride. $C_{17}H_{29}ClN_2O_3$; mol wt 344.88. C 59.20%, H 8.48%, Cl 10.28%, N 8.12%, O 13.92%. Prepn: **Brit. pat. 728,527** (1955 to Novocol Chem.), *C.A.* **50**, 5750 (1956).

Crystals, mp 117-119°. uv max (water): 313 nm ($A_{1cm}^{1\%}$ 6.70). Freely sol in water, ethanol. Moderately sol in acetone, chloroform. Very slightly sol in ether. pH of a 2% aq soln about 5.6.

THERAP CAT: Local anesthetic.

5823. Metachrome Yellow. *2-Hydroxy-5-[(3-nitrophenyl)azo]benzoic acid monosodium salt; C.I. Mordant Yellow 1;* C.I. 14025; alizarine yellow GG; salicyl yellow; sodium *m*-nitrobenzeneazosalicylate. $C_{13}H_8N_3NaO_5$; mol wt 309.22. C 50.49%, H 2.61%, N 13.59%, Na 7.44%, O 25.87%. Prepd from diazotized *m*-nitroaniline and salicylic acid in alkaline medium: R. Nietzki, **U.S. pat. 424,019;** *Colour Index* **vol. 4** (3rd ed., 1971) p 4058.

Yellow powder. Slightly sol in cold water; more sol in hot water.

USE: Biological stain and acid-base indicator. pH: 10.2 colorless, 12.0 yellow.

5824. Metaclazepam. *7-Bromo-5-(2-chlorophenyl)-2,3-dihydro-2-(methoxymethyl)-1-methyl-1H-1,4-benzodiazepine;* brometazepam; metuclazepam; Ka-2547; KC-2547. $C_{18}H_{18}BrClN_2O$; mol wt 393.71. C 54.91%, H 4.61%, Br 20.30%, Cl 9.00%, N 7.12%, O 4.06%. Benzodiazepine deriv with alkoxymethyl group replacing carbonyl in parent compd. Prepn: W. Milkowski *et al.,* **Ger. pat. 2,221,558;** *eidem,* **U.S. pat. 4,098,706** (1973, 1978 to Kali-Chemie). Effect on psychomotor performance and memory in volunteers: Z. Subhan, I. Hindmarch, *Drugs Exp. Clin. Res.* **9**, 567 (1983). Interaction with ethanol: H. R. Musch, M. Ruhland, *Arzneimittel-Forsch.* **32**, 567 (1982); H. J. Mallach *et al., Blut-alkohol* **20**, 196 (1983); V. Schmidt *et al., Med. Welt* **35**, 32 (1984). Clinical studies: G. Laakman *et al., Arzneimittel-Forsch.* **30**, 1233 (1980).

Hydrochloride, $C_{18}H_{19}BrCl_2N_2O$, *Talis.* Crystals from ethanol, mp 193-196°. LD$_{50}$ orally in NMRI white mice: 1578 mg/kg (Milkowski).

THERAP CAT: Anxiolytic.

5825. [2.2]Metacyclophane. *Tricyclo[9.3.1.1^{4,8}]hexadeca-1(15),4,6,8(16),11,13-hexaene;* di-*m*-xylylene; *m*-dixylylene. $C_{16}H_{16}$; mol wt 208.29. C 92.26%, H 7.74%. Prepd by the action of phenyllithium on *m*-xylene dibromide: Allinger *et al., J. Am. Chem. Soc.* **83**, 1974 (1961); *eidem, J. Org. Chem.* **32**, 2272 (1967).

Orthorhombic prisms from ether, mp 132.5°. bp$_{760}$ 290°; bp$_{12}$ 170°. Sparingly sol in alc. More sol in ether, benzene.

5826. Metalaxyl. *N-(2,6-Dimethylphenyl)-N-(methoxyacetyl)-DL-alanine methyl ester;* methyl *N-(2-methoxyacetyl)-N-(2,6-xylyl)-DL-alaninate;* metaxanin; CGA 48988; Ridomil; Subdue. $C_{15}H_{21}NO_4$; mol wt 279.35. C 64.50%, H 7.58%, N 5.01%, O 22.91%. Acylalanine fungicide systemically active against phytopathogens of the order *Peronosporales.* Prepn: A. Hubele, **Ger. pat. 2,515,091** corresp to **U.S. pat. 4,151,299** (1975, 1979 to Ciba-Geigy). Activity: A. Kerkenaar, A. K. Sijpesteijn, *Pestic. Biochem. Physiol.* **15**, 71 (1981). In potato blight: H. W. Platt, *Can. J. Plant Pathol.* **5**, 38 (1983). Enzyme-linked immunosorbent assay of residues in foods: W. H. Newsome, *J. Agr. Food Chem.* **33**, 528 (1985). Comprehensive description: P. A. Urech, F. J. Schwinn, *Phytiatr.-Phytopharm.* **27**, 239-247 (1978).

Whitish crystals, mp 71-72°. Vapor press at 20°: 2.2 × 10^{-6} mm Hg. Soly in water at 20°: 7.1 g/l. Readily sol in most organic solvents. LD$_{50}$ in rats (mg/kg): 669 orally (Urech).

USE: Agricultural fungicide.

5827. Metaldehyde. Metacetaldehyde. A polymer of acetaldehyde. $(C_2H_4O)_n$. Prepd by polymerization of acetaldehyde in the presence of HCl, H_2SO_4 at low temp: Kekule, Zincke, *Ann.* **162**, 125 (1872); in the presence of pyridine and HBr: Wilder, **U.S. pat. 2,426,961** (1947 to Publicker Ind.).

Prisms, mp 246° in sealed tube. Sublimes at about 112°, but dec with partial regeneration of aldehyde above 80°. Practically insol in water; sol in benzene, chloroform; sparingly sol in alcohol, ether. LD$_{50}$ orally in rats: 630 mg/kg, *Toxic Substances List,* H. E. Christensen, Ed. (1973) p 587.

USE: In compressed form as a fuel instead of alcohol; slug and snail poison. *Caution:* Ingestion may cause severe abdominal pain, nausea, vomiting, diarrhea, marked rise in body temp, convulsions, coma.

5828. Metamfepramone. *2-(Dimethylamino)-1-phenyl-*

propanone; 2-(dimethylamino)propiophenone; benzoyl-α-(dimethylamino)ethane; α-(dimethylamino)propiophenone; metamfepyramone; *N*-methylephedrone. $C_{11}H_{15}NO$; mol wt 177.24. C 74.54%, H 8.53%, N 7.90%, O 9.03%. Prepn from 2-bromopropiophenone and dimethylamine: Thomson, Stevens, *J. Chem. Soc.* **1932**, 1932; from *N,N*-dimethyl-2-(dimethylamino)propionamide and phenylmagnesium bromide: Eidebenz, *Arch. Pharm.* **280**, 49 (1942); from α-acetylbenzyl alcohol and dimethylamine: Iwao *et al., J. Pharm. Soc. Japan* **74**, 551 (1954). Prepn of *d*-form: Freudenberg, Nikolai, *Ann.* **510**, 223 (1934). Resolution of *dl*-form: Takamatsu, *J. Pharm. Soc. Japan* **76**, 1219 (1956).

$$C_6H_5-COCHCH_3 \quad \overset{\displaystyle N(CH_3)_2}{|}$$

dl-Form, bp_{13} 126°.

dl-Form hydrochloride, $C_{11}H_{16}ClNO$, *Effilone*. Crystals, mp 202-204° (Iwao) also reported as dec 201-202° (Eidebenz).

l-Form, $[\alpha]_D^{26}$ −60°.

l-Form hydrochloride, prisms, mp 197-199°. $[\alpha]_D^{26}$ −52.5°.

THERAP CAT: Anorexic.

5829. Metamivam. *3-Ethoxy-N,N-diethyl-4-hydroxybenzamide;* 3-ethoxy-4-hydroxybenzoic acid diethylamide; Anacardiol. $C_{13}H_{19}NO_3$; mol wt 237.29. C 65.80%, H 8.07%, N 5.90%, O 20.23%. Prepn: Canonica *et al., Ann. Chim. (Rome)* **45**, 205 (1955); U.S. pat. **3,019,257** (1962 to Ist. Biochim. Ital.).

Crystals from petr ether, mp 92.5°.

THERAP CAT: Cardiac and respiratory stimulant.

5830. Metampicillin. *3,3-Dimethyl-6-[[(methyleneamino)phenylacetyl]amino]-7-oxo-4-thia-1-azabicyclo[3.2.0]-heptane-2-carboxylic acid;* D-6-[α-(methyleneamino)phenyl-acetamido]penicillanic acid; methampicillin; Bonopen; Fedacilina; Micinovo; Pravacilin; Ruticina; Suvipen; Tampilen; Viderpen. $C_{17}H_{19}N_3O_4S$; mol wt 361.42. C 56.49%, H 5.30%, N 11.62%, O 17.71%, S 8.87%. Semi-synthetic antibiotic related to penicillin. Prepn: **Belg.** pat. **661,232,** *C.A.* **65**, 3884e (1966); Gradnik, **Brit.** pat. **1,081,093,** *C.A.* **68**, 114595g (1968), (1965 and 1967, both to E.R.A.S.M.E.). Synthesis: Gradnik *et al., Farmaco, Ed. Sci.* **26**, 20 (1971). Antibacterial activity studies: Sutherland *et al., Chemotherapy* **17**, 145 (1972). Pharmacokinetics: Fleischmann *et al., Farmaco, Ed. Prat.* **26**, 106 (1971). Clinical studies: Farina, *Minerva Med.* **60**, 1999 (1969); Ginocchi, *ibid.* 2003; Cardinale, Arrotta, *ibid.* 2011.

Sodium salt, $C_{17}H_{18}N_3NaO_4$, *Ocelina, Magnipen, Venzoquimpe.*

THERAP CAT: Antibacterial.

5831. Metanephrine. *4-Hydroxy-3-methoxy-α-(methylaminomethyl)benzenemethanol;* α-(methylaminomethyl)-vanillyl alcohol; 3-O-methylepinephrine; 3-O-methyladrenaline; 1-(4-hydroxy-3-methoxyphenyl)-2-methylaminoethanol. $C_{10}H_{15}NO_3$; mol wt 197.23. C 60.89%, H 7.67%, N 7.10%, O 24.34%. A naturally occurring derivative of epinephrine, found in the urine and in certain tissues. Various methods of prepn: Külz, Hornung, **Ger.** pat. **682,394** (1939); *Chem. Zentr.* **1940,** I, 1078, *C.A.* **36**, 3011 (1942);

Axelrod *et al., J. Biol. Chem.* **233**, 697 (1958); Heacock, Hutzinger, *Chem. & Ind. (London)* **1961**, 595.

dl-Form hydrochloride, $C_{10}H_{15}NO_3$.HCl, prisms from ethanol + ether (also reported as crystals from dil acetone). Dec 175°. uv max (ethanol): 231, 280 nm (ε 7600, 3100).

5832. Metanilic Acid. *3-Aminobenzenesulfonic acid; m*-sulfanilic acid; aniline-*m*-sulfonic acid. $C_6H_7NO_3S$; mol wt 173.18. C 41.61%, H 4.07%, N 8.09%, O 27.72%, S 18.51%. Usually obtained by reduction of 3-nitrobenzenesulfonic acid: Fierz-David, Blangey, *Fundamental Processes of Dye Chemistry* (Interscience, New York, 1949) pp 120-123; A. I. Vogel, *Practical Organic Chemistry* (Longmans, London, 3rd ed., 1959) p 589. Large-scale process: *FIAT Final Rept.* **1313** (I), 187-191 (1948). The industrial reduction with iron filings and dil acid gives up to 90% yields, in the lab it seldom exceeds 55%. Better yields with small amounts are claimed for a hydrazine-Raney nickel reduction (about 75%): Gialdi *et al., Farmaco Ed. Sci.* **14**, 765 (1959) or by using WS_2 (about 94%): Ehrmann, **Fr.** pat. **1,336,648** (1963 to BASF), *C.A.* **60**, 2846d (1964).

Anhydrous, orthorhombic needles from water. d 1.69. Very slow crystn yields triclinic prisms of the sesquihydrate. Crystallographic data for both forms: Hall, Maslen, *Acta Cryst.* **18**, 301-306 (1965). Photomicrograph of the sesquihydrate: *Helv. Chim. Acta* **12** (1929), facing page 666. Both forms dec on heating without melting. K at 25° = 2.00 × 10^{-4}. Soly of the sesquihydrate in water (16.8°): 2.37% (w/w). The soly of the anhydr form in water is given as 0.79% (w/w) at 0° and as 6.50% (w/w) at 85°. Sparingly sol in methanol, ethanol.

Sodium salt, $H_2NC_6H_4SO_3Na$, minute crystals from water, dec 302-304°.

USE: The sodium salt in the manuf of azo dyes. In the synthesis of certain sulfa drugs.

5833. Metanil Yellow. *3-[[4-(Phenylamino)phenyl]azo]-benzenesulfonic acid monosodium salt; C.I. Acid Yellow 36; m*-[(*p*-anilinophenyl)azo]benzenesulfonic acid sodium salt; sodium salt of metanilylazodiphenylamine; C.I. 13065; Tropaeolin G; Ext. D & C Yellow No. 1. $C_{18}H_{14}N_3NaO_3S$; mol wt 375.38. C 57.59%, H 3.76%, N 11.20%, Na 6.13%, O 12.79%, S 8.54%. Prepn: Welcher, *Organic Analytical Reagents,* **vol. 4** (Van Nostrand, 1948) p 516; *Colour Index* **vol. 4** (3rd ed., 1971) p 4045.

Brownish-yellow powder. Sol in water, alc; moderately sol in benzene, ether; slightly sol in acetone.

USE: As indicator in 0.1% soln, of which 2 drops are required for 10 ml liquid. pH: 1.2 red to 2.3 yellow. *Caution:* The provisional approval for external use in drugs and cosmetics was terminated by the FDA, see *FDA Consumer,* February, 1979, p 5.

5834. Metaphanine. *8,10-Epoxy-8-hydroxy-3,4-dimethoxy-17-methylhasubanan-7-one.* $C_{19}H_{23}NO_5$; mol wt 345.38. C 66.07%, H 6.71%, N 4.06%, O 23.16%. A member of the hasubanan alkaloids; also possesses an intramolecular hemiketal ring. Isolated from stems of *Stephania japonica* Miers, *Menispermaceae* from which the alkaloids *stephanine* and protostephanine, *q.v.* are also obtained: Kondo, Sanada, *J. Pharm. Soc. Japan* **514**, 5 (1924); *Yakugaku Zasshi* **44**, 5, 1034 (1924); **48**, 177, 930 (1927); Kondo, Watanabe, *ibid.* **58**, 268 (1938). Structure and stereochemistry: Tomita *et al., Tetrahedron Letters* **1964**, 3605; *Chem. Pharm. Bull.* **13**, 695 (1965). Synthesis of *dl*-form: Ibuka *et al., Tetrahedron Letters* **1972**, 1393; *eidem, Chem. Pharm. Bull.* **22**, 907 (1974).

Needles, mp 232°. Also reported as colorless prisms, mp 205-206° for the *dl*-form (Ibuka). pKa 6.03. Sol in water, alcohol.

Dihydrometaphanine, $C_{19}H_{25}NO_5$, crystals, mp 211°. $[\alpha]_D^{20}$ +72° (in chloroform). pKa 6.76.

5835. Metapramine. *10,11-Dihydro-N,5-dimethyl-5H-dibenz[b,f]azepin-10-amine; 10,11-dihydro-5-methyl-10-(methylamino)-5H-dibenz[b,f]azepine;* RP 19560; Timaxel. $C_{16}H_{18}N_2$; mol wt 238.33. C 80.64%, H 7.61%, N 11.75%. Psychotropic agent, related structurally to imipramine, *q.v.* Prepn: J. C. Fouche, C. G. A. Gueremy, **S. Afr. pat. 68 00345** corresp to **U.S. pat. 3,622,565** (1968, 1971 both to Rhone-Poulenc). Determn in plasma by GC: A. R. Viala *et al., Anal. Chem.* **49**, 2354 (1977). HPLC method: J. P. Sommadossi *et al., J. Chromatog.* **228**, 205 (1982). Pharmacological and clinical effects: P. Dick, *Encephale* **4**, 41 (1978). Clinical studies: L. F. Gayral *et al., ibid.* 365; E. J. Caille, J.-P. Brun, *Psychol. Med.* **13**, 1879 (1981). Biotransformation in animals and man: B. Decouvelaere *et al., Therapie* **37**, 249 (1982).

Hydrochloride, $C_{16}H_{19}ClN_2$, cryst from isopropanol/ether, mp 238-240°. Injectable formulation. The fumarate is used for tablet formulations.

THERAP CAT: Antidepressant.

5836. Metaproterenol. *5-[1-Hydroxy-2-[(1-methylethyl)amino]ethyl]-1,3-benzenediol; 3,5-dihydroxy-α-[(isopropylamino)methyl]benzyl alcohol;* 1-(3,5-dihydroxyphenyl)-2-isopropylaminoethanol; 1-(3,5-dihydroxyphenyl)-1-hydroxy-2-isopropylaminoethane; orciprenaline. $C_{11}H_{17}NO_3$; mol wt 211.27. C 62.54%, H 8.11% N 6.63% O 22.72%. Prepn and resolution of *d*- and *l*-forms: **Belg. pat. 611,502** (1961 to Boehringer, Ing.), *C.A.* **57**, 13678i (1962). Pharmacology: Pelz, *Am. J. Med. Sci.* **253**, 321 (1967). Metabolism: Tatsumi *et al., Yakugaku Zasshi* **90**, 639 (1970); **91**, 680 (1971). Toxicity: E. I. Goldenthal, *Toxicol. Appl. Pharmacol.* **18**, 185 (1971).

Crystals, mp 100°.

Sulfate, $C_{22}H_{36}N_2O_{10}S$, *Th 152, Alotec, Alupent, Metaprel, Metsol, Novasmasol.* Crystals from 90% ethanol, mp 202-203°. LD_{50} orally in rats: 42 mg/kg (Goldenthal).

Hydrochloride, $C_{11}H_{18}ClNO_3$, crystals from methanol +

ether, mp 147°. *d*-Form: Crystals, mp 212-213°. $[\alpha]_D$ +45.2° (c = 2 in methanol). *l*-Form: Crystals, mp 212°. $[\alpha]_D$ −45°.

THERAP CAT: Bronchodilator.

5837. Metaraminol. *α-(1-Aminoethyl)-3-hydroxybenzenemethanol;* (−)-*α-(1-aminoethyl)-m-hydroxybenzyl alcohol; m*-hydroxynorephedrine; *m*-hydroxypropadrine; *m*-hydroxyphenylpropanolamine; *m*-hydroxy-α-(1-aminoethyl)benzyl alcohol; 2-amino-1-(m-hydroxyphenyl)-1-propanol; 1-(m-hydroxyphenyl)-2-amino-1-propanol; α-(m-hydroxyphenyl)-β-aminopropanol; metadrine; Pressonex. $C_9H_{13}NO_2$; mol wt 167.20. C 64.65%, H 7.84%, N 8.38%,O 19.14%. Prepn from *m*-hydroxyisonitrosopropiophenone: **Brit. pat. 353,361** (1930 to I. G. Farben); *l*-form: **Brit. pat. 396,951** (1932 to I. G. Farben); from *l-m*-hydroxyphenylacetylcarbinol: **Swiss pat. 162,367** (1931 to I. G. Farben); *see also* **U.S. pats. 1,948,162 and 1,951,302;** Hartung, **U.S. pat. 1,995,709** (1935 to Sharp & Dohme). Pharmacology: A. Cession-Fossion, *Arch. Int. Pharmacodyn. Ther.* **172**, 421 (1968). Toxicity data: O. H. Siegmund *et al., J. Pharmacol. Exp. Ther.* **92**, 207 (1948).

Bitartrate (hydrogen L-tartrate), $C_{13}H_{19}NO_8$, *Aramine, Icoral B, Pressorol.* Crystals, mp 176-177°. Freely sol in water. pH of 1% aq soln about 3.5.

Hydrochloride, $C_9H_{13}NO_2 \cdot HCl$, hygroscopic crystals, $[\alpha]_D^{20}$ −19.75°. Freely sol in water. LD_{50} i.p. in mice: 440 mg/kg (Siegmund).

Oxalate dihydrate, $C_9H_{13}NO_2 \cdot C_2H_2O_4 \cdot 2H_2O$, crystals, mp 190°. $[\alpha]_D^{20}$ −21.66°. Sol in water.

THERAP CAT: Adrenergic.

5838. Metaxalone. *5-(3,5-Dimethylphenoxymethyl)-2-oxazolidinone;* AHR 438; Skelaxin. $C_{12}H_{15}NO_3$; mol wt 221.25. C 65.14%, H 6.83%, N 6.33%, O 21.70%. Prepn: Lunsford *et al., J. Am. Chem. Soc.* **82**, 1166 (1960); Lunsford, **U.S. pat. 3,062,827** (1962 to A. H. Robins).

Crystals from ethyl acetate, mp 121.5-123°.

THERAP CAT: Skeletal muscle relaxant.

5839. Metazocine. *1,2,3,4,5,6-Hexahydro-3,6,11-trimethyl-2,6-methano-3-benzazocin-8-ol;* 2'-hydroxy-2,5,9-trimethyl-6,7-benzomorphan; methobenzmorphan. $C_{15}H_{21}NO$; mol wt 231.33. C 77.88%, H 9.15%, N 6.05%, O 6.92%. Prepd from 3,4-lutidine methiodide: May, Fry, *J. Org. Chem.* **22**, 1366 (1957); May, Ager, *ibid.* **24**, 1432 (1959).

Plates from dil methanol, mp 232-235°.

Hydrochloride monohydrate, $C_{15}H_{22}ClNO \cdot H_2O$, rods from abs alcohol + ether, mp 194-196°.

Caution: May be habit forming. This is a controlled substance (opiate) listed in the U.S. Code of Federal Regulations, Title 21 Part 1308.12 (1987).

THERAP CAT: Narcotic analgesic.

5840. Metcaraphen. *1-(3',4'-Dimethylphenyl)-1-cyclopentanecarboxylic acid 2-diethylaminoethyl ester;* 2-diethylaminoethyl 1-(3',4'-dimethylphenyl)-1-cyclopentanecarb-oxylate; Netrin. $C_{20}H_{31}NO_2$; mol wt 317.46. C 75.67%, H 9.84%, N 4.41%, O 10.08%. Prepd by treating the parent acid chloride with 2-diethylaminoethanol: Martin, Häfliger, U.S. pat. **2,404,588** (1946 to Geigy).

[chemical structure]

Liquid. Faint amine odor, $bp_{0.05}$ 126-128°.
Hydrochloride, $C_{20}H_{31}NO_2 \cdot HCl$, crystals from alcohol + ethyl acetate. Sol in water.
THERAP CAT: Anticholinergic.

5841. Meteloidine. *2-Methyl-2-butenoic acid 6α,7α-di-hydroxy-8-methyl-8-azabicyclo[3.2.1]oct-3β(E)-yl ester; (E)-1αH,5αH-tropane-3α,6β,7β-triol 3-(2-methylcrotonate); 3-(3,6,7-tropanetriol) tiglate; 6,7-dihydroxytropinetiglic acid ester; 6,7-dihydroxy-3-tigloyloxytropane.* $C_{13}H_{21}NO_4$; mol wt 255.31. C 61.15%, H 8.29%, N 5.48%, O 25.07%. From leaves of *Datura meteloides* DC., *Solanaceae*: Pyman, Reynolds, *J. Chem. Soc.* **93**, 2077 (1908); King, *ibid.* **115**, 487 (1919); from *D. ferox* L.: Evans, Wellendorf, *ibid.* **1959**, 1406. Stereochemistry: Heusner, *Z. Naturforsch.* **9b**, 683 (1954). Synthesis: Zeile, Heusner, *Arch. Pharm.* **292**, 238 (1959).

[chemical structure]

Flat needles from benzene, mp 141-142°. uv max: 217 nm (ε 12,200). Freely sol in alcohol, chloroform, acetone; sparingly sol in water, ether, benzene.
Hydrobromide dihydrate, $C_{13}H_{22}BrNO_4 \cdot 2H_2O$, chisel-shaped needles, when anhydr, mp 250°.

5842. Metepa. *1,1',1''-Phosphinylidynetris[2-methyl-aziridine]; tris(2-methyl-1-aziridinyl)phosphine oxide; tris(1-methylethylene)phosphoric triamide; methyl aphoxide; methapoxide; MAPO.* $C_9H_{18}N_3OP$; mol wt 215.24. C 50.22%, H 8.43%, N 19.52%, O 7.43%, P 14.39%. Prepd by treatment of 2-methylethyleneimine with $POCl_3$ in alk medium: Parke *et al.*, U.S. pat. **2,606,902** (1952 to Am. Cyanamid).

[chemical structure]

Liquid, bp 90-92° (0.15-0.3 mm). LD_{50} orally in male, female rats: 136, 213 mg/kg, Gaines, *Toxicol. Appl. Pharmacol.* **14**, 515 (1969).
USE: Chemosterilant; in creaseproofing and flameproofing textiles.

5843. Metergoline. *[[(8β)-1,6-Dimethylergolin-8-yl]-methyl]carbamic acid phenylmethyl ester; D-8β-[(carbobenz-oxyamino)methyl]-1,6-dimethyl-10α-ergoline; D-N-carbo-benzoxydihydro-1-methyllysergamine I; D-8β-[(carboxy-amino)methyl]-1,6-dimethylergoline I benzyl ester; D-N-carboxydihydro-1-methyllysergamine I benzyl ester; D-[(4,6,6a,7,8,9,10,10a-octahydro-4,7-dimethyl-10α-indolo-[4,3-fg]quinolin-9β-yl)methyl]carbamic acid benzyl ester; methergoline; Liserdol.* $C_{25}H_{29}N_3O_2$; mol wt 403.51. C 74.41%, H 7.24%, N 10.41%, O 7.93%. Prepn: Bernardi *et al.*, *Gazz. Chim. Ital.* **94**, 936 (1964); Camerino *et al.*, U.S. pat. **3,238,211** (1966 to Farmitalia). Pharmacology: Camer-

ino *et al.*, *loc. cit.* Metabolic studies: Arcamone *et al.*, *Boll. Chim. Farm.* **110**, 704 (1971).

[chemical structure]

Crystals from benzene + ether, mp 146-149°. $[α]_D^{28}$ −7 ± 2°. uv max: 291 nm ($E_{1cm}^{1\%}$ 165). Very sol in pyridine; sol in alc, acetone, chloroform. Practically insol in benzene, ether, water. LD_{50} orally in mice: 430 mg/kg.
THERAP CAT: Prolactin inhibitor.

5844. Metescufylline. *7-[2-(Diethylamino)ethyl]theo-phylline compound with [(7-hydroxy-4-methyl-2-oxo-2H-1-benzopyran-6-yl)oxy]acetic acid;* etamiphylline methescule-tol; methescufylline; theophylline methesculetol; Veinartan; Venarterin. $C_{25}H_{31}N_5O_8$; mol wt 529.54. C 56.70%, H 5.90%, N 13.23%, O 24.17%. Prepn: P. Chabrier *et al.*, *Fr.* pat. **M1234** (1962 to Lab. Dausse), *C.A.* **57**, 16769f (1962).

[chemical structure]

Crystals, mp 124°. Slightly sol in cold water; freely sol in hot water; practically insoluble in acetone. LD_{50} i.v. in mice 260 mg/kg, P. Chabrier *et al.*, *loc. cit.*
THERAP CAT: Capillary protectant.

5845. Metformin. *N,N-Dimethylimidodicarbonimidic diamide; 1,1-dimethylbiguanide; N,N-dimethyldiguanide; N'-dimethylguanylguanidine; DMGG; La 6023; Diabeto-san; Diabex; Fluamine; Flumamine; Gliguanid.* $C_4H_{11}N_5$; mol wt 129.17. C 37.19%, H 8.58%, N 54.22%. Oral hypo-glycemic agent. Prepn: Werner, Bell, *J. Chem. Soc.* **121**, 1790 (1922); Shapiro *et al.*, *J. Am. Chem. Soc.* **81**, 3728 (1959). Use as antidiabetic: J. J. Sterne, U.S. pat. **3,174,901** (1965 to Jan Marcel Didier Aron-Samuel). Toxicity: *Rx Bulletin* **3**, 25 (1972). Pharmacokinetics in man: G. T. Tucker *et al.*, *Brit. J. Clin. Pharmacol.* **12**, 235 (1981). Review of pharmacology: L. S. Hermann, *Diabete Metab.* **5**, 233-245 (1979).

[chemical structure]

Hydrochloride, $C_4H_{12}ClN_5$, *Diabefagos, Glucophage, Haurymellin, Meguan, Metiguanide.* Prisms from water, mp 232° (Werner, Bell); crystals from propanol, mp 218-220° (uncorr.) (Shapiro *et al.*). Sol in water, 95% alcohol. Practi-cally insol in ether, chloroform. LD_{50} in rats (mg/kg): 1000 orally, 300 s.c. (*Rx Bulletin*).
p-Chlorophenoxyacetate(salt), $C_{12}H_{18}ClN_5O_3$, *Glucinan.* Embonate, $C_{31}H_{48}N_{10}O_6$, *metformin pamoate, Stagid.*
THERAP CAT: Antidiabetic.

5846. Methabenzthiazuron. *N-2-Benzothiazolyl-N,N'-dimethylurea; 1-(2-benzothiazolyl)-1,3-dimethylurea; meta-benzthiazuron; MBU; Bayer 5633; Bayer 74283; Tribunil.* $C_{10}H_{11}N_3OS$; mol wt 221.28. C 54.28%, H 5.01%, N 18.98%, O 7.23%, S 14.49%. Derivative of urea. Prepn and use as pre-emergence herbicide: N. E. Searle, U.S. pat. **2,756,135** (1956 to du Pont). Use as pre- and post-emergence herbi-cide in wheat and barley: H. Hack *et al.*, *Brit. pat.* **1,085,-**

430 (1967 to Bayer). Herbicidal properties: H. Hack, *Pflanzenschutz-Nachr.* **22,** 331 (1969). Toxicity studies: G. Kimmerle, E. Löser, *ibid.* 351. Use in winter cereals: D. C. Clark *et al., Proc. 12th Brit. Weed Control Conf.* 163 (1974). Mode of action: G. F. Collet, *Weed Res.* **9,** 340 (1969). Long-term effect on soil: P. L. Huge, *Pflanzenschutz-Nachr.* **34,** 97 (1981). Brief review: P. Lours, *Def. Veg.* **24,** 91 (1970).

White crystals from benzene, mp 119-120.5°. Soly in water at 20°: 59 ppm. Sol in organic solvents. Vapor pressure at 20°: $< 10^{-6}$ mm Hg. LD_{50} in mice (mg/kg): > 1000 orally; in male, female rats (mg/kg): $> 2500, > 2500$ orally; 540, 315 i.p. (Kimmerle, Löser).

USE: Selective herbicide.

5847. Methacholine Chloride. *2-(Acetyloxy)-N,N,N-trimethyl-1-propanaminium chloride;* acetyl-β-methylcholine chloride; *O*-acetyl-β-methylcholine chloride; (2-hydroxypropyl)trimethylammonium chloride acetate; (2-acetoxypropyl)trimethylammonium chloride; trimethyl-β-acetoxypropylammonium chloride; Amechol; Provocholine. $C_8H_{18}ClNO_2$; mol wt 195.69. C 49.10%, H 9.27%, Cl 18.12%, N 7.16%, O 16.35%. Parasympathomimetic bronchoconstrictor. Prepn from trimethylacetonylammonium chloride: R. T. Major, J. K. Cline, U.S. pat. 2,040,146 (1936 to Merck & Co.). Mechanism of ganglionic blockade in cats: R. L. Volle, *J. Pharmacol. Exp. Ther.* **158,** 66 (1967). Clinical diagnostic efficacy in bronchial asthma: S. L. Spector, R. S. Farr, *J. Allergy Clin. Immunol.* **56,** 308 (1975); J. G. Easton, I. Hirata, *Ann. Allergy* **50,** 171 (1983).

White, hygroscopic needles from ether, mp 172-173°. Slight odor of dead fish. Insol in ether. Freely sol in water, alcohol, chloroform. Aq solns are neutral to litmus. Rapidly dec by alkalies and slowly by water. Should not be handled in very moist atmosphere. The bromide is less hygroscopic.

Antidote: Atropine.

THERAP CAT: Cholinergic. Diagnostic aid (bronchial asthma).

5848. Methacrifos. *(E)-3-[(Dimethoxyphosphinothioyl)-oxy]-2-methyl-2-propenoic acid methyl ester; 3-hydroxy-2-methylacrylic acid methyl ester, O-ester with O,O-dimethyl phosphothioate;* methyl (*E*)-3-(dimethoxyphosphinothioyl-oxy)-2-methylacrylate; CGA 20168; Damfin. $C_7H_{13}O_5PS$; mol wt 240.21. C 35.00%, H 5.45%, O 33.30%, P 12.89%, S 13.35%. Organophosphorus insecticide effective against arthropod pests in stored grains. Prepn: E. Beriger, L. Pinter, S. Afr. pat. 67 04,184 corresp to U.S. pat. 3,594,454 (1967, 1971 both to Ciba); *eidem,* Belg. pat. 766,000 corresp to U.S. pat. 3,923,932 (1971, 1975 both to Ciba-Geigy). GLC determn of residues in stored grain: J. Desmarchelier *et al., Pestic. Sci.* **8,** 473 (1977). Efficacy and long-term stability: R. L. Kirkpatrick *et al., J. Econ. Entomol.* **75,** 277 (1982). Comparative field trial in stored sorghum: M. Bongston *et al., Pestic. Sci.* **14,** 385 (1983). Comprehensive description: R. Wyniger *et al., Proc. Brit. Crop Prot. Conf.—Pests Dis.* **1977,** 1033.

Oil, bp$_{0.01}$ 90°. Vapor pressure (20°): 8.3×10^{-4} mm Hg.

Soly in water (20°): $\sim$400 ppm. Readily sol in many organic solvents. Highly toxic to rainbow trout. LD_{50} in rats (mg/kg): 678 orally; > 3100 dermally (Wyniger).

USE: Grain protectant, insecticide, acaricide.

5849. Methacrylic Acid. *2-Methylpropenoic acid;* α-methylacrylic acid. $C_4H_6O_2$; mol wt 86.09. C 55.80%, H 7.03%, O 37.17%. Occurs in oil from Roman chamomile. Prepd by dehydration of α-hydroxyisobutyric acid: Crawford, U.S. pat. 2,143,941 (1939 to I.C.I.); by hypochlorite oxidation of methyl α-alkylvinyl ketone: Meitzner, U.S. pat. 2,192,142 (1940 to Rohm & Haas); by hydrolysis of acetone cyanohydrin: Crawford, **Brit.** pat. 405,699 (1932 to I.C.I.); by oxidation of methacrolein: Bauer, U.S. pat. 2,153,406 (1939 to Rohm & Haas). *Review:* J. W. Nemec, L. S. Kirch in Kirk-Othmer *Encyclopedia of Chemical Technology* vol. 15 (Wiley-Interscience, New York, 3rd ed., 1981) pp 346-376.

Long prisms, mp 16°, forming a corrosive liquid. Acrid, repulsive odor. d_4^{20} 1.0153. bp$_{760}$ 163°; bp$_{30}$ 81°; bp$_{12}$ 63°. n_D^{20} 1.43143. Flash pt, open cup: 170°F (76°C). Sol in warm water; miscible with alc, ether. Polymerizes easily, especially on heating or in the presence of traces of HCl. The polymer forms a ceramic-looking mass, sol in abs alc, from which it is precipitated by ether.

Methyl ester, **methyl methacrylate,** polymerizes easily, forming a clear plastic known as *Lucite, Plexiglas, Perspex.* Sol in methyl ethyl ketone, tetrahydrofuran, esters, aromatic and chlorinated hydrocarbons. LD_{50} orally in rats: 8.4 g/kg, W. Deichmann, *J. Ind. Hyg. Toxicol.* **23,** 343 (1941).

USE: In the manuf of methacrylate resins and plastics. *Caution:* May act as strong irritant.

5850. Methacrylonitrile. *2-Methyl-2-propenenitrile;* isopropenylnitrile; α-methylacrylonitrile; isopropene cyanide. C_4H_5N; mol wt 67.09. C 71.61%, H 7.51%, N 20.88%. $CH_2=C(CH_3)C\equiv N$. Prepd by vapor-phase catalytic oxidation of methallylamine: Peters *et al., Ind. Eng. Chem.* **40,** 2046 (1948); U.S. pat. 2,375,016 (1945 to Shell); by the dehydration of methacrylamide: Kung, U.S. pat. 2,373,190 (1945 to Goodrich); from isopropylene oxide and ammonia: Spillane, Kayser, U.S. pat. 2,557,703 (1951 to Allied Chem.).

Liquid. d_4^{20} 0.8001; d_4^{30} 0.7896. mp $-35.8°$. bp$_{760}$ 90.3°. n_D^{20} 1.4007; n_D^{30} 1.3954. Flash pt, open cup: 55°F (13°C). Viscosity at 20°, 0.392 cp. Surface tension at 20°, 24.4 dynes/cm. Soly in water at 20° = 2.57 wt-%, at 50° = 2.69 wt-%. Soly of water in methacrylonitrile at 20° = 1.62 wt-%, at 50° = 2.83 wt-%. Misc with acetone, octane, toluene at 20-25°. LD_{50} orally in rats: 0.25 ml/kg, Smyth *et al., Am. Ind. Hyg. Assoc. J.* **23,** 95 (1962).

USE: In prepn of homopolymers and copolymers; as an intermediate in the prepn of acids, amides, amines, esters, nitriles. *Caution:* Lacrimator, insidious poison, delayed skin reaction.

5851. Methacycline. *[4S-(4α,4aα,5α,5aα,12aα)]-4-Dimethylamino-1,4,4a,5,5a,6,11,12a-octahydro-3,5,10,12,12a-pentahydroxy-6-methylene-1,11-dioxo-2-naphthacenecarboxamide;* 6-methyleneoxytetracycline; 6-methylene-5-hydroxytetracycline; metacycline; Bialatan. $C_{22}H_{22}N_2O_8$; mol wt 442.41. C 59.72%, H 5.01%, N 6.33%, O 28.93%. Broad spectrum, semi-synthetic antibiotic related to tetracycline, *q.v.* Prepn from oxytetracycline, *q.v.*: R. K. Blackwood *et al., J. Am. Chem. Soc.* **83,** 2773 (1961); **85,** 3943 (1963); R. K. Blackwood, U.S. pat. 3,026,354 (1962 to Pfizer). Solubility data: Marsh, Weiss, *J. Assoc. Offic. Anal. Chem.* **50,** 257 (1967). Toxicity data: Goldenthal, *Toxicol. Appl. Pharmacol.* **18,** 185 (1971). Comparative clinical study with ampicillin, *q.v.,* in chronic bronchitis: S. Chodosh *et al., Chest* **69,** 587 (1976).

Hydrochloride, $C_{22}H_{23}ClN_2O_8$, *Adriamicina, Ciclobiotic, Germiciclin, Globociclina, Megamycine, Metadomus, Metilenbiotic, Londomycin, Optimycin, Physiomycine, Rindex, Rondomycin.* Obtained as crystals containing 0.5 mol water and 0.5 mol methanol from methanol + acetone + concd HCl + ether. Yellow, crystalline powder, dec ~205°. Bitter taste. Soluble in water, sparingly sol in alcohol. Practically insol in ether, chloroform. uv max (methanol + 0.01N HCl): 253, 345 nm (log ϵ 4.37, 4.19). LD_{50} in rats, mice (mg/kg): 252, 288 i.p. (Goldenthal).

THERAP CAT: Antibacterial.

5852. Methadone Hydrochloride. *6-Dimethylamino-4,4-diphenyl-3-heptanone hydrochloride; 1,1-diphenyl-1-(2-dimethylaminopropyl)-2-butanone hydrochloride; 4,4-diphenyl-6-dimethylamino-3-heptanone hydrochloride.* $C_{21}H_{28}ClNO$; mol wt 345.90. C 72.91%, H 8.16%, Cl 10.25%, N 4.05%, O 4.63%. Prepn: Eisleb, *Office of Publication Board, Department of Commerce,* Report no. **PB-981,** 96A; Schultz *et al., J. Am. Chem. Soc.* **69,** 2454 (1947); Easton *et al., ibid.* 2941 (1947). Resolution: Larsen *et al., ibid.* **70,** 4194 (1948); Howe, Sletzinger, *ibid.* **71,** 2935 (1949); Brode, Hill, *J. Org. Chem.* **13,** 191 (1948); Howe, Tishler, U.S. pat. **2,644,010** (1953 to Merck & Co.); Zaugg, U.S. pat. **2,983,-757** (1961 to Abbott). Pharmacokinetics in patients with chronic pain: C. E. Inturrisi *et al., Clin. Pharmacol. Ther.* **41,** 392 (1987). Toxicity data: Finnegan *et al., J. Pharmacol. Exp. Ther.* **92,** 269 (1948); Winter, Flataker, *ibid.* **98,** 305 (1950). Review: Eddy, *J. Am. Pharm. Assoc., Pract. Pharm. Ed.* **8,** 536 (1947). Comprehensive description: R. H. Bishara in *Analytical Profiles of Drug Substances* vol. 3, K. Florey, Ed. (Academic Press, New York, 1974) pp 365-439. Review of use in opioid dependence: E. C. Senay, *Int. J. Addict.* **20,** 803-821 (1985).

dl-Form, *Adanon hydrochloride, Algidon, Algolysin, Amidon hydrochloride, AN-148, Butalgin, Depridol, Diaminon hydrochloride, Dolophine hydrochloride, Fenadone, Heptadon hydrochloride, Heptanon, Hoechst 10820, Ketalgin hydrochloride, Mecodin, Mephenon, Miadone, Moheptan, Phenadone hydrochloride, Physeptone, Tussal.* Platelets from alcohol + ether, mp 235°. Bitter taste. uv max: 292 nm. Soly (g/100 ml): water 12; alcohol 8; isopropanol 2.4. Practically insol in ether, glycerol. The pH of a 1% aq soln: 4.5-5.6. The free base (mp 78°) is pptd from solns of pH >6. Aq solns can be autoclaved at 120° for one hour without loss of potency. LD_{50} orally in rats: 95 mg/kg (Finnegan).

l-Form, *Levadone, Levothyl,* L-*Polamidon.* Crystals, mp 241°. $[\alpha]_D^{20}$ −145° (c = 2.5); $[\alpha]_D^{20}$ −169° (c = 2.1 in alc). LD_{50} s.c. in rats: 44 mg/kg (Winter, Flataker).

Caution: May be habit forming. This is a controlled substance (opiate) listed in the U.S. Code of Federal Regulations, Title 21 Part 1308.12 (1985).

THERAP CAT: Narcotic analgesic.

5853. Methadyl Acetate. *β-[2-(Dimethylamino)propyl]-α-ethyl-β-phenylbenzeneethanol acetate (ester); 6-(dimethylamino)-4,4-diphenyl-3-heptanol acetate (ester); O-acetyl-6-dimethylamino-4,4-diphenyl-3-heptanol; 3-acetoxy-6-dimethylamino-4,4-diphenylheptane; 5-acetoxy-2-dimethylamino-4,4-diphenylheptane; acetylmethadol; acemethadone; amidolacetate; race-acetylmethadol.* $C_{23}H_{31}NO_2$; mol wt 353.49. C 78.14%, H 8.84%, N 3.96%, O 9.05%. Congener of methadone, *q.v.* Prepn of *β-dl*-form: M. Bockmühl, G. Ehrhart, *Ann.* **561,** 52 (1948); of *α-dl*-form: M. E. Speeter *et al., J. Am. Chem. Soc.* **71,** 57 (1949); R. L. Clark, U.S. pats. **2,565,592; 2,668,814** (1951, 1954 to Merck & Co.); M. E.

Speeter, U.S. pat. **2,649,445** (1953 to Bristol); of *α-d* and *α-l*-forms: A. Pohland *et al., J. Am. Chem. Soc.* **71,** 460 (1949). The *α-dl*-form is more active and less toxic than the *β-dl*-form. The *α-l*-form (levomethadyl acetate, *q.v.*) is less active than the *α-d*-form but is longer acting. Metabolism of the *α-dl*-form: R. E. McMahon *et al., J. Pharmacol. Exp. Ther.* **149,** 436 (1965). Analgesic, depressant activity of *α-dl*-form in mice, rats: N. B. Eddy *et al., ibid.* **98,** 121 (1950); of enantiomers: N. B. Eddy *et al., J. Org. Chem.* **17,** 321 (1952). Comparison of methadone with *α-dl*-form in the treatment of heroin addiction: J. H. Jaffe *et al., J. Am. Med. Assoc.* **211,** 1834 (1970).

α-dl-Form hydrochloride, $C_{23}H_{32}ClNO_2$, crystals from ethyl acetate, mp 213-214°. Sol in water. pH 4-5. LD_{50} in mice (mg/kg): 61.0 s.c., 118.3 orally (Eddy, 1952).

α-dl-Form hydrobromide, $C_{23}H_{32}BrNO_2$, crystals, mp 193-194.5°. Sol in water.

α-d-Form, *alphacetylmethadol.*

α-d-Form hydrochloride, crystals from ethanol-ether, mp 215°. $[\alpha]_D^{25}$ +61.2° (c = 0.2). Sol in water. LD_{50} in mice (mg/kg): 72.2 s.c., 130.4 orally (Eddy, 1952).

α-l-Form, *See* levomethadyl acetate.

β-l-Form, *betacetylmethadol.*

β-dl-Form hydrochloride, crystals, mp 215-217°. Sol in water. LD_{50} in mice (mg/kg): 42.0 s.c., 80.2 orally (Eddy, 1952).

Caution: May be habit forming. This is a controlled substance (opiate) listed in the U.S. Code of Federal Regulations, Title 21 Part 1308.11 (1985).

5854. Methafurylene. *N-(2-Furanylmethyl)-N′,N′-dimethyl-N-2-pyridinyl-1,2-ethanediamine; 2-[(2-dimethylaminoethyl)furfurylamino]pyridine; N,N-dimethyl-N′-(2-pyridyl)-N′-furfurylethylenediamine; N′-(2-furylmethyl)-N′-(2-pyridyl)-N,N-dimethylethylenediamine.* $C_{14}H_{19}N_3O$; mol wt 245.32. C 68.54%, H 7.80%, N 17.13%, O 6.52%. Prepn: Vaughan, Anderson, *J. Am. Chem. Soc.* **70,** 2607 (1948); Kyrides, Zienty, *ibid.* **71,** 1122 (1949); Horclois, U.S. pat **2,502,151** (1950 to Rhône-Poulenc).

Light yellow oil, $bp_{0.2}$ 117.5-118°.
Hydrochloride, $C_{14}H_{20}ClN_3O$, needles from ethyl acetate, mp 117-119°.
Dihydrogen citrate, $C_{20}H_{27}N_3O_7$, crystals from methanol-ether or from methyl ethyl ketone, mp 95-97°.
Fumarate, $C_{18}H_{23}N_3O_5$, *F-151, Foralamin.*

THERAP CAT: Antihistaminic.

5855. Methallatal. *5-Ethyldihydro-5-(2-methyl-2-propenyl)-2-thioxo-4,6(1H,5H)-pyrimidinedione; 5-ethyl-5-(2-methylallyl)-2-thiobarbituric acid; ethyl-β-methylallylthiobarbituric acid; V-12; Mosidal.* $C_{10}H_{14}N_2O_2S$; mol wt 226.29. C 53.07%, H 6.24%, N 12.38%, O 14.14%, S 14.17%. Prepn: Volwiler, Tabern, U.S. pat. **2,153,729** (1939 to Abbott).

Crystals, mp 160-161°. Insol in water. Forms an alkaline sodium salt which is sol in water, alcohol.

Note: This is a controlled substance (depressant) listed in the U.S. Code of Federal Regulations, Title 21, Part 1308.13 (1987).

THERAP CAT: Antiemetic.

5856. Methallenestril. *β-Ethyl-6-methoxy-α,α-dimethyl-2-naphthalenepropionic acid; β-(6-methoxy-2-naphthyl)-α,α-dimethylvaleric acid; 3-(6-methoxy-2-naphthyl)-2,2-dimethylpentanoic acid; Vallestril.* $C_{18}H_{22}O_3$; mol wt 286.36. C 75.49%, H 7.74%, O 16.76%. Prepn: Horeau, Jacques, *Compt. Rend.* **224**, 862 (1947); *Bull. Soc. Chim. France* **1948**, 711; Gay, Horeau, *ibid.* **1955**, 955.

Crystals from dil methanol, mp 139-140°. Sol in ether, vegetable oils.

THERAP CAT: Estrogen.

5857. Methallibure. *N-Methyl-N'-(1-methyl-2-propenyl)-1,2-hydrazinedicarbothioamide; 1-methyl-6-(1-methylallyl)-2,5-dithiobiurea; N-methylthiocarbamoyl-N'-[(1-methylallyl)thiocarbamoyl]hydrazine; metallibure; ICI 33828; Aimax.* $C_7H_{14}N_4S_2$; mol wt 218.35. C 38.51%, H 6.46%, N 25.66%, S 29.37%. Prepn: Paget *et al.*, **Brit. pat.** **878,177** (1961 to I.C.I.). Inhibition of pituitary gonadotropic function in animals: Paget *et al.*, *Nature* **192**, 1191 (1961). Prevention of fetus nidation when given orally to rats: Harper, *J. Reprod. Fert.* **7**, 211 (1964).

Crystals, dec 198-200°.

THERAP CAT (VET): Anterior pituitary activator (for swine).

5858. Methamidophos. *Phosphoramidothioic acid 0,S-dimethyl ester; O,S-dimethyl phosphoramidothioate; Bayer 71628; ENT 27396; Ortho 9006; SRA 5172; Monitor; Tamaron.* $C_2H_8NO_2PS$; mol wt 141.12. C 17.02%, H 5.71%, N 9.93%, O 22.67%, P 21.95%, S 22.72%. Prepn: W. Lorenz *et al.*, **Belg. pat.** **666,143** (1965 to Bayer), *C.A.* **65**, 16864f (1966); **Neth. pat. Appl.** **6,602,588** corresp to P. S. Magee, **U.S. pat.** **3,309,266** (both 1967 to Chevron). Activity: I. Hammann, *Pflanzenschutz-Nachr.* **23**, 140 (1970). Acute toxicity: T. B. Gaines, R. E. Linder, *Fundam. Appl. Toxicol.* **7**, 299 (1986). Review of properties and metabolism: A. M. A. Khasawinah *et al.*, *Pestic. Biochem. Physiol.* **9**, 211-221 (1978).

Crystals from ether, mp 54°. n_D^{40} 1.5092; $d^{44.5}$ 1.31. Vapor press. at 30°: 3×10^{-4} mm Hg. Readily sol in water, ethanol. LD_{50} in adult male, female rats (mg/kg): 25, 27 orally (Gaines, Linder).

USE: Insecticide; acaricide.

5859. Methamphetamine. *N,α-Dimethylbenzeneethanamine; d-N,α-dimethylphenethylamine; d-N-methylamphetamine; d-deoxyephedrine; d-desoxyephedrine; 1-phenyl-2-methylaminopropane; d-phenylisopropylmethylamine; methyl-β-phenylisopropylamine; Norodin.* $C_{10}H_{15}N$; mol wt 149.24. C 80.48%, H 10.13%, N 9.39%. Central stimulant. Can be prepd by reducing ephedrine or pseudoephedrine: Emde, *Helv. Chim. Acta* **12**, 365 (1929). Prepn by reducing the condensation product of benzyl methyl ketone and methylamine: A. Ogata, *J. Pharm. Soc. Japan* **451**, 751

(1919), *C.A.* **14**, 745 (1920). Synthesis from D-phenylalanine: D. B. Repke *et al.*, *J. Pharm. Sci.* **67**, 1167 (1978). Stereochemistry-pharmacology aspects: Patil *et al.*, *J. Pharmacol. Exp. Ther.* **155**, 1, 13 (1967). Toxicity data: A. M. Lands *et al.*, *J. Pharmacol. Exp. Ther.* **89**, 382 (1947). Review of clinical trials in bulimia: H. G. Pope, Jr., J. I. Hudson, *J. Clin. Psychiatry* **47**, 339 (1986).

Hydrochloride, $C_{10}H_{16}ClN$, *"speed", "meth", Amphedroxyn, Desfedrin, Desoxyfed, Desoxyn, Destim, Dexoval, D-O-E, Doxephrin, Drinalfa, Efroxine, Gerobit, Hiropon, Isophen, Madrine, Methampex, Methedrine, Methylisomyn, Pervitin, Semoxydrine, Soxysympamine, Syndrox, Tonedron.* Crystals, mp 170-175°. Bitter taste. $[\alpha]_D^{25}$ +14 to +20°. Sol in water, alcohol, chloroform. Practically insol in ether. A 1% aq soln is neutral or slightly acid to litmus. LD_{50} i.p. in mice: 70 mg/kg (Lands).

Caution: Excessive use may lead to tolerance and physical dependence. This is a controlled substance (stimulant) listed in the U.S. Code of Federal Regulations, Title 21 Part 1308.12 (1985).

THERAP CAT: Anorexic. In attention deficit disorder with hyperactivity.

THERAP CAT (VET): Sympathomimetic, CNS stimulant.

5860. Metham Sodium. *Methylcarbamodithioic acid sodium salt; methyldithiocarbamic acid sodium salt;* sodium methyldithiocarbamate; *N-methylaminodithioformic acid sodium salt;* sodium *N-methylaminodithioformate; N-methylaminomethanethionothiolic acid sodium salt;* sodium *N-methylaminomethanethionothiolate;* sodium metham; *metam sodium; sodium metam; SMDC; carbathione; trimaton; VPM.* $C_2H_4NNaS_2$; mol wt 129.18. C 18.59%, H 3.12%, N 10.85%, Na 17.80%, S 49.64%. $CH_3NHCSSNa$. Prepd from methylamine, carbon disulfide, and NaOH: Compin, *Bull. Soc. Chim. France* [4] **27**, 464 (1920); from methyl isothiocyanate and NaSH: Iliceto, D'Angeli, *Gazz. Chim. Ital.* **89**, 1950 (1959). Use as soil fumigant: S. C. Dorman, A. B. Lindquist, **U.S. pat.** **2,766,554** (1956 to Stauffer). Conversion to methyl isothiocyanate, *q.v.*, in soil: N. J. Turner, M. E. Corden, *Phytopathology* **53**, 1388 (1963); J. H. Smelt, M. Leistra, *Pestic. Sc.* **5**, 401 (1974).

Dihydrate, *N-869, Maposol, Sistan, Vapam.* Crystals, anhyd at 130°. Unpleasant odor, similar to that of carbon disulfide. Non-flammable. Soly in water at 20° = 72.2 g/100 ml H_2O. Moderately sol in alc. Sparingly sol in other solvents. Concd aq solns are stable. Dec when in dil soln and in the presence of acids and heavy metal salts. LD_{50} orally in mice, rats: 285, 820 mg/kg, *RTECS vol. I*, R. J. Lewis, R. L. Tatken, Eds. (1980) p 451.

USE: Soil fumigant to control weeds and weed seeds, nematodes, fungi, and soil insects. *Caution:* Irritating to skin, mucous membranes.

5861. Methandriol. *17α-Methyl-5-androstene-3β,17β-diol;* methylandrostenediol; MAD; mestenediol; Masdiol; Metocryst; Metildiolo; Androdiol; Metidione; Nabadial; Neosteron; Diolandrone; Stenediol; Protandren; Neostene; Crestabolic; Diolostene; Metendiol; Metandiol; Methandiol; Methanabol; Methostan; Neutrormone; Neutrosteron; Androteston-M; Megabion (Japanese); Notandron. $C_{20}H_{32}O_2$; mol wt 304.46. C 78.89%, H 10.59%, O 10.51%. Prepd by the action of methylmagnesium iodide on 3β-hydroxy-5-androsten-17-one, also called 5,6-dehydroandrosterone: Ruzicka *et al.*, *Helv. Chim. Acta* **18**, 1487 (1935); Miescher, Klarer, *ibid.* **22**, 962 (1939). Absorption spectrum in H_2SO_4: Bernstein, Lenhard, *J. Org. Chem.* **18**, 1153 (1953). NMR: Hampel, Kraemer, *Tetrahedron* **22**, 1601 (1966).

Crystals from ethyl acetate, mp 205.5-206.5°. $[\alpha]_D^{20}$ −73° (alc). Insol in water. Slightly sol in some organic solvents. Diacetate, $C_{24}H_{36}O_4$, crystals from hexane, mp 145-146°. $[\alpha]_D^{21}$ −59° (c = 0.984 in alc). Dipropionate, $C_{26}H_{40}O_4$, *Probolin*.

THERAP CAT: Anabolic.

5862. Methandrostenolone. *17-Hydroxy-17-methylandrosta-1,4-dien-3-one;* 17α-methyl-17β-hydroxyandrosta-1,4-dien-3-one; 1-dehydro-17α-methyltestosterone; methandienone; Danabol; Nerobol; Nabolin; Stenolon; Dianabol. $C_{20}H_{28}O_2$; mol wt 300.42. C 79.95%, H 9.39%, O 10.65%. Anabolic steroid. Prepn by microbial dehydrogenation of 17α-methyltestosterone: Vischer *et al., Helv. Chim. Acta* **38**, 1502 (1955); by reduction of 17α-methyltestosterone with selenium dioxide: Meystre *et al., ibid.* **39**, 734 (1956); Wettstein *et al.,* U.S. pat. **2,900,398** (1959 to Ciba).

Crystals from acetone + ether. mp 163-164°. $[\alpha]_D^{26}$ 0° (c = 1.15 in chloroform). uv max: 245 nm (ε 15600).

THERAP CAT: Androgen.
THERAP CAT (VET): Anabolic.

5863. Methane. Marsh gas; methyl hydride. CH_4; mol wt 16.04. C 74.87%, H 25.13%. Widely distributed in nature. American natural gas is about 85% methane. The earth's atm contains 0.00022% by vol. Major constituent of the atm of the outer planets (Jupiter, Saturn, Uranus, Neptune), exact figures in *Landolt-Börnstein,* **vol. III** (Springer, 6th ed., 1952) p 59; G. P. Kuiper, *The Atmospheres of the Earth and the Planets* (University of Chicago Press, 1949). Pure carbon combines directly with pure hydrogen at temperatures above 1100° forming methane. Above 1500° amount of methane formed increases with temperature: Pring, *J. Chem. Soc.* **97**, 498 (1910). Can be prepd from sodium acetate and sodium hydroxide, or from aluminum carbide and water: Matthews, *J. Am. Chem. Soc.* **21**, 647 (1899); Carroll, *J. Phys. Chem.* **22**, 148 (1918). Prepd commercially from natural gas or by fermentation of cellulose and sewage sludge: Cost, U.S. pat. **2,583,090** (1952 to Elliott Co.); Le Paige, de Dommartin, Fr. pat. **994,032** (1951), *C.A.* **51**, 10836i (1957); Oswald, Golueke, *Mech. Eng.* **86**, 40 (1964).

Colorless, odorless, non-poisonous, flammable gas. Burns with a pale, faintly luminous flame. d_4^0 0.554 (air = 1) or 0.7168 g/liter. mp −182.6°. bp −161.4°. Crit temp −82.25°; crit pressure 45.8 atm. Heat of combustion 978 Btu/cu ft at 25° (a kilogram of CH_4 yields 13,300 kcal). Forms exposive mixtures with air, the loudest explosions occur when one vol of methane is mixed with 10 vols of air (or 2 vols of oxygen). Air contg less than 5.53% methane no longer explodes. Air contg more than 14% methane burns without noise. Autoignition temp 650°. Soly in water at 17°: 3.5 ml/100 ml H_2O. Sol in alc, ether, other organic solvents.

USE: Constituent of illuminating and cooking gas, in the manuf of hydrogen, hydrogen cyanide, ammonia, acetylene, formaldehyde, in organic syntheses. *Caution:* Simple asphyxiant.

5864. Methanearsonic Acid. *Methylarsonic acid;* methylarsinic acid; monomethylarsinic acid. CH_5AsO_3; mol wt 139.96. C 8.58%, H 3.60%, As 53.53%, O 34.30%. CH_3-$AsO(OH)_2$. Prepd from sodium arsenite and methyl iodide:

Quick, Adams, *J. Am. Chem. Soc.* **44**, 809 (1922). The disodium salt is easily prepd by treating sodium arsenite with dimethyl sulfate at 85°: Uhlinger, Cook, *Ind. Eng. Chem.* **11**, 105 (1919). Other routes are by the reaction of methyl chloride with sodium arsonate under pressure: Miller *et al.,* U.S. pat. **2,442,372** (1948); by the reaction of dimethyl sulfate with a solution of arsenic trioxide in sodium hydroxide: Schwerdle, U.S. pat. **2,889,347** (1959). Acute toxicity: T. B. Gaines, R. E. Linder, *Fundam. Appl. Toxicol.* **7**, 299 (1986).

Monoclinic, spear-shaped plates from abs alcohol. Pleasant acid taste. mp 161°. Strong dibasic acid. Freely sol in water; sol in alcohol.

Monosodium salt, CH_4AsNaO_3, *monosodium methanearsonate, MSMA, Ansar 170, Ansar 529, Bueno, Daconate, Trans-vert, Weed-E-Rad 120, Weed Hoe.* LD_{50} in adult male, female rats (mg/kg): 1105, 1059 orally (Gaines, Linder).

Disodium salt, $CH_3AsNa_2O_3$, *disodium monomethanearsonate, DSMA, Ansar 184, Ansar 8100, Arrhenal, Arsinyl, Arsynal, Cacodyl New, Clout, Crab-E-Rad, Dal-E-Rad, Granular, Neo-Arsycodile, Sodar, Stenosine, Tonarsin, Weed-E-Rad 360, Weedone Crabgrass Killer Granular.* Hydrated crystals contg $5H_2O$ or $6H_2O$. One gram dissolves in about one ml water; slightly sol in alcohol. LD_{50} in adult male, female rats (mg/kg): 928, 821 orally (Gaines, Linder). Disodium methylarsonate combined with an equimolar amount of mercury salicylate is called *Enesol;* white powder, sol in water.

USE: Herbicide.

5865. Methanesulfonic Acid. Methylsulfonic acid. CH_4-O_3S; mol wt 96.10. C 12.50%, H 4.19%, O 49.95%, S 33.37%. CH_3SO_2OH. Prepd from sulfur trioxide and methane: Snyder, Grosse, U.S. pat. **2,493,038** (1950 to Houdry Process); by oxidation of dimethyl disulfide: Johnson, Wolff, U.S. pat. **2,697,722** (1954 to Standard Oil of Indiana); Proell *et al., Ind. Eng. Chem.* **40**, 1129 (1948). Other prepns and chemistry: Suter, *The Organic Chemistry of Sulfur* (Wiley, New York, 1944).

Solid. d_4^{18} 1.4812. mp 20°. bp_{10} 167°; bp_1 122°. Soly at 26-28° in wt %: hexane, 0; methylcyclopentane, 0; benzene, 1.50; toluene, 0.38; o-chlorotoluene, 0.23; ethyl disulfide, 0.47. Thermally stable at moderately elevated temps. Not hydrolyzed by boiling water or hot aq alkali. Corrosive to iron, steel, brass, copper, lead.

Ethyl ester *see* Ethyl Methanesulfonate.
Methyl ester *see* Methyl Methanesulfonate.

USE: As catalyst in polymerization, alkylation and esterification reactions; as a solvent. *Caution:* Strong irritant.

5866. Methanesulfonyl Chloride. CH_3ClO_2S; mol wt 114.55. C 10.48%, H 2.64%, Cl 30.95%, O 27.94%, S 27.99%. CH_3SO_2Cl. Prepd from methanesulfonic acid and thionyl chloride: Hearst, Noller, *Org. Syn.,* **coll. vol. IV,** 571 (1963).

Liquid. d_4^{18} 1.4805. bp_{730} 161°; bp_{18} 62°. n_D^{23} 1.451. Practically insol in water; sol in alcohol, ether.

5867. Methanethiol. Methyl mercaptan; mercaptomethane; thiomethyl alcohol; methyl sulfhydrate. CH_4S; mol wt 48.11. C 24.96%, H 8.38%, S 66.65%. CH_3SH. Occurs in "sour" gas of W. Texas, in coal tar, and in petroleum distillates. Isolated from roots of *Raphanus sativus.* Produced in the intestinal tract by the action of anaerobic bacteria on albumin. Evolved from *Penicillium brevicaule* bread cultures containing disulfides. Prepn from sodium methyl sulfate and KHS: Klason, *Ber.* **20**, 3409 (1887); Arndt, *ibid.* **54**, 2236 (1921); catalytically from methanol and hydrogen sulfide: Kramer, Reid, *J. Am. Chem. Soc.* **43**, 880 (1921); from methyl chloride and sodium hydrosulfide: Scott *et al., Ind. Eng. Chem.* **47**, 876 (1955). Toxicity data: *Handbook of Toxicology* vol. **1**, W. S. Spector, Ed. (Saunders, Philadelphia, 1956) pp 344-345. Review on occurrence, preparation, properties and reactions: E. E. Reid, *Organic Chemistry of Bivalent Sulfur* vol. I (Chemical Publishing Co., New York, 1958) pp 15-261.

Flammable gas; odor of rotten cabbage. mp −123°. bp_{760} 5.95°; d_4^{20} 0.8665; d_4^{25} 0.9600. Critical temp 196.8°. Critical pressure 71.4 atm. Heat capacity (solid at 14.97-146.57°K): 0.773-17.47 cal/deg/mole; (liq at 154.16-271.06°K): 21.27-21.13 cal/deg/mole, Russell *et al., J. Am. Chem. Soc.* **64**, 165 (1942). Azeotrope with isobutane (14.9% methanethiol) bp

−13.00°. Soly in water at 20°: 23.30 g/l. Forms a cryst hydrate. LC for rats in air: 10,000 ppm (Spector).

Sodium salt heminonahydrate, $CH_3SNa.4\frac{1}{2}H_2O$, needles. Freely sol in water, methanol. Practically insol in ether.

Copper salt, CH_3SCu, pale yellow crystals. Practically insol in water, ethanol, ether, benzene.

USE: Intermediate in manuf of jet fuels, pesticides, fungicides, plastics; synthesis of methionine.

5868. Methanol. Methyl alcohol; carbinol; wood spirit; wood alcohol. CH_4O; mol wt 32.04. C 37.48%, H 12.58%, O 49.93%. CH_3OH. Originally obtained by the destructive distillation of wood, now usually manuf from hydrogen and carbon monoxide or carbon dioxide, also by oxidation of hydrocarbons. *Review:* Faith, Keyes & Clark's *Industrial Chemicals*, F. A. Lowenheim, M. K. Moran, Eds. (Wiley-Interscience, New York, 4th ed., 1975) pp 524-529; L. E. Wade *et al.*, in Kirk-Othmer *Encyclopedia of Chemical Technology* vol. **15** (Wiley-Interscience, New York, 3rd ed., 1981) pp 398-415.

Flammable, poisonous, mobile liq. Slight alcoholic odor when pure; crude material may have a repulsive, pungent odor. Burns with a non-luminous, bluish flame. d_4^0 0.8100; d_4^{15} 0.7960; d_4^{20} 0.7915; d_4^{25} 0.7866. mp −97.8°. bp_{760} 64.7°; bp_{400} 49.9°; bp_{200} 34.8°; bp_{100} 21.2°; bp_{60} 12.1°; bp_{40} +5.0°; bp_{20} −6.0°; bp_{10} −16.2°; bp_5 −25.3°; $bp_{1.0}$ −44.0°; n_D^{15} 1.33066; n_D^{20} 1.3292. Vapor density: 1.11 (air = 1). Flash pt, closed cup: 54°F (12°C). Ignition temp 470° (878°F). Explosive limits (%-vol in air): 6.0 to 36.5. Crit temp 240.0°; crit pressure 78.5 atm. Specific heat at 20-25° = 0.595 to 0.605. Dipole moment 1.69. Miscible with water, ethanol, ether, benzene, ketones, and most other organic solvents. Forms azeotropes with many compds. *Density, freezing and boiling point data* of *methanol-water mixtures:* 10% methanol by vol (d_4^{25}, fp, bp): 0.9836, −5°, 92.8°; 20% methanol: 0.9695, −12°, 87.8°; 30% methanol: 0.9572, −21°, 84.0°; 40% methanol: 0.9423, −33°, 80.9°; 50% methanol: 0.9259, −47°, 78.3°; 60% methanol: 0.9082, −57°, 75.9°. Methanol usually is a better solvent than ethanol, dissolves many inorganic salts, *e.g.*, sodium iodide 43%, calcium chloride 22%, ammonium nitrate 14%, copper sulfate 13%, silver nitrate 4%, ammonium chloride 3.2%, sodium chloride 1.4%.

Caution: Poisoning may occur from ingestion, inhalation or percutaneous absorption. *Acute Effects:* Headache, fatigue, nausea, visual impairment or complete blindness (may be permanent), acidosis, convulsions, mydriasis, circulatory collapse, respiratory failure, death. Death from ingestion of less than 30 ml has been reported. Usual fatal dose 100-250 ml. *Chronic:* Visual impairment, *cf.* Patty's *Industrial Hygiene and Toxicology* vol. **2C**, G. D. Clayton, F. E. Clayton, Eds. (Wiley-Interscience, New York, 3rd ed., 1982) pp 4528-4541.

USE: Industrial solvent. Raw material for making formaldehyde and methyl esters of organic and inorganic acids. Antifreeze for automotive radiators and air brakes; ingredient of gasoline and diesel oil antifreezes. Octane booster in gasoline. As fuel for picnic stoves and soldering torches. Extractant for animal and vegetable oils. To denature ethanol. Softening agent for pyroxylin plastics. Solvent and solvent adjuvant for polymers. Solvent in the manuf of cholesterol, streptomycin, vitamins, hormones, and other pharmaceuticals.

5869. Methantheline Bromide. *N,N-Diethyl-N-methyl-2-[(9H-xanthen-9-ylcarbonyl)oxy]ethanaminium bromide; diethyl(2-hydroxyethyl)methylammonium bromide xanthene-9-carboxylate;* β-diethylaminoethyl 9-xanthenecarboxylate methobromide; MTB 51; SC 2910; Banthine Bromide; Avagal; Uldumont; Vagantin; Metaxan; Methanide; Xanteline; Gastron; Gastrosedan; Methanthine Bromide; Vagamin;

Metanyl; Doladene; Asabaine. $C_{21}H_{26}BrNO_3$; mol wt 420.36. C 60.00%, H 6.24%, Br 19.01%, N 3.33%, O 11.42%. Prepn from 9-xanthenecarboxylic acid and 2-diethylaminoethanol: Cusic, Robinson, *J. Org. Chem.* **16,** 1921 (1951).

Crystals from isopropanol, mp 175-176°. Bitter taste. Very slightly hygroscopic. Freely sol in water, alcohol. Practically insol in ether. pH (2% aq soln): 5.0-5.5. Aq solns tend to hydrolyze after a few days. The corresponding chloride is very hygroscopic. uv max (alc): 246, 282 nm ($E_{1cm}^{1\%}$ 135, 69).

THERAP CAT: Anticholinergic.

THERAP CAT (VET): Anticholinergic, antispasmodic, antisecretory agent.

5870. Methaphenilene. *N,N-Dimethyl-N'-phenyl-N'-(2-thienylmethyl)-1,2-ethanediamine; N,N-dimethyl-N'-phenyl-N'-(2-thenyl)ethylenediamine;* 00836; RP 2740; W-50 base; Diatrin base. $C_{15}H_{20}N_2S$; mol wt 260.39. C 69.18%, H 7.74%, N 10.76%, S 12.31%. Prepd by a sodamide condensation of N,N-dimethyl-N'-phenylethylenediamine with 2-thenyl chloride in dry toluene: Leonard, Solmssen, *J. Am. Chem. Soc.* **70,** 2066 (1948).

Dark yellow oil. bp_7 183-185°. n_D^{25} 1.5902.

Hydrochloride, $C_{15}H_{21}ClN_2S$, Diatrin, Enstamine, Nilhistin. Crystals from ethanol, mp 186-187°. Freely soluble in water. Soluble in alc. LD_{50} i.p. in mice: 117 mg/kg.

THERAP CAT: Antihistaminic.

THERAP CAT (VET): Antihistaminic.

5871. Methapyrilene. *N,N-Dimethyl-N'-2-pyridinyl-N'-(2-thienylmethyl)-1,2-ethanediamine; 2-[(2-dimethyl-aminoethyl)-2-thenylamino]pyridine; N,N-dimethyl-N'-(2-pyridyl)-N'-(2-thenyl)ethylenediamine; N,N-dimethyl-N'-(α-pyridyl)-N'-(2-methylthienyl)ethylenediamine;* thenylpyramine; AH-42; Thenylene; Pyrathyn; Semikon; Thionylan; Tenalin; Dormin; Restryl; Rest-On; Sleepwell; Paradormalene; Pyrinistab; Pyrinistol; Lullamin. $C_{14}H_{19}N_3S$; mol wt 261.38. C 64.32%, H 7.33%, N 16.08%, S 12.27%. Prepd by heating a 2-thenyl halide with an alkali metal salt of N,N-dimethyl-N'-(2-pyridyl)ethylenediamine: Kyrides, U.S. pat. **2,581,868** (1952 to Monsanto). Alternate syntheses: Weston, *J. Am. Chem. Soc.* **69,** 980 (1947); Clapp *et al., ibid.* 1549. Carcinogenicity study: W. Lijinsky *et al., Science* **209,** 817 (1980).

Liquid. $bp_{0.45}$ 125-135°; bp_3 173-175°. n_D^{25} 1.5842 (also reported as 1.5835). LD_{50} orally in guinea pigs: 375 mg/kg.

Hydrochloride, $C_{14}H_{20}ClN_3S$. Bitter crystals, mp 162°. uv max: 238 nm ($E_{1cm}^{1\%}$ 623); min: 272 nm. One gram dissolves in about 0.5 ml water, in 5 ml alcohol, in 3 ml chloroform. Practically insol in ether, benzene.

Fumarate, $C_{40}H_{50}N_6O_{12}S_2$, crystals, mp 135-136°. Prepn: Meyer, *Brit. pat.* **694,805** (1953 to Monsanto).

THERAP CAT: Antihistaminic.

THERAP CAT (VET): Antihistaminic.

5872. Methaqualone. *2-Methyl-3-(2-methylphenyl)-4(3H)-quinazolinone; 2-methyl-3-o-tolyl-4(3H)-quinazolinone; 3,4-dihydro-2-methyl-4-oxo-3-o-tolylquinazoline;* metolquizolone; MAOA; MTQ; ortonal; QZ-2; RIC 272; Rorer 148; TR 495; Cateudyl; Citexal; Dormigoa; Dormogen; Dormutil; Dorsedin; Fadormir; Holodorm; Hyminal; Hypcol; Hyptor; Ipnofil; Melsomin; Mequin; Mollinox; Motolon; Nobedorm; Noctilene; Normi-Nox; Omnyl; Optinoxan; Parminal; Parest; Pro-Dorm; Quaalude; Roulone; Rouqualone; Sindesvel; Somnafac; Sonal; Somberol; Somnomed; Sopor; Soverin; Torinal; Tuazole (Strasenburgh); Tuazolone. $C_{16}H_{14}N_2O$; mol wt 250.29. C 76.78%, H 5.64%, N 11.19%, O 6.39%. Prepn: Kacker, Zaheer, *J. Indian*

Chem. Soc. **28**, 344 (1951); Lab. Toraude, **Brit.** pat. **843,073** (1960); Klosa, *J. Prakt. Chem.* **14**, 84 (1961); **20**, 283 (1963). Improved synthesis: M. S. Manhas *et al., Synthesis* **5**, 309 (1977). Estimation in biol. materials: Akagi *et al., Chem. Pharm. Bull.* **11**, 62 (1963). Metabolism: R. Bonnichsen *et al., Clin. Chim. Acta* **60**, 67 (1975); W.G. Stillwell *et al., Drug Metab. Dispos.* **3**, 287 (1975). Toxicity data: E. I. Goldenthal, *Toxicol. Appl. Pharmacol.* **18**, 185 (1971). Comprehensive description: D. M. Patel *et al.,* in *Analytical Profiles of Drug Substances* **vol. 4**, K. Florey, Ed. (Academic Press, New York, 1975) pp 245-267.

Crystals, mp 120°; also given as mp 114-116°. uv max (ethanol): 225, 263, 304, 316 nm; (0.01N HCl): about 234, 269 nm. Sol in ethanol, ether, chloroform. Practically insol in water. LD$_{50}$ orally in rats: 255 mg/kg (Goldenthal).

Hydrochloride, $C_{16}H_{15}ClN_2O$, *Melsedin, Mequelon, Metadorm, Methased, Optimil, Paxidorm, Revonal, Riporest, Sedaquin, Sleepinal, Somnium, Toquilone, Toraflon.* Crystals, mp 255-265°. Sol in ether, ethanol. Practically insol in water.

One of the ingredients of *Akalon-T, Biphetamine T* (obsolete), *Drastinetten, Gammagrippyl, Mandrax, Merprodem, Spasmipront.*

Caution: May be habit forming. This is a controlled substance (depressant) listed in the U.S. Code of Federal Regulations, Title 21 Part 1308.11 (1985).

THERAP CAT: Hypnotic. Sedative.

5873. Metharbital. *5,5-Diethyl-1-methyl-2,4,6(1H,3H,-5H)-pyrimidinetrione; 5,5-diethyl-1-methylbarbituric acid;* Gemonil. $C_9H_{14}N_2O_3$; mol wt 198.22. C 54.53%, H 7.12%, N 14.13%, O 24.21%. Prepn: Halpern, Jones, *J. Am. Pharm. Assoc.* **38**, 352 (1949); Snyder, Link, *J. Am. Chem. Soc.* **75**, 1881 (1953).

Crystals from benzene + petr ether, mp 155°. Solubility (g/100 ml soln) in water 0.12; in alcohol 4.3; in ether 2.6. pH of satd aq solns: 5.6-5.7. pK 8.45: Fox, Shugar, *Bull. Soc. Chim. Belg.* **61**, 44 (1952).

Caution: May be habit forming. This is a controlled substance (depressant) listed in the U.S. Code of Federal Regulations, Title 21 Parts 329.1 and 1308.13 (1987).

THERAP CAT: Anticonvulsant.

5874. Methargen. *3,3'-Methylenebis(2-naphthalenesulfonic acid) disilver(1+) salt; 2,2'-dinaphthylmethane-3,3'-disulfonic acid disilver salt;* silver methylenebis(2-naphthyl-3-sulfonate); methylenedinaphthalenesulfonic acid disilver salt; silver dinaphthylmethane disulfonate; Viacutan. $C_{21}H_{14}Ag_2O_6S_2$; mol wt 642.23. C 39.27%, H 2.20%, Ag 33.60%, O 14.95%, S 9.99%. Prepn of similar compds: Craven, Pritchard, U.S. pat. **2,539,728** (1951 to Ward, Blenkinsop).

THERAP CAT: Topical antiseptic.

5875. Methazolamide. *N-[5-(Aminosulfonyl)-3-methyl-*

1,3,4-triadiazol-2(3H)-ylidene]acetamide; N-(4-methyl-2-sulfamoyl-Δ²-1,3,4-thiadiazolin-5-ylidene)acetamide; 5-acetylamino-4-methyl-Δ²-1,3,4-thiadiazoline-2-sulfonamide; Neptazane. $C_5H_8N_4O_3S_2$; mol wt 236.27. C 25.42%, H 3.41%, N 23.71%, O 20.32%, S 27.14%. Carbonic anhydrase inhibitor. Prepn: Young *et al., J. Am. Chem. Soc.* **78**, 4649 (1956); U.S. pat. **2,783,241** (1957 to Am. Cyanamid); Pala, *Farmaco Ed. Sci.* **13**, 650 (1958).

Crystals from water, mp 213-214°. pKa 7.30. uv max (95% ethanol): 254 nm (log ε 3.66); (0.1N NaOH): 247 nm (log ε 3.61).

THERAP CAT: Diuretic.

5876. Methazole. *2-(3,4-Dichlorophenyl)-4-methyl-1,2,4-oxadiazolidine-3,5-dione;* oxydiazol; VCS 438; Paxilon; Probe. $C_9H_6Cl_2N_2O_3$; mol wt 261.05. C 41.40%, H 2.32%, Cl 27.16%, N 10.73%, O 18.39%. Pre- and post-emergence herbicide. Prepn: J. Krenzer, S. Afr. pat. **67 01,543**; *idem,* U.S. pat. **3,437,664** (1968, 1969 to Velsicol). Metabolism: H. W. Dorough *et al., Drug Metab. Dispos.* **2**, 129 (1974). Persistence in the soil: W. Bond, H. A. Roberts, *Weed Res.* **16**, 23 (1976); A. Walker, *Pestic. Sci.* **9**, 326 (1978). Mechanism of phytotoxicity: J. Verity *et al., Weed Res.* **21**, 243, 307, 317 (1981). Acute toxicity: T. B. Gaines, R. E. Linder, *Fundam. Appl. Toxicol.* **7**, 299 (1986).

Light tan solid, mp 123-124°. Soly at 25°: water 1.5 ppm; xylene 55 g/l. LD$_{50}$ in adult male, female rats (mg/kg): 777, 925 orally (Gaines, Linder).

USE: Herbicide.

5877. Methdilazine. *10-[(1-Methyl-3-pyrrolidinyl)-methyl]-10H-phenothiazine.* $C_{18}H_{20}N_2S$; mol wt 296.43. C 72.93%, H 6.80%, N 9.45%, S 10.82%. Prepn: Feldkamp, Wu, U.S. pat. **2,945,855** (1960 to Mead Johnson). Toxicity data: E. I. Goldenthal, *Toxicol. Appl. Pharmacol.* **18**, 185 (1971).

Crystals, mp 87-88°.

Hydrochloride, $C_{18}H_{21}ClN_2S$, *Dilosyn, Disyncran, Tacaryl.* Crystals from isopropyl alc, mp 187.5-189°. LD$_{50}$ orally in rats: 320 mg/kg (Goldenthal).

THERAP CAT: Antipruritic.

5878. Methemoglobin. Hemiglobin; ferrihemoglobin; met Hb. Oxidation product of the normal blood pigment, hemoglobin, in which the iron is present in the ferric state. Methemoglobin may be formed through the direct action of oxidants, through the coaction of hydrogen donors and atmospheric oxygen, or through the autoxidation of hemoglobin which occurs to a small extent in normal blood. Methemoglobin does not have the capacity to combine with molecular oxygen. *Reviews:* Bodansky, *Pharmacol. Rev.* **3**, 144 (1951); Jaffé in *The Red Blood Cell,* Bishop, Surgenor, Eds. (Academic Press, 1964) pp 397-422.

Cleaved by acids and bases to yield globin and ferriproto-porphyrin.

5879. Methenamine. *1,3,5,7-Tetraazatricyclo[3.3.1.1³,⁷]-decane; hexamethylenetetramine;* HMT; HMTA; hexamine; 1,3,5,7-tetraazaadamantane; hexamethylenamine; Amino-form; Ammoform; Cystamin; Cystogen; Formin; Uritone; Urotropin. $C_6H_{12}N_4$; mol wt 140.19. C 51.40%, H 8.63%, N 39.97%. From formaldehyde and ammonia: Meissner, Schwiedessen, U.S. pat. **2,762,800** (1956); *see also* U.S. pat. **2,762,799**; Faith, Keyes & Clark's *Industrial Chemicals*, F. A. Lowenheim, M. K. Moran, Eds. (Wiley-Interscience, New York, 4th ed., 1975) pp 445-448. Review of chemistry: E. M. Smolin, L. Rapoport, *s-Triazines and Derivatives* (Interscience, New York, 1959) Chapter X, pp 545-596. Comprehensive review: J. F. Walker, *Formaldehyde* (Reinhold, New York, 3rd ed., 1964) Chapter 19, pp 511-551.

Crystals or granules or powder; odorless. Sublimes at about 263° without melting and with partial decompn; somewhat volatile at lower temp. In contact with a flame it readily burns with smokeless flame. One gram dissolves in 1.5 ml water, 12.5 ml alc, 320 ml ether, 10 ml chloroform. pH of 0.2 molar aq sol 8.4.

USE: In adhesives, coatings, and sealing compounds; in the chemical detection of metals; in the preservation of hides; as cross-linking agent for hardening phenol-formaldehyde resin and vulcanizing rubber; as corrosion inhibitor for steel; as dye fixative; as fuel tablets for camping stoves; as stabilizer for lubricating and insulating oils; with sodium phenate and NaOH as absorber of poisonous gases; the parent substance for the manufacture of explosive compounds (*see* Cyclonite).

THERAP CAT: Antibacterial (urinary).
THERAP CAT (VET): Urinary antiseptic.

5880. Methenamine Allyl Iodide. Allyl iodide hexamine; allyl iodide hexamethylenetetramine; allyliodourotropine. $C_9H_{17}IN_4$; mol wt 308.18. C 35.07%, H 5.56%, I 41.19%, N 18.18%. $C_6H_{12}N_4 \cdot C_3H_5I$. Prepd from allyl iodide and methenamine: F. J. Welcher, *Org. Anal. Reagents* (Van Nostrand, New York, 1st ed., 1947) p 121; Hurd, Evans, *Ind. Eng. Chem., Anal. Ed.* **5**, 16 (1933).

Cryst substance. mp about 148° with decompn. Very sol in water; practically insol in chloroform, ether.

USE: Reagent for detection and determination of cadmium, with which it forms a characteristic cryst ppt in neutral or slightly acid soln.

5881. Methenamine Anhydromethylenecitrate. Hexamethylenetetramine anhydromethylenecitrate; β-(hydroxymethoxy)tricarballylic acid γ-lactone hexamethylenetetramine salt; Helmitol; Formanol; Citramin; Uropurgol; Urotropin New. $C_{13}H_{20}N_4O_7$; mol wt 344.32. C 45.34%, H 5.85%, N 16.27%, O 32.53%. $C_7H_8O_7 \cdot (CH_2)_6N_4$. Prepn: Settimj, *Ann. Chim. Applicata* **39**, 393 (1949), *C.A.* **46**, 1965 (1952).

White, cryst powder. mp about 175° with decompn. Sol in 10 parts water; very slightly sol in alcohol, ether. Dec by acids or alkalies with liberation of formaldehyde. *Incompat:* Acids, alkalies, hot water.

THERAP CAT: Antibacterial (urinary).

5882. Methenamine Hippurate. Hexamethylenetetramine hippurate; Haiprex; Hippramine; Hiprex; Urex [Riker]; Urotractan; Viapta. $C_{15}H_{21}N_5O_3$; mol wt 319.37. C 56.41%, H 6.63%, N 21.93%, O 15.03%. $C_6H_{12}N_4 \cdot C_9H_9NO_3$. Prepd by refluxing one mole hexamethylenetetramine with one mole hippuric acid in methanol: Galat, U.S. pat. **3,004,026** (1959).

Crystals, mp 105-110°. Freely sol in water and alcohol.
THERAP CAT: Antibacterial (urinary).

5883. Methenamine Mandelate. Mandelamine; hexydaline; hexamethylenetetramine phenylglycolate; hexamethylenamine mandelate; Mandastat; Mandacon; Purerin; Uro-

Cedulamin; Mandamina; Mandurin; Reflux; Renelate; Mandoz; Cedulamin; Mantropine; Uromandelin; Uronamin. $C_{14}H_{20}N_4O_3$; mol wt 292.35. C 57.52%, H 6.90%, N 19.17%, O 16.42%. $C_6H_5CHOHCOOH \cdot (CH_2)_6N_4$.

Crystals or white, cryst powder; odorless or practically so, but on long keeping may develop a slight odor. mp 128-130°. Solns have a pH about 4. Very sol in water; sol in alcohol, chloroform; slightly in acetone, ether.

THERAP CAT: Antibacterial (urinary).

5884. Methenamine Salicylate. Saliformin; Formin salicylate. $C_{13}H_{18}N_4O_3$; mol wt 278.31. C 56.10%, H 6.52%, N 20.13%, O 17.25%. $(CH_2)_6N_4 \cdot C_6H_4(OH)COOH$.

White, cryst powder; acidulous taste. Freely sol in water or alcohol.

5885. Methenamine Sulfosalicylate. Hexalet; Hexal; Sulfhexet. $C_{13}H_{18}N_4O_6S$; mol wt 358.37. C 43.57%, H 5.06%, N 15.64%, O 26.79%, S 8.95%.

Monohydrate, odorless, cryst powder. mp about 190° with decompn. Sol in 8 parts water; slightly sol in alcohol.

THERAP CAT: Antibacterial (urinary).

5886. Methenamine Tetraiodine. Hexamethylenetetramine tetraiodide; methenamine tetraiodide; Siomine; Iodoformine; Mirion. $C_6H_{12}I_4N_4$; mol wt 647.87. C 11.12%, H 1.87%, I 78.36%, N 8.65%. $(CH_2)_6N_4I_4$. Prepd by adding potassium mercuric iodide to an aq soln of methenamine: U.S. pat. **1,226,394** (1917).

Reddish powder. Slight odor and taste. Deflagrates at 138°. Almost insol in water. Slightly sol in alcohol, chloroform, ether, carbon disulfide. Sol in acetone, aq solns of sodium or potassium iodides, sodium thiosulfate and in dil HCl. Decompn is liable to occur when in aq soln.

THERAP CAT: Antiseptic, iodide source.
THERAP CAT (VET): Antiseptic, iodine therapy.

5887. Methenolone. *17β-Hydroxy-1β-methyl-5α-androst-1-en-3-one;* 1-methyl-Δ¹-androsten-17β-ol-3-one; méténolone. $C_{20}H_{30}O_2$; mol wt 302.44. C 79.42%, H 10.00%, O 10.58%. Prepn: Wiechert, Kaspar, *Ber.* **93**, 1710 (1960); Popper, Ger. pat. **1,023,764** (1958 to Schering, AG).

Crystals from isopropyl ether, mp 149.5-152°. mp 160-161° (Popper). $[\alpha]_D$ +58.9°.

17-Acetate, *Primobolan Tablets, Primonabol.* Crystals from isopropyl ether, mp 138-139°. uv max (methanol): 240 nm (ε 13,300). Sol in methanol, ether, chloroform.

17-Enanthate, *methenolone enanthate, 17β-heptanoyloxy-1-methyl-5α-androst-1-en-3-one, Primobolan-Depot, Primonabol Depot.* Used as repository form.

THERAP CAT: Anabolic.

5888. Methestrol. *4,4'-(1,2-Diethyl-1,2-ethanediyl)bis-[2-methylphenol]; 4,4'-(1,2-diethylethylene)di-o-cresol;* 3,4-bis(3-methyl-4-hydroxyphenyl)hexane; dimethylhexestrol; promethestrol; γ-promethestrol. $C_{20}H_{26}O_2$; mol wt 298.41. C 80.49%, H 8.78%, O 10.72%. Prepd by hydrolysis of the dipropionate: Niederl *et al., J. Am. Chem. Soc.* **70**, 508 (1948). Chromatographic identification and determn: R. W. Roos, *J. Pharm. Sci.* **63**, 594 (1974). Pharmacologic study: P. H. Jellinck, A. M. Newcombe, *Biochem. Pharmacol.* **29**, 3031 (1980). Pharmaco-therapeutic study: C. B. Hammond, W. S. Maxson, *Fertil. Steril.* **37**, 5 (1982).

Crystals from dil acetic acid, mp 145°.

Dipropionate, $C_{26}H_{34}O_4$, **Meprane Dipropionate.** Cryst, mp 115°. Freely sol in ether, ethyl acetate, benzene; slightly sol in alc. Practically insol in water, dil acids.

THERAP CAT: Estrogen.

5889. Methetoin. *5-Ethyl-1-methyl-5-phenyl-2,4-imid-azolidinedione; 5-ethyl-1-methyl-5-phenylhydantoin;* 1-methylnirvanol; 1-methyl-5-ethyl-5-phenylhydantoin; Deltoin. $C_{12}H_{14}N_2O_2$; mol wt 218.25. C 66.03%, H 6.47%, N 12.84%, O 14.66%. Prepn: **Ger.** pat. **611,057** (1935 to Sandoz); Long *et al., J. Am. Chem. Soc.* **70,** 900 (1948).

Crystals, mp 210°.

THERAP CAT: Anticonvulsant.

5890. Methicillin Sodium. *6-[(2,6-Dimethoxybenzoyl)-amino]-3,3-dimethyl-7-oxo-4-thia-1-azabicyclo[3.2.0]hept-ane-2-carboxylic acid monosodium salt;* 6-(2,6-dimethoxy-benzamido)penicillanic acid sodium salt; sodium 2,6-dimethoxyphenylpenicillin; 2,6-dimethoxyphenylpenicillin sodium salt; sodium 6-(2,6-dimethoxybenzamido)penicillin-ate; 2,6-dimethoxybenzoylpenin sodium salt; dimethoxy-phenecillin sodium; sodium methicillin; BRL 1241; X-1497; Azapen; Belfacillin; Celpillina; Celbenin; Cinopenil; Dimocillin; Flabelline; Penistaph; Staphcillin. $C_{17}H_{19}N_2NaO_6S$; mol wt 402.42. C 50.74%, H 4.76%, N 6.96%, Na 5.71%, O 23.86%, S 7.97%. Semi-synthetic antibiotic related to penicillin. Prepn starting with 6-aminopenicillanic acid: Doyle *et al., J. Chem. Soc.* **1962,** 1457; Doyle *et al.,* **U.S.** pat. **2,951,839** (1960); Glombitza, *Ann.* **673,** 166 (1964). Physical-chemical properties: Cotta-Ramusino, Intonti, *Farmaco, Ed. Prat.* **16,** 227 (1961).

Monohydrate, crystals from acetone, mp 196-197° (dec). $[\alpha]_D^{20}$ +230° (c = 5); +225° (c = 1). uv max: 281 nm ($E_{1cm}^{1\%}$ 55); min 264 nm. Solubilities (mg/ml) at 20°: water > 300; ethanol 40; ether < 0.03; acetone 0.35; chloroform 0.06; isooctane < 0.03.

THERAP CAT: Antibacterial.

THERAP CAT (VET): Antimicrobial.

5891. Methidathion. *Phosphorodithioic acid S-[(5-methoxy-2-oxo-1,3,4-thiadiazol-3(2H)-yl)methyl] O,O-di-methyl ester; phosphorodithioic acid O,O-dimethyl ester S-ester with 4-(mercaptomethyl)-2-methoxy-Δ^2-1,3,4-thiadia-zolin-5-one;* dithiophosphoric acid *O,O'*-dimethyl-*S*-[(2-methoxy-1,3,4-thiadiazol-5(4H)-on-4-yl)methyl] ester; dithiophosphoric acid *O,O'*-dimethyl-*S*-[(5-methoxy-1,3,4-thiadiazol-2(3H)-one-3-yl)methyl] ester; *O,O'*-dimethyl-*S*-[(2-methoxy-1,3,4-thiadiazole-5(4H)-one-4-yl)methyl] di-thiophosphate; GS 13005; Supracide; Ultracid. $C_6H_{11}N_2O_4$-PS_3; mol wt 302.31. C 23.84%, H 3.67%, N 9.26%, O 21.17%, P 10.24%, S 31.81%. Synthesis and degradation products: Rüfenacht, *Helv. Chim. Acta* **51,** 518 (1968); **Fr.** pat. **1,335,755** (1963 to Geigy), *C.A.* **60,** 1764g (1964). Metabolism in rats: Esser *et al., Helv. Chim. Acta* **51,** 513 (1968). Acute toxicity: T. B. Gaines, R. E. Linder, *Fundam. Appl. Toxicol.* **7,** 299 (1986).

Crystals from methanol, mp 39-40°. Soly in water < 1%. Readily sol in benzene, acetone, methanol, xylene and other org solvents. LD_{50} in adult male, female rats (mg/kg): 31, 32 orally (Gaines, Linder).

USE: Insecticide, acaricide.

5892. Methimazole. *1,3-Dihydro-1-methyl-2H-imidaz-ole-2-thione; 1-methylimidazole-2-thiol;* 1-methyl-2-mer-captoimidazole; mercazolyl; thiamazole; Basolan; Dananti-zol; Favistan; Frentirox; Mercazole; Metazolo; Tapazole; Thacapzol; Thycapsol; Strumazol. $C_4H_6N_2S$; mol wt 114.17. C 42.08%, H 5.30%, N 24.54%, S 28.09%. Prepd by treating aminoacetaldehyde diethyl acetal with methyl isothiocya-nate: Wohl, Marckwald, *Ber.* **22,** 1354 (1889); from thiocy-anic acid and *N*-substituted amino acetals: Jones *et al., J. Am. Chem. Soc.* **71,** 4000 (1949). Metabolism: D. S. Sitar, D. P. Thornhill, *J. Pharmacol. Exp. Ther.* **184,** 432 (1973). Comprehensive description: H. Y. Aboul-Enein, A. A. Al-Badr in *Analytical Profiles of Drug Substances* **vol. 8,** K. Florey, Ed. (Academic Press, New York, 1979) pp 351-370. Review of pharmacology and clinical experience: D. S. Cooper, *N. Engl. J. Med.* **311,** 1353-1362 (1984).

Leaflets from alc, mp 146-148°. bp 280° (some decompn). uv max (0.1N H_2SO_4): 211, 251.5 nm ($E_{1cm}^{1\%}$ 593, 1528). Freely sol in water. Sol in alcohol, chloroform. Sparingly sol in ether, petr ether, benzene.

Methiodide, Jomezol.

USE: In cyanide-free silver electroplating.

THERAP CAT: Antihyperthyroid.

5893. Methiocarb. *3,5-Dimethyl-4-(methylthio)phenol methylcarbamate; methylcarbamic acid 4-(methylthio)-3,5-xylyl ester;* 4-(methylthio)-3,5-xylyl methylcarbamate; 4-methylthio-3,5-dimethylphenyl *N*-methylcarbamate; mer-captodimethur; metmercapturon; Bayer 37344; H 321; Draza; Mesurol. $C_{11}H_{15}NO_2S$; mol wt 225.31. C 58.63%, H 6.74%, N 6.21%, O 14.20%, S 14.22%. Prepn: E. Schegk *et al.,* **Brit.** pat. **912,895** corresp to **U.S.** pat. **3,313,684** (1962, 1967, both to Bayer); E. E. Gilbert, J. A. Otto, **U.S.** pat. **3,358,012** (1967 to Allied). Molluscicidal activity: H. H. Crowell, *J. Econ. Entomol.* **60,** 1048 (1967). Bird repellent properties: E.W. Schafer, R. B. Brunton, *J. Wildl. Manage.* **35,** 569 (1971).

White crystalline powder, mp 121.5°. Insol in water; sol in organic solvents. Unstable in alk media. LD_{50} orally in male, female rats: 70, 60 mg/kg, T. B. Gaines, *Toxicol. Appl. Pharmacol.* **14,** 515 (1969).

USE: Insecticide; molluscicide; bird repellent.

5894. Methiodal Sodium. *Iodomethanesulfonic acid sodium salt;* sodium iodomethanesulfonate; Skiodan; Abro-dil; Radiographol; Segosin; Diagnorenol. CH_2INaO_3S; mol wt 244.01. C 4.92%, H 0.83%, I 52.04%, Na 9.42%, O 19.67%, S 13.14%. CH_2ISO_3Na. Prepd by the action of sodium sulfite on methylene iodide at 70° in water-alcohol soln: Ossenbeck *et al.,* **U.S.** pat. **1,842,626** (1932 to Winthrop); from iodoform + sodium sulfite: Allardt, **U.S.** pat. **1,867,793** (1932 to Schering-Kahlbaum AG).

Crystals. Slightly saline taste followed by sweetish after-taste. Freely sol in water (70 g/100 ml); slightly sol in alcohol (2.5 g/100 ml), benzene, ether, acetone.

THERAP CAT: Diagnostic aid (radiopaque medium—urographic).

5895. Methionic Acid. *Methanedisulfonic acid.* CH_4O_6-S_2; mol wt 176.17. C 6.82%, H 2.29%, O 54.49%, S 36.40%. $CH_2(SO_3H)_2$. Prepn from methane + sulfur trioxide: Snyder, Grosse, **U.S.** pat. **2,493,038** (1950 to Houdry Process); by H_2SO_4 oxidation of acetic acid: Schwab, Neuwirth, *Ber.* **90**, 567 (1957); from $MeSO_3H + SO_3$: Crowder, Gilbert, **U.S.** pat. **2,842,589** (1958 to Allied Chem.). Prepn of aluminum salt: Christian, Jenkins, *J. Am. Pharm. Assoc.* **39**, 633 (1950); **U.S.** pat. **2,504,107** (1950 to Purdue Res. Found.).

Crystals, mp 96-100°.

Aluminum salt, $C_3H_6Al_2O_{18}S_6$, crystals from water + alcohol. Hygroscopic. Soly in water at 27°: 69 w/v. pH of 5% aq soln = 3.5.

USE: Antiperspirant.

THERAP CAT: Aluminum salt as topical astringent.

5896. Methionine. Met (IUPAC abbrev.); 2-amino-4-(methylthio)butyric acid; α-amino-γ-methylmercaptobutyric acid; 2-amino-4-methylthiobutanoic acid; γ-methylthio-α-aminobutyric acid; Meonine; Methilanin; Neston; Lobamine; Mertionin; Neo-Methidin; Thiomedon; Cynaron; Dyprin; Metione; Banthionine; Acimetion. $C_5H_{11}NO_2S$; mol wt 149.21. C 40.25%, H 7.43%, O 21.45%, N 9.39%, S 21.49%. $CH_3SCH_2CH_2CH(NH_2)COOH$. An amino acid classified as essential for maintaining growth in rats. Essential component in human nutrition, not synthesized by the human body. Plays an important role in biological methylations. Both L- and D-forms are effective in the rat and in man. The D-form undergoes deamination, and the keto compd is then transaminated with resultant inversion to the natural L-form. Isolation of natural L-methionine from casein hydrolyzate by extraction with butanol: Pirie, *Biochem. J.* **26**, 1270 (1932); by digesting casein with pancreatin: du Vigneaud, Meyer, *J. Biol. Chem.* **94**, 641 (1931/2). *See also* Toennies, Kolb, *ibid.* **126**, 367 (1938); **128**, 399 (1939); Toennies, Callan, *ibid.* **129**, 481 (1939). Phthalimidomalonic ester synthesis: Barger, Weichselbaum, *Org. Syn.* **coll. vol. II,** 384 (1943). Other syntheses: Hill, Robson, *Biochem. J.* **30**, 248 (1936); Snyder *et al., J. Am. Chem Soc.* **64**, 2082 (1942). The industrial synthesis usually starts with β-methylmercaptopropionaldehyde: Pierson *et al., J. Am. Chem. Soc.* **70**, 1450 (1948); Pierson, Tishler, **U.S.** pat. **2,584,496** (1952 to Merck & Co.); Weiss, **U.S.** pat. **2,732,400** (1956 to Am. Cyanamid). Synthesis of labeled methionine: Tarver, Schmidt, *J. Appl. Phys.* **12**, 323 (1941); *J. Biol. Chem.* **146**, 69 (1942).

L-Methionine, minute hexagonal plates from dil alc, mp 280-282° (decomp, sealed capillary). $[\alpha]_D^{25}$ −8.2°. Sol in water, but the crystals are somewhat water-repellent at first. Sol in warm, dil alcohol; insol in abs alcohol, ether, petr ether, benzene, acetone.

DL-Methionine, *racemethionine, Pedameth, Amurex, Urimeth.* Platelets from alc, mp 281° (decompn). d 1.340. pK_1 2.28; pK_2 9.21. pH of 1% aq soln 5.6-6.1. R_f value 0.77. Soly in water (g/l) at 0°: 18.18; at 25°: 33.81; at 50°: 60.70; at 75°: 105.2; at 100°: 176.0. Sol in dil acids, alkalies. Very slightly sol in 95% alcohol; insol in ether.

THERAP CAT: Lipotropic.

THERAP CAT (VET): Essential nutrient, lipotropic agent, has been used to regulate urinary pH.

5897. Methionine Hydroxy Analog. *2-Hydroxy-4-(methylthio)-butanoic acid; 2-hydroxy-4-(methylthio)butyric acid; 2-hydroxy-4-(methylmercapto)butyric acid; Alimet.* $C_5H_{10}O_3S$; mol wt 150.19. C 39.98%, H 6.71%, O 31.96%, S 21.35%. $CH_3SCH_2CH_2CH(OH)COOH$. Prepn (not claimed) and use as poultry feed additive: E. S. Blake, R. J. Wineman, **U.S.** pat. **2,745,745** (1956 to Monsanto). HPLC determn in feeds: A. Baudichau *et al., J. Sci. Food Agr.* **38**, 1 (1987). Use as a feed additive for livestock: A. Papas *et al., J. Nutr.* **104**, 653 (1974); A. K. Clark, A. H. Rakes, *J. Dairy Sci.* **65**, 1493 (1982); D. H. Reifsnyder *et al., J. Nutr.* **114**,

1705 (1984). Efficacy of calcium salt vs free acid: K. P. Boebel, D. H. Baker, *Poultry Sci.* **61**, 1167 (1982).

Calcium salt, $C_{10}H_{18}CaO_6S_2$, MHA.

USE: Dietary supplement in livestock.

5898. Methioprim. *4-Amino-2-methylthio-5-pyrimidinemethanol;* 4-amino-2-methylmercapto-5-pyrimidinemethanol; 4-amino-5-hydroxymethyl-2-methylthiopyrimidine. $C_6H_9N_3OS$; mol wt 171.22. C 42.09%, H 5.30%, N 24.54%, O 9.34%, S 18.73%. Synthesis starting with ethyl ethoxyethylenecyanoacetate: Ulbricht, Price, *J. Org. Chem.* **21**, 567 (1956).

Needle-like prisms from benzene, mp 126-127°.

USE: Tumor antagonist in mice.

5899. Methiotriazamine. *1,6-Dihydro-6,6-dimethyl-1-[4-(methylthio)phenyl]-1,3,5-triazine-2,4-diamine; 4,6-diamino-1,2-dihydro-2,2-dimethyl-1-[p-(methylthio)phenyl]-s-triazine;* 4,6-diamino-1-(4-methylmercaptophenyl)-1,2-dihydro-2,2-dimethyl-1,3,5-triazine. $C_{12}H_{17}N_5S$; mol wt 263.38. C 54.72%, H 6.51%, N 26.59%, S 12.18%. Prepn: Schalit, Cutler, *J. Org. Chem.* **24**, 573 (1959).

White prisms, mp 149-152°.

Hydrochloride, needles from water, mp 204-207°.

Note: Ingredient of the coccidiostat *Trithiadol* which contains bithionol 50%, methiotriazamine 10%, and calcium sulfate.

5900. Methisazone. *2-(1,2-Dihydro-1-methyl-2-oxo-3H-indol-3-ylidene)hydrazinecarbothioamide; 1-methylindole-2,3-dione 3-thiosemicarbazone;* N-methylisatin 3-thiosemicarbazone; 33T57; Marboran; Viruzona. $C_{10}H_{10}N_4OS$; mol wt 234.29. C 51.26%, H 4.30%, N 23.91%, O 6.83%, S 13.69%. Prepn: Bauer, Sadler, *Brit. J. Pharmacol.* **15**, 101 (1960); **Brit.** pat. **975,357** (1964 to Wellcome Found.).

Crystals from butanol, mp 245°.

THERAP CAT: Antiviral.

5901. Methitural. *Dihydro-5-(1-methylbutyl)-5-[2-(methylthio)ethyl]-2-thioxo-4,6(1H,5H)-pyrimidinedione monosodium salt; 5-(1-methylbutyl)-5-[2-(methylthio)ethyl]-2-thiobarbituric acid sodium salt; 5-(2-methylthio)-5-(1-methylbutyl)-2-thiobarbituric acid sodium salt; 5-(2-methylthioethyl)-5-(2-pentyl)-2-thiobarbituric acid sodium salt; methioturiate; Am 109; Neraval; Thiogenal.* $C_{12}H_{19}N_2$-NaO_2S_2; mol wt 310.42. C 46.43%, H 6.17%, N 9.03%, Na 7.41%, O 10.31%, S 20.66%. Prepn: Zima, Von Werder, **U.S.** pat. **2,802,827** (1957 to E. Merck).

Very hygroscopic, yellow crystals. Slight odor of mercaptans. Freely sol in water. pH of a 10% aq soln about 9.5. Water solns are unstable as evidenced by a deepening of color and formation of a cloudy precipitate upon autoclaving and exposure to light. The addition of sodium carbonate has a stabilizing effect and prevents precipitation for about 24 hrs: Irwin *et al.*, *J. Pharmacol. Exp. Ther.* **116**, 317 (1956).

Caution: May be habit forming. This is a controlled substance (depressant) listed in the U.S. Code of Federal Regulations, Title 21 Parts 329.1 and 1308.13 (1987).

THERAP CAT: Sedative, hypnotic.

5902. Methixene. *1-Methyl-3-(9H-thioxanthen-9-ylmethyl)piperidine;* 9-(N-methyl-3-piperidylmethyl)thioxanthene. $C_{20}H_{23}NS$; mol wt 309.47. C 77.62%, H 7.49%, N 4.53%, S 10.36%. Smooth muscle relaxant. Prepn: Caviezel *et al.*, *Pharm. Acta Helv.* **33**, 447 (1958); Schmutz, U.S. pat. **2,905,590** (1959 to Wander).

Slightly yellow viscous liquid, bp$_{0.07}$ 171-175°.

Hydrochloride, $C_{20}H_{24}ClNS$, *Tremoquil, Methixart, Trest, Tremonil, Tremaril, Tremaril, Cholinfall, Methyloxan.* Flakes from ether, mp 215-217°. uv max (dil HCl): 268 nm (ε 10,250). Sol in water, alcohol, chloroform.

THERAP CAT: Anticholinergic.

5903. Methocarbamol. *3-(2-Methoxyphenoxy)-1,2-propanediol 1-carbamate;* 3-(o-methoxyphenoxy)-2-hydroxypropyl 1-carbamate; 2-hydroxy-3-(o-methoxyphenoxy)-propyl 1-carbamate; guaiacol glyceryl ether carbamate; AHR-85; Neuraxin; Miolaxene; Lumirelax; Etroflex; Delaxin; Robamol; Traumacut; Tresortil; Relestrid; Robaxin. $C_{11}H_{15}NO_5$; mol wt 241.24. C 54.76%, H 6.27%, N 5.81%, O 33.16%. Prepn from 3-(o-methoxyphenoxy)-2-hydroxypropyl chlorocarbonate: Murphey, U.S. pat. **2,770,649** (1956 to A. H. Robins).

Crystals from benzene, mp 92-94°. Soly in water at 20°: 2.5 g/100 ml. Sol in alcohol, propylene glycol.

THERAP CAT: Skeletal muscle relaxant.
THERAP CAT (VET): Skeletal muscle relaxant.

5904. Methohexital Sodium. *1-Methyl-5-(1-methyl-2-pentynyl)-5-(2-propenyl)-2,4,6(1H,3H,5H)-pyrimidinetrione sodium salt;* 5-allyl-1-methyl-5-(1-methyl-2-pentynyl)barbituric acid sodium salt; α-dl-1-methyl-5-(1-methyl-2-pentynyl)-5-allylbarbituric acid sodium salt; α-dl-1-methyl-5-allyl-5-(1-methyl-2-pentynyl)barbituric acid sodium salt; methohexitone sodium; Brevital; Brevital Sodium; Brevimytal Sodium; Brietal Sodium. $C_{14}H_{17}N_2NaO_3$; mol wt 284.30. C 59.15%, H 6.03%, N 9.86%, Na 8.07%, O 16.88%. Prepn: Doran, U.S. pat **2,872,448** (1959 to Lilly).

Minute crystals. Soluble in water.

Caution: May be habit forming. This is a controlled substance (depressant) listed in the U.S. Code of Federal Regulations, Title 21 Parts 329.1 and 1308.14 (1987).

THERAP CAT: Anesthetic (intravenous).
THERAP CAT (VET): Ultra-short acting anesthetic.

5905. Methomyl. *N-[[(Methylamino)carbonyl]oxy]ethanimidothioic acid methyl ester; N-[(methylcarbamoyl)oxy]thioacetimidic acid methyl ester; S-methyl N-[(methylcarbamoyl)oxy]thioacetimidate; methyl O-(methylcarbamoyl)thiolacetohydroxamate; Insecticide 1179; Lannate; Nudrin.* $C_5H_{10}N_2O_2S$; mol wt 162.20. C 37.02%, H 6.21%, N 17.27%, O 19.73%, S 19.76%. Prepn: Jelinek, S. Afr. pat. **68 00,093** corresp to U.S. pat. **3,506,698** (1968, 1970 both to du Pont). Metabolism studies: J. Harvey *et al.*, *J. Agr. Food Chem.* **21**, 769, 775, 781 (1973). Crystal structure: Sim, Waite, *J. Chem. Soc. (B)* **1971**, 752. Acetylcholinesterase activity: Pickering, Pickering, *Arch. Toxicol.* **27**, 292 (1971). Toxicity studies: A. M. Kaplan, H. Sherman, *Toxicol. Appl. Pharmacol.* **40**, 1 (1977).

Crystals, mp 78-79°. d_4^{24} 1.2946. Soly at 25° (w/w): water 5.8; methanol 100; ethanol 42; isopropanol 22; acetone 73. LD$_{50}$ orally in rats (male): 17 mg/kg, A. M. Kaplan, H. Sherman, *loc. cit.*

USE: Insecticide.

5906. Methoprene. *11-Methoxy-3,7,11-trimethyl-2,4-dodecadienoic acid 1-methylethyl ester;* isopropyl 11-methoxy-3,7,11-trimethyldodeca-*trans*-2,*trans*-4-dienoate; manta; ENT 70460; ZR-515; Altosid SR-10. $C_{19}H_{34}O_3$; mol wt 310.48. C 73.50%, H 11.04%, O 15.46%. Juvenile hormone mimic. Prepd and claimed in Belg. pat. **778,242** (1971 to Zoecon). Prepd but not claimed in Henrick, U.S. pats. **3,818,047; 3,865,874** (1974, 1975, both to Zoecon).

Amber liquid, bp$_{0.05}$ 100°. Vapor press at 25°: 2.37 × 10^{-5} mm Hg. Soly in water 1.39 ppm. Sol in most organic solvents.

USE: Insecticide; insect growth regulator.

5907. Methopterin. *N-[4-[[(2-Amino-1,4-dihydro-4-oxo-6-pteridinyl)methyl]methylamino]benzoyl]-L-glutamic acid; N-[4-[[(2-amino-4-hydroxy-6-pteridinyl)methyl]methylamino]benzoyl]glutamic acid;* 10-methylpteroylglutamic acid; N^{10}-methylfolic acid. $C_{20}H_{21}N_7O_6$; mol wt 455.45. C 52.74%, H 4.65%, N 21.53%, O 21.08%. Prepn: Cosulich, Smith, *J. Am. Chem. Soc.* **70**, 1922 (1948); Cosulich, U.S. pat. **2,563,707** (1951 to Am. Cyanamid).

Monohydrate, yellow spherulites. uv max (0.1N NaOH): 255, 302, 368 nm (ε × 10^{-3} 26, 27, 9); (0.1N HCl): 307 nm (ε × 10^{-3} 25).

5908. Methotrexate. *N-[4-[[(2,4-Diamino-6-pteridin-yl)methyl]methylamino]benzoyl]-L-glutamic acid;* 4-amino-*N¹⁰*-methylpteroylglutamic acid; 4-amino-10-methylfolic acid; methylaminopterin; amethopterin; MTX; Cl-14377; Emtexate; A-Methopterin; Rheumatrex. $C_{20}H_{22}N_8O_5$; mol wt 454.46. C 52.86%, H 4.88%, N 24.66%, O 17.60%. A folic acid antagonist. Prepn: Seeger *et al., J. Am. Chem. Soc.* **71**, 1753 (1949). Metabolism: Freeman, *J. Pharmacol. Exp. Ther.* **122**, 154 (1958); Henderson *et al., Cancer Res.* **25**, 1008, 1018 (1965). Toxicity studies: Condit *et al., Cancer* **13**, 222-249 (1960); *ibid.* **23**, 126 (1969). Pharmacokinetic models: Bischoff *et al., J. Pharm. Sci.* **59**, 149 (1970); *eidem, ibid.* **60**, 1128 (1971). Comparative study in treatment of acute lymphocytic leukemia in children: A. I. Freemon *et al., N. Engl. J. Med.* **308**, 477 (1983). Short-term efficacy in rheumatoid arthritis: M. E. Weinblatt *et al., ibid.* **312**, 818 (1985). Toxicity data: H. R. Scherf *et al., Arzneimittel-Forsch.* **20**, 1467 (1970). Comprehensive description: A. R. Chamberlin *et al.,* in *Analytical Profiles of Drug Substances* vol. 5, K. Florey, Ed. (Academic Press, New York, 1976) pp 283-306. Review of metabolism and pharmacokinetics: W. E. Evans, *Appl. Pharmacokinet.* **1980**, 518-548. Review of clinical pharmacology: J. R. Bertino, *Cancer Chemother.* **3**, 359-375 (1981); J. Jolivet *et al., N. Engl. J. Med.* **309**, 1094-1104 (1983). Symposium on clinical experience in rheumatoid arthritis: *J. Rheumatol.* **12**, Suppl. 12, 1-44 (1985).

Monohydrate, yellow crystals from dil HCl, dec 185-204° (bath preheated to 160°). uv max (0.1*N* HCl): 244, 307 nm; (0.1*N* NaOH): 257, 302, 370 nm. Soluble in alkaline solns with decompn. LD_{50} i.v. in rats: 14 mg/kg (Scherf).
Disodium salt, $C_{20}H_{20}N_8Na_2O_5$, *Folex, Mexate.*
THERAP CAT: Antineoplastic; antirheumatic.

5909. Methotrimeprazine. *2-Methoxy-N,N,β-trimethyl-10H-phenothiazine-10-propanamine;* 10-(3-dimethylamino-2-methylpropyl)-2-methoxyphenothiazine; 2-methoxy-10-(2-methyl-3-dimethylaminopropyl)phenothiazine; 3-methoxy-10-(3-dimethylamino-2-methylpropyl)phenothiazine; levomepromazine; 2-methoxytrimeprazine; levomeprazine; RP 7044; Sinogan-Debil; Tisercin; Neozine; Nirvan; Nozinan; Levoprome. $C_{19}H_{24}N_2OS$; mol wt 328.46. C 69.47%, H 7.36%, N 8.53%, O 4.87%, S 9.76%. Prepn: Courvoisier *et al., C.R. Soc. Biol.* **151**, 1378 (1957); Jacob, Robert, U.S. pat. **2,837,518** (1958 to Rhône-Poulenc).

Maleate, $C_{23}H_{28}N_2O_5S$, *Minozinan, Milezin, Neuractil, Neurocil, Sofmin, Veractil.* Crystals, darkened by light. Dec about 190°. The free base is levorotatory: $[\alpha]_D^{20} -17°$ (c = 5 in chloroform). The maleate is sparingly sol in water (0.3% at 20°) and in ethanol (0.4%). pH of a 0.3% aq soln is 4.3.
THERAP CAT: Analgesic.

5910. Methoxamine Hydrochloride. *α-(1-Aminoethyl)-2,5-dimethoxybenzenemethanol hydrochloride; α-(1-amino-ethyl)-2,5-dimethoxybenzyl alcohol hydrochloride;* 2-amino-1-(2,5-dimethoxyphenyl)-1-propanol hydrochloride; β-hydroxy-β-(2,5-dimethoxyphenyl)isopropylamine hydrochloride; β-(2,5-dimethoxyphenyl)-β-hydroxyisopropylamine hydrochloride; 2,5-dimethoxynorephedrine hydrochloride; Pressomin Hydrochloride; Vasoxine Hydrochloride; Vasoxyl Hydrochloride; Vasylox Hydrochloride. $C_{11}H_{18}ClNO_3$; mol wt 247.71. C 53.33%, H 7.32%, N 5.66%, Cl 14.31%, O

19.38%. Prepn: Baltzly *et al.,* U.S. pat. **2,359,707** (1944 to Burroughs Wellcome).

Crystals, mp 212-216°. Very sol in water: One gram dissolves in 2.5 ml water, in 12 ml ethanol. Practically insol in ether, benzene, chloroform. pH of a 2% aq soln between 4.5 and 5.5.
THERAP CAT: Adrenergic (vasopressor).

5911. Methoxsalen. *9-Methoxy-7H-furo[3,2-g][1]-benzopyran-7-one; 6-hydroxy-7-methoxy-5-benzofuranacryl-ic acid δ-lactone;* 9-methoxypsoralen; 8-methoxy-4',5':6,7-furocoumarin; 8-methoxy[furano-3',2':6,7-coumarin]; ammoidin; xanthotoxin; 8-methoxypsoralen; 8-MOP; 8-MP; Meladinine; Meloxine; Methoxa-Dome; Oxsoralen Ultra; Psoralon-MOP. $C_{12}H_8O_4$; mol wt 216.18. C 66.67%, H 3.73%, O 29.60%. Naturally occurring analog of psoralen, *q.v.,* found in spp. of *Leguminosae, Umbelliferae,* and *Rutaceae.* Isolation: Priess, *Ber. Deut. Pharm. Ges.* **21**, 227 (1911); Thoms, *Ber.* **44**, 3325 (1911); **45**, 3705 (1912); Jois *et al., J. Indian Chem. Soc.* **10**, 41 (1933); Späth *et al., Ber.* **73**, 1361 (1933); Schonberg, Sina, *Nature* **160**, 468 (1947); **161**, 481 (1948); *J. Am. Chem. Soc.* **72**, 4826 (1950). Synthesis: Späth, Pailer, *Ber.* **69**, 767 (1936); Lagercrantz, *Acta Chem. Scand.* **10**, 647 (1956); Stanley, Vannier, U.S. pat. **2,889,337** (1959 to USDA); P. Nore, E. Honkanen, *J. Heterocycl. Chem.* **17**, 985 (1980). Use in treatment of psoriasis and mycosis fungoides: J. A. Parrish *et al., Int. J. Dermatol.* **19**, 379 (1980). Acute toxicity data: Hakim *et al., J. Pharmacol. Exp. Ther.* **131**, 394 (1961). Phototoxicity study: A. Korn-hauser *et al., Science* **217**, 733 (1982). Comprehensive description: M. A. Loutfy, M. A. Hassan, in *Analytical Profiles of Drug Substances* vol. 9, K. Florey, Ed. (Academic Press, New York, 1980) pp 427-454. Review of use in photochemotherapy: T. F. Anderson, J. J. Voorhees, *Ann. Rev. Pharmacol. Toxicol.* **20**, 235-258 (1980); *Acta Derm. Venereol.* Suppl. **106**, 9-42 (1982).

Silky needles from hot water or benzene + petr ether, long rhombic prisms from alcohol + ether. mp 148°. Odorless. Bitter taste followed by tingling sensation. pH 5.5. uv max: 219, 249, 300 nm (log ∈ 4.32, 4.35, 4.06). Practically insol in cold water; sparingly sol in boiling water, liq petrolatum, ether. Sol in boiling alcohol, acetone, acetic acid, vegetable fixed oils, propylene glycol, benzene. Freely sol in chloroform. Sol in aq alkalies with ring cleavage, but is reconstituted upon neutralization. LD_{50} i.p. in rats: 470 ± 30 mg/kg (Hakim).
THERAP CAT: Pigmentation agent.

5912. Methoxyamine. *O-Methylhydroxylamine;* methoxylamine; α-methylhydroxylamine; hydroxylamine methyl ether. CH_5NO; mol wt 47.06. C 25.52%, H 10.71%, N 29.77%, O 34.00%. CH_3ONH_2. Prepd by treating hydroxylamine disulfonic acid with methyl sulfate: Goldfarb, *J. Am. Chem. Soc.* **67**, 1852 (1945). May also be prepd from hydroxyurethan.
Free base. *Highly poisonous!* Mobile liquid. Fishy, amine odor. bp_{760} 49-50°. Miscible with water, alcohol, ether.
Hydrochloride, $CH_3ONH_2.HCl$, nacreous scales from alcohol + ether, mp 149-151°. Sol in water, alcohol.
Picrate, mp 175°.
USE: Analytical reagent for aldehydes and ketones. The addition of methoxyamine to α,β-unsaturated ketones and the rearrangement of β-methoxyamino ketones is described

by Blatt, *J. Am. Chem. Soc.* **61**, 3494 (1939). *Caution:* Strong irritant.

5913. Methoxychlor. *1,1'-(2,2,2-Trichloroethylidene)-bis[4-methoxybenzene];* 1,1,1-trichloro-2,2-bis(*p*-methoxyphenyl)ethane; 2,2-di-*p*-anisyl-1,1,1-trichloroethane; DMDT; methoxy-DDT; Marlate. $C_{16}H_{15}Cl_3O_2$; mol wt 345.65. C 55.59%, H 4.37%, Cl 30.77%, O 9.26%. Commercial prepn by the condensation of anisole with chloral in the presence of sulfuric acid: Schneller, Smith, *Ind. Eng. Chem.* **41**, 1027 (1949). Activity: P. Läuger *et al., Helv. Chim. Acta* **27**, 892 (1944).

Dimorphic crystals, mp 78-78.2° or 86-88°. Practically insol in water. Soluble in alc. The solubilities are approx those of DDT, *q.v.* LD_{50} orally in rats: 5.0 g/kg, H. C. Hodge *et al., J. Pharmacol. Exp. Ther.* **99**, 140 (1950).

Human Toxicity: Slightly irritating to, but not absorbed appreciably through, skin. Estimated fatal oral dose 7.5 g/kg body wt. Continued ingestion over long periods may cause kidney damage. *See also* DDT.

USE: Insecticide.

THERAP CAT (VET): Ectoparasiticide.

5914. Methoxyflurane. *2,2-Dichloro-1,1-difluoro-1-methoxyethane;* 2,2-dichloro-1,1-difluoroethyl methyl ether; 1,1-difluoro-2,2-dichloroethyl methyl ether; DA 759; Metofane; Penthrane; Pentrane. $C_3H_4Cl_2F_2O$; mol wt 164.97. C 21.84%, H 2.44%, Cl 42.98%, F 23.03%, O 9.70%. CH_3-OCF_2CHCl_2. Prepn and properties: Miller *et al., J. Am. Chem. Soc.* **70**, 431 (1948); Park *et al., ibid.* **73**, 861 (1951); **Brit.** pat. 754,976 (1956 to Standard Tele. & Cable).

Liquid, bp 105°; bp_{100} 51°. mp −35°. n_D^{20} 1.3861; n_D^{25} 1.3839. d_4^{20} 1.4262 (Miller); d_4^{20} 1.4226 (Park).

Note: The medicinal grade contains butylated hydroxytoluene as antioxidant.

THERAP CAT: Anesthetic (inhalation).

THERAP CAT (VET): Anesthetic.

5915. 10-Methoxyharmalan. *4,9-Dihydro-6-methoxy-1-methyl-3H-pyrido[3,4-b]indole;* 1-methyl-6-methoxy-3,4-dihydro-2-carboline; 3,4-dihydromethoxyharman. $C_{13}H_{14}$-N_2O; mol wt 214.26. C 72.87%, H 6.59%, N 13.08%, O 7.47%. Prepn from 5-methoxytryptamine: Späth, Lederer, *Ber.* **63**, 2102 (1930); Petrova *et al., Zh. Obshch. Khim.* **33**, 1333 (1963), *C.A.* **59**, 10149b (1963). Weak serotonin inhibitor: M. M. Airaksinen *et al., Acta Pharmacol. Toxicol.* **46**, 308 (1980).

Crystals, mp 208-209°. R_f 0.74 in 4:1:5 butanol, acetic acid, water.

5916. 3-Methoxy-4-hydroxyphenylglycol. *1-(4-Hydroxy-3-methoxyphenyl)-1,2-ethanediol;* (4-hydroxy-3-methoxyphenyl)ethylene glycol; vanylglycol; MHPG. $C_9H_{12}O_4$; mol wt 184.19. C 58.68%, H 6.57%, O 34.75%. Major metabolite of norepinephrine, *q.v.* Isoln from rat urine: J. Axelrod *et al., Biochim. Biophys. Acta* **36**, 576 (1959); from cat liver and kidney: M. Goldstein, S. B. Gertner, *Nature* **187**, 147 (1960). Metabolism studies on urinary MHPG levels in patients with affective disorders: K. Greenspan *et al., J. Psychiat. Res.* **7**, 171 (1970); J. W. Maas *et al., Biochem. Pharmacol.* **1974**, Suppl., pt. 2, 907; J. J. Schildkraut *et al., Biol. Markers Psychiatry Neurol., Proc. Conf.* **1981**, E. Usdin, I. Hanin, Eds. (Pergamon, Oxford, 1982) pp 23-33. Studies on MHPG levels in cerebrospinal fluid (CSF): E. Garelis *et al., Brain Res.* **79**, 1 (1974); P. Frattini *et al., Clin. Chim. Acta* **125**, 97 (1982); in CSF of patients with CNS disorders: T. N. Chase *et al., J. Neurochem.* **21**, 581 (1973);

M. G. Ziegler *et al., Am. J. Psychiatry* **134**, 565 (1977). Direct method to estimate rate of prodn of MHPG by human brain *in vivo:* J. W. Maas *et al., Science* **205**, 1025 (1979). Review of origin and distribution in body fluids: E. M. DeMet, A. E. Halaris, *Biochem. Pharmacol.* **28**, 3043-3050 (1979). Use as biochemical marker to identify mental disorders: T. H. Maugh, *Science* **214**, 39 (1981).

USE: As biochemical tool.

5917. 2-(Methoxymethyl)-5-nitrofuran. 5-Nitro-2-furfurylmethyl ether; Furbenal; Furaspor. $C_6H_7NO_4$; mol wt 157.12. C 45.86%, H 4.49%, N 8.92%, O 40.73%. Prepd by nitration of furfuryl methyl ether in acetic anhydride: Gilman, Burtner, *Iowa State Coll. J. Sci.* **6**, 389 (1932), *C.A.* **27**, 288 (1933).

Light yellow, oily liquid. d_{20}^{20} 1.283; bp_3 104-105°. b_4 114-117°; n_D^{20} 1.5325-1.5343. Miscible with ethanol. Soly in water 11 g/l; peanut oil 170 g/l; Skellysolve B 24 g/l. Unstable to alkali and direct sunlight.

THERAP CAT: Antifungal.

5918. 2-Methoxynaphthalene. Methyl β-naphthyl ether; nerolin "old"; Yara yara. $C_{11}H_{10}O$; mol wt 158.19. C 83.51%, H 6.37%, O 10.11%. Prepn: Hiers, Hager, *Org. Syn. coll. vol. I*, 59 (1941).

Leaflets from ether, mp 72°. bp 272°. Practically insol in water; sparingly sol in alc; sol in ether, carbon disulfide, benzene.

5919. Methoxyphenamine. *2-Methoxy-N,α-dimethyl-benzeneethanamime; o-methoxy-N,α-dimethylphenethyl-amine;* β-(*o*-methoxyphenyl)isopropylmethylamine; α-(2-methoxyphenyl)-β-methylaminopropane; Proasma; Orthoxine; Ortodrinex. $C_{11}H_{17}NO$; mol wt 179.25. C 73.70%, H 9.56%, N 7.81%, O 8.93%. β-Adrenergic agonist. Prepn: Woodruff *et al., J. Am. Chem. Soc.* **62**, 922 (1940); Heinzelman, *ibid.* **75**, 921 (1953); Morishita, **Japan.** pat. **61 2921** (1961 to Nippon Shinyaku), *C.A.* **55**, 24677f (1961).

Oil, bp_2 97-99°.
Used as the hydrochloride, $C_{11}H_{17}NO.HCl$, bitter crystals from ether + alcohol, mp 129-131°. Freely sol in water, alcohol, chloroform. Slightly sol in ether, benzene. pH 5.3-5.7 (5% aq soln).

THERAP CAT: Bronchodilator.

5920. 4-Methoxy-m-phenylenediamine. *4-Methoxy-1,3-benzenediamine;* 2,4-diaminoanisole; 2,4-DAA; 4 MMPD; C.I. 76050; C.I. Oxidation Base 12. $C_7H_{10}N_2O$; mol wt 138.16. C 60.85%, H 7.30%, N 20.27%, O 11.58%. Prepd by reduction of 2,4-dinitroanisole with iron and acetic acid: **Ger.** pat. 258,653 (1912 to BASF), *Frdl.* **11**, 392 (1912-14). Alternate prepn: K. Fries, *Ann.* **454**, 147 (1927). Prepn of hydrochloride: F. Kehrmann, *Ber.* **50**, 562 (1917).

Mutagenicity studies: B. N. Ames *et al., Proc. Nat. Acad. Sci. USA* **72**, 2423 (1975); D. J. N. Hossack, J. C. Richardson, *Experientia* **33**, 377 (1977). Toxicity data: G. K. Lloyd *et al., Food Cosmet. Toxicol.* **15**, 607 (1977); C. Burnett *et al., J. Toxicol. Environ. Health* **2**, 657 (1977).

Needles from ether, mp 67-68°. Darkens on exposure to light. LD_{50} of an aq soln containing 0.05% Na_2SO_3: 460 mg/kg orally in rats (Lloyd).

Sulfate, *4-MMPDS.* Irritant. Toxicity studies: C. Burnett *et al., Food Cosmet. Toxicol.* **13**, 353 (1975); *eidem, J. Toxicol. Environ. Health* **1**, 1027 (1976). Mutagenicity: W. G. H. Blijleven, *Mutat. Res.-Genetic Toxicol. Testing* **48**, 181 (1977). LD_{50} in rats (mg/kg): 372 i.p.; >4000 orally (Burnett).

Note: This substance may reasonably be anticipated to be a carcinogen: *Fourth Annual Report on Carcinogens* (NTP 85-002, 1985) p 68.

USE: In prepn of dyes, esp hair dyes; as corrosion inhibitor for steel.

5921. 3-(o-Methoxyphenyl)-2-phenylacrylic Acid. α-*[(2-Methoxyphenyl)methylene]benzeneacetic acid; β-(o-methoxyphenyl)-α-phenylacrylic acid; α-phenyl-o-methoxycinnamic acid.* $C_{16}H_{14}O_3$; mol wt 254.27. C 75.57%, H 5.55%, O 18.88%. Prepn: Buckles *et al., J. Am. Chem. Soc.* **73**, 4972 (1951); Crawford, Moore, *J. Chem. Soc.* **1955**, 3445; Alexander, Barthel, *J. Org. Chem.* **23**, 389 (1958).

trans-Form, plates from benzene, mp 184°.
cis-Form, prisms from benzene, mp 131.5°.
Magnesium salt, $C_{32}H_{26}MgO_6$, *AN 1022, Bilcrine.*
THERAP CAT: Choleretic.

5922. N-(p-Methoxyphenyl)-p-phenylenediamine. *N-(4-Methoxyphenyl)-1,4-benzenediamine; 4-amino-4'-methoxydiphenylamine; 4-methoxy-4'-aminodiphenylamine; Variamine Blue base.* $C_{13}H_{14}N_2O$; mol wt 214.26. C 72.87%, H 6.59%, N 13.08%, O 7.47%. Prepd from the corresp nitroso compd: Willstätter, Kubli, *Ber.* **42**, 4139 (1909).

Needles from ligroin, mp 102°. bp_{12} 238°. Freely sol in alcohol, ether, benzene. Slightly sol in water, petr ether.
Sulfate, $(C_{13}H_{14}N_2O)_2.H_2SO_4$, crystals, sol in water. Addition of ferric chloride produces intensely blue solns.

5923. Methoxypromazine. *2-Methoxy-N,N-dimethyl-10H-phenothiazine-10-propanamine; 10-(3-dimethylaminopropyl)-2-methoxyphenothiazine; 3-methoxy-10-(3'-dimethylaminopropyl)phenothiazine; 2-methoxy-10-(3'-dimethylaminopropyl)phenothiazine; methopromazine; RP 4632.* $C_{18}H_{22}N_2OS$; mol wt 314.46. C 68.75%, H 7.05%, N 8.91%, O 5.09%, S 10.20%. Prepd by condensing 3-methoxyphenothiazine with 1-dimethylamino-3-chloropropane in the presence of sodamide: Charpentier *et al., Compt. Rend.* **235**, 59 (1952). Alternate route: **Brit. pat. 789,276** (1958 to May & Baker).

Crystals, mp 44-48°.
Maleate, $C_{22}H_{26}N_2O_5S$, *Mopazine, Tentone, Vetomazin.* Crystals, mp 141-145°. Should be protected from light during storage. Soly at 25° (g/100 ml) in water: 0.3; in methanol: 4; in chloroform: 12.5; in dimethylformamide: 100. Very slightly sol in ethanol, benzene. Practically insol in ether. pH of satd aq soln about 5.
Oxalate, $C_{18}H_{22}N_2OS.C_2H_2O_4$, crystals, dec 178-179°.
Picrate, $C_{18}H_{22}N_2OS.C_6H_3N_3O_7$, reddish crystals, mp 141-142°.
THERAP CAT: Neuroleptic.

5924. 6-Methoxy-α-tetralone. *3,4-Dihydro-6-methoxy-1(2H)-naphthalenone; 6-methoxy-3,4-dihydro-1(2H)-naphthalenone; 1-keto-6-methoxy-1,2,3,4-tetrahydronaphthalene; 5-keto-2-methoxy-5,6,7,8-tetrahydronaphthalene.* $C_{11}H_{12}O_2$; mol wt 176.21. C 74.97%, H 6.86%, O 18.16%. Prepn: Papa, *J. Am. Chem. Soc.* **71**, 3246 (1949); Ananchenko *et al., Tetrahedron* **18**, 1355 (1962).

Crystals from methanol or ligroin, mp 80°. bp_1 135-139°, bp_{11} 171°.
USE: In the synthesis of derivatives of estrane and 19-norsteroids (Ananchenko *et al., loc. cit.*).

5925. 1-Methoxy-3-(trimethylsilyloxy)-1,3-butadiene. *[(3-Methoxy-1-methylene-2-propenyl)oxy]trimethylsilane;* Danishefsky's diene. $C_8H_{16}O_2Si$; mol wt 172.30. C 55.77%, H 9.36%, O 18.57%, Si 16.30%. A silyl enol ether used in Diels-Alder reactions; highly reactive and nucleophilic diene. Prepn and ^{1}H-NMR: S. Danishefsky, T. Kitahana, *J. Am. Chem. Soc.* **96**, 7807 (1974); S. Danishefsky *et al., Org. Syn.* **61**, 147 (1983). In synthesis of vernolepin, *q.v.*: *eidem, J. Am. Chem. Soc.* **99**, 6066 (1977); of pentalenolactone: *eidem, ibid.* **100**, 6536 (1978). Regiospecific addition to juglone, *q.v.*: R. K. Boeckman *et al., ibid.* 7098. Review: S. Danishefsky, *Accounts Chem. Res.* **14**, 400 (1981).

Oil. bp_{23} 78-81°.
USE: In organic syntheses.

5926. 5-Methoxytryptamine. *5-Methoxy-1H-indole-3-ethanamine; 3-(2-aminoethyl)-5-methoxyindole; meksamin; mexamine.* $C_{11}H_{14}N_2O$; mol wt 190.24. C 69.44%, H 7.42%, N 14.73%, O 8.41%. Prepn: Supniewski *et al., C.A.* **55**, 15458 (1961). Proposed as potentiator for hypnotics, sedatives. Claimed to be more active than serotonin: Mashkovsky, Arutyunyan, *Farmakol. Toksikol.* **26** (no. 1), 10 (1963).

Crystals from ethanol, mp 121-122°.
Hydrochloride, $C_{11}H_{14}N_2O.HCl$, crystals, dec 248°.

5927. Methscopolamine Bromide. *[7(S)-(1α,2β,4β,5α,-7β)]-7-(3-Hydroxy-1-oxo-2-phenylpropoxy)-9,9-dimethyl-3-oxa-9-azoniatricyclo[3.3.1.0²,⁴]nonane bromide; 6β,7β-epoxy-*

3α-*hydroxy-8-methyl-1αH,5αH-tropanium bromide tropate (ester); N*-methylscopolammonium bromide; hyoscine methyl bromide; scopolamine methobromide; scopolamine methyl bromide; epoxymethamine bromide; Diopal; Holopon; Mescopil; Neo-Avagal; Pamine bromide; Proscomide; Restropin. $C_{18}H_{24}BrNO_4$; mol wt 398.31. C 54.28%, H 6.07%, Br 20.06%, N 3.52%, O 16.07%. Prepd by the action of methyl bromide on scopolamine base: Visscher, **U.S. pat. 2,753,288** (1956 to Upjohn). Treatment of duodenal ulcer by transdermally administered methscopolamine bromide: R. P. Walt *et al., Brit. Med. J.* **284,** 1736 (1982).

Crystals from ethanol, dec 214-217°. Freely sol in water, in dil ethanol. Slightly sol in abs ethanol.

THERAP CAT: Anticholinergic.

5928. Methsuximide. *1,3-Dimethyl-3-phenyl-2,5-pyrrolidinedione; N,2-dimethyl-2-phenylsuccinimide;* 1,3-dimethyl-3-phenyl-2,5-dioxopyrrolidine; mesuximide; Celontin; Petinutin. $C_{12}H_{13}NO_2$; mol wt 203.23. C 70.91%, H 6.45%, N 6.89%, O 15.75%. Prepd by the action of methylamine on α-methyl-α-phenylsuccinic acid: Miller, Long, *J. Am. Chem. Soc.* **73,** 4895 (1951); *eidem,* **U.S.** pat. **2,643,257** (1953 to Parke, Davis).

Crystals from dil alc, mp 52-53°. bp$_{0.1}$ 121-122°. Freely sol in methanol, alc. LD$_{50}$ orally in mice: 1.55 g/kg.

THERAP CAT: Anticonvulsant.

5929. Methyclothiazide. *6-Chloro-3-(chloromethyl)-3,4-dihydro-2-methyl-2H-1,2,4-benzothiadiazine-7-sulfonamide 1,1-dioxide; 6-chloro-3-chloromethyl-3,4-dihydro-2-methyl-7-sulfamoyl-1,2,4-benzothiadiazine 1,1-dioxide;* 6-chloro-3-chloromethyl-2-methyl-7-sulfamyl-3,4-dihydro-1,2,4-benzothiadiazine 1,1-dioxide; Aquatensen; Duretic; Enduron; Enduronum; Naturon. $C_9H_{11}Cl_2N_3O_4S_2$; mol wt 360.25. C 30.00%, H 3.08%, Cl 19.68%, N 11.66%, O 17.77%, S 17.80%. Prepn: Close *et al., J. Am. Chem. Soc.* **82,** 1132 (1960). Description: Ford, *Curr. Ther. Res.* **2,** 422 (1960). Comprehensive description: J. A. Raihle in *Analytical Profiles of Drug Substances* **vol. 5,** K. Florey, Ed. (Academic Press, New York, 1976) pp 307-326.

Crystals from alcohol + water, mp 225°. uv max (methanol): 226, 267, 311 nm (ε 39300, 21250, 3300). pKa 9.4. Very sol in acetone, pyridine; slightly sol in methanol, ethanol. Almost insol in water, chloroform, benzene.

THERAP CAT: Diuretic; antihypertensive.

5930. Methyl Abietate. Abalyn. $C_{21}H_{32}O_2$; mol wt 316.47. C 79.69%, H 10.19%, O 10.11%. $C_{19}H_{29}COOCH_3$.

The article of commerce is a mixture of the methyl esters of the rosin acids.

Colorless to yellow, almost odorless, thick liquid. d$_{20}^{20}$ 1.040. bp 360-365° with decompn. n_D^{20} 1.530. Flash pt 180-218°. Insol in water; miscible with usual organic solvents, also with aliphatic hydrocarbons. Dissolves ester gums, rosin, many synthetic resins as well as ethyl cellulose, rubber, etc.

USE: As a solvent for ester gums, rosin, many synthetic resins, ethyl cellulose, rubber, etc.; in the manuf of varnish resins; as ingredient in adhesives.

5931. N-Methylacetanilide. *N-Methyl-N-phenylacetamide;* acetomethylanilide; Exalgin. $C_9H_{11}NO$; mol wt 149.19. C 72.45%, H 7.43%, N 9.39%, O 10.72%. Made by the action of acetic anhydride on methylaniline.

Orthorhombic rods from alcohol, plates from ether or petr ether. mp 102-104°. bp$_{712}$ 253°. One gram dissolves in 60 ml water, 2 ml boiling water, 2 ml alcohol, 1.5 ml chloroform, 10 ml ether.

5932. Methyl Acetate. $C_3H_6O_2$; mol wt 74.08. C 48.64%, H 8.16%, O 43.20%. CH_3COOCH_3.
Colorless liquid; pleasant odor. mp −98°. bp 56.9°. d$_4^{20}$ 0.9342; d$_4^{25}$ 0.9279. n_D^{20} 1.3614: Mumford, Phillips, *J. Chem. Soc.* **1950,** 75. Flash pt, closed cup: 14°F (−10°C). Sol in water; miscible with alcohol, ether.

USE: Solvent for nitrocellulose, acetylcellulose, and many resins and oils; manuf artificial leather. *Caution:* Irritating to respiratory tract and, in high concns, narcotic.

5933. Methyl Acetoacetate. *3-Oxobutanoic acid methyl ester.* $C_5H_8O_3$; mol wt 116.11. C 51.72%, H 6.95%, O 41.34%. $CH_3COCH_2COOCH_3$. Toxicity data: H. F. Smyth, C. P. Carpenter, *J. Ind. Hyg. Toxicol.* **30,** 63 (1948).
Liquid. d 1.078-1.080. bp 169-171° (slight dec). mp −80°. n_D^{20} 1.418. Flash pt 82°. Sol in 2 parts water; miscible with alc, ether; gives a deep red color with FeCl$_3$. LD$_{50}$ orally in rats: 3.0 g/kg (Smyth, Carpenter).

5934. Methyl Acetylsalicylate. *2-(Acetyloxy)benzoic acid methyl ester;* acetylsalicylic acid methyl ester; methylaspirin; Methylrodin; Methylrhodine. $C_{10}H_{10}O_4$; mol wt 194.18. C 61.85%, H 5.19%, O 32.96%. Prepd from acetylsalicylic acid and diazomethane: Herzig, Tichatschek, *Ber.* **39,** 1559 (1906); from methyl salicylate and acetic anhydride: Erdmann, *J. Prakt. Chem.* [2] **56,** 154 (1897); Thorp, **U.S.** pat. **1,255,950** (1918).

Plates from petr ether. mp 51-52°. bp$_9$ 134-136°. Very sparingly sol in water; sol in alc, ether, chloroform, glycerol, propylene glycol, oils.

USE: Fixative for perfumes.

5935. Methyl Acrylate. *2-Propenoic acid methyl ester;* acrylic acid methyl ester. $C_4H_6O_2$; mol wt 86.09. C 55.80%, H 7.03%, O 37.17%. $CH_2=CHCOOCH_3$. Convenient prepn from ethylene chlorohydrin: **Ger. pat. 571,123** (1928 to Rohm & Haas). For direct syntheses from acetylene and carbon monoxide *see* W. Reppe, *Chemie & Technik der Acetylen-Druck-Reaktionen* (Weinheim, 2nd ed., 1952).
Monomer: Liquid. Acrid odor. *Lacrimator.* d$_{20}^{20}$ 0.9561; d$_{20}^{20}$ 0.9574; d$_4^0$ 0.9702; d$_4^{-5}$ 0.9868; d$_4^{-10}$ 0.9929. Weighs 8.0 lbs/gal. mp −76.5°. bp$_{608}$ 70°; bp$_{428}$ 60°; bp$_{298}$ 50°; bp$_{200}$ 40°; bp$_{88}$ 20°; bp$_{54}$ 10°; bp$_{41.5}$ 5°; bp$_{32}$ 0°; bp$_{24.5}$ −5°; bp$_{18.5}$ −10°. n_D^{20} 1.401. Sp heat at −60° = 0.444 cal/g/°C; heat of vaporization 8.25 kcal/mol; heat of combustion 502.88 kcal/mol. Soly in water at 20° = 6 g/100 ml, at 40° = 5 g/100 ml.

Soly of water in methyl acrylate at 20° = 1.8 ml/100 g. Sol in alc, ether. Azeotropes: 9.5% water = bp 73°; 49.0% methanol = bp 61°. Easily polymerizes on standing. The polymerization process can be speeded up by heat, light, and peroxides. If pure, the monomer can be stored below +10° without incurring polymerization. LD_{50} orally in rats: 0.3 g/kg, H. F. Smyth, C. P. Carpenter, *J. Ind. Hyg. Toxicol.* **30**, 63 (1948).

Polymer: Transparent, elastic substance. Practically no odor. Little adhesive power. Resists the usual solvents.

USE: The monomer in manuf leather finish resins, textile and paper coatings, and plastic films. Produces the hardest resin of the acrylate ester series. *Caution:* The monomer is highly irritating to eyes, skin, mucous membranes. Lethargy and convulsions may occur if vapors of the monomer are inhaled in high concns.

5936. Methylal. *Dimethoxymethane;* formal; formaldehyde dimethyl acetal. $C_3H_8O_2$; mol wt 76.09. C 47.35%, H 10.59%, O 42.05%. $CH_2(OCH_3)_2$. Prepn by catalytic vapor-phase oxidation of methanol in the presence of small amts of HCl; from methanol and formaldehyde: Frevel, Hedelund, U.S. pats. **2,663,742** and **2,691,684** (1953, 1954, both to Dow); from paraformaldehyde + methanol in the presence of $CaCl_2$ + HCl: R. Rambaud, D. Besserre, *Bull. Soc. Chim. France* **1955**, 45.

Colorless, clear, volatile, flammable liq; chloroform odor; pungent taste. bp_{760} 41.6°; bp_{754} 41.5°. mp −105°. d_4^{44} 0.8669; d_4^{20} 0.8593. n_D^{18} 1.3589. Flash pt, closed cup: 0°F (−18°C). Sol in 3 parts water; miscible with alcohol, ether, oils.

USE: In perfumery; manuf artificial resins; reaction medium for Grignard and Reppe reactions.

5937. Methyl Allyl Trisulfide. *Methyl-2-propenyl trisulfide; allyl methyl trisulfide;* methyl 2-propenyl trisulfane; MATS. $C_4H_8S_3$; mol wt 152.29. C 31.55%, H 5.29%, S 63.16%. $CH_3SSSCH_2CH=CH_2$. Component of garlic oil which inhibits platelet aggregation. Synthesis, identification by GC: D. M. Oaks *et al., Anal. Chem.* **36**, 1560 (1964). Isoln from garlic oil, effect on platelet-rich plasma: T. Ariga *et al., Lancet* **1**, 150 (1981). Improved synthesis: A. W. Mott, G. Barany, *Synthesis* **1984**, 657. Use as antithrombotic agent: **Japan. Kokai 82 209,218** (1982 to Alcon), *C.A.* **98**, 95692n (1983).

Oil, $bp_{0.05}$ 28-30°.

5938. Methylamine. *Methanamine;* monomethylamine; aminomethane. CH_5N; mol wt 31.06. C 38.67%, H 16.23%, N 45.10%. CH_3NH_2. Occurs in herring brine, in urine of dogs after eating meat, in certain plants such as *Mentha aquatica,* in crude methanol together with di- and trimethylamine. Made by heating methyl alcohol, ammonium chloride, and zinc chloride to about 300°; by heating ammonium chloride and formaldehyde: Marvel, Jenkins, *Org. Syn.* **3**, 67 (1923); from methanol + ammonia: Smith, U.S. pat. **2,456,599** (1948 to Comm. Solvents).

Flammable gas at ordinary temp and pressure. Fuming liq when liquefied by cooling in ice and salt mixture. $d_4^{-10.8}$ 0.699. mp −93.5°. bp_{760} −6.3°; bp_{400} −19.7°; bp_{200} −32.4°; bp_{100} −43.7°; bp_{10} −73.8°. Flash pt 32.5°F (0°C). Stronger base than ammonia: K at 25° = $4.42 × 10^{-4}$. Absorption spectrum: Bielecki, Henri, *Compt. Rend.* **156**, 1861 (1913). One vol of water at 12.5° dissolves 1154 vols of the gas, and 959 vols at 25°. 100 ml of a benzene soln satd at 25° contain 10.5 g methylamine. Sol in alc; miscible with ether. Good solvent for many organic substances. Marketed in liquefied form or as a 33% aq soln. LD s.c. in mice: 2.5 g/kg.

Hydrochloride, deliquescent tetragonal tablets from alcohol, mp 227-228° with sublimation. bp_{15} 225-230°. Sol in water, in abs alcohol. 100 g of boiling abs alcohol dissolve 23.01 g; insol in chloroform, acetone, ether, ethyl acetate. *Keep well closed.*

Picrate, $CH_5N.C_6H_3N_3O_7$, mp 215°.

USE: Methylamine is used in tanning. Methylamine and methylamine hydrochloride are used in organic synthesis for introducing the methylamino group. *Caution:* Irritating to eyes skin, respiratory tract.

5939. 2-Methylaminoethanol. Methyl(β-hydroxyethyl)-amine; methylethylolamine. C_3H_9NO; mol wt 75.11. C 47.96%, H 12.08%, N 18.65%, O 21.30%. CH_3NHCH_2-CH_2OH. The *in vivo* precursor of choline. Has been isolated from a mutant of *Neurospora crassa* which has lost its ability to synthesize choline: Horowitz, *J. Biol. Chem.* **162**, 413 (1946). Prepd by mixing ethylene oxide with concd methylamine soln with external cooling: Knorr, Matthes, *Ber.* **31**, 1069 (1898); Lowe *et al., Brit. pat.* **763,434** (1956 to Oxirane Ltd.); Nikolaev *et al., Zh. Obshch. Khim.* **33**, 391 (1963).

Viscous liquid. Fishy odor. d^{20} 0.937. bp_{760} 155-156°; bp_{12} 64-65°. n_D^{20} 1.4385. Miscible with water, alcohol, ether. Corrosive to metals, cork, metals. Strong base. Forms a deliquescent salt with HCl. LD_{50} orally in rats: 2.34 g/kg, *Toxic Substances List,* H. E. Christensen, Ed. (1974) p 339.

Picrate, yellow crystals, mp 148-150°.

Caution: Irritating to skin, eyes, mucous membranes.

5940. p-Methylaminophenol Sulfate. Monomethyl-*p*-aminophenol sulfate; *p*-hydroxymethylaniline sulfate; Photol; Verol; Rhodol; Armol; Elon; Genol; Graphol; Photo-Rex; Pictol; Planetol; Metol. $C_{14}H_{20}N_2O_6S$; mol wt 344.38. C 48.82%, H 5.85%, N 8.14%, O 27.87%, S 9.31%. $(HOC_6H_4NHCH_3)_2.H_2SO_4$.

Crystals. Discolors in air. mp about 260° with decompn. Sol in 20 parts cold, 6 parts boiling water; slightly sol in alc; insol in ether. *Keep well closed and protected from light.*

USE: Photographic developer; dyeing furs.

5941. Methylaniline. *N-Methylbenzenamine; monomethylaniline.* C_7H_9N; mol wt 107.15. C 78.46%, H 8.47%, N 13.07%. $C_6H_5NHCH_3$. Made by heating aniline chloride and methyl alcohol under pressure.

Colorless or slightly yellow liquid; becomes brown on exposure to air. d_4^{20} 0.989. mp −57°. bp 194-196°. $n_D^{21.2}$ 1.5702. Slightly sol in water; sol in alc, ether. LD orally in rabbits: 280 mg/kg.

5942. Methyl Anthranilate. *2-Aminobenzoic acid methyl ester;* methyl 2-aminobenzoate; neroli oil, artificial. C_8H_9-NO_2; mol wt 151.16. C 63.56%, H 6.00%, N 9.27%, O 21.17%. Occurs in neroli, ylang-ylang, bergamot, jasmine, other essential oils and in grape juice; also obtained synthetically by esterifying anthranilic acid with CH_3OH in presence of HCl.

Crystals. d 1.168. mp 24-25°. bp_{15} 135.5°. Slightly sol in water; freely sol in alcohol or ether. LD_{50} orally in rats, mice: 2910, 3900 mg/kg, P. M. Jenner *et al., Food Cosmet. Toxicol.* **2**, 327 (1964).

USE: As perfume for ointments; manuf synthetic perfumes.

5943. 2-Methylanthraquinone. *2-Methyl-9,10-anthracenedione;* β-methylanthraquinone. $C_{15}H_{10}O_2$; mol wt 222.23. C 81.06%, H 4.54%, O 14.40%. Occurs in teakwood. Prepd by oxidation of 2-methylanthracene; by formation from phthalic anhydride and toluene: Fieser, *Org. Syn.* **4**, 43 (1925); from 6- and 7-methylanthraquinone-1-carboxylic acids with powdered copper in quinoline: Fieser, Martin, *J. Am. Chem. Soc.* **58**, 1443 (1936); from 1,4-naphthoquinone and isoprene: Carothers, Berchet, *ibid.* **55**, 2813 (1933).

Needles from alcohol, mp 177°. Sublimes. Very sol in benzene, toluene, xylene; sol in alcohol, ether, glacial acetic acid, concd H_2SO_4; insol in water.

5944. Methylarbutin. *4-Methoxyphenyl-β-D-glucopyranoside;* methylarbutoside. $C_{13}H_{18}O_7$; mol wt 286.28. C 54.54%, H 6.34%, O 39.12%. Occurs together with arbutin in leaves of *Arctostaphylos uva-ursi* (L.) Spreng. and allied *Ericaceae*. Prepn and physical properties: Lindpainter, *Arch. Pharm.* **277**, 398 (1939).

Monohydrate, bitter needles, mp 158-160° and again at 175°. [α]$_D$ −63° (water). Soluble in hot water or alcohol; slightly sol in ether.

5945. 3-Methylarsacetin. *[4-(Acetylamino)-3-methyl-phenyl]arsonic acid; N-acetyl-3-methylarsanilic acid; 3-methyl-4-acetaminophenylarsanilic acid; 3-methyl-4-acetylaminophenylarsonate; Orsudan.* $C_9H_{12}AsNO_4$; mol wt 273.13. C 39.58%, H 4.43%, As 27.43%, N 5.13%, O 23.43%. Prepn: Benda, Kahn, *Ber.* **41**, 1672 (1908).

Needles or rods from water, dec 306°. Sol in alkaline hydroxide and carbonate; slightly sol in hot water, methyl alcohol; very slightly sol in ethyl alcohol and other organic solvents; practically insol in dil acid.

Sodium salt, white, cryst powder. Sol in 4 parts cold water, 2.5 parts hot water.

THERAP CAT: Antimalarial.

5946. Methylbenzethonium Chloride. *N,N-Dimethyl-N-[2-[2-[methyl-4-(1,1,3,3-tetramethylbutyl)phenoxy]ethoxy]ethyl]benzenemethanaminium chloride;* benzyldimethyl[2-[2-(p-1,1,3,3-tetramethylbutylcresoxy)ethoxy]ethyl]ammonium chloride; [2-[2-(p-octylcresoxy)ethoxy]ethyl]dimethylbenzylammonium chloride; benzyldimethyl[2-[2-[4-(1,1,3,3-tetramethylbutyl)tolyloxy]ethoxy]ethyl]ammonium chloride; Diaparene chloride; Hyamine 10X. $C_{28}H_{44}ClNO_2$; mol wt 462.12. C 72.77%, H 9.60%, Cl 7.67%, N 3.03%, O 6.92%.

Monohydrate, crystals. Bitter taste. mp 161-163°. Freely sol in water, alcohol, Cellosolve, chloroform, hot benzene.

THERAP CAT: Topical anti-infective.

5947. Methyl Benzoate. *Benzoic acid methyl ester;* essence of Niobe; oil of Niobe. $C_8H_8O_2$; mol wt 136.14. C 70.57%, H 5.92%, O 23.50%. $C_6H_5COOCH_3$. Toxicity data: Smyth *et al., Arch. Ind. Hyg. Occup. Med.* **10**, 61 (1954).

Colorless, transparent liquid; pleasant odor. d$_4^{15}$ 1.094. mp ~−15°. bp 198-200°. n$_D^{15}$ 1.5205. Flash pt, closed cup: 181°F (82°C). Insol in water; misc with alcohol, ether, methanol. LD$_{50}$ orally in rats: 3.43 g/kg (Smyth).

USE: In perfumes (peau d'Espagne).

5948. Methyl Benzoylsalicylate. *2-(Benzoyloxy)benzoic acid methyl ester;* benzosalin. $C_{15}H_{12}O_4$; mol wt 256.25. C 70.30%, H 4.72%, O 24.97%.

Crystals. mp 85°. bp 385°. Insol in water. One gram dissolves in 35 ml alcohol; sol in benzene, chloroform, ether.

5949. α-Methylbenzylamine. *α-Methylbenzenemethanamine;* α-phenylethylamine; 1-phenylethylamine; α-aminoethylbenzene. $C_8H_{11}N$; mol wt 121.18. C 79.29%, H 9.15%, N 11.56%. Prepn from acetophenone and ammonium formate: A. W. Ingersoll *et al., J. Am. Chem. Soc.* **58**, 1808 (1936); by reduction of acetophenone in liquid ammonia: J. C. Robinson, H. R. Snyder, *Org. Syn.* coll. vol. III, 717

(1955). Stereospecific synthesis of *R*-(+)-form: H. Takahashi *et al., Chem. Pharm. Bull.* **29**, 3387 (1981). Resolution of *(dl)*-methylbenzylamine: W. Theikecker, H.-G. Winkler, *Ber.* **87**, 690 (1954). Use of *(R)*-(+)-methylbenzylamine as resolving agent: E. J. Corey, J. Mann, *J. Am. Chem. Soc.* **95**, 6832 (1973). Use as chiral intermediate in synthesis of α-methyl-α-amino nitriles: K. Weinges *et al., Ber.* **110**, 2098 (1977); of β-amino acids: M. Furukawa *et al., Chem. Pharm. Bull.* **26**, 260 (1978). Toxicity: H. F. Smyth *et al., Arch. Ind. Hyg. Occup. Med.* **4**, 119 (1951).

dl-Form, liquid. Aromatic odor. Absorbs CO_2 from air. d$_4^{15}$ 0.9395. bp$_{18}$ 80-81°. Strong base. Soly in water at 20° about 4.2%. Misc with alcohol, ether. LD$_{50}$ orally in rats: 0.94 g/kg (Smyth).

(+)-Form, liquid, bp 184-186°. d$_4^{22}$ 0.950. [α]$_D^{22}$ +40.3° (neat).

(−)-Form, liquid, bp$_{12}$ 73°. d$_4^{22}$ 0.950. [α]$_D^{22}$ −40.3° (neat).

USE: As resolving agent; chiral intermediate.

5950. Methyl Blue. *[[4-[Bis[4-[(sulfophenyl)amino]-phenyl]methylene]-2,5-cyclohexadien-1-ylidene]amino]benzenesulfonic acid disodium salt;* sodium triphenyl-p-rosanilinetrisulfate; brilliant cotton blue; Helvetia blue; C.I. Acid Blue 93; C.I. 42780. $C_{37}H_{27}N_3Na_2O_9S_3$; mol wt 799.80. C 55.56%, H 3.40%, N 5.25%, Na 5.75%, O 18.00%, S 12.03%. *Do not confuse with methylene blue.* Prepn: *Colour Index* **vol. 4** (3rd ed., 1971) p 4403.

Dark blue powder. Sol in water. Absorption max about 607 nm.

USE: As a coloring; dye for cotton and silk; biological stain.

THERAP CAT: Antiseptic.

5951. Methyl Bromide. *Bromomethane;* monobromomethane; Embafume. CH_3Br; mol wt 94.95. C 12.65%, H 3.18%, Br 84.17%. Prepd industrially by the action of hydrobromic acid on methanol. Several modifications of the process, *e.g.*, sulfuric acid is added to sodium bromide and methanol, methyl bromide being removed by distillation.

Colorless gas. Usually odorless; sweetish, chloroform-like odor at high concns. Burning taste. Non-flammable in air, but burns in oxygen. mp −93.66°; bp 3.56°: Egan, Kemp, *J. Am. Chem. Soc.* **60**, 2097 (1938). Vapor press. (20°): 1420 mm Hg. d$_4^0$ 1.730. d$_{gas}^{20}$ 3.974 g/l; n$_D^{-20}$ 1.4432. Viscosity at 0°: 0.397 cp. Spec. heat at −96.6°: 0.165 cal/g/°C; at −13.0°: 1.97 cal/g/°C; at 25°: 0.107cal/g/°C. Crit temp 194°. Soly in water (20°, 748 mm): 1.75 g/100 g of soln. Forms a cryst hydrate, $CH_3Br.20H_2O$, below 4°. Freely sol in alcohol, chloroform, ether, carbon disulfide, carbon tetrachloride, benzene. Lethal concn for rats in air (6 hrs): 514 ppm, D. D. Irish *et al., J. Ind. Hyg. Toxicol.* **22**, 218 (1940).

Human Toxicity: Inhalation causes dizziness, headache, nausea, vomiting, abdominal pain, mental confusion, tremors, convulsions, pulmonary edema, coma; death from respiratory or circulatory collapse may occur. Chronic exposure can cause CNS depression or kidney injury, *Clinical Toxicology of Commercial Products,* R. E. Gosselin *et al.,* Eds. (Williams & Wilkins, Baltimore, 4th ed., 1976) Section III, pp 233-237.

USE: In ionization chambers. For degreasing wool. Extracting oils from nuts, seeds, flowers. Insect fumigant for mills, warehouses, vaults, ships, freight cars, also as soil fumigant. Because of its toxicity, there has been considerable concern against distributing this gas in small fire extinguishers.

5952. 2-Methyl-1-butanol. Active amyl alcohol; *dl-sec*-butyl carbinol. $C_5H_{12}O$; mol wt 88.15. C 68.13%, H 13.72%, O 18.15%. $CH_3CH_2CH(CH_3)CH_2OH$. One of the major components of fusel oil, *q.v.* Prepn: Hawthorne, *J. Org. Chem.* **23**, 1788 (1958); **Brit. pat. 883,375** (1961 to Continental Oil). Isoln of (−)-form by fractional distillation of fusel oil: Milburn, Truter, *J. Chem. Soc.* **1954**, 3344. Prepn of (+)-form: Carnmalm, *Chem. & Ind. (London)* **1956**, 1093. Review of manuf by fractionation of fusel oil and *via* chlorination of pentanes, and properties: *Industrial Chemicals*, W. L. Faith *et al.*, Eds. (John Wiley, New York, 2nd ed., 1957) pp 107-114.

Liquid, bp 128°. d_{20}^{20} 0.816. n_D^{25} 1.4104. Flash pt, open cup: 122°F (50°C). Slightly sol in water (3.6 g/100 g at 30°); misc with alcohol, ether. LD_{50} orally in rats: 4.92 ml/kg, H. F. Smyth *et al.*, *Am. Ind. Hyg. Assoc. J.* **23**, 95 (1962).

(−)-Form, $[\alpha]_D^{20}$ −4.75°.
(+)-Form, d_{20}^{20} 0.826. n_D^{20} 1.411.

5953. 3-Methyl-2-butanol. *dl-sec*-Isoamyl alcohol; *sec*-isopentyl alcohol; isopropyl methyl carbinol. $C_5H_{12}O$; mol wt 88.15. C 68.13%, H 13.72%, O 18.15%. $(CH_3)_2CHCH(OH)CH_3$. Prepn: Brown, Subba Rao, *J. Am. Chem. Soc.* **81**, 6423 (1959); Cook, *J. Org. Chem.* **27**, 3873 (1962). Review of manuf by fractionation of fusel oil and *via* chlorination of pentanes, and properties: *Industrial Chemicals*, W. L. Faith *et al.*, Eds. (John Wiley, New York, 2nd ed., 1957) pp 107-114.

Liquid, bp 113-114°. bp_{742} 109.5-110.5°. d^{19} 0.819. n_D^{20} 1.4091. Flash pt, closed cup: 103°F (39°C); open cup: 95°F (35°C). Slightly sol in water (2.8 g/100 g at 30°); miscible with alcohol, ether.

5954. Methyl *tert*-Butyl Ether. *2-Methoxy-2-methylpropane; tert-butyl methyl ether;* MTBE. $C_5H_{12}O$; mol wt 88.15. C 68.12%, H 13.73%, O 18.15%. $(CH_3)_3C(OCH_3)$. Prepn: L. Henry, *Rec. Trav. Chim.* **23**, 324 (1904); from methanol and *t*-butyl alcohol: J. F. Norris, G. W. Rigby, *J. Am. Chem. Soc.* **54**, 2088 (1932); from methanol and isobutylene: K. R. Edlund, T. W. Evans, U.S. pat. **1,968,601** (1934 to Shell); *eidem, Ind. & Eng. Chem.* **28**, 1186 (1936); R. D. Morin, A. E. Bearse, *ibid.* **43**, 1596 (1951); from *t*-butyl alcohol and diazomethane: M. Neeman *et al., Tetrahedron* **6**, 36 (1959). Use as chromatographic eluent: C. J. Little *et al., J. Chromatog.* **169**, 381 (1979). Experimental use to dissolve cholesterol gallstones *in vivo*: M. J. Allen *et al., Gastroenterology* **88**, 122 (1985); *eidem, N. Engl. J. Med.* **312**, 217 (1985). Method for therapeutic use in dissolving cholesterol calculi: J. L. Thistle, M. J. Allen, U.S. pat. **4,758,596** (1988 to Research Corp.). Acute toxicity: D. F. Marsh, C. D. Leake, *Anesthesiology* **11**, 455 (1950).

Liquid, bp 55.2°. mp −109°. d_4^{20} 0.7404, n_D 1.3689. Vapor pressure at 25°: 245 mm Hg. Flash pt: −28°C. Ignition temp: 224°C. Soly in water: 4.8 g/100 g. Soly of water in methyl *t*-butyl ether: 1.5 g/100 g. Unstable in acid soln. LC_{50} in mice (15 min): 1.6 mmol/liter of atmosphere (Marsh).

USE: Octane booster in gasoline. Chromatographic eluent esp in HPLC.

THERAP CAT: Cholelitholytic agent.

5955. Methyl Butyl Ketone. *2-Hexanone.* $C_6H_{12}O$; mol wt 100.16. C 71.95%, H 12.08%, O 15.97%. $CH_3COC_4H_9$. Colorless liquid. d 0.830. bp 127°. Slightly sol in water; sol in alcohol, ether. LD_{50} orally in rats: 2.59 g/kg, Smyth *et al., Arch. Ind. Hyg. Occup. Med.* **10**, 61 (1954).

5956. 2-Methyl-3-butyn-2-ol. 2-Hydroxy-2-methyl-3-butyne. C_5H_8O; mol wt 84.11. C 71.39%, H 9.59%, O 19.02%. Toxicity data: W. Keil *et al., Arzneimittel-Forsch.* **4**, 477 (1954); K. Soehring *et al., ibid.* **5**, 161 (1955).

$$CH_3\text{—}\underset{\underset{OH}{|}}{\overset{\overset{CH_3}{|}}{C}}\text{—}C\equiv CH$$

Liquid. d_{20}^{20} 0.8672. 7.24 lbs/gal. mp +2.6°. bp_{760} 104-105°; bp_{80} 52°; bp_{12} 20°. Flash pt 77°F (25°C). n_D^{20} 1.4211. Surface tension at 25° = 23.8 dynes/cm; 5% aq soln = 41.7 dynes/cm. Miscible with water, acetone, benzene, carbon tetrachloride, Cellosolve, cyclohexanone, diethylene glycol, ethyl acetate, kerosine, methyl ethyl ketone, mineral spirits, monoethanolamine, neatsfoot oil, petr ether, soybean oil, Stoddard solvent. Azeotrope with water, bp 90.7°, contains 28.4% H_2O. LD_{50} in mice (mg/kg): 1800 orally (Keil); 2340 s.c. (Soehring).

5957. Methyl Butyrate. *Butanoic acid methyl ester.* $C_5H_{10}O_2$; mol wt 102.13. C 58.80%, H 9.87%, O 31.33%. $CH_3(CH_2)_2COOCH_3$. Colorless liquid. d_{20}^{20} 0.898. mp about −95°. bp 102°. Flash pt 14°. n_D^{20} 1.3879. Sol in about 60 parts water; miscible with alcohol, ether.

USE: Manuf artificial rum and fruit essences.

5958. Methyl Carbamate. Urethylane; methylurethane. $C_2H_5NO_2$; mol wt 75.07. C 32.00%, H 6.71%, N 18.66%, O 42.63%. $H_2NCOOCH_3$. Prepn from silver or mercuricyanate with H_2S and methanol: Birkenbach, Kolb, *Ber.* **68**, 901 (1935). From urea and methanol: **Ger. pat. 753,127** (1940 to I.G. Farbenind.).

White crystals, mp 52-54°. bp 177°. Freely sublimes even at room temp. Freely sol in water, alcohol.

5959. Methyl Carbitol®. *2-(2-Methoxyethoxy)ethanol;* diethylene glycol monomethyl ether; methyl digol. $C_5H_{12}O_3$; mol wt 120.15. C 49.98%, H 10.07%, O 39.95%. $CH_3OCH_2CH_2OCH_2CH_2OH$. Toxicity data: Smyth *et al., J. Ind. Hyg. Toxicol.* **23**, 259 (1941).

Liquid. d_4^{20} 1.035. mp < −84°. bp 193°. n_D^{27} 1.4264. Flash pt, open cup: 200°F (93°C). Miscible with water, alc, glycerol, ether, acetone, DMF. LD_{50} orally in rats: 9.21 g/kg (Smyth).

USE: Used in the same way as 2-ethoxyethanol where a solvent with a higher boiling point is required.

5960. Methyl Carbonate. *Carbonic acid dimethyl ester.* $C_3H_6O_3$; mol wt 90.08. C 40.00%, H 6.71%, O 53.29%. $CO(OCH_3)_2$. Made from phosgene and methyl alcohol.

Liquid. d_4^{17} 1.065. mp 0.5°. bp 90-91°; also stated as 89.7°. n_D^{20} 1.3687. Insol in water. Miscible with alc, ether.

5961. Methyl Cellosolve®. *2-Methoxyethanol;* ethylene glycol monomethyl ether. $C_3H_8O_2$; mol wt 76.09. C 47.35%, H 10.60%, O 42.06%. $HOCH_2CH_2OCH_3$. Prepn from ethylene oxide + methanol: Finch, Hagemeyer, U.S. pat. **2,748,171** (1956 to Kodak); from ethylene glycol + diazomethane: Hesse, Majumdar, *Ber.* **93**, 1129 (1960).

Liquid. *Poisonous!* bp_{760} 124.43°; bp_{20} 34-41°. d_4^{20} 0.9663. Flash pt 115°F. n_D^{20} 1.4028. Miscible with water, alcohol ether, glycerol, acetone, dimethylformamide. LD_{50} orally in rats: 2.46 g/kg; lethal concn for mice in air: 1500 ppm.

Caution: May cause anemia, macrocytosis, appearance of young granulocytes in blood; also CNS symptoms. Readily absorbed through skin. *See* Patty's *Industrial Hygiene and Toxicology* vol. 2C, G. D. Clayton, F. E. Clayton, Eds. (Wiley-Interscience, New York, 3rd ed., 1982) pp 3911-3919.

USE: Solvent for low-viscosity cellulose acetate, natural resins, some synthetic resins and some alcohol-soluble dyes; in dyeing leather, sealing moistureproof cellophane; in nail polishes, quick-drying varnishes and enamels, wood stains. In modified Karl Fischer reagent: Peters, Jungnickel, *Anal. Chem.* **27**, 450 (1955).

5962. Methyl Cellosolve® Acetate. *2-Methoxyethanol acetate;* ethylene glycol monomethyl ether acetate. $C_5H_{10}O_3$; mol wt 118.13. C 50.83%, H 8.53%, O 40.63%. $CH_3OCH_2CH_2OOCCH_3$. Toxicity data: H. F. Smyth *et al., J. Ind. Hyg. Toxicol.* **23**, 259 (1941). Brief review: E. S. Brown *et al.*, "Glycols (Ethylene and Propylene)" in Kirk-Othmer *Encyclopedia of Chemical Technology* vol. 11 (Wiley-Interscience, New York, 3rd ed., 1980) pp 942-946.

Colorless liquid. d_{20}^{20} 1.0067. bp 145°. mp −65.1°. n_D^{20} 1.4019. Flash pt (open cup): 55.6°. Miscible with water, most organic solvents, oils; dissolves gums, resins. LD_{50} orally in rats: 3.4 g/kg (Smyth).

Caution: Toxic symptoms similar to those for Methyl Cellosolve. *See* Rowe in *Industrial Hygiene and Toxicology* vol. 2, F. A. Patty, Ed. (Interscience, New York, 2nd ed., 1962) pp 1587-1588.

USE: Industrial solvent.

5963. Methylcellulose. *Cellulose methyl ether;* Methocel; Cellothyl; Syncelose; Bagolax; Celevac; Cellucon; Cethylose; Cethytin; Cologel; Cellumeth; Hydrolose; Nicel; Tearisol; Tylose. Prepd from wood pulp or chemical cotton by treatment with alkali and methylation of the alkali cellulose with methyl chloride. *Review and bibliography:* Ott, *Cellulose and Cellulose Derivatives* (Wiley-Interscience, New York, 2nd ed., 1954/55); Greminger, Savage, in *Industrial Gums,* R. L. Whistler, Ed. (Academic Press, New York, 2nd ed., 1973) pp 619-647.

White granules. Odorless, tasteless. Sol in cold water. Insol in hot water. An aq soln is best prepd by dispersing the granules in hot (but not boiling) water with stirring and chilling to +5°. The soln is then stable at room temp. Presence of inorganic salts increases the viscosity. The soly is dependent upon degree of substitution. Commercial methylcellulose has a methoxyl content of 29% (degree of substitution about 1.8). Clear films may be cast from aq soln.

USE: As a substitute for water-soluble gums; to render paper greaseproof, in adhesives, as thickening agent in cosmetics, as protective colloid in emulsions, as binder and stabilizer in foods. As bulk producer in the formulation of dietetic foods. Pharmaceutic aid (suspending agent).

THERAP CAT (VET): Laxative.

5964. Methyl Chloride. *Chloromethane.* CH_3Cl; mol wt 50.49. C 23.79%, H 5.99%, Cl 70.22%. Review of mfg processes: Faith, Keyes & Clark's *Industrial Chemicals,* F. A. Lowenheim, M. K. Moran, Eds. (Wiley-Interscience, New York, 4th ed., 1975) pp 530-538.

Colorless gas; compresses to a colorless liq of ethereal odor and sweet taste. *Poisonous!* Burns with a smoky flame. mp −97°; bp −23.7°; n_D (liq at −23.7°) 1.3712: McGovern, *Ind. Eng. Chem.* **35,** 1230 (1943). Slightly sol in water; misc with chloroform, ether, glacial acetic acid; sol in alcohol. Soly at 20° (ml/100 ml): water 303; benzene 4723; carbon tetrachloride 3756; glacial acetic acid 3679; ethanol 3740. Explosive limits in air 8.1 to 17.2% v/v. Lethal concn for mice in air: 3146 ppm, *RTECS* **Vol. II,** R. J. Lewis, R. L. Tatken, Eds. (1980) p 57.

Human Toxicity: Injury to liver, kidneys, CNS may occur, *cf.* Patty's *Industrial Hygiene and Toxicology* **vol. 2B,** G. D. Clayton, F. E. Clayton, Eds. (Wiley-Interscience, New York, 3rd ed., 1981) pp 3436-3442.

USE: As a refrigerant.

THERAP CAT: Local anesthetic.

5965. Methyl Chloroacetate. *Chloroacetic acid methyl ester.* $C_3H_5ClO_2$; mol wt 108.53. C 33.20%, H 4.64%, Cl 32.67%, O 29.48%. $CH_2ClCOOCH_3$.

Colorless liquid. d_{20}^{20} 1.238. mp −33°. bp 130-132°. Insol in water; miscible with alcohol, ether.

USE: As solvent.

5966. Methyl Chlorocarbonate. *Carbonochloridic acid methyl ester;* methyl chloroformate. $C_2H_3ClO_2$; mol wt 94.50. C 25.42%, H 3.20%, Cl 37.52%, O 33.86%. $ClCO_2$-CH_3. Made from phosgene and methyl alcohol.

Clear liquid. d_4^{20} 1.223. bp 71°. Slightly sol in water and gradually dec by it; miscible with alcohol, benzene, chloroform, ether. *Caution:* Vapors strongly irritating to eyes.

5967. 3-Methylcholanthrene. *1,2-Dihydro-3-methylbenz[j]aceanthrylene;* 20-methylcholanthrene; 3-MECA; 3-MC. $C_{21}H_{16}$; mol wt 268.34. C 93.99%, H 6.01%. From desoxycholic acid: Wieland, Schlichting, *Z. Physiol. Chem.* **150,** 267 (1925); Wieland, Wiedersheim, *ibid.* **186,** 229 (1930); Wieland, Dane, *ibid.* **219,** 240 (1933). From cholic acid: Cook, Haslewood, *J. Chem. Soc.* **1934,** 428; Fieser, Seligman; *J. Am. Chem. Soc.* **58,** 2482 (1936); *cf.* Bachmann, *J. Org. Chem.* **1,** 347 (1936). Methylcholanthrene has been produced also by an unusual pyrolytic degradation of cholesterol derivatives. Total synthesis: Buchta, Güllich, *Angew. Chem.* **70,** 190 (1958); P. W. Tang, C. A. Maggiulli, *J. Org. Chem.* **46,** 3429 (1981); S. A. Jacobs, R. G. Harvey, *Tetrahedron Letters* **22,** 1093 (1981). *Review:* E. Clar, *Polycyclic Hydrocarbons* (Academic Press, 1964) 2 vols.

Pale yellow, slender prisms from benzene + ether, mp 179-180°. bp_{80} 280°; d^{20} 1.28. Absorption spectrum: Fieser, Hershberg, *J. Am. Chem. Soc.* **60,** 940 (1938). Sol in benzene, xylene, toluene; slightly sol in amyl alcohol; insol in water.

USE: Exptlly in cancer research.

5968. Methylconiine. *1-Methyl-2-propylpiperidine.* $C_9H_{19}N$; mol wt 141.25. C 76.52%, H 13.56%, N 9.91%. The *d*-form is stated to occur in hemlock in small quantities. The *l*-form can be isolated from residues left in the prepn of coniine by crystn of the hydrobromides. Isoln: Ahrens, *Ber.* **35,** 1330 (1902); Hess, Eichel, *ibid.* **50,** 1386 (1917). Prepns: Leonard, Barthel, *J. Am. Chem. Soc.* **71,** 3098 (1949); Gaumeton, Glacet, *Bull. Soc. Chim. France* **1959,** 1501; Lukes, Cerny, *Coll. Czech. Chem. Commun.* **24,** 1287 (1959); Cervinka, *ibid.* **26,** 673 (1961).

dl-Form, $bp_{10.5}$ 56.6°.
d-Form, oily liq, bp 173-174°. d^{24} 0.8318. $[\alpha]_D^{24}$ +81°. n_D^{13} 1.4538. Slightly sol in cold water; sol in organic solvents.
l-Form has essentially same characteristics. $[\alpha]_D^{24}$ −84°.

5969. 3'-Methyl-1,2-cyclopentenophenanthrene. *16,17-Dihydro-17-methyl-15H-cyclopenta[a]phenanthrene;* Diels' hydrocarbon. $C_{18}H_{16}$; mol wt 232.33. C 93.06%, H 6.94%. Prepd from cholesterol, cholic acid, or suitable sapogenins: Diels, Rickert, *Ber.* **68,** 267 (1935); starting with 2-acetylphenanthrene: Gamble *et al., J. Chem. Soc.* **1935,** 443, 644.

Crystals from acetic acid, mp 126-127°. Absorption spectrum: Hillemann, *Ber.* **69,** 2610 (1936).

5970. 5-Methylcytosine. *4-Amino-5-methyl-2(1H)-pyrimidinone.* $C_5H_7N_3O$; mol wt 125.13. C 47.99%, H 5.64%, N 33.58%, O 12.79%. Occurs in a nucleic acid obtained from tubercle bacillus. Isoln: Johnson, Coghill, *J. Am. Chem. Soc.* **47,** 2838 (1925). Synthesis from ethyl bromide addition product of thiourea and from ethyl formyl propionate: Wheeler, Johnson, *Am. Chem. J.* **29,** 492 (1903).

Prisms from water, may contain ½ mol H_2O. mp 270° (effervescence). One gram dissolves in 29 ml water.

Picrate, $C_5H_7N_3O \cdot C_6H_3N_3O_7$, minute yellow orthorhombic prisms from water, darkens at 250°, dec 286°.

5971. Methyl Demeton. *Phosphorothioic acid O-[2-(ethylthio)ethyl] O,O-dimethyl ester mixt with S-[2-(ethylthio)-*

ethyl] O,O-dimethyl phosphorothioate; demeton-methyl; methyl-mercaptophos; methyl systox; Bayer 21/116; Meta-Systox. $C_6H_{15}O_3PS_2$; mol wt 230.30. C 31.29%, H 6.57%, O 20.84%, P 13.45%, S 27.85%. Isomeric mixture consisting of demeton-*O*-methyl and demeton-*S*-methyl (*O*(and *S*)-[2-(ethylthio)ethyl] *O,O*-dimethyl phosphorothioate). *Ref:* Henglein, Schrader, *Z. Naturforsch.* **10b**, 12 (1955). Prepn: **Brit.** pat. **814,332** (1959 to Bayer).

Commercial product is a light yellow liquid. Hydrolyzed by alkali. LD_{50} orally in rats: 65 mg/kg.
Demeton-*O*-methyl, colorless oil, $bp_{0.15}$ 74°. d_4^{20} 1.190. n_D^{20} 1.5063. Soly in water at room temp: 330 ppm. Soluble in organic solvents. LD_{50} orally in rats: 180 mg/kg.
Demeton-*S*-methyl, pale yellow oil, $bp_{0.15}$ 89°. d_4^{20} 1.207. n_D^{20} 1.5065. Soly in water at room temp: 3,300 ppm. Sol in organic solvents. LD_{50} orally in rats: 40 mg/kg.
USE: Insecticide. *Caution:* Cholinesterase inhibitor. Absorbed readily through skin. For symptoms *see* Parathion.

5972. p-Methyldiphenhydramine. *N,N-Dimethyl-2-[(4-methylphenyl)phenylmethoxy]ethanamine; N,N-dimethyl-2-[(p-methyl-α-phenylbenzyl)oxy]ethylamine; β-dimethylaminoethyl 4-methylbenzhydryl ether; p-methylbenzhydryl 2-dimethylaminoethyl ether.* Neo-Benodine; Toladryl. $C_{18}H_{23}NO$; mol wt 269.37. C 80.25%, H 8.61%, N 5.20%, O 5.94%. Prepn: Rieveschl, U.S. pat. **2,527,962** (1950 to Parke, Davis); **Brit.** pat. **683,483** (1952 to Fabriken Brocades-Stheeman & Pharmacia).

Hydrochloride, $C_{18}H_{23}NO \cdot HCl$, crystals from acetone, mp 150-152° (free base $bp_{0.1}$ 143°).
THERAP CAT: Antihistaminic.

5973. Methyldiphenylamine. *N-Methyl-N-phenylbenzenamine;* diphenylmethylamine. $C_{13}H_{13}N$; mol wt 183.24. C 85.20%, H 7.15%, N 7.64%. $(C_6H_5)_2NCH_3$. Prepd by heating diphenylamine, methyl alcohol and HCl.
Liquid. d_4^{20} 1.0476. mp −7.6°. bp 296-297°; also given as 282-290°, 293°. n_D^{20} 1.6193. Insol in water; sol in alc, ether.
USE: Manuf dyes; as reagent similar to diphenylamine.

5974. Methyldopa. *3-Hydroxy-α-methyl-L-tyrosine; L-3-(3,4-dihydroxyphenyl)-2-methylalanine; L-α-methyl-3,4-dihydroxyphenylalanine; L-2-amino-2-methyl-3-(3,4-dihydroxyphenyl)propionic acid; α-methyldopa; alpha-methyldopa; AMD; MK-351; Aldomet; Aldometil; Aldomine; Dopamet; Dopegyt; Dopatec; Elanpres; Equibar; Lederdopa; Medomet; Medopa; Medopren; Methoplain; Sembrina; Presinol.* $C_{10}H_{13}NO_4$; mol wt 211.21. C 56.86%, H 6.20%, N 6.63%, O 30.30%. Exists as the sesquihydrate. Prepn: Pfister, Stein, U.S. pat. **2,868,818** (1959 to Merck & Co.). Resolution: Jones *et al.,* U.S. pat. **3,158,648** (1964 to Merck & Co.); *cf.* Slates *et al., J. Org. Chem.* **29**, 1424 (1964). Resolution and configuration: Tristram, *ibid.* 2053. Synthesis from asymmetric intermediates: Reinhold *et al., J. Org. Chem.* **33**, 1209 (1968). Prepn of the ethyl ester hydrochloride: **Fr.** pat. **M2153** (1963 to Merck & Co.); of pharmaceutical dosage forms: Marcus, U.S. pat. **3,230,143** (1966 to Merck & Co.). Mode of action: Finch, Haeusler, *Brit. J. Pharmacol.* **45**, 167p (1972); Day *et al., ibid.* 168p. Review of pharmacology: A. Scriabine, Ed. in *Pharmacology of Antihypertensive Drugs* (Raven, New York, 1980) pp 43-54.

L-Form sesquihydrate, crystals from water. Minute anhydr crystals from methanol, considerably hygroscopic, dec ~300°. $[\alpha]_D^{23}$ −4.0° ± 0.5° (c = 1 in 0.1*N* HCl). uv max: 281 nm (ε 2780). Soly in water at 25°: about 10 mg/ml. pH of satd aq soln about 5.0. (Soly of D-form in water at 25°: ~10 mg/ml, of DL-form: ~18 mg/ml.) Practically insol in the common organic solvents. Sol in dil mineral acids.
Ethyl ester hydrochloride, $C_{12}H_{18}ClNO_4$, *methyldopate hydrochloride.* After drying, mol wt 275.7 (1 mg = 0.77 mg of the free acid). Soluble in water to the extent of 10-300 mg/ml at pH 2-7. May be stabilized with sodium bisulfite and/or disodium EDTA.
THERAP CAT: Antihypertensive.

5975. 3-O-Methyldopa. *3-Methoxy-L-tyrosine;* 3-(4-hydroxy-3-methoxyphenyl)-L-alanine; L-3-methoxy-4-hydroxyphenylalanine; OM-dopa; L-3-MTO; OMD. $C_{10}H_{13}NO_4$; mol wt 211.22. C 56.86%, H 6.20%, N 6.63%, O 30.30%. A major metabolite of L-dopa in man and animals, which has a considerably longer biological half-life than L-dopa: Pletscher *et al., Brain Res.* **4**, 106 (1967); Bartholini, Pletscher, *J. Pharmacol. Exp. Ther.* **161**, 14 (1968); Kuruma *et al., Eur. J. Pharmacol.* **10**, 189 (1970). Proposed as a precursor of dopamine *via* its partial demethylation in the organism: Bartholini *et al., Nature* **230**, 533 (1971); Chalmers *et al., Brit. J. Pharmacol.* **43**, 455p (1971); Bartholini *et al., Life Sci.* **14**, 323 (1974). *See also* Carlsson, Waldeck, *Arch. Pharmacol.* **272**, 441 (1972); Bartholini, Pletscher, *ibid.* **274**, 404 (1972). Metabolic pathways: Bartholini *et al., J. Pharmacol. Exp. Ther.* **183**, 65 (1972). Parkinsonism treatment studies: Calne *et al., Clin. Pharmacol. Ther.* **14**, 386 (1973).

5976. Methyleneaminoacetonitrile. α-Hydroformamine cyanide. $C_3H_4N_2$; mol wt 68.07. C 52.93%, H 5.92%, N 41.15%. $CH_2=NCH_2CN$. The actual molecular formula is $C_9H_{12}N_6$: Johnson, Rinehart, *J. Am. Chem. Soc.* **46**, 768, 1653 (1924). Prepd by the action of formaldehyde on a mixture of ammonium chloride, potassium cyanide and acetic acid: Adams, Langley, *Org. Syn. coll. vol. I* (2nd ed., 1941) p 355.
Trimeric form, orthorhombic crystals from alc or acetone, mp 129°. Sol in hot water, alc; slightly sol in benzene.

5977. Methylene Azure. Azure I. A somewhat variable mixture obtained by oxidation of methylene blue, contg primarily azure A and azure B, *q.v.* Prepn: Bernthsen, *Ann.* **230**, 169 (1885); Kehrmann, *Ber.* **39**, 1804 (1906). *Review:* H. J. Conn's *Biological Stains,* R. D. Lillie, Ed. (Williams & Wilkins, Baltimore, 9th ed., 1977) pp 493-502.
Green glistening crystals. Forms blue soln in water; very sparingly sol in alcohol with a reddish brown fluorescence.
Mixture with methylene blue, *azure II, azure blue II, methylene azure II.* Deep-green powder. Sol in water with blue color, less sol in alcohol, slightly sol in chloroform. Insol in ether.
Mixture with methylene blue and eosin, *azure II eosin.* Chief ingredient of Giemsa stain. Green powder. Slightly sol in water; sol in alcohol, methanol, glycerol.
USE: Biological stain.

5978. 4,4'-Methylenebis[2-chloroaniline]. *4,4'-Methylenebis[2-chlorobenzenamine];* 4,4'-diamino-3,3'-dichlorodiphenylmethane; di-(4-amino-3-chlorophenyl)methane; methylenebis(*o*-chloroaniline); MOCA; DACPM. $C_{13}H_{12}Cl_2N_2$; mol wt 267.15. C 58.45%, H 4.53%, Cl 26.54%, N 10.48%. Prepn: Finger, *J. Prakt. Chem.* [2] **79**, 493 (1909); Mayer, *Ber.* **47**, 1161 (1914). Acute toxicity: G. Ya. Kel'man *et al., Kauch. Rezina* **26**(9), 28 (1967), *C.A.* **67**, 120013m (1967). Review and evaluation of studies of carcinogenicity in laboratory animals: *IARC Monographs* **vol. 4**, 65-71 (1974).

Flakes from alcohol, mp 110°. Slightly sol in water; sol in dil acids, ether, alcohol. LD_{50} orally in mice: 880 mg/kg (Kel'man).

Note: This substance may reasonably be anticipated to be a carcinogen: *Fourth Annual Report on Carcinogens* (NTP 85-002, 1985) p 128.

USE: Curing agent for polyurethane and epoxy resins.

5979. Methylene Blue. *3,7-Bis(dimethylamino)phenothiazin-5-ium chloride; C.I. Basic Blue 9;* methylthioninium chloride; tetramethylthionine chloride; 3,7-bis(dimethylamino)phenazathionium chloride; Swiss blue; *C.I. 52015;* solvent blue 8; urolene blue. $C_{16}H_{18}ClN_3S$; mol wt 319.85. C 60.08%, H 5.67%, Cl 11.08%, N 13.14%, S 10.02%. May crystallize with 3, 4 and 5 mols H_2O. First prepd by Caro in 1876. Now usually prepd from dimethylaniline and thiosulfuric acid: H. E. Fierz-David, L. Blangey, *Fundamental Processes of Dye Chemistry* (Interscience, New York, 1949) p 311. *See also Colour Index* vol. 4 (3rd ed., 1971) p 4470.

Trihydrate, dark green, odorless crystals with bronze luster or cryst powder. Absorption max: 668, 609 nm. One gram dissolves in about 25 ml water, about 65 ml alcohol; sol in chloroform. Insol in ether. In aq soln it is decolorized by Zn dust and dil H_2SO_4, but color is restored on exposure to air and more rapidly upon addn of NH_4OH. Forms double salts with many inorganic salts. *Incompat:* Caustic alkali, dichromates, alkali iodides, reducing agents.

USE: Stain in bacteriology; reagent for several chemicals; in mixed indicators; as oxidation-reduction indicator.

THERAP CAT: Antimethemoglobinemic; antidote to cyanide.

THERAP CAT (VET): Antiseptic; disinfectant; antidote to cyanide and nitrate.

5980. Methylene Bromide. *Dibromomethane.* CH_2Br_2; mol wt 173.86. C 6.91%, H 1.16%, Br 91.93%. Prepn: Hartman, Dreger, *Org. Syn. coll. vol. I,* 357 (1941). Manuf along with bromochloromethane, from dichloromethane: Lake, Asadorian, U.S. pat. **2,553,518** (1951 to Dow).

Liquid, bp 97°. mp −52.7°. d_4^{20} 2.4956. n_D^{20} 1.5419. Soly (g/1000 g water): 11.70 (15°); 11.93 (30°). Miscible with alcohol, ether, acetone.

5981. α-Methylene Butyrolactone. *4,5-Dihydro-3-methylene-2(3H)-furanone.* $C_5H_6O_2$; mol wt 98.10. C 61.21%, H 6.17%, O 32.62%. Isoln from aq extracts of *Erythronium americanum* Ker., Liliaceae (dogtooth violet): Cavallito, Haskell, *J. Am. Chem. Soc.* **68**, 2332 (1946).

Liquid. Polymerizes easily when heated above 70°. bp_2 57-60°. Sol in water. pH 4 (150 mg/ml). The lactone is present in all parts of the plant and can be extracted easily with water, but not with ethanol, ether, ethyl acetate or chloroform. *Caution:* Irritating to the skin.

5982. Methylene Chloride. *Dichloromethane;* methylene dichloride; methylene bichloride. CH_2Cl_2; mol wt 84.94. C 14.14%, H 2.37%, Cl 83.49%. Prepn by chlorination of methane: Lukes *et al.,* U.S. pat. **2,792,435** (1957 to Diamond Alkali); Pitt, Bender, U.S. pat. **2,979,541** (1961 to Stauffer); Burks, Obrecht, U.S. pat. **3,126,419** (1964 to

Stauffer). Review of mfg processes: Faith, Keyes & Clark's *Industrial Chemicals,* F. A. Lowenheim, M. K. Moran, Eds. (Wiley-Interscience, New York, 4th ed., 1975) pp 530-538. Toxicity data: E. T. Kimura *et al., Toxicol. Appl. Pharmacol.* **19**, 699 (1971).

Colorless liquid; vapor is not flammable and when mixed with air is not explosive. bp_{760} 39.75°. mp −95°. d_4^0 1.36174; d_4^{15} 1.33479; d_4^{20} 1.3255; d_4^{30} 1.30777. n_D^{20} 1.4244. Sol in about 50 parts water; miscible with alc, ether, DMF. LD_{50} orally in young adult rats: 1.6 ml/kg (Kimura).

Human Toxicity: Narcotic in high concns.

USE: Solvent for cellulose acetate; degreasing and cleaning fluids; as solvent in food processing. Pharmaceutic aid (solvent).

5983. Methylenedigallic Acid. *2,2′-Methylenebis[3,4,5-trihydroxybenzoic acid].* $C_{15}H_{12}O_{10}$; mol wt 352.25. C 51.14%, H 3.43%, O 45.42%. Made by heating a mixture of gallic acid, formaldehyde and dil HCl on the water bath.

Crystals. Insol in cold, sparingly in hot water; sol in alcohol, alkalies.

5984. 5,5′-Methylenedisalicylic Acid. *3,3′-Methylene-bis[6-hydroxybenzoic acid];* 4,4′-dihydroxydiphenylmethane-3,3′-dicarboxylic acid. $C_{15}H_{12}O_6$; mol wt 288.25. C 62.50%, H 4.20%, O 33.30%. Prepd from salicylic acid and formaldehyde in the presence of sulfuric acid: Clemmensen, Heitman, *J. Am. Chem. Soc.* **33**, 737 (1911).

Wedge-like cryst from acetone + benzene. Bitter taste. Dec 238° (higher-melting material may be impure). Turns red at 180° and starts giving off CO_2. Freely sol in methanol, ethanol, ether, acetone, glacial acetic acid. Very slightly sol in hot water. Practically insol in benzene, chloroform, carbon disulfide, petr ether.

Bacitracin salt, *see* Bacitracin Methylenedisalicylic Acid.

Diacetyl deriv, $C_{15}H_{10}O_6(CH_3CO)_2$, white powder, mp 142°, practically insol in water, sol in acetone, ethanol.

5985. Methylene Iodide. *Diiodomethane.* CH_2I_2; mol wt 267.87. C 4.48%, H 0.75%, I 94.76%. Prepd by the reduction of iodoform with sodium arsenite: Adams, Marvel, *Org. Syn.* vol. **1**, p 57 (1921). Also by heating iodoform with sodium acetate in alcohol: Bagnara, *Eng. Mining J. Press* **116**, 51 (1923).

Very heavy, highly refractive liq. Darkens on exposure to light, air, and moisture. d_4^{20} 3.32537. Solidifies in leaflets at 5.2° or in thin needles at 5.7°. Usually cooling to 0° is necessary to start crystn. mp 6.0°. bp_{760} 181°; bp_{70} 107°; bp_{11} 68°. $n_D^{10.5}$ 1.7559; n_D^{15} 1.7425. Viscosity in centipoises: 3.35 (10°); 2.80 (20°); 2.39 (30°). Absorption spectrum: Lowry, Sass, *J. Chem. Soc.* **1926**, 624; Stepanov, *Acta Physicochim. URSS* **20**, 174 (1945). Sol in about 70 parts water; miscible with alcohol, propanol, isopropanol, hexane, cyclohexane, ether, chloroform, benzene. CH_2I_2 dissolves sulfur and phosphorus (more than 1:1 at 25°).

USE: In separating mixtures of minerals. In determining the specific gravity of minerals and other substances. In the manufacture of x-ray contrast media.

5986. Methylenomycins. Members of a family of cyclopentenoid antibiotics related structurally to sarkomycins, *q.v.,* and having *in vitro* activity vs gram-positive and gram-negative organisms. Isoln from *Streptomyces violaceoruber,* physical, chemical, biological properties: M. Arai *et al.,*

Japan. *Kokai* 73 19796 (1973 to Sankyo), *C.A.* **78**, 157861 (1973); T. Haneishi *et al.*, *J. Antibiot.* **27**, 386 (1974). Structures of methylenomycins A and B: *eidem, ibid.* 393. Abs config of methylenomycin A: K. Sakai *et al.*, *Tetrahedron Letters* **1979**, 2365. Crystal and molecular structure of (±)-A: B. H. Toder, A. B. Smith, *J. Cryst. Mol. Struct.* **8**, 1 (1979). Stereospecific total synthesis of (±)-A: R. M. Scarborough *et al.*, *J. Am. Chem. Soc.* **99**, 7085 (1977); *eidem, ibid.* **102**, 3904 (1980); R. M. Scarborough, *Diss. Abstr. B* **40**, 4829 (1980); Y. Takahashi *et al.*, *Chem. Commun.* **1981**, 714. Stereospecific total synthesis and abs config of (+)-methylenomycin A: J. Jernow *et al.*, *J. Org. Chem.* **44**, 4210 (1979). Revised structure and total synthesis of methylenomycin B: *eidem, ibid.* 4212. Additional syntheses of B: K. Tonari *et al.*, *Agr. Biol. Chem.* **45**, 295 (1981); R. F. Newton *et al.*, *Syn. Commun.* **11**, 527 (1981); T. Siwapinyoyos, Y. Thebtaranonth, *J. Org. Chem.* **47**, 598 (1982); Y. Takahashi *et al.*, *Chem. Commun.* **1982**, 496. Prepn of analogs of A and structure-activity correlations: T. Haneishi *et al.*, *J. Antibiot.* **27**, 400 (1974). Toxicity: *eidem, ibid.* 386. Methylenomycin A is the first example of an antibiotic in which all information required for synthesis is carried by a plasmid, SCP1: L. F. Wright, D. A. Hopwood, *J. Gen. Microbiol.* **95**, 96 (1976). Review of biosynthesis: U. Hornemann, D. A. Hopwood, *Antibiotics* **vol. IV**, J. W. Corcoran, Ed. (Springer-Verlag, New York, 1981) pp 123-131. General review: A. Terehara *et al.*, *Heterocycles* **13**, 353-371 (1979).

methylenomycin A   methylenomycin B

Methylenomycin A, $C_9H_{10}O_4$, *(1S)-1α,5α-dimethyl-3-methylene-4-oxo-6-oxabicyclo[3.1.0]hexane-2α-carboxylic acid.* Colorless cryst from chloroform/carbon tetrachloride, mp 115° (dec). mp of the (±)-form: 88.5-89°; after subl (70-75°, 0.025 mm Hg), 107.5-108°. $[\alpha]_D^{20}$ +42.3° (c = 1 in chloroform). uv max (methanol): 224 nm (ε 6300). Sol in benzene, chloroform, ethyl acetate, acetone, methanol, water. Slightly sol in *n*-hexane, CCl₄. pKa' 3.65. LD₅₀ in mice (mg/kg): 1500 orally, 75 i.p. (Haneishi).

Methylenomycin B, $C_8H_{10}O$, *2,3-dimethyl-5-methylene-2-cyclopenten-1-one.* Neutral colorless oil. uv max (methanol): 240 nm (ε 7650). Sol in ether, benzene, chloroform, ethyl acetate, acetone, alcs. Slightly sol in *n*-hexane, petr ether. LD₅₀ in mice (mg/kg): 260 orally, 245 i.p. (Haneishi).

5987. N-Methylephedrine. *α-[1-(Dimethylamino)ethyl]-benzenemethanol; erythro-α-[1-(dimethylamino)ethyl]benzyl alcohol; 2-dimethylamino-1-phenylpropanol; N,N-dimethylnorephedrine.* $C_{11}H_{17}NO$; mol wt 179.25. C 73.70%, H 9.56%, N 7.81%, O 8.93%. Isoln of *l*-form from *Ephedra distachya* L. (*E. vulgaris* Rich.), and allied *Gnetaceae*: Smith, *J. Chem. Soc.* **1927**, 2056; Wolfes, *Arch. Pharm.* **268**, 327 (1930). Prepn of *l*-form: Smith, *loc. cit.*; of *d*-, *l*-, and *dl*-forms, and resolution of racemic mixture: Nagai, Kanao, *Ann.* **470**, 157 (1929); of *dl*-form: Pfanz, Müller, *Arch. Pharm.* **288**, 11 (1955).

dl-Form, crystals from petr ether or methanol, mp 63.5-64.5°. Readily sol in the usual solvents.

dl-Form hydrochloride, *Metheph.* Crystals from acetone, mp 207-208°.

d-Form, crystals, mp 87-87.5°. $[\alpha]_D^{20}$ +29.2° (c = 4 in methanol).

d-Form hydrochloride, crystals from ethyl acetate, mp 192°. $[\alpha]_D^{20}$ +30.1°.

l-Form, crystals from petr ether, mp 87-88°. $[\alpha]_D$ −29.5° (c = 4.5 in methanol).

l-Form hydrochloride, crystals from ethyl acetate or alcohol, mp 192°. $[\alpha]_D^{20}$ −29.8° (c = 4.6). Readily sol in water; less sol in alcohol; sparingly sol in acetone.

β-Camphorsulfonate, *N-methylephedrine camsylate, Tybraine.*

THERAP CAT: Analeptic.

5988. N-Methylepinephrine. *4-[2-(Dimethylamino)-1-hydroxyethyl]-1,2-benzenediol; α-[(dimethylamino)methyl]-3,4-dihydroxybenzyl alcohol; α-(3,4-dihydroxyphenyl)-2-dimethylaminoethanol; α-(3,4-dihydroxyphenyl)-α-hydroxy-β-dimethylaminoethane; dimethylaminomethyl-(3,4-dihydroxyphenyl) carbinol; α-(dimethylaminomethyl)protocatechuyl alcohol; N-methyladrenaline.* $C_{10}H_{15}NO_3$; mol wt 197.23. C 60.89%, H 7.67%, N 7.10%, O 24.34%. Prepn and resolution of racemic mixture: Manna, Campiglio, *Farmaco Ed. Sci.* **14**, 317 (1959). Configuration: Manna, Ghislandi, *ibid.* **19**, 377 (1964).

DL-Form, *Methadren(e)*. Crystals from alcohol + ethyl acetate, mp 142-143°.

D(−)-Form, crystals from ethyl acetate, mp 149-150°. $[\alpha]_D^{18}$ −65.1° (c = 1.41 in 0.5N HCl).

L(+)-Form, crystals, mp 149-150°. $[\alpha]_D^{18}$ +62.3° (c = 1.4).

THERAP CAT: Adrenergic.

5989. Methylergonovine. *9,10-Didehydro-N-[1-(hydroxymethyl)propyl]-6-methylergoline-8-carboxamide; N-[α-(hydroxymethyl)propyl]-D-lysergamide; D-lysergic acid (+)-butanolamide-(2); d-lysergic acid-dl-hydroxybutylamide-2; methylergometrine; methylergobasine.* $C_{20}H_{25}N_3O_2$; mol wt 339.42. C 70.77%, H 7.42%, N 12.38%, O 9.43%. A homolog of ergonovine contg one more CH_2 group. Prepd by condensing isolysergic acid azide with *d*-2-amino-1-butanol and subjecting the reaction product to an isomerization treatment: Stoll, Hofmann, U.S. pat. **2,265,207** (1941 to Sandoz).

Shiny crystals from benzene, mp 172° (some decompn). $[\alpha]_D^{20}$ −45° (c = 0.4 in pyridine). Sparingly sol in water. Freely sol in alcohol, acetone.

Maleate, $C_{24}H_{29}N_3O_6$, *Basofortina, Methergin, Methergine, Metenarin, Methylergobrevin, Ryegonovin, Partergin, Spametrin-M.* White to pinkish-tan microcryst powder; odorless; bitter taste. Slightly sol in water, alcohol; very slightly sol in chloroform, ether.

Tartrate, $C_{44}H_{56}N_6O_{10}$, clusters of needles, usually with 2 mols of methanol of crystn. Easily sol in water, alcohol, very slightly sol in ether, chloroform. pH of aq solns 5.0-5.8.

THERAP CAT: Oxytocic.

5990. Methyl Ether. *Oxybismethane;* dimethyl ether. C_2H_6O; mol wt 46.07. C 52.14%, H 13.13%, O 34.73%. CH_3OCH_3.
Colorless gas; ethereal odor; burns with a slightly luminous flame. d 1.617 (air = 1). bp −23.6°. Flash pt −41°. One vol water takes up 37 vols gas; far more sol in alcohol.

USE: In refrigeration.

5991. Methyl Ethyl Ketone. *2-Butanone;* ethyl methyl

ketone; MEK; 2-oxobutane. C_4H_8O; mol wt 72.10. C 66.63%, H 11.18%, O 22.19%. $CH_3COCH_2CH_3$. Prepn from ethyl 2-methylacetoacetate: J. Schramm, *Ann.* **398**, 242 (1913). Manuf by dehydration of 2-butanol and by catalytic oxidation of *n*-butenes: A. J. Papa, P. D. Sherman, Jr. in Kirk-Othmer *Encyclopedia of Chemical Technology* vol. **13** (Wiley-Interscience, New York, 3rd ed., 1981) pp 903-907. Toxicity: H. F. Smyth *et al.*, *Am. Ind. Hyg. Assoc. J.* **23**, 95 (1962).

Flammable liquid; acetone-like odor. d_4^{20} 0.805. mp $-86°$. bp 79.6°. Flash pt, closed cup: 21°F ($-6°C$). n_D^{15} 1.3814. Sol in about 4 parts water (27.5%); less sol at higher temp; miscible with alcohol, ether, benzene. Constant boiling mixture with water, bp 73.4°, contains 88.7% methyl ethyl ketone. Soly of water in methyl ethyl ketone: 12.5% at 25°. LD_{50} orally in rats: 6.86 ml/kg (Smyth).

USE: As solvent; in the surface coating industry; manuf smokeless powder; colorless synthetic resins.

5992. α-Methylfentanyl. *N-[1-(1-Methyl-2-phenylethyl)-4-piperidinyl]-N-phenylpropanamide; N-[1-(α-methylphenethyl)-4-piperidyl]propionanilide;* 1-(1-methyl-2-phenylethyl)-4-(*N*-propanilido)piperidine. $C_{23}H_{30}N_2O$; mol wt 350.50. C 78.82%, H 8.63%, N 7.99%, O 4.56%. Potent derivative of fentanyl, *q.v.* Prepn: P. A. J. Janssen, Fr. pat. **M2430**; *idem*, U.S. pat. **3,164,600** (1964, 1965 both to Janssen). This substance has erroneously been referred to as "China White", the street term for very pure Southeast Asian heroin. Initial identification of "China White" as α-methylfentanyl: T. C. Kram *et al.*, *Anal. Chem.* **53**, 1379 A (1981). Molecular structure determn using tandem mass spectrometry: M. T. Cheng *et al.*, *ibid.* **54**, 2204 (1982). Identification and quantification in tissue: T. J. Gillespie *et al.*, *J. Anal. Toxicol.* **6**, 139 (1982). Confirmation of identity of "China White": S. Suzuki *et al.*, *Chem. Pharm. Bull.* **34**, 1340 (1986). Immunoassay for detection of use in racehorses: J. McDonald *et al.*, *Res. Commun. Chem. Pathol. Pharmacol.* **57**, 389 (1987).

CH₃CH₂CON(C₆H₅)—⟨piperidine⟩—N—CHCH₂—C₆H₅ (CH₃)

Hydrochloride, $C_{23}H_{31}ClN_2O$, crystals from isopropanol, mp 272.8-273.6°.

Note: "China White" was first thought to be *3-methylfentanyl*, see S. Stinson, *Chem. & Eng. News* **59**, 71 (Jan. 19, 1981).

Caution: This is a controlled substance (opiate) listed in the U.S. Code of Federal Regulations, Title 21 Part 1308.11 (1987).

5993. Methyl Fluorosulfonate. *Fluorosulfuric acid methyl ester;* methyl fluorosulfate; methyl fluosulfonate; Magic Methyl. CH_3FO_3S; mol wt 114.09. C 10.53%, H 2.65%, F 16.65%, O 42.07%, S 28.10%. CH_3OSO_2F. Prepn from dimethyl ether and fluosulfonic acid: J. Meyer, G. Schramm, *Z. Anorg. Allgem. Chem.* **206**, 24 (1932); from dimethyl sulfate and fluosulfonic acid: R. W. Alder, *Chem. & Ind. (London)* **1973**, 983. Electrochemical prepn: J. P. Coleman, D. Pletcher, *Tetrahedron Letters* **1974**, 147. Powerful methylating agent: M. G. Ahmed *et al.*, *Chem. Commun.* **1968**, 1533. Extremely toxic to humans: D. M. W. vanden Ham, D. van der Meer, *Chem. & Eng. News* **54**, 5 (Aug. 30, 1976); *eidem*, *Chem. & Ind. (London)* **1976**, 782. Volatile liq, bp 92-94°, mp $-95°$. *Severe Poison!* d 1.412. n_D^{20} 1.3326. Good solvent for most organic compounds. Proton NMR absorption at tau 5.88. LD_{50} orally in mice: <112 mg/kg; LC_{50} 1 hr for rats: 5-6 ppm, M. Hite *et al.*, *Am. Ind. Hyg. Assoc. J.* **40**, 601 (1979).

Caution: Exposure can cause fatal pulmonary edema.

USE: In organic synthesis as methylating agent.

5994. Methyl Formate. *Formic acid methyl ester.* $C_2H_4O_2$; mol wt 60.05. C 40.00%, H 6.71%, O 53.29%. HCOOCH₃.

Colorless flammable liquid, agreeable odor. d_{15}^{15} 0.987. bp 31.5°. n_D^{20} 1.3440. Flash pt, closed cup: $-2°F$ ($-19°C$). mp

$\sim -100°$. Sol in about 3.3 parts water; miscible with alcohol.

USE: Fumigant and larvicide for tobacco and food crops. Fire hazard is avoided by use with CO_2. *Caution:* Inhalation of vapor produces nasal and conjunctival irritation, retching, narcosis, death from pulmonary effects, Patty's *Industrial Hygiene and Toxicology* vol. **2A**, G. D. Clayton, F. E. Clayton, Eds. (Wiley-Interscience, New York, 3rd ed., 1981) p 2263.

5995. N-Methylglucamine. *1-Deoxy-1-(methylamino)-D-glucitol;* *N*-methyl-D-glucamine; meglumine. $C_7H_{17}NO_5$; mol wt 195.22. C 43.06%, H 8.78%, N 7.18%, O 40.98%. Prepd from D-glucose and methylamine: Karrer, Herkenrath, *Helv. Chim. Acta* **20**, 83 (1937). Efficacy and toxicity in treatment of leishmaniasis: R. G. Muller *et al.*, *Arch. Inst. Cardiol. Mex.* **52**, 155 (1982); P. Bouree *et al.*, *Pathol. Biol.* **33**, 607 (1985). Antileishmanial activity in dogs: W. L. Chapman *et al.*, *Am. J. Vet. Res.* **45**, 1028 (1984).

CH₂NHCH₃
|
HCOH
|
HOCH
|
HCOH
|
HCOH
|
CH₂OH

Crystals from methanol, mp 128-129°. Does not polymerize or dehydrate unless heated above 150° for prolonged periods. $[\alpha]_D^{18}$ $-18.5°$ (Karrer); $[\alpha]_D^{20}$ $-23°$ (Rhône-Poulenc data sheet). Soly (g/100 ml): water at 25°: ~ 100; alcohol at 25°: 1.2; alcohol at 70°: 21. pH of 1% aq soln: 10.5. Forms salts with acids and complexes with metals. Salts with alkyl aryl sulfonic acids act as detergents.

Antimonate, $C_7H_{18}NO_8Sb$, *RP 2168, Glucantim, Glucantime, Protostib.* Powder. Soly in water about 35% ww. Practically insol in alcohol, ether, chloroform. pH of aq solns 6-7.

USE: In the synthesis of surface active agents, pharmaceuticals, dyes. *Human Toxicity:* Mild irritant.

THERAP CAT: Antimonate as antiprotozoal (Leishmania).

5996. N-Methyl-α-L-glucosamine. *2-(Methylamino)-2-deoxy-α-L-glucopyranose.* $C_7H_{15}NO_5$; mol wt 193.20. C 43.50%, H 7.83%, N 7.25%, O 41.41%. Together with streptose forms the streptobiosamine moiety of streptomycin: Kuehl *et al.*, *J. Am. Chem. Soc.* **69**, 3032 (1947). Prepn from D-glucose by *Streptomyces griseus:* Silverman, Rieder, *J. Biol. Chem.* **235**, 1251 (1960); from the antibiotic, bluensomycin: Bannister, Argoudelis, *J. Am. Chem. Soc.* **85**, 34 (1963). *Review:* Lemieux, Wolfrom, *Advan. Carbohyd. Chem.* **3**, 337 (1948).

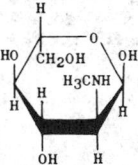

Glass. $[\alpha]_D^{25}$ $-62°$ (c = 1 in methanol).

Hydrochloride, $C_7H_{15}NO_5$·HCl, needles from ethanol, mp 160-163°. Freely sol in water. Shows mutarotation. $[\alpha]_D^{25}$ $-103°$ → $-88°$ (c = 0.6).

N-acetyl deriv, mp 165-166°. $[\alpha]_D^{25}$ $-51°$ (c = 0.4).

Pentaacetyl deriv, $C_{17}H_{25}NO_{10}$, mp 160.5-161.5°. $[\alpha]_D^{25}$ $-100°$ (c = 0.7 in chloroform).

5997. α-Methylglucoside. *Methyl-α-D-glucopyranoside.* $C_7H_{14}O_6$; mol wt 194.18. C 43.30%, H 7.27%, O 49.44%. Prepd by refluxing finely powdered glucose with methanol-HCl: Fischer, *Ber.* **26**, 2405; **27**, 2987; **28**, 1151 (1895); Helferich, Schäfer, *Org. Syn.* **coll. vol. I**, (2nd ed., 1941) 364. Enzymatic synthesis by means of α-glucosidase from yeast: Bourquelot *et al.*, *Compt. Rend.* **156**, 491 (1913). Prepd industrially by reacting glucose with methanol in the presence of a cation exchange material: *Chem. Eng. News* **33**, 4592 (1955). Monograph: G. N. Bollenback, *Methyl Glucoside, Preparation, Physical Constants, Derivatives* (Academic Press, New York, 1958).

Orthorhombic bisphenoidal crystals, d_4^{30} 1.46. mp 168°. $bp_{0.2}$ 200°. $[\alpha]_D^{20}$ +158.9° (p = 10). Ka at 25° = 1.97 × 10^{-14}. Soly at 17° in water 63% (w/w); in 80% alcohol 7.3%; in 90% alcohol 1.6%; practically insol in ether. Soly also reported as 108 g/100 g H_2O at 20° and as 5.2 g/100 g methanol at 20°.

USE: Manuf reconstituted and upgraded drying oils; tall oil esters and varnishes; fatty acid esters; plasticizers; nonionic surface active agents; fast and hard drying alkyd resins, and so-called plasticizing alkyds. In making glucoside hydroxypropyl ethers for polyurethan foam production.

5998. Methyl Green. *4-[[4-(Dimethylamino)phenyl][4-(dimethylimino)-2,5-cyclohexadien-1-ylidene]methyl]-N-ethyl-N,N-dimethylbenzenaminium bromide chloride; [4-[p-(dimethylamino)phenyl]-α-[p-(dimethylamino)phenyl]benzylidene]-2,5-cyclohexadien-1-ylidene]dimethylammonium chloride ethobromide;* C.I. 42590. $C_{27}H_{35}BrClN_3$; mol wt 516.98. C 62.73%, H 6.82%, Br 15.46%, Cl 6.86%, N 8.13%. Usually sold as the double salt with $ZnCl_2$. Prepn: *Colour Index* vol. **4** (3rd ed., 1971) p 4394.

Green powder. Sol in water. Yellow soln in conc H_2SO_4, turning green on dilution.

Note: The term methyl green also applies to the trimethyl analog.

USE: Dyeing and printing textiles; as biological stain.

5999. Methylguanidine. $C_2H_7N_3$; mol wt 73.10. C 32.86%, H 9.65%, N 57.48%. A product of putrefaction. Prepn: Erlenmeyer, *Ber.* **3**, 896 (1870); Werner, Bell, *J. Chem. Soc.* **121**, 1793 (1922); Phillip, Clarke, *J. Am. Chem. Soc.* **45**, 1755 (1923); Philippi, Morsch, *Ber.* **60**, 2120 (1927); Davis, Rosenquist, *J. Am. Chem. Soc.* **59**, 2114 (1937).

Colorless, deliquesc, strongly alkaline mass. *Poisonous!* Very sol in water; sol in alcohol. Reduces permanganate. MLD s.c. in rats: 250 mg/kg.

Sulfate, $2C_2H_7N_3.H_2SO_4$, crystals, mp 239-240°.
Nitrate, $C_2H_7N_3.HNO_3$, crystals, mp 149-150°.

6000. Methylhexaneamine. *4-Methyl-2-hexanamine;* 2-amino-4-methylhexane; 1,3-dimethylamylamine; Forthane; $C_7H_{17}N$; mol wt 115.21. C 72.97%, H 14.87%, N 12.16%. Prepn: U.S. pats. **2,350,318; 2,386,273** (1944, 1945, both to Lilly).

Liquid, amine odor. d 0.7620-0.7655. n_D^{25} 1.4150-1.4175.

bp_{760} 130-135°. Very slightly sol in water. Freely sol in alc, chloroform, ether, dil acids. LD_{50} i.p. in mice: 185 mg/kg.

THERAP CAT: Adrenergic.

6001. Methylhydrazine. Monomethylhydrazine; MMH. CH_6N_2; mol wt 46.07. C 26.07%, H 13.13%, N 60.80%. CH_3NHNH_2. Early prepns: *Beilstein* **4**, 546; 1st suppl., 560. Prepn of sulfate: Hatt, *Org. Syn.* **coll. vol. I**, 395 (1943); Audrieth, Diamond, *J. Am. Chem. Soc.* **76**, 4869 (1954). Manuf and properties: Knight, *Hydrocarbon Process. Petrol. Refiner* **41**, 179 (1962). Toxicity and metabolism: Witkin, *Arch. Ind. Health* **13**, 34 (1956); Dost *et al., Biochem. Pharmacol.* **15**, 1325 (1966); Gregory *et al., Clin. Toxicol.* **4**, 435 (1971); Magee *et al.* in *Proceedings of the Fifth International Congress on Pharmacology, San Francisco, 1972*, vol. 2, T. A. Loomis, Ed. (Karger, New York, 1973) pp 140-149.

Clear liquid, odor characteristic of short chain, organic amines. d^{25} 0.874. mp −52.4°. bp 87.5°. Heat capacity (25°): 32.25 cal/mole/°C. Ignition temp: 196°. Flash pt (Cleveland open cup): 70°. Flammability limits in air (%-vol): 2.5 to 97 ± 2%. Miscible with water, hydrazine, low mol wt monohydric alcohols. Sol in hydrocarbons. Mildly alkaline base. Strong reducing agent. Ignites spontaneously on contact with strong oxidizing agents such as fluorine, chlorine trifluoride, nitrogen tetroxide, fuming nitric acid. LD_{50} orally in mice, rats: 33.0, 32.5 mg/kg (Witkin); orally in rats: 70.7 mg/kg (Gregory).

Sulfate, $CH_3NHNH_2.H_2SO_4$, white plates from 80% ethanol. mp 141-142°.

USE: In rocket fuel; intermediate in chemical syntheses.

6002. Methyl Iodide. Iodomethane. CH_3I; mol wt 141.95. C 8.46%, H 2.13%, I 89.41%. Prepd from methyl alcohol, iodine and red phosphorus: King, *Org. Syn.* **coll. vol. II**, 399 (1943); from potassium iodide and methyl sulfate: Hartman, *ibid.* 404. Toxicity: Irish in *Industrial Hygiene and Toxicology* **vol. 2**, F.A. Patty, Ed. (Interscience, New York, 2nd ed., 1962) pp 1255-1256; Appel *et al., Ann. Intern. Med.* **82**, 534 (1975); Kutob, Plaa, *Toxicol. Appl. Pharmacol.* **4**, 354 (1962).

Colorless, transparent liquid; turns brown on exposure to light. *Poisonous!* d_4^{20} 2.28. mp −66.5°. bp 42.5°; dec at 270°. n_D^{21} 1.5293. Sol in about 50 parts water; miscible with alcohol, ether. *Protect from light.* LD_{50} s.c. in mice: 0.78 mmoles/kg (Kutob, Plaa).

Caution: Massive exposure has led to pulmonary edema; prolonged or repeated exposure may cause CNS effects; prolonged contact may cause skin burns: *Clinical Toxicology of Commercial Products*, R. E. Gosselin *et al.,* Eds. (Williams & Wilkins, Baltimore, 5th ed., 1984) Section II, pp 158-159. This substance may reasonably be anticipated to be a carcinogen: *Fourth Annual Report on Carcinogens* (NTP 85-002, 1985) p 132.

USE: In methylations; in microscopy because of its high refractive index; as imbedding material for examining diatoms; also in testing for pyridine.

6003. Methyl Isobutyrate. *2-Methylpropanoic acid methyl ester.* $C_5H_{10}O_2$; mol wt 102.13. C 58.80%, H 9.87%, O 31.33%. $(CH_3)_2CHCOOCH_3$.

Colorless, mobile liquid. d_D^{20} 0.891. mp −84° to −85°. bp 93°. n_D^{20} 1.3840. Slightly sol in water; miscible with alcohol, ether.

6004. Methyl Isocyanate. *Isocyanatomethane; isocyanic acid methyl ester;* MIC. C_2H_3NO; mol wt 57.05. C 42.10%, H 5.29%, N 24.56%, O 28.05%. $CH_3N=C=O$. Intermediate in the manufacture of insecticides and herbicides including carbaryl, *q.v.* Prepn via Curtius rearrangement: J. W. Boehmer, *Rec. Trav. Chim.* **55**, 379 (1933); by heating *N,N*-diphenyl-*N'*-methylurea: W. Siefken, *Ann.* **562**, 75 (1949); by phosgenation of bis(trimethylsilyl)methylamine: V. F. Mironov *et al., Zh. Obshch. Khim.* **39**, 2598 (1969). Used in prepn of α-aryl-β-methylureas: J. W. Boehmer, *loc. cit.*; of semicarbazides: Ch. C. P. Pacilly, *Rec. Trav. Chim.* **55**, 101 (1936); in conversion of aldoximes to nitriles: J. A. Albright, M. L. Alexander, *Org. Prep. Proced. Int.* **4**, 215 (1972). Toxicity of vapor to rats, mice, humans; physical properties: G. Kimmerle, A. Eben, *Arch. Toxicol.* **20**, 235 (1960). Uptake and distribution studies in animals: J. S.

Ferguson *et al.*, *Toxicol. Appl. Pharmacol.* **94**, 104 (1988). Acute oral toxicity: E. H. Vernot *et al.*, *ibid.* **42**, 417 (1977).

Liquid, bp 39-40°. d^{20} 0.96. Vapor press. at 4.2°: 200 torr; at 13.5°: 300 torr; at 20.6°: 400 torr; at 31.2°: 600 torr. LD_{50} in male rats (mg/kg): 140 single oral dose (Vernot). LC_{50} in rats (4 hours exposure to vapor): 5 ppm (Kimmerle, Eben).

Note: An industrial accident during the manufacture of carbaryl in Bhopal, India on December 3, 1984 resulted in the leakage of an unknown amount of methyl isocyanate into the air. Over 2,000 people died and an estimated 200,000 were exposed to the vapor: *Chem. & Eng. News* **62**, 6 (Dec. 10, 1984); *ibid.* **63**, 14 (Feb. 11, 1985). Series of articles on follow-up studies on survivors: *Indian J. Exp. Biol.* **26**, 149-176, 201-204 (1988).

Caution: Highly volatile. Exposure to 2 ppm for 1-5 min produced tears and irritation of the nose and throat (Kimmerle, Eben).

USE: In organic synthesis; in manuf of carbamate pesticides.

6005. Methyl Isothiocyanate. *Isothiocyanatomethane;* methyl mustard oil; Trapex. C_2H_3NS; mol wt 73.12. C 32.85%, H 4.14%, N 19.16%, S 43.86%. $CH_3N=CS$. Obtained by the action of CS_2 on methylamine.

Crystals, mp 35-36°. bp 119°. n_D^{37} 1.5258. Slightly sol in water; freely sol in alcohol, ether. LD_{50} orally in rats: 305 mg/kg, *Toxic Substances List*, H. E. Christensen, Ed. (1974) p 477.

USE: Pesticide. *Caution:* Highly irritating.

6006. Methyl Isovalerate. *3-Methylbutanoic acid methyl ester.* $C_6H_{12}O_2$; mol wt 116.16. C 62.04%, H 10.41%, O 27.55%. $(CH_3)_2CHCH_2COOCH_3$.

Liquid; valerian odor. d_4^{20} 0.881. bp 116-117°. Slightly sol in water; miscible with alcohol, ether.

6007. Methyl Lactate. *2-Hydroxypropanoic acid methyl ester.* $C_4H_8O_3$; mol wt 104.10. C 46.15%, H 7.75%, O 46.11%. $CH_3CH(OH)COOCH_3$. Prepd by heating 1 mole lactic acid condensation polymer with 2.5-5 moles of methanol and a small quantity of H_2SO_4 at 100° for 1-4 hours in a heavy-walled bottle: Rehberg, *Org. Syn.* coll. vol. III, 47-48 (1955).

Colorless, transparent liquid. d^{19} 1.09. bp 144-145°. n_D^{16} 1.4156. Soluble in alcohol, ether; dec by water.

USE: Cellulose acetate solvent. *Caution:* Mild irritant.

6008. Methyl Linoleate. *9,12-Octadecadienoic acid methyl ester.* $C_{19}H_{34}O_2$; mol wt 294.46. C 77.49%, H 11.64%, O 10.87%. $CH_3(CH_2)_4CH=CHCH_2CH=CH-(CH_2)_7COOCH_3$. Prepn from safflower-seed oil: Parker *et al.*, *Biochem. Prepn.* **4**, 88 (1955).

Colorless oil. d_4^{18} 0.8886. mp -35°. bp_{16} 212°; bp_4 192°. n_D^{25} 1.4593. Iodine value 172.4. More stable to air oxidation than linoleic acid. Miscible with dimethylformamide, fat solvents, oils.

USE: In the vitamin industry.

6009. Methyl Malonate. *Propanedioic acid dimethyl ester;* dimethyl malonate. $C_5H_8O_4$; mol wt 132.11. C 45.45%, H 6.10%, O 48.44%. $CH_3OOCCH_2COOCH_3$.

Liquid. d_4^{20} 1.154. mp -62°. bp 180-181°. n_D^{17} 1.4149. Slightly sol in water; miscible with alcohol, ether, oils.

6010. Methyl Methanesulfonate. *Methanesulfonic acid methyl ester;* methyl methanesulfonic acid; *as*-dimethyl sulfite; methyl mesylate; MMS. $C_2H_6O_3S$; mol wt 110.13. C 21.81%, H 5.49%, O 43.58%, S 29.11%. $CH_3SO_2OCH_3$. Prepn: A. Arbusow, P. Pischtschimuka, *Chem. Zentr.* **II**, 685 (1909); W. E. Bissenger *et al.*, *J. Am. Chem. Soc.* **70**, 3940 (1948). Metabolism: E. A. Barnsley, *Biochem. J.* **106**, 18p (1968). Mutagenicity studies: U. H. Ehling *et al.*, *Mutat. Res.* **5**, 417 (1968); J. Moutschen, *ibid.* **8**, 581 (1969). Review of carcinogenicity studies: *IARC Monographs* **7**, 253-260 (1974). Review of comparative mutagenicity of MMS and ethyl methanesulfonate, *q.v.*: S. Kondo, *Environ. Sci. Res.* **24**, 743-785 (1981).

Liquid, bp_{753} 203°, $bp_{0.6}$ 59°. d_4^{20} 1.2943; n_D^{20} 1.4140. Soly in water about 1:5, in DMF about 1:1. Slightly sol in nonpolar solvents.

USE: Exptlly as mutagen.

6011. 3-Methyl-6,7-methylenedioxy-1-piperonylisoquinoline. *5-(1,3-Benzodioxol-5-ylmethyl)-7-methyl-1,3-dioxolo[4,5-g]isoquinoline.* $C_{19}H_{15}NO_4$; mol wt 321.32. C 71.02%, H 4.71%, N 4.36%, O 19.92%. Papaverine homolog made by synthesis: *Merck's Jahresbericht* **44**, 15 (1930).

Crystals from methanol, mp 141°. Almost insol in water; sol in benzene, chloroform, in hot alcohol, hot methanol, hot ether. A weak base.

Hydrochloride monohydrate, crystals from water, mp about 158°. Anhydr from abs alcohol, dec 253°. One gram dissolves in 100 ml water at 25°; freely sol in hot water; moderately sol in 95% alcohol. Aq solns are acid to litmus.

Note: This drug was formerly known as *Eupaverin, Syntaverin*.

6012. Methyl *N*-Methylnipecotate. *1-Methyl-3-piperidinecarboxylic acid methyl ester; N-methylnipecotic acid methyl ester;* dihydroarecoline. $C_8H_{15}NO_2$; mol wt 157.21. C 61.12%, H 9.62%, N 8.91%, O 20.35%. Prepn: Stoll *et al.*, *Helv. Chim. Acta* **33**, 375 (1950); Sperber *et al.*, *J. Am. Chem. Soc.* **81**, 704 (1959). Electron distribution and dissociation constants: E. Mutschler *et al.*, *Arzneimittel-Forsch.* **17**, 833 (1967). Pharmacologic studies: O. Nieschulz, *ibid.* **22**, 804 (1972). Structure-conformation activity study: G. Lambrecht, E. Mutschler, *ibid.* **24**, 1725 (1974).

Oil, bp_{12} 83-84°. n_D^{25} 1.4520. pKa (20°): 8.66.
Methobromide, $C_9H_{18}BrNO_2$, *Adipsin*.

THERAP CAT: Sialagogue.

6013. *N*-Methylmyosmine. *3-(4,5-Dihydro-1-methyl-1H-pyrrol-2-yl)pyridine; N-methyl-2-(3-pyridyl)-2-pyrroline; 4,5-dihydro-β-nicotyrine.* $C_{10}H_{12}N_2$; mol wt 160.21. C 74.96%, H 7.55%, N 17.49%. Preparation: Korte, Schulze-Steinen, *Ber.* **95**, 2444 (1962).

Light yellow oil, $bp_{0.1}$ 60°. uv max (methanol): 212, 243 nm (log ε 3.70, 3.92).

Picrate, $C_{10}H_{12}N_2 \cdot C_6H_3N_3O_7$, yellow needles from water, mp 156°.

6014. Methyl Nicotinate. *3-Pyridinecarboxylic acid methyl ester;* nicotinic acid methyl ester; Midalgan. $C_7H_7NO_2$; mol wt 137.13. C 61.31%, H 5.15%, N 10.21%, O 23.33%. Obtained by passing HCl gas into a hot methanol soln of nicotinic acid: Engler, *Ber.* **27**, 1787 (1894); by treatment with tetramethylammonium hydroxide in methanol: Prelog, Piantanida, *Z. Physiol. Chem.* **244**, 56 (1936); by heating a mixture of sulfuric acid, selenium and quinoline, followed by refluxing with methanol: Kaufman, *J. Am. Chem. Soc.* **67**, 497 (1945); from nicotinyl chloride hydrochloride and methanol in pyridine: Charonnat *et al.*, *Bull. Soc. Chim. France* **1948**, 1014.

Crystals, mp 39°. bp$_{760}$ 209°; bp$_3$ 70-72°. Soluble in water, alcohol, benzene.

THERAP CAT: Rubefacient.

6015. Methyl Nitrate. *Nitric acid methyl ester.* CH$_3$-NO$_3$; mol wt 77.04. C 15.59%, H 3.92%, N 18.18%, O 62.30%. CH$_3$ONO$_2$. Prepn from methanol and nitric acid in the presence of sulfuric acid: Black, Babers, *Org. Syn. coll. vol.* **II**, 412 (1943).

Liquid. Solid at −83°. bp$_{760}$ 64.6° (explodes). d$_4^{20}$ 1.2075. n$_D^{20}$ 1.3748. Slightly sol in water. Sol in alcohol, ether.

USE: Has been used as rocket propellant. Does not need external oxygen for combustion.

6016. Methyl *p*-Nitrobenzenesulfonate. *p-Nitrobenzenesulfonic acid methyl ester.* C$_7$H$_7$NO$_5$S; mol wt 217.20. C 38.71%, H 3.25%, N 6.45%, O 36.83%, S 14.76%. Reagent for selective methylation of cysteine residues in chemical modification of proteins: Nakagawa, Bender, *J. Am. Chem. Soc.* **91**, 1566 (1969). Procedure: Heinrikson, *Biochem. Biophys. Res. Commun.* **41**, 967 (1970). Prepd by the general method of reacting *p*-nitrobenzenesulfonyl chloride with alcohols. See Morgan, Cretcher, *J. Am. Chem. Soc.* **70**, 375 (1948).

White to cream crystals, mp 93-95°.

USE: Methylating agent.

6017. N-Methyl-N'-nitro-N-nitrosoguanidine. *N-Methyl-N-nitroso-N'-nitroguanidine;* MNNG. C$_2$H$_5$N$_5$O$_3$; mol wt 147.10. C 16.33%, H 3.43%, N 47.61%, O 32.63%. Prepn: A. F. McKay, G. F. Wright, *J. Am. Chem. Soc.* **69**, 3028 (1947). Reviews of carcinogenicity and mutagenicity: *IARC Monographs* **4**, 183-195 (1974); U. Sinha, B. B. Chattoo, *J. Sci. Ind. Res.* **3**, 499-505 (1975).

Yellow crystals from methanol, mp 118° (dec). Reacts with aq KOH to form diazomethane: A. F. McKay, *J. Am. Chem. Soc.* **70**, 1974 (1948). Reacts at acid pH to give methylnitroguanidine.

USE: Experimentally as carcinogen and mutagen. Formerly in prepn of diazomethane.

6018. Methylol Riboflavine. Hyflavin. Mixture of methylol (CH$_2$OH) derivatives of riboflavine formed by the action of formaldehyde on riboflavine in weakly alkaline soln. The number of methylol groups in the ribityl moiety varies from 1 to 3. Prepn: Schoen, Gordon, *Arch. Biochem.* **22**, 149 (1949); U.S. pat. **2,587,533** (1952 to Endo Prod.).

X = H or CH$_2$OH

Orange to yellow, hygroscopic powder. May have a slight odor of formaldehyde. Sol in water. Practically insol in alcohol, benzene, chloroform, ether. Dextrorotatory. The pH of a 10% aq soln is between 6.7 and 7.9. The dry powder is unstable and loses its biological activity in the course of several months with the liberation of formaldehyde and the partial formation of products practically insol in water.

THERAP CAT: Enzyme co-factor vitamin source.

6019. Methyl Orange. *4-[[(4-Dimethylamino)phenyl]-azo]benzenesulfonic acid sodium salt;* sodium *p*-dimethylaminoazobenzenesulfonate; helianthine B; C.I. Acid Orange 52; C.I. 13025; Orange III; Gold Orange; Tropaeolin D. C$_{14}$H$_{14}$N$_3$NaO$_3$S; mol wt 327.33. C 51.37%, H 4.31%, N 12.84%, Na 7.02%, O 14.66%, S 9.79%. Prepn from sulfanilic acid sodium nitrite + dimethylaniline: L. Gattermann, *Die Praxis des organischen Chemikers* (de Gruyter, Berlin, 40th ed., 1961) pp 260-261. *See also Colour Index* vol. **4** (3rd ed., 1971) p 4043.

Orange-yellow powder or cryst scales. Sol in 500 parts water; more sol in hot water; practically insol in alcohol.

USE: As indicator in 0.1% aq soln. pH: 3.1 red, 4.4 yellow. Employed for titrating most mineral acids, strong bases, estimating alkalinity of waters; useless for organic acids. In dyeing and printing of textiles.

6020. Methyl Oxalate. *Ethanedioic acid dimethyl ester;* dimethyl oxalate. C$_4$H$_6$O$_4$; mol wt 118.09. C 40.68%, H 5.12%, O 54.20%. CH$_3$OOCCOOCH$_3$.

Colorless crystals. d^{54} 1.148. mp 54°. bp 163-164°. n$_D^{82.1}$ 1.379. Sol in 17 parts water, in alcohol, ether.

6021. Methylparaben. *4-Hydroxybenzoic acid methyl ester;* methyl *p*-hydroxybenzoate; Nipagin M; Tegosept M; Methyl Chemosept; Methyl Parasept. C$_8$H$_8$O$_3$; mol wt 152.14. C 63.15%, H 5.30%, O 31.55%. Prepn: Ladenburg, Fitz, *Ann.* **141**, 247 (1867); Zbarskii, *C.A.* **33**, 9312^3 (1939). Identification in the vaginal secretions of female dogs in estrus: M. Goodwin *et al.*, *Science* **203**, 559 (1979).

White needles, mp 131°. bp 270-280° (dec). One gram dissolves in 400 ml water, 40 ml warm oil, about 70 ml warm glycerol; freely sol in alcohol, acetone, ether. The soly in water is also given as 0.25% w/w at 20°, and as 0.30% w/w at 25°.

USE: As preservative in foods, beverages and cosmetics.

6022. Methyl Parathion. *Phosphorothioic acid O,O-dimethyl O-(4-nitrophenyl) ester;* O,O-dimethyl O-*p*-nitrophenyl phosphorothioate; O,O-dimethyl O-*p*-nitrophenyl thiophosphate; dimethyl parathion; parathion-methyl; metaphos; E 601; ENT-17292; Dalf (obsolete); Folidol-M; Metacide; Metron; Nitrox 80; Penncap M. C$_8$H$_{10}$NO$_5$PS; mol wt 263.23. C 36.50%, H 3.83%, N 5.32%, O 30.39%, P 11.77%, S 12.18%. Prepn: Fletcher *et al.*, *J. Am. Chem. Soc.* **72**, 2461 (1950). Manuf: Faith, Keyes & Clark's *Industrial Chemicals*, F. A. Lowenheim, M. K. Moran, Eds. (Wiley-Interscience, New York, 4th ed., 1975) pp 552-555. Toxicity data: T. B. Gaines, *Toxicol. Appl. Pharmacol.* **14**, 515 (1969).

Crystals from cold methanol, mp 37-38°. n$_D^{25}$ 1.5367. d$_4^{20}$ 1.358. Soly in water 50 ppm; sol in most organic solvents.

LD$_{50}$ in male, female rats (mg/kg): 14, 24 orally; 67, 67 dermally (Gaines).
USE: Insecticide. *Caution:* Cholinesterase inhibitor.

6023. 5-Methyl-5-(3-phenanthryl)hydantoin. *5-Methyl-5-(3-phenanthrenyl)-2,4-imidazolidinedione;* 5-(3'-phenanthryl)-5-methylhydantoin. C$_{18}$H$_{14}$N$_2$O$_2$; mol wt 290.31. C 74.47%, H 4.86%, N 9.65%, O 11.02%. Prepn: Nitz *et al., Arzneimittel-Forsch.* **5,** 337 (1955); **Brit. pat.** **774,394** (1957 to Cassella Farbw.).

Crystals from ethanol, mp 236-237°.
Sodium salt, *Bagrosin-Natrium.*
THERAP CAT: Anticonvulsant.

6024. N-Methylphenazonium Methosulfate. *5-Methylphenazinium methyl sulfate;* phenazine methosulfate. C$_{14}$H$_{14}$N$_2$O$_4$S; mol wt 306.34. C 54.89%, H 4.61%, N 9.15%, O 20.89%, S 10.47%. Prepn: Kehrmann, *Ber.* **46,** 341 (1913); Hillemann, *ibid.* **71,** 34 (1938); Dickens, McIlwain, *Biochem. J.* **32,** 1615 (1938). Mutagenicity study: S. Venitt, C. Crofton-Sleigh, *Mutat Res.* **68,** 107 (1979).

Flat yellow to brown parallelepipeds from alc, mp 155-157° (Hillemann); mp 167° (Dickens). Oxidation-reduction potential at 30° and pH 7: E$_h$ = +0.080 v.
USE: As an electron carrier in place of the flavine enzyme of Warburg in the hexosemonophosphate system: Dickens, *loc. cit.* In the prepn of succinic dehydrogenase: Green *et al., J. Biol. Chem.* **217,** 551 (1955).

6025. Methylphenidate. *α-Phenyl-2-piperidineacetic acid methyl ester;* methyl phenidylacetate; α-phenyl-α-(2-piperidyl)acetic acid methyl ester; methyl α-phenyl-α-(2-piperidyl)acetate; methylphenidan; 4311/b Ciba; Ritalin; Centedrin; Phenidylate. C$_{14}$H$_{19}$NO$_2$; mol wt 233.30. C 72.07%, H 8.21%, N 6.00%, O 13.72%. Prepn: Panizzon, *Helv. Chim. Acta* **27,** 1748 (1944). Separation of isomers: Rometsch, **U.S. pat.** **2,957,880** (1960 to Ciba). Toxicity data: E. N. Greenblatt, A. C. Osterberg, *J. Pharmacol. Exp. Ther.* **131,** 115 (1961). Comprehensive description: G. R. Padmanabhan in *Analytical Profiles of Drug Substances* vol. **10,** K. Florey, Ed. (Academic Press, New York, 1981) pp 473-497.

Crystals from 50% alc, mp 74-75°. Sol in alcohol, ethyl acetate, ether. Practically insol in water, petr ether.
Hydrochloride, C$_{14}$H$_{20}$ClNO$_2$, crystals, mp 224-226°. pKa 8.9. Sol in water, alc, chloroform. A 5% aq soln is neutral to litmus. LD$_{50}$ orally in mice: 190 mg/kg (Greenblatt, Osterberg).
Note: This is a controlled substance (stimulant) listed in the U.S. Code of Federal Regulations, Title 21 Part 1308.12 (1987).
THERAP CAT: CNS stimulant.

6026. 3-Methyl-5-phenylhydantoin. *3-Methyl-5-phen-*

yl-2,4-imidazolidinedione; Norantoin; Nuvarone. C$_{10}$H$_{10}$N$_2$O$_2$; mol wt 190.20. C 63.15%, H 5.30%, N 14.73%, O 16.82%. Prepn: Pinner, *Ber.* **21,** 2325 (1888); Kjaer, *Acta Chem. Scand.* **4,** 892 (1950); Klosa, *Arch. Pharm.* **285,** 274 (1952).

Fine needles from aq ethanol, mp 162-163°.
THERAP CAT: Anticonvulsant.

6027. 3'-Methylphthalanilic Acid. *2-[[(3-Methylphenyl)amino]carbonyl]benzoic acid;* N-m-tolylphthalamic acid; Duraset; Duraset 20W; Tomaset. C$_{15}$H$_{13}$NO$_3$; mol wt 255.26. C 70.58%, H 5.13%, N 5.49%, O 18.80%. Preparation: Smith, Hoffmann, **U.S. pat.** **2,556,665** (1951 to U.S. Rubber).

Crystals, mp 149-151°. Slightly sol in water; sol in polar solvents with rapid decompn to the imide. LD$_{50}$ orally in rats: 5230 mg/kg.
USE: As a fruit set, *i.e.* to prevent premature drop of apples, pears, cherries, peaches, and of certain vegetable crops, such as tomatoes and peppers. Antishock treatment for plants. *Caution:* On ingestion is hydrolyzed to produce *m*-toluidine. This may cause kidney injury, anemia, methemoglobinemia.

6028. Methylprednisolone. *11,17,21-Trihydroxy-6-methyl-1,4-pregnadiene-3,20-dione;* 1-dehydro-6α-methylhydrocortisone; Δ^1-6α-methylhydrocortisone; 6α-methyl-11β,17α,21-triol-1,4-pregnadiene-3,20-dione; Artisone-Wyeth; Medrate; Medrol; Medrone; Metastab; Metrisone; Promacortine; Suprametil; Urbason. C$_{22}$H$_{30}$O$_5$; mol wt 374.46. C 70.56%, H 8.08%, O 21.36%. Prepn: Spero *et al., J. Am. Chem. Soc.* **78,** 6213 (1956); Fried, *ibid.* **81,** 1235 (1959); Sebek, Spero, **U.S. pat.** **2,897,218** (1959 to Upjohn); Gould, **U.S. pat.** **3,053,832** (1962 to Schering).

Crystals, mp 228-237°. [α]$_D^{20}$ +83° (dioxane). uv max (95% ethanol): 243 nm (α$_M$ 14,875).
21-Acetate, *Depo-Medrate, Depo-Medrol, Depo-Medrone, Mepred.* Crystals, mp 205-208°. [α]$_D^{20}$ +101° (dioxane). uv max (95% ethanol): 243 nm (α$_M$ 14,825). Practically insol in water.
21-Phosphate disodium salt, C$_{22}$H$_{29}$Na$_2$O$_8$P, *Medrol Stabisol.*
21-Succinate sodium salt, C$_{26}$H$_{33}$NaO$_8$, *Urbason-Solubile, Solu-Medrol.* Prepn: Sebek, Spero, *loc. cit.* Sol in water; forms buffered soln suitable for i.v. injection.
THERAP CAT: Glucocorticoid.
THERAP CAT (VET): Glucocorticoid.

6029. p-(2-Methylpropenyl)phenol Acetate. *p-Acetoxyphenylisobutene;* p-isobutenylphenol acetate; AO-12; Isotyl. C$_{12}$H$_{14}$O$_2$; mol wt 190.23. C 75.76%, H 7.42%, O 16.82%. Prepn: v. Braun, *Ann.* **472,** 71 (1929).

Liquid, odor of anise, bp$_{15}$ 148°.

6030. Methyl Propionate. *Propanoic acid methyl ester.* C$_4$H$_8$O$_2$; mol wt 88.10. C 54.53%, H 9.15%, O 36.32%. CH$_3$CH$_2$COOCH$_3$.

Colorless liquid. d$_4^{20}$ 0.915. mp −87°. bp 79.7°. Flash pt −2°. n$_D^{19}$ 1.3769. Sol in 16 parts water; miscible with alcohol, ether.

USE: In organic synthesis.

6031. Methyl Propyl Ether. *1-Methoxypropane;* Neothyl. C$_4$H$_{10}$O; mol wt 74.12. C 64.81%, H 13.60%, O 21.59%. CH$_3$CH$_2$CH$_2$OCH$_3$. Prepd by passing a mixture of methyl and propyl alcohol through a hot layer of β-naphthalenesulfonic or benzenesulfonic acids: Krafft, *Ber.* **26**, 2832 (1893); **Ger. pat. 69,115;** *Frdl.* **III**, 11. Laboratory prepn from sodium propylate and methyl iodide: Chancel, *Ann.* **151**, 305 (1869); Michael, Wilson, *Ber.* **39**, 2573 (1906); Bennett, Phillip, *J. Chem. Soc.* **1928**, 1931.

Mobile, flammable liquid. d$_4^0$ 0.7494; d$_4^{13}$ 0.7356. bp$_{761}$ 38.8°. n$_D^{14.3}$ 1.36019. Soly in water (25°): 5 ml/100 ml H$_2$O. Miscible with alc, ether. Absorbs some water. Oil/water coefficient 10 ± 1. Anesthetic index 2.5. Methyl propyl ether used for anesthesia usually contains 0.002 diphenylamine as stabilizer.

THERAP CAT: Anesthetic (inhalation).

6032. Methyl Propyl Ketone. *2-Pentanone;* ethyl acetone. C$_5$H$_{10}$O; mol wt 86.13. C 69.72%, H 11.70%, O 18.58%. CH$_3$CH$_2$CH$_2$COCH$_3$.

Colorless liquid. d$_4^{20}$ 0.809. mp −78°. bp 102°. n$_D^{20}$ 1.3895. Almost insol in water; misc with alcohol, ether. LD$_{50}$ orally in rats: 3.73 g/kg, H. F. Smyth *et al., Am. Ind. Hyg. Assoc. J.* **23**, 95 (1962).

USE: Solvent.

6033. 5-Methylpyrazole-3-carboxylic Acid. C$_5$H$_6$N$_2$O$_2$; mol wt 126.11. C 47.62%, H 4.80%, N 22.22%, O 25.37%. Prepn: Knorr, MacDonald, *Ann.* **279**, 219 (1894); Lehninger, *J. Am. Chem. Soc.* **64**, 2507 (1942). Hypoglycemic metabolite of 3,5-dimethylpyrazole: Smith *et al., J. Med. Chem.* **8**, 350 (1965).

Crystals from water, mp 236-237°.
Methyl ester, C$_6$H$_8$N$_2$O$_2$, crystals, mp 75-78°.
Ethyl ester, C$_7$H$_{10}$N$_2$O$_2$, crystals, mp 83°.
Toxicity: A powerful hypoglycemic agent in rats (characteristic of pyrazoles).

6034. Methyl Pyridyl Ketone. *1-(3-Pyridinyl)ethanone;* 3-acetylpyridine; β-acetylpyridine; methyl 3-pyridyl ketone; methyl β-pyridyl ketone. C$_7$H$_7$NO; mol wt 121.13. C 69.40%, H 5.83%, N 11.56%, O 13.21%. A nicotinic acid antagonist. Prepd in dry distillation of Ca-nicotinate with Ca-acetate. The ketone is isolated through the phenylhydrazone: Engler, Kiby, *Ber.* **22**, 597 (1889).

Liquid, bp 220°. Freely sol in acids.
Ketoxime, NC$_5$H$_4$C(CH$_3$)=NOH, crystals from alcohol or benzene, mp 112°. Hydrochloride of ketoxime, crystals from alcohol, mp 204°.

Phenylhydrazone, NC$_5$H$_4$C(CH$_3$)=NNHC$_6$H$_5$, yellow needles from alcohol, mp 137°.

6035. Methyl 4-Pyridyl Ketone Thiosemicarbazone. *2-[1-(4-Pyridinyl)ethylidene]hydrazinecarbothioamide.* C$_8$H$_{10}$N$_4$S; mol wt 194.27. C 49.46%, H 5.19%, N 28.84%, S 16.51%. Prepn: Schäfer *et al.,* U.S. pat. **2,744,905** (1956 to Schenley).

Crystals from methanol, dec 229-231°.
Hydrochloride, C$_8$H$_{11}$ClN$_4$S, *Depreton.*
THERAP CAT: Antihypertensive.

6036. N-Methylpyrroline. *N-Methyl-2,5-dihydropyrrole.* C$_5$H$_9$N; mol wt 83.13. C 72.24%, H 10.91%, N 16.85%. In leaves of *Nicotiana tabacum* L. and *Atropa belladonna* L. By reducing N-methylpyrrole with zinc granules in dil acetic or hydrochloric acid.

Almost colorless liquid. Unpleasant, ammonia-like odor. Fumes in air. bp 79-80°. Strong base. Miscible with water; sol in alcohol, ether, chloroform.

6037. Methyl Red. *2-[[4-(Dimethylamino)phenyl]azo]-benzoic acid; C.I. Acid Red 2;* C.I. 13020. C$_{15}$H$_{15}$N$_3$O$_2$; mol wt 269.29. C 66.90%, H 5.61%, N 15.61%, O 11.88%. Prepn by diazotization of anthranilic acid and coupling with dimethylaniline: Clarke, Kirner, *Org. Syn.* **2**, 47 (1922). *See also Colour Index* vol. 4 (3rd ed., 1971) p 4043.

Glistening, violet crystals from toluene. mp 181-182°. pKa$_1$ 2.5; pKa$_2$ 9.5; pKb 4.8. Almost insol in water; sol in alcohol and in acetic acid.

USE: As indicator in 0.1% alcoholic soln; pH: 4.4 red, 6.2 yellow. Used for titrating NH$_3$, weak organic bases, *e.g.,* alkaloids; not suitable for organic acids, except oxalic and picric acid. Methyl red is easily reduced, thereby losing its color, and readings should be made promptly. It is gradually being replaced by sulfonphthalein indicators, such as bromcresol green, which are more stable and exhibit a sharper change in color.

6038. Methyl Salicylate. *2-Hydroxybenzoic acid methyl ester;* wintergreen oil; betula oil; sweet birch oil; teaberry oil. C$_8$H$_8$O$_3$; mol wt 152.14. C 63.15%, H 5.30%, O 31.55%. Present in leaves of *Gaultheria procumbens* L., Ericaceae, in the bark of *Betula lenta* L., Betulaceae, but mostly prepd by esterification of salicylic acid with methanol. The product of commerce is about 99% pure.

Colorless, yellowish or reddish, oily liq; odor and taste of gaultheria. mp −8.6°. bp 220-224°. d$_{25}^{25}$ 1.184. d of the natural ester is about 1.180. n$_D^{20}$ 1.535-1.538. Flash pt, closed cup: 210°F (99°C). Slightly sol in water: one gram in about 1500 ml; sol in chloroform, ether. Misc with alcohol, glacial acetic acid. LD$_{50}$ orally in rats: 887 mg/kg, P. M. Jenner *et al., Food Cosmet. Toxicol.* **2**, 327 (1964).

Human Toxicity: Ingestion of relatively small amounts may cause severe poisoning and death (average lethal dose: 10 ml in children, 30 ml in adults). *Symptoms of poisoning:* nausea, vomiting, acidosis, pulmonary edema, pneumonia, convulsions, death, cf., *Clinical Toxicology of Commercial Products,* R. E. Gosselin *et al.,* Eds. (Williams & Wilkins, Baltimore, 4th ed., 1976) Section III, pp 295-303.

USE: In perfumery; for flavoring candies, etc.

THERAP CAT: Counterirritant.

THERAP CAT (VET): Counterirritant.

6039. Methyl Silicone Resins. Tympanol. Polymeric methyl silicon oxides, probably of a cross-linked siloxane structure: Rochow, Gilliam, *J. Am. Chem. Soc.* **63**, 798 (1941). Prepd by hydrolyzing dimethyldichlorosilane or its esters followed by oxidation with air and a catalyst to the desired CH$_3$/Si ratio: Hyde, DeLong, *ibid.* 1194; by hydrolysis of dimethyldichlorosilane mixed with methyltrichlorosilane or silicon tetrachloride followed by cocondensation of the products: Rochow, Gilliam, *loc. cit.*; by partially methylating silicon tetrachloride to the desired CH$_3$/Si ratio and hydrolyzing the reaction mixture: U.S. pat. **2,258,218** (1941).

Characterized by excellent thermal stability and good resistance to oxidation. Oxidize slowly in air at 300° to silica. Can be heated to 550° *in vacuo* or to 500° in hydrogen without dec or melting.

CH$_3$/Si = 1.2. Solid. d 1.20. n_D^{20} 1.425. Hardens in 2 hours at 100°.

CH$_3$/Si = 1.3. Solid. d 1.15. n_D^{20} 1.422. Hardens in 1.5 hours at 120°.

CH$_3$/Si = 1.4. Solid. d 1.08. n_D^{20} 1.421. Hardens in 4 hours at 141°.

CH$_3$/Si = 1.5. Solid. d 1.06. n_D^{20} 1.418. Hardens in 24 hours at 100°.

CH$_3$/Si < 1.2. Sticky syrups. Harden at room temp.

CH$_3$/Si > 1.5. Oily liquids. Set to a soft gel after several days or weeks at 200°.

USE: In electrical insulation; in heat-resistant paints and varnishes; in protective and decorative finishes.

6040. 4′-(Methylsulfamoyl)sulfanilanilide. *4-Amino-N-[4-[(methylamino)sulfonyl]phenyl]benzenesulfonamide;* N^1-methyl-N^4-sulfanilylsulfanilamide; DB87; Diseptal B; Neo-Uliron. C$_{13}$H$_{15}$N$_3$O$_4$S$_2$; mol wt 341.41. C 45.73%, H 4.43%, N 12.31%, O 18.75%, S 18.78%. Prepn: Fr. pat. **817,034** (1937 to I. G. Farbenind.), *C.A.* **32**, 1715 (1938).

Crystals, mp 141°. Slightly sol in water; more sol in alcohol or acetone; freely sol in aq solns of sodium carbonate and bicarbonate.

THERAP CAT: Antibacterial.

6041. Methyl Sulfate. *Sulfuric acid monomethyl ester;* methylsulfuric acid; methyl hydrogen sulfate; monomethyl sulfate; acid methyl sulfate; "Methylsäure". CH$_4$O$_4$S; mol wt 112.10. C 10.71%, H 3.60%, O 57.09%, S 28.61%. CH$_3$-OSO$_2$OH. Prepn from methanol and methyl chlorosulfonate: Levaillant, Simon, *Compt. Rend.* **169**, 855 (1919); from methanol and sulfuric acid: Dumas, Peligot, *Ann.* **15**, 40 (1835); from sulfur trioxide and methanol below 0°: Chamberlain *et al., BIOS Final Rept.* no. **1482**, pp 6 and 8 (1946). *Review:* C. M. Suter, *The Organic Chemistry of Sulfur* (John Wiley, 1944) pp 18-23.

Oily liquid; does not solidify at −30°. d$_4^{15}$ 1.45-1.47. Freely sol in water, less sol in alcohol. Miscible with ether. *See also* Dimethyl Sulfate.

Barium salt, C$_2$H$_6$BaO$_8$S$_2$, *barium methyl sulfate.* Dihydrate, efflorescent crystals. *Poisonous!* Sol in water, alc.

Calcium salt, C$_2$H$_6$CaO$_8$S$_2$, *calcium methyl sulfate, calcium sulfomethylate.* Crystals. Soluble in water.

USE: In sulfonation; as solvent in bromination of indigo dyes.

6042. Methyl Sulfide. *Thiobismethane.* C$_2$H$_6$S; mol wt 62.13. C 38.66%, H 9.73%, S 51.61%. (CH$_3$)$_2$S. Obtained by distilling potassium methyl sulfate with concn aq K$_2$S. Liquid; disagreeable odor. d$_4^{21}$ 0.846. mp −83°. bp 36.2°; also stated as 38°. Insol in water; sol in alcohol, ether.

USE: As solvent for anhydr mineral salts.

6043. N-Methyltaurine. *2-Methylaminoethanesulfonic acid;* β-methylaminoethane-α-sulfonic acid. C$_3$H$_9$NO$_3$S; mol wt 139.18. C 25.89%, H 6.52%, N 10.07%, O 34.49%, S 23.04%. CH$_3$NHCH$_2$CH$_2$SO$_3$H. Prepd by the reaction of sodium isethionate with methylamine at 270-290° and 200 atm press.: Hoyt, *PB Report* **3868,** Office of Technical Services, Dept. of Commerce, Washington, D.C.

Prisms, mp 241-242°. Freely sol in water. Insol in alcohol, ether.

Sodium salt, *N-methylaminoethane sodium sulfonate, sodium N-methyltaurate.* Usually sold as an aq soln contg 35% of N-methyltaurine sodium salt. d$_4^{25}$ 1.21. At its freezing point (−28°) the soln becomes a suspension of crystals. Also available as a 65% soln.

USE: Intermediate in the manuf of surface-active agents.

6044. 17-Methyltestosterone. *17-Hydroxy-17-methylandrost-4-en-3-one;* 17α-methyl-Δ^4-androsten-17β-ol-3-one; Androsan; Anertan (tabl); Glosso-Sterandryl; Homandren (tabl); Malestrone (tabl); Metandren; Neo-Hombreol-M; Nu-Man; Orchisterone-M; Oreton-M; Perandren (lozenges); Synandrets; Synandrotabs; Testhormona; Testosid (tabl); Testoviron (tabl); Testred. C$_{20}$H$_{30}$O$_2$; mol wt 302.44. C 79.42%, H 10.00%, O 10.58%. Prepn from 17-methyl-Δ^{5,6}-androstene-3,17-diol: Oppenauer, *Rec. Trav. Chim.* **56**, 137 (1937); U.S. pat. **2,384,335** (1945 to Alien Property Custodian); from 17-methyl-Δ^4-androsten-17-ol with CrO$_3$: Miescher, Wettstein, U.S. pats. **2,374,370; 2,374,369** (1945 to Ciba); from androstenedione 3-enol ethyl ether: Miescher, U.S. pat. **2,386,331** (1945 to Ciba); from dehydroandrosterone: Julian *et al.,* U.S. pat. **2,435,013** (1945 to Glidden).

Crystals from hexane. mp 161-166°. Stable in air. $[\alpha]_D^{25}$ +69° to +75° (dioxane). Soluble in alcohol, methanol, ether and in other organic solvents. Sparingly sol in vegetable oils. Practically insol in water.

THERAP CAT: Androgen.

THERAP CAT (VET): Has been used as an androgenic agent.

6045. 17α-Methyltestosterone 3-Cyclopentyl Enol Ether. *(17β)-3-(Cyclopentyloxy)-17-methylandrosta-3,5-dien-17-ol;* RP 12222; Pandrocine; Penmestrol. C$_{25}$H$_{38}$O$_2$; mol wt 370.55. C 81.03%, H 10.34%, O 8.64%. Prepn: Ercoli, Gardi, *J. Am. Chem. Soc.* **82**, 746 (1960); Ercoli, U.S. pat. **3,019,241** (1962).

Crystals, mp 148-152°. $[\alpha]_D$ −150° (in dioxane).
THERAP CAT: Androgen.

6046. 4-Methyl-5-thiazoleethanol. *5-Hydroxyethyl-4-methylthiazole;* 5-(2-hydroxyethyl)-4-methylthiazole; 4-methyl-5-(β-hydroxyethyl)thiazole. C_6H_9NOS; mol wt 143.21. C 50.32%, H 6.33%, N 9.78%, O 11.17%, S 22.39%. The thiazole moiety of thiamine (vitamin B_1). Several syntheses, *e.g.*, by condensing thioformamide with bromoacetopropanol: Buchman, *J. Am. Chem. Soc.* **58**, 1803 (1936); or with γ,γ-dichloro-γ,γ-diacetodipropyl ether: Stein, Stevens, **Fr. pat. 945,198** and **Brit. pat. 641,426** (1947 and 1950 to Merck & Co.); Londergan, Schmitz, **U.S. pat. 2,654,760** (1953 to du Pont).

HOCH$_2$CH$_2$ — S
CH$_3$ — N

Oily, viscous liquid. d_4^{24} 1.196. Characteristic disagreeable odor of thiazole compds, becoming somewhat pleasant at extreme dilutions and imparting a nut-like flavor. bp_7 135°; bp_3 123-124°; bp_1 103°. Sol in alcohol, ether, benzene, chloroform. Very sol in water.
Picrate, $C_{12}H_{12}N_4O_8S$, needles from alcohol, mp 163-164°. Hydrochloride, $C_6H_9NOS.HCl$, hygroscopic crystals, sol in water, alcohol.
USE: Intermediate in the synthesis of vitamin B_1.
THERAP CAT: Sedative, hypnotic.

6047. Methyl Thiocyanate. Methyl sulfocyanate. C_2H_3-NS; mol wt 73.12. C 32.85%, H 4.14%, N 19.16%, S 43.86%. $CH_3SC\equiv N$. Prepd from dimethyl sulfate and aq barium thiocyanate.
Colorless liquid; onion odor. d^{20} 1.068. mp −51°. bp 130-133°. n_D^{20} 1.4697. Very slightly sol in water; miscible with alcohol, ether. LD orally in cats: 8.5 mg/kg.

6048. Methylthiouracil. *2,3-Dihydro-6-methyl-2-thioxo-4(1H)-pyrimidinone;* 6-methyl-2-thiouracil; 4-methyl-2-thiouracil; MTU; Alkiron; Antibason; Basecil; Basethyrin; Methiacil; Methicil; Methiocil; Muracil; Prostrumyl; Strumacil; Thimecil; Thyreostat I. $C_5H_6N_2OS$; mol wt 142.18. C 42.24%, H 4.25%, N 19.70%, S 22.55%, O 11.25%. Prepn: List, *Ann.* **236**, 1 (1886).

CH$_3$ — N(H) — S
NH
O

Crystals, bitter taste, dec 326-331°. Sublimes readily. Very slightly sol in cold water, ether. One part dissolves in about 150 parts boiling water. Slightly sol in alcohol, acetone. Practically insol in benzene, chloroform. Freely sol in aq solns of ammonia and alkali hydroxides. A satd aq soln is neutral or slightly acid to litmus. MLD orally in rabbits: 2.5 g/kg.
THERAP CAT: Thyroid inhibitor.
THERAP CAT (VET): Antithyroid substance to promote growth and finishing.

6049. Methyltrienolone. *17β-Hydroxy-17-methylestra-4,9,11-trien-3-one;* 17β-hydroxy-17-methyl-19-norandrosta-4,9,11-trien-3-one; 17α-methyl-4,9,11-estratrien-17β-ol-3-one; R 1881. $C_{19}H_{24}O_2$; mol wt 284.38. C 80.24%, H 8.51%, O 11.25%. Prepn: Velluz *et al.*, *Compt. Rend.* **257**, 569 (1963); **Neth. pat. appl. 6,401,555** (1964 to Roussel-UCLAF), *C.A.* **62**, 10498d (1965).

Crystals from diisopropyl ether, mp 170°. $[\alpha]_D^{20}$ −58.7° (c = 0.5 in ethanol).
THERAP CAT: Anabolic.

6050. α-Methyl-*m*-tyrosine. *3-Hydroxy-α-methylphenylalanine;* α-methyl-3-(*m*-hydroxyphenyl)alanine; α-MMT. $C_{10}H_{13}NO_3$; mol wt 195.21. C 61.52%, H 6.71%, N 7.18%, O 24.59%. Prepn of DL-form: Stein *et al.*, *J. Am. Chem. Soc.* **77**, 700 (1955); Pfister, Stein, **U.S. pat. 2,868,818** (1959 to Merck & Co.).

CH$_3$
CH$_2$CCOOH
NH$_2$
OH

DL-Form, crystals from methanol, dec 296-297°.
THERAP CAT: Inhibitor of catecholamine synthesis.

6051. 6-Methyluracil. *6-Methyl-2,4(1H,3H)-pyrimidinedione;* 4-methyluracil. $C_5H_6N_2O_2$; mol wt 126.11. C 47.62%, H 4.80%, N 22.22%, O 25.37%. Prepn: Boese, **U.S. pat. 2,138,756** (1938 to Carbide and Carbon Chem.); Gleason, **U.S. pat. 2,174,239** (1939 to Standard Oil). Synthesis from urea and ethyl acetoacetate: J. J. Donleavy, M. A. Kise, *Org. Syn.* coll. vol. II, 422 (1943); by isomerization of 2-hydroxy-4-methoxy-6-methylpyrimidine: McOmie *et al.*, *J. Chem. Soc.* **1957**, 1830; from urea and diketene: Khromov-Borisov, Karlinskaya, *J. Gen. Chem. USSR* **26**, 1728 (1956); **U.S.S.R. pat. 101,690** (1955), *C.A.* **51**, 12989f (1957).

CH$_3$ — N(H) — O
NH
O

Crystals from glacial acetic acid, dec above 300°. uv max: 277 nm (log ϵ 3.83). pH of aq soln 13.

6052. Methyl Vinyl Ketone. *3-Buten-2-one;* Δ^3-2-butenone; methylene acetone; acetyl ethylene; δ-oxo-α-butylene. C_4H_6O; mol wt 70.09. C 68.55%, H 8.63%, O 22.83%. $CH_3COCH=CH_2$. Prepn by condensation of acetone and formaldehyde to 3-ketobutanol and dehydration to methyl vinyl ketone: Merling, Köhler, *J. Soc. Chem. Ind.* **29**, 1037 (1910); White, Haward, *J. Chem. Soc.* **1943**, 25. Prepn from vinylacetylene: Conaway, **U.S. pat. 1,967,225** (1934 to du Pont). For review and polymerization characteristics *see* "Vinyl Ketone Polymers" in *Encyclopedia of Polymer Science and Technology* vol. 14 (Interscience, New York, 1971) pp 617-636.
Liquid with pungent odor. bp_{760} 81.4°. n_D^{20} 1.4086. d_4^{20} 0.8636; d_4^{25} 0.8407. Easily soluble in water, methanol, ethanol, ether, acetone, glacial acetic acid. Slightly sol in hydrocarbons. Forms a binary azeotrope with water, bp_{760} 75° (12% water). uv spectrum and electric moments: Rogers, *J. Am. Chem. Soc.* **69**, 2544 (1947). Polymerizes on standing. LD_{50} in mice and rats: 35 mg/kg, *C.A.* **72**, 124809b (1970).
USE: Alkylating agent; commercial starting material for plastics; as intermediate in the synthesis of steroids and vitamin A. *Caution:* Readily absorbed through skin causing general poisoning of the organism. Irritating to mucous membranes and respiratory tract.

6053. Methymycin. $C_{25}H_{43}NO_7$; mol wt 469.60. C 63.94%, H 9.23%, N 2.98%, O 23.85%. Antibiotic substance produced by a streptomycete from soil near Oswego, N.Y. Has a macrolide structure (12-membered lactone ring, *compare* Picromycin). Isoln and antibacterial activity: Donin *et al.*, *Antibiot. Ann.* **1**, 179 (1953-4). Production using *Streptomyces venezuelae* cultures: Dutcher *et al.*, **U.S. pat. 2,916,483** (1959 to Olin Mathieson). Structure: C. Djerassi, J. A. Zderic, *J. Am. Chem. Soc.* **78**, 6390 (1956). Absolute configuration: Rickards, Smith, *Tetrahedron Letters* **1970**, 1025; Manwaring *et al.*, *ibid.* 1029. Biosynthesis: Birch *et al.*, *J. Chem. Soc.* **1964**, 5274. Synthesis: Masamune *et al.*, *J. Am. Chem. Soc.* **97**, 3512 (1975). Possible mechanism of action: Wilhelm *et al., Antimicrob. Ag. Chemother.* **1967,** 236.

Prisms from abs ethanol, mp 195.5-197°. $[\alpha]_D^{22}$ +61° (c = 0.7 in methanol), +74° (c = 1.1 in chloroform). Basic substance, pKb' 5.7. uv max (methanol): 223, 322 nm (ϵ 10,500, 47). Very slightly sol in water, hexane. Sol in methanol, acetone, chloroform, dil acids. Moderately sol in ethanol, ether.

Sulfate, $2C_{25}H_{43}NO_7 \cdot H_2SO_4$, crystals from methanol + acetone. Sol in water, methanol.

Acid sulfate, $C_{25}H_{43}NO_7 \cdot H_2SO_4$, plates from acetone. Sol in water, acetone.

6054. Methyprylon. *3,3-Diethyl-5-methyl-2,4-piperidinedione;* 2,4-dioxo-3,3-diethyl-5-methylpiperidine; 3,3-diethyl-2,4-dioxo-5-methylpiperidine; Noctan; Dimerin; Noludar. $C_{10}H_{17}NO_2$; mol wt 183.26. C 65.54%, H 9.35%, N 7.65%, O 17.46%. Prepn: Frick, Lutz, **U.S.** pat. **2,680,-116** (1954 to Hoffmann-La Roche). *cf.* **Ger.** pat. **930,206** (1953). Comprehensive description: B. C. Rudy, B. Z. Senkowski, in *Analytical Profiles of Drug Substances* **vol. 2,** K. Florey, Ed. (Academic Press, New York, 1973) pp 363-382.

Crystals. Bitter taste. mp 74-77°. uv max (alcohol): 295 nm ($A_{1cm}^{1\%}$ 2.0). Sol in water, alcohol, benzene, chloroform.

Caution: May be habit forming. This is a controlled substance (depressant) listed in the U.S. Code of Federal Regulations, Title 21 Part 1308.13 (1985).

THERAP CAT: Sedative. Hypnotic.

6055. Methysergid(e). *9,10-Didehydro-N-[1-(hydroxymethyl)propyl]-1,6-dimethylergoline-8-carboxamide; N-[1-(hydroxymethyl)propyl]-1-methyl-d-(+)-lysergamide; N-[α-(hydroxymethyl)propyl]-1-methyl-D-lysergamide;* 1-methylmethylergonovine; 1-methyl-*d*-lysergic acid butanolamide; 1-methyl-*d*-lysergic acid (+)-1-hydroxy-2-butylamide; UML 491. $C_{21}H_{27}N_3O_2$; mol wt 353.45. C 71.36%, H 7.70%, N 11.89%, O 9.05%. Prepn: **Brit.** pat. **854,569** (1960); Hofmann,Troxler, **U.S.** pat. **3,218,324** (1965 to Sandoz). Comparative pharmacology: Z. Votava *et al., Arzneimittel-Forsch.* **16,** 220 (1966); P. N. Chambers, P. B. Marshall, *J. Pharm. Pharmacol.* **19,** 65 (1967). Mechanism of action: D. A. Curran *et al., Res. Clin. Stud. Headache* **1,** 74 (1967); J. E. Hardebo *et al., Neurology* **28,** 64 (1978); S. W. J. Lamberts, R. M. Mac Leod, *Endocrinology* **103,** 287 (1978).

Crystals, mp 194-196°. $[\alpha]_D^{20}$ −45° (c = 0.5 in pyridine). Tartrate, $C_{46}H_{60}N_6O_{10}$, crystals, sparingly sol in water. Dimaleate, dec above 165°. Sol in methanol, less sol in water (1:250). Practically insol in abs ethanol.

Hydrogen maleate, $C_{25}H_{31}N_3O_6$, **methysergide maleate,** *Deseril, Désernil-Sandoz, Sansert.*

THERAP CAT: Antimigraine.

6056. Methysticin. *6-[2-(1,3-Benzodioxol-5-yl)ethenyl]-5,6-dihydro-4-methoxy-2H-pyran-2-one; 5-hydroxy-3-methoxy-7-[3,4-(methylenedioxy)phenyl]-2,6-heptadienoic acid δ-lactone;* 4-methoxy-6-[β-(3',4'-methylenedioxyphenyl)-vinyl]-5,6-dihydro-α-pyrone; 6-(3',4'-methylenedioxystyryl)-4-methoxy-5,6-dihydro-2H-pyran-2-one; kavahin; kavatin. $C_{15}H_{14}O_5$; mol wt 274.26. C 65.69%, H 5.15%, O 29.17%. From root of *Piper methysticum* Forst., *Piperaceae* (kava): Pomeranz, *Monatsh.* **10,** 783 (1889); Borsche *et al., Ber.* **54,** 2229 (1921); Hänsel, Beiersdorff, *Arzneimittel-Forsch.* **9,** 581 (1955). Structure: Borsche *et al., Ber.* **60,** 2113 (1927). Synthesis: Klohs *et al., J. Org. Chem.* **24,** 1829 (1959). Absolute config: Snatzke, Hänsel, *Tetrahedron Letters* **1968,** 1797. Molecular and crystal structure: P. Engel, W. Nowacki, *Z. Kristallogr., Kristallgeom., Kristallphys., Kristallchem.* **136,** 437 (1972). Anticonvulsive activity studies: Kretzschmar, Meyer, *Arch. Int. Pharmacodyn. Ther.* **177,** 261 (1969).

Crystals from methanol, mp 132-134°. uv max (alcohol): 226, 267, 306 nm (log ϵ 4.40, 4.14, 3.93). Practically insol in water; sol in alcohol, ether, acetone.

Dihydromethysticin, $C_{15}H_{16}O_5$, prisms from methanol, mp 118°. uv max (methanol): 232, 288 nm (log ϵ 4.18, 3.56).

6057. Metiazinic Acid. *10-Methyl-10H-phenothiazine-2-acetic acid;* (10-methyl-2-phenothiazinyl)acetic acid; *N*-methyl-3-phenothiazinylacetic acid; methiazic acid; methiazinic acid; metiazic acid; RP 16091; Ambrunate; Soridermal; Soripal. $C_{15}H_{13}NO_2S$; mol wt 271.34. C 66.40%, H 4.83%, N 5.16%, O 11.79%, S 11.81%. Prepn from 2-acetyl-10-methylphenothiazine by a Willgerodt reaction: **Fr.** pat. **M4163,** *C.A.* **68,** 69021d (1968); Farge *et al.,* **Neth.** pat. **Appl. 6,614,516; U.S.** pat. **3,424,748** (1966, 1967, 1969 all to Rhône-Poulenc); Messer *et al., Compt. Rend., Ser. C* **265** (14), 758 (1967). Series of articles on prepn, pharmacology, toxicology and metabolism: *Arzneimittel-Forsch.* **19,** 1193-1221 (1969). Toxicity: L. Joulou *et al., ibid.* 1198.

Crystals from benzene, mp 146°. Sol in acetone, ether, chloroform. Forms a water-sol sodium salt. uv max (0.1*N* NaOH): 253, 305 nm; min 280 nm. LD_{50} in mice, rats (mg/kg): 800, ~500 orally (Joulou).

THERAP CAT: Anti-inflammatory.

6058. Meticrane. *3,4-Dihydro-6-methyl-2H-1-benzothiopyran-7-sulfonamide 1,1-dioxide; 6-methylthiochroman-7-sulfonamide 1,1-dioxide;* 6-methyl-7-sulfamoylthiochroman 1,1-dioxide; Arresten; Fontilix. $C_{10}H_{13}NO_4S_2$; mol wt 275.34. C 43.62%, H 4.75%, N 5.09%, O 23.24%, S 23.29%. Prepn and use as diuretic: Boissier, Malen, **Fr.** pats. **M2790** and **1,365,504** (1962, 1963 to Soc. Ind. Fabric. Antibiot.).

Crystals from methyl Cellosolve, mp 236-237°.
THERAP CAT: Diuretic.

6059. Metipranolol. *4-[2-Hydroxy-3-[(1-methylethyl)-amino]propoxy]-2,3,6-trimethylphenol 1-acetate; 1-(4-hydroxy-2,3,5-trimethylphenoxy)-3-(isopropylamino)-2-propanol 4-acetate;* methypranol; trimepranol; VUFB-6453; Betanol; Disorat; Glauline; Glausyn; Turoptin. $C_{17}H_{27}NO_4$; mol wt 309.42. C 65.99%, H 8.80%, N 4.53%, O 20.68%. β-Adrenergic blocker. Prepn: L. Blaha *et al.*, **Czech. pat. 128,471** (1968), *C.A.* **71**, 3129a (1969). Pharmacology: W. Bartsch *et al.*, *Arzneimittel-Forsch.* **27**, 1022 (1977). Plasma levels and pharmacodynamics in man: O. Mayer *et al.*, *Int. J. Clin. Pharmacol., Ther. Toxicol.* **18**, 113 (1980). Metabolic effects: R. P. Faupel, R. Gotzen, *Med. Klin.* **74**, 929 (1979). Comparative effect in hypertension: K. Hayduk *et al.*, *Therapiewoche* **29**, 7528, 7530 (1979).

Crystals from cyclohexane, mp 105-107°. LD_{50} in mice (mg/kg): 31 i.v. (Bartsch).
THERAP CAT: Antihypertensive, antiarrhythmic, antiglaucoma agent.

6060. Metizoline. *4,5-Dihydro-2-[(2-methylbenzo[b]-thien-3-yl)methyl]-1H-imidazole; 2-[(2-methylbenzo[b]-thien-3-yl)methyl]-2-imidazoline;* 2-methyl-3-(Δ^2-imidazolinylmethyl)benzo[b]thiophene; benazoline (obsolete). $C_{13}H_{14}N_2S$; mol wt 230.32. C 67.79%, H 6.12%, N 12.16%, S 13.92%. Prepn: **Fr. pats. M1614; 1,355,049** (1963, 1964, both to E. Merck), *C.A.* **58**, 12574e (1963); **61**, 4146e (1964). Pharmacology and toxicity: G. F. Rosati, M. G. Poletto, *Farmaco Ed. Prat.* **21**, 204 (1966).

Crystals, mp 156-157°.
Hydrochloride, $C_{13}H_{15}ClN_2S$, *H 1032, Elsyl.* Crystalline powder, mp 244-246°. Slightly sol in water; very sol in ethanol, methanol. Practically insol in chloroform. LD_{50} in mice (mg/kg): 49 i.p.; 9.1 i.v.; 155 orally; in rats (mg/kg): 74 orally (Rosati, Poletto).
THERAP CAT: Adrenergic (vasoconstrictor); nasal decongestant.

6061. Metobromuron. *3-(p-Bromophenyl)-1-methoxy-1-methylurea;* Ciba 3126; Pattonex; Patoran. $C_9H_{11}BrN_2O_2$; mol wt 259.10. C 41.72%, H 4.28%, Br 30.84%, N 10.81%, O 12.35%. Prepd by bromination of *N*-phenyl-*N'*-methoxy-*N'*-methylurea: H. Martin *et al.*, **Swiss pat. 405,821** corresp to **U.S. pat. 3,288,851** (both 1966 to Ciba).

Crystals from cyclohexane, mp 95-96°. Vapor press at 20°: 3×10^{-6} mm Hg. Soly in water at 20° = 320 ppm. Sol in methanol, ethanol, acetone, chloroform. LD_{50} orally in rats: 3.875 g/kg, Novakova, Dinoeva, *C.A.* **80**, 56258j (1974).
USE: Herbicide.

6062. Metochalcone. *1-(2,4-Dimethoxyphenyl)-3-(4-methoxyphenyl)-2-propen-1-one; 2',4,4'-trimethoxychalcone; 3'-(4-methoxyphenyl)-1-(2,4-dimethoxyphenyl)-2-propen-1-one;* CB 1314; Lesidrin; Vesidril; Vesidryl. C_{18}-

$H_{18}O_4$; mol wt 298.32. C 72.46%, H 6.08%, O 21.45%. Prepn from *p*-methoxyacetophenone and 2,4-dimethoxybenzaldehyde: Freudenberg *et al.*, *Ber.* **90**, 957 (1957).

Crystals from aq methanol, mp 97°.
THERAP CAT: Choleretic; diuretic.

6063. Metoclopramide. *4-Amino-5-chloro-N-[(2-diethylamino)ethyl]-2-methoxybenzamide; 4-amino-5-chloro-N-[2-(diethylamino)ethyl]-o-anisamide;* 4-amino-5-chloro-2-methoxy-N-(β-diethylaminoethyl)benzamide; DEL 1267; Elieten; Maltyl; Metoclol; Metox; Moriperan; Peraprin; Plasmil; Pramiel; Reliveran. $C_{14}H_{22}ClN_3O_2$; mol wt 299.81. C 56.09%, H 7.40%, Cl 11.83%, N 14.02%, O 10.67%. Substituted benzamide with neuroleptic activity. Prepn: **Belg. pat. 620,543;** Thominet, **U.S. pat. 3,177,252** (1962, 1965 to Soc. d'Etudes Sci. et Ind. de l'Ile-de-France); R. Pakula *et al.*, *Arch. Pharm.* **313**, 297 (1980). Colorimetric determn in plasma and solubility: G. Pitel, T. Luce, *Ann. Pharm. Franc.* **23**, 673 (1965). Dopamine D_2 receptor antagonist (adenylate cyclase independent receptors): J. W. Kebabian, D. B. Calne, *Nature* **277**, 93 (1979). Pharmacology: M. A. Smith, F. J. Salter, *Drug Intell. Clin. Pharm.* **14**, 169 (1980). Pharmacokinetics: H. Vergin *et al.*, *Arzneimittel-Forsch.* **33**, 458 (1983). Review of efficacy in psychiatric disorders: E. D. Peselow, M. Stanley, *Adv. Biochem. Psychopharmacol.* **35**, 163-194 (1982); of pharmacology and clinical use: R. A. Harrington *et al.*, *Drugs* **25**, 451-494 (1983); *ibid.*, Suppl. 1, 1-88. Comprehensive description: D. Pitré, R. Stradi in *Analytical Profiles of Drug Substances* vol. 16, K. Florey, Ed. (Academic Press, New York, 1987) pp 327-361.

mp 146.5-148°. Soly at 25° (g/100 ml): water 0.02; 95% ethanol 2.30; abs ethanol 1.90; benzene 0.10; chloroform 6.60.
Dihydrochloride monohydrate, $C_{14}H_{24}Cl_3N_3O_2 \cdot H_2O$, *Gastronerton, Imperan, Sorbipéran.* Crystals, dec 145°. Soly at 25° (g/100 ml): water 48; ethanol (95%) 9; abs ethanol 6; benzene 0.10; chloroform 0.10. Stable in acids solns. Unstable in strongly alkaline solns.
Monohydrochloride monohydrate, $C_{14}H_{23}Cl_2N_3O_2 \cdot H_2O$, *AHR-3070-C, Cerucal, Clopromate, Duraclamid, Emperal, Eucil, Gastrese, Gastrobid, Gastromax, Gastrosil, Gastro-Tablinen, Gastrotem, Gastro-Timelets, Maxeran, Maxolon, MCP-ratiopharm, Meclopran, Metamide, Metocobil, Metramid, Mygdalon, Parmid, Paspertin, Plasil, Primperan, Reglan.* White crystalline powder, mp 182.5-184°. Sol in water.
THERAP CAT: Anti-emetic.

6064. Metocurine Iodide. *6,6',7',12'-Tetramethoxy-2,-2,2',2'-tetramethyltubocuraranium diiodide;* (+)-*O,O'-dimethylchondrocurarine diiodide;* *d*-tubocurarine iodide dimethyl ether; dimethyl tubocurarine iodide; tubocurarine dimethyl ether iodide; Metubine Iodide. $C_{40}H_{48}I_2N_2O_6$; mol wt 906.63. C 52.99%, H 5.34%, I 27.99%, N 3.09%, O 10.59%. Prepd by methylation of *d*-tubocurarine with methyl iodide.

Crystals, dec 257-267°. $[\alpha]_D^{22}$ +148 to +158° (c = 0.25). uv max: 280 nm. (E$_{1cm}^{1\%}$ 74). Slightly soluble in water (about 300 mg/100 ml), in dil HCl, in dil NaOH. Very slightly sol in alcohol; practically insol in benzene, chloroform, ether.

Trihydrate, large prisms from water, appreciably sol in methanol.

THERAP CAT: Skeletal muscle relaxant.

6065. Metofenazate. *3,4,5-Trimethoxybenzoic acid 2-[4-[3-(2-chloro-10H-phenothiazin-10-yl)propyl]-1-piperazinyl]ethyl ester;* 2-[4-[3-(2-chlorophenothiazin-10-yl)propyl]-1-piperazinyl]ethyl 3,4,5-trimethoxybenzoate; *N-[β-(3,4,5-trimethoxybenzoyloxy)ethyl]-N'-[γ-(3-chloro-10-phenothiazinyl)propyl]piperazine;* methophenazine; perphenazine 3,4,5-trimethoxybenzoate. $C_{31}H_{36}ClN_3O_5S$; mol wt 598.18. C 62.25%, H 6.07%, Cl 5.93%, N 7.02%, O 13.37%, S 5.36%. Prepn: *Magy. Kém. Lapja* **17**, 169 (1962), *C.A.* **57**, 11314d (1962); Toldy *et al., Acta Chim. Acad. Sci. Hung.* **42**, 351 (1964).

Crystals from ethanol, mp 102-107°.
Difumarate, $C_{39}H_{44}ClN_3O_{13}S$, *Frenolon.*
THERAP CAT: Antipsychotic.

6066. Metofoline. *1-(p-Chlorophenethyl)-1,2,3,4-tetrahydro-6,7-dimethoxy-2-methylisoquinoline;* 1-(p-chlorophenethyl)-2-methyl-6,7-dimethoxy-1,2,3,4-tetrahydroisoquinoline; methopholine; Ro 4-1778/1. $C_{20}H_{24}ClNO_2$; mol wt 345.88. C 69.45%, H 6.99%, Cl 10.25%, N 4.05%, O 9.25%. Prepn: Brossi *et al., Helv. Chim. Acta* **43**, 1459 (1960); Besendorf *et al.,* U.S. pats. **3,067,203; 3,146,266** (1962, 1964, both to Hoffmann-La Roche); Brossi *et al.,* in *Analgetics,* de Stevens, Ed. (Med. Chem. Ser. vol. 5, 1965) p 281. Metabolism: Sciorelli, *Experientia* **23**, 934 (1967).

Colorless leaflets from aqueous methanol, mp 110-111°.
LD$_{50}$ s.c. in mice: 180 mg/kg.
Hydrochloride, crystals, mp 105-106°. uv max (ethanol): 283, 287 nm.
THERAP CAT: Analgesic.

6067. Metolachlor. *2-Chloro-N-(2-ethyl-6-methylphenyl)-N-(2-methoxy-1-methylethyl)acetamide;* 2-chloro-6'-ethyl-N-(2-methoxy-1-methylethyl)acet-o-toluidide; α-chloro-2'-ethyl-6'-methyl-N-(1-methyl-2-methoxyethyl)-acetanilide; metelilachlor; CGA 24705; Dual. $C_{15}H_{22}ClNO_2$; mol wt 283.81. C 63.48%, H 7.83%, Cl 12.49%, N 4.93%, O 11.27%. Selective pre-emergence herbicide. Prepn: C. Vogel, R. Aebi, Ger. pat. **2,328,340** corresp to U.S. pat. **3,937,-**

730 (1973, 1976 to Ciba-Geigy). Metabolism: L. L. McGahen, J. M. Tiedje, *J. Agr. Food Chem.* **26**, 414 (1978).

Clear, odorless liquid. bp$_{0.001}$ 100°. n_D^{20} 1.5301. Vapor press. at 20°: 1.3×10^{-5} mm Hg. Soly in water at 20°: 530 ppm. Sol in most organic solvents. LD$_{50}$ orally in rats: 2780 mg/kg, *RTECS* **vol. 1**, R. J. Lewis, Sr., R. L. Tatken, Eds. (1979) p 50.
USE: Herbicide.

6068. Metolazone. *7-Chloro-1,2,3,4-tetrahydro-2-methyl-3-(2-methylphenyl)-4-oxo-6-quinazolinesulfonamide; 7-chloro-1,2,3,4-tetrahydro-2-methyl-4-oxo-3-o-tolyl-6-quinazolinesulfonamide;* 2-methyl-3-o-tolyl-6-sulfamyl-7-chloro-1,2,3,4-tetrahydro-4-quinazolinone; SR 720-22; Diulo; Metenix; Microx; Mykrox; Oldren; Zaroxolyn. $C_{16}H_{16}ClN_3O_3S$; mol wt 365.84. C 52.53%, H 4.41%, Cl 9.69%, N 11.48%, O 13.12%, S 8.76%. Prepn and activity: B. V. Shetty, U.S. pats. **3,360,518** (1967 to Wallace & Tiernan) and **3,557,111** (1971); B. V. Shetty *et al., J. Med. Chem.* **13**, 886 (1970). Pharmacology: Belair *et al., Arch. Int. Pharmacodyn. Ther.* **177**, 71 (1969); Belair, *Res. Comm. Chem. Pathol. Pharmacol.* **2**, 98 (1971). Clinical studies: Schoonees *et al., N.Y. State J. Med.* **71**, 566 (1971).

Crystals from ethanol, mp 253-259°. Also reported as mp 246-250° (U.S. pat. **3,557,111**). Previously reported as a polymorphic solid with melting range of 218-259°. LD$_{50}$ in mice (mg/kg): > 5000 orally; > 1500 i.p. (Shetty).
THERAP CAT: Diuretic; antihypertensive.

6069. Metomidate. *1-(1-Phenylethyl)-1H-imidazole-5-carboxylic acid methyl ester; 1-(α-methylbenzyl)imidazole-5-carboxylic acid methyl ester;* methyl 1-(α-methylbenzyl)imidazole-5-carboxylate; methoxymol; methomidate. $C_{13}H_{14}N_2O_2$; mol wt 230.27. C 67.81%, H 6.13%, N 12.16%, O 13.90%. Injectable hypnotic related to etomidate, *q.v.* Prepn: **Belg. pat. 662,474;** E. F. Godefroi, C. A. M. van der Eijcken, U.S. pat. **3,354,173** (1965, 1967 both to Janssen); E. F. Godefroi *et al., J. Med. Chem.* **8**, 220 (1965). Anaesthetic effect of combination with azaperone, *q.v.:* J. F. F. Callear, J. F. E. Van Gestel, *Vet. Rec.* **92**, 284 (1973); of combination with fentanyl, *q.v.:* W. Erhardt, *J. Small Animal Pract.* **19**, 401 (1978); C. J. Green, *Lab. Animal* **15**, 171 (1981).

Hydrochloride, $C_{13}H_{15}ClN_2O_2$, R-7315, *Hypnodil.* Crystals from methanol-ether, mp 173-174°. LD$_{50}$ i.v. in rats: 50 mg/kg (Godefroi).
THERAP CAT (VET): Hypnotic.

6070. Metopimazine. *1-[3-[2-(Methylsulfonyl)-10H-phenothiazin-10-yl]propyl]-4-piperidinecarboxamide; 1-[3-[2-(methylsulfonyl)phenothiazin-10-yl]propyl]isonipecotamide;* 10-[3-(4-carbamoylpiperidino)propyl]-2-(methanesulfonyl)phenothiazine; 1-[3-[2-(methylsulfonyl)phenothiazin-10-yl]propyl]-4-piperidinecarboxamide; EXP 999; RP 9965; Vogalene. $C_{22}H_{27}N_3O_3S_2$; mol wt 445.61. C 59.30%, H 6.11%, N 9.43%, O 10.77%, S 14.39%. Prepn: Jacob, Robert, **Ger.** pat. **1,092,476** (to Rhône-Poulenc), *C.A.* **56**, 8723h (1962).

Solid, mp 170-171°. LD_{50} in male rats (mg/kg): 976 orally, 1080 s.c., E. I. Goldenthal, *Toxicol. Appl. Pharmacol.* **18**, 185 (1971).

THERAP CAT: Anti-emetic.

6071. Metopon. *4,5-Epoxy-3-hydroxy-5,17-dimethyl-morphinan-6-one;* methyldihydromorphinone. $C_{18}H_{21}NO_3$; mol wt 299.36. C 72.22%, H 7.07%, N 4.68%, O 16.03%. Prepn from thebaine: Small *et al., J. Am. Chem. Soc.* **58**, 1457 (1936). From dihydrocodeinone enol acetate: Small *et al., J. Org. Chem.* **3**, 204 (1938). Structure: Stork, Bauer, *J. Am. Chem. Soc.* **75**, 4373 (1953).

Needles from alc; sinters 235°, mp 243-245° (evac tube). $[\alpha]_D^{24}$ -141° (c = 1 in alc). Slightly sol in organic solvents.
Hydrochloride, $C_{18}H_{21}NO_3$·HCl, crystals from alc, dec 315-318° (evac tube). $[\alpha]_D^{24}$ -105° (c = 1). Freely sol in water. Sparingly sol in alc. Slightly sol in chloroform. Very slightly sol in ether. Insol in benzene. A 1% aq soln has a pH of about 5.0.
Caution: May be habit forming. This is a controlled substance (opiate) listed in the U.S. Code of Federal Regulations, Title 21 Part 1308.12 (1985).

THERAP CAT: Narcotic analgesic.

6072. Metoprolol. *1-[4-(2-Methoxyethyl)phenoxy]-3-[(1-methylethyl)amino]-2-propanol;* (±)-1-(isopropylamino)-3-[p-(β-methoxyethyl)phenoxy]-2-propanol; CGP 2175; H 93/26. $C_{15}H_{25}NO_3$; mol wt 267.38. C 67.38%, H 9.43%, N 5.24%, O 17.95%. β_1-Adrenergic blocker lacking intrinsic sympathomimetic activity. Prepn: A. E. Brandstrom *et al.,* **Ger.** pat. **2,106,209;** *eidem,* **U.S.** pat. **3,873,600** (1971, 1975 to AB Hässle). Pharmacology: B. Ablad *et al., Life Sci.* **12**, 107 (1973); Johansson, *Eur. J. Pharmacol.* **24**, 194 (1973). Effect on mortality in acute myocardial infarction: A. Hjalmarson *et al., Lancet* **2**, 823 (1981). Clinical trials in acute myocardial infarction: L. Rydén *et al., N. Engl. J. Med.* **308**, 614 (1983); *Drugs* **29**, Suppl. 1, 2-8 (1985). Toxicology: N. O. Bodin *et al., Acta Pharmacol. Toxicol.* **36**, Suppl. V, 96 (1975). Review of pharmacology, therapeutic efficacy: P. Benfield *et al., Drugs* **31**, 376-429 (1986); B. Ablad *et al.,* in *Pharmacology of Antihypertensive Drugs,* A. Scriabine, Ed. (Raven, New York, 1980) pp 247-262. Comprehensive description: J. R. Luch in *Analytical Profiles of Drug Substances* **vol. 12**, K. Florey, Ed. (Academic Press, New York, 1983) pp 325-356.

Tartrate, $C_{34}H_{56}N_2O_{12}$, *Beloc, Betaloc, Lopressor, Lopresor, Metoros, Prelis, Seloken, Selopral.* uv max (water): 223 nm

(ϵ 23400). Soly (mg/ml) at 25°: water > 1000; methanol > 500; chloroform 496; acetone 1.1; acetonitrile 0.89; hexane 0.001. LD_{50} in female mice, male rats (mg/kg): 118, ~90 i.v.; 2090, 3090 orally (Bodin).

THERAP CAT: Antihypertensive; anti-anginal. Treatment of myocardial infarction.

6073. Metoquinone. $C_{20}H_{24}N_2O_4$; mol wt 356.41. C 67.39%, H 6.79%, N 7.86%, O 17.96%. $[C_6H_4(OH)NH-CH_3]_2 \cdot C_6H_4(OH)_2$. A salt-like compound of 1 mol hydroquinone and 2 mols p-monomethylaminophenol.
White crystals. mp about 135° with decompn and blackening. Sol in 100 parts water; less sol in alcohol, acetone; very slightly sol in benzene, chloroform, ether.

USE: As a photographic developer.

6074. Metoserpate. *11,17α,18α-Trimethoxy-3β,20α-yohimban-16β-carboxylic acid methyl ester; O-methyl-18-epi-reserpic acid methyl ester;* 18-epi-O-methylreserpic acid methyl ester; methyl 18-epi-O-methylreserpate; methyl O-methyl-18-epireserpate. $C_{24}H_{32}N_2O_5$; mol wt 428.51. C 67.27%, H 7.53%, N 6.54%, O 18.67%. Prepn: Robison *et al., Experientia* **17**, 14 (1961), *J. Am. Chem. Soc.* **83**, 2694 (1961); Lucas *et al.;* Robison, Lucas; and Ziegler, **U.S.** pats. **3,120,532; 3,126,390** and **3,151,117** (1964) and **3,264,303** (1966) (all to Ciba).

Crystals from benzene + cyclohexane, dec 248.5-251.5°, also reported as 239-241°. $[\alpha]_D^{26}$ -37.5° (CHCl$_3$).
Hydrochloride, $C_{24}H_{33}ClN_2O_5$, *SU-9064, Avicalm, Pacitran.* Solid, dec 240-242°. Sol in water.

THERAP CAT (VET): Hydrochloride as sedative.

6075. Metralindole. *2,4,5,6-Tetrahydro-9-methoxy-4-methyl-1H-3,4,6a-triazafluoranthene;* 3-methyl-8-methoxy-3H-1,2,5,6-tetrahydropyrazino[1,2,3-ab]-β-carboline. $C_{15}H_{17}N_3O$; mol wt 255.32. C 70.56%, H 6.71%, N 16.46%, O 6.27%. Heterocyclic β-carboline derivative used as an antidepressant. Prepn: R. G. Glushkov *et al.,* **Ger.** pat. **2,357,-320;** *eidem,* **U.S.** pat. **4,088,647** (1974, 1978 both to All Union Khim.-Farm Instit. USSR); *eidem, Khim.-Farm. Zh.* **16**, 1054 (1982), *C.A.* **98**, 53831b (1983). Psychotropic activity, toxicity: M. D. Mashkovsky *et al.,* **Ger.** pat. **2,356,-091;** *eidem,* **U.S.** pat. **3,959,470** (1974, 1976 both to All Union Khim.-Farm. Instit. USSR). Pharmacology: N. I. Andreeva, M. D. Mashkovsky, *Farmakol. Toxicol.* **43**, 133 (1980). Effect on the EEG of the brain cortex: L. F. Roshchina, *ibid.* 349. Effect on adrenergic neurotransmission: L. V. Panasiuk *et al., ibid.* **45**, 13 (1982). Clinical studies: M. D. Mashkovsky, N. I. Andreeva, *Zh. Nevropatol. Psikhiatr.* **84**, 410 (1984).

Crystals, mp 164-165°.
Hydrochloride, $C_{15}H_{18}ClN_3O$, *Incazan.* Off-white odorless powder with bitter taste. mp 305-308°. Sol in water. LD_{50} orally in mice: 445 mg/kg (Mashkovsky, 1976).

THERAP CAT: Antidepressant.

6076. Metribuzin. *4-Amino-6-(1,1-dimethylethyl)-3-(methylthio)-1,2,4-triazin-5(4H)-one; 4-amino-6-tert-butyl-3-(methylthio)-as-triazin-5(4H)-one;* Bay 94337; Lexone; Sencor; Sencoral. $C_8H_{14}N_4OS$; mol wt 214.28. C 44.84%, H

6.58%, N 26.15%, O 7.47%, S 14.96%. Pre- and post-emergence triazone herbicide. Prepn: **Belg.** pat. **697,083;** K. Westphal *et al.,* **U.S.** pat. **3,671,523** (1967, 1972 both to Bayer AG). Physicochemical properties, toxicology and herbicidal activity: L. Eue, *Pflanzenschutz-Nachr.* **25,** 175 (1972). Toxicity data: E. Loeser, G. Kimmerle, *ibid.* 186. HPLC determn in plant tissues: C. E. Parker, G. H. Degen, *J. Liq. Chromatog.* **6,** 725 (1983).

White crystalline solid, mp 125-126.5°. d_4^{20} 1.28. Vapor pressure at 20°: $< 10^{-5}$ mm Hg. Sol in methanol, ethanol. Soly in water: 1200 ppm. LD_{50} orally in rats: 2200 mg/kg; LC_{50} in rainbow trout: > 10 ppm (Loeser, Kimmerle).

USE: Herbicide.

6077. Metrizamide. *2-[[3-(Acetylamino)-5-(acetylmethylamino)-2,4,6-triiodobenzoyl]amino]-2-deoxy-D-glucose; 2-[3-acetamido-2,4,6-triiodo-5-(N-methylacetamido)benzamido]-2-deoxy-D-glucose;* 2-[3-acetamido-2,4,6-triiodo-5-(N-methylacetamido)benzamido]-2-deoxy-D-glucopyranose; WIN 39103; Amipaque. $C_{18}H_{22}I_3N_3O_8$; mol wt 789.10. C 27.40%, H 2.81%, I 48.24%, N 5.33%, O 16.22%. X-ray contrast agent related to metrizoic acid, *q.v.* Prepn: H. O. Torsten *et al.,* **Ger.** pat. **2,031,724;** *eidem,* **U.S.** pat. **3,701,-771** (1971, 1972 both to Nyegaard). Metabolism: J. G. Johansen, S. Kolmannskog, *Invest. Radiol.* **13,** 93 (1978). Pharmacology: T. W. Morris *et al., ibid.* 74; G. F. Dibona, *Proc. Soc. Exp. Biol. Med.* **157,** 453 (1978). Series of articles on clinical use: *Ann. Radiol.* **22,** 195-206 (1979). Toxicity studies: S. Salvesen, *Radiology* **123,** 241 (1977); M. Sovak *et al., ibid.* 242; P. Aspelin, *Invest. Radiol.* **11,** 309 (1976). *Review:* L. Hol *et al.,* in *Pharmacological and Biochemical Properties of Drug Substances* vol. 1, M. E. Goldberg, Ed. (Am. Pharm. Assoc., Washington, DC, 1977) pp 387-412.

White crystals from isopropyl alcohol, mp 230° (dec), after drying in vacuo at room temp and 70°. $[\alpha]_D^{20}$ +18.0° (c = 10% in 0.1N HCl; equilibrated with respect to mutarotation). Very sol in water (50% w/v) at room temp. LD_{50} i.v. in mice: 15 g/kg (Torsten); 18.6 g/kg (Salveson); 11.5 g/kg (Sovak); 17.3 g/kg (Aspelin).

THERAP CAT: Diagnostic aid (radiopaque medium).

6078. Metrizoic Acid. *3-(Acetylamino)-5-(acetylmethylamino)-2,4,6-triiodobenzoic acid; 3-acetamido-2,4,6-triiodo-5-(N-methylacetamido)benzoic acid;* N-methyl-3,5-diacetamido-2,4,6-triiodobenzoic acid. $C_{12}H_{11}I_3N_2O_4$; mol wt 627.97. C 22.95%, H 1.77%, I 60.63%, N 4.46%, O 10.19%. Prepn: Pitré, Fumigalli, *Farmaco Ed. Sci.* **17,** 340 (1962); **Brit.** pat. **973,881** (1964 to Nyegaard), *C.A.* **62,** 16139g (1965). Series of articles on use in separation of leukocytes from blood and bone marrow: A. Boyum, *Scand. J. Clin. Lab. Invest.* **21,** Suppl. 97, 7-106 (1968).

Crystals, mp 281-282°.
Sodium salt, *Metrizoate Sodium, Isopaque, Triosil.*
THERAP CAT: Diagnostic aid (radiopaque medium).

6079. Metronidazole. *2-Methyl-5-nitroimidazole-1-ethanol;* 1-(2-hydroxyethyl)-2-methyl-5-nitroimidazole; 1-(β-ethylol)-2-methyl-5-nitro-3-azapyrrole; Bayer 5630; RP 8823; Arilin; Clont; Cont; Danizol; Deflamon; Elyzol; Flagyl; Fossyol; Gineflavir; Klion; MetroGel; Metrolag; Metrolyl; Nidazol; Orvagil; Rathimed N; Sanatrichom; Trichazol; Trichocide; Tricho Cordes; Tricho-Gynaedron; Tricocet; Trivazol; Vagilen; Vagimid; Zadstat. $C_6H_9N_3O_3$; mol wt 171.16. C 42.10%, H 5.30%, N 24.55%, O 28.05%. Prepn: Jacob *et al.,* **U.S.** pat. **2,944,061** (1960 to Rhône-Poulenc); Cossar *et al., Arzneimittel-Forsch.* **16,** 23 (1966). Activity studies: Bock, *ibid.* **11,** 587 (1961). Metabolism: Ings *et al., Biochem. Pharmacol.* **15,** 515 (1966). Comprehensive description: L. L. Wearley, G. D. Anthony, in *Analytical Profiles of Drug Substances* vol. 5, K. Florey, Ed. (Academic Press, New York, 1976) pp 327-344. Review of activity, pharmacokinetics and use in anaerobic infection: R. N. Brogden *et al., Drugs* **16,** 387-417 (1978). Symposium: *Surgery* **93,** 123-234 (1983).

Cream-colored crystals, mp 158-160°. Soly at 20° (g/100 ml): water 1.0; ethanol 0.5; ether <0.05; chloroform <0.05. Sparingly sol in DMF. Sol in dil acids. pH of satd aq soln: 5.8.

Hydrochloride, $C_6H_{10}ClN_3O_3$, SC-32642, *Flagyl I.V.*
Note: This substance may reasonably be anticipated to be a carcinogen: *Fourth Annual Report on Carcinogens* (NTP 85-002, 1985) p 133.

THERAP CAT: Antiprotozoal (Trichomonas).
THERAP CAT (VET): Antiprotozoal. Treponemicide.

6080. Metron S. *6-Methyl-N-(1-methylethyl)-2-heptanamine;* N-isopropyl-1,5-dimethylhexylamine; 2-isopropylamino-6-methylheptane; Metron (Japanese). $C_{11}H_{25}N$; mol wt 171.32. C 77.11%, H 14.71%, N 8.18%. Prepn: Ota *et al., Yakugaku Zasshi* **80,** 1153 (1960), *C.A.* **55,** 3825d (1961).

Oil, bp$_{23}$ 84-85°.
THERAP CAT: Antihistaminic.

6081. Metsulfuron Methyl. *2-[[[[(4-Methoxy-6-methyl-1,3,5-triazin-2-yl)amino]carbonyl]amino]sulfonyl]benzoic acid methyl ester;* N-[(4-methoxy-6-methyl-1,3,5-triazin-2-yl)aminocarbonyl]-2-methoxycarbonylbenzenesulfonamide; DPX T6376; Ally; Allie. $C_{14}H_{15}N_5O_6S$; mol wt 381.36. C 44.09%, H 3.96%, N 18.36%, O 25.17%, S 8.41%. Sulfonyl urea used as pre- and postemergence herbicide. Prepn and herbicidal activity: G. Levitt, **Eur.** pat. **Appl. 7687** (1980 to Dupont); *idem,* **U.S.** pat. **4,383,113** (1983 to Dupont). Comparison with other herbicides in cereals: A. Aamisepp, *Weeds Weed Control* **26,** 26 (1985). Efficacy in control of gorse: A. I. Popay *et al., Proc. 38th N.Z. Weed Pest Control Conf.* **1985,** 94. Review of properties, herbicidal activity: P. Lefebvre, *Def. Veg.* **38,** 331 (1984).

White crystals, mp 163-166°. Vapor pressure at 25°: 5.8 × 10⁻⁵ mm Hg. Soly in water: 0.27 g/l at pH 4.59; 9.5 g/l at pH 6.11. LD₅₀ in male and female rats, in male and female rabbits (mg/kg): > 5000, > 2000 orally. LC₅₀ (96 hour) in American perch, rainbow trout: 150 ppm. LD₅₀ in wild ducks: 2510 mg/kg (Lefebvre).
USE: Herbicide.

6082. Meturedepa. *[Bis(2,2-dimethyl-1-aziridinyl)phosphinyl]carbamic acid ethyl ester;* ethyl [bis(2,2-dimethyl-1-aziridinyl)phosphinyl]carbamate; ethyl *N*-[bis(2,2-dimethyl-ethylenimido)phosphoro]carbamate; AB-132; Turloc. C₁₁H₂₂N₃O₃P; mol wt 275.30. C 47.99%, H 8.06%, N 15.26%, O 17.44%, P 11.25%. Prepn: **Brit.** pat. **911,764** (1962 to Armour), *C.A.* **60**, 1701h (1964).

Crystals, mp 119-121°.
THERAP CAT: Antineoplastic.

6083. Metyrapone. *2-Methyl-1,2-di-3-pyridyl-1-propanone;* 2-methyl-1,2-bis(3-pyridyl)-1-propanone; methopyrapone; mepyrapone; metopyrone; methbipyranone; Su 4885; Metopirone; Metroprine. C₁₄H₁₄N₂O; mol wt 226.27. C 74.31%, H 6.24%, N 12.38%, O 7.07%. Prepn: Chart *et al., Experientia* **14**, 151 (1958); Bencze, Allen, *J. Am. Chem. Soc.* **81**, 4015 (1959); **U.S.** pat. **2,923,710** (1960 to Ciba).

Crystals from ether and pentane, mp 50-51°. Also used as the ditartrate, crystals.
THERAP CAT: Diagnostic aid (pituitary function).

6084. Metyridine. *2-(β-Methoxyethyl)pyridine;* methyridine; Dekelmin; Promintic. C₈H₁₁NO; mol wt 137.18. C 70.04%, H 8.08%, N 10.21%, O 11.66%. Prepn: Arnall, Greenhalgh, **Brit.** pat. **889,748** (1962 to ICI and Midland Tar Distillers); **U.S.** pat. **3,223,710** (1965).

Sweet smelling liquid, bp₁₇ 94-96°. d²⁰ 0.988. n²⁰_D 1.4975. pKa = 5.5. Very sol in water and common solvents.
Sulfate, Mintic.
Hydrochloride, mp 104-105°.
THERAP CAT (VET): Anthelmintic.

6085. Metyrosine. *α-Methyl-L-tyrosine;* α-methyl-*p*-tyrosine; α-methyltyrosine; 4-hydroxy-α-methylphenylalanine; α-methyl-3-(*p*-hydroxyphenyl)alanine; metirosine; L-α-MT; α-MPT; MK-781; Demser. C₁₀H₁₃NO₃; mol wt 195.21. C 61.52%, H 6.71%, N 7.18%, O 24.59%. An inhibitor of the first and rate-limiting reaction in catecholamine biosynthesis, the hydroxylation of tyrosine to dopa. Prepn: **Neth.** pat. **Appl.** **6,607,757** (1966 to Merck & Co.), *C.A.* **67**, 91108p (1967). Prepn of DL-form: Stein *et al., J. Am. Chem. Soc.* **77**, 700 (1955); Potts, *J. Chem. Soc.* **1955**, 1632; Pfister, Stein, **U.S.** pat. **2,868,818** (1959 to Merck & Co.); Saari, *J.*

Org. Chem. **32**, 4074 (1967). Metabolism and biochemical and pharmacologic effects in man: Engelman *et al., J. Clin. Invest.* **47**, 568, 577 (1968). Review of pharmacology and clinical use: R. N. Brogden *et al., Drugs* **21**, 81-89 (1981).

Crystals, mp 310-315°.
DL-Form, crystals from water, dec 320° (Stein *et al., loc. cit.*), also reported as dec 330-332° (Potts, *loc. cit.*). Soly in water at room temp: 0.57 mg/ml.
THERAP CAT: Tyrosine hydroxylase inhibitor; as antihypertensive in pheochromocytoma.

6086. Mevaldic Acid. *3-Hydroxy-3-methyl-5-oxopentanoic acid;* 3-hydroxy-3-methylglutaraldehydic acid. C₆H₁₀O₄; mol wt 146.14. C 49.31%, H 6.90%, O 43.79%. Prepn by acid hydrolysis of 3-hydroxy-3-methyl-5,5-dimethoxypentanoic acid: Shunk *et al., J. Am. Chem. Soc.* **79**, 3294 (1957). Exists in equilibrium with *β,δ-dihydroxy-β-methyl-δ-valerolactone.*

Unstable and reactive compd. Isolated in soln only.

6087. Mevalonic Acid. *3,5-Dihydroxy-3-methylpentanoic acid; 3,5-dihydroxy-3-methylvaleric acid; β,δ-dihydroxy-β-methylvaleric acid;* hiochic acid. C₆H₁₂O₄; mol wt 148.16. C 48.64%, H 8.16%, O 43.20%. Precursor in the biosynthesis of cholesterol. Occurs in equilibrium with the δ-lactone. Isoln from distillers' solns: Wright *et al., J. Am. Chem. Soc.* **78**, 5273 (1956). Synthesis: Wolf *et al., ibid.* **79**, 1486 (1957); Hoffmann *et al., ibid.* 2316; Eggerer *et al., Ann.* **608**, 71 (1957); **U.S.** pats. **2,915,398; 2,915,531-3; 2,915,551** (1959 to Merck & Co.); Shunk *et al.,* **U.S.** pat. **3,014,963** (1961 to Merck & Co.); Hulcher, Hosick, **U.S.** pat. **3,119,-842** (1964); F. C. Huang *et al., J. Am. Chem. Soc.* **97**, 4144 (1975). Resolution: Shunk *et al., ibid.* **79**, 3294 (1957); **U.S.** pat. **2,945,059** (1960 to Merck & Co.). Synthesis of the δ-lactone: R. H. Cornforth *et al., Tetrahedron* **18**, 1351 (1962); E. L. Eliel, K. Soai, *Tetrahedron Letters* **22**, 2859 (1981); A. Banerji, G. P. Kalena, *Syn. Commun.* **12**, 225 (1982). Review of organic and biochemistry: Wagner, Folkers, *Endeavour* **20**, 177-187 (Oct. 1961). Prepn of 5-phosphate: Robinson, Wittreich, **U.S.** pat. **3,014,057** (1961 to Merck & Co.). Review of role of mevalonic acid in sterol biosynthesis: G. J. Schroepfer, Jr., *Ann. Rev. Biochem.* **50**, 585-621 (1981); E. Caspi, *Tetrahedron* **42**, 3-50 (1986).

Oily liquid. Very sol in water, but also sol in organic solvents, especially polar organic solvents.
δ-Lactone, C₆H₁₀O₃, *mevalolactone, mevalonic lactone, β-hydroxy-β-methyl-δ-valerolactone.* Hygroscopic crystals, mp 28°.
Benzhydrylamide, C₁₉H₂₃NO₃, crystals from benzene + petr ether, mp 96-97°. [α]²⁰_D −2.0° (c = 2 in ethanol).

6088. Mevastatin. *[1S-[1α(R*),7β,8β(2S*,4S*),8aβ]]-2-Methylbutanoic acid 1,2,3,7,8,8a-hexahydro-7-methyl-8-[2-(tetrahydro-4-hydroxy-6-oxo-2H-pyran-2-yl)ethyl]-1-naphthalenyl ester;* 7-[1,2,6,7,8,8a-hexahydro-2-methyl-8-(methylbutyryloxy)naphthyl]-3-hydroxyheptan-5-olide; 2β-methyl-8α-(2-methyl-1-oxobutoxy)mevinic acid lactone; compactin; 6-demethylmevinolin; CS 500; ML 236 B. C₂₃H₃₄O₅; mol wt 390.52. C 70.74%, H 8.77%, O 20.49%. Fungal metabolite which is a potent inhibitor of HMG-CoA reductase, the rate controlling enzyme in cholesterol biosynthesis. Isoln from *Penicillium citrinum:* A. Endo *et al.,* **Ger.** pat. **2,524,355** corresp to **U.S.** pat. **3,983,140** (1975, 1976 to Sankyo). Isoln from *P. brevicompactum,* crystal and molecular structure: A. G. Brown *et al., J. Chem. Soc. Perkin Trans I* **1976,** 1165. Inhibition of HMG-CoA reductase activity: A. Endo *et al., FEBS Letters* **72,** 323 (1976); M. S. Brown *et al., J. Biol. Chem.* **253,** 1121 (1978). Therapeutic effects in primary hypercholesterolemia: A. Yamamoto *et al., Atherosclerosis* **35,** 259 (1980). Total synthesis: N. Y. Wang *et al., J. Am. Chem. Soc.* **103,** 6538 (1981); M. Hirama, M. Uei, *ibid.* **104,** 4251 (1982); N. N. Girotra, N. L. Wendler, *Tetrahedron Letters* **23,** 5501 (1982); C.-T. Hsu *et al., J. Am. Chem. Soc.* **105,** 593 (1983); P. A. Grieco *et al., ibid.* 1403; D. L. J. Clive *et al., J. Am. Chem. Soc.* **110,** 6914 (1988). Review of syntheses: T. Rosen, C. H. Heathcock, *Tetrahedron* **42,** 4909-4951 (1986). Review of mevastatin and related compounds: A. Endo, *J. Med. Chem.* **28,** 401-405 (1985).

Crystals from aq ethanol, mp 152°. $[\alpha]_D^{22}$ +283° (c = 0.48 in acetone). uv max: 230, 237, 246 nm (log ϵ 4.28, 4.30, 4.11).

6089. Mevinphos. *3-[(Dimethoxyphosphinyl)oxy]-2-butenoic acid methyl ester; 3-hydroxycrotonic acid methyl ester dimethyl phosphate;* 1-methoxycarbonyl-1-propen-2-yl dimethyl phosphate; methyl 3-(dimethoxyphosphinyloxy)crotonate; *O,O*-dimethyl 1-carbomethoxy-1-propen-2-yl phosphate; 2-carbomethoxy-1-methylvinyl dimethyl phosphate; ENT 22374; OS-2046; Phosdrin. $C_7H_{13}O_6P$; mol wt 224.16. C 37.51%, H 5.84%, O 42.83%, P 13.82%. Prepn: A. R. Stiles, **U.S.** pat. **2,685,552** (1954 to Shell). Activity: R. A. Corey *et al., J. Econ. Entomol.* **46,** 386 (1953). Configuration of isomers: Fukuto *et al., J. Org. Chem.* **26,** 4620 (1961). Metabolism and degradn: K. I. Beynon *et al., Residue Rev.* **47,** 55 (1973).

Commercial product is a yellow liq mixture of *cis-* and *trans-* isomers, the *cis*-isomer being about 100 times more potent in insecticidal activity than the *trans*-isomer: Casida *et al., J. Agr. Food Chem.* **4,** 236 (1956). d_4^{20} 1.25. n_D^{20} 1.4494. bp_1 106-107.5°. Misc with water, acetone, benzene, carbon tetrachloride, chloroform, ethyl and isopropyl alcohols, toluene and xylene. One gram dissolves in 20 ml carbon disulfide and 20 ml kerosene. Practically insol in hexane. LD_{50} in female, male rats: 3.7, 6.1 mg/kg orally; 4.2, 4.7 mg/kg dermally, T. B. Gaines, *Toxicol. Appl. Pharmacol.* **14,** 515 (1969).

USE: Insecticide. *Caution:* Cholinesterase inhibitor. *See* Parathion.

6090. Mexacarbate. *4-(Dimethylamino)-3,5-dimethylphenol methylcarbamate (ester); methylcarbamic acid 4-(dimethylamino)-3,5-xylyl ester;* 4-dimethylamino-3,5-xylyl methylcarbamate; Zectran. $C_{12}H_{18}N_2O_2$; mol wt 222.29. C 64.84%, H 8.16%, N 12.60%, O 14.40%. Outline of prepn starting with 3,5-xylenol: Marquardt in *Analyt. Methods for Pesticides* **2** (Academic Press, New York, 1964) pp 581-596. Metabolic studies: Roberts *et al., J. Agr. Food Chem.* **17,** 107 (1969). Toxicity studies in insects: El-Aziz *et al., J.*

Econ. Entomol. **62,** 318 (1969); in birds, other wildlife: Tucker, Crabtree, *ibid.* 1307; Tucker, Haegel, *Toxicol. Appl. Pharmacol.* **20,** 57 (1971).

Crystals, mp 85°. Vapor pressure <0.1 mm at 139°. Practically insol in water (0.01% at 25°). Freely sol in acetone (162.3% w/w), acetonitrile (142% w/w), methylene chloride (120.6% w/w), ethanol (116.0% w/w), benzene (102.0% w/w). Sparingly sol in petroleum solvents (1.05% w/w). LD_{50} orally in male, female rats: 37, 25 mg/kg, Gaines, *Toxicol. Appl. Pharmacol.* **14,** 515 (1969).

USE: Pesticide for use in the control of snails and slugs.

6091. Mexazolam. *10-Chloro-11b-(2-chlorophenyl)-2,3,7,11b-tetrahydro-3-methyloxazolo[3,2-d][1,4]benzodiazepin-6(5H)-one;* 10-chloro-11b-(o-chlorophenyl)-2,3,7,11b-tetrahydro-3-methyloxazolo[3,2-d][1,4]benzodiazepin-6(5H)-one; CS 386; Melex. $C_{18}H_{16}Cl_2N_2O_2$; mol wt 363.25. C 59.52%, H 4.44%, Cl 19.52%, N 7.71%, O 8.81%. Prepn from 2-amino-2',5-dichlorobenzophenone: R. Tachikawa *et al.,* **Ger.** pat. **1,812,252;** *eidem,* **U.S.** pat. **3,772,371** (1969, 1973 to Sankyo). Synthesis via Schiff base, physical properties: T. Miyadera *et al., J. Med. Chem.* **14,** 520 (1971). Acid-base equilibrium: M. Ikeda, T. Nagai, *Chem. Pharm. Bull.* **30,** 3810 (1982). Effect on socially induced suppression in monkeys: T. Kamioka *et al., Psychopharmacology (Berlin)* **52,** 17 (1977); on gastric contraction in cats: N. Iwata *et al., J. Pharmacobio-Dyn.* **3,** 413 (1980). Toxicity studies: H. Masuda *et al., Sankyo Kenkyusho Nempo* **30,** 175 (1978), *C.A.* **90,** 180217d.

Crystals, mp 172-175°. pKa 6.69. LD_{50} in male, female mice; male, female rats: 4687, 4571, 810, 4500 orally; >6000, >6000, >4000, >4000 i.p. or s.c. (Masuda).

THERAP CAT: Anxiolytic

6092. Mexenone. *(2-Hydroxy-4-methoxyphenyl)(4-methylphenyl)methanone; 2-hydroxy-4-methoxy-4'-methylbenzophenone;* benzophenone-10; Uvistat. $C_{15}H_{14}O_3$; mol wt 242.26. C 74.36%, H 5.83%, O 19.81%. Prepn: Hardy, Forster, **U.S.** pat. **2,773,903** (1956 to Am. Cyanamid).

THERAP CAT: Ultraviolet screen.

6093. Mexicain. A cysteine proteinase obtained from the fruit of *Pileus mexicanus* probably identical with *Leucopremna mexicana* (A.DC.) Standl., *Caricaceae:* Castaneda-Agullo *et al., Science* **96,** 365 (1942); *J. Biol. Chem.* **159,** 751 (1945). Immunological studies: Estrada-Parra *et al., Rev. Latinoamer. Microbiol. Parasitol.* **1969,** 145, *C.A.* **72,** 1744u (1970). Mol wt about 31,100; lysine and alanine believed to be the *N*- and *C*-terminal amino acids resp: Soriano *et al., Rev. Latinoamer. Quim.* **1975,** 143, *C.A.* **84,** 56489t (1976). Similar to but not identical with papain. Mexicain presents some advantage over papain, since it is more stable in solns and does not require cysteine. Isoelectric pt 9.12.

6094. Mexiletine. *1-(2,6-Dimethylphenoxy)-2-propanamine;* 1-(2,6-xylyloxy)-2-propylamine; 1-(2',6'-dimethylphenoxy)-2-aminopropane; 1-methyl-2-(2,6-xylyloxy)ethylamine. $C_{11}H_{17}NO$; mol wt 179.27. C 73.70%, H 9.56%, N 7.81%, O 8.93%. Prepn: **Fr. pat. 1,551,055;** H. Köppe *et al.,* **U.S. pat. 3,659,019** (1968, 1972 to Boehringer, Ing.); *eidem,* **S. Afr. pat. 69 03,772** (1970 to Boehringer, Ing.), *C.A.* **73,** 120307j (1970). Pharmacology: B. N. Singh, E. M. Vaughan Williams, *Brit. J. Pharmacol.* **44,** 1 (1972); J. D. Allen *et al., ibid.* **45,** 561 (1972). Characterization of metabolites by mass spec and ^{13}C-NMR: K. N. Scott *et al., Drug Metab. Dispos.* **1,** 506 (1973). Review of pharmacology and therapeutic efficacy: C. Y. C. Chew *et al., Drugs* **17,** 161-181 (1979).

Hydrochloride, $C_{11}H_{18}ClNO$, *Kö 1173, Katen, Mexitil, Ritalmex.* Crystals from ethanol-ether, mp 203-205°.
THERAP CAT: Anti-arrhythmic.

6095. Mezereum. Mezereon; olive spurge; dwarf bay; magell; paradise plant; spurge flax; wild pepper. Dried bark (also seed, though not official) from aerial portions of *Daphne mezereum* L., *D. gnidium* L., and *D. laureola* L., *Thymelaeaceae.* *Habit.* New England, Canada, mountainous Europe, Siberia. *Constit.* Bark: Mezerein and acrid resin, daphnin, umbelliferone, fixed oil. Seed: Fixed oil, acrid resin.
THERAP CAT: Vesicant.

6096. Mezlocillin. *3,3-Dimethyl-6-[[[[[3-(methylsulfonyl)-2-oxo-1-imidazolidinyl]carbonyl]amino]phenylacetyl]amino]-7-oxo-4-thia-1-azabicyclo[3.2.0]heptane-2-carboxylic acid;* 6R-[2-[3-(methylsulfonyl)-2-oxo-1-imidazolidine carboxamido]-2-phenylacetamido]penicillanic acid; Bay-f 1353. $C_{21}H_{25}N_5O_8S_2$; mol wt 539.59. C 46.74%, H 4.67%, N 12.98%, O 23.72%, S 11.88%. Semisynthetic, broad-spectrum antibiotic related to penicillin and azlocillin, *q.q.v.* Prepn: W. Schroeck *et al.,* **Ger. pat. 2,318,955** (1974 to Bayer), *C.A.* **82,** 31313b (1975). *In vitro* study: G. P. Bodey, T. Pan, *Antimicrob. Ag. Chemother.* **11,** 74 (1977). Pharmacokinetics: H. Lode *et al., Infection* **5,** 163 (1977); S. J. Pancoast, H. C. Neu, *Clin. Pharmacol. Ther.* **24,** 108 (1978). Clinical pharmacology: B. F. Issell *et al., Antimicrob. Ag. Chemother.* **13,** 180 (1978). Tissue concentration and efficacy: E. Helwing *et al., Med. Klin.* **74,** 112 (1979). Series of articles on antibacterial activity, pharmacology, and clinical trials: *Arzneimittel-Forsch.* **29,** 1915-2032 (1979); *Infection* **10,** Suppl. 3, S121-S266 (1982); *J. Antimicrob. Chemother.* **11,** Suppl. C, 1-108 (1983).

Sodium salt monohydrate, $C_{21}H_{24}N_5NaO_8S_2.H_2O$, *Baycipen, Baypen, Mezlin.* Pale yellow cryst; sol in water, methanol, DMF. Insol in acetone and ethanol.
THERAP CAT: Antibacterial.

6097. Mianserin. *1,2,3,4,10,14b-Hexahydro-2-methyldibenzo[c,f]pyrazino[1,2-a]azepine;* 2-methyl-1,2,3,4,10,14b-hexahydro-2H-pyrazino[1,2-f]morphanthridine. $C_{18}H_{20}N_2$; mol wt 264.37. C 81.78%, H 7.63%, N 10.59%. Serotonin receptor antagonist. Prepn: **Neth. pat. Appl. 6,603,256;** van der Burg, Delobelle, **U.S. pat. 3,534,041** (1967, 1970, both to Organon); van der Burg *et al., J. Med. Chem.* **13,** 35 (1970). Pharmacology: Saxena *et al., Eur. J. Pharmacol.* **13,**

295 (1971); Vargaftig *et al., ibid.* **16,** 336 (1971); Van Riezen, *Arch. Int. Pharmacodyn. Ther.* **198,** 256 (1972). Molecular and crystal structure: van Rij, Feil, *Tetrahedron* **29,** 1891 (1973). Review of pharmacology and therapeutic efficacy in depressive illness: R. N. Brogden *et al., Drugs* **16,** 273-301 (1978). 5HT-receptor binding study: B. S. Alexander, M. D. Wood, *J. Pharm. Pharmacol.* **39,** 664 (1987). *Review:* H. van Riezen *et al.,* in *Pharmacological and Biochemical Properties of Drug Substances* vol. 3, M. E. Goldberg, Ed. (Am. Pharm. Assoc., Washington, DC, 1981) pp 56-93; *Brit. J. Clin. Pharmacol.* **15,** Suppl. 2, S141-S342 (1983).

Hydrochloride, $C_{18}H_{21}ClN_2$, *GB 94, Org GB 94, Athymil, Bolvidon, Lantanon, Norval, Tetramide, Tolvin, Tolvon.* mp 282-284°. LD_{50} in male, female mice (mg/kg): 365, 390 orally; 32.5, 31.0 i.v. (van Riezen).
THERAP CAT: Antidepressant.

6098. Mibolerone. *17-Hydroxy-7,17-dimethylestr-4-en-3-one;* 7α,17α-dimethyl-19-nortestosterone; U-10997; Cheque; Matenon. $C_{20}H_{30}O_2$; mol wt 302.46. C 79.42%, H 10.00%, O 10.58%. Synthetic anabolic steroid related to testosterone, *q.v.* Prepn: **Belg. pat. 610,385** corresp to J. C. Babcock, J. A. Campbell, **U.S. pat. 3,341,557** (1962, 1967, both to Upjohn). Biological properties: J. A. Campbell *et al., Steroids* **1,** 317 (1963). Dissolution and solubility rates: W. E. Hamlin *et al., J. Pharm. Sci.* **54,** 1651 (1965). Pharmacodynamics: J. H. Sokolowski, C. W. Kasson, *Am. J. Vet. Res.* **39,** 837 (1978). Efficacy evaluation in dogs: J. H. Sokolowski, *J. Am. Vet. Med. Assoc.* **173,** 983 (1978).

Cryst solid. Soly in deionized water: 0.0454 mg/ml at 37°.
THERAP CAT (VET): Anabolic; androgen.

6099. Michler's Base. *4,4'-Methylenebis(N,N-dimethylbenzenamine); 4,4'-methylenebis(N,N-dimethylaniline);* bis(p-dimethylaminophenyl)methane; 4,4'-tetramethyldiaminodiphenylmethane; tetra-base. $C_{17}H_{22}N_2$; mol wt 254.36. C 80.27%, H 8.72%, N 11.01%. Prepd by heating dimethylaniline with 40% formaldehyde and concd HCl: Mekel, **Ger. pat. 1,026,322** (1958 to BASF); Hey, Sanderson, *J. Chem. Soc.* **1960,** 3203; from dimethylaniline + diacetyl-peroxide: Horner *et al., Ann.* **626,** 1 (1959); from dimethylaniline + *tert*-butylperbenzoate: Sosnovsky, Yang, *J. Org. Chem.* **25,** 899 (1960).

Lustrous leaflets, mp 90-91°. Sublimes without decompn. bp 390°; $bp_{0.1}$ 155-157°. Insol in water. Sol in benzene, ether, carbon disulfide, acids; slightly sol in cold alcohol; more sol in hot alcohol.
Note: This substance may reasonably be anticipated to be a carcinogen: *Fourth Annual Report on Carcinogens* (NTP 85-002, 1985) p 130.
USE: In the form of the hydrochloride as reagent for lead. Has been used in manufacture of dyes.

6100. Michler's Ketone. *Bis[4-(dimethylamino)phenyl]methanone; 4,4'-bis(dimethylamino)benzophenone;* tetramethyldiaminobenzophenone. $C_{17}H_{20}N_2O$; mol wt 268.35. C 76.08%, H 7.51%, N 10.44%, O 5.96%. Prepn from dimethylaniline + phosgene: Michler, *Ber.* **9**, 716 (1876); Michler, Dupertius, *ibid.* 1900; from dimethylaniline, $AlCl_3$ and CCl_4: Fierz, Koechlin, *Helv. Chim. Acta* **1**, 218 (1918).

$(CH_3)_2N$ —〈 〉— CO —〈 〉— $N(CH_3)_2$

White to greenish leaflets, mp 172°. bp above 360° with decompn. Practically insol in water. Sol in alcohol, warm benzene; very slightly sol in ether.

Note: This substance may reasonably be anticipated to be a carcinogen: *Fourth Annual Report on Carcinogens* (NTP 85-002, 1985) p 134.

USE: In manuf of dyes.

6101. Miconazole. *1-[2-(2,4-Dichlorophenyl)-2-[(2,4-dichlorophenyl)methoxy]ethyl]-1H-imidazole; 1-[2,4-dichloro-β-[(2,4-dichlorobenzyl)oxy]phenethyl]imidazole;* Fungisdin. $C_{18}H_{14}Cl_4N_2O$; mol wt 416.12. C 51.96%, H 3.39%, Cl 34.08%, N 6.73%, O 3.84%. Prepn: E. F. Godefroi *et al., J. Med. Chem.* **12**, 784 (1969); E. F. Godefroi, J. Heeres, **Ger.** pat. **1,940,388;** *eidem,* **U.S.** pat. **3,717,655** (1970, 1973 to Janssen). Clinical evaluation: Brugmans *et al., Arch. Dermatol.* **102**, 428 (1970); Godts *et al., Arzneimittel-Forsch.* **21**, 256 (1971). *Review:* P. Janssen, W. Van Bever, in *Pharmacological and Biochemical Properties of Drug Substances* vol. **2**, M. E. Goldberg, Ed. (Am. Pharm. Assoc., Washington, DC, 1979) pp 333-354; R. C. Heel *et al., Drugs* **19**, 7-30 (1980).

Nitrate, $C_{18}H_{15}Cl_4N_3O_4$, *R-14889, Aflorix, Albistat, Andergin, Brentan, Conoderm, Conofite, Daktar, Daktarin, Deralbine, Dermonistat, Epi-Monistat, Florid, Gyno-Daktarin, Gyno-Monistat, Micatin, Miconal Ecobi, Monistat, Prilagin, Vodol.* Crystals, mp 170.5° (Godefroi, Heeres, 1970); 184-185° (Godefroi).

(+)-Form nitrate, mp 135.3°. $[\alpha]_D^{20}$ +59° (methanol).
(−)-Form nitrate, mp 135°. $[\alpha]_D^{20}$ −58° (methanol).

THERAP CAT: Topical antifungal.
THERAP CAT (VET): Topical antifungal.

6102. Micranthine. *6',7-Epoxy-6-methoxy-2-methyloxyacanthan-12'-ol.* $C_{34}H_{32}N_2O_5$; mol wt 548.64. C 74.43%, H 5.88%, N 5.11%, O 14.58%. Minor alkaloid from bark of *Daphnandra micrantha* (Tul.) Benth., *Monimiaceae.* Isoln: Pyman, *J. Chem. Soc.* **105**, 1679 (1914). Structure: Bick, Todd, *ibid.* **1950**, 1606. Revised structure and stereo-

chemistry: Bick *et al., Tetrahedron Letters* **1972**, 33. Biogenesis: I. Ralph, C. Bick, *Heterocycles* **16**, 2105 (1981).

Needles from ethyl acetate or methanol, mp 194-196°. $[\alpha]_D^{22}$ −231° (c = 0.7 in chloroform). Sol in chloroform, ethanol; sparingly sol in methanol, ethyl acetate, benzene; practically insol in ether.

6103. Micrococcin P. Micrococcin. Antibiotic substance produced by a species of *Micrococcus: cf. Brit. J. Exp. Pathol.* **29**, 473 (1948). Isoln of micrococcin P from *B. pumilus:* Heatley, Doery, *Biochem. J.* **50**, 247 (1951). Probable identity of micrococcin and micrococcin P: Abraham *et al., Nature* **178**, 44 (1956). Structure studies: Brookes *et al., J. Chem. Soc.* **1957**, 689; Brookes *et al., ibid.* **1960**, 916; Dean *et al., ibid.* **1961**, 3394; James, Watson, *ibid.* C, **1966**, 1361. The antibiotic is a mixture of two components, P_1 (major) and P_2. Proposed total structure of P_1: J. Walker *et al., Chem. Commun.* **1977**, 706. Revised structure of P_1 and structure of P_2: B. W. Bycroft, M. S. Gowland, *ibid.* **1978**, 256. Mechanism of action: E. Cundliffe, J. Thompson, *Eur. J. Biochem.* **118**, 47 (1981). *Review:* Pestka in *Antibiotics* vol. **3**, J. W. Corcoran, F. E. Hahn, Eds. (Springer-Verlag, New York, 1975) pp 480-486.

micrococcin P_1

Needles from ethanol, dec 222-228°. $[\alpha]_D^{21}$ +116° (c = 0.10 in 90% ethanol). uv max (ethanol): 345 nm. Sol in ethanol, chloroform, acetone, glacial acetic acid, pyridine. Insol in ether, benzene, amyl acetate, glycerol. Very slightly sol in water.

6104. Micronomicin. *O-2-Amino-2,3,4,6-tetradeoxy-6-(methylamino)-α-D-erythro-hexopyranosyl-(1 → 4)-O-[3-deoxy-4-C-methyl-3-(methylamino)-β-L-arabinopyranosyl-(1 → 6)]-2-deoxy-D-streptamine;* 6'-N-methylgentamicin C_{1a}; gentamicin C_{2b}; sagamicin (formerly); antibiotic KW-1062; KW-1062; XK-62-2. $C_{20}H_{41}N_5O_7$; mol wt 463.59. C 51.82%, H 8.91%, N 15.11%, O 24.16%. Gentamicin C complex antibiotic produced by *Micromonospora sagamiensis* var. *nonreducans:* T. Nara *et al.,* **Ger.** pat. **2,326,781** (1973 to Kyowa), *C.A.* **80**, 58389b (1974); *eidem,* **U.S.** pat. **4,045,298** (1977 to Abbott). Isoln, physicochemical and antibacterial properties: R. Okachi *et al., J. Antibiot.* **27**, 793 (1974). Structure: R. S. Egan *et al., ibid.* **28**, 29 (1975). Identity with gentamicin C_{2b}: P. J. L. Daniels *et al., ibid.* **35**. Pharmacology: T. Hashimoto *et al., Japan. J. Antibiot.* **30**, 362 (1977), *C.A.* **89**, 17113z (1978). Toxicity studies: T. Hara *et al., ibid.* 386-449, *C.A.* **89**, 209205c-8f (1978). Biosynthesis: H. Kase *et al., J. Antibiot.* **35**, 1 (1982); *eidem, Agr. Biol. Chem.* **46**, 515 (1982). Series of articles on pharmacology, metabolism and clinical studies: *Chemotherapy (Tokyo)* **25**, 1801-2287 (1977). *See also* Gentamicin.

Consult the cross index before using this section.

White amorphous powder, mp 260° (dec). $[\alpha]_D^{20}$ +116° (c = 1 in water). Sol in water, methanol. Insol in chloroform, ethyl acetate, benzene, petr ether. LD_{50} i.v. in mice: 93 mg/kg (Okachi).

Sulfate, $C_{40}H_{92}N_{10}O_{34}S_5$, *6'-N-methylgentamicin C_{1a} hemipentasulfate, Sagamicin, Santemycin.*

THERAP CAT: Antibacterial.

6105. Midazolam. *8-Chloro-6-(2-fluorophenyl)-1-methyl-4H-imidazo[1,5-a][1,4]benzodiazepine.* $C_{18}H_{13}$-$ClFN_3$; mol wt 325.77. C 66.37%, H 4.02%, Cl 10.88%, F 5.83%, N 12.90%. Short-acting deriv of diazepam, *q.v.* Prepn: R. I. Fryer, A. Walser, **Ger.** pat. **2,540,522** (1976 to Hoffmann-La Roche), *C.A.* **85,** 21497 (1976); A. Walser *et al., J. Org. Chem.* **43,** 936 (1978). Determn in blood and urine: C. V. Puglisi *et al., J. Chromatog.* **145,** 81 (1978). Clinical, electroencephalographic, pharmacokinetic studies: C. R. Brown *et al., Anesthesiology* **50,** 467 (1979). Pharmacological study in dogs: D. J. Jones *et al., ibid.* **51,** 430 (1979). Bioassay: F. H. Sarnquist *et al., ibid.* **52,** 149 (1980). Toxicity data: L. Pieri *et al., Arzneimittel-Forsch.* **31,** 2180 (1981). Series of articles on pharmacology, metabolism, pharmacokinetics, clinical experience: *ibid.* 2177-2288; *Brit. J. Clin. Pharmacol.* **16,** Suppl. 1, 1S-199S (1983). Review of pharmacology and therapeutic use: J. W. Dundee *et al., Drugs* **28,** 519-543 (1984).

Colorless crystals from ether/methylene chloride/hexane, mp 158-160°. uv max (2-propanol): 220 nm (ε 30000). Sol in water.

Maleate, $C_{22}H_{17}ClFN_3O_4$, *Ro 21-3981/001, Dormicum, Flormidal.* Crystals from ethanol/ether, mp 114-117° (solvated). LD_{50} in male mice (mg/kg): 760 orally; 86 i.v. (Pieri).

Hydrochloride, $C_{18}H_{14}Cl_2FN_3$, *Ro 21-3981/003, Hypnovel, Versed.*

Note: This is a controlled substance (depressant) listed in the U.S. Code of Federal Regulations, Title 21 Part 1308.14 (1987).

THERAP CAT: Anesthetic (intravenous).

6106. Midecamycins. Macrolide antibiotic complex produced by *Streptomyces mycarofaciens* nov. sp. Isoln and characterization of the major component, midecamycin A_1: T. Tsuruoka *et al.,* **Japan.** pat. **28834('71**) corresp to U.S. pat. **3,761,588** (1973, both to Meiji); *eidem, J. Antibiot.* **24,** 319, 452 (1971). Structure: I. Shigeharu *et al., ibid.* 460. Isoln, properties, structures of the minor components, midecamycins A_2, A_3, A_4: T. Tsuruoka *et al., ibid.* 476, 526. Enzymatic interconversion of A_1 and A_3: Y. Matsuhashi *et al., ibid.* **32,** 777 (1979). Metabolism: T. Shomura *et al., Chem. Pharm. Bull.* **22,** 2427 (1974). Pharmacokinetics: K. Fukaya *et al., Chemotherapy (Tokyo)* **21,** 692 (1973). Pharmacology: K. Mashimo *et al., ibid.* 729. Antibacterial activity: T. Watanabe *et al., ibid.* **25,** 1624 (1977). Toxicity

and teratogenicity study in mice: M. Moriguchi *et al., Japan. J. Antibiot.* **15,** 187, 193 (1972), *C.A.* **78,** 11652a, 11653b (1973).

Midecamycin A_1, $C_{41}H_{67}NO_{15}$, *leucomycin V 3,4^B-dipropanoate, espinomycin A, mydecamycin, turimycin P3, antibiotic SF 837, antibiotic YL 704B1, SF 837, YL 704B1, Aboren, Medemycin, Midécacine, Midecin, Momicine, Myoxam, Normicina, Rubimycin.* Colorless needles from benzene, mp 155-156° (after drying at 80° for 8 hrs in vacuo). $[\alpha]_D^{23}$ −67° (c = 1 in ethanol). uv max (ethanol): 232 nm ($E_{1cm}^{1\%}$ 325). Weakly basic; pKa' 6.9 in 50% aq ethanol. Sol in methanol, ethanol, acetone, chloroform, ethyl acetate, benzene, ethyl ether, acidic water. Practically insol in *n*-hexane, petr ether, water.

Midecamycin A_3, $C_{41}H_{65}NO_{15}$, *9-deoxy-9-oxoleucomycin V 3,4^B-dipropanoate, antibiotic SF 837A3.* White powder, mp 122-125°. $[\alpha]_D^{22}$ −44° (c = 1 in ethanol). uv max (ethanol): 280 nm ($E_{1cm}^{1\%}$ 295). pKa' 7.0 in 50% aq ethanol. Solubilities similar to midecamycin A_1.

THERAP CAT: Antibacterial.

6107. Midodrine. *2-Amino-N-[2-(2,5-dimethoxyphenyl)-2-hydroxyethyl]acetamide; 1-(2',5'-dimethoxyphenyl)-2-glycinamidoethanol.* $C_{12}H_{18}N_2O_4$; mol wt 254.28. C 56.68%, H 7.13%, N 11.02%, O 25.17%. Prepn: K. Wismayr *et al.,* **Austrian** pat. **241,435**; *eidem,* U.S. pat. **3,340,298** (1965, 1967 both to OSSW). Hemodynamic effects: G. Hitzenberger *et al., Int. J. Clin. Pharmacol. Ther. Toxicol.* **7,** 323 (1973). Pharmacodynamic actions: H. Pittner *et al., Arzneimittel-Forsch.* **26,** 2145 (1976). Plasma levels: N. Kolassa *et al., Arch. Int. Pharmacodyn. Ther.* **238,** 96 (1979). Use in ejaculation disorders: D. Jonas *et al., Eur. Urol.* **5,** 184 (1979); in circulatory disorders: H. Sazovsky, H. Pittner, *Fortschr. Med.* **97,** 733 (1979). Synthesis of potential metabolites: T. Kappe, W. Witoszynaky, *Arch. Pharm.* **308,** 339 (1975).

Hydrochloride, $C_{12}H_{19}ClN_2O_4$, *ST 1085, Alphamine, Amatine, Gutron, Hipertan, Metligine, Midamine.* Crystals, mp 192-193°.

THERAP CAT: α-Adrenergic (vasoconstrictor). Antihypotensive.

6108. Mifentidine. *N-[4-(1H-Imidazol-4-yl)phenyl]-N'-(1-methylethyl)methanimidamide; N-(p-imidazol-4-ylphenyl)-N'-isopropylformamidine.* $C_{13}H_{16}N_4$; mol wt 228.30. C 68.39%, H 7.06%, N 24.54%. Histamine H_2-receptor antagonist. Prepn: E. Cereda *et al.,* **Eur.** pat. **Appl. 53,407**; *eidem,* **U.S.** pat. **4,386,099** (1982, 1983 both to De Angeli); A. Donetti *et al., J. Med. Chem.* **27,** 380 (1984). Pharmacology: A. Giachetti *et al., Agents Actions* **16,** 173 (1985); F. Pagani *et al., Arzneimittel-Forsch.* **35,** 451 (1985); G. Francalanza *et al., ibid.* 456. Receptor binding studies: A. Donetti *et al., ibid.* 306; E. E. J. Haaksma *et al., J. Med. Chem.* **30,** 208 (1987). Effect on gastric acid secretion in cats: C. Scarpignato *et al., Arch. Int. Pharmacodyn.* **276,** 142

(1985); in humans: M. Lazzaroni *et al.*, *Int. J. Clin. Pharmacol. Ther. Toxicol.* **25**, 218 (1987).

Dihydrochloride, $C_{13}H_{18}Cl_2N_4$, *DA-4577.* Crystals from isopropanol, mp 220-223°.

6109. Mifepristone.
(11β,17β)-11-[4-(Dimethylamino)-phenyl]-17-hydroxy-17-(1-propynyl)estra-4,9-dien-3-one; 11β-[4-(*N,N*-dimethylamino)phenyl]-17α-(prop-1-ynyl)-Δ⁴,⁹-estradiene-17β-ol-3-one; RU 486; RU 38486; Mifegyne. $C_{29}H_{35}NO_2$; mol wt 429.60. C 81.08%, H 8.21%, N 3.26%, O 7.45%. Progesterone receptor antagonist. Prepn: J. G. Teutsch *et al.*, *Eur.* pat. **Appl.** 57,115; *eidem,* U.S. pat. **4,386,085** (1982, 1983 both to Roussel-UCLAF). Pharmacology: W. Herrmann *et al.*, *Compt. Rend. Ser. III* **294**, 933 (1982). Pituitary and adrenal responses in primates: D. L. Healy *et al.*, *J. Clin. Endocrinol. Metab.* **57**, 836 (1983). Induction of menstruation in primates: *eidem, Fertil. Steril.* **40**, 253 (1983). Pharmacokinetics in rabbits: G. Wang *et al.*, *Arzneimittel-Forsch.* **36**, 936 (1986). Mechanism of action study: M. Rauch *et al.*, *Eur. J. Biochem.* **148**, 213 (1985). Clinical trial as abortifacient: B. Couzinet *et al.*, *N. Engl. J. Med.* **315**, 1565 (1986); as oral contraceptive: L. K. Nieman *et al.*, *ibid.* **316**, 187 (1987). Symposium on clinical efficacy: *Contraception* **36**, Suppl., 1-42 (1987).

mp 150°. $[\alpha]_D^{20}$ +138.5° (c = 0.5 in chloroform).
THERAP CAT: Abortifacient.

6110. Miglitol.
1,5-Dideoxy-1,5-[(2-hydroxyethyl)imino]-D-glucitol; *N*-(2-hydroxyethyl)moranoline; (2*R*,3*R*,4*R*,-5*S*)-1-(2-hydroxyethyl)-2-(hydroxymethyl)-3,4,5-piperidinetriol; *N*-(β-hydroxyethyl)-1-deoxynojirimycin; BAY m 1099. $C_8H_{17}NO_5$; mol wt 207.23. C 46.37%, H 8.27%, N 6.76%, O 38.60%. α-Glucosidase inhibitor; hydroxyethyl derivative of 1-deoxynojirimycin, *q.v.* Prepn: B. Junge *et al.*, **Ger.** pat. **2,758,025**; *eidem,* U.S. pat. **4,639,436** (1979, 1987 both to Bayer AG). Inhibition of α-glucosidase: B. Lembcke *et al.*, *Digestion* **31**, 120 (1985); and hypoglycemic activity: Y. Yoshikuni *et al.*, *J. Pharmacobio-Dyn.* **11**, 356 (1988). Clinical pharmacology: F. P. Kennedy *et al.*, *Clin. Exp. Pharmacol. Physiol.* **14**, 633 (1987). Clinical evaluations in non-insulin dependent diabetes: N. Katsilambros *et al.*, *Arzneimittel-Forsch.* **36**, 1136 (1986); A. R. Scott, R. B. Tattersall, *Diabetic Med.* **5**, 42 (1988); in insulin dependent diabetes: J. Gerard *et al.*, *Int. J. Clin. Pharmacol. Ther. Toxicol.* **25**, 483 (1987); F. P. Kennedy, J. E. Gerich, *Clin. Pharmacol. Ther.* **42**, 455 (1987).

Crystals from ethanol, mp 114°.
THERAP CAT: Antidiabetic.

6111. Mikamycin.
Antibiotic complex isolated from *Streptomyces mitakaensis* found in the soil at Mitaka City, Japan. Belongs to the streptogramin family of antibiotics. Originally considered a single substance, it was later found to contain several active components, the main ones being designated mikamycin A and B. Isoln and characterization: Arai, Nakamura *et al.*, *J. Antibiot.* **9A**, 193 (1956); Arai, Karasawa *et al.*, *ibid.* **11A**, 14 (1958); Arai, Okabe *et al.*, *ibid.* 21; Okabe, *ibid.* **12A**, 86 (1959). Both components exhibit synergistic action *in vitro* and *in vivo*: Tanaka *et al.*, *ibid.* **12A**, 290 (1959); Watanabe, *ibid.* **13A**, 62 (1960). Production of mikamycin A and B: **Fr.** pat. **1,349,946**; Wantanabe, U.S. pat. **3,137,640** (both 1964 to Kanegafuchi). Nomenclature: P. Crooy, R. De Neys, *J. Antibiot.* **25**, 371 (1972). *Review:* N. Tanaka in *Antibiotics* vol. 3, J. W. Corcoran, F. E. Hahn, Eds. (Springer-Verlag, New York, 1975) pp 487-497.

mikamycin B

Mikamycin A, see Virginiamycin M₁.
Mikamycin B, $C_{45}H_{54}N_8O_{10}$, *4-[4-(dimethylamino)-N-methyl-L-phenylalanine]virginamycin S₁,* mikamycin I_A, ostreogrycin B, streptogramin B, pristinamycin I_A, vernamycin B_α. Isoln and properties: Watanabe *et al.*, *J. Antibiot.* **12A**, 112 (1959); **13A**, 57 (1960). Structure: *ibid.* **13A**, 291, 293 (1960); **14A**, 14 (1961); Cox *et al.*, *Chem. Commun.* **1970**, 1623. Monohydrate, platelets from methanol, mp 160°. Anhydr form obtained by drying at 150° under vacuum for 16 hrs, dec 262-263°. uv max (methanol): 209, 260, 305 nm ($E_{1cm}^{1\%}$ 605, 220, 101). $[\alpha]_D^{20}$ −60.3° (c = 1.0 in methanol). Sol in methanol, ethanol, acetone, butyl acetate, benzene, chloroform; practically insol in water, petr ether, hexane. Stable at neutral and acid pH, unstable at alkaline pH.
THERAP CAT: Antibacterial.

6112. Milbemycins.
Antibiotic B-41. A family of novel macrolide antibiotics with insecticidal and acaricidal activity; structurally related to avermectin, *q.v.*, aglycones. Isoln of nine components from *Streptomyces hygroscopicus* subsp. *aureolacrimosus* and properties: A. Aoki *et al.*, **Ger.** pat. **2,329,486**; *eidem,* U.S. pat. **3,950,360** (1973, 1976 both to Sankyo); of α₁-α₁₀, β₁-β₃, D-H: Y. Takiguchi *et al.*, *J. Antibiot.* **33**, 1120 (1980). Isoln and anthelmintic and acaricidal activity of milbemycin D: **Neth.** pat. **Appl.** 8,004,791; Y. Takiguchi *et al.*, U.S. pat. **4,346,171** (1980, 1982 both to Sankyo); isoln and properties of D-H: Y. Takiguchi *et al.*, *J. Antibiot.* **36**, 502 (1983); of J and K: M. Ono *et al.*, *ibid.* 509. Structures of milbemycins α₁-α₁₀, β₁: M. Kurabayashi *et al.*, *18th Symp. Chem. Natural Products* (Kyoto, 1974) nos. 309-316; of β₁-β₃: H. Mishima *et al.*, *Tetrahedron Letters* **1975**, 711; of D-H, J, K and absolute configuration of D: H. Mishima *et al.*, *J. Antibiot.* **36**, 980 (1983). Model studies on milbemycin β₃ synthesis: S. V. Attwood *et al.*, *Chem. Commun.* **1981**, 556. Total synthesis of (±)-milbemycin β₃: A. B. Smith *et al.*, *J. Am. Chem. Soc.* **104**, 4015 (1982); of the (+)-form: D. R. Williams *et al.*, *ibid.* 4708. Prepn of milbemycin D in *Dirofilaria immitis* infection in dogs: M. Tagawa *et al.*, *Japan. J. Vet. Sci.* **47**, 787 (1985). Toxicity data: N. Matsunuma *et al.*, *Sankyo Kenkyusho Nempo* **35**, 71 (1983), *C.A.* **101**, 32870d (1984).

milbemycin D

Sol in *n*-hexane, benzene, acetone, ethanol, methanol, chloroform; very slightly sol in water.

Milbemycin α_1, $C_{31}H_{44}O_7$, *(6R,25R)-5-O-demethyl-28-deoxy-6,28-epoxy-25-methylmilbemycin B, antibiotic B-41A3.* Cryst from acetone/water, mp 212-215°. $[\alpha]_D^{20}$ +106° (c = 0.25 in acetone). uv max (ethanol): 238, 244 nm (ε 27800, 30500).

Milbemycin β_1, $C_{32}H_{48}O_7$, *25α-methylmilbemycin B, antibiotic B-41A1.* Amorphous solid. $[\alpha]_D^{20}$ +160° (c = 0.25 in acetone). uv max (ethanol): 245 nm (ε 26500).

Milbemycin D, $C_{33}H_{48}O_7$, *(6R,25R)-5-O-demethyl-28-deoxy-6,28-epoxy-25-(1-methylethyl)milbemycin B, antibiotic B41D.* Colorless needles from hexane-ethyl acetate (20:1), mp 186-188°. uv max (ethanol): 244 nm (31000). $[\alpha]_D^{27}$ +107° (c = 0.25 in acetone).

THERAP CAT (VET): Milbemycin D as antiparasitic.

6113. Mildiomycin. *4-Amino-1-[4-[(2-amino-3-hydroxy-1-oxopropyl)amino]-9-[(aminoiminomethyl)amino]-6-C-carboxy-2,3,4,7,9-pentadeoxy-α-L-talo-non-2-enopyranosyl]-5-(hydroxymethyl)-2(1H)-pyrimidinone.* $C_{19}H_{30}N_8O_9$; mol wt 514.45. C 44.36%, H 5.88%, N 21.77%, O 27.99%. A nucleoside antibiotic with anti-mildew activity and low toxicity in mammals and fish. Isoln from *Streptoverticillium rimofaciens* B-98891 and characterization: S. Harada, T. Kishi, *J. Antibiot.* **31**, 519 (1978). Taxonomy and fermentation study: T. Iwasa *et al., ibid.* 511. Structure: S. Harada *et al., J. Am. Chem. Soc.* **100**, 4895 (1978); *Tetrahedron* **37**, 1317 (1981). Anti-mildew spectrum: K. Suetomi, T. Kusaka, *Nippon Noyaku Gakkaishi* **4**, 349 (1979), *C.A.* **92**, 17012c (1980).

Hygroscopic solid, mp > 300° (monohydrate). $[\alpha]_D^{20}$ +100° (c = 0.5 in water); +78.5° (c = 0.5 in 0.1N HCl). pKa': 2.8, 4.3, 7.2, > 12. uv max (pH 7 and 0.1N NaOH): 271 nm ($E_{1cm}^{1\%}$ 164); (0.1N HCl): 280 nm ($E_{1cm}^{1\%}$ 247). LD_{50} in rats and mice: approx 500-1000 mg/kg i.v.; approx 2.5-5.0 g/kg orally, S. Harada, T. Kishi, *loc. cit.*

6114. Milk. Cow's milk. Compn and physical properties (a study of samples representing 8 million quarts of milk from 8 cities): Dahlberg, *Proc. Ann. Conv. Milk Ind. Foundation, Lab. Sec.* **46**, 44 (1953). *Review:* Patton, *Sci. Am.* **221** (1), 58 (July 1969).

White liquid. d 1.032. mp −0.54°. pH 6.62. Fat 3.82%; total solids 12.43%; protein 3.25%; lactose 4.64%; ash 0.73%.

Vitamins and minerals in 100 g: Vitamin A 152 I.U.; riboflavin 0.156 mg; thiamine 0.043 mg; ascorbic acid 1.14 mg; vitamin B_{12} 0.56 γ; vitamin D 2.4 I.U.; vitamin E 0.1 mg; nicotinic acid 0.085 mg; pantothenic acid 0.35 mg; pyridoxine 0.048 mg; biotin 0.0035 mg; folic acid 0.2 γ; inositol 13 mg; choline 13 mg; essential fatty acids 99 mg; Ca 113 mg; P 90 mg; Fe 0.032 mg. Caloric value: 65 cal/100 g.

6115. Milorganite®. Activated sewage sludge prepd by microbial treatment of sewage, useful as fertilizer and as a source of vitamin B_{12}: Schendel, Johnson, *J. Agr. Food. Chem.* **2**, 23 (1954). Isoln of B_{12}-active product: Miner, Wolnak, U.S. pat. **2,646,386** (1953 to Sewerage Commission of Milwaukee); U.S. pat. **2,941,933** (1960 to UCLAF).

Caution: The dust can cause irritation of eyes, respiratory tract.

6116. Miloxacin. *5,8-Dihydro-5-methoxy-8-oxo-1,3-dioxolo[4,5-g]quinoline-7-carboxylic acid; 6,7-methylenedioxy-1-methoxy-4-oxo-1,4-dihydroquinoline-3-carboxylic acid; antibiotic AB 206; AB 206; Fuldazin.* $C_{12}H_9NO_6$; mol wt 263.21. C 54.76%, H 3.45%, N 5.32%, O 36.47%. Quinolone antibacterial; analog of oxolinic acid, *q.v.* Prepn: T. Nakagome *et al.,* Ger. pat. **2,134,451;** *eidem,* U.S. pat. **3,799,930** (1972, 1974 both to Sumitomo). Mode of action: T. Nagate *et al., Antimicrob. Ag. Chemother.* **17**, 763 (1980). Series of articles on antibacterial activity, absorption and excretion in animals, determn of metabolites by HPLC: *ibid.* **18**, 37-49 (1980). Reproduction studies in rats: T. Yamada *et al., Oyo Yakuri* **19**, 651, 815, 833 (1980), *C.A.* **93**, 197897x, 215570s, 215571t (1980).

Colorless prisms from DMF, mp 264° (dec).
THERAP CAT: Antibacterial.

6117. Milrinone. *1,6-Dihydro-2-methyl-6-oxo-(3,4'-bipyridine)-5-carbonitrile; 1,2-dihydro-6-methyl-2-oxo-5-(4-pyridinyl)nicotinonitrile; WIN 47203; Corotrope; Primacor.* $C_{12}H_9N_3O$; mol wt 211.22. C 68.24%, H 4.29%, N 19.89%, O 7.58%. Selective phosphodiesterase inhibitor with vasodilating and positive inotropic activity. Prepn: G. Y. Lesher *et al.,* Belg. pat. **886,336;** G. Y. Lesher, R. E. Philion, U.S. pat. **4,313,951** (1981, 1982 both to Sterling). Improved process: B. Singh, U.S. pat. **4,413,127** (1983 to Sterling); *idem, Heterocycles* **23**, 1479 (1985). Pharmacological profile: A. A. Alousi *et al., J. Cardiovasc. Pharmacol.* **5**, 792, 804 (1983). HPLC determn in body fluids: J. Edelson *et al., J. Chromatog.* **276**, 456 (1983). Clinical studies in congestive heart failure: C. S. Maskin *et al., Circulation* **67**, 1065 (1983); L. S. Sinoway *et al., J. Am. Coll. Cardiol.* **2**, 327 (1983); D. S. Baim *et al., N. Engl. J. Med.* **309**, 748 (1983).

Crystals from DMF + water or from ethanol, mp > 300°.
THERAP CAT: Cardiotonic.

6118. Mimosine. *α-Amino-3-hydroxy-4-oxo-1(4H)-pyridinepropanoic acid; 3-hydroxy-4-oxo-1(4H)-pyridinealanine; β-[N-(3-hydroxy-4-pyridone)]-α-aminopropionic acid; leucenol; leucenine; leucaenine; leucaenol.* $C_8H_{10}N_2O_4$; mol wt 198.18. C 48.48%, H 5.09%, N 14.14%, O 32.29%. A naturally occurring amino acid found in large quantities in the seeds and foliage of the legume genera *Mimosa* and *Leucena.* Isoln from *Leucena glauca* (Willd.) Benth., Leguminosae: Renz, *Z. Physiol. Chem.* **244**, 153 (1936); Mascré, *Compt. Rend.* **204**, 890 (1937); improved large-scale isoln: N. K. Hart *et al.*, *Heterocycles* **7**, 265 (1977). Structure: Adams, Jones, *J. Am. Chem. Soc.* **69**, 1803 (1947). Synthesis: Adams, Johnson, *ibid.* **71**, 705 (1949). Identity of mimosine and leucenol: Kleipool, Wibaut, *Rec. Trav. Chim.* **69**, 37 (1950); Wibaut, Schuhmacher, *ibid.* **71**, 1017 (1952). Crystal structure: A. Mostad *et al.*, *Acta Chem. Scand.* **27**, 164 (1973). Shown to cause inhibition of hair growth and loss of hair in mice: Crounse *et al.*, *Nature* **194**, 694 (1962). Explanation of the toxicity in animals: J. F. Thompson *et al.*, *Ann. Rev. Biochem.* **38**, 137 (1969).

dl-Form, crystals from water, mp 235-236°. Slightly sol in water, much less sol in methanol and ethanol; practically insol in the higher alcohols, in dioxane, ethyl acetate, ether, benzene, chloroform, glacial acetic acid, pyridine, Cellosolve. Sol in dil acids or bases and may be recovered from these solns by adjusting the pH so that it is just acid to bromcresol green. uv max: 282 nm (log ϵ 4.23).

dl-Hemihydrate, crystals, darkens at 215-226°, mp 227-228°.

l-Form, crystals from water, mp 225°. $[\alpha]_D^{22}$ −20°.

Hydrochloride, $C_8H_{11}ClN_2O_4$, dec 175°, sol in water.

Hydrobromide, $C_8H_{11}BrN_2O_4$, dec 179.5°, sol in water.

USE: Depilatory agent, Hegarty *et al.*, *Aust. J. Agr. Res.* **15**, 153 (1964), *C.A.* **60**, 16405e (1964).

6119. Minaprine. *N-(4-Methyl-6-phenyl-3-pyridazinyl)-4-morpholineethanamine; 4-[2-[(4-methyl-6-phenyl-3-pyridazinyl)amino]ethyl]morpholine.* $C_{17}H_{22}N_4O$; mol wt 298.40. C 68.43%, H 7.43%, N 18.78%, O 5.36%. Prepn: H. Laborit, **Ger. pat.** 2,229,215; *idem,* **U.S. pat.** 4,169,158 (1973, 1979 to Centre Etudes Exper. Clin. Physio-Biologie); C.-G. Wermuth, A. Exinger, *Agressologie* **13**, 285 (1972). Pharmacology: H. Laborit *et al.*, *ibid.* 291. Metabolism of ^{14}C-minaprine: A. G. Rico *et al.*, *J. Pharmacol.* **9**, 170 (1978). Pharmacokinetics: J. P. Jeanniot *et al.*, *ibid.* 169. Clinical evaluation: A. Garcia-Maffla, M. H. de Garcia, *Pharmatherapeutica* **2**, 265 (1979). Pharmacological evaluation in depression: K. Bizière *et al.*, *Arzneimittel-Forsch.* **32**, 824 (1982). Toxicologic study: A. G. Mazure *et al.*, *Agressologie* **13**, 319 (1972).

Fine buff-colored needles from isopropanol, mp 122°. Insol in water. Slightly sol in cold ethanol, sol in hot ethanol, chloroform.

Dihydrochloride, $C_{17}H_{24}Cl_2N_4O$, *Agr 1240, CB-30038, Brantur, Cantor.* Crystals from abs ethanol, mp 182°.

THERAP CAT: Psychotropic.

6120. Mineral Spirits. *Petroleum spirits;* white spirits; turpentine substitutes. Name applied to various types of hydrocarbon solvents, primarily petroleum distillates, which have flash points above 100°F (38°C) and distillation ranges between 300°F (149°C) and 415°F (213°C). *See:* A.S.T.M. *Standard Specifications* D 235-83, 71-73 (1983).

Type I, *Stoddard solvent, Texsolve S, Varsol 1.* Regular mineral spirits. Clear liquid. Flash pt (min): 38°C (100°F). $d_{15.6}^{15.6}$ 0.754-0.820. Distillation range: initial bp (min), 149°C (300°F); 50% recovered (max), 182°C (360°F); dry pt (max), 208°C (407°F).

Type II. High flash point mineral spirits. Flash pt (min): 60°C (140°F). $d_{15.6}^{15.6}$ 0.768-0.820. Distillation range: initial bp (min), 177°C (350°F); 50% recovered (max), 196°C (385°F); dry pt (max), 211°C (412°F).

Type III. Odorless mineral spirits. Flash pt (min): 38°C (100°F). $d_{15.6}^{15.6}$ 0.775 (max). Distillation range: initial bp (min), 149°C (300°F); 50% recovered (max), 196°C (385°F); dry pt (max), 213°C (415°F).

Type IV, *Texsolve S-2, Varsol 3.* Low dry point mineral spirits. Flash pt (min): 38°C (100°F). $d_{15.6}^{15.6}$ 0.754-0.800. Distillation range: initial bp (min), 149°C (300°F); 50% recovered (max), 174°C (345°F); dry pt (max), 185°C (365°F).

USE: Solvent; paint thinner. In the coatings and dry cleaning industries.

6121. Minocycline. *4,7-Bis(dimethylamino)-1,4,4a,5,-5a,6,11,12a-octahydro-3,10,12,12a-tetrahydroxy-1,11-dioxo-2-naphthacenecarboxamide;* 7-dimethylamino-6-demethyl-6-deoxytetracycline; Minocyn. $C_{23}H_{27}N_3O_7$; mol wt 457.49. C 60.38%, H 5.95%, N 9.18%, O 24.48%. Semi-synthetic antibiotic effective against tetracycline-resistant staphylococci. Prepn: Boothe, Petisi, **U.S. pats.** 3,148,212 and 3,226,436 (1964 and 1965 to Am. Cyanamid). Synthesis: Martell, Boothe, *J. Med. Chem.* **10**, 44 (1967); Church *et al.*, *J. Org. Chem.* **36**, 723 (1971); Bernardi *et al.*, *Farmaco Ed. Sci.* **30**, 736 (1975). Activity data: Kradolfer *et al.*, *Antimicrob. Ag. Chemother.* **1966**, 359. Metabolism: Kelly, Kanegis, *Toxicol. Appl. Pharmacol.* **11**, 171 (1967). Toxicity studies: Noble *et al.*, *ibid.* 128. Clinical evaluation: Frisk, Tunevall, *Antimicrob. Ag. Chemother.* **1968**, 335; Cappel, Klastersky, *Curr. Ther. Res.* **13**, 227 (1971). Comprehensive description: V. Zbinovsky, G. P. Chrekian in *Analytical Profiles of Drug Substances* vol. 6, K. Florey, Ed. (Academic Press, New York, 1977) pp 323-339.

Bright yellow-orange amorphous solid. $[\alpha]_D^{25}$ −166° (c = 0.524). uv max (0.1N HCl): 352, 263 nm (log ϵ 4.16, 4.23); (0.1N NaOH): 380, 243 nm (log ϵ 4.30, 4.38).

Hydrochloride, $C_{23}H_{28}ClN_3O_7$, *Minocin, Klinomycin, Minomycin, Vectrin.* Yellow crystalline powder. Slightly hygroscopic; sensitive to light and to surface oxidation.

THERAP CAT: Antibacterial.

6122. Minoxidil. *6-(1-Piperidinyl)-2,4-pyrimidinediamine 3-oxide;* 6-amino-1,2-dihydro-1-hydroxy-2-imino-4-piperidinopyrimidine; 2,3-dihydro-3-hydroxy-2-imino-6-(1-piperidinyl)-4-pyrimidinamine; 2,4-diamino-6-piperidinopyrimidine 3-oxide; 6-piperidino-2,4-diaminopyrimidine 3-oxide; PDP; U-10,858; Alopexil; Alostil; Loniten; Lonolox; Minoxinen; Prexidil; Regaine; Rogaine; Tricoxidil. $C_9H_{15}N_5O$; mol wt 209.25. C 51.66%, H 7.22%, N 33.47%, O 7.65%. Prepn: **Neth. pat. Appl.** 6,615,385; W. C. Anthony *et al.*, **U.S. pat.** 3,382,247 (1967, 1968 both to Upjohn); J. M. McCall *et al.*, *J. Org. Chem.* **40**, 3304 (1975). *See also* W. C. Anthony, **U.S. pat.** 3,644,364 (1972 to Upjohn). Metabolism: R. C. Thomas *et al.*, *J. Pharm. Sci.* **64**, 1360 (1975). Pharmacology and pharmacokinetics: D. T. Lowenthal *et al.*, *J. Clin. Pharmacol.* **18**, 500 (1978). Percutaneous absorption and excretion: T. J. Franz, *Arch. Dermatol.* **121**, 203 (1985). Clinical studies: O. Andersson, R. Sivertsson, *Acta Med. Scand.* **205**, 213 (1979); M. Moser, *Advan. Cardiol.* **26**, 38 (1979). Clinical trial in early male pattern baldness: E. A. Olsen *et al.*, *J. Am. Acad. Dermatol.* **13**, 185 (1985). Toxicology: R. G. Carlson, E. S. Feenstra, *Toxicol. Appl. Pharmacol.* **39**, 1 (1977). Review of pharmacology and therapeutic use: V. M. Campese, *Drugs* **22**, 257-278 (1981). Review of topical application in baldness: E. Novak *et al.*, *Int. J. Dermatol.* **24**, 82 (1985). Comprehensive description: D. K. J. Gorecki in *Analytical Profiles of Drug Substances* vol. 17, K. Florey, Ed. (Academic Press, New York, 1988) 185-219.

Crystals from methanol-acetonitrile, mp 248°, dec 259-261° (Anthony, 1972). pKa 4.61. uv max (ethanol): 230, 261, 285 nm (ε 35210, 11210, 11790); (0.01N H_2SO_4): 232, 280 nm (ε 26350, 23850); (0.01N KOH): 231, 261.5, 285 nm (ε 36100, 11400, 12040). Soly (mg/ml): propylene glycol 75, methanol 44, ethanol 29, 2-propanol 6.7, dimethylsulfoxide 6.5, water 2.2, chloroform 0.5, acetone <0.5, ethylacetate <0.5, diethyl ether <0.5, benzene <0.5, acetonitrile <0.5. LD_{50} in rats, mice (mg/kg): 49, 51 i.v. (Carlson, Feenstra).

THERAP CAT: Antihypertensive. Antialopecia agent.

6123. Miokamycin. *Leucomycin V 3^B,9-diacetate 3,4^B-dipropanoate;* 9,3''-diacetylmidecamycin; MOM; ponsinomycin; Miocamycin. $C_{45}H_{71}NO_{17}$; mol wt 898.08. C 60.18%, H 7.97%, N 1.56%, O 30.29%. Macrolide antibiotic deriv of midecamycin A_1. Prepn: S. Inoue *et al.,* **Japan. Kokai 74 124087** (1974 to Meiji), *C.A.* **83**, 10735w (1975); S. Omoto *et al., J. Antibiot.* **29**, 536 (1976); T. Nakamura *et al., Chem. Letters* **1978**, 1293. Physico-pharmaceutical properties of crystalline and non-crystalline miokamycin: T. Sato *et al., Chem. Pharm. Bull.* **29**, 2675 (1981). *In vitro* and *in vivo* antibacterial activity: K. Kawaharajo *et al., J. Antibiot.* **34**, 436 (1981). Metabolism: T. Shomura *et al., Chem. Pharm. Bull.* **29**, 2413 (1981). General pharmacological studies: U. Shibata *et al., Japan. J. Antibiot.* **34**, 734 (1981), *C.A.* **96**, 62631x (1982). Laboratory and clinical pediatric studies: Y. Toyonaga *et al., ibid.* **35**, 1475 (1982), *C.A.* **98**, 216y (1983). Mutagenicity study: F. Hirano, U. Takeda, *J. Antibiot.* **34**, 443 (1981).

Tasteless crystals from isopropanol, mp ~ 220° (with coloration). $[\alpha]_D^{25}$ −53° (c = 1.0 in chloroform). $[\alpha]_D^{20}$ −74° (c = 1 in methanol): 231 nm ($E_{1cm}^{1\%}$ 342). Sol in methanol, acetone, chloroform. Very slightly sol in water. The non-crystalline form converts gradually to the crystalline form in aqueous suspension. In 0.2% hydroxypropylmethylcellulose soln, the non-crystalline solid was at least 10 times more sol than the crystalline form.

THERAP CAT: Antibacterial.

6124. Mipafox. *N,N'-Bis(1-methylethyl)phosphorodiamidic fluoride; N,N'-diisopropylphosphorodiamidic fluoride;* bis(isopropylamino)fluorophosphine oxide; Isopestox; Pestox XV. $C_6H_{16}FN_2OP$; mol wt 182.20. C 39.56%, H 8.85%, F 10.43%, N 15.38%, O 8.78%, P 17.00%. Prepn: Pound *et al.,* **Brit. pat. 688,787** (1953 to Fisons); Heath, *J. Chem. Soc.* **1956**, 3796.

Crystals, mp 65°.
USE: Insecticide. *Caution:* Cholinesterase inhibitor.

6125. Miraculin. Taste-modifying protein; Miralin. A

basic glycoprotein of probable mol wt 44,000. Responsible for the taste-changing properties of *Miracle fruit (agbayun),* the red berries of *Synsepalum dulcificum* (Schum.) Daniell, *Sapotaceae* [alternate name: *Richardella dulcifica* (Schum. and Thonn) Baehni], a shrub indigenous to tropical W. Africa. Sour materials taste sweet after the tasteless mucilaginous pulp of the fruit has been applied to the tongue. First description: Daniell, *Pharm. J.* **11**, 445 (1852). Isolation studies: Inglett *et al., J. Agr. Food Chem.* **13**, 284 (1965); Kurihara, Beidler, *Science* **161**, 1241 (1968). Isoln and purif of miraculin: Brouwer *et al., Nature* **220**, 373 (1968); **Neth. pat. Appl. 6,911,954** corresp to Brouwer *et al.,* U.S. pat. **3,682,880** (1970, 1972 to Lever Bros.). Max sweetening effect occurs when 4 × 10^{-7} M (in 0.02 M citric acid) is held in the mouth for 3 mins. The protein probably binds to receptors of the taste buds and modifies their function. Mechanism of action studies: Kurihara, Beidler, *Nature,* **222**, 1176 (1969). *Reviews:* Henning, *Pharm. Weekbl.* **106**, 271 (1971); Cagan, *Science* **181**, 32 (1973).

USE: Sweetening agent.

6126. Mirex. *1,1a,2,2,3,3a,4,5,5,5a,5b,6-Dodecachloro-octahydro-1,3,4-metheno-1H-cyclobuta[cd]pentalene;* perchloropentacyclo[5.2.1.0^{2,6}.0^{3,9}.0^{5,8}]decane; hexachloropentadiene dimer; CG-1283; ENT 25719; Dechlorane. $C_{10}Cl_{12}$; mol wt 545.59. C 22.01%, Cl 77.99%. Prepn: Prins, *Rec. Trav. Chim.* **65**, 455 (1946); Newcomer, McBee, *J. Am. Chem. Soc.* **71**, 952 (1949); Gilbert, U.S. pat. **2,702,305** (1955 to Allied Chem.); Johnson, U.S. pat. **2,724,730** (1955 to Hooker Electrochem.). Structure: McBee *et al., J. Am. Chem. Soc.* **78**, 1511 (1956). Degradation in the environment to Kepone and related compds: D. A. Carlson *et al., Science* **194**, 939 (1976). Carcinogenicity study: B. M. Ullard *et al., J. Nat. Cancer Inst.* **58**, 133 (1977). Toxicity data: T. B. Gaines, *Toxicol. Appl. Pharmacol.* **14**, 515 (1969). *Reviews:* Ungnade, McBee, *Chem. Rev.* **58**, 249-320 (1958); Alley, *J. Environ. Qual.* **2**, 52-61 (1973); E. M. Waters *et al., Environ. Res.* **14**, 212-222 (1977); K. Kaiser, *Environ. Sci. Technol.* **12**, 520-528 (1978).

Snow-white odorless crystals from benzene, dec 485°. Supports combustion. Practically insol in water. Soly at room temp: dioxane 15.3%; xylene 14.3%; benzene 12.2%; carbon tetrachloride 7.2%; methyl ethyl ketone 5.6%. Practically non-corrosive to metals. LD_{50} orally in female rats (corn oil suspension): 600 mg/kg (Gaines).

Note: This substance may reasonably be anticipated to be a carcinogen: *Fourth Annual Report on Carcinogens* (NTP 85-002, 1985) p 135.

USE: Insecticide; fire retardant for plastics, rubber, paint, paper, electrical goods.

6127. Miroprofen. *4-Imidazo[1,2-a]pyridin-2-yl-α-methylbenzeneacetic acid;* p-imidazo[1,2-a]pyridin-2-ylhydratropic acid; 2-[p-(2-imidazo[1,2-a]pyridyl)phenyl]propionic acid; Y-9213; Antopen. $C_{16}H_{14}N_2O_2$; mol wt 266.30; C 72.16%, H 5.30%, N 10.52%, O 12.02%. Prepn and anti-inflammatory activity: M. Nakanishi *et al.,* **Belg. pat. 817,-247;** *eidem.,* U.S. pat. **3,978,071** (1974, 1976 both to Yoshitomi). Physical properties: J. Sakai *et al., Iyakuhin Kenkyu* **12**, 790 (1981), *C.A.* **95**, 175692r. Metabolism in man: Y. Kato *et al., ibid.* 852. Mode of action: K. Goto *et al., Japan. J. Pharmacol.* **28**, 433 (1978); K. Goto *et al., ibid.* **29**, 67 (1979). Inhibitory effect on platelet aggregation: H. Mikashima, K. Goto, *Yakugaku Zasshi* **102**, 99 (1982). Analgesic and antipyretic activity in mice, dogs: Y. Maruyama *et al., Arzneimittel-Forsch.* **28**, 2102 (1978). Comparison with other anti-inflammatories in laboratory animals: Y. Maruyama *et al., ibid.* **31**, 1111 (1981). Efficacy in acute upper respiratory tract infection: M. Katsu *et al., Kansen-shogaku Zasshi* **56,** 434 (1982).

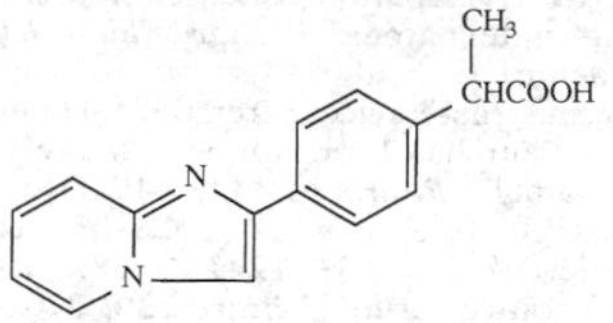

Colorless crystals from methanol, mp 244-246° (dec). Stable against heat and moisture, yellows slightly on exposure to sunlight. Soly in water, ethanol, benzene (w/v): 0.01%, 0.28%, 0.0006%. LD_{50} in male, female mice, male, female rats (mg/kg): 570, 784, 371, 292 orally; 798, 696, 419, 365 i.p.; 687, 878, 396, 372 s.c. (Maruyama, 1978).
Hydrochloride, $C_{16}H_{15}ClN_2O_2$, mp 210-211°.
Ethyl ester, $C_{18}H_{18}N_2O_2$. Colorless scales from isopropyl ether, mp 110-112°.
THERAP CAT: Anti-inflammatory. Analgesic.

6128. Misch Metal. Mixed rare earth metal. Prepd by electrolysis of fused rare earth chlorides. Misch metal contains 94-99% rare earth metals plus traces of aluminum, calcium, carbon, silicon, and iron. A typical composition is 50-60% cerium, 25-30% lanthanum, 15-17% neodymium, 4-6% praseodymium.
USE: In the manuf of flints for gas and cigarette lighters. In metallurgy: decreases microporosity and improves stress resistance, thermomalleability and heat exchange in refined steel.

6129. Misoprostol. *(11α,13E)-(±)-11,16-Dihydroxy-16-methyl-9-oxoprost-13-en-1-oic acid methyl ester; (±)-methyl-(1R,2R,3R)-3-hydroxy-2-[(E)-(4RS)-4-hydroxy-4-methyl-1-octenyl]-5-oxocyclopentaneheptanoate; (±)-15-deoxy-(16RS)-16-hydroxy-16-methyl-PGE1 methyl ester;* SC 29333; Cytotec. $C_{22}H_{38}O_5$; mol wt 382.54. C 69.08%, H 10.01%, O 20.91%. Cytoprotective prostaglandin PGE_1 analog comprised of four stereoisomers in approximately equal proportions (the (+)- and (−)-enantiomers of the 16R- and 16S-forms). Prepn: P. W. Collins, R. Pappo, **Belg.** pat. **827,127**; *eidem,* **U.S.** pat. **3,965,143** (1975, 1976 both to Searle); P. W. Collins *et al., Tetrahedron Letters* **48,** 4217 (1975); H. C. Kluender *et al.,* **U.S.** pat. **4,132,738** (1979 to Miles). Prepn, activity, NMR data: P. Collins *et al., J. Med. Chem.* **20,** 1152 (1977). Mechanism of gastric secretory inhibition: D. G. Colton *et al., Arch. Int. Pharmacodyn. Ther.* **236,** 86 (1978). Protection of gastric mucosa in dogs: K. R. Larsen *et al., Prostaglandins* **21,** Suppl., 119 (1981); in man: C. J. Fimmel *et al., Scand. J. Gastroenterol.* **92,** 184 (1984). Toxicology study: F. N. Kotsonis *et al., Dig. Dis. Sci.* **30,** Suppl. 11, 142S (1985). Symposium on pharmacology and clinical efficacy: *ibid.* 114S-205S. Multicenter clinical trial in prevention of NSAID-induced gastric ulcer: D. Y. Graham *et al., Lancet* **2,** 1277 (1988).

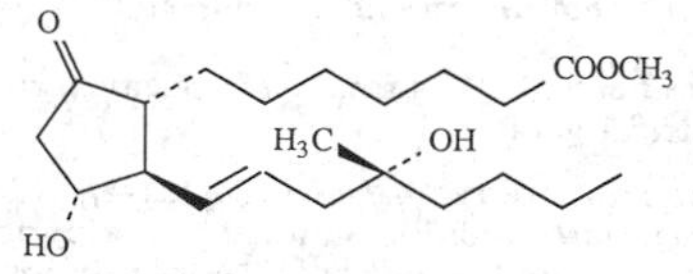

(±)-S-form

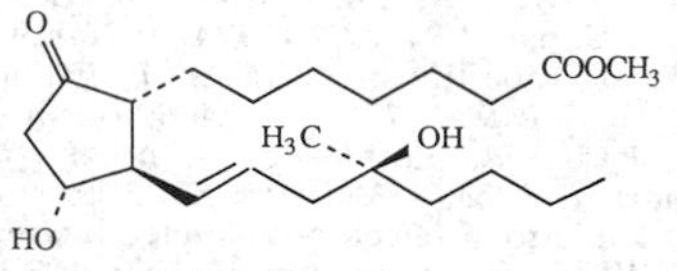

(±)-R-form

Light yellow oil. LD_{50} in rats, mice (mg/kg): 40-62, 70-160 i.p.; 81-100, 27-138 orally (Kotsonis).

THERAP CAT: Anti-ulcerative.

6130. Mitobronitol. *1,6-Dibromo-1,6-dideoxy-D-mannitol;* 1,6-dideoxy-1,6-dibromo-D-mannitol; dibromomannitol; DBM; Myebrol; Myelobromol. $C_6H_{12}Br_2O_4$; mol wt 308.00. C 23.40%, H 3.93%, Br 51.89%, O 20.78%. Cytostatic agent; the first dibromohexitol introduced into clinical use. Manuf: **Brit.** pat. **959,407** (1964 to Chinoin). Pharmaco-biochemical studies: J. Szabo *et al., Neoplasma* **20,** 13 (1973). Clinical studies: D. F. Chiuten *et al., Cancer* **47,** 442 (1981). Comparative study with busulfan, *q.v.:* R. T. Silver *et al., Cancer* **60,** 1442 (1987).

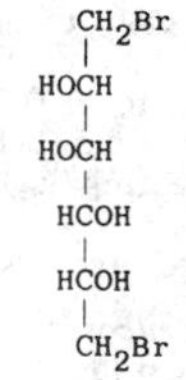

Crystals from methanol + dichloroethane, mp 176-178°.
THERAP CAT: Antineoplastic.

6131. Mitoguazone. *2,2'-(1-Methyl-1,2-ethanediylidene)bis[hydrazinecarboximidamide]; 1,1'-[(methylethanediylidene)dinitrilo]diguanidine;* methylglyoxal bis(guanylhydrazone); pyruvaldehyde bis(amidinohydrazone). $C_5H_{12}N_8$; mol wt 184.21. C 32.60%, H 6.57%, N 60.83%. Prepn: Baiocchi *et al., J. Med. Chem.* **6,** 431 (1963); Oliverio, Denham, *J. Pharm. Sci.* **52,** 202 (1963). Experimentally effective against myelogenous leukemia: Freireich *et al., Cancer Chemother. Repts.* **16,** 183 (1962).

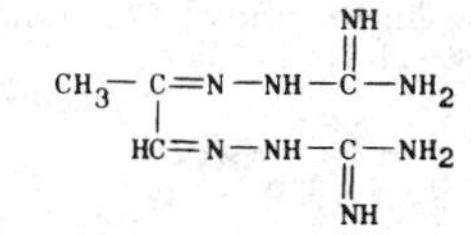

Hemihydrate, crystals from isopropyl alcohol, dec 225°. uv max at pH 1: 283 nm (ε 38,400); at pH 11: 325 nm (ε 33,500).
Dihydrochloride, $C_5H_{14}Cl_2N_8$, *NSC-32946, Methyl-GAG.* Crystals from acetone, dec 256-257°.
Diacetate, *NSC-30689.*
THERAP CAT: Antineoplastic.

6132. Mitolactol. *1,6-Dibromo-1,6-dideoxygalactitol;* 1,6-dibromo-1,6-dideoxydulcitol; dibromodulcitol; dibromodulcit; DBD; NSC-104800; Elobromol; Mitolac. $C_6H_{12}Br_2O_4$; mol wt 307.98. C 23.40%, H 3.93%, Br 51.89%, O 20.78%. Diastereoisomer of mitobronitol, *q.v.,* with myelosuppressive effects. Prepn and anticancer activity: **Neth.** pat. **Appl. 6,600,395** corresp to P. Horváth-Lengyel *et al.,* **U.S.** pat. **3,993,781** (1966, 1976 both to Chinoin); B. Kellner *et al., Nature* **213,** 402 (1967). Structure-activity correlations: L. Institoris *et al., Arzneimittel-Forsch.* **17,** 145 (1967). Pharmacological study: T. H. Corbett *et al., Cancer* **40,** Suppl. 5, 2660 (1977). Metabolism, pharmacokinetics: I. P. Horváth *et al., Eur. J. Cancer* **15,** 337 (1979). Clinical studies: W. Medina, J. M. Kirkwood, *Cancer Treat. Rep.* **66,** 195 (1982); D. C. Tormey *et al., Am. J. Clin. Oncol.* **5,** 33 (1982).

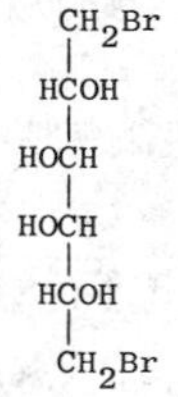

Cryst from methanol, aq alc, or aq acetone, mp 187-188°

(dec). LD$_{50}$ in rats (mg/kg): 1400 orally; 470 i.p., B. Kellner *et al.*, *loc. cit.*

THERAP CAT: Antineoplastic.

6133. Mitomycins. A group of antitumor antibiotics produced by *Streptomyces caespitosus (griseovinaceseus)*. Isoln of mitomycins A and B: T. Hata *et al.*, *J. Antibiot.* **9A**, 141 (1956). Isoln of C: S. Wakaki *et al.*, *Antibiot. Chemother.* **8**, 228 (1958). Production of the complex: Yamamoto, Umezawa, **Japan.** pat. **2898** (1956), *C.A.* **51**, 9100 (1957); Kenkyusho, **Brit.** pat. **830,874** (1960 to Kyowa), *C.A.* **54**, 20071 (1960); Gourevitch *et al.*, U.S. pat. **3,042,582** (1962 to Bristol-Myers). Isoln from *S. verticillatus*, constituents: D. V. Lefemine *et al.*, *J. Am. Chem. Soc.* **84**, 3184 (1962). Structures: J. S. Webb *et al.*, *ibid.* 3185, 3187. Total synthesis of mitomycins A and C: T. Fukuyama *et al.*, *Tetrahedron Letters* **1977**, 4295. Toxicity: T. Hata *et al.*, *loc. cit.*; S. Kinoshita *et al.*, *J. Med. Chem.* **14**, 13 (1971). Review and evaluation of studies of carcinogenic action of mitomycin C in laboratory animals: *IARC Monographs* **10**, 171-179 (1976). Review of syntheses: Y. Kishi, *J. Nat. Prod.* **42**, 551-568 (1979). General review: R. W. Franck, *Fortschr. Chem. Org. Naturst.* **38**, 1-41 (1979).

mitomycin C

Mitomycin A, C$_{16}$H$_{19}$N$_3$O$_6$. The 6-methoxy analog of mitomycin C. X-ray crystallographic study: A. Tulinsky, *J. Am. Chem. Soc.* **84**, 3188 (1962); A. Tulinsky, J. H. van den Hende, *ibid.* **89**, 2905 (1967). Red-violet crystals from acetone + carbon tetrachloride, dec 159-161°. Absorption max (water): 215, 318, 530 nm (E$_{1cm}^{1\%}$ 234, 122, 118.8). Sol in water, benzene, toluene, trichloroethylene, nitrobenzene and many organic solvents. Practically insol in xylene, carbon tetrachloride, carbon disulfide, petr ether, ligroin, cyclohexane. LD$_{50}$ i.v. in mice: 2 mg/kg (Hata; Kinoshita).

Mitomycin B, C$_{16}$H$_{19}$N$_3$O$_6$. Molecular structure: R. Yahashi, I. Matsubara, *J. Antibiot.* **29**, 104 (1976); *ibid.* **31**(6), correction (1978). Violet crystals from acetone + carbon tetrachloride, dec 182-184°. Absorption max (water): 220, 320, 550 nm (E$_{1cm}^{1\%}$ 117.5, 55.0, 9.9). Sol in water and many organic solvents. Practically insol in xylene, carbon tetrachloride, carbon disulfide, petr ether, ligroin, cyclohexane, benzene, toluene, trichloroethylene, nitrobenzene. LD$_{50}$ i.v. in mice: 10 mg/kg (Hata); also reported as 3 mg/kg (Kinoshita).

Mitomycin C, C$_{15}$H$_{18}$N$_4$O$_5$, *[1aS-(1aα,8β,8aα,8bα)]*-6-amino-8-[[(aminocarbonyl)oxy]methyl]-1,1a,2,8,8a,8b-hexahydro-8a-methoxy-5-methylazirino[2',3':3,4]pyrrolo[1,2-a]-indole-4,7-dione, *Ametycine*, *MMC*, *Mitocin-C*, *Mutamycin*. Chemistry and structure: Stevens *et al.*, *J. Med. Chem.* **8**, 1 (1965). Crystal and molecular structure: K. Ogawa *et al.*, *Bull. Chem. Soc. Japan* **52**, 2334 (1979). Revised absolute configuration: K. Shirahata, N. Hirayama, *J. Am. Chem. Soc.* **105**, 7199 (1983). Mechanism of action: C. Rodigheriero, *Farmaco Ed. Sci.* **33**, 651 (1978). Comprehensive description: J. H. Beijnen *et al.*, in *Analytical Profiles of Drug Substances* vol. **16**, K. Florey, Ed. (Academic Press, New York, 1986) pp 361-401. Blue-violet crystals, does not melt below 360°. Absorption max (methanol): 216, 360, 560 nm (E$_{1cm}^{1\%}$ 742, 742, 0.06). Sol in water, methanol, acetone, butyl acetate and cyclohexanone; slightly sol in benzene, carbon tetrachloride, ether. Practically insol in petr ether. LD$_{50}$ i.v. in mice: 5 mg/kg (Wakaki); also reported as 9 mg/kg (Kinoshita). Mitomycin C exhibits delayed toxicity to mice; injected mice die from 2 days to 2 weeks after injection.

N-Methylmitomycin C, see *Porfiromycin*.

THERAP CAT: Mitomycin C as antineoplastic.

6134. Mitotane. *1-Chloro-2-[2,2-dichloro-1-(4-chlorophenyl)ethyl]benzene; 1,1-dichloro-2-(o-chlorophenyl)-2-(p-chlorophenyl)ethane; 2-(2-chlorophenyl)-2-(4-chlorophenyl)-1,1-dichloroethane; 2,4'-dichlorodiphenyldichloroethane; 2,2-bis(2-chlorophenyl-4-chlorophenyl)-1,1-dichloroethane; o,p'-DDD; CB 313; Lysodren.* C$_{14}$H$_{10}$Cl$_4$; mol wt 320.05. C 52.54%, H 3.15%, Cl 44.31%. Constituent of commercial DDD which contains about 10% of this *o,p'*-isomer. Prepn from 2,2-dichloro-1-(o-chlorophenyl)ethanol with chlorobenzene in presence of H$_2$SO$_4$: Haller *et al.*, *J. Am. Chem. Soc.* **67**, 1600 (1945). Isoln from technical grade DDD: Cueto, Brown, *Endocrinology* **62**, 326 (1958).

Crystals from pentane or methanol, mp 76-78°. Sol in ethanol, isooctane, carbon tetrachloride.

THERAP CAT: Antineoplastic.

6135. Mitoxantrone. *1,4-Dihydroxy-5,8-bis[[2-[(2-hydroxyethyl)amino]ethyl]amino]-9,10-anthracenedione; 1,4-dihydroxy-5,8-bis[[2-[(2-hydroxyethyl)amino]ethyl]amino]-9,10-anthraquinone; DHAQ; NSC-279836.* C$_{22}$H$_{28}$N$_4$O$_6$; mol wt 444.09. C 59.45%, H 6.35%, N 12.60%, O 21.60%. Cytostatic deriv of anthraquinone, *q.v.* Prepn: R. K. Y. Zee-Cheng, C. C. Cheng, *J. Med. Chem.* **21**, 291 (1978); K. C. Murdock *et al.*, *ibid.* **22**, 1024 (1979); *eidem*, **Ger.** pat. **2,835,661**, *C.A.* **91**, 56706 (1979) and K. C. Murdock, F. E. Durr, U.S. pat. **4,197,249** (1979, 1980 both to Am. Cyanamid). Activity vs exptl tumors in mice: R. E. Wallace *et al.*, *Cancer Res.* **39**, 1570 (1979). Pharmacological study: K. Lu, T. L. Loo, *ibid.* **40**, 1427 (1980). HPLC studies: R. F. Taylor, L. A. Gaudio, *J. Chromatog.* **187**, 212 (1980); D. L. Reynolds *et al.*, *ibid.* **222**, 225 (1981). Plasma levels in humans: *eidem*, *Int. J. Pharm.* **9**, 67 (1981). Clinical studies: W. R. Wynert *et al.*, *Cancer Treat. Rep.* **66**, 1303 (1982); J. A. Stewart *et al.*, *ibid.* 1327; J. D. Cowan *et al.*, *ibid.* 1779. Safety assessment: B. M. Henderson *et al.*, *ibid.* 1139; B. M. Sparano *et al.*, *ibid.* 1145. Cardiotoxicity study: G. Zbinden, A. K. Beilstein, *Toxicol. Letters* **11**, 289 (1982). Mutagenicity study: A. Nishio *et al.*, *Mutat. Res.* **101**, 77 (1982). *Reviews:* F. Traganos, *Pharmacol. Ther.* **22**, 199-214 (1983); M. McDonald *et al.*, *Drugs Exp. Clin. Res.* **10**, 745-752 (1984). Comprehensive description: J. H. Beijnen *et al.*, in *Analytical Profiles of Drug Substances* vol. **17**, K. Florey, Ed. (Academic Press, New York, 1988) pp 221-258.

Crystals from ethanol/hexane, mp 160-162°. Abs max (ethanol): 244, 279, 525, 620, 660 nm (log ε 4.64, 4.31, 3.70, 4.37, 4.38). Sparingly sol in water, slightly sol in methanol. Practically insol in acetonitrile, chloroform, acetone.

Dihydrochloride, C$_{22}$H$_{30}$Cl$_2$N$_4$O$_6$, *CI-232315*, *NSC-301739*, *DHAD*, *Novantrone*. Hygroscopic blue-black solid from water/ethanol, mp 203-205°. Abs max (water): 241, 273, 608, 658 nm (ε 41000, 12000, 19200, 20900). Sparingly sol in water, slightly sol in methanol. Practically insol in acetonitrile, chloroform, acetone.

THERAP CAT: Antineoplastic.

6136. Mitragynine. *(16E,20β)-16,17-Didehydro-9,17-dimethoxycorynan-16-carboxylic acid methyl ester; (E)-16,17-didehydro-9,17-dimethoxy-17,18-seco-20α-yohimban-16-carboxylic acid methyl ester; 9-methoxycorynantheidine.* C$_{23}$H$_{30}$N$_2$O$_4$; mol wt 398.49. C 69.32%, H 7.59%, N 7.03%, O 16.06%. Major alkaloid of *Mitragyna speciosa* Korth., *Rubiaceae*: Field, *J. Chem. Soc.* **119**, 887 (1921); Ing, Raison, *ibid.* **1939**, 986. Structural studies: Hendrickson, *Chem. & Ind.* *(London)* **1961**, 713; Joshi *et al.*, *ibid.* **1963**, 573. Revised structure: Zacharias *et al.*, *Acta Cryst.* **18**,

1039 (1965). Pharmacology: Macko *et al., Arch. Int. Pharmacodyn. Ther.* **198**, 145 (1972).

White, amorphous powder, mp 102-106°. bp_5 230-240°. $[\alpha]_D$ +39° (chloroform). uv max: 226, 292 nm (ϵ 41,150, 6600). Soluble in alcohol, chloroform, acetic acid.

6137. Mizoribine. *5-Hydroxy-1-β-D-ribofuranosyl-1H-imidazole-4-carboxamide;* 4-carbamoyl-1-β-D-ribofuranosylimidazolium-5-olate; HE-69; Bredinin. $C_9H_{13}N_3O_6$; mol wt 259.22. C 41.70%, H 5.05%, N 16.21%, O 37.03%. Nucleoside antibiotic produced by *Eupenicillium brefedianum* with cytotoxic and immunosuppressive activity. Isoln: K. Mizuno *et al.,* **Belg.** pat. 799,805; *eidem,* **U.S.** pat. 3,888,843 (1973, 1975 to Toyo Jozo); *eidem, J. Antibiot.* **27**, 775 (1974). Synthesis: M. Hayashi *et al., Chem. Pharm. Bull.* **23**, 245 (1975); K. Fukukuwa *et al., ibid.* **32**, 1644 (1984). Prepn of the aglycone: T. Atsumi *et al.,* **Japan. Kokai 76 88,965** (1976 to Sumitomo), *C.A.* **86**, 106582g (1977). HPLC determn in human serum: K. Takada *et al., J. Chromatog.* **222**, 156 (1981). Anti-arthritic activity: H. Iwata *et al., Experientia* **33**, 502 (1977). Cytotoxic effect and comparison with aglycone: K. Sakaguchi *et al., J. Antibiot.* **28**, 798 (1975). Antitumor spectrum of aglycone: N. Yoshida *et al., Cancer Res.* **43**, 5851 (1983). Pharmacokinetics: K. Takada, *Eur. J. Pharmacol.* **24**, 457 (1983). Clinical trials in renal transplantation: T. Inou *et al., Transplantation Proc.* **12**, 526 (1980); *eidem, ibid.* **13**, 315 (1981); A. Tajima *et al., Transplantation* **38**, 116 (1984).

Crystals from methanol, mp >200° (dec). $[\alpha]_D^{27}$ −35° (c = 0.8 in H_2O). pKa 6.75. uv max (H_2O): 245, 279 nm (E 250, 580). Sol in water. Slightly sol in methanol, ethanol. Insol in most organic solvents. LD_{50} in mice (g/kg): >1.5 i.v., >2.4 i.p. (Mizuno, 1975).

Aglycone, $C_4H_5N_3O_2$, *5-hydroxy-1H-imidazole-4-carboxamide;* 4-carbamoyl-5-hydroxyimidazole; 4-carbamoylimidazolium-5-olate, SM-108.

THERAP CAT: Immunosuppressant.

6138. Mocimycin. Antibiotic MYC 8003; delvomycin; kirromycin; MYC 8003. $C_{43}H_{60}N_2O_{12}$; mol wt 796.96. C 64.81%, H 7.59%, N 3.51%, O 24.09%. Antibiotic produced by *Streptomyces ramocissimus.* Isoln and properties: C. Vos, J. Den Admirant, **Ger.** pat. 2,140,674 corresp to U.S. pats. 3,927,211 and 3,923,981 (1972, 1975, 1975, all to Gist-Brocades). Structural studies: C. Vos, P. E. J. Verwiel, *Tetrahedron Letters* **1973**, 2823, 5173. Total structure and identity with kirromycin: H. Maehr *et al., J. Am. Chem. Soc.* **95**, 8449 (1973). Absolute stereochemistry: *eidem, ibid.* **96**, 4034 (1974). Conversion to aurodox: H. Maehr *et al., J. Antibiot.* **32**, 361 (1979). Biological studies: H. Wolf *et al., Proc. Nat. Acad. Sci. USA* **71**, 4910 (1974); J. A. M. Van De Klundeit *et al., FEBS Letters* **81**, 303 (1977). Review of

chemistry: H. Maehr *et al., Can. J. Chem.* **58**, 501-526 (1980). Review of properties and actions: A. Parmeggiani, G. Sander, *Topics in Antibiotic Chemistry* **vol. 5**, P. G. Sammes, Ed. (Halsted Press, New York, 1980) pp 159-221.

Yellow solid. Weakly acidic. $[\alpha]_D^{22}$ −60° (c = 1 in methanol). uv max (methanol/water): 233, 276, 286, 327 nm. Sol in chloroform, methyl isobutyl ketone, butyl acetate, ethyl acetate, acetone, methanol, alkaline soln. Slightly sol in CCl_4, benzene. Insol in diethyl ether, petr ether, water, acid soln. Stable for 4 hrs in 50% aq methanol at pH 3-12. The solid antibiotic stored at 25° and 37° and low relative humidity shows no activity loss for 5 mos. Stable for 3 mos at 25°, 100% relative humidity; for 2 mos at 37°, 100% relative humidity. LD_{50} in mice: >1000 mg/kg i.p., C. Vos, J. Den Admirant, *loc. cit.*

1-Methyl deriv, see Aurodox.

5,6-Dihydromocimycin, $C_{43}H_{62}N_2O_{12}$. Isoln from *S. ramocissimus* and properties: H. Jongsma *et al., Ger.* pat. **2,658,-977** (1977 to Gist-Brocades), *C.A.* **87**, 135078y (1977); *see also* C. Vos *et al.,* U.S. pat. 4,062,948 (1977 to Gist-Brocades). Pale yellow solid, dec begins at 123°. $[\alpha]_D^{20}$ −85° (1% methanolic soln). uv max (methanol/water): 233.5, 267, 291, 333 nm (ϵ 63,000, 23,000, 19,000, 18,000). Solubility similar to mocimycin in most solvents.

USE: Animal growth promotant.

6139. Moclobemide. *4-Chloro-N-[2-(4-morpholinyl)-ethyl]benzamide;* p-chloro-*N*-(2-morpholinoethyl)benzamide; Ro 11-1163; Aurorix. $C_{13}H_{17}ClN_2O_2$; mol wt 268.74. C 58.10%, H 6.38%, Cl 13.19%, N 10.42%, O 11.91%. Reversible monoamine oxidase A (MAO-A) inhibitor. Prepn: W. Burkard, P.-C. Wyss, **Ger.** pat. 2,706,179; **U.S.** pat. 4,210,-754 (1977, 1980 both to Hoffmann-La Roche). Gas chromatography: K. P. Maguire *et al., J. Chromatog.* **278**, 429 (1983). Human pharmacokinetics: F. A. Wiesel *et al., Eur. J. Clin. Pharmacol.* **28**, 89 (1985). Clinical study: J. K. Larsen *et al., Acta Psychiat. Scand.* **70**, 254 (1984). Review of pharmacology: W. P. Burkard, *J. Pharmacol. Exp. Ther.* **248**, 391-399 (1989); of neurochemical profile: M. DaPrada *et al., ibid.* 400-414.

Crystals from isopropanol, mp 137°. LD_{50} in rats (mg/kg): 707 orally (Burkard, Wyss).

Hydrochloride. $C_{13}H_{18}Cl_2N_2O_2$, crystals from isopropanol, mp 208°.

THERAP CAT: Antidepressant.

6140. Modacrylic Fibers. A generic term for acrylic fibers composed of a long chain synthetic polymer containing 35-85% of acrylonitrile units such as *Dynel* and *Verel.* See: R. W. Moncrieff, *Man-Made Fibres* (Wiley, New York, 4th ed., 1963) p 9. *Review:* Kennedy in *Encyclopedia of Polymer Science and Technology* **vol. 8**, N. M. Bikales, Ed. (Interscience, New York, 1968) pp 812-839.

USE: In pile fabrics, carpeting, paint rollers.

6141. Mofebutazone. *4-Butyl-1-phenyl-3,5-pyrazolidinedione;* 4-butyl-1-phenyl-3,5-dioxopyrazolidine; 2-phenyl-3,5-dihydroxy-4-butylpyrazolidine; monophenylbutazone; Arcomonol Tablets; Mobutazon; Mobuzon; Mofesal; Monazan; Monobutyl; Monorheumetten; Reumatox. C_{13}-$H_{16}N_2O_2$; mol wt 232.27. C 67.22%, H 6.94%, N 12.06%, O 13.78%. Prepn: Büchi *et al., Helv. Chim. Acta* **36**, 75 (1953); **Brit.** pat. 839,057 (1960 to Comm. Farm. Milanese). Toxic-

Consult the cross index before using this section.

ity data: Schoetensack, *Arch. Exp. Pathol. Pharmakol.* **233**, 365 (1958).

Crystals from ethanol + water, mp 102-103°. uv max (ethanol): 240, 275 nm ($E_{1cm}^{1\%}$ 443, 245). LD_{50} i.v. in mice: 600 mg/kg (Schoetensack).

THERAP CAT: Anti-inflammatory.

6142. Molindone. *3-Ethyl-1,5,6,7-tetrahydro-2-methyl-5-(4-morpholinylmethyl)-4H-indol-4-one;* 3-ethyl-6,7-dihydro-2-methyl-5-(morpholinomethyl)indol-4(5H)-one. $C_{16}H_{24}N_2O_2$; mol wt 276.37. C 69.53%, H 8.75%, N 10.14%, O 11.58%. Prepn: **Belg. pat.** 670,798 (1966 to Endo), *C.A.* **65**, 7148f (1966). Pharmacology: Sugerman, Herrmann, *Clin. Pharmacol. Ther.* **8**, 261 (1967); Claghorn, *Curr. Ther. Res.* **11**, 524 (1969); Guerrero-Figueroa *et al., ibid.* **15**, 508 (1973).

Crystals, mp 180-181°.
Hydrochloride, *EN 1733 A, Lidone, Moban.* LD_{50} orally in rats: 261 mg/kg, E. I. Goldenthal, *Toxicol. Appl. Pharmacol.* **18**, 185 (1971).

THERAP CAT: Antipsychotic.

6143. Molsidomine. *N-(Ethoxycarbonyl)-3-(4-morpholinyl)sydnone imine;* N-carboxy-3-morpholinosydnonimine ethyl ester; morsydomine; SIN-10; Corvaton; Corvasal; Molsidolat; Morial; Motazomin. $C_9H_{14}N_4O_4$; mol wt 242.23. C 44.62%, H 5.83%, N 23.13%, O 26.42%. Coronary vasodilator; member of a class of non-benzene aromatic, heterocyclic and mesoionic type of compounds previously unknown in the pharmaceutical industry. Developmental work on sydnone imines: Brookes, Walker, *J. Chem. Soc.* **1957**, 4409. Prepn: K. Masuda *et al.,* **Japan.** pat. 6,265-('70); *eidem,* **U.S.** pat. 3,769,283 (1970, 1973 both to Takeda); *eidem, Chem. Pharm. Bull.* **19**, 72 (1971). Stability studies: Asahi *et al., ibid.* **19**, 1079 (1971). Pharmacological studies: Kikuchi *et al., Japan. J. Pharmacol.* **20**, 102, 187, 253 (1970); Hashimoto *et al., Arzneimittel-Forsch.* **21**, 1329 (1971); T. Nakaguchi *et al., Toho Igakkai Zasshi* **17**, 26 (1970). Metabolism: S. Tanayama *et al., Xenobiotica* **4**, 175 (1974). Pharmacokinetics: R. Bergstrand *et al., Eur. J. Clin. Pharmacol.* **27**, 203 (1984). Clinical trial in angina pectoris: P. A. Majid *et al., N. Engl. J. Med.* **302**, 1 (1980); in hypertension: J. Melei *et al., Eur. J. Clin. Pharmacol.* **18**, 231 (1980). Synergism with penbutolol in angina pectoris: A. E. Balestrini *et al., ibid.* **27**, 1 (1984). Review of pharmacology: R.-E. Nitz, V. B. Fiedler, *Pharmacotherapy* **7**, 28 (1987). *Review: Japan. Med. Gaz.* **8**(9), 10 (1971); E. Bassenge, W. R. Kukovetz in *New Drugs Annual: Cardiovascular Drugs,* Vol. 2, A. Scriabine, Ed. (Raven Press, New York, 1984) pp 177-191.

Colorless crystals or white cryst powder, practically tasteless and odorless, mp 140-141° (toluene). Freely sol in $CHCl_3$. Sol in dil HCl, ethanol, ethyl acetate, methanol; sparingly sol in water, acetone, benzene. Very slightly sol in ether, petr ether. pK (100°) 3.0 ± 0.1. Most stable in aq solns pH 5-7; least stable in very alkaline solns. uv max ($CHCl_3$): 326 nm. Sensitive to light of λ < 320 mμ. LD_{50} in

male, female mice, male, female rats (mg/kg): 780, 750, 1380, 1350 s.c.; 860, 800, 830, 760 i.v.; 700, 760, 1250, 1250 i.p.; 830, 840, 1050, 1200 orally (Nakaguchi).

THERAP CAT: Antianginal.

6144. Molybdenum. Mo; at. wt 95.94; at. no. 42; valences 2,3,4,5,6. Naturally occurring isotopes: 98 (23.75%); 96 (16.5%); 95 (15.7%); 92 (15.86%); 94 (9.12%); 100 (9.62%); 97 (9.45%); artificial radioactive isotopes: 88-91; 93; 99; 101-105. Its most important ores are molybdenite, MoS_2, and wulfenite, $PbMoO_4$. Occurrence in the earth's crust: 1-1.5 ppm. Discovered in 1778 by Scheele; isolated in 1782 by Hjelm. Methods of preparation: L. Northcott, *Molybdenum* (Academic Press, New York, 1956) 222 pp; Hein, Herzog, in *Handbook of Preparative Inorganic Chemistry* **vol. 2**, G. Brauer, Ed. (Academic Press, New York, 2nd ed., 1965) pp 1401-1402. Important trace element; participates in biochemical redox reactions such as N_2-fixation: Spence, *Coord. Chem. Rev.* **4**, 475 (1969). Physical properties: Worthing, *Phys. Rev.* [2] **25**, 846 (1925); D. R. Stoll, G. C. Sinke, *Thermodynamic Properties of the Elements,* Advances in Chemistry Series **18**, (American Chemical Society, Washington, 1956) pp 23, 130-131. Review of molybdenum and its compds: Rollinson, "Chromium, Molybdenum and Tungsten" in *Comprehensive Inorganic Chemistry* **vol. 3**, J. C. Bailar Jr. *et al.,* Eds. (Pergamon Press, Oxford, 1973) pp 622-623, 700-742; R. Q. Barr in Kirk-Othmer *Encyclopedia of Chemical Technology* **vol. 15** (Wiley-Interscience, New York, 3rd ed., 1981) pp 670-682. Biochemical review: *Bioinorganic Chemistry* **II**, K. N. Raymond, Ed. (A.C.S., Washington, 1977) pp 353-430. Symposium on the chemistry and uses of molybdenum and its cmpds: *Polyhedron* **5**, 1-606 (1986).

Dark-gray or black powder with metallic luster or coherent mass of silver-white color; body-centered cubic structure. mp 2622° (Worthing). bp about 4825°. d 10.28. Spec heat 5.68 cal/g-atom/deg; heat of fusion: 6.6 kcal/g-atom; heat of vaporization: 142 kcal/g-atom (Stoll, Sinke). Fairly stable at ordinary temp; oxidized to the trioxide at a red heat; slowly oxidized by steam. Not attacked by water, by dil acids or by concd hydrochloric acid. Practically insol in alkali hydroxides or fused alkalies. Reacts with nitric acid, hot concd sulfuric acid, fused potassium chlorate or nitrate. Attacked by fluorine at ordinary temp, by chlorine or bromine at a red heat.

Human Toxicity: Limited data suggest low order of toxicity. *See* E. Browning, *Toxicity of Industrial Metals* (Appleton-Century-Crofts, New York, 1969) pp 243-248.

USE: In the form of ferromolybdenum for manufg special steels for tools, boiler plate, rifle barrels, propeller shafts; electrical contacts, spark plugs, x-ray tubes, filaments, screens and grids for radio tubes; in the production of tungsten; glass-to-metal seals; nonferrous alloys; in colloidal form as lubricant additive.

6145. Molybdenum Disulfide. MoS_2; mol wt 160.08. Mo 59.94%, S 40.06%. Occurs as the mineral *molybdenite*, which is the principal source of molybdenum. Lab prepn: Bell, Herfert, *J. Am. Chem. Soc.* **79**, 3351 (1957).

Lead-gray, lustrous powder; the artificially prepd sulfide is black and lustrous. d_{15}^{15} 5.06; mp 2375°. Begins to sublime at 450°. Insol in water or dil acids.

USE: Dry lubricant and lubricant additive. Hydrogenation catalyst.

6146. Molybdenum Hexafluoride. F_6Mo; mol wt 209.95. F 54.30%, Mo 45.70%. MoF_6. Prepd by direct fluorination of powdered molybdenum: Ruff, Ascher, *Z. Anorg. Allgem. Chem.* **196**, 418 (1931); from MoO_3 and SF_4: Oppengard *et al., J. Am. Chem. Soc.* **82**, 3825 (1960).

Volatile, white, cubic crystals. Very hygroscopic. d_{liq}^{19} 2.543. mp 17.5°. bp 35.0°. Hydrolyzed by water. Forms blue-white clouds in moist air. Soly in anhydr HF: 1.5 moles/1000 g HF, Frlec, Hyman, *Inorg. Chem.* **6**, 1596 (1967). Should be stored in quartz ampuls.

6147. Molybdenum Sesquioxide. Mo_2O_3; mol wt 239.90. Mo 79.99%, O 20.01%.

Grayish-black powder. Very slightly sol in acids.
Combination with ferrous sulfate *Mol-Iron* (obsolete).

THERAP CAT: Hematinic (combination with ferrous sulfate).

6148. Molybdenum Trioxide. Molybdic anhydride. MoO_3; mol wt 143.95. Mo 66.66%, O 33.34%. Prepn from ammonium molybdate: Schumb, Hartford, *J. Am. Chem. Soc.* **56**, 2613 (1934). Toxicity studies: L. T. Fairhall *et al.*, "The Toxicity of Molybdenum," *U. S. Public Health Service Bulletin No. 293*, Washington (1945).

White or slightly yellow to slightly bluish powder or granules. d_4^{26} 4.696. mp 795°. Melts to dark-yellow liquid which solidifies to a yellowish-white cryst mass; sublimes at higher temp. bp 1155°. Sol in water (28°) 0.490 g/liter. Sol in concd mineral acids, in solns of alkali hydroxides, ammonia or potassium bitartrate; after strong ignition it is very slightly sol in acids.

USE: Chiefly as a reagent for chemical analysis.

6149. Molybdic(VI) Acid. H_2MoO_4; mol wt 161.96. H 1.24%, Mo 59.24%, O 39.52%. Prepd as the monohydrate from ammonium molybdate and nitric acid: Hein, Herzog in *Handbook of Preparative Inorganic Chemistry*, vol. 2, G. Brauer, Ed. (Academic Press, New York, 2nd ed., 1965) pp 1412-1413.

Monohydrate, white powder. d 3.1. Sparingly sol in cold, more in hot water; sol in NH_4OH, H_2SO_4, fixed alkalies.

6150. Molybdic Acid, 85%. Consists largely of an ammonium molybdate. Contains 84-86% MoO_3. It is the most widely used form of "molybdic acid".

White or slightly yellow powder. Insol or partly sol in water; sol in ammonia or solns of fixed alkali hydroxides.

USE: Principally for the determination of phosphorus or phosphate and lead; also in glazes for ceramics.

6151. Mometasone Furoate. *(11β,16α)-9,21-Dichloro-17-[(2-furanylcarbonyl)oxy]-11-hydroxy-16-methylpregna-1,4-diene-3,20-dione;* 9,21-dichloro-11β,17-dihydroxy-16α-methylpregna-1,4-diene-3,20-dione 17-(2-furoate); Sch-32088; Elocon. $C_{27}H_{30}Cl_2O_6$; mol wt 521.44. C 62.19%, H 5.80%, Cl 13.60%, O 18.41%. Topical corticosteroid. Prepn: E. L. Shapiro, *Eur.* pat. **Appl.** 57,401; *idem*, **U.S.** pat. **4,472,393** (1982, 1984 both to Schering Corp.); E. L. Shapiro *et al.*, *J. Med. Chem.* **30**, 1581 (1987). Clinical trial in psoriasis: R. S. Medansky *et al.*, *Semin. Dermatol.* **6**, 94 (1987).

Crystals from aq methanol, mp 218-220°. $[\alpha]_D^{26}$ +58.3° (dioxane). uv max (methanol) 247 nm (ϵ 26300).

THERAP CAT: Topical anti-inflammatory.

6152. Monacetin. *1,2,3-Propanetriol monoacetate;* acetin; monoacetin; glyceryl monoacetate. $C_5H_{10}O_4$; mol wt 134.13. C 44.77%, H 7.52%, O 47.71%. $C_3H_7O_2$.OOCCH$_3$. Obtained along with the di- and triacetates by heating glycerol with glacial acetic acid.

Colorless, very hygroscopic liquid; characteristic odor. The commercial product is pale yellow. d_4^{20} 1.206. bp$_{17}$ about 158°; bp$_3$ 129-131°. Soluble in water, alcohol, slightly in ether; insol in benzene. LD$_{50}$ s.c. in rats 6.6 g/kg.

USE: In manuf of smokeless powder and dynamite; as a solvent for basic dyes; in tanning leather.

6153. Monarda. American horsemint; wild bergamot. Leaves of *Monarda punctata* L., *Labiatae.* Habit. New York to Florida, west to Texas and Wisconsin. *Constit.* Volatile oils: thymol, carvacrol, neryl formate, geranyl formate, cineole, γ-terpinene, α- and β-pinene. Gas chromatographic analysis of the *oil of monarda*: R. W. Scora, *J. Chromatog.* **19**, 601 (1965).

USE: Source for thymol.

THERAP CAT: Aromatic stimulant, carminative.

6154. Monardein Chloride. Monardaein chloride; pelargonidin-3-(6-*p*-coumaroyl)glucosido-5-glucoside chloride. $C_{36}H_{37}ClO_{17}$; mol wt 777.16. C 55.64%, H 4.80%, Cl 4.56%, O 35.00%. Coloring matter of *Monarda didyma* L. ("golden balm"), *Labiatae:* Karrer, Widmer, *Helv. Chim. Acta* **10**, 67 (1927); Karrer, *ibid.* **11**, 837 (1928). Structure: Birkofer *et al.*, *Z. Naturforsch.* **20b**, 424 (1965).

Red powder. $[\alpha]_D$ −374.5° (0.1% HCl). Absorption max (methanolic HCl): 286, 313, 507 nm. Sol in water and methanol. The aq soln is scarlet in color, changed to violet by Na_2CO_3 and to light brown by excess NaOH.

6155. Monel®. A line of nickel-copper corrosion-resistant alloys available in wrought and cast forms.

USE: For acid-resistant apparatus, filter cloth, screens.

6156. Monellins. Intensely sweet principle from the fruit of the tropical plant *Dioscoreophyllum cumminsii* Diels, *Menispermaceae* ("Serendipity Berry"). First believed to consist of a single polypeptide chain of approx 91 amino acids, mol wt about 10,700. Preliminary data and extraction: Inglett, May, *J. Food Sci.* **34**, 408 (1969); Van der Wel, *FEBS Letters* **21**, 88 (1971); Van der Wel, Loeve, *ibid.* **29**, 181 (1973). Purification: Morris, Cagan, *Biochim. Biophys. Acta* **261**, 114 (1972). Characterized by scientists at the Univ. of Pennsylvania's Monell Chemical Senses Center as the first protein to elicit a sweet taste in man, a "chemostimulatory protein". *See* Morris *et al.*, *J. Biol. Chem.* **248**, 534 (1973); Cagan, *Science* **181**, 32 (1973). Monellin is composed of two nonidentical subunits: Z. Bohak, S. Li, *Biochim. Biophys. Acta* **427**, 153 (1976); the presence of both subunits is necessary for the sweet taste to occur: B. Jirgenson, *ibid.* **446**, 255 (1976). Structure of subunit I: G. Hudson, K. Biemann, *Biochem. Biophys. Res. Commun.* **71**, 212 (1976). Complete amino acid sequence of subunits A and B consisting of 44 and 50 amino acids, resp.: G. Frank, H. Zuber, *Z. Physiol. Chem.* **357**, 585 (1976). Crystallographic studies: A. Wlodawer, K. Hodgson, *Proc. Nat. Acad. Sci. USA* **72**, 398 (1975); G. E. Tomlinson, S. H. Kim, *J. Biol. Chem.* **256**, 12476 (1981).

Water soluble. Approximately 3000 times sweeter than sucrose on a weight basis. Essentially free from carbohydrate: <5 μg/mg protein. uv max (0.01M sodium phosphate buffer): 277 nm.

USE: Potential low-calorie sweetener.

6157. Monensin. *2-[5-Ethyltetrahydro-5-[tetrahydro-3-methyl-5-[tetrahydro-6-hydroxy-6-(hydroxymethyl)-3,5-dimethyl-2H-pyran-2-yl]-2-furyl]-2-furyl]-9-hydroxy-β-methoxy-α,γ,2,8-tetramethyl-1,6-dioxaspiro[4.5]decane-7-butyric acid;* monensic acid (obsolete); A 3823A. $C_{36}H_{62}O_{11}$; mol wt 670.90. C 64.45%, H 9.32%, O 26.23%. Polyether antibiotic. Major factor in antibiotic complex isolated from *Streptomyces cinnamonensis.* Discovery and isolation: Haney, Hoehn, *Antimicrob. Ag. Chemother.* **1967**, 349. Production: Haney, Hoehn, U.S. pat. **3,501,568** (1970 to Lilly).

Structure: Agtarap *et al., J. Am. Chem. Soc.* **89**, 5737 (1967). Crystal structure studies: Lutz *et al., Helv. Chim. Acta* **53**, 1732 (1970); *ibid.* **54**, 1103 (1971). Fermentation studies: Stark *et al., Antimicrob. Ag. Chemother.* **1967**, 353. Chemistry: Agtarap, Chamberlin, *ibid.* 359. Stereocontrolled total synthesis: T. Fukuyama *et al., J. Am. Chem. Soc.* **101**, 262 (1979); D. B. Collum *et al., ibid.* **102**, 2117, 2118, 2120 (1980). ^{13}C-NMR study: J. A. Robinson, D. L. Turner, *Chem. Commun.* **1982**, 148. Biosynthesis: Day *et al., Antimicrob. Ag. Chemother.* **4**, 410 (1973). *Review:* Stark, "Monensin, A New Biologically Active Compound Produced by a Fermentation Process", in *Fermentation Advances*, Pap. Int. Ferment. Symp., 3rd, **1968**, D. Perlman, Ed. (Academic Press, New York, 1969) pp 517-540.

Crystals, mp 103-105° (monohydrate). $[\alpha]_D$ +47.7°. pKa 6.6 (in 66% DMF). Very stable under alkaline conditions. LD_{50} of monensin complex in mice, chicks (mg/kg): 43.8 ± 5.2, 284 ± 47 orally (Haney, Hoehn).

Sodium salt, $C_{36}H_{61}NaO_{11}$, *Coban, Romensin, Rumensin.* mp 267-269°. $[\alpha]_D$ +57.3° (methanol).

Monensin and its sodium salt are slightly sol in water; more sol in hydrocarbons; very sol in other organic solvents.

THERAP CAT (VET): Antiprotozoal, antibacterial, antifungal. Coccidiostat in chickens. Feed additive.

6158. Monoamine Oxidase. *Amine oxidase;* adrenaline oxidase; tyraminase; MAO. Since its first description by Hare in 1928 as an enzyme which catalyzed the oxidative deamination of tyramine, it has been found to be widely distributed in animals. This enzyme, which catalyzes the oxidative deamination of a variety of biogenic amines, such as serotonin, norepinephrine, epinephrine, tyramine and dopamine, deaminates compounds in which the amine group is attached to the terminal carbon group. Enzymes other than the classical amine oxidase which catalyze the oxidative deamination of amines, have been found in animals, plants and bacteria. In contrast to the classical amine oxidase, they are inhibited by carbonyl reagents and do not act on *N*-substituted amines. Enzymes of this group include: *spermine oxidase, benzylamine oxidase, mescaline oxidase* (of rabbit liver), *plant amine oxidase,* oxidases in microorganisms, and histaminase, *q.v.* Structural studies: Achee, *Biochemistry* **7**, 4329 (1968). Multiple forms of rat brain monoamine oxidase: M. B. H. Youdim *et al., Nature* **223**, 626 (1969). *Review:* H. Blaschko "Amine Oxidase" in *The Enzymes,* vol. **8**, P. D. Boyer *et al.,* Eds. (Academic Press, New York, 2nd ed., 1963) pp 337-351. Book: *Monoamine Oxidase: Structure, Function and Altered Functions,* T. P. Singer *et al.,* Eds. (Academic Press, New York, 1980) 557 pp.

6159. Monobenzone. *4-(Phenylmethoxy)phenol; p-(benzyloxy)phenol;* hydroquinone monobenzyl ether; hydroquinone benzyl ether; *p*-hydroxyphenyl benzyl ether; benzyl hydroquinone; monobenzyl hydroquinone; Benoquin; Depigman; Pigmex; Benzoquin; Agerite. $C_{13}H_{12}O_2$; mol wt 200.23. C 77.98%, H 6.04%, O 15.98%. Prepd from hydroquinone and benzyl bromide in alcoholic KOH: Schiff, Pellizzari, *Ann.* **221**, 370 (1883).

Lustrous leaflets from water, mp 122.5°. Practically insol in cold water. Soly in boiling water about 1.0 g/100 ml. Sol in alcohol, ether, benzene. LD_{50} i.p. in rats: 4.5 g/kg.
THERAP CAT: Depigmentor.

6160. Monocrotaline. *14,19-Dihydro-12,13-dihydroxy-20-norcrotolanan-11,15-dione;* crotaline; NSC 28693; NCI-C56462. $C_{16}H_{23}NO_6$; mol wt 325.35. C 59.06%, H 7.13%, N 4.31%, O 29.51%. Pyrrolizidine alkaloid, the major toxic constituent of *Crotalaria spectabilis* Roth, *Leguminosae:* Adams, Rogers, *J. Am. Chem. Soc.* **61**, 2815 (1939); N. J. Leonard in *The Alkaloids* vol. I, R. H. F. Manske, H. L. Holmes, Eds. (Academic Press, New York, 1950) p 116; Tinker, Lauter, *Econ. Bot.* **10**, 254 (1956); from *C. crispata* F. Muell. ex Benth.: Culvenor, Smith, *Aust. J. Chem.* **16**, 239 (1963). Structure: Adams *et al., J. Am. Chem. Soc.* **74**, 5612 (1952). Configuration: Adams, Fles, *ibid.* **81**, 5803 (1959); Robin, Crout, *J. Chem. Soc. (C)* **1969**, 1386. Crystal structure: H. Stoekli-Evans, *Acta Crystallogr.* **B35**, 231 (1979). Toxicology of monocrotaline and other pyrrolizidine alkaloids: McLean, *Pharmacol. Rev.* **22**, 429-483 (1970); Allen *et al., Toxicol. Appl. Pharmacol.* **23**, 470 (1972); R. A. Roth *et al., ibid.* **60**, 193 (1981). Review and evaluation of studies of carcinogenic and toxic action in laboratory animals: *IARC Monographs* **10**, 291-302, 333-342 (1976). Comprehensive reviews: L. Bull *et al., The Pyrrolizidine Alkaloids* (North-Holland, Amsterdam, 1968) 293 pp; F. L. Warren in *The Alkaloids* vol. **12**, R. H. F. Manske, Ed. (Academic Press, New York, 1970) pp 245-331; D. J. Robins, *Fortschr. Chem. Org. Naturst.* **41**, 115-203 (1982).

Prisms from abs ethanol, mp 197-198° (dec). $[\alpha]_D^{26}$ — 54.7° (c = 5.054 in chloroform). uv max (96% ethanol): 217 nm (log ϵ 3.32), Simánek *et al., Coll. Czech. Chem. Commun.* **34**, 1832 (1969). LD_{50} orally in rats: 71 mg/kg, P. M. Newberne *et al., Toxicol. Appl. Pharmacol.* **18**, 387 (1971).

Hydrochloride, $C_{16}H_{23}NO_6 \cdot HCl$, mp 184° (dec). $[\alpha]_D^{28}$ —38.4° (c = 5.2 in water).

Methiodide, $C_{16}H_{23}NO_6 \cdot CH_3I$, mp 205° (dec). $[\alpha]_D^{28}$ +23.4° (c = 3.1 in methanol).

6161. Monocrotophos. *Phosphoric acid dimethyl [1-methyl-3-(methylamino)-3-oxo-1-propenyl] ester; (E)-phosphoric acid dimethyl ester, ester with 3-hydroxy-N-methylcrotonamide;* 3-(dimethoxyphosphinyloxy)-*N*-methyl-*cis*-crotonamide; dimethyl 2-methylcarbamoyl-1-methylvinyl phosphate; dimethyl phosphate of 3-hydroxy-*N*-methyl-*cis*-crotonamide; C 1414; ENT 27129; SD 9129; Azodrin; Monocron; Nuvacron. $C_7H_{14}NO_5P$; mol wt 223.16. C 37.67%, H 6.32%, N 6.28%, O 35.85%, P 13.88%. General prepn: Whetstone, Stiles, U.S. pat. **2,802,855**; *see also* Ward, Morales, U.S. pat. **3,400,177** (1957, 1968, both to Shell). Metabolite of dicrotophos, *q.v.:* Menzer, Casida, *J. Agr. Food Chem.* **13**, 102 (1965).

Crystals, mp 54-55°. Commercial product is a reddish-brown solid, mp 25-30°. Vapor press. at 20: 7 × 10⁶ mm Hg. Misc with water; sol in acetone, ethanol; practically insol in diesel oils, kerosine. LD_{50} in male, female rats: 17, 20 mg/kg orally; 126, 112 mg/kg dermally, T. B. Gaines, *Toxicol. Appl. Pharmacol.* **14**, 515 (1969).
USE: Insecticide.

6162. Monooctanoin. *Octanoic acid monoester with 1,2,-3-propanetriol; α*-monocaprylin; caprylic acid monoglyceride; glyceryl monocaprylate; octanoic acid glycerol ester; DL-glyceryl-1-mono-octanoate; Capmul 8210; Moctanin. $C_{11}H_{22}O_4$; mol wt 218.29. C 60.52%, H 10.16%, O 29.32%. Cholesterol solvent used to dissolve gallstones by direct biliary infusion. Prepn: A. Heiduschka, H. Schuster, *J. Prakt. Chem.* [2] **120**, 145 (1929); L. Hartman, *J. Chem. Soc.* **1959**, 4134. Prepn of optically active L-isomer: E. Baer, H. O. L. Fischer, *J. Am. Chem. Soc.* **67**, 2031 (1945). Antimicrobial activity: A. J. Conley, J. J. Kabara, *Antimicrob. Ag. Chemother.* **4**, 501 (1973). Clinical dissolution of retained gallstones after surgery: L. N. Jarret *et al.*, *Lancet* **1**, 68 (1981); R. M. Steinhagen, D. Pertsemlidis, *Am. J. Gastroenterol.* **78**, 756 (1983). Effect of reducing agents on gallstone dissolution: B. F. Smith, J. T. LaMont, *J. Clin. Invest.* **76**, 439 (1985). *Review:* G. D. Bell, *Pharmacol. Ther.* **23**, 79-108 (1983).

$$CH_2OOC(CH_2)_6CH_3$$
$$|$$
$$CHOH$$
$$|$$
$$CH_2OH$$

Crystals from light petroleum, mp 39.5-40.5°. L-Form, mp 28-30°. n_D^{22} 1.4548. $[\alpha]_D$ −6.6° (c = 10.2 in pyridine).

THERAP CAT: Cholelitholytic agent.

6163. Monophen®. *2-[(4-Hydroxy-3,5-diiodophenyl)methyl]cyclohexanecarboxylic acid;* 2-(3,5-diiodo-4-hydroxybenzyl)cyclohexanecarboxylic acid. $C_{14}H_{16}I_2O_3$; mol wt 486.09. C 34.59%, H 3.32%, I 52.22%, O 9.87%. Synthesis: Natelson *et al.*, U.S. pats. **2,400,433; 2,496,064** (1946, 1950 to Jewish Hospital of Brooklyn Res. Found.).

Crystals, mp 181-182°. Insol in water, sol in alcohol, strong alkali.

THERAP CAT: Diagnostic aid (radiopaque medium—cholecystographic).

6164. Monorden. *8-Chloro-1a,14,15,15a-tetrahydro-9,11-dihydroxy-14-methyl-6H-oxireno[e][2]benzoxacyclotetradecin-6,12(7H)-dione;* 5-chloro-6-(7,8-epoxy-10-hydroxy-2-oxo-3,5-undecadienyl)-β-resorcylic acid μ-lactone; radicicol. $C_{18}H_{17}ClO_6$; mol wt 364.79. C 59.27%, H 4.70%, Cl 9.72%, O 26.32%. Antibiotic substance from *Monosporium bonorden:* Delmotte, Delmotte-Plaquée, *Nature* **171**, 344 (1953). Identity with radicicol: Mirrington *et al.*, *Tetrahedron Letters* **1964**, 365. Structure: McCapra, Scott, *ibid.* **1964**, 869; Mirrington *et al.*, *Aust. J. Chem.* **19**, 1265 (1966). Isoln: **Neth. pat. Appl. 6,506,173** (1965 to Sandoz).

Crystals from chloroform, alcohol or benzene, mp 193.5°. $[\alpha]_D^{20}$ +203° (chloroform). uv max (ethanol): 264, 272 nm (ε 13200, 13100).

Diacetate, $C_{22}H_{21}ClO_8$, crystals, mp 189°. uv max: 275 nm (ε 13000).

6165. Monosodium Glutamate. *Glutamic acid monosodium salt monohydrate;* monosodium L-glutamate monohydrate; sodium glutamate monohydrate; Chinese seasoning; MSG; RL-50; Accent; Ajinomoto; Glutacyl; Glutavene; Vetsin. $C_5H_8NNaO_4 \cdot H_2O$; mol wt 187.13. C 32.09%, H 5.39%, N 7.49%, Na 12.28%, O 42.75%. HOOCCH(NH₂)·CH₂CH₂COONa.H₂O. The monosodium salt of the naturally occurring L-form of glutamic acid. Manufactured by fermentation of carbohydrate sources such as sugar beet molasses. Production by hydrolysis of vegetable proteins (*see also* Glutamic Acid): Ikeda, Suzuki, **Brit.** pat. **9440**, *C.A.* **5**, 836 (1910); U.S. pat. **1,015,891**, *C.A.* **6**, 717 (1912); from Steffens waste from beet-sugar molasses by acid hydrolysis: Ikeda, U.S. pat. **1,721,820**, *C.A.* **23**, 4591 (1929); by the action of *Micrococcus glutamicus* upon a carbohydrate and subsequent partial neutralization: Kinoshita *et al.*, U.S. pats. **3,002,889** and **3,003,925** (both 1961 to Kyowa Ferment. Ind.). Prepn of cryst Na-glutamate: Shildneck, U.S. pat. **2,306,646**, *C.A.* **37**, 3107 (1943). As a rule, sugar beet products are used in the U.S. and in Europe; other carbohydrate sources such as sugar cane and tapioca are often used in the Orient. Mfg methods: Faith, Keyes, Clark, *Industrial Chemicals* (Wiley, New York, 2nd ed., 1957) p 522. *Review:* T. Yoshida, "L-Monosodium Glutamate (MSG)" in Kirk-Othmer *Encyclopedia of Chemical Technology*, **Vol. 2** (Wiley, New York, 3rd ed., 1978) pp 410-421. *Book:* *Glutamic Acid: Advances in Biochemistry and Physiology*, L. J. Filer *et al.*, Eds. (Raven Press, New York, 1979) 416 pp.

White, practically odorless, free flowing crystals or crystalline powder. Forms rhombic prisms when crystallized from water. Below −8°, it crystallizes as a pentahydrate, which, after filtration and exposure to air, loses water of crystallization and becomes the monohydrate. Commercially preferable crystals are grown in the presence of amino acid contaminants. The optimum concentration is from 0.2 to 0.9% in normally salted food. Slightly levorotatory in water, but dextrorotatory in acid solns (the free L-acid is dextrorotatory). $[\alpha]_D^{25}$ +24.2° to +25.5° (c = 8.0 in 1.0*N* HCl). pH of 0.2% soln = 7.0. Very sol in water; sparingly sol in alcohol.

USE: Flavor enhancer; swine food additive; with sugar, to improve palatability of bitter drugs.

THERAP CAT: Treatment of hepatic coma.

6166. Monotropein. *1-(β-D-Glucopyranosyloxy)-1,4a,7,-7a-tetrahydro-7-hydroxy-7-(hydroxymethyl)cyclopenta[c]pyran-4-carboxylic acid.* $C_{16}H_{22}O_{11}$; mol wt 390.34. C 49.23%, H 5.68%, O 45.09%. From herb of *Monotropa hypopitys* L., Ericaceae: Bridel, *Compt. Rend.* **1742** (1923); *Bull. Soc. Chim. Biol.* **5**, 722 (1923). Structure: Inouye *et al.*, *Tetrahedron Letters* **1963**, 1031; Bobbitt *et al.*, *Chem. & Ind. (London)* **1964**, 931. Stereochemistry: Norio *et al.*, *Tetrahedron Letters* **1967**, 2367. GC-mass spec studies: H. Inouye *et al.*, *J. Chromatog.* **118**, 201 (1976). Purgative activity: *eidem*, *Planta Med.* **25**, 285 (1976).

Monohydrate, needles from water, dec 170-173°. $[\alpha]_D^{18}$ −127.7° (water). uv max: 235 nm (log ε 3.98). Soluble in water, alcohol; practically insol in ethyl acetate.

Pentaacetate, $C_{26}H_{32}O_{16}$, crystals, mp 173-174.5°. $[\alpha]_D^{18}$ −82.5° (ethanol).

6167. Montan Wax. Lignite wax. Obtained by extraction from lignite. Asphalt and resin content and physical properties vary with source of lignite. Brief review: C. S. Letcher in Kirk-Othmer *Encyclopedia of Chemical Technology* **vol. 24** (Wiley-Interscience, New York, 3rd ed., 1984) pp 471-472.

Dark-brown lumps; white or nearly white when bleached. mp 80-86°. Saponification number 88-112. Insol in water; sol in benzene, chloroform, carbon tetrachloride, hot petr ether; incompletely sol in hot ether or boiling alcohol.

USE: Electric cable insulators; in polishes instead of carnauba wax; manuf candles, waterproof paints and varnishes.

6168. Montmorillonite. A clay forming the principal constituent of bentonite and fuller's earth. Approximate formula: $R^+_{0.33}(Al,Mg)_2Si_4O_{10}(OH)_2.nH_2O$ where R^+, in natural material, includes one or more of the cations Na^+, K^+, Mg^{2+}, and Ca^{2+}, and possibly others.

USE: In industrial chromatographic technique, *e.g.*, **U.S. pat. 2,626,888** and **Brit. pat. 697,060** (1953 to Merck & Co.), describing the elution of vitamin B_{12}-active substances adsorbed on montmorillonite adsorbents. Widely used in the petroleum industry. Also as catalyst carrier. *Caution:* Inhalation of the dust can cause respiratory irritation.

6169. Monuron. *N′-(4-Chlorophenyl)-N,N-dimethylurea;* 1,1-dimethyl-3-(*p*-chlorophenyl)urea; CMU; Karmex Monuron Herbicide; Telvar. $C_9H_{11}ClN_2O$; mol wt 198.65. C 54.41%, H 5.58%, Cl 17.85%, N 14.10%, O 8.05%. Prepd by reacting *p*-chlorophenyl isocyanate with dimethylamine: Bucha, Todd, *Science* **114**, 493 (1951); *see also* **U.S. pats. 2,655,444-7.** Review: McCall, *Agricultural Chemicals* **7**, 40 (1952).

Cl—⟨benzene ring⟩—NHCON(CH₃)₂

Thin rectangular prisms from methanol, mp 170.5-171.5° (the commercial product melts at 176-177°). Vapor press at 25°: 5×10^{-7} mm; at 100°: 178×10^{-5} mm. Slight odor. Stable toward oxygen and moisture under ordinary conditions at neutral pH; elevated temps and more acid or alkaline conditions appreciably raise rate of hydrolysis. Very slightly sol in water and in no. 3 Diesel oil: About 230 ppm at 25°. pH of satd aq soln 6.26. Moderately sol in methanol, ethanol, acetone. Practically insol in hydrocarbon solvents. LD_{50} orally in rats: 3700 mg/kg, G. W. Bailey, J. L. White, *Residue Rev.* **10**, 97 (1965).

Trichloroacetate, $C_{11}H_{12}Cl_4N_2O_3$, *monuron TCA, Urox.* mp 78-81°. Prepn: Gilbert *et al.*, **U.S. pat. 2,782,112** (1957 to Allied Chem.). Effective in control of both weeds and grasses.

USE: Herbicide. *Caution:* Anemia and methemoglobinemia have been produced in experimental animals.

6170. Moperone. *1-(4-Fluorophenyl)-4-[hydroxy-4-(4-methylphenyl)-1-piperidinyl]-1-butanone; p-fluoro-4-(4′-hydroxy-4′-p-tolylpiperidino)butyrophenone; p-fluoro-4-(4′-hydroxy-4′-p-methylphenylpiperidino)butyrophenone; 1-(3′-p-fluorobenzoylpropyl)-4-hydroxy-4-p-tolylpiperidine; ω-(4-hydroxy-4-p-tolylpiperidino)-p-fluorobutyrophenone; methylperidol;* R 1658. $C_{22}H_{26}FNO_2$; mol wt 355.46. C 74.34%, H 7.37%, F 5.35%, N 3.94%, O 9.00%. Prepn: Janssen *et al.*, *J. Med. Pharm. Chem.* **1**, 281 (1959); Janssen, **Brit. pat. 881,893** (1961).

F—⟨benzene ring⟩—COCH₂CH₂CH₂—N⟨piperidine ring with OH and tolyl with CH₃⟩

Crystals, mp 118-119.5°. uv max: 246.5 nm (ε 12200). Hydrochloride, $C_{22}H_{27}ClFNO_2$, *Luvatren.* Crystals from isopropyl ether, mp 216-218°.

THERAP CAT: Antipsychotic.

6171. Mopidamol. *2,2′,2″,2‴-[[4-(1-Piperidinyl)-pyrimido[5,4-d]pyrimidine-2,6-diyl]dinitrilo]tetrakisethanol; 2,2′,2″,2‴-[(4-piperidinopyrimido[5,4-d]-pyrimidine-2,6-diyl)dinitrilo]tetraethanol;* 2,6-bis(diethanolamino)-4-piperidinopyrimido[5,4-d]pyrimidine; RA 233; Rapenton. $C_{19}H_{31}N_7O_4$; mol wt 421.51. C 54.14%, H 7.41%, N 23.26%, O 15.18%. An inhibitor of platelet aggregation with antimetastatic properties. Prepn: **Brit. pat. 1,051,218** corresp to J. Roch, H. Scheffler, **U.S. pat. 3,322,755** (1966, 1967 both to Boehringer, Ing.). Effect on tumor cell metastasis: J. L.

Ambrus *et al.*, *J. Med.* **9**, 183 (1978); on inhibition of adenosine uptake into platelets: J. P. Lips *et al.*, *Biochem. Pharmacol.* **29**, 43 (1980). Therapeutic use: R. L. Clark, *Cancer* **43**, 790 (1979); I. L. Bonta *et al.*, *Agents Actions (Suppl.)* **4**, 278 (1979).

(HOCH₂CH₂)₂N—⟨pyrimido-pyrimidine ring system⟩—N(CH₂CH₂OH)₂ with piperidine substituent

Deep yellow crystals, mp 157-158°.
THERAP CAT: Antineoplastic.

6172. Moprolol. *1-(2-Methoxyphenoxy)-3-[(1-methylethyl)amino]-2-propanol; 1-(isopropylamino)-3-(o-methoxyphenoxy)-2-propanol.* $C_{13}H_{21}NO_3$; mol wt 239.32. C 65.24%, H 8.85%, N 5.85%, O 20.06%. β-Adrenergic blocker. Prepn: C. D. Lunsford *et al.*, *J. Am. Chem. Soc.* **82**, 1166 (1960); **Neth. pat. Appl. 301,580** corresp to A. F. Crowther, L. H. Smith, **U.S. pat. 3,501,769** (1965, 1970 both to ICI); A. F. Crowther *et al.*, *J. Med. Chem.* **12**, 638 (1969). Pharmacology: G. Croce *et al.*, *Arzneimittel-Forsch.* **20**, 1074 (1970). Anti-anginal and hemodynamic effects: M. Radice *et al.*, *ibid.* **28**, 2160 (1978). Resolution of enantiomers: G. Ferrari, V. Vecchietti, **Eur. pat. Appl. 15,418**; *eidem*, **U.S. pat. 4,683,245** (1980, 1987 both to Simes). Synthesis of (−)-form: K. Kan *et al.*, *Agr. Biol. Chem.* **49**, 207 (1985). Bioavailability of (−)-form in dog: A. Marzo, J. P. Desager, *Curr. Ther. Res.* **30**, 442 (1981). Pharmacokinetics of racemate and (−)-form: C. Harvengt, J. P. Desager, *Int. J. Clin. Pharmacol. Ther. Toxicol.* **20**, 57 (1982). Clinical comparison of antihypertensive activity of enantiomers: P. Ghirardi, F. Grosso, *J. Cardiovasc. Pharmacol.* **2**, 471 (1980). Clinical studies of (−)-form in hypertension: G. Vrebos *et al.*, *Curr. Ther. Res.* **29**, 654 (1981); H. Hoeffkes, *ibid.* **30**, 88 (1981).

⟨benzene ring with OCH₃⟩—O—CH₂CHCH₂NHCH(CH₃)₂ with OH

Crystals from cyclohexane, mp 82-83°.
Hydrochloride, $C_{13}H_{22}ClNO_3$, *SD-1601, Omeral.* Crystals, mp 110-112°.
l-Form, *levomoprolol.* Crystals from ethyl acetate, mp 78-80°. [α] −5.5 ±0.2 (c = 5.0 in ethanol).
l-Form hydrochloride, *Levotensin.* Crystals from ethyl acetate/95% ethanol, mp 121-123°. $[\alpha]_D^{25}$ −16.3° (c = 5.0 in ethanol).
THERAP CAT: Antihypertensive.

6173. MOPS. *4-Morpholinepropanesulfonic acid;* 3-(N-morpholino)propanesulfonic acid; N-(3-sulfopropyl)morpholine. $C_7H_{15}NO_4S$; mol wt 209.26. C 40.18%, H 7.23%, N 6.69%, O 30.58%, S 15.32%. One of the zwitterionic amino acids known as "Good" buffers; active in the pH range of 6.5-8.0. Prepn: C. F. H. Allen *et al.*, *Anal. Chem.* **37**, 156 (1965); N. E. Good, S. Izawa, "Hydrogen Ion Buffers," in *Methods of Enzymology* vol. 24, A. San Pietro, Ed. (Academic Press, New York, 1972) pp 53-65. Temperature effects on pK: M. Sankar, R. G. Bates, *Anal. Chem.* **50**, 1922 (1978); and pH: R. N. Roy *et al.*, *Cryo-Letters* **6**, 139 (1985). Effects on "Lowry" protein determn: H. M. Himmel, W. Heller, *J. Clin. Chem. Clin. Biochem.* **25**, 909 (1987). Uses as a biological buffer: E. P. Paques *et al.*, *Eur. J. Biochem.* **107**, 447 (1980); L. A. Sonna *et al.*, *J. Biol. Chem.* **263**, 6625 (1988).

O⟨morpholine ring⟩N⁺HCH₂CH₂CH₂SO₃⁻

White powder, mp 283.5-284.5°. pKa 7.15. pKa$_2$ (25°): 7.184; (37°): 7.041. ΔpKa/°C −0.013. Is not completely stable when autoclaved in the presence of glucose.

USE: Biological buffer.

6174. Moquizone. *2,3-Dihydro-1-[(4-morpholinyl)acetyl]-3-phenyl-4(1H)-quinazolinone;* Rec 14-0127. C$_{20}$H$_{21}$N$_3$O$_3$; mol wt 351.41. C 68.36%, H 6.02%, N 11.96%, O 13.66%. Prepn: Bonola *et al., J. Med. Chem.* **11**, 1136 (1968); *eidem,* S. Afr. pat. **68 03,948,** *C.A.* **71,** 81406c (1969) and **Swiss** pat. **474,524** corresp to U.S. pat. **3,637,681** (1968, 1969, 1972, all three to SECEPH). Pharmacology: Setnikar *et al., Arzneimittel-Forsch.* **22,** 1894 (1972). Toxicology: Setnikar, De Fina, *Toxicol. Appl. Pharmacol.* **16,** 571 (1970). Review of pharmacological and clinical aspects: Grossi, Spada, *Clin. Ter.* **50,** 203 (1969).

Crystals from isopropanol, mp 135-137°. Crystals from ethyl acetate or benzene-petr ether, mp 128-130°.

Hydrochloride, C$_{20}$H$_{22}$ClN$_3$O$_3$, *Peristil.* Crystals from ethanol-ether, mp 210-214°. Solubility in water (25°): 12.2 g/100 ml. LD$_{50}$ in mice, rats: 1155, 2135 mg/kg orally; 237, 146 mg/kg i.v.

THERAP CAT: Choleretic.

6175. Morantel. *1,4,5,6-Tetrahydro-1-methyl-2-[2-(3-methyl-2-thienyl)ethenyl]pyrimidine.* C$_{12}$H$_{16}$N$_2$S; mol wt 220.33. C 65.41%, H 7.32%, N 12.71%, S 14.55%. Prepn: Austin *et al.,* **Brit.** pat. **1,120,587** (1968 to Pfizer).

Crystals from chloroform + benzene, mp 239-241°. Tartrate, C$_{16}$H$_{22}$N$_2$O$_6$S, *CP 12009-18, Paratect, Suiminth.* THERAP CAT (VET): Anthelmintic.

6176. Morazone. *1,2-Dihydro-1,5-dimethyl-4-[(3-methyl-2-phenyl-4-morpholinyl)methyl]-2-phenyl-3H-pyrazol-3-one; 4-[(3-methyl-2-phenylmorpholino)methyl]antipyrine;* 1-phenyl-2,3-dimethyl-4-(2'-phenyl-3'-methylmorpholinomethyl)-5-pyrazolone; R 445; Novartrina; Tarugan. C$_{23}$H$_{27}$N$_3$O$_2$; mol wt 377.47. C 73.18%, H 7.21%, N 11.13%, O 8.48%. Prepn: Hengen *et al., Arzneimittel-Forsch.* **8,** 421 (1958); Siemer, Doppstadt, U.S. pats. **2,943,022; 3,005,818** (1960, 1961 both to Ravensberg Chem.). Metabolism: Cartoni *et al., J. Chromatog.* **84,** 419 (1973).

Crystals, mp 149-150°. Freely sol in chloroform; sol in methanol, acetone; slightly sol in ether.

Hydrochloride, C$_{23}$H$_{28}$ClN$_3$O$_2$, crystals from methanol + acetone, mp 171-172° (dec). Sol in water.

Lactate, *Rosimon-Neu.*

THERAP CAT: Analgesic; anti-inflammatory; antipyretic.

6177. Morclofone. *(4-Chlorophenyl)[3,5-dimethoxy-4-[2-(4-morpholinyl)ethoxy]phenyl]methanone; 4'-chloro-3,5-dimethoxy-4-(2-morpholinoethoxy)benzophenone;* dimeclophenone; K 3712; Medicil; Nitux; Plausitin. C$_{21}$H$_{24}$ClNO$_5$;

mol wt 405.88. C 62.14%, H 5.96%, Cl 8.73%, N 3.45%, O 19.71%. Prepn: F. Lauria *et al.,* **Ger.** pat. **2,016,707** corresp to U.S. pat. **3,708,482** (1971, 1973 both to Carlo Erba). Antitussive activity in humans: E. Bosisio *et al., Farmaco Ed. Prat.* **26,** 356 (1971).

Cryst from cyclohexane, mp 91-92°. LD$_{50}$ orally in mice: 552 mg/kg (Lauria).

THERAP CAT: Antitussive.

6178. Moricizine. *[10-[3-(4-Morpholinyl)-1-oxopropyl]-10H-phenothiazin-2-yl]carbamic acid ethyl ester; 10-(3-morpholinopropionyl)phenothiazine-2-carbamic acid ethyl ester;* ethmosine; ethmozin; etmozin; moracizine; EN-313. C$_{22}$H$_{25}$N$_3$O$_4$S; mol wt 427.52. C 61.81.%, H 5.89%, N 9.83%, O 14.97%, S 7.50%. Novel anti-arrhythmic agent with mode of action similar to quinidine, *q.v.* Prepn: A. Gritsenko *et al.,* **Ger.** pat. **2,014,201** (1971 to Acad. Med. Sci., USSR), *C.A.* **76,** 34273a (1972); *eidem,* U.S. pats. **3,740,395** and **3,864,487** (1973, 1975). Pharmacology and toxicity: N. V. Kaverina *et al., Russian Pharmacol. Toxicol.* **35,** 74 (1972). Mode of action: P. N. Shenoy *et al., Clin. Res.* **25,** 276A (1977). Metabolism: Y. I. Vikhlyaev *et al., Farmakol. Toksicol.* **40,** 19 (1977), *C.A.* **86,** 133296y (1977). Effects in dogs: P. Danilo *et al., Eur. J. Pharmacol.* **45,** 127 (1977). Clinical and exptl study of anti-arrhythmic properties: N. V. Kaverina *et al., Kardiologiya* **18,** 67 (1978), *C.A.* **89,** 84861e (1978). Clinical study: P. J. Podrid *et al., Circulation* **61,** 450 (1980). HPLC determn in plasma: C. C. Whitney *et al., J. Pharm. Sci.* **70,** 462 (1981). Symposium on pharmacology, pharmacokinetics and clinical studies: *Am. J. Cardiol.* **60**(11), 1F-89F (1987).

Crystals, mp 156-157°. Hydrochloride, C$_{22}$H$_{26}$ClN$_3$O$_4$S, *Ethmozine, Etmozin.* Crystals from dichloroethane, mp 189° (dec). Sol in water, alcohol. LD$_{50}$ in mice, rats (mg/kg): 36, 12 i.v.; in mice (mg/kg): 131 i.p. (Kaverina).

THERAP CAT: Anti-arrhythmic.

6179. Morin. *2-(2,4-Dihydroxyphenyl)-3,5,7-trihydroxy-4H-1-benzopyran-4-one; 2',3,4',5,7-pentahydroxyflavone;* 2'-hydroxypelargidenolon 1522; C.I. Natural Yellow 8; C.I. Natural Yellow 11; C.I. 75660. C$_{15}$H$_{10}$O$_7$; mol wt 302.23. C 59.61%, H 3.34%, O 37.06%. In wood of old fustic (*Chlorophora tinctoria* (L.) Gaud., *Moraceae*), also called Cuba wood, or yellow Brazil wood. The wood of the Osage orange tree also contains morin and maclurin. Isoln: Perkin, Pate, *J. Chem. Soc.* **67,** 649 (1895); Rolland, *Teintex* **3,** 460 (1938). Synthesis by condensing phloracetophenone dimethyl ether with 2,4-dimethoxybenzaldehyde: von Kostanecki *et al., Ber.* **39,** 625, 4014, 4022 (1906). Comprehensive list of refs in *Colour Index* **vol. 4** (3rd ed., 1971) p 4636.

Crystallizes with 1 or 2 mols water. Anhydr needles from

Consult the cross index before using this section.

abs alcohol, dec 285-290°. One gram dissolves in 4 liters water at 20°, in 1060 ml boiling water. Freely sol in alcohol; slighly sol in ether, acetic acid. Sol in aq alkaline solns with intense yellow color which turns brown on exposure to air. Absorpion spectrum: Grinbaumowna, Marchlewski, *Biochem. Z.* **290**, 261 (1937).

Calico yellow, obtained by the action of bisulfite on fustic extract consists mainly of the bisulfite compound of morin, $C_{15}H_{10}O_7$ + $NaHSO_3$, and is used in calico printing. The calico yellow of commerce is a good source of morin for the laboratory. The bisulfite is removed with HCl.

USE: As a spot test reagent for Al, Be, Zn, Ga, In, and Sc salts: Feigl, *Spot Tests* (New York, 1954); Beck, *Mikrochim. Acta* **2**, 287 (1937). As luminescence indicator: Kocsis, Zádor, *Z Anal. Chem.* **124**, 42-45 (1942). As a textile dye, morin dyes wool mordanted with chromium: olive yellow; mordanted with aluminum: yellow; mordanted with tin: lemon yellow; mordanted with iron: deep olive brown.

6180. Morindin. *1,5-Dihydroxy-2-methyl-6-[(6-O-β-D-xylopyranosyl-β-D-glucopyranosyl)oxy]-9,10-anthracenedione; 2-[[6-O-(6-deoxy-α-L-mannopyranosyl)-β-D-glucopyranosyl]oxy]-1,5-dihydroxy-6-methyl-9,10-anthracenedione.* $C_{27}H_{30}O_{14}$; mol wt 578.54. C 56.05%, H 5.23%, O 38.72%. From bark of *Coprosma australis* Forst. (*C. grandifolia* Hook.), *Rubiaceae*: Briggs, Dacre, *J. Chem. Soc.* **1948**, 564. Structure: Briggs, Le Quesne, *ibid.* **1963**, 3471. On acid hydrolysis yields the aglucone *morindone*. Synthesis and structure proof: B. Vermes *et al.*, *Phytochemistry* **19**, 119 (1980).

Yellow needles from glacial acetic acid, mp 169-171°. $[α]_D^{20}$ −90.0° (c = 0.054 in dioxane). Absorption max (ethanol): 230, 261, 448 nm (log ε 4.15, 3.93, 3.57). Sol in cold water but on boiling forms a water insol form, *β-morindin*. Sol in dioxane, pyridine, acetone, methanol; slightly sol in ethanol, glacial acetic acid; practically insol in ether, chloroform, benzene, petr ether.

6181. Moroxydine. *N-(Aminoiminomethyl)-4-morpholinecarboximidamide; 4-morpholinecarboximidoylguanidine; N¹,N¹-anhydrobis(β-hydroxyethyl)biguanide; abitilguanide; abitylguanide; ABOB; Bioxine; Vironil; Virusmin; Virustat.* $C_6H_{13}N_5O$; mol wt 171.20. C 42.09%, H 7.65%, N 40.91%, O 9.35%. Prepn from $NH_2C(=NH)NHCN$ and morpholine: **Brit. pat.** 776,176 (1957 to AB Kabi). Proposed as influenza suppressant: Melander, *Antibiot. & Chemother.* **10**, 39 (1960).

Hydrochloride, $C_6H_{14}ClN_5O$, *Assur.* Crystals. The acetate, maleate, and nitrate have also been prepd. A very low toxicity in mice is reported. Constituent of *Virugon; Albaton; Spenitol; Flumidin.*

THERAP CAT: Antiviral.

6182. Morphazinamide. *N-(4-Morpholinylmethyl)pyrazinecarboxamide; N-morpholinomethylpyrazinamide; morfazinamide; morinamide; B 2310.* $C_{10}H_{14}N_4O_2$; mol wt 222.24. C 54.04%, H 6.35%, N 25.21%, O 14.40%. Prepn from pyrazinamide, morpholine, and formalin: Felder, Tiepolo, **Ger. pat.** 1,129,492 (1962 to Bracco).

Crystals, mp 118.5-119.5°. uv max (ethanol): 269, 317 nm (log ε 3.95, 2.77). One gram dissolves in 3 ml water, 30 ml ethanol, 30 ml benzene, 2.5 ml chloroform. uv max (ethanol): 269, 317 nm (log ε 3.95, 2.77).

Hydrochloride, $C_{10}H_{15}ClN_4O_2$, *B 2311, Piazofolina, Piazolin.* Crystals, mp 196°. One gram dissolves in 2 ml water, 350 ml ethanol, 2000 ml chloroform.

THERAP CAT: Antibacterial (tuberculostatic).

6183. Morphenol. *Phenanthro[4,5-bcd]furan-3-ol; 3-hydroxy-4,5-oxidophenanthrene; 3-hydroxy-4,5-phenanthrylene oxide.* $C_{14}H_8O_2$; mol wt 208.20. C 80.76%, H 3.87%, O 15.37%. First described by Vongerichten, *Ber.* **30**, 2439 (1897). Prepn from morphine: Mosettig, Meitzner, *J. Am. Chem. Soc.* **56**, 2738 (1934). Approach to synthesis: Dendy *et al.*, *J. Chem. Soc.* **1963**, 4040.

Needles from benzene, mp 145°. Sol in methanol, ethanol, ether. Sol in concd H_2SO_4 giving a yellow soln with green fluorescence; also sol in aq NaOH giving a yellow soln with blue fluorescence.

Methyl ether, $C_{15}H_{10}O_2$, mp 68°.

6184. Morpheridine. *1-[2-(Morpholinyl)ethyl]-4-phenyl-4-piperidinecarboxylic acid ethyl ester; 1-(2-morpholinoethyl)-4-phenylisonipecotic acid ethyl ester; ethyl 1-(2'-morpholinoethyl)-4-phenylpiperidine-4-carboxylate; morpholinoethyl norpethidine.* $C_{20}H_{30}N_2O_3$; mol wt 346.46. C 69.33%, H 8.73%, N 8.09%, O 13.85%. Prepn: Anderson *et al.*, *J. Chem. Soc.* **1956**, 4088; Stern, Anderson, U.S. pat. **2,795,581** (1957 to J. F. Macfarlan). HPLC determn: I. Jane, A. McKinnon, *J. Chromatog.* **323**, 191 (1985). Analgesic activity in rats: R. A. Millar, R. P. Stephenson, *Brit. J. Pharmacol.* **11**, 27 (1956). Pharmacology: A. F. Green, N. B. Ward, *ibid.* **32**.

Liquid. $bp_{0.5}$ 188-192°. n_D^{18} 1.5276.

Dihydrochloride, $C_{20}H_{32}Cl_2N_2O_3$, *TA 1.* Crystals from dil ethanol, mp 264-266° (dec). LD_{50} in mice (mg/kg): 45 i.v. (Green, Ward), 118 i.p. (Millar, Stephenson).

Picrate, $C_{20}H_{30}N_2O_3.2C_6H_3N_3O_7$, dec 247-248°.

Caution: May be habit forming. This is a controlled substance (opiate) listed in the U.S. Code of Federal Regulations, Title 21 Part 1308.11 (1985).

6185. Morphinan. *(4aR)-1,3,4,9,10,10aα-Hexahydro-2H-10α,4aα-(iminoethano)phenanthrene; 1,2,3,9,10,10a-hexahydro-10,4a(4H)-iminoethanophenanthrene.* $C_{16}H_{21}N$; mol wt 227.34. C 84.53%, H 9.31%, N 6.16%. Parent substance of the morphine alkaloids, codeine, thebaine, etc. Prepd from 1-benzyl-1,2,3,4,5,6,7,8-octahydro-2-methylisoquinoline: Grewe, Mondon, *Ber.* **81**, 279 (1948); from β-cyclohexenyl-N-phenacetylethylamine: Grewe *et al.*, *Ber.* **84**, 527 (1951).

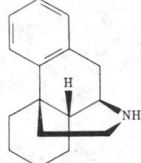

Oily liq, bp$_{0.05}$ 115°.
Hydrochloride, crystals from acetone + ether, dec 229°.
Sulfate, crystals from alcohol + ether, dec 195°.
N-Methylmorphinan, crystals, mp 61°.

6186. Morphine. (5α,6α)-7,8-Didehydro-4,5-epoxy-17-methylmorphinan-3,6-diol; morphium; morphia; Duromorph; Morphina; Nepenthe. $C_{17}H_{19}NO_3$; mol wt 285.33. C 71.56%, H 6.71%, N 4.91%, O 16.82%. The most important alkaloid of opium. A good grade of opium contains between 9 and 14% of anhydr morphine. Extraction procedures: Schwyzer, *Die Fabrikation pharmaceutischer und chemisch-technischer Produkte*, Berlin, 1931; W. R. Heumann, *Bulletin on Narcotics* **IX**, 34 (1957). Modern extraction process using ion-exchange chromatography: Achor, Geiling, *Anal. Chem.* **26**, 1061 (1954); Leete, *J. Am. Chem. Soc.* **81**, 3950 (1959); Mehltretter, Weakly, U.S. pat. **2,740,-787** (1956 to U.S.A.); McGuire et al., *J. Am. Pharm. Assoc.* **46**, 247 (1957). Structure: Knorr, *Ber.* **22**, 1113 (1889); Knorr, Hörlein, *Ber.* **40**, 2032, 3341, 4889 (1907); Gulland, Robinson, *J. Chem. Soc.* **123**, 980 (1925). Total synthesis: Gates, Tschudi, *J. Am. Chem. Soc.* **74**, 1109 (1952); **78**, 1380 (1956); Ginsburg et al., *J. Chem. Soc.* **1951**, 936; **1953**, 1524, 2664; **1954**, 3052. Prepn of (+)-morphine and racemate: Goto, Yamamoto, *Proc. Japan. Acad.* **30**, 769 (1954), *C.A.* **50**, 1052h (1956). Improved synthesis of (+)-form: I. Iijima et al., *J. Org. Chem.* **43**, 1462 (1978); synthesis of (−)-form: E. J. Bijsterveld, H. J. Sinnige, *Rec. Trav. Chim.* **95**, 24 (1976); H. C. Beyerman et al., ibid. **97**, 127 (1978). Biogenesis: Leete, *J. Am. Chem. Soc.* **81**, 3948 (1959). Configuration: G. Stork in *The Alkaloids*, vol. II, Manske, Holmes, Eds. (Academic Press, New York, 1952) pp 171-189; Bick, *Nature* **169**, 755 (1952); Rapoport, Lavigne, *J. Am. Chem. Soc.* **75**, 5329 (1953); Bose, *Chem. & Ind. (London)* **1954**, 130; Mackay, Hodgkin, *J. Chem. Soc.* **1955**, 3261; Kalvoda et al., *Helv. Chim. Acta* **38**, 1847 (1955). Infrared and polarographic data: Seagers et al., *J. Am. Pharm. Assoc.,* *Sci. Ed.* **41**, 640 (1952). Toxicity data: M. E. Buchwald, G. S. Eadie, *J. Pharm. Exp. Ther.* **71**, 197 (1941). Review and bibliography: Small, Lutz, *Chemistry of the Opium Alkaloids* (Suppl. No. 103 to the Public Health Reports) Washington, 1932; K. W. Bentley, *The Chemistry of the Morphine Alkaloids* (Oxford, 1954) 433 pp. Comprehensive description: F. J. Muhtadi in *Analytical Profiles of Drug Substances* vol. 17, K. Florey, Ed. (Academic Press, New York, 1988) pp 259-366.

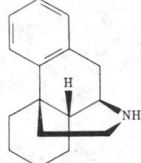

Short, orthorhombic, columnar prisms from anisole, dec 254°, also a metastable phase, mp 197°. High melting form sublimes at 190-200° (0.2 mm pressure at 2 mm distance).
Monohydrate, $C_{17}H_{19}NO_3.H_2O$, usually found in the crystalline form, but the amorphous or freshly pptd form is more sol. The following data are for the usual cryst form: orthorhombic, sphenoidal prisms, needles from methanol. Dec 254-256° with rapid heating. Darkens on exposure to light. Loses water at 130°. d$_4^{20}$ 1.32. [α]$_D^{20}$ −132° (methanol). pKb at 20° = 6.13; pKa 9.85. pH of satd soln, 8.5. uv max in acid: 285 nm; in alkali: 298 nm. One gram dissolves in about 5000 ml water, 1100 ml boiling water, 210 ml alcohol, 98 ml boiling alc, 1220 ml chloroform, 6250 ml ether, 114 ml amyl alc, 10 ml boiling methanol, 525 ml ethyl acetate; freely sol in solns of fixed alkali and alkaline earth hydroxides, in phenol, cresols; moderately sol in mixtures of chloroform with alcohols; slightly sol in ammonia, benzene.
Acetate trihydrate, $C_{19}H_{23}NO_5.3H_2O$, yellowish-white powder; slight acetic odor. Dec with age, losing acetic acid and discoloring. [α]$_D^{15}$ −77° (water). One gram dissolves in 2.25 ml water, 2 ml boiling water, 22 ml alc, 2 ml alc at 60°, 4.5 ml glycerol, 4.75 ml chloroform. Practically insol in ether. *Keep well closed and protected from light.*
Hydriodide dihydrate, $C_{17}H_{20}INO_3.2H_2O$, pale yellow orthorhombic needles; prone to discolor. d 1.66. Sol in hot, slightly in cold water. *Protect from light.*
Lactate, $C_{20}H_{25}NO_6$, cryst powder. Sol in about 10 parts water; slightly sol in alcohol.
Meconate pentahydrate, $(C_{17}H_{19}NO_3)_2.C_7H_4O_7.5H_2O$. The combination in which morphine chiefly exists in opium. Yellowish-white powder, dec 110°. Sol in 20 parts water, in alcohol.
Nitrate, $C_{17}H_{20}N_2O_6$, needles. Sol in 1.5 parts water. *Protect from light.*
Monobasic phosphate hemihydrate, $C_{17}H_{22}NO_7P.½H_2O$. Prepd from equimolar amounts of morphine and phosphoric acid: Homeyer et al., U.S. pat. **2,665,227** (1954 to Mallinckrodt). Stout monoclinic crystals from water + organic solvent. One gram dissolves in 5 ml water. pH of a 1% aq soln at 25° = 4.6.
Polybasic phosphate heptahydrate, approx. $3C_{17}H_{19}NO_3.2H_3PO_4.7H_2O$. Prepn: Pettenkofer, *Buchner's Repertorium der Pharmazie* **4**, 45 (1855). Odorless, needle-like crystals. Darkens in light. Slightly sol in water, dil acids, fixed alkali hydroxide solns. *Protect from light.*
Phthalate pentahydrate, $(C_{17}H_{19}NO_3)_2.C_6H_4(COOH)_2.5H_2O$, yellowish powder. Loses $4H_2O$ over H_2SO_4, the remaining H_2O at 100°.
Tartrate trihydrate, $(C_{17}H_{19}NO_3)_2.C_4H_6O_6.3H_2O$, cryst powder. Sol in 11 parts water; slightly sol in alcohol; practically insol in chloroform, ether, carbon disulfide.
Valerate, $C_{22}H_{29}NO_5$, crystals; valeric acid odor. Dec with age and quickly darkens on exposure to air and light. Soluble in about 5 parts water, alcohol. *Keep tightly closed and protected from light.*
6-Methyl ether, $C_{18}H_{21}NO_3$, *heterocodeine.*
Caution: May be habit forming. This is a controlled substance (opiate) listed in the U.S. Code of Federal Regulations, Title 21 Parts 329.1 and 1308.12 (1987).
THERAP CAT: Narcotic analgesic.
THERAP CAT (VET): Narcotic analgesic, preanesthetic, antitussive, antiperistaltic. Contraindicated in cats; unreliable in horses.

6187. Morphine Hydrobromide. $C_{17}H_{20}BrNO_3$; mol wt 366.27. C 55.75%, H 5.50%, Br 21.82%, N 3.82%, O 13.11%. $C_{17}H_{19}NO_3.HBr$.
Dihydrate, orthorhombic needles, d 1.46. Loses its water at 100° and becomes yellowish. [α]$_D^{15}$ −100.4° in 2.5% aq soln. One gram dissolves in 25 ml water, 1 ml boiling water, 50 ml alcohol, 10 ml boiling alcohol. *Protect from light.*
Caution: May be habit forming. This is a controlled substance (opium derivative) listed in the U.S. Code of Federal Regulations, Title 21 Parts 329.1 and 1308.12 (1987).
THERAP CAT: Narcotic analgesic.

6188. Morphine Hydrochloride. Epimor. $C_{17}H_{20}ClNO_3$; mol wt 321.81. C 63.45%, H 6.26%, Cl 11.02%, N 4.35%, O 14.92%. Toxicity data: M. E. Buchwald, G. S. Eadie, *J. Pharm. Exp. Ther.* **71**, 197 (1941).
Trihydrate, white flakes or crystalline powder; bitter taste. Loses its water of crystn at about 100° and usually becomes yellowish. mp about 200° (dec). [α]$_D^{15}$ −113.5° (c = 2.2 in H_2O, anhydr basis). One gram dissolves in 17.5 ml water, 0.5 ml boiling water, 52 ml alc, 6 ml alc at 60°; slowly sol in glycerol. Insol in chloroform, ether. pH about 5. LD$_{50}$ in mice (mg/kg): 226-318 i.v. (Buchwald, Eadie). *Protect from light.*
Caution: May be habit forming. This is a controlled substance (opium derivative) listed in the U.S. Code of Federal Regulations, Title 21 Parts 329.1 and 1308.12 (1987).
THERAP CAT: Narcotic analgesic.
THERAP CAT (VET): Narcotic analgesic, preanesthetic, anti-

tussive, antiperistaltic. Contraindicated in cats; unreliable in horses.

6189. Morphine Methylbromide. Morphosan. $C_{18}H_{22}$-BrNO$_3$; mol wt 380.28. C 56.85%, H 5.83%, Br 21.02%, N 3.68%, O 12.62%. $C_{17}H_{19}NO_3 \cdot CH_3Br$.

Needles, mp 265-266°. Moderately sol in cold water, freely in hot water; slightly sol in alcohol, chloroform. Insol in ether. *Protect from light.*

Caution: May be habit forming. This is a controlled substance (opium derivative) listed in the U.S. Code of Federal Regulations, Title 21 Parts 329.1 and 1308.11 (1987).

THERAP CAT: Narcotic analgesic.

THERAP CAT (VET): Narcotic analgesic, preanesthetic, antitussive, antiperistaltic. Contraindicated in cats, unreliable in horses.

6190. Morphine Mucate. *Galactaric acid compd with (5α,6α)-7,8-didehydro-4,5-epoxy-17-methylmorphinan-3,6-diol (1:1);* morphine hyperduric. $C_{23}H_{29}NO_{11}$; mol wt 495.48. C 55.75%, H 5.90%, N 2.83%, O 35.52%. The mucic acid salt of morphine. Has a prolonged action upon injection. *Ref:* Eastland, *Nature* **154**, 829 (1944); Heron, *Can. Med. Assoc. J.* **72**, 302 (1955).

Caution: May be habit forming. This is a controlled substance (opium derivative) listed in the U.S. Code of Federal Regulations, Title 21 Parts 329.1 and 1308.12 (1987).

THERAP CAT: Narcotic analgesic.

6191. Morphine Oleate, 20%. Contains 20% hydrated morphine = 18.8% anhydr.

Brown, oily liquid. Insoluble in water. Sol in alcohol. *Note:* Morphine oleate appears to deteriorate rapidly, hence failure to relieve pain may be due to its having been kept too long.

Caution: May be habit forming. This is a controlled substance (opium derivative) listed in the U.S. Code of Federal Regulations, Title 21 Parts 329.1 and 1308.12 (1987).

THERAP CAT: Narcotic analgesic.

6192. Morphine N-Oxide. Genomorphine; morphine oxide. $C_{17}H_{19}NO_4$; mol wt 301.33. C 67.76%, H 6.36%, N 4.65%, O 21.24%. Also said to be bimolecular: K. W. Bentley, *The Chemistry of the Morphine Alkaloids* (Oxford, 1954) p 25. Prepn: Freud, Speyer, *Ber.* **43**, 3310 (1910); **48**, 499 (1915); Chang *et al., J. Org. Chem.* **15**, 634 (1950).

Small prisms from 50% alc, mp 274-275°. Freely sol in ammonia water; slightly sol in water and alcohol. Practically insol in acetone, benzene and chloroform. Most of its salts are insol but the quinate is sol.

Caution: May be habit forming. This is a controlled substance (opium derivative) listed in the U.S. Code of Federal Regulations, Title 21 Part 1308.11 (1987).

THERAP CAT: Narcotic analgesic.

6193. Morphine Sulfate. *7,8-Didehydro-4,5-epoxy-17-methylmorphinan-3,6-diol sulfate salt (2:1);* Moscontin; MS Contin; MST-1 Continus; Oramorph. $C_{34}H_{40}N_2O_{10}S$; mol wt 668.76. C 61.06%, H 6.03%, N 4.19%, O 23.93%, S 4.79%.

Pentahydrate, *MST 10 Mundipharma, MST 30 Mundipharma.* White, fine odorless crystals or powder, or cubical masses. Loses some H$_2$O at ordinary temp; about 3H$_2$O (7.1%) at 100°, the remainder at 130°. Discolors on exposure to light. mp about 250° with decompn when anhydr. $[\alpha]_D^{25}$ −108.7° (c = 4 in H$_2$O, anhydr basis). One gram dissolves in 15.5 ml water at 25°, 0.7 ml water at 80°, 565 ml alcohol, 240 ml alcohol at 60°. Insol in chloroform, ether. pH about 4.8. *Keep well closed and protected from light.* Even ampuled aq solns of morphine sulfate will turn brown on storage. No loss of analgesic potency and no increase in toxicity has ever been demonstrated for such discolored solns: *J. Am. Med. Assoc.* **155**, 28 (1954); *J. Pharm. Pharmacol.* **2**, 673 (1950).

Incompat (of all salts of morphine): Alkalies, tannic acid, iodic acid, potassium permanganate, borax, chlorates, ferric chloride, iodides, lead acetate, magnesia, spirit nitrous ether, mercury bichloride, gold salts.

Caution: May be habit forming. This is a controlled substance (opium derivative) listed in the U.S. Code of Federal Regulations, Title 21 Parts 329.1 and 1308.12 (1987).

THERAP CAT: Narcotic analgesic.

THERAP CAT (VET): Narcotic analgesic, sedative, preanesthetic, gastric sedative. Contraindicated in cats; unreliable in horses.

6194. Morpholine. Tetrahydro-2H-1,4-oxazine; tetrahydro-1,4-oxazine; diethylene oximide; diethylene imidoxide. C_4H_9NO; mol wt 87.12. C 55.14%, H 10.41%, N 16.08%, O 18.36%. Prepd by dehydrating diethanolamine: Knorr, *Ann.* **301**, 1 (1898); Jones, Burns, *J. Am. Chem. Soc.* **47**, 2966 (1925); Hampton, Pollard, *ibid.* **58**, 2338 (1936). Review of morpholine and derivatives: A. L. Wilson, *Ind. Eng. Chem.* **27**, 867-871 (1935). *Monograph:* Morpholine Technical Bulletin, Jefferson Chemical Co. (New York, 1953).

Mobile, hygroscopic liquid. Characteristic amine odor. mp −4.9°. bp$_{760}$ 128.9°; bp$_6$ 20.0°. d_4^{20} 1.007. n_D^{20} 1.4540. Volatile with steam. Does not form an azeotrope with water. Flash pt, open cup: 100°F (38°C). Surface tension at 20° = 37.5 dynes/cm. Viscosity at 20° = 2.23 cp. Dipole moment 1.58. Strong base, pKb 5.6. *Corrosive to human skin.* Miscible with water with evolution of some heat. Immiscible with concd NaOH solns. Also miscible with acetone, benzene, ether, castor oil, methanol, ethanol, ethylene glycol, 2-hexanone, linseed oil, turpentine, pine oil. Will dissolve 109% dimethylamine; 34% trimethylamine; 33% methylamine; > 5% naphtha; < 1% paraffin oil; < 5% sulfur. LD$_{50}$ orally in female rats: 1.05 g/kg, Smyth *et al., Arch. Ind. Hyg. Occup. Med.* **10**, 61 (1954).

Hydrochloride, $C_4H_9NO \cdot HCl$, crystals, dec 175-176°. Sol in water.

N-Methylmorpholine, $C_5H_{11}NO$, mobile liquid, characteristic ammonia odor. bp$_{764}$ 116-117°; bp$_{760}$ 115.4°. d_4^{20} 0.9168. n_D^{20} 1.4332. Miscible with water, alcohol, ether.

Caution: Irritating to eyes, skin, mucous membranes.

USE: Cheap solvent for resins, waxes, casein, dyes. Morpholine fatty acid salts are used as surface-active agents and emulsifiers. Other morpholine compds are used as corrosion inhibitors, antioxidants, plasticizers, viscosity improvers, insecticides, fungicides, herbicides, local anesthetics and antiseptics.

6195. Morpholine Salicylate. *2-Hydroxybenzoic acid compd with morpholine (1:1);* 1,4-tetrahydrooxazine orthoxybenzoate; tetrahydro-1,4-oxazine salicylate; Retarcyl; Desposal; Depot-Salicyl; DSF. $C_{11}H_{15}NO_4$; mol wt 225.24. C 58.65%, H 6.71%, N 6.22%, O 28.41%. Prepn: Henri, Ledrut, U.S. pat. **2,551,682** (1951 to Heyden Pharmacal).

Crystals from ethanol, mp 110-111°. Soluble in water, methanol, ethanol, ethyl acetate, acetone, benzene, chloroform. Practically insol in toluene, xylene, petr ether, ether, carbon tetrachloride.

THERAP CAT: Analgesic; antipyretic; anti-inflammatory.

6196. 7-Morpholinomethyltheophylline. *3,7-Dihydro-1,3-dimethyl-7-(4-morpholinylmethyl)-1H-purine-2,6-dione;* 1,3-dimethyl-7-morpholinomethylxanthine; Xanturil. $C_{12}H_{17}N_5O_3$; mol wt 279.30. C 51.60%, H 6.14%, N 25.08%, O 17.19%. Prepn: Burckhalter, Dill, *J. Org. Chem.* **24**, 562 (1959).

Crystals from ethanol, mp 177°.
THERAP CAT: Diuretic.

6197. Morphothebaine. *5,6,6a,7-Tetrahydro-10-meth-oxy-6-methyl-4H-dibenzo[de,g]quinoline-2,11-diol; 10-methoxyaporphine-2,11-diol;* 2,11-dihydroxy-10-methoxy-aporphine. $C_{18}H_{19}NO_3$; mol wt 297.34. C 72.70%, H 6.44%, N 4.71%, O 16.14%. Prepd by heating thebaine with HCl. Formation from bulbocapnine: Ayer, Taylor, *J. Chem. Soc.* **1956,** 472. *Review:* Small, Lutz, "Chemistry of the Opium Alkaloids", *U.S. Public Health Reports,* Suppl. No. 103, Washington (1932); R. H. F. Manske, H. L. Holmes, *The Alkaloids,* **vol. II** (Academic Press, New York, 1952) pp 112-116.

Rhombic crystals, dec 197°. $[\alpha]_D^{15}$ — 130° (c = 0.88 in alc). Sol in alcohol, methanol, benzene, dil alkalies. Absorption spectra: Farmilo, *Bull. Narcotics U.N. Dept. Social Affairs* **6,** No. 3/4, 18 (1954).
Hydrobromide, $C_{18}H_{19}NO_3$.HBr, crystals from water, mp 270-275°. Also an acid hydrobromide, (B.HBr)$_2$.HBr.
Hydrochloride, $C_{18}H_{19}NO_3$.HCl, needles from water, mp 256-260°. Also an acid hydrochloride, (B.HCl)$_2$.HCl. Absorption spectrum: Csokán, *Z. Anal. Chem.* **124,** 344 (1942).
Hydriodide, $C_{18}H_{19}NO_3$.HI, clusters from dil alc, mp 243-244° (with preliminary softening at 237°).
Methiodide, $C_{18}H_{19}NO_3$.CH$_3$I, crystals from acetic acid, mp 221-222°.

6198. Motilin. A gastrointestinal peptide hormone whose existence was postulated during experimental chemi-cal stimulation of canine duodenum: J. C. Brown *et al., Gastroenterology* **50,** 333 (1966). Motilin stimulates gastric antrum, duodenum, and colon; it has been reported to be about 50 times more powerful than acetylcholine in its ef-fects on sensitive tissue: U. Strunz *et al., ibid.* **68,** 1485 (1975). In man it is found primarily in the upper small in-testine. Although its effects are not known completely, it plays a role in increasing gastric motility and stimulating pepsin output and may also be important in regulating the interdigestive myoelectric complex.

```
1                         7
H-Phe-Val-Pro-Ile-Phe-Thr-Tyr-

8                        15
Gly-Glu-Leu-Gln-Arg-Met-Gln-Glu-

16
Lys-Glu-Arg-Asn-Lys-Gly-Gln-OH

porcine motilin
```

Porcine motilin, which contains 22 amino acids, has been isolated from mucosa of hog small intestine: J. C. Brown *et al., Can. J. Physiol. Pharmacol.* **49,** 399 (1971). Purification, amino acid composition, C-terminal residue: *eidem, Gastroenterology* **62,** 401 (1972). Amino acid sequence: *eidem, Can. J. Biochem.* **51,** 533 (1973); revised sequence: H. Schu-bert, J. C. Brown, *ibid.* **52,** 7 (1974). First synthetic work:

E. Wünsch *et al., Z. Naturforsch.* **28C,** 235 (1973). Synthesis of the complete docosapeptide corresponding to porcine motilin: H. Yajima *et al., Chem. Commun.* **1975,** 159. Al-ternate syntheses: Y. Kai *et al., Chem. Pharm. Bull.* **23,** 2346 (1975); S. Yamada *et al., J. Am. Chem. Soc.* **97,** 7174 (1975); E. Izeboud, H. C. Beyerman, *Rec. Trav. Chim.* **99,** 124 (1980). Solid phase synthesis: N. Ikota *et al., Chem. Pharm. Bull.* **28,** 3347 (1980); D. H. Coy *et al., Peptides* **3,** 137 (1982). Radioimmunoassay: J. R. Dryburgh, J. C. Brown, *Gastroenterology* **68,** 1169 (1975). Identification of motilin endocrine cells: J. M. Polak *et al., Gut* **16,** 225 (1975). Rodent and canine motilin differ from porcine; human motilin has not yet been purified, although its immu-nologic properties suggest strong similarity to porcine motil-in. The presence of a larger motilin has also been detected immunologically, *see* J. M. Polak, A. M. J. Buchan, *Gastro-enterology* **76,** 1065 (1979). *Reviews:* N. D. Christofides *et al.,* "Physiology of Motilin" in *Gut Hormones,* S. R. Bloom, Ed. (Churchill Livingston, Edinburgh, 1978) pp 343-350; S. R. Bloom, J. M. Polak, *Clin. Endocrinol. Metab.* **8,** 401-411 (1979); V. Mutt, *Vitam. Horm.* **39,** 231-426 (1982).

6199. Motretinide. *N-Ethyl-9-(4-methoxy-2,3,6-tri-methylphenyl)-3,7-dimethyl-2,4,6,8-nonatetraenamide; 9-(4-methoxy-2,3,6-trimethylphenyl)-3,7-dimethylnona-2,4,6,8-tetraen-1-oic acid ethyl amide;* Ro 11-1430; Tasmaderm. $C_{23}H_{31}NO_2$; mol wt 353.51. C 78.15%, H 8.84%, N 3.96%, O 9.05%. Aromatic analog of retinoic acid, *q.v.* Prepn: W. Bollag *et al.,* **Ger.** pat. **2,414,619** corresp to **U.S.** pat. **4,105,-681** (1974, 1978, both to Hoffmann -La Roche). Description of properties: *idem, Chemotherapy* **21,** 236 (1975). ^{13}C-NMR study: G. Englert, *Helv. Chim. Acta* **58,** 2367 (1975). Pharmacology: L. J. Wilkoff *et al., Cancer Res.* **36,** 964 (1976); S. S. Shapiro *et al., ibid.* 3702. Inhibitory effect on radiation-induced neoplasms: L. Harisiades *et al., Nature* **274,** 486 (1978). Use in treatment of acne: K. Nordin *et al., Dermatologica* **162,** 104 (1981).

Crystals from ethanol, mp 179-180°.
THERAP CAT: Anti-acne.

6200. Moveltipril. *[R-(R*,S*)]-1-[3-[[2-[(Cyclohexyl-carbonyl)amino]-1-oxopropyl]thio]-2-methyl-1-oxopropyl]-L-proline; N-[3-(N-cyclohexanecarbonyl-D-alanylthio)-2-methylpropanoyl]-L-proline; (-)-N-[(S)-[3-(N-cyclohexyl-carbonyl-D-alanyl)thio]-2-methylpropionyl]-L-proline;* al-tiopril. $C_{19}H_{30}N_2O_5S$; mol wt 398.52. C 57.26%, H 7.59%, N 7.03%, O 20.07%, S 8.04%. Angiotensin-converting en-zyme inhibitor; structural analog of captopril. Prepn: S. Tanaka *et al.,* **Belg.** pat. **893,553** (1982 to Chugai), *C.A.* **98,** 215995n (1983). Pharmacology: K. Sakai *et al., Tohoku J. Exp. Med.* **152,** 363 (1987). Hemodynamic effects in dogs: K. Noguchi *et al., Japan. J. Pharmacol.* **40,** 373 (1986). Antihypertensive activity: J. Aono *et al., Arch. Int. Pharma-codyn.* **292,** 203 (1988). Tissue specific ACE inhibitory ac-tivity: K. Sakai *et al., ibid.* **294,** 228 (1988). Determn by radioimmunoassay: Y. Hinohara *et al., J. Pharmacobio-Dyn.* **11,** 411 (1988).

Solid, mp 113-116°. $[\alpha]_D$ 14.2° (c = 1.05 in methanol). Calcium salt, $C_{38}H_{58}CaN_4O_{10}S_2$, *MC 838, Lowpres.* Prac-tically white powder, mp ~190°. Bitter taste. $[\alpha]_D^{20}$ —48 to —52° (c = 1 in methanol). Freely sol in water, methanol; sol in ethanol, chloroform. Practically insol in acetone, ethyl acetate. 10% aq soln has pH 5.5-6.5. Stable as aq soln and powder at room temp. LD_{50} in male, female mice, male, female rats (g/kg): all > 10.0 orally; 2.1, 2.3, 1.3, 1.3 i.p.; 3.0, 3.8, 3.4, 3.9 s.c.; and in male, female dogs (g/kg): > 6.0, > 6.0 orally (Sakai).

THERAP CAT: Antihypertensive.

6201. Moxalactam. *7-[[Carboxy(4-hydroxyphenyl)acet-yl]amino]-7-methoxy-3-[[(1-methyl-1H-tetrazol-5-yl)thio]methyl]-8-oxo-5-oxa-1-azabicyclo[4.2.0]oct-2-ene-2-carboxylic acid; N-[(6R,7R)-2-carboxy-7-methoxy-3-[[(1-methyl-1H-tetrazol-5-yl)thio]methyl]-8-oxo-5-oxa-1-azabicyclo-[4.2.0]oct-2-en-7-yl]-2-(p-hydroxyphenyl)malonamic acid; 7β-[2-carboxy-2-(4-hydroxyphenyl)acetamido]-7α-meth-oxy-3-[[(1-methyl-1H-tetrazol-5-yl)thio]methyl]-1-oxa-1-dethia-3-cephem-4-carboxylic acid; lamoxactam; latamoxef.* $C_{20}H_{20}N_6O_9S$; mol wt 520.48. C 46.15%, H 3.87%, N 16.15%, O 27.67%, S 6.16%. Oxa-substituted third genera-tion cephalosporin antibiotic (oxacephalosporin). Prepn: M. Narisada, W. Nagata, *Ger. pat.* **2,713,370;** *eidem,* U.S. pat. **4,138,486** (1977, 1979 both to Shionogi); M. Narisada *et al., J. Med. Chem.* **22,** 757 (1979). Laboratory evaluation: T. Yoshida *et al., Antimicrob. Ag. Chemother.* **17,** 302 (1980). Mechanism of action study: Y. Komatsu, T. Nishikawa, *ibid.* 316. Pharmacokinetics: R. Wise *et al., ibid.* **18,** 369 (1980). Comparative *in vitro* activity: T. O. Kurtz *et al., ibid.* 645. Series of articles on pharmacology, teratology, toxicity studies: *Chemotherapy (Tokyo)* **28,** Suppl. 7, 1002-1235 (1980). Clinical evaluation in neonates and infants: U. B. Schaad *et al., J. Pediatr.* **98,** 129 (1981). *Review:* Rev. Infect. Dis. **4,** Suppl., S489-S726 (1982).

Colorless powder, mp 117-122° (dec). $[\alpha]_D^{25}$ −15.3 ±2.6° (c = 0.216 in methanol). uv max (methanol): 276 nm (ε 10200).

Disodium salt, $C_{20}H_{18}N_6Na_2O_9S$, *Ly-12735, S-6059, Festa-moxin, Moxalactam, Moxam, Shiomarin.* $[\alpha]_D^{22}$ −45° (water). uv max (water): 270 nm (ε 12000).

THERAP CAT: Antibacterial.

6202. Moxaverine. *3-Ethyl-6,7-dimethoxy-1-(phenyl-methyl)isoquinoline; 1-benzyl-3-ethyl-6,7-dimethoxyisoquin-oline.* $C_{20}H_{21}NO_2$; mol wt 307.38. C 78.14%, H 6.89%, N 4.56%, O 10.41%. Prepn of hydrochloride: **Fr. pat. 1,362,-765** (1964 to Orgamol, SA), *C.A.* **62,** 536f (1965), corresp to **Brit. pat. 1,030,022.**

Crystals, mp 78-79°.

Hydrochloride, $C_{20}H_{22}ClNO_2$, *Eupaverin, Eupaverina, Kollateral-forte, Sorbosan.* Crystals, mp 214°; also reported as dec 208-210°. Sol in hot water, hot alc, and many other organic solvents; very sparingly sol in cold water.

Note: The name Eupaverin was formerly used to designate 3-methyl-6,7-methylenedioxy-1-piperonylisoquinoline hy-drochloride.

THERAP CAT: Antispasmodic.

6203. Moxestrol. *11β-Methoxy-19-nor-17α-pregna-1,-3,5(10)-trien-20-yne-3,17-diol; 3,17-dihydroxy-11β-meth-oxy-19-nor-17α-pregna-1,3,5-trien-20-yne; 17α-ethynyl-11β-methoxyestra-1,3,5(10)-triene-3,17β-diol; 11β-meth-oxy-17α-ethynyl-Δ^{1,3,5(10)}-estratriene-3,17β-diol; 11β-meth-oxy-17α-ethynylestradiol; R-2858; Surestryl.* $C_{21}H_{26}O_3$; mol wt 326.44. C 77.27%, H 8.03%, O 14.70%. Prepn: Bertin, Pierdet, **S. Afr. pat. 67 03,381** corresp to **U.S. pat. 3,579,545** and **Fr. pat. M6182,** *C.A.* **72,** 90739m (1970), (1968, 1971

and 1968, all to Roussel-UCLAF); Azadian-Boulanger, Bertin, *Chim. Ther.* **8,** 451 (1973). Activity studies: Ray-naud, *Steroids* **21,** 249 (1973); Raynaud *et al., Mol. Pharma-col.* **9,** 520 (1973).

Crystals, mp 280°. $[\alpha]_D^{20}$ +29° (c = 0.6 in ethanol). uv max (ethanol): 280 nm ($E_{1cm}^{1\%}$ 58.4).

THERAP CAT: Estrogen.

6204. Moxisylyte. *4-[2-(Dimethylamino)ethoxy]-2-methyl-5-(1-methylethyl)phenol acetate (ester); 5-(2-dimeth-ylaminoethoxy)carvacrol acetate;* 6-acetoxythymol 2-(di-methylamino)ethyl ether; (6-acetoxythymoxy)ethyldimeth-ylamine; 4-(2-dimethylaminoethoxy)-5-isopropyl-2-meth-ylphenyl acetate; 4-(2-dimethylaminoethoxy)-2-methyl-5-isopropylphenyl acetate; *thymoxamine; Arlytene; Moxyl; Sympal; Uralpha; Vasoklin.* $C_{16}H_{25}NO_3$; mol wt 279.37. C 68.78%, H 9.02%, N 5.01%, O 17.18%. α-Adrenergic block-er. Prepn: Pahlicke, **Ger. pat. 905,738** (1954 to Diwag), *C.A.* **52,** 16294i (1958); **Brit. pat. 745,070** (1956 to Veritas Drug), *C.A.* **51,** 471i (1957); A. Buzas *et al., Bull. Soc. Chim. France* **1959,** 839. Pharmacology: K. Greeff, H. J. Schü-mann, *Arzneimittel-Forsch.* **3,** 341 (1953); J. Mercier *et al., Therapie* **26,** 785 (1971). Toxicology: J. Roquebert, J. Can-ellas, *ibid.* 775. Clinical-pharmacological studies: A. Wasi-lewski, *Arzneimittel-Forsch.* **21,** 1183 (1971). α-Adrenergic antagonist activity: A. T. Birmingham, J. Szolcsanyi, *J. Pharm. Pharmacol.* **17,** 449 (1965); G. M. Drew, *Eur. J. Pharmacol.* **36,** 313 (1976). Metabolism in rats: C. Feniou *et al., J. Pharm. Pharmacol.* **32,** 104 (1980). Series of articles on metabolism: ; K.-O. Vollmer *et al., Eur. J. Drug Metab. Pharmacokinet.* **10,** 61, 71, 139 (1985). Pharmacokinetics: F. Nielsen-Kudsk *et al., Acta Pharmacol. Toxicol.* **47,** 11 (1980). GLC determn of metabolites in plasma: H. Hengy *et al., Eur. J. Drug Metab. Pharmacokinet.* **10,** 203 (1985). HPLC determn: P. Duchene *et al., J. Chromatog.* **424,** 205 (1988). Clinical trial in Raynaud's disease and chilblains: G. V. Jaffe, J. J. Grimshaw, *Brit. J. Clin. Prac.* **34,** 343 (1980). Clinical study in miosis: F. Grehn *et al., Graefe's Arch. Clin. Exp. Ophthalmol.* **224,** 174 (1986); F. Grehn, *Am. J. Ophthalmol.* **103,** 709 (1987). *Review:* M. Wand, W. M. Grant, *Survey Ophthalmol.* **25,** 75-84 (1980).

Hydrochloride, $C_{16}H_{26}ClNO_3$, *Arlitene, Carlytene, Opilon.* Shiny beige needles from ethyl acetate/methanol, mp 208-210°. LD_{50} in mice, rats (mg/kg): 265 ±19, 740 ±51 oral-ly; 200 ±15, 190 ±19 s.c. (Roquebert, Canellas).

THERAP CAT: Vasodilator (peripheral).

6205. MPTP. *1,2,3,6-Tetrahydro-1-methyl-4-phenylpy-ridine; 1-methyl-4-phenyl-1,2,3,6-tetrahydropyridine.* $C_{12}H_{15}N$; mol wt 173.26. C 83.19%, H 8.73%, N 8.08%. Piperi-dine derivative which causes irreversible symptoms of par-kinsonism in humans, monkeys. Prepn as hydrochloride by Grignard reaction: A. Ziering *et al., J. Org. Chem.* **12,** 894 (1947). Alternative prepn: C. J. Schmide, R. C. Mansfield, *J. Am. Chem. Soc.* **78,** 425 (1956). Identification as impurity in "synthetic heroin" and effect on drug users: J. W. Langs-ton *et al., Science* **219,** 979 (1983). Selective destruction of dopaminergic neurons in primates: R. S. Burns *et al., Proc. Nat. Acad. Sci. USA* **80,** 4546 (1983). *In vitro* metabolism by rat brain monoamine oxidase to 1-methyl-4-phenylpyridinium ion (*MPP*+): K. Chiba *et al.*, *Biochem. Biophys. Res. Commun.* **120**, 574 (1984). Studies on mechanism of neurotoxicity: J. W. Langston *et al.*, *Science* **225**, 1480 (1984); S. P. Markey *et al.*, *Nature* **311**, 464 (1984). Binding studies in rat brain: C. M. Wieczorek *et al.*, *Eur. J. Pharmacol.* **98**, 453 (1984); B. Parsons, T. C. Rainbow, *ibid.* **102**, 375 (1984); in rat, human brain: J. A. Javitch *et al.*, *Proc. Nat. Acad. Sci. USA* **81**, 4591 (1984). Comparison of idiopathic and MPTP-induced parkinsonism in humans: R. S. Burns *et al.*, *N. Engl. J. Med.* **312**, 1418 (1985). *Review:* T. P. Singer *et al.*, *Trends Biochem. Sci.* **12**, 266 (1987).

Crystals from heptane, mp 40-42°. $bp_{0.8}$ 85-90°.

6206. MSH. *Melanotropin;* melanophore-affecting hormone; melanocyte-stimulating hormone; melanophore hormone; melanophore dilating hormone; melanophore expanding hormone; melanophore-stimulating hormone; melanotropic hormone; chromatophorotropic hormone; melanosome-dispersing hormone; pigmentation hormone. First known as *Intermedin(e)* or *B hormone*, a pituitary factor causing color changes in fish and amphibia: B. Zondek, H. Kron, *Klin. Wochenschr.* **11**, 849 (1932). Subsequently, the hormone was identified as two linear peptides designated as α- and β-MSH. γ-MSH, a third peptide has also been identified but has no significant melanocyte stimulating activity. The peptides are secreted by the pars intermedia of the pituitary gland. The brain may be a secondary site of synthesis. α-MSH derives from ACTH and β-MSH from β-lipotrophic hormone, *q.q.v.* These MSH precursors derive from two sections of pro-opiomelanocortin, *q.v.* A third portion of pro-opiomelanocortin may serve as the precursor of γ-MSH. Biosynthesis: B. G. Jenks *et al.*, *J. Endocrinol.* **98**, 19 (1983); G. J. M. Martens *et al.*, *Gen. Comp. Endocrinol.* **49**, 73 (1983). Distribution and biosynthesis in the brain: D. F. Swaab *et al.*, "The Distribution of MSH and ACTH in the Rat and Human Brain and its Relation to Pituitary Stores," in *Endogenous Peptides and Learning and Memory Processes,* J. L. Martinez, Jr. *et al.*, Eds. (Academic Press, New York, 1981) pp 7-36. The mechanism of control of MSH release from the pituitary is still being investigated. There is evidence for inhibitory control by the brain via dopaminergic neurons, inhibitory control by the peptide melanotropin (MIF-1), *q.v.* and stimulatory control by β-adrenergic agents and the peptide CRF, *q.v.*: G. Schmitt *et al.*, *Neuroendocrinology* **33**, 306 (1981); P. G. Smelik *et al.*, "The Role of Catecholamines in the Control of the Secretion of Pro-Opiocortin-Derived Peptides from the Anterior and Intermediate Lobes and its Implications in the Response to Stress," in *The Anterior Pituitary Gland,* A. S. Bhatnagar, Ed. (Raven Press, New York, 1983) pp 113-125; L. Proulx-Ferland *et al.*, *J. Steroid Biochem.* **19**, 439 (1983). Although the role of MSH in adaptive color change in lower vertebrates in well known, its physiological significance in mammals appears to be extrapituitary. Effects on reproduction: M. E. Celis, M. Volosin, *Prog. Clin. Biol. Res.* **87**, 113 (1982). Reviews of behavioral and neurochemical effects: P. C. Datta, M. G. King, *Neurosci. Biobehav. Rev.* **6**, 297-310 (1982); B. E. Beckwith, C. A. Sandman, *Peptides* **3**, 411-420 (1982). General reviews: Li, *Advan. Protein Chem.* **12**, 270-295 (1957); Novales, *Neuroendocrinology* **2**, 241 (1967); R. Schwyzer, *Proc. Roy. Soc. London [Biol.]* **210**, 5-20 (1980); F. L. Strand, C. M. Smith, *Pharmacol. Ther.* **11**, 509-533 (1980). Book: A. J. Thody, *The MSH Peptides* (Academic Press, New York, 1980) 162 pp.

α-*Melanotropin.* Contains 13 amino acids. Mammalian α-MSH is the most potent melanocyte-stimulating peptide known. Mechanism of action: T. K. Sawyer *et al.*, *Am. Zool.* **23**, 1983. Isoln from hog pituitary gland: T. H. Lee, A. B. Lerner, *J. Biol. Chem.* **221**, 943 (1956); from pig pituitary gland: Harris, Lerner, *Nature* **179**, 1346 (1957); Harris, *Biochem. J.* **71**, 451 (1959); from horse pituitary gland: Dixon, Li, *J. Am. Chem. Soc.* **82**, 4568 (1960). Structure: Lee *et al.*, *J. Biol. Chem.* **236**, 1390 (1961). Synthesis: Schwyzer *et al.*, *Helv. Chim. Acta* **46**, 870

(1963). Synthesis and biological activity of dogfish α-MSH: A. Eberle *et al.*, *Helv. Chim. Acta* **61**, 2360 (1978). Activity as centrally administered antipyretic agent: M. T. Murphy *et al.*, *Science* **221**, 192 (1983). Cardiovascular effects: D. J. Diz, D. M. Jacobowitz, *Brain Res.* **270**, 265 (1983). Review of the α-melanotropinergic system and its role in the CNS: T. L. O'Donohue, D. M. Jacobowitz in *Polypeptide Hormones,* R. F. Beers, E. G. Bassett, Eds. (Raven Press, New York, 1980) pp 203-222. $[\alpha]_D^{25}$ $-58.5°$ ± $2.5°$ (c = 0.38 in 10% acetic acid).

β-*Melanotropin.* Contains 18-22 amino acids differing slightly from one species to another: Lee *et al.*, *J. Biol. Chem.* **236**, 1390 (1961). Isoln from hog pituitary gland: T. H. Lee, A. B. Lerner, *loc. cit.;* from pig pituitary gland: Geschwind *et al.*, *J. Am. Chem. Soc.* **78**, 4494 (1956); **79**, 620 (1957); Harris, Roos, *Biochem. J.* **71**, 434 (1959); from bovine pituitary gland: Geschwind *et al.*, *J. Am. Chem. Soc.* **79**, 6394 (1957); from human pituitary gland: Dixon, *Biochim. Biophys. Acta* **37**, 38 (1960); Harris, *Nature* **184**, 167 (1959). Synthesis of derivatives: Hofmann, *Ann. N.Y. Acad. Sci.* **88**, art. 3, p 689 (1960). Synthesis of bovine β-melanotropin: Schwyzer *et al.*, *Helv. Chim. Acta* **46**, 1975 (1963); S. Lemaire *et al.*, *J. Med. Chem.* **19**, 373 (1976). Solid-phase synthesis of human and monkey β-melanotropins: Wang *et al.*, *Int. J. Peptide Protein Res.* **5**, 33 (1973); of camel β_{c1}-MSH: Li *et al.*, *Biochemistry* **14**, 953 (1975); of camel β_{c2}-MSH: S. Lemaire *et al.*, *loc. cit.;* of dogfish β-MSH: H. Yajima *et al.*, *Chem. Pharm. Bull.* **26**, 571 (1978); of equine β-MSH: J. Izdebski *et al.*, *Int. J. Peptide Res.* **19**, 327 (1982). It is probable that the human pituitary does not produce β-MSH. The substance reported as human β-MSH in early studies seems to have been an artifact formed by enzymatic degradation of β-LPH during the extraction procedure: K. Tanaka *et al.*, *J. Clin. Invest.* **62**, 94 (1978); A. J. Thody, *loc. cit.*

γ-*Melanotropin.* Contains 12 amino acids. Discovered during nucleotide sequencing of cloned DNA for bovine pro-opiomelanocortin: S. Nakanishi *et al.*, *Nature* **278**, 423 (1979). May be an antagonist of β-endorphin or a partial agonist/antagonist of ACTH. Synthesis and biological activity: N. Ling *et al.*, *Life Sci.* **25**, 1773 (1979); K. Okamoto *et al.*, *Chem. Pharm. Bull.* **28**, 2839 (1980); W. A. Bijl *et al.*, *Rec. Trav. Chim.* **100**, 123 (1981). Radioimmunoassay: T. Shibasaki *et al.*, *Life Sci.* **26**, 1781 (1980). Behavioral profile in rats: J. M. Van Ree *et al.*, *ibid.* **28**, 2875 (1981). Comparison with α-MSH: T. L. O'Donohue *et al.*, *Peptdes* **2**, 101 (1981). *Review:* J. M. Van Ree *et al.*, *Ciba Found. Symp.* **81**, 263-276 (1981). $[\alpha]_D^{20}$ $-38.0°$ (c 0.5 in 10% acetic acid). $[\alpha]_D^{23}$ $-33.4°$ (c = 1.0 in 1% acetic acid).

6207. Mucins. High molecular weight glycoproteins, major constituents of saliva, gastric juice, intestinal juice, and other secretions: F. Haurowitz, *Chemistry and Biology of Proteins* (Academic Press, New York, 1950) p 199. Mucins are capable of forming viscous solutions and thereby act as lubricants or protectants in cavities of the body or on body surfaces. For prepn (*e.g.* from snails) and their biochemical role see P. A. Levene, *The Hexosamines and Mucoproteins* (Longmans, Green, London, 1925). Ovine submaxillary mucin consists of a single polypeptide chain with about 800 disaccharide units attached, each of which is N-acetylneuraminyl$(2 \to 6)$ N-acetylgalactosamine. Carbohydrate accounts for about 45% of the molecular weight. Structure studies: Pigman, Tanaka, *148th Am. Chem. Soc. Meet.* (Chicago, Aug.-Sept., 1964), Abstracts of Papers, p 11C; Ozeki, Yosizawa, *Arch. Biochem. Biophys.* **142**, 177 (1971); Huser *et al.*, *Z. Physiol. Chem.* **354**, 749 (1973). Glycoproteins in bovine cervical mucus: F. A. Meyer *et al.*, *Adv. Exp. Med. Biol.* **89**, 239 (1977). Review on physical and chemical properties, biosynthesis and function of mucin: several authors in *Mucus in Health and Disease,* M. Elstein, D. V. Parke, Eds. (Plenum Press, New York, 1977) pp 171-311.

Mucins are obtained as greenish-gray or yellow powders forming very viscous solns in water; generally sol in dil alkalies. Insol in acetic acid. Acidic substances with an isoelectric point between pH 3 and 5.

USE: Demulcent; adsorbent.

6208. Mucochloric Acid. *2,3-Dichloro-4-oxo-2-butenoic*

acid; *dichloromalealdehydic acid;* α,β-dichloro-β-formyl-acrylic acid; 2,3-dichloromaleic aldehyde acid. $C_4H_2Cl_2O_3$; mol wt 168.97. C 28.43%, H 1.19%, Cl 41.97%, O 28.41%. Prepn from β,γ-dichloropyromucic acid and bromine water or dil nitric acid: Hill, Jackson, *Am. Chem. J.* **12**, 43 (1890); by heating furfural with manganese dioxide and hydrochloric acid: Simonis, *Ber.* **32**, 2085 (1899); Beattie *et al., J. Chem. Soc.* **1932**, 264; Mowry, *J. Am. Chem. Soc.* **72**, 2535 (1950); by chlorination of butyne-1,4-diol: Dury, *Angew. Chem.* **72**, 864 (1960).

Monoclinic prisms from ether and ligroin, mp 127°. Slightly sol in cold water; sol in hot water; hot benzene, alc. LD_{50} in rats: 0.5-1.0 g/kg orally; 10-25 mg/kg i.p., Fassett in *Industrial Hygiene and Toxicology* vol. II, F. A. Patty, Ed. (Interscience, New York, 2nd ed., 1962) p 1977.

Human Toxicity: Strong irritant to skin and eyes; a potent skin sensitizer.

6209. Mucochloric Anhydride. *5,5'-Oxybis[3,4-dichloro-2(5H)-furanone];* bis[3,4-dichloro-2(5)-furanonyl] ether; GC 2466. $C_8H_2Cl_4O_5$; mol wt 319.92. C 30.03%, H 0.63%, Cl 44.33%, O 25.01%. Prepd by refluxing mucochloric acid with benzenesulfonic acid in a mixture of benzene and dioxane: Mowry, *J. Am. Chem. Soc.* **72**, 2535 (1950).

Crystals from benzene + dioxane, α-isomer (racemate): mp 141-143°, β-isomer (*meso* form): mp 180°. Substantially insol in water. Sol in many organic solvents, such as acetone, xylene, cyclohexanone, methylnaphthalenes.

USE: Fungicide. *Ref:* Gilbert, U.S. pat. **2,861,919** (1958 to Allied Chem.).

6210. Muconic Acid. *2,4-Hexadienedioic acid;* 1,3-butadiene-1,4-dicarboxylic acid. $C_6H_6O_4$; mol wt 142.11. C 50.71%, H 4.26%, O 45.03%. HOOCCH=CHCH=CH-COOH. Prepd by oxidation of phenol and peracetic acid: Boeseken, Engelberts, *C.A.* **26**, 2970 (1932); by treatment of ethyl 1,4-dibromoadipate with alcoholic potassium hydroxide: Guha, Sankaran, *Org. Syn.* **26**, 57 (1946); by isomerization of 3-hydroxy-4-carbomethoxybut-1-ene-1-carboxylic acid lactone: Elvidge *et al., J. Chem. Soc.* **1950**, 2235; by carbonylation of acetylene: Tsuji *et al., J. Am. Chem. Soc.* **86**, 2095 (1964). Configuration and separation of isomers: Boeseken, Kerkhoven, *Rec. Trav. Chim.* **51**, 964 (1932); Elvidge *et al., J. Chem. Soc.* **1953**, 708.

trans,trans-Form, prisms from water, mp 301°. uv max (0.1N NaOH): 251, 259, 264 nm (ε 25,600, 29,100, 25,600). One gram dissolves in about 5 liters water at 15°. Quite sol in hot alcohol and glacial acetic acid.

Dimethyl ester, $C_8H_{10}O_4$, crystals from alcohol, mp 159°. Insol in ether.

Diethyl ester, $C_{10}H_{14}O_4$, needles from alcohol, mp 64°; d_4^{100} 0.9829. n_{He}^{99} 1.4675.

cis,cis-Form, prisms from ethanol, mp 194-195°. uv max (0.1N NaOH): 251, 258, 264 nm (ε 15,600, 17,000, 15,300). Freely sol in boiling water; sparingly sol in ether.

Dimethyl ester, $C_8H_{10}O_4$, needles from petr ether, mp 75°. Soluble in ether.

Diethyl ester, $C_{10}H_{14}O_4$, minute crystals from alcohol at −25°; mp 13°. Changes to the *trans-trans*-form on storage at room temp.

cis,trans-Form, needles from hot water, mp 190-191°. uv max (0.1N NaOH): 251, 259, 265 nm (ε 23,400, 25,600, 23,400).

Dimethyl ester, $C_8H_{10}O_4$, needles from aqueous methanol, mp 75°.

6211. Muira Puama. Potency wood. Wood of *Liriosma ovata* Miers, *Oleaceae,* or according to Rebourgeon, *Acan-*

thea virilis Wehmer, *Acanthaceae.* *Habit.* Brazil. *Constit.* Aromatic resin, muirapuamine, fat. Studies of components: Auterhoff, Pankow, *Arch. Pharm. (Weinheim)* **301**, 481 (1968); **302**, 209 (1969). *Reviews:* Gaebler, *Deut. Apoth.* **22**, 94 (1970); Steinmetz, *Quart. J. Crude Drug Res.* **11**, 1787 (1971).

6212. Mupirocin. *[2S-[2α(E),3β,4β,5α[2R*,3R*(1R*,-2R*)]]]-9-[[3-Methyl-1-oxo-4-[tetrahydro-3,4-dihydroxy-5-[[3-(2-hydroxy-1-methylpropyl)oxiranyl]methyl]-2H-pyran-2-yl]-2-butenyl]oxy]nonanoic acid;* pseudomonic acid A; *trans*-pseudomonic acid; BRL-4910A; Bactoderm; Bactroban; Eismycin. $C_{26}H_{44}O_9$; mol wt 500.63. C 62.38%, H 8.86%, O 28.76%. Major component of the pseudomonic acids, *q.v.,* an antibiotic complex produced by *Pseudomonas fluorescens* NCIB 10586. Isoln and characterization: A. T. Fuller *et al., Nature* **234**, 416 (1971); K. D. Barrow, G. Mellows, Ger. pat. **2,227,739**; *eidem,* U.S. pat. **3,977,943**; *eidem,* U.S. pat. **4,071,536** (1973, 1976, 1978 all to Beecham). Purification: P. J. O'Hanlon *et al.,* Ger. pat. **2,842,-358**; *eidem,* U.S. pat. **4,222,942** (1979, 1980 both to Beecham). Structure: E. B. Chain, G. Mellows, *Chem. Commun.* **1974**, 847; *eidem, J. Chem. Soc. Perkin Trans. I* **1977**, 294. Absolute configuration: R. G. Alexander *et al., ibid.* **1978**, 561. Prepn from methyl pseudomonate C: *eidem, Tetrahedron Letters* **22**, 2059 (1981). Total syntheses of (±)-form: B. B. Snider, G. B. Phillips, *J. Am. Chem. Soc.* **104**, 1113 (1982); B. B. Snider *et al., J. Org. Chem.* **48**, 3003 (1983). Biosynthesis: T. C. Feline *et al., J. Chem. Soc. Perkin Trans. I* **1977**, 309. Effect of pH on chemical stability and antibacterial activity: J. P. Clayton *et al., ibid.* **1979**, 838. Inhibition of bacterial protein synthesis: J. Hughes, G. Mellows, *J. Antibiot.* **31**, 330 (1978); of isoleucyl-tRNA synthetase: *eidem, Biochem. J.* **176**, 305 (1978); *eidem, ibid.* **191**, 209 (1980). *In vitro* antibacterial spectrum: R. Sutherland *et al., Antimicrob. Ag. Chemother.* **27**, 495 (1985); M. W. Casewell, R. L. R. Hill, *J. Antimicrob. Chemother.* **15**, 523 (1985). Antimycoplasmal activity *in vitro:* R. M. Banks *et al., J. Antibiot.* **41**, 609 (1988). Clinical evaluations: G. D. Reilly, R. C. Spencer, *ibid.* **13**, 295 (1984); M. W. Casewell, R. L. R. Hill, *ibid.* **17**, 365 (1986). *Reviews:* A. Ward, D. M. Campoli-Richards, *Drugs* **32**, 425-444 (1986); M. W. Casewell, R. L. R. Hill, *J. Antimicrob. Chemother.* **19**, 1-5 (1987).

Crystals from ether, mp 77-78°. $[α]_D^{20}$ −19.3° (c = 1 in methanol). uv max (ethanol): 222 nm (ε 14500).

THERAP CAT: Topical antibacterial.
THERAP CAT (VET): Topical antibacterial.

6213. Muramic Acid. *2-Amino-3-O-(1-carboxyethyl)-2-deoxy-D-glucose;* 2-amino-3-O-(D-1'-carboxyethyl)-2-deoxy-D-glucose; 3-O-α-carboxyethyl-D-glucosamine. $C_9H_{17}NO_7$; mol wt 251.23. C 43.02%, H 6.82%, N 5.58%, O 44.58%. Amino sugar found in the cell walls of bacteria. Discovered by Park, *J. Biol. Chem.* **194**, 885 (1952). Isoln from spores of *Bacillus megatherium,* identification and chemical synthesis: Strange, Kent, *Biochem. J.* **71**, 333 (1959); *Methods Carbohyd. Chem.* **1**, 250-257 (1962). Synthesis: Kent, *Biochem. J.* **67**, 5p (1957); Lambert, Zilliken, *Ber.* **93**, 2915 (1960); Gigg *et al., J. Chem. Soc.* **1965**, 2975. Stereospecific synthesis: Matsushima, Park, *J. Org. Chem.*

27, 3581 (1962); *Biochem. Prepn.* **10**, 109 (1963); Osawa, Jeanloz, *J. Org. Chem.* **30**, 448 (1965).

Crystals from water, mp 152-154°. Hydrated crystals from dil alc. $[\alpha]_D^{20}$ +109° (c = 2); $[\alpha]_D^{25}$ +103° (c = 0.26).

6214. Muramyl Dipeptide. N^2-[*N*-(*N*-Acetylmuramoyl)-L-*alanyl*]-D-*α*-*glutamine*; *N*-acetylmuramyl-L-alanyl-D-iso-glutamine; 2-acetamido-2-deoxy-3-*O*-(D-2-propionyl-L-alanyl-D-isoglutamine)-D-glucopyranose; MDP. $C_{19}H_{32}$-N_4O_{11}; mol wt 492.48. C 46.34%, H 6.55%, N 11.38%, O 35.73%. Synthetic immunoadjuvant corresponding to the smallest immunologically active glycopeptide subunit of the bacterial cell wall. Prepn: A. Adam *et al.*, Ger. pat. **2,450,-355**; *eidem*, U.S. pat. **4,235,771** (1975, 1980 both to Agen. Nat. Valor. Recher.); C. Merser *et al.*, *Biochem. Biophys. Res. Commun.* **66**, 1316 (1975); P. Lefrancier *et al.*, *Int. J. Pep. Prot. Res.* **9**, 249 (1977). Structure-activity relationships: F. Ellouz *et al.*, *Biochem. Biophys. Res. Commun.* **59**, 1317 (1974); A. Adam *et al.*, *ibid.* **72**, 339 (1976). Review of immunoregulating activity: L. Chedid *et al.*, *Prog. Allergy* **25**, 63-105 (1978). Tissue distribution in mice: M. Parant *et al.*, *Int. J. Immunopharmacol.* **1**, 35 (1979). Potentiation of antitumor immunity: S. Sone *et al.*, *J. Biol. Resp. Mod.* **3**, 185 (1984); A. E. Eggers, *ibid.* **7**, 229 (1988).

Crystals from methanol-acetone-ether. $[\alpha]_D^{25}$ +44° (acetic acid).

USE: Immunological adjuvant.

6215. Murexide. 5-[(*Hexahydro*-2,4,6-*trioxo*-5-*pyrimi-dinyl*)*imino*]-2,4,6(1H,3H,5H)-*pyrimidinetrione monoammonium salt*; 5,5′-*nitrilodibarbituric acid monoammonium salt*; acid ammonium purpurate; ammonium purpurate. C_8-$H_8N_6O_6$; mol wt 284.19. C 33.81%, H 2.84%, N 29.57%, O 33.78%. Prepn from alloxan + NH_3: Hartley, *J. Chem. Soc.* **87**, 1791 (1905); Schwartz, Handritschk, **East Ger.** pat. **17,589** (1959), *C.A.* **55**, 3630a (1961); from alloxantin + ammonium acetate: Davidson, *J. Am. Chem. Soc.* **58**, 1821 (1936). Structure: Hartley, *J. Chem. Soc.* **87**, 1796 (1905); Winslow, *J. Am. Chem. Soc.* **61**, 2089 (1939); Schreiber, *Mitt. Deut. Pharm. Ges.* **28**, 20 (1958).

Purple-red crystals with green metallic luster. Absorption max in water: 520 nm. Sparingly sol in cold water, more in hot water; practically insol in alcohol, ether. The H_2O soln is deep purple and the aq NaOH soln is deep blue.

USE: Indicator for complexometric titrations.

6216. Murexine. 2-[[3-(1H-*Imidazol*-4-*yl*)-1-*oxo*-2-*propenyl*]*oxy*]-N,N,N-*trimethylethanaminium*; *β*-(4-imidazolyl)acrylcholine; urocanylcholine. $[C_{11}H_{18}N_3O_2]^+$; mol wt 224.28. Neuromuscular blocker found, often in very large quantities, in the median zone of the hypobranchial body of *Murex trunculus* and of other related species of mollusks: Erspamer, Dordoni, *Ric. Sci. Ricostr.* **16**, 1114 (1946); *Arch. Int. Pharmacodyn.* **74**, 263 (1947); Erspamer, *ibid.* **76**, 308 (1948). Isoln: *idem, Experientia* **4**, 226 (1948); M. Roseghini *et al.*, *Eur. J. Biochem.* **12**, 468 (1970); J. E. Blankenship *et al.*, *Comp. Biochem. Physiol.* **51C**, 129 (1975). Synthesis: Pasini *et al.*, *Ann.* **578**, 6 (1952); Pasini, Coda, U.S.

pat. **2,956,061** (1960 to Società Farmaceutici). Structure: V. Erspamer, O. Benati, *Science* **117**, 161 (1953). Effect on vertebrate and invertebrate muscles: J. C. de Freitas, *Comp. Biochem. Physiol.* **56C**, 57 (1977).

The base is instantly hydrolyzed by water; unstable both in acid or alkaline media.

Chloride, extremely hygroscopic crystals. uv max (pH 4.5): 280-282 nm.

Chloride hydrochloride, hygroscopic microcrystalline powder, mp 219-221° (dec); shows the same uv max as the chloride.

6217. Muroctasin. N^2-[N^2-[*N*-(*N*-Acetylmuramoyl*)-L-*alanyl*]-D-*α*-*glutaminyl*]-N^6-(1-*oxooctadecyl*)-L-*lysine*; N^2-(*N*-acetylmuramoyl-L-alanyl-D-isoglutaminyl)-N^6-stearoyl-L-lysine; MDP-Lys(L18); DJ-7041. $C_{43}H_{78}N_6O_{13}$; mol wt 887.12. C 58.22%, H 8.86%, N 9.47%, O 23.45%. Derivative of muramyl dipeptide, *q.v.*, with immunostimulating activity. Prepn: T. Shiba *et al.*, Eur. pat. Appl. **21,367**; *eidem*, U.S. pat. **4,317,771** (1981, 1982 both to Daiichi Seiyaku). Immunopotentiating effect *in vitro*: Y. Osada *et al.*, *Infect. Immun.* **38**, 848 (1982); *in vivo*: K. Matsumoto *et al.*, *ibid.* **39**, 1029 (1983); M. Akasaki *et al.*, *Agents Actions* **22**, 144 (1987). Enzyme immunoassay in plasma: H. Masayasu *et al.*, *Chem. Pharm. Bull.* **33**, 5522 (1985). Supplement on pharmacology, toxicology and clinical studies: *Arzneimittel-Forsch.* **38**, Suppl. 7A, 951-1074 (1988). Acute toxicity: Y. Ono *et al.*, *ibid.* 1022.

Odorless, tasteless white powder, mp 176-178°. $[\alpha]_D^{20}$ +24.9° (c = 1 in methanol). pKa 5.70. Soly at 20°: DMF 1 g/72 ml; methanol 1 g/190 ml; ethanol 1 g/330 ml. Practically insol in acetone, acetonitrile, chloroform, water. LD_{50} in male, female mice, male, female rats, dogs (mg/kg): 436, 625, 761, 801, >200 s.c. (Ono).

THERAP CAT: Immunostimulant.

6218. Muscalure. 9-*Tricosene*; *cis*-*tricos*-9-*ene*. $C_{23}H_{46}$; mol wt 322.62. C 85.63%, H 14.37%. Sex pheromone of the female common house fly, *Musca domestica* L. Isoln and synthesis: Carlson *et al.*, *Science* **174**, 76 (1971). Alternate syntheses: Eiter, *Naturwiss.* **59**, 468 (1972); Cargill, Rosenblum, *J. Org. Chem.* **37**, 3971 (1972); Gribble *et al.*, *Chem. Commun.* **1973**, 735; Ho, Wong, *Can. J. Chem.* **52**, 1923 (1974); K. Abe *et al.*, *Bull. Chem. Soc. Japan* **50**, 2792 (1977); V. N. Odinokov *et al.*, *Tetrahedron Letters* **23**, 1371 (1982).

$bp_{0.1}$ 157-158°. n_D^{26} 1.4517.

USE: Insect attractant.

6219. Muscarine. *Tetrahydro*-4-*hydroxy*-N,N,N,5-*tetramethyl*-2-*furanmethanaminium*. $[C_9H_{20}NO_2]^+$; mol wt 174.26. Alkaloid from the red variety of *Amanita muscaria* (L.) Pers., *Agaricaceae*, the fly fungus, a poisonous mushroom. Also found in some other fungi: *Inocybe patouillardi*;

I. fastigiata, I. umbrina; I. rimosa. Isoln procedure for the naturally occurring L-(+)-form: Kuehl *et al., J. Am. Chem. Soc.* **77**, 6663 (1955); Eugster, *Helv. Chim. Acta* **39**, 1002 (1956). Structure and synthesis of racemate: Kögl *et al., Rec. Trav. Chim.* **76**, 109 (1957); Kögl *et al., Experientia* **13**, 137, 138 (1957); Cox *et al., Helv. Chim. Acta* **41**, 229 (1958). Alternate syntheses: **Brit.** pat. **828,395** (1960 to Hoffmann-La Roche); Matsumoto *et al., Tetrahedron* **25**, 5889 (1969); W. C. Still, J. A. Schneider, *J. Org. Chem.* **45**, 3375 (1980). Synthesis of muscarine: Whiting *et al., Can. J. Chem.* **50**, 3322 (1972); A. M. Mubarak, D. M. Brown, *Tetrahedron Letters* **21**, 2453 (1980); *eidem, J. Chem. Soc. Perkin Trans I* **1982**, 809; S. Pochet, T. Huynhdinh, *J. Org. Chem.* **47**, 193 (1982). Chemistry and pharmacology: Waser, *Pharmacol. Rev.* **13**, 465-515 (1961). *Reviews:* C. H. Eugster, *Advances in Organic Chemistry* **2**, 427-455 (1960); S. Wilkinson, *Quarterly Revs. (London)* **15**, 153-171 (1961). Configurational relationship in the muscarine series: Bollinger, Eugster, *Helv. Chim. Acta* **54**, 2704 (1971).

Chloride, $C_9H_{20}ClNO_2$, stout prisms from ethanol + acetone, mp 180-181°. Extremely hygroscopic. $[\alpha]_D^{25}$ +8.1° (c = 3.5 in ethanol). Very sol in water, ethanol. Slightly sol in chloroform, ether, acetone. Aq solns are stable. LD_{50} i.v. in mice: 0.23 mg/kg, P. J. Fraser, *Brit. J. Pharmacol.* **12**, 47 (1957).

Human Toxicity: Ingestion may cause sialorrhea, lacrimation, diaphoresis, nausea, vomiting, diarrhea, miosis, bradycardia, circulatory collapse. May progress to convulsions, coma and death in a few hours. *Antidote:* Atropine sulfate.

THERAP CAT: Cholinergic.

6220. Muscazone. *α-Amino-2,3-dihydro-2-oxo-5-oxazoleacetic acid; α-amino-2-oxo-4-oxazoline-5-acetic acid.* $C_5H_6N_2O_4$; mol wt 158.11. C 37.98%, H 3.83%, N 17.72%, O 40.48%. From *Amanita muscaria* (L.) Fr., *Agaricaceae:* Eugster *et al., Tetrahedron Letters* **1965**, 1813. Structure: Fritz *et al., ibid.* **1965**, 2075; Reiner, Eugster, *Helv. Chim. Acta* **50**, 128 (1967). Synthesis: Göth *et al., ibid.* 137.

Crystals, decomp above 190°. uv max at pH 2-7: 212 nm (ε 8700); at pH 12: 220 nm (ε 7500).

6221. Muscimol. *5-(Aminomethyl)-3(2H)-isoxazolone; 5-(aminomethyl)-3-isoxazolol;* 5-aminomethyl-3-hydroxy-isoxazole; 3-hydroxy-5-aminomethylisoxazole; agarin; pantherine. $C_4H_6N_2O_2$; mol wt 114.10. C 42.10%, H 5.30%, N 24.55%, O 28.05%. Potent CNS depressant and GABA agonist isolated from *Amanita muscaria* (L.) Fr., *Agaricaceae.* Structural identity with the insecticidal substances, pantherine and agarin: Onda *et al., Chem. Pharm. Bull.* **12**, 751 (1964); Eugster *et al., Tetrahedron Letters* **1965**, 1813; Bowden, Drysdale, *ibid.* 727. Synthesis: Gagneux *et al., ibid.* 2077; Göth *et al., Helv. Chim. Acta* **50**, 137 (1967); Bowden *et al., J. Chem. Soc. (C)* **1968**, 172. Improved syntheses: Nakamura, *Chem. Pharm. Bull.* **19**, 46 (1971); P. Krogsgaard-Larsen, S. B. Christensen, *Acta Chem. Scand. Ser. B* **B30**, 281 (1976); A. Barco *et al., J. Chem. Res. (S)* **1979**, 176; B. E. McCarry, M. Savard, *Tetrahedron Letters* **22**, 5153 (1981); V. Jager, M. Frey, *Ann.* **1982**, 817. Conformational structure: L. Brehm *et al., ibid.* **26**, 1298 (1972). Industrial patents: Hafliger, Gagneux, U.S. pats. **3,242,190,** and **3,397,209** (1966, 1968 both to Geigy). Pharmacology: Theobald *et al., Arzneimittel-Forsch.* **18**, 311 (1968). Review of pharmacology: F. V. DeFeudis, *Neurochem. Res.* **5**, 1047-1068 (1980).

Crystals, mp 175° (dec). LD_{50} in mice: 3.8 mg/kg s.c., 2.5 mg/kg i.p.; in rats: 4.5 mg/kg i.v., 45 mg/kg orally, Theobald *et al., loc. cit.*

USE: As a molecular probe to study GABA receptors.

6222. Muscone. *3-Methylcyclopentadecanone;* muskone; methylexaltone. $C_{16}H_{30}O$; mol wt 238.40. C 80.60%, H 12.68%, O 6.71%. The odorous principle of musk. Characterization of naturally occurring (−)-form: Ruzicka, *Helv. Chim. Acta* **9**, 715 (1926). Structure: *ibid.* 1008. Synthesis of (±)-form: Ziegler, Weber, *Ann.* **512**, 164 (1934); Ruzicka, Stoll, *Helv. Chim. Acta* **17**, 1308 (1934); Hunsdiecker, *Ber.* **75B**, 1197 (1942). Review of syntheses: S. Abe *et al., Cosmetics and Perfumery* **88**, 67 (1973). Recent syntheses: Stork, Macdonald, *J. Am. Chem. Soc.* **97**, 1264 (1975); H. G. Fliri *et al., Monatsh. Chem.* **110**, 245 (1979); C. Fehr *et al., Helv. Chim. Acta* **62**, 2655 (1979); K. H. Schulte-Elte *et al., ibid.* 2673; G. Cantoni *et al., J. Org. Chem.* **45**, 1906 (1980). Chiral synthesis of (R)-(−)- and (S)-(+)-forms: Q. Branca, A. Fischli, *Helv. Chim. Acta* **60**, 925 (1977).

(−)-Form, oily liquid. Musk odor. bp 328°; $bp_{0.5}$ 130°. d_4^{17} 0.9221. n_D^{17} 1.4802. $[\alpha]_D^{17}$ −13°. Very slightly sol in water; miscible with alc.

(±)-Form, oily liquid. $bp_{0.01}$ 90°. n_D^{26} 1.4767. uv max (ethanol): 285 nm (log ε 1.54).

6223. Musks. Fixatives for perfumes having a characteristic persistent musky aroma. Obtained from plant and animal sources and by chemical synthesis. Natural sources of musk include the musk glands of the male musk deer, *Moschus moschiferus,* the civet cat, *Viverra civetta,* the Louisiana muskrat, *Abelmoschus moschatus,* and angelica roots, *Angelica archangelica.* Musk odor is exhibited by several categories of chemical compounds. Natural musks are macrocyclic ketones or lactones having approximately fifteen carbons in their ring structures. Synthetic musks are of greater industrial importance and include nitro and non-nitro benzenes, indans, and tetralins; derivatives of hydrindacene, isochroman, naphthindan, and coumarin. Purification of natural and synthetic musks: Thomas, Stephens, U.S. pat. **3,415,813** (1968 to Pfizer). Review of macrocyclic musks: Berends, *Am. Perfum. Cosmet.* **80**, 35 (1965). Review of chemical studies of synthetic musks: T. F. Wood, "Chemistry of the Aromatic Musks," (Givaudan Corp., company literature).

USE: In perfumery.

6224. Mustard, Black. Brown mustard; red mustard. Dried ripe seeds of *Brassica nigra* (L.) Koch, or of *B. juncea* (L.) Cosson, and of varieties of these species *(Cruciferae).* *Habit.* Europe, Asia, naturalized in U.S. *Constit.* Sinigrin (potassium myronate), myrosin, sinapine sulfocyanate, fixed oil; erucic, behenic, and sinapolic acids.

USE: Source of volatile oil of mustard.

THERAP CAT: Emetic; counterirritant.

THERAP CAT (VET): Counterirritant. Has been used as emetic, carminative.

6225. Mustard Gas. *1,1'-Thiobis[2-chloroethane]; bis-(2-chloroethyl)sulfide; β,β'-dichloroethyl sulfide; 2,2'-dichlorodiethyl sulfide; bis(β-chloroethyl)sulfide; 1-chloro-2-(β-chloroethylthio)ethane;* sulfur mustard; yellow cross liquid; Kampfstoff "Lost"; Yperite. $C_4H_8Cl_2S$; mol wt 159.08. C 30.20%, H 5.07%, Cl 44.58%, S 20.16%. $(ClCH_2CH_2)_2S$. War gas prepd by treating ethylene with sulfur chloride (Levinstein process): Mann, Pope, *J. Chem. Soc.* **121**, 594 (1922); by treating β,β'-dihydroxyethyl sulfide with HCl gas (German process): Meyer, *Ber.* **19**, 3260 (1886); *Ann.* **240**,

310 (1887); Gomberg, *J. Am. Chem. Soc.* **41**, 1427 (1919). Reactions and derivatives: Helfrich, Reid, *ibid.* **42**, 1208 (1920). Toxicity: Anslow *et al., J. Pharmacol. Exp. Ther.* **93**, 1 (1948). Review of carcinogenicity studies: *IARC Monographs* **9**, 181-192 (1975).

Oily liquid. *Deadly vesicant.* Weak, sweet, agreeable odor. On cooling it forms prisms, mp 13-14°. d^{13} 1.338 (solid); d_4^{20} 1.2741 (liq). bp_{760} 215-217°; bp_{10} 98°. Volatile with steam. n_D^{20} 1.53125. Very sparingly sol in water; sol in fat solvents, other common organic solvents. High lipid soly. Vapor pressure at 0° = 0.025 mm; at 30° = 0.090 mm. Hydrolyzed by alkalies. Recommended neutralizing agent and inactivator: Bleaching powder, sodium hypochlorite. LD_{50} in rats, mice (mg/kg): 3.3, 8.6 i.v. (Anslow).

Human Toxicity: Conjunctivitis, blindness. Produces delayed effects. In 1-12 hrs cough, edema of eyelids, erythema of skin, severe pruritus. May cause edema, ulceration, necrosis of respiratory tract and exposed skin. Ingestion of contaminated material may cause nausea and vomiting. Permanent eye damage, severe respiratory impairment may result. This substance has been listed as a known carcinogen: *Fourth Annual Report on Carcinogens* (NTP 85-002, 1985) p 136.

USE: In chemical warfare.

6226. Mustard, White. Yellow mustard. Dried ripe seeds of *Brassica alba* (L.) Boiss. (*Sinapis alba* L.), Cruciferae. *Habit.* Europe, Asia, adventitious in U.S. *Constit. and Use:* As of mustard, black.

THERAP CAT: Emetic; counterirritant.
THERAP CAT (VET): *See* Mustard, Black.

6227. Muzolimine. *5-Amino-2-[1-(3,4-dichlorophenyl)-ethyl]-2,4-dihydro-3H-pyrazol-3-one;* 3-amino-1-(3,4-dichloro-α-methylbenzyl)-2-pyrazolin-5-one; Bay g 2821; Edrul. $C_{11}H_{11}Cl_2N_3O$; mol wt 272.13. C 48.55%, H 4.08%, Cl 26.05%, N 15.44%, O 5.88%. High-ceiling loop diuretic. Prepn: E. Möller *et al.,* Ger. pat. **2,319,278** corresp to U.S. pat. **4,018,890** (1974, 1977 both to Bayer); *eidem, Experientia* **33**, 382 (1977). *In vitro* effects: H. J. Kramer, *Pharmatherapeutica* **1**, 353 (1977). Mechanism of action: M. Mussche, N. Lameire, *Curr. Med. Res. Opin.* **4**, 462 (1977). Pharmacology: K. Meng *et al., ibid.* 555. Pharmacokinetics: O. Broers *et al., Eur. J. Clin. Pharmacol.* **15**, 105 (1979); *eidem, Curr. Med. Res. Opin.* **6**, 431 (1980). Use in advanced renal disease: A. D. Canton *et al., Brit. Med. J.* **282**, 595 (1981); P. Schmidt *et al., Eur. J. Clin. Pharmacol.* **20**, 23 (1981). Toxicological studies: D. Lorke, P. Mürmann, *Curr. Med. Res. Opin.* **4**, 716 (1977). Symposium on pharmacology, pharmacodynamics, toxicology, clinical studies: *Clin. Nephrol.* **19**, Suppl. 1, S1-S117 (1983).

Cryst from methanol, mp 127-129°. LD_{50} in mice, rats, dogs, rabbits (mg/kg): 1794, 1559, 2000, 1250 orally (Lorke, Mürmann).

THERAP CAT: Diuretic; antihypertensive.

6228. Mycaminose. *3,6-Dideoxy-3-(dimethylamino)-D-glucose.* $C_8H_{17}NO_4$; mol wt 191.22. C 50.25%, H 8.96%, N 7.33%, O 33.47%. Part of the carbomycin molecule. Synthesis and stereochemistry: Richardson, *Proc. Chem. Soc.* **1961**, 430; Foster *et al., Chem. & Ind.* (London) **1962**, 142. Stereochemistry: Grisebach, Hofheinz, *Angew. Chem.* **74**, 499 (1962).

Hydrochloride monohydrate, crystals from moist isopropanol, mp 116-118°. $[\alpha]_D$ +31° (equilib value, c = 0.96).

6229. Mycarose. *2,6-Dideoxy-3-C-methyl-L-ribo-hexose;* 2,6-didesoxy-3-C-methyl-L-ribo-hexose; 2-(4,5-cis)-trihydroxy-4,6-dimethyltetrahydropyran. $C_7H_{14}O_4$; mol wt 162.18. C 51.84%, H 8.70%, O 39.46%. Constituent of the macrolides carbomycin A and B, spiramycin A, B and C and tylosin: Woodward, *Angew. Chem.* **69**, 50 (1957); Paul, Tchelitcheff, *Bull. Soc. Chim. France* **1957**, 443; **1960**, 150; Hamill *et al., Antibiot. & Chemother.* **11**, 328 (1961). Part of carbomycin molecule: Regna *et al., J. Am. Chem. Soc.* **75**, 4625 (1953). Stereochemistry: Grisebach, Hofheinz, *Angew. Chem.* **74**, 499 (1962); Foster *et al., Proc. Chem. Soc.* **1962**, 254; Hofheinz *et al., Tetrahedron* **18**, 1265 (1962). Synthesis of DL-mycarose: Korte *et al., ibid.* 1257; Grisebach *et al., Ber.* **96**, 1823 (1963). Synthesis of L-mycarose: Lemal *et al., Tetrahedron* **18**, 1275 (1962); Koto *et al., Bull. Chem. Soc. Japan* **45**, 532 (1972). Synthesis of D-mycarose: Flaherty *et al., J. Chem. Soc.* (C) **1966**, 398. Isoln from aureolic acid: Berlin *et al., Khim. Prir. Soedin.* **8**, 535 (1972), *C.A.* **78**, 4432a (1973).

DL-Form, crystals from acetone + petr ether, mp 110-111°.

L-Form, needles from boiling acetone + chloroform, mp 128.5-130.5°. $[\alpha]_D^{25}$ −31.1° (c = 4). Sol in water. Exhibits only end absorption in the ultraviolet.

3-O-Methylmycarose, $C_8H_{16}O_4$, Cladinose. Constituent of erythromycin A and B. Liquid, $bp_{0.25}$ at bath temp 120-132°. $[\alpha]_D^{25}$ −23.1° (c = 2.6 in water). Sol in water, alcohol, acetone, ether, benzene, chloroform, carbon tetrachloride; slightly sol in petr ether. Dec by strong acids.

6230. Mycelianamide. *3-[[4-[(3,7-Dimethyl-2,6-octadienyl)oxy]phenyl]methylene]-1,4-dihydroxy-6-methyl-2,5-piperazinedione.* $C_{22}H_{28}N_2O_5$; mol wt 400.46. C 65.98%, H 7.05%, N 7.00%, O 19.98%. Antibiotic substance found in the mycelium of *Penicillium griseofulvum:* Anslow, Raistrick, *Biochem. J.* **25**, 39 (1931); Oxford *et al., ibid.* **29**, 1102 (1935); **33**, 240 (1939); Oxford, Raistrick, *ibid.* **42**, 323 (1948). Structure: Birch *et al., J. Chem. Soc.* **1956**, 3717; Bates, Schauble, *Tetrahedron Letters* **1963**, 1683. Configuration: Gallina *et al., Gazz. Chim. Ital.* **94**, 1301 (1964). Partial synthesis: Brown, Meehan, *Aust. J. Chem.* **21**, 1581 (1968). Total synthesis: N. Shinmon *et al., Chem. Commun.* **1980**, 1020.

Crystals from ethyl acetate, dec 170-172°. $[\alpha]_{546}^{19}$ −217°; $[\alpha]_{579}^{19}$ −182°. Weak acid. Freely sol in acetone, dioxane; sparingly sol in other organic solvents; sol in aq sodium carbonate soln, but not in sodium bicarbonate soln. Quickly destroyed by acids and alkalies.

6231. Mycetins. Antibiotic substances produced by

Actinomyces violaceus: Fainshmidt, Koreniako, *Biokhimiya* **9**, 147 (1944); Krassilnikov, Koreniako, *Microbiologiya* **14**, 80 (1945). Chromatographic and structure studies: Yakubov *et al., ibid.* **31**, 526 (1962); Blinov *et al., J. Chromatog.* **8**, 522 (1962); Yakubov *et al., Antibiotiki* **10**, 771 (1965).

6232. Myclobutanil. α-*Butyl*-α-*(4-chlorophenyl)-1H-1,2,4-triazole-1-propanenitrile;* 2-(4-chlorophenyl)-2-(1*H*-1,2,4-triazol-1-ylmethyl)hexanenitrile; RH-3866; Nova; Rally; Systhane. $C_{15}H_{17}ClN_4$; mol wt 288.78. C 62.39%, H 5.93%, Cl 12.28%, N 19.40%. Broad spectrum systemic triazole fungicide; ergosterol-biosynthesis inhibitor. Prepn: T. T. Fujimoto, **Eur.** pat. **Appl.** 145,294 (1985 to Rohm & Haas), *C.A.* **103**, 160519z (1985). Physical properties and field trials: C. Orpin *et al., Proc. Brit. Crop Prot. Conf.—Pests Dis.* **1986**, 55; A. Perrot, *Défense des Végétaux* **40**, 9 (1986). Antifungal activity: J. A. Quinn *et al., Pestic. Sci.* **17**, 357 (1986).

Light yellow crystals, mp 63-68°. $bp_{1.0}$ 202-208°. Vapor pressure at 25°: 1.6×10^{-6} Torr. Soly at 25°: water 142 ppm. Sol in common organic solvents such as ketones, esters, alcohols, aromatic hydrocarbons. Insol in aliphatic hydrocarbons. LD_{50} in male, female rats (mg/kg): 1600, 2229 orally; LD_{50} in rabbits (mg/kg): 7500 dermally (Orpin).

USE: Systemic fungicide.

6233. Mycobacidin. *4-Oxo-2-thiazolidinehexanoic acid;* ε-[2-(4-thiazolidone)]hexanoic acid; 4-thiazolidone-2-caproic acid; 2-(5-carboxypentyl)-4-thiazolidone; cinnamonin; actithiazic acid; acidomycin. $C_9H_{15}NO_3S$; mol wt 217.28. C 49.75%, H 6.96%, N 6.45%, O 22.09%, S 14.75%. Antibiotic substance produced by *Streptomyces lavendulae, Streptomyces virginiae* and several other *Streptomyces* spp. Isoln: Tejera *et al., Antibiot. & Chemother.* **2**, 333 (1952). Production by fermentation: Grundy *et al.,* **U.S.** pat. **2,678,929** (1954 to Abbott). Synthesis: McLamore *et al., J. Am. Chem. Soc.* **75**, 105 (1953); Nienburg, Friedrichsen, **Ger.** pat. **931,651** (1955 to BASF). *Review:* Caltrider, *Antibiotics,* **vol. I**, D. Gottlieb, P. D. Shaw, Eds. (Springer-Verlag, New York, 1967) pp 666-668.

l-Form, needles from water, ethanol or ethyl acetate, mp 139-140°. $[\alpha]_D^{25}$ −54° (c = 1 in methanol); $[\alpha]_D^{25}$ −60° (c = 1 in ethanol). pK 5.1. Soluble in water, acetone, alcohol, ethylene dichloride, glacial acetic acid. Aq solns are stable over a wide pH range at room temp. Rapidly loses optical activity in dil alkali to give the racemate. Has *in vitro* activity against *Mycobacterium tuberculosis, M. ranae* and *M. phlei.* LD_{50} in mice: 3.5 g/kg i.v., 2 g/kg s.c., Miyake *et al., Pharm. Bull.* **1**, 84, 89 (1953).
Methyl ester, $C_{10}H_{17}NO_3S$, needles from ether + hexane, mp 53-54°. $[\alpha]_D^{25}$ −50.9° (methanol).
d-Form, crystals, mp 138-139°. $[\alpha]_D^{25}$ +57° (methanol).
dl-Form, crystals from water, mp 123°.

6234. Mycobacillin. $C_{65}H_{85}N_{13}O_{30}$; mol wt 1528.50. C 51.08%, H 5.60%, N 11.91%, O 31.40%. Antifungal polypeptide antibiotic isolated from culture filtrates of *Bacillus subtilis:* Majumdar, Bose, *Nature* **181**, 134 (1958). Structural studies: *eidem, Biochem. J.* **74**, 596 (1960); *Arch. Biochem. Biophys.* **90**, 154 (1960). Amino acid configuration: Banerjee, Bose, *Nature* **200**, 471 (1963). Nature of the peptide linkages: Sengupta *et al., Biochem. J.* **121**, 839 (1971). *Review:* Banerjee *et al., Antibiotics* **vol. II**, D. Gottlieb, P. D. Shaw, Eds. (Springer-Verlag, New York, 1967) pp 271-275, 445-446.

Needles, mp 235-240°. uv max 277 nm in a solution of *n*-butanol-water-acetic acid-95% ethanol (10:10:2:2).

6235. Mycobactins. Growth factors for *Mycobacterium paratuberculosis (M. johnei)*, the organism responsible for Johne's disease in cattle. Mycobactins are structurally complex *siderophores* (microbial iron chelators). At least nine mycobactins, designated with the letters A, F, H, M, N, P, R, S, and T, have been isolated from various non-pathogenic species of *Mycobacterium*. Earlier studies were actually on *mycobactin P; see* isoln from *M. phlei:* Francis *et al., Nature* **163**, 365 (1949); *eidem, Biochem. J.* **55**, 596 (1953); structure: Snow, *ibid.* **94**, 160 (1965). Separation and identification of mycobactins: White, Snow, *ibid.* **108**, 593 (1968); **111**, 785 (1969). Chemical and biological properties: *eidem, ibid.* **115**, 1031 (1969). On saponification, mycobactins splits into *mycobactic acid* and *cobactin.* Synthetic studies: Black *et al., Aust. J. Chem.* **25**, 2155 (1972). Synthesis of mycobactin S2: P. J. Maurer, M. J. Miller, *J. Am. Chem. Soc.* **105**, 240 (1983). Review of early studies: Rose, Snow, "Mycobactin: A Growth Factor for Acid-Fast Bacilli" in G. E. W. Wolstenholme, M. P. Cameron, C. M. O'Connor, *Ciba Foundation Symposium on Exptl. Tuberculosis Bacillus and Host* **1955**, 41-54. Comprehensive review: Snow, *Bacteriol. Rev.* **34**, 99 (1970).

R^2, R^3, R^5 = H or CH_3
R^1, R^4 = CH_3, CH_2CH_3 or long chain

Able to chelate with metals, showing strong selectivity toward iron. Isolated as octahedral ferric complexes in the form of red-brown glasses; purified to metal-free mycobactins as white microcrystalline powders. Characterizations, esp mass spectral studies, carried out on the aluminum complexes. Mycobactin solids have definite melting points, are stable to air and heat up to about 100°, and show apple-green fluorescence in uv light. Characteristic uv max (methanol): 250, 311 nm for mycobactins having a methyl substituent on the benzene ring; 243, 249, 304 nm for mycobactins without the methyl substituent. Very sol in chloroform; fairly sol in ethanol; less sol in other alcoholic solvents. Slightly sol in benzene, ether, aliphatic hydrocarbons. Stable to acids; easily broken down under alk. conditions.

USE: In the taxonomy of mycobacteria; in development of iron chelators for clinical use.

6236. Mycolic Acids. High mol wt α-branched, β-hydroxy fatty acids, components of the cell envelopes of all *Mycobacteria*. All known mycolic acids have the basic structure $R^2CH(OH)CHR^1COOH$ where R^1 is a C_{20} to C_{24} linear alkane and R^2 is a more complex structure of 30 to 60 carbon atoms that may contain various numbers of carbon-carbon double bonds and/or cyclopropane rings, methyl branches or oxygen functions such as $C=O$, $CH_3OCH=$, COOH. The structure of mycolic acids varies by families and species. Three principal categories are known: (1) *cory-nomycolic acids* ranging from C_{28} to C_{40} found mostly in *Corynebacteria*; (2) *nocardic acids*, also called *nocardomycol-ic acids*, ranging from C_{40} to C_{60}, produced by strains of *Nocardia*; and (3) mycobacterial mycolic acids ranging from C_{60} to C_{90}. First obtained from a human strain of *Mycobac-terium tuberculosis* and studied in an impure form called "unsaponifiable wax": R. J. Anderson, *J. Biol. Chem.* **85**, 351 (1929). Isoln of the first representative of this group of acids from the human tubercle bacillus and derivation of the name "mycolic acids": F. H. Stodola *et al.*, *ibid.* **126**, 505 (1938). Historical reviews: R. J. Anderson, *Fortschr. Chem. Org. Naturst.* **3**, 145-202 (1939); J. Asselineau, E. Lederer, *ibid.* **10**, 170-273 (1953); J. Asselineau, *The Bacterial Lipids* (Hermann-Holden Day, Paris-New York, 1962) pp 82-148; E. Lederer, *Pure and Appl. Chem.* **25**, 135 (1971). Major mycolic acids of *Mycobacterium smegmatis*: M. Y. H. Wong *et al.*, *J. Biol. Chem.* **254**, 5734 (1979); M. Y. H. Wong, G. R. Gray, *ibid.* 5741. Total synthesis of mycolic acids from *M. smegmatis*: H. C. Huang *et al.*, *J. Org. Chem.* **47**, 4018 (1982). Evaluation as immunotherapeutic agents in the treatment of cancer in animal models: M. V. Pimm *et al.*, *Int. J. Cancer* **24**, 780 (1979). Clinical evaluation in humans: G. J. Vosika, *Cancer* **44**, 495 (1979). Mycolic acids are acid fast. They are always isolated in the form of more or less complex mixtures. They occur in nature esterified with carbohydrates: with arabinose in the cell wall, with trehalose in cord factor, *q.v.*

6237. Mycomycin. *3,5,7,8-Tridecatetraene-10,12-diyn-oic acid.* $C_{13}H_{10}O_2$; mol wt 198.22. C 78.77%, H 5.09%, O 16.14%. $HC{\equiv}C-C{\equiv}CCH=C=CHCH=CHCH=CH-CH_2COOH$. Antibiotic substance produced by *Nocardia acidophilus*: Johnson, Burdon, *J. Bacteriol.* **54**, 281 (1947); Johnson, *Soc. Am. Bacteriologists*, 49th General Meeting (May, 1949), *Abstracts of Papers*, p 68; Celmer, Solomons, 121st *A.C.S.* Meeting (Milwaukee, 1952), *Abstracts of Papers*, p 17-J; *J. Am. Chem. Soc.* **74**, 1870, 2245 (1952); **75**, 1372 (1953). Synthesis: Bohlmann, Sucrow, *Angew. Chem.* **76**, 611 (1964).

Needles from methylene chloride at $-40°$. Very thermo-labile. Complete retention of activity is possible only by storage at $-40°$ or lower. mp $+75°$ (deflagrates). $[\alpha]_D^{25} -130°$ (c = 0.4 in ethanol). uv max (ether): 267, 281 nm (ϵ 61,000, 67,000). Sol in alcohol, ether, amyl acetate, methylene chloride. Forms a water-sol sodium salt. The acid itself is sparingly sol in water and is relatively stable in ex-tremely dil aq solns (approx 0.005 mg/ml), half-life inverse-ly related to concn. Undergoes an unusual rearrangement in normal aq KOH at 27°, involving an allene to acetylene isomerization accompanied by migration of the existing acetylenic bonds. The rearranged acid, *isomycomycin*, has been assigned the structure *3,5-tridecadiene-7,9,11-triynoic acid*, $CH_3-C{\equiv}C-C{\equiv}C-C{\equiv}C-CH=CHCH=CH-CH_2COOH$.

6238. Mycophenolic Acid. *6-(1,3-Dihydro-4-hydroxy-6-methoxy-7-methyl-3-oxo-5-isobenzofuranyl)-4-methyl-4-hexenoic acid; 6-(4-hydroxy-6-methoxy-7-methyl-3-oxo-5-phthalanyl)-4-methyl-4-hexenoic acid;* Melbex. $C_{17}H_{20}O_6$; mol wt 320.35. C 63.74%, H 6.29%, O 29.97%. Antibiotic substance produced by *Penicillium brevi-compactum; P. sto-loniferum* and related spp. Isoln: Oxford, Raistrick, *Bio-chem. J.* **26**, 1902 (1932); **27**, 1176, 1473 (1933); Clutterbuck *et al.*, *ibid.* **26**, 1441 (1932); Florey *et al.*, *Lancet* **I**, 46 (1946); Williams *et al.*, *Antimicrob. Ag. Chemother.* **1968**, 229. Structure: Birkinshaw *et al.*, *Biochem. J.* **50**, 630 (1952); Logan, Newbold, *J. Chem. Soc.* **1957**, 1946. Total syntheses: Birch, Wright, *Aust. J. Chem.* **22**, 2635 (1969); *eidem, Chem.*

Commun. **1969**, 788; Canonica *et al.*, *Tetrahedron Letters* **1971**, 2689. Biosyntheses: Canonica *et al.*, *Chem. Commun.* **1970**, 1357; **1971**, 257; Bedford *et al.*, *ibid.* **1971**, 323; Cano-nica *et al.*, *J. Chem. Soc. Perkins Trans. I* **1972**, 2639. Indus-trial patents: Borrow *et al.*; Carter; Jefferys *et al.*, **Brit.** pats. **1,157,099; 1,157,100; 1,158,387** (all 1969 to ICI). Activity studies: Williams *et al.*, *J. Antibiot.* **21**, 463 (1968); Ando *et al.*, *ibid.* 649; Suzuki *et al.*, *ibid.* **22**, 297 (1969); Carter *et al.*, *Nature* **223**, 848 (1969). Metabolic studies: Noto *et al.*, *J. Antibiot.* **22**, 165 (1969); Nery, Nice, *J. Pharm. Pharmacol.* **23**, 842 (1971). *Review:* Wilson, "Miscellaneous *Penicillium* Toxins," in *Microbial Toxins* vol. VI, A. Ciegler *et al.*, Eds. (Academic Press, New York, 1971) pp 460-470.

Needles from hot water, mp 141°. Weak dibasic acid. Stable. Almost insol in cold water; freely sol in alcohol; moderately sol in ether, chloroform; sparingly sol in benz-ene, toluene. LD_{50} in mice, rats: 2500, 700 mg/kg orally; 550, 450 mg/kg i.v., Wilson, *loc. cit.*

THERAP CAT: Antineoplastic.

6239. Mycosamine. *3-Amino-3,6-dideoxymannose;* 3-amino-3-deoxyrhamnose; 3,6-dideoxy-3-amino-D-manno-pyranose; 3-amino-3,6-dideoxy-D-aldohexose. $C_6H_{13}NO_4$; mol wt 163.17. C 44.16%, H 8.03%, N 8.58%, O 39.22%. An amino sugar which represents the nitrogen-contg moiety of the antifungal antibiotics nystatin, *q.v.*, and amphotericin B, *q.v.* Isoln and structure: Dutcher *et al.*, *Antibiot. Ann.* **1956-1957**, 866. Walters *et al.*, *J. Am. Chem. Soc.* **79**, 5076 (1957); Dutcher *et al.*, *J. Org. Chem.* **28**, 995 (1963). Stereo-chemistry: von Saltza *et al.*, *J. Am. Chem. Soc.* **83**, 2875 (1961); *eidem, J. Org. Chem.* **28**, 999 (1963).

N-Acetylmycosamine, $C_8H_{15}NO_5$, needles from methanol + acetone, mp 195-197°. $[\alpha]_D^{22} -46°$ (ethanol).
Triacetylmycosamine, $C_{12}H_{19}NO_7$, prisms from acetone or ethanol, mp 185-187°. $[\alpha]_D^{22} +85°$ (ethanol).
Tetraacetylmycosamine, $C_{14}H_{21}NO_8$, needles from ben-zene, mp 159-161°. $[\alpha]_D^{23} +39.3°$ (ethanol).
Hydrochloride, $C_6H_{13}NO_4\cdot HCl$, prismatic rods from etha-nol+ ether, mp 162°. $[\alpha]_D^{24} -11.5°$.

6240. Mylabris. Chinese cantharides; Chinese blistering flies. Dried insect, *Mylabris sidae (Phalerata)* Pallas, *Coleop-tera. Habit.* China and East India. *Constit.* Cantharidin, said to be present in larger proportions than in cantharides (1-1.2%).

Black, cylindrical body, rounded above, flattish below; smaller than cantharides; black wing cases marked with a spot at point of insertion and has three tawny bands. *Cau-tion:* Highly toxic! *See* Cantharides.
USE: Source for cantharidin.

6241. Myoral. *Calcium [mercaptoacetato(2−)-O,S]au-rate(1−) (1:2); [(carboxymethyl)thio]gold calcium salt;* calci-um aurothioglycolate; mercaptoacetic acid calcium salt gold derivative. $C_4H_4Au_2CaO_4S_2$; mol wt 614.28. C 7.82%, H 0.66%, Au 64.14%, Ca 6.52%, O 10.42%, S 10.44%. (AuS-$CH_2COO)_2Ca$. Prepn: Denko, Anderson, *J. Am. Chem. Soc.* **67**, 2241 (1945). Description: Lespagnol, *Pharmacie Chi-mique* (Paris, 3rd ed., 1950) p 941.

Reddish-brown powder. Stable to air. Practically insol in water. Employed as a suspension in oil.
THERAP CAT: Antirheumatic.

6242. Myosin. Mol wt approx 500,000. The main protein in muscle, which, by hydrolyzing ATP, provides the energy necessary for muscular contraction. Individual molecules are rod-shaped and consist of two globular heads attached flexibly to a tail of approx 1500Å length. All muscle myosins have a similar subunit structure composed of two heavy chains and two distinct pairs of light chains. One of the light chains is involved in a Ca^{2+}-dependent regulatory process in the vertebrate skeletal muscle system. Biological prepn and review: A. G. Szent-Györgyi, *Chemistry of Muscular Contraction* (Academic Press, New York, 2nd ed., 1951) pp 38-57, 146-148. Other reviews: *idem, Advan. Enzymol. Relat. Subj. Biochem.* **16**, 313-360 (1955); A. G. Szent-Györgyi *et al.*, in *Sulfur in Proteins,* R. Benesch *et al.,* Eds. (Academic Press, New York, 1959) pp 291-295; J. Gergely, *Biochemistry of Muscle Contraction* (Little, Brown, Boston, 1964) pp 3-115; several authors in *Contractile Proteins and Muscle,* K. Laki, Ed. (Dekker, New York, 1971); several authors in *Methods Enzymol.* vol. 85, Part B, D. W. Frederiksen, L. W. Cunningham, Eds. (Academic Press, New York, 1982) p 55-130. Review on structure and function of myosin: W. F. Harrington, "Contractile Proteins of Muscle" in *The Proteins,* vol. 4, H. Neurath, R. L. Hill, Eds. (Academic Press, New York, 1979) p 245-409. Polymorphism of mammalian myosins and distribution of isoforms in different muscle fibers reviewed by A. G. Weeds in *Plasticity of Muscle,* D. Pette, Ed. (DeGruyter, Berlin, 1980) pp 55-68. Role of myosin light chains in calcium regulation: Kendrick-Jones, *Nature* **249**, 631 (1974). *See also* Tropomyosin.

6243. β-Myrcene. *7-Methyl-3-methylene-1,6-octadiene;* 2-methyl-6-methylene-2,7-octadiene. $C_{10}H_{16}$; mol wt 136.23. C 88.16%, H 11.84%. Found in oil of bay, verbena, hop, and others. Isoln: Power, Kleber, *Pharm. Rundschau* **13**, 60 (1895); *see also* Ruzicka, Stoll, *Helv. Chim. Acta* **7**, 272 (1924); Goulding, Roberts, *J. Chem. Soc.* **105**, 2614 (1914); Booker *et al., ibid.* **1940**, 1453; Kugler, Kováts, *Helv. Chim. Acta* **46**, 1480 (1963). Obtained by pyrolysis of β-pinene: Goldblatt, Palkin, *J. Am. Chem. Soc.* **63**, 3517 (1941); Rummelsburg, U.S. pat. **2,444,790** (1948 to Hercules Powder); Houlihan *et al., J. Am. Chem. Soc.* **81**, 4692 (1959). Separation of isomers: Ohloff *et al., Ann.* **675**, 83 (1964). Synthesis: O. P. Vig *et al., Indian J. Chem.* **7**, 450 (1969); **11**, 104 (1973); **13**, 1244 (1975); T. Mandai *et al., Tetrahedron Letters* **22**, 763 (1981). Formation from isoprene: K. Takabe *et al., Synthesis* **1977**, 307.

Oil. Pleasant odor. d_4^{20} 0.794. n_D^{20} 1.4709. uv max (ethanol): 226 nm (ε 16,100). Practically insol in water; sol in alcohol, chloroform, ether, glacial acetic acid.

α-Myrcene. *2-Methyl-6-methylene-1,7-octadiene.* bp_{10} 44°. n_D^{25} 1.4661. d_{25}^{25} 0.7959. uv max (isooctane): 224.5 nm (ε 18,600). Not found in nature. Prepn: Mitzner *et al., J. Org. Chem.* **30**, 646 (1965); Vig *et al., J. Indian Chem. Soc.* **50**, 329 (1973).

USE: β-Myrcene as an intermediate in the manuf of perfume chemicals.

6244. Myricetin. *3,5,7-Trihydroxy-2-(3,4,5-trihydroxyphenyl)-4H-1-benzopyran-4-one; 3,3',4',5,5',7-hexahydroxyflavone;* cannabiscetin; delphidenolon 1575. $C_{15}H_{10}O_8$; mol wt 318.23. C 56.61%, H 3.17%, O 40.22%. From the bark of *Myrica nagi* Thunb., *Myriaceae:* Perkin, Hummel, *J. Chem. Soc.* **69**, 1287 (1896). Structure: Perkin, *ibid.* **81**, 203 (1902). Identity with cannabiscetin: Seshadri, Venkateswarlu, *Proc. Indian Acad. Sci.* **23A**, 296 (1946); *C.A.* **40**, 6447 (1946). Synthesis: Kalff, Robinson, *J. Chem. Soc.* **127**, 181 (1925); Rao *et al., J. Sci. Ind. Res.* **8B**, No. 6, 113 (1949). Occurrence in *Hamamelidaceae* and *Anacardiaceae:* Reznek,

Egger, *Z. Naturforsch.* **15b**, 247 (1960). Metabolism: Smith, Griffiths, *Biochem. J.* **118**, 53p (1970).

Yellow needles from dil alc, mp 357°. uv max (ethanol): 375, 255 nm. Sparingly sol in boiling water; sol in alcohol; practically insol in chloroform, acetic acid.

Hexaacetate, $C_{27}H_{22}O_{14}$, crystals, mp 213°.
Hexaethyl ether, $C_{27}H_{34}O_8$, needles from alcohol, mp 149-151°.
3-Rhamnoside, $C_{21}H_{20}O_{12}$, *myricitrin.* Structure: Hattori, Hayashi, *Acta Phytochim.* **5**, 213 (1931), *C.A.* **26**, 990[8] (1932). Pale yellow leaflets from water, mp 199-200°. uv max (alc): 262, 352 nm. Sparingly sol in water, abs alcohol.

6245. Myristica. Nutmeg; nux moschata; nuces (semen) nucistae. Dried ripe seed of *Myristica fragrans* Houtt., *Myristicaceae,* deprived of its seed coat and with or without a thin coating of lime. *Habit.* Southern Asia, Moluccas, cultivated in many tropical countries. *Constit.* 25-35% fixed oil and 5-15% volatile oils. The myristicin fraction of the latter, together with elemicin (> 25% content) is supposedly responsible for the purported hallucinogenic properties of nutmeg seed: Shulgin, *Nature* **197**, 4865 (1963); Weil, *Econ. Bot.* **19**, 194 (1965). Review of botany, history, chemistry and psychopharmacology: *Public Health Service Publ. No. 1645,* D. H. Efron, Ed., 183-229 (1967).

USE: As a spice in food; as source of myristica oil. *Caution:* Ingestion of large quantities causes drowsiness, stupor, death.

THERAP CAT: Carminative.

6246. Myristic Acid. *Tetradecanoic acid.* $C_{14}H_{28}O_2$; mol wt 228.36. C 73.63%, H 12.36%, O 14.01%. $CH_3(CH_2)_{12}$-COOH. Occurs in nutmeg butter (*Myristica fragrans* Houtt.) to the extent of 70-80%; predominates in the fats of the *Myristicaceae;* in palm seed fats it may comprise 20% of the total fatty acids; in milk fats between 8-12% of the total acids. Occurs in most animal and vegetable fats; has been found in considerable amounts (up to 15%) in sperm whale oil, *see* Markley, *Fatty Acids* (New York, 1947). Prepn: G. D. Beal, *Org. Syn.* coll. vol. I, 379 (2nd ed., 1941). Prepn from tall-oil fatty acids: Segessmann, Molnar, U.S. pat. **2,481,356** (1949); from 9-ketotetradecanoic acid: Ames *et al., J. Chem. Soc.* **1950**, 174; by electrolysis of methyl hydrogen adipate + decanoic acid: Greaves *et al., ibid.* **1950**, 3326; by Maurer oxidation of myristyl alc: Langenbeck, Richter, *Ber.* **89**, 202 (1956); from cetanol: Selwitz, U.S. pat. **2,969,380** (1961 to Gulf). Wide-line NMR spectrum: A. V. Bailey, R. A. Pittman, *J. Am. Oil Chem. Soc.* **48**, 775 (1971).

Crystals from methanol, mp 58.5°. bp_{100} 250.5°; bp_{16} 199°; bp_4 184°. d_4^{54} 0.8622. d_4^{70} 0.8528. d_4^{90} 0.8394. n_D^{60} 1.4305; n_D^{70} 1.4273. Soluble in abs alcohol, methanol, ether, petr ether, benzene, chloroform. Neutralization value: 245.68. Absorption spectrum: Markley, *op. cit.* LD_{50} i.v. in mice: 43±2.6 mg/kg: L. Orö, A. Wretlind, *Acta Pharmacol. Toxicol.* **18**, 141 (1961).

Ethyl ester, $C_{16}H_{32}O_2$, *ethyl myristate.* Liquid, mp 12°; bp 295°; bp_{30} 195°. d_4^{20} 0.856. Insol in water; sol in alcohol; slightly sol in ether.

USE: As ingredient in soaps and shaving creams; in lubricants; in coatings for anodized aluminum.

6247. Myristicin. *4-Methoxy-6-(2-propenyl)-1,3-benzodioxole; 5-allyl-1-methoxy-2,3-(methylenedioxy)benzene.* $C_{11}H_{12}O_3$; mol wt 192.22. C 68.73%, H 6.29%, O 24.97%. Aromatic ether extracted from nutmeg, parsley, carrots. Isoln from nutmeg, *Myristica fragrans* Houtt., *Myristicaceae* and characterization: F. B. Power, A. H. Salway, *J. Chem. Soc.* **91**, 2037 (1907); A. T. Shulgin, *Nature* **197**, 379 (1963); from carrots, *Daucus carota* L., *Umbelliferae:* S. G. Yates, R. E. England, *J. Agr. Food Chem.* **30**, 317 (1982); from parsley, *Petroselinium hortense* Hoffman, *Umbelliferae:* M. Balogh *et al., Herba Hung.* **17**, 39 (1978), *C.A.* **91**, 2543e (1979). Synthesis: V. M. Trikojus, D. E. White, *Nature* **144**, 1016 (1939); F. Dallacker, R. Sluysmans, *Monatsh.* **100**, 560 (1969). HPLC sepn: L. W. Wulf *et al., J. Chromatog.* **161**, 271 (1978); A. W. Archer, *ibid.* **438**, 117 (1988). GC-MS study: G. M. Sammy, W. W. Nawar, *Chem. Ind.* **1968**, 1279. MS-MS study of nutmeg extract: D. V. Davis, R. G. Cooks, *J. Agr. Food Chem.* **30**, 495 (1982). Possible psychotropic properties: A. T. Shulgin, *Nature* **210**, 380 (1966).

Colorless oil, bp_{40} 173°. n_D^{20} 1.54032. d_{20}^{20} 1.1437.

6248. Myristyl Alcohol. *1-Tetradecanol;* tetradecyl alcohol. $C_{14}H_{30}O$; mol wt 214.38. C 78.43%, H 14.11%, O 7.46%. $CH_3(CH_2)_{12}CH_2OH$. Prepn by sodium reduction of fatty acid esters: Hansley, *Ind. Eng. Chem.* **39**, 55 (1947); by $LiAlH_4$ reduction of fatty acids: Watanabe, *Bull. Chem. Soc. Japan* **34**, 398 (1961); from acetaldehyde + dimethylamine: Langenbeck *et al., J. Prakt. Chem.* **8**, 112 (1959).

White crystals, mp 38°. d 0.824. bp_{15} 167°. Practically insol in water; sol in ether, slightly in alcohol.

USE: As emollient for cold creams, etc., also for making the sulfated alcohol whose sodium salt is applicable as a "wetter" in textiles.

6249. Myristyltrimethylammonium Bromide. *N,N,N-Trimethyl-1-tetradecanaminium bromide; trimethyltetradecylammonium bromide;* tetradonium bromide; Morpan T; Mytab. $C_{17}H_{38}BrN$; mol wt 336.41. C 60.70%, H 11.39%, Br 23.75%, N 4.16%. $[CH_3(CH_2)_{13}N(CH_3)_3]^+Br^-$. Cationic germicidal detergent. Prepn and antibacterial activity: R. S. Shelton *et al., J. Am. Chem. Soc.* **68**, 753 (1946). Toxicity and pharmacology: B. Isomaa, K. Bjondahl, *Acta Pharmacol. Toxicol.* **47**, 17 (1980).

White powder, mp 245-250°. Sol in 5 parts water. *Caution: Corrosive!* LD_{50} i.v. in mice, rats: 12.0, 15.0 mg/kg, B. Issomaa, K. Bjondahl, *loc. cit.*

USE: Disinfectant; deodorant; laboratory reagent.

6250. Myrophine. *7,8-Didehydro-4,5-epoxy-17-methyl-3-(phenylmethoxy)morphinan-6-ol tetradecanoate(ester);* 3-benzoylmorphine 6-myristate; benzylmorphine myristic acid ester; benzylmorphinyl myristate; myristyl benzylmorphine; 3-benzyloxy-6-myristoyloxy-N-methyl-4,5-epoxy-7-morphinene; 3-benzyloxy-6-myristoyloxymorphinene; myrocodine; myrophinium; Leucodinine; Myricodine; Peronine myristate. $C_{38}H_{51}NO_4$; mol wt 585.80. C 77.91%, H 8.78%, N 2.39%, O 10.93%. Prepn: Jeanson, U.S. pat. **2,802,828** (1957 to Laborec).

Solid, mp 41°.

Hydrochloride, $C_{38}H_{51}NO_4 \cdot HCl$, solid, mp 198-199°.

Caution: May be habit forming. This is a controlled substance (opium derivative) listed in the U.S. Code of Federal Regulations, Title 21 Part 1308.11 (1985).

THERAP CAT: Narcotic analgesic.

6251. Myrrh. Gum-resin myrrh. From *Commiphora abyssinica* (Berg) Eng. or from other species of *Commiphora, Burseraceae.* Yields not less than 30% alcohol-soluble extract. *Habit.* Nubia, Somaliland, Arabia. *Constit.* 20-25% resin, 57-61% gum, 7-17% volatile oil and a bitter principle.

THERAP CAT: Carminative; astringent.

6252. Myrtecaine. *2-[2-(6,6-Dimethylbicyclo[3.1.1]hept-2-en-2-yl)ethoxy]-N,N-diethylethanamine; 2-[2-(6,6-dimethyl-2-norpinen-2-yl)ethoxy]triethylamine;* 2-homomyrtenyloxy-1-(diethylamino)ethane; homomyrtenyl β-(diethylamino)ethyl ether; Nopoxamine. $C_{17}H_{31}NO$; mol wt 265.43. C 76.92%, H 11.77%, N 5.28%, O 6.03%. Prepn: Goudin, **Brit. pat. 861,900** (1961).

Liquid, $bp_{2.3}$ 135-140°. n_D^{20} 1.477.

THERAP CAT: Local anesthetic.

6253. Myrtol. Gelomyrtol. The fraction of the volatile oil from *Myrtus communis* L., *Myrtaceae* distilling between 166-180° and consisting chiefly of eucalyptol and *dextro-*pinene with a small quantity of an undefined camphor.

Colorless or slightly yellow, strongly dextrorotatory liquid (due to *dextro-*pinene); pleasant odor resembling that of turpentine and eucalyptol. d about 0.895. $[\alpha]_D^{20}$ +10°. n_D^{20} about 1.465. Freely sol in alcohol, ether.

6254. Mytatrienediol. *3-Methoxy-16-methyl-1,3,5(10)-estratriene-16β,17β-diol;* SC 6924; Anvene; Manvene. $C_{20}H_{28}O_3$; mol wt 316.42. C 75.91%, H 8.92%, O 15.17%. Prepn: Tyner, U.S. pat. **2,949,476** (1960 to Searle).

Crystals from benzene + acetone, mp 179-181°. $[\alpha]_D^{20}$ +71° in dioxane.

THERAP CAT: Estrogen; antilipemic; hypocalciuric.

6255. Myxin. *6-Methoxy-1-phenazinol 5,10-dioxide;* 1-hydroxy-6-methoxyphenazine 5,10-dioxide; 3C antibiotic. $C_{13}H_{10}N_2O_4$; mol wt 258.23. C 60.46%, H 3.90%, N 10.85%, O 24.78%. Potent broad-spectrum antibiotic from a *Sorangium* species (strain 3C), a soil-borne myxobacter: Peterson *et al., Can. J. Microbiol.* **12**, 221 (1966). Proposed structure: Edwards, Gillespie, *Tetrahedron Letters* **1966**, 4867. Revised structure and synthesis: Weigele, Leimgruber, *ibid.* **1967**, 715; Sigg, Toth, *Helv. Chim. Acta* **50**, 716 (1967). Activity: Grunberg *et al., Chemotherapia* **12**, 272 (1967). Crystal structure: Hanson, *Acta Cryst., Sect. B* **24**, 1084 (1968). *Review:* Edwards, Gillespie, *Proc. Int. Sym. Drug Res.* **1967**, 236; Leimgruber, Grunberg, *ibid.* 240; Baker, Vezina, *ibid.* 242.

Red needles from acetone, mp 120-130°; also given as mp 149° (dec) (Sigg, Toth). Strongly exothermic at 149°. May explode on drying with heat: Rachlin, *Chem. Eng. News* **45**, 32 (Sept. 4, 1967). Absorption max (0.1*N* HCl): 283, 340, 505 mn (ε 97,000; 5400; 6500). Stable in water after 6 hours at 37° in 0.05*N* acetate buffer at pH 5.0, in 0.05*N* phosphate at pH 6.0 and 7.0, or in 0.005*N* tris-HCl at pH 8.0 and 9.0. Stable at 70° except in phosphate buffer. Acetate, methanol, or aq tris buffer solns remain stable indefinitely at 4°. LD_{50} i.p. in mice: ~40 mg/kg. Capable of inhibiting a wide range of gram-positive and gram-negative bacteria, various fungi and yeasts.

Copper(II) complex, $C_{26}H_{18}CuN_4O_8$, *RO 7-4488/1, cuprimyxin, UNITOP*. Prepn: Leimgruber *et al.*, **Ger.** pat. **1,931,-466** corresp to U.S. pat. **3,586,674** (1970, 1971 to Hoffmann-La Roche). Activity studies: Maestrone *et al.*, *Am. J.* *Vet. Res.* **33**, 185 (1972). Dark green, fine crystals. Absorption max (DMSO): 287, 300, 356, 408, 610 nm (ϵ 68,500, 63,200, 10,000, 10,400, 9500).

THERAP CAT (VET): Antibacterial, antifungal.

N

6256. Nabam. *1,2-Ethanediylbiscarbamodithioic acid disodium salt; ethylenebis[dithiocarbamic acid] disodium salt;* disodium ethylenebis[dithiocarbamate]; Dithane D-14; Parzate Liquid. $C_4H_6N_2Na_2S_4$; mol wt 256.35. C 18.74%, H 2.36%, N 10.93%, Na 17.94%, S 50.03%. Prepd by treating ethylenediamine with CS_2 in the presence of NaOH: Hester, U.S. pat. 2,317,765 (1943 to Rohm & Haas); Flenner, U.S. pat. 2,504,404 (1950 to du Pont).

$$\underset{NaSCNH-CH_2CH_2-NHCSNa}{\overset{\displaystyle S \qquad\qquad S}{\overset{\displaystyle \| \qquad\qquad \|}{}}}$$

Hexahydrate, crystals from alc. Moderately sol in water. Forms a continuous film on plant surfaces, which is said to become insol in water. LD_{50} orally in rats: 395 mg/kg. USE: Agricultural fungicide. *Caution:* Irritating to skin, mucous membranes, and, in high concns, narcotic. In presence of alc can cause violent vomiting.

6257. Nabilone. *3-(1,1-Dimethylheptyl)-6,6a,7,8,10,10a-hexahydro-1-hydroxy-6,6-dimethyl-9H-dibenzo[b,d]pyran-9-one;* Compound 109514; LY-109514; Cesamet. $C_{24}H_{36}O_3$; mol wt 372.55. C 77.37%, H 9.74%, O 12.88%. Synthetic cannabinoid with antiemetic, antiglaucoma, and CNS activity. Prepn: R. A. Archer, **Ger. pat. 2,451,934** corresp to U.S. pat. 3,968,125 (1975, 1976 to Lilly); R. A. Archer *et al., J. Org. Chem.* 42, 2277 (1977). Pharmacology: P. Stark, R. A. Archer, *Pharmacologist* 17, 210 (1975); R. M. Orzelek-O'Neil *et al., Toxicol. Appl. Pharmacol.* 54, 493 (1980). Physiological disposition: A. Rubin *et al., Clin. Pharmacol. Ther.* 22, 85 (1977). Pharmacokinetics: H. R. Sullivan *et al., Biomed. Mass Spectrom.* 5, 296 (1978). Behavioral effects: P. Stark, P. B. Dews, *J. Pharmacol. Exp. Ther.* 214, 124 (1980). Clinical studies as antiemetic in cancer patients: N. Steele *et al., Cancer Treat. Rep.* 64, 219 (1980); C. J. Williams *et al., Cancer Clin. Trials* 3, 363 (1980). Clinical evaluation of anxiolytic effects: L. F. Fabre *et al., J. Clin. Pharmacol.* 21, Suppl 8-9, 377S (1981); R. M. Glass *et al., ibid.* 383S. Review of pharmacology and efficacy as anti-emetic: A. Ward, B. Holmes, *Drugs* 30, 127-144 (1985). Toxicity study in neonatal rats: C. L. Moss *et al., Toxicol. Appl. Pharmacol.* 48, A120 (1979). Comprehensive description: R. W. Souter in *Analytical Profiles of Drug Subtances* vol. 10, K. Florey, Ed. (Academic Press, New York, 1981) pp 499-512.

White crystals from ethyl acetate/hexane, mp 159-160°. uv max (ethanol): 207, 280 nm (ϵ 47000, 250). pKa in 66% DMF: 13.5.
Note: This is a controlled substance (hallucinogen) listed in the U.S. Code of Federal Regulations, Title 21 Part 1308.11 (1987).
THERAP CAT: Antiemetic.

6258. Nabumetone. *4-(6-Methoxy-2-naphthalenyl)-2-butanone;* 4-(6-methoxy-2-naphthyl)-butan-2-one; BRL 14777; Arthaxan; Consolan; Diosmal; Nabuser; Relafen; Relifen; Relifex. $C_{15}H_{16}O_2$; mol wt 228.29. C 78.92%, H 7.06%, O 14.02%. Non-steroidal anti-inflammatory agent. Prepn: A. W. Lake, C. J. Rose, **Ger. pat. 2,442,305;** *eidem,* U.S. pat. 4,061,779 (1975, 1977 both to Beecham); A. C. Goudie *et al., J. Med. Chem.* 21, 1260 (1978). Pharmacology: E. A. Boyle *et al., J. Pharm. Pharmacol.* 34, 562 (1982).

Effect on human platelet activity: B. Nunn, P. D. Chamberlain, *ibid.* 576; on arachidonic acid induced hypotension: T. C. Hamilton, S. D. Longman, *ibid.* 35, 458 (1983). In treatment of rheumatoid arthritis: G. Fostiropoulos, E. A. P. Croydon, *J. Int. Med. Res.* 10, 204 (1982); of osteoarthritis: F. Ginsberg, J. P. Famaey, *ibid.* 209. Metabolism: R. E. Haddock *et al., Xenobiotica* 14, 327 (1984). Pharmacokinetics and bioavailability: H. W. von Schrader *et al., Int. J. Clin. Pharmacol. Ther. Toxicol.* 22, 672 (1984). HPLC determn in plasma: J. E. Ray, R. O. Day, *J. Chromatog.* 336, 234 (1984). Brief review of pharmacology: F. R. Mangan, *Int. Congr. Symp. Ser. - Roy. Soc. Med.* 69, 5-14 (1985). Review of pharmacodynamics, pharmacokinetics and clinical efficacy in rheumatic diseases: H. A. Fridel, P. A. Todd, *Drugs* 35, 504-524 (1988).

Crystals from ethanol, mp 80°.
THERAP CAT: Anti-inflammatory.

6259. Nadide. *Adenosine 5'-(trihydrogen diphosphate) 5' → 5'-ester with 3-(aminocarbonyl)-1-β-D-ribofuranosylpyridinium, hydroxide, inner salt; 3-carbamoyl-1-β-D-ribofuranosylpyridinium hydroxide 5'-ester with adenosine 5'-pyrophosphate inner salt;* nicotinamide-adenine dinucleotide; diphosphopyridine nucleotide; DPN; adenine-D-ribose-phosphate-phosphate-D-ribose-nicotinamide; ARPPRN; NAD; cozymase; coenzyme I; Co I; codehydrogenase; Enzopride. $C_{21}H_{27}N_7O_{14}P_2$; mol wt 663.44. C 38.02%, H 4.10%, N 14.78%, O 33.76%, P 9.34%. The coenzyme of apozymase, necessary for the alcoholic fermentation of glucose. Isolation from fresh baker's yeast (10 kg of wet yeast yield about 250 mg): Ohlmeyer, *Biochem. Z.* 297, 66 (1938); v. Euler, Schlenk, *Z. Physiol. Chem.* 246, 64 (1937); Williamson, Green, *J. Biol. Chem.* 135, 345 (1940); Jandorf, *ibid.* 138, 305 (1941); LePage, *ibid.* 168, 623 (1947); *Biochem. Prepn.* 1, 28 (1949). Isoln from rabbit muscle: Ochoa, *Biochem. Z.* 292, 68 (1937); from erythrocytes: Warburg, Christian, *ibid.* 287, 291 (1936). Manuf by fermentation: **Brit. pat. 1,190,079** (1970 to Kyowa), *C.A.* 73, 54631g (1970). Synthetic approach: Todd *et al., J. Chem. Soc.* 1950, 303; 1957, 3727, 3733. Stereochemistry: Lumieux, *Can. J. Chem.* 41, 889 (1963); Oppenheimer *et al., Proc. Nat. Acad. Sci. USA* 68, 3200 (1971); Sarma, Mynott, *Chem. Commun.* 1972, 977. Nomenclature: Dixon, *Science* 132, 1548 (1960). NMR studies: Blumenstein, Raftery, *Biochemistry* 11, 1643 (1972).

Very hygroscopic white powder. Freely sol in water without residue. uv max: 340 nm (after reduction with Na hydrosulfite in alkaline soln). Monobasic acid. A 1% soln has a pH of about 2. Aq solns are stable for about a week. When neutralized, they are stable for about 2 weeks at 0°. Boiling results in decompn especially at acid pH.
Also precipitated as silver salt, barium salt, lead salt, copper salt, or phosphotungstate. Purification by Al_2O_3 adsorption may result in the compound $C_{21}H_{27}N_7O_{14}P_2 \cdot Al(OH)_3 \cdot H_2O$.
THERAP CAT: Narcotic antagonist; alcohol deterrent.

6260. Nadolol. *5-[3-[(1,1-Dimethylethyl)amino]-2-hydroxypropoxy]-1,2,3,4-tetrahydro-2,3-naphthalenediol;* 1-(*tert*-butylamino)-3-[(5,6,7,8-tetrahydro-*cis*-6,7-dihydroxy-1-naphthyl)oxy]-2-propanol; (2R,3S)-5-[3-(*tert*-butylami-

Consult the cross index before using this section.

no)-2-hydroxypropoxy]-1,2,3,4-tetrahydronaphthalene-2,3-diol; 2,3-cis-1,2,3,4-tetrahydro-5-[2-hydroxy-3-(tert-butylamino)propoxy]-2,3-naphthalenediol; SQ 11725; Anabet; Corgard; Solgol. $C_{17}H_{27}NO_4$; mol wt 309.42. C 65.99%, H 8.80%, N 4.53%, O 20.68%. β-Adrenergic blocker. Prepn: F. P. Hauck et al., **Ger. pat. 2,258,995** corresp to U.S. pat. **3,935,267** (1973, 1976, both to Squibb). Resolution of isomers: F. P. Hauck, J. E. Sundeen, **Ger. pat. 2,421,549** (1974 to Squibb), *C.A.* **82**, 57481e (1975). Metabolism: J. Dreyfuss et al., *J. Clin. Pharmacol.* **17**, 300 (1977). Toxicology: P. L. Sibley et al., *Toxicol. Appl. Pharmacol.* **44**, 379 (1978). Review of pharmacology: R. C. Heel et al., *Drugs* **1**, 1-23 (1980); M. J. Antonaccio, D. B. Evans, in *Pharmacology of Antihypertensive Drugs*, A. Scriabine, Ed. (Raven Press, New York, 1980) pp 295-301. Comprehensive description: L. Slusarek, K. Florey in *Analytical Profiles of Drug Substances* vol. 9, K. Florey, Ed. (Academic Press, New York, 1980) pp 455-485. Book: *International Experience with Nadolol, a Long-Acting β-Blocking Agent*, F. Gross, Ed. (Grune & Stratton, New York, 1981) 229 pp.

Crystalline powder, mp 124-136°. uv max (methanol): 270, 278 nm ($E^{1\%}_{1cm}$ 37.5, 39.1). pKa 9.67. Freely sol in alc, propylene glycol. Slightly sol in chloroform. Insol in acetone, benzene, ether, hexane. LD_{50} in mice, rats (mg/kg): 4500, 5300 orally (Antonaccio, Evans).

THERAP CAT; Antihypertensive, antianginal.

6261. Nadoxolol. *N,3-Dihydroxy-4-(1-naphthalenyloxy)butanimidamide*; 3-hydroxy-4-(1-naphthyloxy)butyramidoxime; 4-(α-naphthyloxy)-3-hydroxybutyramidoxime. $C_{14}H_{16}N_2O_3$; mol wt 260.29. C 64.60%, H 6.20%, N 10.76%, O 18.44%. β-Adrenergic blocker. Prepn: Lafon, **Ger. pat. 2,132,113** corresp to U.S. pat. **3,819,702** (1972, 1974 to Orsymonde). Pharmacology: Duteil et al., *Therapie* **28**, 703 (1973). Clinical studies: Tricot et al., *ibid.* 721.

Hydrochloride, $C_{14}H_{17}ClN_2O_3$, *LL 1530, Bradyl*. White crystalline powder from ethyl ether-methanol, mp 188°. Sol in water, methanol, ethanol. Insol in ethyl ether. LD_{50} in mice (mg/kg): 180 i.v.; 1000 orally (Lafon).

THERAP CAT: Antiarrhythmic.

6262. NADP. *Adenosine 5'-(trihydrogen diphosphate) 2'-(dihydrogen phosphate)* 5' → 5'-ester with 3-(aminocarbonyl)-1-β-D-ribofuranosylpyridinium hydroxide inner salt; 3-carbamoyl-1-β-D-ribofuranosylpyridinium hydroxide 5' → 5'-ester with adenosine 2'-(dihydrogen phosphate) 5'-(trihydrogen pyrophosphate) inner salt; nicotinamide-adenine dinucleotide phosphate; triphosphopyridine nucleotide; TPN; coenzyme II; Co II; phosphocozymase; codehydrase II; codehydrogenase II. $C_{21}H_{28}N_7O_{17}P_3$; mol wt 743.44. C 33.93%, H 3.80%, N 13.19%, O 36.59%, P 12.50%. A coenzyme widely distributed in living matter; present in concns of 0.01 to 0.1 mg per gram of tissue except liver where concn is greater. Present chiefly in the reduced state. Acts as a hydrogen carrier in anaerobic and aerobic oxidations and fermentations. Seems to be reduced by fewer enzymes than NAD. It is a cofactor (without undergoing oxidation) in the hydroxylation of aromatic and steroid rings by enzymes. The structure is described as: adenine—D-ribose (phosphate)—phosphate—phosphate—D-ribose—nicotinamide. Isoln from horse blood: Warburg et al., *Biochem. Z.* **282**,

157 (1935); from hog liver by adsorption on charcoal: LePage, Mueller, *J. Biol. Chem.* **180**, 775 (1949); by anion-exchange chromatography: Cohn, *J. Am. Chem. Soc.* **72**, 1471 (1950); from sheep liver by extraction with hot water: Kornberg, Horecker, *Biochem. Prepn.* **3**, 23 (1953). Review: Sund, "The Pyridine Nucleotide Coenzymes" in *Biological Oxidations*, T. Singer, Ed. (Interscience, New York, 1968) pp 603-639.

Grayish-white powder. Sol in water, methanol. Much less sol in ethanol. Practically insol in ether, ethyl acetate. pKa_1 3.9; pKa_2 6.1. Molar extinction coefficient of reduced NADP is 6.22×10^6 cm² at 340 nm; of oxidized NADP = 18.0×10^6 cm² at 260 nm. Forms a sodium salt tetrahydrate, also forms insol salts with heavy metals.

6263. Naepaine. *2-(Pentylamino)ethanol 4-aminobenzoate (ester)*; p-aminobenzoic acid 2-n-pentylaminoethyl ester; p-aminobenzoic acid 2-n-amylaminoethyl ester; 2-n-amylaminoethyl p-aminobenzoate; 2-n-pentylaminoethyl p-aminobenzoate; Amylsine. $C_{14}H_{22}N_2O_2$; mol wt 250.33. C 67.17%, H 8.86%, N 11.19%, O 12.78%. Prepn: Goldberg, Whitmore, *J. Am. Chem. Soc.* **59**, 2280 (1937). Prepn of dimorphic cryst forms: Kreider, Menotti, *ibid.* **64**, 1227 (1942).

Solid, mp 66°.
Hydrochloride, $C_{14}H_{23}ClN_2O_2$, dimorphic crystals. Rectangular plates, mp 153.5°; rods, mp 176°. Bitter taste followed by a sense of numbness. Sol in water; sparingly sol in alcohol; insol in ether, chloroform, benzene.

THERAP CAT: Hydrochloride as local anesthetic.
THERAP CAT (VET): Topical anesthetic (ocular).

6264. Nafamostat. *4-[(Aminoiminomethyl)amino]benzoic acid 6-(aminoiminomethyl)-2-naphthalenyl ester*; 6-amidino-2-naphthyl-4-guanidinobenzoate; nafamstat. $C_{19}H_{17}N_5O_2$; mol wt 347.38. C 65.70%, H 4.93%, N 20.16%, O 9.21%. Non-peptide protease inhibitor. Prepn: S. Fujii et al., **Eur. pat. Appl. 48,433;** eidem, U.S. pat. **4,454,338** (1982, 1984 both to Torii & Co.); T. Aoyama et al., *Chem. Pharm. Bull.* **33**, 1458 (1985). Inhibitory effects on trypsin, thrombin, kallikrein, plasmin and complement-mediated hemolysis: S. Fujii, Y. Hitomi, *Biochim. Biophys. Acta* **661**, 342 (1981); T. Aoyama et al., *Japan. J. Pharmacol.* **35**, 203 (1984). Effect in experimental acute pancreatitis in rats: M. Iwaki et al., *ibid.* **41**, 155 (1986). Spectrofluorometric determn in biological material: T. Aoyama et al., *Chem. Pharm. Bull.* **33**, 2142 (1985). Clinical anti-complement activity: Y. Miyamoto et al., *Trans. Am. Soc. Artif. Intern. Organs* **31**, 508 (1985).

Dimethanesulfonate, $C_{21}H_{25}N_5O_8S_2$, *nafamostat mesilate, FUT-175, Futhan.* Crystals from ethyl ether, mp 217-220° (Fujii). Also reported as colorless powder from water, mp 260° (dec) (Aoyama).

Dihydrochloride, $C_{19}H_{19}Cl_2N_5O_2$, crystals from DMF-HCl, mp 264-266°.

THERAP CAT: Enzyme inhibitor (protease).

6265. Nafarelin. *6-[3-(2-Naphthalenyl)-D-alanine]luteinizing hormone-releasing factor(pig);* 5-oxo-L-prolyl-L-histidyl-L-tryptophyl-L-seryl-L-tyrosyl-3-(2-naphthyl)-D-alanyl-L-leucyl-L-arginyl-L-prolylglycinamide; [6-[3-(2-naphthyl)-D-alanine]] LHRH; D-nal(2)⁶-LHRH; NAG. $C_{66}H_{83}N_{17}O_{13}$; mol wt 1322.49. C 59.94%, H 6.33%, N 18.00%, O 15.73%. Synthetic peptide agonist analog of LH-RH, *q.v.* Prepn: J. J. Nestor *et al.,* U.S. pat. **4,234,571** (1980 to Syntex); *eidem, J. Med. Chem.* **25,** 795 (1982). Suppression of luteal and placental function in baboons: B. H. Vickery *et al., Fertil. Steril.* **36,** 664 (1981). Inhibition of ovulation in humans: J. A. Gudmundsson *et al., Contraception* **30,** 107 (1984). Radioimmunoassay in plasma and serum: C. Nerenberg *et al., Anal. Biochem.* **141,** 10 (1984). Kinetics and tissue disposition in animals: N. I. Chu *et al., Drug Metab. Dispos.* **13,** 560 (1985). Clinical evaluation in benign prostatic hyperplasia: C. A. Peters, P. C. Walsh, *N. Engl. J. Med.* **317,** 599 (1987); in endometriosis: M. R. Henzl *et al., ibid.* **318,** 485 (1988). Review of clinical studies: P. G. Hoffman *et al., J. Androl.* **8,** Suppl., S17-S22 (1987).

5-oxoPro-His-Trp-Ser-Tyr-N---C---C—Leu-Arg-Pro-Gly-NH₂

Monoacetate, $C_{68}H_{87}N_{17}O_{15}$, mp 188-190°. $[\alpha]_D^{25}$ —27.4° (c = 0.9 in acetic acid).

Acetate hydrate, $C_{66}H_{83}N_{17}O_{13}.xC_2H_4O_2.yH_2O$, *RS-94991-298, Synarel.*

THERAP CAT: Treatment of endometriosis.

6266. Nafcillin Sodium. *6-[[(2-Ethoxy-1-naphthalenyl)carbonyl]amino]-3,3-dimethyl-7-oxo-4-thia-1-azabicyclo[3.2.0]heptane-2-carboxylic acid monosodium salt;* sodium 6-(2-ethoxy-1-naphthamido)penicillanate; 6-(2-ethoxy-1-naphthamido)penicillin sodium salt; Nafcil; Naftopen; Unipen. $C_{21}H_{21}N_2NaO_5S$; mol wt 436.46. C 57.79%, H 4.85%, N 6.42%, Na 5.26%, O 18.32%, S 7.35%. Semi-synthetic antibiotic related to penicillin. Prepn: Doyle *et al.,* Brit. pat. **880,400** (1961 to Beecham).

THERAP CAT: Antibacterial.
THERAP CAT (VET): Antibacterial.

6267. Nafiverine. *α-Methyl-1-naphthaleneacetic acid 1,4-piperazinediyldi-2,1-ethanediyl ester;* 1,4-piperazinediethanol bis(α-methyl-1-naphthaleneacetate); N,N'-bis[[2-(1-naphthyl)propionyloxy]ethyl]piperazine; N,N'-di[α-(1-naphthyl)propionyloxy-2-ethyl]piperazine; Naftidan. $C_{34}H_{38}N_2O_4$; mol wt 538.66. C 75.81%, H 7.11%, O 11.88%. Prepn: Pala, **Brit.** pat. **1,016,968** (1966 to de Angeli). Isoln and characterization of isomers: Pala, Montegani, *Farmaco Ed. Sci.* **22,** 3 (1967).

Dihydrochloride, $C_{34}H_{40}Cl_2N_2O_4$, crystals, mp 220-221°.
THERAP CAT: Antispasmodic.

6268. Nafronyl. *Tetrahydro-α-(1-naphthalenylmethyl)-2-furanpropanoic acid 2-(diethylamino)ethyl ester; tetrahydro-α-(1-naphthylmethyl)-2-furanpropionic acid 2-(diethylamino)ethyl ester;* 3-(1-naphthyl)-2-tetrahydrofurfurylpropionic acid 2-(diethylamino)ethyl ester; α-tetrahydrofurfuryl-1-naphthalenepropionic acid 2-(diethylamino)ethyl ester; N-diethylaminoethyl β-(1-naphthyl)-β-tetrahydrofuryl isobutyrate; naftidrofuryl; Dubimax; Gevatran; Tridus. $C_{24}H_{33}NO_3$; mol wt 383.53. C 75.16%, H 8.67%, N 3.65%, O 12.52%. Prepn: Szarvasi, Bayssat, **Fr.** pat. **1,363,948** corresp to U.S. pat. **3,334,096** (1964, 1967 both to LIPHA); Szarvasi *et al., Compt. Rend.* **260,** 3095 (1965); *eidem, Bull. Soc. Chim. France* **1966,** 1838. Activity studies: Fontaine *et al., Compt. Rend. Ser. D* **262,** 719 (1966). Metabolism and toxicology: Fontaine *et al., Chim. Ther.* **4,** 44 (1969); Fontaine *et al., J. Eur. Toxicol.* **11,** 40 (1969). Acute toxicity: P. Bessin *et al., Eur. J. Med. Chem.* **10,** 291 (1975). Clinical studies: C. A. Clyne *et al., Brit. J. Surg.* **67,** 347 (1980); R. M. Greenhalgh, *ibid.* **68,** 265 (1981).

bp₀.₅ 190°. d_4^{31} 1.0465. n_D^{29} 1.5513. LD₅₀ in mice (mg/kg): 365 orally (Bessin).

Acid oxalate, $C_{26}H_{35}NO_7$, *EU-1806, LS-121, Citoxid, Di-Actane, Dusodril, Praxilene.* Crystals from ethyl acetate, mp 110-111°. Slightly hygroscopic, sol in water.
THERAP CAT: Vasodilator.

6269. Naftalofos. *2-[(Diethoxyphosphinyl)oxy]-1H-benz-[de]isoquinoline-1,3(2H)-dione; N-hydroxynaphthalimide diethyl phosphate; O,O-diethyl O-naphthaloximide phosphate;* naphthalophos; phthalophos; BAY 9002; ENT-25567; S-940; Maretin; Rametin. $C_{16}H_{16}NO_6P$; mol wt 349.29. C 55.02%, H 4.62%, N 4.01%, O 27.48%, P 8.87%. Prepn: Lorenz, Wegler, **Ger.** pat. **962,608** and **Austrian** pat. **193,894** (both 1957 to Bayer).

Minute crystals. The commercial product has a brown to tan color. mp 174-179°. Practically insol in water. Sparingly sol in the usual organic solvents. Fairly sol in methylene chloride.
THERAP CAT (VET): Anthelmintic.

6270. Naftifine. *(E)-N-Methyl-N-(3-phenyl-2-propenyl)-1-naphthalenemethanamine; (E)-N-cinnamyl-N-methyl-*

1-naphthalenemethylamine; *N*-methyl-*N*-(1-naphthylmeth-yl)-3-phenylpropen-1-amine; naftifungin. $C_{21}H_{21}N$; mol wt 287.40. C 87.76%, H 7.37%, N 4.87%. Antimycotic allyl-amine. Prepn: D. Berney, **Belg.** pat. **853,976**; *idem,* **U.S.** pat. **4,282,251** (1977, 1981 both to Sandoz); A. Blade Fond, E. Mendoza-Villela, **Span.** pat. **504,432** (1982 to Lab. Frumtost-Prem S.A.), *C.A.* **99,** 38173r (1983); H. Loibner *et al., Tetrahedron Letters* **25,** 2535 (1984). In vitro activity against dermatophytes and *Candida* spp: A. Georgopoulos *et al., Antimicrob. Ag. Chemother.* **19,** 386 (1981). Morpho-logical changes induced in *Trichophyton mentagrophytes*: J. C. Meingassner *et al., J. Invest. Dermatol.* **77,** 444 (1981). Specific inhibitor of squalene epoxidase, a key enzyme in fungal ergosterol biosynthesis: F. Paltauf *et al., Biochim. Biophys. Acta* **712,** 268 (1982); N. S. Ryder, *Antimicrob. Ag. Chemother.* **25,** 483 (1984). HPLC determn in human plas-ma and urine: F. Schatz, H. Haberl, *Arzneimittel-Forsch.* **36,** 1850 (1986). Clinical trials in dermatophytosis: U. Ganzin-ger *et al., Clin. Trials J.* **19,** 342 (1982).

Colorless, viscous oil, bp$_{0.015 torr}$ 162-167°.
Hydrochloride, $C_{21}H_{22}ClN$, *AW 105 843, SN 105-843, Exoderil, Naftin.* mp 177° (from propanol).
THERAP CAT: Topical antifungal.

6271. Nalbuphine. *17-(Cyclobutylmethyl)-4,5-epoxy-morphinan-3,6,14-triol;* N-cyclobutylmethyl-14-hydroxydi-hydronormorphine. $C_{21}H_{27}NO_4$; mol wt 357.46. C 70.56%, H 7.61%, N 3.92%, O 17.90%. Mixed opioid agonist-antag-onist. Prepn: **Brit.** pat. **1,119,270** corresp to I. J. Pachter, Z. Matossian, **U.S.** pat. **3,393,197** (both 1968 to Endo). Pharmacology: S. G. Holtzman, *Probl. Drug Depend.* **1976,** 374; H. E. Shannon, S. G. Holtzman, *J. Pharmacol. Exp. Ther.* **201,** 55 (1977); G. J. Schaefer, S. G. Holtzman, *ibid.* **205,** 291 (1978). Plasma determn: S. H. Weinstein *et al., J. Pharm. Sci.* **67,** 547 (1978). Clinical study: T. Tammisto, I. Tigerstedt, *Acta Anaesthesiol. Scand.* **21,** 390 (1977). Review of pharmacology and therapeutic efficacy: J. K. Errick, R. C. Heel, *Drugs* **26,** 191-211 (1983).

Cryst from methanol-acetone, mp 230.5°.
Hydrochloride, $C_{21}H_{28}ClNO_4$, *EN-2234A, Nubain.*
THERAP CAT: Narcotic analgesic.

6272. Naled. *Phosphoric acid 1,2-dibromo-2,2-dichloro-ethyl dimethyl ester;* dimethyl 1,2-dibromo-2,2-dichloro-ethyl phosphate; bromchlophos; ENT 24988; RE 4355; Bromex; Dibrom; Ortho-Dibrom. $C_4H_7Br_2Cl_2O_4P$; mol wt 380.79. C 12.62%, H 1.85%, Br 41.97%, Cl 18.62%, O 16.80%, P 8.13%. Prepn: **Brit.** pat. **855,157** corresp to Ospenson, Kohn, **U.S.** pat. **2,971,882** (1960, 1961 both to California Res. Corp.).

Usually obtained as a liquid, slightly pungent odor, bp$_{0.5}$ 110°. d$_4^{25}$ 1.96. Has been crystallized, mp 26.5-27.5°. Vapor

press. at 20° about 2×10^{-3} mm Hg. Practically insol in water, but completely hydrolyzed by water within 48 hrs. Freely sol in aromatic and chlorinated hydrocarbons, ke-tones, alcohols. Sparingly sol in petroleum solvents and mineral oils. LD$_{50}$ in male rats: 250 mg/kg orally; 800 mg/kg dermally, T. B. Gaines, *Toxicol. Appl. Pharmacol.* **14,** 515 (1969).
USE: Insecticide; acaricide.

6273. Nalidixic Acid. *1-Ethyl-1,4-dihydro-7-methyl-4-oxo-1,8-naphthyridine-3-carboxylic acid;* 3-carboxy-1-eth-yl-7-methyl-1,8-naphthyridin-4-one; 1-ethyl-7-methyl-1,8-naphthyridin-4-one-3-carboxylic acid; Win 18320; Betaxina; Cybis; Dixiben; Eucisten; Innoxalon; Kusnarin; Nalidicron; Nalitucsan; Narigix; NegGram; Negram; Nevi-gramon; Nicelate; Nogram; Poleon; Specifin; Uriben; Uri-clar; Uralgin; Urodixin; Uroman; Uroneg; Uropan; Winto-mylon. $C_{12}H_{12}N_2O_3$; mol wt 232.23. C 62.06%, H 5.21%, N 12.06%, O 20.67%. Prepn: Lesher *et al., J. Med. Pharm. Chem.* **5,** 1063 (1962); **S. Afr.** pat. **R61/3073;** Lesher, Gruett, **Belg.** pat. **612,258** corresp to **U.S.** pat. **3,149,104** (1962, 1962, 1964 all to Sterling Drug). Synthesis of inter-mediates: Lappin, *J. Am. Chem. Soc.* **70,** 3348 (1948). Mechanism of action studies: Bourguignon *et al., Antimi-crob. Ag. Chemother.* **4,** 479 (1973); Goss, Cook, *Antibiotics,* **vol. 3,** J. W. Corcoran, F. E. Hahn, Eds. (Springer-Verlag, New York, 1975) pp 174-196; A. M. Pedrini, *ibid.* **vol. 5**(pt. 1), F. E. Hahn, Ed. (1979) pp 154-175; H. T. Wright *et al., Science* **213,** 455 (1981). Comprehensive description: P. E. Grubb in *Analytical Profiles of Drug Substances* vol. **8,** K. Florey, Ed. (Academic Press, New York, 1979) pp 371-397.

Pale buff, cryst powder, mp 229-230°. Soly in mg/ml at 23°: chloroform 35; toluene 1.6; methanol 1.3; ethanol 0.9; water 0.1; ether 0.1. LD$_{50}$ in mice (mg/kg): 3300 orally; 500 s.c.; 176 i.v., Lesher *et al., loc. cit.*
THERAP CAT: Antibacterial.
THERAP CAT (VET): Antibacterial; has been used in urinary tract infections.

6274. Nalmefene. *(5α)-17-(Cyclopropylmethyl)-4,5-epoxy-6-methylenemorphinan-3,14-diol;* 6-desoxy-6-meth-ylenenaltrexone; nalmetrene; ORF 11676. $C_{21}H_{25}NO_3$; mol wt 339.43. C 74.31%, H 7.42%, N 4.13%, O 14.14%. Struc-tural analog of naltrexone, *q.v.,* with opiate antagonist activ-ity. Prepn: J. Fishman, **U.S.** pat. **3,814,768** (1974 to M. J. Lewenstein); E. F. Hahn *et al., J. Med. Chem.* **18,** 259 (1975). Improved synthesis: P. C. Meltzer, J. W. Coe, **Eur.** pat. **Appl. 140,367;** *idem,* **U.S.** pat. **4,535,157** (both 1985 to Key Pharm.). Comparative evaluation of efficacy in mice: R. D. Heilman *et al., Res. Commun. Chem. Pathol. Pharma-col.* **13,** 635 (1976). HPLC determn in human plasma: J. Hsiao, R. Dixon, *ibid.* **42,** 449 (1983). Receptor binding activity in rat brain: M. E. Michel *et al., Meth. Find. Exp. Clin. Pharmacol.* **7,** 175 (1985). Clinical evaluation of nar-cotic antagonism: T. J. Gal, C. A. DiFazio, *Anesthesiology* **64,** 175 (1986). Pharmacokinetics in humans: R. Dixon *et al., Clin. Pharmacol. Ther.* **39,** 49 (1986); *idem, J. Clin. Pharmacol.* **27,** 233 (1987). Use as antagonist to opioid-induced immobilization in large animals: T. J. Kreeger *et al., J. Wildl. Dis.* **23,** 619 (1987).

Crystals from ethyl acetate, mp 188-190°.
THERAP CAT: Narcotic antagonist.

6275. Nalorphine. 7,8-Didehydro-4,5-epoxy-17-(2-propenyl)morphinan-3,6-diol; N-allylnormorphine; allorphine; antorphine; NANM; Nalline; Norfin; Anarcon. $C_{19}H_{21}NO_3$; mol wt 311.39. C 73.29%, H 6.80%, N 4.50%, O 15.41%. Prepd from normorphine: McCawley et al., J. Am. Chem. Soc. **63**, 314 (1941); amended procedure: Hart, McCawley, J. Pharmacol. **82**, 339 (1944); Weijlard, Erickson, J. Am. Chem. Soc. **64**, 869 (1942); U.S. pat. **2,364,833** (1944); Weijlard, U.S. pat. **2,891,954** (1959 to Merck & Co.). Toxicity data: E. I. Goldenthal, Toxicol. Appl. Pharmacol. **18**, 185 (1971).

Crystals from ether, mp 208-209° (not 92-93° as first given by McCawley). $[\alpha]_D^{25}$ −155.3° (c = 3 in methanol). Sparingly sol in water, ether. Sol in alc, acetone, chloroform, dil alkalies. uv max (in acid): 285 nm; (in alkali): 298 nm.
Hydrobromide, $C_{19}H_{22}BrNO_3$, Lethidrone. Crystals from alc, dec 258-259°. Sol in water.
Hydrochloride, $C_{19}H_{22}ClNO_3$, Miromorfalil. Crystals from alc, mp 260-263°. Sol in water. Moderately sol in alc. uv max (water): 285 nm. IR and polarographic data: Seagers et al., J. Am. Pharm. Assoc., Sci. Ed. **41**, 640 (1952). pH of 0.5% aq soln: 5.0. LD_{50} s.c. in rats: 1460 mg/kg (Goldenthal).
Caution: Violent withdrawal symptoms may occur in narcotic addicts. This is a controlled substance listed in the U.S. Code of Federal Regulations, Title 21 Part 1308.13 (1987).
THERAP CAT: Narcotic antagonist.
THERAP CAT (VET): Narcotic antagonist.

6276. Nalorphine Dinicotinate. 7,8-Didehydro-4,5-epoxy-17-(2-propenyl)morphinan-3,6-diol bis(3-pyridinecarboxylate) (ester); N-allylnormorphine dinicotinate; bis(nicotinic acid) diester of N-allylnormorphine; N-allylnormorphine bis(pyridine-3-carboxylic acid) ester; nalorphine bis(nicotinate); Nimelan. $C_{31}H_{27}N_3O_5$; mol wt 521.55. C 71.39%, H 5.22%, N 8.06%, O 15.34%. Prepn: Pongratz, Zirm, Monatsh. **91**, 396 (1960).

Hydrochloride, $C_{31}H_{28}ClN_3O_5$, crystals from ethanol, mp 238-239°.
THERAP CAT: Narcotic antagonist.

6277. Naloxone. 4,5-Epoxy-3,14-dihydroxy-17-(2-propenyl)morphinan-6-one; 1-N-allyl-7,8-dihydro-14-hydroxynormorphinone; 12-allyl-7,7a,8,9-tetrahydro-3,7a-dihydroxy-4aH-8,9c-iminoethanophenanthro[4,5-bcd]furan-5(6H)-one; 1-N-allyl-14-hydroxynordihydromorphinone; N-allylnoroxymorphone; Nalone; Narcan. $C_{19}H_{21}NO_4$; mol wt 327.37. C 69.70%, H 6.47%, N 4.28%, O 19.55%. Specific opiate antagonist. Prepn: Brit. pat. **939,287** (1963 to Sankyo); Lewenstein, Fishman, U.S. pat. **3,254,088** (1966); R. A. Olofson et al., Tetrahedron Letters **1977**, 1567. Activity studies in rodents: Collier, Schneider, Nature **224**, 610 (1969). Human pharmacology: Jasinki et al., J. Pharmacol. Exp. Ther. **157**, 420 (1967). Metabolism: Fujimoto, ibid.

168, 180 (1969); Proc. Soc. Exp. Biol. Med. **113**, 317 (1970). Prepn and properties of (+)-naloxone: I. Iijima et al., J. Med. Chem. **21**, 398 (1978). Potential use in aging and senile dementias: E. Roberts, Ann. N.Y. Acad. Sci. **396**, 165 (1982). Clinical trial in senile dementia of the Alzheimer's type: B. Reisberg et al., N. Engl. J. Med. **308**, 721 (1983). Review of role in rehabilitation of opiate addicts: A. Schecter, Am. J. Drug Alcohol Abuse **7**, 1-18 (1980); of preclinical research: A. L. Misra, Nat. Inst. Drug Abuse Res. Monograph Ser. **28**, 132-146 (1981); of antagonism of discriminative opioid stimulus: E. C. Krimmer, H. Barry, Fed. Proc. **41**, 2319-2322 (1982); of therapeutic uses: L. F. McNicholas, W. R. Martin, Drugs **27**, 81 (1984). Comprehensive description: M. M. A. Hasson et al., in Analytical Profiles of Drug Substances vol. **14**, K. Florey, Ed. (Academic Press, New York, 1985) pp 453-489.

Crystals from ethyl acetate, mp 184° (Lewenstein), 177-178° (Sankyo Co.). $[\alpha]_D^{20}$ −194.5° (c = 0.93 in $CHCl_3$). Sol in chloroform. Practically insol in petr ether.
Hydrochloride, $C_{19}H_{22}ClNO_4$, EN 1530, Narcan. Crystals from ethanol + ether, mp 200-205°. Sol in water, alcohol. Practically insol in ether.
Hydrochloride dihydrate, Narcanti.
THERAP CAT: Narcotic antagonist.
THERAP CAT (VET): Narcotic antagonist.

6278. Naltrexone. 17-(Cyclopropylmethyl)-4,5-epoxy-3,14-dihydroxymorphinan-6-one; N-cyclopropylmethyl-14-hydroxydihydromorphinone; UM-792; Celupan; Trexan. $C_{20}H_{23}NO_4$; mol wt 341.41. C 70.36%, H 6.79%, N 4.10%, O 18.75%. Congener of naloxone, q.v., with opiate-blocking activity. Prepn: H. Blumberg et al., U.S. pat. **3,332,950** (1967 to Endo). Plasma kinetics in dogs, monkeys: R. H. Reuning et al., J. Pharm. Sci. **68**, 411 (1979). Metabolism study: E. J. Cone, Tetrahedron Letters **1973**, 2607. Biosynthesis, isoln, identification of major human metabolite: E. J. Cone et al., J. Pharm. Sci. **64**, 618 (1975). Physiological and psychological effects of single doses in man: E. R. Gritz et al., Clin. Pharmacol. Ther. **19**, 773 (1976). Disposition, metabolism, effects after acute and chronic administration: K. Verebey et al., ibid. **20**, 315 (1976). Urinary excretion profile: H. E. Dayton, C. E. Inturrisi, Drug. Metab. Dispos. **4**, 474 (1976). Enhanced analgesic effect of stress after chronic administration: S. Amir, Z. Amit, Eur. J. Pharmacol. **59**, 137 (1979). Effects on brain opiate receptor supersensitivity: R. S. Zukin et al., Brain Res. **245**, 285 (1982); A. Tempel et al., Life Sci. **31**, 1402 (1982). Operant analysis of effects on human self-administration of heroin: N. K. Mello et al., J. Pharmacol. Exp. Ther. **216**, 45 (1981). Clinical study in chronic schizophrenia: M. Ragheb et al., Int. Pharmacopsychiatry **15**, 1 (1980); in heroin addiction: B. A. Judson et al., Drug Alcohol Dependence **7**, 325 (1981). Preliminary evaluation in infantile autism: Psychopharmacol. Bull. **24**, 135 (1988). Toxicity data: R. P. Maickel et al., Ann. N.Y. Acad. Sci. **281**, 321 (1976). Review: M. S. Gold et al., Med. Res. Rev. **2**, 211-246 (1982).

Crystals from acetone, mp 168-170°. LD$_{50}$ in mice (mg/kg): 586 s.c. (Maickel).

Hydrochloride, C$_{20}$H$_{24}$ClNO$_4$, *EN-1639A, Antaxone, Nalorex.* Crystals from methanol, mp 274-276°.

THERAP CAT: Narcotic antagonist.

6279. Nandinine. *(S)-5,8,13,13a-Tetrahydro-10-methoxy-6H-benzo[g]-1,3-benzdioxolo[5,6-a]quinolizin-9-ol; 10-methoxy-2,3-(methylenedioxy)berbin-9-ol;* 5,6,13,13a-tetrahydro-9-hydroxy-10-methoxy-2,3-(methylenedioxy)-8H-dibenzo[a,g]quinolizine; tetrahydroberberrubine. C$_{19}$-H$_{19}$NO$_4$; mol wt 325.35. C 70.14%, H 5.89%, N 4.31%, O 19.67%. From root bark of *Nandina domestica* Thunb., *Berberidaceae:* Eijkman, *Ber.* **17**, 441 (1884); Kitasato, *J. Pharm. Soc. Japan* **522**, 1 (1925). Structure: *idem, Acta Phytochim. (Japan)* **3**, 175 (1927), *C.A.* **22**, 1779 (1928); Späth, Leithe, *Ber.* **63**, 3007 (1930). Configuration: Corrodi, Hardegger, *Helv. Chim. Acta* **39**, 889 (1956). Synthesis: Kametani *et al., J. Chem. Soc. (C)* **1969**, 2036; *eidem, J. Chem. Soc. Perkin I* **1977**, 1151.

dl-Form, colorless needles from ethanol, mp 183-185°. *l*-Form, crystals from ether. mp 195-196°. [α]$_D^{15}$ − 304° (c = 0.513 in chloroform). *d*-Form, crystals from chloroform + ether, mp 195-196°.

6280. Nandrolone. *17-Hydroxyestr-4-en-3-one;* 17β-hydroxy-4-estren-3-one; 4-estren-17β-ol-3-one; 17β-hydroxy-19-nor-4-androsten-3-one; 19-nortestosterone. C$_{18}$-H$_{26}$O$_2$; mol wt 274.39. C 78.79%, H 9.55%, O 11.66%. Prepd from alkyl ethers of estradiol: Birch, *Quart. Rev.* **4**, 69 (1950); Wilds, Nelson, *J. Am. Chem. Soc.* **75**, 5366 (1953). Alternate syntheses: H. Ueberwasser *et al., Helv. Chim. Acta* **46**, 344 (1963); I. Shimizu *et al., Tetrahedron Letters* **21**, 487 (1980). Benzoate: Hicks, U.S. pat. **2,698,855** (1955 to Organics, Inc.). Hexahydrobenzoate: Muller, U.S. pat. **2,891,973** (1959 to Lab. Français de Chimiother.).

Dimorphic crystals, mp 112° and 124°. [α]$_D^{22}$ +55° (c = 0.93 in chloroform). uv max (ethanol): 241 nm (ε 17,000). Sol in alcohol, ether, chloroform.

Cyclohexanepropionate, C$_{27}$H$_{40}$O$_3$, *19-nortestosterone cyclohexylpropionate, Sanabolicum.*

Benzoate, C$_{25}$H$_{30}$O$_3$, needles from 75% alc, mp 174-175°. [α]$_D^{20}$ +104.5° (alc).

Cyclohexanecarboxylate, C$_{25}$H$_{36}$O$_3$, *19-nortestosterone hexahydrobenzoate, Norlongandron, Nor-Durandron.* Small, elongated prisms from petr ether, mp 88-89°, [α]$_D^{20}$ +50° (c = 0.5 in chloroform).

Dodecanoate, C$_{30}$H$_{48}$O$_3$, *nandrolone laurate, Laurabolin V.*

Furylpropionate, C$_{25}$H$_{32}$O$_4$, *19-nortestosterone furylpropionate, NFP, Demelon.*

THERAP CAT: Anabolic.

6281. Nandrolone Decanoate. *17β-[(1-Oxodecyl)oxy]-estr-4-en-3-one;* 17β-hydroxyestr-4-en-3-one 17-decanoate; 17β-hydroxy-19-norandrost-4-en-3-one 17-decanoate; 19-nortestosterone decanoate; norandrostenolone decanoate; Deca-Durabolin; Deca-Durabol; Deca-Hybolin; Hybolin Decanoate; Retabolil. C$_{28}$H$_{44}$O$_3$; mol wt 428.63. C 78.45%,

H 10.35%, O 11.20%. Prepn: DeWitt, Overbeek, U.S. pat. **2,998,423** (1961 to Organon).

White to yellow crystals, mp 32-35°. Practically insol in water; freely sol in ethanol, ether, acetone, chloroform, oils.

THERAP CAT: Anabolic.

6282. Nandrolone p-Hexyloxyphenylpropionate. *17β-[3-[4-(Hexyloxy)phenyl]-1-oxopropoxy]estr-4-en-3-one;* 17β-hydroxyestr-4-en-3-one p-(hexyloxy)hydrocinnamate; 17β-hydroxyestr-4-en-3-one p-hexyloxyphenylpropionate; 17β-hydroxy-19-norandrost-4-en-3-one p-hexyloxyphenylpropionate; 19-nortestosterone-3-(p-hexyloxyphenyl)propionate; nortestosterone hexoxyphenylpropionate; Anador; Anadur. C$_{33}$H$_{46}$O$_4$; mol wt 506.73. C 78.21%, H 9.15%, O 12.63%. Prepn and pharmacological properties: Diczfalusy *et al., Acta Chem. Scand.* **17**, 2536 (1963). Claimed but not described in: Diczfalusy *et al.,* U.S. pat. **2,904,562** (1959 to AB Leo).

Crystals, mp 53-55°. [α]$_D$ +45° (c = 1.0 in dioxane).

THERAP CAT: Anabolic.

6283. Nandrolone Phenpropionate. *(17β)-(1-Oxo-3-phenylpropoxy)estr-4-en-3-one;* 17β-hydroxyestr-4-en-3-one 3-phenylpropionate; 17β-hydroxy-4-estren-3-one hydrocinnamate; 19-nor-17β-hydroxy-3-ketoandrostene 17-phenylpropionate; 19-nor-Δ^4-androsten-17β-ol-3-one β-phenylpropionate; 19-nortestosterone β-phenylpropionate; Activin; Durabolin; Durabol; Strabolene; Superanabolon; Nandrolin. C$_{27}$H$_{34}$O$_3$; mol wt 406.54. C 79.76%, H 8.43%, O 11.81%. Prepn: Brit. pat. **826,028** (1959 to Organon); Donia, Ott, U.S. pat. **2,868,809** (1959 to Upjohn). Crystals, mp 95-96°. [α]$_D$ +58° (chloroform).

THERAP CAT: Anabolic.

THERAP CAT (VET): Anabolic steroid.

6284. Nandrolone Propionate. *17β-(1-Oxopropoxy)estr-4-en-3-one;* 17β-hydroxyestr-4-en-3-one propionate; 19-nor-Δ^4-androsten-17β-ol-3-one propionate; 19-nortestosterone propionate; Norybol-19; Nortesto. C$_{21}$H$_{30}$O$_3$; mol wt 330.45. C 76.32%, H 9.15%, O 14.53%. Prepn: Djerassi *et al.,* U.S. pat. **2,756,244** (1956 to Syntex); Donia, Ott, U.S. pat. **2,798,879** (1957 to Upjohn); Rao, *J. Org. Chem.* **25**, 1058 (1960).

Crystals from aqueous methanol or isopropyl ether, mp 55-60°. [α]$_D^{23}$ +41° in chloroform (Donia, Ott); [α]$_D^{23.5}$ +58° in chloroform (Rao). uv max: 240 nm (ε 16,650).

THERAP CAT: Anabolic.

6285. Napalm. A coprecipitated aluminum soap from naphthenic acids and the fatty acids of coconut oil developed early in 1942 (Fieser, Harris, Hershberg, Morgana, Novello, Putnam) for prepn of gasoline gels for incendiary munitions: U.S. pats. **2,606,107** (1952); Herron, U.S. pat. **2,684,339** (1954 to Safety Fuel & Chem. Corp.). The name was derived from the *naphthenic* and *palmitic* acids which are its major constituents. Structure and mfg problems: *Chem. & Eng. News* **32**, 2690 (1954). Historical account: L. F. Fieser, *The Scientific Method* (Reinhold, New York, 1964); Bruce, "Chemical Warfare—Flame" in Kirk-Othmer *Encyclopedia of Chemical Technology* (Interscience, New York, 1964) p 888.

USE: Gasoline thickener. In chemical warfare (fire bombs, flame throwers, flame land mines).

6286. Napelline. *21-Ethyl-4-methyl-16-methylene-7,20-cycloveatchane-1,12-15-triol;* luciculine. C$_{22}$H$_{33}$NO$_3$; mol wt 359.49. C 73.50%, H 9.25%, N 3.90%, O 13.35%. Isoln from *Aconitum napellus* L., *Ranunculaceae:* Freudenberg, Rogers, *J. Am. Chem. Soc.* **59**, 2572 (1937); Jacobs, Craig, *J. Biol. Chem.* **143**, 611 (1942). The product obtained by these isoln procedures is a mixture of at least three compounds. Structure: Wiesner, Valenta, *Fortschr. Chem. Org. Naturst.* **16**, 53 (1958). Related stereochemical studies: Okamoto et al., *Chem. Pharm. Bull.* **13**, 1270 (1963). Total synthesis of racemic napelline: Wiesner et al., *Can. J. Chem.* **52**, 2353, 2355 (1974); S. P. Sethi et al., *ibid.* **58**, 1889 (1980).

Rectangular plates from ether + petr ether. Hydrochloride, $C_{22}H_{34}ClNO_3$, solvated crystals, dec 200-222°. $[\alpha]_D^{22}$ —93.9° (c = 5). Sol in water.

6287. Naphazoline. *4,5-Dihydro-2-(1-naphthalenyl-methyl)-1H-imidazole; 2-(1-naphthylmethyl)imidazoline.* $C_{14}H_{14}N_2$; mol wt 210.27. C 79.97%, H 6.71%, N 13.32%. α-Adrenergic agonist. Prepd by reacting the acetic acid anhydride of naphthoimidic acid with ethylenediamine: U.S. pat. **2,161,938** (1939); by reacting naphthylthioacetamide with ethylenediamine: **Dan.** pat. **62,889** (1944), *C.A.* **40**, 4398 (1946). Toxicity data: J. Gylfe et al., *Fed. Proc.* **9**, 280 (1950).

Hydrochloride, $C_{14}H_{15}ClN_2$, *Ak-Con, Clera, Coldan, Naphcon, Niazol, Opcon, Privine hydrochloride, Rhinantin, Rhinoperd, Sanorin, Sanorin-Spofa, Strictylon, Vasocon.* Bitter crystals, mp 255-260°. Freely sol in water (40 g dissolve in 100 ml) and in alcohol. Slightly sol in chloroform. Insol in benzene, ether. A 1% aq soln has a pH of about 6.2. LD_{50} s.c. in rats: 385 mg/kg (Gylfe).

THERAP CAT: Adrenergic (vasoconstrictor); decongestant.

6288. Naphthacene. Tetracene; 2,3-benzanthracene; rubene; chrysogen. $C_{18}H_{12}$; mol wt 228.28. C 94.70%, H 5.30%. Occurs in coal tar. Contaminates commercial anthracene to which it imparts a yellow color. Isoln: Cook et al., *Proc. Roy. Soc. London* **B111**, 455 (1932). Isoln from crude anthracene by chromatography: Winterstein et al., *Z. Physiol. Chem.* **230**, 159 (1934). Synthesis by condensing succinic acid and phthalic anhydride in the presence of sodium acetate: Gabriel, Michael, *Ber.* **10**, 1559, 2207 (1877); **11**, 1682 (1878); Roser, *Ber.* **17**, 2744 (1884); Nathanson, *Ber.* **26**, 2582 (1893); Gabriel, Leupold, *Ber.* **31**, 1159, 1272 (1898); Wanag, *Ber.* **70**, 274 (1937); from 1-naphthol and phthalic anhydride: Deichler, Weizmann, *Ber.* **36**, 547, 719 (1903); **Ger.** pat. **298,345** (1916); from 1,5-dihydroxynaphthalene and phthalic anhydride: Bentley et al., *J. Chem. Soc.* **91**, 411, 1588 (1907); from tetralin and phthalic anhydride: Schroeter, *Ber.* **54**, 2242 (1921); **Ger.** pat. **346,673** (1918); cf. Fieser, *J. Am. Chem. Soc.* **53**, 2329 (1931). Other syntheses: Coulson, *J. Am. Chem. Soc.* **1935**, 77; Weizmann et al., *ibid.* **1939**, 398.

Orange leaflets from xylene. d 1.35. Sublimes *in vacuo.* mp 341° (open capillary tube), mp 357° (copper block). Absorption spectrum: Clar, *Ber.* **69**, 607 (1936). Fluorescence maxima: Krishman, Seshan, *Z. Krist.* **89**, 538 (1934).

Difficultly sol in most solvents. Solns show slight green fluoresence in daylight. Does not form a picrate.

6289. Naphthalene. Naphthalin; naphthene; tar camphor. $C_{10}H_8$; mol wt 128.16. C 93.71%, H 6.29%. Most abundant single constituent of coal tar. Dry coal tar contains about 11%. Crystallizes from the middle or "carbolic oil" fraction of the distilled tar. Purified by hot pressing, which may be followed by washing with H_2SO_4, NaOH, and water, then by fractional distillation or by sublimation. Manuf: Faith, Keyes & Clark's *Industrial Chemicals*, F. A. Lowenheim, M. K. Moran, Eds. (Wiley-Interscience, New York, 4th ed., 1975) pp 556-562. *Review:* R. M. Gaydos in Kirk-Othmer *Encyclopedia of Chemical Technology* **vol. 15** (Wiley-Interscience, New York, 3rd ed., 1981) pp 698-719.

Monoclinic prismatic plates from ether or by sublimation; also sold as white scales, powder, balls, or cakes, mp 80.2°. Odor of moth balls. Volatilizes appreciably at room temp. d_4^{20} 1.162. d_4^{100} 0.9628. Sublimes appreciably at temps above mp; volatile with steam. bp_{760} 217.9°; bp_{400} 193.2°; bp_{200} 167.7°; bp_{100} 145.5°; bp_{60} 130.2°; bp_{40} 119.3°; bp_{20} 101.7°; bp_{10} 85.8°. Flash pt, open cup 174°F (79°C); closed cup 190°F (88°C). Autoignition temp 1053°F (567°C). n_D^{100} 1.58212. Purple fluorescence in Hg light (petr ether soln). Ultraviolet absorption: Several characteristic bands between 217.5 and 320 nm in hexane. Insol in water. One gram dissolves in 13 ml methanol or ethanol, in 3.5 ml benzene or toluene, in 8 ml olive oil or turpentine, in 2 ml chloroform or carbon tetrachloride, in 1.2 ml carbon disulfide. Very sol in ether, hydronaphthalenes, in fixed and volatile oils.

Human Toxicity: Poisoning may occur by ingestion of large doses, inhalation, or skin absorption. *Symptoms and signs:* nausea, vomiting, headache, diaphoresis, hematuria, hemolytic anemia, fever, hepatic necrosis, convulsions, coma, *Clinical Toxicology of Commercial Products*, R. E. Gosselin et al., Eds. (Williams & Wilkins, Baltimore, 4th ed., 1976) Section III, pp 242-246.

USE: Manuf phthalic and anthranilic acids which are used in making indigo, indanthrene, and triphenylmethane dyes. Manuf of hydroxyl (naphthols), amino (naphthylamines), sulfonic acid and similar compds used in the dye industries. Manuf of synthetic resins, celluloid, lampblack, smokeless powder. Manuf of hydronaphthalenes (Tetralin, Decalin) which are used as solvents, in lubricants, and in motor fuels. The use of naphthalene as a moth repellent and insecticide is decreasing due to the introduction of chlorinated compds such as *p*-dichlorobenzene.

THERAP CAT: Has been used as antiseptic (topical and intestinal); anthelmintic (Cestodes).

THERAP CAT (VET): Has been used in dusting powders, as an insecticide and internally as an intestinal antiseptic and vermicide.

6290. 1-Naphthaleneacetic Acid. α-Naphthaleneacetic acid; naphthylacetic acid; NAA; Fruitone-N; Phyomone; Planofix; Rootone; Tre-Hold. $C_{12}H_{10}O_2$; mol wt 186.20. C 77.40%, H 5.41%, O 17.18%. Prepn from naphthalene + chloroacetic acid: Ogata, Ishiguro, *J. Am. Chem. Soc.* **76**, 4302 (1950); Southwick et al., *ibid.* **76**, 754 (1954); U.S. pat. **2,655,531** (1953 to FMC); from naphthylacetonitrile: Wenner, U.S. pat. **2,489,348** (1949 to Hoffmann-La Roche); *J. Org. Chem.* **15**, 548 (1950). Activity: F. E. Gardiner et al., *Science* **90**, 208 (1939). Crystal structure: S. S. Rajan, *Acta Crystallogr.* **B34**, 998 (1978).

Needles from water, mp 134.5-135.5°. Sol in about 30 parts alcohol; freely sol in acetone, ether, chloroform. Soly in water at 17°: 0.38 g/l. LD_{50} orally in rats: 1000 mg/kg, G. W. Bailey, J. L. White, *Residue Rev.* **10**, 97 (1965).

USE: Plant growth regulator.

6291. 1,8-Naphthalenediamine. 1,8-Diaminonaphthalene. $C_{10}H_{10}N_2$; mol wt 158.20. C 75.92%, H 6.37%, N 17.71%. Prepd by reducing 1,8-dinitronaphthalene with phosphorus triiodide: Meyer, Müller, *Ber.* **30**, 775 (1897).

Crystals from dil alc, mp 66.5°. bp_{12} 205°; $n_D^{99.4}$ 1.6828; $d_4^{99.4}$ 1.1265. Sublimable. Turns brown on standing. Soluble in alcohol or ether; slightly sol in water or chloroform.

Dihydrochloride, $C_{10}H_{12}Cl_2N_2$, leaflets, mp 280°.

USE: Antioxidant for lubricating oils. Detection of selenium and nitrites.

6292. 1,6-Naphthalenedisulfonic Acid. Ewer-Pick acid. $C_{10}H_8O_6S_2$; mol wt 288.30. C 41.66%, H 2.80%, O 33.30%, S 22.24%. Prepn: Fierz-David, Hasler, *Helv. Chim. Acta* **6**, 1134 (1923).

Crystals. Very sol in water; sol in alcohol; practically insol in ether.

6293. 2,6-Naphthalenedisulfonic Acid. Ebert-Merz β-acid. $C_{10}H_8O_6S_2$; mol wt 288.30. C 41.66%, H 2.80%, O 33.30%, S 22.24%. Prepn: Fierz-David, Richter, *Helv. Chim. Acta* **28**, 257 (1945).

Deliquesc crystals. Very sol in water, alcohol; practically insol in ether.

6294. 2,7-Naphthalenedisulfonic Acid. Ebert-Merz α-acid. $C_{10}H_8O_6S_2$; mol wt 288.30. C 41.66%, H 2.80%, O 33.30%, S 22.24%. Prepn from naphthalene and concd H_2SO_4: Fierz-David, Blangey, *Fundamental Processes of Dye Chemistry* (Interscience, New York, 1949) p 209.

Very deliquesc crystals. Very sol in water, alcohol; practically insol in ether.

USE: In dye chemistry.

6295. 1-Naphthalenesulfonic Acid. α-Naphthalenesulfonic acid. $C_{10}H_8O_3S$; mol wt 208.23. C 57.68%, H 3.87%, O 23.05%, S 15.40%. Made by sulfonating naphthalene with H_2SO_4 at 0°.

Crystals. mp 90° (dihydrate). Freely sol in water or alcohol, slightly in ether.

USE: Manuf α-naphthol. The sodium salt is used for rendering phenols sol in water.

6296. 2-Naphthalenesulfonic Acid. β-Naphthalenesulfonic acid. $C_{10}H_8O_3S$; mol wt 208.23. C 57.68%, H 3.87%, O 23.05%, S 15.40%. Made by sulfonating naphthalene with H_2SO_4 at 160°.

Monohydrate, white to slightly brownish, cryst leaflets; very hygroscopic. mp 124-125°. mp 91° when anhydr. Freely sol in water.

USE: Manuf β-naphthol and intermediates.

6297. 1-Naphthalenethiol. α-Thionaphthol; 1-thionaphthol; 1-mercaptonaphthalene; 1-naphthyl mercaptan. $C_{10}H_8S$; mol wt 160.23. C 74.95%, H 5.03%, S 20.01%. Prepd by catalytic sulfurization of naphthalene with S, S_2Cl_2, etc., followed by hydrogenation: Lazier *et al.*; Signaigo, U.S. pats. 2,402,645; 2,402,686 (both 1946 to du Pont).

Liquid, strong mercaptan odor. Solidif on cooling. d_4^0 1.1729; d_4^{20} 1.607; d_4^{23} 1.1549. bp_{760} 285°; $bp_{10.3}$ 144.8°; bp_2 138-140°. n_D^{20} 1.6802. Sol in ethanol, ether; sparingly sol in aq alkalies. Volatile with steam.

6298. 2-Naphthalenethiol. β-Thionaphthol; 2-thionaphthol; 2-mercaptonapthalene; 2-naphthyl mercaptan. $C_{10}H_8S$; mol wt 160.23. C 74.95%, H 5.03%, S 20.01%. Prepd by catalytic hydrogenation of a sulfonic acid derivative of naphthalene: Lazier, Signaigo, U.S. pat. 2,402,641 (1946 to du Pont); by reduction of naphthalenesulfonyl chloride with zinc: Holt, U.S. pat. 2,216,840 (1940 to du Pont).

Crystals from ethanol, disagreeable odor. mp 81°. bp_{760} 286°; $bp_{10.3}$ 146.3°. Very sol in ethanol, ether, petr ether; sparingly sol in water. Slightly volatile with steam.

6299. Naphthalic Acid. *1,8-Naphthalenedicarboxylic acid.* $C_{12}H_8O_4$; mol wt 216.18. C 66.67%, H 3.73%, O 29.60%. Made by the oxidation of acenaphthene with chromic acid mixture, etc.: Graebe, Gfeller, *Ber.* **25**, 652 (1895); Ogilvie, Wilder, U.S. pat. 2,379,032 (1945 to Allied Chem.); Karishin, Fedorenko, *Zhur. Priklad. Khim.* **29**, 955 (1956), *C.A.* **50**, 14677b (1956).

Crystals, mp 270°. Practically insol in water; freely sol in warm alcohol, slightly in ether.

6300. Naphthenes. A term used in petroleum chemistry to denote certain saturated hydrocarbons, specifically five- and six-carbon cycloparaffins and their alkyl derivatives, found in crude petroleum. Sometimes used to include polycyclic members found in the higher-boiling fractions.

6301. 1-Naphthoic Acid. *1-Naphthalenecarboxylic acid;* α-naphthoic acid. $C_{11}H_8O_2$; mol wt 172.17. C 76.73%, H 4.68%, O 18.58%. Prepn from α-bromonaphthalene by Grignard reaction: Gilman *et al., Org. Syn.* **coll. vol. II**, 425

(1943); from 4-aminonaphthalenesulfonic acid by Sand-meyer reaction: Bassilios, *Bull. Soc. Chim. France* **1950**, 757; from β-nitronaphthalene by von Richter reaction: Bunnett, Rauhut, *J. Org. Chem.* **21**, 934, 944 (1956); by oxidation of 1-methylnaphthalene: Aries, U.S. pat. **2,930,802** (1960); Barker, U.S. pat. **2,963,508** (1960 to Mid-Century).

Crystals from hot toluene, mp 160.5-162°. bp 300°. uv max (ethanol): 293 nm (log ϵ_m 3.9). Slightly sol in hot wa-ter; freely sol in hot alcohol, ether.
Caution: Moderately irritating to skin, mucous mem-branes.

6302. 2-Naphthoic Acid. *2-Naphthalenecarboxylic acid;* β-naphthoic acid; isonaphthoic acid. $C_{11}H_8O_2$; mol wt 172.17. C 76.73%, H 4.68%, O 18.58%. Prepn from β-acet-ylnaphthalene + NaOCl: Newman, Holmes, *Org. Syn.* coll. vol. **II**, 428 (1943); from naphthalene + octylsodium: Mor-ton *et al., J. Am. Chem. Soc.* **64**, 2250 (1942); from 2-amino-naphthalenesulfonic acid by Sandmeyer reaction: Wahl, Bassilios, *Bull. Soc. Chim. France* **1947**, 482; by carboxyla-tion of naphthalene: Prichard, U.S. pat. **2,729,673** (1956 to du Pont).

Crystals from 95% alcohol, mp 184-185°. bp above 300°. uv max (ethanol): 235, 280, 335 nm (log ϵ_m 4.7, 3.8, 3.1). Slightly sol in hot water; sol in alcohol, ether.
Caution: Moderately irritating to skin, mucous mem-branes.

6303. 1-Naphthol. *1-Naphthalenol;* α-naphthol; alpha-naphthol; α-hydroxynaphthalene. $C_{10}H_8O$; mol wt 144.16. C 83.31%, H 5.59%, O 11.10%. Prepd by fusing the sodium salt of α-naphthalenesulfonic acid with NaOH: Tyrer, U.S. pats. **2,407,044, 2,407,055** and **2,451,996** (1946 and 1948); by oxidation of naphthalene: Loeb, U.S. pat. **3,033,903** (1962 to Union Carbide).

Prisms, mp 96°. Phenolic odor; disagreeable, burning taste. bp 288°; bp_{40} 184°. $d_4^{98.7}$ 1.0954. uv max: 297, 310, 324 nm. Sublimable; volatile with steam. Darkens in light; reduces ammoniacal silver nitrate. Slightly sol in water, freely in alcohol, benzene, chloroform, ether, alkali hydrox-ide solns. *Protect from light.* LD_{50} orally in rats: 2.59 g/kg, H. F. Smyth *et al., Am. Ind. Hyg. Assoc. J.* **23**, 95 (1962).
USE: Manuf dyes, intermediates, synthetic perfumes; also in microscopy. *Human Toxicity: See* 2-Naphthol.

6304. 2-Naphthol. *2-Naphthalenol;* β-naphthol; beta-naphthol; β-hydroxynaphthalene; isonaphthol; C.I. Azoic Coupling Component 1; C.I. Developer 5; C.I. 37500. C_{10}-H_8O; mol wt 144.16. C 83.31%, H 5.59%, O 11.10%. Prepn from sodium naphthalene-2-sulfonate: Schoeffel, Barton, U.S. pat. **2,760,992** (1956 to Sterling); Stevens, Harris, U.S. pat. **2,831,895** (1958 to Dow); Fr. pat. **1,326,175** (1963 to Ciba); by oxidation of naphthalene: Simons, U.S. pat. **2,530,369** (1950 to Phillips Petroleum); from 2-bromonaph-thalene + *tert*-butyl hydroperoxide: Lawesson, Yang, *J. Am. Chem. Soc.* **81**, 4230 (1959).

Crystals, mp 121-123°. bp 285-286°. d 1.22. Flash pt 161°. Slight phenolic odor. Darkens with age on exposure to light. Sublimes when heated, distillable *in vacuo*; volatile with vapors of alcohol or water; reduces ammoniacal silver nitrate. uv max (95% ethanol): 226, 265, 275, 286, 320, 331 nm (ε 91,194, 3911, 4559, 3301, 1861, 2163). One gram dis-solves in 1000 ml water, 80 ml boiling water, 0.8 ml alcohol, 17 ml chloroform, 1.3 ml ether; sol in glycerol, olive oil, solns of alkali hydroxides. *Protect from light. Incompat.:* Antipyrine, camphor, phenol, ferric salts, menthol, potas-sium permanganate and other oxidizing agents, urethane.
Sodium salt, $C_{10}H_7NaO$, *sodium* β-*naptholate, sodium naphthol, Microcidin.* Grayish-white powder; becomes reddish or brownish on exposure to light and air. Soluble in 3 parts water. *Keep well closed and protected from light.*
Human Toxicity: Ingestion of large quantities may cause nephritis, lens opacity, vomiting and diarrhea, abdominal pain, circulatory collapse, convulsions, hemolytic anemia, death. Fatal poisoning from external application has been reported. *Caution:* Local action may produce peeling of the skin which may be followed by persistent pigmentation, *cf. Clinical Toxicology of Commercial Products,* R. E. Gosselin *et al.,* Eds. (Williams & Wilkins, Baltimore, 4th ed., 1976) Section II, p 126.
USE: Manuf medicinal organics, dyes, perfumes; the larg-est single use is probably in making antioxidants for the synthetic rubber industry.
THERAP CAT: Formerly as anthelmintic (Nematodes).
THERAP CAT (VET): Has been used as antiseptic, anthelmin-tic and counter-irritant in alopecia.

6305. 1-Naphthol-8-amino-3,6-disulfonic Acid. *4-Ami-no-5-hydroxy-2,7-naphthalenedisulfonic acid;* 8-amino-1-naphthol-3,6-disulfonic acid; H acid. $C_{10}H_9NO_7S_2$; mol wt 319.31. C 37.61%, H 2.84%, N 4.39%, O 35.07%, S 20.08%. Prepn: Willard, *Color Trade J.* **15**, 40 (1924); Mow, U.S. pat. **2,272,272** (1942 to Allied Chem.); Hayashi, *Yamaguchi J. Sci.* **2**, 67 (1951), *C.A.* **49**, 2390f (1955); Roos *et al.,* U.S. pat. **2,875,243** (1959 to Bayer).

White crystals or gray powder. Slightly sol in water, alco-hol, ether; sol in alkalies.
USE: Prepn of azo dyes.

6306. 1-Naphthol-4,8-disulfonic Acid. *4-Hydroxy-1,5-naphthalenedisulfonic acid;* α-naphtholdisulfonic acid S; Schöllkopf's acid. $C_{10}H_8O_7S_2$; mol wt 304.30. C 39.47%, H 2.65%, O 36.81%, S 21.08%. Prepd by sulfonation of 1-naphthol-8-sulfonic acid sultone: I. G. Farbenindustrie, **PB 74197**, frames 828-829.

Crystals. Soluble in water. The sodium salt is very solu-ble in water.

6307. 2-Naphthol-3,6-disulfonic Acid. *3-Hydroxy-2,7-naphthalenedisulfonic acid;* R acid. $C_{10}H_8O_7S_2$; mol wt 304.30. C 39.47%, H 2.65%, O 36.81%, S 21.08%. Prepn from β-naphthol and concd H_2SO_4: Fierz-David, Blangey, *Fundamental Processes of Dye Chemistry* (Interscience, New York, 1949) p 197.

Deliquesc needles. Very sol in water, alcohol; practically insol in ether. The sodium salt is known as "R Salt."

USE: In dye chemistry.

6308. 2-Naphthol-6,8-disulfonic Acid. *7-Hydroxy-1,3-naphthalenedisulfonic acid;* 2-hydroxynaphthalene-6,8-disulfonic acid; G acid. $C_{10}H_8O_7S_2$; mol wt 304.30. C 39.47%, H 2.65%, O 36.81%, S 21.08%. Prepd by heating β-naphthol with concd H_2SO_4, 2-naphthol-3,6-disulfonic acid (R acid) being obtained as a byproduct: Fierz-David, Blangey, *Fundamental Processes of Dye Chemistry* (Interscience, New York, 1949) p 197; *see also* Forster, Keyworth, *J. Soc. Chem. Ind. (London)* **46**, 27T (1927).

Sodium salt, *G salt.* Platelets or prisms. Freely sol in water, sol in dil alcohol.

Barium salt octahydrate, $C_{10}H_6BaO_7S_2.8H_2O$, minute prisms. Freely sol in water. Practically insol in alcohol, even when much dil with water.

USE: In the manuf of azo dyes, *see* Crossley, Resenvelt, *Ind. Eng. Chem.* **16**, 271 (1924).

6309. α-Naphtholphthalein. *3,3-Bis(4-hydroxynaphthalenyl)-1(3H)-isobenzofuranone; 3,3-bis(4-hydroxy-1-naphthyl)phthalide;* di-p-α-naphtholphthalein; p-α-naphtholphthalein. $C_{28}H_{18}O_4$; mol wt 418.42. C 80.37%, H 4.34%, O 15.29%. Prepn: Werner, *J. Chem. Soc.* **113**, 20 (1918). Commercial prepn contains a small amount of the isomeric 3,3-bis(2-hydroxy-1-naphthyl)phthalide.

Colorless powder when pure; usually grayish-red. mp 253-255°. Practically insol in water; sol in alcohol.

USE: As indicator in 0.1% or 0.04% soln in alc. pH 7.3 colorless to reddish; 8.7 greenish to blue. Particularly adapted for weak acids in strong alcoholic soln.

6310. 1-Naphthol-2-sulfonic Acid. *1-Hydroxy-2-naphthalenesulfonic acid;* Baum's acid; Schaeffer's α-acid. $C_{10}H_8O_4S$; mol wt 224.23. C 53.56%, H 3.60%, O 28.54%, S 14.30%. Prepd by sulfonation of α-naphthol: Hodgson, Hathaway, *J. Soc. Dyers Colourists* **63**, 109 (1947).

Deliquesc crystals. Slightly sol in cold water; sol in boiling water; practically insol in ether.

6311. 1-Naphthol-4-sulfonic Acid. *4-Hydroxy-1-naphthalenesulfonic acid;* Nevile and Winther's acid. $C_{10}H_8O_4S$; mol wt 224.23. C 53.56%, H 3.60%, O 28.54%, S 14.30%. Prepn from 1,4-diazonaphthalenesulfonic acid + sulfuric acid: Erdmann, *Ann.* **247**, 341 (1888); by sulfonation of α-naphthylcarbonate: Reverdin, *Ber.* **27**, 3458 (1894); from

chlorosulfonic acid + α-naphthol: Gebauer-Fülnegg, *Monatsh.* **49**, 195 (1928); Baddiley *et al.*, U.S. pat. **1,452,481** (1923); from sodium naphthionate + SO_2: Binns, Lurie, U.S. pat. **1,880,701** (1933 to Virginia Smelting).

Crystals, dec 170°. K_{OH} 3 × 10⁻⁹. Very sol in water. Salt solns give a blue color with ferric chloride.

USE: Preparation of azo dyes.

6312. 2-Naphthol-6-sulfonic Acid. *6-Hydroxy-2-naphthalenesulfonic acid;* Schaeffer's β-acid. $C_{10}H_8O_4S$; mol wt 224.23. C 53.56%, H 3.60%, O 28.54%, S 14.30%. Prepn from β-naphthol and concd H_2SO_4: Fierz-David, Blangey, *Fundamental Processes of Dye Chemistry* (Interscience, New York, 1949) p 194.

Leaflets. mp 125°. Very sol in water, alcohol; practically insol in ether.

Sodium salt, $C_{10}H_7NaO_4S$, *Schaeffer's salt.* Light yellow to pink powder. Freely sol in water; slightly sol in alcohol.

USE: In dye chemistry.

6313. Naphthol Yellow S. *8-Hydroxy-5,7-dinitro-2-naphthalenesulfonic acid disodium salt;* Citronin A; Sulfur Yellow S; Acid Yellow S; FD & C Yellow no. 1; Ext. D & C Yellow no. 7; C.I. 10316; C.I. Acid Yellow 1. $C_{10}H_4N_2-Na_2O_8S$; mol wt 358.19. C 33.53%, H 1.12%, N 7.82%, Na 12.84%, O 35.74%, S 8.95%. Prepn: *Colour Index*, vol. **4** (3rd ed., 1971) p 4004.

Greenish yellow powder, sol in water to a yellow soln. The free acid is known as flavianic acid, *q.v.*

USE: Dye for wool, silk. *Caution:* Approved by FDA for external use only.

6314. 1,2-Naphthoquinone. *1,2-Naphthalenedione;* β-naphthoquinone. $C_{10}H_6O_2$; mol wt 158.15. C 75.94%, H 3.82%, O 20.23%. Prepd by oxidation of 1-amino-2-naphthol with ferric chloride: Fieser, *Org. Syn.* **17**, 68 (1937); by oxidation of naphthalene: Milas, U.S. pat. **2,395,638** (1946 to Research Corp.); by bacterial dissimilation of naphthalene: Murphy, Stone, *Can. J. Microbiol.* **1**, 579 (1955); from tetrachloro-o-benzoquinone: Horner, Dürckheimer, *Z. Naturforsch.* **14b**, 741 (1959). Structure: Hodgson, Hathaway, *Trans. Faraday Soc.* **41**, 115 (1945). Spectroscopic properties: Oliver *et al.*, *Tetrahedron* **24**, 4067 (1968).

Golden yellow needles, dec 145-147°. uv max (abs alc): 250, 340, 405 nm (log ε 4.35, 3.40, 3.40). Sol in alc, benzene, ether, 5% NaOH, 5% $NaHCO_3$, concd H_2SO_4 with green color. Practically insol in water.

2-Semicarbazone, $C_{11}H_9N_3O_2$, naftazone, Haemostop Injection, Mediaven, Karbinon.

USE: As reagent for resorcinol and thalline.

THERAP CAT: 2-Semicarbazone as hemostatic.

6315. 1,4-Naphthoquinone. *1,4-Naphthalenedione;* α-naphthoquinone; 1,4-dihydro-1,4-diketonaphthalene. $C_{10}H_6O_2$; mol wt 158.15. C 75.94%, H 3.82%, O 20.23%. Prepn: Fieser, *Org. Syn.* **coll. vol. I**, 383 (2nd ed., 1941); Braude, Fawcett, *ibid.*, **coll. vol. IV**, 698 (1963). Substituted 1,4-naphthoquinones occur in nature, *e.g.*, phthiocol, *q.v.*, various pigments, and the vitamins K.

Yellow triclinic needles from alcohol or petr ether. Odor like that of benzoquinone. d 1.422. mp 126°. Begins to sublime below 100°. Sublimation rates: Kempf, *J. Prakt. Chem.* [2] **78**, 236, 257. Easily volatile with steam. uv spectrum: Purvis, *J. Chem. Soc.* **101**, 1318 (1912). Sparingly sol in cold water, slightly in petr ether, freely in hot alcohol, ether, benzene, chloroform, carbon bisulfide, acetic acid; sol in alkali hydroxide solns giving a reddish-brown soln.

6316. Naphthoresorcinol. *1,3-Naphthalenediol;* 1,3-dihydroxynaphthalene. $C_{10}H_8O_2$; mol wt 160.16. C 74.99%, H 5.03%, O 19.98%. Prepn by cyclization of ethyl phenylacetoacetate: Soliman, West, *J. Chem. Soc.* **1944**, 53; by heating 1-amino-3-hydroxy-4-naphthalenesulfonic acid in acidic soln: Meyer, Bloch, *Org. Syn.* **25**, 73 (1945); from 1,3-naphtholsulfonic acid: Kozlov, Odintsov, *J. Appl. Chem. (USSR)* **17**, 219 (1944); Kozlov et al., *Zh. Prikl. Khim.* **35**, 880 (1962).

Leaflets, mp 124-125°. Freely sol in water, alcohol, ether.

USE: Reagent for sugars, oils, and for glucuronic acid in urine: Forsyth, *Nature* **161**, 239 (1948); Heyns, Kelch, *Z. Anal. Chem.* **139**, 339 (1953).

6317. 2-Naphthoxyacetic Acid. *2-Naphthalenyloxyacetic acid;* β-naphthoxyacetic acid; O-(2-naphthyl)glycolic acid. $C_{12}H_{10}O_3$; mol wt 202.20. C 71.28%, H 4.99%, O 23.74%. Prepd by treating β-naphthol with chloroacetic acid in an alkaline medium: Spitzer, *Ber.* **34**, 3192 (1901); Shirley, *Organic Intermediates* (New York, 1951) p 209.

Prisms from hot water or benzene, mp 156°. Moderately sol in hot water. Soluble in alcohol, ether, acetic acid.

Ethyl ester, $C_{14}H_{14}O_3$, leaflets from alcohol, mp 49°. Sol in alcohol, ether.

USE: As plant hormone, to promote growth of roots on clippings, to prevent fruit from falling prematurely; causes stunted growth when used in excess.

6318. 1-Naphthylamine. *1-Naphthalenamine;* 1-aminonaphthalene; α-naphthylamine; naphthalidine. $C_{10}H_9N$; mol wt 143.18. C 83.88%, H 6.34%, N 9.78%. Prepd by reducing α-nitronaphthalene with Fe and HCl: West, *J. Chem. Soc.* **127**, 494 (1925); from 1-naphthalenecarboxylic acid and hydroxylamine: Snyder et al., *J. Am. Chem. Soc.* **75**, 2014 (1953). Review and evaluation of studies of carcinogenicity in laboratory animals and in humans: *IARC Monographs* **4**, 87-96 (1974).

Needles, becoming red on exposure to air, or a reddish, cryst mass; unpleasant odor. Sublimes; volatile with steam. d 1.13. mp 50°. bp 301°. Flash pt 157°. Sol in 590 parts water; freely sol in alc, ether. Reduces warm ammoniacal silver nitrate. *Keep well closed and protected from light.*

Hydrochloride, $C_{10}H_9N.HCl$, cryst powder; becomes bluish on exposure to air and light. Sol in about 27 parts water; sol in alcohol, ether. *Protect from light.*

Sulfate dihydrate, $(C_{10}H_9N)_2.H_2SO_4.2H_2O$, white to yellowish cryst powder. Slightly sol in water or alcohol.

Caution: Harmful dust and vapor. This substance has been listed as a carcinogen by OSHA: *Fed. Reg.* **39**, 3757 (1974).

USE: Manuf dyes; toning prints made with cerium salts; the hydrochloride with sulfanilic acid is a reagent for nitrate.

6319. 2-Naphthylamine. *2-Naphthalenamine;* 2-aminonaphthalene; β-naphthylamine. $C_{10}H_9N$; mol wt 143.18. C 83.88%, H 6.34%, N 9.78%. Prepd by heating β-naphthol with ammonium sulfite and NH_4OH at 150°: Ger. pat. 117,471 (1900); *Frdl.* **6**, 190; from 2-naphthalenecarboxylic acid and hydroxylamine: Snyder et al., *J. Am. Chem. Soc.* **75**, 2014 (1953). Review and evaluation of studies of carcinogenicity in laboratory animals and in humans: *IARC Monographs* **4**, 97-111 (1974).

White to reddish crystals; volatile with steam. d_4^{98} 1.061. mp 111-113°. bp 306°; also stated as 294°. Sol in hot water, alcohol, ether. Reduces warm ammoniacal silver nitrate.

Acetate, $C_{10}H_9N.CH_3COOH$, white to yellowish scales or flakes; slight odor of acetic acid. Very sol in water or alcohol; sol in ether. *Keep well closed and protected from light.*

Note: This substance has been listed as a known carcinogen: *Fourth Annual Report on Carcinogens* (NTP 85-002, 1985) p 137.

USE: Manuf dyes.

6320. 1-Naphthylamine-2,7-disulfonic Acid. *1-Amino-2,7-naphthalenedisulfonic acid;* Kalle's acid. $C_{10}H_9NO_6S_2$; mol wt 303.32. C 39.60%, H 2.99%, N 4.62%, O 31.65%, S 21.14%. Prepn from 1-naphthylamine-2,4,7-trisulfonic acid: Ger. pat. 62,634 (1892 to Kalle); *Frdl.* **3**, 431; from 1-naphthylamine-2,5,7-trisulfonic acid: Ger. pat. 255,724 (1912 to Bayer); *Frdl.* **11**, 217.

Needles.

Sodium salt, prisms from dil alc. Solns display a blue-green fluorescence.

Barium salt, needles. Very slightly sol in hot water.

6321. 1-Naphthylamine-4,6-disulfonic Acid. *4-Amino-1,7-naphthalenedisulfonic acid;* Dahl's acid II. $C_{10}H_9NO_6S_2$; mol wt 303.31. C 39.60%, H 2.99%, N 4.62%, O 31.65%, S 21.14%. Prepn by sulfonation of α-naphthylamine and separation from mixture of sulfonic acids obtained: Erdmann, *Ann.* **275**, 192 (1893).

Crystals. Sol in 6 parts cold water; very sol in hot water; sol in alcohol.

6322. 1-Naphthylamine-4,7-disulfonic Acid. *4-Amino-1,6-naphthalenedisulfonic acid;* Dahl's acid III. Empirical formula and prepn, *see* 1-Naphthylamine-4,6-disulfonic Acid.

Crystals. Sol in 145 parts cold water, in 20 parts boiling water; practically insol in alcohol.

6323. 1-Naphthylamine-4-sulfonic Acid. *4-Amino-1-naphthalenesulfonic acid;* naphthionic acid; Piria's acid. $C_{10}H_9NO_3S$; mol wt 223.26. C 53.80%, H 4.06%, N 6.27%, O 21.50%, S 14.36%. Prepd on a large-scale by heating α-naphthylamine with equimolar amounts of concd H_2SO_4 at 180-200° or by heating α-naphthylamine acid sulfate: Fierz-David, Blangey, *Fundamental Processes of Dye Chemistry* (Interscience, New York, 1949).

Sesquihydrate, shiny needles from water. d_4^{25} 1.673. Dec on heating without melting (the amide mp 206°). K at 25° = 2.1 × 10⁻³. One gram dissolves in 3.45 l of water at 10°, in 3.22 l at 20°, in 1.69 l at 50°, in 438.5 ml at 100°. Very sparingly sol in alcohol, ether. Practically insol in acetic acid and acetic anhydride, but becomes sol when some pyridine is added. In dil solns of alkali hydroxides or carbonates with a blue fluorescence.

Sodium salt tetrahydrate, $C_{10}H_8NNaO_3S.4H_2O$, *101-E, naphthionine.* Ref: Piria, *Ann.* **78,** 41 (1851). Large monoclinic prisms from water, loses 3½ mols water at 80°, anhydr at 130°. Sweet aftertaste. Freely sol in water with a blue fluorescence. Also sol in 95% alc. Practically insol in ether. Sparingly sol in concd aq and alcoholic caustic solns. pH of a 1% aq soln 6.8. Even ampuled aq solns discolor under the influence of light.

Sodium salt glucoside, $C_{16}H_{20}NNaO_8S$, *101-G, naphthionine N-glucoside.* Ref: Estève *et al., Ann. Pharm. Franc.* **10,** 680 (1952). White powder, bitter taste. Dec 255°. $[\alpha]_D^{23.5}$ −113° (c = 10). Sol in water; slightly sol in alcohol. Practically insol in ether.

USE: The sodium salt is an important dye intermediate in the manuf of Congo Red, Fast Red A, Azo Rubine, and similar azo dyes.

THERAP CAT: Sodium salt as hemostatic.

6324. 1-Naphthylamine-5-sulfonic Acid. *5-Amino-1-naphthalenesulfonic acid;* Laurent's acid. $C_{10}H_9NO_3S$; mol wt 223.26. C 53.80%, H 4.06%, N 6.27%, O 21.50%, S 14.36%. Prepn from 5-chloro-1-naphthalenesulfonic acid + ammonia: Oehler, **Ger.** pat. **72,336** (1893); *Frdl.* **3,** 514; from α-naphthylamine + concd sulfuric acid: Erdmann, *Ann.* **275,** 193, 200 (1893); Sixma, *Rec. Trav. Chim.* **73,** 235 (1954); from 2-naphthylamine-8-sulfonic acid: Blangey, *Helv. Chim. Acta* **39,** 977 (1956).

White crystals. Sol in 950 parts cold water; more sol in hot water.

USE: In dye chemistry.

6325. 1-Naphthylamine-8-sulfonic Acid. *8-Amino-1-naphthalenesulfonic acid;* Peri acid. $C_{10}H_9NO_3S$; mol wt 223.25. C 53.80%, H 4.06%, N 6.27%, O 21.50%, S 14.36%. Prepn: Fierz, Weissenbach, *Helv. Chim. Acta* **3,** 310 (1920); Martin, *Z. Farben-Ind.* **20,** 9, 76 (1928); *C.A.* **22,** 2665 (1928); Shebuev *et al.,* USSR pat. **165,746** (1964 to State Scientific Res. Inst. of Org. Intermediates & Dyes), *C.A.* **62,** 10392c (1965).

Needles. Sol in 4800 parts cold, 240 parts boiling water; freely sol in glacial acetic acid.

Caution: Irritating to skin.

6326. 2-Naphthylamine-1-sulfonic Acid. *2-Amino-1-naphthalenesulfonic acid;* Tobias acid. $C_{10}H_9NO_3S$; mol wt 223.25. C 53.80%, H 4.06%, N 6.27%, O 21.50%, S 14.36%. Prepd by sulfonation of 2-naphthylamine: Tinker, Hansen, U.S. pat. **1,969,189** (1934 to du Pont). Utilization in viral biology study: N. Sakota *et al., J. Ferment. Technol.* **56,** 53 (1978). TLC study: J. Franc, V. Koudelkova, *J. Chromatog.* **170,** 89 (1979).

Anhyr scales from hot water; hydrated needles from cold water. Slightly sol in cold water; more sol in hot water; very slightly sol in alcohol, ether.

Diethylammonium salt, $C_{10}H_9NO_3S.(C_2H_5)_2NH$, mp 180°.

USE: Dyestuff intermediate.

6327. 2-Naphthylamine-5-sulfonic Acid. *6-Amino-1-naphthalenesulfonic acid;* Dahl's acid. $C_{10}H_9NO_3S$; mol wt 223.25. C 53.80%, H 4.06%, N 6.27%, O 21.50%, S 14.36%. Prepn by reduction of 6-nitro-1-naphthalenesulfonic acid: Kappeler, *Ber.* **45,** 633 (1912); by sulfonation of β-naphthylamine and separation from mixt of sulfonic acids obtained: Green, Vakil, *J. Chem. Soc.* **113,** 35 (1918).

Needles from water. Sol in 3025 parts water at 20°; almost insol in alcohol.

USE: Dyestuff intermediate.

6328. 2-Naphthyl Benzoate. *2-Naphthalenol benzoate;* betanaphthol benzoate; benzonaphthol; benzoylnaphthol; Lintrin; Haertolan. $C_{17}H_{12}O_2$; mol wt 248.27. C 82.24%, H 4.87%, O 12.89%. $C_6H_5COOC_{10}H_7$.

White, crystalline powder; darkens with age. mp 107-110°. Almost insol in water. Freely sol in hot alcohol, chloroform, glycerol, oils; slightly sol in ether.

USE: Hardening agent for paraffin.

THERAP CAT: Intestinal antiseptic.

6329. N-(1-Naphthyl)ethylenediamine. *N-1-Naphthalenyl-1,2-ethanediamine;* 1-amino-2-(α-naphthylamino)-ethane. $C_{12}H_{14}N_2$; mol wt 186.25. C 77.38%, H 7.58%, N 15.04%. Prepn from α-naphthylamine and bromoethylphthalimide: Bratton, Marshall, *J. Biol. Chem.* **128,** 537 (1939).

Straw-yellow, viscous liquid; bp_9 204°; bp_{760} ~320° (dec); n_D^{25} 1.6648; d_4^{25} 1.114. The soly in water is about 0.2 g in 100 ml at 25°; more sol in hot water. pH of satd aq soln: 10.5. Readily sol in common organic solvents except petr ether.

Dihydrochloride, $C_{12}H_{14}N_2.2HCl$, long hexagonal prisms; mp 188-190°. Readily sol in 95% alcohol, dil hydrochloric acid, hot water; slightly sol in cold water, acetone, abs alc.

USE: Dihydrochloride in determination of sulfanilamide in body fluids; also in determination of potassium, nitrites, and sulfates.

6330. 1-Naphthylisocyanate. *1-Isocyanatonaphthalene.* $C_{11}H_7NO$; mol wt 169.17. C 78.09%, H 4.17%, N 8.28%, O 9.46%. $C_{10}H_7N=CO$.

Colorless liquid; almost odorless at ordinary temp; vapors have a pungent odor characteristic of isocyanates (*e.g.,* phenylisocyanate). d 1.181. bp 270°. Sol in chloroform, ether, petr ether, alcohol. *Keep well closed and protected from light.*

6331. 1-Naphthylisothiocyanate. *1-Isothiocyanatonaphthalene;* ANIT. $C_{11}H_7NS$; mol wt 185.24. C 71.32%, H 3.81%, N 7.56%, S 17.31%. $C_{10}H_7N=C=S$. Prepn: Cymerman-Craig *et al., Org. Syn. coll. vol.* IV, 700 (1963); Jochims, *Ber.* **101**, 1746 (1968). Has been used with pyrethrum as insecticide. Toxicity studies: Schwartz, Warren, *U.S. Pub. Health Repts.* **54**, 1426 (1939); Ambrose, Miller, *Fed. Proc.* **2**, 74 (1943); Capizzo, Roberts, *Toxicol. Appl. Pharmacol.* **17**, 262 (1970).

White needles; practically odorless and tasteless. mp 58°. Insol in water; freely sol in ether, benzene, hot alcohol, acetone, carbon tetrachloride, olive oil, petr ether. LD_{50} orally in mice: 245 mg/kg, Becker, Plaa, *Toxicol. Appl. Pharmacol.* **7**, 804 (1965).

Caution: Hepatotoxic; may cause dermatitis.

6332. 2-Naphthyl Lactate. *2-Hydroxypropanoic acid 2-naphthalenyl ester;* lactic acid β-naphthyl ester; Lactol; Lactonaphthol. $C_{13}H_{12}O_3$; mol wt 216.23. C 72.21%, H 5.59%, O 22.20%. $CH_3CHOHCOOC_{10}H_7$. Prepd by heating equimolar amounts of β-naphthol sodium and sodium lactate to 125-130° in the presence of phosphorus oxychloride: P. Lebeau, *Traité de Pharmacie Chimique* vol. II (Masson, Paris, 1955/56) p 964.

Crystals from alcohol. Moderately sol in alcohol. Insol in water, ether.

THERAP CAT: Intestinal antiseptic.

6333. 2-(2-Naphthyloxy)ethanol. *2-(2-Naphthalenyloxy)ethanol;* β-naphthoxyethanol; betanaphthoxyethanol; β-hydroxyethyl 2-naphthyl ether; 2-(β-hydroxyethoxy)-naphthalene; ethylene glycol mono-2-naphthyl ether; Anavenol. $C_{12}H_{12}O_2$; mol wt 188.22. C 76.57%, H 6.43%, O 17.00%. Prepd by the condensation of β-naphthol with ethylene chlorohydrin: Rindfusz *et al., J. Am. Chem. Soc.* **42**, 157, 164, 165 (1920); Kirner, Richter, *ibid.* **51**, 3409 (1929); by heating β-naphthol with ethylene oxide and sodium ethoxide in alcohol: Boyd, Marle, *J. Chem. Soc.* **105**, 2117 (1914).

Crystals from benzene + petr ether, mp 76.7°. Insoluble in water. One gram dissolves in 4 g of 95% alcohol, in 2 g acetone; also sol in ether, chloroform. uv max (0.004% in chloroform): 273, 328 nm (E ~1.00, ~0.395).

THERAP CAT (VET): As an adjunct to anesthesia, sedative.

6334. 1-Naphthyl Salicylate. *2-Hydroxybenzoic acid 1-naphthalenyl ester;* α-naphthyl salicylate; α-naphthol salicylate; Alphol. $C_{17}H_{12}O_3$; mol wt 264.27. C 77.26%, H 4.58%, O 18.16%.

Crystalline powder. mp 83°. Insol in water. Freely sol in alcohol, ether, oils.

THERAP CAT: Anti-infective; anti-inflammatory.

6335. 2-Naphthyl Salicylate. *2-Hydroxybenzoic acid 2-naphthalenyl ester;* β-naphthol salicylate; Betol; Naphthalol; Naphthosalol; Salinaphthol. $C_{17}H_{12}O_3$; mol wt 264.27. C 77.26%, H 4.58%, O 18.16%.

White, odorless, tasteless, crystalline powder. mp 95°. Insol in water or glycerol. Sparingly sol in cold alcohol; freely sol in boiling alcohol, in benzene, ether.

THERAP CAT: Antiseptic.

6336. Napropamide. *N,N-Diethyl-2-(1-naphthalenyloxy)propanamide; N,N-diethyl-2-(1-naphthyloxy)propionamide;* 2-(α-naphthoxy)-*N,N*-diethylpropionamide; R 7465; Devrinol. $C_{17}H_{21}NO_2$; mol wt 271.37. C 75.24%, H 7.80%, N 5.16%, O 11.79%. Prepn: H. Filles *et al., U.S.* pat. **3,480,671** (1969 to Stauffer). Characteristics in the soil: C. H. Wu *et al., Weed Sci.* **23**, 54 (1975).

Light brown solid from *n*-pentane, mp 63-64° (tech, 69.5°). Solubility in water at 20°: 70 ppm. LD_{50} orally in mice: > 5 g/kg, T. Kawada *et al., C.A.* **80**, 56257h (1974).

USE: Herbicide.

6337. Naproxen. *(S)-6-Methoxy-α-methyl-2-naphthaleneacetic acid; d-2-(6-methoxy-2-naphthyl)propionic acid;* MNPA; RS-3540; Axer; Bonyl; Calosen; Diocodal; Dysmenalgit N; Equiproxen; Floginax; Laraflex; Laser; Naixan; Napren E; Naprium; Naprius; Naprosine; Naprosyn; Naprosyne; Naprux; Naxen; Prexan; Primeral; Proxen; Reuxen; Veradol; Xenar. $C_{14}H_{14}O_3$; mol wt 230.26. C 73.03%, H 6.13%, O 20.84%. Prepn of naproxen and sodium salt: J. H. Fried, I. T. Harrison, **S. Afr.** pat. **67 07,597;** *eidem,* **U.S.** pat. **3,904,682;** *eidem,* U.S. pat. **4,009,197** (1968, 1975, 1977 all to Syntex); I. T. Harrison *et al., J. Med. Chem.* **13**, 203 (1970). Pharmacology: Roszkowski *et al., J. Pharmacol. Exp. Ther.* **179**, 114 (1971). Activity may be due to the ability to inhibit prostaglandin biosynthesis. Mode of action studies: Tomlinson *et al., Biochem. Biophys. Res. Commun.* **46**, 552 (1972). Metabolism: Runkel *et al., J. Pharm. Sci.* **61**, 703 (1972). Clinical studies: Katona *et al., Clin. Trials J.* **8**, 3 (1972); Runkel, *Chem. Pharm. Bull.* **20**, 1457 (1972). *Review: Arzneimittel-Forsch.* **25**, 278-332 (1975). Review of pharmacology and therapeutic efficacy: R. N. Brogden *et al., Drugs* **18**, 241-277 (1979).

Crystals from acetone-hexane, mp 152-154°. $[\alpha]_D$ +66° (in chloroform). LD_{50} in mice (mg/kg): 435 i.v.; 1234 orally; in rats (mg/kg): 575 i.p.; 534 orally (Roszkowski).

Piperazine salt, $C_{20}H_{28}N_2O_4$, piproxen; Numidan.

Sodium salt, $C_{14}H_{13}NaO_3$, RS-3650, Anaprox, Apranax, Flanax, Miranax, Synflex. Crystals from acetone, mp 244-246°. $[\alpha]_D$ −11° (in methanol).

THERAP CAT: Anti-inflammatory; analgesic; antipyretic.

THERAP CAT (VET): Anti-inflammatory.

6338. Naptalam. *2-[(1-Naphthalenylamino)carbonyl]benzoic acid; N-1-naphthylphthalamic acid;* α-naphthylphthalamic acid. $C_{18}H_{13}NO_3$; mol wt 291.31. C 74.21%, H 4.50%, N 4.81% O 16.48%. Selective pre-emergence herbicide. Prepn: A. E. Smith, O. L. Hoffmann, U.S. pat. **2,556,665** (1951 to U.S. Rubber). Activity: O. L. Hoffmann,

A. E. Smith, *Science* **109**, 588 (1949). Mobility in soil: A. E. Smith *et al.*, *J. Agr. Food Chem.* **5**, 748 (1957).

Crystals from ethanol, d_4^{20} 1.40. mp 203° (technical grade, mp 175-180°). Soly: <0.02g/100 ml of water. Sol in alkaline solns, but dec above pH 9.5. Slightly sol in ethanol, acetone, benzene. Hydrolyzed by strong acids and bases.

Sodium salt, $C_{18}H_{12}NNaO_3$, *ACP 332, NPA-3, Alanap*.

USE: Herbicide; as analytical reagent for thorium and zirconium.

6339. Narasin. *(4S)-4-Methylsalinomycin;* narasin A; Compound 79891; Antibiotic A-28086 factor A; C-7819B; Monteban. $C_{43}H_{72}O_{11}$; mol wt 765.05. C 67.51%, H 9.49%, O 23.00%. Main component of a polyether antibiotic complex produced by *Streptomyces aureofaciens* NRRL 5758 & NRRL 8092. Production: D. H. Berg *et al.*, **Ger. pat. 2,-525,095** corresp to U.S. pat. **4,038,384** (1975, 1977 to Lilly); L. D. Boeck *et al.*, *Dev. Ind. Microbiol.* **18**, 471 (1976). Isoln and characterization: D. H. Berg, R. L. Hamill, *J. Antibiot.* **31**, 1 (1978). Biosynthetic studies using ^{13}C-NMR study: D. E. Dorman *et al.*, *Helv. Chim. Acta* **59**, 2625 (1976). Structure: J. L. Occolowitz *et al.*, *Biomed. Mass. Spectrom.* **3**, 272 (1976); H. Seto *et al.*, *J. Antibiot.* **30**, 530 (1977). Anticoccidial activity: M. D. Ruff *et al.*, *Poultry Sci.* **59**, 2008 (1980). Total synthesis: Y. Kishi *et al.*, *Front. Chem., Plenary Keynote Lect. IUPAC Congr., 28th* **1981**, K. J. Laidler, Ed. (Pergamon, Oxford, 1982) pp 287-304. HPLC determn in animal feeds: M. R. LaPointe, H. Cohen, *J. Assoc. Offic. Anal. Chem.* **71**, 480 (1988).

Crystals from acetone-water, mp 98-100°; resolidif and remelts at 198-200°. uv max (ethanol): 285 nm (ε 58). $[\alpha]_D^{25}$ −54° (c = 0.2 in methanol). pKa 7.9 (80% aq DMF). Sol in alcohols, acetone, DMF, DMSO, benzene, chloroform, ethyl acetate. Insol in water. LD_{50} i.p. in mice: 7.15 mg/kg (Berg).

THERAP CAT (VET): Coccidiostat; growth stimulant.

6340. Narbomycin. *12-Deoxypicromycin.* $C_{28}H_{47}NO_7$; mol wt 509.66. C 65.98%, H 9.30%, N 2.75%, O 21.97%. Antibiotic substance produced by *Streptomyces narbonensis* from soil near Cannes, France: Corbaz *et al.*, *Helv. Chim. Acta* **38**, 935 (1955). Structure: Prelog *et al.*, *ibid.* **45**, 4 (1962). Stereochemical studies: Rickards, Smith, *Tetrahedron Letters* **1970**, 1025; H. Ogura *et al.*, *J. Am. Chem. Soc.* **97**, 1930 (1975); *eidem, Tetrahedron* **37**, Suppl. 1, 165 (1981). Isoln and structure of the aglycone *narbonolide:* Hori *et al.*, *Chem. Commun.* **1971**, 304. Synthesis of narbonolide: T. Kaiho *et al.*, *J. Org. Chem.* **47**, 1612 (1982).

Crystals from ether + petr ether, mp 113.5-115°. $[\alpha]_D^{20}$ +68.5° (c = 1.35 in chloroform). uv max (abs ethanol): 225, 286 nm (log ε 4.06, 2.23). LD_{50} s.c. in mice: 500 mg/kg, Corbaz *et al., loc. cit.*

6341. Narceine. *6-[[6-[2-(Dimethylamino)ethyl]-4-methoxy-1,3-benzodioxol-5-yl]acetyl]-2,3-dimethoxybenzoic acid; 6-[[6-[2-(dimethylamino)ethyl]-2-methoxy-3,4-(methylenedioxy)phenyl]acetyl]-o-veratric acid.* $C_{23}H_{27}NO_8$; mol wt 445.45. C 62.01%, H 6.11%, N 3.14%, O 28.73%. Occurs in opium to the extent of 0.1-0.5%. The separation from morphine mother liquors is tedious: Merck, *Chem. Ztg.* **13**, 525 (1889). Prepn from narcotine or gnoscopine: Roser, *Ann.* **247**, 167 (1888); Frankforter, Keller, *Ann. Chem. J.* **22**, 61 (1899); Frerichs, *Arch. Pharm.* **241**, 259 (1903); Hope, Robinson, *J. Chem. Soc.* **105**, 2100 (1914). Structure: Freund, Frankforter, *Ann.* **277**, 20 (1893); Freund, Michaels, *Ann.* **286**, 248 (1895); Freund, *Ber.* **40**, 194 (1907); Freund, Oppenheim, *Ber.* **42**, 1084 (1909); Addinall, Major, *J. Am. Chem. Soc.* **55**, 1202, 2153 (1933).

The anhydr material is very hygroscopic, mp 138°. uv max (ethanol): 270 nm (log ε 3.98). Usually the alkaloid is obtained as the trihydrate. Clusters of silky, prismatic needles from water, mp 176°. pKb at 20° = 10.7, Kb = 2 × 10^{-11}; pKa = 9.3, Ka = 5 × 10^{-10}. pH of satd soln = 5.8. One gram dissolves in 770 ml water, 220 ml boiling water. Moderately sol in hot alcohol, nearly insol in benzene, chloroform, ether, petr ether; sol in alkali hydroxide solns forming salts, also in dil mineral acids.

Ethylnarceine hydrochloride, $C_{25}H_{32}ClNO_8$, *Narcyl.* Plates from water, mp 208-210°. Slightly sol in cold water, sol in hot water, alcohol, chloroform; insol in ether.

THERAP CAT: Ethylnarceine hydrochloride as narcotic analgesic; antitussive.

6342. Narcobarbital. *5-(2-Bromo-2-propenyl)-1-methyl-5-(1-methylethyl)-2,4,6(1H,3H,5H)-pyrimidinetrione; 5-(2-bromoallyl)-5-isopropyl-1-methylbarbituric acid; N-methyl-β-bromoallylisopropylbarbituric acid; 5-isopropyl-5-(β-bromoallyl)-N-methylbarbituric acid; enibomal.* $C_{11}H_{15}BrN_2O_3$; mol wt 303.16. C 43.58%, H 4.99%, Br 26.36%, N 9.24%, O 15.83%. Prepd by heating 5-isopropyl-1-methylbarbituric acid with 2,3-dibromo-1-propene in alkaline soln: Boedecker, Gruber, **Ger. pat. 613,403** (1937 to Riedel de Haen). *See also* **Ger. pat. 627,380**; U.S. pat. **2,080,071**.

Crystals from dilute ethanol, mp 115°. Bitter taste. Very sparingly sol in water. Sol in methanol, ethanol, pyridine, alkaline aq solns.

Sodium salt, *Eunarcon, Narcotal,* alkaline, water-soluble.

Caution: May be habit forming. This is a controlled substance (depressant) listed in the U.S. Code of Federal Regulations, Title 21 Parts 329.1 and 1308.13 (1987).

THERAP CAT: Sedative, hypnotic; anticonvulsant.

6343. Narcotoline. *6,7-Dimethoxy-3-(5,6,7,8-tetrahydro-4-hydroxy-6-methyl-1,3-dioxolo[4,5-g]isoquinolin-5-yl)-1(3H)-isobenzofuranone;* desmethylnarcotine. $C_{21}H_{21}NO_7$; mol wt 399.39. C 63.15%, H 5.30%, N 3.51%, O 28.04%.

Found in the shell of ripe poppyseed capsules: Wrede, *Arch. Pharmacol.* **184**, 331 (1937); Baumgarten, Christ, *Pharmazie* **5**, 80 (1950); Pfeifer, Weiss, *ibid.* **10**, 658 (1955); from opium prepns: Pfeifer, *Arch. Pharm.* **290**, 209 (1957). Preliminary stereochemical studies: Battersby, Spenser, *J. Chem. Soc.* **1965**, 1087. Revised stereochemistry: Blaha *et al., Coll. Czech. Chem. Commun.* **29**, 2328 (1964); Snatzke *et al., Tetrahedron* **25**, 5059 (1969).

Rectangular rods from dil methanol, mp 202°. $[\alpha]_D^{20}$ −189° (0.1 g in 25 ml chloroform, 20 cm tube). $[\alpha]_D^{20}$ +5.8° (0.065 g in 5 ml 0.1N HCl in 20 cm tube). Very sparingly sol in water. Moderately sol in warm alc and ether. Freely sol in chloroform, in dil acids and in dil aq solns of KOH and NaOH. Sparingly sol in dil aq solns of Na_2CO_3.

6344. Naringenin. *2,3-Dihydro-5,7-dihydroxy-2-(4-hydroxyphenyl)-4H-1-benzopyran-4-one; 4′,5,7-trihydroxyflavanone;* naringetol; salipurpol; pelargidanon 1602. C_{15}-$H_{12}O_5$; mol wt 272.25. C 66.17%, H 4.44%, O 29.38%. The aglucon of naringin. Prepn by hydrolysis of naringin: Asahina, Inubuse, *Ber.* **61**, 1514 (1928); Haley, Bassin, *J. Am. Pharm. Assoc.* **40**, 111 (1951). From kino of *Eucalyptus maculata* Hook, *Myrtaceae:* Gell *et al., Aust. J. Chem.* **11**, 372 (1958). Synthesis: Rosenmund, Rosenmund, *Ber.* **61**, 2608 (1928); Zemplén, Bognár, *Ber.* **75**, 648 (1942). Absolute configuration: Gaffield, Waiss, *Chem. Commun.* **1968**, 29.

Needles from dilute alc, mp 251°. uv max: 226, 292 nm. Sol in alcohol, ether, benzene; almost insol in water.

Triacetylnaringenin, $C_{15}H_9(CH_3CO)_3O_5$, crystals, mp 55°.

6345. Naringin. *7-[[2-O-(6-Deoxy-α-L-mannopyranosyl)-β-D-glucopyranosyl]oxy]-2,3-dihydro-5-hydroxy-2-(4-hydroxyphenyl)-4H-1-benzopyran-4-one; 4′,5,7-trihydroxyflavanone 7-rhamnoglucoside;* naringenin-7-rhamnoglucoside; aurantiin. $C_{27}H_{32}O_{14}$; mol wt 580.53. C 55.86%, H 5.56%, O 38.58%. In the flowers of grapefruit trees (*Citrus paradisi* Macfad, *Rutaceae*), also in fruit and rind. Most abundant in immature fruit; main bitter component of grapefruit juice. Extraction from grapefruit peel: Zoller, *Ind. Eng. Chem.* **10**, 364 (1918); Pulley, von Loesecke, *J. Am. Chem. Soc.* **61**, 175 (1939); U.S. pats. **2,421,062/3** (1947 to California Fruit Growers Exchange). Structure: Asahina, Inubuse, *C.A.* **23**, 3475 (1929); *cf.* also *Ber.* **61**, 1514 (1928); Horowitz, Gentili, *Tetrahedron* **19**, 773 (1963). Solubility data: G. N. Pulley, *Ind. Eng. Chem., Anal. Ed.* **8**, 360 (1936). *Review:* Kesterson, Hendrickson, *Naringin, A Bitter Principle of Grapefruit*, Univ. Florida Agric. Expt. Station, Bulletin no. 511 (Jan. 1953).

When crystallized from water, it contains 6 to 8 mols H_2O. mp ~83°. After drying at 110° to constant weight, it contains 2 mols H_2O, mp 171°. Bitter taste (1:10,000 H_2O can be tasted). $[\alpha]_D^{19}$ −82° (alcohol). One gram dissolves in 1000 ml water at 40°. At 75° one gram dissolves in 10 ml water. Sol in acetone, alcohol, warm acetic acid.

6346. Natamycin. *Pimaricin;* antibiotic A 5283; tennecetin; CL 12625; Mycophyt; Myprozine; Natacyn; Pimafucin; Synogil. $C_{33}H_{47}NO_{13}$; mol wt 665.75. C 59.54%, H 7.12%, N 2.10%, O 31.24%. Polyene antifungal antibiotic produced by *Streptomyces natalensis* from soil near Pietermaritzburg, South Africa and by *S. chattanoogensis.* Isoln: Struyk *et al., Antibiot. Ann.* **1957-1958**, 878. Prodn: **Brit.** pat. **844,289** (1960 to Koninklijke Nederlandsche Gist en Spiritus-Fabriek); **Brit.** pat. **846,933** (1960 to Am. Cyanamid). Identity with tennecetin: Divekar *et al., Antibiot. & Chemother.* **11**, 377 (1961). Structure: Golding *et al., Tetrahedron Letters* **1966**, 3551; Meyer, *Chem. Commun.* **1968**, 470. Configuration: Gaudiano *et al., Chim. Ind. (Milan)* **48**, 1327 (1966). Revised structure: R. C. Pandey, K. L. Rinehart, *J. Antibiot.* **29**, 1035 (1976). Soly data: Marsh, Weiss, *J. Assoc. Offic. Anal. Chem.* **50**, 457 (1967). Toxicology: G. J. Levinskas *et al., Toxicol. Appl. Pharmacol.* **8**, 97 (1966). Comprehensive description: H. Brik in *Analytical Profiles of Drug Substances* vol. 10, K. Florey, Ed. (Academic Press, New York, 1981) pp 513-561.

Crystals from methanol + water, darkens ~200°, dec 280-300°. $[\alpha]_D^{20}$ +278° (c = 1 in CH_3COOH). Sensitive to light, but otherwise very stable in the dry state. uv max (methanol + 0.1% CH_3COOH): 220, 280, 290, 303, 318 nm (ε 21300, 26630, 52930, 83220, 76230). Practically insol in higher alcohols, ether, esters, aromatic or aliphatic hydrocarbons, chlorinated hydrocarbons, ketones, dioxane, cyclohexanol, oils. LD_{50} orally in male, female rats (g/kg): 2.73, 4.67 (Levinskas).

THERAP CAT: Antibacterial.

6347. Natural Gas. American natural gas consists of about 85% methane, 9% ethane, 3% propane, 2% nitrogen, 1% butane. Northern Texas gas contains more nitrogen, also sufficient helium to warrant commercial extraction.

USE: Fuel gas, in the manuf of hydrogen, methane, ammonia. *Caution:* Narcotic in high concns. Incomplete combustion can result in production of carbon monoxide.

6348. Nauheim Salts (Artificial). Bad Nauheim salts (artificial). The composition used in European pharmaceutical dispensing practice is as follows: three lbs $CaCl_2.2H_2O$, two lbs NaCl, 1½ lbs $Na_2CO_3.10H_2O$ and ½ lb $NaHCO_3$. This mixture is dissolved in a bath of 50 gallons of water. Then one lb of $NaHSO_4.H_2O$ is slowly added.

6349. Naxagolide. *(4aR)-trans-3,4,4a,5,6,10b-Hexahydro-4-propyl-2H-naphth[1,2-b]-1,4-oxazin-9-ol;* (+)-*trans-1a,2,3,4a,5,6-hexahydro-9-hydroxy-4-propyl-4H-naphth-[1,2-b]-1,4-oxazine;* (+)-*4-propyl-9-hydroxynaphthoxazine;* dopazinol; (+)-PHNO; L 647,339; MK-458; N-0500. $C_{15}H_{21}NO_2$; mol wt 247.34. C 72.84%, H 8.56%, N 5.66%, O 12.94%. Dopamine D_2-receptor agonist. Prepn: J. H. Jones *et al., Eur. pat. Appl. 80,115;* J. H. Jones, U.S. pat. **4,420,-480** (both 1983 to Merck & Co.); J. H. Jones *et al., J. Med. Chem.* **27,** 1607 (1984). Total synthesis and pharmacology: D. Dykstra *et al., Eur. J. Med. Chem.-Chim. Ther.* **20,** 247 (1985). Enantioselective synthesis: D. G. Melillo *et al., J. Org. Chem.* **52,** 5143 (1987). Pharmacology: G. E. Martin *et al., J. Pharmacol. Exp. Ther.* **230,** 569 (1984); A. S. Horn *et al., J. Pharm. Pharmacol.* **36,** 639 (1984). Receptor binding studies: G. E. Martin *et al., J. Pharmacol. Exp. Ther.* **233,** 395 (1985). Dopaminergic activity in rats: M. T. Martin-Iverson *et al., Psychopharmacology* **95,** 534 (1988). GC determn in plasma: D. G. Musson *et al., Biomed. Environ. Mass Spectrom.* **17,** 293 (1988). Efficacy in animal models of Parkinsonism: W. Koller *et al., Movement Dis.* **2,** 193 (1987); C. E. Clarke *et al., Arch. Pharmacol.* **338,** 35 (1988). Clinical evaluation in Parkinson's disease: M. D. Muenter *et al., Neurology* **38,** 1541 (1988).

mp 158-160°. $[\alpha]_D$ +59.54° (c = 0.0964 in ethanol). Hydrochloride, $C_{15}H_{22}ClNO_2$, white crystalline solid, mp 303-305°. $[\alpha]_{509}$ +55.9° (c = 1.0 in 0.01*M* HCl in methanol).

THERAP CAT: Antiparkinsonian.

6350. Nealbarbital. *5-(2,2-Dimethylpropyl)-5-(2-propenyl)-2,4,6(1H,3H,5H)-pyrimidinetrione; 5-allyl-5-neopentylbarbituric acid;* nealbarbitone; neallymal; Nevental; Censedal. $C_{12}H_{18}N_2O_3$; mol wt 238.28. C 60.48%, H 7.61%, N 11.76%, O 20.14%. Synthesis: Brandstrom, *Acta Chem. Scand.* **13,** 613, 615, 619 (1959); **Brit.** pat. **797,017** and U.S. pat. **2,899,435** (1958, 1959, both to Pharmacia). Metabolism: J. N. T. Gilbert *et al., J. Pharm. Pharmacol.* **26,** 119 (1974); J. N. T. Gilbert, J. W. Powell, *Eur. J. Drug Metab. Pharmacokinet.* **3,** 195 (1978).

Slightly bitter crystals, mp 155-157°. Practically insol in water, petr ether. Moderately sol in chloroform; freely sol in alcohol, ether, acetone. Also sol in aq alkaline solns.

Note: This is a controlled substance listed in the U.S. Code of Federal Regulations, Title 21 Part 1308.13 (1987).

THERAP CAT: Sedative, hypnotic.

6351. Neamine. *2-Deoxy-4-O-(2,6-diamino-2,6-dideoxy-α-D-glucopyranosyl)-D-streptamine;* neomycin A. $C_{12}H_{26}N_4O_6$; mol wt 322.36. C 44.71%, H 8.12%, N 17.38%, O 29.78%. Component of the antibiotic complex neomycin, *q.v.* Isoln: Dutcher *et al., J. Am. Chem. Soc.* **73,** 1384 (1951). Identity of neomycin A and neamine: Leach, Teeters, *ibid.* **74,** 3187 (1952). Prepn from neomycin B: Peck *et al., U.S. pat.* **2,691,675** (1954 to Merck & Co.). Structure: Carter *et al., J. Am. Chem. Soc.* **83,** 3723 (1961); Hichens, Rinehart, *ibid.* **85,** 1547 (1963). Synthesis: Umezawa *et al., J. Antibiot.* **20A,** 53 (1967); H. Kohno *et al., Agr. Biol. Chem.*

39, 1091 (1975); A. Harayama *et al., Bull. Chem. Soc. Japan* **52,** 3626 (1979).

Crystals from water or aq alc, dec 225-226°, $[\alpha]_D^{25}$+112.8° (c = 1). IR spectrum: Leach, Teeters, *J. Am. Chem. Soc.* **73,** 2794 (1951).

Hydrochloride, $C_{12}H_{26}N_4O_6.4HCl$, amorphous, dec 250-260°, $[\alpha]_D^{25}$ +83° (c = 1).

N-Acetyl deriv, $C_{12}H_{26}N_4O_6.(CH_3CO)_4$, crystals from methanol, mp 334-336°. $[\alpha]_D^{25}$ +87° (c = 1).

6352. Neatsfoot Oil. Fixed oil from feet of neat (bovine) cattle.

Pale yellow liquid; peculiar odor. d 0.915. Solidif 0° to −10°. n_D^{20} 1.4695-1.4708. Sapon no. 192-203. Iodine no. 44-73.2.

USE: Waterproofing and softening leather; lubricant; oiling wool, etc.

6353. Nebularine. *9-β-D-Ribofuranosyl-9H-purine.* $C_{10}H_{12}N_4O_4$; mol wt 252.23. C 47.62%, H 4.80%, N 22.22%, O 25.37%. Isoln from the mushroom *Clitocybe nebularis* (Batsch.) Quel., *Agaricaceae:* Ehrenberg *et al., Svensk Kem. Tidskr.* **58,** 269 (1946); Löfgren *et al., Acta Chem. Scand.* **8,** 670 (1954); from a streptomycete: Isono, Suzuki, *J. Antibiot.* **13A,** 270 (1960). *In vitro* toxicity towards sarcoma 180 cells, mouse embryonic fibroblasts and epithelial cells: J. J. Biescle *et al., Cancer* **8,** 87 (1955). Synthesis: Brown, Weliky, *J. Biol. Chem.* **204,** 1019 (1953); Fox *et al., J. Am. Chem. Soc.* **80,** 1669 (1958); Hashizume, Iwamura, *Tetrahedron Letters* **1966,** 643; *eidem, J. Org. Chem.* **33,** 1796 (1968). Crystal structure: T. Takeda, *Acta Crystallogr.* **30B,** 825 (1974). Alternate syntheses: V. Nair, S. G. Richardson, *Tetrahedron Letters* **1979,** 1181; P. K. Gupta, D. S. Bhakuni, *Indian J. Chem.* **20B,** 534 (1981).

Small rhombohedra from ethyl methyl ketone + methanol, mp 181-182°; needles from methanol, mp 182-183°. $[\alpha]_D^{25}$ −48.6° (H_2O). uv max (0.1*N* HCl): 262 nm ($E_{1cm}^{1\%}$ 232); (0.1*N* NaOH): 263 nm ($E_{1cm}^{1\%}$ 361). Considerably sol in water (about 10%). Slightly sol in cold ethanol. Very slightly sol in acetone, ether, chloroform. Aq solns may be sterilized by boiling without decompn. LD_{50} s.c. in rats, guinea pigs: 220, 15 mg/kg, Truant, D'Amato, *Fed. Proc.* **14,** 391 (1955).

6354. Neburon. *N-Butyl-N'-(3,4-dichlorophenyl)-N-methylurea;* 3-(3,4-dichlorophenyl)-1-methyl-1-*n*-butylurea; Kloben Neburon. $C_{12}H_{16}Cl_2N_2O$; mol wt 275.18. C 52.37%, H 5.86%, Cl 25.77%, N 10.18%, O 5.81%. Selective pre-emergence herbicide. Prepn: Todd, U.S. pats. **2,655,444-7** (1953 to du Pont).

Crystals from dioxane + water, mp 101.5-103°. Soly in water at 24°: 48 ppm; sparingly sol in hydrocarbon solvents.

Stable toward oxidation and water at ordinary temperatures. LD_{50} orally in rats: > 11,000 mg/kg, G. W. Bailey, J. L. White, *Residue Rev.* **10**, 97 (1965).

USE: Herbicide.

6355. Nedocromil. *9-Ethyl-6,9-dihydro-4,6-dioxo-10-propyl-4H-pyrano[3,2-g]quinoline-2,8-dicarboxylic acid; 4,6-dioxo-1-ethyl-10-propyl-4H,6H-pyrano[3,2-g]quinoline-2,8-dicarboxylic acid;* FPL 59002. $C_{19}H_{17}NO_7$; mol wt 270.55. C 84.35%, H 6.33%, N 5.18%, O 4.14%. Anti-allergic compound with the ability to stabilize both mucosal and connective tissue mast cells. Prepn and pharmacology: H. Cairns, D. Cox, **Belg.** pat. **866,622;** *eidem,* **U.S.** pat. **4,474,-787** (1978, 1984 both to Fisons); H. Cairns *et al., J. Med. Chem.* **28**, 1832 (1985). Prevention of histamine release from mast cells, inhibition of bronchoconstriction in monkeys: R. P. Eady *et al., Brit. J. Pharmacol.* **85**, 323 (1985). Early clinical trial in bronchial asthma: S. Lal *et al., Thorax* **39**, 809 (1984).

Yellow powder, mp 298-300° (dec).
Disodium salt, $C_{19}H_{17}NNa_2O_7$, *FPL 59002KP, Tilade.* Pale yellow powder.

THERAP CAT: Anti-allergic; anti-asthmatic.

6356. Neem. Nim. Subtropical shade tree, *Azadirachta indica,* A. Juss. *(Melia azadirachta L.), Meliceae,* native to the arid regions of India, Pakistan, parts of Africa. Most important constituent is the limonoid, azadirachtin, *q.v.* Known for centuries as being free of insects, disease, nematodes. All parts of the tree, but esp the seeds, are resistant. The bark, leaves and fruit have been used in traditional medicinal remedies: R. N. Chopra *et al., Indigenous Drugs of India* (U. N. Dhur & Sons, Calcutta, 2nd ed., 1958) pp 360-363. Antifeedant activity of neem extracts against locusts: J. H. Butterworth, E. D. Morgan, *Chem. Commun.* **1968,** 23; against Carolina grasshopper, walkingstick, field cricket: V. E. Adler, E. C. Uebel, *J. Environ. Sci. Health* **A19,** 393 (1984); against bengalgram podborer: P. C. Sundaru Babu, B. Rajasekaran, *Pesticides* **18,** 58 (1984). Series of articles on chemistry and activity of neem extracts: *Natural Pesticides from the Neem Tree,* Proc. 1st Int. Neem Conf., 1980, H. Schmutterer *et al.,* Eds. (German Agency for Technical Cooperation, Eschborn, 1981) 291 pp.

Note: Margosan-O, a neem formulation, has been approved by the EPA (1985) for limited use as a pesticide on non-food crops: *Chem. & Eng. News* **63,** 51 (May 27, 1985).

USE: Oil, crushed seeds as insect repellent, antifeedant.

6357. Neem Oil. Nim oil; margosa oil. A fixed oil expressed from seed-kernels of the Indian neem tree, *Azadirachta indica* A. Juss. *(Melia azadirachta L.), Meliaceae.* The seed-kernels yield about 10% of the oil. *Constit.* Mostly glycerides and limonoids including azadirachtin, nimbin, nimbiol, *q.q.v.,* salannin; about 2% bitters. Seed extracts as feeding deterrents for Japanese beetles: T. L. Ladd, Jr. *et al., J. Econ. Entomol.* **71,** 810 (1978). Antifeedant activity of oil: R. S. Saxena *et al.,* in *Natural Pesticides of the Neem Tree,* Proc. 1st Int. Neem Conf., 1980, H. Schmutterer *et al.,* Eds. (German Agency for Technical Cooperation, Eschborn, 1981) pp 171-204. Has been used as a traditional medicinal remedy. Cases of poisoning have been reported: D. Sinniah, G. Baskaran, *Lancet* **1,** 487 (1981).

Yellow oil. Odor of garlic. Bitter taste. d_4^{15} 0.925. Sol in ether, chloroform. Practically insol in alcohol, water.

USE: Agricultural insect repellent, antifeedant.

6358. Nefopam. *3,4,5,6-Tetrahydro-5-methyl-1-phenyl-1H-2,5-benzoxazocine; 5-methyl-1-phenyl-1,3,4,6-tetrahydro-5H-benz[f]-2,5-oxazocine.* $C_{17}H_{19}NO$; mol wt 253.35. C 80.60%, H 7.56%, N 5.53%, O 6.31%. A cyclized analog of orphenadrine and diphenhydramine, *q.q.v.;* representative of a new class of centrally acting skeletal muscle relaxants,

the benzoxazocines. Prepn: **Neth.** pat. **Appl. 6,606,390** (1966), *C.A.* **66,** 55535w (1967); Baltes, **U.S.** pat. **3,487,153** (1969) (both to Rexall). Pharmacology: Bassett *et al., Brit. J. Pharmacol.* **37,** 69 (1969); Klohs *et al., Arzneimittel-Forsch.* **22,** 132 (1972). Review of pharmacology and therapeutic efficacy: R. C. Heel *et al., Drugs* **19,** 249-267 (1980).

Hydrochloride, $C_{17}H_{20}ClNO$, *fenazoxine, R 738, Acupan, Ajan, Sinalgico.* mp 238-242°. LD_{50} in mice, rats (mg/kg): 119, 178 orally; 44.5, 28 i.v. (Baltes).

THERAP CAT: Analgesic. Antidepressant.

6359. Negamycin. *3,6-Diamino-2,3,4,6-tetradeoxy-L-threo-hexonic acid 2-(carboxymethyl)-2-methylhydrazide; 3,6-diamino-5-hydroxyhexanoic acid 2-(carboxymethyl)-2-methylhydrazide; [2-(3,6-diamino-5-hydroxy-1-oxohexyl)-1-methylhydrazino]acetic acid.* $C_9H_{20}N_4O_4$; mol wt 248.28. C 43.53%, H 8.12%, N 22.57%, O 25.78%. Hydrazide antibiotic isolated from strains related to *Streptomyces purpeofuscus:* Hamada *et al., J. Antibiot.* **23,** 170 (1970). Prepn: *idem,* **Japan.** pat. **71 28,835** (1971 to Microbiochem. Res. Found.), *C.A.* **75,** 139416g (1971). Structure and partial synthesis: Kondo *et al., J. Am. Chem. Soc.* **93,** 6305 (1971). Total synthesis of negamycin and its antipode: Shibahara *et al., ibid.* **94,** 4353 (1972). Synthesis of racemic negamycin: W. Streicher *et al., J. Antibiot.* **31,** 725 (1978); G. Pasquet *et al., Tetrahedron Letters* **21,** 931 (1980); A. Pierdet *et al., Tetrahedron* **36,** 1763 (1980). Stereocontrolled synthesis of (+)-negamycin: Y. F. Wang *et al., J. Am. Chem. Soc.* **104,** 6465 (1982); S. DeBernardo *et al., Tetrahedron Letters* **29,** 4077 (1988). Mechanism of action: Mizuno *et al., J. Antibiot.* **23,** 581 (1970); Y. Uehara *et al., Biochim. Biophys. Acta* **442,** 251 (1976).

Colorless powder, mp 110-120° (dec). $[\alpha]_D^{29}$ +2.5° (c = 2). Amphoteric compound. pKa values after treatment with HCl-methanol: 3.55, 8.10, 9.75. Sol in water. Practically insol in methanol, ethanol, butanol, ethyl acetate, chloroform and benzene. LD_{50} i.v. in mice: 400-500 mg/kg (Hameda).

THERAP CAT: Antibacterial.

6360. Negatol®. *Hydroxymethylbenzenesulfonic acid polymer with formaldehyde;* methylenebis(hydroxytoluenesulfonic acid) polymer; dihydroxydimethyldiphenylmethanedisulfonic acid polymer; Albocresil; Albothyl; Negatan. Colloidal prod of relatively high mol wt prepd by reacting sulfonated *m*-cresol and formaldehyde, e.g. the product obtained by Thuau, **Can.** pat. **417,272** (1943 to Lilly), *C.A.* **38,** 1077 (1944). Shown to be a mixture of six components: Pasich, Stawinska, *Farm. Polska* **20,** 829 (1964), *C.A.* **62,** 12976h (1965).

Sol in water, forming colloidal solns. Very acid reaction. pH of 5% w/v soln about 1.0.

THERAP CAT: Antiseptic.

THERAP CAT (VET): Antiseptic.

6361. Neoarsphenamine. *Sulfoxylic acid mono[[[5-[(3-amino-4-hydroxyphenyl)diarsenyl]-2-hydroxyphenyl]amino]methyl] ester monosodium salt; [5-[(3-amino-4-hydroxyphenyl)arseno]-2-hydroxyanilino]methanol sulfoxylate sodium;* arsphenamine methylenesulfoxylate sodium; 3,3'-diamino-4,4'-dihydroxyarsenobenzenemethylenesulfoxylate sodium; Neosalvarsan; Collunovar; N.A.B.; Neo-Arsoluin; Vetarsenobillon; Novarsenobillon; Arsevan; Novarsan; Novarsenobenzol; Miarsenol. $C_{13}H_{13}As_2N_2$-NaO_4S; mol wt 466.13. C 33.49%, H 2.81%, As 32.14%, N

6.01%, Na 4.93%, O 13.73%, S 6.88%. The usual medicinal grade contains a small amount of inert inorganic salts and some solvent. *The National Formulary* requires 19+% As. Prepn from arsphenamine + sodium formaldehydesulfoxyl-ate: Krumwiede, *J. Am. Pharm. Assoc.* **8**, 795 (1919); Heyl, Miller, *ibid.* **11**, 432 (1922); Dohr, U.S. pat. **1,549,465** (1925); Kober, U.S. pat. **1,564,859** (1926); Kraft *et al., Russian* pat. **158,388** (1962).

Yellow powder; odorless or slight odor. Oxidizes in air, becoming darker and more toxic; higher temps accelerate the oxidation; hence marketed in air-evacuated ampuls or filled with a nonoxidizing gas. Very sol in water; sol in glycerol. Slightly sol in alcohol or acetone. Practically insol in chloroform, ether. Its aq soln is practically neutral, unlike arsphenamine, which is acid. LD_{50} i.v. in rats: 300 mg/kg.

Caution: Neoarsphenamine does not require neutralization. Injections should be made with freshly prepd soln. The soln should not be shaken.

THERAP CAT (VET): Has been used in contagious pleuropneumonia, babesiasis, equine petechial fever, eperythrozoonosis.

6362. Neocembrene. *1,5,9-Trimethyl-12-(1-methylethenyl)-1,5,9-cyclotetradecatriene;* cembrene A; neocembrene A. $C_{20}H_{32}$; mol wt 272.48. C 88.16%, H 11.84%. Termite trail pheromone with all-*trans* configuration isolated from *Nasutitermes exitiosus* (Hill): B. P. Moore, *Nature* **211**, 746 (1966). Structure: A. J. Birch *et al., J. Chem. Soc. Perkin Trans. I* **1972**, 2653; V. D. Patil *et al., Tetrahedron* **29**, 341 (1973). Synthesis: M. Kodama *et al., Tetrahedron Letters* **1975**, 3065; Y. Kitahara *et al., Chem. Letters* **1976**, 219; H. Takayanagi *et al., Chem. Commun.* **1978**, 359. Synthesis of neocembrene and geometric isomers: T. Kato *et al., J. Org. Chem.* **45**, 1126 (1980).

Colorless oil, faint wax-like odor. $bp_{0.8}$ 150-152°. n_D^{30} 1.5102.

6363. Neo-cupferron. *N-Hydroxy-N-nitroso-1-naphthalenamine ammonium salt; N-1-naphthyl-N-nitrosohydroxylamine ammonium derivative;* ammonium α-naphthyl-nitrosohydroxylamine. $C_{10}H_{11}N_3O_2$; mol wt 205.21. C 58.52%, H 5.40%, N 20.48%, O 15.59%. $C_{10}H_7N(NO)O-NH_4$.
White to light buff crystals; sensitive to light. mp 125-126° with decompn. Soluble in water, methanol; insol in ether. The aq soln is unstable.

USE: Reagent for iron and copper. *See also* Cupferron.

6364. Neocuproine. *2,9-Dimethyl-1,10-phenanthroline;* 2,9-dimethyl-o-phenanthroline. $C_{14}H_{12}N_2$; mol wt 208.25. C 80.74%, H 5.81%, N 13.45%. Prepn: Case, *J. Am. Chem. Soc.* **70**, 3994 (1948); O'Reilly, Plowman, *Aust. J. Chem.* **13**, 145 (1960).

Hemihydrate, $C_{14}H_{12}N_2 \cdot \frac{1}{2}H_2O$, crystals from water or ligroin, mp 159-160°.
Dihydrate, needles from water. Loses water over P_2O_5 or at 80°.

USE: Clinical reagent (blood glucose assay). In spectrophotometric determination of copper, Nebesar, *Anal. Chem.* **36**, 1961 (1964).

6365. Neodymium. Nd; at. wt 144.24; at. no. 60; valences 2, 3, 4. A lanthanide; belongs to the cerium group of rare earths. Seven naturally occurring isotopes: 142 (27.13%); 143 (12.20%); 144 (23.87%); 145 (8.29%); 146 (17.18%); 148 (5.72%); 150 (5.60%); ^{144}Nd is radioactive, $T_{1/2}$ 2.4×10^{15} years; α emission. Artificial, radioactive isotopes: 138-141; 147; 149; 151. Abundance in earth's crust: 12-24 ppm. Found in cerium cores: cerite, monazite sand, gadolinite etc. Discovered by von Welsbach in 1885. Sepn from Pr: Spedding *et al., J. Am. Chem. Soc.* **69**, 2786 (1947); **76**, 2557 (1954); Kauffman, Blank, *J. Chem. Ed.* **37**, 156 (1960); from other rare earths: Spedding *et al., J. Am. Chem. Soc.* **69**, 2812 (1947). Metal prepd by electrolysis of the chloride at high current density, using a finely divided carbon cathode: Matignon, *Compt. Rend.* **131**, 891 (1900); Sieverts, Roell, *Z. Anorg. Chem.* **150**, 261 (1926); by reduction of the chloride with potassium *in vacuo:* Bommer, Hohmann, *ibid.* **248**, 357 (1941). Spectrum: Joye, *Arch. Sci. Phys. Nat. Geneve* **36**, 41 (1913); Alberston *et al., Phys. Rev.* **61**, 167 (1942). Reviews of prepn, properties and compds of neodymium and other rare earths: *The Rare Earths,* F. H. Spedding, A. H. Daane, Eds. (Krieger, Huntington, N.Y., 1971, reprint of 1961 ed.) 641 pp; Hulet, Bode, "Separation Chemistry of the Lanthanides and Transplutonium Actinides" in *MTP Int. Rev. Sci.: Inorg. Chem.., Ser. One,* Vol. 7, K. W. Bagnall, Ed. (University Park Press, Baltimore, 1972) pp 1-45; Moeller, "The Lanthanides" in *Comprehensive Inorganic Chemistry* vol. 4, J. C. Bailar Jr. *et al.,* Eds. (Pergamon Press, Oxford, 1973) pp 1-101.

Silver-white metal, becomes yellowish on exposure to air. Hexagonal structure at room temp, d 7.003; body centered cubic structure above 868°, d 6.80. mp about 1024°. bp about 3000°. E°(aq) Nd^{3+}/Nd −2.44 V (calc). Experimental reduction potentials (referred to a normal calomel electrode): −1.870, −1.960 V: Noddack, Brukl, *Angew. Chem.* **50**, 362 (1937). Reacts slowly with cold water; rapidly on heating.

Oxide, Nd_2O_3, blue powder, hexagonal structure, exhibits a slightly red fluorescence. Prepd by heating the hydroxide, carbonate, nitrate or oxalate. Very stable. Sol in dilute acids; soly in water: 5.7×10^{-6} g-mol/l at 29°.

Hydroxide, $Nd(OH)_3$, bluish or pink precipitate; on heating at 300-350° is converted into $2Nd_2O_3 \cdot 3H_2O$, grayish-brown; on further increase in temp is converted into $Nd_2O_3 \cdot H_2O$.

Chloride, $NdCl_3$, large purple prisms. Sol in water, in alcohol. Forms addition compds with ammonia. A hexahydrate is obtained from the aq soln; the hydrate is very sol in water (2.46 parts per 1 part of water); mp 124°. LD_{50} in mice: 348; 600 mg/kg i.p.; 5.25 g/kg orally; in guinea pigs: 140 mg/kg i.p., Haley, *J. Pharm. Sci.* **54**, 663 (1965).

Sulfate, $Nd_2(SO_4)_3$, pinkish needles; prepd by heating the oxide with concd sulfuric acid. Heat of formation 57.2 kcal. Sol in water; heat of soln 36.5 kcal. Is dec at 700-800°. Forms several double salts. A penta-, an octa-, and a pentadecahydrate have been prepared.

Nitrate, $Nd(NO_3)_3$, penta- and hexahydrates; prepd by adding the oxide to nitric acid. LD_{50} (hexahydrate) in rats: 270 mg/kg i.p.; 2.75 g/kg orally, Haley, *loc. cit.*

USE: Oxide used in glass filter plate laminated on color TV tubes to improve contrast and brightness; in lasers: Pings, *Colo. Sch. Mines, Miner. Ind. Bull.* **12** (no. 2), (1969) 19 pp.

6366. Neoergosterol. *19-Norergosta-5,7,9,22-tetraen-3β-ol.* $C_{27}H_{40}O$; mol wt 380.59. C 85.20%, H 10.59%, O 4.20%. Prepd from ergosterol: Inhoffen, *Ann.* **497**, 130 (1932); Windaus, Deppe, *Ber.* **70**, 76 (1937); Mosettig, Scheer, *J. Org. Chem.* **17**, 764 (1952). Stereochemistry: Steele *et al., J. Am. Chem. Soc.* **85**, 1134 (1963).

Needles from methanol, mp 152-154°. $[\alpha]_D^{17}$ −11° (c = 2 in chloroform). Practically insol in water; sol in organic solvents. Is precipitated by digitonin.

Methyl ether, $C_{28}H_{42}O$, mp 94°. $[\alpha]_D^{20}$ −5° (c = 1.2 in chloroform).

6367. Neohesperidin Dihydrochalcone. *1-[4-[[2-O-(6-Deoxy-α-L-mannopyranosyl)-β-D-glucopyranosyl]oxy]-2,6-dihydroxyphenyl]-3-(3-hydroxy-4-methoxyphenyl)-1-propanone; 3,5-dihydroxy-4-(3-hydroxy-4-methoxyhydrocinnamoyl)phenyl-2-O-(6-deoxy-α-L-mannopyranosyl)-β-D-glucopyranoside;* neohesperidin DHC; NHDC; Sukor. $C_{28}H_{36}O_{15}$; mol wt 612.60. C 54.90%, H 5.92%, O 39.18%. Prepn from naringin, a flavanone glycoside occurring naturally in grapefruit: Horowitz, Gentile, U.S. pats. **3,087,821** and **3,375,242** (1963, 1968, both to U.S. Secy. Agr.); Robertson *et al., Ind. Eng. Chem. Prod. Res. Develop.* **13**, 125 (1974). Sweetening effect: Inglett *et al., J. Food Sci.* **34**, 101 (1969). Chromatography: Gentile, Horowitz, *J. Chromatog.* **63**, 467 (1971); J. F. Fisher, *J. Agr. Food Chem.* **25**, 682 (1977). *Review:* P. J. Pratter, *Perfum. Flavor.* **5**, 12-18 (1981).

Crystals from acetone, mp 156-158°. 1000 to 1500 times sweeter than sucrose; 20 times sweeter than saccharin.
USE: Sweetening agent, esp in chewing gum and dentifrices.

6368. Neomethymycin. *10-Deoxy-12-hydroxymethymycin.* $C_{25}H_{43}NO_7$; mol wt 469.60. C 63.94%, H 9.23%, N 2.98%, O 23.85%. Macrolide antibiotic found in the mother liquors from methymycin, *q.v.* Differs from methymycin only in the location of one hydroxyl group, which results in marked changes in chemical behavior: Djerassi, Halpern, *J. Am. Chem. Soc.* **79**, 2022, 3926 (1957); *Tetrahedron* **3**, 255 (1958). Production from *Streptomyces venezuelae* cultures: Dutcher *et al.,* U.S. pat. **2,916,486** (1959 to Olin Mathieson). Synthesis of the aglycone, *neomethynolide:* J. Inanago *et al., Chem. Letters* **1981**, 1415.

Crystals from ether + hexane, mp 156-158°. $[\alpha]_D^{25}$ +93°. uv max (ethanol): 227.5 nm (log ε 4.10). Solvates easily and when crystallized from dilute acetone yields 1 cm long crystals of the acetone solvate. Soluble in alcohol, acetone, chloroform, benzene, ethyl acetate, ether. Slightly sol in water, dibutyl ether. Practically insol in hexane, aliphatic hydrocarbons. *Compare also* Picromycin.

6369. Neomycin. Mycifradin; Myacyne; Fradiomycin; Neomin; Neolate; Neomas; Pimavecort; Vonamycin Powder V. Antibiotic complex composed of neomycins A, B and C. Produced by *Streptomyces fradiae:* Waksman, Lechevalier,

Science **109**, 305 (1949); Waksman *et al., J. Clin. Invest.* **28**, 934 (1949); Swart *et al., ibid.* 1045; Waksman, Lechevalier, U.S. pat. **2,799,620** (1957). Purification: Jackson, U.S. pat. **2,848,365** (1958 to Upjohn); Haak, U.S. pat. **3,108,996** (1963 to Upjohn). Recovery: Miller, U.S. pat. **3,005,815** (1961 to Merck & Co.); Moses, U.S. pat. **3,022,228** (1962 to Penick); **Brit.** pat. **945,475** (1964 to O.W.G. Chemie). Characterization: Peck *et al., J. Am. Chem. Soc.* **71**, 2590 (1949); Regna, Murphy, *ibid.* **72**, 1045 (1950); Dutcher *et al., ibid.* **73**, 1384 (1951). Structure of neomycins B and C: K. L. Rinehart *et al., ibid.* **79**, 4567, 4568 (1957); *ibid.* **84**, 3216, 3218 (1962). Abs config of neomycin C: M. Hichens, K. L. Rinehart, *ibid.* **85**, 1547 (1963). Total synthesis of neomycin C: S. Umewaza, Y. Nishimura, *J. Antibiot.* **30**, 189 (1977); S. Umewaza *et al., Bull. Chem. Soc. Japan* **53**, 3259 (1980); of neomycin B: T. Usui, S. Umezawa, *J. Antibiot.* **40**, 1464 (1987). Monographs: S. A. Waksman, *Neomycin* (Rutgers Univ. Press, New Brunswick, N. J., 1953) 219 pp; K. L. Rinehart, Jr., *The Neomycins and Related Antibiotics* (Wiley, New York, 1964) 137 pp. Comprehensive description: W. F. Heyes in *Analytical Profiles of Drug Substances* **vol. 8**, K. Florey, Ed. (Academic Press, New York, 1979) pp 399-488.

Neomycin B

Neomycin complex is an amorphous base sol in water, methanol and acidified alcohol. Practically insol in common organic solvents. Solns up to 250 mg/ml H_2O may be prepared.

Neomycin A: *see* Neamine.

Neomycin B, $C_{23}H_{46}N_6O_{13}$, *Framycetin, Enterfram, Framygen, Soframycin, Actilin, antibiotique EF 185.* Identity of neomycin B and framycetin: K. L. Rinehart *et al., ibid.* **82**, 3938 (1960). Yields on hydrolysis neamine and *neobiosamine B.* Structure of neobiosamine B: K. L. Rinehart *et al., J. Am. Chem. Soc.* **82**, 2970 (1960).

Neomycin B hydrochloride, amorphous white powder. $[\alpha]_D^{20}$ +57° (H_2O). Soly in mg/ml at about 28°: water 15.0; methanol 5.7; ethanol 0.65; isopropanol 0.05; isoamyl alcohol 0.33; cyclohexane 0.06; benzene 0.03. Practically insol in acetone, ether, other organic solvents. For additional soly data *see* Weiss *et al., Antibiot. & Chemother.* **7**, 374 (1957).

Neomycin B sulfate, *Bykomycin, Endomixin, Fraquinol, Myacine, Neosulf, Neomix, Neobrettin, Nivemycin, Tuttomycin.* Amorphous white powder. Practically tasteless. $[\alpha]_D^{20}$ +54° (c = 2 in H_2O). Soly in mg/ml at about 28°: water 6.3; methanol 0.225; ethanol 0.095; isopropanol 0.082; isoamyl alcohol, 0.247; cyclohexane 0.08; benzene 0.05. Practically insol in acetone, ether, chloroform. Aq solns are fairly stable at pH 2 to 9. Highly purified prepns are very stable to alkali and unstable to acids. Refluxing with barium hydroxide for 18 hrs showed no loss of activity. Boiling with mineral acids yields an aldehyde, characterized as furfural, and an organic base.

Neomycin C, $C_{23}H_{46}N_6O_{13}$. Yields on hydrolysis neamine and *neobiosamine C.* Structure of neobiosamine C: K. L. Rinehart, P. Woo, *J. Am. Chem. Soc.* **80**, 6463 (1958).

THERAP CAT: Antibacterial.
THERAP CAT (VET): Antibacterial.

6370. Neomycin Undecylenate. Neodecyllin. 10-Undecenoate salt of the antibiotic neomycin, *q.v.* Commercial prepn contains the equivalent of not less than 30% neomycin base, calculated on the dried basis. Prepn: van De Griendt, U.S. pat. 3,022,286 (1962 to Penick).

THERAP CAT: Antibacterial; antifungal.

6371. Neon. Ne; at. wt 20.179; at. no. 10. Three stable isotopes: 20 (90.92%); 21 (0.26%); 22 (8.82%); artificial, short-lived isotopes: 17-19; 23; 24. Abundance in earth's crust including the atmosphere: 5×10^{-7}; concentration in air: 18.2 ppm by vol. An element of the 0 group: discovered in 1898 by Ramsay and Travers: Travers, *The Discovery of the Rare Gases*, London, 1928. Monograph: *Argon, Helium and the Rare Gases*, vols. 1, 2, G. A. Cook, Ed. (Interscience, New York, 1961) 818 pp; Cockett, Smith, "The Monatomic Gases" in *Comprehensive Inorganic Chemistry* vol. 1, J. C. Bailar, Jr. *et al.*, Eds. (Pergamon Press, Oxford, 1973) pp 139-211.

Inert, odorless gas; does not condense at the temp of liq air; solid at the temp of liq hydrogen; the solid form, face-centered cubic crystals. d^0 (gas) 0.89994 g/l; d (liq at bp): 1.207 g/cc. bp $-246.08°$; triple pt $-248.6°$; crit temp $-228.7°$; crit pressure 26.9 atm.

USE: In neon light tubes; ingredient of gaseous fillers for antifog devices, warning signals, electrical current detectors, high-voltage indicators for high-tension electric lines, lightning arresters, wave-meter tubes; in lasers.

6372. Neopentane. *2,2-Dimethylpropane;* tetramethylmethane. C_5H_{12}; mol wt 72.15. C 83.23%, H 16.77%. Found in petr naphtha. Prepn from *tert*-butyl iodide and dimethylzinc: Lwow, *Z. Chemie* **1870**, 520; *cf. J. Am. Chem. Soc.* **54**, 3460 (1932); from butylmagnesium iodide and dimethyl sulfate in ether: Ferrario, Fagetti, *Gaz. Chim.* **38**, II, 663 (1908); from amylene vapor and hydrogen in presence of an electric discharge: *ibid.* **62**, 621 (1932); by passing pentanes over aluminum chloride at 140°, *J. Gen. Chem. USSR* **14**, 343 (1944), *C.A.* **39**, 3783 (1945); from neopentylmagnesium chloride and water: Whitmore *et al.*, *J. Am. Chem. Soc.* **56**, 749 (1934).

$$H_3C - \underset{\underset{CH_3}{|}}{\overset{\overset{CH_3}{|}}{C}} - CH_3$$

Liquid or gas. d^0_0 0.613 (liq). Solidifies to form tetragonal crystals, mp $-19.8°$. bp 9.5°. Insol in water.

6373. Neopentyl Alcohol. *2,2-Dimethyl-1-propanol;* *tert*-butyl carbinol; neoamyl alcohol; neopentanol. $C_5H_{12}O$; mol wt 88.15. C 68.13%, H 13.72%, O 18.15%. $(CH_3)_3C-CH_2OH$. Prepn: Conant *et al.*, *J. Am. Chem. Soc.* **51**, 1249 (1929). Review of manuf (by fractionation of fusel oil and via chlorination of pentanes) and properties: *Industrial Chemicals*, W. L. Faith *et al.*, Eds. (Wiley & Sons, New York, 2nd ed., 1957) pp 107-114.

Volatile crystals, peppermint odor. bp 114°. mp 53°. d_4^{20} 0.812. Vapor pressure: 16 mm at 20°; 50.5 mm at 46°; 533.5 mm at 100°. Slightly sol in water (about 3.5% at 25°); miscible with alcohol, ether.

6374. Neopentyl Glycol. *2,2-Dimethyl-1,3-propanediol;* dimethyltrimethylene glycol. $C_5H_{12}O_2$; mol wt 104.15. C 57.66%, H 11.61%, O 30.73%. Prepd from isobutyraldehyde.

$$HOCH_2 - \underset{\underset{CH_3}{|}}{\overset{\overset{CH_3}{|}}{C}} - CH_2OH$$

Needles from benzene, mp 127°. bp_{760} 208°. Solubility in water about 65% w/w. Freely sol in alcohol, ether.

USE: In the manuf of plasticizers, polyesters, as modifier of alkyd resins.

6375. Neophyl Chloride. *(2-Chloro-1,1-dimethylethyl)-benzene; 1-chloro-2-methyl-2-phenylpropane;* (β-chloro-*tert*-butyl)benzene. $C_{10}H_{13}Cl$; mol wt 168.67. C 71.21%, H 7.77%, Cl 21.02%. Prepn (68% yield) from benzene and methallyl chloride in the presence of concd sulfuric acid:

Whitmore *et al.*, *J. Am. Chem. Soc.* **65**, 1469 (1943); Smith, Sellas, *Org. Syn. coll. vol. IV*, 702 (1963).

$$C_6H_5 - \underset{\underset{CH_3}{|}}{\overset{\overset{CH_3}{|}}{C}} - CH_2Cl$$

Liquid. d_4^{25} 1.0379. bp_{741} 221° (dec); bp_{90} 111°; bp_{30} 120°; bp_{20} 105°; bp_{18} 104°; bp_{13} 97°; bp_{10} 95°; $bp_{1.0}$ 53°. n_D^{20} 1.5249; n_D^{25} 1.5228. Considerably less reactive with sodium than neopentyl chloride. Forms a Grignard reagent readily and in good yield. Salt effect in acetolysis: Fainberg, Winstein, *J. Am. Chem. Soc.* **78**, 2763 (1956).

6376. Neopine. *8,14-Didehydro-4,5-epoxy-3-methoxy-17-methylmorphinan-6-ol;* β-codeine. $C_{18}H_{21}NO_3$; mol wt 299.36. C 72.22%, H 7.07%, N 4.68%, O 16.03%. An alkaloid from opium, isomeric with codeine. Prepn from 14-bromocodeinone: Conroy, U.S. pat. **2,797,222** (1957 to Merck & Co.); from thebaine: S. Makleit *et al.*, *Acta Chim. Acad. Sci. Hung.* **94**, 161 (1977).

Long needles from petr ether, mp 127.5°. $[\alpha]_D^{23} -28°$ (c = 7.5 in chloroform). The absorption spectrum seems identical with that of codeine: Dobbie, Lauder, *J. Chem. Soc.* **99**, 34 (1911). Solubilities about the same as those of codeine, *q.v.*

Hydrobromide, $C_{18}H_{21}NO_3$.HBr, prismatic crystals from water, darkens about 240°, dec 283°. $[\alpha]_D^{23} +17°$ (c = 3.7). Relatively insol in water, making possible a separation from other more sol hydrobromides of various opium alkaloids.

Note: Incorrectly called *hydroxycodeine,* *cf.* L. J. Sargent, U. Weiss, *J. Org. Chem.* **25**, 987 (1960).

6377. Neoprene. Duprene; GR-M; polychloroprene; poly(2-chloro-1,3-butadiene). Mol wt range of 100,000-300,000. An oil-resistant synthetic rubber, with predominantly *trans* configuration made by polymerization of chloroprene. Prepn: W. H. Carothers *et al.*, *J. Am. Chem. Soc.* **53**, 4203 (1931). *Review:* P. J. Johnson in Kirk-Othmer *Encyclopedia of Chemical Technology* vol. 8 (Wiley-Interscience, New York, 3rd ed., 1979) pp 515-534.

d 1.23-1.35. n 1.55-1.56. Brittle point: $-35°$. Softens ca. 80°. Has high tensile strength, resilience and abrasion resistance.

USE: In mechanical and automotive products; as wire and cable jackets, gaskets, roof coatings, binder for fibers.

6378. Neopterin. *2-Amino-6-(1,2,3-trihydroxypropyl)-4(3H)-pteridinone; 1-(2-amino-4-hydroxy-6-pteridinyl)-1,2,3-propanetriol; 6-(1',2',3'-trihydroxy)pterin;* Crithidia factor. $C_9H_{11}N_5O_4$; mol wt 253.22. C 42.69%, H 4.38%, N 27.66%, O 25.27%. Precursor in the biosynthesis of biopterin, *q.v.* Of the four possible isomers, D-*erythro*, L-*erythro*, D-*threo*, L-*threo*, two have been found in nature: the D-*erythro* form, to which the term neopterin originally referred; first isolated from the pupae of bees: Rembold, Buschmann, *Ann.* **662**, 72 (1963); the L-*threo* form, found to be the growth factor for the protozoan *Crithidia fasciculata* and isolatable from cell-free extracts of *Serratia indica:* Kobashi, Iwai, *Agr. Biol. Chem.* **35**, 47 (1971); **36**, 1685, 1695 (1972).

Both natural forms found in human urine: Fukushima, Shiota, *J. Biol. Chem.* **247**, 4549 (1972). Early synthetic studies and structure: Rembold, Buschmann, *Ber.* **96**, 1406 (1963). Synthesis of L-(−)-form: Viscontini, Provenzale, *Helv. Chim. Acta* **51**, 1495 (1968). Synthesis of natural D-neopterin: Viscontini *et al.*, *ibid.* **53**, 1202 (1970).

D-*erythro*-Form, $[\alpha]_D^{25}$ +45° ± 3° (c = 0.3 in 0.1*N* HCl).
L-*erythro*-Form, $[\alpha]_D^{25}$ −44° ± 3° (c = 0.3 in 0.1*N* HCl).
D-*threo*-Form, $[\alpha]_D^{25}$ −92° ± 3° (c = 0.3 in 0.1*N* HCl).
L-*threo*-Form, pale yellow, spiny crystals. $[\alpha]_D^{25}$ +97° ± 3° (c = 0.3 in 0.1*N* HCl). uv max in 0.1*N* NaOH: 255, 363 nm (ε 20,900, 7050); in 0.1*N* HCl: 248, 323 nm (ε 11,200, 7880). Strong blue fluorescence in neutral or alkaline soln; weak fluorescence in acidic soln.

6379. Neoquassin. *16-Hydroxy-2,12-dimethoxypicrasa-2,12-diene-1,11-dione; 3a,4,5,5a,6,7,7a,8,11a,11b,11c-deca-hydro-5-hydroxy-2,10-dimethoxy-3,8,11a,11c-tetramethyl-phenanthro[10,1-bc]pyran-1,11-dione.* $C_{22}H_{30}O_6$; mol wt 390.46. C 67.67%, H 7.74%, O 24.59%. Found together with the quassin in the mixture of bitter constituents of the wood of *Quassia amara* L., *Simaroubaceae*, known in commerce as Surinam quassia. Forms quassin on oxidation. Isoln: London *et al.*, *J. Chem. Soc.* **1950**, 3431. Structure: Valenta *et al.*, *Tetrahedron Letters* **1960**(20), 25; Carman, Ward, *ibid.* **1961**, 317; Valenta *et al.*, *Tetrahedron* **15**, 100 (1961).

Very bitter polymorphous crystals. Stable form, thick prisms, mp 227.5-228.5°; unstable form, long narrow plates, mp 213°. $[\alpha]_D^{20}$ +41.0° (c = 4.98 in chloroform). uv max ~255 nm (ε ~11,650). Sol in cold acetone, chloroform, pyridine, acetic acid, in warm ethyl acetate, benzene, alcohol. Sparingly sol in ether, petr ether.

6380. Neostigmine. *3-[[(Dimethylamino)carbonyl]oxy]-N,N,N-trimethylbenzenaminium;* (3-dimethylcarbamoxy-phenyl) trimethylammonium; synstigmin; proserine; Juvastigmin; Neostigmin; Normastigmin; Prostigmin. $[C_{12}H_{19}N_2O_2]^+$; mol wt 209.29. Synthesis: Aeschlimann, U.S. pat. **1,905,990** (1933 to Hoffmann-La Roche); Yanagisawa, **Japan.** pat. **51 3071**, *C.A.* **47**, 4908g (1953). Pharmacokinetics and pharmacological effect: T. N. Calvey *et al.*, *Brit. J. Clin. Pharmacol.* **7**, 149 (1979). Synergism of neostigmine toxicity by lithium: W. M. Davis, N. S. Hatoum, *Toxicology* **17**, 1 (1980). Toxicity: L. O. Randall, G. Lehmann, *J. Pharmacol. Exp. Ther.* **99**, 16 (1950). Comprehensive description: A. A. Al-Badr, M. Tariq in *Analytical Profiles of Drug Substances* vol. **16**, K. Florey, Ed. (Academic Press, New York, 1987) pp 403-444.

Bromide, $C_{12}H_{19}BrN_2O_2$, *Neoesserin (tabl.).* Crystals from

alcohol + ether, dec about 167°. One gram dissolves in about 1 ml water. Sol in alc.
Methyl sulfate, $C_{13}H_{22}N_2O_6S$, *Intrastigmina, Metastigmin, Neoesserin (amp.).* Crystals from alc, mp 142-145°. One gram dissolves in about 10 ml water, less sol in alcohol. LD$_{50}$ in mice (mg/kg): 0.16 i.v.; 0.42 s.c.; 7.5 orally, (Randall, Lehmann).
THERAP CAT: Cholinergic.
THERAP CAT (VET): Parasympathomimetic, has been used in myasthenia gravis (dog), to counteract the action of tubo-curarine chloride.

6381. Neotetrazolium Chloride. *3,3'-[1,1'-Biphenyl]-4,4'-diylbis[2,5-diphenyl-2H-tetrazolium] dichloride; 3,3'-[4,4'-biphenylene]bis[2,5-diphenyl-2H-tetrazolium] dichloride;* 2,2',5,5'-tetraphenyl-3,3'-(p-diphenylene)ditetrazolium chloride; 2,2'-(p-diphenylene)-bis[3,5-diphenyl]ditetrazolium chloride; neotetrazolium blue; neo-T; TP; NTC. $C_{38}H_{28}Cl_2N_8$; mol wt 667.60. C 68.37%, H 4.23%, Cl 10.62%, N 16.78%. Prepn: L. J. Pannone, J. B. Rust, U.S. pat. **2,713,581** (1955 to Montclair Res. Corp. and Ellis-Foster Co.). Purification: G. R. N. Jones, *Histochem. J.* **1**, 59 (1968). *See also* Triphenyltetrazolium chloride.

Light yellow powder, mp 297° (dec).
USE: Staining of microorganisms and plant and animal tissues. In determn of dehydrogenases in histochemical and cytochemical studies: K. Neumann, G. Koch, *Z. Physiol. Chem.* **295**, 35 (1953). In amylase determn.

6382. Neovitamin A. *5-cis-Vitamin A.* $C_{20}H_{30}O$; mol wt 286.44. C 83.86%, H 10.56%, O 5.59%. A naturally occurring isomer of vitamin A. Isolated in cryst form from soupfin shark liver oil. Also present in cod, dogfish, halibut, and California jewfish liver oils to the extent of approx 35% of the total vitamin A content. Synthetic vitamin A concentrates also contain neovitamin A in the same proportion. The biological potencies of neovitamin A and vitamin A, as measured by the U.S.P. growth method in cats, proved to be identical. Isoln: Robeson, Baxter, *J. Am. Chem. Soc.* **69**, 136 (1947); Robeson, U.S. pat. **2,552,908** (1951 to Eastman Kodak). Discussion of stereoisomerism: Zechmeister, *Chem. Rev.* **34**, 267 (1944); Zechmeister, *Vitam. Horm. (New York)* **7**, 57-81 (1949); Vogel-Knobloch, *Chemie und Technik der Vitamine* vol. I (Stuttgart, 3rd ed., 1950) p 102. Additional characteristics: Meunier, Jouannetau, *Bull. Soc. Chim. Biol.* **30**, 260 (1948).
Pale yellow needles from ethyl formate, mp 58-60°. More resistant to atmospheric oxidation than vitamin A. uv max: 328 nm. Neovitamin A and its esters react with maleic anhydride at a much slower rate than vitamin A. This reaction forms a basis of assaying neovitamin A in fish liver oils and concentrates.

6383. Nepetalactone. *5,6,7,7a-Tetrahydro-4,7-dimethyl-cyclopenta[c]pyran-1-(4aH)-one.* $C_{10}H_{14}O_2$; mol wt 166.21. C 72.26%, H 8.49%, O 19.25%. The first fully characterized methylcyclopentane monoterpenoid. Isolated from the volatile oil of catnip produced by *Nepeta cataria* L., *Labiatae*: S. M. McElvain *et al.*, *J. Am. Chem. Soc.* **63**, 1558 (1941); J. Meinwald, *ibid.* **76**, 4571 (1954). First believed to be a single entity, it was later shown to be a mixture of the *cis-trans* and *trans-cis* isomers, the *cis-trans* isomer comprising 70-99% of the oil. Structure: S. M. McElvain, E. J. Eisenbraun, *J. Am. Chem. Soc.* **77**, 1599 (1955). Configuration of the *cis-trans* and *trans-cis* isomers: R. B. Bates *et al.*, *ibid.* **80**, 3420 (1958). Separation and properties: R. B. Bates, C. W. Siegel, *Experientia* **19**, 564 (1963). Biosynthesis: F. E. Regnier *et al.*, *Phytochemistry* **7**, 221 (1968). Metabolism in cats: G. R. Waller *et al.*, *Science* **164**, 1281 (1969). Behavioral and toxicological study: J. W. Harney *et al.*, *Lloydia* **41**, 367

(1978). Structural study of isomeric nepetalactones: E. J. Eisenbraun et al., J. Org. Chem. **45**, 3811 (1980).

cis-trans-nepetalactone

Oil. Odor very attractive to cats. d_4^{25} 1.0663. $bp_{0.05}$ 71-72°. $[\alpha]_D^{23}$+3.6°. n_D^{25} 1.4859. Soluble in ether, carbon tetrachloride. When shaken with 10% NaOH yields *nepetalic acid*, crystals, mp 74°, $[\alpha]_D^{30}$ +47.6° (chloroform).

cis-trans-Form, $C_{10}H_{14}O_2$, *2-(2-hydroxy-1-methylethenyl)-5-methylcyclopentanecarboxylic acid delta lactone.* Oil, $[\alpha]_D^{21}$ +3.7° (c = 27 in chloroform). $[\alpha]_D^{27.5}$ +11.11° (chloroform). n_D^{25} 1.4878.

trans-cis-Form, mp 37-39°, $[\alpha]_D^{21}$ — 24.4° (c = 6.15 in chloroform), E. J. Eisenbraun et al., *loc. cit.*, also reported as $[\alpha]^{27.5}$ +21.9° (chloroform), R. B. Bates, C. W. Siegel, *loc. cit.* n_D^{25} 1.4878.

6384. Neptunium. Np; at. wt 237 (most stable isotope); at. no. 93; valence 3, 4, 5, 6, 7. [239]Np was discovered in 1940: McMillan, Abelson, *Phys. Rev.* **57**, 1185 (1940); [237]Np was discovered in 1942: Wahl, Seaborg, *ibid.* **73**, 940 (1948). Presence in nature: Seaborg, Perlman, *J. Am. Chem. Soc.* **70**, 1571 (1948). Chemical properties: Seaborg, Wahl, *ibid.* 1128. Isoln of the element and determination of the half-life of [237]Np: Magnusson, LaChapelle, *ibid.* 3534; redetermn of half-life: Brauer et al., J. Inorg. Nucl. Chem. **12**, 234 (1960). Prepn of solid neptunium compds: Fried, Davidson, *J. Am. Chem. Soc.* **70**, 3539 (1948). *Reviews:* J. J. Katz, G. T. Seaborg, *The Chemistry of the Actinide Elements* (John Wiley, New York, 1957) pp 204-238; Hindman, *J. Chem. Ed.* **36**, 22-26 (1959); Keller, *Fortschr. Chem. Forsch.* **13**, 1-124 (1969); C. Keller, *The Chemistry of the Transactinide Elements* (Verlag Chemie, Weinheim, English Ed., 1971) pp 253-332; W. W. Schulz, G. E. Benedict, *Neptunium-237; Production and Recovery,* AEC Critical Review Series (USAEC, Washington, D.C., 1972) 85 pp; *Comprehensive Inorganic Chemistry* vol. **5**, J. C. Bailar, Jr. et al., Eds. (Pergamon Press, Oxford, 1973) *passim*.

[237]Np, α-emitting isotope, $T_{1/2}$ 2.14 × 10⁶ years. Silvery metal; orthorhombic structure. d 20.45. mp 640°. Neptunium has been obtained in its five oxidation states in soln; the most stable is the pentavalent state. Tetravalent neptunium is readily oxidized to the hexavalent state by permanganate in the cold, or by strong oxidizing agents; on electrolytic reduction in an atmosphere of nitrogen, the trivalent form is obtained.

6385. Nequinate. *6-Butyl-1,4-dihydro-4-oxo-7-(phenylmethoxy)-3-quinolinecarboxylic acid methyl ester; 7-(benzyloxy)-6-n-butyl-1,4-dihydro-4-oxo-3-quinolinecarboxylic acid methyl ester;* 3-methoxycarbonyl-6-n-butyl-7-benzyloxy-4-oxoquinoline; 7-(benzyloxy)-6-n-butyl-4-hydroxy-3-quinolinecarboxylic acid methyl ester; 7-(benzyloxy)-6-n-butyl-3-methoxycarbonyl-4-quinoline; methyl benzoquate; ICI 55052; Statyl. $C_{22}H_{23}NO_4$; mol wt 365.43. C 72.31%, H 6.34%, N 3.83%, O 17.51%. Prepn: **Belg.** pat. 677,592 and **Brit.** pat. 1,070,223 (1966 to I.C.I.), C.A. **68**, 68899j (1968). Coccidiostatic activity: Bowie et al., *Nature* **214**, 1349 (1967). HPLC determn in poultry feedstuffs: G. H. J. Merson et al., *Analyst* **110**, 761 (1985).

Crystals, mp 287-288°.
THERAP CAT (VET): Coccidiostat.

6386. Neriifolin. *(3β,5β)-3-[(6-Deoxy-3-O-methyl-α-L-*

glucopyranosyl)oxy]-14-hydroxycard-20(22)-enolide. $C_{30}H_{46}O_8$; mol wt 534.70. C 67.39%, H 8.67%, O 23.94%. Cardiac glycoside isolated from *Thevetia neriifolia* Juss.: M. Frèrejacque, *Compt. Rend.* **221**, 645 (1945); from *Cerbera odollam* Gaertn.: S. Rangaswami, E. V. Rao, *J. Sci. Ind. Res.* **16B**, 209 (1957); from *Thevetia thevetioides* seeds: J. L. McLaughlin et al., *J. Econ. Entomol.* **73**, 39 (1980). Pharmacology: K. Mezey, *Arch. Int. Pharmacodyn. Ther.* **84**, 367 (1950); K. K. Chen et al., *ibid.* 81.

Rhombic plates from methanol, mp 218-225°. Also reported as mp 208° (Frèrejacque). $[\alpha]_D^{23}$ — 50.2° (CH₃OH). uv max (CH₃OH): 217 nm (log ε 4.1).

2′-Acetate, $C_{32}H_{48}O_9$, *cerberin, veneniferin, monoacetylneriifolin.* Identity of cerberin with monoacetylneriifolin: M. Frèrejacque, *Compt. Rend.* **226**, 835 (1948); of cerberin with veneniferin: M. Frèrejacque, Durgeat, *ibid.* **228**, 1310 (1949). Coarse prisms from methanol + water, mp 212-215°. $[\alpha]_D^{19}$ — 82° (CHCl₃). Sol in alc, CHCl₃, acetic acid, ether. Practically insol in water.
THERAP CAT: Cardiotonic.

6387. Nerol. *3,7-Dimethyl-2,6-octadien-1-ol;* 2,6-dimethyl-2,6-octadien-8-ol. $C_{10}H_{18}O$; mol wt 154.24. C 77.86%, H 11.76%, O 10.37%. The *cis*-isomer of geraniol; found in many essential oils. Readily loses water and cyclizes forming dipentene. Isoln from neroli oil: Hesse, Zeitschel, *J. Prakt. Chem.* **66**, 502 (1902). Structure: Verley, *Bull. Soc. Chim. France* **25**, 68 (1919); J. L. Simonsen, *The Terpenes* vol. **I** (University Press, Cambridge, 2nd ed., 1947) pp 52-54. Stereochemistry: Burrell et al., *Proc. Chem. Soc.* **1959**, 263; Bates et al., *J. Org. Chem.* **28**, 1086 (1963). Synthesis: Yukawa et al., *Bull. Chem. Soc. Japan* **37**, 158 (1964). Stereochemistry and synthesis: Burrell et al., *J. Chem. Soc. (C)* **1966**, 2144; K. Takabe et al., *Chem. Letters* **1977**, 1025.

Liquid. Odor of sweet rose. bp_{745} 224-225°; bp_{25} 125°. d^{15} 0.8813. Optically inactive. uv max: 189-194 nm (ε 18,000). Sol in abs alc.

Tetrabromide, $C_{10}H_{18}Br_4O$, crystals, mp 118°.
Allophanate, $C_{12}H_{20}N_2O_3$, needles from petr ether, mp 84-86°.
USE: Base for manuf of perfumes.

6388. Nerolidol. *3,7,11-Trimethyl-1,6,10-dodecatrien-3-ol; peruviol.* $C_{15}H_{26}O$; mol wt 222.36. C 81.02%, H 11.79%, O 7.20%. Found in essential oils from many flowers such as *Melaleuca viridiflora* Soland., *Myrtaceae* and *Myroxylon pereirae* (Royle) Klotzsch, *Leguminosae*: Naves, *Compt. Rend.* **251**, 900 (1960). Isoln: Hellyer, McKern, *J. Proc. Roy. Soc. N.S. Wales* **89**, (Pt. 4), 188 (1955); Sutherland et al., *Aust. J. Chem.* **13**, 357 (1960). Synthesis: Nazarov et al., *Zh. Obshch. Khim.* **28**, 1444 (1958); Ofner et al., *Helv. Chim. Acta* **42**, 2577 (1959); Shvarts, Petrov, *Zh. Obshch. Khim.* **30**,

3598 (1960). Exists in two stereoisomeric forms: Bates *et al., J. Org. Chem.* **28**, 1086 (1963).

trans-Form, liquid. $bp_{0.15}$ 78°. n_D^{25} 1.4792.
cis-Form, liquid. $bp_{0.10}$ 70°. n_D^{25} 1.4775.
cis/trans-Form, liquid. bp_3 122°. n_D^{25} 1.4769. d_4^{25} 0.8720.
Sol in abs alc (still clearly sol in 3 parts of 70% alc).

6389. Netilmicin. *O-3-Deoxy-4-C-methyl-3-(methylamino)-β-L-arabinopyranosyl-(1 → 6)-O-[2,6-diamino-2,3,-4,6-tetradeoxy-α-D-glycero-hex-4-enopyranosyl-(1 → 4)]-2-deoxy-N¹-ethyl-D-streptamine; (2S-cis)-4-O-[3-amino-6-(aminomethyl)-3,4-dihydro-2H-pyran-2-yl]-2-deoxy-6-O-[3-deoxy-4-C-methyl-3-(methylamino)-β-L-arabinopyranosyl]-N¹-ethyl-D-streptamine;* 1-N-ethylsisomicin; Sch 20569. $C_{21}H_{41}N_5O_7$; mol wt 475.60. C 53.03%, H 8.69%, N 14.73%, O 23.55%. Broad spectrum semi-synthetic aminoglycoside antibiotic, related to sisomicin, *q.v.* Prepn: M. J. Weinstein *et al.,* **Ger.** pat.**2,437,160** (1975 to Sherico); J. J. Wright, **U.S.** pat. **4,029,882** (1977 to Schering); J. J. Wright *et al.,* **U.S.** pat. **4,002,742** (1977 to Schering). Structure and synthesis: J. J. Wright, *Chem. Commun.* **1976**, 206. Biological activity: G. H. Miller *et al., Antimicrob. Ag. Chemother.* **10**, 827 (1976). Radioimmunoassay: A. Broughton *et al., Clin. Chem.* **24**, 717 (1978). Pharmacology: W. Raab, *Adv. Clin. Pharmacol.* **15**, 91 (1978); I. Trestman *et al., Antimicrob. Ag. Chemother.* **13**, 832 (1978). Metabolism and pharmacokinetics: J. C. Pechere *et al., Clin. Pharmacol. Ther.* **23**, 677 (1978); R. E. Brummett *et al., Arch. Otolaryngol.* **104**, 579 (1978). Clinical studies: J. Klastersky *et al., Antimicrob. Ag. Chemother.* **12**, 503 (1977); A. P. Panwalker *et al., ibid.* **13**, 170 (1978). Toxicity studies: L. Albiero *et al., Arch. Int. Pharmacodyn. Ther.* **233**, 343 (1978); F. C. Lust, *J. Int. Med. Res.* **6**, 286 (1978). *Review:* P. Noone, *Drugs* **27**, 548-578 (1984).

$[\alpha]_D^{26}$ +164° (c = 3 in water). LD_{50} in mice (mg/kg): 40 i.v.; 125 i.p.; 175 s.c. (Miller).
Sulfate, $C_{42}H_{92}N_{10}O_{34}S_5$, *Certomycin, Netillin, Netilyn, Netromicine, Netromycin, Nettacin, Vectacin, Zetamicin.*
THERAP CAT: Antibacterial.

6390. Netobimin. *2-[[[(Methoxycarbonyl)amino][[2-nitro-5-(propylthio)phenyl]amino]methylene]amino]ethanesulfonic acid;* 2-[[[(methoxycarbonyl)amino][[2-nitro-5-(propylthio)phenyl]imino]methyl]amino]ethanesulfonic acid; methyl [N'-[2-nitro-5-(propylthio)phenyl]-N-(2-sulfoethyl)amidino]carbamate; N-methoxycarbonyl-N'-[2-nitro-5-(propylthio)phenyl]-N''-2-(ethylsulfonic acid)guanidine; totabin; Sch 32481; Hapadex. $C_{14}H_{20}N_4O_7S_2$; mol wt 420.46. C 39.99%, H 4.79%, N 13.33%, O 26.64%, S 15.25%. Phenylguanidine anthelmintic. Prepn: M. M. Nafissi-Varchei, **Eur.** pat. **Appl.** **50,286**; *idem,* **U.S.** pat. **4,406,893** (1982, 1983 both to Schering). Metabolism in animals: P. Delatour *et al., J. Vet. Pharmacol. Therap.* **9**, 230 (1986). Anthelmintic efficacy in cattle: J. C. Williams *et al., Am. J. Vet. Res.* **46**, 2188 (1985).

mp 215° (dec).
Sodium salt, $C_{14}H_{20}N_4NaO_7S_2$, crystals, mp 150° (dec) (Nafissi-Varchei); also reported as yellow powder, mp 160° (Delatour). uv max (methanol): 225, 347 nm. Sol in water, acetone, methanol, dimethylformamide, DMSO. Insol in ether.
THERAP CAT (VET): Anthelmintic.

6391. Netropsin. *4-[[[(Aminoiminomethyl)amino]acetyl]amino]-N-[5-[[(3-amino-3-iminopropyl)amino]carbonyl]-1-methyl-1H-pyrrol-3-yl]-1-methyl-1H-pyrrole-2-carboxamide; N'-(2-amidinoethyl)-4-(2-guanidinoacetamido)-1,1'-dimethyl-N,4'-bi[pyrrole-2-carboxamide];* sinanomycin; congocidine; T-1384. $C_{18}H_{26}N_{10}O_3$; mol wt 430.47. C 50.22%, H 6.09%, N 32.54%, O 11.15%. Basic oligopeptide antibiotic produced by *Streptomyces netropsis* with wide range of antimicrobial activity. Prepn and antibacterial spectrum: A. C. Finlay *et al., J. Am. Chem. Soc.* **73**, 341 (1951); and trypanocidal activity: C. Cosar *et al., Compt. Rend.* **234**, 1498 (1952). Antiviral activity: F. M. Schabel *et al., Proc. Soc. Exp. Biol. Med.* **83**, 1 (1953). Identity with sinanomycin: K. Watanabe, *J. Antibiot.* **9A**, 102 (1956). Structural studies: M. Julia, N. Joseph, *Compt. Rend.* **243**, 961 (1956). Structure: C. W. Waller *et al., J. Am. Chem. Soc.* **79**, 1265 (1957). Revised structure: M. Julia, N. Préau-Joseph, *Compt. Rend.* **257**, 1115 (1963); *eidem, Bull. Soc. Chim. France* **1967**, 4348. Specific binding to DNA: Ch. Zimmer *et al., J. Mol. Biol.* **58**, 329 (1971); K. Zakrzewska *et al., Nucleic Acids Res.* **11**, 8825, 8841 (1983). Use in gradient separation of DNA: K. M. Tatti *et al., Anal. Biochem.* **89**, 561 (1978). Selective blocking of DNAase I cleavage: Ch. Zimmer *et al., Nucleic Acids Res.* **8**, 2999 (1980). Review of activity: F. E. Hahn in *Antibiotics* **vol. 3**, J. W. Corcoran, F. E. Hahn, Eds. (Springer-Verlag, New York, 1975) pp 79-100. Review of effect on DNA structure and function: Ch. Zimmer, *Prog. Nucleic Acid Res. Mol. Biol.* **15**, 285-318 (1975).

Disulfate, $C_{18}H_{26}N_{10}O_3 \cdot 2H_2SO_4$, needles, mp 288°. Soly in water about 30 mg/ml at 80°, less than 0.5 mg/ml at 25°. Practically insol in the common organic solvents.
Dihydrochloride, $C_{18}H_{26}N_{10}O_3 \cdot 2HCl$, prisms, mp 228°. LD_{50} in mice (mg/kg): 17 i.v.; 70 s.c.; > 300 orally (Finlay).
USE: Non-intercalative DNA binding agent.

6392. Neuraminic Acid. Prehemataminic acid. $C_9H_{17}NO_8$; mol wt 267.24. C 40.45%, H 6.41%, N 5.24%, O 47.90%. Parent acid of a family of amino sugars contg 9 or more carbon atoms (sialic acids or nonulosaminic acids). May be regarded as the aldol condensation product of pyruvic acid and N-acetyl-D-mannosamine. Neuraminic acid has not been isolated and characterized as such, but several N-and O-substituted derivatives are widely distributed in nature. *Review:* Whelan, *Ann. Repts. Progr. Chem.* **54**, 319 (1958); A. Gottschalk, *The Chemistry and Biology of Sialic Acids and Related Substances* (Cambridge Univ. Press, 1960). *See also* Sialic Acids.

6393. Neurine. *N,N,N-Trimethylethenaminium hydroxide; trimethylvinylammonium hydroxide.* $C_5H_{13}NO$; mol wt 103.16. C 58.21%, H 12.70%, N 13.58%, O 15.51%. $CH_2=CHN(CH_3)_3OH$. Found in egg yolk, brain, bile, in cadavers. Formed during putrefaction by dehydration of choline: Hofmann, *Compt. Rend.* **47**, 559 (1858); Renshaw, Ware, *J. Am. Chem. Soc.* **47**, 2992 (1925). Prepn: Meyer, Hopff, *Ber.* **54**, 2277 (1921); from aq trimethylamine and acetylene: Gardner *et al., J. Chem. Soc.* **1949**, 789. Separation and identification in biological fluids: E. Merlevede *et al., Arch. Int. Pharmacodyn.* **122**, 474 (1959). Stereoelectronic config: *Intern. Kongr. Entomol. Verhandl., 11th, Vienna,* **1960**, No. 3, pp 87-93, *C.A.* **57**, 7749b (1962).

Syrupy liq. Fishy odor. *Poisonous!* Forms a cryst trihydrate. Alkaline reaction. Readily absorbs CO_2 from the air. Sol in water, alcohol. Dec readily, forming trimethylamine. Forms an HCl salt. LD s.c. in mice: 46 mg/kg, Hunt, *J. Pharmacol. Exp. Ther.* **28**, 267 (1926).

6394. Neurotensin. A basic tridecapeptide, found in mammalian brain and gut, having a wide variety of hormone-like activities. Neurotensin has been shown to induce hypotension in the rat, to stimulate contraction of guinea pig ileum and rat uterus, and to cause relaxation of rat duodenum. There is also evidence that it acts as a CNS neurotransmitter. Isoln from bovine hypothalamus: R. Carraway, S. Leeman, *J. Biol. Chem.* **248**, 6854 (1973). Amino acid sequence of bovine neurotensin: *eidem, ibid.* **250**, 1907 (1975). Synthesis: *eidem, ibid.* 1912; K. Kitagawa *et al., Chem. Pharm. Bull.* **24**, 2692 (1976); H. Yajima *et al., Int. J. Peptide Protein Res.* **14**, 169 (1979). Isoln, amino acid sequence of chicken neurotensin: R. Carraway, Y. M. Bhatnagar, *Peptides* **1**, 167 (1980). Synthesis: H. Yajima *et al., Chem. Pharm. Bull.* **29**, 2587 (1981). Radioimmunoassay and differential distribution of bovine neurotensin in CNS, small intestine, and stomach: R. Carraway, S. Leeman, *J. Biol. Chem.* **251**, 7035, 7045 (1976). CNS effects: C. B. Nemeroff *et al., Brain Res.* **128**, 485 (1977). Receptor binding in brain: G. R. Uhl *et al., ibid.* **130**, 299 (1977). Pharmacokinetics and effect on gastrointestinal and pituitary hormones in man: A. M. Blackburn *et al., J. Clin. Endocrinol. Metab.* **51**, 1257 (1980). Reviews of biological activity: G. R. Uhl, S. H. Snyder, *Advan. Biochem. Psychopharmacol.* **28**, 87-106 (1981); D. R. Brown, R. J. Miller, *Ann. Rep. Med. Chem.* **17**, 271-280 (1982). Book: *Ann. N.Y. Acad. Sci.* **400**, entitled "Neurotensin, a Brain and Gastrointestinal Peptide", C. B. Nemeroff, A. J. Prange, Eds. (1982) pp 1-444.

```
pyro-Glu-Leu-Tyr-Glu-Asn-Lys-Pro-Arg-Arg-Pro-Tyr-Ile-Leu-OH

              bovine neurotensin
```

Bovine neurotensin triacetate hexahydrate, $C_{84}H_{133}N_{21}O_{26}$.6H_2O, *neurotensin(ox) triacetate(salt).* Fluffy white powder. $[\alpha]_D^{23}$ $-65.6°$ (c = 0.5 in water).

6395. Neutral Red. $N^8,N^8,3$-*Trimethyl-2,8-phenazinediamine monohydrochloride; C.I. Basic Red 5;* 3-amino-7-dimethylamino-2-methylphenazine hydrochloride; toluylene red; neutral red chloride; nuclear fast red; aminodimethylaminotoluaminozine hydrochloride; C.I. 50040; Kernechtrot; Michrome no. 226. $C_{15}H_{17}ClN_4$; mol wt 288.78. C 62.38%, H 5.93%, Cl 12.28%, N 19.40%. Prepn from *N,N*-dimethyl-*p*-nitrosoaniline hydrochloride + toluene-2,4-diamine: Witt, *Ber.* **12**, 931 (1879); Ger. pat. **15,272** (1880), *Frdl.* **1**, 247 (1888); Bernthsen, Schweitzer, *Ann.* **236**, 332 (1886); Pokorny, *J. Soc. Dyers Colour.* **42**, 347 (1926). Toxicity data: F. Stolarsky, T. J. Haley, *Fed. Proc.* **10**, 337 (1951). Brief

review: H. J. Conn's *Biological Stains,* R. D. Lillie, Ed. (Williams & Wilkins, Baltimore, 9th ed., 1977) pp 377-380, 596. *See also: Colour Index* vol. **4** (3rd ed., 1971) p 4446.

Dark green powder. Absorption max (50% alc): 533 nm. pKa 6.7. Soluble in water or alcohol with red color. Soly in water 4.0%, in abs alcohol 1.8%, in Cellosolve 3.75%, in ethylene glycol 3.0%. Practically insol in xylene. LD_{50} in mice, rats, rabbits (mg/kg): 142, 112, 97 i.v. (Stolarsky, Haley).

USE: As indicator (0.1% in 60% alcohol) for alkalinity of water, urea, nitrite, etc. pH: 6.8 red; 8.0 yellow. Also used for preparing neutral-red paper. As biological stain (study of the Golgi apparatus in cells).

6396. Neutral Spirits. Distilled from suitable raw materials, is 95% ethanol (v/v) meaning that it is at least 190 proof when distilled. Used for blending with straight whisky (*see* Whisky) and for making gin, cordials, liqueurs and vodka. It may be made from grain or molasses or by redistillation of other beverages, such as brandy or rum, and is almost colorless and contains no extraneous flavoring.

6397. NGF. *Nerve growth factor.* Potent stimulator of growth and differentiation of peripheral, sympathetic, and sensory ganglia: Levi-Montalcini, *Ann. N.Y. Acad. Sci.* **55**, 330 (1952). First observed in certain mouse sarcomas. Found in trace amounts in a variety of mammalian tissues and body fluids, and in higher concentrations in certain snake venoms and mouse submaxillary glands: Levi-Montalcini, Cohen, *Proc. Nat. Acad. Sci. USA* **42**, 695 (1956); *eidem, Ann. N.Y. Acad. Sci.* **85**, 324 (1960). Composed of acidic, basic, and neutral components called the α-, β-, and γ-subunits, of which only the β-subunit shows nerve growth stimulating activity: Bocchini, Angeletti, *Proc. Nat. Acad. Sci. USA* **64**, 787 (1969). Structure of the active component of mouse submaxillary NGF is a dimeric polypeptide containing identical subunits of 118 amino acid residues of mol wt about 13,000: Angeletti *et al., Biochemistry* **10**, 463 (1971); **12**, 100 (1973). Amino acid sequence: Angeletti, Bradshaw, *Proc. Nat. Acad. Sci. USA* **68**, 2417 (1971). Reviews: Levi-Montalcini, Angeletti, *Physiol. Rev.* **48**, 534 (1968); Angeletti *et al., Advan. Enzymol. Relat. Areas Mol. Biol.* **31**, 51 (1968); Vernon *et al.,* in *Homeostatic Regulators, Ciba Found. Symp.,* G. E. W. Wolstenholme, J. Knight, Eds. (J. & A. Churchill Ltd., London, 1969) pp 57-74. Review and comparison with insulin: Bradshaw *et al., Recent Progr. Horm. Res.* **30**, 575 (1974).

6398. Niacinamide. *3-Pyridinecarboxamide;* nicotinic acid amide; nicotinamide; vitamin PP; Benicot; Aminicotin; Vi-Nicotyl; Dipegyl; Nicamindon; Nicotamide; Nicotilamide; Pelonin Amide; Amide PP; Nicofort; Niozymin; Pelmine. $C_6H_6N_2O$; mol wt 122.12. C 59.01%, H 4.95%, N 22.94%, O 13.10%. Occurs in plants and animals, usually in conjugated form (in enzyme systems). Isolation: Euler *et al., Z. Physiol. Chem.* **258**, 212 (1939); Karrer, Keller, *Helv. Chim. Acta* **22**, 1292 (1939). Prepd by the action of thionyl chloride on nicotinic acid, and treating the resulting acid chloride with ammonia. Alternately prepd by passing NH_3 gas into molten nicotinic acid: Truchan, Davidson, U.S. pat. **2,993,051** (1961 to Cowles Chem.). From 3-cyanopyridine: Gasson, Hadley, U.S. pat. **2,904,552** (1959 to Distillers). Toxicity data: Brazda, Coulson, *Proc. Soc. Exp. Biol. Med.* **62**, 19 (1946).

Needles from benzene, mp 128-131°. Distills at 150-160° at 5×10^{-4} mm Hg. Absorption spectrum: Kuhn, Vetter,

Ber. **68**, 2374 (1935). One gram dissolves in about one ml water, in about 1.5 ml alcohol, in 10 ml glycerol. A 10% w/v soln in water is neutral to litmus. Forms cryst salts with acids. LD_{50} s.c. in rats: 1.68 g/kg (Brazda, Coulson). *Note:* Niacinamide and nicotinic acid have been referred to as *Vitamin B₃:* Lecoq *et al., Compt. Rend.* **222**, 414 (1946), and *Vitamin B₅:* Cheldelin in *The Vitamins* vol. 3, W. H. Sebrell, Jr., R. S. Harris, Eds. (Academic Press, New York, 1954) p 596, 598. *See also* Pantothenic Acid.

THERAP CAT: Vitamin (enzyme cofactor).

THERAP CAT (VET): Nutritional factor in nicotinic acid deficiency in dog, pig.

6399. Nialamide. *4-Pyridinecarboxylic acid 2-[3-oxo-3-[(phenylmethyl)amino]propyl]hydrazide; isonicotinic acid 2-[2-(benzylcarbamoyl)ethyl]hydrazide;* 1-[2-(benzylcarbamoyl)ethyl]-2-isonicotinoylhydrazide; *N*-[2-(benzylcarbamyl)ethylamino]isonicotinamide; *N*-benzyl-β-(isonicotinoylhydrazino)propionamide; *N*-isonicotinoyl-*N'*-[β-(*N*-benzylcarboxamido)ethyl]hydrazide; Espril; Niamid; Niamidal; Niaquitil; Nuredal; Nyazin. $C_{16}H_{18}N_4O_2$; mol wt 298.34. C 64.41%, H 6.08%, N 18.78%, O 10.73%. Prepn: Bloom, Carnahan, U.S. pats. **2,894,972; 3,040,061** (1959, 1962, both to Pfizer).

CONHNHCH₂CH₂CONH — CH₂C₆H₅

Slightly bitter crystals from ethyl acetate, mp 152-153°. Sparingly sol in water; freely sol in acidic solvents. LD_{50} in mice: 590 mg/kg orally; 435 mg/kg i.p.; 120 mg/kg i.v., *Toxic Substances List,* H. E. Christensen, Ed. (1974) p 436.

THERAP CAT: Antidepressant.

6400. Niaprazine. *N-[3-[4-(4-Fluorophenyl)-1-piperazinyl]-1-methylpropyl]-3-pyridinecarboxamide; N-[3-[4-(p-fluorophenyl)-1-piperazinyl]-1-methylpropyl]nicotinamide;* Nopron. $C_{20}H_{25}FN_4O$; mol wt 356.45. C 67.39%, H 7.09%, F 5.33%, N 15.72%, O 4.49%. Prepn: R. Y. Mauvernay, **Ger. pat. 1,957,371** corresp to **U.S. pat. 3,712,893** (1970, 1973 both to CERM). Pharmacology: P. Duchene-Marrullaz *et al., Therapie* **26**, 1203 (1971). Psycholeptic effects in mice: J. Hache, J. Tachon, *J. Pharmacol.* **7**, 469 (1976). Brain catecholamine depletion: P. E. Keane, M. S. Benedetti, *Neuropharmacology* **18**, 595 (1979).

CONHCH—(CH₂)₂—N   N— F
     CH₃

Cryst from ethyl acetate, mp 131°. Photosensitive. LD_{50} in mice (mg/kg): 890 orally, 145 i.v. (Mauvernay).

THERAP CAT: Sedative, hypnotic.

6401. Nicametate. *3-Pyridinecarboxylic acid 2-(diethylamino)ethyl ester; nicotinic acid 2-diethylaminoethyl ester;* diethylaminoethyl 3-pyridinecarboxylate; diethylaminoethyl nicotinate; Eucast. $C_{12}H_{18}N_2O_2$; mol wt 222.28. C 64.84%, H 8.16%, N 12.60%, O 14.40%. Prepd from nicotinoyl chloride and diethylaminoethanol or from nicotinic acid and diethylaminoethyl chloride: Blicke, Jenner, *J. Am. Chem. Soc.* **64**, 1722 (1942). Clinical trial in vascular pathology: P. Langeron, *J. Sci. Med. Lille* **90**, 179 (1972). Use of the citrate in cardiovascular surgery: C. Stankowiak *et al., Lille Med.* **18**, 1039 (1973).

COOC₂H₄N(C₂H₅)₂

Liquid. bp₁₀ 155-157°; bp₂ 120-125°.
Hydrochloride, $C_{12}H_{19}ClN_2O_2$, crystals from acetone, ethanol or isopropyl alcohol, mp 128-129°.
Citrate monohydrate, $C_{18}H_{26}N_2O_9 \cdot H_2O$, *Euclidan, Nutrin, Soclidan.*

THERAP CAT: Vasodilator.

6402. Nicarbazin. *N,N'-Bis(4-nitrophenyl)urea, compd with 4,6-dimethyl-2(1H)-pyrimidinone (1:1); 4,4'-dinitrocarbanilide compd with 4,6-dimethyl-2-pyrimidinol (1:1);* Nicarb; Nicoxin; Nicrazin. $C_{19}H_{18}N_6O_6$; mol wt 426.38. C 53.52%, H 4.26%, N 19.71%, O 22.51%. Prepn: Cuckler *et al., Science* **122**, 244 (1955); Basso, O'Neill, U.S. pats. **2,-731,382-4** (all 1956 to Merck & Co.). Effect on nutritional encephalopathy in chicks: I. Bartov, P. Budowski, *Poultry Sci.* **58**, 597 (1979).

O₂N— —NHCONH— —NO₂

H₃C—⟨ ⟩—OH
     CH₃

Crystals, dec 265-275°. uv max (concd H_2SO_4): 298 nm ($A^{1\%}_{1cm}$ 670). Practically insol in water. Complex is slowly dec by trituration with water, or more rapidly by dil aq acids. The dry crystals are strongly electrostatic and present some dry-mixing problems.

THERAP CAT (VET): Coccidiostat.

6403. Nicardipine. *1,4-Dihydro-2,6-dimethyl-4-(3-nitrophenyl)-3,5-pyridinedicarboxylic acid methyl 2-[methyl-(phenylmethyl)amino]ethyl ester.* $C_{26}H_{29}N_3O_6$; mol wt 479.54. C 65.12%, H 6.10%, N 8.76%, O 20.02%. Dihydropyridine calcium channel blocker. Prepn: M. Iwanami *et al.,* **Japan. Kokai 74 109,384** (1974 to Yamanouchi), *C.A.* **82**, 170692n (1975); of the hydrochloride: M. Murakami *et al.,* **Ger. pat. 2,407,115** (1974 to Yamanouchi), *C.A.* **82**, 4131j (1975); M. Iwanami *et al., Chem. Pharm. Bull.* **27**, 1426 (1979). Vasodilator profile: T. Takenaka *et al., Arzneimittel-Forsch.* **26**, 2172 (1976). Absorption, excretion, metabolism: S. Higuchi *et al., Xenobiotica* **7**, 469 (1977). Clinical pharmacology: T. Seki, T. Takenaka, *Int. J. Clin. Pharmacol. Biopharm.* **15**, 267 (1977). Mechanism of action: K. Satoh *et al., Clin. Exp. Pharmacol. Physiol.* **7**, 249 (1980). Acute toxicity study: Y. Odani, T. Sado, *Oyo Yakuri* **18**, 301 (1979), *C.A.* **92**, 104384u (1980). Symposium: *Brit. J. Clin. Pharmacol.* **20**, Suppl. 1, 1S-208S (1985). Clinical synergy with captopril, *q.v.:* K. A. Conrad *et al., Clin. Pharmacol. Ther.* **42**, 113 (1987). Review of pharmacodynamics, pharmacokinetics and therapeutic efficacy: E. M. Sorkin, S. P. Clissold, *Drugs* **33**, 296-345 (1987).

H₃C   H   CH₃
   CH₃
H₃COOC   COOCH₂CH₂N—CH₂—C₆H₅

NO₂

Hydrochloride, $C_{26}H_{30}ClN_3O_6$, RS-69216, YC-93, Barizin, Bionicard, Cardene, Dacarel, Lecibral, Lescodil, Loxen, Nerdipina, Nicardal, Nicarpin, Nicodel, Nimicor, Perdipina, Perdipine, Ranvil, Ridene, Roxen, Rycarden, Rydene, Vasodin, Vasonase. Isolated in two crystalline forms from methanol/acetone. α-form: mp 179-181°; β-form: mp 168-170°. The forms also have different IR and x-ray diffraction patterns. LD_{50} in male, female rats (mg/kg): 634, 557 orally; 18.1, 25.0 i.v.; in male, female mice: 634, 650 orally; 20.7, 19.9 i.v. (Odani, Sado).

THERAP CAT: Anti-anginal; antihypertensive.

6404. Nicergoline. *(8β)-10-Methoxy-1,6-dimethylergoline-8-methanol 5-bromo-3-pyridinecarboxylate (ester);* 1-methyllumilysergol 8-(5-bromonicotinate) 10-methyl ether; 4,6,6a,7,8,9,10,10a-octahydro-10aα-methoxy-4,7-dimethylindolo[4,3-fg]quinoline-9-methanol 5-bromonicotinate; 8β-[(5-bromonicotinoyloxy)methyl]-1,6-dimethyl-10α-methoxyergoline; nicoterguline; nimergoline; MNE; FI 6714; Duracebrol; Memoq; Nargoline; Sermion; Vasospan. $C_{24}H_{26}BrN_3O_3$; mol wt 484.40. C 59.51%, H 5.41%, Br 16.50%,

N 8.67%, O 9.91%. Prepn: Bernardi *et al.,* U.S. pat. **3,228,-943**; Temperilli, **Ger.** pat. **2,112,273** (1966, 1971, both to Farmitalia); Arcari *et al., Experientia* **28**, 819 (1972). Series of articles on pharmacology, clinical studies, tolerability: *Arzneimittel-Forsch.* **29**, 1213-1316 (1979). Toxicity study: B. W. Neumann, F. Lauschner, *ibid.* 1206. Hemodynamic effects in the dog: *ibid.* **31**, 1693 (1981). Use in acute myocardial infarction with diastolic hypertension: E. Triulzi *et al., Farmaco Ed. Prat.* **36**, 449 (1981).

mp 136-138°. LD_{50} in male mice, rats (mg/kg): 860, 2800 orally; 46, 43 i.v. (Neumann, Lauschner).

THERAP CAT: Vasodilator.

6405. Niceritrol. *3-Pyridinecarboxylic acid 2,2-bis[[(3-pyridinylcarbonyl)oxy]methyl]-1,3-propanediyl ester; nicotinic acid neopentanetetrayl ester;* pentaerythritol tetranicotinate; 8 AL; Perycit; Bufor; Cardiolipol. $C_{29}H_{24}N_4O_8$; mol wt 556.54. C 62.58%, H 4.34%, N 10.07%, O 23.00%. Prepn: Fr. pat. M2046 (1963 to Societé d'Etudes et de Recherches Pharmacotechniques), *C.A.* **60**, 10656e (1964); **Brit.** pats. **1,022,880; 1,053,689** (1966, 1967 to Aktiebolag Bofors), *C.A.* **64**, 19708d (1966), *C.A.* **66**, 104907e (1967); Kuriyama, Kudo, **Japan.** pat. 67 2359 (to Yoshitomi), *C.A.* **66**, 115935p (1967). Acute toxicity data: T. Sugawara *et al., Oyo Yakuri* **14**, 741 (1977), *C.A.* **88**, 131005v (1977).

Crystals, mp reported as 160-162° and as 163-164°. LD_{50} in mice, rats (g/kg): >20, >20 orally; >5, >5 s.c.; >5, >5 i.p.; in rabbits (g/kg): >10 orally; >5 i.p. (Sugawara).

THERAP CAT: Antihyperlipoproteinemic.

6406. Nickel. Ni; at. wt 58.69; at. no. 28; valence 2; seldom 1, 3, 4. Five naturally occurring isotopes: 58 (67.76%); 60 (26.16%); 61 (1.25%); 62 (3.66%); 64 (1.16%); artificial, radioactive isotopes: 56; 57; 59; 63; 65-67. Abundance in earth's crust 0.018%. Its elementary nature was recognized by Cronstedt in 1754: Cronstedt, *Mineralogie* (Stockholm, 1758) p 218. Isolated by Berthier, *Ann. Chim. Phys.* [2] **14**, 52 (1820); **25**, 94 (1824). Occurs free in meteorites. Found in many ores as sulfides, arsenides, antimonides and oxides or silicates; chief sources include chalcopyrite, *q.v.,* pyrrhotite, *pentlandite* [(Fe,Ni)$_9$S$_8$] and *garnierite* [3(Mg,Ni)O.-2SiO$_2$.2H$_2$O]; other ores include *niccolite* (NiAs) and *millerite* (NiS). Methods of extraction and purification: Mackiw, *Can. J. Chem. Eng.* **46**, 3 (1968); Houot, *Ann. Mines* **1969** (April), 9; Queneau, *J. Metals* **22**, 44-48 (1970). Prepn of high purity nickel: Wise, Schaefer, *Metals Alloys* **16**, 424 (1924); from NiO and H$_2$: Glemser in *Handbook of Preparative Inorganic Chemistry* vol. 2, G. Brauer, Ed. (Academic Press, New York, 2nd ed., 1965) pp 1543-1544; by electrolysis: Vu Quang Kinh, Nardin, *Compt. Rend. Ser. C* **266**, 307

(1968). Comprehensive reviews: *Gmelin's, Nickel* (8th ed.) **57**, 5 vols, about 3500 pp (1965-1967); Nicholls in *Comprehensive Inorganic Chemistry* vol. **3**, J. C. Bailar, Jr. *et al.,* Eds. (Pergamon Press, Oxford, 1973) pp 1109-1161; J. K. Tien, T. E. Howson in Kirk-Othmer *Encyclopedia of Chemical Technology* vol. **15** (Wiley-Interscience, New York, 3rd ed., 1981) pp 787-801. Book: *Nickel Toxicology*, S. S. Brown, F. W. Sunderman, Eds. (Academic Press, New York, 1980) 193 pp. Review of carcinogenicity studies: *IARC Monographs* **11**, 75-112 (1976).

Lustrous white, hard, ferromagnetic metal; face-centered cubic crystals. mp 1555°. bp (calc) 2837° (3110°K): D. R. Stull, G. C. Sinke, *Thermodynamic Properties of the Elements,* Advances in Chemistry Series **18** (A.C.S., Washington, 1956). d 8.90. Heat capacity (25°) 6.23 cal/g-atom/°C. Mohs' hardness 3.8. Latent heat of fusion 73 cal/g. Electrical resistivity (20°): 6.844 μohms-cm. E°(aq) Ni/Ni^{2+} 0.250 V. Stable in air at ordinary temp; burns in oxygen, forming NiO; not affected by water; decomposes steam at a red heat. Slowly attacked by dil hydrochloric or sulfuric acid; readily attacked by nitric acid. Not attacked by fused alkali hydroxides.

Caution: May cause dermatitis in sensitive individuals. Ingestion of soluble salts causes nausea, vomiting, diarrhea: E. Browning, *Toxicity of Industrial Metals* (Appleton-Century-Crofts, New York, 2nd ed., 1969) pp 249-260. This substance and certain nickel compounds may reasonably be anticipated to be carcinogens: *Fourth Annual Report on Carcinogens* (NTP 85-002, 1985) p 138.

USE: Nickel-plating; for various alloys such as new silver, Chinese silver, German silver; for coins, electrotypes, storage batteries; magnets, lightning-rod tips, electrical contacts and electrodes, spark plugs, machinery parts; catalyst for hydrogenation of oils and other organic substances. *See also* Raney nickel. Probably the largest use of nickel is in the manuf of Monel metal, stainless steels, and nickel-chrome resistance wire; in alloys for electronic and space applications.

6407. Nickel Acetate. C$_4$H$_6$NiO$_4$; mol wt 176.80. C 27.17%, H 3.42%, Ni 33.21%, O 36.20%. Ni(CH$_3$CO$_2$)$_2$.

Tetrahydrate, green, cryst mass or powder; acetic odor. d 1.744. Sol in 6 parts water, in alcohol. *Keep well closed.*

6408. Nickel Acetylacetonate. *Bis(2,4-pentanedionato-O,O')nickel;* bisacetylacetonatonickel(II); bis(2,4-pentanediono)nickel(II); 2,4-pentanedione nickel complex. C$_{10}$H$_{14}$NiO$_4$; mol wt 256.93. C 46.75%, H 5.49%, Ni 22.85%, O 24.91%. Ni(CH$_3$COCHCOCH$_3$)$_2$. Also Ni(acac)$_2$ or Ni(AA)$_2$. Prepn from acetylacetone and Ni(OH)$_2$: Gach, *Monatsh.* **21**, 103 (1900); from acetylacetone and NiCl$_2$.-6H$_2$O: Charles, Pawlikowski, *J. Phys. Chem.* **62**, 440 (1958); from 4-diethylamino-3-pentene-2-one and NiCl$_2$.6H$_2$O: Gash, *Can. J. Chem.* **45**, 2109 (1967). *See also* Fernelius, Bryant, *Inorg. Syn.* **5**, 105 (1957). Exists as a trimer in the solid state: Bullen, *Nature* **177**, 537 (1956); Bullen *et al., Inorg. Chem.* **4**, 456 (1965); as a monomer in the vapor phase: Fackler *et al., J. Phys. Chem.* **72**, 4631 (1972). Structure of dihydrate: Montgomery, Lingafelter, *Acta Cryst.* **17**, 1481 (1964).

Emerald-green orthorhombic crystals. mp 229-230°. bp$_{11}$ 220-235°. d^{17} 1.455. uv max (10^{-4}M in CHCl$_3$): 298, 265 nm (log ϵ 4.34, 4.44). Sol in water, alcohol, chloroform, benzene. Insol in ether, ligroin.

USE: Catalyst.

6409. Nickel Bromide. Nickel dibromide. Br$_2$Ni; mol wt 218.53. Br 73.13%, Ni 26.87%. NiBr$_2$.

Trihydrate, yellowish-green, very deliquesc crystals; loses its water at about 200°, the anhyd salt is a golden-yellow color and sublimable in absence of air. Sol in one part water, in alcohol. *Keep well closed.*

6410. Nickel Carbonate Hydroxide. CH$_4$Ni$_3$O$_7$; mol wt 304.17. C 3.95%, H 1.32%, Ni 57.91%, O 36.82%. NiCO$_3$.-2Ni(OH)$_2$. Tetrahydrate occurs in nature as the mineral *zaratite.*

Tetrahydrate, green, odorless powder. Insol in water; sol in ammonia and in dil acids with effervescence.

USE: Nickel-plating; catalyst for hardening of fats; in ceramic colors and glazes.

6411. Nickel Carbonyl. Nickel tetracarbonyl. C_4NiO_4; mol wt 170.73. C 28.13%, Ni 34.38%, O 37.48%. $Ni(CO)_4$. Intermediate in nickel refining. Made by passing carbon monoxide over finely divided nickel: Mond et al., J. Chem. Soc. **57**, 749 (1890); Gilliland, Blanchard, Inorg. Syn. **2**, 234 (1946). Use of nickel carbonyl in organic synthesis: G. Wilke et al., Angew. Chem. Int. Ed. **5**, 151 (1966); M. F. Semmelhack in Organic Reactions vol. **19** (Wiley, New York, 1972) p 115; E. J. Corey, H. A. Kirst, J. Am. Chem. Soc. **94**, 667 (1972). Kinetic studies: D. H. Stedman et al., Science **208**, 1029 (1980). Toxicity study: Hackett, Sunderman, Arch. Environ. Health **14**, 604 (1967). Review: Nicholls in Comprehensive Inorganic Chemistry vol. **3**, J. C. Bailar, Jr. et al., Eds. (Pergamon Press, Oxford, 1973) pp 1115-1119.

Colorless, volatile liquid. Poisonous! Oxidizes in the air: explodes at about 60°. d^{17} 1.318. bp 43°. mp −19.3°. Crit temp about 200°. Crit pressure about 30 atm. Sol in about 5000 parts water free from air; sol in alcohol, benzene, chloroform, acetone, carbon tetrachloride. LD_{50} in rats (mg/kg): 39 i.p.; 63 s.c.; 66 i.v. (Hackett, Sunderman).

Caution: One of the most toxic chemicals encountered industrially. Inhalation of vapors may cause pulmonary edema with focal hemorrhage: Clinical Toxicology of Commercial Products, R. E. Gosselin et al., Eds. (Williams & Wilkins, Baltimore, 5th ed., 1984) Section II, p. 145. This substance may reasonably be anticipated to be a carcinogen: Fourth Annual Report on Carcinogens (NTP 85-002, 1985) p 138.

USE: In organic synthesis.

6412. Nickel Chloride. Nickel dichloride. Cl_2Ni; mol wt 129.61. Cl 54.70%, Ni 45.30%. $NiCl_2$.

Hexahydrate, green, deliquesc crystals or cryst powder. Monoclinic. Structure reported to be trans-$[NiCl_2(H_2O)_4]$.-$2H_2O$: Mizuno, J. Phys. Soc. Japan **16**, 1574 (1960), C.A. **55**, 26605g (1961). Sol in about one part water, in alcohol. The anhydr salt is golden-yellow, sublimable in absence of air and readily absorbs NH_3. The aq soln is acid; pH about 4. Keep well closed. LD i.v. in dogs: 40-80 mg/kg, Handbook of Toxicology, vol. **1**, W. S. Spector, Ed. (Saunders, Philadelphia, 1956) pp 212-213.

USE: For nickel-plating cast zinc, manuf sympathetic ink. The anhydr salt as absorbent for NH_3 in gas masks.

6413. Nickel Cyanide. C_2N_2Ni; mol wt 110.74. C 21.69%, N 25.30%, Ni 53.00%. $Ni(CN)_2$. Prepn of yellow-brown anhydr salt: Aynsley, Campbell, J. Chem. Soc. **1958**, 1723. (The commercial salt usually contains 20-25% water.)

Tetrahydrate, apple-green powder. Poison! Insol in water; slightly sol in dil acids, freely in alkali cyanides, in ammonia, and in ammonium carbonate.

USE: In nickel-plating.

6414. Nickel Dimethylglyoxime. Bis[(2,3-butanedione dioximato)(1−)-N,N']nickel; bis(dimethylglyoximato)nickel. $C_8H_{14}N_4NiO_4$; mol wt 288.91. C 33.26%, H 4.88%, N 19.39%, Ni 20.31%, O 22.15%. Prepn: Banks et al., J. Am. Chem. Soc. **77**, 324 (1955); F. J. Welcher, Organic Analytical Reagents vol. **3** (Van Nostrand, New York, 1947) pp 165-179; Thabet et al., Inorg. Nucl. Chem. Letters **8**, 211 (1972). Structure: Godycki, Rundle, Acta Cryst. **6**, 487 (1953); Merritt, Anal. Chem. **25**, 718 (1953).

Scarlet-red, cryst powder. Sublimes at 250°. Insol in water, acetic acid, ammonia; sol in dil mineral acids and appreciably sol in abs alcohol.

USE: As sun-fast pigment in paints, lacquers, cellulose compounds and cosmetics.

6415. Nickel Fluoride. Nickel difluoride; nickelous fluoride. F_2Ni; mol wt 96.69. Ni 60.71%, F 39.29%. NiF_2. Prepn: Henkel, Klemm, Z. Anorg. Allgem. Chem. **222**, 74 (1935); Priest, Inorg. Syn. **3**, 173 (1950); Rochow, Kukin, J. Am. Chem. Soc. **74**, 1615 (1952); Haendler et al., ibid. 3167.

Yellowish to green tetragonal crystals (rutile type). d 4.72. Sublimes in HF stream above 1000°. Slightly sol in water (4 g/100 ml at 25°). Aq solns are dec by boiling. Insol in alcohol, ether. LD_{50} in mice: 130 mg/kg i.v., Toxic Substances List, H. E. Christensen, Ed. (1973) p 652.

Caution: Chronic exposure may cause mottling of teeth, changes in bones.

6416. Nickel Formate. $C_2H_2NiO_4$; mol wt 148.73. C 16.15%, H 1.36%, Ni 39.46%, O 43.03%. $Ni(HCOO)_2$. Prepd by reaction of formic acid with Ni: Johnson, U.S. pat. **2,576,072** (1951 to Harshaw Chemical); with $NiCO_3$: Bircumshaw, Edwards, J. Chem. Soc. **1950**, 1800.

Dihydrate, fine, green, monoclinic crystals. Becomes anhydr on careful heating to 130-140°; decomposes at 180-200° yielding Ni, CO, CO_2, H_2, H_2O, CH_4. $d^{20.2}$ 2.154. Moderately sol in water; practically insol in alc, formic acid.

USE: Manuf of Ni; prepn of Ni catalysts for organic reactions, particularly hydrogenation catalysts.

6417. Nickel Hydroxide. "Green nickel oxide". H_2NiO_2; mol wt 92.72. H 2.17%, Ni 63.32%, O 34.51%. $Ni(OH)_2$.

Monohydrate, apple-green powder. Decomp above 200° to form NiO and H_2O. Insol in water; sol in dil acids, in ammonia.

6418. Nickel Iodide. Nickel diiodide. I_2Ni; mol wt 312.51. I 81.21%, Ni 18.79%. NiI_2.

Iron-black color. Sublimes in absence of air. Hexahydrate, bluish-green very deliquesc crystals. Very sol in water or alcohol. Keep well closed.

6419. Nickel Monoxide. Nickelous oxide; nickel protoxide. NiO; mol wt 74.69. Ni 78.58%, O 21.42%. Occurs as the mineral bunsenite.

Green powder; yellow when hot. Insol in water; sol in acids.

USE: Painting on porcelain.

6420. Nickel Nitrate. N_2NiO_6; mol wt 182.72. N 15.33%, Ni 32.13%, O 52.54%. $Ni(NO_3)_2$.

Hexahydrate, green, deliquesc crystals. d 2.05. mp 56.7°. bp 137°. Sol in 0.4 part water, in alcohol. The aq soln is acid; pH about 4. Keep well closed. LD_{50} orally in rats: 1.62 g/kg, H. F. Smyth et al., Am. Ind. Hyg. Assoc. J. **30**, 470 (1969).

USE: Nickel-plating; manuf brown ceramic colors.

6421. Nickel Oxalate. C_2NiO_4; mol wt 146.73. C 16.37%, Ni 40.00%, O 43.62%. NiC_2O_4.

Dihydrate, light green powder. Insol in water; sol in mineral acids, in solns of ammonium chloride, nitrate, or sulfate.

6422. Nickel Phosphate. $Ni_3O_8P_2$; mol wt 366.07. Ni 48.11%, O 34.97%, P 16.92%. $Ni_3(PO_4)_2$.

Octahydrate, light green powder. Insol in water; sol in acids, ammonia.

USE: On ignition yields "nickel yellow"—a pigment used in oil and water colors.

6423. Nickel Sesquioxide. Nickelic oxide; black nickel oxide. Ni_2O_3; mol wt 165.42. Ni 70.98%, O 29.02%. (Contains a variable quantity of water.)

Gray-black powder. Dec at about 600° into NiO and oxygen. Insol in water; very slightly sol in cold acid; dissolved by hot HCl with evolution of Cl, and by hot H_2SO_4 or HNO_3 with evolution of oxygen.

6424. Nickel Sulfate. NiO_4S; mol wt 154.77. Ni 37.93%, O 41.35%, S 20.71%. $NiSO_4$.

Hexahydrate, two known phases. α-Form blue to blue-green tetragonal crystals; transition to β-form at 53.3°. β-Form, green transparent crystals; stable at 40°; becomes blue and opaque at room temp. Sweet astringent taste; somewhat efflorescent. Loses $5H_2O$ at about 100°. Greenish-yellow anhydr salt formed at 280°. Sol in 1.4 parts water; sparingly sol in alcohol, more in methanol. The aq soln is acid; pH about 4.5. LD s.c. in guinea pigs: 62 mg/kg, Handbook of Toxicology vol. **1**, W. S. Spector, Ed. (Saunders, Philadelphia, 1956) pp 212-213.

Heptahydrate, *single nickel salts.*
USE: In nickel-plating; as mordant in dyeing and printing fabrics; blackening zinc and brass.

6425. Niclosamide. *5-Chloro-N-(2-chloro-4-nitrophenyl)-2-hydroxybenzamide; 2′,5-dichloro-4′-nitrosalicylanilide;* 5-chloro-N-(2′-chloro-4′-nitrophenyl)salicylamide; 5-chlorosalicyloyl-(o-chloro-p-nitranilide); N-(2′-chloro-4′-nitrophenyl)-5-chlorosalicylamide; Bayer 2353; Cestocid; Fenasal; Lintex; Mansonil (vet feed-mix); Nasemo; Niclocide; Ruby; Sulqui; Tredemine; Vermitid; Yomesan. $C_{13}H_8Cl_2N_2O_4$; mol wt 327.13. C 47.73%, H 2.46%, Cl 21.68%, N 8.57%, O 19.56%. Prepn: **Brit.** pat. **824,345**; Schraufstätter, Gönnert, **U.S.** pat. **3,079,297** (1959, 1963 both to Bayer); Bekhli *et al., Med. Prom. SSSR* **1965**, 25.

Pale yellow crystals, mp 225-230°. Practically insol in water. Sparingly sol in ethanol, chloroform, ether. Ethanolamine salt, *Bayluscid.*
USE: The ethanolamine salt as a molluscicide.
THERAP CAT: Anthelmintic (Cestodes).
THERAP CAT (VET): Anthelmintic, teniacide.

6426. Nicoclonate. *3-Pyridinecarboxylic acid 1-(4-chlorophenyl)-2-methylpropyl ester; nicotinic acid p-chloro-α-isopropylbenzyl ester;* 1-(p-chlorophenyl)isobutyl nicotinate; p-chlorophenylisopropylcarbinol nicotinate; 1-(p-chlorophenyl)-1-(nicotinoyloxy)-2-methylpropane; S 486; Lipidium. $C_{16}H_{16}ClNO_2$; mol wt 289.77. C 66.32%, H 5.57%, Cl 12.23%, N 4.83%, O 11.04%. Prepn: J. Nordmann, H. B. Swierkot, **Fr.** pat. **M3454**; *eidem*, **U.S.** pat. **3,367,939** (1965, 1968 both to Kuhlmann).

Base, white needles from methanol and water, mp 61-62° (Maquenne block); 55-56.5° (Culatti block). Very soluble in lipids; sol in alcohols, benzene, toluene, ether and acetone. Practically insol in water.
Hydrochloride, $C_{16}H_{17}Cl_2NO_2$, crystals from methanol, mp 124-127°C. LD_{50} i.p. in mice: 2.27 g/kg (Nordmann, Swierkot).
THERAP CAT: Antilipemic.

6427. Nicofibrate. *2-(4-Chlorophenoxy)-2-methylpropanoic acid 3-pyridinylmethyl ester; 2-(p-chlorophenoxy)-2-methylpropionic acid 3-pyridylmethyl ester;* 3-hydroxymethylpyridine p-chlorophenoxyisobutyrate; clofenpyride. $C_{16}H_{16}ClNO_3$; mol wt 305.77. C 62.85%, H 5.28%, Cl 11.59%, N 4.58%, O 15.70%. Prepn: **Neth.** pat. **Appl. 6,610,738** corresp to Bolhofer, **U.S.** pat. **3,369,025** (1967, 1968, both to Merck & Co.); Nakanishi *et al.,* **Japan.** pat. **14,466('68)** to Yoshitomi); Garzia, **Ger.** pat. **1,948,320** (1970 to Ist. Chem. Ital.). Pharmacology: Marmo *et al., Farmaco Ed. Prat.* **26**, 557 (1971).

mp 48-49°. bp$_{0.4}$ 180°.
Hydrochloride, $C_{16}H_{17}Cl_2NO_3$, *Arterium V.* Crystals from ethyl acetate, mp 115.5-118.5°.
THERAP CAT: Antihyperlipoproteinemic.

6428. Nicofuranose. *β-D-Fructofuranose 1,3,4,6-tetra-3-pyridinecarboxylate; fructose 1,3,4,6-tetranicotinate;* 1,3,4,6-tetranicotinoylfructofuranose; 1,3,4,6-tetranicotinoyl-D-fructose; Vasperdil; Bradilan. $C_{30}H_{24}N_4O_{10}$; mol wt 600.52. C 60.00%, H 4.03%, N 9.33%, O 26.64%. Prepn: Suter, Kuendig, **Swiss.** pat. **366,523** (1963 to Eprova).

Crystals from ethyl acetate, mp 140-142°. $[\alpha]_D^{18}$ −8.5° (after 4-6 hours in $CHCl_3$).
THERAP CAT: Peripheral vasodilator.

6429. Nicomol. *3-Pyridinecarboxylic acid (2-hydroxy-1,3-cyclohexanediylidene)tetrakis(methylene) ester; nicotinic acid 1,1,3,3-tetraester with 2-hydroxy-1,1,3,3-cyclohexanetetramethanol;* 2-hydroxycyclohexane-1,1,3,3-tetramethanol tetraester with nicotinic acid; tetranicotinic acid 2-hydroxycyclohexa-1,1,3,3-tetramethyl ester; 2,2,6,6-tetrakis(hydroxymethyl)-1-cyclohexanol 2,2,6,6-tetranicotinate; nicotinic acid 2-hydroxy-1,3-cyclohexanediylidenetetrakismethylene ester; 2,2,6,6-tetrakis(nicotinoyloxymethyl)cyclohexanol; K-31; Cholexamin. $C_{34}H_{32}N_4O_9$; mol wt 640.66. C 63.74%, H 5.03%, N 8.75%, O 22.48%. Prepn: **Belg.** pat. **658,422** corresp to Irikura *et al.,* **U.S.** pat. **3,299,077** (1965 and 1967 to Kyorin). Hypocholesterolemic action studies: Aso *et al., J. Atheroscler. Res.* **10**, 391 (1969). Acute toxicity data: T. Sugawara *et al., Oyo Yakuri* **14**, 741 (1977), *C.A.* **88**, 131005v (1977). *Review: Japan. Med. Gaz.* **8**(8), 9 (1971).

Crystals from dilute acetic acid, mp 177-180°. Practically odorless and tasteless. One gram is sol in 3 ml 1N HCl and 10 ml chloroform. Slightly sol in water, ethanol and ether. LD_{50} in mice, male and female rats (g/kg): >10, 15, <10 orally; >5, >2.5, >2.5 i.p. (Sugawara).
THERAP CAT: Anticholesteremic.

6430. Nicomorphine. *7,8-Didehydro-4,5-epoxy-17-methylmorphinan-3,6-diol di-3-pyridinecarboxylate (ester);* morphine ester with nicotinic acid; morphine bis(nicotinate); morphine bis(pyridine-3-carboxylate); nicotinic acid morphine ester; morphine dinicotinate; Gewalan; Vilan. $C_{29}H_{25}N_3O_5$; mol wt 495.51. C 70.29%, H 5.09%, N 8.48%, O 16.14%. Prepn: Pongratz, Zirm, *Monatsh.* **88**, 330 (1957); **Brit.** pat. **807,115** (1959 to Lannacher Heilmittel).

Crystals, mp 178-178.5° (corr.). Practically insol in water. Sol in ethanol.

Hydrochloride, $C_{29}H_{25}N_3O_5$.HCl, dec 248°. Sol in water with neutral reaction.

Caution: May be habit forming. This is a controlled substance (opium derivative) listed in the U.S. Code of Federal Regulations, Title 21 Part 1308.11 (1987).

THERAP CAT: Narcotic analgesic.

6431. Nicorandil. *N-[2-(Nitrooxy)ethyl]-3-pyridinecarboxamide;* N-(2-hydroxyethyl)nicotinamide nitrate (ester); SG-75; Perisalol; Sigmart. $C_8H_9N_3O_4$; mol wt 211.17. C 45.50%, H 4.29%, N 19.90%, O 30.31%. Deriv of nicotinamide, *q.v.* Prepn: H. Nagano *et al.,* **Ger. pat. 2,714,713** corresp to U.S. pat. **4,200,640** (1977, 1980 both to Chugai). Pharmacological profile: N. Taira *et al., Clin. Exp. Pharmacol. Physiol.* **6**, 301 (1979). Hemodynamic effects of long-term administration: K. Sakai *et al., ibid.* **8**, 557 (1981). Comparative cardiovascular effects: *eidem, J. Cardiovasc. Pharmacol.* **3**, 139 (1981).

Colorless needles from ether/ethanol, mp 92-93°. LD_{50} in SD rats (mg/kg): 1200-1300 orally; 800-1000 i.v. (Nagano).

THERAP CAT: Coronary vasodilator.

6432. Nicotelline. *3,2':4',3''-Terpyridine;* 2,4-di(3-pyridyl)pyridine; 2,4-di(β-pyridyl)pyridine. $C_{15}H_{11}N_3$; mol wt 233.26. C 77.23%, H 4.75%, N 18.02%. From tobacco leaf: Pictet, Rotschy, **Ber. 34**, 696 (1901); Noga, *Fach. Mitt. Tabakregie* **1914**, Nos. 1 and 2. Structure: Kuffner, Faderl, *Monatsh.* **87**, 71 (1956). Synthesis: Thesing, Müller, *Angew. Chem.* **68**, 577 (1956); *Ber.* **90**, 711 (1957).

Prismatic needles from warm dil alcohol, hot water or chloroform + petr ether, mp 147-148°. bp above 300°. Sol in hot water, chloroform, alcohol, benzene; slightly sol in cold water, ether, petr ether.
Picrate, $C_{21}H_{14}N_6O_7$, yellow crystals, mp 216-217°.

6433. Nicotinamide Ascorbate. L-*Ascorbic acid mixt. with 3-pyridinecarboxamide;* niacinamide ascorbate; ascorbic acid nicotinamide complex; merpress; nicoscorbine; Nicastubin. $C_{12}H_{14}N_2O_7$; mol wt 298.25. C 48.32%, H 4.73%, N 9.39%, O 37.55%. Prepd from one mole of nicotinamide and one mole of ascorbic acid by warming in alcohol, acetone, water: Milhorat, *Proc. Soc. Exp. Biol. Med.* **55**, 52 (1944); Fox, Opferman, U.S. pat. **2,433,688** (1947 to Gelatin Prod.); Bailey *et al., J. Am. Chem. Soc.* **67**, 1184 (1945); Wenner, *J. Org. Chem.* **14**, 22 (1949); Najer, Guépet, *Ann. Pharm. Franc.* **12**, 712 (1954); Runti, *Il Farmaco, Ed. Sci.* **10**, 424 (1955); Genot, **Brit. pat. 771,317** (1957).

Yellow crystals, mp 141-145°, practically odorless. $[\alpha]_D^{20}$ +27.5° (c = 8 in H_2O). pH of 5% aq soln about 4.0. Soly at 20° in water 40%, abs ethanol 2.4%, methanol 10%. Sparingly sol in acetone. Practically insol in benzene, ether. A 45% to 50% w/w soln can be prepd by dissolving in equal volumes of water and propylene glycol.

THERAP CAT: Vitamin.

6434. Nicotine. *3-(1-Methyl-2-pyrrolidinyl)pyridine;* 1-methyl-2-(3-pyridyl)pyrrolidine; β-pyridyl-α-N-methylpyrrolidine; $C_{10}H_{14}N_2$; mol wt 162.23. C 74.03%, H 8.70%, N 17.27%. From the dried leaves of *Nicotiana tabacum* and *N. rustica* where it occurs to the extent of 2 to 8%, combined with citric and malic acids. Commercial nicotine is entirely a byproduct of the tobacco industry. Extraction procedure: Gattermann, Wieland, *Laboratory Methods of Organic Chemistry* (New York, 24th ed., 1937); Schwyzer,

Die Fabrikation pharmazeutischer und chemisch-technischer Produkte (Berlin, 1931). Purification through the zinc chloride double salt: Ratz, *Monatsh.* **26**, 1241 (1905). Structure and synthesis: Pinner, *Ber.* **26**, 294 (1893); Pictet, Rotschy, *Ber.* **37**, 1225 (1904); Craig, *J. Am. Chem. Soc.* **55**, 2854 (1933). Recent synthesis: M. Nakane, C. R. Hutchinson, *J. Org. Chem.* **43**, 3922 (1978). Conformation in soln: T. P. Pitner *et al., J. Am. Chem. Soc.* **100**, 246 (1978). HPLC determn in plasma: M. Harlharan *et al., Clin. Chem.* **34**, 724 (1988). Toxicity data: R. B. Barlow, L. J. McLeod, *Brit. J. Pharmacol.* **35**, 161 (1969). Review and bibliography: Jackson, *Chem. Rev.* **29**, 123 (1941). Review of pharmacology: R. W. Ryall in *Neuropoisons: Their Pathophysiological Actions* vol. 2, L. L. Simpson, D. R. Curtis, Eds. (Plenum, New York, 1974) pp 61-97.

Colorless to pale yellow, oily liq; very hygroscopic; turns brown on exposure to air or light. Acrid burning taste. Develops odor of pyridine. bp_{745} 247° (partial decompn); bp_{17} 123-125°. Volatile with steam. n_D^{20} 1.5282. d^{20} 1.0097. $[\alpha]_D^{20}$ −169°. pK_1 (15°) 6.16; pK_2 10.96. pH of 0.05M soln: 10.2. Forms salts with almost any acid and double salts with many metals and acids. Absorption spectrum: Purvis, *J. Chem. Soc.* **97**, 1035 (1910); Dobbie, Fox, *ibid.* **103**, 1194 (1913). Misc with water below 60°; on mixing nicotine with water the volume contracts. Very sol in alc, chloroform, ether, petr ether, kerosene, oils. Distribution of nicotine between water and petroleum oils: Norton, *Ind. Eng. Chem., Ind. Ed.* **32**, 241 (1940). LD_{50} in mice (mg/kg): 0.3 i.v.; 9.5 i.p.; 230 orally (Barlow, McLeod).

Hydrochloride, $C_{10}H_{14}N_2$.HCl, deliquesc crystals. $[\alpha]_D^{20}$ +104° (p = 10).

Dihydrochloride, $C_{10}H_{14}N_2$.2HCl, deliquesc crystals, very sol in water and alcohol. Nearly insol in ether.

Sulfate (neutral sulfate), $(C_{10}H_{14}N_2)_2$.H_2SO_4, six-sided tablets. $[\alpha]_D^{20}$ +88° (p = 70). Sol in water, alcohol.

Tartrate (acid tartrate, bitartrate), $C_{10}H_{14}N_2$.$2C_4H_6O_6$, dihydrate, crystals. mp 90°. $[\alpha]_D^{20}$ +26° (c = 10). Very sol in water or alcohol.

Zinc chloride double salt monohydrate, $C_{10}H_{16}Cl_4N_2$-Zn.H_2O, also with $4H_2O$. Very sol in water; sparingly sol in abs alcohol and ether.

Salicylate, $C_{17}H_{20}N_2O_3$, *Eudermol.* Six-sided plates, mp 118°. $[\alpha]_D^{20}$ +13° (c = 9). Freely sol in water or alcohol.

Polacrilex, *Nicorette.*

USE: Insecticide; fumigant. In the U.S. a 40% soln of nicotine sulfate, *Black Leaf 40,* was the commonly used form. As a contact poison it is most effective as soap, *i.e.,* as the laurate, oleate, or naphthenate. As a stomach poison a combination with bentonite has come into use. *Caution:* Highly toxic. Symptoms include extreme nausea, vomiting, evacuation of bowel and bladder, mental confusion, twitching, convulsions. Base is readily absorbed through mucous membranes and intact skin, but the salts are not. Treatment must be instituted quickly in order to be effective. *See: Clinical Toxicology of Commercial Products,* R. E. Gosselin *et al.,* Eds. (Williams & Wilkins, Baltimore, 4th ed., 1976) pp 246-249.

THERAP CAT (VET): Ectoparasiticide. Has been used as an anthelmintic.

6435. Nicotinic Acid. *3-Pyridinecarboxylic acid;* pyridine-β-carboxylic acid; P. P. factor; pellagra preventive factor; antipellagra vitamin; niacin; Nicacid; Nicagin; Nicobid; Niconacid; Nico-Span; Nicotene; Nicotinipca; Nicyl; Akotin; Daskil; Tinic; Nicolar; Wampocap. $C_6H_5NO_2$; mol wt 123.11. C 58.53%, H 4.09%, N 11.38%, O 25.99%. Minute amounts occur in all living cells, appreciable amounts are found in liver, yeast, milk, adrenal glands, white meat, alfalfa, legumes, whole cereals, corn. Whole wheat flour contains about 60 μg/g and unenriched patent flour about 16 μg/g. Prepd by oxidation of nicotine with concd HNO_3: McElvain, *Org. Syn.* **4**, 49 (1925); from nicotinonitrile: Woodward *et al., Ind. Eng. Chem.* **36**, 540 (1944); by oxidation of alkyl β-substituted pyridines: Ladenburg, *Ann.* **301**, 152 (1898). Total synthesis from pyridine: McElvain, Goese, *J. Am. Chem. Soc.* **63**, 2283 (1941); Rafikov *et al., Dokl. Akad. Nauk SSSR* **126**, 1286 (1959).

Needles from alcohol or water, mp 236.6°. Nonhygroscopic and stable in air. Sublimes without decompn. K at 25° = 1.4 × 10^{-5}; pK 4.85; pH of satd aq soln 2.7. uv max: 263 nm, Hünecke, *Ber.* **60**, 1451 (1927). One gram dissolves in 60 ml water. Freely sol in boiling water and in boiling alcohol, in alkali hydroxides and carbonates; sol in propylene glycol. Insol in ether (nicotinamide is sol in ether and can be extracted from water solns containing both the free acid and its amide). LD$_{50}$ s.c. in rats: 5 g/kg, Brazda, Coulson, *Proc. Soc. Exp. Biol. Med.* **62**, 19 (1946).

Pharmaceutical Incompatibilities: Nicotinic acid must not be formulated with sodium nitrite: *C.A.* **46**, 7707 (1952).

N-Oxide, see Oxiriacic Acid.

Sodium salt sesquihydrate, $C_6H_4NNaO_2$, *Nicoside, Naotin.* White crystals or cryst powder; stable in air. One gram dissolves in about 1.4 ml water, in about 60 ml alc, in 10 ml glycerol. Insol in ether. pH of aq soln: ~7.

THERAP CAT: Vitamin (enzyme cofactor).

THERAP CAT (VET): For prevention and treatment of a pellagra-like disease in dogs.

6436. Nicotinic Acid Benzyl Ester. *3-Pyridinecarboxylic acid phenylmethyl ester;* pyridine-β-carboxylic acid benzyl ester; benzyl nicotinate; Rubriment; Pycaril; Pykaryl. $C_{13}H_{11}NO_2$; mol wt 213.23. C 73.22%, H 5.20%, N 6.57%, O 15.00%. Prepn: **Brit.** pat. **817,103** (1959 to Nordmark-Werke).

Liquid, bp$_{3.4}$ 170°.

THERAP CAT: Rubefacient.

6437. Nicotinic Acid Monoethanolamine Salt. *3-Pyridinecarboxylic acid compd. with 2-aminoethanol (1:1);* monoethanolamine nicotinate; Nicatol; Nicamin. $C_8H_{12}N_2O_3$; mol wt 184.19. C 52.16%, H 6.57%, N 15.21%, O 26.06%. Prepn: Moore, **U.S.** pat. **2,233,419** (1941 to Abbott).

Oily liquid. Sol in water, ethanol; slightly sol in ether.

THERAP CAT: Vitamin (enzyme cofactor).

6438. Nicotinyl Alcohol. *3-Pyridinemethanol.* β-pyridylcarbinol; 3-hydroxymethylpyridine; nicotinic alcohol; Nu-2121; Roniacol; Ronicol. C_6H_7NO; mol wt 109.12. C 66.03%, H 6.47%, N 12.84%, O 14.66%. Prepd by catalytic hydrogenation of 3-pyridinecarboxaldehyde: Panizzon, *Helv. Chim. Acta* **24**, supplemental issue in honor of Gadient Engi, p 26E (1941); by lithium aluminum hydride reduction of ethyl nicotinate: Rosenmund, Zymalkowski, *Ber.* **85**, 156 (1952); Cohen, **U.S.** pat. **2,520,037** (1950); Mosher, Tessieri, *J. Am. Chem. Soc.* **73**, 4926 (1951); of methyl nicotinate: Bohlmann, *Ber.* **86**, 1423 (1953). From 3-cyanopyridine: Chase, **U.S.** pat. **2,615,896** (1952 to Hoffmann-La Roche); from 3-aminomethylpyridine: Schläpfer, **U.S.** pat. **2,547,048** (1951 to Hoffmann-La Roche); from thionicotinic acid S-methyl ester: Ruzicka, Prelog, **U.S.** pat. **2,509,171** (1950 to Ciba).

Very hygroscopic liquid. bp$_{28}$ 154°; bp$_{16}$ 144-145°; bp$_{12}$ 114°; bp$_{0.1}$ 110°. Freely sol in water, ether. Sparingly sol in petr ether.

d-Tartrate, $C_{10}H_{13}NO_7$, *Roniacol Tartrate, Radecol, Niltuvin.* Crystals, sour taste, mp 147-148°. Sol in ether, freely sol in water, alcohol.

USE: Solubilizer for riboflavin: **U.S.** pat. **2,458,430** (1949).

THERAP CAT: Vasodilator (peripheral).

6439. Nidroxyzone. *1-(2-Hydroxyethyl)-2-[(5-nitro-2-furanyl)methylene]hydrazinecarboxamide; 5-nitro-2-furaldehyde 2-(2-hydroxyethyl)semicarbazone;* NF 67; Furadroxyl. $C_8H_{10}N_4O_5$; mol wt 242.19. C 39.67%, H 4.16%, N 23.14%, O 33.03%. Prepn: Stillman, Scott, **U.S.** pat. **2,416,234** (1947 to Eaton).

Bright orange plates from alc, mp 214-216° (dec). Soly in water 1:2000.

THERAP CAT: Anti-infective.

6440. Nidulin. *2,4,7-Trichloro-3-hydroxy-8-methoxy-1,-9-dimethyl-6-(1-methyl-1-propenyl)-11H-dibenzo[b,e][1,4]-dioxepin-11-one;* methyllustin; ustin methyl ether. $C_{20}H_{17}Cl_3O_5$; mol wt 443.73. C 54.14%, H 3.86%, Cl 23.97%, O 18.03%. Antibiotic substance produced by a strain of *Aspergillus nidulans* (Eidam) Wint.: Dean *et al., Nature* **172**, 344 (1953); Dean *et al., J. Chem. Soc.* **1954**, 1432. Structure: Dean *et al., ibid.* **1960**, 4829; Bycroft *et al., ibid.* **1963**, 5148.

Slender, shiny rods from petr ether, mp 180°. Practically insol in water. Sol in aq solns of sodium hydroxide and sodium bicarbonate. Freely sol in chloroform. Sparingly sol in 95% ethanol and in benzene. Very sparingly sol in hot petr ether. A 4% soln in chloroform shows no optical activity. Completely inhibits *in vitro* the growth of *Mycobacterium tuberculosis* for four weeks at a diln of between 1 in 5000 and 1 in 10,000 when tested in the presence of serum by modified Long's medium and the floating-pellicle method. A 0.1% soln also inhibits the growth of the human parasitic fungi *Trichophyton tonsurans* and *Microsporum audouini,* but shows little or no activity towards a wide range of other microorganisms. Inactive against bacteriophage.

6441. Nifedipine. *1,4-Dihydro-2,6-dimethyl-4-(2-nitrophenyl)-3,5-pyridinedicarboxylic acid dimethyl ester;* 4-(2'-nitrophenyl)-2,6-dimethyl-3,5-dicarbomethoxy-1,4-dihydropyridine; BAY a 1040; Adalat(e); Adapress; Aldipin; Alfadat; Anifed; Aprical; Bonacid; Camont; Citilat; Coracten; Cordicant; Corotrend; Duranifin; Ecodipin; Hexadilat; Introcar; Kordafen; Nifedicor; Nifedin; Nifelan; Nifelat; Orix; Oxcord; Pidilat; Procardia; Sepamit; Tibricol; Zenusin. $C_{17}H_{18}N_2O_6$; mol wt 346.34. C 58.95%, H 5.24%, N 8.09%, O 27.72%. Dihydropyridine calcium channel blocker. Prepn: Bossert, Vater, **S. Afr.** pat. **68 01,482**; *eidem,* **U.S.** pat. **3,485,847** (1968, 1969 both to Bayer). Series of articles on pharmacology, pharmacokinetics, biotransformation, clinical studies: *Arzneimittel-Forsch.* **22**, 1-56, 330-388 (1972). Toxicity data: Vater *et al., ibid.* 1. Comparative study in variant angina: D. D. Waters *et al., Am. J. Cardiol.*

47, 179 (1981). Antihypertensive effects: K. Maeda *et al.*, *Arzneimittel-Forsch.* **32**, 267 (1982). Controlled clinical trial in treatment of Raynaud's phenomenon: R. J. Rodeheffer *et al.*, *N. Engl. J. Med.* **308**, 880 (1983). Review of mechanism of action: T. H. Swanson, C. L. Green, *Gen. Pharmacol.* **17**, 255-260 (1986).

Yellow crystals, mp 172-174°. Easily sol in acetone, chloroform; less sol in ethanol. Practically insol in water. Very light sensitive in soln. LD_{50} in mice, rats (mg/kg): 494, 1022 orally; 4.2, 15.5 i.v. (Vater).
THERAP CAT: Antianginal; antihypertensive.

6442. Nifenalol. ($\pm$)-α-*[[(1-Methylethyl)amino]methyl]-4-nitrobenzenemethanol;* α-*[(isopropylamino)methyl]-p-nitrobenzyl alcohol;* 1-(4-nitrophenyl)-2-(isopropylamino)-ethanol; isophenethanol; INPEA. $C_{11}H_{16}N_2O_3$; mol wt 224.26. C 58.91%, H 7.19%, N 12.49%, O 21.40%. β-Adrenergic blocker. Prepn: Teotino *et al.*, *Farmaco Ed. Sci.* **17**, 252 (1962); *eidem*, **Brit.** pat. **950,682** (1964 to Selvi), *C.A.* **60**, 13184h (1964). Pharmacology: Somani, Lum, *J. Pharmacol. Exp. Ther.* **147**, 194 (1965); Almirante *et al.*, *Farmaco Ed. Sci.* **21**, 637 (1966). Prepn of the optical isomers: Almirante, Murmann, *J. Med. Chem.* **9**, 650 (1966); *eidem*, **Belg.** pat. **673,273** (1966 to Selvi), *C.A.* **66**, 65278d (1967).

mp 98°.
Hydrochloride, $C_{11}H_{17}ClN_2O_3$, *Inpea.* mp 181°. Ingredient of *Beta-Intensain.*
THERAP CAT: Antianginal, antiarrhythmic.

6443. Nifenazone. *N-(2,3-Dihydro-1,5-dimethyl-3-oxo-2-phenyl-1H-pyrazol-4-yl)-3-pyridinecarboxamide; N-antipyrinylnicotinamide;* 2,3-dimethyl-1-phenyl-4-(3-pyridinecarboxamido)-5-pyrazolone; 2,3-dimethyl-4-nicotinamido-1-phenyl-5-pyrazolone; 4-(3-pyridinecarboxamido)-2,3-dimethyl-1-phenyl-5-pyrazolone; 4-nicotinamido-1,5-dimethyl-2-phenyl-3-pyrazolone; 4-(*N*-nicotinoylamino)-1-phenyl-2,3-dimethyl-5-pyrazolone; 4-nicotinamido-2-phenyl-1,5-dimethyl-3-pyrazolone; *N*-nicotinoylaminoantipyrine; Piralgo; Reupiron; Thylin. $C_{17}H_{16}N_4O_2$; mol wt 308.33. C 66.22%, H 5.23%, N 18.17%, O 10.38%. Prepn: Heid, **Ger.** pat. **897,407** (1953); Trommsdorff, **Belg.** pat. **635,597** (1963); Pongraz, Zirm, *Monatsh.* **88**, 330 (1957); Zorn, Schmidt, *Pharmazie* **12**, 396 (1957).

Crystals from alc, mp 252-253° (also reported as mp 256-258°). Sparingly sol in water, acetone, ethyl acetate, ether; sol in hot water, alcohol, chloroform, dil acids.
THERAP CAT: Analgesic, antipyretic.

6444. Niflumic Acid. *2-[[3-(Trifluoromethyl)phenyl]-amino]-3-pyridinecarboxylic acid; 2-(α,α,α-trifluoro-m-toluidino)nicotinic acid;* 2-[3-(trifluoromethyl)anilino]nicotinic acid; UP 83; Actol; Forenol; Landruma; Nifluril. $C_{13}H_9F_3N_2O_2$; mol wt 282.23. C 55.32%, H 3.21%, F 20.20%, N 9.93%, O 11.34%. Prepn: **Neth.** pat. **Appl. 6,414,717**; C. Hoffmann, A. Faure, **U.S.** pat. **3,415,834** (1965, 1968 both

to Lab. U.P.S.A.); *eidem, Bull. Soc. Chim. France* **1966**, 2316. Pharmacological and metabolic studies: Glasson *et al.*, *Biochem. Pharmacol.* **18**, 633 (1969); Boissier *et al.*, *Therapie* **26**, 211 (1971). Determn in human plasma by GLC: G. Houin *et al.*, *J. Chromatog.* **223**, 351 (1981).

Crystals from ethanol, mp 204°.
β-Morpholinoethyl ester, $C_{19}H_{20}F_3N_3O_3$, *Nifluril Suppositories.* Prepn of the ester HCl: Hoffmann, **Ger.** pat. **1,802,777;** *idem*, **U.S.** pat. **3,708,481** (1969, 1973 both to Hexachimie).
Phthalidyl ester, *see* talniflumate.
THERAP CAT: Anti-inflammatory.

6445. Nifuradene. *1-[[(5-Nitro-2-furanyl)methylene]-amino]-2-imidazolidinone;* oxafuradene (rescinded); NF-246; NSC 6470; Renafur. $C_8H_8N_4O_4$; mol wt 224.17. C 42.86%, H 3.59%, N 24.99%, O 28.55%. Prepn: Gever, Michels, **U.S.** pat. **2,746,960** (1956 to Norwich Pharm.); *eidem, J. Am. Chem. Soc.* **78**, 5349 (1956); Michels, **U.S.** pat. **2,776,979** (1957 to Norwich Pharm.). Toxicity data: E. I. Goldenthal, *Toxicol. Appl. Pharmacol.* **18**, 185 (1971).

Lemon-yellow solid from nitromethane, mp 261.5-263° (dec). Turns orange on standing, on washing with ethanol or on heating at 75-85°. uv max (water): 387.5, 273 nm (ϵ 17550, 13200). Soly in water: 88-109 mg/l. LD_{50} orally in rats: 1681 mg/kg (Goldenthal).
THERAP CAT: Antibacterial.

6446. Nifuraldezone. *Aminooxoacetic acid [(5-nitro-2-furanyl)methylene]hydrazide;* 5-nitro-2-furaldehyde semioxamazone; Furamazone. $C_7H_6N_4O_5$; mol wt 226.15. C 37.17%, H 2.68%, N 24.78%, O 35.37%. Prepn: Stillman, Scott, **U.S.** pat. **2,416,238** (1947 to Eaton Labs.).

Yellow cryst powder, mp 270° (dec).
THERAP CAT (VET): Antibacterial.

6447. Nifuratel. *5-[(Methylthio)methyl]-3-[[(5-nitro-2-furanyl)methylene]amino]-2-oxazolidinone; 5-[(methylthio)-methyl]-3-[(5-nitrofurfurylidene)amino]-2-oxazolidinone;* 5-(methylmercaptomethyl)-3-(5-nitro-2-furfurylideneamino)-2-oxazolidinone; methylmercadone; Inimur; Macmiror; Magmilor; Omnes; Polmiror; Tydantil. $C_{10}H_{11}N_3O_5S$; mol wt 285.29. C 42.10%, H 3.89%, N 14.73%, O 28.04%, S 11.24%. Prepn: **Belg.** pat. **635,608** (1963 to Polichimica SAP), *C.A.* **61**, 16069c (1964), corresp to **Brit.** pat. **969,126.**

Crystals from acetic acid, mp 182°.
THERAP CAT: Antibacterial; antifungal; antiprotozoal (Trichomonas).

6448. Nifurfoline. *3-(4-Morpholinylmethyl)-1-[[(5-nitro-2-furanyl)methylene]amino]-2,4-imidazolidinedione; 3-(morpholinomethyl)-1-[(5-nitrofurfurylidene)amino]hydantoin;* Urbac. $C_{13}H_{15}N_5O_6$; mol wt 337.30. C 46.29%, H 4.48%, N 20.76%, O 28.46%. Prepn: J. A. Bofill-Auge, J.

M. Espinos-Taya, **Span.** pat. **297,087** (1964), *C.A.* **63,** 11572e (1966); **Brit.** pat. **1,245,095** (1971 to Esteve), *C.A.* **75,** 121382h (1971).

Yellow cryst, mp 206°, also reported as mp 194-196°, J. Klosa, H. Starke, **E. Ger.** pat. **508,615** (1966), *C.A.* **66,** 65501w (1967).

THERAP CAT: Antibacterial.

6449. Nifuroquine. *4-(5-Nitro-2-furanyl)-2-quinoline-carboxylic acid 1-oxide;* 2-carboxy-4-[2'-(5'-nitrofuryl)]-quinoline 1-oxide; 4-(5-nitro-2-furyl)quinaldic acid 1-oxide; quinaldofur; Abimasten 100. $C_{14}H_8N_2O_6$; mol wt 300.22. C 56.01%, H 2.68%, N 9.33%, O 31.98%. Prepn: R. R. G. Haber, E. Schoenberger, **S. Afr.** pat. **67 03,320** corresp to U.S. pats. **4,217,456** and **4,224,448** (1967, 1980, 1980). Use in bovine mastitis: R. R. G. Haber *et al.,* **Ger.** pat. **2,612,-250** corresp to U.S. pat. **4,070,469** (1977, 1978 both to ABIC). *In vitro* study: G. Ziv *et al., Zentralbl. Veterinaermed. (B)* **23,** 301 (1976). Pharmacology: G. Ziv, A. Saran, *ibid.* 310. Activity under anaerobic conditions: G. M. Maluszynska, L. Bassalik-Chabielska, *Pr. Mater. Zootech.* **23,** 63 (1980), *C.A.* **93,** 143793 (1980).

Yellow cryst powder, mp 190° (dec). Practically insol in water.

THERAP CAT (VET): Antibacterial.

6450. Nifuroxazide. *4-Hydroxybenzoic acid [(5-nitro-2-furanyl)methylene]hydrazide; p-hydroxybenzoic acid (5-nitrofurfurylidene)hydrazide;* 1-(p-hydroxybenzoyl)-2-(5-nitrofurfurylidene)hydrazine; 5-nitro-2-furaldehyde p-hydroxybenzoylhydrazone; RC-27109; Adral; Bacifurane; Diarlidan; Dicoferin; Ercefurol; Ercefuryl; Pentofuryl. $C_{12}H_9$-N_3O_5; mol wt 275.22. C 52.37%, H 3.30%, N 15.27%, O 29.07%. Prepn: **Fr.** pat. **1,327,840** (1963 to Robert et Carriere), *C.A.* **59,** 12763b (1963); M. C. E. Carron, **Brit.** pat. **962,706;** *idem,* **U.S.** pat. **3,290,213** (1964, 1966 both to Robert et Carriere). Antiseptic and antibacterial properties: M. C. E. Carron *et al., Ann. Pharm. Franc.* **21,** 287 (1963). *In vitro* study of activity spectrum: A. Thabaut, J. L. Durosoir, *Gaz. Med. Fr.* **85,** 4516 (1978), *C.A.* **90,** 1487 (1979).

Crystals from pyridine, mp 298°. Practically insol in water.

THERAP CAT: Intestinal antiseptic.

6451. Nifuroxime. *5-Nitro-2-furancarboxaldehyde oxime; anti-*5-nitro-2-furaldoxime; Micofur. $C_5H_4N_2O_4$; mol wt 156.10. C 38.47%, H 2.58%, N 17.95%, O 41.00%. Prepd by treating 5-nitrofurfural with hydroxylamine in alcohol: Ikeda, *C.A.* **50,** 10701 (1956). Claimed in Stillman *et al.,* **U.S.** pat. **2,319,481** (1943 to Norwich Pharmacal).

Tasteless, pale yellow or greenish crystals from ethanol. Darkens on exposure to light. mp 156° (Ikeda); mp 163-164° (Stillman *et al.*). Soly at 25° in water: ~1 g/l, in methanol: 89.0 g/l, in 95% ethanol: 39.0 g/l.

THERAP CAT: Topical anti-infective; antiprotozoal (Trichomonas).

6452. Nifurpirinol. *6-[2-(5-Nitro-2-furanyl)ethenyl]-2-pyridinemethanol; 6-[2-(5-nitro-2-furyl)vinyl]-2-pyridinemethanol;* 6-(hydroxymethyl)-2-[2-(5-nitro-2-furyl)vinyl]-pyridine; furpirinol; furpyrinol; P-7138; Furanace. $C_{12}H_{10}$-N_2O_4; mol wt 246.22. C 58.54%, H 4.09%, N 11.38%, O 25.99%. Prepn: Fujita *et al., J. Pharm. Soc. Japan* **86,** 1014 (1966); **Brit.** pat. **1,053,730** (1967 to Dainippon), *C.A.* **66,** 115605f (1967); Murakami, Iwanami, **Japan.** pat. **14,072**-('70) (to Yamanouchi), *C.A.* **73,** 45364v (1970). Antibacterial activity: Shimizu, Takase, *Progr. Antimicrob. Anticancer Chemother., Proc. 6th Int. Congr. Chemother.* Tokyo, 1969 (Univ. Park Press, Baltimore, 1970) **2,** p 388; Amend, Ross, *Prog. Fish Cult.* **32,** 19 (1970); Ross, *ibid.* **34,** 18 (1972); Takase *et al., Chem. Pharm. Bull.* **21,** 144 (1973).

Yellow needles from acetone or methanol, mp 170-171°C. LD_{50} orally in eels: 1780 mg/kg.

THERAP CAT: Antibacterial.

THERAP CAT (VET): Antibacterial in fish diseases.

6453. Nifurprazine. *6-[2-(5-Nitro-2-furanyl)ethenyl]-3-pyridazinamine; 3-amino-6-[2-(5-nitro-2-furyl)vinyl]pyridazine;* 1-(5-nitro-2-furyl)-2-(6-amino-3-pyridazyl)ethylene; furenapyridazin; Furenazin. $C_{10}H_8N_4O_3$; mol wt 232.20. C 51.72%, H 3.47%, N 24.13%, O 20.67%. Prepn: **Belg.** pat. **630,163** (1963 to Boehringer, Mann.), *C.A.* **60,** 14516a (1964) corresp to **Brit.** pat. **966,832.**

Hydrochloride, $C_{10}H_9ClN_4O_3$, Carofur. Crystals, mp 290°.

THERAP CAT: Topical antibacterial.

6454. Nifurtimox. *3-Methyl-N-[(5-nitro-2-furanyl)-methylene]-4-thiomorpholinamine 1,1-dioxide; 4-[(5-nitrofurfurylidene)amino]-3-methylthiomorpholine-1,1-dioxide;* tetrahydro-3-methyl-4-[(5-nitrofurfurylidene)amino]-2H-1,4-thiazine 1,1-dioxide; 1-[(5-nitrofurfurylidene)amino]-2-methyltetrahydro-1,4-thiazine 4,4-dioxide; BAY 2502; Lampit. $C_{10}H_{13}N_3O_5S$; mol wt 287.29. C 41.81%, H 4.56%, N 14.62%, O 27.85%, S 11.16%. Prepn from 5-nitrofurfural and 4-amino-3-methyltetrahydro-1,4-thiazine 1,1-dioxide: Herlinger *et al.,* **Ger.** pat. **1,170,957** corresp to U.S. pat. **3,262,930** (1964 and 1966 to Bayer). Series of articles on pharmacology and clinical findings: *Arzneimittel-Forsch.* **22,** 1563-1642 (1972). Toxicity data: K. Hoffmann, *ibid.* 1590.

Orange-red crystals from dil acetic acid, mp 180-182°. LD_{50} in mice, rats (mg/kg): 3720, 4050 by gavage (Hoffmann).

THERAP CAT: Antiprotozoal (Trypanosoma).

6455. Nifurtoinol. *3-(Hydroxymethyl)-1-[[(5-nitro-2-furanyl)methylene]amino]-2,4-imidazolidinedione; 3-(hydroxymethyl)-1-[(5-nitrofurfurylidene)amino]hydantoin;* 1-

[(5-nitrofurfurylidene)amino]-3-(hydroxymethyl)hydantoin; Urfadyn. $C_9H_8N_4O_6$; mol wt 268.19. C 40.30%, H 3.01%, N 20.89%, O 35.80%. Prepn: Michels, **Belg. pat. 611,941** corresp to U.S. pat. **3,446,802** (1962, 1969 to Norwich Pharmacal); Spencer, Michels, *J. Org. Chem.* **29**, 3416 (1964).

Yellow crystals from aq formaldehyde. When heated on a melting-point block, loses formaldehyde, then further decomposes at same temp as nitrofurantoin. uv max (2% in dimethylformamide): 367.5, 265 nm (ϵ 17,900, 12,800).

THERAP CAT: Antibacterial.

6456. Nifurzide. *5-Nitro-2-thiophenecarboxylic acid [3-(5-nitro-2-furanyl)-2-propenylidene]hydrazide; N^1-(5-nitro-2-furylacrylidene)-N^2-(5-nitro-2-thenoyl)hydrazine; 5-nitro-2-thiophenecarboxylic acid [3-(5-nitro-2-furyl)allylidene]hydrazide;* Ricridene. $C_{12}H_8N_4O_6S$; mol wt 336.28. C 42.86%, H 2.40%, N 16.66%, O 28.55%, S 9.53%. Bactericidal agent related to nitrofurazone, *q.v.* Prepn: E. Szarvasi, L. Fontaine, **Ger. pat. 2,200,375** corresp to U.S. pat. **3,847,911** (1972, 1974 both to LIPHA); E. Szarvasi *et al., J. Med. Chem.* **16**, 281 (1973). Mode of action study: A. Delsarte *et al., Antimicrob. Ag. Chemother.* **19**, 477 (1981).

Yellow cryst from DMF/ether, mp 235-236°. LD_{50} in mice: 3,200 mg/kg orally, E. Szarvasi, L. Fontaine, *loc. cit.*

THERAP CAT: Anti-infective.

6457. Nigericin. Antibiotic K 178; antibiotic X-464; azalomycin M; helixin C; polyetherin A. $C_{40}H_{68}O_{11}$; mol wt 724.98. C 66.27%, H 9.45%, O 24.28%. Polyether antibiotic which affects ion transport and ATPase activity in mitochondria; produced by *Streptomyces hygroscopicus* E-749 and structurally related to monensin, *q.v.* Isoln, characterization, production: R. L. Harned *et al., Antibiot. & Chemother.* **1**, 594 (1951); J. Berger *et al., J. Am. Chem. Soc.* **73**, 5295 (1951); J. Shoji *et al., J. Antibiot.* **21**, 402 (1968). Structure: L. K. Steinrauf *et al., Biochem. Biophys. Res. Commun.* **33**, 29 (1968); T. Kubota *et al., Chem. Commun.* **1968**, 1541; T. Kubota, S. Matsutani, *J. Chem. Soc. (C)* **1970**, 695. Use in coccidiosis: M. Gorman, R. L. Hamill, U.S. pat. **3,555,150** (1971 to Lilly). Effect on calcium uptake and membrane potential in mitochondria: H. Rottenberg, A. Scarpa, *Biochemistry* **13**, 4811 (1974). Stimulation of ATPase activity: H. Sze, *Proc. Nat. Acad. Sci. USA* **77**, 5904 (1980). Approach to synthesis: C. P. Holmes, P. A. Bartlett, *J. Org. Chem.* **54**, 98 (1989).

Colorless needles, mp 183.5-185°. $[\alpha]_D^{24}$ +36.2° (c = 0.842 in $CHCl_3$). Sol in alcohols, acetone, ethyl acetate, chloroform, benzene, ether; slightly sol in satd hydrocarbons. Practically insol in water. LD_{50} in mice: 10-15 mg/kg i.p. (Shoji), also reported as 2.5 mg/kg (Harned).

Sodium salt, $C_{40}H_{67}NaO_{11}$, crystals, mp 245-255° (dec). Sol in chloroform. Practically insol in water.

6458. Nihydrazone. *Acetic acid [(5-nitro-2-furanyl)-methylene]hydrazide; acetic acid 5-(nitrofurfurylidene)hydrazide; 5-nitro-2-furaldehyde acetylhydrazone; N-acetyl-5-nitro-2-furaldehyde hydrazide; 1-(5-nitro-2-furfurylidene)-*

2-acetylhydrazine; 1-acetyl-2-(5-nitro-2-furfurylidene)-hydrazide; NF 64; HC-064; Furiton; Nidrafur. $C_7H_7N_3O_4$; mol wt 197.15. C 42.64%, H 3.58%, N 21.32%, O 32.46%. Prepn: Stillman, Scott, U.S. pats. **2,416,234** and **2,416,236** (1947 to Eaton Labs).

Yellow crystals from acetic acid + ethanol, dec 230-235°. uv max (water): 253, 364 nm (log ϵ 4.11, 4.23). Soly in water 1:20,000.

THERAP CAT (VET): Antibacterial, antiprotozoal.

6459. Nikethamide. *N,N-Diethyl-3-pyridinecarboxamide; N,N-diethylnicotinamide;* pyridine-3-carboxylic acid diethylamide; nicotinic acid diethylamide; Anacardone; Astrocar; Carbamidal; Cardamine; Cardiamid; Cardimon; Coracon; Coractiv N; Coramine; Cordiamin; Corediol; Cormed; Cormid; Corvitol; Corvotone; Dynacoryl; Eucoran; Inicardio; Niamine; Nicamide; Nicor; Nicorine; Nikardin; Pyricardyl; Salvacard; Stimulin; Ventramine. $C_{10}H_{14}N_2O$; mol wt 178.23. C 67.39%, H 7.92%, N 15.72%, O 8.98%. Prepd by the action of thionyl chloride on nicotinic acid, followed by treatment with diethylamine hydrochloride; also formed by heating nicotinic acid or quinolinic acid anhydride with diethylamine: Hartmann, Seiberth, U.S. pat. **1,403,117** (1922); from nicotinic acid + benzenesulfonyldiethyl amide: Oxley *et al., J. Chem. Soc.* **1946**, 763. Toxicity data: E. I. Goldenthal, *Toxicol. Appl. Pharmacol.* **18**, 185 (1971).

Slightly viscous liquid or cryst solid. Faintly bitter taste followed by a faint sensation of warmth. d_4^{25} 1.058-1.066 (liq). mp 24-26°. bp_{760} 296-300° (some decompn); bp_{10} 158-159°; bp_3 128-129°; $bp_{0.4}$ 115°. n_D^{20} 1.525-1.526; n_D^{25} 1.522-1.524. Misc with water, ether, chloroform, acetone, alc. Usually marketed as a 25% w/v aq soln, pH 6.0-6.5. Incompatible with Na_2CO_3 solns which cause precipitation; tannates produce an amorph ppt. LD_{50} i.p. in rats: 272 mg/kg (Goldenthal).

THERAP CAT: Central and respiratory stimulant.

THERAP CAT (VET): Respiratory stimulant.

6460. Nilutamide. *5,5-Dimethyl-3-[4-nitro-3-(trifluoromethyl)phenyl]-2,4-imidazolidinedione; 1-(3'-trifluoromethyl-4'-nitrophenyl)-4,4-dimethylimidazoline-2,5-dione;* RU-23908; Anandron. $C_{12}H_{10}F_3N_3O_4$; mol wt 317.22. C 45.43%, H 3.18%, F 17.97%, N 13.25%, O 20.17%. Prepn: J. Perronnet *et al.,* Ger. pat. **2,649,925**; *eidem,* U.S. pat. **4,097,578** (1977, 1978 both to Roussel-UCLAF). Pharmacology: J.-P. Raynaud *et al., J. Steroid Biochem.* **11**, 93 (1979); M. Moguilewsky *et al., ibid.* **24**, 139 (1986). Kinetics, metabolism and review of clinical studies: J.-P. Raynaud *et al., Prog. Clin. Biol. Res.* **185A**, 99-120 (1985). Effect on ocular response: C. Harnois *et al., Brit. J. Ophthalmol.* **70**, 471 (1986). Review of pharmacology, mechanism of action and effect on LHRH: J.-P. Raynaud *et al., Prostate* **5**, 299-311 (1984). Clinical study of combination with LHRH agonist in prostate cancer: F. Labrie *et al., Int. Congr. Ser.-Excerpta Med.* **655**, 450 (1984).

Crystals from ethanol, mp 149°.

THERAP CAT: Anti-androgen.

6461. Nilvadipine. *2-Cyano-1,4-dihydro-6-methyl-4-(3-nitrophenyl)-3,5-pyridinedicarboxylic acid 3-methyl 5-(1-*

methylethyl) ester; 5-isopropyl-3-methyl-2-cyano-1,4-dihy dro-6-methyl-4-(*m*-nitrophenyl)-3,5-pyridinedicarboxylate; isopropyl 6-cyano-5-methoxycarbonyl-2-methyl-4-(3-nitrophenyl)-1,4-dihydropyridine-3-carboxylate; nivadipine; nivaldipine; FR 34235; FK 235; SK&F 102362; Nivadil. $C_{19}H_{19}N_3O_6$; mol wt 385.38. C 59.22%, H 4.97%, N 10.90%, O 24.91%. Dihydropyridine calcium channel blocker. Prepn: Y. Sato, **Belg.** pat. **879,263**; *idem,* **U.S.** pat. **4,338,-322** (1980, 1982 both to Fujisawa). Structural studies: A. Miyamae *et al., Chem. Pharm. Bull.* **34**, 3071 (1986). Determn in plasma and urine: Y. Tokuma *et al., J. Chromatog.* **415**, 156 (1987). Preliminary pharmacokinetics and resolution of enantiomers: *eidem, J. Pharm. Sci.* **76**, 310 (1987). Pharmacokinetics in rabbits: Y. Nezasa *et al., Kankyo Igaku Kenkyusho Nenpo* **38**, 200 (1987); *C.A.* **107**, 89273q (1987). Mode of action: P. A. Molyvdas, N. Sperelakis, *J. Cardiovasc. Pharmacol.* **8**, 449 (1986). Cardiovascular effects: G. J. Gross *et al., Gen. Pharmacol.* **14**, 677 (1983). Clinical evaluation in hypertension: K. Mizuno *et al., Res. Comm. Chem. Pathol. Pharmacol.* **52**, 3 (1986).

Yellow prisms from ethanol, mp 148-150°.
(+)-Form, $[\alpha]_D^{20}$ +222.42° (c = 1 in methanol).
(—)-Form, $[\alpha]_D^{20}$ —219.62° (c = 1 in methanol).
THERAP CAT: Antihypertensive; anti-anginal.

6462. Nimbin. *(4α,5α,6α,7α,15β,17α)-6-(Acetyloxy)-7,15:21,23-diepoxy-4,8-dimethyl-1-oxo-18,24-dinor-11,12-secochola-2,13,20,22-tetraene-4,11-dicarboxylic acid dimethyl ester;* 5-(acetyloxy)-2-(3-furanyl)-3,3a,4a,5,5a,6,9,9a,10,10a-decahydro-6-(methoxycarbonyl)-1,6,9a,10a-tetramethyl-9-oxo-2H-cyclopenta[b]naphtho[2,3-d]furan-10-acetic acid. $C_{30}H_{36}O_9$; mol wt 540.59. C 66.65%, H 6.71%, O 26.64%. Bitter principle from various parts of the neem tree, *Azadirachta indica* A. Juss. *(Melia azadirachta* L.), *Meliaceae:* Siddiqui, *Current Sci. (India)* **11**, 278 (1942). Structure: C. R. Narayanan *et al., Chem. & Ind. (London)* **1964**, 322; C. R. Narayanan, R. V. Pachapurkar, *Tetrahedron Letters* **1965**, 4333. Stereochemistry: *eidem, J. Org. Chem.* **31**, 2691 (1966); Harris *et al., Tetrahedron* **24**, 1517 (1968). Crystal structure of dihydronimbin: C. R. Narayanan *et al., Acta Crystallogr.* **B36**, 486 (1980).

Crystals from methanol, mp 205°. $[\alpha]_D^{24}$ +170° (abs ethanol). uv max (95% ethanol): 210, 330 nm (ϵ 32700; 66). Practically insol in water. Sol in ether, abs alcohol.
Dihydronimbin, $C_{30}H_{38}O_9$, crystals from methanol, mp 215-216°. $[\alpha]_D$ +167.5° (chloroform). uv max (95% ethanol): 210, 298 nm (ϵ 16200; 29).

6463. Nimbiol. *2,3,4,4a,10,10a-Hexahydro-6-hydroxy-1,1,4a,7-tetramethyl-9(1H)-phenanthrenone;* 6-hydroxy-7-methyl-9-oxopodocarpane. $C_{18}H_{24}O_2$; mol wt 272.37. C 79.37%, H 8.88%, O 11.75%. (+)-Form found in the bark of the neem tree, *Azadirachta indica* A. Juss. *(Melia azadirachta* L.), *Meliaceae:* P. Sengupta *et al., Chem. & Ind. (London)* **1958**, 861. Structure: P. Sengupta *et al., Tetrahedron* **10**, 45 (1960). Also obtained by conversion of podocarpic acid: Bible, *ibid.* **11**, 22 (1960); Wenkert *et al., J. Am. Chem. Soc.* **83**, 2320 (1961). Total synthesis of (±)-form:

W. L. Meyer *et al., J. Org. Chem.* **40**, 3686 (1975). Synthesis of (±)-methyl ether: Ramachandran, Dutta, *J. Chem. Soc.* **1960**, 4766; Delobelle, Fétizon, *Bull. Soc. Chim. France* **1961**, 1900; Nasipuri, Roy, *J. Indian Chem. Soc.* **40**, 327 (1963).

Crystals from dilute methanol or platelets by high vac sublimation. mp 248-252°. $[\alpha]_D^{25}$ +33° (chloroform). uv max (abs alcohol): 234, 283 nm (log ϵ 4.13, 4.10).
Methyl ether, $C_{19}H_{26}O_2$, crystals, mp 143°. $[\alpha]_D^{25}$ +43.7° (chloroform). uv max (alcohol): 207, 232, 279 nm (log ϵ 4.18, 4.15, 4.12).
(±)-Form, white needles from CH_3OH, mp 237.0-237.5°.
(±)-Methyl ether, $C_{19}H_{26}O_2$. Crystals from hexane. mp 112-113°. uv max (alcohol): 207, 232, 279 nm (log ϵ 4.13, 4.13, 4.11) (Delobelle, Fétizon). Also reported as needles from CH_3OH, mp 117-118° (Nasipuri, Roy).

6464. Nimesulide. *N-(4-Nitro-2-phenoxyphenyl)methanesulfonamide;* 4-nitro-2-phenoxymethanesulfonanilide; R 805; Aulin; Flogovital; Mesulid; Nimed; Sulidene. $C_{13}H_{12}N_2O_5S$; mol wt 308.31. C 50.64%, H 3.92%, N 9.09%, O 25.95%, S 10.40%. Prostaglandin synthetase and platelet aggregation inhibitor. Prepn and anti-inflammatory activity: **Belg.** pat. **801,812**; G. G. I. Moore, J. K. Harrington, **U.S.** pat. **3,840,597** (both 1974 to Riker). Activity, comparison with other non-steroidal anti-inflammatories: K. F. Swingle *et al., Arch. Int. Pharmacodyn. Ther.* **221**, 132 (1976). Mechanism of action: R. L. Vigdahl, R. H. Tukey, *Biochem. Pharmacol.* **26**, 307 (1977). Chromatographic determn in plasma: S. F. Chang *et al., J. Pharm. Sci.* **66**, 1700 (1977). Pharmacology: K. F. Swingle, G. G. I. Moore, *Drugs Exp. Clin. Res.* **10**, 587 (1984). Clinical trials in rheumatic disorders: R. Weissenbach, *J. Int. Med. Res.* **9**, 349 (1981); in osteoarthritis: M. Reiner, *ibid.* **10**, 92 (1982); in acute inflammation: J. M. Pais, F. M. Rosteiro, *ibid.* **11**, 149 (1983); C. Milvio, *ibid.* **12**, 327 (1984); M. E. Nouri, *Clin. Ther.* **6**, 142 (1984).

Light tan crystals from ethanol, mp 143-144.5°. LD_{50} orally in rats: 324 mg/kg (Swingle, Moore).
THERAP CAT: Anti-inflammatory.
THERAP CAT (VET): Anti-inflammatory.

6465. Nimetazepam. *1,3-Dihydro-1-methyl-7-nitro-5-phenyl-2H-1,4-benzodiazepin-2-one;* 1-methyl-5-phenyl-7-nitro-1,3-dihydro-2H-1,4-benzodiazepin-2-one; 1-methylnitrazepam; S 1530; Elimin; Hypnon. $C_{16}H_{13}N_3O_3$; mol wt 295.30. C 65.08%, H 4.44%, N 14.23%, O 16.25%. The desmethyl derivative of nitrazepam, *q.v.* Prepn: L. H. Sternbach *et al., J. Med. Chem.* **6**, 261 (1963); E. Reeder, L. H. Sternbach, **U.S.** pats. **3,109,843; 3,144,439; 3,141,890; 3,243,427** (1963, 1964, 1964, 1966 all to Hoffmann-La Roche); O. Keller *et al.,* **U.S.** pats. **3,121,114; 3,203,990** (1964, 1965 to Hoffmann-La Roche); Sorrentino, **S. Afr.** pat. **67 06,791** (1968 to Dumex), *C.A.* **70**, 57916c (1969); Yamamoto *et al.,* **Ger.** pat. **1,816,046** (1970 to Sumitomo), *C.A.* **73**, 120690d (1970); Inaba *et al., Chem. Pharm. Bull.* **19**, 722 (1971). Pharmacology: S. Sakai *et al., Arzneimittel-Forsch.* **22**, 534 (1972).

Pale yellow plates from ethanol, mp 156.5-157.5°. uv max (methanol): 259, 308 nm (ε 15800, 9600). LD_{50} in male, female mice, rats (mg/kg): 910, 750, 1150, 970 orally; 970, 840, 970, 980 i.p.; 1500, 1500, 1000, 1000 s.c. (Sakai).

Note: This is a controlled substance (depressant) listed in the U.S. Code of Federal Regulations, Title 21 Part 1308.14 (1985).

THERAP CAT: Anticonvulsant. Muscle relaxant (skeletal).

6466. Nimidane. *4-Chloro-N-1,3-dithietan-2-ylidene-2-methylbenzeneamine;* cyclic methylene (4-chloro-o-tolyl)-dithioimidocarbonate; AC 84633; ENT 29106; Abequito. $C_9H_8ClNS_2$; mol wt 229.73. C 47.05%, H 3.51%, Cl 15.43%, N 6.10%, S 27.91%. Prepn: R. W. Addor, S. Kantor, **Ger. pat. 2,305,517** (1973 to Am. Cyanamid), *C.A.* **79,** 115553f (1973); W. W. Brand, M. W. Bullock, **U.S. pat. 3,842,096** (1974 to Am. Cyanamid). Tickicidal activity: N. K. Amaral *et al., J. Econ. Entomol.* **67,** 387 (1974).

White solid, mp 43-46°.
THERAP CAT (VET): Acaricide.

6467. Nimodipine. *1,4-Dihydro-2,6-dimethyl-4-(3-nitrophenyl)-3,5-pyridinedicarboxylic acid 2-methoxyethyl 1-methylethyl ester;* 2-methoxyethyl 1,4-dihydro-5-(isoprop-oxycarbonyl)-2,6-dimethyl-4-(3-nitrophenyl)-3-pyridine-carboxylate; isopropyl 2-methoxyethyl 1,4-dihydro-2,6-dimethyl-4-(3-nitrophenyl)-3,5-pyridinedicarboxylate; 2,6-dimethyl-4-(3'-nitrophenyl)-1,4-dihydropyridine-3,5-dicarboxylic acid 3-β-methoxyethyl ester 5-isopropyl ester; Bay e 9736; Nimotop; Periplum. $C_{21}H_{26}N_2O_7$; mol wt 418.45. C 60.28%, H 6.26%, N 6.69%, O 26.76%. Dihydropyridine calcium channel blocker. Prepn: H. Meyer *et al.,* **Ger. pat. 2,117,571;** *eidem,* **U.S. pat. 3,799,934** (1972, 1974 to Bayer). Pharmacology: R. Towart, S. Kazda, *Brit. J. Pharmacol.* **67,** 409P (1979); K. Tanaka *et al., Arzneimittel-Forsch.* **30,** 1494 (1980); L. M. Auer, *ibid.* **31,** 1423 (1981). Prepn of isomers and pharmacological comparison with racemate: R. Towart *et al., ibid.* **32,** 338 (1982). Use as cerebral vasodilator: H. Meyer *et al.,* **Brit. pat. 2,018,134;** *eidem,* **U.S. pat. 4,406,906** (1979, 1983 to Bayer). Effect on associative learning in aging rabbits: R. A. Deyo *et al., Science* **243,** 809 (1989). GC and LC determns in biological fluids: G. J. Krol *et al., J. Chromatog.* **305,** 105 (1984). Clinical trial in prophylaxis of cerebral vasospasm: G. S. Allen *et al., N. Engl. J. Med.* **308,** 619 (1983). Toxicology: H. Schlüter, *Arzneimittel-Forsch.* **36,** 1733 (1986). Series of articles on clinical pharmacology and therapeutic use: *Am. J. Cardiol.* **55**(3), 139B-153B (1985).

Crystals from petr ether/acetic ester, mp 125°. LD_{50} in mice, rats (mg/kg): 3562, 6599 orally; 33, 16 i.v. (Schlüter).
(+)-Form, $[\alpha]_D^{20}$ +7.9° (c = 0.439 in dioxane).
(−)-Form, $[\alpha]_D^{20}$ −7.93° (c = 0.374 in dioxane).

THERAP CAT: Cerebral vasodilator.

6468. Nimorazole. *4-[2-(5-Nitro-1H-imidazol-1-yl)eth-yl]morpholine;* N-2-morpholinoethyl-5-nitroimidazole; 1-(2-N-morpholinylethyl)-5-nitroimidazole; nitrimidazine; K 1900; Acterol; Esclama; Naxofem; Naxogin; Nulogyl; Sirledi. $C_9H_{14}N_4O_3$; mol wt 226.23. C 47.78%, H 6.24%, N 24.76%, O 21.22%. Prepn: **Belg. pat. 667,262;** Giraldi, Mariotti, **U.S. pat. 3,399,193** (1965, 1968 both to Carlo Erba); **Neth. pat. Appls. 6,609,552; 6,609,553;** Gal and Carlson *et al.,* **U.S. pats. 3,458,528** and **3,646,027** (1967, 1967, 1969, 1972, all to Merck & Co.). Synthesis and biological activity studies: Giraldi *et al., Arzneimittel-Forsch.* **20,** 52 (1970). Pharmacology and toxicology: de Carneri *et al., Progr. Antimicrob. Anticancer Chemother., Proc. 6th Int. Congr. Chemother.* Tokyo, 1969 (Univ. of Tokyo Press, 1970, Tokyo) vol. I, pp 149-154. Clinical results: Emanueli, de Carneri, *ibid* vol. II, pp 369-372; Evans, Catterall, *Brit. Med. J.* **IV,** 146 (1971). Metabolic studies: Giraldi, *Biochem. Pharmacol.* **20,** 339 (1971). Acute toxicity data: B. Cavalleri *et al., J. Med. Chem.* **21,** 781 (1978).

Crystals from water, mp 110-111°. Slightly sol in water at room temp; sol in alcohols, acetone, chloroform. LD_{50} orally in mice: 1530 mg/kg (Cavalleri).
THERAP CAT: Antiprotozoal (Trichomonas).

6469. Nimustine. *N'-[(4-Amino-2-methyl-5-pyrimidin-yl)methyl]-N-(2-chloroethyl)-N-nitrosourea.* $C_9H_{13}ClN_6O_2$; mol wt 272.68. C 39.64%, H 4.80%, Cl 13.00%, N 30.82%, O 11.74%. Chloroethylnitrosourea derivative with antitumor activity. Similar to carmustine, chlorozotocin, lomustine, ranimustine, q.q.v. Prepn: H. Nakao *et al.,* **Ger. pat. 2,257,360** (1973 to Sankyo), *C.A.* **79,** 53186c (1973); *eidem, Yakugaku Zasshi* **94,** 1932 (1974), *C.A.* **82,** 43263y (1975). Effects on macrophage cytostatic activity in rats: N. Saijo *et al., Brit. J. Cancer* **42,** 162 (1980). Distribution, excretion, metabolism in mice: M. Tanaka *et al., Cancer Treat. Rep.* **64,** 575 (1980). Exptl and clinical effect: N. Saijo *et al., Cancer Chemother. Pharmacol.* **4,** 165 (1980). Acute toxicity: H. Masuda *et al., Sankyo Kenkyusho Nempo* **29,** 118 (1977), *C.A.* **88,** 146298s (1978).

Hydrochloride, $C_9H_{14}Cl_2N_6O_2$, *NSC-245,382, ACNU, Nidran.* White to light yellow cryst powder. uv max (0.04N HCl): 245 nm ($E_{1cm}^{1\%}$ 480-510). Sol in methanol, slightly sol in abs ethanol, n-butanol. Practically insol in ethyl acetate, ether, chloroform, benzene, n-hexane. Gradually develops greenish yellow color in light; decomposes slowly in humid air. LD_{50} in mice, rats (mg/kg): 62, 46 i.v. (Masuda).
THERAP CAT: Antineoplastic.

6470. Ninhydrin. *2,2-Dihydroxy-1H-indene-1,3(2H)-dione; 2,2-dihydroxy-1,3-indanedione;* 1,2,3-indantrione monohydrate; triketohydrinden hydrate. $C_9H_6O_4$; mol wt 178.14. C 67.50%, H 2.52%, O 29.98%. Prepn: Ruhemann, *J. Chem. Soc.* **97,** 1438 (1910); Teeters, Shriner, *J. Am. Chem. Soc.* **55,** 3026 (1933); Wanag, Lode, *Ber.* **71,** 1267 (1938). Improved syntheses: L. F. Fieser, *Experiments in Organic Chemistry* (Boston, 3rd ed., 1955) p 123; Becker, Russell, *J. Org. Chem.* **28,** 1896 (1963). Poisonous action on bacteria, mice, guinea pigs: O. Loew, *Biochem. Z.* **69,** 111 (1915). Acute toxicity study: J. C. Breton *et al., Compt. Rend. Soc. Biol.* **151,** 719 (1957). *Reviews:* J. C. Breton, *Etudes Chimiques et Biologiques sur la Ninhydrine, Réactif des Aminoacides* (Imprimerie E. Drouillard, Bordeaux, 1958)

285 pp; McCaldin, *Chem. Rev.* **60**, 39 (1960); A. Schoenberg, E. Singer, *Tetrahedron* **34**, 1285-1300 (1978).

Monohydrate, pale yellow prisms from water or alcohol. Reddens at 125°, swells at 139°, dec 241°. Absorption spectrum: Polonovski *et al.*, *Bull. Soc. Chim.* [5] **6**, 1557 (1939). Freely sol in water. LD_{50} i.p. in mice: 78 mg/kg, J. C. Breton *et al.*, *loc. cit.* (1957).

USE: Reagent for the detection of free amino and carboxyl groups in proteins and peptides, yielding a blue color under the proper conditions. Monograph: Henri Plagnol, *Influence de la Structure des Composés Aminés dans leur Reaction avec la Ninhydrine* (Bordeaux, 1962) 156 pp.

6471. Ninopterin. *N-[p-[[1-(2-Amino-4-hydroxy-6-pteridinyl)ethyl]amino]benzoyl]glutamic acid;* 9-methylpteroylglutamic acid; 9-methylfolic acid; Bremfol. $C_{20}H_{21}N_7O_6$; mol wt 455.45. C 52.74%, H 4.65%, N 21.53%, O 21.08%. Folic acid analog. Prepn: Hultquist *et al.*, *J. Am. Chem. Soc.* **71**, 619 (1949); Hultquist, Smith, **Brit.** pat. **667,098** (1952 to Am. Cyanamid). Clinical study in neoplastic diseases: J. C. Wright *et al.*, *J. Nat. Med. Assn.* **43**, 211 (1951). *In vitro* evaluation as folic acid antagonist: S. Waxman *et al.*, *Chemotherapy* (Basel) **28**, 402 (1982).

uv max in 0.1N NaOH: 255, 283, 365 nm (ϵ 26, 30, 8 × 10^3); in 0.1N HCl: 297 nm (ϵ 18 × 10^3).

THERAP CAT: Experimental antineoplastic.

6472. Niobium. Columbium. Nb; at. wt 92.9064; at. no. 41; valence 2, 3, 4, 5; usually pentavalent. One naturally occurring isotope: ^{93}Nb; artificial, radioactive isotopes: 88-92; 94-101. Approximately as abundant as nickel. Occurs in nature together with tantalum in the minerals *columbite* [(Fe,Mn)(Nb,Ta)$_2O_6$], *pyrochlore* (NaCaNb$_2O_6$F) and tantalite *(q.v.)*. Discovered by Hatchett in 1801, isolated by Blomstrand in 1866, named after Niobe, daughter of Tantalos. Extracted from columbite which is mined largely in Nigeria and Zaire. Less than 10% of niobium-bearing ores come from the U.S.A., Canada, and Norway. Reviews of niobium and its compds: *Technology of Columbium (Niobium)* B. W. Gonser, E. M. Sherwood, Eds. (Wiley, New York, 1958); G. L. Miller, *Tantalum and Niobium* (Academic Press, New York, 1959) 767 pp; Brown, "The Chemistry of Niobium and Tantalum" in *Comprehensive Inorganic Chemistry,* Vol. 3, J. C. Bailar, Jr. *et al.*, Eds. (Pergamon Press, Oxford, 1973) pp 553-622; P. H. Payton in *Kirk-Othmer Encyclopedia of Chemical Technology* vol. 15 (Wiley-Interscience, New York, 3rd ed., 1981) pp 820-840.

Steel-gray, lustrous metal. Ductile and malleable when pure. Lattice structure: body-centered cube, lattice constant: 3.294 Å. d 8.57. mp 2468°. bp 4927°. Sp ht: 6.012 cal/g-atom/°C. Heat of sublimation: 170.9 kcal/g-atom; heat of combustion: 2379 cal/g. Coefficient of linear expansion per °C: 7.1 × 10^{-6}. Electrical resistivity (20°): 13.2 μohm-cm. Temp coefficient of electrical resistivity per °C: 0.00395. Electron work function: 4.01 ev. Ionization potential: 6.77 V. Inert toward HCl, HNO_3 or aqua regia, but attacked by fusion with akali hydroxides or oxidizing agents.

USE: In ferrous metallurgy: Ferroniobium (produced by silicon reduction of columbite) is used to alloy stainless steels and metals for welding rods. In niobium base alloys for high temps and nuclear reactions. Niobium has some use as a getter in electronic vacuum tubes.

6473. Niobium Pentachloride. Columbium pentachloride. Cl_5Nb; mol wt 270.20. Cl 65.61%, Nb 34.39%. $NbCl_5$.

Prepn: Epperson *et al.*, *Inorg. Syn.* **7**, 163 (1963); *cf.* Rolsten, *J. Am. Chem. Soc.* **80**, 2952 (1958). Review of niobium halides: Fairbrother in *Halogen Chemistry*, **Vol. 3**, V. Gutmann, Ed. (Academic Press, New York, 1967) pp 123-178.

Yellow, very deliquesc, monoclinic crystals; dec in moist air with evolution of HCl. d 2.75. mp 204.7-209.5°; bp approx 250°, but begins to sublime at 125°. Sol in HCl, carbon tetrachloride. LD_{50} in mice: 830 mg/kg orally, *Toxic Substances List,* H. E. Christensen, Ed. (1972) p 368.

6474. Niobium Pentafluoride. Columbium pentafluoride. F_5Nb; mol wt 187.91. F 50.56%, Nb 49.44%. NbF_5. Prepn from $NbCl_5$ + HF: Ruff, Zedner, *Ber.* **42**, 492 (1909); Ruff, Schiller, *Z. Anorg. Allgem. Chem.* **72**, 329 (1911); Kwasnik in *Handbook of Preparative Inorganic Chemistry,* **Vol. 1**, G. Brauer, Ed. (Academic Press, New York, 2nd ed., 1963) p 254. Prepn from the elements: Fairbrother, Frith, *J. Chem. Soc.* **1951**, 3051; Junkins *et al.*, *J. Am. Chem. Soc.* **74**, 3464 (1952). Review of transition metal pentafluorides: Peacock, *Advan. Fluorine Chem.* **7**, 113-145 (1973).

Strongly refractive, deliquescent, monoclinic crystals, d_4^{80} 2.6955. mp 80.0°; bp 234.9°. Appreciable vapor pressure at 50°. Hydrolyzes in water, alc, caustic solns. Sparingly sol in carbon disulfide, chloroform. More sol in concd H_2SO_4 than TaF_5. Forms complexes with Lewis bases.

6475. Niobium Pentoxide. Columbium pentoxide. Nb_2O_5; mol wt 265.82. Nb 69.90%, O 30.10%. Separation: Münchow, *Chem. Ztg.* **84**, 490, 527 (1960). Review of polymorphism: Schäfer *et al.*, *Angew. Chem. Int. Ed.* **5**, 40-52 (1966).

White, orthorhombic crystals. Becomes yellow on heating. d 4.6. mp 1520°. Insol in water; sol in HF, hot H_2SO_4.

6476. Niobium Potassium Oxypentafluoride. Columbium potassium oxypentafluoride. F_7K_2NbO; mol wt 320.10. F 41.55%, K 24.43%, Nb 29.02%, O 5.00%. K_2NbOF_5. Exists as monohydrate. Prepn: Hoard, Martin, *J. Am. Chem. Soc.* **63**, 11 (1941).

Monohydrate, lustrous monoclinic leaflets; fatty to touch. Freely sol in hot water. Soly in cold water ~1 part in 13.

6477. Nioxime®. *1,2-Cyclohexanedione dioxime.* $C_6H_{10}N_2O_2$; mol wt 142.16. C 50.69%, H 7.09%, N 19.71%, O 22.51%. Prepn: Hach *et al.*, *Org. Syn.* **32**, 35 (1952); Boyer, Ellzey, *J. Am. Chem. Soc.* **82**, 2525 (1960); Borger *et al.*, *Bull. Soc. Chem. Belges* **73**, 73 (1964); Arthur, **Fr.** pat. **1,339,224** (1963 to du Pont).

Crystals, mp 185-188° (darkening at 170°).
USE: Analytical reagent for nickel and palladium.

6478. Nipecotic Acid. *3-Piperidinecarboxylic acid;* hexahydronicotinic acid. $C_6H_{11}NO_2$; mol wt 129.16. C 55.79%, H 8.58%, N 10.85%, O 24.78%. Prepd by hydrogenation of nicotinic acid with platinum oxide catalyst: McElvain, Adams, *J. Am. Chem. Soc.* **45**, 2738 (1923); with rhodium on alumina catalyst: Freifelder, *J. Org. Chem.* **28**, 1135 (1963); *idem*, **U.S. pat. 3,159,639** (1964 to Abbott).

Crystals, mp 261° (dec). Freely sol in water. Practically insol in abs alcohol and ether.

Hydrochloride, $C_6H_{11}NO_2$·HCl, mp 240-242°. Very sol in water; slightly sol in alc (also reported as freely sol in alc); slightly sol in chloroform. Practically insol in ether, benzene, and acetone.

6479. Nipradilol. *3,4-Dihydro-8-[2-hydroxy-3-[(1-methylethyl)amino]propoxy]-2H-1-benzopyran-3-ol 3-nitrate;* 3,4-dihydro-8-(2-hydroxy-3-isopropylamino)propoxy-3-nitroxy-2*H*-1-benzopyran; 8-[2-hydroxy-3-(iso-propylamino)propoxy]-3-chromanol 3-nitrate; nipradolol; K-351; Hypadil. $C_{15}H_{22}N_2O_6$; mol wt 326.35. C 55.21%, H 6.79%, N 8.58%, O 29.42%. β-Adrenergic blocker with vasodilating activity. Prepn: M. Shiratsuchi *et al.*, **Eur.** pat. **Appl. 42,299;** *eidem,* **U.S.** pat. **4,394,382** (1981, 1983 both to Kowa). Synthesis: M. Shiratsuchi *et al., Chem. Pharm. Bull.* **35,** 632 (1987). Resolution of isomers: M. Shiratsuchi *et al.,* **Eur.** pat. **Appl. 154,511;** *eidem,* **U.S.** pat. **4,727,085** (1985, 1988 both to Kowa). Synthesis and activity of isomers: M. Shiratsuchi *et al., Chem. Pharm. Bull.* **35,** 3691 (1987). Pharmacology: Y. Uchida *et al., Arch. Int. Pharmacodyn.* **262,** 132 (1983). Hemodynamic effects in dogs: H. Hisa *et al., ibid.* **271,** 169 (1984). Effects on cardiac function in dogs: M. Sakanashi *et al., ibid.* **274,** 47 (1985); M. Fujii *et al., Japan. Heart J.* **27,** 233 (1986). Pharmacokinetics and metabolism in rats: S. Kabuto *et al., Arzneimittel-Forsch.* **35,** 1674 (1985); H. Kimata *et al., ibid.* 1680. Metabolic fate in dog and man: M. Yoshimura *et al., Chem. Pharm. Bull.* **33,** 3456 (1985). Clinical evaluation in angina: H. Kishida *et al., Japan. Heart J.* **29,** 309 (1988).

Colorless needles, mp 107-116° (Shiratsuchi, 1983); also reported as mp 110-122° (Shiratsuchi, 1987). LD_{50} i.v. in mice, rats: 74.0, 73.0 mg/kg; orally in mice: 540 mg/kg (Shiratsuchi, 1988).

THERAP CAT: Antianginal; antihypertensive.

6480. Niridazole. *1-(5-Nitro-2-thiazolyl)-2-imidazoli-dinone;* 1-(5-nitro-2-thiazolyl)-2-oxotetrahydroimidazole; nitrothiamidazol; Ba-32644; CIBA 32644-Ba; Ambilhar. $C_6H_6N_4O_3S$; mol wt 214.22. C 33.64%, H 2.82%, N 26.16%, O 22.41%, S 14.97%. Prepn: **Belg.** pat. **632,989** (1963 to Ciba), *C.A.* **61,** 1873d (1964), corresp to **Brit.** pat. **986,562;** Lambert *et al., Experientia* **20,** 452 (1964).

Yellow crystals from dimethylformamide, mp 260-262°. THERAP CAT: Anthelmintic (Schistosoma).

6481. Nisin. $C_{143}H_{230}N_{42}O_{37}S_7$; mol wt 3354.25. C 51.21%, H 6.91%, N 17.54%, O 17.65%, S 6.69%. Polypeptide antibiotic structurally similar to subtilin, *q.v.*, but containing no tryptophan. Produced by *Streptococcus lactis:* A. T. R. Mattick, A. Hirsch, *Nature* **154,** 551 (1944); *eidem, Lancet* **250,** 417 (1946); **253,** 5 (1947). Purification and nature of nisin: Berridge *et al., Biochem. J.* **52,** 529 (1952). Production: Hawley, Hall, **U.S.** pat. **2,935,503** (1960 to Aplin & Barrett). Structure contains 34 amino acid residues, eight of which are rarely found in nature, including lanthionine (two alanines bonded to sulfur at the β-carbons) and β-methyllanthionine. Structure: E. Gross, J. L. Morrell, *J. Am. Chem. Soc.* **93,** 4634 (1971). Confirmation of structure of nisin and its major degradation product: M. Barber *et al., Experientia* **44,** 266 (1988). Partial synthesis: K. Fukase *et al., Bull. Chem. Soc. Japan* **59,** 2505 (1986). Total synthesis: K. Fukase *et al., Tetrahedron Letters* **29,** 795 (1988). Biosynthetic study: G. W. Buchman *et al., J. Biol. Chem.* **263,** 16260 (1988). *Review:* A. Hurst, *Adv. Appl. Microbiol.* **27,** 85-123 (1981).

Abu = α-aminobutyric acid
Dha = dehydroalanine
Dhb = dehydrobutyrine

Crystals from ethanol. Soluble in dilute acids. Stable to boiling in acid soln.
USE: In food processing as a preservative, esp. for cheese and canned fruits and vegetables.

6482. Nisoldipine. *1,4-Dihydro-2,6-dimethyl-4-(2-nitrophenyl)-3,5-pyridinedicarboxylic acid methyl 2-methylpropyl ester;* isobutyl methyl 1,4-dihydro-2,6-dimethyl-4-(o-nitrophenyl)-3,5-pyridinedicarboxylate; isobutyl 1,4-dihydro-5-methoxycarbonyl-2,6-dimethyl-4-(2-nitrophenyl)-3-pyridinecarboxylate; 2,6-dimethyl-3-carbomethoxy-4-(2-nitrophenyl)-5-carbisobutoxy-1,4-dihydropyridine; Bay k 5552; Baymycard; Syscor. $C_{20}H_{24}N_2O_6$; mol wt 388.42. C 61.85%, H 6.23%, N 7.21%, O 24.71%. Dihydropyridine calcium channel blocker. Prepn: E. Wehinger *et al.,* **Ger.** pat. **2,549,568;** *eidem,* **U.S.** pat. **4,154,839** (1977, 1979 both to Bayer). Pharmacology: S. Kazda *et al., Arzneimittel-Forsch.* **30,** 2144 (1980). Preferential effect on vascular smooth muscle in humans: A. Vogt *et al., ibid.* 2162. Pharmacodynamics and pharmacokinetics: F. Pasanisi *et al., Eur. J. Clin. Pharmacol.* **29,** 21 (1985). Series of articles on pharmacokinetics in animals, humans: *Arzneimittel-Forsch.* **38,** 1093-1110 (1988). Comparison of hemodynamic effects in humans with nifedipine, nitrendipine, q.q.v.: N. M. G. Debbas *et al., Eur. J. Clin. Pharmacol.* **30,** 393 (1986). Clinical trial in angina: J. Lam *et al., J. Am. Coll. Cardiol.* **6,** 447 (1985).

Crystals from ethanol, mp 151-152°.
THERAP CAT: Antihypertensive; antianginal.

6483. Nitarsone. *4-Nitrophenylarsonic acid; p-nitrobenzenearsonic acid.* $C_6H_6AsNO_5$; mol wt 247.04. C 29.17%, H 2.45%, As 30.32%, N 5.67%, O 32.38%. Prepn: Jacobs *et al., J. Am. Chem. Soc.* **40,** 1580 (1918); Doak, Freedman, **U.S.** pat. **2,653,160** (1953 to the U.S.A. as represented by the Administrator of the Federal Security Agency); Ruddy, Starkey, *Org. Syn.* **coll. vol. III,** 665 (1955).

Pale yellow leaflets from water, dec 298-300°. Very slightly sol in cold water, cold alcohol; readily sol in warm water, warm alcohol.

THERAP CAT (VET): Antihistomonad.

6484. Nithiazide. *N-Ethyl-N'-(5-nitro-2-thiazolyl)urea;* Hepzide. $C_6H_8N_4O_3S$; mol wt 216.23. C 33.33%, H 3.73%, N 25.91%, O 22.20%, S 14.83%. Prepd by the condensation of ethyl isocyanate with 2-amino-5-nitrothiazole in toluene: Cuckler *et al., Proc. Soc. Exp. Biol. Med.* **92,** 483 (1956); O'Neill *et al.,* U.S. pat. **2,755,285** (1956 to Merck & Co.).

Crystals, dec 228°. pKa 7.3. Practically insol in water: 3 mg/100 ml.

Sodium salt, orange plates. Soly in water: 8 g/100 ml. Aq solns are alkaline and unstable.

Potassium salt, crystals. Soly in water: 3 g/100 ml. Aq solns are more stable than those of the Na salt.

THERAP CAT (VET): Antiprotozoal.

6485. Nitracrine. *N,N-Dimethyl-N'-(1-nitro-9-acridinyl)-1,3-propanediamine; 9-[[3-(dimethylamino)propyl]-amino]-1-nitroacridine.* $C_{18}H_{20}N_4O_2$; mol wt 324.39. C 66.65%, H 6.21%, N 17.29%, O 9.86%. Deriv of acridine, *q.v.,* with cytostatic and cytotoxic properties. Prepn: **Fr. pat. 1,458,183** (1966 to Polfa), *C.A.* **68,** 39493s (1968); A. Ledochowski, B. Stefanska, *Roczniki Chem.* **40,** 301 (1966), *C.A.* **65,** 2219b (1966). Pharmacological studies: J. Gieldanowski *et al., Arch. Immunol. Ther. Exp.* **20,** 399 (1972); *eidem, ibid.* 419. Mechanism of action: J. Konopa *et al., Mater. Med. Pol.* **8,** 258 (1976). Cytotoxicity study: J. Szumiel, M. Walicka, *Neoplasma* **27,** 697 (1980). DNA binding activity: L. Szmigiero, M. Gniazdowski, *Arzneimittel-Forsch.* **31,** 1875 (1981). Comprehensive review: M. Gniazdowski *et al.* in *Antibiotics* **Vol. V,** part 2, F. E. Hahn, Ed. (Springer-Verlag, New York, 1979) pp 275-297.

Cryst from benzene/petr ether, mp 134-135°. Practically insol in water, sol in most organic solvents. pKa_1 6.45; pKa_2 8.8.

Dihydrochloride monohydrate, $C_{18}H_{22}Cl_2N_4O_2 \cdot H_2O$, *C-283, Ledakrin.* Orange cryst, mp 223-224°. Sol in water, methanol, ethanol, slightly sol in benzene, diethyl ether. Conc water solns are acidic (pH 4). LD_{50} in rats, mice (mg/kg): 1, 0.72 i.v.; 34, 26 i.g., M. Gniazdowski *et al., loc. cit.* (1979).

THERAP CAT: Antineoplastic.

6486. Nitralin. *4-(Methylsulfonyl)-2,6-dinitro-N,N-dipropylbenzenamine; 4-(methylsulfonyl)-2,6-dinitro-N,N-propylaniline;* SD 11831; Planavin. $C_{13}H_{19}N_3O_6S$; mol wt 345.37. C 45.21%, H 5.54%, N 12.17%, O 27.80%, S 9.28%. Prepn: S. B. Soloway, K. D. Swahlen, U.S. pats. **3,227,734** and **3,321,292** (1966, 1967, both to Shell). Persistence in the soil: J. H. Miller *et al., Weed Sci.* **23,** 211 (1975).

Golden crystals, mp 150-151°. Vapor pressure at 20°: 9.3 × 10⁻⁹ mm Hg; at 30°: 3.3 × 10⁻⁸ mm Hg. Soly at 22°: water, 0.6 ppm; acetone, 36 g/100 ml; DMSO, 33 g/100 ml. Poorly soluble in common hydrocarbon and aromatic solvents and alcohol. LD_{50} orally in rats: > 2 g/kg.

USE: Herbicide.

6487. Nitramide. $H_2N_2O_2$; mol wt 62.03. H 3.25%, N 45.16%, O 51.59%. H_2NNO_2. Prepd from potassium nitrocarbamate: Thiele, Lachmann, *Ber.* **27,** 1909 (1894); *Ann.* **288,** 267 (1895); Marlies, La Mer, *J. Am. Chem. Soc.* **57,** 2008 (1935); Marlies *et al., Inorg. Syn.* **1,** 68 (1939).

Unstable, shiny, white leaflets from ether + petr ether, mp 72-75° (dec). Dipole moment 3.7. Should be freshly prepd when needed. Does not explode under ordinary conditions. *Handle in glass or platinum!* Sol in ether, alcohol, acetone, water. Slightly sol in benzene. Practically insol in petr ether, chloroform.

6488. Nitramine. *N-Methyl-N,2,4,6-tetranitrobenzenamine; N-methyl-N,2,4,6-tetranitroaniline;* picrylmethylnitramine; picrylnitromethylamine; Tetralite; Tetryl. $C_7H_5N_5O_8$; mol wt 287.15. C 29.28%, H 1.76%, N 24.39%, O 44.57%. $(NO_2)_3C_6H_2N(CH_3)NO_2$.

Yellow crystals. d 1.57. mp 130-132°; explodes at about 180-190°. Insol in water; sol in alc, ether, benzene, glacial acetic acid.

USE: As indicator, 0.1 g in 60 ml alcohol with water to make 100 ml. pH: 10.8 colorless, 13.0 reddish-brown. One to five drops of soln required for 10 ml liquid. Salt error said to be small. Also used in explosives. *Keep soln in the dark. Caution:* Irritating to skin and mucous membranes. Causes yellow staining to skin and hair.

6489. Nitranilic Acid. *2,5-Dihydroxy-3,6-dinitro-2,5-cyclohexadiene-1,4-dione; 2,5-dihydroxy-3,6-dinitro-p-benzoquinone;* 2,5-dihydroxy-3,6-dinitroquinone. $C_6H_2N_2O_8$; mol wt 230.09. C 31.32%, H 0.88%, N 12.16%, O 55.63%. Prepd by treating hydroquinone diacetate with fuming nitric and concd sulfuric acids: Nietzki, Benckiser, *Ber.* **18,** 499 (1885); Nietzki, *Ber.* **43,** 3458 (1910); Meyer, *Ber.* **57,** 326 (1924); Town, *Biochem. J.* **30,** 1833 (1936).

Solvated, golden-yellow plates from a little water + nitric acid. Effloresces at 100°, deflagrates at 170° without melting. Yellow prisms from ethyl acetate. Absorption max: 460 nm. Stable when kept away from moisture. Freely sol in water, alcohol; insol in ether. Aq solns dec forming HCN and oxalic acid.

6490. Nitrapyrin. *2-Chloro-6-(trichloromethyl)pyridine;* α,α,α,6-tetrachloro-2-picoline; N-Serve. $C_6H_3Cl_4N$; mol wt 230.93. C 31.21%, H 1.31%, Cl 61.42%, N 6.07%. Prepn: Johnston *et al.,* Brit. pat. **957,276,** Belg. pat. **624,800** (1964, 1963 to Dow), *C.A.* **61,** 1841b (1964).

Crystals, mp 62.5-62.9° (CH_2Cl_2-pentane). bp_{11} 136-137.5°.

USE: Fertilizer additive to control nitrification and prevent loss of soil nitrogen.

6491. Nitrazepam. *1,3-Dihydro-7-nitro-5-phenyl-2H-1,4-benzodiazepin-2-one;* LA-1; Ro 4-5360; Benzalin; Calsmin; Eatan; Eunoctin; Imeson; Insomin; Ipersed; Mogadan;

Mogadon; Nelbon; Neuchlonic; Nitrados; Nitrenpax; Noctesed; Paxisyn; Pelson; Radedorm; Relact; Remnos; Somnased; Somnibel; Somnite; Sonebon; Sonnolin; Surem; Unisomnia. $C_{15}H_{11}N_3O_3$; mol wt 281.26. C 64.05%, H 3.94%, N 14.94%, O 17.07%. Prepn: Sternbach et al., J. Med. Chem. **6**, 261 (1963); Reeder, Sternbach; Kariss, Newmark, U.S. pats. **3,109,843; 3,116,203** and **3,123,529** (1963, 1963 and 1964, all to Hoffmann-La Roche). Crystal structure: G. Gilli et al., Acta Crystallogr. **33B**, 2664 (1977). Pharmacokinetic studies: Rieder, Arzneimittel-Forsch. **23**, 212 (1973); L. Kangas, Acta Pharmacol. Toxicol. **45**, 16 (1979). Toxicity data: Randall et al., Schweiz. Med. Wochenschr. **95**, 334 (1965). Review: Rieder, Wendt in Benzodiazepines, S. Garattini et al., Eds. (Raven Press, New York, 1973) pp 99-127. Comprehensive description: H. Y. Aboul-Enein et al., in Analytical Profiles of Drug Substances vol. 9, K. Florey, Ed. (Academic Press, New York, 1980) pp 487-517.

Yellow crystals from ethanol, mp 224-226°. uv max (0.1N H_2SO_4): 277.5 nm ($E_{1cm}^{1\%}$ 1500). Sol in alc, acetone, chloroform, ethyl acetate. Practically insol in water, ether, benzene, hexane. LD_{50} orally in rats: 825 ± 80 mg/kg (Randall).

Note: This is a controlled substance (depressant) listed in the U.S. Code of Federal Regulations, Title 21 Part 1308.14 (1985).

THERAP CAT: Anticonvulsant. Hypnotic.

6492. Nitrefazole. *2-Methyl-4-nitro-1-(4-nitrophenyl)-1H-imidazole;* 1-(4-nitrophenyl)-2-methyl-4-nitroimidazole; EMD 15700; Altimol. $C_{10}H_8N_4O_4$; mol wt 248.18. C 48.40%, H 3.25%, N 22.57%, O 25.78%. Aldehyde dehydrogenase inhibitor. Prepn: R. Klink, L. Hepding, **Brit. pat.** **1,133,408** corresp to U.S. pat. **3,491,105** (1968, 1970 to E. Merck). Use as alcohol deterrent: A. Behpour et al., **Ger. pat. 2,645,709** corresp to U.S. pat. **4,182,770** (1978, 1980 to E. Merck). Effect on enzyme activity: C. A. Seyfried et al., Alcohol. Clin. Exp. Res. **6**, 313 (1982). Effect on ethanol consumption: D. E. McMillan, Drug Dev. Res. **3**, 193 (1983). Clinical studies: R. Dusing et al., Arzneimittel-Forsch. **32**, 903 (1982); D. Roscher, W. Poser, ibid. 905.

Crystals from acetone, mp 185-187°.
THERAP CAT: Alcohol deterrent.

6493. Nitrendipine. *1,4-Dihydro-2,6-dimethyl-4-(3-nitrophenyl)-3,5-pyridinedicarboxylic acid ethyl methyl ester;* ethyl 1,4-dihydro-5-(acetoxycarbonyl)-2,6-dimethyl-4-(3-nitrophenyl)-3-pyridinecarboxylate; 3-ethyl-5-methyl-1,4-dihydro-2,6-dimethyl-4-(3-nitrophenyl)-3,5-pyridinedicarboxylate; BAY e 5009; Bayotensin; Baypress; Nidrel. $C_{18}H_{20}N_2O_6$; mol wt 360.37. C 59.99%, H 5.59%, N 7.77%, O 26.64%. Dihydropyridine calcium channel blocker. Prepn: H. Meyer et al., **Ger. pat. 2,117,571**; eidem, **U.S. pat. 3,799,- 934** (1972, 1974 both to Bayer); eidem, Arzneimittel-Forsch. **31**, 407 (1981). Series of articles on pharmacology: ibid. 2056-2067. Hemodynamic effects: H. O. Ventura et al., Am. J. Cardiol. **51**, 783 (1983). Pharmacokinetics: L. Hansson et al., Hypertension **5**, Suppl. II, II-25 (1983). HPLC determn in human serum: R. A. Janis et al., J. Clin. Pharmacol. **23**, 266 (1983). Double-blind, controlled clinical trial: U. Brugmann et al., Hertz **10**, 53 (1985). Symposium on pharmacology and clinical efficacy: J. Cardiovasc. Pharmacol. **6** Suppl. 7, S929-S1113 (1984). Review: A. Scriabine et al., "Nitrendipine" in New Drugs Annual: Cardiovascular

Drugs Vol. 2, A. Scriabine, Ed. (Raven Press, New York, 1984) pp 37-49.

Crystals from ethanol, mp 158°. LD_{50} orally in mice: 3000 mg/kg (Meyer, 1974).
THERAP CAT: Antihypertensive.

6494. Nitrenes. Molecular fragments with six electrons on the nitrogen. Nitrenes are nitrogen analogs of carbenes, q.v. They are formed thermally or photochemically from hydrazoic acid or organic azides. As a result of hydrogen shifts, nitrenes are capable of isomerization to imines, of hydrogen abstraction from neighboring molecules affording primary amines, and of ring formation by internal dehydrogenation (insertion reactions). Survey: Horner, Christmann, Angew. Chem. Int. Ed. **2**, 599 (1963). Books: T. L. Gilchrist, C. W. Rees, Carbenes, Nitrenes and Arynes (Plenum Press, New York, 1969) 131 pp; W. Lwowski, Nitrenes (Wiley-Interscience, New York, 1970). Reviews: idem, Reactive Intermediates vol. 1, M. Jones, R. A. Moss, Eds. (Wiley, New York, 1978) pp 197-227; vol. 2 (1981) pp 315-334.

6495. Nitric Acid. Aquafortis; Salpetersäure (German). HNO_3; mol wt 63.02. H 1.60%, N 22.23%, O 76.17%. Usually produced by the catalytic oxidation of ammonia. Reviews of industrial processes: F. D. Miles, Nitric Acid, Manufacture and Uses (Oxford Univ. Press, 1961, 1963); W. Sommer in Ullmann's Encyklopädie der technischen Chemie vol. 15, pp 3-67 (3rd ed., 1964). Purification by distln: Briner et al., Helv. Chim. Acta **18**, 376 (1935); Ward et al., Inorg. Syn. **3**, 13 (1950); Kaplan, Schechter, ibid. **4**, 52 (1953). Reviews: Several authors in Mellor's Vol. VIII, supplement II, Nitrogen (part 2), 278-352 (1967); Jones in Comprehensive Inorganic Chemistry vol. 2, J. C. Bailar, Jr. et al., Eds. (Pergamon Press, Oxford, 1973) pp 375-388; D. J. Newman in Kirk-Othmer Encyclopedia of Chemical Technology vol. 15 (Wiley-Interscience, New York, 1981) pp 853-871. See also Nitric Acid, Anhydrous.

Colorless liq. Fumes in moist air. Characteristic choking odor. d_4^{25} 1.50269. mp −41.59°. Monohydrate, mp −37.68°. Trihydrate, mp −18.47°. Heat of fusion: 2.503 kcal/mole. Heat of infinite dilution (298.1°K): −7971 cal/mole. bp$_{760}$ 83°. With water it forms a negative azeotrope (min vapor press. max bp), so-called constant boiling acid at 68% HNO_3, bp 120.5°, d_4^{20} 1.41. The oxide-free acid, diluted with an equal vol of water, does not discolor drops of 0.1N $KMnO_4$. Colorless 100% acid cannot be stored in the presence of light without formation of NO_2 which produces the discoloration. Stains woolen fabrics and animal tissue a bright yellow. In the presence of traces of oxides it attacks all base metals except Al and special Cr steels which become passivated. Strong, monobasic acid. Oxidizing agent. Reacts violently with alcohol, turpentine, charcoal, organic refuse. Nitric acid concentrated is a water soln contg 70-71% HNO_3. Density of aq solns: d_4^{20} 1.0036 (1% HNO_3 w/w); 1.0543 (10%); 1.1150 (20%); 1.1800 (30%); 1.2463 (40%); 1.3100 (50%); 1.3667 (60%); 1.4134 (70%); 1.4521 (80%); 1.4826 (90%); 1.5129 (100%): International Critical Tables **III**, 58 (1928). See also Nitric Acid, Fuming.

USE: Manufacture of inorganic and organic nitrates and nitro compounds for fertilizers, dye intermediates, explosives and many different organic chemicals. Pharmaceutic aid (acidifier). Caution: Ingestion causes burning and corrosion of mouth, esophagus, stomach; abdominal tenderness, shock, death. Continued exposure to vapor may cause chronic bronchitis; chemical pneumonitis may occur. Carbonated alkalies must not be used as antidote in case of poisoning by ingestion: Clinical Toxicology of Commercial

Compounds, R. E. Gosselin *et al.*, Eds. (Williams & Wilkins, Baltimore, 5th ed., 1984) Section III, pp 8-11.

THERAP CAT (VET): Cauterizing agent (for warts).

6496. Nitric Acid, Anhydrous. Prepared by distillation of concd nitric acid with concd sulfuric acid; by treating sodium or potassium nitrate with 100% H_2SO_4 and removing HNO_3 by distillation; by fractional crystallization of concd HNO_3. Review of preparation and properties of pure HNO_3: Stern *et al.*, *Chem. Rev.* **60**, 185-207 (1960). *See also* Nitric Acid.

Liquid; decomposes above freezing point to form NO_2, H_2O and O_2. Develops yellow color due to nitrogen oxide formation. mp $-41.59°$. d^0 (liq) 1.5492. Forms white, monoclinic crystals; $d^{-41.6}$ (solid) 1.895.

6497. Nitric Acid, Fuming. Defined as concentrated nitric acid containing dissolved nitrogen dioxide. May be prepared from concd nitric acid by passing nitrogen dioxide into it or by adding a small amount of organic reducing agent, such as formaldehyde.

Yellow to brownish-red, clear, strongly fuming, very corrosive liq; evolves suffocating, poisonous, yellowish-red fumes of nitrogen dioxide and nitrogen tetroxide. The density increases as the free NO_2 content increases: concd HNO_3 with 7.5% NO_2 added $d_4^{20} = 1.526$; with 12.7% NO_2 $d_4^{20} = 1.544$. Fuming nitric acid A.C.S. reagent has a density of approx 1.5 and assays not less than 90% HNO_3 by titration with NaOH using methyl orange indicator. Miscible with water. *Handle with extreme care.*

USE: In those organic reactions where nitric acid acts more as an oxidizing agent than as a source of hydrogen ions.

6498. Nitric Oxide. Mononitrogen monoxide; nitrogen monoxide. NO; mol wt 30.01. N 46.68%, O 53.32%. Prepd industrially by passing air through an electric arc (basis of atmospheric nitrogen fixation) or by oxidation of ammonia over platinum gauze. Convenient lab procedure using sodium nitrite and ferrous sulfate: Blanchard, *Inorg. Syn.* **2**, 126 (1946). Alternate lab methods: Schenk in *Handbook of Preparative Inorganic Chemistry* **vol. 1**, G. Brauer, Ed. (Academic Press, New York, 2nd ed., 1963) pp 485-487. *Reviews:* Beattie, "Nitric Oxide" in *Mellor's* Vol. VIII, supplement II, *Nitrogen* (part 2) 216-240 (1967); Jones in *Comprehensive Inorganic Chemistry* **vol. 2**, J. C. Bailar, Jr. *et al.*, Eds. (Pergamon Press, Oxford, 1973) pp 323-334.

Colorless gas. Burns only when heated with hydrogen. Deep blue when liquid. Bluish-white snow when solid. mp $-163.6°$. bp $-151.7°$. $d^{-150.2}$ (liq) 1.27. d (gas) 1.04 (air = 1). Trouton constant 27.1. Contains an odd number of electrons and is paramagnetic. Crit temp $-94°$. Crit press. 65 atm. Heat of formation (18°): -21.5 kcal/mole. Heat of vaporization (bp): 3.293 kcal/mole. Solubility in water (ml/100 ml; 1 atm): 4.6 (20°); 2.37 (60°). Combines with oxygen to form NO_2 (a brown gas) and with chlorine and bromine to form the nitrosyl halides, such as NOCl, see N. V. Sidgwick, *Chemical Elements and Their Compounds* **vol. I** (Oxford, 1950) p 683. Usually shipped in steel cylinders as compressed gas at 480 lb/sq in.

USE: In large quanities in the manuf of nitric acid, also in the bleaching of rayon, as stabilizer (to prevent free-radical decompn) for propylene, methyl ether, etc. *Caution:* Immediately on contact with air, nitric oxide is converted to the highly poisonous nitrogen dioxide, nitrogen tetroxide, or both. Gas masks plus adequate ventilation are mandatory when handling even small amounts in the laboratory. *See also* Nitrogen Dioxide.

6499. Nitrilotriacetic Acid. *N,N-Bis(carboxymethyl)glycine*; triglycollamic acid; α,α',α''-trimethylaminetricarboxylic acid; tri(carboxymethyl)amine; triglycine; NTA. C_6-H_9NO_6; mol wt 191.14. C 37.70%, H 4.74%, N 7.33%, O 50.22%. $N(CH_2COOH)_3$. Prepn: Heintz, *Ann.* **122**, 260 (1862); Michaelis, Schubert, *J. Biol. Chem.* **106**, 331 (1934); Martell, Bersworth, *J. Org. Chem.* **15**, 46 (1950); Singer, Weisberg, U.S. pats. **2,855,428** and **3,061,628** (1958 and 1962, both to Hampshire Chem.). IR studies: Nakamoto *et al.*, *J. Am. Chem. Soc.* **84**, 2081 (1962); Chapman *et al.*, *Proc. Chem. Soc.* **1962**, 336. Solubility data: Bird, *J. Soc. Dyers Col.* **56**, 473 (1940). Toxicity data: *Soap Chem. Spec.* **42**, 58 (1966). pK data: Schwarzenbach *et al.*, *Helv. Chim. Acta*

28, 828 (1945). *Review:* Souchay, Graizon, *Bull. Soc. Chim. France* **1952**, 34.

Prismatic crystals from hot water, mp 230-235° dec (Michaelis, Schubert). mp 241.5° (dec). 1.28 g dissolves in 1 liter of water at 22.5°, pH of satd aq soln is 2.3 (Bird). At 20° pK is 3.03, pK_2 is 3.07, pK_3 is 10.70.

Sodium salt, $C_6H_6NNa_3O_6$, *NTANa₃, Trilon A.* MLD orally in rats $>4,000$ mg/kg (*Soap Chem. Spec.*).

Note: This substance may reasonably be anticipated to be a carcinogen: *Fourth Annual Report on Carcinogens* (NTP 85-002, 1985) p 140.

USE: Chelating and sequestering agent; builder in synthetic detergents.

6500. Nitrin. *2-Aminobenzaldehyde phenylhydrazone.* $C_{13}H_{13}N_3$; mol wt 211.26. C 73.90%, H 6.20%, N 19.89%. Prepd by refluxing 2-nitrobenzaldehyde with phenylhydrazine: Knöpfer, *Monatsh.* **31**, 97 (1910).

Needles from acetone, mp 227-229° (dec). Sol in acetone. Sparingly sol in cold alcohol, ether, chloroform, benzene. A soln in alcohol or acetone develops a red color with nitrites upon addn of acid.

USE: Detection of nitrites, colibacilli in urine: Pfeiffer, *Münch. Med. Wochenschr.* **92**, 1315 (1950).

6501. Nitroacetanilide. $C_8H_8N_2O_3$; mol wt 180.16. C 53.33%, H 4.48%, N 15.55%, O 26.64%.

m-Nitroacetanilide, *N-(3-nitrophenyl)acetamide*, leaflets. mp 151-153°. Sparingly sol in hot water; freely sol in chloroform, nitrobenzene; practically insol in ether.

o-Nitroacetanilide, yellow leaflets, d 1.42. mp 93-94°. Moderately sol in cold water, freely in boiling water, in cold fixed alkali hydroxide solns; sol in chloroform, alcohol, ether.

p-Nitroacetanilide, prisms. mp 214-216°. Almost insol in cold water; sol in hot water, alcohol, ether; sol in KOH with orange color.

6502. Nitroakridin 3582. *1-Diethylamino-3-[(2,3-dimethoxy-6-nitro-9-acridinyl)amino]-2-propanol;* 5-(γ-diethylamino-β-hydroxypropyl)amino-2-nitro-7,8-dimethoxyacridine; 9-(3-diethylamino-2-hydroxypropylamino)-6,7-dimethoxy-3-nitroacridine; nitroacridine 3582; W 1889; Entozon. $C_{22}H_{28}N_4O_5$; mol wt 428.48. C 61.66%, H 6.59%, N 13.08%, O 18.67%. Nitroacridine dye with antimicrobial activity. Prepn: Bockmühl, Fehrle, U.S. pat. **2,040,070** (1936 to Winthrop); *BIOS rept.* no. 766 (1946); Miller, Wagner, *J. Org. Chem.* **13**, 891 (1948); Steck *et al.*, *J. Am. Chem. Soc.* **79**, 4414 (1957). Exptl use for *in vitro* identification of hypoxic cells: A. C. Begg *et al.*, *Brit. J. Radiol.* **56**, 970 (1983).

Crystals from acetone, mp 168-169°.

Complex with glycarsamide, *q.v.*, *Rutenol. See:* **Brit.** pat. **405,629** (1934 to I.G. Farbenindustrie).

Dihydrochloride, $C_{22}H_{30}Cl_2N_4O_5$, orange crystals, dec 219-220°.

THERAP CAT: Antiseptic.

6503. *m*-Nitroaniline. *3-Nitrobenzenamine; m*-nitrani-
line. $C_6H_6N_2O_2$; mol wt 138.12. C 52.17%, H 4.38%, N
20.28%, O 23.17%. Prepn by nitration of aniline: Holleman
et al., Ber. **44,** 704 (1911); by reduction of *m*-dinitrobenzene:
Brady *et al., J. Chem. Soc.* **1929,** 2266; Kubota *et al., J.
Pharm. Soc. Japan* **76,** 801 (1956); Kuhn, **U.S.** pat **2,768,209**
(1956 to Ringwood); from *m*-nitrobenzoic acid: Snyder *et
al., J. Am. Chem. Soc.* **75,** 2014 (1953).

Yellow crystals from water, mp 114°. Calc bp 306°. d_4^{25}
0.9011. One gram dissolves in 880 ml water, about 20 ml alcohol, 18 ml ether, 11.5 ml methanol. Forms water-sol salts
with mineral acids.
USE: Dyestuff intermediate. *Caution:* Highly toxic; absorbed through skin. Avoid breathing dust; avoid contact
with skin, eyes, clothing. In case of contact, immediately
flush skin or eyes with plenty of water for at least 15 minutes; for eyes, get medical attention. Wash clothing before
reusing. Acute exposure can cause methemoglobinemia,
cyanosis. Chronic exposure may cause liver damage.

6504. *o*-Nitroaniline. *o*-Nitraniline. Empirical formula
etc., see *meta*-isomer. Prepn from *o*-dinitrobenzene: Meisenheimer, Hesse, *Ber.* **52,** 1166 (1919); from *o*-nitroaniline-
p-sulfonic acid: Ehrenfeld, Puterbaugh, *Org. Syn.* **coll. vol.
I,** 388 (1941); from ψ-*o*-dinitrosobenzene: Boyer *et al., J.
Am. Chem. Soc.* **77,** 5688 (1955).
Orange-yellow crystals from boiling water, mp 69-71°.
Calc bp 284°. d_4^{25} 0.9015. Slightly sol in cold water; sol in
hot water, alcohol, chloroform. Forms water soluble salts
with mineral acids.
USE: Dyestuff intermediate. *Caution: See m*-Nitroaniline.

6505. *p*-Nitroaniline. *p*-Nitraniline. Empirical formula
etc., see *meta*-isomer. Prepn from acetanilide: Bart, **U.S.**
pat. **2,406,578** (1946 to Am. Cyanamid); by Schmidt reaction: Stockel, Hall, *Nature* **197,** 787 (1963).
Bright yellow powder, mp 146°. Calc bp 332°. One gram
dissolves in 1250 ml water, 45 ml boiling water, 25 ml alcohol, 30 ml ether; sol in benzene, methanol. Forms water
soluble salts with mineral acids.
USE: Dyestuff intermediate. *Caution: See m*-Nitroaniline.

6506. Nitroanisole. *Methoxynitrobenzene;* nitrophenyl
methyl ether. $C_7H_7NO_3$; mol wt 153.13. C 54.90%, H
4.61%, N 9.15%, O 31.34%.

m-Nitroanisole. Crystals. d 1.373. mp 38-39°. bp 258°.
Volatile with steam. Insol in water; sol in alc.
o-Nitroanisole. Colorless to yellowish liquid. d_4^{20} 1.254.
mp +9.4°. bp 277°; also stated as 272-273°. n_D^{20} 1.5620.
Volatile with steam. Insol in water; sol in alc, ether.
p-Nitroanisole. Crystals. d 1.233. mp 54°. bp 260°; also
stated as 274°. Insol in water; freely sol in alc, ether, boiling
petr ether; slightly sol in cold petr ether.
USE: *o*-Isomer as dye intermediate; in organic syntheses.

6507. 5-Nitrobarbituric Acid. *5-Nitro-2,4,6(1H,3H,5H)-
pyrimidinetrione;* dilituric acid. $C_4H_3N_3O_5$; mol wt 173.08.
C 27.76%, H 1.74%, N 24.28%, O 46.22%.

Trihydrate, prisms and leaflets from water. mp 176° with
decompn, when anhydr. Soluble in about 1200 parts cold
water; more sol in hot water; sol in alcohol, in sodium hydroxide soln; insol in ether.
USE: As a microreagent for potassium with which it forms
a characteristic precipitate.

6508. Nitrobenzaldehyde. $C_7H_5NO_3$; mol wt 151.12. C
55.63%, H 3.34%, N 9.27%, O 31.76%.

m-Nitrobenzaldehyde. Yellowish, cryst powder. mp 58°.
bp_{23} 164°. Volatile with steam. Almost insol in water; sol in
alcohol, chloroform, ether.
o-Nitrobenzaldehyde. Light yellow needles. mp 42-44°.
bp_{23} 153°. Volatile with steam. Slightly sol in water, freely
in alcohol, benzene, ether.
p-Nitrobenzaldehyde. White to yellow crystals, mp 106-
107°. Sublimes; slightly volatile with steam. Slightly sol in
water or ether; sol in alc, benzene, glacial acetic acid.
USE: *o*-Nitrobenzaldehyde as reagent for isopropyl alcohol
and acetone.

6509. Nitrobenzene. Nitrobenzol; essence of mirbane; oil
of mirbane. $C_6H_5NO_2$; mol wt 123.11. C 58.53%, H 4.09%,
N 11.38%, O 25.99%. Produced by treating benzene with a
mixture of HNO_3 and H_2SO_4. *Review:* Faith, Keyes &
Clark's *Industrial Chemicals,* F. A. Lowenheim, M. K.
Moran, Eds. (Wiley-Interscience, New York, 4th ed., 1975)
pp 571-574; K. L. Dunlap in Kirk-Othmer *Encyclopedia of
Chemical Technology* vol. **15** (Wiley-Interscience, New York,
3rd ed., 1981) pp 916-932.

Colorless to pale yellow, oily liquid; odor of volatile oil
almond. *Poisonous!* d_4^{15} 1.205. mp +6°. bp 210-211°.
Flash pt, closed cup: 190°F (88°C). n_D^{20} 1.5529. Volatile
with steam. Sol in about 500 parts water; freely sol in alcohol, benzene, ether, oils. LD_{50} orally in rats: 640 mg/kg,
RTECS Vol. 1, R. J. Lewis, R. L. Tatken, Eds. (1979) p 217.
Caution: Rapidly absorbed through skin, vapor hazardous.
Do not get in eyes, on skin, on clothing. Avoid breathing
vapor. Use only with adequate ventilation. In case of contact, immediately remove all contaminated clothing, including shoes, and flush skin or eyes with plenty of water for at
least 15 min; for eyes, get medical attention. Wash clothing
before reusing. May cause headaches, drowsiness, nausea,
vomiting, methemoglobinemia with cyanosis: E. Browning,
Toxicity and Metabolism of Industrial Solvents (Elsevier, New
York, 1965) pp 298-303.
USE: For the manuf of aniline; in soaps, shoe polishes; for
refining lubricating oils; manuf pyroxylin compds.

6510. Nitrobenzoic Acid. $C_7H_5NO_4$; mol wt 167.12. C
50.31%, H 3.02%, N 8.38%, O 38.29%. Prepn of *m*-isomer:
O. Kamm, J. B. Segur, *Org. Syn.* coll. vol. I (2nd ed., 1964) p
391; of *p*-isomer: O. Kamm, A. O. Matthews, *ibid.* p 392.

m-Nitrobenzoic acid, monoclinic leaflets from water. Bitter taste. d 1.494. mp 142°. Melts in hot water. K (25°) =
3.48 × 10⁻⁴. Absorption spectrum: Purvis, *J. Chem. Soc.*
107, 971 (1915). One gram dissolves in 320 ml water, 3 ml
alc, 4 ml ether, 18 ml chloroform, about 2 ml methanol, 2.5

ml acetone; very slightly sol in benzene, carbon disulfide, petr ether.

o-Nitrobenzoic acid, yellowish-white, intensely sweet crystals. d 1.58. mp 147-148°. One gram dissolves in 146 ml water, 3 ml alcohol, 220 ml chloroform, 4.5 ml ether, 2.5 ml acetone, 2.5 ml methanol; very slightly sol in benzene, carbon disulfide, petr ether.

p-Nitrobenzoic acid, monoclinic leaflets, plates from benzene. d 1.58. mp 242.4°. Sublimes. One gram dissolves in 2380 ml water, 110 ml alcohol, 12 ml methanol, 150 ml chloroform, 45 ml ether, 20 ml acetone; slightly sol in benzene, carbon disulfide; insol in petr ether.

USE: Manuf intermediates; also as a reagent for alkaloids and thorium.

6511. 4-Nitrobenzoyl Chloride. $C_7H_4ClNO_3$; mol wt 185.57. C 45.30%, H 2.17%, Cl 19.11%, N 7.55%, O 25.87%. $NO_2C_6H_4COCl$. Prepd by the action of phosphorus pentachloride on *p*-nitrobenzoic acid: Adams, Jenkins, *Org. Syn.* **3**, 75 (1923).

Bright yellow needles from petr ether. Pungent odor. mp 75°. bp_{105} 205°; bp_{12} 154°. Decomposed by water and alcohol; sol in ether. *Keep well closed.*

6512. p-Nitrobenzyl Cyanide. *4-Nitrobenzeneacetonitrile; p-nitrophenylacetonitrile; p-nitro-α-toluinitrile.* C_8H_6-N_2O_2; mol wt 162.14. C 59.26%, H 3.73%, N 17.28%, O 19.73%. $NO_2C_6H_4CH_2CN$. Prepd by the action of concd nitric acid on benzyl cyanide: Robertson, *Org. Syn.* **2**, 57 (1922).

Elongated prisms from alcohol. mp 117°. Insol in water; sol in alcohol, ether.

USE: In the prepn of *p*-nitrophenylacetic acid.

6513. o-Nitrobiphenyl. *2-Nitro-1,1'-biphenyl; o-nitrodiphenyl; ONB.* $C_{12}H_9NO_2$; mol wt 199.20. C 72.35%, H 4.55%, N 7.03%, O 16.06%. Prepn from diazotized o-nitroaniline and benzene by a modified Gomberg reaction: Elks *et al.*, *J. Chem. Soc.* **1940**, 1284. Prepd industrially by direct nitration of biphenyl (yields about 50% para, 35% ortho).

Orthorhombic bipyramidal plates from ethanol. Characteristic sweetish odor. mp 36.7°. d_4^{25} 1.44. $d_{15.5}^{40}$ (liq) 1.189. The technical liq weighs about 10 lbs/gal. bp_{760} 325°; bp_{30} 205°; bp_{13} 170°; bp_4 166°. n_D^{25} (tech liq) 1.613. Flash pt 179° (354°F). uv max: 325 nm. Dipole moment: 3.79. Viscosity at 45°: 12 cps; at 25°: 38 cps (tech liq). Insol in water. Sol in methanol, ethanol, tetrahydrofurfuryl alcohol, acetone, dimethylformamide, carbon tetrachloride, perchlorethylene, mineral spirits, turpentine, glacial acetic acid.

USE: Plasticizer for resins, cellulose acetate and nitrate, polystyrenes; fungicide for textiles; wood preservative; dye intermediate.

6514. p-Nitrobiphenyl. *4-Nitro-1,1'-biphenyl; 4-nitrobiphenyl; p-nitrodiphenyl; PNB.* $C_{12}H_9NO_2$; mol wt 199.20. C 72.35%, H 4.55%, N 7.03%, O 16.06%. Prepn: G. Schultz, *Ann.* **174**, 210 (1874); O. Kühling, *Ber.* **28**, 42 (1895); E. Bamberger, *ibid.* 404; R. L. Jenkins *et al.*, *Ind. Eng. Chem.* **22**(1), 31 (1930); K. Dimroth *et al.*, *Ber.* **101**, 2215 (1968). Crystal structure: M. Prasad *et al.*, *J. Indian Chem. Soc.* **13**, 519 (1936), *C.A.* **31**, 1271[5] (1937). Review and evaluation of studies of carcinogenicity in laboratory animals: *IARC Monographs* **vol. 4**, 113-117 (1974).

Needles from alcohol, mp 113.7°. bp_{760} 340°; bp_{30} 223.7-224.1°. Insol in water. Slightly sol in cold alc; more readily sol in hot alc; sol in chloroform, ether. uv spectrum: D. F. DeTor, H. J. Scheifele, *J. Am. Chem. Soc.* **73**, 1442 (1951).

Note: This substance has been listed as a carcinogen by OSHA: *Fed. Reg.* **39**, 3757 (1974).

USE: Formerly in prepn of *p*-biphenylamine, *q.v.*

6515. 2-Nitro-1,1-bis(p-chlorophenyl)propane. *1,1'-(2-Nitropropylidene)bis[4-chlorobenzene]; 1,1-bis(p-chlorophenyl)-2-nitropropane;* Prolan. $C_{15}H_{13}Cl_2NO_2$; mol wt 310.19. C 58.08%, H 4.22%, Cl 22.86%, N 4.52%, O 10.32%. Prepn: Hass, Blickenstaff, U.S. pat. **2,516,186** (1950 to Purdue Res. Found.); Neher, Blickenstaff, *Ind. Eng. Chem.* **43**, 2875 (1951); Hodge, *J. Am. Chem. Soc.* **73**, 2341 (1951). Toxicity data: Lehman, *Assoc. Food Drug Officials U.S. Quart. Bull.* **14**, 82 (1951), *C.A.* **45**, 3517h (1951).

Viscous oil, $bp_{0.16}$ 180°. Crystals from abs ethanol, mp 80.5-81.5°. LD_{50} orally in rats: 750 mg/kg (Lehman). Combination with Bulan, *Dilan.*

USE: Insecticide.

6516. m-Nitrocinnamic Acid. *3-(3-Nitrophenyl)-2-propenoic acid.* $C_9H_7NO_4$; mol wt 193.15. C 55.96%, H 3.65%, N 7.25%, O 33.13%. $NO_2C_6H_4CH=CHCOOH$. Prepd by heating a mixture of m-nitrobenzaldehyde, sodium acetate and acetic anhydride: Thayer, *Org. Syn.* **5**, 83 (1925). From m-nitrobenzaldehyde and malonic acid in the presence of glycine in a little water at 100°: Dakin, *J. Biol. Chem.* **7**, 54 (1909-1910).

The above methods yield the *trans*-form. White needles from benzene or alcohol. mp 200-201°. Can be sublimed or distilled at atmospheric press. with very little decompn. Absorption spectrum: Purvis, *J. Chem. Soc.* **107**, 971 (1906). One gram dissolves in about 100 ml alcohol at 25°. Ultraviolet irradiation of a soln in water-alcohol-ammonia gives the *cis*-form in 22% yield: Wollring, *Ber.* **47**, 112 (1914). Minute needles from toluene + methanol, mp 158°. By reirradiation of a chloroform soln containing a trace of bromine, the more stable *trans*-form is regenerated.

6517. Nitrodan. *3-Methyl-5-[(4-nitrophenyl)azo]-2-thioxo-4-thiazolidinone; 3-methyl-5-[(p-nitrophenyl)azo]rhodanine; 3-methyl-5-[(p-nitrophenyl)azo]-2-thio-2,4-thiazolidinedione;* CTR 6110; Nidanthel; Everfree. $C_{10}H_8N_4O_3S_2$; mol wt 296.33. C 40.53%, H 2.72%, N 18.91%, O 16.20%, S 21.64%. Prepn: Benghiat, Howard, U.S. pat. **2,952,673** (1960 to Stauffer); **Belg.** pat. **637,282** (1964 to Cooper, Tinsley Labs.).

Yellow powder, mp 267-268°. Sparingly sol in water.
THERAP CAT (VET): Anthelmintic.

6518. Nitroethane. $C_2H_5NO_2$; mol wt 75.07. C 32.00%, H 6.71%, N 18.66%, O 42.63%. $CH_3CH_2NO_2$. Obtained by vapor phase nitration of ethane with HNO_3: Riedel, *Oil Gas J.* **54**, no. 36, 110-114 (1956). Laboratory prepn by treating 1.5 moles sodium nitrite with 1 mole sodium ethyl sulfate at 125-130° in the presence of 0.0625 moles potassium carbonate: Desseigne, Giral, *Mem. Poudres* **34**, 49-53 (1952), *C.A.* **49**, 836 (1955). Similar procedure according to the equation $EtOSO_2OEt + NaNO_2 \rightarrow EtOSO_2ONa + EtNO_2$: McCombie *et al.*, *J. Chem. Soc.* **1944**, 24.

Oily liquid; pleasant odor. d_{25}^{25} 1.041; d_{20}^{20} 1.052. Flash pt, open cup: 106°F (41.11°C). mp about −50°. bp 114-115°. Undergoes thermal decompn at 335-382°. Heating value (liq): 7,720 Btu/lb. Lower limit of flammability in air = 4.0% by volume. Viscosity at 25° = 0.661 centipoise. n_D^{20}

1.3917; $n_D^{24.3}$ 1.39007. Soly in water: 4.5 ml/100 ml H_2O at 20°. Miscible with methanol, ethanol, ether. Sol in chloroform, aq solns of alkalies. Has high heat of absorption. Sudden absorption of the anhydr liq or gas on activated carbon or Hopcalite may result in flames: *Chem. & Eng. News* **30**, 2344 (1952). LD_{50} orally in rats: 1100 mg/kg, *Toxic Substances List*, H. E. Christensen, Ed. (1974) p 332.

USE: Solvent; in organic syntheses. Experimentally as liq propellant. *Caution:* Irritating to eyes, mucous membranes.

6519. Nitrofen. *2,4-Dichloro-1-(4-nitrophenoxy)benzene; 2,4-dichlorophenyl p-nitrophenyl ether;* nitraphen; nitrophen; nitrofene; FW 925; TOK. $C_{12}H_7Cl_2NO_3$; mol wt 284.10. C 50.73%, H 2.48%, Cl 24.96%, N 4.93%, O 16.90%. Selective pre- and post-emergence herbicide. Prepn: H. F. Wilson, H. M. Dougal, U.S. pat. **3,080,225** (1963 to Rohm & Haas). Toxicity studies: A. M. Ambrose *et al., Toxicol. Appl. Pharmacol.* **19**, 263 (1971). Carcinogenicity studies: J. F. Robens, *Vet. Human Toxicol.* **22**, 328 (1980).

Crystalline solid, mp 70-71°. Vapor pressure at 40°: 8 × 10^{-6} mm Hg. Soly in water at 22°: 0.7-1.2 ppm. LD_{50} orally in rats: 3.58 g/kg (Ambrose).

Note: This substance may reasonably be anticipated to be a carcinogen: *Fourth Annual Report on Carcinogens* (NTP 85-002, 1985) p 143.

USE: Herbicide.

6520. Nitrofurantoin. *1-[[(5-Nitro-2-furanyl)methyleneJamino]-2,4-imidazolidinedione; N-(5-nitro-2-furfurylidene)-1-aminohydantoin;* 1-(5-nitro-2-furfurylideneamino)-hydantoin; Berkfuran; Berkfurin; Chemiofuran; Cyantin; Cystit; Fua-Med; Furachel; Furalan; Furadantin; Furadantine MC; Furadoine; Furadonine; Furantoin; Furobactina; Furophen; Ituran; Macrodantin; Orafuran; Parfuran; Trantoin; Urantoin; Urizept; Urodil; Urodin; Urolong; Uro-Tablinen; Welfurin; Zoofurin. $C_8H_6N_4O_5$; mol wt 238.16. C 40.34%, H 2.54%, N 23.53%, O 33.59%. Prepn from 1-aminohydantoin sulfate and 5-nitro-2-furaldehyde diacetate: Hayes, U.S. pat. **2,610,181** (1952 to Eaton Labs.); Swirska *et al., Przem. Chem.* **11** (34), 306 (1955), *C.A.* **52**, 14079b (1958). Alternate route(s): Michels; Gever, O'Keefe, U.S. pats. **2,898,335; 2,927,110** (1959 and 1960, both to Norwich). Comprehensive description: D. E. Cadwallader, H. W. Jun, in *Analytical Profiles of Drug Substances* vol. **5**, K. Florey, Ed. (Academic Press, New York, 1976) pp 345-373.

Orange-yellow needles from dil acetic acid, dec 270-272°. pKa 7.2. uv max: 370 nm ($E_{1cm}^{1\%}$ 776). Soly (mg/100 ml): water (pH 7) 19.0; 95% ethanol 51.0; acetone 510; DMF 8000; peanut oil 2.1; glycerol 60; polyethylene glycol 1500.

THERAP CAT: Antibacterial.

THERAP CAT (VET): Antibacterial.

6521. Nitrofurazone. *2-[(5-Nitro-2-furanyl)methylene]-hydrazinecarboxamide; 5-nitro-2-furaldehyde semicarbazone;* Amifur; Furacin; Chemofuran; Furesol; Nifuzon; Nitrofural; Nitrozone; Furacinetten; Furacoccid; Furazol W; Mammex; Furaplast; Coxistat; Aldomycin; Nefco; Vabrocid. C_6H_6-N_4O_4; mol wt 198.14. C 36.37%, H 3.05%, N 28.28%, O 32.30%. Prepd from 2-formyl-5-nitrofuran and semicarbazide hydrochloride: Stillman, Scott, U.S. pat. **2,416,234** (1947 to Eaton Labs.). Alternate route: Gever, O'Keefe, U.S. pat. **2,927,110** (1960 to Norwich Pharmacal). Toxicity: J. Godwin *et al., J. Am. Med. Assoc.* **133**, 299 (1947). Tumo-

rigenic activity: Morris *et al., Cancer Res.* **29**, 2145 (1969); Ertürk *et al., ibid.* **30**, 1409 (1970).

Pale yellow needles, dec 236-240°. Bitter aftertaste. Darkens on prolonged exposure to light. uv max: 260, 375 nm. Very slightly sol in water (1:4200); slightly sol in alcohol (1:590), in propylene glycol (1:350). Sol in alkaline solns with dark orange color. Insol in ether. pH of satd water soln 6.0 to 6.5. LD_{50} in rats (g/kg): 0.59 orally; 3.0 s.c. (Godwin).

THERAP CAT: Topical anti-infective.

THERAP CAT (VET): Antimicrobial.

6522. Nitrogen. N; at. wt 14.0067; at. no. 7; valences 3, 5; elemental state: N_2. Two naturally occurring isotopes: 14 (99.635%); 15 (0.365%); five short-lived, artificial, radioactive isotopes: 12; 13; 16-18. Discovered in 1772 by Daniel Rutherford and independently by Scheele and Cavendish. Constitutes about 75.5% by weight or 78.06% by volume of the atmosphere; found frequently in volcanic or mine gases, gases from springs and gases occluded in minerals and rocks; an essential constituent of all living organisms; fixed or combined nitrogen is present in many mineral deposits. Prepn from sodium (and alkaline earth) azides by heating the azide: Tiede, *Ber.* **46**, 4100 (1913); **49**, 1745 (1916); Justi, *Ann. Physik* [5] **10**, 985 (1931). Prepd industrially by fractional distln of liquid air; by removal of oxygen by combustion; by reduction of ammonia. Purification of nitrogen furnished in steel cylinders: Kautsky, Thiele, *Z. Anorg. Allgem. Chem.* **152**, 342 (1926); Kendall, *Science* **73**, 395 (1931); Schenk in *Handbook of Preparative Inorganic Chemistry* vol. **1**, G. Brauer, Ed. (Academic Press, New York, 2nd ed., 1963) pp 458-460. Review of nitrogen and nitrogen compounds: Jones in *Comprehensive Inorganic Chemistry* vol. **2**, J. C. Bailar, Jr. *et al.*, Eds. (Pergamon Press, Oxford, 1973) pp 147-388; R. W. Schroeder in Kirk-Othmer *Encyclopedia of Chemical Technology* vol. **15** (Wiley-Interscience, New York, 3rd ed., 1981) pp 932-941. Books: W. L. Jolly, *The Inorganic Chemistry of Nitrogen* (Benjamin, New York, 1964) 124 pp; *Mellor's* Vol. VIII, Supplements I, II, *Nitrogen*, part 1 (1964) 619 pp; part 2 (1967) 676 pp; M. Sittig, *Nitrogen in Industry* (Van Nostrand, Princeton, 1965) 278 pp.

Odorless gas; condenses to a liq, bp −195.79° (77.36°K); solidifies to a snow-white mass, mp −210.01° (63.14°K). d^{gas} (0°, 1 atm) 1.25046 g/l. Critical temp: −147.1°; critical press: 33.5 atm; critical density: 0.311 g/cm³. Sparingly sol in water: 100 volumes of water absorbs 2.4 volumes of gas at 0°, 1.6 volumes at 20°. Soly in water at 50, 75 and 100° from 25 to 1000 atmospheres: Wiebe *et al., J. Am. Chem. Soc.* **55**, 947 (1933). Soly in liq ammonia: Wiebe *et al., ibid.* 975. Soly in alc: one volume of alcohol dissolves 0.1124 volume of nitrogen at 20°. Liquid oxygen at −195.5° absorbs 50.7% of its weight of gaseous nitrogen. Heat of dissociation of the nitrogen molecule (N_2): 225.1 kcal/mole. Combines with oxygen and hydrogen on sparking, forming nitric oxide and ammonia, respectively. Combines directly with lithium, and at a red heat with calcium, strontium, and barium to form nitrides. Forms cyanides when heated with carbon in presence of alkalies or barium oxide.

USE: In manuf of ammonia, nitric acid, nitrates, cyanides, etc.; in manuf explosives; in filling high-temp thermometers, incandescent bulbs; to form an inert atm for preservation of materials, for use in dry boxes or glove bags. Liquid nitrogen in food-freezing processes; in the laboratory as a coolant. Pharmaceutic aid (air displacement). *Caution:* In high concns it is a simple asphyxiant.

6523. Nitrogen Chloride. Nitrogen trichloride; chlorine nitride; trichlorine nitride; Agene. Cl_3N; mol wt 120.38. Cl 88.36%, N 11.64%. NCl_3. Prepared by the action of chlorine gas or hypochlorous acid on ammonium salts: Dulong, *Ann. Chim. Phys.* [1] **86**, 37 (1813); Balard, *ibid.* [2] **57**, 258 (1834); Hentschel, *Ber.* **30**, 1792, 2642 (1897); Noyes, *J. Am. Chem. Soc.* **42**, 2173 (1920); by reaction of anhydr ammonia and anhydr chlorine: Noyes, Haw, *ibid.* 2167; industrially by electrolyzing an acidified soln of ammonium chloride: U.S. pat. **2,118,903** (1938), **Ger.** pat. **641,816** (1937); **Brit.** pat. **494,188** (1938); for bleaching flour by blowing an air stream containing hydrogen chloride gas through a bed of finely powdered ammonium chloride and potassium hypochlorite distributed on an inert carrier to form a gas mixture of chlorine dioxide and nitrogen chloride: **Brit.** pat. **597,199** (1948). *Review:* Richards in *Mellor's* Vol. **VIII**, supplement II, *Nitrogen* (part 2), 411-415 (1967).

Yellow, thick, oily liq, d 1.653, pungent odor, evaporates rapidly in air, very unstable. Dec in light, vapor pressure at room temp is about 150 mm Hg. Explodes when heated to 93°, subjected to a flash of direct sunlight or magnesium light, sealed in glass containers at 60° after 13 seconds, frozen in liq air and thawed *in vacuo*, in contact with ozone, nitric oxide, grease, and several organic substances. Insol in water. Dec in water after 24 hours, sol in carbon disulfide, phosphorus trichloride, benzene, carbon tetrachloride, chloroform.

USE: Bleaching of flour (prohibited in the U.S.A.), wastage control of citrus fruit.

6524. Nitrogen Dioxide. NO_2; mol wt 46.01. N 30.45%, O 69.55%. Prepd industrially from nitric oxide and air. Convenient lab prepn from lead nitrate: Schenk in *Handbook of Preparative Inorganic Chemistry* vol. 1, G. Brauer, Ed. (Academic Press, New York, 1963) pp 488-489. Ultrapure NO_2 from N_2O_5: Hackspill, Besson, *Bull. Soc. Chim. France, Mem.* [5] **16**, 479 (1949). *Reviews:* Beattie, "Nitrogen Dioxide and Dinitrogen Tetroxide" in *Mellor's* Vol. **VIII**, supplement II, *Nitrogen* (part 2) 246-268 (1967); Jones in *Comprehensive Inorganic Chemistry* vol. 2, J. C. Bailar, Jr. *et al.*, Eds. (Pergamon Press, Oxford, 1973) pp 340-356.

Reddish-brown gas. Liquid below 21.15°. Irritating odor. *Deadly poison!* The commercial brown liq under pressure is called *nitrogen tetroxide*. Actually this is an equilibrium mixture of NO_2 and the colorless N_2O_4. d_4^{20} (liq) 1.448; d (gas) 1.58 (air = 1); $d_{gas}^{21.3}$ 3.3 g/liter. mp −9.3°. bp 21.15°. Crit temp 158.2°. Crit press. 99.96 atm. Heat of vaporization (bp) 9.110 kcal/mole. Does not burn, but supports the combustion of carbon, phosphorus, sulfur. Sol in concd sulfuric and nitric acids. Dec in water forming nitric acid and nitric oxide, reacts with alkalies to form nitrates and nitrites. Corrosive to steel when wet, but may be stored in steel cylinders when moisture content is 0.1% or less.

USE: Intermediate in nitric and sulfuric acid production. Used in the nitration of organic compds and explosives, in the manuf of oxidized cellulose compds (hemostatic cotton). Has been used to bleach flour. Proposed as oxidizing agent in rocket propulsion. *See also:* Sisler, "Reactions in Liquid Dinitrogen Tetroxide," *J. Chem. Ed.* **34**, 555 (1957). *Caution:* One of the most insidious gases. Inflammation of lungs may cause only slight pain or pass unnoticed, but the resulting edema several days later may cause death. 100 ppm is dangerous for even a short exposure, and 200 ppm may be fatal: Y. Henderson, H. W. Haggard, *Noxious Gases*, A.C.S. Monograph Series, no. **35** (Reinhold, New York, 2nd ed., 1943) pp 134-137, 141.

6525. Nitrogen Fluoride. Nitrogen trifluoride. F_3N; mol wt 71.01. F 80.27%, N 19.73%. NF_3. Prepd by electrolysis of melted ammonium acid fluoride, NH_4F_2H: Ruff *et al.*, *Z. Anorg. Allgem. Chem.* **172**, 417 (1928); Ruff, *ibid.* **197**, 273 (1931); Ruff, Staub, *ibid.* **198**, 32 (1931); Kwasnik in *Handbook of Preparative Inorganic Chemistry* vol. 1, G. Brauer, Ed. (Academic Press, New York, 2nd ed., 1963) pp 181-183. Reviews of prepn and chemistry: Hoffman, Neville, *Chem. Rev.* **62**, 1-18 (1962); Kemmitt, Sharp, *Advan. Fluorine Chem.* **4**, 189-190 (1965).

Colorless gas. Moldy odor. mp −208.5°. bp −129°. d (liq at bp) 1.885. Trouton const 19.9. Insoluble in water. Rather inert chemically. Does not attack glass, mercury. Decomposed by electric sparks.

Human Toxicity: Prolonged absorption may cause mottling of teeth, skeletal changes; see Sodium Fluoride (chronic effects).

6526. Nitrogen Pentoxide. Dinitrogen pentoxide; nitric anhydride. N_2O_5; mol wt 108.02. N 25.94%, O 74.06%. Usually prepd by dehydration of nitric acid by means of phosphorus pentoxide: Gruenhut *et al.*, *Inorg. Syn.* **3**, 78

(1950). *Reviews:* Beattie, "Dinitrogen Pentoxide" in *Mellor's* Vol. **VIII**, supplement II, *Nitrogen* (part 2), 269-277 (1967); Jones in *Comprehensive Inorganic Chemistry* vol. 2, J. C. Bailar, Jr. *et al.*, Eds. (Pergamon Press, Oxford, 1973) pp 356-360.

Colorless hexagonal crystals. mp 30°. Sublimes at 32.4° but undergoes moderately fast decompn into O_2 and the NO_2/N_2O_4 equilibrium mixture at temps above −10°. d^{15} 2.05. bp_{760} 47.0°; bp_{100} 7.0°; bp_{10} −20°. Dipole moment 1.39. Freely sol in chloroform without appreciable decomposition. Less sol in carbon tetrachloride. Chloroform solns may be stored at −20° for as long as one week without excessive decompn.

USE: As nitrating agent in chloroform soln.

6527. Nitrogen Selenide. *Selenium nitride.* N_4Se_4; mol wt 371.87. N 15.07%, Se 84.93%. Prepd by passing dry ammonia over selenium tetrachloride: Wöhler, *Bull. Soc. Chim.* [1] **1**, 25 (1859); by treating a dil soln of selenium oxychloride in benzene with dry ammonia and washing the precipitate with water and potassium cyanide: Lenher, Wolesensky, *J. Am. Chem. Soc.* **29**, 215 (1907); by the action of dry ammonia on a dilute soln of selenium monochloride in carbon disulfide: van Valkenburg, Bailar, *ibid.* **47**, 2134 (1925); by the action of dry ammonia on diethyl selenite or dimethyl selenite dissolved in benzene and washing with potassium cyanide: Strecker, Schwarzkopf, *Z. Anorg. Chem.* **221**, 193 (1934); by reacting anhydr, liquid ammonia with selenium dioxide: Jander, Doetsch, *Ber.* **93**, 561 (1960). Crystal structure: Baernighausen *et al.*, *Acta Cryst.* **15**, 615 (1962); **21**, 571 (1966).

Orange-red, amorphous powder or monoclinic crystals. d 4.2. Very hygroscopic. Explosive. Sparingly sol in carbon disulfide, benzene, acetic acid; insol in water, ether, abs alc.

6528. Nitroglycerin. *1,2,3-Propanetriol trinitrate;* glyceryl trinitrate; glycerol nitric acid triester; nitroglycerol; trinitroglycerol; glonoin; trinitrin; blasting gelatin; blasting oil; S.N.G.; Adesitrin; Angibid; Angiolingual; Anginine; Angorin; Aquo-Trinitrosan; Cardamist; Coro-Nitro; Corditrine; Deponit; Diafusor; Gilucor "nitro"; GTN; Klavikordal; Lenitral; Lentonitrina; Millithrol; Minitran; Myoglycerin; Nitradisc; Nitran; Nitriderm; Nitro-Bid; Nitrocine; Nitrocontin; Nitroderm TTS; Nitrodisc; Nitro-Dur; Nitrofortin; Nitrogard; Nitro-Gesanit; Nitroglin; Nitroglyn; Nitrolan; Nitrolande; Nitrolar; Nitro-lent; Nitrolingual; Nitro Mack; Nitromel; Nitronal; Nitrong; Nitro-Pflaster-ratiopharm; NitroPRN; Nitrorectal; Nitroretard; Nitrosigma; Nitrostat; Nitrozell retard; Nysconitrine; Percutol; Perlinganit; Perglottal; Reminitrol; Suscard; Sustac; Sustonit; Transderm-Nitro; Transiderm-Nitro; Tridil; Trinalgon; Trinitrosan; Vasoglyn. $C_3H_5N_3O_9$; mol wt 227.09. C 15.87%, H 2.22%, N 18.50%, O 63.41%. Prepn: Sobrero, *Ann.* **64**, 398 (1847); Williamson, *Ann.* **92**, 305 (1854). Review of the early literature: J. W. Lawrie, *Glycerol and the Glycols* (New York, 1928). Review of chemistry and biochemistry: F. J. DiCarlo, *Drug Metab. Rev.* **4**, 1-38 (1975). Review of mechanism of action: S. F. Vatner, G. R. Heyndrickx, *Handb. Exp. Pharmakol.* **40**, 131-161 (1975). Review of the first hundred years: J. R. Parratt, *J. Pharm. Pharmacol.* **31**, 801-809 (1979). Comprehensive description: E. F. McNiff *et al.*, *Analytical Profiles of Drug Substances,* vol. 9, K. Florey, Ed. (Academic Press, New York, 1980) pp 519-541. Symposium on nitroglycerin therapy, perspectives and mechanisms: *Am. J. Med.* **74**, no. 6B, 1-94 (1983). Review of pharmacology and clinical studies of intravenous administration in heart disease: E. M. Sorkin *et al.*, *Drugs* **27**, 45 (1984).

$$CH_2-ONO_2$$
$$|$$
$$CH-ONO_2$$
$$|$$
$$CH_2-ONO_2$$

Pale yellow, oily liquid. Sweet burning taste. Produces headache on tasting. Explodes on rapid heating or on concussion. Crystallizes in 2 forms: labile form, mp +2.8°; stable form, mp +13.5°. d_{15}^{15} 1.599; d_4^4 1.6144; d_4^{15} 1.6009; d_4^{25} 1.5918. n_D^{15} 1.474. Begins to dec at 50-60°, appreciably volatile at 100°, evolves nitrous yellow vapors at 135°, explodes at 218°. Vapor pressure at 20°: 0.00026 mm; at 93°: 0.31

mm. One gram dissolves in 800 ml water, in 4 g ethanol, in 18 g methanol, in 120 g carbon disulfide. Misc with ether, acetone, glacial acetic acid, ethyl acetate, benzene, nitrobenzene, pyridine, chloroform, ethylene bromide, dichloroethylene. Sparingly sol in petr ether, liq petrolatum, glycerol. Heat of combustion: 1580 cal/g. On explosion harmless gases are produced: $4C_3H_5(ONO_2)_3 \rightarrow 12CO_2 + 10H_2O + 6N_2 + O_2$.

Human Toxicity: Acute poisoning, occurring especially in industrial workers: nausea, vomiting, abdominal cramps, headache, mental confusion, delirium, bradypnea, bradycardia, paralysis, convulsions, methemoglobinemia and cyanosis, circulatory collapse, death. *Chronic poisoning:* severe headache, hallucinations, skin rashes. Alcohol aggravates symptoms. Toxic effects may occur by ingestion, inhalation of dust or absorption through intact skin, *cf.* Patty's *Industrial Hygiene and Toxicology* vol. **2C**, G. D. Clayton, F. E. Clayton, Eds. (Wiley-Interscience, New York, 3rd ed., 1982) pp 4186-4188.

USE: Manuf of *dynamite* (75% nitroglycerol, 24.5% diatomaceous earth, 0.5% sodium carbonate).

THERAP CAT: Antianginal; vasodilator (coronary).

THERAP CAT (VET): Has been used for bronchial asthma in dogs.

6529. Nitroguanidine. $CH_4N_4O_2$; mol wt 104.07. C 11.54%, H 3.87%, N 53.84%, O 30.75%. $H_2NC(NH)NH$-NO_2. Prepd by the action of concd H_2SO_4 on guanidine nitrate: Davis, *Org. Syn.* **7**, 68 (1927).

α-Form, (the usual stable form): Needles, prisms from water, dec 225-250° (depending on speed of heating) giving off NH₃ fumes. Absorption spectrum: Riegel, Buchwald, *J. Am. Chem. Soc.* **51**, 491 (1929). One liter of water dissolves 4.4 g at 25°; 82.5 g at 100°. Slightly soluble in methanol (< 0.5%), alcohol; practically insol in ether. Sol in concd acids from which it is precipitated by water. Sol in cold solns of alkalies (not of carbonates) with slow decompn.

β-Form: Prepd by the nitration of guanidine sulfate with fuming HNO₃. Is converted into the α-form by dissolving in concd H_2SO_4 and pouring into ice water: Davis *et al.*, *J. Am. Chem. Soc.* **47**, 1065 (1925).

USE: Explosive of moderate power. Can be exploded only with a detonator. Intermediate in the synthesis of pharmaceuticals.

6530. Nitrohydrochloric Acid. *Aqua regia;* nitromuriatic acid; chloronitrous acid; chloroazotic acid. Made by mixing 18 ml HNO₃, 82 ml HCl. In addition to the free acids, it contains nitrosyl chloride (NOCl) and some free chlorine.

Yellow, fuming, corrosive, volatile liq; suffocating odor; attacks all metals including gold and platinum. Miscible with water. *Keep in well-closed, glass-stoppered bottles in a cool place and protected from light.*

Human Toxicity: Undiluted form is a strong irritant, corrosive.

6531. Nitromersol. *[2-Methyl-5-nitrophenolato(2 −)-C^6,O^1]mercury; 5-methyl-2-nitro-7-oxa-8-mercurabicyclo-[4.2.0]octa-1,3,5-triene; 3-(hydroxymercuri)-4-nitro-o-cresol inner salt;* Metaphen. $C_7H_5HgNO_3$; mol wt 351.73. C 23.90%, H 1.43%, Hg 57.04%, N 3.98%, O 13.65%. According to the National Formulary this compd is the anhydride of 4-nitro-3-hydroxymercuri-o-cresol. Prepd by treating 4-nitro-o-cresol with mercuric acetate: U.S. pat. **1,544,293**; reissue **17,563** (1925).

Yellow, odorless, tasteless powder or granules. Insol in water; almost insol in acetone, alcohol, ether, aq soln of sodium carbonate. Sol in solns of alkalies and of ammonia by opening the anhydride ring and the formation of a salt. Also sol in boiling glacial acetic acid.

THERAP CAT: Antiseptic.

THERAP CAT (VET): Topical disinfectant.

6532. Nitromethane. Nitrocarbol. CH_3NO_2; mol wt 61.04. C 19.67%, H 4.95%, N 22.95%, O 52.42%. Prepd by the interaction of sodium nitrite and sodium chloroacetate: Whitmore, *Org. Syn.* **3**, 83 (1923). Monograph: *Am. Ind. Hyg. Assoc. J.* **22**, 518 (1961).

Oily liquid with a moderately strong, somewhat disagreeable odor. d_4^{25} 1.1322. One gallon weighs 9.5 lbs. Flash pt 112°F. mp −29°. bp_{760} 101.2°; bp_{100} 46.6°; bp_{40} 27.5°; bp_{20} 14.1°; bp_{10} +2.8°; bp_5 −7.9°; $bp_{1.0}$ −29°. n_D^{22} 1.38056. Absorption spectrum: Hantzsch, Voigt, *Ber.* **45**, 106 (1912). Slightly sol in water (9.5% by vol at 20°); sol in alc, ether, DMF. Water solns are acid to litmus. pH of $0.01M$ aq soln 6.12. Forms an explosive sodium salt which bursts into flame on contact with water. Lethal concn for guinea pigs in air: 1000 ppm. LD_{50} orally in mice: 1.44 g/kg, Weatherby, *Arch. Ind. Health* **11**, 102 (1955).

USE: Rocket fuel; solvent for zein. Used in the coating industry.

6533. Nitromide. *3,5-Dinitrobenzamide.* $C_7H_5N_3O_5$; mol wt 211.13. C 39.82%, H 2.39%, N 19.90%, O 37.89%. Prepd from 3,5-dinitrobenzoyl chloride and ammonium acetate: Finan, Fothergill, *J. Chem. Soc.* **1962**, 2824.

Leaflets from water, mp 183°. Slightly sol in cold water, somewhat more sol in hot water.

Note: An ingredient of *Tristat, Unistat, Unistat-3.*

THERAP CAT (VET): Antibacterial, coccidiostat (for poultry).

6534. Nitron. *1,4-Diphenyl-3-(phenylamino)-1H-1,2,4-triazolium hydroxide, inner salt.* $C_{20}H_{16}N_4$; mol wt 312.36. C 76.90%, H 5.16%, N 17.94%. Prepn: E. Merck, Ger. pat. **161,235** (1905); Frdl. **8**, 1240. Formerly represented as *3,5,-6-triphenyl-2,3,5,6-tetraazabicyclo[2.1.1]hex-1-ene:* Busch, *Ber.* **38**, 858, 4054 (1905). Meso-ionic structure: Baker, Ollis, *Quart. Rev.* **11**, 26 (1957); Olah, *J. Inorg. Nucl. Chem.* **16**, 225 (1961). Use in detection of gold: N. Ganchev, A. Dimitrova, *Mikrochim. Acta* **1971**, 476. In determn of nitrate: A. Hulanicki, M. Maj, *Talanta* **22**, 767 (1975); of molybdenum: T. J. Koralewski, G. A. Parker, *Anal. Chim. Acta* **113**, 389 (1980).

Intensely yellow leaflets from alcohol, solvated needles from chloroform, dec about 189°. Practically insol in water; sol in alcohol, benzene, acetone, chloroform, ethyl acetate, dil acids; slightly sol in ether. The alc soln undergoes partial decompn, indicated by a red color, hence should be protected from light. Dipole moment in benzene: 7.2. Solubility of some Nitron salts in $0.01N$ HCl: Nitrate 0.0099%, perchlorate 0.008%, iodide 0.017%, thiocyanate 0.04%, chromate 0.06%, chlorate 0.12%, nitrite 0.19%, bromide 0.61%.

USE: Detection of nitrate, perchlorate, gold, rhenium. Determination of boron, nitrate, perchlorate, rhenium, tungsten, molybdenum.

6535. 1-Nitronaphthalene. α-Nitronaphthalene. $C_{10}H_7NO_2$; mol wt 173.16. C 69.36%, H 4.07%, N 8.09%, O 18.48%. Prepd from napthalene and a mixture of nitric and sulfuric acids at 50°.

Yellow crystals. d 1.331. mp 59-61°. bp 304°. Insol in water; sol in alcohol, freely in chloroform, ether, carbon disulfide. Gives a dark red soln with concd H_2SO_4.

USE: Deblooming petr oils, 2-3 parts suffice for 1000 parts oil; manuf dyes and intermediates.

6536. 1-Nitro-2-naphthol. *1-Nitro-2-naphthalenol;*

α-nitro-β-naphthol; 1-nitro-2-hydroxynaphthalene. C_{10}-H_7NO_3; mol wt 189.16. C 63.49%, H 3.73%, N 7.41%, O 25.37%.

Yellow needles or platelets, mp 103°. Insol in water; sol in alcohol, ether, glacial acetic acid, in alkali hydroxide solns. USE: As a reagent for the determination of cobalt with which it gives a precipitate.

6537. 3-Nitropentane. $C_5H_{11}NO_2$; mol wt 117.15. C 51.26%, H 9.47%, N 11.96%, O 27.32%. $CH_3CH_2CH(NO_2)$-CH_2CH_3.
Colorless liquid; fusel oil odor. d_4^0 0.957. bp 153-155°. Insol in water; sol in alcohol, ether.

6538. 5-Nitro-*o*-phenetidine. *2-Ethoxy-5-nitrobenzenamine;* 5-nitro-2-ethoxyaniline; 1-ethoxy-2-amino-4-nitrobenzene; 1-nitro-3-amino-4-phenyl ethyl ether; 4-nitro-2-aminophenetole; Neo-Douxan. $C_8H_{10}N_2O_3$; mol wt 182.18. C 52.74%, H 5.53%, N 15.38%, O 26.35%. Prepn: Verkade *et al., Rec. Trav. Chim.* **65**, 346 (1946). Toxicology: Grupp, Bilger, *Arzneimittel-Forsch.* **1**, 326 (1951).

Brown-yellow crystals from benzene, mp 97.5-98.5°. 950 times as sweet as cane sugar, claimed to have no aftertaste. USE: Sweetening agent.

6539. Nitrophenide. *Bis(3-nitrophenyl)disulfide; m,m'-*dinitrodiphenyl disulfide; NP; Megasul. $C_{12}H_8N_2O_4S_2$; mol wt 308.33. C 46.74%, H 2.62%, N 9.09%, O 20.76%, S 20.80%. Prepd by the reduction of *m*-nitrobenzenesulfonyl-chloride with hydriodic acid: Ekbom, *Ber.* **24**, 336 (1891); Foss *et al., J. Am. Chem. Soc.* **60**, 2729 (1938). Anticoccidial activity: Waletzky *et al., Ann. N.Y. Acad. Sci.* **52**, 543 (1949).

Yellow rhomboid crystals, mp 83°. Freely sol in ether, less sol in alcohol. Insol in water.
THERAP CAT (VET): Coccidiostat.

6540. *m*-Nitrophenol. Obtained by boiling diazotized *m*-nitroaniline with water and H_2SO_4: Adams, Wilson, *Org. Syn.* **3**, 87 (1923); Bachmann, Rottschaefer, *ibid.* **18**, 88 (1938); Manske, *ibid.* coll. vol. I, 404 (2nd ed., 1941).
Monoclinic prisms from ether or dil HCl. d_4^{20} 1.485; d_4^{100} 1.2797. mp 97°. bp_{70} 194°. Dec when distilled at ordinary pressure. K at 18° = 4.6 × 10⁻⁹. Absorption spectrum: Baly *et al., J. Chem. Soc.* **97**, 586 (1910); Marchlewski, Moroz, *Bull. Soc. Chim.* [4] **35**, 476 (1924). An aq soln satd at 40° contains 3.02 g in 100 g soln; at 98.7°, 40.9 g: Sidgwick *et al., J. Chem. Soc.* **107**, 1202 (1915). Soly (g/100 g solvent) in acetone: 169.35 (0.2°); 1305.9 (84°); in alcohol: 116.9 (1°); 1105.25 (85°); in ether: 105.9 (0.2°); 1065.8 (83°): Carrick, *J. Phys. Chem.* **25**, 635 (1921). Sol in hot and dil acids, in caustic solns; insol in petr ether. LD₅₀ orally in mice, rats: 1414, 933 mg/kg, K. C. Back *et al., Reclassification of Materials Listed as Transportation Health Hazards* (TSA-20-72-3; PB214-270).
USE: As indicator in 0.3% soln in 50% alc. pH: 6.8 colorless, 8.6 yellow.

6541. *o*-Nitrophenol. $C_6H_5NO_3$; mol wt 139.11. C

51.80%, H 3.62%, N 10.07%, O 34.50%. The *o*- and *p*-nitrophenols are formed by the nitration of phenol; the *o*-form is separated by steam distillation.

Light yellow needles or prisms; peculiar, aromatic odor. d 1.495. mp 44-45°. bp 214-216°. Volatile with steam. Slightly sol in cold water, freely in hot water, in alcohol, benzene, ether, carbon disulfide, alkali hydroxides. LD₅₀ orally in mice, rats: 1.297, 2.828 g/kg, K. C. Back *et al., Reclassification of Materials Listed as Transportation Health Hazards* (TSA-20-72-3; PB214-270).
USE: Manuf of many important compds; as indicator in 2% alc soln. pH: 5.0 colorless, 7.0 yellow, but the color change is not sharp and cannot be used where CO_2 is present; as reagent for glucose.

6542. *p*-Nitrophenol. *Para isomer of o-nitrophenol, q.v.*
Colorless to slightly yellow, odorless crystals; sweetish, then burning taste. d_4^{20} 1.270. mp 113-114°. Sublimes; slightly volatile with steam. Moderately sol in cold water, freely in alcohol, chloroform, ether; also sol in solns of fixed alkali hydroxides and carbonates. LD₅₀ orally in mice, rats: 467, 616 mg/kg, K.C. Back *et al., Reclassification of Materials Listed as Transportation Health Hazards* (TSA-20-72-3; PB214-270).
USE: Manuf of many important compounds; as indicator in 0.1% alcohol soln. pH: 5.6 colorless, 7.6 yellow.

6543. *p*-Nitrophenylacetic Acid. *4-Nitrobenzeneacetic acid;* *p*-nitro-α-toluic acid. $C_8H_7NO_4$; mol wt 181.14. C 53.04%, H 3.90%, N 7.73%, O 35.33%. $NO_2C_6H_4CH_2CO$-OH. Prepd by hydrolysis of *p*-nitrobenzyl cyanide with 50% H_2SO_4: Robertson, *Org. Syn.* **2**, 59 (1922); coll. vol. I, 406 (2nd ed., 1941).
Long, pale yellow needles from water. mp 153°. K at 25° 1.04 × 10⁻⁴. Absorption spectrum: Hewitt *et al., J. Chem. Soc.* **101**, 1774 (1912). Sparingly sol in cold water; sol in alcohol, ether, benzene.
Methyl ester, $C_9H_9NO_4$, mp 54°.
Benzyl ester, $C_{15}H_{13}NO_4$, mp 92°.
Nitrile, see *p*-Nitrobenzyl Cyanide.

6544. 4-Nitro-*o*-phenylenediamine. *4-Nitro-1,2-benzenediamine;* 1,2-diamino-4-nitrobenzene; 4-nitro-1,2-diaminobenzene; 4-nitro-1,2-phenylenediamine. $C_6H_7N_3O_2$; mol wt 153.14. C 47.05%, H 4.61%, N 27.44%, O 20.90%. Prepd by the reduction of 2,4-dinitroaniline using H_2S in ammonia water: Kehrmann, *Ber.* **28**, 1707 (1895); Griffin, Peterson, *Org. Syn.* coll. vol. III, 242 (1955).

Dark red needles from hot water. mp 201°. Sparingly sol in water. Sol in aq solns of hydrochloric acid. LD₅₀ in rats (mg/kg): 3720 orally; > 1600 i.p., C. Burnett *et al., J. Toxicol. Environ. Health* **2**, 657 (1977).
USE: Reagent for α-keto acids: Hockenhull, Floodgate, *Biochem. J.* **52**, 38 (1952); Taylor, Smith, *Analyst* **80**, 607 (1955).

6545. (4-Nitrophenyl)hydrazine. $C_6H_7N_3O_2$; mol wt 153.14. C 47.05%, H 4.61%, N 27.44%, O 20.90%.
Orange-red leaflets or needles. mp about 157° with dec. Slightly sol in cold water; sol in hot water or hot benzene, alcohol, chloroform, ether, ethyl acetate.
USE: As reagent for ketones, aliphatic aldehydes.

6546. *o*-Nitrophenylpropiolic Acid. *3-(2-Nitrophenyl)-*

2-propynoic acid. $C_9H_5NO_4$; mol wt 191.14. C 56.55%, H 2.64%, N 7.33%, O 33.48%.

Yellowish to light brown scales or crystals. mp about 157° with decompn and may explode. Moderately sol in cold water, more in hot water or alcohol, very slightly in chloroform; almost insol in carbon disulfide and petr ether. USE: As a reagent for alkaloids and glucose.

6547. (4-Nitrophenyl)urea. $C_7H_7N_3O_3$; mol wt 181.15. C 46.41%, H 3.90%, N 23.20%, O 26.50%. Prepn from *N*-substituted *S*-phenylthiocarbamates: Crosby, Niemann, *J. Am. Chem. Soc.* **76**, 4458 (1954).

Prisms from abs ethanol, needles from dil ethanol, mp 238°. Practically insol in cold water. Sol in boiling water, methanol, ethanol, dimethylformamide. Sparingly sol in ether, benzene.

6548. 1-Nitropropane. $C_3H_7NO_2$; mol wt 89.09. C 40.44%, H 7.92%, N 15.72%, O 35.92%. $CH_3CH_2CH_2NO_2$. Prepd by vapor-phase nitration of propane: Hass *et al.*, U.S. pat. **1,967,667** (1934 to Purdue Res. Found.); *Ind. Eng. Chem.* **28**, 339 (1936); from natural gas: Bachman, Pollack, *ibid.* **46**, 713 (1954).

Liquid. d_4^{25} 0.9934. wt 8.4 lb/U.S. gal. mp $-108°$. bp_{760} 131.6°; bp_{728} 130°; bp_{401} 110°; bp_{94} 70°; $bp_{7.5}$ 20°; bp_4 10°. n_D^{20} 1.4018. Flash pt 34° (93°F). Heat of formation -40.05 kcal/mole at 25°; heat of combustion 481.33 kcal/mole at 25°. Latent heat of vaporization 10.37 kcal/mole at 25°. Slightly sol in water (1.4 ml/100 ml); 0.5 ml water dissolves in 100 ml of 1-nitropropane. Miscible with many organic solvents.

USE: Solvent for cellulose acetate, vinyl resins, lacquers, synthetic rubbers, fats, oils, dyes, other organic materials; as intermediate, propellant. *Caution:* Irritating to mucous membranes.

6549. 2-Nitropropane. $C_3H_7NO_2$; mol wt 89.09. C 40.44%, H 7.92%, N 15.72%, O 35.92%. $CH_3CH(NO_2)CH_3$. For prepn *see* the ref under 1-Nitropropane. Toxicity and carcinogenicity study: T. R. Lewis *et al.*, *J. Environ Pathol. Toxicol.* **2**, 233 (1979). Health hazard alert: *Am. Ind. Hyg. Assoc. J.* **41**, 18 (1980).

Liquid. d_4^{25} 0.9821. wt 8.4 lb/U.S. gal. mp $-93°$. bp_{760} 120.3°; bp_{564} 110°; bp_{300} 90°; bp_{95} 60°; bp_7 10°. n_D^{20} 1.3944. Flash pt 24° (75°F). Heat of formn -43.78 kcal/mole at 25°; heat of combustion 477.60 kcal/mole at 25°. Latent heat of vaporization 9.88 kcal/mole at 25°. Slightly sol in water (1.7 ml/100 ml); 0.6 ml water dissolves in 100 ml of 2-nitropropane. Misc with many organic solvents.

Note: This substance may reasonably be anticipated to be a carcinogen: *Fourth Annual Report on Carcinogens* (NTP 85-002, 1985) p 145.

USE: *See* 1-Nitropropane.

6550. 5'-Nitro-2'-propoxyacetanilide. *N-(5-Nitro-2-propoxyphenyl)acetamide;* 1-propoxy-2-acetamino-4-nitrobenzene; Falimint. $C_{11}H_{14}N_2O_4$; mol wt 238.24. C 55.45%, H 5.92%, N 11.76%, O 26.86%. Prepd by reacting 1-chloro-2,4-dinitrobenzene with sodium alkoxide, followed by partial reduction and treatment with acetic anhydride: Profft, *Chem. Tech. (Leipzig)* **5**, 393 (1953), *C.A.* **47**, 12757a (1953); by nitration of 1-propoxy-2-acetylaminobenzene: Verkade *et al.*, *Rec. Trav. Chim.* **66**, 374 (1947); also **Brit. pat. 597,-835** (1948 to N. V. Polak & Schwarz's Essencefabrieken).

Crystals from propanol, mp 102.5-103.5°. Flat taste. THERAP CAT: Antipyretic, analgesic.

6551. 5-Nitro-2-propoxyaniline. *5-Nitro-2-propoxybenzenamine;* 2-amino-4-nitro-1-propoxybenzene; 1-propoxy-2-amino-4-nitrobenzene; P-4000; Ultrasüss. $C_9H_{12}N_2O_3$; mol wt 196.20. C 55.09%, H 6.17%, N 14.28%, O 24.46%. A substance which is 4000 times sweeter than sugar, with no bitter aftertaste. Prepn: Verkade *et al.*, *Rec. Trav. Chim.* **65**, 346-360 (1946); **Neth. pat. Appl. 6,514,769** (1967 to N. V. Koffie en Theehandel), *C.A.* **67**, 108430n (1967); A. J. de Koning, *J. Chem. Ed.* **53**, 521 (1976); J. McK. Woollard, *ibid.* **57**, 464 (1980). Used as a sweetener in some European countries but banned in U.S.A. because of possible toxic effects. Toxicity studies: Fitzhugh *et al.*, *J. Am. Pharm. Assoc.* **46**, 583 (1951).

Orange crystals from *n*-propanol + petr ether. mp 47.5-48.5°. Soly in water at 20°: 136 mg/liter. Stable in boiling water and in dil acids.

6552. 5-Nitroquinaldic Acid. *5-Nitro-2-quinolinecarboxylic acid;* 5-nitroquinaldinic acid. $C_{10}H_6N_2O_4$; mol wt 218.17. C 55.05%, H 2.77%, N 12.84%, O 29.33%. Prepn from quinaldic acid: Besthorn, Ibele, *Ber.* **39**, 2333 (1906).

Yellow crystals, mp 275-278° (dec). Slightly sol in water and in most organic solvents. Readily sol in acetic acid. USE: In the determination of zinc.

6553. 3-Nitrosalicylic Acid. *2-Hydroxy-3-nitrobenzoic acid.* $C_7H_5NO_5$; mol wt 183.12. C 45.91%, H 2.75%, N 7.65%, O 43.69%. The 3- and 5-nitrosalicylic acids are formed by the nitration of salicylic acid with H_2SO_4 and HNO_3.

Monohydrate, yellowish crystals. mp 123°; mp 148° when anhydr. Slightly sol in water; freely sol in alcohol, benzene, chloroform, ether. USE: In dye chemistry.

6554. 5-Nitrosalicylic Acid. Anilotic acid. $C_7H_5NO_5$; mol wt 183.12. C 45.91%, H 2.75%, N 7.65%, O 43.69%. Yellowish crystals. d 1.65. mp 228-230°. One gram dissolves in 1475 ml water; more sol in hot water; freely sol in alcohol, ether.

6555. Nitroscanate. *1-Isothiocyanato-4-(4-nitrophenoxy)benzene; isothiocyanic acid p-(p-nitrophenoxy)phenyl ester; p-(p-nitrophenoxy)phenyl isothiocyanate;* 1-(4-isothiocyanatophenoxy)-4-nitrobenzene; 4-nitro-4'-isothiocyanodiphenyl ether; cantrodifene (obsolete); CGA-23654; GS 23654; Lopatol. $C_{13}H_8N_2O_3S$; mol wt 272.27. C 57.35%, H 2.96%, N 10.29%, O 17.63%, S 11.77%. Prepn: **Fr. pat.**

1,491,477 (1967 to Agripat), *C.A.* **69**, 76923y (1968); Mart-von, Antos, *Chem. Zvesti* **23**, 181 (1969); Antos *et al.*, **Ger. pat. 1,932,690** (1970 to Ceskoslovenska Akademie Ved), *C.A.* **72**, 100265v (1970). Activity in dogs: Gemmell, Oude-mans, *Res. Vet. Sci.* **19**, 217 (1975).

O$_2$N—⟨benzene⟩—O—⟨benzene⟩—NCS

Crystals, mp 107-113°, Martvon, Antos, *loc. cit.;* 124-125°, **Fr. pat. 1,491,477.** Insol in water; sol in organic solvents.
THERAP CAT (VET): Anthelmintic.

6556. N-Nitrosodiethanolamine. *2,2'-(Nitrosoimino)-bisethanol; 2,2'-nitrosiminodiethanol;* di-(2-hydroxyethyl)-nitrosamine; NDELA. C$_4$H$_{10}$N$_2$O$_3$; mol wt 134.13. C 35.82%, H 7.51%, N 20.88%, O 35.79%. (HOCH$_2$CH$_2$)$_2$-NNO. Formed by the action of nitrites on di- or triethanol-amine. Prepn: E. R. H. Jones, W. Wilson, *J. Chem. Soc.* **1949**, 547; R. Preussmann, *Ber.* **95**, 1571 (1962); W. Lijinsky *et al.*, *J. Nat. Cancer Inst.* **49**, 1239 (1972). Carcinogenicity study: H. Druckery *et al.*, *Z. Krebsforsch.* **69**, 103 (1967). Impurity in cutting fluids: T. Y. Fan *et al.*, *Science* **196**, 70 (1977); in cosmetics: *eidem, Food Cosmet. Toxicol.* **15**, 423 (1977). Mutagenicity study: A. Hesbert *et al.*, *Mutat. Res.* **68**, 207 (1979).
Light yellow oil, bp$_{0.01}$ 125°. n$_D^{20}$ 1.4849.
Note: This substance may reasonably be anticipated to be a carcinogen: *Fourth Annual Report on Carcinogens* (NTP 85-002, 1985) p 148.

6557. N-Nitrosodiethylamine. *N-Ethyl-N-nitrosoethan-amine;* diethylnitrosamine; DEN; DENA; NDEA. C$_4$H$_{10}$-N$_2$O; mol wt 102.14. C 47.04%, H 9.87%, N 27.43%, O 15.66%. (C$_2$H$_5$)$_2$NNO. Detected in trace amounts in tobac-co smoke: Druckney, Preussman, *Naturwiss.* **49**, 498 (1962) and in various processed foods: Hedler, Marquardt, *Food Cosmet. Toxicol.* **6**, 341 (1968); Friemuth, Glaeser, *Nahrung* **14**, 357 (1970). Formed by the interaction of nitrite with diethylamine and by the action of nitrate-reducing bacteria. Industrial prepn: Reilly, **Ger. pat. 1,085,166** (1960 to du Pont), *C.A.* **56**, 4594h (1962); Levering, Maury, **U.S. pat. 3,090,786** (1963 to Hercules Powder); Minisci, Galli, *Chim. Ind. (Milan)* **46**, 173 (1964). Hepatotoxicity and carcinogen-icity studies: Schmaehl *et al.*, *Naturwiss.* **54**, 341 (1967); Grover, Fischer, *Eur. J. Cancer* **7**, 77 (1971); Bader *et al.*, *Arch. Geschwulstforsch.* **37**, 327 (1971). General review: Magee, Barnes, *Adv. Cancer Res.* **10**, 163-246 (1967).
Slightly yellow liq. d$_4^{20}$ 0.9422. bp 175-177°. bp$_5$ 47°. n$_D^{20}$ 1.4388. Sol in water, alc, ether.
Note: This substance may reasonably be anticipated to be a carcinogen: *Fourth Annual Report on Carcinogens* (NTP 85-002, 1985) p 149.
USE: Gasoline and lubricant additive; antioxidant; sta-bilizer.

6558. N-Nitrosodimethylamine. *N-Methyl-N-nitroso-methanamine;* dimethylnitrosamine; DMN; DMNA. C$_2$H$_6$-N$_2$O; mol wt 74.08. C 32.42%, H 8.16%, N 37.82%, O 21.60%. (CH$_3$)$_2$NNO. Reportedly found in trace amounts in tobacco smoke condensates: Rhoades, Johnson, *Nature* **236**, 307 (1972); in cured meat products, notably bacon: Sen *et al.*, *ibid.* **241**, 473 (1973); in smoked and salted fish: Fazio *et al.*, *J. Agr. Food Chem.* **19**, 250 (1971); Fong, Chan, *Na-ture* **243**, 421 (1973). Formed by the interaction of nitrite with dimethylamine and by the action of nitrate-reducing bacteria. Industrial prepn: **Brit. pat. 772,331** (1957 to Olin Mathieson), *C.A.* **51**, 14783a (1957); Ioffe, *Zh. Obshch. Khim.* **28**, 1296 (1958); Norris, *J. Am. Chem. Soc.* **81**, 3346 (1959); Campbell, **U.S. pat. 2,981,752** (1961 to C.S.C.); Datin, Elliott, **U.S. pat. 3,136,821** (1964 to Allied Chem.). Metabolism and toxicity: Magee, Vandekar; Magee, *Bio-chem. J.* **70**, 600, 606 (1958); Heath, *ibid.* **85**, 72 (1962). Carcinogenicity studies: Magee, Barnes, *Brit J. Cancer* **10**, 114 (1956); *see also eidem, Adv. Cancer Res.* **10**, 163-246 (1967). Chemistry: Layne *et al.*, *J. Am. Chem. Soc.* **85**, 435, 1816 (1963).
Yellow liquid. bp 151-153°. bp$_{40}$ 67.1°. d$_4^{20}$ 1.0048. n$_D^{20}$

1.4368. Very sol in water, alcohol, ether. LD$_{50}$ i.p. in rats: 34 mg/kg (Heath).
Note: This substance may reasonably be anticipated to be a carcinogen: *Fourth Annual Report on Carcinogens* (NTP 85-002, 1985) p 151.
USE: Formerly in the prodn of rocket fuels; antioxidant; additive for lubricants; softener of copolymers.

6559. p-Nitroso-N,N-dimethylaniline. *N,N-Dimethyl-4-nitrosobenzenamine;* Accelerine. C$_8$H$_{10}$N$_2$O; mol wt 150.18. C 63.98%, H 6.71%, N 18.66%, O 10.65%. Prepd in the cold from NaNO$_2$ and a soln of dimethylaniline in HCl.
Green plates or leaflets. mp 92.5-93.5°; also stated as 87-88°. Volatile with steam. Insol in water; sol in alcohol, ether.
USE: Manuf organic compds; accelerator in vulcanizing; in printing fabrics.

6560. p-Nitrosodiphenylamine. *4-Nitroso-N-phenylben-zenamine.* C$_{12}$H$_{10}$N$_2$O; mol wt 198.22. C 72.71%, H 5.09%, N 14.13%, O 8.07%. C$_6$H$_5$NHC$_6$H$_4$NO.
Green plates with bluish luster (from benzene) or steel-blue prisms or plates (from ether + water). mp 144-145°. Slightly sol in water or petr ether; freely sol in alcohol, ether, chloroform, benzene. Dissolves in H$_2$SO$_4$ with a red color, which suddenly changes to violet on warming.
Note: This substance may reasonably be anticipated to be a carcinogen: *Fourth Annual Report on Carcinogens* (NTP 85-002, 1985) p 152.
USE: Accelerator in vulcanizing rubber.

6561. N-Nitrosomorpholine. *4-Nitrosomorpholine;* NMOR. C$_4$H$_8$N$_2$O$_2$; mol wt 116.11. C 41.37%, H 6.94%, N 24.12%, O 27.56%. Prepn: L. Knorr, *Ann.* **301**, 1 (1898); F. Chapman, *J. Chem. Soc.* **1949**, 1631; G. Oláh *et al.*, *Ber.* **89**, 2374 (1956). Carcinogenicity studies: H. Druckrey *et al.*, *Naturwiss.* **48**, 134 (1961); P. Bannasch, H.-A. Müller, *Arz-neimittel-Forsch.* **14**, 805 (1964); *IARC Monographs* **17**, 263 (1978). Metabolism: B. W. Stewart, P. N. Magee, *Biochem. J.* **126**, 21P (1972).

⟨morpholine with N-NO structure⟩

Yellow crystals, mp 29°. bp$_{25}$ 139-140°; bp$_{747}$ 224-224.5°. Sol in water. LD$_{50}$ orally in rats: 282 mg/kg (Druckrey).
Note: This substance may reasonably be anticipated to be a carcinogen: *Fourth Annual Report on Carcinogens* (NTP 85-002, 1985) p 157.

6562. 1-Nitroso-2-naphthol. *1-Nitroso-2-naphthalenol;* nitroso-β-naphthol. C$_{10}$H$_7$NO$_2$; mol wt 173.16. C 69.36%, H 4.07%, N 8.09%, O 18.48%. Prepared by the addition of H$_2$SO$_4$ to a mixture of β-naphthol dissolved in aq NaOH and NaNO$_2$: C. S. Marvel, P. K. Porter, *Org. Syn.* **coll. vol. I**, 411 (2nd ed., 1941).

⟨naphthalene with NO and OH structure⟩

Yellowish-brown needles from petr ether, mp 109-110°. Sol in 1000 parts water, 35 parts alcohol; also sol in hot al-cohol, benzene, ether, carbon disulfide, caustic alkali solns, glacial acetic acid; slightly sol in cold petr ether.
USE: To prevent gum formation in gasoline; in analytical chemistry in the determination of cobalt (to separate it from nickel).

6563. 4-Nitrosophenol. Quinone oxime; quinone mon-oxime. C$_6$H$_5$NO$_2$; mol wt 123.11. C 58.53%, H 4.09%, N 11.38%, O 25.99%.

Pale yellow orthorhombic needles, browns at 126° (dec 144°). *Explodes on contact with concentrated acid, alkali, or fire. A technical grade exploded on storage.* Ka at 25° = 3.3 × 10⁻⁷. Moderately sol in water. Sol in dil alkalies giving green to brownish green solns. Sol in alc, ether, acetone.

6564. N-Nitrosopyrrolidine. *1-Nitrosopyrrolidine;* NPYR; NO-PYR. $C_4H_8N_2O$; mol wt 100.11. C 47.99%, H 8.05%, N 27.98%, O 15.98%. Cyclic nitrosamine that occurs in food products and tobacco smoke. Prepn: F. C. Peterson, *Ber.* **21**, 290 (1888). Carcinogenicity studies: H. Druckrey *et al., Z. Krebsforsch.* **69**, 103 (1967); M. Greenblatt *et al., J. Nat. Cancer Inst.* **48**, 1687 (1972); *IARC Monographs* **17**, 313 (1978). Mechanism of mutagenicity: L. I. Hecker *et al., Mutat. Res.* **62**, 213 (1979). Metabolism: R. C. Cottrell *et al., Adv. Exp. Med. Biol.* **136B**, 1165 (1982).

Yellow liquid, bp_{20} 104-106°. Sol in water. LD_{50} orally in rats: 900 mg/kg (Druckrey).
Note: This substance may reasonably be anticipated to be a carcinogen: *Fourth Annual Report on Carcinogens* (NTP 85-002, 1985) p 160.

6565. Nitroso-R Salt. *3-Hydroxy-4-nitroso-2,7-naphthalenedisulfonic acid disodium salt;* sodium salt of 1-nitroso-2-hydroxynaphthalene-3,6-disulfonic acid. $C_{10}H_5NNa_2$-O_8S_2; mol wt 377.27. C 31.83%, H 1.34%, N 3.71%, Na 12.19%, O 33.93%, S 17.00%.

Golden-yellow, fan-shaped crystals. Sol in about 40 parts water; more sol in hot water; slightly sol in methyl or ethyl alcohol.
USE: As a reagent for cobalt and potassium.

6566. p-Nitrosulfathiazole. *p-Nitro-N-2-thiazolylbenzenesulfonamide;* 2-(p-nitrophenylsulfonamido)thiazole; 2-(4-nitrophenylsulfonamido)thiazole; Nisulfazole. C_9H_7-$N_3O_4S_2$; mol wt 285.30. C 37.89%, H 2.47%, N 14.73%, O 22.43%, S 22.48%. Prepn: Jensen, Thorsteinsson, *Dansk Tids. Farm.* **15**, 41 (1941); *C.A.* **35**, 5109⁵ (1941); Lorenz *et al.*, U.S. pat. **2,443,742** (1948 to Winthrop-Stearns).

Yellow crystals from glacial acetic acid, mp 260°. Very slightly sol in water, chloroform, ether. Somewhat more sol in alc. Practically insol in benzene. Freely sol in alkaline aqueous solns.
THERAP CAT: Antibacterial.

6567. 2-Nitro-4-sulfobenzoic Acid. $C_7H_5NO_7S$; mol wt 247.17. C 34.01%, H 2.04%, N 5.67%, O 45.31%, S 12.97%. Prepn by sulfonation of o-nitrotoluene and oxidation of the resulting 2-nitro-4-toluenesulfonic acid with potassium permanganate: Hart, *Am. Chem. J.* **1**, 352 (1879-80).

Needles from hydrochloric acid. Stable in air under ordinary conditions.
USE: Alkalimetric standard.

6568. Nitrosyl Chloride. ClNO; mol wt 65.47. N 21.40%, O 24.44%, Cl 54.16%. NOCl. Best prepared from nitrosylsulfuric acid and dry HCl: Coleman *et al., Inorg. Syn.* **1**, 55 (1939).
Non-explosive, very corrosive, reddish-yellow gas; liquid at −5.5°; solid at −61.5°. Decomposed by water. Sol in fuming H_2SO_4. Critical temp 167°; crit press. 92.4 atm. The orange color of aqua regia is produced by nitrosyl chloride.
Human Toxicity: Intensely irritating to eyes, skin, mucous membranes. Inhalation may cause pulmonary edema, hemorrhage.

6569. Nitrosyl Fluoride. Nitrogen oxyfluoride. FNO; mol wt 49.01. F 38.77%, N 28.58%, O 32.65%. Preparation: Ruff *et al., Z. Anorg. Allgem. Chem.* **208**, 293 (1932); Balz, Mailänder, *ibid.* **217**, 166 (1934); Faloon, Kenna, *J. Am. Chem. Soc.* **73**, 2937 (1951); Kwasnik in *Handbook of Preparative Inorganic Chemistry* vol. **1**, G. Brauer, Ed. (Academic Press, New York, 2nd ed., 1963) pp 184-185. *Reviews:* Hoffman, Neville, *Chem. Rev.* **62**, 1-18 (1962); Kemmitt, Sharp, *Advan. Fluorine Chem.* **4**, 194-195 (1965); Woolf, *ibid.* **5**, 1-30 (1965); Schmutzler, *Angew. Chem. Int. Ed.* **7**, 440-455 (1968).
Colorless gas. Often bluish because of impurities. Vigorous reaction with glass, corroding action on quartz. May be kept in quartz ampuls if cooled in liq oxygen. mp −132.5°. bp −59.9°. d (liq at bp) 1.326. d (solid) 1.719. Trouton const. 21.3. Reacts with water to form NO, HNO_3 and HF.
Human Toxicity: Highly irritating to skin, eyes, mucous membranes. *See also* Fluorine.
USE: Oxidizer in rocket propellants; stabilizing agent for liquid SO_3; fluorinating agent.

6570. Nitrosylsulfuric Acid. *Sulfuric acid monoanhydride with nitrous acid; nitrosyl sulfate;* chamber crystals; nitrososulfuric acid; nitroxylsulfuric acid; nitrosulfonic acid; nitrosyl hydrogen sulfate; nitro acid sulfite; Nitrose. HN-O_5S; mol wt 127.08. H 0.79%, N 11.02%, O 62.95%, S 25.23%. Formed as an intermediate in the lead chamber process for sulfuric acid by the reaction of sulfur dioxide, nitrogen trioxide, oxygen, and water: Clément, Désormes, *Ann. Chim. Phys.* [1] **59**, 329 (1806); Lunge, *J. Chem. Soc.* **47**, 470 (1885). Prepd from sulfur trioxide, nitrogen oxides and water: Döbereiner, *Schweigger's Journ.* **8**, 239 (1812); de Claubry, *Ann. Chim. Phys.* [2] **45**, 284 (1832); Kuhlmann, *ibid.* [3] **1**, 116 (1843); from silver acid sulfate and nitrosyl bromide: Berl *et al., Z. Anorg. Allgem. Chem.* **209**, 264 (1932). *See also* U.S. pats. **1,909,557** and **1,909,558**. The formation of crystals of nitrosylsulfuric acid may be observed by igniting a mixture of 1 part sulfur and 2 or 3 parts potassium nitrate under a bell jar.

Prisms, dec 73.5°. In moist air the crystals dec with the formation of sulfuric and nitric acids and above 50° nitric oxide and nitrogen dioxide are evolved. Sol in sulfuric acid, dec in water.
USE: For bleaching cereal milling products.

6571. Nitrosyl Tetrafluoroborate. Nitrosonium tetrafluoroborate; nitrosyl borofluoride; nitrosyl fluoborate. BF_4NO; mol wt 116.83. B 9.26%, F 65.05%, N 11.99%, O 13.70%. NOBF₄. Prepd according to the equation: 2 HBF₄ + N_2O_3 → 2 NOBF₄ + H_2O: Wilke-Dörfurt, Balz, *Z. Anorg. Allgem. Chem.* **159**, 219 (1927); Balz, Mailänder, *ibid.*

217, 162 (1934); H. S. Booth, D. R. Martin, *Boron Trifluoride and Its Derivatives* (New York, 1949) p 133 sqq. Review of tetrafluoroborates: Sharp, *Advan. Fluorine Chem.* **1,** 68-128 (1960).

Birefringent, orthorhombic, hygroscopic platelets. d_4^{25} 2.185. Sublimes at 0.01 mm and 250° without decompn. Decomposed by water. May be stored in glass bottles if absolutely dry.

USE: In the prepn of diazonium fluoborates.

6572. Nitrotoluene. *Methylnitrobenzene.* $C_7H_7NO_2$; mol wt 137.13. C 61.31%, H 5.15%, N 10.21%, O 23.33%. Nitration of toluol by a mixture of HNO_3 and H_2SO_4 yields principally *o*- and *p*-nitrotoluol. Prepn of *m*-nitrotoluene from 3-nitro-4-amino-toluene and $NaNO_2$: Clark, Taylor, *Org. Syn.* **3,** 91 (1923).

m-Nitrotoluene. Liquid. d_4^{15} 1.1630; d_4^{20} 1.1581; d_4^{59} 1.124; d_4^{121} 1.063. Solidifies in an ice and salt cooling mixture; melts at 15.5°. bp_{760} 231.9°; bp_{100} 156.9°; bp_{40} 130.7°; bp_{20} 112.8°; bp_{10} 96.0°; bp_5 81.0°; $bp_{1.0}$ 50.2°. n_D^{30} 1.5426. Absorption spectrum: Marchlewski, Mayer, *Bull. Acad. Polon.* [A] **1929,** 188. Soly in water at 30°: 0.498 g/l. Miscible with alcohol and ether. Sol in benzene.

o-Nitrotoluene. Yellowish liquid at ordinary temp. d_{15}^{19} 1.1622. mp −10°. bp 222°. n_D^{20} 1.5472. Almost insol in water; sol in alcohol, benzene, petr ether.

p-Nitrotoluene. Yellowish crystals. d 1.286. mp 53-54°. bp 238°. Flash pt 106°. Almost insol in water; sol in alcohol, benzene, ether, chloroform, acetone.

USE: Manuf of dyes, toluidines, nitrobenzoic acids, etc.

6573. Nitrourea. *N-Nitrocarbamide.* $CH_3N_3O_3$; mol wt 105.06. C 11.43%, H 2.88%, N 40.00%, O 45.69%. NH_2-$CONHNO_2$. Prepd by the action of concd sulfuric acid upon urea nitrate: Thiele, Lachman, *Ann.* **288,** 281 (1895); Ingersoll, Armendt, *Org. Syn.* **5,** 85 (1925). By dropwise addition of HCl to a cooled mixture of silver cyanate and nitramide in water: Davis, Blanchard, *J. Am. Chem. Soc.* **51,** 1794 (1929).

Platelets from alcohol + petr ether dec 158.4-158.8°. K at 20° = 7.0 × 10⁻³. Absorption spectrum: Baly, Desch, *J. Chem. Soc.* **93,** 1753 (1908). Soluble in hot water, but water solns are unstable. Decompn in aq alkaline solns is almost instantaneous. Freely sol in acetone, alcohol, acetic acid. Sparingly sol in petr ether, chloroform, benzene. Stable to oxidizing agents. Can be detonated, but is not sensitive to percussion or heating.

6574. Nitrous Acid. HNO_2; mol wt 47.02. H 2.14%, N 29.79%, O 68.06%. Formed by the action of strong acids on inorganic nitrites. *Review:* Block, "Nitrous Acid, Hyponitrous Acid and their Salts" in *Mellor's* Vol. VIII, supplement II, *Nitrogen* (part 2) 353-408 (1967).

Known only in soln (pale blue in color). Weak acid. K (25°): 4.5 × 10⁻⁴. In water it changes quickly into nitric oxide and nitric acid. Forms stable, water-sol nitrites with Li, Na, K, Ca, Sr, Ba, Ag. Does not form salts with weak polyvalent cations like Al or Be. Forms stable esters with alcohols.

6575. Nitrous Oxide. Dinitrogen monoxide; laughing gas; hyponitrous acid anhydride; factitious air. N_2O; mol wt 44.02. N 63.65%, O 36.35%. Constituent of the earth's atm, about 0.00005% by volume: Slobod, Krogh, *J. Am. Chem. Soc.* **72,** 1175 (1950). Prepd by thermal decompn of ammonium nitrate: E. H. Archibald, *The Preparation of Pure Inorganic Substances* (Wiley, New York, 1932) p 246; Castner, Kirst, U.S. pat. **2,111,276** (1938 to du Pont). Preparation and purification: Schenk in *Handbook of Preparative Inorganic Chemistry* vol. **1,** G. Brauer, Ed. (Academic Press, New York, 2nd ed., 1963) pp 484-485. The chief impurity of the commercial product is N_2, although NO_2, N, O_2, and CO_2 may also be present. Teratogenicity study: G. A. Lane *et al., Science* **210,** 899 (1980). *Reviews:* Beattie, "Nitrous

Oxide" in *Mellor's* Vol. VIII, suppl II, *Nitrogen* (part 2) 189-215 (1967); Jones in *Comprehensive Inorganic Chemistry* vol. **2,** J. C. Bailar, Jr. *et al.,* Eds. (Pergamon Press, Oxford, 1973) pp 316-323.

Colorless gas. *Asphyxiant.* Slightly sweetish odor and taste. Supports combustion. Very stable and rather inert chemically at room temperatures. Dissociation begins above 300° when the gas becomes a strong oxidizing agent. mp −90.81°; bp_{760} −88.46°; Trouton constant 21.4: Hoge, *J. Res. Nat. Bur. Stand.* **34,** 281 (1945). Dipole moment 0.166. d⁻⁸⁹ (liq) 1.226; d(S.T.P.) 1.967; d(gas) 1.53 (air = 1). Critical temp 36.5°; crit press. 71.7 atm. Heat of vaporization (bp): 3.956 kcal/mole. While in the steel cylinder nitrous oxide is compressed to the form of gas over liq and has a pressure of about 800 lbs/sq. in. at room temp. At 20° and 2 atm one liter of the gas dissolves in 1.5 liters of water. Freely sol in sulfuric acid. Sol in alcohol, ether, oils.

USE: To oxidize organic compds at temps above 300°; to make nitrites from alkali metals at their boiling points; in rocket fuel formulations (with carbon disulfide); in the prepn of whipped cream. *Caution:* Narcotic in high concns. Less irritating than other oxides of nitrogen.

THERAP CAT: Anesthetic (inhalation); analgesic.

6576. Nitrovin. *2-[3-(5-Nitro-2-furanyl)-1-[2-(5-nitro-2-furanyl)ethenyl]-2-propenylidene]hydrazinecarboximidamide; [[3-(5-nitro-2-furyl)-1-[2-(5-nitro-2-furyl)vinyl]allylidene]amino]guanidine;* sym-bis(5-nitro-2-furfurylidene)-acetone guanylhydrazone; 1,5-bis(5-nitro-2-furyl)-3-pentadienone guanylhydrazone; 1,5-bis(5-nitro-2-furyl)-3-pentadienone amidinohydrazone; Panazon; Payzone. $C_{14}H_{12}N_6O_6$; mol wt 360.29. C 46.67%, H 3.36%, N 23.32%, O 26.64%. Prepn: Uota *et al.,* **Japan.** pat. **2673**('52) (to Toyama), *C.A.* **48,** 2115h (1954); Uoda, Tanizaki, **Japan.** pat. **4479**('64) (to Fukuju Pharm.), *C.A.* **62,** 10412f (1965).

Blackish violet crystals from ethyl alcohol, mp 217° (dec). Hydrochloride, mp 280° (dec).

THERAP CAT (VET): Growth promoter; antibacterial.

6577. Nitroxoline. *5-Nitro-8-quinolinol;* 5-nitro-8-hydroxyquinoline; Enterocol; Nibiol; Noxibiol; Uritrol; Urocoli. $C_9H_6N_2O_3$; mol wt 190.15. C 56.84%, H 3.18%, N 14.73%, O 25.24%. Prepn: Kostanecki, *Ber.* **24,** 154 (1891); Petrow, Sturgeon, *J. Chem. Soc.* **1954,** 570; Pratt, Duke, *J. Am. Chem. Soc.* **82,** 1155 (1960). *In vitro* antibacterial and antifungal activity: A. Desvignes, P. Leguen, *Ann. Pharm. Franc.* **21,** 803 (1963); M. Medic-Saric *et al., Chemotherapy* **26,** 263 (1980). Toxicological study: O. Angelova *et al., Adv. Antimicrob. Antineoplastic Chemother., Proc. 7th Int. Congr. Chemother.* **1,** 507 (1972). Clinical pharmacokinetics: A. Mrhar *et al., Int. J. Clin. Pharmacol. Biopharm.* **17,** 476 (1979). HPLC determn in plasma and urine: R. H. A. Sorel *et al., J. Chromatog.* **222,** 241 (1981). Clinical evaluation in urinary tract infections: M. R. Jacobs *et al., S. Afr. Med. J.* **54,** 959 (1978); B. Cancet, A. Amgar, *Pathol. Biol.* **35,** 879 (1987).

Yellow needles from alcohol or acetic acid, mp 179.5-181.5°. Freely sol in alkali and hot HCl; sparingly sol in alcohol, ether.

Hydrochloride, $C_9H_7ClN_2O_3$, yellow needles from alcohol, mp 258°.

THERAP CAT: Antibacterial.

6578. Nitroxynil. *4-Hydroxy-3-iodo-5-nitrobenzonitrile;* Dovenix. $C_7H_3IN_2O_3$; mol wt 290.03. C 28.99%, H 1.04%, I

43.76%, N 9.66%, O 16.55%. Preparation: **Neth.** pat. **Appl. 6,516,359** corresp to Collins *et al.*, U.S. pat. **3,331,738** (1966, 1967 both to May & Baker).

Yellow crystals from benzene, mp 137-138°. Sparingly sol in water; moderately sol in most organic solvents.

D-*N*-Methylglucamine salt, $C_{14}H_{20}IN_3O_8$, *nitroxynil meglumine, 4-hydroxy-3-iodo-5-nitrobenzonitrile compd with* D-*1-deoxy-1-(methylamino)glucitol (1:1)*. Solid, mp 85-90°.

N-Ethylglucamine, $C_{15}H_{22}IN_3O_8$, *nitroxynil eglumine, 4-hydroxy-3-iodo-5-nitrobenzonitrile compd with 1-deoxy-1-(ethylamino)glucitol (1:1), Trodax*. Readily sol in water with a yellow, odorless and substantially neutral soln. Aq soln is very stable but contamination with calcium and certain other salts can result in pptn of an insol salt of nitroxynil.

THERAP CAT (VET): Anthelmintic (fasciolicide).

6579. Nitryl Chloride. Nitroxyl chloride. $ClNO_2$; mol wt 81.47. Cl 43.52%, N 17.20%, O 39.28%. NO_2Cl. Conveniently prepd by the addn of chlorosulfonic acid to nitric acid: Dachlauer, *Ger.* pat. **509,405** (1929 to I. G. Farben); Kaplan, Schechter, *Inorg. Syn.* **4**, 52 (1953); Collis *et al.*, *J. Chem. Soc.* **1958**, 438.

Corrosive, toxic, colorless gas. Chlorine-like odor. Vapor density (100°): 2.81 g/l. Dec above 120°. bp −14.3°. mp −145°. d_{liq}^0 1.37; d_{liq}^{16} 1.33. Even the purest liquid may have a pale yellow color. Solns in polar solvents are always yellow. The gas or liquid may attack organic matter with explosive violence.

USE: Nitrating and chlorinating agent in organic syntheses. *Caution:* Strong irritant, corrosive.

6580. Nitryl Fluoride. FNO_2; mol wt 65.01. F 29.23%, N 21.55%, O 49.22%. Credit for original prepn by the spontaneous combustion of nitric oxide in an atm of fluorine is given to Moissan, Lebeau, *Compt. Rend.* **140**, 1573, 1621 (1905); more easily prepd by mixing nitrogen dioxide and fluorine: Ruff *et al.*, *Z. Anorg. Allgem. Chem.* **208**, 298 (1932); Faloon, Kenna, *J. Am. Chem. Soc.* **73**, 2937 (1951). Reviews of prepn and chemistry: Hoffman, Neville, *Chem. Rev.* **62**, 1-18 (1962); Kwasnik in *Handbook of Preparative Inorganic Chemistry* vol. 1, G. Brauer, Ed. (Academic Press, New York, 2nd ed., 1963) pp 186-187; Kemmitt, Sharp, *Advan. Fluorine Chem.* **4**, 195-196 (1965); Woolf, *ibid.* **5**, 1-30 (1965); Schmutzler, *Angew. Chem. Int. Ed.* **7**, 440-455 (1968).

Colorless gas. Pungent odor. Attacks mucous membranes. mp −166.0°. bp −72.4°. d (liq at bp) 1.796. d (solid) 1.924. Trouton const 21.2. May be stored in quartz ampuls if cooled in liq oxygen. Purification can be accomplished by fractional distillation at reduced press. in dry glass or quartz apparatus. Rapidly hydrolyzed in water to form nitric and hydrofluoric acids. Powerful oxidizing agent, with fluorinating powers; slightly weaker than fluorine. Absorbs mercury completely. Spontaneously ignites iodine, selenium, phosphorus (red and white), arsenic, antimony, boron, silicon, thorium, molybdenum. On mild warming attacks lead, bismuth, chromium, manganese, iron, nickel, tungsten, sulfur, charcoal. Does not react readily with hydrogen in the cold. Converts ethanol to ethyl nitrate; benzene to nitrobenzene.

Human Toxicity: See Fluorine.

USE: Oxidizer in rocket propellants; fluorinating agent.

6581. Nivalenol. *12,13-Epoxy-3,4,7,15-tetrahydroxytrichothec-9-en-8-one; 3α,4β,7α,15-tetrahydroxyscirp-9-en-8-one.* $C_{15}H_{20}O_7$; mol wt 312.33. C 57.69%, H 6.45%, O 35.86%. *Trichothecene mycotoxin* isolated from *Fusarium nivale:* T. Tatsuno *et al.*, *Chem. Pharm. Bull.* **16**, 2519 (1968). Structure: *eidem, Tetrahedron Letters* **1969**, 2823. Toxicology: T. Tatsuno, *Cancer Res.* **28**, 2393 (1968). Implicated as a chemical warfare agent in Southeast Asia with

T-2 toxin, *q.v.*: N. Wade, *Science* **214**, 34 (1981); R. T. Rosen, J. D. Rosen, *Biomed. Mass Spectrom.* **9**, 443 (1982).

Crystals, mp 222-223° (dec). $[\alpha]_D^{24}$ +21.54° (c = 1.3 in ethanol). uv max (methanol): 218 nm (ε 6300). Slightly sol in water; sol in polar organic solvents. LD_{50} i.p. in mice: 40 μg/10 g (Tatsuno).

Toxicity: Strong hemorrhagic agent! Causes blisters, necrosis of tissues, dizziness, nausea, vomiting, diarrhea, hemorrhaging; may result in death.

6582. Nizatidine. *N-[2-[[[2-[(Dimethylamino)methyl]-4-thiazolyl]methyl]thio]ethyl]-N'-methyl-2-nitro-1,1-ethenediamine; N-[4-(6-methylamino-7-nitro-2-thia-5-aza-6-heptene-1-yl)-2-thiazolylmethyl]-N,N-dimethylamine; LY 139037; ZE-101; ZL-101; Axid; Calmaxid; Cronizat; Gastrax; Naxidine; Nizax; Nizaxid; Zanizal.* $C_{12}H_{21}N_5O_2S_2$; mol wt 331.45. C 43.49%, H 6.39%, N 21.13%, O 9.65%, S 19.34%. Histamine H_2-receptor antagonist related to ranitidine, *q.v.* Prepn: R. P. Pioch, *Eur.* pat. **Appl. 49,618;** *idem,* U.S. pat. **4,375,547** (1982, 1983 both to Eli Lilly). General pharmacology in animals: K. Bemis *et al.*, *Arzneimittel-Forsch.* **39**, 240 (1989). Pharmacokinetics and gastric acid suppression in humans: J. T. Callaghan *et al.*, *Clin. Pharmacol. Ther.* **37**, 162 (1985). *In vitro* and *in vivo* comparison with structurally related compounds: C. G. Meredith *et al.*, *Toxicol. Appl. Pharmacol.* **77**, 315 (1985). Lack of effect on hepatic drug metabolism in humans: J. W. Secor *et al.*, *Brit. J. Clin. Pharmacol.* **20**, 710 (1985). Disposition and metabolism in humans: M. P. Knadler *et al.*, *Drug. Metab. Dispos.* **14**, 175 (1986). Symposium on pharmacology and clinical studies: *Scand. J. Gastroenterol.* **22**, Suppl. 136, 1-88 (1987).

Crystals from ethanol-ethyl acetate, mp 130-132°. LD_{50} in mice, rats (mg/kg): 265, >300 i.v.; 1685, 1680 orally (Pioch).

THERAP CAT: Antiulcerative.

6583. Nizofenone. *(2-Chlorophenyl)[2-[2-[(diethylamino)methyl]-1H-imidazol-1-yl]-5-nitrophenyl]methanone; 2'-chloro-2-[2-[(diethylamino)methyl]imidazol-1-yl]-5-nitrobenzophenone; 1-[2-(2-chlorobenzoyl)-4-nitrophenyl]-2-(diethylaminomethyl)imidazole.* $C_{21}H_{21}ClN_4O_3$; mol wt 412.88. C 61.09%, H 5.13%, Cl 8.59%, N 13.57%, O 11.62%. Imidazole derivative exhibiting protective activity against cerebral anoxia or ischemia. Prepn: M. Nakanishi *et al.*, *Ger.* pat. **2,403,416;** *idem,* U.S. pat. **3,915,981** (1974, 1975 both to Yoshitomi). Antianoxic effect in animal models: H. Yasuda *et al.*, *Arch. Int. Pharmacodyn.* **233**, 136 (1978). Mechanism of action: *eidem, ibid.* **242**, 77 (1979). Multicenter clinical studies: I. Saito *et al.*, *Neurol. Res.* **5**, 29 (1983); T. Ohta *et al.*, *J. Neurosurg.* **64**, 420 (1986). Toxicity studies: H. Horizoe *et al.*, *Oyo Yakuri* **30**, 627 (1985), *C.A.* **104**, 61754m (1986); K. Okumura *et al.*, *ibid.* 633, *C.A.* **104**, 61755 (1986).

Pale yellow crystals from isopropyl ether, mp 75-76°.
Fumarate, $C_{25}H_{25}ClN_4O_7$, *Y-9179, midafenone, Ekonal.*
Pale yellow crystals from isopropyl ether, mp 157-158°.
LD_{50} in male, female mice, male, female rats (mg/kg): 495,
504, 1711, 1580 orally; 62, 70, 63, 65 i.v.; 270, 278, 1830,
1629 s.c. (Horizoe).
THERAP CAT: Nootropic.

6584. NMDA. *N-Methyl-D-aspartic acid; N*-methyl-D-
aspartate; *N*-methyl-D-asparaginsaure (German). C_5H_9-
NO_4; mol wt 147.13. C 40.82%, H 6.16%, N 9.52%, O
43.50%. Excitotoxic amino acid used to identify a specific
subset of excitatory amino acid receptors; consequently the
receptors are known as NMDA receptors. *See also* quisqualic acid, kainic acid. Prepn: O. Lutz, *Z. Physiol. Chem.*
70, 256 (1909); O. Lutz, Br. Jirgensons, *Ber.* **63**, 448 (1930).
Isoln from the blood shell, *Scapharca broughtonii:* M. Sato
et al., Biochem. J. **241**, 309 (1987). Identification as a neuroexcitant: D. R. Curtis, J. C. Watkins, *J. Physiol.* **166**, 1
(1963); of neurotoxic properties: J. W. Olney *et al., Life Sci.*
25, 537 (1979); R. Zaczek *et al., Neurosci. Letters* **24**, 181
(1981); M. V. Sofroniew, R. C. A. Pearson, *Brain Res.* **339**,
186 (1985). Receptor binding studies: C. K. Mitchell, D. A.
Redburn, *Neurosci. Letters* **28**, 241 (1982); D. T. Monaghan,
C. W. Cotman, *J. Neurosci.* **5**, 2909 (1985); J. M. M. Olson
et al., Neuroscience **22**, 913 (1987). Review of NMDA and
its receptor: J. C. Watkin, R. H. Evans, *Ann. Rev. Pharmacol. Toxicol.* **21**, 165-204 (1981); T. W. Stone, N. R. Burton,
Prog. Neurobiol. **30**, 333-368 (1988).

6585. NMN. *3-(Aminocarbonyl)-1-(5-O-phosphono-β-
D-ribofuranosyl)pyridinium hydroxide inner salt; 3-carbam-
oyl-1-β-D-ribofuranosylpyridinium hydroxide 5'-(dihydrogen phosphate) inner salt;* nicotinamide mononucleotide.
$C_{11}H_{15}N_2O_8P$; mol wt 334.24. C 39.53%, H 4.52%, N 8.38%,
O 38.30%, P 9.27%. Prepd by incubation of NAD with
potato pyrophosphatase in a non-phosphate buffer and in
the presence of fluoride: Plaut, Plaut, *Arch. Biochem. Biophys.* **48**, 189 (1954); *Biochem. Prepn.* **5**, 55 (1957). Enzymatic synthesis from nicotinamide by human erythrocytes
and hemolyzates: Preiss, Handler, *J. Biol. Chem.* **225**, 759
(1957). Prepn by hydrolysis of NAD by a yeast enzyme:
Takei, *Agr. Biol. Chem.* **34**, 23 (1970); by cleavage of NAD
with crude rattlesnake venom: Apps, *FEBS Letters* **15**, 277
(1971); R. Jeck, C. Wönckhaus, *Methods in Enzymol.* **66**, 62
(1980). NMR studies and conformation of α- and β-nico-
tinamide nucleotide: N. J. Oppenheimer, N. O. Kaplan,
Biochemistry **15**, 3981 (1976).

Can be precipitated from slightly acidic aq soln by a large

vol of acetone. Freely sol in water, practically insol in acetone. More stable when stored frozen in soln, than when in
dry form.

6586. Nobelium. Element 102. No; at. wt (longest-lived
known isotope) 259; at. no. 102; valence 2, also 3. Discovery of element 102 first claimed by Fields *et al., Phys. Rev.*
107, 1460 (1957); later experiments failed to confirm these
results. Scientists at the Berkeley and Dubna laboratories
have since prepared nine isotopes with mass numbers 251-
259. Produced by bombarding transuranium elements with
heavy ions (e.g. curium (Cm) isotopes with carbon ions (^{12}C,
^{13}C); plutonium (Pu) isotopes with oxygen ions (^{16}O, ^{18}O)
etc.). ^{259}No ($T_{1/2}$ 58 ± 5 minutes, α-emitter) produced by
bombardment of ^{248}Cm with ^{18}O ions: Silva *et al., Nucl.
Phys. A* **216**, 97 (1973), *C.A.* **80**, 43040g (1974). Reviews of
history, prepn and properties: Seaborg, *J. Chem. Ed.* **36**,
38-44 (1959); Ghiorso, Sikkeland, *Phys. Today* **20**(9), 25-32
(1967); Flerov, *U.S. At. Energy Comm.* JINR-D7-3444
(1967) 36 pp, *C.A.* **68**, 34658q (1968); C. Keller, *The Chemistry of the Transuranium Elements* (Verlag Chemie, Weinheim, English Ed., 1971) pp 601-607; Silva, "Trans-Curium
Elements" in *MTP Int. Rev. Sci.: Inorg. Chem., Ser. One* vol.
8, A. G. Maddock, Ed. (University Park Press, Baltimore,
1972) pp 71-105; *Comprehensive Inorganic Chemistry* vol. **5**,
J. C. Bailar, Jr., *et al.,* Eds. (Pergamon Press, Oxford, 1973)
passim; Ghiorso, *Handb. Exp. Pharmakol.* **36**, 691-715
(1973); Taylor, *ibid.* 717-738 (1973).

6587. Nocardamin. *1,12,23-Trihydroxy-1,6,12,17,23,28-
hexaazacyclotritriacontane-2,5,13,16,24,27-hexone.* $C_{27}H_{48}$-
N_6O_9; mol wt 600.73. C 53.98%, H 8.05%, N 13.99%, O
23.97%. Antibiotic substance produced by a *Nocardia* isolated from old bee-honeycombs: Stoll *et al., Schweiz. Z.
Pathol. u. Bakteriol.* **14**, 225 (1951); *Helv. Chim. Acta* **34**, 862
(1951). Structure: Keller-Schierlein, Prelog, *ibid.* **44**, 1981
(1961).

Needles from hot methanol, mp 192-195°. Optically inactive. Has no basic properties. Soluble in hot water, methanol. An 0.5% aq soln at room temp has a pH of 4.55.
Triacetate, $C_{33}H_{54}N_6O_{12}$, crystals, mp 115-117°.

6588. Nocardicin(s). Monocyclic β-lactam antibiotics
with antimicrobial activity which inhibit bacterial cell wall
biosynthesis. Nocardicins A, B, C, D, E, F, G have been
identified. All are produced by *Nocardia uniformis* subsp.
tsuyamenensis, A being the most important component.
Prodn: H. Aoki *et al.,* Ger. pat. 2,242,699 (1973 to Fujisawa), *C.A.* **78**, 134496k (1973); U.S. pat. 3,923,977 (1975 to
Fujisawa). Isoln and characterization of A: H. Aoki *et al.,
J. Antibiot.* **29**, 492 (1976); of B: M. Kurita *et al., ibid.* 1243.
Nocardicin A has also been produced by *Actinosynnema
mirum:* K. Watanabe *et al., ibid.* **36**, 321 (1983). Structural
determination of A and B: M. Hashimoto *et al., J. Am.
Chem. Soc.* **98**, 3023 (1976), *J. Antibiot.* **29**, 890 (1976).
Isoln, characterization, and structures of C, D, E, F, G: J.
Hosoda *et al., Agric. Biol. Chem.* **41**, 2013 (1977). Total
synthesis of A: G. A. Koppel *et al., J. Am. Chem. Soc.* **100**,
3933 (1978); W. V. Curran *et al., J. Antibiot.* **35**, 329 (1981);
of A and D: T. Kamiya *et al., Tetrahedron* **1979**, 323; of A
and B: H. P. Isenring, W. Hofheinz, *Tetrahedron* **1983**,
2591. Biosynthetic studies of A: J. Hosada *et al., Agric. Biol.*

Chem. **41**, 2007 (1977); C. A. Townsend *et al.*, *J. Am. Chem. Soc.* **105**, 919 (1983). Series of articles on antimicrobial activity of A: *J. Antibiot.* **30**, 917-944 (1977).

Nocardicin A, $C_{23}H_{24}N_4O_9$, *Z-3-[[[4-(3-amino-3-carboxypropoxy)phenyl](hydroxyimino)acetyl]amino]-α-(4-hydroxyphenyl)-2-oxo-1-azetidineacetic acid.* Colorless needles from acidic water, mp 214-216° (dec). $[\alpha]_D^{25}$ −135° (for the sodium salt). uv max (1/15M phosphate buffer): 272 nm ($E_{1cm}^{1\%}$ 310); (0.1N NaOH): 244, 283 nm ($E_{1cm}^{1\%}$ 460, 270). Sol in alkaline solns, slightly sol in methanol. Insol in chloroform, ethyl acetate, ethyl ether. LD_{50} in male mice, rats (mg/kg): > 8000 orally; > 2000 i.v.; 2500, 2600 i.p. (Aoki).

Nocardicin B, the *E*-isomer of nocardicin A. Colorless needles, mp 262-264° (dec). $[\alpha]_D^{25}$ −162° (for the sodium salt). uv max (ethanol/water): 224, 270 nm (ε 24600, 9700); (ethanol/0.1N NaOH): 245, 280 nm (ε 26000, 11100).

6589. Nodakenin. *2-[1-(β-D-Glucopyranosyloxy)-1-methylethyl]-2,3-dihydro-7H-furo[3,2-g][1]benzopyran-7-one;* nodakenetin glucoside. $C_{20}H_{24}O_9$; mol wt 408.39. C 58.82%, H 5.92%, O 35.26%. From root of *Peucedanum decursivum* Maxim., *Umbelliferae:* Arima, *Bull. Chem. Soc. Japan* **4**, 16, 113 (1929); Späth, Kainrath, *Ber.* **69**, 2062 (1936). Structure: Späth, Tyray, *ibid.* **72**, 2089 (1939).

Thin leaflets from alc, dec 218-219°. From dil alc or water it crystallizes with one mol H_2O in form of yellow prisms, melting at 216°. $[\alpha]_D^{20}$ +56.6°. Sol in hot water or alcohol.

Tetraacetate, $C_{28}H_{32}O_{13}$, crystals from methanol, mp 195-196°.

6590. Noformicin. *5-Amino-N-(3-amino-3-iminopropyl)-3,4-dihydro-2H-pyrrole-2-carboxamide; N-(2-amidinoethyl)-5-imino-2-pyrrolidinecarboxamide;* 2-[N-(2-amidinoethyl)carbamoyl]-5-iminopyrrolidine; β-(5-imino-2-pyrrolidinecarboxamido)propamidine; noformycin. $C_8H_{15}N_5O$; mol wt 197.24. C 48.71%, H 7.67%, N 35.51%, O 8.11%. Antiviral agent produced by *Nocardia formica,* culture no. N.R.R.L. 2470: Peck *et al.*, U.S. pat. **2,804,463** (1957 to Merck & Co.); Gray, *Phytopathology* **45**, 281 (1955). Activity studies: Furusawa *et al.*, *Proc. Soc. Exp. Biol. Med.* **116**, 938 (1964); *eidem, Med. Pharmacol. Exp.* **12**, 259 (1965). Synthesis of (+)- and (±)-forms: Diana, *J. Med. Chem.* **16**, 857 (1973).

(+)-Form dihydrochloride, $C_8H_{17}Cl_2N_5O$. The biologically active form. Crystals from methanol, dec 263-264°. $[\alpha]_D^{25}$ +8.8° (methanol). Sol in water. Somewhat sol in organic solvents. pKa 9.4.

(±)-Form dihydrochloride, mp 252-254°.

6591. Nogalamycin. *[2R-(2α,3β,4α,5β,6α,11β,13α,-14α)]-11-[(6-Deoxy-3-C-methyl-2,3,4-tri-O-methyl-α-L-mannopyranosyl)oxy]-4-(dimethylamino)-3,4,5,6,9,11,12,13,-14,16-decahydro-3,5,8,10,13-pentahydroxy-6,13-dimethyl-9,16-dioxo-2,6-epoxy-2H-naphthaceno[1,2-b]oxocin-14-carboxylic acid methyl ester;* U-15167; NSC-70845. $C_{39}H_{49}$NO$_{16}$; mol wt 787.83. C 59.46%, H 6.27%, N 1.78%, O 32.49%. Antitumor antibiotic isolated from *Streptomyces nogalater* var. *nogalater:* B. K. Bhuyan, A. Dietz, *Antimicrob. Ag. Chemother.* **1965**, 836; B. K. Bhuyan *et al.*, U.S. pat. **3,183,157** (1965 to Upjohn). Characterization: P. F. Wiley *et al.*, *Tetrahedron Letters* **1968**, 663. Structure: *eidem, J. Am. Chem. Soc.* **99**, 542 (1977). Stereochemistry: *eidem, J. Org. Chem.* **44**, 4030 (1979). Biosynthesis: *eidem, ibid.* **43**, 3457 (1978). Synthesis of L-*nogalose,* sugar component: L. Valente *et al.*, *Tetrahedron Letters* **1979**, 1153; J. Yoshimura *et al.*, *Chem. Letters* **1979**, 687. Partial stereospecific synthesis: R. P. Joyce *et al.*, *Tetrahedron Letters* **27**, 4885 (1986). Molecular structure, absolute stereochemistry and interactions with DNA: S. K. Arora, *J. Am. Chem. Soc.* **105**, 1328 (1983). Toxicity data: Z. Hadidian, *PB Report* 173334 (1966). Review of pharmacology: B. K. Bhuyan, C. G. Smith, *Handb. Exp. Pharmacol.* **38**(pt. 2), 623-632 (1975).

Orange-red solid from methanol, mp 195-196° (dec). $[\alpha]_D^{25}$ +425° (c = 0.11 in CHCl$_3$). pKa′ 7.45 (60% ethanol). uv max (ethanol): 236, 258, 292 nm (ε 52360, 24755, 9890). Sol in methylene chloride, acetone, chloroform, ethyl acetate. Insol in water, methanol, ethanol. LD_{50} in mice (mg/kg): 11.75 i.v., 4.79 i.p. (Hadidian).

THERAP CAT: Antineoplastic.

6592. Nomifensine. *1,2,3,4-Tetrahydro-2-methyl-4-phenyl-8-isoquinolinamine; 8-amino-1,2,3,4-tetrahydro-2-methyl-4-phenylisoquinoline.* $C_{16}H_{18}N_2$; mol wt 238.33. C 80.64%, H 7.61%, N 11.75%. Novel antidepressant distinguished from existing tricyclic and tetracyclic antidepressants by its bicyclic structure. Prepn: **Brit. pat. 1,164,192** corresp to G. Ehrhart *et al.*, U.S. pat. **3,577,424** (1969, 1971, both to Farbwerke Hoechst); I. Hoffmann *et al.*, *Arzneimittel-Forsch.* **21**, 1045 (1971). Pharmacology: I. Hoffmann, *ibid.* **23**, 45 (1973); P. Hunt *et al.*, *J. Pharm. Pharmacol.* **26**, 370 (1974); B. Costall *et al.*, *Psychopharmacologia* **41**, 153 (1975). Pharmacokinetics: L. Vereczkey *et al.*, *ibid.* **45**, 225 (1975). Study on abuse potential in humans: C. Spyraki, H. C. Fibiger, *Science* **212**, 1167 (1981). *Review:* Brit. J. Clin. Pharmacol. **4**, Suppl. 2 (1977) pp 1S-248S; Int. Pharmacopsychiatry **17**, Suppl. 1 (1982) pp 1-148.

mp 179-181°.

Maleate, $C_{20}H_{22}N_2O_4$, *HOE 984, Alival, Hostalival, Merital, Neurolene, Psicronizer.* Crystals from ethanol, mp 199-201°. LD_{50} in mice, rats (mg/kg): 400, 430 orally; 90, 72 i.v.; in mice (mg/kg): 410 s.c. (Hoffman, 1973).

THERAP CAT: Antidepressant.

6593. Nomilin. *1-(Acetyloxy)-1,2-dihydroobacunoic acid ε-lactone; 1-(3-furyl)decahydro-11-hydroxy-4b,7,7-11a,13a-pentamethyloxireno(4,4a)-2-benzopyrano[6,5-g](2)benzoxepin-3,5,9(3aH,4bH,6H)-trione acetate.* $C_{28}H_{34}O_9$; mol wt 514.57. C 65.36%, H 6.66%, O 27.98%. Bitter principle from citrus juice, seeds. Isoln: O. H. Emerson, *J. Am. Chem. Soc.* **70,** 545 (1948). Structure elucidation: D. L. Dreyer, *Tetrahedron* **21,** 75 (1965); *idem, ibid.* **24,** 3273 (1968). Distribution in citrus seeds: S. Hasegawa *et al., J. Agr. Food Chem.* **28,** 922 (1980); R. L. Rouseff, S. Nagy, *Phytochemistry* **21,** 85 (1982). Identification in grapefruit juice: R. L. Rouseff, *J. Agr. Food Chem.* **30,** 504 (1982). Reduction of nomilin bitterness by *Arthrobacter globiformis* cells: S. Hasegawa, V. A. Pelton, *ibid.* **31,** 178 (1983); by *Cornebacterium fascians:* S. Hasegawa *et al., ibid.* **32,** 457 (1984); *eidem, J. Food Sci.* **50,** 330 (1985). Antifeedant effect against *Spodoptera frugiperda:* M. A. Altieri *et al., Protect. Ecol.* **6,** 91 (1984). Biosynthetic pathways: S. Hasegawa *et al., Phytochemistry* **23,** 1601 (1984).

Needles from methanol, mp 278-279°. Slightly sol in 2-propanol, ethyl acetate. $[\alpha]_D^{23}$ −95.7°.

6594. Nonactin. *2,5,11,14,20,23,29,32-Octamethyl-4,13,-22,31,37,38,39,40-octaoxapentacyclo[32.2.1.1^{7,10}.1^{16,19}.1^{25,28}]-tetracontane-3,12,21,30-tetrone.* $C_{40}H_{64}O_{12}$; mol wt 736.91. C 65.19%, H 8.75%, O 26.06%. Macrotetrolide antibiotic. Produced by several *Streptomyces* spp.: Corbaz *et al., Helv. Chim. Acta* **38,** 1445 (1955); Wallhäusser *et al., Arzneimittel-Forsch.* **14,** 356 (1964). Structure: Dominguez *et al., Helv. Chim. Acta* **45,** 129 (1962). Crystal structure: Dobler, *ibid.* **55,** 1371 (1972). Steroselective syntheses of *nonactic acid,* building block of nonactin: Gerlach, Wetter, *ibid.* **57,** 2306 (1974); R. E. Ireland, J.-P. Vevert, *J. Org. Chem.* **45,** 4260 (1980). Total synthesis of (+)- and (−)-nonactic acids: *eidem, Can. J. Chem.* **59,** 572 (1981). Synthesis of nonactin: Gombos *et al., Monatsh.* **106,** 1043 (1975); *eidem, Tetrahedron Letters* **1975,** 3391; Gerlach *et al., Helv. Chim. Acta* **58,** 2036 (1975); U. Schmidt *et al., Ber.* **109,** 2628 (1976). Three homologs, *monactin,* $C_{41}H_{66}O_{12}$, *dinactin,* $C_{42}H_{68}O_{12}$, and *trinactin,* $C_{43}H_{70}O_{12}$ are known: Beck *et al., Helv. Chim. Acta* **45,** 620 (1962); Gerlach, Prelog, *Ann.* **669,** 121 (1963). Activity of nonactin and its homologues: Meyers *et al., J. Antibiot.* **18A,** 128 (1965).

Needles from methanol, mp 147-148°. $[\alpha]_D^{20}$ 0° ±2° (c =

1.2 in $CHCl_3$). uv max (ethanol): slight peak at 264 nm (log ε 1.5). Remarkably inert to chemical compds.

6595. 2-Nonenal. *trans*-2-Nonenaldehyde. $C_9H_{16}O$; mol wt 140.22. C 77.09%, H 11.50%, O 11.41%. Widely distributed in nature; found in beer, coffee, cucumbers, watermelon, palm oil, potatoes, carrots etc. Prepn: R. Delaby, S. Guillot-Allègre, *Bull. Soc. Chim. France* **53,** 301 (1933); J. Braun *et al., Ber.* **67,** 269 (1934); J. Ficini, H. Normant, *Bull. Soc. Chim. France* **1964,** 1294; J. Ficini *et al., Tetrahedron Letters* **1977,** 3589; M. Wado *et al., Chem. Letters* **1977,** 345. Insecticidal activity: P. M. Guerin, M. F. Ryan, *Experientia* **36,** 1387 (1980). Cockroach repellant activity: T. H. Maugh, *Science* **218,** 278 (1982).

Liquid, bp_{16} 100-102°. n_D^{18} 1.4403.

USE: As flavoring agent.

6596. Nonoxynol. *α-(4-Nonylphenyl)-ω-hydroxypoly-(oxy-1,2-ethanediyl); polyethyleneglycols mono(nonylphenyl) ether;* macrogol nonylphenyl ether; nonoxinol; polyoxyethylene(n)nonylphenyl ether; nonylphenyl polyethyleneglycol ether; nonylphenoxypolyethoxyethanol. Nonionic surfactant mixtures prepd by reacting nonylphenol with ethylene oxide. Prepn of polyoxyethylated alkyl phenols: A. Steindorff *et al.,* U.S. pat. **2,213,477** (1940 to GAF). Average number of ethylene oxide units (*n*) per molecule is indicated by number following nonoxynol (e.g. nonoxynol-15 for n = 15). Trademarks for nonoxynol series include *Conco NI, Dowfax 9N, Igepal CO, Makon, Neutronyx 600's, Nonipol NO, Polytergent B, Renex 600's, Solar NP, Sterox, Surfonic N, T-DET-N, Tergitol NP, Triton N.* Individual members of series indicated by numerical suffixes. Toxicology study: H. F. Smyth Jr., J. C. Calandra, *Toxicol. Appl. Pharmacol.* **14,** 315 (1969). *Review:* C. R. Enyeart, "Polyoxyethylene Alkylphenols" in *Nonionic Surfactants,* M. J. Schick, Ed. (Dekker, New York, 1967) pp 44-85.

Highly stable compounds. Lower adducts (n < 15) are yellow to almost colorless liquids; higher adducts (n > 20) are pale yellow to off-white pastes or waxes. Lower adducts (n < 6) sol in oil; higher ones sol in water. Review of physical properties: Enyeart, *loc. cit.*

Nonoxynol-9, C-Film, Conco NI-90, Dowfax 9N9, Encare, Gynol II, Igepal CO-630, Intercept, Neutronyx 611, Renex 698, Semicid, Staycept, Tergitol TP-9. n = 9; av. mol wt 617. A component of *Conceptrol, Delfen, Gentersal, Ortho-creme.* Almost colorless liquid. For Igepal CO-630: d_4^{25} 1.06; solidif pt 26 ±2°F; pour point 37±2°F; flash point: 535-555°F; cloud point (1% aq soln) 126-133°F; viscosity (25°) 175-250 cp. Sol in water, ethanol, ethylene glycol, ethylene dichloride, xylene, corn oil. Insoluble in Stoddard solvent, deodorized kerosene, low viscosity white mineral oil.

Nonoxynol-11, Duragel, Duracreme.

USE: Nonionic surfactants used as detergents, emulsifiers, wetting agents, dispersants, stabilizers, intermediates in synthesis of anionic surfactants, defoaming agents. Nonoxynols-9, 11 as spermaticide. Nonoxynols-4, 15, 30 as pharmaceutic aids (surfactants).

6597. *n*-Nonyl Acetate. *Acetic acid n-nonyl ester; n-nonyl ethanoate; nonanol acetate; Acetate C-9.* $C_{11}H_{22}O_2$; mol wt 186.29. C 70.92%, H 11.90%, O 17.18%. $CH_3COO-(CH_2)_8CH_3$.

Liquid. Pungent odor, suggestive of mushrooms, resembles odor of gardenias when dil. d_4^{15} 0.8785 (commercial grade, d_{25}^{25} 0.864-0.868). bp 208-212°. n_D^{20} 1.4328 (commercial grade, n_D^{20} 1.422-1.426). Insoluble in water. Freely sol in abs alcohol and ether; 33 ml dissolve in 100 ml of 80% ethanol.

Note: The nonyl acetate of commerce may be the above compound or any of its isomers such as diisobutylcarbinyl acetate.

USE: In perfumery.

6598. *n*-Nonyl Alcohol. *1-Nonanol;* nonalol. $C_9H_{20}O$; mol wt 144.26. C 74.93%, H 13.98%, O 11.09%. CH_3-$(CH_2)_7CH_2OH$. Occurs in oil of orange. Prepn by reduction of ethyl Δ^1-nonylenate: Harding, Weizmann, *J. Chem. Soc.* **97**, 304 (1910); by reduction of pelargonic aldehyde: Tomecko, Adams, *J. Am. Chem. Soc.* **49**, 529 (1927); by oxo process: Russum, Hengstebeck, U.S. pat. **2,638,487** (1953 to Standard Oil of Indiana); by reductive cleavage of oleic acid ozonide: Sousa, Bluhm, *J. Org. Chem.* **25**, 108 (1960); from diborane and propylene trimer: Marshall, Smith, **Brit.** pat. **879,242** (1961 to I.C.I.).

Colorless to yellowish liquid; odor of citronella oil. bp_{760} 215°; bp_{15} 107.5°; $bp_{7.5}$ 95.6°. d_4^{20} 0.8279. n_D^{20} 1.4338. Practically insol in water; miscible with alcohol, ether.

USE: In manufacture of artificial lemon oil.

6599. Nonyl Phenol. Approx mol wt 215. A technical grade mixture of monoalkyl phenols, predominantly *para* substituted. The side chains are isomeric branched-alkyl radicals. Manuf of *p*-nonylphenol: Faith, Keyes & Clark's *Industrial Chemicals,* F. A. Lowenheim, M. K. Moran, Eds. (Wiley-Interscience, New York, 4th ed., 1975) pp 575-578.

Pale yellow liquid. Slight characteristic phenolic odor. Comparatively high viscosity. Hydroxyl no. 253. d_4^{20} 0.950. bp 293-297°. n_D^{20} 1.513. Flash pt (open cup) 300°F. Practically insol in water or dil aq NaOH. Sol in benzene, chlorinated solvents, aniline, heptane, aliphatic alcohols, ethylene glycol.

USE: In the prepn of lubricating oil additives, resins, plasticizers, surface active agents.

6600. (*p*-Nonylphenoxy)acetic Acid. $C_{17}H_{26}O_3$; mol wt 278.83. C 73.34%, H 9.41%, O 17.24%. Prepd by the reaction of sodium chloroacetate with nonyl phenol in alkaline medium: Preston, **Brit.** pat. **812,938** and Preston, Taylor, **Brit.** pat. **831,883** (1959, 1960, to I.C.I.).

Liquid. d_4^{20} 1.010-1.025. Viscosity at 30° = 5200 cp. Flow pt 5°. Practically insol in water. Miscible with benzene, mineral oil, kerosene, petr ether.

USE: Corrosion inhibitor and antifoaming agent in gasoline and cutting oils.

6601. Noprylsulfamide. *1-[[4-(Aminosulfonyl)phenyl]-amino]-3-phenyl-1,3-propanedisulfonic acid disodium salt; 1-phenyl-3-p-sulfamoylanilino-1,3-propanedisulfonic acid disodium salt; N^4-(disodium 1,3-disulfo-3-phenylpropyl)-sulfanilamide;* disodium 1-phenyl-3-*p*-sulfamoylanilino-1,3-propanedisulfonate; disodium *p*-(γ-phenylpropylamino)-benzenesulfonamide-α,γ-disulfonate; RP 40; Solucin; Soluseptasine; Soluseptazine; Solusetazine; Sulphasolucin; Sulphasolutin. $C_{15}H_{16}N_2Na_2O_8S_3$; mol wt 494.49. C 36.43%, H 3.26%, N 5.67%, Na 9.30%, O 25.89%, S 19.45%. Prepd by the action of sodium bisulfite on N^4-cinnamylidenesulfanilamide (prepd from cinnamic aldehyde and sulfanilamide): Despois, U.S. pat. **2,262,544** (1941 to Rhône-Poulenc).

Crystals. One gram dissolves in 5 ml water. Neutral reaction. Breaks down in the body with the liberation of free sulfanilamide.

THERAP CAT: Antibacterial.

6602. Noracymethadol. *α-Ethyl-β-[2-(methylamino)-propyl]-β-phenylbenzeneethanol acetate; α-dl-6-(methylamino)-4,4-diphenyl-3-heptanol acetate; α-dl-3-acetoxy-4,4-diphenyl-6-methylaminoheptane; α-dl-3-acetoxy-6-methyl-* amino-4,4-diphenylheptane; *α-dl*-4,4-diphenyl-6-methyl-amino-3-heptanol acetate. $C_{22}H_{29}NO_2$; mol wt 339.46. C 77.84%, H 8.61%, N 4.13%, O 9.43%. Metabolite of methadyl acetate, *q.v.* Prepn: A. Pohland, U.S. pat. **3,021,360** (1962 to Lilly). Metabolism: G. L. Henderson *et al., Drug Metab. Dispos.* **5**, 321 (1977); M. Man *et al., ibid.* **8**, 55 (1980). HPLC determn: C.-H. Kiang *et al., J. Chromatog.* **222**, 81 (1981).

Hydrochloride, $C_{22}H_{30}ClNO_2$, NIH 7667. Crystals from acetone + ether, mp about 216-217°.

Caution: May be habit forming. This is a controlled substance (opiate) listed in the U.S. Code of Federal Regulations, Title 21 Part 1308.11 (1985).

6603. Norbolethone. *3-Ethyl-17-hydroxy-18,19-dinor-pregn-4-en-3-one; dl-13β,17α-diethyl-17β-hydroxygon-4-en-3-one;* Wy-3475; Genabol. $C_{21}H_{32}O_2$; mol wt 316.47. C 79.70%, H 10.19%, O 10.11%. Prepn of *dl-, d-,* and *l*-forms: Smith *et al., J. Chem. Soc.* **1964**, 4472. Biological activities of *dl-, d-,* and *l*-forms: Edgren *et al., Steroids* **2**, 731 (1963).

Crystals from alcohol, mp 144-145°. uv max: 241 nm (ε 16,500). LD_{50} orally in mice: > 5010 mg/kg, E. I. Goldenthal, *Toxicol. Appl. Pharmacol.* **18**, 185 (1971).

d-Form, crystals from acetone + hexane, mp 175-176°. $[\alpha]_D$ +20.7° (in chloroform).

l-Form, crystals from acetone + hexane, mp 172-175.5°. $[\alpha]_D$ −18.1° (in chloroform).

THERAP CAT: Anabolic.

6604. Norbormide. *3a,4,7,7a-Tetrahydro-5-(hydroxy-phenyl-2-pyridinylmethyl)-8-(phenyl-2-pyridinylmethylene)-4,7-methano-1H-isoindole-1,3(2H)-dione; 5-(α-hydroxy-α-2-pyridylbenzyl)-7-(α-2-pyridylbenzylidene)-5-norbornene-2,3-dicarboximide;* McN 1025; Raticate; Shoxin. $C_{33}H_{25}N_3O_3$; mol wt 511.55. C 77.48%, H 4.93%, N 8.22%, O 9.38%. Prepd by condensation of 2-benzoylpyridine and cyclopentadiene followed by reaction of the obtained fulvene with maleimide: Roszkowski *et al., Science* **144**, 412 (1964). Stereoisomerism: Mohrbacher *et al., J. Org. Chem.* **31**, 2141 (1966); Abrahamsson, Nilsson, *ibid.* 3631. Structure-activity studies: Poos *et al., J. Med. Chem.* **9**, 537 (1966). Toxicity data: Roszkowski, *J. Pharmacol. Exp. Ther.* **149**, 288 (1965).

Crystals from methylene chloride + ether, mp 190-198°. uv max (methanol): 250 nm (ε 17500). LD_{50} in rats (mg/kg): 5.3 orally; 0.65 i.v. (Roszkowski). Practically insol in water unless pH is below 4.

USE: Rodenticide.

6605. Norcarane. *Bicyclo[4.1.0]heptane;* 1,2-methylene-cyclohexane. C_7H_{12}; mol wt 96.17. C 87.42%, H 12.58%. Prepd by distilling the barium salt of norcaranecarboxylic acid with ZnO + BaO: Ebel *et al.,* *Helv. Chim. Acta* **12**, 19 (1929); by the action of methylene iodide and zinc-copper couple on cyclohexene: Smith, Simmons, *Org. Syn.* **41**, 72 (1961); Simmons, U.S. pat. **3,074,984** (1963 to du Pont); from cyclohexene + diazomethane + zinc iodide: Applequist, Badad, *J. Org. Chem.* **27**, 288 (1962); from methylene chloride + cyclohexyl lithium: Closs, *J. Am. Chem. Soc.* **84**, 809 (1962).

Liquid. bp 116-117°. n_D^{25} 1.4546.

6606. Norcholanic Acid. *24-Norcholan-23-oic acid.* $C_{23}H_{38}O_2$; mol wt 346.53. C 79.71%, H 11.05%, O 9.23%. Prepn from ethyl cholanate: Wieland *et al.,* *Z. Physiol. Chem.* **161**, 80 (1926); from 3,7-diketonorcholanic acid: Windaus, van Schoor, *ibid.* **173**, 312 (1928); from 12-keto-norcholanic acid: Cook, Haslewood, *J. Chem. Soc.* **1934**, 428; from 3α,23-dihydroxycholanic acid: Yanuka *et al.,* *Tetrahedron Letters* **1968**, 1725.

Needles from acetic acid, mp 177°.
Methyl ester, $C_{24}H_{40}O_2$, needles from methanol, mp 74°. Ethyl ester, $C_{25}H_{42}O_2$, prisms from alcohol, mp 67°.

6607. Norcodeine. *7,8-Didehydro-4,5-epoxy-3-methoxy-morphinan-6-ol;* *N*-desmethylcodeine; normorphine 3-methyl ether. $C_{17}H_{19}NO_3$; mol wt 285.33. C 71.56%, H 6.71%, N 4.91%, O 16.82%. Prepn from acetylcodeine: von Braun, *Ber.* **47**, 2312 (1914); **Ger. pat. 286,743** (1914); *Chem. Zentr.* **1915**, II, 862; *Frdl.* **12**, 741; *Houben* **4**, 588; *cf.* **Ger. pat. 289,273** (1914); from codeine *N*-oxide: Diels, Fischer, *Ber.* **49**, 1723 (1916); from codeine: Speyer, Walther, *Ber.* **63**, 822 (1930).

Plates or needles from acetone or ethyl acetate, mp 185°. Sparingly sol in water, ether. Moderately sol in acetone. Freely sol in hot methanol, ethanol.
Hydrochloride trihydrate, $C_{17}H_{19}NO_3 \cdot HCl \cdot 3H_2O$, needles from water. When anhydr, dec 309°. Sparingly sol in cold water, more sol in hot water. Freely sol in methanol, ethanol. Almost insol in acetone.
Sulfate, $2C_{17}H_{19}NO_3 \cdot H_2SO_4$, crystals, freely sol in water. Acetate (salt), $C_{17}H_{19}NO_3 \cdot C_2H_4O_2$, crystals, freely sol in water.

6608. Nordazepam. *7-Chloro-1,3-dihydro-5-phenyl-2H-1,4-benzodiazepin-2-one;* desmethyldiazepam; nordiazepam; DMDZ; A-101; Ro 5-2180; Calmday; Madar; Nordaz; Praxadium; Stilny. $C_{15}H_{11}ClN_2O$; mol wt 270.72. C 66.55%, H 4.10%, Cl 13.09%, N 10.35%, O 5.91%. Des-

methyl analog and principal metabolite of diazepam, *q.v.* Prepn: L. H. Sternbach, E. Reeder, *J. Org. Chem.* **26**, 4936 (1961); E. Reeder *et al.,* **Ger. pat. 1,136,709;** E. Reeder, L. Sternbach, U.S. pat. **3,051,701** (both 1962 to Hoffmann-La Roche); A. Stempel, **Belg. pat. 620,020;** *idem,* U.S. pat. **3,202,699** (1963, 1965, both to Hoffmann-La Roche); S. C. Bell *et al.,* *J. Org. Chem.* **27**, 562 (1962); A. Stempel, G. W. Landgraf, *ibid.* 4675. Pharmacology: U. Traversa *et al.,* *J. Pharm. Pharmacol.* **29**, 504 (1977); M. Babbini *et al.,* *Pharmacology* **17**, 121 (178). Metabolism: U. Klotz *et al.,* *Brit. J. Clin. Pharmacol.* **7**, 119 (1979). Pharmacokinetics: M. Konishi, *J. Pharm. Sci.* **67**, 1777 (1978). Clinical study: V. Andreoli *et al.,* *Arzneimittel-Forsch.* **27**, 436 (1977). Toxicity study: L. O. Randall *et al.,* *Curr. Ther. Res.* **7**, 590 (1965). Teratogenicity study: R. P. Miller *et al.,* *Toxicol. Appl. Pharmacol.* **25**, 453 (1973).

White or pale yellow crystalline powder from acetone, mp 216-217°. uv max (chloroform): 313 nm ($E_{1cm}^{1\%}$ 82). Slightly sol in alc, chloroform. Practically insol in water. LD_{50} in mice (mg/kg): 2750 orally, >400 i.p. (Randall). Also reported as LD_{50} in mice, rats (mg/kg): 1300, >5200 orally (company communication).
Note: This is a controlled substance (depressant) listed in te U.S. Code of Federal Regulations, Title 21 Part 1308.14 (1987).
THERAP CAT: Anxiolytic.

6609. Nordefrin Hydrochloride. *4-(2-Amino-1-hydroxypropyl)-1,2-benzenediol hydrochloride;* α-(1-aminoethyl)-3,4-dihydroxybenzyl alcohol hydrochloride; 3,4-dihydroxynorephedrine hydrochloride; 3,4-dihydroxyphenylpropanolamine hydrochloride; 3,4-dihydroxyphenylaminopropanol hydrochloride; α-(α-aminoethyl)protocatechuyl alcohol hydrochloride; α-methylnoradrenaline hydrochloride; α-methylnorepinephrine hydrochloride; norhomoepinephrine hydrochloride; isoadrenaline hydrochloride; aminopropanolpyrocatechol hydrochloride; homoarterenol hydrochloride; Lirotil; Corbasil; Cobefrin Hydrochloride. $C_9H_{14}ClNO_3$; mol wt 219.67. C 49.21%, H 6.42%, Cl 16.14%, N 6.38%, O 21.85%. Prepn: Hartung *et al.,* *J. Am. Chem. Soc.* **53**, 4149 (1931); Bockmühl *et al.,* U.S. pat. **1,948,162** (1934 to Winthrop); Bruckner *et al.,* *Ber.* **76B**, 466 (1943). Configuration: Fodor *et al.,* *Monatsh.* **83**, 1146 (1952). The optical *l*- and *d*-isomers have been prepd. Discussion of their usefulness: Luduena *et al.,* *J. Dental Res.* **37**, 206 (1958).

Crystals, dec 178-179°. One gram dissolves in about 1.5 ml water, 15 ml alcohol; practically insol in ether. Solns are neutral and are easily destroyed by traces of alkali.
l-Isomer, *corbadrine, Levonordefrin.* Prepn: Bockmühl, Gorr, **Ger. pat. 639,126** (1936 to I. G. Farbenind.). Crystals, dec 212-215° (free base). $[\alpha]_D^{25}$ −31.0° (c = 0.5 in 0.01N HCl). uv max (dil HCl): 278 nm. Practically insol in water (free base). Freely sol in aq solns of mineral acids. Very slightly sol in acetone, chloroform, ethanol, ether.
THERAP CAT: Vasoconstrictor.

6610. Nordihydroguaiaretic Acid. *4,4'-(2,3-Dimethyl-1,4-butanediyl)bis[1,2-benzenediol];* 4,4'-(2,3-dimethyltetramethylene)dipyrocatechol; 2,3-bis(3,4-dihydroxybenzyl)butane; β,γ-dimethyl-α,δ-bis(3,4-dihydroxyphenyl)butane;

NDGA. $C_{18}H_{22}O_4$; mol wt 302.36. C 71.50%, H 7.33%, O 21.17%. Occurs in the resinous exudates of many plants. Isoln from the perennial evergreen shrub *Larrea divaricata* (*Covillea tridentata*): Waller, Gisvold, *J. Am. Pharm. Assoc.* **34**, 78 (1945); Gisvold, *ibid.* **37**, 194 (1948). Prepn by hydrogenation and subsequent demethylation of the guaiaretic acid dimethyl ether, a constituent of gum guaiac: Schroeter *et al., Ber.* **51**, 1587 (1918); Haworth, *J. Chem. Soc.* **1934**, 1423. Synthesis: Lieberman *et al., J. Am. Chem. Soc.* **69**, 1540 (1947); Pearl, U.S. pat. **2,644,822** (1953 to Sulphite Products). *Review:* Herrmann, *Pharmazie* **12**, 147 (1957).

Crystals from dil acetic acid, mp 184-185°. Sol in ethanol, methanol, ether, acetone, glycerol, propylene glycol; slightly sol in hot water, chloroform; practically insol in petr ether, benzene, toluene; sol in dil alkalies, developing a deep red color; insol in dil HCl; sol in concd H_2SO_4. Soly in cottonseed oil at 30° = 7.1 mg/g; in lard at 45° = 5.2 mg/g. Lard contg 0.01% NDGA stored at room temp for 19 months in diffused daylight showed no appreciable rancidity or color change. *See* Lundberg *et al., Oil Soap (Chicago)* **21**, 33 (1944); Mattil *et al., ibid.* 160.

USE: As antioxidant in fats and oils.

6611. Norea. *N,N-Dimethyl-N'-(octahydro-4,7-methano-1H-inden-5-yl)urea; 3-(hexahydro-4,7-methanoindan-5-yl)-1,1-dimethylurea;* noruron; Hercules 7531; Herban. $C_{13}H_{22}N_2O$; mol wt 222.32. C 70.23%, H 9.97%, N 12.60%, O 7.20%. Prepn: Diveley, Pombo, U.S. pat. **3,150,179** (1964 to Hercules). Microbial degradn: Murray *et al., Weed Sci.* **17**, 52 (1969).

(CH₃)₂NCONH

Crystals from aq ethanol, mp 176-178°. Sol in acetone, cyclohexanone; slightly sol in benzene, toluene. Practically insol in water, hexane. LD_{50} orally in rats, mice: 2000, 4600 mg/kg, *RTECS* Vol. II, R. J. Lewis, R. L. Tatken, Eds. (1979) p 682.

USE: Herbicide.

6612. Norepinephrine. *4-(2-Amino-1-hydroxyethyl)-1,2-benzenediol; α-(aminomethyl)-3,4-dihydroxybenzyl alcohol; 2-amino-1-(3,4-dihydroxyphenyl)ethanol; 1-(3,4-dihydroxyphenyl)-2-aminoethanol;* noradrenaline. $C_8H_{11}NO_3$; mol wt 169.18. C 56.79%, H 6.55%, N 8.28%, O 28.37%. The *l*-form, the demethylated precursor of epinephrine, occurs in animals and man, and is a sympathomimetic hormone of both adrenal origin and adrenergic orthosympathetic postganglionic origin in man. Physiologic review: Malmejac, *Physiol. Rev.* **44**, 186 (1964). The *l*-form has also been found in plants, e.g., *Portulaca olerocea* L., *Portulacaceae:* Fing *et al., Nature* **191**, 1108 (1961). Synthesis of *dl*-form: Payne, *Ind. Chemist* **37**, 523 (1961). Historic review of synthesis: Loewe, *Arzneimittel-Forsch.* **4**, 583 (1954). Resolution of *dl*-form: Tullar, *J. Am. Chem. Soc.* **70**, 2067 (1948); idem, U.S. pat. **2,774,789** (1956 to Sterling Drug). Configuration: Pratesi *et al., J. Chem. Soc.* **1959**, 4062. Comprehensive description: C. F. Schwender in *Analytical Profiles of Drug Substances* vol. 1, K. Florey, Ed. (Academic Press, New York, 1972) pp 149-173; T. D. Wilson, *ibid.* vol. 11 (1982) pp 555-586.

HOCHCH₂NH₂

dl-Form, crystals, dec 191°. Sparingly sol in water; very slightly sol in alc, ether; readily sol in dilute acids, caustic.

l-Form, **levarterenol, Adrenor, Levophed.** Microcrystals, dec 216.5-218°. $[\alpha]_D^{25}$ −37.3° (c = 5 in water with 1 equiv HCl).

l-Form hydrochloride, $C_8H_{12}ClNO_3$, **Arterenol.** Crystals, mp 145.2-146.4°. $[\alpha]_D^{25}$ −40° (c = 6). Freely sol in water. Solns slowly oxidize under the influence of light and oxygen in a manner comparable to epinephrine hydrochloride.

l-Form *d*-bitartrate, $C_{12}H_{17}NO_9$, **levarterenol bitartrate, Aktamin, Binodrenal.** Obtained as the monohydrate, crystals, mp 102-104°. $[\alpha]_D^{25}$ −10.7° (c = 1.6 in H_2O). When anhydr, mp 158-159° (some decompn). Freely sol in water.

THERAP CAT: *l*-Form as adrenergic (vasopressor); antihypotensive.

THERAP CAT (VET): *l*-Form as a sympathomimetic; vasopressor in shock.

6613. Norethandrolone. *17-Hydroxy-19-norpregn-4-en-3-one; 17α-ethyl-19-nortestosterone; 17α-ethyl-17-hydroxy-4-norandrosten-3-one; 17α-ethyl-17-hydroxy-19-norandrost-4-en-3-one;* Nilevar; Solevar. $C_{20}H_{30}O_2$; mol wt 302.44. C 79.42%, H 10.00%, O 10.58%. Prepn by catalytic hydrogenation of 17α-ethynyl-19-nortestosterone: Colton, U.S. pat. **2,721,871** (1955 to Searle); *J. Am. Chem. Soc.* **79**, 1123 (1957).

Crystals from methanol, mp 140-141°. uv max: 240 nm (ε 16,500). Insol in water; sol in alcohol, benzene, ether, ethyl acetate.

THERAP CAT: Androgen.

6614. Norethindrone. *(17α)-17-Hydroxy-19-norpregn-4-en-20-yn-3-one; 19-nor-17α-ethynyltestosterone; 17α-ethynyl-19-nortestosterone; 17α-ethynyl-17β-hydroxy-4-androsten-3-one; 19-nor-17α-ethynylandrosten-17β-ol-3-one; anhydrohydroxynorprogesterone; 19-norethisterone; norpregneninolone;* "mini-pill"; Conceplan; Conludag; Micronor; Micronett; Mini-Pe; Norcolut; Noriday; Micronovum; Milligynon; Nor-QD; Norluten; Ostro-Primolut; Primolut N; Norluton; Norlutin; Noralutin; Utovlan. $C_{20}H_{26}O_2$; mol wt 298.41. C 80.49%, H 8.78%, O 10.72%. Prepn from 19-nor-4-androstene-3,17-dione: Djerassi *et al., J. Am. Chem. Soc.* **76**, 4092 (1954); U.S. pat. **2,744,122** (1956 to Syntex); De Ruggieri, U.S. pat. **2,849,462** (1958). Prepn of the acetate: O. Engelfried *et al.,* U.S. pat. **2,964,537** (1960 to Schering AG). Pharmacokinetics: H. Singh *et al., Am. J. Obstet. Gynecol.* **135**, 409 (1979); M. Humpel, *Contraception* **26**, 83 (1982). Double-blind, comparative clinical trial: S. Koetsawang *et al., ibid.* **25**, 231 (1982); A. Sheth *et al., ibid.* 243. Clinical trial of enanthate as injectable contraceptive: S. K. Banerjee *et al., ibid.* **30**, 561 (1984). Multicenter trial of combination with ethinyl estradiol: M. Toews *et al., Curr. Ther. Res.* **41**, 509 (1987). Review of carcinogenicity studies: *IARC Monographs* **21**, 441-460 (1979). Comprehensive description: A. P. Schroff, E. S. Moyer in *Analytical Profiles of Drug Substances* vol. 4, K. Florey, Ed. (Academic Press, New York, 1975) pp 268-293.

Crystals from ethyl acetate, mp 203-204°. $[\alpha]_D^{20}$ −31.7° (chloroform). First reported as $[\alpha]_D^{20}$ −25° (chloroform). uv max (ethanol): 240 nm (log ε 4.24). Mixture with ethinyl estradiol, *q.v.*, *Binovum, Brevicon, Brevinor, Modicon, Neocon 1/35, Norimin, Norinyl 1+35, Norquentiel, Ortho-Novum 1/35, Ortho-Novum 7/7/7, Ovcon, Ovysmen, Synphase, Tri-Norinyl, Trinovum.* Mixture with mestranol, *q.v.*, *Norinyl-1, Ortho-Novin 1/50, Ortho-Novum 1/50.*

Acetate, $C_{22}H_{28}O_3$, *Aygestin, Norlutate, Norlutin-A, Primolut-Nor.* Crystals from methylene chloride + hexane, mp 161-162°. uv max: 240 nm (ε 18690). Mixture with ethinyl estradiol, *Etalontin, Primosiston, Anovlar 21, Gynovlar, Loestrin, Minovlar, Norlestrin.*

Enanthate, $C_{27}H_{38}O_3$, *Noristerat.*

Note: This substance may reasonably be anticipated to be a carcinogen: *Fourth Annual Report on Carcinogens* (NTP 85-002, 1985) p 162.

THERAP CAT: Progestogen. Norethindrone and acetate in combination with estrogen as oral contraceptive. Enanthate as injectable contraceptive.

6615. Norethynodrel. *17-Hydroxy-19-norpregn-5(10)-en-20-yn-3-one;* 17α-ethynyl-17-hydroxy-5(10)-estren-3-one; 13-methyl-17-ethynyl-17-hydroxy-1,2,3,4,6,7,8,9,11,-12,13,14,16,17-tetradecahydro-15H-cyclopenta[a]phenanthren-3-one; Enidrel. $C_{20}H_{26}O_2$; mol wt 298.41. C 80.49%, H 8.78%, O 10.72%. Prepn: Colton, U.S. pats. **2,691,028, 2,725,389** (1954 and 1955 to Searle). Metabolism studies in mice: Freudenthal *et al., Toxicol. Appl. Pharmacol.* **24,** 125 (1973).

Crystals from aq methanol, mp 169-170°. [α]$_D$ +108° (1% chloroform). Mixture with mestranol, *q.v.*, *Conovid E, Elan, Enavid, Enovid.*

THERAP CAT: Progestogen. In combination with estrogen as oral contraceptive.

6616. Norfenefrine. *α-(Aminomethyl)-3-hydroxybenzene methanol;* α-(aminomethyl)-m-hydroxybenzyl alcohol; 1-(m-hydroxyphenyl)-2-aminoethanol; m-hydroxyphenylethanolamine; norphenylephrine. $C_8H_{11}NO_2$; mol wt 153.18. C 62.72%, H 7.24%, N 9.14%, O 20.89%. Chem synthesis: Sachs, **Fr.** pat. **866,569** (1941), *C.A.* **43,** 5043c (1949); Legerlotz, U.S. pat. **2,312,916** (1943 to Ciba); Credner, Neugebauer, *Arzneimittel-Forsch.* **3,** 462 (1953); Bretschneider, Hörmann, *Monatsh.* **84,** 1021 (1953). Synthesis and resolution: D'Amico *et al., Chim. Ind. (Milan)* **38,** 93 (1956). Biosynthesis: Hartman *et al., J. Am. Chem. Soc.* **77,** 816 (1955). Activity studies: Gersmeyer, Albrecht, *Med. Welt* **1966,** 657. Metabolism: Barac, *C. R. Soc. Biol.* **155,** 1598 (1961).

dl-Form hydrochloride, $C_8H_{12}ClNO_2$, *Coritat, Depot-Novadral, Energona, Esbuphon, Molycor-R, Novadral, Stagural, Tonolift, Vingsal, Zondel.* Crystals, mp 159-160°. uv max: 274 nm (E$_{1cm}^{1\%}$ 91.21). Freely soluble in water.

THERAP CAT: Adrenergic.

6617. Norfloxacin. *1-Ethyl-6-fluoro-1,4-dihydro-4-oxo-7-(1-piperazinyl)-3-quinolinecarboxylic acid;* AM-715; MK-366; Baccidal; Barazan; Esclebin; Floxacin; Fulgram; Gonorcin; Lexinor; Nolicin; Noracin; Noraxin; Noroxin(e); Norxacin; Sebercim; Uroxacin; Zoroxin. $C_{16}H_{18}FN_3O_3$; mol wt 319.34. C 60.18%, H 5.68%, F 5.95%, N 13.16%, O 15.03%. Fluorinated quinolone antibacterial. Prepn: T. Irikura, **Belg.** pat. **863,429**; *eidem,* U.S. pat. **4,146,719** (1978, 1979 both to Kyorin); M. Pesson, **Ger.** pat. **2,840,910**; ei-

dem, U.S. pat. **4,292,317** (1979, 1981 to Roger Bellon/Dainippon); H. Koga *et al., J. Med. Chem.* **23,** 1358 (1980). Comparative antibacterial activity: K. Hirai *et al., Antimicrob. Ag. Chemother.* **19,** 188 (1981); M. Y. Khan *et al., ibid.* 265; H. H. Gadebusch *et al., Infection* **10,** 41 (1982). Series of articles on *in vivo* and *in vitro* activity, pharmacology, pharmacokinetics, metabolism, early clinical studies, toxicology: *Chemotherapy (Tokyo)* **29,** Suppl. 4, 1-1000 (1981). Acute toxicity data: T. Irikura *et al., ibid.* 783. *In vitro* activity vs gentamicin-resistant *P. aeruginosa:* J. Downs *et al., Antimicrob. Ag. Chemother.* **21,** 670 (1982). Treatment of lower urinary tract infection: H. Giamarellou *et al., Eur. J. Clin. Microbiol.* **2,** 266 (1983); of penicillin-resistant gonorrhea: S. R. Crider *et al., N. Engl. J. Med.* **311,** 137 (1984). Review of pharmacology, antibacterial activity, clinical trials: B. Holmes *et al., Drugs* **30,** 482-513 (1985); R. C. Rowen *et al., Pharmacotherapy* **7,** 92-110 (1987). Symposium on clinical experience: *Scand. J. Infect. Dis.* **Suppl. 48,** 1-91 (1986).

Crystals from methylene chloride/methanol, mp 227-228°. Slightly sol in water. Hygroscopic in air, forms a hemihydrate. LD$_{50}$ in mice, rats (mg/kg): >4000 orally (both species); 1500 s.c. (both species); 470, >500 i.m.; 220, 270 i.v. (Irikura).

THERAP CAT: Antibacterial.

6618. Norflurazon. *4-Chloro-5-(methylamino)-2-[3-(trifluoromethyl)phenyl]-3(2H)-pyridazinone;* 4-chloro-5-(methylamino)-2-(α,α,α-trifluoro-m-tolyl)-3(2H)-pyridazinone; 1-(3-trifluoromethylphenyl)-4-methylamino-5-chloropyridazone; SAN 9789; Evital; Solicam; Zorial. $C_{12}H_9ClF_3$-N_3O; mol wt 303.67. C 47.46%, H 2.99%, Cl 11.67%, F 18.77%, N 13.84%, O 5.27%. Selective pre-emergent herbicide which inhibits carotenoid biosynthesis in susceptible species. Prepn: C. Ebner, M. Schuler, **Belg.** pat. **712,832**; *eidem,* U.S. pat. **3,644,355** (1968, 1972 both to Sandoz). Absorption and metabolism: R. H. Strang, R. L. Rogers, *J. Agr. Food Chem.* **22,** 1119 (1974). Mechanism of action: G. Sandmann *et al., Pest. Biochem. Physiol.* **14,** 185 (1980). HPLC determn in edible crops: W. M. Draper, J. C. Street, *J. Agr. Food Chem.* **29,** 724 (1981). Comprehensive description: E. Ummel *et al., Proc. Brit. Weed Contr. Conf.* **13,** 313 (1976).

Crystals from alcohol, mp 183-185°. Soly in water (25°): 28 ppm. LD$_{50}$ orally in rats: 9300 mg/kg (Ummel).

USE: Herbicide.

6619. Norgesterone. *(17α)-17-Hydroxy-19-norpregna-5(10),20-dien-3-one;* 17β-hydroxy-17α-vinylestr-5(10)-en-3-one; 17α-vinyl-5(10)-estren-17β-ol-3-one; norvinodrel; vinylestrenolone. $C_{20}H_{28}O_2$; mol wt 300.44. C 79.96%, H 9.39%, O 10.65%. Prepn: Ruggieri, Ferrari, U.S. pat. **2,983,735** (1961 to Richter); *eidem,* U.S. pat. **3,062,713** (1962). Synthesis of racemic form: Hiscock, Whitehurst, *J. Chem. Soc.* **1965,** 5772. Biological properties: Ruggieri *et al., Steroids* **5,** 73 (1965).

Crystals from diethyl ether-hexane, mp 142-143°. $[\alpha]_D$ +161° (chloroform).

Mixture with ethynylestradiol, **Vestalin**.

THERAP CAT: Progestogen.

6620. Norgestimate. *(17α)-17-(Acetyloxy)-13-ethyl-18,19-dinorpregn-4-en-20-yn-3-one 3-oxime; (+)-13-ethyl-17-hydroxy-18,19-dinor-17α-pregn-4-en-20-yn-3-one oxime acetate (ester); 17α-acetoxy-13-ethyl-17-ethynylgon-4-en-3-one; dexnorgestrel acetime;* D-138; ORF-10131. $C_{23}H_{31}NO_3$; mol wt 369.50. C 74.76%, H 8.46%, N 3.79%, O 12.99%. Acetate oxime of D-norgestrel, *q.v.* Prepn: A. P. Shroff, **Ger. pat.** 2,633,210; *idem,* **U.S. pat.** 4,027,019 (both 1977 to Ortho). HPLC determn: P. A. Lane *et al., J. Pharm. Sci.* **76,** 44 (1987). Pharmacology: D. W. Hahn *et al., Contraception* **16,** 541 (1977); J. Killinger *et al., ibid.* **32,** 311 (1985). Pharmacodynamics: H. S. Weintraub *et al., J. Pharm. Sci.* **67,** 1406 (1978). Metabolism in female monkeys: S. F. Sisenwine *et al., Contraception* **15,** 25 (1977); in women: K. B. Alton *et al., ibid.* **29,** 19 (1984). Clinical trial as a contraceptive in combination with ethinyl estradiol: B. Rubio-Lotvin, R. Gonzales-Ansorena, *Acta Eur. Fertil.* **9,** 1 (1978).

Crystals from methylene chloride, mp 214-218°. $[\alpha]_D^{25}$ +110°. Mixture with ethinyl estradiol, **Cilest, Lutrapulse, Ortrel, TriCilest**.

THERAP CAT: Progestogen. In combination with estrogen as oral contraceptive.

6621. Norgestrel. *13-Ethyl-17-hydroxy-18,19-dinorpregn-4-en-20-yn-3-one; 13β-ethyl-17α-ethynyl-17β-hydroxygon-4-en-3-one; 17α-ethynyl-18-homo-19-nortestosterone;* Neogest; Ovrette. $C_{21}H_{28}O_2$; mol wt 312.44. C 80.73%, H 9.03%, O 10.24%. Prepn of (±)-form: Hughes *et al., Experientia* **19,** 394 (1963); Smith, **Belg. pat.** 623,844 (1963), *C.A.* **61,** 4427c (1964); Smith *et al., J. Chem. Soc.* **1964,** 4472; Hughes, Smith, **Brit. pat.** 1,041,280 (1966 to Herchel Smith). Prepn of (+)- and (−)-forms: Smith *et al., loc. cit.;* of (−)-form: **Neth. pat.** 6,414,702; G. Amiard, G. Nomine, **U.S. pat.** 3,413,314 (1965, 1968 to Roussel-UCLAF). Crystal structure determn of (+)-form: N. J. DeAngelis *et al., Acta Crystallogr.* **31B,** 2040 (1975). Comprehensive description: A. M. Sopirak, L. F. Cullen, in *Analytical Profiles of Drug Substances* **vol. 4,** K. Florey, Ed. (Academic Press, New York, 1975) pp 294-318. The active enantiomer is levorotatory.

Crystals from ethyl acetate or from methanol, mp 203-206° (Hughes *et al.*), 205-207° (Smith *et al.*). uv max (etha-

nol): 242 nm (ε 16900). Mixture with ethinyl estradiol, *q.v.,* **Lo/Ovral, Ovral, Stediril**.

(−)-Form, *levonorgestrel,* D-*norgestrel, dexnorgestrel (obsolete), Microlut, Microval, Norgeston.* Crystals from methanol + chloroform, mp 239-241°. $[\alpha]_D^{25}$ −42.5° (CHCl₃); $[\alpha]_D^{20}$ −26 ± 5° (c = 0.5 in CHCl₃). Mixture with ethinyl estradiol, **Levlen, Nordette, Tetragynon, Tri-Levlen, Trinordiol, Triphasil**.

(+)-Form, crystals, mp 238-242°. $[\alpha]_D^{25}$ +40.7° (CHCl₃).

THERAP CAT: (±)- and (−)-Forms as progestogens. In combination with estrogen as oral contraceptive.

6622. Norgestrienone. *17-Hydroxy-19-norpregna-4,9,-11-trien-20-yn-3-one; 17α-ethynyl-4,9,11-estratrien-17β-ol-3-one; 17α-ethynyl-17β-hydroxy-Δ⁴,⁹,¹¹-gonatriene-17β-ol-3-one;* Ogyline. $C_{20}H_{22}O_2$; mol wt 294.38. C 81.60%, H 7.53%, O 10.87%. Prepn: **Neth. pat. Appl.** 6,401,555, corresp to Nominé *et al.,* **U.S. pat.** 3,257,278 (1964 and 1966, both to Roussel-UCLAF); Nominé *et al., Compt. Rend.* **260,** 4545 (1965); **Fr. pat.** M3060 and **Neth. pat. Appl.** 6,517,141 (1965 and 1966, both to Roussel-UCLAF), *C.A.* **63,** 8449b (1965) and **65,** 15470b (1966).

Pale yellow needles from diisopropyl ether, mp 169°. $[\alpha]_D^{20}$ + 63° (c = 0.5 in alc). uv max: 342, 238 nm (ε 29100, 5920). Sol in alcohols, ether, acetone, benzene, chloroform; practically insol in water, dil aq acids and alkalies.

THERAP CAT: Progestogen.

6623. Norhyoscyamine. *[3(S)-endo]-α-(Hydroxymethyl)-benzeneacetic acid 8-azabicyclo[3.2.1]oct-3-yl ester; 1αH,-5αH-nortropan-3α-ol (−)-tropate;* 1-tropic acid 3α-nortropanyl ester; pseudohyoscyamine; solandrine. $C_{16}H_{21}NO_3$; mol wt 275.34. C 69.79%, H 7.69%, N 5.09%, O 17.43%. Isoln from *Solanaceae* and structure: Carr, Reynolds, *J. Chem. Soc.* **101,** 946 (1912). Synthesis: Fodor *et al., Ber.* **93,** 2681 (1960). Identity with pseudohyoscyamine: Carr, Reynolds, *loc. cit.;* with solandrine: Petrie, *C.A.* **12,** 2345⁷ (1918).

Crystals, mp 140.5°. Strongly basic. $[\alpha]_D$ −23° (alcohol). Slightly sol in water, ether; sol in alcohol, chloroform.

6624. Norleucine. *2-Aminohexanoic acid;* α-aminocaproic acid; glycoleucine; caprine. $C_6H_{13}NO_2$; mol wt 131.17. C 54.94%, H 9.99%, N 10.68%, O 24.39%. $CH_3(CH_2)_3CH(NH_2)COOH$. An amino acid classified as nonessential with respect to its growth effect in rats. Several syntheses; prepn from α-bromo-*n*-caproic acid by action of 25% ammonia at 50-55°: Marvel, du Vigneaud, *Org. Syn.* **4,** 3 (1925).

DL-Form, lustrous leaflets from water. d 1.172. Dec 327°. pK_1 2.39; pK_2 9.76. Soly in water: 11.49 g/l at 25°, 17.27 g/l at 50°, 28.61 g/l at 75°, 52.0 g/l at 100°. Sparingly sol in alcohol: 0.42 g/100 g at 25°. Sol in acids.

L(+)-Form, slightly sweet, shiny leaflets from water. mp 301° (partial decompn). Sublimes partially at 275-280°. $[M]_D$ +32.1° (5N HCl); +47.9° (glacial acetic acid). $[\alpha]_D^{20}$ +21.3° (c = 4.25 in 6N HCl); +6.26° (c = 0.70 in water).

D(−)-Form, bitter, shiny leaflets from water. mp 301° (partial decompn). Sublimes partially at 275-280°. $[\alpha]_D^{20}$ −22.4° (c = 4.69 in 6N HCl); −4.49° (c = 0.96 in water).

6625. Norlevorphanol. *Morphinan-3-ol;* (−)-3-hydroxymorphinan; 1,3,4,9,10,10a-hexahydro-6-hydroxy-2H-10,4a-iminoethanophenanthrene; 1,3,4,9,10,10a-hexahydro-2H-10,4a-iminoethanophenanthren-6-ol; NIH-7539. $C_{16}H_{21}NO$; mol wt 243.34. C 78.97%, H 8.70%, N 5.76%, O 6.58%. Prepn: Schnider, Grüssner, *Helv. Chim. Acta* **34**, 2211 (1951); Hellerbach, *et al., ibid.* **39**, 429 (1956); **Brit. pat.** **765,920** (1957 to Hoffmann-La Roche).

Crystals from acetone + methanol, mp 270-272°. $[\alpha]_D^{21}$ −42° ±2° (c = 1 in methanol).

Hydrochloride, $C_{16}H_{21}NO.HCl$, crystals from water, mp 320°.

Hydrobromide, $C_{16}H_{21}NO.HBr$, crystals from water, mp 222-224°.

Caution: May be habit forming. This is a controlled substance (opiate) listed in the U.S. Code of Federal Regulations, Title 21 Part 1308.11 (1985).

THERAP CAT: Narcotic analgesic.

6626. Norlobelanine. *2,2'-(2,6-Piperidinediyl)bis[1-phenylethanone];* isolobelanine. $C_{21}H_{23}NO_2$; mol wt 321.42. C 78.47%, H 7.21%, N 4.36%, O 9.96%. In herb of *Lobelia inflata* L., *Lobeliaceae.* Isoln from crude lobelanidine nitrate: Wieland *et al., Ann.* **473**, 118 (1929). Synthesis: Schöpf, Lehmann, *ibid.* **518**, 1 (1935). *Cis-* and *trans-*isomers: Ebnöther, *Helv. Chim. Acta* **41**, 386 (1958).

Crystals, mp 120-121°. Practically insol in water. Sol in alcohol, benzene, chloroform, ether, dil acids. On methylation it is converted to lobelanine.

*trans-*Form, prisms from ethyl acetate + ether, mp 88-91°.

*trans-*Form hydrochloride, $C_{21}H_{23}NO_2.HCl$, crystals from alcohol + ether, dec 163-168°.

6627. Normetanephrine. *α-(Aminomethyl)-4-hydroxy-3-methoxybenzenemethanol;* α-(aminomethyl)vanillyl alcohol; 4-hydroxy-3-methoxy-α-(aminomethyl)benzyl alcohol; 1-(4-hydroxy-3-methoxyphenyl)-2-aminoethanol; 3-O-methylarterenol; 3-O-methylnoradrenaline; 3-O-methylnor-epinephrine. $C_9H_{13}NO_3$; mol wt 183.20. C 59.00%, H 7.15%, N 7.65%, O 26.20%. A naturally occurring derivative of epinephrine, found together with metanephrine in urine and in certain tissues. Prepn: Fodor *et al., Acta Chim. Acad. Sci. Hung.* **1**, 395 (1951), *C.A.* **49**, 897 (1955); Axelrod *et al., J. Biol. Chem.* **233**, 697 (1958); Heacock, Hutzinger, *Chem. & Ind. (London)* **1961**, 595.

*dl-*Form, isolated as the hydrochloride, $C_9H_{13}NO_3.HCl$, prisms from abs ethanol, dec 206-207°. uv max (abs ethanol): 232, 282 nm (ε 7100, 2970).

6628. Normethadone. *6-(Dimethylamino)-4,4-diphenyl-3-hexanone;* 1-dimethylamino-3,3-diphenyl-4-hexanone; 1,1-diphenyl-1-(2-dimethylaminoethyl)-2-butanone; isoamidone I; desmethylmethadone; phenyldimazone; Hoechst 10582. $C_{20}H_{25}NO$; mol wt 295.41. C 81.31%, H 8.53%, N

4.74%, O 5.42%. Prepn from 4-bromo-2,2-diphenylbutane-nitrile and dimethylamine: Bockmühl, Ehrhart, *Ann.* **561**, 72 (1948); Easton *et al., J. Am. Chem. Soc.* **74**, 5772 (1952). Toxicity data: Eddy *et al., J. Pharmacol. Exp. Ther.* **98**, 121 (1950).

Oily liquid, bp$_3$ 164-167°. Alkaline reaction.

Hydrochloride, $C_{20}H_{26}ClNO$, *Ticarda.* Crystals from acetone, mp 174-175°. Sol in water, alcohol. pH of 1% aq soln about 5. LD_{50} s.c. in mice: 90 mg/kg (Eddy).

Caution: May be habit forming. This is a controlled substance (opiate) listed in the U.S. Code of Federal Regulations, Title 21 Part 1308.11 (1985).

THERAP CAT: Narcotic analgesic. Antitussive.

6629. Normethandrone. *17β-Hydroxy-17-methylestr-4-en-3-one;* 17α-methyl-19-nortestosterone; methylestrenolone; normethandrolone; normetandrone; methylnortestosterone; Orgasteron; Metalutin; Methalutin. $C_{19}H_{28}O_2$; mol wt 288.41. C 79.12%, H 9.79%, O 11.10%. Prepn: Djerassi *et al.,* U.S. pats. **2,744,122** and **2,774,777** (1956 to Syntex); *J. Am. Chem. Soc.* **76**, 4092 (1956); De Ruggieri, U.S. pat. **2,849,461** (1958).

Crystals from ether-hexane, mp 156-158° (Kofler). $[\alpha]_D$ +33°. uv max (ethanol): 240 nm (log ε 4.23).

THERAP CAT: Androgen.

6630. Normorphine. *7,8-Didehydro-4,5-epoxymorphinan-3,6-diol;* desmethylmorphine. $C_{16}H_{17}NO_3$; mol wt 271.30. C 70.83%, H 6.32%, N 5.16%, O 17.69%. Prepd by hydrolysis of cyanonormorphine: von Braun, *Ber.* **47**, 2312 (1914); Speyer, Walther, *Ber.* **63**, 852 (1930); Weijlard, Erickson, *J. Am. Chem. Soc.* **64**, 869 (1942); Rapoport, Look, U.S. pat. **2,890,221** (1959 to U.S.A.E.C.).

Sesquihydrate, crystals from water, mp 273°; when anhydr, mp 276-277°. Sparingly sol in hot water, alcohol. Insol in ether, chloroform.

Hydrochloride monohydrate, cryst from water, dec 305°.

Caution: May be habit forming. This is a controlled substance (opium derivative) listed in the U.S. Code of Federal Regulations, Title 21 Part 1308.11 (1985).

THERAP CAT: Narcotic analgesic.

6631. Nornicotine. *3-(2-Pyrrolidinyl)pyridine;* 2-(3-pyridyl)pyrrolidine. $C_9H_{12}N_2$; mol wt 148.20. C 72.93%, H 8.16%, N 18.90%. Occurs in ordinary tobacco, in other species of *Nicotiana,* and also in *Duboisia hopwoodii* F. Muell., *Solanaceae.* Review: Markwood, U.S. Dept. Agr. Bur. Entomol. Plant Quarantine E-561 (1942), Supplement by Roark, E-645 (1945). Synthesis: Mizogruchi, *Chem. Pharm. Bull.* **9**, 818 (1961); M. Nakane, C. R. Hutchinson, *J. Org. Chem.* **43**, 3922 (1978).

Hygroscopic, somewhat viscous liquid, develops a slight amine odor, less pungent than that of nicotine. bp 270°; bp_{11} 131°; bp_3 105-107°. Hardly volatile with steam (difference from nicotine which is readily volatile). d_4^{20} 1.0737. $[\alpha]_D^{22}$ −89° (c = 100). n_D^{20} 1.5378. Miscible with water. Very sol in alcohol, chloroform, ether, petr ether, kerosene, oils. Less volatile and less easily oxidized than nicotine. Commercial nicotine prepns sometimes contain several percent nornicotine. *Cf.* Hoskins, Craig, *Ann. Rev. Biochem.* **15**, 539 (1946). LD i.p. in rats: 23.5 mg/kg; LD i.v. in rabbits: 3 mg/kg.

Dipicrate, $C_9H_{12}N_2 \cdot 2C_6H_3N_3O_7$, yellow crystals, mp 192° (evac tube).

USE: Agricultural or horticultural insecticide. *Caution:* Approx one-third as toxic as nicotine. Faintness, prostration, muscular weakness, severe nausea, vomiting, diarrhea, collapse with or without convulsions.

6632. Nornidulin. *2,4,7-Trichloro-3,8-dihydroxy-1,9-dimethyl-6-(1-methyl-1-propenyl)-11H-dibenzo[b,e][1,4]-dioxepin-11-one;* Ustin. $C_{19}H_{15}Cl_3O_5$; mol wt 429.70. C 53.11%, H 3.52%, Cl 24.75%, O 18.62%. Antibiotic substance produced by *Aspergillus ustus:* Kurung, *Science* **102**, 11 (1945); Hogeboom, Craig, *J. Biol. Chem.* **162**, 363 (1946); Doering *et al., J. Am. Chem. Soc.* **68**, 725 (1946). Structure: Dean *et al., J. Chem. Soc.* **1960**, 4829; Bycroft *et al., ibid.* **1963**, 5148.

Crystals from ether + cyclohexane, mp 185-187°. uv max: 266 nm (ε 8120).

6633. Norpipanone. *4,4-Diphenyl-6-(1-piperidinyl)-3-hexanone;* 1-piperidino-3,3-diphenyl-4-hexanone; 6-(1-piperidyl)-4,4-diphenyl-3-hexanone; Hoechst 10495; Hexalgon. $C_{23}H_{29}NO$; mol wt 335.47. C 82.34%, H 8.71%, N 4.18%, O 4.77%. Prepn: Bockmühl, Ehrhart, *Ann.* **561**, 73 (1949); Dupré *et al., J. Chem. Soc.* **1949**, 500.

Hydrochloride, $C_{23}H_{30}ClNO$, *Orfenso.* Crystals, mp 181-182°. Sol in water, alcohol.
Hydrobromide, $C_{23}H_{30}BrNO$, crystals, mp 192-193°. Sol in water, alcohol.
Caution: May be habit forming. This is a controlled substance (opiate) listed in the U.S. Code of Federal Regulations, Title 21 Part 1308.11 (1987).
THERAP CAT: Narcotic analgesic.

6634. Norpseudoephedrine. *(R*,R*)-α-(1-Aminoethyl)-benzenemethanol; threo-2-amino-1-hydroxy-1-phenylpropane; threo-1-phenyl-1-hydroxy-2-aminopropane; nor-ψ-ephedrine; pseudonorephedrine; ψ-norephedrine; cathine; katine.* $C_9H_{13}NO$; mol wt 151.20. C 71.49%, H 8.67%, N 9.26%, O 10.58%. Occurs naturally as the D-*threo*-form in the leaves of the khat plant, *Catha edulis* Forsk., Celastraceae, an evergreen shrub native to southern Arabia and Ethiopia; also found in smaller amounts in the South American tree *Maytenus krukovii* A. C. Smith, Celastraceae and in mother liquors from Ma Huang after recovery of ephedrine. Isoln from *Catha edulis* leaves: Wolfes, *Arch. Pharm.* **268**,

81 (1930); from Ma Huang: Smith, *J. Chem. Soc.* **1927**, 2056; **1928**, 51; **1929**, 2755. Physical measurements: Gibson, Levin, *ibid.* **1929**, 2754. Synthesis: Nagai, Kanao, *Ann.* **470**, 157 (1929); Pfanz, Muller, *Arch. Pharm.* **288**, 11, 65 (1955); Pfanz, Wieduwilt, *ibid.* 563; Sicher, Pankova, *Coll. Czech. Chem. Commun.* **20**, 1419 (1955). Toxicity data for hydrochloride: H. Hofmann *et al., Arzneimittel-Forsch.* **5**, 367 (1955).

D-Form, free base, plates from methanol, mp 77.5-78°. Strongly alkaline reaction. $[\alpha]_{546}^{20}$ +37.9° (c = 3 in methanol). Sol in alcohol, chloroform, ether, dil acids.
Hydrochloride, $C_9H_{14}ClNO$, *Amorphan, Adiposettin, Fasupond, Fugoa, Minusin, Reduform, Exponcit.* Prisms, mp 180-181°. $[\alpha]_D^{20}$ +43.2° (H_2O). Sol in water. pH of aq soln 5.9-6.1. LD_{50} s.c. in mice: 275 mg/kg (Hofmann).
Sulfate, $C_{18}H_{28}N_2O_6S$, hexagonal plates, mp 298°. $[\alpha]_{546}^{20}$ +48.7° (c = 1.4 in H_2O). Sol in water.
dl-Form hydrochloride, $C_9H_{14}ClNO$, crystals, mp 169-171°.
For other synthetic forms *see* Phenylpropanolamine Hydrochloride.
USE: In the optical resolution of externally compensated acids: Gibson, Levin, *loc. cit.*
THERAP CAT: Anorexic.

6635. Nortriptyline. *3-(10,11-Dihydro-5H-dibenzo[a,-d]cyclohepten-5-ylidene)-N-methyl-1-propanamine;* 10,11-dihydro-N-methyl-5H-dibenzo[a,d]cycloheptene-$\Delta^{5,\gamma}$-propylamine; 5-(α-methylaminopropylidene)dibenzo[a,d]-cyclohepta[1,4]diene; 3-(10,11-dihydro-5H-dibenzo[a,d]-cyclohepten-5-ylidene)-N-methylpropylamine; 10,11-dihydro-5-(3-methylaminopropylidene)-5H-dibenzo[a,d]-[1,4]cycloheptene; desitriptilina; desmethylamitriptyline; Avantyl; Aventyl; Noritren; Ateben; Psychostyl; Sensaval. $C_{19}H_{21}N$; mol wt 263.37. C 86.64%, H 8.04%, N 5.32%. Tricyclic antidepressant. Prepn: Hoffsommer *et al., J. Org. Chem.* **27**, 4134 (1962); Neth. pat. **Appl. 6,408,512**; E. I. Engelhardt, U.S. pat. **3,922,305** (1965, 1975 both to Merck & Co.). Alternative process: N. L. Wendler, U.S. pat. **3,442,949** (1969 to Merck & Co). GC determn in plasma: J. E. Burch, *J. Chromatog.* **308**, 165 (1984). Determn in plasma or serum by radioimmunoassay: J. F. Sayegh, *Neurochem. Res.* **11**, 193 (1986). Pharmacokinetics: E. H. Rubin *et al., J. Clin. Psychiatry* **46**, 418 (1985). Clinical trial in post-stroke depression: J. R. Lipsey *et al., Lancet* **1**, 297 (1984). Anticholinergic effect and clinical efficacy in urinary dysfunction: I. Nissenkorn *et al., Eur. Urol.* **12**, 109 (1986). Comprehensive description: J. L. Hale in *Analytical Profiles of Drug Substances* vol. **1**, K. Florey, Ed. (Academic Press, New York, 1972) pp 233-247.

Hydrochloride, $C_{19}H_{22}ClN$, *Acetexa, Allegron, Altilev, Nortrilen, Norzepine, Pamelor, Sensival, Vividyl.* Crystals from ether + ethanol, mp 213-215°. Sol in ethanol, water, chloroform. Practically insol in ether, acetone, benzene. uv max (methanol): 240 nm (ε 13900).
THERAP CAT: Antidepressant.

6636. Norvaline. *2-Aminovaleric acid;* α-aminovaleric acid; 2-aminopentanoic acid. $C_5H_{11}NO_2$; mol wt 117.15. C 51.26%, H 9.47%, N 11.96%, O 27.32%. $CH_3(CH_2)_2CH$-$(NH_2)COOH$. Prepd by treating butyraldehyde ammonia with HCN and HCl: Slimmer, *Ber.* **35**, 404 (1902); from 2-acetylvaleric acid ethyl ester: Hamlin, Hartung, *J. Biol. Chem.* **145**, 349 (1942); from acetamidomalonic acid diethyl ester: Archer, Albertson, U.S. pat. **2,445,817** (1948 to Win-throp-Stearns); from 1-nitrobutane: Stiles, Finkbeiner, *J. Am. Chem. Soc.* **85**, 616 (1963); U.S. pat. **3,055,936** (1962 to Res. Corp.). Prepn of optically active forms: Abderhalden, Kurton, *Fermentf.* **4**, 328; *Chem. Zentr.* **1921**, III, 296.

DL-Form, minute leaflets from alcohol or water, mp 303° (closed capillary). pK_1' 2.36; pK_2' 9.72. Sublimes without decompn. One gram dissolves in 10 ml water at 18°. Freely sol in hot water; practically insol in alcohol, ether, chloroform, ethyl acetate, petr ether.

L(+)-Form, crystals from dil alc. mp about 305° (closed capillary). $[M]_D$ +29.2° (5N HCl); $[M]_D$ +41.0° (glacial acetic acid). $[\alpha]_D^{20}$ +23.0° (c = 10 in 20% HCl). Freely sol in hot water; insol in alcohol, ether, chloroform, ethyl acetate, petr ether.

D(−)-Form, minute leaflets. mp about 307°. $[\alpha]_D^{20}$ −24.2° (c = 10 in 20% HCl). Freely sol in hot water; insol in alcohol, ether, chloroform, ethyl acetate, petr ether.

6637. Norvinisterone. *(17α)-17-Hydroxy-19-norpregna-4,20-dien-3-one;* 17-hydroxy-17α-vinyl-4-estren-3-one; 17-hydroxy-13-methyl-17α-vinyl-1,2,3,6,7,8,9,10,11,12,-13,14,16,17-tetradecahydro-15*H*-cyclopenta[*a*]phenan-thren-3-one; 17α-vinyl-19-nortestosterone; Nor-Proges-telea. $C_{20}H_{28}O_2$; mol wt 300.42. C 79.95%, H 9.39%, O 10.65%. Prepn: Colton, U.S. pats. **2,655,518; 2,802,015** (both 1953 to Searle).

Crystals from ethyl acetate + petr ether, mp 169-171°. $[\alpha]_D$ +36°.

THERAP CAT: Progestogen.

6638. Noscapine. *[S-(R*,S*)]-6,7-Dimethoxy-3-(5,6,7,-8-tetrahydro-4-methoxy-6-methyl-1,3-dioxolo[4,5-g]isoqui-nolin-5-yl)-1(3H)-isobenzofuranone; narcotine; l-α-narco-tine;* l-α-2-methyl-8-methoxy-6,7-methylenedioxy-1-(6,7-dimethoxy-3-phthalidyl)-1,2,3,4-tetrahydroisoquinoline; narcosine; methoxyhydrastine; opian; opianine; NSC 5366; Coscopin; Coscotabs; Capval; Key-tusscapine; Longatin; Lyobex; Narcompren; Narcotussin; Nectadon; Nicolane; Nipaxon; Noscapal; Noscapalin; Terbenol; Tusscapine; Vadebex. $C_{22}H_{23}NO_7$; mol wt 413.43. C 63.91%, H 5.61%, N 3.39%, O 27.09%. An opium alkaloid, isolated from the plant *Papaver somniferum* L. Papaveraceae. Present in amounts up to 11% depending on season and locality. First isoln: Robiquet, *Ann. Chim. Phys.* [2] **5**, 275 (1817). Extractable from the water-insoluble residue remaining from the processing of opium for the manufacture of morphine. Racemization to gnoscopine: Rabe, McMillian, *Ann.* **377**, 233 (1910); and structural studies: Perkin, Robinson, *J. Chem. Soc.* **99**, 775 (1911); Marshall et al., *ibid.* **1934**, 1318. Preliminary stereochemical studies: Ohta et al., *Tetrahedron Letters* **1963**, 1857; Battersby, Spenser, *J. Chem. Soc.* **1965**, 1087. Revised stereochemistry: Blaha et al., *Coll. Czech. Chem. Commun.* **29**, 2328 (1964); Snatzke et al., *Tetrahedron* **25**, 5059 (1969). Synthesis of racemate: Kerekes, Bognar, *J. Prakt. Chem.* **313**, 923 (1971). Biosynthesis: Battersby, Hirst, *Tetrahedron Letters* **1965**, 669. Metabolism: N. Tsunoda, Y. Yoshimura, *Xenobiotica* **9**, 181 (1979); eidem, *ibid.* **11**, 23 (1981). Pharmacokinetics: B. Dhalstroem et al., *Eur. J. Clin. Pharmacol.* **22**, 535 (1982). Clinical evaluation as antitussive: D. W. Empey et al., *ibid.* **16**, 393 (1979). HPLC determn in serum: K. M. Jensen, *J. Chromatog.* **274**, 381 (1983). Comprehensive description: M. A. Al-Yahya, M. M. A. Hassan in *Analytical Profiles of Drug Substances* vol. **11**, K. Florey, Ed. (Academic Press, New York, 1982) pp 407-461.

Orthorhombic bisphenoidal prisms, tablets from diacetone. Triboluminescent. d 1.395. mp 176°. Sublimes at 150-160° under 11 mm pressure at 2 mm distance. Very weak base forming unstable salts with acids and strong bases. pK 7.8. uv max (ethanol): 209, 291, 309-310 nm (log ε 4.86, 3.60, 3.69). Practically insol in vegetable oils. Slightly sol in NH_4OH, hot solns of KOH and NaOH, forming salts. Salts formed with acids are dextrorotatory and unstable in water.

Hydrochloride, $C_{22}H_{24}ClNO_7$, hemihydrate to tetrahydrate, crystals, very sol in water forming basic salts.

Camphorsulfonate, $C_{32}H_{39}NO_{11}S$, *374 JL, Tulisan.* Contains 35.97% camphosulfonic acid. Prepn: Maillard, U.S. pat. **3,108,106** (1963 to Jacques Logeais). Crystals, mp 188-191°. $[\alpha]_D^{33}$ +32.7° (c = 4.56 in water). Freely sol in water. Sol in methanol, ethanol. Slightly sol in ethyl acetate. Practically insol in ether.

dl-Form, *dl-narcotine, gnoscopine.* Long needles from methanol, mp 232° (dec). pK 7.8. Freely sol in carbon disulfide, hot chloroform; sol in about 1500 parts alcohol; sparingly sol in benzene, water.

THERAP CAT: Antitussive.

6639. Nosiheptide. Multhiomycin; RP 9671; Primofax. $C_{51}H_{43}N_{13}O_{12}S_6$; mol wt 1222.35. C 50.11%, H 3.54%, N 14.90%, O 15.71%, S 15.74%. Polythiazole antibiotic produced by *Streptomyces actuosus.* Isoln and characterization: S. Pinnert et al., Fr. pat. **1,392,453;** eidem, U.S. pat. **3,155,-581** (1961, 1964 both to Rhône-Poulenc); F. Benazet et al., *Experientia* **36**, 414 (1980). NMR determination of mol wt and elemental formula: H. Depaire et al., *Tetrahedron Letters* **1977**, 1397, 1401. Structure and configuration: T. Prange et al., *Nature* **265**, 189 (1977); C. Pascard et al., *J. Am. Chem. Soc.* **99**, 6418 (1977). Biosynthetic study: D. P. Houck et al., *ibid.* **109**, 1250 (1987). Identity with multthio-mycin: T. Endo, H. Yonehara, *J. Antibiot.* **31**, 623 (1978). Mode of action: E. Cundliffe, J. Thompson, *J. Gen. Microbiol.* **126**, 185 (1981). Review: F. Benazet et al., *Experientia* **36**, 414-416 (1980).

Yellow needles, mp 310-320° (dec). $[\alpha]_D^{20}$ +38° (c = 1 in pyridine). uv max (water/DMF): 242, 322 nm ($E_{1cm}^{1\%}$ 525,

229). Sol in chloroform, dioxane, pyridine, DMF, DMSO; slightly sol in methanol, ethanol, ethyl acetate, benzene. Insol in water and petr ether.

THERAP CAT (VET): Antibacterial; growth promotant.

6640. Novembichin. *2-Chloro-N,N-bis(2-chloroethyl)-propanamine hydrochloride;* 2-chloropropyldi(2-chloroethyl)amine hydrochloride; Novoembichin; Embichin 7; Embikhin 7. $C_7H_{15}Cl_4N$; mol wt 255.01. C 32.97%, H 5.93%, Cl 55.61%, N 5.49%. Prepn: Ford-Moore *et al., J. Chem. Soc.* **1946,** 819; E. Wilson, M. Tishler, *J. Am. Chem. Soc.* **73,** 3635 (1951).

$$CH_3CHCH_2N(C_2H_4Cl)_2$$

with Cl substituent on CHCH_2

Crystals from chloroform + ethanol, mp 78-79.4°. Free base, colorless oil, bp 114°.

THERAP CAT: Antineoplastic.

6641. Novobiocin. *N-[7-[[3-O-(Aminocarbonyl)-6-deoxy-5-C-methyl-4-O-methyl-β-L-lyxo-hexopyranosyl]oxy]-4-hydroxy-8-methyl-2-oxo-2H-1-benzopyran-3-yl]-4-hydroxy-3-(3-methyl-2-butenyl)benzamide;* crystallinic acid; streptonivicin; PA 93; U 6591; Albamycin [not albomycin]; Biotexin; Cardelmycin; Cathocin; Cathomycin; Inamycin; Spheromycin; Vulcamicina; Vulcamycin; Vulkamycin. $C_{31}H_{36}N_2O_{11}$; mol wt 612.65. C 60.77%, H 5.92%, N 4.58%, O 28.73%. Antibiotic substance produced by *Streptomyces spheroides*: Kaczka *et al., J. Am. Chem. Soc.* **77,** 6404 (1955); Wolf, U.S. pat. **3,000,873** (1961 to Merck & Co.); Stammer, Miller; Miller; Wallick, U.S. pats. **3,049,475; 3,049,476; 3,049,534** (all 1962 to Merck & Co.). By *Streptomyces niveus*: Hoeksema *et al., J. Am. Chem. Soc.* **77,** 6710 (1955); *Antibiot. & Chemother.* **6,** 143 (1956); French, U.S. pat. **3,068,221** (1962 to Upjohn). Structure: Shunk *et al., J. Am. Chem. Soc.* **78,** 1770 (1956); Hoeksema *et al., ibid.* 2019; Walton *et al., ibid.* **82,** 1489 (1960). Conformation: Golding, Richards, *Chem. & Ind. (London)* **1963,** 1081. Revised configuration: O. Achmatowicz *et al., Tetrahedron* **32,** 1051 (1976). Synthesis: Stammer, U.S. pat. **2,925,411** (1960); Walton, Spencer, U.S. pat. **2,966,484** (1960 to Merck & Co.); Vaterlaus *et al., Helv. Chim. Acta* **47,** 390 (1964). Conversion of *isonovobiocin* to novobiocin: Caron *et al.,* U.S. pat. **2,983,723** (1961 to Upjohn). Antiviral activity: Chang, Weinstein, *Antimicrob. Ag. Chemother.* **1970,** 165. Efficacy in canine respiratory infections: B. W. Maxey, *Vet. Med. Small Anim. Clin.* **75,** 89 (1980). Mechanism of action studies: Smith, Davis, *J. Bacteriol.* **93,** 71 (1967); H. T. Wright *et al., Science* **213,** 455 (1981); I. W. Althaus *et al., J. Antibiot.* **41,** 373 (1988). *Review:* Brock in *Antibiotics,* vol. 1, R. Gottlieb, P. Shaw, Eds. (Springer-Verlag, New York, 1967) pp 651-665; M. J. Ryan, *ibid.* **vol. 5**(pt. 1), F. E. Hahn, Ed. (1979) pp 214-234.

Pale yellow orthorhombic crystals from ethanol. *Sensitive to light.* d 1.3448. Dec at 152-156° (a rarer modification dec 174-178°). Acid reaction: pKa_1 4.3; pKa_2 9.1. $[\alpha]_D^{24}$ −63.0° (c = 1 in ethanol). uv max (0.1N NaOH; 0.1N methanolic HCl; pH 7 phosphate buffer): 307; 324; 390 nm ($E_{1cm}^{1\%}$ 600, 390, 350 resp.). Sol in aq soln above pH 7.5. Practically insol in more acidic solns. Sol in acetone, ethyl acetate, amyl acetate, lower alcohols, pyridine. Additional soly data: Weiss *et al., Antibiot. & Chemother.* **7,** 374 (1957).

Monosodium salt, $C_{31}H_{35}N_2NaO_{11}$, Robiocina. Minute crystals, dec 220°. $[\alpha]_D^{24}$ −38° (c = 2.5 in 95% ethanol); $[\alpha]_D^{24}$ −33° (c = 2.5 in water). Freely sol in water. A 100 mg/ml soln has a pH of 7.5 and a half-life of ~30 days at 25° and several months at 4°. Soly data: Weiss *et al., loc. cit.* Properties: Birlova, Traktenberg, *Antibiotiki* **13,** 997 (1968).

THERAP CAT: Antibacterial.
THERAP CAT (VET): Antimicrobial.

6642. Novoldiamine. *N^1,N^1-Diethyl-1,4-pentanediamine;* 1-diethylamino-4-aminopentane; 4-amino-1-diethylaminopentane; 2-amino-5-diethylaminopentane; δ-diethylamino-α-methylbutylamine; δ-diethylaminoisopentylamine. $C_9H_{22}N_2$; mol wt 158.28. C 68.29%, H 14.01%, N 17.70%. Prepd commercially from 2-diethylaminoethanol and ethyl acetoacetate. 2-Chlorotriethylamine (formed by the action of thionyl chloride on the alcohol) is condensed with the sodium derivative of ethyl acetoacetate to yield an intermediate ester, which is hydrolyzed and decarboxylated to *novol ketone* (5-diethylamino-2-pentanone). This is hydrogenated in the presence of ammonia to yield novoldiamine. Several other prepns, i.e., from 1,4-pentanediol and diethylamine: Kyrides, U.S. pat. **2,365,825** (1944 to Monsanto). Purification procedure: Jones, U.S. pat. **2,400,934** (1946 to Lilly).

$$H_2NCH(CH_2)_3N(C_2H_5)_2$$

with CH_3 on CH

Liquid. Amine odor. d_{26}^{20} 0.819. bp_{753} 200-200.5°. n_D^{26} 1.4403. Sol in water, alcohol, ether.

USE: Manuf quinacrine and other antimalarials having the same basic side chain.

6643. Novonal. *2,2-Diethyl-4-pentenamide;* diethylallyl-acetamide; Epinoval. $C_9H_{17}NO$; mol wt 155.23. C 69.63%, H 11.04%, N 9.02%, O 10.31%. Description: Bockmühl, Schaumann, *Deut. Med. Wochenschr.* **54,** 270 (1928).

$$CH_2=CHCH_2CCONH_2$$

with two C_2H_5 groups

White powder, mp 75-76°. Sol in 120 parts water; freely sol in alcohol, ether.

THERAP CAT: Hypnotic.

6644. Noxiptilin. *10,11-Dihydro-5H-dibenzo[a,d]cyclo-hepten-5-one O-[2-(dimethylamino)ethyl]oxime;* 2-[[(10,11-dihydro-5H-dibenz[a,d]cyclohepten-5-ylidene)amino]oxy]-N,N-dimethylethylamine; 5-(2-dimethylaminoethyloximino)-5H-dibenzo[a,d]cyclohepta-1,4-diene; 5-[β-(dimethylamino)ethoxyimino]-10,11-dihydro-5H-dibenzo[a,d]cyclo-heptene; noxiptyline; dibenzoxin; Sipcar. $C_{19}H_{22}N_2O$; mol wt 294.40. C 77.52%, H 7.53%, N 9.51%, O 5.43%. Prepn: Engelhard *et al.,* Ger. pat. **1,198,353** (1965); Schutz, Hoffmeister, Ger. pat. **1,225,169** corresp to Schutz *et al.,* U.S. pat. **3,505,321** (1966, 1970 to Bayer); Wrigley, Leeming, Brit. pat. **1,045,911** (1966 to Pfizer). Series of articles on pharmacology, metabolism, clinical data: *Arzneimittel-Forsch.* **19,** 458-467; 846-878 (1969).

$$NOCH_2CH_2N(CH_3)_2$$

Free base, $bp_{0.05}$ 160-164°.
Hydrochloride, $C_{19}H_{23}ClN_2O$, Bay 1521, Agedal, Nogédal. Crystals from ethanol + ethyl acetate, mp 185-187°. LD_{50} s.c. in mice: 240 mg/kg, F. Hoffmeister *et al., Arzneimittel-Forsch.* **19,** 846 (1969).

THERAP CAT: Antidepressant.

6645. Noxytiolin. *N-(Hydroxymethyl)-N'-methylthiourea;* 1-(hydroxymethyl)-3-methyl-2-thiourea; noxythiolin; Noxyflex-S. $C_3H_8N_2OS$; mol wt 120.18. C 29.98%, H 6.71%, N 23.32%, O 13.31%, S 26.68%. $HOCH_2NHCSNH-CH_3$. Prepn: Aebi, Hafstetter, Brit. pat. **970,414** (1964 to Ed. Geistlich Söhne), *C.A.* **61,** 15981e (1964). Peritoneal adhesion formation in rats: A. T. Raferty, *Brit. J. Surg.* **66,** 654 (1979). Use in colonic healing: R. D. Rosin *et al., ibid.* **65,** 603 (1978).

Consult the cross index before using this section.

Crystals. mp 84-86°. LD$_{50}$ orally in mice: > 3 g/kg.
THERAP CAT: Antiseptic; disinfectant.

6646. NPA Acid. *4-[[[[[(5-Nitro-2-furanyl)methylene]-hydrazino]carbonyl]amino]phenyl]arsonic acid; N-[3-(5-nitrofurfurylidene)carbazoyl]arsanilic acid; 5-nitro-2-furaldehyde 4-(p-arsonophenyl)semicarbazone.* C$_{12}$H$_{11}$AsN$_4$O$_7$; mol wt 398.15. C 36.20%, H 2.79%, As 18.81%, N 14.07%, O 28.13%. Prepn: Ward, U.S. pat. 2,808,414 (1957 to Norwich Pharmacal).

Crystals from dil DMF, dec 215°. Practically insol in water. Sol in DMF.

Sodium salt, C$_{12}$H$_{10}$AsN$_4$NaO$_7$, *NPA sodium.* Precipitated by alcohol. Soluble in water. Insol in ether. LD$_{50}$ orally in mice: > 1.0 g/kg.

THERAP CAT (VET): Coccidiostat.

6647. Nucleic Acids. Macromolecules (polynucleotides) found in living cells which are of fundamental importance in controlling the reproduction, growth and metabolism of living systems. Nucleic acids occur in close association with proteins to form *nucleoproteins.* First isolated in 1868-1869 by F. Miescher from the nuclei of pus cells and called *nuclein.* The components of nucleic acids are purine and pyrimidine bases (primarily adenine, guanine, cytosine, thymine, uridine), a sugar (D-2-deoxyribose or D-ribose) and phosphoric acid. A purine or pyrimidine base in glycosidic linkage with the sugar forms a *nucleoside* (e.g. adenosine, thymidine, *q.v.*). In naturally occurring *N*-nucleosides, the carbon in the 1'-position of the sugar is attached in β-glycosyl linkage to the nitrogen in the 9 position of the purine or in the 1 position of the pyrimidine. (Some *C*-nucleosides are also known). The phosphate ester of a nucleoside is a *nucleotide* (e.g. 5'-guanylic acid, 3'-cytidylic acid, *q.v.*) which is the smallest repeating unit of nucleic acids. (These terms also apply to synthetic derivatives and analogs of naturally occurring nucleosides and nucleotides). Comprehensive reviews of bases, nucleotides and nucleosides: A. M. Michelson, *The Chemistry of Nucleosides and Nucleotides* (Academic Press, New York, 1963); Ts'o, "Monomeric Units of Nucleic Acids—Bases, Nucleosides and Nucleotides" in *Biological Macromolecules* vol. 4 entitled *Fine Structure of Proteins and Nucleic Acids*, G. D. Fasman, S. N. Timasheff, Eds. (Dekker, New York, 1970) pp 49-190; C. A. Dekker, L. Goodman, "Nucleosides" in *The Carbohydrates* vol. 2A, W. Pigman *et al.*, Eds. (Academic Press, New York, 1970) pp 1-68; G. R. Pettit, *Synthetic Nucleotides* vol. 1 (Van Nostrand Reinhold, New York, 1972) 252 pp.

There are two main types of nucleic acids: deoxyribonucleic acid (DNA) and ribonucleic acid (RNA). Both types are present in all types of cells, whether of plant or animal origin. *See* Deoxyribonucleic Acid and Ribonucleic Acid. Monographs: E. Chargaff, J. N. Davidson, *The Nucleic Acids*, 3 vols. (Academic Press, New York, 1955, 1960); V. R. Potter, *Nucleic Acid Outlines* vol. I (Burgess Publishing, Minneapolis, 1960); D. O. Jordan, *The Chemistry of Nucleic Acids* (Butterworth, London, 1960); R. F. Steiner, R. F. Beers, *Polynucleotides* (Elsevier, New York, 1961); E. Chargaff, *Essays on Nucleic Acids* (Elsevier, New York, 1963) 211 pp; E. Harbers *et al.*, *Introduction to Nucleic Acids* (Reinhold, New York, 1968) 403 pp; *Organic Chemistry of Nucleic Acids* (Parts A, B) N. K. Kochetkov, E. I. Budovskii, Eds. (Plenum, New York, Eng. ed., 1971, 1972) 639 pp; J. N. Davidson, *The Biochemistry of the Nucleic Acids* (Academic Press, New York, 7th ed., 1972) 396 pp; *Basic Principles in Nucleic Acid Chemistry* vols. 1, 2, P. O. P. Ts'o, Ed. (Academic Press, New York, 1974) 636, 519 pp; V. A. Bloomfield *et al.*, *Physical Chemistry of Nucleic Acids* (Harper & Row, New York, 1974) 517 pp; *MTP Int. Rev. Sci.: Biochem. Ser. One* vol. 6 entitled *Biochemistry of Nucleic Acids*, K. Burton, Ed. (University Park Press, Baltimore, 1974) 364 pp. Series of volumes on nucleic acids: *Progress in Nucleic Acid Research and Molecular Biology* (Academic Press, New York, 1963-present).

6648. Nucleocidin. *4'-C-Fluoroadenosine 5'-sulfamate; 9-(4-fluoro-5-O-sulfamoylpentofuranosyl)adenine; 4'-fluoro-5'-O-sulfamoyladenosine.* C$_{10}$H$_{13}$FN$_4$O$_6$S; mol wt 364.31. C 32.97%, H 3.59%, F 5.21%, N 23.07%, O 26.35%, S 8.80%. Antitrypanosomal antibiotic produced by *Streptomyces calvus.* The first naturally occurring derivative of a fluoro sugar. Isoln and activity studies: Thomas *et al.*, Hewitt *et al.*, *Antibiot. Ann.* **1956-1957**, 716, 722. Partial structure: Waller *et al.*, *J. Am. Chem. Soc.* **79**, 1011 (1957). Revised formula and structure: Morton *et al.*, *ibid.* **91**, 1535 (1969). Synthesis: Jenkins *et al.*, *ibid.* **93**, 4324 (1971).

Crystalline, weakly alkaline substance. Monohydrate from water, mp > 190° (dec). uv max (methanol): 259 nm (ε 15,000). LD$_{50}$ i.p. in mice: ~0.2 mg/kg (Thomas *et al.*, *loc. cit.*).

Picrate, mp 145-147°.

6649. Nudic Acids. Antibiotic substances produced by the basidiomycete *Tricholoma nudum.* Isoln: Heatley in Florey *et al.*, *Antibiotics* vol. I (Oxford, 1949) p 358; *cf.* Wilkins, Harris, *Ann. Appl. Biol.* **31**, 261 (1944).

Nudic acid A, C$_{14}$H$_{20}$O$_3$, large glistening needles or plates from hexane, mp 123.5°. pK between 3.5 and 4.0. Optically inactive in alc. Slightly sol in hot water; sol in the usual organic solvents; freely sol in sodium hydroxide and sodium carbonate solns.

Note: Nudic acid B *see* Diatretyne II.

6650. Nupharidine. *4-(3-Furanyl)octahydro-1,7-dimethyl-2H-quinolizine 5-oxide.* C$_{15}$H$_{23}$NO$_2$; mol wt 249.34. C 72.25%, H 9.30%, N 5.62%, O 12.83%. From rhizome of *Nuphar luteum* (L.) Sibth. & Sm. (*Nymphaea lutea* L.), Nymphaeaceae: Goris, Crété, *Bull. Sci. Pharmacol.* **17**, 13 (1910). Isoln from *N. japonicum* DC., Nymphaeaceae and structure: Kotake *et al.*, *Ann.* **606**, 148 (1957). Total synthesis: Kotake *et al.*, *Bull. Chem. Soc. Japan* **35**, 698 (1962). Abs configuration: Kawasaki *et al.*, *ibid.* **41**, 1264 (1968); La Londe *et al.*, *J. Am. Chem. Soc.* **93**, 2501 (1971). Determn of crystal structure, abs config and stereochemistry by x-ray diffraction methods: J. Ohrt *et al.*, *J. Cryst. Mol. Struct.* **3**, 3 (1973).

Crystals, mp 221°. [α]$_D$ +15°. Soluble in alcohol, chloroform, ether, acetone, amyl alcohol, dil acids. The base is tasteless, the salts bitter.

6651. Nutgall. Galla; galls; Aleppo-galls; Turkey-galls; Mecca-galls. Excrescence from the young twigs of *Quercus infectoria* Oliv. and other allied species of *Quercus*, Fagaceae. *Habit.* Asia Minor (Levant). *Constit.* 50-60% tannic acid, 2-4% gallic acid, ellagic acids, resin.

Chinese nutgall is the excrescence on the leaf or leafstalk of *Rhus semialata* Murr., Anacardiaceae. Contains about 70% tannin.

Incompat: Alkalies, alkaloids; salts of Cu, Fe, Pb, Zn; $AgNO_3$; opium in soln.

USE: Manuf tannin and ink; dyeing; tanning. *Caution:* Irritating to mucous membranes.

THERAP CAT: Astringent.

THERAP CAT (VET): Has been used as a topical astringent.

6652. Nux Vomica. Quaker buttons; bachelor's buttons; poison nut; dog buttons; vomit nut. Dried, ripe seeds of *Strychnos nux-vomica* L., *Loganiaceae. Habit.* Southern Asia, Northern Australia. *Constit.* 1-1.4% strychnine, about an equal amount of brucine; strychnicine, loganin, caffeotannic (igasuric) acid, proteins. Nux vomica from Saigon contains 1.6-2% strychnine. *Caution:* Extremely poisonous.

THERAP CAT: Formerly as bitter tonic.

THERAP CAT (VET): Has been used as a bitter tonic.

6653. Nybomycin. *8-(Hydroxymethyl)-6,11-dimethyl-2H,4H-oxazolo[5,4,3-ij]pyrido[3,2-g]quinoline-4,10(11H)-dione;* 6,11-dimethyl-8-(hydroxymethyl)pyrido[3,2-g]oxazolo[5,4,3-ij]quinoline-4,10(2H,11H)-dione. $C_{16}H_{14}N_2O_4$; mol wt 298.29. C 64.42%, H 4.73%, N 9.39%, O 21.46%. Antibiotic substance produced by *Streptomycete* A 717 isolated from Missouri soil: Strelitz *et al., Proc. Nat. Acad. Sci. USA* **41**, 620 (1955); Eble *et al., Antibiot. & Chemother.* **8**, 627 (1958); Brock, Sokolski, *ibid.* 631. Structure: Rinehart, Renfroe, *J. Am. Chem. Soc.* **83**, 3729 (1961). Revised structure: Rinehart *et al., ibid.* **92**, 6994 (1970). Total synthesis of *deoxynybomycin:* Forbis, Rinehart, *ibid.* 6995. Total synthesis of nybomycin: *eidem, J. Antibiot.* **24**, 326 (1971); *eidem, J. Am. Chem. Soc.* **95**, 5003 (1973).

Needles from acetic acid, mp 325-330°. Sublimes at 250° (15 mm). Optically inactive. uv max (ethanol): 266, 285 nm. Soluble in concd acids. Very slightly sol in water, alkalies, and common organic solvents. Shows antiphage and antibacterial properties. LD_{50} i.p. in mice: 650 mg/kg, Brock, Sokolski, *loc. cit.* ^{13}C NMR spectrum: A. M. Nadzan, K. L. Rinehart, *J. Am. Chem. Soc.* **99**, 4647 (1977).

Acetate, $C_{18}H_{16}N_2O_5$, crystals from chloroform + ethanol, mp 236-237°.

Succinate, $C_{20}H_{19}N_2O_7$, crystals from dimethylformamide. Practically insol in water.

6654. Nylidrin. *4-Hydroxy-α-[1-[(1-methyl-3-phenylpropyl)amino]ethyl]benzenemethanol; p-hydroxy-α-[1-[(1-methyl-3-phenylpropyl)amino]ethyl]benzyl alcohol; p-hydroxy-N-(1-methyl-3-phenylpropyl)norephedrine;* buphenine; 1-(p-hydroxyphenyl)-2-(1'-methyl-3'-phenylpropylamino)-1-propanol; phenyl-*sec*-butyl norsuprifen. $C_{19}H_{25}NO_2$; mol wt 299.40. C 76.22%, H 8.42%, N 4.68%, O 10.69%. Prepn: **Fr. pat. 968,273** (1950 to Troponwerke Dinklage); **Brit. pats. 669,574-5** (1952); *Chem. & Eng. News* **33**, 2896 (1955); Külz, Schöpf, **U.S. pats. 2,661,372-3** (1953). Pharmacology: T. T. Yen, D. V. Pearson, *Res. Commun. Chem. Pathol. Pharmacol.* **23**, 11 (1979); B. Fichtl, W. Felix, *Eur. J. Pharmacol.* **65**, 333 (1980).

Crystals from methanol, mp 111-112°.

Hydrochloride, $C_{19}H_{26}ClNO_2$, *SKF-1700-A, Arlidin, Bufedon, Buphedrin, Dilatal, Dilatol, Dilatropon, Dilydrin, Opino, Penitardon, Perdilatal, Rudilin, Rydrin, Tocodilydrin, Tocodrin.* Crystals. Sparingly sol in water; slightly sol in alcohol. Practically insol in ether, chloroform, benzene.

THERAP CAT: Vasodilator (peripheral).

6655. Nylon. Polyamide. Generic term used to describe "a manufactured fiber in which fiber-forming substances are any long-chain synthetic polyamide having recurring poly-amide groups (—CONH—) as an integral part of the polymer chain". Formed from various combinations of diacids, diamines, and amino acids. May be formed also by addition polymerization. The linear polyamides have achieved the greatest commercial success. Shorthand nomenclature of nylons involves the use of numbers: a single numeral indicating the number of carbon atoms in a monomer, e.g. nylon 6; two numbers indicating a polymer formed from diamines and dibasic acids, the first numeral indicating the number of carbon atoms separating the nitrogen atoms of the diamine, the second indicating the number of straight-chain carbon atoms in the dibasic acid, e.g. nylon 6,6. First produced by E. I. du Pont de Nemours & Co. according to patents of W. H. Carothers. The name *nylon* was dedicated to public domain on Oct. 27, 1938 at the Herald Tribune Forum where the product itself was announced. *Reviews:* R. W. Moncrieff, *Man-made Fibres* (John Wiley, New York, 1963) pp 335-355; Snider, Richardson, "Polyamide Fibers" in *Encyclopedia of Polymer Science and Technology* vol. 10 (Interscience, New York, 1969) pp 347-460; J. H. Saunders, "Polyamides (Fibers)" in Kirk-Othmer *Encyclopedia of Chemical Technolgy* vol. 18 (Wiley-Interscience, New York, 3rd ed., 1982) pp 372-405. *Book: Nylon Plastics,* M. I. Kohan, Ed. (Wiley-Interscience, New York, 1973).

$$\left[- CO - R - CO - NH - R' - NH -\right]_n$$

Crystalline solids characterized by low specific gravity, high strength, durability, high flexibility, and high tensile strength. Soluble in phenol, cresols (especially *m*-cresol), xylenol, formic acid. Insoluble in alcohols, esters, ketones, hydrocarbons. Hydrolysis and degradation occur at higher temperatures, esp in the melt. Stable to aqueous alkali. Degrades rapidly in aqueous acids. Undergoes photodegradation.

USE: In production of synthetic fibers for various textile and domestic uses. Surgical aid (nonabsorbable suture).

6656. Nylon 6. *Poly[imino(1-oxo-1,6-hexanediyl)]; poly(iminocarbonylpentamethylene);* Caprolan; Enkalon; Grilon; Kapron; Mirlon; Perlon; Phrilon; Amilan. Linear polymer obtained by polymerization of ε-caprolactam, *q.v.*: Schlack, **U.S. pat. 2,241,321** (1941 to I. G. Farbenind.). The importance of this fiber increased with the discovery that caprolactam can be produced by the nitrosation of cyclohexanecarboxylic acid: Muench *et al.,* **U.S. pats. 3,022,291** and **3,108,096** (1962, 1963, both to Snia Viscosa). *Review:* R. W. Moncrieff, *Man-Made Fibres* (John Wiley & Sons, New York, 1963) pp 335-355; H. K. Reimschuessel, *J. Polym. Sci., Macromol. Rev.* **12**, 65-139 (1977).

$$H \left[NH(CH_2)_5CO \right]_n OH \qquad n = \text{approx } 200$$

Softens at 210°. mp 223°. Can withstand a temp of 100° for long periods of time. d_4^{20} 1.14. Moisture regain is about 4%. Swelling is low; if steeped in water and then centrifuged its volume increases by about 13-14%. Immune to microbiological attack. Resistant to most org chemicals, but dissolved by phenol, cresol, and strong acids.

USE: Tire cord; fishing lines; tow ropes; hose manuf; woven fabrics.

6657. Nylon 46. *Poly[imino-1,4-butanediylimino(1,6-dioxo-1,6-hexanediyl)]; poly(tetramethyleneadipamide);* Stanyl. Symmetrical polyamide with higher melting point and greater tensile strength than nylon 66 or nylon 6, *q.v.* Prepn from 1,4-diaminobutane and adipic acid: W. H. Carothers, **U.S. pat. 2,130,948** (1938 to Du Pont). Improved process: E. H. J. P. Bour, J. M. M. Warnier, **Eur. pat. Appl. 77,106** corresp to **U.S. pat. 4,463,166** (1983, 1984 both to Stamicarbon). Account of prepn by melt polymerization, physical properties: R. J. Gaymans *et al., J. Polym. Sci., Polym.*

Chem. Ed. **15**, 537 (1977). Brief account: D. O'Sullivan, *Chem. & Eng. News* **62**, 33 (May 21, 1984).

H $\left[HN - (CH_2)_4 - NH - CO - (CH_2)_4 - CO \right]_n$ OH

Clear polymer, mol wt 22,000-45,000. mp 283-319°. d 1.20 (solution-cast film). Insol in most solvents. Sol in formic acid. Slightly sol in trifluoroacetic acid.

USE: Industrial fiber; fiber-reinforced rubber products.

6658. Nystatin. Fungicidin; Biofanal; Diastatin; Candex; Candio-Hermal; Mycostatin; Moronal; Multilind; Nystan; Nystavescent; O-V Statin. Polyene antifungal antibiotic complex containing 3 biologically active components, A₁, A₂, A₃. Produced by *Streptomyces noursei, S. aureus* and other *Streptomyces* spp: Hazen, Brown, *Science* **112**, 423 (1950); *Proc. Soc. Exp. Biol. Med.* **76**, 93 (1951); Raubitscheck *et al., Antibiot. & Chemother.* **2**, 179 (1952); Cohen, Webb, *Arch. Pediatrics* **69**, 414 (1952); Dutcher *et al., Antibiot. Ann.* **1953-1954**, 191; *eidem, Therapy of Fungus Diseases* (Little, Brown, Boston, 1955) p 168. Review of early literature: Brown, Hazen, *Trans. N.Y. Acad. Sci.*, Ser. II, **19** (1956-1957) pp 447-456. Purification: Vandeputte, U.S. pat. **2,832,719** (1958 to Olin Mathieson); Renella, U.S. pat. **3,517,100** (1970 to Am. Cyanamid). Chemistry and partial structure: Birch *et al., Tetrahedron Letters* **1964**, 1491; of nystatin A₁ and A₂: Shenin *et al., Antibiotiki* **13**, 387 (1968). Structure of the aglycone: Manwaring *et al., ibid.* **1969**, 5319. Complete structure of A₁: Chong, Rickards, *ibid.* **1970**, 5145; Borowski *et al., ibid.* **1971**, 685. Revised structure: R. C. Pandey, K. L. Rinehart, *J. Antibiot.* **29**, 1035 (1976). Stereochemical study of A₁: J. M. Lancelin *et al., Tetrahedron Letters* **29**, 2827 (1988). Structure of A₃: J. Zielinski *et al., J. Antibiot.* **41**, 1289 (1988). Mechanism of

action: R. W. Holz in *Antibiotics* **vol. 5**, pt. 2, F. E. Hahn, Ed. (Springer-Verlag, New York, 1979) pp 313-340. Toxicity study: H. Seneca, *Antibiot. Ann.* **1955-1956**, 697. Comprehensive description: G. W. Michel in *Analytical Profiles of Drug Substances* **vol. 6**, K. Florey, Ed. (Academic Press, New York, 1977) pp 341-421.

nystatin A₁

Light yellow powder. Gradually decomposes above 160° without melting by 250°. $[\alpha]_D^{25}$ −10° (glacial acetic acid); +21° (pyridine); +12° (DMF); −7° (0.1N HCl in methanol). uv max (ethanol): 290, 307, 322 nm. Exhibits strong reducing properties. Solubilities determined by Weiss *et al., Antibiot. & Chemother.* **7**, 374 (1957) in mg/ml at about 28°: water 4.0; methanol 11.2; ethanol 1.2; carbon tetrachloride 1.23; chloroform 0.48; benzene 0.28; ethylene glycol 8.75. Solns and aq suspensions begin to lose activity soon after prepn. Aq suspensions are stable for 10 minutes on heating to 100° at pH 7.0; also stable in moderately alkaline media, but labile at pH 9 and pH 2. Heat, light, and oxygen accelerate decompn. Activity not diminished by blood or serum. LD_{50} i.p. in mice: ~200 mg/kg (Seneca).

Nystatin A₁, $C_{47}H_{75}NO_{17}$.

THERAP CAT: Antifungal.

THERAP CAT (VET): Antimycotic, growth promotant.

O

6659. Obidoxime Chloride. *1,1'-[Oxybis(methylene)]bis-[4-(hydroxyimino)methyl]pyridinium dichloride; 1,1'-(oxy-dimethylene)bis[4-formylpyridinium]dichloride dioxime; N,N-dimethyleneoxidebis(pyridinium-4-aldoxime) dichloride; bis(4-hydroxyiminomethylpyridinium-1-methyl) ether dichloride; bis(isonicotinaldoxime 1-methyl) ether dichloride; BH 6; LüH6; Toksobidin; Toxogonin.* $C_{14}H_{16}Cl_2N_4O_3$; mol wt 359.22. C 46.81%, H 4.49%, Cl 19.74%, N 15.60%, O 13.36%. Prepn from pyridine aldoxime and α,α'-dichlorodimethyl ether: **Brit. pat. 930,040** corresp. to Lüttringhaus *et al., U.S. pat.* **3,137,702** (1963, 1964 to E. Merck); Lüttringhaus, Hagedorn, *Arzneimittel-Forsch.* **14,** 1 (1964). Pharmacology and toxicology: Erdman, Engelhard, *ibid.* 5; Mayer, Michalek, *Biochem. Pharmacol.* **20,** 3029 (1971); Bajgar *et al., Eur. J. Pharmacol.* **19,** 199 (1972). Hydrolysis studies: Christenson, *Acta Pharm. Suecica* **9,** 309 (1972).

HON=CH—⟨N⟩⁺—CH₂OCH₂—⟨N⟩⁺—CH=NOH   2Cl⁻

Occurs in two interchangeable isomeric forms (*syn* and *anti*). Crystals from HCl contg 70% alcohol, dec 225°. Also reported as *syn*, mp 235-236°; *anti*, mp 218-220° [Leitis *et al., C.A.* **71,** 81098d (1969)]. Freely sol in water, stable in 1-10% aq solns. LD_{50} orally in mice: > 2240 mg/kg.

Dibromide, $C_{14}H_{16}Br_2N_4O_3$, dec 202-203°.

THERAP CAT: Cholinesterase reactivator.

6660. Ochratoxins. Toxic metabolites from *Aspergillus ochraceus* Wilh.: Scott, *Mycopathol. Mycol. Appl.* **25,** 213 (1965); *A. sulphureus* and *A. melleus:* Lai *et al., Appl. Microbiol.* **19,** 542 (1970); and *Penicillium viridicatum* Westling: van Walbeek *et al., Can. J. Microbiol.* **15,** 1281 (1969). As these molds occur widely, some toxins have been found as natural contaminants on corn, peanuts, storage grains, cottonseed and other decaying vegetation. Isolns: van Walbeek *et al., ibid.* **14,** 131 (1968); Shotwell *et al., Appl. Microbiol.* **17,** 765 (1969); Nesheim, *J. Assoc. Offic. Anal. Chem.* **52,** 975 (1969); Natori *et al., Chem. Pharm. Bull.* **18,** 2259 (1970). Structure and stereochemistry: van der Merwe *et al., J. Chem. Soc.* **1965,** 7083; *Nature* **205,** 1112 (1965). Synthesis of ochratoxins, A, B and ester deriv: Steyn, Holzapfel, *Tetrahedron* **23,** 4449 (1967); Nesheim, *loc. cit.;* Roberts, Woolven, *J. Chem. Soc. (C)* **1970,** 278. Facile synthesis of A: G. A. Kraus, *J. Org. Chem.* **46,** 201 (1981). Comprehensive reviews: Steyn, "Ochratoxin and other Dihydroisocoumarins" in *Microbial Toxins* vol. VI, A. Ciegler, *et al.,* Eds. (Academic Press, New York, 1971) p 179-205; F. S. Chu, *Crit. Rev. Toxicol.* **2,** 499-524 (1974); P. Krogh in *Natural Toxins,* D. Eaker, P. Wadström, Eds. (Pergamon, New York, 1980) pp 673-680.

Ochratoxin A, $C_{20}H_{18}ClNO_6$, *(R)-N-[(5-chloro-3,4-dihydro-8-hydroxy-3-methyl-1-oxo-1H-2-benzopyran-7-yl)carbonyl]phenylalanine.* The major ochratoxin component. Crystals from xylene, mp 169°. Exhibits green fluorescence. $[\alpha]_D -118°$ (c = 1.1 in CHCl₃). uv max (ethanol): 215, 333 nm (ε 34,000; 2400) (van der Merwe). Also frequently reported as mp 90° from benzene (one mole of benzene of crystallization). uv max: 213, 332 nm (ε 36,800; 6400) (Steyn, Holzapfel). LD_{50} orally in rats: 20-22 mg/kg, Purchase, Theron, *Food Cosmet. Toxicol.* **6,** 479 (1968).

Ochratoxin B, $C_{20}H_{19}NO_6$, *(R)-N-[(3,4-dihydro-8-hydr-oxy-3-methyl-1-oxo-1H-2-benzopyran-7-yl)carboxyl]phenyl-alanine.* The less toxic dechloro deriv of ochratoxin A. Crystals from methanol, mp 221° (van der Merwe); 208-209° (Nesheim). Exhibits blue fluorescence. $[\alpha]_D -35°$ (c = 0.15 in ethanol). uv max: 218, 318 nm (ε 37,200, 6900).

Ochratoxin C. $C_{22}H_{22}ClNO_6$. The equally toxic amorphous ethyl ester of ochratoxin A: Steyn, Holzapfel, *J. S. Afr. Chem. Inst.* **20,** 186 (1967), *C.A.* **69,** 2797p (1968). uv max (ethanol): 333 nm (ε 6500).

6661. Ocimene. $C_{10}H_{16}$; mol wt 136.23. C 88.16%, H 11.84%. From the leaves of *Ocimum basilicum* L., *Labiatae; Baronia dentigeroides* Cheel, *Rutaceae; Litsea zeylanica* C. & T. Nees, *Lauraceae; Homoranthus flavescens* A. Cunn., *Myrtaceae.* From the fruits of *Evodia rutaecarpa* (Juss.) Hook. f. & Thoms., *Rutaceae.* Exists in two modifications: α-form, *3,7-dimethyl-1,3,7-octatriene;* and β-form, *3,7-dimethyl-1,3,6-octatriene.* *Cis* and *trans* refers to the stereochemistry at the double bond between positions 3 and 4. Isoln and structure: van Romburgh, *Proc. Kon. Ned. Acad. Wetensch.* **3,** 454 (1900); Enklaar, *Rec. Trav. Chim.* **26,** 157 (1907); **27,** 422 (1908); **36,** 215 (1917); **45,** 337 (1926). Structure: Sutherland, *J. Am. Chem. Soc.* **74,** 2688 (1952). Separation of isomers: Ohloff *et al., Ann.* **675,** 83 (1964). Synthesis of *trans-α*-form: O. P. Vig *et al., Indian J. Chem.* **7,** 1111 (1969). Stereospecific synthesis: O. P. Vig *et al., ibid.* **15B,** 25 (1977).

trans-β-ocimene

Oil. Pleasant odor. Mixture of isomers. bp₇₀ 100°. d_4^{20} 0.8006. n_D^{20} 1.4862. uv max (methanol): 233 nm (ε 26,200). Practically insol in water. Sol in alcohol, chloroform, ether, glacial acetic acid.

trans-β-Form, d_4^{20} 0.799. n_D^{20} 1.4893. uv max (ethanol): 232 nm (ε 27,600).

cis-β-Form, d_4^{20} 0.799. n_D^{20} 1.4877. uv max (ethanol): 237.5 nm (ε 21,000).

trans-α-Form, d_4^{20} 0.793. n_D^{20} 1.4802. uv max (ethanol): 231 nm (ε 27,300).

cis-α-Form, d_4^{20} 0.794. n_D^{20} 1.4789. uv max (ethanol): 234.5 nm (ε 21,600).

6662. Octabenzone. *[2-Hydroxy-4-(octyloxy)phenyl]phenylmethanone; 2-hydroxy-4-(octyloxy)benzophenone; benzophenone-12; Spectra-Sorb UV 531.* $C_{21}H_{26}O_3$; mol wt 326.42. C 77.27%, H 8.03%, O 14.71%. Prepn: Armitage *et al., U.S. pat.* **3,098,842** (1963 to du Pont).

Crystals, mp 45-46°.

USE: To stabilize polyethylene against deterioration by ultraviolet light.

THERAP CAT: Ultraviolet screen.

6663. Octacaine. *3-(Diethylamino)-N-phenylbutanamide; 3-diethylaminobutyranilide.* $C_{14}H_{22}N_2O$; mol wt 234.33. C 71.75%, H 9.46%, N 11.96%, O 6.83%. Prepn: Hofstetter, Wilder Smith, *Helv. Chim. Acta* **36,** 1698 (1953); Hofstetter, *U.S. pat.* **2,851,393** (1958 to E. Geistlich Söhne).

CH₃CHCH₂CONHC₆H₅
       |
   N(C₂H₅)₂

Crystals from petr ether, mp 46-47°. bp₁ 200°. Easily sol in stoichiometric amount of HCl, in ether, alc, benzene.

Hydrochloride, $C_{14}H_{23}ClN_2O$, *Amplicain.* mp 132-134°. Sol in water.

THERAP CAT: Local anesthetic.

6664. Octacosanol. *1-Octacosanol;* n-octacosanol; octacosyl alcohol. $C_{28}H_{58}O$; mol wt 410.74. C 81.87%, H 14.23%, O 3.90%. $CH_3(CH_2)_{26}CH_2OH$. Constituent of vegetable waxes. Isoln from the wax found on green blades of wheat: Pollard *et al., Biochem. J.* **27**, 1889 (1933); from carnauba wax: Koonce, Brown, *Oil & Soap* **21**, 231 (1944). Synthesis starting with behenic acid: Bleyberg, Ulrich, *Ber.* **64**, 2504 (1931); Francis *et al., Proc. Roy. Soc. (London)* **A 158**, 691 (1937).

Crystals from much acetone, mp 83.4°. Insol in water. Sol in carbon disulfide, other fat solvents, oils.

6665. Octadecyltrimethylammonium Pentachlorophenate. *N,N,N-Trimethyl-1-octadecanaminium salt with pentachlorophenol;* octadecyltrimethylammonium pentachlorophenoxide; Octriphenate. $C_{27}H_{46}Cl_5NO$; mol wt 577.97. C 56.11%, H 8.02%, Cl 30.67%, N 2.42%, O 2.77%. Prepn: Glarum, Hauk, U.S. pat. **2,541,816** (1951 to Rohm & Haas).

USE: Surface-active germicide.

6666. Octafluorocyclobutane. Perfluorocyclobutane; Freon-C318. C_4F_8; mol wt 200.04. C 24.02%, F 75.98%. Prepd by pyrolysis of chlorodifluoromethane: Downing *et al.,* U.S. pat. **2,384,821** (1945 to Kinetic Chem.); U.S. pats. **2,551,573**; **2,615,926** (1951; 1952 to du Pont). Thermodynamic studies: Furukawa *et al., J. Res. Nat. Bur. Stand.* **52**, 11 (1954); Duus, *Ind. Eng. Chem.* **47**, 1445 (1955).

Nonflammable, nontoxic gas. d_{vapor}^{27} 8.2. mp −41.4°. bp −6.04° (also reported bp −5°). Heat of formation at 25° = 352 kcal. Heat of combustion 1359 ± 39 cal/g.

USE: Refrigerant; heat-transfer medium.

6667. 2,2,3,3,4,4,5,5-Octafluoro-1-pentanol. *1H,1H,-5H-*Octafluoropentanol. $C_5H_4F_8O$; mol wt 232.08. C 25.87%, H 1.74%, F 65.49%, O 6.89%. $CHF_2(CF_2)_3CH_2OH$. Prepd by free radical telomerization of tetrafluoroethylene in methanol: Hanford, Joyce, U.S. pat. **2,562,547** and Joyce, U.S. pat. **2,559,625** (both 1951 to du Pont).

Liquid. d_4^{20} 1.6647. bp₇₆₀ 140-141°. n_D^{20} 1.3178. Surface tension at 20° = 27.6 dynes/cm.

USE: To introduce fluoroalkyl groups into an organic molecule: Kramer, Gilbert, U.S. pat. **2,963,526** (1960 to Esso). Proposed intermediate for plastics, surface-active agents, lubricants, elastomers.

6668. Octamethylcyclotetrasiloxane. $C_8H_{24}O_4Si_4$; mol wt 296.64. C 32.40%, H 8.16%, O 21.57%, Si 37.88%. Isolated from the hydrolysis product of dimethyldichlorosilane: Patnode, Wilcock, *J. Am. Chem. Soc.* **68**, 358 (1946).

Oily liquid. mp 17.5°. bp 175°. bp₂₀ 74°. d 0.9558. n_D^{20} 1.3968.

USE: Preparation of methyl silicon oils.

6669. Octamethyltrisiloxane. $C_8H_{24}O_2Si_3$; mol wt 236.54. C 40.62%, H 10.23%, O 13.53%, Si 35.63%. Prepn: Patnode, Wilcock, *J. Am. Chem. Soc.* **68**, 358 (1946).

Liquid. bp 153°. d 0.8200. n_D^{20} 1.3848. mp ∼−80°. Stable. Inert to most chemical reagents and rubber. Maintains about the same viscosity over a wide temp range. Sol in benzene and the lighter hydrocarbons; slightly sol in alcohol and heavy hydrocarbons.

USE: As a basis for silicone oils or fluids designed to withstand extremes of temp; as a foam suppressant in petroleum lubricating oil.

6670. Octamoxin. *(1-Methylheptyl)hydrazine;* 2-hydrazinooctane; octomoxine; Ximaol. $C_8H_{20}N_2$; mol wt 144.26. C 66.60%, H 13.98%, N 19.42%. Monoamine oxidase inhibitor. Prepd by condensation of methyl hexyl ketone and hydrazine hydrate followed by hydrogenation under pressure: Michel-Ber *et al.,* **Brit.** pat. **899,385** (1962 to Soc. Civile Auguil).

Sulfate, $C_8H_{22}N_2O_4S$, *Nimaol.* Crystals, mp 78-80°.

THERAP CAT: Antidepressant.

6671. Octamylamine. *6-Methyl-N-(3-methylbutyl)-2-heptanamine; N-isopentyl-1,5-dimethylhexylamine; N-*isoamyl-1,5-dimethylhexylamine; 2-methyl-6-(3-methylbutylamino)heptane; 2-isoamylamino-6-methylheptane; *N*-(1,5-dimethylhexyl)isopentylamine; Octinum D; Octisamyl; Neo-Octon; Octometine. $C_{13}H_{29}N$; mol wt 199.38. C 78.31%, H 14.66%, N 7.03%. Prepd by the condensation of 2-methyl-6-heptanone with isoamylamine: **Swiss** pat. **258,-452** (1942 to Knoll).

Oily liquid. Weak aromatic odor. bp₇ 100-101°. Strong base.

Hydrochloride, $C_{13}H_{29}N.HCl$, waxy leaflets, mp 121°. Sol in water, alcohol, ether.

THERAP CAT: Anticholinergic, antispasmodic.

6672. Octane. C_8H_{18}; mol wt 114.22. C 84.12%, H 15.89%. $CH_3(CH_2)_6CH_3$. Found in petroleum.

Flammable liquid. mp −56.8°. d_4^{20} 0.7028. bp₇₆₀ 125.6°. n_D^{20} 1.39764. Flash pt, open cup: 72°F (22°C). Insol in water; slightly sol in alcohol; sol in ether; miscible with benzene, petr ether, gasoline. The term "octane rating" is explained under Gasoline.

Caution: Narcotic in high concns.

6673. Octanohydroxamic Acid. *N-Hydroxyoctanamide;* caprylohydroxamic acid; Oct HA; Taselin. $C_8H_{17}NO_2$; mol wt 159.23. C 60.34%, H 10.76%, N 8.80%, O 20.10%. $CH_3-(CH_2)_6CONHOH$. Synthesis and chemical study: Inoue, Yukawa, *J. Agr. Chem. Soc. Japan.* **16**, 504 (1940); Ichim *et al., Igiena (Bucharest)* **9**, 319 (1960), *C.A.* **55**, 16662a (1961). Activity as a urease inhibitor: Kobashi *et al., Biochim. Biophys. Acta* **65**, 380 (1962); **227**, 429 (1971); as an antimicrobial: Hase *et al., Chem. Pharm. Bull.* **19**, 363 (1971).

White plates from benzene, mp 78.5-79°. Water soluble. Practically insol in petr ether.

THERAP CAT (VET): Antimicrobial; growth promotant.

6674. 1-Octanol. Caprylic alcohol. $C_8H_{18}O$; mol wt 130.22. C 73.78%, H 13.93%, O 12.29%. $CH_3(CH_2)_6CH_2$-

OH. Occurs in the form of esters in some essential oils. Prepd from the esterified products of coconut oil, the methyl caprylate being reduced by Na and alcohol.

Colorless liquid; penetrating, aromatic odor. d_4^{20} 0.827. mp $-16°$ to $-17°$. bp 194-195°. n_D^{20} 1.430. Insol in water; miscible with alcohol, chloroform, ether.

USE: Manuf of perfumes and esters.

6675. 2-Octanol. Secondary caprylic alcohol; methyl hexyl carbinol; hexylmethylcarbinol. $C_8H_{18}O$; mol wt 130.22. C 73.78%, H 13.93%, O 12.29%. $CH_3(CH_2)_5CH-(OH)CH_3$. Prepd by heating sodium ricinoleate with caustic soda in a copper vessel and distilling: Adams, Marvel, *Org. Syn.* **1**, 61 (1921); Kenyon, *ibid.* **6**, 68 (1926); Ellis, Reid, *J. Am. Chem. Soc.* **54**, 1678 (1932).

dl-Form, oily, refractive liquid. Aromatic, yet somewhat unpleasant odor, particularly on heating. d_4^{20} 0.8193. mp $-38.6°$. bp$_{760}$ 178.5°; bp$_{60}$ 107.4°; bp$_{20}$ 83.3°; bp$_{10}$ 70.0°; bp$_5$ 57.6°; bp$_{1.0}$ 32.8°. n_D^{20} 1.42025 [Béhal, *Bull. Soc. Chim.* [4] **25**, 482 (1919)]. Flash pt ~140°F (60°C). Soly in water: 0.096 ml/100 ml. Miscible with alc, ether.

d-Form, bp$_{20}$ 86°. $[\alpha]_D^{17}$ +9.9°. d_4^{20} 0.8216.

l-Form, bp$_{20}$ 86°. $[\alpha]_D^{17}$ -9.9°.

USE: In the manuf of perfumes; in disinfectant soaps. To prevent foaming. Solvent for fats and waxes.

6676. Octaverine. *6,7-Dimethoxy-1-(3,4,5-triethoxyphenyl)isoquinoline;* 1-(3,4,5-triethoxyphenyl)-6,7-dimethoxy-isoquinoline; oktaverine. $C_{23}H_{27}NO_5$; mol wt 397.45. C 69.50%, H 6.85%, N 3.52%, O 20.13%. Prepd by cyclodehydration and subsequent dehydrogenation of N-(3,4,5-triethoxybenzoyl)homoveratrylamine: **Fr. pat.** **760,825** (1934 to Asta Chemische Fabrik); *cf.* Goldberg, Shapero, *J. Pharm. Pharmacol.* **6**, 171 (1954).

Free base. Insol in water.

Hydrochloride, $C_{23}H_{27}NO_5\cdot HCl$, crystals, mp 199-200°. Sparingly sol in water (1:500). Aq solns are acid to litmus.

THERAP CAT: Antispasmodic.

6677. Octhilinone. *2-Octyl-3(2H)-isothiazolone;* 2-octyl-4-isothiazolin-3-one; RH-893; Kathon. $C_{11}H_{19}NOS$; mol wt 213.34. C 61.93%, H 8.97%, N 6.57%, O 7.50%, S 15.03%. Prepn: S. N. Lewis *et al.*, **Fr. pat. 1,555,416** corresp to **U.S. pat. 3,761,488** (1969, 1973 to Rohm & Haas); eidem, *J. Heterocycl. Chem.* **8**, 571 (1971).

Liquid, bp$_{0.01}$ 120°. uv max (methanol): 280 nm (log ϵ 3.88).

USE: Fungicide. Biocide in cooling-tower water, paints, cutting oils, cosmetics and shampoo; for leather preservation.

6678. Octodrine. *6-Methyl-2-heptanamine;* 6-methyl-2-heptylamine; 2-methyl-6-aminoheptane; 6-amino-2-methylheptane; 2-amino-6-methylheptane; α,ϵ-dimethylhexylamine; 1,5-dimethylhexylamine; SKF 51; Vaporpac. $C_8H_{19}N$; mol wt 129.24. C 74.34%, H 14.82%, N 10.84%. α-Adrenergic agonist. Prepd from the corresponding saturated ketone: Rohrmann, Shonle, *J. Am. Chem. Soc.* **66**, 1516 (1944). Pharmacology: E. J. Fellows, *J. Pharmacol. Exp. Ther.* **90**, 351 (1947).

dl-Form, viscous liquid, fishy odor, bp 154-156°. n_D^{24} 1.4200.

Hydrochloride, $C_8H_{20}ClN$, crystals, sol in water. LD_{50} in mice, rats (mg/kg): 59, 41.5 i.p. (Fellows).

Sulfate, $2C_8H_{19}N\cdot H_2SO_4$, crystals, sol in water.

THERAP CAT: Decongestant.

6679. Octopamine. α-*(Aminomethyl)-4-hydroxybenzenemethanol;* α-(aminomethyl)-p-hydroxybenzyl alcohol; 1-(p-hydroxyphenyl)-2-aminoethanol; norsympatol; norsynephrine; p-hydroxyphenylethanolamine; WV 569. $C_8H_{11}NO_2$; mol wt 153.18. C 62.72%, H 7.24%, N 9.14%, O 20.89%. A biogenic amine that is the phenol analog of noradrenaline (norepinephrine, *q.v.*). It is a neurosecretory product found in several vertebrates and invertebrates. Formed by β-hydroxylation of tyramine by the enzyme dopamine β-hydroxylase: Pisano *et al.*, *Biochim. Biophys. Acta* **43**, 566 (1960). Identification: Erspamer, *Nature* **169**, 375 (1952). Found in the salivary glands of *Octopus vulgaris*, *O. macropus*, and of *Eledone moschata*: idem, *Arzneimittel-Forsch.* **2**, 253 (1952); in mammalian nerves: Molinoff, Axelrod, *Science* **164**, 428 (1969); in cockroach nervous system: Nathanson, Greengard, *ibid.* **180**, 308 (1973). Prepd synthetically: Asscher, **U.S. pat. 2,585,988** (1952). The natural D($-$) form is 3 times more potent than the L($+$) form in producing cardiovascular adrenergic responses in anesthetized dogs and cats: Korol, Soffer, *The Pharmacologist* **5**, 247 (1963). Prepn of D- and L-forms: Kappe, Armstrong, *J. Med. Chem.* **7**, 569 (1964). In invertebrate nervous systems octopamine may function as a neurotransmitter: Saavedra *et al.*, *Science* **185**, 364 (1974). Effects on neuromuscular transmission in crustacean muscle: C. A. Breen, H. L. Atwood, *Nature* **303**, 716 (1983).

D($-$)-Form, crystals from hot water which change at about 160° to a compd which melts above 250° (dec). $[\alpha]_D^{25}$ $-56.0°$ (0.1N HCl); $-37.4°$ (H_2O).

DL-Form hydrochloride, $C_8H_{12}ClNO_2$, *Epirenor*, *Norden*, *Norfen*, *Norphen* (ampules). Crystals, dec 170°. Freely sol in water.

THERAP CAT: Adrenergic.

6680. Octotiamine. *6-(Acetylthio)-8-[[2-[[(4-amino-2-methyl-5-pyrimidinyl)methyl]formylamino]-1-(2-hydroxyethyl)-1-propenyl]dithio]octanoic acid methyl ester;* 8-[[2-[N-[(4-amino-2-methyl-5-pyrimidinyl)methyl]formamido]-1-(2-hydroxyethyl)propenyl]dithio]-6-mercaptooctanoic acid methyl ester S(or 6)-acetate; S-(3-acetylthio-7-carbomethoxyheptylthio)thiamine; thiamine 8-(methyl 6-acetyldihydrothioctate) disulfide; Gerostop; Neuvitan; TATD. $C_{23}H_{36}N_4O_5S_3$; mol wt 544.87. C 50.72%, H 6.66%, N 10.28%, O 14.68%, S 17.66%. Prepn: Ohara *et al.*, **U.S. pat. 3,098,856** (1963 to Fujisawa).

Crystals, mp 106-109°. uv max: 234, 277 nm (ϵ 16200, 5820).

Hydrochloride, $C_{23}H_{36}N_4O_5S_3\cdot HCl$, crystals from ether + abs ethanol, mp 134.5-135°. uv max: 233 nm (ϵ 23000).

THERAP CAT: Long-acting oral thiamine source.

6681. Octoxynol. α-*[4-(1,1,3,3,-Tetramethylbutyl)phenyl]-ω-hydroxypoly(oxy-1,2-ethanediyl);* octylphenoxy polyethoxyethanol; polyethylene glycol *p*-isooctylphenyl ether. Prepd by reacting isooctylphenol with ethylene oxide. Refs, nomenclature, *see* nonoxynol. Trademarks for series of octoxynols include *Igepal CA, Polytergent G, Triton X.*

Octoxynol (N.F.), mixture in which *n* ranges from 5 to 15; average comp. (n = 10): $C_{34}H_{62}O_{12}$; av. mol wt 647. Pale yellow, viscous liquid. d_4^{25} 1.0595. n_D^{25} 1.4894. Miscible with water, alcohol, acetone. Sol in benzene, toluene. Insol in petr ether. pH of 5% aq soln: 7-9. *Octoxynol-9* (USAN), average comp. (n = 9): $C_{32}H_{58}O_{11}$. Trademarks of products where n = 9 to 10: *Conco NIX-100, Igepal CA-630, Neutronyx 605, Triton X-100.* Ingredient of *Preceptin.*

USE: Nonionic detergent, emulsifier, dispersing agent. Ingredient of nitrofurazone soln, N.F. Spermaticide.

6682. Octreotide. D-*Phenylalanyl*-L-*cysteinyl*-L-*phenylalanyl*-D-*tryptophyl*-L-*lysyl*-L-*threonyl*-N-*[2-hydroxy-1-(hydroxymethyl)propyl]*-L-*cysteinamide cyclic (2 → 7)-disulfide;* 1,2-dithia-5,8,11,14,17-pentaazacycloeicosane cyclic peptide deriv.; SMS 201-995; Sandostatin. $C_{49}H_{66}N_{10}O_{10}S_2$; mol wt 1019.24. C 57.74%, H 6.53%, N 13.74%, O 15.70%, S 6.29%. Long-acting, octapeptide analog of somatostatin, *q.v.* Prepn: W. Bauer, J. Pless, **Eur. pat. Appl. 29,579;** *eidem,* **U.S. pat. 4,395,403** (1981, 1983 both to Sandoz). Prepn and pharmacology: W. Bauer *et al., Life Sci.* **31,** 1133 (1982). Opiate antagonist properties: R. Maurer *et al., Proc. Nat. Acad. Sci. USA* **79,** 4815 (1982). Inhibitory effect on human gastroenteropancreatic hormone secretion: M. E. Kraenzlin *et al., Experientia* **41,** 738 (1985). Endocrine profile in humans: E. del Pozo *et al., Acta Endocrinol.* **111,** 433 (1986). Clinical evaluation in acromegaly: G. Plewe *et al., Lancet* **2,** 782 (1984); in autonomic neuropathy: R. D. Hoeldtke *et al., ibid.* **2,** 602 (1986). Symposium on chemistry, pharmacology and clinical trials: *Scand. J. Gastroenterol.* **21,** Suppl. 119, 1-274 (1986); on clinical evaluation in gastrointestinal endocrine tumors: *Am. J. Med.* **82,** Suppl. 5B, 1-99 (1987).

D-Phe-Cys-Phe-D-Trp-Lys-Thr-Cys-NHCH(CH$_2$OH)CHOHCH$_3$

$[\alpha]_D^{20}$ −42° (c = 0.5 in 95% acetic acid).
THERAP CAT: Gastric antisecretory agent. Treatment of acromegaly.

6683. Octyl Acetate. *Acetic acid 2-ethylhexyl ester;* 2-ethylhexyl acetate. $C_{10}H_{20}O_2$; mol wt 172.26. C 69.72%, H 11.70%, O 18.58%. $CH_3(CH_2)_3CH(C_2H_5)CH_2OOCCH_3$. Toxicity data: H. F. Smyth, C. P. Carpenter, *J. Ind. Hyg. Toxicol.* **26,** 269 (1944).
Liquid. d_{20}^{20} 0.873. bp 199°. mp ~ −80°. n_D^{20} 1.4204. Flash pt, open cup: 190°F (88°C); closed cup: 56°F (13°C). Very slightly sol in water; misc with alcohol, oils, and other organic liquids. LD$_{50}$ orally in rats: 3.0 g/kg (Smyth, Carpenter).
USE: Solvent for nitrocellulose, some resins, waxes, and oils.

6684. *n*-Octyl Bromide. *1-Bromooctane.* $C_8H_{17}Br$; mol wt 193.13. C 49.75%, H 8.87%, Br 41.38%. $CH_3(CH_2)_6-CH_2Br$. Prepd from hydrobromic acid and *n*-octanol: O. Kamm, C. S. Marvel, *Org. Syn.* **coll. vol. I,** 30 (2nd ed., 1941); *cf.* Norris *et al., J. Am. Chem. Soc.* **38,** 1076 (1916); Whitmore *et al., ibid.* **67,** 2059 (1945); from PBr$_3$ and *n*-octanol: Coulson *et al., J. Chem. Soc.* **1965,** 2364.
Colorless liquid. d_4^{25} 1.108. bp 198-200°. mp −55°. n_D^{25} 1.4503. Insol in water, miscible with alcohol, ether. LD$_{50}$ orally in rats: 4.49 ml/kg, H. F. Smyth *et al., Am. Ind. Hyg. Assoc. J.* **30,** 470 (1969).

USE: In organic syntheses.

6685. *sec*-Octyl Bromide. *2-Bromooctane.* $C_8H_{17}Br$; mol wt 193.13. C 49.75%, H 8.87%, Br 41.38%. $CH_3CHBr-(CH_2)_5CH_3$. Prepd by the action of PBr$_3$ on *sec*-octyl alcohol: Hsueh, Marvel, *J. Am. Chem. Soc.* **50,** 856 (1928); Reynolds, Adkins, *ibid.* **51,** 279 (1929); Shriner, Young, *ibid.* **52,** 3332 (1930); Ellis, Reid, *ibid.* **54,** 1680 (1932); Coulson *et al., J. Chem. Soc.* **1965,** 2364.
dl-Form, d_4^{25} 1.0878. bp$_{14}$ 72°. bp$_6$ 66°. bp$_3$ 61°. n_D^{25} 1.4442. Insol in water, miscible with alcohol, ether.
l-Form, d_4^{25} 1.0982. bp$_{18}$ 73°. bp$_{14}$ 71°. bp$_3$ 60°. n_D^{25} 1.4475. α_D^{20} −44.91°.
d-Form, d_4^{25} 1.0982. bp$_{18}$ 77°. bp$_{14}$ 71°. bp$_3$ 60°. n_D^{25} 1.4475.

6686. *sec*-Octyl Iodide. *2-Iodooctane.* $C_8H_{17}I$; mol wt 240.14. C 40.01%, H 7.14%, I 52.85%. $CH_3CHI(CH_2)_5CH_3$. Prepn: Coulson *et al., J. Chem. Soc.* **1965,** 2364.
dl-Form, oily liquid; discolors in light. d_{15}^{18} 1.318. bp about 210° with decompn; also stated as 190°. *Protect from air and light.*
l-Form, α_D^{20} −64.63°.

6687. Octyl Methoxycinnamate. *3-(4-Methoxyphenyl)-2-propenoic acid 2-ethylhexyl ester;* 2-ethylhexyl *p*-methoxycinnamate; Parsol MCX; Parsol MOX. $C_{18}H_{26}O_3$; mol wt 290.40. C 74.45%, H 9.02%, O 16.53%. UV-B blocker. Manuf process and purification: P. Schudel *et al.,* **U.S. pat. 4,713,473** (1987 to Givaudan). Use in sunscreen preparations: D. H. Liem, L. T. H. Hilderink, *Int. J. Cosmet. Sci.* **1,** 341 (1979). Assessment of photostability: R. Aberturas *et al., Boll. Chim. Farm.* **126,** 208 (1987); A. Deflandre, G. Lang, *Int. J. Cosmet. Sci.* **10,** 53 (1988). HPLC determn in cosmetics: L. Gagliardi *et al., J. Chromatog.* **408,** 409 (1987).

bp at 1 mbar: 185-195°; bp at 0.1 mbar: 140-150°.
THERAP CAT: Ultraviolet screen.

6688. Ofloxacin. (±)-9-*Fluoro-2,3-dihydro-3-methyl-10-(4-methyl-1-piperazinyl)-7-oxo-7H-pyrido[1,2,3-de]-1,4-benzoxazine-6-carboxylic acid;* ofloxacine; DL-8280; HOE 280; Flobacin; Floxin; Oflocet; Oflocin; Oxaldin; Tarivid; Visiren. $C_{18}H_{20}FN_3O_4$; mol wt 361.38. C 59.82%, H 5.58%, F 5.26%, N 11.63%, O 17.71%. Broad spectrum, fluorinated quinolone antibacterial. Prepn: I. Hayakawa *et al.,* **Eur. pat. Appl. 47,005;** *eidem,* **U.S. pat. 4,382,892** (1982, 1983 both to Daiichi). Total synthesis: H. Egawa *et al., Chem. Pharm. Bull.* **34,** 4098 (1986). Comparison of antibacterial activity with other nalidixic acid analogs: G. Siebert *et al., Eur. J. Clin. Microbiol.* **2,** 548 (1983). *In vivo* and *in vitro* antibacterial activity: K. Sato *et al., Antimicrob. Ag. Chemother.* **22,** 548 (1982). Activity vs. clinical isolates of *Mycoplasma pneumoniae:* Y. Osada, H. Ogawa *ibid.* **23,** 509 (1983). Antichlamydial effect: J. M. G. Bailey *et al., ibid.* **26,** 13 (1984). Activity, pharmacology, metabolism, clinical studies, toxicology: *Chemotherapy (Tokyo)* **32,** Suppl. 1, 1-1210 (1984). Toxicity: H. Ohno *et al., ibid.* 1084. Pharmacology and clinical efficacy: *Infection* **14,** Suppl. 1, S1-S109 (1986). Review of antibacterial spectrum, pharmacology, and clinical efficacy: J. P. Monk, D. M. Campoli-Richards, *Drugs* **33,** 346-391 (1987).

Colorless needles from ethanol, mp 250-257° (dec). LD_{50} in male, female mice, male, female rats (mg/kg): 5450, 5290, 3590, 3750 orally; 208, 233, 273, 276 i.v.; > 10000, > 10000, 7070, 9000 s.c. (Ohno).

THERAP CAT: Antibacterial.

6689. Oil of Amber, Rectified. Obtained by the destructive distillation of amber and purified by redistillation. Consists of a mixture of terpenes with resinous, oxygen-containing substances.

Pale yellow to yellowish-brown, volatile oil; penetrating odor; burning acrid taste. d 0.850-0.920. α_D^{20} +22° to +26°. Insol in water. Sol in about 10 vols alcohol; freely sol in chloroform, ether, carbon disulfide, oils.

6690. Oil of Angelica. Volatile oil from root of *Angelica archangelica* L. (*A. officinalis* Moench.), *Umbelliferae*. *Constit.* Phellandrene, valeric acid.

Yellow liquid. d_{15}^{15} 0.857-0.915. α_D^{20} +16° to +32°. n_D^{20} 1.4800. Almost insol in water; sol in 6 vols 90% alcohol. *Keep well closed, cool, and protected from light.*

USE: In manuf liqueurs.

6691. Oil of Anise. Volatile oil from dried ripe fruit of *Pimpinella anisum* L., *Umbelliferae*, or of *Illicium verum* Hook. fil., *Magnoliaceae* (Chinese star anise). *Constit.* 80-90% anethole; methylchavicol, anisaldehyde.

Colorless or pale yellow, refractive liquid. d_{25}^{25} 0.978-0.988. Solidif not below 15°. α_D +1° to −2°. n_D^{20} 1.553-1.560. Slightly sol in water; sol in about 3 vols alcohol, freely in chloroform, ether. *Keep cool in well-closed and well-filled containers, protected from light.*

USE: In manuf of liqueurs; flavor for candies, cookies, dentifrices. Pharmaceutic aid (flavor).

THERAP CAT: Carminative, expectorant.

THERAP CAT (VET): Has been used as a carminative.

6692. Oil Anise, Japanese. Volatile oil from fruit of *Illicium anisatum* L., *Magnoliaceae* (Japanese star anise). *Constit.* Chiefly anethole; also safrol, eugenol.

Colorless to slightly yellow liquid. d about 1.006. Solidifies at −10° to −15°. α_D^{20} about −8°.

THERAP CAT: Carminative; expectorant.

6693. Oil of Asarum. Oil of Canada snakeroot. Volatile oil from rhizome of *Asarum canadense* L., *Aristolochiaceae*. *Constit.* Terpene (pinene), methyleugenol, borneol, linalool, geraniol.

Yellowish-brown liq; aromatic odor and taste. d_{15}^{15} 0.93-0.96. α_D^{20} −1.4 to −3.5°. Practically insol in water; sol in 2 vols 70% alc. *Keep well closed, cool and protected from light.*

6694. Oil of Balm. Oil of melissa balm; oil of lemon balm. Volatile oil from leaves and tops of *Melissa officinalis* L., *Labiatae*. Chiefly citral. Composition studies: Hefendehl, *Arch. Pharm.* **303**, 345 (1970).

Yellow to yellowish-green liquid. d_{15}^{15} 0.89-0.925. Practically insol in water; sol in alcohol. *Keep well closed, cool, and protected from light.*

6695. Oil of Basil. Volatile oil from leaves of *Ocimum basilicum* L., *Labiatae* (sweet basil). *Constit.* Methylchavicol, eucalyptol, linalool, estragol.

Yellowish to greenish liquid; aromatic odor. d_{20}^{20} 0.905-0.930. α_D^{20} −6° to −22°. Almost insol in water; sol in 2 vols 80% alc; miscible with ether, chloroform. *Keep well closed, cool and protected from light.*

6696. Oil of Bay. Oil of Myrcia. Volatile oil distilled from leaves of *Pimenta (Myrcia) acris* Kostel., *Myrtaceae*. *Constit.* 40-55% eugenol; myrcene, chavicol, methyleugenol, methylchavicol, citral, *l*-phellandrene; total phenols, 50-65% by volume.

Yellow to brownish-yellow liq; pleasant odor; sharp, spicy taste; becomes brown on exposure to air. d_{25}^{25} 0.962-0.990. α_D^{25} −3°. n_D^{20} 1.500-1.520. Insoluble in water. Very sol in alcohol, carbon disulfide, glacial acetic acid.

USE: Pharmaceutic aid (aromatic). Manuf bay rum.

6697. Oil Bergamot. Volatile oil expressed from rind of fresh fruit of *Citrus aurantium* L., var. *bergamia* Wight & Arn., *Rutaceae*. *Constit.* 36-45% *l*-linalyl acetate, about 6% *l*-linalool; *d*-limonene, dipentene, bergaptene.

Yellowish-green liquid; agreeable odor. d_{25}^{25} 0.875-0.880. α_D^{25} +8° to +24°. n_D^{20} 1.464-1.467. Acid no. 1-4. Almost insol in water; sol in 0.5 vol 95% alcohol, 2 vols 80% alcohol. *Keep well closed in a cool place, protected from light.*

USE: For masking many disagreeable odors, such as iodoform, naphthalene, etc. In perfumery, hair oils, pomades.

6698. Oil of Bitter Almond. Volatile oil from dried ripe kernels of bitter almonds or from other kernels containing amygdalin, such as apricots, cherries, plums, and especially peaches. Obtained by macerating with water, then steam distilling. *Constit.* Not less than 95% benzaldehyde; 2-4% HCN and phenoxyacetonitrile.

Colorless to yellow, very refractive liq; characteristic odor and taste of benzaldehyde. *Very poisonous!* d_{25}^{25} 1.038-1.060. n_D^{20} 1.5428-1.5439. Slightly sol in water; miscible with alcohol, ether, oils. *Keep cool and protected from light.*

Human Toxicity: Hydrogen cyanide *(q.v.)* component responsible for highly toxic properties.

USE: Only the oil *free from HCN* may be used for liqueurs and foods.

THERAP CAT: Formerly as topical antipruritic.

6699. Oil of Bitter Orange. Volatile oil expressed from fresh peel of *Citrus aurantium* L., *Rutaceae*. *Constit.* About 90% *d*-limonene; citral, decyl aldehyde, methyl anthranilate, linalool, terpineol.

Pale yellow liquid; bitter taste. d_{25}^{25} 0.842-0.848. α_D^{25} +88° to +98°. Very slightly sol in water; miscible with abs alcohol; sol in 4 vols alcohol, in 1 vol glacial acetic acid. *Keep well closed, cool, and protected from light.*

USE: As flavoring; in perfumery.

6700. Oil of Cajeput. Cajuput oil; cajeputi oil. Volatile oil from fresh leaves and twigs of several varieties of *Melaleuca leucadendron* L., and other species of *Melaleuca*, *Myrtaceae*. *Constit.* 50-60% eucalyptol (cineol); *l*-pinene, terpineol; valeric, butyric, benzoic and other aldehydes. Toxicity: P. M. Jenner *et al.*, *Food Cosmet. Toxicol.* **2**, 327 (1964).

Colorless or yellowish liquid, agreeable camphor odor, and bitter aromatic taste. d 0.912-0.925. α_D^{20} < −4°. n_D^{20} 1.4660-1.4710. Very slightly sol in water; sol in 1 vol 80% alcohol. Misc with alcohol, chloroform, ether, carbon disulfide. *Keep well closed, cool, and protected from light.* LD_{50} orally in rats: 3870 mg/kg (Jenner).

THERAP CAT: Expectorant, counterirritant, scabicide.

THERAP CAT (VET): Rubefacient, topical antimycotic.

6701. Oil of Calamus. Oil of sweet flag. Volatile oil from rhizome of *Acorus calamus* L., *Araceae* grown in North America, Europe and Asia. *Constit.* of Jammu, India variety: β-asarone 75.8%; calamene 3.84%; calamol 3.2%; α-asarone 1.32%; camphene 0.92%; β-pinene 0.56%; asaronaldehyde 0.2%, Vashist, Handa, *Soap Perfum. Cosmet.* **37**, 135 (1964).

Yellow to yellowish-brown viscid liquid; aromatic odor; bitter taste. d_{20}^{20} 0.960-0.970. α_D^{20} +9° to +31°. n_D^{20} 1.507-1.515. Sapon no. 16-20. Very slightly sol in water; miscible with alcohol. LD_{50} orally in rats (Jammu variety): 777 mg/kg, J. M. Taylor *et al.*, *Toxicol. Appl. Pharmacol.* **10**, 405 (1967). *Keep well closed, cool and protected from light.*

USE: In perfumery. Formerly as minor ingredient in bitter flavors such as vermouth and flavored wines.

6702. Oil of Camphor, Rectified. Formosa oil of camphor; Japanese oil of camphor; white oil of camphor; light oil of camphor. Volatile oil from *Cinnamomum camphora* T. Nees & Eberm., *Lauraceae*. *Constit.* Safrol, acetaldehyde, camphor, terpineol, eugenol, cineol, *d*-pinene, phellandrene, dipentene, cadinene.

Colorless or yellowish liquid; camphor odor. d_{20}^{20} 0.875-0.900. α_D^{25} +9° to +24°. n_D^{20} 1.465-1.470. Insol in water; sol in chloroform, ether, oils, in about 3 vols alcohol. *Keep well closed, cool, and protected from light.*

USE: As solvent in paint and lacquer industry; in perfuming of soaps and detergents; in technical odor masking.

THERAP CAT: Rubefacient.

6703. Oil of Caraway. Volatile oil distilled from dried ripe fruit of *Carum carvi* L., *Umbelliferae*. *Constit.* 53-63% carvone (by vol), *d*-limonene. Composition studies: Von

Schantz, Ek, *Sci. Pharm.* **39**, 82 (1971). *Review:* Arctander, *Perfume and Flavor Materials of Natural Origin* (S. Arctander, Elizabeth, N.J., 1960) pp 124-125.

Colorless or pale yellow liq; darkens and thickens with age. d_{25}^{25} 0.900-0.910. α_D^{25} +70° to +80°. n_D^{20} 1.485-1.497. Almost insol in water. Sol in 8 vols 80% or in 1 vol 90% alcohol. *Keep well closed, cool, and protected from light.*

USE: In manuf liqueurs and perfuming soaps; flavor for cookies, candies. Pharmaceutic aid (flavor).

THERAP CAT: Carminative.

6704. Oil of Cardamom. One of the oldest known essential oils; distilled from the ripe seeds of *Elettaria cardamomum* Maton, *Zingiberaceae*. Description of distillation processes: Viehoever, Sung, *J. Am. Pharm. Assoc.* **26**, 872 (1937); E. Guenther, *The Essential Oils* vol. V (New York, 1952) pp 85-106. *Constit.* Eucalyptol (cineol), sabinene, *d,α*-terpineol and acetate, borneol, limonene, terpinene, 1-terpinen-4-ol and its formate and acetate. Composition studies: Lewis, *Perfum. Essent. Oil Rec.* **57**, 623 (1966).

Colorless or pale yellow volatile oil. Characteristic, aromatic, agreeable odor and taste. Distills with steam. d_4^{25} 0.917-0.947. α_D^{25} +22° to +44°. n_D^{20} 1.4630-1.4660. Insol in water; miscible with abs ethanol. Sol in ether; one vol dissolves in 7 vols of 70% alcohol. Phenol coefficient 7.5.

USE: For flavoring liqueurs, pharmaceutical syrups, curry sauces, confectionary, baked goods.

THERAP CAT: Carminative.

6705. Oil of Cascarilla. Volatile oil from bark of *Croton eluteria* (L.) Sw., *Euphorbiaceae.* Chromatographic separation of constituents: Claudelafontaine *et al., Bull. Soc. Chim. France* **1973** (part 2), 2866.

Yellow to greenish liquid. d 0.890-0.925. Rotation +2° to +5° in a 100-mm tube at 20°. Very sol in alcohol, ether. *Keep well closed, cool, and protected from light.*

6706. Oil of Cashew Nut Shell. From *Anacardium occidentale* L., *Anacardiaceae* (Southern India). Consists mostly (90%) of anacardic acid, *q.v.* Review and bibliography: Sanyal, Das, *Indian Pharm.* **10**, 272-283 (1955).

Wijs iodine no. over 250. Polymerizes easily when heated with acid.

USE: In insulating varnishes, typewriter rolls, coldsetting cements, floor tile, brake linings; molding compounds from polymers: Harvey, U.S. pat. **2,767,150** (1956 to The Harvel Corp.).

6707. Oil of Cedar Leaf. Cedar leaf oil. Obtained by steam distillation from the fresh leaves of *Thuja occidentalis* L., *Cupressaceae.* Although this oil is known commercially as cedar leaf oil, the tree from whose leaves it is obtained is not a true cedar; it is a coniferous tree popularly known as "arbor vitae," sometimes erroneously called "white cedar." *Habit.* Canada and Northern U.S. *Constit.* α-pinene, *d*-thujone, α-fenchone.

Yellowish volatile oil. d 0.910-0.920. α_D^{25} −10° to −13°. n_D^{20} 1.4560-1.4590. Soly similar to oil of lavender, *q.v.*

Human Toxicity: Ingestion of large quantities causes hypotension, bradycardia, tachypnea, convulsions, death.

USE: Substitute for oil of lavender.

THERAP CAT: Counterirritant.

6708. Oil of Cedar Wood. Volatile oil from wood of *Juniperus virginiana* L., *Cupressaceae,* and other species of cedar. *Constit.* Chiefly cedrene (a terpene), and cedral (cedar camphor).

Colorless or slightly yellow, somewhat viscid liquid. d_{20}^{20} 0.940-0.950. α_D^{20} −25° to −46°. n_D^{20} 1.495-1.510. Insol in water. Sol in 10-20 vols 90% alcohol; sol in ether. *Keep well closed, cool, and protected from light.*

USE: In perfumery; as insect repellent; the thickened oil is used in microscopy as a clearing agent and for use with immersion lenses.

6709. Oil of Celery. Volatile oil from celery seed, *Apium graveolens* L., *Umbelliferae.* *Constit.* *d*-Limonene, phenols, sedanolide, sedanoic acid.

Colorless liquid; celery odor. d 0.870-0.895. α_D^{20} +67° to +79°. Slightly sol in water; very sol in alcohol. *Keep well closed, cool, and protected from light.*

USE: As flavor for soft drinks, unpleasant medicaments.

6710. Oil of Chamomile—German. Hungarian chamomile oil; blue chamomile oil. Volatile oil from flowers of *Matricaria chamomilla* L., *Compositae.*

Bluish, viscid liquid; butter-like consistency when cooled; solid at 0°; agreeable chamomile odor. d_{15}^{15} 0.917-0.957. Slightly dextrorotatory. Sapon no. about 45. Very sol in alcohol, in 6 vols 90% alcohol. *Keep well closed, cool, and protected from light.*

THERAP CAT: Aromatic bitter, emetic.

6711. Oil of Chamomile—Roman. Oil of anthemis. Volatile oil from flowers of *Anthemis nobilis* L., *Compositae.* Principal constituents are esters of angelic acid. Studies and syntheses of constituents: A. F. Thomas *et al., Helv. Chim. Acta* **64**, 1488 (1981); I. Klimes *et al., ibid.* 2338; A. F. Thomas, J. C. Egger, *ibid.* 2393; A. F. Thomas, *ibid.* 2397.

Blue liquid, becomes brownish-yellow on exposure to air and light; strong, pleasant aromatic odor; burning taste. d_{15}^{15} 0.905-0.915. Slightly dextrorotatory. n_D^{20} 1.442-1.448. Sapon no. 260-296. Slightly sol in water; sol in less than one vol of 90%, in 6 vols of 70% alcohol. *Keep well closed, cool, and protected from light.*

USE: In perfumery.

THERAP CAT: Aromatic bitter; emetic.

6712. Oil of Champaca. Volatile oil from flowers of *Michelia champaca* L., *Magnoliaceae.* *Constit.* Esters of benzoic acid, benzaldehyde, benzyl alcohol, isoeugenol.

Light yellow or reddish-yellow to brownish liquid. d_{15}^{15} 0.906-0.935. α_D −12° to −52°. Sapon no. about 77. Slightly sol in water; sol in chloroform, ether; sparingly sol in alcohol. *Keep well closed, cool and protected from light.*

USE: In perfumes.

6713. Oil of Chenopodium. Oil of American wormseed; Chenopodiol; Chenoposan; Chenoposetten. Volatile oil from fresh, aboveground parts of flowering and fruiting plant of *American wormseed, Chenopodium ambrosioides* L. var. *anthelminticum* (L.) Aellen, *Chenopodiaceae* also known as *Mexican tea, Spanish tea, Jerusalem tea, ambrosia. Habit.* Central America, U.S., Canada. Oil contains 60-70% ascaridole, the principal active constituent; *p*-cymene, α-terpinene, *l*-limonene, methadiene.

Colorless or pale yellow liquid; characteristic disagreeable odor and taste. d_{25}^{25} 0.950-0.980. α_D^{20} −4° to −8°. n_D^{20} 1.4723-1.4790. Insoluble in water. Sol in 8 vols 70% alcohol; partly sol in glacial acetic acid. *Keep well closed, cool and protected from light.*

THERAP CAT: Anthelmintic.

THERAP CAT (VET): Anthelmintic.

6714. Oil of Cherry Laurel. Volatile oil from leaves of *Prunus laurocerasus* L., *Rosaceae.* *Constit.* HCN, benzaldehyde, benzaldehyde cyanhydrin, benzyl alcohol.

Pale yellow liquid; odor and taste similar to oil of bitter almond. *Very poisonous!* d_{20}^{20} 1.054-1.066. Slightly sol in water; sol in 2 vols 70% alcohol, in benzene, chloroform, ether. *Keep well closed, cool and protected from light.*

Caution: Hydrogen cyanide (*q.v.*) component responsible for highly toxic properties.

6715. Oil of Cinnamon. Oil of cassia; oil of Chinese cinnamon. Volatile oil from leaves and twigs of *Cinnamomum cassia* Nees, *Lauraceae.* *Constit.* 80-90% cinnamaldehyde; cinnamyl acetate, eugenol.

Yellowish or brownish liquid. Darkens and thickens on exposure to air. d_{25}^{25} 1.045-1.063. $\alpha_D^{}$ −1° to +1°. $n_D^{}$ 1.6020-1.6060. Slightly sol in water; sol in an equal vol of alcohol and of glacial acetic acid. *Keep well closed, cool and protected from light.*

USE: As flavor in foods and in perfumes.

THERAP CAT: Carminative.

6716. Oil of Cinnamon, Ceylon. Volatile oil from bark of Ceylon cinnamon. *Constit.* 50-65% cinnamaldehyde; 4-8% eugenol; phellandrene.

Light yellow liq; gradually becomes reddish; characteristic odor, d 1.000-1.030. α_D 0° to −2°. $n_D^{}$ 1.565-1.582.

THERAP CAT: Carminative.

6717. Oil of Citronella. Volatile oil from fresh grass of *Cymbopogon (Andropogon) nardus* (L.) Rendle, *Gramineae.*

Constit. Ceylon: about 60% geraniol, about 15% citronellal, 10-15% camphene and dipentene, small quantities of linalool, borneol. Java: 25-50% citronellal, 25-45% geraniol.

Almost colorless to pale yellow liq; gradually becomes reddish; pleasant odor. d Ceylon, 0.897-0.912; Java, 0.885-0.900. α_D^{20}: Ceylon, $-6°$ to $-14°$; Java, $-2°$ to $-5°$. n_D^{20} Ceylon 1.479-1.485; Java, 1.468-1.473. Slightly sol in water; sol in 10 vols 80% alcohol. *Keep well closed, cool and protected from light.*

USE: As perfume; insect repellent.

6718. Oil of Clove. Clove oil. Volatile oil from dried flower buds of *Eugenia caryophyllata* Thunb. (*Caryophyllus aromaticus* L.), *Myrtaceae. Constit.* 82-87% eugenol, including about 10% acetyleugenol, caryophyllene, small quantities of furfural, vanillin, methyl amyl ketone.

Colorless to pale yellow liq, becoming darker and thicker with age. d_{25}^{25} 1.038-1.060. $\alpha_D^{25} < -1°10'$. n_D^{20} 1.530. bp about 250°. Insol in water; sol in 2 vols 70% alcohol; very sol in stronger alcohol, in ether, glacial acetic acid. *Keep well closed, cool and protected from light.*

USE: In confectionery, toothpowders; also in microscopy.

THERAP CAT: Local anesthetic (toothache); counterirritant; carminative.

6719. Oil of Copaiba. Volatile oil from copaiba balsam, usually Maracaibo. *Constit.* Chiefly caryophyllene and cadinene.

Colorless or pale yellow liquid. d_{15}^{15} 0.895-0.905, increasing with age. bp 250-275°. α_D $-7°$ to $-35°$. n_D^{20} 1.495-1.500. Insol in water; very sol in alcohol, ether, carbon disulfide. *Keep well closed, cool and protected from light.*

USE: In perfumery.

6720. Oil of Coriander. Volatile oil from dried ripe fruit of *Coriandrum sativum* L., *Umbelliferae. Constit.* Chiefly *d*-linalool and its acetate.

Colorless or pale yellow liquid. d_{25}^{25} 0.863-0.875. α_D^{25} +8° to +15°. n_D^{20} 1.4620-1.4720. Almost insol in water; sol in 3 vols 70% alcohol; more sol in stronger alcohol; very sol in chloroform, ether, glacial acetic acid. *Keep well closed, cool and protected from light.*

USE: Flavoring in foods and alcoholic beverages.

THERAP CAT: Carminative.

6721. Oil of Cubeb. Volatile oil from unripe fruit of *Piper cubeba* L.f., *Piperaceae. Constit.* Dipentene, cadinene, cubeb camphor, the latter in old oils.

Colorless, pale green or yellowish liquid. d_{25}^{25} 0.905-0.925. α_D^{20} $-20°$ to $-40°$. n_D^{20} 1.4800-1.5020. Insol in water; sol in obout 10 vols alcohol; miscible with abs alcohol, chloroform. *Keep well closed, cool and protected from light.*

THERAP CAT: Urinary antiseptic.

6722. Oil of Cumin. Volatile oil from fruit of *Cuminum cyminum* L., *Umbelliferae. Constit.* 30-40% cuminaldehyde; *p*-cymene, β-pinene, dipentene.

Colorless to yellow liquid. d_{25}^{25} 0.900-0.935. α_D^{20} +4° to +8°. n_D^{20} 1.4950-1.5090. Almost insol in water; sol in 10 vols 80% alcohol; more sol in stronger alcohol; very sol in chloroform, ether. *Keep well closed, cool and protected from light.*

USE: Flavoring in Indian curry powder.

6723. Oil of Cypress. Volatile oil from leaves and young branches of *Cupressus sempervirens* L., *Pinaceae. Constit.* Furfural, *d*-pinene, *d*-camphene, cymene, *d*-terpineol, *l*-cadinene, sylvestrene, cypress camphor.

Yellowish liquid. d 0.88-0.89. α_D^{20} +4° to +18°. Slightly sol in water; sol in 2-6 vols 90% alcohol. *Keep well closed, cool and protected from light.*

USE: In perfumery.

6724. Oil of Dill. Dill seed oil; dill fruit oil. Volatile oil from dried ripe fruit of *Anethum graveolens* L., *Umbelliferae. Constit.* About 50% carvone *d*-limonene, phellandrene and other terpenes.

Colorless or pale yellow liquid; characteristic odor. d_{15}^{15} 0.900-0.915. α_D^{20} +70° to +80°. n_D^{20} 1.481-1.492. Insol in water; sol in 1 vol 90% alcohol. *Keep well closed, cool and protected from light.*

THERAP CAT: Aromatic carminative.

6725. Oil of Dwarf Pine Needles. Oil of mountain pine; Pinus Montana oil; Pinus pumilio oil. Volatile oil from fresh leaves of *Pinus montana* Mill. (*P. pumilio* Haenke), *Pinaceae. Constit.* *l*-Pinene, *l*-phellandrene, sylvestrene, dipentene, cadinene, 5-7% bornyl acetate.

Colorless or faintly yellow liq; pleasant odor; bitter taste. d_{25}^{25} 0.853-0.869. α_D^{25} $-5°$ to $-12°$. n_D^{20} 1.4750-1.4800. Insol in water. Sol in 4.5-8 vols 90% alcohol; very sol in chloroform, ether. *Keep well closed, cool and protected from light.*

USE: Pharmaceutic aid (flavor and perfume).

THERAP CAT: Expectorant.

6726. Oil of Eucalyptus. Dinkum oil. Volatile oil from fresh leaves of *Eucalyptus globulus* Labill and of some other species of *Eucalyptus, Myrtaceae.* A dwarf species, called Mallee in Australia, is richest in oil of Eucalyptus. *Constit.* 70-80% cineole (eucalyptol); α-pinene; phellandrene; terpineol; citronellal; geranyl acetate; eudesmol; eudesmyl acetate; piperitone; volatile aldehydes (principally isovaleric): E. Guenther, *The Essential Oils* vol. 4 (van Nostrand, New York, 1950) pp 437-525.

Colorless to pale yellow liquid; characteristic camphoraceous odor; pungent, spicy, cooling taste. d_{25}^{25} 0.905-0.925. Does not solidif below $-15.4°$. α_D $-5°$ to $+5°$. n_D^{20} 1.4580-1.4700. Almost insol in water; sol in 5 vols 70% alcohol; miscible with abs alcohol, oils, fats. *Keep well closed, cool and protected from light.*

THERAP CAT: Expectorant, anthelmintic, local antiseptic.

THERAP CAT (VET): Inhalation expectorant; wound dressing.

6727. Oil of Fennel. Volatile oil from the dried fruit of *Foeniculum vulgare* Mill., *Umbelliferae.* Also from the sweet fennel plant (var. *dulce*). *Constit.* 50-60% anethole, $\sim$20% fenchone, pinene, limonene, dipentene, phellandrene.

Colorless or pale yellow liquid; odor and taste of fennel. d_{25}^{25} 0.953-0.973. α_D^{25} +12° to +24°. n_D^{20} 1.5280-1.5380. Slightly sol in water; sol in 1 vol 90% or in 8 vols 80% alcohol; very sol in chloroform, ether. *Keep well closed, cool and protected from light.*

USE: To cover taste of unpleasant medicines.

THERAP CAT: Carminative.

6728. Oil of Fir. Oil of silver pine; oil of silver fir. Volatile oil from needles and young twigs of *Abies alba* Mill. (*A. picea* Lindl., *A. pectinata* DC.), *Pinaceae, Constit.* *l*-Pinene, *l*-limonene, *l*-bornyl acetate.

Colorless, clear liquid; balsamic odor; terebinthinate taste. d_{15}^{15} 0.869-0.875. α_D^{20} $-20°$ to $-59°$. Insol in water. Sol in 5 vols 90% alc, in ether. *Keep well closed, cool and protected from light.*

USE: Pharmaceutic aid (flavor and perfume).

THERAP CAT: Expectorant.

6729. Oil of Fir—Siberian. Oil of pine; oleum abietis; Siberian pine needle oil. Volatile oil from fresh leaves of *Abies sibirica* Ledeb., *Pinaceae. Constit.* About 40% esters calculated as bornyl acetate; pinene, camphene, dipentene, and phellandrene.

Colorless or pale yellow liquid; aromatic odor; pungent taste. d_{15}^{15} 0.905-0.925. α_D^{20} $-32°$ to $-45°$. n_D^{20} 1.466-1.476. Sol in an equal vol 90% alcohol. *Keep well closed, cool and protected from light.*

USE: Pharmaceutic aid (flavor and perfume).

THERAP CAT: Expectorant.

6730. Oil of Fleabane. Oil of Canada fleabane; oil of erigeron. Volatile oil from fresh flowering herb of *Conyza canadensis* (L.) Cron. (*Erigeron canadensis* L., *Leptilon canadense* (L.)) Britt., *Compositae. Constit.* *d*-Limonene, aldehydes.

Pale yellow liquid; becomes darker and thicker with age and on exposure to air; peculiar odor; aromatic, slightly pungent taste. d_{25}^{25} 0.845-0.865. α_D^{20} about +45°. Slightly sol in water; sol in 1 vol alcohol; very sol in chloroform, ether. *Keep well closed, cool and protected from light.*

6731. Oil of Garlic. Volatile oil from bulb or entire plant *Allium sativum* L., *Liliaceae. Constit.* Allylpropyl disulfide, diallyl disulfide, allyl sulfide, *q.v.*

Yellow liquid; strong garlic odor. d_{15}^{15} 1.046-1.057. Optically inactive.

THERAP CAT: Anthelmintic, rubefacient.

6732. Oil of Geranium. Oil of pelargonium geranium; oil of rose geranium. Volatile oil from leaves of *Pelargonium odoratissimum* Ait. and allied species, *Geraniaceae. Constit.* Geraniol esters, calculated as geranyl tiglate, 21-30% in Algerian or French oil, 25-35% in Bourbon oil; citronellol; some linalool.

Colorless, greenish or brownish liquid. d_{15}^{15} 0.894-0.905. α_D^{20} −7° to −11°. n_D^{20} 1.4650-1.470. Slightly sol in water; sol in 3 vols 10% alcohol; more sol in stronger alcohol; very sol in chloroform, ether. *Keep well closed, cool and protected from light.*

USE: In perfumery; as odorant for tooth and dusting powders, ointments, etc. In manuf of rhodinol, *q.v.*

6733. Oil of Geranium—East Indian. Turkish geranium oil; palmarosa oil; Indian grass oil; rusa oil. Volatile oil from *Andropogon schoenanthus* L., *Gramineae*, and allied species grown in India (not Turkey). *Constit.* 85-95% geraniol; citronellol, dipentene; it is practically devoid of esters.

Colorless or light yellow liquid; pleasant rose odor. d_{15}^{15} 0.885-0.896. α_D +1° 40′ to −2°. n_D^{20} 1.476-1.4085.

6734. Oil of Ginger. Volatile oil from the rhizome *Zingiber officinale* Roscoe, *Zingiberaceae. Constit.* *l*-Zingiberene, *d*-camphene, phellandrene, borneol, cineol, citral.

Yellowish, viscid liquid. d_{15}^{15} 0.875-0.885. α_D −25° to −45°. n_D^{20} 1.4880-1.4950. Almost insol in water; sparingly sol in alcohol; sol in ether, carbon disulfide.

USE: In mouthwashes, ginger beverages, liqueurs, etc.

6735. Oil of Hops. Volatile oil from strobiles of *Humulus lupulus* L., *Moraceae. Constit.* 65-70% humulene, terpenes. Gas chromatography and identification of constituents: Buttery *et al., Nature* **200**, 435 (1963). *See also* review by Stevens, *Chem. Rev.* **67**, 19 (1967).

Light yellow to brownish liquid. d_{15}^{15} 0.855-0.880. Practically inactive optically. n_D^{20} 1.4775. Insoluble in water; slightly sol in alcohol; sol in ether. *Keep well closed, cool and protected from light.*

6736. Oil of Hyssop. Volatile oil from *Hyssopus officinalis* L., *Labiatae. Constit.* About 50% pinene, small quantities of aromatic alcohol, probably also sesquiterpenes.

Colorless or greenish-yellow liquid; sharp, camphor-like taste. d_{15}^{15} 0.925-0.940. α_D^{20} −17° to −23°. bp about 200°. Almost insol in water; sol in 2 to 4 vols 80% alcohol. *Keep well closed and protected from light.*

6737. Oil of Juniper. Juniperberry oil. Volatile oil from dried ripe fruit (berries) of *Juniperus communis* L., *Cupressaceae. Constit.* Pinene, cadinene, camphene, terpineol, juniper camphor. Analysis by GLC of the oil of six western North American species: F. C. Vasec, R. W. Scora, *Am. J. Bot.* **54**, 781 (1967).

Colorless to pale greenish-yellow liquid. Odor and taste as of the berries. d_{25}^{25} 0.854-0.879. $[\alpha]_D^{25}$ 0° to −15°. n_D^{20} 1.4780-1.4840. Almost insol in water. Sol in 4 vols alcohol. Miscible with carbon disulfide, benzene, chloroform, amyl alcohol. *Keep cool in well-closed and well-filled bottles, protected from light.*

USE: Manuf liqueurs; for preserving catgut ligatures. In perfumery.

THERAP CAT: Diuretic.

6738. Oil of Juniper Wood. Steam-distilled oil from wood or branches of *Juniperus communis* L., *Cupressaceae. It is not Juniper tar.*

Colorless to yellowish, turpentine-like liquid; weak juniper odor. Insol in water. Sol in alcohol.

Caution: Not to be used for preserving catgut or silk.

6739. Oil of Lavender. Volatile oil from fresh flowering tops of *Lavandula officinalis* Chaix (*L. vera* DC.), *Labiatae. Constit.* 30-40% esters calculated as linalyl acetate; linalool, pinene, limonene, geraniol, some cineol.

Colorless or yellow liquid. d 0.875-0.888. α_D^{25} −3° to −10°. n_D^{20} 1.459-1.470. Slightly sol in water; sol in 4 vols 70% alcohol; miscible with abs alcohol, carbon disulfide. *Keep well closed, cool and protected from light.*

USE: As perfume and flavor.

THERAP CAT: Aromatic, carminative.

6740. Oil of Lemon. Cedro oil. Volatile oil expressed from fresh peel of *Citrus limonum* (L.) Risso (*C. medica* var. *limon* L.), *Rutaceae. Constit.* About 90% of limonene, terpinene, phellandrene, and pinene combined; 4-6% aldehydes calculated as citral, some citronellal, geranyl acetate, sesquiterpenes.

Pale yellow or greenish-yellow liquid. d_{25}^{25} 0.849-0.855. α_D^{25} +57° to +65.6°. n_D^{20} 1.4742-1.4755. Slightly sol in water; sol in 3 vols alcohol; miscible with carbon disulfide, glacial acetic acid. *Keep cool, in well-filled and well-closed bottles, protected from light.*

USE: For flavoring medicaments; as a flavor in liqueurs, pastry, foods, beverages; also in perfumes.

6741. Oil of Lemon Grass. Indian oil of verbena; Indian melissa oil. Volatile oil from *Cymbopogon (Andropogon) citratus* (DC.) Stapf, and of *C. flexuosus* (Nees) Stapf, *Gramineae. Constit.* 75-85% citral; methylheptenone, citronellal, geraniol, limonene, dipentene.

Reddish-yellow or brownish-red liquid; strong odor of verbena. d_{15}^{15} 0.895-0.908 (0.878-0.882 of West Indian oil). α_D −3°. n_D^{20} 1.483-1.489. Slightly sol in water; sol in 3 vols 70% alcohol; sol in chloroform, ether. *Keep well closed, cool and protected from light.*

USE: As source of citral (synthesis of vitamin A) and in perfumery.

6742. Oil of Levant Wormseed. Volatile oil from the flowers of *Artemisia maritima* var *stechmanniana* Bess (*A. pauciflora* Weber), *Compositae. Constit.* Largely eucalyptol (cineol).

Pale yellow to yellowish-green liquid. d_{20}^{20} 0.915-0.940. Insol in water; sol in alcohol, ether. *Keep well closed, cool and protected from light.*

THERAP CAT: Anthelmintic.

6743. Oil of Linaloe. Volatile oil distilled from a Mexican wood (*Bursera delpechiana* Poiss. and probably other species of *Bursera, Burseraceae). Constit.* Linalool, geraniol, methylheptenone.

Colorless to yellowish liquid; pleasant odor. d_{15}^{15} 0.875-0.890. α_D −5° to −12°. n_D^{20} 1.4638. Slightly sol in water; sol in 2 vols 70% alcohol; sol in ether, chloroform. *Keep well closed, cool and protected from light.*

USE: In perfumery.

6744. Oil of Marjoram. Volatile oil from *Origanum marjorana* L., *Labiatae. Constit.* About 40% terpenes, chiefly terpinene; *d*-terpineol.

Yellow or greenish-yellow liquid. d_{15}^{15} 0.888-0.912. α_D^{20} +13° to +18°. Insol in water; sol in 2 vols 80% alcohol; sol in chloroform, ether.

USE: In perfumes and microscopy.

6745. Oil of Mustard, Expressed. Fixed oil expressed from the mustard seeds of *Brassica alba* (L.) Boiss. and *B. nigra* (L.) Koch, *Cruciferae. Constit.* Chiefly the glycerides of oleic acid and other fatty acids, including arachidic.

Straw-colored or brownish-yellow, or greenish-brown liquid. d_{15}^{15} 0.914-0.916. Solidif −8° to −16°. n_D^{40} 1.4655-1.4670. Sapon no. 170-174. Iodine no. 92-97. Insol in water; slightly sol in alcohol; miscible with chloroform, ether, petr ether.

USE: Manuf oleomargarine, soap; as lubricant; salad oil.

6746. Oil of Myrtle. Volatile oil from leaves of *Myrtus communis* L., *Myrtaceae. Constit.* *d*-Pinene, eucalyptol, dipentene, camphor.

Yellow to greenish liquid; fragrant odor. d_{15}^{15} 0.890-0.915. α_D +10° to +30°. Insol in water; sol in alcohol, chloroform, ether.

6747. Oil of Niaouli. Volatile oil from leaves of *Melaleuca viridiflora* (Soland.) Gaertn., *Myrtaceae. Constit.* About 65% cineol, about 30% terpineol, limonene and *d*-pinene combined.

Slightly yellow liquid; aromatic odor; pungent, refreshing peppermint-like taste. d_{15}^{15} 0.908-0.932. Insol in water or glycerol; sol in alcohol, benzene, ether.

THERAP CAT: Anthelmintic.

6748. Oil of Nutmeg, Expressed. Nutmeg butter; oil of mace. Oil expressed from nutmeg (*Myristica fragrans*

Houtt., *Myristicaceae*). *Constit.* Chiefly trimyristin; some volatile oil.

Orange-red to reddish-brown, soft solid. mp 45-51°. Odor and taste of nutmeg. d 0.990-0.995. Sapon no. 172-179. Iodine no. 40-52. Acid no. 17-23. Partly sol in cold alcohol, almost completely sol in hot alcohol; freely sol in chloroform, ether. LD_{50} orally in rats: 3640 mg/kg, P. M. Jenner *et al., Food Cosmet. Toxicol.* **2**, 327 (1964).

Human Toxicity: Symptoms similar to Oil of Nutmeg, Volatile.

THERAP CAT: Rubefacient.

6749. Oil of Nutmeg, Volatile. Oil of myristica. Steam-distilled oil from dried kernels of ripe seeds of nutmeg *(Myristica fragrans* Houtt., *Myristicaceae*). *Constit.* 60-80% d-Camphene, about 8% d-pinene; dipentene, d-borneol, l-terpineol, about 6% geraniol, safrol, about 4% myristicin. The myristicin fraction together with its more than 25% content of elemicin is supposed to be responsible for the purported hallucinogenic properties of nutmeg seed: Shulgin, *Nature* **197**, 4865 (1963); Weil, *Econ. Botany* **19**, 194-217 (1965).

Colorless or pale yellow liquid; odor and taste of nutmeg. d_{25}^{25} 0.859-0.924. α_D +10° to +30°. n_D^{20} 1.4740-1.4880. Insoluble in water; sol in 1 vol alcohol, in 3 vols 90% alcohol. *Keep well closed, cool and protected from light.* LD_{50} orally in rats: 2620 mg/kg, P. M. Jenner *et al., Food Cosmet. Toxicol.* **2**, 327 (1964).

Human Toxicity: Ingestion of large quantities produces narcosis, delirium, death.

USE: As a flavor.

THERAP CAT: Carminative.

6750. Oil of Orange. Oil sweet orange. Volatile oil expressed from fresh peel of ripe fruit of the orange *(Citrus aurantium* var *sinensis* L., *Rutaceae*). *Constit.* About 90% d-limonene; citral, decyl aldehyde, methyl anthranilate, linalool, terpineol.

Yellow to deep orange liquid; characteristic orange taste and odor. d_{25}^{25} 0.842-0.846. α_D^{25} +94° to +99°. n_D^{20} 1.4723-1.4737. Slightly sol in water; sol in 2 vols 90% alcohol, in 1 vol glacial acetic acid; miscible with abs alcohol, carbon disulfide. *Keep well closed, cool and protected from light.*

USE: Chiefly as flavor and perfume. Pharmaceutic aid (flavor).

THERAP CAT: Expectorant.

6751. Oil of Orange Flowers. Oil of neroli. Volatile oil distilled from fresh orange flowers. *Constit.* Limonene, l-linalool, geraniol, 7-18% linalyl acetate; some methyl anthranilate, nerol and neroli camphor.

Yellowish, fluorescent liquid; very intense and pleasant odor; becomes brown on exposure to light. d_{25}^{25} 0.86-0.88. α_D^{25} +1° 30′ to +9° 8′. n_D^{20} 1.475. Slightly sol in water; sol in 1.5-2 vols 80% alcohol with fine violet fluorescence. *Keep well closed, cool and protected from light.*

USE: As perfume and flavor.

6752. Oil of Origanum. Oil of wild marjoram. Volatile oil from flowering tops of *Origanum vulgare* L., *Labiatae. Constit.* Carvacrol, terpenes.

Light yellow liquid. d_{15}^{15} 0.870-0.910. α_D about −34°. Very slightly sol in water; very sol in alcohol. *Keep well closed, cool and protected from light.*

USE: In perfumery.

6753. Oil of Parsley. Parsley seed oil. Volatile oil from parsley seeds *Petroselinum hortense,* Hoffm. (*P. sativum* Hoffm., *Carum petroselinum* Benth. & Hook.), *Umbelliferae. Constit.* Chiefly apiol; terpene, l-pinene(?).

Colorless or yellow, rather viscid liquid. d_{15}^{15} 1.040-1.100. α_D −5° to −11°. n_D^{20} 1.510-1.519. Very slightly sol in water; sol in 8 vols 80% alcohol; sol in ether.

USE: Flavoring in meat sauces, seasonings, spice blends, pickles.

6754. Oil of Patchouli. Patchouli oil. Essential oil from leaves of several *Labiatae* species. The commercial oil is obtained from the cultivated species, *Pogostemon cablin* (Blanco) Benth. (*P. patchouly* Pellet. var *suavis* Hook. f.), *Labiatae.* Major constituent is patchouli alcohol *(q.v.);* minor constituents include patchoulene, azulene, eugenol, and several unidentified sesquiterpenes: Pfau, Plattner, *Helv.*

Chim. Acta **19**, 874 (1936); Naoko *et al., Bull. Chem. Soc. Japan* **40**, 597 (1967). *Review:* E. Guenther, *The Essential Oils* vol. III (Van Nostrand, New York, 1949) pp 552-575.

Yellowish or greenish to dark brown oil, intense and persistent fragrant odor. Can be stored indefinitely. Odor seems to improve with age. d_{15}^{15} 0.975-0.987. $[\alpha]_D^{20}$ −54° to −65.3°. n_D^{20} 1.5099 to 1.5111. Saponif no. 3.3 to 9.3. Ester no. after acetylation: 17.7 to 22.4. Practically insol in water. Soluble in ether.

USE: In perfume formulations to impart a lasting oriental fragrance, in incense, soaps, cosmetics. To scent fine Indian fabrics and shawls.

6755. Oil of Pennyroyal—American. Oil of hedeoma. Volatile oil from leaves and flowering tops of *Hedeoma pulegioides* (L.) Pers., *Labiatae. Constit.* Chiefly pulegone; 2 ketones; acetic, formic and isoheptoic acids.

Pale yellow liq; aromatic odor. d_{25}^{25} 0.920-0.935. α_D^{20} +18° to +22°. n_D^{20} 1.482. Slightly sol in water; sol in 3 vols 70% alcohol; very sol in chloroform, ether. *Keep well closed, cool and protected from light.*

THERAP CAT: Aromatic carminative.

6756. Oil of Pennyroyal—European. Oil of pulegium. Volatile oil from *Mentha pulegium* L., *Labiatae. Constit.* About 85% pulegone.

Yellow or greenish-yellow liquid; aromatic mint-like odor; aromatic taste. d_{15}^{15} 0.960. α_D^{20} +14° to +28°. n_D^{20} 1.475-1.496.

6757. Oil of Pepper. Volatile oil from unripe fruit of the black pepper. *Constit.* Chiefly l-phellandrene, sesquiterpenes (caryophyllene).

Colorless or yellowish liquid. d_{15}^{15} 0.890-0.900. α_D about −3° to −5°. n_D^{20} 1.4935-1.4977. Insol in water; sol in about 15 vols 90% alcohol.

USE: As condiment.

6758. Oil of Peppermint. Colpermin; Mintec. Steam-distilled, volatile oil from fresh flowering plant *Mentha piperita* L., *Labiatae.* The Japanese oil, also known as oil of Poho, is the liq portion remaining after the separation of menthol from the oil of *Mentha arvensis* L., *Labiatae.* Constit. Not less than 50% total menthol including 5-9% esters calcd as menthyl acetate; menthyl isovalerate, menthone, inactive pinene, l-limonene, cadinene, phellandrene, some acetaldehyde, isovaleric aldehyde, amyl alcohol, dimethyl sulfide.

Colorless to pale yellow liquid; strong, penetrating peppermint odor and pungent taste. d_{25}^{25} 0.896-0.908. α^{25} −18° to −32°. n_D^{20} 1.460-1.471. Very slightly sol in water; sol in 4 vols 90% alcohol.

USE: Pharmaceutic aid (flavor). In flavoring liqueurs.

THERAP CAT: Carminative.

THERAP CAT (VET): Has been used as a carminative.

6759. Oil of Pettigrain. Volatile oil from leaves, twigs and unripe fruit of *Citrus vulgaris* Risso (*C. bigaradia* Loisel.), *Rutaceae. Constit.* 40-80% linalyl acetate; geraniol, geranyl acetate, limonene.

Yellow liquid. d_{15}^{15} 0.887-0.900. α_D^{20} +3° 43′ to −1° 22′. n_D^{20} 1.4623. Slightly sol in water; sol in 2 vols 80% alcohol. *Keep well closed, cool and protected from light.*

USE: In perfumes.

6760. Oil of Pimenta. Oil of allspice; oil of pimento. Volatile oil from fruit of *Pimenta officinalis* Lindl., *Myrtaceae. Constit.* 65-75% by vol of phenols calcd as eugenol and a sesquiterpene.

Colorless, yellow or reddish liquid; odor and taste of allspice; becomes darker with age. d_{25}^{25} 1.018-1.048. α_D^{25} −4°. Very slightly sol in water; sol in glacial acetic acid, in 1 vol 90% alcohol, in 2 vols 70% alcohol. *Keep well closed, cool and protected from light.*

THERAP CAT: Carminative.

6761. Oil of Pine Needles. Oil of Scotch fir; fir-wood oil. Volatile oil from *Pinus sylvestris* L., *Pinaceae. Constit.* Dipentene, pinene, sylvestrene, cadinene, 3-3.5% bornyl acetate.

Yellowish liquid. d_{15}^{15} 0.884-0.886. α_D^{20} +7° 3′ to +10°.

Insol in water. Sol in 10 vols 90% alcohol. *Keep well closed, cool and protected from light.*
USE: Pharmaceutic aid (flavor and perfume).
THERAP CAT: Expectorant.

6762. Oil of Rose. Otto of rose; essence of rose; attar of rose. Volatile oil from fresh flowers of *Rosa gallica* L. and *R. damascena* Mill. and varieties of these species *(Rosaceae).* *Constit.* 70-75% free geraniol and citronellol; small amounts of their esters; terpenes.
Colorless or pale yellow liquid; viscous at 25°; highly fragrant, rose odor. d_{15}^{30} 0.848-0.863. Congeals at 18-22° to a translucent, cryst mass. α_D^{25} -1° to -4°. n_D^{30} 1.457-1.463. Very slightly sol in water, sparingly sol in alcohol; sol in fatty oils, chloroform. *Keep cool in well-closed and well-filled containers and protected from light.*
USE: Largely in perfumery; for flavoring lozenges, ointments, and toilet prepns.

6763. Oil of Rosemary. Volatile oil from fresh flowering tops of *Rosmarinus officinalis* L., *Labiatae. Constit.* Not less than 10% total borneol; not less than 2.5% esters calculated as bornyl acetate; camphor, eucalyptol, pinene, camphene.
Colorless or pale yellow liquid; characteristic rosemary odor; camphoraceous taste. d_{25}^{25} 0.894-0.912. α_D^{25} -5° to $+10$°. n_D^{20} 1.464-1.476. Almost insol in water; sol in 10 vols 80% alcohol. *Keep well closed, cool and protected from light.*
USE: In liniments and in hair lotions.
THERAP CAT: Carminative, rubefacient.

6764. Oil of Rue. Volatile oil from *Ruta graveolens* L., Rutaceae. *Constit.* About 90% methyl nonyl ketone, methyl anthranilate.
Pale yellow liquid; characteristic, sharp, unpleasant odor, but odor is pleasant on dilution. d_{15}^{15} 0.832-0.845. Solidif $+8$° to $+10$°. Optically inactive or slightly dextrorotatory. n_D^{20} 1.430-1.440. Almost insol in water; sol in 3 vols 70% alcohol. *Keep well closed, cool and protected from light.*
Caution: Frequent dermal contact produces erythema, vesication. Ingestion of large quantities causes epigastric pain, nausea, vomiting, confusion, convulsions, death; may cause abortion.
USE: Flavoring agent in food.

6765. Oil of Santal. Sandalwood oil; arheol; East Indian sandalwood oil. Volatile oil from dried heartwood of *Santalum album* L., *Santalaceae. Constit.* At least 90% total alcohols calculated as santalol, about 2-4% santalol as esters.
Pale yellow, somewhat viscid liquid; characteristic sandalwood odor and taste. d_{25}^{25} 0.965-0.980. α^{25} -15° to -20°. n_D^{20} 1.500-1.510. Very slightly sol in water; sol in 5 vols 70% alcohol. *Keep well closed, cool and protected from light.*
Note: West Indian sandalwood oil is derived from *Amyris balsamifera* and is not official. Its rotation is $+24$° to $+29$°. It is less sol in 70% alc than the East Indian.
THERAP CAT: Urinary anti-infective.
THERAP CAT (VET): Formerly as a urinary antiseptic.

6766. Oil of Sassafras. Volatile oil from root of *Sassafras albidum* (Nutt.) Nees (*S. variifolium* (Salisb.) Kuntze; *S. officinale* Nees & Eberm.), *Lauraceae. Constit.* About 80% safrol; small amount of eugenol; pinene, phellandrene, sesquiterpene, *d*-camphor.
Yellow to reddish-yellow liquid; characteristic odor and taste of sassafras. d_{25}^{25} 1.065-1.077. α_D^{20} $+2$° to $+4$°. n_D^{20} 1.5250-1.5350. Very slightly sol in water; sol in 2 vols 90% alcohol. *Keep well closed, cool and protected from light.*
USE: To mask disagreeable odor of medicaments.
THERAP CAT: Topical anti-infective; pediculicide; carminative.

6767. Oil of Savin. Volatile oil from fresh tops of *Juniperus sabina* L., *Cupressaceae. Constit.* 35-55% sabinol, partly free and partly as sabinyl acetate; cadiene, pinene.
Colorless or pale yellow liquid. d_{15}^{15} 0.903-0.923. α_D $+40$° to $+60$°. n_D^{20} 1.470-1.478. Very slightly sol in water; very sol in alcohol. *Keep well closed, cool and protected from light.*
THERAP CAT: Emmenagogue. Anthelmintic.

6768. Oil of Spearmint. Oil of crispmint; oil of curled mint. Volatile oil from the flowering tops of *Mentha spicata*

L. (*M. viridis* L.), *Labiatae. Constit.* At least 50% carvone, *l*-limonene, pinene.
Colorless, yellow or greenish-yellow liq; characteristic spearmint odor and taste. d_{25}^{25} 0.917-0.934. α_D^{20} -48° to -59°. n_D^{20} 1.4820-1.4900. Very slightly sol in water; sol in equal vol 80% alcohol. *Keep well closed, cool and protected from light.*
USE: Pharmaceutic aid (flavor).
THERAP CAT: Carminative.

6769. Oil of Spike. Volatile oil from leaves and tops of *Lavandula spica* L. (*L. latifolia* Vill.), *Labiatae. Constit.* About 35% eucalyptol; camphor, linalool, borneol, terpineol, *d*-camphene, sesquiterpene.
Colorless or pale yellow liq; lavender and eucalyptol odor. d_{15}^{15} 0.900-0.920. α_D -4° to $+6$°. n_D^{20} 1.462-1.469. Practically insol in water; sol in 3 vols 70% alcohol, in 6 vols 65% alcohol. *Keep well closed, cool and protected from light.*

6770. Oil of Sweet Almond. Expressed almond oil. Fixed oil from kernels of varieties of *Prunus amygdalus* Stokes (*Amygdalus communis* L.), *Rosaceae. Constit.* Chiefly glyceryl oleate with small amounts of glycerides of linolic, etc., acids. Stated to contain no stearic acid.
Colorless or pale yellow, almost odorless, oily liq; bland taste. d_{25}^{25} 0.910-0.915. It is clear at -10°; congeals near -20°. n_D^{40} 1.4593-1.4646. Sapon no. 191-200. Iodine no. 93-100. Insol in water. Slightly sol in alcohol; miscible with benzene, chloroform, ether, petr ether. *Keep well closed, cool and protected from light.*
USE: In perfumery, manuf fine soaps; as lubricant for delicate mechanisms such as watches, firearms, etc. Emollient.
THERAP CAT: Cathartic.

6771. Oil of Sweet Bay. Volatile oil of laurel. Volatile oil from leaves of *Laurus nobilis* L., *Lauraceae. Constit.* Eucalyptol, eugenol, methyl chavicol, pinene; also esters of isobutyric and isovaleric acids.
Pale yellow to greenish liquid. d_{15}^{15} 0.92-0.93. α_D^{20} -15° to -18°. Insol in water; sol in alcohol. *Keep well closed, cool and protected from light.*
USE: In perfumes.

6772. Oil of Tansy. Volatile oil from leaves and tops of *Tanacetum vulgare* L., *Compositae. Constit.* Thujone, borneol, camphor.
Yellow liq; becomes brown on exposure to air and light. *Poisonous!* d_{15}^{15} 0.925-0.950. α_D $+30$° to $+45$°. Almost insol in water; sol in alcohol, chloroform, ether. *Keep cool in well-filled and well-closed containers, protected from light.*

6773. Oil of Thyme. Volatile oil distilled from flowering plant *Thymus vulgaris* L., *Labiatae. Constit.* 20-40% by vol of thymol and carvacrol; cymene, pinene, linalool, bornyl acetate.
Colorless to reddish-brown liq; pleasant thymol odor; sharp taste. d_{25}^{25} 0.894-0.930. α_D^{20} < -4°. n_D^{20} 1.4830-1.5100. Very slightly sol in water; sol in 2 vols 80% alcohol. *Keep well closed, cool and protected from light.*
Note: Often misnamed "oil of origanum."
THERAP CAT: Rubefacient, counterirritant, antiseptic, carminative.

6774. Oil of Turpentine. "Spirit of turpentine." Volatile oil distilled from the oleoresin obtained from *Pinus palustris* Mill., *Pinaceae* and other species of *Pinus* yielding only terpene oils.
Colorless liq; characteristic odor and taste, both becoming more pronounced and less agreeable on aging or exposure to air. d_{25}^{25} 0.854-0.868. Greater part distills between 154-170°. n_D^{20} 1.4680-1.4780. Rotation is variable. Insol in water; sol in 5 vols alcohol; miscible with benzene, chloroform, ether, carbon disulfide, petr ether and oils.
Human Toxicity: Vapors can cause eye irritation, headache, dizziness, nausea. Inhalation and ingestion can cause bladder irritation. *See also* Turpentine.
USE: Solvent for oils, resins, varnishes; vehicle for paints.
THERAP CAT: Rubefacient, counterirritant.
THERAP CAT (VET): Externally as rubefacient and counterirritant; internally as antiseptic, carminative and expectorant. Has been used as anthelmintic.

6775. Oil of Turpentine, Rectified. Turpentine oil freed from unpleasant odor and taste by treatment with sodium hydroxide and distillation.

d_{25}^{25} 0.853-0.862; otherwise has the properties of the ordinary oil. *Keep well closed, cool and protected from light.*

THERAP CAT: Inhalation expectorant.

THERAP CAT (VET): Counterirritant and rubefacient. Antiseptic, carminative and expectorant. Bronchitis, acute gastric tympany and flatulent colic in horses, rumen tympanites in cattle. Formerly as an anthelmintic in horses, cattle and sheep.

6776. Oil of Turpentine, Sulfurated. Haarlem oil; Dutch oil; Dutch drops. Mixture of 1 part sulfurated linseed oil with 3 parts oil of turpentine.

Note: Not to be confused with "Dutch liquid," which is ethylene dichloride.

Human Toxicity: Symptoms similar to oil of turpentine.

6777. Oil of Valerian. Volatile oil from rhizome and root of *Valeriana officinalis* L., *Valerianaceae.* *Constit.* Bornyl esters of acetic, formic and isovaleric acids; *l*-pinene, *l*-camphene, *l*-limonene.

Yellowish-green to brownish liquid. d_{15}^{15} 0.93-0.96. α_D −8° to −13°. Sapon no. 100-150. Acid no. 20-50. Slightly sol in water; very sol in alcohol, chloroform, ether. *Keep well closed, cool and protected from light.*

THERAP CAT: Sedative.

6778. Oil of Vetiver. Vetiver oil; oleum Andropogonis muricati; vetyver oil; khas khas oil; khus oil; cus cus oil; vetivert oil; vetiver oil Java; vetiver oil Haiti; vetiver oil Reunion (Bourbon). Distilled from roots of vetiver grass *Vetiveria zizanioides* Stapf., *(Andropogon muricatus* Retz., *Anatherum zizanioides* (L.) Hitch. & Chase, *Gramineae),* grown chiefly in Java, India, Reunion Island and Haiti. The constituents of vetiver oil vary according to the place of origin. Java and Reunion vetiver oils may contain 8-35% (usually 15-27%) sesquiterpene ketones of which α- and β-vetivones, *q.v.* have been isolated. The content of vetivenols (vetiverols, vetiver alcohols) varies from 45 to 65%. Other isolated components are vetivenyl vetivenate, vetivenic acid ($C_{15}H_{22}O_2$), palmitic acid, benzoic acid, and vetivene (C_{15}-H_{24}). Major constituents of Indian vetiver oil are khusol, khusitol, khusinol. Review of chemical studies: Anh, Fetizon, *Am. Perfum. Cosmet.* **80,** 40 (1965). Biogenetically significant components: Kaiser, Naegeli, *Tetrahedron Letters* **1972,** 2009; Paknikar *et al., ibid.* **1975,** 2973. Isoln and synthesis of *zizanal* and *epizizanal,* two insect repellent constituents, from Javanese vetiver oil: S. C. Jain *et al., Tetrahedron Letters* **23,** 4639 (1982).

Brown to reddish-brown viscous oil. Aromatic to harsh, woody odor which improves on aging. d_{15}^{15} 0.990-1.040. n_D^{20} 1.5200-1.5280. $[\alpha]_D^{20}$ +15° to +45°. Ester value after acetylation 110-165. Soluble in 1-3 volumes of 80% alcohol, becoming slightly turbid upon further dilution. Sol in all proportions in most fixed oils, diethyl phthalate, benzyl benzoate, mineral oil (slight turbidity). Practically insol in glycerol and propylene glycol. Fairly stable to dilute acids and weak alkalies; unstable to strong acids and alkalies. Should be stored in a cool place and protected from light. *Refs:* Kretchmar, Pictet, *Chimia* **8,** 123 (1954); Pfau, Plattner, *Helv. Chim. Acta* **22,** 640 (1939); Guenther, *The Essential Oils,* **vol. IV** (Van Nostrand, New York, 1950) p 156.

USE: In soap and perfumery formulations.

6779. Oil of White Cedar. Oil of arbor vitae; oil of thuja. Volatile oil from leaves of *Thuja occidentalis* L., *Cupressaceae.* *Constit.* *d*-Thujone, *l*-fenchone, *d*-pinene.

Colorless or pale yellow or greenish-yellow liq; camphor odor; bitter taste. d_{15}^{15} 0.915-0.935. α_D −5° to −14°. Insol in water. Sol in 4 vols 70% alcohol. *Keep well closed, cool and protected from light.*

USE: Perfume.

THERAP CAT: Counterirritant.

6780. Oil of Wine, "Heavy". An oily liq obtained by distilling alcohol or ether (or both) with sulfuric acid. *Constit.* Etherin and etherol ($C_2H_4)_n$, polymers of ethylene; ethyl sulfate and ethyl sulfovinate.

Colorless to slightly yellow liq; penetrating odor; sharp,

bitter taste. d_{15}^{15} 1.095-1.130. Almost insol in water; miscible with alcohol, ether.

USE: For making ethereal oil which is used in making Hoffman's anodyne; sometimes for flavoring brandy.

6781. Oil of Wormwood. Volatile oil from leaves and tops of *Artemisia absinthium* L., *Compositae.* *Constit.* Thujyl alcohol and acetate; thujone, phellandrene, cadinene; also a blue oil.

Brownish-green liq. d_{15}^{15} 0.925-0.955. n_D^{20} 1.460-1.4741. Very slightly soluble in water; sol in 2 vols 80% alcohol, in ether. *Keep well closed, cool and protected from light.*

USE: In flavoring of vermouth; formerly in absinthe.

THERAP CAT: Anthelmintic, antimalarial.

6782. Oil of Yarrow. Milfoil oil. Volatile oil from leaves and tops of *Achillea millefolium* L., *Compositae.* *Constit.* Cineol.

Blue liquid. d_{20}^{20} 0.905-0.925. Insol in water; very sol in alc, ether. *Keep well closed, cool and protected from light.*

THERAP CAT: Diaphoretic.

6783. Olaquindox. *N-(2-Hydroxyethyl)-3-methyl-2-quinoxalinecarboxamide 1,4-dioxide;* BAY Va 9391; Bayon-ox; Fedan. $C_{12}H_{13}N_3O_4$; mol wt 263.25. C 54.75%, H 4.97%, N 15.96%, O 24.31%. Prepn: Fr. pat. **1,594,628** corresp to K. Ley *et al.,* U.S. pat. **3,908,008** (1970, 1975 to Bayer). As feed additive for pigs: K. Bronsch *et al., Z. Tierphysiol., Tierernaehr. Futtermittelkd.* **36,** 211 (1976), *C.A.* **84,** 149717f (1976); R. S. Barber *et al., Anim. Feed Sci. Technol.* **4,** 117 (1979). HPLC analysis: G. F. Bories, *J. Chromatog.* **172,** 505 (1979).

Pale yellow crystals, mp 209° (dec). Slightly sol in water; insol in most organic solvents.

THERAP CAT (VET): Growth stimulant.

6784. Old Yellow Enzyme. *Reduced nicotinamide adenine dinucleotide phosphate dehydrogenase; dihydronicotinamide adenine dinucleotide phosphate diaphorase;* NADPH₂ diaphorase; OYE. Mol wt 102,000-106,000. Flavoprotein which catalyzes the oxidation of reduced triphosphopyridine nucleotide by oxygen or ferricytochrome c. When the oxidizing agent is molecular oxygen, OYE functions as a true hydrogen carrier, whereas when ferricytochrome c is the oxidizing agent, only electrons are passed on to the cytochrome, the protons being liberated as free hydrogen ions. Shows diaphorase activity as well as low cytochrome c reductase activity. Isoln from dried brewer's bottom yeast: Warburg, Christian, *Biochem. Z.* **266,** 377 (1933); Theorell, Akeson, *Biochem. Prepn.* **6,** 54 (1958). Each molecule contains two flavin mononucleotide groups: Ehrenberg, *Acta Chem. Scand.* **11,** 1257 (1957). *Review:* Akeson *et al., The Enzymes* **vol. 7,** P. D. Boyer *et al.,* Eds. (Academic Press, New York, 1963) pp 477-494.

6785. Oleandomycin. PA 105; Amimycin; Landomycin; Matromycin; Romicil. $C_{35}H_{61}NO_{12}$; mol wt 687.89. C 61.11%, H 8.94%, N 2.04%, O 27.91%. Antibiotic substance produced by *Streptomyces antibioticus* no. ATCC 11891: Sobin *et al.;* Ratajak, Nubel, U.S. pats. **2,757,123; 2,842,481** (1956, 1958 to Pfizer). Structure: Hochstein *et al., J. Am. Chem. Soc.* **82,** 3225 (1960). Absolute configuration: Celmer, *ibid.* **87,** 1797 (1965); Celmer, Hobbs, *Carbohyd. Res.* **1,** 137 (1965); S. Omura *et al., Tetrahedron Letters* **1975,** 2939. Synthetic study: K. Tatsuta *et al., ibid.* **29,** 3975 (1988). Activity: Hahn, *Antibiotics* **1,** 378, 755 (1967). For a review of macrolide antibiotics *see* Keller-Schierlein, *Fortschr. Chem. Org. Naturst.* **30,** 313-460 (1973). Toxicity: H. Sous *et al., Arzneimittel-Forsch.* **8,** 386 (1958).

White amorphous powder. uv max (methanol): 286-289 nm. Moderately sol in water. Sol in dil acids. Freely sol in methanol, ethanol, butanol, acetone. Practically insol in hexane, carbon tetrachloride, dibutyl ether.

Hydrochloride, $C_{35}H_{62}ClNO_{12}$, long needles from ethyl acetate, mp 134-135°. $[\alpha]_D^{25}$ −54° (methanol). Freely sol in water. Forms various cryst hydrates. LD_{50} in mice, rats (mg/kg): 8200, >10,000 orally; 600, 400 i.v. (Sous).

Phosphate, $C_{35}H_{64}NO_{16}P$, *Matromycin*.

Triacetyl deriv *see* Troleandomycin.

THERAP CAT: Antibacterial.

THERAP CAT (VET): Antibacterial.

6786. Oleandrin. *16β-(Acetyloxy)-3β-[(2,6-dideoxy-3-O-methyl-L-arabino-hexopyranosyl)oxy]-14-hydroxycard-20-(22)-enolide;* neriolin; Corrigen; Folinerin. $C_{32}H_{48}O_9$; mol wt 576.70. C 66.64%, H 8.39%, O 24.97%. From the leaves of *Nerium oleander* L., *Apocynaceae* (Laurier rose): Tanret, *Compt. Rend.* **194**, 914 (1932); Neumann, *Ber.* **70**, 1547 (1937). Prepn by enzymic hydrolysis of urechitoxin: Hassall, *J. Chem. Soc.* **1951**, 3193. Structure: Tschesche, *Ber.* **70**, 1554 (1937); Krasso *et al., Helv. Chim. Acta* **46**, 1691 (1963).

Crystals from dil methanol, mp 250°. $[\alpha]_D^{25}$ −48.0° (c = 1.3 in methanol). uv max: 220 nm (log ε 4.20). Practically insol in water; sol in alcohol, chloroform. LD_{50} i.v. in cats: 0.30 mg/kg.

Desacetyloleandrin, $C_{30}H_{46}O_8$, leaflets from alcohol, mp 238-240°. $[\alpha]_D^{18}$ −24.9°.

THERAP CAT: Cardiotonic; diuretic.

6787. Oleanolic Acid. *3-Hydroxyolean-12-en-28-oic acid;* oleanol; caryophyllin. $C_{30}H_{48}O_3$; mol wt 456.71. C 78.90%, H 10.59%, O 10.51%. Occurs in the free state in leaves of *Olea europaea, Oleaceae*, in leaves of *Viscum album* L., *Loranthaceae*, in buds of *Syzygium aromaticum* (L.) Merr. & Perry, *Myrtaceae* (cloves), in *Swertia japonica* (Maxim.) Makino, and in *Centaurium umbellatum* Gilib. (*Erythraea centaurium* (L.) Pers.), *Gentianaceae;* as acetate in birch bark, as glycoside in many saponins. Isoln procedures (from cloves): Winterstein, Stein, *Z. Physiol. Chem.* **202**, 222 (1931); Ruzicka, Hofmann, *Helv. Chim. Acta* **19**, 114 (1936); Picard *et al., J. Chem. Soc.* **1939**, 1047. Structure: Ruzicka *et al., Helv. Chim. Acta* **29**, 210 (1946), and earlier papers.

Review: J. Simonsen, W. C. T. Ross, *The Terpenes* vol. 5 (University Press, Cambridge, 1957) pp 221-285. *Cf.* α- and β-amyrin.

Fine, solvated needles from alc. After drying, mp 310°. $[\alpha]_D^{20}$ +83.3° (c = 0.6 in chloroform). K = 3 × 10⁻⁷. Insol in water. Sol in 65 parts ether, 106 parts 95% alcohol, 35 parts boiling 95% alcohol, 118 parts chloroform, 180 parts acetone, 235 parts methanol.

Acetate, $C_{32}H_{50}O_4$, needles from methanol, mp 268°. $[\alpha]_D^{17}$ +74.5° (c = 0.6 in CHCl₃).

Methyl ester, $C_{31}H_{50}O_3$, mp 201°. $[\alpha]_D^{20}$ +75° (c = 0.6 in CHCl₃).

Acetate of methyl ester, $C_{33}H_{52}O_4$, needles from alcohol, mp 223°. $[\alpha]_D^{20}$ +70° (c = 0.6 in CHCl₃).

6788. Oleic Acid. *(Z)-9-Octadecenoic acid.* $C_{18}H_{34}O_2$; mol wt 282.45. C 76.54%, H 12.13%, O 11.33%. Obtained by the hydrolysis of various animal and vegetable fats and oils. Prepn from olive oil: *Biochem. Prepn.* **2**, 100 (1952). Separation from olive oil by double fractionation via urea adducts: Rubin, Paisley, *Biochem. Prepn.* **9**, 113 (1962). Stereochemistry: Thieme, *Ann.* **343**, 354 (1905). Synthesis: Robinson, Robinson, *J. Chem. Soc.* **127**, 175 (1925). ¹³C-NMR studies: W. Stoffel *et al., Z. Physiol. Chem.* **353**, 1962 (1972); J. G. Batchelor *et al., J. Org. Chem.* **39**, 1698 (1974). Toxicity data: L. Orö, A. Wretlind, *Acta Pharmacol. Toxicol.* **18**, 141 (1961). Exptl use of ¹³¹I-labelled oleic acid in myocardial imaging: F. J. Bonte *et al., Radiology* **108**, 195 (1973). Review of diagnostic use of ³H-oleic acid in pancreatic function: N. T. Pedersen, *Digestion* **37**, Suppl. 1, 25-34 (1987).

$$CH_3(CH_2)_6CH_2 \quad CH_2(CH_2)_6COOH$$

Pure oleic acid is a colorless or nearly colorless liq (above 5-7°). d_{25}^{25} about 0.895. Solidifies to cryst mass, mp 4°. bp_{100} 286°. At atm pressure it dec when heated at 80-100°. n_D^{18} 1.463; n_D^{26} 1.4585. Iodine no. 89.9; acid value 198.6. On exposure to air, especially when impure, it oxidizes and acquires a yellow to brown color and rancid odor. Practically insol in water. Sol in alcohol, benzene, chloroform, ether, fixed and volatile oils. *Keep well closed, protected from light.* LD_{50} i.v. in mice: 230±18 mg/kg (Orö, Wretlind).

Several grades of the acid are available in commerce, varying in color from pale yellow to red-brown and, depending on the amount of saturated acid present, becoming turbid at 8-16°. The acid of commerce usually contains 7-12% saturated acids, e.g., stearic, palmitic; also some linoleic, etc., unsaturated acids.

Methyl ester, $C_{19}H_{36}O_2$, *methyl oleate.* Prepd by refluxing oleic acid with *p*-toluene sulfonic acid in methanol: Rubin, Paisley, *loc. cit.* Iodine no. 85.6. d_4^{18} 0.879. n_D^{20} 1.4510. bp_2 168-170°. Miscible with anhydr ethanol, ether.

Ethyl ester, $C_{20}H_{38}O_2$, *ethyl oleate, (Z)-9-octadecenoic acid ethyl ester.* Yellowish, oily liquid. d 0.87. bp 205-208° (some dec). Insol in water. Misc with alcohol, ether.

Barium salt, $C_{36}H_{66}BaO_4$, *barium oleate.* Yellowish-white, granular masses. *Poisonous!* Practically insol in water. Slightly sol in boiling alcohol.

Sodium salt, $C_{18}H_{33}NaO_2$, *Eunatrol.* White powder, slight tallow-like odor. Sol in ~10 parts water, ~20 parts alcohol. Generally contains small quantities of the sodium salts of stearic, etc. acids. Alkaline in aq solns due to hydrolysis but not in alcohol solns.

USE: Prepn of Turkey red oil, soft soap and other oleates; in polishing compds; waterproofing textiles, oiling wool; manuf driers; thickening lubricating oils. Pharmaceutic aid (solvent). The barium salt in rodent extermination. *Caution:* Mildly irritating to skin, mucous membranes.

THERAP CAT: Diagnostic aid (pancreatic function).

6789. Oleoresin of Aspidium. Oleoresin of male fern. An ether extract of male fern containing not less than 24% crude filicin.

Dark green, thick liq, usually depositing a granular, cryst substance. Bitter, unpleasant taste. d not less than 1.0. Insoluble in water; partly sol in alcohol or chloroform. Almost completely sol in ether; not less than 85% sol in petr ether. *Shake well before dispensing.*

THERAP CAT: Anthelmintic.

THERAP CAT (VET): Teniacide, flukicide.

6790. Oleuropein. $C_{25}H_{32}O_{13}$; mol wt 540.53. C 55.55%, H 5.96%, O 38.48%. Bitter glucoside belonging to a new structural-chemical class of compounds, the *iridoids*. *See* Briggs *et al., Tetrahedron Letters* **1963**, 69. Isolation from olives and the leaves and bark of the olive tree, *Olea europaea* L., *Oleaceae* and structural studies: Panizzi *et al., Gazz. Chim. Ital.* **90**, 1449 (1960); Beyerman *et al., Bull. Soc. Chim. France* **1961**, 1821; Shasha, Leibowitz, *J. Org. Chem.* **26**, 1948 (1961). Isoln from the ripe fruits of *Ligustrum lucidum* and *L. japonicum* Thunb, *Oleaceae:* Inouye, Nishioka, *Tetrahedron* **28**, 4231 (1972). Revised structure and stereo-chemistry: Inouye *et al., Tetrahedron Letters* **1970**, 2459. Pharmacology: Petkov, Manolov, *Arzneimittel-Forsch.* **22**, 1476 (1972).

O-β-D-glucopyranose

Minute crystals from ethyl acetate, mp 87-89°. Hygroscopic. $[\alpha]_D^{20}$ −147° (H₂O, alcohol, or acetone). Shows mutarotation $[\alpha]_D^{20}$ −127° after 9 hrs (H₂O). Freely sol in acetone, ethanol, methanol, pyridine, glacial acetic acid, 5% aq NaOH soln. Moderately sol in water, dioxane, butanol, ethyl acetate, butyl acetate. Practically insol in ether, petr ether, chloroform, benzene, carbon tetrachloride. *Compare* Elenolide.

6791. Oleyl Alcohol. *(Z)-9-Octadecen-1-ol; cis-9-octadecen-1-ol; Ocenol.* $C_{18}H_{36}O$; mol wt 268.47. C 80.52%, H 13.52%, O 5.96%. $CH_3(CH_2)_7CH=CH(CH_2)_7CH_2OH$. Found in fish oils. Usually obtained as a mixture of C_{16} and C_{18} unsaturated alcohols with C_{18} predominating. Prepd from butyl oleate by a Bouveault-Blanc reduction with sodium and butyl alcohol; or from triolein by hydrogenation in the presence of zinc chromite: Noller, Bannerot, *J. Am. Chem. Soc.* **56**, 1563 (1934); Reid *et al., Org. Syn.* **15**, 72 (1935); Adkins, Gillespie, *ibid.* **coll. vol. III**, 671 (1955). Purification by fractional crystallization at −40° from acetone, followed by distillation: Swern *et al., Oil Soap (Chicago)* **21**, 113 (1944); Loev, Dawson, *J. Am. Chem. Soc.* **78**, 1182 (1956).

Oily liquid. Usually pale yellow. Gives off acrid fumes when heated. d_4^{20} 0.850. mp 13-19°. bp_8 195°. $bp_{1.5}$ 182-184°. (Distilling range at 760 mm: 305-370°.) $n_D^{27.5}$ 1.4582. Insol in water; sol in alcohol, ether.

USE: Chiefly in the manuf of its sulfuric esters which are detergents and wetting agents, as an antifoam agent; metal cutting lubricant; in carbon paper, stencil paper, printing ink; as a plasticizer; for softening and lubricating textile fabrics; carrier for medicaments.

6792. Olibanum. Frankincense; gum thus. Gum resin

from *Boswellia carterii* Birdwood and other species of *Boswellia (Burseraceae). Habit.* Ethiopia, Egypt, Arabia, Somaliland. *Constit.* 3-8% volatile oil (pinene, dipentene, etc.); about 60% resins; 20% gum (polysaccharide fraction); 6-8% bassorin; bitter principle. Discussion of polysaccharides present: F. Smith, R. Montgomery, *The Chemistry of Plant Gum and Mucilages* (Reinhold, New York, 1959) pp 312-313.

6793. Oligomycins. Macrolide antibiotic complex produced by an actinomycete similar to *Streptomyces diastatochromogenes:* Smith *et al., Antibiot. & Chemother.* **4**, 962 (1954); McCoy, Peterson, U.S. pat. **2,927,057** (1960 to Wisconsin Alumni Res. Foundation). Complex of several closely related compounds: Masamune *et al., J. Am. Chem. Soc.* **80**, 6092 (1958). Separation of oligomycins A, B and C: Marty, McCoy, *Antibiot. & Chemother.* **9**, 286 (1959). Isoln and activity of D from *Streptomyces rutgersensis:* R. Q. Thompson *et al., Antimicrob. Ag. Chemother.* **1961**, 474. Partial structure of B: Prouty *et al., Biochem. Biophys. Res. Commun.* **44**, 619 (1971); total structure of B: Glehn *et al., FEBS Letters* **20**, 267 (1972). Structures of A and C: G. T. Carter, *Diss. Abstr. Int. B.* **37**, 766 (1976); of D: C. Merienne, T. Staron, *Chem. Commun.* **1978**, 318. *Review:* P. D. Shaw in *Antibiotics* vol. l, D. Gottlieb, P. D. Shaw, Eds. (Springer-Verlag, New York, 1967) pp 585-610.

oligomycin A

Oligomycin A, $C_{45}H_{74}O_{11}$. Two crystalline modifications, mp 140-141° and mp 150-151° (hexagonal crystals, melting with rapid decompn). uv max (abs ethanol): 225 nm (ε about 20,000). Solys at 25° (g/100 ml solvent): water 0.002; ether 28; benzene 6; Skellysolve B 0.02; abs ethanol 25; glacial acetic acid 37.5; acetone 85.

Oligomycin B, $C_{45}H_{72}O_{12}$, *28-oxooligomycin A.* Potent inhibitor of oxidative phosphorylation: Lardy *et al., Arch. Biochem. Biophys.* **78**, 587 (1958).

Oligomycin C, $C_{45}H_{74}O_{10}$, *12-deoxyoligomycin A.*

Oligomycin D, $C_{44}H_{72}O_{11}$, *26-demethyloligomycin A, rutamycin, A 272, RR 32705.* mp 116-119°. $[\alpha]_D^{20}$ −62° (c = 1.36 in CHCl₃).

THERAP CAT: Oligomycin D as antifungal.

6794. Olivacine. *1,5-Dimethyl-6H-pyrido[4,3-b]carbazole;* guatambuinine. $C_{17}H_{14}N_2$; mol wt 246.30. C 82.90%, H 5.73%, N 11.37%. Isolated from the bark and stem of *Aspidosperma olivaceum* Müll. Arg., *A. australe* Müll. Arg., *A. longepetiolatum* Kuhlm., *Apocynaceae:* Schmutz, Hunziger, *Pharm. Acta Helv.* **33**, 341 (1958). Identity with guatambuinine: Marini-Bettolo, Carvalho-Ferreira, *Ann. Chim.* **49**, 869 (1959). Structure: Marini-Bettolo, Schmutz, *Helv. Chim. Acta* **42**, 2146 (1959); Ondetti, Deulofeu, *Tetrahedron* **15**, 160 (1961). Synthesis: Schmutz, Wittwer, *Helv. Chim. Acta* **43**, 793 (1960); Wenkert, Dave, *J. Am. Chem. Soc.* **84**, 94 (1962); Mosher *et al., J. Med. Chem.* **9**, 237 (1966); R. Besselievre, H. Husson, *Tetrahedron Letters* **1976**, 1873; J. Bergman, R. Carlsson, *ibid.* **1978**, 4055; J. P. Kutney *et al., Heterocycles* **16**, 1469 (1981).

Fine yellow needles from dilute methanol, yellow prisms from undiluted methanol, mp 317-325°. Fluoresces in dilute alcoholic soln. uv max (ethanol): 224, 238, 276, 287, 292, 314, 329, 375 nm (log ε 4.39, 4.33, 4.70, 4.85, 4.83, 3.66, 3.80, 3.66). Soly in methanol, acetone, chloroform, carbon tetrachloride, carbon disulfide, tetrahydrofuran, dioxane less than 1%.

6795. Olivanic Acids. A family of naturally occurring carbapenem β-lactamase inhibitors with antibacterial activity, produced by *Streptomyces olivaceus:* D. Butterworth, G. N. Rolinson, **Ger.** pat. **2,146,400** corresp to **U.S.** pat. **3,919,-415** (1972, 1975 both to Beecham). Isoln: A. G. Brown *et al., J. Antibiot.* **29,** 668 (1976). The family contains seven members, all containing the 7-oxo-1-azabicyclo[3.2.0]hept-2-ene ring system. Compounds *MM 4450, MM 13902,* and *MM 17880* are sulfated and are predominant when *S. olivaceus* is grown in sodium sulfate-containing media. *MM 22380, MM 22381, MM 22382, MM 22383* are non-sulfated hydroxy analogs and are produced in sulfate-free media or by mutants of *S. olivaceus* unable to complete the sulfation process. Detection, properties, fermentation of the sulfated members: D. Butterworth *et al., ibid.* **32,** 287 (1979); isoln, characterization: J. D. Hood *et al., ibid.* 295. Fermentation, isoln, characterization of the non-sulfated members: S. J. Box *et al., ibid.* 1239. Structures of MM 4450, MM 13902: A. G. Brown *et al., Chem. Commun.* **1977,** 523; of MM 17880: D. F. Corbett *et al., ibid.* 953. Structures of MM 22380-3: A. G.Brown *et al., J. Antibiot.* **32,** 963 (1979). Total synthesis of racemic MM 22383: R. J. Ponsford, R. Southgate, *Chem. Commun.* **1980,** 1085. Comparative antibacterial activity *in vitro*: M. J. Basker *et al., J. Antibiot.* **33,** 878 (1980). MM 22380, MM 22382, MM 22381, MM 22383, MM 13902, and MM 17880 are identical to *epithienamycins* A through F, respectively, cf. E. O. Stapley *et al., ibid.* **34,** 628 (1981); P. J. Cassidy *et al., ibid.* 637. Synthetic study on epithienamycins: T. Kametani *et al., J. Chem. Soc. Perkin Trans. I* **1981,** 3048.

MM 4550, MM13902, MM 17880 are isolated as disodium salts. Sol in water, methanol, DMF, DMSO. Practically insol in other organic solvents. uv max (water): of MM 4550: 240, 287 nm ($E^{1\%}_{1cm}$ 268); MM 13902: 227, 307 nm ($E^{1\%}_{1cm}$ 356); MM 17880: 298 nm ($E^{1\%}_{1cm}$ 192). The antibiotics are unstable in aq soln outside a narrow pH range; in acids, degradation leads to changes in uv spectra. Addition of hydroxylamine or cysteine to neutral solns results in rapid degradation.

6796. Olive Oil. A fixed oil obtained from ripe olives, the fruit of the cultivated olive tree *Olea europaea* L., *Oleaceae.* Produced almost exclusively in the countries adjoining the Mediterranean Sea, Spain being the largest producer. Whole olives are crushed in edge runner mills and the oil is expressed in open hydraulic presses. *Constit.* Mixed glycerides of oleic acid 83.5%, of palmitic acid 9.4%, of linoleic acid 4.0%, of stearic acid 2.0%, of arachidic acid 0.9%. Minor constitutents are squalene, up to 0.7%, phytosterol and tocopherols about 0.2%. *Reviews and bibliographies:* José M. de Soroa y Pineda, *El aceite de oliva* (Dossat, Madrid, 1944); R. F. Simari, G. B. Martinenghi, *Olivicoltura e Oleificio* (Hoepli, Milano, 1950); P. G. Garoglio, *Technologia*

de los Aceites Vegetales (Mendoza, Argentina, 1951); E. W. Eckey, *Vegetable Fats and Oils* (Reinhold, New York, 1954).

Pale yellow or light greenish-yellow oil with a pleasing delicate flavor. Becomes rancid on exposure to air. Begins to get turbid at +5 to +10°, below 0° it forms a whitish, granular mass. Flash pt 437°F (225°C). Ignition temp 650°F (343°C). d^{15}_{15} 0.914-0.919; d^{25}_{25} 0.909-0.915. n^{25}_D 1.466-1.468; n^{40}_D 1.460-1.464. Titer 17-26°. Acid value 0.2-2.8. Saponification value 187-196. Iodine value 79-90. Thiocyanogen value 75-83. Hydroxyl value 4-12. Reichert-Meissl value 0.2-1.0. Unsaponifiable 0.5-1.3%. Slightly sol in alcohol. Miscible with ether, chloroform, carbon disulfide.

USE: As food in salads, with sardines, for cooking and baking. In the manuf of soaps, textile lubricants, sulfonated oils, cosmetics and pharmaceutical preparations. Emollient.

THERAP CAT (VET): Laxative, emollient.

6797. Olivil. *Tetrahydro-3-hydroxy-2-(4-hydroxy-3-methoxyphenyl)-4-[(4-hydroxy-3-methoxyphenyl)methyl]-3-furanmethanol.* $C_{20}H_{24}O_7$; mol wt 376.41. C 63.82%, H 6.43%, O 29.75%. From gum-resin of *Olea europaea* L., *Oleaceae.* Isoln: Pelletier, *Ann. Chim. Phys.* **51,** 196 (1833). Structure: Traverso, *Gazz. Chim. Ital.* **90,** 792 (1960); Smith, *Tetrahedron Letters* **1963,** 991. Revised structure: Ayres, Mhasalkar, *ibid.* **1964,** 335; *eidem, J. Chem. Soc.* **1965,** 3586.

Monohydrate, crystals. mp 118-120°; when anhydr, mp 142-143°. $[α]^{12}_D$ −127°. Sol in hot water, alcohol, acetic acid, fatty oils.

6798. Olivomycins. A mixture of antibiotics produced by *Streptomyces olivoreticuli:* Gauze *et al., Antibiotiki* (Moscow) **7,** 34 (1962); Brazhnikova *et al., ibid.* 39. Similar to the chromomycins, *q.v.:* Berlin *et al., Tetrahedron Letters* **1966,** 1643. Prepn: Gauze, **Brit.** pat. **1,152,748; Fr.** pat. **1,554,600** (1969, both to Sci. Res. Inst. Antibiot.), *C.A.* **71,** 37490b, 128693m (1969). Composed of olivomycins A (major

olivomycin A

component), B, C, and D. Separation: Berlin *et al., Khim. Prir. Soedin.* **1967**, 331, *C.A.* **68**, 40017w (1968). Structural elucidation of the aglycone, *olivin*, and the carbohydrate moieties: Berlin *et al., Tetrahedron Letters* **1964**, 1323; **1966**, 1431; Berlin *et al., Khim. Prir. Soedin.* **1969**, 567, *C.A.* **73**, 25823r (1970). Stereochemistry: Bakhaeva *et al., Chem. Commun.* **1967**, 10; Berlin *et al., Khim. Prir. Soedin.* **1972**, 519. Revised structure: J. Thiem, B. Meyer, *Tetrahedron* **37**, 551 (1981). Synthetic studies: J. H. Dodd, S. M. Weinreb, *Tetrahedron Letters* **1979**, 3593; J. H. Dodd *et al., J. Org. Chem.* **47**, 4045 (1982). Series of articles on pharmacology: *Antibiotiki (Moscow)* **17** (1972). Toxicity data: M. Slavik, S. K. Carter, *Adv. Pharmacol. Chemother.* **12**, 1 (1975). *Review:* J. D. Skarbek, M. K. Speedie, in *Antitumor Compounds of Natural Origin* vol. **1**, A. Aszalos, Ed.(CRC Press, Boca Raton, 1981) pp 191-235.

    Olivomycin A, $C_{58}H_{84}O_{26}$, 3^B-O-$[2,6$-*dideoxy*-3-C-*methyl*-4-O-$(2$-*methyl*-1-*oxopropyl*$)$-α-L-*arabino-hexopyranosyl*$]$-*olivomycin D.* Formerly known as olivomycin and variant I. Yellow crystals from ethanol-hexane, mp 160-165°. $[\alpha]_D^{20}$ $-36°$ (c = 0.5 in ethanol). uv max (ethanol): 228, 277, 318, 406 nm (log ϵ 4.39, 4.67, 3.81, 4.05). Sol in alc, ether, chloroform. Insol in benzene, carbon tetrachloride, petr ether, water. LD_{50} i.v. in mice: 13.75 mg/kg (Slavik, Carter).
    THERAP CAT: Antineoplastic.

6799. Olsalazine. *3,3'-Azobis(6-hydroxybenzoic acid); C.I. Mordant Yellow 5; 3,3'-dicarboxy-4,4'-dihydroxyazo-benzene; 5,5'-azobis(salicylic acid); azodisal.* $C_{14}H_{10}N_2O_6$; mol wt 302.24. C 55.64%, H 3.33%, N 9.27%, O 31.76%. Dimer of mesalamine, *q.v.*, originally used as a mordant dye. Prepn: **Ger.** pat. **278,613**; C. Mettler, U.S. pat. **1,157,169** (1914, 1915 both to J. R. Geigy). Improved prepn: K. H. Agback, A. S. Nygren, **Eur.** pat. **Appl. 36,636**; *eidem,* **U.S.** pat. **4,528,367** (1981, 1985 both to Pharmacia AB). Use in inflammatory intestinal disorders: K. H. Agback *et al.,* **Eur.** pat. **Appl. 36,637**; *eidem,* U.S. pat. **4,559,330** (1981, 1985 both to Pharmacia AB). Metabolism, distribution in man: C. P. Willoughby *et al., Gut* **23**, 1081 (1982). HPLC determn in serum, urine, feces: R. A. van Hogezand *et al., J. Chromatog.* **305**, 470 (1984). *In vitro* effect on fecal bacteria: H. Sandberg-Gertzen *et al., Scand. J. Gastroenterol.* **20**, 607 (1985). Clinical trials in patients intolerant to sulfasalazine, *q.v.:* H. Sandberg-Gertzen *et al., Gastroenterol.* **90**, 1024 (1986). Comparison with mesalamine in ulcerative colitis: S. S. Rao *et al., Scand. J. Gastroenterol.* **22**, 326 (1987). Brief review: R. A. Levinson, *Am. J. Gastroenterol.* **80**, 203 (1985).

    Disodium salt, $C_{14}H_8N_2Na_2O_6$, *azodisal sodium, disodium azodisalicylate, C.I. 14130, Ph CJ 91B, Dipentum.* Yellow powder. Sol in water, moderately sol in ethanol.
    USE: Mordant dye for wool.
    THERAP CAT: Treatment of ulcerative colitis.

6800. Omeprazole. *5-Methoxy-2-[[(4-methoxy-3,5-dimethyl-2-pyridinyl)methyl]sulfinyl]-1H-benzimidazole;* H 168/68; Antra; Losec; Mopral; Omepral. $C_{17}H_{19}N_3O_3S$; mol wt 345.42. C 59.11%, H 5.54%, N 12.17%, O 13.90%, S 9.28%. Gastric proton-pump inhibitor. Prepn: U. K. Junggren, S. E. Sjostrand, **Eur.** pat. **Appl. 5,129**; *eidem,* U.S. pat. **4,255,431** (1979, 1981 both to AB Hässle). Pharmacology: P. Muller *et al., Arzneimittel-Forsch.* **33**, 1685 (1983). Mechanism of action study: B. Wallmark *et al., Biochim. Biophys. Acta* **778**, 549 (1984). Studies of active metabolite: W. B. Im *et al., J. Biol. Chem.* **260**, 4591 (1985); P. Lindberg *et al., J. Med. Chem.* **29**, 1327 (1986). Chromatographic determn in plasma and urine: P. Lagerstrom, B. Persson, *J. Chromatog.* **309**, 347 (1984). Pharmacokinetics: P. J. Prichard *et al., Gastroenterol.* **88**, 64 (1985). Clinical trial in Zollinger-Ellison syndrome: C. B. H. W. Lamers *et al., N. Engl. J. Med.* **310**, 758 (1984); in duodenal ulcer: K. Lauritsen *et al., ibid.* **312**, 958 (1985); P. J. Prichard *et al., Brit. Med. J.* **290**, 601 (1985). Survey of preclinical data: *Scand. J. Gastroen-*

terol. **20**, Suppl 108, 1-120 (1985). Toxicological studies: L. Ekman *et al., ibid.* 53. Preliminary review of pharmacodynamics, pharmacokinetics and therapeutic use: S. P. Clissold, D. M. Campoli-Richards, *Drugs* **32**, 15-47 (1986).

    Crystals from acetonitrile, mp 156°. LD_{50} in mice, rats (g/kg): 0.08, > 0.05 i.v.; > 4, > 4 orally (Ekman).
    THERAP CAT: Anti-ulcerative; in treatment of Zollinger-Ellison syndrome.

6801. Omoconazole. *(Z)-1-[2-[2-(4-Chlorophenoxy)-ethoxy]-2-(2,4-dichlorophenyl)-1-methylethenyl]-1H-imidazole; (Z)-1-[2,4-dichloro-β-[2-(p-chlorophenoxy)ethoxy]-α-methylstyryl]imidazole;* CM 8282. $C_{20}H_{17}Cl_3N_2O_2$; mol wt 423.73. C 56.69%, H 4.04%, Cl 25.10%, N 6.61%, O 7.55%. Prepn: L. Zirngibl *et al.,* **Ger.** pat. **2,839,388**; *eidem,* U.S. pat. **4,210,657** (both 1980 to Siegfried); and crystal structure: K. Thiele *et al., Helv. Chim. Acta* **70**, 441 (1987). Stereospecific synthesis: L. Zirngibl, K. Thiele, U.S. pat. **4,554,356** (1985 to Siegfried). *In vitro* fungistatic activity: M. Mosse *et al., Pathol. Biol.* **34**, 684 (1986).

    Crystals from ethyl acetate/hexane (1:4), mp 89-90°.
    Nitrate, $C_{20}H_{18}Cl_3N_3O_5$, *Sgd 12878, Fangorex, Fongarex.* Crystals from ethyl acetate/ethanol, mp 118-120° (Büchi), 122.5° (Mettler).
    THERAP CAT: Antifungal (topical).

6802. Ondansetron. *1,2,3,9-Tetrahydro-9-methyl-3-[(2-methyl-1H-imidazol-1-yl)methyl]-4H-carbazol-4-one.* $C_{18}H_{19}N_3O$; mol wt 293.37. C 73.70%, H 6.53%, N 14.32%, O 5.45%. Specific serotonin (5HT$_3$) receptor antagonist. Prepn: I. H. Coates *et al.,* **Eur.** pat. **Appl. 191,562**; *eidem,* U.S. pat. **4,695,578** (1986, 1987 both to Glaxo). Pharmacology: A. Butler *et al., Brit. J. Pharmacol.* **94**, 397 (1988). Anxiolytic activity: R. M. Hagan *et al., Eur. J. Pharmacol.* **138**, 303 (1987); B. J. Jones *et al., Brit. J. Pharmacol.* **93**, 985 (1988). Antiemetic properties: R. Stables *et al., Cancer Treat. Rev.* **14**, 333 (1987). Clinical trials as antiemetic for cancer chemotherapy patients: D. Cunningham *et al., Lancet* **1**, 1461 (1987); M. G. Kris *et al., J. Clin. Oncol.* **6**, 659 (1988).

    Crystals from methanol, mp 231-232°.
    Hydrochloride dihydrate, $C_{18}H_{20}ClN_3O.2H_2O$, *GR 380-32F, GR-C507/75, SN 307.* White crystalline solid from water/isopropanol, mp 178.5-179.5°.
    3S-Form, $[\alpha]_D^{25}$ $-14°$ (c = 0.19 in methanol).
    3R-Form, $[\alpha]_D^{24}$ $+16°$ (c = 0.34 in methanol).
    THERAP CAT: Antiemetic.

6803. Onion Oil. Obtained from crushed onion seeds, the seeds of *Allium cepa* L., *Liliaceae:* Loew, *Ind. Quim.* **10**, 5 (1948); Phadnis *et al., J. Univ. Bombay* **17A**, no. 24, 62

(1948). Contains allylpropyl bisulfide, *S*-(1-propenyl)cysteine sulfoxide and 1-propenylsulfenic acid which is thought to be the lacrimator in onions: Virtanen, Sprare, *Suom. Kemistilehti* **35B**, 28, 29 (1962); *Chem. Ztg.* **86**, 816 (1962).

Pale yellow oil. Strong odor of onions. d_4^{15} 0.9289. Solidifies at $-15°$. n_D^{25} 1.4730. $[\alpha]_D^{20} -1.5°$. Sol in ether, chloroform, carbon disulfide.

6804. Ontianil. *N-(4-Chlorophenyl)-2,6-dioxocyclohexanecarbothioamide;* 4'-chloro-2,6-dioxothiocyclohexanecarboxanilide. $C_{13}H_{12}ClNO_2S$; mol wt 281.76. C 55.42%, H 4.29%, Cl 12.58%, N 4.97%, O 11.36%, S 11.38%. Prepn: H. Ruschig *et al.*, Ger. pat. **2,039,466** corresp to U.S. pat. **3,746,765** (1972, 1973 to Hoechst); of sodium, potassium, calcium salts: *eidem*, Ger. pat. **2,317,579** (1974 to Hoechst), *C.A.* **82**, 64537f (1975).

Cryst from ethanol, mp 113-115°C. LD_{50} in rats (mg/kg): 964 s.c.; 229 i.v., H. Ruschig *et al.*, U.S. pat. **3,746,765**. Potassium salt, $C_{13}H_{12}ClKNO_2S$, *Terondit*.

THERAP CAT (VET): Antifungal.

6805. Oosporein. *2,2',5,5'-Tetrahydroxy-4,4'-dimethyl-[bi-1,4-cyclohexadien-1-yl]-3,3',6,6'-tetrone;* 3,3',6,6'-tetrahydroxy-5,5'-dimethyl-2,2'-bi-*p*-benzoquinone; chaetomidin; iso-oosporein. $C_{14}H_{10}O_8$; mol wt 306.22. C 54.91%, H 3.29%, O 41.80%. Fungal pigment isolated from *Oospora colorans* van Beyma.: Kögl, van Wessem, *Rec. Trav. Chim.* **63**, 5 (1944); from *Chaetonium aureum* Chivers and identity with chaetomidin: Lloyd *et al.*, *J. Chem. Soc.* **1955**, 2163; from *Acremonium* spp: Divekar *et al.*, *Can. J. Chem.* **37**, 2097 (1959); from *Beauveria bassiana* (Bals.) Vuill.: Vining *et al.*, *Can. J. Microbiol.* **8**, 931 (1962). Identity with iso-oosporein: Smith, Thomson, *Tetrahedron* **10**, 148 (1960). Synthesis: J. Kalamar *et al.*, *Helv. Chim. Acta* **57**, 2368 (1974). Biosynthesis: E. Steiner *et al.*, *ibid.* 2377.

Bronze plates from aq methanol, mp 290-295°. uv max (ethanol): 216, 287 nm (log ϵ 3.51, 4.67).

Tetraacetate, $C_{22}H_{28}O_{12}$, yellow needles from methanol, mp 190°. uv max (ethanol): 262 nm (log ϵ 4.41).

Tetramethyl ether, $C_{18}H_{18}O_8$, orange needles from aq methanol, mp 123°. uv max (ethanol): 285.5, 394 nm (log ϵ 4.40, 2.98).

6806. Opianic Acid. *6-Formyl-2,3-dimethoxybenzoic acid;* 5,6-dimethoxyphthalaldehydic acid. $C_{10}H_{10}O_5$; mol wt 210.18. C 57.14%, H 4.80%, O 38.06%. Obtained (together with cotarnine) by heating narcotine with dil HNO_3. Prepn: Wilson *et al.*, *J. Org. Chem.* **16**, 792 (1951); Blair, *J. Chem. Soc.* **1955**, 708. NMR studies: Buu-Hoi *et al.*, *Bull. Soc. Chim. France* **1970**, 137.

Needles from water, mp 150°. Sol in 400 parts cold, 60 parts boiling water; sol in alcohol, ether. uv max: 215, 284 nm (ϵ 20,700, 6500).

6807. Opiniazide. *4-Pyridinecarboxylic acid [(2-carb-oxy-3,4-dimethoxyphenyl)methylene]hydrazide;* 5,6-dimethoxyphthalaldehydic acid isonicotinoylhydrazone; 1-(2-carboxy-3,4-dimethoxybenzylidene)-2-isonicotinoylhydrazine; 2-carboxy-3,4-dimethoxybenzal isonicotinoylhydrazone; saluside; saluzide; saliuzid; saluzid. $C_{16}H_{15}N_3O_5$; mol wt 329.32. C 58.36%, H 4.59%, N 12.76%, O 24.29%. Description: G. N. Pershin, S.A. Vichkanova, *C.A.* **51**, 10747e (1957); *C.A.* **52**, 1485-1486 (1958); V. A. Buskina, *C.A.* **54**, 744f (1960). Bioavailability: A. F. Zaeko, V. G. Perkova, *Farmatsiya (Moscow)* **28**, 28 (1979), *C.A.* **92**, 185805g (1980).

LD_{50} i.v. in guinea pigs: 1.634 g/kg, R. A. Akhundov, *C.A.* **86**, 50678y (1977).

THERAP CAT: Antibacterial (tuberculostatic).

6808. Opipramol. *4-[3-(5H-Dibenz[b,f]azepin-5-yl)propyl]-1-piperazineethanol;* 5-[γ-[4-(β-hydroxyethyl)piperazino]propyl]dibenzo[b,f]azepine; *N*-[3-[4-(2-hydroxyethyl)piperazino]propyl]iminostilbene; 4-[3-(5H-dibenzo[b,f]azepin-5-yl)propyl]-1-(2-hydroxyethyl)piperazine. $C_{23}H_{29}N_3O$; mol wt 363.49. C 75.99%, H 8.04%, N 11.56%, O 4.40%. Prepn: Fr. pat. **M209** (1961 to Rhône-Poulenc), *C.A.* **58**, 3442f (1963), Schindler. Ger. pat. **1,133,729** and Swiss pats. **359,143** and **360,061** (all 1962 to Geigy), *C.A.* **58**, 10219a, 10218f, 10218h (1963).

Crystals, mp 100-101°.

Dihydrochloride, $C_{23}H_{31}Cl_2N_3O_7$, *Dinsidon, Ensidon, Insidon, Nisidana*. Crystals from ethanol, mp 210° (Fr. pat., *loc. cit.*), mp 228-230° (Schindler, *loc. cit.*). Sol in water, alcohol; sparingly sol in acetone.

THERAP CAT: Antidepressant; antipsychotic.

6809. Opium. Gum opium; crude opium. Air-dried, milky exudation from incised, unripe capsules of *Papaver somniferum* L., or *P. album* Mill., *Papaveraceae. Habit.* of the plant: Asia Minor, Persia, China, Africa, India; cultivated in the Balkan States, Hungary, Southern Russia. In Japan the strain cultivated from the production of opium is called "Ikkanshu." Appearance and sources: *Chem. & Eng. News* **32**, 2701 (1954). *Constit.* About 20 alkaloids, constituting about 25% of the opium; meconic acid, some lactic and sulfuric acids, sugar, resinous and waxy-like substances; 12-25% water. Morphine is the most important alkaloid and occurs to the extent of 10-16%, noscapine 4-8%, codeine 0.8-2.5%, papaverine 0.5-2.5%, thebaine 0.5-2%.

Deodorized opium, *"denarcotized" opium*, powdered opium freed from its odor and nauseating substances by treatment with petr ether. Contains 10-10.5% anhydr morphine.

Granulated opium, opium dried at not above 70°, reduced to a 16-50 mesh powder and adjusted with lactose or other inert diluent to contain 10-10.5% anhydr morphine.

Powdered opium, opium dried at a temp not above 70°, finely powdered amd adjusted with lactose or other inert diluent to contain 10-10.5% anhydr morphine.

Caution: May be habit forming. This is a controlled substance (opiate) listed in the U.S. Code of Federal Regulations, Title 21 Parts 329.1 and 1308.12 (1987).

USE: Largely for the manuf of morphine, codeine and other opium alkaloids.

THERAP CAT: Narcotic analgesic. Hypnotic.

THERAP CAT (VET): Narcotic analgesic; antidiarrheal; antitussive.

6810. Opromazine. *2-Chloro-N,N-dimethyl-10H-phenothiazine-10-propanamine 5-oxide;* 2-chloro-10-[3-(dimethylamino)propyl]phenothiazine 5-oxide; 10-(γ-dimethyl-aminopropyl)-3-chlorphenothiazine 9-oxide; 5-oxychlor-promazine; chlorpromazine sulfoxide; Secotil. $C_{17}H_{19}ClN_2$-OS; mol wt 334.89. C 60.97%, H 5.72%, Cl 10.59%, N 8.37%, O 4.78%, S 9.57%. Metabolite of chlorpromazine, q.v.: Salzman, Brodie, J. Pharmacol. Exp. Ther. **118**, 46 (1956). Preparation by enzymatic oxidation of chlorpromazine: Gillette, Kamm, ibid. **130**, 262 (1960). Synthesis: **Fr. pat. 1,167,653** (1958 to Rhône-Poulenc). Toxicity data: Minami, Yoshimoto, Nara Igaku Zasshi **8**, 50 (1957), C.A. **51**, 16966a (1957).

Crystals, mp 115°. LD_{50} s.c. in mice: 102 mg/kg (Minami, Yoshimoto).

6811. Opsins. Broad class of species-specific proteins which form the basis of the visual pigments and of bacterio-rhodopsin, q.v. Structurally integrated into the rods and cones of the retina of the eye or into the photoreceptor membranes of certain bacteria. Each of these cell types produces a genetically specified opsin which has been classified on the basis of cellular source: *scotopsins* (rods), *photopsins* (cones), and *bacteriopsins* (bacteria). Methods for purification, prepn and assay: R. Hubbard et al., "Methodology of Vitamin A and Visual Pigments", in Methods in Enzymology Vol. **18**, D. B. McCormick, L. D. Wright, Eds. (Academic Press, New York, 1971) pp 615-653. Series of articles on photoreceptor biosynthesis: ibid. Vol. **81**, L. Packer, Ed. (Academic Press, New York, 1982) pp 763-815. Review of biosynthetic process: D. S. Papermaster, B. G. Schneider in Cell Biology of the Eye, D. McDevitt, Ed. (Academic Press, New York, 1982) p 475. The photoreceptor activity of visual pigments is due to a carotenoid chromophore, a stereospecific isomer of retinal or 3-dehydroretinal, q.q.v., bound as a protonated Schiff base to a lysine moiety in the opsin portion of the molecule: A. R. Osenoff, R. Callender, Biochemistry **13**, 4243 (1974). Each pigment has unique physiochemical properties. The most significant is the absorption spectrum which is regulated by electrostatic interactions between the chromophore and the charged or dipolar groups on the opsin: R. Hubbard, L. Sperling, Exp. Eye Res. **17**, 581 (1973); B. Honig et al., J. Am. Chem. Soc. **101**, 7084 (1979).
A visual system, a set of pigments spanning the light sensitivity range of a particular species, is generally based on one type of chromophore combined with various opsins. Visual systems based on the chromophore, retinal, are the most widespread in nature. The 11-cis isomer is utilized by rhodopsin, q.v., the most common pigment of rod cells, and by the corresponding trichromatic cone pigments (see Iodopsin). Bacteriorhodopsin is composed of bacterioopsin and trans retinal. Visual systems based on 3-dehydroretinal, exemplified by the pigments porphyropsin and cyanopsin, q.q.v., have been found to occur only in certain fish and amphibians. Visual pigments utilizing both types of chromophore have been found to co-exist in the retina of some of these species: T. E. Reuter et al., J. Gen. Physiol. **58**, 351 (1971). Exposure to light initiates the bleaching of the pigment through a series of distinct intermediates involving the isomerization and ultimate dissociation of the chromophore from the opsin: R. Hubbard, A. Kropf, Proc. Nat. Acad. Sci. USA **44**, 140 (1958); B. Honig et al., ibid. **76**, 2503 (1979). This process initiates the mechanism of energy transduction and visual excitation: T. G. Ebrey, B. Honig, Quart. Rev. Biophys. **8**, 129-184 (1975); B. Honig, Ann. Rev. Phys. Chem. **29**, 31-57 (1978); R. Uhl, E. W. Abrahamson, Chem. Rev. **81**, 291 (1981). Review of energy transduction in invertebrate photoreceptors: P. Hillman et al., Physiol. Rev. **63**, 668-772 (1983); in bacteriorhodopsin: H. V. Westerhoff, Z. Dancshazy, Trends Biochem. Sci. **9**, 112 (1984). General reviews: G. Wald, Science **162**, 230-239 (1968); D.

F. O'Brien, ibid. **218**, 961-966 (1982); A. Maeda, T. Yoshizawa, Photochem. Photobiol. **35**, 891-898 (1983); P. S. Zurer, Chem. & Eng. News **61**, 24-35 (Nov. 28, 1983). See also: Methods in Enzymology Vol. **88**, L. Packer Ed. (Academic Press, New York, 1982) 836 pp.

6812. Orange I. *4-[(4-Hydroxy-1-naphthalenyl)azo]ben-zenesulfonic acid monosodium salt;* Tropaeolin OOO no. 1; C.I. Acid Orange 20; FD & C Orange I; Ext. D & C Orange 3; C.I. 14600; α-naphthol orange; sodium azo-α-naphthol-sulfanilate. $C_{16}H_{11}N_2NaO_4S$; mol wt 350.33. C 54.85%, H 3.17%, Na 6.56%, O 18.27%, N 8.00%, S 9.15%. Prepared by coupling diazotized sulfanilic acid with α-naphthol: Colour Index vol. **4** (3rd ed., 1971) p 4065.

Reddish-brown powder. Sol in water to orange-red soln, in alcohol to orange soln. Acids ppt the aq soln. Sodium hydroxide intensifies the red color of the aq soln. pH: 7.6 brownish-yellow; 8.9 purple.
Caution: Delisted by FDA for use in foods, drugs and cosmetics.

6813. Orange II. *4-[(2-Hydroxy-1-naphthalenyl)azo]-benzenesulfonic acid monosodium salt;* C.I. Acid Orange 7; Betanaphthol orange; D & C Orange No. 4; C.I. 15510; Mandarin G; Tropaeolin OOO no. 2. $C_{16}H_{11}N_2NaO_4S$; mol wt 350.33. C 54.85%, H 3.16%, N 8.00%, Na 6.56%, O 18.27%, S 9.15%. By coupling β-naphthol with diazotized sulfanilic acid in alkaline soln, cf. Org. Syn. coll. vol. **II**, 36 (1943); Colour Index vol. **4** (3rd ed., 1971) p 4078. *Review:* H. J. Conn's Biological Stains, R. D. Lillie, Ed. (Williams & Wilkins, Baltimore, 9th ed., 1977) pp 112-113, 573.

Pentahydrate, orange needles from water. One gram dissolves in 20 ml water. Sol in alc. Absorption max 4844 Å.
USE: One of the most common dyes. Biological stain. Infrequently used as indicator: Amber to orange pH 7.4-8.6; orange to red pH 10.2-11.8. *Caution:* Delisted by the FDA for internal use in 1968. May be used externally without restriction.

6814. Orange B. *4,5-Dihydro-5-oxo-4-[(4-sulfo-1-naphthalenyl)azo]-1-(4-sulfophenyl)-1H-pyrazole-3-carb-oxylic acid 3-ethyl ester disodium salt; 5-oxo-4-[(4-sulfo-1-naphthyl)azo]-1-(p-sulfophenyl)-2-pyrazoline-3-carboxylic acid 3-ethyl ester disodium salt; 5-hydroxy-4-[(4-sulfo-1-naphthalenyl)azo]-1-(4-sulfophenyl)-1H-pyrazole-3-carb-oxylic acid 3-ethyl ester disodium salt; 1-(4-sulfophenyl)-3-ethylcarboxyl-4-(4-sulfonaphthylazo)-5-hydroxypyrazole disodium salt;* C.I. Acid Orange 137; C.I. 19235. $C_{22}H_{16}N_4$-$Na_2O_9S_2$; mol wt 590.50. C 44.75%, H 2.73%, N 9.49%, Na 7.79%, O 24.38%, S 10.86%. Prepn: W. H. Kretlow et al., **U.S. pat. 3,285,906** (1966 to Stange Co.). Review: Fed. Reg. **43**, 45611 (1978). See also Colour Index vol. **4** (3rd ed., 1971) p 4134.

Dull orange crystals. Absorption max (0.04N ammonium acetate): 442 nm. Soly in water at 77°: 220 g/l. Violet soln in conc H_2SO_4, changing to fuchsia then red on dilution. Red soln in conc HCl. Yellowish soln in 10% NaOH, changing to brownish yellow on dilution.

USE: In coloring sausage and frankfurter casings.

6815. Orange Peel, Bitter. Dried rind of unripe fruit of *Citrus aurantium* L., *Rutaceae*. *Habit.* N. India; cultivated near Mediterranean Sea, Spain, W. Indies, Florida, California, etc. *Constit.* Volatile oil, hesperidine, naringin, aurantiamarin, acrid resin, gum, tannin.

USE: Pharmaceutic aid (flavor).

6816. Orange Peel, Sweet. Fresh rind of ripe fruit of *Citrus aurantium* L., var *sinensis* L., *Rutaceae*. *Habit.* As of preceding. *Constit.* Volatile oil, hesperidin, fixed oil, resin, gum, tannin. Closely resembles bitter orange peel, but has an orange yellow color; sweetish, fragrant odor; aromatic and only slightly bitter taste.

USE: Solely as flavor.

6817. Orazamide. *1,2,3,6-Tetrahydro-2,6-dioxo-4-pyrimidinecarboxylic acid compd. with 5-amino-1H-imidazole-4-carboxamide (1:1); 5-aminoimidazole-4-carboxamide orotate;* orotic acid compound with 5(or 4)-aminoimidazole-4(or 5)-carboxamide (1:1); 4-amino-5-imidazolecarboxamide orotate; AICA orotate; Aicamin; Aicorat. $C_9H_{10}N_6O_5$; mol wt 282.22. C 38.30%, H 3.57%, N 29.78%, O 28.35%. Prepn: **Fr. pat. 1,351,141** (1964), corresp to Haraoka, Kamiya, **U.S. pat. 3,271,398** (1966); *eidem*, **Japan. pat. 26,553('64)**, *C.A.* **62**, 10450h (1965) (all three to Fujisawa). Pharmacology and toxicology: Tamura, Shibayama, *Yakugaku Kenkyu* **35**, 94 (1963), *C.A.* **64**, 2652h (1966); Hashimoto, *J. Vitaminol. (Kyoto)* **13**, 9, 19 (1967).

Obtained as the dihydrate, crystals, dec 284-285°. LD_{50} in mice: 0.6 g/kg i.p.; >4.0 g/kg orally.

THERAP CAT: Hepatoprotectant.

6818. Orcein. A dye first prepd from lichens (cudbear, *q.v.*, or *archil*). Prepn by oxidation of orcinol with H_2O_2 in the presence of ammonia water: Zulkowski, Peters, *Monatsh.* **11**, 227 (1890). Can be separated into 14 dyes by distribution chromatography: Musso, *Ber.* **89**, 1659 (1956). Structure studies: Beecken *et al.*, *Angew. Chem.* **73**, 665 (1961). Study of binding by elastin, *q.v.*, and certain collagens termed *collastin:* H. Puchter, S. N. Meloan, *Histochemistry* **64**, 119 (1979). Brief review: H. J. Conn's *Biological Stains*, R. D. Lillie, Ed. (Williams & Wilkins, Baltimore, 9th ed., 1977) pp 400-403. Review of early use as textile dye and histological stain: H. Puchter, S. N. Meloan, *loc. cit.*

Brownish-red microcryst powder. Practically insol in water, benzene, chloroform, ether, carbon disulfide. Sol in alcohol, acetone or acetic acid with red color, in dil aq alkali with bluish-violet color.

α-Aminoorcein, R_1 = H; R_2 = O; R_3 = NH_2.
α-Hydroxyorcein, R_1 = H; R_2 = O; R_3 = OH.
β- and γ-Aminoorcein, R_1 = orcinyl; R_2 = O; R_3 = NH_2.
β- and γ-Hydroxyorcein, R_1 = orcinyl; R_2 = O; R_3 = OH.
β- and γ-Aminoorceimin, R_1 = orcinyl; R_2 = NH; R_3 = NH_2.
β- and γ-Components are *cis-trans* isomers of the same compd. The above eight compds are the major components of orcein.

USE: Biological stain.

6819. Orcinol. *5-Methyl-1,3-benzenediol; 5-methylresorcinol;* orcin; 3,5-dihydroxytoluene. $C_7H_8O_2$; mol wt 124.13. C 67.73%, H 6.50%, O 25.78%. Occurs in many species of lichens: Sastry, Rao, *Curr. Sci.* **10**, 437 (1941). Prepn: Anker, Cook, *J. Chem. Soc.* **1945**, 311; Kisteneva, Rozhdestvenskii, *Zhur. Priklad. Khim.* **22**, 1108 (1949); Stevens, **U.S. pat. 2,603,662** (1952 to Gulf); Zimmer, **U.S. pat. 3,028,410** (1962 to Hooker Chem.).

Monohydrate, crystals; sweet but unpleasant taste; reddens on exposure to air due to oxidation. mp about 58°; 107° when anhydr. bp 290°; bp_{14-20} 165-170°; bp_5 147°. Freely sol in water, alcohol, ether; less sol in benzene; slightly sol in chloroform or carbon disulfide. *Keep well closed and protected from light.* LD_{50} in rats, guinea pigs: 844, 1678 mg/kg orally, Veldre *et al.*, *C.A.* **74**, 51746h (1971).

USE: As a reagent for pentoses, lignin, beet sugar, saccharoses, arabinose, diastase.

6820. Orgotein. Artrolasi; Interceptor; Ormetein (rescinded); Ontosein; Oxinorm; Palosein; Peroxidin; Peroxinorm. Water-soluble protein congeners isolated from red blood cells, liver and other tissues. Molecular weight is about 34,000 (±4 percent) with a compact conformation maintained by about 4 gram atoms of chelated divalent metals, in isolated, substantially pure, injectable form. Presently produced from beef liver as Cu-Zn mixed chelate having superoxide dismutase activity. Isoln from hemolyzed plasma-free bovine erythrocytes: Huber, **S. Afr. pat. 69 02,983**; *idem*, **U.S. pats. 3,579,495; 3,687,927** (1969, 1971, 1972 all to Diagnostic Data). Purification: Huber, **Ger. pat. 2,101,866**; *idem*, **U.S. pat. 3,624,251** (both 1971 to Diagnostic Data). Clinical study: K. Lund-Olesen, K. B. Menander-Huber, *Arzneimittel-Forsch.* **33**, 1199 (1983).

THERAP CAT: Anti-inflammatory.
THERAP CAT (VET): Anti-inflammatory; antirheumatic.

6821. Oripavine. *6,7,8,14-Tetrahydro-4,5-epoxy-6-methoxy-17-methylmorphinan-3-ol; O^3-demethylthebaine.* $C_{18}H_{19}NO_3$; mol wt 297.34. C 72.70%, H 6.44%, N 4.71%, O 16.14%. From *Papaver orientale* L., and *P. bracteatum* Lindl., *Papaveraceae:* Junusov *et al.*, *Ber.* **68**, 2158 (1935); Kiselev, Konovalova, *J. Gen. Chem. USSR* **18**, 142 (1948). Identity of O-methyl derivative with thebaine: *eidem*, *Zhur. Obshch. Khim.* **18**, 855 (1948). Biosynthesis: Gross, Dawson, *Biochemistry* **2**, 186 (1963). *Reviews:* H. L. Holmes in *The Alkaloids*, **vol. II**, Manske, Holmes, Eds. (Academic Press, New York, 1952) p 167; K. W. Bentley, *The Chemistry of the Morphine Alkaloids* (Oxford, 1954) pp 192-196. Synthesis from morphine: Barber, Rapoport, *J. Med. Chem.* **18**, 1074 (1975).

Crystals, mp 200-201°. $[\alpha]_D^{20}$ −211.8°.
Hydrochloride, $C_{18}H_{19}NO_3\cdot HCl$, crystals, dec 244-245°.
Methiodide, $C_{18}H_{19}NO_3\cdot CH_3I$, crystals, dec 207-208°.

6822. Orlon®. Polyacrylonitrile; Fiber A. Obtained by polymerizing acrylonitrile. Review of prepn, properties, and uses: R. W. Moncrieff, *Man-Made Fibres* (John Wiley, New York, 1963) pp 446-467.

White fiber. Sticks at 235°. Ironing temps above 160° may cause yellowing. Sp gr 1.17. Flammability similar to that of rayon and cotton. Generally has very good resistance to mineral acids; excellent resistance to common solvents, oils, greases, neutral salts, sunlight; fairly good resistance to weak alkalies but is degraded by strong alkalies. Resists attack by molds, mildew, insects. 100% polyacrylonitrile fibers rarely used commercially due to difficulty in dyeing.

USE: Fiber suitable for outdoor furnishings (awnings, tents, outdoor furniture), indoor furnishings, anode bags in electroplating, knitwear, rugs.

6823. Ormosinine. *21-Ormosanin-20-yl panamine.* $C_{40}H_{66}N_6$; mol wt 631.00. C 76.14%, H 10.54%, N 13.32%. From seed of *Ormosia dasycarpa* Jacks., *Leguminosae:* Hess, Merck, *Ber.* **52**, 1976 (1919); from *O. panamensis* Benth., *Leguminosae:* Lloyd, Horning, *J. Am. Chem. Soc.* **80**, 1506 (1958). Structural studies: Wilson, *Chem. & Ind. (London)* **1965**, 472; *Tetrahedron* **21**, 2561 (1965). Composed of one molecule of *panamine* and one molecule of *ormosanine*; NMR elucidation of structure: N. S. Bhacca *et al., J. Am. Chem. Soc.* **105**, 2538 (1983).

Needles from ethyl acetate, mp 219-220°. $[\alpha]_{436}^{25}$ +16.0°. $[\alpha]_{589}^{25}$ +8.9° (c = 1.29 in chloroform). Sol in chloroform; slightly sol in ether. Practically insol in water, alcohol. Sublimes to panamine.

6824. Ornidazole. *α-(Chloromethyl)-2-methyl-5-nitro-1H-imidazole-1-ethanol;* 1-(3-chloro-2-hydroxypropyl)-2-methyl-5-nitroimidazole; Ro 7-0207; Avrazor; Madelen; Ornidal; Tiberal. $C_7H_{10}ClN_3O_3$; mol wt 219.62. C 38.28%, H 4.59%, Cl 16.14%, N 19.13%, O 21.86%. Prepn: **Neth. pat. Appl. 6,606,853;** M. Hoffer, **U.S. pats. 3,435,049** and **3,493,582** (1966, 1969 and 1970, all to Hoffmann-La Roche). Activity studies: E. Grunberg *et al., Proc. Soc. Exp. Biol. Med.* **133**, 490 (1970). Synthesis and antiprotozoal activity: M. Hoffer, E. Grunberg, *J. Med. Chem.* **17**, 1019 (1974). Pharmacokinetics: D. E. Schwartz, F. Jeunet, *Chemotherapy* **22**, 19 (1976).

Crystals from toluene, mp 77-78°. uv max (2-propanol): 288, 312 nm (ε 3720, 9150). pKa: 2.4 ± 0.1. LD_{50} in rats, mice (mg/kg): 1780, 1420 orally (Grunberg). Also reported as LD_{50} in mice (mg/kg): > 2000 orally, > 2000 i.p. (Hoffer, Grunberg).

THERAP CAT: Anti-infective.

6825. Ornipressin. *8-L-Ornithinevasopressin;* orn(8)-vasopressin; POR 8 Sandoz. $C_{45}H_{63}N_{13}O_{12}S_2$; mol wt 1042.21. C 51.86%, H 6.09%, N 17.47%, O 18.42%, S 6.15%. Synthetic analog of vasopressin, *q.v.,* in which L-ornithine replaces L-arginine. Prepn: R. L. Huguenin, R. A. Boissonnas, *Helv. Chim. Acta* **46**, 1669 (1963); *eidem,* **Fr. pat. 1,396,607** corresp to **U.S. pat. 3,299,036** (1965, 1967 both to Sandoz). Separation by reversed-phase HPLC: K. Krummen, R. W. Frei, *J. Chromatog.* **132**, 27 (1977). Pharmacologic properties: B. Berde *et al., Experientia* **20**, 42 (1964). Pharmacodynamics: S. Keppens, H. de Wulf, *Biochim. Biophys. Acta* **588**, 63 (1979). Comparison of hemostatic and cardiovascular effects: L. Saarnivaara, P. Leander, *Anaesthetist* **26**, 144 (1977). Clinical study: D. J. Adendorff, D. Davies, *S. Afr. Med. J.* **51**, 131 (1977).

Cys-Tyr-Phe-Gln-Asn-Cys-Pro-Orn-GlyNH$_2$

THERAP CAT: Vasoconstrictor.

6826. Ornithine. *α,δ-Diaminovaleric acid; 2,5-diaminopentanoic acid.* $C_5H_{12}N_2O_2$; mol wt 132.16. C 45.44%, H 9.15%, N 21.20%, O 24.21%. $NH_2(CH_2)_3CH(NH_2)COOH$. An amino acid classified as nonessential for the maintenance of growth in rats. Ornithine has not been isolated from proteins except after hydrolysis with alkali. Under these circumstances DL-ornithine is obtained by secondary decompn and racemization from L-arginine. L-Ornithine has, however, been isolated from tyrocidine: Gordon *et al., Biochem. J.* **37**, 313 (1943), and from gramicidin-S: Synge, *ibid.* **39**, 363 (1945); its presence in bacitracin is suspected. Practical laboratory method for the prepn of L-ornithine from arginine by means of arginase from pig livers: Vickery, Cook, *J. Biol. Chem.* **94**, 393 (1931-2). Prepn of DL-ornithine: Kurtz, *ibid.* **122**, 477 (1938); Kline, **U.S. pat. 3,028,424** (1962 to Lilly); K. Warning *et al., Ann.* **1978**, 1707.

L-Ornithine, microcrystals from alcohol-ether. Softens at 120°, mp 140°. Usually obtained as a syrup. $[\alpha]_D^{25}$ +11.5° (c = 6.5). Aq solns are alkaline. pK_1' 1.94; pK_2' 8.65; pK_3' 10.76. Freely sol in water, alc; sparingly sol in ether. Monohydrochloride, $C_5H_{12}N_2O_2\cdot HCl$, dec 233°. $[\alpha]_D^{23}$ +11.0° (c = 5.5). Soluble in water.
Dihydrochloride, $C_5H_{12}N_2O_2\cdot 2HCl$. $[\alpha]_D^{23}$ +16.7° (c = 5.3). Soluble in water.
Monosulfate, $2C_5H_{12}N_2O_2\cdot H_2SO_4$, dec 234°. $[\alpha]_D^{25}$ +8.4°. Aspartate, *Hepa-Merz, Orparan.*

DL-Ornithine, crystals from water. Sparingly sol in alcohol. Monohydrochloride dec 233°; monosulfate dec 234°.

THERAP CAT: Anticholesteremic.

6827. Ornoprostil. *(11α,13E,15S,17S)-11,15-Dihydroxy-17,20-dimethyl-6,9-dioxoprost-13-en-1-oic acid methyl ester;* methyl (−)-(1R,2R,3R)-3-hydroxy-2-[(E)-(3S,5S)-3-hydroxy-5-methyl-1-nonenyl]-ε,5-dioxocyclopentaneheptanoate; 17S,20-dimethyl-6-oxoprostaglandin E$_1$ methyl ester; 6-oxo-17S,20-dimethyl-PGE$_1$ methyl ester; ronoprost; ONO-1308; OU-1308; Alloca; Ronok. $C_{23}H_{38}O_6$; mol wt 410.55. C 67.29%, H 9.33%, O 23.38%. Analog of prostaglandin E$_1$, *q.v.* Prepn: M. Hayashi *et al.,* **Ger. pat. 2,840,032;** *eidem,* **U.S. pat. 4,278,688** (1979, 1981 both to Ono). Effect on gastric mucosal lesions in rats: H. Kuwata *et al., Nippon Shokakibyo Gakkai Zasshi* **82**, 1858 (1985), *C.A.* **104**, 15709c (1986); *eidem, Curr. Clin. Pract. Ser.* **36**, 243 (1986); on experimental esophagitis: S. Inoue *et al., Gendai Iyro* **19**, 1348 (1987), *C.A.* **107**, 626m (1987).

Consult the cross index before using this section.

THERAP CAT: Anti-ulcerative.

6828. Orotic Acid. *1,2,3,6-Tetrahydro-2,6-dioxo-4-pyrimidinecarboxylic acid;* uracil-6-carboxylic acid; whey factor; animal galactose factor; Oropur; Orotyl. $C_5H_4N_2O_4$; mol wt 156.10. C 38.47%, H 2.58%, N 17.95%, O 41.00%. A pyrimidine precursor in animal organisms, found in milk: Bachstez, *Ber.* **64**, 2683 (1931); Hilbert, *J. Am. Chem. Soc.* **54**, 2082 (1932); Johnson, Schroeder, *ibid.* 2942. Synthesis from aspartic acid: Nye, Mitchell, *ibid.* **69**, 1382 (1947). Prepn by condensation of urea with monoethyl ester of oxalacetic acid in methanol: Scriabine, U.S. pat. 2,937,175 (1960 to Rhône-Poulenc). Older syntheses from urea and oxalacetic ester: Müller, *J. Prakt. Chem.* **56**, 488 (1897); Behrend, Struve, *Ann.* **378**, 165 (1910). Microbial process by a pyrimidine-requiring *Micrococcus glutamicus* mutant: Kinoshita *et al.*, U.S. pat. 3,086,917 (1963 to Kyowa). Classed as a vitamin: Moruzzi *et al.*, *Biochem. Z.* **333**, 318 (1960). Relationship to *vitamin B₁₃*: Manna, Hauge, *J. Biol. Chem.* **202**, 91 (1953).

Crystals from water, mp 345-346°.
Monohydrate, *Lactinium, Oroturic.* Crystals, mp 334°. uv max: 282 nm; min: 255 nm. Sol in water about 1.7 mg/ml.
Methyl ester, $C_6H_6N_2O_4$, crystals, mp 249°.
Ethyl ester, $C_7H_8N_2O_4$, crystals, mp 188-189°.
Choline ester, $C_{10}H_{17}N_3O_5$, *choline orotate, Cholergol.*
USE: Has been proposed as feed supplement in combination with methionine to aid growth of calves.
THERAP CAT: Uricosuric.

6829. Orotidine. *1,2,3,6-Tetrahydro-2,6-dioxo-3-β-D-ribofuranosyl-4-pyrimidinecarboxylic acid;* 3-β-D-ribofuranosylorotic acid; 6-carboxyuridine. $C_{10}H_{12}N_2O_8$; mol wt 288.22. C 41.67%, H 4.20%, N 9.72%, O 44.41%. An orotic acid riboside obtained from cultures of *Neurospora crassa* mutants: Michelson *et al.*, *Proc. Nat. Acad. Sci. USA* **37**, 396 (1951). Isoln: Mitchell, Michelson, U.S. pat. 2,788,346 (1957 to California Inst. Res. Found.). Structure: Fox *et al.*, *Biochim. Biophys. Acta* **23**, 295 (1957). Synthesis: Curran, Angier, *J. Org. Chem.* **31**, 201 (1966); U.S. pat. 3,282,919 (1966 to Am. Cyanamid). Synthesis of orotidine 5'-phosphate: Moffatt, *J. Am. Chem. Soc.* **85**, 1118 (1963).

Needles from methanol + benzene. Turned brown near 200° but failed to melt at 400°. uv max (methanol): 268 nm (ε 8900); in 0.1N HCl: 267 nm (ε 9570); in 0.1N methanolic NaOH: 265 nm (ε 8960). Soluble in hot water, lower aliphatic alcohols and aq solns of such alcohols.

Cyclohexamine salt, $C_{16}H_{25}N_3O_8$, crystals from ethanol + benzene, mp 183-084°. $[\alpha]_D$ +15° (c = 1). Also isolated as the lead salt.
Orotidine 5'-phosphate trisodium salt trihydrate, $C_{10}H_{10}N_2O_{11}PNa_3.3H_2O$, uv max (0.1N HCl): 267 nm (ε 9430). Soluble in water.
N^3-Methylorotidine methyl ester, $C_{12}H_{16}N_2O_8$, stout crystals from isopropanol, mp 135-137°. uv max (methanol): 271 nm (ε 7620).

6830. Oroxylin A. *5,7-Dihydroxy-6-methoxy-2-phenyl-4H-1-benzopyran-4-one;* 5,7-dihydroxy-6-methoxyflavone; oroxylin. $C_{16}H_{12}O_5$; mol wt 284.26. C 67.60%, H 4.26%, O 28.14%. From root bark of *Oroxylum indicum* Vent., *Bignoniaceae:* Naylor, Chaplin, *Pharm. J.* **20**, 257 (1890); Naylor, Dyer, *J. Chem. Soc.* **79**, 954 (1901); Row *et al.*, *Proc. Indian Acad. Sci.* **28A**, 189 (1948). Structure: Shah *et al.*, *J. Chem. Soc.* **1936**, 591; **1938**, 1555. Synthesis: Murti, Seshadri, *Proc. Indian Acad. Sci.* **29A**, 1 (1949); Sarin, Seshadri, *J. Sci. Ind. Res. (India)* **19B**, 117 (1960); Molho, Gerphagnon, *Bull. Soc. Chim. France* **1963**, 607; Varady, *Tetrahedron Letters* **1965**, 4281.

Yellow plates from ethanol, mp 231-232°. Sol in alc, acetone, hot benzene, ether, alkalies, glacial acetic acid; sparingly sol in chloroform; practically insol in water.
Diacetyl oroxylin, $C_{20}H_{16}O_7$, needles from alcohol, mp 131-132°.

6831. Orphenadrine. *N,N-Dimethyl-2-[(2-methylphenyl)phenylmethoxy]ethanamine; N,N-dimethyl-2-(o-methyl-α-phenylbenzyloxy)ethylamine;* o-methyldiphenhydramine; o-monomethyldiphenhydramine; 2-(phenyl-o-tolylmethoxy)ethyldimethylamine; phenyl-o-tolylmethyl dimethylaminoethyl ether; β-dimethylaminoethyl 2-methylbenzhydryl ether; BS 5930; Biorphen; Brocasipal; Disipal. $C_{18}H_{23}$-NO; mol wt 269.37. C 80.25%, H 8.61%, N 5.20%, O 5.94%. Prepd by the action of 2-methylbenzhydryl chloride on dimethylaminoethanol: Bijlsma *et al.*, *Arzneimittel-Forsch.* **5**, 72 (1955); Harms, Nauta, *J. Med. Pharm. Chem.* **2**, 57 (1960). Covered, but not described in Rieveschl, U.S. pat. 2,567,351 (1951 to Parke, Davis). Resolution of optical isomers: van der Stelt *et al.*, *Arzneimittel-Forsch.* **19**, 2010 (1969). Synthesis and pharmacological study of metabolites: Den Besten *et al.*, *ibid.* **20**, 538 (1970). Prepn of dosage forms: Harms, U.S. pat. 2,991,225 (1961 to Brocades-Stheeman).

Liquid. bp₁₂ 195°.
Hydrochloride, $C_{18}H_{24}ClNO$, *Mephenamin.* Crystals, mp 156-157°. Sol in water, alcohol, chloroform. Sparingly sol in acetone, benzene. Practically insol in ether. pH of aq soln about 5.5.
Citrate, $C_{24}H_{31}NO_8$, *Banflex, Norflex, X-Otag.*
THERAP CAT: Relaxant (skeletal muscle); antihistaminic.

6832. Orris. White flag. Rhizome of *Iris florentina* L., *I. pallida* Lam. or *I. germanica* L., *Iridaceae. Habit.* Northern Italy, Germany, France. *Constit.* Iridin, irone, ionone, resin, starch, volatile oil (butter of orris).
USE: As dusting powder. Principally in dentifrices, perfumes, and cosmetics.

6833. Orris Root Oil. *Constit.* About 85% myristic acid; the odorous pinciple irone; methyl myristate, oleic aldehyde.

Yellowish-white to yellow, semisolid fatty substance; intense odor of dried orris root. mp 44-50°. Slightly dextrorotatory. Acid no. 213-222. Sapon no. 2-6.

USE: In perfumes.

6834. *o*-Orsellinic Acid. *2,4-Dihydroxy-6-methylbenzoic acid; 6-methyl-β-resorcylic acid;* 4,6-dihydroxy-*o*-toluic acid; 2,4-dihydroxy-6-methylbenzenecarboxylic acid; orcinolcarboxylic acid. $C_8H_8O_4$; mol wt 168.14. C 57.14%, H 4.80%, O 38.06%. Found in conjugated form or in depside form in *Roccella* and *Lecanora* lichens, postulated to arise by autocondensation of acetoacetic acid. Isoln from the fungus *Chaetomium cochliodes:* Mosback, *Z. Naturforsch.* **14B**, 69 (1959). Synthesis: Sonn, *Ber.* **61**, 926 (1928); Kloss, Clayton, *J. Org. Chem.* **30**, 3566 (1965).

Needles from acetone, mp 176° (effervescence). K (25°) = 1.27 × 10⁻⁴. uv max (0.1*N* HCl): 214, 260, 296 nm; in 0.1*N* NaOH: 272 nm. Sol in water, alcohol, glycerol. Soly in ether at 20° = 15.7%. Slightly sol in benzene.
Monohydrate, needles from water, mp 186-189°.
Methyl ester, $C_9H_{10}O_4$, crystals, mp 140°.
Ethyl ester, $C_{10}H_{12}O_4$, crystals, mp 132°.

6835. Orthanilic Acid. *2-Aminobenzenesulfonic acid; o-sulfanilic acid; o-anilinesulfonic acid.* $C_6H_7NO_3S$; mol wt 173.18. C 41.61%, H 4.07%, N 8.09%, O 27.72%, S 18.51%. Prepn from *o*-nitrobenzenesulfonyl chloride: Wertheim, *Org. Syn.,* **coll. vol. II,** 271 (1943).

Minute hexagonal plates, dec ~325°. Slow crystallization from water below 13.5° may yield a hemihydrate. K (25°) = 3.3 × 10⁻³. Slowly and sparingly sol in water.

6836. Orthocaine. *3-Amino-4-hydroxybenzoic acid methyl ester;* methyl 3-amino-4-hydroxybenzoate; Orthoform; Orthoform New. $C_8H_9NO_3$; mol wt 167.16. C 57.48%, H 5.43%, N 8.38%, O 28.71%. Prepd by dissolving 3-amino-4-hydroxybenzoic acid in methanol, saturating with HCl gas: Einhorn, Pfyl, *Ann.* **311**, 46 (1900); **Ger. pat.** **97,333;** *Chem. Zentr.* **1898,** II, 525. Also prepd by reduction of methyl 3-nitro-4-hydroxybenzoate with aluminum amalgam: Auwers, Röhrig, *Ber.* **30**, 991 (1897); with tin and HCl or with stannous chloride and alcoholic HCl: **Ger. pat.** **97,334;** *Chem. Zentr.* **1898,** II, 526.

Needles from benzene, mp 143°. When crystallized from chloroform it sometimes assumes an allotropic form, mp 111°, which on melting changes to the normal form and after solidifying always melts at 143°. Odorless and tasteless. Neutral reaction. Almost insol in cold water. Moderately sol in hot water with gradual decompn forming 3-amino-4-hydroxybenzoic acid and methanol. One gram dissolves in 6 ml alcohol, 50 ml ether; readily dissolves in aq NaOH. Forms water sol salts with HCl and HBr.
Note: The name Orthoform was applied originally to the methyl ester of 4-amino-3-hydroxybenzoic acid, mp 121°, which was also used as a local anesthetic. This compd now is designated as Orthoform Old.

THERAP CAT: Topical anesthetic.
THERAP CAT (VET): Topical anesthetic.

6837. Orthoformic Acid. *Methanetriol;* trihydroxymethane. Hypothetical compd, illustrated for nomenclature purposes only. The hydrogens of the hydroxyl groups are replaceable with alkyl groups giving rise to orthoformic esters, which are real compds (produced by various syntheses, e.g. from sodium alcoholates and chloroform). Monograph: H. W. Post, *The Chemistry of the Aliphatic Orthoesters* (Reinhold, New York, 1943), 188 pp.

Trimethyl ester, $C_4H_{10}O_3$, *methyl orthoformate, trimethoxymethane.* Prepd from chloroform and methanol in presence of sodium: Sah, Ma, *J. Am. Chem. Soc.* **54**, 2965 (1932). Liquid. d_4^{20} 0.9676; d_4^{25} 0.9623. bp_{760} 100.6°. n_D^{25} 1.3773.
Triethyl ester, $C_7H_{16}O_3$, *ethyl orthoformate, 1,1',1'-[methylidynetris(oxy)]tris[ethane], triethoxymethane, triethyl orthoformate, Aethon.* Prepd from chloroform and ethanol in presence of sodium: Chu, Shen, *C.A.* **38**, 2930⁷ (1944). Liquid, sweetish odor resembling that of pine needles. d_4^{20} 0.8909; d_4^{25} 0.8858. bp_{765} 143°. mp below −18°. n_D^{25} 1.3900. Slightly sol in water with decompn; misc with alc, ether. LD_{50} orally in rats: 7.06 g/kg, H. F. Smyth *et al., Arch. Ind. Hyg. Occup. Med.* **4**, 119 (1951).

6838. Orthovaleric Acid. Hypothetical compd, illustrated for nomenclature purposes only. The hydrogens of the hydroxyl groups are replaceable with alkyl groups giving rise to orthovaleric esters, which are described compds produced by various syntheses. Monograph: H. W. Post, *The Chemistry of the Aliphatic Orthoesters* (Reinhold, New York, 1943) 188 pp.

Trimethyl ester, *methyl orthovalerate.* Prepd from methyl imidovalerate: McElvain *et al., J. Am. Chem. Soc.* **68**, 1922 (1946). Liquid. d_4^{27} 0.9413. bp_{760} 165°. n_D^{24} 1.4090.

6839. Oryzacidin. $C_8H_{13}NO_5$; mol wt 203.19. C 47.29%, H 6.45%, N 6.89%, O 39.37%. Antibiotic substance produced by a strain of *Aspergillus oryzae:* Shimoda, *J. Agr. Chem. Soc. Japan* **25**, 254 (1951), *C.A.* **46**, 10284 (1952); *idem,* **Japan.** pat. 1594('52), *C.A.* **47**, 6097 (1953). Production of oryzacidin A by *Streptomyces: idem,* **Japan.** pat. **13,198**('65), *C.A.* **63**, 12284c (1965).
Hygroscopic needles, dec 162-163°. $[\alpha]_D^{13.5}$ −133°. Sol in water, methanol, ethanol, isopropanol, hot acetone. Slightly sol in butanol, ethyl acetate. Practically insol in ether, acetone, chloroform, carbon tetrachloride.
Quinine salt, tabular crystals, dec 210°.
Oryzacidin A, $C_{22}H_{34}N_4O_{12}$. Orange, amorphous powder, mp 231°. Soluble in ether, benzene, carbon tetrachloride, methanol, ethanol, butyl acetate, chloroform, pyridine, dioxane, acetic acid, toluene, xylene. Slightly sol or insol in water, acetone, carbon disulfide, isopropyl ether, petr ether, hexane. pH of ethanolic soln, 6.6. uv max: 243 nm ($E_{1cm}^{1\%}$ 122.5). Stable in direct sunlight.

6840. Oryzalin. *4-(Dipropylamino)-3,5-dinitrobenzenesulfonamide; 3,5-dinitro-N^4,N^4-dipropylsulfanilamide;* EL 119; Dirimal; Ryzelan; Surflan. $C_{12}H_{18}N_4O_6S$; mol wt 346.36. C 41.61%, H 5.24%, N 16.18%, O 27.72%, S 9.25%. Selective pre-emergence herbicide. Prepn: Q. F. Soper, **U.S. pat.** **3,367,949** (1968 to Lilly). Soil degradation: T. Golab *et al., Pestic. Biochem. Physiol.* **5**, 196 (1975); E. W. Stoller, L. M. Wax, *J. Environ. Qual.* **6**, 124 (1977). Review: O. D. Decker, W. S. Johnson, *Anal. Methods Pestic. Plant Growth Regul.* **8**, 433 (1976).

Yellow-orange crystals, mp 137-138° (tech, 141-142°). Soly in water at 25°: 2.5 ppm. Sol in acetone, ethanol, methanol, acetonitrile; slightly sol in benzene; insol in hexane. Vapor press at 30°: $< 1 \times 10^{-7}$ mm Hg. LD_{50} orally in rats: > 10 g/kg, O. D. Decker, W. S. Johnson, *loc. cit.*

USE: Herbicide.

6841. γ-Oryzanol. OZ; γ-OZ; γ-orizanol; Caclate; Gammajust 50; Gamma-Oz; Gammariza; Gammatsul; Guntrin; Hi-Z; Maspiron; Oliver; Oryvita; Oryzaal; Thiaminogen. Mixture of ferulic acid esters of sterols (campestrol, stigmasterol, β-sitosterol) and triterpene alcohols (cycloartanol, cycloartenol, 24-methylenecycloartanol, cyclobranol) extracted from rice bran, corn and barley oils. Isoln from rice bran oil: R. Kaneko, T. Tsuchiya, *J. Chem. Soc. Japan, Ind. Chem. Sect.* **57**, 526 (1954); T. Tsuchiya *et al.*, **Japan. pat. 56 7182** (1956 to Agcy Ind. Sci. Technol.), *C.A.* **52**, 15848b (1958); *eidem*, **Japan. pat. 57 4895** (1957 to Bureau Ind. Technics), *C.A.* **52**, 5758i (1958); from corn oil: T. Tsuchiya, O. Okubo, **Japan. pat. 60 2945** (1960 to Bureau Ind. Technics), *C.A.* **54**, 25609g (1960). Extraction from rice bran, corn and barley oils and purification: T. Yamamoto, **Ger. pat. 1,301,002** (1969 to Toyo Koatsu), *C.A.* **71**, 128704r (1969). Separation of 3 major components designated oryzanols A, B and C: M. Shimizu *et al.*, *Chem. Pharm. Bull.* **5**, 36 (1957). Structure of A: G. Ohta, M. Shimizu, *ibid.* 40; of C: *eidem, ibid.* **6**, 325 (1958); G. Ohta, *ibid.* **8**, 5, 9 (1960). Oryzanol B was subsequently found to be a mixture of A and C: M. Shimizu, G. Ohta, *ibid.* 108. Separation and properties of the seven components of γ-oryzanol: T. Endo *et al., Yukagaku* **17**, 344 (1968), *C.A.* **69**, 37319m (1968); *eidem, ibid.* **18**, 63 (1969), *C.A.* **70**, 89042f (1969). General pharmacology: Y. Yamaji *et al., Oyo Yakuri* **25**, 947 (1983), *C.A.* **99**, 151730h (1983). Antioxidant activity in cytochrome model systems: K. Tajima *et al., Biochem. Biophys. Res. Commun.* **115**, 1002 (1983). Effect on lipid metabolism in rats: M. Shinomiya, *Tohoku J. Exp. Med.* **141**, 191 (1983). Antiulcerative effect on gastric lesions in mice: Y. Ichimaru *et al., Nippon Yakurigaku Zasshi* **84**, 537 (1984), *C.A.* **102**, 55922g (1985). Clinical evaluation in chronic gastritis: T. Arai, *Kitakanto Igaku* **30**, 71, 85 (1980); in hyperlipidemia: G. Yoshino *et al., Curr. Ther. Res.* **45**, 543 (1989).

oryzanol A

White or slightly yellowish, tasteless powder with little or no odor. Crystals from acetone, mp 135-137°. uv max (heptane): 216, 231, 291, 315 nm.

Oryzanol A, $C_{40}H_{58}O_4$, (3β)-9,19-cyclolanost-24-en-3-ol 3-(4-hydroxy-3-methoxyphenyl)-2-propenoate, cycloartenyl ferulate. Prepd as the monohydrate, mp 150-151.5°. $[\alpha]_D$ +40° (c = 0.68). uv max (heptane): 231, 290, 315 nm (log ε 4.15, 4.24, 4.34).

Oryzanol C, $C_{41}H_{60}O_4$, (3β)-24-methylene-9,19-cyclolano-

stan-3-ol 3-(4-hydroxy-3-methoxyphenyl)-2-propenoate, 24-methylenecycloartanyl ferulate.

THERAP CAT: Antiulcerative; antihyperlipoproteinemic. Also used in treatment of menopausal syndrome.

6842. Osajin. 5-Hydroxy-3-(4-hydroxyphenyl)-8,8-dimethyl-6-(3-methyl-2-butenyl)-4H,8H-benzo[1,2-b:3,4-b']-dipyran-4-one. $C_{25}H_{24}O_5$; mol wt 404.44. C 74.24%, H 5.98%, O 19.78%. Isoln from osage oranges, *Maclura pomifera* (Raf.) Schneid. (*M. aurantiaca* Nutt.), *Moraceae*: Walter *et al., J. Am. Chem. Soc.* **60**, 574 (1938); Wolfrom, Mahan, *ibid.* **64**, 308 (1942). Structure: Wolfrom *et al., ibid.* **68**, 406 (1946). Synthesis of osajin 4'-methyl ether: Jain *et al., Tetrahedron* **26**, 1977 (1970). Synthesis of osajin: Jain, Sharma, *J. Org. Chem.* **39**, 2215 (1974).

Light yellow crystals from xylene or 95% ethanol, mp 189° (uncorr), 193° (corr). Practically insol in water, petr ether, moderately sol in benzene, alcohol, warm carbon tetrachloride; very sol in chloroform, ether, acetone, pyridine. uv max (abs alc): 274 nm, Horowitz, Jurd, *J. Org. Chem.* **26**, 2446 (1961).

6843. Osalmid. 2-Hydroxy-N-(4-hydroxyphenyl)benzamide; 4'-hydroxysalicylanilide; N-(p-hydroxyphenyl)salicylamide; N-salicoylaminophenol; oksafenamide; oxaphenamide; Bichol; Dribazil; Driol; Driol-Labaz; Enidran; Kanochol; Saryuurin. $C_{13}H_{11}NO_3$; mol wt 229.23. C 68.11%, H 4.84%, N 6.11%, O 20.94%. Prepn: Weizmann *et al., J. Org. Chem.* **13**, 796 (1948). Description: *Subsidia Medica* **8**, 103 (1956).

Crystals, mp 179°. Practically insol in cold water, acetic acid. Slightly sol in warm water, benzene, toluene. Freely sol in methanol, ethanol, ether, acetone.

Diacetate, $C_{17}H_{15}NO_5$, needles from alcohol, mp 151°.

THERAP CAT: Choleretic.

6844. Osmaron B. Composed of benzoates of primary aliphatic fatty amines obtained from palm kernel oil and corn oil as starting materials. Preponderantly dodecylammonium benzoate and tetradecylammonium benzoate. Total nitrogen 4.15% by analysis; average benzoic acid content 27.6% by analysis.

Yellowish-brown, thick oil. mp −3 to −4°. n_D^{40} 1.4855-1.4885. Vapors of Osmaron B turn wet litmus paper blue and turmeric paper brown. Soluble in methanol, ethanol, ethylene glycol, glycerol, acetone, ethyl acetate, carbon disulfide, formic acid, acetic acid, and lactic acid. Forms turbid solns with ether, gasoline, benzene and its homologs, and chlorinated hydrocarbons which become clear upon deposition of the fine precipitate on standing. *Ref:* E. Benk, *Chem. Ztg.* **75**, 351 (1951).

USE: Disinfectant; cationic surface-active agent; in udder ointments used with milking machines.

6845. Osmium. Os; at. wt 190.2; at. no. 76; valences 1-8; most common states 3, 4, 6. Seven naturally occuring isotopes: 184 (0.02%); 186 (1.6%); 187 (1.6%); 188 (13.3%); 189 (16.1%); 190 (26.4%); 192 (41.0%); artificial radioactive isotopes: 181-183; 185; 191; 193-195. Occurrence in earth's crust about 0.001 ppm. Found in the mineral osmiridium and in all platinum ores. Discovered by Tennant in 1804. Prepn: Berzelius *et al.*, cited by Mellor, *A Comprehensive Treatise on Inorganic and Theoretical Chemistry* **15**, 687 (1936). Reviews of prepn, properties and chemistry of osmium and other platinum metals: Gilchrist, *Chem. Rev.* **32**, 277-372 (1943); Beamish *et al.* in *Rare Metals Handbook*, C. A. Hampel, Ed. (Reinhold, New York, 1956) pp 291-328; Griffith, *Quart. Rev.* **19**, 254-273 (1965); *idem, The Chemistry of The Rarer Platinum Metals* (John Wiley, New York, 1967) pp 1-125; Livingstone in *Comprehensive Inorganic Chemistry*, vol. **3**, J. C. Bailar, Jr. *et al.*, Eds. (Pergamon Press, Oxford, 1973) pp 1163-1189, 1209-1233.

Bluish-white, lustrous metal; close-packed hexagonal structure. d_4^{20} 22.61; long believed to be the densest element; x-ray data show it to be slightly less dense than iridium. mp about 2700°. bp about 5500°. Sp heat (0°) 0.0309 cal/g/°C. Hardness 7.0 on Mohs' scale. Electrical resistivity (0°) 8.12 μohms-cm. Stable in air in the cold; when finely divided, is slowly oxidized by air even at ordinary temp to form tetroxide. Attacked by fluorine above 100°; by dry chlorine on heating; not attacked by bromine or iodine. Attacked by aqua regia, by oxidizing acids over a long period of time; barely affected by HCl, H_2SO_4. Burns in vapor of phosphorus to form a phosphide, in vapor of sulfur to form a sulfide. Attacked by molten alkali hydrosulfates, by potassium hydroxide and oxidizing agents. Finely divided osmium absorbs a considerable amount of hydrogen.

Osmarins are high-molecular weight polymers of carbohydrate and osmium. Potential use in the treatment of arthritis: *Chem. & Eng. News* **60**, 8 (April 5, 1982).

USE: As alloy with iridium for pen points and fine machine bearings; as catalyst in the synthesis of ammonia; as catalyst in hydrogenation of organic compounds.

6846. Osmium Hexafluoride. F_6Os; mol wt 304.20. Os 62.52%, F 37.48%. OsF_6. Prepd by fluorination of osmium metal: Weinstock, Malm, *J. Am. Chem. Soc.* **80**, 4466 (1958); Hargreaves, Peacock, *Proc. Chem. Soc. (London)* **1959**, 85. Previously thought to be *osmium octafluoride:* Ruff, Tschirch, *Ber.* **46**, 929 (1913).

Pale yellow, volatile solid. *Highly poisonous! Very corrosive to skin!* mp 32.1°; bp 45.9°. Also reported: mp 33.4°; bp 47.5°, Cady, Hargreaves, *J. Chem. Soc.* **1961**, 1563. Hydrolyzed on exposure to moisture; forms white corrosive fumes which soon turn bluish. May be stored in quartz ampuls.

6847. Osmium Tetrachloride. Cl_4Os; mol wt 332.00. Cl 42.71%, Os 57.29%. $OsCl_4$. Prepn: Kolbin *et al.*, *Russ. J. Inorg. Chem.* **8**, 1270 (1963).

Red crystals. Sublimes at 450°. d_4^{20} 4.38. Soluble in water to give a yellow soln which on standing hydrolyzes into osmium oxides and HCl.

6848. Osmium Tetroxide. Osmic acid. O_4Os; mol wt 254.20. O 25.18%, Os 74.82%. OsO_4. Prepd by heating (at 300-400°) finely divided osmium metal in a stream of air or oxygen. Lab prepn: Grube in *Handbook of Preparative Inorganic Chemistry*, vol. **2**, G. Brauer, Ed. (Academic Press, New York, 2nd ed., 1965) pp 1603-1604. Use in treatment of arthritis: M. Nissilä *et al.*, *Scand. J. Rheumatol.* **5**, 111 (1977); A. S. Hendricson *et al.*, *Acta Orthop. Scand.* **52**, 17 (1982). Review of chemistry and biochemistry: W. P. Griffith, *Platinum Metal Rev.* **18**, 94-96 (1974).

Pale yellow solid; monoclinic crystals. mp 40.6°. d 5.10 (calc): Ueki *et al.*, *Acta Cryst.* **19**, 157 (1965). Acrid, chlorine-like odor. Minimum perceptible concn 0.02 mg/liter of air. bp_{760} 130.0°; bp_{400} 109.3°; bp_{200} 89.5°; bp_{100} 71.5°; bp_{60} 59.4°. Begins to sublime and distil well below the boiling point. Vapor press at 27°: 11 mm. Critical temp 405°; crit press. 170 atm. Sol in benzene. Soly at 25° (g/100 g): water 7.24; carbon tetrachloride 375: Anderson, Yost, *J. Am. Chem. Soc.* **60**, 1822 (1938). Also sol in alc, ether, ammonium hydroxide, phosphorus oxychloride.

Caution: Safeguards necessary when opening container. Vapor poisonous! Irritant; causes damage to eyes, respiratory tract, skin: E. Browning, *Toxicity of Industrial Metals* (Appleton-Century-Crofts, New York, 2nd ed., 1969) pp 261-266.

USE: Oxidizing agent, particularly for converting olefins to glycols. Catalyzes chlorate, peroxide, periodate, and other oxidations: P. N. Rylander, *Organic Syntheses with Noble Metal Catalysts* (Academic Press, New York, 1973) pp 121-144. As fixing and staining agent for cell and tissue studies.

6849. Osthole. *7-Methoxy-8-(3-methyl-2-butenyl)-2H-1-benzopyran-2-one; 7-methoxy-8-(3-methyl-2-butenyl)coumarin;* 8-(3-methyl-2-butenyl)herniarin. $C_{15}H_{16}O_3$; mol wt 244.28. C 73.75%, H 6.60%, O 19.65%. From rhizome of *Peucedanum ostruthium* (L.) Koch *(Imperatoria ostruthium* L.) *Umbelliferae:* Herzog, Krohn, *Arch. Pharm.* **247**, 553 (1909); Butenandt, Marten, *Ann.* **495**, 187 (1932); from *Prangos pabularia* Lindl., *Umbelliferae:* Pigulevskii, Kuznetsova, *Doklady Akad. Nauk. SSSR* **61**, 309 (1948), *C.A.* **43**, 3416d (1949); from *Flindersia bennettiana* F. Muell., *Rutaceae:* Galbraith *et al.*, *Aust. J. Chem.* **13**, 427 (1960). Structure: Späth, Pesta, *Ber.* **66**, 754 (1933). Synthesis: Späth, Holzen, *ibid.* **67**, 264 (1934); Murayama *et al.*, *Chem. Pharm. Bull.* **20**, 741 (1972).

Prisms from ether. mp 83-84°. uv max: 322, 258 nm (ϵ 8000, 4300). Practically insol in water; sol in aq alkalies, alcohol, chloroform, acetone, boiling petr ether.

6850. Ostruthin. *6-(3,7-Dimethyl-2,6-octadienyl)-7-hydroxy-2H-1-benzopyran-2-one; 6-(3,7-dimethyl-2,6-octadienyl)-7-hydroxycoumarin;* 6-(3,7-dimethyl-2,6-octadienyl)-umbelliferone. $C_{19}H_{22}O_3$; mol wt 298.37. C 76.48%, H 7.43%, O 16.09%. From the root of *Peucedanum ostruthium* (L.) Koch *(Imperatoria ostruthium* L.), *Umbelliferae:* Butenandt, Marten, *Ann.* **495**, 187 (1932); from *Eriostemon tomentellus, Rutaceae:* Duffield, Jeffries, *Aust. J. Chem.* **16**, 123 (1963). Structure: Späth, Klager, *Ber.* **67**, 859 (1934); Späth, Kainrath, *ibid.* **70**, 2272 (1937).

Crystals from alcohol, mp 117-119°. Practically insol in water, petr ether; sol in chloroform, ethyl acetate, hot alc. Acetate, $C_{21}H_{24}O_4$, leaflets from alcohol, mp 81°. Practically insol in water; sol in alc, ether, benzene, chloroform.

6851. Ostruthol. *2-Methyl-2-butenoic acid 2-hydroxy-2-methyl-1-[[(7-oxo-7H-furo[3,2-g][1]benzopyran-4-yl)oxy]-methyl]propyl ester.* $C_{21}H_{22}O_7$; mol wt 386.39. C 65.27%, H 5.74%, O 28.98%. From rhizome of *Peucedanum ostruthium* L. Koch *(Imperatoria ostruthium* L.), *Umbelliferae:* Herzog, Koch, *Arch. Pharm.* **247**, 553 (1909). Structure: Späth, Christiani, *Ber.* **66**, 1150 (1933). Synthesis: Chatterjee, Dutta, *Sci. Cult.* **34**, 460 (1968).

Silky needles from benzene, mp 136-137°. $[\alpha]_D^{15}$ −18.3° (pyridine). Slightly sol in water or in ether; sol in hot alcohol, benzene, toluene, pyridine.

6852. Otobain. *9α-(1,3-Benzodioxol-5-yl)-6,7,8,9-tetrahydro-7α,8β-dimethylnaphtho[1,2-d]-1,3-dioxole;* 5,6-methylenedioxy-2,3-dimethyl-4-(3',4'-methylenedioxyphenyl)-1,2,3,4-tetrahydronaphthalene; otobite. $C_{20}H_{20}O_4$; mol wt

324.36. C 74.05%, H 6.22%, O 19.73%. Isolated from the otoba fat of *Dialeyanthera otoba* (Humb. & Bonpl.) Warb. *(Myristica obota* Humb. & Bonpl.), *Myristicaceae:* Baughman *et al., J. Am. Chem. Soc.* **43,** 200 (1921). Structure: Gilchrist *et al., J. Chem. Soc.* **1962,** 1780; Bhacca, Stevenson, *J. Org. Chem.* **28,** 1638 (1963). Synthesis: Maclean, Stevenson, *J. Chem. Soc. (C)* **1966,** 1717; McMurry, Kennedy-Skipton, *Tetrahedron Letters* **1966,** 975.

Needles from ethanol, mp 137-138°. $[\alpha]_D$ —40.5° (c = 3.2 in chloroform). uv max (ethanol): 234, 287 nm (ε 9300, 6700). Sol in ether, hot alcohol; practically insol in water.

8,6'-Dinitrootobain, $C_{20}H_{18}N_2O_8$, yellow needles from chloroform + methanol, mp 234-236°. $[\alpha]_D$ —170° (c = 2.45 in chloroform).

8,6'-Dibromootobain, $C_{20}H_{18}Br_2O_4$, prisms from chloroform + methanol, mp 197-199°. $[\alpha]_D$ +64° (c = 2.7 in chloroform).

Dehydrootobain, $C_{20}H_{16}O_4$, needles from chloroform + methanol, mp 185-187°.

6853. Ouabagenin. *1β,3β,5,11α,14,19-Hexahydroxy-5β-card-20(22)-enolide;* G-strophanthidin. $C_{23}H_{34}O_8$; mol wt 438.50. C 62.99%, H 7.82%, O 29.19%. Prepn from ouabain with HCl in cold acetone: Mannich, Siewert, *Ber.* **75,** 737 (1942). Review of structure: Reichstein, Reich, *Ann. Rev. Biochem.* **15,** 155 (1946). Proof of this structure: Tamm *et al., Helv. Chim. Acta* **40,** 1469 (1957), *cf. Experientia* **13,** 185 (1957); Turner, Meschino, *J. Am. Chem. Soc.* **80,** 4862 (1958); Volpp, Tamm, *Helv. Chim. Acta* **42,** 1408, 1418 (1959).

Monohydrate, clusters of needles from water, mp 235-238°. One gram dissolves in about 10 ml boiling water. At room temp the soly in water is less than one per cent. Also sol in dil alcohol; practically insol in abs alcohol, ether, and chloroform. Becomes anhydr at 100° *in vacuo* over P_2O_5. The anhydr compd is hygroscopic, mp 255-256°. $[\alpha]_D^{17}$ +11.3° (c = 1.27).

Tetraacetylouabagenin trihydrate, $C_{31}H_{42}O_{12}.3H_2O$, needles from 10% alcohol, mp 282-285°.

Dihydroouabagenin, $C_{23}H_{36}O_8.CH_3OH$, solvated crystals from methanol + ether contg 1 mol methanol, mp 261°.

6854. Ouabain. *3-[(6-Deoxy-α-L-mannopyranosyl)oxy]-1,5,11α,14,19-pentahydroxycard-20(22)-enolide;* G-strophanthin; Gratus strophanthin; acocantherin; Gratibain; Astrobain; Purostrophan; Strophoperm; Strodival. $C_{29}H_{44}O_{12}$; mol wt 584.64. C 59.58%, H 7.59%, O 32.84%. Obtained from the seeds of *Strophanthus gratus* (Wall. & Hock.) Baill.; also occurs in *Acokanthera ouabaio* Cathel and other *A.* spp, *Apocynaceae.* Isoln: Schwartze *et al., J. Pharmacol.* **36,** 481 (1929). Hydrolysis yields one mol ouabagenin and one mol rhamnose: Jacobs, Bigelow, *J. Biol. Chem.* **96,** 647 (1932). *See also* Fieser, Fieser, *Steroids* (1959, Reinhold,

New York; Chapman & Hall, London) pp 768, 772 and Reichstein, Reich, *Ann. Rev. Biochem.* **15,** 155 (1946).

Octahydrate, shiny plates (from water) which give up their water of crystn at 130°. When anhydr dec about 190°. $[\alpha]_D^{25}$ —31° to —32.5° (c = 1 calcd as anhydr). Stable in air, but affected by light. One gram dissolves in about 75 ml water, in 5 ml boiling water, in 100 ml alcohol, in 8 ml boiling alcohol. Also sol in amyl alcohol, dioxane. Slightly sol in ether, chloroform, ethyl acetate. Aq solns are neutral to litmus. LD_{50} i.v. in rats: 14 mg/kg, Small *et al., Toxicol. Appl. Pharmacol.* **20,** 44 (1971).

THERAP CAT: Cardiotonic.

THERAP CAT (VET): Cardiotonic, diuretic.

6855. Ovalbumin. Egg albumin. The major protein constituent (75%) of egg white from hen's eggs. Mol wt about 45,000. Produced under hormonal control by the bird oviduct. May be isolated and crystallized readily from the filtrate of an acidified mixture of egg white and an equal volume of satd. ammonium sulfate: Sorensen, Hoyrup, *C. R. Trav. Lab. Carlsberg* **12,** 12 (1917). Alternate method: Kekwick, Cannan, *Biochem. J.* **30,** 227 (1936). Can be separated by electrophoresis and chromatography from about ten other minor components including avidin (q.v.), lysozyme (q.v.), conalbumin (q.v.), and ovomucoid. Structure is a complex protein consisting of a single polypeptide chain of about 400 residues (about half of which are hydrophobic), a maximum of two phosphate residues per mole, and an oligosaccharide side chain composed of only mannose and glucosamine residues. Sequences of *N*- and *C*-terminal segments: Narita, Ishii, *J. Biochem.* **52,** 367 (1962); Thompson *et al., Aust. J. Biol. Sci.* **24,** 525 (1971). *Reviews:* R. C. Warner in Neurath-Bailey, *The Proteins,* **vol. II,** part A (New York, 1954) p 443 sqq; Taborsky, *Advan. Protein Chem.* **28,** 34-50 (1974).

Needles or elongated prisms, frequently forming rosettes. The crystals usually contain 2 mols protein and 3 mols H_2SO_4. $[\alpha]_D^{20}$ —30.7°. Coagulation temp 56°. Sol in electrolyte-free water. Combines with salts, acids, and bases. Denaturation can be induced by heating to 56°, by vigorous shaking, by electric current, and by various chemicals, such as acids, ammonia salts, heavy metal salts, and alcohols. All methods produce complete and irreversible denaturation. Isoelectric point: 4.63.

6856. Ovex. *4-Chlorobenzenesulfonic acid 4-chlorophenyl ester;* *p*-chlorophenyl *p*-chlorobenzenesulfonate; chlorfenson; K 6451; Estonmite; Genite 883; Lethalaire G-58; Mitran; Orthotran; Ovochlor; Ovotox; Ovotran. $C_{12}H_8Cl_2O_3S$; mol wt 303.16. C 47.54%, H 2.66%, Cl 23.39%, O 15.83%, S 10.58%. Prepn: Barna, **Brit.** pat. **747,368** (1956 to Diamond Alkali).

Crystals from benzene, mp 86.5-86.8°. Practically insol in water: Solubilities at 25° in g/100 ml: 95% alcohol 1.4; acetone 130.0; carbon tetrachloride 41.0; cyclohexanone 110.0; ethylene dichlcride 110.0; xylene 78.0; petr oils 2.0-2.7. LD_{50} orally in rats: 2000 mg/kg, *RTECS* **Vol. I,** R. J. Lewis, R. L. Tatken, Eds. (1979) p 220.

USE: Miticide; control of powdery mildew. *Caution:* Irritating to skin.

6857. Oxaceprol. *trans-1-Acetyl-4-hydroxy-L-proline;* 4-hydroxy-*N*-acetylproline; 1-acetyl-4-hydroxy-2-pyrrolidinecarboxylic acid; CO 61; AHP 200; Jonctum; Tejuntivo. $C_7H_{11}NO_4$; mol wt 173.17. C 48.55%, H 6.40%, N 8.09%, O 36.96%. Derivative of hydroxyproline having anti-inflammatory activity. Prepn: R. L. M. Synge, *Biochem. J.* **33,** 1924 (1939); J. J. Kolb, G. Toennies, *J. Biol. Chem.* **144,** 193 (1942); O. Leonardo *et al.,* **Ger.** pat. **2,301,358** corresp to U.S. pat. **3,860,607** (1973, 1975 to Richardson-Merrell). Elucidation of cyclic conformation and *cis/trans* isomerism about the amide bond: T. Prange *et al., Biochem. Biophys. Res. Comm.* **61,** 104 (1974). Crystal structure: M. Hospital *et al., Biopolymers* **18,** 1141 (1979). In crystals, the acetyl group is in the *trans* conformation and the ring is puckered. The *trans* form is more stable than the *cis* form in solutions: W. A. Thomas, M. K. Williams, *J. Chem. Soc. Chem. Comm.* **1972,** 994. Anti-inflammatory and wound-healing activity in animals: P. Coirre, B. Coirre, **Brit.** pat. **1,246,141** (1971) related to P. Coirre *et al.,* U.S. pats. **3,891,765** and **3,932,638** (1975, 1976 both to Franco Chimie). Clinical studies of use in several rheumatic conditions: P. Grellat, *Rheumatologie* **27,** 223 (1975); in degenerative joint disease, R. Schubotz, L. Hausmann, *Therapiewoche* **27,** 4248 (1977); in treatment of burns, tumors and other wounds (applied locally): Y. Privat, *Gaz. Med. de France* **84,** 618 (1977).

Crystals from acetone, mp 133-134° (Synge); also reported as mp 126-128° (Leonardo). $[\alpha]_D^{20} -116.5°$ (c = 3.2); $[\alpha]_D^{18}$ $-119.5°$ (c = 3.75). Very sol in alcohol. Sol in water, methanol. Insol in ether, chloroform.

Monohydrate, $C_7H_{13}NO_5$, crystals from moist ethyl acetate or acetone, mp 74-76°.

Zinc salt, $C_{14}H_{20}N_2O_8Zn$, mp 120°.

THERAP CAT: Anti-inflammatory; vulnerary.

6858. Oxacillin. *[[(5-Methyl-3-phenyl-4-isoxazolyl)carbonyl]amino]-3,3-dimethyl-6-7-oxo-4-thia-1-azabicyclo-[3.2.0]heptane-2-carboxylic acid;* 5-methyl-3-phenyl-4-isoxazolylpenicillin; 6-(5-methyl-3-phenyl-2-isoxazoline-4-carboxamido)penicillanic acid; oxazocilline. $C_{19}H_{19}N_3O_5S$; mol wt 401.44. C 56.85%, H 4.77%, N 10.46%, O 19.93%, S 7.98%. Semi-synthetic antibiotic related to penicillin. Prepn: Doyle, Nayler, U.S. pat. **2,996,501** (1961); Doyle *et al., Nature* **192,** 1183 (1961). Toxicity: E. I. Goldenthal, *Toxicol. Appl. Pharmacol.* **18,** 185 (1971).

Sodium salt monohydrate, $C_{19}H_{18}N_3NaO_5S.H_2O$, *penicillin P-12, sodium oxacillin, BRL 1400, Bactocill, Bristopen, Cryptocillin, Micropenin, Oxabel, Penstapho, Penstaphocid, Prostaphlin, Resistopen, Stapenor.* Crystals from isopropanol mp 188° (dec). $[\alpha]_D^{20} +201°$ (c = 1 in water). LD_{50} orally in rats: > 8000 mg/kg (Goldenthal).

THERAP CAT: Antibacterial.

THERAP CAT (VET): Antibacterial.

6859. 1,3,4-Oxadiazole. $C_2H_2N_2O$; mol wt 70.06. C

34.28%, H 2.88%, N 39.99%, O 22.84%. Prepn: Ainsworth, *J. Am. Chem. Soc.* **87,** 5800 (1965).

Liquid. bp 150°. Thermally stable.

6860. Oxadiazon. *3-[2,4-Dichloro-5-(1-methylethoxy)-phenyl]-5-(1,1-dimethylethyl)-1,3,4-oxadiazol-2(3H)-one; 2-tert-butyl-4-(2,4-dichloro-5-isopropoxyphenyl)-Δ^2-1,3,4-oxadiazolin-5-one;* 5-*tert*-butyl-3-(2,4-dichloro-5-isopropoxyphenyl)-1,3,4-oxadiazolin-2-one; RP 17623; Ronstar. $C_{15}H_{18}Cl_2N_2O_3$; mol wt 345.22. C 52.19%, H 5.25%, Cl 20.54%, N 8.11%, O 13.90%. Selective pre-emergence herbicide. Prepn: **Brit.** pat. **1,110,500** corresp to J. Metivier, R. Boesch, **U.S.** pat. **3,385,862** (both 1968 to Rhône Poulenc). Metabolism: A. Guardigli *et al., Arch. Environ. Contam. Toxicol.* **4,** 145 (1976). Persistence in soil: D. Ambrosi *et al., J. Agr. Food Chem.* **25,** 868 (1977).

Crystals, mp 88-90°. Vapor press at 20°: $< 1 \times 10^{-6}$ mm Hg. Soly in water at 20°: 0.7 mg/l. Soly at 20° (g/l): ethanol 100; acetone 600; benzene 1000. LD_{50} orally in rats: 3500 mg/kg, *RTECS* **Vol. II,** R. J. Lewis, R. L. Tatken, Eds. (1979) p 182.

USE: Herbicide.

6861. Oxaflozane. *4-(1-Methylethyl)-2-[3-(trifluoromethyl)phenyl]morpholine;* 4-isopropyl-2-(α,α,α-trifluoro-m-tolyl)morpholine; tetrahydro-4-isopropyl-2-[3-(trifluoromethyl)phenyl]-2H-oxazine. $C_{14}H_{18}F_3NO$; mol wt 273.30. C 61.52%, H 6.64%, F 20.86%, N 5.13%, O 5.85%. Non-tricyclic antidepressant agent. Prepn: R. Y. Mauvernay *et al.,* **Fr.** pat. **1,564,792** corresp to U.S. pat. **3,637,680** (1969, 1972 both to CERM); N. Busch *et al., Eur. J. Med. Chem.-Chem. Ther.* **11,** 201 (1976). Improved synthesis: P. M. Weintraub *et al., J. Org. Chem.* **45,** 4989 (1980). Pharmacological profile: J. Hache *et al., Therapie* **29,** 81 (1974). Urinary metabolism: M. Constantin, J. F. Pognat, *Arzneimit-tel-Forsch.* **29,** 109 (1979). Preliminary clinical study: A. Rascol *et al., Therapie* **29,** 95 (1974).

Oil, $bp_{0.005}$ 52°; bp_3 99°. n_D^{24} 1.4751.

Hydrochloride, $C_{14}H_{19}ClF_3NO$, *CERM-1766, Conflictan.* Cryst from ethyl acetate/ethanol, mp 164°. LD_{50} in mice (mg/kg): 365 orally, 90 i.v., R. Y. Mauvernay *et al., loc. cit.* Also reported as: 420 orally, 80 i.v., J. Hache *et al., loc. cit.*

THERAP CAT: Antidepressant.

6862. Oxaflumazine. *10-[3-[4-[2-(1,3-Dioxan-2-yl)eth-yl]-1-piperazinyl]propyl]-2-(trifluoromethyl)-10H-phenothi-azine.* $C_{26}H_{32}F_3N_3O_2S$; mol wt 507.63. C 61.52%, H 6.35%, F 11.23%, N 8.28% O 6.30%, S 6.32%. Prepn: Ratouis, Boissier, *Bull. Soc. Chim. France* **1966,** 2963; eidem, **Ger.** pat. **1,911,719** (1969 to S.I.F.A.), *C.A.* **72,** 55480y (1970). Pharmacology: Boissier, Dumont, *Thérapie* **26,** 481 (1971). Metabolism: Boissier *et al., Ann. Pharm. Franc.* **30,** 851 (1972). Clinical studies: Deniker *et al., Thérapie* **26,** 227 (1971).

Disuccinate, $C_{34}H_{44}F_3N_3O_{10}S$, *SD 270-31, Oxaflumine.* Crystals from acetonitrile, mp 136-138°. uv max (methanol): 259, 310 nm (log ϵ 4.53, 3.58). LD_{50} in mice: 94 mg/kg i.v.; 175 mg/kg i.p.; 919 mg/kg orally.

THERAP CAT: Antipsychotic.

6863. Oxalacetic Acid. *Oxobutanedioic acid;* oxosuccinic acid; ketosuccinic acid. $C_4H_4O_5$; mol wt 132.07. C 36.37%, H 3.05%, O 60.57%. HOOCCOCH₂COOH. Prepd by hydrolyzing sodium diethyloxalacetate with concd HCl: Heidelberger, Hurlbert, *J. Am. Chem. Soc.* **72**, 4704 (1950). Prepn of *cis*-enol form: Heidelberger, *Biochem. Prepn.* **3**, 59 (1953).

Diethyl ester, *see* Ethyl Oxalacetate.

trans-enol Form, **hydroxyfumaric acid,** crystals from acetone + benzene, mp 184°. K at 17° = 2.76 × 10⁻³. Soluble in water, ethanol, ether. Can be converted into the lower-melting *cis*-form, mp 152°, by dissolving the acid in water and re-isolating it as rapidly as possible.

cis-enol Form, **hydroxymaleic acid,** crystals from ethyl acetate + carbon tetrachloride, mp 152°. K at 17° = 2.505 × 10⁻³. Sol in ethanol, acetone, ethyl acetate. Sparingly sol in ether. Insol in benzene, chloroform, ligroin.

6864. Oxalenediuramidoxime. *N,N''-Bis(aminocarbonyl)-N',N'''-dihydroxyethanediimidamide.* $C_4H_8N_6O_4$; mol wt 204.15. C 23.53%, H 3.95%, N 41.17%, O 31.35%.

Needles from dil alcohol. mp 191-192° (dec). Insol in cold water, in ether, chloroform, benzene, petr ether. Very sol in alcohol, acids, alkalies.

USE: As a reagent for the determination of nickel, with which it gives an orange ppt in dil ammoniacal soln.

6865. Oxalic Acid. *Ethanedioic acid.* $C_2H_2O_4$; mol wt 90.04. C 26.68%, H 2.24%, O 71.08%. (COOH)₂. Present in many plants and vegetables, notably in those of the *Oxalis* and *Rumex* families, where it occurs in the cell sap of the plant as the potassium or calcium salt. It is a product of the metabolism of many molds. Several species of *Penicillium* and *Aspergillus* convert sugar into calcium oxalate with 90% yields under optimum conditions. Oxalic acid was formerly manuf by fusion of cellulose matter, e.g. sawdust, with NaOH or by oxidation with HNO₃; it is now made by passing carbon monoxide into concd NaOH or by heating sodium formate in the presence of NaOH or Na₂CO₃: Wallace, U.S. pat. **1,602,802** (1926); Beckman, U.S. pat. **2,687,433** (1951 to Allied Chem.). Efficient laboratory prepn of anhydrous form: H. T. Clarke, A. W. Davis, *Org. Syn.* coll. vol. **I**, 421 (2nd ed., 1941). *Review:* Wilson, "Miscellaneous *Aspergillus* Toxins" in *Microbial Toxins,* vol. VI, A. Ciegler *et al.,* Eds. (Academic Press, New York, 1971) pp 268-273; C. A. Bernales in Kirk-Othmer *Encyclopedia of Chemical Technology* vol. **16** (Wiley-Interscience, New York, 3rd ed., 1981) pp 618-636.

Dihydrate, monoclinic tablets, prisms, granules. *Poisonous!* $d_4^{18.5}$ 1.653. mp 101-102° giving off water of crystn and starting to sublime. K_1 5.36 × 10⁻²; K_2 5.3 × 10⁻⁵. pH of 0.1M soln 1.3. One gram dissolves in about 7 ml water, 2 ml boiling water; 2.5 ml alcohol, 1.8 ml boiling alcohol, 100 ml ether, 5.5 ml glycerol; insol in benzene, chloroform, petr ether. $d^{17.5}$ of aq solns: 1% (w/w) 1.0035; 3% 1.0105; 5% 1.0175; 10% 1.0350; 13% 1.0455. Oxalic acid can be dehydrated by careful drying at 100°, but considerable loss occurs through sublimation; this, moreover, is harmful to the oven. LD_{50} orally in rats (5% soln): 9.5 ml/kg, E. H. Vernot *et al.,* *Toxicol. Appl. Pharmacol.* **42**, 417 (1977).

Anhydr oxalic acid, crystallized from glacial acetic acid, is orthorhombic, the crystals being pyramidal or elongated octahedra. Hygroscopic, mp 189.5° (dec). Sublimes best at 157°. At higher temps dec into CO₂, CO, formic acid, and H₂O. d_4^{17} 1.90. 100 g of aq soln satd at 15° contain 6.71 g, at 20° 8.34 g, at 25° 9.81 g.

Human Toxicity: Caustic and corrosive to skin, mucous membranes. Ingestion may cause severe gastroenteritis with vomiting, diarrhea, melena. Renal damage can occur as a result of formation of excessive calcium oxalate. Convulsions, coma, death from cardiovascular collapse may occur, *Clinical Toxicology of Commercial Products,* R. E. Gosselin *et al.,* Ed. (Williams & Wilkins, Baltimore, 4th ed., 1976) Section III, pp 260-263.

USE: As analytical reagent; in calico printing and dyeing; for bleaching straw (hats) and leather; removing paint or varnish, rust or ink stains; cleaning wood; manuf oxalates; blue ink; celluloid; intermediates and dyes; in metal polishes; in indigo dyeing; in purifying methanol; for decolorizing crude glycerol; for stabilizing hydrocyanic acid. As a general reducing agent; in ceramics and pigments; in metallurgy as cleanser; in the paper industry; in photography; in process engraving; in the rubber mfg industry; in making glucose from starch; as condensing agent in organic chemistry.

THERAP CAT (VET): In 5% solution with 5% malonic acid as hemostatic agent.

6866. Oxalomolybdic Acid. *[Ethanedioato(2−)-O,O']-trioxomolybdate(2−) dihydrogen; [ethanedioato(1−)-O]-hydroxydioxomolybdenum; hydrogen trioxooxalatomolybdate-(VI).* $C_2H_2MoO_7$; mol wt 233.99. C 10.27%, H 0.86%, Mo 41.01%, O 47.87%. H₂[MoO₃(C₃O₄)]. Usually contains 1 or 2H₂O.

Colorless crystals, sol in water.

USE: In invisible inks.

6867. Oxalyl Chloride. *Ethanedioyl dichloride.* $C_2Cl_2O_2$; mol wt 126.93. C 18.92%, Cl 55.87%, O 25.21%. (COCl)₂. Obtained from oxalic acid and PCl₅.

Colorless, fuming liquid; penetrating odor. *Poisonous!* d_4^{13} 1.488. mp −12°. bp 63-64°. n_D^{13} 1.4340. Violently dec by water, also by alcohol.

Caution: Severely irritating to skin, eyes, respiratory tract.

6868. Oxamarin. *6,7-Bis[2-(diethylamino)ethoxy]-4-methyl-2H-1-benzopyran-2-one; 6,7-bis[2-(diethylamino)-ethoxy]-4-methylcoumarin.* $C_{22}H_{34}N_2O_4$; mol wt 390.52. C 67.66%, H 8.78%, N 7.17%, O 16.39%. Deriv of α-pyrone, related structurally to sulmarin, *q.v.* Prepn: E. Massarani, *Farmaco Ed. Sci.* **12**, 691 (1957); G. Cavallini, E. Massarani, U.S. pat. **2,895,963** (1959 to Maggioni); E. Massarani *et al.,* *J. Med. Pharm. Chem.* **3**, 231 (1961). Comparative study in treatment of varicose syndrome: G. Frantoli *et al.,* *Panminerva Med.* **14**, 290 (1972), *C.A.* **78**, 92741 (1973). Transcutaneous absorption: G. Ciaceri, P. Marini, *Boll. Chim. Farm.* **111**, 321 (1972). Cardiovascular, intestinal smooth muscle effects: *eidem,* G. Ital. Patol. Sci. Affini **19**, 11 (1972), *C.A.* **81**, 130823 (1974). Anti-inflammatory action: *eidem, ibid.* 1, *C.A.* **81**, 130952 (1974).

Oil, bp₀.₅ 195°.

Dihydrochloride, $C_{22}H_{36}Cl_2N_2O_4$, *MG-652, Idro P₃.* Cryst from ethanol, mp 224-226°, E. Massarani, *loc. cit.* (1957), also reported as mp 234-236°, E. Massarani *et al., loc. cit.*

THERAP CAT: Hemostatic.

6869. Oxametacine. *1-(4-Chlorobenzoyl)-N-hydroxy-5-methoxy-2-methyl-1H-indole-3-acetamide; 1-(p-chlorobenzoyl)-5-methoxy-2-methylindole-3-acetohydroxamic acid;* indoxamic acid; Dinulcid; Flogar. $C_{19}H_{17}ClN_2O_4$; mol wt 372.81. C 61.21%, H 4.60%, Cl 9.51%, N 7.51%, O 17.17%. Deriv of indomethacin, *q.v.* Prepn: R. Aries, **Fr.** pat. **1,599,495** (1969), *C.A.* **73**, 3797h (1970); F. De Martiis *et al.,*

Ger. pat. **2,008,332** corresp to U.S. pat. **3,624,103** (both 1971 to ABC); *eidem, Boll. Chim. Farm.* **114,** 309 (1975). Pharmaco-toxicological evaluation: *eidem, ibid.* 319. Disposition in rats: L. F. Elsom *et al., Arzneimittel-Forsch.* **29,** 1155 (1979). Pharmacokinetics: P. Dittrich *et al., ibid.* **31,** 518 (1981). Effect on prostaglandin biosynthesis: J. S. Franzone *et al., Farmaco. Ed. Sci.* **35,** 498 (1980). Clinical study: J. Polderman, M. Colon, *J. Int. Med. Res.* **8,** 156 (1980).

Cryst from dioxane, mp 181-182° (dec). Sol in most organic solvents at elevated temperatures. In strong alkali it is hydrolyzed quickly to the debenzoylated product. LD_{50} orally in rats: 96 mg/kg (De Martiis, **U.S.** patent).

THERAP CAT: Anti-inflammatory.

6870. Oxamic Acid. *Aminooxoacetic acid;* oxamidic acid. $C_2H_3NO_3$; mol wt 89.05. C 26.97%, H 3.40%, N 15.73%, O 53.90%. $H_2NCOCOOH$. Prepd by heating oxamide in water with ammonia, followed by neutralization of resulting ammonium salt: Toussaint, *Ann.* **120,** 237 (1861).

Cryst powder from water, or prisms from alcohol, dec about 210 or 214°. Sparingly sol in water. Practically insol in abs alcohol, ether.

Ammonium salt, $H_2NCOCOONH_4$, monoclinic prisms, dec about 226°. Very slightly sol in cold water, alcohol.

6871. Oxamide. *Ethanediamide;* oxalamide; oxalic acid diamide; ethanedioic acid diamide. $C_2H_4N_2O_2$; mol wt 88.07. C 27.27%, H 4.58%, N 31.81%, O 36.33%. $H_2NCO-CONH_2$. Prepd from formamide by glow-discharge electrolysis: Brown *et al., J. Org. Chem.* **27,** 3698 (1962). Crystal structure: G. DeWith, S. Harkema, *Acta Crystallogr.* **33B,** 2367 (1977). Metabolized in body to form oxalic acid.

Triclinic needles, dec 350°. d_4^{20} 1.667. Sparingly sol in hot water, alcohol.

6872. Oxamniquine. *1,2,3,4-Tetrahydro-2-[[(1-methylethyl)amino]methyl]-7-nitro-6-quinolinemethanol; 1,2,3,4-tetrahydro-2-[(isopropylamino)methyl]-7-nitro-6-quinolinemethanol;* 6-hydroxymethyl-2-isopropylaminomethyl-7-nitro-1,2,3,4-tetrahydroquinoline; UK 4271; Mansil; Vansil. $C_{14}H_{21}N_3O_3$; mol wt 279.34. C 60.20%, H 7.58%, N 15.04%, O 17.18%. Prepn: H. C. Richards, **S. Afr.** pat. **68 03636;** *idem,* **U.S.** pat. **3,821,228** (1967, 1974 both to Pfizer). Discovery of schistosomicidal activity: H. C. Richards, R. Foster, *Nature* **222,** 581 (1969). Metabolism: N. M. Woolhouse, B. Kaye, *Parasitology* **15,** 111 (1977). Genetic activity: T. M. Ong, *Mutat. Res.* **55,** 43 (1978). *In vitro* activity: C. J. Chavasse *et al., Ann. Trop. Med. Parasitol.* **72,** 293 (1978). Mutagenicity study: R. P. Batzinger, E. Bueding, *J. Pharmacol. Exp. Ther.* **200,** 1 (1977). Pharmacology: R. Foster, *Rev. Inst. Med. Trop. Sao Paulo* **15,** Suppl. 1, 189 (1973), *C.A.* **83,** 188417g (1975). Clinical study: R. J. Pitchford, M. Lewis, *S. Afr. Med. J.* **53,** 677 (1978).

Pale yellow crystals from isopropanol, mp 147-149°. LD_{50} in mice, rabbits (mg/kg): > 2000, > 1000 i.m., 1300, 800 orally (Foster).

THERAP CAT: Anthelmintic (Schistosoma).

6873. Oxamyl. *2-(Dimethylamino)-N-[[(methylamino)-carbonyl]oxy]-2-oxoethanimidothioic acid methyl ester; N',N'-dimethyl-N-[(methylcarbamoyl)oxy]-1-thiooxamimi-*

dic acid methyl ester; N,N-dimethyl-α-methylcarbamoyloxy-imino-α-(methylthio)acetamide; methyl 1-(dimethylcarbamoyl)-N-(methylcarbamoyloxy)thioformimidate; thioxamyl; DPX 1410; Vydate. $C_7H_{13}N_3O_3S$; mol wt 219.25. C 38.34%, H 5.98%, N 19.17%, O 21.89%, S 14.62%. Prepn: J. B. Buchanan, **S. Afr.** pat. **68 03,629** corresp to U.S. pats. **3,530,220, 3,658,870** (1968, 1970, 1972 all to du Pont). Decompn: J. Harvey, Jr., J. Han, *J. Agr. Food Chem.* **26,** 536 (1978). Metabolism: *eidem, ibid.* 902; J. Harvey, Jr. *et al., ibid.* 529.

Cryst solid, slight sulfurous odor. mp 100-102°, changing to a different crystalline form, mp 108-110°. Soly in g/100 ml at 25°: water 28; acetone 67; ethanol 33; 2-propanol 11; methanol 144; toluene 1. Toxic to wildlife, fish, bees. LD_{50} orally in rats: 5 mg/kg, M. Fahmy *et al., J. Agr. Food Chem.* **26,** 550 (1978).

USE: Insecticide, nematocide, acaricide.

6874. Oxanamide. *2-Ethyl-3-propyloxiranecarboxamide; 2,3-epoxy-2-ethylhexanamide;* 2-ethyl-3-propylglycidamide; 2-ethyl-3-propyl-2,3-epoxypropionamide; Quiactin. $C_8H_{15}NO_2$; mol wt 157.21. C 61.12%, H 9.62%, N 8.91%, O 20.35%. Prepn: Wheeler *et al., J. Org. Chem.* **25,** 1021 (1960).

Tasteless, odorless, white crystals from petr ether, mp 90-91°. One part is soluble in 95 parts of water at 30°. LD_{50} i.p. in mice: 720 mg/kg; orally in mice, rats: 1220, 1360 mg/kg, *Toxic Substances List,* H. E. Christensen, Ed. (1974) p 398.

THERAP CAT: Anxiolytic.

6875. Oxandrolone. *17-Hydroxy-17-methyl-2-oxaandrostan-3-one;* dodecahydro-3-hydroxy-6-(hydroxymethyl)-3,3a,6-trimethyl-1H-benz[e]indene-7-acetic acid δ-lactone; Anavar; Lonavar; Provitar; Vasorome. $C_{19}H_{30}O_3$; mol wt 306.43. C 74.47%, H 9.87%, O 15.66%. Prepn: Pappo, Jung, *Tetrahedron Letters* **1962,** 365; Pappo, **U.S.** pat. **3,128,283** (1964 to Searle).

Crystals, mp 235-238°. $[\alpha]_D^{25}$ −23°. Absorption max: 2.87, 5.79 μ.

THERAP CAT: Androgen.

6876. Oxantel. *(E)-3-[2-(1,4,5,6-Tetrahydro-1-methyl-2-pyrimidinyl)ethenyl]phenol; (E)-m-[2-(1,4,5,6-tetrahydro-1-methyl-2-pyrimidinyl)vinyl]phenol;* 1-methyl-1,4,5,6-tetrahydro-2-[2-(3-hydroxyphenyl)vinyl]pyrimidine; CP 14445. $C_{13}H_{16}N_2O$; mol wt 216.28. C 72.19%, H 7.46%, N 12.95%, O 7.40%. Analog of pyrantel, *q.v.,* with activity vs whipworms (*Trichuris* spp.). Prepn: J. W. Mc Farland, **S. Afr.** pat. **68 04589** corresp to U.S. pats. **3,579,510** and **3,708,584** (1968, 1971, 1973 all to Pfizer). Synthesis and evaluation in whipworm control: J. W. Mc Farland, H. L. Howes *J. Med. Chem.* **15,** 365 (1972). Evaluation vs *Trichuris* in dogs: H. L. Howes, *Proc. Soc. Exp. Biol. Med.* **139,** 394 (1972). Efficacy vs *T. trichiuris* in humans: E. L. Lee *et al., Am. J. Trop. Med. Hyg.* **25,** 563 (1976); vs *T. suis* in swine: M. Robinson, *Vet. Parasitol.* **5**, 223 (1979). Anthelmintic effects of combination with pyrantel pamoate: B. Sinniah, D. Sinniah, *Ann. Trop. Med. Parasitol.* **75**, 315 (1981).

OH

CH₃

Hydrochloride, $C_{13}H_{17}ClN_2O$, crystals from ethanol, mp 207-208°. uv max (water): 231, 274 nm (ε 12700, 20100). Pamoate, $C_{49}H_{48}N_4O_8$, **CP 14445-16, oxantel embonate, Telopar.** Mixture with pyrantel pamoate (1:1), Quantrel.
THERAP CAT: Anthelmintic (Nematodes).
THERAP CAT (VET): Anthelmintic (Nematodes).

6877. Oxapropanium Iodide. *N,N,N-Trimethyl-1,3-dioxolane-4-methanaminium iodide; (1,3-dioxolan-4-ylmethyl)trimethylammonium iodide;* 4-(dimethylaminomethyl)-1,3-dioxacyclopentane methiodide; 1-dimethylamino-2,3-dioxamethylenepropane methiodide; 2249 F; Vasodilatateur 2249F; Dilvasene. $C_7H_{16}INO_2$; mol wt 273.13. C 30.78%, H 5.90%, I 46.47%, N 5.13%, O 11.72%. Prepd by the reaction of a cyclic acetal of an α-monohalohydrin of glycerol with dimethylamine which on treatment with methyl iodide results in a quaternary ammonium salt: Fourneau, U.S. pat. 2,445,393 (1948 to Rhône-Poulenc).

Crystals, mp 158-160° (the free base bp₂₁ 68°). Freely sol in water. Sol in boiling alcohol (about 40% w/v). Sparingly sol in cold alcohol, ether, chloroform, benzene.
THERAP CAT: Cholinergic.

6878. Oxaprotiline. *α-[(Methylamino)methyl]-9,10-ethanoanthracene-9(10H)-ethanol;* 9-[2-hydroxy-3-(methylamino)propyl]-9,10-dihydro-9,10-ethanoanthracene; 1-(1-methylamino-2-hydroxy-3-propyl)dibenzo[b,e]bicyclo-[2.2.2]octadiene; hydroxymaprotiline. $C_{20}H_{23}NO$; mol wt 293.41. C 81.87%, H 7.90%, N 4.77% O 5.45%. Hydroxy analog of maprotiline, *q.v.* Prepn of racemate: M. Wilhelm *et al.,* **Ger.** pat. **2,207,097;** *eidem,* **U.S.** pat. **4,017,542;** of *S*(+)-form: A. Storni, **Eur.** pat. **Appl. 14,434;** of *R*(−)-form: *idem,* **Eur.** pat. **Appl. 14,433** (1972, 1977, 1980, 1980 all to Ciba-Geigy). Pharmacology of racemate: P. C. Waldmeier *et al., Eur. J. Pharmacol.* **46**, 387 (1977); and of enantiomers: P. C. Waldmeier *et al., Biochem. Pharmacol.* **31**, 2169 (1982); A. Delini-Stula *et al., Adv. Biochem. Psychopharmacol.* **31**, 265 (1982). Clinical pharmacology of *R*(−)-form and racemate: O. Schmidlin *et al., Brit. J. Clin. Pharmacol.* **14**, 799 (1982). Inverse isotope dilution assay of enantiomers in biological fluids: W. Dieterle, W. Faigle, *J. Chromatog.* **242**, 289 (1982). HPLC determn of racemate in plasma: U. Breyer-Pfaff *et al., ibid.* **309**, 107 (1984). Pharmacokinetics of racemate: W. Dieterle *et al., Xenobiotica* **14**, 303 (1984). Stereospecific metabolism: W. Dieterle *et al., Biopharm. Drug Dispos.* **5**, 377 (1984). Clinical studies of racemate in depression: M. Roffman *et al., Curr. Ther. Res.* **32**, 247 (1982); M. Schmauss *et al., ibid.* **41**, 342 (1987). Pharmacology of *R*(−)-form: A. Delini-Stula, E. Mogilnicka, *J. Neural. Transm.* **71**, 91 (1988); A. Delini-Stula *et al., Neuropharmacology* **27**, 943 (1988). Fluorometric determn of *R*(−)-form in plasma: C. Horne *et al., Arzneimittel-Forsch.* **37**, 1179 (1987).

CH₂CH(OH)CH₂NHCH₃

Hydrochloride, $C_{20}H_{24}ClNO$, **C-49802-B-Ba.** mp 234-235°.
R(−)-Form, $C_{20}H_{23}NO$, **levoprotiline, CGP-12103A.**
R(−)-Form hydrochloride, $C_{20}H_{24}ClNO$, mp 231-232°. $[\alpha]_D^{20}$ −9° (in methanol).
S(+)-Form, $C_{20}H_{23}NO$, **CGP-12104A.**
S(+)-Form hydrochloride, $C_{20}H_{24}ClNO$, mp 231-232°. $[\alpha]_D^{20}$ +9° (in methanol).
THERAP CAT: Antidepressant.

6879. Oxaprozin. *4,5-Diphenyl-2-oxazolepropanoic acid; 4,5-diphenyl-2-oxazolepropionic acid; β-(4,5-diphenyl-oxazol-2-yl)propionic acid;* Wy 21743; Alvo; Durapro; Duraprost; Oxapro. $C_{18}H_{15}NO_3$; mol wt 293.33. C 73.71%, H 5.15%, N 4.78%, O 16.36%. First description of anti-inflammatory properties: K. Brown *et al., Nature* **219**, 164 (1968). Prepn: Fr. pat. **2,001,036** (1969 to Inst. Farm. Serono), *C.A.* **72**, 66930w (1970); K. Brown, **Brit.** pat. **1,206,403** and U.S. pat. **3,578,671** (1970, 1971 both to Wyeth). Biochemical properties: M. W. Whitehouse *et al., Biochem. Pharmacol.* **20**, 2309 (1971). Metabolism: F. W. Janssen *et al., Drug Metab. Dispos.* **6**, 465 (1978). Pharmacology: D. A. Shriver *et al., Toxicol. Appl. Pharmacol.* **42**, 75 (1977); F. Awouters *et al., J. Pharm. Pharmacol.* **30**, 41 (1978). Clinical studies: R. Jamar, J. Dequeker, *Curr. Med. Res. Opin.* **5**, 433 (1978); J. A. Hubsher *et al., J. Int. Med. Res.* **7**, 69 (1979); *eidem, Arthritis Rheum.* **25**, S117 (1982). Protein binding and clearance: C. A. Homon *et al., Agents Actions* **12**, 211 (1982). Symposium on pharmacology and clinical efficacy: *Semin. Arthritis Rheum.* **15**, Suppl. 3, 1-107 (1986).

C₆H₅ — O — CH₂CH₂COOH / N / C₆H₅

Crystals from methanol, mp 160.5-161.5°.
THERAP CAT: Anti-inflammatory.

6880. Oxatomide. *1-[3-[4-(Diphenylmethyl)-1-piperazinyl]propyl]-1,3-dihydro-2H-benzimidazol-2-one;* 1-[3-[4-(diphenylmethyl)-1-piperazinyl]propyl]-2-benzimidazolinone; R 35,443; Celtect; Cobiona; Dasten; Tinset. $C_{27}H_{30}N_4O$; mol wt 426.57. C 76.03%, H 7.09%, N 13.13%, O 3.75%. Orally active anti-allergic agent, related structurally to cinnarizine, *q.v.,* and having a novel biphasic mode of action. Prepn: J. Vandenberk *et al.,* **Ger.** pat. **2,714,437;** *eidem,* **U.S.** pat. **4,200,641** (1977, 1980 both to Janssen). Inhibition of release and effects of allergic mediators: F. Awouters *et al., Experientia* **33**, 1657 (1977). *In vitro* study of inhibition and stimulation of histamine release: M. K. Church *et al., Agents Actions* **10**, 4 (1980). *In vivo* study: S. Gatti *et al., Brit. J. Dermatol.* **103**, 671 (1980). Clinical studies: S. W. Barham, F. Moran, *Brit. J. Clin. Pract.* **34**, 323 (1980); E. F. Juniper *et al., Clin. Allergy* **11**, 61 (1981); W. Peremans *et al., Dermatologia* **162**, 42 (1981). Review of pharmacological and clinical studies: D. M. Richards *et al., Drugs* **27**, 210 (1984).

C₆H₅ — CH—N   NCH₂CH₂CH₂ — N / C₆H₅ / O / NH

White powder, mp 153.6°. LD_{50} in guinea pigs, mice, rats (mg/kg): 320, > 2560, > 2560 orally; 23, 27, 30 i.v. (Richards).

THERAP CAT: Anti-allergic; anti-asthmatic.

6881. Oxazepam. *7-Chloro-1,3-dihydro-3-hydroxy-5-phenyl-2H-1,4-benzodiazepin-2-one;* 7-chloro-3-hydroxy-5-phenyl-1,3-dihydro-2H-1,4-benzodiazepin-2-one; Wy-3498; Abboxapam; Adumbran; Aplakil; Azutranquil; Bonare; Durazepam; Enidrel; Hilong; Isodin; Lederpam; Limbial; Nesontil; Noctazepam; Oxanid; Oxa-Puren; Praxiten; Propax; Quilibrex; Rondar; Serax; Serenal; Serenid; Serepax; Seresta; Sigacalm; Sobril; Tazepam; Uskan; Zaxopam. $C_{15}H_{11}ClN_2O_2$; mol wt 286.74. C 62.83%, H 3.87%, Cl 12.37%, N 9.77%, O 11.16%. Prepn: Bell, Childress, *J. Org. Chem.* **27,** 1691 (1962); E. Reeder, L. Sternbach, U.S. pat. **3,109,-843;** E. Reeder *et al.,* **Belg.** pat. **629,227;** *eidem,* U.S. pat. **3,340,253** (1963, 1967, 1967, all to Hoffmann-La Roche). Crystal and molecular structure: G. Gilli *et al., Acta Crystallogr.* **B34,** 2826 (1978). Metabolic studies: Sisenwine *et al., Arzneimittel-Forsch.* **22,** 682 (1972); Knowles, Ruelius, *ibid.* 687. Pharmacokinetics: D. J. Greenblatt, *Clin. Pharmacokinet.* **6,** 89 (1981). Toxicity data: E. I. Goldenthal, *Toxicol. Appl. Pharmacol.* **18,** 185 (1971). Comprehensive description: C. M. Shearer, C. R. Pilla in *Analytical Profiles of Drug Substances* vol. 3, K. Florey, Ed. (Academic Press, New York, 1974) pp 441-464. *Review: Acta Psychiatr. Scand., Suppl.* **274,** R. Chinnery, A. Sundwall, Eds. (1978) 128 pp.

Crystals from alcohol, mp 205-206°. Sol in alcohol, chloroform, dioxane. Practically insol in water. LD_{50} in mice, rats (mg/kg): > 5010, > 5010 orally (Goldenthal).

Caution: May be habit forming. This is a controlled substance (depressant) listed in the U.S. Code of Federal Regulations, Title 21 Part 1308.14 (1987).

THERAP CAT: Anxiolytic.

6882. Oxazidione. *2-(4-Morpholinylmethyl)-2-phenyl-1H-indene-1,3(2H)-dione;* 2-phenyl-2-morpholinomethylindane-1,3-dione; mofedione; LD 4610; Transidione; Amplidione. $C_{20}H_{19}NO_3$; mol wt 321.38. C 74.75%, H 5.96%, N 4.36%, O 14.93%. Prepd by reacting 2-phenyl-1,3-indandione with 4-(ethoxymethyl)morpholine: Giudicelli, Najer, **S. Afr.** pat. **69 02,814** (1969 to Lab. Dausse), *C.A.* **72,** 111159q (1970).

mp 88° from cyclohexane. LD_{50} orally in mice: 235 mg/kg.

Hydrochloride, mp 150°.

THERAP CAT: Anticoagulant.

6883. Oxazolam. *10-Chloro-2,3,7,11b-tetrahydro-2-methyl-11b-phenyloxazolo[3,2-d][1,4]benzodiazepin-6(5H)-one;* 10-chloro-2,3,5,6,7,11b-hexahydro-2-methyl-11b-phenylbenzo[6,7]-1,4-diazepino[5,4-b]oxazol-6-one; oxazolazepam; Serenal; Tranquit. $C_{18}H_{17}ClN_2O_2$; mol wt 328.80. C 65.75%, H 5.21%, Cl 10.78%, N 8.52%, O 9.73%. Prepn: Tachikawa *et al.,* **Ger.** pat. **1,812,252** corresp to U.S. pat. **3,772,371** (1969, 1973 to Sankyo). Synthesis and pharmacology: Miyadera *et al., J. Med. Chem.* **14,** 520 (1971). Metabolism studies: Shindo *et al., Chem. Pharm. Bull.* **19,** 60 (1971); Yasumura *et al., ibid.* 1929.

Crystals, mp 186-188°. Sol in chloroform; slightly sol in ethanol. Practically insol in water.

Note: This is a controlled substance (depressant) listed in the U.S. Code of Federal Regulations, Title 21 Part 1308.14 (1987).

THERAP CAT: Anxiolytic.

6884. Ox Bile Extract. Purified oxgall; sodium choleate; Bicol; Bi-Ketolan; Bilein; Bilicholan; Cholatol; Colalin; Crescefel; Desicol; Doxychol; Glycotauro; Panoxolin; Plebilin; Valachol. One gram = 8 g ox bile. *Constit.* Chiefly the sodium salts of glycocholic and taurocholic acids; also cholesterol, lecithin, glycocol and choline compounds.

Yellowish-green, soft solid; peculiar odor; partly sweet, partly bitter, disagreeable taste. Very sol in water or alc.

THERAP CAT: Choleretic.

THERAP CAT (VET): Has been used as a choleretic.

6885. Oxeladin. *α,α-Diethylbenzeneacetic acid 2-[2-(diethylamino)ethoxy]ethyl ester; 2-ethyl-2-phenylbutyric acid 2-(2-diethylaminoethoxy)ethyl ester;* 2-(2-diethylaminoethoxy)ethyl α,α-diethylphenylacetate; α,α-diethylphenylacetic acid 2-(2-diethylaminoethoxy)ethyl ester. $C_{20}H_{33}NO_3$; mol wt 335.47. C 71.60%, H 9.92%, N 4.18%, O 14.31%. Prepn: V. Petrow *et al., J. Pharm. Pharmacol.* **10,** 40 (1958); *eidem,* U.S. pat. **2,885,404** (1959 to Brit. Drug Houses).

Yellow oil with an acrid odor and bitter taste. $bp_{0.5}$ 150°; $bp_{0.1}$ 140°. Soluble in dil HCl, ethanol, acetone, ether, toluene; practically insol in water. Stable in acids, unstable in alkalies. Non-hygroscopic.

Citrate, $C_{26}H_{41}NO_{10}$, *Pectamol, Pectamon, Paxeladine, Silopentol.* Small needles from ethyl acetate, mp 90-91°. Sol in water.

THERAP CAT: Antitussive.

6886. Oxendolone. *16-Ethyl-17-hydroxyestr-4-en-3-one;* 16β-ethyl-19-nortestosterone; TSAA 291; Prostetin. $C_{20}H_{30}O_2$; mol wt 302.46. C 79.42%, H 10.00%, O 10.58%. Prepn: K. Hiraga *et al.,* **Ger.** pat. **2,100,319** corresp to U.S. pat. **3,856,829** (1971, 1974 both to Takeda); K. Yoshioka *et al., Chem. Pharm. Bull.* **23,** 3203 (1975). Stereoselective synthesis and NMR study: G. Goto *et al., ibid.* **25,** 1295 (1977). Synthesis and anti-androgenic activity: *eidem, ibid.* **26,** 1718 (1978). Physico-chemical properties and stabilities: K. Itakura *et al., Takeda Kenkyusho Ho* **37,** 297 (1978), *C.A.* **91,** 20879 (1979). Disposition and metabolism: S. Tanayama *et al., Steroids* **33,** 65 (1979). Series of articles on pharmacology, mechanism of action, anti-androgen effects: *Acta Endocrinol.* **92**(Suppl 2), 1-107 (1979).

Cryst from ether, mp 152-153°. $[\alpha]_D$ +41° (c = 1.0 in ethanol). uv max (ethanol): 240 nm (ε 15800). Stable against heat, humidity, indoor light. Converted to the 16α- and 17α-epimers in sunlight. LD_{50} in rats, mice: > 10 g/kg orally; 5-10 g/kg i.m. and i.p., K. Hiraga *et al., loc. cit.*

THERAP CAT: Anti-androgen; in treatment of benign prostatic hypertrophy.

6887. Oxenin. *11,12-Didehydro-7,10-dihydro-10-hydroxyretinol; 3,7-dimethyl-9-(2,6,6-trimethyl-1-cyclohexen-1-yl)nona-2,7-dien-4-yne-1,6-diol.* $C_{20}H_{30}O_2$; mol wt 302.44. C 79.42%, H 10.00%, O 10.58%. Intermediate in the vitamin A synthesis. Prepn: Isler *et al.*, *Helv. Chim. Acta* **30**, 1911 (1947); Isler, U.S. pat. **2,451,739** (1948 to Hoffmann La Roche); Kardys, U.S. pat. **3,046,310** (1962 to Pfizer).

Very viscous yellow oil. d_4^{27} 0.9984; n_D^{22} 1.5344. Has been crystallized, mp 58.5-59°.

6888. Oxethazaine. *2,2'-[(2-Hydroxyethyl)imino]bis[N-(1,1-dimethyl-2-phenylethyl)-N-methylacetamide]; N,N-bis-[N-methyl-N-phenyl-tert-butylacetamido]-β-hydroxyethyl-amine; oxetacaine; oxethazine; WY 806; Mutesa; Muthesa; Oxaine; Stomacain; Tepilta; Topicain.* $C_{28}H_{41}N_3O_3$; mol wt 467.63. C 71.91%, H 8.84%, N 8.99%, O 10.26%. Prepn: Seifter *et al.*, U.S. pat. **2,780,646** (1957 to Am. Home Prods.); Salomon, Israeli pat. **10,062** (1957), *C.A.* **52**, 15569 (1958). Pharmacology: Axerio Agnessetti, *An. Real. Acad. Farm.* **35**, 183 (1969), *C.A.* **71**, 122082q (1969). Toxicity studies: Glassman *et al.*, *Toxicol. Appl. Pharmacol.* **5**, 184 (1963).

Crystals from benzene+ hexane, mp 104-104.5°. Insol in water; sol in dil acids.

Hydrochloride, *Emoren*, mp 146-147°. LD_{50} in mice, rats (calcd as base in mg/kg): 399.9, 625.9 orally; 247.2, 502.3 i.m.; 3.6, 1.3 i.v. LD_{50} in rabbits: 0.54 mg/kg i.v., Glassman *et al., loc. cit.*

Note: Oxaine and Tepilta are suspensions in aluminum hydroxide gel.

THERAP CAT: Anesthetic (topical).

6889. Oxetorone. *3-Benzofuro[3,2-c][1]benzoxepin-6(12H)-ylidene-N,N-dimethyl-1-propanamine; N,N-dimethylbenzofuro[3,2-c][1]benzoxepin-Δ^{6(12H),γ}-propylamine; 6-(3-dimethylaminopropylidene)benzo[b]benzofurano[2,3-e]oxepine.* $C_{21}H_{21}NO_2$; mol wt 319.41. C 78.97%, H 6.63%, N 4.38%, O 10.02%. Novel serotonin and histamine antagonist with antimigraine activity. Prepn: F. Binon, M. Descamps, Ger. pat. **1,963,205** corresp to U.S. pat. **3,651,051** (1970, 1972 both to Labaz). Metabolism *in vitro:* E. Rossi *et al., J. Chromatog.* **152**, 228 (1978). Use in treatment of migraine: J. J. Dufresne, *Praxis* **67**, 1148 (1978); J. Florence, *ibid.* 1323; in chronic headache: U. Thoden, *Therapiewoche* **30**, 492 (1980).

Fumarate, $C_{25}H_{25}NO_6$, *L 6257, Nocertone, Oxedix.* Cryst from isopropanol, mp 160°.

THERAP CAT: Analgesic (specific in migraine).

6890. Oxfendazole. *[5-(Phenylsulfinyl)-1H-benzimidazol-2-yl]carbamic acid methyl ester; methyl-5-(phenylsulfinyl)-2-benzimidazolecarbamate; 5-phenylsulfinyl-2-carbomethoxyaminobenzimidazole; RS-8858; Repidose; Synanthic; Systamex.* $C_{15}H_{13}N_3O_3S$; mol wt 315.35. C 57.13%, H 4.15%, N 13.32%, O 15.22%, S 10.17%. Prepn: C. Beard *et*

al., Ger. pat. **2,363,351** corresp to U.S. pat. **3,929,821** (1974, 1975, both to Syntex); E. A. Averkin *et al., J. Med. Chem.* **18**, 1164 (1975). Metabolism: J. P. Bell, R. V. Tomlinson, *Fed. Proc.* **35**, 487 (1976). Radioimmunoassay: C. Nerenberg *et al., J. Pharm. Sci.* **67**, 1553 (1978). Efficacy evaluations: J. L. Duncan, J. G. Reid, *Vet. Rec.* **103**, 332 (1978); N. F. Baker *et al., Am. J. Vet. Res.* **39**, 1258 (1978).

Crystals from chloroform-methanol, mp 253° (dec). LD_{50} in dogs, rats, mice: > 1600, > 6400, > 6400 mg/kg. (Averkin).

THERAP CAT (VET): Anthelmintic.

6891. Oxibendazole. *(5-Propoxy-1H-benzimidazol-2-yl)carbamic acid methyl ester; 5-propoxy-2-benzimidazolecarbamic acid methyl ester; 5-propoxy-2-(carbomethoxyamino)benzimidazole; SKF 30310; Anthelcide EQ; Equitac.* $C_{12}H_{15}N_3O_3$; mol wt 249.27. C 57.82%, H 6.06%, N 16.86%, O 19.26%. Prepn: Brit. pat. **1,123,317** corresp to P. P. Actor, J. F. Pagano, U.S. pat. **3,574,845** (1968, 1971 both to SK & F). Comparative anthelmintic effects in mice: C. S. Karunakaron, D. A. Denham, *J. Parasitol.* **66**, 929 (1980). Clinical studies in foals and horses: J. H. Drudge *et al., Am. J. Vet. Res.* **42**, 526 (1981); E. T. Lyons *et al., ibid.* 685; D. K. Hass *et al., ibid.* **43**, 534 (1982).

Cryst, mp 230-230.5°.

THERAP CAT (VET): Anthelmintic.

6892. Oxiconazole Nitrate. *(Z)-1-(2,4-Dichlorophenyl)-2-(1H-imidazol-1-yl)ethanone O-[(2,4-dichlorophenyl)methyl]oxime mononitrate; 2',4'-dichloro-2-(imidazol-1-yl)acetophenone O-(2,4-dichlorobenzyl)oxime nitrate; Sgd 301-76; Ro 13-8996; Myfungar; Oceral; Oxistat.* $C_{18}H_{14}Cl_4N_4O_4$; mol wt 492.15. C 43.93%, H 2.87%, Cl 28.81%, N 11.38%, O 13.00%. Broad spectrum topical antimycotic agent. Prepn: G. Mixich *et al.,* Ger. pat. **2,657,578**; *eidem,* U.S. pat. **4,124,767** (1977, 1978 both to Siegfried). Stereospecific synthesis of (E)-, (Z)-isomers, ^1H NMR study: G. Mixich, K. Thiele, *Arzneimittel-Forsch.* **29**, 1510 (1979). GLC determn in plasma: M. Zell, L. Herzfeld, *J. Chromatog.* **229**, 111 (1982). *In vitro* and *in vivo* antifungal activity: A. Polak, *Arzneimittel-Forsch.* **32**, 17 (1982). Study of skin, nail penetration: G. Stueltgen, E. Bauer, *Mykosen* **25**, 74 (1982). Comparison of antifungal imidazoles: W. H. Beggs, *IRCS Med. Sci.* **11**, 677 (1983).

Crystals from ethanol, mp 137-138°.

THERAP CAT: Antifungal.

6893. Oxidimethiin. *2,3-Dihydro-5,6-dimethyl-1,4-dithiin 1,1,4,4-tetraoxide; tetrathiin; UBI-N252; Harvade.* $C_6H_{10}O_4S_2$; mol wt 210.26. C 34.27%, H 4.79%, O 30.44%, S 30.50%. Prepn: A. D. Brewer *et al.,* Fr. pat. **2,228,429** corresp to U.S. pat. **3,920,438** (1974, 1975 to Uniroyal). Activity: R. W. Neidermyer *et al., Proc. Brit. Weed Control Conf.,*

12th **3**, 959 (1974). Crystal structure: S. K. Arora *et al.*, *Acta Crystallogr.* **B34**, 2918 (1978).

Long white needles from water, mp 166-168°.
USE: Plant growth regulator.

6894. Oxidized Cellulose. Absorbable cellulose; cellulosic acid; polyanhydroglucuronic acid; Oxycel; Hemo-Pak. A cellulose of varied carboxyl content retaining the fibrous structure. Prepd by oxidizing cellulose with nitrogen dioxide: Yackel, Kenyon, *J. Am. Chem. Soc.* **64**, 121 (1942). The degree of oxidation is sufficiently high to make the product sol in dil alk solns. Insol in water or acidic solns.
THERAP CAT: Local hemostatic.
THERAP CAT (VET): Local hemostatic.

6895. Oxiniacic Acid. *3-Pyridinecarboxylic acid 1-oxide; nicotinic acid 1-oxide;* 3-carboxypyridine *N*-oxide. $C_6H_5NO_3$; mol wt 139.11. C 51.80%, H 3.62%, N 10.07%, O 34.50%. Prepn: Clemo, Koenig, *J. Chem. Soc.* **1949**, S231; Taylor, Crovetti, *J. Org. Chem.* **19**, 1633 (1954). Mass spectrum: Bild, Hesse, *Helv. Chim. Acta* **50**, 1885 (1967). Therapeutic effect: Debay, Thery, **Belg.** pat. **618,968** (1962), *C.A.* **58**, 12375e (1963).

Needles, mp 254-255° (dec). Slightly soluble in cold water, more sol in hot water, hot glacial acetic acid, hot methanol, less sol in ethanol. Insol in light petroleum, benzene, chloroform. uv max (0.1*N* H_2SO_4): 220, 260 nm (ε 22,400, 10,200).
Ethanolamine ester, $C_8H_{10}N_2O_3$, *ethanolamine oxiniacate*, *Novacyl*.
THERAP CAT: Antihyperlipoproteinemic.

6896. Oxiracetam. *4-Hydroxy-2-oxo-1-pyrrolidineacetamide;* 2-(4-hydroxypyrrolidin-2-on-1-yl)acetamide; hydroxypiracetam; CT-848; ISF-2522; Neuractiv; Neuromet. $C_6H_{10}N_2O_3$; mol wt 158.16. C 45.57%, H 6.37%, N 17.71%, O 30.35%. Analog of piracetam, *q.v.*, with psychostimulant activity. Prepn: G. Pifferi, M. Pinza, **Ger.** pat. **2,635,853**; *eidem*, **U.S.** pat. **4,118,396** (1977, 1978 both to I.S.F.); *eidem*, *Farmaco Ed. Sci.* **32**, 602 (1977). Effect on learning and memory in animals and prepn of enantiomers: S. Banfi *et al.*, *ibid.* **39**, 16 (1984). Effect in animals with cerebral impairment: S. Banfi *et al.*, *Pharmacol. Res. Commun.* **16**, 67 (1984). Pharmacokinetics in humans: E. Perucca *et al.*, *Eur. J. Drug Metabol. Pharmacokinet.* **9**, 267 (1984). Clinical evaluation in organic brain syndrome: B. Saletu *et al.*, *Neuropsychobiology* **13**, 44 (1985).

White, microcrystalline powder from methanol, mp 165-168°.
(R)-Form, crystals from acetone + water, mp 135-136°. $[\alpha]_D$ +36.2° (c = 1.00 in water).
(S)-Form, crystals from acetone + water, mp 135-136°. $[\alpha]_D$ −36.0° (c = 1.00 in water).

THERAP CAT: Nootropic.

6897. Oxitropium Bromide. *9-Ethyl-7-(3-hydroxy-1-oxo-2-phenylpropoxy)-9-methyl-3-oxa-9-azoniatricyclo[3.3.-1.0^{2,4}]nonane bromide;* (8r)-6β,7β-epoxy-8-ethyl-3α-hydroxy-1αH,5αH-tropanium bromide (−)-tropate; (−)-*N*-ethylnorscopolamine methobromide; Ba 253; Ba 253-BR-L; Oxivent; Tersigat; Ventilat. $C_{19}H_{26}BrNO_4$; mol wt 412.34. C 55.34%, H 6.36%, Br 19.38%, N 3.40%, O 15.52%. Anticholinergic bronchodilating agent. Prepn: K. Zeile *et al.*, **U.S.** pat. **3,472,861** (1969 to Boehringer, Ing.). Comparison with fenoterol, *q.v.*: D. Nolte, *Respiration* **36**, 32 (1978); with ipratropium bromide, *q.v.*: J. Lulling *et al.*, *ibid.* **42**, 188 (1981). Efficacy in bronchial asthma: H. M. Beumer *et al.*, *Int. J. Clin. Pharmacol. Ther. Toxicol.* **19**, 168 (1981); in exercise-induced asthma: K. Larsson, *Respiration* **43**, 57 (1982).

Cryst, mp 203-204° (dec). $[\alpha]_D^{21}$ −25° (c = 2.0 in water).
THERAP CAT: Bronchodilator.

6898. Oxolamine. *N,N-Diethyl-3-phenyl-1,2,4-oxadiazole-5-ethanamine;* 5-(2-diethylaminoethyl)-3-phenyl-1,2,4-oxadiazole; 3-phenyl-5-(β-diethylaminoethyl)-1,2,4-oxadiazole; 683 M. $C_{14}H_{19}N_3O$; mol wt 245.32. C 68.54%, H 7.81%, N 17.13%, O 6.52%. Prepn: **Ger.** pat. **1,097,998** (1961 to Angelini Francesco), *C.A.* **56**, 11598h (1962). Teratogenicity studies of oxolamine citrate: L. Nilsson, *Arzneimittel-Forsch.* **17**, 781 (1967).

Liquid, bp$_{0.4}$ 127°.
Hydrochloride, $C_{14}H_{19}N_3O.HCl$, crystals, mp 153-154°.
Citrate, $C_{20}H_{27}N_3O_8$, *AF 438*, *Bredon*, *Broncatar*, *Perebron*, *Prilon*, *Flogobron*, *Oxarmin*. Crystals, slightly sol in water, alcohol. uv max (aq soln): 239 nm (ε 260), 273, 283 nm.
THERAP CAT: In inflammatory conditions of respiratory tract.

6899. Oxolinic Acid. *5-Ethyl-5,8-dihydro-8-oxo-1,3-dioxolo[4,5-g]quinoline-7-carboxylic acid;* 1-ethyl-1,4-dihydro-6,7-methylenedioxy-4-oxo-3-quinolinecarboxylic acid; 1-ethyl-6,7-methylenedioxy-4-quinolone-3-carboxylic acid; W 4565; Emyrenil; Nidantin; Oxoboi; Pietil; Prodoxol; Urinox; Uritrate; Uro-Alvar; Urotrate; Uroxin Von Boch; Uroxol; Utibid. $C_{13}H_{11}NO_5$; mol wt 261.24. C 59.77%, H 4.25%, N 5.36%, O 30.62%. Quinolone antibacterial. Prepn: Kaminsky, Meltzer, **U.S.** pat. **3,287,458** (1966 to Warner-Lambert); *eidem*, *J. Med. Chem.* **11**, 160 (1968). Pharmacology: Turner *et al.*, *Antimicrob. Ag. Chemother.* **1967**, 475 sqq. Metabolism studies: DiCarlo *et al.*, *Arch. Biochem. Biophys.* **127**, 503 (1968); Crew *et al.*, *Xenobiotica* **1**, 193 (1971). Use in urinary tract infections: S. Kalowski *et al.*, *Med. J. Aust.* **21**, 345 (1979); E. Anza *et al.*, *Minerva Med.* **70**, 2333 (1979). Mechanism of action study: H. T. Wright *et al.*, *Science* **213**, 455 (1981). Brief review of therapeutic use: R. Glickman *et al.*, *Am. J. Hosp. Pharm.* **36**, 1077-1079 (1979).

Crystals from DMF, mp 314-316° (dec). LD_{50} in mice, rats (mg/kg): > 6000, > 2000 orally (Turner).

THERAP CAT: Antibacterial.

6900. Oxomemazine. *N,N,β-Trimethyl-10H-phenothi-azine-10-propanamine 5,5-dioxide; 10-(3-dimethylamino-2-methylpropyl)phenothiazine 5,5-dioxide;* 3-(9,9-dioxo-10-phenothiazinyl)-2-methyl-1-dimethylaminopropane; RP 6847; Dysedon. $C_{18}H_{22}N_2O_2S$; mol wt 330.46. C 65.42%, H 6.71%, N 8.48%, O 9.68%, S 9.70%. Prepn: Jacob, Robert, U.S. pat. **2,972,612** (1961 to Rhône-Poulenc).

Crystals from heptane, mp 115°.
Hydrochloride, $C_{18}H_{23}ClN_2O_2S$, *Doxergan, Imakol.* Crystals from ethanol + isopropanol, mp 250°.

6901. Oxonic Acid. *1,4,5,6-Tetrahydro-4,6-dioxo-1,3,5-triazine-2-carboxylic acid; allantoxanic acid;* 5-azaorotic acid; s-triazine-2,4-dione-6-carboxylic acid; 2,4-dioxo-1,2,-3,4-tetrahydro-1,3,5-triazine-6-carboxylic acid. $C_4H_3N_3O_4$; mol wt 157.08. C 30.58%, H 1.92%, N 26.75%, O 40.74%. Formed during alkaline oxidation of uric acid with H_2O_2 or $KMnO_4$. Prepn and structure: Brandenberger, *Helv. Chim. Acta* **37**, 641 (1954); Brandenberger, Brandenberger, *ibid.* **37**, 2207 (1954); Piskala, Gut, *Coll. Czech. Chem. Commun.* **27**, 1572 (1962). Metabolism: Chelbova *et al., Biochem. Pharmacol.* **19**, 2785 (1970); Cihak, Sorm, *ibid.* **21**, 607 (1972).

Free acid is very unstable; can be isolated as the mono-potassium salt. uv max (pH 9): 255 nm (ε 6800); (pH 12): 252 nm (ε 4600).
Oxonic acid amide, $C_4H_4N_4O_3$, microcrystalline substance. Practically insol in organic solvents, sparingly sol in boiling water. Does not melt below 350°, darkens above 300°.

6902. Oxophenarsine Hydrochloride. *2-Amino-4-arse-nosophenol hydrochloride;* 3-amino-4-hydroxyphenyl arsin-oxide hydrochloride; Ehrlich 5; Arseno 39; Arsenoxide; Mapharsen; Mapharside; Mapharsal; Fontarsan; Arsenosan; Oxiarsolan. $C_6H_7AsClNO_2$; mol wt 235.49. C 30.60%, H 3.00%, As 31.81%, Cl 15.06%, N 5.95%, O 13.59%. *Ref:* G. W. Raiziss, J. L. Gavron, *Organic Arsenical Compounds* (Chem. Catalog Co., New York, 1923), pp 136-137; Scott, Tullar, **Can.** pat. **405,532** (1942 to Parke, Davis).

Hygroscopic, white or nearly white, powder contg ½ mole ethanol. Readily sol in water, methanol, ethanol, in

solns of alkali hydroxides and carbonates, in dil mineral acids. Sparingly sol in glacial acetic acid. Practically insol in acetone, ether. The aq soln is acid to litmus or methyl red, but alkaline to Congo red. Solns darken on exposure to air. Usually marketed mixed with buffering agents to render its soln physiologically compatible with human blood.

THERAP CAT: Antiprotozoal (Trypanosoma).
THERAP CAT (VET): Has been used as an antirickettsial.

6903. Oxophenylarsine. Arsenosobenzene; phenyl ar-senoxide; phenylarsine oxide; Arzene. C_6H_5AsO; mol wt 168.02. C 42.89%, H 3.00%, As 44.58%, O 9.52%. Prepn: Blicke, Smith, *J. Am. Chem. Soc.* **51**, 3479 (1929); **52**, 2946 (1930).

Crystals, mp 144-146° (also reported as mp 119-120°).
USE: Pesticide.
THERAP CAT (VET): Coccidiostat (poultry).

6904. Oxotremorine. *1-[4-(1-Pyrrolidinyl)-2-butynyl]-2-pyrrolidinone;* 1-(2-oxo-1-pyrrolidinyl)-4-(1-pyrrolidin-yl)-2-butyne. $C_{12}H_{18}N_2O$; mol wt 206.28. C 69.87%, H 8.80%, N 13.58%, O 7.76%. Active metabolite of tremorine, *q.v.:* Cho *et al., Biochem. Biophys. Res. Commun.* **5**, 276 (1961). Synthesis: Bebbington, Shakeshaft, *J. Med. Chem.* **8**, 274 (1965); Nau, U.S. pat. **3,444,185** (1969 to Soc. Civile Auguil). Cholinergic agent, used experimentally as convul-sant in study of parkinsonism: Cho *et al., Proc. Int. Pharmacol. Meeting, 2nd, Prague* 1963, **2**, 75 (1964).

Pale yellow liquid. $bp_{0.6}$ 150-155°; $bp_{0.1}$ 124°. n_D^{25} 1.5156.
Picrolonate, $C_{22}H_{26}N_6O_6$, crystals from acetone, mp 157-159°.
USE: Pharmacological tool.

6905. Oxprenolol. *1-[(1-Methylethyl)amino]-3-[2-(2-propenyloxy)phenoxy]-2-propanol; 1-[o-(allyloxy)phenoxy]-3-(isopropylamino)-2-propanol;* 1-(isopropylamino)-2-hydr-oxy-3-[o-(allyloxy)phenoxy]propane; Coretal. $C_{15}H_{23}NO_3$; mol wt 265.34. C 67.89%, H 8.74%, N 5.28%, O 18.09%. β-Adrenergic blocker. Prepn: **Belg.** pat. **669,402** (1966 to Ciba), *C.A.* **65**, 5402d (1966), corresp to **Brit.** pat. **1,077,603**. Metabolic studies: Garteiz, *J. Pharmacol. Exp. Ther.* **179**, 354 (1971). Crystal structure of hydrochloride: J. M. Leger *et al., Acta Crystallogr.* **33B**, 2156 (1977). Long-term pre-vention study in coronary heart disease: S. H. Taylor *et al., N. Engl. J. Med.* **307**, 1293 (1982).

Crystals from hexane, mp 78-80°.
Hydrochloride, $C_{15}H_{24}ClNO_3$, *Ba-39089, Laracor, Pari-tane, Slow-Pren, Trasicor, Tracosal, Trasacor.* Crystals, mp 107-109°.
THERAP CAT: Antihypertensive, antianginal, antiarrhyth-mic.

6906. Oxyacanthine. *6,6',7-Trimethoxy-2,2'-dimethyl-oxyacanthan-12'-ol;* vinetine. $C_{37}H_{40}N_2O_6$; mol wt 608.71. C 73.00%, H 6.62%, N 4.60%, O 15.77%. From root of *Berberis vulgaris* L., *Berberidaceae:* Rüdel, *Arch. Pharm.* **229**, 631 (1891); Späth, Kolbe, *Ber.* **58**, 2280 (1925). Structure: v. Bruchhausen *et al., Ann.* **507**, 144 (1933); Fujita, *J. Pharm. Soc. Japan* **72**, 213 (1952), *C.A.* **47**, 6429b (1953).

White, cryst, bitter powder. mp 216-217°. $[\alpha]_D^{20}$ +131.5° (chloroform). Practically insol in water; sol in alcohol, chloroform, ether, dil acids.

Dihydrochloride, $C_{37}H_{40}N_2O_6 \cdot 2HCl$, mp 270-271°. $[\alpha]_D$ +185.5°. Sol in water.

6907. Oxybenzone. *(2-Hydroxy-4-methoxyphenyl)phenylmethanone; 2-hydroxy-4-methoxybenzophenone;* 4-methoxy-2-hydroxybenzophenone; benzophenone-3; MOB; Cyasorb UV 9 (obsolete); Spectra-Sorb UV 9; Uvinul M-40. $C_{14}H_{12}O_3$; mol wt 228.24. C 73.67%, H 5.30%, O 21.03%. Prepn: König, v. Kostanecki, *Ber.* **39,** 4027 (1906); Hardy, Forster, **U.S.** pat. **2,773,903** (1956 to Am. Cyanamid); Stanley *et al.,* **U.S.** pats. **2,861,104/5** and **3,073,866** (1958 and 1963, all to General Aniline & Film).

Crystals from isopropanol, mp 66°. Readily sol in most organic solvents. LD_{50} orally in rats: > 12.8 g/kg, H.-J. Lewerenz *et al., Food Cosmet. Toxicol.* **10,** 41 (1972).

THERAP CAT: Ultraviolet screen.

6908. Oxybutynin Chloride. *α-Cyclohexyl-α-hydroxybenzeneacetic acid 4-(diethylamino)-2-butynyl ester hydrochloride; α-phenylcyclohexaneglycolic acid 4-(diethylamino)-2-butynyl ester hydrochloride;* 4-diethylamino-2-butynyl phenylcyclohexylglycolate hydrochloride; oxibutinina hydrochloride; Ditropan; Dridase; Pollakisu; Tropax. $C_{22}H_{32}ClNO_3$; mol wt 393.97. C 67.07%, H 8.19%, Cl 9.00%, N 3.56%, O 12.18%. Prepn: **Brit.** pat. **940,540** (1963 to Mead Johnson). Toxicity: E. I. Goldenthal, *Toxicol. Appl. Pharmacol.* **18,** 185 (1971).

Crystals, mp 129-130°. LD_{50} orally in rats: 1220 mg/kg (Goldenthal).

THERAP CAT: Anticholinergic. Treatment of symptoms in neurogenic bladder.

6909. Oxychlorosene. Monoxychlorosene; Clorpactin; Clorpactin XCB. $C_{20}H_{35}ClO_4S$; mol wt 407.02. C 59.02%, H 8.67%, Cl 8.71%, O 15.72%, S 7.88%. $C_{20}H_{34}O_3S \cdot HOCl$. Described as a buffered organic hypochlorous acid derivative with slightly acid pH. Contains a long chain hydrocarbon (surface-active agent). The chain (14 carbons) may be branched or straight but actually consists of a constant mixture having predominantly the straight chain. The hydrocarbon chain also has a phenyl substituent which in turn holds a sulfonic acid group. The complex corresponds to a formula: $HO_3S-C_6H_4-(C_{14}H_{29}) \cdot HOCl$.

Aq solns are unstable and should be freshly prepd; the solid may be stored in properly stoppered bottles.

Sodium oxychlorosene; Clorpactin WCS.

THERAP CAT: Antiseptic.

6910. Oxycinchophen. *3-Hydroxy-2-phenyl-4-quinolinecarboxylic acid; 3-hydroxy-2-phenylcinchoninic acid;* 3-hydroxycinchophen; HPC; Fenidrone; Magnofenyl;

Magnophenyl; Oxinofen; Reumalon. $C_{16}H_{11}NO_3$; mol wt 265.26. C 72.44%, H 4.18%, N 5.28%, O 18.09%. Prepn: Berlingozzi, Capuano, *Atti Accad. Lincei* [V] **33,** II, 91 (1924); John, Fränkel, *J. Prakt. Chem.* **133,** 259 (1932); improved procedure: Marshall, Blanchard, *J. Pharmacol.* **95,** 186 (1949); Kreysa, **U.S.** pat. **2,776,290** (1957 to Chemo Puro).

Minute, deep yellow prisms from alc, dec 206-207°. Sol in acetic acid, alkalies, hot alc, benzene. Sparingly sol in water, ether. Forms an alkaline, water-sol sodium salt.

THERAP CAT: Antidiuretic, uricosuric.

6911. Oxyclozanide. *2,3,5-Trichloro-N-(3,5-dichloro-2-hydroxyphenyl)-6-hydroxybenzamide; 3,3',5,5',6-pentachloro-2'-hydroxysalicylanilide;* 3,3',5,5',6-pentachloro-2,2'-dihydroxybenzanilide; Zanil. $C_{13}H_6Cl_5NO_3$; mol wt 401.48. C 38.89%, H 1.51%, Cl 44.16%, N 3.49%, O 11.95%. Prepn: **Neth.** pat. **Appl. 6,409,325** corresp to Broome *et al.,* **U.S.** pat. **3,349,090** (1965, 1967, both to I.C.I.).

Crystals, mp 209-211°.

THERAP CAT (VET): Anthelmintic (flukicide).

6912. Oxycodone. *4,5-Epoxy-14-hydroxy-3-methoxy-17-methylmorphinan-6-one;* dihydrohydroxycodeinone; 14-hydroxydihydrocodeinone; Dihydrone. $C_{18}H_{21}NO_4$; mol wt 315.36. C 68.55%, H 6.71%, N 4.44%, O 20.29%. Prepn by catalytic reduction of hydroxycodeinone, its oxime, or its bromination products or by reduction of hydroxycodeinone with sodium hydrosulfite. Bibliography: Small, Lutz, "*Chemistry of the Opium Alkaloids,*" Suppl. No. 103 to Public Health Reports, Washington (1932); K. W. Bentley, *The Chemistry of the Morphine Alkaloids* (Oxford, 1954). Oxycodone apparently exists in two tautomeric forms which give the same salts: Freund, Speyer, *J. Prakt. Chem.* **94,** 135 (1916).

Long rods from alc, mp 218-220°. Sol in alc, chloroform; nearly insol in ether, water, KOH, NaOH, NH_4OH. Does not reduce ammoniacal silver or Fehling's soln.

Tautomeric form: Strongly refringent scales from alc, mp 219-220°, more sol in alc than the other form.

Hydrochloride, $C_{18}H_{22}ClNO_4$, *Dinarkon, Eubine, Eucodal, Eukodal, Eutagen, Oxikon, Oxycon, Pancodine, Tecodin, Tekodin, Thecodine, Thekodin.* Component of *Percodan.* Long rods from water, dec 270-272°. $[\alpha]_D^{20}$ −125° (c = 2.5). One gram dissolves in 10 ml water. Slightly sol in alc.

Pectinate, *Proladone,* used for prolonged action.

Caution: May be habit forming. This is a controlled substance (opiate) listed in the U.S. Code of Federal Regulations, Title 21 Part 1308.12 (1985).

THERAP CAT: Narcotic analgesic.

6913. 4,4'-Oxydi-2-butanol. *4,4'-Oxybis[2-butanol];* 3,3'-dihydroxydibutyl ether; bis[3-hydroxybutyl] ether;

DHBE; Diskin; Dyskinébyl; Dis-Cinil; Colenormol. C_8H_{18}-O_3; mol wt 162.22. C 59.23%, H 11.18%, O 29.59%. CH_3-$CH(OH)CH_2CH_2OCH_2CH_2CH(OH)CH_3$. Prepn: Joulty, **Fr. pat. 1,267,084** (1960). Action on bile flow: Bornmann, *Arzneimittel-Forsch.* **2**, 122 (1952). Use as choleretic and antispasmodic: Albot, Toulet, *Presse Med.* **67**, 2053 (1959). General pharmacological properties: Fregnan, Porta, *Arzneimittel-Forsch.* **26**, 2116 (1976).

Clear, almost colorless, bitter fluid. Miscible with water.

THERAP CAT: Choleretic.

6914. 10,10'-Oxydiphenoxarsine. *10,10'-Oxybis-10H-phenoxarsine;* bis(phenoxarsin-10-yl)ether; bis(10-phenoxarsyl)oxide; bis(10-phenoxarsinyl)oxide; OBPA; Vinyzene. $C_{24}H_{16}As_2O_3$; mol wt 502.23. C 57.40%, H 3.21%, As 29.83%, O 9.56%. Prepn: W. L. Lewis *et al., J. Am. Chem. Soc.* **43**, 891 (1921). Synthesis: K. D. Shvetsova-Shilovskaya *et al., J. Gen. Chem. USSR* **31**, 776 (1961). Use in control of plant growth: M. L. Joseph, J. L. Hardy, **U.S. pat. 3,069,252** (1962 to Dow Chemical); as bactericide in plastics: C. C. Yeager, **U.S. pat. 3,288,674** (1966 to Scientific Chemicals). Toxicity study: D. G. Clark, *Toxicol. Appl. Pharmacol.* **21**, 315 (1972).

Colorless monoclinic prisms, mp 184-185°. dec. 380°. sp. gr. 1.40-1.42. Sol in alcohol, chloroform, methylene chloride. Practically insol in water (5 ppm at 20°) and alkali. LD_{50} in male rats: 35-50 mg/kg (Ventron, company data sheet).

USE: Primarily for fungicidal and bactericidal protection of plastics.

6915. Oxyfedrine. *[R-(R*,S*)]-3-[(2-Hydroxy-1-methyl-2-phenylethyl)amino]-1-(3-methoxyphenyl)-1-propanone;* L-3-[(β-hydroxy-α-methylphenethyl)amino]-3'-methoxypropiophenone; L-(1-hydroxy-1-phenyl-2-propylamino)-1-(m-methoxyphenyl)-1-propanone; oxyphedrine. $C_{19}H_{23}NO_3$; mol wt 313.38. C 72.82%, H 7.40%, N 4.47%, O 15.32%. Partial β-adrenergic agonist with coronary vasodilating and positive inotropic effects. Prepn: K. Thiele, **Belg. pat. 630,296;** idem, **U.S. pat. 3,225,095** (1963, 1965 both to Degussa). K. Thiele *et al., Arzneimittel-Forsch.* **16**, 1064 (1966). Absolute configuration determined by circular dichroism: J. Engel *et al., Chem. Ztg.* **105**, 85 (1981). TLC determn: Musumarra, *J. Chromatog.* **350**, 151 (1985). Pharmacology: H. Hueller *et al., Pharmazie* **27**, 242 (1972). Mode of action: P. Mentz, W. Forster, *Arzneimittel-Forsch.* **34**, 1739 (1984); N. Sternitzke, *Z. Kardiol.* **73**, 586 (1984). In prevention of experimental myocardial necrosis in rats: S. D. Seth *et al., Arzneimittel-Forsch.* **34**, 678 (1984). Comparison with atenolol, *q.v.,* in angina pectoris: L. Fananapazir, C. Bray, *Brit. J. Clin. Pharmacol.* **20**, 405 (1985).

DL-Form hydrochloride, $C_{19}H_{24}ClNO_3$, mp 173-175°. LD_{50} in mice (mg/kg): 34 i.v. (Hueller).

L-Form hydrochloride, *D563, Ildamen, Modacor.* Crystals from methanol, mp 192-194°. LD_{50} in mice (mg/kg): 29 i.v. (Hueller).

THERAP CAT: Anti-anginal. Treatment of coronary insufficiency.

6916. Oxyfluorfen. *2-Chloro-1-(3-ethoxy-4-nitrophenoxy)-4-(trifluoromethyl)benzene;* 2-chloro-α,α,α-trifluoro-p-

tolyl-3-ethoxy-4-nitrophenyl ether; RH-2915; Goal. C_{15}-$H_{11}ClF_3NO_4$; mol wt 361.72. C 49.81%, H 3.07%, Cl 9.80%, F 15.76%, N 3.87%, O 17.69%. Selective pre- and postemergence herbicide. Prepn: **Neth. pat. 7,303,590** corresp to H. O. Bayer *et al.,* **U.S. pat. 3,798,276** (1973, 1974 to Rohm & Haas). Activity: R. Y. Yih, C. Swithenbank, *J. Agr. Food Chem.* **23**, 592 (1975). Metabolism: I. L. Adler *et al., ibid.* **25**, 1339 (1977).

Orange cryst solid, mp 83-84°. Soly in water: 0.1 ppm. Sol in most organic solvents. LD_{50} orally in male albino rats: >5000 mg/kg, R. Y. Yih, C. Swithenbank, *loc. cit.*

USE: Herbicide.

6917. Oxygen. O; at. wt 15.9994; at. no. 8; valence 2. Occurs normally as the diatomic gas O_2, also as ozone O_3. Atomic oxygen (O) can be prepd. Three naturally occurring isotopes: 16 (99.759%); 17 (0.037%); 18 (0.204%); artificial radioactive isotopes: 13-15; 19; 20. The most abundant element on earth; makes up 46.6% of earth's crust; 20.95% by vol of dry air. Obtained on a large scale by liquefaction of air. First obtained by Scheele in 1771 and independently by Priestley in 1774. Monograph: M. Ardon, *Oxygen—Elementary Forms and Hydrogen Peroxide* (Benjamin, New York, 1965) 106 pp. Review of oxygen and its compounds: Ebsworth *et al.,* in *Comprehensive Inorganic Chemistry* vol. 2, J. C. Bailar, Jr. *et al.,* Eds. (Pergamon Press, Oxford, 1973) pp 685-794; A. H. Taylor in Kirk-Othmer *Encyclopedia of Chemical Technology* vol. **16** (Wiley-Interscience, New York, 3rd ed., 1981) pp 653-673.

Colorless, odorless, tasteless, neutral gas; supports combustion. d^0 (gas) 1.429 g/l; d^{-183} (liquid) 1.14 g/ml. mp −218.4°. bp −182.96°. Critical temp −118.95°. Critical press. 50.14 atm. Heat of vaporization (−183°): 50.9 cal/g. Usually marketed under pressure in metal cylinders. One vol gas dissolves in 32 vols water at 20°; in 7 vols alcohol at 20°; also sol in other organic liquids and usually to a greater extent than in water.

USE: In oxyhydrogen or oxyacetylene flame for welding metals and for lighting (calcium light, etc); submarine work by divers, propellant for rockets. In the production of synthesis gas which can be used in the Fischer-Tropsch process for liquid fuels. *Caution:* Avoid smoking, flames, electric sparks—explosion hazard.

THERAP CAT: Medicinal gas to relieve hypoxia; at hyperbaric pressures in cardiac and other surgery, anaerobic infections, carbon monoxide poisoning; in cryotherapy (liq form).

THERAP CAT (VET): In hypoxia and in conjunction with volatile anesthetics.

6918. Oxymesterone. *4,17-Dihydroxy-17-methylandrost-4-en-3-one;* 4,17β-dihydroxy-17α-methyl-3-oxoandrost-4-ene; 4-hydroxy-17α-methyltestosterone; 17α-methyl-4-androstene-4,17β-diol-3-one; oxymestrone; Anamidol; Oranabol; Theranabol. $C_{20}H_{30}O_3$; mol wt 318.44. C 75.43%, H 9.50%, O 15.07%. Prepn: **Brit. pat. 848,288;** Camerino *et al.,* **U.S. pat. 3,060,201** (1960, 1962 to Farmitalia).

Crystals, mp 169-171°. $[α]_D^{20}$ +69° (ethanol). uv max (ethanol): 278 nm ($E_{1cm}^{1\%}$ 406). Practically insol in water. Sol in chloroform, acetone, alcohol.

THERAP CAT: Androgen; anabolic.

6919. Oxymetazoline. *3-[(4,5-Dihydro-1H-imidazol-2-yl)methyl]-6-(1,1-dimethylethyl)-2,4-dimethylphenol; 6-tert-butyl-3-(2-imidazolin-2-ylmethyl)-2,4-dimethylphenol; 2-(4-tert-butyl-2,6-dimethyl-3-hydroxybenzyl)-2-imidazoline;* H 990; Navisin; Hazol; Rhinofrenol; Rhinolitan; Sinerol; Nezeril. $C_{16}H_{24}N_2O$; mol wt 260.37. C 73.80%, H 9.29%, N 10.76%, O 6.15%. Prepd from (4-tert-butyl-2,6-dimethyl-3-hydroxyphenyl)acetonitrile and ethylenediamine: Fruhstorfer, Mueller-Calgan, **Ger.** pat. **1,117,588** (1961 to E. Merck), *C.A.* **57**, 4674a (1962). Toxicity data: Hotovy *et al., Arzneimittel-Forsch.* **11**, 1016 (1961).

Crystals from benzene, mp 181-183°.
Hydrochloride, $C_{16}H_{25}ClN_2O$, *Afrazine, Afrin, Iliadin, Nafrine, Nasivin, Oxilin, Sinex.* Crystals, dec 300-303°. Freely sol in water, alcohol. Practically insol in ether, chloroform, benzene. LD_{50} orally in mice: 10 mg/kg (Hotovy).
Note: Ingredient of *Drixin.*
THERAP CAT: Adrenergic (vasoconstrictor); nasal decongestant.

6920. Oxymetholone. *17-Hydroxy-2-(hydroxymethylene)-17-methylandrostan-3-one;* 2-hydroxymethylene-17α-methyldihydrotestosterone; 4,5α-dihydro-2-hydroxymethylene-17α-methyltestosterone; 2-hydroxymethylene-17α-methyl-17β-hydroxy-5α-androstan-3-one; 2-hydroxymethylene-17α-methylandrostan-17β-ol-3-one; anasterone; Adroyd; Anapolon; Anadrol; Pardroyd; Plenastril; Protanabol; Nastenon; Synasteron. $C_{21}H_{32}O_3$; mol wt 332.47. C 75.86%, H 9.70%, O 14.44%. Anabolic steroid. Prepn: Ringold *et al., J. Am. Chem. Soc.* **81**, 427 (1959); Ringold, Rosenkranz, **Ger.** pat. **1,070,632** (1959 to Syntex).

Crystals from ethyl acetate, mp 178-180°. $[\alpha]_D +38°$. uv max: 285 nm (log ε 3.99).
Enol acetate, $C_{23}H_{34}O_4$, crystals from hexane, mp 144-148°. $[\alpha]_D +27°$ (ethanol). uv max: 255 nm (log ε 4.09).
Enol propionate, $C_{24}H_{36}O_4$, crystals from hexane, mp 135°. $[\alpha]_D +26°$ (ethanol). uv max: 257 nm (log ε 4.11).
Enol benzoate, $C_{28}H_{36}O_4$, crystals from acetone + water, mp 188-190°. $[\alpha]_D \pm0°$. uv max: 230 nm (log ε 4.19).
Note: This substance may reasonably be anticipated to be a carcinogen: *Fourth Annual Report on Carcinogens* (NTP 85-002, 1985) p 164.
THERAP CAT: Androgen.
THERAP CAT (VET): Anabolic steroid for small animals.

6921. Oxymethurea. *N,N'-Bis(hydroxymethyl)urea; N,N'*-dihydroxymethylurea; Methural. $C_3H_8N_2O_3$; mol wt 120.11. C 30.00%, H 6.71%, N 23.33%, O 39.96%. (HO-$CH_2NH)_2CO$. Prepn: Einhorn, Hamburger, *Ber.* **41**, 26 (1908); Walter, U.S. pat. **1,863,426** (1927); U.S. pat. **2,436,-355** (1946 to du Pont).
Crystals from alcohol, mp 137-139°. Very sol in cold water, hot ethyl alcohol, methanol.
USE: In the textile industry in cotton crease- and shrink-proofing, finishing, drying, sizing; in tanning; pesticides; in photographic developers.
THERAP CAT: Antiseptic.

6922. Oxymorphone. *4,5-Epoxy-3,14-dihydroxy-17-methylmorphinan-6-one;* dihydrohydroxymorphinone; dihydro-14-hydroxymorphinone; 14-hydroxydihydromorphin-one. $C_{17}H_{19}NO_4$; mol wt 301.33. C 67.76%, H 6.36%, N 4.65%, O 21.24%. Obtained from dihydrohydroxycodeinone by boiling with concd aq hydrobromic acid: Weiss, *J. Am. Chem. Soc.* **77**, 5891 (1955); Lewenstein, Weiss, U.S. pat. **2,806,033** (1957).

Crystals from boiling ethanol, ethyl acetate or benzene. mp 248-249° (dec). Levorotatory. Sol in boiling acetone and chloroform; readily sol in aq alkalies; moderately sol in boiling ethanol; sparingly sol in benzene.
Hydrochloride, $C_{17}H_{20}ClNO_4$, *Numorphan.* Crystals.
Caution: May be habit forming. This is a controlled substance (opiate) listed in the U.S. Code of Federal Regulations, Title 21 Part 1308.12 (1987).
THERAP CAT: Narcotic analgesic.
THERAP CAT (VET): Narcotic analgesic.

6923. Oxypendyl. *4-[3-(10H-Pyrido[3,2-b][1,4]benzothiazin-10-yl)propyl]-1-piperazineethanol;* 10-[3-[4-(2-hydroxyethyl)-1-piperazinyl]propyl]-10H-pyrido[3,2-b][1,4]-benzothiazine; 10-[3-(1-hydroxyethyl-4-piperazinyl)propyl]-4-azaphenothiazine; oxypendyl; D 706. $C_{20}H_{26}N_4OS$; mol wt 370.53. C 64.83%, H 7.07%, N 15.12%, O 4.32%, S 8.66%. Prepn: Schuler, Klebe, *Ann.* **653**, 172 (1962).

bp_1 260-265°, bp_6 280-300°.
Dihydrochloride, $C_{20}H_{28}Cl_2N_4OS$, *Pervetral.* Crystals, mp 218-220°.
THERAP CAT: Antiemetic.

6924. Oxypertine. *5,6-Dimethoxy-2-methyl-3-[2-(4-phenyl-1-piperazinyl)ethyl]-1H-indole;* 1-[2-(5,6-dimethoxy-2-methyl-3-indolyl)ethyl]-4-phenylpiperazine; WIN 18501-2; Equipertine; Forit. $C_{23}H_{29}N_3O_2$; mol wt 379.49. C 72.79%, H 7.70%, N 11.07%, O 8.43%. Prepn: Archer *et al., J. Am. Chem. Soc.* **84**, 1306 (1962).

Hydrochloride, *Integrin.*
THERAP CAT: Antidepressant.

6925. Oxyphenbutazone. *4-Butyl-1-(4-hydroxyphenyl)-2-phenyl-3,5-pyrazolidinedione; 4-butyl-2-(p-hydroxyphenyl)-1-phenyl-3,5-pyrazolidinedione;* 1-phenyl-2-(p-hydroxyphenyl)-3,5-dioxo-4-n-butylpyrazolidine; 1-(p-hydroxyphenyl)-2-phenyl-4-butylpyrazolidine-3,5-dione; p-hydroxyphenylbutazone; G 27202; Butazonic; Californit; Crovaril; Flogitolo; Flogoril; Frabel; Neo-Farmadol; Oxalid; Rapostan; Rheumapax; Tandacote; Tanderil; Tandearil; Visubutina. $C_{19}H_{20}N_2O_3$; mol wt 324.37. C 70.35%, H 6.22%, N 8.64%, O 14.80%. Prepn: Hafliger, U.S. pat. **2,745,783** (1956 to Geigy); Pfister, Hafliger, *Helv. Chim. Acta* **40**, 395 (1957).

Monohydrate, *Imbun, Phlogistol, Phlogase, Phlogont.* Crystals, mp 96°. Anhydr crystals from ether + petr ether, mp 124-125°. Acidic reaction. Soluble in ethanol, methanol, chloroform, benzene, ether. Forms a water-soluble sodium salt.

Piperazine monohydrate, $C_{23}H_{30}N_4O_3 \cdot H_2O$, *Difmedol.*

THERAP CAT: Anti-inflammatory.

6926. Oxyphencyclimine. *α-Cyclohexyl-α-hydroxybenzeneacetic acid (1,4,5,6-tetrahydro-1-methyl-2-pyrimidinyl)methyl ester; 1,4,5,6-tetrahydro-1-methyl-2-pyrimidinylmethyl α-phenylcyclohexaneglycolate;* α-phenylcyclohexaneglycolic acid 1-methyl-2-tetrahydroxypyrimidylmethyl ester; 1-methyl-1,4,5,6-tetrahydro-2-pyrimidylmethyl α-cyclohexyl-α-phenylglycolate; Antulcus; Caridan; Daricol; Setrol; Vio-Thene; Daricon; Naridan; Zamanil. $C_{20}H_{28}N_2O_3$; mol wt 344.44. C 69.74%, H 8.19%, N 8.13%, O 13.94%. Prepn: **Brit.** pat. **795,758** (1958 to Pfizer); Faust *et al., J. Am. Chem. Soc.* **81**, 2214 (1959).

Hydrochloride, $C_{20}H_{28}N_2O_3 \cdot HCl$, crystals, dec 231-232°. Solubility in water: 1.2 g/100 ml.

THERAP CAT: Anticholinergic.

6927. Oxyphenisatin Acetate. *3,3-Bis[4-(acetyloxy)phenyl]-1,3-dihydro-2H-indol-2-one; 3,3-bis(p-acetoxyphenyl)oxindole;* acetphenolisatin; endophenolphthalein; diacetyldiphenolisatin; diacetyldihydroxydiphenylisatin; diacetylhydroxydiphenylisatin; diacetoxydiphenylisatin; di(acetoxyphenyl)oxindole; diphesatin; diacetyldioxyphenylisatin; Isacen; Isocrin; Isaphen; Laxo-Isatin; Promassolax; Sanapert; Bydolax; Cirotyl; Lisagal; Contax; Prulet; Purgaceen; Bisatin. $C_{24}H_{19}NO_5$; mol wt 401.42. C 71.81%, H 4.77%, N 3.49%, O 19.93%. Prepd from isatin by treatment with phenol followed by acetylation in the presence of H_2SO_4 at temps below 100°: Baeyer, Lazarus, *Ber.* **18**, 2641 (1885); Hoffmann-La Roche, **Swiss** pat. **100,806; Ger.** pat. **406,210** (1923); **U.S.** pat. **1,624,675** (1927); **Ger.** pat. **482,435** (1929); Mizuno, Turuga, **Japan.** pat. **129,200** (1939).

Tasteless crystals, mp 242°. Practically insol in water, ether, dilute HCl; slightly sol in alcohol.

THERAP CAT: Cathartic.

6928. Oxyphenonium Bromide. *2-[(Cyclohexylhydroxyphenylacetyl)oxy]-N,N-diethyl-N-methylethanaminium bromide; diethyl(2-hydroxyethyl)methylammonium α-phenylcyclohexaneglycolate bromide;* α-phenylcyclohexaneglycolic acid ester diethyl(2-hydroxyethyl)methylammonium bromide; diethylaminoethyl α-phenylcyclohexaneglycolate methylbromide; 2-diethylaminoethyl α-cyclohexyl-α-phenylglycolate methobromide; cyclohexylhydroxyphenylacetic acid diethylmethylaminoethyl ester bromide; phenylcyclo-

hexyloxyacetic acid diethylaminoethyl ester bromomethylate; Ba-5473; C-5473; Antrenyl; Spasmophen. $C_{21}H_{34}Br\cdot NO_3$; mol wt 428.41. C 58.87%, H 8.00%, Br 18.65%, N 3.27%, O 11.20%. Prepn: **Swiss** pat. **259,948** (1949 to Ciba).

Crystals, mp 189-194° from ethyl acetate + alc. Freely sol in water. Sparingly sol in alc. Aq solns are neutral.

THERAP CAT: Anticholinergic.

6929. Oxypinocamphone. *2-Hydroxy-2,6,6-trimethylbicyclo[3.1.1]heptan-3-one; 2-hydroxy-3-pinanone;* 2-hydroxypinocamphone; Camphostene; Oxypinone. $C_{10}H_{16}O_2$; mol wt 168.23. C 71.39%, H 9.59%, O 19.02%. Prepn of *l*-form: Kuwata, *J. Am. Chem. Soc.* **59**, 2509 (1937); Harispe *et al., Bull. Soc. Chim. France* **1964**, 1035; of *dl*-form: Kuwata, *loc. cit.* Stereochemistry: Carlson *et al., Tetrahedron Letters* **1968**, 5941.

dl-Form, crystals, mp 38.5-39.0°.
l-Form, crystals from ether + petr ether, mp 35.5-36.5°. $[\alpha]_D^{25}$ −18.56° (c = 14.44 in alcohol).

THERAP CAT: Respiratory insufficiency.

6930. Oxypolygelatin. OPG. Prepd by treating purified gelatin with glyoxal, followed by oxidation with hydrogen peroxide.

Solutions remain liquid at ice-box temps.

THERAP CAT: Plasma volume extender.

6931. Oxytetracycline. *4-(Dimethylamino)-1,4,4a,5,-5a,6,11,12a-octahydro-3,5,6,10,12,12a-hexahydroxy-6-methyl-1,11-dioxo-2-naphthacenecarboxamide;* glomycin; terrafungine; riomitsin; hydroxytetracycline; Berkmycen; Biostat; Engemycin; Oxacycline; Oxatets; Oxydon; Oxy-Dumocyclin; Oxymycin; Oxypan; Oxytetracid; Ryomycin; Stevacin; Terraject; Terramycin; Tetramel; Tetran; Vendarcin; Vendracin. $C_{22}H_{24}N_2O_9$; mol wt 460.44. C 57.39%, H 5.25%, N 6.08%, O 31.27%. Antibiotic substance isolated from the elaboration products of the actinomycete, *Streptomyces rimosus*, grown on a suitable medium: Finlay *et al., Science* **111**, 85 (1950). Isoln: Regna, Solomons, *Ann. N.Y. Acad. Sci.* **53**, 221 (1950); Regna *et al., J. Am. Chem. Soc.* **73**, 4211 (1951). Production from *Streptomyces rimosus:* Sobin *et al.,* **U.S.** pat. **2,516,080** (1950 to Pfizer). Isoln from *S. xanthophaeus:* Brockmann, Musso, *Naturwiss.* **41**, 451 (1954); Brockmann *et al.,* **Ger.** pat. **913,687** (1954 to Bayer), *C.A.* **53**, 4662h (1959). Solubility data: Weiss *et al., Antibiot. & Chemother.* **7**, 374 (1957). Structure: Hochstein *et al., J. Am. Chem. Soc.* **74**, 3708 (1952). Abs config: Dobrynin *et al., Tetrahedron Letters* **1962**, 901. Stereochemistry: Schach von Wittenau *et al., J. Am. Chem. Soc.* **87**, 134 (1965). Total synthesis of the *dl*-form: H. Muxfeldt *et al., ibid.* **101**, 689 (1979).

Dihydrate, *Abbocin, Clinimycin, Imperacin.* Needles from water or methanol, dec 181-182°. $[\alpha]_D^{25}$ −196.6° (0.1N HCl); $[\alpha]_D^{25}$ −2.1° (0.1N NaOH); $[\alpha]_D^{25}$ +26.5° (methanol). uv max (pH 4.5 phosphate buffer 0.1*M*): 249, 276, 353 nm ($E_{1cm}^{1\%}$ 240, 322, 301). Soly in water at 23° at various pH's: pH 1.2 = 31,400 γ/ml, pH 2.0 = 4600 γ/ml, pH 3.0 = 1400 γ/ml; pH 5.0 = 500 γ/ml, pH 6.0 = 700 γ/ml, pH 7.0 = 1100 γ/ml, pH 9.0 = 38,600 γ/ml. Soly in abs ethanol 12,000 γ/ml, in 95% ethanol 200 γ/ml.

Hydrochloride, *Alamycin, Aquacycline, Arcospectron, Bio-Mycin, Duphacycline, Geomycin, Gynamousse, Macocyn, Macodyn, Occrycetin, Oxlopar, Oxybiocycline, Oxybiotic, Oxycycline, Oxyject, Oxylag, Stecsolin, Tetra-Tablinen, Toxinal.* Yellow platelets from water. Very soluble in water (1 g/ml). Concd aq solns at neutral pH hydrolyze on standing and deposit crystals of oxytetracycline. Soly in abs ethanol: 12,000 γ/ml, in 95% ethanol: 33,000 γ/ml.

Disodium salt dihydrate, $C_{22}H_{22}N_2Na_2O_9.2H_2O$, yellow crystals; darkens on standing. Soly in abs alc: 8,000 γ/ml, in methanol: 1500 γ/ml.

Stability: Oxytetracycline crystals show no loss in potency on heating for 4 days at 100°, the hydrochloride crystals show <5% inactivation after 4 mos at 56°. Aq solns of the hydrochloride at pH 1.0 to 2.5 are stable for at least 30 days at 25°. Solns at pH 3.0 to 9.0 show no detectable loss in potency on storage at 5° for at least 30 days. Half-life in hours of aq oxytetracycline solns at 37°: pH 1.0 = 114; pH 2.5 = 134; pH 4.6 = 45; pH 5.5 = 45; pH 7.0 = 26; pH 8.5 = 33; pH 10.0 = 14.

USE: To treat lethal yellowing in palm trees.

THERAP CAT: Antibacterial.

THERAP CAT (VET): Antimicrobial.

6932. Oxythiamine. *3-[(1,4-Dihydro-2-methyl-4-oxo-5-pyrimidinyl)methyl]-5-(2-hydroxyethyl)-4-methylthiazolium chloride; 5-(2-hydroxyethyl)-3-[(4-hydroxy-2-methyl-5-pyrimidinyl)methyl]-4-methylthiazolium chloride.* $C_{12}H_{16}-ClN_3O_2S$; mol wt 301.81. C 47.76%, H 5.34%, Cl 11.75%, N 13.92%, O 10.60%, S 10.62%. Prepn: F. Bergel, A. R. Todd, *J. Chem. Soc.* **1937**, 1504; M. Soodak, L. R. Cerecedo, *J. Am. Chem. Soc.* **66**, 1988 (1944). Improved prepn: H. N. Rydon, *Biochem. J.* **48**, 383 (1951). Thiamine antagonist activity: A. J. Eusebi, L. R. Cerecedo, *Science* **110**, 162 (1949); L. J. Daniel, L. C. Norris, *Proc. Soc. Exp. Biol. Med.* **72**, 165 (1949); and distribution in tissues: C. J. Gubler, D. S. Murdock, *J. Nutr. Sci. Vitaminol.* **28**, 217 (1982). Proposed mechanism of action: S. A. Strumilo *et al., Biomed. Biochim. Acta* **43**, 159 (1984). Determn by HPLC: B. C. Hemming, C. J. Gubler, *J. Liq. Chromatog.* **3**, 1697 (1980).

Hydrochloride, $C_{12}H_{16}ClN_3O_2S.HCl$, flat needles grouped in rosettes, dec 195°. uv max (acid soln): 265, 258, 228, 223 nm; (alkaline soln): 268, 260, 228, 221 nm. Does not give the thiochrome reaction.

Diphosphate, crystals, mp 127-129°.

Monophosphoric acid ester, hygroscopic crystals, dec 185-186°.

Triphosphoric acid ester, minute crystals, dec 245-255°.

6933. Oxythioquinox. *6-Methyl-1,3-dithiolo[4,5-b]quinoxalin-2-one; dithiocarbonic acid cyclic S,S-(6-methyl-2,3-quinoxalinediyl) ester;* 6-methyl-2,3-quinoxalinedithiol cyclic S,S-dithiocarbonate; 6-methyl-2-oxo-1,3-dithio[4,5-b]-quinoxaline; chinomethionat(e); quinomethionate; Bayer 36,205; Forstan; Morestan. $C_{10}H_6N_2OS_2$; mol wt 234.29. C 51.26%, H 2.58%, N 11.95%, O 6.83%, S 27.37%. Prepn and review (in English) of chemical and fungicidal properties with toxicol. data: F. Grewe, H. Kaspers, *Pflanzenschutz Nachr. Bayer* **18** (1), 1-23 (1965), *C.A.* **64**, 5689h (1966).

Yellow crystals from benzene, mp 172°. Practically insol

in water. Freely sol in DMF. Sol in hot benzene, toluene, dioxane. Slightly sol in methanol, ethanol, acetone. LD_{50} orally in male, female rats: 1800, 1100 mg/kg, T. B. Gaines, *Toxicol. Appl. Pharmacol.* **14**, 515 (1969).

USE: Acaricide, fungicide.

6934. Oxytocin. Alpha-hypophamine; ocytocin; Endopituitrina; Pitocin; Syntocinon; Nobitocin S; Orasthin; Oxystin; Partocon; Synpitan; Piton-S; Uteracon. $C_{43}H_{66}N_{12}-O_{12}S_2$; mol wt 1007.23. C 51.28%, H 6.61%, N 16.69%, O 19.06%, S 6.37%. The principal uterus-contracting and lactation-stimulating hormone of the posterior pituitary gland. Isoln: Pierce *et al., J. Biol. Chem.* **199**, 929 (1952). Structure and synthesis: Tuppy, Michl, *Monatsh.* **84**, 1011 (1953); Tuppy, *Biochim. Biophys. Acta* **11**, 449 (1953); du Vigneaud *et al., J. Am. Chem. Soc.* **75**, 4879 (1953); **76**, 3115 (1954); Bodanszky, du Vigneaud, *ibid.* **81**, 2504 (1959); Cash *et al., J. Med. Pharm. Chem.* **5**, 413 (1962); Sakakibara *et al., Bull. Chem. Soc. Japan* **38**, 120 (1965). Solid phase synthesis: Bayer, Hagenmaier, *Tetrahedron Letters* **1968**, 2037; Ives, *Can. J. Chem.* **46**, 2318 (1968). Synthesis of D-oxytocin: Flouret, du Vigneaud, *J. Am. Chem. Soc.* **87**, 3775 (1965). Description of commercial process: Velluz *et al., U.S. pats.* **2,938,891** and **3,076,797** (1960, 1963, both to Roussel-UCLAF). Radioimmunoassay: T. Chard, *Clin. Biochem. Anal.* **5**, 209 (1977). *Review:* du Vigneaud, *Experientia Suppl.* **II** (14th Intl. Congr. Pure and Appl. Chem.), 9-26 (1955); R. Caldeyro-Barcia, H. Heller, *Proc. of an Intl. Symp. on Oxytocin* (Montevideo 1959) 443 pp; several authors, *Advan. Exp. Med. Biol.* **2**, 53-104 (1968); C. R. W. Edwards in *Hormones in Blood* vol. 2, C. H. Gray, V. James, Ed. (Academic Press, New York, 3rd ed., 1979) pp 401-421. Review of role in parturition: A.-R. Fuchs, F. Fuchs, *Advan. Exp. Med.* **1980**, 403-428. Comprehensive description: F. Nachtmann *et al.,* in *Analytical Profiles of Drug Substances* vol. 10, K. Florey, Ed. (Academic Press, New York, 1981) pp 563-600.

```
Cys-Tyr-Ile-Gln-Asn-Cys-Pro-Leu-GlyNH2
    └───────────────────┘
```

White powder. $[\alpha]_D^{22}$ −26.2° (c = 0.53). Sol in water, 1-butanol, 2-butanol. Shows an activity of 450-500 U.S.P. units/mg when compared with the U.S.P. posterior pituitary reference standards.

Citrate, *Pitocin-Buccal.*

Note: Unlike vasopressin which occurs in at least two forms, oxytocin from beef and hog sources shows no difference in amino acid composition.

THERAP CAT: Oxytocic.

THERAP CAT (VET): Stimulates milk let-down, uterine contraction.

6935. Ozagrel. *(E)-3-[4-(1H-Imidazol-1-ylmethyl)-phenyl]-2-propenoic acid; (E)-4-(imidazol-1-ylmethyl)cin-namic acid;* OKY-046. $C_{13}H_{12}N_2O_2$; mol wt 228.25. C 68.41%, H 5.30%, N 12.27%, O 14.02%. Prepn: K. Iizuka *et al.,* Ger. pat. **2,923,815**; *eidem,* U.S. pat. **4,226,878** (both 1980 to Ono; Kissei). Synthesis and thromboxane synthetase inhibitory activity: K. Iizuka *et al., J. Med. Chem.* **24**, 1139 (1981). Pharmacology: S. Hiraku *et al., Japan. J. Pharmacol.* **41**, 393 (1986). Pulmonary vascular effects: R. Garcia-Szabo *et al., Prostaglandins* **28**, 851 (1984). Metabolism in animals: M. Shimizu *et al., Iyakuhin Kenkyu* **17**, 289 (1986), *C.A.* **105**, 72013q (1986). HPLC determn in biological fluids: *eidem, ibid.* **298**, *C.A.* **105**, 72013r (1986). Clinical pharmacology and evaluation in myocardial infarction: T. Ito *et al., Biomed. Biochim. Acta* **43**, S125 (1984). Clinical evaluation in prevention of cerebral vasospasm: S. Suzuki *et al., Acta Neurochir.* **77**, 133 (1985); in coronary artery disease: M. Shikano *et al., Japan. Heart J.* **28**, 663 (1987). Toxicity data: T. Nishigake *et al., Clin. Rep.* **20**, 2671 (1986). Series of articles on pharmacology: *Pharmacometrics* **31**, 527-565 (1986), *C.A.* **105**, 35349-35352 (1986).

Prisms from ethanol-ether, mp 223-224°.

Hydrochloride, $C_{13}H_{13}ClN_2O_2$, crystals from ethanol-ether, mp 214-217°.

Sodium salt, $C_{13}H_{11}N_2NaO_2$, *Cataclot, Xanbon.* LD_{50} in male, female mice, male, female rats (mg/kg): 1940, 1580, 1150, 1300 i.v.; 3800, 3600, 5900, 5700 orally; 2450, 2100, 2300, 2250 s.c. (Nishigake).

THERAP CAT: Antithrombotic, antianginal.

6936. Ozone. Triatomic oxygen. O_3; mol wt 48.00. Found in the atm in varying proportions (about 0.05 ppm at sea level), since it is produced continuously in the outer layers of the atm by the action of solar uv radiation on the oxygen of the air. So-called sterilizing lamps operate on the same principle. In the laboratory ozone is prepd by passing dry air between two plate electrodes connected to an alternating current source of several thousand volts. The reaction is reversible, and after a little ozone has been produced it is dec at the same rate as it is generated. Obtained in pure form by cooling ozonized air to −180° when it separates as a dark blue liquid. *See also* C. E. Thorp, *Bibliography of Ozone Technology* (Armour Res. Found., Chicago). Lab prepn: *Org. Syn.* **coll. vol. III**, 673 (1955). Monograph: *Ozone Chemistry and Technology*, Advances in Chemistry Series no. 21 (A.C.S., Washington D.C., 1959) 465 pp. *Review:* C. Nebel in Kirk-Othmer *Encyclopedia of Chemical Technology* **vol. 16** (Wiley-Interscience, New York, 3rd ed., 1981) pp 683-713.

Bluish, explosive gas or blue liquid. Pleasant, characteristic odor in concns of less than 2 ppm. Irritating and injurious in higher concns. Powerful oxidizing agent. d^0 (gas): 2.144 g/l; $d^{-195.4}$ (liq) 1.614 g/ml. mp −193°. bp −111.9°. Critical temp −12.1°. Critical press. 53.8 atm. Heat of formation 34.4 kcal/mole at 25°. Intense absorption band beginning at about 290 nm. Unstable. Solutions contg ozone explode on warming. Prepn of ozone solns in liq oxygen: Cook, U.S. pat. 3,008,902 (1961 to Union Carbide). Although the stability of ozone in aq solns decreases as alkalinity rises, this effect is reversed at high concns. For example, the half life of ozone is 2 min in $1N$ NaOH; it is increased to 83 hrs in $20N$ NaOH: Heidt, Landi, *J. Chem. Phys.* **41**, 176 (1964).

USE: As disinfectant for air and water by virtue of its oxidizing power. For bleaching waxes, textiles, oils. In organic syntheses. Forms ozonides which are sometimes useful oxidizing compds. *Caution:* High concns may cause severe irritation of respiratory tract, eyes.

P

6937. Paclobutrazol. (*R**,*R**)-(±)-*β*-[*(4-Chlorophenyl)methyl*]-*α*-(*1,1-dimethylethyl*)-*1H-1,2,4-triazole-1-ethanol;* 1-*tert*-butyl-2-(*p*-chlorobenzyl)-2-(1,2,4-triazol-1-yl)ethanol; (2*RS*,3*RS*)-1-(4-chlorophenyl)-4,4-dimethyl-2-(1*H*-1,2,4-triazol-1-yl)pentan-3-ol; ICI-PP 333; PP 333; Bonzi; Clipper; Cultar; Parlay. $C_{15}H_{20}ClN_3O$; mol wt 293.80. C 61.32%, H 6.86%, Cl 12.07%, N 14.30%, S 5.45%. Plant growth regulator with fungicidal activity. Prepn: B. C. Baldwin *et al.*, **Ger. pat. 2,734,426**; S. Balasubramanyan, M. C. Shephard, **U.S. pat. 4,243,405** (1978, 1981 both to ICI). Physical properties and biological activity: B. G. Lever *et al.*, *Proc. Brit. Crop. Prot. Conf.-Weeds* **1982**, 3. Resolution and activity of isomers: B. Sugavanam, *Pestic. Sci.* **15**, 296 (1984). GC determn in plant tissue: E. A. Stahly, D. A. Buchanan, *HortScience* **21**, 534 (1986). Comparison with daminozide, *q.v.*, of effect on apple trees: G. R. Stinchcombe *et al.*, *J. Hort. Sci.* **59**, 323 (1984).

White crystalline solid, mp 165-166°. d 1.22. Vapor pressure at 20°: 1×10^{-6} Pa. Soly: water 35 mg/l, methanol 15%, propylene glycol 5%, acetone 11%, cyclohexanone 18%, methylene dichloride 10%, hexane 1%, xylene 6%.

USE: Plant growth regulator.

6938. Pactamycin. *2-Hydroxy-6-methylbenzoic acid [5-[(3-acetylphenyl)amino]-4-amino-3-[[(dimethylamino)-carbonyl]amino]-1,2-dihydroxy-3-(1-hydroxyethyl)-2-methylcyclopentyl]methyl ester;* NSC 52947; U 15800. $C_{28}H_{38}$-N_4O_8; mol wt 558.64. C 60.20%, H 6.86%, N 10.03%, O 22.91%. Antitumor antibiotic produced by *Streptomyces pactum* var *pactum*. Discovery and biological properties: Bhuyan *et al.*, *Antimicrob. Ag. Chemother.* **1961**, 184. Isoln and characterization: Argoudelis *et al.*, *ibid.* 191. Manuf: **Brit. pat. 980,346** (1965 to Upjohn), *C.A.* **62**, 11115f (1965). Structure: Wiley *et al.*, *J. Org. Chem.* **35**, 1420 (1970). Revised structure: D. J. Duchamp, *Am. Crystallogr. Assn.* (Winter Mtg., Albuquerque, 1972) p 23. Mechanism of action: T. A. Beerman *et al.*, *Adv. Enzyme Regul.* **14**, 207 (1976). ^{13}C-NMR study: D. D. Weller *et al.*, *J. Antibiot.* **30**, 997 (1978). Biosynthesis: D. D. Weller, K. L. Rinehart, *J. Am. Chem. Soc.* **100**, 6757 (1978). *Review:* Goldberg in *Antibiotics* vol. 3, J. W. Corcoran, F. E. Hahn, Eds. (Springer-Verlag, New York, 1975) pp 498-515.

$[\alpha]_D^{25}$ +79° (ethanol) changing to +23° on standing. [α] changes in acetone on standing from +25° to +76° in 24 hours. Amphoteric. Sol in ethanol, chloroform, methylene chloride, benzene, ether, in solns >pH 5 and >9.5. Insol in Skellysolve B, cyclohexane; insol at isoelectric pt, pH 8.3. Unstable in solution. LD_{50} in mice (mg/kg): 10.7 orally; 15.6 i.v.; in rats (mg/kg): 1.4 i.v. (Bhuyan).

6939. Palitantin. *3-(1,3-Heptadienyl)-5,6-dihydroxy-2-*

(*hydroxymethyl)cyclohexanone.* $C_{14}H_{22}O_4$; mol wt 254.32. C 66.11%, H 8.72%, O 25.17%. Metabolic product of *Penicillium palitans* Westling. Isoln: J. H. Birkinshaw, H. Raistrick, *Biochem. J.* **30**, 801 (1936). Derivs and degradation products: J. H. Birkinshaw, *ibid.* **51**, 271 (1952). Structure: K. Bowden *et al.*, *J. Chem. Soc.* **1959**, 1662. Biosynthesis: P. Chaplen, R. Thomas, *Biochem. J.* **77**, 91 (1960); A. J. Birch, M. Kocor, *J. Chem. Soc.* **1960**, 866. Reactivity studies: A. T. Austin, B. Pearson, *Chem. & Ind. (London)* **1966**, 1228. Stereoselective synthesis of (±)-form: A. Ichihara *et al.*, *Tetrahedron Letters* **1977**, 3473; *eidem, Tetrahedron* **36**, 1547 (1980).

Needles from hot water, mp 165°. $[\alpha]_{5461}^{23}$ +4.4° (c = 0.8 in CHCl$_3$). uv max (ethanol): 323 nm (ϵ 34,000). Sol in hot water, alc, chloroform; slightly sol in cold water, ether.

6940. Palladium. Pd; at. wt 106.42; at. no. 46; valences 2, 4. Six naturally occurring isotopes: 102 (1.0%); 104 (11.0%); 105 (22.2%); 106 (27.3%); 108 (26.7%); 110 (11.8%); artificial, radioactive isotopes: 98-101, 103; 107; 109; 111-115. Abundance in earth's crust 0.001-0.01 ppm. Discovered in 1803 by Wollaston. Belongs to the platinum group of metals. Occurs in nature alloyed with platinum or gold and as a selenide; found in nickel sulfide ores; found in the minerals stibiopalladinite, braggite, porpezite. Isoln: Vauguelin *et al.*, cited by Mellor, *A Comprehensive Treatise on Inorganic and Theoretical Chemistry* **15**, 595 (1936). Reviews of prepn, properties and chemistry of palladium and other platinum metals: Gilchrist, *Chem. Rev.* **32**, 277-372 (1943); Beamish *et al.*, in *Rare Metals Handbook*, C. A. Hampel, Ed. (Reinhold, New York, 1956) pp 291-328; Livingstone in *Comprehensive Inorganic Chemistry* vol. 3, J. C. Bailar, Jr. *et al.*, Eds. (Pergamon Press, Oxford, 1973) pp 1163-1189, 1274-1329. Review of uses: E. M. Wise, *Palladium, Recovery, Properties, Uses* (Academic Press, New York, 1968) 187 pp. Use of organic palladium derivatives in synthesis: P. M. Maitlis, *The Organic Chemistry of Palladium* (Academic Press, New York, 1971); J. Tsuji, *Topics in Current Chemistry* **91**, 29 (1980); B. M. Trost, *Tetrahedron* **33**, 2615 (1977); *idem, Accts. Chem. Res.* **13**, 385 (1980).

Silver-white metal, face-centered cubic structure; occurs also as black powder and as spongy masses which can be compressed to a compact mass. mp 1555°; bp 3167°. d$_4^{20}$ 12.02. Hardness on Mohs' scale 4.8, Brinell hardness 61.0. Spec heat 0.0584 cal/g at 0°C. Electrical resistivity at 0° = 10.0 microohms-cm. Appreciably volatile at high temps. At a red heat is converted into the monoxide. Forms dihalides with fluorine or chlorine at a red heat. Reacts with nitric acid, sulfuric acid, a mixture of hydrochloric and chloric acids. Reacts slightly with concd HCl; more readily in the presence of air or free chlorine. Forms a sulfide when heated with sulfur, a phosphide when heated with phosphorus. Absorbs a considerable amount of hydrogen.

USE: In form of gold, silver, and copper alloys in dentistry; for alloy bearings, springs, balance wheels of watches; for mirrors in astronomical instruments; as catalyzer in manuf of sulfuric acid and in other oxidizing processes; in powder form as catalyst in hydrogenation and in ignition of hydrogen or hydrocarbons with oxygen; the spongy form is used in gas analysis for separating hydrogen from mixtures of gases.

6941. Palladium Chloride. Palladous chloride. Cl_2Pd; mol wt 177.30. Cl 39.99%, Pd 60.01%. $PdCl_2$. Prepn: Krustinsons, *Z. Elektrochem.* **44**, 537 (1938). Review of $PdCl_2$ and other halides: J. H. Canterford, R. Colton, *Halides of the Second and Third Row Transition Metals* (John Wiley, New York, 1968) pp 358-389.

Red crystals. mp 678-680°. Dec at high temp to palladium and chlorine. LD in rabbits: 18.6 mg/kg i.v., Orestano, *Boll. Soc. Ital. Biol. Sper.* **8**, 1154 (1933).

Dihydrate, dark brown crystals. Sol in water, alcohol,

acetone. Reduced in soln by hydrogen or CO to metal. *Keep tightly closed.*

USE: In photography, for preparing pictures to be transferred to porcelain; toning solutions; electroplating parts of clocks and watches; manuf indelible ink; for the prepn of the metal for use as a catalyst; $PdCl_2$ paper is used for detecting CO, to find leaks in buried gas pipes. Prepn of palladium catalysts using $PdCl_2$: Mozingo, *Org. Syn.* coll. vol. III, 685 (1955). *Toxicity:* Does not cause the skin sensitization seen with platinum. Animal expts show low toxicity: E. Browning, *Toxicity of Industrial Metals* (Appleton-Century-Crofts, New York, 2nd ed., 1969) pp 267-269.

6942. Palladium Diacetate. *Acetic acid palladium salt;* bis(acetato)palladium; diacetatopalladium(II); palladium(II) acetate. $C_4H_6O_4Pd$; mol wt 224.49. C 21.40%, H 2.69%, O 28.51%, Pd 47.40%. $(CH_3COO)_2Pd$. Prepn from palladium sponge or palladium nitrate and glacial acetic acid: Morehouse *et al., Chem. & Ind. (London)* **1964**, 544; Stephenson *et al., J. Chem. Soc.* **1965**, 3632; from hydrated palladium oxide and acetic acid: Hausman *et al., Fr. pat.* **1,403,398** corresp to U.S. pat. **3,318,891** (1965, 1967 to Engelhard). Crystal structure of trimeric palladium acetate: Skapski, Smart, *Chem. Commun.* **1970**, 658.

Orange-brown crystals; mp 205° (dec). Shown to be trimeric by osmometric determn of mol wt in benzene (37°); monomeric by ebullioscopic measurement in benzene (80°). Crystals are trimeric. Sol in chloroform, methylene dichloride, acetone, acetonitrile, diethyl ether; sol with dec in aq HCl, aq KI solutions. Insol in water, aq NaCl, $NaNO_3$, NaOAc solns, petroleum, alcohols. Dec when warmed with alcohols.

USE: Catalyst.

6943. Palladium Nitrate. Palladous nitrate. N_2O_6Pd; mol wt 230.42. Pd 46.18%, N 12.16%, O 41.66%. $Pd(NO_3)_2$. Prepn from palladium and nitric acid: *Gmelin's, Palladium* (8th ed.) **65**, 269 (1942). Reported to be the dihydrate: Gatehouse *et al., J. Chem. Soc.* **1957**, 4222.

Brown, deliquesc crystals. Sol in water with turbidity; with much water a brown basic salt precipitates; completely sol in dil HNO_3. *Keep well closed, protected from light.*

USE: Sepn of Cl_2 and I_2; catalyst in organic syntheses.

6944. Palladium Oxide. Palladium monoxide; palladous oxide. OPd; mol wt 122.40. O 13.07%, Pd 86.93%. PdO. Prepn: Shriner, Adams, *J. Am. Chem. Soc.* **46**, 1685 (1924).

Black powder. d 8.3. Insol in water, acids; slightly sol in aqua regia; sol in 48% HBr. Dec when strongly heated; also in the presence of H_2.

USE: Reduction catalyst in synthesis of organic compds.

6945. Palmatine. *5,6-Dihydro-2,3,9,10-tetramethoxydibenzo[a,g]quinolizinium; 7,8,13,13a-tetradehydro-2,3,9,10-tetramethoxyberbinium;* calystigine. $[C_{21}H_{22}NO_4]^+$; mol wt 352.42. Obtained only in form of its salts. First isolated from Calumba root (*Jateorhiza palmata* (DC.) Miers, *Menispermaceae*). Now found in many other genera. Palmatine and tetrahydropalmatine, *q.v.* are probably the most widely distributed Berberis alkaloids. Extraction procedure: Feist, Dschu, *Arch. Pharm.* **263**, 301 (1925). Structure: Feist, Sandstede, *ibid.* **256**, 2, 5 (1918); Späth, Lang, *Ber.* **54**, 3064, 3068 (1921); Späth, Böhm, *Ber.* **55**, 2988 (1922); Späth, Meinhard, *Ber.* **75**, 400 (1942). Synthesis: Späth, Quientensky, *Ber.* **58**, 2267 (1925); R. D. Haworth *et al., J. Chem. Soc.* **1927**, 548; Z. Kiparissides *et al., Can. J. Chem.* **58**, 2770 (1980). Identity with calystigine, alkaloid of the Chinese drug Chi-Kuo-Lan: Huang, Chen, *C.A.* **52**, 15827i (1958).

Palmatine forms addn products with acetone and chloroform, as does berberine.

Iodide, $(C_{21}H_{22}NO_4)I$, yellow needles from water, dec 239°; also a dihydrate. Sparingly sol in hot water and alc.

Nitrate, $(C_{21}H_{22}NO_4)NO_3$, sesqui- or dihydrate, yellow needles, dec 239°. Freely sol in most solvents.

Chloride trihydrate, $(C_{21}H_{22}NO_4)Cl.3H_2O$, yellowish-green needles from water. Freely sol in hot water and alc.

Sulfate pentahydrate, $(C_{21}H_{22}NO_4)_2SO_4.5H_2O$, yellow needles, mp 250°. Very sol in alc, sol in water.

6946. Palmidrol. *N-(2-Hydroxyethyl)hexadecanamide; N-(2-hydroxyethyl)palmitamide.* $C_{18}H_{37}NO_2$; mol wt 299.48. C 72.19%, H 12.45%, N 4.68%, O 10.69%. $CH_3(CH_2)_{14}CONHCH_2CH_2OH$. A naturally occurring anti-inflammatory agent. Isolated from soybean lecithin, egg yolk, and peanut meal: Kuehl *et al., J. Am. Chem. Soc.* **79**, 5577 (1957). Synthesis by refluxing ethanolamine with palmitic acid: Roe *et al., ibid.* **74**, 3442 (1952).

Crystals from 95% ethanol or cyclohexane, mp 98-99°.

6947. Palmitic Acid. *Hexadecanoic acid;* hexadecylic acid; cetylic acid. $C_{16}H_{32}O_2$; mol wt 256.42. C 74.94%, H 12.58%, O 12.48%. $CH_3(CH_2)_{14}COOH$. Occurs as the glyceryl ester in many oils and fats. Obtained from palm oil, Japan wax, or Chinese vegetable tallow. Purification: Magne *et al., U.S. pat.* **2,791,596** (1957 to Secretary of Agriculture).

White, cryst scales. d_4^{62} 0.853. mp 63-64°. bp_{15} 215°. n_D^{80} 1.4273. Insol in water; sparingly sol in cold alcohol or in petr ether; freely sol in hot alcohol, in ether, propyl alcohol, chloroform. LD_{50} i.v. in mice: 57±3.4 mg/kg, L. Orö, A. Wretlind, *Acta Pharmacol. Toxicol.* **18**, 141 (1961).

6948. Palm Oil, From Fruit. Palm butter. Obtained from the fruit *Elaeis guineensis* Jacq., *Palmae. Constit.* Chiefly palmitin; some stearin, linolein.

Reddish-yellow to dark dirty red, fatty mass; faint odor of violet. d 0.920-0.927. mp 27-42.5°. n_D^{40} 1.453-1.459. Sapon. no. 200-205. Iodine no. 53-57.

USE: In manuf of soap, candles; lubricant.

6949. Palm Oil, From Seed. An oil obtained from seed of *Elaeis guineensis* Jacq., *Palmae.*

White to yellowish. d 0.952. mp 26-30°. Sapon. no. about 247. Iodine no. about 15.

USE: Manuf soap; with coconut oil in manuf plant butter; has also been used in liniments and ointment.

6950. Palustric Acid. *[1R-(1α,4aβ,10aα)]-1,2,3,4,4a,5,-6,9,10,10a-Decahydro-1,4a-dimethyl-7-(1-methylethyl)-1-phenanthrenecarboxylic acid; 13-isopropylpodocarpa-8,13-dien-15-oic acid.* $C_{20}H_{30}O_2$; mol wt 302.44. C 79.42%, H 10.00%, O 10.58%. Isoln from gum rosin: Loeblich *et al., J. Am. Chem. Soc.* **77**, 2823 (1955); Joye *et al., J. Org. Chem.* **30**, 654 (1965). Structure: Shuller *et al., J. Am. Chem. Soc.* **82**, 1734 (1960). Prepn of the 4-*epi*-form: Tabacik, Poisson, *Bull. Soc. Chim. France* **1969**, 3264.

Crystals from methanol, mp 162-167°. $[\alpha]_D$ +71.6°. uv max (0.01N NaOH): 265-266 nm.

Methyl ester, $C_{21}H_{32}O_2$, crystals from methanol, mp 25-27°. $[\alpha]_D$ +67.7°. uv max: 265-266 nm.

6951. Palytoxin. *Palytoxin (C51-55 hemiacetal);* PTX. $C_{129}H_{223}N_3O_{54}$; mol wt 2680.22. C 57.81%, H 8.39%, N 1.57%, O 32.23%. Potent toxin isolated from zoanthid coral of the genus *Palythoa* that is the most poisonous non-proteinaceous substance known. Isoln from "Limu-make-o-Hana", the Hawaiian name for the highly toxic coelenterate *Palythoa toxica*: R. E. Moore, P. J. Scheuer, *Science* **172**,

495 (1971). Structure: R. E. Moore, G. Bartolini, *J. Am. Chem. Soc.* **103**, 2491 (1981). Structure of palytoxin from *P. tuberculosa* of Okinawa (differs from palytoxin from *P. toxica* at two positions): D. Uemura *et al.*, *Tetrahedron Letters* **22**, 2781 (1981). Proposed absolute configuration of 60 of the 64 chiral centers: R. E. Moore *et al.*, *J. Am. Chem. Soc.* **104**, 3776 (1982). Structure and stereochemistry: J. K. Cha *et al.*, *ibid.* 7369. Discussion of the structural elucidation, *see* Y. Shimizu, *Nature* **302**, 212 (1983); review: R. E. Moore, *Prog. Chem. Org. Nat. Prod.* **48**, 82-202 (1985). Synthetic studies: Y. Kishi, *Chemica Scripta* **27**, 573 (1987). Pharmacological study: P. N. Kaul *et al.*, *Proc. West. Pharmacol. Soc.* **17**, 294 (1974). Mechanism of action and treatment of palytoxin poisoning: J. A. Vick *et al.*, *Toxicol. Appl. Pharmacol.* **34**, 214 (1975). Mode of contractile action on vascular smooth muscle: K. Ito *et al.*, *Eur. J. Pharmacol.* **46**, 9 (1977). Depolarizing action on frog spinal cord: Y. Kudo, S. Shibata, *Brit. J. Pharmacol.* **71**, 575 (1980). Toxicology and toxicity studies: J. S. Wiles *et al.*, *Toxicon* **12**, 427 (1974); K. Ito *et al.*, *Arch. Int. Pharmacodyn. Ther.* **258**, 146 (1982). Brief review of biology: P. J. Scheuer, *Accounts Chem. Res.* **10**, 33-39 (1977). Review of synthetic studies and conformational analysis: Y. Kishi, *Pure Appl. Chem.* **61**, 313-324 (1989).

White amorphous hygroscopic solid. No definite mp; chars when heated to 300°. $[\alpha]_D^{25}$ +26° (water). Insol in chloroform, ether, acetone. Sparingly sol in methanol, ethanol. Sol in pyridine, DMSO, water. LD_{50} in mice: 0.45 μg/kg i.v. (Wiles); 50-100 ng/kg i.p. (Kaul). Palytoxin is an intense vasoconstrictor. In intact dogs, doses of > 0.06 μg/kg i.v. caused a transient rise in arterial pressure followed by rapid hypotension and resulted in death within 5 minutes (Ito, 1982).

USE: As a physiological tool to evaluate anti-anginal chemotherapeutic agents.

6952. Pamabrom. *8-Bromo-3,7-dihydro-1,3-dimethyl-1H-purine-2,6-dione compound with 2-amino-2-methyl-1-propanol (1:1); 8-bromotheophylline compound with 2-amino-2-methyl-1-propanol (1:1);* 2-amino-2-methyl-2-propanol 8-bromotheophyllinate. $C_{11}H_{18}BrN_5O_3$; mol wt 347.21. C 38.05%, H 4.94%, Br 23.02%, N 20.17%, O 13.82%. Prepn from 2-amino-2-methyl-1-propanol and 8-bromotheophylline: Holpert, Grote, U.S. pat. **2,711,411** (1955 to Chattanooga Med.).

Fine white powder, dec 300°. Soly in water > 30 g/100 ml at 25°. pH of satd aq soln 8.0-8.5. One of the ingredients of *Pamprin, Sunril, Donasil.*

THERAP CAT: Diuretic.

6953. Pamaquine. *N[1],N[1]-Diethyl-N[4]-(6-methoxy-8-quinolinyl)-1,4-pentanediamine; 8-[[4-(diethylamino)-1-methylbutyl]amino]-6-methoxyquinoline;* Plasmochin; Aminoquin; Praequine; Béprochine; Gamefar; Quipenyl; Plasmoquine. $C_{19}H_{29}N_3O$; mol wt 315.45. C 72.34%, H 9.27%, N 13.32%, O 5.07%. Preparation by reductive condensation of 1-diethylaminopentan-4-one with 6-methoxy-8-aminoquinoline: Elderfield *et al.*, *J. Am. Chem. Soc.* **70**, 40 (1948). *Review:* Cooper, *U.S. Pub. Health Repts.* **64**, 717 (1949).

Dark yellow oil. $bp_{0.3}$ 175-180°; bp_1 182-194°.
Pamoate, $C_{42}H_{45}N_3O_7$, *pamaquine embonate, pamaquine naphthoate.* Prepn: **Brit. pat. 295,656** (1927 to I. G. Farbenind). Details and list of German pats. in K. H. Slotta, *Grundriss der modernen Arzneistoff-Synthese* (Stuttgart, 1931). Yellow to orange-yellow odorless, almost tasteless powder. Numbs the tongue. Practically insol in water; sol in alcohol, acetone. *Protect from light.*

THERAP CAT: Naphthoate as antimalarial.

6954. Pamidronic Acid. *(3-Amino-1-hydroxypropylidene)bisphosphonic acid;* 3-amino-1-hydroxypropane-1,1-diphosphonic acid; ADP; AHPrBP. $C_3H_{11}NO_7P_2$; mol wt 235.07. C 15.33%, H 4.72%, N 5.96%, O 47.64%, P 26.35%. Calcium metabolism regulator; structurally similar to etidronic acid, *q.v.* Prepn: F. Krueger *et al.*, **Ger. pat. 2,130,-794** (1973 to Benckiser), *C.A.* **78**, 84528z (1973); K.-H. Worms *et al.*, *Z. Anorg. Allgem. Chem.* **457**, 214 (1979). Improved process: H. Blum, K.-H. Worms, U.S. pat. **4,327,039** (1982 to Henkel). Mechanism of inhibition of osteoclastic bone resorption studies: J.-M. Delaisse *et al.*, *Life Sci.* **37**, 2291 (1985); P. M. Boonekamp *et al.*, *Bone Mineral* **1**, 27 (1986); C. W. G. Lowik *et al.*, *Adv. Exp. Med. Biol.* **208**, 275 (1986). Pharmacokinetics in rodents: F. Wingen, D. Schmähl, *Arzneimittel-Forsch.* **37**, 1037 (1987). Clinical evaluations in Paget's disease: W. B. Frijlink *et al.*, *Lancet* **1**, 799 (1979); G. Heynen *et al.*, *Eur. J. Clin. Invest.* **12**, 29 (1982); in osteolytic bone disease: F. J. M. van Breukelen *et al.*, *Lancet* **1**, 803 (1979). Prepn of the ^{99m}Tc-ADP complex, and use as a bone scanning agent: O. J. Degrossi *et al.*, *J. Nucl. Med.* **26**, 1135 (1985).

Disodium salt, $C_3H_9NNa_2O_7P_2$, *Aminomux.*

THERAP CAT: Antipagetic; inhibitor of tumor-induced hypercalcemia. Technetium complex as diagnostic aid (radioactive imaging agent).

6955. Pamoic Acid. *4,4'-Methylenebis[3-hydroxy-2-*

naphthalenecarboxylic acid]; 4,4'-methylenebis(3-hydroxy-2-naphthoic acid); 4,4'-methylenedi(3-hydroxy-2-naphthoic acid); 2,2'-dihydroxy-1,1'-dinaphthylmethane-3,3'-dicarboxylic acid; embonic acid. $C_{23}H_{16}O_6$; mol wt 388.36. C 71.13%, H 4.15%, O 24.72%. Prepn from 2-hydroxy-3-naphthoic acid + formaldehyde: Strohback, *Ber.* **34**, 4162 (1901); Brass, Sommer, *ibid.* **61**, 993 (1928); Barber, Gaimster, *J. Appl. Chem. (London)* **2**, 565 (1952). Prepn of salts: Puetzer, U.S. pat. **2,397,903** (1946 to Vick Chem.); Barber, U.S. pat. **2,641,610** (1953 to May & Baker).

Crystals from dil pyridine, dec above 280° without melting. Practically insoluble in water, alcohol, ether, benzene, acetic acid; sparingly sol in chloroform; sol in nitrobenzene, pyridine.

Methyl ether, $C_{24}H_{18}O_6$, yellow crystals from dil methanol, dec 277-282°.

6956. Pancreatin. Diastase vera; Creon; Pancrease; Pancrex Vet; Pankrotanon; Zypanar. A substance from the fresh pancreas of the hog or ox, contg the enzymes amylopsin, trypsin, and steapsin. It converts not less than 25 times its wt of starch into sol carbohydrates and not less than 25 times its wt of casein into proteoses within 5 mins. (equiv to 150 times in 30 mins.). Its greatest activity is exhibited in neutral or faintly alkaline media; mineral acids or excess alkali hydroxides or carbonates render it inactive. Method of production from cow or pig pancreas: Hoek *et al.*, U.S. pat. **3,223,594** (1965 to N. American Philips).

Yellowish, amorphous powder. Partly sol in water. Insol in alcohol. *Incompat.* Alcohol, acids.

Note: Pancreatin is destroyed in the stomach, hence should be given in salol-coated pills.

THERAP CAT: Digestive aid.

THERAP CAT (VET): In replacement therapy in pancreatic enzyme deficiency, in preparation of "predigested" protein foods.

6957. Pancrelipase. Accelerase; Cotazym; Ilozyme; Ku-Zyme HP; Pancrease; Viokase. Porcine pancreatic enzyme concentrate containing principally lipase, with amylase and protease. Clinical pharmacology and comparison with pancreatin, *q.v.*: Y. W. Cho, D. M. Aviado, *J. Clin. Pharmacol.* **21**, 224 (1981). Clinical trial in cystic fibrosis: E. H. Mischler *et al.*, *Am. J. Dis. Child.* **136**, 1060 (1982); in adult pancreatic insufficiency: S. K. Dutta, D. K. Tilley, *J. Clin. Gastroenterol.* **5**, 51 (1983).

THERAP CAT: Enzyme (digestive adjunct).

6958. Pancuronium Bromide. 1,1'-[3,17-Bis(acetyloxy)-androstane-2,16-diyl]bis[1-methylpiperidinium] dibromide; 1,1'-(3α,17β-dihydroxy-5α-androstan-2β,16β-ylene)bis[1-methylpiperidinium] dibromide diacetate; 3α,17β-diacetoxy-2β,16β-dipiperidino-5α-androstane dimethobromide; 2β,16β-dipiperidino-5α-androstane-3α,17β-diol diacetate dimethobromide; poncuronium bromide (rescinded USAN); NA 97; Org NA 97; Mioblock; Pavulon. $C_{35}H_{60}Br_2N_2O_4$; mol wt 732.70. C 57.38%, H 8.25%, Br 21.81%, N 3.82%, O 8.73%. Prepn: W. R. Buckett *et al.*, *Chim. Ther.* **2**, 186 (1967); *eidem, J. Med. Chem.* **16**, 1116 (1973). Structural studies: Savage *et al.*, *J. Chem. Soc. (B)* **1971**, 410. Structure-activity correlation: Waser, *Anaesthesist* **20**, 23 (1971). Pharmacology: W. R. Buckett, I. L. Bonta, *Fed. Proc.* **25**, 718 (1966); W. R. Buckett *et al.*, *Brit. J. Pharmacol. Chemother.* **32**, 671 (1968); I. L. Bonta *et al.*, *Eur. J. Pharmacol.* **4**, 83, 303 (1968). Comparative study of neuromuscular blocking and vagolytic effect: S. L. Son *et al.*, *Anesthesiology* **55**, 12 (1981). *Review:* Speight, Avery, *Drugs* **4**, 163-226 (1972).

Odorless crystals with bitter taste, mp 215°. One gram dissolves in 30 parts chloroform, one part water (20°). LD_{50} in mice (mg/kg): 0.047 i.v.; 0.152 i.p.; 0.167 s.c.; 21.9 orally; in rats, rabbits: 0.153, 0.016 i.v., W. R. Buckett *et al., loc. cit.* (1968).

THERAP CAT: Relaxant (skeletal muscle).

6959. Pangamic Acid. Controversial mixture of compounds erroneously labeled as *vitamin B₁₅*. Allegedly isolated from apricot kernel: E. T. Krebs, Sr. *et al., Int. Rec. Med.* **164**, 18 (1951). Originally named pangamic acid because of its supposed ubiquity in seeds. There is no clear chemical identity for pangamic acid. An historical review by J. C. Micheau *et al.*, in *Chim. Ther.* **7**, 103 (1972) contains structural studies, analyses, syntheses and discussion of various components of pangamic acid. Products sold as pangamic acid in the U.S. vary considerably in their composition. Some are mixtures of calcium gluconate and *N,N*-dimethylglycine, *q.v.*; others contain diisopropylamine dichloroacetate, *q.v.*: *FDA Drug Bulletin* vol. **8**(6), Dec. 1978-Jan. 1979. *Ref.:* V. Herbert, *Am. J. Clin. Nutr.* **32**, 1534 (1979); V. Herbert, R. Herbert, in *Controversies in Nutrition*, L. Ellenbogen, Ed. (Churchill-Livingstone, New York, 1981) pp 159-170.

6960. Pankrin. A proteolytic enzyme (proteinase, endopeptidase). Partial purification and characterization: Grant, Robbins, *J. Am. Chem. Soc.* **78**, 5888 (1956). Isoln from hog or beef pancreas: *eidem*, U.S. pat. **2,871,165** (1959 to Armour).

Insol in water, sol in a 60% aq alcohol soln at 25°. Maintains proteolytic activity in aq solns of pH 4.7 at 25° for 8 hrs. Incapable of clotting blood plasma, but clots milk. The proteolytic activity upon urea-denatured hemoglobin and upon acetyl-1-tyrosine ethyl ester is such that the ratio of its hemoglobin to ATEE activity is at least 150.

6961. Pantetheine. 2,4-Dihydroxy-N-[3-[(2-mercaptoethyl)amino]-3-oxopropyl]-3,3-dimethylbutanamide; 2,4-dihydroxy-N-[2-[(2-mercaptoethyl)carbamoyl]ethyl]-3,3-dimethylbutyramide; N-(pantothenyl)-β-aminoethanethiol; α,γ-dihydroxy-β,β-dimethylbutyryl-β-alanyl-β-aminoethanethiol. $C_{11}H_{22}N_2O_4S$; mol wt 278.39. C 47.46%, H 7.97%, N 10.07%, O 22.99%, S 11.52%. Growth factor for *Lactobacillus bulgaricus:* Williams *et al., J. Biol. Chem.* **177**, 933 (1949). Intermediate in the pathway of coenzyme A *(q.v.)* formation in mammalian liver and in some microorganisms. Synthesis: Schwyzer, *Helv. Chim. Acta* **35**, 1903 (1952); Baddiley, Thain, *J. Chem. Soc.* **1952**, 800; King *et al., J. Am. Chem. Soc.* **75**, 1290 (1953); Walton *et al., ibid.* **76**, 1146 (1954); Walton, U.S. pats. **2,744,119** and **2,835,704** (1956, 1958, both to Merck & Co.). *Reviews:* Snell, Brown, *Advan. Enzymol.* **14**, 49 (1953); Snell, Wittle, *Methods Enzymol.* **3**, 918 (1957).

$$\underset{\underset{CH_3}{|}}{\overset{\overset{CH_3}{|}}{HOCH_2C}} - \underset{\underset{H}{|}}{\overset{\overset{OH}{|}}{C}}CONHCH_2CH_2CONHCH_2CH_2SH$$

Syrup or glass. $[\alpha]_D^{20}$ +12.9° (c = 4.5 in water). Microbiological activity: 20,000 LBF units/mg. An LBF unit is that amount of the growth factor contained in one mg of Basamine-Busch, a standard yeast extract manuf by Anheuser-Busch, Inc.

Silver mercaptide, $C_{11}H_{21}AgN_2O_4S$, yellow noncryst solid. $[\alpha]_D^{25}$ +8° (c = 4.27 in 0.9N NaCl). Very sol in water. 11,000 LBF units/mg.

Mercuric mercaptide, $C_{22}H_{42}HgN_4O_8S_2 \cdot C_3H_6O$, crystals from acetone, mp 96-98°. $[\alpha]_D^{27}$ +9.6° (c = 4 in water). 15,500 LBF units/mg. uv max: 260-265 nm (ε 1000). Sol in water, methanol.

S-Benzoylpantetheine, $C_{18}H_{26}N_2O_5S$, crystals from ethyl acetate, mp 116°. $[\alpha]_D^{27}$ +31° (ethanol). Sol in water, ethyl acetate, chloroform.

S-Acetylpantetheine, $C_{13}H_{24}N_2O_5S$, thick syrup. $[\alpha]_D^{27}$ +39° (c = 0.8 in ethanol).

6962. Pantethine. *N,N'-[Dithiobis[2,1-ethanediylimino(3-oxo-3,1-propanediyl)]]bis[2,4-dihydroxy-3,3-dimethylbutanamide]; N,N'-[dithiobis(ethyleneiminocarbonylethylene)]bis(2,4-dihydroxy-3,3-dimethylbutyramide];* D-bis(*N*-pantothenyl-β-aminoethyl) disulfide; Lipodel; Pantetina; Panthecin; Pantomin; Pantosin. $C_{22}H_{42}N_4O_8S_2$; mol wt 554.74. C 47.63%, H 7.63%, N 10.10%, O 23.07%, S 11.56%. $[HOCH_2C(CH_3)_2CHOHCONHCH_2CH_2CONHCH_2CH_2S]_2$. Disulfide dimer of pantetheine, *q.v.* Growth factor for *Lactobacillus bulgaricus:* Williams *et al., J. Biol. Chem.* **177**, 933 (1949). Formed by oxidation of pantetheine: Brown, Snell, *J. Biol. Chem.* **198**, 375 (1952). Structure: Snell *et al., J. Am. Chem. Soc.* **72**, 5349 (1950). Synthesis: Wieland, Bokelmann, *Naturwiss.* **38**, 384 (1951); Wittle *et al., J. Am. Chem. Soc.* **75**, 1694 (1953); Viscontini *et al., Helv. Chim. Acta* **37**, 375 (1954); Bowman, Cavalla, *J. Chem. Soc.* **1954**, 1171; Shimizu *et al., Chem. Pharm. Bull.* **13**, 180 (1965). Clinical trial in hyperlipoproteinemia: A. Gaddi *et al., Atherosclerosis* **50**, 73 (1984).

Glassy, colorless to light yellow substance. $[\alpha]_D^{27}$ +13.5° (c = 3.75 in water). Freely sol in water; less sol in ethanol. Practically insol in ether, acetone, ethyl acetate, benzene, chloroform.

THERAP CAT: Antihyperlipoproteinemic.

6963. Pantolactone. *Dihydro-3-hydroxy-4H-dimethyl-2(3H)-furanone;* pantoic acid γ-lactone; pantoyl lactone; pantoic lactone; 2,4-dihydroxy-3,3-dimethylbutyric acid γ-lactone; α-hydroxy-β,β-dimethyl-γ-butyrolactone. $C_6H_{10}O_3$; mol wt 130.14. C 55.37%, H 7.75%, O 36.88%. A degradation product of pantothenic acid from liver: Williams, Major, *Science* **91**, 246 (1940). Important intermediate in the synthesis of pantothenic acid. May be prepd by condensing isobutyraldehyde with formaldehyde yielding α,α-dimethyl-β-hydroxypropionaldehyde which is condensed with hydrocyanic acid in the presence of calcium chloride to form racemic pantolactone. Various modifications of this procedure exist: Glaser, *Monatsh.* **25**, 46 (1904); Stiller *et al., J. Am. Chem. Soc.* **62**, 1785 (1940); Reichstein, Grüssner, *Helv. Chim. Acta* **23**, 650 (1940); Carter, Ney, *J. Am. Chem. Soc.* **63**, 312 (1941). Vast patent literature, *e.g.*, Beckmann *et al.;* Klein, U.S. pats. **2,967,869** and **3,024,250** (1961, 1962, both to Nopco).

D(−)-Form, hygroscopic crystals from benzene + petr ether, mp 92°. $[\alpha]_D^{25}$ − 50.7° (c = 2.05 in H_2O). Can be purified by microsublimation.

L(+)-Form, hygroscopic crystals from benzene, mp 91°. $[\alpha]_D^{27}$ +50.1° (c = 2 in H_2O).

DL-Form, hygroscopic rosettes or prisms, mp 80°, bp_{18} 130°. Freely sol in water. Sol in ether, benzene, chloroform, alcohol, carbon disulfide.

6964. Pantothenic Acid. *(R)-N-(2,4-Dihydroxy-3,3-dimethyl-1-oxobutyl)-β-alanine;* D(+)-N-(2,4-dihydroxy-3,3-dimethylbutyryl)-β-alanine; chick antidermatitis factor. $C_9H_{17}NO_5$; mol wt 219.23. C 49.30%, H 7.82%, N 6.39%, O 36.49%. A member of the B complex vitamins. Occurs everywhere in animal and plant tissue. The richest common source is liver, but jelly of the queen bee contains 6 times as much as liver. Rice bran and molasses are other good sources. Isoln from liver: R. J. Williams *et al., J. Am. Chem. Soc.* **60**, 2719 (1938). Pantothenic acid derivs sold commercially are prepd by synthesis. May be prepd by the direct condensation of β-alanine with the optically resolved form of the lactone of pantoic acid (2,4-dihydroxy-3,3-di-

methylbutyric acid) which is prepd from isobutyraldehyde. Synthesis: E. T. Stiller *et al., ibid.* **62**, 1785 (1940); Reichstein, Grüssner, *Helv. Chim. Acta* **23**, 650 (1940); Grüssner *et al., ibid.* 1276. Absolute configuration: Hill, Chan, *Biochem. Biophys. Res. Commun.* **38**, 181 (1970). Only the natural, dextrorotatory form has vitamin activity. Review of chemistry, biochemistry and pharmacology: "Pantothenic Acid" in *The Vitamins* vol. 2, W. H. Sebrell, Jr., R. S. Harris, Eds. (Academic Press, New York, 1956) p 591-694.

HOCH_2C——CCONH—CH_2CH_2COOH with CH_3, OH, CH_3, H substituents

Unstable, viscous oil. Extremely hygroscopic. Easily destroyed by acids, bases, heat. $[\alpha]_D^{25}$ +37.5°. Freely sol in water, ethyl acetate, dioxane, glacial acetic acid; moderately sol in ether, amyl alcohol. Practically insol in benzene, chloroform. One gram of pantothenic acid equals 70,000-75,000 chick units.

Sodium salt, $C_9H_{16}NNaO_5$, *Panthoject.* Very hygroscopic crystals. $[\alpha]_D^{25}$ +27.1° (c = 2). Can be stored in sealed ampuls only.

Calcium salt, see Calcium Pantothenate.

Note: Pantothenic acid has been referred to as *vitamin B₃:* Cheldelin in *The Vitamins* vol. 3, W. H. Sebrell, Jr., R. S. Harris, Eds. (Academic Press, New York, 1954) pp 596-598, and *vitamin B₅:* Malgras, Pax, *Ann. Inst. Pasteur* **93**, 792 (1957). *See also* Nicotinamide.

THERAP CAT: Vitamin.

THERAP CAT (VET): Nutritional factor: dietary essential except in horses, ruminants.

6965. Papain. Papayotin; vegetable pepsin; Arbuz; Nematolyt; Caroid; Summetrin; Tromasin; Velardon; Vermizym. First recognized member of the class of proteolytic enzymes that needs a free sulfhydryl group for activity. Isolated from the latex of the green fruit and leaves of *Carica papaya* L., Caricaceae. Initial isolation and crystallization: Balls *et al., Science* **86**, 379 (1937); Balls, Lineweaver, *J. Biol. Chem.* **130**, 669 (1939). Prepn from commercial dried papaya latex and physical properties: Kimmel, Smith, *ibid.* **207**, 515 (1954); see also ibid. 533-573; Becker, *Econ. Bot.* **12**, 62 (1958). Purification: Gibian, Bratfisch, U.S. pat. **2,950,-227** (1960 to Schering AG); Lesuk, U.S. pat. **3,011,952** (1961 to Sterling Drug); Blumberg *et al., Eur. J. Biochem.* **15**, 97 (1970). The papain molecule consists of one folded polypeptide chain of 212 residues, mol wt about 23,400. Complete amino acid sequence: Drenth *et al., Nature* **218**, 929 (1968); Mitchel *et al., J. Biol. Chem.* **245**, 3485 (1970). Mechanism of action studies: Morihara, *J. Biochem.* **62**, 250 (1967). Use in treatment of contact lenses to prolong wearing time in keratoconic patients with papillary conjunctivitis: D. R. Korb *et al., Arch. Ophthalmol.* **101**, 48 (1983). Reviews: Kimmel, Smith in *Advan. Enzymol. Relat. Subj. Biochem.* **19**, 267-334 (1957); Glazer, Smith in *The Enzymes* vol. III, P. D. Boyer, Ed. (Academic Press, New York, 3rd ed., 1971) pp 501-537; see also Drenth *et al., ibid.* 485-498 and *eidem., Advan. Protein Chem.* **25**, 79-115 (1971) for a comprehensive review of the structural elucidation.

White or grayish-white, slightly hygroscopic powder. uv max: 278 nm ($A_{1cm}^{1\%}$ 25.0). Incompletely sol in water, glycerol. Practically insol in most organic solvents. *Keep well closed.* Potency varies according to process of prepn, etc. with the usual grade digesting about 35 times its wt of lean meat. Best grades render sol 200-300 times their wt of coagulated egg albumin in alkaline media. A temp. range of 60-90° is favorable for the digestive process with 65° the optimum point. Best pH is 5.0, but it functions also in neutral or alkaline media. Activated by reduction (HCN, H_2S etc.) and inactivated by oxidation (H_2O_2, iodoacetate).

Note: The term papain is currently applied to both the crude dried latex and the crystalline proteolytic enzyme.

USE: For tenderizing meats; for clearing beverages; for bating skins.

THERAP CAT: Enzyme (proteolytic). Debriding agent; digestive aid. Has been used to prevent adhesions; as anthelmintic (Nematodes).

6966. Papaveraldine. *(6,7-Dimethoxy-1-isoquinolinyl)-(3,4-dimethoxyphenyl)methanone; 6,7-dimethoxy-1-isoquin-olyl 3,4-dimethoxyphenyl ketone; 6,7-dimethoxy-1-veratro-ylisoquinoline; xanthaline.* $C_{20}H_{19}NO_5$; mol wt 353.36. C 67.98%, H 5.42%, N 3.96%, O 22.64%. The old name papaveraldine is retained to avoid confusion. From opium; whether papaveraldine occurs as such in the poppy plant, or is formed during the process of extraction, has not been investigated. Oxidation of papaverine to papaveraldine by SeO_2: Menon, *Proc. Indian Acad. Sci.* **19A**, 21 (1944). For older references *see:* Small, Lutz, "Chemistry of the Opium Alkaloids," *U.S. Public Health Reports,* **Suppl. No. 103,** Washington (1932). Synthesis from Reissert compds: Popp, McEwen, *J. Am. Chem. Soc.* **79**, 3773 (1957).

Crystals from abs ethanol, mp 208-209°. Sol in benzene, chloroform; slightly sol in alcohol, ether, petr ether; nearly insol in water, alkalies or carbonates.

Hydrochloride, $C_{20}H_{19}NO_5 \cdot HCl$, yellow crystals from abs alcohol, mp 200°.

6967. Papaveretum. "Concentrated opium"; Omnopon; Pantopon. A mixture of the hydrochlorides of the opium alkaloids in their approximate natural proportions. Contains approx 50% morphine, 3% codeine, 20% noscapine, and 5% papaverine. Exhibits biological action of morphine and other alkaloids present in opium. Clinical evaluation in intravenous analgesia: J. A. Catling *et al., Brit. Med. J.* **281,** 478 (1980).

Yellowish-gray, cryst powder. Freely sol in water.

Caution: May be habit forming. This is a controlled substance (opium derivative) listed in the U.S. Code of Federal Regulations, Title 21 part 1308.12 (1985).

THERAP CAT: Narcotic analgesic.

6968. Papaverine. *1-[(3,4-Dimethoxyphenyl)methyl]-6,7-dimethoxyisoquinoline; 6,7-dimethoxy-1-veratrylisoquin-oline.* $C_{20}H_{21}NO_4$; mol wt 339.38. C 70.78%, H 6.24%, N 4.13%, O 18.86%. Found in opium (0.8-1.0%). Synthesis: Pictet, Gams, *Compt. Rend.* **149**, 210 (1909); *Ber.* **42**, 2943 (1909). Review of commercial syntheses: Goldberg, *Chem. Prod. Chem. News* **17**, 371 (1954). Improved synthetic procedures: Braz, Chizhov, *Soviet Pharmaceutical Research* **3**, 90-93 (New York, 1958). Biosynthetic studies: Battersby, Harper, *J. Chem. Soc.* **1962**, 3526; Brochmann-Hanssen *et al., J. Pharm. Sci.* **60**, 1672 (1971). Pharmacology and toxicology: Preininger in *The Alkaloids* vol. **15**, R. H. F. Manske, Ed. (Academic Press, New York, 1975) pp 209-223. Toxicity: S. Levis *et al., Arch. Int. Pharmacodyn.* **123**, 264 (1960). Clinical effect on cerebral blood flow: H. L. Karpman, J. J. Sheppard, *Angiology* **26**, 592 (1975). Clinical evaluation in intermittent claudication: Y. Sheino *et al., ibid.* **34**, 257 (1983). Comprehensive description: M. S. Hifnawy, F. J. Muhtadi in *Analytical Profiles of Drug Substances* vol. **17**, K. Florey, Ed. (Academic Press, New York, 1988) p 367-447.

Triboluminescent, orthorhombic prisms from alcohol + ether, mp 147°. Sublimes at 135-140° at 11 mm pressure and 2 mm distance. d_4^{20} 1.337. pK (25°) 8.07. uv max (ethanol): 239, 278-280, 314, 327 nm (log ϵ 4.83, 3.86, 3.60, 3.67). Almost insol in water. Sol in hot benzene, glacial acetic acid, acetone; slightly sol in chloroform, carbon tetrachloride, petr ether. Optimal pH for storage of papaverine solns: 2.0-2.8.

Hydrochloride, $C_{20}H_{22}ClNO_4$, *Artegodan, Cepaverin, Cerebid, Cerespan, Dipav, Dynovas, Lapav, Optenyl, Pameion, Panergon, Papacon, Papital T.R., Pavabid, Pavacap, Pavacen, Pavadel, Pavagen, Pavagrant, Pavakey, Pavased, Pavatest, Spasmo-Nit, Therapav, Vasal, Vasospan.* Monoclinic rods from water, mp 220-225°. uv max (ethanol): 249-250, 280-282, 311 nm (log ϵ 4.69, 3.80, 3.82). One gram dissolves in about 40 ml water. Sol in alcohol and chloroform. Practically insol in ether. pH of 0.05 molar soln 3.9; pH of 2% aq soln 3.3. LD_{50} in mice, rats (mg/kg): 27.5, 20 i.v.; 150, 370 s.c. (Levis).

Nitrite, $C_{20}H_{22}N_2O_6$, light yellow, crystalline powder. Slightly soluble in water or alcohol; freely sol in chloroform, acetone.

THERAP CAT: Smooth muscle relaxant. Cerebral vasodilator.

6969. Papaya. Papaw; Carica; melon tree. Fruit of *Carica papaya* L., *Caricaceae. Habit.* Tropical America and Asia, Florida. *Constit.* Papain, the dried and purified latex of the fruit; carpaine; carposide (a glucoside).

USE: Manufacture of papain.

6970. Parabanic Acid. *Imidazolidinetrione;* imidazoletrione; oxalylurea. $C_3H_2N_2O_3$; mol wt 114.06. C 31.59%, H 1.77%, N 24.56%, O 42.08%. Prepd by the condensation of urea with diethyl oxalate in a methanol soln of sodium methoxide: Murray, *Org. Syn.* **37**, 71 (1957).

Crystals, mp about 230°; also stated as 243° with decomposition. Sublimes at 100°. Soluble in about 20 parts water, in alcohol. Its salts are unstable.

6971. Paraffin. Paraffin wax; hard paraffin. A mixture of solid hydrocarbons having the general formula C_nH_{2n+2}, obtained from petroleum.

Colorless or white, somewhat translucent, odorless mass; greasy feel; burns with a luminous flame. d about 0.90. mp 50-57°; also available with higher and lower melting ranges. Insol in water or alcohol. Sol in benzene, chloroform, ether, carbon disulfide, oils; miscible when melted with wax, spermaceti, fats.

USE: For raising mp of ointments. Manuf paraffin paper and candles (so-called wax paper or candles); for fixing drawings, etc., on muslin; water-proofing wood, cork, paper, leather; manuf varnishes; to render wooden vessels impermeable to water or alcohol; in lubricants; to cover food products; in floor polishes, cosmetics, electrical insulators; for extracting perfumes from flowers. Pharmaceutic aid (stiffening agent).

6972. Paraffin Chlorinated. Chlorcosane; Cereclor. Prepared by chlorinating a liquid paraffin. Contains about 50% Cl.

Light yellow to amber, thick, oily liq; odorless and stable in air. d 1.00-1.07. Insol in water; slightly sol in alc; miscible with benzene, chloroform, ether, carbon tetrachloride.

USE: As solvent for dichloramine-T, dissolving about 8%.

6973. Paraflutizide. *6-Chloro-3-[(4-Fluorophenyl)-methyl]-3,4-dihydro-2H-1,2,4-benzothiadiazine-7-sulfon-amide 1,1-dioxide;* 1,1-dioxo-3-(p-fluorophenylmethyl)-3,4-dihydro-6-chloro-1,2,4-benzothiadiazine-7-sulfonamide; LD 3612. $C_{14}H_{13}ClFN_3O_4S_2$; mol wt 405.86. C 41.43%, H 3.23%, Cl 8.74%, F 4.68%, N 10.35%, O 15.77%, S 15.80%. Prepn: **Belg. pat. 620,829** (1962 to Lab. Dausse), *C.A.* **58**, 12585g (1963), corresponds to **Brit. pat. 961,641; Fr. pat.**

M1374 (1962 to Lab. Dausse), *C.A.* **58**, 2462c (1963). Synthesis: J. Klosa, *J. Prakt. Chem.* **33**, 298 (1966).

Crystals from 50% alcohol, mp 238-240°.
THERAP CAT: Diuretic.

6974. Paraformaldehyde. Polyoxymethylene; Paraform; Formagene. Also erroneously referred to as *Triformol* or as *"trioxymethylene"*. Polymerized formaldehyde, $(CH_2O)_n$. Obtained by concentrating formaldehyde soln. Use in mummifying dental pulp: I. Curson, *Brit. Dent. J.* **121**, 519 (1966); P. Hobson, *ibid.* **128**, 275 (1970).

White, cryst powder, having an odor of formaldehyde. Slowly sol in cold, more readily in hot water, with evolution of formaldehyde; insol in alcohol, ether; sol in fixed alkali hydroxide solns. *Keep tightly closed.*

USE: For disinfecting sickrooms, clothing, linen, and sickroom utensils. Active ingredient of contraceptive creams. Also used as fumigant; in dentistry; in manuf synthetic resins and artificial horn or ivory.

6975. Paraldehyde. *2,4,6-Trimethyl-1,3,5-trioxane;* paracetaldehyde; Paral. $C_6H_{12}O_3$; mol wt 132.16. C 54.52%, H 9.15%, O 36.32%. A polymer of acetaldehyde. Prepd by the polymerization of acetaldehyde catalyzed by HCl and H_2SO_4 at medium to high temp: Kekulé, Zincke, *Ann.* **162**, 125 (1872); Baer, Mahan, U.S. pat. **2,864,827** (1958 to Phillips). Toxicity data: Figot *et al.*, *Acta Pharmacol. Toxicol.* **8**, 290 (1952).

Liquid, characteristic aromatic odor and warm, but disagreeable taste. d_{25}^{25} ~0.994. bp ~124°. mp 12°. n_D^{20} 1.4049. Sol in 8 parts water at 25°, in 17 parts boiling water; miscible with alc, chloroform, ether, oils. Gives acetaldehyde on heating with dil HCl or on warming with several drops concd H_2SO_4. *Incompat.* Alkalies, hydrocyanic acid, iodides, oxidizers. LD_{50} orally in rats: 1.65 g/kg (Figot).

Caution: May be habit forming. This is a controlled substance (depressant) listed in the U.S. Code of Federal Regulations, Title 21 Parts 329.1 and 1308.14 (1987).

USE: Manuf organic compounds.
THERAP CAT: Sedative, hypnotic.

6976. Paramethadione. *5-Ethyl-3,5-dimethyl-2,4-oxazolidinedione;* *3,5-dimethyl-5-ethyloxazolidine-2,4-dione;* Paradione. $C_7H_{11}NO_3$; mol wt 157.17. C 53.49%, H 7.05%, N 8.91%, O 30.54%. Prepn: Spielman, U.S. pat. **2,575,693** (1951 to Abbott).

Liquid. Fruity, esterlike odor. d_4^{25} 1.1180-1.1240. n_D^{25} 1.449. Slightly sol in water. Freely sol in alcohol, benzene, chloroform, ether.
THERAP CAT: Anticonvulsant.

6977. Paramethasone. *6α-Fluoro-11β,17,21-trihydroxy-16α-methylpregna-1,4-diene-3,20-dione;* 6α-fluoro-16α-methylprednisolone; 16α-methyl-6α-fluoroprednisolone; Flumethone; Haldrate; Paramezone; Dilar; Dillar; Cortiden; Alondra; Metilar. $C_{22}H_{29}FO_5$; mol wt 392.45. C 67.32%, H 7.45%, F 4.84%, O 20.38%. Prepn: Edwards *et al.*, *J. Am. Chem. Soc.* **82**, 2318 (1960).

21-Acetate, $C_{24}H_{31}FO_6$, *Monocortin, Haldrone, Sintecort, Stemex.* mp 228-241° (dec). $[α]_D$ +85°. uv max (ethanol): 243 nm (log ∈ 4.16). Sol in ethanol, acetone; slightly sol in water. LD_{50} i.p. in female rats: 392±23 mg/kg, E. I. Goldenthal, *Toxicol. Appl. Pharmacol.* **18**, 185 (1971).

Disodium phosphate, *Monocortin S, Solu-Dilar.*
THERAP CAT: Glucocorticoid.

6978. Paranyline. *4-(9H-Fluoren-9-ylidenemethyl)benzenecarboximidamide;* α-*fluoren-9-ylidene-p-toluamidine;* 9-(p-guanylbenzylidene)fluorene; 9-(p-guanylbenzal)fluorene; MER-27. $C_{21}H_{16}N_2$; mol wt 296.35. C 85.11%, H 5.44%, N 9.45%. Prepn: Allen *et al.*, *J. Am. Chem. Soc.* **80**, 591 (1958); Van Campen *et al.*, U.S. pat. **2,877,269** (1959 to Wm. S. Merrell).

Hydrochloride, $C_{21}H_{16}N_2$·HCl, crystals from methanol + isopropanol, mp 308°.
THERAP CAT: Anti-inflammatory.

6979. Paraoxon. *Phosphoric acid diethyl 4-nitrophenyl ester;* diethyl *p*-nitrophenyl phosphate; phosphacol; E 600; Ester 25; Eticol; Fosfakol; Mintacol; Miotisal A; Soluglaucit. $C_{10}H_{14}NO_6P$; mol wt 275.21. C 43.64%, H 5.13%, N 5.09%, P 11.26%, O 34.88%. Prepd by the action of diethyl chlorophosphate on sodium *p*-nitrophenolate or by nitration of diethyl phenyl phosphate: Schrader, *B.I.O.S. Final Rept.* **714**, 52 (1947); *Angew. Chem.* **62**, 471 (1950); Swiss pat. **257,649** (1949); Fagerlind *et al.*, *Svensk Farm. Tid.* **56**, 303, *C.A.* **46**, 9259. Physical properties: E. F. Williams, *Ind. Eng. Chem.* **43**, 950 (1951). Formation from parathion on citrus foliage and soil surface: R. C. Spear *et al.*, *J. Agric. Food Chem.* **23**, 808 (1975); W. F. Spencer *et al.*, *Bull. Environ. Contam. Toxicol.* **14**, 265 (1975). Brief review: G. Schrader, *Die Entwicklung neuer insektizider Phosphorsäure-Ester* (Verlag Chemie, Weinheim, 1963) pp 259-272.

Oily liq. Slight odor. *Poisonous!* $bp_{1.0}$ 169-170°. d_4^{25} 1.2683. n_D^{20} 1.50959. uv max: 274 nm (∈ 8.9 × 10^3). Soly in water (25°): 2400 μg/mL. Freely sol in ether, other organic solvents. Aq solns are stable up to pH 7. Roughly 300 times more stable to hydrolysis than tetraethyl pyrophosphate. The uncatalyzed reaction with water is very small and the overall velocity constant is K = 0.52[OH] + 1 × 10^{-6} min⁻¹. *See* Coates, *Ann. Appl. Biol.* **36**, 158 (1949). LD_{50} orally in rats: 1.8 mg/kg, W. R. Pickering, J. C. Malone, *Biochem. Pharmacol.* **16**, 1183 (1967) .

Caution: Cholinesterase inhibitor. For symptoms *see* Parathion.

USE: Insecticide.

6980. Paraquat. *1,1'-Dimethyl-4,4'-bipyridinium;* *N,N'*-dimethyl-γ,γ'-dipyridylium; methyl viologen (2+). [$C_{12}H_{14}N_2$]$^{2+}$; mol wt 186.25. Non-selective contact herbicide. Prepn of dichloride and bismethyl sulfate derivs: L. Michaelis, E. S. Hill, *J. Am. Chem. Soc.* **55**, 1481 (1933); R. C. Brian *et al.*, *Brit. pat.* 813,531 (1959 to ICI). Activity: R. F. Homer *et al.*, *J. Sci. Food Agr.* **11**, 309 (1960); A. D. Dodge, *Endeavor* **30**, 130 (1971). Degradation: A. Calderbank, P. Slade, *Outlook Agr.* **5**, 55 (1966); A. Calderbank, T. E. Tomlinson, *ibid.* 252 (1968); A. Calderbank, *ibid.* **6**, 128 (1970). Toxicity studies: D. G. Clark *et al.*, *Brit. J. Ind. Med.* **23**, 126 (1966); J. F. Dasta, *Am. J. Hosp. Pharm.* **35**, 1368 (1978). Review: A. A. Akhavein, D. L. Linscott, *Residue Rev.* **23**, 97-145 (1968); A. Calderbank, P. Slade in *Herbicides: Chemistry, Degradation and Mode of Action*, P. C. Kearney, D. Kaufman, Eds. (Dekker, New York, 2nd ed., 1976) pp 501-540.

H$_3$C—N$^+$⟨⟩⟨⟩N$^+$—CH$_3$

Dichloride, $C_{12}H_{14}Cl_2N_2$, *PP 148, Gramoxone.* Colorless crystals, mp 300° (dec). Very sol in water, slightly sol in lower alcohols. Insol in hydrocarbons. Hydrolyzed by alkali. Inactivated by inert clays and anionic surfactants. Corrosive to metal. Non-volatile. Normal potential at 30°: −0.446 volts. LD$_{50}$ orally in rats: 125 mg/kg, D. M. Conning *et al.*, *Brit. Med. Bull.* **25**, 245 (1969).

Bismethyl sulfate, $C_{14}H_{20}N_2O_8S_2$, *PP 910, Paraquat I.* Yellow solid. LD$_{50}$ orally in rats (male): 100 mg/kg, R. D. Kimbrough, T. B. Gaines, *Toxicol. Appl. Pharmacol.* **17**, 679 (1970).

Danger: Poison! Causes ulceration of digestive tract, diarrhea, vomiting, renal damage, jaundice. Edema, hemorrhage, fibrosis of lung, and death from anoxia may result, J. F. Dasta, *loc. cit.*

Note: Controversial spraying of marijuana plants: R. J. Smith, *Science* **199**, 861 (1978).

USE: Herbicide. Dichloride as biological oxidation-reduction indicator.

6981. Parasorbic Acid. *S-5,6-Dihydro-6-methyl-2H-pyran-2-one; 5-hydroxy-2-hexenoic acid lactone;* δ-Δ^{α,β}-hexenolactone; 2-hexen-5,1-olide; sorbic oil. $C_6H_8O_2$; mol wt 112.12. C 64.27%, H 7.19%, O 28.54%. The sole constituent of "Vogelbeeröl", an oil obtained by steam distillation of the acidified juice of the ripe berries of the mountain ash, *Sorbus aucuparia* L., *Rosaceae:* Hofmann, *Ann.* **110**, 129 (1859); Doebner, *Ber.* **27**, 344 (1894); Kuhn, Jerchel, *Ber.* **76**, 413 (1943). Structure: *eidem, ibid.* Synthesis: Haynes, Jones, *J. Chem. Soc.* **1946**, 954; Lamberti *et al.*, *Rec. Trav. Chim. Pays-Bas* **86**, 504 (1967).

Oily liquid, sweet aromatic odor. bp$_{14}$ 104-105°; bp$_{22}$ 119-123°. n_D^{25} 1.4682. d$_4^{18}$ 1.079. [α]$_D^{18}$ +49.3°; [α]$_D^{19}$ +210° (c = 2 in alc). Soluble in water; freely sol in alcohol, ether. Aq solns are neutral and turn acid on storage. LD$_{50}$ i.p. in mice: 750 mg/kg, *Toxic Substances List*, H. E. Christensen, Ed. (1974) p 670. *Caution: The vapor given off on heating is irritating.*

6982. Parathiazine. 1,4-Thiazine. C_4H_5NS; mol wt 99.15. C 48.45%, H 5.08%, N 14.13%, S 32.34%.

Known only in the form of its derivatives, e.g., its tetrahy-

dro form as thiamorpholine. *Note:* Parathiazine is also a generic name for 10-[2-(1-pyrrolidyl)ethyl]phenothiazine.

6983. Parathion. *Phosphorothioic acid O,O-diethyl O-(4-nitrophenyl) ester; O,O-diethyl O-p-nitrophenyl phosphorothioate; diethyl-p-nitrophenyl monothiophosphate; DNTP;* S.N.P.; E 605; AC 3422; ENT 15108; Alkron; Alleron; Aphamite; Etilon; Folidol; Fosferno; Niran; Paraphos; Rhodiatox; Thiophos. $C_{10}H_{14}NO_5PS$; mol wt 291.27. C 41.23%, H 4.84%, N 4.81%, O 27.47%, P 10.64%, S 11.01%. Non-systemic contact and stomach insecticide and acaricide. Original prepn: Thurston, *FIAT Report 949* (1946); Coates, Topley, *BIOS Final Report 1808* (1947). See also Fletcher *et al.*, *J. Am. Chem. Soc.* **70**, 3943 (1948). Conversion to toxic oxygen analogs: See Paraoxon. *Review:* Hall, *Advances in Chemistry Series* **1**, 150 (1950). Review of industrial syntheses: Chadwick, Watt, "Thiophosphates" in *Phosphorus and its Compounds* vol. 2, J. R. Van Wazer, Ed. (Interscience, New York, 1961) pp 1257-1262. Review of distribution, transport and fate in the environment: M. S. Mulla *et al.*, *Residue Rev.* **81**, 1-159 (1981).

C$_2$H$_5$O—P(=S)—O—⟨⟩—NO$_2$ / C$_2$H$_5$O

Pale yellow liquid. bp$_{760}$ 375°; bp$_{0.6}$ 157-162°. mp 6°. n_D^{25} 1.5370. d$_4^{25}$ 1.26. Vapor press at 20°: 3.78 × 10^{-5} mm Hg. Surface tension at 25°: 39.2 dynes/cm. Viscosity at 25°: 15.30 cp. Absorption spectra: Williams, *Ind. Eng. Chem.* **43**, 950 (1951). Freely sol in alcohols, esters, ethers, ketones, aromatic hydrocarbons. Practically insol in water (20 ppm), petr ether, kerosene, and the usual spray oils. Incompatible with substances having a pH higher than 7.5. LD$_{50}$ in female, male rats: 3.6, 13 mg/kg orally; 6.8, 21 mg/kg dermally, T. B. Gaines, *Toxicol. Appl. Pharmacol.* **14**, 515 (1969).

Human Toxicity: Highly toxic. Acute effects include anorexia, nausea, vomiting, diarrhea, excessive salivation, pupillary constriction, bronchoconstriction, muscle twitching, convulsions, coma, respiratory failure. Cholinesterase inhibitor effects are cumulative, *cf. Clinical Toxicology of Commercial Products*, R. E. Gosselin *et al.*, Eds. (Williams & Wilkins, Baltimore, 4th ed., 1976) Section III, pp 263-271. Special precautions necessary to prevent inhalation and skin contamination.

USE: Insecticide; acaricide.

6984. Parathyroid Hormone. *Parathormone; PTH.* Mol wt about 9,500. Regulatory factor in the homeostatic control of calcium and phosphate metabolism, its principal sites of activity being the skeleton, kidneys, and gastrointestinal tract. Prime function is to raise plasma calcium concns. Acts synergistically with vitamin D$_3$ (*q.v.*) except in the kidneys where the latter causes phosphate retention. Secretion from the parathyroid gland varies inversely with serum Ca^{2+} concentrations, unlike calcitonin (*q.v.*) which is secreted in direct proportion to serum calcium levels. Structure consists of a single-chain polypeptide of 84 amino acid residues. Sequence varies slightly among mammalian species. Sequence of bovine PTH: Niall *et al.*, *Z. Physiol. Chem.* **351**, 1586 (1970); Brewer, Ronan, *Proc. Nat. Acad. Sci. USA* **67**, 1862 (1970). Sequence of porcine PTH: O'Riordan *et al.*, *Proc. Roy. Soc. Med.* **64**, 1263 (1971). Isoln of human PTH from parathyroid adenomas: O'Riordan *et al.*, *Endocrinology* **89**, 234 (1971). Fragment exhibiting full biological activity consists of about 35 amino acid residues from the N-terminal: Potts *et al.* in *Parathyroid Hormone and Thyrocalcitonin (Calcitonin)*, R. V. Talmage, L. F. Belanger, Eds. (Excerpta Medica, New York, 1968) p 44; *see* in entirety for review and special studies. Synthesis of active bovine fragment: Potts *et al.*, *Proc. Nat. Acad. Sci. USA* **68**, 63 (1971); of human PTH (1-38): S. Funakoshi *et al.*, *Peptide Chem.* **18**, 223 (1980). Reviews of early literature: Potts *et al.*, *Recent Progr. Horm. Res.* **22**, 101 (1966); Arnaud *et al.*, *Ann. Rev. Physiol.* **29**, 349 (1967). *Reviews:* Behrens, Grinnan, *Ann. Rev. Biochem.* **38**, 83 (1969); Auerbach *et al.*, *Recent Progr. Horm. Res.* **28**, 353 (1972); Parsons, Potts, "Physiology and Chemistry of Parathyroid Hormone" in *Clinics in*

Endocrinology and Metabolism, I. MacIntyre, Ed. (Saunders, Philadelphia, 1972) pp 33-78. Biosynthetic review: J. F. Habener *et al.*, *Recent Progr. Horm. Res.* **33**, 249 (1977).

Note: Aqueous solns of the active principles of bovine parathyroid gland are in use under the names: *Parathorm, Para-thor-mone, Paroidin.*

THERAP CAT: Blood calcium regulator.

6985. Parbendazole. (*5-Butyl-1H-benzimidazol-2-yl*)-*carbamic acid methyl ester;* methyl 5-butyl-2-benzimidazole-carbamate; 5-butyl-2-(carbomethoxyamino)benzimidazole; SKF 29044; Helmatac; Verminum; Worm Guard. $C_{13}H_{17}N_3O_2$; mol wt 247.29. C 63.14%, H 6.93%, N 16.99%, O 12.94%. Prepd from 4-butyl-*o*-phenylenediamine and carbomethoxycyanamide: Actor *et al.*, *Nature* **215**, 321 (1967); **Brit. pat.** 1,123,317 and Stedman; Actor, Pagano, **U.S. pats.** 3,480,642; 3,574,845 (1968, 1969, 1971, all to SKF). Activity: Ostmann, Scheidy, *Progr. Antimicrob. Anticancer Chemother., Proc. Int. Congr. Chemother., 6th, Tokyo, 1969,* **1**, 159 (1970). Identification of metabolites: Dunn *et al.*, *J. Med. Chem.* **16**, 996 (1973).

Crystals from aq ethanol, mp 225-227° (dec). uv max (95% ethanol/1N HCl): 282, 288 nm (ε 16,200, 20,000). Practically insol in water. LD_{50} orally in mice and rats: >4 g/kg, Actor *et al., loc. cit.*

THERAP CAT (VET): Anthelmintic.

6986. Pareira. Pareira brava. Dried root of *Chondodendron platiphyllum* (A. St. Hil.) Miers, *C. microphyllum* (Eichl.) Mold., and *C. tomentosum* Ruiz et Pavon, *Menispermaceae.* *Habit.* Brazil. *Constit.* Bebeerine, chondrodine, fatty acids, tannin.

THERAP CAT: Diuretic, urinary anti-infective.

6987. Parethoxycaine. *4-Ethoxybenzoic acid 2-(diethylamino)ethyl ester;* β-diethylaminoethyl *p*-ethoxybenzoate; Diethoxin; Intracaine; Maxicaine. $C_{15}H_{23}NO_3$; mol wt 265.34. C 67.89%, H 8.74%, N 5.28%, O 18.09%. Prepn: Christiansen, Harris, **U.S. pat.** 2,404,691 (1946 to Squibb).

Hydrochloride, $C_{15}H_{23}NO_3 \cdot HCl$, crystals mp 172.5-173.5°.

THERAP CAT: Hydrochloride as local anesthetic.

6988. Pargyline. *N-Methyl-N-2-propynylbenzenemethanamine; N-methyl-N-2-propynylbenzylamine;* N-benzyl-N-methyl-2-propynylamine; *N-methyl-N-propargylbenzylamine;* MO-911; A 19120; Eudatin; Supirdyl. $C_{11}H_{13}N$; mol wt 159.22. C 82.97%, H 8.23%, N 8.80%. Monoamine oxidase inhibitor. Prepd from propargyl bromide and benzylmethylamine: **Brit. pat.** 906,245; Martin, **U.S. pat.** 3,155,-584 (1962 and 1964, both to Abbott). Activity as a glucuronyl transferase inducer: Yeh, Mitchell, *Experientia* **28**, 298 (1972).

Free base, bp_{11} 96-97°.

Hydrochloride, $C_{11}H_{14}ClN$, *Eutonyl.* Crystals from ethanol + ether, mp 154-155°. Readily sol in water. Aq solns are unstable.

THERAP CAT: Antihypertensive.

6989. Paromomycin. *O-2-Amino-2-deoxy-α-*D-*glucopyranosyl-(1→4)-O-[O-2,6-diamino-2,6-dideoxy-β-*L-*idopyranosyl-(1→3)-β-*D-*ribofuranosyl-(1→5)]-2-deoxy-*D-*streptamine; O-2,6-diamino-2,6-dideoxy-β-*L-*idopyranosyl-(1→3)-O-β-*D-*ribofuranosyl-(1→5)-O-[2-amino-2-deoxy-α-*D-*glucopyranosyl-(1→4)]-2-deoxystreptamine; paromomy-*

cin I; amminosidin; catenulin; crestomycin; estomycin; hydroxymycin; monomycin A; neomycin E; paucimycin; R 400. $C_{23}H_{45}N_5O_{14}$; mol wt 615.65. C 44.87%, H 7.37%, N 11.38%, O 36.38%. Oligosaccharide-type antibiotic isolated from various *Streptomyces.* From *S. rimosus* forma *paromomycinus:* Frohardt *et al.*, **U.S. pat.** 2,916,485 (1959 to Parke, Davis); from *S. catenulae:* Davisson, Finlay, **U.S. pat.** 2,895,876 (1959 to Pfizer); from *S. chrestomyceticus:* Canevazzi, Scotti, *Giorn. Microbiol.* **7**, 242 (1959); Arcamone *et al., ibid.* 251. Identity of paromomycin, catenulin, hydroxymycin and aminosidine: Schillings, Schaffner, *Antimicrob. Ag. Chemother.* **1961**, 274. Structure: Haskell *et al., J. Am. Chem. Soc.* **81**, 3480, 3482 (1959). Rinehart *et al., ibid.* **84**, 3218 (1962); Hichens, Rinehart, *ibid.* **85**, 1547 (1963). Probable identity with *zygomycin A:* Horii, *J. Antibiot.* **15A**, 187 (1962). Identity with monomycin A: Konstantinova, Brazhnikova, *Antibiotiki* **10**(1), 34 (1965); with neomycin E: Hessler *et al., J. Antibiot.* **23**, 464 (1970). Toxicity data: A. DiMarco, C. Bertazzoli, *Antibiot. & Chemother.* **11**, 2 (1963). Review of antimicrobial activity: G. L. Coffey *et al., ibid.* **9**, 730 (1959). Review of pharmacology: Gasparini, Pignatelli, *Veterinaria (Milan)* **21**, 7 (1972).

Amorphous white powder. $[\alpha]_D^{25}$ +65° ±3°. Soluble in water; moderately sol in methanol; sparingly sol in abs ethanol. LD_{50} in rats, mice (mg/kg): >1625, >2275 orally; >650, 423 s.c.; 156, 90 i.v. (Coffey).

Sulfate, $C_{23}H_{47}N_5O_{18}S$, *1600 Antibiotic, Aminoxidin, Aminosidine, Farmiglucin, Farminosidin, Fi 5853, Gabbromicina, Gabbromycin, Gabbroral, Humagel, Humatin, Pargonyl, Paramicina, Paricina, Sinosid.* $[\alpha]_D^{25}$ +50.5° (c = 1.5 in water pH 6). LD_{50} in mice (mg/kg): ~15,000 orally; 700 s.c.; 110 i.v. (Di Marco, Bertazzoli).

THERAP CAT: Antibacterial; anti-amebic.

THERAP CAT (VET): Anti-amebic.

6990. Parotin. A salivary gland hormone. It is a protein of globulin nature having an isoelectric point of pH 5.7. Produced by the parotid gland. Structure studies: Ito *et al., Endocrinol. Japan* **12**, 249 (1966); Shimasaki *et al., ibid.* **14**, 11 (1967). Generally acts on the mesenchymal tissues, esp the hard and connective tissues, to promote their development and growth. It also has a protein-anabolic function. Shows hypocalcemic and leukocytosis-promoting activities: Ito *et al., ibid.* **12**, 298 (1966). *Reviews:* Yosoji Ito, *J. Japan. Biochem. Soc.* **25**, 143-164 (1953); *Ann. N. Y. Acad. Sci.* **85**, art. 1, 228-310 (1960).

6991. Paroxetine. *trans-(−)-3-[(1,3-Benzodioxol-5-yloxy)methyl]-4-(4-fluorophenyl)piperidine;* (−)-*trans-4-(p-*fluorophenyl)-3-[[3,4-(methylenedioxy)phenoxy]methyl]-piperidine; FG-7051; BRL-29060; Aropax; Seroxat. $C_{19}H_{20}FNO_3$; mol wt 329.37. C 69.29%, H 6.12%, F 5.77%, N 4.25%, O 14.57%. Serotonin (5-HT) uptake inhibitor. Prepn: J. A. Christensen, R. F. Squires, **Ger. pat.** 2,404,113; *eidem*, **U.S. pats.** 3,912,743; 4,007,196 (1974, 1975, 1977 all to Ferrosan); of hydrochloride: R. D. B. Barnes *et al.*, **Eur. pat. Appl.** 223,403 (1987 to Beecham). Characterization of serotonin inhibition: J. Buus Lassen, *Eur. J. Pharmacol.* **47**, 351 (1978). Binding to serotonin transporter complex: E. Habert *et al., ibid.* **118**, 107 (1985); in comparison with imipramine, *q.v.*: S. Z. Langer *et al., J. Recept. Res.* **7**, 499 (1987). Pharmacokinetics: J. Lund *et al., Acta Pharmacol. Toxicol.* **44**, 289 (1979); J. Lund *et al., ibid.* **51**, 351 (1982). HPLC determn in plasma: M. A. Brett *et al., J. Chromatog.*

419, 438 (1987). Clinical pharmacology: S. M. Hassan *et al., Brit. J. Clin. Pharmacol.* **19,** 705 (1985). Comparison with amitriptyline, *q.v.*, in treatment of depression: A. L. Laursen *et al., Acta Psychiat. Scand.* **71,** 249 (1985); R. Battegay *et al., Neuropsychobiology* **13,** 31 (1985).

Hydrochloride, $C_{19}H_{21}ClFNO_3$, crystal platelets, mp 118°C.

Hydrochloride hemihydrate, $C_{19}H_{21}ClFNO_3 \cdot \frac{1}{2}H_2O$, crystals, mp 129-131°.

Maleate, crystals from ethanol-ether, mp 136-138°. $[\alpha]_D$ −87° (c = 5 in ethanol). LD_{50} in mice (mg/kg): 845 s.c.; 500 orally (Christensen, Squires, 1977).

THERAP CAT: Antidepressant.

6992. Paroxypropione. *1-(4-Hydroxyphenyl)-1-propanone; 4'-hydroxypropiophenone; p-hydroxypropiophenone; paraoxypropiophenone; p-propionylphenol; ethyl p-hydroxyphenyl ketone; P.O.P.; B 360; H-365; Profenone; Frenantol; Frenohypon; Paroxon; Possipione; Hypostat.* C_9H_{10}- O_2; mol wt 150.17. C 71.98%, H 6.71%, O 21.31%. Prepn: Perkin, *J. Chem. Soc.* **55,** 546 (1889); Goldzweig, Kaiser, *J. Prakt. Chem.* [2] **43,** 86 (1891); Cox, *J. Am. Chem. Soc.* **49,** 1028 (1927); Hartung *et al., ibid.* **53,** 4153 (1931); Farinholt, *ibid.* **55,** 3386 (1933); Miller, Hartung, *Org. Syn. coll. vol. II* (1943) p 543. Derivatives: Buu-Hoi, *Rec. Trav. Chim.* **68,** 759 (1949).

Needles or prisms from water. mp 149°. One part dissolves in 2896 parts of water at 15°, in 30 parts at 100°. Freely sol in alcohol or ether.

THERAP CAT: Pituitary gonadotropic hormone inhibitor.

6993. Parsalmide. *5-Amino-N-butyl-2-(2-propynyloxy)benzamide; 5-amino-O-propynyl-N-butylsalicylamide; 2-propargyloxy-5-amino-N-butylbenzamide; MY41-6; Sinovial; Parsal.* $C_{14}H_{18}N_2O_2$; mol wt 246.31. C 68.27%, H 7.37%, N 11.37%, O 12.99%. Prepn: B. Gradnik *et al.,* **Ger. pat. 2,029,991;** *eidem,* **U.S. pat. 3,739,030** (1970, 1973 both to Soc. Etude Recher. Appl. Sci. Med.). Synthesis and pharmacology: A. Pedrazzoli *et al., Boll. Chim. Farm.* **115,** 125 (1976). Effect on gastric secretion in laboratory animals: G. Bertaccini *et al., Farmaco Ed. Prat.* **34,** 482 (1979). Inhibition of platelet aggregation: R. Fantasia *et al., Arzneimittel-Forsch.* **32,** 1312 (1982). Clinical comparison with phenylbutazone, *q.v.*, in management of joint pain: A. Bajardi, S. Fantato, *Minerva Med.* **67,** 3371 (1976); S. Menci *et al., ibid.* 3403. Series of articles on anti-inflammatory activity, pharmacology and toxicology: *Boll. Chim. Farm.* **115,** 135-230 (1976), *C.A.* **85,** 13981-13986 (1976).

Crystals from ethanol, mp 83-85°. pKa (aq soln, 20°) 4.6 ±0.03. uv max (abs alcohol): 220 ±1, 327 ±2 nm; (water, pH 2): 284 nm (log ε 3.39); (water, pH 8.5): 313 nm (log ε 3.40). Very soluble in alcohol, chloroform, acetone, ethyl acetate, dioxane; slightly sol in water. Practically insol in petr ether, cyclohexane.

THERAP CAT: Anti-inflammatory; analgesic.

6994. Parsley Seed. Parsley fruit. Dried ripe seed of *Petroselinum hortense* Hoffm. (*P. sativum* Hoffm.; *Carum petroselinum* Benth. & Hook.), *Umbelliferae.* Habit. Russia, France, Germany; cultivated everywhere. *Constit.* Seed: volatile and fixed oils; apiol, apiolin, apiin, tannin.

USE: Source of apiol.

6995. Parthenin. *3,3a,4,5,6,6a,9a,9b-Octahydro-6a-hydroxy-6,9a-dimethyl-3-methyleneazuleno[4,5-b]furan-2,9-dione; 1,6β-dihydroxy-4-oxo-10αH-ambrosa-2,11(13)-dien-12-oic acid γ-lactone;* parthenicin. $C_{15}H_{18}O_4$; mol wt 262.31. C 68.69%, H 6.92%, O 24.40%. From herb of *Parthenium hysterophorus* L., *Compositae.* Parthenin is the substance largely responsible for the allergic contact dermatitis caused by *P. hysterophorus.* Isoln: Arny, *Am. J. Pharm.* **69,** 169 (1897). Isoln and structure: Herz *et al., Tetrahedron Letters* **1961,** 82; Herz *et al., J. Am. Chem. Soc.* **84,** 2601 (1962). Abs config: Emerson *et al., Tetrahedron Letters* **1966,** 6151. Total synthesis of (±)-form: P. Kok *et al., Bull. Soc. Chim. Belg.* **87,** 615 (1978); C. H. Heathcock *et al., J. Am. Chem. Soc.* **104,** 6081 (1982).

Crystals from water, mp 163-166°. $[\alpha]_D^{25}$ +7.02° (c = 2.71 in chloroform). uv max: 215, 340 nm (ε 15,100; 22). Practically insol in water; sol in alcohol, chloroform, ether, ethyl acetate.

6996. Parthenolide. *[1aR-(1aR*,4E,7aS*,10aS*,-10bR*)]-2,3,6,7,7a,8,10a,10b-Octahydro-1a,5-dimethyl-8-methyleneoxireno[9,10]cyclodeca[1,2-b]furan-9(1aH)-one; 4,5α-epoxy-6β-hydroxy-germacra-1(10),11(13)-dien-12-oic acid γ-lactone.* $C_{15}H_{20}O_3$; mol wt 248.32. C 72.55%, H 8.12%, O 19.33%. Sesquiterpene lactone found in feverfew, *q.v.*, and in other plants. Isolation from *Chrysanthemum parthenium* (L.) Bernh. *Compositae* and characterization: V. Herout *et al., Chem. & Ind. (London)* **1959,** 1069; M. Soucek *et al., Coll. Czech. Chem. Comm.* **26,** 803 (1961); from *Magnolia grandiflora* L., *Magnoliaceae:* F. S. El-Feraly, Y.-M. Chan, *J. Pharm. Sci.* **67,** 347 (1978). Revised structure and spectral analysis: T. R. Govindachari *et al., Tetrahedron* **21,** 1509 (1965). Absolute configuration: A. S. Bawdekar *et al., Tetrahedron Letters* **1966,** 1225. Crystal structure: A. Quick, D. Rogers, *J. Chem. Soc. Perkins Trans. II,* **4,** 465 (1976). HPLC determn: D. Strack *et al., Z. Naturforsch.* **35,** 915 (1980). Cytotoxicity: K.-H. Lee *et al., Cancer Res.* **31,** 1649 (1971); L. A. J. O'Neill *et al., Brit. J. Clin. Pharmacol.* **23,** 81 (1987).

Colorless plates, mp 115-116°. $[\alpha]_D^{20}$ −81.4° (c = 1.04 in chloroform); $[\alpha]_D^{22}$ −71.4° (c = 0.220 in CH_2Cl_2). uv max: 214 nm (log ε 4.22).

6997. Partricin. Ayfactin; SPA-S132. Heptaene macrolide antibiotic complex produced by *Streptomyces aureofaciens* NRRL 3878. Isoln: T. Bruzzese, R. Ferrari, **U.S. pat. 3,773,925** (1973 to SPA). Recovery and purification process: S. Magnaghi *et al.,* **Brit. pat. 1,462,442** (1977 to SPA), *C.A.* **87,** 66588a (1977). *In vitro* activity: W. Ritzerfeld,

Farmaco Ed. Sci. **27**, 235 (1972); G. A. Meloni *et al., ibid.* **34**, 183 (1979). Use in treatment of benign prostatic hypertrophy: T. Bruzzese, L. Ferrari, U.S. pat. 4,237,117 (1980 to SPA). Separation, characterization and structure of components A and B: R. C. Tweit *et al., J. Antibiot.* **35**, 997 (1982).

partricin A: R' = CH₃
partricin B: R' = H

Partricin A, $C_{59}H_{86}N_2O_{19} \cdot 4H_2O$. Greenish-yellow powder, mp > 300° (dec). uv max (75% methanol in DMF): 232, 240, 247, 288, 342, 358, 378, 400 nm (ε 33476, 32237, 22936, 15493, 58282, 76883, 102308, 89280). pKa's (70% aq DMF): 6.07, 8.91.

Partricin B, $C_{58}H_{84}N_2O_{19} \cdot 2\frac{1}{2}H_2O$. Brownish-yellow powder, mp > 300° (dec). uv max (75% methanol in DMF): 232, 240, 247, 288, 342, 358, 378, 400 nm (ε 34761, 32826, 23174, 20594, 50207, 73392, 100425, 87558). pKa's (70% aq DMF): 6.31, 8.95. $[\alpha]_D^{26}$ +87.2° (c = 0.06 in DMF).

Complex, amphoteric yellow crystals. uv max (ethanol): 401, 379, 359, 341 nm. Sol in DMF, DMSO, dimethyl acetamide, pyridine. Practically insol in water, common organic solvents. LD_{50} in mice (mg/kg): 300 orally; 0.5 i.p. (Bruzzese).

Methyl ester, *see* Mepartricin.

6998. Parvaquone. *2-Cyclohexyl-3-hydroxy-1,4-naphthalenedione; 2-cyclohexyl-3-hydroxy-1,4-naphthoquinone;* 2-hydroxy-3-cyclohexyl-1,4-naphthoquinone; BW 993C; Clexon. $C_{16}H_{16}O_3$; mol wt 256.30. C 74.98%, H 6.29%, O 18.73%. Prepn: L. F. Fieser, *J. Am. Chem. Soc.* **70**, 3165 (1948); L. F. Fieser, M. T. Leffler, U.S. pat. 2,553,648 (1951 to Research Corp.). Antimalarial activity: L. F. Fieser, A. P. Richardson, *J. Am. Chem. Soc.* **70**, 3156 (1948). Electron transport inhibition: A. L. Tappel, *Biochem. Pharmacol.* **3**, 289 (1960); by uncoupling: J. L. Howland, *Biochim. Biophys. Acta* **131**, 247 (1967). *In vitro* and *in vivo* antiprotozoal activity: A. T. Hudson *et al., Parasitology* **90**, 45 (1985). Bioassay in serum: N. McHardy, J. Mercer, *Kenya Vet.* **8**(2), 9 (1984). Treatment of East Coast fever in cattle: T. T. Dolan *et al., Vet. Parasitol.* **15**, 103 (1984); of *Theileria annulata* in calves: N. McHardy, D. W. Morgan, *Res. Vet. Sci.* **39**, 1 (1985). Field comparison of antitheilarial activity with buparvaquone, *q.v.:* N. McHardy *et al., ibid.* **39**.

Bright yellow needles, mp 135-136°.
THERAP CAT (VET): Antiprotozoal (Theileria).

6999. Pasiniazide. *4-Pyridinecarboxylic acid hydrazide mono(4-amino-2-hydroxybenzoate); isoniazid 4-aminosalicylate;* isonicotinic acid hydrazide *p*-aminosalicylate; *p*-aminosalicylic acid isonicotinyl hydrazide; GEWO 399; Paraniazide; Anazid; Dipasic. $C_{13}H_{14}N_4O_4$; mol wt 290.27. C 53.79%, H 4.86%, N 19.30%, O 22.05%. An equimolecular salt produced by dissolving the components in hot dil alc, followed by cooling and evapn: Charonnat, Boime, *Compt.*

Rend. **236**, 2140 (1953). Prepn: **Swiss** pat. **303,085** (1955 to Hoffmann-La Roche).

Yellow crystals from methanol or ethanol, mp 142-144°. Sparingly sol in water. uv max 272, 303 nm ($E_{1cm}^{1\%}$ 550, 445).
THERAP CAT: Antibacterial (tuberculostatic).

7000. Passiflora. Passion flower; passion vine; May pops. Dried flowering and fruiting tops of *Passiflora incarnata* L., *Passifloraceae. Habit.* Southeastern U.S. *Constit.* Harman.
THERAP CAT: Sedative, analgesic.

7001. Patchouli Alcohol. *Octahydro-4,8a,9,9-tetramethyl-1,6-methanonaphthalen-1(2H)-ol;* patchouli camphor. $C_{15}H_{26}O$; mol wt 222.36. C 81.02%, H 11.79%, O 7.20%. A tricyclic sesquiterpene alcohol isolated from oil of patchouli: Gadamer, Amenomiya, *Arch. Pharm.* **241**, 39 (1903). Proposed structure: Treibs, *Ann.* **564**, 141 (1949). Revised structure: Dobler *et al., Proc. Chem. Soc. (London)* **1963**, 383. Structural studies: Büchi *et al., J. Am. Chem. Soc.* **78**, 1262 (1956); **83**, 927 (1961); **84**, 3205 (1962); **86**, 4438 (1964). Synthesis of *dl*-form: Danishevsky, Dumas, *Chem. Commun.* **1968**, 1287; Mirrington, Schmalzl, *J. Org. Chem.* **37**, 2871 (1972); K. Yamada *et al., Tetrahedron* **35**, 293 (1979). Stereoselective total synthesis of racemic form: F. Näf, G. Ohloff, *Helv. Chim. Acta* **57**, 1868 (1974); of natural, racemic and (+)-forms: F. Näf *et al., ibid.* **64**, 1387 (1981). *Review:* Walker, *Mfg. Chem. Aerosol News* **39**, no. 7, 27 (1968).

Large crystals (hexagonal-trapezohedral) from the higher boiling fractions of oil of patchouli or from petr ether, mp 56°. mp (racemate) 39-40° (Danishevsky, Dumas); also reported as 46-47° (Mirrington, Schmalzl). bp_8 140°. d_4^{20} 1.0284. $[\alpha]_D^{20}$ −97.4° (c = 24 in chloroform). n_D^{65} 1.5029. Practically insol in water. Sol in alcohol, ether, common organic solvents.

7002. Patulin. *4-Hydroxy-4H-furo[3,2-c]pyran-2(6H)-one;* clavacin; clavatin; claviformin; expansine; mycoin C_3; penicidin. $C_7H_6O_4$; mol wt 154.12. C 54.55%, H 3.93%, O 41.52%. An antibiotic derived from the metabolites of a number of fungi, e.g., *Aspergillus clavatus, A. claviforme, A. giganteus, A. terreus, Penicillium patulum, P. expansum, P. melinii, P. leucopus, P. urticae* and *Gymnoascus* spp. Isoln and antibacterial activity: Birkinshaw *et al., Lancet* **245**, 625 (1943); Birkinshaw, Michael, U.S. pat. 2,417,584 (1947 to Therap. Res. Corp. of Great Britain); *cf.* Waksman *et al., Science* **96**, 202 (1942); Brack, *Helv. Chim. Acta* **30**, 1 (1947); Norstadt, McCalla, *Appl. Microbiol.* **17**, 193 (1969). Structure: Birkinshaw, *loc. cit.;* Bergel *et al., J. Chem. Soc.* **1944**, 415; Woodward, Singh, *J. Am. Chem. Soc.* **71**, 758 (1949); Shemyakin, Khokhlov, *Doklady Akad. Nauk SSSR* **75**, 47 (1950). Synthesis: Woodward, Singh, *J. Am. Chem. Soc.* **72**, 1428 (1950). Biosynthesis: Bu'Lock, Ryan, *Proc. Chem. Soc.* **1958**, 222; Tanenbaum, Bassett, *J. Biol. Chem.* **234**, 1861 (1959). Has carcinogenic activity attributable to α,β-unsaturation together with an external conjugated double bond attached to 4 position of the γ-lactone ring: Dickens, *Brit. Med. Bull.* **20**, 96 (1964). Review amd evaluation of studies of carcinogenic action in laboratory animals: *IARC Monographs* **10**, 205-210 (1976). Also inhibits uptake of K⁺ ions in erythrocytes: Kahn, *J. Pharmacol. Exp. Ther.* **121**, 234 (1957). Toxicity: R. Kinosita, T. Shikata, "On Toxic

Moldy Rice" in *Mycotoxins in Foodstuffs*, G. N. Wogan, Ed. (The M.I.T. Press, Cambridge, 1965) pp 117-119. *Review:* Ciegler *et al.*, "Patulin, Penicillic Acid and Other Carcinogenic Lactones" in *Microbial Toxins*, vol. VI, A. Ciegler *et al.*, Eds. (Academic Press, New York, 1971) p 409-414.

Compact prisms or thick plates from ether or chloroform, mp 111.0°. uv max: 276.5 nm (Bergel, *loc. cit.*). Sol in water and the common organic solvents except petr ether; very sol in ethyl or amyl acetate. Unstable in alkali with loss of biological activity. LD_{50} s.c. in mice: 10-15 mg/kg (Kinosita, Shikata).

Acetylpatulin, $C_9H_8O_5$, prisms from 50% alcohol, mp 118-120°.

7003. Pavoninin-5. $(3\beta,15\alpha,25R)$-26-(*Acetyloxy*)-3-*hydroxycholest-5-en-15-yl* 2-(*acetylamino*)-2-*deoxy*-β-D-*glucopyranoside*. $C_{37}H_{61}NO_9$; mol wt 663.89. C 66.94%, H 9.26%, N 2.11%, O 21.69%. Major component of a group of closely related shark repellents isolated from the defense secretion of the Pacific sole, *Pardachirus pavoninus*. The *pavoninins* are a series of ichthyotoxic and hemolytic steroid amine glycosides. Isoln, structure determn and evaluation of pavoninins 1-6 as shark repellents: K. Tachibana *et al.*, *Science* **226**, 703 (1984). ^{13}C, ^1H-NMR studies, chemical correlations and ichthyotoxicity: *eidem*, *Tetrahedron* **41**, 1027 (1985). Brief review: *Chem. & Eng. News* **62**, 36 (Nov. 19, 1984).

Sol in ethyl acetate. $[\alpha]_D^{29}$ +21° (c = 0.7 in ethanol).

7004. Peach Oil, Expressed. Persic oil. Fixed oil from seed of *Prunus persica (L.)* Stokes and allied spp., *Rosaceae*.

Light yellow liquid; closely resembles expressed oil of almond. d_{15}^{15} 0.917-0.921. Solidif not above −15°. n_D^{40} 1.464-1.465. Sapon no. 189-193. Iodine no. 100-110. Acid no. about 8. Sol in chloroform, ether, petr ether, slightly sol in alcohol.

USE: Emollient.

7005. Peanut. Groundnut; earthnut. Ripe, underground pods with seeds of *Arachis hypogaea* L., *Leguminosae*. *Habit.* Parana River Valley in Paraguay, Brazil and Argentina. Cultivated in the reasonably warm regions of all continents, e.g., through the southern U.S.A. *Composition:* pericarp (shell) 21-29%, episperm (skin) 1.95-3.2%, kernel (+ germ) 71-75%. Botanically the peanut is kin to peas and beans, but its constituents are more like those of true nuts. *Constit.* of kernels (roasted with skin): proteins 26.2%, oil 48.7%, water 1.8%, carbohydrates 20.6%, ash 2.7%. The chief proteins are arachin (25% in oil-free meal) and conarchin (8%). Both are globulins of different soly: Johns, Jones, *J. Biol. Chem.* **28**, 77 (1916). The vitamin content of peanuts is moderate, the largest portion being in the episperm. Trace mineral content of kernels: iron 20 mg/kg; manganese 8.51 mg/kg; copper 6.8 mg/kg; zinc 16 mg/kg. *Reviews:* J. Adam, *Les Plantes a Matiere Grasse*, vol. 3, *L'Arachide* (Paris, 1947); N. J. Morris, F. G. Dollear, *Abstract Bibliography of the Chemistry and Technology of Peanuts*, Southern Regional Res. Lab. (New Orleans, 1949); J. G. Woodruff, C. T. Young in Kirk-Othmer *Encyclopedia of Chemical Technology* vol. 14 (Interscience, New York, 1967) p 122.

USE: In peanut butter, candy; as salted peanuts; for fodder and seeding; crushed for oil. Production in the U.S.A. is

∼1.5 × 10⁶ metric tons; world production ∼16-17 × 10⁶ metric tons. Peanut proteins have been used to produce a fiber, *Sarelon*. The shells are used in the manuf of furfural, xylose, cellulose, plastics, mucilage, also in fertilizers and cattle feed.

7006. Peanut Oil. Arachis oil; groundnut oil; earthnut oil; katchung oil. Prepd by pressing the shelled and skinned seeds of *Arachis hypogaea* L., *Leguminosae:* N. J. Morris, F. G. Dollear, *Abstract Bibliography of the Chemistry and Technology of Peanuts* (Southern Regional Res. Lab., New Orleans, 1949); E. W. Eckey, *Vegetable Fats and Oils* (Reinhold, New York, 1954). *Constits.* of cold pressed oil: glycerides of the following fatty acids: palmitic 8.3%, stearic 3.1%, arachidic 2.4%, behenic 3.1%, lignoceric 1.1%, oleic 56.0%, linoleic 26.0%. Traces of capric and lauric acids have been reported. Unsaponifiable matter 0.8% (includes tocopherols 0.022 to 0.059%, sterols 0.19 to 0.25%, squalene 0.027% and other hydrocarbons).

Greenish-yellow or almost colorless oil. Mild, pleasant odor; bland taste. d_{15}^{15} 0.917-0.921. d_{25}^{25} 0.910-0.915. Clouds at low room temp. Solidifies at about −5°. A.S.T.M. cloud point +4.5°; A.S.T.M. pour point +1°. Titer 26 to 32°. Flash pt 540°F (283°C); ignition temp 833°F (443°C). n_D^{25} 1.466-1.470; n_D^{40} 1.4605-1.4645. Acid value 0.08-6; saponification value 188-195; iodine value 84-102; thiocyanogen value 67-73; hydroxyl value 2.5-9.5; Reichert-Meissl value 0.2-1.0; Polenske value 0.2-0.7. Very slowly thickens and becomes rancid on prolonged exposure to air. Miscible with ether, petr ether, chloroform, carbon disulfide. Sol in benzene, carbon tetrachloride, oils; very slightly sol in alc.

USE: Edible oil: for salad oil as is; in hydrogenated state as shortening, in mayonnaise, in confections. For the manuf of margarine, soaps, paints. In pharmacy as vehicle for i.m. medication, in the laboratory as heat transfer medium in melting point apparatus. Pharmaceutic aid (solvent).

THERAP CAT (VET): Has been used in control of pasture bloat.

7007. Pebulate. *Butylethylcarbamothioic acid S-propyl ester;* S-propyl butylethylthiocarbamate; propyl ethyl-*n*-butylthiolcarbamate; PEBC; Stauffer 2061; Tillam. $C_{10}H_{21}$-NOS; mol wt 203.36. C 59.06%, H 10.41%, N 6.89%, O 7.87%, S 15.77%. Prepn: Campbell, Klingman, U.S. pat. **2,983,747** (1961 to Stauffer). Toxicity data: G. W. Bailey, J. L. White, *Residue Rev.* **10**, 97 (1965).

Liquid, bp_{20} 142°. n_D^{30} 1.4752. d_4^{30} 0.9458. Slightly sol in water; miscible with acetone, benzene, isopropanol, methanol, xylene. LD_{50} orally in rats: 1.12 g/kg (Bailey, White).

USE: Selective herbicide.

7008. Pecilocin. 1-(*8-Hydroxy-6-methyl-1-oxo-2,4,6-dodecatrienyl*)-2-*pyrrolidinone;* Supral; Variotin. $C_{17}H_{25}$-NO₃; mol wt 291.38. C 70.07%, H 8.65%, N 4.81%, O 16.47%. Antifungal antibiotic isolated from *Paecilomyces varioti* Bainier var. *antibioticus:* S. Takeuchi *et al.*, *J. Antibiot.* **12A**, 109, 195 (1959). Prepn: Sumiki *et al.*, **Brit.** pat. **866,425** (1961 to Japan. Antibiot. Res. Assoc. and Nippon Kayaku). Structure: S. Takeuchi *et al.*, *J. Antibiot.* **17A**, 267 (1964). Stereochemistry: S. Takeuchi, H. Yonehara, *Tetrahedron Letters* **1966**, 5197. Revised structure and stereochemistry (E,Z,E to E,E,E): *eidem*, *J. Antibiot.* **22**, 179 (1969). Synthesis of the *dl*-form: A. Ishida, T. Mukaiyama, *Chem. Letters* **1977**, 467; *eidem*, *Bull. Chem. Soc. Japan* **51**, 2077 (1978).

Neutral oil with ester-like odor. Does not show definite boiling or dec pt. $[\alpha]_D^{28}$ −5.68° (methanol). Freely sol in methanol, ethanol, acetone, ethyl acetate, benzene, ether, chloroform, pyridine, dioxane, acetic acid; slightly sol in water, petr ether, ligroin. uv max (methanol): *ca.* 318, 324 nm ($E_{1cm}^{1\%}$ 1198). Unstable and gradually loses antifungal activity in desiccator, though it is fairly stable in organic solvents. Unstable under alkaline conditions.

Monohydrate, needles from ethyl acetate + petr ether, mp 41.5-42.5°. uv max: 320 nm (ϵ 46,000).

THERAP CAT: Antifungal.

7009. Pectin. Mol wt 20,000-400,000. Polysaccharide substance present in cell walls of all plant tissues which functions as an intercellular cementing material. One of the richest sources of pectin is lemon or orange rind which contains about 30% of this polysaccharide. Occurs naturally as the partial methyl ester of α-(1→4) linked D-polygalacturonate sequences interrupted with (1→2)-L-rhamnose residues. Neutral sugars: D-galactose, L-arabinose, D-xylose and L-fucose form side chains on the pectin molecule. Structure studies: D. A. Rees, A. W. Wight, *J. Chem. Soc. B,* **1971,** 1366. Secondary and tertiary structure in solution and in gels: D. A. Rees, E. J. Welsh, *Angew. Chem. Int. Ed.* **16,** 214 (1977). Review and bibliography: Towle, Christensen, in *Industrial Gums,* R. L. Whistler, Ed. (Academic Press, New York, 2nd ed., 1973) p 429-461. Book: Z. I. Kertesz, *The Pectic Substances* (Interscience, New York, 1951).

Occurs as a coarse or fine powder, yellowish-white in color, practically odorless, and with a mucilaginous taste. Almost completely sol in 20 parts water, forming a viscous soln contg negatively charged, very much hydrated particles. Acid to litmus. Insol in alcohol or in diluted alcohol, and in other organic solvents. Dissolves more readily in water, if first moistened with alcohol, glycerol or sugar syrup, or if first mixed with 3 or more parts of sucrose. Stable under mildly acidic conditions; more strongly acidic or basic conditions cause depolymerization.

USE: In the preparation of jellies and similar food products: Owens *et al.,* "Factors Influencing Gelation with Pectin" in Advances in Chemistry Series, *Natural Plant Hydrocolloids* (A.C.S., Washington, 1954) pp 10-15.

THERAP CAT (VET): Antidiarrheal.

7010. Pectolinarigenin. *5,7-Dihydroxy-6-methoxy-2-(4-methoxyphenyl)-4H-1-benzopyran-4-one; 5,7-dihydroxy-4',6-dimethoxyflavone;* 6-methoxyacacetin; 6-hydroxypelargidenon-6,4'-dimethyl ether 1467. $C_{17}H_{14}O_6$; mol wt 314.28. C 64.96%, H 4.49%, O 30.55%. From leaves of *Linaria vulgaris* Mill., *Scrophulariaceae:* Schmid, Rumpel, *Monatsh.* **57,** 421 (1931); Merz, Wu, *Arch. Pharm.* **274,** 126 (1936). Structure: Schmid, Rumpel, *Monatsh.* **60,** 8 (1932). Synthesis: Wessely, Moser, *ibid.* **56,** 97 (1930); Zemplén, Farkas, *Ber.* **76,** 937 (1943); Murti, Seshadri, *Proc. Indian Acad. Sci.* **30A,** 78 (1949), *C.A.* **44,** 3987d (1950); Farkas, Strelisky, *Tetrahedron Letters* **1970,** 187.

Yellow needles from methanol, mp 220-223°. uv max (methanol): 275, 335 nm. Sol in alcohol, acetone, ether, ethyl acetate. Practically insol in water, benzene, chloroform, petr ether.

Diacetate, $C_{21}H_{18}O_8$, needles from 96% alcohol, mp 151°. 7-Rutinoside, $C_{29}H_{34}O_{15}$, *pectolinarin, neolinarin.* From leaves of *L. vulgaris* Mill., *Scrophulariaceae:* Klobb, *Compt. Rend.* **145,** 331 (1907); Zemplén *et al., Ber.* **75,** 489 (1942); from *Cirsium oleraceum* Scop., *Compositae:* Wagner *et al., Arch. Pharm.* **293,** 1053 (1960). Structure: Zemplén, Bognár, *Ber.* **74,** 1818 (1941). Yellow crystals from methanol, mp 275°. uv max (methanol): 275, 330 nm (log ϵ 4.256, 4.365).

7011. Pederin. *N-[[6-(2,3-Dimethoxypropyl)tetrahydro-*

4-hydroxy-5,5-dimethyl-2H-pyran-2-yl]methoxymethyl]tetrahydro-α-hydroxy-2-methoxy-5,6-dimethyl-4-methylene-2H-pyran-2-acetamide; N-[[6-(2,3-dimethoxypropyl)tetrahydro-4-hydroxy-5,5-dimethyl-2H-pyran-2-yl]methoxymethyl]tetrahydro-2-methoxy-5,6-dimethyl-4-methylene-2H-pyran-2-glycolamide; pederine; paederine. $C_{25}H_{45}NO_9$; mol wt 503.65. C 59.62%, H 9.01%, N 2.78%, O 28.59%. The toxic principle isolated from blister beetles, *Paederus fuscipes* Curt.: Pavan, Bo, *Physiol. Comparata et Oecol.* **3,** 307 (1953), *C.A.* **48,** 10217g (1954); **Brit. pat. 932,875** (1963 to Farmitalia). Powerful inhibitor of protein biosynthesis and mitosis. Structure: Cardani *et al., Gazz. Chim. Ital.* **96,** 3 (1966). Revised structure: Matsumoto *et al., Tetrahedron Letters* **1968,** 6297. Biosynthesis: Cardani *et al., ibid.* **1973,** 2815. Synthetic approach: K. Issak, P. Kocienski, *Chem. Commun.* **1982,** 460. Total synthesis: F. Matsuda *et al., Tetrahedron Letters* **23,** 4043 (1982); **24,** 1277 (1983).

Crystals from hexane, benzene + hexane, ether + hexane. mp 112-112.5°. Slightly sol in water, hexane. Sol in methanol, ethanol, carbon disulfide, chloroform, carbon tetrachloride, benzene, and acids. Practically insol in petr ether, NH_4OH, NaOH.

7012. Pefloxacin. *1-Ethyl-6-fluoro-1,4-dihydro-7-(4-methyl-1-piperazinyl)-4-oxo-3-quinolinecarboxylic acid;* pefloxacine; EU-5306; 1589RB; AM 725; Abaktal; Peflacine. $C_{17}H_{20}FN_3O_3$; mol wt 333.36. C 61.25%, H 6.07%, F 5.69%, N 12.60%, O 14.39%. Fluorinated quinolone antibacterial; analog of norfloxacin, *q.v.* Prepn: M. Pesson, **Ger. pat. 2,840,910;** *idem,* **U.S. pat. 4,292,317** (1979, 1981 to Roger Bellon/Dainippon). Pharmacology and antibacterial spectrum: Y. Goueffon *et al., Compt. Rend. Ser. III* **292,** 37 (1981). Pharmacokinetics: J. Barre *et al., J. Pharm. Sci.* **73,** 1379 (1984). Bioavailability and metabolism: A. Contrepois *et al., J. Antimicrob. Chemother.* **14,** 51 (1984); G. Montay *et al., Antimicrob. Ag. Chemother.* **25,** 463 (1984). HPLC determn in urine and plasma: *eidem, J. Chromatog.* **272,** 359 (1983).

Crystals from DMF, mp 270-272° (dec). Slightly sol in water; sol in alkaline and acidic solutions. LD_{50} in mice (mg/kg): 225 i.v., 1000 orally; in rats (g/kg): 1.5 i.p., 2.5 orally (Goueffon).

Methanesulfonate dihydrate, $C_{18}H_{24}FN_3O_6S.2H_2O$, *pefloxacine mesylate, 1589mRB.*

THERAP CAT: Antibacterial.

7013. Pelargonic Acid. *Nonanoic acid;* nonylic acid; nonoic acid. $C_9H_{18}O_2$; mol wt 158.23. C 68.31%, H 11.47%, O 20.22%. $CH_3(CH_2)_7COOH$. Occurs as an ester in oil of pelargonium: Redtenbacher, *Ann.* **59,** 41, 52, 54 (1846). Prepn from unsaturated hydrocarbons by the oxo process: Hill, **U.S. pat. 2,815,355** (1957 to Standard Oil of Indiana); from tall oil unsaturated fatty acids: Maggiolo, **U.S. pat. 2,865,937** (1958 to Welsbach); by oxidation of oleic acid: Mackenzie, Morgan, **U.S. pat. 2,820,046** (1958 to Celanese); from rice bran oil fatty acid: Mihara *et al.,* **U.S. pat. 3,060,-211** (1962 to Toya Koatsu Ind.). Purification: Port, Riser, **U.S. pat. 2,890,230** (1959 to U.S.D.A.).

Colorless, oily liquid at ordinary temp; crystallizes when cooled; characteristic odor. d_4^{20} 0.907. mp 12.5°. bp_{756} 252-

253°; bp$_{14}$ 143-145°; bp$_{6.3}$ 132-133°. n$_D^{20}$ 1.4330; n$_D^{40}$ 1.4245. Practically insol in water; sol in alcohol, chloroform, ether. LD$_{50}$ i.v. in mice: 224±4.6 mg/kg, L. Orö, A. Wretlind, *Acta Pharmacol. Toxicol.* **18**, 141 (1961).

USE: In the production of hydrotropic salts (hydrotropic salts form aq solns which dissolve sparingly sol substances to a greater extent than water); in the manuf of lacquers, plastics. *Caution:* Strong irritant.

7014. Pelargonidin. *3,5,7-Trihydroxy-2-(4-hydroxyphenyl)-1-benzopyrylium chloride; 3,4',5,7-tetrahydroxyflavylium chloride; 3,4',5,7-tetrahydroxy-2-phenylbenzopyrylium chloride.* C$_{15}$H$_{11}$ClO$_5$; mol wt 306.70. C 58.74%, H 3.62%, Cl 11.56%, O 26.08%. The aglucone of pelargonin: Willstätter, Bolton, *Ann.* **408**, 42 (1914). Synthesis: Malkin, Robinson, *J. Chem. Soc.* **127**, 1190 (1925); Robertson *et al.*, **1928**, 1533. Prepn from kaempferol: Mirza, Robinson, *Nature* **166**, 997 (1950); King, White, *J. Chem. Soc.* **1957**, 3901.

Reddish-brown prisms from 2% HCl or from alcoholic HCl. Not melted at 350°. Absorption max (ethanol + 0.01% HCl): 530 nm (ε 32,000). Sol in alcohol, methanol; moderately sol in water; slightly sol in chloroform.

3,5-Diglucoside, C$_{27}$H$_{31}$ClO$_{15}$, *3,5-bis(β-D-glucopyranosyloxy)-7-hydroxy-2-(4-hydroxyphenyl)-1-benzopyrylium chloride, pelargonin, monardin, salvinin, punicin.* From flowers of *Pelargonium zonale* Ait. var. meteor, *Geraniaceae:* Willstätter, Bolton, *loc. cit.;* from scarlet roses: Harborne, *Experientia* **17**, 72 (1961). Identity with monardin and salvinin: Robinson, Todd, *J. Chem. Soc.* **1932**, 2488. Identity with punicin: Karrer, Widmer, *Helv. Chim. Acta* **10**, 67 (1927). Structure: Leon *et al.*, *J. Chem. Soc.* **1931**, 2672. Red needles with green luster from methanol + HCl, dec 175-180°. [α]$_D$ −291°. Absorption max (methanol + HCl): 269, 505 nm. Sol in water, alc.

3-Glucoside, C$_{21}$H$_{21}$ClO$_{10}$, *3-(β-D-glucopyranosyloxy)-5,7-dihydroxy-2-(4-hydroxyphenyl)-1-benzopyrylium chloride, callistephin.* From purple-red aster (*Callistephus chinensis* (L.) Nees, *Compositae):* Willstätter, Burdick, *Ann.* **412**, 149 (1916); from strawberries: Sondheimer, Kertesz, *J. Am. Chem. Soc.* **70**, 3476 (1948). Structure and synthesis: Robertson, Robinson, *J. Chem. Soc.* **1928**, 1460. Dark brownish-red needles with bronze luster. Absorption max (ethanol + HCl): 515 nm (ε 13,000). Sol in water, methanol, ethanol, 0.5-7% aq HCl; moderately sol in 10% HCl.

7015. Pelletierine. *1-(2-Piperidinyl)-2-propanone; isopelletierine; 2-acetonylpiperidine; punicine.* C$_8$H$_{15}$NO; mol wt 141.22. C 68.04%, H 10.71%, N 9.92%, O 11.33%. From the rootbark of the pomegranate tree, *Punica granatum* L., *Punicaceae.* Isoln: Tanret, *Compt. Rend.* **86**, 1270 (1878). Structure: Gilman, Marion, *Bull. Soc. Chim. France* **1961**, 1993; Drillien, Viel, *ibid.* **1963**, 2393. Synthesis: Anet *et al.*, *Nature* **164**, 501 (1949); Beyerman, Maat, *Rec. Trav. Chim.* **82**, 1033 (1963); **84**, 385 (1965).

(±)-Form, slightly colored oily liq; bp 195°; bp$_{11}$ 102-107°. d^{20} 0.988. Sol in alc, ether, chloroform. One gram dissolves in 20 ml water. Should be stored under nitrogen. LD i.v. in rabbits: 40 mg/kg.

Hydrochloride, C$_8$H$_{16}$ClNO, crystals from alcohol + ether, mp 145°. Soluble in water and alcohol.

Sulfate, crystals from methanol + ether, mp 135-138°. Active form [α]$_D^{25}$ −29.5° (c = 1.05).

Methyl deriv, C$_9$H$_{17}$NO, *N-methylpelletierine.* Oily liquid. d$_4^{20}$ 0.948. bp$_{13}$ 96-98°. n$_D^{20}$ 1.46737. Sol in water, petr ether.

(−)-Form, crystals from aq ethanol, mp 130-132°. [α]$_D^{27}$ −7.5° (c = 0.88 in ethanol).

(+)-Form, crystals from aq ethanol, mp 130-131°. [α]$_D^{20}$ +11.9° (c = 0.85 in ethanol).

THERAP CAT: Anthelmintic (Cestodes).

THERAP CAT (VET): *See* Pelletierine Tannate.

7016. Pelletierine Tannate. Punicine tannate. A mixture of the tannates of the several alkaloids from pomegranate, pelletierine, methyl- and pseudopelletierine. It contains an amount of the alkaloids equivalent to not less than 20% as the hydrochloride.

Light yellow, amorphous powder; astringent taste. Sol in about 250 parts water; in alcohol, in warm dil acids; slightly in ether. Almost insol in chloroform. A satd aq soln is acid to litmus. *Protect from light.*

THERAP CAT: Anthelmintic (Cestodes).

THERAP CAT (VET): Has been used as a teniacide.

7017. Pellitorine. *(E,E)-N-(2-Methylpropyl)-2,4-decadienamide; E,E-N-isobutyl-2,4-decadienamide.* C$_{14}$H$_{25}$NO; mol wt 223.36. C 75.28%, H 11.28%, N 6.27%, O 7.16%. Pungent principle isolated from *Anacyclus pyrethrum* DC., *Compositae:* J. M. Gulland, G. U. Hopton, *J. Chem. Soc.* **1930**, 6. Structure: M. Jacobson, *J. Am. Chem. Soc.* **71**, 366 (1949). Synthesis: *idem, ibid.* **75**, 2584 (1953); L. Crombie, *J. Chem. Soc.* **1955**, 1007; J. Tsuji *et al., Tetrahedron Letters* **1977**, 1917; J. Nokami *et al., ibid.* **21**, 4455 (1980); T. Mandai *et al., Chemistry Letters* **1980**, 313. Stereoselective synthesis: R. Bloch, D. Hassangonzales, *Tetrahedron* **42**, 4975 (1986). Insecticidal activity: R. T. Lalonde *et al., J. Chem. Ecol.* **6**, 35 (1980).

Needles from petr ether, mp 90°. uv max (abs ethanol): 258 nm (E$_{1cm}^{1\%}$ 1330). Sol in organic solvents; sparingly sol in water. Practically insol in dil acid or alkalies.

7018. Pellotine. *1,2,3,4-Tetrahydro-6,7-dimethoxy-1,2-dimethyl-8-isoquinolinol; N-methylanhalonidine; 8-hydroxy-6,7-dimethoxy-1,2-dimethyl-1,2,3,4-tetrahydroisoquinoline.* C$_{13}$H$_{19}$NO$_3$; mol wt 237.29. C 65.80%, H 8.07%, N 5.90%, O 20.23%. From the mescal buttons (pellote) of *Lophophora williamsii* (Lemaire) Coult., *Cactaceae.* Structure: Späth *et al., Ber.* **65**, 1771 (1932). Synthesis: Brossi *et al., Helv. Chim. Acta* **47**, 2089 (1964); **49**, 403 (1966); Takido *et al., J. Pharm. Sci.* **59**, 271 (1970). Biosynthetic studies: Battersby *et al., Tetrahedron Letters* **1967**, 563; **1968**, 6111.

Plates from petr ether, mp 112°. Alkaline reaction. Freely sol in alc, acetone, ether, chloroform; sparingly sol in water.

Hydrochloride, C$_{13}$H$_{19}$NO$_3$·HCl, prisms, freely sol in water; sparingly in alcohol.

Hydriodide, C$_{13}$H$_{19}$NO$_3$·HI, prisms, mp 130°, sol in water, alcohol, almost insol in ether.

7019. α-Peltatin. *[5R]-5,8,8aα,9-Tetrahydro-10-hydroxy-5α-(4-hydroxy-3,5-dimethoxyphenyl)furo[3',4':6,7]-naphtho[2,3-d]-1,3-dioxol-6(5aH)-one; 8-hydroxy-2-hydroxymethyl-6,7-methylenedioxy-4-(4'-hydroxy-3',5'-dimethoxyphenyl)-1,2,3,4-tetrahydronaphthalene-3-carboxylic acid lactone.* C$_{21}$H$_{20}$O$_8$; mol wt 400.37. C 62.99%, H 5.04%, O 31.97%. Exists as a glucoside in the rhizomes of *Podophyllum peltatum* L., *Berberidaceae:* von Wartburg *et al., Helv. Chim. Acta* **40**, 1331 (1957). Isoln from resin podophyllum: Hartwell, Detty, *J. Am. Chem. Soc.* **72**, 246 (1950).

Prismatic leaflets from abs ethanol. *Irritates the skin.* Begins to sinter at 236°. Dec 242-246°. $[\alpha]_D^{20}$ − 124.8° (c = 0.5 in chloroform). Soly in water at 20° about 30 mg/liter. Fairly sol in chloroform, hot ethanol, acetic acid, acetone, dilute caustic; less sol in benzene, ether, carbon tetrachloride, propylene glycol. Practically insol in petr ether.

α-Peltatin-β-D-glucoside, $C_{27}H_{30}O_{13}$, prismatic needles from acetone, mp 168-171°. $[\alpha]_D^{20}$ − 128.9° (c = 0.590 in methanol); $[\alpha]_D^{20}$ − 174.4° (c = 0.579 in pyridine).

7020. β-Peltatin. *(5R)-5α,8,8aα,9-Tetrahydro-10-hydroxy-5-(3,4,5-trimethoxyphenyl)furo[3',4':6,7]naphtho-[2,3-d]-1,3-dioxol-6(5aβH)-one; 8-hydroxy-2-hydroxy-methyl-6,7-methylenedioxy-4-(3',4',5'-trimethoxyphen-yl)-1,2,3,4-tetrahydronaphthalene-3-carboxylic acid lactone;* β-peltatin A. $C_{22}H_{22}O_8$; mol wt 414.40. C 63.76%, H 5.35%, O 30.89%. Exists as a glucoside in the rhizomes of *Podophyllum peltatum* L., *Berberidaceae:* von Wartburg *et al., Helv. Chim. Acta* **40**, 1331 (1957). Isoln from resin podophyllum: Hartwell, Detty, *J. Am. Chem. Soc.* **70**, 2833 (1948); **72**, 246 (1950); from *Hyptis verticillata* Jacq., *Labiatae:* German, *J. Pharm. Sci.* **60**, 649 (1971).

Prisms from abs ethanol. *Irritates the skin.* Dec 238-241° (slight sintering begins at 234°). $[\alpha]_D^{20}$ − 122.9° (c = 0.578 in chloroform). Somewhat less sol than α-peltatin. Soly in water at 23°: 13 mg/liter. Fairly sol in chloroform, hot ethanol, acetic acid, acetone, dil caustic; less sol in benzene, ether, carbon tetrachloride, propylene glycol. Practically insol in petr ether.

β-Peltatin-β-D-glucoside, $C_{28}H_{32}O_{13}$, white amorph powder from acetone + ether, dec 156-159°. $[\alpha]_D^{20}$ − 122.7° (c = 0.587 in methanol); $[\alpha]_D^{20}$ − 169.2° (c = 0.556 in pyridine).

7021. Pemoline. *2-Amino-5-phenyl-4(5H)-oxazolone;* phenoxazole; phenylisohydantoin; phenylpseudohydantoin; azoxodone; PIO; LA 956; Yh 1; Azoksodon; Cylert; Dantromin; Deltamine; Endolin; Hyton; Kethamed; Nitan; Notair; Pioxol; Pondex; Ronyl; Sigmadyn; Sistral; Sofro; Tradon; Volital. $C_9H_8N_2O_2$; mol wt 176.17. C 61.36%, H 4.58%, N 15.90%, O 18.16%. Prepn: W. Traube, R. Ascher, *Ber.* **46**, 2077 (1913); L. Schmidt, H. Scheffler, U.S. pat. **2,892,753** (1959 to Boehringer, Ing.). Prepn magnesium hydroxide mixture, originally thought to be a chelate complex: W. E. Lange *et al., J. Pharm. Sci.* **51**, 477 (1962); B. H. Candon, M. Chessin, U.S. pat. **3,108,045** (1963 to Purdue Frederick). CNS stimulant activity: L. Schmidt, *Arzneimittel-Forsch.* **6**, 423 (1956). Efficacy in minimal brain dysfunction in hyperkinetic children: C. K.Connors *et al., Psychopharmacologia* **26**, 321 (1972); J. G. Page *et al., J. Learn. Dis.* **7**, 498 (1974). Toxicity data: E. W. Schafer, *Toxicol. Appl. Pharmacol.* **21**, 315 (1972). *Review:* A. T. Dren, R. S.

Janicki, *in Pharmacological and Biochemical Properties of Drug Substances* **vol. 1,** M. E. Goldberg, Ed. (Am. Pharm. Assoc., Washington, DC, 1977) pp 33-65.

Crystals, mp 256-257° (dec). Practically insol in water, ether, acetone, dil hydrochloric acid. Sol in propylene glycol (1%); in hot alcohol. LD_{50} orally in rats: 500 mg/kg (Schafer).

Magnesium hydroxide mixture, *magnesium pemoline, Abbott 30400, Ecylert, Tamilan.*

Ingredient of *Stimul.*

Note: This is a controlled substance (stimulant) listed in the U.S. Code of Federal Regulations, Title 21 Part 1308.14 (1987).

THERAP CAT: CNS stimulant.

7022. Pempidine. *1,2,2,6,6-Pentamethylpiperidine.* $C_{10}H_{21}N$; mol wt 155.28. C 77.35%, H 13.63%, N 9.02%. Prepn by methylation of 2,2,6,6-tetramethylpiperidine: Leonard, Hauck, *J. Am. Chem. Soc.* **79**, 5289 (1957); Hall, *ibid.* 5447. Description: Spinks, Young, *Nature* **181**, 1397 (1958); Lee *et al., ibid.* 1717. Ganglion blocking agent.

Liquid. pK at 30° = 11.25. Extremely basic reaction for a tertiary amine. bp_{760} 147°. n_D^{21} 1.4550.

p-Toluenesulfonate, $C_{17}H_{19}NO_3S$, crystals from ethyl acetate + ethanol, mp 162-163°.

Tartrate (bitartrate, acid tartrate, hydrogen tartrate), $C_{14}H_{27}NO_6$, *M & B 4486, Pempidil, Pempiten, Perolysen, Tenormal, Tensinol, Tensoral, Viotil.* Crystals, mp 160°. Sol in alcohol, moderately sol in water.

Hydrochloride, $C_{10}H_{21}N \cdot HCl$, crystals, sol in water, alc.

THERAP CAT: Tartrate as antihypertensive.

7023. Penaldic Acids. A term used to describe *N*-acyl-aminomalonaldehydic acids which are found in the degradation products of the penicillins.

These acids rapidly lose carbon dioxide, hence are unstable in the free state. Benzylpenaldic acid, better known as (2-phenylacetamido)malonaldehydic acid (R is −$CH_2C_6H_5$) is a degradation product of penicillin G, and has been isolated in one instance as its *N*-benzylamide 2,4-dinitrophenyl-hydrazone. Review and bibliography: E. Chain in *Antibiotics* **vol. 2,** H. W. Florey *et al.,* Eds. (Oxford University Press, New York, 1949) pp 829-839.

7024. Penamecillin. *3,3-Dimethyl-7-oxo-6-[(phenylacetyl)amino]-4-thia-1-azabicyclo[3.2.0]heptane-2-carboxylic acid (acetyloxy)methyl ester; penicillin G hydroxymethyl ester acetate;* acetoxymethyl benzylpenicillinate; benzylpenicillin acetoxymethyl ester; Wy-20788; Havapen. $C_{19}H_{22}N_2O_6S$; mol wt 406.48. C 56.14%, H 5.46%, N 6.89%, O 23.62%, S 7.89%. Semi-synthetic antibiotic related to penicillin. Prepn: Jansen, Russell, *J. Chem. Soc.* **1965**, 2127; *eidem,* **Brit.** pat. **1,003,479** corresp to U.S. pat. **3,250,679** (1965, 1966 to John Wyeth & Brother).

Crystals from isopropanol + ethanol, mp 106-108°. [α]$_D^{20}$ +154°. Not inactivated by gastric acid.

THERAP CAT: Antibacterial.

7025. Penbutolol. (S)-1-(2-Cyclopentylphenoxy)-3-[(1,1-dimethylethyl)amino]-2-propanol; (S)-1-(tert-butyl-amino)-3-(o-cyclopentylphenoxy)-2-propanol; (−)-1-tert-butylamino-2-hydroxy-3-(2'-cyclopentylphenoxy)propane. C$_{18}$H$_{29}$NO$_2$; mol wt 291.44. C 74.18%, H 10.03%, N 4.81%, O 10.98%. β-Adrenergic blocker. Prepn: H. Ruschig et al., S. Afr. pat. 68 07,915; eidem, U.S. pat. 3,551,493 (1969, 1970 both to Hoechst). Preliminary chemistry and pharmacology: G. Härtfelder et al., Arzneimittel-Forsch. 22, 930 (1972). Physicochemical and analytical study: P. Hajdu, D. Damm, ibid. 29, 602 (1979). Pharmacology: J. Kaiser et al., ibid. 30, 420 (1980). Action specificity: J. Kaiser ibid. 427. Crystallographic study: J. M. Leger et al., Mol. Pharmacol. 17, 339 (1980). Hemodynamic effects: P. Lund-Johansen, Eur. J. Clin. Pharmacol. 16, 149 (1979). Clinical study in hypertension: J. L. Cangiano et al., J. Clin. Pharmacol. 19, 384 (1979). HPLC determn in plasma: R. K. Bhamra et al., Biomed. Chromatog. 1, 140 (1986). Review of pharmacology and therapeutic efficacy: R. C. Heel et al., Drugs 22, 1-25 (1981).

Crystals, mp 68-72°. [α]$_D^{20}$ −11.5° (c = 1 in methanol). pKa 9.3 (1.5 mmol/1 in 25% ethanol). Sol in methanol, ethanol, chloroform.

Sulfate, C$_{36}$H$_{60}$N$_2$O$_8$S, HOE 893d, HOE 39-893d, Betapressin, Levatol, Paginol. Crystals, mp 216-218° (dec). [α]$_D^{20}$ −24.6° (c = 1 in methanol).

THERAP CAT: Antihypertensive; antianginal; anti-arrhythmic.

7026. Pendimethalin. N-(1-Ethylpropyl)-3,4-dimethyl-2,6-dinitrobenzenamine; N-(1-ethylpropyl)-2,6-dinitro-3,4-xylidine; N-(1-ethylpropyl)-3,4-dimethyl-2,6-dinitroaniline; penoxalin; AC 92553; Prowl; Herbadox; Stomp. C$_{13}$H$_{19}$N$_3$-O$_4$; mol wt 281.31. C 55.50%, H 6.81%, N 14.94%, O 22.75%. Pre-emergence and pre-planting herbicide. Prepn: R. H. Kupelian, Ger. pat. 2,232,263, C.A. 78, 84010z (1973) and A. W. Lutz, R. E. Diehl, Ger. pat. 2,241,408, C.A. 78, 135858s (1973) (both 1973 to Am. Cyanamid). Activity: H. A. Roberts, W. Bond, Proc. Brit. Weed Control Conf., 12th 1, 387 (1974). Persistence in soil: A. Walker, W. Bond, Pestic. Sci. 8, 359 (1977).

Orange-yellow cryst solid, mp 56-57°. Vapor press at 25°: 3 × 10^{-5} mmHg. Soly in water at 20°: 0.3 mg/l. Sol in most organic solvents.

USE: Herbicide.

7027. Penethamate Hydriodide. 3,3-Dimethyl-7-oxo-6-[(phenylacetyl)amino]-4-thia-1-azabicyclo[3.2.0]heptane-2-carboxylic acid 2-(diethylamino)ethyl ester monohydriodide; penicillin G 2-diethylaminoethyl ester hydriodide; benzylpenicillin β-diethylaminoethyl ester hydriodide; β-diethylaminoethyl benzylpenicillinate hydriodide; ephicillin hydr-

iodide; penethecillin; Alivin; Broncopen; Estopen; Neopenil; Pulmaxil N; Bronchocillin; Pulmo 500; Leocillin. C$_{22}$H$_{32}$-IN$_3$O$_4$S; mol wt 561.50. C 47.06%, H 5.75%, I 22.60%, N 7.48%, O 11.40%, S 5.71%. Refs: Jensen et al., Ugeskrift for Laeger 112, 1043, 1075 (1950); Acta Path. Microbiol. Scand. 28, 407 (1951); Woodard, J. Pharm. Pharmacol. 4, 1019 (1952); Ebelhare, Ferlauto, Antibiot. & Chemother. 3, 873 (1953); Span. pats. 206,653-5 (1953 to Antibiotics S.A.); Villax, Brit. pat. 759,603 (1956). Residue studies in ruminants: J. F. M. Nouws, G. Ziv, Zentralbl. Veterinaermed. A 25, 312 (1978); J. F. M. Nouws et al., Arch. Lebensmittelhyg. 30, 4 (1979), C.A. 91, 37554p (1979).

Crystals, mp 178-179°. Theoretical potency 1058 I.U./mg. Slightly soluble in water (0.96% at 20°). pH of satd aq soln 4.5-5.2. In dry form the compd has a shelf life of at least 2 years. Aq solns are unstable, the ester is hydrolyzed to free penicillin and diethylaminoethanol, the velocity of the reaction increasing with rise in temperature and pH. If stored in a cool place (below 25°) sterile aq suspensions retain their potency for a max of 7 days and for 30 days in a refrigerator at 4°.

THERAP CAT: Antibacterial.

THERAP CAT (VET): Antibacterial.

7028. Penfluridol. 1-[4,4-Bis(4-fluorophenyl)butyl]-4-[4-chloro-3-(trifluoromethyl)phenyl]-4-piperidinol; 1-[4,4-bis(p-fluorophenyl)butyl]-4-(4-chloro-α,α,α-trifluoro-m-tolyl)-4-piperidinol; 1-(4,4-bis(4-fluorophenyl)butyl)-4-hydroxy-4-(3-trifluoromethyl-4-chlorophenyl)piperidine; R-16341; Semap. C$_{28}$H$_{27}$ClF$_5$NO; mol wt 523.99. C 64.18%, H 5.19%, Cl 6.77%, F 18.13%, N 2.67%, O 3.05%. Prepn: Hermans, Niemegeers, Ger. pat. 2,040,231 corresp to U.S. pat. 3,575,990 (both 1971 to Janssen); Sindelár et al., Coll. Czech. Chem. Commun. 38, 3879 (1973). Pharmacology and toxicology: Janssen et al., Eur. J. Pharmacol. 11, 139 (1970). Crystal structure: Koch, Acta Crystallogr. 29B, 1538 (1973).

White, microcrystals, mp 105-107°. Slightly sol in water, dil HCl (< 0.5 mg/ml). LD$_{50}$ orally in mice (day 7): 86.8 mg/kg, Janssen et al., loc. cit.

THERAP CAT: Antipsychotic.

7029. Penicillamine. 3-Mercapto-D-valine; β,β-dimethylcysteine; α-amino-β-methyl-β-mercaptobutyric acid; DMC; β-thiovaline; Cuprenil; Cuprimine; Depamine; Depen; Emtexate; Mercaptyl; Pendramine; Perdolat; Trolovol. C$_5$H$_{11}$NO$_2$S; mol wt 149.21. C 40.25%, H 7.43%, N 9.39%, O 21.45%, S 21.49%. The most characteristic degradn product of penicillin type antibiotics. Prepn by hydrolysis of penicillins: E. P. Abraham et al., Nature 151, 107 (1943). Review of syntheses of racemic penicillamine and enantiomers: H. M. Crooks in The Chemistry of Penicillin, H. T. Clarke et al., Eds. (Princeton Univ. Press, 1949) pp 455-472; W. M. Weigert et al., Angew. Chem. Int. Ed. 14, 330 (1975). Polymorphism of D-form: J. A. G. Vidler, J. Pharm. Pharmacol. 28, 662 (1976). Pharmacokinetics and metabolism of the D- and L-forms: Ruiz-Torres, Arzneimittel-Forsch. 24, 914 (1974). In treatment of rheumatoid arthritis: P. Davis, Clin. & Invest. Med. 7, 41 (1983); of Wilson's Disease: A. Deiss, Ann. Int. Med. 99, 398 (1983); of progressive systemic sclerosis: B. Kang et al., J. Allergy Clin. Immun. 69, 297 (1982). Evaluation in fibrotic lung disease: R. Chapela et al., *Int. J. Clin. Pharmacol. Ther. Tox.* **24**, 16 (1986). Review of assay methods: N. Kucharczyk, S. Shahiniam, *J. Rheumatol.* **8**, Suppl. 7, 28-84 (1981); of metabolism and therapeutic applications: J. C. Crawhall *et al., Biopharm. & Drug Dispos.* **1**, 73-95 (1979); of side effects: R. S. Levy *et al., J. Am. Acad. Dermatol.* **8**, 548-558 (1983). Toxicity data: Veis *et al., Antibiotiki* **14**, 837 (1969). General reviews: I. A. Jaffe in *Pharmacological and Biochemical Properties of Drug Substances* **Vol. 2**, M. E. Goldberg, Ed. (Am. Pharm. Assoc., Washington, DC, 1979) pp 465-478; C. C. Chiu, L. T. Grady in *Analytical Profiles of Drug Substances* **Vol. 10**, K. Florey, Ed. (Academic Press, New York, 1981) pp 601-637.

SH NH$_2$
| |
(CH$_3$)$_2$C—CHCOOH

mp 202-206° (Weigert *et al.*). $[\alpha]_D^{25}$ −63° (c = 0.1 in pyridine). LD$_{50}$ in rats (mg/kg): >10000 orally, >660 i.p. (Jaffe).

Hydrochloride, C$_5$H$_{12}$ClNO$_2$S, *Distamine, Metalcaptase.* Hygroscopic crystals, dec 177.5°. $[\alpha]_D^{25}$ −63° (1N NaOH). Freely sol in water, sol in ethanol. Aq solns are comparatively stable at pH 2-4. LD$_{50}$ i.v. in mice: 2289 mg/kg (Veis).

DL-Form, crystals, dec 201°. pK values: carboxyl 1.8; α-amino 7.9; β-thiol 10.5. LD$_{50}$ orally in rats: 365 mg/kg (Jaffe).

DL-Form hydrochloride, crystals, dec 145-148°.

L-Form, crystals, mp 190-194°. $[\alpha]_D^{25}$ +63° (in N NaOH). LD$_{50}$ i.p. in rats: 350 mg/kg (Jaffe).

USE: L-Form as exptl inducer of convulsions in rats, *C.A.* **84**, 159420t (1976).

THERAP CAT: Chelating agent (copper) in Wilson's disease. Antirheumatic.

7030. Penicillamine Cysteine Disulfide. *3-[(2-Amino-2-carboxyethyl)dithio]-D-valine; 3,3-dimethyl-3-3'-dithiodialanine;* 1,6-diamino-5-5-dimethyl-3,4-dithiahexane-1,6-dicarboxylic acid. C$_8$H$_{16}$N$_2$O$_4$S$_2$; mol wt 268.37. C 35.81%, H 6.01%, N 10.44%, O 23.85%, S 23.90%. Prepn: Tabachnick *et al., Nature* **174**, 701 (1954); Schöberl *et al., Angew. Chem.* **68**, 213 (1956); Schöberl, Grafje, *Ann.* **617**, 71 (1958); Levine, *Nature* **187**, 940 (1960).

NH$_2$
|
SCH$_2$CHCOOH
|
S   NH$_2$
| |
(CH$_3$)$_2$C—CHCOOH

Crystals, mp 195°.

7031. Penicillamine Disulfide. *3,3,3',3'-Tetramethylcystine; 3,3'-dithiodivaline;* 3,3'-dithiobis[valine]. C$_{10}$H$_{20}$N$_2$O$_4$S$_2$; mol wt 296.42. C 40.52%, H 6.80%, N 9.45%, O 21.59%, S 21.63%. Prepn: Süs, *Ann.* **561**, 31 (1948); Berg *et al.,* **Brit.** pat. **621,915** (1949 to Merck & Co.); Butenandt *et al., Z. Physiol. Chem.* **285**, 238 (1950).

(CH$_3$)$_2$C—CHCOOH
| |
S   NH$_2$
|
S   NH$_2$
| |
(CH$_3$)$_2$C—CHCOOH

DL-Form, crystals, mp 181-183°.

D-Form, crystals, mp 204-205°. $[\alpha]_D^{23}$ +27° (c = 1.46 in 1N HCl).

L-Form, crystals, mp 207°. $[\alpha]_D^{22}$ −26° (1N HCl).

7032. Penicillanic Acid. *3,3-Dimethyl-7-oxo-4-thia-1-azabicyclo[3.2.0]heptane-2-carboxylic acid.* A building block of penicillin, devoid of significant antibacterial activity: Sheehan *et al., J. Am. Chem. Soc.* **75**, 3292 (1953); **81**, 5838 (1959). Separation from its esters and other penicillins by

gas chromatog: Evard *et al., Nature* **201**, 1124 (1964). *See also* 6-Aminopenicillanic Acid.

7033. Penicillic Acid. *3-Methoxy-5-methyl-4-oxo-2,5-hexadienoic acid;* γ-keto-β-methoxy-δ-methylene-Δ$^\alpha$-hexenoic acid. C$_8$H$_{10}$O$_4$; mol wt 170.16. C 56.46%, H 5.93%, O 37.61%. Antibiotic substance produced by the following fungi: *Penicillium puberulum, P. cyclopium, P. thomii, P. suaveolens, P. baarnense, Aspergillus ochraceus, A. melleus.* Isoln: Alsberg, Black, *U.S. Dept. Agric. Bur. Plant Industry,* bull. no. **270**, (1913); Birkinshaw *et al., Biochem. J.* **30**, 394 (1936); Oxford *et al., Chem. & Ind. (London)* **20**, 22 (1942); Karow *et al., Arch. Biochem.* **5**, 279 (1944); Burton, *Nature* **165**, 274 (1950); Natori *et al., Chem. Pharm. Bull.* **18**, 2259 (1970). Acid in tautomeric equilibrium with its lactone. Structure: Birkinshaw *et al., loc. cit.* Physical properties: Kovac, Solcaniova, *Tetrahedron* **25**, 3617 (1969). Synthesis: Raphael, *Nature* **160**, 261 (1947); *J. Chem. Soc.* **1948**, 1508; C. L. Yeh *et al., Tetrahedron Letters* **1978**, 3987. Biosynthesis: Birch *et al., J. Chem. Soc.* **1958**, 4582; Bentley, Keil, *J. Biol. Chem.* **237**, 867 (1962). Has carcinogenic activity attributable to α,β-unsaturation together with an external conjugated double bond attached to the 4 position of the γ-lactone ring: Dickens, *Brit. Med. Bull.* **20**, 96 (1964). Review and evaluation of studies of carcinogenic action in laboratory animals: *IARC Monographs* **10**, 211-216 (1976). Activity studies: Suzuki *et al., Agr. Biol. Chem.* **35**, 287 (1971). *Review:* Ciegler *et al.,* "Patulin, Penicillic Acid and Other Carcinogenic Lactones" in *Microbial Toxins,* **vol. VI**, A. Ciegler *et al.,* Eds. (Academic Press, New York, 1971) p 414.

Needles from petr ether, mp 83-84°. uv max: about 220 nm. Acid reaction, turns Congo red paper blue. Moderately sol in cold water (2 g/100 ml); freely sol in hot water, alcohol, ether, benzene, chloroform; slightly sol in hot petr ether. Practically insol in pentane-hexane. LD$_{50}$ s.c. in mice: 100 mg/kg.

Monohydrate, large transparent monoclinic or triclinic, rhombic crystals from water, mp 58-64°.

7034. Penicillinase. β-Lactamase; Neutrapen. Mol wt about 50,000. Enzymes found in many bacteria which destroy penicillins and cephalosporins by catalyzing the hydrolysis of the amide bond in the β-lactam ring. Good penicillinase producers are *Bact. coli,* the *Bacillus subtilismesentericus* group, *Bacillus anthracis* and *Staphylococci.* Both intra- and extracellular penicillinase are of protein nature. There are probably as many different penicillinases as there are bacteria producing them. Ion-exchange procedures for the purification of penicillinase: Puetzer, Boschetti, U.S. pat. **2,982,696** (1961 to Schenley). Amino acid sequence studies: Ambler, Meadway, *Nature* **222**, 24 (1969). *Reviews:* Chain *et al.* in *Antibiotics* **vol. 2**, Flory *et al.,* Eds. (Oxford, 1949) p 1090; Rothe, *Pharmazie* **5**, 25 (1950); Citri, Pollock, *Advan. Enzymol.* **28**, 237 (1966); Citri in *The Enzymes,* P. D. Boyer, Ed. (Academic Press, New York, 3rd ed., 1971) pp 23-46.

USE: In culture media to antagonize antibacterial activity of penicillin.

THERAP CAT: Enzyme (penicillin inactivating).

7035. Penicillin BT. *6-[[(Butylthio)acetyl]amino]-3,3-dimethyl-7-oxo-4-thia-1-azabicyclo[3.2.0]heptane-2-carboxylic acid;* butylmercaptomethylpenicillin; butylthiomethylpenicillin. C$_{14}$H$_{22}$N$_2$O$_4$S$_2$; mol wt 346.48. C 48.53%, H 6.40%, N 8.09%, O 18.47%, S 18.51%. Biosynthesis: Behrens *et al.*, *J. Biol. Chem.* **175**, 793 (1948); Rhodehamel; Behrens *et al.*, U.S. pats. **2,528,175** and **2,623,876** (1951, 1952, both to Lilly); Umezawa, *C.A.* **45**, 9128a (1951); Suami *et al.*, *C.A.* **48**, 3631d (1954).

Procaine salt, $C_{27}H_{42}N_4O_6S_2$, crystals from ethanol, mp 110°.

7036. Penicillin G Benethamine. *3,3-Dimethyl-7-oxo-6-[(phenylacetyl)amino]-4-thia-1-azabicyclo[3.2.0]heptane-2-carboxylic acid compd with N-(phenylmethyl)benzeneethanamine (1:1);* benzylpenicillinic acid *N-benzyl-β-phenylethylamine salt; N-benzyl-2-phenylethylamine salt of benzylpenicillin;* benethamine penicillin G; Benapen; Betapen; Benetolin. $C_{31}H_{35}N_3O_4S$; mol wt 545.71. C 68.23%, H 6.46%, N 7.70%, O 11.73%, S 5.88%. Semi-synthetic antibiotic prepd from penicillin G and *N*-benzylphenethylamine: Jansen, Hems, **Brit.** pat. **732,559** (1955 to Glaxo). Clinical studies: Nelson *et al.*, *Brit. Med. J.* **1954**, II, 339; Boger *et al.*, *Antibiot. Ann.* **1954-55**, 123.

Crystals, mp 146-147°. Slight, characteristic, amine taste. Potency ~1100 units/mg (as compared to 1670 units/mg of penicillin G sodium). Very slightly sol in water (0.1 w/v at 40°). Solutions of 1000 units/ml approach the limits of aq soly. This degree of soly may be compared with that of procaine penicillin, approx 4000 units/ml and that of benzathine penicillin, approx 200 units/ml.

THERAP CAT: Antibacterial.
THERAP CAT (VET): Antibacterial.

7037. Penicillin G Benzathine. *[2S-(2α,5α,6β)]-3,3-Dimethyl-7-oxo-6-[(phenylacetyl)amino]-4-thia-1-azabicyclo-[3.2.0]heptane-2-carboxylic acid compd with N,N'-bis(phenylmethyl)-1,2-ethanediamine; penicillin G compound with N,N'-dibenzylethylenediamine (2:1); N,N'-dibenzylethylenediamine bis[benzylpenicillin]; dibenzylethylenediamine dipenicillin G;* penicillin G salt of *N,N'-dibenzylethylenediamine;* benzethacil; benzathine penicillin G; diamine penicillin; Beacillin; Megacillin suspension; Bicillin; Cillenta; Permapen; Duropenin; Dibencillin; DBED Penicillin; Penidural; Tardocillin; Dibencil; Lentopenil; Vicin Neolin; Pen-Di-Ben; Penidure; Moldamin; Extencilline; Longacilina; Longicil; Penadur; Penditan; Cepacilina. $C_{48}H_{56}N_6O_8S_2$; mol wt 909.11. C 63.41%, H 6.21%, N 9.25%, O 14.08%, S 7.05%. Semi-synthetic antibiotic prepd by adding an aq soln of one mole *N,N'*-dibenzylethylenediamine diacetate to an aq soln of 2 moles sodium penicillin G: Szabo *et al.*, *Antibiot. & Chemother.* **1**, 499 (1951); Szabo, Bruce, U.S. pat. **2,627,491** (1953 to Wyeth). Comprehensive description: F. Kreuzig in *Analytical Profiles of Drug Substances* vol. **11**, K. Florey, Ed. (Academic Press, New York, 1982) pp 463-482.

Hydrated, tasteless crystals. mp 123-124°. $[\alpha]_D^{25}$ +206° (c = 0.105 in formamide). Activity not less than 1050 units/mg (theory: 1307 units/mg). Soly (mg/ml): water 0.15 (about 200 units/ml); benzene 0.38; alc 5.2; acetone 1.5; formamide 28.0. pH of satd aq soln about 6. Solubilities determined by Weiss *et al.*, *Antibiot. & Chemother.* **7**, 374 (1957) in mg/ml: water 0.315; methanol 16.9; ethanol 15.4.

THERAP CAT: Antibacterial.
THERAP CAT (VET): Antibacterial.

7038. Penicillin G Benzhydrylamine. *[2S-(2α,5α,6β)]-3,3-Dimethyl-7-oxo-6-[(phenylacetyl)amino]-4-thia-1-azabicyclo[3.2.0]heptane-2-carboxylic acid compd with α-phenylbenzenemethanamine (1:1); 1,1-diphenylmethylamine benzylpenicillinate;* benzylpenicillin benzhydrylamine; benzhydrylamine penicillin; penicillin aminodiphenylmethane salt;

benzylpenicillinic acid benzhydrylamine salt; Orencil; Penidryl. $C_{29}H_{31}N_3O_4S$; mol wt 517.63. C 67.29%, H 6.04%, N 8.12%, O 12.36%, S 6.20%. $C_{16}H_{18}N_2O_4S \cdot C_{13}H_{13}N$. Semisynthetic antibiotic. Prepd from potassium penicillin G and benzhydrylamine HCl in dil acetone: Hagemann *et al.*, *Ann. Pharm. Franc.* **12**, 565 (1954); Goldsmith *et al.*, *Antibiot. & Chemother.* **5**, 648 (1955); Penau, Hagemann, U.S. pat. **2,710,863** (1955 to Labs. Français de Chimiothérapie).

Crystals from dil acetone, mp 159°. $[\alpha]_D^{20}$ +206° (c = 1). Activity 1150 units/mg. Soly in water at 24°: 0.7%. Insol in vegetable oils. The solid is stable for 2 years at 24° and for 2 days at 100°.

THERAP CAT: Antibacterial.

7039. Penicillin G Calcium. *[2S-(2α,5α,6β)]-3,3-Dimethyl-7-oxo-6-[(phenylacetyl)amino]-4-thia-1-azabicyclo-[3.2.0]heptane-2-carboxylic acid calcium salt (2:1);* benzylpenicillin calcium; calcium penicillin G; calcium benzylpenicillinate; benzylpenicillinic acid calcium salt. $C_{32}H_{34}CaN_4O_8S_2$; mol wt 706.84. C 54.37%, H 4.85%, Ca 5.67%, N 7.93%, O 18.11%, S 9.07%. $(C_{16}H_{17}N_2O_4S)_2Ca$. Antibiotic obtained by extraction of an ether soln of the free acid (prepared from the sodium salt) with satd calcium hydroxide soln to pH 6.5, followed by lyophilizing: Clarke *et al.*, *Chemistry of Penicillin* (Princeton, 1949) p 88.

White powder, granules, scales. Freely sol in water, in isotonic sodium chloride soln, in glucose solns. Also sol in alcohol, glycerol, acetone, ethyl acetate, chloroform. Insol in fixed oils, liquid petrolatum. pH of an aq soln contg 20,000 units per ml is 5.0 to 6.5.

THERAP CAT: Antibacterial.
THERAP CAT (VET): Antibacterial.

7040. Penicillin G Hydrabamine. *N,N'-Bis(dehydroabietyl)ethylenediamine dipenicillin G; benzylpenicillinic acid N,N'-bis(dehydroabietyl)ethylenediamine double salt;* hydrabamine penicillin G; Compocillin. $C_{74}H_{100}N_6O_8S_2$; mol wt 1265.79. C 70.22%, H 7.96%, N 6.64%, O 10.11%, S 5.07%. Semi-synthetic antibiotic. Pharmacology: MacCorquodale *et al.*, *Antibiot. Ann.* **1954-1955**, 133; Lepper, Spies, *ibid.* 137; DeRose *et al.*, *Antibiot. & Chemother.* **5**, 315, 324 (1955). Purification: DeRose, U.S pat. **2,812,326** (1956 to Abbott).

Tasteless crystals from dil methanol, dec 171-173°. $[\alpha]_D^{25}$ +115.3° (c = 10 in chloroform). Potency 939 units/mg. Soly in water 9 units/ml (bioassay). Solubilities determined by Weiss *et al.*, *Antibiot. & Chemother.* **7**, 374 (1957) in mg/ml at about 28°: Water 0.075; methanol 7.3; ethanol 5.2; isopropanol 1.7; isoamyl alcohol 3.1; cyclohexane 0.115; benzene 0.60; toluene 0.39; petr ether 0.10; isooctane 0.055; carbon tetrachloride 0.50; ethyl acetate 1.65; isoamyl acetate 1.4; acetone 3.4; methyl ethyl ketone 3.65; ether 0.70.

THERAP CAT: Antibacterial.

7041. Penicillin G Potassium. *[2S-(2α,5α,6β)]-3,3-Dimethyl-7-oxo-6-[(phenylacetyl)amino]-4-thia-1-azabicyclo-[3.2.0]heptane-2-carboxylic acid monopotassium salt;* benzylpenicillin potassium; potassium penicillin G; potassium benzylpenicillinate; benzylpenicillinic acid potassium salt; Notaral; Crystapen; Hipercilina; Pentid; Tabilin; Eskacillin; Forpen; Hylenta; Cosmopen; Falapen; Hyasorb; Cristapen; M-Cillin; Monopen; Megacillin tablets; Scotcil. $C_{16}H_{17}KN_2O_4S$; mol wt 372.47. C 51.59%, H 4.60%, K 10.50%, N 7.52%, O 17.18%, S 8.61%. For general prepn and references *see* Benzylpenicillin Sodium. Rapid synthesis: A. N. Voldeng *et al.*, *J. Heterocycl. Chem.* **16**, 621 (1979). HPLC determn: J. M. Blaha *et al., J. Pharm. Sci.* **64**, 1384 (1975). Comprehensive description: J. Kirschbaum in *Analytical Profiles of Drug Substances* **vol. 15**, K. Florey, Ed. (Academic Press, New York, 1986) pp 427-507.

Crystalline antibiotic, moderately hygroscopic. Dec 214-217°. $[\alpha]_D^{22}$ +285-310° (c = 0.7). Freely sol in water, in isotonic sodium chloride soln, and in dextrose solns. Sol in methanol, ethanol, glycerol. pH of 3% aq soln 5.0 to 7.5. Solutions stored at refrigerator temps remain stable for several days. They are rapidly inactivated by acids, alkali hydroxides, and by oxidizing agents. One milligram of benzylpenicillin potassium is equivalent to 1595 units. One international or U.S.P. penicillin unit is equivalent to 0.6 micrograms of benzylpenicillin sodium.

THERAP CAT: Antibacterial.

THERAP CAT (VET): Antibacterial.

7042. Penicillin G Procaine. *[2S-(2α,5α,6β)]-3,3-Dimethyl-7-oxo-6-[(phenylacetyl)amino]-4-thia-1-azabicyclo-[3.2.0]heptane-2-carboxylic acid compd with 2-(diethylamino)ethyl 4-aminobenzoate (1:1) monohydrate; penicillin G compd with 2-(diethylamino)ethyl p-aminobenzoate monohydrate;* benzylpenicillin procaine; procaine benzylpenicillinate; procaine penicillin G; Abbocillin-DC; Afsillin; Ampinpenicillin; Aquacillin; Aquasuspen; Avloprocil; Cilicaine; Crysticillin; Despacilina; Depocillin; Distaquaine; Dorsallin "A.R."; Duracillin; Flo-Cillin Aqueous; Hydracillin; Ilcocillin P; Kabipenin; Ledercillin; Lenticillin; Megapen; Mylipen; Neoproc; Penaquacaine G; Pen-Fifty; Premocillin; Procanodia; Pro-Pen; Wycillin. $C_{29}H_{38}N_4O_6S.H_2O$; mol wt 588.72. C 59.16%, H 6.85%, N 9.52%, O 19.02%, S 5.45%. Semisynthetic antibiotic. Prepn: N. P. Sullivan *et al., Science* **107**, 169 (1948); C. J. Sullivan *et al., J. Am. Chem. Soc.* **70**, 1287 (1948); S. L. Ruskin, U.S. pat. **2,676,961** (1954 to Physiological Chem.). Additional processes: O. R. Sumner, T. C. Grenfell, U.S. pat. **2,725,336** (1955 to Pfizer); M. P. Bardolph, U.S. pat. **2,739,962** (1956 to Comm. Solvents Corp.). Crystal structure: Rose, *Anal. Chem.* **27**, 1841 (1955). Toxicity: K. Soehring *et al., Arzneimittel-Forsch.* **1**, 28 (1951). Pharmacokinetics in horses: S. M. Stover *et al., Am. J. Vet. Res.* **42**, 629 (1981); in humans: B. T. Goh *et al., Brit. J. Vener. Dis.* **60**, 371 (1984). Review of use in syphilis: M. W. Adler, *Brit. Med. J.* **288**, 551-553 (1984).

Monoclinic hemimorphic crystals from methanol-water, mp 106-110° (with decompn). d 1.255-1.256. Not appreciably affected by air or light. Aq solns are dextrorotatory. The pH of a satd aq soln is between 5 and 7.5. One gram dissolves in 250 ml water, in 30 ml alcohol, in 60 ml chloroform (*U.S.P.*). Soly profile: Weiss *et al., Antibiot. & Chemother.* **7**, 374 (1957). Soly in mg/ml at about 28°: water 6.8; methanol > 20; isopropanol 6.5; benzene 0.075; toluene 1.05; petr ether 0.12; isooctane 0.0; carbon tetrachloride 0.12; ethyl acetate 3.35. Rapidly inactivated by acids, alkali hydroxides, oxidizing agents. Penicillin potency: approx 1000 units/mg. LD_{50} s.c. in mice: 2.3 g/kg (Soehring).

THERAP CAT: Antibacterial.

THERAP CAT (VET): Antibacterial.

7043. Penicillin N. *[2S-[2α,5α,6β(S*)]]-6-[(5-Amino-5-carboxy-1-oxopentyl)amino]-3,3-dimethyl-7-oxo-4-thia-1-azabicyclo[3.2.0]heptane-2-carboxylic acid; 6-(D-5-amino-5-carboxyvaleramido)-3,3-dimethyl-7-oxo-4-thia-1-azabicyclo[3.2.0]heptane-2-carboxylic acid;* (D-4-amino-4-carboxybutyl)penicillinic acid; cephalosporin N; adicillin; Synnematin B. $C_{14}H_{21}N_3O_6S$; mol wt 359.40. C 46.79%, H 5.89%, N 11.69%, O 26.71%, S 8.92%. Antibiotic substance produced by *Cephalosporium* spp found in sewage outpours: Gottshall *et al., Proc. Soc. Exp. Biol. Med.* **76**, 307 (1951); Abraham *et al., Nature* **171**, 343 (1953); **176**, 551 (1955). Produced also by *Paecilomyces persicimus:* Pisano *et al., Antimicrobial Agents Annual* **1960** (Plenum Press, New York, 1961) pp 41, 48; by *Penicillium chrysogenum:* Flynn *et al., J. Am. Chem. Soc.* **84**, 4594 (1962). Structure: Abraham, Newton, *Bio-*

chem. J. **58**, 103 (1954). Production: Miller *et al.*, U.S. pat. **2,831,797** (1958). Purification: Goodall, Sutcliffe, U.S. pat. **2,899,425** (1959 to ICI).

Soluble in water. Dextrorotatory. Inactivated by penicillinase as is penicillin G, but differs from the common penicillin by its antibacterial activity and hydrophilic character. When an aq soln is kept at pH 2.7 and 37° for 2 hrs, there is a loss of antibacterial activity and an increase in dextrorotation. Active against *Sarcina lutea, Proteus vulgaris, Salmonella typhimurium, Diplococcus pneumoniae.* Shows practically no activity against *B. subtilis* and *Staph. aureus.* The toxicity is somewhat less than that of penicillin G, although penicillin N is excreted more slowly.

Barium salt, white powder. $[\alpha]_D^{20}$ +187° (c = 0.6). Freely sol in water, sparingly sol in methanol. Practically insol in ethanol.

THERAP CAT: Antibacterial.

7044. Penicillin O. *3,3-Dimethyl-7-oxo-6-[[(2-propenylthio)acetyl]amino]-4-thia-1-azabicyclo[3.2.0]heptane-2-carboxylic acid; [(allylthio)methyl] penicillin;* allylmercaptomethylpenicillin; allylmercaptomethylpenicillinic acid; penicillin AT. $C_{13}H_{18}N_2O_4S_2$; mol wt 330.43. C 47.25%, H 5.49%, N 8.48%, O 19.37%, S 19.41%. Antibiotic produced by *Penicillium chrysogenum.* Biosynthesis of salts: Behrens *et al., J. Biol. Chem.* **175**, 793 (1948); Rhodehamel, Behrens *et al.*, U.S. pats. **2,528,175** and **2,623,876** (1950, 1952, both to Lilly); Ford *et al., Antibiot. & Chemother.* **3**, 1149 (1953); Ford, U.S. pat. **2,647,894** (1953 to Upjohn); Palecková, Slechta, *C.A.* **50**, 17309g (1956).

2-Chloroprocaine salt monohydrate, $C_{26}H_{37}ClN_4O_6S_2.H_2O$, *chloroprocaine penicillin O, penicillin O 2-chloroprocaine, Depo-Cer-O-Cillin Chloroprocaine.* Slender needles from hot water, mp 79-81°. Practically insol in cold water. Stable in dry form at room temp. Aq suspensions are stable at room temp for 1 week, at refrigerator temps for 3 weeks. Calculated activity: 949 units/mg. Solubilities: Weiss *et al., Antibiot. & Chemother.* **7**, 374 (1957).

Potassium salt, $C_{13}H_{17}KN_2O_4S_2$, *potassium penicillin O, penicillin O potassium.* Crystals from acetone. Soluble in water. Stable in dry form at room temp for at least 3 years. Requires no refrigeration when dry. Aq solns may be kept for 3 days at +10° without significant loss of activity. Behrens' prepn assayed 1630 units/mg. Less toxic than benzylpenicillin in exptl animals.

Procaine salt, $C_{26}H_{38}N_4O_6S_2$, crystals from water.

Sodium salt, $C_{13}H_{17}N_2NaO_4S_2$, *Cer-O-Cillin Sodium.* Crystals from acetone.

THERAP CAT: Antibacterial.

7045. Penicillin S Potassium. *6-[[[(3-Chloro-2-butenyl)thio]acetyl]amino]-3,3-dimethyl-7-oxo-4-thia-1-azabicyclo[3.2.0]heptane-2-carboxylic acid monopotassium salt; γ-chlorocrotylmercaptomethylpenicillin potassium;* γ-chlorocrotylmercaptomethylpenicillinic acid potassium salt; potassium γ-chlorocrotylmercaptomethylpenicillinate; potassium penicillin S. $C_{14}H_{18}ClKN_2O_4S_2$; mol wt 417.00. C 40.32%, H 4.35%, Cl 8.50%, K 9.38%, N 6.72%, O 15.35%, S 15.38%. Obtained by growing a penicillin-producing mold in a culture medium contg γ-chlorocrotylmercaptoacetic acid: Ford *et al., J. Am. Chem. Soc.* **70**, 3522 (1948); Ford *et al., Antibiot. & Chemother.* **3**, 1149 (1953).

Crystals from acetone. Disagreeable odor and taste. Sol in water. Bioassay: 1900 units/mg.

7046. Penicillin V. *3,3-Dimethyl-7-oxo-6-[(phenoxyacetyl)amino]-4-thia-1-azabicyclo[3.2.0]heptane-2-carboxylic acid; 6-phenoxyacetamidopenicillanic acid;* penicillin phenoxymethyl; phenoxymethylpenicillin; phenoxymethylpenicillinic acid; Acipen-V; Apopen; Distaquaine V; Eskacillin V; Fenacilin; Fenospen; Meropenin; Oracillin; Oratren; Orocillin; Ospen; Pencompren; Pen-Oral; Pen-Vee; Phenopenicillin; P-Mega-Tablinen; Stabicillin; V-Cil; V-Cillin; Vebecillin. $C_{16}H_{18}N_2O_5S$; mol wt 350.38. C 54.84%, H 5.18%, N 8.00%, O 22.83%, S 9.15%. Obtained by adding 2-phenoxyethanol to the *Penicillium* culture using yeast autolyzate as source of nitrogen: Brandl *et al., Wien. Med. Wochenschr.* **1953**, 602; Brandl, Margreiter, *Oesterr. Chem.-Ztg.* **55**, 11-21 (1954), *C.A.* **48**, 10296 (1954). Purification: Parker *et al., J. Pharm. Pharmacol.* **7**, 683 (1953). Total synthesis: Sheehan, Henery-Logan, *J. Am. Chem. Soc.* **79**, 1262 (1957); **81**, 3089 (1959); **84**, 2983 (1962). Prepn from 6-aminopenicillanic acid: Glambitza, *Ann.* **673**, 166 (1964). Soly data: Weiss *et al., Antibiot. & Chemother.* **7**, 374 (1957). The biologically active form is the dextrorotatory D-form; the DL-form is half as active. L-Penicillin V has little, if any, antibiotic activity. Toxicity data: E. I. Goldenthal, *Toxicol. Appl. Pharmacol.* **18**, 185 (1971). Comprehensive description of the potassium salt: J. M. Dunham in *Analytical Profiles of Drug Substances* **vol. 1**, K. Florey, Ed. (Academic Press, New York, 1972) pp 249-300; D. H. Sieh, *ibid.*, **vol. 17** (1988) pp 677-748.

Crystals, dec 120-128°. Stable in air up to 37°; relatively stable to acid. uv max: 268, 274 nm (ε 1330, 1100). Soln in water at pH 1.8 (acidified with HCl) = 25 mg/100 ml. Sol in polar organic solvents. Practically insol in vegetable oils and in liq petrolatum.

Potassium salt, $C_{16}H_{17}KN_2O_5S$, *Antibiocin, Apsin VK, Arcacil, Arcasin, Beromycin, Beromycin 400, Betapen VK, Calciopen K, Cliacil, Compocillin VK, Distakaps V-K, Distaquaine V-K, Dowpen V-K, Fenoxypen, Icipen, Isocillin, Ispenoral, Ledercillin VK, Megacillin oral, Oracil-VK, Orapen, Ospeneff, Pedipen, Penagen, Pencompren, Pen-Vee K, Penvikal, Pfizerpen VK, PVK, SK-Penicillin VK, Stabillin V-K, Sumapen VK, Suspen, Uticillin VK, V-Cil-K, V-Cillin K, Veetids, Vepen.* Sol in water. $[\alpha]_D^{25}$ +223° (c = 0.2). LD_{50} orally in rats: > 1040 mg/kg (Goldenthal).

Sodium salt, $C_{16}H_{17}N_2NaO_5S$, soluble in water.

Calcium salt, $C_{32}H_{34}CaN_4O_{10}S_2$, *Bantogen, Calcipen-V, Penavlon V, Penicals, Septocillin.*

THERAP CAT: Antibacterial.

THERAP CAT (VET): Antibacterial.

7047. Penicillin V Benzathine. *[2S-(2α,5α,6β)]-3,3-Dimethyl-7-oxo-6-[(phenoxyacetyl)amino]-4-thia-1-azabicyclo[3.2.0]heptane-2-carboxylic acid compd with N,N'-bis-(phenylmethyl)-1,2-ethanediamine (2:1);* penicillin V DBED; benzathine penicillin V; benzathine benzylpenicillin; Ospen; Ostrocilline; Bicillin V; Falcopen V. Semi-synthetic antibiotic prepd in a manner analogous to benzathine penicillin G from one mole *N,N'*-dibenzylethylenediamine and 2 moles phenoxymethyl penicillin.

Solubilities determined by Weiss *et al., Antibiot. & Chemother.* **7**, 374 (1957) in mg/ml at about 28°: water 0.321; ethanol 14.6.

THERAP CAT: Antibacterial.

7048. Penicillin V Hydrabamine. *N,N'*-Bis(dehydroabietyl)ethylenediamine bis(phenoxymethylpenicillin); hydrabamine phenoxymethyl penicillin; hydrabamine penicillin

V; Abbocillin V; Compocillin-V. $C_{74}H_{100}N_6O_{10}S_2$; mol wt 1297.79. C 68.49%, H 7.77%, N 6.48%, O 12.33%, S 4.92%. Semi-synthetic antibiotic. Prepn: DeRose, U.S. pat. **2,812,-326** (1956 to Abbott).

Solubilities determined by Weiss *et al., Antibiot. & Chemother.* **7**, 374 (1957) in mg/ml at about 28°: Water 0.05; methanol 11.05; ethanol 5.8; isopropanol 1.75; isoamyl alcohol 6.85; cyclohexane 0.12; benzene 1.4; toluene 1.07; petr ether 0.06; isooctane 0.065; carbon tetrachloride 3.30; ethyl acetate 4.0; isoamyl acetate 4.9; acetone 10.2; methyl ethyl ketone 13.7; ether 0.095; ethylene chloride > 20; dioxane 7.5.

THERAP CAT: Antibacterial.

7049. Penicilloic Acids. Substances obtained by the alkaline cleavage of the different penicillins. Penicilloic acids have the general formula:

Synthetic penicilloic acids can be derived from D-, DL-, or L-penicillamine and are designated accordingly. The four stereoisomers which can be derived from D-penicillamine are arbitrarily designated as α, β, γ, δ. Review and bibliography: Chain in *Antibiotics*, vol. 2, H. W. Florey *et al.*, Eds. (Oxford University Press, New York, 1949) p 839.

7050. Penicilloyl Polylysine. Benzylpenicilloyl polylysine; PPL; Cilligen; Pre-Pen; Testarpen. Prepn from polylysine and a penicillenic acid: Parker *et al., J. Exp. Med.* **115**, 803 (1962). Prepn and use as diagnostic aid: M. A. Stahmann, S. S. Wagle, **Brit.** pat. **1,226,773;** *eidem,* **U.S.** pat. **3,979,508** (1971, 1976 both to Kremers-Urban Co.). Intradermal test for penicillin sensitivity: Brown *et al., J. Am. Med. Assoc.* **189**, 599 (1964); Van Arsdale, *ibid.* **191**, 238 (1965); T. J. Sullivan, *J. Allergy Clin. Immunol.* **68**, 171 (1981).

THERAP CAT: Diagnostic aid (penicillin sensitivity).

7051. Penillic Acids. 2,3,7,7a-Tetrahydro-2,2-dimethylimidazo[5,1-b]thiazole-3,7-dicarboxylic acid derivatives. Degradation products of penicillins obtained by acid hydrolysis at pH 2. Review and bibliography: Abrahams in *Antibiotics* vol. 2, H. W. Florey *et al.*, Eds. (Oxford University Press, New York, 1949) p 889.

7052. Penimepicycline. *[2S-(2α,5α,6β)]-3,3-Dimethyl-7-oxo-6-[(phenoxyacetyl)amino]-4-thia-1-azabicyclo[3.2.0]heptane-2-carboxylic acid compd with [4S-(4α,4aα,5aα,6β,-12aα)]-4-(dimethylamino)-1,4,4a,5,5a,6,11,12a-octahydro-3,6,10,12,12a-pentahydroxy-N-[[4-(2-hydroxyethyl)-1-piperazinyl]methyl]-6-methyl-1,11-dioxo-2-naphthacenecarboxamide (1:1);* N-[4-(β-hydroxyethyl)diethylenediaminomethyl]tetracycline phenoxymethylpenicillinate; mepicycline phenoxymethylpenicillinate; penicillin V compd with mepicycline; penimepiciclina; Criseocil; Geotricyn; Mepenicycline; Olimpen; Penetracyne; Peniltetra; Prestociclina. $C_{45}H_{56}N_6O_{14}S$; mol wt 937.07. C 57.68%, H 6.02%, N 8.97%, O 23.90%, S 3.42%. $C_{16}H_{18}N_2O_5S \cdot C_{29}H_{38}N_4O_9$. Semi-synthetic antibiotic. Prepn: Pedrazzoli *et al., Boll. Chim. Farm.* **98**, 516 (1959), *C.A.* **54**, 3856a (1960); Gradnik, Pedrazzoli, **Brit.** pat. **897,826** (1962 to E.R.A.S.M.E.), *C.A.* **57**, 15034e (1962). Properties: Gradnik *et al., Pharm. Acta Helv.* **35**, 529 (1960).

Yellowish-white cryst powder, dec above 143° (not corr). Sensitive to light, heat, air. Slightly bitter taste. pH of 2% aq soln, 5.5-5.7. Soly in water at 20° = 1 g/0.7 ml. $[\alpha]_D^{20}$ −50.5° (c = 2 in methanol).

THERAP CAT: Antibacterial.

7053. Pennyroyal. Hedeoma; squaw mint; mosquito plant. Dried leaves and flowering tops of *Hedeoma pule-*

gioides (L.) Pers., *Labiatae.* *Habit.* Canada to Florida and west to Nebraska. *Constit.* Volatile oil (1%), bitter principles, tannin.

THERAP CAT: Diaphoretic, emmenagogue.

7054. Pentaborane(9). Pentaboron nonahydride. B_5H_9; mol wt 63.17. B 85.64%, H 14.36%. Prepd from diborane: Stock, Mathing, *Ber.* **69B**, 1456 (1936); Schlesinger, Burg, *J. Am. Chem. Soc.* **53**, 4321 (1931); **55**, 4009 (1933). Molecular structure determination by rotational spectroscopy: D. Schwoch *et al., Inorg. Chem.* **16**, 3219 (1977). Review of toxicity: see Decaborane(14). *Review:* Greenwood in *Comprehensive Inorganic Chemistry* vol. 1, J. C. Bailar, Jr. *et al.,* Eds. (Pergamon Press, Oxford, 1973) pp 792-801.

Liquid. mp $-46.6°$; bp $60°$. d_4^0 0.61; vp 66 mm Hg at $0°$. Dec very slowly at $150°$. Ignites spontaneously in air. Hydrolyzes in water after long heating. Reacts with ammonia to form a diammoniate.

7055. Pentaborane(11). Dihydropentaborane(9); pentaboron undecahydride. B_5H_{11}; mol wt 65.19. B 82.99%, H 17.01%. Prepd from diborane: Burg, Schlesinger, *J. Am. Chem. Soc.* **55**, 4009 (1933).

Liquid. mp $-123°$. bp $63°$; bp_{53} $0°$. Unstable. When heated or allowed to stand for long periods of time, it produces diborane, tetraborane, hydrogen, pentaborane, decaborane and brown nonvolatile liqs and solids. Ignites spontaneously in air. Hydrolyzes in water to boric acid and hydrogen. Reacts with ammonia to form a tetraammoniate.

7056. Pentabromoacetone. *1,1,1,3,3-Pentabromo-2-propanone.* C_3HBr_5O; mol wt 452.62. C 7.96%, H 0.22%, Br 88.28%, O 3.53%. $CBr_3COCHBr_2$. Prepd by the addition of 12 parts bromine to 1 part acetone: Mulder, *Jahresbericht über die Fortschritte der Chemie* 1864, 330.

Orthorhombic needles or prisms from alcohol or ether. Penetrating odor. mp $76°$ (sublimes above mp). Volatile with steam. Practically insol in water. Freely sol in organic solvents. Forms bromoform under the influence of alkalies.

7057. Pentacene. Benzo[*b*]naphthacene; 2,3,6,7-dibenzoanthracene; *lin*-naphthoanthracene. $C_{22}H_{14}$; mol wt 278.35. C 94.93%, H 5.07%. Synthesis from *m*-xylene or from 4-benzyl-1,3-dimethylbenzene and benzoyl chloride in presence of aluminum chloride or from terephthalyl chloride and *o*-tolylmagnesium bromide: Clar, John, *Ber.* **62**, 940 (1929); **63**, 2967 (1930); **64**, 981 (1931). Synthesis by reduction of 6,13-pentacenequinone: Bruckner *et al., Tetrahedron Letters* **1960**, no. 1, 5; Bruckner, Tomasz, *Acta Chim. Acad. Sci. Hung.* **28**(4), 405 (1961). Structure: Campbell *et al., Acta Cryst.* **14**, 705 (1961).

Deep blue needles with violet luster from hot nitrobenzene. Sublimes in CO_2 stream under reduced pressure at about $300°$ (Clar, John, *loc. cit.*). In presence of air dec above $300°$. Practically insol in water; sparingly sol in organic solvents.

Note: Anthracene (3 linear rings) is colorless; naphthacene (4 linear rings) is orange; pentacene (5 linear rings) is blue; hexacene (6 linear rings) is green.

7058. Pentachloroethane. Pentalin. C_2HCl_5; mol wt 202.31. C 11.87%, H 0.50%, Cl 87.63%. CCl_3CHCl_2. Toxicity data: G. S. Barsoum, K. Saad, *Quart. J. Pharm. Pharmacol.* **7**, 205 (1934).

Liquid; chloroform-like odor. d_4^{25} 1.6712; bp 161-162°. mp $-29°$. n_D^{15} 1.5054. Insol in water. Miscible with alcohol, ether. MLD (mg/kg) in dogs: 1750 orally; 100 i.v.; in rabbits: 700 s.c. (Barsoum, Saad).

Caution: Irritant; narcotic.

7059. Pentachlorophenol. Penta; PCP; penchlorol; Santophen 20. C_6HCl_5O; mol wt 266.35. C 27.05%, H 0.38%, Cl 66.56%, O 6.01%. Prepd by the chlorination of phenol in the presence of a catalyst.

Needle-like crystals, mp 190-191°; bp about 309-310° (dec). d_4^{22} 1.978. Very pungent odor only when hot. Sublimes in needles. Almost insol in water (8 mg in 100 ml); freely sol in alc, ether; sol in benzene; slightly sol in cold petr ether. LD_{50} orally in male, female rats: 146, 175 mg/kg, T. B. Gaines, *Toxicol. Appl. Pharmacol.* **14**, 515 (1969).

Forms a sodium salt, sodium pentachlorophenate; *sodium pentachlorophenoxide, Santobrite, Dowicide G,* which is sol in water.

USE: Insecticide for termite control; pre-harvest defoliant; general herbicide. Has been recommended for use in the preservation of wood, wood products, starches, dextrins, glues. *Caution:* Ingestion causes increase then decrease of respiration, blood pressure, urinary output; fever; increased bowel action; motor weakness; collapse with convulsions and death. Causes lung, liver, kidney damage; contact dermatitis. May be absorbed through skin. More toxic in organic solvents. Dust causes sneezing. *See:* Patty's *Industrial Hygiene and Toxicology,* vol. 2A, G. D. Clayton, F. E. Clayton, Eds. (Wiley-Interscience, New York, 1981) pp 2604-2612.

7060. Pentacynium Bis(methyl sulfate). *4-[2-[(5-Cyano-5,5-diphenylpentyl)dimethylammonio]ethyl]-4-methylmorpholinium bis(methyl sulfate); N^1-(5-cyano-5,5-diphenylpentyl)-N^1,N^1,N^2-trimethylethylene-1-ammonium-2-morpholinium bis(methyl sulfate); ethylene-1-[(5-cyano-5,5-phenylpentyl)dimethylammonium]-(4-methylmorpholinium) bis(methyl sulfate); pentacyone mesylate; Presidal.* $C_{29}H_{45}$-$N_3O_9S_2$; mol wt 643.83. C 54.10%, H 7.05%, N 6.53%, O 22.37%, S 9.96%. Prepn: Billinghurst, U.S. pat. **2,851,458** (1958 to Burroughs Wellcome). Ganglionic blocker.

Crystals from ethanol, mp 173-175°. Sol in water.

THERAP CAT: Antihypertensive.

7061. 3-Pentadecylcatechol. *3-Pentadecyl-1,2-benzenediol; 3-pentadecylpyrocatechol;* tetrahydrourushiol; hydrourushiol; dihydrorhengol; 3-PDC. $C_{21}H_{36}O_2$; mol wt 320.50. C 78.69%, H 11.32%, O 9.98%. Constituent of the irritant oil of poison ivy (*Toxicodendron radicans* (L.) Kuntze) and other *Toxicodendron* spp. (*Anacardiaceae*). Prepn by hydrogenation of extracts from fruits of *Semecarpus heterophylla*: Backer, Haack, *Rec. Trav. Chim.* **57**, 225 (1938). Synthesis from 2,3-dimethoxybenzaldehyde and tetradecyl chloride: Mason, *J. Am. Chem. Soc.* **67**, 1538 (1945); from *o*-veratraldehyde: Backer, Haack, *loc. cit.;* Dawson *et al., J. Am. Chem. Soc.* **68**, 534 (1946); Keil *et al.,* U.S. pat. **2,451,955** (1948); from 2,3-dibenzyloxybenzaldehyde: Loev, Dawson, *J. Org. Chem.* **24**, 980 (1959); from furan derivs: Boehme, *J. Am. Chem. Soc.* **82**, 499 (1960); from catechol: Hanafusa, Yukawa, *Chem. & Ind. (London)* **1961**, 23. Criticism of reported syntheses and synthesis of dimethyl deriv: Byck, Dawson, *J. Org. Chem.* **32**, 1084 (1967). Total synthesis: E. Wenkert *et al., J. Am. Chem. Soc.* **105**, 2021 (1983). Evaluation of diagnostic patch test: M. V. Dahl *et al., Arch. Dermatol.* **120**, 1022 (1984).

Short needles from toluene or petr ether, mp 59-60°. Can be purified by molecular distn. uv max: 277 nm. Sol in alc, ether, benzene, toluene. Sparingly sol in petr ether.

THERAP CAT: Diagnostic aid (contact allergen).

7062. Pentaerythritol. *2,2-Bis(hydroxymethyl)-1,3-propanediol; tetrakis(hydroxymethyl)methane;* tetramethylol-methane; Metab-Auxil; Penetek; Pentek. $C_5H_{12}O_4$; mol wt 136.15. C 44.11%, H 8.88%, O 47.01%. Prepd by treating acetaldehyde with formaldehyde in an aq soln of calcium hydroxide: H. B. J. Schurink, *Org. Syn. coll. vol. I,* 425 (2nd ed., 1941); Fieser, Fieser, *Organic Chemistry* (2nd ed, 1950) p 133. Review of mfg processes: P. W. Sherwood, *Petroleum Refiner,* **Nov. 1956,** p 171-179; Faith, Keyes & Clark's *Industrial Chemicals,* F. A. Lowenheim, M. K. Moran, Eds. (Wiley-Interscience, New York, 4th ed., 1975) pp 598-603. Monograph: E. Berlow *et al., The Pentaerythritols,* ACS Monograph Series no. **136** (Reinhold, New York, 1958).

$$HOCH_2 - \overset{\overset{\displaystyle CH_2OH}{|}}{\underset{\underset{\displaystyle CH_2OH}{|}}{C}} - CH_2OH$$

Ditetragonal crystals from dil HCl, mp 260°. One gram dissolves in 18 ml water at 15°. Sol in ethanol, glycerol, ethylene glycol, formamide. Insol in acetone, benzene, paraffin, ether, carbon tetrachloride.

Trinitrate, *see* pentrinitrol.

USE: In synthetic resins, in paint and varnish industries.

7063. Pentaerythritol Chloral. *1,1'-[2,2-Bis[(2,2,2-trichloro-1-hydroxyethoxy)methyl]-1,3-propanediylbis(oxy)]bis-[2,2,2-trichloroethanol];* Petrichloral; Periclor. $C_{13}H_{16}Cl_{12}O_8$; mol wt 725.76. C 21.51%, H 2.22%, Cl 58.63%, O 17.64%. Prepn: Bruce, U.S. pat. **2,784,237** (1957 to Am. Home Prod.).

$$Cl_3CCHOCH_2 - \overset{\overset{\displaystyle CH_2OCHCCl_3}{|}{\overset{\displaystyle OH}{|}}}{\underset{\underset{\displaystyle CH_2OCHCCl_3}{|}{\underset{\displaystyle OH}{|}}}{\underset{\displaystyle OH}{C}}} - CH_2OCHCCl_3$$

Yellow glass (easily powdered), mp 52-54°. Less odor and aftertaste than choral hydrate. Sol in water. Sol to the extent of 30 or more g/100 ml in ethyl acetate, formamide, ethanol, chloroform, benzene; to the extent of 5 g/100 ml in tetrachloroethylene and isopropanol.

Caution: May be habit forming. This is a controlled substance (depressant) listed in the U.S. Code of Federal Regulations, Title 21 Part 1308.14 (1985).

THERAP CAT: Hypnotic. Sedative.

7064. Pentaerythritol Dichlorohydrin. *2,2-Bis(chloromethyl)-1,3-propanediol;* Dispranol. $C_5H_{10}Cl_2O_2$; mol wt 173.05. C 34.71%, H 5.82%, Cl 40.98%, O 18.49%. $(HOCH_2)_2C(CH_2Cl)_2$. Prepn: Fecht, *Ber.* **40,** 3883 (1907); Rapoport, *J. Am. Chem. Soc.* **68,** 341 (1946); Dee, U.S. pat. **2,763,679** (1956 to Heyden Chem.).

Crystals from water or benzene, mp 79-80° (uncorr), 83° (corr), also reported as 65° and 95°, *see* Rapoport, *loc. cit.* bp$_{12}$ 158.5-160°. Sublimes in high vacuum.

THERAP CAT: Tranquilizer; muscle relaxant.

7065. Pentaerythritol Tetraacetate. *2,2-Bis[(acetyloxy)methyl]-1,3-propanediol diacetate;* tetra-O-acetylpentaerythritol; pentaerythrityl tetraacetate; Normosterol. $C_{13}H_{20}O_8$; mol wt 304.29. C 51.31%, H 6.63%, O 42.07%. Prepn: Wolfrom *et al., J. Am. Chem. Soc.* **73,** 874 (1951); Bonner *et al., J. Chem. Soc.* **1960,** 2914.

$$CH_3COOCH_2 - \overset{\overset{\displaystyle CH_2OOCCH_3}{|}}{\underset{\underset{\displaystyle CH_2OOCCH_3}{|}}{C}} - CH_2OOCCH_3$$

Crystals, mp 83-84°.

THERAP CAT: Antilipemic.

7066. Pentaerythritol Tetranitrate. *2,2-Bis[(nitrooxy)methyl]-1,3-propanediol dinitrate (ester);* pentaerythrityl tetranitrate; 2,2-bisdihydroxymethyl-1,3-propanediol tetranitrate; PETN; nitropentaerythritol; penthrit; niperyt; Lentrat; Hasethrol; Peritrate; Mycardol; Nitropenton; Pentral 80; Dipentrate; Dilcoran-80; Terpate; Pentrite; Perityl; Pentritol; Pentanitrine; Prevangor; Subicard; Pentryate; Vasodiatol; Neo-Corovas; Pentafin; Quintrate; Pergitral; Pentitrate; Metranil; Cardiacap; Angitet; Nitropenta. $C_5H_8N_4O_{12}$; mol wt 316.15. C 18.99%, H 2.55%, N 17.72%, O 60.73%. Prepd by nitration of pentaerythritol: Acken, Vyverberg, U.S. pat. **2,370,437** (1945 to du Pont).

$$O_2NOCH_2 - \overset{\overset{\displaystyle CH_2ONO_2}{|}}{\underset{\underset{\displaystyle CH_2ONO_2}{|}}{C}} - CH_2ONO_2$$

Tetragonal holohedra from acetone + alcohol, mp 140°. d_4^{20} 1.773. Soluble in acetone. Practically insoluble in water (1.5 γ/ml). Sparingly sol in alcohol, ether. Does not reduce Fehling's soln (difference from erythrityl tetranitrate). *Caution:* Explodes on percussion. More sensitive to shock than TNT. For medicinal purposes it is dil with an inert ingredient, usually lactose, to prevent accidental explosions.

USE: Mainly in the manuf of detonating fuse (Primacord), a waterproof textile filled with powdered PETN.

THERAP CAT: Vasodilator.

7067. Pentagastrin. *N-[(1,1-Dimethylethoxy)carbonyl]-β-alanyl-L-tryptophyl-L-methionyl-L-α-aspartyl-L-phenylalaninamide; N-carboxy-β-alanyl-L-tryptophyl-L-methionyl-L-aspartylphenyl-L-alaninamide N-tert-butyl ester;* N-(α-carbamoylphenethyl)-3-[2-[2-[3-(carboxyamino)propionamido]-3-indol-3-ylpropionamido]-4-(methylthio)butyramido]succinamic acid *N-tert-butyl ester; N-[N-[N-[N-(N-tert-butoxycarbonyl-β-alanyl)-L-tryptophanyl]-L-methionyl]-L-aspartyl]-L-phenylalaninamide; Boc-β-Ala-Try-Met-Asp-Phe(NH₂);* AY-6608; ICI 50123; Gastrodiagnost; Peptavlon. $C_{37}H_{49}N_7O_9S$; mol wt 767.93. C 57.87%, H 6.43%, N 12.77%, O 18.75%, S 4.18%. Prepn: Hardy *et al.,* **Belg. pat. 665,591** corresp to U.S. pat. **3,896,103** (1965, 1975 to I.C.I.); Davey *et al., J. Chem. Soc. (C)* **1966,** 555; Sakakibara *et al., Bull. Chem. Soc. Japan* **41,** 438 (1968). Structure-function relationship studies: Morley *et al., Nature* **207,** 1356 (1965). Formulation and pharmacological studies: Wai *et al., J. Pharm. Pharmacol.* **22,** 923 (1970). *Reviews:* Makhlouf, *Fed. Proc.* **27,** 1322 (1968); Sanders, Schimmel, *Am. J. Med.* **49,** 380 (1970).

Fine, colorless needle-shaped crystals from 2-ethoxyethanol + water; mp 229-230° (dec). $[\alpha]_D^{22}$ −28.8° ±0.5° (DMF). uv max (2N ammonium hydroxide): 280, 289 nm (ε 5340, 4590). Sol in DMF, DMSO. Almost insol in water, ethanol, ether, benzene.

THERAP CAT: Diagnostic aid (gastric secretion stimulant).

7068. Pentagestrone. *3-(Cyclopentyloxy)-17-hydroxypregna-3,5-dien-20-one;* 17α-hydroxyprogesterone 3-cyclopentyl enol ether. $C_{26}H_{38}O_3$; mol wt 398.56. C 78.35%, H 9.61%, O 12.04%. Prepn of acetate: Ercoli, Gardi, *J. Am. Chem. Soc.* **82,** 746 (1960); of free alcohol and acetate: Ercoli, U.S. pat. **3,019,241** (1962).

Consult the cross index before using this section.

Solid, mp 184.5-186.5°. $[\alpha]_D$ −115° (dioxane).

Acetate, $C_{28}H_{40}O_4$, *17α-acetoxyprogesterone 3-cyclopentyl enol ether, Gestovis.* Solid, mp 137-138°, also reported as 157-158°, *see* Ercoli, **Brit.** pat. **893,315** (1962 to Vismara). $[\alpha]_D$ −147° (dioxane).

THERAP CAT: Progestogen.

7069. Pentahomoserine. *5-Hydroxynorvaline; α*-amino-*δ*-hydroxy-*n*-valeric acid; 2-amino-5-hydroxypentanoic acid. $C_5H_{11}NO_3$; mol wt 133.15. C 45.10%, H 8.33%, N 10.52%, O 36.05%. Synthesis: Sörensen, *Bull. Soc. Chim. France* **33**, 1052 (1904); Plieninger, *Ber.* **83**, 271 (1950); Gaudry, *Can. J. Chem.* **29**, 544 (1951). Enzymatic resolution: Berlinguet, Gaudry, *J. Biol. Chem.* **198**, 765 (1952).

L-Form, crystals. $[M]_D$ +38.3° (5N HCl); $[M]_D$ +46.6° (glacial acetic acid). $[\alpha]_D^{23}$ −28.8° (c = 2 in 6N HCl).

D-Form, crystals from dil alcohol. $[\alpha]_D^{23}$ −28.0° (c = 2 in 6N HCl).

DL-Form, crystals from dil alcohol, mp 218°; 218-220°; 223-224°, depending on rate of heating.

7070. Pentamethonium Bromide. *N,N,N,N′,N′,N′-Hex-amethyl-1,5-pentanediaminium bromide; pentamethylenebis-[trimethylammonium bromide]; α,ω-bis(trimethylammoni-um)pentane dibromide; 1,5-dibromohexamethylpentaneam-monium; C-5; Penthonium; Lytensium.* $C_{11}H_{28}Br_2N_2$; mol wt 348.18. C 37.94%, H 8.11%, Br 45.90%, N 8.05%. Ganglion blocking agent.

$$\left[(CH_3)_3 \overset{+}{N} - (CH_2)_5 - \overset{+}{N}(CH_3)_3 \right] \; 2Br^-$$

Hygroscopic crystals, slight fishy odor, slightly bitter, saline taste. Aq solns are stable, almost neutral; can be sterilized by autoclaving. The compd is almost completely ionized in aq solns. It contains 54.10% of pentamethonium ion.

THERAP CAT: Antihypertensive.

7071. Pentamidine. *4,4′-[1,5-Pentanediylbis(oxy)]bis-benzenecarboximidamide; 4,4′-(pentamethylenedioxy)dibenz-amidine; 4,4′-diamidino-α,ω-diphenoxypentane;* Pneumo-pent. $C_{19}H_{24}N_4O_2$; mol wt 340.43. C 67.04%, H 7.10%, N 16.46%, O 9.40%. Prepn: A. J. Ewins, **Brit.** pat. **507,565** (1939); J. N. Ashley *et al., J. Chem. Soc.* **1942**, 103; of isethi-onate: G. Newbery, A. P. T. Easson, **U.S.** pat. **2,394,003** (1946 to May & Baker). Trypanocidal activity: E. M. Lourie, W. Yorke, *Ann. Trop. Med. Parasitol.* **33**, 289 (1939). Preliminary pharmacological studies in animals: R. Wien, *ibid.* **37**, 1 (1943). Activity in fibrinolytic systems: J. D. Geratz, *Thromb. Diath. Haemorrh.* **29**, 154 (1973). Pharma-codynamics in men and mice: T. P. Waalkes *et al., Clin. Pharmacol. Ther.* **11**, 505 (1970). *In vitro* activity against *Pneumocystis carinii*: E. L. Pesanti, C. Cox, *Infec. Immun.* **34**, 908 (1981). Uptake and distribution of aerosolized form in animals: R. J. Debs *et al., Am. Rev. Respir. Dis.* **135**, 731 (1987). *In vivo* efficacy of aerosolized form in rats: *eidem, Antimicrob. Ag. Chemother.* **31**, 37 (1987). Determn in plasma, urine and tissues: T. P. Waalkes, V. T. DeVita, *J. Lab. Clin. Med.* **75**, 871 (1970); by HPLC: C. M. Dickinson *et al., J. Chromatog.* **345**, 91 (1985). Preliminary clinical evaluation in *P. carinii* pneumonia: V. T. DeVita *et al., N. Engl. J. Med.* **280**, 287 (1968). Comparison with sulfamethoxazole-trimethoprim mixture in *P. carinii* pneumonia in AIDS: J. M. Wharton *et al., Ann. Int. Med.* **105**, 37 (1986). Early

review of pharmacology, mode of action and clinical appli-cations: E. B. Schoenbach, E. M. Greenspan, *Medicine* **27**, 327-377 (1948). *Review:* S. Drake *et al., Clin. Pharm.* **4**, 507-516 (1985); M. Sands *et al., Rev. Infec. Dis.* **7**, 625-634 (1985); of activity, pharmacokinetics and therapeutic use: K. L. Goa, D. M. Campoli-Richards, *Drugs* **33**, 242-258 (1987).

Crystallizes as colorless plates from water. Dec 186°.

Dihydrochloride, $C_{19}H_{26}Cl_2N_4O_2$, fine needles from dil HCl, mp 232-234°. LD_{50} in mice (mg/g): 0.028 i.v.; 0.064 s.c. (Wein).

Isethionate, $C_{23}H_{36}N_4O_{10}S_2$, *M & B 800, RP 2512, Aero-pent, Benambax, Pentacarinat, Pentam 300.* Hygroscopic, very bitter crystals, mp about 180°. Slight butyric odor. Sol in water (about 1 in 10 at 25°, about 1 in 4 at 100°); sol in glycerol, more readily on warming; slightly sol in alcohol. Insol in ether, acetone, chloroform, liq petr. pH of a 5% w/v soln in water: 4.5 to 6.5.

Dimethanesulfonate, $C_{21}H_{32}N_4O_8S_2$, *pentamidine mesylate, Lomidine.* White powder.

THERAP CAT: Antiprotozoal (Trypanosoma, Leishmania); antipneumocystis.

THERAP CAT (VET): Antiprotozoal (Babesia, Leishmania).

7072. Pentane. *n*-Pentane. C_5H_{12}; mol wt 72.15. C 83.23%, H 16.77%. $CH_3CH_2CH_2CH_2CH_3$. Occurs in petro-leum; it is a constituent of gasoline. Sepn from natural gasoline: Love, *Petrol. Eng.* **12**, no. 10, 130 sqq (1941), *C.A.* **35**, 7162 (1941); from virgin naphthas: Tongberg *et al., Ind. Eng. Chem.* **30**, 166 (1938). Prepd by dehydration and sub-sequent hydrogenation of 2- and 3-pentanol: Mair, *Bur. Stand. J. Res.* **9**, 457 (1932); from 2-bromopentane by Gri-gnard reaction: Noller, *Org. Syn.* **11**, 84 (1931).

Flammable liq. d_0^0 0.64529; d_4^{20} 0.62638; d_4^{30} 0.6163; mp −129.7°. bp_{760} 36.1°; bp_{400} 18.5°; bp_{200} +1.9°; bp_{100} −12.6°; bp_{60} −22.2°; bp_{40} −29.2°; bp_{20} −40.2°; bp_{10} −50.1°; bp_5 −62.5°; $bp_{1.0}$ −76.6°; n_D^{20} 1.35768. Flash pt, closed cup: < −40°F (−40°C). Explosive limits, % by vol in air: lower 1.4; upper 8.0. Autoignition temp +588°F (+309°C). Soly in water at 16°: 0.36 g/l. Miscible with alc, ether, many organic solvents. Lethal concn for mice in air: 128,200 ppm, *Handbook of Toxicology* vol. 1, W. S. Spector, Ed. (Saunders, Philadelphia, 1956) pp 346-347.

Caution: Narcotic in high concns.

7073. 1,5-Pentanediol. Pentamethylene glycol; 1,5-di-hydroxypentane. $C_5H_{12}O_2$; mol wt 104.15. C 57.66%, H 11.61%, O 30.72%. $HOCH_2(CH_2)_3CH_2OH$. Prepd by hy-drogenolysis of tetrahydrofurfuryl alcohol in the presence of copper chromite: Connor, Adkins, *J. Am. Chem. Soc.* **54**, 4678 (1932); D. Kaufman, W. Reeve, *Org. Syn.* **coll. vol. III**, 693 (1955).

Viscous, oily liquid. Bitter taste. d^{20} 0.9941. mp −18°. bp_{760} 239°, $bp_{3.0}$ 120°. n_D^{20} 1.4499. Flash pt 125° (275°F). Miscible with water, methanol, alc, acetone, ethyl acetate. Soly in ether (25°): 11% w/w. Limited soly in benzene, tri-chloroethylene, methylene chloride, petr ether, heptane. LD_{50} orally in rats: 5.89 g/kg, H. F. Smyth *et al., Am. Ind. Hyg. Assoc. J.* **23**, 95 (1962).

USE: As plasticizer in cellulose products and adhesives, in brake fluid compositions. Forms esters and polyesters which can be used as plasticizers, emulsifying agents and resin intermediates.

7074. 1-Pentanol. Pentyl alcohol; *n*-amyl alcohol; *n*-bu-tyl carbinol. $C_5H_{12}O$; mol wt 88.15. C 68.13%, H 13.72%, O 18.15%. $CH_3(CH_2)_4OH$. Prepn from 1-pentene: Brown, Rao, *J. Am. Chem. Soc.* **81**, 6434 (1959); Brown, **U.S.** pat. **2,925,437** (1960). Review of manuf by fractionation of fusel oil and via chlorination of pentanes, and properties: *Indus-trial Chemicals*, W. L. Faith *et al.*, Eds. (John Wiley, New York, 2nd ed., 1957) pp 107-114.

Liquid, mild characteristic odor. bp 137.5°. mp −79°. d_4^{20} 0.8146; d_4^{25} 0.8110. n_D^{20} 1.4103; Mumford, Phillips. *J.*

Chem. Soc. **1950,** 75. Flash pt, closed cup: 100°F (38°C). Slightly sol in water (2.7 g/100 ml at 22°); misc with alc, ether. LD_{50} orally in rats: 3030 mg/kg, P. M. Jenner *et al., Food Cosmet. Toxicol.* **2,** 327 (1964).

USE: In organic syntheses; as solvent. *Toxicity:* Irritating to eyes, respiratory passages. Narcotic: E. Browning, *Toxicity and Metabolism of Industrial Solvents* (Elsevier, New York, 1965) pp 356-367.

7075. 2-Pentanol. *dl-sec*-Amyl alcohol; methyl propyl carbinol. $C_5H_{12}O$; mol wt 88.15. C 68.13%, H 13.72%, O 18.15%. $CH_3CH_2CH_2CH(OH)CH_3$. Prepn: Brown, Nakagawa, *J. Am. Chem. Soc.* **77,** 3614 (1955); Brown, Wheeler, *ibid.* **78,** 2199 (1956). Sepn of optical isomers: Adembri, *Ann. Chim. (Rome)* **46,** 62 (1956). Review of manuf by fractionation of fusel oil and via chlorination of pentanes, and properties: *Industrial Chemicals,* W. L. Faith *et al.,* Eds. (John Wiley, New York, 2nd ed., 1957) pp 107-114.

Liquid, characteristic odor. bp 119.3°. bp_{745} 118°. d_4^{20} 0.8098. n_D^{25} 1.4041, n_D^{20} 1.406. Flash pt (open cup) 105°F. Slightly sol in water (16.6 g/100 ml at 20°). Miscible with alcohol, ether.

Caution: See 1-Pentanol.

7076. 3-Pentanol. Diethyl carbinol. $C_5H_{12}O$; mol wt 88.15. C 68.13%, H 13.72%, O 18.15%. $CH_3CH_2CH(OH)$-CH_2CH_3. Prepn: Brown, Nakagawa, *J. Am. Chem. Soc.* **77,** 3614 (1955). Review of manuf by fractionation of fusel oil and via chlorination of pentanes, and properties: *Industrial Chemicals,* W. L. Faith *et al.,* Eds. (John Wiley, New York, 2nd ed., 1957) pp 107-114.

Liquid, characteristic odor. bp 115.6°, bp_{738} 113.5-113.7°. d_4^{25} 0.815; n_D^{25} 1.4077, n_D^{20} 1.4097. Slightly sol in water (5.5 g/100 g at 30°); sol in alcohol, ether. LD_{50} orally in rats: 1.87 g/kg, Smyth *et al., Arch. Ind. Hyg. Occup. Med.* **10,** 61 (1954).

USE: As flotation agent, as solvent, in organic synthesis. *Caution: See 1-Pentanol.*

7077. Pentapiperide. *α-(1-Methylpropyl)benzeneacetic acid 1-methyl-4-piperidinyl ester; 3-methyl-2-phenylvaleric acid 1-methyl-4-piperidyl ester;* 2-phenyl-3-methylpentanoic acid 1-methyl-4-piperidyl ester; 1-methyl-4-piperidyl 3-methyl-2-phenylvalerate. $C_{18}H_{27}NO_2$; mol wt 289.40. C 74.70%, H 9.40%, N 4.84%, O 11.06%. Prepn: **Brit.** pat. **781,382** and Martin, Habicht, **U.S.** pat. **2,987,517** (1957, 1961 to Cilag).

Fumarate, $C_{22}H_{31}NO_6$, *C 4675.* Crystals, mp 91-93°.

Methosulfate, $C_{20}H_{33}NO_6S$, *pentapiperium methylsulfate, pentapiperide methylsulfate, Crylène, Hycholin, Quilene.* Crystals, mp 110-12°.

THERAP CAT: Anticholinergic; antispasmodic.

THERAP CAT (VET): Anticholinergic.

7078. Pentazocine. *(2α,6α,11R*)-1,2,3,4,5,6-Hexahydro-6,11-dimethyl-3-(3-methyl-2-butenyl)-2,6-methano-3-benzazocin-8-ol;* 2-dimethylallyl-5,9-dimethyl-2'-hydroxybenzomorphan; 3-(3-methyl-2-butenyl)-1,2,3,4,5,6-hexahydro-6,11-dimethyl-2,6-methano-3-benzazocin-8-ol; NSC-107430; Win 20228; Fortalin; Fortalgesic; Fortral; Liticon; Pentagin; Pentalgina; Sosigon; Talwin. $C_{19}H_{27}NO$; mol wt 285.44. C 79.95%, H 9.54%, N 4.91%, O 5.60%. Mixed opioid agonist-antagonist. Prepn: Archer, **Belg.** pat. **611,000** (1962 to Sterling Drug), *C.A.* **58,** 2440b (1963); Archer *et al., J. Med. Chem.* **7,** 123 (1964). Review of pharmacology: Brogden *et al., Drugs* **5,** 6-91 (1973). Clinical study in post-operative pain: F. Camu, *Eur. J. Clin. Pharmacol.* **19,** 259 (1981). HPLC analysis in blood and plasma: R. D. Anderson *et al., J. Chromatog.* **227,** 239 (1982). Toxicity data: E. I. Goldenthal, *Toxicol. Appl. Pharmacol.* **18,** 185 (1971).

Crystals from methanol + water, mp 145.4-147.2°. LD_{50} s.c. in male rats: 175±36 mg/kg (Goldenthal).

Hydrochloride, $C_{19}H_{28}ClNO$, *Algopent.*

Caution: May be habit forming. This is a controlled substance listed in the U.S. Code of Federal Regulations, Title 21 Part 1308.14 (1987).

THERAP CAT: Narcotic analgesic.

7079. 1-Pentene. Propylethylene; *α-n*-amylene. C_5H_{10}; mol wt 70.13. C 85.63%, H 14.37%. $CH_3(CH_2)_2CH=CH_2$. Occurs in coal tar. Prepd from allyl bromide and ethylmagnesium bromide in ether or better in dipropyl ether: Meisenheimer, Casper, *Ber.* **54,** 1663 (1921); Norris, Joubert, *J. Am. Chem. Soc.* **49,** 885 (1927); Kirrmann, *Compt. Rend.* **184,** 1178 (1927).

Liquid. d_4^{20} 0.6429; bp_{760} 30.1°; n_D^{20} 1.3714. Insol in water; miscible with alcohol, ether, benzene.

7080. 2-Pentene. *β-n*-Amylene; *sym*-methylethylethylene. C_5H_{10}; mol wt 70.13. C 85.63%, H 14.37%. CH_3CH_2-$CH=CHCH_3$. Prepn of mixture of *cis* and *trans* isomers by dehydration of 2-pentanol: Norris, *Org. Syn.* **coll. vol. I,** (2nd ed., 1941) p 430. Prepn of the geometrical isomers from *cis*- and *trans-α*-methyl-*β*-ethylacrylic acids: Lucas, Prater, *J. Am. Chem. Soc.* **59,** 1682 (1937). Prepn of the *cis* form from 1-ethyl-2-iodobutyric acid with quinoline, of the *trans* form from 1-ethyl-2-iodobutyric acid with sodium carbonate: Sherrill, Matlak, *ibid.* 2134; prepn of both isomers from *sec*-amyl alcohol with H_2SO_4 and diatomaceous earth at 90-110° for 3 hrs: Lucas *et al., ibid.* **63,** 22 (1941). Absorption spectra: Carr, Stücklen, *ibid.* **59,** 2138 (1937).

cis-Form, liquid. d_4^{20} 0.6503; d_4^{80} 0.5824; d_4^{30} 0.6392; d_4^0 0.6710. mp −180 to −178°. bp_{760} 37.0°. n_D^{20} 1.38130.

trans-Form, liquid. d_4^{20} 0.6482; d_4^{80} 0.5814; d_4^{30} 0.6381; d_4^0 0.6675. mp −136 to −135°. bp_{760} 35.85°. n_D^{20} 1.37921.

7081. 2-Pentenylpenicillin Sodium. *3,3-Dimethyl-7-oxo-6-[(1-oxo-3-hexenyl)amino]-4-thia-1-azabicyclo[3.2.0]-heptane-2-carboxylic acid monosodium salt; Δ-β,γ-pentenyl-penicillin sodium; sodium penicillin F; sodium penicillin I; sodium 2-pentenylpenicillinate.* $C_{14}H_{19}N_2NaO_4S$; mol wt 334.37. C 50.29%, H 5.73%, N 8.38%, Na 6.88%, O 19.14%, S 9.59%. 2-Pentenylpenicillin is produced by certain strains of *Penicillium chrysogenum* and *P. notatum.* Separation and purification by partition chromatography, *cf.* monograph by Boon and Carrington in *Chemistry of Penicillin* (Princeton, 1949). Prepn of cryst penicillin F: Wintersteiner, Adler, **U.S.** pat. **2,485,227** (1949 to Squibb).

Crystals. A trihydrate and a sesquihydrate have been obtained. The anhydr form exists in two cryst modifications. Usually long, fine, blunt-ended needles from water + butanol. After drying *in vacuo* over P_2O_5 and then at 55-60° in high vacuum: mp 204-205° (with decompn, open capillary in block preheated to 200°). $[α]_D^{25}$ +316° (c = 0.88). pK in water at 5° = 2.83; at 25° = 2.87. Very sol in water, in isotonic sodium chloride soln and in glucose solns; also sol in alcohol, but is inactivated by this solvent, likewise by glycerol and other primary alcohols; insol in benzene, carbon tetrachloride, liquid petrolatum.

7082. Pentetate Calcium Trisodium. *[N,N-Bis[2-[bis-(carboxymethyl)amino]ethyl]glycinato(5−)]calciate(3−)tri-sodium;* calcium trisodium pentetate; sodium[[[(carboxymethyl)imino]bis(ethylenenitrilo)]tetraacetato]calciate;

[[(carboxymethyl)imino]bis(ethylenenitrilo)]tetraacetic acid calcium complex trisodium salt; trisodium calcium diethylenetriaminepentaacetate; pentacin; Calcium Chel 330; Ditripentat; Penthamil. $C_{14}H_{18}CaN_3Na_3O_{10}$; mol wt 497.36. C 33.81%, H 3.65%, Ca 8.06%, N 8.45%, Na 13.87%, O 32.17%. Prepn: Rubin, Dexter, U.S. pat. **3,062,719** (1962 to Geigy). Effects in uranium poisoning in rats: R. Dagirmanjian et al., J. Pharmacol. Exp. Ther. **117**, 20 (1956); in mixed fusion products poisoning: K. Kostial et al., J. Appl. Toxicol. **3**, 291 (1983).

Solid. Sol in water. Practically insol in alcohol. Used in aq soln. LD_{50} i.p. in rats: 3.8 g/kg (Dagirmanjian).

THERAP CAT: Chelating agent (plutonium and other transuranium elements).

7083. Pentetic Acid. *N,N-Bis[2-[bis(carboxymethyl)-amino]ethyl]glycine; [[(carboxymethyl)imino]bis(ethylenenitrilo)]tetraacetic acid;* pentacarboxymethyl diethylenetriamine; diethylenetriamine pentaacetic acid; DTPA. $C_{14}H_{23}N_3O_{10}$; mol wt 393.35. C 42.74%, H 5.89%, N 10.68%, O 40.68%. Prepn of the acid: Curme et al., U.S. pat. **2,384,816** (1945 to Carbide and Carbon Chem.); Samoilova, Yashunskii, U.S.S.R. pat. **144,479** (1962), C.A. **57**, 12326d (1962); of Na salt: Brit. pats. **601,816/7** (1948 to Carbide and Carbon Chem.). Structure and infrared spectrum: Nakamoto et al., J. Am. Chem. Soc. **85**, 309 (1963).

$$HOOCCH_2\!\!\diagdown\!\!N\!\!-\!CH_2CH_2NCH_2CH_2N\!\!\diagup\!\!CH_2COOH$$
(with CH₂COOH branch and CH₂COOH groups)

Trisodium calcium salt, *see* Pentetate Calcium Trisodium.

USE: Chelating agent.

THERAP CAT: Chelating agent (iron).

7084. Penthienate Bromide. *2-[(Cyclopentylhydroxy-2-thienylacetyl)oxy]-N,N-diethyl-N-methylethanaminium bromide; diethyl(2-hydroxyethyl)methylammonium bromide* α-*cyclopentyl-2-thiopheneglycolate;* diethyl-2-[1-hydroxy-1-cyclopentyl-1-(2-thienyl)acetoxy]ethylmethylammonium bromide; 2-diethylaminoethyl-2-cyclopentyl-2-(2-thienyl)-hydroxyacetate methobromide; 2-diethylamino-α-cyclopentyl-2-thiopheneglycolate methobromide; Win 4369; Monodral bromide. $C_{18}H_{30}BrNO_3S$; mol wt 420.41. C 51.42%, H 7.19%, Br 19.01%, N 3.33%, O 11.42%, S 7.63%. Prepn: Blicke, Tsao, J. Am. Chem. Soc. **66**, 1645 (1944); Blicke, U.S. pat. **2,541,634** (1951).

Pale yellow crystals, mp 124.6°. uv max 238 nm ($A_{1cm}^{1\%}$ 189). One gram dissolves in 5 ml water. Freely sol in alc, chloroform. Slightly sol in acetone. Insol in ether. pH of a 1% aq soln 5.5 to 7.0.

THERAP CAT: Anticholinergic.

7085. Pentifylline. *1-Hexyl-3,7-dihydro-3,7-dimethyl-1H-purine-2,6-dione;* 1-hexyltheobromine; 1-hexyl-3,7-dimethylxanthine; 3,7-dimethyl-1-hexyl-1H,3H-purin-2,6-dione; SK-7. $C_{13}H_{20}N_4O_2$; mol wt 264.32. C 59.07%, H 7.63%, N 21.20%, O 12.11%. Prepn from theobromine and hexyl halide: Eidebenz, Schuh, Ger. pat. **860,217** (1952 to Chem. Werke Albert); Serchi, Chimica **40**, 451 (1964); Chkhikvadze et al., U.S.S.R. pat. **202,152** (1967), C.A. **69**, 19368x (1968). Pharmacology: Cugurra, Echinard-Garin, Arch. Int. Pharmacodyn. Ther. **123**, 481 (1960); Ramos et al., ibid. **153**, 430 (1965); Mohler et al., Arzneimittel-Forsch. **16**, 1524 (1966). Metabolic studies: Mohler et al., Arch. Pharm. **299**, 448 (1966).

Crystals, mp 82-83°.

Principal ingredient of *Cosaldon* (also contains nicotinic acid): Brit. pat. **815,969** (1959 to Chem. Werke Albert).

USE: Stabilizer of vitamin preparations: Nook, Eidebenz, Ger. pat. **1,810,705** (1970 to Chem. Werke Albert).

THERAP CAT: Vasodilator.

7086. Pentigetide. *N^2-[1-[N-(N-L-α-Aspartyl-L-seryl)-*L-α-*aspartyl]-L-prolyl]-L-arginine;* human IgE pentapeptide; HEPP; Pentyde. $C_{22}H_{36}N_8O_{11}$; mol wt 588.57. C 44.90%, H 6.16%, N 19.04%, O 29.90%. Synthetic pentapeptide with antiallergic activity. Structure corresponds to amino acids 320-324 of the epsilon chain of human immunoglobulin E (IgE). Prepn: R. N. Hamburger, Ger. pat. **2,602,443**; idem, U.S. pat. **4,171,299** (1976, 1979 both to Univ. California). Inhibition of IgE-mediated immediate hypersensitivity response: idem, Science **189**, 389 (1975). Stimulation of peritoneal macrophages: E. Tzehoval et al., Proc. Nat. Acad. Sci. USA **75**, 3400 (1978). Mechanism of action study: R. N. Hamburger, Immunology **38**, 781 (1979). Clinical evaluations in allergic rhinitis: G. A. Cohen et al., Ann. Allergy **52**, 83 (1984); B. M. Prenner, ibid. **58**, 332 (1987).

Asp-Ser-Asp-Pro-Arg

$[\alpha]_D^{20}$ −78.6° (c = 1 in water).

THERAP CAT: Antiallergic.

7087. Pentobarbital Sodium. *5-Ethyl-5-(1-methylbutyl)-2,4,6(1H,3H,5H)-pyrimidinetrione monosodium salt; sodium 5-ethyl-5-(1-methylbutyl)barbiturate;* soluble pentobarbital; pentobarbitone sodium; 844; Nembutal; Embutal; Pentyl; Pentone; Praecicalm; Sopental; Carbrital; Continal; Euthatal; Mebumal Sodium; Sagatal; Penbar; Somnopentyl; Sonistan. $C_{11}H_{17}N_2NaO_3$; mol wt 248.26. C 53.22%, H 6.90%, N 11.28%, Na 9.26%, O 19.33%. Toxicity data: E. W. Schafer, Toxicol. Appl. Pharmacol. **21**, 315 (1972).

Crystalline granules or white powder. Slightly bitter taste. Dec ∼127°. Freely sol in water, alc. Practically insol in ether. Aq solns are unstable and dec on storage. They must not be sterilized by boiling. LD_{50} orally in rats: 118 mg/kg (Schafer).

Acid, $C_{11}H_{18}N_2O_3$, **Barpental, Sotyl, Neodorm**.

Calcium salt, $C_{22}H_{34}CaN_4O_6$, **pentobarbital calcium, Nembutal Calcium, Repocal**.

Caution: May be habit forming. This is a controlled substance (depressant) listed in the U.S. Code of Federal Regulations, Title 21 Parts 329.1, 1308.12 and 1308.13 (1987).

THERAP CAT: Sedative, hypnotic.

THERAP CAT (VET): Anesthetic (intravenous); for euthanasia.

7088. 1-Pentol. *3-Methyl-2-penten-4-yn-1-ol;* 1-hydroxy-3-methyl-2-penten-4-yne. C_6H_8O; mol wt 96.12. C 74.97%, H 8.39%, O 16.65%. Prepd by allylic rearrangement of methylvinylethynyl carbinol with moderately strong acid at 80°: Oroshnik, J. Am. Chem. Soc. **78**, 2651 (1956).

$$HC\!\equiv\!C\!-\!\overset{\overset{\displaystyle CH_3}{|}}{C}\!=\!CHCH_2OH$$

cis-Isomer [1'-pentol], oily liquid. $bp_{9.4}$ 65°. n_D^{20} 1.4820. uv max (ethanol): 223 nm (ϵ 11,000).

p-Nitrobenzoate, mp 61-62°.

trans-Isomer [1''-pentol], oily liquid. $bp_{9.4}$ 73°. n_D^{20} 1.4934. uv max (ethanol): 224 nm (ϵ 13,100).

p-Nitrobenzoate, mp 63-64°.

Caution: Both isomers tend to polymerize and will explode when heated above 120° in a sealed bomb tube.

USE: Intermediate in vitamin A synthesis.

7089. Pentolinium Tartrate. *1,1'-(1,5-Pentanediyl)bis-[1-methylpyrrolidinium] salt with [R-(R*,R*)]-2,3-dihydroxybutanedioic acid (1:2);* pentamethylene-1,5-bis(1-methylpyrrolidinium) hydrogen tartrate; pentapyrrolidinium bitartrate; pentolonium bitartrate; M & B 2050A; Ansolysen Tartrate; Ansolysen Bitartrate; Pentilium. $C_{23}H_{42}N_2O_{12}$; mol wt 538.58. C 51.29%, H 7.86%, N 5.20%, O 35.65%. General procedure of prepn: Libman *et al., J. Chem. Soc.* **1952,** 2305. Ganglionic blocking agent.

Crystals, dec 203°. Non-hygroscopic. Acid taste. Freely sol in water or in 25% aq polyvinylpyrrolidone soln. One gram dissolves in 0.4 ml water, in 810 ml ethanol. Insol in ether, chloroform. A 10% aq soln has a pH of about 3.5.

THERAP CAT: Antihypertensive.

7090. Pentosan Polysulfate. *Xylan hydrogen sulfate;* xylan polysulfate; CB 8061; Fibrase; Hémoclar. Semisynthetic sulfated polyanion composed of β-D-xylopyranose residues with properties similar to heparin, *q.v.* Mol wt ranges from 1500 to 5000. Prepn: **Swiss pat. 293,566** (1953 to Wander), *C.A.* **49,** 1787h (1955). Anticoagulant activity: B. Paramelle, *Therapie* **17,** 719 (1962); J.-B., Dureux *et al., ibid.* **19,** 879 (1964). Pharmacokinetics: R. Taugner *et al., Arch. Int. Pharmacodyn.* **189,** 250 (1971). Effects on aggregation of human blood platelets: G. Kindness *et al., Thromb. Res.* **16,** 97 (1979). Comparison of pentosan polysulfate with heparin: C. Soria *et al., ibid.* **19,** 455 (1980); M. Ryde *et al., ibid.* **23,** 435 (1981); A.-M. Fischer *et al., Thromb. Haemostasis* **47,** 104, 109 (1982).

Sodium salt, *sodium pentosan polysulfate, sodium xylan polysulfate, SP 54, Thrombocid.* White odorless powder, slightly hygroscopic. $[\alpha]_D^{20}$ −57°. pH of 10% aq soln ∼6.0. n_D^{20} of 10% aq soln: 1.344. Sol in water.

THERAP CAT: Anticoagulant.

7091. Pentostatin. *(R)-3-(2-Deoxy-β-D-erythro-pentofuranosyl)-3,6,7,8-tetrahydroimidazo[4,5-d][1,3]diazepin-8-ol;* 2'-deoxycoformycin; DCF; 2'-dCF; Cl 67310465; NSC-218321; CI-825. $C_{11}H_{16}N_4O_4$; mol wt 268.27. C 49.25%, H 6.01%, N 20.88%, O 23.86%. Potent adenosine deaminase inhibitor. Isolation from *Streptomyces antibioticus* and structure determn: P. W. K. Woo *et al., J. Heterocycl. Chem.* **11,** 641 (1974). Isolation and purification: A. Ryder *et al.,* **Ger. pat. 2,517,596;** *eidem,* **U.S. pat. 3,923,785** (both 1975 to Parke-Davis). Total synthesis: E. Chan *et al., J. Org. Chem.* **47,** 3457 (1982). Preliminary biosynthetic study: J. C. Hanvey *et al., Biochemistry* **27,** 5790 (1988). Inhibition of deaminases *in vitro:* T. Rogler-Brown *et al., Biochem. Pharmacol.* **27,** 2289 (1978); C. Frieden *et al., Biochem. Biophys. Res. Commun.* **91,** 278 (1979); *in vivo:* W. Plunkett *et*

al., Biochem. Pharmacol. **28,** 201 (1979). Pharmacokinetics in mice: W. R. McConnell *et al., Drug Metab. Dispos.* **7,** 11 (1979). Toxicology study: J. F. Smyth *et al., Cancer Chemother. Pharmacol.* **1,** 49 (1978). Clinical pharmacology: J. F. Smyth *et al., ibid.* **5,** 93 (1980); P. P. Major *et al., Blood* **58,** 91 (1981). Enzymatic determn in biological fluids: M. M. Chassin *et al., Biochem. Pharmacol.* **28,** 1849 (1979); A. E. Staubus *et al., ibid.* **33,** 1633 (1984). Efficacy in cancer patients: F. J. Cummings *et al., Clin. Pharmacol. Ther.* **44,** 501 (1988). Clinical evaluation in hairy cell leukemia: J. B. Johnston *et al., J. Nat. Cancer Inst.* **80,** 765 (1988). Brief review: P. O'Dwyer *et al., Ann. Int. Med.* **108,** 733 (1988).

White crystals from methanol/water, mp 220-225° (Woo), also reported as 204-209.5° with darkening at > 150° (Chan). uv mx (water, pH 7): 282 nm (ϵ 8000); (pH 11): 283 nm (ϵ 7970); (pH 2) : 273 nm (ϵ 7570 initially, 3143 after 6.5 hrs). $[\alpha]_D^{25}$ +76.4° (c = 1 in water); $[\alpha]_D^{23}$ +73.0° (c = 1, pH 7 buffer). pKa 5.2 in water.

5'-Phosphate, $C_{11}H_{17}N_4O_7P$.

THERAP CAT: Antineoplastic.

7092. Pentoxifylline. *3,7-Dihydro-3,7-dimethyl-1-(5-oxohexyl)-1H-purine-2,6-dione; 1-(5-oxohexyl)theobromine;* 1-(5-oxohexyl)-3,7-dimethylxanthine; 3,7-dimethyl-1-(5-oxohexyl)-1H,3H-purin-2,6-dione; oxpentifylline; vazofirin; BL 191; Azupentat; Durapental; Rentylin; Torental; Trental; Xiphen. $C_{13}H_{18}N_4O_3$; mol wt 278.31. C 56.10%, H 6.52%, N 20.13%, O 17.25%. Identity as a metabolite of pentifylline, *q.v.:* Mohler *et al., Arch. Pharm.* **299,** 448 (1966); Mohler *et al., Arzneimittel-Forsch.* **16,** 1524 (1966). Prepn: **Neth. pat. Appl. 6,511,581;** Mohler *et al.,* **Ger. pat. 1,235,-320;** *eidem,* **U.S. pat. 3,422,107** (1966, 1967, 1969 all to Chem. Werke Albert). Chemistry and synthesis data: W. Mohler, A. Söder, *Arzneimittel-Forsch.* **21,** 1159 (1971). Pharmacology: K. Popendiker *et al., ibid.* 1160; I. Boksay *et al., ibid.* 1174; U. J. Jovanovic, *ibid.* **22,** 994 (1972). Clinical trials: F. Lehrach, R. Müller, *ibid.* **21,** 1171 (1971); L. Wilbert, *ibid.* **22,** 751 (1972). *Review:* R. Müller, F. Lehrach, *Curr. Res. Med. Opin.* **7,** 253-263 (1981); of pharmacodynamics, pharmacokinetics and therapeutic efficacy: A. Ward, S. P. Clissold, *Drugs* **34,** 50-97 (1987).

Bitter tasting, colorless needles from methanol, mp 105°. uv max: 273, 208 nm ($E_{1cm}^{1\%}$ 365, 935). Soly in water: 77 mg/ml at 25°, 191 mg/ml at 37°; in benzene: 11 g/100 ml. LD_{50} orally in mice: 1385 mg/kg (Popendiker).

THERAP CAT: Vasodilator.

7093. Pentoxyl. *5-Hydroxymethyl-6-methyl-2,4(1H,-3H)-pyrimidinedione; 5-hydroxymethyl-6-methyluracil;* 5-hydroxymethyl-4-methyluracil; 4-methyl-5-hydroxymethyluracil. $C_6H_8N_2O_3$; mol wt 156.14. C 46.15%, H 5.16%, N 17.94%, O 30.74%. Prepn: Endicott, Johnson, *J. Am. Chem. Soc.* **63,** 2063 (1941).

Crystals from boiling water, dec 314-315°.

THERAP CAT: Leukopoietic stimulant.

7094. Pentrinitrol. *2,2-Bis[(nitrooxy)methyl]-1,3-propanediol mononitrate (ester); pentaerythritol trinitrate;* W 2197; Petrin. $C_5H_9N_3O_{10}$; mol wt 271.14. C 22.15%, H 3.34%, N 15.50%, O 59.01%. Coronary vasodilator related to pentaerythritol tetranitrate, *q.v.* Prepn (no data given): **Ger. pats.** 638,422-3 (1936 to Westfälisch-Anhaltisch Sprengstoff), *C.A.* **31,** 1212²(1937); N. J. Marans *et al., J. Am. Chem. Soc.* **76,** 1304 (1954); A. T. Camp *et al., ibid* **77,** 751 (1955); J. Simecek, *Coll. Czech. Chem. Commun.* **27,** 362 (1962). Use in coronary insufficiency: F. J. DiCarlo, **U.S. pat.** 3,419,571 (1968 to Warner-Lambert). Electrical moment and molecular rotation: A. R. Lawrence, A. J. Matuszko, *J. Phys. Chem.* **65,** 1903 (1961). Chromatographic determn: I. W. F. Davidson *et al., J. Chromatog.* **57,** 345 (1971). Metabolism: F. J. DiCarlo *et al., Clin. Pharmacol. Ther.* **22,** 309 (1977). Comparative duration of action: J. Vohra *et al., Aust. N. Z. J. Med.* **9,** 554 (1979).

Viscous liq. n_D^{20} 1.4941. d_4^{20} 1.554. Viscosity: 166.8 centistokes at 40°; 77.8 at 50°. Soly in water: 0.705 g/100 ml at 20°; in benzene: 21.40 g/100 ml at 20°. Very sol in ethanol, ether. Forms a cryst hydrate, mp 32°, when washed with water and allowed to stand overnight at 20°; returns to unhydrated form when allowed to stand at 60° for 2 hrs. When used medicinally it is usually diluted with lactose or mannitol to reduce explosive liability, *see* pentaerythritol tetranitrate.

THERAP CAT: Vasodilator (coronary).

7095. Pentryl®. *2-[Nitro(2,4,6-trinitrophenyl)amino]-ethanol nitrate (ester); 2-(N,2,4,6-tetranitroanilino)ethanol nitrate; sym-trinitrophenylnitroaminoethyl nitrate.* C_8H_6-N_6O_{11}; mol wt 362.19. C 26.53%, H 1.67%, N 23.21%, O 48.59%. Prepd by nitration of 2,4-dinitrophenylaminoethanol: Clark, *Ind. Eng. Chem.* **25,** 1385 (1933); Desseigne, *Mém. poudres* **33,** 255 (1951); B. T. Fedoroff *et al., Encyclopedia of Explosives and Related Items,* **vol. I** (Picatinny Arsenal, Dover, New Jersey, 1960) pp A425-429.

Small, cream-colored crystals from chloroform, mp 129° (slight decompn). d 1.82. Explodes when heated to 235°. Heat of combustion: 911.1 kcal/mole; heat of explosion: 372.4 kcal/mole; heat of formation: 43.4 kcal/mole. Soly (w/w) at 25° in benzene: 0.70; carbon tetrachloride: trace; chloroform: 0.07; ethanol: 0.11; ether: 0.16; ethylene dichloride: 0.72; methanol: 0.67; nitroglycerol: freely sol; toluene: 0.63; water: trace.

USE: High explosive. Base charge in detonators.

7096. tert-Pentyl Alcohol. *2-Methyl-2-butanol; tert-amyl alcohol; dimethyl ethyl carbinol; ethyl dimethyl carbinol; tert-pentanol; amylene hydrate.* $C_5H_{12}O$; mol wt 88.15. C 68.13%, H 13.72%, O 18.15%. $(CH_3)_2C(OH)CH_2CH_3$. Prepd from 2-methyl-2-butene alone or mixed with 2-methyl-1-butene: Fenske, Jones, **U.S. pat.** 2,858,331 (1958 to Esso); Odioso *et al., Ind. Eng. Chem.* **53,** 209 (1961).

Review of manuf by fractionation of fusel oil and *via* chlorination of pentanes, and properties: *Industrial Chemicals,* W. L. Faith *et al.,* Eds. (John Wiley, New York, 2nd ed., 1957) pp 107-114. Toxicity: Schaffarzick, Brown, *Science* **116,** 663 (1952).

Volatile liquid; characteristic odor, burning taste. bp_{765} 102.5°. mp −9.0°. d^{20} 0.8084: Costello, Bowden, *Rec. Trav. Chim.* **77,** 36 (1958). n_D^{20} 1.4052. Flash pt, closed cup: 67°F (19°C); open cup: 70°F (21°C). Sol in 8 parts water; miscible with alcohol, ether, benzene, chloroform, glycerol, oils. *Keep tightly closed and protected from light.* LD_{50} orally in rats: 1.0 g/kg (Schaffarzick, Brown).

Human Toxicity: Moderately irritating to mucous membranes. Narcotic in high concns. *See also* 1-Pentanol.

THERAP CAT: Hypnotic.

7097. Pentylenetetrazole. *6,7,8,9-Tetrahydro-5H-tetra-zolo[1,5-a]azepine; α,β-cyclopentamethylenetetrazole; 1,5-pentamethylenetetrazole; 6,7,8,9-tetrahydro-5-azepotetrazole; 1,2,3,3a-tetrazacyclohepta-8a,2-cyclopentadiene; 7,8,-9,10-tetrazabicyclo[5.3.0]-8,10-decadiene;* Cardiazol; Cenalene-M; Cenazol; Coranormol; Corazole; Corvasol; Deumacard; Gewazol; Korazole; Leptazol; Metrazole; Pentetrazole; Phrenazol; Ventrazol. $C_6H_{10}N_4$; mol wt 138.17. C 52.15%, H 7.30%, N 40.55%. Prepd by the addition of cyclohexanone to a benzene or Tetralin solution of hydrazoic acid: Schmidt, *Ber.* **57,** 704 (1924); **U.S. pats.** 1,564,631 (1925); 1,599,493 (1926); Prochazka, **Czech.** pat. 92,456 (1959), *C.A.* **56,** 4776e (1962). Synthesis from caprolactam: Glushkov, Golovchinskaya, *Zh. Prikl. Khim.* **32,** 920 (1959); *Med. Prom. SSSR* **14,** no. 1, 12 (1960).

Slightly pungent, bitter crystals, mp 57-60°. Freely sol in water, most organic solvents. Aq solns are neutral to litmus. Very stable, not easily attacked by other substances. LD_{50} orally in rats (mg/kg): 85±2 s.c., 62 i.p., E. I. Goldenthal, *Toxicol. Appl. Pharmacol.* **18,** 185 (1971).

THERAP CAT: Central stimulant.

THERAP CAT (VET): CNS stimulant, narcotic antagonist.

7098. p-tert-Pentylphenol. *4-(1,1-Dimethylpropyl)phenol; p-tert-amylphenol; 2-methyl-2-p-hydroxyphenylbutane;* Pentaphen. $C_{11}H_{16}O$; mol wt 164.24. C 80.44%, H 9.82%, O 9.74%. Prepd by condensation of *tert*-pentanol or 2-methyl-3-butanol with phenol in the presence of aluminum chloride: Huston, Hsieh, *J. Am. Chem. Soc.* **58,** 439 (1936); Huston *et al., ibid.* **67,** 899 (1945). Physical properties: Pardee, Weinrich, *Ind. Eng. Chem.* **36,** 595 (1944).

Crystals, mp 94-95°. bp 262.5°. bp_{740} 248-250°, bp_{15} 138.5°, bp_3 112-120°. d_4^{20} 0.962. Practically insol in water; sol in alcohol, ether, benzene, chloroform. LD_{50} orally in rats: 3.08 g/kg, H. F. Smyth *et al., Am. Ind. Hyg. Assoc. J.* **23,** 95 (1962).

USE: In the manuf of oil-soluble resins; has been recommended as a germicide and fumigant; intermediate for organic mercury germicides, for pesticides, for chemicals used in rubber and petroleum industries.

7099. Peonidin. *3,5,7-Trihydroxy-2-(4-hydroxy-3-methoxyphenyl)-1-benzopyrylium chloride; 3,4',5,7-tetrahydroxy-3'-methoxyflavylium chloride; 3,4',5,7-tetrahydroxy-3'-methoxy-2-phenylbenzopyrylium chloride.* $C_{16}H_{13}ClO_7$; mol wt 336.73. C 57.07%, H 3.89%, Cl 10.53%, O 28.51%. The aglucon of peonin: R. Willstätter, T. J. Nolan, *Ann.*

408, 136 (1915). Structure and synthesis: T. J. Nolan *et al., J. Chem. Soc.* **1926,** 1968; S. Murakami, R. Robinson, *ibid.* **1928,** 1537.

Monohydrate, reddish-brown needles from aq 20% hydrochloric acid soln. Moderately sol in water giving a reddish-brown soln. Sol in alc giving a purplish-red soln.

3,5-Diglucoside, $C_{28}H_{33}ClO_{16}$, *3,5-bis(β-D-glucopyranosyloxy)-7-hydroxy-2-(4-hydroxy-3-methoxyphenyl)-1-benzopyrylium chloride, peonin.* From deep violet-red peonies: R. Willstätter, T. J. Nolan, *loc. cit.* Synthesis: R. Robinson, A. R. Todd, *J. Chem. Soc.* **1932,** 2488. Deep purple needles with water of crystn from dil HCl, mp 165-167° (dec).

7100. Peplomycin. *N¹-[3-[(1-Phenylethyl)amino]propyl]bleomycinamide; N¹-[3-[[(S)-α-methylbenzyl]amino]-propyl]bleomycinamide; pepleomycin; NK-631.* $C_{61}H_{88}N_{18}$-$O_{21}S_2$; mol wt 1473.62. C 49.72%, H 6.02%, N 17.11%, O 22.80%, S 4.35%. Deriv of bleomycin, *q.v.* with cytostatic activity and less pulmonary toxicity than the natural bleomycin mixture. Prepn of the *(R,S)*-form: H. Umezawa *et al.,* U.S. pat. **3,846,400** (1974 to Microbiochem. Res. Found.); of the *(S)*-form: T. Takita *et al.,* **Ger. pat. 2,828,-933** corresp to U. S. pat. **4,195,018** (1979, 1980 to Nippon Kayaku); W. Tanaka *et al., Heterocycles* **13,** 469 (1979). Biological study of degradation products: K. Takahashi *et al., J. Antibiot.* **32,** 36 (1979). General pharmacology: Y. Ishii *et al., Japan. J. Antibiot.* **31,** 886 (1978), *C.A.* **91,** 215c (1979). Absorption, distribution, excretion, metabolism: H. Takayama *et al., ibid.* 895, *C.A.* **91,** 32580j (1979). Properties and stability: A. Fuji *et al., Iyakuhin Kenkyu* **10,** 197 (1979), *C.A.* **91,** 9408a (1979). Effect in prostatic cancer: T. Niijima, K. Koiso, *Scand. J. Respir. Dis.* **Suppl. 55,** 177 (1980). Review of clinical studies: S. Oko, *Recent Results Cancer Res.* **74,** 163 (1980). Relative pulmonary toxicity: B. I. Sikic *et al., Cancer Treat. Rep.* **64,** 659 (1980). Acute toxicity study: K. Ito *et al., Japan. J. Antibiot.* **31,** 719 (1978), *C.A.* **91,** 68461k (1979).

Sulfate salt, $C_{61}H_{90}N_{18}O_{25}S_3$, *Pepleo Injection.* Pale yellow amorphous powder, mp 196-198°. $[\alpha]_{436}^{25}$ —2.0° (c = 1 in water). pKa 2.9, 4.8, 7.4, 9.0. Sol in water, methanol, acetic acid, DMSO, DMF. Slightly sol in dioxane. Insol in ethyl acetate, acetone, ether, benzene. Stable at 37° for 3 months, 50° for 6 weeks, room temp for 30 months in a sealed container. LD50 in male rats, mice (mg/kg): 234, 88 s.c.; 208, 85 i.p.; 245, 51 i.v., K. Ito *et al., loc. cit.*

THERAP CAT: Antineoplastic.

7101. Pepper. Black pepper. Dried, unripe fruit of *Piper nigrum* L., *Piperaceae.* White pepper is the decorticated ripe fruit of black pepper. *Habit.* India, Malabar coast, Philippines, Sumatra, Java, Ceylon, etc. Predominant pungent principle in black pepper is piperine, *q.v.* Other pungent substances present in small quantities are chavicine, piperidine, *q.q.v.* and piperettine. Isoln of constituents: R. Grewe *et al., Ber.* **103,** 3752 (1970). HPLC determn of constituents: M. Verzele *et al., J. Chromatog.* **172,** 493 (1979); M. Rathnawathie, K. A. Buckle, *ibid.* **264,** 316 (1983); D. E. Games, N. J. Alcock, *ibid.* **294,** 269 (1984). Insecticidal properties of black pepper: H. C. F. Su, *J. Econ. Entomol.* **70,** 18 (1977); W. P. Scott, G. H. McKibben, *ibid.* **71,** 343 (1978); H. C. F. Su, R. Horvat, *J. Agr. Food Chem.* **29,** 115 (1981).
USE: Chiefly as a spice.

7102. Peppermint. Brandy mint; lamb mint. Dried leaves and flowering tops of *Mentha piperita* L., *Labiatae. Habit.* Asia, Europe, North America; cultivated in gardens. *Constit.* Volatile oil, tannin, resin, gum.
USE: Flavoring agent. Peppermint oil and peppermint water as pharmaceutic aids (flavor).

7103. Pepsin. Puerzym. Mol wt 34,500. Principal digestive enzyme of gastric juice; controls the degradation of proteins to proteoses and peptones. Distinctive among enzymes for having a very low isoelectric point and a very low pH optimum. Hydrolyzes only peptide linkages. The N.F. grade digests not less than 3000 or more than 3500 times its wt of freshly coagulated and disintegrated egg albumin in 2½ hours at 52° in water acidulated with HCl. Isoln from gastric juice of swine and beef: Northrup, *J. Gen. Physiol.* **13,** 739 (1930); **16,** 615 (1933); from salmon and tuna: Norris, Elam, *J. Biol. Chem.* **134,** 443 (1940); **204,** 673 (1953). Amino acid composition: Brand in *Crystalline Enzymes,* J. H. Northrup *et al.* (Columbia Univ. Press, New York, 2nd ed., 1948) p 26. Active site: Herriott, *J. Gen. Physiol.* **45,** Suppl., 57 (1962). Review: Bovey, Yanari in *The Enzymes* **vol. 4,** P. D. Boyer *et al.,* Eds. (Academic Press, New York, 1960) pp 63-92.

White or yellowish-white translucent scales or granules, or an amorphous slightly hygroscopic powder, or spongy masses. It has a slightly acid or saline taste. $[\alpha]_D^{26}$ —64.5° (water pH 4.6). Isoelectric point < pH 1.0. Freely sol in water with more or less opalescence. Practically insol in alc, chloroform, ether. Very stable to acid. The activity of pepsin in solns is destroyed by heating above 70°, or by alkalies; dry pepsin is not injured by heating to 100°. Incompat. Alc, tannin, alkaline substances and salts of heavy metals.

Note: Pepsin of an activity of 4000, 5000, 6000, 10,000, 15,000, and 20,000 is likewise available, and it has also been obtained in pure crystalline form.
THERAP CAT: Enzyme (digestive).
THERAP CAT (VET): Has been used as a digestive aid in deficiency of pepsin secretion.

7104. Pepstatin. *N-[(3-Methyl-1-oxobutyl)-L-valyl-L-valyl-4-amino-3-hydroxy-6-methylheptanoyl-L-alanyl]-4-amino-3-hydroxy-6-methylheptanoic acid; N-isovaleryl-L-valyl-L-valyl-3-hydroxy-6-methyl-γ-aminoheptanoyl-L-alanyl-3-hydroxy-6-methyl-γ-aminoheptanoic acid; pepstatin A.* $C_{34}H_{63}N_5O_9$; mol wt 685.91. C 59.54%, H 9.26%, N 10.21%, O 20.99%. A pentapeptide pepsin inhibitor, isolated from cultured broths of *Streptomyces testaceus* Hamada et Okami and *Streptomyces argenteolus* var. *toyonakensis:* H. Umezawa *et al., J. Antibiot.* **23,** 259 (1970). Prepn: H. Umezawa *et al.,* **Ger. pat. 2,028,403** (1971 to Microbiochemical Res. Found.), *C.A.* **74,** 75201c (1971). Structure: H. Morishima *et al., J. Antibiot.* **23,** 263 (1970). Synthesis: H. Morishima *et al., ibid.* **25,** 551 (1972). Biological properties: T. Aoyagi *et al., ibid.* **24,** 687 (1971). *N-n*-Caproyl and *N-iso*-caproyl derivatives, *pepstatin B* and *pepstatin C,* also isolated from pepstatin-producing *Streptomyces,* as minor components of crude preparations: T. Miyano *et al., ibid.* **25,** 489 (1972). Mechanism of pepsin inhibition: S. Kunimoto *et al., ibid.* 251; **27,** 413 (1974); J. Marciniszyn *et al., Adv. Exp. Med. Biol.* **95,** 199 (1977); D. H. Rich, E. T. O. Sun, *Biochem. Pharmacol.* **29,** 2205 (1980); of other acid protease inhibition: D. H. Rich *et al., Biochemistry* **24,** 3165

(1985). Tissue distribution in rats: D. A. W. Grant *et al.*, *Biochem. Pharmacol.* **31**, 2302 (1982). Effect on gastric ulcers in man: O. Bonnevie *et al.*, *Gut* **20**, 624 (1979); L. B. Svendsen *et al.*, *Scand. J. Gastroenterol.* **14**, 929 (1979).

Colorless needles, mp 228-229° (dec). $[\alpha]_D^{27}$ −90.3° (c = 0.288 in methanol). Sol in methanol, ethanol, acetic acid DMSO. Practically insol in benzene, chloroform, ether, and water. LD_{50} in mice, rats, rabbits, dogs (mg/kg): 1090, 875, 820, 450 i.p.; all >2000 orally (Umezawa, 1970).

7105. Peptide T. $C_{35}H_{55}N_9O_{16}$; mol wt 857.87. C 49.00%, H 6.46%, N 14.69%, O 29.84%. Octapeptide segment of the human immunodeficiency virus (HIV) envelope glycoprotein (gp 120); named peptide T because of its high threonine content. Has been reported to block the *in vitro* binding of HIV envelope to human leukocyte receptor CD4. Isolation, neuropharmacology and anti-HIV activity of peptide and analogs: C. B. Pert *et al.*, *Proc. Nat. Acad. Sci. USA* **83**, 9254 (1986). Characterization of active core structure, T[4-8], and chemotactic effects: C. B. Pert, M. R. Ruff, *Clin. Neuropharmacol.* **9**, Suppl. 4, 482 (1986). Chemotactic effects and structural homology with vasoactive intestinal peptide (VIP), *q.v.*: M. R. Ruff *et al.*, *FEBS Letters* **211**, 17 (1987). Competitive binding studies at VIP receptor: T. D. Nguyen, *Peptides* **9**, 425 (1988). Structural homology with thymosin α_1, *q.v.*: T. D. Nguyen, L. A. Scheving, *Biochem. Biophys. Res. Commun.* **145**, 884 (1987). Evaluation of anti-HIV activity: J. Sodroski *et al.*, *Lancet* **1**, 1428 (1987). Clinical evaluation in treatment of AIDS: L. Wetterberg *et al.*, *ibid.* 159. Chromatographic purification of [D-Ala¹] peptide T amide, a metabolically stable and more potent analog of peptide T: T. R. Burke, M. Knight, *J. Chromatog.* **411**, 431 (1987). Blood to brain transport of the amide in mice: C. M. Barrera *et al.*, *Brain Res. Bull.* **19**, 629 (1987). Brief review of debate over effectiveness of peptide T: D. M. Barnes, *Science* **237**, 128 (1987).

Ala-Ser-Thr-Thr-Thr-Asn-Tyr-Thr

7106. Peptonized Iron. Iron peptonized; Saferon. A compd of iron oxide and peptone, rendered sol by the presence of Na citrate, contg 16-18% Fe. The iron is in nonionic form, hence not detectable by the usual reactions and is assumed to be more readily assimilated. Prepn: *U.S.D.*, 25th ed., p 1800.

Dark brown, lustrous granules or brown powder. Characteristic odor and taste. Affected by light. Freely sol in water, yielding neutral or alkaline solns. Practically insol in alcohol. *Protect from light.*

THERAP CAT: Hematinic.

7107. Peracetic Acid. *Ethaneperoxoic acid;* peroxyacetic acid; acetyl hydroperoxide. $C_2H_4O_3$; mol wt 76.05. C 31.58%, H 5.30%, O 63.11%. CH_3COOOH. Prepd from acetaldehyde and oxygen in the presence of cobalt acetate: **Ger.** pats. **269,937; 272,738** (1914); *Frdl.* **11**, 73; by the autooxidation of acetaldehyde: Wallace; Golding, **U.S.** pats. **2,833,813-4** (1958 to du Pont). A 50% soln may be obtained from acetic anhydride, hydrogen peroxide, and sulfuric acid: D'Ans, Frey, *Ber.* **45**, 1848 (1912); Erlenmeyer, *Helv. Chim. Acta* **8**, 795 (1925).

Liquid, acrid odor. Explodes violently on heating to 110°. Freely sol in water, alcohol, ether, H_2SO_4. Stable in dil aq soln. Strong oxidizing agent.

Caution: Strongly irritating to skin and eyes.

7108. Perazine. *10-[3-(4-Methyl-1-piperazinyl)propyl]-10H-phenothiazine;* N-methylpiperazinyl-N'-propylphenothiazine; 10-(γ-methylpiperazinopropyl)phenothiazine; P 725; Taxilan. $C_{20}H_{25}N_3S$; mol wt 339.49. C 70.75%, H 7.42%, N 12.38%, S 9.45%. Prepn: Hromatka *et al.*, *Monatsh.* **88**, 56, 193 (1957); **91**, 107 (1960); Horclois, **Brit.** pat. **780,193** (1957 to Rhône-Poulenc).

Crystals, mp 51-53°. $bp_{0.001}$ 160-170° (air bath temp). Dihydrochloride, $C_{20}H_{27}Cl_2N_3S$, hygroscopic needles, dec 228-230°.

Dihydrochloride hemihydrate, platelets from ethanol, mp 225-227°.

Dimaleate, $C_{28}H_{33}N_3O_8S$, crystals from water, mp 210°.

THERAP CAT: Antipsychotic.

7109. Perbenzoic Acid. *Benzenecarboperoxoic acid; peroxybenzoic acid;* benzoyl hydroperoxide. $C_7H_6O_3$; mol wt 138.12. C 60.87%, H 4.38%, O 34.75%. C_6H_5COOOH. Prepd from dibenzoyl peroxide by treatment with a soln of sodium methoxide in methanol at 0°: Braun, *Org. Syn.* **13**, 86 (1933); *cf.* Bergmann, Witte, **Ger.** pat. **409,779**; *Chem. Zentr.* **1925**, I, 1911.

Leaflets from benzene. Acrid odor. mp 41-43°. Very volatile. Sublimes in desiccator. bp_{15} 100-110° (partial decomposition). Volatile with steam. Very sparingly sol in water, but turns liquid upon contact with water; sparingly sol in petr ether; freely sol in other organic solvents. Stability of solns with chloroform, carbon tetrachloride, ether, benzene: Prileshajew, *Chem. Zentr.* **1911**, I, 1280; Kolthoff *et al.*, *J. Polymer Sci.* **2**, 199 (1947), *C.A.* **41**, 4960 (1947). Forms an unstable acid sodium salt, and a somewhat more stable neutral sodium salt.

USE: To convert ethylenic compounds into oxides; in analysis of unsatd compounds, to determine the number of double bonds.

7110. Perchloric Acid. $ClHO_4$; mol wt 100.47. Cl 35.29%, H 1.01%, O 63.70%. $HClO_4$. Prepd from potassium perchlorate and sulfuric acid: Schmeisser in *Handbook of Preparative Inorganic Chemistry* **vol. 1**, G. Brauer, Ed. (Academic Press, New York, 2nd ed., 1963) pp 318-320. Comprehensive monograph: J. C. Schumacher, *Perchlorates* (Reinhold, New York, 1960).

The anhyd acid is a colorless, volatile, very hygroscopic liquid. d^{22} 1.768; bp_{11} 19°. Dec when distilled at atmospheric pressure, sometimes with explosive violence. mp −112°. Combines vigorously with water with evolution of heat. Undergoes spontaneous and explosive decompn, hence it is marketed only in mixture with water contg 60-70% $HClO_4$, density 1.5 and 1.6, respectively. The aq acid is very caustic and may deflagrate in contact with oxidizable substances. Density of aq solns at 15°: 1% = 1.0050; 10% = 1.0597; 20% = 1.1279; 30% = 1.2067; 40% = 1.2991; 50% = 1.4103; 60% = 1.5389; 70% = 1.6736. Density of aq solns at 25°: 65.0% = 1.597; 70.0% = 1.664; 75.0% = 1.731.

USE: The acid in analytical chemistry as an oxidizer and for separation of potassium from sodium. Its salts for explosives and for plating of metals. *Caution:* Corrosive to skin, mucous membranes.

7111. Perchloryl Fluoride. $ClFO_3$; mol wt 102.46. Cl 34.61%, F 18.54%, O 46.85%. *Reviews:* Downs, Adams in *Comprehensive Inorganic Chemistry* **vol. 2**, J. C. Bailar, Jr. *et al.*, Eds. (Pergamon Press, Oxford, 1973) pp 1391-1393; Christe, Schack in *Advan. Inorg. Chem. Radiochem.* **18**, 319-398.

Usually stored in cylinders as liq under pressure. mp −147.7°. bp −46.7°. d^{20} (liq) 1.434. Critical temp 95.2°. Crit pressure 53 atm. Crit density: 0.637. Heat of vaporization 4.6 kcal/mol. Trouton constant: 20.4. Dipole moment 0.03. Heat of formation of gas at 25° −5.12 kcal/mol. Sp heat of liq: 0.229 cal/g/°C at −40°; 0.290 at +50°. Does not corrode base metals when anhydr. Shows the greatest resistance to electrical breakdown known for any gas.

USE: In organic synthesis to introduce fluorine atoms into organic molecules. As oxidizing agent. As insulator for high voltage systems. *Handle with care.* Do not bring in direct contact with reducing agents, alcohols, etc.: *Chem. & Eng. News* **37**, 60 (1959). Explosions have occurred. *Caution:* Causes methemoglobinemia and cyanosis. Can be absorbed through skin.

7112. Perezone. *2-(1,5-Dimethyl-4-hexenyl)-3-hydroxy-5-methyl-2,5-cyclohexadiene-1,4-dione; 2-(1,5-dimethyl-4-hexenyl)-3-hydroxy-5-methyl-p-benzoquinone;* pipitzahoic acid. $C_{15}H_{20}O_3$; mol wt 248.31. C 72.55%, H 8.12%, O 19.33%. From roots of *Trixis pipitzahuac* Shaffner *(Perezia adnata* Gr., *Compositae):* Weld, *Ann.* **95**, 188 (1855); from roots of *Radix pereziae:* Mylius, *Ber.* **18**, 463, 480, 937 (1885); Anschütz, *Ber.* **18**, 709 (1885); Fichter *et al., Ann.* **395**, 1, 15 (1913). Structure: Archer, Thomson, *Chem. Commun.* **1965**, 354; Bates *et al., Chem. & Ind. (London)* **1965**, 1793.

Yellow plates from water or alcohol, mp 103-104°. $[\alpha]_D^{20}$ −17° (ether).

7113. Perfluidone. *1,1,1-Trifluoro-N-[2-methyl-4-(phenylsulfonyl)phenyl]methanesulfonamide; 1,1,1-trifluoro-4'-(phenylsulfonyl)methanesulfono-o-toluidide; 2-methyl-4-phenylsulfonyltrifluoromethanesulfonanilide;* MBR-8251; Destun. $C_{14}H_{12}F_3NO_4S_2$; mol wt 379.38. C 44.32%, H 3.19%, F 15.02%, N 3.69%, O 16.87%, S 16.90%. Selective pre-emergence herbicide. Prepn: Moore, Harrington, **Ger. pat. 2,118,190** (1972 to Minnesota Mining & Mfg.), *C.A.* **77**, 19402u (1972). Activity: W. A. Gentner, *Weed Sci.* **21**, 122 (1973).

Cryst solid, mp 142-144°. pKa 2.5. Vapor press. at 25°: $<1 \times 10^{-5}$ mm Hg. Soly in water at 22°: 60 mg/l. Soly (g/l): acetone 750; benzene 11; dichloromethane 162; methanol 595.

USE: Herbicide.

7114. Performic Acid. *Methaneperoxoic acid; peroxyformic acid;* permethanoic acid; formyl hydroperoxide. CH_2O_3; mol wt 62.03. C 19.36%, H 3.25%, O 77.39%. HCOOOH. A strong oxidizing agent. A 90% soln is obtained when a mixture of 20 g formic acid, 25 g 100% H_2O_2 and 6.5 g H_2SO_4 is allowed to interact for 2 hrs and is then distilled: D'Ans, Kneip, *Ber.* **48**, 1137 (1915); Greenspan, *J. Am. Chem. Soc.* **68**, 907 (1946).

The 90% soln is a colorless liq. *Prone to explode on contact with metals, their oxides, reducing substances, or on distillation.* Has lower vapor pressure than formic acid. Miscible with water, alc, ether. Sol in benzene, chloroform. Solns are unstable, gassing being noticeable after a few hours, and the effective concn showing a definite decline in 2 hrs.

USE: For oxidation, epoxidation and hydroxylation reactions. *Caution:* Irritant.

7115. Pergolide. *8-[(Methylthio)methyl]-6-propylergoline;* D-6-n-propyl-8β-methylmercaptomethylergoline; Ly-141B. $C_{19}H_{26}N_2S$; mol wt 314.49. C 72.57%, H 8.33%, N 8.91%, S 10.19%. Dopaminergic agonist that also decreases plasma prolactin concentrations. Prepn: E. C. Kornfeld, N. J. Bach, **U.S. pat. 4,166,182** (1979 to Lilly). Dopaminergic effects in rats: R. W. Fuller *et al., Life Sci.* **24**, 375 (1979);

T. T. Yen *et al., ibid.* **25**, 209 (1979). Clinical pharmacology: L. Lemberger, R. E. Crabtree, *Science* **205**, 1151 (1979). Pharmacological evaluation as antiparkinson agent: W. C. Koller, *Neuropharmacology* **19**, 831 (1980). Clinical study in galactorrhea: J. T. Callaghan *et al., Life Sci.* **28**, 95 (1981); in hyperprolactinemia: S. Francks *et al., Lancet* **2**, 659 (1981); in treatment of pituitary tumors secreting prolactin or growth hormone: D. L. Kleinberg *et al., N. Engl. J. Med.* **309**, 704 (1983). Clinical studies in Parkinson's disease: J. Jankovic, J. Orman, *Adv. Neurol.* **45**, 551 (1986); C. W. Olanow, M. J. Alberts, *ibid.* 555.

Solid, mp 206-209°.
Mesylate, $C_{20}H_{30}N_2O_3S_2$, *Ly-127,809; Permax.* Crystals, mp 225° (dec).

THERAP CAT: Antiparkinsonian.

7116. Perhexiline. *2-(2,2-Dicyclohexylethyl)piperidine; 1,1-dicyclohexyl-2-(2-piperidyl)ethane;* perhexilene. $C_{19}H_{35}N$; mol wt 277.50. C 82.24%, H 12.71%, N 5.05%. Calcium blocking agent. Prepn: **Brit. pat. 1,025,578** (1966 to Richardson-Merrell), *C.A.* **65**, 2229f (1966). Pharmacology: Hudak *et al., J. Pharmacol. Exp. Ther.* **173**, 371 (1970). Clinical studies: Winsor, *Clin. Pharmacol. Ther.* **11**, 85 (1970). Series of articles: *Postgrad. Med. J. Suppl.* **49**, 8-132 (1973). Book: *Antiarrhythmic Action and the Puzzle of Perhexiline,* E. M. Vaughan Williams, Ed. (Academic Press, New York, 1980) 143 pp.

Maleate, $C_{23}H_{39}NO_4$, *Pexid.* mp 188.5-191°. LD_{50} in rats, mice: >7, 4.37 g/kg orally, Causa, Perri, *Arzneimittel-Forsch.* **21**, 114 (1971).
Hydrochloride, $C_{19}H_{35}N.HCl$, white crystalline powder, mp 243-245.5° (dec).

THERAP CAT: Vasodilator (coronary); diuretic.

7117. Pericyazine. *10-[3-(4-Hydroxy-1-piperidinyl)-propyl]-10H-phenothiazine-2-carbonitrile; 2-cyano-10-[3-(4-hydroxypiperidino)propyl]phenothiazine;* periciazine; propericiazine; RP 8908; Aolept; Neulactil; Neuleptil. $C_{21}H_{23}N_3OS$; mol wt 365.50. C 69.00%, H 6.34%, N 11.50%, O 4.38%, S 8.77%. Psychotherapeutic phenothiazine. Prepn: Robert, Jacques, **Fr. pat 1,212,031** (1960 to Rhône-Poulenc). Hypotensive action in dogs: K. P. Singh *et al., Ind. J. Med. Res.* **58**, 1467 (1970). GLC determn in human urine: A. P. De Leenheer, *J. Pharm. Sci.* **63**, 389 (1974). Clinical studies in psychiatric patients: U. Spiegelberg, G. Kleu, *Arzneimittel-Forsch.* **17**, 159 (1967); J. C. Barker, M. Miller, *Brit. J. Psychiat.* **115**, 169 (1969). Toxicity and metabolism: L. Julou *et al., Proc. Eur. Soc. Study Drug Toxic.* **9**, 11 (1968). Toxicity data: Schafer, *Toxicol. Appl. Pharmacol.* **21**, 315 (1972). Use in spectrophotometry: H. Gowda *et al., Anal. Chem.* **55**, 1816 (1983); A. T. Gowda *et al., Anal. Chim. Acta* **154**, 347 (1983).

Crystals, mp 116-117°. uv max: 232.5, 271.5 nm (log ε 4319, 4503). LD_{50} orally in rats: 395 mg/kg (Schafer).
USE: Spectrophotometric reagent for palladium and ruthenium.

THERAP CAT: Antipsychotic.

7118. Perilla Ketone. *1-(3-Furanyl)-4-methyl-1-penta-none; 1-(3-furyl)-4-methyl-1-pentanone; β-furyl isoamyl ketone.* $C_{10}H_{14}O_2$; mol wt 166.22. C 72.26%, H 8.49%, O 19.25%. Potent pulmonary edematogenic agent in animals. Isoln from the mint plant, *Perilla frutescens* Britton, *Labiatae,* and structure: R. Goto, *J. Pharm. Soc. Japan* **57,** 77 (1937). Synthesis: T. Matsuura, *Bull. Chem. Soc. Japan* **30,** 430 (1957); K. Kondo, M. Matsumoto, *Tetrahedron Letters* **1976,** 4363; T. Kitamura *et al., Synth. Commun.* **7,** 521 (1977); K. Inomata *et al., Chem. Letters* **1979,** 709. Potent lung toxin implicated in emphysema of grazing cattle: B. J. Wilson *et al., Science* **197,** 573 (1977).

Colorless oil, bp 196°. n_D^{20} 1.4781; d_{15}^{15} 0.9920. uv max (ethanol): 207, 253 nm (ε 14100, 5800). Sensitive to oxygen; becomes reddish-orange on standing. LD_{50} i.p. in female and male mice, male rats: 2.5, 5, 6, 10 mg/kg, B. J. Wilson *et al., loc. cit.*

7119. Perillaldehyde. *4-(1-Methylethenyl)-1-cyclohex-ene-1-carboxaldehyde; 4-isopropenyl-1-cyclohexene-1-carb-oxaldehyde.* $C_{10}H_{14}O$; mol wt 150.21. C 79.95%, H 9.39%, O 10.65%. Isoln from *Perilla arguta* Benth., *Labiatae:* Semmler, Zaar, *Ber.* **44,** 52, 815 (1911); from essential oil of *Sium latifolium* L., *Umbelliferae:* Parczewski, *Dissertationes Pharm.* **12,** 223 (1960), *C.A.* **55,** 7765c (1961); from mandarin peel oil *(Citrus reticulata* Blanco, *Rutaceae):* Kugler, Kováts, *Helv. Chim. Acta* **46,** 1480 (1963). Prepn by chromic oxidation of perilla alcohol: Naves, *ibid.* **29,** 553 (1946); Ritter, Ginsburg, *J. Am. Chem. Soc.* **72,** 2381 (1950); Kergomard, Philibert-Bigou, *Bull. Soc. Chim. France* **1958,** 393, 1174; Naves, Grampoloff, *ibid.* **1960,** 37.

d-Form, liquid. bp_{745} 237°; bp_7 98-100°. d_4^{20} 0.953. n_D^{20} 1.5058. $[α]_D^{20}$ +127° (c = 13.1 in carbon tetrachloride).
l-Form, liquid. bp_{10} 104-105°. d_4^{20} 0.9645. n_D^{20} 1.5069. $[α]_D^{20}$ −146°.
Oxime, $C_{10}H_{15}NO$, *l-Perillaldehyde α-syn-oxime,* previously referred to as *l-perillaldehyde α-anti-oxime, perillartine, "perilla sugar".* Synthesis: Andô *et al., Science* (Tokyo) **17,** 241 (1947), *C.A.* **45,** 1976d (1951). Clarification of structure: Acton *et al., Experientia* **26,** 473 (1970). Needles, mp 102°. uv max (alc): 232 nm (ε 21,800). About 2000 times as sweet as sucrose: Furukawa, *Koryo* No. **11,** 11, 40 (1950), *C.A.* **44,** 6083g (1950). LD_{50} orally in rats: 2500 mg/kg, *RTECS* Vol. I, R. J. Lewis, R. L. Tatken, Eds. (1980) p 568.
USE: The oxime is used as sweetening agent in Japan.

7120. Perimethazine. *1-[3-(2-Methoxy-10H-phenothia-zin-10-yl)-2-methylpropyl]-4-piperidinol;* 2-methoxy-10-[2-

methyl-3-(4-hydroxypiperidino)propyl]phenothiazine; 3-methoxy-10-[3-(4-hydroxypiperidyl)-2-methylpropyl]-phenothiazine; perimetazine; RP 9159; AN 1317; Leptryl. $C_{22}H_{28}N_2O_2S$; mol wt 384.53. C 68.72%, H 7.34%, N 7.28%, O 8.32%, S 8.34%. Prepn: **Brit. pat. 904,210;** Jacob, Robert, **U.S. pat. 3,075,976** (1962, 1963 both to Rhône-Poulenc). Pharmacology: L. Julon *et al., Compt. Rend. Soc. Biol.* **160,** 1852 (1966). Clinical evaluation as antipsychotic: M. Bourgeois, A.-M. Sicart, *Bord. Med.* **3,** 515 (1970).

Crystalline powder from benzene:cyclohexane (15:85), mp 137-138°.
Hydrochloride, $C_{22}H_{29}ClN_2O_2S$. LD_{50} in mice (mg/kg): 115 i.v.; 140 i.p.; 330 s.c.; 310 orally (Julon).

THERAP CAT: Antipsychotic.

7121. Perimycin. Aminomycin; fungimycin; WX 2412; NC 1968. Polyene antifungal antibiotic complex produced by *Streptomyces coelicolor* var. *aminophilus* NRRL 2390; Wooldridge, **U.S. pat. 2,956,925** (1960); Mohan *et al., Antimicrob. Ag. Chemother.* **1963,** 462; McDaniel *et al.,* **U.S. pat. 3,182,004** (1965 to Warner-Lambert). Purification: Borowski *et al., Antimicrob. Ag. Ann.* **1960,** 532. Found to be a mixture of three active components, perimycin A (major), B, and C. Structural studies: C. H. Lee, C. P. Schaffner, *Tetrahedron Letters* **1966,** 5837; *eidem, Tetrahedron* **25,** 2229 (1969). Isoln of perimycins A, B, C and structural elucidation of A: P. Kolodziejczyk *et al., Tetrahedron Letters* **1976,** 3603. Description of unique biological properties: E. Borowski, B. Cybulska, *Nature* **213,** 1034 (1967).

perimycin A

Amorphous, golden-yellow solid. Has no definite mp but dec slowly with darkening upon heating. uv max (methanol): 383 nm ($E_{1cm}^{1\%}$ 1000). Sol in the following solvents in the presence of water: lower alcohols, pyridine, tetrahydrofuran, acetone, dioxane. Sol in warm methanol, DMF, dimethylsulfoxide, and in the lower fatty acids. Practically insol in water, petr ether, ethyl acetate, benzene.
Perimycin A. $C_{59}H_{88}O_{17}N_2$. The first member of the aromatic subgroup of heptaene macrolides for which the complete structure has been elucidated.

THERAP CAT: Antifungal.

7122. Perindopril. *[2S-[1-[R*,(R*)],2α,3aβ,7aβ]]-1-[2-[[1-(Ethoxycarbonyl)butyl]amino]-1-oxopropyl]octahydro-1H-indole-2-carboxylic acid;* (2S,3aS,7aS)-1-[(S)-N-[(S)-1-carboxybutyl]alanyl]hexahydro-2-indolinecarboxylic acid 1-ethyl ester; (2S)-2-[(1S)-1-carbethoxybutylamino]-1-oxopropyl-(2S,3aS,7aS)-perhydroindole-2-carboxylic acid; S-9490; S-9490-3; Coversyl; Electan; Procaptan. $C_{19}H_{32}N_2O_5$; mol wt 368.47. C 61.93%, H 8.75%, N 7.60%, O 21.71%. Angiotensin-converting enzyme inhibitor. Hydrolyzed *in vivo* to the active diacid metabolite. Prepn: G. Remond *et al., Eur. pat. Appl.* **49,658** (1982 to Sci. Union et

Cie. - Soc. Franc. Rech. Med.); M. Vincent *et al.*, U.S. pat. **4,508,729** (1985 to ADIR). Stereoselective synthesis: M. Vincent *et al.*, *Tetrahedron Letters* **23**, 1677 (1982). NMR study: N. Platzer *et al.*, *Magnet. Res. Chem.* **26**, 296 (1988). Hemodynamic effects in humans: K. R. Lees, J. L. Reid, *Brit. J. Clin. Pharmacol.* **23**, 159 (1987). Clinical evaluation in essential hypertension: T. Morgan *et al.*, *J. Cardiovasc. Pharmacol.* **10**, Suppl. 7, S116 (1987). Pharmacokinetics, pharmacodynamics of the diacid: K. R. Lees, J. L. Reid, *ibid.* **10**, 129 (1987).

Diacid form, $C_{17}H_{28}N_2O_5$, *perindoprilat, S-9780.*
THERAP CAT: Antihypertensive.

7123. Periodic Acid. H_5IO_6; mol wt 227.96. H 2.21%, I 55.68%, O 42.11%. Prepd by electrolytic oxidation of iodic acid or from barium periodate and nitric acid: Willard, *Inorg. Syn.* **1**, 172 (1939). Chemistry of periodic acid and periodates: H. Siebert, *Fortschr. Chem. Forsch.* **8**, 470 (1967). Periodic acid and periodates in organic and bioorganic chemistry: A. J. Fatiadi, *Synthesis* **1974**, 229. Book: G. Dryhurst, *Periodate Oxidation of Diol and Other Functional Groups* (Pergamon Press, New York, 1970).

Monoclinic, hygroscopic crystals. mp 122°; dec 130-140° forming I_2O_5, H_2O, and O. Freely sol in water. A 38% w/w soln had d_4^{17} 1.3875. Sol in alcohol, slightly in ether. Soly in nitric acid (d 1.42) at 26° = 7.82 g/100 ml. When aq solns are evapd at room temp the "ortho" acid H_5IO_6 crystallizes out. If this is heated in a vacuum at 100° and 12 mm it loses water and is converted to the "meta" acid HIO_4, an intermediate $H_4I_2O_9$ being formed at 80°. Periodic acid is a dibasic acid, much weaker than perchloric acid. K_1 at 25° = 2.3×10^{-2}; $K_2 = 2 \times 10^{-6}$. Very easily reduced to iodic acid by nitrous or sulfurous acids and even by hydrochloric and sulfuric acids. Oxidizes organic material.
USE: In organic synthesis.

7124. Periodyl. *12-Hydroxy-9,10-diiodo-9-octadecenoic acid;* diiodoricinstearolic acid; ricinstearolic acid diiodide; 8,9-diiodo-11-hydroxy-8-heptadecene-1-carboxylic acid; Diiodyl; Joristen. $C_{18}H_{32}I_2O_3$; mol wt 550.28. C 39.29%, H 5.86%, I 46.13%, O 8.72%. $CH_3(CH_2)_5CHOHCH_2CI=CI-(CH_2)_7COOH$. Prepn: Mühle, *Ber.* **46**, 2091 (1913); Ger. pat. **296,495** (to Riedel).

Tasteless needles from dil alcohol, mp 62°. Practically insol in water, acids; sol in dil alkalies, alcohol, ether, chloroform; slightly sol in benzene.
THERAP CAT: Iodine source.

7125. Periplanones. Sex pheromones of the American cockroach, *Periplaneta americana* L., found primarily in the alimentary tract and excreta of the insect. They act as close proximity sex-excitants and function over relatively short distances compared to long range insect sex-attractants. Isoln: D. R. A. Wharton *et al.*, *Science* **137**, 1062 (1962). Improved isolns, identification and proposed structures of periplanone B: C. J. Persoons *et al.*, *Tetrahedron Letters* **1976**, 2055; E. Talman *et al.*, *Isr. J. Chem.* **17**, 227 (1978); C. J. Persoons *et al.*, *J. Chem. Ecol.* **5**, 221 (1979). Total synthesis and structure of (±) periplanone B: W. C. Still, *J. Am. Chem. Soc.* **101**, 2493 (1979); H. Hauptmann, G. Muhlbauer, *Tetrahedron Letters* **27**, 1315 (1986); of the (−)-form: T. Kihara *et al.*, *ibid.* 1343. Absolute configuration of B: M. A. Adams *et al.*, *J. Am. Chem. Soc.* **101**, 2495 (1979). Structural and stereochemical studies of periplanone B: C. J. Persoons *et al.*, *J. Chem. Ecol.* **8**, 439 (1982); H. Hauptmann, G. Muhlbauer, *Tetrahedron Letters* **27**, 6189 (1986); Y. Shizuri *et al.*, *ibid.* **28**, 1791, 1795 (1987); L. MacDonald *et al.*, *Heterocycles* **25**, 305 (1987). Short synthesis of periplanone A: Y. Shizuri *et al.*, *Tetrahedron Letters* **29**, 1971 (1988). Use in insecticidal formulations: W. J. Bell *et al.*, *Environ.*

Entomol. **13**, 448 (1984); W. J. Bell *et al.*, *Pest Control* **54**, 40 (1986).

periplanone B

Periplanone A, $C_{15}H_{20}O_2$. Found primarily in excreta of *P. americana*, it is unstable and gradually rearranges to a biologically inactive compound. uv max (hexane): 220 nm (log ε 4.18).

Periplanone B, $C_{15}H_{20}O_3$, *[1R*,2R*,5S*,6E,10R*]-(±)-8-methylene-5-(1-methylethyl)spiro(11-oxabicyclo[8.1.0]undec-6-ene-2,2'-oxiran)-3-one.* More stable and more active than periplanone A, it is found in both the alimentary tract and excreta of *P. americana*. The ratio of A to B in excreta is 1:10. Crystals, mp 48-50°. uv max (hexane): 226 nm.

(−)-*Periplanone B*, crystals, mp 47-50°. $[\alpha]_D^{22}$ −667° (c = 0.13 in *n*-hexane).

7126. Periplocin. *3β-[(2,6-Dideoxy-4-O-β-D-glucopyranosyl-β-D-ribo-hexopyranosyl)oxy]-5β,14-dihydroxycard-20(22)-enolide;* glucoperiplocymarin; periplocoside. $C_{36}H_{56}O_{13}$; mol wt 696.84. C 62.05%, H 8.10%, O 29.85%. From *Periploca graeca* L., *Asclepiadaceae*: Lehmann, *Arch. Pharm.* **235**, 157 (1897); Jacobs, Hoffmann, *J. Biol. Chem.* **79**, 519 (1928); from *Strophanthus preussii* Engl. and Pax., *Apocynaceae*: Ruppol, Trukovic, *J. Pharm. Belg.* **10**, 221 (1955), *C.A.* **50**, 12089d (1956). Structure: Stoll, Renz, *Helv. Chim. Acta* **22**, 1193 (1939). Acid hydrolysis yields periplogenin, *q.v.*, and periplobiose (cymarose + glucose, $C_{13}H_{24}O_9$, $[\alpha]_D^{20}$ +32°). Enzymatic hydrolysis with strophanthobiase splits off glucose, yielding periplocymarin. The biose is attached to the hydroxyl group at C-3 of the aglycon.

Dihydrate, fine needles from water. Becomes anhydr after drying for one hour in high vacuum at 105°. mp 224° when bath is preheated to 200°. $[\alpha]_D^{20}$ +23° (c = 0.7 in alcohol). One gram dissolves in about 20 ml boiling water, at 25° the soly is about 1:2500. Freely sol in alcohol; almost insol in ether, chloroform. LD s.c. in rabbits: 10 mg/kg, *Handbook of Toxicology*, **vol. 1**, W. S. Spector, Ed. (Saunders, Philadelphia, 1956) pp 226-227.

Tetraacetylperiplocin, $C_{44}H_{64}O_{17}$, six-sided prisms from alcohol, mp 195°. $[\alpha]_D^{20}$ +20° (c = 0.5 in alcohol). Sol in alcohol, chloroform; very slightly sol in water.

7127. Periplocymarin. *(2,6-Dideoxy-3-O-methyl-β-D-ribo-hexopyranosyl)oxy-5,14-dihydroxy-card-20(22)-enolide.* $C_{30}H_{46}O_8$; mol wt 534.67. C 67.39%, H 8.67%, O 23.94%. By extracting bark and wood of *Periploca graeca* L., *Asclepiadaceae* with 70% alcohol and treating with the enzyme strophanthobiase from seeds of *Strophanthus courmonti* Sacleux, *Apocynaceae*: Jacobs, Hoffmann, *J. Biol. Chem.* **79**, 519 (1928); Katz, Reichstein, *Pharm. Acta Helv.* **19**, 231 (1944); from *S. hypoleucus* Stapf: von Euw, Reichstein, *Helv. Chim. Acta* **33**, 544 (1950); from *S. ledienii* Stein: Lichti *et al.*, *ibid.* **39**, 1914 (1956); from *S. eminii* Asch. et Pax.: Zelnik, Schindler, *ibid.* **40**, 2110 (1957). Structure: von Euw, Reichstein, *ibid.* **31**, 883 (1948). Acid hydrolysis yields one mol periplogenin *(q.v.)* and one mol cymarose *(q.v.)*. The cymarose is attached to the hydroxyl group at C-3 of the aglycon.

Lustrous needles from methanol contg approx one mol CH_3OH of crystn. Bitter taste, but not nearly so marked as that of cymarin. Becomes solvent-free at 100° *in vacuo*. Sinters at 138°. mp 148°. $[\alpha]_D^{27}$ +29° (c = 0.94 in 95% alcohol for the anhydr substance). Readily sol in alcohol, chloroform, acetone; less readily in methanol; very sparingly sol in water and practically insol in ether. Readily hydrolyzed in the cold.

7128. Periplogenin. *3β,5,14-Trihydroxy-5β-card-20(22)-enolide;* desoxostrophanthidin. $C_{23}H_{34}O_5$; mol wt 390.53. C 70.74%, H 8.78%, O 20.49%. The aglycon of periplocin and periplocymarin; accompanies strophanthidin in *Strophanthus eminii* Asch. et Pax., *Apocynaceae*. Isoln: Lehmann, *Arch. Pharm.* **235**, 157 (1897); Jacobs, Hoffmann, *J. Biol. Chem.* **79**, 519 (1928); Stoll, Renz, *Helv. Chim. Acta* **22**, 1193 (1939); Lardon, *ibid.* **33**, 639 (1950). Structure: Speiser, Reichstein, *Experientia* **3**, 323 (1947); *eidem, Helv. Chim. Acta* **30**, 2143 (1947); **31**, 622 (1948). Synthesis: Deghenghi, Gaudry, *Tetrahedron Letters* **1963**, 2045; Kamano *et al., J. Org. Chem.* **39**, 2319 (1974).

Solvated prisms from methanol which contain an undetermined amount of methanol, sinter at 140°, mp 235°. $[\alpha]_D^{27}$ +31.5° (c = 1.04 in alcohol). Sol in alcohol and chloroform; slightly sol in ether, water (1:2500); practically insol in benzene, petr ether. pH of aq solns about 7. Strong positive Legal test.

Monobenzoate, $C_{30}H_{38}O_6$, stout glistening wedges from 95% alcohol, mp 235°.

Dihydroperiplogenin, $C_{23}H_{36}O_5$, stout prisms from 25% alcohol, mp 204° (slight preliminary softening).

3-*O*-Acetylperiplogenin, $C_{25}H_{36}O_6$, crystals, mp 242-244°, $[\alpha]_D^{22}$ +46.9° (c = 0.32 in chloroform).

7129. Perisoxal. *α-(5-Phenyl-3-isoxazolyl)-1-piperidineethanol;* 3-(2-piperidino-1-hydroxyethyl)-5-phenylisoxazole. $C_{16}H_{20}N_2O_2$; mol wt 272.35. C 70.56%, H 7.40%, N 10.28%, O 11.75%. Prepn: **Neth. pat. Appl. 6,408,694** corresp to H. Kano *et al.* **U.S. pat. 3,321,475** (1965, 1967 both to Shionogi); *eidem, J. Med. Chem.* **10**, 411 (1967). Series of articles on pharmacological properties: *Oyo Yakuri* **6**, 1285-1338 (1972), *C.A.* **79**, 49177h-49179k (1973). Studies on physical dependence in rats: K. Yamamoto *et al., ibid.* **16**, 871 (1978), *C.A.* **90**, 180250w (1979). Absorption, excretion, distribution, metabolism: J. Watanabe *et al., Chem. Pharm. Bull.* **27**, 1075 (1979).

Colorless plates from 50% aq ethanol, mp 107-108°.

Citrate, $C_{38}H_{48}N_4O_{11}$, *31252-S, Isoxal.* Colorless prisms from ethanol/acetone, mp 143-145°. LD_{50} in mice: 416 mg/kg s.c., H. Kano *et al., J. Med. Chem.* **10**, 411 (1967).

THERAP CAT: Anti-inflammatory; analgesic.

7130. Perivine. *4-Demethyl-3-oxovobasan-17-oic acid methyl ester.* $C_{20}H_{22}N_2O_3$; mol wt 338.39. C 70.98%, H 6.55%, H 8.28%, O 14.18%. Indole alkaloid from *Vinca rosea* Linn., *Apocynaceae*: G. Svoboda, *J. Am. Pharm. Assoc.* **47**, 834 (1958); G. Svoboda *et al., ibid.* **48**, 659 (1959). Structure: Gorman, Sweeny, *Tetrahedron Letters* **1964**, 3105. Cytotoxicity: D. G. I. Kingston, *J. Pharm. Sci.* **67**, 272 (1978).

Prisms from methanol, dec 218-221°. Also reported as

180-181° (Svoboda 1959). $[\alpha]_D^{26}$ −121.4° (chloroform). pKa in 66% DMF: 7.5. uv max (ethanol): 314 nm ($E_{1cm}^{1\%}$ 2.67).

7131. Perlapine. *6-(4-Methyl-1-piperazinyl)-11H-dibenz[b,e]azepine; 6-(4-methyl-1-piperazinyl)morphanthridine;* AW 14'2333; HF-2333; MP-11; Hypnodin. $C_{19}H_{21}N_3$; mol wt 291.40. C 78.32%, H 7.26%, N 14.42%. Prepn: **Brit. pat. 1,006,156** corresp to Schmutz, Hunziker, **U.S. pat. 3,389,139** (1965, 1968, both to Wander); Hunziker *et al., Helv. Chim. Acta* **49**, 1433 (1966). Pharmacology: Take *et al., Takeda Kenkyusho Ho* **29**, 416 (1970); Yokotani *et al., ibid.* 441; Stille *et al., Psychopharmacologia* **28**, 325 (1973).

Yellow, prismatic crystals from acetone-petrol ether, mp 136-138°. LD_{50} orally in mice: 415 mg/kg.

THERAP CAT: Hypnotic.

7132. Permethrin. *3-(2,2-Dichloroethenyl)-2,2-dimethylcyclopropanecarboxylic acid (3-phenoxyphenyl)methyl ester;* 3-(phenoxyphenyl)methyl (±)-*cis,trans*-3-(2,2-dichloroethenyl)-2,2-dimethylcyclopropanecarboxylate; *m*-phenoxybenzyl (±)-*cis,trans*-3-(2,2-dichlorovinyl)-2,2-dimethylcyclopropanecarboxylate; FMC 33297; NIA 33297; NRDC 143; PP557; SBP 1513; S 3151; Ambush; Corsair; Dragnet; Ectiban; Eksmin; Expar; Nix; Perigen; Pounce; Pynosect; Ridect Pour-On. $C_{21}H_{20}Cl_2O_3$; mol wt 391.29. C 64.46%, H 5.15%, Cl 18.12%, O 12.27%. Synthetic pyrethroid insecticide, more stable to light and at least as active as the natural pyrethrins and with low mammalian toxicity: M. Elliott *et al., Nature* **246**, 169 (1973). Of the four possible isomers, the (1R,*trans*)- and the (1R,*cis*)-isomers are the two esters primarily responsible for insecticidal activity: P. E. Burt *et al., Pestic. Sci.* **5**, 791 (1974). Prepn of the racemic mixture: T. Mizutani *et al.,* **Ger. pat. 2,437,882** (1975 to Sumitomo); F. Mori *et al.,* **Ger. pat. 2,544,150;** *eidem,* **U.S. pat. 4,113,-968** (1976, 1978, both to Kuraray). Metabolism: M. Elliott *et al., J. Agr. Food Chem.* **24**, 270 (1976); L. C. Gaughan *et al., ibid.* **26**, 613 (1978). Photodecompn: R. L. Holmstead *et al., ibid.* 590. Toxicity studies in mammals: L. Metker *et al., U.S. NTIS, AD REP.* **1977**, AD-AO47284, 70 pp. Field evaluation to control flies and ticks in cows: S. S. Quisenberry, D. R. Strohbehn, *J. Econ. Entomol.* **77**, 422 (1984); R. B. Davey, E. H. Ahrens, *Am. J. Vet. Res.* **45**, 1008 (1984). Clinical trial in pediatric pediculosis: E. Ares Mazas *et al., Int. J. Dermatol.* **24**, 603 (1985). *Review:* C. N. E. Ruscoe, *Pestic. Sci.* **8**, 236 (1977).

Technical material is a mixture of approx 60% *trans*- and 40% *cis*-isomers: Colorless crystals to a pale yellow viscous liquid, mp ~35°. $bp_{0.05}$ 220°. d^{20} 1.190-1.272. Vapor pressure at 50° <1 × 10⁻⁶ mm Hg. Soly in water: <1 ppm. Sol or miscible with org solvents except ethylene glycol. LD_{50} orally in female rats: 3800 mg/kg (Metker). Toxic to bees and fish. *Caution:* Mild irritant to skin and eyes.

USE: Insecticide.

THERAP CAT (VET): Ectoparasiticide.

7133. Pernambuco. Fernambuco; Brazil wood; Nicaragua wood; Lima wood; redwood. Wood of *Caesalpinia echinata* Lam., *Leguminosae. Habit.* Brazil. *Constit.* Brazilin.

USE: Dyeing red; manuf of a red lake pigment. With alkalies = purplish-red; with acids = yellow.

7134. Peroxidases. Enzymes which catalyze the oxidation of certain compounds by peroxide. They occur in many plants as well as animals, fungi, and bacteria. The richest plant sources are the sap of the fig tree and the root of horseradish. Review and bibliography: Paul, "Peroxidases", in *The Enzymes*, **vol. 8**, P. D. Boyer *et al.*, Eds. (Academic Press, New York, 1963) pp 227-274; B. C. Saunders *et al.*, *Peroxidase* (Butterworths, Washington, D.C., 1964) 271 pp.

7135. Perphenazine. *4-[3-(2-Chloro-10H-phenothia-zin-10-yl)propyl]-1-piperazineethanol;* 2-chloro-10-[3-[1-(2-hydroxyethyl)-4-piperazinyl]propyl]phenothiazine; 1-(2-hydroxyethyl)-4-[3-(2-chloro-10-phenothiazinyl)propyl]piperazine; chlorpiprazine; chlorpiprozine; PZC; Sch 3940; Trilafon; Trilifan; Decentan; Fentazin; Perphenan. $C_{21}H_{26}$-ClN_3OS; mol wt 403.97. C 62.43%, H 6.49%, Cl 8.78%, N 10.40%, O 3.96%, S 7.94%. Prepn: Cusic, U.S. pat. **2,766,-235** (1956); Sherlock, Sperber, U.S. pat. **2,860,138** (1958 to Schering). Metabolism: U. Breyer, H. J. Gaertner, *Adv. Biochem. Psychopharmacol.* **9**, 167 (1974); H. J. Gaertner *et al.*, *Drug Metab. Dispos.* **3**, 437 (1975). Crystal structure: J. J. H. McDowell, *Acta Crystallogr.* **B34**, 686 (1978).

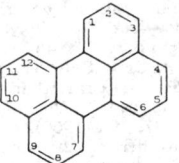

Crystals. Sensitive to light. mp 94-100°. $bp_{0.15}$ 214-218°; bp_1 278-281°. Practically insol in water. Soly (mg/ml): ethanol 153; acetone 82. Practically insol in sesame oil.
Dihydrochloride, $C_{21}H_{28}Cl_3N_3OS$, crystals from alcohol, mp 225-226°.
THERAP CAT: Antipsychotic.
THERAP CAT (VET): Tranquilizer, pre-anesthetic agent.

7136. Peruvoside. *(3β,5β)-3-[(6-Deoxy-3-O-methyl-α-L-glucopyranosyl)oxy]-14-hydroxy-19-oxocard-20(22)-enolide;* cannogenin α-L-thevetoside; Encordin. $C_{30}H_{44}O_9$; mol wt 548.65. C 65.67%, H 8.08%, O 26.25%. Cardenolide (cardiac glycoside) found in seeds of the tropical garden plant *Thevetia peruviana* (Pers.) K. Schum. (*Th. neriifolia* Juss.), *Apocynaceae* (named trumpet flower in Florida). Isoln: Rangaswami, Venkata Rao, *J. Sci. Ind. Res.* **17B**, 331 (1958). Structure: Bloch *et al.*, *Helv. Chim. Acta* **43**, 652 (1960). Pharmacological studies: *Arzneimittel-Forsch.* **18**, 1582-1608 (1968).

Spears from methanol + ether, mp 161-164°. $[\alpha]_D^{22}$ -71.7° (c = 1.54 in methanol). One gram dissolves in about 2500 ml water. Freely soluble in chloroform and acetone, sparingly soluble in methanol and ethanol.
THERAP CAT: In cardiac insufficiency.

7137. Perylene. Dibenz[de,kl]anthracene; *peri*-dinaphthalene. $C_{20}H_{12}$; mol wt 252.30. C 95.20%, H 4.79%. Occurs in coal tar. Isoln from pitch distillate: Cook *et al.*, *J. Chem. Soc.* **1933**, 395. From 2,2'-dihydroxy-1,1'-dinaphthyl: Zinke *et al.*, *Monatsh.* **64**, 415 (1934). From the reaction of phenanthrene with acrolein in anhydr HF: Weinmayr, U.S. pat. **2,145,905** (1939); *cf. J. Am. Chem. Soc.* **61**, 949 (1939).

Yellow to colorless crystals from toluene. mp 273-274°. Sublimes 350-400°. d 1.35. Absorption spectrum: Clar, *Ber.* **65**, 848 (1932). Freely sol in CS_2, chloroform; moderately sol in benzene; slightly in ether, alcohol, acetone; very sparingly sol in petr ether. Insol in water.
Monopicrate, $C_{26}H_{15}N_3O_7$, dark violet-blue needles, mp 223-225°.

7138. Petrolatum. Petroleum jelly; paraffin jelly; vasoliment; Filtrolatum; Filtrosoft; Kremoline; Pureline; Sherolatum; Stanolind; Vaseline; Saxoline; Cosmoline. Purified mixture of semisolid hydrocarbons, chiefly of the methane series of the general formula C_nH_{2n+2}. Actually, petrolatum is a colloidal system of nonstraight-chain solid hydrocarbons and high-boiling liq hydrocarbons, in which most of the liq hydrocarbons are held inside the micelles. Detailed historical account including chemistry and modern mfg methods: Schindler, *Drug Cosmet. Ind.* **89**, 36-37, 76, 78-80, 82 (1961).
Yellowish to light amber or white, semisolid, unctuous mass; practically odorless and tasteless. d_{25}^{60} 0.820-0.865. mp 38-54°. n_D^{60} 1.460-1.474. White petrolatum is transparent in thin layers even at 0°. Practically insol in water, glycerol, alcohol. Sol in benzene, chloroform, ether, petr ether, carbon disulfide, oils.
Premium white petrolatum, *Stanolene.*
USE: As ointment base in pharmaceuticals and cosmetics. Lubricating firearms and machinery, leather grease, shoe polish, rust preventives, modeling clays.

7139. Petrolatum, Liquid. Liquid paraffin; mineral oil; white mineral oil; paraffin oil; Clearteck; Drakeol; Hevyteck; Filtrawhite; Frigol; Kremol; Kaydol; Alboline; Nujol; Paroleine; Saxol; Adepsine oil; Glymol. A mixture of liquid hydrocarbons from petroleum.
Colorless, oily liquid; practically tasteless and odorless even when warmed. The density of the "light" oil is usually 0.83-0.860; the "heavy" 0.875-0.905. Surface tension at 25° slightly below 35 dynes/cm. Insol in water, alc. Sol in benzene, chloroform, ether, carbon disulfide, petr ether, oils.
THERAP CAT: Cathartic.
THERAP CAT (VET): Laxative, externally as a protectant, lubricant.

7140. Petroleum. Crude oil; mineral oil; rock oil; coal oil; seneca oil. Consists of a mixture of hydrocarbons from C_2H_6 and up—chiefly of the paraffins, cycloparaffins, or of cyclic aromatic hydrocarbons, with small amounts of benzene hydrocarbons, sulfur, and oxygenated compounds. *Occurrence:* U.S., Mexico, Iran, Russia, Roumania, Poland, Dutch East Indies, etc.
Dark yellow to brown or greenish-black, oily liquid. Insol in water and only a small portion of it may dissolve in alcohol; sol in benzene, chloroform, ether.
USE: Source of gasoline, petr ether, liq and solid petrolatum, fuel and lubricating oils, butane, isopropyl alcohol, etc.

7141. Petroleum Benzin. *Naphtha;* benzin; petroleum naphtha. Term that has been applied to low boiling fractions of petroleum, consisting chiefly of hydrocarbons of the methane series—principally pentanes and hexanes.
Clear, colorless, nonfluorescent, *highly flammable,* volatile liq; characteristic odor; does not solidify in the cold. The vapors mixed with air explode if ignited. d 0.625-0.660; bp between 35-80°. Flash pt about -40°. Insol in water. Miscible with abs alc, benzene, chloroform, ether, carbon disulfide, carbon tetrachloride, and oils except castor oil. *Caution: Keep tightly closed in a cool place and away from fire.*
Human Toxicity: See Kerosene.
USE: Pharmaceutic aid (solvent).
THERAP CAT: Counterirritant.

7142. Petroselinic Acid. *cis-6-Octadecenoic acid;* petroselic acid; 5-heptadecylene-1-carboxylic acid; Δ^5-octadecylenic acid. $C_{18}H_{34}O_2$; mol wt 282.45. C 76.54%, H 12.13%, O 11.33%. $CH_3(CH_2)_{10}CH=CH(CH_2)_4COOH$. Isoln from parsley seed oil, the oil extracted from dried ripe seed of *Petroselinum hortense* Hoffm., *Umbelliferae:* Fore *et al., J. Am. Oil Chem. Soc.* **37,** 490 (1960).

Leaflets from petr ether, mp 29.5-30.1°. bp_{18} 237-238°; d_4^{40} 0.8700. n_D^{40} 1.4533. Low temp solubilities: Heptane at $-10° = 0.50$ g/100 g solution; methanol at $-20° = 0.48$ g/100 g; ethyl acetate at $-20° = 0.73$ g/100 g; ether at $-20° = 3.52$ g/100 g. Ozonolysis yields 85% of adipic acid. Neutralization equivalent: 282.45; iodine value 89.87%.

Methyl ester, $C_{19}H_{36}O_2$, liq, d_4^{20} 0.8767; n_D^{20} 1.4501; bp_{10} 208-210°.

Glyceryl triester, $C_{57}H_{104}O_6$, *glyceryl tripetroselinate, tripetroselin.* Solidifies at 16.5°. n_D^{40} 1.4619.

Amide, $C_{18}H_{35}NO$, needles, mp 76°.

7143. Petunidin. 2-(3,4-Dihydroxy-5-methoxyphenyl)-3,5,7-trihydroxy-1-benzopyrylium chloride; *3,3',4',5,7-pentahydroxy-5'-methoxyflavylium chloride;* petunidol. $C_{16}H_{13}$-ClO_7; mol wt 352.74. C 54.48%, H 3.71%, Cl 10.05%, O 31.75%. The aglucone of petunin: Willstätter, Burdick, *Ann.* **412,** 217 (1917). Synthesis: Bradley *et al., J. Chem. Soc.* **1930,** 793; Robinson, Robinson, *Biochem. J.* **25,** 1687 (1931). Chromatographic separation: Spaeth, Rosenblatt, *Anal. Chem.* **22,** 1321 (1950).

Gray-brown leaflets or prisms from dil HCl.

3,5-Diglucoside, $C_{28}H_{33}ClO_{17}$, 2-(3,4-dihydroxy-5-methoxyphenyl)-3,5-bis(β-D-glucopyranosyloxy)-7-hydroxy-1-benzopyrylium chloride, petunin. From *Petunia hybrida* Hort., *Solanaceae:* Willstätter, Burdick, *loc. cit.* Synthesis: Bell, Robinson, *J. Chem. Soc.* **1934,** 1604. Violet plates with a coppery luster from dil HCl, mp about 178°. Absorption max (methanolic HCl): 540 nm.

7144. Peucedanin. *3-Methoxy-2-(1-methylethyl)-7H-furo[3,2-g][1]benzopyran-7-one; 6-hydroxy-2-isopropyl-3-methoxy-5-benzofuranacrylic acid δ-lactone;* 4-methoxy-5-isopropylfuro[2,3:6,7]coumarin; oreoselone methyl ether. $C_{15}H_{14}O_4$; mol wt 258.26. C 69.76%, H 5.46%, O 24.78%. Coumarin deriv obtained from rhizome of *Peucedanum officinale* L., *Umbelliferae:* Schlatter, *Ann.* **5,** 201 (1833); Hlasiwetz, Weidel, *ibid.* **174,** 67 (1874); A. Jassoy, P. Haensel, *Arch. Pharm.* **236,** 662 (1898); Popper, *Monatsh.* **19,** 268 (1898); from *Peucedanum morisonii,* Bess., *Umbelliferae:* G. K. Nikonov, A. I. Ivashenko, *Zh. Obshch. Khim.* **33,** 2740 (1963). Fluorescence spectrum: R. H. Goodwin, F. Kavanagh, *Arch. Biochem.* **27,** 182 (1950). Antitumor activity and toxicity studies: E. M. Vermel, S. A. Kruglyak-Syrkina, *Voprosy Onkol.* **5,** 43 (1959), *C.A.* **53,** 19162h (1959). Structure: E. Späth *et al., Ber.* **64,** 2203 (1931); E. Späth, K. Klager, *ibid.* **66,** 749 (1933). Synthesis: H. Schmid, A. Ebnöther, *Helv. Chim. Acta* **34,** 1982 (1951).

Colorless needles from ether/petr ether, mp 84-87°; also reported as mp 95-97° (Schmid, Ebnöther); from ligroin, mp 102.5° (Nikonov, Ivashenko). uv max (methanol): 255, 295, 340 nm (log ϵ 4.40, 4.05, 3.70). Practically insol in water. Freely sol in chloroform, carbon disulfide; sol in hot alcohol,

ether, acetic acid; sparingly sol in benzene, petr ether. LD_{50} orally in mice: 315 mg/kg (Vermel, Kruglyak-Syrkina).

7145. Peyonine. *1-[2-(3,4,5-Trimethoxyphenyl)ethyl]-1H-pyrrole-2-carboxylic acid.* $C_{16}H_{19}NO_5$; mol wt 305.32. C 62.94%, H 6.27%, N 4.59%, O 26.20%. A peyote alkaloid, isolated from *Lophophora williamsii* (Lem.) Coult. Isoln and synthesis: Kapadia, Shah, *Lloydia* **30,** 287 (1967). Structure studies: Kapadia, Highet, *J. Pharm. Sci.* **57,** 191 (1968).

Crystals. mp 131-133.5°. uv max (methanol): 261 nm (ϵ 10,000). IR and NMR data: Kapadia, Highet, *loc. cit.*

7146. Pfeiffer's Substance. 4-(Dimethylamino)antipyrine compd with 5,5-diethylbarbituric acid (1:1); corps de Pfeiffer. $C_{21}H_{29}N_5O_4$; mol wt 415.48. C 60.70%, H 7.04%, N 16.86%, O 15.40%. A molecular compd of aminopyrine and barbital, prepd by repeated crystn of stoichiometric amounts of the components from a minimum of hot water: Pfeiffer, *Z. Physiol. Chem.* **146,** 98 (1925).

Silky needles, mp 113-115°. At 115° the melt is turbid, but turns into a clear liquid at 140°. Freely sol in water. The aminopyrine moiety dissolves in benzene, while the barbital moiety precipitates out. Ingredient of *Veramon.*

7147. Phalloidin. Phalloidine. $C_{35}H_{48}N_8O_{11}S$; mol wt 788.89. C 53.29%, H 6.13%, N 14.20%, O 22.31%, S 4.06%. Best known of the toxins isolated from the poisonous green fungus *Amanita phalloides* (Fr.) Seer., *Agaricaceae,* known as "the green death cap" or "deadly agaric": Lynen, Wieland, *Ann.* **533,** 93 (1938). Structure: Wieland, Schön, *ibid.* **593,** 157 (1955); Wieland, Schöpf, *ibid.* **626,** 174 (1959); Wieland, Schnabel, *ibid.* **657,** 225 (1962). Differs from amanitin in rapidity of action; at high dose levels, death of mice or rats occurs within 1 or 2 hours. Phalloidin acts by binding actin, *q.v.,* an essential internal structural protein. Ultrastructural pathology: M. A. Russo *et al., Am. J. Pathol.* **109,** 133 (1982). Review of the chemistry and toxicology of the toxins of *Amanita phalloides:* Wieland, Wieland, *Pharmacol. Rev.* **11,** 87-107 (1959); *see also* T. Wieland, *Fortschr. Chem. Org. Naturst.* **25,** 214-250 (1967); T. Wieland, H. Faulstich, *Crit. Rev. Biochem.* **5,** 185-260 (1978).

Hexahydrate, needles from water, mp 280-282°. uv max (water): 295 nm ($E_{1cm}^{1\%}$ 0.597). Soly in water (0°): 0.5%; much more sol in hot water; freely sol in methanol, ethanol, butanol, pyridine. LD_{50} in albino mice: 3.3 μg/g i.m., Vogt, *Arch. Exp. Pathol. Pharmakol.* **190,** 406 (1938); 2 mg/kg i.p., Wieland, Wieland, *loc. cit.*

Human Toxicity: See Amanitin.

7148. Phanquinone. *4,7-Phenanthroline-5,6-dione;* 4,7-phenanthroline-5,6-quinone; 5,6-dioxo-5,6-dihydro-4,7-phenanthroline; phanchinone; phanquone; Ciba 11925; Entobex. $C_{12}H_6N_2O_2$; mol wt 210.18. C 68.57%, H 2.88%, N 13.33%, O 15.23%. Prepd from 5,(6)-methoxy-4,7-phenanthroline: Druey, Schmidt, *Helv. Chim. Acta* **33,** 1080 (1950); **Brit.** pat. **688,802** (1953 to Ciba).

Crystals from methanol, mp 295° (dec). Sparingly sol in water, alcohol; sol in dil mineral acids.

THERAP CAT: Anti-amebic.

7149. Phaseolin. *(6bR-cis)-6b,12b-Dihydro-3,3-dimethyl-3H,7H,furo[3,2-c:5,4-f']bis[1]benzopyan-10-ol;* phaseollin. $C_{20}H_{18}O_4$; mol wt 322.34. C 74.52%, H 5.63%, O 19.85%. Antifungal phytoalexin isolated from French bean *(Phaseolus vulgaris L., Leguminosae):* I. A. M. Cruickshank, D. R. Perrin, *Life Sci.* **2,** 680 (1963). Structure: D. R. Perrin, *Tetrahedron Letters* **1964,** 29; D. R. Perrin *et al., ibid.* **1972,** 1673. Crystal structure: C. DeMartinis *et al., Tetrahedron* **34,** 1849 (1978). Biosynthesis: S. L. Hess *et al., Phytopathology* **61,** 79 (1971); P. M. Dewick, M. J. Steele, *Phytochemistry* **21,** 1599 (1982). Total synthesis: S. E. N. Mohamed *et al., J. Chem. Soc. Perkin Trans. I* **1987,** 431. Antifungal properties: I. A. M. Cruickshank, D. R. Perrin, *Phytopathol.* **70,** 209 (1971); M. A. Gordon *et al., Antimicrob. Ag. Chemother.* **17,** 120 (1980). Mode of action: F. D.Van Etten, D. F. Bateman, *Phytopathology* **61,** 1363 (1971).

Crystals, mp 177-178°. $[\alpha]_{578}$ −145°. pKa 9.13. uv max (ethanol): 207, 230, 280, 286 nm (log ε 4.68, 4.40, 3.97, 3.90).

7150. Phasin. Poisonous agglutinin from beans. A polypeptide composed of glutamic acid, aspartic acid, serine, alanine, tyrosine, lysine and arginine: Piekarski, *Dissertationes Pharm.* **9,** 255 (1957), *C.A.* **52,** 6456i (1958). Prepn: Wienhaus, *Biochem. Z.* **18,** 228 (1909), *C.A.* **3,** 2471 (1909); Piekarski, *loc. cit.*

Amorphous powder. Its toxicity and agglutination properties are destroyed by heating.

7151. α-Phellandrene. *2-Methyl-5-(1-methylethyl)-1,3-cyclohexadiene; p-mentha-1,5-diene;* 5-isopropyl-2-methyl-1,3-cyclohexadiene; 4-isopropyl-1-methyl-1,5-cyclohexadiene. $C_{10}H_{16}$; mol wt 136.23. C 88.16%, H 11.84%. Isoln of *l*-form from essential oils of *Eucalyptus dives* Schau. and *E. phellandra* Baker & Smith, *Myrtaceae:* Smith *et al., J. Chem. Soc.* **123,** 1657 (1923). Isoln of *d*-form from oil of bitter fennel *(Foeniculum vulgare* Mill., *Umbelliferae):* Wallach, *Ann.* **336,** 9 (1904). Structure: Semmler, *Ber.* **36,** 1749 (1903). Synthesis of *dl*-form: Read, Storey, *J. Chem. Soc.* **1930,** 2770; B. Singaram, J. Verghese, *Ind. J. Chem.* **14B,** 1003 (1976). Configuration: Burgstahler *et al., J. Am. Chem. Soc.* **83,** 4660 (1961). Circular dichroism: G. Snatzke *et al., Tetrahedron Letters* **1966,** 4551. *Review:* J. L. Simonsen, *The Terpenes* **vol. 1** (Univ. Press, Cambridge, 2nd ed., 1947) pp 193-204; B. Singaram, J. Verghese, *Perfum. Flavor.* **2,** 33-38 (1978).

l-Form, mobile oil. bp$_{758}$ 171-172°; bp$_{16}$ 58-59°. d$_4^{20}$ 0.8410. n$_D^{20}$ 1.4709. $[\alpha]_D^{20}$ −217°. Practically insol in water; sol in ether.

d-Form, mobile oil. bp$_{16}$ 66-68°. d$_4^{25}$ 0.8463. n$_D^{25}$ 1.4777. $[\alpha]_D^{16}$ +86.4°. Practically insol in water; sol in ether.

Caution: Can be irritating to, and absorbed through, skin. Ingestion can cause vomiting, diarrhea.

USE: In fragrances.

7152. β-Phellandrene. *3-Methylene-6-(1-methylethyl)cyclohexene; p-mentha-1(7),2-diene;* 3-isopropyl-6-methylene-1-cyclohexene; 4-isopropyl-1-methylene-2-cyclohexene. $C_{10}H_{16}$; mol wt 136.23. C 88.16%, H 11.84%. Isoln of *d*-form from oil of water fennel *(Phellandrium aquaticum L., Umbelliferae):* Berry *et al., J. Chem. Soc.* **1937,** 1448. Isoln of *l*-form from Canada balsam oil: Macbeth *et al., ibid.* **1938,** 119. Structure: Wallach, *Ann.* **343,** 28 (1905). Synthesis of *dl*-form: Deorha, Sareen, *Rec. Trav. Chim.* **82,** 137 (1965); B. Singaram, J. Verghese, *Ind. J. Chem.* **14B,** 1003 (1976).

d-Form, mobile oil. bp$_{760}$ 171-172°; bp$_{11}$ 57°. d$_4^{20}$ 0.8520. n$_D^{20}$ 1.4788. $[\alpha]_D^{20}$ +65.2°. Practically insol in water; sol in ether.

l-Form, mobile oil. bp$_{758}$ 178-179°; bp$_{12}$ 53°. d$_{15}^{15}$ 0.8497. n$_D^{20}$ 1.4800. $[\alpha]_D^{20}$ −51.9°. Practically insol in water, alcohol; sol in ether.

7153. Phenacaine Hydrochloride. *N,N'-Bis(4-ethoxyphenyl)ethanimidamide monohydrochloride; N^1,N^2-bis(p-ethoxyphenyl)acetamidine hydrochloride;* Holocaine Hydrochloride. $C_{18}H_{23}ClN_2O_2$; mol wt 334.84. C 64.56%, H 6.92%, Cl 10.59%, N 8.37%, O 9.56%. Prepn: *Ger.* pats. **79,868; 80,568.**

Monohydrate, faintly bitter crystals producing transient numbness of the tongue. Stable in air. When anhydr mp 190-192°. One gram dissolves in 50 ml water. Freely sol in boiling water, alcohol, chloroform. Insol in ether. Incompatible with alkalies and their carbonates and the usual alkaloidal reagents. Aq solns are stable and are not dec by boiling.

THERAP CAT: Topical anesthetic (ophthalmic).

THERAP CAT (VET): Ocular anesthetic.

7154. Phenacemide. *N-(Aminocarbonyl)benzeneacetamide; (phenylacetyl)urea;* phenacetylurea; Epiclase; Phacetur; Phenurone; Phetyl ureum. $C_9H_{10}N_2O_2$; mol wt 178.19. C 60.66%, H 5.66%, N 15.72%, O 17.96%. $C_6H_5CH_2CONH$-$CONH_2$. Prepd by the action of aq NH_3 on phenacetylurethan: Basterfield, Greig, *Can. J. Res.* **8,** 454 (1933); of phenacetyl chloride on urea: Spielman *et al., J. Am. Chem. Soc.* **70,** 4189 (1948).

Crystals from alcohol, mp 212-216°. Very slightly sol in water; slightly sol in alcohol, benzene, chloroform, ether. LD_{50} orally in mice, rats: 5.54, > 10 mmol/kg, K. Nakamura *et al., Arzneimittel-Forsch.* **18,** 524 (1968).

THERAP CAT: Anticonvulsant.

7155. Phenacetin. *N-(4-Ethoxyphenyl)acetamide; p-acetophenetidide; p-ethoxyacetanilide; acetophenetidin; para-acetphenetidin; acet-p-phenetidin; p-acetophenetide.* $C_{10}H_{13}NO_2$; mol wt 179.21. C 67.02%, H 7.31%, N 7.82%, O 17.85%. Prepn: *Beilstein* **vol. XIII,** 461. Improved process: Eaker, Campbell, **U.S.** pat. **2,887,513** (1959 to Monsanto). *Monograph:* P. K. Smith, *Acetophenetidin* (Interscience, New York, 1958) 180 pp. Toxicity: Boyd, *Toxicol. Appl. Phar-*

macol. **1**, 240 (1959). Review of toxicity and metabolism studies: L. Fishbein, *IARC Sci. Publ.* **40**, 287-310 (1981). Epidemiologic study of renal morbidity and mortality: U. C. Dubach *et al.*, *N. Engl. J. Med.* **308**, 357 (1983). Evaluation of renal effects: D. P. Sandler *et al.*, *ibid.* **320**, 1238 (1989).

Slightly bitter, cryst scales or powder. mp 134-135°. One gram dissolves in 1310 ml cold water, 82 ml boiling water; 15 ml cold alcohol, 2.8 ml boiling alcohol; 14 ml chloroform, 90 ml ether; sol in glycerol. Gives a pasty mass with phenol, chloral hydrate or pyrocatechol, strong acids or alkalies, salicylic acid, oxidizers, iodine, spirit nitrous ether. LD$_{50}$ orally in rats: 1.65 g/kg (Boyd).

Note: Component of *APC* tablets, also containing aspirin and caffeine. Phenacetin may reasonably be anticipated to be a carcinogen: *Fourth Annual Report on Carcinogens* (NTP 85-002, 1985) p 165.

THERAP CAT: Analgesic, antipyretic.

THERAP CAT (VET): Analgesic, antipyretic.

7156. Phenacetolin. Degener's indicator. Reaction product of concd H$_2$SO$_4$ and glacial acetic acid on phenol. Yellowish-brown powder. Slightly sol in water; sol in alc.

USE: Has been employed as indicator, particularly for mixtures of alkali hydroxides and carbonates; the yellow color changes to red when hydroxide is neutralized, and again to yellow when carbonate is fully dec by acid; was also used for determining alkalinity of water. The indicator soln is prepd by digesting 1 g with 100 ml warm alcohol and filtering after cooling; 2-3 drops used for 100 ml liquid.

7157. Phenacridane Chloride. *9-[4-(Hexyloxy)phenyl]-10-methylacridinium chloride;* 9-chloro-9-[*p*-(hexyloxy)-phenyl]-10-methylacridan; 9-(*p*-hexyloxyphenyl)-10-methyl-9-acridanyl chloride; Acrizane chloride; Micridium chloride. C$_{26}$H$_{28}$ClNO; mol wt 405.98. C 76.92%, H 6.95%, Cl 8.73%, N 3.45%, O 3.94%. Prepn: Tabern, U.S. pat. **2,645,-594** (1953 to Abbott). Appears to be a tautomeric structure with the major portion in the acridinium form.

Hygroscopic yellow powder, mp 184.5-185.5°. Freely sol in water, alcohol. The prepns of choice are an aq soln and a tincture contg 500 ppm. It stains the skin intensely yellow, which color cannot be removed easily. *Ref: Am. Surgeon* **19**, 792 (1953).

THERAP CAT: Topical anti-infective.

7158. Phenactropinium Chloride. *3-[(Hydroxyphenyl-acetyl)oxy]-8-methyl-8-(2-oxo-2-phenylethyl)-8-azoniabicy-clo[3.2.1]octane chloride; 8-benzoylmethyl-3α-hydroxy-1αH,5αH-tropanium chloride mandelate;* 8-phenacylhomatropinium chloride; Trophenium. C$_{24}$H$_{28}$ClNO$_4$; mol wt 429.95. C 67.05%, H 6.56%, Cl 8.25%, N 3.26%, O 14.89%. Prepn: Johnston, Spencer, U.S. pat. **2,828,312** (1958 to T. & H. Smith). Ganglionic blocking agent.

Crystals from hot alcohol, mp 195-197°.

THERAP CAT: Antihypertensive.

7159. Phenacylamine. *2-Amino-1-phenylethanone; 2-aminoacetophenone;* α-aminoacetophenone; ω-aminoacetophenone. C$_8$H$_9$NO; mol wt 135.16. C 71.09%, H 6.71%, N 10.36%, O 11.84%. Prepn from α-phenylethylamine: H. E. Baumgarten, J. M. Petersen, *J. Am. Chem. Soc.* **82**, 459 (1960); *eidem, Org. Syn.* **coll. vol.** V, 909 (1973).

Hydrochloride, C$_8$H$_9$NO.HCl, crystals from isopropanol + HCl, dec 188.5°.

7160. Phenadoxone. *6-(4-Morpholinyl)-4,4-diphenyl-3-heptanone;* 6-tetrahydrooxazine-4,4-diphenyl-3-heptanone; heptazone; morphodone. C$_{23}$H$_{29}$NO$_2$; mol wt 351.49. C 78.59%, H 8.32%, N 3.99%, O 9.10%. Prepn: M. Bockmühl, G. Ehrhart, *Ann.* **561**, 52 (1948). Comparison with other analgesics: C. A. Winter, L. Flataker, *J. Pharmacol. Exp. Ther.* **98**, 305 (1950). HPLC determn: I. Jane *et al.*, *J. Chromatog.* **323**, 191 (1985).

mp 75-76°.

Hydrochloride, C$_{23}$H$_{30}$ClNO$_2$, *Hoechst 10600, CB 11, Hepagin, Heptalgin, Heptalin, Heptone, Supralgin.* Crystals, mp 224-225° (some decompn). pKa at 25° = 6.7 in 40% alcohol. pH of 1% aq soln 4.1; of 5% aq soln 3.7. The free base begins to precipitate when the pH is raised to 5.5. uv max (abs alcohol): 260, 295 nm. Soly (g/100 ml) in water: 10 at 25°; 80 at the bp; in methanol: 27; in abs ethanol: 24; in chloroform: 50. Slightly sol in acetone. Almost insol in benzene, ethyl acetate. LD$_{50}$ i.v. in mice: 47.5 mg/kg (Winter, Flataker).

Caution: May be habit forming. This is a controlled substance (opiate) listed in the U.S. Code of Federal Regulations, Title 21 Part 1308.11 (1985).

THERAP CAT: Narcotic analgesic.

7161. Phenaglycodol. *2-(4-Chlorophenyl)-3-methyl-2,3-butanediol;* Acalo; Acalmid; Sinforil; Ultran. C$_{11}$H$_{15}$-ClO$_2$; mol wt 214.70. C 61.54%, H 7.04%, Cl 16.51%, O 14.90%. Prepn: Mills, U.S. pat. **2,812,363** and **Brit.** pat. **788,896** (1957, 1958, both to Lilly). Acute toxicity: E. W. Schafer, *Toxicol. Appl. Pharmacol.* **21**, 315 (1972).

Crystals from hexane or benzene + petr ether, mp 77-78°. Practically insol in water. Quite sol in many alcohols and oils. LD$_{50}$ orally in rats: 832 mg/kg (Schafer).

THERAP CAT: Anxiolytic.

7162. Phenallymal. *5-Phenyl-5-(2-propenyl)-2,4,6(1H,-*

3H,5H)-pyrimidinetrione; 5-allyl-5-phenylbarbituric acid;
5-phenyl-5-allylbarbituric acid; phenallymalum; alphenal;
prophenal; Alphenate; Allofenyl; Fenallymal. $C_{13}H_{12}N_2O_3$;
mol wt 244.24. C 63.92%, H 4.95%, N 11.47%, O 19.65%.
Its methods of prepn are similar to those used for phenobarbital: Slotta, *Grundriss der modernen Arzneistoff-Synthese*
(Stuttgart, 1931) p 45. Metabolism: D. J. Harvey *et al.*,
Res. Commun. Chem. Pathol. Pharmacol. **4**, 247 (1972);
eidem, Drug Metab. Dispos. **5**, 527 (1977).

Crystals, mp 156-157.5°. Bitter taste. Readily sol in alcohol, chloroform. One gram dissolves in 580 ml water, in 10
ml ether, in 500 ml benzene, in 4000 ml carbon tetrachloride,
in 17,500 ml petr ether.
Caution: May be habit forming. This is a controlled substance (depressant) listed in the U.S. Code of Federal Regulations, Title 21 Part 1308.13 (1987).
THERAP CAT: Sedative, hypnotic.

7163. Phenamacide Hydrochloride. *(±)-α-Aminoben-
zeneacetic acid 3-methylbutyl ester hydrochloride; 2-phenyl-
glycine isopentyl ester hydrochloride;* isopentyl 2-phenylgly-
cinate hydrochloride; phenylaminoacetic acid isoamyl ester
hydrochloride; isoamyl phenylaminoacetate hydrochloride;
Aklonin. $C_{13}H_{20}ClNO_2$; mol wt 257.77. C 60.57%, H
7.82%, Cl 13.76%, N 5.44%, O 12.41%. Prepn: Klosa, *Arch.
Pharm.* **285**, 332 (1952). Pharmacology: Schultz *et al.*,
Pharmazie **25**, 560 (1970).

Crystals, mp 154°.
THERAP CAT: Antispasmodic.

7164. Phenamet. *N-[[4-[Bis(2-chloroethyl)amino]phen-
yl]acetyl]-L-methionine ethyl ester.* $C_{19}H_{28}Cl_2N_2O_3S$; mol wt
435.43. C 52.41%, H 6.48%, Cl 16.28%, N 6.43%, O 11.02%,
S 7.37%. Prepn: Golubeva *et al., Dokl. Akad. Nauk SSSR*
119, 83 (1958); Knunyants *et al., ibid.* **132**, 836 (1960).

Crystals from ethyl acetate + petr ether, mp 93-95°.
THERAP CAT: Antineoplastic.

7165. Phenamidine. *4,4'-Oxybisbenzenecarboximidam-
ide; 4,4'-oxydibenzamidine; 4,4'-diamidinodiphenyl ether;*
$C_{14}H_{14}N_4O$; mol wt 254.29. C 66.13%, H 5.55%, N 22.03%,
O 6.29%. Prepn: A. J. Ewins *et al.*, **Brit.** pat. **507,565** (1939
to May & Baker); J. N. Ashley *et al., J. Chem. Soc.* **1942**,
103; of isethionate: G. Newbery, A. P. T. Easson, U.S. pat.
2,410,796 (1946 to May & Baker). Trypanocidal activity:
E. M. Lourie, W. Yorke, *Ann. Trop. Med. Parasitol.* **33**, 289
(1939). Preliminary pharmacology: R. Wien, *ibid.* **37**, 1
(1943). Use in treatment of Babesia infections in dogs: M.
D. Ruff *et al., Am. J. Vet. Res.* **34**, 641 (1973); R. Gothe *et
al., Kleintierpraxis,* **32**, 97 (1987). Early review of pharmacology, mode of action and clinical applications: E. M.
Schoenbach, E. M. Greenspan, *Medicine* **27**, 327-377 (1948).

Irregular plates from water, mp 215-216°.
Isethionate, $C_{18}H_{26}N_4O_8S_2$, **M&B 736**, *Lomadine.* Crystals, mp 225° (dec). One part is sol in 1.4 parts water; 300
parts alcohol. Practically insol in ether, chloroform.
THERAP CAT (VET): Antiprotozoal (Babesia).

7166. Phenampromid(e). *N-[1-Methyl-2-(1-piperidin-
yl)ethyl]-N-phenylpropanamide; N-(1-methyl-2-piperidino-
ethyl)propionanilide;* 1-piperidino-2-(N-propionylanilino)-
propane. $C_{17}H_{26}N_2O$; mol wt 274.39. C 74.41%, H 9.55%,
N 10.21%, O 5.83%. Synthesis and analgesic activity: W. B.
Wright, Jr. *et al., J. Am. Chem. Soc.* **81**, 1518 (1959); W. B.
Wright, Jr., H. J. Brabander, U.S. pat. **3,016,382** (1962 to
Am. Cyanamid). (−)-Phenampromid(e), the more active
enantiomer, has the *R*-configuration. Absolute configuration: Portoghese, *J. Med. Chem.* **8**, 147 (1965). Pharmacology: Kikuchi *et al., Nippon Yakurigaku Zasshi* **57**, 585
(1961), *C.A.* **59**, 2082b (1963). Determn by HPLC: I. Jane,
A. McKinnon, *J. Chromatog.* **323**, 191 (1985).

(−)−phenampromid(e)

Liquid, $bp_{0.2}$ 124-128°. n_D^{28} 1.518.
(−)-Hydrochloride, $C_{17}H_{27}ClN_2O$, levorotatory crystals
from alcoholic HCl + ether, mp 201-202°.
Caution: May be habit forming. This is a controlled substance (opiate) listed in the U.S. Code of Federal Regulations, Title 21 Part 1208.11 (1985).

7167. Phenanthrene. $C_{14}H_{10}$; mol wt 178.22. C 94.34%,
H 5.66%. Isomeric with anthracene. Occurs in coal tar.
Isoln: Ostermayer, Fittig, *Ber.* **5**, 933 (1872); Glaser, *ibid.*
982. Purification from contaminating carbazole and anthracene): Clar, *Ber.* **65**, 852 (1932). Formation from toluene, bibenzil, 9-methylfluorene or stilbene by passage
through red-hot tube: Graebe, *Ber.* **7**, 48 (1874); *Ann.* **167**,
161 (1879); *Ber.* **37**, 4145 (1904). Also from coumarone and
benzene: Kraemer, Spilker, *Ber.* **23**, 85 (1890). Pschorr synthesis from o-nitrobenzaldehyde and phenylacetic acid: *Ber.*
29, 500 (1896). From diphenylethylene: Cook, Hewett, *J.
Chem. Soc.* **1933**, 1098. Diene synthesis from 1-vinylnaph-
thalene and maleic anhydride: Cohen, Warren, *ibid.* **1937**,
1315. From o-phenylbenzoic acid: Schönberg, Warren,
Chem. & Ind. (London) **58**, 199 (1939). By irradiation of
stilbene: Mallory *et al., J. Am. Chem. Soc.* **84**, 4361 (1962).
Synthesis by double succinoylation of benzene: Rahman *et
al., J. Org. Chem.* **28**, 3571 (1963). Structure: Trotter, *Acta
Cryst.* **16**, 605 (1963).

Monoclinic plates from alcohol. d^{25} 1.179; mp 100°; bp
340°. Sublimes in high vacuum. Absorption spectrum:
Clar, Lombardi, *Ber.* **65**, 1412 (1932); Mayneord, Roe, *Proc.
Roy. Soc. London A* **152**, 317 (1935). Practically insol in
water; sol in organic solvents, especially in aromatic hydrocarbons. One gram dissolves in 60 ml cold, 10 ml boiling
95% alcohol, 25 ml abs alcohol, 2.4 ml toluene or carbon
tetrachloride, 2 ml benzene, 1 ml carbon disulfide, 3.3 ml
anhydr ether. Soluble in glacial acetic acid. Solns exhibit a

blue fluorescence. Forms molecular compds with picric acid, picryl chloride, dinitrobenzene and similar nitro compounds. LD_{50} orally in mice: 700 mg/kg, *Toxic Substances List*, H. E. Christensen, Ed. (1973) p 708.

Caution: Can cause photosensitization of skin.

7168. Phenanthrenequinone. *9,10-Phenanthrenedione;* 9,10-phenanthraquinone. $C_{14}H_8O_2$; mol wt 208.20. C 80.76%, H 3.87%, O 15.37%. Prepn by oxidation of phenanthrene: Curtis *et al.*, **U.S. pat. 2,956,065** (1960 to U.S. Steel); Binder, Koch, **Ger. pat. 1,166,176** (1964).

Orange-red crysts. mp 206-207°. bp about 360°. d 1.405. Sublimes. Practically insol in water; sol in benzene, ether, glacial acetic acid, hot alc. With concd H_2SO_4 gives a dark green color.

7169. o-Phenanthroline. *1,10-Phenanthroline;* 4,5-phenanthroline. $C_{12}H_8N_2$; mol wt 180.20. C 79.98%, H 4.48%, N 15.55%. Made by heating o-phenylenediamine with glycerol, nitrobenzene and concd H_2SO_4, or in the same manner from 8-aminoquinoline. Prepn: Madeja, *J. Prakt. Chem.* **17**, 104 (1962). Crystal and molecular structure: S. Nishigaki *et al.*, *Acta Crystallogr.* **B34**, 875 (1978). Review of analytical uses: Vydra, Kopanica, *Chemist-Analyst* **52**, 88-94 (1963).

Monohydrate, white, cryst powder. mp 93-94°; anhydr, 117°. Soluble in about 300 parts water, 70 parts benzene; sol in alcohol, acetone.

USE: Forms a complex compound with ferrous ions which is used under the name of "*Ferroin*" as an indicator in oxidation reduction systems, *e.g.*, titration of ferrous salts. Also used in determination of nickel, ruthenium, silver and other metals.

7170. Phenarsazine Chloride. *10-Chloro-5,10-dihydrophenarsazine;* 5-aza-10-arsenaanthracene chloride; 10-chloro-5,10-dihydroarsacridine; diphenylaminechlorarsine; phenazarsine chloride; adamsite; DM. $C_{12}H_9AsClN$; mol wt 277.57. C 51.92%, H 3.27%, As 26.99%, Cl 12.77%, N 5.05%. Prepd by heating diphenylamine with arsenic trichloride: **Ger. pat. 281,049** (1914 to I. G. Farben.); Wieland, Reinheimer, *Ann.* **423**, 12 (1921); Lewis, Hamilton, *J. Am. Chem. Soc.* **43**, 2222 (1921); Burton, Gibson, *J. Chem. Soc.* **1926**, 450.

Canary-yellow crystals from carbon tetrachloride. *Poisonous.* Dimorphous. The stable form occurs as orthorhombic crystals; d 1.65; mp 195°; bp 410° (decompn). Sublimes readily. Vapor press. at 20° = 2×10^{-13} mm; volatility 0.02 mg/cu m; heat of volatilization 54.8 cal; spec heat 0.268 cal. Practically insol in water. Slightly sol in benzene, xylene, carbon tetrachloride. Corrodes iron, bronze, brass. [The metastable form melts at 186° if monoclinic, and at 182° if triclinic.]

USE: As war gas, dispersed in air in the form of minute particles. For riots in combination with tear gas (chloroacetophenone). In the formulation of wood-treating solns,

against marine borers and similar pests. *Caution:* Irritating to skin and respiratory tract. Causes profuse watery nasal discharge; severe pain in nose, sinuses, chest; sneezing, coughing, nausea, vomiting, marked depression, weakness. Sensory disturbances may occur later.

7171. Phenarsone Sulfoxylate. *[(5-Arsono-2-hydroxyphenyl)amino]methanesulfinic acid disodium salt; 4-hydroxy-N-(sulfinomethyl)-m-arsanilic acid disodium salt;* Aldarsone. $C_7H_8AsNNa_2O_6S$; mol wt 355.10. C 23.67%, H 2.27%, As 21.10%, N 3.95%, Na 12.95%, O 27.03%, S 9.03%. Prepn: Raiziss, Kremens, **U.S. pat. 2,074,757** (1937 to Abbott).

Amorphous, white powder. Sol in water, dil acids, alkali and alkali carbonates; slightly sol in methanol; practically insol in ether, chloroform, ethanol. pH of 5% aq soln: 7.0-7.5.

THERAP CAT: Anti-amebic.

7172. Phenatine. *N-(1-Methyl-2-phenylethyl)-3-pyridinecarboxamide; N-(α-methylphenethyl)nicotinamide;* nicotinic acid β-phenylisopropylamide; nicotinoyl-β-phenylisopropylamine; fenatin; perviton. $C_{15}H_{16}N_2O$; mol wt 240.29. C 74.97%, H 6.71%, N 11.66%, O 6.66%. Synthesis from nicotinoyl chloride and $C_6H_5CH_2CH(CH_3)NH_2$: Arbuzov *et al.*, *Sb. Statei Obshch. Khim., Akad. Nauk SSSR* **1**, 714 (1953), *C.A.* **49**, 1072 (1955). Pharmacology: Arbuzov, *Farmakol. Toksikol.* **31**, 373 (1968). Analysis: P. P. Suprun, *Farm. Zh. (Kiev)* **30**, 49 (1975); O. N. Shcherbina *et al.*, *ibid.* **1979**, 49.

Crystals from benzene, mp 99-100°.

Phosphate, $C_{15}H_{16}N_2O.2H_3PO_4$, crystals from alcohol + ether, mp 162°. Sol in water, warm alcohol. Insol in ether. Aq solns may be boiled for 3 hrs without decompn.

7173. Phenazine. Dibenzopyrazine; dibenzoparadiazine; azophenylene. $C_{12}H_8N_2$; mol wt 180.20. C 79.98%, H 4.48%, N 15.55%. Obtained (with other products) by passing aniline vapor through a red-hot tube: Bernthsen, *Ber.* **19**, 3257 (1886); by heating aniline with nitrobenzene and sodium hydroxide to 140°: Wohl, Aue, *Ber.* **34**, 2446 (1901); Wohl, *Ber.* **36**, 4135 (1903); by heating o-phenylenediamine with pyrocatechol in sealed tube: Ris, *Ber.* **19**, 2206 (1886); Hinsberg, Garfunkel, *Ann.* **292**, 258 (1896); upon distilling 2-aminodiphenylamine with PbO: Fischer, Heiler, *Ber.* **26**, 383 (1893); by heating 2-aminodiphenylamine with 2-nitrodiphenylamine in the presence of sodium acetate: Kehrmann, Havas, *Ber.* **46**, 342 (1913); by heating nitrobenzene with barium oxide: Zerewitinoff, Ostromisslensky, *Ber.* **44**, 2402 (1911); by boiling 2,2'-dinitrodiphenylamine with stannous chloride in hydrochloric and acetic acids, followed by oxidation with hydrogen peroxide: Eckert, Steiner, *Monatsh.* **35**, 1154 (1914).

Pale yellow needles from alcohol or by sublimation. Colorless needles from dilute alcohol. mp 171°; bp above 360°. Practically insol in water. One part dissolves in 50 parts alcohol. Moderately sol in ether, benzene; sol in mineral acids giving yellow to red solns.

7174. Phenazocine. *1,2,3,4,5,6-Hexahydro-6,11-dimeth-*

yl-3-(2-phenethyl)-2,6-methano-3-benzazocin-8-ol; 2'-hydroxy-2-(N,β-phenethyl)-5,9-dimethyl-6,7-benzomorphan; 2'-hydroxy-5,9-dimethyl-2-phenethyl-6,7-benzmorphan; phenethylazocine; phenobenzorphan. $C_{22}H_{27}NO$; mol wt 321.44. C 82.20%, H 8.47%, N 4.36%, O 4.98%. Prepn: E. L. May, N. B. Eddy, *J. Org. Chem.* **24**, 294, 1435 (1959); Ager, May, *ibid.* **25**, 984 (1960); Gordon *et al.*, U.S. pat. **2,959,594** (1960 to SK&F).

(±)-Form, rods from methanol, mp 181-182°.
(−)-Form, needles from methanol, mp 159-159.5°. $[\alpha]_D^{20}$ −122° (c = 0.74 in 95% ethanol).
(±)-Hydrobromide, $C_{22}H_{28}BrNO$, *NIH 7519, SKF 6574, Narphen, Prinadol, Xenagol.* Rods from acetone or abs alc + ether, mp 166-170°. LD_{50} s.c. in mice: 332 mg/kg (May, Eddy).
(−)-Hydrobromide, crystals, mp 284-287°. $[\alpha]_D^{20}$ −84.1° (c = 1.12 in 95% ethanol). LD_{50} s.c. in mice: 147 mg/kg (May, Eddy).
Caution: May be habit forming. This is a controlled substance (opiate) listed in the U.S. Code of Federal Regulations, Title 21 Part 1308.12 (1985).
THERAP CAT: Narcotic analgesic.

7175. Phenazopyridine Hydrochloride. *3-(Phenylazo)-2,6-pyridinediamine monohydrochloride; 2,6-diamino-3-phenylazopyridine hydrochloride;* β-phenylazo-α,α'-diaminopyridine hydrochloride; Pyridium; Pyripyridium; Bisteril; Pyridiate; Pyridacil; Mallophene; Sedural; Uridinal; Urodine; Phenazodine. $C_{11}H_{12}ClN_5$; mol wt 249.70. C 52.91%, H 4.84%, Cl 14.20%, N 28.05%. The commercial product may contain some β,β'-bis(phenylazo)-α,α'-diaminopyridine and is therefore identified simply as phenylazo-α,α-diaminopyridine monohydrochloride. Prepd by coupling diazotized aniline with α,α-diaminopyridine: Chichibabin, Zeide, *J. Russ. Phys. Chem. Soc.* **46**, 1216 (1914); *Chem. Zentr.* **1915**, I, 1064; Chichibabin, Ossetrowa, *J. Am. Chem. Soc.* **56**, 1711 (1934); Ostromislensky, *ibid.* 1713; Ger. pat. **515,781** (1927 to Boehringer); U.S. pats. **1,680,108; 1,680,-109; 1,680,110; 1,680,111** (all 1928 to Pyridium); Shreve *et al.*, *J. Am. Chem. Soc.* **65**, 2241 (1943). Physical characteristics: Collins, *J. Am. Pharm. Assoc.* **20**, 455 (1931). Toxicity data: B. A. Becker, J. G. Swift, *Toxicol. Appl. Pharmacol.* **1**, 42 (1959). Comprehensive description: K. W. Blessel *et al.*, in *Analytical Profiles of Drug Substances* vol. 3, K. Florey, Ed. (Academic Press, New York, 1974) pp 465-482.

Brick-red microcrystals, slight violet luster. Slightly bitter taste. (The free base mp 139°.) Slightly sol in cold water, about 1 part in 300; sol in boiling water, about 1 part in 20. Forms supersatd solns easily. Will precipitate out of a 2% soln at 25° after about 2 days, out of a 1% soln only after months. Aq solns may be stabilized by the addn of 10% glucose. One part is sol in about 100 parts glycerol U.S.P. Sol in ethylene and propylene glycols. Slightly sol in alc, lanolin; sol in acetic acid. Insol in acetone, benzene, chloroform, ether, toluene. Aq solns are yellow to brick-red and slightly acid. Pyridium stains may be removed by soaking stained fabric (cotton, wool, rayon, nylon, Dacron) in 0.25% sodium hydrosulfite soln. LD_{50} orally in rats: 403 mg/kg (Becker, Swift).
Note: This substance may reasonably be anticipated to be a carcinogen: *Fourth Annual Report on Carcinogens* (NTP 85-002, 1985) p 166.
THERAP CAT: Analgesic (urinary tract).
THERAP CAT (VET): Urinary analgesic.

7176. Phenbenzamine. *N,N-Dimethyl-N'-phenyl-N'-phenylmethyl-1,2-ethanediamine; N-benzyl-N',N'-dimethyl-N-phenylethylenediamine; N-(2-dimethylaminoethyl)-N-benzylaniline;* RP 2339; Antergan; Bridal; Lergitin. $C_{17}H_{22}N_2$; mol wt 254.36. C 80.27%, H 8.72%, N 11.01%. Prepn: Mosnier, Fr. pat. **913,161** (1946 to Rhône-Poulenc); Miescher, Klarer, U.S. pat. **2,505,133** (1950 to Ciba); Kyrides, Zienty, U.S. pat. **2,634,293** (1953 to Monsanto).

Liquid. bp_7 179-180°; $bp_{0.03}$ 195-196°. d_{25}^{25} 1.019. n_D^{25} 1.5794.
Hydrochloride, $C_{17}H_{23}ClN_2$, *Antergan Hydrochloride.* Crystals, mp 210-211°. Freely sol in water; sol in alc. pKb 6.27. pH of soln contg 25 mg/ml 5.72. *Cf.* Huttrer, *Enzymologia* **12**, 284 (1948). LD_{50} orally in rats: 300 mg/kg, *Handbook of Toxicology* vol. 1, W. S. Spector, Ed. (Saunders, Philadelphia, 1956) pp 26-27.
Citrate, crystals, mp 94°.
Picrate, red crystals, dec 194-195°.
THERAP CAT: Antihistaminic.

7177. Phenbutamide. *N-[(Butylamino)carbonyl]benzenesulfonamide; 1-butyl-3-(phenylsulfonyl)urea; N-benzenesulfonyl-N'-butylurea;* fenbutamide; AL 132; R 131; Diaperos; Ipoglon. $C_{11}H_{16}N_2O_3S$; mol wt 256.34. C 51.54%, H 6.29%, N 10.93%, O 18.73%, S 12.51%. $C_6H_5SO_2NH-CONHC_4H_9$. Prepn: Brit. pat. **808,071** (1959 to Hoechst), *C.A.* **53**, 14056c (1959); Hokfelt, Joenssen, *J. Med. Pharm. Chem.* **5**, 231 (1962); Brzozowaki, *Rocz. Chem.* **43**, 1761 (1969). Activity studies: Tonse *et al.*, *Nature* **193**, 891 (1962); Khurana *et al.*, *Indian J. Med. Res.* **55**, 1084 (1967).
Crystals, mp 130-132°.
THERAP CAT: Antidiabetic.

7178. Phencarbamide. *Diphenylcarbamothioic acid S-[2-(diethylamino)ethyl] ester; diphenylthiocarbamic acid S-[2-(diethylamino)ethyl] ester; S-[2-(diethylamino)ethyl] diphenylthiocarbamate;* fencarbamide. $C_{19}H_{24}N_2OS$; mol wt 328.49. C 69.47%, H 7.36%, N 8.53%, O 4.87%, S 9.76%. Prepd from diphenylcarbamyl chloride and β-diethylaminoethanethiol: Brit. pat. **871,774** corresp to K. H. Risse *et al.*, U.S. pat. **3,228,949** (1961, 1966 to Bayer). Pharmacology: W. Wirth *et al.*, *Arch. Int. Pharmacodyn.* **151**, 515 (1964); W. Wirth, R. Goesswald, *ibid.* **155**, 393 (1965); B. R. Madan *et al.*, *ibid.* **185**, 53 (1970).

Crystals, mp 48-49°, $bp_{0.01}$ 120-126°. Practically insol in water. Freely sol in methanol, ether, chloroform. Sol in petr ether.
Hydrochloride, $C_{19}H_{25}ClN_2OS$, *Ba 1355, Escorpal.* Crystals, mp 180-181°. Sol in water. LD_{50} orally in rats: 410 mg/kg, W. Wirth, R. Goesswald, *loc. cit.*
1,5-Naphthalenedisulfonate, $C_{29}H_{32}N_2O_7S_3$, *phencarbamide napadisilate, Gelosedine.* LD_{50} orally in rats: 920 mg/kg, W. Wirth, R. Goesswald, *loc. cit.*
Component of *Spasmo-Dolviran.*
THERAP CAT: Anticholinergic.

7179. Phencyclidine. *1-(1-Phenylcyclohexyl)piperidine;* angel dust; HOG; PCP; CI-395. $C_{17}H_{25}N$; mol wt 243.38. C 83.89%, H 10.35%, N 5.76%. Prepn: Brit. pat. **836,083** and Godefroi *et al.*, U.S. pat. **3,097,136** (1960, 1963 to Parke, Davis); V. H. Maddox *et al.*, *J. Med. Chem.* **8**, 230 (1965). Pharmacology: G. Chen *et al.*, *J. Pharmacol. Exp. Ther.* **127**, 241 (1959); J. C. Munch, *Bull. Narcotics* **26**, 9 (1974). Human metabolism: L. K. Wong, K. Biemann, *Biomed. Mass Spectrom.* **2**, 204 (1975). Toxicity: K. Bailey *et al.*, *J. Pharm. Pharmacol.* **28**, 713 (1976). Extensive bibliography: R. L. Balster, R. S. Pross, *J. Psychedelic Drugs* **10**, 1-15 (1978). *Review:* R. E. Garey, *ibid.* **11**, 265-275 (1979).

Colorless crystals, mp 46-46.5°. bp$_{1.0}$ 135-137°. uv max (0.1N HCl): 252, 257.5, 262, 268.5 nm (E$_{1cm}^{1\%}$ 7.9, 11.2, 13.0, 9.7).

Hydrochloride, C$_{17}$H$_{26}$ClN, *Sernyl, Sernylan*. Crystals from 2-propanol, mp 233-235°. uv max (ethanol): 254, 258, 262.5, 269 nm (ε$_{1cm}^{1\%}$ 7.9, 10.8, 12.7, 10.0). LD$_{50}$ orally in mice: 76.5 mg/kg (Bailey).

Hydrobromide, crystals, mp 214-218°.

Caution: This is a controlled substance (depressant) listed in the U.S. Code of Federal Regulations, Title 21 Part 1308.12 (1987). The ethylamine, pyrrolidine and thiophene analogs are listed as hallucinogens, Title 21 Part 1308.11.

THERAP CAT: Anesthetic (intravenous).

THERAP CAT (VET): Analgesic. Anesthetic.

7180. Phendimetrazine. *3,4-Dimethyl-2-phenylmorpholine; d-2-phenyl-3,4-dimethylmorpholine; 3,4-dimethyl-2-phenyltetrahydro-1,4-oxazine;* Antapentan; Sedafamen. C$_{12}$H$_{17}$NO; mol wt 191.26. C 75.35%, H 8.96%, N 7.32%, O 8.37%. Prepn from ethylene chlorohydrin and *l*-ephedrine: Otto, *Angew. Chem.* **68**, 181 (1956); from 2-phenyl-2-hydroxy-3,4-dimethylmorpholine: Boehringer *et al.*, **Brit. pats.** 791,416 (1958); 862,198 (1961). Stereochemistry: Dvornik, Schilling, *J. Med. Chem.* **8**, 466 (1965). Toxicity data: Stegen *et al., Toxicol. Appl. Pharmacol.* **2**, 589 (1960).

bp$_8$ 122-124°; bp$_{12}$ 134-135°.

Bitartrate, *Dietrol, Plegine, Adphen, Reducto, Bacarate, Phenazine, Limit, Minus, Trimtabs, Obepar, Statobex, Hourbese, Symetra, Trimstat, Neo-Nilorex.*

Hydrochloride, C$_{12}$H$_{18}$ClNO, mp 191° (Otto), 208° (Boehringer). [α]$_D^{20}$ +35.7°. LD$_{50}$ in rats (mg/kg): 455 orally; 245 i.p. (Stegen).

Pamoate, C$_{47}$H$_{50}$N$_2$O$_8$, *Fringanor.*

Caution: Excessive use may lead to tolerance and physical dependence. This is a controlled substance (stimulant) listed in the U.S. Code of Federal Regulations, Title 21 Part 1308.13 (1985).

THERAP CAT: Anorexic.

7181. Phenelzine. *(2-Phenethyl)hydrazine; β-phenylethylhydrazine;* phenalzine. C$_8$H$_{12}$N$_2$; mol wt 136.19. C 70.55%, H 8.88%, N 20.57%. C$_6$H$_5$CH$_2$CH$_2$NHNH$_2$. Monoamine oxidase inhibitor. Prepn: Votocek, Leminger, *Coll. Czech. Chem. Commun.* **4**, 271 (1932), *C.A.* **26**, 5294 (1932); Biel *et al., J. Am. Chem. Soc.* **81**, 2805 (1959); Biel, U.S. pat. **3,000,903** (1959 to Lakeside). Toxicity: J. A. Gylys *et al., Ann. N.Y. Acad. Sci.* **107**, 899 (1963). Comprehensive description: R. E. Daly in *Analytical Profiles of Drug Substances* **vol. 2**, K. Florey, Ed. (Academic Press, New York, 1973) pp 383-407.

Liquid. bp$_{0.1}$ 74°. n$_D^{20}$ 1.5494.

Acid sulfate, C$_8$H$_{14}$N$_2$O$_4$S, *W-1544a, Nardil, Estinerval, Kalgan, Nardelzine, Phenodyn, Stinerval.* White powder, sol in water. LD$_{50}$ orally in mice: 156 mg/kg (Gylys).

Hydrochloride, C$_8$H$_{13}$ClN$_2$, crystals, mp 174°.

THERAP CAT: Antidepressant.

7182. Phenesterine. *Cholest-5-en-3β-ol 4-[bis(2-chloroethyl)amino]benzeneacetate; p-[bis(2-chloroethyl)amino]phenylacetic acid cholesteryl ester;* cholesteryl *p*-bis(2-chloroethyl)aminophenylacetate; Fenesterin; Fenesterol. C$_{39}$H$_{59}$Cl$_2$-NO$_2$; mol wt 644.82. C 72.65%, H 9.22%, Cl 11.00%, N 2.17%, O 4.96%. Prepd from *p*-bis(2-chloroethyl)amino-

phenylacetic acid chloride and cholesterol: Shkodinskaya *et al., Zh. Obshch. Khim.* **32**, 959 (1962), *C.A.* **58**, 2502h (1963).

Crystals, mp 90-90.5°. [α]$_D$ —13.7°.

THERAP CAT: Antineoplastic.

7183. Phenetharbital. *5,5-Diethyl-1-phenyl-2,4,6(1H,-3H,5H)-pyrimidinetrione; 5,5-diethyl-1-phenylbarbituric acid;* 5,5-diethyl-2,4,6-trioxo-1-phenylhexahydropyrimidine; 1-phenyl-5,5-diethylbarbituric acid; *N*-phenylbarbital; phenidiemal; Fedibaretta; Pyrictal. C$_{14}$H$_{16}$N$_2$O$_3$; mol wt 260.28. C 64.60%, H 6.20%, N 10.76%, O 18.44%. Prepd from ethyl diethylmalonate and phenylurea: Fischer, Dilthey, *Ann.* **335**, 334 (1904); Buck, *J. Am. Chem. Soc.* **58**, 1284 (1936).

Small, thick, glittering plates, mp 178°. Freely sol in hot alcohol, alkali.

Note: This is a controlled substance (depressant) listed in the U.S. Code of Federal Regulations, Title 21 Part 1308.13 (1987).

THERAP CAT: Anticonvulsant.

7184. Phenethicillin Potassium. *3,3-Dimethyl-7-oxo-6-[(1-oxo-2-phenoxypropyl)amino]-4-thia-1-azabicyclo[3.2.0]-heptane-2-carboxylic acid potassium salt;* (α-phenoxyethyl)-penicillin potassium; α-phenoxyethylpenicillinic acid potassium salt; 6-(α-phenoxypropionamido)penicillanic acid potassium salt; penicillin-152; penicillin MV; penicillin-152 potassium; potassium phenethicillin; Alfacillin; Alpen (obsolete); Alticina; Bendralan; Brocsil; Broxil; Chemipen; CVK; Darcil; Dramcillin-S; Feneticilline; Maxipen; Optipen; α-Oracillin; Oralopen; Pen-200; Penemve; Peniplus; Penorale; Penova; Pensig; Rocillin; Semopen; Synapen; Syncillin; Synerpenin; Synthecilline; Synthepen Tabl.; Triospen. C$_{17}$-H$_{19}$KN$_2$O$_5$S; mol wt 402.53. C 50.73%, H 4.76%, K 9.71%, N 6.96%, O 19.87%, S 7.97%. Semi-synthetic antibiotic related to penicillin. Prepn from 6-aminopenicillanic acid: Perron *et al., Antibiot. Ann.* **1959-1960**, 107; *J. Am. Chem. Soc.* **82**, 3934 (1960); Glombitza, *Ann.* **673**, 166 (1964).

dl-Form, crystals from acetone, dec 230-232°. Much less hygroscopic than benzylpenicillin sodium. Freely sol in water.

l-Form, *epiphenethicillin potassium, l-Maxipen potassium.* Crystals from 20% butanol, dec 238-239°. [α]$_D^{24}$ +218° (c = 0.01 in water).

THERAP CAT: Antibacterial.

7185. Phenethyl Alcohol. *Benzeneethanol; 2-phenylethanol; β-phenylethyl alcohol;* benzyl carbinol; *β-hydroxyethylbenzene.* C$_8$H$_{10}$O; mol wt 122.16. C 78.65%, H 8.25%, O 13.10%. Found in a number of natural essential oils, such as rose, carnation, hyacinth, Aleppo pine, orange blossom, geranium Bourbon, neroli and in the essential oil of champaca. Prepd by reduction of ethyl phenylacetate with sodium in abs alcohol: Bouveault, Blanc, *Bull. Soc. Chim.* [3]

31, 672 (1904); Leonard, *J. Am. Chem. Soc.* **47**, 1778 (1925); by hydrogenation of phenylacetaldehyde in the presence of nickel catalyst: Skita, Ritter, *Ber.* **43**, 3398 (1910); v. Braun, Kochendörfer, *Ber.* **56**, 2176 (1923); Milligan, Reid, *J. Am. Chem. Soc.* **44**, 204 (1922). Isoln from the fungus *Gibberella fujikuroi*: Cross et al., *J. Chem. Soc.* **1963**, 2937. Antibacterial activity: R. M. E. Richards et al., *J. Pharm. Pharmacol.* **21**, 681 (1969). Use as preservative in ophthalmic solutions: R. M. E. Richards, R. J. McBride, *ibid.* **24**, 145 (1972). Toxicity: Jenner et al., *Food Cosmet. Toxicol.* **2**, 327 (1964).

Liquid. Floral odor, rose character. mp $-27°$. d_{25}^{25} 1.017 to 1.019. bp_{750} 219-221°; bp_{14} 104°; bp_{12} 98-100°. n_D^{20} 1.530 to 1.533. Two ml dissolve in 100 ml water after thorough shaking. One part is clearly sol in 1 part of 50% alc. Miscible with alcohol, ether. LD_{50} orally in rats: 1790 mg/kg (Jenner).

USE: Pharmaceutic aid (antimicrobial). In flavors and perfumery (esp rose perfumes).

7186. Phenethylamine. *Benzeneethanamine; β-phenylethylamine;* 1-amino-2-phenylethane; β-aminoethylbenzene; PEA. $C_8H_{11}N$; mol wt 121.18. C 79.29%, H 9.15%, N 11.56%. $C_6H_5CH_2CH_2NH_2$. Endogenous amine related structurally and pharmacologically to amphetamine: P. Mantegazza, M. Riva, *J. Pharm. Pharmacol.* **15**, 472 (1963). Present in oil of bitter almonds. Found in normal human urine (about 30 μg/liter). Prepn: Johnson, Guest, *Am. Chem. J.* **42**, 346 (1909); Robinson, Snyder, *Org. Syn.* **coll. vol. III**, 720 (1955). Elevated levels in urine due to paranoid chronic schizophrenia: S. G. Potkin et al., *Science* **206**, 470 (1979); due to stress: M. A. Paulos, R. E. Tessel, *ibid.* **215**, 1127 (1982).

Liquid. Fishy odor. Absorbs CO_2 from air. Does not solidify when cooled in an ice-salt mixture. Strong base. d_4^{25} 0.9640; bp 194.5-195°. Sol in water. Freely sol in alc, ether.
Hydrochloride, $C_8H_{11}N.HCl$, orthorhombic bipyramidal platelets from abs alcohol, mp 217°. Freely sol in water (100 parts H_2O will dissolve 80 parts at 15°). Sol in alc. Insol in ether. LD_{50} s.c. in mice: 470 mg/kg, P. Mantegazza, M. Riva, *loc. cit.*
Caution: Skin irritant and possible sensitizer.

7187. o-Phenetidine. *2-Ethoxybenzenamine;* 2-amino-phenetole. $C_8H_{11}NO$; mol wt 137.18. C 70.04%, H 8.08%, N 10.21%, O 11.66%.
Oily liquid; rapidly becomes brown on exposure to light and air. mp $<-20°$. bp 228-230°. Insol in water; sol in alcohol.
USE: Manuf of dyes. *Caution:* Absorbed through skin; vapor hazardous.

7188. p-Phenetidine. *4-Ethoxybenzenamine;* 4-amino-phenetole; 4-ethoxyaniline; *p-*aminophenyl ethyl ether. $C_8H_{11}NO$; mol wt 137.18. C 70.04%, H 8.08%, N 10.21%, O 11.66%. Prepn by reduction of *p*-nitrophenetole: West, *J. Chem. Soc.* **127**, 494 (1925). Prepn from nitrobenzene, ethanol, Mg, and sulfuric acid: Yukawa, *J. Chem. Soc. Japan* **71**, 547 (1950).

Colorless liquid; becomes red to brown on exposure to air and light. d_4^{16} 1.0652; mp about 3°; bp 253-255°. Practically insol in water; sol in alcohol. *Keep well closed and protected from light.*
Citrate, $C_{14}H_{19}NO_8$, *Citrophen.* Crystalline powder, mp 186-188°. Slightly acid taste. Sol in about 40 parts water.
USE: Manuf acetophenetidine, phenocoll, dulcin, dyes. *Caution:* Absorbed through skin; vapor hazardous.

7189. Phenetole. *Ethoxybenzene;* ethyl phenyl ether; phenyl ethyl ether. $C_8H_{10}O$; mol wt 122.16. C 78.65%, H 8.25%, O 13.10%. $C_6H_5OC_2H_5$. Prepd from phenol or its salts by the use of the following ethylating agents: ethyl chloride, Wohl, *Ber.* **39**, 1953 (1906); ethyl bromide, White et al., *J. Am. Chem. Soc.* **46**, 965 (1924); ethyl *p*-toluenesulfonate, Finzi, *Ann. Chim. Applicata* **15**, 41 (1925); diethyl sulfate or triethyl phosphate, Noller, Dutton, *J. Am. Chem. Soc.* **55**, 424 (1933).
Oily liquid. d_4^{20} 0.967; mp $-30°$; bp 171-173°; n_D^{20} 1.507. Practically insol in water; freely sol in alcohol, ether. MLD s.c. in rats: 4.0 g/kg.

7190. Pheneturide. *N-(Aminocarbonyl)-α-ethylbenzene-acetamide;* (2-phenylbutyryl)urea; α-phenyl-α-ethylacetyl-urea; N-(α-phenylbutyryl)urea; α-ethyl-α-phenylacetylurea; ethylphenacemide; EPA; PBU; Benuride. $C_{11}H_{14}N_2O_2$; mol wt 206.24. C 64.06%, H 6.84%, N 13.58%, O 15.52%. Prepd by heating α-phenylbutyric acid chloride with urea: **Ger. pat. 249,241** (1912 to Bayer); *Chem. Zentr.* **1912**, II, 396; *Frdl.* **10**, 1165. From α-phenylbutyric acid chloride, urea, and antipyrine: Gold-Aubert, *Helv. Chim. Acta* **41**, 1512 (1958); **Swiss pat. 374,644** (1964 to Labs. Sapos). Pharmacokinetics: R. L. Galeazzi et al., *J. Pharmacokinet. Biopharm.* **7**, 453 (1979). Toxicity: M. J. Orloff et al., *Neurology* **1**, 377 (1951). Clinical studies: J. C. Bowe, *Brit. J. Clin. Pract.* **27**, 174 (1973); F. B. Gibberd et al., *J. Neurol. Neurosurg. Psychiat.* **45**, 1113 (1982).

dl-Form, needles from ethanol, mp 149-150°.
d-Form, needles from ethanol, mp 168-169°. $[\alpha]_D^{17}$ +54.0° (c = 1 in ethanol); $[\alpha]_D^{22}$ +53.8° (c = 1 in acetone); +48.2° (c = 1 in dioxane).
l-Form, crystals from 50% ethanol, mp 162-163°. $[\alpha]_D^{30}$ $-51.6°$ (c = 1 in ethanol).
THERAP CAT: Anticonvulsant.

7191. Phenformin. *N-(2-Phenylethyl)imidodicarbon-imidic diamide;* 1-phenethylbiguanide; phenethyldiguanide; *N'*-β-phenethylformamidinyliminourea; fenformin; fenormin; DBI; β-PEBG; PEDG; Azucaps; D Bretard; Debeone; Debinyl; Diabis; Glucopostin; Retardo; Dibein; Dibiraf; Dibotin; Feguanide; Glyphen; Insoral; Lentobetic; Normoglucina. $C_{10}H_{15}N_5$; mol wt 205.27. C 58.51%, H 7.37%, N 34.12%. Prepn: Shapiro et al., *J. Am. Chem. Soc.* **81**, 2220 (1959); Shapiro, Freedman, **U.S. pats. 2,961,377 and 3,057,780** (1960 and 1962, both to U.S.V.). Metabolism: R. Beckmann, *Diabetologia* **3**, 368 (1967). Experimental model of phenformin-induced lactic acidosis in rats: R. Assan et al., *ibid.* **14**, 261 (1978). Comprehensive description: J. E. Moody in *Analytical Profiles of Drug Substances* **vol. 4**, K. Florey, Ed. (Academic Press, New York, 1975) pp 319-332.

Hydrochloride, $C_{19}H_{16}ClN_5$, *DBI-TD, Dipar, Meltrol.* Crystals from isopropanol, mp 175-178°. Sol in water; pH (0.1M aq soln): 6.7. LD_{50} in mice: 450 mg/kg orally, 19 mg/kg i.v., G. Proske et al., *Arzneimittel-Forsch.* **12**, 314 (1962).
Human Toxicity: Causes lactic acidosis! FDA has declared this substance an imminent hazard to public health: *Fed. Reg.* **43**, 54995 (1978).
THERAP CAT: Antidiabetic.

7192. Phenglutarimide. *3-[2-(Diethylamino)ethyl]-3-phenyl-2,6-piperidinedione; 2-(2-diethylaminoethyl)-2-phenylglutarimide; α-phenyl-α-(diethylaminoethyl)glutarimide; 3-phenyl-3-(β-diethylaminoethyl)-2,6-dioxopiperidine.* $C_{17}H_{24}N_2O_2$; mol wt 288.38. C 70.80%, H 8.39%, N 9.72%, O 11.10%. Prepn: Tagmann *et al., Helv. Chim. Acta* **35**, 1235 (1952); **37**, 185 (1954); Hoffmann, Tagmann, U.S. pat. **2,664,424** (1953 to Ciba).

Crystals from isopropanol, mp 125-127°. Can also be crystallized from a mixture of ethyl acetate and ligroin.
Hydrochloride, $C_{17}H_{25}ClN_2O_2$, *Aturban(e), Aturbal.* Crystals from methanol + ethyl acetate, dec 168-172°, also 176-177°. Freely sol in water. Dec by alkali.
THERAP CAT: Anticholinergic.

7193. Phenicarbazide. *2-Phenylhydrazinecarboxamide; 1-phenylsemicarbazide;* 1-carbamyl-2-phenylhydrazine; Cryogenine; Kryogenin. $C_7H_9N_3O$; mol wt 151.17. C 55.61%, H 6.00%, N 27.80%, O 10.58%. $C_6H_5NHNHCO\cdot NH_2$. Prepd by mixing 50 g phenylhydrazine with 37 g 90% acetic acid and 250 g water, then slowly adding a soln of 45 g potassium cyanate in 150 g water: Wildman, *Ber.* **26**, 2613 (1893).
Leaflets from water, mp 172°. Very sol in hot water, alcohol, methanol, and acetone; difficultly sol in cold water, ether, benzene, ligroin.
THERAP CAT: Antipyretic.

7194. Phenicin. *2,2'-Dihydroxy-4,4'-dimethyl[bi-1,4-cyclohexadien-1-yl]-3,3',6,6'-tetrone; 3,3'-dihydroxy-5,5'-dimethyl-2,2'-bi-p-benzoquinone;* phoenicin. $C_{14}H_{10}O_6$; mol wt 274.22. C 61.32%, H 3.68%, O 35.01%. Fungal pigment produced by *Penicillium phoeniceum* and *P. rubrum:* Friedheim, *Helv. Chim. Acta* **21**, 1464 (1938); Charollais *et al., Arch. Sci.* (Geneva) **16**, 474 (1963). Synthesis: Posternak *et al., Helv. Chim. Acta* **26**, 2031 (1943); Musso, Beecken, *Ber.* **92**, 1416 (1959); J. Kalamar *et al., Helv. Chim. Acta* **57**, 2368 (1974). Biosynthesis: E. Steiner *et al., ibid.* 2377.

Yellowish-brown crystals from alc, mp 230-231°. uv max (chloroform): 268, 406 nm (log ε 4.52, 3.36). Sparingly sol in water; freely sol in chloroform, acetic acid, hot alcohol. Very acidic solns are yellow, turning red at pH 1.6-3.6 and violet at pH 4.9-6.

7195. Phenindamine. *2,3,4,9-Tetrahydro-2-methyl-9-phenyl-1H-indeno[2,1-c]pyridine;* 2-methyl-9-phenyl-2,3,-4,9-tetrahydro-1-pyridindene; 1,2,3,4-tetrahydro-2-methyl-9-phenyl-2-azafluorene; Nu-1504; Thephorin. $C_{19}H_{19}N$; mol wt 261.35. C 87.31%, H 7.33%, N 5.36%. Prepn: Plati, Wenner, U.S. pat. **2,470,108** (1949 to Hoffmann-La Roche).

Crystals, mp 91°, d 1.17.
Hydrogen tartrate, $C_{23}H_{25}NO_6$, *Pernovin.* Crystals, mp 165-167°. Soly: about 2.5% in water; sparingly sol in propylene glycol. Insol in 95% ethanol, glycerol, ether. LD_{50} orally in rats: 280 mg/kg.
Note: Thephorin was formerly used to designate theobromine sodium formate.

THERAP CAT: Tartrate as antihistaminic.

7196. Phenindione. *2-Phenyl-1H-indene-1,3(2H)-dione; 2-phenyl-1,3-indandione;* 2-phenyl-1,3-diketohydrindene; Pindione; Bindan; Danilone; Athrombon; Hedulin; Dindevan; Dineval; Diadilan; Hemolidione; Indon; Indema; Fenilin; Fenhydren; Rectadione; Diophindane; Cronodione; Thrombasal; PID. $C_{15}H_{10}O_2$; mol wt 222.23. C 81.06%, H 4.54%, O 14.40%. By heating phthalide with benzaldehyde and sodium ethylate soln: Dieckmann, *Ber.* **47**, 1439 (1914).

Leaflets from alcohol, mp 149-151°. Practically insol in cold water. Slightly sol in warm water. pH of satd soln at 25° = 4.5. Readily sol in alkaline solns. Soluble in methanol, ethanol, ether, acetone, benzene, chloroform. Solns in alkalies are red, in concd H_2SO_4 blue.
THERAP CAT: Anticoagulant.

7197. Pheniprazine. *(1-Methyl-2-phenylethyl)hydrazine; (α-methylphenethyl)hydrazine;* (1-benzylethyl)hydrazine; 1-phenyl-2-hydrazinopropane; β-phenylisopropylhydrazine; PIH. $C_9H_{14}N_2$; mol wt 150.22. C 71.95%, H 9.39%, N 18.65%. $C_6H_5CH_2CH(CH_3)NHNH_2$. Prepn of DL-form: Biel *et al., J. Am. Chem. Soc.* **80**, 1519 (1958); **81**, 2805 (1959); Biel, U.S. pats. **2,978,461** and **3,000,903** (both to Lakeside). Resolution of stereoisomers: Bernstein *et al., J. Am. Chem. Soc.* **81**, 4433 (1959). Synthesis: Biel *et al., ibid.* 4995.
DL-Form, liquid, $bp_{0.5}$ 82-86°. n_D^{20} 1.5401.
DL-Form hydrochloride, $C_9H_{15}ClN_2$, *JB 516, Catral, Catron, Catroniazid, Cavodil.* Crystals, mp 124-125°.
D-Form, bp_{10} 135-138°. n_D^{20} 1.5385. $[\alpha]_D^{25}$ +4.5° (c = 5 in methanol).
D-Form hydrochloride, crystals, mp 152-154°. $[\alpha]_D^{25}$ +12.8° (c = 5 in water); also reported as mp 148-149°, $[\alpha]_D^{25}$ +13.8° (c = 1 in water).
L-Form, bp_{10} 135-138°. n_D^{20} 1.5385. $[\alpha]_D^{25}$ −4.5° (c = 5 in methanol).
L-Form hydrochloride, crystals, mp 152-154°. $[\alpha]_D^{25}$ −12.5° (c = 5 in water); also reported as mp 148-149°, $[\alpha]_D^{25}$ −14.0° (c = 1.0 in water).
THERAP CAT: Antihypertensive.

7198. Pheniramine. *N,N-Dimethyl-γ-phenyl-2-pyridinepropanamine; 2-[α-(2-dimethylaminoethyl)benzyl]pyridine;* 1-phenyl-1-(2-pyridyl)-3-dimethylaminopropane; 3-phenyl-3-(2-pyridyl)-N,N-dimethylpropylamine; prophenpyridamine; propheniramine; Avil; Tripoton. $C_{16}H_{20}N_2$; mol wt 240.34. C 79.95%, H 8.39%, N 11.66%. Synthesis: Sperber *et al.,* U.S. pats. **2,567,245** and **2,676,964** (1951, 1954, both to Schering).

Oily liq. Slightly yellow color. Characteristic amine-like odor. d 1.0081; bp_{13} 181°; bp_2 142°; $bp_{0.5}$ 135°; n_D^{25} 1.5519 to 1.5521. Insol in water. Sol in dil acids, alcohol, benzene, chloroform, ether.
Maleate, $C_{20}H_{24}N_2O_4$, *Daneral, Inhiston, Trimeton (formerly).* Crystals from amyl alcohol, faint amine-like odor, mp 107°. Freely sol in water, alcohol. Slightly sol in ether, benzene. pH of 1% aq soln between 4.3 and 4.9.
p-Aminosalicylate, $C_{16}H_{20}N_2 \cdot C_7H_7NO_3$, small needles from acetone + ethyl acetate, large octahedra from water, dec 142°. One gram dissolves in 10 ml water, freely sol in alcohol. Sparingly sol in ethyl acetate, ether, acetone.
THERAP CAT: Antihistaminic.

7199. Phenmedipham. *3-(Methylphenyl)carbamic acid 3-[(methoxycarbonyl)amino]phenyl ester; m-hydroxycarban-*

ilic acid methyl ester m-methylcarbanilate; methyl 3-(*m*-tolylcarbamoyloxy)phenylcarbamate; Schering 38584; Betanal. $C_{16}H_{16}N_2O_4$; mol wt 300.32. C 63.99%, H 5.37%, N 9.33%, O 21.31%. Prepn: **Neth. pat. Appl. 6,604,363** (1966 to Schering AG), *C.A.* **66**, 104813w (1967). Persistence in the soil: K. Kossmann, *Weed Res.* **10**, 349 (1970).

Colorless crystals, mp 139-142 (tech, 143-144°). Soly in water at room temp: < 10 ppm. LD_{50} orally in rats: > 8 g/kg.
USE: Herbicide.

7200. Phenmetrazine. *3-Methyl-2-phenylmorpholine;* 3-methyl-2-phenyltetrahydro-2*H*-1,4-oxazine; 2-phenyl-3-methyltetrahydro-1,4-oxazine; A 66; Preludin; Probese-P; Psychamine A 66. $C_{11}H_{15}NO$; mol wt 177.24. C 74.54%, H 8.53%, N 7.90%, O 9.03%. Prepn: Thomae, Wick, **U.S. pat. 2,835,669** (1958 to Boehringer, Ing.); Siemer, Hengen, **U.S. pat. 3,018,222** (1962 to Ravensberg Chem.); Clark, *J. Org. Chem.* **27**, 3251 (1962); Klosa, *J. Prakt. Chem.* **21**, 12 (1963).

Liquid. bp_{12} 138-140°; bp_1 104°.
Hydrochloride, $C_{11}H_{16}ClNO$, *Marsin, Neo-Zine.* Crystals from ethanol + ether, mp 182°. One gram dissolves in 0.4 ml water, in 2.0 ml 95% alc, in 2.0 ml chloroform. Sparingly sol in ether.
Caution: Excessive use may lead to tolerance and physical dependence. This is a controlled substance (stimulant) listed in the U.S. Code of Federal Regulations, Title 21 Part 1308.12 (1985).
THERAP CAT: Anorexic.

7201. Phenobarbital. *5-Ethyl-5-phenyl-2,4,6(1H,3H,-5H)-pyrimidinetrione;* 5-ethyl-5-phenylbarbituric acid; phenylethylmalonylurea; phenobarbitone; Luminal; Gardenal; Barbenyl; Barbiphenyl; Dormiral; Euneryl; Neurobarb; Barbipil; Lubrokal; Lubergal; Phenyral; Cratecil; Nunol; Phenonyl; Phenobal; Noptil; Agrypnal; Eskabarb; Etilfen; Gardepanyl; Somonal. $C_{12}H_{12}N_2O_3$; mol wt 232.23. C 62.06%, H 5.21%, N 12.06%, O 20.67%. Prepn: **Ger. pat. 247,952** (1911 to Bayer); Hoerlein, **U.S. pat. 1,025,872** (1912), *Frdl.* **11**, 926; *Chem. Zentr.* **1912**, II, 212; Chamberlain *et al., J. Am. Chem. Soc.* **57**, 352 (1935); Inman, Bitler, **U.S. pat. 2,358,072** (1944 to Kay-Fries Chem.); J. T. Pinhey, B. A. Rowe, *Tetrahedron Letters* **21**, 965 (1980). Toxicity data: E. I. Goldenthal, *Toxicol. Appl. Pharmacol.* **18**, 185 (1971). Comprehensive description: M. K. C. Chao *et al.,* in *Analytical Profiles of Drug Substances* vol. 7, K. Florey, Ed. (Academic Press, New York, 1978) pp 359-399.

Crystals (3 different phases), mp 174-178°. uv max (pH 10 buffer): 240 nm ($A_{1cm}^{1\%}$ 431); (0.1*N* NaOH): 256 nm ($A_{1cm}^{1\%}$ 314). Slightly bitter taste. One gram dissolves in about one liter of water, 8 ml alc, 40 ml chloroform, 13 ml ether, about 700 ml benzene. Soluble in alkali hydroxides and carbonates. A satd aq soln is acid to litmus. pK_1 7.3, pK_2 11.8. LD_{50} orally in rats: 162 ±14 mg/kg (Goldenthal).
Caution: May be habit forming. This is a controlled substance (depressant) listed in the U.S. Code of Federal Regulations, Title 21 Parts 329.1 and 1308.14 (1987).

THERAP CAT: Anticonvulsant; sedative, hypnotic.

7202. Phenobarbital Sodium. *5-Ethyl-5-phenyl-2,4,6-(1H,3H,5H)-pyrimidinetrione monosodium salt; sodium 5-ethyl-5-phenylbarbiturate;* sol phenobarbital; sol phenobarbitone; Luminal sodium; Gardenal sodium. $C_{12}H_{11}N_2NaO_3$; mol wt 254.22. C 56.69%, H 4.36%, N 11.02%, Na 9.05%, O 18.88%. Acute toxicity: Schaffarzick, Brown, *Science* **116**, 663 (1952).

Bitter, slightly hygroscopic crystals or white powder. One gram dissolves in about 1 ml water, about 10 ml alcohol. Insol in ether or chloroform. Aq solns are alkaline to litmus and phenolphthalein, pH about 9.3. Aq solns are unstable, but may be kept at 10° for a few days. LD_{50} orally in rats: 660 mg/kg (Schaffarzick, Brown).
Incompat. Insol phenobarbital is precipitated by acidic drugs such as thiamine chloride. Alkali sensitive drugs are readily destroyed by the high alkalinity of the sodium phenobarbital. Likewise ammonia is liberated from ammonium salts, and chloroform from chloral hydrate.
Caution: May be habit forming. This is a controlled substance (depressant) listed in the U.S. Code of Federal Regulations, Title 21 Parts 329.1 and 1308.14 (1987).
THERAP CAT: Sedative, hypnotic. Anticonvulsant.
THERAP CAT (VET): Sedative, long-acting anesthetic, anticonvulsant.

7203. Phenobutiodil. *2-(2,4,6-Triiodophenoxy)butanoic acid; 2-(2,4,6-triiodophenoxy)butyric acid;* α-(2,4,6-triiodophenoxy)butyric acid; 1-(2,4,6-triiodophenoxy)propanecarboxylic acid; 4114 Th; Baygnostil; Trijobil; Vésipaque. $C_{10}H_9I_3O_3$; mol wt 557.91. C 21.53%, H 1.63%, I 68.24%, O 8.60%. Prepn: Redel *et al., Bull. Soc. Chim. France* **1954**, 342; Redel, Maillard, **U.S. pat. 2,796,432** (1957 to Chimie et Atomistique). Pharmacology: H. Bekker *et al., Pharmazie* **27**, 411 (1972).

Crystals from ether + petr ether, mp 124-125°. LD_{50} in mice (mg/kg): 1800 orally; 265 i.v. (Bekker).
THERAP CAT: Diagnostic aid (radiopaque medium—cholecystographic).

7204. Phenocoll. *2-Amino-N-(4-ethoxyphenyl)acetamide; 2-amino-p-acetophenetidide;* α-amino-p-acetophenetide; aminoacetophenetidine; 4-glycylaminophenol ethyl ether; glycine p-phenetidide; 4-glycylaminophenetol; glycocoll-p-phenetidide; phenokoll; Phenamine. $C_{10}H_{14}N_2O_2$; mol wt 194.23. C 61.83%, H 7.27%, N 14.42%, O 16.48%. Prepn: Majert, **Ger. pats. 59,121; 59,874** (1891 to Schering), *Frdl.* **3**, 915, 918; **Ger. pat. 346,809** (1921 to Schering), *Frdl.* **13**, 1066; Karrer, Haebler, *Helv. Chim. Acta* **7**, 534 (1924).

Crystals, mp 100.5°.
Monohydrate, crystals, mp 95°.
Hydrochloride, $C_{10}H_{14}N_2O_2$·HCl, cryst powder. Sol in 20 parts water, in alcohol. More sol in hot water; slightly sol in benzene, chloroform, ether. Aq soln is neutral. *Incompat.* Alkali hydroxides or carbonates.
Salicylate, $C_{17}H_{18}N_2O_5$, *Salocoll.* Fine needles. Sweetish taste. Sol in 200 parts cold, 20 parts hot water. *Incompat.* As for the hydrochloride; with substances incompat with salicylates.

THERAP CAT: Antipyretic, analgesic.

7205. Phenoctide. *N,N-Diethyl-N-[2-[4-(1,1,3,3-tetra-methylbutyl)phenoxy]ethyl]benzenemethanaminium chloride; benzyldiethyl-2-[(p-1,1,3,3-tetramethylbutyl)phenoxy]ethyl-ammonium chloride; β-p-tert-octylphenoxyethyldiethylben-zylammonium chloride; Octaphen.* $C_{27}H_{42}ClNO$; mol wt 432.11. C 75.05%, H 9.80%, Cl 8.21%, N 3.24%, O 3.70%. Prepn: Goldberg, Besly, **Brit.** pat. **703,477** (1954 to Ward Blenkinsop); Erekaev, *C.A.* **54**, 12483i (1960).

Crystals from ethyl acetate. mp 112-114° (Goldberg, Besly); mp 95° (Erekaev).
USE: Orthophosphate as lubricant: Semmens, Summers-Smith, **Brit.** pat. **790,056** (1958 to I.C.I.).
THERAP CAT: Topical anti-infective.

7206. Phenol. Carbolic acid; phenic acid; phenylic acid; phenyl hydroxide; hydroxybenzene; oxybenzene. C_6H_6O; mol wt 94.11. C 76.57%, H 6.43%, O 17.00%. Obtained from coal tar, or made by fusing sodium benzenesulfonate with NaOH, or by heating monochlorobenzene with aq NaOH under high pressure. The crystalline article of commerce contains at least 98% phenol. Review of mfg processes: A. Dierichs, R. Kubicka, *Phenole und Basen, Vorkommen und Gewinnung* (Akademie-Verlag, Berlin, 1958) 472 pp; Faith, Keyes & Clark's *Industrial Chemicals,* F. A. Lowenheim, M. K. Moran, Eds. (Wiley-Interscience, New York, 4th ed., 1975) pp 612-623. Use in treatment of spasticity: D. E. Garland *et al., Clin. Orthop.* **165,** 217 (1982); *eidem, Arch. Phys. Med. Rehab.* **65,** 243 (1984). Review of use in pain relief: K. M. Wood, *Pain* **5,** 205-229 (1978). Review of toxicology: H. Babich, D. L. Davis, *Regul. Toxicol. Pharmacol.* **1,** 90-109 (1981). Toxicity: W. B. Deichmann, S. Witherup, *J. Pharmacol. Exp. Ther.* **80,** 233 (1944). *Review:* C. Thurman in Kirk-Othmer *Encyclopedia of Chemical Technology* vol. **17** (Wiley-Interscience, New York, 3rd ed., 1982) pp 373-384.

Colorless, acicular crystals or white, crystalline mass; characteristic odor. *Poisonous and caustic!* Prone to redden on exposure to air and light, hastened by presence of alkalinity. d 1.071. When free from water and cresols it congeals at 41° and melts at 43°. Ultrapure material mp 40.85°. The commercial product contains an impurity which raises the mp to 182°. Flash pt, closed cup: 175°F (79°C). n_D^{41} 1.5425. pKa at 25° = 10.0. pH of aq solns about 6.0. It is liquefied by mixing with about 8% water. One gram dissolves in about 15 ml water, 12 ml benzene; very sol in alcohol, chloroform, ether, glycerol, carbon disulfide, petrolatum, volatile and fixed oils, aq alkali hydroxides. Almost insol in petr ether. LD_{50} orally in rats: 530 mg/kg (Deichmann, Witherup). *Keep well closed and protected from light. Do not handle with bare hands.*
Incompat. Phenol coagulates collodion; liquefies or becomes semiliquid when triturated with acetanilide, butyl-chloral hydrate, camphor, monobromated camphor, chloral hydrate, diuretin, lead acetate, menthol, naphthalene, naphthol, acetophenetidin, pyrogallol, resorcinol, salol, sodium phosphate, thymol, urethane, chloralamide, terpin hydrate.
Ammonium salt, C_6H_9NO, *ammonium phenate, ammonium carbolate.* White to pink cryst masses. Sol in water.
Sodium salt, C_6H_5NaO, *phenolate sodium.*
Aqueous solution with phenolate sodium, *Chloraseptic.*
Human Toxicity: Ingestion of even small amounts may cause nausea, vomiting, circulatory collapse, tachypnea,

paralysis, convulsions, coma, greenish or smoky-colored urine, necrosis of mouth and G.I. tract, icterus, death from respiratory failure, sometimes from cardiac arrest. Average fatal dose is 15 g but death from as little as one gram has been reported. Fatal poisoning may also occur by skin absorption following application to large areas. Chronic poisoning with renal and hepatic damage may occur from industrial contact. *See* C. J. Polson, R. N. Tattersall, *Clinical Toxicology* (Lippincott, Philadelphia, 1969) pp 51-62; Deichmann, Keplinger in Patty's *Industrial Hygiene and Toxicology* vol. **2A,** G. D. Clayton, F. E. Clayton, Eds. (Wiley-Interscience, New York, 3rd ed., 1981) pp 2567-2584.
USE: As a general disinfectant, either in soln or mixed with slaked lime, etc.; for toilets, stables, cesspools, floors, drains, etc.; for the manuf of colorless or light-colored artificial resins, many medical and industrial organic compds and dyes; as a reagent in chemical analysis. Pharmaceutic aid (preservative).
THERAP CAT: Aqueous soln as topical anesthetic; topical antiseptic; topical antipruritic.
THERAP CAT (VET): Antiseptic caustic. Topical anesthetic in pruritic skin conditions. Has been used internally and externally as an antiseptic.

7207. Phenoldisulfonic Acid. *4-Hydroxy-1,3-benzenedisulfonic acid; 4-hydroxy-m-benzenedisulfonic acid;* 1-phenol-2,4-disulfonic acid. $C_6H_6O_7S_2$; mol wt 254.23. C 28.34%, H 2.38%, O 44.05%, S 25.22%. Conveniently prepd by hydrolysis of the dichloride which is obtained by the action of chlorosulfonic acid on phenol at room temp: Pollak *et al., Monatsh.* **46,** 395 (1925). Monograph: E. E. Gilbert, *Sulfonation and Related Reactions* (Interscience, New York, 1965) 529 pp.

Deliquescent needles, vague melting range from 89° to 100°. Decomp above 100°. Freely sol in water, methanol. Practically insol in ether, petr ether.
USE: In the manuf of aminophenoldisulfonic acids which are intermediates in the dye industry.

7208. Phenolphthalein. *3,3-Bis(4-hydroxyphenyl)-1-(3H)-isobenzofuranone; 3,3-bis(p-hydroxyphenyl)phthalide; α-(p-hydroxyphenyl)-α-(4-oxo-2,5-cyclohexadien-1-ylid-ene)-o-toluic acid; Chocolax; Darmol; Laxin.* $C_{20}H_{14}O_4$; mol wt 318.31. C 75.46%, H 4.43%, O 20.10%. Prepd by condensing phenol with phthalic anhydride: Baeyer, *Ann.* **202,** 69 (1880); Herzog, *Chem. Ztg.* **51,** 84 (1927); Hubacher, **U.S.** pat. **2,192,485** (1940 to Ex Lax); Gamrath, **U.S.** pat. **2,522,939** (1950 to Monsanto).

Minute, triclinic crystals, often twinned, mp 258-262°. d 1.299. Color: White or yellowish-white, *see also* Yellow Phenolphthalein. Almost insoluble in water. One gram dissolves in 12 ml alcohol, in about 100 ml ether. Very slightly sol in chloroform.
USE: A 1% alcoholic soln as an indicator in titrations of mineral and organic acids and most alkalies. Not suitable for ammonia. Very sensitive to CO_2, and in estimating carbonates the liq must be boiled. Borax can be titrated with

phenolphthalein as an indicator only when glycerol is present, because the color gradually fades away as the acid is added. Usable with a few alkaloids. Colorless to pH 8.5; pink to deep-red, above pH 9.

THERAP CAT: Cathartic.

THERAP CAT (VET): Has been used as a laxative.

7209. Phenolphthalein Sodium. *3,3-Bis(4-hydroxyphenyl)-1(3H)-isobenzofuranone disodium salt.* $C_{20}H_{12}Na_2O_4$; mol wt 362.29. C 66.30%, H 3.34%, Na 12.70%, O 17.67%. Prepd from an alcoholic soln of 1 mol phenolphthalein by the addition of 2 mols NaOH. Structure: Buu-Hoi, *Bull. Soc. Chim.* [5] **8**, 165 (1941).

Reddish-brown granular mass with coppery luster, or pale red powder. Dec in air. Soluble in water, giving a deep-red soln. Absorption max: 550-555 nm. The unstable enol form is colorless, as is the trisodium salt.

7210. Phenolphthalin. *2-[Bis(4-hydroxyphenyl)methyl]-benzoic acid;* decolorized phenolphthalein; phthalin; 4',4''-dihydroxytriphenylmethane-2-carboxylic acid. $C_{20}H_{16}O_4$; mol wt 320.33. C 74.99%, H 5.04%, O 19.98%. Made by boiling phenolphthalein with zinc dust in alkaline soln. Called "*Kastle-Meyer reagent*" when in soln. Prepd by dissolving 2 g phenolphthalin + 20 g KOH in a desired amt of doubly distd H_2O and diluting with an equal vol of 95% ethanol. Detailed directions for prepn of test soln and its use: Lecoq, *Bull. Soc. Chim. Belge* **54**, 186-202 (1945), *C.A.* **41**, 2346i (1947).

Colorless crystals, mp 237°. Insol in water; sol in alcohol, ether, aq alkali. The alkaline solns gradually become pink on exposure to air or other oxidizing substances.

USE: As a reagent for oxidases, blood, HCN, peroxides, copper.

7211. Phenolphthalol. *2-[Bis(4-hydroxyphenyl)methyl]-benzenemethanol;* o-[bis(p-hydroxyphenyl)methyl]benzyl alcohol; dihydroxyphenylmethenylbenzyl alcohol; 2-(4,4'-dihydroxybenzhydryl)benzyl alcohol; bis(4-hydroxyphenyl)-(2-hydroxymethylphenyl)methane; Egmol; Regolax. $C_{20}H_{18}O_3$; mol wt 306.34. C 78.41%, H 5.92%, O 15.67%. Prepn: Hubacher, *J. Am. Chem. Soc.* **74**, 5216 (1952); Schultz, Geller, *Arch. Pharm.* **288**, 234 (1955); Bulcsu, **Ger.** pat. 1,141,293 (1962 to Iromedica), *C.A.* **59**, 1535b (1963).

Crystals from dil alcohol, mp 201-202°.

Monoacetate, $C_{22}H_{20}O_4$, crystals from benzene or chloroform, mp 171-174°. Soluble in dil NaOH.

Triacetate, $C_{26}H_{24}O_6$, crystals from methanol or ethanol, mp 104-106°. Practically insol in dil alkali.

THERAP CAT: Cathartic.

7212. p-Phenolsulfonic Acid. *4-Hydroxybenzenesulfonic acid;* sulfocarbolic acid. $C_6H_6O_4S$; mol wt 174.17. C 41.37%, H 3.47%, O 36.74%, S 18.41%. Commercially available as a 65% soln. Prepn by hydrolysis of p-chloro- or p-bromobenzenesulfonic acid: Zollinger, Roehling, **U.S.** pat. **1,321,271** (1920); by sulfonation of phenol: Davidson, Byrne, **Brit.** pat. **820,659** (1959 to Hardman & Holden). Prepn of ammonium salt: Oxley *et al.*, *J. Chem. Soc.* **1948**, 303.

Deliquescent needles. Miscible with water, alcohol.

Ammonium salt, $C_6H_9NO_4S$, *ammonium p-phenolsulfonate*. Plates from water, dec 270-271°.

Barium salt, $C_{12}H_{10}BaO_8S_2$, *barium p-phenolsulfonate*. Monohydrate, powder. *Poisonous!* Sol in water; slightly sol in alcohol.

Caution: Irritating to skin.

USE: Intermediate in mfg of pharmaceuticals, dyestuffs. In the Ferrostan process of tin plating (U.S. Steel).

7213. Phenolsulfonphthalein. *4,4'-(3H-2,1-Benzoxathiol-3-ylidene)bisphenol S,S-dioxide;* α-hydroxy-α,α-bis(p-hydroxyphenyl)-o-toluenesulfonic acid γ-sultone; 3,3-bis(p-hydroxyphenyl)-3H-2,1-benzoxathiole 1,1-dioxide; phenol red; P.S.P.; Sulfonphthal. $C_{19}H_{14}O_5S$; mol wt 354.37. C 64.39%, H 3.98%, O 22.57%, S 9.05%. Prepd by the action of o-sulfobenzoic anhydride or of o-sulfobenzoyl chloride on phenol: Kekulé, Barbaglia, *Ber.* **5**, 876 (1872); Kekulé, *Ber.* **6**, 943 (1873); Heumann, Kochlin, *Ber.* **15**, 1118 (1882); **Ger.** pat. **142,116**; *Chem. Zentr.* II, 79 (1903); Orndorff, Sherwood, *J. Am. Chem. Soc.* **45**, 486 (1923). Diagnostic use: G. Dunea, P. Freedman, *J. Am. Med. Assoc.* **204**, 159 (1968); R. D. Wilkes *et al.*, *Vet. Med. Small Anim. Clin.* **76**, 289 (1981).

Bright red to dark red crystals. Stable in air. pK = 7.9. One gram dissolves in about 1300 ml water, in about 350 ml alc, in 500 ml acetone. Almost insol in chloroform, ether. Readily sol in aq alkali hydroxides or carbonates with red color, which is discharged by boiling with zinc dust.

USE: As indicator in 0.02-0.05% alcohol soln. pH 6.8 yellow, 8.4 red.

THERAP CAT: Diagnostic aid (renal function).

THERAP CAT (VET): Diagnostic aid (renal function).

7214. Phenoltetrachlorophthalein. *4,5,6,7-Tetrachloro-3,3-bis(4-hydroxyphenyl)-1(3H)-isobenzofuranone; 3,4,5,6-tetrachlorophenolphthalein.* $C_{20}H_{10}Cl_4O_4$; mol wt 456.11. C 52.66%, H 2.21%, Cl 31.09%, O 14.03%. Prepd by condensation of phenol and tetrachlorophthalic acid or its anhydride: W. R. Orndorff, J. A. Black, *Am. Chem. J.* **41**, 349 (1909); Zalkind, Belikova, *Zh. Prikl. Khim. (Leningrad)* **8**, 1210 (1935). Metabolism and hepatic chromosecretion of phthalein dyes: P. Hykes, J. Jirsa, *Acta Univ. Carol. Med.* **25**, 135 (1979), *C.A.* **95**, 111282z (1981).

White powder, dec above 300°. Almost insol in water, chloroform, benzene. Sol in alcohol, ether, acetone, glacial acetic acid; sol in aq alkali hydroxides or carbonates with deep purple color when concd, violet-red when dil, and bluish when very dil.

Disodium salt, $C_{20}H_8Cl_4Na_2O_4$, *Chlor-Tetragnost*. Dihydrate, violet crystals. Freely sol in water. Dec on exposure to air.

THERAP CAT: Cathartic; disodium salt as diagnostic aid (hepatic function).

7215. Phenomorphan. *17-(2-Phenylethyl)morphinan-3-ol; 3-hydroxy-N-(2-phenylethyl)morphinan;* (−)-1,3,4,9,10,10a-hexahydro-11-phenethyl-2*H*-10,4a-iminoethanophenanthren-6-ol; NIH-7274. $C_{24}H_{29}NO$; mol wt 347.48. C 82.95%, H 8.41%, N 4.03%, O 4.60%. Prepn: Grüssner *et al., Helv. Chim. Acta* **40**, 1232 (1957).

Crystals from dimethylformamide, mp 243-245°.

Hydrobromide, $C_{24}H_{30}BrNO$, crystals from dil alc, mp 300-301°. $[\alpha]_D^{20}$ −63.12° (c = 3.27 in alc).

D-Tartrate monohydrate, $C_{28}H_{35}NO_7 \cdot H_2O$, crystals from isopropanol, mp 125-126°. $[\alpha]_D^{20}$ −42.75° (c = 0.983 in water).

Methyl bromide, $C_{25}H_{32}BrNO$, crystals from alcohol, mp 239-240°. $[\alpha]_D^{20}$ −42.81° (c = 1.155 in methanol).

Caution: May be habit forming. This is a controlled substance (opiate) listed in the U.S. Code of Federal Regulations, Title 21 Part 1308.11 (1985).

7216. Phenoperidine. *1-(3-Hydroxy-3-phenylpropyl)-4-phenyl-4-piperidinecarboxylic acid ethyl ester; 1-(3-hydroxy-3-phenylpropyl)-4-phenylisonipecotic acid ethyl ester;* 1-[γ-hydroxy-γ-phenylpropyl]-4-phenyl-4-carbethoxypiperidine; 3-(4-carbethoxy-4-phenylpiperidino)-1-phenyl-1-propanol; 1-phenyl-3-[(4'-phenyl-4'-carbethoxy)piperidino]-1-propanol; Phenoperidin; Lealgin. $C_{23}H_{29}NO_3$; mol wt 367.47. C 75.17%, H 7.95%, N 3.81%, O 13.06%. Prepn: Janssen, Eddy, *J. Med. Pharm. Chem.* **2**, 31 (1960); Pohland, U.S. pat. **2,951,080** (1960 to Lilly); Cutler, Fisher, U.S. pat. **2,962,501** (1960 to Merck & Co.).

Hydrochloride, $C_{23}H_{30}ClNO_3$, *Operidine*. Crystals from ethyl acetate + methanol, or ethanol, mp 200-202°. Soluble in water.

Caution: May be habit forming. This is a controlled substance (opiate) listed in the U.S. Code of Federal Regulations, Title 21 Part 1308.11 (1987).

THERAP CAT: Narcotic analgesic.

7217. Phenopyrazone. *1,4-Diphenyl-3,5-pyrazolidinedione;* 1,4-diphenyl-3,5-dioxopyrazolidine. $C_{15}H_{12}N_2O_2$;

mol wt 252.26. C 71.41%, H 4.80%, N 11.11%, O 12.69%. Prepn: Kraft, U.S. pat. **2,909,465** (1959 to Knoll).

Crystals, mp 233-234°. May be recrystallized from dioxane or ethyl acetate.

2-Diethylaminoethanol salt, component of *Osadrin*.

THERAP CAT: Analgesic, antipyretic.

7218. Phenosafranin. *3,7-Diamino-5-phenylphenazinium chloride;* C.I. 50200; Safranin B Extra. $C_{18}H_{15}ClN_4$; mol wt 322.79. C 66.97%, H 4.68%, Cl 10.98%, N 17.36%. Prepd by dichromate oxidation of 1 mol *p*-phenylenediamine hydrochloride and 2 mols aniline hydrochloride: Witt, *Ber.* **12**, 939 (1879); **19**, 3121 (1886). *See also Colour Index* vol. **4** (3rd ed., 1971) p 4449.

Green, lustrous needles from dil HCl. Freely sol in water, also sol in alc. Aq and alcoholic solns are purplish-red and have a greenish-yellow fluorescence. Absorption max about 530 nm.

USE: Biological stain.

7219. Phenosulfazole. *4-Hydroxy-N-2-thiazolylbenzenesulfonamide; N-(2-thiazolyl)-1-phenol-4-sulfonamide;* Virazene. $C_9H_8N_2O_3S_2$; mol wt 256.30. C 42.17%, H 3.15%, N 10.93%, O 18.73%, S 25.02%. *Refs:* Sanders *et al., Tex. Rep. Biol. Med.* **6**, 385 (1948); *Chem. & Eng. News* **26**, 2710 (1948); Jungeblut, *Proc. Soc. Exp. Biol. Med.* **70**, 371 (1949); LoGrippo *et al., ibid.* 528; Cox *et al., ibid.* 530; Weil, Warren, *ibid.* 534; Francis, Brown, *ibid.* 535.

Crystals, insol in water. Forms a water-sol sodium salt.

THERAP CAT: Formerly in poliomyelitis.

7220. Phenothiazine. *10H-Phenothiazine;* thiodiphenylamine; dibenzothiazine; Phenoxur; Contaverm; Fenoverm; Padophène; Phénégic; Lethelmin; AFI-Tiazin; Antiverm; Biverm; Fentiazin; Helmetina; Nemazine; Orimon; Phenoverm; Reconox; Souframine; Vermitin; Phenovis. $C_{12}H_9NS$; mol wt 199.26. C 72.32%, H 4.55%, N 7.03%, S 16.09%. Prepd by fusing diphenylamine with sulfur: Bernthsen, *Ber.* **16**, 2896 (1883); *Ann.* **230**, 73 (1885); *Ger.* pat. **25,150** (1883), *Frdl.* **1**, 252. Improved yields with iodine as catalyst: Knoevenagel, *J. Prakt. Chem.* [2] **89**, 11 (1914); Mitchell, Webb, U.S. pat. **2,415,363** (1947 to Koppers). Purification: Vierling, U.S. pat. **2,887,482** (1959); Rigby, U.S. pat. **3,000,887** (1961 to Shell Oil). Crystal structure: J. D. Bell *et al., Chem. Commun.* **1968**, 1656.

Yellow, rhombic leaflets or diamond-shaped plates from

toluene or butanol, mp 185.1°. Sublimes at 130° at 1 mm. bp$_{760}$ 371°; bp$_{40}$ 290°. Freely sol in benzene; sol in ether, in hot acetic acid; slightly sol in alcohol and in mineral oils. Practically insol in petr ether, chloroform, water. Readily oxidized by sunlight or when in presence of a finely divided inert carrier, acquiring a greenish-brown tint. This can be prevented by the admixture of 0.3-1.0% methenamine.

USE: Insecticide; manuf pharmaceuticals.

THERAP CAT (VET): Anthelmintic.

7221. Phenothrin. *2,2-Dimethyl-3-(2-methyl-1-propenyl)cyclopropanecarboxylic acid (3-phenoxyphenyl)methyl ester; 2,2-dimethyl-3-(2-methylpropenyl)cyclopropanecarboxylic acid m-phenoxybenzyl ester;* 3-phenoxybenzyl *cis,trans*-chrysanthemate; S-2539; Sumithrin. $C_{23}H_{26}O_3$; mol wt 350.46. C 78.82%, H 7.48%, O 13.70%. Potent synthetic pyrethroid insecticide. Prepn of racemic mixture: N. Itaya *et al.,* **Ger. pat. 1,926,433** corresp to **U.S. pat. 3,666,789** (1969, 1972 to Sumitomo). Comparative activity of isomers: K. Fujimoto *et al., Agr. Biol. Chem.* **37,** 2681 (1973); Y. Okuno *et al.,* **Ger. pat. 2,348,930** corresp to **U.S. pat. 3,934,-023** (1973, 1976 to Sumitomo).

The commercial product is a mixture of isomers. Colorless liquid. d$_{25}^{25}$ 1.06; n$_D^{25}$ 1.5483. Sol in acetone, xylene; insol in water.

USE: Insecticide.

7222. Phenoxazine. Phenazoxine. $C_{12}H_9NO$; mol wt 183.20. C 78.67%, H 4.95%, N 7.65%, O 8.73%. Prepn from 2-aminophenol and 2-aminophenol·HCl: Kehrmann, Neil, *Ber.* **47,** 3107 (1903); de Antoni, *Bull. Soc. Chim. France* **1963,** 2871; by heating 2-amino-2'-hydroxydiphenyl ether in a sealed tube at 270-280° for 40 hrs: Cullinane *et al., J. Chem. Soc.* **1934,** 718.

Leaflets from alc, mp 156°. Freely sol in abs methanol, ethanol, ether, CHCl$_3$, benzene. Sparingly sol in petr ether.

10-Acetylphenoxazine, $C_{14}H_{11}NO_2$, prisms, mp 142°. Sparingly sol in hot water; sol in alc; freely sol in glacial acetic acid.

7223. Phenoxyacetic Acid. Phenoxyethanoic acid; O-phenylglycolic acid; phenyl ether glycolic acid; Phenylium. $C_8H_8O_3$; mol wt 152.14. C 63.15%, H 5.30%, O 31.55%. Prepd from phenol and monochloroacetic acid: Giacosa, *J. Prakt. Chem.* [2] **19,** 396 (1879); van Alphen, *Rec. Trav. Chim.* **46,** 148 (1927).

Needles from water, mp 98°; bp 285° (some decompn). K at 25° = 7.54 × 10^{-4}. One gram dissolves in about 75 ml water. Freely sol in alcohol, ether, benzene, carbon disulfide, glacial acetic acid.

USE: Fungicide; keratin exfoliative (to relieve and to soften calluses, corns, and other hard skin surfaces; applied as plasters, pads or in liquids). *Caution:* Mild irritant.

7224. Phenoxyacetyl Cellulose. *Cellulose phenoxyacetate;* Enzorb-A. Support used to immobilize proteins, enzymes and microsomes by hydrophobic adsorption. The attractive forces between the hydrophobic phenoxyacetate groups of the support and the hydrophobic surface areas of protein molecules arise from the common repulsion of the aqueous medium. Prepn and properties: L. G. Butler, *Arch. Biochem. Biophys.* **171,** 645 (1975). Applications: J. Dixon *et al., Biotechnol. Bioeng.* **21,** 2113 (1979). Stable in dry form; slowly hydrolyzed in aq media above pH 8. Resistant to enzymatic hydrolysis.

USE: In enzyme immobilization.

7225. Phenoxybenzamine. *N-(2-Chloroethyl)-N-(1-methyl-2-phenoxyethyl)benzenemethanamine; N-(2-chloroethyl)-N-(1-methyl-2-phenoxyethyl)benzylamine; N-phenoxyisopropyl-N-benzyl-β-chloroethylamine;* bensylyt; 688-A; $C_{18}H_{22}ClNO$; mol wt 303.84. C 71.16%, H 7.30%, Cl 11.67%, N 4.61%, O 5.27%. Prepn: Kerwin, Ullyot, **U.S. pat. 2,599,000** (1952 to SK & F).

Crystals from petr ether, mp 38-40°. Sol in benzene.

Hydrochloride, $C_{18}H_{23}Cl_2NO$, *Dibenyline, Dibenzyline, Dibenzyran.* Crystals from alcohol + ether, mp 137.5-140°. Sol in alcohol, propylene glycol. Very sparingly sol in water. Solutions in propylene glycol should be sterilized by filtration.

THERAP CAT: Antihypertensive.

7226. 2-Phenoxyethanol. 1-Hydroxy-2-phenoxyethane; ethylene glycol monophenyl ether; β-hydroxyethyl phenyl ether; Phenyl Cellosolve; Phenoxethol; Phenoxetol. $C_8H_{10}O_2$; mol wt 138.16. C 69.54%, H 7.30%, O 23.16%. $C_6H_5OCH_2CH_2OH$. Obtained by treating phenol with ethylene oxide in an alkaline medium: Becker, Barthell, *Monatsh.* **77,** 80 (1947); *see also* Roithner, *ibid.* **15,** 674, 678 (1894); Rindfusz, *J. Am. Chem. Soc.* **41,** 669 (1919).

Oily liquid. Faint aromatic odor. Burning taste. d$_{20}^{20}$ 1.1094; d$_4^{22}$ 1.102; mp 14°; bp$_{760}$ 245.2°; bp$_{80}$ 165°; bp$_{25}$ 137°; bp$_{20}$ 128-130°. n$_D^{20}$ 1.534. Flash pt 250°F. Soly in water: 2.67 g/100 ml. Freely sol in alcohol, ether, NaOH solns. LD$_{50}$ orally in rats: 1.26 g/kg, H. F. Smyth *et al., J. Ind. Hyg. Toxicol.* **23,** 259 (1941).

Acetate, $C_{10}H_{12}O_3$, liquid, bp 243°.

USE: Fixative for perfumes, in org synthesis; as bactericide in conjunction with quaternary ammonium compds; as insect repellent.

THERAP CAT: Topical antiseptic.

7227. Phenoxypropazine. *(1-Methyl-2-phenoxyethyl)hydrazine;* fenoxypropazine. $C_9H_{14}N_2O$; mol wt 166.22. C 65.03%, H 8.49%, N 16.85%, O 9.63%. Prepn: Drain *et al., J. Med. Chem.* **6,** 63 (1963).

bp$_{0.2}$ 98-102°.

Hydrogen maleate, $C_{13}H_{18}N_2O_5$, *Drazine.* Prismatic needles from isopropanol, mp 107-108°. Very sol in water; less sol in ethanol, isopropanol.

THERAP CAT: Formerly as monoamine oxidase inhibitor.

7228. Phenpentermine. *α,α,β-Trimethylbenzeneethanamine; α,α,β-trimethylphenethylamine;* 1,1-dimethyl-2-phenylpropylamine; 2-phenyl-3-methyl-3-butylamine; pentorex. $C_{11}H_{17}N$; mol wt 163.25. C 80.92%, H 10.50%, N 8.58%. Prepn of DL-form, DL-form hydrochloride and DL-, D- and L-forms hydrogen D-tartrate: **Fr. pat. M2594** (1964 to Nordmark-Werke), *C.A.* **61,** 16013f (1964).

DL-Form, bp$_{20}$ 109-111°.

DL-Form hydrochloride, C$_{11}$H$_{17}$N.HCl, crystals, mp 164-166°.

DL-Form hydrogen D-tartrate, C$_{15}$H$_{23}$NO$_6$, *Modatrop.* Crystals, mp 160-162°. $[\alpha]_D^{20}$ +13.4° (c = 0.8 in water).

D-Form hydrogen D-tartrate, crystals, mp 167-169°. $[\alpha]_D^{20}$ +17.9°.

L-Form hydrogen D-tartrate, crystals, mp 164-166°. $[\alpha]_D^{20}$ +3.45°.

THERAP CAT: Anorexic.

7229. Phenprobamate. *Benzenepropanol carbamate; carbamic acid 3-phenylpropyl ester;* 1-carbamoyloxy-3-phenylpropane; γ-phenylpropyl carbamate; proformiphen; MH 532; Extacol; Palmita; Spantol; Ansepron; Gamaquil; Quamaquil. C$_{10}$H$_{13}$NO$_2$; mol wt 179.21. C 67.02%, H 7.31%, N 7.82%, O 17.86%. H$_2$NCOOCH$_2$CH$_2$CH$_2$C$_6$H$_5$. Prepn: **Brit. pat. 837,718** (1960 to Siegfried AG). Pharmacology: G. Stille, *Arzneimittel-Forsch.* **12,** 340 (1962).

Shiny leaflets from aq ethanol, mp 101-104°. Slightly bitter, somewhat burning taste. Numbs the tongue slightly afterwards. Soluble in abs ethanol, chloroform, propylene glycol, ethylenediamine, dimethylformamide; sparingly sol in ether, 50% ethanol. Practically insol in water. LD$_{50}$ orally in mice: 840 mg/kg (Stille).

THERAP CAT: Anxiolytic, muscle relaxant (skeletal).

7230. Phenprocoumon. *4-Hydroxy-3-(1-phenylpropyl)-2H-1-benzopyran-2-one;* 3-(α-ethylbenzyl)-4-hydroxycoumarin; 3-(1-phenylpropyl)-4-hydroxycoumarin; Marcumar; Liquamar. C$_{18}$H$_{16}$O$_3$; mol wt 280.31. C 77.12%, H 5.75%, O 17.12%. Prepn: Grüssner, Balthasar, **U.S. pat. 2,723,276** (1955 to Hoffmann-La Roche); Junek, Ziegler, *Monatsh.* **87,** 218 (1956); Schroeder, Link, *J. Am. Chem. Soc.* **79,** 3291 (1957); **U.S. pat. 2,872,457** (1959 to Wisconsin Alumni Res. Found.); **Brit. pat. 805,748** (1958 to Geigy); L. R. Pohl, *J. Med. Chem.* **18,** 513 (1975). Resolution: Preis *et al.,* **U.S. pat. 3,239,529** (1966 to Wisconsin Alumni Res. Found.). Conformation in soln: E. J. Valente *et al., J. Med. Chem.* **21,** 141, 231 (1978).

Crystals or prisms from dil methanol, mp 179-180°.

THERAP CAT: Anticoagulant.

7231. Phensuximide. *1-Methyl-3-phenyl-2,5-pyrrolidinedione; N-methyl-2-phenylsuccinimide;* N-methyl-α-phenylsuccinimide; Lifène; Succitimal; Milontin; Mirontin. C$_{11}$H$_{11}$NO$_2$; mol wt 189.21. C 69.82%, H 5.86%, N 7.40%, O 16.91%. Prepd by the action of methylamine on phenylsuccinic acid: Miller, Long, *J. Am. Chem. Soc.* **73,** 4895 (1951); **U.S. pat. 2,643,258** (1953 to Parke, Davis).

Fine crystals from hot 95% ethanol, mp 71-73°. Slightly sol in water (about 4.2 mg/ml at 25°). Readily sol in methanol, ethanol. Aq solns are fairly stable at pH 2-8, but hydrolysis sets in under more alkaline conditions. LD$_{50}$ orally in mice: 960 mg/kg, Chen, Ensor, *J. Lab. Clin. Med.* **41,** 78 (1953).

THERAP CAT: Anticonvulsant.

7232. Phentermine. *α,α-Dimethylbenzeneethanamine;*

α,α-*dimethylphenethylamine;* phenyl-*tert*-butylamine; α-benzylisopropylamine. C$_{10}$H$_{15}$N; mol wt 149.23. C 80.48%, H 10.13%, N 9.39%. Prepn: Shelton, Van Campen, **U.S. pat. 2,408,345** (1946 to Wm. S. Merrell); Abell *et al.,* **U.S. pat. 2,590,079** (1952 to Wyeth).

Oily liquid. bp$_{750}$ 205°; bp$_{21}$ 100°.

Hydrochloride, C$_{10}$H$_{16}$ClN, *Adipex-P, Fastin, Wilpo.* Crystals, mp 198°.

Ingredient of *Ionamin, Mirapront, Linyl, Omnibex, Duromine* (phentermine resins).

Caution: Excessive use may lead to tolerance and physical dependence. This is a controlled substance (stimulant) listed in the U.S. Code of Federal Regulations, Title 21 Part 1308.14 (1985).

THERAP CAT: Anorexic.

7233. Phentetiothalein Sodium. *4,5,6,7-Tetraiodophenolphthalein sodium;* 3,3-bis(4-hydroxyphenyl)-4,5,6,7-tetraiodo-1(3H)-isobenzofuranone disodium salt; 3,4,5,6-tetraiodophenolphthalein disodium salt; phenoltetraiodophthalein sodium; Iso-Iodeikon. C$_{20}$H$_8$I$_4$Na$_2$O$_4$; mol wt 865.94. C 27.74%, H 0.93%, I 58.63%, Na 5.31%, O 7.39%. Prepd by the condensation of phenol and tetraiodophthalic acid or its anhydride: Orndorff, Black, *Am. Chem. J.* **41,** 349 (1909); Zalkind, Belikova, *Zh. Prikl. Khim. (Leningrad)* **8,** 1210 (1935). Isomeric with iodophthalein sodium, *q.v.*

Bronze-purple, odorless, slightly hygroscopic granules. Dec on exposure becoming incompletely sol. Sol in water, alcohol. *Keep tightly closed.*

THERAP CAT: Diagnostic aid (radiopaque medium—cholecystography and hepatic function).

7234. Phentolamine. *3-[[(4,5-Dihydro-1H-imidazol-2-yl)methyl](4-methylphenyl)amino]phenol; 2-[N-(m-hydroxyphenyl)-p-toluidinomethyl]imidazoline;* 2-(m-hydroxy-N-p-tolylanilinomethyl)-2-imidazoline; 2-(N'-p-tolyl-N'-m-hydroxyphenylaminomethyl)-2-imidazoline; C 7337. C$_{17}$H$_{19}$-N$_3$O; mol wt 281.35. C 72.57%, H 6.81%, N 14.94%, O 5.69%. α-Adrenergic blocker. Prepn: K. Miescher *et al.,* **U.S. pat. 2,503,059** (1950 to Ciba); E. Urech *et al., Helv. Chim. Acta* **33,** 1386 (1950). Pharmacology and toxicity: R. Meier *et al., Proc. Soc. Exp. Biol. Med.* **71,** 70 (1949). HPLC determn in biological samples: B. D. Kerger *et al., Anal. Biochem.* **170,** 145 (1988). Diagnostic use: G. Spergel *et al., J. Am. Med. Assoc.* **211,** 266 (1970). Review of pharmacology and therapeutic applications: L. Gould, C. V. R. Reddy, *Am. Heart J.* **92,** 397-402 (1976).

Crystals, mp 174-175°.

Hydrochloride, C$_{17}$H$_{20}$ClN$_3$O, bitter crystals, mp 239-

240°. One gram dissolves in 50 ml water, in 70 ml alcohol. Very slightly sol in chloroform. Practically insol in acetone, ethyl acetate. pH of 1% aq soln 4.5-5.5. Aq solns cannot be stored. LD_{50} in rats (mg/kg): 75 i.v.; 275 s.c.; 1250 orally (Meier).

Methanesulfonate, $C_{18}H_{23}N_3O_4S$, *Regitine, Rogitine.* Crystals, mp 177-181°. One gram dissolves in 50 ml water, 23 ml alcohol, 660 ml chloroform. pH of 1% aq soln 4.5-5.5. Aq solns cannot be stored.

THERAP CAT: Antihypertensive; treatment of pheochromocytoma; diagnostic aid (pheochromocytoma).

7235. Phentydrone. *1,2,3,4-Tetrahydro-9-fluorenone;* THF. $C_{13}H_{12}O$; mol wt 184.23. C 84.75%, H 6.57%, O 8.68%. Prepn: Beyerman,Veer, U.S. pat. **2,692,898** (1954 to Organon); House *et al., J. Am. Chem. Soc.* **82,** 1457 (1960).

Light yellow needles or prisms from pentane, mp 81-82° (Beyerman); golden yellow prisms from petr ether, mp 41.5-42.5° (House). $bp_{0.05}$ 139-140°. uv max (alcoholic molar soln): 265, 232 nm (log ε 4.48, 3.62); (carbon tetrachloride): 238, 245, 317 nm (ε 39,700, 46,900, 960). LD_{50} s.c. in mice: 1-1.3 g/kg.

7236. Phenylacetaldehyde. *Benzeneacetaldehyde;* α-toluic aldehyde; α-tolualdehyde; Hyacinthin. C_8H_8O; mol wt 120.14. C 79.97%, H 6.71%, O 13.32%. $C_6H_5CH_2CHO$. Prepd by oxidizing phenylethyl alcohol with chromic acid. High-yield synthesis from styrene oxide or styrene glycol: G. Paparatto, G. Gregorio, *Tetrahedron Letters* **29,** 1471 (1988).

Oily, colorless liq which polymerizes and grows more viscous on standing. Odor reminiscent of lilac and hyacinth. (Has been crystallized, mp 33-34°.) d_{25}^{25} 1.023-1.030; bp_{760} 195°; bp_{18} 88°; bp_{10} 78°; n_D^{20} 1.524-1.528. Slightly sol in water. Soluble in alcohol, ether. One part is sol in 2 parts of 80% alc forming a clear solution.

USE: In perfumery; intermediate in organic synthesis.

7237. α-Phenylacetamide. *Benzeneacetamide;* α-toluamide. C_8H_9NO; mol wt 135.16. C 71.09%, H 6.71%, N 10.36%, O 11.84%. The amide of phenylacetic acid, not to be confused with N-phenylacetamide which is acetanilide. Prepd by Willgerodt reaction from acetophenone or from styrene. Review and lab procedures: Carmack, Spielman in *Org. Reactions* **II** (New York, 1946) p 83. Has also been prepd by heating the ammonium salt of phenylacetic acid: Menschutkin, *Ber.* **31,** 1429 (1898). Lab procedure from benzyl cyanide: Wenner, *Org. Syn.* **coll. vol. IV,** 760 (1963).

Bimorphous plates, leaflets. mp 155°. Distils *in vacuo.* Freely sol in alc. Slightly sol in water, ether, benzene.

USE: In manuf of penicillin G.

7238. Phenyl Acetate. *Acetic acid phenyl ester;* acetylphenol. $C_8H_8O_2$; mol wt 136.14. C 70.57%, H 5.92%, O 23.50%. $C_6H_5OOCCH_3$. Prepd from phenol and acetyl chloride.

Colorless, mobile, highly refractive liquid; phenolic odor. d_4^{20} 1.073; bp 195-196°. n_D^{20} 1.5030 (almost the same as glass). Practically insol in water; miscible with alc, chloroform, ether; sol in glacial acetic acid. LD_{50} orally in rats: 1.63 ml/kg, H. F. Smyth *et al., Am. Ind. Hyg. Assoc. J.* **30,** 470 (1969).

7239. Phenylacetic Acid. *Benzeneacetic acid;* α-toluic acid. $C_8H_8O_2$; mol wt 136.14. C 70.57%, H 5.92%, O 23.50%. $C_6H_5CH_2COOH$. Made by refluxing benzyl cyanide with dil H_2SO_4 or HCl: Adams, Thal, *Org. Syn.* **2,** 63

(1922). Absorption spectrum: Baly, Tryhorn, *J. Chem. Soc.* **107,** 1065 (1915).

Leaflets on distillation *in vacuo;* plates, tablets from petr ether; mp 76.5°. bp_{760} 265.5°; bp_{100} 198.2°; bp_{40} 173.6°; bp_5 127°; $bp_{1.0}$ 97°. d_4^{77} 1.091. K at 25° = 5.56 × 10⁻⁵. Slightly sol in cold, freely in hot water. The aq soln satd at 25° is 0.131N. Sol in alcohol, ether. Soly at 25° in chloroform (moles/l): 4.422; in carbon tetrachloride: 1.842; in acetylene tetrachloride: 4.513; in trichlorethylene: 3.299; in tetrachlorethylene: 1.558; in pentachloroethane: 3.252.

Methyl ester, $C_9H_{10}O_2$, liquid, bp 215°.

Ethyl ester, *see* Ethyl phenylacetate.

USE: Starting material in manuf synthetic perfumes, condensation products with aldehydes.

7240. Phenylacetone. *l-Phenyl-2-propanone;* benzyl methyl ketone. $C_9H_{10}O$; mol wt 134.18. C 80.56%, H 7.51%, O 11.92%. $C_6H_5CH_2COCH_3$. Synthesis from phenylacetic and acetic acids: R. H. Pickard, J. Kenyon, *J. Chem. Soc.* **105,** 1124 (1914); R. M. Herbst, R. H. Manske, *Org. Syn.* **coll. vol. II,** 389 (1943); from α-phenylacetoacetonitrile: P. L. Julian, J. J. Oliver, *ibid.* 391; from α-methyl-α-phenylethylene oxide: S. Danilow, E. Venus-Danilowa, *Ber.* **60,** 1050 (1927); from diethyl malonate: H. G. Walker, C. R. Hauser, *J. Am. Chem. Soc.* **68,** 1386 (1946). Conformational calculations: M. Hirota *et al., Tetrahedron* **39,** 3091 (1983). Use as prochiral ketone in enantioselective hydrosilylation: H. Brenner *et al., Ber.* **117,** 1330 (1984).

Oil, mp −16° to −15°. bp_{760} 214°; bp_{14} 100-101°. d_4^{20} 1.0157. n_D 1.5174. uv max (ethanol): 258, 283 nm (ε 255, 150).

Note: This is a controlled substance listed in the U.S. Code of Federal Regulations, Title 21 Part 1308.12 (1987).

USE: In organic synthesis; production of benzyl radicals by photolysis.

7241. Phenyl Acetylsalicylate. *2-(Acetyloxy)benzoic acid phenyl ester;* acetylphenylsalicylate; Acetylsalol; Spiroform; Vesipyrin. $C_{15}H_{12}O_4$; mol wt 256.25. C 70.30%, H 4.72%, O 24.97%. $CH_3COOC_6H_4COOC_6H_5$.

White, crystalline powder. mp 97°; bp_{11} 198°. Insol in water. Sol in alcohol, ether.

THERAP CAT: Analgesic, antipyretic, anti-inflammatory.

7242. Phenylalanine. Phe (IUPAC abbrev.); β-phenylalanine; α-aminohydrocinnamic acid; α-amino-β-phenylpropionic acid. $C_9H_{11}NO_2$; mol wt 165.19. C 65.43%, H 6.71%, N 8.48%, O 19.37%. $C_6H_5CH_2CH(NH_2)COOH$. An amino acid classified as essential with respect to its growth effect in rats. Essential component in human nutrition, not synthesized by the human body. Recommended intake for normal adult male 2.2 g/day (L-phenylalanine). Whole egg contains 5.4% and skim milk 5.1% of L-phenylalanine. D-Phenylalanine is utilized very little by the human organism. L-Phenylalanine is isolated commercially from proteins such as ovalbumin, lactalbumin, zein, and fibrin. Synthesis of L-phenylalanine from L-tyrosine: Coffey *et al., J. Chem. Soc.* **1959,** 4100; V. Viswanatha, V. J. Hruby, *J. Org. Chem.* **45,** 2010 (1980). Microbial synthesis: Huang, U.S. pat. **2,973,304** (1961 to Pfizer). Synthesis of DL-phenylalanine: Fischer, *Ber.* **37,** 3064 (1904); *Org. Syn.* **coll. vol. II,** 489 (1943); **coll. vol. III,** 705-708 (1955). Evaluation of D-form in multiple sclerosis: A. Winter, *Neurol. Orthopaed. J. Med. Surg.* **5,** 39 (1984). Reviews of metabolism of L-form and role in phenylketonuria: S. P. Bessman, *Nutr. Rev.* **37,** 209-220 (1979); F. Güttler, *Acta Paedr. Scand.* **73,** 705-716 (1984).

L-Phenylalanine, monoclinic plates, leaflets from warm concd aq solns. Hydrated needles from dil solns. Dec 283°. Sublimes *in vacuo.* $[α]_D^{25}$ −35.1° (c = 1.94). Soly in water (g/l): 19.8 at 0°; 29.6 at 25°; 44.3 at 50°; 66.2 at 75°; 99.0 at 100°. Very slightly sol in methanol, ethanol.

D-Phenylalanine (from the DL-form by partial fermentation with yeast and sugar), *Sabiden.* Leaflets from water, dec 285°. $[α]_D^{20}$ +35.0° (c = 2.04); $[α]_D^{20}$ +7.1° (c = 3.8 in 18% HCl). One gram dissolves in 35.5 ml water at 16°. Sparingly sol in methanol.

DL-Phenylalanine, monoclinic leaflets or prisms from water or alcohol, sweetish taste. Dec 271-273°. Sublimes *in vacuo.* pK_1 2.58; pK_2 9.24. Soly in water (g/l): 9.97 at 0°; 14.11 at 25°; 21.87 at 50°; 37.08 at 75°; 68.9 at 100°.

THERAP CAT: Nutrient.

7243. Phenyl Aminosalicylate. *4-Amino-2-hydroxybenzoic acid phenyl ester; p-aminosalicylic acid phenyl ester;* phenyl p-aminosalicylic acid; p-aminosalol; fenamisal; Phenyl PAS; Pheny-PAS-Tebamin; Tebamin; Tebanyl. $C_{13}H_{11}$-NO_3; mol wt 229.23. C 68.11%, H 4.84%, N 6.11%, O 20.94%. Description of properties: Meyer, *Antibiot. Ann.* **1957-1958**, 614. Prepd by Raney nickel reduction of the corresp nitro ester in ethyl acetate: Friere, U.S. pat. **2,604,-488** (1952 to Rhône-Poulenc).

Crystals from isopropanol, mp 153°. Soly in water: 0.7 mg/100 ml; in serum: 12 mg/100 ml. One gram of phenyl PAS is equivalent to 0.67 g PAS.

THERAP CAT: Antibacterial (tuberculostatic).

7244. N-Phenylanthranilic Acid. *2-(Phenylamino)benzoic acid;* 2-anilinobenzoic acid; diphenylamine-2-carboxylic acid. $C_{13}H_{11}NO_2$; mol wt 213.23. C 73.22%, H 5.20%, N 6.57%, O 15.01%. Prepd from o-chlorobenzoic acid and aniline: Ullmann, Dieterle, *Ann.* **365,** 322 (1907); Ullmann, *Ber.* **36,** 2382 (1907); from 2-iodobenzoic acid and phenylhydroxylamine: Wieland, Roseeu, *Ber.* **48,** 1120 (1915).

Leaflets from alcohol, dec 183-184°. Sol in hot alcohol. Very slightly sol in hot water, hot benzene, ether.

USE: Detection of vanadium in steel: Adamovich, Zagorulko, *Zavodskaya Lab.* **9,** 465 (1940), *C.A.* **37,** 1346 (1943).

7245. 2-Phenyl-1H-benzimidazole. *N,N'-Benzenyl-o-phenylenediamine;* phenzidole; Gainex. $C_{13}H_{10}N_2$; mol wt 194.23. C 80.38%, H 5.19%, N 14.42%. Prepn from N-benzoyl-o-phenylenediamine: v. Auwers, Frese, *Ber.* **59,** 548 (1926); from o-phenylenediamine and benzonitrile: Hölljes, Wagner, *J. Org. Chem.* **9,** 31 (1944). Anthelmintic in sheep: Forsyth, *Australian Vet. J.* **38,** 398 (1962).

Needles from benzene, plates from water, mp 291°. Sol in methanol, abs ethanol. Sparingly sol in water, benzene, chloroform.

Hydrochloride, $C_{13}H_{11}ClN_2$, needles from alcohol, dec 328°. Freely sol in water.

THERAP CAT (VET): Has been used as an anthelmintic.

7246. Phenyl Benzoate. *Benzoic acid phenyl ester.* C_{13}-$H_{10}O_2$; mol wt 198.21. C 78.77%, H 5.09%, O 16.14%. $C_6H_5COOC_6H_5$.

Monoclinic prisms; geranium odor. d 1.235; mp 70°; bp 314°. Insoluble in water; freely sol in hot alcohol; slightly sol in cold alcohol or ether.

7247. Phenyl Biguanide. *N-Phenylimidodicarbonimidic diamide;* phenyl diguanide; 1-phenylbiguanide; N-phenyl-N'-guanylguanidine. $C_8H_{11}N_5$; mol wt 177.21. C 54.22%, H 6.26%, N 39.52%. Prepd by reacting aniline hydrochloride and dicyandiamide in water: Cohn, *J. Prakt. Chem.* [2] **84,** 396; from aniline, dicyandiamide in pyridine contg HCl: Jacobs, Jolles, Brit. pat. **587,907** (1947 to ICI).

$$C_6H_5NHCNHCNH_2$$
(with NH NH groups above)

Crystals from water or toluene. Sharp, somewhat bitter taste. mp 144-146°. pK_1 10.76, pK_2 2.13. Freely sol in water and in alcohol.

Hydrochloride, $C_8H_{11}N_5$·HCl, prisms, mp 237°. Sol in about 115 parts of water at 32°.

Nitrate, $C_8H_{11}N_5$·HNO_3, crystals, mp 208-209°.

Picrate, $C_8H_{11}N_5$·$C_6H_3N_3O_7$, yellow needles from alc, mp 176-179°; very sparingly sol in water; very sol in hot alc.

7248. Phenylbutazone. *4-Butyl-1,2-diphenyl-3,5-pyrazolidinedione;* 4-butyl-1,2-diphenyl-3,5-dioxopyrazolidine; 3,5-dioxo-1,2-diphenyl-4-n-butylpyrazolidine; flexazone; diphebuzol; fenibutazona; G 13871; Alindor; Antadol; Anuspiramin; Artrizin; Azolid; Benzone; Betazed; Bizolin; Bizolin 200; Bunetzone; Butacote; Butadion; Butapirazol; Butazolidin; Butidiona; Butoz; Buzon; Ecobutazone; Elmedal; Equipalazone; Exrheudon N; Fenibutol; Fenilbutina; Fenotone; Ia-But; Intrabutazone; Intrazone; Mephabutazon; Phebuzine; Phenyzone; Pirarreumol "B"; Praecirheumin; R-3-ZON; Reudo; Reudox; Robizone-V; Schemergen; Spondyril; Tevcodyne; Ticinil; Uzone; Zolaphen. $C_{19}H_{20}N_2$-O_2; mol wt 308.37. C 74.00%, H 6.54%, N 9.09%, O 10.38%. Prepn: Stenzl, U.S. pat. **2,562,830** (1951 to Geigy); cf. Brit. pat. **812,449** (1959 to Geigy). Review of synthesis: *Ullmann's Encyklopädie der Technischen Chemie,* **13,** 298 (1962). Physical properties and pharmacology: v. Rechenberg, *Phenylbutazone* (Edward Arnold, London, 1962) 197 pp. Acute toxicity: T. B. Gaines, R. E. Linder, *Fundam. Appl. Toxicol.* **7,** 299 (1986). Soly data: Pulver *et al., Schweiz. Med. Wochenschr.* **86,** 1080 (1956). Comprehensive description: S. L. Ali in *Analytical Profiles of Drug Substances* vol. **11,** K. Florey, Ed. (Academic Press, New York, 1982) pp 483-521. Review of hematological effects: G. A. Faich, *Pharmacotherapy* **7,** 25 (1987).

Crystals from ethanol, mp 105°. Soly in water at 22.5°: 0.7 mg/ml (also reported as 2.2 mg/ml). pK 4.5 (from uv in water), pK 4.89 (titration in 50% ethanol), pK 5.25 (titration in 80% 2-methoxyethanol). uv max (acid methanol): 239.5 nm (log ε 4.19).

Sodium salt, crystals, freely sol in water. pH of aq solns about 8.2.

Calcium salt, $C_{38}H_{36}CaN_4O_4$, Pyrazon. LD_{50} in adult male, female rats (mg/kg): 1311, 647 orally (Gaines, Linder).

Piperazine salt, $C_{23}H_{30}N_4O_2$, pyrazinobutazone, pyrasanone, Carudol, Dartranol, Ranoroc. mp 140-141° (solidifies and remelts at ∼180°).

2-Amino-2-thiazoline salt, see Thiazolinobutazone.

THERAP CAT: Anti-inflammatory.

THERAP CAT (VET): Analgesic; anti-inflammatory.

7249. α-Phenylbutyramide. *α-Ethylbenzeneacetamide;* 2-phenylbutanamide; α-ethyl-α-phenylacetamide; α-phenyl-α-ethylacetamide; TH 4128; Hyposterol; Natol; Phenetamid; Redusterol. $C_{10}H_{13}NO$; mol wt 163.21. C 73.59%, H 8.03%, N 8.58%, O 9.80%. Prepn from phenylethylacetylurea: Ger. pat. **249,241** (1912 to Bayer); *Chem. Zentr.* **1912,** II, 396; *Frdl.* **10,** 1164.

Practically tasteless crystals, mp 86°; also given as mp 83-84°. Very sparingly sol in cold water. Soly in boiling water: about 1:50; in ethanol at 25°: 1:10. Sol in many other organic solvents.

THERAP CAT: In hypercholesterolemia.

7250. Phenyl Carbonate. *Carbonic acid diphenyl ester;* diphenyl carbonate. $C_{13}H_{10}O_3$; mol wt 214.22. C 72.89%, H 4.71%, O 22.41%. $C_6H_5OCOOC_6H_5$. Prepn: Eckenroth, *Ber.* **27**, 3410 (1894); Bischoff, *ibid.* **35**, 3434 (1902); Gomberg, Snow, *J. Am. Chem. Soc.* **47**, 198 (1925).

Lustrous needles, mp 80-81°. bp 302-306°; bp_{15} 168°. Practically insol in water; sol in hot alcohol, benzene, ether, glacial acetic acid.

USE: In the molten state as solvent for nitrocellulose.

7251. 2-Phenyl-6-chlorophenol. *3-Chloro[1,1'-biphenyl]-2-ol;* 6-chlororthoxenol. $C_{12}H_9ClO$; mol wt 204.65. C 70.42%, H 4.43%, Cl 17.33%, O 7.82%.

Pale yellow, viscous liquid. Slight, characteristic odor. d_4^{25} about 1.24; mp 6°; bp 317-318° with decompn. n_D^{30} 1.6237. Insoluble in water; sol in fixed alkali hydroxide solns and in most organic solvents.

USE: Germicide; fungicide.

7252. 4-Phenyl-2-chlorophenol. *3-Chloro-(1,1'-biphenyl)-4-ol.* Pale yellow crystals. mp about 77°. bp_7 about 160-162°. Soly and use as for 2-phenyl-6-chlorophenol, *q.v.*

7253. α-Phenylcinnamic Acid. *α-(Phenylmethylene)benzeneacetic acid; 2,3-diphenylacrylic acid;* 2,3-diphenylpropenoic acid; stilbene-α-carboxylic acid. $C_{15}H_{12}O_2$; mol wt 224.25. C 80.33%, H 5.39%, O 14.27%. Prepn of *cis*-form: Buckles, Hausman, *J. Am. Chem. Soc.* **70**, 415 (1948); Buckles, *J. Chem. Ed.* **27**, 210 (1950); Buckles, Bremer, *Org. Syn.* **coll. vol. IV**, 777 (1963); B. H. Patwardhan, G. Bagavant, *Indian J. Chem.* **10**, 59 (1972). Prepn of *cis-* and *trans*-forms: L. F. Fieser, *Experiments in Organic Chemistry* (Heath, Boston, 3rd ed., 1955) p 182. Isomerization studies: S. V. Kessar *et al.*, *Indian J. Chem.* **20B**, 4 (1981).

cis-form

cis-Form, silky needles from ether + petr ether, mp 174. pKa 6.1 in 60% ethanol. uv max (ethanol): 223, 280 nm (ε 32,100, 19,500). Sol in hot water, methanol, ethanol, isopropanol, ether, benzene.

trans-Form, stout prisms from ether + petr ether, mp 138-139°. Less stable than the *cis*-form. pKa 4.8 in 60% ethanol. uv max (ethanol): 222, 289 nm (ε 14,500, 22,500). More sol than the *cis*-form.

7254. m-Phenylenediamine. *1,3-Benzenediamine;* m-diaminobenzene. $C_6H_8N_2$; mol wt 108.14. C 66.64%, H 7.46%, N 25.91%. Prepd by the reduction of *m*-dinitrobenzene: Kuhn, U.S. pat. **2,768,209** (1956 to Ringwood); Faust, *J. Prakt. Chem.* **6**, 14 (1958); Neilson *et al.*, *J. Chem. Soc.* **1962**, 371; Tallee, Peltier, *Compt. Rend.* **259**, 400 (1964).

White crystals becoming red on exposure to air. d 1.139; mp 62-63°; bp 284-287°; dipole moment 1.79. Fire point: 175°. Sol in water, methanol, ethanol, chloroform, acetone,

dimethylformamide, methyl ethyl ketone, dioxane. Slightly sol in ether, carbon tetrachloride, isopropanol, dibutyl phthalate. Very slightly sol in benzene, toluene, xylene, butanol. *Keep well closed and protected from light.* LD_{50} in rats (mg/kg): 650 orally; 283 i.p., C. Burnett *et al.*, *J. Toxicol. Environ. Health* **2**, 657 (1977).

Hydrochloride, white or slightly red, cryst powder; becomes darker on exposure to air. Freely sol in water; sol in alcohol.

USE: Manuf dyes; rubber curing agents, ion exchange resins, decolorizing resins, formaldehyde condensates, resinous polyamides, block polymers, textile fibers, urethanes, petroleum additives, rubber chemicals, corrosion inhibitors; in photography; as reagent for gold and bromine. The hydrochloride chiefly as a reagent for nitrite.

7255. o-Phenylenediamine. *1,2-Benzenediamine;* o-diaminobenzene. $C_6H_8N_2$; mol wt 108.14. C 66.64%, H 7.46%, N 25.91%. Made by reducing *o*-nitroaniline with Zn and NaOH.

Brownish-yellow crystals. mp 103-104°. bp 256-258°. Slightly sol in water; freely sol in alcohol, chloroform, ether. LD_{50} in rats (mg/kg): 1070 orally; 516 i.p., C. Burnett *et al.*, *J. Toxicol. Environ. Health* **2**, 657 (1977).

USE: Manufacture of dyes.

7256. p-Phenylenediamine. *1,4-Benzenediamine;* p-diaminobenzene; p-aminoaniline; orsin; C.I. 76076; Ursol D. $C_6H_8N_2$; mol wt 108.14. C 66.64%, H 7.46%, N 25.91%. Prepn: A. Rinne, T. Zincke, *Ber.* **7**, 869 (1874); Ger. pat. **202,170** (1907 to BASF), *C.A.* **3**, 382 (1909); A. J. Quick, *J. Am. Chem. Soc.* **42**, 1033 (1920); J. F. Norris, E. O. Cummings, *Ind. & Eng. Chem.* **17**, 305 (1925). *See also:* Beilstein **XIII**, 61 (1930). Crystal structure: A. Domenicano *et al.*, *Acta Crystallogr.* **B33**, 1664 (1977). Mutagenicity studies: B. N. Ames *et al.*, *Proc. Nat. Acad. Sci. USA* **72**, 2423 (1975); W. G. H. Blijleven, *Mutat. Res.* **48**, 181 (1977).

White to slightly red crystals; darkens on exposure to air. mp 145-147°; bp 267°; sol in 100 parts cold water; sol in alcohol, chloroform, ether. A black color is developed with 3% H_2O_2; brown with 5% $FeCl_3$ soln. *Keep well closed and protected from light.* LD_{50} in rats: 80 mg/kg orally, 37 mg/kg i.p., C. Burnett *et al.*, *J. Toxicol. Environ. Health* **2**, 657 (1977).

Hydrochloride, white to slightly reddish crystals. Freely sol in water, slightly in alcohol, ether.

USE: Dyeing furs; also in photochemical measurements, accelerating vulcanization; manuf azo dyes, etc. The hydrochloride as reagent for blood, H_2S, amyl alcohol; in testing of milk. *Caution:* Pure compd or intermediate oxidation products may produce eczematoid contact dermatitis, bronchial asthma.

7257. Phenylephrine Hydrochloride. *(R)-3-Hydroxy-α-[(methylamino)methyl]benzenemethanol hydrochloride; l-m-hydroxy-α-[(methylamino)methyl]benzyl alcohol hydrochloride; l-1-(m-hydroxyphenyl)-2-methylaminoethanol hydrochloride; l-α-hydroxy-β-methylamino-1-ethylbenzene hydrochloride; m-methylaminoethanolphenol hydrochloride;* metaoxedrin; Adrianol; Ak-Dilate; Ak-Nefrin; Fenilfar; Isophrin Hydrochloride; Lexatol; Meta-Synephrine hydrochloride; Meta Sympatol; *m*-Sympatol; Mezaton; Neophryn; Neo-Synephrine hydrochloride; Oftalfrine; Pyracort D; Prefrin; Mydfrin. $C_9H_{14}ClNO_2$; mol wt

203.67. C 53.07%, H 6.93%, Cl 17.41%, N 6.88%, O 15.71%. α-Adrenergic agonist. Prepn from m-hydroxy-ω-chloroacetophenone and methylamine: U.S. pats. **1,932,347; 1,954,-389** (1934). Synthesis from m-benzyloxybenzaldehyde: Bergmann, Sulzbacher, *J. Org. Chem.* **16**, 84 (1951). Toxicity data: M. R. Warren, H. W. Werner, *J. Pharmacol. Exp. Ther.* **86**, 284 (1946). Comprehensive description: C. A. Gaglia in *Analytical Profiles of Drug Substances* vol. 3, K. Florey, Ed. (Academic Press, New York, 1974) pp 483-512.

$$\text{HOCHCH}_2\text{NHCH}_3 \cdot \text{HCl}$$

Bitter crystals, mp 140-145°. (free base mp 169-172°: *U.S.P. XVI*). $[\alpha]_D^{25}$ −46.2° to −47.2°. Freely sol in water or alcohol. The aq soln is neutral to litmus paper. LD_{50} in rats (mg/kg): 17 ±1.1 i.p.; 33 ±2.0 s.c. (Warren, Werner).

THERAP CAT: Mydriatic; decongestant.

7258. Phenylethanolamine. α-*(Aminomethyl)benzenemethanol;* α-*(aminomethyl)benzyl alcohol;* β-hydroxyphenethylamine. $C_8H_{11}NO$; mol wt 137.18. C 70.04%, H 8.08%, N 10.21%, O 11.66%. Prepn: Dornow, Theidel, *Ber.* **88**, 1267 (1955).

$$\text{HOCHCH}_2\text{NH}_2$$

Pale yellow crystals, mp 56-57°. bp_{17} 157-160°. Freely sol in water. Gives an alkaline reaction in water and forms salts with acids under mild conditions.

Sulfate, $C_{16}H_{24}N_2O_6S$, *dl*-α-*phenyl*-β-*aminoethanol sulfate, Apophedrin.* Prepn: Roth, *Arch. Pharm.* **292**, 76 (1959). Crystals, mp 275-276° (Kofler). Soluble in water.

USE: Free base as a stopping agent during polymerization of styrene-butadiene rubber; in the hardening of waxes. Intermediate in the manuf of pressor amines.

THERAP CAT: Sulfate as topical vasoconstrictor.

7259. Phenyl Ether. *1,1'-Oxybisbenzene;* diphenyl ether; diphenyl oxide. $C_{12}H_{10}O$; mol wt 170.20. C 84.68%, H 5.92%, O 9.40%. $C_6H_5OC_6H_5$. Prepd by heating sodium phenolate with chlorobenzene.

Liquid; characteristic odor. d^{20} 1.075 (liq); mp 28°, and remains liq at lower temp; bp 259°. Flash pt 115°. Insol in water; sol in alcohol, benzene, ether, glacial acetic acid.

USE: As heat transfer medium; in perfuming soaps; in organic syntheses.

7260. 5-(α-Phenylethyl)semioxamazide. *Oxo-[(1-phenylethyl)amino]acetic acid hydrazide; 5-(α-methylbenzyl)semioxamazide.* $C_{10}H_{13}N_3O_2$; mol wt 207.23. C 57.96%, H 6.32%, N 20.28%, O 15.44%. Prepd from α-phenylethylamine, ethyl oxalate, and hydrazine: Leonard, Boyer, *J. Org. Chem.* **15**, 42 (1950).

$$\text{CH}_3$$
$$\text{CHNHCOCONHNH}_2$$

dl-Form, fine needles from ethanol, mp 157°.
d-Form, crystals, mp 167-168°. $[\alpha]_D^{25}$ +102.0° (c = 1.04 in chloroform).
l-Form, crystals, mp 167-168°. $[\alpha]_D^{25}$ −102.5° (c = 0.625 in chloroform).

USE: The racemic form as reagent for the characterization of aldehydes and ketones. The optically active forms as resolving agents for carbonyl compds possessing asymmetric carbon atoms.

7261. Phenylglyceryl Ether. *3-Phenoxy-1,2-propanediol;* Antodyne. $C_9H_{12}O_3$; mol wt 168.19. C 64.27%, H 7.19%, O 28.54%. $C_6H_5OCH_2CH(OH)CH_2OH$. Prepd from phenol

and 3-chloro-1,2-propylene carbonate: Smith, U.S. pat. **2,967,892** (1961 to Dow).

Crystals, mp 50-52°, $bp_{0.6}$ 129-142°. Sol in water, alcohol. LD_{50} i.p. in mice: 1280 mg/kg, Hine *et al.*, *J. Pharmacol. Exp. Ther.* **97**, 414 (1949).

7262. N-Phenylglycine. Anilinoacetic acid. $C_8H_9NO_2$; mol wt 151.16. C 63.56%, H 6.00%, N 9.27%, O 21.17%. $C_6H_5NHCH_2COOH$. Prepd from aniline and chloroacetic acid: Curtius, *J. Prakt. Chem.* **38**, 436 (1888); Thorpe, Wood, *J. Chem. Soc.* **103**, 1606 (1913).

Crystals, mp 127-128°. $K = 3.8 \times 10^{-6}$ at 25°. Moderately sol in water. Less sol in alc. Sparingly sol in ether. Forms water-sol salts with alkali hydroxides.

Ethyl ester, $C_{10}H_{13}NO_2$, leaflets, mp 58°. Sol in alc, ether.

USE: Intermediate in the Heumann synthesis of indigo.

7263. α-Phenylglycine. α-*Aminobenzeneacetic acid;* α-aminophenylacetic acid; α-amino-α-toluic acid; *C*-phenylglycine. $C_8H_9NO_2$; mol wt 151.16. C 63.56%, H 6.00%, N 9.27%, O 21.17%. $C_6H_5CH(NH_2)COOH$. Prepd by hydrolysis of α-aminophenylacetonitrile with dil hydrochloric acid: Marvel, Noyes, *J. Am. Chem. Soc.* **42**, 2264 (1920); R. E. Steiger, *Org. Syn. coll. vol. III*, 84 (1955).

dl-Form, lustrous platelets. Sublimes without melting at about 255°. Slightly sol in the usual organic solvents. Sol in alkalies.

Methyl ester, $C_9H_{11}NO_2$, needles from petr ether, mp 32°. Sol in alcohol, ether, benzene.

Ethyl ester, $C_{10}H_{13}NO_2$, liquid, bp 257°; bp_{16} 149°; bp_5 114-115°; n_D^{25} 1.500.

l-Form, needles from dil alc, mp 305-310°. $[\alpha]_D^{20}$ −157.8° (dil HCl).

d-Form, crystals. $[\alpha]_D^{20}$ +156° (dil HCl).

7264. Phenylhydrazine. Hydrazinobenzene. $C_6H_8N_2$; mol wt 108.14. C 66.64%, H 7.46%, N 25.91%. C_6H_5-NHNH₂. Prepd by diazotizing aniline with $NaNO_2$ and HCl, then treating the soln with Na_2SO_3 followed by NaOH: Fischer, *Anleitung zur Darstellung organischer Präparate* (Braunschweig, 10th ed., 1922) p 23; Coleman, *Org. Syn.* **2**, 71 (1922), **coll. vol. I** (2nd ed., 1941) p 442. Mechanism of oxidative hemolysis of erythrocytes: B. Goldberg, A. Stern, *Mol. Pharmacol.* **13**, 832 (1977); B. Vilsen, H. Nielsen, *Biochem. Pharmacol.* **33**, 2739 (1984). Review of interaction with hemoglobin: M. D. Shetlar, H. A. O. Hill, *Environ. Health Perspec.* **64**, 265-281 (1985).

Monoclinic prisms or oil. Turns yellow to dark red on exposure to air and light. Faint aromatic odor. Forms a hemihydrate, mp 24°. Data for the anhydr compd: d_4^{20} 1.0978; mp 19.5°; bp_{760} 243.5° (decompn); bp_{100} 173.5°; bp_{40} 148.2°; bp_{20} 131.5°; bp_{10} 115.8°; bp_5 101.6°; $bp_{1.0}$ 71.8°; $n_D^{20.3}$ 1.60813. Weak base, pK (15°) 8.79. Miscible with alcohol, ether, chloroform, benzene. Sparingly sol in water, petr ether; sol in dil acids. *Keep well closed and protected from light.*

Caution: May cause contact dermatitis, hemolytic anemia, and liver and kidney injury: *Patty's Industrial Hygiene and Toxicology* Vol. 2A, G. D. Clayton, R. E. Clayton, Eds. (Wiley-Interscience, New York, 3rd ed., 1981) pp 2792-2795, 2804-2805.

USE: Manuf dyes, antipyrine, nitron (a stabilizer for explosives); reagent for sugars, aldehydes, ketones.

THERAP CAT: Hemolytic.

7265. Phenylhydrazine Hydrochloride. $C_6H_9ClN_2$; mol wt 144.59. C 49.84%, H 6.27%, N 19.37%, Cl 24.52%. Prepd by neutralizing the base with HCl: Fischer, *Ann.* **190**, 83 (1878); or by heating it with CCl_4: Brunner, Eiermann, *Ber.* **31**, 1406 (1898); by reduction of benzenediazonium chloride: Rüetschi, Trümpler, *Helv. Chim. Acta* **36**, 1649 (1953); Hupfer, **Ger.** pat. **1,143,825** (1963 to Hoechst).

Leaflets from alc, mp 243-246° (slight browning). Sublimes. Freely sol in water; sol in alcohol; practically insol in ether. Is precipitated from aq soln by concd HCl.

THERAP CAT: Hemolytic.

7266. Phenylhydroxylamine. *N-Hydroxybenzenamine;* β-phenylhydroxylamine; *N*-phenylhydroxylamine. C_6H_7-NO; mol wt 109.12. C 66.04%, H 6.47%, N 12.84%, O 14.66%. C_6H_5NHOH. Prepd by zinc reduction of nitrobenzene in ammonium chloride soln: Kamm, *Org. Syn.* **4**, 57 (1925).

Needles from satd NaCl soln. mp 82°. *Deteriorates on storage and should be used promptly.* Sol in 50 parts cold, in 10 parts hot water. Freely sol in alcohol, ether, carbon disulfide, chloroform, hot benzene, dil mineral acids, acetic acid. Slightly sol in petr ether. The oxalate is more stable.

USE: Manufacture of cupferron.

7267. Phenyl Isocyanate. *Isocyanatobenzene;* carbanil; phenylcarbimide. C_7H_5NO; mol wt 119.12. C 70.58%, H 4.23%, N 11.76%, O 13.43%. $C_6H_5N=C=O$. Prepd by passing carbonyl chloride into a hot soln of aniline in toluene, saturated with HCl: Hardy, *J. Chem. Soc.* **1934**, 2011.

Liquid, with acrid odor. $d_4^{11.6}$ 1.101; d_4^{15} 1.092; $d_4^{19.6}$ 1.0956; $d_4^{25.9}$ 1.08870. bp 158-168°; bp_{18-20} 58.2-59.5°; bp_{13} 55°. $n_D^{19.6}$ 1.53684; $n_D^{25.9}$ 1.53412.

Human Toxicity: Irritating to eyes.

7268. Phenyl Isothiocyanate. *Isothiocyanatobenzene;* isothiocyanic acid phenyl ester; phenyl mustard oil; thiocarbanil. C_7H_5NS; mol wt 135.18. C 62.19%, H 3.73%, N 10.37%, S 23.72%. $C_6H_5N=C=S$. Prepd from ammonium phenyldithiocarbamate by the action of lead nitrate: Dains *et al., Org. Syn.* **coll. vol. I** (2nd ed., 1941) p 447.

Liquid. mp −21°. d_4^{25} 1.1288; d_4^{35} 1.1202; d_4^{50} 1.1061. bp_{760} 221°; bp_{33} 117.1°; bp_{12} 95°. Distills with water without dec. $n_D^{23.4}$ 1.64918. Insol in water. Sol in alcohol, ether.

USE: Derivatizing agent for primary, secondary amines. In sequencing of peptides by Edman degradation. In amino acid analyses by HPLC (Pico-Tag).

7269. Phenylmagnesium Chloride. *Chlorophenylmagnesium.* C_6H_5ClMg; mol wt 136.88. C 52.64%, H 3.68%, Mg 17.77%, Cl 25.91%. C_6H_5MgCl. One of the Grignard reagents. Prepd by reacting chlorobenzene with magnesium at reflux temperatures in the presence of catalytic amounts of an organic nitrate: Ramsden, U.S. pat. **2,816,937** (1957 to M. & T. Corp.).

Sol in ether (ethyl ether, other ethers may be used as solvents). A solution which is about 3 molar has an approx strength of 48%, and a d_4^{20} of about 1.15. Reacts with water, steam or acids to produce toxic and flammable vapors with evolution of heat.

USE: In organic synthesis, especially in the production of hydrocarbons, alcohols, ketones, organic acids, amines, silicones, borane.

7270. N-Phenylmaleimide. *1-Phenyl-1H-pyrrole-2,5-dione.* $C_{10}H_7NO_2$; mol wt 173.16. C 69.36%, H 4.07%, N 8.09%, O 18.48%. Prepn from maleanilic acid: Cava *et al., Org. Syn.* **41**, 93 (1961).

Canary-yellow needles from cyclohexane, mp 89-89.8°.

USE: Dienophile in the Diels-Alder reaction. Usually gives cryst adducts.

7271. Phenylmercuric Acetate. *(Acetato)phenylmercury;* acetoxyphenylmercury; phenylmercury acetate; PMA; PMAC; PMAS; Ceresan Slaked Lime; Gallotox; Liquiphene; Phix; Mersolite; Tag Fungicide; Tag HL-331; Nylmerate; Scutl; Riogen. $C_8H_8HgO_2$; mol wt 336.75. C 28.53%, H 2.39%, Hg 59.57%, O 9.50%. $C_6H_5HgOOCCH_3$. Made by heating benzene with mercuric acetate: Maynard, *J. Am. Chem. Soc.* **46**, 1510 (1925); Brit. pat. **325,846** (1928 to I. G. Farbenind.); Renschler, U.S. pat. **2,050,018** (1936 to Hamilton Labs.); Grave *et al., J. Am. Pharm. Assoc.* **25**, 752 (1936). *Review:* F. C. Whitmore, *Organic Compounds of Mercury* (Chemical Catalog Co., New York, 1921) pp 175-176.

Small, lustrous prisms from ethanol, mp 149°. Sol in about 600 parts water; sol in alcohol, benzene, acetone. LD_{50} orally in rats: 22 mg/kg, *RTECS* Vol. II, R. J. Lewis, R. L. Tatken, Eds. (1980) p 46.

USE: Herbicide; fungicide. An ingredient of *Agrosan GN:* Gates, *Ann. Appl. Biol.* **47**, 502 (1959); and of *Ceresan Dry:*

Bayer Pflanzenschutz Compendium vol. **1** (Bayer, Leverkusen, 1962) pp 24-27.

7272. Phenylmercuric Chloride. *Chlorophenylmercury;* phenylmercury chloride; Stopspot. C_6H_5ClHg; mol wt 313.18. C 23.01%, H 1.61%, Cl 11.32%, Hg 64.06%. C_6H_5-HgCl. Antibacterial activity: H. L. Friedman, *Ann. N.Y. Acad. Sci.* **65**, 461 (1957). Use as fungicide: U. Prota, *Phytopath. Mediter.* **8**, 87 (1970), *C.A.* **72**, 42138t (1970); M. Makes, Z. Lokaj, *Agrochemia* **16**, 181 (1976), *C.A.* **85**, 187602t (1976). Toxicity: M. Umeda *et al., Japan. J. Exp. Med.* **39**, 47 (1969).

White satiny leaflets, mp 250-252°. Sol in about 20,000 parts cold water, in benzene, ether, pyridine; slightly in hot alcohol. LD_{50} s.c. in rats: 30 mg/kg (Umeda).

USE: Agricultural fungicide.

7273. Phenylmercuric Nitrate, Basic. *(Nitrato-O)phenylmercury;* merphenyl nitrate; Phermernite; Phenmerzyl nitrate. $C_{12}H_{11}Hg_2NO_4$; mol wt 634.45. C 22.72%, H 1.75%, Hg 63.24%, N 2.21%, O 10.09%. $C_6H_5HgOH.C_6H_5HgNO_3$. Made by boiling benzene with mercuric acetate, then treating the resulting acetate with an alkali nitrate: Pyman, Stevenson, *Pharm. J.* **133**, 269 (1934); T. B. Grave *et al., J. Am. Pharm. Assoc.* **25**, 752 (1936); Brit. pat. **446,703** (1936 to Schering-Kahlbaum). Toxicology: R. Wien, *Quart. J. Pharm. Pharmacol.* **12**, 212 (1939).

Pearly lustrous scales, mp 187-190° (dec). Phenol coefficient stated to be about 600 and relatively unaffected by organic matter. Soluble in about 1250 parts water; slightly sol in alcohol, moderately in glycerol. Practically insol in other usual organic solvents. LD_{50} in mice (mg/g): 0.045 s.c.; 0.027 i.v. (Wien).

Human Toxicity: Local application of the more concd solns may cause irritation.

USE: Pharmaceutic aid (antimicrobial). Antiseptic, germicide, fungicide esp for tree wounds.

THERAP CAT (VET): Antibacterial, antifungal; chiefly used topically, may also be injected.

7274. Phenylmercury Borate. *[Orthoborato(3−)-O]-phenylmercurate(2−) dihydrogen; (dihydrogen borato)-phenylmercury;* phenylmercuric borate; Famosept; Gyne-Merfen; Merfen. $C_6H_7BHgO_3$; mol wt 338.56. C 21.29%, H 2.09%, B 3.20%, Hg 59.25%, O 14.18%. Prepn: Christiansen, U.S. pat. **2,196,384** (1950 to Lever Bros.).

Crystalline powder, mp 112-113°. Sol in water; alcohol, glycerol.

Note: Product of commerce may be an equimolar composition of phenylmercuriborate and phenylmercurihydroxide.

THERAP CAT: Topical antiseptic.

7275. Phenylmethylbarbituric Acid. *5-Methyl-5-phenyl-2,4,6(1H,3H,5H)-pyrimidinetrione; 5-methyl-5-phenylbarbituric acid;* 2,4,6-trioxo-5-methyl-5-phenylhexahydropyrimidine; Eudan; Rutonal. $C_{11}H_{10}N_2O_3$; mol wt 218.21. C 60.54%, H 4.62%, N 12.84%, O 22.00%. Prepn: Ger. pat. **247,952** (1912 to Bayer); Dvornik *et al., C.A.* **50**, 312f (1956).

Slightly bitter crystals, mp 226°. Practically insol in water. Sol in alcohol, ether, aq solns of the alkalies. Forms a water-sol sodium salt.

Note: Often confused with mephobarbital, but it is not the same compd.

Caution: May be habit forming. This is a controlled substance (depressant) listed in the U.S. Code of Federal Regulations, Title 21 Parts 329.1 and 1308.13 (1987).

THERAP CAT: Anticonvulsant; sedative, hypnotic.

7276. o-Phenylphenol. *(1,1'-Biphenyl)-2-ol; 2-biphenylol;* orthoxenol; o-hydroxydiphenyl; 2-hydroxydiphenyl; Dowicide 1. $C_{12}H_{10}O$; mol wt 170.20. C 84.68%, H 5.92%, O 9.40%. Prepn from phenyl ether: Lüttringhaus, Sääf, *Ann.* **542,** 241 (1939); from dibenzofuran: Gilman, Esmay, *J. Am. Chem. Soc.* **75,** 2947 (1953); Müller, U.S. pat. **2,862,-035** (1958 to Bayer). Purification: Widiger, U.S. pat. **3,087,969** (1963 to Dow). Toxicity data: Hodge *et al., J. Pharmacol. Exp. Ther.* **104,** 202 (1952).

White, flaky crystals. Mild, characteristic odor. mp 55.5-57.5°; bp 280-284°; bp$_{15}$ 152-154°. Practically insol in water; sol in fixed alkali hydroxide solns and most organic solvents. LD$_{50}$ orally in rats: 2.48 g/kg (Hodge).
Sodium salt tetrahydrate, $C_{12}H_9NaO.4H_2O$, *Dowicide A, Natriphene.* White flakes. Soly in g/100 g of solvent: water 122; acetone 156; methanol 138; propylene glycol 28. Practically insol in petroleum fractions, pine oil. pH of satd water soln at 25° = 12.0-13.5.
USE: In the rubber industry; agricultural fungicide; disinfectant. *Caution:* Toxic symptoms similar to phenol, *q.v.*

7277. p-Phenylphenol. *[1,1'-Biphenyl]-4-ol;* 4-hydroxydiphenyl; Paraxenol.
mp 164-165°; bp 305-308°. Possesses approx the soly of the ortho compd.
USE: As intermediate in the manuf of resins; also in the rubber industry.

7278. Phenyl Phthalate. *1,2-Benzenedicarboxylic acid diphenyl ester;* diphenyl phthalate. $C_{20}H_{14}O_4$; mol wt 318.31. C 75.46%, H 4.43%, O 20.10%. $C_6H_4(COOC_6H_5)_2$. Prepd from phenol and phthalic anhydride.
White odorless crystals. d^{74} 1.572; mp 70-73°; bp$_{14}$ about 255°. Insol in water; sol in acetone and other ketones, in liquid esters and chlorinated hydrocarbons.
USE: Plasticizer in nitrocellulose lacquers.

7279. Phenylpropanolamine Hydrochloride. *(R*,S*)-(±)-α-(1-Aminoethyl)benzenemethanol hydrochloride; α-(1-aminoethyl)benzyl alcohol hydrochloride; dl-norephedrine hydrochloride;* 2-amino-1-phenyl-1-propanol hydrochloride; α-hydroxy-β-aminopropylbenzene hydrochloride; 1-phenyl-2-amino-1-propanol; Kontexin; Monydrin; Mucorama; Mydriatine; Obestat; Propadrine hydrochloride; Temporinolo. $C_9H_{14}ClNO$; mol wt 187.67. C 57.60%, H 7.52%, Cl 18.89%, N 7.46%, O 8.53%. Sympathomimetic amine. Prepn: Vanderbilt, Hass, *Ind. Eng. Chem.* **32,** 34 (1940); *cf.* Nagai, Kanao, *Ann.* **470,** 157 (1929); Kamlet, U.S. pat. **2,151,517** (1939); Hoover, Hass, *J. Org. Chem.* **12,** 506 (1947). Industrial prepn: Jenkins, Hartung, *Chemistry of Organic Medicinal Products* (New York, 1949). Improved process: Wilbert, Sosis, U.S. pat. **3,028,429** (1962 to Nepera Chem.). Prepn of optically active compds: Rusch *et al.,* Ger. pat. **1,014,553** (1957 to Knoll). Separation of enantiomers: S. Kanao, *J. Pharm. Soc. Japan* **48,** 947 (1928); C. Jarowski, W. H. Hartung, *J. Org. Chem.* **8,** 564 (1943); C. Pettersson, H. W. Stuurman, *J. Chromatog. Sci.* **22,** 441 (1984). Crystal structure: A. Podder *et al., Indian J. Phys.* **53A,** 652 (1979). HPLC determn: V. Das Gupta, A. G. Ghanekar, *J. Pharm. Sci.* **66,** 895 (1977); N. Muhammad, J. A. Bodnar, *J. Liquid Chromatog.* **3,** 113 (1980). Clinical studies of decongestant activity: O. K. Haugeto *et al., J. Otolaryngol.* **10,** 359 (1981); M. Bende *et al., Rhinology* **23,** 43 (1985); in weight loss: S. Altschuler *et al., Int. J. Obesity* **6,** 549 (1982); in stress incontinence: E. Fossberg *et al., Urologia Int.* **38,** 293 (1983). Toxicity: E. I. Goldenthal, *Toxicol. Appl. Pharmacol.* **18,** 185 (1971); P. Pentel, *J. Am. Med. Assoc.* **252,** 1898 (1984). Comprehensive description: I. Kanfer *et al.,* in *Analytical Profiles of Drug Substances* **vol. 12,** K. Florey, Ed. (Academic Press, New York, 1983) pp 357-383.

Crystals, mp 190-194°. Odor resembling that of crude benzoic acid. pKa 9.44±0.04. Freely sol in water, alcohol. Practically insol in ether, chloroform, benzene. The aq soln is neutral to litmus. LD$_{50}$ orally in rats: 1490 mg/kg (Goldenthal).
(±)-Form base, mp 101-101.5°.
(+)-Form hydrochloride, mp 171-172°. $[\alpha]_D^{25}$ +32° (water).
D-*threo*-Form, *see* Norpseudoephedrine.
THERAP CAT: Decongestant. Anorexic.
THERAP CAT (VET): Bronchodilator. Nasal decongestant.

7280. Phenylpropylmethylamine. *N,β-Dimethylbenzeneethanamine; dl-N,β-dimethylphenethylamine; dl-N-methyl-2-phenylpropylamine;* 1-methylamino-2-phenylpropane; phenpromethamine; 1-methylamino-2-methyl-2-phenylethane; Vonedrine. $C_{10}H_{15}N$; mol wt 149.23. C 80.48%, H 10.13%, N 9.39%. Synthesis starting with the condensation of chlorobenzene with allyl chloride, followed by ammonolysis: Patrick *et al., J. Am. Chem. Soc.* **68,** 1009 (1946).

Volatile liquid. d$_4^{25}$ 0.915-0.925; bp$_{760}$ 205-210°; bp$_{15}$ 95-96°; n$_D^{20}$ 1.5102. Slightly sol in water (1.2 g/100 ml). Freely sol in alcohol, ether, benzene. Aq solns are strongly alkaline; pH of a soln of 2 drops (about 0.1 ml) dil with 10 ml H_2O is about 10.5.
Hydrochloride, $C_{10}H_{15}N.HCl$, mp 144-148°.
THERAP CAT: Adrenergic.

7281. 1-Phenyl-3-pyrazolidinone. 1-Phenyl-3-pyrazolidone; Phenidone. $C_9H_{10}N_2O$; mol wt 162.19. C 66.65%, H 6.22%, N 17.27%, O 9.86%. Prepd by heating phenylhydrazine with β-chloropropionic acid: Ger. pat. **53,834** (1889); by acid hydrolysis of 3-amino-1-phenylpyrazoline: Kendall, Duffin, Brit. pats. **650,911** and **669,591** (1951, 1952, both to Ilford). *Review:* Kendall, *Brit. J. Photogr.* **100,** 56 (1953).

Leaflets or needles from benzene, mp 121°. One gram dissolves in 10 ml boiling water, in 10 ml hot alcohol, in 37.5 ml boiling benzene. Practically insol in ether, petr ether. Freely sol in dil aq solns of acids and alkalies. Pharmacological tests indicate low oral toxicity. No cases of dermatitis have occurred.
USE: Non-staining, high contrast photographic developer. The amount required is about one-fifth to one-tenth that of Metol.

7282. Phenyl Salicylate. *2-Hydroxybenzoic acid phenyl ester;* Salol. $C_{13}H_{10}O_3$; mol wt 214.21. C 72.89%, H 4.71%, O 22.41%. Made by the action of phosphorus oxychloride on a mixture of phenol and salicylic acid.

White, small crystals or crystalline powder; pleasant aromatic odor and taste. d 1.25; mp 41-43°; bp$_{12}$ 173°. One gram dissolves in 6670 ml water, 6 ml alcohol, 1.5 ml benzene, 5 ml amyl alcohol, 10 ml liq paraffin, 4 ml almond oil. Sol in acetone, chloroform, ether, oils; very slightly sol in glycerol. Solubility at 25° in g/100 g: absolute ethanol 53; ethyl acetate 470; methyl ethyl ketone 620; toluene 460; Stoddard solvent 88; water less than 0.1%.

Incompat. Bromine water, ferric salts; camphor, monobromated camphor, phenol, chloral hydrate, thymol or urethan in trituration.

USE: In the manuf of various polymers for the plastics industry, also in lacquers, adhesives, waxes, polishes. In suntan oils and cremes. As light absorber to prevent discoloration of plastics. Has some plasticizer properties.

THERAP CAT: Analgesic, antipyretic, anti-inflammatory.

THERAP CAT (VET): Has been used externally as a disinfectant, internally as an intestinal antiseptic and antipyretic.

7283. 4-Phenylsemicarbazide. *N-Phenylhydrazinecarb-oxamide;* anilinoformylhydrazine. C$_7$H$_9$N$_3$O; mol wt 151.17. C 55.61%, H 6.00%, N 27.80%, O 10.58%. C$_6$H$_5$NHCO-NHNH$_2$. Prepd by the action of hydrazine hydrate on phenylurea: Curtius, *J. Prakt. Chem.* [2] **58**, 216 (1898); Wheeler, *Org. Syn.* coll. vol. **I** (2nd ed., 1941) p 450.

Orthorhombic plates from water, mp 122°. Insol in ether, difficultly sol in hot water; freely sol in alcohol, chloroform, dil acids and alkalies.

Hydrochloride, C$_7$H$_9$N$_3$O.HCl, prisms, mp 215°; freely sol in water and alcohol.

7284. N-Phenylsulfanilic Acid. *4-(Phenylamino)benz-enesulfonic acid;* 4-diphenylaminesulfonic acid; *p*-anilinobenzenesulfonic acid. C$_{12}$H$_{11}$NO$_3$S; mol wt 249.28. C 57.81%, H 4.45%, N 5.62%, S 12.86%, O 19.25%. Prepd by acetylation followed by sulfonation of diphenylamine: Merz, Weith, *Ber.* **6**, 1512 (1873); Gnehm, Werdenberg, *Z. Angew. Chem.* **12**, 1027 (1899); Sarver, Kolthoff, *J. Am. Chem. Soc.* **53**, 1902 (1931).

Leaves. Becomes blue on exposure to light. Soluble in water, in alcohol; insol in ether. Dec into diphenylamine and sulfuric acid when heated above 200° with water contg hydrochloric acid.

Sodium salt, C$_{12}$H$_{10}$NNaO$_3$S, very sol in water.

Potassium salt, C$_{12}$H$_{10}$KNO$_3$S, leaves; slightly sol in alcohol, very sol in water.

Barium salt, (C$_{12}$H$_{10}$NO$_3$S)$_2$Ba, leaflets; slightly sol in water. *Poisonous!*

USE: Colorimetric determination of nitrates; oxidation-reduction indicator; detection of oxidizing substances.

7285. Phenyl Sulfide. *1,1'-Thiobis[benzene];* diphenyl-sulfide. C$_{12}$H$_{10}$S; mol wt 186.27. C 77.37%, H 5.41%, S 17.22%. (C$_6$H$_5$)$_2$S. Willard, Hall, *J. Am. Chem. Soc.* **44**, 2219 (1922).

Colorless, almost odorless liquid. d$_{15}^{15}$ 1.118; mp about −40°; bp 295-297°; n$_D^{18}$ 1.6350. Insol in water; sol in hot alcohol; misc with benzene, ether, carbon disulfide. LD$_{50}$ orally in rats: 0.49 ml/kg, H. F. Smyth *et al.*, *Am. Ind. Hyg. Assoc. J.* **23**, 95 (1962).

7286. Phenylthiourea. Phenylthiocarbamide. C$_7$H$_8$N$_2$S; mol wt 152.22. C 55.23%, H 5.30%, N 18.41%, S 21.07%. C$_6$H$_5$NHCSNH$_2$. Prepd by evaporating an aq soln of aniline hydrochloride and ammonium thiocyanate and carefully heating the residue.

Bitter or tasteless needles, depending upon heredity of taster. d 1.3. mp 154°. Soluble in 400 parts cold water, 17 parts boiling water; soluble in alcohol. LD$_{50}$ orally in rats, rabbits: 3, 40 mg/kg, Scheline *et al.*, *J. Med. Pharm. Chem.* **4**, 109 (1961).

USE: In medical genetics.

7287. Phenyltoloxamine. *N,N-Dimethyl-2-[2-(phenyl-methyl)phenoxy]ethanamine; N,N-dimethyl-2-(α-phenyl-o-tolyloxy)ethylamine;* N,N-dimethyl-2-(α-phenyl-o-toloxy)-ethylamine; 2-(2-dimethylaminoethoxy)diphenylmethane;

2-benzhydryl β-dimethylaminoethyl ether; 2-benzylphenyl β-dimethylaminoethyl ether; PRN; bistrimin; C 5581H; Antin; Phenoxadrine; Bristamin. C$_{17}$H$_{21}$NO; mol wt 255.35. C 79.96%, H 8.29%, N 5.49%, O 6.27%. Prepn: Cheney *et al.*, *J. Am. Chem. Soc.* **71**, 60 (1949); Binkley, Cheney, U.S. pat. 2,703,324 (1955 to Bristol Labs.).

Oily liquid, bp 141-144° (at < 0.1 mm Hg).

Dihydrogen citrate, C$_{17}$H$_{21}$NO.C$_6$H$_8$O$_7$, crystals from water or methanol, mp 138-140°. Soluble in water.

Hydrochloride, C$_{17}$H$_{21}$NO.HCl, crystals from methyl isobutyl ketone, mp 119-121°. Soluble in water.

Phenyltoloxamine adsorbed on a resin is marketed as *Histionex.*

THERAP CAT: Dihydrogen citrate as antihistaminic.

7288. Phenyl Tolyl Ketone. C$_{14}$H$_{12}$O; mol wt 196.24. C 85.68%, H 6.16%, O 8.15%. *o*-Isomer made from benzene and *o*-toluic acid chloride in the presence of AlCl$_3$. *p*-Isomer made from benzoyl chloride and toluene in presence of AlCl$_3$.

o-Phenyl tolyl ketone, *(2-methylphenyl)phenylmethanone, 2-methylbenzophenone.* Oily liquid. mp below −18°. bp 309-311°. Insol in water. Freely sol in alcohol, oils, most organic solvents.

p-Phenyl tolyl ketone, *(4-methylphenyl)phenylmethanone, 4-methylbenzophenone.* Crystals, mp 59-60°. bp$_{720}$ 311-312°. Insol in water. Sol in alcohol; easily sol in benzene, ether, oils.

USE: As fixative in perfumery.

7289. Phenyltrimethylammonium Iodide. *N,N,N-Tri-methylbenzaminium iodide.* C$_9$H$_{14}$IN; mol wt 263.13. C 41.08%, H 5.36%, N 5.32%, I 48.23%. C$_6$H$_5$(CH$_3$)$_3$NI. Ref: Pass, Ward, *Analyst* **58**, 667 (1933).

White, cryst powder. mp 175°. Sol in water or alcohol.

USE: For the detection and determination of cadmium.

7290. Phenylurea. Phenylcarbamide. C$_7$H$_8$N$_2$O; mol wt 136.15. C 61.75%, H 5.92%, N 20.58%, O 11.75%. C$_6$H$_5$-NHCONH$_2$. Obtained with carbanilide as byproduct by refluxing aniline hydrochloride and urea in water: Davis, Blanchard, *Org. Syn.* **3**, 95 (1923).

Monoclinic prisms from water or alcohol. d 1.302; mp 147° (dec); bp 238°. On slow cooling separates in needles several centimeters in length. Soluble in hot water, hot alcohol, ether, ethyl acetate, glacial acetic acid.

7291. Phenylurethan(e). *Phenylcarbamic acid ethyl ester;* ethyl phenylcarbamate; ethyl carbanilate. C$_9$H$_{11}$NO$_2$; mol wt 165.19. C 65.43%, H 6.71%, N 8.48%, O 19.37%. C$_6$H$_5$-NHCOOC$_2$H$_5$. Made by the action of aniline on ethyl chloroformate.

White, acicular crystals. d 1.106; mp 52-53°; bp 238° with slight decompn; n$_D^{20}$ 1.5376. Slightly sol in water, freely in alcohol, ether; scarcely attacked by boiling for a short time with HCl or NaOH.

7292. Phenyramidol. *α-[(2-Pyridinylamino)methyl]ben-zenemethanol;* 2-(β-hydroxyphenethylamino)pyridine; fenyramidol; IN 511; MJ 505; NSC-17777; Elan; Abbolexin; Miodar; Evasprin. C$_{13}$H$_{14}$N$_2$O; mol wt 214.26. C 72.87%, H 6.59%, N 13.08%, O 7.47%. Prepn from 2-mandelamidopy-ridine or 2-aminopyridine: Gray *et al.*, *J. Am. Chem. Soc.* **81**, 4347, 4351 (1959).

Crystals from dil methanol, mp 82-85°. uv max (95% eth-anol): 243, 303 nm (log ε 4.24, 3.63). pKa 5.85.

Hydrochloride, $C_{13}H_{15}ClN_2O$, *Anabloc, Analexin, Cabral, Bonapar.* Crystals from ethanol + ether, mp 140-142°. Soluble in water. Aq solns are slightly acid and stable in ampuls.

Methiodide, $C_{14}H_{17}IN_2O$, crystals from ethanol + ether, mp 164-166°.

THERAP CAT: Hydrochloride as analgesic, relaxant (skeletal muscle).

7293. Phenytoin. *5,5-Diphenyl-2,4-imidazolidinedione;* diphenylhydantoin; Difhydan; Dihycon; Di-Hydan; Di-Lan; Dilabid; Dilantin; Ekko; Hydantol; Lehydan; Phenhydan (tabl); Zentropil. $C_{15}H_{12}N_2O_2$; mol wt 252.26. C 71.41%, H 4.80%, N 11.11%, O 12.68%. Prepn from benzophenone: Henze, U.S. pat. **2,409,754** (1946 to Parke, Davis); of sodium salt from benzil, urea and NaOH: Biltz, *Ber.* **41,** 1391 (1908); **44,** 411 (1911). Pharmacology: Gillis *et al., J. Pharmacol. Exp. Ther.* **179,** 599 (1971). Toxicity of base: G. Stille, I. Brunckow, *Arzneimittel-Forsch.* **4,** 723 (1954); of sodium salt: G. B. Fink, E. A. Swinyard, *J. Pharmacol. Exp. Ther.* **127,** 318 (1959). Reviews: Damato, *Progr. Cardiov. Dis.* **12,** 1-15 (1969); Dreifus, Watanabe, *Am. Heart J.* **80,** 709-713 (1970). Review of carcinogenicity studies: *IARC Monographs* **13,** 201-225 (1977). Comprehensive description: J. Philip *et al.,* in *Analytical Profiles of Drug Substances* vol. 13, K. Florey, Ed. (Academic Press, New York, 1984) pp 417-445.

Powder, mp 295-298°. Practically insol in water. One gram dissolves in about 60 ml alcohol, about 30 ml acetone. Sol in alkali hydroxides. LD_{50} in mice (mg/kg): 92 i.v.; 110 s.c. (Stille, Brunckow).

Sodium salt, $C_{15}H_{11}N_2NaO_2$, *phenytoin soluble, sodium 5,5-diphenyl hydantoinate, Dilantin Sodium, Dihydan soluble, Di-Len, Denyl Sodium, Danten, Citrullamon, Diphenylan Sodium, Diphenine Sodium, Lepitoin Sodium, Antisacer, Alepsin, Epanutin, Derizene, Eptoin, Hidantal, Minetoin, Diphentoin, Phenhydan (inj), Tacosal, Thilophenyt, Solantoin, Solantyl.* White powder; bitter, soapy taste. Somewhat hygroscopic. Easily dissociated even by weak acids (incl CO_2 absorbed on exposure to air) regenerating phenytoin. One gram dissolves in 10.5 ml alc; in ~66 ml water (aq soln turbid unless pH adjusted to >11.7, the pH of a satd soln). Insol in ether, chloroform. LD_{50} orally in mice: 490 mg/kg (Fink, Swinyard).

Note: This substance may reasonably be anticipated to be a carcinogen: *Fourth Annual Report on Carcinogens* (NTP 85-002, 1985) p 167.

THERAP CAT: Anticonvulsant; anti-epileptic.

THERAP CAT (VET): Sodium salt has been recommended in epileptiform seizures in dogs.

7294. Pheromones. Chemical substances used for communication between individual organisms of the same species; intraspecific chemical messengers. Perceived primarily by the olfactory sense and to a lesser extent the gustatory sense. From the Greek *pherein* (to transfer) and *hormon* (to excite, stimulate), coined to replace *ectohormone:* Karlson, Lüscher, *Naturwiss.* **46,** 63 (1959); *Nature* **183,** 55 (1959); Karlson, Butenandt, *Annu. Rev. Entomol.* **4,** 39 (1959). Evidence of pheromone communication has been found in mammals incl some primates, in reptiles, amphibians, fish, earthworms, fungi, molds and esp in insects. There are two classes of pheromones, the releasers and the primers, differentiated by the types of response they elicit. Releasers generally have an immediately obvious, direct and reversible effect on social or sexual behavior, i.e. alarm pheromones,

trail and territorial markers, sex attractants and aphrodisiacs. The primer pheromones, less well-known than the releasers, bring about long-term physiological changes in the target organism or population. One example is the queen substance of the honeybee which inhibits ovarian development in workers after mating and thereby prevents the appearance of new queens. *Reviews:* MacConnell, Silverstein, *Chemistry* **44**(7), 6-9 (1971); M. Jacobson, *Insect Sex Pheromones* (Academic Press, New York, 1972) 382 pp.

Allomones and *kairomones* are interspecific chemical messengers, used for communication between different species. Allomones cause a reaction in another species which is favorable to the organism secreting the substance, whereas, kairomones are either nonadaptive or actually detrimental to the organisms producing them. See Brown *et al., BioScience* **20,** 21 (1970).

USE: Exptlly in insect pest control.

7295. Phethenylate Sodium. *5-Phenyl-5-(2-thienyl)-2,4-imidazolidinedione monosodium salt; 5-phenyl-5-(2-thienyl)hydantoin sodium;* sodium 5-phenyl-5-(2-thienyl)hydantoinate; Thiantoin(e) Sodium. $C_{13}H_9N_2NaO_2S$; mol wt 280.25. C 55.71%, H 3.24%, N 10.00%, Na 8.20%, O 11.42%, S 11.44%. Prepn: Spurlock, U.S. pat. **2,366,221** (1942).

Base, mp 256-257°.

Minute, hygroscopic crystals. Sol in water, alc. Aq solns are alkaline and dec easily.

THERAP CAT: Anticonvulsant.

7296. Phillyrin. *4-[4-(3,4-Dimethoxyphenyl)tetrahydro-1H,3H-furo[3,4-c]furan-1-yl]-2-methoxyphenyl-β-D-glucopyranoside;* phyllyrin; phillyroside; forsythin. $C_{27}H_{34}O_{11}$; mol wt 534.54. C 60.66%, H 6.41%, O 32.92%. From bark of *Phillyrea latifolia* L., and allied *Oleaceae:* Campona, *Ann.* **24,** 242 (1837); Eijkman, *Rec. Trav. Chim.* **5,** 127 (1886); Kramer, *Compt. Rend.* **196,** 814 (1933); Sosa, *Bull. Soc. Chim. Biol.* **29,** 918 (1947). Structure: Kaku *et al., J. Pharm. Soc. Japan* **59,** 248 (1939); M. Chiba *et al., Chem. Pharm. Bull.* **25,** 3435 (1977).

α-Form, needles from dil alc, mp 154-155°. $[α]_D$ +48.4° (alcohol).

β-Form, needles, mp 184-185°. $[α]_D$ +48.5° (alcohol).

7297. Phloionic Acid. (±)-*9,10-Dihydroxyoctadecanedioic acid;* floionic acid. $C_{18}H_{34}O_6$; mol wt 346.45. C 62.40%, H 9.89%, O 27.71%. Isoln from cork: Guillemonat, Cesaire, *Bull. Soc. Chim. France* **1949,** 792; Ribas, Seoane, *An. Real Soc. Espan. Fis. Quim. (Madrid)* **50B,** 963 (1954); *C.A.* **50,** 806f (1956); Brown, Rosen, U.S. pat. **2,872,464** (1959 to Crown Cork). Structure: Duhamel, *Bull. Soc. Chim. France* **1965,** 399. Synthesis: Ruzicka *et al., Helv. Chim. Acta* **25,** 1086 (1942); Gensler, Schlein, *J. Am. Chem. Soc.* **77,** 4846 (1955). Resolution of isomers: Gender, Mahadevan, *ibid.* **78,** 169 (1956); Alvarez-Varquez, Ribas-Marques, *An. Quim.* **64,** 783 (1968). Synthesis of (±), (+), and (−) forms: McGhie *et al., Chem. & Ind. (London)* **1972,** 536.

Crystals from ethanol + water, mp 126°.

Dimethyl ester, $C_{20}H_{38}O_6$, crystals from ethanol, mp 77.5-78°.

7298. Phloionolic Acid. *(R*,R*)-9,10,18-Trihydroxy-octadecanoic acid;* 9,10,18-trihydroxystearic acid. $C_{18}H_{36}O_5$; mol wt 332.47. C 65.02%, H 10.92%, O 24.06%. Isolation from cork: Seoane, Ribas, *An. Real Soc. Espan. Fis. Quim. (Madrid)* **47B**, 61 (1951), *C.A.* **46**, 89i (1952); **51**, 8651a (1957); Ribas, Seoane, *ibid.* **50B**, 963 (1954), *C.A.* **50**, 806f (1956); Brown, Rosen, U.S. pat. 2,872,464 (1959 to Crown Cork). Structure: Seoane *et al., Chem. & Ind. (London)* **1957**, 490; *An. Real Soc. Espan. Fis. Quim. (Madrid)* **55B**, 839 (1959); Duhamel, *Bull. Soc. Chim. France* **1965**, 399.

HOCH$_2$(CH$_2$)$_7$CH—CH(CH$_2$)$_7$COOH
(OH OH)

Crystals from ethanol + water, mp 101-102°.

Methyl ester, $C_{19}H_{38}O_5$, crystals from benzene + petr ether, mp 80.5-81.5°.

7299. Phloretin. *3-(4-Hydroxyphenyl)-1-(2,4,6-trihydroxyphenyl)-1-propanone; 2',4',6'-trihydroxy-3-(p-hydroxyphenyl)propiophenone;* β-(p-hydroxyphenyl)phloropropiophenone; β-(p-hydroxyphenyl)-2,4,6-trihydroxypropiophenone. $C_{15}H_{14}O_5$; mol wt 274.26. C 65.69%, H 5.15%, O 29.17%. The aglucon of phlorizin, *q.v.* From root bark of apple trees: Rochleder, *J. Prakt. Chem.* **98**, 205 (1866). Structure: Seshadri, *Ann. Rev. Biochem.* **20**, 495 (1951). Synthesis: Fischer, Nouri, *Ber.* **50**, 611 (1917); Rosenmund, Rosenmund, *ibid.* **61B**, 2608 (1928); Shinoda *et al., J. Pharm. Soc. Japan* **49**, 797 (1929); Johnston, U.S. pat. 2,789,995 (1957 to Union Carbide).

Needles from dil alcohol, dec 262°. Absorption spectrum: Lambrechts, *Arch. Int. Physiol.* **44**, Suppl., 1-39 (1937). Freely sol in alc, methanol, acetone; sol in alkalies in hot glacial acetic acid; very sparingly sol in benzene, chloroform; practically insol in water, ether, the soly being increased by addition of alcohol.

7300. Phloridzin. *1-[2-(β-D-Glucopyranosyloxy)-4,6-dihydroxyphenyl]-3-(4-hydroxyphenyl)-1-propanone;* phlorhizin; phlorizin; phlorrhizen; phloretin-2'-β-glucoside; 4,6-dihydroxy-2-(β-D-glucosido)-β-(p-hydroxyphenyl)propiophenone. $C_{21}H_{24}O_{10}$; mol wt 436.40. C 57.79%, H 5.54%, O 36.66%. A dihydrochalcone occurring in all parts of the apple tree except the mature fruit. Once thought to occur in pear, cherry trees and other *Rosaceae*: A. H. Williams in *Comparative Phytochemistry*, T. Swain, Ed. (Academic Press, New York, 1966) pp 297-307. Isoln from root bark: De Koninck, *Ann.* **15**, 75, 258 (1835); Stass, *Ann.* **30**, 192 (1839); Bridel, Kramer, *Bull. Soc. Chim. Biol.* **15**, 544 (1933). Hydrolysis by dil mineral acids yields phloretin and glucose. Procedure for acid hydrolysis: Wessely, Sturm, *Monatsh.* **53-54**, 557 (1929); Müller, Robertson, *J. Chem. Soc.* **1933**, 1170. The energy of activation required for the hydrolysis is much less than for other glucosides and approaches that for γ-fructosides (sucrose, raffinose). Is hydrolyzed by saccharase at pH 4.45, by enzymes of *Aspergillus niger* and other enzymes found in invertebrates: Kobert, *Pflüger's Arch. Ges. Physiol.* **99**, 116 (1903); Moelwyn-Hughes, *J. Gen. Physiol.* **13**, 807 (1930); *cf. Trans. Faraday Soc.* **25**, 81 (1929). Synthesis: Zemplen, Bognár, *Ber.* **75B**, 1040 (1942); *cf. ibid.* 645 and **76B**, 386 (1943).

Dihydrate, long needles from water, mp 110°. Sweet, with bitter aftertaste. $[\alpha]_D^{25}$ −52° (0.16 g in 5 ml of 96% alcohol). One gram dissolves in about one liter of water at 22°, in 64 ml at 60°, in 22 ml at 70°. Freely sol in boiling water; in about 4 parts alcohol, in methanol, amyl alcohol, acetone, ethyl acetate, pyridine, aniline, quinoline and other organic bases; in aq alkaline solns and in glacial acetic acid. Practically insol in ether, chloroform, benzene.

USE: Experimentally to produce glycosuria in animals.

7301. Phloroglucinol. *1,3,5-Benzenetriol;* 1,3,5-trihydroxybenzene; phloroglucin; Dilospan S; Spasfon-Lyoc. $C_6H_6O_3$; mol wt 126.11. C 57.14%, H 4.80%, O 38.06%. Prepn by reducing trinitrobenzoic acid or trinitrobenzene: Pascoe, *Chem. Products* **18**, 454 (1955); from TNT: Kastens, Kaplan, *Ind. Eng. Chem.* **42**, 402 (1950); from picryl chloride: Heertjes, *Rec. Trav. Chim.* **78**, 452 (1959); from m-isopropylresorcinyl diacetate: Zimmer, U.S. pat. 3,028,410 (1962 to Hooker Chem.); from tribromobenzene: McKillop *et al., Syn. Commun.* **4**, 35 (1974).

White rhombic crystals from water, mp 218°. Sweet taste. Sublimes with decompn. Discolors in light. Soluble in 100 parts water, 10 parts alcohol, 0.5 part pyridine; sol in ether. *Protect from light.* LD_{50} in mice (mg/kg): 4550 orally; 4050 i.p.; 5520 s.c., *RTECS* Vol. II, R. J. Lewis, R. L. Tatken, Eds. (1980) p 304.

USE: In diazo-type printing and textile dyeing. As reagent for pentoses and pentosans, aldehydes, lignin, HCl, methanol, chloral hydrate, turpentine oil, lignified cell tissue, free acid (HCl) in gastric juice; in microscopy, as an excellent decalcifier of bone specimens.

THERAP CAT: Antispasmodic.

7302. Phlorol. *2-Ethylphenol.* $C_8H_{10}O$; mol wt 122.16. C 78.65%, H 8.25%, O 13.10%. Prepn by heating ethylene and phenol with phosphoric acid at 200°: Ipatieff, *J. Am. Chem. Soc.* **60**, 1162 (1939); by heating ethanol + phenol over thoria + alumina at 350°: Hansch, Robertson, *ibid.* **72**, 4810 (1950); from phenol, ethylene chlorohydrin + sodium: Carlton, Bradbury, *ibid.* **78**, 1069 (1956). Purification: Biddiscombe *et al., J. Chem. Soc.* **1963**, 5764.

Colorless liq; phenol odor. Cooling the liq to −30° gives a metastable cryst form, mp about −28°; when kept for 24-48 hrs, this changes to the stable cryst form, mp −3.4°. bp_{760} 204.52°. d_{25} 1.01459. Practically insol in water. Freely sol in alcohol, benzene, glacial acetic acid.

7303. Pholcodine. *7,8-Didehydro-4,5-epoxy-17-methyl-3-[2-(4-morpholinyl)ethoxy]morphinan-6-ol;* 3-[2-(4-morpholinyl)ethyl]morphine; tetrahydro-1,4-oxazinylmethylcodeine; 3-(2-morpholinoethyl)morphine; β-morpholinylethylmorphine; homocodeine; Dia-Tuss; Ethnine; Galenphol; Galphol; Glycodine; Memine; Codylin; Hibernyl; Pectolin; Prodromine; Weifacodine. $C_{23}H_{30}N_2O_4$; mol wt 398.49. C 69.32%, H 7.59%, N 7.03%, O 16.06%. Prepn: Chabrier *et* *al.*, U.S. pat. **2,619,485** (1952 to Dausse). Description: *Ann. Pharm. Franc.* **8**, 261 (1950). Toxicity data: B. Kelentey *et al.*, *Arzneimittel-Forsch.* **8**, 325 (1958).

Monohydrate, crystals, mp 91°. Bitter taste. $[\alpha]_D^{20}$ −95.3° (c = 2 in ethanol). Slightly sol in water (2% w/v), ether. Sol in alc (1:3), chloroform, benzene. pH of 2% aq soln 9.5 to 9.8. LD_{50} s.c. in mice: 0.540 g/kg (Kelentey).

Caution: May be habit forming. This is a controlled substance (opium derivative) listed in the U.S. Code of Federal Regulations, Title 21 Part 1308.11 (1985).

THERAP CAT: Antitussive.

7304. Pholedrine. *4-[2-(Methylamino)propyl]phenol*; *p*-hydroxy-*N,α*-dimethylphenethylamine; *β*-(*p*-hydroxyphenyl)isopropylmethylamine; *α*-(*p*-hydroxyphenyl)-*β*-methylaminopropane; *p*-hydroxy-*N*-methylbenzedrine; Knoll H$_{75}$; Paredrinol; Pulsotyl; Veritol. $C_{10}H_{15}NO$; mol wt 165.23. C 72.69%, H 9.15%, N 8.48%, O 9.68%. Prepd from *p*-methoxybenzyl methyl ketone: **Fr. pat. 822,422, Brit. pat. 482,-414, Ger.** pat. **674,753** (1937, 1938 and 1939 to Knoll); Hildebrandt and Hildebrandt, Freese, **Ger.** pats. **665,793** and **767,161** (1938 and 1951 to Knoll). Synthesis: Savitskii, Makhnenko, *J. Gen. Chem. USSR* **10**, 1819 (1940); Buzas, Dufour, *Bull. Soc. Chim. France* **1950**, 139. Toxicity data: Lindner, *Arch. Exp. Pathol. Pharmacol.* **188**, 675 (1938).

Crystals from methanol, mp 162-163°. Acrid, burning taste. Alkaline reactions. Slightly sol in water. Sol in alcohol, ether; readily sol in dil acids.

Sulfate, $2C_{10}H_{15}NO.H_2SO_4$, crystals, dec 320-323°. Sol in water. LD s.c. in rats: 500 mg/kg (Lindner).

THERAP CAT: Adrenergic; vasopressor.

THERAP CAT (VET): Sympathomimetic. Circulatory stimulant.

7305. Phorate. *Phosphorodithioic acid O,O-diethyl S-[(ethylthio)methyl] ester*; *O,O*-diethyl *S*-(ethylthio)methyl phosphorodithioate; *O,O*-diethyl *S*-ethylmercaptomethyl dithiophosphate; American Cyanamid 3911; EI 3911; ENT 24042; Thimet. $C_7H_{17}O_2PS_3$; mol wt 260.40. C 32.29%, H 6.58%, O 12.29%, P 11.90%, S 36.94%. Prepn: Schrader, Lorenz, **U.S.** pat. **2,759,010** and **Brit.** pat. **797,307** (1956, 1958, both to Bayer). Metabolism: J. B. Bowman, J. E. Casida, *J. Agr. Food Chem.* **5**, 192 (1957). Persistence in soil: D. L. Suett, *Pestic. Sci.* **6**, 385 (1975).

Clear liquid, $bp_{0.1}$ 75-78°, $bp_{0.8}$ 118-120°, $bp_{2.0}$ 125-127°. d_4^{25} 1.156. n_D^{25} 1.5329. Vapor press. at 20°: 8.4×10^{-4} mm Hg. Soly: 50 ppm in water. Misc with xylene, carbon tetrachloride, dioxane, methyl Cellosolve, dibutyl phthalate, vegetable oils. Stable at room temp. Hydrolyzed in the presence of water and alkali. LD_{50} in female, male rats: 1.1,

2.3 mg/kg orally; 2.5, 6.2 mg/kg dermally, T. B. Gaines, *Toxicol. Appl. Pharmacol.* **14**, 515 (1969).

USE: Insecticide. *Caution:* Cholinesterase inhibitor.

7306. Phorbol. *1,1a,1b,4,4a,7a,7b,8,9,9a-Decahydro-4a,7b,9,9a-tetrahydroxy-3-(hydroxymethyl)-1,1,6,8-tetramethyl-5H-cyclopropa[3,4]benz[1,2-e]azulen-5-one.* $C_{20}H_{28}O_6$; mol wt 364.44. C 65.91%, H 7.74%, O 26.34%. Parent alcohol of the tumor promoting compounds in croton oil, *q.v.*, the oil expressed from the seeds of *Croton tiglium* L., Euphorbiaceae. Phorbol has a structural skeleton based on cyclopropabenzazulene. Isoln: Flaschenträger, Wigner, *Helv. Chim. Acta* **25**, 569 (1942); Kauffmann, Neumann, *Ber.* **92**, 1715 (1959); S. Tseng *et al.*, *J. Org. Chem.* **42**, 3645 (1977). Structure and stereochemistry: Hecker *et al.*, *Tetrahedron Letters* **1967**, 3165; Pettersen, Ferguson, *Chem. Commun.* **1967**, 716. Unlike its diesters, phorbol does not appear to be co-carcinogenic or enhance chemically-induced mutagenesis: C. J. Soper, F. J. Evans, *Cancer Res.* **37**, 2487 (1977). Mechanism of action study on phorbol esters: A. S. Kraft, W. B. Anderson, *Nature* **301**, 621 (1983). Comprehensive review of phorbol and its esters: Hecker, Schmidt, *Fortschr. Chem. Org. Naturst.* **31**, 377-467 (1974); P. M. Blumberg, *Crit. Rev. Toxicol.* **8**, 199-234 (1981).

Anhydr crystals, dec 250-251°. Two forms of solvated crystals from ethyl acetate: mp 162-163° and 233-234°. Solvated crystals from methanol or ethanol; mp 249-250°. $[\alpha]_D^{24}$ +102° (water). $[\alpha]_D^{20}$ +118° (c = 0.4 in dioxane). uv max (ethanol): 235, 334 nm (ε 5200, 70). Quite sol in polar solvents, including water.

12-Myristate 13-acetate diester, $C_{36}H_{56}O_8$, *12-O-tetradecanoylphorbol-13-acetate*, TPA, croton oil factor A_1. uv max (ethanol): 232, 333 nm (ε 5400, 73). Differentiation of human leukemia cells: J. B. Weinberg, *Science* **213**, 655 (1981).

Caution: Phorbol diesters are potent co-carcinogens.

7307. Phorone. *2,6-Dimethyl-2,5-heptadien-4-one*; diisopropylideneacetone. $C_9H_{14}O$; mol wt 138.20. C 78.21%, H 10.21%, O 11.58%. $(CH_3)_2C=CHCOCH=C(CH_3)_2$. Prepn from isobutenyllithium + CO_2: Braude, Timmons, *J. Chem. Soc.* **1950**, 2000; from acetone: Dolgov, Samsonova, *Zh. Obshch. Khim.* **22**, 632 (1952); Joseph, Blumenthal, *J. Org. Chem.* **24**, 1371 (1959); Tsmur, *Zh. Prikl. Khim.* **34**, 1628 (1961); M. Konieczny, G. Sosnovsky, *Z. Naturforsch.* **33B**, 454 (1978).

Yellow liquid or yellowish-green prisms, mp 28°. bp 198-199°; bp_{17} 88°. d_4^{20} 0.885. n_D^{21} 1.4968.

7308. Phosalone. *Phosphorodithioic acid S-[(6-chloro-2-oxo-3(2H)-benzoxazolyl)methyl] O,O-diethyl ester*; *phosphorodithioic acid, O,O-diethyl ester, S-ester with 6-chloro-3-(mercaptomethyl)-2-benzoxazolinone*; 3-(*O,O*-diethyldithiophosphorylmethyl)-6-chlorobenzoxazolinone; 6-chloro-3-(*O,O*-diethyldithiophosphorylmethyl)benzoxazolone; *S*-(6-chloro-2-oxobenzoxazolin-3-yl)methyl diethyl phosphorothiolothionate; RP 11974; Zolone. $C_{12}H_{15}ClNO_4PS_2$; mol wt 367.80. C 39.19%, H 4.11%, Cl 9.64%, N 3.81%, O 17.40%, P 8.42%, S 17.43%. Prepn: **Brit.** pat. **1,005,372** (1965 to Rhône-Poulenc). Toxicology, metabolism, insecticidal properties: Colinese, Terry, *Chem. & Ind. (London)* **1968**, 1507. Use as molluscicide: Cessac, **Fr.** pat. **1,453,989** (1966 to Rhône-Poulenc).

Crystals, mp 47.5-48°. Sol in ketones, alcohols and most aromatic solvents. Practically insol in water and aliphatic hydrocarbons. LD$_{50}$ orally in mice, rats: 180-205, 120-175 mg/kg.

USE: Insecticide, acaricide.

7309. Phosfolan. *1,3-Dithiolan-2-ylidenephosphoramidic acid diethyl ester; phosphonodithioimidocarbonic acid cyclic ethylene P,P-diethyl ester;* diethyl 1,3-dithiolan-2-ylidenephosphoramidate; 2-(diethoxyphosphinylimino)-1,3-dithiolane; pholan; AC 47031; CL 47031; EI 47031; ENT 25830; Cyolane; Cylan. C$_7$H$_{14}$NO$_3$PS$_2$; mol wt 255.28. C 32.93%, H 5.53%, N 5.49%, O 18.80%, P 12.13%, S 25.12%. Prepn: R. W. Addor, J. B. Lovell, **Belg. pat.** *618,-* *155*, *C.A.* **59**, 10066h (1963) and R. W. Addor, **U.S. pat.** *3,197,481* (1962 and 1965 to Am. Cyanamid). Activity: D. L. Bull *et al., J. Econ. Entomol.* **57**, 112 (1964); R. L. Ridgway *et al., ibid.* **59**, 149 (1966). Fate in soil: G. Tantawy *et al., Alexandria J. Agr. Res.* **22**, 315 (1974). Metabolism: I. P. Kapoor, R. C. Blinn, *J. Agr. Food Chem.* **25**, 413 (1977).

Solid, mp 36.5°, bp$_{0.001}$ 115-118°. Sol in water, acetone, benzene, ethanol, cyclohexane, toluene. Slightly sol in ether; sparingly sol in hexane. LD$_{50}$ orally in rats: 29 mg/kg, *RTECS* Vol. I, R. J. Lewis, R. L. Tatken, Eds. (1979) p 789.

USE: Insecticide.

7310. Phosgene. *Carbonic dichloride;* carbonyl chloride; chloroformyl chloride. CCl$_2$O; mol wt 98.92. C 12.14%, O 16.17%, Cl 71.69%. Cl$_2$C=O. Prepn from chlorine + carbon monoxide: Whitehouse, **U.S. pat.** *1,231,226* (1917); Peacock, **U.S. pat.** *1,360,312* (1921); Bradner, **U.S. pat.** *1,457,493* (1923); Douthitt, **U.S. pat.** *2,847,470* (1958 to Texas Co.); from carbon monoxide + nitrosyl chloride: Williams, **U.S. pat.** *1,746,506* (1930 to du Pont Ammonia Corp.); from carbon tetrachloride + oleum: Murphy, Reuter, *Aust. Chem. Inst. J. & Proc.* **15**, 144 (1948). Toxicology: S. A. Cucinell, *Arch. Environ. Health* **28**, 272 (1974). Review: E. E. Hardy in Kirk-Othmer *Encyclopedia of Chemical Technology* vol. 17 (Wiley-Interscience, New York, 3rd ed., 1982) pp 416-425.

Colorless, highly toxic gas; suffocating odor; when much diluted with air there is an odor reminiscent of moldy hay. Condenses at about 0° to a clear, colorless, fuming liquid. d$_4^0$ 1.432. mp −118°. bp$_{760}$ 8.2°. Vapor press at 20°: 1215 mm. Slightly sol in water and slowly hydrolyzed by it; freely sol in benzene, toluene, glacial acetic acid and most liq hydrocarbons.

Human Toxicity: Insidious poison as it is not irritating immediately, even when fatal concns are inhaled. May cause severe pulmonary edema (may be quickly fatal) or pneumonia. Inhalation of high concns causes choking, constricted feeling in chest, coughing, painful breathing, bloody sputum. Vapors strongly irritating to the eyes, *cf.* Patty's *Industrial Hygiene and Toxicology* vol. 2C, G. D. Clayton, F. E. Clayton, Eds. (Wiley-Interscience, New York, 3rd ed., 1982) pp 4126-4128.

USE: For the prepn of many organic chemicals; as a war gas.

Warning: Paper soaked in alcoholic or carbon tetrachloride soln contg 10% of a mixture of equal parts of *p*-dimethylaminobenzaldehyde and colorless diphenylamine, then dried, will turn from yellow to deep orange in the presence of approx the max allowable concn of phosgene, and should always be used where the generation of this gas is possible or suspected.

7311. Phosmet. *Phosphorodithioic acid S-[(1,3-dihydro-1,3-dioxo-2H-isoindol-2-yl)methyl] O,O-dimethyl ester; phosphorodithioic acid O,O-dimethyl ester S-ester with N-(mercaptomethyl)phthalimide; O,O-dimethyl S-phthalimidomethyl phosphorothionate; N-(mercaptomethyl)phthalimide S-(O,O-dimethyl phosphorodithioate);* phthalophos (USSR); ENT 25705; R-1504; Imidan; Prolate. C$_{11}$H$_{12}$NO$_4$PS$_2$; mol wt 317.32. C 41.64%, H 3.81%, N 4.41%, O 20.17%, P 9.76%, S 20.21%. Prepn: Fancher, **U.S. pat.** *2,767,194* (1956 to Stauffer). Activity: B. A. Butt, J. C. Keller, *J. Econ. Entomol.* **54**, 813 (1961).

Off-white, cryst solid, mp 71.9°. Tech product (95-98% pure), mp 66.5-69.5°. Dec below its boiling point. Vapor press. at 50°: 1 × 10^{-3} mm Hg. Soly in water at 25°: 25 ppm. LD$_{50}$ orally in male, female rats: 113, 160 mg/kg, T. B. Gaines, *Toxicol. Appl. Pharmacol.* **14**, 515 (1969).

USE: Insecticide, acaricide.

7312. Phosphamidon. *Phosphoric acid 2-chloro-3-(diethylamino)-1-methyl-3-oxo-1-propenyl dimethyl ester; phosphoric acid dimethyl ester, ester with 2-chloro-N,N-diethyl-3-hydroxycrotonamide;* 2-chloro-2-diethylcarbamoyl-1-methylvinyl dimethyl phosphate; Ciba 570; ENT 25515; Dimecron. C$_{10}$H$_{19}$ClNO$_5$P; mol wt 299.69. C 40.08%, H 6.39%, Cl 11.83%, N 4.67%, O 26.69%, P 10.33%. Prepn: Beriger, Sallmann, **U.S. pat.** *2,908,605* (1959 to Ciba); Anliker *et al., Helv. Chim. Acta* **44**, 1622 (1961). Toxicity data: T. B. Gaines, *Toxicol. Appl. Pharmacol.* **14**, 515 (1969).

Oil. d$_4^{25}$ 1.2132. n$_D^{25}$ 1.4718. bp$_{1.5}$ 162°; bp$_{0.001}$ 120°. mp −45°. Vapor pressure at 20°: 2.5 × 10^{-5} mm Hg. Absorption spectra: Anliker *et al., loc. cit.* Misc with water and most organic solvents except saturated hydrocarbons. One gram dissolves in about 30 g hexane. Stable in neutral or acid media; hydrolyzed by alkali. LD$_{50}$ orally in rats: 24 mg/kg (Gaines).

USE: Insecticide. *Caution:* Cholinesterase inhibitor. *See* Parathion.

7313. Phosphine. H$_3$P; mol wt 34.00. H 8.89%, P 91.11%. PH$_3$. Formed in small quantity in the putrefaction of organic matter contg phosphorus. Prepd from white phosphorus and aq alkali hydroxide; also by treatment of PH$_4$I with KOH: Klement in *Handbook of Preparative Inorganic Chemistry* vol. 1, G. Brauer, Ed. (Academic Press, New York, 2nd ed., 1963) pp 525-530; by pyrolysis of phosphorous acid: Gokhale, Jolly, *Inorg. Syn.* **9**, 56 (1967); by hydrolysis of a metal phosphide such as calcium phosphide: Klement, *loc. cit.;* Baudler *et al., Z. Anorg. Allgem. Chem.* **353**, 122 (1967).

Gas. *Poisonous!* Odor of decaying fish; bp −87.7°. mp −133°. Spontaneously flammable in air if there is a trace of P$_2$H$_4$ present; burns with a luminous flame. Slightly sol in water (0.26 vol. at 20°). Combines violently with oxygen and the halogens. Liberates hydrogen and forms the phosphide when passed over heated metal. Forms phosphonium salts when brought in contact with the halogen acids. Lowest published lethal concn for hamsters (inhalation): 8 ppm, *Toxic Substances List*, H. E. Christensen, Ed. (1973) p 737.

Human Toxicity: Pain in region of diaphragm, a feeling of coldness. Weakness, vertigo, dyspnea, bronchitis, edema, lung damage, convulsions, coma, death: L. T. Fairhall, *Industrial Toxicology* (Hafner, New York, 2nd ed., 1969) pp 91-92.

7314. Phosphinothricin. *2-Amino-4-(hydroxymethylphosphinyl)butanoic acid;* (3-amino-3-carboxypropyl)methylphosphinic acid; homoalanin-4-yl(methyl)phosphinic acid; 2-amino-4-methylphosphinobutyric acid; glufosinate. $C_5H_{12}NO_4P$; mol wt 181.13. C 33.16%, H 6.68%, N 7.73%, O 35.33%, P 17.10%. First reported naturally occurring amino acid containing a phosphinic acid group. Glutamine synthetase inhibitor with herbicidal activity. Active component of antibiotic peptides produced by several *Streptomyces spp.* The naturally occurring form is the L-isomer. Isoln from *S. viridochromogenes* and synthesis of DL-form: E. Bayer *et al., Helv. Chim. Acta* **55**, 224 (1972). Prepn by hydrolysis of *SF-1293*, also known as *bialaphos*, an antibiotic produced by *S. hydroscopicus:* Y. Ogawa *et al., Japan.* **Kokai 73 85,538** (1973 to Meiji Seika Kaisha), *C.A.* **80**, 60035b (1974); *eidem, Meiji Seika Kenkyo Nempo* **13**, 42 (1973), *C.A.* **81**, 37806r (1974). Alternate synthesis: H. Gross, T. Gnauk, *J. Prakt. Chem.* **318**, 157 (1976). Crystal structure and absolute configuration of L-phosphinothricin: E. F. Paulus, S. Grabley, *Z. Kristallogr.* **160**, 63 (1982). Biosynthetic studies: H. Seto *et al., J. Antibiot.* **35**, 1719 (1982); *eidem, ibid.* **36**, 96 (1983). Asymmetric synthesis of enantiomers: N. Minowa *et al., Tetrahedron Letters* **25**, 1147 (1984). Activity and use as herbicide: W. Rupp *et al.,* **Belg. pat. 854,753;** U.S. pat. **4,168,963** (1977, 1979 both to Hoechst AG); P. Langelüddeke *et al., Meded. Fac. Landbouww. Rijksuniv. Ghent* **47**, 95 (1982), *C.A.* **98**, 48584v (1982). Comparative phytotoxicity: K. L. Carlson, O. C. Burnside, *Weed Sci.* **32**, 841 (1984). Mechanism of action study: R. R. Bellinger *et al., ibid.* **33**, 779 (1985).

$$\underset{\underset{OH}{|}\quad\underset{NH_2}{|}}{H_3CPCH_2CH_2CHCOOH}$$
$$\overset{O}{\overset{\|}{}}$$

L-Form, $[\alpha]_D^{25}$ +13.4° (c = 1 in water).
D-Form, $[\alpha]_D^{25}$ −12.4° (c = 1 in water).
DL-Form, mp 229-231° (dec).
DL-Form hydrochloride, $C_5H_{13}ClNO_4P$, crystals from water + ethanol, mp 195-198°.
DL-Form monoammonium salt, $C_5H_{15}N_2O_4P$, *glufosinate-ammonium, Hoe 661, Hoe 39866, Basta, Total.* Sol in water. LD_{50} in male, female mice, male, female rats (mg/kg): 431, 416, 2000, 1620 orally (Langelüddeke).
USE: DL-Form monoammonium salt as nonselective postemergent herbicide.

7315. Phosphocreatine. *N-[Imino(phosphonoamino)methyl]-N-methylglycine; N-(phosphonoamidino)sarcosine;* creatine phosphate; creatinephosphoric acid; PC. $C_4H_{10}N_3O_5P$; mol wt 211.11. C 22.76%, H 4.77%, N 19.90%, O 37.89%, P 14.67%. The distribution of phosphocreatine (PC) in various tissues of a number of vertebrates has been determined; skeletal muscle contains much more than other tissues. Although PC is characteristic of the vertebrates, it also occurs in certain invertebrates. Isoln from frog muscle: Eggleton, *Biochem. J.* **21**, 190 (1927); from cat muscle: Fiske, Subbarow, *J. Biol. Chem.* **81**, 629 (1929). Synthesis by phosphorylation of creatine: Zeile, Fawaz, *Z. Physiol. Chem.* **256**, 193 (1938); Ennor, Stocken, *Biochem. J.* **43**, 190 (1948). Enzymic determn in muscle: K. Yoshikawa *et al., Anal. Biochem.* **159**, 303 (1986). Clinical trial as cardioprotective: M. L. Semenovsky *et al., J. Thorac. Cardiovasc. Surg.* **94**, 762 (1987); as anti-arrhythmic: M. Y. Ruda *et al., Am. Heart J.* **116**, 393 (1988). *Review:* Ennor, Morrison, *Physiol. Rev.* **38**, 631 (1958).

$$\underset{\underset{HO}{}}{\overset{HO}{}}\underset{\underset{NH}{\|}}{\overset{O}{\overset{\|}{P}}}-NHC\underset{}{N}CH_2COOH$$

Sodium salt hexahydrate, $C_4H_8N_3Na_2O_5P\cdot6H_2O$, *Creatergyl, Neoton.* Platelets from water + ethanol. Very sol in water. (Free acid, pKa_2 is 4.6).
Calcium salt tetrahydrate, $C_4H_8CaN_3O_5P\cdot4H_2O$, hygroscopic crystals. Soluble in water; sparingly sol in alcohol.
Note: The term *phosphagen,* originally a synonym for phosphocreatine, has since been used to describe any naturally occurring phosphorylated guanidine compd.
THERAP CAT: Cardioprotective.

7316. Phosphomolybdic Acid. *Molybdophosphoric acid;* dodecamolybdophosphoric acid. Formula approx $24MoO_3\cdot P_2O_5\cdot xH_2O$. Prepn: Wu, *J. Biol. Chem.* **43**, 189 (1920); Hastings, Frediani, *Anal. Chem.* **20**, 382 (1948). Formula also reported to be $20MoO_3\cdot P_2O_5\cdot51H_2O$: *U.S.P.* **XXI**, 1398. Review of phosphomolybdic acids: *Mellor's,* Vol. **XI**, pp 659-672 (1931).
Bright yellow crystals. Sol in less than 0.4 part water; very sol in alcohol, ether.
USE: Weighting silks; as reagent for alkaloids, uric acid, xanthine, creatinine, some metals, with hematoxylin as nerve stain in microscopy.

7317. Phosphonium Iodide. H_4IP; mol wt 161.93. P 19.13%, H 2.49%, I 78.38%. PH_4I. Prepd by hydrolysis of a mixture of diphosphorus tetraiodide and white phosphorus: Work, *Inorg. Syn.* **2**, 141 (1946). Improved apparatus for its prepn: Beredjick, *ibid.* **6**, 91 (1960).
Large, transparent, colorless crystals (usually cubes). Tetragonal system. Sublimes at room temp. *Store in sealed ampuls in refrigerator!* Vapor pressure: 50 mm at 20°; 760 mm at 62.5°. mp 18.5° under its own vapor pressure. Heat of fusion 12,680 cal/mol. *Caution:* Heat or traces of moisture or alcohol cause decompn into PH_3 and HI. Will detonate if heated rapidly.
USE: In the laboratory prepn of phosphine.

7318. Phosphoric Acid. Orthophosphoric acid. H_3O_4P; mol wt 98.00. H 3.09%, O 65.31%, P 31.61%. H_3PO_4. Obtained commercially from phosphate rock deposits in Florida, Tennessee, and the Western United States. Phosphate rock is essentially tricalcium phosphate and one of the large scale processes is based on the equation: $Ca_3(PO_4)_2 + 3H_2SO_4 + 6H_2O \rightarrow 2H_3PO_4 + 3(CaSO_4\cdot2H_2O)$. Description of various processes: W. H. Waggaman, *Phosphoric Acid, Phosphates and Phosphatic Fertilizers* (Reinhold, New York, 1952); *Phosphoric Acid,* Vol. **1**, parts I, II, A. V. Slack, Ed. (Dekker, New York, 1968) 1159 pp; Faith, Keyes & Clark's *Industrial Chemicals,* F. A. Lowenheim, M. K. Moran, Eds. (Wiley-Interscience, New York, 4th ed., 1975) pp 628-639. Prepn of ultrapure, cryst H_3PO_4: Simon, Schulze, *Z. Anorg. Allgem. Chem.* **242**, 322 (1939); Weber, King, *Inorg. Syn.* **1**, 101 (1939). *Reviews:* J. R. Van Wazer, *Phosphorus and its Compounds,* Vol. **1**, Chemistry (Interscience, New York, 1958) pp 479-491; R. B. Hudson *et al.,* "Phosphoric Acids and Phosphates" in Kirk-Othmer *Encyclopedia of Chemical Technology,* Vol. **17** (Wiley-Interscience, New York, 3rd ed., 1982) pp 426-472.

$$\underset{\underset{HO}{}}{\overset{HO}{}}\overset{O}{\overset{\|}{P}}-OH$$

Unstable, orthorhombic crystals, mp 42.35°, or clear, syrupy liquid; easily supercooled into a glass. Pleasing acid taste when suitably diluted. An acid containing about 88% H_3PO_4 will frequently crystallize on prolonged cooling; forms hemihydrate, mp 29.32°. Becomes anhydr at 150°; gradually changes to pyrophosphoric acid at about 200°, and changes to metaphosphoric acid when heated above 300°. The hot concd acid attacks porcelain and granite ware. May be stored in suitable stainless steel containers. Tribasic acid: $K_1 = 7.107 \times 10^{-3}$; $K_2 = 7.99 \times 10^{-8}$; $K_3 = 4.8 \times 10^{-13}$. Other reported values of dissociation consts discussed by Van Wazer, *loc. cit.* The pH of a 0.1N aq soln is 1.5. Heat of formation (cryst): −306.2 kcal/mole. Heat of soln: +2.79 kcal/mole. Sol in water, alc; sol in 8 vols of a 3:1 ether:alcohol mixture. Properties of phosphoric acid solns. d^{25} 1.8741 (100% soln); 1.6850 (85% soln); 1.3334 (50% soln); 1.0523 (10% soln). Density measurements: Christensen, Reed, *Ind. Eng. Chem.* **47**, 1277 (1950); Egan, Luff, *ibid.* 1280. $n_D^{17.5}$ 1.34203 (10% soln); 1.35032 (20% soln); 1.35846 (30% soln). Spec heat (21.3°): 0.4359 (88% soln). Viscosity data: Van Wazer, *loc. cit.*

Human Toxicity: Concd solns are irritating to skin, mucous membranes.

USE: In the manuf of superphosphates for fertilizers, other phosphate salts, polyphosphates, detergents. Acid catalyst in making ethylene, purifying hydrogen peroxide. As acidulant and flavor, synergistic antioxidant and sequestrant in food. Pharmaceutic aid (solvent). In dental cements; process engraving; rustproofing of metals before painting; coagulating rubber latex; as analytical reagent.

THERAP CAT (VET): Has been used in lead poisoning.

7319. Phosphoric Acid, Meta. *Metaphosphoric acid;* glacial phosphoric acid. $(HPO_3)_n$; H 1.26%, O 60.01%, P 38.73%.

Transparent, glass-like solid or soft silky masses; hygroscopic. Volatilizes at red heat. Very slowly sol in cold water, slowly changing to H_3PO_4, the change is hastened by boiling; sol in alcohol. *Keep tightly closed.* For greater convenience it is also marketed in form of rods made by the addition of sodium phosphate.

USE: In dentistry for making zinc oxyphosphate cement; as reagent in chemical analysis.

7320. Phosphorous Acid. *Phosphonic acid.* H_3O_3P; mol wt 82.00. H 3.69%, O 58.54%, P 37.78%. Prepd by hydrolysis of PCl_3 according to the equation $PCl_3 + 3H_2O \rightarrow H_3PO_3 + 3HCl$, the rather violent reaction can be slowed down by the initial presence of concd HCl: Milobedzki, Friedmann, *Chem. Pol.* **15**, 76 (1917); *Chem. Z.* **1918**, I, 933; Simon, Fehér, *Z. Anorg. Allgem. Chem.* **230**, 298 (1937). Alternate procedure carrying out the reaction in carbon tetrachloride: Voight, Gallais, *Inorg. Syn.* **4**, 55 (1953). *Review:* Ohashi, "Lower Oxo Acids of Phosphorus and Their Salts" in *Topics in Phosphorus Chemistry,* Vol. 1, M. Grayson, E. J. Griffith, Eds. (Interscience, New York, 1964) pp 113-187.

$$\underset{HO}{\overset{HO\quad O}{\diagdown}}P-H$$

White, very hygroscopic and deliquesc, crystalline mass; garlic-like taste; slowly oxidized by oxygen (air) to H_3PO_4. d_4^{21} 1.65; d_4^{76} liq 1.597. mp about 73°; above 180° is dec into PH_3 and H_3PO_4. Very sol in water, alcohol. pK_1 1.29; pK_2 6.74. Usually marketed as a 20% aq soln.

7321. Phosphorus. P; at. wt 30.97376; at. no. 15; valences 3, 5. One naturally occurring isotope: ^{31}P; artificial, radioactive isotopes: 28-30; 32-34. Abundance in earth's crust: about 0.12%. Does not occur free in nature; found in the form of phosphates in the minerals *chlorapatite* [$3Ca_2$-$(PO_4)_2 \cdot CaCl_2$], *fluorapatite* [$3Ca(PO_4)_2 \cdot CaF_2$], vivianite, wavellite and "phosphate rock" or phosphorite; occurs in small quantities in granite rocks; occurs in all fertile soil; an essential constituent of protoplasm, nervous tissue and bones. Discovered in 1669 by Brandt. Prepn: Ullmann, *Enzyklopädie der Technischen Chemie* **8**, 362 (1931); DeWitt, Skolnik, *J. Am. Chem. Soc.* **68**, 2305 (1946); Skolnik *et al.*, *ibid.* 2310. Lab prepn and purification: Klement in *Handbook of Preparative Inorganic Chemistry* vol. 1, G. Brauer, Ed. (Academic Press, New York, 2nd ed., 1963) pp 518-525. *Reviews:* J. R. Van Waser, *Phosphorus and Its Compounds,* 2 vols. (Interscience, New York, 1958, 1961) 2046 pp; Corbridge, "The Structural Chemistry of Phosphorus Compounds" in *Topics in Phosphorus Chemistry,* Vol. 3, E. J. Griffith, M. Grayson, Eds. (Interscience, New York, 1966) pp 57-394; Toy, "Phosphorus" in *Comprehensive Inorganic Chemistry,* Vol. 2, J. C. Bailar, Jr. *et al.*, Eds. (Pergamon Press, Oxford, 1973) pp 389-545; J. R. Van Wazer in Kirk-Othmer *Encyclopedia of Chemical Technology* vol. 17 (Wiley-Interscience, New York, 3rd ed., 1982) pp 473-490.

Phosphorus exists in three main allotropic forms: white, black, and red. The same liquid is obtained on melting. Phosphorus atoms exist as symmetrical, tetrahedral P_4 molecules in the liquid phase and in the vapor phase below 800°; molecules dissociate to P_2 above 800°.

White phosphorus. Colorless or white, transparent, cryst solid; waxy appearance; darkens on exposure to light.

Sometimes called *yellow phosphorus;* color due to impurities. Two allotropic modifications: α-form exists at room temp; cubic crystals containing P_4 molecules; d 1.83. β-Form prepd by conversion of α-form at $-79.6°$; hexagonal crystals; d 1.88. mp 44.1° (vapor press. 0.181 mm); bp 280°. Volatile; sublimes *in vacuo* at ordinary temp when exposed to light. When exposed to air in the dark, emits a greenish light and gives off white fumes. Solubilities in water: one part/300,000 parts water; in abs alc: one g/400 ml; in abs ether: one g/102 ml; in $CHCl_3$: one g/40 ml; in benzene: one g/35 ml; in CS_2: one g/0.8 ml. Soly in oils: one gram phosphorus dissolves in 80 ml olive oil, 60 ml oil of turpentine, about 100 ml almond oil. Ignites at about 30° in moist air; the ignition temp is higher when the air is dry. *Caution: handle with forceps. Keep under water.* The fumes and the element itself are poisonous. Combines directly with the halogens to form tri- or pentahalides; combines with sulfur to form sulfides. Reacts with several metals to form phosphides. Yields orthophosphoric acid when treated with nitric acid. Reacts with alkali hydroxides with formation of phosphine and sodium hypophosphite. *Incompat.* Sulfur, iodine, oil of turpentine, potassium chlorate.

Black Phosphorus. Polymorphic. Orthorhombic crystalline form: stable in air; resembles graphite in texture; produced from the white modification under high pressures: Bridgman, *J. Am. Chem. Soc.* **36**, 1344 (1914); Jacobs, *J. Chem. Phys.* **5**, 945 (1937); Krebs, *Inorg. Syn.* **7**, 60 (1963). d 2.691. Does not catch fire spontaneously. Insol in organic solvents. Amorphous form prepd at lower pressures: Jacobs, *loc. cit.* At higher pressure the orthorhombic form undergoes reversible transition to a rhombohedral structure, d 3.56, and a cubic structure, d 3.83: Jamieson, *Science* **139**, 1291 (1963).

Red phosphorus. Red to violet powder; polymorphism: Roth *et al.*, *J. Am. Chem. Soc.* **69**, 2881 (1947); Corbridge, *loc. cit.* Crystal structure of one form, *Hittorf's phosphorus:* Thurn, Krebs, *Acta Cryst.* **25B**, 125 (1969). The properties of red phosphorus are intermediate between those of the white and black forms. Sublimes at 416°, triple point 589.5° under 43.1 atm. d 2.34. Insol in organic solvents. Sol in phosphorus tribromide. Less active than the white form; reacts only at high temp. Yields the white modification when distilled at 290°. Catches fire when heated in air to about 260° and burns with formation of the pentoxide. Burns when heated in an atmosphere of chlorine.

Caution: Avoid contact with $KClO_3$, $KMnO_4$, peroxides and other oxidizing agents; explosions may result on contact or friction.

Human Toxicity: Ingestion of even small amounts of white phosphorus may produce severe G.I. irritation, bloody diarrhea, liver damage, skin eruptions, oliguria, circulatory collapse, coma, convulsions, death. The approx fatal dose is 50 to 100 mg. External contact may cause severe burns. Chronic poisoning (from ingestion or inhalation) is characterized by bony necrosis, especially of the mandible, spontaneous fractures, anemia, weight loss. Red phosphorus is relatively nontoxic unless it contains the white form as an impurity: *Clinical Toxicology of Commercial Products,* M. N. Gleason *et al.*, Eds. (Williams & Wilkins, Baltimore, 1969) Section III, pp 192-195.

USE: White phosphorus: manuf rat poisons; for smoke screens, gas analysis. Red phosphorus: pyrotechnics; manuf safety matches; in organic synthesis; manuf phosphoric acid, phosphine, phosphoric anhydride, phosphorus pentachloride, phosphorus trichloride; manuf fertilizers, pesticides, incendiary shells, smoke bombs, tracer bullets.

7322. Phosphorus Hemitriselenide. P_2Se_3; mol wt 298.84. P 20.73%, Se 79.27%. Prepd by warming a mixture of phosphorus and selenium in the correct proportions: Hahn, *J. Prakt. Chem.* [1] **93**, 430 (1864); Muthmann, Clever, *Z. Anorg. Chem.* **13**, 191 (1897); Meyer, *ibid.* **30**, 258 (1902); **61B**, 1807 (1928).

Dark red mass. Dec by heat. Dec in moist air and water. Sol in potassium hydroxide; insol in carbon disulfide.

7323. Phosphorus Oxybromide. *Phosphoryl bromide;* phosphoryl tribromide. Br_3OP; mol wt 286.72. Br 83.62%, O 5.58%, P 10.80%. $POBr_3$. Prepd according to the equation: $3PBr_5 + P_2O_5 \rightarrow 5POBr_3$: Hönigschmid, Hirschbold-

Wittner, *Z. Anorg. Allgem. Chem.* **243**, 355 (1940); Johnson, Nunn, *J. Am. Chem. Soc.* **63**, 141 (1941); Booth, Seegmiller, *Inorg. Syn.* **2**, 151 (1946).

Thin plates, faint orange tint, d 2.822. mp 56° (in hot water). bp_{758} 193° (dec). Slowly hydrolyzes in water forming H_3PO_4 and HBr. Sol in ether, benzene, chloroform, carbon disulfide, concd H_2SO_4. *Store in sealed glass ampuls!*

7324. Phosphorus Oxychloride.
Phosphoryl chloride; phosphorus chloride. Cl_3OP; mol wt 153.35. Cl 69.36%, O 10.43%, P 20.20%. $POCl_3$. Manuf: Faith, Keyes & Clark's *Industrial Chemicals*, F. A. Lowenheim, M. K. Moran, Eds. (Wiley-Interscience, New York, 4th ed., 1975) pp 646-649.

Colorless, clear, strongly fuming liquid; pungent odor. d^{25} 1.645. bp 105.8°; mp 1.25°: J. R. Van Wazer, *Phosphorus and its Compounds—Chemistry*, vol. 1 (Interscience, New York, 1958) p 254. Reacts exothermically with water, alc. *Keep in tightly closed containers.*

USE: As chlorinating agent, especially to replace oxygen in organic compounds; as solvent in cryoscopy. *Caution:* Intensely irritating to skin, eyes, mucous membranes. Inhalation may cause pulmonary edema.

7325. Phosphorus Pentabromide.
Phosphoric bromide; phosphorus perbromide. Br_5P; mol wt 430.56. Br 92.80%, P 7.20%. PBr_5. *Review:* Payne, "The Chemistry of Phosphorus Halides" in *Topics in Phosphorus Chemistry, Vol. 4*, M. Grayson, E. J. Griffith, Eds. (Interscience, New York, 1967) pp 85-155.

Yellow, cryst mass. mp above 100° with decompn. Dec by water or alcohol; sol in carbon disulfide or tetrachloride. *Keep tightly closed. Caution:* corrosive.

USE: Brominating agent for converting organic acids to acyl bromides.

7326. Phosphorus Pentachloride.
Phosphoric chloride; phosphorus perchloride. Cl_5P; mol wt 208.27. Cl 85.12%, P 14.88%. PCl_5. Prepn: Maxson, *Inorg. Syn.* **1**, 99 (1939). *Review:* Payne, "The Chemiistry of Phosphorus Halides" in *Topics in Phosphorus Chemistry, Vol. 4*, M. Grayson, E. J. Griffith, Eds. (Interscience, New York, 1967) pp 85-155.

White to pale yellow, fuming, deliquesc, cryst mass; pungent, unpleasant odor; attacks the eyes and mucous membranes. mp 148° under pressure; sublimes at about 100° without melting. bp 160°. Hydrolyzed by water to form phosphoric acid and hydrogen chloride. Reacts with alcohols (ROH) to form the corresponding chloride (RCl). Sol in carbon disulfide or tetrachloride. *Keep in tightly closed containers and handle with caution. Corrosive.*

USE: As catalyst in manuf acetylcellulose; for replacing hydroxyl groups by Cl, particularly for converting acids into acid chlorides.

7327. Phosphorus Pentafluoride.
F_5P; mol wt 125.98. F 75.41%, P 24.59%. PF_5. Prepd by treating PF_3 with bromine to form PF_3Br which then disproportionates to PF_5 and PBr_5: Moissan, *Compt Rend.* **100**, 1348; **101**, 1490 (1885); by heating P_2O_5 with CaF_2: Lucas, Ewing, *J. Am. Chem. Soc.* **49**, 1270 (1927); Booth, Bozarth, *ibid.* **55**, 3890 (1933); from PCl_5 and AsF_3: Thorpe, *Ann.* **182**, 201 (1876); *Proc. Roy. Soc.* **25**, 122 (1877); O. Ruff, *Die Chemie des Fluors* (Berlin, 1920), p 29; Kwasnik in *Handbook of Preparative Inorganic Chemistry*, vol. 1, G. Brauer, Ed. (Academic Press, New York, 2nd ed., 1963) pp 190-191; from PCl_3 and CaF_2: Muetterties *et al., J. Inorg. Nucl. Chem.* **16**, 52 (1960); from phosphoryl fluoride, hydrogen fluoride and sulfur trioxide: Wiesboeck, U.S. pats. **3,584,999; 3,592,594** (1971 to U.S. Steel). *Reviews:* Burg in *Fluorine Chemistry*, vol. 1, J. H. Simons, Ed. (Academic Press, New York, 1950) pp 97-98; Kemmitt, Sharp, *Advan. Fluorine Chem.* **4**, 197-198 (1965); Schmutzler, *ibid.* **5**, 32-285 (1965).

Colorless gas. Fumes strongly in air. d (gas) 5.805 g/l. mp −93.8°. bp −84.6°. Trouton constant 21.8. Dipole moment: zero. High thermal stability. Does not attack dry glass even at 250°, but a slight trace of moisture leads to formation of POF_3 and HF. Water hydrolysis ultimately yields phosphoric acid; intermediates are oxyfluophosphates. Lewis acid; forms complexes with amines, ethers, nitrates, sulfoxides, organic bases. Forms a cryst addn product PF_5.-NO_2 at −10° which dissociates on warming. May be stored in steel cylinders.

Caution: Intensely irritating to skin, eyes, mucous membranes. Inhalation may cause pulmonary edema.

USE: Catalyst in ionic polymerization reactions.

7328. Phosphorus Pentaselenide.
P_2Se_5; mol wt 456.76. P 13.57%, Se 86.43%. Prepd by melting a mixture of phosphorus and selenium in an atm of CO_2 or N_2: Carius, Bogen, *Ann.* **124**, 57 (1862); Kudchadker *et al., Can. J. Chem.* **46**, 1415 (1968).

Amorphous, glass, black-purple solid. Dec in steam and boiling water. Reacts with CCl_4; insol in carbon disulfide.

7329. Phosphorus Pentasulfide.
Phosphoric sulfide; thiophosphoric anhydride; phosphorus persulfide. P_2S_5; mol wt 222.29. P 27.87%, S 72.13%. Exists as P_4S_{10}. Conveniently prepd by fusing red phosphorus with sulfur: Stock, Herscovici, *Ber.* **43**, 1223 (1910); Klement in *Handbook of Preparative Inorganic Chemistry*, vol. 1, G. Brauer, Ed. (Academic Press, New York, 2nd ed., 1963) p 567. Manuf: Faith, Keyes & Clark's *Industrial Chemicals*, F. A. Lowenheim, M. K. Moran, Eds. (Wiley-Interscience, New York, 4th ed., 1975) pp 650-653.

Light yellow, triclinic crystals; peculiar odor. d 2.09. mp 286-290°; bp 513-515°. Dec by water forming H_3PO_4 and H_2S; sol in carbon disulfide, in aq soln of alkali hydroxides. *Keep tightly closed.*

USE: In manuf of lube oil additives and pesticides. Manuf safety matches, ignition compds, and for introducing sulfur into organic compds.

7330. Phosphorus Pentoxide.
Phosphoric anhydride; diphosphorus pentoxide. O_5P_2; mol wt 141.96. O 56.36%, P 43.64%. P_2O_5. Exists as P_4O_{10}. Prepd commercially by burning phosphorus in a current of dry air. Purification: Manley, *J. Chem. Soc.* **121**, 331 (1922); de Decker, McGillavry, *Rec. Trav. Chim.* **60**, 153, 413 (1941). Review of phosphorus oxides: J. R. Van Waser, *Phosphorus and its Compounds—Chemistry*, vol. 1 (Interscience, New York, 1958) pp 267-286.

Very deliquescent crystals. *Corrosive.* Several cryst and amorphous modifications. Commercial form; hexagonal; d 2.30. mp 340°. Sublimation temp 360°. Nonflammable. Does not support combustion. Heat of formation: −365.83 kcal/mole. Sp heat: 0.170 cal/g/°C. Heat of fusion: 8.2 kcal/mole. Heat of volatilization: 22.7 kcal/mole of P_4O_{10}. Readily absorbs moisture from the air. Exothermic hydrolysis by water to form phosphoric acid. The reaction with alcohol is similar.

USE: Drying and dehydrating agent. Condensing agent in organic synthesis. *Caution:* Strong irritant; corrosive to skin, mucous membranes and eyes.

7331. Phosphorus Sulfochloride.
Thiophosphoryl chloride. Cl_3PS; mol wt 169.41. Cl 62.79%, P 18.28%, S 18.92%. $PSCl_3$. Prepd from P_2S_5 + PCl_5: Martin, Duvall, *Inorg. Syn.* **4**, 73 (1953). Alternate procedure from PCl_3, $AlCl_3$ and S: Moeller *et al., ibid.* 71.

Fuming liq, crystallizes as α-form at −40.8° or as β-form at −36.2°. bp_{760} 125°. n_D^{25} 1.635. Sol in benzene, carbon tetrachloride, carbon disulfide, chloroform. Hydrolyzes slowly in water, rapidly in alkaline solns. In water the hydrolysis products are orthophosphoric acid, hydrochloric acid, and hydrogen sulfide.

Caution: Strong irritant.

7332. Phosphorus Tribromide.
Phosphorous bromide. Br_3P; mol wt 270.73. Br 88.56%, P 11.44%. PBr_3. Prepn: Gray, Maxson, *Inorg. Syn.* **2**, 147 (1946). *Review:* Payne, "The Chemistry of Phosphorus Halides" in *Topics in Phosphorus Chemistry*, vol. 4, M. Grayson, E. J. Griffith, Eds. (Interscience, New York, 1967) pp 85-155.

Colorless, fuming liquid; very penetrating odor. d^{15} 2.85. mp −41.5°. bp 173.2°. Vapor pressure: 10 mm (48°). Dissolved and dec by water or alc; sol in acetone, carbon disulfide. *Keep tightly closed. Caution:* Corrosive.

7333. Phosphorus Trichloride.
Phosphorous chloride. Cl_3P; mol wt 137.35. Cl 77.45%, P 22.55%. PCl_3. Prepd from red phosphorus and dry chlorine in the presence of refluxing PCl_3: Forbes *et al., Inorg. Syn.* **2**, 145 (1946). Manuf: Faith, Keyes & Clark's *Industrial Chemicals*, F. A. Lowenheim, M. K. Moran, Eds. (Wiley-Interscience, New

York, 4th ed., 1975) pp 654-657. *Review:* Payne, "Chemistry of Phosphorus Halides" in *Topics in Phosphorus Chemistry,* vol. 4, M. Grayson, E. J. Griffith, Eds. (Interscience, New York, 1967) pp 85-155.

Colorless, clear, fuming liquid. d_4^{21} 1.574. mp —112°. bp 76°. Vapor pressure: 100 mm (21°). Decomposed by water or alc. Sol in benzene, chloroform, ether, carbon disulfide. *Keep in tightly closed containers and handle with caution.*

USE: As of phosphorus oxychloride; manuf $POCl_3$, PCl_5; producing iridescent metallic deposits. *Caution:* Highly irritating and corrosive to skin, mucous membranes.

7334. Phosphorus Trifluoride. F_3P; mol wt 87.98. F 64.79%, P 35.21%. PF_3. Prepd by halogen exchange between PCl_3 and AsF_3: Moissan, *Compt. Rend.* **100,** 272 (1885); Hoffman, *Inorg. Syn.* **4,** 149 (1953); between PCl_3 and ZnF_2: Williams, *ibid.* **5,** 95 (1957); between PCl_3 and CaF_2 or SbF_3: Booth, Bozarth, *J. Am. Chem. Soc.* **61,** 2927 (1939); Muetterties *et al., J. Inorg. Nucl. Chem.* **16,** 52 (1960); between PCl_3 and HF: Kwasnik in *Handbook of Preparative Inorganic Chemistry,* vol. 1, G. Brauer, Ed. (Academic Press, New York, 2nd ed., 1963) pp 189-190. *Reviews:* Burg in *Fluorine Chemistry,* vol. 1, J. H. Simons, Ed. (Academic Press, New York, 1950) pp 98-100; Kemmitt, Sharp, *Advan. Fluorine Chem.* **4,** 198-199 (1965); Schmutzler, *ibid.* **5,** 32-285 (1965).

Colorless gas. Does not fume in air. Does not attack glass except at high temps. *Poisonous!* d (gas) 3.907 g/l. mp —151.30°. bp —101.38°. Critical temp —2.05°; critical press. 42.69 atm. May be stored in steel cylinders or in glass, also in a gasometer over Hg. Slowly hydrolyzed by water. Absorbed by aq bases at a rate increasing with pH, producing a fluophosphite which boiling aq HNO_3 does not convert to phosphate. Anhydr KOH may be used to dry PF_3 with little loss. Dry NH_3 forms a solid addn product. Aq oxidizers such as chromic acid, permanganate, or bromine rapidly destroy PF_3, and alcohols convert it to alkyl phosphite. Phosphides and fluorides are formed upon reaction with hot metals.

Caution: See Phosphorus Pentafluoride.

7335. Phosphorus Trioxide. Diphosphorus trioxide. O_3P_2; mol wt 109.95. O 43.66%, P 56.34%. P_2O_3. Exists as P_4O_6. Prepd by treating PCl_3 with tetramethylammonium sulfite in liq SO_2: Jander *et al., Ber.* **77,** 689 (1944). Alternate prepn from the elements: Thorpe, Tutton, *J. Chem. Soc.* **57,** 545 (1890); Miller, *ibid.* **1928,** 1847; **1929,** 1823; Wolf, Schmager, *Ber.* **62,** 779 (1929). Chemistry: Riess, Van Wazer, *Inorg. Chem.* **5,** 178 (1966).

Transparent monoclinic crystals or colorless liquid. *Very poisonous!* d_4^4 2.135. mp 23.8°. bp 173.1° in nitrogen atm. Disproportionates into red P and P_2O_4 when heated above 210°. Sol in benzene, carbon disulfide. When placed in cold water, H_3PO_3 is formed slowly. Hot water produces a violent reaction with the formation of red phosphorus, phosphine, and H_3PO_4.

7336. Phosphorus Triselenide. Tetraphosphorus triselenide. P_4Se_3; mol wt 360.80. P 34.35%, Se 65.65%. Prepn: Meyer, *Z. Anorg. Chem.* **30,** 258 (1902); **61B,** 1807 (1928); Irgolic *et al., Inorg. Chem.* **4,** 1421 (1965).

Orange-red crystals. Irritating odor. mp 245-246°. d 1.31. bp 360-400°. Phosphoresces at 160°. Flammable when heated in air. Dec in moist air. Sol in carbon tetrachloride, carbon disulfide, chloroform, benzene, toluene, acetone, acetylene dichloride, acetylene trichloride.

7337. Phosphorylcholine. *N,N,N-Trimethyl-2-(phosphonooxy)ethanaminium chloride; choline chloride dihydrogen phosphate;* (2-hydroxyethyl)trimethylammonium chloride phosphate; phosphorylcholine chloride; choline phosphate chloride; choline chloride phosphate; choline phosphoric acid ester (chloride). $C_5H_{15}ClNO_4P$; mol wt 219.61. C 27.34%, H 6.88%, Cl 16.15%, N 6.38%, O 29.14%, P 14.10%. $[(CH_3)_3{}^+NCH_2CH_2OPO_3H_2]Cl^-$. Prepd by phosphorylation of choline chloride with diphenylphosphoryl chloride: Baer, McArthur, *J. Biol. Chem.* **154,** 451 (1944); Baer, *J. Am. Chem. Soc.* **69,** 1253 (1947); *Biochem. Prepn.* **2,** 96 (1952); by heating choline or choline chloride with pyrophosphoric or polyphosphoric acid: Cherbuliez, Rabinowitz, *Helv. Chim. Acta* **42,** 1154 (1959).

Barium salt, $C_5H_{13}BaClNO_4P$, hydrated glistening leaflets from water + alcohol. Must be dried *in vacuo* (0.5 mm) over P_2O_5 at 100° for 16 hrs. Freely sol in water, practically insol in abs ethanol, ether.

Potassium salt, prepn by Baer, McArthur, *loc. cit.*

Calcium salt, $C_5H_{13}CaClNO_4P$, *Colifos, Epafosforil, Fosfocolina, Isocolin.* Prepn: Cherbuliez, *loc. cit.*

Magnesium salt, *Heparexine.*

THERAP CAT: Hepatobiliary dysfunction.

7338. Phosphoserine. *Serine dihydrogen phosphate (ester);* serine phosphate; *O*-phospho-DL-serine. $C_3H_8NO_6P$; mol wt 185.08. C 19.47%, H 4.35%, N 7.57%, O 51.87%, P 16.74%. Prepn: Neuhaus, Korkes, *Biochem. Prepn.* **6,** 75 (1958); D. M. Theodoropoulos, *J. Chem. Soc.* **1960,** 5257. Other methods of prepn and isoln from protein hydrolysates: Greenstein, Winitz, *Chemistry of the Amino Acids* (Wiley, New York, 1961), *passim.* Crystal structure: G. H. McCallum, *Nature* **184,** 1863 (1959).

Crystals from ethanol + ether, mp 166-167° (dec).

Note: Used in combination with L-glutamine and vitamin B_{12} as a roborant: *Vitasprint B_{12}, Fosforina B_{12}.*

7339. Phosphotungstic Acid. *Tungstophosphoric acid.* Approx composition $24WO_3.2H_3PO_4.48H_2O$. The H_2O content may vary appreciably. Prepn: Bailar, *Inorg. Syn.* **1,** 132 (1939).

White or slightly yellowish-green, slightly efflorescent crystals or cryst powder. Sol in about 0.5 part water. Also sol in alcohol and in ether.

USE: As reagent for alkaloids and many other nitrogen bases, for phenols, albumin, peptone, amino acids, uric acid, urea, blood, carbohydrates; as biological stain.

7340. Phosvitin. A phosphoprotein of mol wt about 40,000. Isoln from egg yolk: Mecham, Olcott, *J. Am. Chem. Soc.* **71,** 3670 (1949); *Biochem. Prepn.* **2,** 15 (1952); Joubert, Cook, *Can. J. Biochem. Physiol.* **36,** 399 (1958); from roe: Mano, Yoshida, *J. Biochem. (Tokyo)* **66,** 105 (1969). Represents about 7% of the total yolk protein; high in phosphorus content and a ratio of serine to all other amino acid residues of 6:4. *In vitro* synthesis in hens liver: Heald, McLachlan, *Biochem. J.* **94,** 32 (1965); Rudack, Wallace, *Biochim. Biophys. Acta* **155,** 299 (1968). Estradiol-induced synthesis in roosters: Veuving, Gruber, *ibid.* **232,** 524, 529 (1971); Jailkhani, Talwar, *Nature New Biol.* **236,** 239 (1972). Amino acid sequence: Belitz, *Angew. Chem.* **76,** 574 (1964); Shainkin, Perlmann, *J. Biol. Chem.* **246,** 2278 (1971). Conformation: Grizzuti, Perlmann, *ibid.* **245,** 2573 (1970). Anticlotting activity studies: Sato *et al., Am. J. Physiol.* **203,** 1170 (1962). *Review:* Taborsky, *Advan. Protein Chem.* **28,** 50-78 (1974).

THERAP CAT: Anticoagulant.

7341. Phoxim. *4-Ethoxy-7-phenyl-3,5-dioxa-6-aza-4-phosphaoct-6-ene-8-nitrile 4-sulfide; phenylglyoxylonitrile oxime O,O-diethyl phosphorothioate; O,O*-diethyl *O*-(α-cyanobenzylideneamino)phosphorothioate; α-[[(diethoxyphosphinothioyl)oxy]imino]benzeneacetonitrile; Bay 5621; Bay 77488; Baythion; Sebacil; Volaton. $C_{12}H_{15}N_2O_3PS$; mol wt 298.29. C 48.32%, H 5.07%, N 9.39%, O 16.09%, P 10.38%, S 10.75%. Prepn: Neth. pat. Appl. 6,605,907 corresp to W. Lorenz *et al., U.S. pat. 3,591,662* (1966,1971 to Bayer). Activity: C. R. Harris, *J. Econ. Entomol.* **63,** 782 (1970); D. C. Read, *ibid.* **69,** 429 (1976). Toxicity studies: J. H. Vinopal, T. R. Fukuto, *Pestic. Biochem. Physiol.* **1,** 44 (1971). Degradation in soil: G. Dräger, *Pflanzenschutz-Nachr.* **30,** 28 (1977).

Pale yellow oil, bp$_{0.01}$ 102°. n_D^{20} 1.5405. d_4^{20} 1.176. Sol in alc, ketones, aromatic hydrocarbons. LD$_{50}$ orally in mice: >2000 mg/kg, J. H. Vinopal,T. R. Fukuto, *loc. cit.*

USE: Insecticide.

7342. Phrenosin. *N-[1-[(β-D-Galactopyranosyloxy)meth-yl]-2-hydroxy-3-heptadecenyl]-2-hydroxytetracosanamide;* Cerebron. C$_{48}$H$_{93}$NO$_9$; mol wt 828.23. C 69.60%, H 11.32%, N 1.69%, O 17.39%. Cerebroside easily separated from a commercially available beef spinal cord lipid concentrate or from nerve tissue. A large proportion of its fatty acid content (25%) is 2-hydroxystearic acid, the rest is cerebronic (2-hydroxylignoceric) acid. Isoln: Radin *et al., J. Biol. Chem.* **219**, 977 (1956); Skipski *et al., Arch. Biochem. Biophys.* **82**, 487 (1959). Synthesis: Shapiro, Flowers, *J. Am. Chem. Soc.* **83**, 3327 (1961).

Crystals from pyridine + acetone, methanol + chloroform, toluene + alcohol (1:1) or chloroform + alcohol + water, [α] +4.5° (c = 2 in pyridine). Absorption spectrum: Schwarz *et al., Ann. N.Y. Acad. Sci.* **69**, 118 (1957). Soluble in hot dioxane, 1,1-dichloro-1-nitroethane, butanol and acetonitrile + chloroform.

7343. Phthalamide. *1,2-Benzenedicarboxamide;* phthalic acid diamide. C$_8$H$_8$N$_2$O$_2$; mol wt 164.16. C 58.53%, H 4.91%, N 17.07%, O 19.49%. Prepd from phthalic anhydride and anhydr ammonia in benzene: Dominikiewicz, *Arch. Chem. Farm.* **3**, 141 (1937).

Minute crystals, mp 228° (commercial grades mp 221-223°). Melts with decomposition to NH$_3$ and phthalimide. Slightly sol in cold water, methanol. More sol in hot solvents. Heat of combustion 921.7 cal/mol. Boiling of aq or alcoholic solns produces NH$_3$ and phthalimide. The presence of acids accelerates decompn.

7344. Phthalazine. 2,3-Benzodiazine; benzo[d]pyridazine; β-phenodiazine. C$_8$H$_6$N$_2$; mol wt 130.14. C 73.83%, H 4.65%, N 21.53%. Prepd by reduction of 1-chlorophthalazine with hydriodic acid and red phosphorus: Gabriel, Eschenbach, *Ber.* **30**, 3024 (1897); Paul, *Ber.* **32**, 2015 (1899); by redn of 1-hydrazinophthalazine: Armarego, *J. Appl. Chem.* **11**, 70 (1961); from o-phthalaldehyde + hydrazine sulfate: Smith, Otremba, *J. Org. Chem.* **27**, 879 (1962); from 1,2-benzodinitrile: Carter, Cheeseman, *Org. Prep. Proced. Int.* **6**, 67 (1974).

Pale yellow needles from ether, mp 90-91°. bp 315-317° (dec); bp$_{29}$ 189°; bp$_{17}$ 175°. uv max (water): 218, 261, 292, 305 nm (log ε 4.83, 3.53, 3.18, 3.11). Freely sol in water; sol in ethanol, methanol, benzene, ethyl acetate; less sol in ether. Practically insol in ligroin.

Hydrochloride, C$_8$H$_6$N$_2$.HCl, needles from ether, mp 231° (efferv).

7345. Phthalic Acid. *1,2-Benzenedicarboxylic acid.* C$_8$-H$_6$O$_4$; mol wt 166.13. C 57.83%, H 3.64%, O 38.52%. Manufactured by catalytic oxidation of o-toluic acid and oxidation of xylene: Taylor, Dean, U.S. pat. **3,064,046** (1962 to I.C.I.); Cier, U.S. pat. **3,088,974** (1963 to Esso Res. & Eng.). Isoln from the fungus, *Gibberella fujikuroi:* Cross *et al., J. Chem. Soc.* **1963**, 2937.

Crystals, mp about 230° when rapidly heated, forming phthalic anhydride and water. One gram dissolves in 160 ml water, 10 ml alcohol, 205 ml ether, 5.3 ml methanol; practically insol in chloroform. LD$_{50}$ orally in rats: 7.9 g/kg, C. B. Shaffer *et al., J. Ind. Hyg. Toxicol.* **27**, 130 (1945).

Ethyl ester, C$_{12}$H$_{14}$O$_4$, *ethyl phthalate, diethyl phthalate, Neantine, Palatinol A.* Colorless, practically odorless, oily liq; bitter disagreeable taste. d$_4^{14}$ 1.232. bp 295°. Flash pt 140°. n_D^{14} 1.5049. Insol in water; misc with alcohol, ether and many other organic solvents. LD$_{50}$ i.p. in rats: 5.06 ml/kg, A. R. Singh *et al., J. Pharm. Sci.* **61**, 59 (1972).

USE: Ethyl phthalate used in manuf celluloid; solvent for cellulose acetate in manuf varnishes and dopes; fixative for perfumes; denaturing alc. *Caution:* Irritating to mucous membranes, and, in high concns, narcotic.

7346. Phthalic Anhydride. *1,3-Isobenzofurandione.* C$_8$-H$_4$O$_3$; mol wt 148.11. C 64.87%, H 2.72%, O 32.41%. Prepd from naphthalene by oxidation with a mixture of HgSO$_4$ and CuSO$_4$ in presence of H$_2$SO$_4$; by passing naphthalene and oxygen over a suitable catalyst at 400-500°. Review of mfg processes: Faith, Keyes & Clark's *Industrial Chemicals,* F. A. Lowenheim, M. K. Moran, Eds (Wiley-Interscience, New York, 4th ed., 1975) pp 658-665.

White, lustrous needles. d 1.53; mp 130.8°; bp 295°. Sublimes. Sol in 162 parts water, more in hot water with conversion into phthalic acid, in 125 parts carbon disulfide; sol in alcohol, sparingly in ether.

USE: Manuf phthaleins, phthalates, benzoic acid, synthetic indigo, artificial resins (glyptal).

7347. Phthalimide. *1H-Isoindole-1,3(2H)-dione.* C$_8$H$_5$-NO$_2$; mol wt 147.13. C 65.30%, H 3.43%, N 9.52%, O 21.75%. Prepd from phthalic anhydride and NH$_4$OH or (NH$_4$)$_2$CO$_3$: W. A. Noyes, P. K. Porter, *Org. Syn.* coll. vol. I, 457 (2nd ed., 1941). Teratogenicity study: K. Fickentscher *et al., Pharmazie* **31**, 172 (1976).

Monoclinic prisms from water or by sublimation. mp 238°. Has slightly acidic properties, Ka = 5 × 10^{-9}. Absorption spectrum: Hartley, Hedley, *J. Chem. Soc.* **91**, 317 (1907). Slightly soluble in water. A satd aq soln contains 0.036 g at 25°; 0.07 g at 40°; about 0.4 g at the bp. 100 g of boiling alcohol dissolves 5 g phthalimide. Almost insol in benzene, petr ether; fairly sol in boiling acetic acid; freely sol in aq alkali hydroxides.

7348. Phthalofyne. *1,2-Benzenedicarboxylic acid mono-(1-ethyl-1-methyl-2-propynyl) ester; phthalic acid 1-ethyl-1-methyl-2-propynyl ester;* 1-ethyl-1-methyl-2-propynyl acid phthalate; 3-methyl-1-pentyn-3-yl acid phthalate; ftalofyne; NSC-25614; Whipcide. $C_{14}H_{14}O_4$; mol wt 246.25. C 68.28%, H 5.73%, O 25.99%. Prepn: Sugimoto, Okumura, **Japan.** pat. **1833('54)** (Tanabe), *C.A.* **49**, 11711e (1955); **Brit.** pat. **736,993** (1955 to Schering).

Crystals from benzene or hexane, mp 96-98°. Weak acid. Slightly sol in water. Unstable in strong alkali.
THERAP CAT (VET): Anthelmintic.

7349. Phthaloyl Chloride. *1,2-Benzenedicarbonyl dichloride.* $C_8H_4Cl_2O_2$; mol wt 203.02. C 47.33%, H 1.99%, Cl 34.93%, O 15.76%. Obtained by the action of PCl_5 on phthalic anhydride.

Colorless, oily liquid. d^{20} 1.409. Solidif +12°; mp 15-16°; bp 280-282°; n_D^{20} 1.5692. Dec by water or alcohol; sol in ether. *Keep tightly closed.*

7350. Phthalylsulfacetamide. *2-[[[4-[(Acetylamino)sulfonyl]phenyl]amino]carbonyl]benzoic acid; 4'-(acetylsulfamyl)phthalanilic acid;* phthalylsulfacetimide; N^1-acetyl-N^4-phthaloylsulfanilamide; N^1-acetyl-N^4-phthalylsulfanilamide; phthaloylsulfacetimide; ftalicetimida; N-[p-(o-carboxybenzamido)benzenesulfonyl]acetamide; N-(o-carboxybenzoyl)-sulfacetamide; Enterocid; Sulfacyl; Enterosulfamid; Enterosulfon; Talsutin; Talecid; Talicetimida; Thalocid; Thalamyd; Tamid; Rabalan; Sterathal. $C_{16}H_{14}N_2O_6S$; mol wt 362.36. C 53.03%, H 3.89%, N 7.73%, O 26.49%, S 8.85%. Prepd by refluxing N^1-acetylsulfanilamide and phthalic anhydride in alcohol: Basu, *J. Indian Chem. Soc.* **26**, 130 (1949).

Needles from dil alcohol, mp 196°. Soluble in alcohol. Very sparingly sol in water. Forms a water-sol sodium salt. *Note:* The name Sulfacyl is also applied to N^1-acetylsulfanilamide, *q.v.*
THERAP CAT: Antibacterial.
THERAP CAT (VET): Antibacterial.

7351. Phthalylsulfathiazole. *2-[[[4-[(2-Thiazolylamino)sulfonyl]phenyl]amino]carbonyl]benzoic acid; 4'-(2-thiazolylsulfamyl)phthalanilic acid;* 2-(N^4-phthalylaminobenzenesulfonamido)thiazole; 2-(N^4-phthalylsulfanilamido)thiazole; phthalylsulfonazole; Sulfathalidine; Enteramida; Entexidin; Entexidina; AFI-Ftalyl; Entero-Sulfina; Ftalazol; Intestiazol; Sulfacetil; Sulftalyl; Taleudron; Talidine; Thalistatyl; Ultratiazol; Thalazole. $C_{17}H_{13}N_3O_5S_2$; mol wt 403.43. C 50.61%, H 3.25%, N 10.42%, O 19.83%, S 15.90%. May be prepd by condensing sulfathiazole with phthalic anhydride: M. L. Moore, **U.S.** pats. **2,324,013; 2,324,015** (1943 to Sharp & Dohme). Prepn of 8-hydroxyquinoline salt: **Ger.** pat. **1,008,296** (1957 to Geistlich Söhne). Toxicity data: P. A. Mattis *et al., J. Pharmacol. Exp. Ther.* **81**, 116 (1944).

Crystals. Slightly bitter taste. Darkens on prolonged exposure to light. Effervesces at 244° to 250°. mp 272-277° (dec) when the melting point bath is preheated to 220-225°. Slightly sol in alcohol; very slightly sol in ether; readily sol in NaOH or KOH soln, ammonia water and concd HCl. Practically insol in chloroform and water. LD_{50} i.p. in mice: 920 mg/kg (Mattis).
8-Hydroxyquinoline salt, *Ilentazol, Colitiazolo.* Yellow crystals from dil methanol, dec above 220°. Split into its components by dil acids.
THERAP CAT: Antibacterial.
THERAP CAT (VET): Antibacterial.

7352. Phthiocol. *2-Hydroxy-3-methyl-1,4-naphthalenedione; 2-hydroxy-3-methyl-1,4-naphthoquinone.* $C_{11}H_8O_3$; mol wt 188.18. C 70.21%, H 4.28%, O 25.51%. Antibiotic substance produced by *Mycobacterium tuberculosis:* Terni, *Boll. Soc. Ital. Biol. Sper.* **25**, 60 (1949), *C.A.* **45**, 2054f (1951). Possesses some vitamin K activity. Isoln and synthesis: Anderson, Newman, *J. Biol. Chem.* **101**, 773 (1933); **103**, 197, 405 (1933); L. F. Fieser, *ibid.* **133**, 391 (1940); Tarbell *et al., J. Am. Chem. Soc.* **72**, 379 (1950); Burton, Praill, *J. Chem. Soc.* **1952**, 755; Eistert, Müller, *Ber.* **92**, 2071 (1959); H. Kallmayer, *Arch. Pharm.* **307**, 806 (1974); K. Maruyama, S. Arakawa, *J. Org. Chem.* **42**, 3793 (1977).

Yellow prisms from ether-petr ether, mp 173-174°. Sublimes. Volatile with steam. Slightly sol in water; sol in the usual organic solvents except petr ether. Forms deep red water-sol salts. E_0^{alc} 0.299 volts.
Acetate, $C_{13}H_{10}O_4$, pale yellow crystals, mp 101-102°.

7353. Phycobiliproteins. Deeply colored, highly fluorescent photoreceptor pigments found in blue-green, red and cryptomonad algae that contain a linear tetrapyrrole as the prosthetic group. They are composed of a bile pigment or *phycobilin* and an apoprotein. Phycobiliproteins are classified according to uv-vis absorption maxima as *phycocyanins* (blue pigment), *phycoerythrins* (red pigment), and *allophycocyanins* (pale blue pigment). Phycocyanins and phycoerythrins occur as large mol wt aggregates called *phycobilisomes* that are attached to the photosynthetic membranes. They are closely linked to the chlorophyll containing system for efficient energy transfer. *Phytochrome* is a similar biliprotein that functions in plant photomorphogenesis. It is widely distributed in plants but only in trace amounts. It exists in two forms that are interconverted upon alternate exposure to red and far-red light. *Reviews:* H. W. Siegelman *et al., Biochem. Soc. Symp.* **28**, 107-120 (1968); P. O'Carra, C. O'hEocha in *Chemistry and Biochemistry of Plant Pigments* **vol. 1,** T. W. Goodwin, Ed. (Academic, New York, 2nd ed., 1976) pp 328-376; A. Bennett, H. W. Siegelman in *Porphyrins* **vol. 6,** D. Dolphin, Ed. (Academic, New York, 1979) pp 493-520.

7354. Physalaemin. *Physalemin.* $C_{58}H_{84}N_{14}O_{16}S$; mol wt 1265.49. C 55.05%, H 6.69%, N 15.50%, O 20.23%, S 2.53%. An undecapeptide belonging to the group of proteins named tachykinins. Found in skin of the amphibian *Physalaemus fuscumaculatus:* Erspamer *et al., Experientia* **18**, 562 (1962). Structure: Erspamer *et al., ibid.* **20**, 489 (1964); Anastasi *et al., Arch. Biochem. Biophys.* **108**, 341 (1964). Synthesis: Bernardi *et al., Experientia* **20**, 490 (1964); Nakamura *et al.*, **Japan.** pat. **25,384('71)** (to Dainippon), *C.A.* **75**, 152083r (1971). Solid-phase synthesis: W. Voelter *et al., Tetrahedron* **28**, 5963 (1972). Biological activities similar to the tachykinins, eledoisin and substance P, *q.q.v.* Exerts a powerful hypotensive action, stimulates salivary secretion, intestinal contraction, and vasodilation. Occurrence in other *Physalaemus* spp. and pharmacology: G. Bertaccini *et al., Brit. J. Pharmacol.* **25**, 363 (1965); G. Bertaccini, *Pharmacol. Rev.* **28**, 127 (1976). Differentiation of physalaemin and substance P: Geipert *et al., Arch. Pharmacol.* **265**, 225

(1969). Immunoreactivity study in human lung small-cell carcinoma: L. H. Lazarus *et al.*, *Science* **219**, 79 (1983).

```
5-oxo-Pro-Ala-Asp-Pro-Asn-Lys-Phe-Tyr-Gly-Leu-Met-NH₂
```

Trifluoroacetate dihydrate, $C_{58}H_{84}N_{14}O_{16}S.CF_3COOH$. - $2H_2O$, dec 180°. $[\alpha]_D^{20}$ −56° (c = 0.2 in ethanol). uv max: 278 nm (ϵ 1780). Slowly loses activity when incubated in blood. Inactivated by liver and kidney homogenates.

Hydrochloride trihydrate, $C_{58}H_{84}N_{14}O_{16}S.HCl.3H_2O$, dec about 185°. $[\alpha]_D^{25}$ −43° (c = 1 in 95% acetic acid).

7355. Physodic Acid. *3,8-Dihydroxy-11-oxo-1-(2-oxo-heptyl)-6-pentyl-11H-dibenzo[b,e][1,4]dioxepin-7-carboxylic acid; 4,4′,6′-trihydroxy-6-(2-oxoheptyl)-2′-pentyl-2,3′-oxydibenzoic acid-ϵ-lactone; physodalin.* $C_{26}H_{30}O_8$; mol wt 470.50. C 66.37%, H 6.43%, O 27.20%. From the lichens *Parmelia physodes* (L.) Ach.: Hesse, *Ber.* **30**, 1983 (1897); Klosa, *Pharm. Ind.* **15**, 46 (1953); and *Cetraria ciliaris* Ach.: Culberson, *Science* **143**, 255 (1964). Structure: Asahina, Nogami, *Ber.* **67**, 805 (1934); **68**, 77 (1935).

Needles from methanol, mp 205°. uv max (95% ethanol): 256 nm (log ϵ 4.2). Sol in ether, acetone, hot methanol; somewhat sol in cold ethanol, methanol, hot chloroform; practically insol in benzene, petr ether, carbon disulfide.

Diacetate, $C_{30}H_{34}O_{10}$, plates from acetone + carbon disulfide, mp 153-155.5°.

Methyl ester, $C_{27}H_{32}O_8$, prisms from 80% alcohol, mp 156-157°.

7356. Physostigma. Calabar bean; ordeal bean; chop nut; split nut. Dried ripe seed of *Physostigma venenosum* Balf., *Leguminosae. Poisonous! Habit.* West Africa (near mouths of Niger and Old Calabar Rivers); introduced into India and Brazil. *Constit.* 0.15-0.3% alkaloids consisting of physostigmine (eserine), physovenine, eseridine, eseramine, phytosterol.

7357. Physostigmine. *(3aS-cis)-1,2,3,3a,8,8a-Hexahydro-1,3a,8-trimethylpyrrolo[2,3-b]indol-5-ol methylcarbamate (ester);* Eserine; Physostol. $C_{15}H_{21}N_3O_2$; mol wt 275.34. C 65.43%, H 7.69%, N 15.26%, O 11.62%. From Calabar beans, the seeds of the vine *Physostigma venenosum* Balf., *Leguminosae:* Jobst, Hesse, *Ann.* **129**, 115 (1864); Hesse, *ibid.* **141**, 82 (1867). Extraction procedure: Schwyzer, *Die Fabrikation pharmazeutischer und chemisch-technischer Produkte* (Berlin, 1931) p 338; Chemnitius, *J. Prakt. Chem.* **116**, 59 (1927). Structure: Stedman, Barger, *J. Chem. Soc.* **127**, 247 (1925). Abs config: R. B. Longmore, B. Robinson, *Chem. & Ind. (London)* **1969**, 622. Crystal structure: Petcher, Pauling, *Nature* **241**, 277 (1973). Synthesis: Julian, Pikl, *J. Am. Chem. Soc.* **57**, 755 (1935); Harley-Mason, Jackson, *J. Chem. Soc.* **1954**, 3651; J. Wijnberg, W. N. Speckamp, *Tetrahedron* **34**, 2399 (1978). Cholinesterase inhibitor: Engelhart, Loewi, *Arch. Exp. Pathol. Pharmakol.* **150**, 1 (1930); Matthes, *J. Physiol.* **70**, 338 (1930). Toxicity data: W. T. Lynch, J. M. Coon, *Toxicol. Appl. Pharmacol.* **21**, 53 (1972); R D. Sofia, L. C. Knabloch, *ibid.* **28**, 227 (1974). Improvement of long-term memory: K. L. Davis *et al.*, *Science* **201**, 272 (1978). Memory enhancement in Alzheimer's disease: L. J. Thal *et al.*, *N. Engl. J. Med.* **308**, 720 (1983); *eidem, Ann. Neurol.* **13**, 491 (1983); L. Gustafson *et al., Psychopharmacol.* **93**, 31 (1987). Review: Robinson in *The Alkaloids* Vol. X, R. H. F. Manske, Ed. (Academic Press, New York, 1968) pp 383-388.

Orthorhombic sphenoidal prisms or clusters of leaflets from ether or benzene. mp 105-106° (also an unstable, low-melting form, mp 86-87°). $[\alpha]_D^{17}$ −76° (c = 1.3 in chloroform); $[\alpha]_D^{25}$ −120° (benzene). K_1 7.6 × 10^{-7}; K_2 5.7 × 10^{-13}. Slightly sol in water; sol in alc, benzene, chloroform, oils. Solid and solns turn red on exposure to heat, light, air, and on contact with traces of metals. Under certain conditions the oxidation may proceed to yield eserine blue, $C_{17}H_{23}N_3O_2$. LD_{50} orally in mice: 4.5 mg/kg (Lynch, Coon).

Salicylate, $C_{22}H_{27}N_3O_5$, *Antilirium.* Acicular crystals, mp 185-187°. One gram dissolves in 75 ml water at 25° (pH of 0.5% aq soln 5.8); in 16 ml water at 80°; in 16 ml alcohol, 5 ml boiling alcohol, in 6 ml chloroform, 250 ml ether. *Solns should be kept well closed in light-resistant, alkali-free glass containers and used within a week.* Turns red and loses effectiveness on exposure to heat, light, air. LD_{50} i.p. in mice: 0.64 mg/kg (Sofia, Knabloch).

Sulfate, $C_{30}H_{44}N_6O_8S$, deliquescent scales, mp 140° (after drying at 100°). One gram dissolves in 0.4 ml alcohol, 4 ml water (pH of 0.05M soln 4.7), 1200 ml ether. The solns are more prone to change color than those of the salicylate.

Sulfite, $C_{30}H_{44}N_6O_7S$, white powder, freely sol in water, alcohol. The aq soln is stated to remain colorless for a long time.

THERAP CAT: Cholinergic (anticholinesterase); miotic.

THERAP CAT (VET): Cholinergic; miotic. In atony of gastrointestinal tract.

7358. Physovenine. *3,3a,8,8a-Tetrahydro-3a,8-dimethyl-2H-furo[2,3-b]indol-5-ol methylcarbamate (ester).* $C_{14}H_{18}N_2O_3$; mol wt 262.30. C 64.10%, H 6.92%, N 10.68%, O 18.30%. From seeds of *Physostigma venenosum* Balf., *Leguminosae:* Salway, *J. Chem. Soc.* **99**, 2148 (1911). Structure: Robinson, *ibid.* **1964**, 1503. Synthesis: Longmore, Robinson, *Chem. & Ind. (London)* **1965**, 1297. Abs config: *eidem, ibid.* **1969**, 622.

Plates from ether, mp 124-125°. $[\alpha]_D^{22.5}$ −92° (ethanol). uv max (ethanol): 252, 310 nm (ϵ 13,200, 3300). Sol in alcohol, benzene, chloroform; slightly sol in ether. Practically insol in water, petr ether.

7359. Phytic Acid. *myo-Inositol hexakis(dihydrogen phosphate);* inositolhexaphosphoric acid; 1,2,3,4,5,6-cyclohexanehexolphosphoric acid; cyclohexanehexyl hexaphosphate; Alkalovert. $C_6H_{18}O_{24}P_6$; mol wt 660.08. C 10.92%, H 2.75%, O 58.18%, P 28.16%. $C_6H_6[OPO(OH)_2]_6$. Major phosphorus compound in plants; particularly abundant in oil seeds, legumes and cereal grains. Forms insoluble complexes with di- and trivalent cations. Prepn: S. Posternak, U.S. pat. **1,313,014** (1919 to Ciba); *idem, Helv. Chim. Acta* **4**, 155 (1921); M. J. Thomas, U.S. pat. **2,718,523** (1955 to Staley Mfg.); D. S. Bolley *et al.*, U.S. pat. **2,732,395** (1956 to Natl. Lead Co.); A. R. Baldwin *et al.*, U.S. pat. **2,815,360** (1957 to Corn Prods. Ref.). HPLC determn in plants: E. Graf, F. R. Dintzis, *Anal. Biochem.* **119**, 413 (1982); in foods and biological samples: B.E. Knuckles *et al., J. Food Sci.* **47**, 1257 (1982). Review of effect of dietary phytate on zinc metabolism: *Nutr. Rev.* **41**, 64-66 (1983); on mineral bioavailability: J. L. Kelsay, *Am. J. Gastoenterol.* **82**, 983-986 (1987).

Syrupy, straw-colored liquid. Dec on heating. Acid reaction; the pH of a 10% aq soln has been reported as 0.86. Miscible with water, 95% alc, glycerol; sol in water contg alcohol-ether mixtures; very slightly sol in abs alc, methanol. Practically insol in anhydr ether, benzene, chloroform.

Calcium iron salt, *Calciphos.* Obtained in the processing of corn. Grayish powder contg 20% Ca, 14% P, and 2% Fe. Sparingly sol in water, dil mineral acids.

Calcium magnesium salt, *phytin.* White, odorless powder. Poor soly in water. Sol in dil acids.

Sodium salt, $C_6H_9Na_9O_{24}P_6$, *sodium phytate, Phytat D.B., Rencal.* Soluble in water with neutral reaction.

USE: Complexing agent for the removal of traces of heavy metal ions (also employed as the sodium salt). Starting material in manufacture of inositol. Fermentation nutrient.

THERAP CAT: Sodium salt as hypocalcemic.

7360. Phytochlorin. *18-Carboxy-20-(carboxymethyl)-8-ethenyl-13-ethyl-2,3-dihydro-3,7,12,17-tetramethyl-21H,-23H-porphine-2-propanoic acid;* chlorine e_6; phytochlorin(e) e. $C_{34}H_{36}N_4O_6$; mol wt 596.66. C 68.44%, H 6.08%, N 9.39%, O 16.09%. Prepn from broccoli-leaf extract: Wall, U.S. pat. **2,555,583** (1951 to U.S.D.A.). *Review:* H. Fischer, A. Stern, *Die Chemie des Pyrrols* vol. II, 2 (Leipzig, 1940) pp 91-94; *The Chlorophylls,* L. P. Vernon, G. R. Seely, Eds. (Academic Press, New York, 1966) pp 9, 12, 71, 90, 95, 98, 142.

Greenish-brown rectangular plates. May contain one mol water. $[\alpha]_D^{20}$ −141° in acetone. Ether solns are olive-green with deep red fluorescence. Sparingly sol in ethanol, ether, acetone. Freely sol in pyridine.

7361. Phytofluene. *7,7',8,8',11,12-Hexahydro-ψ,ψ-carotene;* 5,6,7,8,9,10,10',9',8',7',6',5'-dodecahydrolycopene. $C_{40}H_{68}$; mol wt 548.94. C 87.51%, H 12.49%. Polyene hydrocarbon widespread in the vegetable kingdom where it has been observed in chlorophyll-free tissues which contain considerable amounts of carotenoid pigments. Extracted from *Diospyros kaki* L.f., *Ebenaceae, Arbutus unedo* L., *Ericaceae, Pyracantha angustifolia* (Franch.) Schneid., *Rosaceae,* and tomatoes: Zechmeister, Sandonal, *J. Am. Chem. Soc.* **68,** 197 (1946); Wallace, Porter, *Arch. Biochem. Biophys.* **36,** 468 (1952); from the basidiomycete, *Dacromyces stillatus:* Goodwin, *Biochem. J.* **53,** 538 (1953). Structure: Porter, Lincoln, *Arch. Biochem. Biophys.* **27,** 390 (1950); Zechmeister, *Experientia* **10,** 1 (1954).

Pale orange, viscous oil which solidifies upon cooling, forming a glassy mass without apparent crystal structure. bp$_{0.0001}$ 140-185° (bath temp). uv max (petr ether): 367, 348, 332 nm. Freely sol in petr ether, ether, benzene. Practically insol in water, methanol, ethanol. Strong green fluorescence in ultraviolet spectra: Wallace, Porter, *loc. cit.*

7362. Phytol. *3,7,11,15-Tetramethyl-2-hexadecen-1-ol;* 2,6,10,14-tetramethylhexadec-14-en-16-ol. $C_{20}H_{40}O$; mol wt 296.52. C 81.01%, H 13.60%, O 5.40%. Decompn product of chlorophyll: Willstätter, *Ann.* **354,** 205 (1907);

371, 1 (1909); **378,** 1, 73 (1911); **418,** 121 (1918). Synthesis: Fischer, Löwenberg, *Ann.* **475,** 183 (1929); Karrer, Ringier, *Helv. Chim. Acta* **22,** 610 (1939); Karrer *et al., ibid.* **26,** 1741 (1943); from ethyl levulinate: Lukes, Zobacova, *Chem. Listy* **51,** 330 (1957); from acetone: Sato *et al., J. Org. Chem.* **32,** 177 (1967). Stereochemistry: Burrell *et al., Proc. Chem. Soc.* **1959,** 263. Abs config: Crabbe *et al., ibid.* **1959,** 264. Stereochemistry and synthesis: Burrell *et al., J. Chem. Soc. (C)* **1966,** 2144. Stereoselective total synthesis of natural phytol: T. Fujisawa *et al., Tetrahedron Letters* **22,** 4823 (1981); M. Schmid *et al., Helv. Chim. Acta* **65,** 684 (1982). *Review:* J. Simonsen, D. H. R. Barton, *The Terpenes,* vol. III (Cambridge University Press, Cambridge, 1952) pp 345-349.

Oily liquid. d_4^{25} 0.8497. n_D^{25} 1.4595. bp$_{10}$ 203-204°; bp$_{0.03}$ 145°. uv max (abs alcohol): 212 nm (log ε 3.04), Bader, *Helv. Chim. Acta* **34,** 1632 (1951). Practically insol in water; sol in the usual organic solvents.

USE: Preparation of vitamins E and K₁.

7363. Phytolacca. Poke root; garget; pocan; American nightshade root; scoke. Dried root of *Phytolacca americana* L. (*P. decandra* L.), *Phytolaccaceae* (berries also used). *Habit.* North America; naturalized in Southern Europe. *Constit.* Resin, tannin, about 10% sugar; phytolaccine, phytolaccic acid, asparagine.

THERAP CAT: Antirheumatic; emetic; ectoparasiticide.

7364. Phytosterols. A generic name for sterols obtained from higher plants. They differ from sterols occurring in the animal kingdom in having C_1 or C_2 residues at C-24 and/or a double bond at C-22. Examples of phytosterols are: stigmasterol, sitosterol, fucosterol, brassicasterol, campesterol. First isolated from Calabar beans and identified by O. Hesse, *Ann.* **192,** 175 (1878). Reviews of biosynthesis of phytosterols: E. Lederer, *Biochem. J.* **93,** 449 (1964); *idem, Quart. Revs.* **23,** 453 (1969); T. W. Goodwin, *Biochem. J.* **123,** 293 (1971); B. A. Knights, *Chem. Brit.* **9,** 106 (1973). Synthesis of *monofluorophytosterols* and pro-insecticidal activity: T. B. Kline, G. D. Prestwich, *Tetrahedron Letters* **23,** 3043 (1982); G. D. Prestwich *et al., Bio/Technol.* **1,** 62 (1983).

7365. Piberaline. *1-(Phenylmethyl)-4-(2-pyridinylcarbonyl)piperazine;* 1-benzyl-4-picolinoylpiperazine; EGYT-475; Trelibet. $C_{17}H_{19}N_3O$; mol wt 281.36. C 72.57%, H 6.81%, N 14.93%, O 5.69%. Prepn: J. Körösi *et al.,* **Belg.** pat. **781,494;** *eidem,* **U.S.** pat. **3,865,828** (1972, 1975 both to EGYT). Alternate prepns: Z. Budai *et al.,* Ger. pat. **2,828,-888,** *C.A.* **90,** 168643u (1979); *eidem,* **Hung.** pat. **17,182,** *C.A.* **92,** 215463p (1980) (both 1979 to EGYT); *eidem, Acta Chim. Acad. Sci. Hung.* **105,** 241 (1980), *C.A.* **94,** 192275d (1981). HPLC separation: J. Borda *et al., J. Chromatog.* **258,** 271 (1983). Effect on avoidance behavior in rats: G. Telegdy *et al., Arch. Int. Pharmacodyn. Ther.* **266,** 50 (1983). Metabolism in animals and humans: K. Magyar, *Pol. J. Pharmacol. Pharm.* **39,** 107 (1987). Mechanism of action study: K. Tekes *et al., ibid.* 203.

Dihydrochloride, $C_{17}H_{21}ClN_3O$, crystals from ethanol, mp 214-215°C.

Maleate, $C_{21}H_{23}N_3O_5$, crystals, mp 169-170°.

Fumarate, $C_{21}H_{23}N_3O_5$, crystals, mp 165°.

THERAP CAT: Antidepressant.

7366. Picadex. *1-Piperazinecarbodithioic acid;* 1-piperazinecarbodithioic acid betaine; 1-piperazinecarbodithioic acid inner salt; piperazine carbon disulfide complex; piperazine dithiocarbamate; Choisine; Elmifarma; Parvex; Safersan. $C_5H_{10}N_2S_2$; mol wt 162.29. C 37.00%, H 6.21%, N 17.27%, S 39.52%. Prepn: Schmidt, Wichmann, *Ber.* **24**, 3237 (1891); Herz, *ibid.* **30**, 1584 (1897). As anthelmintic: Leiper, Watkins, U.S. pat. **2,814,583** (1957 to Boots Pure Drug).

Melting pt depends on method of determination: in an open-end capillary tube in a heating bath—sublimes 225-228°; in a sealed capillary decomp 235°. Practically insol in ether, benzene, chloroform, water.

THERAP CAT (VET): Anthelmintic.

7367. Picein. *1-[4-(β-D-Glucopyranosyloxy)phenyl]ethanone; p-(acetylphenyl)-β-D-glucopyranoside; 4'-(β-D-glucopyranosyloxy)acetophenone; p-hydroxyacetophenone-D-glucoside; piceoside; salinigrin; salicinerein; ameliaroside.* $C_{14}H_{18}O_7$; mol wt 298.28. C 56.37%, H 6.08%, O 37.55%. In needles and sprouts of *Pinus picea* L., *Picea excelsa* Link., *Picea glehnii* Mast., *Coniferae:* Tanret, *Bull. Soc. Chim.* [3] **11**, 944 (1894); Kariyone *et al.*, *Yakugaku Zasshi* **79**, 394 (1959), *C.A.* **53**, 14096i (1953). In various willow barks, especially in bark of *Salix discolor* Muhl., *S. nigra* Marsh., *Salicaceae:* Jowett, *J. Chem. Soc.* **77**, 707 (1900); Nonomura, *J. Pharm. Soc. Japan* **75**, 80 (1955). In English mistletoe, *Amelanchier vulgaris* Moench., *Rosaceae:* Bridel *et al.*, *Compt. Rend.* **187**, 56 (1928). Synthesis: Mauthner, *J. Prakt. Chem.* [2] **85**, 564 (1912).

Needles or prisms from methanol; mp 195-196°. $[\alpha]_D^{23}$ $-88°$ (c = 1). One gram dissolves in 50 ml water at 15°, in 1 ml boiling water, in about 650 ml abs alc at 15°, in about 40 ml boiling abs alc, in 140 ml ethyl acetate at 15°. Sol in hot glacial acetic acid. Practically insol in ether, chloroform. Hydrolysis with dil mineral acids or with emulsin yields D-glucose and p-hydroxyacetophenone (*Piceol*). Alkaline hydrolysis yields levoglucosan and p-hydroxyacetophenone: Montgomery *et al.*, *J. Org. Chem.* **10**, 194 (1945).

7368. Picene. *3,4-Benzchrysene; 1,2,7,8-dibenzphenanthrene; dibenzo[a,i]phenanthrene; β,β-binaphthyleneethene.* $C_{22}H_{14}$; mol wt 278.33. C 94.93%, H 5.07%. Found in tar oils from soft coal: Burg, *Ber.* **13**, 1834 (1880); Lang *et al.*, *ibid.* **97**, 494 (1964); in petroleum residues from the cracking process: Meyer, Hofmann, *Monatsh.* **37**, 681 (1916). Synthesis from cholic acid by selenium dehydrogenation: Ruzicka *et al.*, *Helv. Chim. Acta* **17**, 200 (1934); from o-xylylenedicyanide with o-nitrobenzaldehyde by a double Pschorr synthesis: Waldmann, Pitschak, *Ann.* **527**, 183 (1937); by a diene synthesis with tetrahydrodinaphthyl and maleic anhydride: Weidlich, *Ber.* **71**, 1203 (1938); from 9,10-dihydrophenanthrene: Phillips, *J. Am. Chem. Soc.* **75**, 3223 (1953); from ethyl 4,6-dioxoheptane-1,5-dicarboxylate: Nasipuri, *Chem. & Ind. (London)* **1956**, 795.

Fluorescent plates from ethyl acetate, mp 366-367°. bp 518-520°. Absorption spectrum: Mayneord, Roe, *Proc. Roy. Soc. London* **A152**, 319 (1935). Difficultly sol in most solvents. Somewhat sol in boiling benzene, chloroform, glacial acetic acid; more sol in cumene (isopropylbenzene).

7369. Picilorex. *3-(4-Chlorophenyl)-5-cyclopropyl-2-methylpyrrolidine;* UP-507-04. $C_{14}H_{18}ClN$; mol wt 235.75. C 71.33%, H 7.69%, Cl 15.04%, N 5.94%. Prepn: J. M. Teulon, **Ger.** pat. **2,446,317**; *idem*, **U.S.** pat. **4,005,103** (1975, 1977 both to Hexachemie). Anorexic activity in animals: G. Dumeur *et al.*, *Brit. J. Pharmacol.* **58**, 437 (1976).

Hydrochloride, $C_{14}H_{19}Cl_2N$, *Roxenan.* Crystals from isopropanol, mp 191°.

THERAP CAT: Anorexic.

7370. Picloram. *4-Amino-3,5,6-trichloro-2-pyridinecarboxylic acid; 4-amino-3,5,6-trichloropicolinic acid;* Tordon. $C_6H_3Cl_3N_2O_2$; mol wt 241.48. C 29.85%, H 1.25%, Cl 44.05%, N 11.60%, O 13.25%. Prepn: H. Johnston, M. S. Tomita, **Belg.** pat. **628,487**, *C.A.* **61**, 1838d (1964), and *eidem*, **U.S.** pat. **3,285,925** (1963, 1966 both to Dow). Activity: J. W. Hamaker *et al.*, *Science* **141**, 363 (1963). Carcinogenicity study: M. D. Reuber, *J. Toxicol. Environ. Health* **7**, 207 (1981).

Crystals, mp 218-219°. Vapor press. at 35°: 6.16×10^{-7} mm Hg. pKa 3.6. Soly in (g/l) at 25°: water 0.43; acetone 19.8; 2-propanol 5.5; dichloromethane 0.6. LD_{50} orally in rats, mice, rabbits: 3.75, 1.5, 2.0 g/kg, *RTECS* **Vol. II**, R. J. Lewis, R. L. Tatken, Eds. (1979) p 305.

USE: Herbicide.

7371. Picloxydine. *N,N''-Bis[[(4-chlorophenyl)amino]iminomethyl]-1,4-piperazine dicarboximidamide; N,N''-bis-[(p-chlorophenyl)amidino]-1,4-piperazinedicarboxamidine; 1,4-bis(N^4-p-chlorophenylamidinoamidinyl)piperazine; 1,1'-[1,4-piperazinediylbis(imidocarbonyl)]bis[3-(p-chlorophenyl)guanidine].* $C_{20}H_{24}Cl_2N_{10}$; mol wt 475.38. C 50.53%, H 5.09%, Cl 14.92%, N 29.46%. Heterocyclic biguanide with antibacterial activity. Prepn: J. W. James, L. F. Wiggins, **Brit.** pat. **855,017**; *eidem*, **U.S.** pat. **3,101,336** (1960, 1963 both to Aspro-Nicholas). Prepn and antibacterial spectrum: J. W. James *et al.*, *J. Med. Chem.* **11**, 942 (1968). Bactericidal effect in disinfectant formulations: A. M. Gordon, *J. Clin. Pathol.* **22**, 496 (1969). Mode of action study: B. D. Rawal, J. V. Hardy, *Microbios* **14**, 135 (1974). *In vitro* activity vs *Chlamydia trachomatis:* D. Thomas *et al.*, *Pathol. Biol.* **32**, 544 (1984).

Dihydrochloride, $C_{20}H_{26}Cl_4N_{10}$, *Vitabact.* Crystals from water, mp 274°. LD_{50} in mice (mg/kg): 150 i.p. (James).
THERAP CAT: Topical antibacterial.

7372. α-Picoline. *2-Methylpyridine.* C_6H_7N; mol wt 93.12. C 77.38%, H 7.58%, N 15.04%. Found in coal tar and in bone oil. Synthesis (40-50% yield) from cyclohexylamine with excess ammonia and zinc chloride at 350°: Nordt, *PB Report 704* (1941). Prepn from ethylene-mercuric acetate adduct and ammonia water (70% yield): Gumboldt, Feichtinger, *Z. Naturforsch.* **4b,** 123 (1949). Review of the various methods of condensing aldehydes, ketones, ethylene, butadiene, etc., with ammonia: F. Brody, P. R. Ruby in A. Weissberger, *The Chemistry of Heterocyclic Compounds,* **vol. 14,** part I (New York, 1960) pp 99-589. Contains 1851 references.
Colorless liquid; strong unpleasant odor. d_4^{15} 0.950; mp −70°; bp 128-129°; n_D^{20} 1.501. Addnl physical constants: Biddiscombe *et al., J. Chem. Soc.* **1954,** 1957. Freely sol in water; miscible with alcohol, ether. LD_{50} orally in rats: 1.41 g/kg, H. F. Smyth *et al., Arch. Ind. Hyg. Occup. Med.* **4,** 119 (1951).
USE: Solvent; intermediate in the dye and resins industries. *Caution:* Irritating to respiratory tract.

7373. β-Picoline. *3-Methylpyridine.* C_6H_7N; mol wt 93.12. C 77.38%, H 7.58%, N 15.04%.
Colorless liq; sweetish, not unpleasant odor. d_4^{15} 0.9613; bp 143-144°; n_D^{20} 1.5043. Miscible with water, alcohol, ether.
USE: Solvent; intermediate in the dye and resins industries; in the manuf of insecticides, waterproofing agents, niacin, and niacinamide.

7374. γ-Picoline. *4-Methylpyridine;* 4-picoline. C_6H_7N; mol wt 93.12. C 77.38%, H 7.58%, N 15.04%. Found in coal tar, in bone oil, in urine of horses. Isoln from technical picolines: Flaschner, *J. Chem. Soc.* **95,** 669 (1909).

Flammable liquid. Obnoxious, sweetish odor. Turns brown if not very pure. d_4^{15} 0.9571. bp_{760} 145°. n_D^{17} 1.5064. Kb at 25° = 1.1 × 10⁻⁸. Sol in water, alcohol, ether. LD_{50} orally in rats: 1.29 g/kg, H. F. Smyth *et al., Arch. Ind. Hyg. Occup. Med.* **10,** 61 (1954).
USE: Manuf isonicotinic acid and derivatives. In waterproofing agents for fabrics. As solvent for resins.

7375. Picolinic Acid. *2-Pyridinecarboxylic acid;* o-pyridinecarboxylic acid. $C_6H_5NO_2$; mol wt 123.11. C 58.53%, H 4.09%, N 11.38%, O 25.99%. An isomer of nicotinic acid. Prepn: Singer, McElvain, *Org. Syn. coll. vol.* III, 740 (1955). *Review:* Oliveto, "Pyridinecarboxylic Acids" in *Pyridine and Its Derivatives,* **Part 3,** E. Klingsberg, Ed. (Interscience, New York, 1962) p 179.

Needles from water, alcohol or benzene, mp 134-136°. pK 5.4. Sublimes. Very sol in glacial acetic acid; practically insol in ether, chloroform, carbon disulfide.
Hydrochloride, crystals from abs ethanol/ether, mp 210-212° (slow heating).

7376. Picoperine. *N-Phenyl-N-[2-(1-piperidinyl)ethyl]-2-pyridinemethanamine; 1-[2-[N-(2-pyridylmethyl)anilino]-ethyl]piperidine;* N-(2-pyridylmethyl)-N-phenyl-N-2-(piperidinoethyl)amine; 1-[2-[phenyl(2-pyridylmethyl)amino]ethyl]piperidine; N-(2-picolyl)-N-phenyl-N-(2-piperidinoethyl)amine; N-(2-piperidinoethyl)-N-(2-pyridylmethyl)aniline; N-phenyl-N-(2-pyridylmethyl)-2-piperidinoethylamine; picoperidamine; TAT-3. $C_{19}H_{25}N_3$; mol wt 295.43. C 77.24%, H 8.53%, N 14.22%. Prepn: **Fr. pat.** 1,511,398 and

U.S. pat. **3,471,501** (1968 and 1969 to Takeda); Miyano *et al., J. Med. Chem.* **13,** 704 (1970). Prepn of salts: Sawa *et al., Takeda Kenkyusho Ho* 29(2), 275 (1970), *C.A.* **73,** 98758h (1970); Masuda *et al.,* **Ger. pat.** **1,935,172** (1970 to Takeda), *C.A.* **73,** 98818c (1970). Pharmacology: Y. Kasé *et al., Arzneimittel-Forsch.* **19,** 1916 (1969); eidem, *ibid.* **20,** 37 (1970). *Review: Japan. Med. Gaz.* 8(11), 9 (1971).

Pale yellow liquid, bp_4 195-196°.
Hydrochloride, $C_{19}H_{26}ClN_3$, *Coben.* Water-soluble white crystalline powder, odorless with slightly bitter taste, mp 183-185°. Freely sol in ethanol and chloroform. Slightly sol in acetone, dioxane, benzene. Practically insol in hexane, ether. LD_{50} in mice (mg/kg): 210 s.c.; 85 i.p.; 17 i.v.; 240 orally (Kasé, 1969).
Tripalmitate, $C_{67}H_{121}N_3O_6$, *Coben P.* White, tasteless, odorless crystalline powder, mp 57-60°. Freely sol in ethanol, chloroform; slightly sol in dioxane, acetone, benzene. Practically insol in hexane, water. LD_{50} orally in mice: 1900 mg/kg (Kasé, 1969).
THERAP CAT: Antitussive.

7377. Picosulfate Sodium. *4,4'-(2-Pyridinylmethylene)-bisphenol bis(hydrogen sulfate) (ester) disodium salt; 4,4'-(2-pyridylmethylene)diphenolbis(hydrogen sulfate) (ester) disodium salt; 4,4'-(2-picolylidene)bis(phenylsulfuric acid) disodium salt; 2-picolylidenebis(p-phenyl sodium sulfate); disodium 4,4'-disulfoxydiphenyl-(2-pyridyl)methane; picosulfol; sodium picosulfate; Evanol; Guttalax; Laxidogol; Laxoberal; Laxoberon; Neopax; Pico-Salax.* $C_{18}H_{13}NNa_2O_8S_2$; mol wt 481.41. C 44.91%, H 2.72%, N 2.91%, Na 9.55%, O 26.59%, S 13.32%. Prepn: Pala, **Fr. pat. M5832** and U.S. pats. **3,528,986, 3,558,643** (1968, 1970, 1971 all to Ist. de Angeli); Seeger, Machleidt, **Ger. pat.** **1,904,322** (1970 to Thomae). Synthesis and physical-chemical data: Pala *et al., Helv. Chim. Acta* **51,** 1164 (1968). Pharmacology: Pala *et al., Arch. Int. Pharmacodyn. Ther.* **164,** 356 (1966). Metabolism: Perego *et al., Arzneimittel-Forsch.* **19,** 1889 (1969).

White crystalline solid from ethanol or methanol, mp 272-275° (dec). uv max (H_2O): 218, 262 nm (ε 20450, 6075). Readily sol in water; slightly sol in alcohol. Practically insol in most organic solvents.
THERAP CAT: Cathartic.

7378. Picotamide. *4-Methoxy-N,N'-bis(3-pyridinylmethyl)-1,3-benzenedicarboxamide; 4-methoxy-N,N'-bis(3-pyridylmethyl)isophthalamide; N,N'-bis(3-picolyl)-4-methoxyisophthalamide; G-137.* $C_{21}H_{20}N_4O_3$; mol wt 376.20. C 66.99%, H 5.36%, N 14.89%, O 12.76%. Prepn: Orzalesi *et al., Chim. Ther.* **6,** 203 (1971); G. Orzalesi, R. Selleri, **Fr. pat.** **2,100,850,** *C.A.* **77,** 164502f (1972); prepn and use as inhibitor of blood platelet aggregation: eidem, **U.S. pat.** **3,973,026** (1972, 1976 both to Soc. Italo-Brit. L. Manetti-H. Roberts). Crystal structure of monohydrate: E. Foresti *et al., Acta Crystallogr. Sect. C* **42,** 220 (1986). Mechanism of action study: M. Berrettini *et al., Boll. Soc. Ital. Biol. Sper.* **59,** 309 (1983). Clinical studies: R. Schmutzler *et al., Age Ageing* **7,** 246 (1978); V. Coto *et al., Minerva Cardioangiol.* **34,** 601 (1986).

Crystalline powder from benzene, mp 124°. LD_{50} i.p. in male mice: 1205 mg/kg (Orzalesi, Selleri, 1976).

Monohydrate, $C_{21}H_{20}N_4O_3 \cdot H_2O$, *Plactidil*.

THERAP CAT: Antithrombotic; fibrinolytic; anticoagulant.

7379. Picramic Acid. *2-Amino-4,6-dinitrophenol;* 4,6-dinitro-2-aminophenol. $C_6H_5N_3O_5$; mol wt 199.12. C 36.19%, H 2.53%, N 21.10%, O 40.17%. Prepd from picric acid, concd NH_4OH, and H_2S followed by acetic acid neutralization of ammonium salt: Egerer, *J. Biol. Chem.* **35**, 565 (1918). Prepn of ammonium salt from picric acid, aqueous NH_3-soln, and ammonium sulfide: Dehn, U.S. pat. **1,472,-791** (1924).

Dark red needles from alcohol, prisms from chloroform. mp 169-170°. At 22-25°, 0.065 g dissolves in 100 ml H_2O; not much more sol in hot water. Sparingly sol in ether, chloroform; moderately sol in alcohol; sol in benzene, glacial acetic acid, aniline. Flashes at 210°; in contact with open flame in glass tube or beaker, ignites rapidly and burns relatively fast: Blinov, *Khim. Prom.* **1959**, 419, *C.A.* **55**, 6866d (1961).

USE: Manuf of azo dyes; rarely as indicator (yellow with acids, red with alkalies); reagent for albumin. *Caution:* Toxic symptoms similar to 2,4-dinitrophenol, *q.v.*

7380. Picric Acid. *2,4,6-Trinitrophenol;* picronitric acid; carbazotic acid; nitroxanthic acid. $C_6H_3N_3O_7$; mol wt 229.11. C 31.45%, H 1.32%, N 18.34%, O 48.88%. Prepd by sulfonating phenol then treating with nitric acid: Olsen, Goldstein, *Ind. Eng. Chem.* **16**, 66 (1924); by treating benzene with nitric acid and mercuric nitrate: Teeters, Mueller, U.S. pat. **2,455,322** (1948 to Allied Chem.); by nitration of 2-*tert*-butyl-4,6-dinitrophenol: Ley, Müller, *Ber.* **89**, 1402 (1956). Crystal structure: E. N. Duesler *et al., Cryst. Struct. Commun.* **7**, 449 (1978).

Pale yellow, odorless, intensely bitter crystals. d 1.763. mp 122-123°. Explodes above 300°. One gram dissolves in 78 ml water, 15 ml boiling water, 12 ml alc, 10 ml benzene, 35 ml chloroform, 65 ml ether. *Keep in a cool place and remote from fire. Explodes when rapidly heated or by percussion! Incompat:* All oxidizable substances, albumin, gelatin, alkaloids.

Note: For safety in transportation, 10-20% water is usually added.

Human Toxicity: Local or generalized allergic reactions may occur following topical use. Ingestion or percutaneous absorption may cause nausea, vomiting, diarrhea, abdominal pain, oliguria, anuria, yellow staining of skin (not icterus), pruritus, skin eruptions, stupor, convulsions, death, *cf. Clinical Toxicology of Commercial Products,* R. E. Gosselin *et al.,*

Eds. (Williams & Wilkins, Baltimore, 4th ed., 1976) Section II, p 133.

USE: Explosives; matches; in leather industry; electric batteries; etching copper; manuf colored glass; textile mordant; also as reagent.

7381. Picrocrocin. *4-(β-D-Glucopyranosyloxy)-2,6,6-tri-methyl-1-cyclohexene-1-carboxaldehyde;* saffron-bitter. $C_{16}H_{26}O_7$; mol wt 330.37. C 58.17%, H 7.93%, O 33.90%. From stigmas of *Crocus sativus* L., *Iridaceae.* Isoln: Kayser, *Ber.* **17**, 2228 (1884). Structure: Kuhn, Winterstein, *Ber.* **67**, 344 (1934). Exerts sex-determining influences in the plant organism: Kuhn, *Angew. Chem.* **53**, 1 (1940). Its moieties are glucose and safranal, *q.v.* Abs config: Buchecker, Eugster, *Helv. Chim. Acta* **56**, 1121 (1973). Synthesis: H. Mayer, J.-M. Santer, *Helv. Chim. Acta* **63**, 1463 (1980).

Crystals, mp 154-156°. $[\alpha]_D^{20}$ −58° (c = 0.6). Bitter taste. Alkali unstable. Sol in water, alcohol; slightly sol in chloroform, ether. Practically insol in petr ether, benzene.

7382. Picrolichenic Acid. *6-Hydroxy-2'-methoxy-2,4'-dioxo-4,6'-dipentylspiro[benzofuran-3(2H),1'-[2,5]cyclo-hexadiene]-5-carboxylic acid;* picrolichenin. $C_{25}H_{30}O_7$; mol wt 442.49. C 67.85%, H 6.83%, O 25.31%. Bitter acidic principle of crustose lichen, *Pertusaria amara* (Ach.) Nyl., *Pertusariaceae.* Isoln: Alms, *Ann. Pharm.* **1**, 61 (1832). Identity of picrolichenin and picrolichenic acid: Zopf, *Ann.* **321**, 38 (1902). Isoln and structure: Wachmeister, *Acta Chem. Scand.* **12**, 147 (1958). Synthesis: Davidson, Scott, *J. Chem. Soc.* **1961**, 4075.

Prisms from benzene, dec 184-187°; prisms from aqueous acetic acid, decomp 187-190°. Intense bitter taste similar to that of quinine. Optically inactive. uv max: 245, 270-277 nm (ε 24,000, 7800). Readily sol in most organic solvents except benzene and light petroleum.

Methyl picrolichenate, $C_{26}H_{32}O_7$, needles from methanol, mp 102-103.5°. Soluble in cold 0.5N NaOH.

Methyl O-methylpicrolichenate, $C_{27}H_{34}O_7$, needles from hexane, mp 80-82°. Practically insol in cold 0.1N NaOH.

7383. Picrolonic Acid. *2,4-Dihydro-5-methyl-4-nitro-2-(4-nitrophenyl)-3H-pyrazol-3-one;* 3-methyl-4-nitro-1-(p-nitrophenyl)-2-pyrazolin-5-one. $C_{10}H_8N_4O_5$; mol wt 264.21. C 45.46%, H 3.05%, N 21.21%, O 30.28%.

Yellow leaflets, mp 116-117°; dec at 125°. Sparingly sol in water; sol in alcohol.

USE: As reagent for alkaloids, tryptophan, phenylalanine, and for the detection and estimation of calcium.

7384. Picromycin. *(E)-14-Ethyl-13-hydroxy-3,5,7,9,-13-pentamethyl-6-[[3,4,6-trideoxy-3-(dimethylamino)-β-D-xylo-hexopyranoside]oxy]oxacyclotetradec-11-ene-2,4,10-*

trione; pikromycin; albomycetin; amaromycin. $C_{28}H_{47}NO_8$; mol wt 525.70. C 63.97%, H 9.01%, N 2.67%, O 24.35%. First macrolide antibiotic isolated. Isoln from *Actinomyces* spp by Lindenbein and Bauer, *see* Brockmann, Henkel, *Naturwiss.* **37**, 138 (1950); *Ber.* **84**, 284 (1951); Brockmann, Bohne, U.S. pat. **2,693,433** (1954 to Schenley). Structure: Brockmann, Oster, *Ber.* **90**, 605 (1957); Anliker, Gubler, *Helv. Chim. Acta* **40**, 119, 1768 (1957). Studies in stereochemistry: Djerassi, Halpern, *J. Am. Chem. Soc.* **79**, 3926 (1957); Ogura *et al., ibid.* **97**, 1930 (1975); *eidem, Tetrahedron* **37**, Suppl. 1, 165 (1981). Revised structure: Rickards, Smith, *Chem. Commun.* **1968**, 1049; Muxfeldt *et al., J. Am. Chem. Soc.* **90**, 4748 (1968). Synthesis from narbonolide: Maezawa *et al., J. Antibiot.* **26**, 771 (1973).

Very bitter, rectangular platelets from methanol, mp 169.5-170°. Stable to heat. $[\alpha]_D^{24}$ +8.2° (c = 3.5 in ethanol); $[\alpha]_D^{20}$ −33.5° (c = 2.07 in chloroform); $[\alpha]_D^{24}$ −50.2° (c = 6.3 in chloroform). uv max (ethanol): 225 nm (log ϵ 3.97). Rotary dispersion data: Djerassi, Halpern, *Tetrahedron* **3**, 268 (1958). Very sparingly sol in water, petr ether, carbon disulfide. Soly in ethanol: 3.5 g/100 ml at 20°. Freely sol in acetone, benzene, chloroform, ethyl acetate, dioxane. Moderately sol in ether, methanol.

7385. Picropodophyllin. *[5R]-5,8,8aα,9-Tetrahydro-9α-hydroxy-5α-(3,4,5-trimethoxyphenyl)furo[3',4':6,7]naphtho-[2,3-d]-1,3-dioxol-6(5aH)-one;* picropodophyllinic acid lactone. $C_{22}H_{22}O_8$; mol wt 414.40. C 63.76%, H 5.35%, O 30.89%. Found in resin podophyllum, a dried alcoholic extract of *Podophyllum peltatum* L., *Berberidaceae*. Isomer of podophyllotoxin *q.v.*; picropodophyllin does not exist in the fresh plant, but is formed from podophyllotoxin β-D-glucoside during the extraction process. Structure: Schrecker, Hartwell, *J. Am. Chem. Soc.* **76**, 752 (1954); *J. Org. Chem.* **21**, 381 (1956); von Wartburg *et al., Helv. Chim. Acta* **40**, 1331 (1957). Isoln from resin podophyllum: Späth *et al., Ber.* **65**, 1545 (1932). Isoln as the glucoside: von Wartburg *et al., loc. cit.* Synthesis: Gensler, Wang, *J. Am. Chem. Soc.* **76**, 5890 (1954); Gensler *et al., ibid.* **82**, 1714 (1960).

Crystals from acetone. Irritates the skin. mp 214°. Higher-melting modification from abs ethanol, mp 228°. $[\alpha]_D^{20}$ +9.4° (c = 0.7 in chloroform). Soly in water at 25° about 100 mg/l. Fairly sol in chloroform, hot ethanol, acetic acid, acetone, dil caustic; less sol in benzene, ether, carbon tetrachloride, propylene glycol. Practically insol in petr ether. Picropodophyllin-β-D-glucoside, $C_{28}H_{32}O_{13}$, crystals, mp 231-233°; 252-254°. $[\alpha]_D^{20}$ −10.9° (c = 0.549 in pyridine).

7386. Picrorhiza. Dried rhizome of *Picrorhiza kurroa* Royle, *Scrophulariaceae*. *Habit.* Himalayas. *Constit.* Picrorhizin—a glucoside.

7387. Picrotin. *[1aR-(1aα,2aβ,3β,6β,6aβ,8aS*,8bβ,-9S*)]-Hexahydro-2a-hydroxy-9-(1-hydroxy-1-methylethyl)-8b-methyl-3,6-methano-8H-1,5,7-trioxacyclopenta[ij]cyclo-prop[a]azulene-4,8(3H)-dione.* $C_{15}H_{18}O_7$; mol wt 310.29. C 58.06%, H 5.85%, O 36.09%. Nontoxic component of picrotoxin. Prepn from picrotoxin: Meyer, Bruges, *Ber.* **31**, 2958 (1898). Structure studies: Bensted *et al., J. Chem. Soc.* **1952**, 1042; Slater, Wilson, *ibid.* 1597; Holker *et al., ibid.* **1958**, 2987. Total synthesis: E. J. Corey, H. L. Pearce, *Tetrahedron Letters* **21**, 1823 (1980). Toxicity data: C. H. Jarboe *et al., J. Med. Chem.* **11**, 729 (1968). *Review:* Porter, *Chem. Rev.* **67**, 441 (1967).

Fine needles from water; thick, shiny prisms from aq alc soln, mp 248-250°. $[\alpha]_D$ −64.7° (c = 2.31 in abs alc). Freely sol in abs alc, acetic acid, boiling water. Slightly sol in cold water. Practically insol in ether, chloroform, benzene. LD_{50} i.p. in mice: 135 mg/kg (Jarboe).

7388. Picrotoxin. Cocculin. $C_{30}H_{34}O_{13}$; mol wt 602.57. C 59.79%, H 5.69%, O 34.52%. Bitter principle isolated from the seed of *Anamirta cocculus* L. Wight & Arn., *Menispermaceae*, also found in *Tinomiscium philippinense* Diels. Extraction procedure: Clark, *J. Am. Chem. Soc.* **57**, 1111 (1935). For a complete chemical bibliography up to April 1949 *see Helv. Chim. Acta* **32**, 1859 (1949). Picrotoxin is a molecular compd of one mole picrotoxinin, *q.v.*, and one mole picrotin, *q.v.*, into which it is readily separated. Crystal and molecular structure: L. Dupont *et al., Acta Crystallogr.* **B32**, 2987 (1976).

Intensely bitter and very poisonous shiny rhomboid leaflets, mp 203°. $[\alpha]_D^{16}$ −29.3° (c = 4 in abs ethanol). One gram dissolves in about 350 ml water, in about 5 ml boiling water, in 13.5 ml 95% ethanol, in about 3 ml boiling alcohol. Sparingly sol in ether, chloroform. Readily sol in strong ammonia water, in aq solns of NaOH. Highly toxic to fish. LD_{50} i.p. in mice: 7.2 mg/kg, I. Setnikar *et al., J. Pharmacol. Exp. Ther.* **128**, 176 (1960).

THERAP CAT: Central and respiratory stimulant.

THERAP CAT (VET): CNS stimulant, antidote to barbiturates.

7389. Picrotoxinin. *[1aR-(1aα,2aβ,3β,6β,6aβ,8aS*,8bβ,-9R*)]-Hexahydro-2a-hydroxy-8b-methyl-9-(1-methylethen-yl)-3,6-methano-8H-1,5,7-trioxacyclopenta[ij]cycloprop[a]-azulene-4,8(3H)-dione.* $C_{15}H_{16}O_6$; mol wt 292.28. C 61.64%, H 5.52%, O 32.85%. Toxic component of picrotoxin. Prepn from picrotoxin: Horrmann, *Ber.* **45**, 2090 (1912). Structure: Conroy, *J. Am. Chem. Soc.* **73**, 1889 (1951); **79**, 5550 (1957); Craven, *Tetrahedron Letters* no. 19, 21 (1960). Total synthesis: E. J. Corey, H. L. Pearce, *J. Am. Chem. Soc.* **101**, 5841 (1979). Structure-activity relationship: C. H. Jarboe *et al., J. Med. Chem.* **11**, 729 (1968). *Review:* Porter, *Chem. Rev.* **67**, 441 (1967).

Very bitter large prisms or small crystals contg water, mp 209.5°. $[\alpha]_D^{17}$ +4.4° (c = 4.28 in abs alc), +3.49° (c = 7.57 in acetone). Soluble in hot common organic solvents, in cold alcohol and chloroform. LD_{50} i.p. in mice: 3 mg/kg (Jarboe).

7390. Picryl Chloride. *2-Chloro-1,3,5-trinitrobenzene.* $C_6H_2ClN_3O_6$; mol wt 247.56. C 29.11%, H 0.81%, Cl 14.32%, N 16.98%, O 38.78%.

Almost white needles. d 1.797. mp 83°. Insol in water; freely sol in benzene, hot chloroform, boiling alcohol, slightly in ether or petr ether.

7391. Pifarnine. *1-(1,3-Benzodioxol-5-ylmethyl)-4-(3,-7,11-trimethyl-2,6,10-dodecatrienyl)piperazine;* 1-piperonyl-4-(3,7,11-trimethyl-2,6,10-dodecatrienyl)piperazine; U-27; Pifazin; $C_{27}H_{40}N_2O_2$; mol wt 424.63. C 76.37%, H 9.49%, N 6.60%, O 7.54%. Non-anticholinergic gastric anti-secretory agent. Prepn: S. Tricerri *et al.*, **Ger. pat. 2,310,044** corresp to **U.S. pat. 3,875,163** (1973, 1975 both to Pierrel); *eidem, Eur. J. Med. Chem.* **9,** 555 (1974). Pifarnine is a mixture of four stereoisomers: *ZZ, EZ, ZE, EE.* Separation does not give compounds with significantly different activity/toxicity ratios from the mixture. Stereochemistry and pharmacology of the four isomers: G. Guadagnini *et al., ibid.* **10,** 585 (1975). Pharmacological study of pifarnine: A. Bianchetti *et al., Arzneimittel-Forsch.* **25,** 580 (1975). Physical-chemical and analytical studies: G. Guadagnini *et al., Pharm. Ind.* **38,** 296 (1976). Absorption, distribution, excretion: M. Riva *et al., Farmaco Ed. Prat.* **34,** 542 (1979). Clinical studies: A. Porro *et al., ibid.* 85; M. Petrillo *et al., Curr. Ther. Res.* **25,** 457 (1979).

Light yellow viscous liq with a slight odor and bitter taste. uv max (ethanol): 287 nm ($E_{1cm}^{1\%}$ 94.6). n_D^{20} 1.5235. Relative density at 20°: 1.013-1.015. pK_1 4.10; pK_2 3.25. Readily sol in most org solvents. Slightly sol in aq solns of organic acids. Practically insol in alkali, water. LD_{50} in mice, rats (mg/kg): 2175, 2610 orally; 40.6, 33.3 i.v. (Bianchetti). Approx LD_{50} i.p. in mice: 500 mg/kg (Tricerri, 1974).

THERAP CAT: Anti-ulcerative.

7392. Pifoxime. *1-[[4-[1-(Hydroxyimino)ethyl]phen-oxy]acetyl]piperidine; 1-[(p-acetylphenoxy)acetyl]piperidine p-oxime;* 1-[p-(1-oximidoethyl)phenoxyacetyl]piperidine; 4'-(1-piperidylcarbonylmethoxy)acetophenone oxime; N-[p-(1-isonitrosoethyl)phenoxyacetyl]piperidine; pixifenide; SW 77; Flamanil. $C_{15}H_{20}N_2O_3$; mol wt 276.34. C 65.20%, H 7.29%, N 10.14%, O 17.37%. Prepn: Mieville, **Ger. pat. 2,003,430** (1970 to Orchimed) corresp to **U.S. pat. 3,907,792** (1975); **Fr. pat. 2,068,443** (1971 to Lab. Fournier), *C.A.* **76,** 15377y (1972). Synthesis and pharmacology: De Cointet *et al., Chim. Ther.* **8,** 574 (1973). Crystal and molecular structure: Tran Qui Duc *et al., Acta Crystallogr.* **30B,** 2237 (1974).

White, odorless, tasteless powder, mp 168°. uv max: 210, 258 nm (ε 23,200, 18,900). Insol in water; partially sol in abs ethanol. LD_{50} orally in mice: 1000 mg/kg, De Cointet *et al., loc. cit.*

THERAP CAT: Anti-inflammatory.

7393. Piketoprofen. *3-Benzoyl-α-methyl-N-(4-methyl-2-pyridinyl)benzeneacetamide;* m-benzoyl-N-(4-methyl-2-pyridyl)hydratropamide; 2-(3-benzoylphenyl)-N-(4-methyl-2-pyridyl)propionamide; Calmatel (aerosol). $C_{22}H_{20}N_2O_2$; mol wt 344.41. C 76.72%, H 5.85%, N 8.13%, O 9.29%. Derivative of ketoprofen, *q.v.* used topically as cream or aerosol. Prepn: R. G. W. Spickett *et al.,* **Brit. pat. 1,436,-502** (1976 to Antonio Gallardo, S.A.). Pharmacology and clinical efficacy: E. Tarrus *et al., 2nd Eur. Congr. Biopharm. Pharmacokinet.* **1,** 483 (1984).

Oil, sol in methylene chloride, ethanol. Insol in water. Hydrochloride, $C_{22}H_{21}ClN_2O_2$, Calmatel (cream), mp 180-182°.

THERAP CAT: Topical anti-inflammatory.

7394. Pildralazine. *6-[(2-Hydroxypropyl)methylamino]-3(2H)-pyridazinone hydrazone;* 3-hydrazino-6-[(2-hydroxy-propyl)methylamino]pyridazine; (±)-1-[(6-hydrazino-3-pyridazinyl)methylamino]-2-propanol; propyldazine; propildazine. $C_8H_{15}N_5O$; mol wt 197.24. C 48.72%, H 7.66%, N 35.51%, O 8.11%. Peripheral vasodilator with hypotensive activity; related to hydralazine, *q.v.* Prepn: G. Pifferi, **Ger. pat. 2,154,245;** *idem,* **U.S. pat. 3,769,278** (1972, 1973 both to I.S.F.); G. Pifferi *et al., J. Med. Chem.* **18,** 741 (1975). Pharmacology in animals: L. Dorigotti *et al., Pharmacol. Res. Commun.* **8,** 295 (1976); L. Dorigotti *et al., Arzneimittel-Forsch.* **34,** 876 (1984); in humans: L. Terzoli *et al., Boll. Soc. Ital. Cardiol.* **22,** 1053 (1977). GLC determn in plasma: P. Ventura *et al., J. Chromatog.* **161,** 237 (1978). Clinical evaluation in essential hypertension: R. Pellegrini, G. Abbondati, *Farmaco Ed. Prat.* **32,** 19 (1977). Pharmacokinetics and tissue binding: E. Noack *et al., Arzneimittel-Forsch.* **37,** 407 (1987).

Dihydrochloride, $C_8H_{17}Cl_2N_5O$, ISF 2123, Atensil. Crystals from ethanol, mp 206-209° (dec). LD_{50} in mice, rats (mg/kg): 357, 355 i.p.; 1170, 1230 orally (Dorigotti, 1976).

THERAP CAT: Antihypertensive.

7395. Pilocarpine. *3-Ethyldihydro-4-[(1-methyl-1H-imidazol-5-yl)methyl]-2(3H)-furanone;* Chibro Pilocarpine; Ocusert Pilo. $C_{11}H_{16}N_2O_2$; mol wt 208.25. C 63.44%, H 7.74%, N 13.45%, O 15.37%. Cholinergic principle from *Pilocarpus jaborandi* Holmes, *Rutaceae.* Isoln: Petit, Polanovski, *Bull. Soc. Chim.* [3] **17,** 557, 702 (1897). Structure: Jowett, *J. Chem. Soc.* **77,** 473, 851 (1900); **83,** 438 (1903). Stereoisomeric with isopilocarpine: Polonovski, Polonovski, *Bull. Soc. Chim.* [4] **31,** 1314 (1922). Has the *cis* configuration; isopilocarpine is *trans:* Zav'yalov, *Doklady Akad. Nauk SSSR* **82,** 257 (1952). Absolute configuration: Hill, Barcza, *Tetrahedron* **22,** 2889 (1966). Synthesis: Preobrashenski *et al., Ber.* **66,** 1187 (1933); Samokhvalov, *Med. Prom. SSSR* **11,** no. 2, 10 (1957); DeGraw, *Tetrahedron* **28,** 967 (1972); Link, Bernauer, *Helv. Chim. Acta* **55,** 1053 (1972). Stereoselective synthesis: A. Noordam *et al., Rec. Trav. Chim.* **98,** 467 (1979). *Review:* Langenbeck, *Angew. Chem.* **60,** 297 (1948); van Rossum *et al., Experientia* **16,** 373 (1960). Toxicity studies: Beccari, *Boll. Chim. Farm.* **106,** 8 (1967). Comprehensive description: A. A. Al-Badr, H. Y. Aboul-Enein, in *Analytical Profiles of Drug Substances* **vol. 12,** K. Florey, Ed. (Academic Press, New York, 1983) pp 385-432.

Pilocarpine, oil or crystals, mp 34°. bp$_5$ 260° (partial conversion to isopilocarpine). [α]$_D^{18}$ +106° (c = 2). pK$_1$ (20°) 7.15; pK$_2$ (20°) 12.57. Sol in water, alcohol, chloroform; sparingly sol in ether, benzene. Almost insol in petr ether.

Pilocarpine hydrochloride, C$_{11}$H$_{17}$ClN$_2$O$_2$, *Akarpine, Almocarpin, Pilomiotin, Isopto Carpine, Pilopine HS, Pilovisc.* Hygroscopic crystals from alcohol, mp 204-205°. [α]$_D^{18}$ +91° (c = 2). Freely sol in water, alcohol. Practically insol in ether, chloroform. *Keep well closed and protected from light.*

Pilocarpine nitrate, C$_{11}$H$_{17}$N$_3$O$_5$, *Licarpin, Pilofrin, Pilagan.* mp 173.5-174.0° (dec). *Poisonous!* [α]$_D$ +77° to +83° (c = 10). One gram dissolves in 4 ml water, 75 ml alcohol. Insol in chloroform, ether.

Incompat. Silver nitrate, mercury bichloride, iodides, gold salts, tannin, calomel, KMnO$_4$, alkalies.

Isopilocarpine, β-pilocarpine. Hygroscopic oily liquid or prisms. bp$_{10}$ 261°. [α]$_D^{18}$ +50° (c = 2). pK$_1$ (18°) 7.17. Miscible with water and alcohol; very sol in chloroform; less sol in benzene, ether. Almost insol in petr ether.

Isopilocarpine hydrochloride hemihydrate, C$_{11}$H$_{17}$ClN$_2$O$_2$.½H$_2$O, scales from alcohol + ether, mp 127°; when anhydr, mp 161°. [α]$_D^{18}$ +39° (c = 5). Sol in 0.27 part water; 2.1 parts alcohol.

Isopilocarpine nitrate, C$_{11}$H$_{17}$N$_3$O$_5$, prisms from water, scales from alcohol, mp 159°. [α]$_D^{18}$ +39° (c = 2). Sol in 8.4 parts water, in 350 parts abs alcohol.

THERAP CAT: Antiglaucoma agent; miotic.

THERAP CAT (VET): Parasympathomimetic; miotic; gastric secretory stimulant.

7396. Pilocarpus. Jaborandi. Leaves of *Pilocarpus jaborandi* Holmes (Pernambuco jaborandi), or of *P. microphyllus* Stapf (Maranhao jaborandi), *Rutaceae.* Habit. Brazil, Paraguay. *Constit.* About 1% alkaloids of which about 0.5% is pilocarpine; pilocarpidine, isopilocarpine, jaborine, jaboridine, jaboric acid, pilocarpic acid.

THERAP CAT: Sudorific; miotic.

7397. Pilocereine. C$_{45}$H$_{65}$N$_3$O$_6$; mol wt 744.04. C 72.64%, H 8.81%, N 5.65%, O 12.90%. From *Pilocereus sargentianus* Orcutt, *Lophocereus schottii* Britt. and Rose, *Cactaceae.* Isoln: Heyl, *Arch. Pharm.* **239**, 451 (1901); Djerassi et al., *J. Am. Chem. Soc.* **75**, 3632 (1953). Structure: *eidem, ibid.* **84**, 3210 (1962). Biosynthesis: Schütte, Seelig, *Ann.* **730**, 186 (1969); O'Donovan et al., *J. Chem. Soc. (C)* **1971**, 2398.

Needles from ethanol, mp 176.5-177°. uv max (abs ethanol): 284 nm (log ε 3.72). Sol in alcohol, benzene, chloroform, ether, petr ether. Practically insol in water.

Methyl ether, C$_{46}$H$_{67}$N$_3$O$_2$, crystals from hexane, mp 103-105°, solidifies and remelts at 153-155°. uv max: 282, 320 nm (log ε 3.68, 2.05).

Ethyl ether, C$_{47}$H$_{69}$N$_3$O$_6$, crystals from hexane, mp 90-95°, solidifies and remelts at 152-153°.

Acetate, C$_{47}$H$_{67}$N$_3$O$_7$, crystals from ether + acetone, mp 186-186.5°. uv max (abs ethanol): 284 nm (log ε 3.84).

7398. Pimaric Acid. *7-Ethenyl-1,2,3,4,4a,4b,5,6,7,9,10,-10a-dodecahydro-1,4a,7-trimethyl-1-phenanthrenecarboxylic acid; 13α-methyl-13-vinylpodocarp-8(14)-ene-15-oic acid;* dextropimaric acid; *d*-pimaric acid; α-pimaric acid. C$_{20}$H$_{30}$O$_2$; mol wt 302.44. C 79.42%, H 10.00%, O 10.58%. Isoln from American rosin: Rimbach, *Ber. Deut. Pharm. Ges.* **6**, 61 (1896); from French galipot from *Pinus maritima* Mill, *Pinaceae:* Ruzicka, Balas, *Helv. Chim. Acta* **6**, 677 (1923); Ruzicka et al., *ibid.* **15**, 915 (1932). Structure: Ruzicka, Sternbach, *ibid.* **23**, 124 (1940); Fleck, *J. Am. Chem. Soc.* **62**, 2044 (1940); Harris, Sanderson, *ibid.* **70**, 2081 (1948). Stereochemistry: Wenkert, Chamberlin, *ibid.* **81**, 688 (1959).

Orthorhombic, bisphenoidal crystals from acetone or acetic acid, mp 217-219°; bp$_{18}$ 282°; [α]$_D^{18}$ +74.7° (c = 0.4 in chloroform); [α]$_D^{20}$ +87.3° (in chloroform).

Hydrochloride, C$_{20}$H$_{31}$O$_2$Cl, dec 184°. [α]$_D^{20}$ +13.6° (c = 0.5 in alcohol). Can be reconverted by heating with quinoline at 250°.

Quinidine salt, C$_{40}$H$_{54}$N$_2$O$_4$, mp 90°.

Methyl ester, C$_{21}$H$_{32}$O$_2$, mp 69°.

7399. Pimeclone. *2-(1-Piperidinylmethyl)cyclohexanone;* 2-(piperidinomethyl)cyclohexanone; Nu-582; NA 66; Spiractin; Karion. C$_{12}$H$_{21}$NO; mol wt 195.30. C 73.79%, H 10.84%, N 7.17%, O 8.19%. Prepn: Mannich, Honig, *Arch. Pharm.* **265**, 598 (1927); Dimroth et al., *Ber.* **73**, 1399 (1940).

Liquid, bp$_{14}$ 118-120°; bp$_6$ 124°.

Hydrochloride, C$_{12}$H$_{21}$NO.HCl, crystals from isopropanol, mp 161-165°.

THERAP CAT: Respiratory stimulant.

7400. Pimefylline. *3,7-Dihydro-1,3-dimethyl-7-[2-[(3-pyridinylmethyl)amino]ethyl]-1H-purine-2,6-dione; 7-[2-[(3-pyridylmethyl)amino]ethyl]theophylline;* 7-(β-3'-picolylaminoethyl)theophylline; pimephylline; ES 771. C$_{15}$H$_{18}$N$_6$O$_2$; mol wt 314.35. C 57.32%, H 5.77%, N 26.73%, O 10.18%. Prepn: **Neth.** pat. **Appl. 6,600,250** corresp to Suter, Zutter, **U.S. pat. 3,350,400** (1966, 1967, both to Eprova). Structure determination, properties and chemistry: Suter, Zutter, *Pharm. Acta Helv.* **48**, 133 (1973). Pharmacology: Ciaceri, Attaguile, *Gazz. Med. Ital.* **132**, 36, 108 (1973). Metabolism: Pitrè, *Farmaco Ed. Prat.* **29**, 46 (1974).

Crystals from isopropyl acetate, mp 111-112°. uv max (water): 270 nm. Readily sol in cold water, chloroform; sol in acetone, ethanol. LD$_{50}$ orally, i.v. in mice: 1900, 402 mg/kg, Suter, Zutter, *loc. cit.* (1973).

Nicotinate, $C_{21}H_{23}N_7O_4$, *ES 902, Teonicon.* Colorless, odorless powder from ethanol, mp 159-160°. uv max (water): 267 nm. Very sol in water (40% soln can be prepd); slightly sol in methanol; very slightly sol in acetone, chloroform. LD$_{50}$ orally, i.v. in mice: 2530, 470 mg/kg, Suter, Zutter, *loc. cit.* (1973).

THERAP CAT: Vasodilator (coronary).

7401. Pimelic Acid. *Heptanedioic acid;* 1,5-pentanedicarboxylic acid. $C_7H_{12}O_4$; mol wt 160.17. C 52.49%, H 7.55%, O 39.96%. HOOC$(CH_2)_5$COOH. Prepn starting with cyclohexanone and ethyl oxalate: Snyder *et al., Org. Syn. coll. vol. II*, p 531 (1943); by the action of sodium and isoamyl alcohol on salicylic acid: Müller, *ibid.* p 535.

Monoclinic prisms from benzene, mp 105.7-105.8°. Tends to sublime. bp$_{100}$ 272°; bp$_{50}$ 251.5°; bp$_{15}$ 223°; bp$_{10}$ 212°. 2.52 parts dissolve in 100 parts water at 13.5°; 5 parts dissolve in 100 parts water at 20°. Freely sol in alcohol, ether. Practically insol in cold benzene.

Diethyl ester, $C_{11}H_{20}O_4$, liquid, d$_4^{20}$ 0.99448; bp$_{748}$ 254°; bp$_{24}$ 155°; bp$_{15}$ 140°. Sol in alcohol, ether, ethyl acetate.

7402. Pimenta. Allspice; Jamaica pepper; semen Amomi. Dried, nearly ripe fruit of *Pimenta officinalis* Lindl., *Myrtaceae. Habit.* East and West Indies, Central and South America. *Constit.* 3-4.5% volatile oil containing about 70% eugenol and small quantities eugenol methyl ether; cineol, *l-α*-phellandrene and caryophyllene; resin, tannin, fixed oil, sugar, gum.

7403. Piminodine. *4-Phenyl-1-[3-(phenylamino)propyl]-4-piperidinecarboxylic acid ethyl ester; 1-(3-anilinopropyl)-4-phenylisonipecotic acid ethyl ester; ethyl 4-phenyl-1-[(3-phenylamino)propyl]-4-piperidinecarboxylate; ethyl 1-(3-anilinopropyl)-4-phenylisonipecotate.* $C_{23}H_{30}N_2O_2$; mol wt 366.49. C 75.37%, H 8.25%, N 7.64%, O 8.73%. Prepn: Elpern *et al., J. Am. Chem. Soc.* **81**, 3784 (1959).

$C_6H_5NHCH_2CH_2CH_2$—N⟨ ⟩COOC$_2H_5$ / C$_6H_5$

Dihydrochloride, $C_{23}H_{32}Cl_2N_2O_2$, crystals, mp 219.4-222.2°.

Ethanesulfonate, $C_{25}H_{36}N_2O_5S$, *NIH 7590, WIN-14098, Alvodine, Anopridine, Cimadon, Pimadin.*

Caution: May be habit forming. This is a controlled substance (opiate) listed in the U.S. Code of Federal Regulations, Title 21 Part 1308.12 (1987).

THERAP CAT: Narcotic analgesic.

7404. Pimozide. *1-[1-[4,4-Bis(4-fluorophenyl)butyl]-4-piperidinyl]-1,3-dihydro-2H-benzimidazol-2-one; 1-[1-[4,4-Bis(p-fluorophenyl)butyl]-4-piperidyl]-2-benzimidazolinone; 1-[4,4-di-(4-fluorophenyl)butyl]-4-(2-oxo-1-benzimidazolinyl)piperidine; R 6238; Orap; Opiran.* $C_{28}H_{29}F_2N_3O$; mol wt 461.56. C 72.86%, H 6.33%, F 8.23%, N 9.10%, O 3.47%. Prepn: Janssen, *Fr. pat.* M3695 (1965 to Janssen), *C.A.* **66**, 115709t (1967). Pharmacological studies: Janssen *et al., Arzneimittel-Forsch.* **18**, 261, 279, 282 (1968). Metabolism in rats: Soudijn, Wijngaarden, *Life Sci.* **8**, 291 (1969). Clinical experience: Poldinger, *Curr. Ther. Res.* **13**, 23 (1971). Pimozide blocks establishment but not expression of amphetamine-produced environment-specific conditioning: R. J. Beninger, B. L. Hahn, *Science* **220**, 1304 (1983). Use in treatment of acute schizophrenia: B. Shopsin, G. Selzer, *Curr. Ther. Res.* **21**, 755 (1977); in treatment of chronic schizophrenia: F. Kline *et al., ibid.* 768. In management of Gilles de la Tourette's syndrome: M. S. Ross, H. Moldofsky, *Lancet* **1**, 103 (1977); A. K. Shapiro, E. Shapiro, *Am. J. Psychiat.* **140**, 1235 (1983); *eidem, J. Am. Acad. Child Psychiat.* **23**, 161 (1984).

Microcrystals, mp 214-218°. Weak base, pKa 7.32. Almost insol in water (< 0.01 mg/ml), very slightly sol in dil aq solns of organic and mineral acids (< 5 mg/ml).

THERAP CAT: Antipsychotic.

7405. Pimpinella. Pimpernel; brunet saxifrage; small saxifrage. Root of *Pimpinella saxifraga* L. or *P. magna* L. *Umbelliferae. Habit.* Europe; adventitious in U.S. *Constit.* Volatile oil, resin, benzoic acid, pimpinellin.

THERAP CAT: Aromatic carminative.

7406. Pimpinellin. *5,6-Dimethoxy-2H-furo[2,3-h]-1-benzopyran-2-one;* 4-hydroxy-6,7-dimethoxy-5-benzofuranacrylic acid δ-lactone. $C_{13}H_{10}O_5$; mol wt 246.21. C 63.41%, H 4.09%, O 32.49%. Found in *Pimpinella saxifraga* L., *Heracleum spondylium* L., *H. lanatum* Michx., and *H. panaces* L., *Umbelliferae.* Isoln: Heut, *Arch. Pharm.* **236**, 162 (1898); Herzog, Hancu, *ibid.* **246**, 402 (1908); Wessely, Kallab, *Monatsh.* **59**, 161 (1932); Späth, Simon, *ibid.* **67**, 344 (1936); Fujita, Furuya, *J. Pharm. Soc. Japan* **74**, 795 (1954); **76**, 535 (1956); Svendsen, Ottestad, *Pharm. Acta Helv.* **32**, 457 (1957); Svendsen, *C.A.* **52**, 2173g (1958); Svendsen *et al., Planta Med.* **7**, 113 (1959); *Pharm. Acta Helv.* **34**, 33 (1959); Jastrzebski, *C.A.* **54**, 1205b (1960). Synthesis: M. W. Reed, H. W. Moore, *J. Org. Chem.* **53**, 4166 (1988).

Off-white needles from methylene chloride/hexane, mp 119°. Practically insol in water. Sol in alcohol. Absorption spectrum: Wessely, Koltan, *Monatsh.* **86**, 430 (1955).

7407. Pinacidil. *(±)-N-Cyano-N'-4-pyridinyl-N''-(1,2,2-trimethylpropyl)guanidine monohydrate;* P-1134; Pindac. $C_{13}H_{19}N_5$.H$_2$O; mol wt 263.34. C 59.29%, H 8.04%, N 26.59%, O 6.08%. Potassium channel opening vasodilator. Prepn: H. J. Petersen, *Ger. pat.* 2,557,438; *idem, U.S. pat.* 4,057,636 (1976, 1977 both to Leo Pharm.); H. J. Petersen *et al., J. Med. Chem.* **21**, 773 (1978). Mechanism of action: E. Arrigoni-Martelli *et al., Experientia* **36**, 445 (1980); K. M. Bray *et al., Brit. J. Pharmacol.* **91**, 421 (1987). Metabolism: E. Eilertsen *et al., Xenobiotica* **12**, 187 (1982). Bioavailability: *eidem, ibid.* 177. Determn in plasma: M. Hamilton *et al., J. Chromatog.* **375**, 359 (1986). Pharmacokinetics and hypotensive effects: J. W. Ward *et al., Eur. J. Clin. Pharmacol.* **26**, 603 (1984). Clinical comparison with hydralazine, *q.v.*: R. L. Byyny *et al., Clin. Pharmacol. Ther.* **42**, 50 (1987). Review of pharmacology and mechanism of action: M. L. Cohen: *Drug Develop. Res.* **9**, 249-258 (1986).

Crystals, mp 164-165°. LD$_{50}$ in mice, rats (mg/kg): 600, 570 orally (Petersen, 1978).

THERAP CAT: Antihypertensive.

7408. Pinacol. *2,3-Dimethyl-2,3-butanediol;* pinacone; tetramethylethylene glycol. $C_6H_{14}O_2$; mol wt 118.17. C 60.98%, H 11.94%, O 27.08%. Prepd by the reduction of acetone: Holleman, *Rec. Trav. Chim.* **25**, 206 (1906); R.

Adams, E. W. Adams, *Org. Syn.* **coll. vol. I**, 459 (2nd ed., 1941); **Ger. pat. 233,894** in *Frdl.* **10**, 1000. Convenient lab procedure: L. F. Fieser, *Experiments in Organic Chemistry* (Boston, 3rd ed., 1955) p 101. Crystal structure: G. A. Jeffrey, A. Robbins, *Acta Crystallogr.* **B34**, 3817 (1978).

$$H_3C \quad CH_3$$
$$CH_3C \mathrm{-} CCH_3$$
$$HO \quad OH$$

Hexahydrate, $C_6H_{14}O_2.6H_2O$, four-sided plates from water. mp 45.4°; d^{15} 0.967 (supercooled liquid). The anhydr compd crystallizes in needles from alc or ether, mp 41.1°; bp_{760} 174.4°. Freely sol in hot water, in alc, in ether; slightly sol in cold water, carbon disulfide.

Dimethyl ether, $C_8H_{18}O_2$, liquid. Agreeable odor. bp 144°.

7409. Pinacolone. *3,3-Dimethyl-2-butanone; tert*-butyl methyl ketone; pinacolin. $C_6H_{12}O$; mol wt 100.16. C 71.94%, H 12.08%, O 15.97%. Prepd by distillation of pinacol hydrate with dil H_2SO_4: Fittig, *Ann.* **114**, 56 (1860); Hill, Flosdorf, *Org. Syn.* **coll. vol. I** (2nd ed., 1941) p 462; L. F. Fieser, K. L. Williamson, *Organic Experiments* (D. C. Heath and Co., Lexington, Mass., 5th ed., 1983) p 332. Almost quantitative yields are obtained when 56.5 g pinacol hydrate is boiled for 3 hrs with 0.5 liter 25% H_2SO_4 and the product is steam distilled: Boeseken, van Tonningen, *Rec. Trav. Chim.* **39**, 189 (1920).

Liquid. Peppermint or camphor-like odor. d_{25}^{25} 0.7250. mp −52.5°. bp_{760} 106.2°. n_D^{25} 1.3939. Volatile in steam. Moderately sol in water (2.44% at 15°). Soluble in alcohol, ether, acetone.

Oxime, $C_6H_{13}NO$, needles from aq alcohol. mp 78°. bp_{748} 171.6°. Sol in alc, ether, petr ether, benzene, chloroform.

7410. Pinaverium Bromide. *4-[(2-Bromo-4,5-dimethoxyphenyl)methyl]-4-[2-[2-(6,6-dimethylbicyclo[3.1.1]hept-2-yl)ethoxy]ethyl]morpholinium bromide;* 4-(6-bromoveratryl)-4-[2-[2-(6,6-dimethyl-2-norpinyl)ethoxy]ethyl]morpholinium bromide; 1717; Dicetel. $C_{26}H_{41}Br_2NO_4$; mol wt 591.45. C 52.80%, H 6.99%, Br 27.02%, N 2.37%, O 10.82%. Spasmolytic agent with low incidence of anticholinergic effects. Prepn: **Belg. pat. 769,469** corresp to R. Baronnet, U.S. pat. **3,845,048** (1971, 1974 both to Societe Berri-Balzac). Synthesis and pharmacology: R. Baronnet *et al., Eur. J. Med. Chem.* **9**, 182 (1974). Pharmacodynamics: J. Bretaudeau *et al., Therapie* **30**, 919 (1975). Synthesis of ^{14}C pinaverium bromide and pharmacokinetics: C. Jacquot *et al., Eur. J. Med. Chem.* **13**, 61 (1978). Mechanism of action: J. Bretaudeau, O. Foussard-Blanpin, *J. Pharmacol.* **11**, 233 (1980). Inhibition of gastrointestinal contractile activity in dogs: Z. Itoh, T. Takahashi, *Arzneimittel-Forsch.* **31**, 1450 (1981). Effects in humans on gastric emptying and transit time: J. Bertrand *et al., Therapie* **36**, 555 (1981).

Cryst from methyl ethyl ketone, mp 181°. LD_{50} in mice: 37±2.4 mg/kg i.v., R. Baronnet *et al., loc. cit.* Also reported as mp ca. 170°; LD_{50} in mice (mg/kg): 1400 orally; 66 i.v., R. Baronnet, U.S. pat. **3,845,048**.

THERAP CAT: Spasmolytic.

7411. Pinazepam. *7-Chloro-1,3-dihydro-5-phenyl-1-(2-propynyl)-2H-1,4-benzodiazepin-2-one;* 7-chloro-1-propar-

gyl-5-phenyl-3H-1,4-benzodiazepin-2(1H)-one; Z-905; Domar. $C_{18}H_{13}ClN_2O$; mol wt 308.76. C 70.02%, H 4.24%, Cl 11.48%, N 9.07%, O 5.18%. Prepn: F. Tenconi *et al.,* **Ger. pat. 2,339,790** (1974 to Zambeletti), *C.A.* **80**, 133492k (1974); C. Podesva, K. Vagi, U.S. pat. **3,842,094** (1974 to Delmar). Metabolism study: A. Trebbi *et al., J. Chromatog.* **110**, 309 (1975). Pharmacological and toxicological studies: F. Scrollini *et al., Arzneimittel-Forsch.* **25**, 934 (1975). Pharmacokinetics: P. M. Boselli, F. Scrollini, *Boll. Chim. Farm.* **116**, 363 (1977). Psychopharmacological properties: F. Scrollini *et al., Arzneimittel-Forsch.* **28**, 423 (1978). Physicochemical profile: G. Filipi, A Trebbi, *Boll. Chim. Farm.* **118**, 105 (1979). Clinical study: V. Bertoncelli *et al., Clin. Ter.* **94**, 641 (1980).

Cryst from methanol/water, mp 140-142°. LD_{50} in mice, rats (mg/kg): 1355, 5819 orally; 266, 622 i.p. (Scrollini, 1975). LD_{50} also reported as 670 mg/kg orally in mice (Podesva, Vagi).

Note: This is a controlled substance (depressant) listed in the U.S. Code of Federal Regulations, Title 21 Part 1308.14 (1985).

THERAP CAT: Anxiolytic.

7412. Pindolol. *1-(1H-Indol-4-yloxy)-3-[(1-methylethyl)amino]-2-propanol;* 4-[2-hydroxy-3-(isopropylamino)propoxy]indole; prinodolol; LB 46; Betapindol; Blocklin L; Calvisken; Decreten; Durapindol; Glauco-Viskin; Pectobloc; Pinbetol; Pynastin; Visken. $C_{14}H_{20}N_2O_2$; mol wt 248.32. C 67.71%, H 8.12%, N 11.28%, O 12.88%. β-Adrenergic blocker with partial agonist activity. Prepn: Troxler, Swiss pats. **469,002** and **472,404** (both 1969 to Sandoz). HPLC determn in plasma: H. Smith, *J. Chromatog.* **415**, 95 (1987). Symposium on clinical studies: *Am. Heart J.* **104**, Suppl. 2, pt. 2, 333-520 (1982); *Brit. J. Clin. Pharmacol.* **13**, Suppl. 2, 143S-450S (1982). Brief review: W. H. Frishman, *N. Engl. J. Med.* **308**, 940-944 (1983).

Crystals from ethanol, mp 171-173°.

THERAP CAT: Antihypertensive, antianginal, antiarrhythmic, antiglaucoma agent.

7413. Pindone. *2-(2,2-Dimethyl-1-oxopropyl)-1H-indene-1,3(2H)-dione;* 2-pivaloyl-1,3-indandione; 2-pivalyl-1,3-indandione; 2-trimethylacetyl-1,3-indandione; pivalyl indandione; pivaldione; Pival; Pivalyl Valone; Tri-Ban. $C_{14}H_{14}O_3$; mol wt 230.25. C 73.02%, H 6.13%, O 20.85%. Prepn: Kilgore *et al., Ind. Eng. Chem.* **34**, 494 (1942); Zelmens, Vanags, *C.A.* **53**, 21830d (1959).

Bright yellow crystals from ethanol, mp 108-110°. LD_{50} orally in male rats: 280 mg/kg, T. B. Gaines, *Toxicol. Appl. Pharmacol.* **14**, 515 (1969).

Sodium salt, *Pivalyn.* Bright yellow crystals, mp 205-

210°. Schwarz, U.S pat. **2,880,132** (1959 to Morton Chem.). Sol in water.

USE: Rodenticide; insecticide. *Caution:* Reduces blood coagulation. Symptoms resemble warfarin, *q.v.*

7414. α-Pinene. *2,6,6-Trimethylbicyclo[3.1.1]hept-2-ene; 2-pinene; pinene.* $C_{10}H_{16}$; mol wt 136.23. C 88.16%, H 11.84%. Obtained from oil of turpentine which contains 58-65% α-pinene along with 30% β-pinene, *q.v.*: E. Gildemeister, F. Hoffmann, *Die ätherischen Oele* **Band IV** (Akademie-Verlag, Berlin, 4th ed., 1956) p 39. α-Pinene in North American oils is dextrorotatory, in most European oils it is levorotatory. Constituent of many volatile oils. Isoln of *d*-α-pinene from Port Oxford cedar wood oil *(Chamaecyparis lawsoniana* Parl., *Pinaceae):* Thurber, Roll, *Ind. Eng. Chem.* **19**, 739 (1927). Isoln of *l*-α-pinene from mandarin peel oil *(Citrus reticulata* Blando, *Rutaceae):* Kugler, Kováts, *Helv. Chim. Acta* **46**, 1480 (1963). Total synthesis of α- and β-forms: Komppa, *Ann. Acad. Sci. Fennicae* **A59**, 3 (1943), *C.A.* **41**, 425 (1947); Thomas, Fallis, *Tetrahedron Letters* **1973**, 4687; *eidem, J. Am. Chem. Soc.* **98**, 1227 (1976). *Review:* Palmer, *Ind. Eng. Chem.* **34**, 1028 (1942); J. L. Simonsen, *The Terpenes* vol. II (Cambridge Univ. Press, 2nd ed., 1949) pp 105-191; D. V. Banthorpe, D. Whittaker, *Chem. Rev.* **66**, 643-654 (1966); *Food Cosmet. Toxicol.* **16**, Suppl. I, 853-857 (1978).

dl-Form, liquid, characteristic odor of turpentine. bp_{760} 155-156°; bp_{20} 52.5°. d_4^{20} 0.8592. n_D^{20} 1.4664. Practically insol in water; sol in alc, chloroform, ether, glacial acetic acid.

Hydrochloride, $C_{10}H_{17}Cl$, mp 132°.

d-Form, bp_{760} 155-156°. d_4^{20} 0.8591. n_D^{20} 1.4663. $[\alpha]_D^{20}$ +51.14°.

Hydrochloride, mp 132°. $[\alpha]_D^{20}$ +33.52° (alcohol).

l-Form, bp_{760} 155-156°. d_4^{20} 0.8590. n_D^{20} 1.4662. $[\alpha]_D^{20}$ -51.28°.

Hydrochloride, mp 132°. $[\alpha]_D^{20}$ -33.24° (alcohol).

USE: Manufacture of camphor, insecticides, solvents, plasticizers, perfume bases, synthetic pine oil. *Caution:* Toxic effects similar to turpentine, *q.v.*

7415. β-Pinene. *6,6-Dimethyl-2-methylenebicyclo-[3.1.1]heptane; nopinene.* $C_{10}H_{16}$; mol wt 136.23. C 88.16%, H 11.84%. Found in most essential oils which contain α-pinene, but in far smaller proportions; the *l*-form occurs most commonly. Initial identification: A. von Baeyer, *Ber.* **29**, 25 (1896). Isoln of the *d*-form from *Ferula galbaniflua* Boiss. et Buhse, *Umbelliferae:* B. N. Rutovski, I. V. Vinogradova, *J. Prakt. Chem.* **120**, 41 (1928); from *Cynomara-thrum nuttallii* A. Gray, *Umbelliferae:* E. K. Nelson, *J. Am. Chem. Soc.* **55**, 3400 (1933). Irreversible isomerization of β-pinene to α-pinene occurs on shaking with platinum black satd with hydrogen: F. Richter, W. Wolff, *Ber.* **59**, 1733 (1926). Synthesis: G. Bonnet, *Bull. Inst. Pin* **1938**, 217; **1939**, 1, *C.A.* **33**, 4223⁴ (1939); K. J. Crowley, *Proc. Chem. Soc. (London)* **1962**, 245; *Tetrahedron* **21**, 1001 (1965); L. M. Harwood, M. Julia, *Synthesis* **1980**, 456. For general refs *see* α-pinene.

dl-Form, bp_{760} 165-166°.
d-Form, bp_{760} 164-166°. d_{20}^{20} 0.8654. n_D^{20} 1.4739. $[\alpha]_D$

+28.59° (Nelson). Also reported as bp_{760} 162-163°. d_{20}^{20} 0.8662. n_D^{20} 1.4745. $[\alpha]_D$ +20.75° (Rutovski, Vinogradova). *l*-Form, bp_{760} 162-163°. d^{15} 0.874. n_D^{15} 1.4872, $[\alpha]_D$ -22.4°.

7416. Pine Oil. Yarmor. An oil from *Pinus palustris* Mill. and certain other species of pines, *Pinaceae.* It is obtained from pitch-soaked pine wood by steam distillation or solvent extraction followed by steam distillation and also by destructive distillation. It consists mainly of isomeric tertiary and secondary, cyclic terpene alcohols.

Colorless to pale yellow liquid, turpentine-like odor. d about 0.9; bp 200-220°. Insol in water. Sol in the usual organic solvents.

Human Toxicity: Irritating to skin, mucous membranes. Large doses may cause CNS depression.

USE: Pharmaceutic aid (flavor and perfume). Manuf terpin hydrate and other terpin products; as a solvent, disinfectant and deodorant; in textile scouring; for flotation of lead and zinc ores.

7417. Pine Tar. A product obtained by destructive distillation of wood of *Pinus palustris* Mill., or other species of pine, *Pinaceae.*

Blackish-brown, viscous liquid; heavier than water; empyreumatic odor and sharp taste. Slightly sol in water; sol in alc, chloroform, ether, acetone, glacial acetic acid, fixed and volatile oils, and in solns of caustic alkalies. Principal constituents: turpentine, resin, guaiacol, creosol, methylcreosol, phenol, phlorol, toluene, xylene, and other hydrocarbons.

THERAP CAT: Topical antieczematic; rubefacient.

THERAP CAT (VET): Mild irritant, antiseptic in chronic skin conditions. Expectorant.

7418. Pinguinain. A proteolytic enzyme obtained from the juice of the fruit of *Bromelia pinguin* Plum. ex L., *Bromeliaceae* (pineapple family): Asenjo, *Science* **95**, 48 (1942); Bloch, Messing, U.S. pat. **2,977,287** (1961 to Ethicon). Immunochemical studies: E. Toro-Goyco, I. Rodriguez-Costas, *Arch. Biochem. Biophys.* **175**, 359 (1976). Structure studies: E. Toro-Goyco *et al., Biochim. Biophys. Acta* **622**, 151 (1980).

Aq solns are capable of digesting necrotic tissue, but do not attack viable tissue. Optimum proteolytic activity is at pH 5.2 to 5.5 and also at pH 7.3. Inactivated at temps above 80°.

7419. Pinosylvin. *(E)-5-(2-Phenylethenyl)-1,3-benzenediol; E-3,5-stilbenediol;* 5-styrylresorcinol; *trans-3,5-dihydroxystilbene.* $C_{14}H_{12}O_2$; mol wt 212.24. C 79.22%, H 5.70%, O 15.08%. Occurs together with its monomethyl and dimethyl ethers in the heartwood of pine and other woody plants. Naturally occurring pinosylvins have the *trans* configuration. Isoln from *Pinus sylvestris* L., *Pinaceae:* H. Erdtman, *Ann.* **539**, 116 (1939); from other *Pinus* species: G. Lindstedt, *Acta Chem. Scand.* **3**, 755-772 (1949); J. C. Alvarez-Novoa *et al., ibid.* **4**, 444 (1950); from *Alnus sieboldiana, Betulaceae:* Y. Asakawa, *Bull. Chem. Soc. Japan* **44**, 2761 (1971); from *Polygonum nodosum, Polygonaceae:* M. Kuroyanagi *et al., Chem. Pharm. Bull.* **30**, 1602 (1982). Synthesis of pinosylvin: E. Späth, F. Liebherr, *Ber.* **74**, 869 (1941); of monomethyl ether: E. Späth, K. Kromp, *ibid.* 1424; of dimethyl ether: G. Aulin-Erdtman, H. Erdtman, *ibid.* 50; of pinosylvin and derivatives: A. A. Loman, L. R. Snowdon, *Can. J. Chem.* **48**, 1554 (1970). Biosynthesis: Birch, *Fortschr. Chem. Org. Naturst.* **14**, 186 (1957). Toxicological study: K. O. Frykholm, *Nature* **155**, 454 (1945). Use as antimicrobial agent: E. H. Sheers, *Ger.* pat. **1,952,451**; *idem,* U.S. pat. **3,577,230** (1970, 1971 both to Arizona Chem. Co.). Deterrent to feeding behavior of snowshoe hare: J. P. Bryant *et al., Science* **222**, 1023 (1983).

Fine needles from glacial acetic acid, mp 155.5-156°. uv max (ethanol): 305 nm (log ε 4.49). Practically insol in water. Sol in benzene, acetone, chloroform, glacial acetic acid.

Monomethyl ether, $C_{15}H_{14}O_2$, crystals, mp 122-123°. uv max (ethanol): 303 nm (log ε 4.26). More sol in benzene than pinosylvin. Also sol in methanol, glacial acetic acid.

Dimethyl ether, $C_{16}H_{16}O_2$, crystals from methanol-water, mp 55-56°. uv max (ethanol): 305 nm (log ε 4.39).

7420. Pipacycline. *4-(Dimethylamino)-1,4,4a,5,5a,6,-11,12a-octahydro-3,6,10,12,12a-pentahydroxy-N-[[4-(2-hydroxyethyl)-1-piperazinyl]methyl]-6-methyl-1,11-dioxo-2-naphthacenecarboxamide;* N-[[4-(2-hydroxyethyl)-1-piper-azinyl]methyl]tetracycline; N-[4-(β-hydroxyethyl)diethyl-enediamino-1-methyl]tetracycline; mepicycline; mepiciclina; Ambraveine; Ambra-vena; Sieromicin; Valtomicina; Valto-mycin. $C_{29}H_{38}N_4O_9$; mol wt 586.63. C 59.37%, H 6.53%, N 9.55%, O 24.55%. Semi-synthetic broad spectrum antibiotic related to tetracycline. Prepn: Pedrazzoli et al., *Boll. Chim. Farm.* **98**, 516 (1959), *C.A.* **54**, 3856a (1960); Gradnik et al., **Brit. pat. 888,968** corresp to **U.S. pat. 3,149,114** (1962 and 1964 to E.R.A.S.M.E.). Properties: eidem, *Pharm. Acta Helv.* **35**, 529 (1960). Pharmacokinetic studies: A. Scalvini, A. Delmonte, *Gazz. Med. Ital.* **131**, 1 (1972).

Yellow cryst powder, dec 162-163°. [α]$_D^{20}$ −195° (c = 0.5). [α]$_D^{20}$ −175° (c = 0.5 in methanol). uv max (10 γ/ml 0.1N HCl): 286, 355 nm. pH of 2% aq soln, 7.2-7.4. Freely sol in water, methanol, formamide; slightly sol in ethanol, isopropanol; practically insol in ether, benzene, chloroform. Sensitive to light, heat, and air. LD$_{50}$ i.v. in white mice: 188 mg/kg (Scalvini, Delmonte).

THERAP CAT: Antibacterial.

7421. Pipamazine. *1-[3-(2-Chloro-10H-phenothiazin-10-yl)propyl]-4-piperidinecarboxamide;* 1-[3-(2-chlorophe-nothiazin-10-yl)propyl]isonipecotamide; 10-[3-(4-carbamo-ylpiperidin-1-yl)propyl]-2-chlorophenothiazine; 2-chloro-10-[3-(4-carbamoylpiperidinyl)propyl]phenothiazine; 10-[3-(4-carbamoylpiperidino)propyl]-2-chlorophenothiazine; SC 9387; Nometine; Nausidol; Mornidine. $C_{21}H_{24}ClN_3OS$; mol wt 401.97. C 62.75%, H 6.02%, Cl 8.82%, N 10.45%, O 3.98%, S 7.98%. Prepn: Cusic et al., **U.S. pat. 2,957,870** (1960 to Searle).

Crystals from 2-propanol + petr ether, mp about 139°. Hydrochloride, crystals, mp about 196-197° with forma-tion of bubbles.

THERAP CAT: Anti-emetic.

7422. Pipamperone. *1'-[4-(4-Fluorophenyl)-4-oxobut-yl]-[1,4'-bipiperidine]-4'-carboxamide;* 1'-[3-(p-fluoro-benzoyl)propyl]-[1,4'-bipiperidine]-4'-carboxamide; 1-(p-fluorophenyl)-4-(4-piperidino-4-carbamoylpiperidino)-1-butanone; 1-[γ-(4-fluorobenzoyl)propyl]-4-piperidinopiper-idine-4-carboxamide; 4'-fluoro-4-[N-[4-(N-piperidino)-4-carbamido]piperidino]butyrophenone; floropipamide; R 3345; Dipiperon; Piperonyl; Propitan. $C_{21}H_{30}FN_3O_2$; mol wt 375.49. C 67.17%, H 8.05%, F 5.06%, N 11.19%, O 8.52%. Prepn of the dihydrochloride by reaction of γ-chloro-4-fluorobutyrophenone and 4-piperidinopiperidine-4-carbox-

amide: Janssen, **Belg. pat. 610,830** (1962 to Janssen), *C.A.* **57**, 13740b (1962).

Dihydrochloride, $C_{21}H_{30}FN_3O_2$.2HCl. Crystals, mp 124.5-126.0°.

THERAP CAT: Antipsychotic.

7423. Pipazethate. *10H-Pyrido[3,2-b][1,4]benzothiadi-azine-10-carboxylic acid 2-(2-piperidinoethoxy)ethyl ester;* 2-(2-piperidinoethoxy)ethyl 10H-pyrido[3,2-b][1,4]benzo-thiadiazine-10-carboxylate; thiophenylpyridylamino-10-carboxylic acid piperidinoethoxyethyl ester; 2-(2-piperidino-ethoxy)ethyl 10-thia-1,9-diazaanthracene-10-carboxylate; 1-azaphenothiazine-10-carboxylic acid 2-(2-piperidinoeth-oxy)ethyl ester; D 254. $C_{21}H_{25}N_3O_3S$; mol wt 399.52. C 63.13%, H 6.31%, N 10.52%, O 12.01%, S 8.03%. Prepn: Schuler et al., *Ann.* **673**, 102 (1964); Schuler, **U.S. pat. 2,989,529** (1961 to Degussa).

Hydrochloride, $C_{21}H_{26}ClN_3O_3S$, *Lenopect, Selvigon, Theratuss.* Pale yellow crystals from isopropanol, mp 160-161°. Soluble in water, methanol. Practically insol in ace-tone, petr ether. LD$_{50}$ orally in rats: 560 mg/kg (Schuler).

THERAP CAT: Antitussive.

7424. Pipebuzone. *4-Butyl-4-[(4-methyl-1-piperazinyl)-methyl]-1,2-diphenyl-3,5-pyrazolidinedione;* 1,2-diphenyl-3,5-dioxo-4-[(4-methyl-1-piperazinyl)methyl]-4-butylpyra-zolidine; LD 4644; Elarzone. $C_{25}H_{32}N_4O_2$; mol wt 420.57. C 71.40%, H 7.67%, N 13.32%, O 7.61%. The 4-[(4-methyl-1-piperazinyl)methyl]- deriv of phenylbutazone, *q.v.* Prepn: **Ger. pat. 1,958,722** and **Fr. pat. M7914** (both 1970 to Dausse), *C.A.* **73**, 25460v (1970); **76**, 140797j (1972).

Crystals from isopropanol, mp 129°.

THERAP CAT: Anti-inflammatory; antipyretic; analgesic.

7425. Pipecolic Acid. *2-Piperidinecarboxylic acid;* pipe-colinic acid; hexahydropicolinic acid; homoproline; dihydro-baikiaine. $C_6H_{11}NO_2$; mol wt 129.16. C 55.79%, H 8.58%, N 10.85%, O 24.78%. The l-form occurs in plants: Phillips, *Chem. & Ind. (London)* 1953, 127. Prepn: A. Ladenburg, *Ber.* **24**, 640 (1891); Stevens, Ellman, *J. Biol. Chem.* **182**, 75 (1950); V. Asher et al., *Tetrahedron Letters* **22**, 141 (1981). Synthesis of L-pipecolic acid from L-lysine: Fujii, Miyoshi, *Bull. Chem. Soc. Japan* **48**, 1341 (1975).

l-Form, needles by sublimation, mp 270°. $[\alpha]_D^{25}$ −34.9°. Sol in water, dil alcohol. Sparingly sol in abs alcohol, acetone, chloroform. Insol in ether.

d-Form, platelets from alcohol, mp 270°. $[\alpha]_D^{25}$ +35.7°. Sol in water. Somewhat sol in alcohol.

dl-Form, leaflets from water, mp 264°. Sol in water, boiling alcohol.

dl-Hydrochloride, $C_6H_{11}NO_2 \cdot HCl$, crystals (warts) from alcohol + benzene, mp 258-262°.

7426. Pipecurium Bromide. *4,4'-[3,17-Bis(acetyloxy)-androstane-2,16-diyl]bis(1,1-dimethylpiperazinium)dibromide dihydrate;* 4,4'-(3α,17β-dihydroxy-5α-androstan-2β,-16β-ylene)bis(1,1-dimethylpiperazinium)dibromide diacetate dihydrate; pipecuronium bromide; RGH 1106; Arduan. $C_{35}H_{62}Br_2N_4O_4 \cdot 2H_2O$; mol wt 787.77. C 52.63%, H 8.33%, Br 20.01%, N 7.01%, O 12.02%. Non-depolarizing curare-like agent; deriv of androstane, *q.v.*, related structurally to pancuronium bromide, *q.v.* Prepn: Z. Tuba *et al.*, Ger. pat. **2,337,882** (1974 to Gedeon Richter), *C.A.* **80**, 121210d (1974); Z. Tuba, *Arzneimittel-Forsch.* **30**, 342 (1980). Series of articles on pharmacology, properties, disposition, pharmacokinetics, safety tests, clinical studies: *ibid.* 346-394. Determn in human serum: G. Szabo, E. Tassonyi, *ibid.* **31**, 1013 (1981).

Cryst from methylene dichloride/acetone, mp 262-264° (dec). $[\alpha]_D^{25}$ +8.1° (c = 1 in water). LD_{50} in mice, rats (mg/kg): 29.7, 172.6 i.v.; 70.6, 449.6 i.p.; 60.5, 455.8 s.c., E. Kárpáti, K. Biro, *Arzneimittel-Forsch.* **30**, 346 (1980).

THERAP CAT: Relaxant (skeletal muscle).

7427. Pipemidic Acid. *8-Ethyl-5,8-dihydro-5-oxo-2-(1-piperazinyl)pyrido[2,3-d]pyrimidine-6-carboxylic acid;* piperamic acid; 1489 RB; Dolcol; Filtrax; Memento 400; Pipeacid; Pipedac; Pipedase; Pipemid; Pipram; Pipurin; Tractur; Uropimid; Urosten; Uroval. $C_{14}H_{17}N_5O_3$; mol wt 303.33. C 55.43%, H 5.65%, N 23.09%, O 15.82%. Quinolone antibacterial. Prepn: S. Minami *et al.*, **Ger. pat. 2,341,146;** *eidem,* **U.S. pat. 3,887,557** (1974, 1975 both to Dainippon Pharm.); Pesson *et al.*, *C.R. Acad. Sci., Ser. C* **278**, 1169 (1974); De Lajudie *et al., ibid. Ser. D* **279**, 1931 (1974); Matsumoto, Minami, *J. Med. Chem.* **18**, 74 (1975). Characterization: Ficicchia, *Farmaco Ed. Prat.* **30**, 207 (1974). Antibacterial activity: Ficicchia, De Lajudie, *ibid.* 252. Pharmacology: Shimizu *et al., Antimicrob. Ag. Chemother.* **7**, 441 (1975).

Yellowish-white, odorless, bitter-tasting crystals, mp 253-255°. Hygroscopic. Yellows slowly in light. Sol in acid soln and alkaline. Very slightly sol in water, alcohol; slightly sol in chloroform (0.5%), methanol (0.4%). Practically insol in ether, benzene. LD_{50} in mice (mg/kg): 4000 orally; 1000 i.p., 50 i.v. (Ficicchia).

Trihydrate, *Deblaston, Solupemid.* Nearly colorless needles, mp 253-255°.

THERAP CAT: Antibacterial.

7428. Pipenzolate Bromide. *1-Ethyl-3-[(hydroxydiphenylacetyl)oxy]-1-methylpiperidinium bromide; 1-ethyl-3-hydroxy-1-methylpiperidinium bromide benzilate;* benzilic acid, 1-ethyl-3-piperidyl ester methyl bromide; *N*-ethyl-3-

piperidyl benzilate methobromide; pipenzolate methylbromide; pipenzolone bromide; JB 323; Piptal. $C_{22}H_{28}BrNO_3$; mol wt 434.38. C 60.83%, H 6.50%, Br 18.40%, N 3.22%, O 11.05%. Synthesis starting with *N*-ethyl-3-chloropiperidine and benzilic acid: Biel *et al., J. Am. Chem. Soc.* **74**, 1485 (1952); U.S. pat. **2,918,406** (1959 to Lakeside).

Crystals from methyl ethyl ketone, mp 179-180°. Soluble in water.

Note: *N*-Ethyl-3-piperidyl benzilate is a controlled substance (hallucinogen) listed in the U.S. Code of Federal Regulations, Title 21, Part 1308.11 (1987).

THERAP CAT: Anticholinergic.

7429. Piperacetazine. *1-[10-[3-[4-(2-Hydroxyethyl)-1-piperidinyl]propyl]-10H-phenothiazin-2-yl]ethanone; 10-[3-[4-(2-hydroxyethyl)piperidino]propyl]phenothiazin-2-yl methyl ketone;* 2-acetyl-10-[3-[4-(β-hydroxyethyl)piperidino]propyl]phenothiazine; 2-acetyl-10-[3-[γ-(2-hydroxyethyl)piperidino]propyl]phenothiazine; PC-1421; Psymod; Quide. $C_{24}H_{30}N_2O_2S$; mol wt 410.59. C 70.21%, H 7.37%, N 6.82%, O 7.79%, S 7.81%. Prepn: **Brit. pat. 861,807** (1961 to Searle).

Hydrochloride, mp 100-110°.

THERAP CAT: Antipsychotic.

THERAP CAT (VET): Tranquilizer.

7430. Piperacillin. *6-[[[[(4-Ethyl-2,3-dioxo-1-piperazinyl)carbonyl]amino]phenylacetyl]amino]-3,3-dimethyl-7-oxo-4-thia-1-azabicyclo[3.2.0]heptane-2-carboxylic acid;* (2S,5R,6R)-6-[(R)-2-(4-ethyl-2,3-dioxo-1-piperazinecarboxamido)-2-phenylacetamido]-3,3-dimethyl-7-oxo-4-thia-1-azabicyclo[3.2.0]heptane-2-carboxylic acid; 6-D-(-)-α-(4-ethyl-2,3-dioxo-1-piperazinylcarbonylamino)-α-phenylacetamidopenicillanic acid; 4-ethyl-2,3-dioxopiperazine carbonyl ampicillin. $C_{23}H_{27}N_5O_7S$; mol wt 517.57. C 53.37%, H 5.26%, N 13.53%, O 21.64%, S 6.19%. Broad spectrum semi-synthetic antibiotic related to pencillin. Prepn: I. Saikawa *et al.*, **Ger. pat. 2,519,400;** *eidem,* **U.S. pat. 4,087,424** (1976, 1978 both to Toyama). *In vitro* studies: G. P. Bodey, B. LeBlanc, *Antimicrob. Ag. Chemother.* **14**, 78 (1978); R. Wise *et al., ibid.* 549. *In vitro* and *in vivo*: N. A. Kuck, G. S. Redin, *J. Antibiot.* **31**, 1175 (1978). Pharmacokinetics: M. A. Evans *et al., J. Antimicrob. Chemother.* **4**, 255 (1978). Metabolism: K. Iida *et al., Antimicrob. Ag. Chemother.* **14**, 257 (1978). Clinical study: T. Saito, Y. Yamada, *Japan. J. Antibiot.* **30**, 835 (1977). Toxicity study: A. Takai *et al., Chemother. (Tokyo)* **25**, 816 (1977). Review of antibacterial activity, pharmacokinetics: G. L. Mandell, *Clin. Ther.* **7**, Suppl. B, 28-35 (1985). Review of clinical experience in urinary tract infections: S. J. Childs, *ibid.* 36-45. Series of articles on antibacterial activity and clinical use: *Chemotherapy (Tokyo)* **36**, Suppl. 7, 1-85 (1988).

Sodium salt, $C_{23}H_{26}N_5NaO_7S$, *Cl 227193, T-1220, Isipen, Pentcillin, Pipracil, Pipril.* mp 183-185° (dec). LD_{50} in mice, rats, dogs, monkeys (g/kg): 5, 2.7, >6, >4 i.v. (Takai).

THERAP CAT: Antibacterial.

7431. Piperazine. Hexahydropyrazine; piperazidine; diethylenediamine; Lumbrical; Worm-Away; Wurmirazin. $C_4H_{10}N_2$; mol wt 86.14. C 55.77%, H 11.70%, N 32.52%. Prepd by the action of alcoholic ammonia on ethylene chloride: Cloez, *Jahresber.* **1853,** 468; by the reduction of pyrazine with sodium in alcohol: Wolff, *Ber.* **26,** 724 (1893); by catalytic deamination of diethylenetriamine and of ethylenediamine: Kyrides, U.S. pat. **2,267,686** (1941); Martin, Martell, *J. Am. Chem. Soc.* **70,** 1817 (1948); MacKenzie, Turbin, U.S. pat. **2,901,482** (1959 to Dow); Moss, Godfrey, U.S. pat. **3,037,023** (1962 to Jefferson Chem.).

Leaflets from alc, mp 106°. Salty taste. bp_{760} 146°. Strong base: pKa = 4.19, absorbs water and CO_2 from air. Absorption spectrum: Purvis, *J. Chem. Soc.* **103,** 2286, 2293 (1913). Freely sol in water, glycerol, glycols; one gram dissolves in 2 ml of 95% alcohol. Insol in ether. pH of a 10% aq soln 10.8-11.8. Forms a solid compd with theophylline. *Keep tightly closed and protect from light.*

Hexahydrate, $C_4H_{10}N_2.6H_2O$, *Anthalazine, Arpezine, Arthriticine, Ascaril, Dispermin, Eraverm (syrup), Helmifren, Parid, Piavetrin, Tasnon (elixir), Upixon, Vermicompren (syrup), Vermisol.* Crystals from water (contg 44.34% anhydr piperazine), mp 44°. bp 125-130° The piperazine of commerce is usually this hydrate. Freely sol in water; sol in alc (about 1:2). Practically insol in ether. pH of a 10% aq soln 10.8-11.8.

Phosphate, $C_4H_{13}N_2O_4P$, *Eraverm (tabl.), Pincets, Pinsirup, Piperaverm (tabl.), Piperazate, Pripsen, Tasnon (tabl.), Uvilon (tabl.).* Minute crystals. Very slightly sol in water. pH of saturated aq soln, 6.5.

N^1,N^4-Dibenzoylpiperazine, $C_{18}H_{18}N_2O_2$, prepd from piperazine and benzoyl chloride in dil NaOH. Crystals from alc, mp 191°.

THERAP CAT: Anthelmintic (Nematodes).

THERAP CAT (VET): Anthelmintic (Nematodes).

7432. Piperazine Adipate. *Hexanedioic acid compd with piperazine (1:1);* Dietelmin; Entacyl; Oxyzin (tabl.); Vermicompren (tabl.); Nometan; Vermilass; Oxypaat; Pipadox; Oxurasin; Adiprazine. $C_{10}H_{20}N_2O_4$; mol wt 232.28. C 51.70%, H 8.68%, N 12.06%, O 27.55%. $C_4H_{10}N_2.C_6H_{10}O_4$. The neutral salt of piperazine and adipic acid: Davies *et al., J. Pharm. Pharmacol.* **6,** 707 (1954). Prepn in methanolic medium: Forrest, Petrow, U.S. pat. **2,799,617** (1957 to British Drug Houses); Brit. pat. **767,826.** Pharmacology: B. G. Cross *et al., J. Pharm. Pharmacol.* **6,** 711 (1954).

Prisms, mp 256-257°. Stable to heat and air. Pleasant, slightly acid taste. Dissolves slowly. Soly in 100 ml water at 20°: 5.53 g, at 30°: 6.61 g, at 56.3°: 10.14 g; in 100 g methanol at 25°: 0.02 g. Practically insol in abs ethanol, isopropanol, dioxane. Aq solns of 0.2-0.01M have pH 5.45; this pH is only slightly affected by increases in ionic strength caused by the addition of simple neutral salts. LD_{50} in mice, rats (g/kg): 11.4, 7.9 orally (Cross).

THERAP CAT: Anthelmintic (Nematodes).

THERAP CAT (VET): Anthelmintic (Nematodes).

7433. Piperazine Citrate. *Piperazine 2-hydroxy-1,2,3-propanetricarboxylate (3:2);* tripiperazine dicitrate; Antepar; Multifuge; Oxucide; (3P-2C); Piperaverm (syrup); Pipizan Citrate; Pinrou; Piptelate; Ta-Verm; Exelmin; Rhomex; Nemadital; Pinozan; Pipracid (syrup); Exopin; Uvilon (syrup); Oxyzin (syrup); Parazine; Helmezine; Arpezine. $C_{24}H_{46}N_6O_{14}$; mol wt 642.68. C 44.85%, H 7.21%, N 13.08%, O 34.85%. $3C_4H_{10}N_2.2C_6H_8O_7$. Formed from 3 moles piperazine and 2 moles citric acid: Hefferren *et al., J. Am. Pharm. Assoc.* **44,** 678 (1955).

Crystals, dec 182-187°. Freely sol in water. Practically insol in alcohol, ether, chloroform. pH of a 10% aq soln 5.0-6.0.

THERAP CAT: Anthelmintic (Nematodes).

THERAP CAT (VET): Anthelmintic.

7434. 2,5-Piperazinedione. Glycine anhydride; cycloglycylglycine; α,γ-diacipiperazine; glycylglycine lactam; diglycolyldiamide; 2,5-diketopiperazine; 2,5-dioxopiperazine. $C_4H_6N_2O_2$; mol wt 114.10. C 42.10%, H 5.30%, N 24.55%, O 28.04%. Can exist in 5 enol forms: Richardson *et al., J. Am. Chem. Soc.* **51,** 3074 (1929). Prepn from glycine ethyl ester hydrochloride: Fischer, *Ber.* **39,** 2930 (1906).

Platelets from water, needles by sublimation. Dec 311-312° (begins to sublime at 260° and sinters at 305°). Sparingly sol in water; sol in HCl (d 1.19) from which it can be precipitated by the addition of alcohol. Acts as a weak base. Hydrolyzed to glycylglycine by acids and alkalies.

Hydrochloride, $C_4H_6N_2O_2.HCl$, crystals, mp 129-130°.

7435. Piperazine Edetate Calcium. *[[N,N'-1,2-Ethanediylbis[N-(carboxymethyl)glycinato]](4−)-N,N',O,O',O^N,-O^N]-(OC-6-21)calciate (2−) dihydrogen compd with piperazine (1:1);* (ethylenedinitrilo)tetraacetic acid piperazine calcium salt; piperazine calcium edetate; piperazine calcium edathamil; Justelmin; Perin. $C_{14}H_{24}CaN_4O_8$; mol wt 416.45. C 40.38%, H 5.81%, Ca 9.62%, N 13.45%, O 30.74%. Prepd by the action of ethylenediaminetetraacetic acid on calcium carbonate and piperazine: Schlesinger *et al.,* U.S. pat. **2,-834,782** (1958 to Endo).

Dihydrate, crystals; slightly salty taste. Freely sol in water. Very slightly sol in alcohol, chloroform. Practically insol in ether. pH of 20% aq soln 4.3-5.4.

THERAP CAT: Anthelmintic (Nematodes).

7436. Piperazine Tartrate. *[R-(R*,R*)]-Piperazine-2,3-dihydroxybutanedioate (1:1);* piperate; Veroxil; Noxiurotan; Paravermin; (1P-1T). $C_8H_{16}N_2O_6$; mol wt 236.23. C 40.67%, H 6.83%, N 11.86%, O 40.64%. $C_4H_{10}N_2.C_4H_6O_6$. Formed from one mole piperazine and one mole tartaric acid: Hefferren *et al., J. Am. Pharm. Assoc.* **44,** 678 (1955).

Crystals, dec 258-263°. Soly in g/100 ml at 25°: water 26; alcohol 0.01; chloroform 0.01. pH of 1% soln: 4.8.

THERAP CAT: Anthelmintic (Nematodes).

7437. Piperic Acid. 5-(1,3-Benzodioxol-5-yl)-2,4-pentadienoic acid; 5-(3,4-methylenedioxyphenyl)-2,4-pentadienoic acid. $C_{12}H_{10}O_4$; mol wt 218.20. C 66.05%, H 4.62%, O 29.33%. Early literature treats the title compound without specifying stereochemistry; however, four isomers exist: piperinic acid *(E,E)*; isochavicinic acid *(E,Z)*; isopiperinic acid *(Z,E)*; chavicinic acid *(Z,Z)*. Physical constants given in early references for piperic acid agree with those of piperinic acid. *See* Beilstein **19,** 281; suppl. II, 300. Synthesis of *(E,E)*-form: L. von Babo, E. Keller, *J. Prakt. Chem.* **72,** 53 (1857); A. Ladenburg, M. Sholtz, *Ber.* **27,** 2958 (1894); of *(E,Z)*-form: H. Lohaus, H. Gall, *Ann.* **517,** 278 (1935); of *(Z,E)*-form: E. Ott, F. Eichler, *Ber.* **55,** 2653 (1922); of

(Z,Z)-form: *idem, loc. cit.;* H. Lohaus, H. Gall, *loc. cit.;* of all isomers: R. Grewe *et al., Ber.* **103,** 3752 (1970); R. De Cleyn, A. Verzele, *Bull. Soc. Chim. Belg.* **81,** 529 (1972).

(E,E)-*Form, **piperinic acid, piperinsäure (German).*** Needles from alc. Colorless when freshly prepd, rapidly turns yellow on exposure to light. mp 216-217°. Sublimes as yellow needles with partial decompn. uv max (methanol): 340 nm (ε 28,800). Sol in 50 parts boiling alcohol, 275 parts abs alcohol at 25°. Practically insol in water, ether, benzene, carbon disulfide.

(E,Z)-*Form, **isochavicinic acid.*** Yellow crystals from methanol/water, sublimes as needles. mp 134-136° (Lohaus, Gall); 143° (Grewe). uv max (methanol): 335 nm (ε 14,500). Sol in methanol, benzene.

(Z,E)-*Form, **isopiperinic acid.*** Needles from benzene. mp 145° (Ott, Eichler); 153° (Grewe); 138° (DeCleyn, Verzele). uv max (methanol): 328 nm (ε 22,000). Sol in benzene.

(Z,Z)-*Form, **chavicinic acid.*** Amorphous yellow granules from benzene. Wide disparities in mp have been reported: 200-202° (Ott, Eichler); 130° (Grewe); 120° (De Cleyn, Verzele). uv max (methanol): 335 nm (ε 17,500). 0.55 g sol in 16 g boiling 95% alcohol, and in 65 g boiling benzene.

7438. Piperidine. Hexahydropyridine. $C_5H_{11}N$; mol wt 85.15. C 70.52%, H 13.02%, N 16.45%. Found in small quantities in *Piper nigrum* L., *Piperaceae* (black pepper). May be obtained from piperine by heating with alcoholic KOH, or from 1,5-diaminopentane hydrochloride by cyclization. Usually prepd by electrolytic reduction of pyridine. Forms complexes with salts of heavy metals. Because of its reactivity, piperidine is useful in the prepn of cryst derivatives of aromatic nitro compds contg nuclear halogen atoms: Seikel, *J. Am. Chem. Soc.* **62,** 750 (1940). Review of physical constants of piperidine and *N*-alkyl piperidines: Magnusson, Schierz, *Univ. Wyoming Pub.* **7,** 1 (1940).

Liquid. Characteristic odor. Soapy feel. Solidifies −13° to −17°; mp −7°; bp₇₆₀ 106°; bp₂₀ 18°; d_4^{20} 0.8622; n_D^{20} 1.4534. Infrared absorption spectrum: Freymann, *Compt. Rend.* **205,** 852 (1937); *Ann. Chim.* **11,** 11 (1939). Ultraviolet and Raman spectra: Lecomte, *Compt. Rend.* **207,** 395 (1938). Strong base: pK at 25° = 2.80; K = 1.6 × 10⁻³. Misc with water. Sol in alcohol, benzene, chloroform. LD₅₀ orally in rats: 0.52 ml/kg, H. F. Smyth *et al., Am. Ind. Hyg. Assoc. J.* **23,** 95 (1962).

Hydrochloride, $C_5H_{11}N$·HCl, orthorhombic prisms from alcohol, mp 247°. Freely sol in water, alcohol.

Nitrate, $C_5H_{11}N$·HNO₃, hygroscopic plates, sublimes (10 mm) 75°, mp 110°. Freely sol in water, alcohol, ether. Absorption spectrum: Harper, Macbech, *J. Chem. Soc.* **107,** 91 (1915).

Bitartrate, $C_5H_{11}N$·$C_4H_6O_6$, crystals, freely sol in water.

Aurichloride, $C_5H_{11}N$·HAuCl₄, yellow crystals, mp 219°.

Platinichloride, $2C_5H_{11}N$·H_2PtCl_6, yellow monoclinic prisms from water, mp 202°. Freely sol in water, slightly in alcohol.

N-Benzoylpiperidine, long needles, mp 44-48°. *Ref: Org. Syn.* **coll. vol. I** (2nd ed., 1941) p 99.

Phosphate, crystals, mp 204-206°. Prepn from piperidine + phosphoric acid: Abood, U.S. pat. **3,035,977** (1962).

7439. Piperidione. *3,3-Diethyl-2,4-piperidinedione; 3,3-diethyl-2,4-dioxopiperidine;* dihyprylone; Sedulon; Tusseval. $C_9H_{15}NO_2$; mol wt 169.22. C 63.88%, H 8.94%, N 8.27%,

O 18.91%. Prepn: Tsukita, *J. Pharm. Soc. Japan* **69,** 194 (1949).

Crystals, mp 102-107°. Bitter taste. Freely sol in water, alcohol, chloroform.

THERAP CAT: Sedative, antitussive.

7440. Piperidolate. α-*Phenylbenzeneacetic acid 1-ethyl-3-piperidinyl ester; diphenylacetic acid 1-ethyl-3-piperidyl ester; N-ethyl-3-piperidyl diphenylacetate; JB 305.* $C_{21}H_{25}NO_2$; mol wt 323.44. C 77.98%, H 7.79%, N 4.33%, O 9.89%. Prepd from diphenylacetyl chloride and *N*-ethyl-3-hydroxypiperidine: Biel *et al., J. Am. Chem. Soc.* **74,** 1485 (1952); U.S. pat. **2,918,407** (1959 to Lakeside Labs.).

Liquid. bp₀.₁₈ 191-192°.
Hydrochloride, $C_{21}H_{26}ClNO_2$, *Crapinon; Dactil.* Crystals, mp 195-196°. Soluble in water.
THERAP CAT: Anticholinergic.

7441. Piperilate. α-*Hydroxy-α-phenylbenzeneacetic acid 2-(1-piperidinyl)ethyl ester; β-piperidylethyl benzilate; pipethanate; benzilic acid 1-piperidineethanol ester; 1-piperidineethanol benzilate; 2-(1-piperidino)ethyl benzilate.* $C_{21}H_{25}NO_3$; mol wt 339.44. C 74.31%, H 7.42%, N 4.13%, O 14.14%. Prepn: Ford-Moore, Ing, *J. Chem. Soc.* **1947,** 55.

Hydrochloride, $C_{21}H_{26}ClNO_3$, *Daipisate, Norticon, Panpurol (tabl.), Pensanate, Pipenale, Sycotrol.* Crystals from acetone or ethanol, mp 170-171°. Ingredient of *Modutrol.*
Ethyl bromide, $C_{23}H_{30}BrNO_3$, *Panpurol (inj.).*
THERAP CAT: Anticholinergic; antispasmodic.

7442. Piperine. *1-[5-(1,3-Benzodioxol-5-yl)-1-oxo-2,4-pentadienyl]piperidine; 1-piperoylpiperidine.* $C_{17}H_{19}NO_3$; mol wt 285.33. C 71.55%, H 6.71%, N 4.91%, O 16.82%. Isolated from black pepper (*Piper nigrum* L.); also in *P. longum* L., *P. retrofractum* Vahl. (*P. officinarum* C.D.C.), and *P. clusii* C.D.C.; in root bark of *Piper geniculatum* Sw., *Piperaceae.* Extraction procedure: Cazeneuve, Caillot, *Bull. Soc. Chim.* [2] **27,** 291 (1877). Synthesis: Rugheimer, *Ber.* **15,** 1390 (1882); Newman, *Chem. Products* **16,** 379 (1953); Normant, Feugeas, *Compt. Rend.* **258,** 2846 (1964). Spectroscopic structural elucidation and preparative separation of piperine and its stereoisomers isopiperine, isochavicine and chavicine, *q.v.:* R. De Cleyn, M. Verzele, *Bull. Soc. Chim. Belg.* **84,** 435 (1975). Synthesis of isomers: R. Grewe *et al., Ber.* **103,** 3752 (1970); of piperine and isochavicine: S. Tsuboi *et al., Tetrahedron Letters* **1979,** 1043. Stereoselective synthesis of piperine: R. A. Olsen, G. O. Spessard, *J. Agr. Food Chem.* **29,** 942 (1981). More toxic to houseflies than pyrethrum: Harvill *et al., Contrib. Boyce Thompson Inst.* **13,** 87 (1943).

Monoclinic prisms from alcohol, mp 130°. Tasteless at first, but burning aftertaste. Neutral to litmus. pK (18°): 12.22; K = 6 × 10⁻¹³. Almost insol in water (40 mg/liter at 18°), in petr ether. One gram dissolves in 15 ml alcohol, 1.7 ml chloroform, 36 ml ether. Sol in benzene, acetic acid.

(E,Z)-Form, *isochavicine.* Crystals from chloroform/-hexane, mp 89° (Grewe), 103° (De Cleyn). uv max (methanol): 333 nm (ε 16300).

(Z,E)-Form, *isopiperine.* Crystals from chloroform/-hexane, mp 110° (Grewe), 86° (De Cleyn). uv max (methanol): 332 nm (ε 21800).

USE: To impart pungent taste to brandy. As insecticide.

7443. Piperitone. *3-Methyl-6-(1-methylethyl)-2-cyclohexen-1-one; p-menth-1-en-3-one;* 4-isopropyl-1-methyl-1-cyclohexen-3-one. $C_{10}H_{16}O$; mol wt 152.23. C 78.89%, H 10.59%, O 10.51%. Isoln of *d*-form from oil of *Cymbopogon sennaarensis* Chiov., *Gramineae:* Roberts, *J. Chem. Soc.* **107**, 1465 (1915); from oil of *Andropogon iwarancusa* Jones: Simonsen, *ibid.* **119**, 1644 (1921); from oil of *Mentha* spp., *Labiatae:* Reitsma, *J. Am. Pharm. Assoc.* **47**, 265, 267 (1958). Isoln of *l*-form from Sitka spruce oil: von Rudloff, *Can. J. Chem.* **42**, 1057 (1964). Isoln of *dl*-form from oil of *Eucalyptus dives* Schau., *Myrtaceae:* Read, Smith, *J. Chem. Soc.* **119**, 779 (1921); from peppermint oil: Katsuragi, *Koryo No. 24*, 16 (1953). Synthesis of *dl*-form: Misrock, Church, *Ind. Eng. Chem.* **49**, 822 (1957); Bain *et al.*, U.S. pat. **2,972,-632** (1961 to Glidden); Wiemann, Dubois, *Bull. Soc. Chim. France* **1962**, 1813; Stepanov, Myrsina, *Zh. Obshch. Khim.* **34**, 3092 (1964).

d-Form, liquid. Peppermint odor. bp 232-235°; bp₂₀ 116-118.5°. d_4^{20} 0.9344. $[α]_D^{20}$ +49.13°. n_D^{20} 1.4848.

l-Form, liquid. bp₁₅ 109.5-110.5°. d_4^{20} 0.9324. $[α]_D^{20}$ -15.9°. n_D^{20} 1.4823. Practically insol in water; sol in alcohol, oils.

dl-Form, liquid. bp₇₆₉ 232-233°; bp₁₆ 116-118°. d_4^{20} 0.9331. n_D^{24} 1.4823. uv max (ethanol): 232.5 nm (ε 13,350).

USE: In masking odors in dentifrices.

7444. Piperocaine. *2-Methyl-1-piperidinepropanol benzoate;* γ-(2-methylpiperidyl)propyl benzoate; (2-methylpiperidino)propyl benzoate; 3-benzoxy-1-(2-methylpiperidino)propane; benzoyl-γ-(2-methylpiperidino)propanol; Neothesin; Metycaine. $C_{16}H_{23}NO_2$; mol wt 261.37. C 73.53%, H 8.87%, N 5.36%, O 12.24%. Prepn of the hydrochloride from 2-methylpiperidine and γ-chloropropyl benzoate: U.S. pat. **1,784,903** (1930).

Hydrochloride, $C_{16}H_{24}ClNO_2$, crystals, mp 172-175°. Slightly bitter taste followed by numbness of the mouth. Stable in air. One gram dissolves in 1 ml water. Sol in alc, chloroform. Insol in ether, olive oil. A 1:10 aq soln is faintly acid to litmus. Alkali carbonates and hydroxides precipitate piperocaine base as an oil from aq solns of the hydrochloride. LD₅₀ in rats: 1.3 g/kg s.c.; 20 mg/kg i.v.

THERAP CAT: Local anesthetic.

THERAP CAT (VET): Local anesthetic.

7445. Piperonal. *1,3-Benzodioxole-5-carboxaldehyde;* 3,4-(methylenedioxy)benzaldehyde; heliotropin; piperonylaldehyde; dioxymethyleneprotocatechuic aldehyde. $C_8H_6O_3$; mol wt 150.13. C 64.00%, H 4.03%, O 31.97%. Prepn: Blair, U.S. pat. **2,916,499** (1959 to Welsbach Corp.); Holum, *J. Org. Chem.* **26**, 4814 (1961); Feugeas, *Bull. Soc. Chim. France* **1964**, 1892. Toxicity: Hagan *et al.*, *Toxicol. Appl. Pharmacol.* **7**, 18 (1965).

Colorless, lustrous crystals, mp 37°. Heliotrope odor. bp about 263°; bp₀.₅ 88°. Sol in 500 parts water; freely sol in alcohol, ether. *Keep in cool place protected from light.* LD₅₀ orally in rats: 2700 mg/kg (Hagan).

USE: In perfumery, in cherry and vanilla flavors, in organic syntheses.

THERAP CAT: Pediculicide.

7446. Piperonyl Butoxide. *5-[[2-(2-Butoxyethoxy)ethoxy]methyl]-6-propyl-1,3-benzodioxole; α-[2-(2-butoxyethoxy)ethoxy]-4,5-methylenedioxy-2-propyltoluene;* [3,4-(methylenedioxy)-6-propylbenzyl] butyl diethyleneglycol ether; 6-propylpiperonyl butyl diethylene glycol ether; butylcarbityl (6-propylpiperonyl) ether; ENT 14250; Butacide. $C_{19}H_{30}O_5$; mol wt 338.43. C 67.43%, H 8.94%, O 23.64%. Prepn: H. Wachs, U.S. pats. **2,485,681; 2,550,737** (1949, 1951, both to U.S. Industrial Chemicals); *idem, Science* **105**, 530 (1947). Toxicity: T. B. Gaines, *Toxicol. Appl. Pharmacol.* **14**, 515 (1969).

Liquid. d 1.04-1.07. bp₁.₀ 180°. n_D^{20} 1.50. Flash pt 340°F. Miscible with methanol, ethanol, benzene, Freons, Geons, other organic solvents, oils. LD₅₀ orally in female, male rats: 6150, 7500 mg/kg (Gaines).

Available in various mixtures with synthetic pyrethroids, *e.g., Derringer, Duracide, Grovex, Prentox, Scourge;* also in combination with rotenone, *q.v.,* **PB-NOX, Chem-Fish, Rotacide.**

Toxicity: Large doses have caused vomiting, diarrhea: Sarles *et al., Am. J. Trop. Med.* **29**, 151 (1949).

USE: Insecticide synergist, especially for pyrethroids and rotenone. Ingredient of insecticidal oil solutions, aerosols, dusts, wettable powders, slurries.

7447. Piperonylic Acid. *1,3-Benzodioxole-5-carboxylic acid;* 3,4-methylenedioxybenzoic acid; protocatechuic acid methylene ether. $C_8H_6O_4$; mol wt 166.13. C 57.83%, H 3.64%, O 38.52%. Occurs in Paracoto bark. Prepd by permanganate oxidation of piperonal: Shriner, Kleiderer, *Org. Syn.* **10**, 82 (1930).

Prisms (by sublimation), needles from alc, feathery crystals from water. mp 229°. Sublimes around 210°. Slightly sol in water, chloroform, cold alcohol, ether. Absorption spectrum: Dobbie, Lauder, *J. Chem. Soc.* **83**, 621 (1903). Methyl ester, $C_9H_8O_4$, mp 53°. Sublimes easily. Volatile in steam. Freely sol in alcohol, ether.

7448. Piperoxan. *1-[(2,3-Dihydro-1,4-benzodioxin-2-yl)methyl]piperidine; 2-piperidinomethyl-1,4-benzodioxan;* 2-(1-piperidylmethyl)-1,4-benzodioxan; benzodioxane; 933F; Fourneau 933; Benodaine. $C_{14}H_{19}NO_2$; mol wt 233.30. C 72.07%, H 8.21%, N 6.00%, O 13.72%. α-Adrenergic blocker. Prepn: Fourneau, U.S. pat. **2,056,046** (1936 to Rhône-Poulenc).

bp₁₇ 193°.

dl-Form hydrochloride, $C_{14}H_{20}ClNO_2$, crystals, mp 232-234° (darkens at 220°). uv max: 275 nm ($E_{1cm}^{1\%}$ 82); min: 240

nm ($E_{1cm}^{1\%}$ < 5). Freely sol in water, pH of 1% soln ~5. Alkalies liberate the water-insol base. Soluble in acid solns. Soly in isopropanol: about 10.8 mg/g at 25°. Crystals are not hygroscopic and are stable to light, air, and normal storage temps. An aq soln at its own pH (5) is stable to autoclaving and to many months of storage at room temp.

THERAP CAT: Antihypertensive. Diagnostic aid (pheochromocytoma).

7449. Piperylone. *4-Ethyl-1,2-dihydro-2-(1-methyl-4-piperidinyl)-5-phenyl-3H-pyrazol-3-one; 4-ethyl-1-(1-methyl-4-piperidyl)-3-phenyl-3-pyrazolin-5-one; 1-(N-methylpiperid-4-yl)-3-phenyl-4-ethylpyrazol-5-one;* PR 66. $C_{17}H_{23}N_3O$; mol wt 285.38. C 71.54%, H 8.12%, N 14.73%, O 5.61%. Prepn: Ebnother et al., Helv. Chim. Acta **42**, 1201 (1959); Jucker et al., U.S. pat. 2,903,460 (1959 to Sandoz).

Crystals from methanol + ether, dec 162-163°. pK$_1$ 5.90; pK$_2$ 9.15. Very sparingly sol in water. Sol in ethanol, isopropanol, acetone, ethyl acetate.

Ingredient of *Palerol (Pelerol)*.

Hydrobromide, $C_{17}H_{23}N_3O \cdot HBr$, crystals from methanol + ether, dec 194-197°. Sol in water.

Tartrate, crystals from methanol, mp 145-147°.

THERAP CAT: Analgesic.

7450. PIPES. *1,4-Piperazinediethanesulfonic acid;* piperazine-N,N'-bis(2-ethanesulfonic acid); 1,4-piperazinebis-(ethanesulfonic acid). $C_8H_{18}N_2O_6S_2$; mol wt 302.36. C 31.78%, H 6.00%, N 9.26%, O 31.75%, S 21.21%. One of the zwitterionic N-substituted aminosulfonic acids known as "Good" buffers; active in the pH range 6-8.5. Prepn: N. Good et al., Biochemistry **5**, 467 (1966). Interference with Lowry protein determn: H. M. Himmel, W. Heller, J. Clin. Chem. Clin. Biochem. **25**, 909 (1987). Use as biological buffer: R. Salema, I. Brandao, J. Submicros. Cytol. **5**, 79 (1973); S. Haviernick et al., J. Micros. **135**, 83 (1984).

Monosodium salt, $C_8H_{17}N_2NaO_6S_2$, crystals from water and alcohol, mp > 300 (dec). pKa$_1$ ~3, pKa$_2$ (20°): 6.82 (0.1M); 6.82 (0.2M); 6.96 (0.01M). ΔpKa/°C −0.0085.

Disodium salt, $C_8H_{16}N_2Na_2O_6S_2$, *sodium pipesate.*

USE: Biological buffer.

7451. Pipobroman. *1,4-Bis-(3-bromo-1-oxopropyl)piperazine; 1,4-bis-(3-bromopropionyl)piperazine;* A-8103; NSC-25154; Amedel; Vercyte. $C_{10}H_{16}Br_2N_2O_2$; mol wt 356.09. C 33.73%, H 4.53%, Br 44.88%, N 7.87%, O 8.99%. Prepn: Horrom, Carbon, **Ger.** pat. 1,138,781 (1962 to Abbott), C.A. **58**, 7955c (1963); Groszkowski, Roczniki Chem. **38**, 229 (1964), C.A. **60**, 14506a (1964). Structure-cytostatic effect studies: Groszkowski et al., J. Med. Chem. **11**, 621 (1968); Oteleanu, Retezeanu, Farmacia (Bucharest) **16**, 279 (1968), C.A. **69**, 33591w (1968).

Crystals, mp 106-107°.

THERAP CAT: Antineoplastic.

7452. Piposulfan. *1,4-Bis[3-[(methylsulfonyl)oxy]-1-oxopropyl]piperazine; 1,4-dihydracryloylpiperazine dimethanesulfonate;* 1,4-bis(3-hydroxypropionyl)piperazine dimethanesulfonate; N,N'-bis(3-methanesulfonyloxypropionyl)piperazine; N,N'-bis(3-methylsulfonyloxypropionyl)piperazine; A-20968; NSC-47774; Ancyte. $C_{12}H_{22}N_2O_8S_2$; mol wt 386.46. C 37.29%, H 6.99%, N 8.83%, O 33.12%, S 16.59%. Prepn: Horrom, Carbon, **Ger.** pat. 1,177,162 (1964 to Abbott), C.A. **61**, 13329a (1964).

Crystals from water, mp 175-177°.

THERAP CAT: Antineoplastic.

7453. Pipotiazine. *10-[3-[4-(2-Hydroxyethyl)-1-piperidinyl]propyl]-N,N-dimethyl-10H-phenothiazine-2-sulfonamide;* 2-[1-[3-[2-[(dimethylamino)sulfonyl]-10H-phenothiazin-10-yl]propyl]-4-piperidinyl]ethanol; pipothiazine; RP 19366; Piportil. $C_{24}H_{33}N_3O_3S_2$; mol wt 475.66. C 60.60%, H 6.99%, N 8.83%, O 10.09%, S 13.48%. Prepn: **Fr.** pat. M7835 (1970 to Rhône-Poulenc), C.A. **78**, 43499x (1973). Pharmacology and metabolism of pipotiazine and its esters: Julou et al., Acta Psychiat. Scand., Suppl. **241**, 9 sqq (1973); eidem, Thérapie **28**, 475 (1973). Pharmacokinetics: P. J. DeSchepper et al., Arzneimittel-Forsch. **29**, 1056 (1979).

Palmitic ester, $C_{40}H_{63}N_3O_4S_2$, *pipotiazine palmitate, RP 19552, Piportil L4.*

Undecylenic ester, $C_{40}H_{63}N_3O_4S_2$, *pipotiazine undecylenate, RP 19551, Piportil M2.*

THERAP CAT: Antipsychotic.

7454. Pipoxolan Hydrochloride. *5,5-Diphenyl-2-[2-(1-piperidinyl)ethyl]-1,3-dioxolan-4-one hydrochloride;* 2-(β-N-piperidylethyl)-4,4-diphenyl-1,3-dioxolan-5-one hydrochloride; Rowapraxin. $C_{22}H_{26}ClNO_3$; mol wt 387.91. C 68.12%, H 6.76%, Cl 9.14%, N 3.61%, O 12.37%. Prepn: Pailer et al., Monatsh. **99**, 892 (1968); **Brit.** pat. 1,109,959 and **Belg.** pat. 719,230 (1968, 1969 to Rowa-Wagner).

Crystals from isopropyl alc, mp 207-209°. Soluble in water; stable to mild attack by acids or bases. LD$_{50}$ orally in rats, mice: 1500, 700 mg/kg.

THERAP CAT: Antispasmodic.

7455. Pipradrol. *α,α-Diphenyl-2-piperidinemethanol;* α-(2-piperidyl)benzhydrol; alpha-pipradrol; MRD-108; Detaril; Gerodyl; Meratran (formerly); Pipradol. $C_{18}H_{21}NO$; mol wt 267.36. C 80.86%, H 7.92%, N 5.24%, O 5.98%. Prepd by hydrogenation of α,α-diphenyl-2-pyridinemethanol: Tilford et al., J. Am. Chem. Soc. **70**, 4001 (1948); Werner, Tilford, U.S. pat. 2,624,739 (1953 to Merrell).

Hydrochloride monohydrate, $C_{18}H_{22}ClNO.H_2O$, *Meratonic, Metadin, Stimolag Fortis*. Crystals from butanone, dec 308-309°. Slightly bitter taste. One gram dissolves in 60 ml of hot water.

Note: This is a controlled substance (stimulant) listed in the U.S. Code of Federal Regulations, Title 21 Part 1308.14 (1987).

THERAP CAT: CNS stimulant.

7456. Piprinhydrinate. *8-Chloro-3,7-dihydro-1,3-dimethyl-1H-purine-2,6-dione compd. with 4-(diphenylmethoxy)-1-methylpiperidine (1:1);* N-methylpiperidyl 4-benzhydryl ether 8-chlorotheophyllinate; diphenylpyraline 8-chlorotheophyllinate; Kolton (tabl.); Koltonal; Mepedyl. $C_{26}H_{30}ClN_5O_3$; mol wt 496.39. C 62.91%, H 6.09%, Cl 7.14%, N 14.18%, O 9.67%. Prepn: Schuler, **Ger.** pat. **934,890** (1955 to Promonta). *See also* Diphenylpyraline.

Minute crystals, mp 151°. Sparingly sol in water. Freely sol in alcohol.

THERAP CAT: Antihistaminic. Anti-emetic.

7457. Piprozolin. *[3-Ethyl-4-oxo-5-(1-piperidinyl)-2-thiazolidinylidene]acetic acid ethyl ester; 3-ethyl-4-oxo-5-piperidino-$\Delta^{2,\alpha}$-thiazolidineacetic acid ethyl ester;* Gö 919; W 3699; Coleflux; Epsyl; Probilin; Secrebil. $C_{14}H_{22}N_2O_3S$; mol wt 298.40. C 56.35%, H 7.43%, N 9.39%, O 16.09%, S 10.74%. Prepn: G. Satzinger, *Ann.* **665,** 150 (1963); G. Satzinger *et al.,* **Ger.** pat. **2,414,345** (1975 to Goedecke), corresp to **U.S.** pat. **3,971,794** (1976 to Warner-Lambert). Metabolism and pharmacokinetics: K. O. Vollmer, F. W. Koss, *Arch. Int. Pharmacodyn. Ther.* **198,** 312 (1972). Mechanism of action: F. W. Koss *et al., ibid.* 333. Series of articles on synthesis, pharmacology, toxicology and clinical studies: *Arzneimittel-Forsch.* **27,** 463-526 (1977). Toxicity data: M. Herrmann *et al., ibid.* 467.

Colorless cryst, mp 86-87°. Practically insol in water. Sol in dil aq acids and most organic solvents. uv max (methanol): 245, 285 nm (ε 8200, 20000) (Vollmer). LD_{50} in mice, rats (mg/kg): 1070, 3256 orally (Herrmann).

THERAP CAT: Choleretic.

7458. Pipsyl Chloride. *4-Iodobenzenesulfonyl chloride;* p-iodophenyl sulfonyl chloride. Usually prepd from iodide ion and p-diazobenzenesulfonic acid, followed by treatment with phosphorus pentachloride. A 5-10-fold excess reacts quantitatively with amino acids as indicated by the disappearance of amino nitrogen. Radioactive pipsyl chloride is used in the analysis of proteins: Keston *et al., J. Am. Chem. Soc.* **68,** 1390 (1946).

7459. Piracetam. *2-Oxo-1-pyrrolidineacetamide;* 2-pyrrolidoneacetamide; 2-pyrrolidinoneacetamide; 2-ketopyrrolidine-1-ylacetamide; 1-acetamido-2-pyrrolidinone; UCB 6215; Avigilen; Cerebroforte; Encetrop; Euvifor; Gabacet; Genogris; Nootron; Nootrop; Nootropyl; Norzetam; Normabraïn; Pirroxil. $C_6H_{10}N_2O_2$; mol wt 142.15. C 50.69%, H 7.09%, N 19.70%, O 22.51%. Prepn: H. Morren, **Neth.** pat.

Appl. 6,509,994; *eidem,* **U.S.** pat. **3,459,738** (1966, 1969 both to U.C.B.). Pharmacology: Giurgea *et al., Arch. Int. Pharmacodyn. Ther.* **166,** 238 (1967); Giurgea, Moyersoons, *ibid.* **188,** 401 (1970); Giurgea *et al., Psychopharmacologia* **20,** 160 (1971). Metabolism and biochemical studies: Gobert, *J. Pharm. Belg.* **27,** 281 (1972). Clinical studies: W. J. Oosterveld, *Arzneimittel-Forsch.* **30,** 1947 (1980); G. Chouinard *et al., Psychopharmacol. Bull.* **17,** 129 (1981); in dyslexia: M. Di Ianni *et al., J. Clin. Psychopharmacol.* **5,** 272 (1985).

Crystals from isopropanol, mp 151.5-152.5°.

THERAP CAT: Nootropic.

7460. Pirarubicin. *[8S-[8α,10α(S*)]]-10-[[3-Amino-2,3,6-trideoxy-4-O-(tetrahydro-2H-pyran-2-yl)-α-L-lyxo-hexopyranosyl]oxy]-7,8,9,10-tetrahydro-6,8,11-trihydroxy-8-(hydroxyacetyl)-1-methoxy-5,12-naphthacenedione;* 4'-O-tetrahydropyranyl doxorubicin; (2''R)-4'-O-tetrahydropyranyladriamycin; (8S,10S)-10-[[3-amino-2,3,6-trideoxy-4-O-(tetrahydro-2H-pyran-2-yl)-α-L-lyxo-hexopyranosyl]oxy]-8-glycoloyl-7,8,9,10-tetrahydro-6,8,11-trihydroxy-1-methoxy-5,12-naphthacenedione; tepirubicin; THP; THP-ADM; THP-adriamycin; 1609RB; Pinorubicin; Theprubicin; Therarubicin. $C_{32}H_{37}NO_{12}$; mol wt 627.65. C 61.24%, H 5.94%, N 2.23%, O 30.59%. Structural analog of doxorubicin, *q.v.* Prepn of (2''R) and (2''S)-diastereomers: H. Umezawa *et al., J. Antibiot.* **32,** 1082 (1979); H. Umezawa *et al.,* **Eur.** pat. **Appl. 14,853;** *eidem,* **U.S.** pat. **4,303,785** (1980, 1981 both to Microbiochem. Res. Found. Japan). Absolute configuration: H. Umezawa *et al., J. Antibiot.* **37,** 1094 (1984). HPLC determn in biological fluids: Y. Matsushita *et al., J. Antibiot.* **36,** 880 (1983). Preliminary toxicology: D. Dantchev *et al., ibid.* **32,** 1085 (1979). *In vitro* antitumor activity: T. Tsuruo *et al., Cancer Res.* **42,** 1462 (1982); *in vivo* activity: T. Hisamatsu *et al., Japan. J. Cancer Res.* **76,** 1008 (1985). Cellular uptake and inhibition of DNA synthesis: S. Kunimoto *et al., J. Antibiot.* **36,** 312 (1983); and cytostatic activity: S. Kunimoto *et al., ibid.* **37,** 1697 (1984). Pharmacokinetics and tissue distribution: H. Iguchi *et al., Cancer Chemother. Pharmacol.* **15,** 132 (1985). Clinical pharmacology: A. A. Miller, C. G. Schmidt, *Cancer Res.* **47,** 1461 (1987). Clinical studies in breast cancer: G. Mathe *et al., Biomed. Pharmacother.* **40,** 376 (1986); in leukemia and malignant lymphoma: R. Ohno *et al., Cancer Chemother. Pharmacol.* **20,** 230 (1987). Brief review of early clinical evaluations: H. Majima, K. Ohta, *Biomed. Pharmacother.* **41,** 237-243 (1987).

Red solid, mp 188-192° (dec). $[\alpha]_D^{25}$ 175 ±25° (c = 0.2 in $CHCl_3$). uv and visible max (methanol): 234, 252, 290, 498, 531.5, 580 nm ($E_{1cm}^{1\%}$ 480, 350, 110, 140, 100, 45). Sol in ethyl acetate, chloroform, and ethanol; slightly sol in water, n-hexane, petr ether. Ethanolic and acidic solutions are red in color; give a positive ninhydrin reaction and do not reduce Fehling's solution. LD_{50} i.v. in mice: 27.8 mg/kg (Umezawa, 1979).

THERAP CAT: Antineoplastic.

7461. Pirbuterol. α^6-[[(1,1-Dimethylethyl)amino]meth-

yl]-3-hydroxy-2,6-pyridinedimethanol; 2-hydroxymethyl-3-hydroxy-6-(1-hydroxy-2-*tert*-butylaminoethyl)pyridine. $C_{12}H_{20}N_2O_3$; mol wt 240.30. C 59.98%, H 8.39%, N 11.66%, O 19.97%. Analog of albuterol, *q.v.*, with β_2-adrenergic stimulating activity. Prepn: W. E. Barth, **Ger.** pat. **2,204,-195;** *idem.* **U.S.** pat. **3,700,681** (both 1972 to Pfizer). Stability study: P. C. Bansal, D. C. Monkhouse, *J. Pharm. Sci.* **66,** 819 (1977). Biotransformation: H. M. McIlhenny, M. S. D. Ghaly, *Fed. Proc.* **38,** 1130 (1979). Pharmacokinetics and cardiopulmonary effects in dogs: J. W. Constantine *et al., J. Pharmacol. Exp. Ther.* **208,** 371 (1979). Comparative study in respiratory disease: A. J. Dyson, A. D. Mackay, *Brit. J. Dis. Chest* **74,** 70 (1980). Use in treatment of cardiac failure: N. A. Awan *et al., Clin. Res.* **28,** 17A (1980); W. S. Colucci *et al., N. Engl. J. Med.* **305,** 185 (1981); G. I. Nelson *et al., Eur. Heart J.* **3,** 238 (1982); K. T. Weber *et al., Circulation* **66,** 1262 (1982). Review of pharmacology and efficacy in bronchospastic disease: D. M. Richards, R. N. Brogden, *Drugs* **30,** 6-21 (1985).

(CH₃)₃CNCH₂CH ... CH₂OH structure

Dihydrochloride, $C_{12}H_{22}Cl_2N_2O_3$, *CP-24,314-1,* **Broncocor, Broncocur, Exirel.** Crystals from ethanol/isopropyl ether, mp 182° (dec).
Monoacetate, $C_{14}H_{24}N_2O_5$, *Maxair.*
THERAP CAT: Bronchodilator.

7462. Pirenoxine. *1-Hydroxy-5-oxo-5H-pyrido[3,2-a]-phenoxazine-3-carboxylic acid;* 1-hydroxy-5*H*-pyrido-[3,2-*a*]phenoxazin-5-one-3-carboxylic acid; 1-hydroxy-3-carboxy-5*H*-pyrido[3,2-*a*]phenoxazin-5-one; pirfenoxone; Catalin. $C_{16}H_8N_2O_5$; mol wt 308.24. C 62.34%, H 2.62%, N 9.09%, O 25.95%. Prepn: S. Ogino, **Japan.** pat. **2227('59),** *C.A.* **54,** 11058i (1960); S. Ishii, **Japan.** pat. **1782('61),** *C.A.* **55,** 21494e (1961) (both to Chizu Drug). Prepn of the sodium salt: S. Ishii, K. Ogata, **Japan.** pat. **73 02,672** (1973 to Senju), *C.A.* **80,** 6959t (1974). Pharmacological studies: F. Ikemoto *et al., Oyo Yakuri* **8,** 937 (1974), *C.A.* **83,** 71663t (1975). Toxicological studies: *eidem, ibid.* 911, 923, *C.A.* **82,** 51617g, 68262k (1975). Influence on carbohydrate metabolism in the lens: I. Korte *et al., Ophthalmic Res.* **7,** 282 (1975). Effect on NADH, NADPH: *eidem, ibid.* 440. Effect on senile cataracts: T. Murata, *Folia Ophthalmol. Japan* **31,** 1217 (1980). Clinical trial in treatment of cataracts: S. K. Angra *et al., Indian J. Ophthalmol.* **31,** 5 (1983).

Orange-yellow powder, mp 247-248° (dec).
Sodium salt, $C_{16}H_7N_2NaO_5$, *Clarvisan.* Very sol in water. LD_{50} in mice (mg/kg): > 10,000 orally; > 5,000 s.c.; 2120-2250 i.p. LD_{50} in rats (mg/kg) i.p.: 2400 (males); 1460 (females) (Ikemoto).
THERAP CAT: Treatment of cataracts.

7463. Pirenzepine. *5,11-Dihydro-11-[(4-methyl-1-piperazinyl)acetyl]-6H-pyrido[2,3-b][1,4]benzodiazepin-6-one;* L-S 519. $C_{19}H_{21}N_5O_2$; mol wt 351.42. C 64.94%, H 6.02%, N 19.93%, O 9.11%. Tricyclic gastric-acid inhibitor. Prepn: **Fr.** pat. **1,505,795** (1967 to Thomae), *C.A.* **70,** 4154w (1969). Pharmacology: W. Eberlein *et al., Arzneimittel-Forsch.* **27,** 356 (1977). Pharmacokinetics: R. Hammes *et al., ibid.* 928. Mechanism of action: G. Heller *et al., Verh. Deut. Ges. Inn. Med.* **84,** 991 (1978), *C.A.* **90,** 132984s (1979). Human pharmacology: H. Brunner *et al., Arzneimittel-Forsch.* **27,** 684 (1977). Multicenter controlled clinical trial: *Scand. J. Gastroenterol.* **17,** Suppl. 81, 1-42 (1982). Radioimmunoassay determn in human plasma and urine: C. A. Homon *et al., Ther. Drug Monit.* **9,** 236 (1987). Symposium: *ibid.*

Suppl. 72, 1-273. Review of pharmacology and therapeutic efficacy: A. A. Carmine, R. N. Brogden, *Drugs* **30,** 85-126 (1985). Comprehensive description: H. A. El-Obeid *et al.,* in *Analytical Profiles of Drug Substances* **vol. 16,** K. Florey, Ed. (Academic Press, New York, 1987) pp 445-506.

Dihydrochloride, $C_{19}H_{23}Cl_2N_5O_2$, *L-S 519-C12,* **Bisvanil, Duogastral, Durapirenz, Gasteril, Gastrozepin, Leblon, Maghen, Pirefar, Tabe, Ulcoforton, Ulcosan.** Sol in water, slightly sol in methanol. Practically insol in ether.
THERAP CAT: Antiulcerative.

7464. Piretanide. *3-(Aminosulfonyl)-4-phenoxy-5-(1-pyrrolidinyl)benzoic acid;* 4-phenoxy-3-(1-pyrrolidinyl)-5-sulfamoylbenzoic acid; HOE 118; S 73 4118; Arelix; Diumax; Tauliz. $C_{17}H_{18}N_2O_5S$; mol wt 362.40. C 56.34%, H 5.01%, N 7.73%, O 22.07%, S 8.85%. High-ceiling loop diuretic, structurally related to bumetanide, *q.v.* Prepn: D. Bormann *et al.,* **Ger.** pat. **2,419,970;** *eidem,* **U.S.** pat. **4,010,-273** (1975, 1977 both to Hoechst). Chemistry and pharmacology: W. Merkel *et al., Eur. J. Med. Chem.* **11,** 399 (1976). Diuretic activity in man: N. Pozet *et al., Brit. J. Clin. Pharmacol.* **9,** 577 (1980); T. Saruta, E. Kato, *Arzneimittel-Forsch.* **30,** 1807 (1980). Vascular effects of piretanide: E. Klaus *et al., ibid.* **33,** 1273 (1983); renal effects: M. Omosu *et al., ibid.* 1277. Review of pharmacology and therapeutic efficacy: S. P. Clissold, R. N. Brogden, *Drugs* **29,** 489-530 (1985).

Pale yellow platelets from methanol/water, mp 225-227°. Exhibits intense light blue fluorescence at 366 nm. LD_{50} in rats, mice (mg/kg): 5601, 3672 orally (Merkel).
THERAP CAT: Diuretic.

7465. Piribedil. *2-[4-(1,3-Benzodioxol-5-ylmethyl)-1-piperazinyl]pyrimidine;* 2-(4-piperonyl-1-piperazinyl)pyrimidine; 2-[4-(3,4-methylenedioxybenzyl)piperazino]pyrimidine; 1-(2-pyrimidyl)-4-piperonylpiperazine; 1-(2''-pyrimidyl)-4-(methylene-3',4'-dioxybenzyl)piperazine; ET 495; EU 4200; Trivastal. $C_{16}H_{18}N_4O_2$; mol wt 298.35. C 64.41%, H 6.08%, N 18.78%, O 10.73%. Central dopaminergic agonist. Prepn: **Neth.** pat. **Appl. 6,413,349** corresp to G. Regnier *et al.,* **U.S.** pat. **3,299,067;** *eidem,* **Brit.** pat. **1,101,425** (1965, 1967, 1968, all three to Sci. Union et Cie-Soc. Franc. Rech. Med.); G. Regnier *et al., J. Med. Chem.* **11,** 1151 (1968). Activity in man: R. Royer, *Proc. Int. Pharmacol. Meet., 3rd, 1966,* **3,** R. K. Richards, Ed. (Pergamon Press, New York, 1968) pp 45-55. Pharmacology: M. Laubie *et al., Eur. J. Pharmacol.* **6,** 75 (1969). Metabolism: D. B. Campbell *et al., Advan. Neurol.* **3,** 199 (1973).

Crystals from anhydr ethanol, mp 98°. LD_{50} i.p. in mice: 690.3 mg/kg.
THERAP CAT: Vasodilator (peripheral).

7466. Piridocaine. *2-Piperidineethanol 2-aminobenzoate (ester);* anthranilic acid 2-(2-piperidyl)ethyl ester; 2-piperidineethanol anthranilate; β-(2-piperidyl)ethyl *o*-aminobenzoate; PT-14; Lucaine. $C_{14}H_{20}N_2O_2$; mol wt 248.32. C 67.71%, H 8.12%, N 11.28%, O 12.89%. Prepn: Walter, Fosbinder, *J. Am. Chem. Soc.* **61**, 1713 (1939); U.S. pat. **2,229,533** (1941 to Maltbie Chem.).

Hydrochloride, $C_{14}H_{21}ClN_2O_2$, crystals from abs alcohol, mp 209-211°.

THERAP CAT: Local anesthetic.

7467. Pirifibrate. *2-(4-Chlorophenoxy)-2-methylpropanoic acid [6-(hydroxymethyl)-2-pyridinyl]methyl ester;* 2,6-pyridinedimethanol mono-*p*-chlorophenoxyisobutyrate. $C_{17}H_{18}ClNO_4$; mol wt 335.79. C 60.81%, H 5.40%, Cl 10.56%, N 4.17%, O 19.06%. Hypolipemic agent related structurally to clofibrate, *q.v.* Prepn: D. Humbert, R. Ratouis, **Ger. pat. 2,432,322** corresp to **U.S. pat. 3,971,798** (1975, 1976 both to Roussel). Multicenter study in hyperlipoproteinemias: A. J. Domingo *et al., Clin. Ther.* **3**, 219 (1980).

Cryst from isopropyl ether, mp 46°. LD_{50} in mice: 915-1098 mg/kg i.p., D. Humbert, R. Ratouis, *loc. cit.*

Hydrochloride, $C_{17}H_{19}Cl_2NO_4$, *EL-466, Bratenol.* Cryst, mp 110°. LD_{50} in mice: 1098-1281 mg/kg i.p. (expressed as free base), D. Humbert, R. Ratouis, *loc. cit.*

THERAP CAT: Antihyperlipoproteinemic.

7468. Pirimicarb. *Dimethylcarbamic acid 2-(dimethylamino)-5,6-dimethyl-4-pyrimidinyl ester;* 2-(dimethylamino)-5,6-dimethyl-4-pyrimidinyl dimethylcarbamate; 5,6-dimethyl-2-dimethylamino-4-dimethylcarbamoyloxypyrimidine; PP-062; ENT 27766; Aphox; Fernos; Pirimor. $C_{11}H_{18}N_4O_2$; mol wt 238.29. C 55.45%, H 7.61%, N 23.51%, O 13.43%. Selective aphicide. Prepn: F. L. C. Baranyovits *et al.*, **S. Afr. pat. 67 01,588** corresp to **U.S. pat. 3,493,574** (1968, 1970 to ICI). Activity: F. L. C. Baranyovits, R. Ghosh, *Chem. & Ind. (London)* **1969**, 1018. Degradn in soil: I. R. Hill, *ACS Symp. Ser.* **29**, 358 (1976). GLC analysis: P. D. Bland, *J. Assoc. Off. Anal. Chem.* **64**, 1315 (1981).

Cryst solid, mp 90.5°. Vapor press at 30°: 3×10^{-5} mm Hg. Soly in water at 25°: 2.7 g/l. Sol in most organic solvents. Dec by prolonged boiling with acids or alkali. Aq solns are unstable to light. LD_{50} orally in female rats: 147 mg/kg, F. L. C. Baranyovits, R. Ghosh, *loc. cit.*

USE: Insecticide.

7469. Pirimiphos-ethyl. *O-[2-(Diethylamino)-6-methyl-4-pyrimidinyl]phosphorothioic acid O,O-diethyl ester;* O,O-diethyl O-[2-(diethylamino)-6-methyl-4-pyrimidinyl]phosphorothioate; PP-211; Fernex; Primicid. $C_{13}H_{24}N_3O_3PS$; mol wt 333.38. C 46.83%, H 7.26%, N 12.60%, O 14.40%, P 9.29%, S 9.62%. Broad spectrum contact and fumigant insecticide. Prepn: B. K. Snell, S. P. Sharpe, **Brit. pat. 1,205,000** (1970 to ICI); *C.A.* **74**, 52482f (1971). Persistence and degradn in soil: D. L. Suett, *Pestic. Sci.* **6**, 385

(1975). Analysis: D. J. W. Bullock, *Anal. Methods Pestic. Plant Growth Regul.* **8**, 171 (1976).

Straw-colored liquid, dec > 130°. d^{20} 1.14; n_D^{25} 1.520. Vapor pressure at 25°: 2.9×10^{-4} mm Hg. Soly in water at 30°: < 1 mg/l. Miscible with most organic solvents. LD_{50} orally in rats: 140 mg/kg, *RTECS* vol. II, R. J. Lewis, R. L. Tatken, Eds. (1980) p 338.

O,O-Dimethyl analog, $C_{11}H_{20}N_3O_3PS$, *pirimiphos-methyl, PP-511, Actellic, Actellifog, Blex, Silo San, Tomahawk.* Straw-colored liquid, d^{30} 1.157; n_D^{25} 1.527. Vapor pressure at 30°: $\sim 1 \times 10^{-4}$ mm Hg. Soly in water at 30°: 5 mg/l.

USE: Insecticide.

7470. Piritramide. *1'-(3-Cyano-3,3-diphenylpropyl)-[1,4'-bipiperidine]-4'-carboxamide;* 1-(3,3-diphenyl-3-cyanopropyl)-4-piperidino-4-piperidinecarboxamide; 2,2-diphenyl-4-(4-piperidino-4-carbamoylpiperidino)butyronitrile; pirinitramide; A65; R 3365; Dipidolor; Piridolan. $C_{27}H_{34}N_4O$; mol wt 430.57. C 75.31%, H 7.96%, N 13.01%, O 3.72%. Prepn: **Belg. pat. 606,850**; P. A. J. Janssen, **U.S. pat. 3,080,366** (1961, 1963 both to Janssen). Crystal structure: C. Humblet *et al., Acta Crystallogr.* **B33**, 1615 (1977).

Crystals from acetone, mp 149-150°.

Caution: May be habit forming. This is a controlled substance (opiate) listed in the U.S. Code of Federal Regulations, Title 21 Part 1308.11 (1987).

THERAP CAT: Narcotic analgesic.

7471. Pirmenol. *cis-(±)-α-[3-(2,6-Dimethyl-1-piperidinyl)propyl]-α-phenyl-2-pyridinemethanol;* (±)-*cis*-2,6-dimethyl-α-phenyl-α-2-pyridyl-1-piperidinebutanol; (±)-1-phenyl-1-(2-pyridyl)-4-(*cis*-2,6-dimethyl-1-piperidyl)butanol. $C_{22}H_{30}N_2O$; mol wt 338.49. C 78.06%, H 8.93%, N 8.28%, O 4.73%. Prepn: R. W. Fleming, **Ger. pat. 2,806,654;** idem, **U.S. pat. 4,112,103** (both 1978 to Parke, Davis). Anti-arrhythmic profile in dogs: T. E. Mertz, T. J. Steffe, *J. Cardiovasc. Pharmacol.* **2**, 527 (1980). Toxicology study: J. L. Schardein *et al., Toxicol. Appl. Pharmacol.* **56**, 294 (1980). HPLC determn in biological fluids: E. L. Johnson, L. A. Pachla, *J. Pharm. Sci.* **73**, 754 (1984). Pharmacokinetics in humans: S. C. Hammill *et al., Circulation* **65**, 369 (1982); S. W. Sanders *et al., J. Clin. Pharmacol.* **23**, 113 (1983). Hemodynamic effects in cardiac patients: M. S. Nieminen *et al., Eur. Heart J.* **7**, 150 (1986). Clinical evaluations in ventricular arrhythmias: L. K. Toivonen *et al., J. Cardiovasc. Pharmacol.* **8**, 156 (1986); E. M. Hampton *et al., Eur. J. Clin. Pharmacol.* **31**, 15 (1986). Symposium on pharmacology and clinical efficacy: *Am. J. Cardiol.* **59**, Suppl., 1H-57H (1987).

Crystals from petroleum ether, mp 70-71°.

Monohydrochloride, $C_{22}H_{31}ClN_2O$, *CI-845.* mp 171-172°. LD_{50} in mice, rats, dogs (mg/kg): 20.8, 23.6, > 7.0 i.v.; 215.5, 359.9, > 40.0 orally (Schardein).

THERAP CAT: Anti-arrhythmic.

7472. Piroctone. *1-Hydroxy-4-methyl-6-(2,4,4-trimeth-ylpentyl)-2(1H)-pyridinone.* $C_{14}H_{23}NO_2$; mol wt 237.34. C 70.85%, H 9.77%, N 5.90%, O 13.48%. Pyridone deriv related structurally to ciclopirox, *q.v.* Prepn: G. Lohaus, W. Dittmar, **Ger. pat. 2, 214,608** corresp to U.S. pat. **3,972,888** (1973, 1976 both to Hoechst); *eidem, Arzneimittel-Forsch.* **31**, 1311 (1981). Evaluation of efficacy as anti-dandruff agent: E. Futterer, *J. Soc. Cosmet. Chem.* **32**, 327 (1981).

Cryst, mp 108°.
Ethanolamine salt (1:1), $C_{16}H_{30}N_2O_3$, *Octopirox.*
THERAP CAT: Antiseborrheic.

7473. Piroheptine. *3-(10,11-Dihydro-5H-dibenzo[a,d]cyclohepten-5-ylidene)-1-ethyl-2-methylpyrrolidine.* $C_{22}H_{25}$-N; mol wt 303.45. C 87.08%, H 8.30%, N 4.62%. Prepn: **Neth. pat. Appl. 6,609,280** corresp to Y. Deguchi *et al.,* U.S. pat. **3,454,595** (1967, 1969 both to Fujisawa). Crystal structure: Y. Tokuma *et al., Bull. Chem. Soc. Japan* **44**, 2665 (1971). Pharmacological studies: M. Hitomi *et al., Arzneimittel-Forsch.* **22**, 953, 961 (1972). *In vitro* study: T. Ohashi *et al., ibid.* 966. Metabolism: Y. Tokuma *et al., Bull. Chem. Soc. Japan* **48**, 294 (1975). Clinical pharmacology: A. Barbeau, *Ann. Rev. Pharmacol.* **14**, 91 (1974).

Liquid, bp_4 167°. uv max (95% ethanol): 240 nm (ϵ 12,100).
Hydrochloride, $C_{22}H_{26}ClN$, *Trimol.* Cryst, mp 250-253°. LD_{50} in male mice, rats (mg/kg): 153, 600 orally; 19, 17 i.v.; 95, 110 i.p.; 109, 330 s.c., M. Hitomi *et al., Arzneimittel-Forsch.* **22**, 961 (1972).
THERAP CAT: Antiparkinsonian.

7474. Piromen. Pyromen; desacchromin dispersion. Pyrogenic *pseudomonas* polysaccharide-nucleic acid complex. Prepd by proteolyzing the bacterial organism and separating the complex from inactive material by dialyzing: N. M. Nesset, L. G. Ginger, **Brit. pat. 699,663** (1953 to Baxter Labs.), *C.A.* **48**, 6083 (1954). Contains deoxyribonucleic acid, ribonucleic acid, and hexosamine, the pyrogenic reducing sugar: N. M. Nesset *et al., J. Am. Pharm. Assoc.* **39**, 456 (1950). Use in detn of bone marrow granulocyte reserves: B. C. Korbitz *et al., Curr. Ther. Res., Clin. Exp.* **11**, 491 (1969). Effects on spinal cord regeneration: M. A. Matthews *et al., Neuropathol. Appl. Neurobiol.* **5**, 161 (1979); *eidem, Acta Neurobiol. Exp.* **40**, 489 (1980).

7475. Piromidic Acid. *8-Ethyl-5,8-dihydro-5-oxo-2-(1-pyrrolidinyl)pyrido[2,3-d]pyrimidine-6-carboxylic acid;* 5,8-dihydro-8-ethyl-5-oxo-2-pyrrolidinopyrido[2,3-d]pyrimidine-6-carboxylic acid; PD 93; Bactramyl; Enterol; Gasturol; Panacid; Pirodal; Purim; Reelon; Septural; Uropir. $C_{14}H_{16}N_4O_3$; mol wt 288.31. C 58.32%, H 5.59%, N 19.43%, O 16.65%. Quinolone antibacterial. Prepn: S. Minami *et al.,* **Japan. pat. 67 25,912;** *eidem,* **Brit. pat. 1,129,358** (1967, 1968 both to Dainippon Pharm.); *eidem, Chem. Pharm. Bull.* **19**, 1426 (1971). Activity studies: M. Shimizu *et al., Antimicrob. Ag. Chemother.* **1970**, 117. Metabolism: *eidem, ibid.* 123.

Crystals from ethanol-chloroform, mp 314-316°. LD_{50} in male, female mice, male, female rats (mg/kg): 287, 268, 177, 158 i.v.; all > 4000 orally, s.c. and i.p. (Shimizu).
THERAP CAT: Antibacterial.

7476. Piroxicam. *4-Hydroxy-2-methyl-N-2-pyridinyl-2H-1,2-benzothiazine-3-carboxamide 1,1-dioxide;* 3,4-dihydro-2-methyl-4-oxo-N-2-pyridyl-2H-1,2-benzothiazine-3-carboxamide 1,1-dioxide; CP 16171; Artroxicam; Baxo; Bruxicam; Caliment; Erazon; Feldene; Flogobene; Larapam; Piroflex; Reudene; Riacen; Roxicam; Roxiden; Solocalm; Unicam; Zunden. $C_{15}H_{13}N_3O_4S$; mol wt 331.35. C 54.37%, H 3.95%, N 12.68%, O 19.31%, S 9.68%. Non-steroidal anti-inflammatory with long half-life. Prepn (keto form): J. Lombardino, **Ger. pat. 1,943,265;** *idem,* **U.S. pat. 3,591,584** (1970, 1971 to Pfizer). Synthesis and biological properties: J. Lombardino, E. Wiseman, *J. Med. Chem.* **15**, 848 (1972); J. Lombardino *et al., ibid.* **16**, 493 (1973). Pharmacology: E. Wiseman *et al., Arzneimittel-Forsch.* **26**, 1300 (1976). Evaluation of ulcerogenic effects in comparison with droxicam, *q.v.:* G. Palacios *et al., Method Find. Exp. Clin. Pharmacol.* **9**, 353 (1987). Clinical pharmacology: L. Martinez *et al., ibid.* **10**, 729 (1988). *Review: eidem,* in *Pharmacological and Biochemical Properties of Drug Substances* vol. 3, M. E. Goldberg, Ed. (Am. Pharm. Assoc., Washington, DC, 1981) pp 324-346. Review of pharmacology and therapeutic efficacy: R. N. Brogden *et al., Drugs* **22**, 165-187 (1981); *eidem, ibid.* **28**, 292-323 (1984). Symposium on clinical efficacy and safety: *Am. J. Med.* **81**, Suppl. 5B, 1-55 (1986). Comprehensive description: M. Mihalic *et al.* in *Analytical Profiles of Drug Substances* vol. 15, K. Florey, Ed. (Academic Press, New York, 1986) pp 509-531.

Crystals from methanol, mp 198-200°. pKa 6.3 (2:1 dioxane-water). LD_{50} orally in mice: 360 mg/kg (Wiseman).
Cinnamic acid ester, $C_{24}H_{19}N_3O_5S$, *piroxicam cinnamate, cinnoxicam, SPA-S510, Sinartrol.*
Compd with β-cyclodextrin (7:1), $C_{57}H_{83}N_3O_{39}S$, *Brexin.*
THERAP CAT: Anti-inflammatory.

7477. Pirozadil. *3,4,5-Trimethoxybenzoic acid 2,6-pyridinediylbis(methylene)ester;* 2,6-pyridinedimethanol bis-(3,4,5-trimethoxybenzoate); 722 D; Pemix. $C_{27}H_{29}NO_{10}$; mol wt 527.54. C 61.47%, H 5.54%, N 2.66%, O 30.33%. Hypolipidemic agent that inhibits platelet aggregation. Prepn: J. P. Cochs, **Ger. pat. 2,411,902** (1974 to Inst. Int. Ter.), *C.A.* **82**, 4136q (1975). Effect on cerebral metabolic blood flow in rabbits: J. Balasch, L. Palacios, *Arch. Farmacol. Toxicol.* **3**, 137 (1977). Efficacy in exptl atherosclerosis: M. R. Parwaresch *et al., Atherosclerosis* **31**, 395 (1978). Toxicological and histopathological study: J. Roca *et al., Arch. Farmacol. Toxicol.* **6**, 41 (1980). Clinical trials in hyperlipoproteinemia: M. Shinomiya *et al., Arzneimittel-Forsch.* **37**, 1069 (1987); R. Tapounet, I. Marti Ragué, *Drugs Exp. Clin. Res.* **13**, 447 (1987).

White cryst powder, mp 119-126°. Very sol in chloroform; sol in dioxane, acetonitrile. Practically insol in ether, water.

THERAP CAT: Antihyperlipoproteinemic.

7478. Pirprofen. *3-Chloro-4-(2,5-dihydro-1H-pyrrol-1-yl)-α-methylbenzeneacetic acid; 3-chloro-4-(3-pyrrolin-1-yl)hydratropic acid;* SU 21524; Rangasil 400; Rengasil; Seflenyl. $C_{13}H_{14}ClNO_2$; mol wt 251.71. C 62.03%, H 5.61%, Cl 14.08%, N 5.57%, O 12.71%. Non-steroidal anti-inflammatory. Prepn: R. W. Carney, G. DeStevens, U.S. pat. **3,641,040** (1972 to Ciba); R. W. Carney et al., *Experientia* **29**, 938 (1973). Metabolism: R. C. Luders et al., *Clin. Pharmacol. Ther.* **21**, 721 (1977). Pharmacology: G. Wilhelmi, *Pharmacology* **16**, 268 (1978). Pharmacokinetics: M. Beth et al., *J. Clin. Pharmacol.* **15**, 563 (1975). Comparative efficacy: R. J. Saykaly et al., *ibid.* **19**, 59 (1979). Review of pharmacology and therapeutic efficacy: P. A. Todd, R. Beresford, *Drugs* **32**, 509-537 (1986).

Crystals from benzene-hexane, mp 98-100°.
THERAP CAT: Anti-inflammatory.

7479. Piscidic Acid. *2,3-Dihydroxy-2-[(4-hydroxyphenyl)methyl]butanedioic acid; (p-hydroxybenzyl)tartaric acid.* $C_{11}H_{12}O_7$; mol wt 256.21. C 51.56%, H 4.72%, O 43.71%. Isoln from root bark of *Piscidia piscipula* (L.) Sarg. (*P. erythrina* L.), *Leguminosae* (Jamaica dogwood): Freer, Clover, *Am. Chem. J.* **25**, 390 (1901); from *Narcissus poeticus* L., *Amaryllidaceae*: Smeby et al., *J. Am. Chem. Soc.* **76**, 6127 (1954). Structure: Bridge et al., *J. Chem. Soc.* **1948**, 257. Synthesis of *p-O*-methyl deriv: Buckle et al., *ibid.* **1954**, 3981.

Elongated prisms. mp 186-187°. $[\alpha]_D^{20}$ +41.02° (c = 2.65). Sol in water, ethyl acetate; practically insol in $CHCl_3$. *p-O*-Methyl derivative, $C_{12}H_{14}O_7$, *(p-methoxybenzyl)tartaric acid.* Crystals, mp 169-170°. $[\alpha]_D^{23}$ +44.01° ±5.0 (c = 1.262).

7480. Pithecolobine. *19-Heptyl-10-hydroxy-1,5,10,14-tetraazacyclononadecan-15-one.* $C_{22}H_{46}N_4O_2$; mol wt 398.62. C 66.28%, H 11.63%, N 14.06%, O 8.03%. Occurs in the bark of *Samanea saman* Merr. (formerly *Pithecolobium saman* Benth.), bark and seed of *P. bigeminum* Mart. and *P. lobatum* Benth., *Leguminosae*. Isoln: Greshoff, *Ber.* **23**, 3541 (1890); K. Wiesner et al., *Can. J. Chem.* **30**, 761 (1952). Structure: *eidem, J. Am. Chem. Soc.* **75**, 6348 (1953); D. E. Orr, K. Wiesner, *Chem. & Ind. (London)* **1959**, 672; K. Wiesner, D. E. Orr, *Tetrahedron Letters* no. 16, 11 (1960); K. Wiesner et al., *Can. J. Chem.* **46**, 1886, 3617 (1968).

Crystals, mp 67-69° or oily liquid, $bp_{0.007}$ 230°. Sublimes at 135° at 0.007 mm pressure. Sol in water, alcohol, chloroform, ether, petr ether.

7481. Pituitary, Posterior. Pituamin; Di-Sipidin; Pituitrin. Desiccated hypophysis. The cleaned, dried, and powdered posterior lobe of pituitary body of domesticated animals used for food by man. Contains both oxytocin and vasopressin, *q.q.v.*
Yellowish or grayish, amorphous powder; characteristic odor. Partially sol in water. *Keep well closed and in a cool place.*
THERAP CAT: Oxytocic; antidiuretic.
THERAP CAT (VET): Oxytocic; antidiuretic.

7482. Pivalic Acid. *2,2-Dimethylpropanoic acid; α,α-dimethylpropionic acid; trimethylacetic acid.* $C_5H_{10}O_2$; mol wt 102.13. C 58.80%, H 9.87%, O 31.33%. Prepd by the reaction of *tert*-butylmagnesium chloride and carbon dioxide: Bouveault, *Compt. Rend.* **138**, 1108 (1904); Puntambeker, Zoellner, *Org. Syn.* **8**, 104 (1928); other methods: *ibid.* 108. Forms higher esters (e.g., isobutyl ester) only with difficulty.

Needles, mp 35.5°. bp_{760} 163.8°; d^{50} 0.905; $n_D^{36.5}$ 1.3931. K at 25° = 9.76 × 10^{-6}. One gram dissolves in 40 ml water. Freely sol in alcohol, ether.
Ethyl ester, $C_7H_{14}O_2$, liquid. d_4^{18} 0.8580; bp 118.2°; n_D^{18} 1.3922.

7483. Pivalylbenzhydrazine. *2,2-Dimethylpropanoic acid 2-(phenylmethyl)hydrazide; pivalic acid 2-benzylhydrazide; N^1-pivaloyl-N^2-benzylhydrazine; 1-benzyl-2-trimethylacethydrazine; 1-benzyl-2-pivaloylhydrazine; 1-(trimethylacetyl)-2-benzylhydrazine; pivhydrazine; Ro 4-1634; Tersavid.* $C_{12}H_{18}N_2O$; mol wt 206.28. C 69.87%, H 8.80%, N 13.58%, O 7.76%. $C_6H_5CH_2NHNHOCC(CH_3)_3$. Prepn: **Brit.** pat. **883,379** (1961 to Hoffmann-La Roche).
Crystals, mp 68-69°.
THERAP CAT: Monoamine oxidase inhibitor.

7484. Pivampicillin. *6-[(Aminophenylacetyl)amino]-3,3-dimethyl-7-oxo-4-thia-1-azabicyclo[3.2.0]heptane-2-carboxylic acid (2,2-dimethyl-1-oxopropoxy)methyl ester; hydroxymethyl 6-(2-amino-2-phenylacetamido)-3,3-dimethyl-7-oxo-4-thia-1-azabicyclo[3.2.0]heptane-2-carboxylate pivalate (ester);* 6-[D-α-aminophenylacetamido]penicillanic acid pivaloyloxymethyl ester; pivaloyloxymethyl D-α-aminobenzylpenicillinate; ampicillin pivaloyloxymethyl ester; pivaloyloxymethyl ampicillin; MK-191. $C_{22}H_{29}N_3O_6S$; mol wt 463.55. C 57.00%, H 6.30%, N 9.06%, O 20.71%, S 6.92%. Semi-synthetic antibiotic related to penicillin. Prepn: E. K. Frederiksen, W. O. Godtfredsen, **S. Afr.** pat. **68 05,952** corresp to U.S. pat. **3,660,575** (1969, 1972 to Leo Pharm.); von Daehne et al., *J. Med. Chem.* **13**, 607 (1970). Pharmacology: *eidem, loc. cit.;* Jordan et al., *Antimicrob. Ag. Chemother.* **1970**, 438; Foltz et al., *ibid.* 442. Pharmacokinetics in man: M. Ehrnebo et al., *J. Pharmacokinet. Biopharm.* **7**, 429 (1979). Toxicity: von Daehne et al., *Antimicrob. Ag. Chemother.* **1970**, 430.

Hydrochloride, $C_{22}H_{30}ClN_3O_6S$, *Alphacilina, Alphacillin, Berocillin, Centurina, Devonium, Diancina, Inacilin, Maxifen, Pivatil, Pondocil, Pondocillin, Pondocillina, Sanguicillin.* Microcrystalline powder, mp 155-156° (dec). $[\alpha]_D^{20}$ +196° (c = 1 in water). Weak uv max (water): 268, 262, 256 nm ($E_{1cm}^{1\%}$ ~3.9, 5.7, 6.3). pKa ~7.0. Relatively stable in acid soln; ester hydrolyzes slowly in neutral soln. pH of 0.5 g/100 ml water: ~ 4.5. Very sol in water and chloroform; freely sol in ethanol; sparingly sol in *n*-propanol, *tert*-butanol, and ethyl ether. LD_{50} in mice, rats (g/kg): 3.34, 5.00 orally; 3.60, 4.50 s.c. (von Daehne).

THERAP CAT: Antibacterial.

7485. Pivcefalexin. *7-[(Aminophenylacetyl)amino]-3-methyl-8-oxo-5-thia-1-azabicyclo[4.2.0]oct-2-ene-2-carboxylic acid (2,2-dimethyl-1-oxopropoxy)methyl ester; 7-(D-2-amino-2-phenylacetamido)desacetoxycephalosporanic acid pivaloyloxymethyl ester; pivcephalexin; Cefalex.* $C_{22}H_{27}N_3O_6S$; mol wt 461.54. C 57.25%, H 5.90%, N 9.10%, O 20.80%, S 6.95%. Orally active semi-synthetic cephalosporin antibiotic; deriv of cephalexin, *q.v.* Prepn: W. O. Godtfredsen, E. T. Binderup, Ger. pat. **1,951,012** (1970 to Leo Pharm.), *C.A.* **72**, 132761v (1970). Pharmacology: P. Foresta *et al., Arzneimittel-Forsch.* **27**, 819 (1977). Absorption and excretion: E. Trabucchi *et al., Clin. Ther.* **81**, 299 (1977). Clinical trial: C. Vittorini *et al., Arzneimittel-Forsch.* **31**, 1163 (1981).

Hydrochloride, $C_{22}H_{28}ClN_3O_6S$, *ST-21, Bencef, Pivacef, Sigmacef.*

THERAP CAT: Antibacterial.

7486. Pizotyline. *4-(9,10-Dihydro-4H-benzo[4,5]cyclohepta[1,2-b]thien-4-ylidene)-1-methylpiperidine; 4-(1-methyl-4-piperidylidene)-9,10-dihydro-4H-benzo[4,5]cyclohepta[1,2-b]thiophene; pizotifen; pizotifan; BC 105; Litec.* $C_{19}H_{21}NS$; mol wt 295.45. C 77.24%, H 7.16%, N 4.74%, S 10.85%. A benzocycloheptathiophene deriv with a basic side chain resembling that of cyproheptadine. Prepn: Jucker *et al., Belg. pat.* **636,717** and U.S. pat. **3,272,826** (1964, 1966 to Sandoz); Bastian *et al., Helv. Chim. Acta* **49**, 214 (1966). Toxicity data: Bagdon, Dorado, *Pharmacologist* **12**, No. 2, 297 (1970). Review of pharmacological properties and therapeutic efficacy: Speight, Avery, *Drugs* **3**, 159-203 (1972).

Hydrochloride, crystals from isopropanol-ether, mp 261-263° (dec).

Malate, $C_{23}H_{27}NO_5S$, *Sandomigran, Sanmigran, Sanomigran, Moségor.* Crystals from methanol, mp 185-186° (dec).

THERAP CAT: Anabolic; antidepressant; serotonin inhibitor (specific in migraine).

7487. Placenta. Lioplacentyl; Tissulina; Inplacen; Placentafil; Tissural. Human placenta prepd for injection or topical use. Has been used to stimulate regeneration of damaged tissue.

7488. Plafibride. *2-(4-Chlorophenoxy)-2-methyl-N-[[(4-morpholinylmethyl)amino]carbonyl]propanamide; N-2-(p-chlorophenoxy)isobutyryl-N'-morpholinomethylurea; ITA-104; Idonor; Perifunal.* $C_{16}H_{22}ClN_3O_4$; mol wt 355.83. C 54.01%, H 6.23%, Cl 9.96%, N 11.81%, O 17.99%. Analog

of clofibrate, *q.v.* Prepn: J. Iniesta-Pons, Ger. pat. **2,716,374** (1977 to Investigacion Tecnica y Aplicada), *C.A.* **88**, 37810g (1978). Series of articles on antiplatelet aggregation activity, hypolipemic activity, pharmacology, and toxicology: *Arch. Farmacol. Toxicol.* **4**, 132-142 (1978).

Cryst, mp 100-102°. Sol in acetone. Slightly sol in alcohol. Practically insol in water, petr ether. LD_{50} in mice, rats, guinea pigs (mg/kg): 3569, >4000, 2168 orally, J. Zapatero, L. Brugeghini, *Arch. Farmacol. Toxicol.* **4**, 137 (1978).

THERAP CAT: Antithrombotic.

7489. Plantago Seed. Psyllium seed; flea seed. Seed from *Plantago psyllium* L. or *P. arenaria* Waldst. & Kit. (*P. ramosa* Asch.) (Spanish or French psyllium seed) or of *P. ovata* Forsk. (blond or Indian plantago seed), *Plantaginaceae. Review:* BeMiller in *Industrial Gums*, R. L. Whistler, Ed. (Academic Press, New York, 2nd ed., 1973) pp 345-354.

Small, dark reddish-brown, odorless, almost tasteless seeds. Mixed with an equal bulk of water, forms a mucilaginous mass.

THERAP CAT: Cathartic.

7490. Plantisul. Banabins. Orally effective antidiabetic principle contained in aq extracts from the leaves and fruits of the banaba tree, *Lagerstroemia speciosa (L.) Pers., Lythraceae.* Extraction procedure: Faustino Garcia, *Philippine J. Sci.* **76**, no. 3, 3-21 (1944); *J. Philippine Med. Assoc.* **31**, 216-224 and 276-282 (1955). *Constit.* of banaba extracts: Carew, Chin, *Nature* **190**, 1108 (1961).

7491. Plasma (in Physics). In physics and in nuclear chemistry plasma means highly ionized gases, i.e., a mixture of ions, electrons and at times neutral particles at high energy levels. Books: J. G. Linhart, *Plasma Physics* (North Holland Publ., Amsterdam, 1960); *Topics in Current Chemistry*, S. Veprek, M. Venugopalan, Eds., **Vols. 89, 90** and **94**, entitled *Plasma Chemistry I, II* and *III* (Springer-Verlag, New York, 1980) 143 pp, 121 pp and 125 pp, resp.

7492. Plasmalogens. Aldehydogenic lipids characteristic of the animal kingdom; references to their existence in plants and bacteria are rare. Plasmalogens contain an aldehydogenic chain linked to glycerol as an α,β-unsaturated ether. Although most are aldehydogenic phosphatides, nonphosphatide or neutral plasmalogens have been detected in animal tissues. "Native plasmalogens", when deacylated, yield lysoderivatives, *lysoplasmalogens*, contg the aldehydogenic chain linked to glycerol as an α,β-unsaturated ether. The nomenclature "*phosphatidal ethanolamine*", "*phosphatidal choline*", etc. has been proposed for native plasmalogen phosphatides, and "*lysophosphatidal ethanolamine*", "*lysophosphatidal choline*", etc. for corresponding deacylated derivatives. This nomenclature minimizes the confusion arising from less precise terms, such as "plasmalogen" or "*ethanolamine plasmalogen*", "*choline plasmalogen*", etc. to designate either native compds, lysoderivatives, or other structures which may occur (molecules with two α,β-unsaturated ether chains or with one saturated ether and one α,β-unsaturated ether chain, and true cyclic glyceryl acetal derivatives). First isoln of a pure native plasmalogen (phosphatidal choline): Gottfried, Rapport, *Fed. Proc.* **20**, 278 (1961); *eidem, J. Biol. Chem.* **237**, 329 (1962). *Ref:* Rapport, Norton, *Ann. Rev. Biochem.* **31**, 103 (1962). Synthesis: Piantadosi *et al., J. Org. Chem.* **28**, 2425 (1963); Chacko *et al., ibid.* **32**, 3698 (1967); Slotboom *et al., Chem. Phys. Lipids* **1**, 192 (1967); Gigg, Gigg, *J. Chem. Soc. C* **1968**, 16, 2030; Vtorov *et al., Tetrahedron Letters* **1971**, 4605. *Reviews:* E. Klenk, H. Debuch, "Plasmalogens" in Holman *et al., Progr. Chem. Fats Lipids* vol. 6, (Macmillan, New York, 1963) pp 1-29; Piantadosi, Snyder, *J. Pharm. Sci.* **59**, 283-297 (1970).

native plasmalogen phosphatide

7493. Plasmin. Fibrinolysin; serum tryptase; Actase; Thrombolysin. Mol wt about 90,000. Trypsin-like proteolytic enzyme which cleaves fibrin, fibrinogen, q.q.v., and other plasma proteins. Component of the mammalian fibrinolytic system specifically responsible for the dissolution of fibrin clots. Exists in plasma as an inactive precursor, plasminogen, q.v. Converted to the active enzyme at the clot site by tissue plasminogen activator, q.v. Also activated by streptokinase and urokinase, q.q.v. Rapidly inactivated in plasma by α_2-antiplasmin, a glyceroprotein with high specific binding affinity for plasmin. Prepn: Christensen, MacLeod, J. Gen. Physiol. **28**, 599 (1945); E. C. Loomis, U.S. pat. **2,624,691** (1953 to Parke, Davis); J. H. Hink, J. K. McDonald, U.S. pat. **3,234,106** (1966 to Cutter Labs); K. C. Robbins, L. Summaria, J. Biol. Chem. **238**, 952 (1963). Converted from plasminogen by proteolysis of a single arg-val bond: K. C. Robbins et al., ibid. **242**, 2333, 4279 (1967). Structure consists of two polypeptide chains connected by two disulfide bonds. The heavy (A) chain (MW ~65,000) originates from the amino-terminal portion of plasminogen and contains the binding sites. The light (B) chain (MW ~25,000) originates from the carboxy-terminus and contains the active site. Isolation and characterization of heavy and light chains: L. Summaria et al., J. Biol. Chem. **242**, 5046 (1967); eidem, ibid. **246**, 2143 (1971). Amino acid sequence studies: W. R. Groskopf et al., ibid. **244**, 3590 (1969); Hartley, Phil. Trans. Roy. Soc. London Ser. B **257**, 77 (1970); S. Nagasawa, T. Suzuki, Biochem. Biophys. Res. Commun. **41**, 562 (1970). Amino acid sequence of active site: K. C. Robbins et al., J. Biol. Chem. **248**, 1631 (1973). Review of structural studies: F. J. Castellino, Semin. Throm. Hemostasis. **10**, 18-23 (1984). Enzyme specificity for lysine and arginine peptide bonds: W. Troll et al., J. Biol. Chem. **208**, 85 (1954). Review of biochemistry: D. Ogston, J. Clin. Pathol. **33**, Suppl. 14, 5-9 (1980). Reviews: D. Collen, Thromb. Haemostasis **43**, 77-89 (1980); K. C. Robbins et al., Methods Enzymol. **80**, 379-387 (1981); K. K. Kane, Ann. Clin. Lab. Sci. **14**, 443-449 (1984).
THERAP CAT: Thrombolytic enzyme.

7494. Plasminogen. Profibrinolysin; plasma trypsinogen. Mol wt ~90,000. The circulating plasma precursor (zymogen) or inactive form of plasmin, q.v. Prepn from blood plasma: Loomis et al., Arch. Biochem. **12**, 1 (1947); Oncley et al., J. Am. Chem. Soc. **71**, 541 (1949). Purification: Christensen, Smith, Proc. Soc. Exp. Biol. Med. **74**, 840 (1950); Kline, J. Biol. Chem. **204**, 949 (1953); Hagan et al., ibid. **235**, 1005 (1960); Hagan et al., U.S. pat. **3,066,079** (1962 to Am. Cyanamid); Derechin et al., Biochem. J. **84**, 336 (1962); Mertz, Chan, Can. J. Biochem. Physiol. **41**, 1811 (1963); D. G. Deutsch, E. T. Mertz, Science **170**, 1095 (1970). Physical properties: Davies, Englert, J. Biol. Chem. **235**, 1011 (1960). Converted to plasmin by natural activators such as streptokinase, urokinase, or tissue plasminogen activator, q.q.v., through the cleavage of a single arg-val bond. Mechanism of activation: K. C. Robbins et al., ibid. **242**, 2333, 4279 (1967). Native plasminogen is a single chain glycopeptide. Two major molecular forms exist, differing in carbohydrate content and separable by affinity chromatography. Mol wt studies: Barlow et al., ibid. **244**, 1138 (1969). Sequence studies: K. C. Robbins et al., ibid. **247**, 6757 (1972); F. J. Castellino et al., Biochem. Biophys. Res. Commun. **53**, 845 (1973). Carbohydrate comp: M. L. Hayes, F. J. Castellino, J. Biol. Chem. **254**, 8768, 8772, 8777 (1979). Amino acid sequence and biochemistry: F. J. Castellino, Semin. Thromb. Hemostasis **10**, 18 (1984). Biosynthesis and secretion by cultured hepatocytes: J. F. Bohmfalk, G. M. Fuller, Science **209**, 408 (1980). Therapeutic use with streptokinase: I. D. Walker et al., Thromb. Haemostasis **51**, 204 (1984). Review of fibrinolytic system: D. Collen, ibid. **43**, 77-89 (1980). Reviews: Fibrinolysis, D. L. Kline, K. N. N.

Reddy, Eds. (CRC, Boca Raton, 1980) 256 pp; F. J. Castellino, Chem. Rev. **81**, 431-446 (1981).
Soluble below pH 5 and above pH 9 and only sparingly sol at intermediate pH values. Resistant to heat below pH 5. Displays maximum stability at pH 2-3.

7495. Plasmocid. N,N-Diethyl-N'-(6-methoxy-8-quinolinyl)-1,3-propanediamine; 8-(3-diethylaminopropylamino)-6-methoxyquinoline; 6-methoxy-8-(3-diethylaminopropylamino)quinoline; 710 F; SN 3115; Antimalarine; Fourneau 710; Rhodoquine. $C_{17}H_{25}N_3O$; mol wt 287.39. C 71.04%, H 8.77%, N 14.62%, O 5.57%. Prepn: Fourneau et al., Ann. Inst. Pasteur **44**, 503 (1930); **46**, 514 (1931); **50**, 731 (1933); Magidson, Strukov, Arch. Pharm. **271**, 359 (1933); Strukov, Russian pat. **39,105** (1934), C.A. **30**, 3445 (1936); Magidson et al., J. Appl. Chem. USSR **9**, 304 (1936); Giral, Ciencia (Mexico) **6**, 253 (1945), C.A. **40**, 3563 (1946); Yanko et al., J. Am. Chem. Soc. **67**, 664 (1945).

Oily liquid, d_4^{24} 1.0569. $bp_{1.0}$ 182°. n_D^{24} 1.5855.
Dihydrochloride, $C_{17}H_{27}Cl_2N_3O$, yellow crystals, mp 218-220°. Slightly sol in water, alcohol.
Diphosphate, yellow crystals, mp 169-171°. Slightly sol in water, alcohol.
Note: The name Rhodoquine was first applied to 8-(3-dimethylaminopropylamino)-6-methoxyquinoline.
THERAP CAT: Antimalarial.

7496. Platelet Activating Factor. PAF; 1-O-alkyl-2-acetyl-sn-glycero-3-phosphorylcholine; acetyl glyceryl ether phosphorylcholine; AGEPC; PAF-acether; antihypertensive polar renomedullary lipid; APRL. Phospholipid mediator of platelet aggregation, inflammation, and anaphylaxis. Produced in response to specific stimuli by a variety of cell types, including neutrophils, basophils, platelets, and endothelial cells. Several molecular species of PAF have been identified which vary in the length of the O-alkyl side chain. In vitro identification of a soluble platelet activating factor produced by antigen stimulated leukocytes: P. M. Henson, J. Exp. Med. **131**, 287 (1970). Isoln from rabbit basophils, characterization of PAF response and potential role in immune complex disease: J. Benveniste et al., ibid. **136**, 1356 (1972). Isoln from human leukocytes: J. Benveniste, Nature **249**, 581 (1974). Release of PAF in vivo during anaphylaxis: R. N. Pinckard et al., J. Immunol. **123**, 1847 (1979). Structural study: J. Benveniste et al., Nature **269**, 170 (1977); and synthetic approaches: J. Benveniste et al., Compt. Rend. Ser. D **289**, 1037 (1979); C.A. Demopoulos et al., J. Biol. Chem. **254**, 9355 (1979). Identity with APRL and antihypertensive activity: M. L. Blank et al., Biochem. Biophys. Res. Commun. **90**, 1194 (1979). Structures of 1-O-octadecyl- and 1-O-hexadecyl-PAF, two of the predominant molecular forms: D. J. Hanahan et al., ibid. **255**, 5514 (1980). Total synthesis of 1-O-octadecyl-PAF: J. Godfroid et al., FEBS Letters **116**, 161 (1980). Enantiomeric synthesis of C_{16}- and C_{18}-PAF: M. Ohno et al., Chem. Pharm. Bull. **33**, 572 (1985). Molecular heterogeneity of naturally occurring PAF: R. N. Pinckard et al., Biochem. Biophys. Res. Commun. **122**, 325 (1984); H. W. Mueller et al., J. Biol. Chem. **259**, 14554 (1984). Ulcerogenic effects: A.-C. Rosam et al., Nature **319**, 54 (1986). Effects on human pulmonary and cardiovascular function: F. M. Cuss et al., Lancet **2**, 189 (1986). Review of isoln and analytical methods: D. J. Hanahan, S. T. Weintraub, Meth. Biochem. Anal. **31**, 195-219 (1985); of chemical and biochemical characteristics: D. J. Hanahan, R. Kumar, Prog. Lipid Res. **26**, 1-28 (1987). Reviews of biosynthesis, biological activities and PAF antagonists: P. Braquet et al., Pharmacol. Rev. **39**, 97-145 (1987); P. J. Barnes et al., J. Allergy Clin. Immunol. **81**, 919-934 (1988). Books: Platelet-Activating Factor and Related Lipid Mediators, F. Snyder, Ed. (Plenum Press, New York, 1987) 471 pp; Prog. Biochem. Pharmacol. **22** entitled

"Biologically Active Ether Lipids" by P. Braquet *et al.*, Eds. (Karger, Basel, 1988).

$$CH_3COCH \begin{matrix} CH_2O(CH_2)_nCH_3 \\ | \\ | \\ O \\ \| \\ CH_2OPOCH_2CH_2N(CH_3)_3 \\ | \\ O^- \end{matrix}$$

n usually represents 15 or 17

1-*O*-Hexadecyl PAF, $C_{26}H_{54}NO_7P$, *(R)-7-(acetyloxy)-4-hydroxy-N,N,N-trimethyl-3,5,9-trioxa-4-phosphapentacosan-1-aminium hydroxide inner salt 4-oxide*, C_{16}-*PAF*. White, amorphous solid, mp 247° (dec). $[\alpha]_D^{21}$ —3.66° (c = 0.71 in chloroform).

1-*O*-Octadecyl-PAF, $C_{28}H_{58}NO_7P$, *(R)-7-(acetyloxy)-4-hydroxy-N,N,N-trimethyl-3,5,9-trioxa-4-phosphaheptacosan-1-aminium hydroxide inner salt 4-oxide*, C_{18}-*PAF*. mp 212-215° (dec). $[\alpha]_D^{20}$ —4.00° (c = 0.71 in chloroform).

7497. Platelet-Derived Growth Factor. PDGF. Cationic glycoprotein mitogen for fibroblasts, smooth muscle cells and glial cells. Mol wt ~30,000 daltons. Principal factor in serum required for the growth and proliferation of mesenchymal derived cells in tissue culture. Originally isolated from blood platelets, PDGF-like peptides have subsequently been found to be synthesized and secreted by a number of normal and neoplastic cell types. Identification of platelet-derived mitogen for smooth muscle cells: R. Ross *et al.*, *Proc. Nat. Acad. Sci. USA* **71**, 1207 (1974); for fibroblasts: N. Kohler, A. Lipton, *Exp. Cell Res.* **87**, 297 (1974); for glial cells: B. Westermark, A. Wasteson, *Adv. Metabol. Dis.* **8**, 85 (1975). Isoln of *human serum growth factor:* H. N. Antoniades *et al.*, *Proc. Nat. Acad. Sci. USA* **72**, 2635 (1975); and immunological identity with PDGF: H. N. Antoniades, C. D. Scher, *ibid.* **74**, 1973 (1977). Purification: H. N. Antoniades *et al.*, *ibid.* **76**, 1809 (1979); C.-H. Heldin *et al.*, *ibid.* 3722. PDGF is stored in platelet α-granules and released at sites of tissue injury upon platelet activation. Intracellular localization and potential role in wound repair: D. R. Kaplan *et al.*, *Blood* **53**, 1043 (1979). Chemotactic effect on smooth muscle cells: G. R. Grotendorst *et al.*, *Proc. Nat. Acad. Sci. USA* **78**, 3669 (1981); on monocytes and neutrophils: T. F. Deuel *et al.*, *J. Clin. Invest.* **69**, 1046 (1982). PDGF is a disulfide-linked dimer of two polypeptide chains, designated A and B, which may be assembled as a heterodimer and/or as AA or BB homodimers. Proposed structure as AB heterodimer: C.-H. Heldin *et al.*, *Biochem. J.* **193**, 907 (1981); A. Johnsson *et al.*, *Biochem. Biophys. Res. Commun.* **104**, 66 (1982). PDGF-B is structurally homologous to *p28^sis*, the transforming protein of the simian sarcoma virus: M. D. Waterfield *et al.*, *Nature* **304**, 35 (1983); R. F. Doolittle *et al.*, *Science* **221**, 275 (1983). *Osteosarcoma-derived growth factor* or *ODGF*, a growth factor isolated from human osteosarcoma cells, is structurally similar to a PDGF-A homodimer: C.-H. Heldin *et al.*, *Nature* **319**, 511 (1986). Comparison of biological properties and transforming potential of A and B chains: M. P. Beckmann *et al.*, *Science* **241**, 1346 (1988). Identification of PDGF receptor: C.-H. Heldin *et al.*, *Proc. Nat. Acad. Sci. USA* **78**, 3664 (1981). Correlation of receptor binding with mitogenic activity: D. F. Bowen-Pope, R. Ross, *J. Biol. Chem.* **257**, 5161 (1982). Molecular mechanisms of action: S. R. Coughlin *et al.*, *Prog. Clin. Biol. Res.* **266**, 39 (1988); L. T. Williams, *Science*, **243**, 1564 (1989). Review of purification, properties and biological activities: T. F. Deuel, J. S. Huang, *Prog. Hematol.* **13**, 201-221 (1983). Review of relationship to oncogenes and malignant transformation: H. N. Antoniades, *Biochem. Pharmacol.* **33**, 2823-2828 (1984); C.-H. Heldin *et al.*, *J. Cell. Sci.* **Suppl. 3**, 65-76 (1985); of role in atherosclerosis: R. Ross, *N. Engl. J. Med.* **314**, 488-500 (1986); in arteriosclerosis: H. R. Baumgartner, M. Hosang, *Experientia* **44**, 109-112 (1988). *Reviews:* C.-H. Heldin *et al.*, *Mol. Cell. Endocrinol.* **39**, 169-187 (1985); R. Ross *et al.*, *Cell* **46**, 155-169 (1986); T. F. Deuel *et al.*, *J. Cell*

Physiol. **Suppl. 5**, 95-99 (1987); R. Ross, *Ann. Rev. Med.* **38**, 71-79 (1987).

7498. Platinic Chloride. *Hydrogen hexachloroplatinate-(IV); acid platinic chloride;* hexachloroplatinic(IV) acid; chloroplatinic acid. Cl_6H_2Pt; mol wt 409.80. Cl 51.90%, H 0.49%, Pt 47.61%. H_2PtCl_6. Exists as hexahydrate.

Hexahydrate, brownish-yellow, very deliquesc, cryst mass. d 2.431; mp 60°. Easily sol in water, alcohol. *Keep tightly closed, protected from light.*

USE: In platinum plating, photography, platinum mirrors, platinum luster on glass and porcelain, platinized carbon for acetic acid manuf; platinizing pumice stone or asbestos, as catalyst in manuf SO_3; indelible ink; relief etching of zinc for artistic and commercial purposes; fixing microscopic prepns, etc. *Caution:* May cause asthma or dermatitis, E. Browning, *Toxicity of Industrial Metals* (Appleton-Century-Crofts, New York, 2nd ed., 1969) pp 270-275.

7499. Platinic Iodide. I_4Pt; mol wt 702.73. I 72.23%, Pt 27.77%. PtI_4. Prepn: Argue, Banewicz, *J. Inorg. Nucl. Chem.* **25**, 923 (1963).

Brownish-black to black powder. Sol in water.

7500. Platinic Oxide. Platinum dioxide; Adams' catalyst. O_2Pt; mol wt 227.09. O 14.09%, Pt 85.92%. PtO_2. Prepd by the interaction of platinic chloride and excess NaOH: Wöhler, *Z. Anorg. Chem.* **40**, 434 (1904); Bellucci *ibid.* **44**, 171 (1905); by the reduction of chloroplatinic acid with formaldehyde: Willstätter, Waldschmidt-Leitz, *Ber.* **54**, 113 (1921); Feulgen, *ibid.* 360; by fusing chloroplatinic acid with sodium nitrate: Adams *et al.*, *Org. Syn.* **8**, 92 (1928). Determination of orthorhombic $(CaCl_2)$ crystal structure of β-PtO_2: Siegel *et al.*, *J. Inorg. Nucl. Chem.* **31**, 3803 (1969).

Several hydrates have been reported; the compd generally used for catalysis is the monohydrate. Black powder. When freshly pptd it is sol in concd acids, also in dil H_3PO_4, esp when warmed. Easily sol in dilute solns of potassium hydroxide.

USE: As catalyst in hydrogenations. The actual catalyst is platinum black which is formed *in situ* by reduction of the PtO_2 by the hydrogen used for the hydrogenation. Especially useful for reduction at room temp and hydrogen pressures up to 4 atmospheres. Suitable for the reduction of double and triple bonds, aromatic nuclei, carbonyl groups, nitro groups, and nitriles.

7501. Platinous Chloride. Cl_2Pt; mol wt 266.00. Cl 26.64%, Pt 73.36%. $PtCl_2$. Prepn from hexachloroplatinic acid, $H_2PtCl_6.6H_2O$: Cohen, *Inorg. Syn.* **6**, 209 (1960).

Grayish-green to brown powder. d 5.87. Insol in water, alcohol, ether; sol in hydrochloric acid. Combines with PCl_3 to form a compd sol in benzene or chloroform.

7502. Platinous Iodide. Platinum diiodide. I_2Pt; mol wt 448.91. I 56.53%, Pt 43.47%. PtI_2. Prepn: Argue, Banewicz, *J. Inorg. Nucl. Chem.* **25**, 923 (1963).

Heavy, black powder. Dec. 325°. Insol in water or alkali iodides.

7503. Platinum. Pt; at. wt 195.08; at. no. 78; valences 2,4; seldom 1, 5, 6. Six naturally occurring isotopes: 190 (0.01%); 192 (0.8%); 194 (32.9%); 195 (33.8%); 196 (25.2%); 198 (7.2%); 190 is radioactive: $T_{1/2}$ 6.9 × 10^{11} years. Artificial, radioactive isotopes: 173-189; 191; 193; 197; 199-201. Abundance in earth's crust about 0.01 ppm. Believed to be mentioned by Pliny under the name "alutiae". Has been known and used in South America as "platina del Pinto". Reported by Ulloa in 1735; brought to Europe by Wood, and described by Watson in 1741. Occurs native alloyed with one or more members of its group (iridium, osmium, palladium, rhodium, and ruthenium) in gravels and sands. Prepn: Wichers *et al.*, *Trans. Amer. Inst. Min. Met. Eng.* **76**, 602 (1928). Reviews of prepn, properties and chemistry of platinum and other platinum metals: Gilchrist, *Chem. Rev.* **32**, 277-372 (1943); Beamish *et al.*, in *Rare Metals Handbook*, C. A. Hampel, Ed. (Reinhold, New York, 1956) pp 291-328; Livingstone in *Comprehensive Inorganic Chemistry*, **Vol. 3**, J. C. Bailar, Jr. *et al.*, Eds. (Pergamon Press, Oxford, 1973) pp 1163-1189, 1330-1370; F. R. Hartley, *The Chemis-*

try of *Platinum and Palladium with Particular Reference to Complexes of the Elements* (Halsted Press, New York, 1973).

Silver-gray, lustrous, malleable and ductile metal; face-centered cubic structure. Also prepd in the form of a black powder (platinum black) and as spongy masses (platinum sponge). mp 1773.5 ± 1°; Roeser *et al.*, *Nat. Bur. Stand. J. Res.* **6**, 1119 (1931); bp about 3827°. d 21.447 (calcd). Brinell hardness: 55. Sp heat 0.0314 cal/g at 0°. Electrical resistivity (20°) 10.6 μohm-cm. Does not tarnish on exposure to air. Absorbs hydrogen at a red heat and retains it tenaciously at ord. temp, gives off the gas at a red heat *in vacuo*. Occludes carbon monoxide, carbon dioxide, nitrogen. Volatilizes considerably when heated in air at 1500°. The heated metal absorbs oxygen; gives it off on cooling. Not affected by water or by single mineral acids. Reacts with boiling aqua regia with formation of chloroplatinic acid, also with molten alkali cyanides. Attacked by halogens, by fusion with caustic alkalies, alkali nitrates, alkali peroxides; by arsenates and phosphates in presence of reducing agents.

Human Toxicity: Inhalation of the dust of sol platinum salts is irritating and may cause rhinorrhea, sneezing, tightness of chest, shortness of breath, cyanosis, wheezing, cough. May cause dermatitis: E. Browning, *Toxicity of Industrial Metals* (Appleton-Century-Crofts, New York, 2nd ed., 1969) pp 270-275.

USE: Manuf apparatus for laboratory and industrial use, thermocouples, platinum resistance thermometers, acidproof containers, electrodes, etc. In dentistry; jewelry; electroplating. As oxidation catalyst in manuf acetic acid, nitric acid from ammonia, manuf sulfuric acid; control of automotive emissions.

7504. Platonin. *2,2'-[3-[(3-Heptyl-4-methyl-2(3H)-thiazolylidene)ethylidene]-1-propene-1,3-diyl]bis(3-heptyl-4-methylthiazolium) diiodide; 4,4',4''-trimethyl-3,3',3''-triheptyl-8-(2''-thiazolyl)-2,2'-pentamethinethiazolocyanine 3,3''-diiodide; 3,3',3''-triheptyl-4,4',4''-trimethyl-7-(2''-methylthiazolyl-2,2'-trimethine)thiazolcyanine 3,3''-diiodide; platonin J; NK-19; Kankohso 101; Photosensitizer 101.* $C_{38}H_{61}I_2N_3S_3$; mol wt 909.91. C 50.16%, H 6.76%, I 27.89%, N 4.62%, S 10.57%. Photosensitizing cyanine dye with antimicrobial and immunomodulating activities. General prepn and use in rheumatoid arthritis: I. Yamamoto, **Belg. pat. 894,635**; *idem*, **U.S. pat. 4,464,383** (1983, 1984). Antifungal properties: K. Ito, P. K. Kuroda, *Bull. Pharm. Res. Inst. Osaka Med. Coll.* **3**, 20 (1952), *C.A.* **47**, 11335f (1953). Effect on experimentally induced leukopenia: H. Iwata *et al.*, *Folia Pharmacol. Japon.* **50**, 169 (1954), *C.A.* **49**, 10518b (1955). Immunomodulating effect in animals: H. Ichihashi, T. Kondo, *Gann* **58**, 529 (1967); Y. Oyanagui, *Arch. Int. Pharmacodyn. Ther.* **266**, 162 (1983). Enzyme immunoassay in plasma: I. Yamamoto, K. Morishita, *K. Immunoassay* **7**, 17 (1986). Toxicity study: T. Kimoto, K. Nishitani, *Kanko Shikiso* **85**, 43 (1977), *C.A.* **86**, 84374m (1977).

Brilliant green crystalline powder, mp 204°. Sol in water, alcohol. LD_{50} in male, female mice (mg/kg): 46.9, 50.5 i.p.; in male, female rats (mg/kg): 1539, 1571 orally (Kimoto, Nishitani).

THERAP CAT: Immunomodulator.

7505. Platyphylline. *1,2-Dihydro-12-hydroxysenecionan-11,16-dione; platifillin.* $C_{18}H_{27}NO_5$; mol wt 337.40. C 64.07%, H 8.06%, N 4.15%, O 23.71%. Toxic pyrrolizidine alkaloid from *Senecio platyphyllus* DC. and other *Senecio*

spp., *Compositae:* Orékhov, Tiedebel, *Ber.* **68**, 650 (1935); Orékhov *et al.*, *ibid.* 1886; Konovalova, Orékhov, *ibid.* **69**, 1908 (1936); *Bull. Soc. Chim. France* [5] **4**, 2037 (1937); Allen, *Am. J. Pharm.* **117**, 110 (1945). Structure: Dry *et al.*, *J. Chem. Soc.* **1955**, 63. Comprehensive reviews of platyphylline and other pyrrolizidine alkaloids: L. Bull *et al.*, *The Pyrrolizidine Alkaloids* (North-Holland, Amsterdam, 1968) 293 pp; F. L. Warren in *The Alkaloids*, **vol. 12**, R. H. F. Manske, Ed. (Academic Press, New York, 1970) pp 245-331.

Crystals, mp 129°. $[\alpha]_D^{20}$ −56.4° (c = 0.7 in chloroform). Practically insol in water. Sol in alcohol, chloroform, ether, dil acids.

Bitartrate, $C_{18}H_{27}NO_5 \cdot C_4H_6O_6$, crystals, mp 199°. $[\alpha]_D^{25}$ −40° (c = 2). Sol in 10 parts of water, in 5 parts of hot water. Sparingly sol in alc, in 42 parts of boiling alc. Nearly insol in chloroform, ether.

7506. Plaunotol. *(Z,E,E)-2-(4,8-Dimethyl-3,7-nonadienyl)-6-methyl-2,6-octadiene-1,8-diol; (E,Z,E)-7-hydroxymethyl-3,11,15-trimethyl-2,6,10,14-hexadecatetraen-1-ol; CS-684; Kelnac.* $C_{20}H_{34}O_2$; mol wt 306.49. C 78.38%, H 11.18%, O 10.44%. Acyclic diterpene alcohol isolated from a Thai medicinal plant identified as *Croton sublyratus* Kurz, *Euphorbiaceae.* Isoln, synthesis and anti-ulcer activity: H. Mishima *et al.*, **Japan. Kokai 77 62,213**; *eidem*, **U.S. pat. 4,059,641** (both 1977 to Sankyo); A. Ogiso *et al.*, *Chem. Pharm. Bull.* **26**, 3117 (1978). Pharmacology: S. Kobayashi *et al.*, *Oyo Yakuri* **24**, 599 (1982).

Light yellow oil. Aromatic odor, bitter taste. Sol in benzene, acetone, alcohols, ethers. Insol in water. LD_{50} in male, female mice, rats (μl/kg): 8800, 8100, 10900, 11200 orally (Kobayashi).

THERAP CAT: Anti-ulcerative.

7507. Plegatil. *N,N-Diethyl-N-methyl-2-[2-(trimethylammonio)ethoxy]-1-propanaminium diiodide; N,N-diethyl-N,N',N',N'-tetramethyl-N,N'-(2-methyl-3-oxapentamethylene)bis[ammonium iodide]; 2-(2-dimethylaminoethoxy)-N,N-diethylpropylamine dimethiodide; β-dimethylaminoethyl β'-diethylamino-α'-methylethyl ether dimethiodide; 2-dimethylaminoethyl 2'-diethylaminoisopropyl ether bismethiodide; HL 8731.* $C_{13}H_{32}I_2N_2O$; mol wt 486.24. C 32.11%, H 6.63%, I 52.20%, N 5.76%, O 3.29%. Prepn: Pakleppa, Lenke, **East Ger. pat. 11,363** (1956) (not covered). *Ref:* Lenke, *Arzneimittel-Forsch.* **7**, 616 (1957); *ibid.* **11**, 571 (1961). Ganglionic blocking agent.

Crystals, mp 178-180°. LD_{50} i.v. in mice: 39.1 mg/kg, Lenke, *loc. cit.* (1957).

7508. Pleuromutilin. *Hydroxyacetic acid 6-ethenyldeca-*

hydro-5-hydroxy-4,6,9,10-tetramethyl-1-oxo-3a,9-propano-3aH-cyclopentacycloocten-8-yl ester; glycolic acid 8-ester with octahydro-5,8-dihydroxy-4,6,9,10-tetramethyl-6-vinyl-3a,9-propano-3aH-cyclopentacycloocten-1(4H)-one; drosophilin B; BC 757. $C_{22}H_{34}O_5$; mol wt 378.49. C 69.81%, H 9.05%, O 21.14%. Antibiotic substance produced by the basidio-mycetes *Pleurotus mutilus* (Fr.) Sacc. and *P. passeckerianus* Pilat.: Kavanagh *et al., Proc. Nat. Acad. Sci. USA* **37**, 570 (1951); **38**, 555 (1952). Isoln procedure: Anchel, *J. Biol. Chem.* **199**, 133 (1952). Structure: D. Arigoni, *Gazz. Chim. Ital.* **92**, 884 (1962). A. J. Birch *et al., Chem. & Ind. (London)* **1963**, 374; *Tetrahedron* **Suppl. 8**, 359 (1966). Stereo-chemistry: D. Arigoni, *Pure Appl. Chem.* **17**, 331 (1968). Fermentation and biosynthetic study: F. Knauseder, E. Brandl, *J. Antibiot.* **29**, 125 (1976). Mechanism of action: G. Hoegenauer, *Antibiotics* vol. 5, pt. 1, F. E. Hahn, Ed. (Springer-Verlag, New York, 1979) p 344. Synthetic stud-ies: E. G. Gibbons, *J. Org. Chem.* **45**, 1540 (1980); M. Kahn, *Tetrahedron Letters* **21**, 4547 (1980). Total synthesis: E. G. Gibbons, *J. Am. Chem. Soc.* **104**, 1767 (1982).

Crystals from ethyl acetate + Skellysolve B, mp 170-171°. $[\alpha]_D^{24}$ +20° (c = 3 in abs ethanol). uv max (5 mg/ml in 95% ethanol): 290 nm. Soluble in methanol, ethanol, ethyl ace-tate, chloroform. LD_{50} in mice: > 60 mg/kg, Kavanagh *et al., loc. cit.* (1952).
Diacetate, $C_{26}H_{38}O_7$, crystals from abs ethanol, mp 145.5°.

7509. Pleurotin(e). *(2aα,4aβ,5β,6β,8aα,12bβ,12cβ,-12dβ)-(−)-2a,3,4,4a,5,6,7,8a,12b,12c-Decahydro-6-methyl-2H-5,12d-ethanofuro[4',3',2':4,10]anthra[9,1-bc]oxepin-2,9,12-trione.* $C_{21}H_{22}O_5$; mol wt 354.41. C 71.17%, H 6.26%, O 22.57%. Antibiotic substance produced by the fungus *Pleurotus griseus:* Robbins *et al., Bull. Torrey Bot. Club* **72**, 165 (1945). Isoln: Robbins *et al., Proc. Nat. Acad. Sci. USA* **33**, 171 (1947). Partial structure: Huls, *Publ. Univ. Congo Elisabethville* **4**, 109 (1962), *C.A.* 3894f (1965). Struc-ture and stereochemistry: J. Grandjean, R. Huls, *Tetrahe-dron Letters* **1974**, 1893. Synthesis of racemate: D. J. Hart, H.-C. Huang, *J. Am. Chem. Soc.* **110**, 1634 (1988).

Amber crystals from ether + chloroform, mp 200-215°. $[\alpha]_D^{23}$ −20° (chloroform). Neutral reaction. Sparingly sol in water which inactivates it. More stable at acid pH, but still loses ~50% of its activity when boiled for 10 min at pH 3. Aq solns also are inactivated on exposure to light. Modera-tely sol in alcohol, ether. Freely sol in chloroform.

7510. Plicamycin. Mithramycin; aureolic acid; mitra-mycin; aurelic acid; antibiotic LA-7017; Mithracin. $C_{52}H_{76}$-O_{24}; mol wt 1085.18. C 57.55%, H 7.06%, O 35.39%. Oligo-saccharide antibiotic produced by *Streptomyces argillaceus* n. sp. and *S. tanashiensis.* Isoln: Grundy *et al., Antibiot. & Chemother.* **3**, 1215 (1953); Philip, Schenck, *ibid.* 1218; Rao *et al., ibid.* **12**, 182 (1962). Identity of aureolic acid with mithramycin: Berlin *et al., Nature* **218**, 193 (1968). Prepn: Gado *et al.,* **Hung. pat.** 155,679 (1969 to Gyogyszerkutato Intezet), *C.A.* **70**, 118093f (1969). Similar to the chromomy-

cins and olivomycins, *q.q.v.* Structural elucidation of the aglycone, *chromomycinone,* and the carbohydrate moieties: Bakhaeva *et al., Tetrahedron Letters* **1968**, 3595; Berlin *et al., Khim. Prir. Soedin.* **1972**, 537, 542, *C.A.* **78**, 16436t, 16443t (1973). Revised structure: J. Thiem, B. Meyer, *Tetrahedron* **37**, 551 (1981). Metabolic studies: Kennedy *et al., Cancer Res.* **27**, 1534 (1967). Mode of action studies: Kushch *et al., Antibiotiki* **17**, 504 (1972). *Review:* J. D. Skarbek, M. K. Speedie in *Antitumor Compounds of Natural Origin* vol. 1, A. Aszalos, Ed. (CRC Press, Boca Raton, 1981) pp 191-235.

Yellow solid, mp 180-183° (Bakhaeva). $[\alpha]_D^{20}$ −51° (c = 0.4 in ethanol). Sol in lower alcohols, acetone, ethyl acetate, water. Moderately sol in chloroform. Slightly sol in ether, benzene. LD_{50} in mice, rats (mg/kg): 2.14, 1.74 i.v., M. Slavik, S. K. Carter, *Adv. Pharmacol. Chemother.* **12**, 1 (1975).

THERAP CAT: Antineoplastic.

7511. Plumbagin. *5-Hydroxy-2-methyl-1,4-naphtha-lenedione; 5-hydroxy-2-methyl-1,4-naphthoquinone.* C_{11}-H_8O_3; mol wt 188.17. C 70.21%, H 4.29%, O 25.51%. Found in the roots of *Plumbago europaea* L., *P. zeylanica, P. rosea* L., *Plumbaginaceae.* Isoln: Dulong d'Astafort, *J. Pharm. Chim.* **14**, 441 (1828); Wefers-Bettink, *Rec. Trav. Chim.* **8**, 319 (1889); Ray, Dutt, *J. Indian Chem. Soc.* **5**, 419 (1928). Structure and synthesis: Fieser, Dunn, *J. Am. Chem. Soc.* **58**, 572 (1936). Biosynthesis: Durand, Zenk, *Tetrahedron Letters* **1971**, 3009. Chemotherapeutic proper-ties: Vichkanova *et al., C.A.* **78**, 66906s (1973). Efficient syntheses: A. Ichihara *et al., Agr. Biol. Chem.* **44**, 211 (1980); G. Wurm *et al., Arch. Pharm.* **314**, 861, 1055 (1981); H. Möhrle, H. Foltmann, *ibid.* **321**, 259 (1988). Toxicity data: M. Debray *et al., Plant Med. Phytother.* **7**, 77 (1973), *C.A.* **79**, 14255e (1973).

Yellow needles from dil alcohol, mp 78-79°. Sublimes. Volatile with steam. Slightly sol in hot water; sol in alc, ace-tone, chloroform, benzene, acetic acid. LD_{50} i.p. in mice: ~0.015 g/kg (Debray).

7512. Plumericin. *[3aS-(3E,3aα,4aβ,7aβ,9aR*,9bβ)]-3-Ethylidene-3,3a,7a,9b-tetrahydro-2-oxo-2H,4aH-1,4,5-tri-oxadicyclopent[a,hi]indene-7-carboxylic acid methyl ester.* $C_{15}H_{14}O_6$; mol wt 290.26. C 62.06%, H 4.86%, O 33.07%. A sesquiterpenoid, exhibiting *in vitro* activity against fungi, some bacteria including *Mycobacterium tuberculosis* 607. Isoln from roots of *Plumeria multiflora* Muell.-Arg., *Apo-cynaceae,* also from roots of *P. rubra* var. *alba:* J. E. Little, D. B. Johnstone, *Arch. Biochem.* **30**, 445 (1951); from *Alla-manda cathartica* Linn., *Apocynaceae:* B. R. Pai *et al., Indi-an J. Chem.* **8**, 851 (1970). Structure, physical properties and IR, ^{1}H-NMR spectra: G. Albers-Schönberg, H. Schmid, *Helv. Chim. Acta* **44**, 1447 (1961). Synthetic approach: J. K. Whitesell *et al.*, *Syn. Commun.* **7**, 355 (1977). Biomimetic total synthesis of (±)-form: B. M. Trost *et al.*, *J. Am. Chem. Soc.* **105**, 6755 (1983); *eidem, ibid.* **108**, 4974 (1986). Alternate synthesis of (±)-form: *eidem, ibid.* 4965; K. E. B. Parkes, G. Pattenden, *J. Chem. Soc. Perkin Trans. I* **1988**, 1119.

Narrow, rectangular plates from alc, toluene or methylene chloride + ether, dec 211.5-212.5°. Sublimes in high vacuum at 160-180°. $[\alpha]_D^{25}$ +197.5° ±2 (c = 0.982 in chloroform). uv max (ethanol): 214-215 nm (log ϵ 4.24). Sol in chloroform; slightly sol in methanol, alc, ether, acetone, benzene. Practically insol in petr ether, water.

7513. Plumieride. *1-(β-D-Glucopyranosyloxy)-4a,7a-dihydro-4'-(1-hydroxyethyl)-5'-oxospiro[cyclopenta[c]pyran-7(1H),2'(5'H)-furan]-4-carboxylic acid methyl ester;* agoniadin. $C_{21}H_{26}O_{12}$; mol wt 470.42. C 53.61%, H 5.57%, O 40.81%. Found in bark of *Plumeria lancifolia* Muell.-Arg., *Apocynaceae*, also in *P. acutifolia* and *P. rubra* var. *alba*: Peckolt, *Arch. Pharm.* [2] **142**, 40 (1870); *Merck's Jahresber.* **1895**, 11; Boorsma, *Mededeel. Lands Plant.* **13**, 27 (1894); **31**, 132 (1899); Franchimont, *Rec. Trav. Chim.* **18**, 334, 477 (1899); **19**, 350 (1900); Halpern, Schmid, *Helv. Chim. Acta* **41**, 1109 (1958). Structure: Albers-Schönberg, Schmid, *ibid.* **44**, 1447 (1961). Biosynthesis: D. A. Yeowell, H. Schmid, *Experientia* **20**, 250 (1964); K. Inoue *et al.*, *Chem. Pharm. Bull.* **27**, 3115 (1979).

Monohydrate, bitter crystals, mp 156-158°. When anhydrous, mp 224-225°. $[\alpha]_D^{20}$ −114°; $[\alpha]_D^{20}$ −80° (methanol). Broad uv absorption: 217-238 nm. Sol in water, alcohol, ethyl acetate.

7514. Plutonium. Pu; at. wt (most stable known isotopes) 242, 244; at. no. 94; valences 3, 4, 5, 6, 7. Occurrence in the earth's crust: 10^{-22}%. First isotope, ^{238}Pu, produced in 1940: Seaborg *et al.*, *Phys. Rev.* **69**, 366, 367 (1946). Presence of ^{239}Pu in pitchblende: Seaborg, Perlman, *J. Am. Chem. Soc.* **70**, 1571 (1948). Known isotopes (mass number): 232-246; all are radioactive. Two most useful isotopes: ^{238}Pu ($T_{1/2}$ 86.4 years; α-emitter); ^{239}Pu ($T_{1/2}$ 24,390 years; α-emitter). Chemical properties: Seaborg, Wahl, *ibid.* 1128; Harvey *et al.*, *J. Chem. Soc.* **1947**, 1010. Reviews: J. J. Katz, G. T. Seaborg, *Chemistry of the Actinide Elements* (John Wiley, New York, 1957) pp 239-330; *J. Chem. Ed.* **36**, 22-26 (1959); J. M. Cleveland, *The Chemistry of Plutonium* (Gordon & Breach, New York, 1970) 653 pp; C. Keller, *The Chemistry of the Transuranium Elements* (Verlag Chemie, Weinheim, English Ed., 1971) pp 333-484; *Comprehensive Inorganic Chemistry* vol. 5, J. C. Bailar, Jr. *et al.*, Eds. (Pergamon Press, Oxford, 1973) *passim*; several authors in *Handb. Exp. Pharmakol.* **36**, 307-688 (1973); F. Weigel in Kirk-Othmer *Encyclopedia of Chemical Technology* vol. 18 (Wiley-Interscience, New York, 3rd ed., 1982) pp 278-301; *Plutonium Chemistry*, W. T. Carnall, G. R. Choppin, Eds.

(Am. Chem. Soc., Washington, D.C., 1983) 484 pp. Review of biology and toxicology of plutonium: W. J. Bair, R. C. Thompson, *Science* **183**, 715-722 (1974).

Silvery-white metal. Highly reactive. mp 639.5°. Six allotropic forms. d_4^{20} 19.816.

Trivalent plutonium is a weak reducing agent. Stable in soln in absence of air. Slowly oxidized to the tetravalent plutonium by atmospheric oxygen, by permanganate in acid soln in the cold; oxidized to the hexavalent form by permanganate at 60°. Trivalent salts are blue; form complexes very readily; form a series of double sulfates. Crystal structure of the complex and double salts: Zachariasen, *J. Am. Chem. Soc.* **70**, 2147 (1948).

Tetravalent plutonium in aq soln is reduced to the trivalent form by sulfur dioxide, hydroxylamine hydrochloride, hydrazine hydrochloride, the uranous ion, the iodide ion; by shaking with mercury in chloride soln; electrolytically at a platinum cathode. Tetravalent salts are pink or greenish; form complexes very readily.

Hexavalent plutonium is obtained by the action of strong oxidizing agents (ceric salts, dichromates, permanganates, or hot bromate soln contg nitric acid) on the tri- or tetravalent form. Reduced to tri- or tetravalent plutonium by sulfur dioxide or ferrocyanide.

USE: ^{238}Pu as heat source; as radioisotope thermoelectric generator; in radionuclide batteries for pacemakers; with Be as neutron source. ^{239}Pu in atomic weapons; in power reactors. *Caution:* Radiation hazard; concentrates in bone. Max permissible concn of ^{238}Pu in air: 7×10^{-13} μCurie/cc; of ^{239}Pu in air: 6×10^{-13} μCurie/cc: *National Bureau of Standards Handbook* **69**, 87 (1959). *See also Handb. Exp. Pharmakol., loc. cit.*

7515. PMSG. Pregnant Mare Serum Gonadotrop(h)in; equine cyonin; equine gonadotrop(h)in; serum gonadotrop(h)in; Gormon; Eleagol; Lobulantina; Priatin; Serotropin; Gestyl; Antostab; Gonadotraphon F.S.H.; Serogan; Seragon; Seragonin; Predalon-S; Gonadyl; P.M.S.; Anteron; Antex-490; Primantron. Glycoprotein found in blood of pregnant mares which stimulates the growth of the ovarian follicles and the formation of the corpus luteum. Discovery: Cole, Hart, *Am. J. Physiol.* **93**, 57 (1930). Isoln procedures: Goss, Cole, *Endocrinology* **26**, 244 (1940); Rimington, Rowlands, *Biochem. J.* **35**, 736 (1941); **38**, 54 (1944). Contains ~30% protein and 45% carbohydrate; rich in sialic acids. Physico-chemical studies: Harbon-Chabbat *et al.*, *Bull. Soc. Chim. Biol.* **43**, 1339 (1961); Shams, Papkoff, *Biochim. Biophys. Acta* **263**, 139 (1972); Gospodarowicz, *Endocrinology* **91**, 101 (1972).

Powder. Loses its biological activity even in the dry state. Isoelec pt pH 2.60-2.65. Soluble in water, dil alcohol, dil glycerol and glycols. The hormone seems most stable in neutral or slightly alkaline soln. A water soln contg 0.1 mg/ml (no buffers added) remained unchanged for 2 days at 37°, while solns buffered with sodium acetate to pH 4.5 lost 50% of their activity. For a definition of International Units of PMSG consult Thayer, *Vitamins and Hormones*, vol. IV, 312 (1946). Usually 0.25 mg = 1 I.U.; 1000 I.U. = 50 rat units.

THERAP CAT: Gonadotropic hormone.

THERAP CAT (VET): Gonadotropic hormone.

7516. Podocarpic Acid. *1,2,3,4,4a,9,10,10a-Octahydro-6-hydroxy-1,4a-dimethyl-1-phenanthrenecarboxylic acid; 12-hydroxypodocarpa-8,11,13-trien-16-oic acid.* $C_{17}H_{22}O_3$; mol wt 274.35. C 74.42%, H 8.08%, O 17.50%. Chief acidic constituent of the resin of the Javanese *Podocarpus cupressina* (L'Hérit.) Pers., *Coniferae*, esp. in var. *imbricata*. Also in New Zealand kahikatea resin (from *Podocarpus dacrydioides*) and in rimu resin (*Dacrydium cupressinum*). A related component, "nimbiol", *q.v.*, occurs in the nim tree, *Azadirachta indica* Juss. (*Melia azadirachta* L.), *Meliaceae*. Isoln from *Podocarpus cupressina*: A. C. Oudemans, *Ber.* **6**, 1122 (1873); from *Dacrydium cupressinum*: I. R. Sherwood, W. F. Short, *J. Chem. Soc.* **1938**, 1008. Structure: Campbell, Todd, *J. Am. Chem. Soc.* **64**, 928 (1942). Conversion to nimbiol as proof of structure: Bible, *Tetrahedron* **11**, 22 (1960). Synthesis from desoxypodocarpic acid: Wenkert, Jackson, *J. Am. Chem. Soc.* **80**, 217 (1958). Resolution of dl-desoxypodocarpic acid: Wenkert *et al.*, *ibid.* **86**, 2038

(1964). Stereoselective total synthesis of *dl*-form: Meyer, Maheshwari, *Tetrahedron Letters* **1964**, 2175; Spencer *et al.*, *J. Org. Chem.* **33**, 719 (1968); S. C. Welch *et al.*, *ibid.* **42**, 2879 (1977); P. R. Kanjilal *et al.*, *Syn. Commun.* **11**, 795 (1981). Synthesis of *d*-form: Pelletier *et al.*, *Tetrahedron Letters* **1971**, 4179.

Platelets from dil alcohol, mp 193.5°. $[\alpha]_{546}^{20}$ +165° (c = 4 in abs ethanol). Sol in methanol, ethanol, ether, acetic acid. Practically insol in water, chloroform, benzene, carbon disulfide. Methyl ester, $C_{18}H_{24}O_3$, crystals from alcohol, mp 208°.

7517. Pododacric Acid. *1,2,3,4,4a,9,10,10a-Octahydro-6-hydroxy-7-[2-hydroxy-1-(hydroxymethyl)ethyl]-1,4a-dimethyl-1-phenanthrenecarboxylic acid; 12-hydroxy-13-[2-hydroxy-1-(hydroxymethyl)ethyl]podocarpa-8,11,13-trien-16-oic acid.* $C_{20}H_{28}O_5$; mol wt 348.44. C 68.94%, H 8.10%, O 22.96%. From heartwood of *Podocarpus dacrydioides*; *P. totara*: Briggs *et al.*, *Tetrahedron* **7**, 270 (1959); Cambie, Mander, *ibid.* **18**, 465 (1962). Structure: Cambie, Mathai, *Chem. Commun.* **1971**, 154.

Colorless needles from 30% methanol, mp 213-214°. $[\alpha]_D^{25}$ +118° (c = 0.9 in 5:1 chloroform-methanol). uv max: 225, 284 nm (ε 5500, 3100). pK 8.44 (in methyl cellosolve system).

7518. Podophyllic Acids. $C_{22}H_{24}O_9$; mol wt 432.41. C 61.10%, H 5.59%, O 33.30%. Since the prepn of a "podophyllic acid", mp 145-150° (crude), 163-165° (pure), by Borsche, Niemann, *Ann.* **494**, 126 (1932), fairly extensive studies have been made of two isomers: the 2,3-*trans*-hydroxy acid and the 2,3-*cis*-hydroxy acid. Prepn of the *trans*-acid: Kuhn, Wartburg, *Experientia* **19**, 391 (1963); **Neth. pat. Appl. 6,405,480** (1964 to Sandoz); *C.A.* **62**, 9083c (1965); Renz *et al.*, *Ann.* **681**, 207 (1965). Prepn of the *cis*-acid: Renz *et al.*, *loc. cit.* Prepn of a DL-stereoisomer: Gensler *et al.*, *J. Am. Chem. Soc.* **76**, 315 (1954). Configuration and nomenclature: Rutschmann, Renz, *Helv. Chim. Acta* **42**, 890 (1959); Renz *et al.*, *loc. cit.*

2,3-*trans*-Hydroxy acid, R = α-COOH, *podophyllinic acid.* Crystals from acetone + ether, mp 164-168°. $[\alpha]_D^{20}$ −199.9° (c = 0.463 in ethanol), −292.6° (c = 0.671 in pyridine).
2,3-*cis*-Hydroxy acid, R = β-COOH, *picropodophyllinic acid.* Needles from methanol + ether, double mp 150-155° and 200-232°. $[\alpha]_D^{21}$ −100.4° (c = 0.615 in ethanol), −185° (c = 0.746 in pyridine).
2,3-*trans*-Hydroxy acid hydrazide, R = α-CONHNH₂, *podophyllinic acid hydrazide.* Prisms from methanol, mp 198-199°. $[\alpha]_D$ −202° (c = 0.4 in ethanol). Prepn and

properties: Rutschmann, Renz, *loc. cit.;* Rutschmann, **U.S. pat. 2,977,359** (1961 to Sandoz).
2,3-*cis*-Hydroxy acid hydrazide, R = β-CONHNH₂, *picropodophyllinic acid hydrazide, picropodophyllic acid hydrazide.* Rhombic scales, mp 152-154°. $[\alpha]_D$ −106° (c = 0.4 in ethanol). Prepn and properties: Rutschmann, Renz, *loc. cit.;* Rutschmann, *loc. cit.*

7519. Podophyllinic Acid 2-Ethylhydrazide. *5,6,7,8-Tetrahydro-8-hydroxy-7-(hydroxymethyl)-5-(3,4,5-trimethoxyphenyl)naphtho[2,3,d]-1,3-dioxole-6-carboxylic acid 2-ethyl hydrazide;* 1-ethyl-2-podophyllinic acid hydrazide; mitopodozide; SP-I; Proresid. $C_{24}H_{30}N_2O_8$; mol wt 474.50. C 60.75%, H 6.37%, N 5.90%, O 26.98%. Prepn: Rutschmann, **U.S. pat. 3,054,802** (1962 to Sandoz). *Note:* Patent also claims prepn of *picropodophyllinic acid 2-ethylhydrazide* or *1-ethyl-2-picropodophyllinic acid hydrazide*.

Amorphous powder, pptd from chloroform + petr ether. $[\alpha]_D$ −154° (c = 0.5 in chloroform).
THERAP CAT: Antineoplastic.

7520. Podophyllotoxin. *5,8,8a,9-Tetrahydro-9-hydroxy-5-(3,4,5-trimethoxyphenyl)furo[3',4':6,7]naphtho[2,3-d]-1,3-dioxol-6(5aH)-one;* 1-hydroxy-2-hydroxymethyl-6,7-methylenedioxy-4-(3',4',5'-trimethoxyphenyl)-1,2,3,4-tetrahydronaphthalene-3-carboxylic acid lactone; podophyllinic acid lactone; podofilox; Condyline; Condylox; Martec; Warticon. $C_{22}H_{22}O_8$; mol wt 414.40. C 63.76%, H 5.35%, O 30.89%. Antineoplastic glucoside. Found in the rhizomes of North American *Podophyllum peltatum* L., *Podophyllaceae:* V. Podwyssotski, *Arch. Exp. Pathol. Pharmakol.* **13**, 29 (1880); also in *P. emodi* Wall.: von Wartburg *et al.*, *Helv. Chim. Acta* **40**, 1331 (1957); in *Juniperus virginiana* L., *Cupressaceae:* Kupchan *et al.*, *J. Pharm. Sci.* **54**, 659 (1965). Structure and absolute configuration: Schrecker, Hartwell, *J. Org. Chem.* **21**, 381 (1956); *cf. Helv. Chim. Acta* **37**, 1541 (1954). Crystal structure: T. J. Petcher *et al.*, *J. Chem. Soc., Perkin Trans. II* **1973**, 288. Synthesis: Gensler, Gatsonis, *J. Am. Chem. Soc.* **84**, 1748 (1962); *J. Org. Chem.* **31**, 4004 (1966); W. S. Murphy, S. Wattanasin, *Chem. Commun.* **1980**, 262; D. Rajapaksa, R. Rodrigo, *J. Am. Chem. Soc.* **103**, 6208 (1981); T. Kaneko, H. Wong, *Tetrahedron Letters* **28**, 517 (1987). Prepn of derivs and analogs: Gensler *et al.*, *J. Am. Chem. Soc.* **82**, 6074 (1960); Schreier, *Helv. Chim. Acta* **47**, 1529 (1964). Study of diastereomers: Aiyar, Chang, *J. Org. Chem.* **40**, 2384 (1975). Asymmetric total synthesis of (−)-form: R. C. Andrews *et al.*, *J. Am. Chem. Soc.* **110**, 7854 (1988). Toxicity data: F. S. Phillips *et al.*, *Fed. Proc.* **7**, 249 (1948). Clinical evaluation for treatment of genital warts: A. Lassus, *Lancet* **2**, 513 (1987); K. R. Bentner *et al.*, *ibid.* **1**, 831 (1989). Review of early literature: Hartwell, Schrecker in *Fortschr. Chem. Org. Naturst.* **15**,

98-121 (1958). Review of chemistry and antineoplastic activity: I. Jardine, *Anticancer Agents Based on Natural Product Models,* J. M. Cassady, J. D. Douros, Eds. (Academic Press, New York, 1980) pp 319-351.

Solvated crystals. mp 114-118° (effervescence). Several polymorphic modifications. After drying: mp 183.3-184.0°. $[\alpha]_D^{20}$ —132.7° (chloroform). Soly in water at 23°: 120 mg/l. Sol in alcohol, chloroform, acetone, warm benzene, glacial acetic acid. *Caution: Irritates the skin.* LD_{50} in rats (mg/kg): 8.7 i.v.; 15 i.p. (Philips).

Podophyllotoxin-β-D-glucoside, $C_{28}H_{32}O_{13}$, hygroscopic amorphous white flakes, mp 152-154°. $[\alpha]_D^{20}$ —76.4° (c = 0.576 in methanol); $[\alpha]_D^{20}$ —117.0° (c = 0.668 in pyridine).

THERAP CAT: Topical antiviral.

7521. Podophyllum. May apple; mandrake root; Indian apple; vegetable calomel. Dried rhizome and roots of *Podophyllum peltatum* L., *Berberidaceae. Habit.* North America. *Constit.* 3-6% resin, 0.2-1% podophyllotoxin, picropodophyllin, quercetin, peltatins. Sixteen well-characterized compds have been isolated. They fall into two chemical classes: lignans and flavonol pigments. History, isoln procedures, structures: Hartwell, Schrecker, *Fortschr. Chem. Org. Naturst.* **15,** 83-166 (1958), a review with 227 refs.

THERAP CAT: Cathartic.

THERAP CAT (VET): Has been used as a purgative.

7522. Podophyllum, Indian. Rhizome and roots of *Podophyllum emodi* Wall., *Berberidaceae. Habit.* Himalaya Mountains, Kashmir. *Constit.* 6-12% resin, 1-4% podophyllotoxin; berberine.

THERAP CAT: Cathartic.

7523. Podophyllum Resin. Podophyllin. A dry, alcoholic extract of the rhizomes and roots of *Podophyllum peltatum* L., *Berberidaceae:* "Extract powdered podophyllum (1 kilo) by slow percolation until it is exhausted of its resin, using alcohol as the menstruum. Concentrate the percolate by evaporation until the residue has the consistency of a thin syrup, and pour this with constant stirring, into one liter of water contg 10 ml concd HCl and previously cooled to a temp below 10°. Allow the precipitate to settle, decant the clear liq, and wash the precipitate with two 1000 ml portions of cold water. Dry the resin and powder it." *Constit.* Podophyllotoxin (up to 20%), α- and β-peltatin, desoxypodophyllotoxin, dehydropodophyllotoxin. These constits are originally present in the plant as β-D-glucosides. The picropodophyllin often reported in the literature is not present in the fresh plant, but forms during isoln of the resin by epimerization of podophyllotoxin and may be regarded as the more stable *cis*-form of the natural *trans*-podophyllotoxin: von Wartburg *et al., Helv. Chim. Acta* **40,** 1331 (1957). *Review:* Kelly, Hartwell, *J. Natl. Cancer Inst.* **14,** 967 (1954).

Light brown to greenish-yellow powder, or small, yellowish, bulky, fragile lumps becoming darker on exposure to light or heat (> 25°); faint odor; acrid, bitter taste. Sol in alcohol, usually with a slight opalescence, also sol in dil caustic. Not less than 75% is sol in ether and not less than 65% in chloroform

Human Toxicity: Extremely caustic to the skin and mucous membranes; dust inflames the eyes.

THERAP CAT: Caustic.

THERAP CAT (VET): Externally as a caustic for warts. Has been used internally as a purgative.

7524. Poi. The root of the taro plant, *Colocasia esculenta* (L.) Schott, *Araceae.* Used as a cereal substitute, particularly in allergy cases and potentially allergic infants.

THERAP CAT: Nutrient.

7525. Poison Ivy. Poison vine; markweed. *Toxicodendron radicans* (L.) Kuntze, *Anacardiaceae.* Erroneously called poison oak. Vigorous woody vine, shrub, or subshrub with trifoliate, alternate leaves. *Habit.* All states of the United States east of the Cascade Mountains, the Great Basin, and the Mojave Desert (absent in Nevada); all states of Mexico except in the Yucatan Peninsula and northern Baja California. Its southern limit is Huehuetenango Department of Guatemala and its northern limit is the 52nd parallel of latitude. It is found in Bermuda and several islands in the Bahamas, in Japan, and in the middle elevations in the mountains of Taiwan and of central and western China.

Review: Gillis in J. M. Kingsbury, *Poisonous Plants of the United States and Canada* (Prentice-Hall, Englewood Cliffs, N. J., 1964) pp 209-214. *Constit.* Toxic constituent: urushiol *q.v.:* Dawson, *Trans. N.Y. Acad. Sci.* [2] **18,** 427 (1956).

Human Toxicity: Can cause severe allergic dermatitis.

THERAP CAT: Extract as ivy poisoning counteractant.

7526. Poison Oak. Western poison oak. *Toxicodendron diversilobum* (T. & G.) Greene, *Anacardiaceae.* Similar to poison ivy in that it has three leaflets, grows as a shrub or vine and produces dermatitis in man. Interbreeds with poison ivy. *Habit.* Western North America from southern British Columbia to northern Baja California. *Constit.* Poisons presumably closely related if not identical to those in poison ivy.

Eastern poison oak; Toxicodendron quercifolium (Michx.) Greene, *Anacardiaceae.* Differs significantly from poison ivy. Never climbs or produces aerial roots and rarely hybridizes with poison ivy. *Habit.* Southern New Jersey to Florida, west to eastern Texas and Kansas. *Constit.* Poisons may be the same as in poison ivy or Western poison oak.

Extract of *T. quercifolium, Anergex.*

Human Toxicity: Can cause severe allergic dermatitis.

THERAP CAT: Extract as allergy inhibitor.

7527. Poison Sumac. Poison elder. *Toxicodendron vernix* (L.) Kuntze, *Anacardiaceae.* Tall, rangy shrub with leaves compounded with seven to eleven leaflets. Rachis is bright red and leaflets have no teeth on their margins, thus differing from nonpoisonous sumacs. *Habit.* Southern Quebec to central Florida predominantly east of Mississippi River. Found only in bogs, swamps and wet bottom lands.

Human Toxicity: Can cause severe allergic dermatitis.

THERAP CAT: Extracts for prophylaxis of poison sumac dermatitis.

7528. Polar® Yellow. *5-Chloro-2-[4,5-dihydro-3-methyl-4-[[4-[[(4-methylphenyl)sulfonyl]oxy]phenyl]azo]-5-oxo-1H-pyrazol-1-yl]benzenesulfonic acid monosodium salt;* 4-*p*-hydroxybenzeneazo-1-*p*-chloro-*o*-sulfophenyl-3-methyl-5-hydroxypyrazole toluene-*p*-sulfonyl ester sodium salt; C.I. Acid Yellow 40; C.I. 18950. $C_{23}H_{18}ClN_4NaO_7S_2$; mol wt 584.99. C 47.22%, H 3.10%, Cl 6.06%, N 9.58%, Na 3.93%, O 19.15%, S 10.96%. Prepn from diazotized *p*-aminophenol and 1-(*o*-sulfo-*p*-chlorophenyl)-3-methyl-5-pyrazolone: B. Richard, U.S. pat. 1,067,881 (Geigy). *See also: Colour Index* vol. 4 (3rd ed., 1971) p 4128.

Yellowish-brown powder. Sol in water giving a pure yellow soln; in alcohol (yellow color); in alkalies; in concd sulfuric acid. Slightly sol in ethanol, acetone. Insol in other org solvents. Precipitated by dil acids.

USE: To dye wool. In the isoln of polymyxin.

7529. Poldine Methylsulfate. *2-[[(Hydroxydiphenylacetyl)oxy]methyl]-1,1-dimethylpyrrolidinium methyl sulfate (salt); 2-hydroxymethyl-1,1-dimethylpyrrolidinium methyl sulfate benzilate;* 2-benziloyloxymethyl-1,1-dimethylpyrrolidinium methyl sulfate; (1-methyl-2-pyrrolidinyl)methyl benzilate methylmethosulfate; IS 499; McN-R-726-47; Nacton; Nactate. $C_{22}H_{29}NO_7S$; mol wt 451.55. C 58.52%, H 6.47%, N 3.10%, O 24.80%, S 7.10%. Prepn: Doyle *et al., J. Chem. Soc.* **1958,** 4458. Prepn of the base: Blicke, Lu, *J. Am. Chem. Soc.* **79,** 29 (1955).

Needles from methyl ethyl ketone + ethanol + ether, mp 154-155°. Soluble in water.

THERAP CAT: Anticholinergic.

7530. Pole Reagent Paper. Unsized paper impregnated with a soln of 1-2 g phenolphthalein in 10 ml alcohol and 110 ml water and then passed, while still wet, through a soln of 20 g Na_2SO_4 (cryst) in 100 ml water; the paper is then dried.

USE: For recognizing the negative poles of batteries (the reaction at the negative pole coloring the paper red). The ends of the wires are placed about ½ inch apart on the moistened paper.

7531. Polidexide. *Sephadex 2-(diethylamino)ethyl ether;* dextran 2-(diethylamino)ethyl 2-[[2-(diethylamino)ethyl]-diethylammonio]ethyl ether chloride hydrochloride epichlorohydrin crosslinked; poly[2-(diethylamino)ethyl] polyglycerylene dextran hydrochloride; PDX chloride; DEAE-Sephadex; Secholex. An anion exchange resin containing quaternary ammonium groups, which reduces serum cholesterol by binding the bile acids in the intestine, thus facilitating their increased faecal excretion. This depletion of the bile acids is countered by synthesis of more bile acids from cholesterol. Prepn: **Brit. pat. 1,013,585** (1965 to Pharmacia AB). Activity studies: Parkinson, *J. Lipid Res.* **8**, 24 (1967); Howard, Hyams, *Brit. Med. J.* **3**, 25 (1971); Evans *et al.*, *Angiology* **24**, 22 (1973); Ritland *et al.*, *Scand. J. Gastroenterol.* **10**, 791 (1975).

Tasteless powder; forms gel in water.

USE: Ion exchange resin.

THERAP CAT: Anticholesteremic.

7532. Polidocanol. α-*Dodecyl-ω-hydroxypoly(oxy-1,2-ethanediyl); polyethylene glycol monododecyl ether;* dodecyl alcohol polyoxyethylene ether; hydroxypolyethoxydodecane; polyoxyethylene lauryl ether; Aetoxisclerol; Atlas G-2133; Atlas G-3705; Cimagel; Lubrol PX; Mergital LM 11; Simulsol 330M; Texofor B 9; Thesit. $C_{12}H_{25}(OCH_2CH_2)_xOH$. Contains an average of nine ethylene oxide units and has an average mol wt of about 600. Prepd by reaction of ethylene oxide and dodecyl alcohol: Pertsemlides, Soehring, *Arzneimittel-Forsch.* **10**, 990 (1960). Sol in lipids and water.

USE: Solvent; non-ionic emulsifier.

THERAP CAT: Topical anesthetic; antipruritic; sclerosing agent.

7533. Polonium. Po; at. wt of naturally occurring isotope 210; at. no. 84; valence 4, occasionally 2, rarely 6. Isotopes range in mass number from 193-218; all are radioac-

tive. The first radioactive substance discovered by Mme. Curie in 1898. A product of disintegration of radium; one gram is contained in about 25,000 tons of pitchblende or in 7.5 kg radium that is more than 30 years old. Separated in form of a deposit on a bismuth plate immersed in a soln of the chloride: Marckwald, *Ber.* **35**, 2285 (1902); using a silver, gold or nickel plate: Curie, Joliot, *J. Chim. Phys.* **28**, 201 (1931); Haissinsky, *ibid.* **33**, 97 (1936); Rollier, *Gazz. Chim. Ital.* **66**, 797 (1936); Ziv, *C. R. Acad. Sci. URSS* **25**, 743 (1939). Obtained in the metallic form by volatilization from nickel on a collodion film: Rollier *et al.*, *J. Chim. Phys.* **4**, 648 (1936). The only readily accessible isotope is the penultimate member of the radium decay series, $^{210}_{84}Po$, also called *Radium F* (Ra-F). Decays by α-emission ($T_{1/2}$ 138.4 days) to $^{206}_{82}Pb$. Comprehensive reviews: K. W. Bagnall, *Chemistry of the Rare Radioelements* (Butterworths, London, 1957); idem, *Endeavour* **22**(86), 61 (May 1963); idem, "Selenium, Tellurium and Polonium" in *Comprehensive Inorganic Chemistry* vol. 2, J. C. Bailar, Jr., et al., Eds. (Pergamon Press, Oxford, 1973) pp 935-1008.

Two allotropic forms; coexist between 18° and 54°; d (α-form) 9.196; d (β-form) 9.398. mp 254°; bp 962°. Latent heat of vaporization: 24.597 kcal/mole. Resistivity: α-Po = 42 μohm-cm at 0°; β-Po = 44 μohm-cm at 0°. Physical properties: Maxwell, *J. Chem. Phys.* **17**, 1288 (1949). Chemically resembles tellurium and bismuth. Forms a volatile, unstable hydride, PoH_2. Forms a polonide, Na_2Po; a carbonyl PoCO; a hydroxide $Po(OH)_4$.

Caution: Radiation hazard; alpha emitter. Sol compds hazard to kidneys, spleen; insol, airborne compds hazard to lungs. Max permissible concn of insol $^{210}_{84}Po$ in air: 7 × 10^{-11} μ-Curie/cc, *National Bureau of Standards Handbook* **69**, 79 (1959).

7534. Polonium Dioxide. O_2Po; mol wt 241.00. O 13.28%, Po 86.72%. PoO_2. Formed from the elements at 250°: Martin, *J. Phys. Chem.* **58**, 911 (1954).

Two crystal modifications: yellow, low temperature form, face-centered cubic symmetry; red, high-temperature form, tetragonal symmetry. Darkens in color on heating, chocolate brown at sublimation temp of 885°. Dec into the elements at 500° under vacuum; slowly reduced to the metal in hydrogen at 200°. Heat of formation ~30 kcal/mole. Sol in aq solns of ammonium carbonate, phosphoric acid.

7535. Polonium Tetrachloride. Cl_4Po; mol wt 350.80. Cl 40.42%, Po 59.58%. $PoCl_4$. May be prepd by dissolving metallic polonium in hydrochloric acid; by heating polonium dioxide in carbon tetrachloride vapor at 200°; by heating the metal in dry chlorine at 200°; also by heating the dioxide in dry hydrogen chloride, thionyl chloride vapor, or with phosphorus pentachloride: *Review:* K. W. Bagnall, *Chemistry of the Rare Radioelements* (Butterworths, London, 1957) p 64.

Hygroscopic, bright yellow, cryst solid. Monoclinic or triclinic. Hydrolyzed by moist air, forming a white solid of indefinite composition. mp ca 300° (in chlorine), turns scarlet red at 350°. bp 390°. Vapors are purple-brown, turning blue-green above 500°. Sol in water (fairly slow hydrolysis). Freely sol in hydochloric acid, thionyl chloride. Moderately sol in ethanol, acetone. Dec by dil nitric acid. Forms a complex with two moles of tributyl phosphate.

7536. Poloxalene. *Methyloxirane polymer with oxirane.* Poly(oxyethylene)-poly(oxypropylene)-poly(oxyethylene) polymer; bis[hydroxyethyl poly(ethyleneoxy)ethyl]polypropyleneglycol; dipolyoxyethylated polypropyleneglycol ether; oxyethylene oxypropylene polymer; SK & F 18667; Bloat Guard; Therabloat. A block polymer of ethylene oxide and propylene oxide, with a mol wt of approx 3000. Example of prepn: Lundsted, **U.S. pat. 2,674,619** (1954 to Wyandotte Chem.). Field trial in wheat pasture bloat of cattle: E. E. Bartley *et al.*, *J. Animal Sci.* **41**, 752 (1975). Absorption and excretion in rats: J. B. Rogers *et al.*, *Drug Metab. Dispos.* **12**, 631 (1984). Hypocholesterolemic effect in rabbits: idem, *J. Clin. Invest.* **71**, 1490 (1983); idem, *Atherosclerosis (Shannon, Ire.)* **64**, 37 (1987). *Review:* Stanton, *Am. Perfumer Aromat.* **72**, No. 4, 54, 56, 58 (1958). *See also* Poloxamers.

$$HO(CH_2CH_2O)_a[CH(CH_3)CH_2O]_b(CH_2CH_2O)_cH$$

average values for a, b, c are:

a = 12; b = 34; c = 12

A liquid nonionic surfactant polymer.

USE: Pharmaceutic aid (surfactant).

THERAP CAT (VET): Prevention of bloat in cattle.

7537. Poloxamers. *Methyl oxirane polymers, polymer with oxirane; polyethylenepolypropylene glycols, polymers; α-hydro-ω-hydroxypoly(oxyethylene)poly(oxypropylene)-poly(oxyethylene) block copolymers; Pluronics.* Series of nonionic surfactants with the structure $HO(CH_2CH_2O)_a$-$(CH(CH_3)CH_2OH)_b(CH_2CH_2O)_cH$ where b is at least 15 and $(CH_2CH_2O)_{a+c}$ is varied from 20 to 90% by weight. Mol wt ranges from 1000 to greater than 16,000. The poly(oxypropylene) segment is hydrophobic; the poly(oxyethylene) segment hydrophilic. Comprehensive reviews: I. R. Schmolka, *Am. Perfumer Cosmet.* **82**(7), 25-30 (1967); *idem* "Polyalkylene Oxide Block Copolymers" in *Nonionic Surfactants,* M. Schick, Ed. (Dekker, New York, 1967) pp 300-371. *See also* Poloxalene.

Mobile liquids, pastes or flakeable solids. Relatively nonhygroscopic. Vary from water-insol to very water-sol compds; more sol in cold than hot water. In general, sol in aromatic solvents (benzene, toluene, xylene), chlorinated solvents, acetone, alc, propylene or hexylene glycol, butyl cellosolve, butyl carbitol, methyl ethyl ketone, cyclohexanone. Insol in ethylene glycol, kerosene, mineral oil. Low-foaming properties. Stable to acids, alkalies, metallic ions.

Poloxamer 182LF, Pluronic L62LF. a = 7, b = 30, c = 7; av. mol wt 2450. Low-foaming liquid. d_{25}^{25} 1.035. Brookfield viscosity (25°) 375 cp; cloud pt (10% aq soln): 22°.

Poloxamer 188, poloxalkol (obsolete), Exocorpol, Pluronic F68. a = 75, b = 30, c = 75; av. mol wt 8350. Flakeable solid. mp 50° (minimum); cloud pt (10% aq soln): > 100°.

Poloxamer 331, Pluronic L101. a = 7, b = 54, c = 7; av. mol wt 3800. Liquid. d_{25}^{25} 1.018; Brookfield viscosity (25°) 756 cp; cloud pt (10% aq soln): 11°.

USE: Food additives; defoamers; antistatic agents; demulsifiers; detergents; wetting agents, gelling agents; emulsifiers; foam controllers; dispersants; dye levelers.

THERAP CAT: Poloxamer 182LF as pharmaceutic aid; 188 as cathartic.

7538. Polyamine-Methylene Resin. *Resinat; Exorbin.* Phenol condensation product with polyamines. An ion-exchange resin specially purified for medicinal use.

Light amber, granular, free-flowing powder. Insol in water, alcohol, ether, aq solns of acids and alkalies. Under the conditions of the old N.N.R. assay for acid-consuming capacity, not less than 50 ml 0.1*N* hydrochloric acid is consumed by 0.2 g of the resin.

THERAP CAT: Antacid.

7539. Polybenzarsol. *(4-Hydroxyphenyl)arsonic acid polymer with formaldehyde; Benzodol.* A polymeric mixture obtained by adding formaldehyde (40%) (0.116 mole) over a 3-hr period to *p*-hydroxybenzenearsonic acid (0.209 mole) in 180 g of 90% H_2SO_4 at 0-5° and keeping it cold for 21 hrs. Dilution of the mixture with H_2O precipitates the product:

Faith, *J. Am. Chem. Soc.* **72**, 837 (1950). Description: Jones *et al., Antibiot. & Chemother.* **8**, 400 (1958).

White powder. Somewhat sol in water; sol in alcoholic NaOH. LD_{50} i.p. in mice: 235 mg/kg. No deaths after 4 g/kg i.g. in mice.

THERAP CAT: Antiamebic.

7540. Polybrominated Biphenyls. PBB's; brominated biphenyls; polybromobiphenyls. Mixtures with structures similar to polychlorinated biphenyls, *q.v.,* where each X = H or Br. Once widely used commercially. Prepn: H. Hahn *et al., Ger. pat.* **1,161,547** (1964 to Chem. Fabrik Kalb); G. A. Burk, U.S. pat. **3,733,366** (1973 to Dow); L. C. Mitchell, D. R. Breckenridge, U.S. pats. **3,763,248** and **3,833,674** (1973, 1974 both to Ethyl Corp.). Persistence in soils: L. W. Jacobs *et al., J. Agr. Food Chem.* **24**, 1198 (1976). Photo-degradation: L. O. Ruzo *et al., ibid.* 1062. Review of environmental hazards: K. Kay, *Environ. Res.* **13**, 74-93 (1977); F. J. DiCarlo *et al., Environ. Health Perspect.* **23**, 351-365 (1978).

Firemaster BP-6, major component is *2,2',4,4',5,5'-hexabromobiphenyl.* Softens at 72°, dec above 300°. Low vapor pressure; degraded by uv light. Very sol in benzene, toluene. Insol in water.

Note: The 1973 "Michigan Incident" in which BP-6 was accidentally added to animal feed, and resulted in widespread destruction of contaminated farm animals, led to the removal of BP-6 from the market: L. J. Carter, *Science* **192**, 240 (1976). This substance may reasonably be anticipated to be a carcinogen: *Fourth Annual Report on Carcinogens* (NTP 85-002, 1985) p 169.

USE: Flame retardant.

7541. Polychlorinated Biphenyls. PCBs; chlorinated biphenyls; chlorobiphenyls; Aroclor; Clophen; Fenclor; Kaneclor; Phenoclor; Pyralene; Santotherm. Once widely used industrial chemicals whose high stability contributed to both their commercial usefulness and their long-term deleterious environmental and health effects. Early synthesis: H. Schmidt, G. Schulz, *Ann.* **207**, 338 (1881). Commercially available since 1930: C. Penning, *Ind. Eng. Chem.* **22**, 1180 (1930). Commercial PCBs are mixtures. The Aroclors are characterized by four digit numbers. The first two digits indicate that the mixture contains biphenyls (12), triphenyls (54) or both (25, 44); the last two digits give the weight percent of chlorine in the mixture (e.g. Aroclor 1242 contains biphenyls with approx 42% chlorine). Reviews of environmental impact and toxicity: L. Fishbein, *Ann. Rev. Pharmacol.* **14**, 139-156 (1974); R. D. Kimbrough, *CRC Crit. Rev. Toxicol.* **2**, 445-498 (1974); *National Conference on Polychlorinated Biphenyls,* Nov. 19-21, 1975 (EPA-560/6-75-004, 1976) 487 pp. Accumulation of airborne PCBs in foliage: E. H. Buckley, *Science* **216**, 520 (1982). *Reviews:* H. L. Hubbard in *Kirk-Othmer Encyclopedia of Chemical Technology* **vol. 5** (Interscience, New York, 2nd ed., 1964) pp 289-297; O. Hutzinger *et al., The Chemistry of PCB's* (CRC Press, Cleveland, Ohio, 1974) 269 pp; J. W. Lloyd *et al., J. Occup. Med.* **18**, 109-113 (1976). Review of carcinogenicity studies: *IARC Monographs* **18**, 43-103 (1978).

X = H or Cl

Aroclor 1242, clear, mobile liquid; av. number Cl/molecule: 3.10. d_4^{25} 1.381, $d_4^{15.5}$ 1.392. Distillation range 325-366°. Flash point (open cup) 348-356°F. n_D^{20} 1.627-1.629. Dielectric constant (1000 cycles) 5.6 (25°), 4.9 (100°).

Aroclor 1254, light yellow, viscous liquid; av. number Cl/molecule: 4.96. d_4^{65} 1.495; $d_4^{15.5}$ 1.505. Distillation range 365-390°. No open cup flash point to boiling. n_D^{20} 1.629-1.641. Dielectric constant (1000 cycles) 5.0 (25°), 4.3 (100°). LD_{50} orally in weanling rats: 1295 mg/kg (Kimbrough).

Aroclor 1260, light yellow, soft, sticky resin; av. number

Cl/molecule: 6.30. d_4^{90} 1.555; $d_4^{15.5}$ 1.566. Distillation range 385-420. No open cup flash point to boiling. n_D^{20} 1.647-1.649. Dielectric constant (1000 cycles) 4.3 (25°); 3.7 (100°). LD_{50} orally in weanling rats: 1315 mg/kg (Kimbrough).

Caution: Toxic effects in humans include chloracne, pigmentation of skin and nails, excessive eye discharge, swelling of eyelids, distinctive hair follicles, gastrointestinal disturbances. In Japan, accidental contamination of rice bran oil with Kanechlor 400 led to an outbreak of what became known as "Yusho disease", *see* M. Kuratsune *et al.,* in EPA-560/6-75-004, *loc. cit.,* p 14. Toxic symptoms in animals include hepatocellular carcinoma, hypertrophy of the liver, adenofibrosis, weight and hair loss, mouth and eyelid edema, acneform lesions, decreased hemoglobin + hematocrit, gastric mucosal ulceration and reduced ability to reproduce. These substances may reasonably be anticipated to be carcinogens: *Fourth Annual Report on Carcinogens* (NTP 85-002, 1985) p 170.

USE: In electrical capacitors, electrical transformers, vacuum pumps, gas-transmission turbines. Formerly used in U.S. as hydraulic fluids, plasticizers, adhesives, fire retardants, wax extenders, dedusting agents, pesticide extenders, inks, lubricants, cutting oils, in heat transfer systems, carbonless reproducing paper.

7542. Polydatin. *3-Hydroxy-5-[2-(4-hydroxyphenyl)ethenyl]phenyl-β-D-glucopyranoside; 3-hydroxy-5-(p-hydroxystyryl)phenyl glucoside; 3,4',5-trihydroxystilbene-3-β-D-glucoside; resveratrol-3-β-mono-D-glucoside; piceid.* C_{20}-$H_{22}O_8$; mol wt 390.40. C 61.53%, H 5.68%, O 32.79%. Isoln from fresh root of *Polygonum cuspidatum* Sieb. & Zucc., *Polygonaceae,* and structure: Nonomura *et al., Yakugaku Zasshi* **83,** 988 (1963).

Trihydrate, crystals, mp 225-226°. $[\alpha]_D^{27}$ −74.9° (c = 1.709 in ethanol).

7543. Polyestradiol Phosphate. *Estradiol phosphate polymer;* PEP; Estradurin. Polymeric ester of phosphoric acid and estradiol. Mol wt ~26,000. Prepn: Diczfalusy, *Endocrinology* **54,** 471 (1954); Diczfalusy *et al.,* U.S. pat. **2,928,849** (1960 to AB Leo); Fernö *et al., Acta Chem. Scand.* **12,** 1675 (1958). Proposed structure: Fernö *et al., loc. cit.*

(—ORO— is the estradiol radical and n is about 80)

Solid, mp 195-202°. Very sol in aq pyridine; sol in aq alkali; very slightly sol in ethanol, ethanol + water (1:1), water, dioxane, acetone, chloroform. Intrinsic viscosity [η] in 0.25M NaCl soln at pH 7.5 = 0.04.

THERAP CAT: Estrogen (used in prostatic carcinoma).

7544. Polyethylene. *Ethene homopolymer;* Agilene; Alathon; Alkathene; Courlene; Lupolen; Platilon; Polythene; Pylen; Reevon. Mol wt about 1500-100,000. C 85.7%, H 14.3%. Prepd by polymerization of liq ethylene at high temps and high or low pressure. *Reviews:* Aggarwal, Sweeting, *Chem. Rev.* **57,** 665-742 (1957); Raff, Allison, *Polyethylene,* vol. XI of High Polymers series (Interscience, New York, 1956); Faith *et al., Industrial Chemicals* (Wiley, New York, 3rd ed., 1965) pp 624-630.

Plastic solid of milky transparency. d_4^{20} 0.92. Tough and flexible at room temps, mp 85-110°. Breaks with cryst fracture at −50°. Good electrical insulator. Surface resistivity: 10^{14} ohms. Will burn, but hardly supports combustion. Stable to water, non-oxidizing acids and alkalies, alcohols, ethers, ketones, esters at ordinary temps. Attacked by oxidizing acids such as nitric acid and perchloric acid, free halogens, benzene, petr ether, gasoline and lubricating oils, aromatic and chlorinated hydrocarbons.

USE: Laboratory tubing; in making prostheses; electrical insulation; packaging materials; kitchenware; tank and pipe linings; paper coatings; textile stiffeners.

7545. Polyethylene Glycol. *α-Hydro-ω-hydroxypoly-(oxy-1,2-ethanediyl);* macrogol; PEG; Carbowax; Jeffox; Nycoline; Pluracol E; Poly-G; Polyglycol E; Solbase. Liquid and solid polymers of the general formula $H(OCH_2$-$CH_2)_nOH$, where n is greater than or equal to 4. In general, each PEG is followed by a number which corresponds to its average mol wt. Synthesis: Fordyce, Hibbert, *J. Am. Chem. Soc.* **61,** 1905, 1910 (1939). *Reviews:* Glycols, G. O. Curme, Jr., F. Johnston, Eds., A.C.S. Monograph Series no. **114** (Reinhold, New York, 1952) pp 176-202; Kastens in *High Polymers,* H. Mark *et al.,* Eds., vol. 13 entitled *Polyethers,* part 1 (Interscience, New York, 1963) pp 169-189, 274-291; G. M. Powell, III in *Handbook of Water-Soluble Gums & Resins,* R. L. Davidson, Ed. (McGraw-Hill, New York, 1980) pp 18/1-18/31.

Clear, viscous liquids or white solids which dissolve in water forming transparent solns. Sol in many organic solvents. Readily sol in aromatic hydrocarbons. Only slightly sol in aliphatic hydrocarbons. Do not hydrolyze or deteriorate on storage, will not support mold growth. Polyethylene glycols are compds of low toxicity: Smyth *et al., J. Am. Pharm. Assoc., Sci. Ed.* **39,** 349 (1950). Toxicity data (PEG 400): W. Bartsch *et al., Arzneimittel-Forsch.* **26,** 1581 (1976).

Polyethylene glycol 200, average value of n is 4, mol wt range 190-210. Viscous, hygroscopic liq; slight characteristic odor; d_{25}^{25} 1.127. Viscosity (210°F): 4.3 centistokes. Supercools upon freezing.

Polyethylene glycol 400, average value of n between 8.2 and 9.1, mol wt range 380-420. Viscous, slightly hygroscopic liq; slight characteristic odor; d_{25}^{25} 1.128. mp 4-8°. Viscosity (210°F): 7.3 centistokes. LD_{50} orally in rats: 30 ml/kg (Bartsch).

Polyethylene glycol 600, average value of n between 12.5 and 13.9, mol wt range 570-630. Viscous, slightly hygroscopic liq; characteristic odor; d_{25}^{25} 1.128. mp 20-25°. Viscosity (210°F): 10.5 centistokes.

Polyethylene glycol 1500, average value of n between 29 and 36, mol wt range 1300-1600. White, free-flowing powder; d_{25}^{25} 1.210. mp 44-48°. Viscosity (210°F): 25-32 centistokes.

Polyethylene glycol 4000, average value of n between 68 and 84, mol wt range 3000-3700. White, free-flowing powder or creamy-white flakes; d_{25}^{25} 1.212. mp 54-58°. Viscosity (210°F): 76-110 centistokes. LD_{50} orally in rats (divided doses): 59 g/kg (Smyth).

Polyethylene glycol 6000, average value of n between 158 and 204, mol wt range 7000-9000. Powder or creamy-white flakes; d_{25}^{25} 1.21. mp 56-63°. Viscosity (210°F): 470-900 centistokes. LD_{50} orally in rats: > 50 g/kg (Smyth).

USE: As water-soluble lubricants for rubber molds, textile fibers, and metal-forming operations. In food and food packaging. In hair prepns, in cosmetics in general. Pharmaceutic aid (ointment and suppository base). As a stationary phase in gas chromatography. Also in water paints, paper coatings, polishes and in the ceramics industry. *Caution: Solvent action on some plastics!*

THERAP CAT (VET): Ointment base.

7546. Polyethylene Terephthalates. PET. Fiber forming polyesters prepd from terephthalic acid, *q.v.* or its esters and ethylene glycol: Whinfield, Dickson, U.S. pat. **2,465,-319** (1949 to du Pont). Review of structures, definition of trade names: R. W. Moncrieff, *Man-Made Fibres* (John Wiley & Sons, New York, 4th ed., 1963) pp 361-389, 707-723.

R—OC—⟨benzene ring⟩—COOCH$_2$CH$_2$O—H]$_n$

R = OH, **Dacron, Amilar, Fiber V.** Solid, dec at approx 250°. Sp gr 1.38. Sol in hot *m*-cresol, trifluoroacetic acid, *o*-chlorophenol, a mixture of 7 parts of trichlorophenol and 10 parts (by wt) of phenol, a mixture of 2 parts of tetrachloroethane and 3 parts (by wt) of phenol. Fiber has good resistance to weak acids even at boiling temp, to strong acids in the cold, to weak alkalies, to bleaches, to most alcohols, ketones, soaps, detergents, and dry cleaning agents. Fabric has good resistance to creasing, abrasion, heat aging, and sunlight when behind glass. When "heat-set", fabric will not shrink in either boiling water or boiling drycleaning solvent. Fabric burns, but local melting generally prevents spread of fire. Insects cannot thrive on the fiber, but some can cut through it. Molds, mildew, and fungi may grow on applied finishes, but do not attack fiber.

R = OCH$_3$, **Terylene.** For physical properties, *see* Dacron above. Other similar products: **Diolen, Enkalene, Fortrel, Tergal, Terital, Terlenka, Trevira, Mylar.**

USE: In fabric manufacture; as films; as base for magnetic coatings. Surgical aid (arterial grafts).

7547. Polyferose. Jefron. An iron carbohydrate chelate contg approx 45% iron in which the metallic (Fe) ion is sequestered within a polymerized carbohydrate derived from sucrose.

THERAP CAT: Hematinic.

7548. Polygodial. [*1R-(1α,4aβ,8aα)*]*-1,4,4a,5,6,7,8,8a-Octahydro-5,5,8a-trimethyl-1,2-naphthalenedicarboxaldehyde*; tadeonal. C$_{15}$H$_{22}$O$_2$; mol wt 234.34. C 76.88%, H 9.46%, O 13.66%. Widely distributed drimane sesquiterpene with insect antifeedant properties; naturally occurring as the (−)-form. Isoln from *Polygonum hydropiper* L., *Polygonaceae* (Australia) and structure: C. S. Barnes, J. W. Loder, *Aust. J. Chem.* **15**, 322 (1962); from the bark of *Warburgia stuhlmanni* Engl. or *W. ugandensis, Canellacceae* (E. Africa): I. Kubo *et al., Chem. Commun.* **1976**, 1013; from nudibranch *Dendrodons limbata* (Mediterranean): G. Cimino *et al., Science* **219**, 1237 (1983); from nudibranchs *D. nigra, D. tuberculosa* (Hawaii) and *D. krebsii* (Mexico): R. K. Okuda *et al., J. Org. Chem.* **48**, 1866 (1983). Relationship between structure and antifeedant-activity: K. Nakanishi, I. Kubo, *Isr. J. Chem.* **16**, 28 (1977); M. D'Ischia *et al., Tetrahedron Letters* **1982**, 3295. Partial synthesis: M. J. Cortes *et al., Chem. & Ind. (London)* **1985**, 735. Synthesis of (±)-form: T. Kato *et al., Tetrahedron Letters* **1971**, 1961; S. P. Tanis, K. Nakanishi, *J. Am. Chem. Soc.* **101**, 4398 (1979); M. Jalali-Naini *et al., Tetrahedron Letters* **1981**, 2995; S. C. Howell *et al., Chem. Commun.* **1981**, 507.

Colorless needles from petroleum (40-60°), mp 57° (Barnes, Loder). [α]$_D^{24}$ −131° (c = 0.96 in ethanol). uv max (ethanol): 231, 295 nm (ε 11800, 76).

(±)-Form, mp 93-94° (Tanis, Nakanishi).

7549. Polylysine. A lysine polypeptide or homopolymer, the chain length of which varies with the method of prepn. Prepn: Katchalski *et al., J. Am. Chem. Soc.* **69**, 2564 (1947); **70**, 2094 (1948); Fasman *et al., ibid.* **83**, 709 (1961); Sela *et al., Biopolymers* **1**, 517 (1963); Strojny, White, U.S. pat. **3,215,684** (1965 to Dow).

H—[NHCHCO]—OH
  |
 (CH$_2$)$_4$
  |
 NH$_2$]$_n$

L-Form hydriodide, average dp (or n) = 32. Transparent, solid, film-like polymer. Readily sol in water; practically insol in the usual organic solvents. Transition of high-mol wt poly-L-lysine (dp 1500) in aq soln from a helical to a randomly coiled conformation under the influence of decreasing pH or increasing temp: Applequist, Doty, *C.A.* **58**, 6925b (1963).

7550. Polymyxin. Antibiotic complex produced by *Bacillus polymyxa:* Brownlee, Jones, *Biochem. J.* **43**, XXV (1948). Prepn: Ainsworth, Pope, U.S. pat. **2,565,057** (1951 to Burroughs Wellcome); Petty, U.S. pat. **2,595,605** (1952 to Am. Cyanamid); Benedict, Stodola, U.S. pat. **2,771,397** (1956 to USDA). Purification: Hastings *et al.,* **Brit.** pat. **782,926** (1957 to Distillers Co.). Polymyxins A, B, C, D, E, F, K, M, P, S and T have been identified. Isoln of polymyxins A, B, C and E: Few, Schulman, *Biochem. J.* **54**, 171 (1953); of *polymyxin D*: Stansley *et al., Bull. Johns Hopkins Hosp.* **81**, 43 (1947); of *polymyxin F*: W. L. Parker *et al, J. Antibiot.* **30**, 767 (1977); of *polymyxin K*: Kimura, **Japan.** pat. **16,152('71)**, *C.A.* **75**, 62180r (1971); of *polymyxin M*: Khokhlov *et al., C.A.* **55**, 5653h (1961); of *polymyxin P*: Kimura *et al., J. Antibiot.* **22**, 449 (1969); of *polymyxin S$_1$* and *polymyxin T$_1$*: J. Shoji *et al., ibid.* **30**, 1029 (1977). Resolution of polymyxin B into B$_1$ and B$_2$: Hausmann, Craig, *J. Am. Chem. Soc.* **76**, 4892 (1954). Structure and synthesis of polymyxin B$_1$: Wilkinson, Lowe, *Nature* **202**, 1211 (1964); Vogler *et al., Helv. Chim. Acta* **48**, 1161 (1965). Structure of polymyxin B$_2$: Wilkinson, Lowe, *Nature* **204**, 993 (1964). Separation of polymyxin D into D$_1$ and D$_2$ and structures: Hayashi *et al., Experientia* **22**, 354 (1966). Structure of polymyxin S$_1$: J. Shoji *et al., J. Antibiot.* **30**, 1035 (1977); of polymyxin T$_1$: *eidem, ibid.* 1042. *Review:* Vogler, Studer, *Experientia* **22**, 345-354 (1966); Paulus, "Polymyxins" in *Antibiotics* II, D. Gottlieb, P. Shaw, Eds. (Springer-Verlag, New York, 1967) pp 254-267.

γ-NH$_2$
|
L-DAB → D-X ──→ L-Y
=| |
R → L-DAB → L-Thr → Z → L-DAB
=| =   ↑
γ-NH$_2$ L-Thr ← L-DAB ← L-DAB
         =| =| =|
       γ-NH$_2$ γ-NH$_2$

DAB = α,γ-diaminobutyric acid

Obtained as the hydrochloride. Nearly colorless powder, dec 228-230°. [α]$_D^{23}$ −40° (c = 1.05). The hydrochloride is very sol (more than 40%) in water and methanol. The solv decreases in the higher alcs. Practically insol in the usual ethers, esters, ketones, hydrocarbons, and the chlorinated solvents. Forms water insol salts with a number of ppts such as picric acid, helianthic acid, Reinecke salt. The free base is slightly sol in water; almost insol in alc.

Polymyxin B. Mixture of polymyxins B$_1$ and B$_2$. [α]$_{5461}$ −106.3° (1N HCl).

Polymyxin B sulfate, *Aerosporin.* Solubilities: Weiss *et al., Antibiot. & Chemother.* **7**, 374 (1957). Mixture with trimethoprim, *Polytrim.*

Polymyxin B$_1$. C$_{56}$H$_{98}$N$_{16}$O$_{13}$. R = (+)-6-methyloctanoyl; X = phenylalanine; Y = leucine; Z = L-DAB.

Polymyxin B$_1$ pentahydrochloride, C$_{56}$H$_{103}$Cl$_5$N$_{16}$O$_{13}$, white powder. [α]$_D^{25}$ −85.11° (c = 2.33 in 75% ethanol).

Polymyxin B$_2$. C$_{55}$H$_{96}$N$_{16}$O$_{13}$. R = 6-methylheptanoyl; X = phenylalanine; Y = leucine; Z = L-DAB. [α]$_{5461}^{22}$ −112.4° (2% acetic acid).

Polymyxin D$_1$. C$_{50}$H$_{93}$N$_{15}$O$_{15}$. R = (+)-6-methyloctanoyl; X = leucine; Y = threonine; Z = D-serine.

Polymyxin D$_2$. C$_{49}$H$_{91}$N$_{15}$O$_{15}$. R = 6-methylheptanoyl; X = leucine; Y = threonine; Z = D-serine.

Polymyxin E see Colistin.
THERAP CAT: Antibacterial.
THERAP CAT (VET): Antibacterial.

7551. Polymyxin B-Methanesulfonic Acid. *Polymyxin B N-sulfomethyl deriv sodium salt;* polymyxin B-*N,N,N,N,N*-pentakis(methanesulfonic acid). Prepn of sodium salt from polymyxin B sulfate, formaldehyde, and sodium bisulfite: Wilkinson, U.S. pat. **3,044,934** (1962 to Burroughs Wellcome).
Sodium salt, *Thiosporin.*
B₁ Analog pentasodium salt, *sulfomyxin, Dynamyxin.*
THERAP CAT: Antibacterial.
THERAP CAT (VET): Antibacterial.

7552. Polynoxylin. *Urea polymer with formaldehyde;* poly[methylenedi(hydroxymethyl)urea]; oxymethyleneurea; polynoxyline; polyoxymethyleneurea; Anaflex; Larex; Ponoxylan. A condensation product of formaldehyde and urea, substantially linear chains, possibly lightly cross-linked. Prepn: Haler, Aebi, *Nature* **190,** 734 (1961); **Brit.** pat. **905,-195** (1962 to Ed. Geistlich Soehne).

$$\left[\begin{array}{c} CH_2OH \quad CH_2OH \\ \\ -N-C-N-CH_2- \\ \| \\ O \end{array}\right]_n$$

Amorphous powder, dec without melting. Soly in water: 0.28-0.31%.
THERAP CAT: Topical antibacterial.

7553. Polyoxins. Agricultural antifungal antibiotic complex produced by *Streptomyces cacaoi* var *asoensis* and *S. piomogenus.* Polyoxins A through O are known, all except C and I having specific activity against phytopathogenic fungi by inhibiting cell wall chitin synthesis. Isoln and characterization of polyoxin A: S. Suzuki *et al., J. Antibiot.* **18A,** 131 (1965); of A and B: K. Isono *et al., Agr. Biol. Chem.* **29,** 848 (1965); S. Suzuki *et al., Japan.* pat. **15,520('66);** of C through L: K. Isono *et al., Agr. Biol. Chem.* **30,** 817 (1966); **31,** 190 (1967); **32,** 792 (1968); of M: K. Isono, S. Suzuki, *Tetrahedron Letters* **1970,** 425. Isoln of polyoxins N and O: S. Suzuki *et al., Japan.* **Kokai 72 23,596** (to Inst. Phys. Chem. Res., and Hokko Chem. Ind.), *C.A.* **78,** 41566t (1973); production of N and O: *eidem,* **Japan.** pat. **20,555-** ('77) (to Inst. Phys. Chem. Res.), *C.A.* **87,** 150183x (1977). Isoln and structure of M: N. Uramoto *et al., Nucleic Acids Res., Spec. Publ.* **5,** 327 (1978). Structural elucidation of polyoxins A through L: K. Isono *et al., J. Am. Chem. Soc.* **91,** 7490 (1969). Total synthesis of J: H. Kuzuhara *et al., Tetrahedron Letters* **1973,** 5055. Biosynthetic studies: K. Isono *et al., Biochemistry* **14,** 2992 (1975); S. Funayama, K. Isono, *ibid.* 5568; K. Isono, R. J. Suhadolnick, *Arch. Biochem. Biophys.* **173,** 141 (1976); S. Funayama, K. Isono, *Biochemistry* **16,** 3121 (1977); K. Isono *et al., J. Am. Chem. Soc.* **100,** 3937 (1978). Mode of action: N. Ohta *et al., Agr. Biol. Chem.* **34,** 1224 (1970); M. Hori *et al., ibid.* **35,** 1280 (1971); **38,** 691, 699 (1974). *Review:* R. J. Suhadolnick, *Nucleoside Antibiotics* (Wiley-Interscience, New York, 1970) pp 218-234; K. Isono, S. Suzuki, *Heterocycles* **13,** 333-351 (1979).

Polyoxin A, C₂₃H₃₂N₆O₁₄, *[S-(E)]-1-[5-[[2-amino-5-O-*

(aminocarbonyl)-2-deoxy-L-xylonoyl]amino]-1,5-dideoxy-1-[3,4-dihydro-5-(hydroxymethyl)2,4-dioxo-1(2H)-pyrimidin-yl]-β-D-allofuranuronoyl]-3-ethylidene-2-azetidinecarboxyl-ic acid. Colorless needles from aq ethanol, dec >180°. [α]²⁰_D −30°. uv max (0.05*N* HCl): 262 nm (log ε 3.94); (0.05*N* NaOH): 264 nm (log ε 3.80).
Polyoxin B, C₁₇H₂₅N₅O₁₃, *5-[[2-amino-5-O-(aminocarb-onyl)-2-deoxy-L-xylonoyl]amino]-1,5-dideoxy-1-[3,4-di-hydro-5-(hydroxymethyl)-2,4-dioxo-1(2H)-pyrimidinyl]-β-D-allofuranuronic acid, Polyoxin AL.* Amorphous powder from aq ethanol. [α]²⁰_D +34°. uv max (0.05*N* HCl): 262 nm (log ε 3.94); (0.05*N* NaOH): 264 nm (log ε 3.82). LD₅₀ in mice: 800 mg/kg orally, *RTECS* Vol. II, R. J. Lewis, R. L. Tatken, Eds. (1980) p 409. *See also* Suzuki, *loc. cit.* (1965).
USE: Fungicide, esp for *Alternaria* leaf spot of many agricultural products.

7554. Polyoxyethylene Alcohols. Polyethylene glycol fatty alcohol ethers; ethoxylated fatty alcohols; macrogol fatty alcohol ethers. R(OCH₂CH₂)ₙOH where R is a long chain alkyl group or mixture of alkyl groups. Nonionic surfactants prepared by ethoxylation of fatty alcohols with ethylene oxide. Prepn and reaction mechanism: J. D. Malkemus, *J. Am. Oil Chem. Soc.* **33,** 571 (1956); W. Satkowski, C. G. Hsu, *Ind. Eng. Chem.* **49,** 1875 (1957); R. D. Fine, *J. Am. Oil Chem. Soc.* **35,** 542 (1958). Compounds with a broad range of properties can be prepared by varying the fatty alcohol (lipophile) used and the degree of polymerization of the polyethylene glycol (hydrophile) segment. CFTA-assigned names based on fatty alcohol segment include *ceteth* (cetyl alc), *laureth* (lauryl alc), *myreth* (myristyl alc), *oleth* (oleyl alc), *steareth* (stearyl alc), *trideth* (tridecyl alc). The average number of ethylene oxide units in the polyethylene glycol segment is indicated by an appended number (e.g. ceteth-20). Additional products prepared from fatty alcohol mixtures include *ceteareth* (cetyl/stearyl alcs), *laneth* (lanlin alcs). Trademarks for some commercially available series of compounds: *Alfonic ethoxylates, Bio Soft EA, Brij, Dehydol, Emulphogene BC, Ethosperse, Eumulgin C, Ethoxyl, Lipal* (CA, CSA, LA, MA, OA, TD), *Lipocol, Macol, Polychol, Poly-Tergent J, Siponic* (E, L, Y), *Tergitol* (L, S), *Trycol, Volpo.* *Review:* W. B. Satkowski *et al.,* "Polyoxyethylene Alcohols" in *Nonionic Surfactants* M. J. Schick, Ed. (Dekker, New York, 1967) pp 86-141.
Liquids to waxy solids. Compds with one to five moles ethylene oxide are sol in oil and many hydrocarbons. Water soly increases with increasing ethylene oxide content. Properties of Brij surfactants: G. King, *Drug Cosmet. Ind.* **90,** 24 (1962).
Laureth 9, C₁₂H₂₅(OCH₂CH₂)ₙOH where the av. value of *n* is 9, *Lipal 9LA, lauromacrogol 400.* Component of *Lanettes.*
Cetomacrogol 1000, CH₃(CH₂)ₘ(OCH₂CH₂)ₙOH where *m* may be 15 or 17 and *n* may be 20 to 24, *ceteth-20, polyeth-yleneglycol 1000 monocetyl ether, Brij 58, Lipocol C-20, Texo-for A1P.* Prepd from cetyl or cetyl/stearyl alcohol. Waxy, unctuous mass, mp 39°. n⁶⁰_D 1.448-1.452. Sol in water, alc, acetone. Insol in petr ether. May be sterilized by heating to 150° for one hour.
USE: Used as emulsifiers, wetting agents, antistats, solubilizers, defoamers, detergents, lubricants in pharmaceutical, cosmetic and other industrial applications. Laureth 9 as spermaticide; pharmaceutic aid (surfactant).

7555. Polyoxyethylene Fatty Acid Esters. Polyethylene glycol esters of fatty acids; PEG fatty acid esters; POE fatty acid esters; ethoxylated fatty acid esters; macrogol fatty acid esters. RCOO(CH₂CH₂O)ₙH or RCOO(CH₂CH₂O)ₙOCR where R is a long chain alkyl group or mixture of alkyl groups. Nonionic surfactants prepared commercially by esterification of fatty acids with ethylene oxide or with polyethylene glycol. Tall oil, *q.v.,* frequently used as source of fatty acids. Wide range of properties achieved by changing the hydrophobic fatty acid segment and/or varying the degree of polymerization of the hydrophilic polyoxyethylene segment. Trademarks for some commercially available series of compounds include *Emerest 2600 series, Emulphor* (VN, VT), *Emulsynt, Ethofat, Lipal* (DL, DS, L, MS, OL, SR, W), *Lipopeg, Mapeg, Myrj, Nonisol, Nopalcol, Pegosperse, Renex 20, Sterox CD, Varonic 400.* Review of prepn, properties and uses: W. B. Satkowski *et al.,* "Polyoxyethylene

Esters of Fatty Acids" in *Nonionic Surfactants*, M. J. Schick, Ed. (Dekker, New York, 1967) pp 142-174.

Liquids, soft solids, solids or flakes. Solubility properties depend on length of polyoxyethylene (POE) segment added to a specific fatty acid. Compounds are insol in water. Sol in oil and hydrocarbon solvents when < 8 ethylene oxide units are added. Soly in water begins when 12-15 ethylene oxide units are added. Specific gravity and viscosity increase with increasing ethylene oxide content.

Polyoxyl 8 stearate, polyethylene glycol 400 monostearate, Cerasynt 660, Emerest 2640, Hodag 40-S, Lipal 400S, Lipopeg 4-S, Myrj 45, Mapeg 400 MS, Pegosperse 400-MS. Mixture of unesterified polyethylene glycols and mono- and distearates of polyethylene glycols: R. L. Birkmeier, J. D. Brander, *J. Agr. Food. Chem.* **6**, 471 (1958).

Polyoxyl 40 stearate, Emerest 2672, Lipal 39S, Mapeg S 40, Myrj 52, Myrj 52S, Pegosperse 1750-MS. Mixture of mono- and distearate esters of polyoxyethylene and of free POE. The average number of oxyethylene units is 40. Waxy, white to light tan solid; odorless or faint fat-like odor. Sol in water, alc, ether, acetone. Insol in mineral oil, vegetable oils.

USE: As antistats, emulsifiers, defoamers, wetting agents, solubilizers, conditioning agents, lubricants, detergents. Have wide range of cosmetic, pharmaceutical and other industrial applications. Polyoxyl 8 and 40 stearates as pharmaceutic aids (surfactant).

7556. Polyphosphazenes. High polymers containing an inorganic backbone of alternating nitrogen and phosphorus atoms. They exist as glasses, flexible solids, or rubbery solids with a low tendency for crystallization; non-flammable and more elastic than carbon-backbone polymers. Prepn of the first polyphosphazene, hexachlorocyclotriphosphazene: J. Liebig, *Ann.* **11**, 139 (1834). Improved synthesis and basis of modern mfg methods: R. Schenck, G. Römer, *Ber.* **57B**, 1343 (1924). Review of syntheses, properties, chemistry and applications: H. R. Allcock, *Science,* **193**, 1214-1219 (1976); idem, *Angew. Chem. Int. Ed.* **16**, 147-156 (1977); E. N. Peters, "Inorganic High Polymers", in Kirk-Othmer *Encyclopedia of Chemical Technology* **vol. 13** (Wiley-Interscience, New York, 3rd ed., 1981) pp 398-413.

$$\left[-\underset{\underset{X}{|}}{\overset{\overset{X}{|}}{P}}=N- \right]_n$$

USE: In waterproofing; as flame retardants; in gaskets, o-rings, hydrocarbon fuel hoses.

7557. Polyphosphoric Acid. Phospholeum; tetraphosphoric acid. May be prepd by heating H_3PO_4 with sufficient phosphoric anhydride to give the resulting product an 82-85% P_2O_5 content: Bell, *Ind. Eng. Chem.* **40**, 1464 (1949); Van Wazer, Holst, *J. Am. Chem. Soc.* **72**, 639 (1950); Kennard, *Org. Chem. Bull.* **29**, no. 1 (1957). Consists of about 55% tripolyphosphoric acid, the remainder being H_3PO_4 and other polyphosphoric acids. Typical analysis: 83.0% P_2O_5; ortho equivalent 115.0%.

Viscous liquid at room temps. Conveniently fluid at 60°. Solidifies to a glass at low temps. Sol in water with evolution of heat and hydrolysis to H_3PO_4.

USE: In organic synthesis for cyclizations and acylations. *Caution:* In strong concns moderately irritating to skin, mucous membranes.

7558. Polypropylene. *1-Propene homopolymer;* propylene polymer. Three forms are possible. *Isotactic* (fiber-forming): methyl groups are all on same side of plane of zig-zag carbon atom chain. *Syndiotactic:* methyl groups are on alternate sides of plane of carbon atom chain. *Atactic* (not fiber-forming, amorphous): methyl groups are in a random arrangement with respect to plane of carbon atom chain. Early synthesis of isotactic form with Ziegler catalyst and comparison with atactic form: Natta *et al., J. Chem. Soc.* **77**, 1708 (1955); Natta, *J. Polymer Sci.* **16**, 143 (1955). *Reviews:* N. G. Gaylord, H. F. Mark, *Linear and Stereoregular Addition Polymers* (Interscience, New York, 1959) pp

54-65; R. W. Moncrieff, *Man-Made Fibres* (Wiley, New York, 4th ed., 1963) pp 500-510; J. G. Cook, *Handbook of Textile Fibres* (Merrow Publishing Co., England, 3rd ed., 1964) pp 369-379; G. Crespi, L. Luciani, "Olefin Polymers (Polypropylene)" in Kirk-Othmer *Encyclopedia of Chemical Technology* **Vol. 16** (Wiley-Interscience, New York, 3rd ed., 1981) pp 453-469.

$$\left[-CH(CH_3)CH_2- \right]_n$$

Isotactic form, *Amco, Amerfil, Beamette, Courlene PY, DLP, Gerfil, Herculon, Lambeth, Meraklon, Moplen, Olane, Prolene, Tuff-Lite, Ulstron.* Solid material, softens at ~155°, mp ~165°. Low flammability comparable to that of wool. Keeps strength down to -100°. d 0.90-0.92. Practically insol in cold org solvents; sol in hot decalin, hot tetralin, boiling tetrachloroethane. Shrinks in boiling trichloroethylene. Resistant to acids, alkalies; attacked by strong oxidizing agents, *e.g.,* hydrogen peroxide. Good resistance to abrasion ("pilling"). Tendency to develop static charges. Unstabilized material has poor resistance to sunlight. Difficult to dye, lacks dye-attracting polar groups in structure.

USE: Isotactic form: for fishing gear, ropes, filter cloths, laundry bags, protective clothing, blankets, fabrics, carpets, yarns, etc.

7559. Polysorbate 80. *Sorbitan mono-9-octadecenoate poly(oxy-1,2-ethanediyl) derivs;* polyoxyethylene (20) sorbitan mono-oleate; sorethytan (20) mono-oleate; polyethylene oxide sorbitan mono-oleate; Sorbitan mono-oleate polyoxyethylene; Sorlate; Tween 80; Monitan; Olothorb. An oleate ester of sorbitol and its anhydrides copolymerized with approx 20 moles of ethylene oxide for each mole of sorbitol and sorbitol anhydrides. *See also* Span.

$$HO(C_2H_4O)_w \quad (OC_2H_4)_x OH$$
$$CH(OC_2H_4)_y OH$$
$$H_2C(OC_2H_4)_z R$$

[Sum of w, x, y, z is 20;
R is $(C_{17}H_{33})COO$]

Amber-colored, viscous liquid. d 1.06-1.10. Viscosity 270-430 centistokes. Very sol in water; sol in alcohol, cottonseed oil, corn oil, ethyl acetate, methanol, toluene. Insol in mineral oil. pH of 5% aq soln between 5 and 7.

USE: Pharmaceutic aid (surfactant); as emulsifier and dispersing agent in medicinal products; as defoamer and emulsifier in foods.

7560. Polytetrafluoroethylene. *Tetrafluoroethene homopolymer; tetrafluoroethylene polymer;* polytetrafluoroethylene resin; polytef; PTFE; Fluon; Fluoroflex; Teflon. A highly stable thermoplastic tetrafluoroethylene homopolymer. Composed of at least 20,000 C_2F_4 monomer units linked into very long unbranched chains. Prepd by polymerization of tetrafluoroethylene: Plunkett, U.S. pat. **2,230,654** (1941 to Kinetic Chem.); Brubaker, U.S. pat. **2,393,967**; Joyce, U.S. pat. **2,394,243** (both 1946 to du Pont); Hanford, Joyce, *J. Am. Chem. Soc.* **68**, 2082 (1946); Renfrew, Lewis, *Ind. Eng. Chem.* **38**, 870 (1946); Renfrew, U.S. pat. **2,534,058** (1950 to du Pont); C. E. Schildknecht, *Vinyl and Related Polymers* (Wiley, New York, 1952) pp 483-494. Account of discovery by Roy J. Plunkett: A. B. Garrett, *J. Chem. Ed.* **39**, 288 (1962). *Reviews:* R. W. Moncrieff, *Man-Made Fibres* (John Wiley, New York, 4th ed., 1963) pp 512-517; McCane in *Encyclopedia of Polymer Science and Technology* **vol. 13**, N. M. Bikales, Ed. (Interscience, New York, 1970) pp 623-654; S. V. Gangal in Kirk-Othmer *Encyclopedia of Chemical Technology* **vol. 11** (Wiley-Interscience, New York, 3rd ed., 1980) pp 1-24.

n ≥ 20,000

Nonflammable, high polymer. White translucent to opaque solid (depending on thickness). Very inert chemically. Useful temp range from cryogenic to +260°. Melts to an extremely viscous gel at 327° and reverts to the gaseous monomer at temperatures above 400°. d 2.2. Shore hardness 55-56. Tensile strength 3500-4500 psi. Flexural modulus ~ 80,000-90,000 psi at room temp. Brittle point below −80°. Dielectric constant (at 60 to 3 × 10⁹ cycles) 2.0-2.05. Not affected by water, aqua regia, chlorosulfonic acid, acetyl chloride, boron fluoride, hot nitric acid, boiling solns of sodium hydroxide, and organic solvents. Not wetted by water. No substance has been found which will dissolve the polymer at moderate temperatures, but prolonged contact with fluorine, hot plasticizers and polymeric waxes is not recommended. Is subject to cold flow at high pressure. Because of its high melt viscosity molding and sintering techniques similar to those used in powder metallurgy are normally used for fabrication.

Caution: The finished compound is inert under ordinary conditions. There have been reports of polymer fume fever, the symptoms of which resemble those of an attack of influenza, in humans exposed to the heated polymer under conditions of inadequate ventilation. Contamination of smoking tobacco with polytetrafluoroethylene, even in small amounts, is to be avoided.

USE: As tubing and sheets for chemical laboratory and process work; for lining reaction vessels; for gaskets and pump packings, sometimes mixed with graphite or glass fibers; as electrical insulator esp in high frequency applications; filtration fabrics; protective clothing. Prosthetic aid.

7561. Polythiazide. *6-Chloro-3,4-dihydro-2-methyl-3-[[(2,2,2-trifluoroethyl)thio]methyl]-2H-1,2,4-benzothiadiazine-7-sulfonamide 1,1-dioxide;* 2-methyl-3-(β,β,β-trifluoroethylthiomethyl)-6-chloro-7-sulfamyl-3,4-dihydro-1,2,4-benzothiadiazine 1,1-dioxide; 6-chloro-3,4-dihydro-2-methyl-7-sulphamoyl-3-(2,2,2-trifluoroethylthiomethyl)-2H-benzo-1,2,4-thiadiazine 1,1-dioxide; Drenusil; Nephril; Renese. $C_{11}H_{13}ClF_3N_3O_4S_3$; mol wt 439.90. C 30.03%, H 2.98%, Cl 8.06%, F 12.96%, N 9.55%, O 14.55%, S 21.87%. Prepn: McManus, U.S. pat. **3,009,911** (1961 to Pfizer).

Crystals from isopropanol, mp 202.5°. Practically insol in water. Sol in aq solns made alkaline with carbonates or hydroxides of the alkali metals. Rate of decompn increases with increase in pH.

THERAP CAT: Diuretic, antihypertensive.

7562. Polyvinyl Alcohol. *Ethenol homopolymer;* PVA; Akwa Tears; Alvyl; Elvanol; Gelvatol; Liquifilm; Moviol; Polyviol; Resistoflex; Rhodoviol; Sno Tears; Solvar; Vinarol; Vinol. A polymer prepd from polyvinyl acetates by replacement of the acetate groups with hydroxyl groups. The alcoholysis proceeds most rapidly in a methanol + methyl acetate mixture in the presence of catalytic amounts of alkali or mineral acids: Hermann, Haehnel, *Ber.* **60,** 1658 (1927). *Monograph:* C. E. Schildknecht, *Vinyl and Related Polymers* (Wiley, New York, 1952). The head-to-tail or 1,3-glycol structure is favored: Staudinger *et al., Ber.* **60,** 1782 (1927); *J. Prakt. Chem.* **155,** 261 (1940); Marvel, Denoon, *J. Am. Chem. Soc.* **60,** 1045 (1938); McDowell, Kenyon, *ibid.* **62,** 415 (1940); Marvel, Inskeep, *ibid.* **65,** 1710 (1943). *Reviews:* M. Leeds in Kirk-Othmer *Encyclopedia of Chemical Tech-*

nology **vol. 21** (Wiley-Interscience, New York, 2nd ed., 1970) pp 353-368; *Polyvinyl Alcohol,* A. C. Finch, Ed. (Wiley, New York, 1973) 640 pp; A. S. Dunn, *Chem. & Ind. (London)* **1980,** 801-806. Vinyl alcohol monomer is incapable of existence.

Dry, unplasticized polyvinyl alcohol powders are white to cream colored, soften at about 200° with decompn. Commercial polyvinyl alcohols have different contents of residual acetyl groups and therefore different viscosity characteristics. The first code number following the trade name indicates the degree of hydrolysis, while the second set of numbers indicates the approx viscosity in centipoises (4% aq soln at 20°). Polyvinyl alcohols are essentially sol in hot and cold water, but those coded 20-105 require alcohol-water mixtures. Aq solns are colloidal and compatible with lower alcohols. Pure aq solns are neutral or faintly acid and subject to mold growth. Insol in petroleum solvents.

USE: In the plastics industry in molding compds, surface coatings, films resistant to gasoline, textile sizes and finishing compositions; can be compounded to yield elastomers to be used in manuf artificial sponges, fuel hoses, etc., also in printing inks for plastics and glass, in pharmaceutical finishing, cosmetics, water-sol film and sheeting. Pharmaceutic aid (viscosity increasing agent); ophthalmic lubricant.

7563. Polyvinyl Chloride. *Chloroethene homopolymer; chloroethylene polymer;* PVC; Vybak; Geon; Breon; Welvic; Movyl; Tevilon; Deckor; Vinacort; Ultron; Vinylite V; Koroseal; Marvinol. Polyvinyl chloride fibers are marketed under the names: *Rhovyl, Fibravyl, Thermovyl, Isovyl, Retractyl, Crinovyl, Envilon, Nip.* Average mol wt, about 60,000 to 150,000. Prepn: Baumann, *Ann.* **163,** 308 (1872); Schoenfeld, U.S. pat. **2,168,808** (1937 to B. F. Goodrich). Structure: Natta, Rigamonti, *Atti accad. Lincei Classe sci. fis., mat. nat.* **24,** 381 (1936); *C.A.* **31,** 4563⁹ (1937); Marvel *et al., J. Am. Chem. Soc.* **61,** 3241 (1939). *Reviews:* C. E. Schildknecht, *Vinyl and Related Polymers* (Wiley, New York, 1952) pp 392-442. Technology: W. S. Penn, *PVC Technology* (Maclaren, London, 1962); J. A. Davidson, K. L. Gardner, "Vinyl Polymers (PVC)" in Kirk-Othmer *Encyclopedia of Chemical Technology* vol. 23 (Wiley-Interscience, New York, 3rd ed., 1983) pp 886-936. Book: Sarvetnick, *Polyvinyl Chloride* (Van Nostrand, Reinhold, New York, 1969).

Plastic solid. d 1.406; n 1.54. Stabilizers are necessary to prevent discoloration from exposure to light or heat. Solvents for unmodified polyvinyl chloride of high mol wt: cyclohexanone, methyl cyclohexanone, dimethyl formamide, nitrobenzene, tetrahydrofuran, isophorone, mesityl oxide. Solvents for lower polymers: dipropyl ketone, methyl amyl ketone, methyl isobutyl ketone, acetonylacetone, methyl ethyl ketone, dioxane, methylene chloride.

USE: Rubber substitutes, electric wire and cable-coverings, pliable thin sheeting, film finishes for textiles, non-flammable upholstery, raincoats, tubing, belting, gaskets, shoe soles.

7564. Pomegranate. Granatum. Dried bark of stem or root of *Punica granatum* L., *Punicaceae.* Habit. Mediterranean region; Eastern, Western, and Southern Asia; cultivated in subtropical countries. *Constit.* 0.5-1% alkaloids consisting of pelletierine, methylpelletierine, pseudopelletierine (granatonine), and isopelletierine; mannite, about 20% tannin. Rind of fruit contains about 30% tannin.

THERAP CAT: Formerly as teniafuge.

7565. Ponalrestat. *3-[(4-Bromo-2-fluorophenyl)methyl]-3,4-dihydro-4-oxo-1-phthalazineacetic acid;* 3-(4-bromo-2-fluorobenzyl)-4-oxo-3H-phthalazin-1-ylacetic acid; 2-(2-fluoro-4-bromobenzyl)-1,2-dihydro-1-oxophthalazin-4-ylacetic acid; ICI-128436; MK-538; Prodiax; Statil. $C_{17}H_{12}BrFN_2O_3$; mol wt 391.20. C 52.19%, H 3.09%, Br 20.43%, F 4.86%, N 7.16%, O 12.27%. Aldose reductase inhibitor. Prepn: D. R. Brittain, R. Wood, **Eur.** pat. **Appl. 2,895;** *eidem,* **U.S.** pat. **4,251,528** (1979, 1981 both to ICI). Effect on cataract development and nerve conduction in diabetic rats: D. Stribling *et al., Metabolism* **34,** 336 (1985). Effect on peripheral nerve edema in galactosemic rats: A. P. Mizisin *et al., J. Neurol. Sci.* **74,** 35 (1986). *In vitro* inhibition of aldose reductase from bovine retina: R. Poulsom, *Biochem. Pharmacol.* **35,** 2955 (1986); from human retina: *idem, Curr. Eye Res.* **6,** 427 (1987). Clinical evaluation in diabetic neuropathy: G. Sundkvist *et al., Acta Med. Scand.* **221,** 445 (1987).

Crystals from methanol + water, mp 184-185°; crystals from dry methanol, mp 180-182°.

THERAP CAT: Treatment of diabetic neuropathy and retinopathy.

7566. Ponasterone A. *(2β,3β,5β,22R)-2,3,14,20,22-Pentahydroxycholest-7-en-6-one;* 25-deoxyecdysterone. $C_{27}H_{44}O_6$; mol wt 464.65. C 69.79%, H 9.54%, O 20.66%. Polyhydroxylated steroid with strong moulting hormone activity; first phytoecdysone to be isolated. Isoln with ponasterones B, C, D, from *Podocarpus nakaii* Hay., *Podocarpaceae* and structure determn: K. Nakanishi *et al., Chem. Commun.* **1966,** 915; isoln from various Japanese ferns: T. Takemoto *et al., Chem. Pharm. Bull.* **21,** 2336 (1973). Stereochemical elucidation: H. Moriyama, K. Nakanishi, *Tetrahedron Letters* **1968,** 1111; using ^{13}C-NMR: H. Hikino *et al., Chem. Pharm. Bull.* **23,** 125 (1975). Chromatographic sepn: M. Hori, *Steroids* **14,** 33 (1969). HPLC analysis: I. D. Wilson *et al., J. Chromatog.* **238,** 97 (1982); R. E. Isaac *et al., ibid.* **246,** 317 (1982). Synthesis: G. Hüppi, J. B. Siddall, *Tetrahedron Letters* **1968,** 1113. Moulting hormone activity on houseflies, silkworms: M. Kobayashi *et al., Steroids* **9,** 529 (1967); on various insects: W. E. Robbins *et al., ibid.* **16,** 105 (1970). Hormonal regulation to increase yield from silkworms: T. Okauchi *et al.,* **U.S.** pat. **3,941,879** (1976 to Takeda). Used to characterize ecdysteroid receptors in *Drosophilia* cells: P. Maroy *et al., Proc. Natl. Acad. Sci. USA* **75,** 6035 (1978); B. A. Sage *et al., J. Biol. Chem.* **257,** 6373 (1982). *Review of* **ponasterones:** K. Nakanishi, *Bull. Soc. Chim. France* **1969,** 3475.

Crystals from ethanol, mp 259-260° (dec). $[\alpha]_D^{15}$ +90° (methanol). uv max (methanol): 244, 326 nm (ϵ 12400, 130).

7567. Ponceau 3R. *3-Hydroxy-4-[(2,4,5-trimethylphenyl)azo]-2,7-naphthalenedisulfonic acid disodium salt;* **C.I. Food Red 6;** sodium cumeneazo-β-naphthol disulfonate; **C.I. 16155; FD & C Red no. 1; Ext. D & C Red no. 15.** $C_{19}H_{16}N_2Na_2O_7S_2$; mol wt 494.46. C 46.15%, H 3.26%, N 5.67%, Na 9.30%, O 22.65%, S 12.97%. Prepn: *Colour Index* **vol. 4** (3rd ed., 1971) p 4092. Toxicity studies: Hansen *et al., Toxicol. Appl. Pharmacol.* **5,** 105 (1963).

Dark red powder. Soluble in water with cherry-red color, slightly in alcohol. Addition of HCl to its aq soln does not change the color, but NaOH produces a yellow ppt. It dissolves in concd H_2SO_4 to a cherry-red soln which does not change color on dilution.

USE: Dyeing wool. *Caution:* Delisted by the FDA for use in foods, drugs, and cosmetics.

7568. Ponceau SX. *3-[(2,4-Dimethyl-5-sulfophenyl)azo]-4-hydroxy-1-naphthalenesulfonic acid disodium salt;* **FD & C Red No. 4; C.I. Food Red 1; C.I. 14700.** $C_{18}H_{14}N_2Na_2O_7S_2$; mol wt 480.42. C 45.00%, H 2.94%, N 5.83%, Na 9.57%, O 23.31%, S 13.35%. Prepn: E. Nölting, O. Kohn, *Ber.* **19,** 137 (1886). Metabolism: J. L. Radomski, T. J. Mellinger, *J. Pharmacol. Exp. Ther.* **136,** 259 (1962). Chronic toxicity study: K. J. Davis *et al., Toxicol. Appl. Pharmacol.* **8,** 306 (1966). *Review: IARC Monographs* **8,** 207-214 (1975). *See also: Colour Index* **vol. 4** (3rd ed., 1971) p 4068.

Red crystals. Absorption max (0.02 N CH_3COONH_4): 500 nm. Sol in water; slightly sol in ethanol; insol in vegetable oils. Deep red soln in conc H_2SO_4, changing to red ppt on dilution. Orange soln in conc HNO_3, turning yellow. LD_{50} orally in rats: > 2 g/kg, F. C. Lu, A. Lavalle, *Can. Pharm. J.* **97,** 30 (1964).

USE: In externally applied drugs and cosmetics; formerly in maraschino cherries. Delisted by FDA for use in food: *Fed. Reg.* **41,** 41852 (1976).

7569. Poppy Capsules. Poppy heads. Fully grown, dried capsules of *Papaver somniferum* L., *Papaveraceae. Habit.* Europe, Asia. *Constit.* Capsules: 0.15-0.5% morphine and small amounts of other opium alkaloids. Seed: Fixed oil (poppy oil), albuminoids.

USE: For preparing emulsions—only the white seeds should be used. Poppy seeds are chiefly used for making the oil, and in baking; the bluish-black variety is generally used in the U.S.A. for baking.

7570. Poppy Oil. Poppy-seed oil. Expressed from poppy seeds. Contains no morphine or other opium alkaloids. Pale yellow, drying oil; pleasant odor and taste. d 0.924-0.927. Solidif about −18°. n_D^{20} 1.4766-1.4774. Sapon. no. 189-197. Iodine no. 133-158.

USE: Manuf paints, varnishes, and soaps. Edible grades are marketed in Europe and Asia.

See also Ethiodized Oil; Iodized Oil.

7571. Populin. *2-(Hydroxymethyl)phenyl-β-D-glucopyranoside 6-benzoate;* populoside; salicin benzoate. $C_{20}H_{22}O_8$; mol wt 390.38. C 61.53%, H 5.68%, O 32.79%. In bark and leaves of *Populus tremula* L., *P. nigra* L., *P. nigra* L. var. *italica* Duroi, *P. canadensis* Moench., *P. grandidentate* Michx., and *P. tremuloides* Michx., *Salicaceae,* perhaps also in *Salix helix, Salix purpureae* L. var *helix* (L.) Koch. Isoln: Pearl *et al., J. Org. Chem.* **27,** 2685 (1962). May be made from salicin by melting with benzoic anhydride, or from salicin and benzoyl chloride in presence of potassium hydroxide: Richtmyer, Yeakel, *J. Am. Chem. Soc.* **56,** 2495 (1934). Alkaline cleavage produces benzoic acid and salicin. Enzymatic hydrolysis using taka-diastase *(q.v.)* gives salicyl alcohol and benzoyl glucose: Kitasato, *Biochem. Z.* **190**, 109 (1927).

Dihydrate, needles from water, sweet taste, like licorice. Becomes anhydr at 100°. mp 179°. $[\alpha]_D^{20}$ −2.0° (c = 5 in pyridine); $[\alpha]_D^{25}$ −29.7° (c = 5 in acetone). One gram dissolves in about 2 liters water, in 42 ml boiling water, in about 100 ml alcohol. Practically insol in ether.

7572. Porfiromycin. *6-Amino-8-[[(aminocarbonyl)oxy]-methyl]-1,1a,2,8,8a,8b-hexahydro-8a-methoxy-1,5-dimethyl-azirino[2',3':3,4]pyrrolo[1,2-a]indole-4,7-dione;* N-methylmitomycin C; U-14743. $C_{16}H_{20}N_4O_5$; mol wt 348.35. C 55.16%, H 5.79%, N 16.08%, O 22.97%. Antibiotic substance isolated from a *Streptomyces ardus* fermentation broth: Herr *et al., Antimicrob. Ag. Ann.* **1960**, 23. Isoln from *S. verticillatus* and structure: Webb *et al., J. Am. Chem. Soc.* **84**, 3185, 3187 (1962). Production process: Bohonos *et al.,* U.S. pat. **3,219,530** (1965 to Am. Cyanamid). Synthesis of (±)-form: F. Nakatsubo *et al., J. Am. Chem. Soc.* **99**, 8115 (1977). For stereochemistry and other synthetic studies, *see* Mitomycins.

Dark purple triclinic crystals, dec 201-201.5°. $[\alpha]_D^{25}$ +275° ±55° (c = 0.1% in methanol); $[\alpha]_D^{25}$ +242° ±100° (c = 0.045% in methanol). uv max (methanol): 217, 360, 555 nm (ε 24600, 23000, 209). Slightly sol in water, moderately sol in polar organic solvents, practically insol in hydrocarbon solvents.

THERAP CAT: Antibacterial; antineoplastic.

7573. Porofor® BSH. *Benzenesulfonic acid hydrazide;* benzenesulfohydrazide; phenylsulfohydrazide. $C_6H_8N_2O_2S$; mol wt 172.21. C 41.85%, H 4.68%, N 16.27%, O 18.58%, S 18.61%. $C_6H_5SO_2NHNH_2$. *Ref:* Lober, *Angew. Chem.* **64**, 65 (1952).

Crystals, dec 103-104° with the evolution of nitrogen. May be stored indefinitely at temps up to 80°. Sensitive to moist oxidizing agents.

USE: Gas generating agent for use in making foam rubber and foam plastics.

7574. Porphine. *21H,23H-Porphine;* porphin. $C_{20}H_{14}$-N_4; mol wt 310.34. C 77.40%, H 4.55%, N 18.06%. Parent substance of the *porphyrins,* a group of compounds found in all living matter which are the basis of respiratory pigments in animals and plants. In porphyrins, side chains are substituted for the hydrogens in the porphine pyrrole rings. *Chlorins* are dihydroporphyrins. *See also:* chlorophyll, hemoglobin, vitamin B_{12}, hematin. Prepn of porphine: Fischer, Gleim, *Ann.* **521**, 157 (1935); Rothemund, *J. Am. Chem. Soc.* **58**, 625 (1936); Krol, *J. Org. Chem.* **24**, 2065 (1959). Study of porphyrin analogs: C. L. Honeybourne *et al., Tetrahedron* **36**, 1833 (1980). Review of biosyntheses of porphyrins and chlorins: A. R. Battersby, E. McDonald in *Porphyrins Metalloporphyrins,* K. M. Smith, Ed. (Elsevier, New York, 1975) pp 61-122. Review of porphyrin syntheses: R. P. Evstigneeva, *Pure Appl. Chem.* **53**, 1129-1140 (1981). Comprehensive seven volume treatise: *The Porphyrins,* D. Dolphin, Ed. (Academic Press, New York, 1978).

Dark red, shiny leaflets from chloroform-methanol. Darkens at 360° but does not melt. The absorption bands are those characteristic of the substituted porphyrins, details in Fischer-Orth, *Die Chemie des Pyrrols* **vol. II**, 1, 175 (1937). Sol in pyridine, dioxane, and phenol; slightly sol in chloroform, bromoform, glacial acetic acid. Almost insol in acetone, methanol, ether. HCl number: 1.7 (Fischer); 3.3 (Rothemund).

Iron salt, $C_{20}H_{12}N_4 \cdot FeCl$, brown cubes from ether.
Magnesium salt, $C_{20}H_{12}MgN_4$, red needles.
Copper salt, $C_{20}H_{12}N_4Cu$, brown needles.

7575. Porphobilinogen. *5-(Aminomethyl)-4-(carboxymethyl)-1H-pyrrole-3-propanoic acid;* 2-aminomethylpyrrol-3-acetic acid 4-propionic acid. $C_{10}H_{14}N_2O_4$; mol wt 226.23. C 53.09%, H 6.24%, N 12.38%, O 28.29%. An intermediate in the biosynthesis of heme, *q.v.,* found in the urine of patients with acute porphyria: Westall, *Nature* **170**, 614 (1952). Isoln and structure: Cookson, Rimington, *Biochem. J.* **57**, 476 (1954). Synthesis: Jackson, MacDonald, *Can. J. Chem.* **35**, 715 (1957). Chemistry of conversion into porphyrins: Mathewson, Corwin, *J. Am. Chem. Soc.* **83**, 135 (1961).

Monohydrate, minute pink crystals from dil ammonium acetate soln at pH 4. Dec 172-175°. pK' = 3.70; 4.95; 10.1. Slightly sol in water. Converted to uroporphyrins by hot dil HCl or by blood hemolyzates.

Hydrochloride monohydrate, fine triclinic needles from dil HCl, dec 165-170°. Soluble in water.

7576. Porphyrillic Acid. *1,3-Dihydro-4,10-dihydroxy-8-methyl-3-oxoisobenzofuro[5,4-b]benzofuran-7-carboxylic acid;* 1,7-dihydroxy-9-hydroxymethyl-3-methyl-4,8-dibenzofurandicarboxylic acid (8,9)-γ-lactone. $C_{16}H_{10}O_7$; mol wt 314.27. C 61.15%, H 3.21%, O 35.64%. Constituent of lichens. Isoln from *Haematomma coccineum* Dicks., *Lecanoreae,* and structure studies: Wachtmeister, *Acta Chem. Scand.* **8**, 1433 (1954); **10**, 1404 (1956).

Needles from ethanol or dioxane, mp 280-283° (dec) after darkening from 270°. Soluble in concd sulfuric acid, sodium carbonate, sodium hydroxide; sparingly sol in sodium bicarbonate.

7577. Porphyropsin. Photoreceptor protein found in the retinal rod cells of fresh water and migrating fish, lampreys, and certain amphibians. Absorption maximum approx 520-530 nm. Composed of the chromophore, 11-*cis*-3-dehydroretinal, *q.v.,* bound to scotopsin, the specific protein component of rod pigments (see Opsins). Biological activity is similar to that of rhodopsin, *q.v.* Isoln from retinas: G. Wald, *Nature* **139**, 1017 (1937); *idem, J. Gen. Physiol.* **22**, 775 (1939). Prepn from 3-dehydroretinal and opsin: *idem, Ann. Rev. Biochem.* **22**, 497 (1953). Methods of purification, prepn and assay: R. Hubbard *et al.,* "Methodology of Vitamin A and Visual Pigments", in *Methods in Enzymology* **vol. 18,** D. B. McCormick, L. D. Wright, Eds. (Academic Press, New York, 1971) pp 615-653. Exposure to light initiates the conversion of porphyropsin through a series of distinct intermediates to yield opsin and *trans*-3-dehydroretinal. Bleaching kinetics and photochemistry: T. Yoshizawa in *Handbook of Sensory Physiology* Vol. VII(2), H. J. A. Dartnall, Ed. (Springer-Verlag, New York, 1972) pp 146-179. Porphyropsin may co-exist with rhodopsin in the retinas of certain amphibians and fish: T. E. Reuter *et al., J. Gen. Physiol.* **58,** 351 (1971). Environmental effect on visual pigment formation and interconversion: P. Witkovsky *et al., ibid.* **72,** 821 (1978); A. T. C. Tsin, D. D. Beatty, *Exp. Eye Res.* **30,** 143 (1980). *Review:* G. Wald, *Science* **162,** 230-239 (1968).

7578. Potash Sulfurated. Liver of sulfur; sulfurated potassa; hepar sulfuris. Mixture of potassium polysulfides, chiefly trisulfide, and potassium thiosulfate, containing not less than 12.8% sulfur as sulfide. *U.S.D.* (5th ed., 1955) p 1092. Has been incorrectly called "potassium sulfide".

Yellowish-brown lumps; fresh fracture is liver-brown, slight H_2S odor. Dec on exposure to air, forming free sulfur and potassium carbonate and becomes yellow to gray. Freely sol in water, usually with slight residue; partially sol in alcohol. *Keep tightly closed.*

Incompat. Acids, alcohol, carbonated waters, acid salts. *Caution:* Avoid metal bathtubs, metal spoons, and water containing much CO_2.

USE: Producing color effects on brass, bronze, nickel, gunmetal.

THERAP CAT: Sulfide source (skin diseases).

THERAP CAT (VET): Externally for chronic and parasitic skin diseases.

7579. Potassium. Kalium. K; at. wt 39.0983; at. no. 19; valence 1. Alkali metal. Occurrence in earth's crust: 2.59% by wt. Natural isotopes: 39 (93.22%); 40 (0.012%); 41 (6.77%); ^{40}K is radioactive: $T_{1/2}$ 1.26 $\times$ 19^9 years; isotopes range in mass number from 37 to 47. Found mainly as the chloride (sylvite); also in the aluminosilicates *orthoclase,* and *microcline* ($KAlSi_3O_8$), and as *carnallite* ($KCl.MgCl_2.6H_2O$). First prepd in free form by Davy in 1807 by electrolysis of fused potassium hydroxide. Prepns: Hackspill, *Helv. Chim. Acta* **11,** 1003 (1928); Jackson, Werner, U.S. pat. **2,480,655** (1949 to Mine Safety Appliances Co.). NMR spectrum of potassium anion (K$^-$): P. P. Edwards *et al., Nature* **317,** 242 (1985). *Review:* Whaley, "Sodium, Potassium, Rubidium, Cesium, and Francium" in *Comprehensive Inorganic Chemistry* **vol. 1,** J. C. Bailar, Jr. *et al.,* Eds. (Pergamon Press, Oxford, 1973) pp 369-529; J. S. Greer *et al.,* in Kirk-Othmer *Encyclopedia of Chemical Technology* **vol. 18** (Wiley-Interscience, New York, 3rd ed., 1982) pp 912-920.

Soft, silvery-white metal; body-centered cubic structure; tarnishes on exposure to air; becomes brittle at low temps; mp 63.2°. bp 765.5°. d^{20} 0.856. Specific heat (0°): 0.176 cal/g deg. Thermal conductivity (cal/sec °C cm): 0.23 (21°); 0.956 (400°). Sol in liq ammonia, ethylenediamine, aniline; sol in several metals, forming alloys. One of the most active metals; E^0 (aq) K/K$^+$ 2.922 V. Reacts vigorously with oxygen; with water even at $-100°$; with acids; with the halogens, igniting with bromine and iodine. Molten metal reacts with sulfur; with hydrogen sulfide. Reacts with hydrogen slowly at 200°, rapidly at 350-400°. Reacts slowly with anhyd hydrogen halides at room temp; molten metal ignites in the reaction. Reduces silicates, sulfates, nitrates, carbonates, phosphates, oxides and hydroxides of the heavy metals, often with the separation of the metal. Reacts with organic compds containing active groups. Inert to saturated aliphatic and to aromatic hydrocarbons. *Keep under liquid containing no oxygen,* e.g., liquid petrolatum, petroleum, etc.

USE: In synthesis of inorganic potassium compds; in organic syntheses involving condensation, dehalogenation, reduction, and polymerization reactions. As heat transfer medium together with sodium: *Chem. & Eng. News* **33,** 648 (1955).

7580. Potassium Acetate. $C_2H_3KO_2$; mol wt 98.14. C 24.48%, H 3.08%, K 39.85%, O 32.60%. CH_3COOK.

Colorless, lustrous, rapidly deliquesc crystals or white cryst powder or flakes. d^{25} 1.57. mp 292°. One gram dis-

solves in 0.5 ml water, 0.2 ml boiling water, 2.9 ml alcohol. The aq soln is alkaline to litmus. pH of 0.1 molar aq soln 9.7. *Keep tightly closed.* LD_{50} orally in rats: 3.25 g/kg, H. F. Smyth *et al., Am. Ind. Hyg. Assoc. J.* **30,** 470 (1969).

THERAP CAT: Alkalizer.

THERAP CAT (VET): Has been used in cardiac arrhythmias, as an expectorant. Diuretic.

7581. Potassium Aluminate. *Aluminum potassium oxide.* $Al_2K_2O_4$; mol wt 196.16. Al 27.51%, K 39.86%, O 32.63%. Trihydrate, hard, lustrous crystals. Freely sol in water; insol in alcohol. The aq soln is strongly alkaline.

7582. Potassium *p*-Aminobenzoate. *p-Aminobenzoic acid potassium salt;* potassium *para*-aminobenzoate; KPABA; Potaba. $C_7H_6KNO_2$; mol wt 175.23. C 47.98%, H 3.45%, K 22.31%, N 8.00%, O 18.26%. Prepn: E. A. Meyers *et al., J. Am. Chem. Soc.* **89,** 3565 (1967). Crystal structure: G. S. Ciminago, *Compt. Rend.* Ser. C **267,** 1402 (1968). Antifibrotic efficacy in idiopathic pulmonary fibrosis: U. H. Cegla *et al., Pneumonologie* **152,** 75 (1975); in Peyronie's disease: A. J. Riley, *Brit. J. Sex. Med.* **6,** 29 (1970); G. Williams, N. A. Green, *Brit. J. Urol.* **52,** 392 (1980).

Crystals from alcohol. Saline taste. Slightly alkaline to litmus. pH of 1% soln about 7. Very freely sol in water, less sol in alc. Practically insol in ether. Reported to cause less gastric irritation than the free acid or the sodium salt.

USE: Catalyst in the manuf of condensation polymers of polyglycol ethers.

THERAP CAT: Antifibrotic.

7583. Potassium Arsenate. Potassium acid arsenate; potassium dihydrogen arsenate; Macquer's salt. AsH_2KO_4; mol wt 180.02. As 41.61%, H 1.12%, K 21.72%, O 35.55%. KH_2AsO_4.

Colorless crystals or white, cryst mass or powder. *Poisonous!* d 2.8. Sol in 5.5 parts cold, more sol in hot water, slowly in 1.6 parts glycerol; insol in alcohol.

USE: In the textile, tanning, and paper industries. In insecticidal formulations (especially fly paper).

7584. Potassium Arsenite. Somewhat variable composition; the article of commerce is of a composition corresponding to approx $KAsO_2.HAsO_2$.

White, hygroscopic powder; gradually dec on exposure to air (by the CO_2). *Very poisonous!* Sol in water. *Keep well closed.* LD_{50} orally in rats: 14 mg/kg, *Handbook of Toxicology* vol. 1, W. S. Spector, Ed. (Saunders, Philadelphia, 1956) pp 240-241.

USE: In manuf of mirrors to reduce the silver salt to metallic silver.

7585. Potassium Arsenite Solution. Fowler's soln; arsenical soln. Made from arsenic trioxide 10 g; potassium bicarbonate 7.6 g; alcohol 30 ml; distilled water to one liter. *Very poisonous!*

Colorless and odorless, clear liquid. It may mold. *Incompat.* Alkaloidal salts, hypophosphites and sulfites in acid soln; salts of Fe and of most other heavy metals; tannic acid.

THERAP CAT: Antineoplastic; dermatologic.

THERAP CAT (VET): Has been used as an alterative or tonic, in pulmonary emphysema, chronic coughs, anemias, chronic skin diseases.

7586. Potassium Bicarbonate. Potassium acid carbonate; Kaylox; K-Lyte. $CHKO_3$; mol wt 100.11. C 11.99%, H 1.01%, K 39.05%, O 47.94%. $KHCO_3$. Contains not less than 99% $KHCO_3$.

Colorless, transparent crystals, white granules or powder. Sol in 2.8 parts water, 2 parts water at 50°. Almost insol in alcohol. pH: 8.2 (in 0.1 molar concn).

USE: In baking powders, effervescent salts.

THERAP CAT: Potassium supplement.

7587. Potassium Bifluoride. Potassium acid fluoride; potassium hydrogen fluoride. F_2HK; mol wt 78.11. F 48.65%, H 1.29%, K 50.06%. $K\bar{F}.HF$. Prepd according to the eq $KOH + 2HF = KHF_2 + H_2O$: Lange, Eichler, *Z. Physik. Chem.* **129**, 285 (1927); Kwasnik in *Handbook of Preparative Inorganic Chemistry* vol. **1**, G. Brauer, Ed. (Academic Press, New York, 2nd ed., 1963) p 237. Made commercially from potassium carbonate and hydrofluoric acid. Tetragonal crystals. *Poisonous!* d 2.37. mp 238.7°. Transformn pt 195°. Soly in water (g/100 ml): 30.1 (10°); 39.2 (20°); 114.0 (80°). Sol in dil alc. Insol in abs alc.

USE: In the prepn of pure potassium fluoride; as an electrolyte in the manuf of fluorine; frosting glass; treating coal to prevent slag formation; flux for silver solders; catalyst in the alkylation of benzene with olefins. *Caution:* Corrosive and irritating to skin, mucous membranes.

7588. Potassium Binoxalate. Potassium acid oxalate; salt of sorrel; sal acetosella. C_2HKO_4; mol wt 128.13. C 18.75%, H 0.79%, K 30.51%, O 49.95%. KOOCCOOH. Incorrectly "*salt of lemon*". The same synonyms apply to potassium tetraoxalate.

Monohydrate, white, odorless crystals. *Poisonous!* d 2.0. Sol in 40 parts cold, 6 parts boiling water, slightly in alcohol. pH of 0.1 molar aq soln: 2.7.

USE: Removing ink stains, scouring metals, cleaning wood; photography; as mordant in dyeing; bleaching stearic acids.

7589. Potassium Biphthalate. *Phthalic acid potassium acid salt;* potassium acid phthalate; potassium hydrogen phthalate; acid potassium phthalate. $C_8H_5KO_4$; mol wt 204.22. C 47.05%, H 2.47%, K 19.14%, O 31.34%. HOOC-C_6H_4COOK. Prepd by half-neutralization of a phthalic anhydride soln: F. J. Welcher, *Organic Analytical Reagents* vol. **2** (Van Nostrand, New York, 1947) pp 75-79.

Orthorhombic crystals, stable in air. d_4^{25} 1.636. Acid reaction; pH of 0.05*M* aq soln at 25° = 4.005 (glass electrode). Sol in about 12 parts cold water, 3 parts boiling water; slightly sol in alcohol.

USE: As primary standard for preparing volumetric alkali solns, also as a buffer in pH determinations.

7590. Potassium Biselenite. Potassium hydroselenite. HKO_3Se; mol wt 167.06. H 0.60%, K 23.40%, O 28.73%, Se 47.26%. $KHSeO_3$. Prepd by boiling a soln of potassium carbonate and selenious acid and evaporating over sulfuric acid: Nilsen, *Bull. Soc. Chim.* [2] **21**, 253 (1874); *ibid.* [2] **23**, 262 (1875).

Orthorhombic prisms. Slowly loses water at 100°; at higher temps selenium oxide is liberated. Very deliquescent. Sol in water; slightly sol in alcohol.

7591. Potassium Bisulfate. Potassium acid sulfate; potassium hydrogen sulfate; sal enixum. HKO_4S; mol wt 136.17. H 0.74%, K 28.71%, O 47.00%, S 23.55%. $KHSO_4$.

White, deliquesc crystals, pieces, or granules. d 2.24. mp 197°; at higher temp loses water and is converted into pyrosulfate. Sol in 1.8 parts water, 0.85 part boiling water. *Keep well closed.*

USE: As flux in analysis of ores and siliceous compds.

THERAP CAT: Cathartic.

7592. Potassium Bisulfide. Potassium hydrosulfide; potassium hydrogen sulfide; potassium sulfhydrate. HKS; mol wt 72.17. H 1.40%, K 54.18%, S 44.42%. KHS. Prepd industrially from $Ca(HS)_2$ and K_2SO_4: Hene, **Ger.** pat. 380,-385 (1922); from H_2S and K_2S: Bassett, **U.S.** pat. **1,662,735** (1925); Strosacker, Jones, **U.S.** pat. **1,771,384** (1926 to Dow). Prepn of pure material by the action of dry H_2S upon potassium metal dissolved in abs ethanol: Rule, *J. Chem. Soc.* **99**, 558, 564 (1911); West, *Z. Kryst.* **88**, 102 (1934).

Colorless, deliquescent crystals or white, strongly hygroscopic, cryst mass. Usually present as the hemihydrate. Trigonal system. d 1.70. Rapidly becomes yellow on exposure to air with formation of polysulfides and H_2S. Becomes anhydr at 175-200°. mp 450-510° forming a dark red liquid. Heat of formation: +62.5 kcal. Heat of soln at 17°: +0.77 kcal, for hemihydrate at 16°: +0.62 kcal. Freely sol in water, alcohol.

7593. Potassium Bitartrate. Potassium acid tartrate; acid potassium tartrate; potassium hydrogen tartrate; cream of tartar; cremor tartari; faecula; faecla. $C_4H_5KO_6$; mol wt 188.18. C 25.53%, H 2.68%, K 20.78%, O 51.01%. KHC_4-H_4O_6. Obtained from the sediments in the manuf of wine, known as argols or wine lees. The salt is at least 99.5% pure. *See also* Argol and Tartaric Acid.

Colorless crystals or white, cryst powder; pleasant acidulous taste. One gram dissolves in 162 ml water, in 16 ml boiling water, 8820 ml alcohol; readily sol in dil mineral acids, in solns of alkalies or borax. Soly in water also given as about 0.4% at 10° to about 6% at 100°.

USE: Largely in baking powders; coloring metals, galvanic tinning of metals; reducer of CrO_3 in mordants for wool.

THERAP CAT: Cathartic.

THERAP CAT (VET): Laxative, diuretic.

7594. Potassium Borohydride. *Potassium tetrahydroborate.* BH_4K; mol wt 53.95. B 20.06%, H 7.47%, K 72.47%. KBH_4. Prepd from potassium tetramethoxyborohydride and diborane at low temps: Schlesinger *et al., J. Am. Chem. Soc.* **75**, 199 (1953). Commercial process from sodium borohydride and NaOH in water: Banus *et al., ibid.* **76**, 3848 (1954). Review of potassium and other metal tetrahydroborates: James, Wallbridge, *Prog. Inorg. Chem.* **11**, 99-231 (1970).

Non-hygroscopic cryst (difference from $NaBH_4$). Stable to air. d 1.11. Dec at about 500° without melting. Supports combustion. Thermally more stable and less reactive to oxidation than $NaBH_4$. Negative heat of soln in H_2O = 6.3 kcal/mol. Soly (w/w) in water at 25°: 19%; liq ammonia at 25°: 20%; ethylenediamine at 75°: 3.9%; methanol at 20°: 0.7%; DMF at 20°: 15.0%. Soly in a 4:1 water-methanol mixture: 13 g/100 g. Alkaline aq solns are stable.

USE: As of sodium borohydride.

7595. Potassium Borotartrate. Potassium tartratoborate; soluble cream of tartar; borated cream of tartar; potassium sodium borotartrate. Made by evaporating a soln of 2 parts borax and 7 parts potassium bitartrate.

White, odorless powder. Freely sol in water.

USE: Has been used in photography as a retarder for alkaline developers.

7596. Potassium Bromate. $BrKO_3$; mol wt 167.01. Br 47.85%, K 23.41%, O 28.74%. $KBrO_3$.

White crystals or granules. d 3.27. mp about 350°, decomposing at about 370° with evolution of oxygen. Sol in 12.5 parts water, 2 parts boiling water; almost insol in alc.

USE: Bread- and flour-improving agent; in analytical chemistry. *Caution:* Ingestion may cause vomiting, diarrhea, methemoglobinemia, renal injury.

7597. Potassium Bromide. BrK; mol wt 119.01. Br 67.15%, K 32.85%. KBr. Continuous electrolytic process of prepn: Maylott, Elkins, **U.S.** pat. **2,989,450** (1961 to Dow).

Colorless crystals or white granules or powder. d 2.75. mp 730°. One gram dissolves in 1.5 ml water, 1 ml boiling water, 250 ml alc, 4.6 ml glycerol. The aq soln is neutral.

Human Toxicity: Large doses cause CNS depression. Prolonged intake may cause mental deterioration, acneform skin eruptions.

USE: Manuf photographic papers and plates; process engraving.

THERAP CAT: Sedative, anticonvulsant.

THERAP CAT (VET): Sedative.

7598. Potassium Carbacrylic Resin. A generic name for the potassium form of cross-linked polyacrylic polycarboxylic ion exchange resins.

7599. Potassium Carbonate. Salt of tartar; pearl ash. CK_2O_3; mol wt 138.20. C 8.69%, K 56.58%, O 34.73%. K_2CO_3.

Hygroscopic, odorless granules or granular powder. d 2.29; mp 891°. Sol in 1 part cold, 0.7 part boiling water; practically insol in alcohol. Its aq soln is strongly alkaline. pH 11.6. *Keep tightly closed.* LD_{50} orally in rats: 1.87 g/kg, H. F. Smyth *et al., Am. Ind. Hyg. Assoc. J.* **30**, 470 (1969).

Sesquihydrate, small granular crystals. When it contains the full amount of water (16.36%) it is not hygroscopic. Sol in less than 1 part water; practically insol in alcohol. The aq soln is strongly alkaline.

USE: Manuf soap, glass, pottery, smalts and many potassium salts; in process engraving and lithography; tanning and finishing leather; liq shampoos; for removal of water from organic liqs; in anal. chemistry. *Caution:* Irritant, caustic.

THERAP CAT: Alkalizer, diuretic.

7600. Potassium Chlorate. Potcrate. $ClKO_3$; mol wt 122.55. Cl 28.93%, K 31.91%, O 39.17%. $KClO_3$. Contains at least 99% $KClO_3$.

Colorless, lustrous crystals, or white granules or powder. d 2.32. mp 368°; above this temp it dec into perchlorate and oxygen. One gram dissolves slowly in 16.5 ml water, 1.8 ml boiling water, about 50 ml glycerol; almost insol in alcohol. *Keep out of contact with organic matter or other oxidizable substances. Caution:* Explodes with sulfuric acid; inflames with explosion if triturated with any organic substances, sulfur, phosphorus, sulfite, hypophosphite, and other oxidizable substances. *Incompat.* Iodides, tartaric acid.

Human Toxicity: Irritating to G.I. tract, kidney; can cause hemolysis of red blood cells and methemoglobinemia: *Clinical Toxicology of Commercial Products,* R. E. Gosselin *et al.,* Eds. (Williams & Wilkins, Baltimore, 5th ed., 1984) Section II, p 112; Section III, pp 74-77.

USE: Explosives; fireworks; matches; printing and dyeing cotton and wool black; manuf aniline black and other dyes; source of oxygen; in chemical analyses.

THERAP CAT: Formerly as topical antiseptic.

THERAP CAT (VET): In dilute soln as antiseptic mouthwash.

7601. Potassium Chloride. Camcopot; Chloropotassuril; Chlorvescent; Diffu-K; Enseal potassium chloride; Kaleorid; Kalitabs; Kalium-Duriles; Kaon-Cl; Kaskay; Kayback; Kay-Cee-L; K-Contin; Klor-Con; K-Norm; K-Tab; Lento-Kalium; Leo K; Micro K; Nu-K; Peter-Kal; Pfiklor; Potavescent; Rekawan; Repone K; Slow-K; Span-K. ClK; mol wt 74.55. Cl 47.56%, K 52.44%. KCl. Occurs in nature as the mineral *sylvine* or *sylvite.* Industrial prepns: Faith, Keyes & Clark's *Industrial Chemicals,* F. A. Lowenheim, M. K. Moran, Eds. (Wiley-Interscience, New York, 4th ed., 1975) pp 666-673.

White crystals or crystalline powder. d 1.98. mp 773°. One gram dissolves in 2.8 ml water, 1.8 ml boiling water, 14 ml glycerol, about 250 ml alcohol. Insol in ether, acetone. Hydrochloric acid, sodium or magnesium chlorides diminish its soly in water. d of saturated aq soln at 15° is 1.172. pH: about 7.

Human Toxicity: Large doses by mouth can cause G.I. irritation, purging, weakness and circulatory disturbances.

USE: In photography. In buffer solns, electrode cells.

THERAP CAT: Electrolyte replenisher.

7602. Potassium Chromate(VI). Neutral potassium chromate. CrK_2O_4; mol wt 194.20. Cr 26.78%, K 40.26%, O 32.96%. K_2CrO_4.

Lemon-yellow crystals; d 2.73; mp 975°. Sol in 1.6 parts cold, 1.2 parts boiling water; insol in alcohol. The aq soln is alkaline to litmus or phenolphthalein.

USE: Has a limited application in enamels, finishing leather, rustproofing of metals, being replaced by the sodium salt; as reagent in analytical chemistry.

7603. Potassium Citrate. Urocit-K. $C_6H_5K_3O_7$; mol wt 306.40. C 23.52%, H 1.64%, K 38.28%, O 36.55%. $K_3C_6H_5O_7$. It is at least 99% pure.

Monohydrate, white crystals, granules or powder. Loses its water at 180°. One gram dissolves in 0.65 ml water; very slowly in 2.5 ml glycerol. Practically insol in alcohol. The aq soln is alkaline to litmus; pH about 8.5.

THERAP CAT: Antiurolithic. Antacid.

THERAP CAT (VET): Diuretic.

7604. Potassium Citrate, Monobasic. Monopotassium citrate. $C_6H_7KO_7$; mol wt 230.21. C 31.30%, H 3.07%, K 16.98%, O 48.65%. $KH_2C_6H_5O_7$.

White, cryst powder. Sol in water; the soln is subject to molding.

USE: A 0.05 molal solution as standard for pH scale (pH at 25° 3.776): Staples, Bates, *J. Res. Nat. Bur. Stand.* 73A, 37 (1969).

7605. Potassium Cobaltous Selenate. $CoK_2O_8Se_2$; mol wt 423.05. Co 13.93%, K 18.48%, O 30.26%, Se 37.33%. $K_2Co(SeO_4)_2$. Prepd by evaporating a soln of the component salts: von Hauer, *Sitzungsber. Akad. Wien* 39, 839 (1860).

Hexahydrate, garnet-red monoclinic crystals. d 2.514. Stable in air.

7606. Potassium Cyanate. CKNO; mol wt 81.11. C 14.81%, K 48.20%, N 17.27%, O 19.72%. Inhibitor of the sickling of erythrocytes *in vitro:* Cerami, Manning, *Proc. Nat. Acad. Sci. USA* 68, 1180 (1971). *See also* Sodium Cyanate. Pharmacology: Cerami *et al., J. Pharmacol. Exp. Ther.* 185, 653 (1973).

White powder. d 2.05. Sol in water, very slightly in alcohol. LD_{50} i.p. in mice: 320 mg/kg.

7607. Potassium Cyanide. CKN; mol wt 65.11. C 18.44%, K 60.05%, N 21.51%. KCN. The article of commerce contains about 95% KCN.

White, deliquesc, granular powder or fused pieces; odor of HCN. *Violent poison!* On exposure to air it is gradually dec by CO_2 and moisture. d 1.52; mp 634°. Sol in 2 parts cold, 1 part boiling water, 2 parts glycerol, 100 parts alcohol, 25 parts methanol. The aq soln is strongly alkaline and rapidly dec. pH of 0.1N aq soln: 11.0. *Keep tightly closed and protected from light. Incompat.* Acids and acid syrups; alkaloids, chloral hydrate, iodine, metallic salts, permanganates, chlorates, peroxides. LD_{50} orally in rats: 10 mg/kg, Hayes, *Toxicol. Appl. Pharmacol.* 11, 327 (1967).

Human Toxicity: Poisoning may occur by ingestion, absorption through injured skin or inhalation of hydrogen cyanide, liberated by action of carbon dioxide or other acids. Strong solns are corrosive to skin. For symptoms *see* Hydrogen Cyanide.

USE: Similar to sodium cyanide.

7608. Potassium Dichromate(VI). Potassium bichromate. $Cr_2K_2O_7$; mol wt 294.21. Cr 35.36%, K 26.58%, O 38.07%. $K_2Cr_2O_7$. In the U.S.A. it is usually prepared by the reaction of potassium chloride on sodium dichromate: Vetter in Kirk-Othmer *Encyclopedia of Chemical Technology* vol. 3 (Interscience, New York, 1949) p 951; Hartford, Copson, *ibid.* vol. 5 (2nd ed., 1964) pp 484-488. In Germany it is obtained from potassium chromate produced by roasting the chrome ore with KOH. *Ref:* Müller, Glissmann in *Ullmann's Encyklopädie der Technischen Chemie,* vol. 5 (Munich, 3rd ed., 1954) p 580.

Bright orange-red crystals. Not hygroscopic or deliquescent (difference from sodium dichromate). Crystal habit: prismatic. Crystal system: triclinic pinacoidal, transition to monoclinic at 241.6°. d_4^{25} 2.676. Bulk density: 100 lbs/cu ft. mp 398°. Dec at about 500°. Heat of fusion 29.8 cal/g. Heat of soln −62.5 cal/g. Specific heat 0.186 at 16° −98°. Soluble in water. A satd aq soln contains at 0°: 4.3%, at 20°: 11.7%, at 40°: 20.9%, at 60°: 31.3%, at 80°: 42.0%, at 100°: 50.2%. Acid reaction: A 1% aq soln has a pH of 4.04 and a 10% soln has a pH of 3.57.

Human Toxicity: Intern. a corrosive poison. Industrial contact may result in ulceration of hands, destruction of mucous membranes and perforation of nasal septum. See E. Browning, *Toxicity of Industrial Metals* (Appleton-Century-Crofts, New York, 2nd ed., 1969) pp 119-131. *See also* Chromium.

USE: In tanning leather, dyeing, painting, decorating porcelain, printing, photolithography, pigment-prints, staining wood, pyrotechnics, safety matches; for bleaching palm oil, wax, and sponges; waterproofing fabrics; as oxidizer in the manuf of organic chemicals; in electric batteries; as depolarizer for dry cells. As corrosion inhibitor in preference to sodium dichromate where lower soly is advantageous. Pharmaceutic aid (oxidizing agent).

THERAP CAT (VET): Caustic.

7609. Potassium Dicyanoaurate(I). Gold potassium cyanide; potassium aurocyanide. C_2AuKN_2; mol wt 288.13. C 8.34%, Au 68.37%, K 13.57%, N 9.72%. $KAu(CN)_2$. Prepd by electrolysis of Au in KCN: Glassford, Napier, *Phil. Mag.* 25, 61 (1844).

Dihydrate, cryst powder. One gram dissolves in 7 ml water, 0.5 ml boiling water; slightly sol in alcohol; practically insol in ether.

USE: For electroplating.

7610. Potassium Ethyl Sulfate. Potassium sulfovinate. $C_2H_5KO_4S$; mol wt 164.22. C 14.63%, H 3.07%, K 23.81%, O 38.97%, S 19.53%. $KC_2H_5SO_4$.
White crystals. Freely sol in water or alcohol.

7611. Potassium Ferricyanide. *Tripotassium hexakis-(cyano-C)ferrate(3−); potassium hexacyanoferrate(III);* red prussiate of potash. $C_6FeK_3N_6$; mol wt 329.25. C 21.89%, Fe 16.96%, K 35.62%, N 25.53%. $K_3Fe(CN)_6$.
Ruby-red crystals. d 1.89. Slowly sol in 2.5 parts cold water, in 1.3 parts boiling water; slightly sol in alc; dec by acids. The aq soln dec slowly on standing. *Protect from light.*
USE: Chiefly for blueprints; in photography; also for stain-ing wood, dyeing wool, calico printing, as etching liquid (Mercer's liquor), tempering iron and steel; in electroplat-ing; as a mild oxidizing agent in organic synthesis; in analyt-ical chemistry.

7612. Potassium Ferrocyanide. *Tetrapotassium hexakis-(cyano-C)ferrate(4−); potassium hexacyanoferrate(II);* yel-low prussiate of potash. $C_6FeK_4N_6$; mol wt 368.34. C 19.56%, Fe 15.16%, K 42.46%, N 22.82%. $K_4Fe(CN)_6$. Re-view of properties, chemistry and syntheses: *The Chemistry of Ferrocyanides,* American Cyanamid Co. (Beacon Press, New York, 1953) 112 p.
Trihydrate, soft, slightly efflorescent crystals; begins to lose water at 60°, becomes anhyd at 100°. d 1.85.

7613. Potassium Fluoride. FK; mol wt 58.10. F 32.70%, K 67.30%. KF. Prepd by thermal decompn of KHF_2 or by neutralizing HF with K_2CO_3: Lange, Eichler, *Z. Physik. Chem.* **129**, 285, 286 (1927); Kwasnik in *Handbook of Pre-parative Inorganic Chemistry* vol. 1, G. Brauer, Ed. (Aca-demic Press, New York, 2nd ed., 1963) p 236.
Cubic crystals (NaCl lattice). Usually obtained as white, deliquesc powder or solid. *Poisonous!* d 2.481. mp 859.9°. bp 1505°. Soly in water (g/100 ml): 92.3 (18°); 96.4 (21°). Very freely sol in boiling water. Also sol in aq HF, liq NH_3. Insol in alcohol unless water is present. May be stored in aluminum containers. *Caution:* Attracts moisture from the air. Aq solns corrode glass and porcelain. MLD orally in guinea pigs: 250 mg/kg.
Dihydrate, monoclinic crystals, mp 41°. Soly in water (18°) : 349.3 g/100 ml.
Tetrahydrate, crystals, mp 19.3°.
USE: In the fluorination of organic compds; in flux for hard solder; to prevent unwanted fermentations; in insecti-cide formulations; for frosting glass. *Caution:* Irritating to skin, eyes, mucous membranes.

7614. Potassium Formate. $CHKO_2$; mol wt 84.10. C 14.28%, H 1.20%, K 46.48%, O 38.04%. HCOOK.
Colorless, deliquesc granules. d 1.91. mp 167°, dec at higher temp with evolution of H_2. Sol in 0.4 part water. The aq soln is practically neutral. *Keep tightly closed.*

7615. Potassium Gluconate. *Gluconic acid potassium salt;* Gluconsan K; Kalimozan; Kaon; Katorin; Potasoral; Potassuril; K-IAO. $C_6H_{11}KO_7$; mol wt 234.24. C 30.76%, H 4.73%, K 16.69%, O 47.81%. The normal potassium salt of gluconic acid.

```
            H   H  OH H
            |   |   |  |
HOCH2 - C - C - C - C - COOK
            |   |   |  |
           OH  OH  H  OH
```

Yellowish-white crystals. Stable in air. Mild, slightly sa-line taste. Dec 180°. Freely sol in water. Practically insol in abs alcohol, ether, benzene, chloroform. Aq solns are alka-line to litmus and have a pH of 7.5-8.5.
THERAP CAT: Replenisher (electrolyte).

7616. Potassium Glycerophosphate. $C_3H_7K_2O_6P$; mol wt 302.31. C 14.51%, H 2.84%, K 31.50%, O 38.67%, P 12.48%. $K_2C_3H_7PO_6$.
Trihydrate, colorless to slightly yellow, viscous mass. Very sol in water. Alkaline to phenolphthalein. Usually

marketed as a 50-75% aq soln of more or less syrupy consis-tency and colorless to pale yellow.
THERAP CAT: Tonic.

7617. Potassium Guaiacolsulfonate. *3-Hydroxy-4-meth-oxybenzenesulfonic acid monopotassium salt mixt with mono-potassium 4-hydroxy-3-methoxybenzenesulfonate;* Thiocol; Sulfoguaiacol; Orthocoll. $C_7H_7KO_5S$; mol wt 242.29. C 34.70%, H 2.91%, K 16.14%, O 33.02%, S 13.23%. A mix-ture of the potassium salts of 4- and 5-guaiacolsulfonic acid.
White, odorless crystals or cryst powder; faint bitter taste. Gradually turns pink on exposure to air and light. Soluble in 7.5 parts water; almost insol in alcohol; insol in ether. The aq soln is neutral to litmus. *Keep well closed and pro-tected from light.*
THERAP CAT: Expectorant.

7618. Potassium Hexachloroosmate(IV). Osmium po-tassium chloride. Cl_6K_2Os; mol wt 481.13. Cl 44.22%, K 16.25%, Os 39.53%. $K_2[OsCl_6]$. Preparation: Turner *et al., Analyt. Chem.* **30**, 1708 (1958).
Dark red to almost black, cubic crystals. Freely sol in water; sparingly sol in alcohol.

7619. Potassium Hexachloroplatinate(IV). Platinic po-tassium chloride; potassium platinic chloride; potassium chloroplatinate. Cl_6K_2Pt; mol wt 486.03. Cl 43.76%, K 16.08%, Pt 40.16%. $K_2[PtCl_6]$.
Orange-yellow crystals or yellow powder. d 3.50. Slight-ly sol in cold water; sol in hot water; almost insol in alcohol.
USE: In photography.

7620. Potassium Hexacyanocobaltate(III). *Tripotassium hexakis(cyano-C)cobaltate(3−);* potassium cyanocobaltate-(III); potassium cobalticyanide; potassium cobaltihexacya-nide; cobalt potassium cyanide. $C_6CoK_3N_6$; mol wt 332.34. C 21.68%, Co 17.73%, K 35.29%, N 25.29%. $K_3[Co(CN)_6]$. Prepd by reacting cobalt acetate or cyanide with potassium cyanide, followed by air-oxidation of the cobaltocyanide formed: Biltz, Biltz, *Z. Anorg. Allgem. Chem.* **50,** 108 (1906); Grube, *Z. Elektrochem.* **32,** 561 (1926); Benedetti-Pichler, *Z. Anal. Chem.* **70,** 258 (1927); Bigelow, *Inorg. Syn.* **2,** 225 (1946).
Faintly yellow, monoclinic crystals from water. d 1.906. Melts with decompn forming an olive-green mass. Freely sol in water, acetic acid solns. Insol in alcohol. Very slight-ly sol in liq ammonia. Dec by strong mineral acids. Aq solns are slightly yellow and are regarded as non-poisonous when freshly prepd: Zwenger, *Ann.* **62,** 178 (1847). An odor of HCN develops after about one week's standing in the dark. Exposure to light causes rapid development of a yellow color, but hydrolysis is slight.

7621. Potassium Hexafluoromanganate(IV). Potassium manganese hexafluoride; potassium hexafluomanganite. F_6K_2Mn; mol wt 247.11. F 46.13%, K 31.64%, Mn 22.22%. $K_2[MnF_6]$. Prepn: Jenssen, Bandte, *Angew. Chem.* **65,** 304 (1953).
Golden-yellow, hexagonal platelets; become brown on heating, regain the original color on cooling. Hydrolyzed by water with precipitation of MnO_2.

7622. Potassium Hexafluorosilicate. Potassium fluosili-cate; potassium silicofluoride. F_6K_2Si; mol wt 220.25. F 51.76%, K 35.50%, Si 12.75%. $K_2[SiF_6]$.
White, fine powder or crystals. d 2.27. Slightly sol in cold water; insol in alcohol. In hot water it hydrolyzes to KF, HF, and silicic acid. pH of 1% aq soln 3.4.
USE: In the manuf of opalescent glass, in porcelain enam-els, in insecticides. Also used in aluminum metallurgy. *Caution:* Strong irritant. Ingestion can cause vomiting, diarrhea.

7623. Potassium Hexafluorozirconate(IV). Potassium zirconium fluoride; zirconium potassium fluoride. F_6K_2Zr; mol wt 283.41. F 40.22%, K 27.59%, Zr 32.19%. $K_2[ZrF_6]$.
Crystals. Slightly sol in cold water, sol in hot water.
USE: For manuf of metallic zirconium.

7624. Potassium Hexathiocyanatoplatinate(IV). *Dipo-tassium hexakis(thiocyanato-S)platinate(2−);* platinic potas-sium thiocyanate; potassium platinic thiocyanate. $C_6K_2N_6$-

PtS_6; mol wt 621.90. C 11.59%, K 12.57%, N 13.51%, Pt 31.39%, S 30.93%. $K_2[Pt(SCN)_6]$.

Carmine-red crystals. Sol in water.

7625. Potassium Hydroxide. Potassium hydrate; caustic potash; potassa. HKO; mol wt 56.10. H 1.80%, K 69.69%, O 28.52%. KOH. Prepd industrially by electrolysis of potassium chloride: Faith, Keyes & Clark's *Industrial Chemicals*, F. A. Lowenheim, M. K. Moran, Eds. (Wiley-Interscience, New York, 4th ed., 1975) pp 674-678. Toxicity: H. F. Smyth *et al.*, *Am. Ind. Hyg. Assoc. J.* **30**, 470 (1969).

White or slightly yellow lumps, rods, pellets. *Very caustic to tissue.* Rapidly absorbs moisture and CO_2 from the air and deliquesces. mp about 360°; mp 380° when anhydr. Sol in 0.9 part water, about 0.6 part boiling water, 3 parts alcohol, 2.5 parts glycerol. When dissolved in water or alcohol or when the soln is treated with an acid, much heat is generated. *Keep tightly closed and do not handle with bare hands.* A 0.1M aq soln has a pH of 13.5. LD_{50} orally in rats: 1.23 g/kg (Smyth).

Human Toxicity: Extremely corrosive. Ingestion may produce violent pain in throat and epigastrium, hematemesis, collapse. If not immediately fatal, stricture of esophagus may develop, *cf. Clinical Toxicology of Commercial Products*, R. E. Gosselin *et al.*, Eds. (Williams & Wilkins, Baltimore, 4th ed., 1976) Section III, pp 206-212.

USE: Manuf liq soap; mordant for wood; absorbing CO_2; mercerizing cotton; paint and varnish removers; electroplating, photoengraving and lithography; printing inks; in analytical chemistry and in organic syntheses. Pharmaceutic aid (alkalizer).

THERAP CAT (VET): Caustic. In disbudding calves' horns. In aq solution to dissolve scales and hair in skin scrapings.

7626. Potassium Hypophosphite. H_2KO_2P; mol wt 104.08. H 1.93%, K 37.57%, O 30.74%, P 29.76%. KH_2PO_2.

White crystals or granular, deliquesc powder. Odorless; pungent, saline taste. When strongly heated it dec with evolution of phosphine which ignites spontaneously in air. *It explodes when triturated with chlorates or other oxidizing agents.* One gram dissolves in 0.6 ml water, 9 ml alc, 5 ml boiling alc. The aq soln is neutral or slightly alkaline. *Keep well closed.*

7627. Potassium Iodate. IKO_3; mol wt 214.02. I 59.30%, K 18.27%, O 22.43%. KIO_3.

White, odorless crystals or cryst powder. d 3.89. mp 560° with partial decompn. Slowly sol in 12 parts water, in 3.1 parts boiling water; insol in alcohol.

USE: Oxidizing agent in volumetric chemical analysis; as maturing agent and dough conditioner.

THERAP CAT: Topical antiseptic (mucous membrane).

THERAP CAT (VET): In feeds as a source of iodine.

7628. Potassium Iodide. Jodid; KI-N; Knollide. IK; mol wt 166.02. I 76.45%, K 23.55%. KI. Potassium iodide of commerce contains about 99.5% KI. Prepd from HI and $KHCO_3$. Purification by melting in dry hydrogen: Lingane, Kolthoff, *Inorg. Syn.* **1**, 163 (1939). Continuous electrolytic process for large scale industrial prepn: Morylott, Elkins, U.S. pat. 2,989,450 (1961 to Dow). Toxicity data: Hildebrandt, *Arch. Exp. Pathol. Pharmakol.* **96**, 292 (1923). Use in the treatment of radiation poisoning resulting from a nuclear accident: W. K. Waterfall, *Brit. Med. J.* **281**, 988 (1980); *Bull. N.Y. Acad. Med.* **57**, 395 (1981).

Colorless or white, cubical crystals, white granules, or powder. Slightly deliquescent in moist air; on long exposure to air becomes yellow due to liberation of iodine, and small quantities of iodate may be formed; light and moisture accelerate the decompn. Aq solns also become yellow in time due to oxidation, but a small amount of alkali prevents it. d 3.12. mp 680° (volatilizes at higher temp). One gram dissolves in 0.7 ml water, 0.5 ml boiling water, 22 ml alcohol, 8 ml boiling alcohol, 51 ml abs alcohol, 8 ml methanol, 75 ml acetone, 2 ml glycerol, about 2.5 ml glycol. Potassium iodide solns readily dissolve elemental iodine. The aq soln is neutral or, usually, slightly alkaline. pH: 7-9. 30 g KI with 21 ml water gives 30 ml of a saturated soln at 25°. Approx LD i.v. in rats: 285 mg/kg (Hildebrandt).

Incompat. Alkaloidal salts, chloral hydrate, tartaric and other acids, calomel, potassium chlorate, metallic salts.

USE: Manuf photographic emulsions; in animal and poultry feeds to the extent of 10-30 parts per million; in table salt as a source of iodine and in some drinking water; also in anal. chemistry.

THERAP CAT: Antifungal; expectorant; iodine supplement.

THERAP CAT (VET): In actinobacillosis, actinomycosis. For simple goiter. As expectorant. In iodine deficiency and in chronic poisoning with lead or mercury. Orally only, not by injection. Externally for treatment of bursal enlargements.

7629. Potassium Manganate(VI). K_2MnO_4; mol wt 197.12. K 39.67%, Mn 27.87%, O 32.47%. Prepn: Scholder, Waterstradt, *Z. Anorg. Allgem. Chem.* **277**, 172 (1954).

Dark green crystals; dec at 190°. Sol in water. Sol and stable in KOH solns. It is an oxidizing agent. With HCl it gives free chlorine.

7630. Potassium Metabisulfite. *Potassium pyrosulfite.* $K_2O_5S_2$; mol wt 222.32. K 35.17%, O 35.98%, S 28.85%. $K_2S_2O_5$. The article of commerce contains ~95% $K_2S_2O_5$.

White crystals or cryst powder; sulfur dioxide odor; acid reaction; liberates SO_2 with acids; oxidizes in air to sulfate, more readily in presence of moisture. It may catch fire if much heat develops in powdering it. Freely sol in water; insol in alcohol. *Keep dry and well closed.*

USE: As antifermentative in breweries and wineries; bleaching straw; preservative for fruits and vegetables.

7631. Potassium Metaphosphate. Potassium Kurrol's salt; potassium polymetaphosphate; potassium polyphosphate. $(KPO_3)_x$. High mol wt polymer; degree of polymerization dependent upon preparative conditions. Prepd by dehydration of KH_2PO_4: Pfansteil, Iler, *J. Am. Chem. Soc.* **74**, 6059 (1952). Structural studies: Jost, *Acta Cryst.* **16**, 623 (1963); Jost, Schulze, *ibid.* **25B**, 1110 (1969); *eidem, ibid.* **27B**, 1345 (1971). Reviews of metaphosphates: J. R. Van Wazer, *Phosphorus and its Compounds* vol **1** (Interscience, New York, 1958) pp 601-678; Thilo, *Advan. Inorg. Chem. Radiochem.* **4**, 1-75 (1962).

White, monoclinic crystals. d^{20} 2.45. Insol in pure water. Sol in aq solns of alkali metal (except potassium) salts.

7632. Potassium Methyl Sulfate. CH_3KO_4S; mol wt 150.19. C 8.00%, H 2.01%, K 26.03%, O 42.61%, S 21.35%. KCH_3SO_4.

Hemihydrate, white crystals. Sol in water, alcohol.

USE: In organic syntheses.

7633. Potassium Molybdate(VI). K_2MoO_4; mol wt 238.14. K 32.84%, Mo 40.29%, O 26.87%.

Pentahydrate, white, deliquesc, cryst powder. d 2.3; mp 919°. Sol in 0.6 part water; insol in alc. *Keep well closed.*

7634. Potassium Nitrate. Saltpeter; niter. KNO_3; mol wt 101.10. K 38.67%, N 13.86%, O 47.47%. Contains about 99.5% KNO_3.

Colorless transparent prisms, white granular or cryst powder; cooling, saline, pungent taste. d 2.11; mp 333°; dec at 400° with evolution of O_2. One gram dissolves in 2.8 ml water, 0.5 ml boiling water, 620 ml alc. Sol in glycerol; insol in abs alc. Dissolves in water with a lowering of the temp. pH about 7. LD_{50} orally in rabbits: 1.166 g anion/kg, Dollahite, Rowe, *Southwest Vet.* **27**, 246 (1974).

Human Toxicity: Ingestion of large quantities may cause violent gastroenteritis. Prolonged exposure to small amts may produce anemia, methemoglobinemia, nephritis.

USE: In fireworks, fluxes, pickling meats; manuf glass, matches, gunpowder, blasting powders; freezing mixtures; impregnating candle wicks; treating tobacco to make it burn evenly; tempering steel.

THERAP CAT: Diuretic.

7635. Potassium Nitrite. KNO_2; mol wt 85.10. K 45.94%, N 16.46%, O 37.60%. The nitrite of commerce usually contains about 85% KNO_2, the remainder consisting chiefly of nitrate.

White or slightly yellow, deliquesc granules or rods. Dec even by weak acids with evolution of brown fumes of nitrous anhydride. d 1.915; mp 441° (decompn starts at 350°). Sol in 0.35 part water, slightly in alc. The aq soln is alkaline. *Keep well closed.* LD_{50} orally in rabbits: 108 mg anion/kg, Dollahite, Rowe, *Southwest Vet.* **27**, 246 (1974).

USE: In analytical chemistry.
THERAP CAT: Vasodilator; antidote (cyanide poisoning).

7636. Potassium Nitroprusside. *Dipotassium pentakis-(cyano-C)nitrosylferrate(2—); potassium nitrosylpentacyano-ferrate(II).* $C_5FeK_2N_6O$; mol wt 294.14. C 20.42%, Fe 18.99%, K 26.58%, N 28.57%, O 5.44%. $K_2Fe(CN)_5(NO)$.
Dihydrate, garnet-red, hygroscopic crystals. Sol in 1 part water, in alcohol. *Keep well closed.*

7637. Potassium Oleate. *Oleic acid potassium salt.* Approx $C_{18}H_{33}KO_2$.
Yellowish or brownish, soft mass. Sol in water, alc. The aq soln is alkaline to phenolphthalein.
USE: Detergent.

7638. Potassium Osmate(VI). K_2O_4Os; mol wt 332.40. K 23.53%, O 19.25%, Os 57.22%. K_2OsO_4.
Dihydrate, violet, hygroscopic crystals. *Poisonous!* Sol in water; insol in alcohol, ether. Slowly dec in aq solns with formation of the tetroxide. *Keep well closed.*

7639. Potassium Oxalate. $C_2K_2O_4$; mol wt 166.22. C 14.45%, K 47.05%, O 38.50%. $K_2C_2O_4$.
Monohydrate, colorless, odorless crystals; efflorescent in warm dry air. *Poisonous!* d 2.13. Loses its water at about 160°; when ignited is converted into carbonate without appreciable charring. Sol in 3 parts water.
USE: Cleaning and bleaching straw, removing stains in photography; in examination of blood to prevent its coagulation; also in analytical chemistry.

7640. Potassium Percarbonate. $C_2K_2O_6$; mol wt 198.22. C 12.12%, K 39.45%, O 48.43%. $K_2C_2O_6$. Prepn of practically anhydr compd: Partington, Fathallah, *J. Chem. Soc.* **1950**, 1934.
Monohydrate, white, granular mass. Sol in water with evolution of oxygen. One part potassium percarbonate is sol in 15 parts of cold water; dec in boiling water; 100 parts water dissolve 6.5 parts potassium percarbonate at ordinary temp. *Keep dry and protected from light.*
USE: Has been used in microscopy for detecting tubercle bacilli stained with fuchsin in smears; in photography under the name *Anti-hypo*, to remove last traces of sodium thiosulfate; also as oxidizing agent in chem analyses, but is no longer favored. *Caution:* Strong irritant. Causes vomiting if swallowed. Large quantities can be fatal.

7641. Potassium Perchlorate. Peroidin; Perchloracap. $ClKO_4$; mol wt 138.55. Cl 25.59%, K 28.22%, O 46.19%. $KClO_4$.
Colorless crystals or white, cryst powder. Dec at 400°; also dec by organic matter, oxidizable substances and on concussion, but is less reactive than the chlorate. d 2.52. Sol in 65 parts cold water, 15 parts boiling water; practically insol in alcohol.
USE: In explosives, pyrotechnics and photography, in analytical chemistry.

7642. Potassium Periodate. Potassium metaperiodate. IKO_4; mol wt 230.01. I 55.18%, K 17.00%, O 27.82%. KIO_4. Prepd by oxidizing potassium iodate with chlorine in alkaline soln: Hill, *J. Am. Chem. Soc.* **50**, 2678 (1928); *Inorg. Syn.* **1**, 171 (1939).
Colorless tetragonal crystals, d_4^{15} 3.618. mp 582°. Soly in water (g/100 g H_2O): 0.168 at 0°; 0.42 at 20°; 0.93 at 40°; 2.16 at 60°; 4.44 at 80°; 7.87 at 100°; also given as 0.66 at 13°. Sparingly sol in aq KOH.
USE: Powerful oxidizer in acid soln, oxidizing manganese compds to permanganate; used for this purpose in analytical chemistry (colorimetric estimation of Mn); also for the oxidation of some organic compds. *Caution:* Highly irritating to skin, eyes, mucous membranes.

7643. Potassium Permanganate. *Permanganic acid potassium salt;* chameleon mineral. $KMnO_4$; mol wt 158.03. K 24.74%, Mn 34.76%, O 40.50%. Prepn from manganese ore by electrolytic oxidation: Faith, Keyes & Clark's *Industrial Chemicals,* F. A. Lowenheim, M. K. Moran, Eds. (Wiley-Interscience, New York, 4th ed., 1975) pp 679-683.
Dark purple or bronze-like, odorless crystals. Almost opaque by transmitted light and of a blue metallic luster by reflected light. Sweet with astringent aftertaste; stable in air.

Dec about 240° with evolution of oxygen. d 2.7. Soluble in 14.2 parts cold, 3.5 parts boiling water. Dec by alc and many other organic solvents, also by concd acids with liberation of oxygen; with HCl, chlorine is liberated. Readily dec by many reducing substances, such as ferrous salts, iodides, oxalates, etc., especially in the presence of an acid. *Caution:* Take great care in handling as explosions may occur if it is brought into contact with organic or other readily oxidizable substances, either in soln or in the dry state. LD_{50} orally in rats: 1.09 g/kg, H. F. Smyth *et al., Am. Ind. Hyg. Assoc. J.* **30**, 470 (1969).
Incompat. Alcohol, arsenites, bromides, iodides, hydrochloric acid, charcoal; organic substances generally; ferrous or mercurous salts, hypophosphites, hyposulfites, sulfites, peroxides, oxalates.
Human Toxicity: Dilute solns are mildly irritating and high concns are caustic.
USE: Bleaching resins, waxes, fats, oils, straw, cotton, silk and other fibers and chamois skins; dyeing wood brown; printing fabrics; washing CO_2 in manuf mineral waters; exterminating *Oidium tuckeri;* photography; tanning leathers; purifying water; with formaldehyde soln to expel formaldehyde gas for disinfecting; as an important reagent in analytical and synthetic organic chemistry.
THERAP CAT: Topical anti-infective.
THERAP CAT (VET): Topical antiseptic, astringent, deodorant.

7644. Potassium Persulfate. $K_2O_8S_2$; mol wt 270.32. K 28.93%, O 47.35%, S 23.72%. $K_2S_2O_8$. The article of commerce contains 93-97% $K_2S_2O_8$.
Colorless or white, odorless crystals. Gradually dec, losing available oxygen; dec more quickly at higher temps; completely dec ~100°. A powerful oxidizing agent. Sol in about 50 parts water, 25 parts water at 40°; insol in alc; the aq soln dec at ordinary temp and more rapidly on warming. The aq soln is acid. *Keep well closed, in a cool place.*
USE: Bleaching fabrics, soaps; in photography under the name *Anthion* to remove last traces of thiosulfate from plates and paper; in analytical chemistry.

7645. Potassium Phenolsulfonate. *Hydroxybenzenesulfonic acid potassium salt;* potassium sulfocarbolate. $C_6H_5-KO_4S$; mol wt 212.26. C 33.95%, H 2.37%, K 18.42%, O 30.15%, S 15.10%. $C_6H_4(OH)SO_3K$.
Monohydrate, white, odorless crystals. d 1.87. Sol in water, alcohol. The aq soln is practically neutral to litmus.

7646. Potassium Phenoxide. *Phenol potassium salt;* potassium phenate; potassium phenylate; potassium carbolate. C_6H_5KO; mol wt 132.20. C 54.51%, H 3.81%, K 29.59%, O 12.10%. C_6H_5OK. Prepd from phenol and KOH in dil methanol: Kornblum, Lurie, *J. Am. Chem. Soc.* **81**, 2710 (1959).
White to reddish, hygroscopic, cryst lumps. Very sol in water; sol in alcohol. The aq soln is strongly alkaline. *Keep tightly closed.*

7647. Potassium Phosphate, Dibasic. Dipotassium phosphate; dikalium phosphate; DKP; dipotassium hydrogen phosphate. HK_2O_4P; mol wt 174.18. H 0.58%, K 44.89%, O 36.74%, P 17.79%. K_2HPO_4.
White, somewhat hygroscopic granules. Very sol in water, slightly in alcohol. 100 g will dissolve rapidly and completely in 67 g of cold water. Converted into pyrophosphate by ignition. The aq soln is slightly alkaline to phenolphthalein. *Keep well closed.*
USE: Buffering agent in antifreeze solns; nutrient in the culturing of antibiotics; ingredient of instant fertilizers; as sequestrant in the prepn of non-dairy powdered coffee creams.
THERAP CAT: Cathartic.

7648. Potassium Phosphate, Monobasic. Potassium biphosphate; potassium acid phosphate; potassium dihydrogen phosphate; monopotassium phosphate; Sörensen's potassium phosphate. H_2KO_4P; mol wt 136.09. H 1.48%, K 28.73%, O 47.03%, P 22.76%. KH_2PO_4.
Colorless crystals or white, granular powder; permanent in air; at 400° loses H_2O, forming metaphosphate. d 2.34. Sol in about 4.5 parts water. Insol in alcohol. pH 4.4-4.7.

USE: In buffers for determination of pH. Pharmaceutic aid (buffering agent).

7649. Potassium Phosphate, Tribasic. Tripotassium phosphate. K_3O_4P; mol wt 212.27. K 55.25%, O 30.15%, P 14.59%. K_3PO_4. Purification: Jänecke, Z. Physik. Chem. **127**, 75 (1927); Simon, Schulze, Z. Anorg. Allgem. Chem. **242**, 331 (1939).
Deliquescent, orthorhombic crystals. d_4^{17} 2.564. mp 1340°. Soly in water: 43.7% at 0°; 50.8% at 25°; 59.7% at 45.1°. Insol in alcohol. Aq solns are strongly alkaline. Octahydrate, flat, rectangular platelets, mp 45.1°.

7650. Potassium Phosphite. HK_2O_3P; mol wt 158.18. H 0.64%, K 49.43%, O 30.34%, P 19.59%. K_2HPO_3.
White, deliquesc powder. Slowly oxidizes in the air to phosphate; dec by heat. Very sol in water; insol in alcohol. *Keep well closed.*

7651. Potassium Picrate. *2,4,6-Trinitrophenol potassium salt;* potassium trinitrophenolate. $C_6H_2KN_3O_7$; mol wt 267.20. C 26.97%, H 0.75%, N 15.73%, O 41.92%, K 14.63%.
Yellow, reddish or greenish, lustrous needles. *Explodes when struck or heated.* Sol in 200 parts cold water, 4 parts boiling water.

7652. Potassium Pyroantimonate, Acid. *Potassium antimonate(V).* Approx $K_2H_2Sb_2O_7.4H_2O$.
White granules or cryst powder. Moderately sol in cold water; insol in alcohol.
USE: Formerly used for detection of sodium.

7653. Potassium Pyrophosphate. *Diphosphoric acid tetrapotassium salt.* $K_4O_7P_2$; mol wt 330.34. K 47.35%, O 33.90%, P 18.75%. $K_4P_2O_7$. Manuf: Faith, Keyes & Clark's *Industrial Chemicals,* F. A. Lowenheim, M. K. Moran, Eds. (Wiley-Interscience, New York, 4th ed., 1975) pp 684-687.
Trihydrate, colorless, deliquesc granules or cryst mass; freely sol in water; insol in alcohol. The aq soln is alkaline.
USE: In detergents and surfactants; in water treatment; in drilling muds as a clay thinner.

7654. Potassium Pyrosulfate. *Disulfuric acid dipotassium salt;* "anhydrous" potassium acid sulfate. $K_2O_7S_2$; mol wt 254.32. K 30.75%, O 44.04%, S 25.22%. $K_2S_2O_7$.
Colorless, fused pieces. d 2.28. mp about 325°. Sol in water. The aq soln is strongly acid.

7655. Potassium Salicylate. $C_7H_5KO_3$; mol wt 176.21. C 47.71%, H 2.86%, K 22.19%, O 27.24%. HOC_6H_4COOK.
White, odorless powder. Becomes pink on exposure to light. Very sol in water, alcohol. One gram dissolves in 0.85 ml H_2O. A satd aq soln contains 55.82% w/w at 28.5°. *See* Sidgwick, Ewbank, J. Chem. Soc. **121**, 1847, 1850 (1922). The aq soln is neutral or slightly acid to litmus. *Keep well closed and protected from light.*

7656. Potassium Selenate. K_2O_4Se; mol wt 221.15. K 35.36%, O 28.94%, Se 35.70%. K_2SeO_4.
Colorless crystals or white powder. d 3.07. Sol in about one part water.
USE: Reagent.

7657. Potassium Selenide. K_2Se; mol wt 157.15. K 49.76%, Se 50.24%. Prepd by heating selenium with an excess of potassium: Fonzes-Diacon, *Contribution a l'Etudes des Séléniures Métalliques,* Montpellier (1901); by adding selenium to a soln of potassium in liq ammonia: Hugot, *Recherches sur l'Action du Sodammonium et du Potassammonium sur Quelque Métalloides,* Paris (1900); Feher in *Handbook of Preparative Inorganic Chemistry* vol. 1, G. Brauer, Ed. (Academic Press, New York, 2nd ed., 1963) p 421; by igniting potassium selenite or selenate in an atm of hydrogen: Berzelius, cited in *Mellor's* vol. **10**, 767 (1930).
Crystalline mass. d 2.29. Reddens on exposure to air. Turns brownish-black when heated. Deliquescent. Forms potassium selenite with selenious acid. Forms potassium hydroselenide with hydrogen selenide. Sol in water; insol in ammonia.
Nonahydrate, needle-like cryst. Freely sol in water.

7658. Potassium Silicate. Soluble potash glass; soluble potash water glass. Variable composition: $K_2Si_2O_5$ to K_2-Si_3O_7; may also contain water.
Colorless or yellowish, translucent to transparent, hygroscopic, glass-like pieces; strong alkaline reaction. Usually very slowly sol in cold water, or depending on the composition, almost insol. More readily sol in water when heated with it under pressure. Insol in alcohol; dec by acids with precipitation of silica. *Keep well closed.*
USE: As binder (*e.g.,* in carbon electrodes, lead pencils, protective coatings, insol pigments); detergent, in glass and ceramics manuf.

7659. Potassium Silver Cyanide. *Potassium bis(cyano-C)argentate(1−);* potassium dicyanoargentate(I); silver potassium cyanide. C_2AgKN_2; mol wt 199.01. C 12.07%, Ag 54.21%, K 19.65%, N 14.08%. $KAg(CN)_2$.
White crystals; sensitive to light. *Poisonous!* Sol in water; acids ppt silver cyanide from the soln. *Protect from light.*
USE: In silver plating; as bactericide.

7660. Potassium Sodium Tartrate. Rochelle salt; Seignette salt. $C_4H_4KNaO_6$; mol wt 210.16. C 22.86%, H 1.92%, K 18.60%, Na 10.94%, O 45.68%. $KNaC_4H_4O_6$.
Tetrahydrate, translucent crystals or white, cryst powder; cooling saline taste. Slightly effloresces in warm air. d 1.79; mp 70-80°; at 100° loses $3H_2O$; becomes anhyd at 130-140°; at 220° begins to dec. Sol in 0.9 part water; almost insol in alcohol. The aq soln is slightly alkaline to litmus. pH 7-8. *Incompat.* Acids, calcium or lead salts, magnesium sulfate, silver nitrate.
USE: Manuf of mirrors; as a constituent of Fehling's soln. For the control of radio frequencies, and wherever piezoelectric crystals are used.
THERAP CAT: Cathartic.

7661. Potassium Sorbate. *2,4-Hexadienoic acid potassium salt; sorbic acid potassium salt;* potassium 2,4-hexadienoate; BB powder. $C_6H_7KO_2$; mol wt 150.22. C 47.97%, H 4.70%, K 26.03%, O 21.30%. $CH_3CH=CHCH=CH-COOK$. Prepd from sorbic acid and KOH: Probst, Oehme, U.S. pat. **3,173,948** (1965 to Hoechst).
Crystals, d_{20}^{25} 1.363. Dec above 270°. Soly in water at 20°: 58.2%; in alc: 6.5%.
USE: As mold and yeast inhibitor, like sorbic acid, esp where greater soly in water is desirable.

7662. Potassium Stannate(IV). K_2O_3Sn; mol wt 244.89. K 31.93%, O 19.60%, Sn 48.47%.
Trihydrate, colorless crystals. d 3.197. Sol in one part water; insol in alcohol. The aq soln is alkaline.
USE: In textile dyeing and printing, in tin plating.

7663. Potassium Stannosulfate. Marignac's salt. K_2O_8-S_2Sn; mol wt 389.02. K 20.10%, O 32.90%, S 16.49%, Sn 30.51%. $K_2Sn(SO_4)_2$.
White crystals. Partly dec by water; sol in dil fixed alkali hydroxide solns.

7664. Potassium Stearate. *Stearic acid potassium salt.* Approx $KC_{18}H_{35}O_2$. The article of commerce contains a considerable proportion of palmitate.
White powder; usually with slight odor of fat. Slowly sol in cold, readily in hot water or alcohol. The aq soln is strongly alkaline to litmus or phenolphthalein, but the alcoholic soln is only slightly alkaline to phenolphthalein.
USE: In the manuf of textile softeners.

7665. Potassium Sulfate. Sal polychrestum; arcanum duplicatum; tartarus vitriolatus. K_2O_4S; mol wt 174.26. K 44.87%, O 36.73%, S 18.40%. K_2SO_4. Contains at least 99% K_2SO_4.
Colorless or white, odorless, hard, bitter crystals, or white granules or powder; permanent in air. d 2.66; mp 1067°. One gram dissolves in 8.3 ml water, 4 ml boiling water, 75 ml glycerol; insol in alcohol. Its soly in water is decreased by KCl or $(NH_4)_2SO_4$ and is practically insol in a saturated soln of the latter. The aq soln is neutral. pH about 7.
Human Toxicity: Swallowing large doses causes severe G.I. irritation.
USE: Technical grades are used in fertilizers for manuf of potassium alum, potassium carbonate and glass; the reagent grade is used in the Kjeldahl determination of nitrogen.

THERAP CAT: Cathartic.

7666. Potassium Sulfide. Potassium monosulfide. K_2S; mol wt 110.26. K 70.92%, S 29.08%. Best prepd from the elements in liq ammonia: Klemm et al., *Z. Anorg. Allgem. Chem.* **241**, 281 (1939); Feher in *Handbook of Preparative Inorganic Chemistry* vol. **1**, G. Brauer, Ed. (Academic Press, New York, 2nd ed., 1963) pp 360-361.

White, cubic crystals or fused plates. Discolors in air. Very hygroscopic. Unstable. May explode on percussion or rapid heating. d 1.74. mp 912°. Freely sol in water. Aq solns are strongly alkaline.

Pentahydrate, colorless rhombs. Odor of hydrogen sulfide. Discolors upon exposure to light and air (yellow to yellowish red). mp 60°. Freely sol in water, alc, glycerol. Insol in ether. Densities of aq solutions at 25° (calcd as % $K_2S.5H_2O$): 1.82%: 1.009; 10.90%: 1.049; 21.80%: 1.100; 29.07%: 1.136; 39.97%: 1.192; 50.88%: 1.250; 81.77%: 1.432. Aq solns are very caustic and prone to decompn. *See* the hydrolysis reactions *under* Sodium Sulfide.

7667. Potassium Sulfite. K_2O_3S; mol wt 158.26. K 49.41%, O 30.33%, S 20.26%. K_2SO_3. The sulfite of commerce is 90-95% pure.

Dihydrate, white crystals or cryst powder; gradually oxidizes in the air to sulfate. Soluble in about 3.5 parts water, slightly in alcohol; dec by dil acids with evolution of sulfur dioxide. pH about 8. *Keep well closed in a cool place.*
USE: In photographic developers.
THERAP CAT: Cathartic, diuretic.

7668. Potassium Sulfobenzoate. 2-Sulfobenzoic acid dipotassium salt. $C_7H_4K_2O_5S$; mol wt 278.36. C 30.20%, H 1.45%, K 28.09%, O 28.74%, S 11.52%. $KSO_3C_6H_4COOK$.
White powder. Freely sol in water or alcohol.

7669. Potassium Tartrate. Soluble tartar. $C_4H_4K_2O_6$; mol wt 226.27. C 21.23%, H 1.78%, K 34.56%, O 42.43%.
Hemihydrate, white crystals or granular powder; loses its water at about 150°. d 1.98. Sol in about 0.5 part water; almost insol in alc. The aq soln is slightly alkaline to litmus; pH 7-8.
THERAP CAT: Cathartic.

7670. Potassium Tellurate(IV). Potassium tellurite. K_2O_3Te; mol wt 253.80. K 30.81%, O 18.91%, Te 50.28%. K_2TeO_3.
White, granular powder. Sol in water to an alkaline soln.
USE: In 1:50,000 soln for detecting presence of living pathogenic bacteria in sera and vaccines, such organisms forming characteristic black reduction compds.

7671. Potassium Tellurate(VI). K_2O_4Te; mol wt 269.80. K 28.98%, O 23.72%, Te 47.29%. K_2TeO_4. Crystallizes with 3 or $5H_2O$.
Trihydrate, white, cryst powder. Sol in 4 parts water.
USE: Anhidrotic.

7672. Potassium Tetraborate. Potassium biborate; potassium borate. $B_4K_2O_7$; mol wt 233.44. B 18.52%, K 33.50%, O 47.98%. $K_2B_4O_7$.
Pentahydrate, white, cryst powder. Sol in 4 parts water, slightly in alc.

7673. Potassium Tetrabromoaurate(III). Gold potassium bromide; potassium auribromide. $AuBr_4K$; mol wt 555.74. Au 35.45%, Br 57.52%, K 7.03%. Prepn: Block, *Inorg. Syn.* **4**, 14 (1953).
Dihydrate, violet crystals. Sol in water, alcohol. *Protect from light.*

7674. Potassium Tetrachloroaurate(III). Gold potassium chloride; potassium aurichloride. $AuCl_4K$; mol wt 377.90. Au 52.13%, Cl 37.52%, K 10.35%.
Dihydrate, yellow, monoclinic crystals. Sol in water. *Protect from light.*
USE: Photography, painting on porcelain and glass; prepn of several other gold compds.

7675. Potassium Tetrachloroplatinate(II). Platinous potassium chloride; potassium platinochloride; potassium chloroplatinate. Cl_4K_2Pt; mol wt 415.26. Cl 34.15%, K 18.83%, Pt 47.01%. $K_2[PtCl_4]$.

Ruby-red crystals. Sol in water.
USE: In photography, in acid toning baths.

7676. Potassium Tetracyanomercurate(II). *Dipotassium tetrakis(cyano-C)mercurate(2−)*; mercuric potassium cyanide. $C_4HgK_2N_4$; mol wt 382.87. C 12.55%, Hg 52.40%, K 20.42%, N 14.63%. $K_2[Hg(CN)_4]$.
Colorless or white crystals. Sol in water. *Violent poison!*
USE: In manuf of mirrors to prevent the silver coating from becoming yellow; as reagent in testing for free acids.

7677. Potassium Tetracyanonickelate(II). *Dipotassium tetrakis(cyano-C)nickelate(2−)*; nickel potassium cyanide. $C_4K_2N_4Ni$; mol wt 240.96. C 19.94%, K 32.45%, N 23.25%, Ni 24.36%. $K_2[Ni(CN)_4]$.
Monohydrate, orange-yellow, cryst powder. *Poison!* Loses H_2O at about 100°. Sol in water.

7678. Potassium Tetracyanoplatinate(II). *Dipotassium tetrakis(cyano-C)platinate(2−)*; platinous potassium cyanide; potassium platinocyanide. $C_4K_2N_4Pt$; mol wt 377.36. C 12.73%, K 20.72%, N 14.85%, Pt 51.70%. $K_2[Pt(CN)_4]$.
Crystallizes as a tri- or a dodecahydrate.
Trihydrate, almost colorless, rhombic prisms; blue in direction of principal axis. Sol in hot water.

7679. Potassium Tetracyanozincate. *Dipotassium tetrakis(cyano-C)zincate(2−)*; zinc potassium cyanide. $C_4K_2N_4Zn$; mol wt 247.60. C 19.40%, K 31.58%, N 22.63%, Zn 26.41%. $K_2[Zn(CN)_4]$.
Cryst powder. *Poison!* Freely sol in water.

7680. Potassium Tetrafluoroborate. Potassium borofluoride; potassium fluoborate; avogadrite. BF_4K; mol wt 125.92. B 8.59%, F 60.36%, K 31.05%. KBF_4. Prepared according to the eq $H_3BO_3 + 4HF + KOH = KBF_4 + 4H_2O$: Vorländer et al., *Ber.* **65**, 535 (1932); Kwasnik in *Handbook of Preparative Inorganic Chemistry* vol. **1**, G. Brauer, Ed. (Academic Press, New York, 2nd ed., 1963) p 223. For other methods of prepn *see* H. S. Booth, D. R. Martin, *Boron Trifluoride and its Derivatives* (New York, 1949) pp 99-106.
Orthorhombic bipyramidal or cubic crystals. d_4^{20} 2.505. mp 530°. Soly in water (g/100 g): 0.3 (3°); 0.448 (20°); 0.55 (25°); 1.4 (40°); 6.27 (100°). Index of refraction of solns is lower than that of water. A satd soln (0.6%) on heating gives color effects, if excess crystals are added. With a 10% aq soln a transparent blue color appears at 100°, turning green at 90°, and yellow at 60°. In a concd fluoboric acid soln these phenomena occur with variations in room temp. Aq solns of KBF_4 are at first neutral to litmus, but upon standing, diluting, or heating become acidic without etching their glass containers. Slightly sol in boiling alc.
USE: Has been proposed as a flux for soldering and brazing; filler in resin-bonded grinding wheels.

7681. Potassium Tetraiodoaurate(III). Gold potassium iodide; potassium auric iodide. AuI_4K; mol wt 743.74. Au 26.51%, I 68.24%, K 5.26%.
Black, lustrous crystals. Sol in water with decompn and liberation of iodine. *Protect from light.*

7682. Potassium Tetraiodocadmate. *Dipotassium tetraiodocadmate(2−)*; cadmium potassium iodide; potassium cadmium iodide; potassium iodocadmate; Marme's reagent. CdI_4K_2; mol wt 698.21. Cd 16.10%, I 72.70%, K 11.20%. $K_2[CdI_4]$. Dihydrate prepd from aq solns of stoichiometric amounts of CdI_2 and KI; must be recrystallized from water at low temps: Croft, *Phil. Mag.* **21**, 355 (1842); *J. Prakt. Chem.* **58**, 399 (1856); Hittorf, *Pogg. Ann.* **106**, 525 (1859); Eder, *Photogr. Mitt.* **13**, 67 (1867); Rimbach, *Ber.* **38**, 1562 (1905).
Dihydrate, large, water-clear, somewhat distorted octahedra. Deliquesc upon standing in air. Turns yellow with age. d_4^{21} 3.359. One part (w/w) dissolves at 15° in 0.73 part water, in 1.4 part alcohol, in 24.5 parts ether, in 4.5 parts of a 1:1 alcohol-ether mixture; also sol in ethyl acetate.
USE: In testing for alkaloids, amines, imines, heterocyclic nitrogens, peroxides in ether. In chelatometric quantitative organic analysis: *Chim. Anal. (Paris)* **43**, 449 (1961).

7683. Potassium Tetraiodomercurate(II). Mercuric po-

tassium iodide. HgI_4K_2; mol wt 786.48. Hg 25.51%, I 64.55%, K 9.94%. Also crystallizes with 1-2 mols H_2O.
Sulfur-yellow crystals; deliquesc in moist air. *Poison!* Very sol in water; sol in alcohol, ether, acetone. *Keep well closed.*
THERAP CAT: Topical anti-infective, disinfectant.
THERAP CAT (VET): Topical antiseptic, disinfectant.

7684. Potassium Tetroxalate. Potassium quadroxalate; sal acetosella; salt of sorrel. $C_4H_3KO_8$; mol wt 218.16. C 22.02%, H 1.39%, K 17.92%, O 58.67%. $KHC_2O_4 \cdot H_2C_2O_4$. Incorrectly called *"essential salt of lemons"*.
Dihydrate, colorless or white crystals; permanent in air. Sol in 60 parts cold, 12 parts boiling water; slightly sol in alcohol.
USE: Removing rust and ink spots; in metal polishes; in analytical chemistry.

7685. Potassium Thioantimonate(V). Potassium sulfantimonate. K_3S_4Sb; mol wt 367.29. K 31.94%, S 34.91%, Sb 33.15%. K_3SbS_4.
Heminonahydrate, colorless to yellowish crystals. Sol in water; insol in alcohol. The aq soln is strongly alkaline. *Keep well closed in a cool place.*

7686. Potassium Thiocarbonate. Potassium sulfocarbonate; potassium trithiocarbonate. CK_2S_3; mol wt 186.41. C 6.44%, K 41.95%, S 51.61%. K_2CS_3.
Yellowish-red, deliquesc granules or crystals. Very sol in water. The aq soln is strongly alkaline. *Keep tightly closed.*
USE: In analytical chemistry. Monograph: K. N. Johri, *Chemical Analysis without H_2S using Potassium Trithiocarbonate* (Asia Pub., New York, 1963) 107 pp. *Caution:* Irritant, caustic.

7687. Potassium Thiocyanate. Potassium sulfocyanate; potassium rhodanide; Rhocya. CKNS; mol wt 97.18. C 12.36%, K 40.23%, N 14.41%, S 33.00%. KSCN.
Colorless, deliquesc crystals. d 1.89; mp about 173°, the fused salt turning brown, then green, blue, and white again on cooling. One gram dissolves in 0.5 ml acetone, 12 ml alcohol, 8 ml boiling alcohol. When dissolved in its own wt of water, the temp drops about 30°. The aq soln is neutral. *Keep well closed.* LD_{50} orally in mice, rats: 594, 854 mg/kg, Andersen, Chen, *J. Am. Pharm. Assoc.* **29**, 152 (1940).
Human Toxicity: May cause skin eruptions, psychosis.
USE: Manuf artificial mustard oil; printing and dyeing textiles; in photography as intensifier; in analytical chemistry. The sodium salt now is replacing it for most of these uses.
THERAP CAT: Hypotensive.

7688. Potassium Thiosulfate. Potassium hyposulfite. $K_2O_3S_2$; mol wt 190.32. K 41.08%, O 25.22%, S 33.70%. $K_2S_2O_3$. May crystallize with 0.33 to 1.5 mol H_2O.
Colorless, hygroscopic crystals. Sol in water; insol in alcohol. *Keep well closed.*

7689. Potassium Titanyl Oxalate. *Dipotassium bis[ethanedioato(2−)-O,O′]oxotitanate(2−);* titanium potassium oxalate; titanyl potassium oxalate. $C_4K_2O_9Ti$; mol wt 318.14. C 15.10%, K 24.58%, O 45.26%, Ti 15.06%. $K_2TiO(C_2O_4)_2$.
Crystals or cryst powder. Very sol in water.
Dihydrate, *potassium oxodioxalatodiaquotitanate(IV).*
USE: As mordant in dyeing cotton and leather.

7690. Potassium Triiodide. I_3K; mol wt 419.80. I 90.69%, K 9.31%. Monohydrate prepd by adding the stoichiometric amount of iodine to a hot soln of KI and cooling to 0°: Wells, Wheeler, *Z. Anorg. Allgem. Chem.* **1**, 453 (1892); Foote, Chalker, *J. Am. Chem. Soc.* **39**, 565 (1908).
Reasonably stable only as the monohydrate, dark brown, hygroscopic, monoclinic prisms. d_4^{15} 3.498. mp 38° (closed tube). Gives off iodine at 225°, leaving KI. "Incongruently" sol in water, sol (with partial decompn) in alcohol, ether.

7691. Potassium Triiodomercurate(II) Solution. Mercuric potassium iodide soln; potassium mercuriiodide soln; soln potassium iodohydrargyrate; Channing's soln; Thoulet's soln. Prepd by dissolving 1 g HgI_2 and 0.8 g KI in water to make 100 ml. *Poison!*
USE: As reagent for alkaloids.
THERAP CAT: Topical antiseptic, disinfectant.

7692. Potassium Triiodozincate. Potassium zinc iodide; zinc potassium iodide. I_3KZn; mol wt 485.24. K 8.06%, I 78.47%, Zn 13.47%. $KZnI_3$.
Very hygroscopic crystals. Very sol in water. *Keep well closed.*

7693. Potassium Tungstate(VI). K_2O_4W; mol wt 326.06. K 23.98%, O 19.63%, W 56.40%. K_2WO_4. Crystallizes also with $2H_2O$.
Heavy, deliquesc, cryst powder. d 3.12. mp 921°. Sol in about 2 parts cold, about 0.7 part boiling water; insol in alcohol. *Keep well closed.*

7694. Potassium Uranate(VI). Potassium diuranate; uranium oxide orange. $K_2O_7U_2$; mol wt 666.33. K 11.74%, O 16.81%, U 71.46%. $K_2U_2O_7$.
Orange powder. Insol in water; sol in acids.
USE: Painting on porcelain.

7695. Potassium Uranyl Nitrate. Uranyl potassium nitrate. $KN_3O_{11}U$; mol wt 495.19. K 7.90%, N 8.49%, O 35.54%, U 48.08%. $K(UO_2)(NO_3)_3$.
Greenish-yellow, cryst powder. Sol in about 1 part water.

7696. Potassium Uranyl Sulfate. Uranyl potassium sulfate. $K_2O_{10}S_2U$; mol wt 540.35. K 14.47%, O 29.61%, S 11.87%, U 44.05%. $K_2(UO_2)(SO_4)_2$.
Dihydrate, greenish-yellow, cryst powder. Freely sol in water.

7697. Potassium Xanthogenate. *Carbonodithioic acid O-ethyl ester potassium salt; ethylxanthic acid potassium salt;* potassium ethyldithiocarbonate; potassium ethylxanthogenate; potassium ethylxanthate. $C_3H_5KOS_2$; mol wt 160.30. C 22.48%, H 3.14%, K 24.39%, O 9.98%, S 40.01%. $C_2H_5-OCS_2K$. Made by treating an alcoholic soln of CS_2 with alcoholic KOH. Usually contains 8-10% H_2O.
White to pale yellow crystals or cryst powder. Very sol in water; sol in alcohol. The aq soln is strongly alkaline. *Keep well closed and protected from light.*
USE: As reagent in analytical chemistry.

7698. Potassium Zinc Sulfate. Zinc potassium sulfate. $K_2O_8S_2Zn$; mol wt 335.71. K 23.29%, O 38.13%, S 19.10%, Zn 19.48%. $K_2Zn(SO_4)_2$.
Hexahydrate, crystals. Sol in water.

7699. Potassium Zirconium Sulfate. Zirconium potassium sulfate. $K_4O_{16}S_4Zr$; mol wt 631.88. K 24.75%, O 40.51%, S 20.30%, Zr 14.44%. $K_4Zr(SO_4)_4$.
Trihydrate, $K_4Zr(SO_4)_4 \cdot 3H_2O$, cryst powder. Sparingly sol in water; the aq soln is acid to litmus.

7700. Povidone. *1-Ethenyl-2-pyrrolidinone polymers; 1-vinyl-2-pyrrolidinone polymers;* poly[1-(2-oxo-1-pyrrolidinyl)ethylene]; polyvinylpyrrolidone; polyvidone; P.V.P.; RP 143; Kollidon; Peregal ST; Periston; Plasdone; Plasmosan; Protagent; Subtosan; Vinisil. Solns are known as *Haemodyn.* Produced commercially as a series of products having mean mol wts ranging from about 10,000 to 700,000. Prepared by Reppe's process: 1,4-butanediol obtained in the Reppe butadiene synthesis is dehydrogenated over copper at 200° forming γ-butyrolactone; reaction with ammonia yields pyrrolidone. Subsequent treatment with acetylene gives the vinyl pyrrolidone monomer. Polymerization is carried out by heating in the presence of H_2O_2 and NH_3. Cf. P. B. reports **163; 1288**; also DeBell *et al., German Plastics Practice* (Springfield, 1946); Hecht, Weese, *Münch. Med. Wochenschr.* **1943,** 11; Weese, *Naturforschung & Medizin* **62**, 224 (Wiesbaden, 1948), and the corresp vol. of *FIAT Review of German Science.* Monographs: General Aniline and Film Corp., *PVP* (New York, 1951); W. Reppe, *Polyvinylpyrrolidon* (Monographie zu *"Angewandte Chemie"* no. 66, Weinheim/Bergstr., 1954).

Faintly yellow solid resembling albumin, but does not give the reactions of albumin. Sol in water giving a colloidal soln. Also sol in alcohol, chloroform. Practically insol in ether. A 3.5% soln develops an osmotic pressure of about 400 mm water.

USE: Pharmaceutic aid (dispersing and suspending agent). Proposed as clarifying agent in wines.

7701. Povidone-Iodine. *1-Ethenyl-2-pyrrolidinone homopolymer compd with iodine; 1-vinyl-2-pyrrolidinone polymers, iodine complex;* iodine-polyvinylpyrrolidone complex; polyvinylpyrrolidone-iodine complex; PVP-1; Betadine; Betaisodona; Braunol; Braunosan H; Disadine D.P.; Disphex; Efo-Dine; Inadine; Isodine; Proviodine; Traumasept; Videne. An iodophor, *q.v.,* prepd by Beller, Hosmer, U.S. pat. **2,706,701**; Hosmer, U.S. pat. **2,826,532**; Siggia, U.S. pat. **2,900,305** (1955, 1958, and 1959, all to GAF). Prepn, history and use: Shelanski, Shelanski, *J. Int. Coll. Surgeons* **25**, 727 (1956).

Yellowish-brown, amorphous powder with slight characteristic odor. Aq solns have a pH near 2 and may be made more neutral (but less stable) by the addition of sodium bicarbonate. Sol in alc, water. Practically insol in chloroform, carbon tetrachloride, ether, solvent hexane, acetone. Solns do not give the familiar starch test when freshly prepared.

THERAP CAT: Topical anti-infective.

7702. PPACK. D-*Phenylalanyl-N-[4-[(aminoiminomethyl)amino]-1-(chloroacetyl)butyl]-L-prolinamide;* D-phenylalanyl-prolyl-arginine chloromethyl ketone; FPRMeCl. $C_{21}H_{31}ClN_6O_3$; mol wt 450.97. C 55.93%, H 6.93%, Cl 7.86%, N 18.64%, O 10.64%. Selective irreversible inhibitor of thrombin: C. A. Kettner, E. N. Shaw, *Thromb. Res.* **14**, 969 (1979). Synthesis: E. N. Shaw, C. A. Kettner, U.S. pat. **4,318,904** (1982 to Research Corp.). Anticoagulant and antithrombotic activity: J. Hauptman, F. Markwardt, *Thromb. Res.* **20**, 347 (1980). *In vivo* studies: D. Collen *et al., J. Lab. Clin. Med.* **99**, 76 (1982).

$$\underline{\text{D-Phe-L-Pro-L-Arg-CH}_2\text{Cl}}$$

Note: Commercially available as the dihydrochloride.
USE: As a research tool; as a diagnostic tool to measure the level of thrombin in blood.

7703. Practolol. *N-[4-[2-Hydroxy-3-[(1-methylethyl)-amino]propoxy]phenyl]acetamide; 4'-[2-hydroxy-3-(isopropylamino)propoxy]acetanilide;* 1-(4-acetamidophenoxy)-3-isopropylamino-2-propanol; AY 21011; ICI 50172; Dalzic; Eraldin. $C_{14}H_{22}N_2O_3$; mol wt 266.34. C 63.13%, H 8.33%, N 10.52%, O 18.02%. β-Adrenergic blocker. Prepn: Howe, Smith, **Neth. pat. Appl. 6,512,676** corresp to U.S. pat. **3,408,387** (1966, 1968, to I.C.I.). Synthesis of *R(+)*-form: Danilewicz, Kemp, *J. Med. Chem.* **16**, 168 (1973). Pharmacological studies: Dunlop, Shanks, *Brit. J. Pharmacol. Chemother.* **32**, 201 (1968); Barrett, *Postgrad. Med. J., Suppl.* **47**, 7 (1971). Metabolic studies: Scales, Cosgrove, *J. Pharmacol. Exp. Ther.* **175**, 338 (1970).

mp 134-136° (BuOAc). Sol in warm isopropanol. Hydrochloride monohydrate, mp 140-142°. *R(+)*-Form, crystals from dioxane, mp 130-131.5°. $[\alpha]_{365}^{25}$ +4.3°, $[\alpha]_{578}^{25}$ +3.5° (ethanol). *R(+)*-Hydrochloride, $[\alpha]_{436}^{25}$+26.0°, $[\alpha]_{578}^{25}$ +14.0°. THERAP CAT: Antiarrhythmic.

7704. Prajmaline. *17,21-Dihydroxy-4-propylajmalanium; N⁴-propylajmalinium;* prajmalium; N-propylajmaline. $[C_{23}H_{33}N_2O_2]^+$; mol wt 369.52. Prepn: Keck, *Z. Naturforsch.* **18b**, 177 (1963); Bonati, Bombardelli, *Farmaco Ed. Sci.* **18**, 851 (1963); Keck, Ger. pat. **1,154,120** (1963 to Thomae) corresp to U.S. pat. **3,414,577** (1968 to Boehringer, Ing.). Pharmacology: Koch, *Arzneimittel-Forsch.* **22**, 2079 (1972); Mertens *et al., ibid.* **23**, 642 (1973); Diederich, Boyk-

sen, *ibid.* 1302. Toxicology: Von Philipsborn, Stalder, *ibid.* **22**, 2085 (1972). Pharmacokinetics: A. T. Trompler *et al., ibid.* **33**, 436 (1983).

Hydrogen tartrate, $C_{27}H_{38}N_2O_8$, *prajmaline bitartrate, NPA, NPAB, GT-1012, Neo-Gilurytmal.* White crystals from ethanol + ether, mp 149-152°. LD_{50} in mice (mg/kg): 43 orally; 1.7 i.v. (Von Philipsborn, Stalder).
THERAP CAT: Anti-arrhythmic.

7705. Pralidoxime Chloride. *2-[(Hydroxyimino)methyl]-1-methylpyridinium chloride; 2-formyl-1-methylpyridinium chloride oxime;* 1-methyl-2-formylpyridinium chloride oxime; N-methylpyridinium-2-aldoxime chloride; 2-pyridine aldoxime methyl chloride; 2-PAM chloride; Protopam chloride. $C_7H_9ClN_2O$; mol wt 172.63. C 48.71%, H 5.25%, Cl 20.54%, N 16.23%, O 9.27%. Prepn of salts: I. B. Wilson *et al.,* U.S. pat. **2,816,113** (1957 to U.S. Sec'y. of Army); A. A. Kondritzer *et al., J. Pharm. Sci.* **50**, 109 (1961). Manufacturing processes: R. I. Ellin *et al.,* U.S. pat. **3,140,289** (1964 to U.S. Dept. of the Army); McDowell, U.S. pat. **3,155,674** (1964 to Olin Mathieson). Commercial prepn: R. I. Ellin, *Ind. Eng. Chem. Prod. Res. Develop.* 3 (1), 20 (1964). Toxicology: Christensen, Richter, *Arch. Environ. Health* **15**, 599 (1967). Toxicity data: Fleisher *et al., Toxicol. Appl. Pharmacol.* **16**, 40 (1970); R. I. Ellin, J. H. Wills, *J. Pharm. Sci.* **53**, 1143 (1964). Comprehensive description: U. V. Banakar, U. N. Patel in *Analytical Profiles of Drug Substances* **vol. 17**, K. Florey, Ed. (Academic Press, New York, 1988) pp 533-569.

Crystals from alcohol + ether, mp 235-238° (dec). Soly (g/100 ml) 25°: acetone 0, isopropanol 0.09, ethanol 0.89, methanol 8.5, water 65.5. LD_{50} in rats (mg/kg): 96 i.v. (Fleisher). LD_{50} in rabbits (mg/kg): 95 i.v.; LD_{50} in mice (mg/kg): 115 i.v., 205 i.p., 4100 orally (Ellin, Wills).

Pralidoxime iodide, $C_7H_9IN_2O$, *2-pyridine aldoxime methiodide, 2-PAM, Protopam iodide.* Yellow crystals from alcohol, mp 225-226°. Soly at 25°: 48 mg/ml. Very sol in water, fairly sol in hot alcohol, poorly sol in cold alcohol. Insol in ether, acetone. LD_{50} in mice (mg/kg): 140-178 i.v., 136-260 i.p., 290-340 s.c., 1500-4000 orally (Ellin, Wills).

Pralidoxime mesylate, $C_8H_{12}N_2O_4S$, *Protopam methanesulfonate, Contrathion, P2S.* Very hygroscopic crystals from ethanol, mp 155°. pKa 8.0. Soly in water: 1 g in 2 ml. LD_{50} in mice (mg/kg): 118-122 i.v., 216 i.p., 3700 orally; in rats (mg/kg): 109 i.v., 262 i.v. (Ellis, Wills).

THERAP CAT: Cholinesterase reactivator. Antidote (nerve gases and cholinesterase-inhibitor type insecticides.)

7706. Pramiverin. *N-(1-Methylethyl)-4,4-diphenylcyclohexanamine; N-isopropyl-4,4-diphenylcyclohexylamine;* pramiverine; primaverine; primaverin; propaminodiphen. $C_{21}H_{27}N$; mol wt 293.46. C 85.95%, H 9.28%, N 4.77%. Prepn: **Neth. pat. Appl. 6,515,046** corresp to R. Unger *et al.,* U.S. pat. **3,376,-312** (1966, 1968 both to E. Merck). Series of articles on analysis, metabolism, distribution, kinetics, pharmacology, toxicology, clinical and exptl studies: *Arzneimittel-Forsch.* **26**, 686-752 (1976). Use in management of labor: G. LoDico *et al., Minerva Ginecol.* **31**, 683 (1979).

Liq, $bp_{0.05}$ 164-165°; also reported as a solid, mp 70° [H. J. Enenkel *et al.*, *Arzneimittel-Forsch.* **26**, 690 (1976)].

Hydrochloride, $C_{21}H_{28}ClN$, *EMD 9806, HSP 2986, Monoverin, Sistalgin.* Off-white cryst powder, mp 230°; also reported as 234-237° (H. J. Enenkel *et al.*, *loc. cit.*). LD_{50} (after 14 days) in mice, rats (mg/kg): 346, 623 orally, 25, 26 i.v., M. Von Eberstein *et al.*, *Arzneimittel-Forsch.* **26**, 703 (1976).

THERAP CAT: Antispasmodic.

7707. Pramoxine. *4-[3-(4-Butoxyphenoxy)propyl]morpholine; p-butoxyphenyl γ-morpholinopropyl ether; γ-morpholinopropyl 4-n-butoxyphenyl ether; pramocaine; proxazocain; Tronothane.* $C_{17}H_{27}NO_3$; mol wt 293.39. C 69.59%, H 9.28%, N 4.77%, O 16.36%. Prepn: Wright, Moore, *J. Am. Chem. Soc.* **73**, 2281 (1951); **76**, 4396 (1954); U.S. pat. 2,870,151 (1959 to Abbott).

bp_6 196°; $bp_{2.8}$ 183-184°.

Hydrochloride, $C_{17}H_{28}ClNO_3$, *Proctofoam, Tronolane.* Crystals, mp 181-183°. Sol in water. LD_{50} i.v. in mice: 79.5 mg/kg.

THERAP CAT: Hydrochloride as topical anesthetic.

7708. Pranoprofen. *α-Methyl-5H-[1]benzopyrano[2,3-b]pyridine-7-acetic acid; 2-(5H-[1]benzopyrano[2,3-b]pyridin-7-yl)propionic acid; Y-8004; Niflan; Nifran.* $C_{15}H_{13}NO_3$; mol wt 255.28. C 70.58%, H 5.13%, N 5.49%, O 18.80%. Prepn: M. Nakanishi *et al.*, *Ger. pat.* 2,337,052; *eidem*, U.S. pat. 3,931,205 (1974, 1976 both to Yoshitomi); *eidem*, *Yakugaku Zasshi* **96**, 99 (1976), *C.A.* **84**, 135515b (1976). Physicochemical properties: M. Nobutoki, Y. Ota, *Iyakuhin Kenkyu* **7**, 200 (1976), *C.A.* **88**, 110468h (1978). Absorption, metabolism, distribution, excretion: Y. Kato *et al.*, *Yakuri To Chirgo* **4**, 1473 (1976), *C.A.* **88**, 130710j (1978). Pharmacological study: Y. Maruyama *et al.*, *Nippon Yakurigaku Zasshi* **73**, 113 (1977), *C.A.* **88**, 276m (1978). Toxicity study: M. Edanaga *et al.*, *Iyakuhin Kenkyu* **7**, 211 (1976), *C.A.* **88**, 131015y (1978).

Cryst from aq dioxane, mp 182-183°. LD_{50} in male mice, rats (mg/kg): 447.3, 87.3 orally (Edanaga).

THERAP CAT: Anti-inflammatory.

7709. Praseodymium. Pr; at. wt 140.9077; at. no. 59; valences 3, 4. A lanthanide; belongs to the cerium group of rare earths. One naturally occurring isotope: ^{141}Pr; artificial, radioactive isotopes: 134-140; 142-149. Radioactive isotopes produced by bombardment: Seaborg, *Chem. Rev.* **27**, 199 (1940); Huber *et al.*, *Helv. Phys. Acta* **18**, 221 (1945). Abundance in earth's crust: 3.5-5.5 ppm. Found in rare earth minerals. Reported in 1843 by Mosander as *didymium* which was a mixture of praseodymium and neodymium; separated in 1885 by von Welsbach. Separated from other rare earth elements by fractional crystn. Also by ion exchange: Spedding *et al.*, *J. Am. Chem. Soc.* **69**, 2786, 2812 (1947). Prepn of the metal by electrolysis of the fused chloride at 1000°: Muthmann, Weiss, *Ann.* **331**, 1 (1904); by electrolysis of a eutectic mixture of praseodymium chloride, potassium chloride, and sodium chloride, using a graphite anode and tungsten cathode: Canneri, Rossi, *Gazz. Chim. Ital.* **62**, 1160 (1932); Mazza, *Atti X Congr. Int. Chim.* **3**, 604 (1939); by electrolysis of fused rare earth with alkali metal

salts in presence of Mg and Cd: Trombe, Mahn, *Compt. Rend.* **220**, 778 (1945). Reviews of prepn, properties and compds of praseodymium and other lanthanides: *The Rare Earths,* F. H. Spedding, A. H. Daane, Eds. (Krieger, Huntington, N.Y., 1971, reprint of 1961 ed.) 641 pp; Hulet, Bode, "Separation Chemistry of the Lanthanides and Transplutonium Actinides" in *MTP Int. Rev. Sci.: Inorg. Chem.*, *Ser. One*, Vol. 7, K. W. Bagnall, Ed. (Univ. Park Press, Baltimore, 1972) pp 1-45; Moeller, "The Lanthanides", *Comprehensive Inorganic Chemistry*, Vol. 4, J. C. Bailar, Jr. *et al.*, Eds. (Pergamon Press, Oxford, 1973) pp 1-101.

Yellowish metal, forms oxide film on exposure to moist air. Two cryst forms: α-form, hexagonal close-packed lattice, d 6.77; β-form, body-centered cubic lattice, forms at 800°, d 6.64. mp 935°, bp 3290° ± 90°. E^0 (aq) Pr^{3+}/Pr −2.47 V (calc). Experimental reduction potentials (referred to a normal calomel electrode): −1.875, −1.990 V: Noddack, Brukl, *Angew. Chem.* **50**, 362 (1937).

Oxide, Pr_2O_3, d 7.07, prepd by reducing with hydrogen the oxides PrO_2 or Pr_6O_{11}. Oxidizes to Pr_6O_{11} on heating in air; forms the dioxide, PrO_2, on fusing with potassium chlorate.

Hydroxide, $Pr(OH)_3$, a gelatinous pale green precipitate, obtained by the action of alkali hydroxide on a soln of a praseodymium salt; a purple powder when obtained by the action of water on praseodymium carbide.

Chloride, $PrCl_3$, heptahydrate, green crystals; mp in its water of crystn at 111°; on partial dehydration yields the hexa-, tri-, or monohydrate; on heating to 180-200° in a stream of hydrogen chloride yields the anhydr chloride, green needles, mp 769-782°, sol in water, alcohol. LD_{50} in mice: 600 mg/kg i.p.; 4.5 g/kg orally, Haley *et al.*, *Toxicol. Appl. Pharmacol.* **6**, 614 (1964).

Sulfate, $Pr_2(SO_4)_3$, light green crystals, prepd by evaporating a soln of the oxide in H_2SO_4 and cooling the resulting product over phosphorus pentoxide. Several hydrates and double sulfates are known.

7710. Prasterone. *3-Hydroxyandrost-5-en-17-one; dehydroepiandrosterone; dehydroisoandrosterone; trans-dehydroandrosterone; Δ⁵-androsten-3β-ol-17-one; 17-Hormoforin; Astenile; Deandros; Diandron; Psicosterone.* $C_{19}H_{28}O_2$; mol wt 288.41. C 79.12%, H 9.79%, O 11.09%. Isoln from male urine: Butenandt, Tscherning, *Z. Physiol. Chem.* **229**, 167 (1934); Butenandt, Dannenbaum, *ibid.* 192. Prepn from cholesterol: Butenandt *et al.*, *ibid.* **237**, 57 (1935); Ruzicka, Wettstein, *Helv. Chim. Acta* **18**, 986 (1935); Wallis, Fernholz, *J. Am. Chem. Soc.* **57**, 1379, 1504 (1935); Schoeller *et al.*, *Naturwiss.* **23**, 337 (1935); from sitosterol: Oppenauer, *Nature* **135**, 1039 (1935). High yield prepn: H. Hosoda *et al.*, *J. Org. Chem.* **38**, 4209 (1973). Metabolism study: P. Knapstein *et al. Acta Endocrinol.* **58**, 261 (1968). Study of the sulfate as an indicator of adrenal androgen function: R. A. Lobo *et al.*, *Obstet. Gynecol.* **57**, 69 (1981). Toxicity study of the sulfate: M. Yahara *et al.*, *J. Toxicol. Sci.* **2**, 161 (1977).

Dimorphous, needles, mp 140-141°; leaflets, mp 152-153°. $[\alpha]_D^{18}$ +10.9° (c = 0.4 in alc). Pptd by digitonin. Sol in benzene, alcohol, ether; sparingly sol in chloroform, petr ether.

Sodium sulfate, $C_{19}H_{27}NaO_5S$, *sodium dehydroepiandrosterone sulfate, DHA-S, Mylis.* White cryst powder, mp 154° (dec). Sol in methanol, slightly sol in water, abs ethanol. Practically insol in acetone, chloroform, benzene. LD_{50} in male mice, rats (mg/kg): > 10,000 orally; 899, 1005 s.c.; 460, 559 i.p.; 274, 468 i.v., *Japan. Med. Gaz.* **18**(5), 10 (1981).

THERAP CAT: Androgen.

7711. Pratensein. *5,7-Dihydroxy-3-(3-hydroxy-4-methoxyphenyl)-4H-1-benzopyran-4-one; 3',5,7-trihydroxy-4'-methoxyisoflavone.* $C_{16}H_{12}O_6$; mol wt 300.26. C 64.00%, H

4.03%, O 31.97%. Isoln from red clover (*Trifolium pratense* L., *Leguminosae*), structure and synthesis: E. Wong, *J. Org. Chem.* **28**, 2336 (1963). Alternate syntheses: A. C. Jain, P. K. Bambab, *Indian J. Chem.* **26B**, 488 (1987).

Needles from ethanol, mp 272-273°. uv max (ethanol): 263 nm (log ε 4.53).

Triacetylpratensein, $C_{22}H_{18}O_9$, white needles from chloroform, mp 175-177°.

7712. Pravastatin Sodium. *[1S-[1α(βS*,δS*),2α,6α,8β-(R*),8aα]]-1,2,6,7,8,8a-Hexahydro-β,δ,6-trihydroxy-2-methyl-8-(2-methyl-1-oxobutoxy)-1-naphthaleneheptanoic acid monosodium salt;* sodium (+)-(3R,5R)-3,5-dihydroxy-7-[(1S,2S,6S,8S,8aR)-6-hydroxy-2-methyl-8-[(S)-2-methyl-butyryloxy]-1,2,6,7,8,8a-hexahydro-1-naphthyl]heptanoate; eptastatin; 3β-hydroxycompactin sodium salt; CS-514; SQ-31000; Mevalotin; Pravachol. $C_{23}H_{35}NaO_7$; mol wt 446.52. C 61.87%, H 7.90%, Na 5.15%, O 25.08%. HMG-CoA reductase inhibitor; bioactive metabolite of mevastatin, *q.v.* Prepn by microbial hydroxylation: A. Terahara, M. Tanaka, *Ger.* pat. 3,122,499; *eidem*, *U.S.* pat. 4,346,227 (1981, 1982 both to Sankyo); N. Serizawa *et al.*, *J. Antibiot.* **36**, 604 (1983). Structure elucidation: H. Haruyama *et al.*, *Chem. Pharm. Bull.* **34**, 1459 (1986). Inhibition of sterol synthesis and hypolipidemic effect in animals: Y. Tsujita *et al.*, *Biochim. Biophys. Acta* **877**, 50 (1986); in humans: N. Nakaya *et al.*, *Atherosclerosis* **61**, 125 (1986). Preliminary clinical trial in primary hypercholesterolemia: T. Kazumi *et al.*, *Horm. Metabol. Res.* **18**, 654 (1986). Clinical comparison with probucol, *q.v.*: G. Yoshino *et al.*, *Lancet* **2**, 740 (1986).

Amorphous powder. uv max (methanol): 230, 237, 245 nm.

Free lactone, $C_{23}H_{34}O_6$, colorless plate crystals, mp 138-142°. $[α]_D^{22} +194.0°$ (c = 0.51 in methanol). uv max (methanol): 230, 237, 245 nm.

THERAP CAT: Antihyperlipoproteinemic.

7713. Prazepam. *7-Chloro-1-(cyclopropylmethyl)-1,3-dihydro-5-phenyl-2H-1,4-benzodiazepin-2-one;* 1-(cyclopropylmethyl)-5-phenyl-7-chloro-1H-1,4-benzodiazepin-2-(3H)-one; W 4020; Centrax; Demetrin; Lysanxia; Prazene; Sedapran; Settima; Trepidan; Verstran. $C_{19}H_{17}ClN_2O$; mol wt 324.83. C 70.26%, H 5.28%, Cl 10.92%, N 8.63%, O 4.93%. Prepn: McMillan, Pattison, *Fr.* pat. 1,394,287, corresp to *U.S.* pat. 3,192,199 (both 1965 to Warner-Lambert); Wuest, *U.S.* pat. 3,192,200 (1965); Inaba *et al.*, *Chem. Pharm. Bull.* **17**, 1263 (1969). Pharmacology: Robichaud *et al.*, *Arch. Int. Pharmacodyn. Ther.* **185**, 213 (1970). Metabolism: DiCarlo *et al.*, *J. Pharm. Sci.* **58**, 960 (1969).

Crystals from methanol, mp 145-146°.

Note: This is a controlled substance (depressant) listed in the U.S. Code of Federal Regulations, Title 21 Part 1308.14 (1987).

THERAP CAT: Anxiolytic.

7714. Praziquantel. *2-(Cyclohexylcarbonyl)-1,2,3,6,7,-11b-hexahydro-4H-pyrazino[2,1-a]isoquinolin-4-one;* EMBAY 8440; Biltricide; Cesol; Droncit. $C_{19}H_{24}N_2O_2$; mol wt 312.41. C 73.05%, H 7.74%, N 8.97%, O 10.24%. Prepn: J. Seubert *et al.*, *Ger.* pat. 2,362,539 corresp to *U.S.* pat. 4,001,411 (1975, 1977, both to E. Merck, W. Ger.); D. Frehel *et al.*, *Heterocycles* **20**, 1731 (1983). Properties: J. Seubert *et al.*, *Experientia* **33**, 1036 (1977). *In vitro* study: C. J. Chavasse *et al.*, *Z. Parasitenk.* **58**, 169 (1979). Effect on ultrastructure of trematodes: H. Mehlhorn *et al.*, *Arzneimittel-Forsch.* **33**, 91 (1983). Pharmacodynamics: G. C. Coles, *J. Helminthol.* **53**, 31 (1979). Ovicidal activity in *Echinococcus*: A. S. Thakur *et al.*, *Exp. Parasitol.* **47**, 131 (1979). Clinical pharmacology: G. Leopold *et al.*, *Eur. J. Clin. Pharmacol.* **14**, 281 (1978). Efficacy in dogs: F. L. Anderson *et al.*, *Am. J. Vet. Res.* **40**, 700 (1979). Mutagenicity study: H. Bartsch *et al.*, *Mutat. Res.* **58**, 133 (1978). Tolerance study: P. Muermann *et al.*, *Vet. Med. Rev.* **1976**(2), 142. Symposium on African schistosomiasis: *Arzneimittel-Forsch.* **31**, 535-618 (1981). Clinical studies (human schistosomiasis): T. E. Nash *et al.*, *Am. J. Trop. Med. Hyg.* **31**, 977 (1982); R. N. H. Pugh, C. H. Teesdale, *Brit. Med. J.* **286**, 429 (1983); C. H. Schutte *et al.*, *S. Afr. Med. J.* **64**, 7 (1983). Clinical trial in human parenchymal brain cysticercosis: J. Sotelo *et al.*, *N. Engl. J. Med.* **310**, 1001 (1984).

Crystals, mp 136-138°. Soly (g/100 ml): ethanol 9.7; chloroform 56.7; water 0.04. LD_{50} in mice and rats (mg/kg): 2,000 to 3,000 orally; > 3,000 s.c. (Muermann).

THERAP CAT: Anthelmintic (Schistosoma).

THERAP CAT (VET): Anthelmintic.

7715. Prazosin. *1-(4-Amino-6,7-dimethoxy-2-quinazolinyl)-4-(2-furanylcarbonyl)piperazine;* 2-[4-(2-furoyl)piperazin-1-yl]-4-amino-6,7-dimethoxyquinazoline; furazosin. $C_{19}H_{21}N_5O_4$; mol wt 383.41. C 59.52%, H 5.52%, N 18.27%, O 16.69%. Prepn: **Brit.** pat. 1,156,973; Hess, **U.S.** pat. 3,511,836 (1969, 1970 to Pfizer); **Neth.** pat. Appl. 7,206,067 (1972 to Brocades-Stheeman), *C.A.* **78**, 72180s (1973); E. Honkanen *et al.*, *J. Heterocycl. Chem.* **17**, 797 (1980). Pharmacology and clinical data: Scriabine *et al.*, *Experientia* **24**, 1150 (1968); Cohen, *J. Clin. Pharmacol. J. New Drugs* **10**, 408 (1970). Pharmacokinetics: P. Jaillon, *Clin. Pharmacokinet.* **5**, 365 (1980). Review of pharmacology: I. Cavero, A. G. Roach, *Life Sci.* **27**, 1525-1540 (1980); A. Scriabine, Ed. in *Pharmacology of Antihypertensive Drugs* (Raven, New York, 1980) pp 151-160. Book: *Prazosin: Pharmacology, Hypertension and Congestive Heart Failure*, M. D. Rawlins, Ed. (Grune & Stratton, New York, 1981) 143 pp. Review of pharmacology and therapeutic use: W. F. Stanaszek *et al.*, *Drugs* **25**, 339-384 (1983); J. L. Reid, J. Vincent, *Cardiology* **73**, 164-174 (1986).

Crystals, mp 278-280°.

Hydrochloride, $C_{19}H_{22}ClN_5O_4$, *CP-12299-1*, *Alpress, Duramipress, Eurex, Hypovase, Minipress, Peripress, Sinetens.* THERAP CAT: Antihypertensive. α_1-Adrenergic blocker.

7716. Precocenes. Anti-JH. Anti-juvenile hormones found in plants that induce reversible precocious metamorphosis and sterilization in insects by suppressing the function of the corpora allata gland: W. S. Bowers *et al., Science* **193,** 542 (1976); W. S. Bowers, R. Martinez-Pardo, *ibid.* **197,** 1369 (1977). Isoln from *Ageratum mexicanum* Sw., *Compositae* and structure: A. R. Alertsen, *Acta Chem. Scand.* **9,** 1725 (1955). Synthesis of precocene I: R. Livingstone, R. B. Watson, *J. Chem. Soc.* **1957,** 1509; of precocene II: R. Huls, *Bull. Soc. Chim. Belg.* **66,** 409 (1957); **67,** 22 (1958); of I and II: J. R. Hlubucek *et al., Aust. J. Chem.* **24,** 2347 (1971); A. Banerji, N. C. Goomer, *Indian J. Chem.* **20B,** 144 (1981); V. K. Ahluwalia *et al., Chem. & Ind. (London)* **1982,** 369. Biosynthesis: A. V. Vyas, N. B. Mulchandani, *Phytochemistry* **19,** 2597 (1980). Metabolism studies: D. M. Soderlund *et al., J. Agr. Food Chem.* **1980,** 724; B. J. Bergot *et al., Pestic Biochem. Physiol.* **13,** 95 (1980). Mechanism of action: W. S. Bowers, *Am. Zool.* **21,** 737 (1981).

Precocene I, $C_{12}H_{14}O_2$, *7-methoxy-2,2-dimethyl-2H-1-benzopyran, 7-methoxy-2,2-dimethylchromene, 6-demethoxy-ageratochromene.* R = H. Oil, bp$_6$ 120°. n_D^{21} 1.5548. uv max (ethanol): 280, 304 nm (ϵ 6644, 5844).

Precocene II, $C_{13}H_{16}O_3$, *6,7-dimethoxy-2,2-dimethyl-2H-1-benzopyran, 6,7-dimethoxy-2,2-dimethylchromene, ageratochromene.* R = OCH$_3$. Crystals, mp 47.5°, bp$_6$ 136°. uv max (ethanol): 280, 323 nm (ϵ 5500, 9200).

7717. Prednicarbate. *(11β)-17-[(Ethoxycarbonyl)oxy]-11-hydroxy-21-(1-oxopropoxy)pregna-1,4-diene-3,20-dione; 11β,17,21-trihydroxypregna-1,4-diene-3,20-dione 17-(ethyl carbonate) 21-propionate; prednisolone 17-ethylcarbonate 21-propionate; HOE 777; S 77 0777; EsCort; Dermatop; Regenit.* $C_{27}H_{36}O_8$; mol wt 488.58. C 66.37%, H 7.43%, O 26.20%. Non-halogenated glucocorticoid. Prepn: U. Stache *et al., Ger. pat.* **2,735,110;** *eidem,* U.S. pat. **4,242,334** (1979, 1980 both to Hoechst AG); U. Stache *et al., Arzneimittel-Forsch.* **35,** 1753 (1985). Topical and systemic activity in rats: H. G. Alpermann *et al., ibid.* **32,** 633 (1982). Series of articles on pharmacology, pharmacokinetics and clinical efficacy: Z. *Hautkr.* **61,** Suppl. 1, 1-96 (1986).

Crystals from ethanol + diethylether, mp 110-112°; second crystalline form, mp 183°. uv max (ethanol): 241 nm (ϵ 15000). $[\alpha]_D^{20}$ +63° (c = 0.1 in ethanol).
THERAP CAT: Topical anti-inflammatory.

7718. Prednimustine. *21-[4-[4-[Bis(2-chloroethyl)amino]phenyl]-1-oxobutoxy]-11,17-dihydroxypregna-1,4-diene-3,20-dione; 11β,17,21-trihydroxypregna-1,4-diene-3,20-dione 21-[4-[p-[bis(2-chloroethyl)amino]phenyl]butyrate]; pred-*

nisolone 21-[4-[p-[bis(2-chloroethyl)amino]phenyl]butyrate]; Leo 1031; Sterecyt; Stereocyt. $C_{35}H_{45}Cl_2NO_6$; mol wt 646.66. C 65.01%, H 7.01%, Cl 10.96%, N 2.17%, O 14.85%. Prednisolone ester of chlorambucil, *q.v.* Prepn: H. F. Fex *et al., Ger. pat.* **2,001,305** corresp to U.S. pat. **3,732,260** (1970, 1973 both to AB Leo). *In vitro* study: A. H. Evenaar *et al., Eur. J. Cancer* **9,** 773 (1973). Metabolism: R. Y. Kirdani *et al., Oncology* **35,** 47 (1978). Clinical studies: L. Hakarsson *et al., ibid.* 103; J. E. Johnsson *et al., Cancer Treat. Rep.* **63,** 421 (1970). Toxicity and antitumor activity study: K. R. Harrap *et al., Eur. J. Cancer* **13,** 873 (1977). Series of articles on pharmacology, toxicology and clinical efficacy: *Semin. Oncol.* **13,** Suppl. 1, 1-44 (1986).

Crystals from methanol-water, mp 163-164°. $[\alpha]_D^{24}$ +92.9° (c = 1.06 in chloroform).
THERAP CAT: Antineoplastic.

7719. Prednisolone. *(11β)-11,17,21-Trihydroxypregna-1,4-diene-3,20-dione; 1,4-pregnadiene-3,20-dione-11β,17α,21-triol; 1,4-pregnadiene-11β,17α,21-triol-3,20-dione; 3,20-dioxo-11β,17α,21-trihydroxy-1,4-pregnadiene; metacortandralone; delta F; Δ^1-dehydrocortisol; Δ^1-hydrocortisone; Δ^1-dehydrohydrocortisone; hydroretrocortine; Decaprednil; Meticortelone; Ultracortene-H; Precortilon; PreCortisyl; Precortancyl; Cortalone; Predniretard; Predniliderm; Solone; Prednicen; Dicortol; Hefasolon; Hydeltra; Klismacort; Delta-Cortef; Deltastab; Paracortol; Deltisolone; Codelcortone; Hydrodeltisone; Prenolone; Hydrodeltalone; Sterolone; Sterane; Prednelan; Decortin H; Deltacortril; Scherisolon; Di-Adreson-F; Hostacortin H; Predonine; Nisolone; Ropredlone; Flamasone.* $C_{21}H_{28}O_5$; mol wt 360.44. C 69.97%, H 7.83%, O 22.20%. Synthetic corticosteroid. Microbiological prepn by dehydrogenation of hydrocortisone: Nobile *et al., J. Am. Chem. Soc.* **77,** 4184 (1955); Nobile, U.S. pats. **2,837,464** and **3,134,718** (1958, 1964, both to Schering); Herzog *et al., Tetrahedron* **18,** 581 (1962). Prepn: Oliveto, Gould, U.S. pat. **2,897,216** (1959 to Schering). Structure: *eidem, Science* **121,** 176 (1955). Anti-inflammatory and immunosuppressant activity in rabbits: I. M. Hunneyball, *Agents and Actions* **11,** 49 (1981). Tissue distribution and metabolic interconversion with prednisone, *q.v.:* N. Khalafallah, W. J. Jusko, *J. Pharmacol. Exp. Ther.* **229,** 719 (1984). Clinical studies in asthma: R. F. Willey *et al., Thorax* **39,** 340 (1984); in breast cancer: A. M. Blackburn *et al., Cancer Treat. Rep.* **68,** 1447 (1984). Review of clinical pharmacokinetics, bioavailability: J. G. Gambertoglio *et al., J. Pharmacokinet. Biopharm.* **8,** 1-52 (1980). Use in cattle: P. L. Toutain *et al., Am. J. Vet. Res.* **46,** 719 (1985).

Crystals, dec 240-241°. $[\alpha]_D^{25}$ +102° (dioxane). uv max (methanol): 242 nm (ϵ 15,000; $A_{1cm}^{1\%}$ 414). Very slightly sol in water. One gram dissolves in about 30 ml of alcohol, in about 180 ml of chloroform, in about 50 ml of acetone. Sol in methanol, dioxane.

21-Acetate, $C_{23}H_{30}O_6$, *Ak-Tate, Deltacortenolo, Hydrocortidelt, Inflanefran, Pred Forte, Pred Mild, Pricortin.* Crystals, decomp 237-239°. $[\alpha]_D^{25}$ +116° (dioxane).

21-*m*-Sulfobenzoate sodium salt, $C_{28}H_{32}NaO_9S$, *Corti-Clyss, Predenema, Solupred.*
Hemisuccinate, *Fiasone.*
THERAP CAT: Glucocorticoid.
THERAP CAT (VET): Anti-inflammatory.

7720. Prednisolone 21-Diethylaminoacetate. *N,N-Di-ethylglycine 11β,17-dihydroxy-3,20-dioxopregna-1,4-dien-21-yl ester; 11β,17,21-trihydroxypregna-1,4-diene-3,20-dione 21-(N,N-diethylglycine) ester;* prednisolone 21-*N,N*-diethyl-glycinate; prednisolamate. $C_{27}H_{39}NO_6$; mol wt 473.59. C 68.47%, H 8.30%, N 2.96%, O 20.27%. $ROCCH_2N(C_2H_5)_2$ where R is the 21-prednisolone radical. Prepared from prednisolone and N,N-diethylaminoacetyl chloride hydro-chloride: Pancrazio, Sbarigia, *Farmaco Ed. Prat.* **16,** 190 (1961); from prednisolone and chloroacetyl chloride fol-lowed by reaction with diethylamine: **Brit. pat. 862,370** (1961 to Pfizer).
Crystals from acetone + hexane, mp 175-177°.
Hydrochloride, $C_{27}H_{40}ClNO_6$, *Deltacortril DA.* Crystals from ethanol, mp 237.4-239.8°. $[\alpha]_D$ +120.7° (water).
THERAP CAT: Glucocorticoid.

7721. Prednisolone Sodium Phosphate. *11,17-Dihydr-oxy-21-(phosphonooxy)pregna-1,4-diene-3,20-dione disodi-um salt; 11β,17,21-trihydroxypregna-1,4-diene-3,20-dione 21-(dihydrogen phosphate) disodium salt;* 21-prednisolone-phosphoric acid disodium salt; prednisolone phosphate di-sodium; disodium prednisolone 21-phosphate; Ak-Pred; Codelsol; Hydeltrasol; Inflamase; Metreton; Minims; Pred-nesol; Predsol; Predsolan; Solucort; Solu-Predalone. $C_{21}H_{27}Na_2O_8P$; mol wt 484.39. C 52.07%, H 5.62%, Na 9.49%, O 26.42%, P 6.40%. $RPO(ONa)_2$ where R is the 21-predni-solone radical. Prepn: Sarett, **U.S. pat. 2,789,117** (1957 to Merck & Co.). Alternate synthesis: Poos *et al., Chem. & Ind. (London)* **1958,** 1260; Elks, Phillips, **U.S. pat. 2,936,-313** (1960 to Glaxo).
White powder. Slightly hygroscopic. Stable at room temps. $[\alpha]_D^{25}$ +102.5°. uv max (methanol): 243 nm ($A_{1cm}^{1\%}$ 308). Sol in water, methanol, ethanol. pH of 1% aq soln 7.5 to 8.5.
THERAP CAT: Glucocorticoid.
THERAP CAT (VET): Glucocorticoid.

7722. Prednisolone Sodium Succinate. *21-(3-Carboxy-1-oxopropoxy)-11β,17-dihydroxypregna-1,4-diene-3,20-dione monosodium salt;* Solu-Decortin-H; Di-Adreson-F-aquo-sum; Meticortelone Soluble. $C_{25}H_{31}NaO_8$; mol wt 482.53. C 62.23%, H 6.48%, Na 4.76%, O 26.53%. $ROCCH_2CH_2$-COONa where R is the 21-prednisolone radical. Prepn: Shull, Kita, **Ger. pat. 1,045,400** (1958 to Pfizer), *C.A.* **55,** 2746f (1961).
THERAP CAT: Glucocorticoid.
THERAP CAT (VET): Glucocorticoid.

7723. Prednisolone Sodium 21-*m*-Sulfobenzoate. *11,-17-Dihydroxy-21-[(3-sulfobenzoyl)oxy]pregna-1,4-diene-3,20-dione monosodium salt;* Cortico-Sol; Predfoam; Solu-pred. $C_{28}H_{31}NaO_9S$; mol wt 566.62. C 59.35%, H 5.52%, Na 4.06%, O 25.41%, S 5.66%. Prepd from prednisolone and *m*-carboxybenzenesulfonyl chloride pyridine complex in tri-ethylamine: Allais, Girault, **U.S. pat. 3,032,568** and from prednisolone 21-methanesulfonate and sodium *m*-sulfoben-zoate: Joly, Warnant, **U.S. pat. 3,037,034** (both 1962 to Roussel-UCLAF).

(R = 21-prednisolone radical)

Crystals from water, dec 293-295°. $[\alpha]_D^{20}$ +170° (water).
THERAP CAT: Glucocorticoid.

7724. Prednisolone 21-Stearoylglycolate. *11β,17-Di-hydroxy-21-[[[(1-oxooctadecyl)oxy]acetyl]oxy]pregna-1,4-diene-3,20-dione; 11β,17,21-trihydroxypregna-1,4-diene-3,20-dione 21-ester with glycolic acid stearate;* prednisolone

steaglate; Sintisone. $C_{41}H_{64}O_8$; mol wt 684.92. C 71.89%, H 9.42%, O 18.69%. $ROCCH_2OOC(CH_2)_{16}CH_3$ where R is the 21-prednisolone radical. Prepn: Giraldi, Nannini, **U.S. pat. 3,171,846** (1965 to Carlo Erba); Giraldi *et al., Arzneimittel-Forsch.* **16,** 162 (1966).
Crystals from dil alcohol or butyl ether, mp 105-107°. uv max (methanol): 242 nm ($E_{1cm}^{1\%}$ 212 ±10). $[\alpha]_D^{20}$ +57-63°.
THERAP CAT: Glucocorticoid.

7725. Prednisolone Tebutate. *21-(3,3-Dimethyl-1-oxo-butoxy)-11,17-dihydroxypregna-1,4-diene-3,20-dione; 11β,-17,21-trihydroxypregna-1,4-diene-3,20-dione 21-(3,3-di-methylbutyrate);* prednisolone 21-*tert*-butylacetate; Codel-cortone-T.B.A.; Hydeltra-T.B.A.; Hydeltrone-T.B.A.; Predalone T.B.A. $C_{27}H_{38}O_6$; mol wt 458.57. C 70.71%, H 8.35%, O 20.93%. $ROCCH_2C(CH_3)_3$ where the R is the 21-prednisolone radical. Prepd from prednisolone and 3,3-dimethylbutyryl chloride: Sarett, **U.S. pat. 2,736,734** and **Ger. pat. 1,135,904** (1956, 1962, both to Merck & Co.).
Crystals from ethanol, mp 266-273°.
THERAP CAT: Glucocorticoid.
THERAP CAT (VET): Glucocorticoid.

7726. Prednisolone 21-Trimethylacetate. *(11B)-21-(2,2-Dimethyl-1-oxopropoxy)-11,17-dihydroxypregna-1,4-diene-3,20-dione; 11β,17,21-trihydroxypregna-1,4-diene-3,20-dione 21-pivalate;* Ultracortenol; Ultracorterenol. C_{26}-$H_{36}O_6$; mol wt 444.55. C 70.24%, H 8.16%, O 21.59%. $ROCC(CH_3)_3$ where R is the 21-prednisolone radical. Prepn: Vischer *et al., Helv. Chim. Acta* **38,** 1502 (1955); Joly, Warnant, **U.S. pat. 3,037,034** (1962 to Roussel-UCLAF).
Crystals from acetone, mp 233-236°. $[\alpha]_D^{26}$ +103° (c = 1.208 in chloroform), $[\alpha]_D^{20}$ +97.5° (c = 1 in chloroform). uv max: 244 nm (ε 14,700).
THERAP CAT: Glucocorticoid.
THERAP CAT (VET): Adrenocortical steroid. Glucocorticoid, anti-inflammatory.

7727. Prednisone. *17,21-Dihydroxypregna-1,4-diene-3,11,20-trione;* 1,4-pregnadiene-17α,21-diol-3,11,20-trione; $Δ^1$-dehydrocortisone; $Δ^1$-cortisone; deltacortisone; delta E; metacortandracin; retrocortine; NSC 10023; Juvason; Meti-corten; Deltasone; Decortin; Decortisyl; Dekortin; Di-Adreson; Dacortin; Hostacortin; Deltisone; Colisone; Para-cort; Ancortone; Cortancyl; Decortancyl; Ultracorten; Bi-cortone; Delta-Cortelan; Rectodelt; Ultracortene; Deltra; Deltacortone; Delta Prenovis; Fernisone; Nurison; Predni-longa; Lisacort; Orasone; Sone; Servisone; Pronison; Encor-ton; Keteocort; Precortal. $C_{21}H_{26}O_5$; mol wt 358.44. C 70.37%, H 7.31%, O 22.32%. Prepn: Oliveto, Gould, **U.S. pat. 2,897,216** (1959 to Schering). Microbiological prepn: Nobile *et al., J. Am. Chem. Soc.* **77,** 4184 (1955); Nobile, **U.S. pat. 2,837,464** and **3,134,718** (1958, 1964, both to Schering); Herzog *et al., Tetrahedron* **18,** 581 (1962). Struc-ture: Herzog *et al., Science* **121,** 176 (1955); *cf.* Djerassi *et al.,* **U.S. pat. 2,579,479** (1951 to Syntex).

Crystals, dec 233-235°. $[\alpha]_D^{25}$ +172° (dioxane). uv max (methanol): 238 nm (ε 15,500). Very slightly sol in water. One gram dissolves in about 150 ml alcohol, in about 200 ml chloroform. Slightly sol in methanol, dioxane.
21-Acetate, $C_{23}H_{28}O_6$, *Delta-Corlin, Delcortin.* Crystals, dec 226-232°. $[\alpha]_D^{25}$ +186° (dioxane). uv max (ethanol): 238 nm (ε 16,100).
THERAP CAT: Glucocorticoid.
THERAP CAT (VET): Adrenocortical steroid. Glucocorticoid, anti-inflammatory.

7728. Prednival. *11,21-Dihydroxy-17-[(1-oxopentyl)-oxy]pregna-1,4-diene-3,20-dione; 11β,17,21-trihydroxypreg-*

na-1,4-diene-3,20-dione 17-valerate; prednisolone-17-valerate; W 4869. $C_{26}H_{36}O_6$; mol wt 444.55. C 70.24%, H 8.16%, O 21.59%. Prepn: Vitali, Ercoli, *Tetrahedron Letters* **1961**, 448; Ercoli, Gardi, U.S. pat. **3,152,154** (1964 to Vismara).

Crystals from aq methanol, mp 210-213°. $[\alpha]_D$ +3.5° (dioxane).
21-Acetate, $C_{28}H_{38}O_7$, *Acepreval*.
THERAP CAT: Glucocorticoid.

7729. Prednylidene. *(11β)-11,17,21-Trihydroxy-16-methylenepregna-1,4-diene-3,20-dione;* $\Delta^{1,4}$-pregnadiene-16-methylene-11β,17α,21-triol-3,20-dione; 16-methylene-11β,17α,21-trihydroxypregna-1,4-diene-3,20-dione; 16-methyleneprednisolone; Dacortilen; Decortilen; Sterocort. $C_{22}H_{28}O_5$; mol wt 372.44. C 70.94%, H 7.58%, O 21.48%. Prepn: Mannhardt *et al., Tetrahedron Letters* no. 16, 21 (1960).

Solid, mp 233-235°. $[\alpha]_D^{23}$ +31° (dioxane). uv max: 243 nm (ϵ 15900).
THERAP CAT: Glucocorticoid; anti-inflammatory.

7730. Prednylidene 21-Diethylaminoacetate. *N,N-Diethylglycine 11β,17-dihydroxy-16-methylene-3,20-dioxopregna-1,4-dien-21-yl ester;* 16-methyleneprednisolone 21-diethylaminoacetate. $C_{28}H_{39}NO_6$; mol wt 485.60. C 69.25%, H 8.10%, N 2.88%, O 19.77%. $ROCCH_2N(C_2H_5)_2$ where R is the 21-prednylidene radical. Prepn: Mannhardt *et al.,* Ger. pat. **1,134,074** (1962 to E. Merck), *C.A.* **58**, 568a (1963).
Solid, mp 195-196°. $[\alpha]_D$ +32° (dioxane).
Hydrochloride, $C_{28}H_{40}ClNO_6$, *Decortilen soluble.* Solid, mp 245-246°. $[\alpha]_D$ +45° (water). uv max (water): 246-247 nm ($E_{1cm}^{1\%}$ 300).
THERAP CAT: Glucocorticoid.

7731. Pregnane. *5β-Pregnane;* 17β-ethyletiocholane. $C_{21}H_{36}$; mol wt 288.52. C 87.42%, H 12.58%. Prepd by reduction of etiocholyl methyl ketone or of pregnanedione: Butenandt, *Ber.* **64**, 2529 (1931); Marker *et al., J. Am. Chem. Soc.* **60**, 1067 (1938); Steiger, Reichstein, *Helv. Chim. Acta* **21**, 161 (1938).

Monoclinic scales, plates from methanol; mp 83.5°. d_4^{15} 1.032. $[\alpha]_D^{19}$ +20° (c = 2 in chloroform).

7732. Pregnanediol. *(3α,5β,20S)-Pregnane-3,20-diol.* $C_{21}H_{36}O_2$; mol wt 320.50. C 78.69%, H 11.32%, O 9.98%. A metabolite of progesterone, present in large amounts in pregnancy urine. Isoln from pregnancy urine of women: Marrian, *Biochem. J.* **23**, 1090 (1929); Butenandt, *Ber.* **63**, 659 (1930); of cows, mares, and chimpanzees: Fish *et al., J. Biol. Chem.* **143**, 716 (1942). Prepn by reduction of pregn-16-ene-3,20-dione: Marker *et al.,* U.S. pat. **2,352,852** (1944 to Parke, Davis). Conversion to progesterone: Butenandt, Schmidt, *Ber.* **67**, 1893, 1901 (1934). Conversion to 3α-hydroxypregnan-20-one: Marker, U.S. pat. **2,223,377** (1940 to Parke, Davis). Prepn of the 3-acetate: Hirschmann, *J. Biol. Chem.* **140**, 797 (1941); Ralls *et al., ibid.* **210**, 709 (1954). Prepn of the 20-acetate: Hirschmann, *loc. cit.* Prepn of the diacetate: Johnson *et al., J. Chem. Soc.* **1954**, 1302. Crystal structure: Haner, Norton, *Acta Cryst.* **16**, 707 (1963). Review of metabolism, bioactivity and assay during pregnancy: P. J. Keller, *Contrib Gynecol. Obstet.* **2**, 75-91 (1976).

Crystals from acetone or ethanol, mp 239°. $[\alpha]_D^{20}$ +27.4° (c = 0.7 in alc). Sparingly sol in organic solvents. Not precipitated by digitonin.
3-Acetate, $C_{23}H_{38}O_3$, *3α-acetoxypregnan-20α-ol.* Crystals, mp 132°. $[\alpha]_D^{25}$ +45° ($CHCl_3$).
20-Acetate, $C_{23}H_{38}O_3$, *20α-acetoxypregnan-3α-ol.* Crystals, mp 174°.
Diacetate, $C_{25}H_{40}O_4$, *3α,20α-diacetoxypregnane.* Crystals from light petroleum, mp 180°, also reported as mp 182-183°. $[\alpha]_D^{15}$ +35° (c = 1.1 in $CHCl_3$).
USE: In manuf of progesterone.

7733. 3,20-Pregnanedione. *(5β)-Pregnane-3,20-dione.* $C_{21}H_{32}O_2$; mol wt 316.47. C 79.69%, H 10.19%, O 10.11%. From pregnancy urine of mares. Prepn from other steroids: Butenandt, *Ber.* **63**, 659 (1930); Butenandt, Fleischer, *Ber.* **68**, 2094 (1935); Marker *et al., J. Am. Chem. Soc.* **59**, 1595 (1937); Shoppee, Reichstein, *Helv. Chim. Acta* **24**, 356 (1941); U.S. pats. **2,160,719; 2,352,852; 2,323,276; 2,229,818.**

Needles from dil alc, mp 123°. Insol in water. Freely sol in the usual organic solvents.
Dioxime, $C_{21}H_{34}N_2O_2$, dec above 250°.

7734. Pregnan-3α-ol-20-one. *(3α,5β)-3-Hydroxypregnan-3-one.* $C_{21}H_{34}O_2$; mol wt 318.48. C 79.19%, H 10.76%, O 10.05%. Found in pregnancy urine of women and sows. Formed by partial hydrogenation of 3,20-pregnanedione in neutral medium: Marker, Kamm, *J. Am. Chem. Soc.* **59**, 1373 (1937); Marker *et al., ibid.* 1841; U.S. pats. **2,223,377; 2,231,017/18; 2,156,275.** *See also* Marrian, Gough, *Biochem. J.* **40**, 376 (1946); Lieberman *et al., J. Biol. Chem.* **172**, 263 (1948).

Rosettes of needles from hexane, mp 149.5°. $[\alpha]_D^{13}$ +109° (c = 0.98 in ethanol). Not precipitated by digitonin.

Acetate, $C_{23}H_{36}O_3$, mp 112° (Marker); mp 99° (Marrian, Lieberman).

3β-Analog, *pregnan-3β-ol-20-one*. Crystals from dilute alcohol, mp 149°. Precipitated by digitonin.

7735. 4-Pregnene-20,21-diol-3,11-dione. *20,21-Dihydroxypregn-4-ene-3,11-dione;* Reichstein's substance T. $C_{21}H_{30}O_4$; mol wt 346.45. C 72.80%, H 8.73%, O 18.47%. Isoln from adrenal glands: Reichstein, von Euw, *Helv. Chim. Acta* **22**, 1222 (1939).

Crystals, mp about 210°. $[\alpha]_D^{20}$ +176° (acetone). Diacetate, $C_{25}H_{34}O_6$, crystals, mp 212-213°.

7736. 4-Pregnene-11β,17α,20β,21-tetrol-3-one. *11β,-17,20β,21-Tetrahydroxypregn-4-ene-3-one;* 17-(1,2-dihydroxyethyl)androsten-3-one-11,17-diol; Reichstein's substance E. $C_{21}H_{32}O_5$; mol wt 364.47. C 69.20%, H 8.85%, O 21.95%. Occurs in adrenal cortex. Isoln: Reichstein, *Helv. Chim. Acta* **19**, 29 (1936); **20**, 953 (1937); Reichstein, von Euw, *ibid.* **24**, 247E (1941).

Hydrated crystals from dil acetone, dec 125°. $[\alpha]_D^{20}$ +87° (alc). $[M]_D$ +317°. uv max: 240 nm.

20,21-Diacetate, crystals, dec 229-230°. $[\alpha]_D^{20}$ +162.7°; $[M]_D$ +730° (acetone).

7737. 4-Pregnene-17α,20β,21-triol-3,11-dione. *17,20,-21-Trihydroxypregn-4-ene-3,11-dione;* Reichstein's substance U. $C_{21}H_{30}O_5$; mol wt 362.45. C 69.58%, H 8.34%, O 22.07%. Isoln from adrenal glands: Reichstein, von Euw, *Helv. Chim. Acta* **24**, 247E (1941).

Clusters of needles from acetone + ether, mp 208-209°. 20,21-Diacetate, $C_{25}H_{34}O_7$, pointed needles from acetone

+ ether or chloroform + ether, mp 253°. $[\alpha]_D^{21}$ +178.5° (c = 0.924 in acetone). uv max (alc): 239 nm (log ε 4.1). Less sol than the 20,21-diacetate of 4-pregnene-11β,17α,-20,21-tetrol-3-one.

7738. 4-Pregnene-17α,20β,21-triol-3-one. *17α,20β,21-Trihydroxypregn-4-en-3-one;* 17-(1,2-dihydroxyethyl)-Δ⁴-androsten-3-on-17α-ol; 17α-pregnenetriolone. $C_{21}H_{32}O_4$; mol wt 348.47. C 72.38%, H 9.26%, O 18.37%. Prepd from 17-ethynyltestosterone by hydrogenation with palladium in pyridine, allylic rearrangement, and hydroxylation with osmium tetroxide: Ruzicka, Müller, *Helv. Chim. Acta* **22**, 755 (1939); Logemann, *Naturwiss.* **27**, 196 (1939); from a 3-enol ester of a 17,21-diacyloxyprogesterone by reduction followed by saponification: **Swiss pat. 207,496** (1940), *C.A.* **36**, 3636 (1942).

Crystals from methanol. mp 190°. Sol in dioxane, chloroform, methanol. $[\alpha]_D$ +63° (c = 1 in dioxane). uv max: 240 nm (log ε 4.1).

20,21-Diacetate, $C_{25}H_{36}O_6$, crystals from acetone + ether. Polymorphic; mp 170° and 194°; $[\alpha]_D^{20}$ +125° (dioxane). Reaction with zinc in toluene yields 17-isodesoxycorticosterone acetate.

7739. Pregnenolone. *3-Hydroxypregn-5-en-20-one;* Δ⁵-pregnen-3β-ol-20-one; 17β-(1-ketoethyl)-Δ⁵-androsten-3β-ol. $C_{21}H_{32}O_2$; mol wt 316.47. C 79.69%, H 10.19%, O 10.11%. Prepn from stigmasterol: Butenandt *et al.*, *Ber.* **67**, 1611 (1934); Butenandt, Fleischer, *Ber.* **70**, 96 (1937); from androstenolone: H. Butenandt, J. Schmidt-thome, *Ber.* **72**, 182 (1939); S. Danishefsky *et al.*, *J. Org. Chem.* **40**, 1989 (1975); from Δ⁵-3-acetoxyetiocholenic acid chloride: Wettstein, *Helv. Chim. Acta* **23**, 1373 (1940); from diosgenin: Marker, Krueger, *J. Am. Chem. Soc.* **62**, 3349 (1940); Marker *et al.*, *ibid.* **69**, 2173 (1947); from nologenin: *eidem, ibid.* 2395; by treating a 21-halo-Δ⁵-pregnen-3-ol-20-one with a reducing agent: **Swiss pat. 215,139** (1941), *C.A.* **42**, 3144 (1948). Crystal structure: J. Bordner *et al.*, *Cryst. Struct. Commun.* **7**, 513 (1978).

Needles from dil alc, mp 193°. $[\alpha]_D^{20}$ +28° (alc). Very sparingly sol in water. Soly (g/100 ml of soln): carbon tetrachloride 0.5; petr ether 0.1; ethyl acetate 1.1; acetone 0.6; chloroform 17.0; ethanol 1.9; benzene 0.9; isopropanol 1.5. Soly (g/100 ml of solvent): propylene glycol 0.1; dioxane 3.1; benzyl alcohol 8.1. On refluxing with methyl alcohol yields the 17-isopregnenolone, mp 172-173°, $[\alpha]_D^{20}$ −140.5° (alcohol).

Acetate, $C_{23}H_{34}O_3$, needles from alcohol, mp 149-151°. $[\alpha]_D^{20}$ +22° (alcohol). Soly (g/100 ml of soln): carbon tetrachloride 5.0; petr ether 1.0; ethyl acetate 7.9; acetone 2.7; chloroform 55.0; ethanol 2.5; benzene 26.0; isopropanol 2.0. Soly (g/100 ml of solvent): propylene glycol 20.2; benzyl alcohol 11.1; benzyl benzoate 9.1.

7740. Pregnenolone Methyl Ether. *3β-Methoxypregn-5-en-20-one.* $C_{22}H_{34}O_2$; mol wt 330.49. C 79.95%, H 10.37%, O 9.68%. Prepn: Butenandt, Grosse, *Ber.* **70**, 1446 (1937);

Riegel, Meyer, *J. Am. Chem. Soc.* **68**, 1097 (1946); Huffman, Sadler, *J. Org. Chem.* **18**, 919 (1953).

Crystals from abs or dil methanol, mp 123.5°. [α]$_D^{18}$ +18° (c = 1.085 in chloroform).

7741. Prelog-Djerassi Lactone. *Tetrahydro-α,3,5-trimethyl-6-oxo-2H-pyran-2-acetic acid;* 6-(1-carboxyethyl)-3,4,5,6-tetrahydro-3,5-dimethyl-2-pyranone. $C_{10}H_{16}O_4$; mol wt 200.24. C 59.98%, H 8.06%, O 31.96%. Intermediate in the synthesis of macrolide antibiotics. Isoln of (+)-form as degradation product of narbomycin, q.v.: R. Anliker *et al., Helv. Chim. Acta* **39**, 1785 (1956); of methymycin, q.v.: C. Djerassi, J. A. Zderic, *J. Am. Chem. Soc.* **78**, 6390 (1956). Abs config: R. W. Rickards, R. M. Smith, *Tetrahedron Letters* **1970**, 1025. Synthesis of (±)-form: S. Masamune *et al., J. Am. Chem. Soc.* **97**, 3512 (1975); M. Hirami *et al., Tetrahedron Letters* **1979**, 3937; P. A. Bartlett, J. L. Adams, *J. Am. Chem. Soc.* **102**, 337 (1980). Synthesis of (+)-form: P. A. Grieco *et al., ibid.* **101**, 4749 (1979); M. Isobe *et al., Tetrahedron Letters* **22**, 4287 (1981); D. A. Evans, J. Bartroli, *ibid.* **23**, 807 (1982).

(+)-Form, cryst, mp 124-125°. [α]$_D$ +33° (c = 0.797 in CHCl$_3$).

7742. Prenalterol. *(S)-4-[2-Hydroxy-3-[(1-methylethyl)amino]propoxy]phenol;* (−)-(S)-1-(p-hydroxyphenoxy)-3-(isopropylamino)-2-propanol. $C_{12}H_{19}NO_3$; mol wt 225.29. C 63.97%, H 8.50%, N 6.22%, O 21.31%. A β$_1$-adrenergic agonist. Prepn: K. A. Jaeggi *et al.,* Ger. pat. **2,503,968** corresp to U.S. pats. **3,978,041** and **4,049,797** (1974, 1976, 1977, all to Ciba-Geigy). Pharmacologic study: E. Carlsson *et al., Arch. Pharmacol.* **300**, 101 (1977). Metabolism, hemodynamic effects, pharmacokinetics in man: O. Rönn *et al., Eur. J. Clin. Pharmacol.* **17**, 81 (1980). Cardiovascular effects: D. H. Scott *et al., Brit. J. Clin. Pharmacol.* **7**, 365 (1979). Clinical study in coronary heart disease: I. Hutton *et al., Brit. Heart J.* **43**, 134 (1980). See also T. P. Kenakin, D. Beek, *J. Pharmacol. Exp. Ther.* **213**, 406 (1980) for a discussion of the selectivity of action. Prepns of the racemic mixture: **Neth. pat. Appl. 6,409,883** corresp to H. Köppe *et al.,* U.S. pat. **3,637,852** (1965, 1972 both to Boehringer, Ing.); **Neth. pat. Appl. 301,580** corresp to A. F. Crowther, L. H. Smith, U.S. pat. **3,501,769** (1965, 1970 both to ICI); A. F. Crowther *et al., J. Med. Chem.* **12**, 638 (1969). Symposium: *Acta. Med. Scand.,* suppl. 659, 1-325 (1982).

Crystals from ethyl acetate, mp 127-128°. [α]$_D^{20}$ −1° ±1°; [α]$_{Hg}^{20}$ +2° ±1° (c = 0.940 in methanol). Hydrochloride, $C_{12}H_{20}ClNO_3$, *H 133/22, CGP 7760B,* (−)H 80/62, Hyprenan, Varbian.

THERAP CAT: Cardiotonic.

7743. Prenoxdiazine Hydrochloride. *1-[2-[3-(2,2-Diphenylethyl)-1,2,4-oxadiazol-5-yl]ethyl]piperidine hydro-*

chloride; 3-(β,β-diphenylethyl)-5-(β-piperidinoethyl)-1,2,4-oxadiazole hydrochloride; HK-256; Libexin; Lomapect; Tibexin. $C_{23}H_{28}ClN_3O$; mol wt 397.95. C 69.42%, H 7.09%, Cl 8.91%, N 10.56%, O 4.02%. Prepn: K. Harsanyi *et al.,* **Hung.** pat. **151,748** (1964 to Chinoin), *C.A.* **62**, 11821f (1965). Pharmacology: L. Tardos, I. Erdély, *Arzneimittel-Forsch.* **16**, 617 (1966). Stability study: E. Pandula *et al., Acta Pharm. Hung.* **38**, 68 (1968). Review of pharmacology and clinical studies: K. Harsanyi *et al., Boll. Chim. Farm.* **112**, 691 (1973).

Crystals from ethanol, mp 192-193°. LD$_{50}$ in mice, rats (mg/kg): 920, > 2000 orally; 34, 32 i.v., L. Tardos, I. Erdély, *loc. cit.*

THERAP CAT: Antitussive.

7744. Prenylamine. *N-(1-Methyl-2-phenylethyl)-γ-phenylbenzenepropanamine;* N-(3,3-diphenylpropyl)-α-methylphenethylamine; N-(3'-phenyl-2'-propyl)-1,1-diphenyl-3-propylamine; 1-phenyl-2-[1',1'-diphenylpropyl-3'-amino]propane; B-436; Bismethin; Carditin; Corpax; Elecor; Hostaginan; Valecor. $C_{24}H_{27}N$; mol wt 329.46. C 87.49%, H 8.26%, N 4.25%. Prepn: G. Ehrhart *et al.,* Ger. pat. **1,100,031**, *C.A.* **56**, 3413h (1962) and Ger. pat. **1,111,642** corresp to U.S. pat. **3,152,173** (1961, 1961, 1964 all to Hoechst); G. Erhart, *Arch. Pharm.* **295**, 196 (1962). Series of articles on pharmacology and chemistry: *Arzneimittel-Forsch.* **10**, 569-588 (1960). Metabolism: M. Volz, *ibid.* **21**, 1320 (1971). *Review:* J. E. Murphy, *J. Int. Med. Res.* **1**, 204-209 (1973).

mp 36.5-37.5°.
Lactate, $C_{27}H_{33}NO_3$, *Agozol, Angormin, Coredamin, Corontin, Crepasin, Daxauten, Incoran, Irroin, Lactamine, Plactamin, Reocorin, Roinin, Seccidin, Sedolatan, Segontin, Synadrin.* mp 140-142°. Sparingly sol in water (~ 0.5%); sol in organic solvents. uv max (chloroform): 260 nm (E$_{1cm}^{1\%}$ 170).

THERAP CAT: Coronary vasodilator.

7745. Prephenic Acid. *1-Carboxy-4-hydroxy-α-oxo-2,5-cyclohexadiene-1-propanoic acid;* 1-carboxy-4-hydroxy-2,5-cyclohexadiene-1-pyruvic acid. $C_{10}H_{10}O_6$; mol wt 226.18. C 53.10%, H 4.46%, O 42.44%. Non-aromatic biosynthetic intermediate that represents a secondary branch-point in the pathway from chorismic acid to phenylalanine and tyrosine, q.q.v., in many organisms. Isoln from cultures of mutant *Escherichia coli:* B. D. Davis, *Science* **118**, 251 (1953). Characterized as cryst Ba salt: U. Weiss *et al., ibid.* **119**, 774 (1954); R. L. Metzenberg, H. K. Mitchell, *Arch. Biochem. Biophys.* **64**, 51 (1956). Structure: H. Plieninger, G. Keilich, *Z. Naturforsch.* **16b**, 81 (1961). Synthetic approaches: H. Plieninger, *Angew. Chem. Int. Ed.* **1**, 367 (1962). Total synthesis of disodium salt: S. Danishevsky, M. Hirama, *J. Am. Chem. Soc.* **99**, 7740 (1977); S. Danishevsky *et al., ibid.* **101**, 7013 (1979); W. Gramlich, H. Plieninger, *Ber.* **112**, 1571 (1979). *Review:* U. Weiss, J. M. Edwards, *The Biosynthesis of Aromatic Compounds* (Wiley, New York, 1980) pp 144-184. See also shikimic acid.

Barium salt monohydrate, $C_{10}H_8BaO_6 \cdot H_2O$, crystals from water + methanol or water + pyridine.

7746. Pretilachlor. *2-Chloro-N-(2,6-diethylphenyl)-N-*

(2-propoxyethyl)acetamide; N-propoxyethyl-N-chloroacetyl-2,6-diethylaniline; 2,6-diethyl-N-(2'-n-propoxyethyl)chloroacetanilide; CGA 26 423; CG 113; Rifit; Solnet. $C_{17}H_{26}ClNO_2$; mol wt 311.85. C 65.48%, H 8.40%, Cl 11.37%, N 4.49%, O 10.26%. Selective herbicide for control of perennial weeds in transplanted rice. Prepn, herbicidal properties: **Neth. pat. Appl. 7,307,584;** C. Vogel, R. Aebi, **U.S. pat. 4,168,965** (1973, 1979 both to Ciba-Geigy). Residue analysis, stability: H. Egli, *J. Agr. Food Chem.* **30**, 861 (1982). Effect on leaf elongation in rice, use in combination with a safening agent (CGA 123,407): R. A. Christ, *Weed Res.* **25**, 193 (1985). Brief account: J. Rufener, M. Quadranti, *Proc. 10th Conf. Int. Congr. Plant Prot.* **1**, 332 (1983).

Colorless liquid, $bp_{0.001}$ 135°. Vapor pressure at 20°: 1×10^{-6} mm Hg. Almost insol in water (50 mg/l at 20°). Sol in most organic solvents. n_D^{20} 1.5204. LD_{50} in rats (mg/kg): 6099 orally; >3100 dermally (Rufener, Quadranti). LC_{50} in rainbow trout, carp, catfish: 3.0, 3.0, 2.6 ppm (Vogel, Aebi). USE: Herbicide for use in rice paddys.

7747. Pridinol. *α,α-Diphenyl-1-piperidinepropanol;* 1,1-diphenyl-3-piperidino-1-propanol; 3-piperidino-1,1-diphenyl-1-propanol; 1,1-diphenyl-3-(1-piperidyl)-1-propanol; 3-(N-piperidyl)-1,1-diphenyl-1-propanol; ridinol; C-238; Nonplesin. $C_{20}H_{25}NO$; mol wt 295.41. C 81.31%, H 8.53%, N 4.74%, O 5.42%. May be prepd from ethyl 1-piperidine-propionate and phenylmagnesium bromide: Adamson, **Brit. pat. 624,118** (1949 to Wellcome Found.).

Crystals, mp 120-121°. Soluble in acetone.
Hydrochloride, $C_{20}H_{26}ClNO$, *Par KS-12, Mitanoline.* Crystals, dec 238°. Sol in alc. LD_{50} in mice, rats (mg/kg): 35, 33 i.v.; 131, 91 i.p., R. W. Cunningham *et al., J. Pharmacol. Exp. Ther.* **96**, 151 (1949).
Methanesulfonate, $C_{21}H_{29}NO_4S$, *Konlax, Loxeen, Lyseen.* Crystals, mp 152.5-155.0°. Sparingly sol in water.
THERAP CAT: Antiparkinsonian; anticholinergic.

7748. Prifinium Bromide. *3-(Diphenylmethylene)-1,1-diethyl-2-methylpyrrolidinium bromide;* 3-(diphenylmethylene)-1-ethyl-2-methylpyrrolidine ethyl bromide; pyrodifenium bromide; Padrin; Riabal. $C_{22}H_{28}BrN$; mol wt 386.39. C 68.39%, H 7.30%, Br 20.68%, N 3.63%. Prepn: Sadao Oki, **Japan. pat. 22,462('65),** *C.A.* **64**, 3489a (1966); of the free base: *idem,* **Japan. pat. 17,015('65),** *C.A.* **63**, 18034b (1965). Ganglion blocking agent. Metabolism: Nakai *et al., Arzneimittel-Forsch.* **20**, 1112 (1970). Toxicity studies: Kumada *et al., ibid.* **22**, 706 (1972).

Crystals, mp 216-218°. (Free base, $bp_{0.15}$ 183-185°). LD_{50} in male mice: 11 mg/kg i.v.; 43 mg/kg i.p.; 30 mg/kg s.c.; 330 mg/kg orally.
THERAP CAT: Antispasmodic.

7749. Prilocaine. *N-(2-Methylphenyl)-2-(propylamino)-propanamide; 2-(propylamino)-o-propionotoluidide; N-(α-propylaminopropionyl)-o-toluidine; α-propylamino-2-*

methylpropionanilide; propitocaine; L 67. $C_{13}H_{20}N_2O$; mol wt 220.31. C 70.87%, H 9.15%, N 12.72%, O 7.26%. Prepn: Löfgren, Tegner, *Acta Chem. Scand.* **14**, 486 (1960); **Brit. pat. 839,943** (1960 to Astra).

Needles, mp 37-38°. $bp_{0.1}$ 159-162°. n_D^{20} 1.5299.
Hydrochloride, $C_{13}H_{21}ClN_2O$, *Citanest, Xylonest.* Crystals from ethanol + isopropyl ether, mp 167-168°. Readily sol in water.
THERAP CAT: Local anesthetic.

7750. Primaperone. *1-(4-Fluorophenyl)-4-(1-piperidinyl)-1-butanone;* 4'-fluoro-4-piperidinobutyrophenone; 4-piperidino-4'-fluorobutyrophenone. $C_{15}H_{20}FNO$; mol wt 249.34. C 72.26%, H 8.08%, F 7.62%, N 5.62%, O 6.42%. Prepn: Beregi *et al.,* **Fr. pats. 1,301,863** and **M1459** (both 1962 to Science Union).

Oil, $bp_{0.5}$ 127-130°.
Hydrochloride, $C_{15}H_{21}ClFNO$, crystals from isopropyl alcohol, mp 180-181°. Main constituent of *Diviator.*
THERAP CAT: Antihypertensive.

7751. Primaquine. *N^4-(6-Methoxy-8-quinolinyl)-1,4-pentanediamine; 8-(4-amino-1-methylbutylamino)-6-methoxyquinoline;* SN 13272. $C_{15}H_{21}N_3O$; mol wt 259.34. C 69.46%, H 8.16%, N 16.20%, O 6.17%. Prepn following the synthesis of pamaquine: Elderfield *et al., J. Am. Chem. Soc.* **68**, 1525 (1946); improved procedure: Elderfield *et al., ibid.* **77**, 4816 (1955). *Review:* Olenick in *Antibiotics* **vol. 3**, J. W. Corcoran, F. E. Hahn, Eds. (Springer-Verlag, New York, 1975) pp 516-520.

Viscous liquid, $bp_{0.2}$ 175-179°. Soluble in ether.
Diphosphate, $C_{15}H_{21}N_3O.2H_3PO_4$, yellow crystals from 90% ethanol, mp 197-198°. Moderately sol in water.
Oxalate, $C_{15}H_{21}N_3O.C_2H_2O_4$, yellow crystals from 80% ethanol, mp 182.5-185°.
THERAP CAT: Antimalarial.

7752. Primeverose. *6-O-β-D-Xylopyranosyl-D-glucose;* 6-(β-D-xylosido)-D-glucose. $C_{11}H_{20}O_{10}$; mol wt 312.27. C 42.31%, H 6.46%, O 51.24%. By enzymatic hydrolysis of gaultherin (monotropitoside), primeverin, rhamnicoside and other glycosides: Goris, Vischniac, *Compt. Rend.* **169**, 871 (1919); Bridel, *ibid.* **179**, 991 (1924); Bridel, Charaux, *ibid.* **180**, 857 (1925); Richter, *J. Chem. Soc.* **1936**, 1701. Synthesis: Helferich, Rauch, *Ber.* **59**, 2655 (1926); *Ann.* **455**, 168 (1927); McCloskey, Coleman, *J. Am. Chem. Soc.* **65**, 1778 (1943).

Crystals from methanol or 80% alcohol. Darkens at 190°, mp 209-210° (copper block). Shows mutarotation. $[\alpha]_D^{20}$ +23° → −3.2° (c = 5 in water). Sol in water, methanol, 80% alcohol. Is hydrolyzed to 1 mol D-glucose and 1 mol D-xylose by boiling 2% H_2SO_4 for 5 hrs. Reduces Fehling's soln slowly in the cold, instantly when hot.

7753. Primidone. *5-Ethyldihydro-5-phenyl-4,6(1H,5H)-pyrimidinedione;* 5-ethyl-5-phenylhexahydropyrimidine-4,6-dione; 5-phenyl-5-ethylhexahydropyrimidine-4,6-dione; 2-desoxyphenobarbital; Lepsiral; Liskantin; Mylepsin; Mylepsinum; Mysoline; Primaclone; Primoline; Resimatil; Sertan. $C_{12}H_{14}N_2O_2$; mol wt 218.25. C 66.03%, H 6.47%, N 12.84%, O 14.66%. Prepd by electrolytic reduction of phenobarbital or by catalytic desulfuration of the corresp 2-thiobarbituric acid: Boon *et al., Brit.* pat. **666,027** (1952 to I.C.I.); Bogue, Carrington, *Brit. J. Pharmacol.* **8**, 230 (1953). Comprehensive description: R. D. Daley in *Analytical Profiles of Drug Substances* vol. 2, K. Florey, Ed. (Academic Press, New York, 1973) pp 409-437; A. A. Al-Badr, H. A. El-Obeid, *ibid.* vol. 17 (1988) pp 749-795.

Crystals, mp 281-282°. Practically tasteless. Has no acidic properties. Sparingly sol in water (0.6 g/l at 37°); in most organic solvents.

THERAP CAT: Anticonvulsant.

THERAP CAT (VET): Anticonvulsant. Chiefly to control epileptiform seizures.

7754. Primocarcin. *5-(Acetylamino)-4-oxo-5-hexenamide;* 4-acetamido-4-penten-3-one-1-carboxamide. $C_8H_{12}N_2O_3$; mol wt 184.19. C 52.16%, H 6.57%, N 15.21%, O 26.06%. Antineoplastic antibiotic produced by the actinomycete, *Nocardia fukana:* Y. Sumiki *et al., J. Antibiot.* **13A**, 416 (1960); Isono, Suzuki, *ibid.* **15A**, 77 (1962). Isoln from *Streptomyces diastatochromogens* var. *luteus:* H. Abe *et al., ibid.* **20A**, 167 (1967). Structure: Isono, *ibid.* **15A**, 80 (1962). Synthesis: Bowman *et al., J. Chem. Soc.* **1965**, 470.

Needles from hot methanol, mp 130-131°. uv max (water): 253 nm (ε 3420); in 0.1N NaOH: 215, 347 nm (ε 3230, 304). Sol in water, lower alcohols, acetone, pyridine; slightly sol in most other organic solvents. Aq soln stable from pH 5-8. LD_{50} in mice (mg/kg): 50-60 i.p.; 300 i.v. (Isono, Suzuki).

Dihydroprimocarcin, $C_8H_{14}N_2O_3$, plates from acetone, mp 137-141°.

Tetrahydroprimocarcin, $C_8H_{16}N_2O_3$, needles from methanol+ acetone, mp 183°.

7755. Primulaverin. $C_{20}H_{28}O_{13}$; mol wt 476.42. C 50.42%, H 5.92%, O 43.66%. Isoln from species of *Primulaceae:* Thieme, Winkler, *Pharmazie* **26**, 434 (1971). Synthesis: Chaudhury *et al., J. Chem. Soc.* **1948**, 2220.

Dihydrate, star-shaped clusters from ethyl acetate, needles from alc, mp 163°. $[\alpha]_D^{20}$ −67°. Sol in water, alc, acetone.

7756. Primycin. Macrolide antibiotic complex of more than 20 components produced by actinomycetes found in the intestinal tract of the wax moth *(Galeria melonella).* Nine primary components, in 3 major groups designated A, B, and C, represent 90% of the total material. Originally

thought to be a single entity which was subseqently identified as the major component, primycin A_1. Isoln from cultures of *Streptomyces primycini:* Válvi-Nagy *et al., Nature* **174**, 1105 (1954); Hung. pats. **146,332** and **151,197** (1962, 1964 both to Hung. Acad. Sci.). Production by *Thermopolyspora galeriensis* (also designated *Micromonospora galeriensis):* T. Válvi-Nagy *et al.,* Hung. pat. **153,593;** *eidem,* U.S. pat. **3,498,884** (1967, 1970 both to Chinoin). Structure of A_1: J. Aberhart *et al., J. Am. Chem. Soc.* **92**, 5816 (1970); *eidem, J. Chem. Soc. Perkin Trans. I* **1974**, 816, 836. TLC sepn of components: I. Szilagyi *et al., J. Chromatog.* **295**, 141 (1984). Structures of the nine primary components: J. Frank *et al., Tetrahedron Letters* **28**, 2759 (1987). Antibacterial activity and toxicity: T. Válvi-Nagy *et al., Pharmazie* **11**, 304 (1956). Antifungal activity *in vitro:* J. V. Uri, P. Actor, *J. Antibiot.* **32**, 1207 (1979). Mode of action: I. Horvath *et al., Arch. Microbiol.* **121**, 135 (1979). Clinical studies in dermatological infections: C. Mészáros, K. Vezekényi, *Ther. Hung.* **35**, 77 (1987); J. Biro, V. Várkonyi, *ibid.* 136; G. Bálint, *ibid.* 140. *Review:* J. V. Uri, *Acta Microbiol. Hung.* **33**, 141 (1986).

Primycin A_1

Sulfate, *Ebrimycin.* mp 192-195° (dec) (Aberhart, 1974); also reported as 202-206° (dec) (Uri, 1986). Infrared and ultraviolet spectra: Szilágyi *et al., Nature* **193**, 243 (1962). Fairly soluble in methanol (2.5%), less sol in the higher alcohols, sparingly sol in pyridine, glacial acetic acid, water. LD_{50} in mice, guinea pigs, rats, rabbits (mg/kg): 2.5, 5.0, 10.0, 10.0 i.p. (Válvi-Nagy, 1956).

THERAP CAT: Topical antibacterial.

7757. Pristane. *2,6,10,14-Tetramethylpentadecane;* norphytane; Robuoy. $C_{19}H_{40}$; mol wt 268.51. C 84.98%, H 15.02%. Isoprenoid alkane obtained from the unsaponifiable fraction of shark liver oil where it occurs to an extent of 14%: Tsujimoto, *J. Soc. Chem. Ind.* **51**, 317T (1932); Sörensen, Mehllum, *Acta Chem. Scand.* **2**, 140 (1948). Identity with norphytane: Pliva, Sörensen, *ibid.* **4**, 846 (1950). Isoln from petroleum crude oils: Bendoraitis *et al., Anal. Chem.* **34**, 49 (1962); from wool wax: Mold *et al., Nature* **199**, 283 (1963). Synthesis from phytol: Sörensen, Sörensen, *Acta Chem. Scand.* **3**, 939 (1949). Metabolism: McKenna, Kallio, *Proc. Nat. Acad. Sci. USA* **68**, 1552 (1971).

Mobile, transparent, stable liq. d_4^{20} 0.78267. Congealing point −100°. bp_{760} 296°; bp_{10} 158°; $bp_{0.001}$ 68° (bath temp). n_D^{20} 1.43848. Acid no. 0-5. Iodine no. 0-7.5. Sapon no. 0-5. Viscosity at 25°: 5 cps. Soluble in ether, petr ether, benzene, chloroform, carbon tetrachloride.

USE: Lubricant; transformer oil. Anti-corrosion agent. Biological marker.

7758. Pristinamycin. RP 7293; Pyostacine; Stapyocine. An antibiotic mixture produced by *Streptomyces pristinaespiralis* (NRRL 2958). Isoln: Mancy *et al.,* Fr. pat. **1,301,857** corresp to U.S. pat. **3,154,475** (1962, 1964 both to Rhône-

Poulenc); Preud'homme *et al.*, *Compt. Rend.* **260**, 1309 (1965). Consists of pristinamycins I$_A$, I$_B$, I$_C$, which are identical to vernamycins B$_α$, B$_β$, B$_γ$ (*q.v.*) respectively, and pristinamycins II$_A$, II$_B$, which are identical to virginiamycin M$_1$, *q.v.*, and 26,27-dihydrovirginiamycin M$_1$. Separation of components and structure: Preud'homme *et al.*, *Bull. Soc. Chim. France* **1968**, 585. Nomenclature: Crooy, De Neys, *J. Antibiot.* **25**, 371 (1972).

THERAP CAT: Antibacterial.

7759. Probarbital. *5-Ethyl-5-(1-methylethyl)-2,4,6-(1H,3H,5H)-pyrimidinetrione; 5-ethyl-5-isopropylbarbituric acid;* Ipral. C$_9$H$_{14}$N$_2$O$_3$; mol wt 198.22. C 54.53%, H 7.12%, N 14.13%, O 24.22%. Prepn: Thorp, U.S. pats. **1,255,951**; **1,576,014** (1918, 1926). Toxicity data: R. H. Fitch, A. L. Tatum, *J. Pharmacol. Exp. Ther.* **44**, 325 (1932).

Needles, mp 197-198°. Very slightly sol in cold water, more readily sol in hot water; easily sol in alcohol, ether. Calcium salt trihydrate, C$_{18}$H$_{26}$CaN$_4$O$_6$.3H$_2$O, crystals. Slightly bitter taste. One gram dissolves in about 40 ml water. Practically insol in alcohol. Aq solns are alkaline to litmus. Solns are unstable and pptn occurs on boiling. LD$_{50}$ in rats, rabbits (mg/kg): 110, 110 i.p. (Fitch, Tatum).

Sodium salt, C$_9$H$_{13}$N$_2$NaO$_3$, hygroscopic powder. Freely sol in water; slightly sol in alc. Practically insol in ether, chloroform. Aq solns are alkaline to litmus. Solns are unstable and pptn occurs on boiling.

Caution: May be habit forming. This is a controlled substance (depressant) listed in the U.S. Code of Federal Regulations, Title 21 Part 1308.13 (1987).

THERAP CAT: Sedative, hypnotic.

THERAP CAT (VET): Sedative, hypnotic.

7760. Probenecid. *4-[(Dipropylamino)sulfonyl]benzoic acid; p-(dipropylsulfamoyl)benzoic acid; p-*(dipropylsulfamyl)benzoic acid; Benemid; Probecid; Proben. C$_{13}$H$_{19}$NO$_4$S; mol wt 285.36. C 54.72%, H 6.71%, N 4.91%, O 22.43%, S 11.23%. Prepn from *p*-carboxybenzenesulfonyl chloride and dipropylamine: C. S. Miller, U.S. pat. **2,608,507** (1952 to Sharp & Dohme). Study of metabolites: Z. H. Israili *et al.*, *J. Med. Chem.* **15**, 709 (1972). Pharmacokinetics in man: R. F. Cunningham *et al.*, *Clin. Pharmacokinet.* **6**, 135 (1981). Comprehensive description: A. A. Al-Badr, H. A. El-Obeid in *Analytical Profiles of Drug Substances* vol. **10**, K. Florey, Ed. (Academic Press, New York, 1981) pp 639-663.

Crystals from dilute alcohol, mp 194-196°. uv max (0.1*N* NaOH): 242.5 nm. pKa 5.8. Slightly bitter taste, pleasant aftertaste. Sol in chloroform, in dil solns of NaOH buffered to pH 7.4. Nearly insol in water. LD$_{50}$ orally in rats: 1.6 g/kg.

THERAP CAT: Uricosuric.

7761. Probucol. *4,4'-[(1-Methylethylidene)bis(thio)]bis-[2,6-bis(1,1-dimethylethyl)phenol]; 4,4'-(isopropylidenedithio)bis[2,6-di-tert-butylphenol];* acetone bis(3,5-di-*tert*-butyl-4-hydroxyphenyl)mercaptole; DH-581; Biphenabid; Bisbid; Bisphenabid; Lorelco; Lurselle; Sinlestal. C$_{31}$H$_{48}$-O$_2$S$_2$; mol wt 516.84. C 72.04%, H 9.36%, O 6.19%, S 12.41%. Prepn: M. B. Neuworth, Fr. pat. **1,561,853**; *eidem,* U.S. pat. **3,576,883** (1969, 1971 both to Consolidation Coal Co.); and use as a cholesterol-lowering agent: J. W. Barnhart, P. J. Shea, U.S. pat. **3,862,332** (1975 to Dow). Prepn and activity studies: M. B. Neuworth *et al.*, *J. Med. Chem.* **13**, 722 (1970). Pharmacological studies: J. W. Barnhart *et al.*, *Am. J. Clin. Nutr.* **23**, 1229 (1970); Drake *et al.*, *Circulation* **40**, Suppl. 3, 73 (1969). Clinical studies: *eidem, Metab. Clin. Exp.* **18**, 916 (1969); Kalams *et al.*, *Curr. Ther. Res.* **13**,

692 (1971). Review of pharmacology and therapeutic use: R. C. Heel *et al.*, *Drugs* **15**, 409-428 (1978). Symposium on mechanism of action, clinical efficacy and safety: *Am. J. Cardiol.* **57**, 1H-54H (1986).

White crystalline solid from ethanol, mp 124.5-126°; fine, yellow crystals from isopropanol, mp 125-126.5°.

THERAP CAT: Antihyperlipoproteinemic.

7762. Procainamide Hydrochloride. *4-Amino-N-[2-(diethylamino)ethyl]benzamide monohydrochloride;* procaine amide hydrochloride; Amidoprocain; Amisalin; Novocamid; Novocainamid; Procamide; Procan-SR; Procapan; Procardyl; Promide; Pronestyl hydrochloride; Supicaine Amide hydrochloride. C$_{13}$H$_{22}$ClN$_3$O; mol wt 271.79. C 57.45%, H 8.16%, Cl 13.05%, N 15.46%, O 5.89%. Prepn: M. Yamazaki *et al.*, *J. Pharm. Soc. Japan* **73**, 294 (1953); Y. Tashika, M. Kuranari, *ibid.* 1069. Comprehensive description: R. B. Poet, H. Kadin in *Analytical Profiles of Drug Substances* vol. **4**, K. Florey, Ed. (Academic Press, New York, 1975) pp 333-383.

Crystals, mp 165-169°. uv max: 278 nm. Freely sol in water. Sol in alc; slightly sol in chloroform; very sparingly sol in benzene, ether. The pH of a 10% aq soln is 5.5. Commercially available aq solns are preserved with 0.9% benzyl alcohol and 0.09% sodium bisulfite.

THERAP CAT: Cardiac depressant (antiarrhythmic).

7763. Procaine. *4-Aminobenzoic acid 2-(diethylamino)ethyl ester; p-*aminobenzoyldiethylaminoethanol; 2-diethylaminoethyl *p-*aminobenzoate. C$_{13}$H$_{20}$N$_2$O$_2$; mol wt 236.30. C 66.07%, H 8.53%, N 11.86%, O 13.54%. Benzoic acid derivative with anesthetic activity. Prepn: A. Einhorn, U.S. pat. **812,554** (1906); *idem, Ann.* **371**, 125 (1909); A. Einhorn, E. Uhlfelder, *ibid.* 131. CNS effects: C. G. Peterson, *Anesthesiology* **16**, 976 (1955). Intravenous pharmacokinetics in humans: A. B. Seifen *et al.*, *Anesth. Analg. (Cleveland)* **58**, 382 (1979). Clinical evaluation as anti-arrhythmic and cough suppressant during anesthesia: D. S. Thompson *et al.*, *Am. J. Surg.* **138**, 798 (1979). Stabilization of vascular smooth muscle *in vitro*: K. Kitamura *et al.*, *Drugs Exp. Clin. Res.* **12**, 773 (1986). Toxicity data: W. C. North, K. F. Urbach, *J. Am. Pharm. Assoc. Sci. Ed.* **45**, 382 (1956); E. I. Goldenthal, *Toxicol. Appl. Pharmacol.* **18**, 185 (1971).

Hygroscopic, anhydr plates, tablets from ligroin or ether, mp 61°. When freshly precipitated, one gram dissolves in 200 ml water. Sol in alc, ether, benzene, chloroform. LD$_{50}$ in mice (mg/kg): 195 i.p.; 45 i.v. (North, Urbach).

Dihydrate, needles from aq alc, mp 51°. Slightly bitter taste; applied to the tongue causes transitory numbing sensation.

Nitrate, C$_{13}$H$_{21}$N$_3$O$_5$, crystals, mp 100-102°. Sol in water, alcohol. The aq soln is neutral. Particularly useful with silver nitrate because no precipitate forms.

Butyrate, C$_{17}$H$_{28}$N$_2$O$_4$, *Probutylin.* Hygroscopic crystals. Soluble in water, alcohol, vegetable oils.

Borate, C$_{13}$H$_{25}$B$_5$N$_2$O$_{12}$, *Borocaine.* Small, monoclinic, tabular crystals, mp 165-166°. One gram dissolves in 4 ml water. Sol in alcohol. Insol in benzene, chloroform, ether. Aq solns are alkaline, may be sterilized by brief boiling.

Hydrochloride, $C_{13}H_{21}ClN_2O_2$, *Allocaine, Alocaine, Aminocaine, Anesthesol, Anestil, Atoxicocaine, Bernacaine, Cetain, Ethocaine, Eugerase, Irocaine, Isocaine-Asid, Isocaine-Heisler, Jenacaine, Juvocaine, Kerocaine, Medaject, Naucaine, Neocaine, Novocain, Paracain, Planocaine, Scurocaine, Sevicaine, Syncaine, Topokain, Westocaine.* Crystals. Six-sided plates, monoclinic or triclinic. mp 153-156°. Numbing taste. Stable in air. One gram dissolves in 1 ml water and in 30 ml alcohol. Slightly sol in chloroform. Almost insol in ether. The pH of a 0.1M aq soln is 6.0. Aq solns may be sterilized by boiling. LD$_{50}$ in mice (mg/kg): 660 ± 60 s.c. (Goldenthal).

THERAP CAT: Local anesthetic.

THERAP CAT (VET): Local anesthetic.

7764. Procarbazine. *N-(1-Methylethyl)-4-[(2-methylhydrazino)methyl]benzamide; N-isopropyl-α-(2-methylhydrazino)-p-toluamide; N-4-isopropylcarbamoylbenzyl-N′-methylhydrazine; p-(N^1-methylhydrazinomethyl)-N-isopropylbenzamide; ibenzmethyzin; MIH; Ro 4-6467.* $C_{12}H_{19}N_3O$; mol wt 221.30. C 65.12%, H 8.65%, N 18.99%, O 7.23%. Prepn: **Belg.** pat. **618,638** corresp to W. Bollag *et al.*, **U.S.** pat. **3,520,926** (1962, 1970 to Hoffmann-La Roche). Toxicity data: E. I. Goldenthal, *Toxicol. Appl. Pharmacol.* **18**, 185 (1971). Comprehensive description: R. J. Rucki in *Analytical Profiles of Drug Substances* **vol.** 5, K. Florey, Ed. (Academic Press, New York, 1976) pp 403-427.

CH$_3$NHNHCH$_2$—⟨ ⟩—CONHCH(CH$_3$)$_2$

Hydrochloride, $C_{12}H_{20}ClN_3O$, *Matulane, Natulan.* Crystals from methanol, mp 223-226°. LD$_{50}$ orally in rats: 785 ±34 mg/kg (Goldenthal).

Hydrobromide, $C_{12}H_{19}N_3O \cdot HBr$, crystals from methanol + ether, dec 216-217°.

Note: This substance may reasonably be anticipated to be a carcinogen: *Fourth Annual Report on Carcinogens* (NTP 85-002, 1985) p 172.

THERAP CAT: Antineoplastic.

7765. Procaterol. *8-Hydroxy-5-[1-hydroxy-2-[(1-methylethyl)amino]butyl]-2(1H)-quinolinone; (±)-erythro-8-hydroxy-5-[1-hydroxy-2-(isopropylamino)butyl]carbostyril.* $C_{16}H_{22}N_2O_3$; mol wt 290.37. C 66.18%, H 7.64%, N 9.65%, O 16.53%. Sympathomimetic amine with selective β$_2$-adrenergic agonist activity. Prepn: K. Nakagawa *et al.*, **Belg.** pat. **823,841;** *eidem*, **U.S.** pat. **4,026,897** (1975, 1977 both to Otsuka); S. Yoshizaki *et al.*, *J. Med. Chem.* **19**, 1138 (1976). Prepn of optical isomers: *eidem, ibid.* **20**, 1103 (1977). Assessment of selective action: H. Himori, N. Taira, *Brit. J. Pharmacol.* **61**, 9 (1977). Metabolism: Y. Yasuda *et al.*, *Arzneimittel-Forsch.* **29**, 261 (1979). Pharmacokinetics: M. Ishigami *et al., ibid.* 266. Series of reproduction studies: *Iyakuhin Kenkyu* **10**, 68-111 (1979), *C.A.* **90**, 197728f-30a (1979). Antigenicity test: N. Nakagiri, S. Tei, *Oyo Yakuri* **17**, 363 (1979), *C.A.* **91**, 186577a (1979).

Hydrochloride hemihydrate, $C_{16}H_{23}ClN_2O_3 \cdot \frac{1}{2}H_2O$, *OPC-2009, Masacin, Meptin, Onsukil, Pro-Air, Procadil, Promaxol, Propulm.* Off-white cryst powder, mp 193-197° (dec). Sol in methanol; slightly sol in ethanol. Practically insol in acetone, ether, ethyl acetate, chloroform, benzene. Colors on exposure to light. LD$_{50}$ of the hydrochloride in male rats (mg/kg): 2600 orally, 80 i.v. (Nakagawa).

THERAP CAT: Bronchodilator.

7766. Procerin. *2-Hydroxy-5-(3-methyl-2-butenyl)-4-(1-methylethenyl)-2,4,6-cycloheptatrien-1-one.* $C_{15}H_{18}O_2$;

mol wt 230.29. C 78.23%, H 7.88%, O 13.90%. A tropolone isolated from *Juniperus procera* Hochst., *Cupressaceae,* obtained in Kenya: Petterson, Runeberg, *Acta Chem. Scand.* **15**, 713 (1961). Structure: Runeberg, *ibid.* 645. Synthesis: Kitahara, Kato, *Bull. Chem. Soc. Japan* **37**, 895 (1964).

Crystals from dil alc, mp 71-72°. Sol in alc, petr ether.

7767. Prochloraz. *N-Propyl-N-[2-(2,4,6-trichlorophenoxy)ethyl]-1H-imidazole-1-carboxamide; 1-[N-propyl-N-[2-(2,4,6-trichlorophenoxy)ethyl]carbamoyl]imidazole; BTS 40542; Sportak.* $C_{15}H_{16}Cl_3N_3O_2$; mol wt 376.67. C 47.83%, H 4.28%, Cl 28.24%, N 11.16%, O 8.49%. Broad spectrum fungicide for use on cereal crops, fruit and vegetables. Prepn from imidazole: R. F. Brookes *et al.*, **Ger.** pat. **2,429,523;** *eidem*, **U.S.** pat. **3,991,071** (1975, 1976 both to Boots). Fungicidal properties: R. J. Birchmore *et al.*, *Proc. Brit. Crop. Prot. Conf.-Pests Dis.* **1977**, 593; R. G. Harris *et al., ibid.* **1979**, 53. Efficacy against cereal powdery mildew: D. M. Weighton *et al., ibid.* **1977**, 25. Protective effect against blotch in winter wheat and barley: R. G. Harris, G. Barnes, *ibid.* **1981**, 267. Effect on cytochrome P-450 and sterol biosynthesis in Japanese quail: J.-L. Riviere *et al., Pestic. Sci.* **15**, 317 (1984). *Review:* A. de Saint-Blanquat, J. My, *Def. Veg.* **37**, 121 (1983).

White crystalline solid, mp 38.5-41°. Technical product is pale yellow viscous oil. Vapor pressure at 20°: 0.57 × 10^{-9} torr. Almost insol in water (0.0055 g/l). Soly in chloroform, diethyl ether, toluene, xylene: 2500 g/l; in acetone: 3500 g/l. LD$_{50}$ in rats (mg/kg): 1600 orally; > 5000 s.c.; 400-800 i.p.; LC$_{50}$ (96 hour) in rainbow trout: 1 mg/l; in bluegill: 2.2 mg/l (de Saint-Blanquat, My). Relatively non-toxic to bees.

USE: Fungicide.

7768. Prochlorperazine. *2-Chloro-10-[3-(4-methyl-1-piperazinyl)propyl]-10H-phenothiazine; 3-chloro-10-[3-(4-methyl-1-piperazinyl)propyl]phenothiazine; 2-chloro-10-[3-(1-methyl-4-piperazinyl)propyl]phenothiazine; N-[γ-(4′-methylpiperazinyl-1′)propyl]-3-chlorophenothiazine; chlormeprazine; prochlorpemazine; proclorperazine; Bayer A 173; RP 6140; SKF 4657; Nipodal; Tementil; Kronocin; Capazine; Emelent; Meterazine.* $C_{20}H_{24}ClN_3S$; mol wt 373.94. C 64.24%, H 6.47%, Cl 9.48%, N 11.24%, S 8.57%. Prepn: R. J. Horclois, **Brit.** pat. **780,193;** **Fr.** pat. **1,167,627;** **U.S.** pat. **2,902,484** (1957, 1958, 1959, all to Rhône-Poulenc). Toxicity data: S. Courvoisier *et al.*, *C.R. Soc. Biol.* **152**, 1371 (1958).

Dimaleate, $C_{28}H_{32}ClN_3O_8S$, *Buccastem, Compazine, Emetiral, Nibromin A, Stemetil, Vertigon.* Minute crystals, mp 228°. Very slightly sol in water (less than 0.1% at 20°). Slightly sol in methanol, ethanol. Practically insol in ether, benzene, chloroform. LD_{50} in mice (mg/kg): 400 s.c.; 120 i.p.; 90 i.v., 400 orally (Courvoisier).

THERAP CAT: Anti-emetic. Antipsychotic. Treatment of vertigo.

THERAP CAT (VET): Anti-emetic.

7769. Procodazole. *1H-Benzimidazole-2-propanoic acid;* β-*(2-benzimidazolyl)propionic acid;* 2-(2-carboxyethyl)benzimidazole; propazol; AL-1241; Estimulocel. $C_{10}H_{10}$-N_2O_2; mol wt 190.20. C 63.15%, H 5.30%, N 14.73%, O 16.82%. Non-specific, active immunoprotective agent against viral and bacterial infections. Prepn: J. Maier, *Ann.* **327,** 17 (1903); R. Meyers, H. Lüders, *ibid.* **415,** 29 (1918); B. Chatterjee, *J. Chem. Soc.* **1929,** 2966; Span. pat. **407,882** (1972 to Lafarquim). Pharmacological studies: C. Fernández *et al., Rev. Clin. Esp.* **135,** 539 (1974); *eidem, ibid.* **141,** 51 (1976). Clinical studies: M. Pérez Tascon, J. M. Monturio, *Med. Klin.* **181,** 78 (1976).

Silky white needles from water, mp 228° (dec). Sol in alc, warm water. Practically insol in ether, benzene. The water soln is extremely sweet-tasting.

THERAP CAT: Immunopotentiator (non-specific).

7770. Procyclidine. α-*Cyclohexyl*-α-*phenyl*-1-*pyrrolidinepropanol;* 1-cyclohexyl-1-phenyl-3-(1-pyrrolidinyl)-1-propanol; 1-cyclohexyl-1-phenyl-3-pyrrolidino-1-propanol. $C_{19}H_{29}NO$; mol wt 287.43. C 79.39%, H 10.17%, N 4.87%, O 5.57%. Prepn of free base and hydrochloride: Adamson *et al., J. Chem. Soc.* **1951,** 52; Adamson, U.S. pat. **2,891,890** (1959 to Burroughs Wellcome). Prepn of methochloride: Bottorff, U.S. pat. **2,826,590** (1958 to Lilly); Harfenist, Magnien, U.S. pat. **2,842,555** (1958 to Burroughs Wellcome).

Crystals from petr ether, mp 85.5-86.5°. uv max (0.17% in ethanol): 258.5 nm (ε 233).

Hydrochloride, $C_{19}H_{30}ClNO$, *Arpicolin, Kemadrin, Osnervan.* Crystals from ethanol + ethyl acetate, dec 226-227°. Moderately sol in water (about 3.0 g/100 ml); more sol in alcohol, chloroform; very slightly sol in ether.

Methochloride, $C_{20}H_{32}ClNO$, *1-(3-cyclohexyl-3-hydroxy-3-phenylpropyl)-1-methylpyrrolidinium chloride, tricyclamol chloride, Elorine chloride, Lergine chloride, Tricoloid chloride.* Crystals from nitroethane, mp 159-164°. Moderately sol in water, alcohol. Practically insol in ether.

Methosulfate, $C_{21}H_{35}NO_5S$, *Elorine sulfate, Vagosin sulfate.* Crystals, mp about 100°. Soluble in water (about 2% at 25°), alcohol. Practically insol in ether.

THERAP CAT: Anticholinergic.

7771. Procymate. α-*Ethylcyclohexanemethanol carbamate; carbamic acid 1-cyclohexylpropyl ester;* 1-cyclohexylpropyl carbamate; T 3033; Equipax. $C_{10}H_{19}NO_2$; mol wt 185.26. C 64.83%, H 10.34%, N 7.56%, O 17.27%. Prepn: Swierkot, Brit. pat. **979,236** (1960 to Comp. Franc. Matières Colorantes).

Crystals, mp 128-129°.

THERAP CAT: Tranquilizer.

7772. Procymidone. *3-(3,5-Dichlorophenyl)-1,5-dimethyl-3-azabicyclo[3.1.0]hexane-2,4-dione;* N-(3,5-dichlorophenyl)-1,2-dimethyl-1,2-cyclopropanedicarboximide; dicyclidine; S-7131; Sumisclex; Sumilex. $C_{13}H_{11}Cl_2NO_2$; mol wt 284.14. C 54.95%, H 3.90%, Cl 24.95%, N 4.93%, O 11.26%. Dicarboximide fungicide systemically active against *Botrytis* and *Sclerotinia spp.* on fruits and vegetables. Prepn: Neth. pat. Appl. **7,003,836;** A. Fujinami *et al.,* U.S. pat. **3,903,090** (1970, 1975 both to Sumitomo). Antimicrobial spectrum and systemic activity: Y. Hisada *et al., J. Pestic. Sci.* **1,** 145 (1976). Mechanism of action: A. C. Pappas, D. J. Fisher, *Pestic. Sci.* **10,** 239 (1979). Field evaluation: I. F. Jackson, B. N. Smith, *Proc. 32nd N.Z. Weed Pest Control Conf.* 278 (1979). HPLC determn in wine musts and extracts: P. Cabras *et al., J. Chromatog.* **256,** 176 (1983). *Review:* N. Mikami, J. Miyamoto, *Rev. Plant Protect. Res.* **14,** 85-95 (1981).

Crystalline solid, mp 165-167°. d^{25} 1.42-1.46. Vapor pressure: 1.32×10^{-4} mm Hg. uv max: 207.5, 275 nm (ε 4.2×10^4, 4.1×10^2). Soly in water (25°): 4.5 ppm. Highly sol in acetonitrile, acetone, ether, chloroform. Moderately sol in benzene, toluene. Stable in solvents. Unstable in alkaline media. LD_{50} in male rats (mg/kg): 6800 orally, > 10,000 dermally (Jackson). Also reported as LD_{50} in male, female rats (g/kg): 7.8, 9.1 orally (Mikami).

USE: Systemic agricultural fungicide.

7773. Prodiamine. *2,4-Dinitro-N^3,N^3-dipropyl-6-(trifluoromethyl)-1,3-benzenediamine;* α,α,α-*trifluoro-3,5-dinitro-N^4,N^4-dipropyltoluene-2,4-diamine;* 5-dipropylamino-α,α,α-trifluoro-4,6-dinitro-*o*-toluidine; 2,6-dinitro-N^1,N^1-dipropyl-4-trifluoromethyl-*m*-phenylenediamine; N^3,N^3-di-*n*-propyl-2,4-dinitro-6-trifluoromethyl-1,3-phenylenediamine; USB-3153; CN-11-2936; Marathon; Blockade; Endurance; Rydex. $C_{13}H_{17}F_3N_4O_4$; mol wt 350.30. C 44.57%, H 4.89%, F 16.27%, N 15.99%, O 18.27%. Broad-spectrum, pre-emergence dinitroaniline herbicide. Prepn: D. L. Hunter *et al.,* Ger. pat. **2,013,510;** *eidem,* U.S. pat. **3,764,623** (1970, 1973 both to U.S. Borax and Chemical Corp.). Weed control in container grown plants: T. A. Fretz, W. J. Sheppard, *HortScience* **15,** 489 (1980); S. A. Duray, F. T. Davies, *J. Environ. Hort.* **5,** 82 (1987). Field studies: M. G. Sybouts, *Proc. West. Soc. Weed Sci.* **40,** 169 (1987). Evaluation of herbicidal activity: W. Bond, *Crop Prot.* 7, 75 (1988). Review of herbicidal activity: S. J. Bowe, *Proc. West. Soc. Weed Sci.* **39,** 216-218 (1986).

Orange needles from 95% ethanol, mp 124-125°.

USE: Herbicide.

7774. Prodigiosin. *4-Methoxy-5-[(5-methyl-4-pentyl-2H-pyrrol-2-ylidene)methyl]-2,2'-bi-1H-pyrrole;* 2,2'-[3-methoxy-4'-amyl-5'-methyl-5-(2''-pyrryl)]dipyrrylmethene; prodigiosine. $C_{20}H_{25}N_3O$; mol wt 323.40. C 74.27%, H 7.79%, N 12.99%, O 4.95%. Antibiotic pigment produced by *Chromobacterium prodigiosum (Serratia marcescens).* Exhibits antimicrobial and cytotoxic properties. Isoln: Wrede, Hettche, *Ber.* **62,** 2678 (1929); Lasseur, Georges, *Trav. Lab. Microbiol. Nancy* **9,** 47 (1936); Lasseur, Melcion, *ibid.* **13,** 192 (1944). Purification: Morgan, Tanner, *J. Chem. Soc.* **1955,** 3305. Structure: Wrede, Rothhaas, *Z. Physiol. Chem.*

226, 95 (1934). Revised structure and synthesis: H. H. Wasserman *et al., J. Am. Chem. Soc.* **82,** 506 (1960); H. Rapoport, K. G. Holden, *ibid.* 5510; *ibid.* **84,** 635 (1962); A. J. Castro *et al., J. Org. Chem.* **28,** 857 (1963). Total synthesis and *in vitro* cytotoxic activity: D. L. Boger, M. Patel, *Tetrahedron Letters* **28,** 2499 (1987); *eidem, J. Org. Chem.* **53,** 1405 (1988). NMR studies: R. J. Cushley *et al., Can. J. Chem.* **53,** 148 (1975). Antimalarial activity: A. J. Castro, *Nature* **213,** 903 (1967). *Review:* R. P. Williams, W. R. Hearn in *Antibiotics* vol. 2, D. Gottlieb, P. D. Shaw, Eds. (Springer-Verlag, New York, 1967) pp 410-432, 449-451. Review of synthesis: A. H. Jackson, K. M. Smith, in *The Total Synthesis of Natural Products,* **vol. 1,** J. ApSimon, Ed. (Wiley-Interscience, New York, 1973) pp 227-232; of biological activity and production: V. Alonzo, *Ig. Mod.* **81,** 557-564 (1984), *C.A.* **101,** 126398h (1984).

Lustrous square pyramids (dark red with green reflex) from petr ether, mp 151-152°. Almost insol in water. Moderately sol in alcohol, ether; freely sol in chloroform, bromoform, benzene. Alkaline or neutral solns are orange-yellow, acid solns are red. Absorption max (isopropanol): 466 nm (ε 43000); 336, 280 nm.

Hydrochloride, $C_{20}H_{26}ClN_3O$, magenta crystals from benzene + petr ether, dec 148.5-150°. Absorption max (isopropanol): 540, 294 nm (ε 70700, 10800).

7775. Prodilidine. *1,2-Dimethyl-3-phenyl-3-pyrrolidinol propanoate (ester); propionic acid 1,2-dimethyl-3-phenyl-3-pyrrolidinyl ester;* 1,2-dimethyl-3-phenyl-3-pyrrolidyl propionate; 1,2-dimethyl-3-phenyl-3-propionoxypyrrolidine; 1,2-dimethyl-3-phenyl-3-propionyloxypyrrolidine. $C_{15}H_{21}NO_2$; mol wt 247.33. C 72.84%, H 8.56%, N 5.66%, O 12.94%. Prepn: J. F. Cavalla *et al., J. Med. Pharm. Chem.* **4,** 1 (1961); J. F. Cavalla, J. Davoll, **Brit.** pat. **862,513** corresp to **U.S.** pat. **3,256,297** (1961, 1966 to Parke, Davis). Crystal structure: C. Humblet *et al., Acta Crystallogr.* **B33,** 1618 (1977).

dl-Form, liq. n_D^{20} 1.5164. $bp_{1.1}$ 126-128°; $bp_{1.5}$ 117-130°.
dl-Form hydrochloride, $C_{15}H_{22}ClNO_2$, *Cogesic.* mp 194-195°. LD_{50} i.p. in rats: 123 mg/kg, E. I. Goldenthal, *Toxicol. Appl. Pharmacol.* **18,** 185 (1971).

THERAP CAT: Analgesic.

7776. Prodipine. *1-(1-Methylethyl)-4,4-diphenylpiperidine; 1-isopropyl-4,4-diphenylpiperidine.* $C_{20}H_{25}N$; mol wt 279.43. C 85.97%, H 9.02%, N 5.01%. CNS stimulating agent with antidepressant and antiparkinson properties. Prepn: H. G. Menge, J. Klosa, **Ger.** pat. **1,936,452;** *eidem,* **U.S.** pat. **4,016,280** (1971, 1977 both to Byk Gulden). Evaluation of MAO-inhibitory properties *in vivo* and *in vitro*: G. Planz *et al., Arzneimittel-Forsch.* **23,** 281 (1973). Effects on 5-hydroxytryptamine uptake in platelets: M. Eltze *et al., ibid.* **30,** 1129 (1980). Comparative study of effects in Parkinson's disease: E. Gründig, F. Gerstenbrand, *Wien Klin. Wochenschr.* **92,** 868 (1980).

Oil, $bp_{0.01}$ 117-125°.
Hydrochloride, $C_{20}H_{26}ClN$, *Anthen.* Crystals from isopropanol, mp 267°.

THERAP CAT: Antiparkinsonian.

7777. Prodlure. *9,11-Tetradecadien-1-ol acetate; cis-9,-trans-11-tetradecadienyl acetate; cis-9,trans-11-TDDA;* Ferodin SL; Litlure A. $C_{16}H_{28}O_2$; mol wt 252.40. C 76.14%, H 11.18%, O 12.68%. Major component of sex pheromone of female *spodoptera litura* (F.) and Egyptian cotton leafworm, *S. littoralis* (Boisd.). Isoln and prepn: Y. Tamaki *et al., Appl. Entomol. Zool.* **8,** 200 (1973); B. F. Nesbitt *et al., Nature New Biol.* **244,** 208 (1973). Prepn: T. Yushimo *et al.,* **Ger.** pat. **2,406,259** (1974 to Natl. Inst. Agr. Sci.), *C.A.* **82,** 97681b (1975). Stereoselective synthesis: D. R. Hall *et al., Chem. & Ind. (London)* **1975,** 216; G. Goto *et al., Chem. Letters* **1975,** 103; G. Decodts *et al., Synthesis* **7,** 510 (1979). Activity studies in presence of minor component, *(Z,E)-9,12-tetradecadien-1-ol acetate:* S. Neumark *et al., Environ. Letters* **6,** 219 (1974); Y. Tamaki, T. Yushima *et al., J. Insect. Physiol.* **20,** 1005 (1974); M. Kehat *et al., Appl. Entomol. Zool.* **11,** 45 (1976).

Colorless liquid, $bp_{0.2}$ 147-148°, $bp_{0.003}$ 85-86°. uv max (hexane): 232 nm (ε 27,300).

USE: Insect sex attractant.

7778. Pro-Drone®. *1-(8-Methoxy-4,8-dimethylnonyl)-4-(1-methylethyl)benzene;* 2-methoxy-9-(*p*-isopropylphenyl)-2,6-dimethylnonane; MV-678; AI 3-36206. $C_{21}H_{36}O$; mol wt 304.52. C 82.83%, H 11.92%, O 5.25%. Synthetic analog of juvenile hormones, *q.v.* Prepn: M. Schwarz *et al.,* **Ger.** pat. **2,536,298** corresp to **U.S.** pat. **4,002,769** (1976, 1977 to Stauffer). Insect maturation inhibiting activity: *eidem, J. Econ. Entomol.* **67,** 598 (1974); J. E. Wright *et al., ibid.* **69,** 79 (1976); C. H. Schaefer *et al., ibid.* 119.

USE: Insecticide.

7779. Producer Gas. Blow gas. Obtained by blowing air and steam through incandescent coke. *Composition:* 14% CO_2; 10% CO; 75% N_2; 1% Ar.

USE: In the manuf of ammonia as source of nitrogen. *cf.* Water gas. *Caution:* Asphyxiant. See Carbon Monoxide.

7780. Proflavine. *3,6-Acridinediamine;* 3,6-diaminoacridine; 2,8-diaminoacridine. $C_{13}H_{11}N_3$; mol wt 209.25. C 74.62%, H 5.30%, N 20.08%. Prepn: M. Schöpff, *Ber.* **27,** 2320 (1894); **Ger.** pat. **230,412** (1910 to Cassella), *C.A.* **5,** 2734 (1911); W. P. Thompson, **Brit.** pat. **137,214** (1919 to Poulenc Frères), *C.A.* **14,** 1445 (1920); A. Albert, *J. Chem. Soc.* **1941,** 121, 484; **1947,** 244. *Review: The Acridines,* A. Albert, Ed. (St. Martin's Press, New York, 2nd ed., 1966) pp 300-302; *Acridines,* R. M. Acheson, Ed. (Interscience, New York, 1956) pp 341-344.

Yellow needles from alc, mp 281° (Schöpff), mp 288° (Albert). Sol in water, ethanol; insol in benzene, ether. pKa 9.7. Solns are brownish and when diluted are fluorescent. *Solns should be discarded when they become turbid.* LD$_{50}$ s.c. in mice: 0.14 g/kg, S. D. Rubbo, *Brit. J. Exp. Pathol.* **28**, 1 (1947).

Dihydrochloride, $C_{13}H_{13}Cl_2N_3$, orange-yellow needles. pH of 0.1% soln 2.5-3.0.

Sulfate, $C_{13}H_{13}N_3O_4S$, red needles. pH of 0.1% soln is 2.5.

Sulfate dihydrate, $C_{13}H_{13}N_3O_4S$, *proflavine hemisulfate, neutral proflavine sulfate.* Orange-red bitter crystals, hygroscopic. Sol in 300 parts cold water, in 1 part boiling water; slightly sol in alc; insol in ether, CHCl$_3$. pH of satd soln is 6-8.

Mixture with the hydrochloride of proflavine methochloride: *See* Acriflavine.

THERAP CAT: Topical antiseptic.

THERAP CAT (VET): Topical antiseptic.

7781. Profluralin. *N-(Cyclopropylmethyl)-2,6-dinitro-N-propyl-4-(trifluoromethyl)benzenamine; N-(cyclopropylmethyl)-α,α,α-trifluoro-2,6-dinitro-N-propyl-p-toluidine;* N-cyclopropylmethyl-*N*-n-propyl-4-trifluoromethyl-2,6-dinitroaniline; CGA 10832; ER 5461; B 4576; Tolban; Pregard. $C_{14}H_{16}F_3N_3O_4$; mol wt 347.30. C 48.42%, H 4.64%, F 16.41%, N 12.10%, O 18.43%. Selective pre-planting herbicide. Prepn: **Neth. pat. Appl. 6,911,565** corresp to L. L. Maravetz, **U.S. pat. 3,546,295** (1969, 1970 to Esso). Persistence in soil: J. M. Kennedy *et al., Weed Sci.* **25**, 373 (1977).

Yellow-orange crystals or liquid, mp 33-36°. Soly in water at 20°: 0.1 mg/l. Readily sol in most organic solvents. LD$_{50}$ orally in rats: 10,000 mg/kg, *RTECS Vol. II*, R. J. Lewis, R. L. Tatken, Eds. (1979) p 628.

USE: Herbicide.

7782. Progabide. *4-[[(4-Chlorophenyl)(5-fluoro-2-hydroxyphenyl)methylene]amino]butanamide;* 4-[[α-(p-chlorophenyl)-5-fluorosalicylidene]amino]butyramide; 4-[[α-(p-chlorophenyl)-5-fluoro-2-hydroxybenzylidene]amino]butyramide; halogabide; SL 76002; Gabren(e). $C_{17}H_{16}ClFN_2O_2$; mol wt 334.78. C 60.99%, H 4.82%, Cl 10.59%, F 5.67%, N 8.37%, O 9.56%. Gamma-aminobutyric acid (GABA) antagonist with anti-epileptic activity. Prepn: J.-P. Kaplan *et al.,* **Ger.** pat. **2,634,288;** *eidem,* **U.S.** pat. **4,094,992** (1977, 1978 to Synthelabo); *eidem, J. Med. Chem.* **23**, 702 (1980). Use as analgesic: J.-P. Kaplan, U.S. pat. **4,361,583** (1982 to Synthelabo). Pharmacokinetics: I. Johno *et al., J. Pharm. Sci.* **71**, 633 (1982). HPLC determn in biological fluids: P. Padovani *et al., J. Chromatog.* **308**, 229 (1984). Doubleblind clinical trial in therapy resistant epilepsy: P. Loiseau *et al., Epilepsia* **24**, 703 (1983); in spasticity: K. Mondrup, E. Pedersen, *Acta Neurol. Scand.* **69**, 200 (1984).

Crystals from cyclohexane and toluene, mp 133-135°; possible second crystalline form, mp 142.5° (Kaplan, *J. Med. Chem.*). uv max in methanol: 332, 250, 210 (ε 4200, 10800, 24000). LD$_{50}$ i.p. in mice: 900 mg/kg (Kaplan).

THERAP CAT: Anticonvulsant.

7783. Progesterone. *Pregn-4-ene-3,20-dione;* Δ^4-pregnene-3,20-dione; corpus luteum hormone; luteohormone;

Corlutin; Corlutina; Corluvite; Cyclogest; Flavolutan; Fologenon; Gestone; Gestormone; Gestron; Hormoflaveine; Hormoluton; Lipo-Lutin; Lucorteum Sol; Luteodyn; Luteogan; Luteol; Luteosan; Luteovis; Lutex; Lutidon; Lutocyclin M; Lutocylin; Lutoform; Lutogyl (Inj); Lutren; Lutromone; Nalutron; Percutacrine Luteinique; Primolut; Progekan; Progestasert; Progesterol; Progestin; Progestogel; Progestol; Progestone; Prolidon (Inj); Proluton; Syngesterone; Utrogestan. $C_{21}H_{30}O_2$; mol wt 314.45. C 80.21%, H 9.62%, O 10.18%. Active principle of the corpus luteum, secreted during the latter half of the menstrual cycle. If pregnancy ensues, secretion continues. Exerts an antiovulatory effect when administered during days 5 to 25 of the normal menstrual cycle. Isoln from corpus luteum of pregnant sows: Butenandt, Westphal, *Ber.* **67**, 1440 (1934); Wintersteiner, Allen, *J. Biol. Chem.* **107**, 321 (1934). Structure: Butenandt *et al., Ber.* **67**, 1611 (1934). Synthesis of DL-form: Johnson *et al., J. Am. Chem. Soc.* **93**, 4332 (1971). Prepn from other steroids in review by W. H. Strain in Gilman's *Organic Chemistry* vol. **II** (Wiley, New York, 2nd ed., 1943) pp 1487-1489. Numerous patents, e.g., U.S. pats. **2,379,832; 2,232,438; 2,314,185.** Anesthetic effect and toxicity: H. Selye, *Proc. Soc. Exp. Biol. Med.* **46**, 116 (1941). Review of physiology: Csapo, *Sci. Am.* **198**, 40-46 (April, 1958); Rothchild, *Vitam. Horm. (New York)* **23**, 209-327 (1965). Book: *Progesterone and Progestins,* C. W. Bardin *et al.,* Eds. (Raven Press, New York, 1982) 462 pp.

Exists in two cryst forms of equal physiologic activity and which are readily interconverted. The α-form is orthorhombic (prisms from dil alc) with a:b:c = 0.750:1:0.905. Crystals show (011), (110), (010). Poor (011) cleavage. d^{23} 1.166. mp 127-131°. The β-form is orthorhombic (needles) with a:b:c = 0.563:1:0.275. Crystals acicular with parallel extinction and negative elongation. Cleavage on (001) and (110). d^{20} 1.171. mp 121°. [α]$_D^{20}$ +172° to +182° (c = 2 in dioxane). uv max: 240 nm. Insol in water. Sol in alcohol, acetone, dioxane, concd H$_2$SO$_4$. Sparingly sol in vegetable oils. 1 mg = 1 I.U.

Note: This substance may reasonably be anticipated to be a carcinogen: *Fourth Annual Report on Carcinogens* (NTP 85-002, 1985) p 173.

THERAP CAT: Progestogen.

THERAP CAT (VET): Progestational hormone. Has been used to control habitual abortion, to suppress or synchronize estrus.

7784. Proglumetacin. *1-(4-Chlorobenzoyl)-5-methoxy-2-methyl-1H-indole-3-acetic acid 2-[4-[3-[[4-(benzoylamino)-5-(dipropylamino)-1,5-dioxopentyl]oxy]propyl]-1-piperazinyl]ethyl ester;* (±)-N-[2-[1-(p-chlorobenzoyl)-5-methoxy-2-methyl-3-indolylacetoxy]ethyl]-N'-[3-(N-benzoyl-N',N'-di-n-propyl-DL-isoglutaminoyl)oxypropyl]piperazine. $C_{46}H_{58}ClN_5O_8$; mol wt 844.46. C 65.43%, H 6.92%, Cl 4.20%, N 8.29%, O 15.16%. Deriv of indomethacin, *q.v.* Prepn: F. Makovec *et al.,* **Ger. pat. 2,535,799** corresp to U.S. pat. **3,985,878** (both 1976 to Rotta). Series of articles on pharmacology, mechanism of action, safety: *Arzneimittel-Forsch.* **29**, 1116-1129 (1979). Bioavailability study: A. A. Bignamini, P. L. Casula, *Curr. Med. Res. Opin.* **6**, 299 (1979). Clinical evaluation: J. Münzenberg, S. Tachibana, *Pharmatherapeutica* **2**, 279 (1980); P. Loizzi *et al., ibid.* 285. Mutagenicity study: R. Vidal y Plana *et al., Farmaco Ed. Prat.* **33**, 543 (1978). Toxicity data for dimaleate: A. L. Rovati *et al., Arzneimittel-Forsch.* **29**, 1116 (1979).

Dimaleate, $C_{54}H_{66}ClN_5O_{16}$, *CR-604, protacine, Afloxan, Miridacin, Protaxit, Protaxon, Proxil.* Crystals from ethanol, mp 146-148°. LD_{50} in male mice, rats (mg/kg): 262, 450 orally (Rovati).

THERAP CAT: Anti-inflammatory.

7785. Proglumide. *4-(Benzoylamino)-5-(dipropylamino)-5-oxopentanoic acid; DL-4-benzamido-N,N-dipropylglutaramic acid; N-benzoyl-N',N'-di-n-propyl-DL-isoglutamine;* xylamide; CR 242; W-5219; Milid; Gastridene; Midelid; Milide; Promid. $C_{18}H_{26}N_2O_4$; mol wt 334.42. C 64.65%, H 7.83%, N 8.37%, O 19.14%. Prepn: **Neth. pat. Appl. 6,510,006** and **S. Afr. pat. 65/4065** (both 1966 to Rotta), C.A. **65**, 3793b (1966). Pharmacological activity: A. L. Rovati *et al., Minerva Med.* **58**, 3653 (1967); T. Umetzu *et al., Eur. J. Pharmacol.* **64**, 69 (1980). Pharmacokinetic study: A. A. Bignamini *et al., Arzneimittel-Forsch.* **29**, 639 (1979). Clinical study in duodenal ulcer: W. Bergemann *et al., Med. Klin.* **76**, 226 (1981). Cholecystokinin receptor antagonist activity: W. F. Hahne *et al., Proc. Nat. Acad. Sci. USA* **78**, 6304 (1981). Selective blockade of cholecystokinin CNS effects: L. A. Chiodo, B. S. Bunney, *Science* **219**, 1449 (1983).

$$(CH_3CH_2CH_2)_2NCCHCH_2CH_2COOH$$

Crystals, mp 142-145°. LD_{50} i.v., orally in mice: 2211-2649; 7350-8861 mg/kg, A. L. Rovati *et al., loc. cit.*

THERAP CAT: Anticholinergic.

7786. Proheptazine. *Hexahydro-1,3-dimethyl-4-phenyl-1H-azepin-4-ol propanoate (ester); DL-α-1,3-dimethyl-4-phenyl-4-propionoxyazacycloheptane; 1,3-dimethyl-4-phenyl-4-propionoxyhexamethylenimine; 4-phenyl-4-propionoxy-1,3-dimethylazacycloheptane; 4-propionoxy-1,3-dimethyl-4-phenylhexamethylenimine;* Wy 757; Dimephepramine. $C_{17}H_{25}NO_2$; mol wt 275.38. C 74.14%, H 9.15%, N 5.09%, O 11.62%. Prepn: Diamond, Bruce, U.S. pat. **2,775,589** (1956 to American Home Prod.). Structure and synthesis: Diamond *et al., J. Med. Chem.* **7**, 57 (1964).

Liquid, $bp_{0.3}$ 126°. n_D^{28} 1.5182. n_D^{21} 1.5215. Hydrobromide, $C_{17}H_{26}BrNO_2$, crystals from acetone + methanol, dec 207-207.5°. Soly in water at 25°: 1-2%. Hydrochloride, $C_{17}H_{26}ClNO_2$, crystals, dec 207°.

Caution: May be habit forming. This is a controlled substance (opiate) listed in the U.S. Code of Federal Regulations, Title 21 Part 1308.11 (1987).

THERAP CAT: Narcotic analgesic.

7787. Proinsulin. Single chain insulin precursor consisting of the insulin A and B chains and a connecting polypeptide (C-peptide), which contains 30-35 amino acids; the number and sequence of these amino acids are species dependent. Its presence was discovered in a human islet cell adenoma: D. F. Steiner, P. E. Oyer, *Proc. Nat. Acad. Sci. USA* **57**, 473 (1967). Conversion of proinsulin to insulin has a half-time of about 1 hour in rat islets *in vitro;* it is postulated that proteolytic enzymes cleave proinsulin at the sites

where two amino acids connect the C-peptide to the insulin chain.This cleavage results in the production of insulin and the C-peptide, both of which are retained in the secretory granules of the beta cells and discharged in equimolar amounts during exocytosis of the granules. *See* W. Kemmler, D. F. Steiner, *Biochem. Biophys. Res. Commun.* **41**, 1223 (1970). *In vitro* conversion of proinsulin to insulin with trypsin and carboxypeptidase B: W. Kemmler *et al., J. Biol. Chem.* **246**, 6786(1971), **248**, 4544 (1973). Isoln of bovine proinsulin: D. F. Steiner *et al., Diabetes* **17**, 725 (1968). Structural studies on mammalian proinsulin: R. E. Chance *et al., Science* **161**, 165 (1968) (porcine); C. Nolan *et al., J. Biol. Chem.* **246**, 2780 (1971) (bovine); P. E. Oyer *et al., ibid.* 1365 (human, monkey); J. D. Petersen *et al., ibid.* **247**, 4866 (1972) (monkey, sheep, dog); H. S. Tager, D. F. Steiner, *ibid.* 7936 (rat, horse); D . E. Massey, D. G. Smyth, *ibid.* **250**, 628 (1975) (guinea pig). Proposed three-dimensional structure: C. R. Snell, D. G. Smyth, *ibid.* 6291. Syntheses of C-peptides and human proinsulin: N. Yanaihara *et al., Diabetes* 27(Suppl. 1), 149 (1978). Synthesis of rat proinsulin in bacteria: L. Villa-Komaroff *et al.,* in *Polypeptide Hormones,* R. F. Beers, E. G. Bassett, Eds. (Raven Press, New York, 1980) pp 49-65. *Reviews:* D. F. Steiner *et al.* in *Diabetes, 8th Proc. Congr. Int. Diabetes Fed.,* W. J. Malaisse, J. Pirart, Eds. (Excerpta Med., Amsterdam, 1974) pp 119-133; A. H. Rubenstein *et al.,* in *Rec. Progr. Horm. Res.* **33**, R. O. Greep, Ed. (Academic Press, New York, 1977) pp 435-475; A. E. Kitabshi, *Metabolism* **26**, 547-587 (1977). Book: *Proinsulin, Insulin, C-Peptides,* S. Baba *et al.,* Eds. (Excerpta Medica, Amsterdam, 1979) 468 pp.

7788. Prolactin. *Pituitary lactogenic hormone;* adenohypophysial luteotropin; anterior pituitary luteotropin; lactogen; galactin; mammotropin; luteotropic hormone; luteotropin; LTH; Ferolactan. Polypeptide hormone of mol wt about 23,000; active principle of adenohypophysial gland essential in the induction of lactation in mammals at parturition. Its synergistic action with estrogen, *q.v.* promotes mammary gland proliferation. Also brings about the release of progesterone, *q.v.* from lutein cells which renders the uterine mucosa suited for the imbedding of the ovum, should fertilization occur. Isoln procedures from adenohypophyseal tissue or whole pituitary glands of ox, sheep, and swine: Lyons, *Cold Spring Harbor Symposia Quant. Biol.* **5**, 198 (1937); Li *et al., J. Biol. Chem.* **146**, 627 (1942); White *et al., ibid.* **143**, 447 (1942). Isoln from sheep pituitaries: Reisfeld *et al., J. Am. Chem. Soc.* **83**, 3719 (1961); from other mammalian pituitaries: Nelson; Eppstein, U.S. pats. **3,265,580** and **3,317,392** (1966, 1967, both to Upjohn); from human pituitaries: Lewis *et al., Biochem. Biophys. Res. Commun.* **44**, 1169 (1971). Amino acid sequence of ovine prolactin: Li *et al., Nature* **224**, 695 (1969); Li, Dixon, *Arch. Biochem. Biophys.* **146**, 233 (1971). Review of structural studies of human prolactin and relationship to somatotropin, *q.v.*: H. D. Niall *et al., Recent Progr. Horm. Res.* **29**, 387 (1973). Amino acid sequence of human prolactin: B. Shome, A. F. Parlow, *J. Clin. Endocrinol. Metab.* **45**, 1112 (1977). Effects of prolactin on the murine immune system: E. Nagy *et al., Acta Endocrinol.* **102**, 351 (1983). Symposium on clinical endocrinology: *Horm. Res.* **22**, 129-252 (1985). General reviews: Li, Evans, *Hormones* **1**, 631 (1948); White, *Vitam. Horm. (New York)* **7**, 253 (1949); Voss, *Arzneimittel-Forsch.* **4**, 467 (1954); Apostalakis, *Vitam. Horm. (New York)* **26**, 197 (1968). Books: *Prolactin* vols. **1-8**, D. F. Horrobin, Ed. (Eden Press, Quebec, 1973-1981); *Prolactin: Physiology, Pharmacology and Clinical Findings,* O. Hutzinger *et al.,* Eds. (Springer-Verlag, New York, 1982) 224 pp; *Prolactin and Prolactinomas,* G. Tolis *et al.,* Eds. (Raven Press, New York, 1983) 478 pp.

Crystals. Isoelectric point 5.73. $[\alpha]_D^{25}$ −40.5° (c = 1 in phosphate buffer of pH 7). Practically insol in water (0.102 g/l) unless electrolytes are present. Forms a water-soluble hydrochloride. Sol in abs methanol or ethanol, if a small amount of acid is present. These data apply to prolactin obtained from ox glands. Prolactin from sheep glands is slightly different: In 0.357M NaCl at pH 2.25 the sheep hormone has a soly of 0.506 g/l, while the soly of the ox hormone is only 0.316 g. In citrate buffer (1M, pH 6.36) and in alcohol the ox protein is more sol than the sheep protein. In the absence of salt, prolactin shows little loss of potency after boiling for one hour at pH 8.0 or at 60° for 5 hours; in the presence of salts complete destruction may occur. An 0.04% soln was stable in a boiling water bath for 15 minutes at pH 1 to 9, but lost activity rapidly at pH 11 and 13. As a rule the hormone is more stable in acid than in alkaline solns. 1 mg = 30 I.U.

THERAP CAT: Lactation stimulating hormone.

7789. Prolamins. A group of simple proteins. Yield only amino acids upon cleavage by enzymes or acids. Examples: gliadin, hordein, zein from grains.

Insoluble in water or neutral salt solvents. Sol in dil acids and alkalies and in 70-90% alcohol.

7790. Proline. Pro (IUPAC abbrev.); 2-pyrrolidinecarboxylic acid. $C_5H_9NO_2$; mol wt 115.13. C 52.16%, H 7.88%, O 27.79%, N 12.17%. An amino acid classified as nonessential with respect to its growth effect in rats. Isoln from wheat gliadin or from gelatin: Cox, King, *J. Biol. Chem.* **84**, 533 (1929); Town, *Biochem. J.* **22**, 1083 (1928); **30**, 1837 (1936); Bergmann, *J. Biol. Chem.* **110**, 471 (1935); Mayeda, *Bull. Inst. Phys. Chem. Res. (Tokyo)* **19**, 261, 271 (1940). Synthesis starting with γ-bromopropylmalonic ester: Willstätter, *Ber.* **33**, 1160 (1900); Willstätter, Ettlinger, *Ann.* **326**, 91 (1903); Leuchs, *Ber.* **44**, 1507 (1911); from γ-phthalimidopropylmalonic ester: Fischer, *Ber.* **34**, 454 (1901); Sorensen, Andersen, *Z. Physiol. Chem.* **56**, 236 (1908); from arylpiperidines: Fischer, Zemplén, *Ber.* **42**, 2989 (1909); from α-piperidone: Heymons, *Ber.* **66**, 846 (1933); from α-pyrrolidonecarboxylic ester: Fischer, Boehner, *Ber.* **44**, 1332 (1911); Signaigo, Adkins, *J. Am. Chem. Soc.* **58**, 709, 1122 (1936); from cyclopentanone: C. L. A. Schmidt, *Chemistry of Amino Acids and Proteins* (Springfield, Ill., 2nd ed., 1944) p 89; from L-pyroglutamic acid: Monteiro, *Synthesis* **1974**, 137. Stereospecific synthesis: S. L. Titouani *et al.*, *Tetrahedron* **36**, 2961 (1980).

L-Form, the naturally occurring form. Flat needles from alcohol + ether, prisms from water, dec 220-222°. $[\alpha]_D^{23.4}$ −85.0°; $[\alpha]_D^{20}$ −52.6° (c = 0.58 in 0.50N HCl); $[\alpha]_D^{20}$ −93.0° (c = 2.4 in 0.6N KOH). pK$_1$ 1.99; pK$_2$ 10.60. Soly in 100 ml water: 127 g at 0°; 162 g at 25°; 206.7 g at 50°; 239 g at 65°. One gram of L-proline is sol in 1.5 ml abs alc at 19°. Insol in ether, butanol, isopropanol.

DL-Form, the synthetic form. Monohydrate, crystals, mp 190° (when anhydr, dec 205°). Sol in water, alc; sparingly sol in acetone, chloroform, benzene; insol in ether.

7791. Prolintane. *1-[1-(Phenylmethyl)butyl]pyrrolidine;* *1-(α-propylphenethyl)pyrrolidine;* 1-phenyl-2-pyrrolidylpentane; phenylpyrrolidinopentane; Sp 732. $C_{15}H_{23}N$; mol wt 217.34. C 82.89%, H 10.67%, N 6.45%. Prepn: **Brit.** pat. **807,835** and Seeger, Kottler, **Ger.** pat. **1,093,799** (1959, 1960, both to Thomae); *C.A.* **55**, 19950c (1961). Pharmacological studies: R. Kadatz, E. Pötzsch, *Arzneimittel Forsch.* **7**, 344 (1957); K. Takagi *et al.*, *Oyo Yakuri* **5**, 5 (1971), *C.A.* **76**, 94575k (1972).

bp$_{0.5}$ 105°; bp$_{16}$ 153°.

Hydrochloride, $C_{15}H_{24}ClN$, *Promotil.* Crystals from methyl ether, mp 133-134°. LD$_{50}$ orally in mice: 257 mg/kg, R. Kadatz, E. Pötzsch, *loc. cit.*

Mixture with vitamins, *Katovit, Catovit, Villescon.*

THERAP CAT: Central stimulant; antidepressant.

7792. Prolonium Iodide. *2-Hydroxy-N,N,N,N′,N′,N′-hexamethyl-1,3-propanediaminium diiodide; (2-hydroxytrimethylene)bis[trimethylammonium] iodide;* 1,3-bis(trimethylamino)-2-propanol diiodide; hexamethyldiaminoisopropa-

nol diiodide; di(iodohexamethyl)diaminoisopropanol; iodisan; Endojodin; Entodon; Esoiodine; Hexayodina; Propiodal; Yodanodia. $C_9H_{24}I_2N_2O$; mol wt 430.14. C 25.13%, H 5.62%, I 59.01%, N 6.51%, O 3.72%. Prepn: Callsen, **U.S.** pat. **1,526,627** (1925).

White, crystalline powder. mp about 275° with decompn, but becomes brown at 240°. Freely sol in water, slightly in alcohol. Practically insol in ether, acetone.

THERAP CAT: Iodine source.

7793. Promazine. *N,N-Dimethyl-10H-phenothiazine-10-propanamine; 10-(3-dimethylaminopropyl)phenothiazine;* RP 3276; Wy 1094; A 145; Liranol; Promwill; Sinophenin; Tomil; Neo-Hibernex; Ampazine; Esparin. $C_{17}H_{20}N_2S$; mol wt 284.41. C 71.79%, H 7.09%, N 9.85%, S 11.27%. Prepd by heating a xylene soln of phenothiazine and 3-dimethylamino-1-chloropropane in the presence of sodamide: Charpentier, **U.S.** pat. **2,519,886** (1950 to Rhône-Poulenc).

Oily liq. Amine odor. Alkaline reaction. bp$_{0.3}$ 203-210°. Hydrochloride, $C_{17}H_{21}ClN_2S$, *Centractil, Centractyl, Prazine, Protactyl, Sparine, Starazine, Talofen, Verophene.* White to slightly yellow crystals, dec 181° (microblock). Oxidizes upon prolonged exposure to air and acquires a blue or pink color. Hygroscopic. One gram dissolves in about 3 ml water. Sol in methanol, ethanol, chloroform. Practically insol in ether, benzene. Aq solns are slightly acid to litmus. Incompatible with alkalies, oxidizing agents, heavy metals.

THERAP CAT: Antipsychotic.

THERAP CAT (VET): Tranquilizer.

7794. Promecarb. *3-Methyl-5-(1-methylethyl)phenol methylcarbamate; methylcarbamic acid m-cym-5-yl ester;* m-cym-5-yl methylcarbamate; 3-methyl-5-isopropyl N-methylcarbamate; Schering 34615; Carbamult; Minacide. $C_{12}H_{17}NO_2$; mol wt 207.28. C 69.53%, H 8.27%, N 6.76%, O 15.44%. Prepn: **Brit.** pat. **913,707**, *C.A.* **58**, 7316c (1963) corresp to **U.S.** pat. **3,167,472** (1962, 1965, both to Schering). *Review:* Jaeger, *Z. Angew. Entomol.* **58**, 188 (1966).

Colorless crystals, mp 87.0-87.5°. bp$_{0.1}$ 117°. Sol in polar organic solvents; insol in water. Hydrolyzed in alkalies. LD$_{50}$ in mice, rats: 39.5, 60 mg/kg orally, *C.A.* **78**, 119931e (1973).

USE: Insecticide. *Caution:* Cholinesterase inhibitor.

7795. Promedol. *1,2,5-Trimethyl-4-phenyl-4-piperidinol propanoate (ester);* 1,2,5-trimethyl-4-phenyl-4-propionyloxypiperidine; 1,2,5-trimethyl-4-phenyl-4-piperidyl propionate; dimethylmeperidine. $C_{17}H_{25}NO_2$; mol wt 275.38. C 74.14%, H 9.15%, N 5.09%, O 11.62%. Preparation: Nazarov *et al.*, *J. Gen. Chem. USSR* **26**, 3117 (1956); Nazarov, Shvestov, *ibid.* 3533. Conformation studies: Prostakov, Mikheeva, *ibid.* **31**, 108 (1961); **33**, 2931 (1963); *eidem, Russ. Chem. Rev.* **31**, 556 (1962). Of the four possible isomers α and γ are shown:

α-isomer γ-isomer

α-Isomer hydrochloride, $C_{17}H_{26}ClNO_2$, α-*promedol*. Crystals from benzene, mp 153-154°. Has been also reported as mp 106-107° or 126-131°, Prostakov, Mikheeva, *loc. cit.* (1961).

β-Isomer hydrochloride, *isopromedol*. Crystals, mp 183-184°, Nazarov, Shvestov, *Bull. Acad. Sci. USSR Phys. Ser.* **1959**, 2059.

γ-Isomer hydrochloride, *trimeperidine*, γ-*promedol*. Crystals from acetone, mp 222-223°.

Caution: May be habit forming. This is a controlled substance (opiate) listed in the U.S. Code of Federal Regulations, Title 21 Part 1308.11 (1987).

THERAP CAT: Narcotic analgesic.

7796. Promegestone. *(17β)-17-Methyl-17-(1-oxopropyl)estra-4,9-dien-3-one; 17α-methyl-17-propionylestra-4,9-dien-3-one; 17α-methyl-17β-propionyl-19-nor-4,9-androstadien-3-one; 17α,21-dimethyl-19-norpregna-4,9-diene-3,20-dione; R 5020; RU 5020; Surgestone.* $C_{22}H_{30}O_2$; mol wt 326.48. C 80.94%, H 9.26%, O 9.80%. Synthetic progestin with no androgenic activity and with high affinity for the progesterone receptor. Prepn: J. Warnant, A. Farcilli, **Belg. pat. 763,099;** *eidem,* **U.S. pats. 3,679,714, 3,761,591** (1971, 1972, 1973 all to Roussel-UCLAF). Inhibition of gonadotropin secretion and lack of androgenic activity: F. Labrie *et al., Fert. Steril.* **28**, 1104 (1977). Binding studies in mouse uterus: D. Philibert, J.-P. Raynaud, *Steroids* **22**, 89 (1973); *eidem, Endocrinology* **94**, 627 (1974). Binding to human endometrium: *eidem, Contraception* **10**, 457 (1974); M. Haukkamaa, T. Luukkainen, *J. Steroid Biochem.* **5**, 447 (1974). Review and possible use in treatment of hormone-dependent breast cancer: J.-P. Raynaud, T. Ojasoo, *J. Gynecol. Obstet. Biol. Reprod.* **12**, 697 (1983).

Colorless crystals from isopropyl ether, mp 152°. Sol in acetone, benzene. Insol in water. $[\alpha]_D^{20}$ −262° (c = 0.5 in ethanol). uv max in ethanol: 215, 305 nm ($E_{1cm}^{1\%}$ 202, 648).

USE: As radioligand for the progestin receptor.

THERAP CAT: Progestogen.

7797. Promethazine. *N,N,α-Trimethyl-10H-phenothiazine-10-ethanamine; 10-(2-dimethylaminopropyl)phenothiazine; 10-(2-dimethylamino-2-methylethyl)phenothiazine; N-(2'-dimethylamino-2'-methyl)ethylphenothiazine; proazamine; RP 3277; RP 3389; Hiberna; Vallergine; Dimapp.* $C_{17}H_{20}N_2S$; mol wt 284.41. C 71.79%, H 7.09%, N 9.85%, S 11.27%. Prepn from 10-phenothiazinepropyl chloride and dimethylamine in presence of Cu: Charpentier, *Compt. Rend.* **225**, 306 (1947); **U.S. pat. 2,530,451** (1950 to Rhône-Poulenc); from Grignard complexes of dimethylaminopropyl halide and phenothiazine: Berg, Ashley, **U.S. pat. 2,607,773** (1952 to Rhône-Poulenc). Structure and isomerism: Edge, Wragg, *J. Pharm. Pharmacol.* **5**, 279 (1953). Metabolic studies: Huang *et al., J. Pharm. Sci.* **59**, 772 (1970). Comprehensive description: C. M. Shearer, S. M. Miller in *Analytical Profiles of Drug Substances* vol. **5**, K. Florey, Ed. (Academic Press, New York, 1976) pp 429-465. Toxicity: Rajsner, *Coll. Czech. Chem. Commun.* **34**, 1019 (1969).

Crystals, mp 60°. bp₃ 190-192°.

Hydrochloride, $C_{17}H_{21}ClN_2S$, *Atosil, Duplamin, Fargan, Fenazil, Provigan, Prorex, Diphergan, Ganphen, Fellozine, Remsed, Dorme, Fenergan, Lergigan, Phencen, Phenergan, Promantine, Promethegan, Promine, Protazine, Prothazin, Thiergan.* Crystals from ethylene dichloride, mp 230-232° (some dec). Turns blue on prolonged exposure to air and moisture. Lower mp reported in the literature are caused by admixture with isopromethazine, *q.v.* uv max (water): 249, 297 nm (ε 28770, 3400). pH of 10% aq soln 5.3. Freely sol in water. Sol in alcohol, chloroform. Practically insol in acetone, ether, ethyl acetate. LD₅₀ i.v. in mice: 55.0 mg/kg (Rajsner).

8-Chlorotheophylline salt, *Avomine.*

THERAP CAT: Antihistaminic.

THERAP CAT (VET): Antihistaminic, antiemetic, CNS depressant.

7798. Promethium. Pm; at. wt of best known isotope 147; at. no. 61; valence 3. All known isotopes are radioactive; mass numbers: 140-154. ¹⁴⁷Pm, a β-emitter, $T_{1/2}$ 2.62 years. The discovery of element 61 in rare earth concentrate was claimed by Harris and Hopkins, *J. Am. Chem. Soc.* **48**, 1585 (1926) and by Rolla and Fernandez, *Gazz. Chim. Ital.* **56**, 435 (1926). Evidence of existence in nature is inconclusive: Yost *et al., The Rare Earth Elements and Their Compounds* (John Wiley, New York, 1947). First obtained synthetically by irradiating neodymium and praseodymium with neutrons, deuterons, and alpha particles: Law *et al., Phys. Rev.* **59**, 936 (1941). Positive identification by ion-exchange chromatography: Marinsky *et al., J. Am. Chem. Soc.* **69**, 2781 (1947). Can be isolated in quantitative yields from the uranium fission-product mixtures and from the waste products of radiochemical processing plants. Metallic Pm first prepd by reduction of PmF₃ by lithium *in vacuo:* Weigel, *Angew. Chem. Int. Ed.* **2**, 326 (1963); prepd by reduction of PmCl₃ by calcium: E. J. Wheelwright, *J. Phys. Chem.* **73**, 2867 (1969). *Reviews:* Boyd, *J. Chem. Ed.* **36**, 3-14 (1959); Weigel, *Fortschr. Chem. Forsch.* **12**, 539-621 (1969). Review of chemistry, toxicology and industrial uses: *Promethium Technology,* E. J. Wheelwright, Ed. (American Nuclear Society, Hindsdale, IL, 1973) 395 pp.

¹⁴⁷Pm, metallic solid; d 7.22. mp 1080° (Weigel); 1169° (Wheelwright). A number of salts have been prepared including the trihalides (PmX₃), the sesquioxide (Pm₂O₃), the hydroxide [Pm(OH)₃] and the nitrate [Pm(NO₃)₃.xH₂O]. *See* review by Weigel, *loc. cit.*

USE: As β-particle source for thickness gauges; in the prepn of self-luminous compds; in the construction of miniature atomic batteries.

7799. Prometon. *6-Methoxy-N,N'-bis(1-methylethyl)-1,3,5-triazine-2,4-diamine; 2,4-bis(isopropylamino)-6-methoxy-s-triazine; 2-methoxy-4,6-bis(isopropylamino)-s-triazine; methoxypropazine; prometone; G 31435; Gesafram; Primatol; Pramitol.* $C_{10}H_{19}N_5O$; mol wt 225.29. C 53.31%, H 8.50%, N 31.09%, O 7.10%. Nonselective pre- and post-emergent herbicide for use on noncrop land. General preparative information and use as herbicide: H. Gysin, E. Knüsli, **U.S. pat. 2,909,420** (1959 to Geigy); *see also:* M. A. Priola, **U.S. pat. 3,713,806** (1973 to Ciba-Geigy). Alternate process: H. V. Lemaster, **U.S. pat. 3,663,542** (1972 to Geigy). Comparison with other triazine herbicides: H. Gysin, E. Knüsli, *Adv. Pest Contr. Res.* **3**, 289 (1960). Metabolism in rats: J. E. Bakke *et al., J. Agr. Food Chem.* **15**, 628 (1967). GC determn in human urine: *ibid.* **27**, 740 (1979). Toxicology study in sheep and cattle: A. E. Johnson *et al., Am. J. Vet. Res.* **33**, 1433 (1972). Review of properties and analytical methods: B. G. Tweedy, R. A. Kahrs, *Anal. Methods Pestic. Plant Growth Regul.* **10**, 493 (1978).

Crystalline solid, mp 91-92°. Soly in water (20°): 750 ppm. Very sol in organic solvents. Vapor pressure at 20°: 2.3×10^{-6} Torr. LD_{50} in mice, rats (mg/kg): 2160, 2980 orally (Gysin, Knüsli, *Adv. Pest Contr. Res.*). LC_{50} (96 hour) in bluegill sunfish, rainbow trout (ppm): >32, 20 (Tweedy, Kahrs).

USE: Nonselective herbicide.

7800. Prometryn. *N,N'-Bis(1-methylethyl)-6-methyl-thio-1,3,5-triazine-2,4-diamine; 2,4-bis(isopropylamino)-6-(methylthio)-s-triazine;* 2-methylthio-4,6-bis(isopropylamino)-s-triazine; G 34161; Gesagard; Caparol. $C_{10}H_{19}N_5S$; mol wt 241.37. C 49.76%, H 7.93%, N 29.02%, S 13.29%. Selective pre- and post-emergence herbicide. General preparative information and use as herbicide: H. Gysin, E. Knüsli, **Swiss** pat. **337,019;** *eidem,* U.S. pat. **2,909,420** (both 1959 to Geigy). See also: M. A. Priola, U.S. pat. **3,713,806** (1973 to Ciba-Geigy). Alternative process: **Fr.** pat. **1,372,-089;** E. Knüsli, W. Stammbach, U.S. pat. **3,207,756** (1964, 1965 both to Geigy). Activity: H. Gysin, *Chem. & Ind. (London)* **1962,** 1393. Metabolism in rats and rabbits: C. Böhme, F. Bär, *Food Cosmet. Toxicol.* **5,** 23 (1967). Toxicity: Geigy, *Toxicology Data on Prometryne,* July, 1964. Review of properties and analytical methods: B. G. Tweedy, R. A. Kahrs, *Anal. Methods Pestic. Plant Growth Regul.* **10,** 493 (1978).

Crystals, mp 118-120°. Vapor pressure at 20°: 1×10^{-6} mm Hg. Soly in water at 20°: 48 ppm. Readily sol in org solvents. Stable in neutral or slightly acid or alkaline media. Hydrolyzed under stronger acidic or basic conditions. LD_{50} orally in rats: 3.75 g/kg (Geigy). LC_{50} (96 hour) in bluegill sunfish, rainbow trout (ppm): 10.0, 2.5 (Tweedy, Kahrs).

USE: Herbicide.

7801. Promoxolane. *2,2-Bis(1-methylethyl)-1,3-dioxo-lane-4-methanol; 2,2-diisopropyl-1,3-dioxolane-4-methanol;* 2,2-diisopropyl-4-hydroxymethyl-1,3-dioxolane; Dimethyl-yn; Dimethylane. $C_{10}H_{20}O_3$; mol wt 188.26. C 63.79%, H 10.71%, O 25.50%. Prepn: Boekelheide *et al., J. Am. Chem. Soc.* **71,** 3303 (1949).

Liquid. d_4^{21} 0.995. bp 115°. n_D^{21} 1.4502.

THERAP CAT: Skeletal muscle relaxant.

7802. Pronethalol. *α-[(Isopropylamino)methyl]-2-naph-thalenemethanol;* [2-hydroxy-2-(2-naphthyl)ethyl]isopropyl-amine; 2-isopropylamino-1-(2-naphthyl)ethanol; 1-(2'-naphthyl)-2-isopropylaminoethanol; ICI 38174; Nethalide; Alderlin. $C_{15}H_{19}NO$; mol wt 229.31. C 78.56%, H 8.35%, N 6.11%, O 6.98%. β-Adrenergic blocker. Prepd by reduction of 2-naphthacyl bromide with $NaBH_4$ followed by reaction with isopropylamine: Stephenson, **Brit.** pat. **909,357** (1962 to I.C.I.). Pharmacology: Black *et al., Brit. J. Pharmacol. Chemother.* **25,** 577 (1965). Carcinogenicity in mice: R. Howe, *Nature* **207,** 594 (1965). Metabolism: W. G. Stillwell, M. G. Horning, *Res. Commun. Chem. Pathol. Pharmacol.* **9,** 601 (1974).

Crystals, mp 108°.

Hydrochloride, $C_{15}H_{20}ClNO$, crystals, mp 184°. LD_{50} in mice (mg/kg): 512 orally, 46 i.v. (Black).

THERAP CAT: Antianginal; antiarrhythmic; antihypertensive.

7803. Pro-opiomelanocortin. *Proopiocortin;* ACTH-β-lipotropin common precursor; precursor to ACTH-LPH-β-endorphin; 31K-precursor; POMC. A precursor protein of mol wt about 30,000, synthesized in the hypothalamus, pituitary gland, brain, and several peripheral tissues that incorporates the amino acid sequences of the pituitary hormones ACTH and β-lipotropin. These two hormones, in turn, contain biologically active component peptides: α-MSH, *corticotropin-like intermediate lobe peptide (CLIP),* α-LPH, β-MSH, endorphins, and met-enkephalin. First description of the presence of ACTH and LPH in the same molecule: P. J. Lowry *et al., Int. Congr. Ser.-Excerpta Med.* **402,** 71 (1976). Confirmation of the common precursor by radioimmunoassay: R. E. Mains *et al., Proc. Nat. Acad. Sci. USA* **74,** 3014 (1977); J. L. Roberts, E. Herbert, *ibid.* 4826, 5300. Isoln of rat pro-opiomelanocortin: M. Rubenstein *et al., ibid.* **75,** 669 (1978). Partial sequence analysis using cDNA from mouse: J. L. Roberts *et al., ibid.* **76,** 2153 (1979). Complete sequence of bovine POMC using cloned cDNA: S. Nakanishi *et al., Nature* **278,** 423 (1979). The precursor has also been found in human ACTH ectopic tumor: X. Bertagna *et al., Proc. Nat. Acad. Sci. USA* **75,** 5160 (1978). Primary structure of the NH_2-terminal glycopeptide of human pituitary POMC: N. G. Seidah *et al., J. Biol. Chem.* **256,** 7977 (1981). *Reviews:* M. Chrétien, N. G. Seidah, *Mol. Cell. Biochem.* **34,** 101-127 (1981); D. De Wied, J. Jolles, *Physiol. Rev.* **62,** 976-1059 (1982); Y. P. Loh *et al., Peptides* **3,** 397-404 (1982).

7804. Propacetamol. *N,N-Diethylglycine 4-(acetylami-no)phenyl ester; N,N-*diethylglycine ester with 4'-hydroxy-acetanilide; 4-acetamidophenyl (diethylamino)acetate. $C_{14}H_{20}N_2O_3$; mol wt 264.32. C 63.61%, H 7.63%, N 10.60%, O 18.16%. Injectable prodrug of acetaminophen, *q.v.* Prepn and pharmacology: J. C. Cognacq, **Belg.** pat. **854,376;** *idem,* U.S. pat. **4,127,671** (1977, 1978 both to Hexachimie). Clinical comparison with injectable aspirin: R. De Marneffe, L. Mokassa, *Compt. Rend. Ther. Pharm. Clin.* **3,** 23 (1985). Clinical studies: P. Delacroix *et al., Sem. Hop.* **61,** 2739 (1985); J. Modai, *ibid.* **62,** 587 (1986).

Thick oil.

Hydrochloride, $C_{14}H_{21}ClN_2O_3$, *UP-34101,* Pro-Dafalgan. mp 228°. Sol in water.

THERAP CAT: Analgesic; antipyretic.

7805. Propachlor. *2-Chloro-N-(1-methylethyl)-N-phen-ylacetamide; 2-chloro-N-isopropylacetanilide;* N-isopropyl-α-chloroacetanilide; CP 31393; Bexton; Prolex; Ramrod. $C_{11}H_{14}ClNO$; mol wt 211.69. C 62.41%, H 6.66%, Cl 16.75%, N 6.62%, O 7.56%. Selective pre-emergence herbicide. Prepn: P. C. Hamm, A. J. Speziale, U.S. pat. **2,863,-752** (1958 to Monsanto). Activity studies: Duke, *Diss. Abstr. B* **28,** 1315 (1967); Jaworski, *J. Agr. Food Chem.* **17,** 165 (1969); Lamoureaux *et al., ibid.* **19,** 346 (1971); Dhillon, Anderson, *Weed Res.* **12,** 182 (1972). Metabolism: J. E. Bakke *et al., Science* **210,** 433 (1980). Toxicity data: E. E. Kenaga, *Down Earth* **35,** 25 (1979).

Consult the cross index before using this section.

Light tan solid, mp 67-76°. Vapor pressure at 110°: 0.03 mm Hg. Soly in water at 20°: 700 mg/l. Sol in common organic solvents except aliphatic hydrocarbons. LD_{50} orally in rats: 710 mg/kg (Kenaga).

USE: Herbicide.

7806. Propafenone. *1-[2-[2-Hydroxy-3-(propylamino)-propoxy]phenyl]-3-phenyl-1-propanone; 2'-[2-hydroxy-3-(propylamino)propoxy]-3-phenylpropiophenone;* SA 79. $C_{21}H_{27}NO_3$; mol wt 341.46. C 73.87%, H 7.97%, N 4.10%, O 14.06%. Prepn: R. Sachse, **Ger.** pat. **2,001,431** (1971 to Helopharm), *C.A.* **75,** 151538f (1971). Pharmacology: H.-J. Hapke, E. Prigge, *Arzneimittel-Forsch.* **26,** 1849 (1976). Use in arrhythmia: O. A. Beck, H. Hochrein, *Deut. Med. Wochenschr.* **103,** 1261 (1978). Pharmacokinetics: H. Blanke et al., ibid. **104,** 587 (1979); M. Hollmann et al., *Arzneimittel-Forsch.* **33,** 763 (1983). Brief review: M. Fischer, *Med. Klin.* **75,** 39 (1980).

Hydrochloride, $C_{21}H_{28}ClNO_3$, *Pulonon, Rythmol, Rytmonorm.* Fine white crystals. Slightly bitter taste. Sol in basic lower aliphatic alcohols, CCl_4, hot water; slightly sol in cold water. Insol in ether. LD_{50} in rats (mg/kg): 18.8 i.v.; 700 orally (Hapke, Prigge).

THERAP CAT: Antiarrhythmic.

7807. Propallylonal. *5-(2-Bromo-2-propenyl)-5-(1-methylethyl)-2,4,6(1H,3H,5H)pyrimidinetrione; 5-(2-bromoallyl)-5-isopropylbarbituric acid;* Noctal. $C_{10}H_{13}BrN_2O_3$; mol wt 289.13. C 41.54%, H 4.53%, Br 27.64%, N 9.69%, O 16.60%. Prepn: **U.S.** pat. **1,622,129** (1927); cf. Herzog, *Arch. Pharm.* **263,** 216 (1925). Acute toxicity: Maloney, *J. Pharmacol. Exp. Ther.* **42,** 267 (1931).

Crystals. mp 177-179°. Slightly bitter taste. Slightly sol in water; freely sol in alcohol, glacial acetic acid, acetone, alkalies. Sparingly sol in ether, chloroform, benzene. MLD orally in rabbits: 300-350 mg/kg (Maloney).

Caution: May be habit forming. This is a controlled substance (depressant) listed in the U.S. Code of Federal Regulations, Title 21 Parts 329.1 and 1308.13 (1987).

THERAP CAT: Sedative, hypnotic.

7808. Propamidine. *4,4'-[1,3-Propanediylbis(oxy)]bis-benzenecarboximidamide; 4,4'-(trimethylenedioxy)dibenzamidine; 4,4'-diamidino-α,ω-diphenoxypropane.* $C_{17}H_{20}N_4O_2$; mol wt 312.37. C 65.37%, H 6.45%, N 17.94%, O 10.24%. Prepn: A. J. Ewins et al., **Brit.** pat. **507,565** (1939 to May & Baker); J. N. Ashley et al., *J. Chem. Soc.* **1942,** 103; of isethionate: G. Newbery, A. P. T. Easson: **U.S.** pat. **2,394,003** (1946 to May & Baker). Trypanocidal activity: E. M. Lourie, W. Yorke, *Ann. Trop. Med. Parasitol.* **33,** 289 (1939). Preliminary pharmacology: R. Wien, ibid. **37,** 1 (1943). Determn in biological fluids: D. P. Jackson et al., *J. Biol. Chem.* **167,** 377 (1947). Mode of action: M. J. Pine, *Biochem. Pharmacol.* **17,** 75 (1968). Activity in fibrinolytic systems: J. D. Geratz, *Thromb. Diath. Haemorrh.* **29,** 154 (1973). Clinical use in treatment of *Acanthamoeba* keratitis: D. L. Easty, *Brit. Med. J.* **296,** 228 (1988); J. J. Wiens, W. B. Jackson, *Can. J. Ophthalmol.* **23,** 107 (1988). Early review of pharmacology, mode of action and clinical applications: E.

B. Schoenbach, E. M. Greenspan, *Medicine* **27,** 327-377 (1948).

Isethionate, $C_{21}H_{32}N_4O_{10}S_2$, *M & B 782, Broline drops.* Hygroscopic, very bitter crystals or granular powder, mp ~235°. Soluble in water (about 1 in 5), glycerol, 95% alcohol (about 1 in 33). Practically insol in ether, chloroform, fixed oils, liq petrolatum. pH of a 5% w/v soln in water = 4.5 to 6.5.

THERAP CAT: Antiprotozoal (Trypanosoma); antiamebic.

THERAP CAT (VET): Topical anti-infective. Formerly used as antiprotozoal (Trypanosoma, Babesia).

7809. Propane. Dimethylmethane; propyl hydride. C_3H_8; mol wt 44.09. C 81.72%, H 18.29%. $CH_3CH_2CH_3$. Constituent of natural gas and of crude petroleum. Obtained by the so-called "stabilization process" using fractional distillation under pressure: Francis, Robbins, *J. Am. Chem. Soc.* **55,** 4339 (1933). Many syntheses, e.g., by using butyronitrile and sodium: Timmermans, *J. Chim. Phys.* **18,** 133 (1920).

Gas. Odorless when pure. Burns with a luminous, smoky flame. Explosive limits, % by vol in air: 2.37-9.5. Heavier than air. One liter weighs 2.0200 g at 0° and 760 mm; 1.8324 g at 25° and 760 mm. Liquefies at −42°; solid at −187.7°. bp (1 atm) −42.1°; bp (2 atm) −25.6°; bp (5 atm) +1.4°; bp (10 atm) 26.9°; bp (20 atm) 58.1°; bp (30 atm) 78.7°; bp (40 atm) 94.8°. Crit temp 96.81°; crit press. 42.01 atm. Heat of combustion (const vol) 528.4 cal, (const pressure) 553.5 cal. 100 vols water dissolve 6.5 vols at 17.8° and 753 mm pressure; 100 vols abs alc dissolve 790 vols at 16.6° and 754 mm pressure; 100 vols ether dissolve 926 vols at 16.6° and 757 mm pressure; 100 vols chloroform dissolve 1299 vols at 21.6° and 757 mm pressure; 100 vols benzene dissolve 1452 vols at 21.5° and 757 mm pressure; 100 vols turpentine dissolve 1587 vols at 17.7° and 757 mm pressure.

USE: As fuel gas, sometimes mixed with butane. In organic syntheses. As refrigerant. *Caution:* May be narcotic in high concns.

7810. 1-Propanearsonic Acid. *n-Propylarsonic acid.* $C_3H_9AsO_3$; mol wt 168.01. C 21.44%, H 5.40%, As 44.58%, O 28.57%. $C_3H_7AsO(OH)_2$.
White needles, mp 125°. Freely sol in water; sol in alcohol; insol in ether.

USE: For the determination of zirconium.

7811. 1,3-Propanedithiol. 1,3-Dimercaptopropane; trimethylenedithioglycol; dithiotrimethyleneglycol; trimethylenedimercaptan. $C_3H_8S_2$; mol wt 108.23. C 33.29%, H 7.46%, S 59.25%. $HSCH_2CH_2CH_2SH$. Prepd by alkaline hydrolysis of propylene-1,3-diisothiuronium dihydrochloride: Grogan et al., *J. Org. Chem.* **20,** 50 (1955).

Oil. Disagreeable odor. d_4^{20} 1.0772. bp_{760} 169-170°; bp_{759} 170-171°; bp_{56} 92-98°; n_D^{20} 1.5392. Volatile with steam. Slightly sol in water. Miscible with alcohol, ether, chloroform and benzene.

7812. Propanethial S-Oxide. *Thiopropionaldehyde S-oxide;* thiopropanal S-oxide. C_3H_6OS; mol wt 90.14. C 39.97%, H 6.71%, O 17.75%, S 35.57%. $CH_3CH_2CH{=}SO$. Lachrymatory factor of the onion, *Allium cepa L.,* found as a 95% cis- and 5% trans- mixture. Early structure studies: W. D. Niegisch, W. H. Stahl, *Food Res.* **21,** 657 (1956); C. G. Spare, A. I. Virtanen, *Acta Chem. Scand.* **17,** 641 (1963). Structure: M. H. Brodnitz, J. V. Pascale, *J. Agr. Food Chem.* **19,** 269 (1971). Stereochemistry: E. Block et al., *J. Am. Chem. Soc.* **101,** 2200 (1979); eidem, *Tetrahedron Letters* **21,** 1277 (1980). Chemistry: eidem, *J. Am. Chem. Soc.* **102,** 2490 (1980).

7813. Propanidid. *4-[2-(Diethylamino)-2-oxoethoxy]-3-methoxybenzeneacetic acid propyl ester; [4-[(diethylcarbamoyl)methoxy]-3-methoxyphenyl]acetic acid propyl ester; [3-methoxy-4-[(N,N-diethylcarbamido)methoxy]phenyl]acetic* acid *n*-propyl ester; propyl [4-[(diethylcarbamoyl)methoxy]-3-methoxyphenyl]acetate; Bayer 1420; FBA 1420; Epontol; Sombrevin. $C_{18}H_{27}NO_5$; mol wt 337.40. C 64.07%, H 8.06%, N 4.15%, O 23.71%. Prepn: Hiltmann *et al.*, **Ger. pat. 1,134,981** corresp to **U.S. pat. 3,086,978** (1962, 1963 to Bayer).

Pale yellow oil, $bp_{0.7}$ 210-212°. Practically insol in water; sol in alcohol, chloroform. LD_{50} orally in rats: > 10,000 mg/kg, E. I. Goldenthal, *Toxicol. Appl. Pharmacol.* **18**, 185 (1971).

THERAP CAT: Anesthetic (intravenous).

7814. Propanil. *N-(3,4-Dichlorophenyl)propanamide; 3',4'-dichloropropionanilide;* N-(3,4-dichlorophenyl)propionamide; DPA; FW-734; Stam; Stampede; Rogue; Chem Rice; Surcopur. $C_9H_9Cl_2NO$; mol wt 218.09. C 49.57%, H 4.16%, Cl 32.51%, N 6.42%, O 7.34%. Selective contact herbicide. Prepn: W. Schäfer *et al.*, **Ger. pat. 1,039,779** (1958 to Bayer), *C.A.* **54**, 20060i (1960); Huffman, Allen, *J. Agr. Food Chem.* **8**, 298 (1960).

White crystalline solid, mp 91-93°. Soly in water at room temp: 225 ppm. LD_{50} orally in rats: 1384 mg/kg, G. W. Bailey, J. L. White, *Residue Rev.* **10**, 97 (1965).

USE: Herbicide. In nematocide formulations: Fielding, Stoddard, **U.S. pat. 3,108,038** (1963 to du Pont).

7815. Propanocaine. *α-[2-(Diethylamino)ethyl]benzenemethanol benzoate (ester); α-[2-(diethylamino)ethyl]benzyl alcohol benzoate;* 3-diethylamino-1-phenylpropyl benzoate; α-(2-diethylaminoethyl)benzyl benzoate; 1-benzoyloxy-3-diethylamino-1-phenylpropane; Detraine. $C_{20}H_{25}NO_2$; mol wt 311.41. C 77.13%, H 8.09%, N 4.50%, O 10.28%. Prepd from acetophenone and trioxymethylene in the presence of diethylamine hydrochloride followed by catalytic hydrogenation and reaction with benzoyl chloride: Detrie, **Fr. pat. 1,234,701** (1960 to Clin-Byla), *C.A.* **55**, 24682g (1961).

Yellow oil.
THERAP CAT: Local anesthetic.

7816. Propantheline Bromide. *N-Methyl-N-(1-methylethyl)-N-[2-[(9H-xanthen-9-ylcarbonyl)oxy]ethyl]-2-propanaminium bromide; (2-hydroxyethyl)diisopropylmethylammonium bromide xanthene-9-carboxylate; β-diisopropylaminoethyl 9-xanthenecarboxylate methobromide;* Corrigast; Ercotina; Giquel; Pro-Banthine; Prodixamon; Ketaman; Neo-Metantyl; Pantheline. $C_{23}H_{30}BrNO_3$; mol wt 448.42. C 61.61%, H 6.74%, Br 17.82%, N 3.12%, O 10.70%. Prepn: Cusic, Robinson, *J. Org. Chem.* **16**, 1921 (1951); **U.S. pat. 2,659,732** (1953 to Searle). Metabolic studies: Beermann *et al.*, *Clin. Pharmacol. Ther.* **13**, 212 (1972).

Crystals from isopropanol + ether, mp 159-161°. Very soluble in water, alcohol, chloroform. Practically insol in ether, benzene.

THERAP CAT: Anticholinergic.

7817. Proparacaine. *3-Amino-4-propoxybenzoic acid 2-(diethylamino)ethyl ester;* 2-(diethylamino)ethyl 3-amino-4-propoxybenzoate; proxymetacaine. $C_{16}H_{26}N_2O_3$; mol wt 294.38. C 65.28%, H 8.90%, N 9.52%, O 16.31%. Prepn of the hydrochloride: Clinton *et al.*, *J. Am. Chem. Soc.* **74**, 592 (1952). Pharmacology: McIntyre, Sievers, *J. Pharmacol. Exp. Ther.* **63**, 369 (1938). Comprehensive description: D. B. Whigan in *Analytical Profiles of Drug Substances* vol. 6, K. Florey, Ed. (Academic Press, New York, 1977) pp 423-456.

Hydrochloride, $C_{16}H_{27}ClN_2O_3$, *Ak-Taine, Alcaine, Ophthaine, Ophthetic.* Prisms from abs alcohol + ethyl acetate, mp 182.0-183.3°. uv max (methanol): 225, 270, 300 nm. Sol in water, warm alcohol, methanol. Insol in ether, benzene. Solns are neutral to litmus. pKa 3.22.

THERAP CAT: Topical anesthetic (ophthalmic).

7818. Propargite. *Sulfurous acid 2-[4-(1,1-dimethylethyl)phenoxy]cyclohexyl 2-propynyl ester; sulfurous acid 2-(p-tert-butylphenoxy)cyclohexyl 2-propynyl ester;* 2-(p-tert-butylphenoxy)cyclohexyl propargyl sulfite; cyclosulfyne; propargil; BPPS; ENT 27226; DO 14; Omite; Comite. $C_{19}H_{26}O_4S$; mol wt 350.47. C 65.11%, H 7.48%, O 18.26%, S 9.15%. Acaricide for control of mites on crops. Prepn and insecticidal properties: R. A. Covey *et al.*, **U.S. pat. 3,272,854** (1966 to U.S. Rubber). Field tests, comparison with other acaricides in control of Pacific spider mite: E. M. Stafford, *J. Econ. Entomol.* **61**, 1641 (1968). Efficacy against citrus red mite: L. R. Jeppson *et al.*, *ibid.* **62**, 531 (1969). Residue determn: J. M. Devine, H. R. Sisken, *J. Agr. Food Chem.* **20**, 59 (1972). Brief account of properties, analysis: G. M. Stone, *Anal. Methods Pestic. Plant Growth Regul.* **7**, 355 (1973). Toxicity studies: T. B. Gaines, *Toxicol. Appl. Pharmacol.* **14**, 515 (1969).

Viscous liquid, distils with decomp, stabilized by adding 0.5-1.0% propylene oxide. Sol in most organic solvents. Practically insol in water (10.5 ppm). LD_{50} in male, female rats (mg/kg): 1480, 1480 orally; 250, 680 dermally (Gaines). LC_{50} in rainbow trout, bluegill sunfish (ppb): > 100, 31 (Stone).

USE: Acaricide.

7819. Propargyl Alcohol. *2-Propyn-1-ol.* C_3H_4O; mol wt 56.06. C 64.27%, H 7.19%, O 28.54%. $HC \equiv CCH_2OH$. Prepd by heating β-bromoallyl alcohol with conc KOH: Henry, *Ber.* **5**, 456, 569 (1872); **6**, 729 (1873); **14**, 404 (1881); from formaldehyde and sodium acetylide: Henne, Greenlee, *J. Am. Chem. Soc.* **67**, 484 (1945); from epichlorohydrin +

Consult the cross index before using this section.

sodium: Eglinton *et al., J. Chem. Soc.* **1952**, 2873; from acetylene and formaldehyde: Reppe, *Ann.* **596**, 1 (1955).

Moderately volatile liquid. Mild geranium odor. d_4^{20} 0.9715. mp $-48°$ to $-52°$. bp_{760} 114-115°; $bp_{490.3}$ 100°; $bp_{147.6}$ 70°; $bp_{35.4}$ 40°; $bp_{20.6}$ 30°; $bp_{11.6}$ 20°. n_D^{20} 1.43064. Viscosity at 20° = 1.68 cp. Flash pt 33°C. Spec heat: 0.616 cal/g. Misc with water, benzene, chloroform, 1,2-dichloroethane, ethanol, ether, acetone, dioxane, tetrahydrofuran, pyridine. Appreciable heat is evolved on mixing with pyridine. Moderately sol in carbon tetrachloride. Immiscible with aliphatic hydrocarbons. With water, propargyl alcohol forms an azeotrope, bp 97.5°. This mixture has a composition of 39.5 parts by weight of propargyl alcohol and 60.5 parts of water. Polymerized by heat or caustic. Acidified aq solns are resistant to polymerization. LD_{50} orally in rats: 0.07 g/kg, company information, G.A.F. Corp.

Caution: Irritating to skin, mucous membranes.

7820. Propargyl Chloride. *3-Chloro-1-propyne.* C_3H_3Cl; mol wt 74.51. C 48.36%, H 4.06%, Cl 47.59%. Prepd by treating propargyl alcohol with phosphorus trichloride: Henry, *Ber.* **7**, 761 (1874); **8**, 398 (1875); Kirsmann, *Bull. Soc. Chim. France* [4] **39**, 698 (1926).

Liquid. d_4^{25} 1.0306. mp $-78°$. bp_{760} 57°. Flash pt $< 60°$. Practically insol in water, glycerol. Miscible with benzene, carbon tetrachloride, ethanol, ethylene glycol, ether, ethyl acetate. Reacts with hydroxy compds to form ethers; with sulfides, ammonia, amines or metal hypoiodites to give the corresp propargyl compds; with aldehydes and ketones to give β-acetylenic alcohols. Undergoes isomerization.

USE: Intermediate in organic synthesis.

7821. Propatyl Nitrate. *2-Ethyl-2-[(nitrooxy)methyl]-1,3-propanediol dinitrate (ester); 2-ethyl-2-(hydroxymethyl)-1,3-propanediol trinitrate; 1,1,1-tris(nitratomethyl)propane; ethyltrimethylolmethane trinitrate; trimethylolethylmethane trinitrate; 2,2-bis(hydroxymethyl)-1-butanol trinitrate; 1-hydroxy-2,2-bis(hydroxymethyl)butane trinitrate; ETTN; WIN 9317; Atrilon 5; Etradil; Etrynit; Ettriol trinitrate; Gina; Ginapect; Vasangor.* $C_6H_{11}N_3O_9$; mol wt 269.18. C 26.77%, H 4.12%, N 15.61%, O 53.50%. Prepn: Médard, *Mém. poudres* **35**, 113 (1953); Bourjol, *ibid.* **36**, 79 (1954); Hensinger, **Fr.** pat. **1,103,113** (1955).

$$O_2NOCH_2-\overset{\overset{\displaystyle CH_2CH_3}{|}}{\underset{\underset{\displaystyle CH_2ONO_2}{|}}{C}}-CH_2ONO_2$$

White powder, mp 51-52°. d 1.49. Readily sol in acetone, alcohol; practically insol in water. Lowest explosive temp: 220°. Heat of combustion 829.2 kcal/mol. Explosive but only slightly sensitive to shock.

THERAP CAT: Coronary vasodilator.

7822. Propazine. *6-Chloro-N,N'-bis(1-methylethyl)-1,3,5-triazine-2,4-diamine; 2-chloro-4,6-bis(isopropylamino)-s-triazine; 2,4-bis(isopropylamino)-6-chloro-s-triazine;* G-30028; Gesamil; Milogard; Prozinex. $C_9H_{16}ClN_5$; mol wt 230.09. C 46.98%, H 7.01%, Cl 15.41%, N 30.60%. Selective pre-emergence herbicide. Prepn: Gysin, Knüsli, **Swiss** pats. **342,784-5** (both 1960 to Geigy).

Crystals, mp 213°. Soly in water at 20° = 8.6 ppm; difficultly sol in organic solvents. LD_{50} orally in rats: > 5000 mg/kg, G. W. Bailey, J. L. White, *Residue Rev.* **10**, 97 (1965).

USE: Herbicide.

7823. Propentofylline. *3,7-Dihydro-3-methyl-1-(5-oxohexyl)-7-propyl-1H-purine-2,6-dione; 3-methyl-1-(5-oxohexyl)-7-propylxanthine; 1-(5'-oxohexyl)-3-methyl-7-*

propylxanthine; HWA 285; Albert-285; Hoe-285; Hextol; Karsivan. $C_{15}H_{22}N_4O_3$; mol wt 306.36. C 58.81%, H 7.24%, N 18.29%, O 15.66%. Peripheral vasodilator which inhibits cyclic AMP phosphodiesterase. Prepn: W. Mohler *et al.,* **Ger.** pat. **2,330,742;** *eidem,* **U.S.** pat. **4,289,776** (1975, 1981 both to Hoechst). Cardiovascular effects in animals: O. Hudlicka *et al., Brit. J. Pharmacol.* **72**, 723 (1981). Effect on cAMP phosphodiesterase: K. Nagata *et al., Arzneimittel-Forsch.* **35**, 1034 (1985). Inhibition of adenosine uptake: B. B. Fredholm, K. Lindström, *Acta Pharmacol. Toxicol.* **58**, 187 (1986). Cerebrovascular effects in animals: J. J. Grome *et al., Drug Dev. Res.* **5**, 111 (1985). Effect on postischemic brain edema: B. B. Mrsulja *et al., ibid.* **6**, 339 (1985). Effect on cognitive function in humans: I. Hindmarch, Z. Subhan, *ibid.* **5**, 379 (1985). Acute toxicity: M. Inazu *et al., Oyo Yakuri* **31**, 357 (1986), *C.A.* **104**, 218960a (1986).

Crystals from diisopropyl ether, mp 69-70°. Soly in water at 25°: 3.2%; in ethanol: > 10%; in DMSO: > 10%. LD_{50} in male, female mice, male, female rats (mg/kg): 900, 780, 1150, 940 orally; 168, 170, 180, 195 i.v.; 375, 346, 199, 196 i.p.; 450, 508, 400, 338 s.c. (Inazu).

THERAP CAT: Cognition activator.

THERAP CAT (VET): Peripheral and cerebral vasodilator.

7824. Propenzolate. *α-Cyclohexyl-α-hydroxybenzeneacetic acid 1-methyl-3-piperidinyl ester; α-phenylcyclohexaneglycolic acid 1-methyl-3-piperidyl ester; 1-methyl-3-piperidyl α-phenylcyclohexaneglycolate; N-methyl-3-piperidyl phenylcyclohexylglycolate.* $C_{20}H_{29}NO_3$; mol wt 331.55. C 72.45%, H 8.84%, N 4.23%, O 14.48%. Prepn: Biel *et al., J. Org. Chem.* **26**, 4096 (1961); Biel, **U.S.** pat. **2,995,492** (1961 to Lakeside).

Hydrochloride, $C_{20}H_{30}ClNO_3$, *NDR 263, Delinal.* Crystals, mp 222°.

Note: The (+)-1-methyl-3-piperidinyl ester is designated as *oxyclipine* or *oxiclipine.*

THERAP CAT: Anticholinergic.

7825. Properdin. Mol wt 223,000. A highly basic serum protein believed to be a factor in natural immunity against germ and virus diseases, perhaps also against cancer: L. Pillemer *et al., Science* **120**, 279 (1954). Combines with zymosan, *q.v.,* the insol cell wall residue from yeast: L. Pillemer, O. A. Ross, *ibid.* **121**, 732 (1955). Participates also in a nonspecific manner in a variety of immunological reactions of normal serum. Isoln from human serum: L. Pillemer *et al., J. Exp. Med.* **103**, 1 (1956). Purification: Spicer *et al.,* **U.S.** pat. **3,038,838** (1962 to Merck & Co.). Characterization of highly purified human properdin: J. Pensky *et al., J. Immunol.* **100**, 142 (1968). Activity studies: O. Götze, H. J. Müller-Eberhard, *J. Exp. Med.* **139**, 44 (1974). The human properdin system has been studied extensively; it was the subject of long years of scientific controversy. Components of the human properdin system that have been identified and characterized are: properdin, factor B, factor D and C3. Comprehensive review of L. Pillemer's work and history of the properdin system discovery: I. H. Lepow, *J. Immunol.* **125**, 471 (1980). The rabbit properdin system: G. B. Naff, *ibid.* **124**, 2625 (1980).

7826. Properidine. *1-Methyl-4-phenyl-4-piperidinecar-*

boxylic acid 1-methylethyl ester; 1-methyl-4-phenylisonipecotic acid isopropyl ester; isopropyl 1-methyl-4-phenylpiperidine-4-carboxylate; isopropyl 1-methyl-4-phenylisonipecotate; ipropethidine; isopedine; Gevelina; Spasmodolisina. $C_{16}H_{23}NO_2$; mol wt 261.35. C 73.53%, H 8.87%, N 5.36%, O 12.24%. Prepn: Bergel et al., J. Chem. Soc. **1944**, 265.

Hydrochloride, $C_{16}H_{23}NO_2 \cdot HCl$, crystals, mp 192-195°. Soluble in water.

Caution: May be habit forming. This is a controlled substance (opiate) listed in the U.S. Code of Federal Regulations, Title 21 Part 1308.11 (1985).

THERAP CAT: Hydrochloride as narcotic analgesic; antispasmodic.

7827. Propetamphos. *3-[[(Ethylamino)methoxyphosphinothioyl]oxy]-2-butenoic acid 1-methylethyl ester; (E)-3-hydroxycrotonic acid isopropyl ester O-ester with O-methyl ethylphosphoramidothioate; (E)-1-methylethyl 3-[[(ethylamino)methoxyphosphinothioyl]oxy]-2-butenoate; (E)-O-2-isopropoxycarbonyl-1-methylvinyl O-methyl ethylphosphoramidothioate;* SAN 322I; Blotic; Safrotin. $C_{10}H_{20}NO_4PS$; mol wt 281.31. C 42.70%, H 7.16%, N 4.98%, O 22.75%, P 11.01%, S 11.40%. Prepn: J. P. Leber, K. Lutz, **Ger.** pat. 2,035,103 corresp to U.S. pat. 3,758,645 (1971, 1973 to Sandoz). Activity: W. Berg, R. Gothe, Proc. Brit. Crop Prot. Conf.-Pests Dis. **1979**, 517.

Yellowish liquid, $bp_{0.005}$ 87-89°. n_D^{20} 1.495. d_4^{20} 1.1294. Soly in water at 24°: 110 mg/l. Sol in most organic solvents. LD_{50} orally in male rats: 82 mg/kg (Berg, Gothe).

USE: Insecticide.

THERAP CAT (VET): Ectoparasiticide.

7828. Propham. *Phenylcarbamic acid 1-methylethyl ester; carbanilic acid isopropyl ester; N-phenyl isopropyl carbamate; isopropyl carbanilate; O-isopropyl N-phenyl carbamate;* INPC; IPC; IsoPPC. $C_{10}H_{13}NO_2$; mol wt 179.21. C 67.02%, H 7.31%, N 7.82%, O 17.86%. Prepn: Allen, U.S. pat. 2,615,916 (1952 to Columbia-Southern Chem. Corp.); Kovalenko, Zh. Obshch. Khim. **24**, 1041 (1954); J. L. Hermanson, R. Olson, Trans. Kansas Acad. Sci. **64**, 231 (1961). Acute toxicity: T. B. Gaines, R. E. Linder, Fundam. Appl. Toxicol. **7**, 299 (1986).

Crystals, mp 90°. Practically insol in water. Sol in most organic solvents. LD_{50} in male, female rats (mg/kg): 3724, 4315 orally (Gaines, Linder).

USE: Herbicide, applied as a spray to the soil.

7829. Propicillin. *[2S-(2α,5α,6β)]-3,3-Dimethyl-7-oxo-6-[(1-oxo-2-phenoxybutyl)amino]-4-thia-1-azabicyclo[3.2.0]heptane-2-carboxylic acid; 6-(α-phenoxybutyramido)penicillanic acid; α-phenoxypropylpenicillin; levopropylcillin.* $C_{18}H_{22}N_2O_5S$; mol wt 378.44. C 57.13%, H 5.86%, N 7.40%, O 21.14%, S 8.47%. Semi-synthetic antibiotic related to penicillin. Prepn from 6-aminopenicillanic acid: **Brit.** pat. 877,120 (1961 to Beecham); of salts: Perron et al., J.

Am. Chem. Soc. **82**, 3934 (1960); Glombitza, Ann. **673**, 166 (1964). Metabolism in humans: M. Cole et al., Antimicrob. Ag. Chemother. **3**, 463 (1973). TLC determn: S. Hendrickx et al., J. Chromatog. **291**, 211 (1984).

Potassium salt, $C_{18}H_{21}KN_2O_5S$, **BRL 284, PA 248, Baycillin, Brocillin, Cetacillin, Oricillin, Trescillin, Ultrapen.** Crystals, dec 195-197°. Soluble at 20° in 1.2 parts water, 23 parts 95% (w/v) alcohol. pH of 1% w/v soln: 5-7.5.

THERAP CAT: Antibacterial.

7830. Propiconazole. *1-[[2-(2,4-Dichlorophenyl)-4-propyl-1,3-dioxolan-2-yl]methyl]-1H-1,2,4-triazole;* proconazole; CGA 64250; Banner; Desmel; Orbit; Radar; Tilt. $C_{15}H_{17}Cl_2N_3O_2$; mol wt 342.22. C 52.64%, H 5.01%, Cl 20.72%, N 12.28%, O 9.35%. Systemic foliar fungicide. Prepn: G. Van Reet et al., **Ger.** pat. 2,551,560; eidem, **U.S.** pat. 4,079,062 (1976, 1978 both to Janssen). Physicochemical properties, toxicity and antifungal activity: P. A. Urech et al., Proc. Brit. Crop Prot. Conf. - Pests Dis. **1979**, 508. Efficacy vs cereal diseases: G. Eyries, Phytiatr. Phytopharm. **30**, 37 (1981). GC determn in soil, water, plant material: B. Büttler, J. Agr. Food Chem. **31**, 762 (1983).

Yellowish, viscous liquid. $bp_{0.1mm}$ 180°. Vapor pressure at 20°: $< 3 \times 10^{-6}$ mm Hg. Soly in water at 20°: 110 mg/l. Sol in most organic solvents: LD_{50} orally in rats: 1517 mg/kg (Urech).

USE: Agricultural fungicide.

7831. Propineb. *[[(1-Methyl-1,2-ethanediyl)bis[carbamodithioato]](2−)]zinc; [propylenebis(dithiocarbamato)]zinc;* zinc 1,2-propylene bisdithiocarbamate; mezineb; methylzineb; Bayer 46131; Antracol. $C_5H_8N_2S_4Zn$; mol wt 289.73. C 20.72%, H 2.78%, N 9.67%, S 44.26%, Zn 22.56%. Prepn and use as fungicide: **Ital.** pat. 611,046 (1960 to Montecatini); Lehmann et al., **Belg.** pats. 611,960 and 628,114 (1962 and 1963 to Bayer). Chemistry, biological activity and toxicology: Grewe, Pflanzenschutz-Nachr. **20**, 581 (1967). Fungicidal activity: A. G. Channon, Ann. Appl. Biol. **65**, 481 (1965); E. J. S. Reddy et al., Pesticides **19**, 67 (1985). Distribution and degradation in soil: W. Mittelstaedt, F. Fuhr, Landwirtsch. Forsch. **30**, 221 (1977); in fruit: K. Vogeler et al., Pflanzenschutz-Nachr. **30**, 72 (1977).

White to yellowish powder, practically odorless. Decomposes and turns brown above 160°. Decomposes in strongly acid or alkaline media. Practically insol in all conventional solvents. LD_{50} in male rats, rabbits, cats (mg/kg): 8500, > 2500, > 2500 orally (Grewe).

USE: Agricultural fungicide.

7832. β-Propiolactone. *2-Oxetanone; hydracrylic acid β-lactone; β-propionolactone;* propanolide; NSC-21626; Betaprone. $C_3H_4O_2$; mol wt 72.06. C 50.00%, H 5.60%, O 44.41%. Prepd by the condensation of ketene with formaldehyde: Küng, U.S. pat. 2,356,459 (1941). Purification:

Gresham, Jansen, U.S. pat. **2,602,802** (1952 to B. F. Goodrich). Use in Diels-Alder diene synthesis: Gresham *et al.*, *J. Am. Chem. Soc.* **76**, 609 (1954). Review and evaluation of studies of carcinogenicity in laboratory animals: *IARC Monographs* **4**, 259-269 (1974).

Liquid. d_4^{20} 1.1460; d_4^{25} 1.1420; d_{20}^{20} 1.1490. mp $-33.4°$. bp_{760} 162° (dec); bp_{750} 150° (dec); bp_{20} 61°; bp_{10} 51°. Flash pt 70° (158°F). n_D^{20} 1.4131; n_D^{25} 1.4110. Dipole moment 3.8. Slowly hydrolyzed to hydracrylic acid. Stable when stored at 5° in glass containers. Soly in water: 37% v/v. Misc with alcohol, acetone, ether, chloroform.

Note: This substance may reasonably be anticipated to be a carcinogen: *Fourth Annual Report on Carcinogens* (NTP 85-002, 1985) p 176.

USE: Versatile intermediate in organic synthesis.

THERAP CAT: Disinfectant.

7833. Propiolic Acid. *2-Propynoic acid;* acetylenecarboxylic acid; propargylic acid. $C_3H_2O_2$; mol wt 70.05. C 51.44%, H 2.88%, O 45.68%. CH≡CCOOH. Prepn: Wilson, Wenzke, *J. Am. Chem. Soc.* **57**, 1265 (1935); Owen, Sultanbawa, *J. Chem. Soc.* **1949**, 3109; Wolf, *Ber.* **86**, 735 (1953); **87**, 668 (1954). Manuf: Wolf, U.S. pat. **2,786,022** (1957); Pachter, U.S. pat. **2,799,703** (1957 to Ethyl Corp.). Review on its occurrence and biological activity: Reisch, *Pharmazie* **20**, 194 (1965).

Liquid at room temp (crystals from carbon disulfide, mp 9°). bp 144° (dec), $bp_{10.5}$ 54-55°, bp_{50} 70-75°. d_4^{15} 1.1435, d_4^{20} 1.1380, d_4^{25} 1.1325. $n_D^{20.4}$ 1.4302. Dipole moment at 25°, 2.08D (in dioxane): Wilson, Wenzke, *loc. cit.*

7834. Propiomazine. *1-[10-[2-(Dimethylamino)propyl]-10H-phenothiazin-2-yl]-1-propanone;* 10-dimethylaminoisopropyl-2-propionylphenothiazine; 3-propionyl-10-dimethylaminoisopropylphenothiazine; propionylpromethazine; Wy-1359; Phenoctyl. $C_{20}H_{24}N_2OS$; mol wt 340.50. C 70.55%, H 7.10%, N 8.23%, O 4.70%, S 9.42%. Prepn: Schmitt *et al.*, *Bull. Soc. Chim. France* **1957**, 1474; Schmitt, **Fr. pat. addn. 71,342** (addn. to Fr. pat. **1,176,919** to Clin-Byla). Comprehensive description of the hydrochloride: K. B. Crombie, L. F. Cullen, in *Analytical Profiles of Drug Substances* **vol. 2**, K. Florey, Ed. (Academic Press, New York, 1973) pp 439-466. GC determn in plasma and urine: U. Ahs, G. Wickström, *J. Chromatog.* **183**, 229 (1980). Clinical use as sedative: W. F. Powell *et al.*, *Anesth. Analg.* **49**, 132 (1970); as hypnotic: M. Viukari, P. Miettinen, *Neuropsychobiology* **12**, 134 (1984).

$bp_{0.5}$ 235-245°.

Maleate, $C_{24}H_{28}N_2O_5S$, *1678 CB, Dorevane, Indorm, Propavan.* Crystals from isopropanol, mp 160-161°.

Hydrochloride, $C_{20}H_{25}ClN_2OS$, *Largon.* Yellow, practically odorless powder. Very soluble in water; freely sol in alcohol. Insol in benzene.

Methobromide, $C_{18}H_{21}BrN_2OS$, *Secergan.*

THERAP CAT: Sedative, hypnotic.

THERAP CAT (VET): Tranquilizer.

7835. Propionaldehyde. *Propanal;* methylacetaldehyde; propylaldehyde. C_3H_6O; mol wt 58.08. C 62.04%, H 10.41%, O 27.55%. CH_3CH_2CHO. Prepd by treating propyl alcohol with a bichromate oxidizing mixture: C. D. Hurd, R. N. Meinert, *Org. Syn. coll. vol. II*, 541 (1943); by passing propyl alcohol vapor over copper at high temp: Sabatier, Senderens, *Compt. Rend.* **136**, 923 (1903); Willimott, *Analyst* **50**, 13; *Chem. Zentr.* **1925, I,** 2097.

Liq. Suffocating odor. d_4^0 0.8432; $d_4^{3.7}$ 0.8192; d_4^{24} 0.8071;

d_4^{33} 0.7898. mp $-81°$. bp_{760} 49°; bp_{740} 47°; bp_{687} 45°. $n_{580}^{16.6}$ 1.3695; n_D^{19} 1.36460. Flash pt, open cup: $<20°F$ ($<-6°C$). Absorption spectrum: Kwiecinski, Marchlewski, *Bull. Soc. Chim.* [4] **45**, 608 (1929). Sol in 5 vols water at 20°. Misc with alc and ether. LD_{50} orally in rats: 1.4 g/kg; lethal concn for rats in air: 8000 ppm, H. F. Smyth *et al.*, *Arch. Ind. Hyg. Occup. Med.* **4**, 119 (1951).

7836. Propionamide. *Propanamide;* propionic acid amide. C_3H_7NO; mol wt 73.09. C 49.30%, H 9.65%, N 19.17%, O 21.89%. Prepd by heating ammonium propionate under pressure: Hofmann, *Ber.* **15**, 981 (1882); by dropping propionyl chloride into cooled ammonia water: Aschan, *Ber.* **31**, 2347 (1898).

Orthorhombic platelets from benzene. d_4^{20} 1.0335; d_4^{80} 0.9597; mp 79°; bp 222.2°. Volatile with steam. n_D^{110} 1.4160. Freely sol in water, alcohol, ether, chloroform. Lethal concn for rats in air: 8000 ppm.

7837. Propionic Acid. *Propanoic acid;* methylacetic acid; ethylformic acid. $C_3H_6O_2$; mol wt 74.08. C 48.64%, H 8.16%, O 43.20%. CH_3CH_2COOH. Occurs in dairy products in small amounts. Can be obtained from wood pulp waste liquor by a fermentation process using bacteria of the genus *Propionibacterium:* Wayman *et al.*, U.S. pat. **3,067,107** (1962 to Columbia Cellulose). Prepn from ethylene, carbon monoxide and steam: Reppe, *Angew. Chem.* **1956**, 46; Larson, U.S. pat. **2,448,375** (1948 to du Pont); from ethanol and carbon monoxide using a boron trifluoride catalyst: Loder, U.S. pats. **2,135,448; 2,135,451; 2,135,453** (1939 to du Pont); by oxidation of propionaldehyde: Hasche, U.S. pat. **2,294,984** (1942 to Kodak); from natural gas by the Fischer-Tropsch process; as a byproduct in the pyrolysis of wood; by the action of microorganisms on a variety of materials in small yields. Very pure propionic acid can be obtained from propionitrile.

Oily liquid. Slightly pungent, disagreeable, rancid odor. d_4^{20} 0.99336. mp $-21.5°$. bp_{760} 141.1°; bp_{400} 122.0°; bp_{100} 85.8°; $bp_{1.0}$ 4.6°. n_D^{25} 1.3848. Flash pt, open cup: 136°F (58°C). Viscosity (cp) at 15°: 1.175; at 25°: 1.020; at 30°: 0.956; at 60°: 0.668; at 90°: 0.495. Surface tension in dynes/cm at 15°: 27.21. Ka at 25°: 1.34×10^{-5}. Misc with water. Can be salted out of water solns by the addn of $CaCl_2$ or other salts. Sol in alcohol, ether, chloroform. Azeotrope with water, bp 99.98°, contains 17.7% acid; with toluene, bp 110.45°, contains 3% acid; with o-xylene, bp 135.4°, contains 43% acid; with ethylbenzene, bp 131.1°, contains 28% acid. LD_{50} orally in rats: 4.29 g/kg, H. F. Smyth *et al.*, *Am. Ind. Hyg. Assoc. J.* **23**, 95 (1962).

Barium salt monohydrate, $C_6H_{10}BaO_4 \cdot H_2O$, *barium propionate.* Powder, usually with slight odor. *Poisonous!* Freely sol in water; slightly sol in alcohol.

USE: Esterifying agent; in the production of cellulose propionate (thermoplastic) and other propionates, e.g., calcium propionate, used as mold inhibitors and preservatives; in the manuf of ester solvents, fruit flavors, and perfume bases. β-Propiolactone is useful in organic syntheses and to preserve arterial grafts.

THERAP CAT: Antifungal.

7838. Propionic Anhydride. *Propanoic acid anhydride;* propanoic anhydride; methylacetic anhydride. $C_6H_{10}O_3$; mol wt 130.14. C 55.37%, H 7.75%, O 36.88%. $(CH_3CH_2CO)_2O$. Obtained by dehydration of the acid or by carbonylation of its esters: Reppe, Friederich, U.S. pat **2,730,546** (1956 to Badische Anilin- & Soda-Fabrik); from propionaldehyde by air oxidation in the presence of cobalt and copper acetate catalysts: McFarlane, U.S. pat. **2,491,572** (1949 to Celanese). Other syntheses, e.g., from ethanol and carbon monoxide.

Liquid. Odor more pungent than that of the acid. d_4^0 1.0336; d_4^{15} 1.0169; d_4^{20} 1.0125; d_4^{40} 0.98974; d_4^{50} 0.97913. 8.4 lbs/gal at 20°. mp $-45°$. bp_{760} 167.0°; bp_{400} 146.0°; bp_{200} 127.8°; bp_{100} 107.2°; bp_{60} 94.5°; bp_{40} 85.6°; bp_{20} 70.4°; bp_{10} 57.7°; bp_5 45.3°; $bp_{1.0}$ 20.6°. Flash pt, open cup: 165°F (74°C). n_D^{17} 1.4041; n_D^{20} 1.4038. Viscosity (cp): 1.144 at 20°; 0.978 at 30°; 0.853 at 40°; 0.7511 at 50°. Dec by water. Sol in methanol, ethanol, ether, chloroform. LD_{50} orally in rats: 2.36 g/kg, H. F. Smyth *et al.*, *Arch. Ind. Hyg. Occup. Med.* **10**, 61 (1954).

USE: Esterifying agent for certain perfume oils, fats, oils, and especially cellulose. In the production of alkyd resins, dyestuffs and drugs. Has been used as a dehydrating agent in some sulfonations and nitrations.

7839. Propionitrile. *Propanenitrile;* ethyl cyanide. C_3H_5N; mol wt 55.08. C 65.42%, H 9.15%, N 25.43%. CH_3-CH_2CN. May be prepd by dehydration of propionamide (or propionic acid + NH_3) or by distilling ethyl sulfate and concd aq KCN, also by reduction of acrylonitrile. Has been shown to be a duodenal ulcerogen in rats: S. Szabo *et al., Res. Commun. Chem. Pathol. Pharmacol.* **16**, 311 (1977); L. M. Lichtenberger *et al., Gastroenterology* **73**, 1305 (1977).
Liquid. *Poisonous when heated to decompn or on contact with acids.* Pleasant, ethereal, sweetish odor. d_4^0 0.8020; d_4^{20} 0.7818; d_4^{30} 0.7716; d_4^{56} 0.7515; $d_4^{70.2}$ 0.7291. mp −91.8°. bp_{760} 97.2°; bp_{400} 77.7°; bp_{200} 58.2°; bp_{100} 41.4°; bp_{60} 30.1°; bp_{40} 22.0°; bp_{20} +8.8°; bp_{10} −3.0°; bp_5 −13.6°; $bp_{1.0}$ −35.0°. n_D^{15} 1.36812; n_D^{20} 1.36585; n_D^{30} 1.36132. Soly in water at 40° = 11.9 g/100 g H_2O; at 100° = 29 g/100 g H_2O. Misc with alcohol, ether, DMF. LD_{50} orally in rats: 39 mg/kg, H. F. Smyth *et al., Arch. Ind. Hyg. Occup. Med.* **4**, 119 (1951).

7840. Propionyl Chloride. *Propanoyl chloride.* C_3H_5Cl-O; mol wt 92.53. C 38.94%, H 5.45%, Cl 38.32%, O 17.29%.
Liquid, pungent odor. d_4^{20} 1.065; mp −94°; bp 80°; n_D^{20} 1.4051. Vigorously dec and dissolved by water or alcohol.

7841. Propionylpromazine. *1-[10-[3-(Dimethylamino)propyl]-10H-phenothiazin-2-yl]-1-propanone;* 3-propionyl-10-(γ-dimethylaminopropyl)phenothiazine; 10-(3-dimethyl-aminopropyl)-2-propionylphenothiazine; propiopromazine; 1497 CB. $C_{20}H_{24}N_2OS$; mol wt 340.55. C 70.55%, H 7.10%, N 8.23%, O 4.70%, S 9.42%. Prepn: Schmitt *et al., Bull. Soc. Chim. France* **1957**, 938, 1474; Schmitt, **Fr. pat. addn. 71,342** (1956 to Clin-Byla), *C.A.* **56**, 3486h (1962).

Crystals, mp 69-70°.
Hydrochloride, $C_{20}H_{25}ClN_2OS$, *Combelen, Tranvet, Tranvex.*
Maleate, $C_{20}H_{24}N_2OS.C_4H_4O_4$, crystals from acetone, mp 135°.
Methiodide, $C_{20}H_{24}N_2OS.CH_3I$, yellow crystals from iso-propanol, mp 79-80°.
THERAP CAT (VET): Tranquilizer.

7842. Propiophenone. *1-Phenyl-1-propanone;* ethyl phenyl ketone; propionylbenzene; phenyl ethyl ketone. $C_9H_{10}O$; mol wt 134.17. C 80.56%, H 7.51%, O 11.93%. Prepared from propionyl chloride and benzene in the presence of anhydr aluminum chloride: Read, *J. Am. Chem. Soc.* **44**, 1751 (1922).

Leaflets or tabular crystals. Strong, persistent, agreeable, flowery odor. Usually supplied as a liquid. d_4^0 (solid) 1.157; d_4^{20} (liq) 1.0105; d_{20}^{20} 1.0118; d_{25}^{25} 1.0087; $d_4^{41.8}$ 0.9934; $d_4^{61.2}$ 0.9776; $d_4^{85.5}$ 0.9572. mp 21°. Typical freezing point: 18.6°. bp_{760} 218.0°; bp_{400} 194.2°; bp_{200} 170.2°; bp_{100} 149.3°; bp_{60} 135.0°; bp_{40} 124.3°; bp_{20} 107.6°; bp_{10} 92.2°; bp_5 77.9°. $n_D^{15.9}$ 1.5290; n_D^{20} 1.5269. uv max (hexane): 250, 280, 323 nm. Dipole moment: 2.7. Parachor 328.7. Flash pt 99° (210°F). Misc with methanol, anhydr ethanol, ether, benzene, toluene. Insol in water, glycerol, ethylene glycol, propylene glycol.
USE: In perfumery; in the synthesis of ephedrine and related compds.

7843. Propipocaine. *3-(1-Piperidinyl)-1-(4-propoxyphenyl)-1-propanone;* 3-piperidino-4'-propoxypropiophenone; β-piperidinoethyl 4-propoxyphenyl ketone; 4-n-propoxy-β-(1-piperidyl)propiophenone; 4-propoxyphenyl piperidine

ethyl ketone. $C_{17}H_{25}ClNO_2$; mol wt 275.40. C 74.14%, H 9.15%, N 5.09%, O 11.62%. Prepd from p-propoxyaceto-phenone by condensation with formaldehyde and piperidine HCl: Profft, *Chem. Tech. (Leipzig)* **4**, 241 (1952), *C.A.* **47**, 10531 (1953). *See also:* Dyclonine.

Hydrochloride, $C_{17}H_{26}ClNO_2$, *falicain, Exotancain.* Crystals, mp 166°. Sol in water, alcohol, acetone. Phenol coefficient 3.6. Aq solns should not be sterilized by autoclaving.
THERAP CAT: Local anesthetic.

7844. Propiram. *N-[1-Methyl-2-(1-piperidinyl)ethyl]-N-2-pyridinylpropanamide;* N-(1-methyl-2-piperidinoethyl)-N-2-pyridylpropionamide; N-propionyl-2-(1-piperidinoiso-propyl)aminopyridine; N-propionyl-N-(2-pyridyl)-1-piperi-dino-2-aminopropane. $C_{16}H_{25}N_3O$; mol wt 275.38. C 69.78%, H 9.15%, N 15.26%, O 5.81%. Prepn: Hiltmann *et al.,* **Brit. pat. 939,947** corresp to **U.S. pat. 3,163,654; Fr. pat. M 2283** (1963, 1964, 1964 all to Bayer). Resolution of isomers: **Neth. pat. Appl. 6,610,362;** Wollweber *et al.,* **U.S. pat. 3,594,477** (1967, 1971 both to Bayer). Pharmacology: Jasinski *et al., Clin. Pharmacol. Ther.* **12**, 613 (1971). Series of articles on chemistry, pharmacology, toxicology, pharma-cokinetics, clinical trials: *Arzneimittel-Forsch.* **24**, 581-722 (1974). Toxicity data: D. Tettenborn, *ibid.* 624. Absolute configuration and analgesic activity of enantiomers: W. Wollweber, *Eur. J. Med. Chem.-Chim. Ther.* **17**, 125 (1982).

bp 162-163°.
Fumarate, $C_{20}H_{29}N_3O_5$, *FBA 4503, Bay 4503, Algeril, Dirame.* LD_{50} in mice, rats (mg/kg): 874, 1657 orally (Tettenborn).
Note: This is a controlled substance (opiate) listed in the U.S. Code of Federal Regulations, Title 21 Part 1308.11 (1987).
THERAP CAT: Narcotic analgesic.

7845. Propivane. *α-Propylbenzeneacetic acid 2-(diethyl-amino)ethyl ester hydrochloride;* 2-diethylaminoethyl α-phen-ylvalerate hydrochloride; 2-diethylaminoethanol hydrochloride α-propyltolu-ate hydrochloride; diethylaminoethanol hydrochloride prop-ylphenylacetate; RP 177; Prospasmine. $C_{17}H_{28}ClNO_2$; mol wt 313.86. C 65.06%, H 8.99%, Cl 11.29%, N 4.46%, O 10.20%. Prepn: Viaud, **U.S. pat. 2,219,796** (1941 to Rhône-Poulenc).

Crystals. Bitter taste. mp 109° (free base bp_3 140-144°). Freely sol in water; very sparingly sol in alcohol, ether. A 5% aq soln is neutral to litmus.
THERAP CAT: Antispasmodic.

7846. Propizepine. *6-[2-(Dimethylamino)propyl]-1,6-dihydro-5H-pyrido[2,3-b][1,5]benzodiazepin-5-one;* 6,11-dihydro-6-[2-(dimethylamino)-2-methylethyl]-5H-pyrido-[2,3-b][1,5]benzodiazepin-5-one. $C_{17}H_{20}N_4O$; mol wt 296.38. C 68.89%, H 6.80%, N 18.91%, O 5.40%. Prepn: **Neth. pat. Appl. 6,600,065** (1966 to Labs. U.P.S.A.); Hoff-mann, Faure, *Bull. Soc. Chim. France* **1966**, 2316. Psycho-pharmacology: Lwoff *et al., Thérapie* **26**, 451 (1971).

mp 122°.
Hydrochloride, $C_{17}H_{21}ClN_4O$, *UP 106, Depressin, Vagran.* mp 235°.

THERAP CAT: Antidepressant.

7847. Propofol. *2,6-Bis(1-methylethyl)phenol; 2,6-diiso-propylphenol;* disoprofol; ICI 35868; Diprivan; Disoprivan; Rapinovet. $C_{12}H_{18}O$; mol wt 178.27. C 80.85%, H 10.18%, O 8.97%. Prepn: A. J. Kolka *et al., J. Org. Chem.* **21**, 712 (1956); **22**, 642 (1957); G. G. Ecke, A. J. Kolka, U.S. pat. **2,831,898** (1958 to Ethyl Corp.); T. J. Kealy, D. D. Coffman, *J. Org. Chem.* **26**, 987 (1961); B. E. Firth, T. J. Rosen, U.S. pat. **4,447,657** (1984 to Universal Oil Products). Chromatographic study: J. K. Carlton, W. C. Bradbury, *J. Am. Chem. Soc.* **78**, 1069 (1956). Animal studies: J. B. Glen, *Brit. J. Anaesth.* **52**, 731 (1980). Pharmacokinetics: H. K. Adam *et al., ibid.* 743; *idem, ibid.* **55**, 97 (1983). Determn in blood: *eidem, J. Chromatog.* **223**, 232 (1981). Comparative studies vs other injectable anesthetics: B. Kay, D. K. Stephenson, *Anaesthesia* **35**, 1182 (1980); D. V. Rutter *et al., ibid.* 1188. Use in i.v. anesthesia: E. Major *et al., ibid.* **37**, 541 (1982). Cardiovascular effects: D. Al-Khudhairi *et al., ibid.* 1007. Pharmacology of emulsion formulation: J. B. Glen, S. C. Hunter, *Brit. J. Anaesth.* **56**, 617 (1984). Series of articles on pharmacology and clinical experience: *Postgrad. Med. J.* **61**, Suppl. 3, 1-169 (1985).

bp$_{30}$ 136°. bp$_{17}$ 126°. mp 19°. n_D^{20} 1.5134. n_D^{25} 1.5111. d$_{20}$ 0.955.

THERAP CAT: Anesthetic (intravenous).
THERAP CAT (VET): Intravenous anesthetic (dogs and cats).

7848. Propolis. Bee bread; hive dross. A resinous substance found in beehives. Collected by bees from buds. Isoln of caffeic acid from propolis: Cizmárik, Matel, *Experientia* **26**, 713 (1970). Antimicrobial constituents of propolis: J. Metzner *et al., Pharmazie* **30**, 799 (1975); E. M. Schneidewind *et al., ibid.* 803. Review on the origin, chemical constituents and therapeutic activity: M. H. Haydak, *State of Iowa, Repts. State Apiarist 1953,* p 74-87; M. Vanhaelen, R. Vanhaelen-Fastre, *J. Pharm. Belge* **34**, 253 (1979).

Greenish-brown, sticky mass. Aromatic odor. d 1.2. mp 64°. Becomes brittle when cooled below 15°. Extraction with alcohol gives *propolis wax.* The residue from the alcohol extraction is called *propolis resin,* yielding *propolis balsam* on extraction with hot petr ether. Propolis balsam has a hyacinth odor and is said to contain 10% cinnamyl alcohol.

7849. Propoxur. *2-(1-Methylethoxy)phenol methylcarbamate; o-isopropoxyphenyl N-methylcarbamate;* aprocarb; BAY 39007; BAY 9010; Baygon; Bifex; Blattanex; Invisi-Gard; Propyon; Suncide; Sendran; Unden. $C_{11}H_{15}NO_3$; mol wt 209.24. C 63.14%, H 7.23%, N 6.69%, O 22.94%. Prepn: U.S. pat. **3,111,539** (1963 to Bayer; Chemagro Corp.). Properties: *Pflanzenschutz Nachr. Bayer* **18**, 53 (1965). Toxicity data: T. B. Gaines, *Toxicol. Appl. Pharmacol.* **14**, 515 (1969). Teratogenicity study: K. D. Courtney *et al., J. Environ. Sci. Health* **B20**, 373 (1985).

Minute crystals, mp 91.5°. Dec at high temp forming methyl isocyanate. Sol in methanol, acetone and many organic solvents, but only slightly sol in cold hydrocarbons. Water soly about 0.2% at 20°. Unstable in highly alkaline media. LD$_{50}$ orally in male, female rats: 83, 86 mg/kg (Gaines).

USE: Insecticide.

7850. Propoxycaine Hydrochloride. *4-Amino-2-propoxybenzoic acid 2-diethylaminoethyl ester hydrochloride; 2-diethylaminoethyl 4-amino-2-propoxybenzoate hydrochloride; 2-diethylaminoethyl 2-propoxy-4-aminobenzoate* hydrochloride; Ravocaine hydrochloride; Pravocaine hydrochloride; Blockaine hydrochloride. $C_{16}H_{27}ClN_2O_3$; mol wt 330.86. C 58.08%, H 8.23%, Cl 10.72%, N 8.47%, O 14.51%. Prepn: Clinton, Laskowski, U.S. pat. **2,689,248** (1954 to Sterling Drug).

White, odorless crystals, mp 148-150°. Discolors upon prolonged exposure to light and to air. Freely sol in water; sol in ethanol, chloroform. Sparingly sol in ether. Practically insol in acetone, chloroform. pH of a 2% aq soln 5.4.

THERAP CAT: Local anesthetic.

7851. Propoxyphene. *[S-(R*,S*)]-α-[2-(Dimethylamino)-1-methylethyl]-α-phenylbenzeneethanol propanoate (ester); α-d-4-dimethylamino-3-methyl-1,2-diphenyl-2-butanol propionate;* (+)-1,2-diphenyl-2-propionoxy-3-methyl-4-dimethylaminobutane; (+)-4-dimethylamino-1,2-diphenyl-3-methyl-2-propionyloxybutane; *d-propoxyphene;* dextropropoxyphene. $C_{22}H_{29}NO_2$; mol wt 339.48. C 77.83%, H 8.61%, N 4.13%, O 9.43%. Prepn of racemate: Pohland, Sullivan, *J. Am. Chem. Soc.* **75**, 4458 (1953); Pohland, U.S. pat. **2,728,779** (1955 to Lilly). Prepn of (+)-form: Pohland, Sullivan, *J. Am. Chem. Soc.* **77**, 3400 (1955). Stereochemistry: Sullivan *et al., J. Org. Chem.* **28**, 2381 (1963); Casy, Myers, *J. Pharm. Pharmacol.* **16**, 455 (1964). Stereospecific synthesis: Pohland *et al., J. Org. Chem.* **28**, 2483 (1963). Metabolism: S. L. Due *et al., Biomed. Mass Spectrom.* **3**, 217 (1976). The α-*dl*- and *d*-diastereoisomers possess marked analgesic activity in contrast to the β-diastereoisomers which are substantially inactive. Toxicity: E. I. Goldenthal, *Toxicol. Appl. Pharmacol.* **18**, 185 (1971); J. L. Emerson *et al., ibid.* **19**, 445 (1971). Comprehensive description: B. McEwan in *Analytical Profiles of Drug Substances* vol. 1, K. Florey, Ed. (Academic Press, New York, 1972) pp 301-318. Symposium on pharmacology, toxicology, and clinical efficacy of propoxyphene alone and in combination with acetaminophen: *Human Toxicol.* **3**, Suppl., 1S-238S (1984).

Crystals from petr ether, mp 75-76°. $[\alpha]_D^{25}$ +67.3° (c = 0.6 in chloroform).

α-d-Hydrochloride, $C_{22}H_{30}ClNO_2$, *Algafan, Antalvic, Darvon, Depromic, Deprancol, Develin, Dolene, Dolocap, Doraphen, Erantin, Femadol, Harmar, Propox, Propoxychel, Proxagesic.* Bitter crystals from methanol + ethyl acetate, mp 163-168.5°. $[\alpha]_D^{25}$ +59.8° (c = 0.6 in water). Sol in water, alc, chloroform, acetone. Practically insol in benzene, ether. LD$_{50}$ in mice, rats (mg/kg): 28, 15 i.v.; 111, 58 i.p.; 211, 134 s.c.; 282, 230 orally (Emerson).

α-d-Form napsylate monohydrate, $C_{32}H_{37}NO_5S \cdot H_2O$, *Darvon-N, Doloxene.* LD$_{50}$ orally in female rats: 990 mg/kg (Goldenthal).

α-*l*-Form, see Levopropoxyphene.

α-dl-Form, *racemic propoxyphene, diméprotane.*
α-dl-Form hydrochloride, crystals from methanol + ethyl acetate, mp 170-171°. Soluble in water, alcohol, chloroform. Practically insol in benzene, ether.
β-dl-Form, crystals from acetone + ether. mp 187-188°. More soluble than the α-form.
Caution: May be habit forming. Bulk dextropropoxyphene (non-dosage forms) is a controlled substance (opiate) listed in the U.S. Code of Federal Regulations, Title 21 Part 1308.12 (1987); dextropropoxyphene is a controlled substance (narcotic), *ibid.,* Title 21 Part 1308.14.
THERAP CAT: Narcotic analgesic.

7852. Propranolol. *1-[(1-Methylethyl)amino]-3-(1-naphthalenyloxy)-2-propanol; 1-(isopropylamino)-3-(1-naphthyloxy)-2-propanol;* Avlocardyl; Euprovasin; Sumial. $C_{16}H_{21}NO_2$; mol wt 259.34. C 74.10%, H 8.16%, N 5.40%, O 12.34%. β-Adrenergic blocker. Prepn: **Belg. pats. 640,312** and **640,313**; Crowther, Smith, **U.S. pats. 3,337,628** and **3,520,919** (1964, 1964, 1967, 1970 all to I.C.I.). Description of optical isomers: Howe, Shanks, *Nature* **210,** 1336 (1966). Biological studies: Barrett, Cullum, *Brit. J. Pharmacol.* **34,** 43 (1968). Metabolism: Bond, *Nature* **213,** 721 (1967). Multi-center clinical trial in myocardial infarction: V. Hansteen *et al., Brit. Med. J.* **284,** 155 (1982). Use as migraine prophylactic: S. Diamond, E. Millstein, *J. Clin. Pharmacol.* **28,** 193 (1988); S. Diamond *et al., Headache* **27,** 70 (1987). Toxicity data: M. Martin, P. Linee, *Eur. J. Med. Chem.* **9,** 563 (1974). Review of pharmacokinetics: P. A. Routledge, D. G. Shand, *Appl. Pharmacokinet.* **1980,** 464-485; of pharmacology: J. D. Fitzgerald in *Pharmacology of Antihypertensive Drugs,* A. Scriabine, Ed. (Raven Press, New York, 1980) pp 195-208.

OH
|
OCH₂CHCH₂NHCH(CH₃)₂

Crystals from cyclohexane, mp 96°.
Hydrochloride, $C_{16}H_{22}ClNO_2$, *AY 64043, ICI 45520, NSC-91523, Angilol, Apsolol, Bedranol, Beprane, Berkolol, Beta-Neg, Beta-Tablinen, Beta-Timelets, Cardinol, Caridorol, Deralin, Dociton, Dumopranol, Duranol, Efektolol, Elbrol, Frekven, Inderal, Inderex, Indobloc, Intermigran, Kemi, Oposim, Prano-Puren, Propahexal, Prophylux, Propranur, Pylapron, Rapynogen, Sagittol, Sloprolol, Tesnol.* Crystals from *n*-propanol, mp 163-164°. Sol in water, alc. Practically insol in ether, benzene, ethyl acetate. LD₅₀ in mice (mg/kg): 565 orally; 22 i.v.; 107 i.p. (Martin, Linee).
THERAP CAT: Antihypertensive; antianginal; antiarrhythmic.

7853. Propyl Acetate. *Acetic acid n-propyl ester.* $C_5H_{10}O_2$; mol wt 102.13. C 58.80%, H 9.87%, O 31.33%. $CH_3COOCH_2CH_2CH_3$. Prepn from acetic acid and *n*-propyl alcohol: Wagner, *J. Chem. Ed.* **27,** 245 (1950). Manuf from acetic acid and mixture of propene + propane in the presence of $ZnCl_2$ catalyst: Biller, **Brit. pat. 872,876** (1961).
Liquid, odor of pears, bp 101.6°. mp −92°. d_4^{20} 0.836, d_{20}^{20} 0.887. n_D^{20} 1.3844. Flash pt, closed cup: 58°F (14°C). Soly in water at 16°: 1.6:100. Misc with alcohol, ether. LD₅₀ orally in rats, mice: 9370, 8300 mg/kg, P. M. Jenner *et al., Food Cosmet. Toxicol.* **2,** 327 (964).
USE: Mfg flavors, perfumes. Solvent for resins, cellulose derivatives, plastics. *Caution:* May be irritating to skin, mucous membranes, and, in high concns, narcotic.

7854. *n*-Propyl Alcohol. *1-Propanol;* propylic alcohol; Optal. C_3H_8O; mol wt 60.09. C 59.96%, H 13.42%, O 26.62%. $CH_3CH_2CH_2OH$. Discovered by Chancel in 1853 in crude fusel oil and obtained therefrom by fractionation; available as a byproduct of the reaction between carbon monoxide and hydrogen.
Liquid; alcoholic and slightly stupefying odor. mp −127°. bp 97.2°. d_4^{20} 0.8053; d_4^{25} 0.8016: Mumford, Phillips, *J. Chem. Soc.* **1950,** 75. Flash pt 22°. n_D^{20} 1.3862. Miscible

with water, alcohol, ether. LD₅₀ orally in rats: 1.87 g/kg, Smyth *et al., Arch. Ind. Hyg. Occup. Med.* **10,** 61 (1954).
USE: As a solvent for resins and cellulose esters, etc. *Caution:* Mildly irritating to eyes, mucous membranes; depressant action similar to ethyl alcohol: E. Browning, *Toxicity and Metabolism of Industrial Solvents* (Elsevier, New York, 1965) pp 332-334, 401-411.

7855. Propylamine. *1-Propanamine;* 1-aminopropane. C_3H_9N; mol wt 59.11. C 60.95%, H 15.35%, N 23.70%. $CH_3CH_2CH_2NH_2$. Prepn from propionaldehyde + ammonia with a Raney nickel catalyst: Olin, Schwoegler, **U.S. pat. 2,373,705** (1945 to Sharples); by low pressure catalytic hydrogenation of nitropropane: Iffland, Cassis, *J. Am. Chem. Soc.* **74,** 6284 (1952); by catalytic hydrogenation of propionitrile: Rylander, Kaplan, **U.S. pat. 3,117,162** (1964 to Englehard).
Colorless, alkaline liq; strong ammonia odor. d_{20}^{20} 0.719; mp −83°; bp 48-49°; n_D^{20} 1.389. Flash pt, closed cup: 10°F (−12°C). Misc with water, alcohol, ether. *Keep tightly closed.* LD₅₀ orally in rats: 0.57 g/kg, H. F. Smyth *et al., Am. Ind. Hyg. Assoc. J.* **23,** 95 (1952).
Hydrochloride, $C_3H_9N.HCl$, deliquesc crystals, mp 157-158°. Soluble in 0.4 part water, 1.5 parts chloroform. *Keep well closed.*
Caution: Strong irritant, possible skin sensitizer.

7856. *n*-Propylbenzene. 1-Phenylpropane. C_9H_{12}; mol wt 120.19. C 89.93%, H 10.06%. $C_6H_5CH_2CH_2CH_3$. Prepd by the action of diethyl sulfate on benzylmagnesium chloride: H. Gilman, W. E. Catlin, *Org. Syn.* coll. **vol. I,** 471 (2nd ed., 1941).
Liquid. d_4^{20} 0.8621. mp −99.2°. bp₇₆₀ 159.2°; bp₄₀₀ 135.7°; bp₂₀₀ 113.5°; bp₁₀₀ 94°; bp₄₀ 71.6°; bp₂₀ 56.8°; bp₁₀ 43.4°; bp₅ 31.3°; bp₁.₀ 6.3°. n_D^{20} 1.4919. Very slightly sol in water (0.06 g/l); sol in alcohol, ether. LD₅₀ orally in rats: 6040 mg/kg, P. M. Jenner *et al., Food Cosmet. Toxicol.* **2,** 327 (1964).
USE: In textile dyeing and printing; as solvent for cellulose acetate.

7857. Propyl Bromide. *1-Bromopropane.* C_3H_7Br; mol wt 123.00. C 29.29%, H 5.74%, Br 64.97%. $CH_3CH_2CH_2Br$.
Colorless liquid. d_{20}^{20} 1.353; mp −110°; bp 71°; n_D^{20} 1.4341. Sol in 400 parts water; miscible with alcohol, etc.

7858. Propyl Butyrate. *Butanoic acid propyl ester.* $C_7H_{14}O_2$; mol wt 130.18. C 64.58%, H 10.84%, O 24.58%. $CH_3CH_2CH_2COOC_3H_7$.
Colorless liquid. d_4^{15} 0.879; mp −95°; bp 143°; n_D^{20} 1.4005. Slightly sol in water; misc with alcohol, ether. LD₅₀ orally in rats: 15,000 mg/kg, P. M. Jenner *et al., Food Cosmet. Toxicol.* **2,** 327 (1964).

7859. Propyl Chloride. *1-Chloropropane.* C_3H_7Cl; mol wt 78.54. C 45.87%, H 8.98%, Cl 45.14%. $CH_3CH_2CH_2Cl$. Prepd from propyl alc and PCl_3 in the presence of $ZnCl_2$.
Colorless liquid. d_{20}^{20} 0.890; mp −122° to −123°; bp 46-47°; n_D^{20} 1.3886. Sol in about 300 parts water; miscible with alcohol, ether.

7860. Propyl Chlorocarbonate. *Carbonochloridic acid propyl ester;* propyl chloroformate. $C_4H_7ClO_2$; mol wt 122.55. C 39.20%, H 5.76%, Cl 28.93%, O 26.11%. C_3H_7O-COCl.
Colorless liquid. d^{20} 1.090; bp 114-116°. Gradually dec by water or alc; miscible with benzene, chloroform, ether.
Caution: Vapors strongly irritating to eyes and mucous membranes.

7861. Propyl Docetrizoate. *3-(Diacetylamino)-2,4,6-triiodobenzoic acid propyl ester;* propyl 3-(diacetylamino)-2,4,-6-triiodobenzoate; docetrizoate propyl; Pulmidol. $C_{14}H_{14}I_3NO_4$; mol wt 641.00. C 26.23%, H 2.20%, I 59.40%, N 2.19%, O 9.98%. Prepd by acylation of propyl 3-acetamido-2,4,6-triiodobenzoate: Ashley, Easson, **Brit. pat. 898,780** (1962 to May & Baker).

Crystals from ethanol, mp 158-160°.
THERAP CAT: Diagnostic aid (radiopaque medium).

7862. Propylene. *1-Propene;* methylethylene; methylethene. C_3H_6; mol wt 42.08. C 85.63%, H 14.37%. $H_2C=CHCH_3$. Obtained from petr oils during the refining of gasoline. Catalytic or thermal cracking of hydrocarbons always yields propylene. If necessary, it can be obtained by catalytic dehydrogenation of propane. *Reviews:* R. F. Goldstein, *The Petroleum Chemicals Industry* (New York-London, 1949) p 114 sqq.; Sherwood, *Ind. Chemist* **1960**, 542-546; *Chim. et Ind.* **1961**, 576-587; Haney, "Ethylene, Propylene and 1-Butene" in *Vinyl and Diene Monomers*, E. C. Leonard, Ed. (Interscience, New York, 1971) pp 577-689; M. R. Schoenberg *et al.*, in Kirk-Othmer *Encyclopedia of Chemical Technology* vol. 19 (Wiley-Interscience, New York, 3rd ed., 1982) pp 228-246.

Flammable gas. Burns with yellow sooty flame. d 1.49 (air = 1.0). mp (triple pt) −185°. bp_{760} −48°. Critical temp 91.8°. Critical pressure 45.6 atm. Heat of fusion 717.6 cal/mol. Liquefies at 7-8 atm. d_4^{20} (liq) 0.5139. Flammable limits in air: 2.4-10.3% (by volume). Latent heat of vaporization at bp: 104.62 cal/g. Dipole moment 0.35. n_D^{40} 1.3567. Surface tension at 90°: 16.70 dynes/cm. Shipped as a liquefied gas in low pressure steel cylinders under its own vapor pressure of about 136 pounds per square inch. Contaminants are propane, butane, carbon dioxide.

Caution: Simple asphyxiant and mild anesthetic at high concentrations: E. E. Sandmeyer in Patty's *Industrial Hygiene and Toxicology* Vol. 2B, G. D. Clayton, F. E. Clayton, Eds. (Wiley-Interscience, New York, 3rd ed., 1981) pp 3199-3201.

USE: In polymerized form as polypropylene plastic. Raw material in the manuf of acetone, isopropylbenzene, isopropanol, isopropyl halides, propylene oxide.

7863. Propylene Chlorohydrin. *2-Chloro-1-propanol;* 2-chloropropyl alcohol. C_3H_7ClO; mol wt 94.54. C 38.11%, H 7.46%, Cl 37.50%, O 16.92%. $CH_3CHClCH_2OH$.

Colorless liquid; pleasant odor. d^{20} 1.103; bp 133-134°; n_D^{20} 1.4362. Sol in water, alcohol, etc. LD_{50} orally in rats: 0.22 ml/kg; by skin penetration in rabbits: 0.48 ml/kg, Smyth *et al.*, *Am. Ind. Hyg. Assoc. J.* **30**, 470 (1969).

USE: In prepn of propylene oxide *(q.v.)*.

7864. sec-Propylene Chlorohydrin. *1-Chloro-2-propanol;* 1-chloroisopropyl alcohol. C_3H_7ClO; mol wt 94.54. C 38.11%, H 7.46%, Cl 37.50%, O 16.92%. $ClCH_2CH(OH)-CH_3$.

Colorless liquid. d^{20} 1.115; bp 126-127°; n_D^{20} 1.4392. Sol in water, alcohol, etc.

7865. Propylenediamine. *1,2-Propanediamine.* $C_3H_{10}N_2$; mol wt 74.13. C 48.61%, H 13.60%, N 37.80%. $CH_3CH-(NH_2)CH_2NH_2$. Prepd from propylene dibromide and alcoholic ammonia at 100°.

Extremely hygroscopic, strongly alkaline liq. Rapidly absorbs moisture to form a hemihydrate. d^{15} 0.878 in anhydr form. bp 119-120°. Very sol in water. *Keep tightly closed.*

USE: In conjunction with cupric sulfate it is a very sensitive reagent for mercury.

7866. Propylene Dibromide. *1,2-Dibromopropane.* $C_3H_6Br_2$; mol wt 201.91. C 17.84%, H 3.00%, Br 79.16%. $CH_3CHBrCH_2Br$. Prepd from propyl bromide and Br in the presence of $AlCl_3$ or $AlBr_3$.

Colorless liquid. mp −55°; bp 140-142°; n_D^{20} 1.5203; d^{20} 1.933. Slightly sol in water; miscible with organic solvents.

7867. Propylene Dichloride. *1,2-Dichloropropane.* $C_3H_6Cl_2$; mol wt 112.99. C 31.89%, H 5.35%, Cl 62.76%. $CH_3CHClCH_2Cl$. Prepd from propyl chloride and Sb_2Cl_5. Toxicity data: H. F. Smyth *et al.*, *Am. Ind. Hyg. Assoc. J.* **30**, 470 (1969).

Flammable, mobile liq. Odor of chloroform. d_{25}^{25} 1.159; bp 95-96°. Solidifies below −70°. n_D^{20} 1.4388. Flash point (ASTM open cup) 21° (70°F). Despite the low flash pt it does not catch fire readily in industrial applications. Fire pt 38°. Slightly sol in water; miscible with organic solvents. LD_{50} orally in rats: 1.19 ml/kg (Smyth).

USE: Oil and fat solvent; in dry cleaning fluids; in degreasing. In insecticidal fumigant mixtures. *Caution:* May be irritating to eyes, mucous membranes, and in high concns, narcotic. Has caused liver, kidney necrosis in exptl animals.

7868. Propylene Glycol. *1,2-Propanediol;* methyl glycol; 1,2-dihydroxypropane. $C_3H_8O_2$; mol wt 76.09. C 47.35%, H 10.60%, O 42.05%. $CH_3CHOHCH_2OH$. Prepn from glycerol: Raschig, Prahl, *Ber.* **61**, 185 (1928). Prepn of levorotatory propylene glycol from hydroxyacetone by yeast reduction: Levene, Walti, *Org. Syn. coll. vol. II*, 545 (1943). Synthesis of S-(+)-form: C. Melchiorre, *Chem. Ind. (London)* **1976**, 218. Manuf from propylene oxide by hydration: Faith, Keyes & Clark's *Industrial Chemicals*, F. A. Lowenheim, M. K. Moran, Eds. (Wiley-Interscience, New York, 4th ed., 1975) pp 688-691. Taken internally, propylene glycol is harmless, probably because its oxidation yields pyruvic and acetic acids, *cf.* Whitmore, *Organic Chemistry* (New York, 1951). Toxicity data: W. Bartsch *et al.*, *Arzneimittel-Forsch.* **26**, 1581 (1976). Review on toxicity, metabolism and biochemistry: Ruddick, *Toxicol. Appl. Pharmacol.* **21**, 102 (1972).

dl-Form, hygroscopic, viscous liquid. Slightly acrid taste. d_4^{25} 1.036. mp −59°. bp_{760} 188.2°; bp_{400} 168.1°; bp_{200} 149.7°; bp_{100} 132.0°; bp_{60} 119.9°; bp_{40} 111.2°; bp_{20} 96.4°; bp_{10} 83.2°; bp_5 70.8°; $bp_{1.0}$ 45.5°. Flash pt, open cup: 210°F (99°C). Miscible with water, acetone, chloroform. Sol in ether. Will dissolve many essential oils, but is immiscible with fixed oils. It is a good solvent for rosin. Under ordinary conditions propylene glycol is stable, but at high temps it tends to oxidize giving rise to products such as propionaldehyde, lactic acid, pyruvic acid and acetic acid. LD_{50} orally in rats: 25 ml/kg (Bartsch).

l-Form, bp_{12} 88-90°, bp_{760} 187-189°. $[\alpha]_D^{20}$ −15.0°.
d-Form, bp_{14} 94-96°. $[\alpha]_D^{20}$ +15.84° (neat). d^{25} 1.04.

USE: As nontoxic antifreeze in breweries and dairy establishments. Substitute for ethylene glycol and glycerol. In the manuf of synthetic resins. As inhibitor of fermentation and mold growth. As mist to disinfect air. As emulsifier in foods. Pharmaceutic aid (humectant; solvent).

THERAP CAT (VET): Glucogenic (orally) in ruminants.

7869. Propylene Oxide. *Methyloxirane;* propene oxide. C_3H_6O; mol wt 58.08. C 62.04%, H 10.41%, O 27.55%. Results from the action of KOH (aq) on propylene chlorohydrin. Toxicity: H. F. Smyth *et al.*, *J. Ind. Hyg. Toxicol.* **23**, 259 (1941). *Reviews:* Holden in *Glycols*, G. O. Curme, F. Johnston, Eds., A.C.S. Monograph Series no. **114** (Reinhold, New York, 1952) pp 250-261; Faith, Keyes & Clark's *Industrial Chemicals*, F. A. Lowenheim, M. K. Moran, Eds. (Wiley-Interscience, New York, 4th ed., 1975) pp 692-697; R. O. Kirk, T. J. Dempsey in Kirk-Othmer *Encyclopedia of Chemical Technology* vol. 19 (Wiley-Interscience, New York, 3rd ed., 1982) pp 246-274.

Colorless ethereal liquid. *Extremely flammable.* d_4^0 0.859. mp −112.13°. bp 34.23°. Flash pt, closed cup: −31°F (−35°C). Soly in water (20°): 40.5% by wt; soly of water in propylene oxide: 12.8% by wt; miscible with alcohol, ether. LD_{50} orally in rats: 1.14 g/kg (Smyth).

USE: Chemical intermediate in prepn of polyethers to form polyurethanes; in prepn of propylene and dipropylene glycols; in prepn of lubricants, surfactants, oil demulsifiers. Also as solvent; fumigant; soil sterilant.

7870. Propyl Ether. *1,1'-Oxybispropane;* dipropyl ether. $C_6H_{14}O$; mol wt 102.17. C 70.53%, H 13.81%, O 15.66%. $C_3H_7OC_3H_7$. Obtained by heating propyl alcohol with benzenesulfonic acid.

Mobile liquid. Extremely flammable. d_4^{20} 0.7360; mp −122°; bp 89-91°; n_D^{20} 1.3807. Flash pt, open cup: −5°F

(−20°C). Slightly sol in water; sol in alcohol, ether. Highly volatile. Tends to form explosive peroxides, esp when anhydr. Do not allow to evaporate to near dryness.

7871. Propyl Formate. *Formic acid propyl ester.* $C_4H_8O_2$; mol wt 88.10. C 54.53%, H 9.15%, O 36.32%. $HCOOC_3H_7$.

Colorless liquid; pleasant odor. d^{20} 0.901; mp −93°; bp 81-82°. Flash pt, closed cup: 27°F (−3°C). n_D^{20} 1.3771. Sol in 45 parts water; misc with alcohol, ether. LD_{50} orally in rats: 3980 mg/kg, P. M. Jenner *et al.*, *Food Cosmet. Toxicol.* **2**, 327 (1964).

7872. Propyl Gallate. *3,4,5-Trihydroxybenzoic acid propyl ester;* *n*-propyl gallate; gallic acid propyl ester; PG; Progallin P; Tenox PG. $C_{10}H_{12}O_5$; mol wt 212.20. C 56.60%, H 5.70%, O 37.70%. Spectrophotometric determn: C. S. Sastry *et al.*, *Talanta* **29**, 917 (1982). Effects on survival of *Saccharomyces cerevisiae:* V. L. Eubanks, L. R. Beuchat, *J. Food Prot.* **46**, 29 (1983). Antioxidant effectiveness: M. A. Augustin, S. K. Berry, *J. Am. Oil Chem. Soc.* **60**, 105 (1983).

Crystals, mp 150°. Soly at 25° in water = 0.35 g/100 ml; in alcohol = 103 g/100 g; in ether = 83 g/100 g. Soly in cottonseed oil at 30° = 1.23 g/100 g; in lard at 45° = 1.14 g/100 g. Darkens in the presence of iron and iron salts. Synergic with acids, BHA, BHT. LD_{50} orally in rats: 3.8 g/kg, *Patty's Industrial Hygiene and Toxicology* **vol. 2A**, G. D. Clayton, F. E. Clayton, Eds. (Wiley-Interscience, New York, 3rd ed., 1981) p 2326.

USE: Antioxidant for foods, fats, oils, ethers, emulsions, waxes, transformer oils.

7873. Propylhexedrine. *N,α-Dimethylcyclohexaneethanamine;* 1-cyclohexyl-2-methylaminopropane; hexahydrodesoxyephedrine; Benzedrex; CHP-Depot. $C_{10}H_{21}N$; mol wt 155.28. C 77.34%, H 13.63%, N 9.02%. Prepd by catalytic hydrogenation of the phenyl analog using Adams platinum catalyst and glacial acetic acid as the solvent: Zenitz *et al.*, *J. Am. Chem. Soc.* **69**, 1117 (1947); Ullyot, U.S. pat. **2,454,-746** (1948 to SK & F).

dl-Form, oily liq, d_4^{25} 0.8501, amine odor, bp$_{760}$ 205°; bp$_{20}$ 92-93°; volatilizes slowly at room temp. n_D^{20} 1.4600. Very slightly sol in water. Absorbs carbon dioxide from air and its aq solns are alkaline to litmus. Miscible with alcohol, chloroform, ether.

dl-Form hydrochloride, $C_{10}H_{22}ClN$, crystals, dec 127-128°. Sol in water.

d-Form, oily liq, bp$_{10}$ 82-83°; n_D^{20} 1.4588.

d-Form hydrochloride, crystals, dec 138-139°; $[\alpha]_D^{26}$ +14.73°. Sol in water.

l-Form, oily liq, bp$_9$ 80-81°. n_D^{20} 1.4590.

l-Form ethylphenylbarbiturate, $C_{22}H_{33}N_3O_3$, *l*-1-cyclohexyl-2-(methylamino)propane ethylphenylbarbiturate, phenobarbital compound with 1-propylhexedrine, barbexaclone, Maliasin.

l-Form hydrochloride, crystals, dec 138-139°; $[\alpha]_D^{26}$ −14.74°. Sol in water.

THERAP CAT: Adrenergic (vasoconstrictor); nasal decongestant.

7874. Propylidene Chloride. *1,1-Dichloropropane.* $C_3H_6Cl_2$; mol wt 112.99. C 31.89%, H 5.35%, Cl 62.76%. $CH_3CH_2CHCl_2$. Obtained by the action of PCl_5 on propionaldehyde.

Liquid. d^{10} 1.143; bp 87°. Very slightly sol in water; sol in many organic solvents. LD_{50} orally in rats: 6.5 g/kg, Smyth *et al.*, *Arch. Ind. Hyg. Occup. Med.* **10**, 61 (1954).

7875. Propyl Iodide. *1-Iodopropane.* C_3H_7I; mol wt 170.01. C 21.19%, H 4.15%, I 74.66%. $CH_3CH_2CH_2I$. Prepd by heating propyl alcohol with iodine and red phosphorus.

Colorless or slightly yellow liquid. d_4^{20} 1.747; mp about −98°; bp 102-103°; n_D^{20} 1.5051. Sol in 575 parts water; miscible with alcohol, ether.

7876. Propyliodone. *3,5-Diiodo-4-oxo-1(4H)-pyridineacetic acid propyl ester;* 3,5-diiodo-4-pyridone-*N*-acetic acid propyl ester; propyl 3,5-diiodo-4-pyridone-*N*-acetate; propiodone; Dionosil. $C_{10}H_{11}I_2NO_3$; mol wt 447.03. C 26.87%, H 2.48%, I 56.78%, N 3.13%, O 10.74%. Prepn: Branscombe, **Brit.** pat. **517,382** (1940 to I.C.I.). Description: Tomich *et al.*, *Brit. J. Pharmacol.* **8**, 166 (1953).

Crystals, dec 186-187°. Slightly sol in saline, serum. Soly in water at 15°: 0.014 g/100 ml; at 35°: 0.020 g/100 ml; at 95°: 0.020 g/100 ml. LD_{50} i.v. in mice: 300 mg/kg, Tomich *et al., loc. cit.*

THERAP CAT: Diagnostic aid (radiopaque medium).

7877. *n*-Propyl Nitrate. *Nitric acid propyl ester.* $C_3H_7NO_3$; mol wt 105.09. C 34.28%, H 6.71%, N 13.33%, O 45.67%. $CH_3CH_2CH_2ONO_2$. Prepd by nitration of propanol with nitric acid, usually in the presence of urea and ammonium nitrate or sulfuric acid: Vogel, *J. Chem. Soc.* **1948**, 1847.

Pale yellow liquid. Sweet, sickly odor. d_4^{20} 1.0538. bp$_{762}$ 110°. *Heating may cause it to explode.* n_D^{20} 1.3979. Dipole moment: 2.98. Azeotrope with water contg 75% $C_3H_7NO_3$, bp$_{760}$ 84.8°. Very slightly sol in water. Sol in alcohol, ether. More physical data (especially concerning combustion and propulsion): Penner, Ducarme, *The Chemistry of Propellants* (Pergamon Press, 1960).

USE: Fuel ignition promoter, in rocket fuel formulations, as organic intermediate.

7878. *n*-Propyl Nitrite. *Nitrous acid n-propyl ester.* $C_3H_7NO_2$; mol wt 89.09. C 40.44%, H 7.92%, N 15.72%, O 35.92%. $CH_3CH_2CH_2ONO$. Prepd from silver nitrite and *n*-propyl bromide: Reynolds, Adkins, *J. Am. Chem. Soc.* **51**, 279 (1929); from *n*-propanol, sodium nitrite, and dil sulfuric acid: Cowley, Partington, *J. Chem. Soc.* **1933**, 1252.

Liquid. d_4^{20} 0.8864. bp$_{760}$ 46-48°. n_D^{20} 1.3592 (n_D^{20} 1.3613). Soluble in alcohol, ether.

USE: Jet propellant. *Caution:* Inhalation causes vasodilation, smooth muscle relaxation, hypotension.

7879. Propylparaben. *4-Hydroxybenzoic acid propyl ester;* propyl *p*-hydroxybenzoate; Nipasol; Chemocide PK; Propyl Chemosept; Solbrol P; Propyl Parasept. $C_{10}H_{12}O_3$; mol wt 180.20. C 66.65%, H 6.71%, O 26.64%. Prepn: Stohmann, *J. Prakt. Chem.* **36**, 368 (1887); L. Nobli, *Giorn. Farm. Chim.* **84**, 168 (1935), *C.A.* **30**, 3423⁹ (1936).

White crystals, mp 96-97°. Sol in 2000 parts water; freely sol in alcohol, ether; slightly sol in boiling water.

USE: Pharmaceutic aid (antifungal). Preservative in foods.

7880. Propyl Propionate. *Propanoic acid propyl ester.* $C_6H_{12}O_2$; mol wt 116.16. C 62.04%, H 10.41%, O 27.55%. $CH_3CH_2COOC_3H_7$.

Liquid. d^{20} 0.883; mp $-76°$; bp 122-124°; n_D^{20} 1.3935. Sol in 200 parts water; miscible with alcohol, ether.

7881. Propyl Sulfide. *1,1'-Thiobispropane;* dipropyl sulfide. $C_6H_{14}S$; mol wt 118.24. C 60.94%, H 11.94%, S 27.12%. $C_3H_7SC_3H_7$. Prepd from propyl bromide and an alcoholic soln of Na_2S.

Colorless liquid. d^{17} 0.814; mp about $-102°$; bp 142°. Insol in water; sol in alcohol, ether.

7882. Propylthiouracil. *2,3-Dihydro-6-propyl-2-thioxo-4(1H)pyrimidinone; 6-propyl-2-thiouracil;* 2-thio-4-oxo-6-propyl-1,3-pyrimidine; 2-thio-6-propyl-1,3-pyrimidin-4-one; Propacil; Propycil; Procasil; Propyl-Thyracil; Prothyran; Thyreostat II. $C_7H_{10}N_2OS$; mol wt 170.23. C 49.39%, H 5.92%, N 16.46%, O 9.40%, S 18.84%. Prepd by the condensation of ethyl β-oxocaproate with thiourea: Anderson *et al., J. Am. Chem. Soc.* **67**, 2197 (1945). Comprehensive description: H. Y. Aboul-Enein in *Analytical Profiles of Drug Substances* vol. 6, K. Florey, Ed. (Academic Press, New York, 1977) pp 457-486. Review of pharmacology and clinical experience: D. S. Cooper *et al., N. Engl. J. Med.* **311**, 1353-1362 (1984).

White, bitter cryst powder of starch-like appearance to the eye and to the touch, mp 219-221°. uv max (methanol): 275, 214 nm (ε 15800, 15600); (methanolic KOH): 315.5, 260, 207.5 nm (ε 10900, 10700, 15400). One part dissolves in about 900 parts water at 20°, in 100 parts boiling water, in 60 parts ethyl alcohol, in 60 parts acetone. Practically insol in ether, chloroform, benzene. Freely sol in aq solns of ammonia and alkali hydroxides. A satd aq soln is neutral or slightly acid to litmus.

Note: This substance may reasonably be anticipated to be a carcinogen: *Fourth Annual Report on Carcinogens* (NTP 85-002, 1985) p 176.

THERAP CAT: Antihyperthyroid.

THERAP CAT (VET): Antihyperthyroid. Has been used to promote fattening.

7883. Propylure. *10-Propyl-5,9-tridecadien-1-ol acetate; trans*-1-acetoxy-10-(*n*-propyl)trideca-5,9-diene; 10-propyl-*trans*-5,9-tridecadienyl acetate. $C_{18}H_{32}O_2$; mol wt 280.44. C 77.09%, H 11.50%, O 11.41%. Once thought to be the sex attractant of the pink bollworm moth, *Pectinophora gossypiella* (Saunders), a destructive cotton pest, H. E. Hummel *et al., Science* **181**, 873 (1973), *cf.* Gossyplure. Reported as the first known natural product possessing di *n*-propyl branching. Isoln, structure, and synthesis: Jones *et al., Science* **152**, 1516 (1966). Improved total syntheses: Pattenden, *J. Chem. Soc. (C)* **1968**, 2385; Meyers, Collington, *Tetrahedron* **27**, 5979 (1971); Vig *et al., J. Indian Chem. Soc.* **50**, 39 (1973); K. Utimoto *et al., Tetrahedron Letters* **1975**, 4233. Photochemical synthesis: Kossanyi *et al., ibid.* **1973**, 3459. Stereoselective synthesis: A. Alexakis *et al., ibid.* **1978**, 2027. *trans*-Propylure is rendered inactive by the presence of > 15% of the *cis*-isomer: Jacobson, *Science* **163**, 190 (1969).

Colorless liquid having no detectable odor. $bp_{0.1}$ 135°. n_D^{25} 1.4635. Strong IR band at 970 cm^{-1}.

USE: Insect sex attractant.

7884. Propyphenazone. *1,2-Dihydro-1,5-dimethyl-4-(1-methylethyl)-2-phenyl-3H-pyrazol-3-one; 4-isopropylantipyrine;* 4-isopropyl-2,3-dimethyl-1-phenyl-3-pyrazolin-5-one; 2,3-dimethyl-1-phenyl-4-isopropylpyrazolone; isopro-

pylphenazone; Baukal; Causyth. $C_{14}H_{18}N_2O$; mol wt 230.30. C 73.01%, H 7.88%, N 12.17%, O 6.95%. Prepn: Stenz, U.S. pat. **1,972,036** (1934 to Hoffmann-La Roche); Sawa, *J. Pharm. Soc. Japan* **57**, 953 (1937), *C.A.* **32**, 2533 (1938).

Slightly bitter crystals, mp 103°. Readily sol in alcohol, ether. Soly in water: 0.24 g/100 ml at 16.5°.

Note: Used mostly as an ingredient in analgesics such as *Irgapyrin, Gardan, Dolibrax.*

THERAP CAT: Analgesic, antipyretic, antiphlogistic.

7885. Propyromazine. *1-Methyl-1-[1-methyl-2-oxo-2-(10H-phenothiazin-10-yl)ethyl]pyrrolidinium bromide; 1-methyl-1-[1-(phenothiazin-10-ylcarbonyl)ethyl]pyrrolidinium bromide;* 10-[2-(1-methyl-1-pyrrolidinyl)propionyl]phenothiazine bromide; 10-[2-(1-pyrrolidinyl)propionyl]phenothiazine methobromide; 10-(2-methyl-1-oxo-2-pyrrolidinylethyl)phenothiazine methobromide; SD 104.19; Diaspasmyl. $C_{20}H_{23}BrN_2OS$; mol wt 419.41. C 57.28%, H 5.53%, Br 19.06%, N 6.68%, O 3.81%, S 7.65%. Prepd by *N*-acylation of phenothiazine: Boissier *et al., Thérapie* **13**, 989 (1958). Prepn of the free base from 10-(α-bromopropionyl)-phenothiazine and pyrrolidine: Dahlbom, Ekstrand, *Acta Chem. Scand.* **5**, 102 (1951); *eidem,* U.S. pat. **2,615,886** (1952 to Astra).

Crystals, mp 228-229°.

Free base, $C_{19}H_{20}N_2OS$, crystals from petr ether, mp 94.5-95.5°.

THERAP CAT: Anticholinergic, antispasmodic.

7886. Propyzamide. *3,5-Dichloro-N-(1,1-dimethyl-2-propynyl)benzamide;* pronamid; RH 315; Kerb. $C_{12}H_{11}Cl_2$-NO; mol wt 256.13. C 56.27%, H 4.33%, Cl 27.68%, N 5.47%, O 6.25%. Selective pre-emergence herbicide. Prepn: B. W. Horrom *et al., S. Afr. pat.* **68 00,090** corresp to U.S. pats. **3,534,098** and **3,640,699** (1969, 1970, 1972 all to Rohm & Haas). Activity: K. L. Viste *et al., Science* **167**, 280 (1970); C. Swithenbank *et al., J. Agr. Food Chem.* **19**, 417 (1971). Metabolism: R. Y. Yih *et al., Weed Sci.* **18**, 604 (1970); R. Y. Yih, C. Swithenbank, *J. Agr. Food Chem.* **19**, 314, 320 (1971); J. D. Fisher, *ibid.* **22**, 606 (1974). Carcinogenicity study: M. D. Reuber, *Environ. Res.* **23**, 1 (1980).

White solid, mp 155-156°. Soly in water at 25°: 15 ppm. Sol in aliphatic, aromatic solvents. Vapor press at 25°: 8.5 × 10^{-5} mm Hg. LD_{50} orally in rats (male, female): 8350, 5620 mg/kg, K. L. Viste, *loc. cit.*

USE: Herbicide.

7887. Proquazone. *7-Methyl-1-(1-methylethyl)-4-phenyl-2(1H)-quinazolinone; 1-isopropyl-7-methyl-4-phenyl-2(1H)-quinazolinone;* RU 43-715; Sandoz 43-715; Arthrex; Biarison; Biarsan. $C_{18}H_{18}N_2O$; mol wt 278.35. C 77.67%, H

6.52%, N 10.06%, O 5.75%. Prepn: H. Ott, M. Denzer, **Ger. pat. 1,805,501** corresp to U.S. pats. **3,845,128** and **3,925,548** (1969, 1974 and 1975, all to Sandoz); R. V. Coombs *et al., J. Med. Chem.* **16**, 1237 (1973). Metabolism: M. B. Zucker, *Proc. Soc. Exp. Biol. Med.* **156**, 209 (1977); H. Ott, J. Meier, *Scand. J. Rheumatol. (Suppl.)* **21**, 12 (1978). Pharmacology: H. U. Gubler, M. Baggiolini, *ibid.* 8; H. Ott, *ibid.* 5. Clinical studies: P. Sfikakis, P. Tsachalos, *Ther. Umsch.* **34**, 730 (1977); series of articles in *Scand. J. Rheumatol. (Suppl.)* **21**, 15-39 (1978). Review of pharmacodynamics, pharmacokinetics and therapeutic efficacy: S. P. Clissold, R. Beresford, *Drugs* **33**, 478-502 (1987).

Yellow crystals from ethyl acetate, mp 137-138°. Sol in chloroform. Insol in water.

THERAP CAT: Anti-inflammatory.

7888. Proscar®. (5α,17β)-(1,1-*Dimethylethyl*)-3-*oxo-4-azaandrost-1-ene-17-carboxamide*; 17β-(N-tert-butylcarbamoyl)-4-aza-5α-androst-1-en-3-one; MK-906. $C_{23}H_{36}$-N_2O_2; mol wt 372.55. C 74.15%, H 9.74%, N 7.52%, O 8.59%. Inhibitor of 5α-reductase, the enzyme which converts testosterone to the more potent androgen, 5α-dihydrotestosterone. Prepn: G. H. Rasmusson, G. F. Reynolds, **Eur. pat. 155,096;** *eidem,* **U.S. pat. 4,760,071** (1985, 1988 both to Merck & Co.); G. H. Rasmusson *et al., J. Med. Chem.* **29**, 2298 (1986); A. Bhattacharya *et al., J. Am. Chem. Soc.* **110**, 3319 (1988). Inhibition of 5α-reductase *in vitro:* T. Liang *et al., Endocrinology* **117**, 571 (1985); *in vivo:* J. R. Brooks *et al., Steroids* **47**, 1 (1986). HPLC determn in plasma and urine and preliminary pharmacokinetics: J. R. Carlin *et al., J. Chromatog.* **427**, 79 (1988).

White to off-white crystalline solid, mp ~257° (dried to anhydrous under N_2). Also reported as mp 252-254° (Rasmusson, 1988). $[\alpha]_D$ −59° (c = 1 in methanol). Freely sol in chloroform, DMSO, ethanol, methanol, n-propanol; sparingly sol in propylene glycol, polyethylene glycol 400; very slightly sol in 0.1N HCl, 0.1N NaOH, water.

THERAP CAT: Treatment of benign prostatic hypertrophy.

7889. Proscillaridin. 3-[(6-*Deoxy-α-L-mannopyranosyl)oxy]-14-hydroxybufa-4,20,22-trienolide*; 14-hydroxy-3β-(rhamnosyloxy)bufa-4,20,22-trienolide; 3β-rhamnosido-14β-hydroxy-Δ^4,20,22-bufatrienolide; proscillaridin A; desglucotransvaaline; scillarenin 3β-rhamnoside; Caradrin; Cardion; Carmazon; Herzo; Procillan; Prostosin; Proszin; Protasin; Purosin-TC; Sandoscill; Scillacrist; Scilla "Didier"; Simeon; Solestril; Stellarid; Talucard; Talusin; Urgilan; Wirnesin. $C_{30}H_{42}O_8$; mol wt 530.64. C 67.90%, H 7.98%, O 24.12%. Prepn by acid cleavage of scillaren A: Stoll *et al., Helv. Chim. Acta* **16**, 703 (1933); by enzymic decompn of glucoscillaren A with strophanthobiase: Stoll *et al., ibid.* **35**, 2495 (1952); from *Urginea burkei* Baker, *Liliaceae:* Zoller, Tamm, *ibid.* **36**, 1744 (1953); from *U. (Scilla) maritima* (L.) Baker, *Liliaceae:* Görlich, *Arzneimittel-Forsch.* **10**, 770 (1960). Structure: Stoll *et al., Helv. Chim. Acta* **35**, 1934 (1952). Pharmacology: Lenke, Brock, *Arzneimittel-Forsch.*

20, 1 (1970). Metabolic studies: Davis *et al., Arch. Int. Pharmacodyn.* **177**, 231 (1969); Nakano *et al., ibid.* **183**, 199 (1970). Clinical studies: Several authors, *Minerva Med.* **58**, 4243-4322 (1967).

Prisms from methanol, mp 219-222°. $[\alpha]_D^{20}$ −91.5° (CH₃OH). LD₅₀ orally in male, female rats: 56, 76 mg/kg, E. I. Goldenthal, *Toxicol. Appl. Pharmacol.* **18**, 185 (1971).
Proscillaridin-4'-methyl ether, $C_{31}H_{44}O_8$, *meproscillarin, Clift.* mp 213-217°. $[\alpha]_D^{20}$ −94° (CH₃OH). uv max (CH₃OH): 297 nm (log ε 3.79), (1N KOH/CH₃OH): 355 nm (log ε 4.65). Sol in methanol, ethanol, THF, dioxane; slightly sol in CHCl₃, CH₂Cl₂, acetone; insol in water, nonpolar organics. Series of articles on prepn, pharmacology, toxicology, pharmacokinetics, metabolism: *Arzneimittel-Forsch.* **28**, 493-573 (1978).

THERAP CAT: Cardiotonic.

7890. Prostacyclin. (5Z,9α,11α,13E,15S)-6,9-*Epoxy-11,15-dihydroxyprosta-5,13-dien-1-oic acid*; (5Z)-9-deoxy-6,9α-epoxy-Δ^5-PGF_{1α}; epoprostenol; prostaglandin I₂; prostaglandin X; PGI₂; PGX; U-53217. $C_{20}H_{32}O_5$; mol wt 352.48. C 68.15%, H 9.15%, O 22.70%. A prostaglandin produced by enzymatic transformation of prostaglandin endoperoxides (*PGG₂, PGH₂*), which dilates blood vessels and is approximately 30 times more potent than prostaglandin E₁, *q.v.*, in inhibiting platelet aggregation. Evidence for its occurrence during biosynthetic conversion of arachidonic acid by rat stomach homogenates: C. Pace-Asciak, L. S. Wolfe, *Biochemistry* **10**, 3657 (1971). Isoln from microsomes of pig and rabbit aorta by J. R. Vane and co-workers: S. Moncada *et al., Nature* **263**, 663 (1976). PGI₂ is also synthesized in bovine coronary arteries as well as human arteries and veins: *eidem, Lancet* **1**, 18 (1977); G. J. Dusting *et al., Prostaglandins* **13**, 3 (1977); by cultured human and bovine endothelial cells: B. B. Weksler *et al., Proc. Nat. Acad. Sci. USA* **74**, 3922 (1977); by pig aortic endothelial cells: D. E. MacIntyre *et al., Nature* **271**, 549 (1978). It has been suggested that endoperoxides released by platelets can be converted to PGI₂ by vascular tissue and that a balance between formation of PGI₂ and release of thromboxane A₂, *q.v.*, which induces platelet aggregation, controls the formation of thrombi in blood vessels. It has also been postulated that PGI₂ acts to stimulate platelet adenylate cyclase and to prevent the action of thrombi on phospholipid breakdown as well as platelet aggregation. Structure: R. A. Johnson *et al., Prostaglandins* **12**, 915 (1976). Synthesis: E. J. Corey *et al., J. Am. Chem. Soc.* **99**, 3006 (1977); of sodium salt and stereochemistry: R. A. Johnson *et al., ibid.* 4182. Additional syntheses: I. Tomoskozi *et al., Tetrahedron Letters* **1977**, 2627; N. Whittaker, *ibid.* 2805; K. Nicolaou, *Chem. Commun.* **1977**, 630. Synthesis of the 5E-isomer: E. J. Corey *et al., Tetrahedron Letters* **1977**, 3529. Chemical stability in aq solns: M. J. Cho, M. A. Allen, *Prostaglandins* **15**, 943 (1978).
Biosynthetic study: V. Tomasi *et al., Nature* **273**, 670 (1978). Biological properties: R. J. Gryglewski *et al., Prostaglandins* **12**, 685 (1976). Preliminary clinical study: A. E. S. Gimson *et al., Lancet* **1**, 173 (1980). Antimetastatic effects: K. V. Honn *et al., Science* **212**, 1270 (1981); *eidem, ibid.* **217** 542 (1982). Preliminary study of effect of PGX infusion in patients with acute myocardial infarction: O. Edhag *et al., N. Engl. J. Med.* **308**, 1032 (1983). Review of biological properties: S. Moncada, J. R. Vane, *Clin. Sci.* **61**,

369-372 (1981); of therapeutic potential: *eidem, Advan. Pharmacol. Ther.* **4**, 215-233 (1982); of physiological role: J. R. Vane *et al., Int. Rev. Exp. Pathol.* **23**, 161-207 (1982). General reviews: S. Moncada, J. R. Vane, *Fed. Proc.* **38**, 66-71 (1979); J. C. McGiff, *Ann. Rev. Pharmacol. Toxicol.* **21**, 479-509 (1981); S. Moncada *et al., Advan. Pharmacol. Ther.* **6**, 39-47 (1982). Books: *Prostacyclin*, J. R. Vane, S. Bergstrom, Eds. (Raven Press, New York, 1979) 453 pp; *Prostaglandins in Cardiovascular and Renal Function*, A. Scriabine *et al.*, Eds. (Spectrum Publications, New York, 1980) 498 pp.

Chemically unstable in aq soln. Hydrolyzes to 6-oxo-PGF$_{1\alpha}$. Half-life at 4° is approx 14.5 min when total phosphate is 0.165 *M*. Anti-aggregating activity disappears within 0.25 min on boiling or within 10 min at 37°.

Sodium salt, C$_{20}$H$_{31}$NaO$_5$, *U-53217A, Cyclo-Prostin, Flolan.* Hygroscopic, free-flowing white powder. Stable for 2 months if kept dry at −30°.

THERAP CAT: Platelet aggregation inhibitor.

7891. Prostaglandin(s). A family of biologically potent lipid acids first discovered in seminal fluid and extracts of accessory genital glands of man and sheep: von Euler, *Arch. Exp. Pathol. Pharmakol.* **175**, 78 (1934); *Klin. Wochenschr.* **14**, 1182 (1935). Isoln: Bergstrom, Sjovall, U.S. pats. 3,-069,322 and 3,598,858 (1962, 1971); Samuelsson, *J. Biol. Chem.* **238**, 3229 (1963). Also found in lower concns in other organs: *idem, Biochim. Biophys. Acta* **84**, 707 (1964). The single non-mammalian source of prostaglandin intermediates, or *syntons*, is the gorgonian sea whip or sea fan, *Plexaura homomalla:* Weinheimer, Spraggins, *Tetrahedron Letters* **1969**, 5185; Schneider *et al., J. Am. Chem. Soc.* **94**, 2122 (1972). Prostaglandins are named as derivatives of **prostanoic acid.** Prostaglandins are divided into the types E, F, A, B, C, and D based on functions in the cyclopentane ring. Numerical subscripts refer to the number of unsaturations in the side chains; α or β subscripts refer to the configuration of substituents in the ring. Six naturally occurring prostaglandins, E$_1$, E$_2$, E$_3$, F$_{1\alpha}$, F$_{2\alpha}$, F$_{3\alpha}$, are considered primary in that no one is derived from another in the living organism. First structural and stereochemical elucidations: Bergstrom *et al., Acta Chem. Scand.* **16**, 501 (1962); *eidem, J. Biol. Chem* **238**, 3555 (1963). Absolute config: Nugteren *et al., Nature* **212**, 38 (1966). First total synthesis of racemic PGE$_1$ and PGF$_{1\alpha}$: Corey *et al., J. Am. Chem. Soc.* **90**, 3245 (1968). Review of synthetic studies: Pike, *Fortschr. Chem. Org. Naturst.* **28**, 313 (1970); Axen *et al.*, in *The Total Synthesis of Natural Products*, vol. 1, J. ApSimon, Ed. (Wiley-Interscience, New York, 1973) pp 81-143; Clarkson in *Progress in Organic Chemistry*, vol. 8, W. Carruthers, J. K. Sutherland, Eds. (Wiley, New York, 1973) pp 1-28. Book: J. S. Bindra, R. Bindra, *Prostaglandin Synthesis* (Academic Press, New York, 1977).

prostanoic acid

Biosynthesis occurs by enzymatic conversion of unsaturated twenty-carbon fatty acids. Review of biosynthetic studies: Samuelsson, *Progr. Biochem. Pharmacol.* **5**, 109 (1969). Review of metabolism: Samuelsson *et al., Ann. N.Y. Acad. Sci.* **180**, 138 (1971). Biological activities include stimulation of smooth muscle, dilation of small arteries, bronchial dila-

tion, lowering of blood pressure, inhibition of gastric secretion, of lipolysis, and of platelet aggregation, induction of labor, abortion, and menstruation, and increase in ocular pressure. Implicated also in dysmenorrhea, inflammatory reactions, nasal vasoconstriction, kidney function, and in autonomic neurotransmission. Reviews of pharmacological and biochemical aspects: Horton, *Experientia* **21**, 113 (1965); Weeks, *Ann. Rev. Pharmacol.* **12**, 317 (1972); Hinman, *Ann. Rev. Biochem.* **41**, 161 (1972). Review of biological activities of synthetic prostaglandins: Ramwell *et al., Nature* **221**, 1251 (1969). Comprehensive review of analytical and preparative techniques: *Methods Enzymol.* **86**, entitled "Prostaglandins and Arachidonate Metabolites", W. E. M. Landis, W. L. Smith, Eds. (Academic Press, New York, 1982) 705 pp. General reviews: Bergstrom, *Science* **157**, 382 (1967); Bergstrom *et al., Pharmacol. Rev.* **20**, 1 (1968); Ramwell, Shaw, *Recent Progr. Horm. Res.* **26**, 139 (1970); Pike, *Sci. Amer.* **225**, 84 (Nov., 1971); Bindra, Bindra, *Progress in Drug Research*, **vol. 17** (Birkhäuser Verlag, Basel, 1973) pp 410-487. Books: *The Prostaglandins*, vols. 1, 2, P. Ramwell, Ed. (Plenum Press, New York, 1973, 1974); *Prostaglandins in Cardiovascular and Renal Function*, A. Scriabine *et al.*, Eds. (Spectrum Publications, New York, 1980) 498 pp; *Cardiovascular Pharmacology of the Prostaglandins*, A. Herman, Ed. (Raven Press, New York, 1982) 472 pp. Three volume book series: *Prostaglandins and Reproduction; Prostaglandins: Chemical and Biochemical Aspects; Prostaglandins: Physiological, Pharmacological and Pathological Aspects*, all edited by S. M. M. Karim (University Park Press, Baltimore, 1975, 1976, 1976).

7892. Prostaglandin E$_1$. *11,15-Dihydroxy-9-oxoprost-13-en-1-oic acid; 3-hydroxy-2-(3-hydroxy-1-octenyl)-5-oxo-cyclopentaneheptanoic acid;* alprostadil; PGE$_1$; U-10136; Liple; Minprog; Palux; Prostandin; Prostin VR. C$_{20}$H$_{34}$O$_5$; mol wt 354.49. C 67.76%, H 9.67%, O 22.57%. A primary prostaglandin; easily crystallized from purified biological extracts. Isoln from sheep seminal vesicle tissue, and structure: Bergstrom *et al., Acta Chem. Scand.* **16**, 501 (1962); *eidem, J. Biol. Chem.* **238**, 3555 (1963). Enzymic conversion from 8,11,14-eicosatrienoic acid: Nugteren *et al., Rec. Trav. Chim.* **85**, 405 (1966). Synthesis of the *dl*-form: Corey *et al., J. Am. Chem. Soc.* **90**, 3245, 3247 (1968); Schneider *et al.*, *ibid.* 5895; **91**, 5372 (1969); Axen *et al., Chem. Commun.* **1969**, 303; Taub *et al., ibid.* **1970**, 1258; Slates *et al., ibid.* **1972**, 304; Kuo *et al., Tetrahedron Letters* **1972**, 5317; Taub *et al., Tetrahedron* **29**, 1447 (1973); Miyano, Stealey, *Chem. Commun.* **1973**, 180; Finch *et al., J. Org. Chem.* **38**, 4412 (1973). Synthesis of natural form: Corey *et al., J. Am. Chem. Soc.* **91**, 535 (1969); **92**, 2586 (1970); Sih *et al., ibid.* **94**, 3643 (1972); **95**, 1676 (1973); Schaaf, Corey, *J. Org. Chem.* **37**, 2921 (1974); Slates *et al., Tetrahedron* **30**, 819 (1974). Metabolism in guinea pigs: Anggard, Samuelsson, *J. Biol. Chem.* **239**, 4097 (1964). Metabolism in humans: Hamberg, Samuelsson, *ibid.* **246**, 6713 (1971). Review of biological activities: Berti *et al., Progr. Biochem. Pharmacol.* **3**, 110 (1967). Comparative pharmacology with respect to other prostaglandins: Weeks, *Ann. Rev. Pharmacol.* **12**, 317 (1972). Clinical use to maintain patency of ductus arteriosus in neonatal cardiac problems: P. M. Olley *et al., Adv. Prostaglandin Thromboxane Res.* **7**, 913 (1980); J. S. Donahoo *et al., J. Thoracic Cardiovasc. Surg.* **81**, 227 (1981); E. D. Silove *et al., Circulation* **63**, 682 (1981). Use in non-atherosclerotic vasculopathy: D. L. Wooster *et al., J. Am. Med. Assoc.* **245**, 1846 (1981). For general refs *see* Prostaglandins.

Crystals from ethyl acetate + heptane, mp 115-116°. [α]$_{578}$ −61.6° (c = 0.56 in tetrahydrofuran). Easily dehydrated in soln at pHs < 4 or > 8.

THERAP CAT: Vasodilator.

7893. Prostaglandin E$_2$. *(5Z,11α,13E,15S)-11,15-Dihydroxy-9-oxoprosta-5,13-dien-1-oic acid; 7-[3-hydroxy-2-(3-*

hydroxy-1-octenyl)-5-oxocyclopentyl]-5-heptenoic acid; dino-
prostone; PGE$_2$; U-12062; Cerviprost; Minprostin E$_2$; Prepidil; Prostarmon-E; Prostin E$_2$. C$_{20}$H$_{32}$O$_5$; mol wt 352.48. C
68.15%, H 9.15%, O 22.70%. The most common and most
biologically potent of mammalian prostaglandins. Isoln
from sheep seminal vesicle tissue: Bergstrom et al., Acta
Chem. Scand. **16**, 501 (1962). Total synthesis of the dl-form:
Schneider, Chem. Commun. **1969**, 304; Corey et al., J. Am.
Chem. Soc. **91**, 5675 (1969); Corey et al., Tetrahedron Letters
1970, 307; Schneider, Ger. pat. **2,011,969** (1970 to Upjohn),
C.A. **74**, 87486n (1971); Fried et al., J. Am. Chem. Soc. **94**,
4342 (1972). Synthesis of naturally occurring form: Corey
et al., ibid. **92**, 397, 2586 (1970); Heather et al., Tetrahedron
Letters **1973**, 2313; from Plexaura homomalla prostaglandin
intermediates: Bundy et al., J. Am. Chem. Soc. **94**, 2123
(1972); Schneider et al., Chem. Commun. **1973**, 254. Biosynthesis: Van Dorp et al., Biochim. Biophys. Acta **90**, 204
(1964); Bergstrom et al., ibid. 207; Neth. pat. **Appl. 6,505,-
799** (1965 to Unilever), C.A. **65**, 7584h (1966). Metabolism:
Anggard, Samuelsson, Mem. Soc. Endocrinol., no. 14, 107
(1966); Hamberg, Samuelsson, J. Biol. Chem. **246**, 6713
(1971). Several reviews in Prostaglandin Symp. Worcester
Found. Exp. Biol., P. Ramwell, Ed. (Interscience, New York,
1968). For general refs see Prostaglandins.

Natural form, colorless crystals. mp 66-68°. [α]$_D^{26}$ −61°
(c = 1 in tetrahydrofuran). Easily dehydrated in soln at
pHs <4 or >8.
THERAP CAT: Oxytocic; abortifacient.

7894. Prostaglandin F$_{2\alpha}$. 9,11,15-Trihydroxyprosta-5,13-
dien-1-oic acid; 7-[3,5-dihydroxy-2-(3-hydroxy-1-octenyl)-
cyclopentyl]-5-heptenoic acid; dinoprost; PGF$_{2\alpha}$; U-14583;
Enzaprost F; Glandin; Prostarmon F; Prostin F$_2$ Alpha.
C$_{20}$H$_{34}$O$_5$; mol wt 354.49. C 67.76%, H 9.67%, O 22.57%.
One of the most biologically studied of the primary prosta-
glandins. Closely related to prostaglandin E$_2$ (PGE$_2$) in that
both prostaglandins are biosynthesized from the same pre-
cursors and that PGF$_{2\alpha}$ is the synthetic reduction product of
PGE$_2$. For refs to synthesis of dl and natural forms see
Prostaglandin E$_2$. Alternate synthesis of natural PGF$_{2\alpha}$:
Schneider, Murray, J. Org. Chem. **38**, 397 (1973); R. B.
Woodward et al., J. Am. Chem. Soc. **95**, 6853 (1973); G.
Stork et al., ibid. **100**, 8272 (1978); K. Kondo et al., Tetrahedron Letters **1978**, 3927; N. R. A. Beeley et al., Tetrahedron
37, Suppl. 9, 411 (1981); R. J. Cave et al., J. Chem. Soc.
Perkin Trans. I **1981**, 646. Causes vasocontraction and
exhibits luteolytic activity; is most commonly associated
with its role in pregnancy: Karim et al., J. Obstet. Gynaecol.
Brit. Commonw. **78**, 172 (1971). Metabolism in female subjects: Granstrom, Samuelsson, J. Biol. Chem. **246**, 5254
(1971). Toxicity data: T. Fujita et al., Iyakuhin Kenkyo **9**,
261 (1978), C.A. **89**, 71399k (1978). For general refs see
Prostaglandins.

Natural form, crystals, mp 25-35°. [α]$_D^{25}$ +23.5° (c = 1 in
tetrahydrofuran). Freely sol in methanol, abs ethanol, ethyl
acetate, chloroform; slightly sol in water. Stable for two
years in light resistant containers at 5-15°. Degrades in one
week when exposed to sunlight or in three months at 40°.
LD$_{50}$ in rabbits (mg/kg): 2.5-5.0 i.v.; 2.5-5.0 i.m. (Fujita).
Tromethamine salt, C$_{24}$H$_{45}$NO$_8$, Lutalyse, Prostin F$_2$ Alpha
Injectable. White or off-white cystalline powder. Readily
sol in water to at least 200 mg/ml.
THERAP CAT: Oxytocic. Abortifacient.
THERAP CAT (VET): Oxytocic.

7895. Prostalene. 9,11,15-Trihydroxy-15-methylprosta-
4,5,13-trien-1-oic acid methyl ester; (±)-methyl 7-[3,5-di-
hydroxy-2-[(E)-3-hydroxy-3-methyl-1-octenyl]cyclopent-
yl]-4,5-heptadienoate; RS-9390; Synchrocept. C$_{22}$H$_{36}$O$_5$;
mol wt 380.53. C 69.44%, H 9.54%, O 21.02%. Synthetic
analog of prostaglandin F$_{2\alpha}$, q.v. Prepn: P. Crabbé, J. H.
Fried, Ger. pat. **2,258,668** corresp to U.S. pat. **3,879,438**
(1973, 1975 both to Syntex). Use in induction of parturition
in sows: W. Holtz et al., J. Anim. Sci. **49**, 367 (1979). Pharmacodynamics in mares: R. G. Loy et al., J. Reprod. Fertil.
(Suppl) **27**, 229 (1979).

THERAP CAT (VET): Luteolytic.

7896. Prosultiamine. N-[(4-Amino-2-methyl-5-pyrimi-
dinyl)methyl]-N-[4-hydroxy-1-methyl-2-(propyldithio)-1-
butenyl]formamide; 2-(2-methyl-4-aminopyrimidin-5-yl)-
methylformamido-5-hydroxy-2-penten-3-yl propyl disulfide; vitamin B$_1$ propyl disulfide; thiamine propyl disulfide;
dithiopropylthiamine; DTPT; TPD; Alinamin; Aneurimec;
Ausovit B$_1$; Betatron; Binova; Ditiovit; Liponeurina; Marineurina; Orobetina; Proneurin; Sintotiamina; Tipidi. C$_{15}$-
H$_{24}$N$_4$O$_2$S$_2$; mol wt 356.51. C 50.53%, H 6.78%, N 15.72%,
O 8.98%, S 17.99%. Synthesis: Matsukawa, Kawasuki, J.
Pharm. Soc. Japan **73**, 216 (1953), C.A. **48**, 2071 (1954);
Matsukawa et al., J. Vitaminol. **1**, 13 (1954); Fujiwara et al.,
U.S. pat. **2,833,768** (1958 to Takeda), cf. Fr. pat. **1,068,459**
(1954 to Takeda). Structural studies: Nishikawa et al.,
Chem. Pharm. Bull. **17**, 932 (1969). Metabolism: Suzuoki-
Ziro et al., J. Biochem. **58**, 279 (1965); Nishikawa et al., J.
Pharmacol. Exp. Ther. **157**, 589 (1967).

Prisms from benzene, dec 128-129°. Sparingly sol in water. Soluble in organic solvents and lipids. Better absorbed
upon oral ingestion by man, than thiamine hydrochloride.
Hydrochloride, C$_{15}$H$_{25}$ClN$_4$O$_2$S$_2$, Neurvit, Nevritar, Supertiamin, Tiotiamina. Crystals, dec 160-161°.
THERAP CAT: Enzyme co-factor vitamin.

7897. Protactinium. Protoactinium. Pa; at. wt (longest
lived isotope) 231; at. no. 91; valence 4, 5. First isotope discovered, ^{234m}Pa (T$_{1/2}$ 1.175 minutes); called brevium, uranium
X$_2$ or UX$_2$: Fajans, Göhring, Naturwiss. **1**, 339 (1913), C.A.
7, 3916 (1913); Göhring, C.A. **10**, 852^1 (1916). Long-lived
isotope, ^{231}Pa (T$_{1/2}$ 3.25 × 10^4 years), discovered in 1918 by
Hahn, Meitner, Phys. Z. **19**, 208 (1918); and independently
by Soddy, Cranston, Proc. Roy. Soc. **94A**, 384 (1918). The
parent compd of actinium. Uranium Z (UZ), ^{234}Pa (T$_{1/2}$ 6.75
hours), isolated by Hahn, Ber. **54B**, 1131 (1921). Review of
discovery: Fajans, Morris, Nature **244**, 137-138 (1973). See
also C. M. Lederer et al., Table of Isotopes (John Wiley, New
York, 6th ed., 1968) pp 140-141. Prepn of ^{231}Pa: Katzin et
al., J. Am. Chem. Soc. **72**, 4815 (1950). Reviews: J. J. Katz,
G. T. Seaborg, The Chemistry of the Actinide Elements (John
Wiley, New York, 1957) pp 67-93; Comprehensive Inorganic
Chemistry vol. 5, J. C. Bailar, Jr. et al., Eds. (Pergamon
Press, Oxford, 1973) passim.
^{231}Pa, shiny malleable metallic mass. Tetragonal crystal
structure. Easily tarnished in air to an undetermined oxide.
d (calc) 15.37 g/cc. mp 1560°. Also reported as 1575°: J. C.
Bailar, Jr. et al., loc. cit. Reacts with H$_2$ at 250-300° to form
PaH$_3$. In dil solns of HF, is deposited (10-96%) on Be, Al,
Mn, Zn, and Pl; gives small deposits on Cr, Ta, Fe, Cd, Ni,
Cu, Hg, W: Camarcat et al., J. Chim. Phys. **46**, 153-157
(1949).
Caution: Radiation hazard; alpha emitter (approx 5 Mev).

Inhalation hazard in insol form; general hazard if absorbed systemically. Max permissible concn of insoluble form in air: 4×10^{-11} μCi/cc; of sol form in air: 4×10^{-13} μCi/cc; *Natl. Bur. Stand. Handbook* **69**, 83 (1959).

7898. Protamines. A group of simple proteins that yield basic amino acids on hydrolysis and that occur combined with nucleic acid in the sperm of fish, e.g., salmine, q.v., from salmon sperm, clupeine, q.v., from herring sperm, *iridine* from trout sperm, *sturine* from sturgeon sperm, and *scombrine* from mackerel sperm. Protamines contain very few kinds of amino acids. All of them contain arginine, alanine, and serine; most of them contain proline and valine; many contain glycine and isoleucine. Individual protamines contain histidine, lysine, threonine, and aspartic and glutamic acids. Tyrosine and tryptophan are absent. Isoln procedures: A. Kossel, *The Protamines and Histones* (Longmans, New York, 1928). Structure studies: Felix, Hashimoto, *Z. Physiol. Chem.* **330**, 205 (1963). Improved fractionation method: Ando, Watanabe, *Int. J. Prot. Res.* **1**, 221 (1969). *Review:* Felix, *Advan. Protein Chem.* **15**, 1 (1960). Protamine (from salmon) has been shown to inhibit angiogenesis (capillary proliferation) observed in embryogenesis, inflammation, certain immune reactions and the growth of solid tumors: S. Taylor, J. Folkman, *Nature* **297**, 307 (1982).

Alkaline reaction. Sol in water, dil acids, and ammonia water. Not coagulated by heat.

Sulfate, *Protamine Sulfate Injection.* Fine white or faintly colored amorphous or cryst powder. Sparingly sol in water. When administered alone it has an anticoagulant effect, but when given in the presence of heparin, a stable salt is formed which results in loss of anticoagulant activity of both substances.

THERAP CAT: Antidote (to heparin).

7899. Protamine Zinc Insulin Suspension. *Insulin protamine zinc;* Protamine Zinc Insulin Injection; Protamine, Zinc and Iletin; Insulin Zinc Protamine; Depo-Insulin; Deposulin; Insulyl-Retard. Sterile suspension, in a buffered water medium, of insulin modified by the addition of zinc chloride and protamine sulfate. The protamine sulfate is prepd from sperm or from mature testes of fish belonging to the genus *Oncorhynchus* Suckley, or *Salmo* Linné *Salmonidae*. Contains 40, 80 or 100 U.S.P. insulin units/ml. Prepn: Scott, Fisher, *J. Pharmacol. Exp. Ther.* **58**, 78 (1936); *eidem,* U.S. pats. **2,143,591** and **2,179,384** (1936 and 1937 to Governors of the Univ. of Toronto).

White or almost white suspension. pH 7.1-7.4. A pH of 7.2 corresp to the isoelectric point at which it exhibits its max insolubility. At any other pH the soly is increased and the intensity and duration of hypoglycemic action is affected: Hagedorn *et al., J. Am. Med. Assoc.* **106**, 177 (1936). Contains 1.4 to 1.8% (w/v) of glycerin and either 0.18 to 0.22% (w/v) of cresol or 0.22 to 0.28% (w/v) of phenol. Contains 0.15 to 0.25% (w/v) of Na_2HPO_4. Also contains 0.20 to 0.25 mg of zinc and 1.0 to 1.5 mg of protamine per 100 U.S.P, insulin units. Onset of action occurs from 4-6 hrs after s.c. injection, reaching its max in 14-24 hrs. Duration of action is 36 hrs. *Caution:* Suspension should not be used if outdated or if ppt is clumped or granular in appearance.

THERAP CAT: Antidiabetic.

7900. Protein Hydrolysates. Amigen; Aminonat; Elamine; Lactamin (obsolete); Neo-Protostan; Parenamine; Protolysate; Aminosol; Protigényl; Travamin; C.P.H.; Hyprotigen; Lacotein; Virex; Trophysan; Aminokrovin. Sterile solution of amino acids and short-chain peptides which represent the approx nutritive equivalent of the casein, lactalbumin, plasma, fibrin or other suitable protein from which it is derived by acid, enzymatic or other method of hydrolysis. It may be modified by partial removal and restoration or addn of one or more amino acids. It may contain alcohol, dextrose, or other carbohydrate suitable for i.v. infusion. Not < 50% of the total nitrogen present is in the form of α-amino nitrogen.

THERAP CAT: Replenisher (fluid and nutrient).
THERAP CAT (VET): Parenteral nutrient.

7901. Protheobromine. *3,7-Dihydro-1-(2-hydroxypropyl)-3,7-dimethyl-1H-purine-2,6-dione; 1-(2-hydroxypropyl)-theobromine;* Cordabromin; Theocor; Tebe; Bonicor. C_{10}-

$H_{14}N_4O_3$; mol wt 238.24. C 50.42%, H 5.92%, N 23.52%, O 20.15%. Prepn: Rojahn, Fegler, *Arch. Pharm.* **268**, 570 (1930); **Brit.** pat. **816,299** (1959 to Chemiewerk Homburg), *cf.* Ger. pat. **1,067,025.** Improved procedure: Carstens, Donat, East Ger. pat. **26757** (1964). Pharmacology: Giertz *et al., Arzneimittel-Forsch.* **6**, 457 (1956).

Crystals from isopropanol, mp 140-142°. Freely sol in water at neutral or slightly acid pH. Sol in chloroform, hot ethanol, warm glycerol. Practically insol in ether. LD_{50} s.c. in mice: 580 mg/kg, R. Taugner *et al., Arzneimittel-Forsch.* **6**, 601 (1956).

THERAP CAT: Diuretic.

7902. Prothipendyl. *N,N-Dimethyl-10H-pyrido[3,2-b]-[1,4]benzothiazine-10-propanamine; 10-(3-dimethylaminopropyl)-10H-pyrido[3,2-b][1,4]benzothiazine; 10-(γ-dimethylaminopropyl)-1-azaphenothiazine; N-(3-dimethylaminopropyl)thiophenylpyridylamine; 2,3-pyridino-(5',6')-5,6-benzo-4-(3''-dimethylaminopropyl)-1,4-thiazine;* D206; Dominal; Phrenotropin; Tolnate; Timovan. $C_{16}H_{19}N_3S$; mol wt 285.42. C 67.33%, H 6.71%, N 14.72%, S 11.24%. Prepn: Yale, Bernstein, **U.S.** pat. **2,943,086** (1960 to Olin Mathieson); von Schlichtegroll, *Proc. Int. Congr. Neuro-Pharm., 1st, Rome,* **1958**, 408 (1959), *C.A.* **54**, 13400g (1960); **Fr.** pat. **1,173,134** (1959 to Rhône-Poulenc).

Liquid, $bp_{0.7}$ 217-219°; $bp_{0.5}$ 195-198°.

Hydrochloride monohydrate, $C_{16}H_{20}ClN_3S.H_2O$, crystals, mp 108-112° (when anhyd, mp 177-178° with sintering about 176°). Freely sol in water, methanol. Practically insol in petr ether, ether.

Dihydrochloride, $C_{16}H_{21}Cl_2N_3S$, crystals from acetonitrile, mp 205-207°.

THERAP CAT: Antipsychotic, antihistaminic.

7903. Prothrombin. *Blood-coagulation factor II;* factor II; prothrombase; serozyme; thrombogen. Mol wt 69,000-74,000. Coagulation proenzyme present in highest concentration in blood. Prothrombin is one of the vitamin-K dependent blood coagulation factors. It is converted to thrombin by the action of factor X_a, factor V and phospholipid in the presence of Ca^{2+} ions. Accounts for < 0.2% of total plasma protein. Prepn of human and bovine prothrombin: Goldstein *et al., J. Biol. Chem.* **234**, 2857 (1959); Lanchantin *et al., ibid.* **238**, 238 (1963). Chemistry of activation: Magnussen, *Biochem. J.* **115**, 2P (1969). Synthesized normally by liver parenchymal cells in a cyclic asynchronous manner. Dicoumarol derivatives halt the synthesis, while vitamin K_1 stimulates synchronized activity of all the liver parenchymal cells: Barnhart, Anderson, *Biochem. Pharmacol.* **9**, 23 (1962). The glycoprotein structure probably consists of a single polypeptide chain containing between 8 and 10% carbohydrate: Magnussen, *Thromb. Diath. Haemorrh.* **suppl. 54**, 31 (1973). *Reviews:* W. H. Seegers, *Prothrombin* (Harvard Univ. Press, 1962) 728 pp; Magnussen "Prothrombin and Thrombin," in *The Enzymes*, Vol. III, P. D. Boyer, Ed. (Academic Press, New York, 3rd ed., 1971) pp 277-321; K. G. Mann, *Methods Enzymol.* **45B**, 123-156 (1976). Review on prothrombin activation: several authors in *Ann. N.Y. Acad. Sci.* **370**, 336-528 (1981).

Most stable within the range pH 4-9.5. Isoelec pt pH 4.2. Very sol in water but pptd at pH 4.2-4.5. Solns tend to activate spontaneously. There is little loss of activity when drying from the frozen state but the dry material alters in a few months. Drying with organic solvents destroys activity.

7904. Protiofate. *3,4-Dihydroxy-2,5-thiophenedicarboxylic acid dipropyl ester;* dipropyl 3,4-dihydroxy-2,5-thiophenedicarboxylate; Atrimycon. $C_{12}H_{16}O_6S$; mol wt 288.31. C 49.99%, H 5.59%, O 33.30%, S 11.12%. Thiophene derivative with antimycotic activity: L. Riviera *et al.*, *Chemioterapia* **3**, 116 (1984). Prepn: **Fr. pat. 2,255,063** (1975 to Bioindustria), *C.A.* **84**, 89986f (1976).

CH$_3$(CH$_2$)$_2$OOC — [thiophene ring] — COO(CH$_2$)$_2$CH$_3$
HO — OH

Crystals from ethanol-water, mp 96-97°.
THERAP CAT: Topical fungicide.

7905. Protionamide. *2-Propyl-4-pyridinecarbothioamide;* *2-propylthioisonicotinamide;* prothionamide; 2-propyl-4-thiocarbamoylpyridine; Ektebin; Peteha; Trevintix. $C_9H_{12}N_2S$; mol wt 180.29. C 59.96%, H 6.71%, N 15.54%, S 17.79%. Prepn: Libermann *et al.*, *Compt. Rend.* **242**, 2409 (1956); **Brit. pat. 800,250** (1958 to Chimie et Atomistique); L. Bernardi, G. Palamidessi, **Fr. pat. 2,091,338** (1972 to Societa Farm. Ital); L. N. Yakhontov *et al.*, *Khim.-Farm. Zh.* **10**, 96 (1976), *C.A.* **85**, 32787h (1976). Pharmacology: A. M. Il'in, *Farmakol. Toksikol.* **38**, 471 (1975).

[pyridine ring structure with CH$_2$CH$_2$CH$_3$ and CSNH$_2$]

Crystals, mp 142°. Sol in ethanol, methanol; slightly sol in ether, chloroform. Practically insol in water. LD_{50} in mice, rats, cats: 1.0, 1.32 mg/kg, > 1 g/kg i.p.
THERAP CAT: Antibacterial (tuberculostatic).

7906. Protizinic Acid. (±)-*7-Methoxy-α,10-dimethyl-10H-phenothiazine-2-acetic acid;* 7-methoxy-α,10-dimethylphenothiazine-3-acetic acid; 2-(7-methoxy-10-methyl-2-phenothiazinyl)propionic acid; Pirocrid; $C_{17}H_{17}NO_3S$; mol wt 315.40. C 64.74%, H 5.43%, N 4.44%, O 15.22%, S 10.16%. Prepn: **Brit. pat. 1,048,680** corresp to Farge *et al.*, **U.S. pat. 3,450,698**; *eidem*, **S. Afr. pat. 68 00,435**, *C.A.* **70**, 68394g (1969), (1966, 1969, 1968 all to Rhône-Poulenc).

[phenothiazine structure with CH$_3$O, N-CH$_3$, and CH$_3$CHCOOH groups]

Crystals from diisopropyl ether or acetonitrile, mp 124-125°.
THERAP CAT: Anti-inflammatory.

7907. Protoanemonin. *5-Methylene-2(5H)-furanone;* *4-hydroxy-2,4-pentadienoic acid γ-lactone;* 5-methylene-2-oxodihydrofuran; Isomycin. $C_5H_4O_2$; mol wt 96.08. C 62.50%, H 4.20%, O 33.30%. An antibacterial principle of *Anemone pulsatilla* L., *Ranunculaceae:* Seegal, Holden, *Science* **101**, 413 (1945). Exists as a glucoside, *ranunculin*, in the intact plant, and is released by an enzymatic process on maceration of the plant tissue: R. Hill, R. van Heyningen, *Biochem. J.* **49**, 332 (1951). Isoln: Asahina, Fujita, *Acta Phytochim. Japan* **1**, 1 (1922); Baer *et al.*, *J. Biol. Chem.* **162**, 65 (1946). Structure and synthesis: Asahina, Fujita, *loc. cit.;* Muskat *et al.*, *J. Am. Chem. Soc.* **52**, 326 (1930); Fox, Jr., *Proc. Soc. Exp. Biol.* **51**, 102 (1942); Shaw, *J. Am. Chem. Soc.* **68**, 2510 (1946). Industrial prepns: **Brit. pat. 759,999** (1956 to Olin Mathieson), *C.A.* **51**, 9678f (1957); Reicheneder *et al.*, **Ger. pat. 1,088,047** (BASF), *C.A.* **56**, 14086i (1962); Sakuma, Hirano, **U.S. pat. 3,203,863** (1965 to Lion Dentifrice).

[furanone structure: H$_2$C, O, O]

Pale yellow oil; volatile with steam. bp$_{1.5}$ 45°. Sol in ethylene dichloride, chloroform. Solubility in water about 1%. Stable in water. When pure the compd turns to a hard polymer which, when ground and extracted with boiling ethyl acetate, yields *anemonin*, mp 149-150°.
THERAP CAT: Antibacterial.

7908. Protocatechualdehyde. *3,4-Dihydroxybenzaldehyde;* 3,4-dihydroxybenzenecarbonal; protocatechuic aldehyde; rancinamycin IV. $C_7H_6O_3$; mol wt 138.12. C 60.87%, H 4.38%, O 34.75%. Prepn from catechol: Reimer, Tiemann, *Ber.* **9**, 1268 (1876); Tiemann, Koppe, *ibid.* **14**, 2015 (1881); from vanillin: Tiemann, Haarmann, *ibid.* **7**, 620 (1874); from veratric aldehyde: Dreyfus, **Ger. pat. 193,958;** *Frdl.* **9**, 161 (1908-10); from piperonal: Hoering, Baum, *Ber.* **41**, 1914 (1908); Barger, *J. Chem. Soc.* **93**, 563 (1908); Buck, Zimmerman, *Org. Syn.* **coll. vol. II**, 549 (1943).

[benzaldehyde structure: CHO, OH, OH]

Platelets from water or toluene. Dimorphic. Dec 153-154°. K at 25° = 2.8 × 10^{-8}. Soly in water (g/100 ml): 5 (20°); 33 (99°); in ethanol: 79 (78°). Freely sol in ether.

7909. Protocatechuic Acid. *3,4-Dihydroxybenzoic acid.* $C_7H_6O_4$; mol wt 154.12. C 54.55%, H 3.92%, O 41.52%. Minute amounts are found in wheat grains, in wheat seedlings, and in many other plants: L. Hörhammer, A. Scherm, *Arch. Pharm.* **288**, 441 (1955). Prepd by the alkaline fusion of vanillin: Tiemann, Haarmann, *Ber.* **7**, 617 (1874); Pearl, *J. Am. Chem. Soc.* **68**, 2180 (1946); *Org. Syn.* **coll. vol. III**, 745 (1955).

[benzoic acid structure: COOH, OH, OH]

White to brownish, cryst powder; discolors in air. d 1.54. mp about 200° with decompn. Sol in 50 parts water; sol in alcohol, ether. *Keep well closed.*

7910. Protokosin. From dried flowers of *Hagenia abyssinica* J. J. Gmel. (*Brayera anthelmintica* Kunth.), *Rosaceae.* Isoln: Leichsenring, *Arch. Pharm.* **232**, 58 (1894); Lobeck, *ibid.* **239**, 682 (1901); Hems, Todd, *J. Chem. Soc.* **1937**, 562. Structural studies: Birch, Todd, *ibid.* **1952**, 3102; Orth, Riedl, *Ann.* **663**, 83 (1963). Exists in several tautomeric forms and is a mixt of isobutyryl, isovaleryl, and 2-methylbutyryl side chain homologues. Structural elucidation: Lounasmaa *et al.*, *Acta Chem. Scand.* **28B**, 1200 (1974).

[complex chemical structure with CH$_3$O, OH, CH$_3$, C—R, CH$_2$, H$_3$C, etc.]

R = —CH(CH$_3$)$_2$, —CH$_2$—CH(CH$_3$)$_2$, —CH(CH$_3$)CH$_2$CH$_3$

Colorless needles from acetone, mp 181-183°. [α]$_D^{25}$

+13.9° (c = 0.610 in chloroform). uv max (cyclohexane): 224, 285 nm (ε 28,200, 36,400). Practically insol in water; slightly sol in alcohol, light petroleum; freely sol in ether, acetone, ethyl acetate, chloroform.

7911. Protokylol. *4-[2-[[2-(1,3-Benzodioxol-5-yl)-1-methylethyl]amino]-1-hydroxyethyl]-1,2-benzenediol; α-[(α-methyl-3,4-methylenedioxyphenethylamino)methyl]protocate-chuyl alcohol;* α-(3,4-dihydroxyphenyl)-β-[2-(3,4-methyl-enedioxyphenyl)isopropylamino]ethanol; 1-(3,4-dihydroxy-phenyl)-2-(α-methyl-3,4-methylenedioxyphenethylamino)-ethanol; N-[β-(3,4-methylenedioxyphenyl)isopropyl]-β-(3,4-dihydroxyphenyl)-β-hydroxyethylamine; N-[2-(3,4-methylenedioxyphenylisopropyl)]norepinephrine. $C_{18}H_{21}$-NO_5; mol wt 331.36. C 65.24%, H 6.39%, N 4.23%, O 24.14%. β-Adrenergic agonist. Prepn: Biel *et al., J. Am. Chem. Soc.* **76,** 3149 (1954); Biel, U.S. pat. **2,900,415** (1959 to Lakeside Labs.). Toxicity: E. I. Goldenthal, *Toxicol. Appl. Pharmacol.* **18,** 185 (1971).

Hydrochloride, $C_{18}H_{22}ClNO_5$, *JB-251, Caytine, Ventaire.* Crystals from isopropanol, mp 126-127°. Sol in water. LD_{50} orally in rats: 938 ±96 mg/kg (Goldenthal).
THERAP CAT: Bronchodilator.
THERAP CAT (VET): Bronchodilator.

7912. Protopine. *4,6,7,14-Tetrahydro-5-methyl-bis[1,3]-benzodioxolo[4,5-c:5',6'-g]azecin-13(5H)-one; 7-methyl-2,3:9,10-bis(methylenedioxy)-7,13a-secoberbin-13a-one;* fu-marine; macleyine. $C_{20}H_{19}NO_5$; mol wt 353.36. C 67.98%, H 5.42%, N 3.96%, O 22.64%. From opium: Hesse, *Ber.* **4,** 693 (1871); also from the herb *Fumaria officinalis* L., *Chelidonium majus* L., and many other *Papaveraceae* and *Fumariaceae:* Manske in *The Alkaloids* vol. IV, R. H. F. Manske, H. L. Holmes, Eds. (Academic Press, New York, 1954) pp 157-159. Structure: Perkin, Jr., *J. Chem. Soc.* **109,** 815 (1916); Gadamer, Bruchhausen, *Arch. Pharm.* **260,** 97 (1922); Mottus *et al., Can. J. Chem.* **31,** 1144 (1953); Anet, Marion, *ibid.* **32,** 452 (1954). Synthesis: Haworth, Perkin, *J. Chem. Soc.* **1926,** 1769. Crystal structure: Hall, Ahmed, *Acta Cryst.* **24B,** 337 (1968).

Monoclinic prisms from alcohol + chloroform, mp 208°. d 1.399 (calc). uv max (95% ethanol): 293 nm (log ε 3.93). Sol in 15 parts chloroform, 900 parts alc, 1000 parts ether. Slightly sol in ethyl acetate, carbon disulfide, benzene, petr ether. Practically insol in water.
Hydrochloride, $C_{20}H_{19}NO_5$.HCl, prisms from alcohol, sol in 143 parts water at 13°, sol in alcohol. Also a hexahydrate, needles from water.
Methiodide, $C_{20}H_{19}NO_5$.CH$_3$I, twinned crystals from methanol, dec 217°.

7913. Protoporphyrin IX. *7,12-Diethenyl-3,8,13,17-tetramethyl-21H,23H-porphine-2,18-dipropanoic acid; 3,7,12,17-tetramethyl-8,13-divinyl-2,18-porphinedipropionic acid; 1,3,5,8-tetramethyl-2,4-divinylporphine-6,7-dipropionic acid; ooporphyrin; Kammerer's porphyrin.* $C_{34}H_{34}N_4O_4$; mol wt 562.64. C 72.58%, H 6.09%, N 9.96%, O 11.37%. Biological precursor of blood and plant pigments. Prepd from hemin: Fischer-Orth, *Die Chemie des Pyrrols* II, 1, 396 (Leipzig, 1937); Ramsey, *Biochem. Prepn.* **3,** 39 (1953).

Structure: Sparatore, Mauzerall, *J. Org. Chem.* **25,** 1073 (1960). Synthesis: Carr *et al., J. Chem. Soc. (C)* **1971,** 487. Crystal and molecular structure: W. S. Caughey, J. A. Ibers, *J. Am. Chem. Soc.* **99,** 6639 (1977). Chelates with metals, esp iron, in the ferrous state to form heme *(q.v.),* in the ferric state to form hematin *(q.v.). Review:* Rimington, Kennedy, in M. Florkin, H. S. Mason, *Comparative Biochemistry* (Academic Press, New York, 1962) pp 557-614.

Monoclinic, brownish-yellow prisms from ether. Absorption max (25% HCl): 602.4, 582.2, 557.2 nm. Freely sol in chloroform, glacial acetic acid, alcohol contg HCl, ether contg some glacial acetic acid, hydrochloric acid. Somewhat sol in dil alkalies, aniline, pyridine. Forms sparingly sol disodium and dipotassium salts.
Disodium salt, $C_{34}H_{32}N_4Na_2O_4$, *Depocolin-S, Dojin-PM.*
Dimethyl ester, $C_{36}H_{38}N_4O_4$, crystals from chloroform + methanol, mp 228-230°. Absorption max (25% HCl): 601, 556, 409 nm. Soluble in chloroform, slightly sol in methanol. Insol in sodium carbonate solns.
THERAP CAT: In liver disease.

7914. Protostephanine. *6,7,8,9-Tetrahydro-2,3,10,12-tetramethoxy-7-methyl-5H-dibenz[d,f]azonine.* $C_{21}H_{25}O_4N$; mol wt 355.44. C 70.96%, H 7.09%, O 18.01%, N 3.94%. A member of the hasubanan alkaloids. First alkaloid known to possess the unique dibenz[d,f]azonine structure. Isoln from *Stephania japonica,* Miers, *Menispermaceae:* H. Kondo, T. Sanada, *J. Pharm. Soc. Japan* **541,** 177 (1927), *C.A.* **21,** 2700⁴ (1927); H. Kondo, T. Watanabe, *ibid.* **58,** 268 (1938), *C.A.* **32,** 5403⁵ (1938). Structure: K. Takeda, *Bull. Agr. Chem. Soc. Japan* **20,** 165 (1956). Synthesis: *idem, C.A.* **60,** 5570f (1964); B. Pecherer, A. Brossi, *J. Org. Chem.* **32,** 1053 (1967); A. R. Battersby *et al., J. Chem. Soc. Perkin Trans. I* **1981,** 2002. Biosynthesis: *eidem, ibid.* 2010, 2016, 2030.

White cryst solid from benzene, mp 84-86° (Pecherer). Also reported as mp 75° (Kondo).

7915. Protoveratrines. From the rhizome of *Veratrum album* L., *Liliaceae:* Salzberger, *Arch Pharm.* **228,** 462 (1890); Poethke, *ibid.* **275,** 357 (1937); Craig, Jacobs, *J. Biol. Chem.* **143,** 427 (1942); **149,** 271 (1943). Mixture of proto-veratrines A and B: Glen *et al., Nature* **170,** 932 (1952); Klohs *et al., J. Am. Chem. Soc.* **74,** 5107 (1952); Nash, Brooker, *ibid.* **75,** 1942 (1953); Stoll, Seebeck, *Helv. Chim. Acta* **36,** 718 (1953); Nash, Brooker, U.S. pat. **2,929,812** (1960 to Allied Labs). Structure of protoveratrines A and B: Kupchan, Ayres, *J. Am. Chem. Soc.* **82,** 2252 (1960).

Slightly bitter, sternutative crystals from alc, dec 266-267°. $[\alpha]_D^{25}$ −38.6° (pyridine). Soluble in chloroform; dil aq acidic solns; slightly sol in ether. Practically insol in water, petr ether.

Maleate, *Provell maleate, Puroverin(e), Veralba.* Sternutative crystals, dec 210-222°. $[\alpha]_D^{25}$ −33.2° to −38.4° (pyridine). uv max: 211 nm ($A_{1cm}^{1\%}$ 148). pH of a 0.2% aq soln 4.1-4.7. Soluble in water, alcohol, chloroform. Practically insol in water, petr ether.

Protoveratrine A, $C_{41}H_{63}NO_{14}$, *4α,9-epoxycevane-3β,4,-6α,7α,14,15α,16β,20-octol 6,7-diacetate 3(S)-(2-hydroxy-2-methylbutanoate) 15(R)-(2-methylbutanoate), Pro-Amid, Protalba.* R = H. Crystals from acetone, dec 267-269°. $[\alpha]_D^{25}$ −40.5° (pyridine); $[\alpha]_D^{25}$ −10.5° (chloroform). Soluble in chloroform, pyridine, hot alcohol.

Protoveratrine B, $C_{41}H_{63}NO_{15}$, *veratetrine, neoprotoveratrine.* R = OH. Crystals from acetone, dec 268-270°. $[\alpha]_D^{25}$ −37° (pyridine); $[\alpha]_D^{25}$ −3.5° (chloroform). Soluble in chloroform, pyridine, hot alcohol.

Note: Tensatrin contains protoveratrine A and B.

THERAP CAT: Antihypertensive.

7916. Protoverine. *4,9-Epoxycevane-3,4,6,7,14,15,16,20-octol.* $C_{27}H_{43}NO_9$; mol wt 525.62. C 61.69%, H 8.25%, N 2.67%, O 27.40%. Obtained by alkaline hydrolysis of protoveratrine A: Poethke, *Arch. Pharm.* **275**, 571 (1937); Stoll, Seebeck, *Helv. Chim. Acta* **36**, 718 (1953). Structure: Kupchan *et al., Chem. & Ind. (London)* **1958**, 1626; *J. Am. Chem. Soc.* **81**, 1009 (1959); **82**, 2242 (1960).

Fine needles from methanol, dec 220-222°. $[\alpha]_D^{20}$ −15.7° (c = 1.1 in pyridine). Sol in CHCl₃, methanol, pyridine.

7917. Protriptyline. *N-Methyl-5H-dibenzo[a,d]cycloheptene-5-propanamine;* 5-(3-methylaminopropyl)-5H-dibenzo[a,d]cycloheptene; 7-(3-methylaminopropyl)-1,2:5,6-dibenzocycloheptatriene; amimetilina. $C_{19}H_{21}N$; mol wt 263.37. C 86.64%, H 8.04%, N 5.32%. Prepn: Engelhardt, Christy, **Belg.** pat. **617,967** (1962 to Merck & Co.), *C.A.* **59**, 517f (1963); Tishler *et al.,* U.S. pats. **3,244,748** and **3,271,-451** (both 1966 to Merck & Co.); Engelhardt *et al., J. Med. Chem.* **11**, 325 (1968). Metabolism in man, pig and dog: Sisenwine *et al., J. Pharmacol. Exp. Ther.* **175**, 51 (1970). Use in treatment of sleep apnea: R. W. Clark *et al., Neurology* **29**, 1287 (1979); L. G. Brownell *et al., N. Engl. J. Med.* **307**, 1037 (1982).

Hydrochloride, $C_{19}H_{22}ClN$, *MK-240, Concordin, Maximed, Triptil, Vivactil.* Crystals from isopropanol-ethyl ether, mp 169-171°. uv max: 290 nm (ε 13311). Freely sol in water. pKa 8.2.

THERAP CAT: Antidepressant.

7918. Pro-Urokinase. *Prourokinase (enzyme-activating);* single-chain urokinase-type plasminogen activator; single-chain pro-urokinase; scu-PA; pro-UK; pro u-PA; PUK; Tomieze; Sandolase. Single-chain proenzyme form of urokinase, *q.v.,* with intrinsic thrombolytic activity. Produced by the kidney; present in urine and blood. Consists of 411 amino acid residues, mol wt ~54,000 daltons. Converted by plasmin and kallikrein to active two-chain urokinase by proteolytic cleavage of the Lys 158 - Ile 159 peptide bond. Identification of proenzyme activity from human kidney cell culture: M. B. Bernik, *J. Clin. Invest.* **52**, 823 (1973). Isolation by affinity chromatography: C. Nolan *et al., Biochim. Biophys. Acta* **496**, 384 (1977). Purification, characterization and comparison with urokinase: T.-C. Wun *et al., J. Biol. Chem.* **257**, 7262 (1982). Extraction from urine or from kidney tissue culture and purification: S. S. Husain *et al.,* U.S. pat. **4,381,346** (1983); *eidem, Arch. Biochem. Biophys.* **220**, 31 (1983). Amino acid sequence: S. Kasai *et al., J. Biol. Chem.* **260**, 12382 (1985). Thrombolytic activity study: V. Gurewich *et al., J. Clin. Invest.* **73**, 1731 (1984). Cloning and expression of human pro-UK gene: W. E. Holmes *et al., Biotechnology* **3**, 923 (1985). Series of articles on thrombolytic activity of recombinant proenzyme: *Thromb. Haemostas.* **52**, 19-33 (1984). Mechanism of action: R. Pannell, V. Gurewich, *Blood* **67**, 1215 (1986); H. R. Lijnen *et al., J. Biol. Chem.* **261**, 1253 (1986). Reaction kinetics in thrombolysis: D. Collen *et al., ibid.* 1259. Comparative *in vitro* study wth tissue plasminogen activator, *q.v.:* V. Gurewich, R. Pannell, *Thromb. Res.* **44**, 217 (1986). Clinical pharmacology and fibrinolytic activity: G. Trubestein *et al., Haemostasis* **17**, 238 (1987). Clinical evalution in acute myocardial infarction: F. Van de Werf *et al., Ann. Int. Med.* **104**, 345 (1986); C. Bode *et al., Am. J. Cardiol.* **61**, 971 (1988). *Reviews:* V. Gurewich, *J. Am. Coll. Cardiol.* **10**, 16B-21B (1987); of biochemical properties: V. Gurewich, R. Pannell, *Sem. Thromb. Hemostas.* **13**, 146-151 (1987); of physicochemical properties: V. Gurewich, *ibid.* **14**, 110-115 (1988).

Recombinant human pro-urokinase, $C_{2031}H_{3145}N_{585}O_{601}S_{31}$, *saruplase.*

THERAP CAT: Thrombolytic.

7919. Proxazole. *N,N-Diethyl-3-(1-phenylpropyl)-1,2,4-oxadiazole-5-ethanamine; 5-[2-(diethylamino)ethyl]-3-(α-ethylbenzyl)-1,2,4-oxadiazole;* propaxoline; Aerbron. $C_{17}H_{25}N_3O$; mol wt 287.39. C 71.04%, H 8.77%, N 14.62%, O 5.57%. Prepn: **Brit.** pat. **924,608** and Palazzo, Silvestrini, U.S. pat. **3,141,019** (1963, 1964 to Angelini Francesco). Pharmacology: Silvestrini, Pozzatti, *Arzneimittel-Forsch.* **13**, 798 (1963). Separation and pharmacology of the enantiomers: De Feo *et al., Farmaco Ed. Sci.* **26**, 370 (1971).

bp$_{0.2}$ 132°.

Citrate, $C_{23}H_{33}N_3O_8$, *AF-634, Flou, Pirecin, Toness.* LD$_{50}$ i.p., orally in rats: 39, 60 mg/kg, Silvestrini, Pozzatti, *loc. cit.*

Nitrate, crystals, mp 127-128°.

THERAP CAT: Relaxant (smooth muscle); analgesic; antiinflammatory.

7920. Proxibarbal. *5-(2-Hydroxypropyl)-5-(2-propen-yl)-2,4,6(1H,3H,5H)-pyrimidinetrione; 5-allyl-5-(2-hydroxy-propyl)barbituric acid; 5-allyl-5-(β-hydroxypropyl)barbi-turic acid; 5-allyl-5-(β-hydroxypropyl)malonylurea; proxi-barbital; HH 184; Axeen; Centralgol; Centralgyl; Ipronal.* $C_{10}H_{14}N_2O_4$; mol wt 226.23. C 53.09%, H 6.24%, N 12.38%, O 28.29%. Prepn: Bobranski *et al., Roczniki Chem.* **30**, 175 (1956), *C.A.* **51**, 438f (1957); **Brit. pat.** 953,387 (1964 to Hommel A.G.), *C.A.* **61**, 3123a (1964); Smissman *et al., J. Med. Chem.* **14**, 853 (1971). Metabolism: Bobranski *et al., Arch. Immunol. Ther. Exp.* **10**, 895 (1962); B. Lambrey *et al., Eur. J. Med. Chem.-Chim. Ther.* **15**, 463 (1980). Pharmaco-kinetics: *eidem, ibid.* **12**, 565 (1977).

Crystals from benzene + ethanol, mp 157-158°; also re-ported as mp 166.5-168.5° from acetone + chloroform (Smissman). Moderately sol in water.
Note: This is a controlled substance (depressant) listed in the U.S. Code of Federal Regulations, Title 21 Part 1308.13 (1987).
THERAP CAT: Sedative, hypnotic.

7921. Proxyphylline. *3,7-Dihydro-7-(2-hydroxypropyl)-1,3-dimethyl-1H-purine-2,6-dione; 7-(2-hydroxypropyl)the-ophylline;* Brontyl; Proxy-Retardoral; Purophyllin; Spantin; Spasmolysin; Thean; Theon. $C_{10}H_{14}N_4O_3$; mol wt 238.24. C 50.42%, H 5.92%, N 23.52%, O 20.15%. Prepd by reflux-ing 1-chloro-2-propanol with theophylline in an alkaline medium: Rice, **U.S. pat.** 2,715,125 (1955 to Gane's Chem. Works).

Crystals from abs ethanol, mp 135-136°. One gram dis-solves in about 1 ml water, in 14 ml abs ethanol. More sol in boiling ethanol. pH of a 5% aq soln 5.5 to 7.0. Solns may be sterilized by heating.
THERAP CAT: Bronchodilator; vasodilator; smooth muscle relaxant.

7922. Prozapine. *1-(3,3-Diphenylpropyl)hexahydro-1H-azepine; 1-(3,3-diphenylpropyl)hexamethylenimine; 1,1-di-phenyl-3-hexamethyleneiminopropane; hexadiphane.* $C_{21}H_{27}N$; mol wt 293.43. C 85.95%, H 9.27%, N 4.77%. Prepn: Janssen, de Jongh, **U.S. pat.** 2,881,165 (1959 to N. V. Ne-derland. Comb. Chem. Ind.).

bp_1 170-174°.
Hydrochloride, $C_{21}H_{28}ClN$, *Norbiline.*
THERAP CAT: Antispasmodic, choleretic.

7923. Prunetin. *5-Hydroxy-3-(4-hydroxyphenyl)-7-methoxy-4H-1-benzopyran-4-one; 4',5-dihydroxy-7-meth-oxyisoflavone;* prunusetin. $C_{16}H_{12}O_5$; mol wt 284.26. C 67.60%, H 4.26%, O 28.14%. Isoln from *Prunus* spp., *Rosa-ceae:* Hasegawa, Shirato, *J. Am. Chem. Soc.* **79**, 450 (1957); Hasegawa, *ibid.* 1738; Goel, Seshadri, *Tetrahedron* **5**, 91 (1959); Plouvier, *Compt. Rend.* **250**, 594 (1960). Identity

with prunusetin: King, Jurd, *J. Chem. Soc.* **1952**, 3211. Structure: Shrimer, Hull, *J. Org. Chem.* **10**, 288 (1945). Synthesis: Bradbury, White, *J. Chem. Soc.* **1953**, 871.

Needles from ethanol, mp 240°.
Diacetate, $C_{20}H_{16}O_7$, rods from methanol, mp 222.5°.
4'-Glucoside, $C_{22}H_{22}O_{10}$, *prunitrin.* Isoln: Finnemore, *Pharm. J.* **31**, 604 (1910). Structure and synthesis: Zem-plén, Farkas, *Ber.* **90**, 836 (1957). Needles, mp 235-236°. Sol in hot water, ethyl acetate.

7924. Pseudoaconitine. *(1α,3α,6α,14α,16β)-20-Ethyl-1,6,16-trimethoxy-4-(methoxymethyl)aconitane-3,8,13,14-tetrol 8-acetate 14-(3,4-dimethoxybenzoate);* acraconitine; feraconitine; nepaline; veratroylaconine; "English" aconitine; "Nepal" aconitine. $C_{36}H_{51}NO_{12}$; mol wt 689.78. C 62.68%, H 7.45%, N 2.03%, O 27.83%. From tubers of *Aconitum ferox* Wall., *Ranunculaceae. Ref:* Wright, Luff, *J. Chem. Soc.* **33**, 151 (1878); Dunstan, Carr, *ibid.* **71**, 350 (1897); Henry, Sharp, *ibid.* **1928**, 1105, 3094; K. Paech, M. V. Tra-cey, *Modern Methods of Plant Analysis* **Vol. IV** (Springer-Verlag, Berlin, 1955) p 375. Structure: Gilman, Marion, *Tetrahedron Letters* **1962**, 923. Stereochemistry: Tsuda, Marion, *Can. J. Chem.* **41**, 1485 (1963).

White crystals, or amorphous, syrupy mass. *Very poison-ous!* mp 214°. $[α]_D^{20}$ +17° (alc); the salts are levorotatory. Practically insol in water. Sol in alcohol, ether.
Hydrobromide trihydrate, $C_{36}H_{52}BrNO_{12}\cdot 3H_2O$, triangular prisms, mp 199°. $[α]_D^{20}$ −18.5° (water).

7925. Pseudobaptigenin. *3-(1,3-Benzodioxol-5-yl)-7-hy-droxy-4H-1-benzopyran-4-one; 7-hydroxy-3',4'-(methyl-enedioxy)isoflavone.* $C_{16}H_{10}O_5$; mol wt 282.24. C 68.08%, H 3.57%, O 28.34%. The aglycon of pseudobaptisin. Prepn by hydrolysis of natural pseudobaptisin: Gorter, *Arch. Pharm.* **235**, 494 (1897). Structure: Späth, Schmidt, *Monatsh.* **53**, 454 (1929). Syntheses: Späth, Lederer, *Ber.* **63**, 743 (1930); Mahal *et al., J. Chem. Soc.* **1934**, 1771; Baker *et al., ibid.* **1937**, 805; **1953**, 1852; Farkas, *Ber.* **91**, 2858 (1958); Dhoub-hadel, Joshi, *J. Indian Chem. Soc.* **52**, 440 (1975).

Long, felted needles from alc, dec 296-298°. Sublimes in high vacuum at ~220°. Sparingly sol in the usual solvents.
7-Rhamnoglucoside trihydrate, $C_{28}H_{30}O_{14}\cdot 3H_2O$, *pseudo-baptisin.* Crystals, becomes anhydr at 120°. mp 148-150° (evac tube). If heating is continued, it resolidif at 180-210° and melts again at 249-251°. $[α]_D^{14}$ −98.1° (c = 1.1 in meth-anol). Freely sol in methanol, hot water, hot acetone.

7926. Pseudococaine. *3-(Benzoyloxy)-8-methyl-8-aza-bicyclo[3.2.1]octane-2-carboxylic acid methyl ester; 3β-hydr-oxy-1αH,5αH-tropane-2α-carboxylic acid methyl ester benzo-ate; 2α-carbomethoxy-3β-benzoxytropane; depsococaine;*

dextrocaine; isococaine; Delcaine. $C_{17}H_{21}NO_4$; mol wt 303.35. C 67.31%, H 6.98%, N 4.62%, O 21.10%. Cocaine diastereomer with greater local anesthetic activity than the natural substance. Synthesis: Einhorn, Marquardt, *Ber.* **23**, 468, 981 (1890); Willstätter, Bode, *Ann.* **326**, 42 (1903); Willstätter, Bommer, *ibid.* **422**, 34 (1921); Willstätter *et al.*, *ibid.* **434**, 138 (1923); **Brit. pat.** 210,050 (1923 to E. Merck). Configuration: S. P. Findlay, *J. Am. Chem. Soc.* **76**, 2855 (1954). Conformational analysis and ^1H, ^{13}C NMR studies: F. I. Carroll *et al.*, *J. Org. Chem.* **1982**, 13. Pharmacokinetics: A. L. Misra *et al.*, *Experientia* **32**, 895 (1976). Interaction with sodium channels: J C. Matthews, A. Collins, *Biochem. Pharm.* **32**, 455 (1983). Pharmacology of cocaine and pseudococaine: G. Schmidt *et al.*, *Arch. Exp. Pathol. Pharmakol.* **240**, 523 (1961).

Prisms, mp 47°. $[\alpha]_D^{20}$ +42° (c = 5 in chloroform). Freely sol in ether, chloroform, benzene, petr ether. Slightly sol in water.

Hydrochloride, $C_{17}H_{21}NO_4 \cdot HCl$, crystals from alcohol, mp 210°. $[\alpha]_D^{20}$ +41° (c = 5), less sol in water than cocaine hydrochloride.

Tartrate, $C_{21}H_{27}NO_{10}$, *Psicaine*. Crystals, mp 139°. $[\alpha]_D^{20}$ +43° (c = 5 in water). Sol in 4 parts water, in alcohol. The aq soln is stable and may be sterilized by boiling, without decompn. pH about 3.7 (2% soln).

Note: Psicaine N, pseudococaine sodium tartrate, is more sol in water than psicaine; esp useful where a neutral soln is desired. *Neopsicaine* is the hydrochloride of the *n*-propyl ester of *d*-benzoylpseudotropinecarboxylic acid.

THERAP CAT: Formerly as topical anesthetic.

7927. Pseudocodeine. *6,7-Didehydro-4,5-epoxy-3-methoxy-17-methylmorphinan-8-ol;* ψ-codeine; neoisocodeine. $C_{18}H_{21}NO_3$; mol wt 299.36. C 72.22%, H 7.07%, N 4.68%, O 16.03%. An isomer of codeine. Review and bibliography: K. W. Bentley, *The Chemistry of the Morphine Alkaloids* (Oxford, 1954).

White needles, mp 181-182°. $[\alpha]_D$ −96.6° (alc). Slightly sol in water; sol in alcohol.

7928. Pseudoconhydrine. *(3R-trans)-6-Propyl-3-piperidinol;* 5-hydroxy-2-propylpiperidine; 5-hydroxyconiine. $C_8H_{17}NO$; mol wt 143.22. C 67.09%, H 11.96%, N 9.78%, O 11.17%. In *Conium maculatum* L., *Umbelliferae* (hemlock). Extraction procedure: Ladenburg *et al.*, *Ber.* **24**, 1671 (1891); Braun, *Ber.* **38**, 3108 (1905); Löffler, *Ber.* **42**, 116 (1909). Structure: Späth *et al.*, *Ber.* **66**, 591 (1933). Stereochemistry: Hill, *J. Am. Chem. Soc.* **80**, 1611 (1958).

Hygroscopic needles from abs ether, mp 106° (monohydrate, scales, mp 60° from moist ether). bp 236°. $[\alpha]_D^{20}$ +11° (c = 10 in alc). K at 18° = 2 × 10⁻⁴. Soluble in water and most organic solvents.

Hydrochloride, $C_8H_{17}NO \cdot HCl$, crystals from alcohol, mp 213°. Freely sol in water, sparingly sol in alcohol, acetone.

7929. Pseudocumene. *1,2,4-Trimethylbenzene;* pseudocumol; asymmetrical trimethylbenzene. C_9H_{12}; mol wt 120.19. C 89.93%, H 10.06%. Occurs in coal tar and in many petroleums. Physical properties: Hirschler, Falconer, *J. Am. Chem. Soc.* **68**, 210 (1946). Toxicity: *Handbook of Toxicology* vol. 1, W. S. Spector, Ed. (Saunders, Philadelphia, 1956) pp 306-307.

Liquid. d_4^{20} 0.8761. bp 169-171°. mp −43.78° (Hirschler, Falconer). n_D^{21} 1.5044. Practically insol in water; sol in alc, benzene, ether. LD i.p. in rats: ~2.0 ml/kg (Spector).

USE: Sterilizing catgut by heating one hour at 160°; manuf dyes, perfumes and resins. Oxidation yields trimellitic anhydride. *Caution:* CNS depressant; respiratory irritant.

7930. Pseudohecogenin. *(3β,5α,25R)-3,26-Dihydroxyfurost-20(22)-en-12-one.* $C_{27}H_{42}O_4$; mol wt 430.61. C 75.30%, H 9.83%, O 14.86%. Prepn from hecogenin: Marker *et al.*, *J. Am. Chem. Soc.* **69**, 2167 (1947); Cameron *et al.*, *J. Chem. Soc.* **1955**, 2807; Wall, Serota, U.S. pat. **2,870,143** (1959 to USDA). Properties of free alcohol and diacetate: Cameron *et al.*, *loc. cit.*

Plates from aq acetone, mp 190-191°. $[\alpha]_D^{20}$ +103° (c = 1.5 in chloroform), +96° (c = 1 in dioxane). uv max (ethanol): 213 nm (ε 6400).

3,26-Diacetate, $C_{31}H_{46}O_6$, plates from methanol, mp 92.5-94.0°. $[\alpha]_D^{20}$ +74° (c = 1.8 in chloroform).

7931. Pseudoionone. *6,10-Dimethyl-3,5,9-undecatriene-2-one;* citrylideneacetone; 2,6-dimethylhendeca-2,6,8-trien-10-one. $C_{13}H_{20}O$; mol wt 192.29. C 81.20%, H 10.48%, O 8.32%. Intermediate in the synthesis of α- and β-ionone. Prepn from citral and acetone: Tiemann, Krüger, *Ber.* **26**, 2692 (1893); Stiehl, *J. Prakt. Chem.* [2] **58**, 84 (1898); Tiemann, *Ber.* **32**, 115 (1899); Hibbert, Cannon, *J. Am. Chem. Soc.* **46**, 119 (1924); A. Russel, R. L. Kenyon, *Org. Syn. coll.* vol. III, 747 (1955). A less pure product is obtained by the treatment of oil of lemon grass and acetone with bleaching powder, cobalt nitrate, and alcohol: Ziegler, *J. Prakt. Chem.* [2] **57**, 493 (1898); Tiemann, *Ber.* **31**, 2313 (1898); Haarmann, Reimer & Co., **Ger. pat.** 73,089; *Frdl.* **3**, 889. Synthesis: T. Onishi *et al.*, *Synthesis* **1980**, 651.

Pale yellow oil, bp_2 114-116°; bp_4 124-126°; bp_{12} 143-145°; d^{20} 0.8984; n_D^{20} 1.53346.

7932. Pseudojervine. *(3β,23β)-17,23-Epoxy-3-(β-D-glucopyranosyloxy)veratraman-11-one.* $C_{33}H_{49}NO_8$; mol wt 587.76. C 67.44%, H 8.40%, N 2.38%, O 21.78%. Glucoside of jervine. Isoln from *Veratrum album* L., *Liliaceae:* Wright, Luff, *J. Chem Soc.* **35**, 405 (1879); Poethke, *Arch.*

Pharm. **276,** 170 (1938); from *V. viride* Ait., *Liliaceae:* Jacobs, Craig, *J. Biol. Chem.* **155,** 565 (1944); from *V. eschscholtzii* Gray, *Liliaceae:* Klohs *et al., J. Am. Chem. Soc.* **75,** 2133 (1953).

Lustrous leaflets. mp 300-301° (dec). $[\alpha]_D^{25}$ −133° (c = 0.48 in 1:3 ethanol-chloroform). Sol in chloroform, benzene; slightly sol in alcohol. Almost insol in ether.

7933. Pseudomonic Acids. A group of antibacterial antibiotics produced by *Pseudomonas fluorescens* NCIB 10586 that have unusual structural features. Four members of the group are known: pseudomonic acid A, the major component, *pseudomonic acid B,* the 3,4,5-trihydroxy analog of A (also referred to as *pseudomonic acid I*), *pseudomonic acid D,* the 4-nonenoic acid analog of A; and *pseudomonic acid C,* in which the epoxide oxygen is replaced by a double bond. Isoln and characterization of A and B: A. T. Fuller *et al., Nature* **234,** 416 (1971); K. D. Barrow, G. Mellows, **Ger.** pat. **2,227,739;** *eidem,* **U.S.** pat. **3,977,943** (1973, 1976 both to Beecham). Structure of B: E. B. Chain, G. Mellows, *J. Chem. Soc. Perkin Trans. I* **1977,** 318. Prepn of C: N. H. Rogers, P. J. O'Hanlon, **Eur.** pat. **Appl. 3069;** *eidem,* **U.S.** pat **4,205,002** (1979, 1980 both to Beecham). Isoln, structure, configuration of C: J. P. Clayton *et al., Tetrahedron Letters* **21,** 881 (1980). Total syntheses of naturally occurring (+)-form of C: A. P. Kozikowski *et al., J. Am. Chem. Soc.* **102,** 6577 (1980); G. E. Keck *et al., J. Org. Chem.* **49,** 1462 (1984); J. C. Barrish *et al., ibid.* **52,** 1372 (1987); of (±)-A and (±)-C: B. B. Snider, G. B. Phillips, *J. Am. Chem. Soc.* **104,** 1113 (1982); B. B. Snider *et al., J. Org. Chem.* **48,** 3003 (1983). Prepn of D: P. J. O'Hanlon, **Eur.** pat. **Appl. 68,680** (1983 to Beecham), *C.A.* **98,** 159135t (1983); P. J. O'Hanlon *et al., J. Chem. Soc. Perkin Trans. I* **1983,** 2655. Antimycoplasmal activity in vitro: R. M. Banks *et al., J. Antibiot.* **41,** 609 (1988).

Pseudomonic Acid C

Isolated as a mixture of sodium salts. Can be stored at 0° for several months with no activity loss. Stable within pH 4-9 at 37° for 24 hrs. Hemolytic; inactivated by serum at conc > 50%.

Pseudomonic Acid A, See Mupirocin.

Pseudomonic Acid C, $C_{26}H_{44}O_8$, *[2S-[2α(E),3β,4β,5α(2E,-4S*,5R*)]]-9-[[3-methyl-1-oxo-4-[tetrahydro-3,4-dihydroxy-5-(5-hydroxy-4-methyl-2-hexenyl)-2H-pyran-2-yl]-2-butenyl]oxy]nonanoic acid.* $[\alpha]_D^{25}$ +7.64° (c = 0.78 in chloroform). uv max (ethanol): 222 nm (ε 14100).

Pseudomonic Acid D, $C_{26}H_{42}O_9$, *[2S-[2α[E(E)],3β,4β,5α-[2R*,3R*(1R*,2R*)]]]-9-[[3-methyl-1-oxo-4-[tetrahydro-3,4-dihydroxy-5-[[3-(2-hydroxy-1-methylpropyl)oxiranyl]methyl]-2H-pyran-2-yl]-2-butenyl]oxy]-4-nonenoic acid.* Oil. uv max (ethanol): 220 nm (ε 15499).

7934. Pseudomorphine. *7,7',8,8'-Tetradehydro-4,5:4',-5'-diepoxy-17,17'-dimethyl-[2,2'-bimorphinan]-3,3',6,6'-tetrol; 2,2'-bimorphine.* $C_{34}H_{36}N_2O_6$; mol wt 568.67. C 71.81%, H 6.38%, N 4.93%, O 16.88%. A dimolecular base formed by the gentle oxidation of morphine in alkaline soln. *Refs:* Broockmann, Polstorff, *Ber.* **13,** 88 (1880); Bentley, Dyke, *J. Chem. Soc.* **1959,** 2574. Review and bibliography: K. W. Bentley, *The Chemistry of the Morphine Alkaloids* (Oxford, 1954) p 54.

Trihydrate, crystalline powder; levorotatory in acid soln. Becomes anhydr at 150° and dec at about 327°. Insol in cold water, alcohol, ether, dil H_2SO_4. Slightly sol in cold, more sol in hot ammonia; sol in fixed alkali hydroxide solns, in pyridine, aniline, benzyl alcohol, guaiacol.

7935. Pseudopederin. *N-[[6-(2,3-Dimethoxypropyl)tetrahydro-3-hydroxy-5,5-dimethyl-2H-pyran-2-yl]methoxymethyl]tetrahydro-α,2-dihydroxy-5,6-dimethyl-4-methylene-2H-pyran-2-acetamide;* ψ-paederine; ψ-pederine; pseudopaederine; pseudopederine. $C_{24}H_{43}NO_9$; mol wt 489.59. C 58.87%, H 8.85%, N 2.86%, O 29.41%. Similar to Pederin, *q.v.* Isoln from blister beetles, *Paederus fuscipes:* Quilico *et al., Chem. Ind. (Milan)* **43,** 1434 (1961), *cf.* Cardani *et al., Tetrahedron Letters* **1965,** 2537; **Brit.** pat. **932,875** (1963 to Farmitalia). Structure: Cardani *et al., loc. cit.*

Crystals from benzene, mp 133°.

7936. Pseudopelletierine. *9-Methyl-9-azabicyclo[3.3.1]-nonan-3-one; 9-methyl-3-granatanone;* pseudopunicine. $C_9H_{15}NO$; mol wt 153.22. C 70.55%, H 9.87%, N 9.14%, O 10.44%. In root bark of *Punica granatum* L., *Punicaceae.* Extraction procedure: Hess, Eichel, *Ber.* **50,** 380, 1391, 1395 (1917); *ibid.* **52,** 1012 (1919). Synthesis under physiological conditions from calcium acetonedicarboxylate, glutardialdehyde, and methylamine: Menzies, Robinson, *J. Chem. Soc.* **1924,** 2163; *cf.* Schöpf, *Angew. Chem.* **50,** 779, 797 (1937). Alternate syntheses: Cope *et al., Org. Syn.* **37,** 73 (1957); Robinson, Hunt, *J. Pharm. Pharmacol.* **22,** 29S (1970); Bottini, Gal, *J. Org. Chem.* **36,** 1718 (1971). Conformation: Chen, LeFevre, *J. Chem. Soc. (B)* **1966,** 539.

Orthorhombic prisms from petr ether, mp 54°; bp 246°. Volatile. Strong base. One gram dissolves in about 2.5 ml water, 10 ml ether. Freely sol in alcohol and chloroform; sparingly sol in petr ether.
Dihydrate, plates from water.
Hydrochloride, $C_9H_{15}NO.HCl$, rhombohedra. One gram dissolves in about 1 ml water.
Sulfate tetrahydrate, $2C_9H_{15}NO.H_2SO_4.4H_2O$, crystals, sol in water.

7937. Pseudotropine. *exo-8-Methyl-8-azabicyclo[3.2.1]-octan-3-ol; 1αH,5αH-tropan-3β-ol; 3β-tropanol; ψ-tropine; 3-pseudotropanol.* $C_8H_{15}NO$; mol wt 141.21. C 68.04%, H 10.71%, N 9.92%, O 11.33%. Stereoisomeric with tropine. Prepn, structure, separation from tropine, and stereochemical configuration: See Tropine. Stereochemical synthesis: J. J. Tufariello *et al., J. Am. Chem. Soc.* **101,** 2435 (1979).

Orthorhombic, bipyramidal crystals from petr ether + benzene. mp 109°; bp 241°; pK at 15° = 3.80. K = 1.6 × 10^{-4}. pH of 0.05 M soln 11.5. Freely sol in water, alcohol, benzene.
Hydrochloride, $C_8H_{15}NO.HCl$, needles from alcohol, dec 282°. Soluble in water, hot alcohol.
Tropate, $C_8H_{15}NO.C_9H_8O_2$, cryst from benzene, mp 132°.

Consult the cross index before using this section.

7938. Pseudoyohimbine. *17α-Hydroxy-3β-yohimban-16α-carboxylic acid methyl ester.* $C_{21}H_{26}N_2O_3$; mol wt 354.46. C 71.16%, H 7.39%, N 7.90%, O 13.54%. Present in the bark of *Corynanthe johimbe* K. Schum., *Rubiaceae.* Obtained from the residues of commercial isoln procedures for the manuf of yohimbine: Karrer, Salomon, *Helv. Chim. Acta* **9**, 1059 (1926). Structure and stereochemistry: Janot *et al., Bull. Soc. Chim. France* **1952**, 1085; **1961**, 637. Synthesis: van Tamelen *et al., J. Am. Chem. Soc.* **80**, 5006 (1958); **91**, 7315 (1969); Stork, Guthikonda, *ibid.* **94**, 5109 (1972).

Rhombic platelets, mp 293° (corr, Maquenne block), mp 268° (open capillary), browns at 250°. Also reported as crystals from methanol, mp 252-256° (van Tamelen). $[\alpha]_D^{19}$ +27° (pyridine). uv max (methanol): 225, 281, 290 nm (log ε 4.54, 3.86, 3.80).

Hydrochloride, $C_{21}H_{26}N_2O_3 \cdot HCl$, needles from alcohol + ether, dec 258°. $[\alpha]_D^{20} - 10°$ (c = 1 in water).

7939. Psicofuranine. *9-β-D-Psicofuranosyl-9H-purin-6-amine; 9β-D-psicofuranosyladenine;* 6-amino-9-D-psicofuranosylpurine; angustmycin C; U-9586. $C_{11}H_{15}N_5O_5$; mol wt 297.27. C 44.44%, H 5.09%, N 23.56%, O 26.91%. Nucleoside antibiotic produced by *Streptomyces hygroscopicus* var. *decoyicus:* Eble *et al., Antibiot. & Chemother.* **9**, 419 (1959); Yüntsen, *J. Antibiot.* **11A**, 244 (1958); Eble, Lewis, U.S. pat. 3,020,274 (1962 to Upjohn). Antibacterial and antitumor activity: N. Tanaka *et al., J. Antibiot.* **14A**, 98 (1961). Structure: Schroeder, Hoeksema, *J. Am. Chem. Soc.* **81**, 1767 (1959); Garrett, *ibid.* **82**, 827 (1960). Synthesis: Farkas, Sorm, *Coll. Czech. Chem. Commun.* **28**, 882 (1963); L. A. Aleksandrova, F. W. Lichtenthaler, *Nucleic Acids Symp. Ser.* **9**, 263 (1981). Biosynthesis: Sugimori, Suhadolnik, *J. Am. Chem. Soc.* **87**, 1136 (1965). Inhibits the conversion of xanthosine-5'-phosphate to guanosine-5'-phosphate: Slechta, *Biochem. Pharmacol.* **5**, 96 (1960).

Crystals, dec 212-214°. $[\alpha]_D^{25} - 68°$ (dimethylformamide). uv max (0.01N acid): 259 nm ($E_{1cm}^{1\%}$ 508); in 0.01N base: 261 nm ($E_{1cm}^{1\%}$ 527). Soly at 25° (mg/ml): water 8; methanol 8; ethanol 6; butanol 2; ethyl acetate 0.23.

7940. D-Psicose. D-*ribo*-2-Ketohexose; D-ribohexulose; D-allulose; D-erythrohexulose; pseudofructose. $C_6H_{12}O_6$; mol wt 180.16. C 40.00%, H 6.71%, O 53.29%. Occurrence and identification as a non-fermentable substance in cane molasses: Zerban, Sattler, *Ind. Eng. Chem.* **34**, 1180 (1942). Structure: Ohle, Just, *Ber.* **68**, 601 (1935). Prepn starting with D-ribono-γ-lactone: Wolfrom *et al., J. Am. Chem. Soc.* **67**, 1793 (1945); from D-allose: Steiger, Reichstein, *Helv. Chim. Acta* **19**, 184 (1937); from D-glucose: Hough *et al., J. Chem. Soc.* **1953**, 2005.

Sweet syrupy liquid. $[\alpha]_D^{25}$ +4.7° (c = 4.3 in water; no detectable mutarotation). Soluble in water, methanol, ethanol; practically insol in acetone.

Phenylosazone, $C_{18}H_{22}N_4O_4$, yellow crystals from water, mp about 162-163° (dec); also reported mp 178°.

7941. Psilocin. *3-[2-(Dimethylamino)ethyl]-1H-indol-4-ol;* 4-hydroxy-*N,N*-dimethyltryptamine; psilocyn. $C_{12}H_{16}N_2O$; mol wt 204.27%. C 70.56%, H 7.90%, N 13.71%, O 7.83%. The minor hallucinogenic component of Teonanácatl, the sacred mushroom of Mexico. Isolated in trace amounts from the fruiting bodies of *Psilocybe mexicana* Heim, *Agaricaceae:* Hofmann *et al., Experientia* **14**, 107 (1958); Heim *et al., Helv. Chim. Acta* **42**, 1557 (1959). Prepn: Heim *et al.,* Ger. pat. **1,087,321** (1960 to Sandoz). Synthetic precursor of psilocybin: Hofmann, Troxler, U.S. pat. 3,075,992 (1963 to Sandoz). Psilocin, the 4-hydroxy analog of psilocybin, is formed by metabolic dephosphorylation of psilocybin and is the active species in the central nervous system: Horita, Weber, *Toxicol Appl. Pharmacol.* **4**, 730 (1962). Crystal structure: T. J. Petcher, H. P. Weber, *J. Chem. Soc. Perkin Trans. II* **1974**, 946. Review: Hofmann, *Bull. Narcotics* **23**, 3 (1971).

Plates from methanol, mp 173-176°. Amphoteric substance. Unstable in soln, esp. akaline soln. Very slightly sol in water. uv max: 222, 260, 267, 283, 293 nm (log ε 4.6, 3.7, 3.8, 3.7, 3.6).

Caution: This is a controlled substance (hallucinogen) listed in the U.S. Code of Federal Regulations, Title 21 Part 1308.11 (1985).

7942. Psilocybin. *3-[2-(Dimethylamino)ethyl]-1H-indol-4-ol dihydrogen phosphate ester;* O-phosphoryl-4-hydroxy-*N,N*-dimethyltryptamine; Indocybin. $C_{12}H_{17}N_2O_4P$; mol wt 284.27. C 50.70%, H 6.03%, N 9.86%, O 22.51%, P 10.90%. The major of two hallucinogenic components of Teonanácatl, the sacred mushroom of Mexico, the other component being psilocin, *q.v.* from the fruiting bodies of *Psilocybe mexicana* Heim, *Agaricaceae:* Hofmann *et al., Experientia* **14**, 107 (1958); Heim *et al., Helv. Chim. Acta* **42**, 1557 (1959); Heim *et al.,* Ger. pat. **1,087,321** (1960 to Sandoz). Structure: Hofmann *et al., Experientia* **14**, 397 (1958). Synthesis: Hofmann, Troxler, U.S. pat. 3,075,992 (1963 to Sandoz). Crystal structure: H. P. Weber, T. J. Petcher, *J. Chem. Soc.,Perkin Trans. II* **1974**, 942. Converted to psilocin *in vivo.* Toxicity data: E. Usdin, D. H. Efron, *Psychotropic Drugs and Related Compounds* (National Institute of Mental Health, Rockville, Md., 2nd ed., 1972) p 138. Reviews: Hofmann, *Proc. 1st Int. Congr. Neuro-Pharm.,* Rome **1958**, 446; Cerletti, *Deut. Med. Wochenschr.* **84**, 2317 (1959); Hofmann, *Bull. Narcotics* **23**, 3 (1971).

Crystals from boiling water, mp 220-228°; from boiling methanol, mp 185-195°. uv max (methanol): 220, 267, 290 nm (log ϵ 4.6, 3.8, 3.6). pH 5.2 in 50% aq ethanol. Sol in 20 parts boiling water, 120 parts boiling methanol; difficultly sol in ethanol. Practically insol in chloroform, benzene. LD_{50} in mice, rats, rabbits (mg/kg): 285, 280, 12.5 i.v. (Usdin, Efron).

Caution: This is a controlled substance (hallucinogen) listed in the U.S. Code of Federal Regulations, Title 21 Part 1308.11 (1985).

THERAP CAT: Psychomimetic.

7943. PSK®. KRESTIN. Protein-bound polysaccharide having immunostimulant and anticancer properties. Isolated from a basidiomycete, *Coriolus versicolor* (Fr.) Quel.: S. Otsuka *et al.*, **Japan.** pat. 73 08,489 (1973 to Kureha), *C.A.* **80**, 41025g (1974). Antitumor activity: S. Tsukagoshi *et al.*, *Prog. Chemother. (Antibacterial, Antiviral, Antineoplast.)*, *Proc. Int. Congr. Chemother.*, *8th, Athens 1973*, vol. **3**, G. K. Daikos, Ed. (Hellen. Soc. Chemother., Athens, 1974) pp 799-803. Efficacy vs mouse sarcoma: S. Tsukagoshi, F. Ohashi, *Gann* **65**, 557 (1974), *C.A.* **82**, 119037 (1975). Immunochemotherapy study: J. Akiyama *et al.*, *Cancer Res.* **37**, 3042 (1977). Comparative *in vitro* study: B. T. Nguyen, S. Stadtsbaeder, *Adv. Exp. Med. Biol.* **121A**, 255 (1979). Effects on immunological parameters and expt'l infections: P. Mayer, J. Drews, *Infection* **8**, 13 (1980). Acute toxicity: *Japan. Med. Gaz.* **13**(9), 11 (1976). *Review:* S. Tsukagoshi *et al.*, *Cancer Treat. Rev.* **11**, 131-155 (1984).

Brownish powder. Tasteless, but has a slight odor. Sol in water. Practically insol in methanol, pyridine, chloroform, benzene, hexane. pH of 1% soln is 6.6-7.2. LD_{50} in mice, rats (mg/kg): > 20,000 orally; > 5,000 i.p.; > 5000 s.c.; > 1300, > 600 i.v. (*Japan. Med. Gaz.*).

THERAP CAT: Antineoplastic.

7944. Psoralen. *7H-Furo[3,2-g][1]benzopyran-7-one;* 6-hydroxy-5-benzofuranacrylic acid δ-lactone; furo[3,2-g]-coumarin; ficusin. $C_{11}H_6O_3$; mol wt 186.16. C 70.97%, H 3.25%, O 25.78%. One of a group of furocoumarins occurring naturally in more than two dozen plant sources, including *Rutaceae* (*e.g.* bergamot, limes, cloves), *Umbelliferae* (*e.g.* celery, parsnips), *Leguminosae* (*e.g. Psoralen coryfolia*), and *Moraceae* (*e.g.* figs). Isoln: H. S. Jois *et al.*, *J. Indian Chem. Soc.* **10**, 41 (1933); A. Stoll *et al.*, *Helv. Chim. Acta* **33**, 1637 (1950); F. E. King *et al.*, *J. Chem. Soc.* **1954**, 1392. Synthesis: E. Späth *et al.*, *Ber.* **69**, 1087 (1936); R. C. Esse, B. E. Christensen, *J. Org. Chem.* **25**, 1565 (1960); O. Dann, D. Volz, *Arch. Pharm.* **308**, 121 (1975); V. K. Ahluwalia *et al.*, *Monatsch.* **111**, 877 (1980). Psoralens are *phytoalexins;* they are used by plants in a defensive response to attacks by fungi and insects: M. Berenbaum, P. Feeny, *Science* **212**, 927 (1981). They have also shown photosensitizing and phototoxic effects in animals and humans and have been used in photochemotherapy for management of vitiligo, psoriasis, and mycosis fungoides, *cf.* T. F. Anderson, J. J. Voorhees, *Ann. Rev. Pharmacol. Toxicol.* **20**, 235 (1980); A. Kornhauser *et al.*, *Science* **217**, 733 (1982). Review of psoralen photochemistry: B. J. Parsons, *Photochem. Photobiol.* **32**, 813-821 (1980). Review of genetic toxicity of psoralen and uv radiation in human cells: *Acta Derm. Venereol.* **Suppl. 104**, 4-40 (1982). *See* Methoxsalen, Trioxsalen, Bergapten for additional refs.

Crystals from ether, mp 163-164°; 169-179° (Späth). Absorption spectra: Wessely, Kaltan, *Monatsh.* **86**, 430 (1955).

USE: As photochemical probe in biological systems: P.-S. Song, C.-N. Ou, *Ann. N.Y. Acad. Sci.* **346**, 355 (1980).

7945. Psoralidin. *3,9-Dihydroxy-2-(3-methyl-2-buten-yl)-6H-benzofuro[3,2-c][1]benzopyran-6-one;* 6-(3-methyl-but-2-enyl)coumestrol. $C_{20}H_{16}O_5$; mol wt 336.33. C 71.42%, H 4.80%, O 23.79%. Extracted from the seed kernel of *Psoralea corylifolia* Linn., *Leguminosae.* Isoln: Chakravarti, *J. Sci. Ind. Res.* **7B**, 24 (1948). Structure: Duttagupta

et al., *Chem. & Ind. (London)* **1960**, 937. Synthesis: Khastgir *et al.*, *Tetrahedron* **14**, 275 (1961).

Crystals from acetone, mp 290-292°. Shows violet fluorescence in dil ethanol soln. Absorption spectra: Khastgir *et al.*, *loc. cit.*

Diacetate, $C_{24}H_{20}O_7$, crystals from ethyl acetate, mp 221-223°.

7946. Psychotrine. *1',15-Didehydro-7',10,11-trimeth-oxyemetan-6'-ol.* $C_{28}H_{36}N_2O_4$; mol wt 464.58. C 72.38%, H 7.81%, N 6.03%, O 13.77%. Minor alkaloid found in ipecac, the ground roots of *Uragoga ipecacuanha* (Brot.) Baill. [*Cephaelis ipecacuanha* (Brot.) A. Rich.], *Rubiaceae:* F. H. Carr, F. L. Pyman, *J. Chem. Soc.* **105**, 1591 (1914); O. Hesse, *Ann.* **405**, 1 (1914). Structure and stereochemistry: A. R. Battersby *et al.*, *J. Chem. Soc.* **1959**, 2704, 3512. Mass spectrum: H. Budzikiewicz *et al.*, *Tetrahedron* **20**, 399 (1964); ^{1}H,^{13}C-nmr of O-methylpsychotrine: T. Fujii *et al.*, *Chem. Pharm. Bull.* **30**, 598 (1982); ^{13}C-nmr of psychotrine: eidem, ibid. **31**, 2583 (1983). Proposed alternate structure with exocyclic double bond: E. E. van Tamelen *et al.*, *J. Am. Chem. Soc.* **79**, 4817 (1957); C. Schuij *et al.*, *J. Chem. Soc. Perkin Trans. I* **1979**, 970. *Review:* M. Janot in Manske, Holmes, *The Alkaloids* vol. **3** (Academic Press, New York, 1953) pp 363-394.

Tetrahydrate, yellow prisms with blue fluorescence from dil acetone or alcohol. *Very bitter taste, produces nausea instantly.* The anhydr material sinters at 120°, becomes transparent at 120-126° and melts completely at 128°. $[\alpha]_D^{15}$ +69.3° (c = 2 in alcohol, calcd as the tetrahydrate). Sparingly sol in water, benzene, petr ether, ether. More sol in alcohol, acetone, chloroform. uv max (0.1N HCl): 240, 288, 306, 356 nm (ϵ 13900, 5700, 6250, 6800).

Sulfate trihydrate, $C_{28}H_{38}N_2O_8S \cdot 3H_2O$, pale yellow scales from water, dec 214-217° when anhydr. $[\alpha]_D^{20}$ +39.2°.

Note: O-Methylpsychotrine has a methoxy group in place of the hydroxyl group. mp 123-124°. $[\alpha]_D^{20}$ +43.2° (alc). uv max (water) pH 1: 241.5, 288.5, 305, 354 (log ϵ 4.26, 3.86, 3.92, 3.91); pH 13: 226, 278.5, 307 (log ϵ 4.43, 3.96, 3.77).

7947. Pteridine. Pyrazino[2,3-d]pyrimidine; pyrimido-[4,5-b]pyrazine; pyrimidine-4',5':2,3-pyrazine; 1,3,5,8-tet-raazanaphthalene; azinepurine; benzotetrazine. $C_6H_4N_4$; mol wt 132.12. C 54.54%, H 3.05%, N 42.41%. Prepn from 4,5-diaminopyrimidine and polyglyoxal or glyoxal sodium bisulfite: Albert *et al.*, *J. Chem. Soc.* **1951**, 474.

Yellow crystals from benzene, mp 138-138.5°. uv max (water): 298, 309 nm (log ϵ 3.87, 3.83). Sol in 7.2 parts water at 20-25°.

Caution: Related to folic acid antagonists.

7948. Pterocarpin. *(6aR-cis)-6a,12a-Dihydro-3-meth-oxy-6H-[1,3]dioxolo[5,6]benzofuro[3,2-c][1]benzopyran.* $C_{17}H_{14}O_5$; mol wt 298.28. C 68.45%, H 4.73%, O 26.82%.

From red sandalwood (*Pterocarpus santalinus* L.f., *Leguminosae*): Cazeneuve, Hugouneng, *Compt. Rend.* **104**, 1722 (1887); **107**, 737 (1888); Leonhardt, Fay, *Arch. Pharm.* **273**, 53 (1935); McGookin *et al.*, *J. Chem. Soc.* **1940**, 787; from *Pterocarpus* spp: King *et al.*, *ibid.* **1953**, 3693. Structure: Bredenberg, Shoolery, *Tetrahedron Letters* **1961**, 285. Synthesis of the *dl*-form: Fukui, Nakayama, *ibid.* **1966**, 1805; *eidem*, *Bull. Chem. Soc. Japan* **42**, 1408 (1969); Uchiyama, Matsui, *Agr. Biol. Chem.* **31**, 1490 (1967). Stereochemistry: Ito *et al.*, *Chem. Commun.* **1965**, 595; Pachler, Underwood, *Tetrahedron* **23**, 1817 (1967).

Crystalline plates from petr ether or from ethanol, mp 164.5°. $[\alpha]_D$ −214.5° and $[\alpha]_{5461}^{20.5}$ −207.5° (c = 0.53 in chloroform). Practically insol in water, cold alcohol or ether. Sol in hot alcohol, chloroform.

7949. Pteroic Acid. 4-[[(2-Amino-1,4-dihydro-4-oxo-6-pteridinyl)methyl]amino]benzoic acid; p-[(2-amino-4-hydroxy-6-pteridylmethyl)amino]benzoic acid. $C_{14}H_{12}N_6O_3$; mol wt 312.28. C 53.84%, H 3.87%, N 26.91%, O 15.37%. Intermediate in the synthesis of folic acids. Prepn: Waller *et al.*, *J. Am. Chem. Soc.* **70**, 19 (1948); *cf.* Hultquist *et al.*, *ibid.* 23; Boothe, U.S. pat. **2,472,462** (1949 to Am. Cyanamid); C. Temple *et al.*, *J. Org. Chem.* **46**, 3666 (1981). Synthesis: M. G. Nair *et al.*, *ibid.* 3152.

Crystals from dil HCl. Sol in aq NaOH solns.

7950. Pteropterin. N-[N-[N-[4-[[(2-Amino-1,4-dihydro-4-oxo-6-pteridinyl)methyl]amino]benzoyl]-L-γ-glutamyl]-L-glutamyl]-L-glutamic acid; N-[N-(N-pteroyl-γ-glutamyl)-γ-glutamyl]glutamic acid; fermentation L. casei Factor; pterolyl-γ-glutamyl-γ-glutamylglutamic acid; pteroyldi-γ-glutamylglutamic acid; pteroyltriglutamic acid; Teropterin; PTGA. $C_{29}H_{33}N_9O_{12}$; mol wt 699.64. C 49.78%, H 4.75%, N 18.02%, O 27.44%. Isoln from aerobic culture of *Corynebacterium*: Hutchings *et al.*, *J. Am. Chem. Soc.* **70**, 1 (1948). Identification, structure and synthesis: Boothe *et al.*, *ibid.* 1099; *ibid.* **71**, 2304 (1949). Prepn: Cosulich, U.S. pat. **2,563,707** (1951 to Am. Cyanamid).

Monohydrate, crystals from water adjusted to pH 2.8 with HCl and contg some NaCl. Has the general properties of a polypeptide. Absorption spectrum, Hutchings, *loc. cit.* Soly in water at 5°: 0.10 mg/ml; at 80°: 3.00 mg/ml. Sol in NaOH solns.

Methyl ester, crystals from methanol contg NaCl. Soly in water at 5°: 0.12 mg/ml; at 80°: 1.00 mg/ml; in methanol at −5°: 0.30 mg/ml; at 60°: 5.00 mg/ml.

Forms a crystalline barium salt.

THERAP CAT: Formerly as antineoplastic.

7951. Pteroylhexaglutamylglutamic Acid. Pteroylheptaglutamic acid; vitamin Bc conjugate; Bc conjugate; PHGA. $C_{49}H_{61}N_{13}O_{24}$; mol wt 1216.13. C 48.39%, H 5.06%, N 14.97%, O 31.58%. One of the naturally occurring polyglutamates of folic acid. Isoln from yeast: Pfiffner *et al.*, *Science* **102**, 228 (1945); *J. Am. Chem. Soc.* **68**, 1392 (1946). Solid phase synthesis: C. L. Krumdieck, C. M.

Baugh, *Biochemistry* **8**, 1568 (1969); *eidem*, *Methods in Enzymology* **66**, 523 (1980). Chain length characterization of pteroylpolyglutamates: D. G. Priest *et al.*, *Anal. Biochem.* **115**, 163 (1981).

Rosettes of minute needles from 5% NaCl soln. Has no definite mp. When heated on the hot stage it begins to darken from about 200° and partially melts at 230-260°, and remains partially melted up to 360°. The ultraviolet absorption curves are practically identical with those of folic acid. Sparingly sol in water contg NaCl. Sol in dil NaOH soln.

7952. Puberulic Acid. 3,4,6-Trihydroxy-5-oxo-1,3,6-cycloheptatriene-1-carboxylic acid. $C_8H_6O_6$; mol wt 198.13. C 48.49%, H 3.05%, O 48.45%. Antibiotic substance produced by *Penicillium puberulum*, *P. aurantiovirens*, *P. johannioli*, *P. cyclopium viridicatum*. Isoln: Birkinshaw, Raistrick, *Biochem. J.* **26**, 441 (1932); Barger, Dorrer, *ibid.* **28**, 11 (1934). Structure: McGowan, *Chem. & Ind. (London)* **1947**, 205; Corbett *et al.*, *ibid.* **1949**, 626; *J. Chem. Soc.* **1950**, 6. Synthesis: Johns *et al.*, *ibid.* **1954**, 198. Biosynthesis: Ferretti, Richards, *Proc. Nat. Acad. Sci. USA* **46**, 1438 (1960). Review: Nozoe, *Fortschr. Chem. Org. Naturst.* **13**, 232-301 (1956).

Cream-colored powder from water, mp 316-318°. Sublimes in high vacuum. Dibasic acid. uv max (water): 270, 350 nm (log ε 4.55, 3.86). Sol in hot water. Forms a sodium salt which is freely sol in water.

7953. Puberulonic Acid. 5,6,7-Trihydroxy-1H-cyclohepta[c]furan-1,3,4-trione; 4,5,6-trihydroxy-7-oxo-1,3,5-cycloheptatriene-1,2-dicarboxylic anhydride. $C_9H_4O_7$; mol wt 224.12. C 48.23%, H 1.80%, O 49.97%. Antibiotic substance produced by *Penicillium puberulum*, *P. aurantiovirens*, *P. johannioli*, *P. cyclopium viridicatum*. Isoln: Birkinshaw, Raistrick, *Biochem. J.* **26**, 441 (1932); Barger, Dorrer, *ibid.* **28**, 11 (1934). Structure: McGowan, *Chem. & Ind. (London)* **1947**, 205; Aulin-Erdtman, *Acta Chem. Scand.* **4**, 1325 (1950). Johnson *et al.*, *J. Chem. Soc.* **1951**, 1139. Synthesis: Nozoe *et al.*, *Bull. Chem. Soc. Japan* **33**, 1071 (1960). Biosynthesis: Ferritti, Richards, *Proc. Nat. Acad. Sci. USA* **46**, 1438 (1960); A. I. Scott, E. Lee, *Chem. Commun.* **1972**, 655. Review: Nozoe, *Fortschr. Chem. Org. Naturst.* **13**, 232-301 (1956).

Yellow plates, dec 296°. Sol in water. The yellow color of an aq soln changes to pink on neutralization and disappears altogether when the soln is made alkaline.

Sodium salt, crystals from ethanol, freely sol in water.

7954. Pukateine. (R)-6,7,7a,8-Tetrahydro-7-methyl-5H-benzo[g]-1,3-benzodioxolo[6,5,4-de]quinolin-12-ol; 1,2-(methylenedioxy)aporphin-11-ol; 1,2-methylenedioxy-11-hydroxyaporphine. $C_{18}H_{17}NO_3$; mol wt 295.32. C 73.20%, H 5.80%, N 4.74%, O 16.25%. From bark of *Laurelia novae-zelandiae* A. Cunn, *Lauraceae*. Isoln: Aston, *J. Chem. Soc.* **97**, 1381 (1910); Bernauer, *Helv. Chim. Acta* **50**, 1583 (1967). Structure: Barger, Giradet, *ibid.* **14**, 481, 504 (1931); Barger, Schlittler, *ibid.* **15**, 381 (1932). Total synthesis: Zymalkowski, Happel, *Tetrahedron Letters* **1969**, 219; *Ber.* **102**, 2959 (1969); Kametani *et al.*, *J. Chem. Soc. Perkin Trans. 1* **1972**, 1435.

dl-Form, crystals from absolute alcohol, mp 232-233°.
l-Form, crystals from alcohol, mp 208-212°. $[\alpha]_D^{25}$ —240° (c = 0.097 in alcohol). Practically insol in water. Sol in alc, ether, chloroform, pyridine; slightly sol in petr ether.

7955. Pulegone. *5-Methyl-2-(1-methylethylidene)cyclohexanone; R-(+)-p-menth-4(8)-en-3-one;* 1-methyl-4-isopropylidene-3-cyclohexanone. $C_{10}H_{16}O$; mol wt 152.23. C 78.89%, H 10.59%, O 10.51%. Found in oils derived from plants of the *Labiatae* family. Readily isolated in quantity from the pennyroyal oils from *Mentha pulegium* L., *M. longifolia* (L.) Huds., and *Hedeoma pulegioides* (L.) Pers., *Labiatae:* Gildemeister, Hoffmann, *Die ätherischen Ole* **Vol. I,** p 560 (1928); Simonsen, *The Terpenes* **Vol. I** (2nd ed, 1947) p 370. Synthesis: Kuhn, Schinz, *Helv. Chim. Acta* **36,** 161 (1953). Synthesis of the (±)-form: Black *et al., J. Chem. Soc.* **1956,** 2971; Wolinsky *et al., J. Org. Chem.* **30,** 3207 (1965); of the S(—)-form: E. J. Corey *et al., ibid.* **41,** 380 (1976). Improved synthesis of the R(+)-form: T. Sato *et al., Tetrahedron Letters* **1980,** 3377. Conversion of the R(+) to the S(—)-form: H. E. Ensley, R. V. C. Carr, *ibid.* **1977,** 513. Biosynthetic study: A. Akhila, D. V. Banthorpe, *Z. Pflanzenphysiol.* **99,** 277 (1980), *C.A.* **94,** 2073r (1981).

Oil. Pleasant odor, midway between peppermint and camphor. d_4^{15} 0.9346. bp_{760} 224°; bp_{100} 151-153°; bp_{17} 103°; bp_6 84°. $[\alpha]_D^{20}$ +21°; $[\alpha]_{546}^{20}$ +28.2°. n_D^{20} 1.4894. Practically insol in water. Miscible with alcohol, ether, chloroform.
(±)-Form. Liquid. bp_7 78-80°. n_D^{16} 1.4856. uv max (alc): 253.3 nm (log ε 3.86).
(—)-Form. Liquid. bp_{20} 104-108°. $[\alpha]_D^{23}$ —22.5° (neat).

7956. Pulsatilla. Pasque flower; wind flower; meadow anemone; Easter flower. Dried herb of *Anemone pulsatilla* L. (*Pulsatilla vulgaris* Mill.), *A. pratensis* L. (*P. pratensis* (L.) Mill.), or *A. patens* L. (*P. patens* (L.) Mill.), *Ranunculaceae. Habit.* Europe, Asia. *Constit.* Anemone camphor, volatile oil, tannin.

7957. Pumice. Light, hard, rough, porous, gray masses or a gritty, gray-colored powder of volcanic origin. Found chiefly in the Lipari islands and in the Greek archipelagos. Consists mainly of complex silicates of Al, K, and Na. It is insol in water and is not attacked by acids.
USE: Abrasive in metal polishes; in fireproofing and insulating compds; also as a carrier for metal catalysts. In pharmacy as filtering medium and dispersant. In cosmetics for removing rough skin.

7958. Pumpkin Seed. Pepo. Dried ripe seeds of cultivated varities of *Cucurbita pepo* L., *Cucurbitaceae. Habit.* Southern Asia, Europe, America. *Constit.* Fixed oil, acrid resin, myosin, vitellin, sugar.
THERAP CAT: Anthelmintic.

7959. 1H-Purine. 7H-Imidazo[4,5-d]pyrimidine. C_5-H_4N_4; mol wt 120.11. C 50.00%, H 3.36%, N 46.65%. Prepd by heating 4,5-diaminopyrimidine with anhydr formic acid in a current of carbon dioxide: Isay, *Ber.* **39,** 251

(1906); by boiling 2,6-diiodopurine with zinc dust and water under carbon dioxide: Fischer, *ibid.* **31,** 2551, 2564 (1898).

Needles from toluene or alcohol, mp 216-217° (partial sublimation on rapid heating). Freely sol in water, hot alcohol. Slightly sol in hot ethyl acetate, acetone. Practically insol in ether, chloroform. Aq solns are neutral to litmus. Forms salts with acids and bases.

7960. Puromycin. *3'-[[2-Amino-3-(4-methoxyphenyl)-1-oxopropyl]amino]-3'-deoxy-N,N-dimethyladenosine; 3'-(α-amino-p-methoxyhydrocinnamamido)-3'-deoxy-N,N-dimethyladenosine;* 6-dimethylamino-9-[3-deoxy-3-(p-methoxy-L-phenylalanylamino)-β,D-ribofuranosyl]-β-purine; 6-dimethylamino-9-[3'-(p-methoxy-L-phenylalanylamino)-3'-deoxy-β,D-ribofuranosyl]purine; Cl 13900; P-638; 3123 L; Stylomycin. $C_{22}H_{29}N_7O_5$; mol wt 471.51. C 56.04%, H 6.20%, N 20.79%, O 16.97%. Antibiotic substance produced by the soil actinomycete *Streptomyces alboniger.* Isoln: Porter *et al., Antibiot. & Chemother.* **2,** 409 (1952); *eidem,* U.S. pat. **2,763,642** and Szumski, Goodman, U.S. pat. **2,797,187** (1956, 1957 both to Am. Cyanamid). Structure: Waller *et al., J. Am. Chem. Soc.* **75,** 2025 (1953); Fryth, Waller, *ibid.* **80,** 2736 (1958). Synthesis: Baker *et al., ibid.* **76,** 4044 (1954); **77,** 12 (1955). Conformation: Jardetzky, *J. Am. Chem. Soc.* **85,** 1823 (1963). Puromycin is known to interfere with protein formation by interfering with the function of RNA in the cells involved. *Review:* Nathans in *Antibiotics* vol. I, D. Gottlieb, P. D. Shaw, Eds. (Springer-Verlag, New York, 1967) pp 259-277.

Crystals, mp 175.5-177°. $[\alpha]_D^{25}$ —11° (ethanol). uv max (0.1N NaOH): 275 nm (ε 20,300); in 0.1N HCl: 267.5 nm (ε 19,500). It is a diacidic base and readily forms a water-sol dihydrochloride or monosulfate. LD_{50} in mice: 350 mg/kg i.v.; 525 mg/kg i.p.; 675 mg/kg orally, Porter *et al., loc. cit.* (1952).
Note: Puromycin was once referred to by the trademark "Achromycin" which has since 1953 been reassigned to tetracycline hydrochloride.
THERAP CAT: Antineoplastic; antiprotozoal (Trypanosoma).

7961. Purothionin. A low mol wt wheat protein composed of α_1-, α_2- and β-purothionins. Contains about 20% cystine, 10% arginine and 10% lysine. Isoln from unbleached patent flour: A. K. Balls, W. S. Hale, *Cereal Chem.* **17,** 243 (1940); A. K. Balls *et al., ibid.* **19,** 279 (1942). Inhibits the growth of bacteria and yeasts: L. S. Stuart, T. H. Harris, *ibid.* 288. Toxic effects in animals: E. J. Coulson *et al., ibid.* 301. Sepn and characterization of α- and β-purothionin: D. G. Redman, N. Fisher, *J. Sci. Food Agr.* **19,** 651 (1968); C. C. Nimmo *et al., ibid.* **25,** 607 (1974). Amino acid sequences of α_1- and β-purothionin: S. Ohtani *et al., Agr. Biol. Chem.* **39,** 2269 (1975); *eidem, Z. Biochem.* **82,** 753 (1977); of β-purothionin: A. S. Mak, B. L. Jones, *Can. J. Chem.* **54,** 835 (1976); of α_1 and α_2-purothionin: B. L. Jones, A. S. Mak, *Cereal Chem.* **54,** 511 (1977). *Review:* D.

D. Kasarda *et al.*, "Wheat Proteins" in *Advances in Cereal Science and Technology* vol. 1 (Am. Cereal Chem., St. Paul, 1976) pp 208-210.

Forms a cryst hydrochloride. Freely sol in water. Easily digested by proteolytic enzymes including cryst chymotrypsin, chymopapain, papain.

α_1-*Purothionin*, $C_{198}H_{340}N_{68}O_{56}S_8$, *5-L-argine-27-glycine-33-L-isoleucine-34-L-serine-42-glycine-purothionin A I (reduced), purothionin A II.*

α_2-*Purothionin*, $C_{202}H_{348}N_{68}O_{59}S_8$, *5-L-arginine-6-L-threonine-18-L-serine-26-L-serine-27-L-threonine-42-glycine-purothionin A I (reduced).*

β-*Purothionin*, $C_{203}H_{339}N_{67}O_{59}S_8$, *purothionin A I.*

7962. Purpurin. *1,2,4-Trihydroxy-9,10-anthracenedione; 1,2,4-trihydroxyanthraquinone;* C.I. Natural Red 8; C.I. Natural Red 16; C.I. 58205; C.I. 75410. $C_{14}H_8O_5$; mol wt 256.20. C 65.63%, H 3.15%, O 31.22%. Occurs as glycoside in the madder root (*Rubia tinctorum* L., *Rubiaceae*) of commerce. Is formed during storage; no appreciable amount in the fresh root: Hill, Richter, *J. Chem. Soc.* **1936**, 1714. Although a dye itself, it is usually considered as an undesirable contaminant of alizarin extracted from madder. May be prepd from alizarin by oxidation with ammonium persulfate: Wacker, *J. Prakt. Chem.* [2] **54**, 90 (1896); also by Friedel-Crafts condensation of hydroxyhydroquinone with phthalic anhydride: Dimroth, Fick, *Ann.* **411**, 321 (1916).

Long orange needles with 1 H_2O from dil alcohol, anhydr at 100°. Anhydr red needles from abs alcohol or by sublimation around 150° in high vacuum (less than 2 mm Hg). mp 257°. Absorption spectrum: Meek, *J. Chem. Soc.* **111**, 969 (1917); Ezaby, *ibid.* (B) **1970**, 1293. More sol in boiling water than alizarin (yellow color with yellowish hue). Freely sol in alcohol (red), in ether (intensely yellow with fluorescence). Soluble in benzene, toluene, xylene (dark yellow), in boiling alum soln (red).

2-Methyl ether, $C_{15}H_{10}O_5$, orange crystals from benzene, mp 240°.

2,4-Dimethyl ether, $C_{16}H_{12}O_5$, orange needles, mp 186-189°.

USE: Forms colored "lakes" with various metal salts and is a fast dye for cotton printing. Now used mostly in the manuf of acid and chrome dyes. Reagent for the detection of boron; for detection of insol calcium salts in the cell contents of histological material and as a nuclear stain.

7963. Purpurogallin. *2,3,4,6-Tetrahydroxy-5H-benzocyclohepten-5-one.* $C_{11}H_8O_5$; mol wt 220.17. C 60.00%, H 3.66%, O 36.33%. The aglycone of several glycosides from various nutgalls. Prepn by oxidation of pyrogallol: Perkin, Steven, *J. Chem. Soc.* **83**, 192 (1903); Perkin, *ibid.* **101**, 803 (1912); Nierenstein, Spiers, *Ber.* **46**, 3151 (1913); Evans, Dehn, *J. Am. Chem. Soc.* **52**, 3647 (1930). Structure: Willstätter, Heiss, *Ann.* **433**, 17 (1923); Barltrop, Nicholson, *J. Chem. Soc.* **1948**, 116. Synthesis: Caunt *et al.*, *ibid.* **1950**, 1631; **1951**, 1313.

Deep red needles from glacial acetic acid, dec 274-275°. Sparingly sol in most solvents.

Diglucoside, $C_{23}H_{28}O_{15}$, *dryophantin.* Coloring matter of "red pea" gall produced by *Dryophanta divisa* Ald. on *Quercus pedunculata* Ehrh., *Fagaceae*: Nierenstein, *J. Chem. Soc.* **115**, 1328 (1919); Nierenstein, Swanton, *Biochem. J.* **38**, 373 (1944). Dark red needles with bronze luster, mp 220-221°.

Slightly sol in water, cold alcohol; sol in boiling alcohol; in methanol, acetone.

USE: As an additive to edible or inedible fats or oils, hydrocarbon fuels or lubricants, retards oxidation or metal contamination: Thompson, U.S. pat. **2,770,545** (1956 to Universal Oil Prod.).

7964. Putrescine. *1,4-Butanediamine;* tetramethylenediamine. $C_4H_{12}N_2$; mol wt 88.15. C 54.50%, H 13.73%, N 31.77%. $NH_2(CH_2)_4NH_2$. Biogenic polyamine and precursor of spermidine, *q.v.*, initially detected in decaying animal tissues, but now known to be present in all cells and certain bacterial cultures. It is essential for both normal and neoplastic tissue growth. Formed via decarboxylation of ornithine or by decarboxylation of arginine, followed by hydrolysis. Prepn: A. Ladenburg, *Ber.* **19**, 780 (1886); G. Ciamician, C. U. Zanetti, *ibid.* **22**, 1970 (1889); R. Willstätter, W. Heubner, *ibid.* **40**, 3871 (1907); of the dihydrochloride: *Org. Syn. coll. vol. IV* (1963) p 819. Role in cell growth processes: C. W. Tabor, H. Tabor, *Ann. Rev. Biochem.* **45**, 285 (1976); J. Janne *et al.*, *Biochim. Biophys. Acta* **473**, 241 (1978). Formation and interconversion of putrescine and spermidine in mammalian cells: A. E. Pegg *et al.*, *Advan. Enzyme Regul.* **19**, 427 (1980). Biosynthetic study in fungi: L. Stevens, *Med. Biol.* **59**, 308 (1981). Regulation of tRNA methyl transferase activity: M. Mach *et al.*, *Biochem. J.* **202**, 153 (1982). Alteration of DNA conformation in rat brain tumor cells by depletion of intracellular putrescine: D. T. Hung *et al.*, *Science* **221**, 368 (1983). Use of labeled putrescine as a positron-emission tomographic tracer in brain tumors: N. Volkow *et al.*, *ibid.* 673. Reviews of early literature: M. Guggenheim, *Die biogenen Amine* (S. Karger, Basel, 1951, 4th ed.) 619 pp; H. Tabor *et al.*, *Ann. Rev. Biochem.* **30**, 579-604 (1961). Review of formation of GABA, the major inhibitory neurotransmitter in vertebrate brains, from putrescine: N. Seiler, *Physiol. Chem. Phys.* **12**, 411-429 (1980). Review of metabolism: T. L. Sourkes, K. Missala, *Agents Actions* **11**, 20-27 (1981). Book: *Polyamines in Biology and Medicine*, D. R. Morris, L. J. Marton, Eds. (Dekker, New York, 1981) 512 pp.

Colorless oil, bp 158-160°. Cryst on cooling, mp 23-24°. Strong piperidine-like odor. Very sol in water.

Dihydrochloride, $C_4H_{14}Cl_2N_2$, cryst from 85% alc, melts above 275°.

USE: As a tool in biochemical research.

7965. Pyocyanine. *1-Hydroxy-5-methylphenazinium hydroxide inner salt.* $C_{13}H_{10}N_2O$; mol wt 210.23. C 74.27%, H 4.80%, N 13.32%, O 7.61%. Produced by *Pseudomonas aeruginosa* (*Ps. pyocyanea, Bacillus pyocyaneus*). Isoln from broth cultures: Wrede, Strack, *Z. Physiol. Chem.* **140**, 1 (1924). Isoln from agar cultures: Schoental, *Brit. J. Exp. Pathol.* **22**, 137 (1941). Synthesis by dimethyl sulfate methylation of α-hydroxyphenazine and alkaline treatment of the resulting methosulfate: Wrede, Strack, *Z. Physiol. Chem.* **181**, 58 (1929); *Ber.* **62**, 2051 (1929); Surrey, *Org. Syn.* **26**, 86 (1946); **coll. vol. III**, 753 (1955).

Dark blue needles from water (usually with 1 H_2O which is lost at 50° over P_2O_5 *in vacuo*). mp 133°. Upon further heating it sublimes with decompn. Absorption spectrum: Nitzsche, *Ber.* **77**, 337 (1944). Freely sol in chloroform. Sol in nitrobenzene, pyridine, phenol, acetic acid, hot water, hot alcohol. Slightly sol in cold water and benzene. The blue water soln, made alkaline with Na_2CO_3, is easily rendered colorless by reduction with glucose or sodium hydrosulfite. Acidic $KMnO_4$ solns are decolorized by pyocyanine. An alkaline water soln of pyocyanine acquires a maroon color upon heating.

Picrate, crystals from alcohol, dec 190°, dec in water.

Aurichloride, dec 185°.

7966. Pyracarbolid. *3,4-Dihydro-6-methyl-N-phenyl-2H-pyran-5-carboxamide; 3,4-dihydro-6-methyl-2H-pyran-5-carboxanilide;* 5,6-dihydro-2-methyl-3-(phenylcarbamoyl)-4*H*-pyran; Hoe 13764; Sicarol. $C_{13}H_{15}NO_2$; mol wt 217.27. C 71.86%, H 6.96%, N 6.45%, O 14.73%. Prepn: **Fr.** pat. **2,000,552**; O. Scherer *et al.,* **U.S.** pat. **3,632,821** (1969, 1972 both to Hoechst). Activity: B. Jank, F. Grossman, *Pestic. Sci.* **2**, 43 (1971).

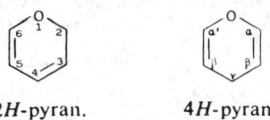

Colorless solid, mp 110-111°. Vapor press: 7.66×10^{-3} mm Hg. Soly in water at 40°: 0.6 g/l. Sol in most organic solvents; slightly sol in benzene, xylene, cyclohexane.
USE: Agricultural fungicide.

7967. Pyran. This information given for nomenclature purposes only.

2*H*-pyran.   4*H*-pyran.

2H-Pyran. Synonyms: *1,2-pyran, α-pyran.*
4H-Pyran. Synonyms: *1,4-pyran, γ-pyran.*

7968. Pyrantel. *(E)-1,4,5,6-Tetrahydro-1-methyl-2-[2-(2-thienyl)ethenyl]pyrimidine.* $C_{11}H_{14}N_2S$; mol wt 206.32. C 64.04%, H 6.84%, N 13.58%, S 15.54%. Prepn: **Belg.** pat. **658,987** (1965 to Pfizer), *C.A.* **64**, 8192c (1966); Austin *et al., Nature* **212**, 1273 (1966); Kasubrick, McFarland, **S. Afr.** pat. **68 00,516**; *eidem,* **U.S.** pat. **3,502,661** (1968, 1970 both to Pfizer). Structure-activity studies: McFarland *et al., J. Med. Chem.* **12**, 1066 (1969). Pharmacology and animal trials: Cornwell, *Vet. Rec.* **79**, 590 (1966); Howes, Lynch, *J. Parasitol.* **53**, 1085 (1967); Eyre, *J. Pharm. Pharmacol.* **22**, 26 (1970).

Crystals from methanol, mp 178-179°.
Tartrate, $C_{15}H_{20}N_2O_6S$, *CP-10423-18, Strongid, Banminth.* White crystals from hot methanol, mp 148-150°. uv max (water): 312 nm (log ε 4.27).
Pamoate, $C_{34}H_{30}N_2O_6S$, *CP-10423-16, pyrantel embonate, Antiminth, Cobantril, Combantrin, Early Bird, Helmex, Piranver.* Tasteless, yellow crystalline powder. Insol in water. Mixture with oxantel pamoate (1:1), *Quantrel.*
THERAP CAT: Anthelmintic (Nematodes).
THERAP CAT (VET): Anthelmintic.

7969. Pyrathiazine. *10-[2-(1-Pyrrolidinyl)ethyl]-10H-phenothiazine;* 10-[2-(1-pyrrolidyl)ethyl]phenothiazine; *N*-(β-pyrrolidinoethyl)phenothiazine; parathiazine. $C_{18}H_{20}N_2S$; mol wt 296.45. C 72.93%, H 6.80%, N 9.45%, S 10.82%. Prepn: Reid *et al., J. Am. Chem. Soc.* **70**, 3100 (1948); Hunter, Reid, **U.S.** pat. **2,483,999** (1949 to Upjohn); Cusic, **U.S.** pat. **2,687,414** (1954 to Searle). Toxicity data: M. J. Vander Brook *et al., J. Pharmacol. Exp. Ther.* **94**, 197 (1948).

Brown oil, bp_2 222-226°; $bp_{0.1}$ 142-146°.
Hydrochloride, $C_{18}H_{21}ClN_2S$, *Pyrrolazote, Pyrrolazote Abergic, Rolazote.* Cubes from isopropanol, mp 200-201°. Sol in water, alcohol. LD_{50} in mice (mg/kg): ~1340 s.c.; ~37 i.v. (Vander Brook).
Combination with acetophenetidin, Orthoxine, aspirin, and caffeine, *Pyrroxate.*
THERAP CAT: Antihistaminic.
THERAP CAT (VET): Antihistaminic.

7970. Pyrazinamide. *Pyrazinecarboxamide;* pyrazinoic acid amide; pyrazine carboxylamide; D-50; Aldinamide; Eprazin; Pezetamid; Pyrafat; Pirilène; Piraldina; Tebrazid; Unipyranamide; Zinamide. $C_5H_5N_3O$; mol wt 123.11. C 48.78%, H 4.09%, N 34.14%, O 13.00%. Prepd by ammonolysis of methyl pyrazinoate (from quinoxaline): Dalmer, Walter, **Ger.** pat. **632,257** (1936 to E. Merck); **Brit.** pat. **451,304**; **U.S.** pat. **2,149,279** (1939 to Merck & Co.); Hall, Spoerri, *J. Am. Chem. Soc.* **62**, 664 (1940); Kushner *et al., ibid.* **74**, 3617 (1952); Williams, Kushner, **U.S.** pat. **2,677,-641** (1954 to Am. Cyanamid); by thermal decarboxylation of pyrazine-2-carboxamide-3-carboxylic acid: Webb, Arlt, **U.S.** pat. **2,780,624** (1957 to Am. Cyanamid). Pharmacology: I. M. Weiner, J. P. Tinker, *J. Pharmacol. Exp. Ther.* **180**, 411 (1972). Comprehensive description: E. Feldner, D. Pitre, in *Analytical Profiles of Drug Substances* vol. 12, K. Florey, Ed. (Academic Press, New York, 1983) pp 433-462.

Crystals from water or alcohol, mp 189-191°. Begins to sublime at 60°. pKa 0.5. uv max: 269 nm ($E^{1\%}_{1cm}$ 660). Soly (mg/ml): water 15 ; methanol 13.8; abs ethanol 5.7; isopropanol 3.8; ether 1.0; isooctane 0.01; chloroform 7.4. Aq solns are neutral.
THERAP CAT: Antibacterial (tuberculostatic).

7971. Pyrazine. 1,4-Diazine; paradiazine. $C_4H_4N_2$; mol wt 80.09. C 59.98%, H 5.03%, N 34.98%. Prepn: S. Gabriel, G. Pinkus, *Ber.* **26**, 2197 (1893); P. Brandes, C. Stoehr, *J. Prakt. Chem.* [2] **54**, 481 (1896). *See also: Beilstein* **XXIII**, 91 (1936). Crystal structure: G. De With *et al., Acta Crystallogr.* **B32**, 3178 (1976).

Crystals or wax-like solid. Strong pyridine-like odor. d_4^{61} 1.031; mp 53°; bp 115-118°; n_D^{61} 1.4953. Freely sol in water, alcohol, ether. Volatile with steam.

7972. 2,3-Pyrazinedicarboxylic Acid. $C_6H_4N_2O_4$; mol wt 168.11. C 42.86%, H 2.40%, N 16.67%, O 38.07%. Prepd by permanganate oxidation of quinoxaline: Gabriel, Sonn, *Ber.* **40**, 4850 (1907); Sausville, Spoerri, *J. Am. Chem. Soc.* **63**, 3153 (1941); Jones, McLaughlin, *Org. Syn.* **30**, 86 (1950).

Dihydrate, prisms from water. Becomes anhydr at 100°, then melts at 183-185° (with evolution of CO_2). The dry dust provokes sneezing. Freely sol in water. Soluble in methanol, acetone, ethyl acetate. Slightly sol in ethanol, ether, chloroform, benzene, petr ether.

Dimethyl ester, $C_8H_8N_2O_4$, crystals, mp 50°. Sol in water, acetone; insol in benzene, petr ether.

Diamide, $C_6H_6N_4O_2$, crystals, dec 240°. Sol in boiling water.

7973. Pyrazinoic Acid. *Pyrazinecarboxylic acid;* pyrazinemonocarboxylic acid. $C_5H_4N_2O_2$; mol wt 124.10. C 48.39%, H 3.25%, N 22.58%, O 25.79%. Prepn by thermal decompn of 2,3-pyrazinedicarboxylic acid: Hall, Spoerri, *J. Am. Chem. Soc.* **62**, 664 (1940); by oxidation of alkylpyrazines: Gainer, *J. Org. Chem.* **24**, 691 (1959).

Fine white needles from water, dec 225-229°. Sublimes. K at 25° = 1.2×10^{-3}. Slightly sol in cold water, more sol in hot water. One gram dissolves in 120 g abs ethanol at 25°. Practically insol in ether, chloroform, benzene.

Methyl ester, crystals from ether, mp 59°.

Amide, *see* Pyrazinamide.

7974. Pyrazole. 1,2-Diazole. $C_3H_4N_2$; mol wt 68.08. C 52.92%, H 5.92%, N 41.15%. Prepn from acetylene and diazomethane: H. v. Pechmann, *Ber.* **31**, 2950 (1898). Two tautomeric forms. ^1H-NMR: J. F. K. Wilshire, *Aust. J. Chem.* **19**, 1935 (1966); ^{13}C-NMR: W. M. Litchman, *J. Am. Chem. Soc.* **101**, 545 (1979); ^{15}N-NMR: I. I. Schuster *et al.*, *J. Org. Chem.* **44**, 1765 (1979). Spectral studies: D. Dumanovic *et al.*, *Talanta* **22**, 819 (1975). Inhibition of alcohol dehydrogenase: W. K. Lelbach, *Experientia* **25**, 816 (1969). Induction of thyroid necrosis: S. Szabo *et al.*, *Science* **199**, 1209 (1979). Toxicity: G. Magnussen *et al.*, *Experientia* **28**, 1198 (1972).

Needles or prisms from petr ether; pyridine-like odor, bitter taste. mp 69.5-70°. $bp_{757.9}$ 186-188°. K at 25° = 3.0×10^{-12}. uv max ($12M$ H_2SO_4): 215 nm (log ϵ 3.76); at pH 6.1: 310 nm (log ϵ 3.61). Sol in water, alcohol, ether, benzene. LD_{50} rats, mice (mmol/kg) 24 hr mortality: 19, 21 i.v.; 21, 22 orally (Magnussen).

USE: Chelating agent; in organic synthesis.

7975. 2-Pyrazoline. *4,5-Dihydro-1H-pyrazole.* $C_3H_6N_2$; mol wt 70.09. C 51.40%, H 8.63%, N 39.97%. Prepn from ethylene and diazomethane: Azzarello, *Atti Acad. Lincei* [5] **14**, II, 285 (1905).

Liquid. Faint amine odor. An odor resembling chocolate has been detected in some prepns. d_4^{17} 1.0200. bp_{760} 144°. Volatile with steam. $n_{587}^{17.2}$ 1.4796. Miscible with water, alc. Picrate, $C_9H_9N_5O_7$, yellow needles from alcohol, mp 130°.

7976. Pyrazophos. *2-[(Diethoxyphosphinothioyl)oxy]-5-methylpyrazolo[1,5-a]pyrimidine-6-carboxylic acid ethyl ester; 2-hydroxy-5-methylpyrazolo[1,5-a]pyrimidine-6-carboxylic acid ethyl ester, O-ester with O,O-diethyl phosphorthioate;* HOE 2873; Afugan; Curamil. $C_{14}H_{20}N_3O_5PS$; mol wt 373.36. C 45.04%, H 5.40%, N 11.25%, O 21.43%, P 8.29%, S 8.59%. Systemic fungicide. Prepn: **Neth. pat. Appl.** **6,602,131;** O. Scherer, H. Mildenberger, **U.S. pat. 3,632,757** (1966, 1972 both to Hoechst). Account of fungicidal properties and toxicity: F. M. Smit, *Meded. Rijksfac. Landbouwwetensch. Gent* **34**, 763 (1969), *C.A.* **73**, 76038w (1970). In control of rusty spot on peaches: R. H. Daines, *Plant Dis. Res.* **58**, 254 (1974). Effectiveness against powdery mildew: D. J. Butt *et al.*, *Ann. Appl. Biol.* **75**, 217 (1973); F. O. Rior-

dain, *Plant Dis. Rep.* **58**, 12 (1974); R. T. Burchill, M. E. Cook, *Plant Pathol.* **24**, 194 (1975). Enzymatic detection by TLC: G. F. Ernst, F. Schuring, *J. Chromatog.* **49**, 325 (1970). Absorption and metabolism by plants: S. Gorbach *et al.*, *Environ. Qual. Saf. Suppl.* **3**, 840 (1975).

Yellow oil, mp 38-40°C. LD_{50} in rats, marmots, rabbits (mg/kg): 140, 184, 435 orally (Smit).

USE: Fungicide.

7977. Pyrene. Benzo[*def*]phenanthrene. $C_{16}H_{10}$; mol wt 202.24. C 95.02%, H 4.98%. Occurs in coal tar. Isoln: Kruber, *Ber.* **64**, 84 (1931). Also obtained by the destructive hydrogenation of hard coal: I. G. Farben., **Ger. pats. 639,-240; 640,580; 654,201.** Purification by chromatography: Winterstein *et al.*, *Z. Physiol. Chem.* **230**, 162 (1934). Synthesis from *o,o'*-ditolyl: Weitzenböck, *Monatsh.* **34**, 193 (1913). From *peri*-trimethylenenaphthalene and malonyl chloride with AlCl₃: Fleischer, Retze, *Ber.* **55**, 3280 (1922). From α-tetralone: Braun, Rath, *Ber.* **61**, 956 (1928). From 4-keto-1,2,3,4-tetrahydrophenanthrene by a Reformatsky reaction: Cook, Hewett, *J. Chem. Soc.* **1934**, 366.

Monoclinic prismatic tablets from alcohol or by sublimation. d^{23} 1.271. Pure pyrene is colorless, the usual contaminant which gives it a yellow color is tetracene. Solid and solns have slight blue fluorescence. mp 156°. bp 404°. Absorption spectrum: Clar, *Ber.* **69**, 1677 (1936); Seshan, *Proc. Indian Acad. Sci.* **A3**, 148 (1936). Fluorescence maxima: Sannié, Poremski, *Bull. Soc. Chim.* [5] **3**, 1139 (1936). Insol in water; fairly sol in organic solvents.

7978. Pyrethrins. Active insecticidal constituents of pyrethrum flowers. Isoln: Staudinger, Ruzicka, *Helv. Chim. Acta* **7**, 177 (1924). Prepn by reconstitution from pyrethrolone and chrysanthemic acid: Elliott, Janes, *Chem. & Ind. (London)* **1969**, 270. Structure: Crombie *et al.*, *J. Chem. Soc.* **1956**, 3963; Godin *et al.*, *ibid.* (C) **1966**, 332. Stereochemistry: Begley *et al.*, *Chem. Commun.* **1972**, 1276. Biosynthesis: Crowley *et al.*, *Biochim. Biophys. Acta* **60**, 312 (1962). *Review:* Crombie, Elliott, *Fortschr. Chem. Org. Naturst.* **19**, 120 (1961).

Pyrethrin I, $C_{21}H_{28}O_3$, *2,2-dimethyl-3-(2-methyl-1-propenyl)cyclopropanecarboxylic acid 2-methyl-4-oxo-3-(2,4-pentadienyl)-2-cyclopenten-1-yl ester, chrysanthemummonocarboxylic acid pyrethrolone ester.* R = CH_3. Viscous liquid. $bp_{0.0005}$ 146-150°. n_D^{20} 1.5242. $[\alpha]_D^{20}$ −14° (isooctane). uv max (95% ethanol): 225 nm (ϵ 36,400). Oxidizes readily and becomes inactive in air. Should be refrigerated and stored in darkness. Stability data of pyrethrins I and II: Godin, *Pyrethrum Post* **9**, 17 (1968), *C.A.* **71**, 12031q (1969). Practically insol in water; sol in alcohol, petr ether, kerosene, carbon tetrachloride, ethylene dichloride, nitromethane. LD_{50} orally in rats: 1.2 g/kg.

Pyrethrin II, C$_{22}$H$_{28}$O$_5$, *3-(3-methoxy-2-methyl-3-oxo-1-propenyl)-2,2-dimethylcyclopropanecarboxylic acid 2-methyl-4-oxo-3-(2,4-pentadienyl)-2-cyclopenten-1-yl ester, chrysanthemumdicarboxylic acid monomethyl ester pyrethrolone ester.* R = COOCH$_3$. Viscous liquid. Oxidizes rapidly and becomes inactive in air. bp$_{0.007}$ 192-193°. n$_D^{20}$ 1.5355. [α]$_D^{19}$ +14.7° (isooctane-ether). uv max (95% ethanol): 229 nm (ε 45,850). Practically insol in water; sol in alc, petr ether (less sol than pyrethrin I), kerosene, carbon tetrachloride, ethylene dichloride, nitromethane. LD$_{50}$ orally in rats: 1.2 g/kg.

USE: Insecticide. *Caution:* Can cause severe allergic dermatitis, systemic allergic reactions. Large amounts may cause nausea, vomiting, tinnitus, headache, and other CNS disturbances.

7979. Pyrethrosin. *6-(Acetyloxy)-1a,3,6,6a,7,9a,10,10a-octahydro-4,10a-dimethyl-7-methyleneoxireno[8,9]cyclodeca-[1,2-b]furan-8(2H)-one; 1,10-epoxy-6,8-dihydroxygermacra-4,11(13)-dien-12-oic acid 12,8-lactone acetate.* C$_{17}$H$_{22}$O$_5$; mol wt 306.35. C 66.65%, H 7.24%,O 26.11%. From flowers of *Chrysanthemum cinerariaefolium* Vis., *Compositae.* Isoln: Thoms, *Pharm. Ztg.* **1891,** 503. Structure: Barton, de Mayo, *J. Chem. Soc.* **1957,** 150; Barton *et al., ibid.* **1960,** 2263. Revised structure: Iriuchijima, Tamura, *Agr. Biol. Chem.* **34,** 204 (1970); Gabe *et al., Chem. Commun.* **1971,** 559.

Crystals from ethanol, ethyl acetate or benzene + light petroleum, mp 198-200°. [α]$_D$ −31° (c = 1.73 in chloroform). uv max: 204, 210, 220, 230 nm (ε 14,600; 12,200; 6000; 1700). Practically insol in water; sol in hot alcohol, chloroform; slightly sol in ether or petr ether.

7980. Pyrethrum Flowers. Dalmatian insect powder; Persian insect powder. Flowers of *Chrysanthemum (Pyrethrum) cinerariaefolium* Vis., and of *C. coccineum* Willd. (*C. roseum* Adam), *Compositae. Habit.* Dalmatia, Montenegro, Western Asia. *Constit.* About 1-1.5% volatile oil; pyrethrin, pyretol, pyrethrotoxic acid, pyrethrosin, chrysanthemine, chrysathemumic acid.

USE: In insecticide prepns.

THERAP CAT: Scabicide.

THERAP CAT (VET): Ectoparaciticide.

7981. Pyridate. *O-(6-Chloro-3-phenyl-4-pyridazinyl)-carbonothioic acid S-octyl ester;* fenpyrate; CL 11.344; Lentagran. C$_{19}$H$_{23}$ClN$_2$O$_2$S; mol wt 378.92. C 60.22%, H 6.12%, Cl 9.36%, N 7.40%, O 8.44%, S 8.46%. Contact herbicide for control of weeds in cereals. Prepn, herbicidal activity: **Belg. pat.** 816,574; R. Schonbeck *et al.,* **U.S. pat.** 3,953,445 (1974, 1976 to Chemie Linz). Brief description of properties, herbicidal activity: A. Diskus *et al., Proc. Brit. Crop. Prot. Conf. - Weeds* **1976,** 717. Account of use on fodder cabbage, toxicity data: M. Chéroux, J.-M. Moncorge, *Def. Veg.* **38,** 310 (1984). Residue analysis by HPLC: W. Lindner, H. Ruckendorfer, *Int. J. Environ. Anal. Chem.* **16,** 205 (1983).

Oily brown liquid, bp$_{0.1}$ 220°. mp 27°. n$_D^{20}$ 1.568. d^{20} 1.555. Flash pt open cup: 200°C. Almost insol in water (90 ppm). Sol in organic solvents. LD$_{50}$ orally in male rats, female rats, mice (mg/kg): 1970, 2400, > 10,000. LC$_{50}$ (96 hour exposure) in carp, trout (mg/l): > 100, 81 (Chéroux, Moncorge).

USE: Herbicide.

7982. Pyridazine. 1,2-Diazine; orthodiazine; oizine. C$_4$H$_4$N$_2$; mol wt 80.09. C 59.98%, H 5.03%, N 34.98%. Prepd from maleic hydrazide: Mizzoni, Spoerri, *J. Am. Chem. Soc.* **73,** 1873 (1951). Resonance energy and position of double bonds: Maccoll, *J. Chem. Soc.* **1946,** 671.

Liquid. d$_4^{18}$ 1.107; d$_4^{23.5}$ 1.1035. mp −8°. bp$_{760}$ 208°; bp$_{14}$ 87°; bp$_{1.0}$ 48°. n$_D^{23.5}$ 1.52311. Dipole moment in dioxane at 35° = 3.94. uv max: 338 nm. Resonance energy: 22 kcal. Miscible with water, benzene, DMF. Freely sol in methanol, ethanol, ether. Practically insol in petr ether.

Hydrochloride, C$_4$H$_4$N$_2$.HCl, yellow solid, mp 161-163°.

Picrate, C$_4$H$_4$N$_2$.C$_6$H$_3$N$_3$O$_7$, yellow needles from water or alcohol, dec 170-175°.

7983. Pyridine. C$_5$H$_5$N; mol wt 79.10. C 75.92%, H 6.37%, N 17.71%. Discovered in coal tar: T. Anderson, *Ann.* **60,** 86 (1846). General review: G. L. Goe in Kirk-Othmer *Encyclopedia of Chemical Technology* **vol. 19** (Wiley-Interscience, New York, 3rd ed., 1982) pp 454-483. Books: H. Maier-Bode, *Das Pyridin und Seine Derivate in Wissenschaft und Technik* (Edwards Bros., Ann Arbor, Mich., 1943); Hill in *Chemistry of Coal Utilization* **vol. 2,** H. H. Lowry, Ed. (Wiley, New York, 1945), Chapter 27; *The Chemistry of Heterocyclic Compounds* **vol. 14,** a series of books entitled "Pyridine and its Derivatives": Parts 1-4, E. Klingsberg, Ed. (1960-1964); Suppl. Parts 1-4, R. A. Abramovitch, Ed. (1974-1975); Part 5, G. R. Newkome, Ed. (1984) (Wiley-Interscience, New York). Physical data: B. A. Middleton, J. R. Partington, *Nature* **141,** 516 (1938). Acute toxicity: H. F. Smyth Jr. *et al., Arch. Ind. Hyg. Occup. Med.* **4,** 119 (1951). Human toxicity: E. Browning, *Toxicity and Metabolism of Industrial Solvents* (Elsevier, New York, 1965) pp 304-309. Brief review of commercial synthesis and uses of pyridine and derivatives: D. J. Berry, *Spec. Chem.* **3,** 13 (1983). Review of occurrence of pyridine and derivatives in foods, tobacco and essential oils: G. Vernin, *Perfum. Flavor.* **7,** 23-26 (1982); of physical, chemical, toxicological properties and environmental fate: A. Jori *et al., Ecotoxicol. Environ. Saf.* **7,** 251 (1983); of use in the study of surface properties of transition metal oxides: M. C. Kung, H. H. Kung, *Catal. Rev. Sci. Eng.* **27,** 425-460 (1985).

Flammable, colorless liq; characteristic disagreeable odor; sharp taste. d$_4^{20}$ 0.98272. Flash pt, closed cup: 68°F (20°C). mp −41.6°. bp 115.2-115.3°. n$_D^{20}$ 1.50920. Dipole moment in benzene: 2.26. Forms an azeotropic mixture with 3 mols water, boiling at 92-93°. Volatile with steam. Miscible with water, alcohol, ether, petr ether, oils and many other organic liquids. Good solvent for many organic and inorganic compds. Weak base; forms salts with strong acids. pKa 5.19. pH of 0.2 molar soln in H$_2$O: 8.5. LD$_{50}$ orally in rats: 1.58 g/kg (Smyth).

Human Toxicity: May cause CNS depression, irritation of skin and respiratory tract. Large doses may produce G.I. disturbances, kidney and liver damage.

USE: As solvent for anhydr mineral salts. Synthetic intermediate in laboratory and industry.

7984. Pyridine 1-Oxide. Pyridine *N*-oxide. C$_5$H$_5$NO; mol wt 95.10. C 63.15%, H 5.30%, N 14.73%, O 16.82%. Prepd industrially by heating pyridine in glacial acetic acid at 80° with 30% hydrogen peroxide. The acid is neutralized and the oxide recovered by distillation: Katritzky *et al., J. Chem. Soc.* **1957,** 1769. Laboratory procedure using peracetic acid: H. S. Mosher *et al., Org. Syn.* **coll. vol. IV,** 828 (1963).

Deliquescent crystals, mp 66°; bp$_{1.0}$ 100-105°.
Hydrochloride, $C_5H_5NO.HCl$, crystals from isopropanol, mp 179.5-181°.
USE: Synthetic intermediate.

7985. Pyridinium Bromide Perbromide. *Hydrogen tribromide compd with pyridine (1:1).* $C_5H_6Br_3N$; mol wt 319.85. C 18.78%, H 1.89%, Br 74.95%, N 4.38%. C_5H_5-$N^+HBr_3^-$. Prepd from pyridine, hydrobromic acid, and bromine: L. F. Fieser, M. Fieser, *Reagents for Organic Chemistry* vol. 1 (John Wiley, New York, 1967) p 967.
Orange-needles from acetic acid. Odorless. Nonvolatile. Sparingly sol in acetic acid. Dissociates in the presence of a bromine acceptor, such as an alkene, to liberate one mole of bromine.
USE: In small-scale brominations, where it is much more convenient and agreeable to measure and use than elemental bromine.

7986. Pyridinium Chlorochromate. *Chlorotrioxochromate(1−), hydrogen, compd with pyridine (1:1);* PCC. C_5-$H_6ClCrNO_3$; mol wt 215.56. C 27.86%, H 2.80%, Cl 16.45%, Cr 24.12%, N 6.50%, O 22.27%. Oxidizing agent for the efficient conversion of primary or secondary alcohols to carbonyl compounds. Prepn from chromyl chloride and pyridine: R. J. Meyer, H. Best, *Z. Anorg. Allgem. Chem.* **22**, 192 (1899); J. Bernard, M. Camelot, *C.R. Acad. Sci.* **258**, 5881 (1964). Prepn from chromium trioxide in HCl and pyridine, use as an oxidant: E. J. Corey, J. W. Suggs, *Tetrahedron Letters* **1975**, 2647. Oxidation, ring enlargement of furans to pyranones: G. Piancatelli *et al., ibid.* **1977**, 2199; oxidation of tertiary allylic alcohols: W. G. Dauben, D. M. Michno, *J. Org. Chem.* **42**, 682 (1977). Role in oxidative cationic cyclization: E. J. Corey, D. L. Boger, *Tetrahedron Letters* **1978**, 2461. *Review:* G. Piancatelli *et al., Synthesis* **1982**, 245-258.

Orange-yellow solid, mp 205° (dec). Insol in dichloromethane.
USE: Oxidizing agent in organic synthesis.

7987. Pyridinol Carbamate. *2,6-Pyridinedimethanol bis(methylcarbamate) (ester);* 2,6-pyridinylenebis[methyl-*N*-methylcarbamate]; pyricarbate; H-3749; Anginin; Angioxine; Aterosan; Atover; Cicloven; Colesterinex; Duvaline; Movecil; Prodectin; Ravenil; Sospitan; Vasagin; Vasapril; Vasocil; Vasoverin; Veranterol. $C_{11}H_{15}N_3O_4$; mol wt 253.25. C 52.17%, H 5.97%, N 16.59%, O 25.27%. Prepd by the reaction of methylamine with 2,6-pyridinedimethanol-[phenylcarbonate] in methanol: Matsumoto, **Japan.** pat. **66 22,185** (1966 to Banyu), *C.A.* **66**, 75907x (1967). Pharmacology and clinical evaluation: T. Shimamoto *et al., Am. Heart J.* **71**, 297 (1966). Metabolism: Mallein *et al., Therapie* **28**, 115 (1973). Effects on lipid metabolism and antiatherogenic effect in animals: V. Orbetzova *et al., Artery* **8**, 560 (1980). HPLC determn in biological fluids: A. Suenaga *et al., Acta Pharm. Suec.* **23**, 245 (1986).

Needles from methanol or acetone, mp 136-137°. uv max (methanol): 264-265 nm. Sparingly sol in cold water; freely sol in hot water. LD$_{50}$ in mice, rats, rabbits, dogs (mg/kg): 4500, 3400, 5200, 1000 orally (Shimamoto).
THERAP CAT: Antiarteriosclerotic.

7988. Pyridofylline. *3,7-Dihydro-1,3-dimethyl-7-[2-(sulfooxy)ethyl]-1H-purine-2,6-dione compd with 5-hydroxy-6-methyl-3,4-pyridinedimethanol (1:1);* 7-(2-hydroxyethyl)-theophylline hydrogen sulfate compound with pyridoxol; β-[1,3-dimethyl-2,6(1H,3H)-purinedion-7-yl]ethyl sulfate of 5-hydroxy-6-methyl-3,4-pyridinedimethanol; pyridoxol compd with 7-(2-hydroxyethyl)theophylline hydrogen sulfate; theophyllineethyl sulfate of pyridoxine; Atherophylline. $C_{17}H_{23}N_5O_9S$; mol wt 473.48. C 43.13%, H 4.90%, N 14.79%, O 30.41%, S 6.77%. Prepn and therapeutic properties: A. Debarge, **Fr.** pat. **M828** (1961), *C.A.* **58**, 9098g (1963).

Crystals from abs alcohol, mp 144-146°. LD$_{50}$ in mice (g/kg): 1 i.v.; 1.6 orally (Debarge).
THERAP CAT: Respiratory stimulant; coronary vasodilator.

7989. Pyridomycin. *[5R-(2Z,5R*,6S*,9S*,10S*,11R*)]-3-Hydroxy-N-[10-hydroxy-5,11-dimethyl-2-(1-methylpropylidene)-3,7,12-trioxo-9-(3-pyridinylmethyl)-1,4-dioxa-8-azacyclododec-6-yl]-2-pyridinecarboxamide.* $C_{27}H_{32}N_4O_8$; mol wt 540.59. C 59.99%, H 5.96%, N 10.36%, O 23.68%. Antimycobacterial antibiotic substance produced by *Streptomyces albidofuscus* Okami et Umezawa, nov sp., renamed *S. pyridomyceticus.* Isoln: Maeda *et al., J. Antibiot.* **6A**, 140 (1953); K. Yagishita, *ibid.* **7A**, 143 (1954). Degradation studies: Maeda, *ibid.* **10A**, 94 (1957). Structure: Koyama *et al., Tetrahedron Letters* **1967**, 3587; H. Ogawara *et al., Chem. Pharm. Bull.* **16**, 679 (1968). Synthetic study: M. Kinoshita, M. Awamura, *Bull. Chem. Soc. Japan* **51**, 869 (1978); M. Kinoshita *et al., ibid.* 3595. Production: **Japan.** pats. **54 7048; 55 1349; 56 9566** (1954, 1955, 1956 all to Nippon Antibiotic Subst. Sci. Assoc.).

Crystals from ethanol, mp 222°. uv max (ethanol): 303 nm (E$_{1cm}^{1\%}$ 209). Soluble in lower alcohols, ethyl or butyl acetate, benzene, acetone, dioxane, tetrahydrofuran. Practically insol in water.
Hydrochloride, crystals, mp 194-196°. Freely sol in water. $[\alpha]_D^{16}$ −53.2° (water). pH of aq solns about 2.0. LD$_{50}$ i.p. in mice: >2100 mg/kg (Yagishita).

7990. Pyridostigmine Bromide. *3-[[(Dimethylamino)-carbonyl]oxy]-1-methylpyridinium bromide;* 3-hydroxy-1-methylpyridinium bromide dimethylcarbamate; 1-methyl-3-hydroxypyridinium bromide dimethylcarbamate; 3-(dimethylcarbamyloxy)-1-methylpyridinium bromide; Ro 1-5130; Mestinon Bromide; Kalymin; Regonol. $C_9H_{13}BrN_2O_2$; mol wt 261.14. C 41.39%, H 5.02%, Br 30.60%, N 10.73%, O 12.25%. Prepn: Urban, **U.S.** pat. **2,572,579** (1951 to Hoffmann-La Roche).

Shiny, hygroscopic crystals from abs ethanol, mp 152-154°. Freely sol in water, alcohol. Practically insol in ether, acetone, benzene. Aq solns may be sterilized by autoclaving with steam.

THERAP CAT: Cholinergic.

7991. Pyridoxal. *3-Hydroxy-5-(hydroxymethyl)-2-methyl-4-pyridinecarboxaldehyde; 3-hydroxy-5-(hydroxymethyl)-2-methylisonicotinaldehyde;* 2-methyl-3-hydroxy-4-formyl-5-hydroxymethylpyridine. $C_8H_9NO_3$; mol wt 167.16. C 57.48%, H 5.43%, N 8.38%, O 28.72%. Synthesis and structure: Harris et al., J. Biol. Chem. **154**, 315 (1944); J. Am. Chem. Soc. **66**, 2088 (1944).

Hydrochloride, $C_8H_9NO_3$·HCl, rhombic crystals, mp approx 165° with decompn. Soluble in water (1 g/2 ml); sol in 95% ethanol (1.7 g/100 ml); pH of 1% water soln = 2.65. The water solns are sensitive to heat. uv max: 292.5 nm (E mol 7600). Can be reduced to pyridoxine hydrochloride (mp 206-208°).

Monoethylacetal hydrochloride, $C_{10}H_{14}ClNO_3$, *1,3-dihydro-1-ethoxy-6-methylfuro[3,4-c]pyridin-7-ol hydochloride.* Crystals from alcohol and ether, mp 142-143°.

7992. Pyridoxal 5-Phosphate. *3-Hydroxy-2-methyl-5-[(phosphonooxy)methyl]-4-pyridinecarboxaldehyde;* pyridoxal 5-monophosphoric acid ester; 3-hydroxy-5-(hydroxymethyl)-2-methylisonicotinaldehyde 5-phosphate; codecarboxylase; Pyromijin; Sechvitan; Vitazechs. $C_8H_{10}NO_6P$; mol wt 247.15. C 38.88%, H 4.08%, N 5.67%, O 38.84%, P 12.54%. Prepn by the action of adenosine triphosphate on pyridoxal: Gunsalus et al., J. Biol. Chem. **155**, 685 (1944); by the action of phosphorus oxychloride on pyridoxal in aq soln: Gunsalus et al., ibid. **161**, 743 (1945); Umbreit et al., Arch. Biochem. **7**, 189 (1945); by phosphorylation of pyridoxamine with 100% H_3PO_4 followed by oxidation: Wilson, Harris, J. Am. Chem. Soc. **73**, 4693 (1951). Alternate route: Schorre, U.S. pat. 3,124,587 (1964 to E. Merck). Isoln in pure form as the oxime and as the O-methyl oxime and structure: Heyl et al., J. Am. Chem. Soc. **73**, 3430 (1951). Activity as a co-transaminase in the synthesis of amino acids: Lichstein et al., J. Biol. Chem. **161**, 311 (1945). Activity of natural and synthetic material: Umbreit et al., loc. cit.

Colorless in acid soln, bright-yellow in alkaline soln. uv max (alkaline soln): 390 nm (E_m 3.7); in acid soln: 295 nm (E_m 5.1). Gives a negative 2,6-dichloroquinone chlorimide test. On oxidation with H_2O_2 in alkaline solution yields [(2-methyl-3,4-dihydroxy-5-pyridyl)methyl]phosphoric acid.

Oxime, $C_8H_{11}N_2O_6P$, crystals, dec 229-230°. Practically insol in water, alcohol, ether.

O-Methyloxime, $C_9H_{13}N_2O_6P$, crystals, dec 212-213°. Practically insol in water, alcohol, ether.

Calcium salt, *Aderoxal*, yellow precipitate. Practically insol in water, alcohol or ether.

THERAP CAT: Enzyme co-factor vitamin.

7993. Pyridoxamine Dihydrochloride. *4-(Aminomethyl)-5-hydroxy-6-methyl-3-pyridinemethanol dihydrochloride;* 2-methyl-3-hydroxy-4-aminomethyl-5-hydroxymethylpyridine dihydrochloride. $C_8H_{14}Cl_2N_2O_2$; mol wt 241.12. C 39.85%, H 5.85%, Cl 29.41%, N 11.62%, O 13.27%. $C_8H_{12}N_2O_2$·2HCl. Synthesis and structure: Harris et al., J. Biol. Chem. **154**, 315 (1944); J. Am. Chem. Soc. **66**, 2088 (1944).

Platelets, mp 226-227° with dec. Soluble in water (approx 1 g/2 ml); sol in 95% alc (0.65 g/100 ml). Reasonably stable at room temp; shows no decompn in a few days at 60°. Liquefies on exposure to 80% relative humidity. pH of a 1% water soln 2.4. uv max (pH 1.94): 287.5 nm (E mol 91,000).

Free base, crystals from alc; mp 193-193.5°; sol in alc.

Pyridoxamine phosphate, $C_8H_{13}N_2O_5P$, 2-methyl-3-hydroxy-4-aminomethyl-5-pyridylmethylphosphoric acid. Prepn in soln by autoclaving codecarboxylase with glutamic acid: McNutt, Snell, J. Biol. Chem. **182**, 557 (1950); prepn by direct phosphorylation of pyridoxamine in aq soln with phosphorus oxychloride: Heyl et al., J. Am. Chem. Soc. **73**, 3430 (1951).

7994. 4-Pyridoxic Acid. *3-Hydroxy-5-(hydroxymethyl)-2-methyl-4-pyridinecarboxylic acid; 3-hydroxy-5-(hydroxymethyl)-2-methylisonicotinic acid;* 2-methyl-3-hydroxy-4-carboxy-5-hydroxymethylpyridine. $C_8H_9NO_4$; mol wt 183.16. C 52.46%, H 4.95%, N 7.65%, O 34.94%. Occurs in urine. It is the chief metabolic product of pyridoxine pyridoxal, and pyridoxamine. Isoln from human urine: Singal, Sydenstricker, Science **78**, 545 (1941). Isoln and synthesis: Huff, Perlzweig, J. Biol. Chem. **155**, 345 (1944).

Wedge-shaped crystals, mp 247-248°. Slightly sol in water, alcohol, pyridine. Insoluble in ether and in aq acid soln, but completely sol in aq alkaline soln. Possesses two acidic groups, one a phenolic and the other a carboxyl having pK values of 9.75 and 5.50, respectively. Characteristic blue fluorescence, max at pH 3 to 4. The fluorescence disappears on reduction with hydrosulfite and is restored to the original intensity with H_2O_2. Is adsorbed on zeolite from aq solns at pH 4 to 5 and can be eluted with 25% KCl; butanol extracts of neutral eluates also show characteristic blue fluorescence which is increased by a trace of acetic acid. Stable to boiling with dil alkali (1N); but upon being heated with 0.5N acid for a few minutes it is converted to the lactone.

Lactone, $C_8H_7NO_3$, *β-pyracine.* mp 263-265°, exhibits much stronger fluorescence. Is more easily followed in the course of biochemical investigations.

7995. Pyridoxine Hydrochloride. *5-Hydroxy-6-methyl-3,4-pyridinedimethanol hydrochloride; pyridoxol hydrochloride;* vitamin B_6 hydrochloride; pyridoxinium chloride; adermine hydrochloride; 2-methyl-3-hydroxy-4,5-bis(hydroxymethyl)pyridine hydrochloride; 5-hydroxy-6-methyl-3,4-pyridinedicarbinol hydrochloride; 3-hydroxy-4,5-dimethylol-α-picoline hydrochloride; Bonasanit; Hexabione hydrochloride; Hexabetalin; Hexavibex; Paxadon; Pyridipca; Pyridox; Bécilan; Benadon; Hexermin; Campoviton 6; Hexobion. $C_8H_{12}ClNO_3$; mol wt 205.64. C 46.72%, H 5.88%, Cl 17.24%, N 6.81%, O 23.34%. One of the vitamins of the B_6 complex; see also Codecarboxylase; Pyridoxal Hydrochloride; Pyridoxamine Dihydrochloride. Present in many foodstuffs; esp good sources are yeast, liver and cereals. Isoln: Keresztesy, Stevens, Proc. Soc. Exp. Biol. Med. **38**, 64 (1938); György, J. Am. Chem. Soc. **60**, 983 (1938); Kuhn, Wendt,

Ber. **71**, 780, 1118 (1938). Structure: E. T. Stiller *et al., J. Am. Chem. Soc.* **61**, 1237 (1939); R. Kuhn *et al., Ber.* **72**, 305 sqq. (1939). uv absorption spectrum: E. A. Peterson, H. A. Sober, *J. Am. Chem. Soc.* **76**, 169 (1954). Synthesis: Harris, Folkers, *J. Am. Chem. Soc.* **61**, 1242, 1245, 3307 (1939); P. G. Stevens, U.S. pats. **2,680,743** and **2,734,063** (1954, 1956 both to Gen. Aniline); P. I. Pollak, U.S. pats. **2,904,551, 3,024,244, 3,024,245** (1959, 1962, 1962 all to Merck & Co.); E. E. Harris *et al., J. Org. Chem.* **27**, 2705 (1962); W. Böll, H. König, *Ann.* **1979**, 1657; T. Shono *et al., Chem. Letters* **1981**, 1121. Biosynthesis: Hiel, Spenser, *Science* **169**, 773 (1970). Review: *The Vitamins* vol. 2, W. H. Sebrell, R. S. Harris, Eds. (Academic, New York, 2nd ed., 1968) pp 1-117.

Platelets or thick, birefringent rods from alcohol + acetone. Reasonably stable to light and air. Dec 205-212°. Sublimes. (Free base, mp 160°.) uv max (0.1*N* HCl): 290 nm (ϵ 8400); (phosphate buffer, pH 7): 253, 325 nm (ϵ 3700, 7100). One gram dissolves in about 4.5 ml water, 90 ml alcohol. Sol in propylene glycol. Sparingly sol in acetone. Insol in ether, chloroform. pH of a 10% w/v soln in water: 3.2. Acidic aq solns are stable and may be heated for 30 min at 120° without decompn.

THERAP CAT: Enzyme co-factor vitamin.
THERAP CAT (VET): Nutritional factor.

7996. Pyrilamine. *N-[(4-Methoxyphenyl)methyl]-N',N'-dimethyl-N-2-pyridinyl-1,2-ethanediamine; 2-[(2-dimethylaminoethyl)(p-methoxybenzyl)amino]pyridine; N-p-methoxybenzyl-N',N'-dimethyl-N-α-pyridylethylenedi-amine;* mepyramine; pyranisamine; RP 2786; Neo-Antergan; Antallergan; Harvamine; Isamin; Pyra; Kriptin; Thylogen; Wait's Green Mountain Antihistamine; Paraminyl; Maranhist; Histapyran; Histasan; Histalon; Histacap; Coradon; Copsamine; Antamine; Afko-Hist; Mepyren; Neo-Bridal; Pyramal; Stangen; Statomin; Nyscaps; Parmal. $C_{17}H_{23}N_3O$; mol wt 285.38. C 71.54%, H 8.12%, N 14.73%, O 5.61%. Prepn: Bovet *et al., C.R. Soc. Biol.* **138**, 99 (1944); Huttrer *et al., J. Am. Chem. Soc.* **68**, 1999 (1946); Viaud, *Prod. Pharm.* **2**, 53 (1947); Horclois, U.S. pat. **2,502,151** (1950 to Rhône-Poulenc). Toxicity: A. von Schlichtergroll, *Arzneimittel-Forsch.* **7**, 237 (1957); of the maleate: F. Hunziker *et al., ibid.* **13**, 324 (1963).

Oily liquid. bp$_5$ 201°; bp$_{0.06}$ 168-172°. n_D^{25} 1.5760-1.5765. LD$_{50}$ orally in mice: 312 mg/kg (Schlichtergroll).

Hydrochloride, $C_{17}H_{23}N_3O$.HCl, crystals, mp 143-143.5°. Very sol in water.

Maleate, $C_{21}H_{27}N_3O_5$, *Anthisan, Dorantamin, Enrumay, Histan, Histatex, Histosol, Paraminyl Maleate, Stamine.* Crystals. Bitter saline taste. Stable in air. mp 100-101°. uv max: 244 nm (E$_{1cm}^{1\%}$ 420). One gram dissolves in about 0.4 ml water, in about 15 ml abs alc. pH of 10% soln ~5.1. On raising the pH to 7.5 or 8.0 pptn of the oily, free base begins. Slightly sol in benzene, ether. LD$_{50}$ orally in mice: 338 mg/kg (Hunziker).

8-Bromotheophyllinate, $C_{24}H_{30}BrN_7O_3$, *SMP 68-40, NSC-14279, Glybrom, Bromth, Sanbromal.* Crystals. Freely sol in alc, chloroform; sol in benzene; slightly sol in water. Almost insol in ether. pH (satd aq soln) ~7.

THERAP CAT: Antihistaminic.
THERAP CAT (VET): Antihistaminic.

7997. Pyrimethamine. *5-(4-Chlorophenyl)-6-ethyl-2,4-pyrimidinediamine; 2,4-diamino-5-(p-chlorophenyl)-6-eth-ylpyrimidine;* RP 4753; Daraprim; Chloridin; Darapram;

Malocide; Tindurin. $C_{12}H_{13}ClN_4$; mol wt 248.71. C 57.94%, H 5.27%, Cl 14.25%, N 22.53%. Prepn: Russel, Hitchings, *J. Am. Chem. Soc.* **73**, 3763 (1951); Logemann *et al., Ber.* **87**, 435 (1954). Commercial process: Chase, Walker, *J. Chem. Soc.* **1953**, 3518; Hitchings *et al., U.S.* pats. **2,576,939; 2,579,259,** and **2,602,794** (1951 and 1952 to Burroughs Wellcome); Jacob, U.S. pat. **2,680,740** (1954 to Rhône-Poulenc). Review: Burchall in *Antibiotics* vol. 3, J. W. Corcoran, F. E. Hahn, Eds. (Springer-Verlag, New York, 1975) pp 312-320. Comprehensive description: M. A. Loutfy, H. Y. Aboul-Enein in *Analytical Profiles of Drug Substances* vol. 12, K. Florey, Ed. (Academic Press, New York, 1983) pp 463-482.

Crystals, mp 233-234° (capillary); mp 240-242° (copper block). Practically insol in water. Slightly sol in ethanol, (about 9 g/l), in dil HCl (about 5 g/l); sol in boiling ethanol (about 25 g/l). Very sparingly sol in propylene glycol and dimethylacetamide at 70°.

THERAP CAT: Antimalarial.
THERAP CAT (VET): Antimicrobial, usually as a sulfonamide synergist.

7998. Pyrimidine. 1,3-Diazine; metadiazine; miazine. $C_4H_4N_2$; mol wt 80.09. C 59.98%, H 5.03%, N 34.98%. Prepd by reducing 2,4,6-trichloropyrimidine with zinc dust: Gabriel, *Ber.* **33**, 3666 (1900); by reducing tetrachloropyrimidine: Emery, *ibid.* **34**, 1480 (1901); by reducing 2,4-dichloropyrimidine with magnesium oxide and palladium-charcoal: Wittaker, *J. Chem. Soc.* **1953**, 1646.

Liquid or cryst mass; penetrating odor. mp 20-22°. bp$_{762}$ 123-124°. uv max (water): 240 nm (ϵ 2400). Soluble in water, alcohol, ether.

Picrate, $C_4H_4N_2.C_6H_3N_3O_7$, yellow needles, mp 156°.
Methiodide, $C_4H_4N_2.CH_3I$, yellow plates from ethanol, mp 136-137°.

7999. Pyriminil. *N-(4-Nitrophenyl)-N'-(3-pyridinyl-methyl)urea; N-3-pyridylmethyl-N'-p-nitrophenylurea;* pyrinuron; RH-787; DLP-787; Vacor. $C_{13}H_{12}N_4O_3$; mol wt 272.27. C 57.35%, H 4.44%, N 20.58%, O 17.63%. Prepn: J. E. Ware *et al.,* **Ger.** pat. **2,409,686** (1974 to Rohm & Haas). Evaluation as rodenticide: D. L. Peardon, *Pest Contr.* **42**, 14, 16, 18, 27 (1974); J. E. Brooks, P. T. Htun, *J. Hyg.* **80**, 401 (1978). Mode of action: D. L. Peardon, J. E. Ware, *Pest Contr.* **45**, 49 (1977).

mp 223-225° (dec). LD$_{50}$ orally in male, female rats: 6.2, 7.2 mg/kg (Brooks, Htun).

USE: Rodenticide.

8000. Pyrimithate. *Phosphorothioic acid O-[2-(dimeth-ylamino)-6-methyl-4-pyrimidinyl] O,O-diethyl ester; O,O-di-ethyl O-[2-(dimethylamino)-4-methyl-6-pyrimidinyl] phos-phorothioate;* ICI 29661; Diothyl. $C_{11}H_{20}N_3O_3PS$; mol wt 305.35. C 43.27%, H 6.60%, N 13.76%, O 15.72%, P 10.15%, S 10.50%. Prepn: McHattie, **Brit.** pat. **1,019,227** corresp to U.S. pat. **3,287,453** (both in 1966 to ICI).

Liquid, bp$_{0.04}$ 128-132°. Sp gr 1.165. Practically insol in water; sol in alcohols, acetone, benzene, solvent naphtha.
USE: Insecticide, acaricide.

8001. Pyrinoline. α-*[3-(Di-2-pyridinylmethylene)-1,4-cyclopentadien-1-yl]-α-2-pyridinyl-2-pyridinemethanol; 3-(di-2-pyridylmethylene)-α,α-di-2-pyridyl-1,4-cyclopentadiene-1-methanol; di-2-pyridyl-(6,6-di-2-pyridylfulven-2-yl)methanol*; McN-1210; Surexin. C$_{27}$H$_{20}$N$_4$O; mol wt 416.46. C 77.86%, H 4.84%, N 13.45%, O 3.84%. Prepn: **Brit.** pat. **1,009,012** (1965 to McNeil).

Crystals, mp 146.5-147.5°.
THERAP CAT: Cardiac depressant (anti-arrhythmic).

8002. Pyrisuccideanol. *Butanedioic acid 2-(dimethylamino)ethyl [5-hydroxy-4-(hydroxymethyl)-6-methyl-3-pyridinyl]methyl ester; pyridoxol 3-[2-(dimethylamino)ethyl succinate]*; 2-(dimethylamino)ethyl [5-hydroxy-4-(hydroxymethyl)-6-methyl-3-pyridyl]succinic acid methyl ester; pirisudanol. C$_{16}$H$_{24}$N$_2$O$_6$; mol wt 340.38. C 56.46%, H 7.11%, N 8.23%, O 28.20%. Prepn: Esanu, **Ger.** pat. **2,102,831** (1971 to Soc. d'Etudes de Prods. Chim.) corresp to **U.S.** pat. **3,717,636** (1973). Activity studies: Hugelin *et al.*, *C.R. Soc. Biol.* **166**, 1435 (1972).

Dimaleate, C$_{24}$H$_{32}$N$_2$O$_{14}$, *Mentium, Nadex, Stivane*, mp 134°.
THERAP CAT: Antidepressant.

8003. Pyrithiamine. *1-[(4-Amino-2-methyl-5-pyrimidinyl)methyl]-3-(2-hydroxyethyl)-2-methylpyridinium bromide monohydrobromide*; 1-(2-methyl-4-amino-5-pyrimidyl)-methyl-2-methyl-3-hydroxyethylpyridinium bromide hydrobromide; neopyrithiamine. C$_{14}$H$_{20}$Br$_2$N$_4$O; mol wt 420.16. C 40.02%, H 4.80%, Br 38.04%, N 13.34%, O 3.81%. Thiamine antagonist prepd by condensation of 2-methyl-3-(β-hydroxyethyl)pyridine with the pyrimidine moiety of vitamin B$_1$: Tracy, Elderfield, *J. Org. Chem.* **6**, 54 (1941); improved procedure: Wilson, Harris, *J. Am. Chem. Soc.* **71**, 2231 (1949); U.S. pat. **2,587,262** (1952 to Merck & Co.). *Cf.* Woolley, *J. Am. Chem. Soc.* **72**, 5763 (1950).

Crystals from acetone, dec 218-220°. uv max (water): 238, 271 nm. Sol in water.
Chloride hydrochloride, C$_{14}$H$_{20}$Cl$_2$N$_4$O, dec 234-236°. uv max (water): 235, 273 nm. Sol in water.

8004. Pyrithione. *1-Hydroxy-2(1H)-pyridinethione;*

2-pyridinethiol 1-oxide; 2-mercaptopyridine 1-oxide; PTO; Omadine. C$_5$H$_5$NOS; mol wt 127.18. C 47.22%, H 3.96%, N 11.02%, O 12.58%, S 25.22%. Prepn: Shaw *et al.*, *J. Am. Chem. Soc.* **72**, 4362 (1950); Semenoff, Dolliver, **U.S.** pat. **2,745,826** (1956 to Olin Mathieson). Prepn of zinc deriv: **Brit.** pat. **761,171** (1956 to Olin Mathieson). Activity of zinc deriv: Karsten *et al.*, and Judge *et al.*, U.S. pats. **3,236,-733** and **3,281,366** (both 1966 to Procter and Gamble).

Sodium salt, C$_5$H$_4$NNaOS, *Fonderma*.
Zinc derivative, C$_{10}$H$_8$N$_2$O$_2$S$_2$Zn, *zinc pyrithione, zinc pyridinethione, bis(2-pyridylthio)zinc 1,1'-dioxide, bis-(1-hydroxy-2(1H)-pyridinethionato-O,S)zinc, De-Squaman, Vancide ZP, Zinc Omadine*. Ingredient in *Head & Shoulders*.
Dimer, C$_{10}$H$_8$N$_2$O$_2$S$_2$, *dipyrithione, 2,2'-dithiobispyridine 1,1'-dioxide, OMDS, Omadine Disulfide*.
USE: Fungicide, bactericide.
THERAP CAT: Zinc deriv as antibacterial; antifungal; antiseborrheic. Dimer as antibacterial; antifungal.

8005. Pyrithyldione. *3,3-Diethyl-2,4-(1H,3H)pyridinedione*; 3,3-diethyl-2,4-dioxotetrahydropyridine; 2,4-dioxo-3,3-diethyltetrahydropyridine; 2,4-dioxo-3,3-diethyl-1,2,-3,4-tetrahydropyridine; Presidon; Persedon; Tetridin; Benedorm. C$_9$H$_{13}$NO$_2$; mol wt 167.20. C 64.65%, H 7.84%, N 8.38%, O 19.14%. Prepn: Schnider, *Festschrift Emil Barell* **1936**, 195; Preiswerk, Schnider, U.S. pat. **2,090,068** (1937 to Hoffmann-La Roche); Strukov *et al.*, *Med. Prom. SSSR* **13**, no. 9, 9 (1959). Alternate procedure: Rechen *et al.*, U.S. pat. **3,019,230** (1962 to Hoffmann-La Roche).

Exists in three cryst modifications: I, mp 92-93°; II, mp 97-98°; III, mp 81-86° [Scheibl, Wachter, *Arch. Toxikol.* **18**, 253 (1960)]. bp$_{14}$ 187-189°. Moderately sol in water; freely sol in the usual organic solvents except petr solvents. Aq solns are just acid to litmus.
THERAP CAT: Hypnotic; sedative.

8006. Pyritinol. *3,3'-[Dithiobis(methylene)]bis[5-hydroxy-6-methyl-4-pyridinemethanol]*; bis(4-hydroxymethyl-5-hydroxy-6-methyl-3-pyridylmethyl) disulfide; bis[(3-hydroxy-4-hydroxymethyl-2-methyl-5-pyridyl)methyl] disulfide; dipyridoxolyldisulfide; pyridoxine-5-disulfide; pyrithioxin; Enbol Base; Bonifen; Epocan. C$_{16}$H$_{20}$N$_2$O$_4$S$_2$; mol wt 368.48. C 52.16%, H 5.47%, N 7.60%, O 17.37%, S 17.40%. Prepn: Zima, Schorre, U.S. pat. **3,010,966** (1961 to E. Merck); Iwanami *et al.*, *Bitamin* **36**, 122 (1967); *J. Vitaminol. (Kyoto)* **14**, 321, 326 (1968). HPLC determn in urine: K. Kitao *et al.*, *Chem. Pharm. Bull.* **25**, 1335 (1977). Pharmacokinetics and metabolism: Darge *et al.*, *Arzneimittel-Forsch.* **19**, 5, 9, (1969); Nowak, Schorre, *ibid.* 11. Clinical trial in dementia: S. Hoyer *et al.*, *ibid.* **27**, 671 (1977); A. J. Cooper, R. V. Magnus, *Pharmacotherapeutica* **2**, 317 (1980); in cerebrovascular disorders: Y. Tazaki *et al.*, *J. Int. Med. Res.* **8**, 118 (1980).

Crystals, mp 218-220°.

Dihydrochloride monohydrate, $C_{16}H_{22}Cl_2N_2O_4S_2 \cdot H_2O$, **Biocefalin, Encephabol, Enerbol, Life.** mp 184°.
Note: Has no vitamin B_6 activity.
THERAP CAT: Nootropic.

8007. Pyrocalciferol. *10α-Ergosta-5,7,22-trien-3β-ol;* 9α-lumisterol; 9α-lumista-5,7,22-trien-3β-ol. $C_{28}H_{44}O$; mol wt 396.63. C 84.78%, H 11.18%, O 4.03%. Differs from ergosterol in the steric configuration at C-9 and C-10. Formation via isopyrocalciferol: Askew *et al., Proc. Roy. Soc. London* **B109**, 488 (1932); Windaus *et al., Nachr. Ges. Wiss. Göttingen, math.-phys. Klasse* **1932**, 150. Dehydrogenation with selenium gives Diel's hydrocarbon (3'-methyl-1,2-cyclopentenophenanthrene). Fieser, Fieser, *Steroids* (New York, 1959) pp 136-143, argue the configuration at C-9 and C-10. *See also* Castells *et al., J. Chem. Soc.* **1959**, 1159. Critical examination and comparison with other unnatural steroids: Castells *et al., ibid.* **1962**, 2907.

Needles from methanol, mp 93-95°. $[\alpha]_D^{20}$ +512°, $[\alpha]_{546}^{20}$ +624° (c = 0.15 in alcohol). uv max: 274, 294 nm (Askew, *loc. cit.*).
Acetate, $C_{30}H_{46}O_2$, mp 81-82°, $[\alpha]_D$ +403° (chloroform). 3,5-Dinitrobenzoate, $C_{35}H_{46}N_2O_6$, mp 168-170°. $[\alpha]_D^{21}$ +195°, $[\alpha]_{546}^{20}$ +249° (both c = 2 in chloroform).

8008. Pyrocarbonic Acid Diethyl Ester. *Dicarbonic acid diethyl ester; oxydiformic acid diethyl ester;* diethyl pyrocarbonate; diethyl oxydiformate; DEPC; Ue 5908; Baycovin. $C_6H_{10}O_5$; mol wt 162.14. C 44.44%, H 6.22%, O 49.34%. $C_2H_5OOCOCOOC_2H_5$. Prepd according to the equation $R_1OOCCl + NaOCOOR_2 \rightarrow R_1OOCOCOOR_2 + NaCl$: Kovalenko, *Zh. Obshch. Khim.* **22**, 1546 (1952); Thoma, Rinke, *Ann.* **624**, 30 (1959); alternate method: Degering *et al., J. Am. Pharm. Assoc. Sci. Ed.* **39**, 626 (1950). *Review: Handbook of Food Additives,* T. A. Furia, Ed. (Chemical Rubber Co., Cleveland, 1968) pp 181-185.
Viscous liquid. Fruity odor. d_4^{20} 1.12. Viscosity at 20° = 1.97 cp. Soluble in alcohols, esters, ketones, hydrocarbons. A 50% w/w soln in ethanol may be prepd. Slowly sol in water with hydrolytic decompn yielding ethanol and CO_2.
USE: Gentle esterifying agent. Preservative for wines, soft drinks, fruit juices. *Caution:* Concd DEPC is irritating to eyes, mucous membranes and skin.

8009. Pyrocatechol. *1,2-Benzenediol;* pyrocatechin; catechol; 1,2-dihydroxybenzene. $C_6H_6O_2$; mol wt 110.11. C 65.44%, H 5.49%, O 29.06%. Prepd by treating salicylaldehyde with hydrogen peroxide, or from its monomethyl ether (guaiacol) by treatment with hydrobromic acid: Dakin, *Org. Syn.* **coll. vol. I**, 149 (2nd ed., 1941). *Review:* J. Varagnat, "Hydroquinone, Resorcinol, and Catechol", in Kirk-Othmer *Encyclopedia of Chemical Technology* vol. 13 (Wiley-Interscience, New York, 3rd ed., 1981) pp 39-69.

Monoclinic tablets, prisms from toluene. Discolors in air and light. d 1.344; mp 105°; bp_{760} 245.5°; bp_{400} 221.5°; bp_{200} 197.7°; bp_{100} 176°; bp_{60} 161.7°; bp_{40} 150.6°; bp_{20} 134°; bp_{10} 118.3°; bp_5 104°. Sublimes. Volatile with steam. K at 18° = 3.3×10^{-10}. Sol in 2.3 parts water, in alcohol, benzene, chloroform, ether; very sol in pyridine, aq alkalies. Its aq solns soon turn brown. LD_{50} in mice (mg/kg): 260 orally; 190 i.p., A. J. Lehman *et al., Adv. Food Res.* **3**, 197 (1951).

Note: Catechol also refers to catechin, *q.v.*
Human Toxicity: Can cause eczematous dermatitis. Systemic effects similar to phenol, *q.v.*
USE: In photography; dyeing fur; as reagent.
THERAP CAT: Topical antiseptic.

8010. Pyrogallol. *1,2,3-Benzenetriol;* 1,2,3-trihydroxybenzene; pyrogallic acid. $C_6H_6O_3$; mol wt 126.11. C 57.14%, H 4.80%, O 38.06%. Observed by Scheele in 1786; prepd by Braconot in 1818. Prepn from gallic acid: Marsh, **Brit.** pat. 144,897 (1919); Rinderknecht, Niemann, *J. Am. Chem. Soc.* **70**, 2605 (1948); from *p-tert*-butylphenol: Stevens, **U.S.** pat. 2,603,662 (1952 to Gulf). Synthesis from aliphatic sources: Shipchandler *et al., J. Chem. Soc. Perkin Trans. I* **1975**, 1400. Isoln from *Penicillium patulum:* Tanenbaum, Bassett, *Biochim. Biophys. Acta* **28**, 21 (1958). Acute toxicity: J. W. Dollahite *et al., Am. J. Vet. Res.* **23**, 1264 (1962).

White, odorless crystals, mp 131-133°. Becomes grayish on exposure to air and light. *Poisonous!* d 1.45; bp 309°. Sublimes when slowly heated. One gram dissolves in 1.7 ml water, 1.3 ml alc, 1.6 ml ether; slightly sol in benzene, chloroform, carbon disulfide. The aq soln darkens on exposure to air, quite rapidly when alkaline. *Keep well closed and protected from light. Incompat.* Alkalies, ammonium hydroxide, antipyrine, camphor, phenol, menthol. LD_{50} orally in rabbits: 1.6 g/kg (Dollahite).
Monoacetate, $C_8H_8O_4$, *eugallol.* White or brownish liquid, bp_{23} ~185°. Sol in water, alcohol, chloroform, ether, acetone and castor oil. Marketed as a 67% soln in acetone.
Triacetate, $C_{12}H_{12}O_6$, *lenigallol, acetpyrogall.* White, crystalline powder, mp 165°. Slightly sol in water; sol in alcohol. Dec by alkali hydroxide solns.
Caution: Ingestion may cause G.I. irritation, renal and hepatic damage, hemolysis, methemoglobinemia, convulsions, circulatory collapse, death. Poisoning and death have occurred from percutaneous absorption. *See: Clinical Toxicology of Commercial Products,* R. E. Gosselin *et al.,* Eds. (Williams & Wilkins, Baltimore, 5th ed., 1981) section II, p 190.
USE: Developer in photography; making colloidal solns of metals; as mordant for wool, staining leather, process engraving; manuf various dyes; dyeing furs, hair, etc.; in analytical chemistry as a reagent for antimony and bismuth; as an active reducer for gold, silver and mercury salts; especially for absorption of oxygen in gas analysis.
THERAP CAT: Has been used as antipsoriatic.

8011. Pyrogens. *Fever-producing substances.* Agents which produce a febrile response when administered parenterally to man and certain animals. A number of pyrogens, primarily of a polysaccharide nature, are produced by various bacteria, molds, viruses, and by yeast. Since pyrogens commonly occur in distilled water, and many pyrogen-producing microorganisms are air-borne, parenterals can readily be contaminated with these substances during prepn. *Reviews:* "Symposium on Pyrogens," *J. Pharm. Pharmacol.* **6**, 302 (1954); A. Berger *et al.,* "Pyrogens" in *Advances in Chemistry Series* **No. 16** (Am. Chem. Soc., Washington, D.C., 1956) pp 168-197. A potent pyrogen used in the study of exptl fever is *etiocholanolone* or *3α-hydroxy-5β-androstan-17-one,* a steroid metabolite of human origin: Kimball *et al., J. Clin. Endocrinol. Metab.* **26**, 222 (1966).

8012. L-Pyroglutamic Acid. *5-Oxo-L-proline; 5-oxo-2-pyrrolidinecarboxylic acid;* 2-pyrrolidone-5-carboxylic acid; glutimic acid; glutiminic acid; α-aminoglutaric acid lactam; glutamic acid lactam; pyroGlu. $C_5H_7NO_3$; mol wt 129.11. C 46.51%, H 5.47%, N 10.85%, O 37.18%. Cyclized internal amide of L-glutamic acid found in vegetables, fruits, grasses, and molasses. Easily prepd from L-glutamic acid by autoclaving with an equal wt of water at 135-140°: Dearborn, Stekol, **U.S.** pat. 2,528,267; purification: Blish, **U.S.** pat. 2,738,353 (1950, 1956, both to International Minerals and

Chem.). *Review:* Orlowski, Meister, in *The Enzymes,* **vol. 4,** P. D. Boyer, Ed. (Academic Press, New York, 3rd ed., 1971) pp 123-151; C. Moret, M. Briley, *Trends Pharmacol. Sci.* **9,** 278-279 (1988).

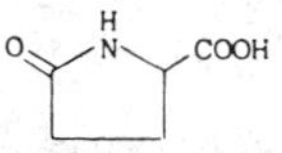

Orthorhombic bisphenoidal crystals from alcohol + petr ether. mp 162-163°. $[\alpha]_D^{20}$ −11.9 (c = 2); $[\alpha]_D^{25}$ −23.6° (c = 5 at pH 7). Sol in water, alcohol, acetone.
USE: In the resolution of racemic amines.

8013. Pyrolan®. *Dimethylcarbamic acid 3-methyl-1-phenyl-1H-pyrazol-5-yl ester;* 1-phenyl-3-methyl-5-pyrazolyl dimethylcarbamate; 3-methyl-1-phenyl-5-pyrazolyl dimethylcarbamate; ENT 17588; G-22008. $C_{13}H_{15}N_3O_2$; mol wt 245.27. C 63.66%, H 6.16%, N 17.13%, O 13.05%. Prepn from 1-phenyl-3-methyl-5-pyrazolone and dimethylcarbamoyl chloride: **Swiss pat.** 279,553 (1952 to Geigy); **Brit.** pat. 681,376. Anticholinesterase activity: Pulver, Domenjoz, *Experientia* **7,** 306 (1951); Ferguson, Alexander, *J. Agr. Food Chem.* **1,** 888 (1953); Müller, Spindler, *Experientia* **10,** 91 (1954).

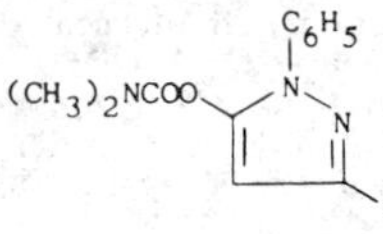

Crystals, mp 50°. $bp_{0.2}$ 160-162°. Possesses water and lipoid solubility. LD_{50} orally in rats: 62 mg/kg, E. W. Schafer, *Toxicol. Appl. Pharmacol.* **21,** 315 (1972).
USE: Insecticide. *Caution:* Cholinesterase inhibitor.

8014. Pyroligneous Acid. Wood vinegar; pyroligneous vinegar. Obtained by destructive distillation of wood.
Yellowish, acid liquid; empyreumatic odor; contains about 6% acetic acid. Miscible with water, alcohol.
USE: Largely for smoking meats.

8015. Pyromellitic Acid. *1,2,4,5-Benzenetetracarboxylic acid.* $C_{10}H_6O_8$; mol wt 254.15. C 47.26%, H 2.38%, O 50.36%. Prepd by heating mellitic (benzenehexacarboxylic) acid with $KHSO_4$ and H_2SO_4: Silberrad, *J. Chem. Soc.* **89,** 1795 (1906) (actually the dianhydride, mp 286°, is obtained by this method). Laboratory prepn from pine or spruce charcoal: Philippi, Thelen, *Org. Syn.* **coll. vol. II,** 551 (1943); by oxidation of benzene derivatives contg substituents in the 1,2,4,5 positions: Jacobsen, *Ber.* **17,** 2516 (1884).

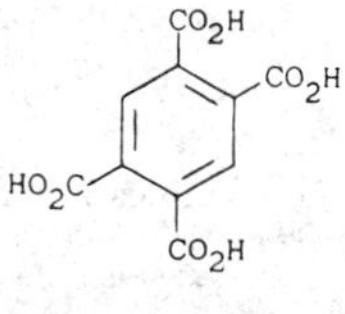

Dihydrate, triclinic plates from water. When anhydr, mp 276°. Distills with anhydride formation. 1.5 g (anhydr) dissolves in 100 ml water. Freely sol in alc.
Tetramethyl ester, $C_{14}H_{14}O_8$, leaflets from alc, mp 141.5°.

8016. Pyronine B. *3,6-Bis(diethylamino)xanthylium chloride;* [6-(diethylamino)-3H-xanthen-3-ylidene]diethyl-ammonium chloride; C.I. 45010. $C_{21}H_{27}ClN_2O$; mol wt 358.91. C 70.28%, H 7.58%, Cl 9.88%, N 7.80%, O 4.46%. Prepn of metallic complex: **Ger.** pat. **54,190** (1889 to Bayer); **Ger.** pat. 59,003 (1891 to A. Leonhardt); Biehringer, *Ber.* **27,** 3299 (1894); *J. Prakt. Chem.* **54,** 217 (1896); Albert, *J. Chem. Soc.* **1947,** 244. Structure: Chamberlin *et al., J. Org. Chem.* **27,** 2263 (1962). The commercial product is a complex with iron as depicted below.

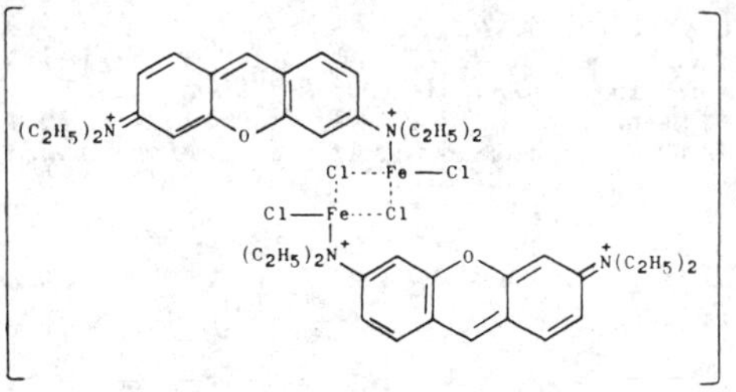

Ferric chloride complex, green metallic needles, mp 176-178°. Absorption max (50% ethanol): 555 nm ($E_{1cm}^{1\%}$ 2324).
USE: Stain for bacteria, molds, ribonucleic acids.

8017. Pyronine Y. *3,6-Bis(dimethylamino)xanthylium chloride;* N-[6-(dimethylamino)-3H-xanthene-3-ylidene]-N-methylmethanaminium chloride; [6-(dimethylamino)-3H-xanthen-3-ylidene]dimethylammonium chloride; tetramethyldiaminoxanthylium chloride; C.I. 45005; pyronine G. $C_{17}H_{19}ClN_2O$; mol wt 302.80. C 67.43%, H 6.32%, Cl 11.71%, N 9.25%, O 5.28%. Prepn from dimethyl-*m*-aminophenol: Mohlau, Koch, *Ber.* **27,** 2887 (1894); **Ger.** pats. **58,955; 59,003; 63,081** (1892 to Leonhardt); *Frdl.* **3,** 92-94. *See also: Colour Index* vol. **4** (3rd ed., 1971) p 4417.

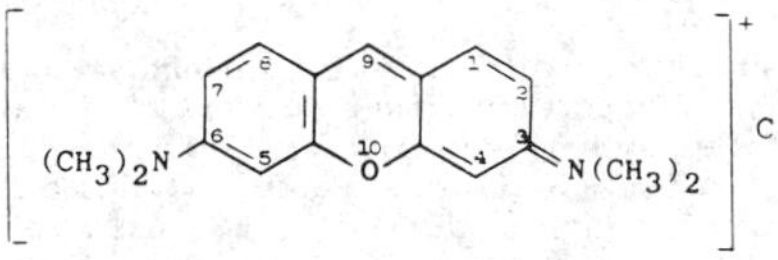

Ferric chloride complex, lustrous green crystals from alc. Soly in water at 26°: 8.96% giving a red soln. Soly in alc at 26°: 0.60%. Solns show a yellowish fluorescence. Absorption max about 552 nm; curves: Stotz *et al., Stain Technol.* **25,** 57 (1950).
USE: Bacterial and biological stain.

8018. Pyrophosphoric Acid. *Diphosphoric acid.* $H_4O_7P_2$; mol wt 177.99. H 2.27%, O 62.93%, P 34.81%. Prepd according to the equation $5H_3PO_4 + POCl_3 \rightarrow 3H_4P_2O_7 + 3HCl$: Geuther, *J. Prakt. Chem.* [2] **8,** 359 (1874); Partington, Wallsom, *Chem. News* **136,** 97 (1928); by heating H_3PO_4: Bell, *Ind. Eng. Chem.* **40,** 1464 (1948).

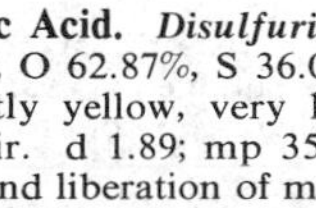

Hygroscopic glass, seldom acicular crystals, mp 61°. K_1 at 18° = 0.14; K_2 = 0.011; K_3 = 2.1 × 10^{-7}; K_4 = 4.1 × 10^{-10}. Forms normal salts, such as $Na_4P_2O_7$, and dihydrogen salts, such as $Na_2H_2P_2O_7$. Solubility in water (23°): 709 g/100 ml. Quickly converted to phosphoric acid when dissolved in hot water. Sol in alcohol and ether.

8019. Pyrosulfuric Acid. *Disulfuric acid.* $H_2O_7S_2$; mol wt 178.15. H 1.13%, O 62.87%, S 36.00%. H_2SO_7.
Colorless to slightly yellow, very hygroscopic crystals, fuming strongly in air. d 1.89; mp 35°. Very sol in water with violent hissing and liberation of much heat. *Keep tightly closed and handle with caution.*

8020. Pyrosulfuryl Chloride. *Disulfuryl chloride;* disulfur pentoxydichloride; chlorosulfonic anhydride. $Cl_2O_5S_2$; mol wt 215.05. Cl 32.98%, O 37.20%, S 29.82%. $S_2O_5Cl_2$.
Colorless, mobile, very refractive, fuming liquid. d 1.837; mp −37°; bp 151°. Violently dec by water into H_2SO_4, and HCl. *Keep tightly closed and handle with caution.*

8021. Pyrovalerone. *1-(4-Methylphenyl)-2-(1-pyrrolidinyl)-1-pentanone;* 4′-methyl-2-(1-pyrrolidinyl)valerophenone; α-pyrrolidino-p-methylvalerophenone; 1-(1-pyrrolidinyl)butyl p-tolyl ketone; 1-(p-tolyl)-1-oxo-2-pyrrolidinon-pentane; 1-(p-tolyl)-2-pyrrolidino-1-pentanone. $C_{16}H_{23}$-

NO; mol wt 245.35. C 78.32%, H 9.45%, N 5.71%, O 6.52%. Prepn: **Brit. pats. 927,475** and **933,507** (both 1963 to Wander and to Thomae); Heffe, *Helv. Chim. Acta* **47**, 1289 (1964). Pharmacology: Stille *et al., Arzneimittel-Forsch.* **13**, 871 (1963). Metabolism: Michaelis *et al., J. Med. Chem.* **13**, 497 (1970).

bp$_{0.08}$ 104°.
Hydrochloride, C$_{16}$H$_{24}$ClNO, **F 1983, Centroton, Thymergix**. Crystals from 2-butanone or from methanol + acetone + diethyl ether, mp 178°. LD$_{50}$ orally in mice: 350 mg/kg, Stille *et al., loc. cit.*
THERAP CAT: Central stimulant.

8022. Pyroxylin. *Cellulose nitrate; nitrocellulose;* collodion cotton; soluble gun cotton; collodion wool; colloxylin; xyloidin; celloidin; Parlodion. Variable mixture which consists chiefly of cellulose tetranitrate. *Review:* R. T. Bogan *et al.* in Kirk-Othmer *Encyclopedia of Chemical Technology* **vol. 5** (Wiley-Interscience, New York, 3rd ed., 1979) pp 129-143.
Yellowish-white, matted mass of filaments, having the appearance of raw cotton. *Highly flammable;* pyroxlin with higher nitrogen content *may explode!* Flash pt 40°F (closed cup); ignites at 160-170°. When kept in well-closed containers and exposed to light it dec. Sol in 25 parts of a mixture of 1 vol alcohol and 3 vols ether; also sol in methanol, acetone, glacial acetic acid, amyl acetate. *Keep loosely packed in cartons and protected from light and moisture.* Can be shipped with safety only when wet with 25-30% water or alcohol.
USE: In manuf of collodions; in lacquer coatings, inks, adhesives. Cellulose hexanitrate is used in explosives and propellants. Celloidin is used for embedding sections in microscopy; in electrotechnics, photography, galvanoplasty.
THERAP CAT: Topical protectant.

8023. Pyrobutamine. *1-[4-(4-Chlorophenyl)-3-phenyl-2-butenyl]pyrrolidine; 1-(γ-p-chlorobenzylcinnamyl)pyrrolidine;* 1-p-chlorophenyl-2-phenyl-4-pyrrolidyl-2-butene; Pyronil. C$_{20}$H$_{22}$ClN; mol wt 311.87. C 77.03%, H 7.11%, Cl 11.37%, N 4.49%. *Ref:* Lee *et al., Proc. Soc. Exp. Biol. Med.* **80**, 458 (1952). Prepn: Mills, **U.S. pat. 2,655,509** (1953 to Eli Lilly).

Oily liquid, bp$_{0.3}$ 190-195°. On standing gives crystals, mp 48-49°.
Diphosphate, C$_{20}$H$_{28}$ClNO$_8$P$_2$, crystals from alcohol + ether, mp 129.5-130°. Soluble in warm water to the extent of 10%. Soly in alcohol at 25° about 5%. Practically insol in chloroform, ether.
Hydrochloride, C$_{20}$H$_{22}$ClN.HCl, crystals from alcohol + ether, mp 227-228°.
Hydrobromide, C$_{20}$H$_{22}$ClN.HBr, crystals from alcohol + ether, mp 228-229°.
THERAP CAT: Antihistaminic.

8024. Pyrrocaine. *N-(2,6-Dimethylphenyl)-1-pyrrolidineacetamide; 1-pyrrolidineaceto-2',6'-xylidide;* 2-(1-pyrrolidinyl)-2',6'-acetoxylidide; 1-pyrrolidinoaceto-2,6-dimethylanilide; EN 1010; NSC-52644; Endocaine; Dynacaine. C$_{14}$H$_{20}$N$_2$O; mol wt 232.32. C 72.38%, H 8.68%, N 12.06%, O 6.89%. Prepn: Schlesinger, Gordon, **U.S. pat. 2,813,861** (1957 to Endo); Löfgren *et al., Acta Chem. Scand.* **11**, 1724 (1957).

Crysts from hexane or petr ether + dibutyl ether, mp 83°. Hydrochloride, C$_{14}$H$_{20}$N$_2$O.HCl, crystals from isopropanol, mp 205°. Soluble in water, alcohol, isopropyl alcohol. Practically insol in chloroform, ether.
THERAP CAT: Anesthetic (local).

8025. 1H-Pyrrole. Azole; imidole; divinylenimine. C$_4$H$_5$N; mol wt 67.09. C 71.60%, H 7.51%, N 20.89%. A constituent of coal tar and bone oil: Runge, *Ann. Phys.* **31**, 67 (1834). Prepd industrially by fractional distillation of bone oil, or by the thermal decompn of ammonium mucate with glycerol or mineral oil: McElvain, Bolliger, *Org. Syn.* **coll. vol. I** (2nd ed., 1941) p 473; Blicke, Powers, *Ind. Eng. Chem.* **19**, 1334 (1927). Also formed on heating of albumin; on heating sheep's wool with aq barium hydroxide soln; by pyrolysis of gelatin. Alternate prepns from acetaldehyde and ammonia: Tschitschibabin, *Chem. Zentr.* **1916**, I, 920; from succindialdehyde with ammonia and acetic acid: Harries, *Ber.* **34**, 1496 (1901); **35**, 1183 (1902); distilling succinimide with zinc or sodium: Bell, Bernthsen, *Ber.* **13**, 877, 1049 (1880). Purification and physical properties: R. V. Helm *et al., J. Phys. Chem.* **62**, 858 (1958). *Review:* Fischer-Orth, *Die Chemie des Pyrrols* (Leipzig, 1934-1940); E. Vittort, L. R. Anderson in Kirk-Othmer *Encyclopedia of Chemical Technology* **vol. 19** (Wiley-Interscience, New York, 3rd ed., 1982) pp 499-520.

Liquid. Agreeable empyreumatic odor resembling that of chloroform. Colorless when freshly distilled, darkens unless every trace of oxygen is removed. d$_4^{20}$ 0.9691. bp$_{760}$ 129.8°. Best distilled *in vacuo.* n$_D^{20}$ 1.5085. Flash pt, closed cup: 102° F (390° C). Absorption spectrum: Menczel, *Phys. Chem.* **125**, 161; *Chem. Zentr.* **1927**, I, 2510. Sparingly sol in water; freely sol in alcohol, benzene, ether. Insol in aq alkalies. Sol in dil acids with decompn. Solns in dil HCl yield pyrrole red, an amorphous, orange-colored substance; also polymerization takes place under the influence of acids and glycols.

8026. Pyrrolidine. Tetrahydropyrrole. C$_4$H$_9$N; mol wt 71.12. C 67.55%, H 12.76%, N 19.70%. Found in tobacco and carrot leaves. Probable biosynthesis from ornithine and putrescine. Usually prepd by reduction of pyrrole.

Almost colorless liquid; unpleasant ammonia-like odor. Fumes in air. bp 88.5-89°. d$_4^{22.5}$ 0.8520. n$_D^{28}$ 1.4402. Strong base. K at 25° = 1.3 × 10^{-3}. Miscible with water. Soluble in alcohol, ether, chloroform.

8027. 2-Pyrrolidone. *2-Pyrrolidinone;* 2-oxopyrrolidine; α-pyrrolidone; 2-ketopyrrolidine. C$_4$H$_7$NO; mol wt 85.10. C 56.45%, H 8.29%, N 16.46%, O 18.80%. Prepd on a large scale from butyrolactone by a Reppe process: **Ger. pat. 1,085,525** (to BASF). Other prepns: Metzger, Seelert, *Angew. Chem.* **75**, 919 (1963); Copenhaver, Ney, **U.S. pat. 3,095,423** (1963 to Minnesota Mining & Manuf); Lidov, **U.S. pat. 3,109,005** (1963 to Halcon International).

Liquid above 25°. bp$_{760}$ 245°; bp$_{9.2}$ 113-114°; bp$_{0.2}$ 76°. d$_4^{25}$ 1.116. Flash pt, open cup: 265°F (129°C). Viscosity at 25° = 13.3 cps. Dipole moment: 2.3. In the presence of the stoichiometric amount of water a cryst monohydrate, mp 30°, can be formed. Non-corrosive. Good chemical stability. Miscible with water, ethanol, ether, chloroform, benzene, ethyl acetate, carbon disulfide.

USE: Intermediate in the manuf of polyvinylpyrrolidone and polypyrrolidone (a polymer, formed in the presence of alkaline catalysts). Also used as high-boiling solvent in petroleum processing, acrylonitrile manuf. Industrial solvent for polymers, chlordane, DDT, sorbitol, glycerol, iodine, sugars. In specialty printers inks. As plasticizer and coalescing agent for acrylic-styrene emulsion-type floor polishes.

8028. 3-Pyrroline. *2,5-Dihydro-1H-pyrrole.* C$_4$H$_7$N; mol wt 69.10. C 69.52%, H 10.21%, N 20.27%. Prepd by reduction of pyrrole with zinc and glacial acetic or hydrochloric acid.

Almost colorless liquid. Unpleasant, ammonia-like odor. Fumes in air. Very hygroscopic, also absorbs CO$_2$. bp$_{748}$ 90-91°; d$_4^{20}$ 0.9097; n$_D^{20}$ 1.4664. Strong base. Miscible with water. Sol in alcohol, ether, chloroform.

8029. Pyrrolnitrin. *3-Chloro-4-(3-chloro-2-nitrophenyl)pyrrole;* 3-chloro-4-(2'-nitro-3'-chlorophenyl)pyrrole; PN; Pyroace. C$_{10}$H$_6$Cl$_2$N$_2$O$_2$; mol wt 257.09. C 46.72%, H 2.35%, Cl 27.58%, N 10.90%, O 12.45%. Antifungal antibiotic isolated from *Pseudomonas pyrrocinia* n. sp.: K. Arima et al., Agr. Biol. Chem. 28, 575 (1964); eidem, J. Antibiot. 18, 201 (1965). Structure: Imanaka et al., ibid. 207. Synthesis: S. Umio et al., Belg. pat. 670,427; eidem, U.S. pat. 3,428,648 (1964, 1969 both to Fujisawa); Nakano et al., Tetrahedron Letters 1966, 737; S. Umio et al. (10 publications) Chem. Pharm. Bull. 17, 559-628 (1969); Gosteli, Helv. Chim. Acta 55, 451 (1972). Pharmacological studies: M. Nishida et al., J. Antibiot. 18, 211 (1965).

Pale yellow crystals from hot cyclohexane, mp 124.5°. Gradually changes to a red or brown color on exposure to sunlight and loses antibiotic activity. uv max: 252 nm (ε 7500). Slightly sol in water, petr ether, cyclohexane; sol in methanol, ethanol, butanol, acetone, ethyl acetate, benzene, ether, chloroform, carbon tetrachloride, pyridine, acetic acid. LD$_{50}$ in rats, rabbits (mg/kg): 68, 105 i.p. (Nishida).
THERAP CAT: Antifungal.

8030. Pyruvaldehyde. *2-Oxopropanal;* methylglyoxal; 2-ketopropionaldehyde; acetylformaldehyde. C$_3$H$_4$O$_2$; mol wt 72.06. C 50.00%, H 5.60%, O 44.40%. CH$_3$COCHO. Obtained by warming isonitrosoacetone with dilute sulfuric acid: v. Pechmann, Ber. 20, 3213 (1887); by distilling a dilute solution of dihydroxyacetone from calcium carbonate: Bernhauer, Görlich, Biochem. Z. 212, 462 (1929). There are several biochemical methods of prepn.

Yellow, mobile liquid, pungent odor. Polymerizes very readily. d^{24} 1.0455. n$_D^{17.5}$ 1.4002. Hygroscopic. bp$_{760}$ 72° forming a yellowish-green vapor. May be kept in a sealed tube for several days. The polymer, a viscous solid, dissolves in water with the evolution of heat, giving monomeric solns. Faintly acid to litmus. Sol in alc, giving a colorless soln. Sol in ether, benzene, giving yellow solns.
Oxime, see Isonitrosoacetone.

8031. Pyruvate Decarboxylase. *Pyruvic decarboxylase;*

pyruvic carboxylase; α-carboxylase; 2-oxo-acid carboxylyase. An enzyme found in yeast, in plants, in wheat germ and in bacteria: Neuberg, Rosenthal, Biochem. Z. 51, 142 (1913); Holzer et al., ibid. 327, 331 (1956); Singer, Pensky, J. Biol. Chem. 196, 375 (1952); King, Chelelin, ibid. 208, 821 (1954); Howells, Lindstrom, J. Bacteriol. 75, 305 (1958). Cofactors: Diphosphothiamine, divalent metal ions. Catalyzes decarboxylation of α-ketoacids to aldehydes + CO$_2$: Juni, J. Biol. Chem. 236, 2302 (1961). Structure and mechanism of action of the yeast enzyme: Schellenberger, Angew. Chem. Int. Ed. 6, 1024 (1967). Review: Utter, The Enzymes vol. 5, P. D. Boyer et al., Eds. (Academic Press, New York, 2nd ed., 1961) pp 320-323.

8032. Pyruvic Acid. *2-Oxopropanoic acid;* α-ketopropionic acid; acetylformic acid; pyroracemic acid; Brenztraubensäure (German). C$_3$H$_4$O$_3$; mol wt 88.06. C 40.92%, H 4.58%, O 54.51%. CH$_3$COCOOH. Intermediate in sugar metabolism and in enzymatic carbohydrate degradation (alcoholic fermentation) where it is converted to acetaldehyde and CO$_2$ by carboxylase, see Nord, Chem. Rev. 26, 423 (1940). In muscle, pyruvic acid (derived from glycogen) is reduced to lactic acid during exertion, which is reoxidized and partially retransformed to glycogen during rest. The liver can convert pyruvic acid to alanine by amination. Pyruvic acid has been isolated from cane sugar fermentation broth by fixation with β-naphthylamine giving α-methyl-β-naphthocinchonic acid: Grab, Biochem. Z. 123, 84 (1921). The most practical method of preparing pyruvic acid is by distillation of tartaric acid in presence of potassium acid sulfate as dehydrating agent. The distillate must be fractionated under reduced pressure: Erlenmeyer, Ber. 14, 321 (1881); Döbner, Ann. 242, 269 (1887); Howard, Fraser, Org. Syn. coll. vol. I (2nd ed., 1941) p 475.

Liquid. Odor resembling that of acetic acid. d$_4^{15}$ 1.267; mp 11.8°; bp$_{760}$ 165° (dec); bp$_{100}$ 106.5°; bp$_{40}$ 85.3°; bp$_{20}$ 70.8°; bp$_{10}$ 57.9°; bp$_5$ 45.8°; bp$_{1.0}$ 21.4°; n$_D^{20}$ 1.4138. Absorption spectrum: Henri, Fromageot, Bull. Soc. Chim. [4] 37, 846 (1925). K at 25°: 3.2 × 10^{-3}. Miscible with water, alcohol, ether. Polymerizes and dec on standing unless pure and kept in container with airtight closure.

Methyl ester, C$_4$H$_6$O$_3$, liquid, bp$_{760}$ 134-137°.
Ethyl ester, C$_5$H$_8$O$_3$, liquid, bp$_{760}$ 155°; bp$_{750}$ 147.5°; bp$_{42}$ 69-71°; bp$_{20}$ 66° also 56°. Prepn from ethyl lactate by oxidation with KMnO$_4$: Cornforth, Org. Syn. coll. vol. IV, 467 (1963).

8033. Pyrvinium Chloride. *6-(Dimethylamino)-2-[2-(2,5-dimethyl-1-phenyl-1H-pyrrol-3-yl)ethenyl]-1-methylquinolinium chloride;* viprynium chloride; SN 4395. C$_{26}$H$_{28}$ClN$_3$; mol wt 417.99. C 74.71%, H 6.75%, Cl 8.48%, N 10.05%. Prepn: Van Lare, Brooker, U.S. pat. 2,515,912 (1950).

Dihydrate, deep-red powder, dec 249-251°. Sparingly sol in water.

8034. Pyrvinium Pamoate. *6-(Dimethylamino)-2-[2-(2,5-dimethyl-1-phenyl-1H-pyrrol-3-yl)ethenyl]-1-methylquinolinium, salt with 4,4'-methylenebis[3-hydroxy-2-naphthalenecarboxylic acid] (2:1); bis[6-(dimethylamino)-2-[2-(2,5-dimethyl-1-phenylpyrrol-3-yl)vinyl]-1-methylquinolinium] 4,4'-methylenebis(3-hydroxy-2-naphthoate); 6-(dimethylamino)-2-[2-(2,5-dimethyl-1-phenyl-3-pyrryl)vinyl]-1-methylquinolinium salt of 2,2'-dihydroxy-1,1'-dinaphthylmethane-3,3'-dicarboxylic acid;* pyrvinium embonate; viprynium embonate; Alnoxin; Molevac; Neo-Oxypaat; Pamovin; Poquil; Povan; Povanyl; Pyrcon; Altolat; Tolapin; Tru; Vanquil; Vanquin; Vermitiber. C$_{75}$H$_{70}$N$_6$O$_6$; mol wt

1151.44. C 78.24%, H 6.13%, N 7.30%, O 8.34%. Prepn:
Elslager, Worth, U.S. pat. **2,925,417** (1960 to Parke, Davis).

Bright orange or orange-red to almost black precipitate,
mp 210-215° (softens at 190°). Stable to heat, light and air.
Slightly soluble in chloroform, methoxyethanol; very slightly
sol in alcohol. Practically insol in water, ether. Absorption
max: 236, 356, 503 nm.

THERAP CAT: Anthelmintic (Nematodes).

Q

8035. Q-Enzyme®. α-*Glucan branching glycosyltrans-ferase;* potato branching enzyme; enzyme Q; branching factor; glucosan transglycosylase. A widely distributed enzyme which transfers part of a 1,4-glucosan chain from a 4- to a 6-position in polysaccharides. Isoln from potatoes: Barker *et al., J. Chem. Soc.* **1949**, 1705. Crystallization: Gilbert, Patrick, *Biochem. J.* **51**, 181 (1952). Improved isoln procedure: Griffen, Wu, *Biochemistry* **7**, 3063 (1968). Although Q-enzyme is unstable in aq soln, it can be stored for long periods without appreciable loss in activity when it is freeze-dried. *Review:* Barker, Bourne, *Quart. Rev. Chem. Soc.* **7**, 65-68 (1953); Dixon, Webb, *Enzymes* (Academic Press, New York, 1958) *passim.*

8036. Quassia. Bitter wood; bitter ash. The wood of *Picrasma excelsa* (Sw.) Planch, or of *Quassia amara* L., *Simaroubaceae.* The first is known in commerce as Jamaica quassia, the second as Surinam quassia. *Habit. Picrasma excelsa* inhabits Jamaica and the Caribbean Islands; *Quassia amara* is a native of Brazil and Guiana and is cultivated in Colombia, Panama, and the West Indies. Quassin and neo-quassin are the bitter principles of Surinam quassia; picrasmin, that of Jamaica quassia. These bitter pinciples are obtained in yields of 0.1-0.2% and appear commercially under the name of quassin.

Unground quassia occurs usually in chips, raspings, or shavings, occasionally in billets; yellowish-white to bright yellow with a few light gray pieces; coarsely grained, fibrous. Slight odor; very bitter taste. The powdered form is pale yellow in color.

USE: The extract is used for fly poison on flypaper; to imitate hops.

THERAP CAT: Anthelmintic.

THERAP CAT (VET): Has been used as a bitter, as an anthelmintic.

8037. Quassin. 2,12-*Dimethoxypicrasa-2,12-diene-1,11,-16-trione; 3aβ,6aβ,7,7aα,8,11a,11bα,11c-octahydro-2,10-di-methoxy-3,8α,11aβ,11cβ-tetramethylphenanthro[10,1-bc]pyran-1,5,11(4H)-trione;* nigakilactone D. $C_{22}H_{28}O_6$; mol wt 388.44. C 68.02%, H 7.27%, O 24.71%. One of the bitter constituents of the wood of *Quassia amara* L., *Simarouba-ceae* known in commerce as Surinam quassia. Obtained by the resolution of the mixture of bitter constituents of quassia wood: E. P. Clark, *J. Am. Chem. Soc.* **59**, 927 (1937); London *et al., J. Chem. Soc.* **1950**, 3431. Structure: Valenta *et al., Tetrahedron Letters* no. **20**, 25 (1960); Carman, Ward, *ibid.* **1961**, 317; Valenta *et al., Tetrahedron* **15**, 100 (1961). Stereochemistry: Valenta *et al., ibid.* **18**, 1433 (1962). Identity with nigakilactone D: Murae *et al., ibid.* **27**, 1545 (1971). Synthetic approach: Stojanac *et al., Can. J. Chem.* **53**, 619 (1975); P. A. Grieco *et al., Tetrahedron Letters* **21**, 1619 (1980). Total synthesis of *dl*-form: *eidem, J. Am. Chem. Soc.* **102**, 7586 (1980).

Very bitter rectangular plates from dilute methanol, mp 222°. $[\alpha]_D^{20}$ +34.5° (c = 5.09 in $CHCl_3$). uv max: ∼255 nm (ε ∼11,650). Sol in benzene, alc, acetone, chloroform, pyridine, acetic acid, hot ethyl acetate. Sparingly sol in ether, petr ether. Bitterness threshold 1:60,000.

8038. Quatrimycin. [4R-(4α,4aβ,5aβ,6α,12aβ)]-4-(Dimethylamino)-1,4,4a,5,5a,6,11,12a-octahydro-3,6,10,12,-12a-pentahydroxy-6-methyl-1,11-dioxo-2-naphthacenecarb-oxamide; epitetracycline. $C_{22}H_{24}N_2O_8$; mol wt 444.43. C

59.45%, H 5.44%, N 6.30%, O 28.80%. Prepn by epimerization of tetracycline: McCormick *et al., J. Am. Chem. Soc.* **79**, 2849 (1957); Kaplan *et al., Antibiot. Chemother. (Basel)* **7**, 569 (1957); Remmers *et al., J. Pharm. Sci.* **52**, 752 (1963). Acute toxicity: L. Kung, H. Sun, *Yao Hsueh Hsueh Pao* **13**, 244 (1966), *C.A.* **65**, 7869e (1966).

Monohydrate, crystals, mp 178° dec. $[\alpha]_D^{25}$ −335° (c = 0.5 in 0.03N HCl). uv max (0.01N H_2SO_4): 216, 255, 270, 355 (ε 13,900, 16,400, 15,200, 14,700). LD_{50} i.v. in mice: 85.8 mg/kg (Kung, Sun).

Ammonium salt monohydrate, $C_{22}H_{27}N_3O_8 \cdot H_2O$, yellow crystals, mp 170° dec. $[\alpha]_D^{25}$ −321° (c = 0.5 in 0.03N HCl). Soly in water > 50 mg/kg.

Hydrochloride, $C_{22}H_{25}ClN_2O_8$, amorphous yellow solid. $[\alpha]_D^{25}$ −325° (c = 0.58 in 0.2N HCl). Soly in water > 100 mg/ml.

Methiodide, $C_{23}H_{27}IN_2O_8$, crystals, mp 161-162°. $[\alpha]_D^{25}$ −265° (c = 0.5 in 0.03N HCl).

8039. Quazepam. 7-*Chloro-5-(2-fluorophenyl)-1,3-di-hydro-1-(2,2,2-trifluoroethyl)-2H-1,4-benzodiazepine-2-thi-one;* Sch-16134; Dormalin; Oniria; Prosedar; Quazium; Selepam. $C_{17}H_{11}ClF_4N_2S$; mol wt 386.79. C 52.79%, H 2.87%, Cl 9.16%, F 19.65%, N 7.24%, S 8.29%. Benzodiazepine hypnotic. Prepn: M. Steinman, **Ger. pat. 2,138,773;** *idem,* **U.S. pat. 3,845,039** (1972, 1974 both to Schering Corp.); M. Steinman *et al., J. Med. Chem.* **16**, 1354 (1973). Pharmacology: A. Barnett *et al., Arzneimittel-Forsch.* **32**, 1452 (1982); E. Ongini *et al., ibid.* 1456. GC determn in plasma: M. Hilbert *et al., J. Pharm. Sci.* **73**, 516 (1984). Pharmacokinetics: M. Chung *et al., Clin. Pharm. Ther.* **35**, 520 (1984). Metabolism: N. Zampaglione *et al., Drug Metabol. Dispos.* **13**, 25 (1985). Clinical trials in insomnia: J. W. Goethe, G. Kader, *Curr. Ther. Res.* **32**, 150 (1982); M. Mamelak *et al., J. Clin. Pharmacol.* **24**, 65 (1984). Toxicology: H. E. Black *et al., Arzneimittel-Forsch.* **37**, 906 (1987).

Crystals from methylene chloride-hexane, mp 137.5-139°. LD_{50} in mice (mg/kg): > 1370 i.v., > 5000 orally (Orgini). LD_{50} in male, female mice, rats (mg/kg): 845, 921, 3072, 2749 i.p. (Black).

Note: This is a controlled substance (depressant) listed in the U.S. Code of Federal Regulations, Title 21 Part 1308.14 (1987).

THERAP CAT: Sedative; hypnotic.

8040. Quebrachamine. 7-*Ethyl-1,4,5,6,7,8,9,10-octahy-dro-2H-3,7-methanoazacycloundecino[5,4-b]indole;* kamassin. $C_{19}H_{26}N_2$; mol wt 282.41. C 80.80%, H 9.28%, N 9.92%. From the bark of *Aspidosperma quebracho blanco* Schlecht., *Apocynaceae:* Hesse, *Ann.* **211**, 249 (1882); Field, *J. Chem. Soc.* **125**, 1444 (1924). Identity with kamassin: Gellert, Witkop, *Helv. Chim. Acta* **35**, 114 (1952). Structure: Witkop, *J. Am. Chem. Soc.* **79**, 3193 (1957); Kny, Witkop, *J. Org. Chem.* **25**, 635 (1960); Biemann, Spiteller, *J. Am. Chem. Soc.* **84**, 4578 (1962). Crystal structure: C. Puglisi *et al., Acta Crystallogr.* **B32**, 1900 (1976). Total synthesis of *dl*-form: Stork, Dolfini, *J. Am. Chem. Soc.* **85**, 2872 (1963);

Ziegler *et al., ibid.* **91,** 2342 (1969); Kutney *et al., ibid.* **92,** 1727 (1970); V. S. Giri *et al., J. Heterocycl. Chem.* **17,** 1133 (1980); S. Takano *et al., Heterocycles* **16,** 247 (1981). Enantioselective synthesis: *eidem, Chem. Commun.* **1980,** 616; **1981,** 1153.

Bitter leaflets, mp 145-147°. $[\alpha]_D^{20}$ −109 to −110° (acetone). uv max (methanol): 230, 287, 293 nm (log ε 4.55, 3.85, 3.84). Sol in acetone, alc, chloroform, ether, dil acids.

8041. Quebracho Colorado. Red quebracho. Wood of *Loxopterygium lorentzii* Griseb., *Anacardiaceae. Habit.* Argentine Republic. *Constit.* Tannin, coloring matter, loxopterygine.
USE: In dyeing and tanning.

8042. Queen Substance. *(E)-9-Oxo-2-decenoic acid.* $C_{10}H_{16}O_3$; mol wt 184.23. C 65.19%, H 8.75%, O 26.05%. Secreted in the mandibular gland of queen honey bees *(Apis mellifera, A. florea, A. cerana, A. dorsata)*; inhibits the development of ovaries in worker bees, prevents queen cell formation and attracts male bees (drones) to virgin queens for the purpose of mating: Butler, *Experientia* **13,** 256 (1957); Sannasi, Rajulu, *Life Sci.* **10,** part 2, 195 (1971). Similarity with the ovary inhibiting hormone of prawns *(Leander serratus):* Carlisle, Butler, *Nature* **177,** 276 (1956). Extraction and purification: Carlisle, Butler, *loc. cit.;* Butler *et al., Nature* **184,** 1871 (1959). Synthesis: Barbier *et al., Compt. Rend.* **251,** 1133 (1960); Jaeger, Robinson, *Tetrahedron* **14,** 320 (1961); B. M. Trost, T. N. Salzman, *J. Org. Chem.* **40,** 148 (1975); J. Tsuji *et al., Tetrahedron Letters* **1977,** 2267; C. S. Subramaniam *et al., Ind. J. Chem.* **16B,** 318 (1978); T. Fujisawa *et al., Chem. Letters* **1982,** 219; Y. Naoshima *et al., Agr. Biol. Chem.* **48,** 2151 (1984).

Transparent elongated plates from ether + petr ether or aq methanol, mp 54.5-55.5°. Stable to heat, acids, less stable to alkalies. Sol in acetone, alcohol. IR spectrum: Butler *et al., loc. cit.*

8043. Quercetagetin. *2-(3,4-Dihydroxyphenyl)-3,5,6,7-tetrahydroxy-4H-1-benzopyran-4-one; 3,3′,4′,5,6,7-hexahydroxyflavone;* 6-hydroxycyanidenolon 1555. $C_{15}H_{10}O_8$; mol wt 318.23. C 56.61%, H 3.17%, O 40.22%. From flowers of French marigold, *Tagetes patula* Linn., *Compositae:* Perkin, *J. Chem. Soc.* **103,** 209 (1913). Synthesis: Baker *et al., ibid.* **1929,** 74; Rao, Seshadri, *Proc. Indian Acad. Sci.* **23A,** 23 (1946), *C.A.* **40,** 5052² (1946).

Dihydrate, pale yellow needles from dil alcohol, mp 318°. uv max (alc): 259, 361 nm (log ε 4.23, 4.34). Sol in hot alcohol; sparingly sol in boiling water.
Hexaacetate, $C_{27}H_{22}O_{14}$, needles from alcohol + acetic acid, mp 209-211°. Sparingly sol in alc.
7-Glucoside, $C_{21}H_{20}O_{13}$, *quercetagitrin.* From flowers of the African marigold, *Tagetes erecta* L., *Compositae:* Rao,

Seshadri, *Proc. Indian Acad. Sci.* **14A,** 289 (1941), *C.A.* **36,** 2555³ (1942); from *Chrysanthemum coronarium* L., *Compositae:* Anyas, Steelink, *Arch. Biochem. Biophys.* **90,** 63 (1960). Structure: Rajagopalan, Seshadri, *Proc. Indian Acad. Sci.* **28A,** 31 (1948), *C.A.* **43,** 4265b (1949). Crystals from aqueous pyridine, dec 236-238°. uv max (95% ethanol): 260, 272, 362 nm.

8044. Quercetin. *2-(3,4-Dihydroxyphenyl)-3,5,7-trihydroxy-4H-1-benzopyran-4-one; 3,3′,4′,5,7-pentahydroxyflavone;* meletin; sophoretin; cyanidenolon 1522. $C_{15}H_{10}O_7$; mol wt 302.23. C 59.61%, H 3.34%, O 37.06%. The aglucon of quercitrin, of rutin, and of other glycosides. Widely distributed in the plant kingdom, esp in rinds and barks, in clover blossoms and in ragweed pollen. Isoln from *Rhododendron cinnabarinum* Hook, *Ericaceae:* Rangaswami *et al., Proc. Indian Acad. Sci.* **56A,** 239 (1962), *C.A.* **58,** 9414a (1963). Structure: Underhill *et al., Can. J. Biochem. Physiol.* **35,** 219 (1957). Biosynthesis: Watkin *et al., ibid.* 229; Grisebach, *Biochem. J.* **85,** 3p (1962); Patschke *et al., Z. Naturforsch.* **21b,** 201 (1966). Synthesis: Shakhova *et al., Zh. Obshch. Khim.* **32,** 390 (1962), *C.A.* **58,** 1426f (1963). Metabolism: Nakagawa *et al., Biochim. Biophys. Acta* **97,** 233 (1965). Toxicity data: M. Sullivan *et al., Proc. Soc. Exp. Biol. Med.* **77,** 269 (1951). *See also* Bioflavonoids.

Dihydrate, yellow needles from dil alcohol. Becomes anhyd at 95-97°. When anhydr dec 314°. uv max (alc): 258, 375 nm (log ε 2.75, 2.75). One gram dissolves in 290 ml abs alc, in 23 ml boiling alc. Soluble in glacial acetic acid; in aq alkaline solns with yellow color. Practically insol in water. Alcoholic solns taste very bitter. LD_{50} orally in mice: 160 mg/kg (Sullivan).
Pentabenzyl ether, $C_{50}H_{40}O_7$, *3,3′,4′,5,7-pentakis(benzyloxy)flavone, penta-O-benzylquercetin, Parietrope.* Prepn: Chopin, Chadenson, *Compt. Rend. Ser. C* **263,** 729 (1966); Binovic, **Ger.** pat. **2,122,514** (1972 to Biosedra), *C.A.* **76,** 113072n (1972). Crystals, mp 123-125°. uv max (chloroform): 249, 343 nm (log ε 4.43, 4.14).
3-D-Galactoside hemipentahydrate, $C_{21}H_{20}O_{12}\cdot2\frac{1}{2}H_2O$, *hyperin, hyperoside.* From *Acacia melanoxylon* R. Br., *Leguminosae:* Falco, de Vries, *Naturwiss.* **51,** 462 (1964). Yellow needles from ethanol, dec 227-230°. $[\alpha]_D^{20}$ −83° (c = 0.2 in pyridine). uv max: 259, 364 nm (log ε 4.31, 4.39).
THERAP CAT: Capillary protectant.

8045. Quercimeritrin. *2-(3,4-Dihydroxyphenyl)-7-(β-D-glucopyranosyloxy)-3,5-dihydroxy-4H-1-benzopyran-4-one;* quercetin-7-D-glucoside; 3,3′,4′,5,7-pentahydroxyflavone-7-D-glucoside. $C_{21}H_{20}O_{12}$; mol wt 464.37. C 54.31%, H 4.34%, O 41.34%. Found in flowers of *Gossypium herbaceum* L., *Malvaceae:* Perkin, *J. Chem. Soc.* **95,** 2181 (1909); from leaves of *Chrysanthemum segetum* L. and *C. coronarium* L., *Compositae:* Geissman, Steelink, *J. Org. Chem.* **22,** 946 (1957); Anyas, Steelink, *Arch. Biochem. Biophys.* **90,** 63 (1960). Structure: Attree, Perkin, *J. Chem. Soc.* **1927,** 234; Rao, Seshadri, *Proc. Indian Acad. Sci.* **9A,** 365 (1939), *C.A.* **34,** 107¹ (1940); Pacheco, Grouiller, *Compt. Rend.* **253,** 1178 (1961).
Trihydrate, yellow plates from aq pyridine. The water of crystn is given up at 100°, the anhydr material is hygroscopic, mp 247-249°. uv max (ethanol): 372, 257 nm (log ε 4.33, 4.38). Practically insol in cold water, more sol in hot water; sol in methanol. Sol in aq alkaline solns with deep yellow color. Is hydrolyzed by 7% H_2SO_4 yielding 1 mol quercetin and 1 mol D-glucose.
In the mother liquor from quercimeritrin the glucosides gossypitrin and isoquercitrin *q.v.,* are also found. *Gossypitrin,* $C_{21}H_{20}O_{13}$, orange-yellow needles melting at 200-202°; slightly sol in alcohol and acetic acid.

8046. d-Quercitol. *2-Deoxy-D-chiro-inositol;* D-1-deoxy-*muco*-inositol; "acorn sugar"; (+)-protoquercitol; 1,2,-3,4,5-cyclohexanepentol. $C_6H_{12}O_5$; mol wt 164.16. C 43.90%, H 7.37%, O 48.73%. Found in the acorns of various spp of *Quercus, Fagaceae*: Prunier, *Ann. Chim. Phys.* **15**, 1 (1878); from leaves of the European palm (*Chaemerops humilis*): Muller, *J. Chem. Soc.* **91**, 1766 (1907). Configuration: Posternak, *Helv. Chim. Acta* **19**, 1007 (1936). Synthesis: G. E. McCasland *et al., J. Org. Chem.* **33**, 4220 (1968).

Sweet crystals. mp 234-235°, $[\alpha]_D^{20}$ +24 to +26°. Sol in water; slightly sol in hot, almost insol in cold alcohol; practically insol in ether.

8047. Quercitrin. *3-[(6-Deoxy-α-L-mannopyranosyl)oxy]-2-(3,4-dihydroxyphenyl)-5,7-dihydroxy-4H-1-benzopyran-4-one;* quercitroside; quercimelin; quercetin-3-L-rhamnoside; thujin. $C_{21}H_{20}O_{11}$; mol wt 448.37. C 56.25%, H 4.50%, O 39.25%. From *Aesculus hippocastanum* L., Hippocastanaceae: Hörhammer *et al., Arch. Pharm.* **292**, 113 (1959). Structure: Zemplén *et al., Ber.* **61**, 2486 (1928); Marchlewski, Skarzynski, *Biochem. Z.* **297**, 56 (1938); Wolfrom, Thompson in R. L. Whistler, M. L. Wolfrom, *Methods in Carbohydrate Chemistry* vol. 1 (Academic Press, New York, 1962) p 202.

Yellow crystals from dil methanol or ethanol, mp 176-179°; from water, mp 167°. uv max (ethanol): 350, 258 nm (log ε 4.18, 4.30). Practically insol in cold water, ether; sol in alc; moderately sol in hot water; sol in aq alkaline solns with intense yellow color which is oxidized by air to brown.

USE: Has been used as textile dye. *Flavine yellow shade* is prepared by extracting quercitron bark with high pressure steam and consists mainly of quercitrin: Tisdale, *Can. Textile J.* **57**, 44 (1941).

8048. Quercus. White oak. Dried inner bark of trunk and branches of *Quercus alba* L., Fagaceae. *Habit.* Eastern U.S. and Canada. *Constit.* Tannic acid, oak-red, resin, pectin, levulin, quercitrin.

THERAP CAT: Astringent.

8049. Quillaic Acid. *3β,16α-Dihydroxy-23-oxoolean-12-en-28-oic acid;* quillaja sapogenin. $C_{30}H_{46}O_5$; mol wt 486.67. C 74.03%, H 9.53%, O 16.44%. Prepn from quillaja saponin: Windaus *et al., Z. Physiol. Chem.* **160**, 301 (1926); Elliott, Kon, *J. Chem. Soc.* **1939**, 1130. Structure: Windaus, *loc. cit.;* Ruzicka, van Veen, *Z. Physiol. Chem.* **184**, 69 (1929); Ruzicka *et al., Coll. Czech. Chem. Commun.* **15**, 893 (1950).

Crystals from ethyl acetate, mp 292-293°; also reported as 258-265° (dec): Kubota *et al., Tetrahedron Letters* **1969**, 771. $[\alpha]_D^{20}$ +56.1° (c = 2.9 in pyridine). Soluble in alcohol, ether, acetone, ethyl acetate, glacial acetic acid.

Diacetate, $C_{34}H_{50}O_7$, needles from glacial acetic acid, mp 180-187°.

Methyl ester, $C_{31}H_{48}O_5$, needles from dil methanol, mp 223-225°; also reported as 200-205° (dec), Kubota *et al., loc. cit.* $[\alpha]_D^{20}$ +40.5° (c = 4 in pyridine).

8050. Quillaja. Soap bark; quillay bark; Panama bark; China bark; Murillo bark. Inner dried bark of *Quillaja saponaria* Molina, *Rosaceae.* *Habit.* South America (Peru, Chile); cultivated in Northern Hindustan. *Constit.* Quillaic acid, quillajasaponin, sucrose, tannin.

USE: Manuf of saponin; in mineral water industry, shampoo liquids, etc.; as foam producer.

8051. Quillaja Saponin. The saponin of quillay bark. Isolation: Kobert, *Arch. Exp. Pathol. Pharmakol.* **23**, 233 (1887); Pachorukow, *Arbeiten des Pharmakologischen Instituts zu Dorpat* **I**, 5 (1888); Cofman-Nicoresti, *Pharm. J.* **111**, 103 (1923), found the total saponin content of quillay bark to be about 9 or 10%. The saponin is built from quillaic acid, a sugar and possibly other substances.

Amorphous, deliquescent powder which causes sneezing when dispersed in the air. Freely sol in dil alc. Foams easily when shaken with water, the foam being relatively stable.

8052. Quinacillin. *6-[[(3-Carboxy-2-quinoxalinyl)carbonyl]amino]-3,3-dimethyl-7-oxo-4-thia-1-azabicyclo-[3.2.0]heptane-2-carboxylic acid;* 3-carboxy-2-quinoxalinyl penicillin. $C_{18}H_{16}N_4O_6S$; mol wt 416.42. C 51.92%, H 3.87%, N 13.46%, O 23.05%, S 7.70%. Semi-synthetic antibiotic related to penicillin. Prepd by condensation of quinoxaline-2,3-dicarboxylic anhydride with 6-aminopenicillanic acid: Richards *et al., Nature* **199**, 354 (1963).

Disodium salt, $C_{18}H_{14}N_4Na_2O_6S$, crystals, dec 261-262°. $[\alpha]_D^{23}$ +183.5° (water). Very hygroscopic. uv max (containing 9.2% water): 242, 326 nm (ε 32,100; 7280). Acquires a bright yellow color on exposure to strong sunlight but is stable at 100° for at least 3 months. Freely sol in water; a 25% aq soln is stable for 2 months at 0°. Antimicrobial activity is highest against *Staphylococcus aureus.*

Bistriethylammonium salt monohydrate, $C_{30}H_{46}N_6O_6S$.H_2O, crystals from acetone, dec 135-137°. $[\alpha]_D^{20}$ +142° (c = 0.376 in water).

THERAP CAT: Antibacterial.

8053. Quinacrine. *N^4-(6-Chloro-2-methoxy-9-acridinyl)-N^1,N^1-diethyl-1,4-pentanediamine;* 6-chloro-9-[[4-(diethylamino)-1-methylbutyl]amino]-2-methoxyacridine; 3-chloro-7-methoxy-9-(1-methyl-4-diethyl aminobutylamino)acridine; 2-chloro-5-(ω-diethylamino-α-methylbutylamino)-7-methoxyacridine; mepacrine. $C_{23}H_{30}ClN_3O$; mol wt 399.96. C 69.07%, H 7.56%, Cl 8.86%, N 10.51%, O 4.00%. Prepd by condensing 1-diethylamino-4-aminopentane with 3,9-dichloro-7-methoxyacridine: F. Mietzsch, H. Mauss, **Ger.** 553,072; 571,499 (1934); *eidem*, U.S. pat. **2,113,357** (1938 to Winthrop Chemical); *cf.* Drosdov, Cherutzov, *J. Gen. Chem. USSR* **5**, 1576, 1736 (1935); **8**, 1192 (1938); Magidson, Grigorovskii, *Khim. Farm. Prom.* **1933**, 187; Jensch, Eisleb, U.S. pat. **1,782,727** (1930); Schulemann *et al.,* U.S. pat. **1,889,704** (1932). Review of quinacrine and other acridines: A. O. Wolfe in *Antibiotics* vol. 3, J. W. Corcoran, F. E. Hahn, Eds. (Springer-Verlag, New York, 1975) pp 203-233.

Dihydrochloride dihydrate, $C_{23}H_{32}Cl_3N_3O.2H_2O$, **RP 866, SN 390, Atabrine dihydrochloride, Atebrin hydrochloride, Chinacrin hydrochloride, Erion, Acriquine, Acrichine, Palacrin, Metoquine, Italchin.** Bitter, bright yellow crystals. Dec 248-250° (mp poorly discernible). One gram dissolves in about 35 ml water. Much more sol in hot water. Slightly sol in ethanol, somewhat more sol in methanol. Insol in ether, benzene, acetone. pH of a 1% aq soln about 4.5. Under uv light the yellow aq solns exhibit a vivid fluorescence which is detectable in a dilution of 1:5,000,000.

Methanesulfonate monohydrate, $C_{25}H_{38}ClN_3O_7S_2.H_2O$, bitter, bright yellow crystals. One part dissolves in 3 parts water at 15.5°, in 36 parts 95% alcohol at 15.5°. pH of 2.0% w/v soln in water 3.0-5.0.

THERAP CAT: Anthelmintic (Cestodes); antimalarial.

THERAP CAT (VET): Antiprotozoal, teniacide.

8054. Quinaldic Acid. *2-Quinolinecarboxylic acid;* quinaldinic acid. $C_{10}H_7NO_2$; mol wt 173.17. C 69.36%, H 4.07%, N 8.09%, O 18.48%.

Dihydrate, crystals. mp 155-157° (anhydr). Moderately sol in water; sol in alcohol and in alkali solns.

USE: For the determination of copper, zinc and uranium with which it forms insol salts.

8055. Quinaldine. *2-Methylquinoline.* $C_{10}H_9N$; mol wt 143.18. C 83.88%, H 6.34%, N 9.78%. Occurs in coal tar; made from aniline, acetaldehyde and HCl. Lab procedure: A. I. Vogel, *Practical Organic Chemistry* (Longmans, London, 3rd ed., 1959) p 831; Gattermann-Wieland, *Praxis des Organischen Chemikers* (de Gruyter, Berlin, 40th ed., 1961) p 318. Pharmacology: Brown *et al.*, *Comp. Biochem. Physiol. A* **42**, 223 (1972). Toxicity data: Smyth *et al.*, *Arch. Ind. Hyg. Occup. Med.* **4**, 119 (1951); L. L. Marking, *Investigations in Fish Control* **23**, 3 (1969).

Colorless, oily liquid; quinoline odor; becomes reddish-brown on exposure to air. d about 1.06. bp 246-247°. Practically insol in water. Sol in chloroform, ether. *Keep tightly closed and protected from light.* LD_{50} orally in rats: 1.23 g/kg (Smyth).

Caution: Highly irritating to mucous membranes.

USE: As anesthetic in transport and handling of fish.

8056. Quinaldine Blue. *1-Ethyl-2-[3-(1-ethyl-2(1H)-quinolinylidene)-1-propenyl]quinolinium chloride;* bis[1-ethylquinoline-(2)]trimethinecyanine chloride; 1,1'-diethyl-2,2'-trimethinequinocyanine chloride; pinacyanol chloride; Vernitest Reagent. $C_{25}H_{25}ClN_2$; mol wt 388.96. C 77.20%, H 6.48%, Cl 9.12%, N 7.20%. Prepn: Fischer, *J. Prakt. Chem.* **98**, 204 (1918). Properties and use as histological stain: F. Proescher, *Proc. Soc. Exp. Biol. Med.* **31**, 79 (1933); H. P. Klinger *et al.*, *Stain Technol.* **46**, 43 (1971); R. K. J. Narayan, *ibid.* **55**, 9 (1980).

Bright blue-green alcohol-containing prisms or needles from alcohol. Drying at 110° expels alcohol of crystn. Dec at about 263°. Moderately sol in water with a violet-red color, in alcohol with a blue color. Solns are dichroic.

USE: Histological stain (chromosomes).

8057. Quinaldine Red. *2-[2-[4-(Dimethylamino)phenyl]ethenyl]-1-ethylquinolinium iodide;* 2-[p-(dimethylamino)styryl]-1-ethylquinolinium iodide; 2-(p-dimethylamino-styryl)quinoline ethiodide; α-(p-dimethylaminophenylethylene)quinoline ethiodide; Eastman no. 1361. $C_{21}H_{23}IN_2$; mol wt 430.33. C 58.61%, H 5.39%, I 29.49%, N 6.51%. Prepd by condensing quinaldine ethyliodide and p-dimethylamino-

benzaldehyde in alc in the presence of piperidine: König, *J. Prakt. Chem.* [2] **86**, 172 (1912); Ger. pat. 294,744; *Frdl.* **14**, 735.

Very dark red powder. Sparingly sol in water. Freely sol in alcohol, giving a dark red soln. Solns are slowly dec by light. pKb 11.25. Colorless at pH 1.4; red at pH 3.2. In the presence of chlorides the transition interval is from pH 1.2 to pH 3.0.

USE: As an indicator.

8058. Quinaldine Sulfate. *2-Methylquinoline sulfate.* $C_{10}H_9N.H_2SO_4$; mol wt 241.27. C 49.78%, H 4.60%, O 5.81%, N 26.52%, S 13.29%. Prepn: S. Hoogewerff, W. A. Van Dorp, *Rec. Trav. Chim.* **3**, 345 (1884); J. L. Allen, J. B. Sills, *Investigations in Fish Control* **47**, 3 (1973). Efficacy as anesthetic in fish: P. A. Gilderhus *et al.*, *ibid.* **49**, 3 (1973); G. C. Blasiola Jr., *J. Fish Biol.* **10**, 113 (1977). Toxicity in fish: L. L. Marking, V. K. Dawson, *Investigations in Fish Control* **48**, 3 (1973).

Light yellow crystalline powder, mp 211-214°. Soly (g/100 ml) in water 104.05, in methanol 7.44, in ethanol 2.27, in acetone 0.08. Practically insol in ether, benzene, hexane. uv max ($0.1N$ H_2SO_4): 236, 317 nm. LC_{50} (96 hour) in largemouth bass: 6.80 mg/l; in carp: 72.5 mg/l (Marking, Dawson).

USE: As anesthetic in transport and handling of fish.

8059. Quinalizarin. *1,2,5,8-Tetrahydroxy-9,10-anthracenedione; 1,2,5,8-tetrahydroxyanthraquinone;* C.I. Mordant Violet 26; C.I. 58500; Alizarinbordeaux. $C_{14}H_8O_6$; mol wt 272.20. C 61.77%, H 2.96%, O 35.27%. From hemipic (hemipinic) acid and hydroquinone with H_2SO_4: Liebermann, Kostanecki, *Ann.* **240**, 245 (1887); Liebermann, Wense, *Ber.* **20**, 862 (1887). From alizarin or quinizarin by treatment with 80% oleum, then boiling with caustic: Gattermann, *J. Prakt. Chem.* [2] **43**, 246 (1891); Schmidt, *ibid.* **237**, 242 (1891); *Bull. Soc. Ind. Mulhouse* **84**, 409 (1914). *See also: Colour Index* vol. 4 (3rd ed., 1971) p 4519.

Red needles with green metallic luster from acetic acid or from nitrobenzene or by sublimation *in vacuo.* mp above 275°. Absorption spectrum: Meek, Watson, *J. Chem. Soc.* **109**, 544 (1916); Meek, *ibid.* **111**, 969 (1917). Insol in water; very slightly sol in most other solvents. Dissolves in aq solns of alkalies with reddish-violet, in acetic acid with yellow, in sulfuric acid with blue-violet color.

USE: Dyes cotton mordanted with Al salts dark red, cotton mordanted with Cr salts bluish-violet. No longer used as a dye, except occasionally in printing cotton.

8060. Quinamine. *8a-(5-Ethenyl-1-azabicyclo[2.2.2]-oct-2-yl)-2,3,8,8a-tetrahydro-3aH-furo[2,3-b]indol-3a-ol.* $C_{19}H_{24}N_2O_2$; mol wt 312.40. C 73.04%, H 7.74%, N 8.97%, O 10.24%. In Cinchona barks, especially those of *Cinchona pubescens* Vahl. (*C. succirubra* Pav.) and *C. officinalis* L. (*C. ledgeriana* Moens), Rubiaceae: Hesse, *Ber.* **5**, 265 (1872); **10**, 2157 (1877); *Ann.* **207**, 288 (1881); Oudemans, *Ann.* **197**, 49 (1879). Isoln from quinine sulfate mother liquors: De Vrij, *Pharm. J.* **4**, 609 (1874); Howard, *ibid.* **5**, 1 (1875); Henry *et al.*, *J. Chem. Soc.* **1945**, 524. Reduction with lithium aluminum hydride yields cinchonamine: Goutarel *et al.*, *Helv. Chim. Acta* **33**, 150 (1950). Structure: Witkop, *J. Am. Chem. Soc.* **72**, 2311 (1950).

Needles from boiling benzene, mp 185-186°. $[\alpha]_D$ +116° or +104° (c = 0.5 in alcohol). uv max (methanol): 242, 301 nm (log ε 3.93, 3.39). One gram dissolves in 1500 ml water, 50 ml ether, about 120 ml 80% alcohol. Sol in abs alc, hot ether, hot benzene, hot petr ether.

Hydrochloride monohydrate, $C_{19}H_{25}ClN_2O_2 \cdot H_2O$, crystals, mp 166-167°. $[\alpha]_D$ +102.8°.

Hydriodide, $C_{19}H_{25}IN_2O_2$, crystals, mp 224°. $[\alpha]_D$ +84.82° (c = 0.5 in alcohol).

Nitrate, $C_{19}H_{25}N_3O_5$, crystals, mp 186-188°. $[\alpha]_D$ +94.9°.

8061. Quinapyramine. *4-Amino-6-[(2-amino-1,6-dimethyl-4(1H)-pyrimidinylidene)amino]-1,2-dimethylquinolinium conjugate monoacid;* 4-amino-6-[(2-amino-6-methyl-4-pyrimidinyl)amino]-1-methylquinaldinium methosalts; 4-amino-6-[(2-amino-1,6-dimethyl-4-pyrimidinyl)amino]-1,2-dimethylquinoline salts; 4-amino-6-(2-amino-6-methyl-4-pyrimidinyl)amino]quinaldine-1,1'-dimethosalts; M 7555; Antrycide. Prepn: F. H. S. Curd, D. G. Davey, *Brit. J. Pharmacol.* **5**, 25 (1950); Curd, **U.S. pat. 2,585,917** (1952 to I.C.I.); Ainley *et al.*, *J. Chem. Soc.* **1953**, 59.

Dimethosulfate, $C_{19}H_{28}N_6O_8S_2$, creamy white crystals from aq methanol, mp 265-266°. Freely sol in water. LD_{50} i.v. in mice: 10-15 mg/kg (Curd, Davey).

Diiodide, pale cream needles, mp 312-313° (dec). Sparingly sol in water.

Dichloride, crystals from water, mp 316-317° (dec). Sparingly sol in water. LD_{50} i.v. in mice: 10-15 mg/kg (Curd, Davey).

THERAP CAT: Antiprotozoal (Trypanosoma).

8062. Quinazoline. 1,3-Benzodiazine; benzo[a]pyrimidine; 5,6-benzopyrimidine; phenmiazine. $C_8H_6N_2$; mol wt 130.14. C 73.83%, H 4.65%, N 21.53%. Prepn from 2-nitrobenzylidenebis(formamide): Riedel, **Ger. pat. 174,941**; *Chem. Zentr.* **1906**, II, 1372; *Frdl.* **8**, 1238; Bogert, McColm, *J. Am. Chem. Soc.* **49**, 2650 (1927).

Leaflets from petr ether; odor of quinoline. Slightly bitter taste. mp 48.0-48.5°. bp$_{772.5}$ 243°; bp$_{764}$ 241.5°. Freely sol in water. Neutral reaction. Sol in many organic solvents. Picrate, red needles, mp 188-190°.

8063. Quinbolone. *17β-(1-Cyclopenten-1-yloxy)androsta-1,4-dien-3-one;* 1-dehydrotestosterone 17-cyclopent-1'-

enyl ether; Anabolicum Vister. $C_{24}H_{32}O_2$; mol wt 352.52. C 81.77%, H 9.15%, O 9.08%. Prepn from 17β-hydroxyandrosta-1,4-dien-3-one: Ercoli *et al.*, *Chem. & Ind. (London)* **1962**, 1284.

Crystals from ethanol, mp 133-135°. $[\alpha]_D^{20}$ +61° (dioxane). uv max: 244-245 nm ($E_{1cm}^{1\%}$ 430-450). Practically insol in water; sol in many organic solvents; sparingly sol in ethanol, hexane. Soly in sesame oil: 40-45 mg/ml.

THERAP CAT: Anabolic.

8064. Quince Seed. Gum quince seed; semen cydonia; golden apple seed; cydonia seed. Seed of *Cydonia oblonga* Mill. (*C. vulgaris* Pers.), *Rosaceae.* **Habit.** Southern Asia, Europe; widely cultivated. *Constit.* Amygdalin, emulsin, about 15% fatty oil, about 20% of a mucilage named cydonin. Brief review of seed and gum uses: BeMiller in *Industrial Gums*, R. L. Whistler, Ed. (Academic Press, New York, 2nd ed., 1973) pp 339-345.

USE: Gum from the seeds as suspending agent, stabilizer; in hair and cosmetic prepns.

8065. Quinestradiol. *(16α,17β)-3-(Cyclopentyloxy)estra-1,3,5(10)-trien-16,17-diol;* quinestradol; estriol 3-cyclopentyl ether; Colpovis; Pentovis. $C_{23}H_{32}O_3$; mol wt 356.49. C 77.49%, H 9.05%, O 13.46%. Prepn from estriol and cyclopentylbromide: Ercoli, **Brit. pat. 909,662** (1962 to Vismara).

Crystals, mp 98-100°.

THERAP CAT: Estrogen.

8066. Quinestrol. *3-(Cyclopentyloxy)-19-nor-17α-pregna-1,3,5(10)-trien-20-yn-17-ol;* 17α-ethinylestradiol 3-cyclopentyl ether; W 3566; Estrovis. $C_{25}H_{32}O_2$; mol wt 364.51. C 82.37%, H 8.85%, O 8.78%. Prepn: Ercoli, Gardi, *Chem. & Ind. (London)* **1961**, 1037; Ercoli, **U.S. pat. 3,159,543**; Ercoli *et al.*, **U.S. pat. 3,231,567** (1964, 1966, both to Vismara).

Crystals, mp 107-108°. $[\alpha]_D^{25}$ +5° (c = 0.5 in dioxane).

THERAP CAT: Estrogen.

8067. Quinethazone. *7-Chloro-2-ethyl-1,2,3,4-tetrahydro-4-oxo-6-quinazolinesulfonamide;* 7-chloro-2-ethyl-6-sulfamoyl-1,2,3,4-tetrahydro-4-quinazolinone; 7-chloro-2-ethyl-1,2,3,4-tetrahydro-4-oxo-6-sulfamoylquinazoline; CL 36010; Hydromox; Aquamox. $C_{10}H_{12}ClN_3O_3S$; mol wt 289.76. C 41.45%, H 4.18%, Cl 12.24%, N 14.50%, O 16.56%, S 11.07%. Prepn: Cohen *et al.*, *J. Am. Chem. Soc.* **82**, 2731 (1960); Cohen, Vaughan, Jr., **U.S. pat. 2,976,289** (1961 to Am. Cyanamid).

Consult the cross index before using this section.

Fibrous crystals from 50% acetone, mp 250-252°. Sol in acetone, alcohol.

THERAP CAT: Diuretic, antihypertensive.

8068. Quinfamide. *2-Furancarboxylic acid 1-(dichloro-acetyl)-1,2,3,4-tetrahydro-6-quinolinyl ester;* 2-furoic acid ester with 1-(dichloroacetyl)-1,2,3,4-tetrahydro-6-quino-linol; 1-(dichloroacetyl)-6-(2-furoyloxy)-1,2,3,4-tetrahy-droquinoline; Win 40014; Amenox; Amenide. $C_{16}H_{13}Cl_2$-NO_4; mol wt 354.19. C 54.26%, H 3.70%, Cl 20.02%, N 3.95%, O 18.07%. Prepn: D. M. Bailey, U.S. pat. **3,997,542** (1976 to Sterling); D. M. Bailey *et al., J. Med. Chem.* **22**, 600 (1979). Amebicidal activity and toxicological evaluation: R. G. Slighter *et al., Parasitology* **81**, 157 (1980). Distribution and metabolism in rats: J. F. Baker, *Arch. Int. Pharmaco-dyn.* **258**, 29 (1982). Clinical evaluation in adults with chronic amebiasis: L. Guevara *et al., Clin. Therap.* **6**, 43 (1983); in children: F. A. Rojas *et al., ibid.* 47.

Crystals from ethyl acetate, mp 150.5-151°.

THERAP CAT: Anti-amebic.

8069. Quingestrone. *3-(Cyclopentyloxy)pregna-3,5-dien-20-one;* progesterone cyclopentyl-3-enol ether; W 3399; Enol Luteovis. $C_{26}H_{38}O_2$; mol wt 382.59. C 81.63%, H 10.01%, O 8.36%. Prepn: Ercoli, U.S. pat. **3,019,241** (1962). Purification and stabilization: McMillan, Ninger, U.S. pat. **3,179,675** (1965 to Warner-Lambert).

Crystals, mp 105-106°. $[\alpha]_D$ −47.5° (dioxane).

THERAP CAT: Progestogen.

8070. Quinhydrone. *2,5-Cyclohexadiene-1,4-dione com-pound with 1,4-benzenediol (1:1);* green hydroquinone. C_{12}-$H_{10}O_4$; mol wt 218.20. C 66.05%, H 4.62%, O 29.33%. An addn compd of one mol hydroquinone and one mol quinone. Prepn from hydroquinone and quinone: Gattermann-Wie-land, *Praxis des Organischen Chemikers* (de Gruyter, Berlin, 40th ed., 1961) p 270. Alternate prepn by the action of fer-ric ammonium sulfate on hydroquinone: A. I. Vogel, *Prac-tical Organic Chemistry*, (Longmans, London, 3rd ed., 1959) p 747.

Green crystals with metallic luster; reddish-brown by transmitted light. d 1.40. mp 171°; sublimes with partial decompn. Slightly sol in cold water; sol in hot water,

ammonia, alc, ether; insol in petr ether. LD_{50} orally in rats: 225 mg/kg, Woodward *et al., Fed. Proc.* **8**, 348 (1949).

USE: In pH determinations (quinhydrone electrode).

8071. Quinic Acid. *1,3,4,5-Tetrahydroxycyclohexanecar-boxylic acid;* chinic acid; kinic acid; hexahydro-1,3,4,5-tetra-hydroxybenzoic acid. $C_7H_{12}O_6$; mol wt 192.17. C 43.75%, H 6.30%, O 49.95%. Found in cinchona bark, particularly in South American barks; also in many other plants, such as tobacco leaves, carrot leaves, apples, peaches, pears, plums, etc. Structure and configuration: Fischer, Dangschat, *Ber.* **65**, 1009 (1932). Total synthesis: Grewe *et al., ibid.* **87**, 793 (1954); Smissman, Oxman, *J. Am. Chem. Soc.* **85**, 2184 (1963). Stereospecific synthesis: Wolinsky *et al., J. Org. Chem.* **29**, 3596 (1964). *Review:* Bohm, *Chem. Rev.* **65**, 435 (1965).

White crystals; strong acid taste. d 1.64. mp 162-163°; at higher temps forms a lactone. $[\alpha]_D^{20}$ −42° to −44° in water. Sol in 2.5 parts water, in alcohol, glacial acetic acid.

8072. Quinidine. *6′-Methoxycinchonan-9-ol;* 6-meth-oxy-α-(5-vinyl-2-quinuclidinyl)-4-quinolinemethanol; α-(6-methoxy-4-quinolyl)-5-vinyl-2-quinuclidinemethanol; conquinine; pitayine; β-quinine. $C_{20}H_{24}N_2O_2$; mol wt 324.41. C 74.04%, H 7.46%, N 8.64%, O 9.86%. A dextro-rotatory stereoisomer of quinine, *q.v.* Present in cinchona barks to the extent of 0.25-3.0%. Found in quinine sulfate mother liquors. Configuration: Prelog, Zalán, *Helv. Chim. Acta* **27**, 535 (1944); Prelog, Häfliger, *ibid.* **33**, 2021 (1950); Roth, *Pharmazie* **16**, 257 (1961). Crystal and molecular structure: R. Doherty *et al., J. Pharm. Sci.* **67**, 1698 (1978). Rotatory dispersion studies: Lyle, Gaffield, *Tetrahedron Letters* **1963**, 1371. Most of the material sold today is made by isomerization of quinine: Doering *et al., J. Am. Chem. Soc.* **69**, 1700 (1949); Dietrich, Stein, **Ger.** pat. **877,611** (1953 to Boehringer, Mann.). Total synthesis: Uskokovic *et al., J. Am. Chem. Soc.* **92**, 203, 204 (1970); Grethe *et al., Helv. Chim. Acta* **56**, 1485 (1973); Gutzwiller, Uskokovic, *ibid.* 1494; *eidem, J. Am. Chem. Soc.* **100**, 576 (1978); G. Grethe *et al., ibid.* 589. Toxicity data: K. Dietmann *et al., Arznei-mittel-Forsch.* **27**, 589 (1977). Review of pharmacology: D. M. Ariado, H. Salem, *J. Clin. Pharmacol.* **15**, 477 (1975).

Triboluminescent. mp 174-175° after drying of solvated crystals. $[\alpha]_D^{15}$ +230° (c = 1.8 in chloroform), $[\alpha]_D^{17}$ +258° (alc), $[\alpha]_D^{17}$ +322° (c = 1.6 in 2M HCl). pK_1 at 20° = 5.4; pK_2 10.0. Blue fluorescence in dil H_2SO_4. The uv absorp-tion spectrum is identical with that of quinine. One gram dissolves in about 2000 ml cold, 800 ml boiling water, 36 ml alcohol, 56 ml ether, 1.6 ml chloroform; very sol in metha-nol. Practically insol in petr ether. LD_{50} in rats (mg/kg): 30 i.v., 263 orally (Dietmann)

Hemipentahydrate, prisms from dil alcohol, loses ½ H_2O in air, mp about 168°.

Acid sulfate tetrahydrate, $C_{20}H_{26}N_2O_6S.4H_2O$, *quinidine bisulfate, Chinidin-Duriles, Kiditard, Kinichron, Kinidin, Kinidrin, Quiniduran.* Rods, sol in 8 parts water with blue fluorescence.

Sulfate, see separate entry.

Hydrochloride monohydrate, $C_{20}H_{24}N_2O_2.HCl.H_2O$, lustrous needles, dec 259° (dry). $[\alpha]_D^{20}$ +200°. Sol in 60 parts water, freely sol in hot water, in alcohol, chloroform, slightly in ether. Neutral reaction.

Dihydrobromide trihydrate, $C_{20}H_{24}N_2O_2 \cdot 2HBr \cdot 3H_2O$, crystals from water, dec 115°. $[\alpha]_D^{15}$ +223° (0.4 molar soln).

Methiodide monohydrate, $C_{20}H_{24}N_2O_2 \cdot CH_3I \cdot H_2O$, crystals from water, mp 248° (dry). Readily sol in hot water, alc.

Gluconate, $C_{26}H_{36}N_2O_9$, *gluconic acid quinidine salt*, *Duraquin, Dura-Tab, Gluquinate, Quinaglute*. Crystals, mp 175-176.5°. Sol in 9 parts water, 60 parts alcohol. Used as a long-acting quinidine preparation.

THERAP CAT: Cardiac depressant (anti-arrhythmic).

8073. Quinidine Polygalacturonate. Cardioquin; Galactoquin; Naticardina; Sineflutter. $(C_{20}H_{24}N_2O_2 \cdot C_6H_{10}O_7 \cdot H_2O)_x$. C 58.18%, H 6.76%, N 5.22%, O 29.84%. Prepn: A. Halpern *et al.*, *Am. J. Pharm.* **130**, 190 (1958); A. Halpern, U.S. pat. 2,878,252 (1959 to Synergistics Inc.). Pharmacology: A. Halpern *et al.*, *Antibiot. Chemother.* **9**, 97 (1959).

Amorphous powder, mp 180° (dec). Anhydr product is insol in methanol, ethanol, chloroform, ether, acetone, dioxane; soly in hot 40% methanol or ethanol: 12%; in water at 25°: about 2%. LD_{50} in rats, mice (mg/kg): 3200 ± 350, 2680 ± 210 orally (Halpern *et al.*, 1959).

THERAP CAT: Antiarrhythmic.

8074. Quinidine Sulfate. *6'-Methoxycinchonan-9-ol sulfate;* Cin-Quin; Quinidex; Quinicardine; Quinitex; Quinora. $C_{40}H_{50}N_4O_8S$; mol wt 746.93. C 64.32%, H 6.75%, N 7.50%, O 17.14%, S 4.29%. Toxicity data: C. Turba *et al.*, *Arzneimittel-Forsch.* **18**, 1127 (1968). Comprehensive description: M. A. Loutfy *et al.*, in *Analytical Profiles of Drug Substances* vol. **12**, K. Florey, Ed. (Academic Press, New York, 1983) pp 483-546.

Dihydrate, white, very bitter, odorless, fine crystals, frequently cohering in masses. Darkens on exposure to light. Does not lose all of its water below 120°. $[\alpha]_D^{25}$ about +212° (95% alcohol); about +260° (dil HCl). pKa 4.2, 8.8. pH (1% aq soln): 6.0-6.8. One gram dissolves in about 90 ml water, 15 ml boiling water, 10 ml alcohol, 3 ml methanol, 12 ml chloroform. Insol in ether, benzene. *Protect from light.* LD_{50} in mice, rats (mg/kg): 700, 455.8 orally; 83, 56 i.v. (Turba).

THERAP CAT: Cardiac depressant (anti-arrhythmic).

THERAP CAT (VET): Has been used in atrial fibrillation.

8075. Quinine. *6'-Methoxycinchonan-9-ol.* $C_{20}H_{24}N_2O_2$; mol wt 324.41. C 74.04%, H 7.46%, N 8.64%, O 9.86%. For chemical names *see* Quinidine. The most important alkaloid of cinchona bark, the bark of *Cinchona officinalis* L. (*C. ledgeriana* Moens), *Rubiaceae*, which contains about 8% quinine, other barks 1 to 4%. Cinchona trees grow wild in South America and are cultivated in Java. Extraction procedure: Schwyzer, *Die Fabrikation der Alkaloide* (Berlin, 1927); Vetter in *Emil Barell Festschrift* (Basel, 1936) p 541; Jucker, Stoll, in *Ullmann's Enzyklopädie der technischen Chemie* **3**, 213-218 (1953). Configuration: Prelog, Zalán, *Helv. Chim. Acta* **27**, 535 (1944); Prelog, Häfliger, *ibid.* **33**, 2021 (1950); Roth, *Pharmazie* **16**, 257 (1961). Synthesis: Woodward, Doering, *J. Am. Chem. Soc.* **66**, 849 (1944); **67**, 860 (1945); Taylor, Martin, *ibid.* **94**, 6218 (1972); Gutzwiller, Uskokovic, *ibid.* **100**, 576 (1978); G. Grethe *et al.*, *ibid.* 589; T. Imanishi *et al.*, *Chem. Pharm. Bull.* **30**, 1925 (1982). *Review:* F. E. Hahn, Ed. in *Antibiotics* vol. **5**(pt. 2) (Springer-Verlag, New York, 1979) pp 353-362.

Triboluminescent, orthorhombic needles from abs alcohol, mp 177° (some decompn). Sublimes in high vacuum at 170-180°. $[\alpha]_D^{15}$ -169° (c = 2 in 97% alcohol), $[\alpha]_D^{17}$ -117° (c =

1.5 in chloroform), $[\alpha]_D^{15}$ -285° (c = 0.4M in 0.1N H_2SO_4). pK_1 at 18° = 5.07; pK_2 9.7. pH of satd aq soln 8.8. Absorption spectra: Dobbie, Lauder, *J. Chem. Soc.* **99**, 1260 (1911); Dobbie, Fox, *ibid.* **101**, 78 (1912). Fluorescence: Rabe, Marschall, *Ann.* **382**, 362 (1911). The blue fluorescence is especially strong in dil H_2SO_4. One gram dissolves in 1900 ml water, 760 ml boiling water, 0.8 ml alcohol, 80 ml benzene (in 18 ml benzene at 50°), in 1.2 ml chloroform; 250 ml dry ether, 20 ml glycerol, 1900 ml of 10% ammonia water. Almost insol in petr ether.

Trihydrate, microcrystalline powder, mp 57°, efflorescent, loses one H_2O in air, two H_2O over H_2SO_4, anhydr at 125°. Tartrate, $(C_{20}H_{24}N_2O_2)_2 \cdot C_4H_6O_6$, crystals from ethanol, mp 211-212.5°. $[\alpha]_D^{25}$ -156.4° (c = 0.97 in methanol).

A standardized method for the assay of quinine in cinchona bark named "Brussels 1949" is published in the journal *De Belgische Chemische Industrie (Ind. Chim. Belge)* **15**, 328-338 (1950).

THERAP CAT: Antimalarial.

THERAP CAT (VET): Bitter stomachic, analgesic, antipyretic.

8076. Quinine Bisulfate. Acid quinine sulfate. $C_{20}H_{26}N_2O_6S$; mol wt 422.50. C 56.85%, H 6.20%, N 6.63%, O 22.72%, S 7.59%. $C_{20}H_{24}N_2O_2 \cdot H_2SO_4$.

Heptahydrate, $C_{20}H_{26}N_2O_6S \cdot 7H_2O$, *Dentojel, Bi-Quinate, Quinbisan*. Very bitter crystals or cryst powder; efflorescent on exposure to air and darkens on exposure to light. One gram dissolves in 9 ml water, 0.7 ml boiling water, 23 ml alcohol, 0.7 ml alcohol at 60°, 625 ml chloroform, 2500 ml ether, 15 ml glycerol. pH: 3.5. *Protect from light.*

THERAP CAT: Antimalarial; oral sclerosing agent.

8077. Quinine Carbonate. Diquinine carbonate; Aristoquin; Aristoquinine; Aristochin. $C_{41}H_{46}N_4O_5$; mol wt 674.81. C 72.97%, H 6.87%, N 8.30%, O 11.85%. $CO(OC_{20}H_{23}N_2O)_2$. Prepd by reacting quinine and phosgene in chloroform: Schwyzer, *Die Fabrikation Pharmazeutischer und Chemisch-technischer Produkte* (Springer Verlag, Berlin, 1931) p 318.

Tasteless powder. mp 189°. $[\alpha]_D^{20}$ -160.5°. Practically insol in water or ether; slightly sol in cold alcohol, more in hot alcohol and hot chloroform; sol in dil acids. Dec by alkali solns. *Protect from light.*

THERAP CAT: Antimalarial.

8078. Quinine Dihydrobromide. Quinine dibromide; quinine acid hydrobromide. $C_{20}H_{26}Br_2N_2O_2$; mol wt 486.26. C 49.40%, H 5.39%, Br 32.87%, N 5.76%, O 6.58%. $C_{20}H_{24}N_2O_2 \cdot 2HBr$.

Trihydrate, white to yellowish, odorless powder. The salt of commerce usually contains less water. Sol in 3 parts water, 10 parts alcohol, the soln being strongly acid. *Keep well closed and protected from light.*

THERAP CAT: Antimalarial.

8079. Quinine Dihydrochloride. Quinine dichloride; quinine bimuriate; acid quinine hydrochloride. $C_{20}H_{26}Cl_2N_2O_2$; mol wt 397.34. C 60.45%, H 6.60%, Cl 17.85%, N 7.05%, O 8.05%.

Very bitter powder or crystals. One gram dissolves in about 0.6 ml water, in about 12 ml alcohol. Slightly sol in chloroform, very slightly sol in ether. Aq solns are strongly acid to litmus paper (pH about 2.6). *Protect from light.*

THERAP CAT: Antimalarial.

THERAP CAT (VET): Bitter stomachic, analgesic, antipyretic.

8080. Quinine Ethylcarbonate. *6'-Methoxycinchonan-9-ol ethyl carbonate (ester);* euquinine; Tasteless Quinine. $C_{23}H_{28}N_2O_4$; mol wt 396.47. C 69.67%, H 7.12%, N 7.07%, O 16.14%. Prepd by reacting quinine and ethyl chlorocarbonate in benzene: Schwyzer, *Die Fabrikation pharmazeutische und chemisch-technischer Produkte* (Springer-Verlag, Berlin, 1931) p 314.

Fine matted needles, almost tasteless, mp 90-92°: Page, *Quart. J. Pharm. Pharmacol.* **7**, 361 (1934). $[\alpha]_D^{20}$ -43.7°. The satd aq soln is alkaline to litmus paper. Slightly sol in water. One gram dissolves in 2 ml alcohol, in 1 ml chloroform, and in 10 ml ether. Readily sol in dil acids, but dec by them. *Protect from light.*

THERAP CAT: Antimalarial.

8081. Quinine Formate. Quinoform. $C_{21}H_{26}N_2O_4$; mol wt 370.43. C 68.09%, H 7.07%, N 7.56%, O 17.28%. $C_{20}H_{24}N_2O_2.HCOOH$.

White, cryst powder. mp about 113°. Sol in about 30 parts water, in alcohol, chloroform, very slightly in ether. The aq soln is neutral. *Protect from light.*

THERAP CAT: Antimalarial.

8082. Quinine Gluconate. Gluconic acid quinine salt. $C_{26}H_{36}N_2O_9$; mol wt 520.56. C 59.99%, H 6.97%, N 5.38%, O 27.66%. $C_{20}H_{24}N_2O_2.HOOCC_5H_6(OH)_5$. Prepn of the dihydrate: Haegland, U.S. pat. **2,049,442** (1936 to Merck & Co.). Prepn of hemihydrate and anhydr compd: Willemot, *Ann. Pharm. Franc.* **23**, No. 3, 203 (1965). Existence of dihydrate questioned: Willemot, *ibid.*

Crystals from methanol, mp 159-161°. Stable in air.

Hemihydrate, crystals from water or aq acetone, mp 146-149° (corr). Very bitter taste. $[\alpha]_D^{20}$ -105° ± 3° (aq soln). pH of a 6% aq soln, 6.20 ± 0.20. Very sol in water, methanol, 95% ethanol; slightly sol in abs alcohol; very slightly sol in anhydr acetone.

Dihydrate, cryst powder. Chars when heated to about 100°; may be heated to about 80° without appreciable discoloration. Sol in about 5 parts of water at 25°, and in about 45 parts by wt of 95% alcohol.

THERAP CAT: Antimalarial.

8083. Quinine Hydriodide. Quinine "iodide". $C_{20}H_{25}IN_2O_2$; mol wt 452.34. C 53.11%, H 5.57%, I 28.06%, N 6.19%, O 7.07%. $C_{20}H_{24}N_2O_2.HI$.

Yellowish, odorless powder; becomes darker on exposure to air and light. Sol in about 200 parts water; more sol in hot water; sol in alcohol, dil acids. *Keep well closed and protected from light.*

THERAP CAT: Antimalarial.

8084. Quinine Hydrobromide. Quinine "bromide". $C_{20}H_{25}BrN_2O_2$; mol wt 405.34. C 59.25%, H 6.25%, Br 19.72%, N 6.91%, O 7.89%. $C_{20}H_{24}N_2O_2.HBr$.

Monohydrate, white, odorless, bitter, silky needles. Darkens in the light. One gram dissolves in 40 ml water, 3 ml boiling water, 1 ml alcohol, 0.6 ml chloroform, 7 ml glycerol; slightly sol in ether. Crystallizes from chloroform or bromoform with 2.5 and 2 mols of the solvent, respectively. pH about 6. *Protect from light.*

8085. Quinine Hydrochloride. *6'-Methoxycinchonan-9-ol monohydrochloride;* quinine chloride; quinine muriate; quinine monohydrochloride. $C_{20}H_{25}ClN_2O_2$; mol wt 360.88. C 66.56%, H 6.98%, Cl 9.83%, N 7.76%, O 8.87%. $C_{20}H_{24}N_2O_2.HCl$. Comprehensive description: F. J. Muhtadi *et al.*, in *Analytical Profiles of Drug Substances* vol. 12, K. Florey, Ed. (Academic Press, New York, 1983) pp 547-621.

Dihydrate, bitter, silky needles. Effloresces on exposure to warm air. Does not lose all its water below 120°. One gram dissolves in 16 ml water, in 0.5 ml boiling water, in 1.0 ml alcohol, in about 7.0 ml glycerol, in about 1 ml chloroform, in about 350 ml ether. pH (1% aq soln): 6.0-7.0. Bitterness threshold 1:30,000. *Protect from light.*

THERAP CAT: Antimalarial.

8086. Quinine Iodosulfate. Herapathite; iodoquinine sulfate. $C_{80}H_{104}I_6N_8O_{20}S_3$; mol wt 2355.35. C 40.79%, H 4.45%, I 32.33%, N 4.76%, O 13.59%, S 4.08%. $4C_{20}H_{24}N_2O_2.3H_2SO_4.2HI.I_4$. Named after its discoverer, Herapath, an English physician.

Hexahydrate, plate-like crystals of pale olive-green color by transmitted light and of a brilliant green to reddish green by reflected light. The crystals polarize light 5 times as much as tourmaline. Loses its water at 100° and becomes red. Almost insol in water; sol in about 1000 parts boiling water; in 800 parts cold, 50 parts boiling alcohol, in 60 parts hot glacial acetic acid.

USE: In the manuf of polarizing glasses and plastics.

8087. Quinine Oleate. Usually with an excess of oleic acid. Contains about 21% anhydr quinine.

Brown, thick liquid. Insol in water; sol in alcohol, ether, oils, petr ether.

USE: Has been credited with opacity to ultraviolet rays and suggested as prophylactic against sunburn, applied as a 5-10% ointment. A soln of the oleate without excess of oleic acid in petr naphtha has been recommended for mothproofing of fabrics.

8088. Quinine Salicylate. $C_{27}H_{30}N_2O_5$; mol wt 462.55. C 70.11%, H 6.54%, N 6.06%, O 17.30%. $C_{20}H_{24}N_2O_2.C_6H_4(OH)COOH$.

Monohydrate, white, odorless, bitter crystals or cryst powder. One gram dissolves in about 1500 ml water, 20 ml alcohol, 25 ml chloroform, 160 ml ether, 15 ml glycerol. *Protect from light.*

THERAP CAT: Antimalarial; analgesic.

8089. Quinine Sulfate. Basic quinine sulfate. $C_{40}H_{50}N_4O_8S$; mol wt 746.93. C 64.32%, H 6.75%, N 7.50%, O 17.14%, S 4.29%. $2C_{20}H_{24}N_2O_2.H_2SO_4$. Quinine sulfate crystallizes with 8 mols water (16.17% H_2O), but it loses one mol very rapidly even at 20°. On exposure to air at ordinary temp it gradually loses 5 more mols of water. Double-blind study in prevention of hemodialysis-induced muscle cramps: D. M. Kaji *et al.*, *Lancet* **2**, 66 (1976); in prevention of nocturnal recumbency muscle cramps: K. Jones, C. M. Castleden, *Age Ageing* **12**, 155 (1983).

Dihydrate, $C_{40}H_{50}N_4O_8S.2H_2O$, *Coco-Quinine, Quinamin, Quinamm, Quine, Quinate, Quinsan.* Dull needles or rods, making a light and readily compressible mass. Becomes brownish on exposure to light. Loses its water of crystn at about 100°. $[\alpha]_D^{15}$ -220° (5% soln in about 0.5N HCl); *see also* the values given for quinine base. One gram dissolves in 810 ml water, 32 ml boiling water, 120 ml alcohol, 10 ml alcohol at 78°. Slightly sol in chloroform, ether, but freely sol in a mixture of 2 vols chloroform and 1 vol abs alcohol. Aq solns are neutral to litmus, pH of satd soln 6.2. *Keep well closed and protected from light.* Incompat: Ammonia, alkalies, limewater, Donovan's soln, tannic acid, iodine, iodides, acetates, citrates, tartrates, benzoates, salicylates.

THERAP CAT: Antimalarial. Muscle relaxant (skeletal).

THERAP CAT (VET): Bitter stomachic. Analgesic. Antipyretic.

8090. Quinine Tannate. Contains about 33% anhydr quinine.

Yellowish-white, amorphous, odorless powder; tasteless or almost tasteless; astringent taste. Slightly sol in water, chloroform or ether; somewhat more sol in alcohol.

THERAP CAT: Antimalarial.

8091. Quinine Urea Hydrochloride. $C_{21}H_{30}Cl_2N_4O_3$; mol wt 457.40. C 55.14%, H 6.61%, Cl 15.50%, N 12.25%, O 10.49%. $C_{20}H_{24}N_2O_2.HCl + CO(NH_2)_2.HCl$. A double salt of quinine and urea hydrochlorides.

Pentahydrate, prisms from water. One gram dissolves in 0.9 ml water, 2.4 ml alcohol. pH of 1:20 aq soln 3.1.

THERAP CAT: Sclerosing agent; antimalarial; local anesthetic.

THERAP CAT (VET): Has been used as a local anesthetic.

8092. Quininic Acid. *6-Methoxy-4-quinolinecarboxylic acid;* 6-methoxycinchoninic acid. $C_{11}H_9NO_3$; mol wt 203.19. C 65.02%, H 4.46%, N 6.89%, O 23.62%.

Pale yellow crystals. mp about 280° with dec. Slightly sol in water, cold alcohol or ether; sol in about 80 parts boiling abs alcohol; sol in aq alkalies.

8093. Quininone. (8α)-6'-Methoxycinchonan-9-one.

Consult the cross index before using this section.

$C_{20}H_{22}N_2O_2$; mol wt 322.39. C 74.51%, H 6.88%, N 8.69%, O 9.93%. By careful oxidation of quinine or quinidine: Rabe, Kuliga, *Ann.* **364**, 346, 349 (1909); Woodward *et al., J. Am. Chem. Soc.* **67**, 1425 (1945). Alternate procedure: Rabe, Kindler, *Ber.* **51**, 466 (1918). Prepn of amorphous epimeric mixture of quininone and quinidinone from quinotoxine: Gutzwiller, Uskokovic, *Helv. Chim. Acta* **56**, 1494 (1973).

Crystals from ether, mp 108° (rapid heating). Shows mutarotation, final $[\alpha]_D^{20}$ +76° (c = 2 in alcohol). Alkaline reaction to litmus. Freely sol in alcohol, ether, chloroform, benzene. Almost insol in water and petr ether.

Hydrochloride, $C_{20}H_{23}ClN_2O_2$, hygroscopic crystals, mp 212°. Final $[\alpha]_D^{18}$ +59° (c = 2 in abs alcohol).

8094. Quinizarin. *1,4-Dihydroxy-9,10-anthracenedione; 1,4-dihydroxyanthraquinone;* C.I. 58050. $C_{14}H_8O_4$; mol wt 240.20. C 70.00%, H 3.36%, O 26.64%. Prepn from *p*-chlorophenol and phthalic anhydride: Reynolds, Bigelow, *J. Am. Chem. Soc.* **48**, 420 (1926); U.S. pat. **1,845,632**, *C.A.* **26**, 2203; *Org. Syn.* **coll. vol. I**, 476 (New York, 1941). Prepn from hydroquinone: Gattermann-Wieland, *Praxis des Organischen Chemikers* (de Gruyter, Berlin, 40th ed., 1961) p 299. Also prepd from diazotized *p*-chloroaniline and phthalic anhydride: **Brit. pat. 373,999**, *C.A.* **27**, 3946 (1933); by treating anthraquinone with ammonium persulfate in sulfuric acid: Wacker, *J. Prakt. Chem.* [2] **54**, 90 (1896). Purification from contaminating purpurin: *Org. Syn.* (loc. cit.). See also: *Colour Index* vol. 4 (3rd ed., 1971) p 4515.

Orange crystals from acetic acid, mp 200-203° (*Org. Syn.*). Orange plates from ether. Deep red needles from alcohol, benzene, toluene, xylene. mp 196°. Sublimes in high vacuum. Absorption spectrum: Meek, Watson, *J. Chem. Soc.* **109**, 544 (1916); Meek, *ibid.* **111**, 969 (1917). K at 18° = 3.1 × 10⁻¹⁰. Moderately sol in alcohol with red color. Sol in ether with brown color and yellow fluorescence. Sol with violet color in aq alkalies and in ammonia. Black precipitate with CO_2. One gram dissolves in about 13 g of boiling glacial acetic acid.

Dimethyl ether, $C_{16}H_{12}O_4$, mp 177°.

8095. Quinizarin Green SS. *1,4-Bis[(4-methylphenyl)-amino]-9,10-anthracenedione; 1,4-di-p-toluidinoanthraquinone;* D & C Green No. 6; C.I. Solvent Green 3; C.I. 61565. $C_{28}H_{22}N_2O_2$; mol wt 418.50. C 80.36%, H 5.30%, N 6.69%, O 7.65%. Discovered by R. E. Schmidt in 1894: *Colour Index* vol. 4 (3rd ed., 1971) p 4541.

Dark violet needles, mp 218°. Blue soln in conc H_2SO_4 giving a blue-green ppt on dilution.

USE: For sutures: *Fed. Regist.* **40**, 18167 (1975). Permitted for use in externally applied drugs and cosmetics: *ibid.* **47**, 14138 (1982).

8096. Quinocide. N^1-*(6-Methoxy-8-quinolinyl)-1,4-pentanediamine; 8-(4-aminopentylamino)-6-methoxyquinoline; 8-[(4-amino-4-methylbutyl)amino]-6-methoxyquinoline; 6-methoxy-8-(4-aminopentylamino)quinoline;* Chinocide; Khinocide. $C_{15}H_{21}N_3O$; mol wt 259.34. C 69.46%, H 8.16%, N 16.21%, O 6.17%. Prepn: Braude, Stavrovskaya, *J. Gen. Chem. USSR* **26**, 999 (1956); modified method: *idem, Med. Prom.* **11**(7), 19 (1957), *C.A.* **52**, 11043 (1958); B. Balkrishen *et al., Chem. & Ind.* (London) **1983**, 899.

Hydrochloride, $C_{15}H_{22}ClN_3O$, crystals from abs alcohol, mp 224-224.5° (base, bp₁ 183-186°; mp 46°).
Dihydrochloride, $C_{15}H_{23}Cl_2N_3O$, mp 227-227.5°.
Diphosphate, $C_{15}H_{27}N_3O_9P_2$, mp 174-176°.
THERAP CAT: Antimalarial.

8097. Quinoline. Leucoline; chinoleine; 1-benzazine; benzo[b]pyridine. C_9H_7N; mol wt 129.15. C 83.69%, H 5.46%, N 10.85%. Occurs in small amounts in coal tar. Prepd by the Skraup synthesis by heating aniline with glycerol and nitrobenzene in presence of sulfuric acid: Clarke, Davis, *Org. Syn.* **coll. vol. I**, 478 (2nd ed., 1941) 478; alternate syntheses: Manske, *Chem. Rev.* **30**, 113 (1942); Bergstrom, *ibid.* **35**, 150 (1944). Process based on the interaction of aniline with acetaldehyde and a formaldehyde hemiacetal: Cislak, Wheeler, U.S. pat. **3,020,281** (1962 to Reilly Tar & Chem.). Toxicity data: Smyth *et al., Arch. Ind. Hyg. Occup. Med.* **4**, 119 (1951).

Hygroscopic liquid. Absorbs as much as 22% water. Darkens on storage in ordinary, stoppered bottle. Penetrating odor, not as offensive as pyridine. Volatile with steam. d_4^{25} 1.0900; mp −15°; bp₇₆₀ 237.7°; bp₁₀₀ 163.2°; bp₄₀ 136.7°; b₂₀ 119.8°; bp₁₀ 103.8°; bp₅ 89.6°; bp₁.₀ 59.7°. n_D^{20} 1.62683. Absorption spectrum: Hantzsch, *Ber.* **44**, 1824 (1911). Weak base (neutral to phenolphthalein), forms water-soluble salts with strong acids: pK 9.5. Difficultly sol in cold water, more easily in hot water. Miscible with alcohol, ether, carbon disulfide. Dissolves sulfur, phosphorus, arsenic trioxide. *Protect from light and moisture.* LD₅₀ orally in rats: 460 mg/kg (Smyth).

Bisulfate, $C_9H_9NO_4S$, white to grayish-white, crystalline powder. mp 163-165°. Freely sol in water; one gram dissolves in 50 ml cold, in 9 ml boiling abs alcohol. *Protect from light.*

Hydrochloride, C_9H_8ClN, white deliquesc crystals. mp 93-94°. Freely sol in water, alcohol, hot benzene, chloroform, sparingly in cold ether. *Keep well closed and protected from light.*

Salicylate, $C_{16}H_{13}NO_3$, reddish-gray crystalline powder. Sol in 80 parts water; freely sol in alcohol, benzene, ether, glycerol, oils. *Protect from light.*

Tartrate, $C_{43}H_{45}N_3O_{24}$, white crystals, pungent odor, sharp taste. mp 125° (dec). Sol in 80 parts water, 150 parts alcohol. Insol in ether.

USE: Manuf dyes; prepn hydroxyquinoline sulfate, *q.v.*, niacin. As preservative for anatomical specimens. Solvent for resins, terpenes.

THERAP CAT: Antimalarial.

8098. 8-Quinolineboronic Acid. $C_9H_8BNO_2$; mol wt

172.98. C 62.49%, H 4.66%, B 6.26%, N 8.10%, O 18.50%. Prepd by reacting 8-bromoquinoline with butyllithium and tributyl borate: Letsinger, Dandegaonker, *J. Am. Chem. Soc.* **81**, 498 (1959).

Crystals from alc, mp > 300°.
Butyl diester, $C_{17}H_{24}BNO_2$, bp$_4$ 180°. n_D^{25} 1.4840.
Chloroethyl diester, $C_{13}H_{14}BCl_2NO_2$, crystals from toluene, mp 193-194°.

8099. 8-Quinolinecarboxylic Acid. $C_{10}H_7NO_2$; mol wt 173.16. C 69.36%, H 4.07%, N 8.09%, O 18.48%. Prepd by heating 2-aminobenzoic acid with 2-nitrobenzoic acid, glycerol and sulfuric acid: Schlosser, Skraup, *Monatsh.* **2**, 530 (1881); by oxidizing 8-quinolylaldehyde with chromic acid: Howitz, *Ber.* **35**, 1275 (1902).

Needles from water. mp 186-187.5°. Sublimes above mp. Slightly sol in cold water, appreciably sol in hot water and in alcohol. Freely sol in acids and alkalies.
USE: Detection of cadmium, copper, iron, lead, mercury, silver, thallium; quantitative determination of copper.

8100. Quinoline Yellow. *C.I. Acid Yellow 3;* C.I. 47005; D&C Yellow No. 10; Acid Yellow 3; Food Yellow 13. Synthetic dye. A mixture of the sodium salts of the mono- and disulfonic acids of quinoline yellow spirit soluble, *q.v.* Principal constituents: sodium salts of *2-(2,3-dihydro-1,3-dioxo-1H-indene-2-yl)-6-quinolinesulfonic acid* and *2-(2,3-dihydro-1,3-dioxo-1H-indene-2-yl)-8-quinolinesulfonic acid*. Prepn of sulfonic acids: M. C. Traub, *Ber.* **16**, 297 (1883); *Colour Index* vol. 4 (3rd ed., 1971) p 4435. HPLC determn: J. Chudy *et al., J. Chromatog.* **154**, 306 (1978). Purification: N. A. Ambrosiano *et al.*, U.S. pat. **4,398,916** (1983 to Sterling Drug). Stability: H. Delonca *et al., Pharm. Acta Helv.* **58**, 332 (1983). ^{13}C-NMR spectrum: M. Gelbcke *et al., Bull. Soc. Chim. Belg.* **91**, 237 (1982). TLC identification: J. A. Steele, *J. Assoc. Off. Anal. Chem.* **67**, 540 (1984). Composition and skin sensitivity: Y. Sata *et al., Contact Dermatitis* **10**, 30 (1984). The unsulfonated quinoline yellow spirit soluble may remain as an impurity and is limited by the FDA: *Code of Federal Regulations*, Title 21, Parts 1-99, Sec. 74.1710(b), p 307 (1985). LC determn of unsulfonated impurity: A. L. Goldberg, *J. Assoc. Off. Anal. Chem.* **68**, 477 (1985). Toxicity: F. C. Lu, A. Lavalle, *Can. Pharm. J.* **97**, 30 (1964).

Bright greenish yellow. Orange in conc H_2SO_4. Sol in water, slightly sol in ethanol. Practically insol in vegetable oils. LD$_{50}$ in rats (g/kg): > 2 orally (Lu, Lavalle).
USE: Textile dye for wool, nylon, silk. Paper dye. Barium salt in printing inks. Color for food, drugs, cosmetics. Approved by the FDA for use in drugs, cosmetics, except for use in the area of the eye: *Fed. Reg.* **48**, 39217 (1983); *ibid.* **49**, 27744 (1984).

8101. Quinoline Yellow Spirit Soluble. *C.I. Solvent Yellow 33;* C.I. 47000; D&C Yellow No. 11; quinoline yellow base; quinoline yellow A. Synthetic dye consisting principally of *quinophthalone (2-(2-quinolinyl)-1H-indene-1,3-(2H)-dione)*. Prepn: M. C. Traub, *Ber.* **16**, 297 (1883); E. Jacobsen, C. L. Reimer, *ibid.* 1082; *Colour Index* vol. 4 (3rd ed., 1971) p 4435. ^{13}C-NMR spectrum: M. Gelbcke *et al., Bull. Soc. Chim. Belg.* **91**, 237 (1982). Composition and skin sensitivity: Y. Sato *et al., Contact Dermatitis* **10**, 30 (1984); S. Kita *et al., ibid.* **11**, 210 (1984). Inhalation toxicity in animals: T. C. Marrs *et al., Human Toxicol.* **3**, 289 (1984).

quinophthalone
Bright greenish yellow. Insol in water. Slightly sol in ethanol (yellow), linseed oil, mineral oil, oleic acid, paraffin wax, stearic acid, turpentine. Sol in acetone, chloroform, benzene, toluene. Yellow brown in conc H_2SO_4, yellow flocculent ppt in dilution.
USE: In spirit lacquers, polystyrenes, polycarbonates, polyamides, and acrylic resins. In colored smokes. Occasionally in hydrocarbon solvents. Approved by the FDA for use in externally applied drugs and cosmetics: *Code of Federal Regulations*, Title 21, Parts 1-99, Secs. 74.1711, 74.3106, pp 307, 315 (1985).

8102. Quinolinic Acid. *2,3-Pyridinedicarboxylic acid.* $C_7H_5NO_4$; mol wt 167.12. C 50.31%, H 3.02%, N 8.38%, O 38.29%. Metabolite of tryptophan, *q.v.* Prepn: S. Hoogewerff, W. A. van Dorp, *Ber.* **12**, 747 (1879); W. Koenigs, *ibid.* 983; O. Fischer, E. Renouf, *ibid.* **17**, 755 (1884); E. Sucharda, *ibid.* **58B**, 1727 (1925); V. Yu. Stiks, S. A. Bulgach, *ibid.* **65B**, 11 (1932); A. F. Lindenstruth, C. A. VanderWerf, *J. Am. Chem. Soc.* **71**, 3020 (1949). Metabolism studies: L. M. Henderson *et al., J. Biol. Chem.* **181**, 667, 677, 687, 731 (1949); H. P. Sarett, *ibid.* **193**, 627 (1951). Neuroexcitatory activity and possible role in neurodegenerative disorders: R. Schwarcz *et al., Science* **219**, 316 (1983).

Odorless crystals. mp 190° when rapidly heated, with decompn into CO_2 and nicotinic acid. Sol in 180 parts water, in alkalies, slightly in alcohol; almost insol in ether or benzene.

8103. Quinone. *2,5-Cyclohexadiene-1,4-dione; p-quinone; 1,4-benzoquinone; 1,4-cyclohexadienedione.* $C_6H_4O_2$; mol wt 108.09. C 66.67%, H 3.73%, O 29.60%. Made by oxidation of aniline with sodium dichromate in presence of sulfuric acid. Laboratory prepn from hydroquinone: Vliet, *Org. Syn.* **2**, 85 (1922); also *Org. Syn.* **coll. vol. I** (2nd ed., 1941). *Cf.* Underwood, Walsh, *ibid.* **16**, 73 (1936).

Yellow monoclinic prisms from water or petr ether. Penetrating odor resembling that of chlorine. Irritating vapors. d$_4^{20}$ 1.318. mp 115.7°. Sublimes [sublimation velocities *in vacuo*: Kempf, *J. Prakt. Chem.* [2] **78**, 236 (1908)]. Volatile with steam. Absorption spectrum: Hantzsch, *Ber.* **49**, 522 (1916). Dipole moment: 0.67. Polemic over correct values: Paoloni, *J. Am. Chem. Soc.* **80**, 3879 (1958). Slightly sol in water; sol in alcohol, ether, hot petr ether, alkalies. LD$_{50}$ orally in rats: 130 mg/kg, Woodard *et al., Fed. Proc.* **8**, 348 (1949).
Picrate, $C_{12}H_7N_3O_9$, yellow crystals, mp 78-79°.
USE: Oxidizing agent; in photography; manuf dyes; manuf hydroquinone; tanning hides; making gelatin insol; strengthening animal fibers; as reagent. *Caution:* Can cause dermatitis with discoloration, erythema, formation of papules and vesicles. In severe cases there can be necrotic changes in the skin. Vapors acting on eye can cause serious disturbances, including conjunctivitis and even corneal ulceration. Discoloration of conjunctiva and cornea have been reported.

8104. Quinovic Acid. *3-Hydroxyurs-12-ene-27,28-dioic acid;* quinovaic acid; chinovic acid; chinova acid. $C_{30}H_{46}O_5$;

mol wt 486.67. C 74.03%, H 9.53%, O 16.44%. Isoln from cinchona bark as the glycoside quinovin: Hlasiwetz, *Ann.* **79**, 129 (1851); **111**, 182 (1859); Votocek, Rác, *Coll. Czech. Chem. Commun.* **1**, 234 (1929); Tschesche *et al.*, *Ann.* **667**, 151 (1963); from *Zygophyllum coccineum* L., *Zygophyllaceae:* Soliman, *J. Chem. Soc.* **1939**, 1760; from bark of *Mitragyna inermis* Kuntze and leaves of *M. ciliata* Aubrev & Pellegr. and *M. rubrostipulacea* Havil., *Rubiaceae:* Badger *et al., ibid.* **1950**, 867. Structure: Brossi *et al., Helv. Chim. Acta* **34**, 244 (1951); Barton, de Mayo, *J. Chem. Soc.* **1953**, 3111. *Review:* J. Simonsen, W. C. J. Ross, *The Terpenes* vol. 5 (University Press, Cambridge, 1957) pp 75-113.

Prisms from dil pyridine, dec 297°. Very bitter taste. $[\alpha]_D^{20}$ +99° (c = 2.48 in pyridine). Practically insol in water and other solvents except pyridine.

O-Acetylquinovic acid, $C_{32}H_{48}O_6$, needles from dil acetone, mp 282-284°.

Dimethyl ester, $C_{32}H_{50}O_5$, needles from dil acetone, mp 175°. $[\alpha]_D^{19}$ +117.4° (c = 2.2 in $CHCl_3$). Sol in organic solvents.

8105. Quinovin. Quinova-bitter; chinovin. Isoln from cinchona bark (*cortex chinae*): Hlasiwetz, *Ann.* **111**, 182 (1859); Liebermann, Giesel, *Ber.* **16**, 926 (1883). A mixture of three glycosides, 60% A, 5% B and 30% C: Tschesche *et al., Ann.* **667**, 151 (1963).

Glycoside A, $C_{36}H_{56}O_9$, *quinovic acid* β-D-*quinovoside.* Crystals from methanol + water, mp 237-238°. $[\alpha]_D^{20}$ +57° ± 2° (ethanol).

Glycoside B, $C_{36}H_{56}O_9$, *cincholic acid* β-D-*quinovoside.* Crystals from methanol + water, mp 193-195°. $[\alpha]_D^{19}$ +78° ± 3° (c = 0.78 in ethanol).

Glycoside C, $C_{36}H_{56}O_{10}$, *quinovic acid* β-D-*glucoside.* Crystals from methanol + water, mp 247-250°. $[\alpha]_D^{20}$ +62° ± 2° (ethanol).

8106. Quinovose. *6-Deoxy*-D-*glucose;* D-glucomethylose; D-isorhamnose; D-epirhamnose; isorhodeose; epifucose; chinovose. $C_6H_{12}O_5$; mol wt 164.16. C 43.90%, H 7.37%, O 48.73%. Isoln from cinchona bark: Freudenerg, *Ber.* **62**, 373 (1929). Identity with D-glucomethylose: Votocek, Rác, *Coll. Czech. Chem. Commun.* **1**, 239 (1929). Structure: Karrer, Boettcher, *Helv. Chim. Acta* **36**, 570 (1953). Synthesis from 6-*O*-*p*-tolylsulfonyl-D-glucose: Schmidt, "6-Deoxy-α-D-glucose" in *Methods in Carbohydrate Chemistry* vol. I, R. L. Whistler, M. L. Wolfrom, Eds. (Academic Press, New York, 1962) pp 198-201. Alternate facile synthesis: V. K. Srivastava, L. M. Lerner, *Carbohydr. Res.* **64**, 263 (1978).

Crystals from ethyl acetate, mp 146°. $[\alpha]_D^{20}$ +73° (5 min) → +30° (3 hr, final; c = 8.3 in water). Sol in water, ethanol; practically insol in ether, acetone.

8107. Quinoxaline. 1,4-Benzodiazine; benzo[*a*]pyrazine; benzoparadiazine; phenpiazine. $C_8H_6N_2$; mol wt 130.14. C 73.83%, H 4.65%, N 21.53%. Prepd from *o*-phenylenediamine and glyoxal or glyoxal sodium bisulfite: Hinsberg,

Ber. **17**, 320 (1884); *Ann.* **237**, 334 (1887); Jones, McLaughlin, *Org. Syn.* **30**, 86 (1950).

Crystals, mp 29-30°. Odor of quinoline when cool, odor of piperidine when hot. d_4^{48} (liq) 1.1334. $bp_{760.3}$ 229.5°. bp_{12} 108-111°. n_D^{48} 1.6231. Very freely sol in water, alc, ether, benzene.

Monohydrate, crystals, mp 37°.

Sulfate, $C_8H_6N_2.H_2SO_4$, leaflets, mp 186-187°. Freely sol in water. Less sol in alcohol.

8108. Quintozene. *Pentachloronitrobenzene;* PCNB; terrachlor; PKhNB; Avicol; Botrilex; Brassicol; Folosan; Terraclor; Tilcarex; Tritisan. $C_6Cl_5NO_2$; mol wt 295.36. C 24.40%, Cl 60.03%, N 4.74%, O 10.83%. Prepd by treating pentachlorobenzene with fuming nitric acid: Jungfleisch, *Ann. Chim.* [4] **15**, 286 (1868); A. Roedig, K. Kiepert, *Ann.* **593**, 71 (1955). Persistence of residues in foods: P. B. Baker, B. Flaherty, *Analyst* **97**, 378 (1972); A. R. P. Paxton, D. Purser, *Pestic. Sci.* **13**, 401 (1982); in soil: J. Beck, K. E. Hansen, *ibid.* **5**, 41 (1974). Metabolism by sheep and goats: P. W. Aschbacher, V. J. Feil, *J. Agr. Food Chem.* **31**, 1150 (1983). Toxicity data: J. K. Finnegan *et al., Arch. Int. Pharmacodyn. Ther.* **114**, 38 (1958). *Review:* WHO *Environmental Health Criteria* **41**, 1-38 (1984). Comparative review of the effects on soil organisms: E. R. Ingham, *Crop. Prot.* **4**, 3-32 (1985).

Fine needles from alcohol, platelets from carbon disulfide. d_4^{25} 1.718. mp 144°. bp_{760} 328° (some dec). Practically insol in water, cold alcohol. Freely sol in carbon disulfide, benzene, chloroform. LD_{50} in male, female rats (g/kg): 1.71 ± 0.20, 1.65 ± 0.17 by gavage (Finnegan).

USE: Fungicide for seed and soil treatment.

8109. Quinuclidine. *1-Azabicyclo[2.2.2]octane;* 1,4-ethylenepiperidine. $C_7H_{13}N$; mol wt 111.18. C 75.61%, H 11.79%, N 12.60%. Prepn: Löffler, Stietzel, *Ber.* **42**, 124 (1909); Meisenheimer *et al., Ann.* **420**, 191 (1920); Clemo, Metcalfe, *J. Chem. Soc.* **1937**, 1989; Prelog, U.S. pat. **2,192,-840** (1940); Wawzonek *et al., J. Am. Chem. Soc.* **73**, 2806 (1951); Leonard, Elkin, *J. Org. Chem.* **27**, 4635 (1962).

Prisms from petr ether. Sublimes. mp 156° (sealed tube). Very sol in water and the usual organic solvents.

Picrate, $C_7H_{13}N.C_6H_3N_3O_7$, yellow prisms from alcohol, dec 275°. Soluble in 35-40 parts hot alcohol.

Hydrochloride, $C_7H_{13}N.HCl$, crystals from abs ethanol, mp 364-365°.

8110. 3-Quinuclidinol. *1-Azabicyclo[2.2.2]octan-3-ol;* 3-hydroxyquinuclidine. $C_7H_{13}NO$; mol wt 127.18. C 66.10%, H 10.30%, N 11.01%, O 12.58%. Prepn: Sternbach, Kaiser, *J. Am. Chem. Soc.* **74**, 2215 (1952); Sternbach, U.S. pat. **2,648,667** (1953 to Hoffmann-La Roche); of free alcohol, acetate and/or benzoate esters: Grob *et al., Helv. Chim. Acta* **40**, 2170 (1957); Mikhalina, Rubtsov, *Zh. Obshch. Khim.* **30**, 163 (1960), *C.A.* **54**, 22632h (1960). Resolution of *dl*-forms via *d*-camphorsulfonates: Sternbach, Kaiser, *loc. cit.*; Sternbach, *loc. cit.* Structure activity of esters: M. D. Mashkovsky in *Proc. 1st Int. Pharmacol. Mtg. Stockholm, 1961* **vol. 7** (Macmillan, New York, 1963) pp 359-366.

dl-Form, crystals from benzene or acetone, mp 221-223° also reported as 225-227°. Sublimes at 120° and 20 mm Hg. Very sol in water.

l-Form, prisms from acetone, mp 220-222°. $[\alpha]_D^{25}$ −2.0° (c = 6.5), −43.8° (c = 3 in 1N HCl).

dl-Form hydrochloride, prisms from methanol + acetone, mp > 300°.

dl-Form acetate (ester), $C_9H_{15}NO_2$, *aceclidine, 3-acetoxyquinuclidine, 3-quinuclidinyl acetate.* Oily liq, $bp_{0.4}$ 73-74°, bp_{11} 113-115°. n_D^{25} 1.4675.

dl-Form acetate hydrochloride, $C_9H_{16}ClNO_2$, *Glaucotat, Glaudin.* Crystals, mp 166°.

dl-Form benzilate (ester), $C_{21}H_{23}NO_3$, *3-quinuclidinyl benzilate, BZ, QNB, Ro 2-3308.* Prepn: Sternbach, Kaiser, *loc. cit.,* 2219. Crystals from acetone-ether, mp 164-165°.

dl-Form benzoate (ester), $C_{14}H_{17}NO_2$, *3-benzoyloxyquinuclidine, 3-quinuclidinyl benzoate.* Oily liquid, $bp_{0.3}$ 148-150°.

dl-Form benzoate hydrochloride, $C_{14}H_{18}ClNO_2$, *oksilidin, oxylidine.* Crystals, mp 238-240°.

USE: Benzilate (ester) has been used as an incapacitating agent in chemical warfare: *Health Aspects of Chemical and Biological Weapons* (WHO, Geneva, 1970) pp 49-51.

THERAP CAT: Hypotensive; acetate (ester) as cholinergic.

8111. Quinupramine. *5-(1-Azabicyclo[2.2.2]oct-3-yl)-10,11-dihydro-5H-dibenz[b,f]azepine;* 10,11-dihydro-5-(3-quinuclidinyl)-5H-dibenz[b,f]azepine; LM-208; Kinupril; Kevopril. $C_{21}H_{24}N_2$; mol wt 304.43. C 82.85%, H 7.95%, N 9.20%. Analog of imipramine, *q.v.* Prepn: C. Gueremy, P. C. Wirth, **Ger. pat. 2,030,492** (1971 to Sogeras), *C.A.* **74,** 141581e (1971). Animal studies: W. Van Dorsser, A. Dresse, *Arch. Int. Pharmacodyn. Ther.* **208,** 373 (1974); *eidem, ibid.* **220,** 164 (1976). Clinical study: R. Volmat *et al., Clin. Neurol. Psychiat.* **239,** 445 (1978).

Cryst, mp 150°.
THERAP CAT: Antidepressant.

8112. Quisqualic Acid. *(S)-α-Amino-3,5-dioxo-1,2,4-oxadiazolidine-2-propanoic acid;* L-quisqualic acid; β-(3,5-dioxo-1,2,4-oxodiazolidin-2-yl)-L-alanine. $C_5H_7N_3O_5$; mol wt 189.13. C 31.75%, H 3.73%, N 22.22%, O 42.30%. Excitatory amino acid (EAA) used to identify a specific subset of EAA receptors; consequently, the receptors are known as quisqualate receptors. *See also* NMDA, kainic acid. Isoln from the seeds of *Quisqualis chinesis* and anthelmintic activity: Y.-C. Tuan *et al., Yao Hsueh Hsueh Pao* **5,** 87 (1957), *C.A.* **56,** 14896b (1958); from *Q. indica:* S.-T. Fang, J.-H. Chu, *Hua Hsueh Hsueh Pao* **30,** 226 (1964), *C.A.* **61,** 7359f (1962); from *Q. fructus:* T. Takemoto *et al., Yakugaku Zasshi* **95,** 176 (1975), *C.A.* **82,** 152211a (1975). Enzymic synthesis: I. Murakoshi *et al., Chem. Pharm. Bull.* **22,** 473 (1974). Total synthesis: J. E. Baldwin *et al., Chem. Commun.* **1985,** 256. Crystal structure: J. L. Flippen, R. D. Gilardi, *Acta Crystallogr.* **B32,** 951 (1976). Identification as a neuroexcitant: H. Shinozaki, I. Shibuya, *Neuropharmacology* **13,** 665 (1974). Receptor binding studies: K. Koshiya, *Life Sci.* **37,** 1373 (1985); J. T. Greenamyre, *J. Pharmacol. Exp. Ther.* **233,** 254 (1985). Review of isolation of quisqualic acid and other EAAs: T. Takemoto, in *Kainic Acid as a Tool in Neurobiology,* R. G. McGeer *et al.,* Eds. (Raven Press, New York, 1978) pp 1-15.

Crystals from water-ethanol, mp 190-191°. $[\alpha]_D^{20}$ +17.0° (c = 2.0 in 6M HCl).

8113. Quizalofop-Ethyl. *2-[4-[(6-Chloro-2-quinoxalinyl)oxy]phenoxy]propanoic acid ethyl ester;* ethyl 2-[4-(6-chloro-2-quinoxalinyloxy)phenoxy]propionate; quinofop-ethyl; DPX-Y 6202; NCI 96683; NC 302; Assure; Targa; Pilot. $C_{19}H_{17}ClN_2O_4$; mol wt 372.81. C 61.21%, H 4.60%, Cl 9.51%, N 7.51%, O 17.17%. Post-emergence herbicide for control of grassy weeds in broad-leaved crops. Prepn and herbicidal activity: Y. Ura *et al.,* **Ger. pat. 3,004,770** (1980 to Nissan), *C.A.* **94,** 103421h (1981). Prepn and structure-activity relationships: G. Sakata *et al., J. Pestic. Sci.* **10,** 61 (1985). Pharmacological effects in laboratory animals: K. Inokuchi *et al., Oyo Yakuri* **30,** 509 (1985), *C.A.* **104,** 30101s (1986). Site of action studies: T. Ikai *et al., Proc. Brit. Crop Prot. Conf.-Weeds* **1985,** 163. Account of properties, toxicity, crop tolerance: G. Sakata *et al., Proc. 10th Conf. Int. Congr. Plant Prot.* **1,** 315 (1983).

White crystals, mp 92-93°. $bp_{0.2}$ 220°. Practically insol in water (0.3 × 10⁻⁶ g/ml at 20°). Soly in acetone, ethanol, benzene, xylene at 20° (g/ml): 0.11, 0.009, 0.29, 0.12. LD_{50} in male, female rats, male, female mice (mg/kg): 1670, 1480, 2350, 2360 orally; all 10,000 dermally. LC_{50} (96 hr) in rainbow trout: 10.7 mg/l (Sakata).

USE: Herbicide.

R

8114. R-11. *1,5a,6,9,9a,9b-Hexahydro-4a(4H)-dibenzo-furancarboxaldehyde;* 2,3,4,5-bis(2-butenylene)tetrahydro-furfural; MGK-11; MGK Repellent 11. $C_{13}H_{16}O_2$; mol wt 204.26. C 76.44%, H 7.90%, O 15.67%. Prepd by heating furfuraldehyde with butadiene and water under pressure: Hillyer, Nicewander, U.S. pat. **2,683,151** (1954 to Phillips Petroleum). Clinical efficacy vs. sand flies: F. P. Fossati, M. Maroli, *Trans. Roy. Soc. Trop. Med. Hyg.* **80**, 771 (1986).

Liquid, bp 307°. d_4^{20} 1.10. mp —80°. n_D^{20} 1.5254. Practically insol in water.
Oxime, mp 97.2°.
Dinitrophenylhydrazone, mp 152.8°.
USE: Insect repellent.

8115. Racefemine. (±)-*α-Methyl-N-(1-methyl-2-phenoxyethyl)benzeneethanamine; dl-threo-α-methyl-N-(1-methyl-2-phenoxyethyl)phenethylamine; dl-threo-α*-methyl-N-(1-phenoxy-2-propyl)phenethylamine; *dl-threo-N-(1-methyl-2-phenylethyl)-N-(2-phenoxy-1-methylethyl)amine;* CB 3697. $C_{18}H_{23}NO$; mol wt 269.37. C 80.25%, H 8.61%, N 5.20%, O 5.94%. Smooth muscle relaxant. Prepn from amphetamine and phenoxyacetone with isomer separation: **Neth.** pat. **Appl. 6,407,309** (1964 to Clin-Byla), *C.A.* **63**, 353c (1965).

Isomers: From *dl*-amphetamine, a yellow liq, $bp_{0.05}$ 132-135°, was obtained which was separated into isomers I and II. *I-Fumarate:* mp 162°. *I-Hydrochloride:* mp 156-157°. *II-Hydrochloride:* mp 167-168°. From *d*-amphetamine, isomers III, $[α]_D^{21}$ +41° (c = 0.01 in ethanol), and IV, $[α]_D^{22}$ +22° (c = 0.005 in ethanol), were obtained. *III-Fumarate:* mp 164-165°, $[α]_D^{20}$ +19° (c = 0.01 in ethanol). *III-Hydrochloride:* mp 186-187°, $[α]_D^{21}$ +22° (c = 0.01 in ethanol). *IV-Hydrochloride:* mp 160-163°, $[α]_D^{20}$ +22° (c = 0.01 in ethanol). From *l*-amphetamine, isomers V, $[α]_D^{24}$ —24° (c = 0.01 in ethanol), and VI, $[α]_D^{21}$ —41° (c = 0.01 in ethanol), were obtained. *V-Hydrochloride:* mp 189-190°, $[α]_D^{21}$ —22° (c = 0.01 in ethanol). *VI-Fumarate:* mp 164-164.5°, $[α]_D^{22}$ —15° (c = 0.01 in ethanol). *VI-Hydrochloride:* mp 186-187°, $[α]_D^{21}$ —21.5° (c = 0.01 in ethanol).
Fumarate, $C_{22}H_{27}NO_5$, *Dysmalgine.*
THERAP CAT: Antispasmodic.

8116. Racemethorphan. (±)-*3-Methoxy-17-methylmorphinan; dl-cis*-1,3,4,9,10,10a-hexahydro-6-methoxy-11-methyl-2H-10,4a-iminoethanophenanthrene; *dl-cis*-1,2,3,9,-10,10a-hexahydro-6-methoxy-11-methyl-4H-10,4a-iminoethanophenanthrene; deoxydihydrothebacodine; methorphan. $C_{18}H_{25}NO$; mol wt 271.41. C 79.66%, H 9.28%, N 5.16%, O 5.90%. Prepn: Schnider, Grüssner, U.S. pat. **2,676,177** (1954 to Hoffmann-La Roche). Prepn of *d*-form: *eidem, ibid.;* Häfliger *et al., Helv. Chim. Acta* **39**, 2053 (1956). Prepn of *l*-form hydrobromide: Corrodi *et al., ibid.* **42**, 215 (1959). Crystal structure and absolute configuration of *d*-form hydrobromide: L. Gylbert, D. Carlstrom, *Acta Crystallogr.* **B**, 2833 (1977).

Hydrobromide, $C_{18}H_{26}BrNO$, *Ro 1-5470.* Crystals, mp 124-126°.
d-Form hydrobromide, **dextromethorphan hydrobromide, demorphan hydrobromide, Ro 1-5470/5, Antisep, Benylin DM, Canfodion, Cosylan, Delsym, Demethocaine, Romilar Hydrobromide, Sacophan, Silentium, Servicof, Supressin, Symptom 1, Testamin, Torfan, Tusilan.** Occurs as the monohydrate, crystals, mp 122- 124°. $[α]_D^{20}$ +27.6° (c = 1.5 in water). Approx soly in water: 1.5% at 25°, 5% at 50°, 25% at 85°. Soly (w/w): 25% in 95% ethanol at room temp, 10% in glycerol. Sol in propylene glycol, chloroform. Practically insol in ether. pH of a 1% aq soln: 5.2-6.5. Long range stability of aq solns obtained by adjusting pH within 4-5.6. Reacts with alkalies forming the free base which is practically insol in water.
l-Form hydrobromide, **levomethorphan hydrobromide, Ro 1-5470/6, Ro 1-7788.** Occurs as the dihydrate, crystals mp 124-126°. $[α]_D^{20}$ —26.3°.
Caution: May be habit forming. This is a controlled substance (opiate) listed in the U.S. Code of Federal Regulations, Title 21 Part 1308.12 (1987).
THERAP CAT: Antitussive.

8117. Radicinin. *2,3-Dihydro-3-hydroxy-2-methyl-7-(1-propenyl)-4H,5H-pyrano[4,3-b]pyran-4,5-dione;* stemphylone. $C_{12}H_{12}O_5$; mol wt 236.22. C 61.01%, H 5.12%, O 33.87%. A mold metabolite from the plant pathogen *Stemphylium radicinum:* Clark, Nord, *Arch. Biochem. Biophys.* **45**, 469 (1953); **59**, 269 (1955). Structure and identity with stemphylone: Grove, *J. Chem. Soc.* **1964**, 3234. Absolute configuration: M. Nukina, S. Marumo, *Tetrahedron Letters* **1977**, 3271. Synthesis of *dl*-form: Kato *et al., Chem. Commun.* **1969**, 95.

Needles from ethanol, dec 238-240°. $[α]_D^{27}$ —217.4° (c = 2.37 in pyridine); $[α]_D^{27}$ —175.7° (c = 0.2 in ethanol); $[α]_D^{27}$ —208° (c = 1.25 in chloroform). uv max: 343, 280, 270 nm (log ε 4.27, 3.62, 3.79). Soluble in alkali. Practically insol in dil acids, sodium bicarbonate, sodium carbonate. Radicinin has no effect on the unsaturation of the fat synthesized by *Fusarium lini* Bolley and causes an increase in the rate of dehydrogenation of isopropanol when incorporated in the growth medium for this organism.
Monoacetate, $C_{14}H_{14}O_6$, crystals from methanol, mp 197°. $[α]_D^{27}$ —267° (c = 0.69 in pyridine).

8118. Radium. Ra; at. wt 226 (mass number of most stable isotope); at. no. 88; valence 2. A radioactive alkaline earth metal. Occurrence in earth's crust: approx 10^{-11}% by wt. Natural isotopes: 223, **actinium X;** 224, **thorium X;** 226; 228, **mesothorium I.** ^{226}Ra is a product of disintegration of uranium and is present in all ores contg uranium. Separated in the form of a salt by P. and M. S. Curie from the pitchblende of Joachimsthal, Bohemia: Curie *et al., Compt. Rend.* **127**, 1215 (1898). Isoln of the element by electrolysis of an aq soln of radium chloride: Curie, Debierne, *ibid.* **151**, 523 (1910). ^{228}Ra ($T_{1/2}$ 6.7 years); produced by disintegration of thorium (^{232}Th); discovered in 1907 by O. Hahn in monazite residues from isolating thorium. Zaire (Congo) is the main producer of radium, Canada next. Clinical evaluation in brachytherapy of tongue: J. Horiuchi *et al.*, *Int. J. Rad. Oncol. Biol. Phys.* **8**, 82 (1982); M. Hoshina *et al.*, *Brit. J. Radiol.* **62**, 59 (1989); in intracavitary radiation of uterus: D. A. Jones, R. Stout, *Clin. Radiol.* **37**, 169 (1986). Review of radiotherapy in cervical carcinoma: P. R. Reddi *et al.*, *Obstet. Gynecol.* **43**, 238-247 (1974); in rectal carcinoma: *Dis. Colon Rectum* **29**, 600-614 (1986), reprint of C. Gordon-Watson, *Brit. J. Surg.* **17**, 649-669 (1930). Comprehensive reviews: K. W. Bagnall, *Chemistry of the Rare Radio-elements* (New York, Academic Press, 1957); Goodenough, Stenger, "Magnesium, Calcium, Strontium, Barium, and Radium" in *Comprehensive Inorganic Chemistry* **vol. 1**, J. C. Bailar, Jr. *et al.*, Eds. (Pergamon Press, Oxford, 1973) pp 591-664.

Brilliant, white metal; body-centered cubic structure; blackens on exposure to air. mp 700°; bp 1737°; d 5.5. $T_{1/2}$ 1600 yrs (^{226}Ra). One gram of radium evolves about 1000 kcal per year. Undergoes spontaneous disintegration with formation of radon. One gram of radium produces about 0.0001 ml of radon per day at normal temp and pressure. Radium is produced and used in the form of its salts: the chloride, bromide, carbonate, sulfate. Its compds closely resemble those of barium; the element itself is more volatile than barium. Radium salts impart a carmine-red color to a flame.

Bromide, Br$_2$Ra. White or slightly brownish crystals. d 5.79. mp 728°. Sublimes at 900°. Sol in water. The salt of commerce is usually a mixture with barium bromide.

Chloride, Cl$_2$Ra. White or slightly brownish crystals. d 4.91. mp 1000°. Sol in water. The salt of commerce is usually a mixture with barium chloride.

Human Toxicity: Inhalation, ingestion or body exposure may result in lung cancer, osteogenic sarcoma, osteitis, blood dyscrasias, skin injury.

USE: In physical research. In radiography of metals because the penetration of gamma rays is more pronounced than that of x-rays. As source of radon. No longer used to make luminous paints.

THERAP CAT: Antineoplastic (radiation source).

8119. Radon. Rn; at. no. 86; mass no. of longest-lived "natural" isotope 222. Obsolete synonyms: *niton*, symbol Nt, and *radium emanation* or *emanation*, symbol Em. The first isotope of this element, ^{220}Rn, unofficial name *Thoron*, symbol Tn, $T_{1/2}$ 55.3 sec, was discovered by Owens in 1899. ^{222}Rn, $T_{1/2}$ 3.823 days, was discovered by Rutherford in 1900 and a third isotope ^{219}Rn, unofficial name *actinon*, symbol An, $T_{1/2}$ 4.0 sec, was discovered by Debièrne and Giesel in 1902. These three isotopes are formed by α-disintegration of radium and its isotopes. At least 18 additional isotopes with mass nos. between 202 and 224 have been prepd by various nuclear reactions. All the isotopes are radioactive with short half-lives. With the exception of ^{223}Rn and ^{224}Rn, β⁻ emitters, they decay by α-emission or α-emission and e-capture in the case of the lightest isotopes. Occurrence in earth's crust 4×10^{-17} wt%. Although considered to be a "noble", chemically inert gas because of its electronic structure, K-, L-, M-, N-shells filled, $5s^25p^65d^{10}6s^26p^2$, the prepn of *radon fluoride* has been reported: Fields *et al.*, *J. Am. Chem. Soc.* **84**, 4164 (1962); Stein, *Science* **168**, 362 (1970). Radon is obtained by pumping the gases off a soln of a radium salt, sparking the gas mixture to combine the H$_2$ and O$_2$, removing the H$_2$O and CO$_2$ by adsorption, and freezing out the radon: Jennings, Russ, *Radon: Its Technique and Use* (Murray, London, 1948) pp 79-95. *Reviews: Argon, Helium and the Rare Gases*, G. A. Cook, Ed. (Interscience, New York, 1961); Haissinsky, Adloff, *Radiochemical Survey of the Elements* (Elsevier, New York, 1965) pp 126-128.

Colorless, odorless, inert gas. bp −62°. d (gas at 1 atm and 0°) 9.73 g/l; d (liq at bp) 4.4. Sol in water (230 cm^3/l at 20°), and organic solvents. Strongly adsorbed on various surfaces. Heat capacity at 25° and 1 atm: 4.9860 cal/g-atom/°K. Critical pressure 62 atm; crit temp 105°. Triple point −71°. Latent heat of vaporization at bp: 4325 cal/g-atom; latent heat of fusion at triple point: 776 cal/g-atom.

Human Toxicity: Toxicity due to ionizing radiation. Max permissible concn of ^{222}Rn in air: 10^{-8} μ-Curie/cc: *National Bureau of Standards Handbook* **69**, 79 (1959). Sources and health effects of indoor radon exposure: E. P. Radford, *Environ. Health Perspect.* **62**, 281 (1985).

USE: To initiate and influence chemical reactions, as a surface label in the study of surface reactions; in the determination of radium or thorium; in the study of the behavior of filters; in combination with Be or other light materials as a source of neutrons.

THERAP CAT: Antineoplastic (radiation source).

8120. Raffinose. β-D-*Fructofuranosyl-O-α-D-galacto-pyranosyl-(1 → 6)-α-D-glucopyranoside;* gossypose; melitose; melitriose. C$_{18}$H$_{32}$O$_{16}$; mol wt 504.46. C 42.86%, H 6.39%, O 50.75%. A trisaccharide built from 1 mol each of D-galactose, D-glucose, and D-fructose which are obtained from it by acid hydrolysis. Invertase splits it into melibiose and saccharose. Occurs in Australian manna (from *Eucalyptus* spp, *Myrtaceae*); in cottonseed meal. Prepn: C. S. Hudson, T. S. Harding, *J. Am. Chem. Soc.* **36**, 2110 (1914); E. P. Clark, *ibid.* **44**, 210 (1922); Harding, *Sugar* **25**, 82 (1923); Hungerford, Nees, *Ind. Eng. Chem.* **26**, 462 (1934). Configuration: Haworth *et al.*, *J. Chem. Soc.* **1923**, 3125; Charlton *et al.*; Haworth *et al.*, *ibid.* **1927**, 1527, 3146. Structure: Hassid, Ballou in W. Pigman, *The Carbohydrates* (Academic Press, New York, 1957) p 517. Synthesis: Suami *et al.*, *Carbohyd. Res.* **26**, 234 (1973). Review: E. B. Rathbone, *Dev. Food Carbohyd.* **2**, 145-185 (1980).

Pentahydrate, crystals in clusters from dil alc. Indifferent taste. d 1.465. mp 80°. Loses water of crystn upon slow heating to 100°. The anhydrous form dec 118-119°. $[α]_D^{20}$ +105.2° (c = 4). One gram dissolves in 7 ml water (soly table: Hungerford, Nees), in 10 ml methanol. Sol in pyridine, slightly sol in alc. Does not form an osazone and does not reduce Fehling's soln.

8121. Rafoxanide. N-*[3-Chloro-4-(4-chlorophenoxy)-phenyl]-2-hydroxy-3,5-diiodobenzamide; 3'-chloro-4'-(p-chlorophenoxy)-3,5-diiodosalicylanilide;* MK-990; Bovanide; Duofas; Flukanide; Ranide. C$_{19}$H$_{11}$Cl$_2$I$_2$NO$_3$; mol wt 626.01. C 36.45%, H 1.77%, Cl 11.32%, I 40.54%, N 2.24%, O 7.67%. Prepn: **Neth. pat. Appl. 6,815,783; Belg. pat. 724,668** (both 1969 to Merck & Co.); Mrozik *et al.*, *Experientia* **25**, 883 (1969). Trials in sheep and cattle: Snijders, Horak, Louw, *J. S. Afr. Vet. Med. Assoc.* **43**, 397 (1972); **44**, 251 (1973); **46**, 265 (1975); Horak, Snijders, *Vet. Rec.* **94**, 12 (1974).

Crystals, mp 168-170°. Moderately sol in acetone and acetonitrile; practically insol in water.

THERAP CAT (VET): Fasciolicide; anthelmintic.

8122. Ramifenazone. 1,2-Dihydro-1,5-dimethyl-4-[(1-methylethyl)amino]-2-phenyl-3H-pyrazol-3-one; *4-isopropylamino-2,3-dimethyl-1-phenyl-3-pyrazolin-5-one; 4-isopropylaminoantipyrine; isopropylaminophenazone; isopyrin.* C$_{14}$H$_{19}$N$_3$O; mol wt 245.32. C 68.54%, H 7.81%, N 17.13%, O 6.52%. Prepn: E. Skita, W. Stühmer, **Ger. pats. 930,328; 932,677** (both 1955), *C.A.* **52**, 16372i, 20200h (1958). Acute toxicity: E. Tubaro *et al.*, *Arzneimittel-Forsch.* **20**, 1024 (1970). Clinical evaluation of combination with phenylbutazone, *q.v.*, in migraine: W. Sacks, *S. Afr. Med. J.* **58**, 444 (1980).

CRYSTALS image area:

Crystals from acetone + glacial acetic acid, mp 80°. LD_{50} in mice (mg/kg): 843 i.p.; 1070 orally (Tubaro). Hydrochloride monohydrate, $C_{14}H_{20}ClN_3O.H_2O$. Mixture with phenylbutazone, **Tomanol**.

THERAP CAT: Analgesic, antipyretic, anti-inflammatory.

8123. Ramipril. *[2S-[1[R*(R*)],2α,3aβ,6aβ]]-1-[2-[[1-(Ethoxycarbonyl)-3-phenylpropyl]amino]-1-oxopropyl]octahydrocyclopenta[b]pyrrole-2-carboxylic acid; N-(1S-carboethoxy-3-phenylpropyl)-S-alanyl-cis,endo-2-azabicyclo-[3.3.0]octane-3S-carboxylic acid; (2S,3aS,6aS)-1-[(S)-N-[(S)-1-carboxy-3-phenylpropyl]alanyl]octahydrocyclopenta[b]pyrrole-2-carboxylic acid 1-ethyl ester;* Hoe 498; Cardace. $C_{23}H_{32}N_2O_5$; mol wt 416.52. C 66.32%, H 7.74%, N 6.73%, O 19.21%. Angiotensin converting enzyme inhibitor. Prepn: V. Teetz *et al, Eur.* **pat. Appl. 79,022** (1983 to Hoechst A.G.), *C.A.* **100,** 52012h (1984); E. H. Gold *et al.,* U.S. pat. **4,587,258** (1986 to Schering Corp.); V. Teetz *et al., Arzneimittel-Forsch.* **34,** 1399 (1984). Converted to active, diacid metabolite, *ramiprilat.* General pharmacology: M. Omosu *et al., ibid.* **38,** 1309 (1988). Series of articles on pharmacology, pharmacokinetics and preliminary clinical studies: *ibid.* **34,** 1402-1454 (1984). Radioimmunoassay determn in human serum and plasma: H. G. Eckert *et al., ibid.* **35,** 1251 (1985). Toxicology: H. H. Donaubauer, D. Mayer, *ibid.* **38,** 14 (1988). Symposium on pharmacology and clinical efficacy: *Am. J. Cardiol.* **59,** 1D-177D (1987).

Felty needles from ether, mp 109°. $[α]_D^{24}$ +33.2° (c = 1 in 0.1N ethanolic HCl). LD_{50} (14 day) in male, female mice, male, female rats (mg/kg): 1194, 1158, 687, 608 i.v.; 10933, 10048, >10000, >10000 orally (Donaubauer, Mayer).

THERAP CAT: Antihypertensive.

8124. Raney Nickel®. Raney nickel catalyst. Prepd by fusing 50 parts nickel with 50 parts aluminum: Raney, U.S. pats. **1,628,190** (1927); **1,915,473** (1933); pulverizing the alloy and dissolving out most of the aluminum with NaOH soln: Covert, Adkins, *J. Am. Chem. Soc.* **54,** 4116 (1932); Ruggli, Preiswerk, *Helv. Chim. Acta* **22,** 494 (1939); Mozingo, *Org. Syn.* **21,** 15 (1941). Absorption studies: Kokes, Emmett, *J. Am. Chem. Soc.* **83,** 29 (1961). The residual aluminum, which amounts to several percent, appears to be necessary for proper catalytic activity. *Review:* J. S. Pizey, *Synthetic Reagents* vol. 2 (John Wiley, New York, 1974) pp 175-311.

Grayish-black powder or cubic crystals. *Ignites on contact with air.* Contains hydrogen and has been attributed the formula Ni_2H. Generally stored under alcohol, although ether, water, methylcyclohexane, and dioxane may be used. Raney nickel loses its hydrogen slowly on storage and becomes inactive. Properly prepd and stored it should remain active for 6 months.

USE: Catalyst for the hydrogenation of organic compds with gaseous hydrogen. Usually from 1 to 10% of the substance to be reduced is employed. Is active at room temp and 1 atm pressure, but is also used at high temps and high pressures, *see* H. B. Adkins, *Reactions of Hydrogen* (Madison, 1937).

8125. Ranimustine. *Methyl 6-[[[(2-chloroethyl)nitrosoamino]carbonyl]amino]-6-deoxy-α-D-glucopyranoside;* methyl N-carbamyl-N'-(2-chloroethyl)-N'-nitroso-6-amino-6-deoxy-α-D-glucopyranoside; 6-[3-(2-chloroethyl)-3-nitrosoureido]-6-deoxy-α-D-glucopyranoside; 3-(methyl-

α-D-glucopyranos-6-yl)-1-(2-chloroethyl)-1-nitrosourea; ranomustine; MCNU; NSC-0270516; Cymerine; Thymerin. $C_{10}H_{18}ClN_3O_7$; mol wt 327.72. C 36.65%, H 5.54%, Cl 10.82%, N 12.82%, O 34.17%. Chloroethylnitrosourea derivative with antitumor activity. Similar to carmustine, chlorozotocin, lomustine, nimustine, *q.q.v.* Prepn: **Neth. pat. Appl. 7,507,973;** G. Kimura, J. Sekine, U.S. pat. **4,057,-684; Neth.** pat. **Appl. 7,800,920;** G. Kimura, U.S. pat. **4,156,777** (1976, 1977, 1978, 1979, all to Tokyo Tanabe). Antitumor activity in rodents, comparison with other nitrosoureas: S. Sekido *et al., Cancer Treat. Rep.* **63,** 961 (1979); S. Fujimoto, M. Ogawa, *Cancer Chemother. Pharmacol.* **9,** 134 (1982); S. Fujimoto *et al., Gann* **75,** 937 (1984). Mechanism of action: R. Kanamaru *et al., Curr. Chemother. Immunother.* **2,** 1377 (1982). Effect of sugar alcohols on toxicity in mice: T. Tashiro *et al., Cancer Chemother. Pharmacol.* **8,** 183 (1982). Series of articles on metabolic fate in rats: Y. Esumi *et al., Iyakuhin Kenkyu* **16,** 381-428 (1985), *C.A.* **103,** 115620q, 115621r, 134369f (1985). Clinical study in hematological malignant diseases: T. Masaoka *et al., Chemotherapy (Tokyo)* **33,** 271 (1985).

Pale yellowish needles from anhydrous ethanol-ethyl ether (1:1), mp 111-112°. $[α]_D^{20}$ +93.2° (c = 0.5 in methanol) (Kimura). Also reported as 101-103° (dec) from isopropanol. $[α]_D^{25}$ +73.2° (c = 0.3 in methanol) (Kimura, Sekine). Solubility in water: 900 mg/ml at 25°. LD_{50} in male rats (mg/kg): 42 i.p., 42 i.v., 50 orally (Kimura).

THERAP CAT: Antineoplastic.

8126. Ranitidine. *N-[2-[[[-5-[(Dimethylamino)methyl]-2-furanyl]methyl]thio]ethyl]-N'-methyl-2-nitro-1,1-ethenediamine.* $C_{13}H_{22}N_4O_3S$; mol wt 314.41. C 49.66%, H 7.05%, N 17.82%, O 15.27%, S 10.20%. Histamine H_2-receptor antagonist which inhibits gastric acid secretion. Prepn: B. J. Price *et al.,* **Fr.** pat. **2,384,765;** *eidem,* U.S. pat. **4,128,658** (both 1978 to Allen & Hanburys). HPLC determn in plasma: P. F. Carey, L. E. Martin, *J. Liq. Chromatog.* **1979,** 1291. Pharmacological studies: J. Bradshaw *et al., Brit. J. Pharmacol.* **66,** 464 (1979); M. J. Daly *et al., Gut* **21,** 408 (1980). Efficacy in treatment of duodenal ulcers: A. Berstad *et al., Scand. J. Gastroenterol.* **15,** 637 (1980); R. P. Walt *et al., Gut* **22,** 49 (1981). Review of pharmacology and therapeutic use: R. N. Brogden *et al., Drugs* **24,** 267-303 (1982). Comprehensive description: M. Hohnjec *et al.* in *Analytical Profiles of Drug Substances* vol. 15, K. Florey, Ed. (Academic Press, New York, 1986) pp 533-561.

Solid, mp 69-70°.
Hydrochloride, $C_{13}H_{23}ClN_4O_3S$, *AH 19065, Azantac, Melfax, Noctone, Raniben, Ranidil, Raniplex, Sostril, Taural, Terposen, Trigger, Ulcex, Ultidine, Zantac, Zantic, Zantidon.* Off-white solid, mp 133-134°. Freely sol in acetic acid and water, sol in methanol, sparingly sol in ethanol. Practically insol in chloroform.

THERAP CAT: Antiulcerative.

8127. Rapeseed Oil. Colza oil. Oil expressed from seeds of *Brassica campestris* L., *Cruciferae.*
Pale yellow, rather viscid liq. d 0.913-0.917; n_D^{20} 1.4720-1.4752. Solidif −2° to −10°. Sapon no. 170-177. Iodine no. 97-105. Sol in chloroform, ether, CS_2.

USE: Lubricant; manuf rubber substitutes, margarine, soft soaps, blown oils; oiling woolens.

8128. Raspberry. Fresh ripe fruit of varieties of *Rubus idaeus* L., or *R. stigosus* Michx., *Rosaceae. Habit.* Europe, Asia, cultivated in Canada and U.S. *Constit.* Sugar, malic and citric acids.

USE: Flavoring. Pharmaceutic aid (flavor).

8129. Raubasine. *16,17-Didehydro-19-methyloxayohimban-16-carboxylic acid methyl ester;* δ-yohimbine; py-tetrahydroserpentine; tetrahydroserpentine; ajmalicine; Circolene; Hydrosarpan; Isoarteril; Lamuran. $C_{21}H_{24}N_2O_3$; mol wt 352.42. C 71.57%, H 6.86%, N 7.95%, O 13.62%. Alkaloid from bark of *Corynanthe johimbe* K. Schum., *Rubiaceae:* H. Heinemann, *Ber.* **67**, 15 (1934); from roots of *Rauwolfia serpentina* (L.) Benth., *Apocynaceae:* S. Siddiqui, R. H. Siddiqui, *J. Indian Chem. Soc.* **8**, 667 (1931); A. H. Popelak *et al., Naturwiss.* **40**, 625 (1953); A. Hofmann, *Helv. Chim. Acta* **37**, 849 (1954); M. W. Klohs *et al., J. Am. Chem. Soc.* **76**, 1332 (1954). Review of early literature: R. E. Woodson *et al., Rauwolfia: Botany, Pharmacognosy, Chemistry and Pharmacology* (Little, Brown and Co., Boston, 1957) 147 pp. Structure: Goutarel, Le Hir, *Bull. Soc. Chim. France* **18**, 909 (1951). Stereochemistry: Wenkert *et al., J. Am. Chem. Soc.* **83**, 5037 (1961); Shamma, Richey, *ibid.* **85**, 2507 (1963). Total synthesis of *dl*-form: van Tamelen, Placeway, *ibid.* **83**, 2594 (1961); van Tamelen *et al., ibid.* **91**, 7359 (1969); J. Gutzwiller *et al., Helv. Chim. Acta* **64**, 1663 (1981); T. Kametani *et al., J. Chem. Soc. Perkin Trans. I* **1981**, 3168. Biosynthesis: N. Nagakura *et al., ibid.* **1979**, 2308; M. Rueffer *et al., Chem. Commun.* **1979**, 1016. Pharmacokinetics: A. Marzo *et al., Farmaco Ed. Prat.* **36**, 173 (1981). Clinical evaluation of platelet anti-aggregant activity: J. Neuman *et al., Arzneimittel-Forsch.* **36**, 1394 (1986). Evaluation of combination with almitrine, *q.v.*, in cerebral ischemia in rats: M. G. Borzeix, J. Cahn, *ibid.* **37**, 491 (1987).

Prisms from methanol, dec 257°. $[\alpha]_D^{20}$ −60° (c = 0.5 in chloroform); $[\alpha]_D^{20}$ −45° (c = 0.5 in pyridine); $[\alpha]_D^{20}$ −39° (c = 0.25 in methanol). uv max (methanol): 227, 292 nm (log ε 4.61, 3.79).

Hydrochloride, $C_{21}H_{25}ClN_2O_3$, leaflets from ethanol, mp 290° (dec). $[\alpha]_D^{20}$ −17° (c = 0.5 in methanol). Sparingly sol in water or dil HCl.

Hydrobromide, $C_{21}H_{25}BrN_2O_3$, diamond-shaped platelets from methanol, mp 295-296°.

THERAP CAT: Antihypertensive, anti-ischemic (cerebral and peripheral).

8130. Raunescine. *17α-Hydroxy-18β-[(3,4,5-trimethoxybenzoyl)oxy]-3β,20α-yohimban-16β-carboxylic acid methyl ester;* 3,4,5-trimethoxybenzoyl methyl raunescate. $C_{31}H_{36}$-N_2O_8; mol wt 564.62. C 65.94%, H 6.43%, N 4.96%, O 22.67%. Isoln from *Rauwolfia canescens* L., *R. tetraphylla* and other *Rauwolfia* spp, *Apocynaceae.* Isoln and structure: Hosansky, Smith, *J. Am. Pharm. Assoc. Sci. Ed.* **44**, 639

(1955); Huebner, Schlittler, *J. Am. Chem. Soc.* **79**, 250 (1957); van Tamelen, Taylor, *ibid.* **79**, 5256 (1957). Extraction: **Brit. pat. 833,149** (1960 to Ciba).

Monohydrate, hexagonal prisms from 90% methanol, mp 160-170°. $[\alpha]_D^{25}$ −74° (chloroform). uv max (U.S.P. alc): 218, 271 nm (log ε 4.77, 4.26).

Nitrate, $C_{31}H_{36}N_2O_8 \cdot HNO_3$, mp 223-225°; (mp 208-210°). $[\alpha]_D^{25}$ −80° (5N acetic acid).

Monoacetate monohydrate, $C_{33}H_{38}N_2O_9 \cdot H_2O$, mp 159-162°. $[\alpha]_D^{25}$ −157° (chloroform).

8131. Rauwolfia serpentina. *Rauwolfia serpentina* (L.) Benth. *(Ophioxylon serpentinum* L.) *Apocynaceae.* A small shrub native to the Orient from India to Sumatra. The crude botanical drug is usually the root. Contains many indole alkaloids, e.g., reserpine, reserpinine, yohimbine, ajmaline, serpentine, serpetinine. Review and bibliography with botanical information: Monachino, *Econ. Bot.* **8**, 349-365 (1954). The genus *Rauwolfia* is large and has a wide distribution in the tropics.

Extracts are sold under trade names, such as *Austrawolf, Egalin, Gendon, Hiwolfia, Koglucoid, Ralfen, Raucolyt, Raudixin, Raulfin, Raurixin, Rauserpol, Rautensine, Rauverid, Rauwiloid, Ra-Valeas, Rauwoldin, Rawotal, Rivadescin, Roxinil, Sarpagan, Serpina, Wolfin.*

THERAP CAT: Antihypertensive.

THERAP CAT (VET): *See* Reserpine.

8132. Rayon. Regenerated cellulose. The Code of Federal Regulations defines rayon as a manufactured fiber composed of regenerated cellulose, as well as manufactured fibers composed of regenerated cellulose in which substituents have replaced not more than 15 percent of the hydrogens of the hydroxyl groups. Now produced almost exclusively by the viscose process. *Review:* J. Lundberg, A. Turbak in Kirk-Othmer *Encyclopedia of Chemical Technology* **Vol. 19** (Wiley-Interscience, New York, 3rd ed., 1982) pp 855-880.

Note: Avisco Rayon is *purified rayon,* a medicinal grade of rayon.

USE: Purified rayon as surgical aid.

8133. Razoxane. *4,4'-(1-Methyl-1,2-ethanediyl)bis-2,6-piperazinedione;* (±)*-4,4'-propylenedi-2,6-piperazinedione;* (±)-(3,5,3',5'-tetraoxo)-1,2-dipiperazinopropane; (±)-1,2-bis(3,5-dioxopiperazinyl)propane; ICI 59118; ICRF 159; NSC 129943; Razoxin. $C_{11}H_{16}N_4O_4$; mol wt 268.28. C 49.25%, H 6.01%, N 20.88%, O 23.86%. Prepn: A. M. Creighton *et al., Nature* **222**, 384 (1969); A. M. Creighton, **Brit. pat. 1,234,935** corresp to U.S. pat. **3,941,790** (1971, 1976 both to Natl. Res. Dev. Corp.). Mode of action: H. B. A. Sharpe *et al., Nature* **226**, 524 (1970). Metabolism: R. E. Bellet *et al., Eur. J. Cancer* **13**, 1293 (1977). Pharmacology: A. Atherton, *ibid.* **11**, 383 (1975). Clinical studies: M. T. Bakowski *et al., Int. J. Radiat. Oncol. Biol. Phys.* **4**, 115 (1978); H. W. Bruckner *et al., Cancer Treat. Rep.* **66**, 1713 (1982). Toxicity study: E. Hassenstein, K. Renner, *Strahlentherapie* **154**, 122 (1978). Mitigating effects on daunomycin-induced cardiomyopathy in mice: V. W. Fischer *et al., Drug Chem. Toxicol.* **5**, 155 (1982). *Review:* M. T. Bakowski, *Cancer Treat. Rev.* **3**, 95-107 (1976).

Pale cream microcrystalline solid, mp 237-239°. 320 mg/kg/day administered orally to beagle dogs was lethal, E. J. Gralla *et al., Cancer Chemother. Rep.* **5**, 1 (1974).

THERAP CAT: Antineoplastic.

8134. Reductic Acid. *2,3-Dihydroxy-2-cyclopenten-1-one;* 2-cyclopenten-2,3-diol-1-one. $C_5H_6O_3$; mol wt 114.10. C 52.63%, H 5.30%, O 42.07%. Antioxidant obtained from pectic substances: Reichstein, Oppenauer, *Helv. Chim. Acta* **16**, 988 (1933); **17**, 390 (1934); Goldstein, U.S. pat. **2,-854,484** (1958). Alternate syntheses: Hesse *et al., Ann.* **563**, 31 (1949); **592**, 137 (1955); **736**, 134 (1970). Crystal structure: D. Semmingsen, *Acta Chem. Scand. B* **31**, 81 (1977). Similar to vitamin C in structure but not activity.

Pale yellow crystals (has been obtained colorless) from ethyl acetate, dec 213-213.5° (also reported as 207.5° and 211°). Soluble in water, methanol, ethanol. Sparingly sol in ether, ethyl acetate, acetone. Insol in benzene.

5-Methylreductic acid, mp 71°. Stronger reducing agent than reductic acid.

USE: As antioxidant, like isoascorbic or ascorbic acid.

8135. Reinecke Salt. *Ammonium diamminetetrakis(thiocyanato-N)chromate(1 −)*; *ammonium reineckate*; ammonium tetrathiocyanodiammonochromate. $C_4H_{10}CrN_7S_4$; mol wt 336.41. C 14.28%, H 3.00%, Cr 15.46%, N 29.14%, S 38.12%. $NH_4[Cr(NH_3)_2(SCN)_4]$. Exists as monohydrate; made by fusing ammonium thiocyanate with ammonium dichromate. Prepn: Dakin, *Org. Syn. coll. vol.* **II**, 555 (1943).

Monohydrate, dark red crystals or red cryst powder. Sparingly sol in cold water; sol in hot water, alcohol. Dec in aq soln with formation of a blue color and free HCN; decompn occurs in about 2 weeks at room temp, rapidly above 65°. A similar decompn takes place in boiling alcohol.

Note: The K-salt is also known as Reinecke salt.

USE: Precipitant for primary and secondary amines, proline, hydroxyproline, and certain amino acids; also a reagent for mercury with which it gives a red color or precipitate.

8136. Reissert Compounds. A term used to designate 2-acyl-1,2-dihydroisoquinaldonitriles and 1-acyl-1,2-dihydroquinaldonitriles: Reissert, *Ber.* **38**, 1603, 3415 (1905); McEwen, Cobb, *Chem. Rev.* **55**, 511 (1955).

8137. Relaxin. Cervilaxin; Releasin (formerly). A polypeptide hormone secreted by the corpora lutea of many mammalian species during pregnancy; it is also produced in several non-mammalians, including the shark. Relaxin facilitates the birth process by causing a softening and lengthening of the pubic symphysis and cervix; it also inhibits contraction of the uterus and may play a role in timing of parturition. First discovered in estrogen-primed guinea pigs: F. L. Hisaw, *Proc. Soc. Exp. Biol. Med.* **23**, 661 (1926). Extraction from corpora lutea: Fevold *et al., J. Am. Chem. Soc.* **52**, 3340 (1930); Albert *et al., Endocrinology* **40**, 370 (1947); from pregnant rabbit serum: Abramowitz *et al., Anat. Rec.* **84**, 456 (1942); cf. Hall, Newton, *J. Physiol.* **106**, 18 (1947). Can be obtained commercially from pregnant sows' ovaries. Isoln procedure and purification by resin chromatography: Lehrman *et al., J. Am. Pharm. Assoc.* **44**, 206 (1955); Kroc, Phillips; Phillips, U.S. pats. **2,852,431-2** (both 1958 to Warner-Lambert). Isoln from ovarian tissue using glacial acetic acid: Cohen, U.S. pat. **2,930,737** (1960 to Princeton Laboratories). Alternate method: Keck, U.S. pat. **3,008,878** (1961 to Thomae). Prepn from hog ovaries: Doczi, U.S. pat. **3,096,246** (1963 to Warner-Lambert). Porcine relaxin has a mol wt of approx 6000 and consists of two peptide chains, A and B, of 22 and 31 residues respectively, linked covalently by one intra- and two inter-chain disulfide bonds. Purification, characterization of porcine relaxin: O. D. Sherwood, E. O'Byrne, *Arch. Biochem. Biophys.* **160**, 185 (1974). Structure of the A chain: C. Schwabe *et al., Biochem. Biophys. Res. Commun.* **70**, 397 (1976); of the B chain: *eidem, ibid.* **75**, 503 (1977). Relaxin is structurally homologous to insulin and related growth factors: R. A. Bradshaw, *Rev. Biochem.* **47**, 191 (1978); S. Bedarkar *et al., Nature* **270**, 449 (1977); N. Isaacs *et al., ibid.* **271**, 278 (1978). Complete amino acid sequence of porcine relaxin: R. James *et al., ibid.* **267**, 544 (1977); of rat: M. J. John *et al., Endocrinology* **108**, 726 (1981). Isoln, characterization of relaxin from the sand tiger shark (*Odontaspis taurus*): J. W. Reinig *et al., ibid.* **109**, 537 (1981); structure: L. K. Gowan *et al., FEBS Letters* **129**, 80 (1981). Demonstration of synthesis of rat relaxin as a preprorelaxin molecule with a connecting peptide of 105 amino acid residues, using molecular

cloning: P. Hudson *et al., Nature* **291**, 127 (1981). Structure of a genomic clone from which the amino acid sequence of biologically active human relaxin was predicted: *eidem, ibid.* **301**, 628 (1983). *Reviews:* C. Schwabe *et al., Recent. Progr. Horm. Res.* **34**, 123-211 (1978); *Ann. N.Y. Acad. Sci.* **380**, entitled "Relaxin: Structure, Function, and Evolution", B. G. Steinetz *et al.,* Eds. (1982) pp 1-244.

Amorphous powder. Diffusion constant, $cm^2/sec = 12.6 \times 10^{-7}$. Sedimentation constant, $sec = 1.6 \times 10^{-13}$. Isoelec pt around pH 7.0. uv max (water pH 7.0): 277.5 nm ($E^{1\%}_{1cm}$ 7.20). Slightly sol in water and 95% alcohol. Sol in acid or alkaline solns. Insol in abs alcohol, ether, acetone, petr ether, benzene. Dec above pH 9. Neutral or acid solns which are free from oxidizing agents, are stable. Ampuled solns at pH 7.0 and 3.2 have been kept in refrigerator for one year without loss of activity.

THERAP CAT: Hormone (ovarian).

8138. Renin. Highly specific aspartyl proteinase of mol wt about 40000 Da, produced and secreted by the kidney. Found also in amniotic fluid. Conc. of renin in the human kidney is about 20 times less than in hog kidney. Purification from hog kidneys: Haas *et al., Arch. Biochem. Biophys.* **42**, 368 (1953); from kidneys of various species: Haas *et al., ibid.* **48**, 256 (1954). Prepn of human renin: Haas *et al., ibid.* **110**, 534 (1965). Exists in α, β and γ configurations: Haas *et al., ibid.* **44**, 79 (1953). Purification on DEAE cellulose columns: Passananti, *Biochim. Biophys. Acta* **34**, 246 (1959); Maier, Morgan, *ibid.* **128**, 193 (1966). Has been observed to exist in high molecular weight inactive forms which upon acid treatment decrease in molecular size and increase in enzymic activity. *See* review of purification studies: E. Haber, E. E. Slater, *Circ. Res.* **40**, Suppl. I, I36 (1977). Renin is secreted by juxtaglomerular cells and acts on the plasma substrate, *angiotensinogen,* to split off the inactive decapeptide angiotensin I, *q.v.,* which is converted to the active pressor agent angiotensin II, *q.v.* Renin itself has no activity. Secretion of renin is stimulated by constriction of a renal artery, blood loss, low sodium levels, adrenal insufficiency. Increased levels of renin shown also in pregnancy. Cloning and sequence analysis of human renin cDNA: T. Imai *et al., Proc. Nat. Acad. Sci. USA* **80**, 7405 (1983). Crystal structure of human recombinant renin: A. R. Sielecki *et al., Science* **243**, 1346 (1989). *Review:* W. S. Peart, "The Renin-Angiotensin System" in *Pharmacol. Rev.* **17**, 143 (1965); *idem, Proc. Roy. Soc. Ser. B* **173**, 317 (1969); Smeby, Bumpus, "Renin" in *Renal Hypertension,* I. Page, J. McCubbin, Eds. (Year Book Medical Publishers, Chicago, 1968) pp 14-61. Books: M. R. Lee, *Renin and Hypertension* (Williams & Wilkins, Baltimore, 1969); *Renin vol.* **5**, S. Oparil *et al.,* Eds. (Eden Press, Quebec, 1980) 368 pp. Review of the renin-angiotensin system: J. C. Romero, F. G. Knox, *Hypertension* **11**, 724-738 (1988).

8139. Rennin. Chymosin; rennase; lab; abomasal enzyme; Lab Ferment (German). Mol wt about 31,000. The predominant milk-clotting enzyme from the true stomach or abomasum of the suckling calf: Tauber, Kleiner, *J. Biol. Chem.* **96**, 745 (1932). Secreted as an inactive precursor called *prorennin* and converted in the acid environment of the stomach to the active enzyme. Obtained by extracting dried strips of calf stomach with 5 to 10% NaCl soln contg boric acid for 5 days; improved process: Keil, U.S. pat. **2,339,931** (1942). Structure consists of a single polypeptide chain with internal disulfide bridges. Amino acid composition: Schwander *et al., Helv. Chim. Acta* **35**, 553 (1952). Determination of *N*-terminal amino acid sequence: Foltmann, *Phil. Trans. Roy. Soc. London* **257B**, 147 (1970). Homologies in amino acid sequences of rennin and pepsin: Tang, Hartley, *Biochem. J.* **118**, 611 (1970). *Reviews:* Berridge, *Advan. Enzymol.* **15**, 423 (1954); Foltmann, *Milk Proteins* **2**, 217 (1971).

Yellowish powder, grains or scales. Peculiar, not unpleasant odor. Characteristic, slightly salty taste. Slightly hygroscopic. Can be purified enough to crystallize. The dry substance (sometimes diluted with NaCl or lactose) is stable, aq solns are not. Partially sol in water, dil alcohol. Insol at pH values near its isoelectric point of pH 4.5. pH of aq soln 5.8 (acceptable range 5.3-6.3). Strongly affected by ultraviolet (sunlight). Causes milk to coagulate, optimum temp 37-43°, optimum pH of milk 5.8. Not active below 15° nor above 55°. When a commercial prepn is labeled 1:10,000 it means that one gram will curdle 10 liters of milk at 35° within 40 minutes. Other strengths are 1:100,000 to 1:7,600,000.

USE: *Rennet* (a dried extract contg rennin) is used in the manuf of cheese and rennet casein, also sold retail for the making of junket or rennet custards.

THERAP CAT: Digestive enzyme.

8140. Repirinast. *5,6-Dihydro-7,8-dimethyl-4,5-dioxo-4H-pyrano[3,2-c]quinoline-2-carboxylic acid 3-methylbutyl ester;* isopentyl 5,6-dihydro-4,5-dioxo-4H-pyrano[3,2-c]quinoline-2-carboxylate; MY-5116; Romet. $C_{20}H_{21}NO_5$; mol wt 355.39. C 67.59%, H 5.96%, N 3.94%, O 22.51%. Prepn: **Belg. pat. 876,751;** Y. Morinaka, K. Takahashi, **U.S. pat. 4,298,610** (1979, 1981 both to Mitsubishi Yuka); Y. Morinaka *et al., Eur. J. Med. Chem. - Chim. Ther.* **16,** 251 (1981). General pharmacology in animals: K. Takahashi *et al., Oyo Yakuri* **32,** 233 (1986), *C.A.* **105,** 218691j (1986). Anti-allergic effects: S. Kono, K. Ohata, *Arerugi* **35,** 1105 (1986), *C.A.* **106,** 165870z (1987). Toxicity: M. Nagase *et al., Iyakuhin Kenkyu* **17,** 545 (1986), *C.A.* **106,** 108121r (1987).

Crystals from chloroform + *n*-hexane, mp 236-241°. LD_{50} in rats, mice (mg/kg): > 5000 orally, > 5000 s.c. both species (Nagase).

THERAP CAT: Anti-allergic.

8141. Reposal. *5-Bicyclo[3.2.1]oct-2-en-3-yl-5-ethyl-2,4,6(1H,3H,5H)-pyrimidinetrione; 5-bicyclo[3.2.1]oct-2-en-3-yl-5-ethylbarbituric acid;* 5-ethyl-5-(bicyclo[3.2.1]octen-yl)barbituric acid; WT 161; Reposamal. $C_{14}H_{18}N_2O_3$; mol wt 262.30. C 64.10%, H 6.92%, N 10.68%, O 18.30%. Prepn: Taub, **U.S. pat. 2,808,408** (1956 to Calanda Stiftung). Metabolism: Nielsen, Tarding, *Acta Pharmacol. Toxicol.* **26,** 521 (1968). Toxicity: Frey, *Arzneimittel-Forsch.* **12,** 389 (1962).

Needles from isopropanol, mp 213°. pKa 7.4-7.5°. LD_{50} i.p. in mice: 175 mg/kg (Frey).

Note: This is a controlled substance (depressant) listed in the U.S. Code of Federal Regulations, Title 21 Part 1308.13 (1987).

THERAP CAT: Sedative, hypnotic.

8142. Reproterol. *7-[3-[[2-(3,5-Dihydroxyphenyl)-2-hydroxyethyl]amino]propyl]-3,7-dihydro-1,3-dimethyl-1H-purine-2,6-dione; 7-[3-[(β,3,5-trihydroxyphenethyl)amino]propyl]theophylline;* D 1959. $C_{18}H_{23}N_5O_5$; mol wt 389.42. C 55.52%, H 5.95%, N 17.99%, O 20.54%. Theophylline deriv which has selective activity on β_2 receptors. Prepn: **Fr. pat. M5969** (1968 to Degussa), *C.A.* **71,** 70644c (1969). *See also* **Fr. pat. Addn. 0308** (1970 to Degussa), *C.A.* **78,** 72216h (1973). Synthesis: K. H. Klingler, *Arzneimittel-Forsch.* **27,** 4 (1977). Toxicology: S. Habersang, *ibid.* 45. Series of articles on metabolism, pharmacology and clinical trials: *ibid.* 15-76. Other clinical studies: D. Nolte *et al., Deut. Med. Wochenschr.* **102,** 619 (1977); H. Budmiger *et al., Schweiz. Med. Wochenschr.* **108,** 1190 (1978).

Hydrochloride, $C_{18}H_{24}ClN_5O_5$, *W-2946M, Asmaterolo, Bronchospasmin, Bronchodil.* Crystals, mp 249-250°. LD_{50} in mice (mg/kg): 148 i.v., > 10,000 orally (Habersang).

THERAP CAT: Bronchodilator.

8143. Resacetophenone. *1-(2,4-Dihydroxyphenyl)ethanone; 2′,4′-dihydroxyacetophenone.* $C_8H_8O_3$; mol wt 152.14. C 63.15%, H 5.30%, O 31.55%. Made by heating resorcinol, glacial acetic acid, and anhydr zinc chloride at 145-150°: Cooper, *Org. Syn.* **21,** 103 (1941); from resorcinol, acetic anhydride and boron trifluoride: Killelea, Lindwall, *J. Am. Chem. Soc.* **70,** 428 (1948).

Needles or leaflets, mp 145-147°. Gradually dec by water; sol in pyridine, in warm alcohol, in glacial acetic acid; practically insol in benzene, ether, chloroform.

USE: Reagent for iron as a 10% alcoholic soln. A red color is obtained with ferric ions in slightly acid soln: Cooper, *Ind. Eng. Chem. Anal. Ed.* **9,** 334 (1937).

8144. Resazurin. *7-Hydroxy-3H-phenoxazin-3-one 10-oxide;* "diazoresorcinol"; Resazoin. $C_{12}H_7NO_4$; mol wt 229.18. C 62.88%, H 3.08%, N 6.11%, O 27.92%. Prepn from resorcinol: Rast *et al., Ger. pat.* **961,829** (1957 to Bayer), *C.A.* **53,** 15098 (1959).

Dark red, small crystals with greenish luster. Insol in water, ether; sparingly sol in alcohol, glacial acetic acid; sol in dil alkali hydroxides.

USE: As indicator, using a soln of 0.1 g in 20 ml 0.1N NaOH, and water to make 500 ml. pH: 3.8 orange, 6.5 dark violet. In the detection of hyposulfite (sulfoxylate). In food research (reductase test).

8145. Rescimetol. *18-[[3-(4-Hydroxy-3-methoxyphenyl)-1-oxo-2-propenyl]oxy]-11,17-dimethoxyyohimban-16-carboxylic acid methyl ester;* methyl O-(4-hydroxy-3-methoxycinnamoyl)reserpate; CD-3400; WHO 4939; Toscarna. $C_{33}H_{38}N_2O_8$; mol wt 590.67. C 67.10%, H 6.48%, N 4.74%, O 21.67%. Analog of rescinnamine, *q.v.* Prepn: **Belg. pat. 782,661** corresp to T. Kametani, **U.S. pat. 3,898,215** (1972, 1975 both to Nippon Chemiphar); T. Kametani *et al., J. Med. Chem.* **15,** 686 (1972). *See also* T. Kametani *et al., J. Chem. Soc. Perkin Trans. I* **1975,** 932. Antihypertensive activity: H. Kato *et al., Japan. J. Pharmacol.* **27,** 87 (1977). Effects on respiration, cardiovascular system, renal function: T. Kamishiro *et al., Oyo Yakuri* **20,** 441 (1980), *C.A.* **96,** 28430w (1982). Toxicity study: T. Fukuda *et al., ibid.* **18,** 71 (1979), *C.A.* **92,** 104392v (1980).

Pale yellow needles from methanol, mp 259-260°.
THERAP CAT: Antihypertensive.

8146. Rescinnamine. *11,17-Dimethoxy-18-[[1-oxo-3-(3,4,5-trimethoxyphenyl)-2-propenyl]oxy]-3,20-yohimban-16-carboxylic acid methyl ester;* 3,4,5-trimethoxycinnamic acid ester of methyl reserpate; methyl 3,4,5-trimethoxycinnamoyl reserpate; reserpinine; Anaprel; Apoterin; Cartric; Cinnaloid; Scinnamina; Rescaloid; Rescisan; Moderil. $C_{35}H_{42}N_2O_9$; mol wt 634.71. C 66.23%, H 6.67%, N 4.41%, O 22.69%. Found in *Rauwolfia serpentina* Benth., *Apocynaceae:* Haack *et al., Naturwiss.* **41**, 214 (1954); Ordway, Guercio, U.S. pat. **2,876,228** (1959 to Pfizer); Klohs *et al.,* U.S. pat. **2,974,144** (1961 to Riker); from *Rauwolfia* spp: Banes *et al., J. Am. Pharm. Assoc.* **47**, 625 (1958). Identity with reserpinine: Haack *et al., Naturwiss.* **42**, 47 (1955). Structure: Klohs *et al., J. Am. Chem. Soc.* **77**, 2241 (1955).

Fine needles from benzene, mp 238-239° (vac). $[\alpha]_D^{24}$ −97° (c = 1 in chloroform). uv max (methanol): 228, 302 nm (log ε 4.79, 4.48). Practically insol in water. Moderately sol in methanol, benzene, chloroform, other organic solvents.
THERAP CAT: Antihypertensive.

8147. Reserpic Acid. *18β-Hydroxy-11,17α-dimethoxy-3β,20α-yohimban-16β-carboxylic acid;* reserpinolic acid. $C_{22}H_{28}N_2O_5$; mol wt 400.46. C 65.98%, H 7.05%, N 7.00%, O 19.98%. Obtained by controlled alkaline hydrolysis of reserpine: Dorfman *et al., Helv. Chim. Acta* **37**, 59 (1954); Schlittler, U.S. pat. **2,824,874** (1958).

Crystals from methanol, mp 241-243°.
Hydrochloride hemihydrate, $C_{22}H_{28}N_2O_5 \cdot HCl \cdot \frac{1}{2}H_2O$, crystals, mp 257-259°. $[\alpha]_D^{23}$ −81°.
Methyl ester, $C_{23}H_{30}N_2O_5$, needles from methanol, mp 235-240°. $[\alpha]_D^{25}$ −106°. Hydrochloride, mp about 228°.

8148. Reserpiline. *16,17-Didehydro-10,11-dimethoxy-19-methyloxayohimban-16-carboxylic acid methyl ester.* $C_{23}H_{28}N_2O_5$; mol wt 412.47. C 66.97%, H 6.84%, N 6.79%, O 19.40%. Isoln from *Rauwolfia serpentina* Benth., *Apocynaceae:* Klohs *et al., Chem. & Ind. (London)* **1954**, 1264; from *R. canescens* L.: Stoll *et al., Helv. Chim. Acta* **38**, 270

(1955). Stereochemistry: Shamma, Richey, *J. Am. Chem. Soc.* **85**, 2507 (1963). Partial synthesis: S.-I. Sakai *et al., Heterocycles* **17**, 99 (1982).

Amorphous powder. $[\alpha]_D^{20}$: −38° (ethanol); −14° (c = 1.5 in pyridine); −12° (c = 1.7 in chloroform). uv max (ethanol): 229, 300-304 nm (log ε 4.57, 4.03). Freely sol in ethanol, acetone, chloroform, benzene.
Hydrochloride, $C_{23}H_{28}N_2O_5 \cdot HCl$, crystals, mp 205-207°. $[\alpha]_D^{24}$ −40° (c = 0.44 in ethanol).
Note: The dimethylaminoethyl ester dihydrochloride of reserpilic acid, $C_{26}H_{35}N_3O_5 \cdot 2HCl$, is known as *Paratensiol, Resporisan.*

8149. Reserpine. *(3β,16β,17α,18β,20α)-11,17-Dimethoxy-18-[(3,4,5-trimethoxybenzoyl)oxy]yohimban-16-carboxylic acid methyl ester;* 3,4,5-trimethoxybenzoyl methyl reserpate; Alserin; Austrapine; Crystoserpine; Eskaserp; Hiserpia; Orticalm; Quiescin; Rau-sed; Reserpex; Reserpoid; Rivasin; Roxinoid; Sandril; Sedaraupin; Serfin; Serolfia; Serpanray; Serpasil; Serpasol; Serpate; Serpen; Serpine; Serpiloid. $C_{33}H_{40}N_2O_9$; mol wt 608.70. C 65.11%, H 6.62%, N 4.60%, O 23.66%. Found in *Rauwolfia* spp. Isoln from the roots of *Rauwolfia serpentina* L. Benth., *Apocynaceae:* J. M. Müller *et al., Experientia* **8**, 338 (1952); Dorfman *et al., Helv. Chim. Acta* **37**, 59 (1954); Schwyzer, Mueller, U.S. pat. **2,833,771** (1958 to Ciba). Structure: Dorfman *et al., loc. cit.;* Neuss *et al., J. Am. Chem. Soc.* **76**, 2463 (1954). Stereochemistry: Diassi *et al., ibid.* **77**, 2028, 4687 (1955); Huebner, Wenkert, *ibid.* 4180; van Tamelen, Hance, *ibid.* 4693; Aldrich *et al., ibid.* **81**, 2481 (1959); Ban, Yonemitsu, *Tetrahedron* **20**, 2877 (1964). Total synthesis: R. B. Woodward *et al., J. Am. Chem. Soc.* **78**, 2023 (1956); *eidem, Tetrahedron* **2**, 1-57 (1958); B. A. Pearlman, *J. Am. Chem. Soc.* **101**, 6404 (1979); P. A. Wender *et al., ibid.* **102**, 6157 (1980). Stereoselective synthesis: G. Stork, *Pure Appl. Chem.* **61**, 439 (1989). Comprehensive description: R. E. Schirmer in *Analytical Profiles of Drug Substances* vol. **4**, K. Florey, Ed. (Academic Press, New York, 1975) pp 384-430. *Review:* A. Scriabine, Ed. in *Pharmacology of Antihypertensive Drugs* (Raven, New York, 1980) pp 119-125. Review of carcinogenicity studies: *IARC Monographs* **10**, 217-229 (1976); R. M. Diener *et al., Toxicol. Pathol.* **8**, 1-21 (1980).

Long prisms from dil acetone, dec 264-265° (dec 277-277.5° in evac tube). $[\alpha]_D^{23}$ −118° (CHCl3); $[\alpha]_D^{26}$ −164° (c = 0.96 in pyridine); $[\alpha]_D^{26}$ −168° (c = 0.624 in DMF). uv max (CHCl3): 216, 267, 295 nm (ε 61700, 17000, 10200). Weak base. pK 6.6. Very sparingly sol in water. Freely sol in chloroform (~1 g/6 ml), methylene chloride, glacial acetic acid. Sol in benzene, ethyl acetate. Slightly sol in acetone, methanol, alcohol (1 g/1800 ml), ether, and in aq solns of acetic and citric acids. Upon standing most solns acquire a yellow color and pronounced fluorescence; esp after the addition of acid or upon exposure to light.

Hydrochloride hydrate, $C_{33}H_{41}ClN_2O_9 \cdot H_2O$, crystals, dec 224°.

Note: This substance may reasonably be anticipated to be a carcinogen: *Fourth Annual Report on Carcinogens* (NTP 85-002, 1985) p 177.

THERAP CAT: Antihypertensive.

THERAP CAT (VET): Hypotensive, tranquilizer. Has been used to prevent aortic rupture in turkeys.

8150. Resibufogenin. *(3β,5β,15β)-14,15-Epoxy-3-hydr-oxy-5-bufa-20,22-dienolide;* Respigon. $C_{24}H_{32}O_4$; mol wt 384.50. C 74.97%, H 8.39%, O 16.65%. Cytotoxic constituent of toad venom. Isoln: Meyer, *Helv. Chim. Acta* **35**, 2444 (1952); Linde, Meyer, *Pharm. Acta Helv.* **33**, 327 (1958). Structure: Thiessen, *Chem. & Ind. (London)* **1958**, 440; Linde, Meyer, *Experientia* **14**, 238 (1958); *eidem, Helv. Chim. Acta* **42**, 807 (1959). Synthesis: Pettit *et al., J. Org. Chem.* **36**, 3736 (1971); Haede *et al., Ann.* **1973**, 5. Stereochemical study of cytotoxic properties: Y. Kamono *et al., J. Chem. Res. (S)* **1977**, 78. Pharmacology: Leigh, Caldwell, *J. Pharm. Pharmacol.* **21**, 708 (1969).

Crystals from acetone + water, mp 113-140°/155-168°. $[\alpha]_D^{22}$ −7.1° (c = 1.259 in chloroform). Also obtained as an amorphous solid, $[\alpha]_D^{16}$ −5.4° (c = 2.030 in chloroform). Hydrochloride, $C_{24}H_{33}ClO_4$, crystals from acetone, dec 230-232°. $[\alpha]_D^{15}$ +15.1° (c = 0.5302 in chloroform). uv max (alc): 298 nm (log ε 3.74).

THERAP CAT: Cardiotonic.

8151. Resilin. Highly amorphous rubberlike protein found in insect cuticle. Isoln and characterization: Weis-Fogh, *J. Exp. Biol.* **37**, 889 (1960). The structure is a cross-linked random network of flexible protein chains, incorporating 15 of the more common amino acids and involving covalent bonds. Cross-links have been identified as di- and trityrosine. Biosynthesis: Coles, *J. Insect Physiol.* **12**, 679 (1966). Similar to elastin *(q.v.)* but does not form fibers; occurs as masses or typically as 2-5 μ layers. Elasticity approaches that of the ideal cross-linked rubber. Responsible for various springlike actions of winged insects. In flea: Rothschild, *Nature* **239**, 45 (1972). *Review:* Andersen, *Acta Physiol. Scand.* **66**, Suppl. 263 (1966); *idem,* "Resilin" in *Comprehensive Biochemistry* vol. **26C**, M. Florkin, E. H. Stotz, Eds. (Elsevier, New York, 1971) pp 633-657.

The colorless material is insol in water < 140° and in all other solvents not rupturing peptide bonds. Swells in water and protein solvents; shrinks and becomes hard in absolute methanol, ethanol, dioxane, acetone. Shows no tendency to crystallize at all. Fluorescent in uv light.

8152. Resin Ipomea. Resin of Mexican scammony. Made by extracting the dried roots of *Ipomoea orizabensis* Ledenois, *Convolvulaceae* with alcohol and pptg with water.

Brown, translucent, brittle fragments or pale brown powder. Insol in water; sol in alc, chloroform, partially in ether.

THERAP CAT: Cathartic.

8153. Resin Jalap. Prepd by extracting the powdered root of *Exogonium purga* (Hayne) Lindl. (*E. jalapa* Baill., *Ipomoea purga* Hayne), *Convolvulaceae* with alcohol and pptg with water.

Yellow to brown, amorphous mass or powder. Insol in water, benzene, carbon disulfide, oils; freely sol in alc.

THERAP CAT: Cathartic.

8154. Resin Scammony. Made by extracting the tubers of *Convolvulus scammonia* L., *Convolvulaceae* with alcohol and pptg with water.

Brown, amorphous mass. Freely sol in alc; not less than 95% sol in ether and in caustic alkalies (with gentle heat).

THERAP CAT: Cathartic.

8155. Resistomycin. *3,5,7,10-Tetrahydroxy-1,1,9-tri-methyl-2H-benzo[cd]pyrene-2,6(1H)-dione;* X-340. $C_{22}H_{16}O_6$; mol wt 376.35. C 70.21%, H 4.29%, O 25.51%. Unusually stable antibiotic substance produced by *Streptomyces resistomycificus:* Brockmann, Schmidt-Kastner, *Naturwiss.* **38**, 479 (1951); *Ber.* **87**, 1460 (1954); Brockmann, Ger. pat. **888,918** (1953). Structure: Brockmann, *Angew. Chem.* **76**, 863 (1964). Revised structure: Bailey *et al., Chem. Commun.* **1968**, 374; Brockmann *et al., Ber.* **102**, 1224 (1969). Synthetic studies: J. F. Kingston, L. Weiler, *Can. J. Chem.* **55**, 785 (1977); K. James, R. A. Raphael, *Tetrahedron Letters* **1979**, 3895. Total synthesis: B. A. Keay, R. Rodrigo, *J. Am. Chem. Soc.* **104**, 4725 (1982). Isoln from *Streptomyces griseoflavus* B71, ^{13}C-NMR and biosynthesis: G. Höfle, H. Wolf, *Ann.* **1983**, 835.

Yellow needles from dioxane, dec 315°. Sublimes at 200-205° (10^{-4} mm) without activity loss. Stable to hot conc H_2SO_4 or hot N KOH. Weakly acid. Absorption max: 268, 290, 320, 337, 366, 457 nm (ε 24000, 23000, 14400, 13900, 11000, 15400). Slight soln in water; fair in ether, benzene, alcohol, acetone, acetic acid. Orange-red in sodium hydroxide and piperidine; red in pyridine. Fluoresces in alcohol, benzene, and acetone solns, also in sulfuric acid.

Note: The name Resistomycin is also used as a trademark for Kanamycin A sulfate.

8156. Resodec. A polycarboxylic cation exchange resin. Almost inert, nonabsorbable powder, intended to aid the dietary restriction of salt by removing sodium from the contents of the intestinal tract.

THERAP CAT: Ion exchange resin (sodium).

8157. Resorantel. *N-(4-Bomophenyl)-2,6-dihydroxy-benzamide;* 4'-bromo-γ-resorcylanilide; 4'-bromo-2,6-di-hydroxybenzanilide; 2-6-dihydroxybenzoic acid 4'-bromo-anilide; resorcylam; Hoe 296V; Terenol. $C_{13}H_{10}BrNO_3$; mol wt 308.14. C 50.67%, H 3.27%, Br 25.93%, N 4.55%, O 15.58%. Prepn: LeMaire *et al., J. Pharm. Sci.* **50**, 831 (1961); **Brit.** pat. **1,124,613** corresp to Ruschig *et al.,* U.S. pat. **3,449,420** (1968, 1969 to Hoechst). Animal studies: Düwel; Behrens, Matschullat; Pfeiffer, *Deut. Tierärztl. Wochenschr.* **77**, 97-107 (1970); Christ *et al., Berlin. München. Tierärztl. Wochenschr.* **83**, 61 (1970).

Colorless powder, mp 229-230°. Also reported as mp 183-186° (LeMaire). Stable at room temp for at least two years. Very resistant to acids and alkalies. Highly sensitive to iron compds which stain it deep red or violet. Very soluble in DMF; slightly sol in lower alcohols. Insol in water, hydrocarbons, vegetable oils.

THERAP CAT (VET): Anthelmintic (sheep and other ruminants).

8158. Resorcinol. *1,3-Benzenediol;* m-dihydroxybenzene; resorcin. $C_6H_6O_2$; mol wt 110.11. C 65.44%, H 5.49%, O 29.06%. Is at least 99.5% pure. Made by fusing m-benzenedisulfonic acid with excess NaOH. *Review:* J. Varagnat in Kirk-Othmer *Encyclopedia of Chemical Technology* vol. **13** (Wiley-Interscience, New York, 3rd ed., 1981) pp 39-69.

White, needle-like crystals; sweetish taste. Becomes pink on exposure to light and air, or by contact with iron. d 1.272. mp 109-111°. bp 280°, but volatilizes at lower temp and is slightly volatile with steam. One gram dissolves in 0.9 ml water, 0.2 ml water at 80°, 0.9 ml alcohol; freely sol in ether, glycerol; slightly sol in chloroform. pH: 5.2. *Protect from light. Incompat.* Acetanilide, albumin, alkalies, antipyrine, camphor, ferric salts, menthol, spirit nitrous ether, urethan.

Monoacetate, $C_8H_8O_3$, *acetylresorcinol, Euresol.* Golden-yellow, thick, syrupy, oily liquid. bp about 283° with decompn. Insol in water. Miscible with alc, benzene, chloroform, acetone. Sol in solns of alkali hydroxides.

Caution: Irritating to skin, mucous membranes. Absorption can cause methemoglobinemia, cyanosis, convulsions, death, *cf.* Patty's *Industrial Hygiene and Toxicology* vol. 2A, G. D. Clayton, F. E. Clayton, Eds. (Wiley-Interscience, New York, 3rd ed.,1981) pp 2586-2589.

USE: In tanning; manuf resins, resin adhesives, hexylresorcinol, *p*-aminosalicylic acid, explosives, dyes; in cosmetics; dyeing and printing textiles; as a reagent for zinc.

THERAP CAT: Keratolytic; antiseborrheic.

THERAP CAT (VET): Topical antipruritic and antiseptic. Has been used as an intestinal antiseptic.

8159. β-Resorcylaldehyde. *2,4-Dihydroxybenzaldehyde;* 2,4-dihydroxybenzenecarbonal. $C_7H_6O_3$; mol wt 138.12. C 60.87%, H 4.38%, O 34.75%. Prepd by saturating with HCl a soln of resorcinol and anhyd hydrocyanic acid in abs ether and boiling the precipitated resorcylaldimide-HCl with water: Gattermann, Köbner, *Ber.* **32**, 278 (1899); **Ger. pat.** 106,508; *Chem. Zentr.* **1900,** I, 742. By adding $POCl_3$ to a soln of resorcinol and formanilide in ether, then boiling with dil NaOH, acidifying with H_2SO_4 and extracting with ether: Dimrath, Zöppritz, *Ber.* **35,** 995 (1902); by passing HCl into a mixture of resorcinol, ether, anhyd zinc chloride and cyanogen bromide, then boiling the precipitate with water: Karrer, *Helv. Chim. Acta* **2,** 92 (1919).

Needles from water or ether + ligroin. mp 135-136°. bp$_{22}$ 226°. Freely sol in water, alcohol, ether, chloroform, glacial acetic acid. Difficultly sol in cold benzene. β-Resorcylaldehyde is easily dec by acids and alkalies. Prolonged exposure to moist air converts it to a brown, amorphous powder which is insol in ether.

8160. β-Resorcylic Acid. *2,4-Dihydroxybenzoic acid;* 2,4-dihydroxybenzenecarboxylic acid; BRA. $C_7H_6O_4$; mol wt 154.12. C 54.55%, H 3.92%, O 41.52%. Prepared from resorcinol and $KHCO_3$ in glycerol or water by heating and passing CO_2 through the mixture: Brunner, *Ann.* **351,** 320 (1907). *Cf.* Bistrycki, Kostanecki, *Ber.* **18,** 1984 (1885); Nierenstein, Clibbens, *Org. Syn. coll. vol.* II, 557 (1943).

Hydrated crystals from water. Becomes anhydr at 100°, mp 213° (rapid heating). K at 25° = 5.00 × 10^{-4}. Sol in hot water, alcohol, ether, olive oil. Boiling with water, acids or salt solns results in loss of CO_2.

USE: Intermediate for dyestuffs and drugs; spot test reagent for iron.

8161. Retamine. *Dodecahydro-7,14-methano-2H,6H-dipyrido[1,2-a:1',2'-e][1,5]diazocin-1-ol.* $C_{15}H_{26}N_2O$; mol wt 250.37. C 71.95%, H 10.47%, N 11.19%, O 6.39%. From bark and young branches of *Genista sphaerocarpa* Lam. (*Retama sphaerocarpa* (Lam.) Boiss.), *Leguminosae.* Isoln: Battandier, Malosse, *Compt. Rend.* **125,** 360, 450 (1897); Ribas *et al., Anales Fis. y Quim.* **42,** 516 (1946), *C.A.* **41,** 4894c (1947). Structure: Bohlmann *et al., Ber.* **98,** 659 (1965). Synthesis: Bohlmann *et al., ibid.* 653. Stereochemistry: Ribas *et al., Tetrahedron Letters* **1965,** 3181.

Bitter crystals from alc, mp 165-166°. $[\alpha]_D^{20}$ +46.2° (c = 1.017 in abs ethanol). Practically insol in water; slightly sol in ether; sol in chloroform, alcohol, methanol; very sol in benzene.

Hydrochloride, mp 272-273°.

Picrate, crystals from acetone, mp 207°.

8162. Retene. *1-Methyl-7-(1-methylethyl)phenanthrene.* $C_{18}H_{18}$; mol wt 234.32. C 92.26%, H 7.74%. Occurs in pine tar, in fossilized pine, in high-boiling tar oils. Formed by Se or Pd dehydrogenation of abietic acid: Ruzicka, Waldmann, *Helv. Chim. Acta* **16,** 842 (1933). Synthesis starting with methyl β-(6-isopropyl-2-naphthoyl)propionate: Bardhan, Sengupta, *J. Chem. Soc.* **1932,** 2520, 2798; *cf.* Bogert, *Science* **77,** 289 (1933). Review and bibliography: Fieser, Fieser, *Natural Products Related to Phenanthrene* (New York, 3rd ed., 1949).

Plates, scales from alc, mp 99°. bp$_{10}$ 208°; bp$_{760}$ 390-394°. Absorption spectrum: Askew, *J. Chem. Soc.* **1935,** 509. Insol in water. Sol in hot alcohol, benzene, hot ether, carbon disulfide. Upon treatment with CrO_3 in acetic acid forms retenequinone, $C_{18}H_{16}O_2$, mp 197.5°.

8163. Reticulin (the Protein). One of the connective tissue proteins occurring wherever connective tissue forms a boundary: Jacobson, *Nature and Structure of Collagen,* J. T. Randall, Ed. (Butterworths, London, 1953) pp 6-13, *C.A.* **48,** 10814e (1954). Found together with collagen and elastin. Particularly abundant in membranes of the glomeruli and tubules of the kidney and in supportive structure of the vasculature and lung alveoli. Isoln from kidney tissue slices: Kramer, Little, *ibid.* pp 33-43, *C.A.* **48,** 10815a (1954). Contains about 85% protein, the amino acid content of which is very similar to that of collagen. Also contains about 4.2% carbohydrate and approx 11% bound lipid which yields, on hydrolysis, 95% myristic acid and 5% palmitic acid: Windrum *et al., Brit. J. Exp. Pathol.* **36,** 49 (1955). Lacks hydroxyproline; glycine content is that of normal proteins: Pras, Glynn, *ibid.* **54,** 449 (1973). Stable in boiling water and in 1N HCl. Can be dispersed in distilled water but precipitated by physiological saline (0.15 mol/l).

8164. Reticuline. *1,2,3,4-Tetrahydro-1-[(3-hydroxy-4-methoxyphenyl)methyl]-6-methoxy-2-methyl-7-isoquinolinol;* 1,2,3,4-tetrahydro-6-methoxy-2-methyl-1-vanillyl-7-isoquinolinol; 2-methyl-7-hydroxy-6-methoxy-1-(3-hydroxy-4-methoxybenzyl)-1,2,3,4-tetrahydroisoquinoline; coclanoline. $C_{19}H_{23}NO_4$; mol wt 329.38. C 69.28%, H 7.04%, N 4.25%, O 19.43%. Precursor of many aporphine and morphine-type alkaloids. Isoln of *d*-form from *Anona reticulata* Linn., *Anonaceae:* Gopinath *et al., Ber.* **92,** 776 (1959); *dl*-form from opium: Brochmann-Hanssen, Furuya, *J. Pharm. Sci.* **53,** 575 (1964); *Planta Med.* **12,** 328 (1964).

Structure: Brochmann-Hanssen, Nielson, *Tetrahedron Letters* **1965**, 1271. Configuration: Battersby, Evans, *ibid.* 1275. Synthesis of amorphous *dl*-form: Gopinath *et al.*, *Ber.* **92**, 1657 (1959); P. Kerekes *et al.*, *Acta Chim. Acad. Sci. Hung.* **98**, 491 (1978). Synthesis of crystalline *dl*-form: Baxter *et al.*, *J. Chem. Soc. (C)* **1965**, 3645; Chan, Maitland, *ibid.* **1966**, 753; K. C. Rice, A. Brossi, *J. Org. Chem.* **45**, 592 (1980); G. Dornyei *et al.*, *Tetrahedron Letters* **1982**, 2913. Asymmetric synthesis of (S)-reticuline: Konda *et al.*, *Chem. Pharm. Bull.* **23**, 1063 (1975). Biosynthesis: D. S. Dewan *et al., J. Chem. Soc. Perkin Trans. I* **1977**, 1662; A. Barrett *et al., ibid.* **1979**, 652.

dl-Form, pink crystals, mp 146°. uv max: 284 nm (log ε 3.85). Sol in aqueous buffer of pH < 7.5 or > 11. Practically insol in water at pH 8-10.

(S)-Form perchlorate, $C_{19}H_{23}NO_4 \cdot HClO_4$, colorless prisms from ethanol, mp 203-204°. $[\alpha]_D^{18}$ +88.3° (c = 0.21 in alc).

8165. Retinal. Retinaldehyde; retinene₁; vitamin A aldehyde; axerophthal. $C_{20}H_{28}O$; mol wt 284.42. C 84.45%, H 9.92%, O 5.63%. Carotenoid component of the visual pigments. Of the 16 possible stereoisomers, 13 have been synthesized. The 11-*cis* isomer is the chromophore of the majority of naturally occurring opsins, *q.v.* Isoln from retinas: G. Wald, *J. Gen. Physiol.* **19**, 351 (1935); recognition as vitamin A aldehyde: R. A. Morton, *Nature* **153**, 69 (1944). Prepd by the oxidation of vitamin A: Ball *et al.*, *Biochem. J.* **42**, 516 (1948); by the oxidation of β-carotene: Wendler *et al.*, *J. Am. Chem. Soc.* **72**, 234 (1950); from β-ionone: Eiter, Truscheit, U.S. pat. **3,060,229** (1962 to Bayer); Jacobs *et al.*, *Rec. Trav. Chim.* **84**, 1113 (1965). Synthesis *in vitro* from vitamin A₁: G. Wald, R. Hubbard, *Proc. Nat. Acad. Sci. USA* **36**, 92 (1950). Synthesis of 11-*cis* isomer: J. M. Dieterle, C. D. Robeson, *Science* **120**, 219 (1954). X-ray crystallography of retinals: R. Gilardi *et al.*, *Nature* **232**, 187 (1971); H. Matsumoto *et al.*, *J. Am. Chem. Soc.* **102**, 4259 (1980). Exposure of opsin-bound retinal to light initiates the *cis-trans* isomerization and resultant dissociation of retinal from the apoprotein. This catalyzes an enzyme cascade which leads to visual excitation: R. Hubbard, A.

11-*cis* retinal

all-*trans* retinal

Kropf, *Proc. Natl. Acad. Sci. USA* **44**, 130 (1958); B. Honig *et al., ibid.* **76**, 2503 (1979). Bioregulation of absorption maxima: B. Honig *et al.*, *J. Am. Chem. Soc.* **101**, 7084 (1979). HPLC determn in biological material: T. Suzuki, M. Makino-Tasaka, *Anal. Biochem.* **129**, 111 (1983). Review of syntheses: V. Balogh-Nair, K. Nakanishi, *Methods in Enzymology* vol. **88**, L. Packer, Ed. (Academic Press, New York, 1982) pp 496-506; R. Liu, A. Asata, *ibid.* pp 506-516. Role in vision: D. F. O'Brien, *Science* **218**, 961-966 (1982); P. S. Zurer, *Chem. & Eng. News* **61**, 24-35 (Nov. 28,1983). Book: *The Retinoids*, Vol. 1-2, M. B. Sporn *et al.*, Eds. (Academic Press, New York, 1984).

trans-Isomer, orange colored crystals from petr ether, mp 61-64°. uv max (hexane): 368 nm (ε 48000). Sol in ethanol, chloroform, cyclohexane, petr ether, oils. Practically insol in water.

11-*cis*-Isomer, orange prisms from petr ether, mp 63.5-64.4°. uv max (hexane): 365 nm (ε 26360).

8166. Retine. Inhibitor of malignant growth, extracted along with *promine*, a cancer growth-promoting and sterilizing substance, from calves' thymus: Szent-Györgyi *et al.*, *Proc. Nat. Acad. Sci. USA* **48**, 1439 (1962). Sterilizing action of promine caused by another substance, *infertine:* Hegyeli *et al., ibid.* **49**, 230 (1963). Prepn of retine from human urine: Hegyeli *et al., Science* **142**, 1571 (1963); from calves' aortas: Szent-Györgyi *et al., ibid.* **140**, 1391 (1963); from both sources: *eidem,* U.S. pat. **3,297,533** (1967 to U.S. Dept. of HEW). Identification studies: Szent-Györgyi, *Proc. Nat. Acad. Sci. USA* **57**, 1642 (1967). Isoln of promine and retine from calf thymus and liver and their identification as ε-*N*-methylated lysines and guanidino-methylated arginines, resp.: E. Tyihak, A. Patthy, *Acta Agron. (Budapest)* **22**, 445 (1973). Polemic on the findings on Tyihak and Patthy: T. Nakajima, *Acta Agron. Acad. Sci. Hung.* **23**, 236 (1974); C. Marmasse, *ibid.* 216.

8167. Retinoic Acid. 3,7-Dimethyl-9-(2,6,6-trimethyl-1-cyclohexen-1-yl)-2,4,6,8-nonatetraenoic acid; vitamin A acid; tretinoin; Aberel; Airol; Aknoten; Cordes Vas; Dermairol; Effederm; Epi-Aberel; Eudyna; Retin-A. $C_{20}H_{28}O_2$; mol wt 300.42. C 79.95%, H 9.39%, O 10.65%. The naturally occurring form is all-*trans*. Prepn: C. D. Robeson *et al.*, *J. Am. Chem. Soc.* **77**, 4111 (1955); Pommer, Sarnecki, U.S. pat. **3,006,939** (1961 to BASF). Prepn by oxidation of retinal: Lakshmanan *et al.*, *Biochem. J.* **90**, 569 (1964). Single-step process for all-*trans* and 13-*cis* isomers: R. Marbet, **Ger.** pat. **2,061,507;** U.S. pat. **3,746,730** (1971, 1973 both to Hoffmann-La Roche). New process for 13-*cis* form: R. Lucci, **Eur.** pat. Appl. **111,325;** U.S. pat. **4,556,518** (1984, 1985 both to Hoffmann-La Roche). Structure: Stam, Mac Gillavry, *Acta Cryst.* **16**, 62 (1963). Inhibition of bladder carcinogenesis in rats: M. B. Sporn *et al.*, *Science* **195**, 487 (1977); M. Tannenbaum *et al.*, *Fed. Proc.* **38**, 1073 (1979). Use of 13-*cis* form in treatment of severe cystic acne: G. L. Peck *et al.*, *N. Engl. J. Med.* **300**, 329 (1979). Toxicology, carcinogenicity and teratogenicity: J. J. Kamm, *J. Am. Acad. Dermatol.* **6**, 652 (1982). Human teratogenicity study: E. J. Lammer *et al.*, *N. Engl. J. Med.* **313**, 837 (1985). Clinical evaluation in treatment of photoaged skin: J. S. Weiss *et al.*, *J. Am. Med. Assoc.* **259**, 527 (1988). Review of pharmacology and therapeutic efficacy of 13-*cis* form in skin disorders: A. Ward *et al.*, *Drugs* **28**, 6-37 (1984); of clinical pharmacology: A. Vahlquist, O. Rollman, *Dermatologica* **175**, Suppl. 1, 20-27 (1987). Review of effect on cellular differentiation and proliferation: F. Frickel, *Pure Appl. Chem.* **57**, 709 (1985). *Reviews:* D. B. Ott, P. A. Lachance, *Am. J. Clin. Nutr.* **32**, 2522-2531 (1979); D. S. Goodman, *N. Engl. J. Med.* **310**, 1023-1031 (1984). Book: *The Retinoids,*

all-*trans*-retinoic acid

Vols. 1-2, M. B. Sporn *et al.*, Eds. (Academic Press, New York, 1984).

Crystals from ethanol, mp 180-182°. uv max (methanol): 351 nm (ε 45000). LD$_{50}$ (10 day) in mice, rats (mg/kg): 790, 790 i.p.; 2200, 2000 orally (Kamm).

9-*cis*-Form, yellow needles from ethanol, mp 189-190°. uv max (methanol): 343 nm (ε 36500).

13-*cis*-Form, *Ro 4-3780, isotretinoin, Accutane, Isotrex, Roaccutane, Teriosal.* Reddish-orange plates from isopropyl alcohol, mp 174-175°. uv max: 354 nm (ε 39800). LD$_{50}$ (20 day) in mice, rats (mg/kg): 904, 901 i.p.; 3389, >4000 orally (Kamm).

THERAP CAT: All-*trans* and 13-*cis* forms as keratolytic.

8168. Retronecine. *(1R-trans)-2,3,5,7a-Tetrahydro-1-hydroxy-1H-pyrrolizine-7-methanol.* C$_8$H$_{13}$NO$_2$; mol wt 155.19. C 61.91%, H 8.44%, N 9.03%, O 20.62%. The most common base portion of pyrrolizidine alkaloids. "*Necine*" *bases* are *1-methylpyrrolizidines* of different stereochemical configurations and degrees of hydroxylation that occur in the form of esters in alkaloids of *Senecio, Crotalaria* and a number of genera of the *Boraginaceae.* Retronecine occurs in nature in the (+)-form. It is the necine base of monocrotaline, senecionine, *q.q.v.*, seneciphylline, retrorsine and numerous other hepatotoxic pyrrolizidine alkaloids. Structure: R. Adams, E. F. Rogers, *J. Am. Chem.* Soc. **63**, 228 (1941). Total synthesis: T. A. Geissman, A. C. Waiss, *J. Org. Chem.* **27**, 139 (1962). Stereospecific synthesis: E. Vedejs, G. R. Martinez, *J. Am. Chem. Soc.* **102**, 7993 (1980); H. Niwa *et al.*, *Tetrahedron Letters* **27**, 4605 (1986). Total synthesis of (−)-form: J. Cooper *et al.*, *Chem. Commun.* **1988**, 509. Toxicity data: P. N. Harris *et al.*, *J. Pharmacol. Exp. Ther.* **75**, 78 (1942). Biosynthesis: D. J. Robins, J. R. Sweeney, *Chem. Commun.* **1979**, 120; G. Grue Sorensen, I. D. Spenser, *J. Am. Chem. Soc.* **103**, 3208 (1981). Comprehensive reviews on retronecine and other necine bases: F. L. Warren, *Fortschr. Chem. Org. Naturst.* **12**, 198-269 (1955); **24**, 329-406 (1966); D. J. Robins, *ibid.* **41**, 115-203 (1982); F. L. Warren in *Alkaloids* Vol. **XII**, R. H. F. Manske, Ed. (Academic Press, New York, 1970) pp 245-331.

Crystals from acetone, mp 119-120° C. [α]$_D^{20}$ +4.95° (c = 0.58 in alc). LD$_{50}$ i.v. in mice: 634.0 ±26.0 mg/kg (Harris). (±)-Form, crystals from acetone, mp 130-131°.

8169. Retrorsine. *12,18-Dihydroxysenecionan-11,16-dione;* β-longilobine. C$_{18}$H$_{25}$NO$_6$; mol wt 351.41. C 61.52%, H 7.17%, N 3.99%, O 27.32%. Hepatotoxic pyrrolizidine alkaloid; common constituent of *Senecio* species. Isoln from *Senecio retrorsus* DC, *Compositae*: R. H. F. Manske, *Can. J. Chem.* **5**, 651 (1931); G. Barger *et al.*, *Chem. Soc.* **1935**, 11; from *Crotalaria usaramoensis* E. G. Baker, *Leguminosae*: C. C. J. Culvenor, L. W. Smith, *Aust. J. Chem.* **19**, 2127 (1966). Structure: S. M. H. Christie *et al.*, *J. Chem. Soc.* **1949**, 1700; E. C. Leisegang, F. L. Warren, *ibid.* **1950**, 702. Identity with β-longilobine: F. L. Warren *et al.*, *J. Am. Chem. Soc.* **72**, 1421 (1950). Review and evaluation of toxicity and carcinogenicity studies: *IARC Monographs* **10**, 303-312, 333-342 (1976). Comprehensive reviews of pyrrolizidine alkaloids: L. B. Bull *et al.*, *The Pyrrolizidine Alkaloids* (North Holland, Amsterdam, 1968) 293 pp; F. L.Warren in *The*

Alkaloids vol. **12**, R. H. F. Manske, Ed. (Academic Press, New York, 1970) pp 245-331.

Crystals from ethyl acetate, mp 212° [Barger *et al.*]; 216-216.5° [Bull *et al.*]. [α]$_D^{18}$ −17.6° (c = 1.99 in ethanol). uv max (water): 217 nm (log ε 3.85). Readily sol in alcohol, chloroform; slightly sol in water, acetone, ethyl acetate; practically insol in ether.

N-Oxide, C$_{18}$H$_{25}$NO$_7$, *isatidine, 12,18-dihydroxysenecionan-11,16-dione-4-oxide, retrorsine N-oxide.* Isolated from *Senecio* species. Crystals from ethanol, mp 140.5-141.5° (Christie *et al.*). Review and evaluation of carcinogenicity and toxicity studies: *IARC Monographs* **10**, 269-273 (1976).

8170. Rhamnetin. *2-(3,4-Dihydroxyphenyl)-3,5-dihydroxy-7-methoxy-4H-1-benzopyran-4-one; 3,3',4',5-tetrahydroxy-7-methoxyflavone;* 7-methylquercetin; β-rhamnocitrin; cyanidenolon-7-methyl ether 1537. C$_{16}$H$_{12}$O$_7$; mol wt 316.26. C 60.76%, H 3.82%, O 35.41%. The aglucon of xanthorhamnin. From fruit of *Rhamnus cathartica* L., *Rhamnaceae*: Krassowski, *J. Russ. Phys. Chem. Soc.* **40**, 1510 (1909); from commercial xanthorhamnin: Nystrom *et al.*, *J. Org. Chem.* **22**, 1272 (1957). Structure: Oesch, Perkin, *J. Chem. Soc.* **105**, 2354 (1914); Jurd, Horowitz, *J. Org. Chem.* **22**, 1618 (1957). Synthesis: Kuhn, Low, *Ber.* **77**, 211 (1944); Jurd, *J. Am. Chem. Soc.* **80**, 5531 (1958); U.S. pat. **2,892,845** (1959 to USDA); Anand *et al.*, *J. Sci. Ind. Res.* **21B**, 322 (1962), *C.A.* **57**, 13712f (1962); Kawano *et al.*, *Chem. Pharm. Bull.* **15**, 711 (1967).

Yellow needles from acetone + methanol, mp 292-294°. uv max (ethanol): 371, 256 nm (log ε 4.41, 4.40). Sol in hot phenol. Slightly sol in hot water, hot alcohol, hot glacial acetic acid, hot acetone. Freely sol in dil alkalies with intense yellow color.

Tetraacetate, C$_{24}$H$_{20}$O$_{11}$, needles from acetone + methanol, mp 189-190°.

USE: Has been used for dyeing wool and cotton.

8171. Rhamnose. *6-Deoxy-L-mannose;* L-rhamnose; L-mannomethylose; isodulcit. C$_6$H$_{12}$O$_5$; mol wt 164.16. C 43.90%, H 7.37%, O 48.73%. Occurs free in poison sumac *(Rhus toxicodendron* L., *Anacardiaceae)*, combined in the form of glycosides of many plants. Preparation: Clark, *J. Biol. Chem.* **38**, 255 (1919). Structure and configuration: Fischer, Morrell, *Ber.* **27**, 384 (1894); Fischer, Zach, *ibid.* **45**, 3762 (1912); Hirst, Macbeth, *J. Chem. Soc.* **1926**, 23; Avery, Hirst, *ibid.* **1929**, 2466. Isoln from walls of gram-negative bacteria: Salton, *Biochim. Biophys. Acta* **45**, 364 (1960); from *Afraegle paniculata* Engl., *Rutaceae*: Torto, *J. Chem. Soc.* **1961**, 5234; from rabbit skin: Malawista, Davidson, *Nature* **192**, 871 (1961); from leaves of *Solanum chacoense* Bitter, *Solanaceae*: Kuhn, Löw, *Ber.* **94**, 1088 (1961).

α-L-rhamnose

α-Form, always obtained by crystn from H$_2$O or EtOH. Monohydrate, holohedric rods from water, hemihedric monoclinic columns from alcohol. Loses water of crystn on heating and partially changes to the β-modification. Very sweet taste. mp 82-92°: Ghosh, *Proc. Roy. Soc. Edinburgh* **36**, 216 (1915/16). Sublimes at 105° and 2 mm Hg. d$_4^{20}$

1.4708. Shows mutarotation. $[\alpha]_D^{20}$ $-7.7°$ → $+8.9°$: Hudson, Yanovsky, *J. Am. Chem. Soc.* **39**, 1032 (1917).

β-Form, prepd by heating α-rhamnose monohydrate on a steam bath; crystallized from anhydr acetone + alcohol. Needles, mp 122-126° (rapid heating). $[\alpha]_D^{20}$ $+31.5°$ (1 min, p = 10). After a short time the rotation adjusts to the same final value as α-rhamnose. The β-form is hygroscopic and changes into crystals of the α-modification upon exposure to moist air.

8172. Rhamnus cathartica. Buckthorn bark. Dried, ripe fruit of *Rhamnus cathartica* L., *Rhamnaceae. Habit.* Europe, Northern Africa to Middle Asia. *Constit.* Rhamnocathartin, rhamnotannic acid, rhamnin, rhamnetin.

THERAP CAT: Cathartic.

8173. Rhapontin. *3-Hydroxy-5-[2-(3-hydroxy-4-methoxyphenyl)ethenyl]phenyl β-D-glucopyranoside; 4'-methoxy-3,3',5-stilbenetriol 3-glucoside;* rhaponticin; ponticin. $C_{21}H_{24}O_9$; mol wt 420.40. C 59.99%, H 5.75%, O 34.25%. From root of *Rheum rhaponticum* L., *Polygonaceae:* Hesse, *J. Prakt. Chem.* **77**, 321 (1908); Schürhoff, Plettner, *Arch. Pharm.* **275**, 281 (1937). On hydrolysis yields rhapontigenin and glucose: Kawamura, *J. Pharm. Soc. Japan* **58**, 405 (1938), *C.A.* **32**, 6655[4] (1938). Constitution of rhapontigenin: Takaoka, *Proc. Imp. Acad. (Tokyo)* **16**, 408 (1940), *C.A.* **35**, 1399[2] (1941).

Crystals, dec 236-237°. $[\alpha]_D^{32}$ $-59.5°$ (acetone). Exhibits bright blue fluorescence. Sol in dil alcohol, hot acetone, hot water; slightly sol in ether, alc, acetone, cold water; practically insol in benzene, petr ether, chloroform.

8174. Rheadine. *(8β)-8-methoxy-16-methyl-2,3:10,11-bis[methylenebis(oxy)]rheadan; rhoeadine.* $C_{21}H_{21}NO_6$; mol wt 383.39. C 65.78%, H 5.52%, N 3.65%, O 25.04%. Isoln from the seed capsules of corn poppy *(Papaver rhoeas* L., *Papaveraceae):* Hesse, *Ann.* **140**, 145 (1866); *Arch. Pharm.* **228**, 7 (1890); Awe, *ibid.* **274**, 439 (1936); Awe, Winkler, *ibid.* **290**, 367 (1957); Nemecková, Santavy, *Coll. Czech. Chem. Commun.* **27**, 1210 (1962). Structure and stereochemistry: Santavy *et al., ibid.* **30**, 3479 (1965); **32**, 4452 (1967). Total synthesis: Klotzer *et al., Helv. Chim. Acta* **54**, 2057 (1971); **55**, 2228 (1972); Irie *et al., J. Chem. Soc. Perkin Trans. I* 1972, 2986.

Crystals from ethanol, mp 252-254°. $[\alpha]_D^{23}$ $+235°$ (c = 1.01 in chloroform); $[\alpha]_D^{22}$ $+174°$ (c = 0.69 in pyridine). uv max (ethanol): 239, 290 nm (ε 9150, 9180). Slightly sol in ethyl acetate, methylene chloride. Practically insol in water, alcohol, chloroform, ether, benzene.

8175. Rhein. *9,10-Dihydro-4,5-dihydroxy-9,10-dioxo-2-anthracenecarboxylic acid;* 1,8-dihydroxyanthraquinone-3-carboxylic acid; 4,5-dihydroxyanthraquinone-2-carboxylic acid; chrysazin-3-carboxylic acid; monorhein; rheic acid; cassic acid; parietic acid; rhubarb yellow. $C_{15}H_8O_6$; mol wt 284.21. C 63.39%, H 2.84%, O 33.78%. Found in the free state and as glucoside in *Rheum* spp, *Polygonaceae* (rhubarb) and in Senna leaves; also in several spp of *Cassia (Legumino-*

sae). Diacetate used as antirheumatic. Isoln from Chinese rhubarb: J. Schlossberger, O. Doepping, *Ann.* **50**, 196 (1844); O. Hesse, *ibid.* **309**, 32 (1899); F. Tutin, H. W. B. Clewer, *J. Chem. Soc.* **99**, 946 (1911); from aloe-emodin, *q.v.:* O. A. Oesterle, *Arch. Pharm.* **241**, 604 (1903); from *Cassia alata* L.: H. Hauptmann, L. L. Nazario, *J. Am. Chem. Soc.* **72**, 1492 (1950); from *C. fistula:* V. K. Murty *et al., Tetrahedron* **23**, 515 (1967). Prepn from chrysophanic diacetate: Fischer *et al., J. Prakt. Chem.* [2] **83**, 208; **84**, 369 (1911). Structure: H. Nawa *et al., J. Org. Chem.* **26**, 979 (1961). Use in arthritis: C. A. Friedmann, Ger. pat. **2,711,493**; *idem,* U.S. pat. **4,244,968** (1977, 1981 both to Proter). Spectrophotometric study: A. A. Habib, N. A. El-Sebakhy, *J. Nat. Prod.* **43**, 452 (1980). Determn by gas chromatography, mass spectroscopy: G. W. Van Eijk, H. J. Roeijmans, *J. Chromatog.* **295**, 497 (1984); HPLC: A. J. J. Van den Berg, R. P. Labadie, *ibid.* **329**, 311 (1985). Pharmacological effects on colonic mucosa: R. Wanitschke, *Pharmacology* **20**, Suppl. 1, 21 (1980); K. Ewe, *ibid.* 27. Metabolism study: J. Lemli, L. Lemmons, *ibid.* 50. In electron transport: P. Egerer *et al., Z. Physiol. Chem.* **363**, 627 (1982). Inhibition of proteases: L. Raimondi *et al., Pharmacol. Res. Commun.* **14**, 103 (1982). Effect on prostaglandin biosynthesis: S. Franchi-Micheli *et al., J. Pharm. Pharmacol.* **35**, 262 (1983).

Yellow needles by sublimation, mp 321-322°, dec 330°. Absorption max (methanol): 229, 258, 435 nm (ε 36,800, 20,100, 11,100). Practically insol in water. Sol in alkalies, pyridine; slightly sol in alc, benzene, chloroform, ether, petr ether. Forms a red potassium salt and a pink sodium salt. Red precips are also obtained with $Ca(OH)_2$ and $Ba(OH)_2$. Diacetate, *See* Diacerein.

8176. Rhenium. Re; at. wt 186.207; at. no. 75; valences 1-7; the heptavalent state is the most stable. Two natural isotopes: 185 (37.07%); 187 (62.93%); the latter is radioactive, $T_{1/2}$ ~10^{11} years. Artificial, radioactive isotopes: 177-184; 186; 188-192. Occurs in gadolinite, molybdenite, columbite, rare earth minerals, and some sulfide ores. Average concn in earth's crust 1×10^{-9} (0.001 ppm). Discovered by Noddack *et al., Naturwiss.* **13**, 567, 571 (1925). Prepn of metallic rhenium by reduction of potassium perrhenate or ammonium perrhenate: Hurd, Brim, *Inorg. Syn.* **1**, 175 (1939). Prepn of high purity rhenium: Rosenbaum *et al., J. Electrochem. Soc.* **103**, 518 (1956). *Reviews:* Melaven in *Rare Metals Handbook,* C. A. Hampel, Ed. (Reinhold, New York, 1954) pp 347-364; Peacock in *Comprehensive Inorganic Chemistry* vol. 3, J. C. Bailar, Jr. *et al.,* Eds. (Pergamon Press, Oxford, 1973) pp 905-978; P. M. Treichel in Kirk-Othmer *Encyclopedia of Chemical Technology* vol. 20 (Wiley-Interscience, New York, 3rd ed., 1982) pp 249-258.

Hexagonal close-packed crystals, black to silver-gray. d 21.02. mp 3180°. bp 5900° (estimated). Specific heat 0-20° 0.03263 cal/g/°C. Specific electrical resistance: 0.21 × 10^{-4} ohm/cm at 20°. Brinell hardness: 250. Latent heat of vaporization 152 kcal/mol. Reacts with oxidizing acids, nitric and concd sulfuric; not with HCl.

Human Toxicity: No toxic manifestations have been reported in exptl animals or in man. *See* E. Browning, *Toxicity of Industrial Metals* (Appleton-Century-Crofts, New York, 1969) pp 276-277.

USE: Electron tube and semiconductor applications, in alloys for electrical contacts, as catalyst; possibly in high temp thermocouples and to improve the workability of tungsten and molybdenum alloys; plating jewelry, medical instruments, high vac equipment, mirror backings.

8177. Rhenium Heptoxide. Dirhenium heptoxide. O_7Re_2; mol wt 484.44. O 23.12%, Re 76.89%. Re_2O_7. Prepd from rhenium metal or its lower oxides in a stream of air or oxygen at 400-425°: Noddack, Noddack, *Naturwiss.* **17**, 93 (1929); *Z. Anorg. Allgem. Chem.* **181**, 1 (1929); **215**, 129

(1933); Briscoe *et al.*, *Nature* **129**, 618 (1932); Melaven *et al.*, *Inorg. Syn.* **3**, 188 (1950). Crystal structure: Krebs *et al.*, *Chem. Commun.* **1968**, 263.

Canary-yellow, very deliquescent crystals. Begins to sublime at 250°. mp 300.3°; bp 360.3°: Smith *et al.*, *J. Am. Chem. Soc.* **74**, 4964 (1952). Freely sol in water, alcohol, ether, ethyl acetate, dioxane, pyridine. Readily absorbs water forming perrhenic acid, $HReO_4$.

8178. Rhenium Hexafluoride. F_6Re; mol wt 300.22. F 37.97%, Re 62.03%. ReF_6. Prepd by direct fluorination of rhenium metal: Ruff, Kwasnik, *Z. Anorg. Allgem. Chem.* **209**, 113 (1932); **219**, 65 (1934); Malm, Selig, *J. Inorg. Nucl. Chem.* **20**, 189 (1961). The product obtained by Ruff was probably a mixture of ReF_6 and ReF_7: *eidem, J. Am. Chem. Soc.* **82**, 1510 (1960).

Yellow, cubic crystals or liquid; colorless gas; extremely hygroscopic. Forms bluish vapors on contact with air, discolors to a dark purple liquid. mp 18.5°. bp 33.7°. Sol in nitric acid. Soly in anhyd HF: 1.75 moles/1000 g HF, Frlec, Hyman, *Inorg. Chem.* **6**, 1596 (1967). Does not attack dried Pyrex glass or silica. May be handled in copper up to 150°.

8179. Rhenium Oxychloride. *Rhenium chloride oxide;* perrhenyl chloride; rhenium trioxide chloride. ClO_3Re; mol wt 269.68. Cl 13.14%, O 17.79%, Re 69.06%. ReO_3Cl. Prepd by direct chlorination of rhenium(VI) oxide: Wolf *et al.*, *J. Am. Chem. Soc.* **79**, 4257 (1957).

Clear, colorless liquid. mp +4.5°. bp$_{760}$ 128°; also reported to be 131°: Bruckl, Ziegler, *Ber.* **65**, 916 (1932). Distills *in vacuo* without decompn. Can be stored in sealed glass ampuls. Any purple color developing is due to residual ReO_3. Sol in CCl_4. Hydrolyzed by water to $HReO_4$ and HCl. Extremely reactive toward mercury, silver and a large number of organic substances.

8180. Rhenium Trioxide. Rhenic anhydride. O_3Re; mol wt 234.22. O 20.49%, Re 79.51%. ReO_3. Prepd by reduction of Re_2O_7 with dioxane: Nechamkin, Hiskey, *Inorg. Syn.* **3**, 186 (1950); with CO: Melaven *et al., ibid.* 187.

Red cubic crystals; green luster by transmitted light; the color of the powder may appear blue (*rhenium blue*). At 400° disproportionates *in vacuo* to Re_2O_7 and ReO_2. bp 750°. d 6.9-7.4. Practically insol in water, alkalies, non-oxidizing acids; oxidized by HNO_3 to $HReO_4$.

8181. Rhizopterin. *4-[[(2-Amino-1,4-dihydro-4-oxo-6-pteridinyl)methyl]formylamino]benzoic acid; p-[N-[(2-amino-4-hydroxy-6-pteridinyl)methyl]formamido]benzoic acid; p-[N-(2-amino-4-hydroxypyrimido[4,5-b]pyrazin-6-ylmethyl)formamido]benzoic acid;* S.L.R. factor; formyl-pteroic acid. $C_{15}H_{12}N_6O_4$; mol wt 340.29. C 52.94%, H 3.55%, N 24.70%, O 18.81%. A folic acid factor that possesses high growth-promoting activity for *Streptococcus lactis R (S. faecalis R)*, but almost none for *Lactobacillus casei*, chick and rat: Keresztesy *et al., Science* **97**, 465 (1943). Isoln from charcoal adsorbate obtained from the purification of *Rhizopus nigricans* fumaric acid fermentation liquors: Rickes *et al., J. Am. Chem. Soc.* **69**, 2749 (1947); Keresztesy, Rickes, U.S. pat. **2,478,404** (1949 to Merck & Co.). Structure: Wolf *et al., J. Am. Chem. Soc.* **69**, 2753 (1947).

Light yellow platelets from aq ammonium acetate soln. Does not melt below 300°. Absorption spectrum: Rickes *et al., J. Am. Chem. Soc.* **69**, 2749, 2751 (1947). Practically insol in water and the usual organic solvents. Sol in ammonia water, in aq solns of alkali hydroxides, and in pyridine.

8182. Rhodamine B. *N-[9-(2-Carboxyphenyl)-6-(diethylamino)-3H-xanthen-3-ylidene]-N-ethylethanaminium chloride;* tetraethylrhodamine; D & C Red No. 19; C.I. Basic Violet 10; C.I. 45170. $C_{28}H_{31}ClN_2O_3$; mol wt 479.00. C 70.20%, H 6.52%, Cl 7.40%, N 5.85%, O 10.02%. Prepn: *Colour Index* **vol. 4** (3rd ed., 1971) p 4420.

Green crystals or reddish-violet powder. Very sol in water with a bluish-red color, dil solns being strongly fluorescent; very sol in alcohol; slightly sol in HCl, NaOH. LD$_{50}$ i.v. in rats: 89.5 mg/kg, J. M. Webb *et al., Toxicol. Appl. Pharmacol.* **3**, 696 (1961).

USE: As a dye, especially for paper; as a reagent for antimony, bismuth, cobalt, niobium, gold, manganese, mercury, molybdenum, tantalum, thallium, tungsten; as biological stain. Provisionally listed for use in drugs and cosmetics.

8183. Rhodanilic Acid. *Hydrogen bis(benzenamine-N)-tetrakis(thiocyanato-N)chromate(1−).* $C_{16}H_{15}CrN_6S_4$; mol wt 471.62. C 40.75%, H 3.21%, Cr 11.03%, N 17.82%, S 27.20%. $[Cr(SCN)_4(C_6H_5NH_2)_2]H$. Prepared from chrome alum, potassium thiocyanate and aniline: Bergmann, *J. Biol. Chem.* **110**, 471 (1935).

Ammonium salt sesquihydrate, $C_{16}H_{18}CrN_7S_4$, *ammonium rhodanilate*. Rose-colored platelets. Sol in methanol.

USE: In the separation of amino acids, *cf.* Reinecke Salt.

8184. Rhodanine. *2-Thioxo-4-thiazolidinone;* 2-thio-4-ketothiazolidine; 4-oxo-2-thionothiazolidine; rhodanic acid. $C_3H_3NOS_2$; mol wt 133.18. C 27.05%, H 2.27%, N 10.52%, O 12.01%, S 48.14%. Prepd by the action of sodium chloroacetate on ammonium dithiocarbamate: Julian, Sturgis, *J. Am. Chem. Soc.* **57**, 1126 (1935); Redemann *et al., Org. Syn.* **coll. vol. III**, 763 (1955).

Pale yellow crystals (long blades) from glacial acetic acid or water. d 0.868. mp 168.5° (capillary); mp 170.5-171° (hot stage). May explode on rapid heating. K at 25° = 3 × 10^{-6}. Freely sol in boiling water, in methanol, ethanol, DMF, ether, alkalies, ammonia, hot acetic acid.

USE: In the synthesis of phenylalanine.

8185. Rhodinal. *3,7-Dimethyl-7-octenal;* α-citronellal. $C_{10}H_{18}O$; mol wt 154.24. C 77.86%, H 11.76%, O 10.37%. An isomer of citronellal. For isoln and structure *see* citronellal.

Liquid. bp$_{14}$ 51°. n_D^{20} 1.4410. $[\alpha]_D^{25}$ +9.75°. *Caution:* Mild irritant.

8186. Rhodinol. *S-(−)-3,7-Dimethyl-7-octen-1-ol;* α-citronellol. $C_{10}H_{20}O$; mol wt 156.26. C 76.86%, H 12.90%, O 10.24%. Name that has been used to describe a group of related chemicals as well as a single chemical. The term, at times, was applied to what was considered the most important alcohol in geranium oil, *l-citronellol. See:* Barbier, Bouveault, *Ber.* **29**, 289 (1896). Prepn in pure form: Sutherland, *J. Am. Chem. Soc.* **73**, 2385 (1951). Synthesis from *cis*-pinane: Rienäcker, *Chimia* **27**, 97 (1973).

Oily liquid, odor of rose. d_4^{20} 0.8549. bp_{12} 114-115°. n_D^{20} 1.4556. $[\alpha]_D^{20}$ —2.88°. uv max: 186-189 nm (ε 9000). Very slightly sol in water, miscible with alcohol, ether.

USE: In perfumery. *See also* β-Citronellol.

8187. Rhodium. Rh; at. wt 102.9055; at. no. 45; valences 1-6; most common states 1, 3. One naturally occurring isotope: 103; artificial radioactive isotopes: 97-102; 104-110. Belongs to the platinum group of metals. One of the rarest elements: constitutes about $1 \times 10^{-7}\%$ of the earth's crust; found in small quantities associated with all native platinum; in the minerals rhodite, sperrylite, iridosmine; in some nickel-copper ores. Discovered in 1803 by Wollaston, *Phil. Trans.* **94**, 419 (1804). Prepn: Vauquelin, etc., cited by Mellor, *A Comprehensive Treatise on Inorganic and Theoretical Chemistry* **15**, 546 (1936). Practical methods of separation: Wichers, Gilchrist, *Trans. Am. Inst. Mining Metallurg. Eng.* **76**, 619 (1928). Reviews of prepn, properties and chemistry of rhodium and other platinum metals: Gilchrist, *Chem. Rev.* **32**, 277-372 (1943); Beamish *et al.*, in *Rare Metals Handbook*, C. A. Hampel, Ed. (Reinhold, New York, 1956) pp 291-328; W. P. Griffith, *The Chemistry of the Rarer Platinum Metals* (John Wiley, New York, 1967) pp 1-41, 313-430; Livingstone in *Comprehensive Inorganic Chemistry*, Vol. 3, J. C. Bailar Jr. *et al.*, Eds. (Pergamon Press, Oxford, 1973) pp 1163-1189, 1233-1253.

Silvery-white, soft, ductile, malleable metal; face-centered cubic structure. mp 1966°: Roeser, Wensel, *Nat. Bur. Stand. J. Res.* **12**, 519 (1934); d^{20} 12.41. Electr resistivity (0°) 4.51 μ-ohms-cm. Brinell hardness: 100. Not attacked by acids, even aqua regia when in compact form; the finely divided metal reacts with aqua regia. Absorbs oxygen when melted; at a red heat is slowly oxidized to the sesquioxide. Converted to the trihalide by chlorine or bromine at a red heat; not attacked by fluorine.

Human Toxicity: No toxic effects have been reported in exptl animals or in man, E. Browning, *Toxicity of Industrial Metals* (Appleton-Century-Crofts, New York, 1969) pp 278-279.

USE: As an alloy with platinum; as a corrosion-resistant electroplate for protecting silverware from tarnishing; for making high-reflectivity mirrors for cinema projectors, searchlights. Spongy or black rhodium is used as a catalyst in various organic hydrogenation and oxidation reactions.

8188. Rhodium Carbonyl Chloride. *Tetracarbonyldi-μ-chlorodirhodium;* dichlorotetracarbonylrhodium; chlorodicarbonylrhodium(I) dimer. $C_4Cl_2O_4Rh_2$; mol wt 388.75. C 12.36%, Cl 18.24%, O 16.46%, Rh 52.94%. $[Rh(CO)_2Cl]_2$. Prepn: Hieber, Lagally, *Z. Anorg. Allgem. Chem.* **251**, 96 (1943); McCleverty, Wilkinson, *Inorg. Syn.* **8**, 211 (1966); Colton *et al.*, *Aust. J. Chem.* **23**, 1351 (1970); Cramer, *Inorg. Syn.* **15**, 14 (1974). Structure: Dahl *et al.*, *J. Am. Chem. Soc.* **83**, 1761 (1961).

Orange-red crystalline solid or red crystalline sublimate. mp 124-125°. Sol in most organic solvents (except aliphatic hydrocarbons). Stable to dry air. Solutions in organic solvents decompose when exposed to air.

USE: Catalyst.

8189. Rhodium Chloride. Rhodium trichloride. Cl_3Rh; mol wt 209.28. Cl 50.85%, Rh 49.17%. $RhCl_3$.

Red powder. Insol in water; sol in alkali hydroxide or cyanide solns. LD_{50} in rats: 198 mg/kg i.v., Landolt *et al.*, *Toxicol. Appl. Pharmacol.* **21**, 589 (1972).

The chloride is obtainable also as a water-sol hydrate with variable amounts of water. Prepn of trihydrate: Anderson, Basolo, *Inorg. Syn.* **7**, 214 (1963). Similar to the chlorides of other platinum group metals, rhodium chloride readily forms double salts with alkali chlorides.

8190. Rhododendrin. *3-(4-Hydroxyphenyl)-1-methyl-*

propyl β-D-*glucopyranoside;* betuloside. $C_{16}H_{24}O_7$; mol wt 328.35. C 58.52%, H 7.37%, O 34.11%. From leaves of *Rhododendron* spp., *Ericaceae:* Archangelski, *Arch. Exp. Pathol. Pharmakol.* **46**, 313 (1901); Kawaguchi *et al.*, *J. Pharm. Soc. Japan* **62**, 4 (1942), *C.A.* **44**, 9634a (1950). Structure: Kim, *ibid.* 455, *C.A.* **45**, 4222h (1951). Identity with betuloside: Kim, *ibid.* **63**, 103 (1943), *C.A.* **45**, 4222i (1951). Chemotaxonomic significance: Thieme, Winkler, *Pharmazie* **24**, 703 (1969).

Bitter crystals, mp 187°. Sol in hot water, alcohol; slightly sol in chloroform, ether.

Pentaacetate, needles, mp 96-97°.

Methyl ether, prisms from warm water, mp 102-103°. $[\alpha]_D^{21}$ —17.1° (96% alcohol).

USE: In systematic classification of *Rhododendron* species.

8191. Rhodomycins. Antibiotic substance produced by *Streptomyces purpurascens* nov. sp.: Brockmann, Bauer, *Naturwiss.* **37**, 492 (1950); Brockmann *et al.*, *Ber.* **84**, 700 (1951); **Brit.** pat. 708,749 (1954 to Bayer). Separation of rhodomycin A and rhodomycin B: Brockmann, Patt, *Ber.* **88**, 1455 (1955). Isoln and structural elucidation of additional rhodomycins produced by *Streptomyces stammes:* Biedermann, Brauniger, *Pharmazie* **27**, 782 (1972); Brockmann *et al.*, *Tetrahedron Letters* **1973**, 3699. Total synthesis of racemic rhodomycins (the aglycones of the rhodomycins): A. S. Kende, Y.-G. Tsay, *Chem. Commun.* **1977**, 140.

Rhodomycin A

Rhodomycin A, $C_{36}H_{48}N_2O_{12}$, β-*rhodomycin II.* Dec on mild acidic hydrolysis into 1 mole β-rhodomycinone and 2 moles rhodosamine: Brockmann, Spohler, *Naturwiss.* **48**, 716 (1961). Structure of β-rhodomycinone: Brockmann, Niemeyer, *ibid.* 570. Structure of rhodosamine: Brockmann, Spohler, *ibid.* **42**, 154 (1955). Total structure: Brockmann *et al.*, *Tetrahedron Letters* **1969**, 415.

Dihydrochloride, $C_{36}H_{48}N_2O_{12} \cdot 2HCl$, red prisms from ethanol+ isopropanol, mp 205°. Freely sol in water, lower alcohols; sparingly sol in benzene, chloroform, practically insol in ether, petr ether.

Perchlorate, $C_{36}H_{48}N_2O_{12} \cdot 2HClO_4$, minute red needles, mp 188°. Freely sol in methanol, acetone; sparingly sol in water, butanol.

Rhodomycin B, $C_{28}H_{31}NO_9$. Dec on mild acidic hydrolysis into 1 mole β-rhodomycinone and 1 mole rhodosamine: Brockmann, Spohler, *ibid.* **48**, 716 (1961).

Hydrochloride, red prisms, mp 180°. Freely sol in lower alcohols; sparingly sol in water, acetone; practically insol in chloroform, benzene.

8192. Rhodopin. *1,2-Dihydro-1-hydroxy-ψ,ψ-carotene; 1,2-dihydro-1-hydroxylycopene.* $C_{40}H_{58}O$; mol wt 554.86. C 86.58%, H 10.54%, O 2.88%. Pigment from *Rhodovibrio* and

Thiocystis bacteria: Karrer, Solmssen, *Helv. Chim. Acta* **18**, 1306 (1935); **19**, 1019 (1936); Karrer *et al., ibid.* **21**, 454 (1938); from *Rhodomicrobium vannielii:* Volk, Pennington, *J. Bacteriol.* **59**, 169 (1950); Ryvarden, Jensen, *Acta Chem. Scand.* **18**, 643 (1964). Structure: Jensen, *ibid.* **13**, 842, 2142 (1959). Synthesis: Bonnett *et al., ibid.* **18**, 1739 (1964); Surmatis *et al., J. Org. Chem.* **31**, 186 (1966). Synthesis of deuterated rhodopin: Johansen, Liaaen-Jensen, *Acta Chem. Scand.* **28B**, 301 (1974).

Aggregates of dark red crystals from carbon disulfide + petr ether, mp 171° (preliminary sintering). Absorption max (chloroform): 521, 486, 453 nm.

8193. Rhodopsin.
Visual purple. Mol wt about 40,000. A photoreceptor protein found in the rods of the retina of the eye. Composed of 11-*cis* retinal, *q.v.*, bound as a protonated Schiff base to a lysine moiety in the apoprotein opsin, *q.v.* Prepn from retinal and opsin: G. Wald, P. K. Brown, *Proc. Natl. Acad. Sci. USA* **36**, 84 (1950); R. Hubbard, G. Wald, *ibid.* **37**, 69 (1951). Purification and amino acid compn of bovine rhodopsin: Shields *et al., Biochim. Biophys. Acta* **147**, 238 (1967). Synthesis of labelled rhodopsin and studies on the active site: Hirtenstein, Akhtar, *Biochem. J.* **119**, 359 (1970). Characterization in synthetic membranes: W. L. Hubbell, *Accounts Chem. Res.* **8**, 85 (1975). Amino acid sequence of bovine rhodopsin: P. A. Hargrave *et al., Biophys. Struct. Mech.* **9**, 235 (1983). Studies on transmembrane molecular structure: E. A. Dratz, P. A. Hargrave, *Trends Biochem. Sci.* **8**, 128 (1983). Isoln and sequence of gene encoding bovine rhodopsin: J. Nathans, D. S. Hogness, *Cell* **34**, 807 (1983); of gene encoding human rhodopsin: *eidem, Proc. Natl. Acad. Sci. USA* **81**, 4851 (1984). Exposure to light initiates the conversion of rhodopsin through a series of distinct intermediates to yield opsin + *trans*-retinal: R. Hubbard, A. Kropf, *Proc. Natl. Acad. Sci. USA* **44**, 130 (1958); R. Hubbard *et al., Nature* **183**, 142 (1959); B. Honig *et al., Proc. Natl. Acad. Sci. USA* **76**, 2503 (1979). Review of visual transduction and bioregulation of absorption maxima: K. Nakanishi, *Pure Appl. Chem.* **57**, 769-776 (1985). *Reviews:* T. G. Ebrey, B. Honig, *Quart. Rev. Biophys.* **8**, 129-184 (1975); D. F. O'Brien, *Science* **218**, 961-966 (1982); A. Maeda, T. Yoshizawa, *Photochem. Photobiol.* **35**, 891-898 (1982); P. S. Zurer, *Chem. & Eng. News* **61**, 24-35 (Nov. 28, 1983). *See also: Methods in Enzymology* **81** and **88**, Lester Packer, Ed. (Academic Press, New York, 1982).

8194. Rhodoquinone.
2-*Amino*-5-(3,7,11,15,19,23,27,-31,35,39-*decamethyl*-2,6,10,14,18,22,26,30,34,38-*tetraconta-decaenyl*)-3-*methoxy*-6-*methyl*-2,5-*cyclohexadiene*-1,4-*dione*; rhodoquinone-10. $C_{58}H_{89}NO_3$; mol wt 848.36. C 82.12%, H 10.57%, N 1.65%, O 5.66%. Isoln from *Rhodospirillum rubrum:* Gloves, Threlfall, *Biochem. J.* **85**, 14P (1962). Structure and synthesis: Moore, Folkers, *J. Am. Chem. Soc.* **87**, 1409 (1965); **88**, 567 (1966); Daves *et al., ibid.* **90**, 5587 (1968).

Violet needles, mp 70-71°. uv max (cyclohexane): 280, 500 nm ($E_{1cm}^{1\%}$ 154, 14).

8195. Rhodoviolascin.
3,3′,4,4′-*Tetradehydro*-1,1′,2,2′-*tetrahydro*-1,1′-*dimethoxy*-ψ,ψ-*carotene*; 3,3′,4,4′-*tetrade-hydro*-1,1′,2,2′-*tetrahydro*-1,1′-*dimethoxylycopene*; spirilloxanthin. $C_{42}H_{60}O_2$; mol wt 596.90. C 84.51%, H 10.13%, O

5.36%. Carotenoid pigment isolated from *Rhodovibrio* and *Thioceptis* bacteria: Karrer, Solmssen, *Helv. Chim. Acta* **18**, 1306 (1935). Identity with spirilloxanthin: Zechmeister *et al., Arch. Biochim.* **5**, 243 (1944). Structure: Karrer, Koenig, *Helv. Chim. Acta* **23**, 460 (1940); Barber *et al., Proc. Chem. Soc.* **1959**, 96. Synthesis: Surmatis, Ofner, *J. Org. Chem.* **28**, 2735 (1963); Surmatis, U.S. pat. **3,160,658** (1964 to Hoffman-La Roche).

Dark red, spindle-shaped crystals from benzene, mp 218°. Absorption max (chloroform): 544, 507, 476 nm. Almost insol in petr ether, ligroin, methanol, somewhat more sol in hot benzene.

8196. Rhodoxanthin.
4′,5′-*Didehydro*-4,5-*retro*-β,β-*carotene*-3,3′-*dione*. $C_{40}H_{50}O_2$; mol wt 562.80. C 85.36%, H 8.95%, O 5.69%. Carotenoid pigment found widely distributed in nature, but in very small amounts only. One of very few carotenoids with retro configuration. Best isolated from the arils of *Taxus baccata* L., *Taxaceae:* Kuhn, Brockmann, *Ber.* **66**, 828 (1933). Occurs also in the feathers of birds such as *Phoenicircus nigricollis, Megaloprepia magnifica.* Isoln: Volker, *Z. Physiol. Chem.* **292**, 75 (1953). Structure: Kuhn, Brockmann, *loc. cit.;* Karrer, Solmssen, *Helv. Chim. Acta* **18**, 477 (1935). Synthesis: Mayer *et al., ibid.* **50**, 1606 (1967); Surmatis, Walser, U.S. pats. **3,466,331; 3,624,105** (1969, 1971 to Hoffmann-La Roche); P. R. Ellis *et al., Helv. Chim. Acta* **64**, 1092 (1981); E. Widmer *et al., ibid.* **65**, 944 (1982). *Review:* Liaaen-Jensen in *Aspects Terpenoid Chem. Biochem., Proc. Phytochem. Soc. Symp.,* 2nd, 1970, T. W. Goodwin, Ed. (Academic Press, London, 1971) p 223-254.

Rosettes of deep purple, lancet-shaped crystals from benzene:methanol (1:4). mp 219° (evacuated tube). Absorption max (chloroform): 546, 510, 482 nm. Freely sol in pyridine; sol in benzene, chloroform. Sparingly sol in ethanol and methanol. Practically insol in hexane, petr ether.

USE: Coloring material for food, beverages, pharmaceuticals and cosmetics.

8197. Rhubarb.
Rhizome and roots of *Rheum officinale* Baill., *R. palmatum* L., or other *Rheum* spp, *Polygonaceae.* *Habit.* Central Asia; cultivated in Europe, Southern Siberia, North America. *Constit.* Chrysophanic acid, emodin, rhein, rheotannic acid, erythroretin, methylchrysophanic acid, rhabarberon, cinnamic and gallic acids, calcium oxalate. Comprehensive studies: Workman, Hiner, *J. Am. Pharm. Assoc.* **49**, 118 (1960). *Incompat.* Mineral acids, iron salts, lead acetate, zinc salts, lime water; infusions of catechu, cinchona, and nutgall; tartar emetic, tannic acid.

THERAP CAT: Cathartic.

THERAP CAT (VET): Has been used as a laxative.

8198. Rhynchophylline.
16,17-*Didehydro*-17-*methoxy*-2-*oxocorynoxan*-16-*carboxylic acid methyl ester;* rhyncophylline; mitrinermine. $C_{22}H_{28}N_2O_4$; mol wt 384.46. C 68.72%, H 7.34%, N 7.29%, O 16.65%. From stems and roots of *Uncaria rynchophylla* Miq. [*Ourouparia rynchophylla* (Miq.)] Matsum, *Rubiaceae:* Kondo *et al., J. Pharm. Soc. Japan* **48**, 54 (1928); from *Mitragyna rotundifolia* (Roxb.), *Rubiaceae:* Barger *et al., J. Org. Chem.* **4**, 418 (1939). Structure: Seaton, Marion, *Can. J. Chem.* **35**, 1102 (1957); **38**, 1035 (1960). Stereochemistry: Nozoye, *Chem. Pharm. Bull.* **6**, 309 (1958); Ban, Oishi, *Tetrahedron Letters* **1961**, 791; *eidem, Chem.*

Pharm. Bull. **11**, 441, 446, 451 (1963). Partial synthesis: Finch, Taylor, J. Am. Chem. Soc. **84**, 1318, 3871 (1962). Total synthesis: Ban et al., Tetrahedron Letters **1972**, 2113.

Crystals from methanol, mp 216°; dl-form reported as colorless pillars from ethyl acetate-ether, mp 197-199° (Ban et al., loc. cit.). $[\alpha]_D^{13}$ — 14.7° (c = 2.5 in chloroform). pKa 6.4. uv max: 245, 280 nm (log ϵ 4.24, 3.15). Sol in chloroform; moderately sol in acetone, alcohol, benzene; sparingly sol in ether, ethyl acetate. Practically insol in petr ether.

8199. Ribavirin. *1-β-D-Ribofuranosyl-1H-1,2,4-triazole-3-carboxamide;* ICN-1229; RTCA; Viramid; Virazid; Virazole. $C_8H_{12}N_4O_5$; mol wt 244.21. C 39.34%, H 4.95%, N 22.94%, O 32.76%. The first synthetic, non-interferon-inducing, broad-spectrum antiviral nucleoside. Preliminary information: Chem. & Eng. News **50**, 26 (April 17, 1972). Synthesis: J. T. Witkowski et al., 163rd Am. Chem. Soc. Meeting (Boston, April 1972), Abstracts of Papers MEDI 19; eidem, J. Med. Chem. **15**, 1150 (1972); eidem, J. Carbohyd. Nucl. Nucl. **5**, 363 (1978). Regioselective synthesis: Y. Ito et al., Tetrahedron Letters **1979**, 2521; R. R. Schmidt, D. Heermann, Ber. **114**, 2825 (1981). Structure and conformation studies: Kreishman et al., J. Am. Chem. Soc. **94**, 5894 (1972); Prusiner, Sundaralingam, Nature New Biol. **244**, 116 (1973). Activity studies: Sidwell et al., Science **177**, 705 (1972); Huffman et al., Antimicrob. Ag. Chemother. **3**, 235 (1973); Sidwell et al., ibid. 242; Khare et al., ibid. 517. In vitro inhibition of HIV-1 (HTLV-III/LAV) virus replication: J. B. McCormick et al., Lancet **2**, 1367 (1984). Toxicity data: J. T. Witkowski et al., J. Med. Chem. **15**, 1150 (1972). Teratogenicity studies: V. H. Ferm et al., Teratology **17**, 93 (1978); D. M. Kochrar et al., Toxicol. Appl. Pharmacol. **52**, 99 (1980). Controlled clinical trial in infants with respiratory syncytial viral infection: C. B. Hall et al., N. Engl. J. Med. **308**, 1443 (1983); in Lassa fever: J. B. McCormick et al., ibid. **314**, 20 (1986). Review: R. Sidewell et al., Pharmacol. Ther. **6**, 123-146 (1979); F. E. Hahn, Ed. in Antibiotics vol. **5**, pt. 2 (Springer-Verlag, New York, 1979) pp 439-458.

Colorless, water-soluble, stable material. Exists in two polymorphic forms: mp 166-168° (aq ethanol); mp 174-176° (ethanol). $[\alpha]_D^{25}$ — 36.5°. LD_{50} i.p. in mice: 1.3 g/kg; orally in rats: 5.3 g/kg (Witkowski).
THERAP CAT: Antiviral.

8200. α-Ribazole. *5,6-Dimethyl-1-α-D-ribofuranosyl-1H-benzimidazole.* $C_{14}H_{18}N_2O_4$; mol wt 278.31. C 60.42%, H 6.52%, N 10.07%, O 22.99%. Nucleoside moiety of vitamin B_{12}, q.v. Isoln by acid hydrolysis of vitamin B_{12}: N. G. Brink et al., J. Am. Chem. Soc. **72**, 1866 (1950); N. G. Brink, K. Folkers, ibid. **74**, 2856 (1952). Syntheses of α- and β-anomers: F. W. Holly et al., ibid. 4521; R. S. Wright et al., ibid. **80**, 2004 (1958); J. D. Stevens et al., J. Org. Chem. **33**, 1806 (1968). Biosynthetic study: J. Hörig, P. Renz, FEBS Letters **80**, 337 (1977). Spectroscopic characterization: K. L. Brown et al., Inorg. Chem. **23**, 1463 (1984).

Crystals from water, mp 198-199°. $[\alpha]_D^{23}$ +14° (c = 0.9 in pyridine).

8201. Riboflavine. Vitamin B_2; lactoflavine; vitamin G; 7,8-dimethyl-10-(D-ribo-2,3,4,5-tetrahydroxypentyl)isoalloxazine; 7,8-dimethyl-10-ribitylisoalloxazine; Beflavine; Flavaxin; Ribipca. $C_{17}H_{20}N_4O_6$; mol wt 376.36. C 54.25%, H 5.36%, N 14.89%, O 25.51%. Nutritional factor found in milk, eggs, malted barley, liver, kidney, heart, leafy vegetables. Richest natural source is yeast. Minute amounts present in all plant and animal cells. Occurs in the free form only in the retina of the eye, in whey, and in urine; principal forms occurring in tissues and cells are *flavine mononucleotide (FMN, riboflavine-5-phosphate)* and *flavin-adenine dinucleotide (FAD)*. First syntheses: Karrer et al., Helv. Chim. Acta **18**, 426, 522 (1935); Kuhn et al., Naturwiss. **23**, 260 (1935). Riboflavin for therapeutic use is produced by synthesis, the most common starting materials being o-xylene, D-ribose, and alloxan. Improved synthetic process: Howe, U.S. pat. **2,807,611** (1957 to Merck & Co.). Additional synthesis: F. Yoneda et al., J. Chem. Soc. Perkin Trans. I, **1978**, 348. Several fermenting organisms, such as Ashbya gossypii and Eremothecium ashbyii, have the capacity to synthesize large quantities of riboflavin. Concentrates for poultry and livestock feeds are produced by the fermentation process. Example of production by microorganisms (Ashbya gossypii): Malzahn et al., U.S. pat. **2,876,169** (1959 to Grain Processing Corp.). Early reviews: Rosenberg in The Vitamins (New York, 1945); Sebrell, Harris, The Vitamins vol. **III** (Academic Press, New York, 1954) pp 299-402; Vogel-Knobloch, Chemie und Technik der Vitamine vol. **II**, part 1 (Enke, Stuttgart, 1955). Review of metabolism: Rivlin, N. Engl. J. Med. **283**, 463 (1970). Review of extraction and assay methods: Pearson, The Vitamins vol. **VII**, P. György, W. N. Pearson, Eds. (Academic Press, New York, 2nd ed., 1967) pp 99-136. Comprehensive review: ibid. vol. **V**, pp 1-96 (1972).

Fine orange-yellow needles from 2N acetic acid, alcohol, water, or pyridine. Dec at 278-282° (darkens at about 240°). Three different crystal forms having different solubilities in water: Dale, U.S. pats. **2,603,633** and **2,797,215.** When dry, it is not appreciably affected by diffused light, but in alkaline soln it deteriorates quite rapidly, the deterioration being accelerated by light. $[\alpha]_D^{25}$ — 112° to — 122° (50 mg in 2 ml 0.1N alcoholic NaOH dil to 10 ml with water). Absorption max: 220-225, 266, 371, 444, 475 nm. One gram dissolves in from 3000 to about 15,000 ml of water, the variations in the soly being due to differences in the crystal structure. Slightly more sol in NaCl solns; less sol in alcohol than in water (0.0045 g/100 ml of abs ethanol at 27.5°), also slightly sol in cyclohexanol, amyl acetate, benzyl alcohol, phenol; insol in ether, chloroform, acetone, benzene; very sol in dil

Consult the cross index before using this section.

alkalies (with decompn). Neutral to litmus. pKa 10.2; pKb
1.7. pH of satd aq soln about 6. Isoelectric pt: pH 6. Aq
solns are yellow showing a green fluoresence with max at
565 nm. Optimal fluorescence is at pH 6-7.

Although sensitive to alkalies, riboflavine is stable to mineral acids in the dark. At 27° the decompn of acidic solns
buffered to maintain the pH at 5.0 was observed to take
place at the rate of 1.2% per month. Visible or ultraviolet
irradiation of alkaline solns causes the formation of lumiflavine, q.v., whereas irradiation of acid or neutral solns gives
rise to the production of the blue, fluorescent substance
lumichrome, q.v., together with varying amounts of lumiflavine. Prepn of forms of riboflavine having increased soly in
water without loss of physiological activity: Stecher, U.S.
pat. 2,480,517 (1949 to Merck & Co.). See also Riboflavine
Phosphate (Sodium). Practically nontoxic. LD$_{50}$ in mice,
rats: 340, 560 mg/kg i.p., Kuhn, Boulanger, Z. Physiol.
Chem. 241, 233 (1936); Unna, Greslin, J. Pharmacol. Exp.
Ther. 76, 75 (1942).

Recommended daily allowance: sedentary 70 kg male:
1.7 mg (National Research Council).

Tetrabutyrate, C$_{33}$H$_{44}$N$_4$O$_{10}$, Bituvitan, Eyekas, Hibon,
Lacflavin, Riboract, Ribovis, Wakaflavin L, Viras.

THERAP CAT: Enzyme co-factor vitamin.

THERAP CAT (VET): Nutritional factor for all species; horses
and ruminants normally need no dietary sources.

8202. Riboflavine Phosphate (Sodium). Riboflavin
5'-(dihydrogen phosphate) monosodium salt; vitamin B$_2$
phosphate (sodium salt); flavine mononucleotide; riboflavine
5'-phosphate ester monosodium salt; cytoflav; coflavinase;
alloxazine mononucleotide; A.M.N.; Hyryl; Ribo. C$_{17}$H$_{20}$-
N$_4$NaO$_9$P; mol wt 478.34. C 42.69%, H 4.22%, N 11.71%,
Na 4.81%, O 30.10%, P 6.47%. One gram equals 0.730 g
riboflavine. Prepd by phosphorylation of riboflavine with
chlorophosphoric acid: Flexser, Farkas, XIIth International
Congress of Pure and Applied Chemistry (New York, 1951)
Abstracts of Papers, p 71; U.S. pats. 2,610,178-9 (both 1952
to Hoffmann-La Roche); with pyrophosphoric acid: Brei-
vogel, Ridge, U.S. pat. 2,535,385 (1950 to White Labs.);
with metaphosphoric acid: Viscontini et al., Helv. Chim.
Acta 35, 457 (1952); U.S. pat. 2,740,775 (1956 to Hoffmann-
La Roche); with pyrocatechol cyclic phosphate: Ukita et al.,
U.S. pat. 3,118,876 (1964 to Takeda).

Dihydrate, yellow crystals. Soly in water at pH 6.9 = 112
mg/ml; at pH 5.6 = 68 mg/ml; at pH 3.8 = 43 mg/ml.
The commercial product may have a pH of 5.0 to 6.0 because of a small amount of the disodium salt. Theoretical
pH of aq soln (1 millimole/150 ml H$_2$O) is 4.5. Fully active
biologically, microbiologically, and enzymatically. The
greater sensitivity of the phosphate ester to destruction by
ultraviolet light necessitates careful protection of dil solns
from exposure: Sleezer et al., Drug & Cosmet. Ind. 74, 196
(1954).

THERAP CAT: Enzyme co-factor vitamin.

THERAP CAT (VET): Nutritional factor.

8203. Ribonuclease. RNase. An enzyme that digests
RNA. First isolated from beef pancreas in 1920. Crystallization from acid extracts of pancreas: Kunitz, J. Gen. Physiol. 24, 15 (1940). Isoln: Kunitz, McDonald, Biochem.
Prepn. 3, 9 (1953). Bovine pancreatic RNase A is a single-
chain peptide of 124 amino acid residues. Primary structure:
C. H. W. Hirs et al., J. Biol. Chem. 235, 633 (1960). Amino
acid sequence: D. H. Spackman et al., ibid. 648. Pictorial
representation of entire structure: Stein, Moore, Sci. Am.
204, no. 2, 81-92 (Feb. 1961). Ribonuclease from plant
leaves has slightly different characteristics: Markham,
Strominger, Biochem. J. 64, 46P (1956). Can be obtained as
a by-product of microbial erythromycin production: Japan.
pat. 26,938('63) (to Shionogi). Chemical synthesis of materials possessing partial RNase enzyme activity: Gutte,
Merrifield, J. Am. Chem. Soc. 91, 501 (1969); Denkewalter,
Hirschmann et al., ibid. 502. Series of articles describing the
total synthesis of a protein having the full enzymic activity
of bovine pancreatic RNase: N. Fujii, H. Yajima, J. Chem.
Soc. Perkin Trans. I 1981, 789-841. Specifically catalyzes
the cleavage of the phosphodiester bond between the 3' and
5' positions of the ribose moieties in RNA with the formation of oligonucleotides terminating in 2',3'-cyclic phos-

phate derivs: Roberts et al., Proc. Nat. Acad. Sci. USA 62,
1151 (1969). Reviews: Anfinsen, White, The Enzymes, vol.
5, P. D. Boyer et al., Eds. (Academic Press, New York, 2nd
ed., 1961) pp 95-122; Richards, Wyckoff, ibid. vol. 4 (3rd
ed., 1971) pp 647-806.

Crystals. [α]$_D^{25}$ (per mg N) −0.47° (c = 5). uv max (0.1M
KCl soln): 277.5 nm (ε 9700). Isoelec pt about pH 8.0.
Freely sol in water. Aq solns of cryst ribonuclease are quite
stable at temps below 25°. The region of maximum stability
is between pH 2 and 4.5. Stable for years stored as refrigerated dry powder or in frozen soln. Aggregates on lyophil-
ization and storage. Shows affinity for glass surfaces. The
optimum temp for digestion of yeast ribonucleic acid is 65°.
The optimum pH is 7.7. Inhibited by heavy metal ions; by
magnesium ions at concns as low as 0.0005M. Competitively inhibited by DNA, the effect of denatured DNA being
greater than that of the native nucleic acid. Ribonuclease is
precipitated by trichloroacetic acid, and does not diffuse
through collodion or cellophane.

8204. Ribonucleic Acid. RNA; yeast nucleic acid. Po-
lynucleotide directly involved in protein synthesis; found in
both the nucleus and the cytoplasm of cells. Description of
components of RNA: see Nucleic Acids. The four primary
nucleosides are adenosine, guanosine, cytidine and uridine;
minor nucleosides are also found. The nucleosides are
linked by phosphate diester bonds from the 3'-hydroxyl of
one D-ribose to the 5'-hydroxyl of the next. The secondary
structure of RNA is that of an incompletely organized sin-
gle-stranded polynucleotide consisting of some areas with
helical structure alternating with nonhelical lengths. Compare Deoxyribonucleic Acid (DNA). Structure: Brown,
Todd in The Nucleic Acids vol. 1, E. Chargaff, J. N. Davidson, Eds. (Academic Press, NewYork, 1955) pp 409-439;
Spirin, Progr. Nucleic Acid Res. 1, 301-345 (1963); J. N.
Davidson, The Biochemistry of Nucleic Acids (Academic
Press, New York, 7th ed., 1972) pp 106-128. Review of
NMR studies: B. R. Reid, Ann. Rev. Biochem. 50, 969-996
(1981). Book: The Ribonucleic Acids, P. R. Stewart, D. S.
Letham, Eds. (Springer-Verlag, New York, 2nd ed., 1977).
Comprehensive reviews and monographs: see Nucleic Acids.

Several types of RNA have been identified. Ribosomal
RNA or rRNA; about 80% of the RNA found in cells; a metabolically stable form; an important component of ribosomes (q.v.) which play a central role in protein synthesis.
Two rRNA species of high molecular weight have been isolated from ribosomes: mol wts approx 0.6 × 10^6 and 1.1 ×
10^6 in bacterial cells; higher in eukaryotic cells. In addition,
at least one rRNA species of low mol wt, 5S RNA, has been
identified; approx 120 nucleotides. Reviews: Attardi, Amal-
di, Ann. Rev. Biochem. 39, 183-226; Möller, Garrett, "The
Ribonucleic Acids of the Bacterial Ribosome" in Protein
Synthesis vol. 1, E. H. McConkey, Ed. (Dekker, New York,
1971) pp 229-272; Craig, "Ribosomal RNA Synthesis in
Eukaryotes and its Regulation" in MTP Int. Rev. Sci.: Biochem., Ser. One vol. 6, K. Burton, Ed. (University Park
Press, Baltimore, 1974) pp 255-288.

Messenger RNA, mRNA, informational RNA. A short-
lived form of high molecular weight; acts as template for
protein synthesis in the cell; complementary to one strand of
DNA. A linear relationship exists between the sequence of
amino acids in a polypeptide and the sequence of nucleotides
in the corresponding mRNA and DNA. The genetic code
has been proposed to explain the necessary translation process. The four purine and pyrimidine bases are treated as
letters which can be combined to form 3-letter words or
codons; 4^3 or 64 codons can be formed. 61 have been found
to code for specific amino acids; the remaining three are
probably chain termination codons. Reviews of mRNA and
the genetic code: Lipmann, "Messenger Ribonucleic Acid"
in Progr. Nucleic Acid Res. 1, 135-161 (1963); Crick, "The
Recent Excitement in the Coding Problem" ibid., pp 163-
217; Woese, "The Present Status of the Genetic Code" in
Progr. Nucleic Acid Res. Mol. Biol. 7, 107-172 (1967); Hadji-
olov, "Ribonucleic Acids and Information Transfer in Animal Cells" ibid. 195-242; Jukes, Gatlin, "Recent Studies
Concerning the Coding Mechanism" ibid. 11, 303-350
(1971); J. N. Davidson, The Biochemistry of the Nucleic Acids
(Academic Press, New York, 7th ed., 1972) pp 290-383;
Mathews, "Mammalian Messenger RNA" in Essays in Bio-

chemistry vol. 9, P. N. Campbell, F. Dickens, Eds. (Academic Press, New York, 1973) pp 59-102; Brawerman, *Ann. Rev. Biochem.* **43**, 621-642 (1974); Watts, Watts, "The Genetic Code" in *MTP Int. Rev. Sci.: Biochem.*, *Ser. One* vol. 7, H. R. V. Arnstein, Ed. (University Park Press, Baltimore, 1975) pp 255-294.

Transfer RNA, tRNA, soluble RNA, sRNA. Low mol wt: 23,000-27,000; approx 75-85 nucleotides. Each tRNA is specific for and binds with a particular amino acid; more than one may exist for each amino acid. Performs three functions during protein synthesis: binds with its specific amino acid; recognizes the corresponding codon on mRNA and places the amino acid in the correct position for attachment to the polypeptide chain being formed; binds the polypeptide to the ribosome. First determination of total structure of a transfer RNA (yeast alanine tRNA): Holley *et al.*, *Science* **147**, 1462 (1965). Reviews of structure and function: Miura, "Specificity in the Structure of Transfer RNA" in *Progr. Nucleic Acid Res. Mol. Biol.* **6**, 39-82 (1967); Cramer, "Three-Dimensional Structure of tRNA", *ibid.* **11**, 391-421 (1971); *Nucleic Acid Sequence Analysis*, S. Mandeles (Columbia University Press, New York, 1972) pp 256-280; Nishimura, "Transfer RNA: Structure and Biosynthesis" in *MTP Int. Rev. Sci.: Biochem.*, *Ser. One* vol. 6, K. Burton, Ed. (University Park Press, Baltimore, 1974) pp 289-322; A. Rich, V. L. Raj Bhandary, *Ann. Rev. Biochem.* **45**, 805-860 (1976); P. F. Agris, *The Modified Nucleosides of Transfer RNA, II* (A. R. Liss, New York, 1983) 220 pp.

8205. D-Ribose. $C_5H_{10}O_5$; mol wt 150.13. C 40.00%, H 6.71%, O 53.29%. Prepd by hydrolysis of yeast-nucleic acid: Levene, Jacobs, *Ber.* **42**, 1201, 3247 (1909); Levene, Clark, *J. Biol. Chem.* **46**, 19 (1921); Bredereck, *Ber.* **71**, 408 (1938); Bredereck *et al.*, *ibid.* **73**, 956 (1940); Phelps, U.S. pat. **2,152,662** (1939 to U.S. Gov't); by ion-exchange resin chromatography: Cohn, *Science* **109**, 377 (1949); *J. Am. Chem. Soc.* **71**, 2275 (1949); **72**, 1471 (1950). From glucose: Karrer, *Helv. Chim. Acta* **18**, 1435 (1935); Austin, Humoller, *J. Am. Chem. Soc.* **56**, 1152 (1934); Kuhn *et al.*, *Ber.* **68**, 1765 (1935); from nucleosides: Laufer, Charney, U.S. pats. **2,379,913-4** (1945 to Schwarz Labs.); from D-erythrose: Sowden, *J. Am. Chem. Soc.* **72**, 808 (1950); from L-glutamic acid: Koga *et al.*, *Tetrahedron Letters* **1971**, 263. Reduction of D-ribonic acid: van Ekenstein, Blanksma, *Chem. Zentr.* **1913**, II, 1562; Steiger, *Helv. Chim. Acta* **19**, 189 (1936). Review: Overend, Stacey, in *Nucleic Acids* vol. I, E. Chargaff, J. N. Davidson, Eds. (Academic Press, New York, 1955) pp 1-80.

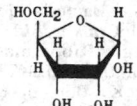

α-D-ribofuranose

Plates from abs alcohol, mp 87°. Shows complex mutarotation: Phelps *et al.*, *J. Am. Chem. Soc.* **56**, 748 (1934). Final $[\alpha]_D^{24}$ −25° (water). Sol in water, slightly sol in alc.

Phenylosazone, $C_{17}H_{20}N_4O_3$, yellow needles from pyridine + water, mp 163-164°.

Methyl-D-riboside, crystals from ethyl acetate, mp 83-84°. $[\alpha]_D^{20}$ −113.6° (p = 3), Minsaas, *Ann.* **512**, 286 (1934).

8206. D-Ribose-5-phosphoric Acid. *5-(Dihydrogen phosphate)*-D-*ribose*; D-ribofuranose-5-phosphoric acid; D-ribose-5-phosphate. $C_5H_{11}O_8P$; mol wt 230.12. C 26.10%, H 4.82%, O 55.62%, P 13.46%. From inosinic acid: Levene, Jacobs, *Ber.* **41**, 2703 (1908); **44**, 746 (1911); Levene, Mori, *J. Biol. Chem.* **81**, 215 (1929); Levene, Stiller, *ibid.* **104**, 299 (1934); LePage, Umbreit, *ibid.* **148**, 255 (1943); Marmur *et al.*, *Arch. Biochem. Biophys.* **34**, 209 (1951). Purification: Groth *et al.*, *ibid.* **199**, 398 (1952). Differentiation between ribose-5-phosphate and ribose-3-phosphate by means of the orcinol-pentose reaction: Albaum, Umbreit, *J. Biol. Chem.* **167**, 369 (1947).

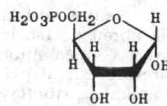

Ba salt hemihendecahydrate, $C_5H_9BaO_8P.5\frac{1}{2}H_2O$, hexagonal plates from water. Sparingly soluble in cold water. Reduces Fehling's soln.

8207. Ribosome. Small, irregularly shaped particle found in all living cells; made up of protein and ribosomal RNA molecules. Molecular weight of bacterial (70 S) ribosome: 2.7×10^6; of ribosome found in yeast, higher plants and animals (80 S): $\sim 4 \times 10^6$. Each ribosome consists of one large and one small subunit: 50 S and 30 S subunits in bacteria; 60 S and 40 S subunits in higher plants and animals. The *polyribosome* or *polysome*, a cluster of two or more ribosomes associated with a strand of messenger RNA, is the active unit in protein synthesis in living cells. The ribosomes move along the mRNA strand while engaged in the process of mfg proteins according to the instructions of mRNA. *See also* Ribonucleic Acid. Reviews of properties, structure and function: M. L. Petermann, *The Physical and Chemical Properties of Ribosomes* (Elsevier, New York, 1964) 258 pp; Schlessinger, *Bacteriol. Rev.* **33**, 445-453 (1969); Nomura, *Science* **179**, 864-873 (1973); *Cold Springs Harbor Monograph Series: Ribosomes*, M. Nomura *et al.*, (1974) 930 pp; Cox, Godwin, "Ribosome Structure and Function" in *MTP Int. Rev. Sci.: Biochem.*, *Ser. One* vol. 7, H. R. V. Arnstein, Ed. (University Park Press, Baltimore, 1975) pp 179-253; C. G. Kurland, *Ann. Rev. Biochem.* **46**, 173-200 (1977); H. G. Wittmann, *ibid.* **51**, 155-183 (1982); A. Liljas, *Prog. Biophys. Mol. Biol.* **40**, 161-228 (1982).

8208. Ribostamycin. *O-2,6-Diamino-2,6-dideoxy-α-*D-*glucopyranosyl-(1 → 4)-O-[β-*D-*ribofuranosyl-(1 → 5)]-2-deoxy-*D-*streptamine*; SF 733 antibiotic. $C_{17}H_{34}N_4O_{10}$; mol wt 454.49. C 44.93%, H 7.54%, N 12.33%, O 35.20%. Aminocyclitol antibiotic belonging to the neomycin, *q.v.*, group of antibiotics. Produced by *Streptomyces ribosidificus* (formerly called *S. thermoflavus*). Taxonomy, isoln and toxicity data: T. Shomura *et al.*, *J. Antibiot.* **23**, 155 (1970). Prepn: *eidem*, Ger. pat. **1,814,735** corresp to U.S. pats. **3,661,892, 3,799,-842** (1969, 1972, 1974 all to Meiji). Structure: E. Akita *et al.*, *J. Antibiot.* **23**, 173 (1970). Synthesis from neamine, *q.v.*: Ito *et al.*, *Antimicrob. Ag. Chemother.* **1970**, 33; *eidem*, *Agr. Biol. Chem.* **34**, 980 (1970); V. Kumar, W. A. Remers, *J. Org. Chem.* **43**, 3327 (1978); **46**, 4298 (1981). Total synthesis: H. Fukami *et al.*, *Tetrahedron Letters* **1976**, 545; *eidem*, *Agr. Biol. Chem.* **41**, 1689 (1977). *In vitro* antibacterial activity: E. Yourassowsky, M. P. Vander Linden, *Arzneimittel-Forsch.* **26**, 184 (1976).

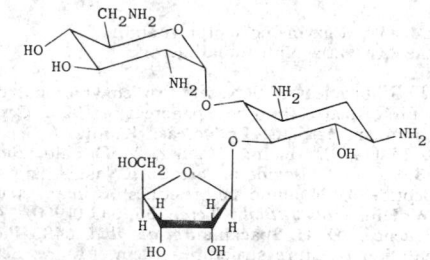

Colorless needles from methanol, mp 192-195°. [Also reported as the monohydrate from water-methanol-ethanol, mp 175-180° (dec)]. $[\alpha]_D^{23}$ +42°. pKa' 7.70. Stable in neutral and alkaline solns; slightly unstable in acidic media. Sol in water; slightly sol in methanol. Practically insol in acetone, *n*-butanol, ethyl acetate, benzene, hexane, ether.

Sulfate, *Ibistacin, Landamycine, Ribostamin, Ribomycine, Vistamycin*. White or yellowish powder. $[\alpha]_D^{20}$ +39° (c = 1). LD_{50} i.v. in mice: 225 mg/kg (Shomura).

THERAP CAT: Antibacterial.

8209. D-Ribulose. D-*erythro-2-Pentulose*; D-adonose; D-*erythro-2*-ketopentose. $C_5H_{10}O_5$; mol wt 150.13. C

40.00%, H 6.71%, O 53.29%. Prepn from D-arabinose: Glatthaar, Reichstein, *Helv. Chim. Acta* **18**, 80 (1935); from 2-keto-D-gluconic acid: Hall *et al., Biochem. J.* **60**, 271 (1955); from D-fructose: P. M. Collins *et al., J. Chem. Soc. Perkin Trans. I* **1980**, 277. Stereoselective synthesis: M. Yamaguchi, T. Mukaiyama, *Chem. Letters* **1981**, 1005; K. Suzuki *et al., ibid.* 1529.

$$CH_2OH$$
$$|$$
$$CO$$
$$|$$
$$HCOH$$
$$|$$
$$HCOH$$
$$|$$
$$CH_2OH$$

Sweet syrup. $[\alpha]_D^{24}$ −15° (c = 0.5 in water).
Diphosphate, $C_5H_{12}O_{11}P_2$. Prepn from D-ribose-5-phosphate: Horecker *et al., Biochem. Prep.* **6**, 83 (1958).

8210. Rice Bran Oil. Oil of rice bran; rice oil. Extracted from bran obtained in making white or polished rice (which is actually the endosperm of the seed of *Oryza sativa* L., *Gramineae*). Rice bran separated by milling of hulled rice consists of bran, aleurone layer and germ, and contains about 15% oil. The oil is usually obtained by solvent extraction, see J. R. Loeb, N. J. Morris: *Abstract Bibliography of the Chemistry, Processing, and Utilization of Rice Bran and Rice Bran Oil,* U.S. Dept. of Agriculture, Ms. no AIC-328 (1952 and suppl.). Rice bran contains an unusually active lipase which raises the free fatty acid content on storage. A variety of methods to inhibit this rise can be found in the literature: Williams *et al., J. Am. Oil Chem. Soc.* **42**, 151 (1965). The free fatty acid content of the oil should not exceed 4 to 5%, but may run as high as 70%. Acceptable rice bran oil contains 15-20% of satd and 80-85% unsatd fatty acids as glycerides: Myristic 0.4-1%, palmitic 12-18%, stearic 1-3%, C_{20}-C_{22} satd 1%, oleic 40-50%, linoleic 29-42%, linolenic trace to 1%, palmitoleic 0.2-0.4%. *See also* Sreenivasan, *J. Am. Oil Chem. Soc.* **45**, 263 (1968).

Golden yellow oil, difficult to bleach. Not affected by temporary heating to 160°. d_{25}^{25} 0.916-0.921. n_D^{25} 1.470-1.473; n_D^{40} 1.465-1.468. Titer (solidification pt) 24-28. Saponification value 181-189. Acid value 4-120 (see above comments on the lipase content of the bran). Iodine value 99-108. Thiocyanogen value 69-76. Hydroxyl value 5-14. Reichert-Meissl value <0.5. Polenské value <0.5. Hehner value 95.3. Unsaponifiable matter 3-5%. The high refractive index is ascribed to the presence of squalene. Also contains other antioxidants such as tocopherols. Miscible with hexane and other fat solvents.

USE: Soaps, hydrogenated shortenings. Edible oil, also suitable for cosmetics and pharmaceuticals when low in free fatty acids and all the lipase has been deactivated.

8211. Ricin. A toxic lectin and hemagglutinin isolated from castor bean, *Ricinus communis* L., *Euphorbiaceae:* Stillmark, *Arb. Pharmak. Inst. Dorpat* **1889**, 59; T. Osborne *et al., Am. J. Physiol.* **14**, 259 (1905); H. L. Craig *et al.,* U.S. pat. **3,060,165** (1962 to U.S. Dept. Army). Purification of toxic ricin protein: M. Ishiguro *et al., J. Biochem.* **55**, 587 (1964). The hemagglutinating activity of ricin was initially believed to be the cause of its high toxicity, but later studies have shown that separate proteins are responsible for the toxicity and hemagglutination: S. Olsnes, A. Pihl, *Biochemistry* **12**, 3121 (1973). "Ricin", the original term for the mixed extract, is now used in various ways. Two ricin agglutinins and two toxins have been identified: *RCL I, RCL II* (agglutinins), *RCL III* or *Ricin D* and *RCL IV* (toxins), *cf.* T. T. Lin, S. S. Li, *Eur. J. Biochem.* **105**, 453 (1980). All four lectins consist of two different polypeptide chains joined by a disulfide bond; the toxins are dimers of an A-chain (30,000 Da) and a B-chain (33,000 Da) and the agglutinins occur as a tetramer composed of two 30,000 and two 33,000 mol wt subunits. Ricin has been shown to have anti-tumor properties: J. Y. Lin *et al., Nature* **227**, 92 (1970); P. E. Thorp *et al., ibid.* **297**, 594 (1982). Synergistic effect with daunorubicin, cisplatin, and vincristine, *q.q.v.*, in systemic L 1210 leukemia: O. Fodstad, A. Pihl, *Cancer Res.*

42, 2152 (1982). Conjugates of ricin and cell-binding antigens or antibodies, called immunotoxins, have been used in cancer therapy: E. S. Vitetta *et al., Science* **219**, 644 (1983). Experimental study with ricin-A chain conjugate in AIDS: M. A. Till *et al., Science* **242**, 1166 (1988). Radioimmunoassay in blood: A. Godal *et al., J. Toxicol. Environ. Health* **8**, 409 (1981). *Reviews:* M. Funatsu, "The Structure and Toxic Function of Ricin" in *Proteins, Structure and Function* vol. 2, M. Funatsu *et al.,* Eds. (Kodansha, Tokyo, Wiley, New York, 1972) pp 103-139; G. A. Balint, *Toxicology* **2**, 77-122 (1974). *See also* Abrin, Lectins.

Isoelectric pt 7.1. uv max: 280 nm (ε 85,000).

Caution: Ricin is among the most toxic compounds known. MLD i.p. in mice at 48 hrs: 0.001 μg ricin nitrogen/g body wt, M. Ishiguro *et al., loc. cit.* Seeds of *R. communis,* if thoroughly masticated, can produce serious poisoning and death: J. M. Kingsbury, *Poisonous Plants of the United States and Canada* (Prentice-Hall, New Jersey, 1964) pp 194-197.

USE: As a tool in studies of cell-surface properties; exptly in cancer research.

8212. Ricinine. *1,2-Dihydro-4-methoxy-1-methyl-2-oxo-3-pyridinecarbonitrile; 1,2-dihydro-4-methoxy-1-methyl-2-oxonicotinonitrile;* ricidine. $C_8H_8N_2O_2$; mol wt 164.16. C 58.53%, H 4.91%, N 17.07%, O 19.49%. From seeds and leaves of the castor plant, *Ricinus communis* L., *Euphorbiaceae.* Extraction procedure: Böttcher, *Ber.* **51**, 673 (1918). Synthesis starting with polymerization of cyanoacetyl chloride: Schroeter *et al., Ber.* **65**, 432 (1932). Biosynthesis: Waller, Henderson, *J. Biol. Chem.* **236**, 1186 (1961); Essery *et al., Can. J. Chem.* **41**, 1142 (1963).

Prisms, needles from alc, mp 201.5°. Sublimes at 170-180° under 20 mm pressure. Sparingly sol in water, alc, chloroform, ether. Neutral to litmus. Does not form salts with acids.

Human Toxicity: Ingestion may cause nausea, vomiting, hemorrhagic gastroenteritis, hepatic and renal damage, convulsions, coma, hypotension, respiratory depression, death.

8213. Ricinoleic Acid. *[R-(Z)]-12-Hydroxy-9-octadecenoic acid; d-12-hydroxyoleic acid.* $C_{18}H_{34}O_3$; mol wt 298.45. C 72.43%, H 11.48%, O 16.08%. $CH_3(CH_2)_5CH(OH)CH_2$-$CH=CH(CH_2)_7COOH$. Found primarily in oils from the seeds of *Ricinus* spp, *Euphorbiaceae.* Accounts for about 90% of the triglyceride fatty acids of castor oil, and up to about 40% of the glyceride fatty acids of ergot oil. Bibliography on its isoln: Ralston, *Fatty Acids* (New York, 1948) p 189. Also isolated from *Linum mucronatum* (flax), *Linaceae:* Kleiman, Spencer, *Lipids* **6**, 962 (1971). Structure: Goldsobel, *Ber.* **27**, 3121 (1894). Mechanism of biosynthesis: Morris, *Biochem. Biophys. Res. Commun.* **29**, 311 (1967).

Liquid. $d_4^{27.4}$ 0.940; mp +5.5°; bp_{10} 245°. $[\alpha]_D^{\circ}$ +6.67°; $[\alpha]_D^{26}$ +7.15 (c = 5 in acetone). n_D^{20} 1.4716. Neutralization value 187.98; iodine value 85.05. Sol in alcohol, acetone, ether, chloroform (*cf.* the solubilities of castor oil).

Acid sulfate, $C_{18}H_{34}O_6S$, *ricinolsulfuric acid.* Obtained by the action of chlorosulfonic acid. Viscous brown liquid with weak blue fluorescence. Sol in water (about 10%), alcohol, ether, chloroform.

Sodium salt, *Soricin, Colidosan.* Sodium salts of the fatty acids from castor oil. White or slightly yellow, odorless or almost odorless powder. Sol in water or alcohol. The aq soln is alkaline.

USE: In textile finishing; sometimes added to Turkey red oil, dry-cleaning soaps.

THERAP CAT: Has been used in contraceptive jellies. The sodium salt has been used as sclerosing agent.

8214. Rifamide. *4-O-[2-(Diethylamino)-2-oxoethyl]-*

rifamycin; 2-[(1,2-dihydro-5,6,17,19,21-pentahydroxy-23-methoxy-2,4,12,16,18,20,22-heptamethyl-1,11-dioxo-2,7-(epoxypentadeca[1,11,13]trienimino)naphtho[2,1-b]furan-9-yl)oxy]-N,N-diethylacetamide; N,N-diethylrifomycin B amide; rifamycin B diethylamide; M-14; Rifocin M; Rifocina M. $C_{43}H_{58}N_2O_{13}$; mol wt 810.95. C 63.69%, H 7.21%, N 3.46%, O 25.65%. Prepn: **Belg. pat. 632,770** corresp to Sensi, Maggi, U.S. pat. **3,313,804** (1963 and 1967 to Lepetit); Sensi *et al., J. Med. Chem.* **7**, 596 (1964). Physical and chemical properties: Maggi *et al., Farmaco Ed. Prat.* **20**, 147 (1965). Activity data: Pallanza *et al., Arzneimittel-Forsch.* **15**, 800 (1965); Monnier, Bourse, *Pathol. Biol.* **16**, 901 (1968). Metabolic studies: Fürész *et al., Arzneimittel-Forsch.* **15**, 802 (1965); Maffii, Shiatti, *Toxicol. Appl. Pharmacol.* **8**, 138 (1966). Toxicology: Dezulian *et al., ibid.* 126.

Yellow-orange ppt, crystallized from benzene + hexane. No definite mp, begins to soften at 140°, melts completely at 170° (dec). $[\alpha]_D^{20}$ −48.7° (c = 0.4 in methanol). Absorption spectrum in phosphate buffer (pH 7.38): 222, 302, 421 nm (ε 42,820, 20,770, 16,200). LD_{50} in mice, rats (mg/kg): 2450, >4000 orally; 640, 2500 s.c.; 320, 535 i.p.; 315, 380 i.v. (Dezulian).

THERAP CAT: Antibacterial.

8215. Rifampin. *3-[[(4-Methyl-1-piperazinyl)imino]methyl]rifamycin; 5,6,9,17,19,21-hexahydroxy-23-methoxy-2,4,12,16,18,20,22-heptamethyl-8-[N-(4-methyl-1-piperazinyl)formimidoyl]-2,7-(epoxypentadeca[1,11,13]trienimino)naphtho[2,1-b]furan-1,11(2H)-dione 21-acetate;* rifampicin; rifaldazine; rifamycin AMP; R/AMP; Abrifam; Dipicin; Eremfat; Rifa; Rifadin; Rifadine; Rifaldin; Rifaprodin; Rifobac; Riforal; Rifoldin; Rifoldine; Rimactan. $C_{43}H_{58}N_4O_{12}$; mol wt 822.96. C 62.75%, H 7.10%, N 6.81%, O 23.33%. Semisynthetic antibiotic obtained by reacting 3-formylrifamycin SV with 1-amino-4-methylpiperazine in tetrahydrofuran. Prepn and structure: Maggi *et al., Chemotherapia* **11**, 285 (1966); Neth. pat. **Appl. 6,509,961**; Maggi, Sensi, U.S. pat. **3,342,810** (1966, 1967 both to Lepetit). Chemical and biological properties: Fürész, *Antibiot. & Chemother. (Basel)* **16**, 316 (1970). Activity studies and clinical survey: Arioli *et al., Arzneimittel-Forsch.* **17**, 523 (1967); Pallanza *et al., ibid.* 529; Bergamini, *ibid.* **20**, 1546 (1970); Dans *et al., Am. J. Med. Sci.* **259**, 120 (1970). Metabolism: Meyer-Brunot *et al., Int. Congr. Chemother. Proc., 5th, Vienna 1967* **1**(2), 763; Fürész *et al., Arzneimittel-Forsch.* **17**, 534 (1967); Maggi *et al., ibid.* **19**, 651 (1969). Inhibition of protein synthesis in mammalian cells: W. C. Buss *et al., Science* **200**, 432 (1978). Comprehensive reviews: Binda *et al., Arzneimittel-Forsch.* **21**, 1907-1978 (1971); Lester, *Ann. Rev. Microbiol.* **26**, 88-

102 (1972). Comprehensive description: G. G. Gallo, P. Radaelli, in *Analytical Profiles of Drug Substances* **vol. 5**, K. Florey, Ed. (Academic Press, New York, 1976) pp 467-513. Symposium on the use of rifampin in the treatment of nontuberculous infections: *Rev. Infect. Dis.* **5**, Suppl. 3, S399-S632 (1983).

Red to orange platelets from acetone, dec 183-188°. Absorption max (pH 7.38): 237, 255, 334, 475 nm (ε 33200, 32100, 27000, 15400). Rifampin is a "zwitterion" with pKa 1.7 related to the 4-hydroxy and pKa 7.9 related to the 3-piperazine nitrogen. Very stable in DMSO; rather stable in water. Freely sol in CH_3Cl, DMSO; sol in ethyl acetate, methanol, tetrahydrofuran; slightly sol in water (pH <6), acetone, carbon tetrachloride. LD_{50} in mice, rats (mg/kg): 885, 1720 orally; 260, 330 i.v.; 640, 550 i.p. (Fürész).

THERAP CAT: Antibacterial (tuberculostatic).

8216. Rifamycins. *Rifomycins.* Group of antibiotics characterized by a natural ansa structure (chromophoric naphthohydroquinone group spanned by a long aliphatic bridge) not previously found in other known antibiotics. Isoln from the fermentation broths of *Streptomyces mediterranei*: P. Sensi *et al., Antibiot. Ann.* **1959-1960**, 262. Among rifamycins, rifamycin B, O, S, SV and X are the more studied members. Prepn of rifamycin B derivs: P. Sensi *et al.,* U.S. pat. **3,313,804** (1967 to Lepetit). Structure: Prelog, *Pure Appl. Chem.* **7**, 551 (1963); *Chemotherapia* **7**, 133 (1963); Oppolzer *et al., Experientia* **20**, 336 (1964); Oppolzer, Prelog, *Helv. Chim. Acta* **56**, 2287 (1973). Total synthesis of rifamycin S: H. Nagaoka *et al., J. Am. Chem. Soc.* **102**, 7962 (1980); H. Iio *et al., ibid.* 7965; H. Nagaoka, Y. Kishi, *Tetrahedron* **37**, 3873 (1981); S. Hanessian *et al., J. Am. Chem. Soc.* **104**, 6164 (1982). Review of chemistry of ansamycin antibiotics: K. L. Rinehart *et al., Fortschr. Chem. Org. Naturst.* **33**, 231-307 (1976). General reviews: P. Sensi, *Research Progress in Organic-Biological & Medicinal Chemistry* **vol. 1**, U. Gallo, L. Santamaria, Eds. (Società Editoriale Farmaceutica, Milan, Italy, 1964) pp 337-421; Wehrli, Staehelin, in *Antibiotics* **vol. 3**, J. W. Corcoran, F. E. Hahn, Eds. (Springer-Verlag, New York, 1975) pp 252-268; C. Gurgo *et al.,* "Rifamycins" in *Handbook of Experimental Pharmacology* **vol. 61**, G. V. R. Born *et al.,* Eds., entitled "Chemotherapy of Viral Infections", P. E. Came, L. A. Caliguiri, Eds. (Springer-Verlag, New York, 1982) pp 519-555.

Rifamycin AMP, see Rifampin.

Rifamycin B, $C_{39}H_{49}NO_{14}$, *4-O-(carboxymethyl)rifamycin, nancimycin.* R = —OCH_2COOH; R' = —OH. Yellow prismatic needles from benzene, mp 300° (dec 160-164°). $[\alpha]_D^{20}$ −11° (methanol). Absorption max (phosphate buffer soln pH 7.3): 223, 304, 425 nm ($E_{1cm}^{1\%}$ 555, 275, 220). Dibasic acid. Very stable. Solubilities: water 0.027% (w/w), methanol 2.62%, ethanol 0.44%. LD_{50} in mice (mg/kg): 2040 i.v., >3000 i.p., s.c., and orally (Sensi, review).

Rifamycin O, $C_{39}H_{47}NO_{14}$, *4-O-(carboxymethyl)-1-deoxy-1,4-dihydro-4-hydroxy-1-oxorifamycin γ-lactone.* R = (1,3-dioxolan-4-on)-2-yl; R' = =O. Prepn: P. Sensi *et al., Farmaco Ed. Sci.* **15**, 228 (1960); Umezawa, Japan. pat. **15,518('66)**, *C.A.* **66**, 1583v (1967). Pale yellow crystals from methanol, mp 300° (dec 160°). Also reported as mp 180-185° (Umezawa). $[\alpha]_D^{20}$ +71.5° (c = 1 in dioxane). uv max (methanol contg 5% acetate buffer soln pH 4.62): 226, 273, 370 nm ($E_{1cm}^{1\%}$ 365, 440, 60). Weak acid. Practically insol in dil acids and water. Slowly sol in alkaline soln with red-violet color. Sol in acetone, tetrahydrofuran; slightly sol in methanol, ethanol, ethyl acetate; practically insol in ether, petr ether.

Rifamycin S, $C_{37}H_{45}NO_{12}$, **1,4-dideoxy-1,4-dihydro-1,4-dioxorifamycin.** R = R′ = =O. Activation product found in solns of rifamycin B and rifamycin O: P. Sensi *et al.*, *Experientia* **16**, 412 (1960). Yellow-orange crystals from methanol, dec 179-181°. $[\alpha]_D^{20}$ +476° (c = 0.1 in methanol). Absorption max (phosphate buffer soln pH 7.3): 317, 525 nm ($E_{1cm}^{1\%}$ 426, 62). LD_{50} in mice (mg/kg): 122 i.v.; 258 i.p.; 3000 orally (Sensi, review).

Rifamycin SV, *see* separate entry.

Note: Rifamycin AG is a condensation product of rifamycin O and aminoguanidine: P. Sensi *et al.*, *Antibiot. & Chemother.* **12**, 448 (1962).

8217. Rifamycin SV. *5,6,9,17,19,21-Hexahydroxy-23-methoxy-2,4,12,16,18,20,22-heptamethyl-2,7-(epoxypentadeca[1,11,13]trienimino)naphtho[2,1-b]furan-1,11(2H)-dione 21-acetate;* rifomycin SV; rifamicine SV; Rifocin; Rifocyn. $C_{37}H_{47}NO_{12}$; mol wt 697.80. C 63.69%, H 6.79%, N 2.01%, O 27.52%. Semi-synthetic antibiotic derived from Rifamycin S. Prepn: Sensi *et al.*, *Experientia* **16**, 412 (1960); *Farmaco, Ed. Sci.* **16**, 165 (1961). Comprehensive review: Bergamini, Fowst, *Arzneimittel-Forsch.* **15**, 951-1002 (1965). For general refs *see* Rifamycins.

Yellow-orange crystals, mp 300° (dec 140°). $[\alpha]_D^{20}$ −4° (methanol). uv max (phosphate buffer soln pH 7.3): 223, 314, 445 nm ($E_{1cm}^{1\%}$ 586, 322, 204). Acid reaction. Slightly sol in water, petr ether; sol in ether, bicarbonate soln; very sol in methanol, ethanol, acetone, ethyl acetate. A reducing substitute, such as ascorbic acid, should be added to aq solns of rifamycin SV to prevent its transformation to rifamycin S. LD_{50} in mice (mg/kg): 550 i.v.; 625 i.p.; 2120 orally (Bergamini, Fowst).

Sodium salt, *Chibro-Rifamycin.* Orange-red crystals. Soly in water pH 7.2: ∼5 g/100 ml.

THERAP CAT: Antibacterial.

8218. Rifaximin. *[2S-(2R*,16Z,18E,20R*,21R*,22S*,-23S*,24S*,25R*,26S*,27R*,28E)]-25-(Acetyloxy)-5,6,21,23-tetrahydroxy-27-methoxy-2,4,11,16,20,22,24,26-octamethyl-2,7,[epoxypentadeca(1,11,13)trienimino]benzofuro[4,5-e]pyrido[1,2-a]benzimidazole-1,15(2H)-dione;* 5,6,21,23,25-pentahydroxy-27-methoxy-2,4,11,16,20,22,24,26-octamethyl-2,7-(epoxypentadeca[1,11,13]trienimino)benzofuro[4,5-e]-pyrido[1,2-a]benzimidazole-1,15(2H)-dione 25-acetate; 4-deoxy-4′-methylpyrido[1′,2′-1,2]imidazo[5,4-c]rifamycin SV; rifaxidin; L 105 (Alfa); L 105SV; α-0817185; Fatroximin; Normix; Rifacol; Ritacol. $C_{43}H_{51}N_3O_{11}$; mol wt 785.89. C 65.72%, H 6.54%, N 5.35%, O 22.39%. Nonabsorbable intestinal disinfectant related to rifamycin SV, *q.v.* Prepn: **Belg.** pat. **888,895;** E. Marchi, L. Montecchi, **U.S.** pat. **4,341,785** (1981, 1982 both to Alfa); E. Marchi *et al.*, *J. Med. Chem.* **28**, 960 (1985); prepn and NMR study: M. Brufani *et al.*, *J. Antibiot.* **37**, 1611 (1984). Structural study: M. Brufani *et al.*, *ibid.* 1623. Antibacterial spectrum: C. Eftimiadi *et al.*, *Drugs Exp. Clin. Res.* **10**, 691 (1984); *in vitro* and *in vivo* antibacterial activity: A. P. Venturini, E. Marchi, *Chemioterapia* **5**, 257 (1986). Pharmacodynamics in rats and dogs: A. P. Venturini, *Chemotherapy (Basel)* **29**, 1 (1983). Absorption profile: A. P. Venturini *et al.*, *Drugs*

Exp. Clin. Res. **13**, 233 (1987). Toxicological study: G. Borelli, D. Bertoli, *Chemioterapia* **4**, 263 (1986). Clinical evaluation in infectious diarrhea: M. Vinci *et al.*, *Curr. Ther. Res.* **36**, 92 (1984); in hepatic encephalopathy: F. De Marco *et al.*, *ibid.* 668; in intestinal infections: V. Alvisi *et al.*, *J. Int. Med. Res.* **15**, 49 (1987).

Red orange powder, mp 200-205° (dec). uv max: 232, 260, 292, 320, 370, 450 nm ($E_{1cm}^{1\%}$ 489, 339, 295, 216, 119, 159). Sol in alcohols, ethyl acetate, chloroform, toluene. Insol in water. LD_{50} orally in rats: >2000 mg/kg (Borelli, Bertoli).

THERAP CAT: Antibacterial.

8219. Rilmazafone. *5-[[(Aminoacetyl)amino]methyl]-1-[4-chloro-2-(2-chlorobenzoyl)phenyl]-N,N-dimethyl-1H-1,2,4-triazole-3-carboxamide;* 1-(2-o-chlorobenzoyl-4-chlorophenyl)-5-glycylaminomethyl-3-dimethylcarbamoyl-1H-1,2,4-triazole; 2′,5-dichloro-2-(3-dimethylcarbamoyl-5-glycylaminomethyl-1H-1,2,4-triazol-1-yl)benzophenone; $C_{21}H_{20}Cl_2N_6O_3$; mol wt 475.33. C 53.06%, H 4.24%, Cl 14.92%, N 17.68%, O 10.10%. Ring-opened triazolobenzo-diazepine derivative. Prepn: K. Hirai, H. Sugimoto, **Ger.** pat. **2,725,164;** *eidem,* **U.S.** pat. **4,159,374** (1977, 1979 both to Shionogi); and CNS activity: K. Hirai *et al.*, *J. Heterocycl. Chem.* **19**, 1363 (1982). Receptor binding study of rilmazafone and its closed ring active metabolites: M. Fujimoto *et al.*, *Biochem. Pharmacol.* **33**, 1645 (1984). Combined HPLC determn and enzyme immunoassay in plasma: G. Kominami *et al.*, *J. Chromatog.* **417**, 216 (1987). Series of articles on teratogenicity studies: *Oyo Yakuri* **30**, 749-833 (1985), *C.A.* **104**, 81885h-81887k; 102328w (1986).

Monohydrochloride dihydrate, $C_{21}H_{21}Cl_3N_6O_3 \cdot 2H_2O$, *450191-S, Rhythmy.* Solid from 95% ethanol, mp 107° LD$_{50}$ orally in mice: >1500 mg/kg (Hirai, 1979).

THERAP CAT: Sedative; hypnotic.

8220. Rilmenidene. *N-(Dicyclopropylmethyl)-4,5-dihydro-2-oxazolamine;* 2-[N-(dicyclopropylmethyl)amino]oxazoline; oxaminozoline; S 3341; Hyperium. $C_{10}H_{16}N_2O$; mol wt 180.25. C 66.63%, H 8.95%, N 15.54%, O 8.88%. α_2-Adrenoceptor agonist. Prepn: C. Malen *et al.*, **Ger.** pat. **2,362,754;** *eidem,* **U.S.** pat. **4,102,890** (1974, 1978 both to Sci. Union et Cie. - Soc. Franc. Recher. Med.). Adrenoceptor binding study: P. Guicheney *et al.*, *J. Pharmacol.* **12**, 255

(1981). Pharmacokinetics in hypertensive subjects: J. Velly *et al., ibid.* **13**, 413 (1982). Animal pharmacology: M. Laubie *et al., ibid.* **16**, 259 (1985); P. A. van Zwieten *et al., Arch. Int. Pharmacodyn.* **279**, 130 (1986). Hypotensive effect in humans: K. Weerasuriya *et al., Eur. J. Clin. Pharmacol.* **27**, 281 (1984).

Crystals from hexane, mp 106-107°.
Fumarate, $C_{14}H_{20}N_2O_5$, crystals from ethanol, mp 170°.
THERAP CAT: Antihypertensive.

8221. Rimantadine. α-*Methyltricyclo[3.3.1.1*^{3,7}*]decane-1-methanamine;* α-*methyl-1-adamantanemethylamine;* re-mantadin(e). $C_{12}H_{21}N$; mol wt 179.31. C 80.38%, H 11.81%, N 7.81%. Deriv of adamantane, *q.v.* Prepn: **Neth. pat. Appl. 6,408,505;** W. W. Prichard, **U.S. pat. 3,352,912** (1965, 1967 both to du Pont). Antiviral activity: A. Tsunoda *et al., Antimicrob. Ag. Chemother.* **1965**, 553. Effects on influenza in mice: J. W. McGahen *et al., Ann. N.Y. Acad. Sci.* **173**, 557 (1970). Mechanism of action study: A. Bukrinskaya *et al., Arch. Virol.* **66**, 275 (1980); *eidem, J. Gen. Virol.* **60**, 49 (1982). Pharmacokinetics in humans: R. J. Wills *et al., Antimicrob. Ag. Chemother.* **31**, 826 (1987). Clinical trial in prophylaxis of influenza A infection: R. Dolin *et al., N. Engl. J. Med.* **307**, 580 (1982). Comparative toxicity of rimantadine and amantadine in healthy adults: F. G. Hayden *et al., Antimicrob. Ag. Chemother.* **19**, 226 (1981). Controlled study of CNS effects: V. M. Millet *et al., ibid.* **21**, 1 (1982). Review of studies in the USSR on exptl and clinical pharmacology: D. M. Zlydnikov *et al., Rev. Infect. Dis.* **3**, 408-421 (1981).

Hydrochloride, $C_{12}H_{22}ClN$, *EXP-126, Flumadine, Meradan(e), Roflual.* White crystals, mp 373-375° (sealed tube).
THERAP CAT: Antiviral.

8222. Rimazolium Metilsulfate. *3-(Ethoxycarbonyl)-6,7,8,9-tetrahydro-1,6-dimethyl-4-oxo-4H-pyrido[1,2-a]pyrimidinium methyl sulfate; 3-carboxy-6,7,8,9-tetrahydro-1,6-dimethyl-4-oxo-4H-pyrido[1,2-a]pyrimidinium methyl sulfate, ethyl ester; 1,6-dimethyl-3-carbethoxy-4-oxo-6,7,8,9-tetrahydrohomopyrimidazole methyl sulfate; MZ 144; MZ 0780; Ro 11-780; Dolcuran; Probon; Probonal; Rimagina.* $C_{14}H_{22}N_2O_7S$; mol wt 362.40. C 46.40%, H 6.12%, N 7.73%, O 30.90%, S 8.85%. Prepn: Z. Mészáros *et al.,* **Hung. pat. Teljes 519** (1970 to Chinoin), *C.A.* **74**, 42374h (1971); *eidem, Arzneimittel-Forsch.* **22**, 815 (1972). Anti-inflammatory and analgesic effects in animals: K. Gyires *et al., Drugs Exptl. Clin. Res.* **11**, 493 (1985). Clinical analgesic effect: H. Graber, *Int. J. Clin. Pharmacol. Ther. Toxicol.* **6**, 354 (1972); M. Haataja *et al., Curr. Ther. Res.* **22**, 784 (1977). Series of articles on pharmacology, pharmacodynamics and toxicology: *Arzneimittel-Forsch.* **21**, 717-738 (1971). Toxicity data: J. Knoll *et al., ibid.* 719.

White crystals, mp 165-166°. uv max: 336, 258 nm (ε 3630, 27500). Readily sol in water. LD_{50} in rats (mg/kg): 220 i.v.; 720 i.p.; 790 s.c.; 1600 orally (Knoll).
THERAP CAT: Analgesic.

8223. Rimiterol. *4-(Hydroxy-2-piperidinylmethyl)-1,2-benzenediol; erythro-α-(3,4-dihydroxyphenyl)-2-piperidine-methanol; erythro-3,4-dihydroxyphenyl-2-piperidinylcarbi-nol.* $C_{12}H_{17}NO_3$; mol wt 223.28. C 64.55%, H 7.68%, N 6.27%, O 21.50%. Prepn: Sankey, Whiting, **Ger. pat. 2,024,049** corresp to **U.S. pat. 3,910,934** (1970, 1975 to 3M); Kaiser, Ross, **Ger. pat. 2,047,937** corresp to **U.S. pat. 3,705,169** (1971, 1972 to SK & F); Sankey, Whiting, *J. Heterocycl. Chem.* **9**, 1049 (1972). Pharmacology: Carney *et al., Arch. Int. Pharmacodyn. Ther.* **194**, 334 (1971); Griffin, Turner, *J. Clin. Pharmacol.* **11**, 280 (1971); Bowman, Rodger, *Brit. J. Pharmacol.* **45**, 574 (1972).

Crystals from ethyl acetate, mp 203-204°.
Hydrobromide, $C_{12}H_{18}BrNO_3$, *R 798, WG 253, Asmaten, Pulmadil.* White powder, mp 220° (dec).
THERAP CAT: Bronchodilator.

8224. Rimocidin. $C_{39}H_{61}NO_{14}$; mol wt 767.93. C 61.00%, H 8.01%, N 1.82%, O 29.17%. Polyene antifungal antibiotic produced by *Streptomyces rimosus* along with Terramycin (oxytetracycline, *q.v.*). Recovered from fermentation broth by extracting the mycelium with butanol. Isoln and antifungal activity: Davisson *et al., Antibiot. & Chemother.* **1**, 289 (1951); Seneca *et al., ibid.* **2**, 435 (1952); Davisson *et al.,* **U.S. pat. 2,963,401** (1960 to Pfizer). Partial structure: Cope *et al., J. Am. Chem. Soc.* **87**, 5452 (1965). Structure: L. Falkowski *et al., J. Antibiot.* **29**, 197 (1976); R. Pandey, K. L. Rinehart, *ibid.* **30**, 146 (1977).

Dec above 110°. $[\alpha]_D^{25}$ +116° (pyridine). uv max (80% methanol): 279, 291, 304, 318 nm. Slightly sol in water, acetone, lower alcohols. A cryst sodium salt was prepd by reaction with NaOH in methanol.
Sulfate heptahydrate, large plates from dil methanol, dec 151°. $[\alpha]_D^{25}$ +75.2° (methanol). Sol in water. LD_{50} i.v. in mice: 20 mg/kg (Seneca).

8225. Rioprostil. *(11α,13E)-1,11,16-Trihydroxy-16-methylprost-13-en-9-one;* *(2R,3R,4R)-4-hydroxy-2-(7-hydroxyheptyl)-3-[(E)-(4RS)-(4-hydroxy-4-methyl-1-octenyl)]cyclopentanone;* 16-methyl-1,11α,16RS-trihydroxyprost-13E-en-9-one; Bay-o-6893; ORF-15927; TR-4698; Rostil. $C_{21}H_{38}O_4$; mol wt 354.53. C 71.15%, H 10.80%, O 18.05%. Prostaglandin E_1 analog with cytoprotective and gastric antisecretory activity. Prepn: H. C. Kluender *et al.*, U.S. pat. **4,132,738** (1979 to Miles). Pharmacology: D. A. Shriver *et al.*, *Arzneimittel-Forsch.* **35**, 839 (1985). Effect on human gastric secretion: P. Demol *et al.*, *ibid.* 861; B. Vaona *et al.*, *Adv. Prostaglandin Thromboxane Leukotriene Res.* **17**, 328 (1987). Clinical evaluation in duodenal ulcer: H. G. Dammann *et al.*, *ibid.* 303.

$[α]_D$ −58.6° (c = 1 in chloroform).

THERAP CAT: Antiulcerative.

8226. Risocaine. *4-Aminobenzoic acid propyl ester;* propyl *p*-aminobenzoate; Propaesin; Propesin; Propazyl; Raythesin. $C_{10}H_{13}NO_2$; mol wt 179.21. C 67.02%, H 7.31%, N 7.82%, O 17.85%. Synthesis and properties: Büchi *et al.*, *Arzneimittel-Forsch.* **18**, 791 (1968). Alternate prepn: Kadaba *et al.*, *J. Pharm. Sci.* **58**, 1422 (1969).

Crystals, mp 75-76°; pKa 2.68. Solubility in water: 1.67 mmol/l. Freely sol in alc, benzene, chloroform, ether; about 7% in oils. uv max (pH 7.4): 286, 219 nm (ε 17,196, 9000). Component of *Orabiotic*.

THERAP CAT: Local anesthetic, antipruritic.

8227. Ristocetin. Ristomycin; Spontin; Riston. Glycopeptide antibiotic complex produced by the actinomycete *Nocardia lurida.* Ristocetin A and the more active ristocetin B differ in the number of glucose, mannose, rhamnose and D-arabinose groups in their side chains. Isoln, crystallizn and chemical properties: J. E. Philip *et al.*, *Antibiot. Ann.* **1956-57**, p 699; *eidem,* U.S. pat. **2,990,329** (1961 to Abbott). Structural studies of ristocetin A: Fehlner *et al.*, *Proc. Nat. Acad. Sci. USA* **69**, 2420 (1972); C. M. Harris *et al.*, *J. Am. Chem. Soc.* **101**, 437 (1979); F. Sztaricskai *et al.*, *ibid.* **102**, 7093 (1980). Total structure determn of ristocetin A: D. H. Williams *et al.*, *Chem. Commun.* **1979**, 906; J. R. Kalman, D. H. Williams, *J. Am. Chem. Soc.* **102**, 897 (1980). Identity of ristocetin A with ristomycin A: D. H. Williams *et al.*, *J. Chem. Soc. Perkin Trans. I* **1979**, 787. ^{13}C-NMR studies: F. Sztaricskai *et al.*, *Tetrahedron Letters* **21**, 2983 (1980); M. P. Williamson, D. H. Williams, *J. Chem. Soc. Perkin Trans. I* **1981**, 1483. Revised configuration: C. M. Harris, T. M. Harris, *J. Am. Chem. Soc.* **104**, 363 (1982). Biosynthesis: S. J. Hammond *et al.*, *Chem. Commun.* **1983**, 116. Review: D. C. Jordan, *Antibiotics* vol. 1, D. Gottlieb, P. Shaw, Eds. (Springer-Verlag, New York, 1967) pp 84-89.

Ristocetin A

Crystalline sulfates. Sol in acidic aq solns; much less sol in the neutral pH range. Generally insol in organic solvents. Both components show good stability in aq acidic solns, but are readily inactivated above pH 7.0. Commercial prepns are mixtures of both with >90% ristocetin A.

Ristocetin A, $C_{95}H_{110}N_8O_{44}$, ristomycin A. Cryst sulfate, $[α]_D$ −120° to −133° (water).

Ristocetin B, *ristomycin B.* Cryst sulfate, $[α]_D$ −144° to −149° (water).

USE: Tool for investigation of platelet aggregation: Howard, Firkin, *Thromb. Diath. Haemorrh.* **26**, 362 (1971).

THERAP CAT: Antibacterial.

8228. Ritodrine. *(R*,S*)-4-Hydroxy-α-[1-[[2-(4-hydroxyphenyl)ethyl]amino]ethyl]benzenemethanol; erythro-p-hydroxy-α-[1-[(p-hydroxyphenethyl)amino]ethyl]benzyl alcohol;* N-[2-(p-hydroxyphenyl)ethyl]-N-[2-(p-hydroxyphenyl)-2-hydroxy-1-methylethyl]amine; 1-(4-hydroxyphenyl)-2-[2-(4-hydroxyphenyl)ethylamino]propanol; N-(p-hydroxyphenylethyl)-4-hydroxynorephedrine. $C_{17}H_{21}NO_3$; mol wt 287.37. C 71.05%, H 7.37%, N 4.87%, O 16.70%. $β_2$-Adrenergic agonist. Prepn: Belg. pat. **660,244** (1965 to N.V. Philips); Claassen *et al.*, U.S. pat. **3,410,944** (1968 to No. Am. Philips). Clinical investigations: Coutinho *et al.*, *Am. J. Obstet. Gynecol.* **104**, 1053 (1969); Landesman *et al.*, *ibid.* **110**, 111 (1971); Wesselius-De Casparis *et al.*, *Brit. Med. J.* **3**, 144 (1971). Clinical efficacy in treatment of preterm labor: J. F. Larsen *et al.*, *Obstet. Gynecol.* **67**, 607 (1986).

Base, resinous mass, mp 88-90°.

Hydrochloride, $C_{17}H_{22}ClNO_3$, *Du 21220, Miolene, Prempar, Pre-Par, Utemarin, Utopar, Yutopar.* mp 193-195° (dec) from ethanol-ether. uv max: 267.5 nm (ε 3310).

THERAP CAT: Tocolytic.

8229. Robenidine. *Bis[(4-chlorophenyl)methylene]carbonimidic dihydrazide; 1,3-bis[(p-chlorobenzylidene)amino]guanidine.* $C_{15}H_{13}Cl_2N_5$; mol wt 334.21. C 53.91%, H 3.92%, Cl 21.21%, N 20.96%. Prepn: Tomcufcik, **Ger.** pat. **1,933,112** (1970 to Am. Cyanamid), *C.A.* **72**, 90113c (1970). Activity studies: Kantor *et al.*, *Science* **168**, 373 (1970); Wong *et al.*, *Biochem. Biophys. Res. Commun.* **46**, 621 (1972). Metabolism: Zulalian *et al.*, 163rd Am. Chem. Soc.

Meeting (Boston, April 1972) *Abstracts of Papers,* PEST 12, 13. Animal studies: Millard, *Res. Vet. Sci.* **11,** 394 (1970); Joyner, Norton, *ibid.* **13,** 279 (1972).

Hydrochloride, $C_{15}H_{14}Cl_3N_5$, *robenzidene, Cycostat, Robenz.* Crystals from ethanol, mp 289-290°.

THERAP CAT (VET): Coccidiostat.

8230. Robinin. *3-[[6-O-(6-Deoxy-α-L-mannopyranosyl)-β-D-galactopyranosyl]oxy]-7-[(6-deoxy-α-L-mannopyranosyl)oxy]-5-hydroxy-2-(4-hydroxyphenyl)-4H-1-benzopyran-4-one;* kaempferol 3-robinoside 7-rhamnoside. $C_{33}H_{40}O_{19}$; mol wt 740.68. C 53.51%, H 5.44%, O 41.04%. Dimorphic flavanoid isolated from the leaves and flowers of *Robinia pseudoacacia* L., *Leguminosae:* C. Zwenger, F. Dronke, *Ann.* **suppl. 1,** 257 (1861); C. Sando, *J. Biol. Chem.* **94,** 675 (1932). Structure: Zemplén, Bognár, *Ber.* **74B,** 1783 (1941). Total synthesis and structure: L. Farkas *et al., Phytochemistry* **15,** 215 (1976).

β-Form. Yellow crystals, mp 250-254° (Farkas); also reported as straw-yellow needles from alc, mp 249-250° (Sando). uv max (ethanol): 352, 368 nm (log ε 4.14, 4.18), Jurd, Horowitz, *J. Org. Chem.* **22,** 1619 (1957). Sol in hot water, hot alc; practically insol in ether. On hydrolysis yields kaempferol, *q.v.*

α-Form. Obtained by crystallization from water and dehydrating, mp 195-197° (Sando). Also reported as hydrate, yellow needles from aq methanol, mp 196-199° (Farkas).

8231. Roccellic Acid. *(S)-2-Dodecyl-3-methylbutanedioic acid; d-2-dodecyl-3-methylsuccinic acid; d-α-dodecyl-β-methylsuccinic acid; d-α-methyl-α'-dodecylsuccinic acid.* $C_{17}H_{32}O_4$; mol wt 300.43. C 67.96%, H 10.74%, O 21.30%. Occurs in lichens. Isoln from *Lecanora sordida* (Pers.) Th. Fries, *Parmeliaceae:* Hesse, *J. Prakt. Chem.* **58,** 497 (1898); Kennedy *et al., Sci. Proc. Roy. Dublin Soc.* **21,** 557 (1937); from *Roccella montagnei, Graphidaceae:* Subbaraya, Seshadri, *Proc. Indian Acad. Sci.* **12A,** 466 (1940); from *Crocynia membranacae* (Dicks.) Zahlbr., *Chrysotrichaceae:* Akermark *et al., Acta Chem. Scand.* **13,** 1855 (1959). Structure: Kennedy *et al., loc. cit.* Absolute configuration: Akermark, *Acta Chem. Scand.* **16,** 599 (1962).

Rectangular rods from acetone, mp 132-133°. $[α]_D^{20}$ +18° (c = 1.84 in ethanol). Practically insol in water. Freely sol in alcohol, ether; sol in aq sodium bicarbonate solns. Forms a water-sol sodium salt.

8232. Rociverine. *1-Hydroxy[1,1'-bicyclohexyl]-2-carboxylic acid 2-(diethylamino)-1-methylethyl ester; 2-(diethylamino)-1-methylethyl cis-1-hydroxy[bicyclohexyl]-2-carboxylate;* LG-30158; Rilaten. $C_{20}H_{37}NO_3$; mol wt 339.53. C 70.75%, H 10.98%, N 4.13%, O 14.14%. Spasmolytic agent with balanced neurotropic and myotropic properties. Prepn: L. Turbanti, S. Afr. pat. **67 05649,** *C.A.* **70,** 47117d (1969) and U.S. pat. **3,700,675** (1968, 1972 both to Guidotti). Antispasmodic activity *in vitro* and *in vivo:* G. Toson *et al., Arzneimittel-Forsch.* **28,** 1130 (1978). Effect in cystitis or bladder spasm: A. Manganelli, *Farmaco Ed. Prat.* **34,** 384 (1979). Clinical studies: M. Petrillo *et al., Curr. Med. Res.*

Opin. **7,** 73 (1980); R. Assisi, S. deStefano, *Acta Ther.* **6,** 353 (1980); F. Marsala, *Minerva Med.* **73,** 2179 (1982).

Oil, $bp_{0.1}$ 148-150°. n_D^{20} 1.4820. Sol in alc, ether, chloroform, benzene, dil mineral acids. Insol in water.

THERAP CAT: Relaxant (smooth muscle).

8233. Rokitamycin. *Leucomycin V 4B-butanoate 3B-propanoate;* 3''-propionylleucomycin A5; 5-[O-(2,6-dideoxyhexopyranosyl)-(1→4)-3,6-dideoxy-3-dimethylamino-β-D-glucopyranosyloxy]-6-formylmethyl-3,9-dihydroxy-4-methoxy-8-methyl-10,12-hexadecadien-15-olide; rikamycin; M-19-Q; TMS-19Q; Ricamycin. $C_{42}H_{69}NO_{15}$; mol wt 828.00. C 60.92%, H 8.40%, N 1.69%, O 28.98%. Macrolide antibiotic active against Mycoplasma and macrolide resistant strains of *Staphylococcus aureus* and *Streptococcus pyogenes.* Prepn: H. Sakakibara *et al.,* **Ger. pat. 2,918,954** corresp to U.S. pat. **4,242,504** (1979, 1980 both to Toyo Jozo); *eidem, J. Antibiot.* **34,** 1001 (1981). Series of articles on antibacterial activity, pharmacology, metabolism, toxicology: *Chemotherapy (Japan)* **32,** Suppl. 6, 1-627 (1984). Toxicity data: K. Matsumoto *et al., ibid.* 138.

mp 116°. $[α]_D^{20}$ −71° (c = 1.0 in chloroform). uv max (ethanol): 232 nm (ε 28000).

THERAP CAT: Antibacterial.

8234. Rolicyprine. *5-Oxo-N-(2-phenylcyclopropyl)-2-pyrrolidinecarboxamide;* D-N-(trans-2-phenylcyclopropyl)-L-5-pyrrolidone-2-carboxamide; rolicyprine; Ex 4883; RMI 83027; Cypromin. $C_{14}H_{16}N_2O_2$; mol wt 244.28. C 68.83%, H 6.60%, N 11.47%, O 13.10%. Prepn: (isomer obtained not specified) **Brit. pat. 961,313** (1964 to Lakeside Labs.); **Brit. pat. 1,000,895** (1965 to Colgate-Palmolive); Biel, U.S. pat. **3,192,229** (1965 to Colgate-Palmolive). Prepn of other stereoisomers: Biel, *loc. cit.*

Crystals from water, mp 144-147°. $[α]_D^{25}$ +104.28° (dimethylformamide). LD_{50} orally in rats: 96±8 mg/kg, E. I. Goldenthal, *Toxicol. Appl. Pharmacol.* **18,** 185 (1971).

THERAP CAT: Antidepressant.

8235. Rolipram. *4-[3-(Cyclopentyloxy)-4-methoxyphenyl]-2-pyrrolidinone;* ZK 62711. $C_{16}H_{21}NO_3$; mol wt 275.35. C 69.79%, H 7.69%, N 5.09%, O 17.43%. Pyrrolidone antidepressant; selective inhibitor of cyclic AMP phosphodiesterase. Prepn: **Belg. 826,923;** R. Schmiechen *et al.,* U.S. pat. **4,193,926** (1975, 1980 both to Schering AG). Selective enzyme inhibition and enhancement of cAMP accumulation in rat brain tissue: U. Schwabe *et al., Mol. Pharmacol.* **12**, 900 (1976). Mechanism of action studies: H. Wachtel, *Neuropharmacol.* **22**, 267 (1983); H. Wachtel, H. H. Schneider, *ibid.* **25**, 1119 (1986). Pharmacokinetics in animals: W. Krause, G. Kühne, *Xenobiotica* **18**, 561 (1988). Clinical evaluation in depression: E. Zeller *et al., Pharmacopsychiatry* **17**, 188 (1984). Comparative clinical trial with amitriptyline, *q.v.*, in severe depression: F. Eckmann *et al., Curr. Ther. Res.* **43**, 291 (1988).

Crystals from ethyl acetate, mp 132°.

THERAP CAT: Antidepressant.

8236. Rolitetracycline. *4-(Dimethylamino)-1,4,4a,5,5a,-6,11,12a-octahydro-3,6,10,12,12a-pentahydroxy-6-methyl-1,11-dioxo-N-(1-pyrrolidinylmethyl)-2-naphthacenecarboxamide;* N-(pyrrolidinomethyl)tetracycline; N-(1-pyrrolidinylmethyl)tetracycline; Bristacin; Reverin; Superciclin; Syntetrex; Syntetrin; Synotodecin; Tetraverin; Transcycline; Velacicline; Velacycline. $C_{27}H_{33}N_3O_8$; mol wt 527.56. C 61.47%, H 6.31%, N 7.97%, O 24.26%. Semi-synthetic antibiotic prepd from tetracycline: Siedel *et al., München. Med. Wochenschr.* **100**, 661 (1958); Lindner *et al., S. Afr. pat.* **3169/57**. Alternate procedure and structure: Gottstein *et al., J. Am. Chem. Soc.* **81**, 1198 (1959); Cheney *et al., U.S. pat.* **3,104,240** (1963 to Bristol-Myers). Toxicity: E. I. Goldenthal, *Toxicol. Appl. Pharmacol.* **18**, 185 (1971).

Fine, pale yellow needles, dec 162-165°. Amphoteric. More sol than tetracycline and tetracycline hydrochloride. Soly in water at 25°: 1.25 g/ml. Freely sol in alc. Sol in dil acids and alkalies.

Nitrate sesquihydrate, $C_{27}H_{34}N_4O_{11}\cdot1\frac{1}{2}H_2O$, *Pyrrocycline-N, Bristacin-A, Tetrex PMT, Tetrim, Tetriv.* LD$_{50}$ i.v. in rats: 91 mg/kg (Goldenthal).

THERAP CAT: Antibacterial.

8237. Ronidazole. *1-Methyl-5-nitroimidazole-2-methanol carbamate (ester);* carbamic acid (1-methyl-5-nitroimidazol-2-yl)methyl ester; 1-methyl-2-[(carbamoyloxy)methyl]-5-nitroimidazole; (1-methyl-5-nitroimidazole-2-yl)methyl carbamate; MCMN; Dugro; Ridzol. $C_6H_8N_4O_4$; mol wt 200.16. C 36.00%, H 4.03%, N 27.99%, O 31.97%. Preparation and antiprotozoal activity: Neth. pats. **6,609,552; 6,609,553** (both 1967 to Merck & Co.), *C.A.* **67**, 54123u, 11487y (1967); Verdi; Kollonitsch, U.S. pats. **3,450,710; 3,459,764** (both 1969 to Merck & Co.).

Pale yellow crystals, mp 167-169°. Soly in water (at pH 6.5) about 2.9 mg/ml at room temp. More sol in acid solns. Unstable in alkaline solns. pKa 1.2. Freely sol in acetone. Sol in methanol, ethanol, chloroform and ethyl acetate.

THERAP CAT (VET): Antimicrobial.

8238. Ronifibrate. *3-Pyridinecarboxylic acid 3-[2-(4-chlorophenoxy)-2-methyl-1-oxopropoxy]propyl ester;* 3-(nicotinoyloxy)propyl *p*-chlorophenoxyisobutyrate; 3-[α-(*p*-chlo-

rophenoxy)isobutyryloxy]propyl nicotinate; I-612; Cloprane. $C_{19}H_{20}ClNO_5$; mol wt 377.82. C 60.40%, H 5.34%, Cl 9.38%, N 3.71%, O 21.17%. Diester of clofibric and nicotinic acids, *q.q.v.* Prepn: Y. Hirata *et al.*, **Japan. Kokai 73 40,777** (1973 to Yamanouchi), *C.A.* **79**, 66180w (1973); H. Shindo *et al.*, **Japan. Kokai 74 30,377** (1974 to Kowa Pharm.), *C.A.* **81**, 135984s (1974). Clinical trial in hyperdyslipidemia: A. Bucalossi *et al., Clin. Ter.* **89**, 127 (1979). Efficacy and tolerability: G. Buzzelli *et al., ibid.* 251.

LD$_{50}$ orally in mice: 4.08 g/kg (Bucalossi).

THERAP CAT: Antihyperlipoproteinemic.

8239. Ronnel. *Phosphorothioic acid, O,O-dimethyl O-(2,4,5-trichlorophenyl)ester;* fenchlorphos; dimethyl trichlorophenyl thiophosphate; Trolene; Etrolene; Nankor; Korlan; Viozene; Ectoral. $C_8H_8Cl_3O_3PS$; mol wt 321.57. C 29.88%, H 2.51%, Cl 33.08%, O 14.93%, P 9.63%, S 9.97%. Prepn: Schrader, Ger. pat. **814,152** (1948 to Bayer); Moyle, U.S. pat. **2,599,516** (1952 to Dow). Description of mfg process: *Can. Dept. Agr. Bull.*, (3rd ed., Oct. 1957) p 136.

White powder, mp 41°. Vapor pressure at 25°: 8×10^{-4} mm Hg. Practically insol in water (0.004 g/100 ml at 25°). Freely sol in acetone, carbon tetrachloride, ether, methylene chloride, toluene, kerosene. LD$_{50}$ orally in male, female rats: 1250, 2630 mg/kg, T. B. Gaines, *Toxicol. Appl. Pharmacol.* **14**, 515 (1969).

USE: Insecticide. *Caution:* Cholinesterase inhibitor.

THERAP CAT (VET): Insecticide (systemic).

8240. Rosaprostol. *2-Hexyl-5-hydroxycyclopentaneheptanoic acid;* 9-hydroxy-19,20-bisnorprostanoic acid; C-83; IBI-C83; Rosal. $C_{18}H_{34}O_3$; mol wt 298.47. C 72.44%, H 11.48%, O 16.08%. Prostaglandin analog. Prepn, hypolipemic, platelet aggregation inhibitory activity: U. Valcavi, Ger. pat. **2,535,343**, *eidem*, U.S. pat. **4,073,938** (1976, 1978 both to Ist. Biochim. Ital.). Alternate process: V. Marotta, G. Zabban, Eur. pat. Appl. **155,392** (1985 to Ist. Biochim. Ital.). Gastric antisecretory, cytoprotective activity: U. Valcavi *et al., Arzneimittel-Forsch.* **32**, 657 (1982). Effect on mucus and gastrin secretion in duodenal ulcer: D. Foschi *et al., Prostaglandins Leukotrienes Med.* **15**, 147 (1984). Comparison with cimetidine, *q.v.: eidem, Drugs Exp. Clin. Res.* **10**, 427 (1984). Clinical evaluation in treatment of ulcers: G. P. Tincani *et al., Minerva Med.* **78**, 847 (1987).

Oil.

Sodium salt, $C_{18}H_{33}NaO_3$, white solid. LD$_{50}$ in mice, rats: ~3000 mg/kg orally (Valcavi, 1978), >5 g/kg orally (Valcavi, 1982).

THERAP CAT: Anti-ulcerative.

8241. Rosaramicin. *4'-Deoxycirramycin A$_1$;* 3-ethyl-7-hydroxy-2,8,12,16-tetramethyl-5,13-dioxo-9-[[3,4,6-trideoxy-3-(dimethylamino)-β-D-*xylo*-hexopyranosyl]oxy]-4,17-dioxabicyclo[14.1.0]heptadec-14-ene-10-acetaldehyde; antibiotic 67-694; juvenimicin A$_3$; rosamicin; M-4365A2; Sch 14947. $C_{31}H_{51}NO_9$; mol wt 581.76. C 64.00%, H 8.84%, N

2.41%, O 24.75%. Macrolide antibiotic isolated from fermentations of *Micromonospora rosaria* NRRL-3718. Isoln and properties: M. J. Weinstein *et al.*, **S. Afr. pat. 71 00,402** and **Fr. pat. 2,081,448** (1971, 1972 both to Sherico), *C.A.* **76**, 139065n (1972); **78**, 109310n (1973); G. H. Wagman *et al.*, *J. Antibiot.* **25**, 641 (1972). Biological studies: J. A. Waitz *et al.*, *ibid.* 647. Structure: H. Reimann, R. S. Jaret, *Chem. Commun.* **1972**, 1270. Crystal structure: A. K. Ganguly *et al.*, *Tetrahedron Letters* **21**, 4699 (1980). Biosynthesis: A. K. Ganguly *et al.*, *J. Antibiot.* **29**, 976 (1976). Identity with juvenimicin A₃: T. Kishi *et al.*, *ibid.* 1171. Isoln from *M. capillata* MCRL 0940 and identity with antibiotic M-4365A2: A. Kinumaki *et al.*, *ibid.* **30**, 450 (1977). *In vitro* activity: S. Feltham *et al.*, *J. Antimicrob. Chemother.* **5**, 731 (1979). Use in experimental pneumococcal meningitis: C. M. Nolan *et al.*, *Antimicrob. Ag. Chemother.* **16**, 776 (1979).

Crystals from chloroform, mp 119-122°. $[\alpha]_D^{26}$ — 35° (ethanol). uv max (methanol): 240 nm (ϵ 14,600). Very sol in methanol, acetone, chloroform, benzene. Sparingly sol in ether. Slightly sol in water. LD_{50} in mice (mg/kg): 625 s.c.; 350 i.p.; 155 i.v., G. H. Wagman *et al.*, *loc. cit.*

THERAP CAT: Antibacterial.

8242. Rose Bengal. *4,5,6,7-Tetrachloro-3',6'-dihydroxy-2',4',5',7'-tetraiodospiro[isobenzofuran-1(3H),9'-[9H]xanthen]-3-one dipotassium salt; 4,5,6,7-tetrachloro-2',4',5',7'-tetraiodofluorescein potassium (or sodium) derivative potassium (or sodium) salt;* C.I. Acid Red 94; Rose Bengale B; C.I. 45440. $C_{20}H_2Cl_4I_4K_2O_5$; mol wt 1049.84. C 22.88%, H 0.19%, Cl 13.51%, I 48.35%, K 7.45%, O 7.62%. Also found as the disodium salt. Discovered by Gnehm 1882; tetraiodinate of 4,5,6,7-tetrachlorofluorescein (from resorcinol and tetrachlorophthalic anhydride) converted to potassium salt: *Colour Index* vol. 4 (3rd ed., 1971) 4428. Labeling with ¹³¹I: Liebster, Andrysek, *Nature* **184**, 913 (1959). See also: *H. J. Conn's Biological Stains*, R. D. Lillie, Ed., (Williams & Wilkins, Baltimore, 9th ed., 1977) p 350. Diagnostic use in hepatic function: K. S. Nijran, D. C. Barber, *Phys. Med. Biol.* **31**, 563 (1986).

Bright bluish pink. Soluble in water (bluish red); brown soln in concd H_2SO_4, on dilution gives flesh pink ppt. ¹³¹I-Labeled sodium salt, *sodium rose bengal I 131, radiorose bengal sodium, Rose Bengal Sodium I 131, Robenogatope.* USE: As a dye, biological stain; in coloring straw, wood chips, and inks; for coloring edible products and cosmetics. THERAP CAT: Diagnostic aid (corneal trauma indicator). Sodium Rose Bengal I 131 as diagnostic aid (hepatic function).

8243. Rose Hips. Hipberries. The fruits or berries from wild rose bushes, notably *Rosa canina* L., *R. gallica* L., *R.*

condita Scop. and *R. rugosa* Thunb., *Rosaceae.* Rich in ascorbic acid: M. Szczyglowa *et al.*, *Roczniki Panstwowego Zakladu Hig.* **1**, 523-532 (1950), *C.A.* **46**, 1177d. Ascorbic acid content of hips of *R. canina* reported as 19-27 mg/g dry weight: Å. Gustafsson, J. Schröderheim, *Nature* **153**, 196 (1944); as 33.1 mg/g dry weight: I. Roubani *et al.*, *J. Hort. Sci.* **51**, 375 (1976). Comparison of methods for determn of ascorbic acid in hips: K. Gliniecki *et al.*, *Pharm. Z.* **127**, 823 (1982).

8244. Rosemary. Garden rosemary. Flowers and leaves of *Rosmarinus officinalis* L., *Labiatae. Habit.* Mediterranean Basin; cultivated in gardens. *Constit.* Flowers: About 1% volatile oil, resin, bitter principle. Leaves: Volatile oil, tannin.

USE: In perfumery; manuf oil rosemary.

THERAP CAT: Emmenagogue.

8245. Rosin. Colophony; yellow resin; abietic anhydride. Residue left after distilling off the volatile oil from the oleoresin obtained from *Pinus palustris* and other species of *Pinus, Pinaceae.* Offered as wood rosin (from Southern pine stumps), gum rosin (the exudate from incisions in the living tree, *P. palustris* and *P. caribaea*), and tall oil rosin, *see* Tall Oil. Rosin is chiefly produced in the U.S.A. *Constit.* About 90% resin acids and 10% neutral matter. Of the resin acids about 90% are isomeric with abietic acid ($C_{20}H_{30}O_2$); the other 10% is a mixture of dihydroabietic acid ($C_{20}H_{32}O_2$) and dehydroabietic acid ($C_{20}H_{28}O_2$).

Pale yellow to amber, translucent fragments; brittle fracture at ordinary temp; slight turpentine odor and taste. Readily fusible when heated. d 1.07-1.09. Acid no. not less than 150. Insol in water. Freely sol in alc, benzene, ether, glacial acetic acid, oils, carbon disulfide; also sol in dil solns or fixed alkali hydroxides.

USE: Manuf varnishes, varnish and paint driers, printing inks, cements, soap, sealing wax, wood polishes, floor coverings, paper, plastics, fireworks, tree wax, sizes, rosin oil; for water-proofing cardboard, walls, etc. Pharmaceutic aid (stiffening agent).

8246. Rosin Oil. Rosinol; retinol. Obtained by dry distillation of rosin.

Yellow, viscid, fluorescent, oily liquid. bp above 280°. Insol in water; sol in ether, oil turpentine and other oils. It dissolves phosphorus, sulfur, camphor, phenols, and many other organic compds.

USE: Manuf of carbon black for lithography and printing inks; in varnishes, retinol colors, lacquers, brewers' pitch, axle greases.

8247. Rosoxacin. *1-Ethyl-1,4-dihydro-4-oxo-7-(4-pyridinyl)-3-quinolinecarboxylic acid;* acrosoxacin; Win 35213; Eracine; Eradacil; Eradacin; Roxadil; Winuron. $C_{17}H_{14}N_2O_3$; mol wt 294.31. C 69.38%, H 4.79%, N 9.52%, O 16.31%. Quinolone antibacterial. Prepn: G. Y. Lesher, P. M. Carabateas, **Ger. pat. 2,224,090**; *eidem*, **U.S. pat. 3,753,-993** (1972, 1973 both to Sterling). HPLC determn in plasma and urine: M. P. Kullberg *et al.*, *J. Chromatog.* **173**, 155 (1979). Pharmacological studies: S. Maigaard *et al.*, *Urol. Res.* **8**, 113 (1980); *eidem*, *Invest. Urol.* **17**, 149 (1979). Clinical study in gonorrhea: B. M. Limson *et al.*, *Curr. Ther. Res.* **26**, 842 (1979).

Yellow crystals from DMF, mp 290°. Stable in dry heat at 70°. Light sensitive. Solns and suspensions stable at pH 2.0-9.5.

THERAP CAT: Antibacterial.

8248. Rotenone. *[2R-(2α,6aα,12aα)]-1,2,12,12a-Tetrahydro-8,9-dimethoxy-2-(1-methylethenyl)-[1]benzopyrano-[3,4-b]furo[2,3-h][1]benzopyran-6(6aH)-one;* Canex. C_{23}-

$H_{22}O_6$; mol wt 394.41. C 70.04%, H 5.62%, O 24.34%. Principal insecticidal constituent of derris root, cubé, etc. Powerful inhibitor of mitochondrial electron transport. Isoln from *Lonchocarpus nicou* (Aubl.) DC., *Leguminosae:* Geoffrey, *Ann. Inst. Colon. Marseille* **2**, 1 (1895). Review of structure: La Forge *et al., Chem. Rev.* **12**, 181 (1933); Butenandt, McCartney, *Ann.* **494**, 17 (1932); King, *Ann. Rep. Progr. Chem. (Chem. Soc. London)* **29**, 186 (1932). Absolute configuration: Büchi *et al., J. Chem. Soc.* **1961**, 2843; Nakazaki, Arakawa, *Bull. Chem. Soc. Japan* **34**, 1246 (1961); Begley *et al., Chem. Commun.* **1975**, 850. NMR spectrum: Crombie, Lown, *ibid.* **1962**, 775. Total synthesis: Miyano *et al., Agr. Biol. Chem.* **25**, 673 (1961); Miyano, *J. Am. Chem. Soc.* **87**, 3958 (1965). Alternate synthesis: Crombie *et al., J. Chem. Soc. Perkin Trans I* **1973**, 1277, *eidem, Chem. Commun.* **1979**, 1142; I. Sasaki, K. Yamashita, *Agr. Biol. Chem.* **43**, 137 (1979). Synthesis of stereoisomers: Unai, Yamamoto, *ibid.* **37**, 897 (1973). Toxicology: Santi, Toth, *Farmaco Ed. Sci.* **20**, 270 (1965). Toxicity data: J.-I. Fukami *et al., Science* **155**, 713 (1967). *Review:* H. Fukami, M. Nakajima in *Naturally Occurring Insecticides,* M. Jacobson, D. G. Crosby, Eds. (Dekker, New York, 1971) pp 71-97; S. B. Soloway, *Environ. Health Persp.* **14**, 109-117 (1976); T. J. Haley, *J. Environ. Pathol. Toxicol.* **1**. 315-337 (1978).

Orthorhombic, six-sided plates from trichloroethylene, mp 165-166° (dimorphic form, mp 185-186°). $[\alpha]_D^{20}$ −228° (c = 2.22 in benzene). uv spectra: Büchi *et al., loc. cit.* Practically insol in water. Sol in alcohol, acetone, carbon tetrachloride, chloroform, ether, and many other organic solvents. Dec upon exposure to light and air. Colorless solns in organic solvents oxidize upon exposure and become yellow, orange and then deep red and may deposit crystals of dehydrorotenone and rotenone which are toxic to insects. LD_{50} i.p. in mice: 2.8 mg/kg (Fukami); in rats (mg/kg): 132 orally; 6 i.v. (Soloway).

Human Toxicity: Inhalation or ingestion of large doses may cause numbness of oral mucous membrane, nausea and vomiting, muscle tremors, tachypnea. With lethal doses respiratory paralysis occurs. Death may be preceded by convulsions. Chronic poisoning may produce fatty changes in liver, kidney. Direct contact occasionally causes mild irritation of skin or conjunctiva. More toxic when inhaled than when ingested. *See Clinical Toxicology of Commercial Products,* R. E. Gosselin *et al.,* Eds. (Williams & Wilkins, Baltimore, 4th ed., 1976) Section III, pp 293-295.

USE: Pesticide.

THERAP CAT (VET): Acaricide, ectoparasiticide.

8249. Rotraxate. *trans-4-[[4-(Aminomethyl)cyclohexyl]carbonyl]benzenepropanoic acid;* 3-[p-(trans-4-aminomethylcyclohexylcarbonyl)phenyl]propionic acid; *p-[[trans-4-(aminomethyl)cyclohexyl]carbonyl]hydrocinnamic acid;* traxaprone. $C_{17}H_{23}NO_3$; mol wt 289.37. C 70.56%, H 8.01%, N 4.84%, O 16.59%. Gastric cytoprotectant structurally similar to cetraxate, *q.v.* Prepn: T. Takeshita *et al., Eur. pat. Appl.* **44,541;** *eidem,* U.S. pat. **4,402,975** (1982, 1983 both to Teijin); T. Takeshita *et al., Chem. Pharm. Bull.* **33**, 5059 (1985). Crystal and molecular structure: K. Koyano *et al., ibid.* **34**, 3900 (1986). Antiulcerative effects in animals: K. Hoshina *et al., Arzneimittel-Forsch.* **35**, 493 (1985). Effects on gastric mucosa: *eidem, ibid.* **37**, 1154, 1284, 1289 (1987). Acute toxicity: Y. Izawa *et al., Yakuri to Chiryo* **16**, 357 (1988), *C.A.* **109**, 16846k (1988). Series of articles on toxicology: *eidem, ibid.* 371-599, *C.A.* **109**, 558g-565g; 16847n (1988).

Hydrochloride, $C_{17}H_{24}ClNO_3$, *TEI-5103, TG-51, Cumelon.* Crystals from water, mp 245° (dec) (Takeshita, 1985); also reported as mp 221-227° (Takeshita, 1983). LD_{50} in male, female rats (mg/kg): 9800, 9800 orally; 862, 835 i.p.; 5000, 5000 s.c. (Izawa).

THERAP CAT: Antiulcerative.

8250. Rottlerin. *1-[6-[(3-Acetyl-2,4,6-trihydroxy-5-methylphenyl)methyl]-5,7-dihydroxy-2,2-dimethyl-2H-1-benzopyran-8-yl]-3-phenyl-2-propen-1-one;* 5,7-dihydroxy-2,2-dimethyl-6-(2,4,6-trihydroxy-3-methyl-5-acetylbenzyl)-8-cinnamoyl-1,2-chromene; mallotoxin. $C_{30}H_{28}O_8$; mol wt 516.52. C 69.75%, H 5.46%, O 24.78%. Most important and powerful toxic component of *Kamala, q.v.* Isoln from *Kamata* or *Mallotus philippinensis* (Lam.) Muell.-Arg. (*Rottlera tinctoria* Roxb.), *Euphorbiaceae:* Telle, *Arch. Pharm.* **244**, 446 (1906); Dutt, *J. Chem. Soc.* **127**, 2044 (1925); McGookin *et al., ibid.* **1937**, 748. Structure: Brockmann, Maier, *Ann.* **535**, 149 (1938); McGookin *et al., J. Chem. Soc.* **1939**, 1579. Synthesis of tetrahydrorottlerin: Backhouse *et al., ibid.* **1948**, 113.

Light yellow prisms from toluene, mp 206-207°. Sol in ether, chloroform, alcohol, benzene, ethyl acetate; slightly sol in glacial acetic acid; practically insol in water.

5,7-Dimethyl ether, $C_{32}H_{32}O_8$, yellow crystals from ethyl acetate + acetone or chloroform + methanol, dec 245-246°. Sol in chloroform, pyridine, hot glacial acetic acid; slightly sol in cold methanol, ethyl acetate, acetone, benzene, ether.

Pentamethyl ether, $C_{35}H_{38}O_8$, crystals from petr ether, methanol or 90% alcohol, mp 144°. Very sol in acetone, glacial acetic acid, benzene, ethyl acetate.

Pentaacetate, $C_{40}H_{38}O_{13}$, leaflets from benzene + alcohol or acetone + alcohol, prisms from ethyl acetate, mp 212°.

Tetrahydrorottlerin, $C_{30}H_{32}O_8$, yellow prisms from alcohol, mp 214°.

8251. Roxarsone. *4-Hydroxy-3-nitrophenylarsonic acid; 4-hydroxy-3-nitrobenzenearsonic acid;* 3-nitro-4-hydroxyphenylarsonic acid; 2-nitro-1-hydroxybenzene-4-arsonic acid; nitrophenolarsonic acid; NSC-2101; Ren-O-sal; Ristat. $C_6H_6AsNO_6$; mol wt 263.03. C 27.40%, H 2.30%, As 28.48%, N 5.33%, O 36.50%. Prepared by treating sodium p-hydroxyphenylarsonate with a mixture of nitric and sulfuric acids at 0°: Benda, Bertheim, *Ber.* **44**, 3446 (1911); *Ger. pat.* **224,953,** *C.A.* **5**, 156 (1911).

Tufts of pale-yellow needles or rhombohedral plates from water. Puffs up and deflagrates on heating. Slightly sol in cold water; sol in about 30 parts boiling water; freely sol in methanol, ethanol, acetic acid, acetone, alkalies; sparingly sol in dil mineral acids; insol in ether, ethyl acetate. Forms mono-, di-, and trisodium salts. LD_{50} in rats, chickens:

155, 110-123 mg/kg orally; 66, 34 mg/kg i.p., Kerr et al., Toxicol. Appl. Pharmacol. **5**, 507 (1963).

USE: Formerly in manuf arsphenamines; as reagent for zirconium.

THERAP CAT: Antibacterial.

THERAP CAT (VET): Control of enteric infections. To improve growth and feed efficiency.

8252. Roxatidine Acetate. 2-(Acetyloxy)-N-[3-[3-(1-piperidinylmethyl)phenoxy]propyl]acetamide; aceroxatidine. $C_{19}H_{28}N_2O_4$; mol wt 348.44. C 65.49%, H 8.10%, N 8.04%, O 18.37%. Histamine H_2-receptor antagonist. Prepn and anti-ulcerative activity: K. Shibata et al., **Eur. pat. Appl. 24,510**; eidem, **U.S. pat. 4,293,557** (both 1981 to Teikoku Hormone). Pharmacology in animals: M. Tarutani et al., Arzneimittel-Forsch. **35**, 703 (1985). Comparison with cimetidine of effect on gastric acid secretion and ulcer formation in animals: eidem, ibid. 844. Metabolism in humans: S. Honma et al., Oyo Yakuri **30**, 555 (1985); C.A. **104**, 61453n (1985).

CH₃COOCH₂CONHCH₂CH₂CH₂O

mp 59-60°. LD_{50} orally in male mice: 1000 mg/kg (Shibata).

Hydrochloride, $C_{19}H_{29}ClN_2O_4$, *pifatidine*, *Hoe-760*, *TZU-0460*, *Altat*, *Roxane*, *Roxin*, *Roxit*, *Xarcin*. mp 145-146°.

THERAP CAT: Anti-ulcerative.

8253. Roxithromycin. Erythromycin 9-[O-[(2-methoxyethoxy)methyl]oxime]; oxacyclotetradecane erythromycin deriv; 9-(2′,5′-dioxahexyloxyimino)erythromycin; RU 28965; RU 965; Rulid. $C_{41}H_{76}N_2O_{15}$; mol wt 837.06. C 58.83%, H 9.15%, N 3.35%, O 28.67%. Semisynthetic erythromycin derivative. Prepn: S. Gouin d'Ambrieres et al., **Fr. pat. 2,473,525**; eidem, **U.S. pat. 4,349,545** (1981, 1982 both to Roussel-UCLAF). In vitro antibacterial spectrum: R. N. Jones et al., Antimicrob. Ag. Chemother. **24**, 209 (1983); Y. J. Drabu et al., Drugs Exptl. Clin. Res. **13**, 201 (1987). Antiprotozoal activity in mice: B. J. Luft, Eur. J. Clin. Microbiol. **6**, 479 (1987); H. R. Chang, J.-C. R. Pechere, Antimicrob. Ag. Chemother. **31**, 1147 (1987). Pharmacokinetics: R. Wise et al., ibid. 1051. Bioassay in biological fluids: A. L. Barry, R. R. Packer, Eur. J. Clin. Microbiol. **5**, 536 (1986). Clinical evaluations in respiratory tract infections: J. M. Lachat et al., Schweiz. Med. Wochenschr. **116**, 1739 (1986); C. Grassi et al., Chemioterapia **6**, 41 (1987). Series of articles on antimicrobial activity, pharmacokinetics and clinical efficacy: J. Antimicrob. Chemother. **20**, Suppl. B, 1-185 (1987).

$[\alpha]_D^{25}$ −77.5 ±2° (c = 0.45 in chloroform).

THERAP CAT: Antibacterial.

8254. Royal Jelly. Queen bee jelly; apilak; Weiselfuttersaft (German); Gelée royale (French). Production from bee hives: Ritschel, Oesterr. Drogisten-Ztg. **12**, 4-7 (1958). Elementary analysis: N 5%, P 0.7%, S 0.4%. Group analysis: Water 24%, proteins 31%, ether-sol substances 15%, carbohydrates 15%, ash 2%. Trace elements present: Fe, Mn, Ni, Co, Si, Cr, Au, Hg, Bi, As. Vitamins: thiamine 1.2-7.4

mg-%, riboflavine 5.2-10 mg-%, niacin 60-150 mg-%, ascorbic acid 12 mg-%, pyridoxine 2.2-10.2 mg-%, B_{12} 0.15 mg-%, pantothenic acid 65-200 mg-%, biotin 0.9-3.7 mg-%, inositol 80-150 mg-%, folic acid 0.2 m-%, vitamin E <0.2 mg-%. A synthetic mixture of the above composition fed to bee larvae maintains life, but does not produce queens. The presence of hormones affecting mammals has not been demonstrated: Hinglais, Gautherie, Compt. Rend. **242**, 2483 (1956). Natural royal jelly fed to adult rats deficient in vitamin E, had no action in restoring spermatogenesis in males or producing normal fertility in females. No effect has been demonstrated in humans, except a tendency to produce wakefulness. Judged to be of no practical utility in human nutrition because of the very large amounts required for any definite effect: Moreaux, Bull. Soc. Sci. Nancy **14**, 49-53 (1955), C.A. **50**, 13214f (1956). Monographs: B. deBelvefer, Royal Jelly (Paris, Librairie Maloine, 1958) 270 pp; H. Rembold, Biologically Active Substances in Royal Jelly in Vitamins and Hormones **23**, 359-382 (1965).

Royal jelly acid, $C_{10}H_{18}O_3$, *trans-10-hydroxy-Δ²-decenoic acid*. HO(CH₂)₇CH=CHCOOH. Constitutes about 10% of the dried royal jelly. Isoln: Townsend, Lucas, Biochem. J. **34**, 1155 (1940); Butenandt, Rembold, Z. Physiol. Chem. **308**, 284 (1957). Synthesis: Fray et al., Tetrahedron Letters **4**, 15 (1960); Smissman et al., J. Org. Chem. **29**, 3517 (1964); Bestmann et al., Ann. **699**, 33 (1966); J. Tsuji et al., Bull. Chem. Soc. Japan **50**, 2507 (1977); T. Fujisawa et al., Chem. Letters **1982**, 219; R. Chiron, J. Chem. Ecol. **8**, 709 (1982). Prevention of leukemia in mice with royal jelly acid: Townsend et al., Nature **183**, 1270 (1959). Prisms from ether + petr ether or methanol + water, mp 64-65°. uv max: 211 nm (ε 12,000).

See also Queen Substance.

8255. Rubber. Caoutchouc; India rubber. $(C_5H_8)_n$. Primarily obtained by coagulating the milk juice (latex) of several tropical trees, chiefly Hevea brasiliensis Muell.-Arg., Euphorbiaceae. Habit. Brazil, East Indies, Java, etc. Rubber also occurs in a number of plants, among which *guayule* (Parthenium argentatum Gray), a shrub which grows primarily in the northern Mexican desert, represents a potentially useful renewable source. Guayule produces rubber which is chemically identical to Hevea rubber in amounts ranging from 10-20% (dry basis) distributed in the roots and stems. See E. Campos-Lopez, J. Polymer Sci., Polym. Lett. Ed. **14**, 649 (1976); J. Polymer Sci., Polym. Chem. Ed. **14**, 1561 (1976); J. D. Johnson, C. W. Hinman, Science **208**, 460 (1980). Possible biosynthesis from acetate through mevalonic acid and isopentenyl pyrophosphate: Archer et al., Nature **184**, 268 (1959). Comprehensive review of chemistry: Ellis, Chem. Ind. (London) **1962**, 1447. Book: L. G. Polhamus, Rubber (Wiley, New York, 1962). Review: Jirgenson's, Natural Organic Macromolecules (Pergamon Press, New York, 1962) pp 115-124; D. R. St. Cyr in Kirk-Othmer Encyclopedia of Chemical Technology vol. **20** (Wiley-Interscience, New York, 3rd ed., 1982) pp 468-491. Natural rubber is defined as a cis-1,4-polyisoprene with a molecular weight varying from 100,000 to one million.

The best grades of raw rubber (pale crepe or smoked sheet) contain about 95% rubber hydrocarbon. The rest consists of proteins (2-3%), acetone-sol resins and fatty acids (2%), small amounts of sugar and a little mineral matter. Vulcanization, which consists of heating rubber with 1-3% of sulfur, introduces cross links between chains to produce a three-dimensional lattice of improved elasticity, strength, and temp sensitivity. Accelerators such as zinc dimethyldithiocarbamate greatly decrease the time or lower the temp required for vulcanization.

Pure rubber is nearly colorless and transparent in thin layers; odorless and tasteless. It is very elastic and lighter than water. Dec at about 120°. Burns with smoky flame. Emits characteristic offensive odor while burning. Practically insol in water, alcohol, dil acids, or alkali; sol in abs ether, chloroform, most fixed and volatile oils, petr ether, carbon disulfide, oil of turpentine.

8256. Rubeanic Acid. *Ethanedithioamide; dithiooxamide.* $C_2H_4N_2S_2$; mol wt 120.20. C 19.98%, H 3.35%, N 23.31%, S 53.35%. NH$_2$CSCSNH$_2$. Review of rubeanic acid and derivs as chelating ligands and analytical reagents: Ray, Xavier, *J. Indian Chem. Soc.* **38**, 535 (1961).

Red crystals or cryst powder; dec at about 200°. Slightly sol in water; sol in alcohols. Insol in ether.

USE: As a reagent for copper, cobalt, and nickel. As a stabilizer of ascorbic acid solns: Smoczkiewicz, Grochmalicka, *Nature* **192**, 16 (1961).

8257. Ruberythric Acid. *1-Hydroxy-2-[(6-O-β-D-xylopyranosyl-β-D-glucopyranosyl)oxy]-9,10-anthracenedione; 1-hydroxy-2-anthraquinonyl 6-O-β-D-xylopyranosyl-β-D-glucopyranoside; β-2-alizarin primeveroside; ruberythrinic acid; rubian; rubianic acid.* $C_{25}H_{26}O_{13}$; mol wt 534.46. C 56.18%, H 4.90%, O 38.92%. Isoln from root of *Rubia tinctorium* L., *Rubiaceae:* Hill, Richter, *Proc. Roy. Soc. (London) Ser. B* **121**, 547 (1937). Structure: Richter, *J. Chem. Soc.* **1936**, 1701. Synthesis: Zemplén, Bognár, *Ber.* **72B**, 913 (1939). Metabolism: Maehner, Dulce, *Z. Klin. Chem. Klin. Biochem.* **6**, 99 (1968).

Golden-yellow, silky, lustrous prisms or long needles. mp 259-261° (from water). Sol in hot water, alkalies; slightly sol in alc, ether. Practically insol in benzene.

8258. Rubiadin. *1,3-Dihydroxy-2-methyl-9,10-anthracenedione; 1,3-dihydroxy-2-methylanthraquinone.* $C_{15}H_{10}O_4$; mol wt 254.23. C 70.86%, H 3.96%, O 25.17%. From *Rubia tinctorum* L., *Coprosma* var., *Morinda citrifolia* Linn, *Rubiaceae:* Schunck, *Ann.* **87**, 344 (1853); Briggs, Nicholls, *J. Chem. Soc.* **1949**, 1241; Briggs *et al.*, ibid. **1952**, 1718; Borvie, Cooke, *Aust. J. Chem.* **15**, 332 (1962). Structure: Marchlewski, *J. Chem. Soc.* **63**, 1137 (1893); Schunck, Marchlewski, ibid. **65**, 182 (1894); Jones, Robertson, ibid. **1930**, 1699. Synthesis: Joshi *et al.*, *J. Sci. Ind. Res.* **14B**, 87 (1955); Hirase, *Chem. Pharm. Bull.* **8**, 417 (1960).

Yellow plates from glacial acetic acid, mp 302°; yellow slender plates from alc, mp 290°. Absorption max (ethanol): 246, 280, 415 nm (log ε 4.39, 4.52, 3.87). Sol in alc, ether, practically insol in water, alkalies.

Diacetate, yellow rods from acetic anhydride, mp 228°. Dimethyl ether, yellow needles from alc, mp 160.5°.

8259. Rubidium. Rb; at. wt 85.4678; at. no. 37; valence 1. Alkali metal. Widely distributed in very small quantities in earth's crust: 0.0034% by wt. Natural isotopes: 85 (72.15%); 87 (27.85%); [87]Rb is radioactive: $T_{1/2}$ 4.8×10^{10} years; isotopes range in mass number from 79 to 95. Found with other alkali metals in rhodizite (borate), lepidolite (aluminosilicate), rubidium carnallite (chloride); in sea water; in mineral springs and salt lakes. Discovered by Bunsen and Kirchhoff in 1861. Prepn: Hackspill, *Helv. Chim. Acta* **11**, 1003 (1928). *Review:* Whaley, "Sodium, Potassium, Rubidium, Cesium and Francium" in *Comprehensive Inorganic Chemistry* vol. I, J. C. Bailar Jr. *et al.*, Eds. (Pergamon Press, Oxford, 1973) pp 369-529; F. B. White, W. G. Lidman in Kirk-Othmer *Encyclopedia of Chemical Technology* vol. 20 (Wiley-Interscience, New York, 3rd ed., 1982) pp 492-499.

Lustrous, silvery-white, soft metal; body-centered cubic structure; rapidly tarnishes on exposure to air; mp 39°; bp

688°; d^{20} 1.532. Sp heat 0.0802 cal/g·deg. One of the most active metals. E° (aq) Rb/Rb$^+$ 2.924 V. In its chemical properties closely resembles potassium. Dec water rapidly, even at −108°. Ignites spontaneously in oxygen; when molten readily takes fire in the air. Reacts vigorously with the halogens. Forms a series of solid solns with potassium, cesium, sodium. Combines vigorously with mercury. *Keep under benzene, petroleum, or other liq not containing oxygen.*

USE: In making rubidium salts; as a reagent in making zeolite catalysts; in photoelectric cells.

8260. Rubidium Bromide. BrRb; mol wt 165.39. Br 48.32%, Rb 51.68%. RbBr.

White, cryst powder; d 3.35; mp 682°; bp 1340°. Sol in about 1 part cold water, 0.5 part boiling water. The aq soln is neutral.

8261. Rubidium Chloride. Rubinorm. ClRb; mol wt 120.94. Cl 29.32%, Rb 70.68%. RbCl. Causes shock-induced aggression in rats: B. S. Eichelman, *Psychopharmacol. Bull.* **9**, 21 (1973). Biochemical, behavioral and metabolic studies in humans: R. R. Fieve *et al.*, *Am. J. Psychiatry* **130**, 55 (1973). Clinical evaluations in depression: C. Paschalis *et al.*, *J. Roy. Soc. Med.* **71**, 343 (1978); G. Placidi *et al.*, *J. Clin. Psychopharmacol.* **8**, 184 (1988).

White, crystalline powder. d 2.76. mp 715°. bp 1390°. One gram dissolves in 1 ml cold, 0.7 ml boiling water, 90 ml methanol, 1650 ml alc. The aq soln is neutral.

USE: As catalyst; as gasoline octane number improver.

THERAP CAT: Antidepressant.

8262. Rubidium Hydroxide. HORb; mol wt 102.49. H 0.98%, O 15.61%, Rb 83.40%. RbOH. Book: L. S. Itkina, *Lithium, Rubidium and Cesium Hydroxide* (Nauka, Moscow, USSR, 1973) 136 pp. Toxicological study: G. T. Johnson *et al.*, *Toxicol. Appl. Pharmacol.* **32**, 239 (1975).

Grayish-white, deliquesc mass. d 3.203. Melts at red heat. Sol in 0.6 part water; sol in alcohol. It is a stronger base than potassium hydroxide. *Keep well closed.*

USE: Catalyst in oxidative chlorination.

8263. Rubidium Iodide. IRb; mol wt 212.40. I 59.76%, Rb 40.24%. RbI.

White crystals or cryst powder; discolors on exposure to air and light; d 3.55; mp 642°; bp 1300°. Sol in 0.66 part water; sol in alcohol. The aq soln is neutral or slightly alkaline. *Keep well closed and protected from light.*

THERAP CAT: Iodine source.

8264. Rubijervine. *Solanid-5-ene-3β,12α-diol;* Δ^5-3β,-12α-dihydroxysolanidene; rubigervine. $C_{27}H_{43}NO_2$; mol wt 413.62. C 78.40%, H 10.48%, O 7.74%, N 3.39%. From various species of *Veratrum, Liliaceae.* Isoln: Wright, Luff, *J. Chem. Soc.* **35**, 405 (1879); Salzberger, *Arch. Pharm.* **228**, 462 (1890); Jacobs, Craig, *J. Biol. Chem.* **148**, 41 (1943); Sato, Jacobs, ibid. **179**, 623 (1949); Pelletier, Locke, *J. Am. Chem. Soc.* **79**, 4531 (1957). Stereochemistry: Höhne *et al.*, *Tetrahedron* **22**, 673 (1966).

Solvated needles from alc, mp 240-246°. $[\alpha]_D^{25}$ +19.0° (ethanol). Very sparingly sol in water. Soluble in ethanol, methanol, benzene, chloroform. Slightly sol in ether, petr ether. Red color in concd H$_2$SO$_4$. Precipitated by digitonin.

Hydrobromide, $C_{27}H_{43}NO_2$.HBr, needles from methanol + acetone, dec 265-270°. LD$_{50}$ i.v. in rats: 70 mg/kg.

THERAP CAT: Antifungal.

8265. Rubixanthin. *(3R)-β,ψ-Caroten-3-ol.* $C_{40}H_{56}O$; mol wt 552.85. C 86.90%, H 10.21%, O 2.89%. Carotenoid isomeric with cryptoxanthin. Isolated from hipberries (rose hips, *Rosa rubiginosa* L., *Rosaceae*): Kuhn, Grundmann,

Ber. **67**, 339, 1133 (1934); from the flowers of *Tagetes patula* L., *Compositae.* Found also in other *Rosa* varieties. Structural studies: Brown, Weedon, *Chem. Commun.* **1968**, 382. Abs config: L. Bartlett *et al., J. Chem. Soc. (C)* **1969**, 2527. Biosynthesis studies: McDermott *et al., Biochem. J.* **134**, 1115 (1973). Has no vitamin A activity.

Deep red needles with metallic luster from benzene + methanol. Orange crystals from benzene + petr ether, mp 160°. Absorption max (chloroform): 509, 474, 439 nm. Sol in benzene, chloroform; slightly sol in petr ether, alcohol.

8266. Rubus. Blackberry bark. Dried bark of rhizome and roots of the section *Eubatus* Focke of the genus *Rubus* L. or *R. nigrobaccus* Bailey, *Rosaceae. Habit.* Eastern U.S. *Constit.* Tannin, gallic acid, villosin (a saponin).
THERAP CAT: Antidiarrheal.

8267. Rufigallol. *1,2,3,5,6,7-Hexahydroxy-9,10-anthra-cenedione; 1,2,3,5,6,7-hexahydroxyanthraquinone;* rufigallic acid; C.I. 58600. $C_{14}H_8O_8$; mol wt 304.20. C 55.27%, H 2.65%, O 42.08%. Prepd by the action of H_2SO_4 on gallic acid: Robiquet, *Ann.* **19**, 204 (1836); by dry distn of gallic acid: Kunz-Krause, Manicke, *Ber.* **53**, 190 (1920). *See also: Colour Index* **vol. 4** (3rd ed., 1971) p 4520.

Red needles. Does not melt, but sublimes with partial decompn upon heating. Absorption spectrum: Treibs, Steinmetz, *Ann.* **506**, 171, 191 (1933). Practically insol in water; freely sol in acetone; slightly sol in alcohol, ether, with yellow color. Sol in alkali hydroxide solns with violet color, but is soon dec by oxidation.
Hexamethyl ether, $C_{20}H_{20}O_8$, yellow needles, mp 240°.
USE: For color reactions with Zr and Hf.

8268. Rugulovasines. $C_{16}H_{16}N_2O_2$; mol wt 268.32. C 71.62%, H 6.01%, N 10.44%, O 11.93%. Mixture of the interconvertible diastereomers, rugulovasine A and rugulovasine B. Isolated as the hydrochlorides from surface cultures of *Penicillium concavo-rugulosum* Abe and other *penicillium* strains: Abe *et al., Agr. Biol. Chem.* **33**, 469 (1969); R. J. Cole *et al., Can. J. Microbiol.* **22**, 741 (1976). Prepn and separation of isomers: Abe, **Ger. pat.** **1,919,255** (1969 to Takeda), *C.A.* **72**, 30311f (1970). Structure: Yamatodani *et al., Agr. Biol. Chem.* **34**, 485 (1970). Abs config and mechanism of interconversion of the isomers: R. J. Cole *et al., Tetrahedron Letters* **1976**, 3849. Total synthesis: J. Rebek, Y. K. Shue, *J. Am. Chem. Soc.* **102**, 5426 (1980). Hypotensive and cardiovascular activity studies of the hydrochloride isomers: Nagaoka *et al., Arzneimittel-Forsch.* **22**, 137, 143 (1972).

Rugulovasine A, cis-(±)-dihydro-4'-methyl-4-(methyl-amino)spiro[benz[cd]indole-5(1H),2'(5'H)-furan]-5'-one. Colorless prisms from benzene, chloroform, or acetonitrile, mp 138° (dec). $[\alpha]_{436}^{22}$ −3.0° (pyridine). uv max (ethanol): 224, 277, 288, 295 nm (log ε 4.37, 3.70, 3.78, 3.77). Sparingly sol in water, petr ether. Moderately sol in ether, chloro-

form, benzene, acetonitrile. Readily sol in ethyl acetate, acetone, methanol, ethanol, pyridine, and dil acids.
Hydrochloride, $C_{16}H_{16}N_2O_2$·HCl, colorless prisms from water, mp 225° (dec).
Rugulovasine B, trans analog of Rugulovasine A. Difficult to crystallize from common solvents. Separated from benzene as colorless resinous oil. $[\alpha]_{436}^{22}$ +1.4° (pyridine). uv max (ethanol): 227, 278, 288, 295 nm (log ε 4.16, 3.68, 3.73, 3.72). Same solubilities as rugulovasine A.
Hydrochloride, $C_{16}H_{16}N_2O_2$·HCl, white prisms from water, mp 187° (dec).

8269. Rumex. Yellow dock; curled dock. Dried root of *Rumex crispus* L., or of *R. obtusifolius* L., *Polygonaceae. Habit.* Europe, North America. *Constit.* Chrysophanic acid, emodin, tannin, calcium oxalate, lapathin.
THERAP CAT: Cathartic, astringent.

8270. Ruscogenin. *(25R)-Spirost-5-ene-1β,3β-diol;* Ruscorectal. $C_{27}H_{42}O_4$; mol wt 430.61. C 75.31%, H 9.83%, O 14.86%. A sapogenin from *Ruscus aculeatus* L., *Liliaceae.* Isoln: Lapin, Sannié, *Bull. Soc. Chim. France* **1955**, 1522; Gonzalez *et al., Tetrahedron* **28**, 128 (1972). Structure: Benn *et al., J. Am. Chem. Soc.* **79**, 3920 (1957); Burn *et al., J. Chem. Soc.* **1958**, 795.

Platelets, mp 205-210° (dried); $[\alpha]_D^{19}$ −127°.
Diacetate, $C_{31}H_{46}O_6$, needles from methanol, mp 194-196°. $[\alpha]_D$ −82° (c = 0.2 in $CHCl_3$); also reported as $[\alpha]_D^{20}$ −68° (c = 0.33 in $CHCl_3$): Burn *et al., loc. cit.*
THERAP CAT: Treatment of hemorrhoids.

8271. Rutecarpine. *8,13-Dihydroindolo[2',3':3,4]pyrido[2,1-b]quinazolin-5(7H)-one;* rutaecarpine. $C_{18}H_{13}N_3O$; mol wt 287.31. C 75.24%, H 4.56%, N 14.63%, O 5.57%. From fruit of *Evodia rutaecarpa* Hook. & Thoms. and *Hortia arborea* Engl., *Rutaceae:* Asahina, Kashiwaki, *J. Pharm. Soc. Japan* **1915**, 1293; Pachter *et al., J. Am. Chem. Soc.* **82**, 5187 (1960). Structure: Y. Asahina, *J. Pharm. Soc. Japan* **1924**, 1. Synthesis: Y. Asahina *et al., J. Chem. Soc.* **1927**, 1708; T. Kametani *et al., J. Am. Chem. Soc.* **99**, 2306 (1977); *eidem, Chem. Pharm. Bull.* **26**, 1922 (1978); H. Möhrle *et al., Arch. Pharm.* **313**, 990 (1980); J. Bergman, S. Bergman, *Heterocycles* **16**, 347 (1981). Simple synthesis: J. Kökösi *et al., Tetrahedron Letters* **22**, 4861 (1981). Synthesis under physiological conditions: Schöpf, *Angew. Chem.* **50**, 779, 797 (1937). Biosynthesis: M. Yamazaki *et al., Tetrahedron Letters* **1966**, 3221; **1967**, 3317. Mass spec.: J. Tamas *et al., Acta Chim. Acad. Sci. Hung.* **89**, 85 (1976).

Needles from ethyl acetate, mp 259.5-260°. uv max (ethanol): 278, 290, 332, 345, 364 nm (log ε 3.83, 3.88, 4.49, 4.54, 4.44). Sol in alc, benzene, chloroform, ether; practically insol in water.

8272. Ruthenium. Ru; at. wt 101.07; at. no. 44; valences 1-8; most common states 2, 3, 4. Naturally occurring isotopes: 96 (5.46%); 98 (1.87%); 99 (12.63%); 100 (12.53%);

101 (17.02%); 102 (31.6%); 104 (18.87%); artificial radioactive isotopes: 93-95; 97; 103; 105-108. Belongs to the platinum group of metals. Found in the minerals osmiridium, laurite, in platinum ores, in some copper-nickel ores. Discovered in 1828 by Osann; prepd in the pure state in 1845 by Klaus. Constitutes about 0.0004 ppm of the crust of the earth. Prepn from osmiridium: Fremy *et al.*, cited by Mellor, *A Comprehensive Treatise on Inorganic and Theoretical Chemistry* **15**, 499 (1936). Reviews of prepn, properties and chemistry of ruthenium and other platinum metals: Gilchrist, *Chem. Rev.* **32**, 277-372 (1943); W. P. Griffith, *The Chemistry of The Rarer Platinum Metals* (John Wiley, New York, 1967) pp 1-41, 126-226; Livingstone in *Comprehensive Inorganic Chemistry* vol. 3, J. C. Bailar, Jr. *et al.*, Eds. (Pergamon Press, Oxford, 1973) pp 1163-1209.

Lustrous, hard metal; hexagonal, close-packed structure. d_4^{20} 12.45. mp about 2450°; bp about 4150°. Sp heat (0°): 0.057 cal/g/°C. Does not react with acids, even aqua regia. Not oxidized by air in the cold; on heating combines readily with oxygen; the powdered metal forms the dioxide on igniting in air. Superficially attacked by concd alkaline hypochlorites. The powdered metal is attacked by chlorine above 200°; by bromine between 300-700°. Oxidized by fused alkali hydroxides. Forms alloys with platinum, palladium, cobalt, nickel, tungsten; forms definite compds with zinc and with tin.

USE: As substitute for platinum in jewelry; for pen nibs; as hardener in electrical contact alloys, electrical filaments; in ceramic colors; catalyst in synthesis of long chain hydrocarbons.

8273. Ruthenium Red. C.I. 77800; ruthenium oxychloride ammoniated. $Cl_6H_{42}N_{14}O_2Ru_3$; mol wt 786.35. Cl 27.05%, H 5.38%, N 24.94%, O 4.07%, Ru 38.56%. $[(NH_3)_5Ru-O-Ru(NH_3)_4-O-Ru(NH_3)_5]Cl_6$. Exists as tetrahydrate. Identification: A. Joly, *C.R. Acad. Sci.* **115**, 1299 (1893). Prepn and structure: Fletcher *et al.*, *J. Chem. Soc.* **1961**, 2000. Mössbauer study: Clausen *et al.*, *Inorg. Nucl. Chem. Letters* **7**, 485 (1971). Alternate structure and composition: Sterling, *Am. J. Bot.* **57**, 172 (1970). Use as microscopic stain: L. Mangin, *C.R. Acad. Sci.* **116**, 653 (1893); in electron microscopy: R. Dierichs, *Histochemistry* **64**, 171 (1979). Differential staining: P. R. Blanquet, *ibid.* **47**, 175 (1976); M. G. Gutierrez-Gonzalvez *et al.*, *J. Microscopy* **145**, 333 (1987). Series of articles on chemistry, mechanism of action, specific uses: *Anat. Rec.* **171**, 347-442 (1971). Conversion *in situ* to *ruthenium violet*: J. R. Luft, *ibid.* 347; M. G. Guiterrez-Gonzalvez *et al.*, *loc. cit.*

Brownish-red powder. Sol in water, ammonia.

USE: In microscopy as a general stain.

8274. Ruthenium Tetroxide. Ruthenium(VIII) oxide. O_4Ru; mol wt 165.70. Ru 61.38%, O 38.62%. RuO_4. Prepd by fusing Ru with $KMnO_4$ and KOH in a proportion of 1:2:20. Lab procedures: Ruff, Vidic, *Z. Anorg. Allgem. Chem.* **136**, 49 (1924); Grube in *Handbook of Preparative Inorganic Chemistry* vol. 2, G. Brauer, Ed. (Academic Press, New York, 2nd ed., 1965) p 1599. Crystal structure: Tréhoux *et al.*, *Compt. Rend. Ser. C* **268**, 246 (1969). Review of use as oxidizing agent: P. N. Rylander, *Organic Syntheses with Noble Metal Catalysts* (Academic Press, New York, 1973) pp 134-144.

Golden-yellow monoclinic prisms. Very volatile, sublimes at room temp. *Handle in hood only. Vapors irritating to eyes and respiratory tract.* mp 25.4°. bp 40°. Sparingly sol in water, about 2.0% (w/v) at 20°. Freely sol in carbon tetrachloride, other chlorinated solvents. Also sol in bromine, liquid SO_2. Reacts explosively with alcohol and filter paper.

USE: Oxidizing agent similar to osmium tetroxide, but more difficult to handle. The solvents normally employed in OsO_4 oxidations (ether, benzene, pyridine) cannot be used because of their violent reaction with RuO_4. Only CCl_4 is recommended: Djerassi, Engle, *J. Am. Chem. Soc.* **75**, 3838 (1953).

8275. Ruthenium Trichloride. Cl_3Ru; mol wt 207.47. Cl 51.27%, Ru 48.73%. $RuCl_3$. Two forms, α and β, are known. Prepn: Fletcher *et al.*, *Nature* **199**, 1089 (1963).

α-$RuCl_3$, black lustrous crystals. Insol in alcohol, water.

β-$RuCl_3$, dark-brown, fluffy, hexagonal crystals. Sol in alcohol.

8276. Rutin. 3-*[[6-O-(6-Deoxy-α-L-mannopyranosyl)-β-D-glucopyranosyl]oxy]-2-(3,4-dihydroxyphenyl)-5,7-dihydroxy-4H-1-benzopyran-4-one;* rutoside; rutinoside; 3,3',4',5,7-pentahydroxyflavone-3-rutinoside; melin; phytomelin; eldrin; ilixathin; sophorin; globularicitrin; paliuroside; osyritrin; osyritin; myrticolorin; violaquercitin; Birutan; Rutabion; Rutozyd; Tanrutin. $C_{27}H_{30}O_{16}$; mol wt 610.51. C 53.11%, H 4.95%, O 41.93%. Identity with ilixanthin: Schindler, Herb, *Arch. Pharm.* **288**, 372 (1955). Found in many plants, especially the buckwheat plant *(Fagopyrum esculentum* Moench., *Polygonaceae)* which contains about 3% (dry basis): Couch *et al.*, *Science* **103**, 197 (1946). From tobacco *(Nicotiana tabacum* L., *Solanaceae)* Couch, Krewson, *U.S. Dept. Agr., Eastern Regional Res. Lab.*, **AIC-52** (1944). In forsythia [*Forsythia suspensa* (Thunb.) Vahl. var. *fortunei* (Lindl.) Rehd., *Oleaceae*], in hydrangea (*Hydrangea paniculata* Sieb., *Saxifragaceae*), in pansies *(Viola* sp., *Violaceae*). General extraction procedure: *Beilstein* **XXXI**, 376. From leaves of *Eucalyptus macroryncha* F. v. Muell., *Myrtaceae:* Attree, Perkin, *J. Chem. Soc.* **1927**, 234. Industrial production from *Eucalyptus* spp.: Humphreys, *Econ. Bot.* **18**, 195 (1964). Structure: Zemplén, Gerecs, *Ber.* **68B**, 1318 (1935). Synthesis: Shakhova *et al.*, *Zh. Obshch. Khim.* **32**, 390 (1962), *C.A.* **58**, 1426e (1963). Rutin is hydrolyzed by rhamnodiastase from the seed of *Rhamnus utilis* Decne, *Rhamnaceae* (Chinese buckthorn); emulsin is not effective: Bridel, Charaux, *Compt. Rend.* **181**, 925 (1925). Toxicity data: Harrison *et al.*, *J. Am. Pharm. Assoc.* **39**, 557 (1950). *Book:* J. Q. Griffith, Jr., *Rutin and Related Flavonoids* (Mack, Easton, Pa., 1955). Comprehensive description: T. I. Khalifa *et al.* in *Analytical Profiles of Drug Substances* vol. 12, K. Florey, Ed. (Academic Press, New York, 1983) pp 623-681.

Pale yellow needles from water, gradual darkening on exposure to light. The crystals contain 3 H_2O and become anhydr after 12 hrs at 110° and 10 mm Hg. Anhydr rutin browns at 125°, becomes plastic at 195-197°, and dec 214-215° (with effervescence). $[\alpha]_D^{23}$ +13.82° (ethanol); $[\alpha]_D^{23}$ −39.43° (pyridine). Anhydr rutin is hygroscopic. One gram dissolves in about 8 liters water, about 200 ml boiling water, 7 ml boiling methanol. Sol in pyridine, formamide and alkaline solns; slightly sol in alcohol, acetone, ethyl acetate; practically insol in chloroform, carbon bisulfide, ether, benzene, petr solvents. Dil solns give green color with ferric chloride. Rutin is colored brown by tobacco enzyme under experimental conditions: Neuberg, Kobel, *Naturwiss.* **23**, 800 (1935). LD_{50} i.v. in mice: 950 mg/kg (propylene glycol soln) (Harrison).

THERAP CAT: Capillary protectant.

8277. Rutinose. 6-O-(6-Deoxy-α-L-*mannopyranosyl)*-D-glucose; 6-O-α-L-rhamnosyl-D-glucose. $C_{12}H_{22}O_{10}$; mol wt 326.30. C 44.17%, H 6.80%, O 49.03%. A disaccharide present in glycosides. Prepn from rutin by enzymatic hydrolysis using rhamnodiastase: Charaux, *Compt. Rend.* **178**, 1312 (1924); *Bull. Soc. Chim. Biol.* **6**, 634 (1924); Zemplén, Gerecs, *Ber.* **68B**, 1320 (1935). Synthesis: *eidem, Ber.* **67B**, 2049 (1934); **68B**, 1318 (1935). Structure: Gorin, Perlin, *Can. J. Chem.* **37**, 1930 (1959).

Very hygroscopic powder from alcohol and ether turning into a syrup on exposure to air. The freshly pptd material loses about 6% of its wt at 100°. The dried rutinose softens around 140°, dec 189-192°. $[\alpha]_D^{20}$ +3.2° → 0.8° (c = 4 in water). $[\alpha]_D^{20}$ −10° (95% alc). Very sol in water. Sol in alc. Hydrolysis with dil HCl yields 1 mol L-rhamnose and 1 mol D-glucose.

Heptaacetylrutinose, mp 169-170°. $[\alpha]_D^{20}$ −27.7° (chloroform).

8278. Ryania. A genus of tropical American shrubs and trees belonging to the *Flacourtiaceae* family. The wood of various species is insecticidal. Ground stem wood of *Ryania speciosa* Vahl., *Flacourtiaceae* is employed in the commercial insecticide formulations **Ryanex, Ryanicide** (formerly). *See* Folkers *et al.*, U.S. pat. **2,400,295** (1946 to Merck & Co.); Pepper, Carruth, *J. Econ. Entomol.* **38**, 59 (1945); Heal, *Agr. Chem.* **4**, 37 (May, 1949). Insecticidal components such as ryanodine, *q.v.*, are extractable by water, chloroform, or methanol. Toxicity data: Kuna, Heal, *J. Pharmacol. Exp. Ther.* **93**, 407 (1948).

LD$_{50}$ orally in rats, mice, rabbits, guinea pigs (mg/kg): 1200, 650, 650, 2500 (Kuna, Heal).

USE: Insecticide. *Toxicity:* In exptl animals causes weakness, deep and slow respiration, vomiting, diarrhea, tremors, convulsions, coma; death may occur.

8279. Ryanodine. *Ryanodol 3-(1H-pyrrole-2-carboxylate).* $C_{25}H_{35}NO_9$; mol wt 493.54. C 60.84%, H 7.15%, N 2.84%, O 29.18%. Insecticidal principle isolated from *Rya-*

nia speciosa Vahl., *Flacourtiaceae:* E. F. Rogers *et al.*, *J. Am. Chem. Soc.* **70**, 3086 (1948); R. B. Kelly *et al.*, *Can. J. Chem.* **29**, 905 (1951). Structure: D. R. Babin *et al.*, *Experientia* **21**, 425 (1965); J. Santroch *et al.*, *ibid.* **21**, 730 (1965); K. Wiesner *et al.*, *Tetrahedron Letters* **1967**, 221; K. Wiesner, *Advan. Org. Chem.* **8**, 295 (1972). Crystal structure: S. N. Srivastava, M. Przybylska, *Can. J. Chem.* **46**, 795 (1968). ^{1}H-NMR: A. L. Waterhouse *et al.*, *Chem. Commun.* **1984**, 1265. Synthetic studies: P. Deslongchamps, *Pure Appl. Chem.* **49**, 1329 (1977). Effect on calcium ion uptake by cardiac sarcoplasmic reticulum vesicles: L. R. Jones *et al.*, *J. Pharmacol. Exp. Ther.* **209**, 48 (1979); L. R. Jones, S. E. Cala, *J. Biol. Chem.* **256**, 11809 (1981).

Crystals, dec 219-220°. $[\alpha]_D^{25}$ +26° (methanol). uv max (alc): 268.5 nm (log ε 4.18). Sol in water, alcohol, acetone, ether, chloroform. Practically insol in benzene, petr ether.
USE: Insecticide.

S

8280. Sabadilla. Cevadilla; caustic barley. The dried ripe seeds of *Schoenocaulon officinale* (Schlecht. & Cham.) A. Gray (*Sabadilla officinarum* Brandt, *Asagraea officinalis* Lindl.), *Liliaceae*. *Habit.* The Andes of Mexico, Guatemala, Venezuela. *Constit.* About 3 to 6% total alkaloids of which cevadine (cryst veratrine) and veratridine (amorphous vera-trine) are the most important; other alkaloids are sabadinine, sabadilline, and sabadine; also contains sabadillic and vera-tric acids; fatty oil, resin. Toxicity data: E. D. Swiss, R. O. Bauer, *Proc. Soc. Exp. Biol. Med.* **76**, 847 (1951).

LD$_{50}$ i.p. in mice: 7.5 mg/kg (Swiss, Bauer).

THERAP CAT: Pediculicide.

THERAP CAT (VET): Has been used as an insecticide.

8281. Sabadine. (*3β,4α,16β*)-*Cevane-3,4,12,14,16,17,20-heptol 3-acetate*; sabatine. C$_{29}$H$_{47}$NO$_8$; mol wt 537.71. C 64.78%, H 8.81%, N 2.60%, O 23.80%. Isolated from Saba-dilla seed: Merck, *Arch. Pharm.* **229**, 164 (1891); Hennig *et al.*, *J. Am. Pharm. Assoc.* **40**, 168 (1951). Structure: Auter-hoff, Schwartz, *Arch. Pharm.* **288**, 549 (1955); Möhrle, Au-terhoff, *ibid.* **292**, 337 (1959); Kupchan *et al.*, *J. Med. Pharm. Chem.* **5**, 690 (1962). NMR data: Ito *et al.*, *Tetrahedron* **20**, 913 (1964).

Prisms, mp 256-258° (dec). [α]$_D^{25}$ −11° (c = 1.93 in alc).

8282. Saccharin. *1,2-Benzisothiazol-3(2H)-one 1,1-di-oxide; 2,3-dihydro-3-oxobenzisosulfonazole*; 1,2-dihydro-2-ketobenzisosulfonazole; saccharin insoluble; benzosulfimide; *o*-sulfobenzimide; benzoic sulfimide; *o*-sulfobenzoic acid imide; Gluside; Glucid; Garantose; Saccharinol; Saccharin-ose; Saccharol; Saxin; Sykose; Hermesetas. C$_7$H$_5$NO$_3$S; mol wt 183.18. C 45.89%, H 2.75%, N 7.65%, O 26.20%, S 17.50%. Prepn from *o*-sulfamoylbenzoic acid: Remsen, Fahlberg, *Ber.* **12**, 470 (1879). History and comparison with other sweetening agents: R. W. Moncrieff, *The Chemical Senses* (Wiley, New York, 1946). Mfg processes: O. Beyer, *Handbuch der Saccharinfabrikation* (Rascher, Zürich, 1923); *FIAT Rep., PB* **901** (1945); *Chem. Eng. (New York)* **61**, 128, 150 (July 1954). Maumee Chemical's process: *Chem. & Eng. News* **41**, 76-78 (Dec. 9, 1963). Stability studies: DeGarmo *et al.*, *J. Am. Pharm. Assoc. (Sci. Ed.)* **41**, 17 (1952). Carcinogenicity study: G. T. Bryan *et al.*, *Science* **168**, 1238 (1970). Evidence for induction of morphine toler-ance in genetically selected rats by chronically elevated sac-charin intake: I. Lieblich *et al.*, *Science* **221**, 871 (1983). Series of articles on toxicology and metabolism: *Food Chem. Toxicol.* **23**, 419-546 (1985). Review of toxicology and car-cinogenicity studies: D. L. Arnold, *Fundam. Appl. Toxicol.* **4**, 674 (1984). *Review:* R. Mazur, "Sweeteners" in Kirk-Othmer *Encyclopedia of Chemical Technology* vol. 22 (Wiley-Interscience, New York, 3rd ed., 1983) pp 448-464.

Monoclinic crystals, mp 228.8-229.7°. Twinning on (001). Perfect 100 cleavage. Acicular crystals by vacuum sublima-tion. In dil aq soln it is 500 times as sweet as sugar; the sweet taste is still detectable in 1:100,000 dilution. Bitter, metallic aftertaste. d 0.828. Heat of combustion at constant volume: 4753.1 cal/g. uv max (0.1*N* NaOH): 267.3 nm (ε 1570). One gram dissolves in 290 ml water, 25 ml boiling water, 31 ml alcohol, 12 ml acetone, about 50 ml glycerol; freely sol in solns of alkali carbonates; slightly sol in chloro-form, ether. pH of 0.35% aq soln 2.0. Hydrolysis products of saccharin are *o*-sulfamoylbenzoic acid (alkaline hydroly-sis) and ammonium *o*-sulfobenzoic acid (acid hydrolysis).

Ammonium salt, C$_7$H$_8$N$_2$O$_3$S, *saccharin ammonium*, *Daramin, Sucline.*

Sodium salt dihydrate, C$_7$H$_4$NNaO$_3$S.2H$_2$O, *soluble sac-charin, saccharin sodium, Kristallose, Crystallose, Dagutan, Sucaryl, Sucromat.* Crystalline powder; effloresces in dry air. In dil aq soln it is 300-500 times as sweet as sugar. One gram dissolves in 1.2 ml water, in about 50 ml alcohol. Aq solns are neutral or alkaline to litmus, but not alkaline to phenolphthalein. Anhydrous form, *Sucrédulcor.*

Note: This substance may reasonably be anticipated to be a carcinogen: *Fourth Annual Report on Carcinogens* (NTP 85-002, 1985) p 179.

USE: Non-caloric sweetener; pharmaceutical aid (flavor). In formulations for electroplating-bath brighteners.

8283. L-Saccharopine. *N-(5-Amino-5-carboxypentyl)-L-glutamic acid*; ε-*N*-(L-glutar-2-yl)-L-lysine. C$_{11}$H$_{20}$N$_2$O$_6$; mol wt 276.29. C 47.82%, H 7.30%, N 10.14%, O 34.75%. A lysine precursor in the aminoadipic acid-lysine pathway in yeast: Darling, Larsen, *Acta Chem. Scand.* **15**, 743 (1961); Kjaer, Larsen, *ibid.* 750; Kuo *et al.*, *Biochem. Biophys. Res. Commun.* **8**, 227 (1962); Trupin, Broquist, *J. Biol. Chem.* **240**, 2524 (1965); Jones, Broquist, *ibid.* **241**, 3430 sqq. (1966). Isoln from mycelium of the yeast *Candida utilis:* Marimoto, Yamano, *Biochem. Z.* **340**, 155 (1964).

Hydrated crystals; loses water of crystn over P$_2$O$_5$ at 100°. Dec 240-248°, when anhydr. [α]$_D^{23}$ +33.6° (c = 1 in 0.5*N* HCl); [M]$_D$ +93°. pK$_2$ 2.6; pK$_3$ 4.1; pK$_4$ 9.2; pK$_5$ 10.3. Sparingly sol in water, ethanol. Readily sol in alkaline solns and in strong acids.

8284. Safflower Oil. The oil from the seed of *Carthamus tinctorius* L., *Compositae*. Milling and extraction procedures: Winter, *J. Am. Oil Chem. Soc.* **27**, 82 (1950). Purification and stabilization: Freedman, Shapiro, U.S. pat. **2,978,381** (1961). Monograph: R. E. Woodward, G. M. Severson, *Industrial Survey of Safflower*, Chemurgy Dept. Report no. S-3 (Agricult. Expt. Sta., Lincoln, Nebraska, 1951). Fatty acids present as glycerides: palmitic 6.4%, stearic 3.1%, arachidic 0.2%, oleic 13.4%, linoleic 76.6-79.0%, linolenic 0.04-0.13%.

Edible drying oil, intermediate between soybean and lin-seed oil. d$_{25}^{25}$ 0.9211-0.9215. Titer: 15-18°. n$_D^{25}$ 1.472-1.475. n$_D^{40}$ 1.4690-1.4692. Acid value 1.0-9.7. Saponif value 188-194. Iodine value 140-150. Thiocyanogen value 82.5-86.0. Reichert-Meissl value below 0.5. Hydroxyl value 2.9-6.0. Unsaponifiable below 1.5%. Thickens and becomes rancid on prolonged exposure to air. Sol in the usual oil and fat solvents.

USE: As linseed oil in paints. For salad oil blends, in hy-drogenated state as shortening.

THERAP CAT: Dietary supplement in hypercholesteremia (and possible prophylaxis and treatment of atherosclerosis).

8285. Saffron. Crocus; Spanish saffron; French saffron. Stigmas of *Crocus sativus* L., *Iridaceae*. *Habit.* Western Asia, Southern Europe. *Constit.* About 1% volatile oil, pi-crocrocin—a bitter glucoside, crocin (polychroit), fixed oil, wax.

Flattish-tubular, almost thread-like stigmas ~1.25 inches (3 cm) long; orange-brown color; strong, peculiar, aromatic odor; bitterish, aromatic taste.

USE: Almost entirely used for coloring and flavoring.

8286. Safranal. *2,6,6-Trimethyl-1,3-cyclohexadiene-1-carboxaldehyde;* 2,2,6-trimethyl-4,6-cyclohexadien-1-aldehyde. $C_{10}H_{14}O$; mol wt 150.21. C 79.95%, H 9.39%, O 10.65%. Prepn from picrocrocin: Kuhn, Winterstein, *Ber.* **67**, 354 (1934). Synthesis: Kuhn, Wendt, *Ber.* **69**, 1549 (1936); Könst *et al.*, *Tetrahedron Letters* **1974**, 3175. Biogenetic synthesis: T. Kametani *et al.*, *Chem. Pharm. Bull.* **29**, 105 (1981).

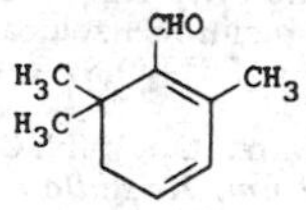

Liquid. d_4^{19} 0.9734. $bp_{1.0}$ 70° (bath temp). n_D^{19} 1.5281. Sol in methanol, ethanol, petr ether, glacial acetic acid.

8287. Safrole. *5-(2-Propenyl)-1,3-benzodioxole; 4-allyl-1,2-methylenedioxybenzene;* allylcatechol methylene ether; allyldioxybenzene methylene ether; *m*-allylpyrocatechin methylene ether. $C_{10}H_{10}O_2$; mol wt 162.18. C 74.05%, H 6.22%, O 19.73%. Constituent of several essential oils, notably of sassafras in which it is present to the extent of about 75%. Has been used as topical antiseptic; pediculicide. Metabolism studies: Oswald *et al.*, *Biochim. Biophys. Acta* **230**, 237 (1971). Review and evaluation of studies of carcinogenic action in laboratory animals: *IARC Monographs* **10**, 231-241 (1976). Toxicity data: Hagan *et al.*, *Toxicol. Appl. Pharmacol.* **7**, 18 (1965).

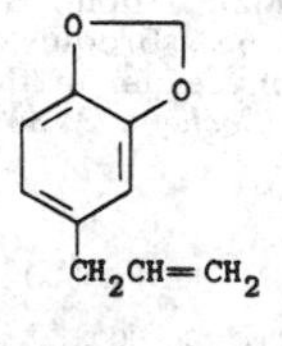

Colorless or slightly yellow liq; sassafras odor. d^{20} 1.096. mp ~11°. bp 232-234°. n_D^{20} 1.5383. Insol in water. Very sol in alcohol. Miscible with chloroform, ether. LD_{50} in rats, mice (mg/kg): 1950, 2350 orally (Hagan).
Note: This substance may reasonably be anticipated to be a carcinogen: *Fourth Annual Report on Carcinogens* (NTP 85-002, 1985) p 180.
USE: In perfumery; denaturing fats in soap manuf; also in manuf of heliotropin.

8288. SAICAR. *N-[[5-Amino-1-(5-O-phosphono-β-D-ribofuranosyl)-1H-imidazol-4-yl]carbonyl]-L-aspartic acid; N-[(5-amino-1-ribofuranosylimidazol-4-yl)carbonyl]aspartic acid 5'-phosphate;* 5-amino-N-(1,2-dicarboxyethyl)-1-ribofuranosylimidazole-4-carboxamide 5'-phosphate; 5-amino-4-imidazole-N-succinocarboxamide ribonucleotide; 5-amino-4-imidazole-N-succinocarboxamide ribotide; N-[5-amino-1-(5'-phosphoribofuranosyl)-4-imidazolecarbonyl]aspartic acid; succino-AICAR. $C_{13}H_{19}N_4O_{12}P$; mol wt 454.29. C 34.37%, H 4.22%, N 12.33%, O 42.26%, P 6.82%. Important purine precursor. Synthesis by enzymic reaction according to the equation AIR + ATP + aspartic acid + CO_2 → SAICAR + ADP + orthophosphate:

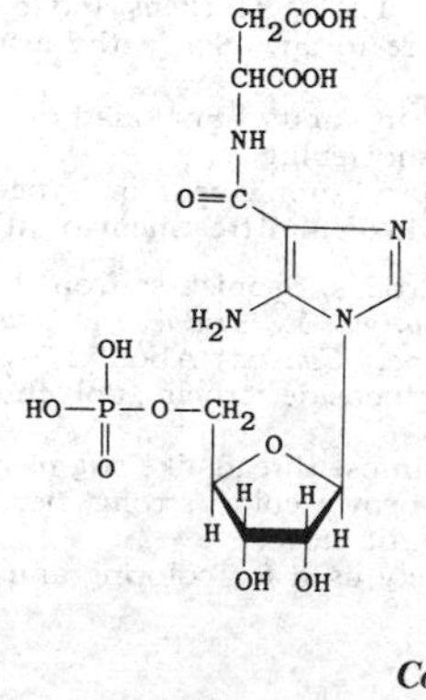

Lukens, Buchanan, *J. Biol. Chem.* **234**, 1791 (1959). Production by adenine-requiring mutants of *Bacillus subtilis:* Ohmura *et al.*, U.S. pat. **3,280,007** (1966 to Takeda). Synthesis of SAICAR and derivatives: Burrows, Shaw, *J. Chem. Soc. (C)* **1967**, 1088; Burrows *et al.*, *ibid.* **1968**, 40.
Dibarium salt, $C_{13}H_{15}Ba_2N_4O_{12}P$. uv max (pH 1): 269, 244 nm (ε 11,850, 9580). If the Bratton-Marshall procedure [*J. Biol. Chem.* **128**, 537 (1939)] is modified by keeping the sample in ice during, and for 10 minutes after the addition of the reagents, SAICAR produces a purple chromophore having max absorption at 550 nm. The diazonium salt of SAICAR is very unstable and dec before it can couple under the usual test conditions.

8289. Sakuranetin. *2,3-Dihydro-5-hydroxy-2-(4-hydroxyphenyl)-7-methoxy-4H-1-benzopyran-4-one; 4',5-dihydroxy-7-methoxyflavanone;* naringen 7-methyl ether. $C_{16}H_{14}O_5$; mol wt 286.27. C 67.12%, H 4.93%, O 27.95%. Isoln by hydrolysis of 5-glucoside obtained from bark of *Prunus yedoensis* Matsum., *Rosaceae:* Asahina, *Arch. Pharm.* **246**, 259 (1908); from wood of various *Prunus* species: Hasegawa, Shirato, *J. Am. Chem. Soc.* **76**, 5559 (1954); **77**, 3557 (1955); Hasegawa, *ibid.* **79**, 1738 (1957). Structure: Asahina *et al.*, *J. Pharm. Soc.* Japan No. **550**, 1007 (1927); Shinoda, Sato, *ibid.* **48**, 220 (1928). Synthesis: Zemplén *et al.*, *Ber.* **75B**, 1432 (1942). Absolute configuration of (+)-isomer: Arakawa, Nakazaki, *Ann.* **636**, 111 (1960).

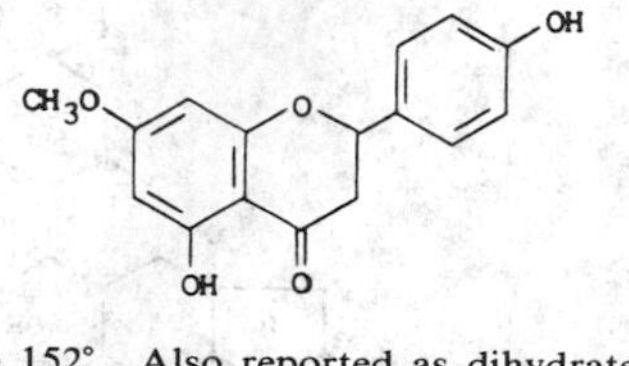

Needles, mp 152°. Also reported as dihydrate, mp 131-132° (Arakawa, Nakazaki, *loc. cit.*). (+)-Isomer, $[\alpha]_D^{16}$ +8.0° (c = 7.92 in acetone); $[\alpha]_D^{12}$+8.4° (c = 6.28 in methanol). Sol in alcohol, ether, chloroform, benzene, ethyl acetate, pyridine; slightly sol in boiling water. Practically insol in cold water.
5-Glucoside, $C_{22}H_{24}O_{10}$, *4',5-dihydroxy-7-methoxyflavanone 5-glucoside, sakuranin.* Tetrahydrate, bitter needles from dil alcohol. Begins to lose water at 100°. mp 212-214°. $[\alpha]_D^{28}$ −106.6° (tetrahydrate in acetone); $[\alpha]_D^{28}$ −123.2° (anhydr compd in acetone); $[\alpha]_D^{16}$ −97.4° (c = 1.58 for anhydr compd in 75% acetone). Very sol in dil alcohol, pyridine; slightly sol in abs alcohol, hot water, ethyl acetate. Practically insol in ether, cold water.

8290. Salacetamide. *N-Acetyl-2-hydroxybenzamide; N-acetylsalicylamide;* acetsalicylamide; Actylamide; Labazyl (obsolete). $C_9H_9NO_3$; mol wt 179.17. C 60.33%, H 5.06%, N 7.82%, O 26.79%. Prepn: Hicks, *J. Chem. Soc.* **97**, 1033 (1910); Anshutz, Mitarbeiter, *Ann.* **442**, 25 (1925); **Belg. pat.** **496,438** (1950 to Soc. Belge de l'Azote Prod. Chim. Marly); Prost, Charlier, *Experientia* **18**, 319 (1962). Metabolism: Rayet *et al.*, *Arch. Int. Pharmacodyn. Ther.* **88**, 159 (1951).

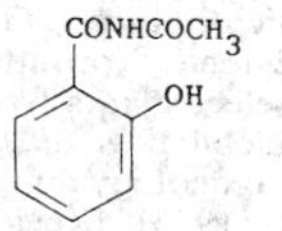

mp 148°.
THERAP CAT: Analgesic, antipyretic, anti-inflammatory.

8291. Salazosulfadimidine. *5-[[4-[[(4,6-Dimethyl-2-pyrimidinyl)amino]sulfonyl]phenyl]azo]-2-hydroxybenzoic acid; 5-[p-[(4,6-dimethyl-2-pyrimidinyl)sulfamoyl]phenylazo]salicylic acid;* 4'-(4,6-dimethylpyrimidin-2-ylsulfamoyl)-4-hydroxyazobenzene-3-carboxylic acid; 5-[p-[(4,6-dimethyl-2-pyrimidinyl)aminosulfonyl]phenylazo]salicylic acid; salicylazosulfadimidine; salicylazosulfamethazine; Azudimidine. $C_{19}H_{17}N_5O_5S$; mol wt 427.45. C 53.38%, H 4.01%, N 16.39%, O 18.72%, S 7.50%. Prepn: Korkuczanski, *Przem. Chem.* **37**, 162 (1958), *C.A.* **52**, 13727

(1958); Malatesta, Lotti, *Ann. Chim. (Rome)* **50**, 114 (1960), *C.A.* **56**, 8613g (1962).

Brown crystals, mp 207°.
THERAP CAT: Antibacterial.

8292. Salep. Dried tubers of several species of *Orchis, Orchidaceae. Habit.* Europe, Asia Minor. *Constit.* Mucilage, starch.
USE: Nutrient, demulcent; as a vehicle for acrid remedies.

8293. Salicin. *2-(Hydroxymethyl)phenyl-β-D-glucopyranoside;* salicoside; salicyl alcohol glucoside; saligenin-β-D-glucopyranoside. $C_{13}H_{18}O_7$; mol wt 286.27. C 54.54%, H 6.34%, O 39.12%. Usually obtained by making hot water extracts from the ground bark of poplar *(Populus)* and willow *(Salix);* also found in the leaves and female flowers of the willow. Isoln from root bark of *Viburnum prunifolium* L., *Caprifoliaceae:* Evans *et al., J. Am. Pharm. Assoc.* **34**, 207 (1945). Structure and synthesis: Irvine, Rose, *J. Chem. Soc.* **89**, 814 (1906); Kunz, *J. Am. Chem. Soc.* **48**, 262 (1926). Hydrolysis by acids: Moelwyn-Hughes, *Trans. Faraday Soc.* **25**, 503 (1929). Enzymatic hydrolysis: Pigman, *J. Res. Nat. Bur. Stand.* **27**, 6 (1941). Enzymatic hydrolysis in heavy water: Stacie, *Z. Phys. Chem.* **28B**, 236 (1935).

Orthorhombic crystals from water, mp 199-202°. $[\alpha]_D^{25}$ −62° to −67° (c = 3). $[\alpha]_D^{20}$ −45.6° (c = 0.6 in abs alc), Brauns, *J. Am. Chem. Soc.* **47**, 1292 (1925). One gram dissolves in 23 ml water, 3 ml boiling water, 90 ml alcohol, 30 ml alcohol at 60°. Soluble in alkalies, pyridine, glacial acetic acid. Practically insol in ether, chloroform. Aq solns are neutral to litmus and have a bitter taste.
USE: Standard substrate in evaluating enzyme prepns contg β-glucosidase: Weidenhagen, *Z. Ver. Zucker-Ind.* **79**, 591 (1929).
THERAP CAT: Analgesic.
THERAP CAT (VET): Has been used as a bitter stomachic, as an antirheumatic and as an analgesic.

8294. Salicyl Alcohol. *2-Hydroxybenzenemethanol;* saligenin; saligenol; o-hydroxybenzyl alcohol. $C_7H_8O_2$; mol wt 124.13. C 67.73%, H 6.50%, O 25.78%. Prepd by the action of emulsin on salicin; by heating phenol with methylene chloride and aq NaOH.

Plates or cryst powder. d 1.16. mp 86-87°; sublimes at 100°. Soluble in 15 parts water; very sol in alcohol, chloroform, ether, sol in benzene. Gives a red color with H_2SO_4.
THERAP CAT: Local anesthetic.

8295. Salicylaldehyde. *2-Hydroxybenzaldehyde;* salicylic aldehyde. $C_7H_6O_2$; mol wt 122.12. C 68.84%, H 4.95%, O 26.20%. Made by heating sodium phenolate and chloroform with NaOH.

Clear, colorless, oily liq; bitter almond-like odor; burning taste. d_4^{20} 1.167; mp −7°; bp 196-197°; n_D^{20} 1.5735. Slightly sol in water; sol in alc, ether. Gives an orange color with H_2SO_4. MLD s.c. in rats: 1.0 g/kg, *Handbook of Toxicology*, W. S. Spector, Ed. (Saunders, Philadelphia, 1956) p 262.
USE: In perfumery.

8296. Salicylaldoxime. *2-Hydroxybenzaldehyde oxime;* Saldox. $C_7H_7NO_2$; mol wt 137.13. C 61.31%, H 5.15%, N 10.21%, O 23.33%. $HOC_6H_4CH=NOH$.
Prisms, mp 57°. Slightly sol in cold water; more sol in hot water; freely sol in alcohol, ether, benzene, dil HCl. Insol in petr ether. On heating it dec into salicylaldehyde and hydroxylamine.
USE: As a reagent for copper and nickel.

8297. Salicylamide. *2-Hydroxybenzamide;* Salamid; Samid; Acket; Algiamida; Cidal; Oramid; Panithal; Salizell; Novecyl; Dolomide; Salipur; Salrin; Salymid; Saliamin; Urtosal; Amid-Sal; Algamon; Benesal. $C_7H_7NO_2$; mol wt 137.13. C 61.31%, H 5.15%, N 10.21%, O 23.33%.

White or slightly pink, crystalline powder, somewhat bitter taste. Sensation of warmth on tongue. mp 140°. Soly in water at 30° = 0.2%, at 47° = 0.8%; in glycerol at 5° = 2.0%, at 39° = 5.0%, at 60° = 10.0%; in propylene glycol at 5° = 10.0%. Sol in hot water, alcohol, chloroform, ether. pH of satd aq soln at 28° about 5. Forms a water-soluble sodium salt at pH 9. LD_{50} orally in mice: 1.4 g/kg, Hart, *J. Pharmacol. Exp. Ther.* **89**, 205 (1947).
THERAP CAT: Analgesic.

8298. Salicylamide O-Acetic Acid. *[2-(Aminocarbonyl)phenoxy]acetic acid;* o-(carbamylphenoxy)acetic acid; α-(2-carbamoylphenoxy)acetic acid. $C_9H_9NO_4$; mol wt 195.17. C 55.38%, H 4.65%, N 7.18%, O 32.79%. Prepn: Merriman, *J. Chem. Soc.* **103**, 1838 (1913); Tzofin, *J. Gen. Chem. USSR* **3**, 17 (1933); Klosa, *Arch. Pharm.* **288**, 389 (1955); Izumi *et al.,* **Japan.** pat. 56 874 (1956 to Yoshitomi).

Crystals, mp 221°. Sol in aq alkali.
Sodium salt, *Salizell (amp.),* Algamon soluble. Crystals, mp 212-215°.
Diethylamine salt, $C_{13}H_{22}N_2O_4$, *Akistin.*
THERAP CAT: Analgesic, anti-inflammatory, antipyretic.

8299. Salicylanilide. *2-Hydroxy-N-phenylbenzamide;* N-phenylsalicylamide; Salinidol; Shirlan Extra. $C_{13}H_{11}NO_2$; mol wt 213.24. C 73.22%, H 5.20%, N 6.57%, O 15.01%. Usually made by the reaction of salicylic acid with aniline in the presence of PCl_3 at an elevated temp. Better yields are obtained by using an inert organic solvent such as toluene or carbon tetrachloride as a reaction diluent. Novel process using ion-exchange resins: Majewski, Skelly, U.S. pat. **3,221,051** (1965). Alternate process: Majewski *et al.,* U.S. pat. **3,231,611** (1966 to Dow).

Odorless leaflets. mp 135.8-136.2°. Slightly sol in water; freely sol in alcohol, ether, chloroform, benzene.
Human Toxicity: In concd form may cause irritation of skin, mucous membranes. *See also* Salicylic Acid.
USE: Anti-mildew, fungicide.

THERAP CAT: Topical antifungal.
THERAP CAT (VET): Topical antifungal.

8300. Salicylhydroxamic Acid. *N,2-Dihydroxybenzamide;* 2-hydroxybenzhydroxamic acid. $C_7H_7NO_3$; mol wt 153.13. C 54.90%, H 4.61%, N 9.15%, O 31.34%. HOC_6H_4-$CONHOH$ or $HOC_6H_4C(OH)=NOH$. Prepn: Jeanrenaud, *Ber.* **22**, 1272 (1889); Urbanski, *Nature* **166**, 267 (1950). Trypanocidal activity: F. Opperdoes *et al., Exp. Parasitol.* **40**, 198 (1976); A. B. Clarkson, F. H. Brohn, *Science* **194**, 204 (1976); A. J. Barnicoat *et al., Experientia* **37**, 1291 (1981).
Needles from acetic acid, mp 168° (slow heating), mp 176-178° (quick heating). Sublimes. K at 25° $= 6.4 \times 10^{-6}$.
Sodium salt, $C_7H_6NNaO_3$, plates. Freely sol in water. pH of 0.1N soln 7.7.
USE: Complexing agent.

8301. Salicylic Acid. *2-Hydroxybenzoic acid;* Keralyt; Verrugon. $C_7H_6O_3$; mol wt 138.12. C 60.87%, H 4.38%, O 34.75%. Occurs in the form of esters in several plants, notably in wintergreen leaves and the bark of sweet birch. Made synthetically by heating sodium phenolate with carbon dioxide under pressure. Large scale process details: Faith *et al., Industrial Chemicals* (Wiley, New York, 3rd ed., 1965) pp 652-655. Novel method by microbial oxidation of naphthalene: Zajic, Dunlap, U.S. pat. **3,274,074** (1966 to Kerr-McGee). Toxicity data: K. Sota *et al., J. Pharm. Soc. Japan* **89**, 1392 (1969).

COOH
OH

Acicular crystals, or crystalline powder, mp 157-159°. Gradually discolors in sunlight. d 1.44. bp_{20} about 211°. Sublimes at 76°; when rapidly heated at atm pressure it dec into phenol and CO_2. One gram dissolves in 460 ml water, 15 ml boiling water, 2.7 ml alcohol, 3 ml acetone, 42 ml chloroform, 3 ml ether, 135 ml benzene, 52 ml oil turpentine, about 60 ml glycerol, about 80 ml fats or oils. Soly in water increased by sodium phosphate, borax, alkali acetates or citrates. pH of satd aq soln 2.4. *Keep protected from light.* LD_{50} i.v. in mice: 500 mg/kg (Sota).
Salicylic acid or its salts are colored reddish even by merest traces of ferric salts. *Incompat:* Iron salts, spirit nitrous ether, lead acetate, iodine.
Lithium salt, $C_7H_5LiO_3$, *lithium salicylate.* White or grayish-white, odorless, sweetish powder. Deliquesc in moist air. Very sol in water, alcohol; the aq soln is slightly acid to litmus. *Keep well closed and protected from light.*
Silver salt, $C_7H_5AgO_3$, *silver salicylate.* White to reddish-white crystals. Slightly sol in water, alc. *Protect from light.*
Human Toxicity: Ingestion of large amounts can cause vomiting, abdominal pain, increased respiration, acidosis, mental disturbances. May cause skin rashes in sensitive individuals. *See: Clinical Toxicology of Commercial Products,* R. E. Gosselin *et al.,* Eds. (Williams & Wilkins, Baltimore, 4th ed., 1976) pp 295-303.
USE: As preservative of food products, but its use for this purpose is forbidden in some countries; manuf methyl salicylate, acetylsalicylic acid and other salicylates, dyes; as reagent in analytical chemistry.
THERAP CAT: Topical keratolytic.
THERAP CAT (VET): Externally as an antiseptic and antifungal agent, and for various skin conditions. Keratolytic. Has been used internally in equine gastric tympany.

8302. 4-Salicyloylmorpholine. *4-(2-Hydroxybenzoyl)-morpholine;* salicyl morpholide; L 1102; Deposal; Retarcyl; Tardisal. $C_{11}H_{13}NO_3$; mol wt 207.22. C 63.75%, H 6.32%, N 6.76%, O 23.16%. Prepn: Carron, *Fr.* pat. **1,096,209** (1955 to Robert & Carrière); Schweitzer, U.S. pat. **2,906,728** (1959 to Dow).

OH
CO-N   O

Crystals, mp 175°. Soly (g/100 ml): water 0.41; alcohol 3.3; ether 0.22. pH of satd aq soln 6.2.
THERAP CAT: Choleretic.

8303. Salicylsulfuric Acid. *2-(Sulfooxy)benzoic acid;* salicylic acid, acid sulfate; salicylic acid sulfuric acid ester. $C_7H_6O_6S$; mol wt 218.18. C 38.53%, H 2.77%, O 44.00%, S 14.70%. Prepd by treating salicylic acid with chlorosulfonic acid in pyridine: Loeper *et al., Compt. Rend. Soc. Biol.* **135**, 917 (1941); *Chim. Ind. (Paris)* **49**, 99 (1943).

COOH
OSO₃H

Monosodium salt, *sodium salicylsulfate, Salcyl, Salcylix.* Fine needles. Sol in water. Insol in organic solvents.
THERAP CAT: Analgesic, anti-inflammatory.

8304. Salinazid. *[(2-Hydroxyphenyl)methylene]hydrazide 4-pyridinecarboxylic acid;* 1-isonicotinoyl-2-salicylidenehydrazine; isonicotinic acid salicylidenehydrazide; saliniazid; *N'*-*o*-hydroxybenzylidenepyridine-4-carboxyhydrazide; *o*-hydroxybenzal isonicotinylhydrazone; Salizid; Nupasal; Acozid; Nupasal-213. $C_{13}H_{11}N_3O_2$; mol wt 241.24. C 64.72%, H 4.60%, N 17.42%, O 13.26%. Prepd by the interaction of isoniazid and salicylaldehyde in water: Hart *et al., Antibiot. & Chemother.* **4**, 803 (1954); *cf.* Yale *et al., J. Am. Chem. Soc.* **75**, 1933 (1953). Tuberculostatic activity and toxicity: Bavin *et al., J. Pharm. Pharmacol.* **7**, 1032 (1955).

N
OH
CONHN==CH

Crystals from ethanol, mp 232-233° (also given as 251°). Soly at 25° in 100 ml water: 0.005 g; in 100 ml abs ethanol: 0.18 g; in 100 ml propylene glycol: 0.212 g. Sol in dil aq acids and alkalies forming a yellow soln.
THERAP CAT: Antibacterial (tuberculostatic).

8305. Salinomycin. Coxistac. $C_{42}H_{70}O_{11}$; mol wt 751.02. C 67.17%, H 9.40%, O 23.43%. Polyether ionophoric antibiotic having a unique tricyclic spiroketal ring system and an unsaturated 6-membered ring in the molecule. Produced by a strain of *Streptomyces albus* (FERM-P No. 419 and ATCC 21838). Production: Y. Miyazaki *et al.,* **Japan. Kokai 72 25392** (1972 to Kaken Chem.), *C.A.* **78**, 41561 (1973). Structure: H. Kinashi *et al., Tetrahedron Letters* **1973**, 4955. Taxonomy, production, isolation and physicochemical and biological properties: Y. Miyazaki *et al., J. Antibiot.* **27**, 814 (1974). Use as a coccidiostat: Y. Tanaka *et al.,* **Ger.** pat. **2,253,031;** *eidem,* U.S. pat. **3,857,948** (1973, 1974 to Kaken Chem.). Anticoccidial efficacy: T. T. Migaki *et al., Poultry Sci.* **58**, 1192 (1979). Total synthesis: Y. Kishi *et al.,* Front. Chem., *Plenary Keynote Lect. IUPAC Congr., 28th* **1981**, K. J. Laidler, Ed. (Pergamon, Oxford, 1982) pp 287-304. HPLC determn in animal feeds: M. R. LaPointe, H. Cohen, *J. Assoc. Offic. Anal. Chem.* **71**, 480 (1988).

HOOC
CH₃   CH₃   OH   CH₃   OH
CH₂
CH₃

mp 112.5-113.5°. pKa' 6.4 (DMF). $[\alpha]_D^{25}$ −63° (c = 1 in ethanol). uv max (ethanol-water, 2:1): 284 nm (ϵ 126). LD_{50} in mice (mg/kg): 18 i.p.; 50 orally (Miyazaki).

Sodium salt, $C_{42}H_{69}NaO_{11}$, *Salocin, Ovicox*. mp 140-142°. $[\alpha]_D^{25}$ −37° (c = 1 in ethanol).

THERAP CAT (VET): Anticoccidial agent.

8306. Salmine Sulfate. A protamine found in the sperm of salmon. Contains arginine, proline, serine, glycine, valine, leucine, alanine, threonine, isoleucine, lysine, histidine, aspartic and glutamic acids. Mol wt 6000 to 7000. N about 25%. Prepn and properties: Fisher, Scott, *J. Pharmacol. Exp. Ther.* **58**, 78 (1936); Felix, *Am. Sci.* **43**, 431 (1955); Callanan *et al., J. Biol. Chem.* **229**, 279 (1957); Carroll *et al., ibid.* **234**, 2314 (1959). Separation of components and amino acid sequence of major component: Ando, Watanabe, *Int. J. Protein Res.* **1**, 221 (1969).

Salmine base is a powder sol in disodium phosphate buffer; the sulfate separates from water in the cold as a clear, immiscible liquid. LD_{50} i.v. in rats: 75 mg/kg, Starbuck *et al., Arch. Int. Pharmacodyn. Ther.* **165**, 374 (1967). *See also* Protamines.

8307. Salsalate. *2-Hydroxybenzoic acid 2-carboxyphenyl ester;* disalicylic acid; salicylic acid bimolecular ester; salicyl-oxysalicylic acid; salicylsalicylic acid; NSC 49171; Arcylate; Diplosal; Disalcid; Disalgesic; Mono-Gesic; Salflex; Salysal. $C_{14}H_{10}O_5$; mol wt 258.22. C 65.12%, H 3.90%, O 30.98%. Nonacetylated aspirin analog. Prepn: **Ger. pats. 211,403** and **214,044** (1909, both to Boehringer, Mann.), *Frdl.* **9**, 928 and *C.A.* **4**, 368 (1910); W. Baker *et al., J. Chem. Soc.* **1951**, 201. Metabolism: S. M. Dromgoole *et al., J. Pharm. Sci.* **73**, 1657 (1984). Clinical evaluation in arthritis: T. C. McPherson, *Clin. Ther.* **6**, 388 (1984). Mechanism of action studies: C. A. Divincenzo, F. R. Venezio, *Curr. Ther. Res.* **42**, 720 (1987). HPLC determn in plasma and urine: L. I. Harrison *et al., J. Pharm. Sci.* **69**, 1268 (1980). Review of chemistry, pharmacokinetics, safety and clinical efficacy: P. T. Singleton, *Clin. Ther.* **3**, 80-102 (1980).

Crystals from benzene. mp 148-149°. Practically insol in water but gradually hydrolyzed by it into 2 mols of salicylic acid. Sol in alc, ether; sparingly sol in benzene.

THERAP CAT: Analgesic, anti-inflammatory.

8308. Salsoline. *1,2,3,4-Tetrahydro-7-methoxy-1-methyl-6-isoquinolinol;* 6-hydroxy-7-methoxy-1-methyl-1,2,3,4-tetrahydroisoquinoline. $C_{11}H_{15}NO_2$; mol wt 193.24. C 68.37%, H 7.82%, N 7.25%, O 16.56%. In *Salsola richteri* Karel., *Chenopodiaceae*. Extraction procedure: Orechoff, Proskurnina, *Ber.* **66**, 841 (1933); Proskurnina, Orekhov, *J. Gen. Chem. USSR* **7**, 1999 (1937); *Bull. Soc. Chim. France* **1937**, 1265. Structure and synthesis: Späth *et al., Ber.* **67**, 1214 (1934); Späth, Dengel, *Ber.* **71**, 113 (1938); Kovács, Fodor, *Ber.* **84**, 795 (1951). Abs config: Battersby, Edwards, *J. Chem. Soc.* **1960**, 1214. Biosynthetic studies: McFarlane, Slayton, *Phytochemistry* **11**, 235 (1972). Synthesis of the natural (+)- and the (−)-form: Teitel *et al., J. Med. Chem.* **17**, 134 (1974).

Crystals from alc, mp 221°. $[\alpha]_D^{20}$ +34.5° (c = 1 in 0.1N HCl). Sol in chloroform, hot alc, dil NaOH. Slightly sol in water, benzene; almost insol in ether, petr ether.

Hydrochloride, mp 174-175°. $[\alpha]_D$ +31.0° (methanol). uv max (isopropanol): 204, 227, 284, 286 nm (ϵ 39,400, 5900, 3540, 3530). Sol in water, hot alcohol. Very sparingly sol in acetone, chloroform. LD_{50} in mice: 140 mg/kg i.v.; > 1000 mg/kg orally, Teitel *et al., loc. cit.*

8309. Salutaridine. *5,6,8,14-Tetradehydro-4-hydroxy-*

3,6-dimethoxy-17-methylmorphinan-7-one; floripavine. $C_{19}H_{21}NO_4$; mol wt 327.39. C 69.70%, H 6.47%, N 4.28%, O 19.55%. Intermediate in morphine synthesis; naturally occurring in (+)-form. Isolated from various species of *Croton* and *Papaver:* Barton *et al., J. Chem. Soc.* **1965**, 2423; Pfeifer, Kühn, *Pharmazie* **23**, 267 (1968). Structure and identity with floripavine: Mndzhoyan *et al., C.A.* **68**, 114787w (1968). Synthesis: T. Kametani *et al., J. Chem. Soc. (C)* **1969**, 2030; G. Horvath, S. Makleit, *Acta Chim. Acad. Sci. Hung.* **106**, 37 (1981); of racemate: T. Kametani *et al., J. Chem. Soc. Perkin Trans. I* **1972**, 1435; C. Szántay *et al., J. Org. Chem.* **47**, 594 (1982); of racemate and (−)-form: W. Ludwig, H. J. Schafer, *Angew. Chem. Int. Ed.* **25**, 1025 (1986). *Review:* Stuart, *Chem. Revs.* **1971**, 47.

Rods from ethyl acetate, mp 197-198°. $[\alpha]_D^{12}$ +114° (c = 0.5 in methanol). uv max (methanol): 236, 279 nm (log ϵ 4.23, 3.76).

(−)-Form, *Sinoacculine*. $[\alpha]_D^{20}$ −98° (c = 0.55 g/100 ml CH_3OH). uv max (methanol): 276, 240 nm (log ϵ 3.76, 4.24).

8310. Salverine. *2-[2-(Diethylamino)ethoxy]-N-phenyl-benzamide; 2-[2-(diethylamino)ethoxy]benzanilide.* $C_{19}H_{24}N_2O_2$; mol wt 312.40. C 73.04%, H 7.74%, N 8.97%, O 10.24%. Prepn: **Austrian pat. 239,779** (1965 to Montavit), *C.A.* **63**, 4211h (1965).

Hydrochloride, $C_{19}H_{25}ClN_2O_2$, crystals, mp 173°. Ingredient in *Montamed*.

THERAP CAT: Analgesic.

8311. Salvia. Sage. Dried leaves of *Salvia officinalis* L., *Labiatae. Habit.* Southern Europe; cultivated in U.S., England, France. *Constit.* 1-2.5% volatile oil; resin, tannin, bitter principles.

USE: Flavoring, condiment.

THERAP CAT: Antisecretory agent.

8312. Samaderins. Antitumor agents isolated from the bark and seed of *Samadera indica* Gaertn, *Simaroubaceae:* van der Marck, *Arch. Pharm.* **239**, 96 (1901). Separable into three components, samaderins A, B and C: Polonsky *et al., Bull. Soc. Chim. France* **1962**, 1715. According to Zylber *et al., ibid.* **1963**, 1322, samaderin C is identical with *samaderoside A* isolated by Mitra, Gregg, *Naturwiss.* **14**, 327 (1962). Structure of B and C: Zylber, Polonsky, *Bull. Soc. Chim. France* **1964**, 2016. Structure of A: M. C. Wani *et al., Chem. Commun.* **1977**, 295. Crystal structure of A: K. D. Onan, A. T. McPhail, *J. Chem. Res.(S)* **1978**, 14.

samaderin A

samaderins B and C

Samaderin A, $C_{18}H_{18}O_6$. mp 253-255°. $[\alpha]_D^{25}$ −31.3° (c = 0.259 in pyridine). uv max (methanol): 288 nm (ϵ 13400).

Samaderin B, $C_{19}H_{22}O_7$. R = O. Prisms from ethyl acetate, mp 235-240°.

Samaderin C, $C_{19}H_{24}O_7$. R = OH. Shiny plates from ethyl acetate+ alcohol, mp 265-268°. Sol in pyridine; practically insol in ether; very sparingly sol in other org solvents.

8313. Samandarine. $1\alpha,4\alpha$-*Epoxy-3-aza-A-homoandrostan-16β-ol.* $C_{19}H_{31}NO_2$; mol wt 305.45. C 74.71%, H 10.23%, N 4.59%, O 10.48%. Constitutes together with samandarone approx 75% of the alkaloid content in the skin glands of the European fire and Alpine salamanders: *Salamandra maculosa* resp. *atra* Laur., *Salamandridae*. Isoln: Schöpf, Braun, *Ann.* **514,** 69 (1934). Structure and configuration: Wölfel *et al., Ber.* **94,** 2361 (1961).

Needles from abs methanol or 50% acetone, mp 187-188°. Solvated crystals from dil methanol. $[\alpha]_D^{17}$ +43.3° (acetone). Freely sol in most organic solvents. Practically insol in water and NaOH solns.

Hydrochloride, $C_{19}H_{31}NO_2 \cdot HCl$, crystals from methanol, mp 312-313°; from water, mp 321-322° (monohydrate).

Samandarone, $C_{19}H_{29}NO_2$, ketone corresponding to the secondary alcohol in samandarine. Synthesis: Hara, Oka, *J. Am. Chem. Soc.* **89,** 1041 (1967). Needles from methyl ethyl ketone, mp 191-192°; from 35% acetone, mp 189-191°. $[\alpha]_D^{21}$ −115.7° (acetone).

8314. Samarium. Sm; at. wt 150.36; at. no. 62; valences 2, 3. A lanthanide; belongs to the cerium group of rare earths. Seven naturally occurring isotopes: 144 (3.09%); 147 (14.97%); 148 (11.24%); 149 (13.83%); 150 (7.44%); 152 (26.72%); 154 (22.71%); three are radioactive: 147, $T_{1/2}$ 1.05 × 10^{11} years, α-emission; 148 and 149, conflicting reports on half-life and emission: C. M. Lederer *et al., Table of Isotopes* (John Wiley, New York, 6th ed., 1967) p 86. Artificial radioactive isotopes (mass numbers): 142; 143; 145; 146; 151; 153; 155-157. Abundance in earth's crust: 4.5-7 ppm. Occurs in samarskite, cerite, orthite, ytterbite, monazite and fluorspar. Discovered in 1879 by Lecoq de Boisbaudran; isolated from samarskite as the oxide: L. de Boisbaudran, *Compt. Rend.* **88,** 322 (1879); **89,** 212 (1880). Sepn by crystn of the nitrates from concd nitric acid: Demarcay, *ibid.* **122,** 728 (1896); by crystn of the simple nitrates and double magnesium nitrates: Feit, Przibylla, *Z. Anorg. Chem.* **43,** 203 (1905); sepn of the metal by electrolysis of the fused chloride in a graphite cell with a graphite anode: Schumacher, Harris, *J. Am. Chem. Soc.* **48,** 3108 (1926); by reduction of the trivalent salt with sodium amalgam: Marsh, *J. Am. Chem. Soc.* **1942,** 398, 523; **1943,** 8. More recent prepns of metal by reduction of salts: Daane *et al., J. Am. Chem. Soc.* **75,** 2272 (1953); Onstott, *ibid.* **75,** 5128 (1953); **77,** 812 (1955). Reviews of prepn, properties and compds of samarium and other rare earths: Prandtl, *Z. Anorg. Allgem. Chem.* **238,** 321-334 (1938); *The Rare Earths,* F. H. Spedding, A. H. Daane (Krieger, Huntington, N.Y., 1971, reprint of 1961 ed) 641 pp; Hulet, Bode, "Separation Chemistry of the Lanthanides and Transplutonium Actinides" in *MTP Int. Rev. Sci: Inorg. Chem., Ser. One,* Vol. 7, K. W. Bagnall, Ed. (University Park Press, Baltimore, 1972) pp 1-45; Moeller, "The Lanthanides" in *Comprehensive Inorganic Chemistry,* Vol. 4, J. C. Bailar Jr. *et al.,* Eds. (Pergamon Press, Oxford, 1973) pp 1-101.

Yellow metal; tarnishes on exposure to air. Hardest metal of the cerium group. Rhombohedral structure at room temp; d 7.536; body-centered cubic above 917°. mp 1072°. E°(aq) Sm^{3+}/Sm −2.41 V (calc).

Oxide, Sm_2O_3, yellowish-white powder. d 8.347.

Hydroxide, $Sm(OH)_3$, gelatinous precipitate.

Trichloride, $SmCl_3$, white-yellowish powder. d 4.465, mp 686°. Forms addition compds with ammonia. Forms a

hexahydrate, $SmCl_3 \cdot 6H_2O$, d 2.382, yellow cryst plates. By reducing the anhydr trichloride at high temps with hydrogen, ammonia or aluminum powder, samarium dichloride is obtained. LD_{50} in mice: 585 mg/kg i.p.; >2000 mg/kg orally, Haley, *J. Pharm. Sci.* **54,** 663 (1965).

Dichloride, $SmCl_2$, dark brown cryst mass. d^{22} 3.687; insol in alcohol; dec by water.

Sulfate, $Sm_2(SO_4)_3$, octahydrate, light yellow crystals, d^{18} 2.930. Sparingly sol in water.

USE: Oxide in control rods of some commercial nuclear power reactors.

8315. Sambucus. Elder; American elder; sweet elder. Flowers of *Sambucus canadensis* L., or of *S. nigra* L., *Caprifoliaceae. Habit.* U.S., east of Rocky Mts. *Constit.* Volatile oil, resin, fat, mucilage, tannin.

THERAP CAT: Diaphoretic, diuretic, cathartic.

8316. Sancycline. *4-(Dimethylamino)-1,4,4a,5,5a,6,11,-12a-octahydro-3,10,12,12a-tetrahydroxy-1,11-dioxo-2-naphthacenecarboxamide;* 6-demethyl-6-deoxytetracycline; norcycline; Bonomycin. $C_{21}H_{22}N_2O_7$; mol wt 414.40. C 60.86%, H 5.35%, N 6.76%, O 27.03%. Semi-synthetic antibiotic related to tetracycline, *q.v.* Prepn: Beereboom *et al., J. Am. Chem. Soc.* **82,** 1003 (1960); McCormick *et al., ibid.* 3381; McCormick, Jensen, U.S. pat. **3,019,260** (1962 to Am. Cyanamid); **Brit.** pat. **901,209** (1962 to Pfizer). Total synthesis: Conover *et al., J. Am. Chem. Soc.* **84,** 3222 (1962).

Hydrochloride hemihydrate, $C_{21}H_{22}N_2O_7 \cdot HCl \cdot \frac{1}{2}H_2O$, crystals from ethanol, dec 215-220°. uv max (MeOH/0.01N HCl): 267, 347 nm (ϵ 19300, 15500); in 0.1N H_2SO_4: 217, 268, 343 nm (ϵ 13400, 19000, 14600).

THERAP CAT: Antibacterial.

8317. Sandalwood, White. White saunders; yellow saunders. Heartwood of *Santalum album* L., *Santalaceae. Habit.* East Indies. *Constit.* Volatile oil, resin, tannin. A source of Oil of Santal, *q.v.*

USE: As incense and fumigant.

8318. Sandarac. Resin from *Callitris quadrivalvis* Vent., *Pinaceae. Habit.* Morocco. *Constit.* About 80% pimaric acid, about 10% callitrolic acid; sandaricinic acid.

Light yellow, brittle, elongated tears; translucent with vitreous fracture; crumbles to powder when masticated. Insol in water, benzene, petr ether. Sol in alcohol, ether, acetone, hot caustic alkalies; partially sol in chloroform, volatile oils, oil turpentine, carbon disulfide.

USE: Tooth cements, lacquers, varnishes; as an incense. Pharmaceutic aid (in ointments and plasters).

8319. Sanguinaria. Bloodroot; red puccoon; red root; puccoon root; tetterwort. Dried rhizome and roots of *Sanguinaria canadensis* L., *Papaveraceae. Habit.* North America. *Constit.* Sanguinarine, chelerythrine, protopine, homochelidonine, resin. An extract of *Sanguinaria canadensis* has been used in cough syrup; used experimentally as toothpaste base, in gingivitis and in periodontal disease: P. A. Ladanyi, U.S. pat. **4,145,412** (1979 to Vipont Chem. Co.).

8320. Sanguinarine. *13-Methyl[1,3]benzodioxolo[5,6-c]-1,3-dioxolo[4,5-i]phenanthridinium;* pseudochelerythrine; ψ-chelerythrine. $[C_{20}H_{14}NO_4]^+$; mol wt 332.34. From the root of *Sanguinaria canadensis* L., and other *Papaveraceae*: Schmidt *et al., Arch. Pharm.* **231,** 145 (1893). Widely distributed in poppy-fumaria species; constituent of argemone oil: S. A. E. Hakim *et al., Nature* **189,** 198 (1961). Identity with pseudochelerythrine: Gadamer, Stichel, *ibid.* **262,** 488 (1924). Structure: Späth, Kuffner, *Ber.* **64,** 370, 2034 (1931); Beke, *Acta Chim. Acad. Sci. Hung.* **17,** 463 (1958). Biosynthesis: Leete, *J. Am. Chem. Soc.* **85,** 473 (1963). Synthesis: Dyke *et al., Tetrahedron Letters* **1968,** 3933;

Sainsbury et al., J. Chem. Soc. (C) 1970, 1797; Onda et al., Chem. Pharm. Bull. 17, 404 (1969); 19, 31 (1971); M. Hanaoka et al., Chem. Letters 22, 739 (1986). Purification procedure: Stipanovic et al., J. Heterocycl. Chem. 9, 1453 (1972). Studies of effect of sanguinarine and argemone oil on intraocular pressure: S. A. E. Hakim, Brit. J. Ophthalmol. 38, 193 (1954); G. C. Dobbie, M. E. Langham, ibid. 45, 81 (1961). Anti-inflammatory activity in rats: J. Lenfeld et al., Planta Med. 43, 161 (1981). Clinical study of sanguinarine oral rinse as antiplaque agent: G. L. Southard et al., J. Am. Dental Assn. 108, 338 (1984). Short-term toxicity study: P. J. Becci et al., J. Toxicol. Environ. Health 20, 199 (1987). Review: M. Shamma, The Isoquinoline Alkaloids (Academic Press, New York, 1972) pp 315-343.

Monohydrate, $C_{20}H_{13}NO_4 \cdot H_2O$, crystals from water, mp 278-280° (Stipanovic). Soluble in alcohol, chloroform, acetone, ethyl acetate.

Chloride, $C_{20}H_{14}ClNO_4$. Viadent. LD_{50} in male mice (mg/kg): 15.9 i.v.; in female mice (mg/kg): 102.0 s.c. (Lenfeld). LD_{50} in rats (mg/kg): 29 i.v.; 1658 orally (Becci).

Chloride dihydrate, $C_{20}H_{14}ClNO_4 \cdot 2H_2O$, fine orange needles, mp 273-274° (dec). uv max (methanol): 234, 283, 325 nm (log ϵ 4.50, 4.52, 4.18).

USE: In mouthwash and toothpaste.

8321. α-Santalol. 5-(2,3-Dimethyltricyclo[2.2.1.0²,⁶]hept-3-yl)-2-methyl-2-penten-1-ol. $C_{15}H_{24}O$; mol wt 220.34. C 81.76%, H 10.98%, O 7.26%. A sesquiterpene alcohol which together with β-santalol comprises about 90% of commercial sandalwood oil: Semmler, Bode, Ber. 40, 1124 (1907); E. Gruenther, The Essential Oils vol. 2 (Van Nostrand, New York, 1949) pp 266-269. Structure: Semmler, Ber. 43, 1893 (1910). Separation of α- and β-santalols of high purity: Bradfield et al., J. Chem. Soc. 1935, 309. Synthesis of α-santalol and conversion to β-santalol: Bhattacharyya, Sci. Cult. 13, 207 (1947); C.A. 43, 5484gh (1949); see also Sathe et al., Indian J. Chem. 4, 393 (1966); Kamat et al., Tetrahedron 23, 4487 (1967). Stereochemistry studies: Brieger, Tetrahedron Letters 1963, 2123. Natural α-santalol has a seqcis-configuration. Total synthesis and geometric configuration: Lewis et al., ibid. 1967, 401. Stereospecific synthesis: Corey et al., J. Am. Chem. Soc. 92, 226, 6314 (1970); Julia, Ward, Bull. Soc. Chim. France 1973, 3065; M. Tamura, G. Suzukamo, Tetrahedron Letters 22, 577 (1981). Synthesis and separation of cis- and trans-isomers: K. Sato et al., Bull. Chem. Soc. Japan 49, 3351 (1976). Review: J. Simonsen, D. H. R. Barton, The Terpenes vol. 3 (Univ. Press, Cambridge, 1952) pp 178-188; Bhati, Flavour Ind. 1, 235-251 (1970).

Liquid. bp_{14} 166-167°. d_{25}^{25} 0.9770. n_D^{25} 1.5017. α_{5461} +10.3°. $[\alpha]_D^{20}$ +17.20° (c = 0.8 in chloroform). Sol in alcohol; slightly sol in propylene glycol, glycerine. Practically insol in water.

USE: In perfumes, soaps and detergents.

8322. β-Santalol. 2-Methyl-5-(2-methyl-3-methylenebicyclo[2.2.1]hept-2-yl)-2-penten-1-ol; 2-methyl-5-(2-methyl-3-methylene-2-norbornyl)-2-penten-1-ol. $C_{15}H_{24}O$; mol wt 220.34. C 81.76%, H 10.98%, O 7.26%. A sesquiterpene alcohol from sandalwood oil. Isoln and purification and additional refs: see α-santalol. Structure: Ruzicka, Thomann, Helv. Chim. Acta 18, 355 (1935). Synthesis: Bhattacharyya, Sci. Cult. 13, 209 (1957), C.A. 43, 5385a (1949); Kretschmar, Erman, U.S. pats. 3,662,008, 3,679,756 (1972

to Procter & Gamble). Natural β-santalol has a seqcis-configuration. Total synthesis and geometric configuration: Kretschmar, Erman, Tetrahedron Letters 1970, 41. Stereoselective synthesis: P. A. Christenson, B. J. Willis, J. Org. Chem. 44, 2012 (1979). Synthesis of dl-form: M. Baumann, W. H. Hoffman, Ann. 1979, 743; K. Sato et al., Chem. Letters 1981, 1183.

Liquid. bp_{17} 177-178°. d_{25}^{25} 0.9717. n_D^{25} 1.5100. α_{5461} −87.1°. Sol in alcohol; practically insol in water.
dl-Form, $bp_{0.1}$ 101-103°.

USE: In perfumes, soaps and detergents.

8323. Santonica. Levant wormseed. Dried, unexpanded flower heads of Artemisia maritima L., sens. lat., Compositae. Habit. Iran, Turkestan. Constit. 2-4% santonin, 2-3% volatile oil; artemisin, resin.

THERAP CAT: Anthelmintic.

8324. Santonic Acid. Octahydro-α,3a,5-trimethyl-6,8-dioxo-1,4-methano-1H-indene-1-acetic acid; hexahydro-α,3a,5-trimethyl-6,8-dioxo-1,4-methanoindan-1-acetic acid. $C_{15}H_{20}O_4$; mol wt 264.31. C 68.16%, H 7.63%, O 24.21%. Prepd from santonin by the action of hot concd aq bases, preferably potassium hydroxide: v. Oettingen, Dissertation, Göttingen, 1913; Woodward et al., J. Am. Chem. Soc. 70, 4216 (1948). Structure: Wedekind, Engel, J. Prakt. Chem. 139, 115 (1934); Woodward et al., loc. cit.; Woodward, Yates, Chem. & Ind. (London) 1954, 1391. Chemistry: Hortmann, Daniel, J. Org. Chem. 37, 4446 (1972).

Crystals from water or alcohol, mp 170-172°; bp_{15} 285°; $[\alpha]_D^{20}$ −74.0° (chloroform). Soluble in 190 parts of water at 17°. Freely sol in alc, chloroform, ether, glacial acetic acid.

8325. α-Santonin. [3S-(3α,3aα,5aβ,9bβ)]-3a,5,5a,9b-Tetrahydro-3,5a,9-trimethylnaphtho[1,2-b]furan-2,8(3H,-4H)-dione; 1,2,3,4,4a,7-hexahydro-1-hydroxy-α,4a,8-trimethyl-7-oxo-2-naphthaleneacetic acid γ-lactone; l-santonin. $C_{15}H_{18}O_3$; mol wt 246.29. C 73.14%, H 7.37%, O 19.49%. Anthelmintic isolated from the dried unexpanded flower heads of Artemisia maritima L., sens. lat., Compositae [Levant wormseed] and other species of Artemisia found principally in Russian and Chinese Turkestan and the Southern Ural region. Structure: Clemo et al., J. Chem. Soc. 1930, 1110; Ruzicka, Steiner, Helv. Chim. Acta 17, 614 (1934). Stereochemistry: Corey, J. Am. Chem. Soc. 77, 1044 (1955); Cocker, McMurry, Tetrahedron 8, 181 (1960); Asher, Sim, Proc. Chem. Soc. 1962, 111, 335; Nakazaki, Arakawa, Bull. Chem. Soc. Japan 37, 464 (1964); Pregosin et al., J. Chem. Soc. Perkin I 1972, 299. Total synthesis: Abe et al., Proc. Japan Acad. 30, 116 (1954); J. Am. Chem. Soc. 78, 1422 (1956); U.S. pat. 2,836,604 (1958 to Takeda); J. A. Marshall, P. G. M. Wuts, J. Org. Chem. 43, 1086 (1978). Mass spectral studies of the santonins and derivs: Woseda et al., Tetrahedron 23, 4623 (1967). Review: C. H. Heathcock in The Total Synthesis of Natural Products vol. 2, J. W. ApSimon, Ed. (Wiley, New York, 1973) pp 315-324.

(−)-Form, tabular crystals, orthorhombic sphenoidal, mp 170-173°. Almost tasteless with bitter aftertaste. $[\alpha]_D^{25}$ −170° to −175° (c = 2 in alc). Becomes yellow on exposure to light. Irritating to mucous membranes. d 1.187. One part dissolves in 5000 parts of cold water, in 250 parts of boiling water, in 280 parts of 50% alcohol at 17°, in 10 parts of boiling 50% alcohol, in 44 parts of cold 90% alcohol, in 3 parts of boiling 90% alcohol, in 125 parts of cold ether, in 72 parts of boiling ether, in 4.3 parts of cold chloroform.

(±)-Form, colorless plates from methanol, mp 181°. uv max (ethanol): 241 nm (log ε 4.10).

(+)-Form, colorless plates from methanol, mp 172°. $[\alpha]_D^{20}$ +165.9° (c = 1.92 in ethanol).

4-Hydroxysantonin, see Artemisin.

THERAP CAT: Anthelmintic (Nematodes).

THERAP CAT (VET): Has been used as an anthelmintic.

8326. Saponaria. Soapwort; soaproot; fuller's herb; bruisewort; bouncing bet. Herb and root of *Saponaria officinalis* L., *Caryophyllaceae*. *Habit.* Europe to Middle Asia; naturalized in U.S. *Constit.* Saponin, sapotoxin, saponarin.

8327. Saponarin. 6-β-D-*Glucopyranosyl*-7-(β-D-*glucopyranosyloxy*)-5-*hydroxy*-2-(4-*hydroxyphenyl*)-4H-1-*benzopyran*-4-*one*. $C_{27}H_{30}O_{16}$; mol wt 610.54. C 53.12%, H 4.95%, O 41.93%. From leaves of *Saponaria officinalis* L., *Caryophyllaceae*: Barger, *J. Chem. Soc.* **89**, 1210 (1906); from *Hibiscus syriacus* L., *Malvaceae*: Nakoaki, *J. Pharm. Soc. Japan* **64**, 304 (1944), *C.A.* **46**, 108d (1952); from *Lemna (Spirodela) oligorrhiza* Kurz., *Lemnaceae*: Jurd *et al.*, *Arch. Biochem. Biophys.* **67**, 284 (1957). Structure: Seikel, Geissman, *ibid.* **71**, 17 (1957). Revised structure: Hörhammer *et al.*, *Tetrahedron Letters* **1965**, 1707. On acidic hydrolysis yields the aglycone, *saponaretin*.

Monohydrate, pale yellow granules, mp 228°. $[\alpha]_D$ −7.9° (aq pyridine). uv max (ethanol): 335, 272 nm. Practically insol in cold water; sparingly sol in hot water, alcohol; sol in alkalies with yellow color and in concd H_2SO_4 with blue fluorescence; sol in pyridine.

8328. Saponins. Sapogenin glycosides. A type of glycoside widely distributed in plants. Each saponin consists of a sapogenin which constitutes the aglucon moiety of the molecule, and a sugar. The sapogenin may be a steroid or a triterpene and the sugar moiety may be glucose, galactose, a pentose, or a methylpentose. Poisonous towards the lower forms of life and used for killing fish by the aborigines of South America. *Review and bibliography:* R. J. McIlroy, *The Plant Glycosides* (Edward Arnold & Co., London, 1951) Chapter IX; Y. Birk, I. Peri in *Toxic Const. Plant Foodst.*, I. E. Liener, Ed. (Academic Press, New York, 2nd ed., 1980) pp 161-182.

Bitter taste. All saponins foam strongly when shaken with water. They form oil-in-water emulsions and act as protective colloids.

Caution: Although practically non-toxic to man upon oral ingestion, they act as powerful hemolytics when injected into the blood stream, dissolving the red corpuscles even at extreme dilutions.

8329. Sapphire. The corundum modification of aluminum oxide, Al_2O_3. Mined in Montana, Siam, Burma, Ceylon. The highest priced variety of the native gem has the

blue color of cornflowers. Easily prepd synthetically. Hydrothermal synthesis: Laudise, Ballman, *J. Am. Chem. Soc.* **80**, 2655 (1958).

USE: In jewelry, as phonograph needles, industrial abrasive, watch and instrument bearings.

8330. Saralasin. *1-(N-Methylglycine)-5-L-valine-8-L-alanineangiotensin II*; *N*-[1-[*N*-[*N*-[*N*-[*N*-[*N²*-(*N*-methylglycyl)-L-arginyl]-L-valyl]-L-tyrosyl]-L-valyl]-L-histidyl]-L-prolyl]-L-alanine; 1-sar-8-ala-angiotensin II. $C_{42}H_{65}$-$N_{13}O_{10}$; mol wt 912.07. C 55.31%, H 7.18%, N 19.96%, O 17.54%. A specific antagonist of angiotensin II. Prepn: Sipos *et al.*, Ger. pat. **2,127,393** corresp to U.S. pat. **3,751,-404** (1972, 1973 to Norwich Pharmacal). Activity studies: Pals *et al.*, *Circ. Res.* **29**, 673 (1971); Pals, Fulton, *Arch. Int. Pharmacodyn. Ther.* **204**, 20 (1973); Solomon, Buckley, *J. Pharm. Sci.* **63**, 1109 (1974); Streeten *et al.*, *N. Engl. J. Med.* **292**, 657 (1975). Comprehensive review of the acetate: C. T. Huang *et al.*, in *Pharmacological and Biochemical Properties of Drug Substances* vol. 3, M. E. Goldberg, Ed. (Am. Pharm. Assoc., Washington, DC, 1981) pp 176-225.

Sar-Arg-Val-Tyr-Val-His-Pro-Ala
1   2   3   4   5   6   7   8

Hydrated acetate, *P-113*, saralasin acetate, *Sarenin*. Described as $C_{42}H_{65}N_{13}O_{10}.xC_2H_4O_2.xH_2O$. Fluffy white powder, mp 256°. Sol in water, 5% aq dextrose, 90-95% aq alcohol. LD_{50} i.v. in male mice: 1171 mg/kg (Huang).

THERAP CAT: Antihypertensive, diagnostic aid (renindependent hypertension).

8331. Sarcosine. *N-Methylglycine*; *N*-methylaminoacetic acid; *N*-methylglycocoll; methylaminoethanoic acid. C_3H_7-NO_2; mol wt 89.09. C 40.44%, H 7.92%, N 15.72%, O 35.92%. CH_3NHCH_2COOH. Has been found in starfish and sea urchins: A. Kossel, S. Edlbacher, *Z. Physiol. Chem.* **94**, 264 (1915); in rock lobsters: L. Novellie, H. M. Schwartz, *Nature* **173**, 450 (1954). Formation from caffeine by dec with barium hydroxide: Paulmann, *Arch. Pharm.* **232**, 601 (1894). Prepd on a large scale from formaldehyde, sodium cyanide, and methylamine: Eschweiler, *Ann.* **279**, 39 (1894); Baumann, *J. Biol. Chem.* **21**, 563 (1915); Caverly, U.S. pat. **2,720,540** (1955 to du Pont); Leake, Brakebill, U.S. pat. **3,009,954** (1961 to Allied Chem.). Synthesis: M. Ebata *et al.*, *Bull. Chem. Soc. Japan* **39**, 2535 (1966). Review of syntheses: T. Shirai, *Synthetic Production and Utilization of Amino Acids*, T. Kaneko *et al.*, Eds. (Wiley, New York, 1974) pp 184-186.

Orthorhombic, deliquescent crystals from dil methanol. Dec 212°. Sweetish taste. pK_1' 2.23; pK_2' 10.01. Sol in water: 5 ml of a satd aq soln contain 2.1412 g sarcosine. Slightly sol in alc.

Forms a sodium salt; usually sold as aq soln, pH 12.0.

Hydrochloride, $C_3H_7NO_2.HCl$, needles from alcohol, dec 171°. Freely sol in water; slightly sol in alcohol, ether.

USE: Intermediate in the synthesis of antienzyme agents for toothpaste.

8332. Sarin. *Methylphosphonofluoridic acid 1-methylethyl ester*; isopropoxymethylphosphoryl fluoride; isopropylmethylphosphonofluoridate; GB. $C_4H_{10}FO_2P$; mol wt 140.09. C 34.29%, H 7.20%, F 13.56%, O 22.84%, P 22.11%. Nerve gas: Holmstedt, *Acta Physiol. Scand.* **25**, suppl. 90, p 106 (1951); *Protar* **14**, 113 (1948); **16**, 131 (1950). *See also* the ref and prepns given *under* Tabun. Additional ref on synthesis: Schrader, *B.I.O.S.* **714**, 41 (1947); B. C. Saunders, *Some Aspects of the Chemistry and Toxic Action of Organic Compounds Containing Phosphorus and Fluorine* (Cambridge, 1957) p 92 sqq; Bryant *et al.*, *J. Chem. Soc.* **1960**, 1553. Brief review: Schrader, *Die Entwicklung neuer insektizider Phosphorsäure-Ester* (Verlag Chemie, Weinheim, 1963) p 4. Abs config: Benschop *et al.*, *Rec. Trav. Chim.* **87**, 387 (1968).

Liquid, d_4^{20} 1.10. mp $-57°$. bp$_{760}$ 147°; bp$_{16}$ 56°. Miscible with and hydrolyzed by water. Rapidly hydrolyzed by dil aq sodium hydroxide or sodium carbonate forming relatively non-toxic products. Water alone removes the fluorine atom producing the non-toxic acid $CH_3PO[OCH(CH_3)_2]OH$. LD$_{50}$ i.p. in mice: 0.42 mg/kg, B. Holmstedt, *Pharmacol. Rev.* **11**, 567 (1959). Lethal dose for man may be as low as 0.01 mg/kg, *Chem. & Eng. News* **31**, 4676 (1953).

Human Toxicity: Extremely active cholinesterase inhibitor. Toxic effects similar to, but more severe than, parathion, *q.v.*

USE: Chemical warfare agent.

8333. Sarkomycin. Sarcomycin. Antibiotic and antitumor substance produced by *Streptomyces erythrochromogenes*, strain W-115-C, from soil found at Kamakura, Japan: Umezawa *et al., J. Antibiot.* **6A**, 101, 147, 153 (1953); *eidem, Antibiot. & Chemother.* **4**, 514 (1954). Structure of sarkomycin A, the active principle: Hooper *et al., ibid.* **5**, 585 (1955). Characterization of sarkomycins A, A' and B: Maeda, Kondo, *J. Antibiot.* **11A**, 37 (1958). Sarkomycin reacts with H_2S to form sarkomycins S$_1$, S$_2$ and S$_3$: Tatsuoka *et al., ibid.* **9B**, 107, 110 (1956). Structure of S$_1$ and S$_2$: *eidem, ibid.* **11B**, 275 (1958). Synthesis of *d-* and *l*-sarkomycin A: Toki, *ibid.* **10A**, 35, 226 (1957). Synthesis of *dl*-sarkomycin A: Toki, *Bull. Chem. Soc. Japan* **30**, 450 (1957); R. K. Boeckman, Jr. *et al., J. Org. Chem.* **45**, 752 (1980); J. N. Marx, G. Minaskanian, *ibid.* **47**, 3306 (1982); B. A. Wexler *et al., ibid.* 3333; A. P. Kozikowski, P. D. Stein, *J. Am. Chem. Soc.* **104**, 4023 (1982). Abs config of A: Sato *et al., Chem. Pharm. Bull.* **11**, 829 (1963). Revised config of A: Hill *et al., J. Org. Chem.* **32**, 2330 (1967). Review: S. A. Waksman, H. A. Lechevalier, *The Actinomycetes* vol. **III** (Williams & Wilkins, Baltimore, 1962) pp 362-364; Sung in *Antibiotics*, vol. **I**, D. Gottlieb, P. Shaw, Eds. (Springer-Verlag, New York, 1967) pp 156-165.

sarkomycin A

Oily liquid. $[\alpha]_D^{15}$ $-32.5°$ (c = 1 in methanol). uv max (water): 230 nm. Sol in water, methanol, ethanol, butanol, ethyl acetate. Sparingly sol in petr ether.

Sarkomycin A, $C_7H_8O_3$, *(R)-2-methylene-3-oxocyclopentanecarboxylic acid*.

8334. Sarmentogenin. *3β,11α,14-Trihydroxy-5β-card-20(22)-enolide.* $C_{23}H_{34}O_5$; mol wt 390.50. C 70.74%, H 8.78%, O 20.49%. An aglucon from sarmentocymarin isolated from the seeds of *Strophanthus sarmentosus* DC. var. *senegambiae* (A. DC.) Monachino, *Apocynaceae:* Jacobs, Heidelberger, *J. Biol. Chem.* **81**, 765 (1929); Katz, *Helv. Chim. Acta* **31**, 993 (1948). Structure: Callow, Taylor, *J. Chem. Soc.* **1952**, 2299; Repke, Klesczewski, *Arch. Exp. Pathol. Pharmakol.* **239**, 131 (1960).

Prisms from methanol or methanol + acetone, mp 278-282°. $[\alpha]_D^{19}$ $+21.1° ± 4°$ (c = 0.521 in methanol). Absorption max (98% H_2SO_4): 230, 415 nm (E$_{1cm}^{1\%}$ 460, 430). Sol in alcohol, methanol, pyridine; sparingly sol in acetone, chloroform; practically insol in benzene, ether. Dissolves in concd H_2SO_4 with a bright golden color which slowly turns green and then indigo. In dil pyridine soln gives a deep red color with alkaline nitroprusside soln. When recrystallized from

pyridine, it forms long microplatelets contg about 1 mol of the solvent and melting at 258° with effervescence. Is not precipitated with digitonin. Yields sarmentogenone on oxidation.

Diacetate, $C_{27}H_{38}O_7$, needles from abs ether, mp 135-155°. Not affected by CrO_3 in acetic acid.

Dibenzoate, $C_{37}H_{42}O_7$, flat hexagonal prisms from acetone, mp 281°. $[\alpha]_D^{20}$ $+14°$ (acetone). Sparingly sol in alc; practically insol in ether. Not affected by CrO_3 in acetic acid.

8335. Sarmentose. *2,6-Dideoxy-3-O-methyl-xylo-hexose.* $C_7H_{14}O_4$; mol wt 162.18. C 51.84%, H 8.70%, O 39.46%. A methyl ether of a 2-desoxyhexomethylose, isomeric with cymarose. Obtained on hydrolysis of sarmentocymarin, a glycoside isolated from seeds of *Strophanthus sarmentosus* DC., *Apocynaceae* by the enzymic method: Jacobs, Heidelberger, *J. Biol. Chem.* **81**, 765 (1929); Jacobs, Bigelow, *ibid.* **96**, 355 (1932). Synthesis: H. Hauenstein, T. Reichstein, *Helv. Chim. Acta* **33**, 446 (1950).

Prisms, plates from ether + petr ether, mp 78-79°. Shows mutarotation. $[\alpha]_D^{20}$ $+12°$ (20 min) → $+15.8°$ (24 hrs c = 1.08): Elderfield, *Advan. Carbohyd. Chem.* **1**, 172 (1945).

8336. Sarpagine. *Sarpagan-10,17-diol;* raupine. $C_{19}H_{22}$-N_2O_2; mol wt 310.38. C 73.52%, H 7.14%, N 9.03%, O 10.31%. Isoln from *Rauwolfia serpentina* (L.) Benth., *Apocynaceae:* Stoll, Hofmann, *Helv. Chim. Acta* **36**, 1143 (1953). Structure: Stauffacher *et al., ibid.* **40**, 508 (1957); Poisson *et al., Bull. Soc. Chim. France* **1957**, 610; Biemann, *J. Am. Chem. Soc.* **83**, 4801 (1961). Stereochemistry: Bartlett *et al., ibid.* **84**, 622 (1962); Ohashi *et al., Tetrahedron* **19**, 2241 (1963).

Needles from methanol or ethanol, long plates from acetone, dec > 320°. $[\alpha]_D^{20}$ $+54°$ (c = 0.5 in pyridine). uv max (ethanol): 230, 278 nm (log ε 4.30, 3.92). One gram dissolves in 60 ml boiling ethanol, 250 ml boiling methanol, 400 ml boiling acetone. Practically insol in chloroform.

Hydrochloride, $C_{19}H_{22}N_2O_2$·HCl, needles from alcohol, dec > 220°.

8337. Sarsaparilla. Dried root of *Smilax aristolochiaefolia* Mill. (*S. medica* Cham. & Schlecht.) (Mexican Sarsap.), or *S. regelii* Killip & Morton (*S. ornata* Hook f.) (Jamaica or Central American Sarsap.), *Liliaceae. Habit.* Honduras, Jamaica, Mexico (Vera Cruz), Brazil, Guatemala. *Constit.* Sarsaponin, smilacin (parillin), paroaparic acid, resin, volatile oil.

USE: As flavor in beverages. Pharmaceutic aid (flavor).

8338. Sarsasapogenin. *(25S)-Spirostan-3β-ol;* parigenin. $C_{27}H_{44}O_3$; mol wt 416.62. C 77.83%, H 10.65%, O 11.52%. Steroid sapogenin from *Smilax ornata* Hooker, fil., *Liliaceae* (Sarsaparilla): Power, Salway, *J. Chem. Soc.* **105**, 201 (1914); Jacobs, Simpson, *J. Biol. Chem.* **105**, 501 (1934); **109**, 573 (1935). Partial structure: Hirschmann *et al., J. Org. Chem.* **20**, 572 (1955). Stereochemistry: Taylor, *Chem. & Ind. (London)* **1954**, 1066; Wall, Serota, *J. Am. Chem. Soc.* **76**, 2850 (1954); Callow, Massey-Beresford, *J. Chem. Soc.* **1957**, 4482; Rosen *et al., J. Am. Chem. Soc.* **81**, 1687 (1959). *Reviews:* Turner, *The Steroidal Sapogenins, Univ. Microfilms* (Ann Arbor, Mich.), Pub. no. 361, 249 pp (1941); Shabica, *Sarsasapogenin and Related Compounds, Univ. Microfilms* (Ann Arbor, Mich.), Pub. no. 549, 89 pp (1943); L. F. Fieser, M. Fieser, *Steroids* (Reinhold, New York, 1959) p 810. *See also* Smilagenin.

Large prismatic needles from acetone, mp 199-199.5°. $[\alpha]_D^{25}$ −75°, $[\alpha]_{546}^{25}$ −89° (c = 0.5 in CHCl₃). uv spectrum: Smith, Eddy, *Anal. Chem.* **31**, 1539 (1959). Sol in alcohol, acetone, benzene, chloroform. Is precipitated by digitonin.

Acetate, $C_{29}H_{46}O_4$, flat needles from methanol, mp 144-145°. $[\alpha]_D^{25}$ −70.2°; $[\alpha]_{546}^{25}$ −83.1° (c = 1.18 in CHCl₃).

USE: In the manuf of compds of the pregnane series.

8339. Sarverogenin. *7β,8-Epoxy-3β,11α,14-trihydroxy-12-oxo-5β-card-20(22)-enolide.* $C_{23}H_{30}O_7$; mol wt 418.49. C 66.01%, H 7.23%, O 26.76%. The aglycone of sarveroside isolated from seeds of *Strophanthus* spp, *Apocynaceae:* von Euw *et al., Festschrift Prof. Paul Casparis* (Zürich, 1949) p 178; Buzas *et al., Helv. Chim. Acta* **33**, 465 (1950); Rothrock *et al., J. Am. Chem. Soc.* **72**, 3827 (1950); von Euw, Reichstein, *Helv. Chim. Acta* **33**, 1006 (1950); Taylor, *J. Chem. Soc.* **1952**, 4832. Structural studies: Taylor, *Chem. & Ind. (London)* **1953**, 62; Schindler, *Helv. Chim. Acta* **39**, 375 (1956); Taylor, *J. Chem. Soc. (C)* **1966**, 790. Revised structure: Fuhrer *et al., Helv. Chim. Acta* **52**, 616 (1969).

Crystallizes in various forms with different melting points, but [according to Taylor, *J. Chem. Soc.* **1952**, 4832] on re-crystn from methanol all of these forms yield the one melting at 232-234°. $[\alpha]_D^{15}$ +44.5° ± 2° (c = 1.1683 in methanol); $[\alpha]_D^{23}$ +49.5° (methanol). uv max: 218, 277.5 nm (log ε 4.2, 1.83). Sol in water, methanol, chloroform.

Diacetate, crystals, mp 254-256° (corr). $[\alpha]_D^{22}$ +24.2° (chloroform).

Dibenzoate, cryst powder from methanol, mp 1889° (Taylor, *loc. cit.*); mp 187-191° (micro-block, Rothrock *et al., loc. cit.*); $[\alpha]_D^{23}$ +30.7° ± 1° (c = 1.301 in acetone). Sol in acetone; slightly sol in methanol.

8340. Sassafras. Saxifrax; ague tree; cinnamon wood; saloop. Bark of root of *Sassafras albidum* (Nutt.) Nees [*S. variifolium* (Salisb.) Kuntze, *S. officinale* Nees & Eberm.], *Lauraceae.* *Habit.* North America. *Constit.* Root: ∼2% volatile oil. Bark of root: 6-9% volatile oil; sassafrid, wax.

USE: Aromatic, flavoring agent.

THERAP CAT: Sudorific.

8341. Sassy Bark. Saucy bark; Mancona bark; ordeal bark; red-water tree bark; casca bark; Saxon bark; doom bark; teli; bondou. Bark of *Erythrophleum guineense* G. Don, *Leguminosae.* *Habit.* Central and Western Africa. *Constit.* Erythrophleine, tannin, resin.

8342. Saunders, Red. Red sandalwood; ruby wood. Heartwood of *Pterocarpus santalinus* L. f., *Leguminosae.* *Habit.* East Indies. *Constit.* Santalin (santalinic acid), santal pterocarpin, tannin.

USE: For coloring tinctures and similar medicinal prepns; formerly used as a dye.

8343. Savin. Young shoots of *Juniperus sabina* L., *Cupressaceae.* *Habit.* Europe, Northern Asia, North America

south to New York, and Montana. *Constit.* 2-4% volatile oil, tannin, savinin, podophyllotoxin. Isoln and structure of *savinin:* Hartwell *et al., J. Am. Chem. Soc.* **75**, 235 (1953); Schrecker, Hartwell, *ibid.* **76**, 4896 (1954).

THERAP CAT: Emmenagogue. Anthelmintic. Antirheumatic.

8344. Saxitoxin. Mussel poison; clam poison; paralytic shellfish poison; gonyaulax toxin; STX. $[C_{10}H_{17}N_7O_4]^{2+}$; mol wt 299.30. Powerful neurotoxin produced by the dinoflagellates *Gonyaulax catenella*, or *G. tamarensis*, the consumption of which causes the California sea mussel *Mytilus californianus*, the Alaskan butterclam *Saxidomus giganteus* and the scallop to become poisonous: Sommer *et al., Arch. Pathol.* **24**, 537, 560 (1937); Schantz *et al., Can. J. Chem.* **39**, 2117 (1961); Ghazarossian *et al., Biochem. Biophys. Res. Commun.* **59**, 1219 (1974). These poisonous shellfish have been connected to instances of toxic "red-tides" where the high concentrations of algae discoloring the water were of the *Gonyaulax* genus. Isoln and partial characterization: Schantz *et al., J. Am. Chem. Soc.* **79**, 5230 (1957); Mold *et al., ibid.* 5235. Purifn and biochemical data: Schantz *et al., Biochemistry* **5**, 1191 (1966); Schantz, *Ann. N.Y. Acad. Sci.* **90**, 843 (1960). pKa values: R. S. Rogers, H. Rapoport, *J. Am. Chem. Soc.* **102**, 7335 (1980). Toxicity: G. S. Wiberg, N. R. Stephenson, *Toxicol. Appl. Pharmacol.* **2**, 607 (1960). Structural studies: Schuett, Rapoport, *J. Am. Chem. Soc.* **84**, 2266 (1962); Wong *et al., ibid.* **93**, 4633, 7344 (1971). Revised structure: Schantz *et al., ibid.* **97**, 1238 (1975); Bordner *et al., ibid.* **97**, 6008 (1975). Pharmacology including study of sodium transport inhibition: Cheymol *et al., Arch. Int. Pharmacodyn. Ther.* **174**, 393 (1968); Evans, *Brit. Med. Bull.* **25**, 263 (1969). Stereospecific total synthesis of dl-saxitoxin: H. Tanino *et al., J. Am. Chem. Soc.* **99**, 2818 (1977); Y. Kishi, *Heterocycles* **14**, 1477 (1980). *Reviews:* Kao, *Pharmacol. Rev.* **18**, 997 (1966); Kao, *Fed. Proc., Fed. Am. Soc. Exp. Biol.* **31**, 1117 (1972); Narahashi, *ibid.* 1124, Evans, *Int. Rev. Neurobiol.* **15**, 83 (1972); Y. Shimizu in *Progress in the Chemistry of Natural Products* vol. **45**, W. Herz *et al.,* Eds. (Springer-Verlag, New York, 1984) pp 239-246. *See also* Brevetoxins.

Dihydrochloride, white, hygroscopic solid. pKa in water: 8.24, 11.60. $[\alpha]_D^{25}$ +130°. Very sol in water, methanol; sparingly sol in ethanol, glacial acetic acid. Practically insol in lipid solvents. Stable in acid solns; decomposes rapidly in alkaline media. Boiling 3-4 hrs at pH 3 causes loss of activity. LD_{50} in mice (μg/kg): 10 i.p.; 263 orally; 3.4 i.v. (Wiberg, Stephenson).

USE: As a tool in neurochemical research.

8345. SBR Rubber. Styrene-butadiene rubber; GR-S; Government Rubber Styrene. General-purpose synthetic rubbers that were originally produced in government owned plants as GR-S. They are copolymers obtained by the polymerization of butadiene and styrene in a ratio of approx 3:1. The chains contain a random sequence of the two monomers. Very similar in composition to *Buna S, Kraton, Solprene, Stereon.* *Review:* P. Schneider *et al.,* in Ullmann, *Encyklopädie der technischen Chemie,* 3 Aufl., Bd. IX (München-Berlin, 1957) p 331; M. Morton, *Introduction to Rubber Technology* (Reinhold, 1959) pp 56 and 256-284; R. G. Bauer in Kirk-Othmer *Encyclopedia of Chemical Technology* vol. **8** (Wiley-Interscience, New York, 3rd ed., 1979) pp 608-625.

USE: In tires, adhesives, chewing gum.

8346. Scabiolide. *2,3-Dihydroxy-2-methylpropanoic acid 6-[(acetyloxy)methyl]-2,3,3a,4,7,8,11,11a-octahydro-10-methyl-3-methylene-2-oxocyclodeca[b]furan-4-yl ester.* $C_{21}H_{28}O_8$; mol wt 408.45. C 61.75%, H 6.91%, O 31.34%. A

Consult the cross index before using this section.

sesquiterpenic lactone, the bitter principle of *Centaurea sca-biosa* (L.) Presl., *Compositae*. Isolation: Suchy *et al., Coll. Czech. Chem. Commun.* **27**, 1905 (1962). Structure: *eidem, ibid.* **27**, 2398 (1962). Revised structure: *eidem, ibid.* **30**, 3473 (1965). Further structural revisions: *eidem, ibid.* **33**, 2238 (1968).

Crystals, mp 120°. $[\alpha]_D^{20}$ +101.0° (c = 3.67 in chloroform). Slightly sol in ether, diisopropyl ether, acetone; practically insol in petr ether.

8347. Scammony Root. Root of *Convolvulus scammonia* L., *Convolvulaceae*, yielding not less than 8% resin. *Habit.* Eastern Mediterranean region, especially Syria. *Constit.* 8-13% resin; dihydroxycinnamic acid, β-methylesculetin, ipuranol, a reducing sugar, starch.
THERAP CAT: Cathartic.

8348. Scandium. Sc; at. wt 44.9559; at. no. 21; valence 3. A rare earth metal. Naturally occurring isotope: [45]Sc; artificial, radioactive isotopes: 41-44; 46-50. Abundance in earth's crust: 5-6 ppm. Widely dispersed in nature. Occurs in the minerals *thortveitite* [(Sc,Y)$_2$Si$_2$O$_7$] and in other rare earth minerals such as davidite, ytterbite, orthite and cerrite; frequently associated with tin or zirconium. Predicted and called "ekaboron" by Mendeleev. Discovered by Nilson: *Ber.* **12**, 551, 554 (1879); **13**, 1430, 1439 (1880). Sepn from wolframite: Lukens, *J. Am. Chem. Soc.* **35**, 1470 (1913). Sepn based on extreme insolubility of double sulfate in satd potassium sulfate soln or on extraction of Sc(CNS)$_3$ with ether: Fischer, Bock, *Z. Anorg. Chem.* **249**, 146 (1942). Review of isolns including ion-exchange techniques: Spedding *et al., J. Electrochem. Soc.* **105**, 683-686 (1958). Review of scandium and its compounds: R. C. Vickery, *The Chemistry of Yttrium and Scandium* (Pergamon Press, New York, 1960) 123 pp; *idem,* "Scandium, Yttrium and Lanthanum" in *Comprehensive Inorganic Chemistry* vol. **1**, J. C. Bailar, Jr. *et al.,* Eds. (Pergamon Press, Oxford, 1973) pp 329-353.

Metal. Reported to be dimorphic. α-form: hexagonal close-packed structure; d 2.985. β-form: evidence of existence is inconclusive; face-centered cubic structure; d 3.19. mp 1538°. E°(aq) Sc^{3+}/Sc −2.08 V (calc). Salts are hydrolyzed in aq soln.

Oxide, O$_3$Sc$_2$, *scandia*, fine white powder, d 3.864. Obtained by igniting the metal or its compds. Readily sol in hot or concd acids.

Hydroxide, Sc(OH)$_3$, white gelatinous precipitate forming a hard, horny mass on exposure to air; obtained by the action of alkalies on solns of the salts; dissolves readily in dil acids.

Chloride, ScCl$_3$, white deliquescent solid. Prepd by the action of a mixture of sulfur chloride and chlorine on the heated oxide; mp 960°; crystallizes with 6 mols of water. Sol in water, insol in alc. LD$_{50}$ in mice: 755 mg/kg i.p.; 4 g/kg orally, Haley *et al., J. Pharm. Sci.* **51**, 1043 (1962).

Sulfate, Sc$_2$(SO$_4$)$_3$, pentahydrate, d 2.519. Most sol of the sulfates of the rare earths (54.6 g/100 ml at 25°). Converted to the dihydrate on heating above 100°.

Nitrate, Sc(NO$_3$)$_3$, crystallizes as the tetrahydrate, prismatic deliquescent crystals, readily dissolves in water or alc.

8349. Scarlet Red. *1-[[2-Methyl-4-[(2-methylphenyl)-azo]phenyl]azo]-2-naphthalenol; C.I. Solvent Red 24; o-tolyl-azo-o-tolylazo-β-naphthol;* 1-(4-o-tolylazo-o-tolylazo)-2-naphthol; C.I. 26105; Biebrich Scarlet Red; Sudan IV; Fat Ponceau R. C$_{24}$H$_{20}$N$_4$O; mol wt 380.43. C 75.77%, H 5.30%, N 14.73%, O 4.21%. Prepd by diazotizing *o*-amino-azotoluene and coupling with β-naphthol: *Colour Index* vol. **4** (3rd ed., 1971) p 4227. Review of carcinogenicity studies: *IARC Monograph* **8**, 217 (1975).

Dark brown powder. Softens at 175°; mp 181-188°; dec completely at 260°. Practically insol in water. One gram dissolves in 15 ml chloroform. Sol in oils, fats, warm petrolatum, paraffin, phenol. Slightly sol in acetone, alcohol, benzene.
USE: Fat stain.
THERAP CAT: Has been used to promote wound healing.
THERAP CAT (VET): Has been used to stimulate healing of wounds.

8350. Schiff Bases. R-CH=N-C$_6$H$_5$. Condensation products of aldehydes and ketones with primary amines. The compounds are stable if there is at least one aryl group on the nitrogen or carbon. H. Schiff, *Ann. Suppl.* **3**, 343 (1864-1865). *Reviews:* N. V. Sidgewick, *Organic Chemistry of Nitrogen* (Clarendon Press, Oxford, 3rd ed., 1966) p 164-166; R. L. Reeves, *Chemistry of the Carbonyl Group,* S. Patai, Ed. (Wiley, New York, 1966) pp 608-614; J. Fastrez, *Ind. Chim. Belg.* **34**, 835 (1969); A. Bruylants, E. Feymantis-de Medicis, *Chemistry of the Carbon-Nitrogen Double Bond,* S. Patai, Ed. (Wiley, New York, 1970) pp 465-502. Metal complexes: S. Yamada, *Coord. Chem. Res.* **1**, 415 (1966); E. Sinn, C. M. Harris, *ibid.* **4**, 391 (1969); L. F. Lindsy, *Quart. Rev.* **25**, 379 (1971).

8351. Schradan. *Octamethyldiphosphoramide; octamethyl pyrophosphoramide; bis[bisdimethylaminophosphonous] anhydride; bis-N,N,N',N'-tetramethylphosphorodiamidic anhydride; OMPA; Pestox III; Sytam.* C$_8$H$_{24}$N$_4$O$_3$P$_2$; mol wt 286.26. C 33.56%, H 8.45%, N 19.57%, O 16.77%, P 21.66%. Synthesis: Schrader, *B.I.O.S. Final Report no.* **1808** (1948); Gardiner, Kilby, *J. Chem. Soc.* **1950**, 1769; *Biochem. J.* **51**, 79 (1952); Hartley *et al., J. Sci. Food Agr.* **2**, 303, 310 (1951); **3**, 60 (1952); Toy, Costello, *U.S. pat.* **2,717,249** (1955 to Victor Chem.); Toy, Walsh, *Inorg. Syn.* **7**, 73 (1963). Has been found to inhibit peripheral cholinesterase without pronounced effects on the central nervous system, see DuBois *et al., J. Pharmacol. Exp. Ther.* **99**, 376 (1950).

Viscous liquid, d$_4^{25}$ 1.09. bp$_{0.5}$ 120-125°. bp$_{2.0}$ 154°. mp 14-20°. n$_D^{25}$ 1.462. Vapor press at 25°: 1 × 10^{-3} mm Hg. Misc with water. Sol in most organic solvents, including ketones, nitriles, esters, aromatic hydrocarbons and alcohols. Practically insol in higher aliphatic hydrocarbons. Hydrolyzed by acids. LD$_{50}$ in male, female rats: 9.1, 42 mg/kg orally; 15, 44 mg/kg dermally, T. B. Gaines, *Toxicol. Appl. Pharmacol.* **14**, 515 (1969).
USE: Insecticide.

8352. Schweizer's Reagent. *Tetraamminecopper dihydroxide.* Approximately [Cu(NH$_3$)$_4$](OH)$_2$. Usually applied in the form of an ammoniacal soln which is easily obtained by dissolving Cu(OH)$_2$ in 20% ammonia water. The soln is an intense azure color. Evaporation yields long, blue needles: Schweizer, *J. Prakt. Chem.* **72**, 109, 344 (1857).
USE: Dissolves cellulose; used in rayon production.

8353. Scillabiose. *6-Deoxy-4-O-β-D-glucopyranosyl-L-mannose;* 4-O-D-glucopyranosyl-L-rhamnose; glucosido-rhamnose. C$_{12}$H$_{22}$O$_{10}$; mol wt 326.30. C 44.16%, H 6.80%, O 49.03%. Obtained by acid hydrolysis of scillaren A: Stoll *et al., Helv. Chim. Acta* **16**, 703 (1933); Zemplén, *C.A.* **33**, 4202 (1939); from glucoscilliphäosid: Stoll *et al., Helv. Chim. Acta* **35**, 2495 (1952). Structure: Stoll, Renz, *Enzymologia* **7**, 362 (1939), *C.A.* **34**, 6298[3] (1940).

Syrup. $[\alpha]_D^{20}$ —24.8° (0.83 g in 25 ml). Soluble in water, pyridine.

Phenylosazone, $C_{24}H_{32}N_4O_8$, crysts from dil alc, mp 165°. Hexacetylscillabiose, $C_{24}H_{34}O_{16}$, needles from alc, mp 97°.

8354. Scillaren. A mixture of glycosides, scillaren A and B in the proportions in which they occur in fresh squill, *Urginea (Scilla) maritima* (L.) Baker, *Liliaceae,* about 2 parts of A to 1 part of B. Isolation of scillaren and separation of A and B: Stoll *et al., Helv. Chim. Acta* **16,** 703 (1933).

Granular, very bitter powder. On drying in high vacuum at 78° for 15 hrs it loses not more than 6% of its wt. $[\alpha]_D^{20}$ —25° to —35° (0.5 g in 25 ml of 75% w/w alcohol). One gram dissolves in 3000 ml water, in 5 ml abs alcohol, in 5 ml methanol; practically insol in ether and chloroform. Aq solns are neutral to litmus. LD i.v. in cats: 0.18-0.62 mg/kg; LD orally in rabbits: 0.9 mg/kg; LD s.c. in rats: 10 mg/kg, *Handbook of Toxicology,* W. S. Spector, Ed. (Saunders, Philadelphia, 1956) p 266.

Scillaren A, $C_{36}H_{52}O_{13}$, *glucoproscillaridin A, transvaalin.* Mol wt 692.78. C 62.41%, H 7.57%, O 30.02%. Characterization and structure: Stoll *et al., Z. Physiol. Chem.* **222,** 24 (1933); *Helv. Chim. Acta* **17,** 641, 1334 (1934); **18,** 82, 120, 401, 644, 1247 (1935); **24,** 1380 (1941); Zoller, Tamm, *ibid.* **36,** 1744 (1953).

Very bitter taste. Two crystal modifications from methanol: prisms, mp 184-186°; leaflets, mp 208-211°. $[\alpha]_D^{23}$ —71.9° (c = 1.011 in methanol) (Zoller, Tamm, *loc. cit.*). Tendency to form solvated crystals. Six-sided plates, tablets contg 1 mol CH_3OH + 1 mol H_2O from dil methanol, mp 230-240°. Anhydr from 85% alcohol, solvent-free: mp 270°; $[\alpha]_D^{20}$ —72° to —78° (c = 1 to 3 in 75% alcohol), $[\alpha]_D^{20}$ —73.8° (Stoll, *primo loc. cit.*). Soluble in 350 parts alcohol, in 80 parts methanol, in 40 parts dil alcohol (4 vols ethanol + 1 vol water). Practically insol in chloroform, ether. Sparingly sol in water. LD i.v. in cats: 0.143 mg/kg.

Scillaren B. A water-sol mixture of glucosides remaining after extraction of scillaren A. Amorphous, granular powder. Very bitter taste. Scillaren B dried in a high vacuum at 78°C for 15 hrs loses not more than 5% of its weight. Completely dried scillaren B contains approx 99.5% active glycosidal substance. $[\alpha]_D^{20}$ +35° to 41° (0.5 g in 25 ml of 75% w/w ethanol). Freely sol in water. Soluble in alcohol, methanol (about 1 in 5); very slightly sol in chloroform (about 1 in 10,000). Practically insol in ether. Aq solns are neutral to litmus.

THERAP CAT: Cardiotonic.

8355. Scillarenin. *3β,14-Dihydroxybufa-4,20,22-trienolide.* $C_{24}H_{32}O_4$; mol wt 384.50. C 74.97%, H 8.39%, O 16.64%. Prepn by adaptive enzymic decompn of proscillaridin A: Stoll *et al., Helv. Chim. Acta* **34,** 2301 (1951); from *Urginea burkei* Baker, *Liliaceae:* Zoller, Tamm, *ibid.* **36,** 1744 (1953). Structure: Stoll *et al., ibid.* **35,** 1934 (1952).

Prisms from methanol, mp 232-238°. $[\alpha]_D^{20}$ —16.8° (c = 0.357 in methanol); $[\alpha]_D^{20}$ +17.9° (c = 0.39 in chloroform). uv max: 300 nm (log ε 3.72). Mean LD i.v. in cats: 0.1567 mg/kg, Chen, Henderson, *J. Pharmacol. Exp. Ther.* **111,** 365 (1954).

3-Acetate, $C_{26}H_{34}O_5$, crystals from methanol, dec 240-243°. $[\alpha]_D^{20}$ —23.4° (c = 1.365 in chloroform).

THERAP CAT: Cardiotonic.

8356. Scilliroside. *(3β,6β)-6-(Acetyloxy)-3-(β-D-glucopyranosyloxy)-8,14-dihydroxybufa-4,20,22-trienolide.* $C_{32}H_{44}O_{12}$; mol wt 620.67. C 61.92%, H 7.15%, O 30.93%. Glycoside from red squill, the red variety of *Urginea maritima* (L.) Baker, *Liliaceae:* Stoll, Renz, *Helv. Chim. Acta* **25,** 43 (1942); Wichtt, Fuchs, *Arch. Pharm.* **295,** 361 (1962). Structure: Stoll *et al., Helv. Chim. Acta* **26,** 648 (1943); von Wartburg, Renz, *ibid.* **42,** 1620 (1959). Toxicity data: F. Dybing *et al., Acta Pharmacol. Toxicol.* **8,** 391 (1952).

Long prisms from dilute methanol. The crystals are solvated and lose about 8% of their wt in high vacuum, but retain ½ mol H_2O. For the hemihydrate, indistinct mp 168-170°, dec 200°. $[\alpha]_D^{20}$ —59° to —60° (methanol). Absorption max (96% ethanol): 300 nm (log ε 3.73); (98% H_2SO_4): 295, 505 nm ($E_{1cm}^{1\%}$ 260, 315). Freely sol in the lower alcohols, in ethylene glycol, dioxane, glacial acetic acid. Slightly sol in water, acetone, chloroform, ethyl acetate. Practically insol in ether, petr ether. LD_{50} in mice (mg/kg): 0.471 s.c.; 0.440 orally (Dybing).

Tetraacetyl scilliroside, $C_{40}H_{52}O_{16}$, rosettes of long needles from methanol, mp 199°. Sol in chloroform. The four additional acetyl groups are on the sugar.

8357. Scoparin. *8-β-D-Glucopyranosyl-5,7-dihydroxy-2-(4-hydroxy-3-methoxyphenyl)-4H-1-benzopyran-4-one;* 8-glycosyl-4',5,7-trihydroxy-3'-methoxyflavone; scoparoside. $C_{22}H_{22}O_{11}$; mol wt 462.40. C 57.14%, H 4.80%, O 38.06%. From leaves and branches of *Spartium scoparium* L., *Leguminosae:* Stenhouse, *Ann.* **78,** 15 (1851); Mascré, Paris, *Compt. Rend.* **204,** 1270 (1937). Structural study: Hörhammer *et al., Naturwiss.* **49,** 393 (1962). Revised structure: Prox, *Tetrahedron* **24,** 3697 (1968).

Needles from 80% methanol, mp 253°. uv max (methanol): 345, 270 nm (log ε 4.27, 4.18). Practically insol in cold water, ether, chloroform, benzene; sol in hot water, ethanol, methanol, acetic acid, ethyl acetate, acetone, pyridine. Heptaacetate, $C_{36}H_{36}O_{18}$, crystals, mp 240-241°.

8358. Scoparius. Broom; green broom; Scotch broom; Irish broom; hogweed; bannal. Dried tops of *Cytisus scoparius* (L.) Link, *Leguminosae*. The seeds and flowers are also used, though not officially. *Habit.* Western Asia, Southern and Western Europe; cultivated in U.S.A., particularly in Oklahoma. *Constit.* Sparteine, scoparin, genisteine, sarothamnine, volatile oil, tannin, fat, wax, sugar.

THERAP CAT: Diuretic, cathartic.

8359. Scoparone. *6,7-Dimethoxy-2H-1-benzopyran-2-one; 6,7-dimethoxycoumarin;* esculetin dimethyl ether. $C_{11}H_{10}O_4$; mol wt 206.19. C 64.07%, H 4.89%, O 31.04%. Isolation from *Zanthoxylum* and *Artemisia* spp: Araki, Miyashita, *J. Pharm. Soc. Japan* **48**, 437 (1928); Sera, Shibuye, *C.A.* **24**, 5742 (1930); Parihar, Dutt, *Proc. Indian Acad. Sci.* **25A**, 153 (1947); King *et al., J. Chem. Soc.* **1954**, 1392; Singh *et al., J. Sci. Ind. Res. (India)* **15B**, 190 (1956); Ban'kovskaya, Ban'kovskaya, *C.A.* **55**, 17776c (1961). Structure and synthesis: Singh *et al., Chem. & Ind. (London)* **1954**, 1294. Improved synthesis: R. S. Mali *et al., Indian J. Chem.* **21B**, 759 (1982).

Crystals, mp 145-146°. Absorption spectra: Cingolani, *Gazz. Chim. Ital.* **89**, 985, 999 (1959).

8360. Scopola. Dried rhizome of *Scopola carniolica* Jacq., *Solanaceae. Habit.* Germany (Bavaria), Hungary, Russia. *Constit.* Scopolamine (hyoscine), atropine, hyoscyamine; total about 0.7% alkaloids.

THERAP CAT: Anticholinergic.

8361. Scopolamine. *[7(S)-(1α,2β,4β,5α,7β)]-α-(Hydroxymethyl)benzeneacetic acid 9-methyl-3-oxa-9-azatricyclo[3.3.1.0²,⁴]non-7-yl ester; 6β,7β-epoxy-1αH,5αH-tropan-3α-ol (−)-tropate; 6β,7β-epoxy-3α-tropanyl S-(−)-tropate; 6,7-epoxytropine tropate;* scopine tropate; tropic acid ester with scopine; hyoscine; *l*-scopolamine; Scop; Scopoderm TTS; Transderm-V. $C_{17}H_{21}NO_4$; mol wt 303.35. C 67.31%, H 6.98%, N 4.62%, O 21.10%. Tropane alkaloid isolated from *Datura metel* L., *Scopola carniolica* Jacq. and other *Solanaceae.* Constituent of impure duboisine from *Duboisia myoporoides* R. Br., pure duboisine is *l*-hyoscyamine, q.v. Isoln: A. Ladenburg, *Ann.* **206**, 274 (1881); E. Schmidt, *Arch. Pharm.* **230**, 207 (1892). Identity with hyoscine: O. Hesse, *Ann.* **271**, 100 (1892); idem, *J. Prakt. Chem.* **66**, 194 (1902). Absorption spectra: J. J. Dobbie, J. J. Fox, *J. Chem. Soc.* **103**, 1194 (1913). Resolution of isomers and review of early literature: H. King, *ibid.* **1919**, 476. Extraction procedure: F. Chemnitius, *J. Prakt. Chem.* **120**, 221 (1928). Structural studies: J. Gadamer, F. Hammer, *Arch. Pharm.* **259**, 122 (1921); K. Hess, O. Wahl, *Ber.* **55**, 1979 (1922); R. Willstätter, E. Berner, *ibid.* **56**, 1079 (1923); W. Steffens, *Arch. Pharm.* **262**, 205 (1924). Configuration: G. Fodor, *Nature* **170**, 278 (1952); J. Meinwald, *J. Chem. Soc.* **1953**, 712. Review of stereochemistry: G. Fodor, *Tetrahedron* **1**, 86 (1957). Absolute configuraton of tropic acid moiety: G. Fodor, G. Csepreghy, *Tetrahedron Letters* **1959**, 16; *eidem, J. Chem. Soc.* **1961**, 3222. Synthesis of DL-form: G. Fodor *et al., Chem. & Ind.* **1956**, 764; P. Dobo *et al., J. Chem. Soc.* **1959**, 3461. Clinical evaluation in motion sickness: J. J. Brand, P. Whittingham, *Lancet* **2**, 232 (1970); in peripheral vertigo: T. Rahko, P. Karma, *J. Larynogol. Otol.* **99**, 653 (1985). Acute toxicity of the hydrobromide: Stockhaus, Wick, *Arch. Int. Pharmacodyn. Ther.* **180**, 155 (1969). Review of CNS effects in humans: D. J. Safer, R. P. Allen, *Biol. Psychiat.* **3**, 347-355 (1971); of use in anesthesia: L. E. Shutt, J. B. Bowles, *Anaesthesia* **34**, 476-490 (1979); of pharmacology and clinical efficacy: S. P. Clissold, R. C. Heel, *Drugs* **29**, 190-207 (1985).

Viscous liquid. $[α]_D^{20}$ −28° (c = 2.7). Sol in 9.5 parts water at 15°. Forms a cryst monohydrate, mp 59°. Freely sol in hot water, in alcohol, ether, chloroform, acetone. Sparingly sol in benzene, petr ether. Easily hydrolyzed by acids or alkalies. Dec on standing.

Hydrobromide trihydrate, $C_{17}H_{22}BrNO_4 \cdot 3H_2O$, *scopolammonium bromide, Scopos.* Orthorhombic sphenoidal crystals from water, slightly efflorescent in dry air. mp 195° (after drying at 105° for 3 hours). $[α]_D^{25}$ −24° to −26° (c = 5, calculated on anhydrous basis). pH of 0.05 molar soln 5.85. One gram dissolves in 1.5 ml water, 20 ml alcohol. Slightly sol in chloroform. Practically insol in ether. LD_{50} in rats (mg/kg): 3800 s.c. (Stockhaus, Wick).

Hydrochloride, $C_{17}H_{22}ClNO_4$, crystals from acetone, mp 200°. Dihydrate, prisms from water, melts in water of crystn at 80°. Very sol in water and alcohol. pH of 0.05 molar soln 5.85.

Methyl bromide, See Methscopolamine Bromide.

Methyl nitrate, $C_{18}H_{24}N_2O_7$, *methscopolamine nitrate, Skopolate, Skopyl.* Crystals. Freely sol in water, dil alcohol; slightly sol in abs alcohol.

DL-Form, *atroscine.* Dihydrate, chisel-shaped prisms from ethanol + water, mp 38-40°. Monohydrate, efflorescent crystals, mp 55-57°. Anhydrous, long prisms, mp 82-83°. Very slightly sol in water; sol in alc, chloroform, ether, oils.

THERAP CAT: Anticholinergic. Treatment of motion sickness.

THERAP CAT (VET): Has been used as a sedative, a preanesthetic and to control motion sickness.

8362. Scopolamine N-Oxide. *[7(S)-(1α,2β,4β,5α,7β)]-α-(Hydroxymethyl)benzeneacetic acid 9-methyl-3-oxa-9-azatricyclo[3.3.1.0²,⁴]non-7-yl ester N-oxide; 6β,7β-epoxy-1αH,5αH-tropan-3α-ol (−)-tropate 8-oxide;* scopolamine aminoxide; Genoscopolamine. $C_{17}H_{21}NO_5$; mol wt 319.35. C 63.93%, H 6.63%, N 4.39%, O 25.05%. Prepn: Polonovski, Polonovski, *Compt. Rend.* **180**, 1755 (1925). Absolute configuration: Huber *et al., Can. J. Chem.* **49**, 3258 (1971).

White powder, mp about 80°. $[α]_D^{20}$ −14° in water.

Hydrobromide, $C_{17}H_{22}BrNO_5$, prisms from water, mp 135-138°. $[α]_D^{25}$ −25° (c = 2). Soly in water about 10 g/100 ml. Slightly sol in alcohol, acetone. pH of 3% aq soln about 3.2.

THERAP CAT: Anticholinergic.

8363. Scopoletin. *7-Hydroxy-6-methoxy-2H-1-benzopyran-2-one; 7-hydroxy-6-methoxycoumarin;* 6-methoxyumbelliferone; *β*-methylesculetin; chrysatropic acid; gelseminic acid. $C_{10}H_8O_4$; mol wt 192.16. C 62.50%, H 4.20%, O 33.30%. The aglucone of scopolin. Occurs in root of *Scopolia japonica* Maxim., *Scopolia carniolica* Jacq., *Atropa belladonna* L., *Solanaceae, Convolvulus scammonia* L., *Convolvulaceae.* Isoln: Eykman, *Ber.* **17** III, 442 (1884). Synthesis: Crosby, *J. Org. Chem.* **26**, 1215 (1961); Desai, Desai, *J. Indian Chem. Soc.* **40**, 456 (1963).

Needles or prisms from chloroform or acetic acid, mp 204°. uv max: 230, 254, 260, 298, 346 nm (log ε 4.11, 3.68, 3.63, 3.68, 4.07), Ballantyne *et al.*, *Tetrahedron* **27**, 871 (1971). Slightly sol in water or cold alcohol; sol in hot alcohol or hot glacial acetic acid; moderately sol in chloroform; practically insol in benzene. Its alcoholic soln has a blue fluorescence. Reduces Fehling's soln.

8364. Scopolin. 7-(β-D-*Glucopyranosyloxy*)-6-methoxy-2H-1-benzopyran-2-one. $C_{16}H_{18}O_9$; mol wt 354.30. C 54.24%, H 5.12%, O 40.64%. Mono-β-glucopyranoside of scopoletin. From root of *Scopolia japonica* Maxim., *S. carniolica* Jacq., *Solanaceae* and *Nerium odorum* Sol., *Apocynaceae*. Isoln: Eykman, *Ber.* **17** III, 442 (1884); Ritter *et al.*, *Helv. Chim. Acta* **36**, 434 (1953). Synthesis: Merz, *Arch. Pharm.* **270**, 476 (1932). Possible identity with *murrayin*: Bose, Mookerjee, *J. Indian Chem. Soc.* **14**, 489 (1937). Chromatography, and spectrum of murrayin: Chakraborty, Bose, *ibid.* **33**, 905 (1956); Chakraborty, Chakraborty, *Trans. Bose Res. Inst. (Calcutta)* **24** (1), 15 (1961), *C.A.* **55**, 24228c (1961).

Needles. mp 218°. Soluble in water, alcohol. Practically insol in chloroform, ether.

Tetraacetate, $C_{24}H_{26}O_{13}$, polymorphous crystals, prisms, mp 166° or flat plates, mp 184-185°.

8365. Scopoline. Hexahydro-4-methyl-2,5-methano-2H-furo[3,2-b]pyrrol-6-ol; 3β,7β-epoxy-1βH,5βH-tropan-6α-ol; 3α,6α-epoxy-7β-hydroxytropane; Oscine. $C_8H_{13}NO_2$; mol wt 155.19. C 61.91%, H 8.44%, N 9.03%, O 20.62%. Decomposition product of scopolamine: Fodor, Kovács, *J. Chem. Soc.* **1953**, 2341. Prepn from teloidine: Zeile, Heusner, *Ber.* **90**, 2800, 2809 (1957); Ger. pats. **1,056,617; 1,062,-703** (1959 to Boehringer, Ing.). Structure: Fodor *et al.*, *J. Chem. Soc.* **1955**, 3504. Stereochemistry: Fodor, *Tetrahedron* **1**, 86 (1957).

Crystals from petr ether, mp 107-109°. bp 248°. Sol in water, alcohol, acetone, ether.

Hydrochloride, $C_8H_{13}NO_2.HCl$, crystals from ethanol, mp 255-257°.

Acetate, $C_{10}H_{15}NO_3$, crystals from petr ether, mp 56-57°.

8366. Scotophobin. L-Seryl-L-aspartyl-L-asparaginyl-L-asparaginyl-L-glutaminyl-L-glutaminylglycyl-L-lysyl-L-seryl-L-alanyl-L-glutaminyl-L-glutaminylglycylglycyl-L-tyrosinamide. $C_{62}H_{97}N_{23}O_{26}$; mol wt 1580.62. C 47.11%, H 6.19%, N 20.38%, O 26.32%. Pentadecapeptide, isolated from the brains of rats with acquired fear of the dark, which induces dark avoidance in untrained mice. (From the Greek *skotos*, dark and *phobos*, fear.) Thought to be the first isolation, identification and synthesis of one of the "chemical code words" that control memory and learning. *See* Ungar *et al.*, *Nature* **238**, 198 (1972). Original training, testing and extraction procedures: Ungar *et al.*, *ibid.* **217**, 1259 (1968); *eidem*, *Proc. West. Pharmacol. Soc.* **13**, 149 (1970). Structural elucidation: Desiderio *et al.*, *Chem. Commun.* **1971**, 432. Synthesis of preliminary structures: Ali *et al.*, *Experientia* **27**, 1138 (1971); *eidem*, *Int. J. Peptide Protein Res.* **4**, 395 (1972). Solid phase synthesis of revised structure: Parr, Holzer, *Z. Physiol. Chem.* **352**, 1043 (1971); *eidem*, *Experientia* **28**, 884 (1972). Passive transfer of dark avoidance to mice: Malin, Guttman, *Science* **178**, 1219 (1972); to goldfish: Guttman *et al.*, *Nature New Biol.* **235**, 126 (1972); Bryant *et al.*, *Science* **177**, 635 (1972). Chemistry and critical analysis of published data on scotophobin: Stewart, *Nature* **238**, 202-210 (1972).

Ser-Asp-Asn-Asn-Gln-Gln-Gly-Lys-Ser-Ala-Gln-Gln-Gly-Gly-TyrNH₂

8367. Scutellarein. 5,6,7-Trihydroxy-2-(4-hydroxyphenyl)-4H-1-benzopyran-4-one; 4',5,6,7-tetrahydroxyflavone; 6-hydroxypelargidenon 1465. $C_{15}H_{10}O_6$; mol wt 286.23. C

62.94%, H 3.52%, O 33.54%. By hydrolysis of scutellarin: Molisck, Goldschmidt, *Monatsh.* **22**, 679 (1901); Marsh, *Biochem. J.* **59**, 58 (1955). From *Digitalis lanata* Ehrh., *Scrophulariaceae*: Rangaswami, Rao, *Proc. Indian Acad. Sci.* **54A**, 51 (1961), *C.A.* **56**, 10076f (1962). Structure: Wessely, Moser, *Monatsh.* **56**, 97 (1930). Synthesis: Sastri, Seshadri, *Proc. Indian Acad. Sci.* **23A**, 262 (1946), *C.A.* **41**, 449a (1947); Zemplén *et al.*, *Acta Chim. Acad. Sci. Hung.* **16**, 445 (1958); Jouanne, Mentzer, *Compt. Rend. Ser. C* **263**, 1022 (1966).

Yellow leaflets from methanol, does not melt below 300°. uv max (ethanol): 286, 339 nm (ε 16,600; 18,300).

Tetraacetate, $C_{23}H_{18}O_{10}$, prisms from ethyl acetate, mp 235-237°.

Glucuronide, $C_{21}H_{18}O_{12}$, scutellarin. From leaves of *Scutellaria altissima* Linn., *Labiatae*: Goldschmiedt, Zerner, *Monatsh.* **31**, 439 (1910); from *Centaurea scabiosa* L., *Compositae*: Charaux, Rabaté, *J. Pharm. Chim.* **9**, 155 (1940); from *Scutellaria* spp.: Marsh, *Biochem. J.* **59**, 58 (1955); *Nature* **183**, 1824 (1959). Synthesis: Farkas *et al.*, *Ber.* **107**, 3878 (1974). Needles from alc, darkens above 230°, does not melt below 300°. $[\alpha]_D^{18}$ −14° (water); $[\alpha]_D^{20}$ −139° (pyridine). uv max (ethanol): 285, 335 nm (ε 20,000; 26,100). Practically insol in water; sol in alkali hydroxides, glacial acetic acid; slightly sol in organic solvents.

8368. Scutellaria. Skullcap; helmet flower. Dried overground portion of *Scutellaria lateriflora* L., *Labiatae*. *Habit.* North America. *Constit.* Scutellarin, volatile oil, tannin.

8369. Sebacic Acid. Decanedioic acid; 1,8-octanedicarboxylic acid. $C_{10}H_{18}O_4$; mol wt 202.24. C 59.38%, H 8.97%, O 31.64%. $HOOC(CH_2)_8COOH$. Obtained on a large scale by heating castor oil or ricinoleic acid with sodium hydroxide. Convenient lab procedures: Topchiev, Pavlov, *C.A.* **47**, 8002h (1953); Dominguez *et al.*, *J. Chem. Ed.* **29**, 446 (1952). Also prepd by ozonization of undecylenic acid: Noorduyn, *Rec. Trav. Chim.* **38**, 326 (1919); Verkade *et al.*, *ibid.* 385. Review and bibliography: Jones, *Chimia* **5**, 169-173 (1951).

Monoclinic prismatic tablets, leaflets from acetone + petr ether (commercial product is a white, free-flowing powder having a mild fatty acid odor). d_4^{20} 1.207. mp 134.5°. bp_{100} 294.5°; bp_{50} 243.5°; bp_{15} 243.5°; bp_{10} 232°. Sp heat (solid) 0.48 cal/g; sp heat (liq) 0.53 cal/g. Sublimes slowly at 750 mm when heated to mp. n_D^{134} 1.422. K_1 = 2.6 × 10⁻⁵; K_2 = 2.6 × 10⁻⁶. Aq solns are neutral to methyl orange. Soly in water: 0.004% at 0°; 0.10% at 20°; 0.42% at 65°; 2.0% at 100°. Freely sol in alcohols, esters, ketones. Soly in ether: 0.1% at 17°. Sparingly sol in hydrocarbons and chlorinated hydrocarbons.

Ethyl ester, $C_{14}H_{26}O_4$, ethyl sebacate, diethyl sebacate. Colorless to yellowish fluid. d_4^{20} 0.965. bp about 307° with some decompn. n_D^{20} 1.4369. Sol in about 700 parts cold, 50 parts boiling water; misc with alc, ether and other organic solvents. LD_{50} orally in rats: 14,470 mg/kg, P. M. Jenner *et al.*, *Food Cosmet. Toxicol.* **2**, 327 (1964).

USE: Raw material in the manuf of synthetic resins of the alkyd or polyester type, non-migrating plasticizers, polyester rubbers, synthetic fibers of the polyamide type.

8370. Sebacil. 1,2-Cyclodecanedione. $C_{10}H_{16}O_2$; mol wt 168.23. C 71.39%, H 9.59%, O 19.02%. Prepd by oxidation of sebacoin with chromium trioxide in acetic acid: Blomquist *et al.*, *J. Am. Chem. Soc.* **74**, 3640 (1952); *cf.* Prelog *et al.*, *Helv. Chim. Acta* **35**, 1610 (1952); Blomquist, Goldstein, *Org. Syn.* coll. vol. **IV**, 838 (1963).

Crystals, mp 40-41°. bp_{18} 120-125°; bp_{10} 104-105°.

8371. Sebacoin. *2-Hydroxycyclodecanone;* 1-cyclodecanol-2-one. $C_{10}H_{18}O_2$; mol wt 170.24. C 70.54%, H 10.66%, O 18.80%. Prepd by cyclization of methyl or ethyl sebacate with sodium metal: Hansley, U.S. pat. **2,228,268** (1941 to du Pont); Prelog *et al., Helv. Chim. Acta* **35**, 1598 (1952); Allinger, *Org. Syn.* **coll. vol. IV**, 840 (1963).

Prisms from ether + petr ether, mp 42°. bp_6 113-114°; $bp_{0.1}$ 77°.

8372. Secalonic Acids. $C_{32}H_{30}O_{14}$; mol wt 638.59. C 60.19%, H 4.74%, O 35.08%. Toxic mold metabolites which are diastereoisomeric dimeric hydroxanthone derivatives manifest as yellow pigments. First isolation studies: Stoll *et al., Helv. Chim. Acta* **35**, 2022 (1952). Seven isomers have been isolated: *secalonic acid A, secalonic acid B,* and *secalonic acid C* from ergot: Aberhart *et al., Tetrahedron* **21**, 1417 (1965); Franck *et al., Ber.* **99**, 3842, 3875 (1966); *secalonic acid D* from *Penicillium oxalicum:* Steyn, *Tetrahedron* **26**, 51 (1970); *secalonic acid E* from *Phoma terrestris:* Hansen, Howard *et al., Chem. Commun.* **1973**, 464; *secalonic acid F* from *Aspergillus aculeatus:* R. Andersen *et al., J. Org. Chem.* **42**, 352 (1977); *secalonic acid G* from *Pyrenochaeta terrestris:* I. Kurobane, L. C. Vining, *Tetrahedron Letters* **1978**, 4633. Secalonic acid A is enantiomeric with D; secalonic acid B is enantiomeric with E. Chromatographic separation of the ergot pigments: Franck, *Angew. Chem. Int. Ed.* **8**, 251 (1969); structures: ApSimon *et al., J. Chem. Soc.* **1965**, 4144. Revised structures of secalonic acids A, B, C, D: Hooper *et al., Chem. Commun.* **1971**, 111; *eidem, J. Chem. Soc. (C)* **1971**, 3580. Crystal and molecular structure of secalonic acid A: C. C. Howard *et al., J. Chem. Soc. Perkin Trans. I* **1976**, 1820. Identity of secalonic acid A with entothein: Yoshioka *et al., Chem. Pharm. Bull.* **16**, 2090 (1968). Biosynthesis of secalonic acid A: I. Kurobane *et al., Tetrahedron Letters* **1978**, 1379. Review of secalonic acids and other ergochromes: Franck, Flasch in *Fortschr. Chem. Org. Naturst.* **30**, 151-206 (1973).

Secalonic acid A, *ergochrome AA(2,2');* 6,6',7,7'-tetrahydro-1,1',5β,5'β,8,8'-hexahydroxy-6α,6'α-dimethyl-9,9'-dioxo-[2,2'-bixanthene]-10αβ,10'αβ(5H,5'H)-dicarboxylic acid dimethyl ester. Fine lemon-yellow needles from dioxane + petr ether + chloroform, mp 246-248° (dec); also reported as mp 260° (Yoshioka *et al., loc. cit.).* $[\alpha]_D$ −73° (CHCl₃). uv max (methanol): 247, 340 nm (ε 18,140, 31,200). pK_{DMF} 8.7. Readily soluble in pyridine, dimethylformamide, dioxane; moderately sol in chloroform, methylene chloride, acetone. Gives a wine-red color reaction with 5% $FeCl_3$ in methanol.

8373. Secnidazole. α,2-Dimethyl-5-nitro-1H-imidazole-1-ethanol; 1-(2-hydroxypropyl)-2-methyl-5-nitroimidazole; 1-(2-methyl-5-nitroimidazol-1-yl)-2-propanol; PM-185184; RP-14539; Flagentyl. $C_7H_{11}N_3O_3$; mol wt 185.18. C 45.40%, H 5.99%, N 22.69%, O 25.92%. Analog

of metronidazole, *q.v.* Prepn: **Fr. pat. M3270** (1965 to Rhône-Poulenc), *C.A.* **63**, 11571d (1965); C. Cosar *et al., Arzneimittel-Forsch.* **16**, 23 (1966). Anti-amebic and trichomonacidal activities: F. Benazet, L. Guillaume, *Bull. Soc. Pathol. Exot. Ses Fil.* **69**, 309 (1976), *C.A.* **90**, 145922v (1979). Serum half-life: J. Symonds, *J. Antimicrob. Chemother.* **5**, 484 (1979). Therapeutic use: D. Videau *et al., Brit. J. Vener. Dis.* **54**, 77 (1978).

Cryst from toluene, mp 76° (Cosar).
THERAP CAT: Anti-amebic. Antiprotozoal (Trichomonas).

8374. Secobarbital Sodium. *5-(1-Methylbutyl)-5-(2-propenyl)-2,4,6(1H,3H,5H)-pyrimidinetrione monosodium salt; sodium 5-allyl-5-(1-methylbutyl)barbiturate;* 5-allyl-5-(1-methylbutyl)barbituric acid sodium salt; 5-allyl-5-(1-methylbutyl)malonylurea sodium salt; meballymal sodium; quinalbarbitone sodium; Barbosec; Bipinal sodium; Immenoctal; Imesonal; Seotal; Pramil; Quinalspan; Sebar; Sedutain; Seconal sodium. $C_{12}H_{17}N_2NaO_3$; mol wt 260.27. C 55.37%, H 6.58%, N 10.76%, Na 8.84%, O 18.44%. Prepn: U.S. pat. **1,954,429** (1934). Comprehensive description: I. Comer in *Analytical Profiles of Drug Substances* vol. 1, K. Florey, Ed. (Academic Press, New York, 1972) pp 343-365. Toxicity data: E. W. Schafer, *Toxicol. Appl. Pharmacol.* **21**, 315 (1972).

White, hygroscopic powder. Bitter taste. Very sol in water. Sol in alc. Practically insol in ether. Aq solns are alkaline to litmus. The soly in elixirs is increased by the presence of methenamine. LD_{50} orally in rats: 125 mg/kg (Schafer).
Free acid, $C_{12}H_{17}N_2O_3$, mp 100°.
Caution: May be habit forming. This is a controlled substance (depressant) listed in the U.S. Code of Federal Regulations, Title 21 Parts 329.1, 1308.12 and 1308.13 (1987).
THERAP CAT: Sedative, hypnotic.
THERAP CAT (VET): Short-acting hypnotic, sedative. Preanesthetic.

8375. Secretin. Secretine; Secretin-Kabi; Sekretolin. A strongly basic polypeptide gastrointestinal hormone, produced in the duodenum and jejunum of many animal species and man. Its primary biological activity is stimulation of exocrine pancreatic secretion. Discovered by W. M. Bayliss and E. H. Starling, *Ergeb. Physiol.* **5**, 664 (1905). Isoln of porcine secretin: J. E. Jorpes, V. Mutt, *Acta Chem. Scand.* **15**, 1790 (1961). Structure: J. E. Jorpes *et al., Biochem. Biophys. Res. Commun.* **9**, 275 (1962); V. Mutt *et al., Biochemistry* **4**, 2358 (1965). Synthesis: M. Bodanszky *et al., J. Am. Chem. Soc.* **89**, 685, 6753 (1967); M. A. Ondetti *et al., ibid.* **90**, 4711 (1968); M. Bodanszky *et al.,* U.S. pat. **3,767,- 639** (1973 to Squibb); E. Wuensch *et al., Ber.* **104**, 2430 (1971); **105**, 2508 (1972); E. Wuensch, *Naturwiss.* **59**, 239 (1972); B. Hemmasi, E. Bayer, *Int. J. Peptide Protein Res.* **9**, 63 (1977); D. H. Coy *et al., Peptides* **3**, 137 (1982). Isoln of bovine secretin and amino acid sequence (identical to porcine): M. Carlquist *et al., FEBS Letters* **127**, 71 (1981). Stability study: H. Beyerman *et al., Life Sci.* **29**, 885 (1981). Comparative effects of synthetic and natural secretin on pancreatic secretion and on secretin, insulin, and glucagon levels in man: C. Beglinger *et al., Dig. Dis. Sci.* **27**, 231 (1982). Review of radioimmunoassay methods: T. M. Chang, W. Y. Chey, *ibid.* **25**, 529-552 (1980). General reviews: W. H. Häcki, *Clin. Gastroenterol.* **9**, 609-632 (1980); K. G. Wormsley, *Scand. J. Gastroenterol.* **15**, 513-517 (1980); V. Mutt, *Vitam. Horm.* **39**, 231-426 (1982).

```
               1                                    9
        His-Ser-Asp-Gly-Thr-Phe-Thr-Ser-Glu-

        10                                   18
        Leu-Ser-Arg-Leu-Arg-Asp-Ser-Ala-Arg-

        19                                   27
        Leu-Gln-Arg-Leu-Leu-Gln-Gly-Leu-Val

                    porcine secretin
```

Porcine secretin, $C_{130}H_{220}N_{44}O_{41}$, *secretin [pig], porcine hormone secretin, secretin [ox].*

THERAP CAT: Hormone (pancreatic).

8376. Securinine. *Securinan-11-one.* $C_{13}H_{15}NO_2$; mol wt 217.26. C 71.86%, H 6.96%, N 6.45%, O 14.73%. From leaves and roots of *Securinega suffruticosa* Rehder, *Euphorbiaceae* found in the Ussuri region: Murav'eva, Ban'kovskii, *C.A.* **50**, 17335c (1956); *eidem, Doklady Akad. Nauk. SSSR* **110**, 998 (1956), *C.A.* **51**, 8121a (1957). Improved extraction process: Kogan *et al.*, **Brit.** pat. **1,169,471** (1959 to All-Union Sci. Res. Inst. of Medicinal Plants), *C.A.* **72**, 47349x (1970) corresp to U.S. pat. **3,538,103** (1970). Structure: Satoda *et al., Tetrahedron Letters* **1962**, 1199; Mukherjee *et al., Naturwiss.* **50**, 155 (1963); Saito *et al., Tetrahedron* **19**, 2085 (1963). Stereochemistry: Nakano *et al., Chem. & Ind. (London)* **1963**, 1034; Parello *et al., Bull. Soc. Chim. France* **1963**, 898; Nakano *et al., J. Org. Chem.* **28**, 2619 (1963); Horii *et al., Tetrahedron* **19**, 2101 (1963); Imado *et al., Chem. & Ind. (London)* **1964**, 1691. Synthesis of the racemate: Saito *et al., Chem. Pharm. Bull.* **14**, 313, 1059 (1966); Horii *et al., Tetrahedron* **23**, 1165 (1967).

Yellow crystals from alcohol, mp 142-143°. $[\alpha]_D^{20}$ −1042° (c = 1 in alc). uv max (alc): 256, 330 nm (log ε 4.27, 3.30). Hydrochloride, mp 230°. $[\alpha]_D^{20}$ −259.2° (alcohol). Nitrate, mp 205°. $[\alpha]_D^{20}$ −312.12° (alcohol).

8377. Sedecamycin. *[1S-(1R*,2S*,3E,5E,7R*,9E,11E,-13R*,15S*,19S*)]-N-[13-(Acetyloxy)-7-hydroxy-1,4,10,19-tetramethyl-17,18-dioxo-16-oxabicyclo[13.2.2]nonadeca-3,5,-9,11-tetraen-2-yl]-2-oxopropanamide; N-(7,13-dihydroxy-1,4,10,19-tetramethyl-17,18-dioxo-16-oxabicyclo[13.2.2]-nonadeca-3,5,9,11-tetraen-2-yl)pyruvamide 13-acetate;* Antibiotic T-2636A; bundlin B; lankacidin A; lankacidin C 14-acetate; T-2636A; Takelan. $C_{27}H_{35}NO_8$; mol wt 501.58. C 64.66%, H 7.03%, N 2.79%, O 25.52%. One of 7 components of antibiotic *T-2636*. First isolated as bundlin B from *Streptomyces griseofuscus:* J. M. J. Sakamoto *et al., J. Antibiot.* **A15**, 98 (1962). Isoln from *S. rochei var volubilis:* **Neth.** pat. **Appl. 6,807,119;** E. Higashide *et al.*, **U.S.** pat. **3,626,055** (1968, 1971 both to Takeda); *eidem, J. Antibiot.* **24**, 1 (1971); S. Harada *et al., ibid.* 13. Similar to lankacidin which is isolated from *S. violaceoniger* with lankamycin, *q.v.* Structure, absolute configuration and identity with bundlin B: H. Harada *et al., Tetrahedron Letters* **1969**, 2239; K. Kamiya *et al., ibid.* 2245; M. Uramoto *et al., ibid.* 2249. Enzymatic conversion to other lankacidins: T. Fugono *et al., Experientia* **26**, 26 (1970). Antibacterial activity: K. Tsuchiya *et al., J. Antibiot.* **24**, 29 (1971). Mass spectra: M. Uramoto *et al., Agr. Biol. Chem.* **38**, 855 (1974). Biosynthetic studies: *eidem, J. Am. Chem. Soc.* **100**, 3616 (1978); K. Kakinima *et al., Tetrahedron Letters* **23**, 5303 (1982). Synthetic study: M. J. Fray, E. J. Thomas, *Tetrahedron* **40**, 673 (1984). Activity against *Treponema hyodysenteriae* (swine dysentery): **Belg.** pat. **891,275;** N. Narukawa *et al.*, **U.S.** pat. **4,425,356** (1982, 1984 both to Takeda).

Neutral lipophilic colorless plates from ether-ethyl acetate, mp 207-210° (dec). $[\alpha]_D^{25}$ −235° (c = 1.0 in ethanol). uv max (ethanol): 228 mu (ε 46900). Slightly sol in n-hexane, petroleum ether. Sol in ethyl acetate, acetone, chloroform, methanol. LD_{50} in mice (g/kg): up to 10 orally, 8-10 i.p. (Harada).

THERAP CAT (VET): Antibacterial. In swine dysentery.

8378. Seidlitz Mixture. A mixture of 3 parts Rochelle salt and 1 part sodium bicarbonate. Ten grams of the mixture are employed with 2.17 g tartaric acid for one Seidlitz powder.

THERAP CAT: Cathartic.

8379. Selagine. *5-Amino-11-ethylidene-5,8,9,10-tetrahydro-7-methyl-5,9-methanocycloocta[b]pyridin-2(1H)-one.* $C_{15}H_{18}N_2O$; mol wt 242.31. C 74.35%, H 7.49%, N 11.56%, O 6.60%. Occurs in stem and leaves of *Lycopodium selago* L., *Lycopodiaceae:* Muszynski, *Quart. J. Pharm. Pharmacol.* **21**, 34 (1948). Structure: Valenta *et al., Tetrahedron Letters* **1960**(10), 26; Yoshimura *et al., ibid.* no. 12, 14. Synthetic studies: A. S. Kende *et al., ibid.* **25**, 1341 (1984); D. Gravel *et al., Can. J. Chem.* **62**, 2945 (1984). Proposed as substitute for pilocarpine and eserine in glaucoma: Miratynska-Ernestowa, *Klinika Oczna* **18**, 437 (1948).

Amorphous base, odor of coniine. mp 224-226°. $[\alpha]_D^{25}$ −99° (methanol). uv max (ethanol): 231, 313 nm (ε 10700, 8500). pKa = 7.18. Forms deliquescent salts.

8380. Selenic Acid. H_2O_4Se; mol wt 144.98. H 1.39%, O 44.15%, Se 54.46%. H_2SeO_4. Prepd by treating lead selenate with hydrogen sulfide and concentrating the filtered soln by evaporation: Mitscherlich, *Pogg. Ann.* **9**, 623 (1827); **11**, 327 (1827); *Ann. Chim. Phys.* [2] **36**, 100 (1827); by treating calcium selenate with cadmium oxalate and hydrogen sulfide: von Hauer, *J. Prakt. Chem.* [1] **80**, 214, 317 (1860); by treating a soln of silver selenite with bromine and evaporating the filtered soln: Thomsen, *Ber.* **2**, 598 (1869); by oxidizing selenium oxide with 30% hydrogen peroxide and removing water: Gilbertson, King, *J. Am. Chem. Soc.* **58**, 180 (1936); *Inorg. Syn.* **3**, 137 (1950).

Hexagonal prisms; mp 58°; d_4^{15} 2.9508; bp 260°. Very sol in water; sol in sulfuric acid; insol in ammonia; dec in alcohol. Very deliquescent. Reduced by hydrobromic acid, hydriodic acid, hydrogen sulfide, hydroxylamine hydrochloride, phenylhydrazine, formic acid, oxalic acid, malonic acid, pyruvic acid, acetyl chloride, and several metals.

8381. Selenious Acid. Monohydrated selenium dioxide; selenous acid. H_2O_3Se; mol wt 128.98. H 1.56%, O 37.22%, Se 61.22%. H_2SeO_3. Prepd by dissolving selenium dioxide in hot water and cooling: Berzelius, *Acad. Handl. Stockholm* **39**, 13 (1818).

Deliquescent hexagonal prisms; d_4^{15} 3.004. Vapor press. (mm Hg): 2 at 15°; 4.5 at 35°; 7 at 40.3°. $K_1 = 0.0024$; $K_2 = 4.8 \times 10^{-9}$. Gives off water upon heating and selenium oxide sublimes. Oxidized to selenic acid by strong oxidizing agents such as ozone, hydrogen dioxide, chlorine. Reduced to selenium by most reducing agents including hydriodic acid, sulfurous acid, sodium hyposulfite, hydroxylamine salts, hydrazine salts, hypophosphorous acid, phosphorous acid. 90 parts dissolve in 100 parts water at 0°, 400 parts at 90°; very sol in alcohol; insol in ammonia.

USE: As a reagent for alkaloids; as oxidizing agent.

8382. Selenium. Se; at. wt 78.96; at. no. 34; valences 2, 4, 6. Six stable isotopes: 74 (0.87%); 76 (9.02%); 77 (7.58%); 78 (23.52%); 80 (49.82%); 82 (9.19%); artificial, radioactive isotopes: 70-73; 75; 79; 81; 83-85; 87. Discovered in 1817 by Berzelius. Constitutes about 0.09 ppm of the earth's crust. Occurs in nature usually in the sulfide ores of the heavy metals; found in small quantities in pyrite; in the minerals *clausthalite* (PbSe), *naumannite* [(Ag,Pb)Se], *tiemannite* (HgSe); in selenosulfur. Prepn: Waitkins *et al.*, *Ind. Eng. Chem.* **34**, 899 (1942). Purification: Nielsen, Heritage, *J. Electrochem. Soc.* **106**, 39 (1959). Purification by distn from a mixture contg a small amount of Mg: Oberbacher, Schlier, U.S. pat. 2,930,678 (1960 to Norddeutsche Raffinerie). Essential trace element in animal diets at levels of approx 0.1 ppm. Toxic element at higher concentrations. Reviews of nutritional and toxic properties of selenium: I. Rosenfeld, O. A. Beath, *Selenium* (Academic Press, New York, 1964) 411 pp; E. Browning, *Toxicity of Industrial Metals* (Appleton-Century-Crofts, New York, 2nd ed., 1969) pp 286-295; E. J. Underwood, *Trace Elements in Human and Animal Nutrition* (Academic Press, New York, 1971) pp 323-368; Frost, *Crit. Rev. Toxicol.* **1**, 467-514 (1972); Allaway, *Cornell Vet.* **63**, 151-170 (1973); Frost, Lish, *Ann. Rev. Pharmacol.* **15**, 259-284 (1975). Review of selenium biochemistry: T. C. Stadtman, *Ann. Rev. Biochem.* **49**, 93-110 (1980); *idem, FASEB J.* **1**, 375-379 (1987). Symposium on organic selenium and tellurium compds: Y. Okamoto, W. H. H. Gunther, Eds., *Ann. N.Y. Acad. Sci.* **192**, 1-225 (1972). Reviews of selenium and its compds: K. W. Bagnall, *The Chemistry of Selenium, Tellurium and Polonium* (Elsevier, New York, 1966) 200 pp; Bagnall in *Comprehensive Inorganic Chemistry* vol. 2, J. C. Bailar, Jr. *et al.*, Eds. (Pergamon Press, Oxford, 1973) pp 935-1008; *Selenium*, R. A. Zingaro, W. C. Cooper, Eds. (Van Nostrand Reinhold, New York, 1974) 835 pp; E. M. Elkin in Kirk-Othmer *Encyclopedia of Chemical Technology* vol. 20 (Wiley-Interscience, New York, 3rd ed., 1982) pp 575-601.

Exists in several allotropic forms: amorphous, cryst or red, and gray or metallic. Liq is a brownish red, boils at 685° forming dark red vapors. Sol in dil aq caustic alkali solns; in aq potassium cyanide soln, in potassium sulfite soln. Burns in air with a bright blue flame forming the dioxide and emitting a characteristic odor resembling rotten horseradish. Combines directly with hydrogen, with the halogens (excluding iodine). Oxidized to selenious acid by nitric acid, to selenic acid by sulfuric acid. Reduces hot aqueous solutions of silver and gold salts with formation of silver selenide and metallic gold, respectively. Reacts with many metals.

Amorphous forms: Vitreous, black selenium; dark red-brown to bluish-black solid; formed when molten Se is cooled rapidly. d 4.28. Softens at 50-60° and becomes elastic at 70°. Red, amorphous form; formed by reduction of selenious acid in water; by condensation of Se vapor on a cold surface. d 4.26. Review of structural studies: Richter, Breitling, *Z. Naturforsch.* **26B**, 1699-1708, 2074-2075 (1971). When freshly precipitated, reacts with water at 50° forming selenious acid and hydrogen. Sol in carbon disulfide, methylene iodide, benzene, or quinoline.

Crystalline or red: Two monoclinic forms; dark red, transparent crystals. α-Form prepd by slow evaporation of CS_2 soln of Se; β-form by rapid evaporation of soln. mp 200°. d (α-form) 4.46. Structure of α-form: Cherin, Unger, *Acta Cryst.* **28B**, 513 (1972); *see also* Bagnall, *loc. cit.* Both forms are metastable and change into gray form on heating.

Gray or metallic: Lustrous gray to black hexagonal crystals. The most stable form. d_4^{20} 4.81. mp 217°. Mohs' hardness: 2.0. Latent heat of fusion 16.4 cal/g. Latent heat

of vaporization 20.6 kcal/mol. Linear coefficient of thermal expansion per degree C = 37×10^{-6}. Specific heat (28°): 0.084 cal/g/°C. Surface tension (217°): 92.5 dynes/cm. Thermal conductivity (25°): 0.0007-0.00183 cal/(cm)(°C)-(sec). Insol in water, alcohol; very slightly sol in carbon disulfide (2 mg/100 ml at ord temp). Sol in ether. Conducts electricity and rectifies alternating current; the conductivity increases up to a thousand times on exposure to light.

USE: As ingredient of toning baths in photography; as pigment in manuf ruby-, pink-, orange-, or red-colored glass; as metallic base in making electrodes for arc lights, electrical instruments and apparatus, as rectifier in radio and television sets; in selenium photocells, in semiconductor fusion mixtures, selenium cells, telephotographic apparatus; as vulcanizing agent in processing of rubber; as catalyst in determination of nitrogen by Kjeldahl method; for dehydrogenation of organic compds. *Caution:* Occupational exposure has caused pallor, nervousness, depression, garlic odor of breath and sweat, G.I. disturbances, dermatitis. Liver injury has been produced in exptl animals.

THERAP CAT (VET): Nutritional factor (interrelationship with vitamin E). Chiefly to prevent muscle degenerative diseases.

8383. Selenium Bromide. Selenium monobromide. Br_2Se_2; mol wt 317.75. Br 50.30%, Se 49.70%. Se_2Br_2. Prepd by mixing bromine and selenium in the correct proportions: Schneider, *Pogg. Ann.* **128**, 327 (1866); **129**, 450, 634 (1866); by treating selenium dioxide with hydrobromic acid: Lenher, Kao, *J. Am. Chem. Soc.* **47**, 772 (1925).

Dark red liquid. Unpleasant odor. d_4^{15} 3.604. Dec when heated. Dec in water and moist air. Sol in chloroform, ethyl bromide, carbon disulfide.

8384. Selenium Chloride. Selenium monochloride. Cl_2Se_2; mol wt 228.83. Cl 30.99%, Se 69.01%. Se_2Cl_2. Prepd from the elements: Sacc *et al.*, cited in *Mellor's* vol. X, 894 (1930); by the action of PCl_5 on selenium, on the selenide of phosphorus or antimony, on SeO_2 or on $SeOCl_2$: Baudrimont, *Ann. Chim. Phys.* [4] **2**, 5 (1864); Michaelis, *Z. Chem.* [2] **6**, 460 (1870); by the action of HCl on a soln of selenium in fuming sulfuric acid: Divers, Shimose, *J. Chem. Soc.* **45**, 194, 198, 201 (1884); from Se, SeO_2 and HCl: Lenher, Kao, *J. Am. Chem. Soc.* **47**, 772 (1925); **48**, 1550 (1926).

Deep red, oily liquid. bp_{733} 127° (dec). d_4^{25} 2.7741. mp −85°. Sol in chloroform, benzene, carbon tetrachloride, carbon disulfide, fuming sulfuric acid; dec in water.

8385. Selenium Hexafluoride. F_6Se; mol wt 192.96. F 59.08%, Se 40.92%. SeF_6. Prepd by passing gaseous fluorine over finely divided selenium in a copper vessel: Klemm, Henkel, *Z. Anorg. Allgem. Chem.* **207**, 74 (1932); Yost, Claussen, *J. Am. Chem. Soc.* **55**, 885 (1933); Yost, *Inorg. Syn.* **1**, 121 (1939). *Review:* Kemmitt, Sharp, *Advan. Fluorine Chem.* **4**, 235 (1965).

Gas. mp −50.8°; sublimes −63.8°. Vapor press. (mm Hg): 651.2 (−48.7°); 213.1 (−64.8°); 30.4 (−87.5°). Insol in water. Reacts with ammonia gas at 200° to give selenium, nitrogen, and hydrogen fluoride. Covalently satd, does not attack glass.

USE: Gaseous electric insulator.

8386. Selenium Oxide. Selenium dioxide; selenious anhydride. O_2Se; mol wt 110.96. O 28.84%, Se 71.16%. SeO_2. Prepd by burning selenium in oxygen: Berzelius, *Acad. Handl. Stockholm* **39**, 13 (1818); *Ann. Chim. Phys.* [2] **9**, 160, 225, 337 (1818); by burning selenium in oxygen and nitrogen dioxide: Lenher, *J. Am. Chem. Soc.* **20**, 559 (1898); by oxidizing selenium with nitric acid: Julien, *Bull. Soc. Chim.* **47**, 1799 (1925). *Review:* Waitkins, Clark, *Chem. Rev.* **36**, 235-289 (1945).

Lustrous, tetragonal needles. Acidic taste; leaves a burning sensation. Its yellowish green vapor has a pungent, sour smell. mp 340°; d_{15}^{15} 3.954. Vapor pressure (mm Hg): 12.5 at 70°; 20.2 at 94°; 39.0 at 181°; 760.0 at 315°; 848.0 at 320°. n_D^{20} < 1.76. Soly (parts/100 parts solvent) in water: 38.4 (14°); in methanol: 10.16 (11.8°); in 93% ethanol: 6.67 (14°); in acetone: 4.35 (15.3°); in acetic acid: 1.11 (13.9°). Sol in concd H_2SO_4. Stable to light and heat. Rapidly absorbs dry hydrogen fluoride, hydrogen chloride, hydrogen bromide, and hydrogen iodide to form the corresponding selenium oxyhalide. Reacts with ammonia to form nitrogen and selenium; in alcohol soln forms ammonium ethyl selenite, $HN_4(C_2H_5)SeO_3$. Yields nitrogen and black amorphous selenium with hydrazine; nitrogen and reddish-brown amorphous selenium with hydroxylamine hydrochloride. Forms selenic acid with nitric acid. Reduced by carbon and other organic substances.

USE: In the manuf of other selenium compounds; as a reagent for alkaloids; as oxidizing agent: L. F. Fieser, M. Fieser, *Reagents for Organic Chemistry* vol. 1 (New York, Wiley, 1967) p 992. *Caution:* Causes intense local irritation of skin, eyes: Cerwenka, Cooper, *Arch. Environ. Health* **3**, 189 (1961).

8387. Selenium Oxybromide. *Seleninyl bromide;* selenyl bromide. Br_2OSe; mol wt 254.79. Br 62.73%, O 6.28%, Se 30.99%. $SeOBr_2$. Prepd by the reaction of selenium dioxide, fused selenium, and fused calcium bromide: Lenher, *J. Am. Chem. Soc.* **44**, 1668 (1922).
Red-yellow solid. mp 41.5-41.7°; d^{50} 3.38; bp_{740} 217° (dec). Dec in water. Chlorine displaces the bromine. Dec by sulfur and hydrogen sulfide. Sol in sulfuric acid, carbon disulfide, chloroform, benzene, toluene, xylene, carbon tetrachloride. Reacts explosively with sodium and potassium.

8388. Selenium Oxychloride. *Seleninyl chloride.* Cl_2OSe; mol wt 165.87. Cl 42.75%, O 9.65%, Se 47.60%. $SeOCl_2$. Prepd by passing chlorine into SeO_2 suspended in carbon tetrachloride: Lenher, *J. Am. Chem. Soc.* **42**, 2498 (1920); Hönigschmid, Görnhardt, *Z. Naturforsch.* **1**, 661 (1946); by dehydration of dichloroselenious acid: Smith, Jackson, *Inorg. Syn.* **3**, 130 (1950); from $SOCl_2$ and SeO_2: Paetzold, Aurich, *Z. Anorg. Allgem. Chem.* **315**, 72 (1962); from $AlCl_3$ and SeO_2: Drago, Whitten, *Inorg. Chem.* **5**, 677 (1966). Acute toxicity: C. G. Wilber, *Clin. Toxicol.* **17**, 171 (1980).
Nearly colorless or yellowish, corrosive liquid; fumes in the air. d_4^{16} 2.44. Solidif about 5°. bp 180°. n_D^{20} 1.651. Dec by water into HCl and selenious acid. Miscible with carbon tetrachloride, chloroform, carbon disulfide, benzene, toluene. It is an excellent solvent for many substances including metals. LD_{50} s.c. in rabbits: 7 mg/kg (Wilber).
Caution: Strong irritant, vesicant: *Clinical Toxicology of Commercial Products,* R. E. Gosselin *et al.,* Eds. (Williams & Wilkins, Baltimore, 5th ed., 1984) Section II, p 129.

8389. Selenium Oxyfluoride. *Seleninyl fluoride;* selenyl fluoride. F_2OSe; mol wt 132.97. F 28.58%, O 12.03%, Se 59.38%. F_2SeO. Prepd from selenium dioxide and fluorine: Aynsley *et al., J. Chem. Soc.* **1952**, 1231; from selenium oxychloride and sodium fluoride: Tullock, Coffman, *J. Org. Chem.* **25**, 2016 (1960); from selenium tetrafluoride and tellurium oxide: J. Carre *et al., J. Fluorine Chem.* **14**, 139 (1979).
Fuming liquid. Ozone-like odor. mp 15.0°; bp 126°. $d^{21.5}$ 2.8. Rapidly attacks glass.
Caution: Strong irritant, vesicant. Can cause fatal pulmonary edema.

8390. Selenium Sulfides. Prepn from selenious acid and H_2S; product may be a mixture of the elements: *Gmelin's Selenium* (8th ed.) **10** (part B), 159-176 (1949). Prepn by fusion of the elements; solid solns of sulfur (S_8), selenium (Se_8) and selenium sulfides (S_nSe_{8-n}; $n = 1$-7) formed: Fergusson *et al., J. Inorg. Nucl. Chem.* **24**, 157 (1962); Hawes, *Nature* **198**, 1267 (1963); Cooper, Culka, *J. Inorg. Nucl. Chem.* **27**, 755 (1965); **29**, 1217 (1967).
Se_4S_4, red tabular crystals from benzene, d 3.20. mp 113° (dec). Sol in CS_2. Solubility in benzene (20°): 0.4 g/l. Se_2S_6, light-orange needles from benzene, d 2.44. mp 121.5°. Sol in CS_2. Solubility in benzene (20°): 12 g/l.
Selenium sulfide detergent prepn, *Exsel, Selenex, Sel-O-Rinse, Selsun.* Prepn: Baldwin, Young, U.S. pat. 2,694,669 (1954 to Abbott). Safety evaluation: Cummins, Kimura, *Toxicol. Appl. Pharmacol.* **20**, 89 (1971). Review of use as antidandruff agent: Matson, *J. Soc. Cosmet. Chem.* **7**, 459-466 (1956).
Note: Selenium sulfide may reasonably be anticipated to be a carcinogen: *Fourth Annual Report on Carcinogens* (NTP 85-002, 1985) p 182.
THERAP CAT: Detergent prepn as topical antiseborrheic.

THERAP CAT (VET): Topically in eczemas and dermatomycoses.

8391. Selenium Tetrabromide. Br_4Se; mol wt 398.62. Br 80.19%, Se 19.81%. $SeBr_4$. Prepd by adding excessive bromine to selenium: Schneider, *Pogg. Ann.* **128**, 327 (1866); **129**, 450, 634 (1866); by dissolving selenium dioxide in hydrobromic acid: Muthmann, Schäfer, *Ber.* **26**, 1008 (1893).
Red-brown, cryst powder. Unpleasant odor. Dec 70-80°. Dec in moist air and water. Sol in carbon disulfide, chloroform, ethyl bromide.

8392. Selenium Tetrachloride. Cl_4Se; mol wt 220.79. Se 35.76%, Cl 64.24%. $SeCl_4$. Prepd by the action of an excess of chlorine on selenium: Berzelius, cited in *Mellor's* vol X, 898 (1930); by the action of thionyl chloride or phosphorus trichloride on selenium oxychloride and by the action of phosphorus pentachloride on selenium oxide: Michaelis, *Z. Chem.* [2] **6**, 460 (1870); by the action of anhydr selenic acid on acetyl chloride at 0°: Lamb, *Am. Chem. J.* **30**, 209 (1903). Crystal structure: Shoemaker, Abrahams, *Acta Cryst.* **18**, 296 (1965).
White to pale yellow crystals. d 2.6. Sublimes when heated. Dec in water and moist air. Insol in liquid bromine. Dec by dry ammonia.

8393. Selenium Tetrafluoride. F_4Se; mol wt 154.96. F 49.04%, Se 50.96%. SeF_4. Fluorinating agent. Prepn from $SeCl_4$ and AgF: Prideaux, Cox, *J. Chem. Soc.* **1927**, 928; **1928**, 1603; from Se and F_2: Aynsley *et al., ibid.* **1952**, 1231; J. Carre *et al., J. Fluorine Chem.* **14**, 139 (1979); from Se and AgF: Glemser *et al., Naturwiss.* **52**, 130 (1965); from Se_2Cl_2 and F_2: Goggin, *J. Inorg. Nucl. Chem.* **28**, 661 (1966); from Se and ClF: Pitts, Jache, U.S. pat. 3,373,000 (1968 to Olin Mathieson); from Se and ClF_3: Olah *et al., J. Am. Chem. Soc.* **96**, 925 (1974).
Colorless liquid; fumes in air. mp −9.5°. bp 106°. d^{18} 2.75. Forms a white hygroscopic solid. Hydrolyzed violently by water. Attacks glass slowly when perfectly dry. Completely miscible with ether, ethanol, iodine pentafluoride, sulfuric acid; sol in chloroform, carbon tetrachloride.
USE: Fluorination of alcohols, carboxylic acids and carbonyl compds: Olah *et al., loc. cit.*

8394. Selenocysteine. *3-Selenyl-L-alanine;* selenium cysteine. $C_3H_7NO_2Se$; mol wt 168.05. C 21.44%, H 4.20%, N 8.33%, O 19.04%, Se 46.98%. Naturally occurring amino acid found in the active site of certain prokaryotic and eukaryotic enzymes. Highly oxidizable making synthesis difficult. Easily prepared from its oxidized form *selenocystine.* Prepn of *dl*-selenocystine: A. Fredga, *Svensk. Kem. Tidskrift.* **48**, 160 (1936); E. P. Painter, *J. Chem. Soc.* **69**, 229 (1947); of *l*-form: H. Tanaka, K. Soda, *Methods in Enzymology* **143**, 240 (1987). Reduction of selenocystine and separation of L-selenocysteine: J. N. Burnell *et al., J. Inorg. Biochem.* **12**, 343 (1980). Chemical properties: R. E. Huber, R. S. Criddle, *Arch. Biochem. Biophys.* **122**, 164 (1967). Determn by HPLC/MS: H. E. Ganther *et al., Methods in Enzymology* **107**, 582 (1984); colorimetric determn: N. Esaki, K. Soda, *ibid.* **143**, 148 (1987). Identification of Se as a component in protein: S.-H. Oh *et al., Biochemistry* **13**, 1825 (1974). Characterization of selenocysteine as the Se containing amino acid moiety in various enzymes: J. E. Cone *et al., Proc. Nat. Acad. Sci. USA* **73**, 2659 (1976); J. B. Jones *et al., Arch. Biochem. Biophys.* **195**, 255 (1979); presence in the catalytic site of mammalian glutathione peroxidase: J. W. Forstrom *et al., Biochemistry* **17**, 2639 (1978); J. J. Zabrowski *et al., Biochem. Biophys. Res. Commun.* **84**, 248 (1978). Metabolism in animals: H. Tanaka *et al., Curr. Topics Cell Reg.* **27**, 487 (1985). Identification of selenocysteine-specific tRNA: W. C. Hawkes *et al., Biochim. Biophys. Acta* **699**, 183 (1982); W. Leinfelder *et al., Nature* **331**, 323 (1988). *Review:* T. C. Stadtman, *FASEB J.* **1**, 375-379 (1987).

```
              COOH
               |
    H₂N ─── C ─── H
               |
              CH₂
               |
              SeH
```

pK 5.2. Unstable to acid hydrolysis.

8395. Selenomethionine. *2-Amino-4-(methylseleno)butanoic acid;* α-amino-γ-(methylseleno)butyric acid. C_5H_{11}-NO_2Se; mol wt 196.11. C 30.62%, H 5.65%, N 7.14%, O 16.32%, Se 40.26%. $CH_3SeCH_2CH_2CH(NH_2)COOH$. Selenium analog of methionine. Prepn of DL-form: E. P. Painter, *J. Am. Chem. Soc.* **69**, 232 (1949); H. J. Klosterman, E. P. Painter, *ibid.* 2009; G. Zdansky, *Ark. Kemi* **19**, 559 (1962), *C.A.* **58**, 6919h (1963). Prepn of L-form: *idem, ibid.,* **29**, 437 (1968); C. S. Pande *et al., J. Org. Chem.* **35**, 1440 (1970). Biosynthesis of radioactive (^{75}Se) form: M. Blau, *Biochim. Biophys. Acta* **49**, 389 (1961). Metabolism: M. Blau, J. F. Holland in *Radioactive Pharmaceuticals,* G. A. Andrews *et al.,* Eds. (USAEC Symposium Series #6, 1976) pp 409-421. Review of use in pancreatic scanning: J. E. Agnew *et al., Brit. J. Radiol.* **49**, 979-995 (1976).

DL-Form, transparent, hexagonal sheets or plates from methanol and water, metallic luster; mp 265° (dec). IR spectrum: Zdansky, *loc. cit.* (1962).

L-Form, crystals from aqueous acetone; mp 266-268° (dec). $[\alpha]_D^{22}$ +21.6° (c = 0.5 in 2N HCl).

^{75}Se-Injection, *Sethotope.*

THERAP CAT: ^{75}Se as diagnostic aid (radioactive imaging agent).

8396. Semicarbazide Hydrochloride. *Hydrazinecarboxamide monohydrochloride;* aminourea hydrochloride; carbamylhydrazine hydrochloride. CH_6ClN_3O; mol wt 111.54. C 10.77%, H 5.42%, Cl 31.78%, N 37.68%, O 14.34%. NH_2-$NHCONH_2.HCl$. Prepd by electrolytic reduction of nitrourea with cathodes of copper, nickel, lead, and mercury in hydrochloric acid solution: Ingersoll *et al., Org. Syn.* **5**, 93 (1925). Commercial prepn from hydrazine hydrate, iron carbonyl, and carbon monoxide: Sampson, U.S. pat. **2,589,-289** (1952 to du Pont).

Prisms from dil alc, dec 175-185°. Freely sol in water with acid reaction. Dissociation consts of free base: Kb at 20° = 29 × 10^{-11}; Ka = 1.6 × 10^{-11}. Very slightly sol in hot alcohol, insol in anhydr ether. Ammonolysis with liq NH_3 gives a 95% yield of the free base; mp 96°: Audrieth, *J. Am. Chem. Soc.* **52**, 1250 (1930).

Acetone semicarbazone, $C_4H_9N_3O$, needles from water, mp 192° (heated stage); mp 188° (open capillary tube).

USE: As a reagent for ketones and aldehydes with which it affords cryst compds having characteristic melting points.

8397. Semioxamazide. *Aminooxoacetic acid hydrazide;* aminooxamide; oxamic acid hydrazide. $C_2H_5N_3O_2$; mol wt 103.08. C 23.30%, H 4.89%, N 40.77%, O 31.04%. H_2NCO-$CONHNH_2$. Made from hydrazine sulfate, ethyl oxamate and KOH. *Ref:* Wilson, Pickering, *J. Chem. Soc.* **123**, 394 (1923); Leonard, Boyer, *J. Org. Chem.* **15**, 42 (1950).

Lustrous leaflets, mp about 220° with decompn. Soluble in 400 parts water; more sol in hot water; easily sol in acids and alkalies with formation of salts; insol in alcohol, ether.

USE: As a reagent for aldehydes and ketones.

8398. Sempervirine. *2,3,4,13-Tetrahydro-1H-benz[g]indolo[2,3-a]quinolizin-6-ium;* 3,4,5,6,14,15,20,21-octadehydroyohimbanium; sempervirene. $C_{19}H_{16}N_2$; mol wt 272.33. C 83.79%, H 5.92%, N 10.29%. Found in rhizome and roots of Carolina jessamine (yellow jessamine) *Gelsemium sempervirens* (L.) Ait. f., *Loganiaceae:* Stevenson, Sayre, *J. Am. Pharm. Assoc.* **4**, 60 (1915); **8**, 708 (1919); Hasenfratz, *Bull. Soc. Chim. France* [4] **53**, 1084 (1933); Forsyth *et al., J. Chem. Soc.* **1945**, 579; Prelog, *Helv. Chim. Acta* **31**, 588 (1948). Believed to exist primarily as inner salt depicted below: Woodward, Witkop, *J. Am. Chem. Soc.* **71**, 379 (1949). Synthesis: Woodward, McLamore, *ibid.* 379; Swan, *J. Chem. Soc.* **1958**, 2039; Ban, Seo, *Tetrahedron* **16**, 11 (1961); Potts, Mattingly, *J. Org. Chem.* **33**, 3985 (1968).

Monohydrate, yellow needles from chloroform, mp 228°. Brownish-yellow leaflets from dil ethanol, mp 258°. uv max (ethanol): 243, 249, 297, 345, 387 nm (log ε 4.58, 4.57, 4.20, 4.26, 4.24). Dipole moment: 8.5 (dioxane); 7.5 (benzene). Soluble in alcohol, chloroform, pyridine. Slightly sol in acetone. Practically insol in ether, benzene.

Methochloride, $C_{20}H_{19}ClN_2$, minute yellow needles from ethanol, mp 330-332°. Soluble in water giving yellow solns with purple fluorescence. uv max: 241, 292, 330, 395 nm (log ε 4.56, 4.20, 4.28, 4.22).

8399. Senecaldehyde. *3-Methyl-2-butenal; 3,3-dimethylacrolein;* 3-methylcrotonaldehyde; senecioaldehyde; β,β-dimethylacrolein. C_5H_8O; mol wt 84.11. C 71.39%, H 9.59%, O 19.02%. $(CH_3)_2C=CHCHO$. Prepn from isoamyl alcohol: Fischer *et al., Ber.* **64**, 30 (1931); Fischer, Löwenberg, *Ann.* **494**, 272 (1932); *cf.* Burkhardt *et al.,* **Brit.** pat. **512,465** (1939); Montavon, Saucy, U.S. pat. **2,902,515** (1959 to Hoffmann-La Roche).

Liquid. Odor more pungent than that of isovaleraldehyde. Autooxidizes easily, but the pure substance can be stored under vacuum. d_4^{20} 0.8722; bp_{730} 133°; n_D^{20} 1.4526. Absorption spectrum: Forbes, Skilton, *J. Org. Chem.* **24**, 436 (1959).

Caution: Irritant to mucous membranes. Narcotic in high concentrations.

8400. Senecic Acid. *5-Ethylidene-2-hydroxy-2,3-dimethylhexanedioic acid.* $C_{10}H_{16}O_5$; mol wt 216.24. C 55.54%, H 7.46%, O 37.00%. One of the most widely studied *necic acids* which are constituent acids of the hepatotoxic pyrrolizidine alkaloids and are isolated from these alkaloids by hydrolysis or by hydrogenolysis. Necic acids are C_5- to C_{10}-acids; many are γ- or δ-hydroxy acids and have structures that may be dissected into isoprene units. Synthesis of senecic acid: C. C. J. Culvenor, T. A. Geissman, *J. Am. Chem. Soc.* **83**, 1647 (1961). uv spectra of senecic acid and its isomers: V. Simanek *et al., Collect. Czech. Chem. Commun.* **34**, 1832 (1969). Biosynthesis: D. H. G. Crout *et al., J. Chem. Soc. Perkin Trans. I* **1972**, 671. Comprehensive reviews on necic acids: F. L. Warren, *Fortschr. Chem. Org. Naturst.* **12**, 198-269 (1955); **24**, 329-406 (1966); D. J. Robins *ibid.* **41**, 115-203 (1982); F. L. Warren in *Alkaloids,* **Vol. XII**, R. H. F. Manske, Ed. (Academic Press, New York, 1970) pp 245-331.

Needles from benzene or ethyl acetate, mp 145-146°. Sublimes readily at ~125°.

8401. Senecio. Golden ragwort; squaw weed; life root. Dried plant of *Senecio aureus* L., *Compositae.* *Habit.* Canada and Eastern U.S. *Constit.* The alkaloids, senecifoline, senecine, etc., resins.

THERAP CAT: Emmenagogue.

8402. Senecionine. *12-Hydroxysenecionan-11,16-dione;* aureine. $C_{18}H_{25}NO_5$; mol wt 335.39. C 64.46%, H 7.51%, N 4.18%, O 23.85%. Hepatotoxic pyrrolizidine alkaloid from whole plant of *Senecio vulgaris* L., *Compositae* and several other *Senecio* spp. Isoln: Barger, Blackie, *J. Chem. Soc.* **1936**, 743. Structure: Adams, Govindachari, *J. Am. Chem. Soc.* **71**, 1953 (1949); Koekmoer, Warren, *J. Chem. Soc.* **1955**, 63. Conformation: Culvenor, *Tetrahedron Letters* **1966**, 1091. Review of toxicology of senecionine and other pyrrolizidine alkaloids: E. K. McLean, *Pharmacol. Rev.* **22**, 429-483 (1970). Comprehensive reviews of pyrrolizidine alkaloids: L. Bull *et al., The Pyrrolizidine Alkaloids* (North-Holland, Amsterdam, 1968) 293 pp; F. L. Warren in *The Alkaloids* vol. **12**, R. H. F. Manske, Ed. (Academic Press, New York, 1970) pp 245-331; D. J. Robins, *Fortschr. Chem. Org. Naturst.* **41**, 115-203 (1982).

Bitter plates, mp 236°. $[\alpha]_D^{25}$ $-55.1°$ (c = 0.034 in chloroform). Practically insol in water; freely sol in chloroform; slightly sol in alcohol, ether. LD_{50} i.v. in mice: 64.12 ±2.24 mg/kg, P. N. Harris *et al., J. Pharmacol. Exp. Ther.* **75**, 69 (1942).

C_{15}-*trans* Isomer, *integerrimine, squalidine, [15(20)E]-12-hydroxysenecionan-11,16,dione.* Isolated from several *Senecio* and *Crotalaria* spp.: L. Bull *et al., loc. cit.*; F. L. Warren, *loc. cit.* Identity of squalidine and integerrimine: A. G. Gonzalez, A. Calero, *Chem. & Ind. (London)* **1958**, 126. mp 172-172.5°; $[\alpha]_D^{28}$ $+4.3°$ (c = 0.8 in methanol).

8403. Seneciphylline. *13,19-Didehydro-12-hydroxysenecionan-11,16-dione;* jacodine; α-longilobine. $C_{18}H_{23}NO_5$; mol wt 333.39. C 64.85%, H 6.95%, N 4.20%, O 24.00%. Hepatotoxic pyrrolizidine alkaloid, common constituent of *Senecio* species. Isoln from *Senecio platyphllus* DC. *Compositae:* A. Orechoff, W. Tiedebel, *Ber.* **68**, 650 (1935); from *S. jacobaea* L.: G. Barger, J. J. Blackie, *J. Chem. Soc.* **1937**, 584; from *Crotalaria juncea* L. *Leguminosae:* R. Adams, M. Gianturco, *J. Am. Chem. Soc.* **78**, 1919 (1956). Identity with α-longilobine: R. Adams, J. H. Looker, *ibid.* **73**, 134 (1951). Identity with jacodine: R. B. Bradbury, C. C. J. Culvenor, *Chem. & Ind. (London)* **1954**, 1021. Structural study: R. Adams *et al., J. Am. Chem. Soc.* **74**, 700 (1952). Revised structure: S. Masume, *Chem. & Ind. (London)* **1959**, 21. Review and evaluation of toxicity and carcinogenicity studies: *IARC Monographs* **10**, 319-325, 333-342 (1976). Comprehensive reviews of seneciphylline and other pyrrolizidine alkaloids: L. Bull *et al., The Pyrrolizidine Alkaloids* (North-Holland, Amsterdam, 1968) 293 pp; F. L. Warren in *The Alkaloids* vol. **12**, R. H. F. Manske, Ed. (Academic Press, New York, 1970) pp 245-331.

Small rhombic platelets from hot alcohol or acetone, mp 217-218°. $[\alpha]_D$ $-128°$ (chloroform). Easily sol in chloroform, ethylene chloride; less sol in alc, acetone. Difficultly sol in ether, ligroin.

8404. Senega. Senega snakeroot; Seneca snakeroot; rattlesnake root. Dried root of *Polygala senega* L., *Polygalaceae. Habit.* North America. *Constit.* Polygalic acid, senegin (polygalin)—a saponin, senegenin *q.v.*, fixed and volatile oils. Quantitative analysis of senega saponins: Kita *et al., Yakugaku Zasshi* **89**, 1111 (1969).

THERAP CAT: Expectorant, emetic.

THERAP CAT (VET): Has been used as expectorant.

8405. Senegenin. *12-(Chloromethyl)-2β,3β-dihydroxy-27-norolean-13-ene-23,28-dioic acid.* $C_{30}H_{45}ClO_6$; mol wt 537.15. C 67.08%, H 8.44%, Cl 6.60%, O 17.87%. From root of *Polygala senega* Linn., *Polygalaceae:* Jacobs, Isler, *J. Biol. Chem.* **119**, 155 (1937). Structure studies: Shamma, Reiff, *Chem. & Ind. (London)* **1960**, 1272; Dugan *et al., Can. J. Chem.* **42**, 491 (1964). Structure: Dugan *et al., Proc.*

Chem. Soc. **1964**, 264. Formation from presenegenin: Pelletier *et al., Chem. Commun.* **1966**, 727.

Needles from dil alc, mp 281-283°. $[\alpha]_D$ $+19°$ (c = 0.70 in ethanol). uv max: 205 nm (ε 6950). Sol in alcohol, acetone, acetic acid; practically insol in chloroform, benzene.

Monoacetate, $C_{32}H_{47}ClO_7$, crystals from methanol + chloroform, mp 209-211°. $[\alpha]_D$ $+23°$ (c = 1.32). uv max (methanol): 205 nm (ε 6400).

Diacetate, $C_{34}H_{49}ClO_8$, dec 257-263°. $[\alpha]_D$ $+29°$ (c = 1.09 in methanol).

8406. Senna. Dried leaflets of *Cassia senna* L. (C. acutifolia* Delile), (Alexandria senna), or of *C. angustifolia* Vahl (India or Tinnevelly senna), *Leguminosae. Habit.* Egypt and neighboring districts, Southern India (Tinnevelly). *Constit.* Sennosides A and B; glucosides of rhein and chrysophanic acid: Becker, *Planta Med.* **7**, 390 (1959); Fairbairne, *ibid.* **12**, 260 (1964). Prepn of senna powder: Friedmann, Ryan, **Brit.** pat. **832,017** (1960 to Westminster Labs.); Corran, **Brit.** pat. **852,343** (1960 to C. E. Fulford); Baetz, **Ger.** pat. **1,158,211** (1963 to Ludwig Heumann).

Incompat: Mineral acids, carbonates, infusion cinchona, lime water, salts of heavy metals, tartar emetic.

THERAP CAT: Cathartic.

THERAP CAT (VET): Has been used as a purgative.

8407. Sennoside A & B. Amyron; Colonorm; Glysennid; Laxenna; Nytilax; Pursennid. $C_{42}H_{38}O_{20}$; mol wt 862.72. C 58.47%, H 4.44%, O 37.09%. Anthraquinone glucosides found in senna in equal amounts and in the rhubarbs where sennoside A predominates. Isoln from Tinnevelly senna *(Cassia angustifolia* Vahl, *Leguminosae):* Stoll *et al., Helv. Chim. Acta* **32**, 1892 (1949); **Brit.** pat. **804,232** (1958 to Byk-Gulden); Khorana, Sanghavi, *J. Pharm. Sci.* **53**, 110 (1964); Menssen *et al., U.S.* pat. **3,517,269** (1970 to Nattermann & Cie). Isoln from the rhyzome of *Rheum palmatum* L.: Zwaving, *Planta Med.* **13**, 474 (1965). Isoln of sennoside A from rhubarb: Miyamoto *et al., Yakugaku Zasshi* **87**, 1040 (1967). Sennosides B, C, D, E, and F have also been isolated from rhubarb, see Oshio *et al., Chem. Pharm. Bull.* **22**, 823 (1974). Structure: Stoll *et al., Helv. Chim. Acta* **33**, 313 (1950). Sennoside A is built up from the dextrorotatory aglucon sennidin A and D-glucose. Sennoside B is built up from the intramolecularly compensated *meso*-sennidin B and D-glucose. *Review:* Stoll, Becker, *Fortschr. Chem. Org. Naturst.* **7**, 248 (1950). Transport and mechanism of action studies: Dobbs *et al., Farmaco Ed. Sci.* **30**, 147 (1975).

Sennoside A, *5,5'-Bis(β-D-glucopyranosyloxy)-9,9',10,-10'-tetrahydro-4,4'-dihydroxy-10,10'-dioxo[9,9'-bianthracene]-2,2'-dicarboxylic acid.* Rectangular yellow plates from dil acetone, dec 200-240°. $[\alpha]_D^{20}$ $-164°$ (c = 0.1 in 60% acetone); $[\alpha]_D^{20}$ $-147°$ (c = 0.1 in 70% acetone); $[\alpha]_D^{20}$ $-24°$

(c = 0.2 in 70% dioxane). Insol in water, benzene, ether, chloroform. Sparingly sol in methanol, Carbitol, acetone, dioxane. The soly increases tremendously in water-miscible organic solvents contg an optimum of 30% w/w of water. Sol in aq solns of sodium bicarbonate. Sennoside A can be slowly isomerized to B in $NaHCO_3$ soln at 80°.

Sennoside B, light yellow prisms from dil acetone; fine needles from water, dec 180-186°. $[\alpha]_D^{20}$ −100° (c = 0.2 in 70% acetone); $[\alpha]_D^{20}$ −67° (c = 0.4 in 70% dioxane). uv max (0.5% $NaHCO_3$): 270, 308, 355 nm. Has the same soly characteristics as sennoside A, but is more sol. Can be recrystallized from large amounts of hot water. *Keep protected from air and light.*

THERAP CAT: Cathartic.

8408. Senociclin. *Succinic acid α-ester with* D(−)*threo-2,2-dichloro-N-[β-hydroxy-α-(hydroxymethyl)-p-nitrophenethyl]acetamide compd with 4-(dimethylamino)-1,4,4a,5,5a,-6,11,12a-octahydro-3,6,10,12,12a-pentahydroxy-6-methyl-1,11-dioxo-N-(1-pyrrolidinylmethyl)-2-napthacenecarboxamide;* chloramphenicol succinate compd with rolitetracycline; rolitetracycline chloramphenicol succinate; cafrolicycline; gradocycline; levocycline; protercycline; Ciclincaf; Clorociclin; Crovicina; Levociclina U.M.; Metilcaf; Proterciclina; Reicaf; Tecaf; Tetrafenicol. $C_{42}H_{49}Cl_2N_5O_{16}$; mol wt 950.79. C 53.06%, H 5.19%, Cl 7.46%, N 7.37%, O 26.92%. Broad spectrum semi-synthetic antibiotic. Prepn: **Belg. pat. 636,234,** *C.A.* **61,** 13258g (1967); and Scevola, **U.S. pat. 3,218,335** (1963 and 1965, both to Laboratori Pro-Ter SpA).

Bright yellow powder, dec 140-144°. Bitter taste. Very sol in water; practically insol in ether, ligroin, hexane.

THERAP CAT: Antibacterial.

8409. Sepia. Cuttle-fish bone. Calcareous substance found under the skin of the back of *Sepia officinalis* L., *Cephalopoda. Habit.* Mediterranean Sea, Atlantic and Pacific Oceans. *Constit.* Calcium carbonate and phosphate, gluten.

USE: As a polishing agent and in tooth powders.

8410. Sepiomelanin. The eumelanin isolated from ink sacs of the cuttlefish, *Sepia officinalis* L., *Cephalopoda.* Present in the raw ink as a suspension of small, dark granules in a colorless plasma. Structure is a macromolecule or probably a mixture of polyacid polymers in which the predominant chemical unit is the indole type. Structure studies: Piattelli, Nicolaus, *Tetrahedron* **15,** 66 (1961); Piattelli *et al., ibid.* **19,** 2061 (1963). *Review:* R. A. Nicolaus, *Melanins* (Hermann, Paris, 1968) pp 68-91.

In the dry form, an amorphous, hygroscopic, black powder. Insol in all solvents.

USE: Pigment.

8411. Serine. Ser (IUPAC abbrev.); 2-amino-3-hydroxypropionic acid; β-hydroxyalanine; 2-amino-3-hydroxypropanoic acid; α-amino-β-hydroxypropionic acid. $C_3H_7NO_3$; mol wt 105.09. C 34.28%, H 6.71%, N 13.33%, O 45.67%. $HOCH_2CH(NH_2)COOH$. An amino acid classed as nonessential for the maintenance of growth in rats. The naturally occurring form is L-serine. Isoln from silk fibroin (sericine): Stein *et al., J. Biol. Chem.* **139,** 481 (1941); Coleman, Howitt, *Proc. Roy. Soc. (London)* **A190,** 145 (1947). Synthesis of DL-serine: Fischer, Leuchs, *Ber.* **35,** 3799 (1902); Leuchs, Geiger, *Ber.* **39,** 2644 (1906); Erlenmeyer, *Ber.* **35,** 3769 (1902); Erlenmeyer, Stoop, *Ann.* **337,** 236 (1904); Schlitz, Carter, *J. Biol. Chem.* **193,** 793 (1936); H. E. Carter, H. D. West, *Org. Syn. coll. vol.* **3,** 774 (1955). Synthetic reviews: K. Toi in *Synth. Prod. Util. Amino Acids,* T.

Kaneko *et al.,* Eds. (Halsted, New York, 1974) pp 187-195; L. Bassignani *et al., Ber.* **112,** 148 (1979).

L-Serine, hexagonal plates or prisms. Sweetish taste, insipid aftertaste. Dec 228°. Sublimes at 150° in high vac (10⁻⁴ mm Hg). $[\alpha]_D^{20}$ −6.83° (1.5 g in 15 g aq soln); $[\alpha]_D^{25}$ +14.45° (c = 0.5 g in 5.6N HCl). Absorption spectrum: *Z. Physiol. Chem.* **176,** 257 (1928). Sol in water; practically insol in abs alcohol, ether.

DL-Serine, monoclinic prismatic leaflets from water. d 1.537. Dec 246° (closed capillary, bath preheated to 225°). Sublimes at 150° in high vac (10⁻⁴ mm Hg). pK_1' 2.21; pK_2' 9.15. Soly in water (g/l) at 0° = 22.04; at 25° = 50.23; at 50° = 103; at 75° = 192; at 100° = 322. Insol in abs alcohol, ether.

8412. Sermorelin. Human growth hormone-releasing factor(1-29)amide; human pancreatic somatoliberin(1-29)-amide; GRF(1-29)NH_2; hpGRF(1-29)NH_2; SM-8144; Geref. $C_{149}H_{246}N_{44}O_{42}S$; mol wt 3357.92. C 53.30%, H 7.38%, N 18.35%, O 20.01%, S 0.95%. Amidated fragment of human somatoliberin, *q.v.* Characterization of *in vitro* activity: J. Rivier *et al., Nature* **300,** 276 (1982). Prepn: N. Ling *et al., Biochem. Biophys. Res. Commun.* **123,** 854 (1984); J. E. F. Rivier *et al.,* **U.S. pat. 4,703,035** (1987 to Salk Inst. Biol. Stud.). Clinical pharmacology: M. Losa *et al., Klin. Wochenschr.* **62,** 1140 (1984). Clinical evaluation in growth hormone deficiency: M. B. Ranke *et al., Eur. J. Pediatr.* **145,** 485 (1986); R. Hümmelink *et al., Acta Paediatr. Scand.* **Suppl. 331,** 48 (1987). Field trial in pigs and dairy heifers: D. Petitclerc *et al., J. Anim. Sci.* **65,** 996 (1987).

```
        1                                        14
Tyr-Ala-Asp-Ala-Ile-Phe-Thr-Asn-Ser-Tyr-Arg-Lys-Val-Leu-Gly-

Gln-Leu-Ser-Ala-Arg-Lys-Leu-Leu-Gln-Asp-Ile-Met-Ser-Arg-NH2
 15                                                        29
```

$[\alpha]_D^{20}$ −63.1° (c = 1 in 30% acetic acid).

THERAP CAT: Growth hormone releasing factor.

8413. Serotonin. *3-(2-Aminoethyl)-1H-indol-5-ol;* 5-hydroxytryptamine; 3-(β-aminoethyl)-5-hydroxyindole; 5-hydroxy-3-(β-aminoethyl)indole; enteramine; thrombocytin; thrombotonin; 5-HT. $C_{10}H_{12}N_2O$; mol wt 176.21. C 68.16%, H 6.86%, N 15.90%, O 9.08%. Vasoactive amine found in tissues and fluids of vertebrates and invertebrates. Extraction of enteramine from rabbit tissue, pharmacology: V. Erspamer, *Arch. Exp. Pathol. Pharmakol.* **196,** 343 (1940). Isoln of 5-HT from beef serum: M. M. Rapport *et al., Science* **108,** 329 (1948); *eidem, J. Biol. Chem.* **176,** 1237, 1243 (1948); M. M. Rapport, *ibid.* **180,** 961 (1949). Identity with enteramine, the enterochromaffin cell hormone: V. Erspamer, B. Asero, *Nature* **169,** 800 (1952). Ultrastructural localization in enterochromaffin cells: R. D. Dey, J. Hoffpauir, *J. Histochem. Cytochem.* **32,** 661 (1984). Synthesis starting with 5-benzyloxyindole: M. E. Speeter *et al., J. Am. Chem. Soc.* **73,** 5514 (1951); K. E. Hamlin, **U.S. pat. 2,715,129** (1955 to Abbott). Alternate routes: R. Justoni, R. Pessina, **U.S. pat. 2,947,757** (1960 to Vismara). Pharmacology: G. Reid, M. Rand, *Nature* **169,** 801 (1952). Existence of multiple serotonin receptors: S. J. Peroutka, S. H. Snyder, *Molec. Pharmacol.* **16,** 687 (1979); G. Engel *et al., Arch. Pharmacol.* **324,** 116 (1983); W. Feniuk *et al., Eur. J. Pharmacol.* **96,** 71 (1983). Extraneural synthesis in CNS: C. Maruki *et al., J. Neurochem.* **43,** 316 (1984). Effect on smooth muscle: W. Feniuk *et al., loc. cit.* Activity as neurotransmitter: S. G. Griffith *et al., Brain Res.* **247,** 388 (1982). Role in hemostasis: F. De Clerck *et al., Agents Actions* **15,** 627 (1984). Series of articles on role in hypertension, Raynaud's phenomenon and platelet activation: *J. Cardiovasc. Pharmacol.* **7,** Suppl. 7, 1-182 (1985). Review of role in migraine: E. T. MacKenzie *et al., Cephalalgia* **5,** 69-78 (1985). *Review:* V. Erspamer in E. Jucker, *Progress in Drug Research* **3,** 151-367 (1961). Books: *Serotonin in Health and Disease* vols. I-V, W. B. Essman, Ed. (Spectrum, New York, 1978); *Adv. Exp. Med.* **133,** entitled "Serotonin: Current Aspects of Neurochemistry and Function," B. Haber, Ed. (1981) 824 pp; *Serotonin and the Cardiovascular System,* P. M. Vanhoutte, Ed. (Raven Press, New York, 1985) 280 pp. *See also* Bufotenine and Ketanserin.

Hydrochloride, $C_{10}H_{12}N_2O \cdot HCl$, hygroscopic crystals, sensitive to light, mp 167-168°. Water sol. Aq solns are stable at pH 2-6.4.

Complex with creatinine sulfate, monohydrate, $C_{14}H_{21}N_5$-$O_6 \cdot S \cdot H_2O$, *Antemovis*. Plates, 215° dec. uv max (water at pH 3.5): 275 nm (ϵ 15000). $pK_1' = 4.9$; $pK_2' = 9.8$. pH of 0.01 molar aq soln: 3.6. Sol in glacial acetic acid. Very sparingly sol in methanol, 95% ethanol. Insol in abs ethanol, acetone, pyridine, chloroform, ethyl acetate, ether, benzene.

8414. Serpentaria. Snakeroot; snakeweed; sangrel; birthwort. Dried rhizome and roots of *Aristolochia serpentaria* L. (Virginia snakeroot), and of *Aristolochia reticulata* Nutt. (Texas snakeroot), *Aristolochiaceae*. Habit. U.S. Constit. Aristolochic acid, volatile oil (about 1%), resin, tannin.

THERAP CAT: Bitter tonic.

8415. Serpentine (Alkaloid). *3,4,5,6,16,17-Hexadehydro-16-(methoxycarbonyl)-19α-methyloxayohimbanium.* $C_{21}H_{20}$-N_2O_3; mol wt 348.39. C 72.39%, H 5.79%, N 8.04%, O 13.78%. From roots of *Rauwolfia serpentina* (L.) Benth., *Apocynaceae:* Schlittler, Schwarz, *Helv. Chim. Acta* **33**, 1463 (1950). Structure: Schlittler *et al., ibid.* **37**, 1912 (1954); Klohs *et al., J. Am. Chem. Soc.* **76**, 1332 (1954); Wenkert, Roychaudhuri, *ibid.* **80**, 1613 (1958). Stereoisomer of alstonine *(q.v.).* Stereochemistry: Fritz, *Ann.* **655**, 148 (1962); Shamma, Richey, *J. Am. Chem. Soc.* **85**, 2507 (1963).

Solvated, yellow rods or leaflets from abs ethanol, mp 158° (air-dried), mp 175° (after drying at 120° in a high vacuum, but still not entirely solvent-free). Dec and turns red on drying at 150°. $[\alpha]_D^{25}$ +292° (c = 0.27 in methanol); $[\alpha]_D^{25}$ +267° (c = 0.21 in ethanol). uv max (ethanol): 252, 308, 370 nm (log ϵ 4.49, 4.30, 3.61). Freely sol in methanol. Sol in 10% acetic acid, in other organic solvents.

Nitrate, $C_{21}H_{20}N_2O_3 \cdot HNO_3$, yellow crystals, mp 170-172°. Only slightly sol in water.

Hydrochloride monohydrate, $C_{21}H_{20}N_2O_3 \cdot HCl \cdot H_2O$, yellow crystals, mp 246-248°. $[\alpha]_D^{23}$ +178°. Sol in water.

Perchlorate, $C_{21}H_{20}N_2O_3 \cdot HClO_4$, yellow crystals, mp 255-256°. $[\alpha]_D^{22}$ +185° (c = 0.5 in acetone). uv max (methanol): 252, 306-308, 365-370 nm (log ϵ 4.54, 4.35, 3.68).

Picrate, $C_{21}H_{20}N_2O_3 \cdot C_6H_3N_3O_7$, reddish crystals, mp 261-262°.

8416. Sertraline. *(1S-cis)-4-(3,4-Dichlorophenyl)-1,2,-3,4-tetrahydro-N-methyl-1-naphthalenamine; (1S,4S)-4-(3,4-dichlorophenyl)-1,2,3,4-tetrahydro-N-methyl-1-naphthylamine.* $C_{17}H_{17}Cl_2N$; mol wt 306.23. C 66.68%, H 5.60%, Cl 23.15%, N 4.57%. Selective serotonin uptake inhibitor. Prepn: W. M. Welch *et al., Eur. pat. Appl.* **30,081**; *eidem*, U.S. pat. **4,536,518** (1981, 1985 both to Pfizer). Prepn and stereospecific activity: W. M. Welch *et al., J. Med. Chem.* **27**, 1508 (1984). Pharmacology in animals: B. K. Koe *et al., J. Pharmacol. Exp. Ther.* **226**, 686 (1983). Psychopharmacology and pharmacodynamics in humans: B. Saletu *et al., J. Neural Transm.* **67**, 241 (1986). GC determn in human plasma: H. G. Fouda *et al., J. Chromatog.* **417**, 197 (1987). Clinical evaluation in depression: F. W. Reimherr *et al., Psychopharmacol. Bull.* **24**, 200 (1988).

Hydrochloride, $C_{17}H_{18}Cl_3N$, *CP-51974-1.* Crystals, mp 243-245°. $[\alpha]_D^{23}$ +37.9° (c = 2 in methanol).

THERAP CAT: Antidepressant.

8417. Serum Albumin. Albumin, from blood; Albuminate (obsolete); Albuminar; Albumisol; Albuspan; Buminate; Pro-Bumin; Proserum. Smallest and most abundant (comprising about 60%) of the plasma proteins. Mol wt of human, about 69,000. Only trace amounts present in normal urine but becomes major protein constit in the condition of proteinuria. Involved in osmotic regulation and the transport of sparingly soluble metabolic products from one tissue to another, esp in the transport of free fatty acids. Structure consists of a carbohydrate-free polypeptide chain connecting four globular segments of unequal size; stabilized by seventeen S-S bridges. Structural studies: Bradshaw, Peters, *J. Biol. Chem.* **244**, 5582 (1969); Swaney, Klotz, *Biochemistry* **9**, 2570 (1970); McMenamy *et al., J. Biol. Chem.* **246**, 4744 (1971). Review: Peters, *Advan. Clin. Chem.* **13**, 37-111 (1970).

Brownish, amorphous lumps, scales, or powder. Net negative charge of 18 at pH 7.4. High solubility: 40% solns readily prepared at pH 7.4. Low viscosity. Stable, even at high temperatures.

^{131}I-labelled form, *iodinated (*^{131}I*) human serum albumin,* ^{131}I *HSA, Albumotope-*^{131}I*, Macroscan-131, Risa-131-H.* Normal human serum albumin mildly iodinated with radioactive iodine (^{131}I) which has a half-life of 8 days, and emits beta and gamma rays. Contains not more than one atom of iodine per molecule of albumin (mol wt 60,000). Physical and chemical properties essentially the same as those of albumin. Stable at 10° for at least 3 weeks. Supplied for injection in aq isotonic sodium chloride soln.

^{123}I-labelled form. Human serum albumin iodinated with ^{123}I which has a half-life of 60 days and emits gamma rays.

THERAP CAT: Plasma volume expander; correction of hypoproteinemia. Iodinated form as diagnostic aid (blood volume determination).

8418. Sesame Oil. Benne oil; teel oil; gingilli oil. From the seeds of cultivated varieties of *Sesamum indicum* L., *Pedaliaceae*. Constit. Olein, stearin, palmitin, myristin, linolein, sesamin, sesamolin. Review: Budowski, Markley, *Chem. Rev.* **48**, 125 (1951).

Pale yellow oil; almost odorless; bland taste. d 0.916-0.920. Solidif about −5°. $[\alpha]_D^{40}$ +1° to +9°. n_D^{40} 1.4650-1.4665. Sapon no. 188-193. Iodine no. 103-122. Soluble in chloroform, ether, petr ether, carbon disulfide; slightly sol in alcohol. Insol in water.

USE: Manuf oleomargarine, cosmetics, iodized oil, etc. Pharmaceutic aid (solvent).

THERAP CAT (VET): Formerly as emollient, pediculicide.

8419. Sesamex. *5-[1-[2-(2-Ethoxyethoxy)ethoxy]ethoxy]-1,3-benzodioxole; acetaldehyde 2-(2-ethoxyethoxy)ethyl 3,4-(methylenedioxy)phenyl acetal; 2-(3,4-methylenedioxyphenoxy)-3,6,9-trioxaundecane; Sesoxane.* $C_{15}H_{22}O_6$; mol wt 298.33. C 60.39%, H 7.43%, O 32.18%. Prepn: M. Beroza, U.S. pat. **2,832,792** (1958 to U.S. Secy. Agr.). Use as a synergist: S. A. El-Aziz *et al., J. Econ. Entomol.* **62**, 318 (1969); H. T. Gordon, L. T. Jao, *ibid.* **64**, 546 (1971).

Straw-colored liquid, $bp_{0.08}$ 137-141°. Sol in kerosene, Freon 11, Freon 12; other organic solvents. n_D^{25} 1.4938.

USE: Insecticide, especially as synergist for pyrethrins, allethrin, methoxychlor.

8420. Sesamin. *5,5′-(Tetrahydro-1H,3H-furo[3,4-c]-furan-1,4-diyl)bis-1,3-benzodioxole; tetrahydro-1,4-bis[3,4-(methylenedioxy)phenyl]-1H,3H-furo[3,4-c]furan; 2,6-bis-(3,4-methylenedioxyphenyl)-3,7-dioxabicyclo[3.3.0]octane.* $C_{20}H_{18}O_6$; mol wt 354.34. C 67.79%, H 5.12%, O 27.09%. Isolated from the bark of various *Fagara* species; from sesame oil: Bertram *et al.*, *Biochem. Z.* **197**, 1 (1928); from fruit of *Piper lowong* Blume, *Piperaceae*: Peinemann, *Arch. Pharm.* **234**, 238 (1896). Structure: Cohen, *Rec. Trav. Chim.* **57**, 653 (1938). Synthesis of *dl*-form: Beroza, Schechter, *J. Am. Chem. Soc.* **78**, 1242 (1956); Freudenberg, *Naturwiss.* **43**, 16 (1956). Identity of *dl*-form with *fagarol*: Carnmalm, *Acta Chem. Scand.* **9**, 1111 (1955); identity of *d*-form with *pseudocubebin*: idem, *ibid.* **10**, 134 (1956). Stereochemistry of the *d*-isomer: Freudenberg, Sidhu, *Ber.* **94**, 851 (1961); Becker, Beroza, *Tetrahedron Letters* **1962**, 157. *Review:* Budowski, *J. Am. Oil Chem. Soc.* **41**, 280 (1964).

(+)-sesamin

d-Form, needles from ethanol, mp 122-123°. $[\alpha]_D^{20}$ +64.5° (c = 1.75 in chloroform).
dl-Form, crystals from ethanol, mp 125-126°. Freely sol in chloroform, benzene, acetic acid, acetone. Practically insol in water, alkaline solns and hydrochloric acid.
USE: Insecticide synergist.

8421. Sesamolin. *5-[4-(1,3-Benzodioxolol-5-yloxy)tetrahydro-1H,3H-furo[3,4-c]furan-1-yl]-1,3-benzodioxole; tetrahydro-1-[3,4-(methylenedioxy)phenoxy]-4-[3,4-(methylenedioxy)phenyl]-1H,3H-furo[3,4-c]furan.* $C_{20}H_{18}O_7$; mol wt 370.34. C 64.86%, H 4.90%, O 30.24%. Constituent of sesame seeds, the seeds of *Sesamum indicum* L., *Pedaliaceae*. Isoln: Canzoneri, Perciabosco, *Gazz. Chim. Ital.* **33**, II, 253 (1907). Structure and synthesis: J. Boseken, W. D. Cohen, *Rec. Trav. Chim.* **55**, 815 (1936); E. Haslam, R. D. Haworth, *J. Chem. Soc.* **1955**, 827. Stereochemistry: E. Haslam, *ibid.* **1970**, 2332.

White plates from ethanol, mp 93-94°. $[\alpha]_D^{20}$ +212° (chloroform).
USE: Synergist for pyrethrum insecticides.

8422. Sesin. *Benzoic acid 2,4-dichlorophenoxyethyl ester; 2-(2,4-dichlorophenoxy)ethanol benzoate; 2,4-dichlorophenoxyethyl benzoate.* $C_{15}H_{12}Cl_2O_3$; mol wt 311.18. C 57.90%, H 3.89%, Cl 22.79%, O 15.42%. Prepn: Lambrech, U.S. pat. **2,765,224** (1956 to Union Carbide).

Crystals, mp 66°. $bp_{1.5}$ 185°.

USE: In weed control.

8423. Setastine. *1-[2-[1-(4-Chlorophenyl)-1-phenylethoxy]ethyl]hexahydro-1H-azepine; 1-[2-[(p-chloro-α-methyl-α-phenylbenzyl)oxy]ethyl]hexahydro-1H-azepine; 1-[2-(4-chloro-α-methyl-α-phenylbenzyloxy)ethyl]azepam; N-[2-(α-methyl-p-chlorobenzhydryloxy)ethyl]hexamethyleneimine; N-[2-[(4-chloro-α-methyl-α-phenylbenzyl)oxy]ethyl]-hexamethylenimine.* $C_{22}H_{28}ClNO$; mol wt 357.92. C 73.83%, H 7.88%, Cl 9.91%, N 3.91%, O 4.47%. Nonsedating-type histamine H_1-receptor antagonist; structurally similar to clemastine, *q.v.* Prepn: I. Beck *et al.*, Ger. pat. **2,528,-194** (1976 to EGYT), *C.A.* **85**, 46439h (1976). *In vitro* antihistaminic activity: A. Németh, Z. Huszti, *Agents Actions* **23**, 194 (1988). Clinical pharmacology: Z. Vezekényi *et al.*, *Acta Med. Hung.* **44**, 55 (1987).

Hydrochloride, $C_{22}H_{29}Cl_2NO$, *EGIS-2062, EGYT-2062, Loderix.*
THERAP CAT: Antihistaminic.

8424. Sethoxydim. *2-[1-(Ethoxyimino)butyl]-5-[2-(ethylthio)propyl]-3-hydroxy-2-cyclohexen-1-one;* BAS 9052; NP 55; Poast. $C_{17}H_{29}NO_3S$; mol wt 327.86. C 62.35%, H 8.92%, N 4.28%, O 14.66%, S 9.79%. Selective post-emergence herbicide. Prepn: **Japan. Kokai 81 75,408** (1981 to Nippon Soda), *C.A.* **95**, 110175e (1981). Activity: C. M. Knott, *Proc. Brit. Crop. Prot. Conf.—Weeds* **1980**, 39; G. H. Ingram *et al.*, *ibid.* 91.

USE: Herbicide.

8425. Shark Liver Oil. Robecote. Oil expressed from the livers of sharks and other *Elasmobranchii* species. *Constit:* Squalene, pristane, vitamins A and D, esters of fatty acids (esp. palmitic and oleic), glycerol ethers, triglycerides, cholesterol and fatty alcohols. Isoln: A. Ehrenreich, **Brit.** pat. **284,657** (1927); C. E. Brown, S. A. Reed, **Brit.** pat. **1,021,542** (1966 to Marfleet Refining Co. Ltd.). Composition: N. A. Sörensen, J. Mehlum, *Acta Chem. Scand.* **2**, 140 (1948); E. Gelpi, J. Oro, *J. Am. Oil Chem. Soc.* **45**, 144 (1968); R. Lombardi *et al.*, *Ann. Pharm. Franc.* **29**, 429 (1971). Comparison with cod liver oil: A. O. Banjo, *J. Food Technol.* **14**, 107 (1979).
Clear, yellow oil. Phys props of liver oil from basking shark: n_D^{20} 1.4784; d^{20} 0.922; sapon no. 140-146; iodine no. 140-145 (Lombardi).
THERAP CAT: Topical protectant.

8426. Shellac. Lacca; lac. A resinous excretion of the insect *Laccifer (Tachardia) lacca* Kerr, order *Homoptera*, family *Coccidae*. Different resiniferous trees of India serve as host trees as the season progresses. The insects suck the juice of the tree and excrete "stick-lac" almost continuously. Whitest shellac is produced while the kusum tree (*Schleichera trijuga*) is the host. Most shellac is produced in the Central and in the United provinces of India and in the states of Bihar and Orissa. The major component of lac is a resin which upon mild hydrolysis gives a complex mixture of aliphatic and alicyclic hydroxy acids and their polyesters. The composition of the hydrolysate depends on the lac source and the time of collection. The major component of the aliphatic fraction is aleuritic acid, *q.v.*; the major component of the alicyclic fraction is shellolic acid, *q.v.*: Yates, Field, *Tetrahedron* **26**, 3135 (1970). Physical and chemical properties of shellac: Cockeran, Levine, *J. Soc. Cosmet. Chem.* **12**, 316 (1961). Trends in shellac chemistry: *eidem, Am. Ink Maker* **39** (7), 26 (1961); E. Hicks, *Shellac, Its Origin and Applications* (Chemical Publ. Co., New York, 1961). *Review:* J. Martin in Kirk-Othmer *Encyclopedia of Chemical Technology* vol. **20** (Wiley-Interscience, New York, 3rd ed., 1982) pp 737-747.

Brittle, yellowish, transparent sheets or crushed pieces or powder. d 1.035-1.140. mp 115-120°. Sapon no. 185-210; iodine no. 10-18. Solubility in alcohol 85-95% w/w (very slowly sol); in ether 13-15%; in benzene 10-20%; in petr ether 2-6%. Sparingly sol in oil of turpentine. Insol in water; sol in aq solns of ethanolamines, of alkalies or borax with slightly purple color.

USE: Chiefly in lacquers and varnishes; also in manuf buttons, grinding wheels, sealing wax, cements, inks, phonograph records, paper; for stiffening hats; in electrical machines; coating confections and medicinal tablets; finishing leather.

8427. Shellolic Acid. *2,3,4,7,8,8a-Hexahydro-4-hydroxy-8-(hydroxymethyl)-8-methyl-1H-3a,7-methanoazulene-3,6-dicarboxylic acid; 10β,13-dihydroxycedr-8-ene-12,15-dioic acid.* $C_{15}H_{20}O_6$; mol wt 296.33. C 60.80%, H 6.80%, O 32.40%. Major component of the alicyclic fraction of shellac hydrolysate of which it usually constitutes 5-8%. Isoln: Harries, Nagel, *Ber.* **55**, 3833 (1922); Nagel, Mertens, *ibid.* **70**, 2173 (1937); Kirk *et al., J. Am. Chem. Soc.* **63**, 1243 (1941). Structure studies: Carruthers *et al., J. Chem. Soc.* **1961**, 5251; Cookson *et al., Tetrahedron* **18**, 547, 1321 (1962); Yates, Field, *ibid.* **26**, 3135 (1970). Stereochemistry: Yates *et al., ibid.* **26**, 3159 (1970).

Crystals, mp 204-207°.
Dimethyl ester, crysts from ethyl acetate, mp 153-154.5°. uv max (ethanol): 231 nm (ε 6200).

8428. Shikimic Acid. *[3R-(3α,4α,5β)]-3,4,5-Trihydroxy-1-cyclohexene-1-carboxylic acid.* $C_7H_{10}O_5$; mol wt 174.15. C 48.27%, H 5.79%, O 45.94%. Naturally occurring (−)-form is a major biosynthetic precursor of phenylalanine, tyrosine, and tryptophan and hence of the majority of plant alkaloids. It is also involved in the biosynthesis of lignin, *q.v.*, flavonoids and other important aromatic compounds. Isoln from the fruit of the oriental plant *Illicium religiosum* Sieb. et Zucc., *Magnoliaceae* (called in Japanese *shikimi-no-ki*): J. F. Eykman, *Rec. Trav. Chim.* **4**, 32 (1885); *idem, Ber.* **24**, 1278 (1891). Structural study: H. O. L. Fischer, G. Dangshat, *Helv. Chim. Acta* **17**, 1200 (1934). Configuration: *eidem, ibid.* **18**, 1206 (1935); **20**, 705 (1937). Conformation in soln: L. D. Hall, *J. Org. Chem.* **29**, 297 (1964). Enzymatic synthesis: P. R. Srinivasan *et al., J. Am. Chem. Soc.* **77**, 4943 (1955). Stereospecific synthesis: R. McCrindle *et al., J. Chem. Soc.* **1960**, 1560; E. E. Smissman *et al., J. Am. Chem. Soc.* **84**, 1040 (1962). Improved synthesis: J. L. Pawlak, G. A. Berchtold, *J. Org. Chem.* **52**, 1765 (1987). Synthesis of the (±)-form: R. Grewe, I. Hinrichs, *Ber.* **97**, 443 (1964); R. Grewe, S. Kersten, *Angew. Chem.* **77**, 859 (1965). Carcinogenicity study: I. A. Evans, M. A. Osman, *Nature* **250**, 348 (1974). *Reviews:* B. A. Bohm, *Chem. Rev.* **65**, 435-466 (1965); B. Ganem, *Tetrahedron* **34**, 3353-3383 (1978); J. B. Harborne, *Biosynthesis* **6**, 40-75 (1980). Books: E. Haslam, *The Shikimate Pathway* (Halsted Press, New York, 1974) 316 pp; *idem,* "Shikimic Acid Metabolites" in *Comprehensive Organic Chemistry* vol. **5** (Pergamon Press, Oxford, 1979) pp 1167-1205; U. Weiss, J. M. Edwards, *The Biosynthesis of Organic Compounds* (Wiley, New York, 1980) 728 pp.

Needles from methanol/ethyl acetate, mp 183-184.5°. Sublimes. $[\alpha]_D$ −161° (c = 0.57 in methanol). $[\alpha]_D^{18}$ −183.8° (c = 4.03 in water). uv max (alcohol): 213 nm (ε 8900). K at 14.1° = 7.1 × 10⁻⁵. Soly in water about 18 g/100 ml. Soly at 23° (g/100 m): 2.25 in abs alcohol: 0.015 in anhydr ether. Practically insol in chloroform, benzene, petr ether. LD_5 i.p. in mice: 1000 mg/kg (Evans, Osman).

(±)-Form, needles from methanol/ethyl acetate, mp 191-192°. uv max (ethanol): 212 nm (ε 8200).

8429. Shionone. *D:A-Friedo-18,19-secolup-19-en-3-one; 3β,5α,8,17aβ-tetramethyl-3-(4-methyl-3-pentenyl)-D-homoandrostan-17-one.* $C_{30}H_{50}O$; mol wt 426.70. C 84.44%, H 11.81%, O 3.75%. From *Aster tartaricus* L., *Compositae:* Nakaoki, *J. Pharm. Soc. Japan* **52**, 499 (1932); *C.A.* **26**, 4821⁴ (1932); Takahashi *et al., ibid.* **79**, 1281 (1959); *C.A.* **54**, 4667g (1960). Structure: Y. Tanahashi *et al., Bull. Soc. Chim. France* **1966**, 1670. Conformation: Y. Moriyama *et al., ibid.* **1968**, 2890.

Crystals from alc, mp 161-162°. $[\alpha]_D^{26.5}$ −56.1° (c = 1.07 in chloroform). uv max (chloroform): 290 nm (log ε 1.47).

8430. Showdomycin. *3-β-D-Ribofuranosyl-1H-pyrrole-2,5-dione; 2-β-D-ribofuranosylmaleimide.* $C_9H_{11}NO_6$; mol wt 229.19. C 47.16%, H 4.84%, N 6.11%, O 41.89%. Nucleoside antibiotic isolated from *Streptomyces showdoensis:* Nishimura *et al., J. Antibiot.* **17A**, 148 (1964); Nishimura, Fr. pat. M2751 corresp to U.S. pat. 3,316,149 (1964, 1967, both to Shionogi). Structure: Nakagawa *et al., Tetrahedron Letters* **1967**, 4105; Darnall *et al., Proc. Nat. Acad. Sci. USA* **57**, 548 (1967). Synthesis: Kalvoda *et al., Tetrahedron Letters* **1970**, 2297; Trummlitz, Moffatt, *J. Org. Chem.* **38**, 1841 (1973); T. Sato *et al., Tetrahedron Letters* **1978**, 1829; J. G. Buchanan *et al., J. Chem. Soc. Perkin Trans. I* **1979**, 225; T. Inoue, I. Kuwajima, *Chem. Commun.* **1980**, 251; A. P. Kozikowski, A. Ames, *J. Am. Chem. Soc.* **103**, 3923 (1981); Y. Araki *et al., Tetrahedron Letters* **29**, 351 (1988). Exhibits antitumor activity: Shinzo *et al., J. Antibiot.* **17A**, 234 (1964). *Review:* D. W. Visser, S. Roy-Burman in *Antibiotics* vol. **5**(pt. 2), F. E. Hahn, Ed. (Springer-Verlag, New York, 1979) pp 363-371.

Leaflets from acetone + benzene, mp 153-154°. $[\alpha]_D^{22.5}$ +49.9° (H_2O). uv max: 220-221 nm ($E_{1cm}^{1\%}$ 422). Sol in water, alcohol, acetone, dioxane; insol in ethyl ether, benzene, petr ether. More stable in acid media than in neutral or alkaline. LD_{50} in mice (mg/kg): 25 i.p.; 18 s.c.; 110 i.v. (Nishimura).

8431. Sialic Acids. Nonulosaminic acids. Family of amino sugars contg 9 or more carbon atoms, N- and O-substituted derivatives of neuraminic acid, $q.v.$ Widely distributed throughout the animal kingdom; appear to be regular components of all types of mucoproteins, mucopolysaccharides and certain mucolipids. Bovine and ovine submaxillary gland sialic acids are diacetyl derivs. Pig submaxillary mucin, and pig, horse, and ox erythrocytes contain sialic acids having an N-glycolyl group. *Reviews:* A. Gottschalk, *The Chemistry and Biology of Sialic Acids and Related Substances* (Cambridge U. Press, 1960); idem, *Rev. Pure Appl. Chem.* **12,** 46 (1962). Books: *Biological Roles of Sialic Acid,* A. Rosenberg, C.-L. Schengrund, Eds. (Plenum Press, New York, 1976) 393 pp; *Sialic Acids: Chemistry, Metabolism & Function,* R. Schauer, Ed. (Springer-Verlag, New York, 1983) 346 pp.

N-acetylneuraminic acid

N-Acetylneuraminic acid; 5-(acetylamino)-3,5-dideoxy-D-*glycero*-D-*galacto*-2-nonulosonic acid, *NAN, NANA, lactaminic acid, O-sialic acid.* $C_{11}H_{19}NO_9$; mol wt 309.28. The most important of the 5 sialic acids. Isoln: Klenk, Faillard, *Z. Physiol. Chem.* **298,** 230 (1954). Prepn from sheep submaxillary mucins: Blix, *Methods Carbohyd. Chem.* **1,** 246 (1962). Synthesis: Cornforth *et al., Biochem. J.* **68,** 57 (1958); Carrol, Cornforth, *Biochim. Biophys. Acta* **39,** 161 (1960); Kuhn, Baschang, *Ann.* **659,** 156 (1962); Rinderknecht, Rebane, *Experientia* **19,** 342 (1963). Improved synthesis: How *et al., Carbohyd. Res.* **11,** 313 (1969); L. Benzing-Nguyen, M. B. Perry, *J. Org. Chem.* **43,** 551 (1978). First synthesis from non-carbohydrate sources: M. P. DeNinno, S. J. Danishevsky, *J. Org. Chem.* **51** 2615 (1986). Configuration: Kuhn, Brossmer, *Angew. Chem. Int. Ed.* **1,** 218 (1962). Dec 185-187°. $[\alpha]_D^{22}$ —32° (water).

8432. Siccanin. *[4aS-(4aα,6aα,11bα,13aR*,13bα)]-1,2,-3,4,4a,5,6,6a,11b,13b-Decahydro-4,4,6a,9-tetramethyl-13H-benzo[a]furo[2,3,4-mn]xanthen-11-ol;* Tackle. $C_{22}H_{30}O_3$; mol wt 342.48. C 77.15%, H 8.83%, O 14.02%. Antibiotic obtained from the cultured broth of *Helminthosporium siccans* Drechsler, a parasitic organism of rye grass (*Lolium multiflorum* Lam). Isoln: K. Ishibashi, **Japan.** pat. **62 3548** (1962 to Sankyo); idem, *J. Antibiot.* **15A,** 161 (1962). The *cis* formation of the decahydronaphthalene system may be the first example of the naturally occurring drimane skeleton. Structure: K. Hirai *et al., Tetrahedron Letters* **1967,** 2177; K. Hirai *et al., Tetrahedron* **27,** 6057 (1971). Approach to synthesis: H. Oida *et al., Chem. Pharm. Bull.* **20,** 2634 (1972); T. Yoshida *et al., ibid.* 2642. Total synthesis of racemate: M. Kato *et al., J. Am. Chem. Soc.* **103,** 2434 (1981); *eidem, Tetrahedron* **43,** 711 (1987). CD spectral study: S. Nozoe *et al., Tetrahedron* **30,** 2773 (1974). *In vitro* activity studies: M. Arai *et al., Antimicrob. Ag. Chemother.* **1969,** 247; M. G. Bellotti, L. Riviera, *Chemioterapia* **4,** 431 (1985); *in vivo* studies: S. Sugarwara, *Antimicrob. Ag. Chemother.* **1969,** 253. Toxicology: Y. Suzuki *et al., Sankyo Kenkyusho Nempo* **21,** 120 (1969), *C.A.* **72,** 130939k (1970). Mode of action: K. Nose, A. Endo, *J. Bacteriol.* **105,** 176 (1971). Biosynthesis: K. Suzuki *et al., Bioorg. Chem.* **3,** 72 (1974). Clinical evaluation in surface mycosis: D. Crippa *et al., G. Ital. Dermatol. Venereol.* **120,** 57 (1985).

White to pale yellow odorless crystals, mp 139-140°. $[\alpha]_D^{20}$ —136° (c = 2 in chloroform). uv max (ethanol): 210, 285 nm (ε 45690, 1717). pK 10.9. Very sol in chloroform, dimethylformamide, benzene; freely sol in acetone, ether, ethyl acetate; sol in alcohol. Practically insol in water. LD_{50} in mice, rats (mg/kg): > 6000, > 1000 orally; > 3000, > 600 s.c. or i.p. (Suzuki).

THERAP CAT: Antifungal.

8433. Siduron. *N-(2-Methylcyclohexyl)-N'-phenylurea;* Tupersan. $C_{14}H_{20}N_2O$; mol wt 232.32. C 72.38%, H 8.68%, N 12.06%, O 6.89%. Prepn: Luckenbaugh, **Belg.** pat. **631,-289** (1963 to du Pont), *C.A.* **60,** 15775h (1964), corresp to **Brit.** pat. **1,028,818.** Prepn of *trans*-form: Hayman *et al., J. Pharm. Pharmacol.* **16,** 538 (1964).

Solid, mp 133-138°. Sol in water (18 ppm); sol to the extent of 10% or more in ethanol, dimethylacetamide, dimethylformamide, methylene chloride.

trans-Form, crystals, mp 157-159°.

USE: Herbicide.

8434. Sikkimotoxin. *3a,4,9,9a-Tetrahydro-4-hydroxy-6,7-dimethoxy-9-(3,4,5-trimethoxyphenyl)naphtho[2,3-c]-furan-1(3H)-one;* 1-hydroxy-2-hydroxymethyl-6,7-dimethoxy-4-(3',4',5'-trimethoxyphenyl)-1,2,3,4-tetrahydronaphthalene-3-carboxylic acid lactone. $C_{23}H_{26}O_8$; mol wt 430.44. C 64.17%, H 6.09%, O 29.74%. Isoln from roots and rhizomes of *Podophyllum sikkimensis* R. Chatterjee et Mukerjee, *Berberidaceae:* Chatterjee, Datta, *Indian J. Physiol. Allied Sci.* **4,** 61 (1950). Structure: Chatterjee, Chakravarti, *J. Am. Pharm. Assoc. Sci. Ed.* **41,** 415 (1952); Schreier, *Helv. Chim. Acta* **47,** 1529 (1964).

Crystals from ethanol + benzene, mp 120°. $[\alpha]_D^{25}$ —91.8°. Freely sol in ethanol, methanol, chloroform. Sparingly sol in benzene. Practically insol in water and petr ether.

8435. Silane. Silicon tetrahydride; silicane; monosilane. H_4Si; mol wt 32.09. H 12.56%, Si 87.44%. SiH_4. Prepd from aluminum silicide and hydrochloric acid or by electrolysis of an aq soln of sodium, ammonium, manganese, or ferrous chloride using an aluminum-silicon alloy as the positive pole: Buff, Wöhler, *Ann.* **102,** 128 (1857); from magnesium silicide and HCl: Wöhler, *ibid.* **106,** 56 (1858); Stock, Somiesky, *Ber.* **49,** 111 (1916); from zinc silicide and acid: Deville, Caron, *Compt. Rend.* **45,** 163 (1857); from lithium silicide and acid: Moissan, *ibid.* **135,** 1284 (1902); from magnesium silicide and ammonium bromide or chloride in liquid ammonia: Johnson, Isenberg, *J. Am. Chem. Soc.* **57,** 1349 (1935); Clasen, **Ger.** pat. **926,069** (1952); *Angew. Chem.* **70,** 179 (1958); from lithium aluminum hydride in ether and silicon tetrachloride: Finholt *et al., J. Am. Chem. Soc.* **69,** 2692 (1947); from silica gel, aluminum oxide and hydrogen: Jackson, U.S. pat. **3,068,069** (1962 to du Pont); from silicon tet-rachloride and sodium aluminum hydride: Mason, Kelly, U.S. pat. **3,043,664** (1962 to Allied Chem.). Purification: Jacob, Trenner, U.S. pat. **3,019,087** (1962 to Merck & Co.); Shoemaker, Sumrell, U.S. pat. **3,041,141** (1962 to Baker Chem.).

Gas, repulsive odor. d^{-185} (liq) 0.68. Solidifies at approx $-200°$. mp $-185°$; bp $-112°$. Stable at ordinary temps; completely dec at approx 400° into silicon and hydrogen; dec by an electric discharge; ignites in an atm of chlorine; ignites in air by raising the temp. Slowly dec in water; practically insol in alc, ether, benzene, chloroform, silicochloroform and silicon tetrachloride. Dec in potassium hydroxide solns.
USE: Source of hyperpure silicon for semiconductors. *Caution:* Irritating to respiratory tract.

8436. Silanols. Silanols are compds contg hydroxyl groups directly bound to the silicon atom: R_3SiOH; $R_2Si-(OH)_2$; $RSi(OH)_3$.

8437. Silicic Acid. Precipitated silica. Approx H_2SiO_3. Occurs in nature as *opal.* Monograph: E. A. Hauser, *Silicic Science* (Van Nostrand, Princeton, 1958). Short review of chemistry: S. A. Greenberg, *J. Chem. Ed.* **36**, 218-219 (1959).

White, amorphous powder. Insol in water or acids except hydrofluoric (*see under* Silicon Dioxide); sol in hot fixed alk hydroxide solns.
Silica gel, Dri-Die. Pptd silicic acid in form of lustrous granules, esp prepd and adapted for absorption of various vapors.
USE: Silica gel as insecticide.

8438. Silicon. Si; at. wt 28.0855; at. no. 14; valence 4, also 2. Three naturally occurring isotopes: 28 (92.18%); 29 (4.71%); 30 (3.12%); artificial isotopes: 25-27; 31; 32. Does not occur free in nature; found as silica (quartz, sand, sand-stone) or as silicate (feldspar or orthoclase, kaolinite, anor-thite, etc.). Constitutes about 27.6% of the earth's crust; second most abundant element on earth, oxygen being first. Prepd industrially by carbon reduction of silica in an electric arc furnace. Purification by zone refining (*see* ref. *under* Germanium). Very pure silicon is obtained by decomp of silicon tetraiodide: Litton, Anderson, *J. Electrochem. Soc.* **101**, 287 (1954); *Chem. & Eng. News* **34**, 5007 (1956). From silicon tetrachloride: Lyon *et al., Trans. Electrochem. Soc.* **96**, 359 (1949); Klyuchnikov, *J. Appl. Chem. USSR* **29**, 139 (1956). By thermal decompn of a chlorosilane: Schering, U.S. pat. **3,041,144** (1962 to Siemens-Schuckertwerke). Review of silicon and its compounds: Rochow, "Silicon" in *Comprehensive Inorganic Chemistry* vol. **1**, J. C. Bailar, Jr. *et al.,* Eds. (Pergamon Press, Oxford, 1973) pp 1323-1467; W. Runyan in Kirk-Othmer *Encyclopedia of Chemical Technology* vol. **20** (Wiley-Interscience, New York, 3rd ed., 1982) pp 826-845. The uses of silicon compounds in organic chemis-try: E. W. Colvin, *Chem. Soc. Rev.* **7**, 15 (1978); I. Fleming, *ibid.* **10**, 83 (1981); L. A. Paquette, *Science* **217**, 793 (1982).

Black to gray, lustrous, needle-like crystals or octahedral platelets (cubic system). The amorphous form is a dark brown powder. Poor conductor of electricity. d_4^{25} 2.33. mp 1410°. Average heat capacity (16-100°): 0.1774 cal/g/°C. Lattice constant (25°): 5.41987×10^{-8} cm. Compressibility (V/V_0) at 25×10^3 kg/cm²: 0.978; at 100×10^3 kg/cm²: 0.940, *Gmelin's Silicon* (8th ed.) **15B** (1959) p 57. Dielectric const: 13. Covalent bond ionization energy at 0°K = 1.2 ev. Band gap: 1.106 ev. Impurity atom ionization energy: $\sim$0.04 ev. Intrinsic resistivity at 300°K = 0.23 megohm. Electron mobility at 300°K: 1500 cm²/volt/sec. Hole mo-bility at 300°K: 500 cm²/volt/sec. Intrinsic charge density at 300°K: 1.5×10^{10}. Electron diffusion constant at 300°K: 38. Hole diffusion constant at 300°K: 13. Practically insol in water. Attacked by hydrofluoric or a mixture of hydro-fluoric and nitric acids (depending upon cryst modifica-tions). Soluble in molten alkali oxides. Burns in fluorine, chlorine.
USE: In making silanes and silicones, the Si—C bond be-ing about as strong as a C—C bond. In the manuf of tran-sistors, silicon diodes and similar semiconductors. For mak-ing alloys such as ferrosilicon, silicon bronze, silicon copper. As a reducing agent like aluminum in high temp reactions.

8439. Silicon Carbide. CSi; mol wt 40.07. C 29.97%, Si 70.03%. SiC. Made by heating coke and sand (and salt as flux) in electric furnace.

Exceedingly hard, green to bluish-black, iridescent, sharp crystals; hardness = 9.5. d 3.23. Dielectric constant: 7.0. Electron mobility: > 100 cm²/volt-sec. Hole mobility: > 20 cm²/volt-sec. Band gap energy: 2.8 ev.
USE: Polishing glass, granite; smoothing bisque ware; in sharpening-stones; abrasive; manuf porcelain, "emery" paper and wheels, shoe soles, refractory brick, furnace linings; an-tiskid pavements; in semiconductor technology.

8440. Silicon Dioxide. *Silica;* silicic anhydride. O_2Si; mol wt 60.09. O 53.25%, Si 46.75%. SiO_2. Occurs in nature as *agate, amethyst, chalcedony, cristobalite, flint, quartz, sand, tridymite.* Reviews: Several authors in Kirk-Othmer *Encyclopedia of Chemical Technology* vol. **18** (Interscience, New York, 2nd ed., 1969) pp 46-111; Rochow in *Compre-hensive Inorganic Chemistry* vol. **1**, J. C. Bailar, Jr. *et al.,* Eds. (Pergamon Press, Oxford, 1973) pp 1388-1402.

Transparent, tasteless crystals, or amorphous powder. d (amorphous) 2.2. d^0 (quartz) 2.65. Melts to a glass. Silica has the lowest coefficient of expansion by heat of any known substance. It is practically insol in water or acids, except hydrofluoric acid in which it readily dissolves forming the gas silicon tetrafluoride; it is also slowly attacked by heating with concd phosphoric acid. The crystallized forms of silica are scarcely attacked by alkalies, while the amorphous is sol, especially when finely divided. *See also* Infusorial Earth.
USE: Manuf glass, water glass, refractories, abrasives, ce-ramics, enamels; decolorizing and purifying oils, petroleum products, etc.; in scouring- and grinding-compounds, ferro-silicon, molds for castings; as anticaking and defoaming agent. *Caution:* Prolonged inhalation of the dust can cause fibrosis of the lung (silicosis), L. T. Fairhall, *Industrial Toxi-cology* (Hafner, New York, 1969) pp 105-107.

8441. Silicon Disulfide. S_2Si; mol wt 92.21. S 69.54%, Si 30.46%. SiS_2. Prepd from SiO_2 and Al_2S_3: Tiede, *Ber.* **59**, 1703 (1926); Zintl, Loosen, *Z. Phys. Chem.* **A174**, 301 (1935).

White, fibrous mass or needles. Yields H_2S on contact with moist air. d 2.02. mp 1090° (sublimes). Burns when held in flame. Sol with decompn in water, alcohol, alkaline solns. Insol in benzene.

8442. Silicones. The name was coined by F. S. Kipping and is now used to designate any organosilicon oxide poly-mer in which the structural unit is usually —$R_2Si—O$—, where R is a monovalent organic radical. Physical proper-ties of silicones depend on the size and type of the radical (methyl, phenyl), the R:Si ratio, and the molecular configu-ration of the polymer (linear, cyclic, degree of crosslinking). Commercial preparation involves: the synthesis of organo-silicon halides by the Grignard reaction, by direct reaction of hydrocarbon halides with silicon, by olefin addition to silanes, or by the sodium method; the hydrolysis of the organosilicon halides; and the condensation of the resulting diols. The products have remarkably high thermal and chemical stability and unusual release from sticking and surface properties.

Silicones are classified as fluids (or oils), rubbers, resins and compounds. Review of chemistry, manuf, and uses: E. G. Rochow, *Chemistry of the Silicones* (John Wiley, New York, 2nd ed., 1950); R. R. McGregor, *Silicones and Their Uses* (McGraw-Hill, New York, 1954); *Silicones,* S. Ford-ham, Ed. (Philosophical Library, New York, 1961) 252 pp; Lichtenwalner, Sprung in *Encyclopedia of Polymer Science and Technology* vol. **12** (Interscience, New York, 1970) pp 464-569.
USE: As lubricants; hydraulic oils, heat-transfer media; dielectric oils; rubber for electrical coatings, molding, and extruding; electrical and nonelectrical adhesives and caulk-ing compounds; release agents for the food industry, for metalworking application, and for rubber, plastics, and ce-ramics; in paints, enamels, and varnishes; water repellents for textiles, paper, masonry, and concrete; damping fluids; electrical insulation, and in cosmetic and pharmaceutical formulations. A specialty product known as "bouncing putty" or "silly putty" has unusual rheological properties

which allow it to be stretched, shattered, molded, or to rebound like rubber when dropped on hard surface.

THERAP CAT: Dermatologic; antiflatulent; bronchial anti-foam.

THERAP CAT (VET): Antifrothing agents: have been used in bloat control and as an aerosol in pulmonary edema.

8443. Silicon Monoxide. OSi; mol wt 44.09. O 36.29%, Si 63.71%. SiO. Prepd by high vacuum sublimation of a mixture of silicon and quartz: Schenk in *Handbook of Preparative Inorganic Chemistry* vol. 1, G. Brauer, Ed. (Academic Press, New York, 2nd ed., 1963) pp 696-697. Existence below 1200° in doubt; may disproportionate to a mixture of Si and SiO_2: Benyon, *Vacuum* **20**, 293 (1970). *Review:* Rochow in *Comprehensive Inorganic Chemistry* vol. 1, J. C. Bailar, Jr. *et al.*, Eds. (Pergamon Press, Oxford, 1973) pp 1353-1355.

Brownish-black scales by sublimation. Has been crystallized. Cubic system. About as hard as silicon. Nonconductor of electricity even at red heat. d 2.18.

8444. Silicon Nitride. Trisilicon tetranitride. N_4Si_3; mol wt 140.28. N 39.94%, Si 60.06%. Si_3N_4. Non-oxide ceramic material of high thermal stability. Manuf by direct nitridation of pure silicon. Prepn by pyrolysis of silicon diimide $(Si(NH_2)_2)$: O. Glemser, P. Naumann, *Z. Anorg. Allgem. Chem.* **298**, 134 (1954); and characterization: K. S. Mazdiyasni, C. M. Cooke, *J. Am. Ceram. Soc.* **56**, 628 (1973); *eidem*, U.S. pat. **3,959,446** (1976 to USA); by pyrolysis of silazanes: D. Seyferth *et al.*, *J. Am. Ceram. Soc.* **66**, C-13 (1983); *eidem*, U.S. pat. **4,397,828** (1983 to M.I.T.). Brief review: D. L. Segal, *Chem. & Ind. (London)* **1985**, 544.

Exists in two crystalline forms, α-Si_3N_4 and β-Si_3N_4. α-Form occurs when powdered Si_3N_4 is heated to 1200° > 4 hr; increasing the temperature to > 1450°C for 2 hr produces the β-form. White when pure but often brown or black powder. mp 2173°K (sublimes). d 3.2 kg/dm³. Coefficient of thermal expansion K^{-1}: 2.5×10^{-6}.

USE: As high temp engineering material for use in gas turbines, diesel engines.

8445. Silicon Tetraacetate. *Acetic acid tetraanhydride with silicic acid.* $C_8H_{12}O_8Si$; mol wt 264.27. C 36.36%, H 4.58%, O 48.44%, Si 10.63%. $Si(CH_3COO)_4$. Prepd from $SiCl_4 + (CH_3CO)_2O$: Balthis, *Inorg. Syn.* **4**, 45 (1953).

Extremely hygroscopic cryst mass. Hydrolyzed rapidly by moist air. mp 110°. Dec at 160-170° with evolution of acetic anhydride. $bp_{6.0}$ 148°. Violent reaction on contact with water. Forms SiO_2 and ethyl acetate on contact with alcohol. Moderately sol in acetone, benzene.

8446. Silicon Tetrabromide. Silicon bromide. Br_4Si; mol wt 347.72. Br 91.93%, Si 8.07%. $SiBr_4$. Prepd from the elements: Gattermann, *Ber.* **22**, 189 (1889).

Colorless, fuming liq; disagreeable odor; becomes yellow on exposure to air. d 2.8. Solidif −12°; mp 5°; bp 153°. Dec by water into silicic acid and HBr with great evolution of heat. Reacts violently with metallic potassium.

8447. Silicon Tetrachloride. Silicon chloride. Cl_4Si; mol wt 169.89. Cl 83.48%, Si 16.52%. $SiCl_4$. Prepd from the elements: Schenk in *Handbook of Preparative Inorganic Chemistry* vol. 1, G. Brauer, Ed. (Academic Press, New York, 2nd ed., 1963) pp 682-683.

Colorless, clear, mobile, fuming liq; suffocating odor. d_4^0 1.52. mp −70°; bp 59°. Dec by water with much heat into silicic acid and HCl. Miscible with benzene, ether, chloroform, petr ether.

USE: Producing smoke screens in warfare. In the prepn of pure silicon. *Caution:* May be irritating to eyes, respiratory tract.

8448. Silicon Tetrafluoride. Silicon fluoride. F_4Si; mol wt 104.06. F 73.03%, Si 26.97%. SiF_4. Commercial prepn from waste gases of phosphate fertilizer production: Molstad, U.S. pat. **2,833,628** (1958 to W. R. Grace). Lab prepns: Booth, Swinehart, *J. Am. Chem. Soc.* **57**, 1337 (1935); Hoffman, Gutowski, *Inorg. Syn.* **4**, 145 (1953); Belf, *Chem. & Ind. (London)* **1955**, 1296.

Colorless gas; very pungent odor similar to that of hydrogen chloride. Forms heavy clouds with moist air. Dec by water into silicic acid and HF. Sublimes −95.7°; mp −90.2°

(under pressure): d (liq; −80°) 1.590. Critical temp −1.5°; critical pressure 50 atm.

8449. Silicotungstic Acid. *Tungstosilicic acid.* H_4O_{40}-SiW_{12}; mol wt 2878.31. H 0.14%, O 22.23%, Si 0.98%, W 76.65%. $H_4[SiO_4\cdot(W_3O_9)_4]$. Prepn of hydrated compd: Grüttner, Jander in *Handbook of Preparative Inorganic Chemistry* vol. 2, G. Brauer, Ed. (Academic Press, New York, 2nd ed., 1965) pp 1717-1718.

Hydrate, white to slightly yellow, deliquesc crystals. Very sol in water, alcohol. *Keep tightly closed.*

USE: Preparing heavy solns for separating minerals; as a mordant for basic aniline dyes; as a reagent for pptn and determination of alkaloids.

8450. Silver. Ag; at. wt 107.868; at. no. 47; valence 1, 2. Occurrence in the earth's crust: 0.1 ppm; also present in seawater: 0.01 ppm. Natural isotopes: 107 (51.35%); 109 (48.65%); artificial isotopes (mass numbers): 100-106, 108, 110-117. One of the earliest known metals. Found native or associated with copper, gold and lead. Principle ores are argentite *q.v.*, *cerargyrite* or *horn silver* (mixture of halides), *proustite* ($3Ag_2S.As_2S_3$), *pyrargyrite* ($Ag_2S.Sb_2S_3$). Extraction from ores: Percy *et al.*, cited by Mellor, *A Comprehensive Treatise on Inorganic and Theoretical Chemistry* **3**, 301 (1928); *Silver, Its Economics, Extraction, Use,* A. Butts, C. D. Coxe, Eds. (Van Nostrand, Princeton, 1967) 480 pp. Reviews of silver and silver compounds: Thompson, "Silver" in *Comprehensive Inorganic Chemistry*, vol. 3, J. C. Bailar, Jr. *et al.*, Eds. (Pergamon Press, Oxford, 1973) pp 79-128; G. H. Sistare and H. B. Lockhart in Kirk-Othmer *Encyclopedia of Chemical Technology* vol. 21 (Wiley-Interscience, New York, 3rd ed., 1983) pp 1-32.

White metal, face-centered cubic structure. More malleable and ductile than any other metal except gold; excellent conductor of heat and electricity. mp 960.5°. bp ~2000°. d^{15} 10.49. Not attacked by water or atmospheric oxygen; blackened by ozone, by hydrogen sulfide, by sulfur. Inert to most acids; readily reacts with dil nitric acid, hot concd sulfuric acid; superficially attacked by hydrochloric acid. Sol in fused alkali hydroxides in presence of air, in fused alkali peroxides, in alkali cyanides in presence of air or oxygen. Most silver salts are light-sensitive.

USE: For coinage, most frequently alloyed with copper or gold; for manuf tableware, mirrors, jewelry, ornaments; for electroplating; for making vessels and apparatus used in manuf medicinal chemicals, in processing foods and beverages, in handling organic acids; as catalyst in hydrogenation and oxidation processes; as ingredient of dental alloys. Has been used for purification of drinking water because of toxicity to bacteria and lower forms of life. Some salts used in photography. *Caution:* Does not cause serious toxic manifestations, but prolonged absorption of silver compds can lead to grayish blue discoloration of skin, known as argyria or argyrosis. Inhalation of dust should be avoided. Many silver salts are irritating to skin, mucous membranes. *See* E. Browning, *Toxicity of Industrial Metals* (Appleton-Century-Crofts, New York, 2nd ed., 1969) pp 296-301.

8451. Silver Acetate. $C_2H_3AgO_2$; mol wt 166.92. C 14.39%, H 1.81%, Ag 64.63%, O 19.17%. CH_3COOAg.

White to slightly grayish, lustrous needles or cryst powder. d 3.26. Sol in 100 parts of cold water, 35 parts of boiling water; freely sol in dil nitric acid. *Protect from light.*

USE: Oxidizing agent for use in liquid ammonia: Kline, Kershner, *Inorg. Chem.* **5**, 932 (1966).

8452. Silver Bromide. AgBr; mol wt 187.80. Ag 57.45%, Br 42.55%.

Yellowish, odorless powder; darkens on exposure to light. d 6.47; mp 432°. Soly in water (25°): 0.135 mg/l. Insol in alcohol, or most acids. Slightly sol in dil ammonia, but moderately sol in concd ammonia. One liter of 10% ammonia dissolves 3.3 g at 12°; sol in solns of alkali cyanides, sparingly sol in solns of thiocyanates or thiosulfates; sol in 220 parts saturated NaCl, in 35 parts saturated KBr solns; slightly sol in ammonium carbonate soln. *Protect from light.*

USE: In photography.

THERAP CAT: Topical anti-infective; astringent.

8453. Silver Carbonate. CAg_2O_3; mol wt 275.77. C 4.36%, Ag 78.24%, O 17.41%. Ag_2CO_3. Prepn from silver nitrate and sol carbonate soln: Poyer *et al., Inorg. Syn.* **5**, 19 (1957).

Light yellow powder when freshly pptd, but becomes darker on drying and on exposure to light. Dec at about 220° into Ag_2O and CO_2 and at higher temp into metal Ag. d 6.08. Sol in 30,000 parts cold water, 2000 parts boiling water; readily sol in dil nitric acid, ammonia, or alkali cyanides. *Protect from light.*

8454. Silver Chlorate. Argentous chlorate. $AgClO_3$; mol wt 191.34. Ag 56.38%, Cl 18.53%, O 25.09%. Prepd from $AgNO_3$ and $NaClO_3$: Nicholson, Holley, Jr., *Inorg. Syn.* **2**, 4 (1946).

White, tetragonal crystals. d_4^{20} 4.430. mp 230°, dec 270° forming silver chloride and oxygen. Solubility in water (g/100 ml): 10 (15°); 20 (27°); 50 (80°). Slightly sol in alc. Darkens upon exposure to light due to slow decompn. *Keep away from light, organic vapors and oxidizable substances.*
USE: In organic synthesis as an effective oxidizing agent for certain organic compds.

8455. Silver Chloride. AgCl; mol wt 143.34. Ag 75.26%, Cl 24.74%.

White powder; darkens on exposure to light. d 5.56; mp 455°; bp 1550°. Soly in water (25°): 1.93 mg/l; hydrochloric acid increases its soly. Sol in 250 parts of concd HCl, in 13 parts of 10% ammonia; more sol in stronger ammonia and also at higher temps. Sol in solns of alkali cyanides, thiosulfates, ammonium carbonates; appreciably sol in concd aq solns of ammonium chloride; mercuric nitrate, and silver nitrate. The freshly pptd chloride dissolves more readily than the dried precipitate. Insol in alcohol, or dil acids. *Protect from light.*
USE: In silver plating, in making antiseptic silver prepns.

8456. Silver Chromate(VI). Ag_2CrO_4; mol wt 331.77. Ag 65.03%, Cr 15.68%, O 19.29%.

Dark brownish-red, cryst powder. d^{25} 5.625. Soly in water (0°): 0.0014%. Sol in nitric acid and ammonia.
USE: Catalyst for formation of aldol from alcohol; formed at end point of Mohr titration of halides.

8457. Silver Citrate. *2-Hydroxy-1,2,3-propanetricarboxylic acid silver salt;* Itrol; Silberol. $C_6H_5Ag_3O_7$; mol wt 512.74. C 14.05%, H 0.98%, Ag 63.12%, O 21.84%.

White, odorless, heavy, crystalline powder; darkens in light. Sol in 3500 parts water, more sol in boiling water; readily sol in dil HNO_3, ammonia. *Protect from light.*
USE: Anti-infective dusting powder.

8458. Silver Cyanide. CAgN; mol wt 133.90. C 8.97%, Ag 80.57%, N 10.46%. AgCN.

White or grayish, odorless powder; stable in dry air; darkens on exposure to light; dec at 320°. *Poisonous!* d 3.95. Insol in water, alcohol, or dil acids; sol in alkali cyanides and in boiling concd nitric acid; converted by dil HCl into hydrocyanic acid and silver chloride; sparingly sol in dil, more in concd ammonia. *Protect from light.*
USE: For silver plating; formerly used for extemporaneous prepn of dil hydrocyanic acid by treatment with HCl.

8459. Silver Difluoride. Argentic fluoride. AgF_2; mol wt 145.88. Ag 73.95%, F 26.05%. Prepd by the action of fluorine on silver: Ebert *et al., J. Am. Chem. Soc.* **55**, 3056 (1933); by the interaction of fluorine and silver halides: Ruff, Giese, *Z. Anorg. Allgem. Chem.* **219**, 143 (1934); v. Wartenberg, *ibid.* **242**, 406 (1939); Struve *et al., Ind. Eng. Chem.* **39**, 353 (1947). Laboratory procedure according to the equation $2AgCl + 2F_2 \rightarrow 2AgF_2 + Cl_2$: Priest, Swinehart, *Inorg. Syn.* **3**, 176 (1950); Kwasnik in *Handbook of Preparative Inorganic Chemistry*, vol. 1, G. Brauer, Ed. (Academic Press, New York, 2nd ed., 1963) pp 241-242. Prepd from silver chloride and chlorine trifluoride: Rochow, Kukin, *J. Am. Chem. Soc.* **74**, 1615 (1952).

White when pure. Usually obtained as a gray-black or brownish, amorphous solid showing a yellow bloom. Sensitive to light. d 4.7. mp 690°. Very hygroscopic, is converted to a greasy black mass on exposure to atmospheric moisture. Violent reaction with water (instant hydrolytic cleavage). Powerful oxidizing agent giving off some ozone when treated with dil acids. Liberates iodine from iodides. May be stored in quartz or iron ampuls.

USE: In the fluorination of hydrocarbons. *Caution:* Symptoms due to fluoride. Highly toxic.

8460. Silver Fluoride. Argentous fluoride; silver monofluoride. AgF; mol wt 126.88. Ag 85.03%, F 14.97%. Prepared according to the equation $Ag_2CO_3 + 2HF \rightarrow 2AgF + CO_2 + H_2O$; also prepd by reduction of AgF_2 with H_2: Ruff, *Die Chemie des Fluors* (Berlin, 1920) p 37; Emeléus in *J. H. Simons' Fluorine Chemistry*, **vol. I** (Academic Press, New York, 1950) p 32; Kwasnik in *Handbook of Preparative Inorganic Chemistry*, **vol. 1**, G. Brauer, Ed. (Academic Press, New York, 2nd ed., 1963) pp 240-241.

Flexible leaflets (cubic, NaCl lattice). Very hygroscopic. Darkens on exposure to light. d 5.852. mp 435°. bp about 1150°. Soly in water when freshly prepd: 182 g/100 ml at 15.5°. In moist air it gradually becomes insol because of basic fluoride formation. Aq solns are neutral and are solvents for silver oxide which renders them alkaline. Also sol in HF, NH_3, CH_3CN. Forms several hydrates. The dihydrate is stable to 39.5° and the tetrahydrate is stable from $-14°$ to $+18.7°$. In addition, an acid fluoride, AgF.3HF can be prepd by cooling a soln of AgF in HF. This loses HF at 0°, forming AgF.HF.
Human Toxicity: Prolonged absorption may cause mottling of teeth, skeletal changes.
USE: To convert organic Br and Cl compds to their fluoro analogs; as antiseptic.
THERAP CAT: Anti-infective.

8461. Silver Iodate. $AgIO_3$; mol wt 282.80. Ag 38.15%, I 44.88%, O 16.97%.

White, cryst powder. d 5.53. mp above 200°. Sol in 1875 parts water at 25°, in about 1000 parts of 35% nitric acid, in 2.5 parts of 10% ammonia. *Protect from light.*
USE: As reagent for determining small quantities of chloride, e.g., in blood: Sendroy, *J. Biol. Chem.* **120**, 335-445 (1937).

8462. Silver Iodide. AgI; mol wt 234.80. Ag 45.95%, I 54.05%. Prepd according to the equation $AgNO_3 + KI \rightarrow AgI + KNO_3$: Kolkmeijer, van Hengel, *Z. Kristallogr.* **A88**, 317 (1934).

Light yellow, odorless powder; slowly darkened by light. Crystals are hexagonal or cubic. d 5.67; mp 552°. Practically insol in water (0.03 mg/l); in acid (except concd HI in which it dissolves on heating); in ammonium carbonate. Freely sol in solns of alkali cyanides or iodides; 35 mg dissolve in a liter of 10% ammonia; appreciably sol in concd solns of alkali bromides, chlorides, thiocyanates, thiosulfates, mercuric and silver nitrates. It is slowly attacked by boiling concd acids, but not affected by hot solns of alkali hydroxides.
USE: In cloud precipitation (rain-making).
THERAP CAT: Local anti-infective.
THERAP CAT (VET): In colloidal suspensions as local antiseptic for mucous membranes.

8463. Silver Lactate. *2-Hydroxypropanoic acid silver salt.* $C_3H_5AgO_3$; mol wt 214.97. C 18.29%, H 2.56%, Ag 54.78%, O 24.37%. $CH_3CH(OH)COOAg$.

Monohydrate, white or slightly gray, cryst powder readily affected by light. Sol in about 15 parts water; slightly sol in alcohol. *Protect from light.*
THERAP CAT: Topical anti-infective; astringent.

8464. Silver Nitrate. $AgNO_3$; mol wt 169.89. Ag 63.50%, N 8.25%, O 28.25%. Silver nitrate of commerce is, with the exception of 0.1-0.2% water, practically 100% pure.

Colorless, odorless, transparent, large crystals or white small crystals. *Poisonous!* Not photosensitive when pure; presence of trace amounts of organic material promotes photoreduction. Reduced by H_2S in the dark. d 4.35. mp 212°, forming a yellowish liq solidifying to a white, cryst mass on cooling. Dec at 440° into metallic silver, nitrogen, oxygen and nitrogen oxides. One gram dissolves in 0.4 ml water, 0.1 ml boiling water, 30 ml alc, 6.5 ml boiling alc, 253 ml acetone; readily sol in ammonia water, slightly in ether. The aq and alc solns are neutral to litmus. pH ~6.
Incompat: Alkalies, antimony salts, arsenites, bromides, carbonates, chlorides, iodides, thiocyanates, ferrous salts,

hypophosphites, morphine salts, oils, creosote, phosphates, tannic acid, tartrates, vegetable decoctions, and extracts.

Human Toxicity: Caustic and irritating to skin, mucous membranes. Swallowing can cause severe gastroenteritis that may be fatal. *See also* Silver.

USE: In photography; manuf of mirrors; other silver salts; silver plating; in sympathetic and indelible inks; dyeing hair; coloring porcelain; etching ivory; as a very important and extensively used reagent in analytical chemistry.

THERAP CAT: Topical anti-infective.

THERAP CAT (VET): Astringent, antiseptic, caustic, styptic. In dilute solution as eye lotion. Intramammary to destroy secretory tissue.

8465. Silver Nitrate, Toughened. Lunar caustic; fused silver nitrate; molded silver nitrate. Consists of about 97-98% silver nitrate, remainder is silver chloride to toughen it.

White or grayish, hard rods or thin small cones. Darkens on exposure to light. *Protect from light.*

THERAP CAT: Caustic.

THERAP CAT (VET): Locally as escharotic.

8466. Silver Nitrite. $AgNO_2$; mol wt 153.89. Ag 70.10%, N 9.10%, O 20.79%.

Pale yellow, odorless needles; non-hygroscopic; becomes gray in light. d 4.45. Dec at 140°. Sol in 300 parts water; more sol in boiling water; less sol in aqueous solns of silver nitrate; partly dec on prolonged boiling with water; insol in alcohol; dec by dil acids. *Protect from light.*

USE: For prepn of standard $NaNO_2$ soln for water analysis; prepn of aliphatic nitro-compds; as reagent for primary, secondary, and tertiary alcohols.

8467. Silver Oxalate. *Ethanedioic acid silver salt.* C_2-Ag_2O_4; mol wt 303.78. C 7.91%, Ag 71.03%, O 21.07%. AgOOCCOOAg.

White, cryst powder; at 140° dec violently. d 5.03. Sol in 24,000 parts water; sol in moderately concd nitric acid, in ammonia.

USE: Exptl photographic emulsion (low sensitivity).

8468. Silver Oxide. Argentous oxide. Ag_2O; mol wt 231.76. Ag 93.10%, O 6.90%. May be prepd by making sol silver salt solns alkaline: Madsen, *Z. Anorg. Allgem. Chem.* **79,** 197 (1913).

Brownish-black, heavy, odorless powder. d_4^{25} 7.22. Begins to dec at about 200° and at 250-300° decompn is rapid; it also breaks up into its constituents in sunlight. Reduced by hydrogen, carbon monoxide and most metals. Moist silver oxide absorbs carbon dioxide. Sol in 40,000 parts water; freely sol in dil nitric acid, in ammonia; somewhat sol in NaOH solns; practically insol in alcohol. *Protect from light.* LD_{50} orally in rats: 2.82 g/kg, H. F. Smyth *et al., Am. Ind. Hyg. Assoc. J.* **30,** 470 (1969).

USE: As catalyst; in the purification of drinking water; in the glass industry (polishing, coloring glass yellow).

THERAP CAT (VET): Has been used as germicide and parasiticide.

8469. Silver(II) Oxide. Argentic oxide; silver peroxide; silver suboxide; Divasil. AgO; mol wt 123.88. Ag 87.08%, O 12.92%. Is actually a silver(I)-silver(III) oxide: McMillan, *Chem. Rev.* **62,** 65 (1962). Prepd by the oxidation of silver nitrate with potassium peroxydisulfate in an alk medium: Hammer, Kleinberg, *Inorg. Syn.* **4,** 12 (1953). Prepn of single crystals: *Chem. & Eng. News* **47,** 32 (Aug 11, 1969).

Charcoal-gray powder. *Avoid contact with skin, organic matter, strong ammonia and alkalies.* Malleable. Cubic or orthorhombic system when cryst. d_4^{25} 7.483. The dry solid is stable at 100° for 18 hrs, dec above 100° into silver and oxygen. Possesses semiconductor properties and is diamagnetic. *Strong oxidizing agent.* Ka 7.9×10^{-13}. Practically insol in water: 27 mg/l at 25° (decompn). Sol in alkalies and NH_4OH (with decompn and evolution of N_2). In dil acids oxygen is evolved immediately, in concd acid intensely colored solns are formed (brown in nitric acid and olive green in sulfuric acid).

USE: In the manuf of silver oxide-zinc alkali batteries. *Caution:* Highly irritating to skin, eyes, mucous membranes, respiratory tract.

8470. Silver Perchlorate. $AgClO_4$; mol wt 207.34. Ag

52.03%, Cl 17.10%, O 30.87%. Prepn from $NOClO_4$ + AgBr: Markowitz *et al., J. Inorg. Nucl. Chem.* **16,** 159 (1960).

Deliquescent crystals, dec 486°. d^{25} 2.806. Freely sol in water: 557 g/100 ml; a satd aq soln contains 84.8% w/w at 25°: Smith, Ring, *J. Am. Chem. Soc.* **59,** 1889 (1937). Much less sol in 60% perchloric acid (5.63% w/w). Forms a monohydrate stable to 43°. Sol in many organic solvents, e.g., aniline, pyridine, benzene, toluene, nitromethane, glycerol, nitrobenzene, and chlorobenzene. Solvated crystals have been obtained contg 6 mols aniline, 4 mols pyridine, 1 mol benzene, 1 mol toluene. These compds explode readily when struck: Brinkley, *J. Am. Chem. Soc.* **62,** 3524 (1940).

USE: In the explosives industry. *Caution:* Irritating to skin, mucous membranes.

8471. Silver Permanganate. Silver manganate(VII). Ag-MnO_4; mol wt 226.81. Ag 47.56%, Mn 24.22%, O 28.22%. Prepn: Büssem, Herrmann, *Z. Kristallogr.* **A74,** 459 (1930).

Violet, cryst powder; dec in light. d 4.49. Soly in water at room temps about 9 g/l. More sol in hot water; dec by alcohol. *Protect from light.*

USE: In gas masks.

8472. Silver Phosphate. Silver orthophosphate. Ag_3O_4P; mol wt 418.62. Ag 77.31%, O 15.29%, P 7.40%. Ag_3PO_4.

Yellow, odorless powder; darkens in light. d 6.37. mp 849°. Reduced by hydrogen. Soluble in 15,500 parts water, the soly in water is decreased by presence of silver nitrate; slightly sol in dil acetic acid; freely sol in dil nitric acid, in ammonia, ammonium carbonate, also in alkali cyanides and thiosulfates. *Protect from light.*

USE: In photography for collodion emulsions.

8473. Silver Picrate. *2,4,6-Trinitrophenol silver salt;* picric acid silver derivative; silver trinitrophenolate; Picragol; Picrotol. $C_6H_2AgN_3O_7$; mol wt 335.97. C 21.45%, H 0.60%, Ag 32.11%, N 12.51%, O 33.33%. $(O_2N)_3C_6H_2OAg$.

Monohydrate, yellow crystals. Sol in about 50 parts water; sparingly sol in alc; slightly sol in acetone or glycerol. Insol in chloroform, ether.

THERAP CAT: Antiprotozoal (Trichomonas).

THERAP CAT (VET): Has been used in bovine granular vaginitis.

8474. Silver Protein, Mild. Mild silver protein; silver nucleate; silver nucleinate; Mild Protargin; Argentum Vitellinatum; Argentum Vitellinum; Argyn; Solargentum; Silvol; Argyrol; Cargentos; Lunargen; Argisal. May be prepd from silver oxide and gelatin in the presence of alkali, or from silver oxide and serum albumin, or by suspending moist silver oxide in a solution of casein, and heating the mixture until no precipitate is obtained on the addition of a solution of sodium chloride, and then evaporating the mixture to dryness. Alternate procedure from peptone and silver oxide: Ger. pat. **105,866** (1897 to Bayer); Brit. pat. **18,478** (1897). It contains not less than 19% nor more than 23% Ag.

Brown, dark-brown, or almost black, odorless, lustrous scales or granules; somewhat hygroscopic. Freely sol in water. Almost insol in alcohol, chloroform, ether. *Keep well closed and protected from light. Solutions should be freshly made and protected from light.*

Mild Silver Protein is differentiated from Strong Silver Protein by the fact that no, or very little, precipitate of silver chloride is formed by addition of NaCl soln to a filtrate obtained by treating an aq soln of the silver protein with solid ammonium sulfate and filtering. Under the same conditions Strong Silver Protein gives a heavy precipitate of silver chloride. *Incompat.* Acids; most neutral and acid salts in concd solns.

THERAP CAT: Topical anti-infective.

THERAP CAT (VET): Topical antiseptic (ointment or solution) for eyes or mucous membranes. Has been used in infectious sinusitis of turkeys.

8475. Silver Protein, Strong. Strong Silver Protein; Strong Protein Silver; Protargol; Strong Protargin; Protargentum; Proganol. A compd of silver and protein, contg 7.5-8.5% silver. Prepd by adding a silver nitrate solution to a solution of peptone. The precipitate is digested with protalbumose until a solution is obtained which is evaporated in a moderate vacuum. Instead of the silver nitrate solution silver oxide may be suspended in the peptone soln. The suspension is shaken until the silver peptone has formed. The digested and dried product is identical with that obtained from silver nitrate: **Ger. pats. 105,866; 118,353; 118,-496**; H. P. Kaufmann, *Arzneimittel-Synthese* (Springer, 1953) p 588. Differentiated from Mild Silver Protein in that its silver, most of which is present in ionic form, is found in the filtrate obtained as described under Mild Silver Protein, and gives a heavy precipitate with NaCl soln: Van Deripe, Konnerth, *Am. J. Pharm.* **111**, 65 (1939). Although Strong Silver Protein contains about one-third as much silver as Mild Silver Protein, it is more irritating and stands midway between silver nitrate and Mild Silver Protein in both germicidal and irritant action.

Pale yellowish-orange to brownish-black, odorless powder. Somewhat hygroscopic. Freely sol in water; almost insol in alcohol, ether, chloroform. *Keep well closed and protect from light. Solns should be freshly made with cold water and be dispensed in amber-colored bottles.*

THERAP CAT: Antiseptic (mucous membrane).

THERAP CAT (VET): Local antiseptic. Used in dilute solutions and in ointments for eye and mucous membranes.

8476. Silver Selenate. Ag_2O_4Se; mol wt 358.72. Ag 60.15%, O 17.84%, Se 22.01%. Ag_2SeO_4. Prepd by treating silver carbonate with selenic acid: Mitscherlich, *Pogg. Ann.* **9**, 623 (1827); *Ann. Chim. Phys.* [2] **36**, 100 (1827).

Orthorhombic crystals. d 5.72. Solubility in water (20°): 1.182 g/l.

8477. Silver Selenide. Ag_2Se; mol wt 294.72. Ag 73.21%, Se 26.79%. Occurs in nature as the mineral naumannite. Prepd by melting silver and selenium together and by passing hydrogen selenide into a soln of silver nitrate: Berzelius, cited by Mellor, *A Comprehensive Treatise on Inorganic and Theoretical Chemistry* **10**, 771 (1930); by combination of ions or elements in solns or melts: Kulifay, U.S. pat. 3,026,175 (1962 to Monsanto).

Gray, hexagonal, microscopic needles. mp 880°. d^{15} 8.216. Two forms exist; transition temp at 133°. Forms metallic silver and selenium oxide when heated in oxygen; transformed by chlorine into silver and selenium chlorides; oxidized to silver selenite by fuming nitric acid. Sol in molten silver or bismuth without chemical change. Practically insol in water.

8478. Silver Selenite. Ag_2O_3Se; mol wt 342.72. Ag 62.96%, O 14.01%, Se 23.04%. Ag_2SeO_3. Prepd by dissolving silver selenide in boiling nitric acid and cooling: Berzelius, cited by Mellor, *A Comprehensive Treatise on Inorganic and Theoretical Chemistry* **10**, 824 (1930); from a soln of silver carbonate in selenious acid: Thomsen, *Ber.* **2**, 598 (1869).

Needles. d 5.9297. Stable to mp: 530°; dec above this temp to silver, selenium oxide and oxygen. Sparingly sol in cold water; freely in hot water; sol in nitric acid.

8479. Silver Subfluoride. Disilver fluoride. Ag_2F; mol wt 234.76. Ag 91.91%, F 8.09%. Prepd by electrolytic reduction of argentous fluoride, AgF: Hettich, *Z. Anorg. Allgem. Chem.* **167**, 67 (1927); Scholder, Traulsen, *ibid.* **197**, 57 (1931). By the reaction of elemental silver with argentous fluoride at 50-90°: Güntz, *Compt. Rend.* **110**, 1337 (1890); **112**, 861 (1891); Poyer *et al., Inorg. Syn.* **5**, 18 (1957).

Hexagonal, bronze-colored crystals with green luster, d 8.57. Turns grayish-black on prolonged exposure to air. Thermal decompn to Ag and AgF begins at 100° and is complete at 200°. Good electrical conductor. On contact with water almost instant hydrolysis with precipitation of Ag powder. Apparently stable to abs ethanol; insensitive to light. When dry may be kept in a stoppered container for weeks without evidence of decompn.

8480. Silver Sulfate. Ag_2O_4S; mol wt 311.83. Ag 69.19%, O 20.52%, S 10.28%. Ag_2SO_4. Conveniently prepd from silver nitrate and sulfuric acid: Richards, Jones, *Z. Anorg. Allgem. Chem.* **55**, 72 (1907).

Crystals or cryst powder; slowly darkens on exposure to light. mp 657°; dec 1085°. d 5.45. Slowly sol in 125 parts water, 71 parts boiling water; sol in nitric acid, ammonia, concd sulfuric acid. Soly in $AgNO_3$ solns: Lietzke, Stough-

ton, *J. Inorg. Nucl. Chem.* **28**, 1877 (1966). *Protect from light.*

8481. Silver Sulfide. Argentous sulfide. Ag_2S; mol wt 247.83. Ag 87.06%, S 12.94%. Occurs in nature as the mineral **argentite**. Usually prepd from the elements: Hönigschmid, Sachtleben, *Z. Anorg. Allgem. Chem.* **195**, 207 (1931).

Grayish-black, heavy powder. The cryst structure is orthorhombic, changing to cubic when heated above 179°. d_4^{20} 7.234. mp 845°. Insol in water; sol in nitric acid, in solns of alkali cyanides.

USE: In ceramics.

8482. Silver Tetraiodomercurate(II). Mercuric silver iodide. Ag_2HgI_4; mol wt 924.05. Ag 23.35%, Hg 21.71%, I 54.94%.

Deep yellow thermochromic powder. Insol in water or dil acids; sol in solns of alkali iodides or cyanides.

USE: Thermal sensor to detect overheating in journal bearings, etc., becomes blood-red at 40-50° and yellow again on cooling.

8483. Silvex. *2-(2,4,5-Trichlorophenoxy)propionic acid;* fenoprop; 2,4,5-TC; Miller Nu Set (Hormone Spray). $C_9H_7Cl_3O_3$; mol wt 269.53. C 40.11%, H 2.62%, Cl 39.47%, O 17.81%. Selective pre- and post-emergence herbicide. Many commercial products are formulated as various esters and salts. Prepn: Toothill *et al., Ann. Appl. Biol.* **44**, 547 (1956). Toxicity: G. W. Bailey, J. W. White, *Residue Rev.* **10**, 97 (1965).

Crystals, mp 181.6°. Soly at 25° in water 0.014%; acetone 15.2%; benzene 0.16%; carbon tetrachloride 0.024%; ether 7.13%; heptane 0.017%; methanol 10.5%. LD_{50} orally in rats: 650 mg/kg (Bailey, White).

Triethanolamine salt, **silvexamine.** Sol in water.

Esters with butyl ethers of mono-, di-, or polypropylene glycols, **Kuron.** Prepn: Williams, U.S. pat. 2,749,360 (1956 to Dow).

USE: Formerly in the control of woody plants on uncropped land. *Caution:* Irritating to eyes, skin, mucous membranes.

Note: In March 1985, the E.P.A. terminated all registrations for the use of this herbicide on rice fields, orchards, sugarcane, rangeland and other noncrop sites. This follows the 1979 emergency action of the agency in banning almost all uses of the pesticide: Chem. & Eng. News **63**, 6 (Mar. 25, 1985).

8484. Silymarin-Group. Apihepar; Laragon; Legalon; Pluropen; Silarin; Silliver. Antihepatotoxic principle isolated from the fruit of *Silybum marianum* (L.) Gaertn. (*Carduus marianus* L.). Comprised mainly of three isomers: silybin (formerly called silymarin), **silidianin** and **silicristin**. Characterized as a new class of substances, flavonolignans, almost certainly produced in the plant by a radical coupling of a flavonoid and coniferyl alcohol: Hänsel *et al., Deut. Apotheker Ztg.* **108**, 1988 (1968). Isoln and chemistry: Wagner *et al., Arzneimittel-Forsch.* **18**, 688 (1968); **24**, 466 (1974). Structural work: Janiak, Hänsel, *Planta Med.* **8**, 71 (1960); Hänsel, Schöpflin, *Tetrahedron Letters* **1967**, 3645; Pelter, Hänsel, *ibid.* **1968**, 2911. Pharmacology and toxicology: Hahn *et al., Arzneimittel-Forsch.* **18**, 698 (1968); H. M. Rauen, H. Schriewer, *ibid.* **23**, 148-170 (1973); Halbach, Trost, *ibid.* **24**, 866 (1974).

Silybin

Silybin, $C_{25}H_{22}O_{10}$, *2-[2,3-dihydro-3-(4-hydroxy-3-meth-oxyphenyl)-2-(hydroxymethyl)-1,4-benzodioxin-6-yl]-2,3-dihydro-3,5,7-trihydroxy-4H-1-benzopyran-4-one, silybum substance E₆, silymarin I, Dura-Silymarin, Silepar, Silirex.* The major component of the group. Revised structure and synthetic studies: R. Hänsel *et al., Chem. Commun.* **1972**, 195; *eidem, Ber.* **110**, 3664 (1977). Total synthesis: L. Merlini *et al., Chem. Commun.* **1979**, 695; *eidem, J. Chem. Soc. Perkin Trans. I* **1980**, 775. Monohydrate, crystals from acetone + petr ether, mp 167°, dec 180°; from methanol + water, mp 180°. (Anhydr subst, mp 158°, dec 180°.) $[\alpha]_D^{20}$ +11° (c = 0.25 in acetone + alcohol). uv max (methanol): 288 nm (log ε 4.33). Sol in acetone, ethyl acetate, methanol, ethanol; sparingly sol in $CHCl_3$. Practically insol in water.
THERAP CAT: Treatment of liver dysfunction.

8485. Simazine. *6-Chloro-N,N'-diethyl-1,3,5-triazine-2,4-diamine; 2-chloro-4,6-bis(ethylamino)-s-triazine;* 2,4-bis(ethylamino)-6-chloro-s-triazine; G-27692; Gesatop; Primatol S; Princep; Simanex. $C_7H_{12}ClN_5$; mol wt 201.67. C 41.69%, H 6.00%, Cl 17.58%, N 34.73%. Selective pre-emergence herbicide. Prepn: Hofmann, *Ber.* **18**, 2776 (1885); Pearlman, Banks, *J. Am. Chem. Soc.* **70**, 3726 (1948); Thurston *et al., ibid.* **73**, 2981 (1951). As herbicide: Gysin, Knüsli, U.S. pat. **2,891,855** (1959 to Geigy).

Crystals from ethanol or methyl Cellosolve, mp 226-227°. Practically insol in water. Slightly sol in dioxane, ethyl Cellosolve. LD_{50} orally in rats: 5000 mg/kg, G. W. Bailey, J. L. White, *Residue Rev.* **10**, 97 (1965).
USE: Herbicide.

8486. Simethicone. Dimethicone; dimethyl polysiloxane; dimeticon; Aeropax; Antifoam A; Baros; Bicolon; Ceolat; Delesan; Endo-Paractol; Infacol; Lefax; Meteorex; MS Antifoam M; Mylicon; Mylocon; Ovol; Phasil; Phazyme; Polysilon; Sab Simplex; Silain; Siligaz. Mixture of dimethyl polysiloxanes and silica gel suitably purified for pharmaceutical use. Prepn of homologous liquid methyl siloxane polymers: Hyde, U.S. pat. **2,441,098** (1948 to Corning Glass). Physical properties: *Drug Standards* **25**, 163 (1957).

n = 200 - 350

Water-white, viscous, oil-like liquid. d 0.965-0.970. n_D^{25} about 1.404. Viscosity at 25° about 60,000 cs. Immiscible with water, alcohol. Miscible with chloroform, ether.
USE: Ointment base ingredient, topical drug vehicle, skin protectant.
THERAP CAT: Antiflatulent.
THERAP CAT (VET): Antibloating agent.

8487. Simetride. *1,4-Bis[(2-methoxy-4-propylphenoxy)-acetyl]piperazine;* AP2; Kyorin AP2. $C_{28}H_{38}N_2O_6$; mol wt 498.60. C 67.44%, H 7.68%, N 5.62%, O 19.25%. Prepn: Fr. pat. M2936 (1964 to Kyorin Pharmaceutical Co., Ltd.);

T. Irikura *et al.*, U.S. pat. **3,184,463** (1965). Pharmacological studies: T. Irikura *et al., Yakugaku Zasshi* **83**, 785 (1963), *C.A.* **60**, 1008b (1964). Analgesic activity: K. Nishino *et al., Yakugaku Zasshi* **85**, 715 (1965), *C.A.* **63**, 16957b (1965).

Crystals, mp 128-130° (Fr. pat.); also reported as mp 138.5-139° (Irikura, 1963). Practically insol in water, ethanol. Sol in $CHCl_3$. LD_{50} i.p. in mice: 15000 mg/kg (Irikura).
THERAP CAT: Analgesic.

8488. Simetryne®. *N,N'-Diethyl-6-(methylthio)-1,3,5-triazine-2,4-diamine; 2,4-bis(ethylamino)-6-(methylthio)-s-triazine;* 2,4-bis(ethylamino)-6-(methylthio)-1,3,5-triazine; 2-methylthio-4,6-bis(monoethylamino)-s-triazine; simetryn; G 32911; Gy-bon. $C_8H_{15}N_5S$; mol wt 213.32. C 45.05%, H 7.09%, N 32.83%, S 15.03%. Prepn: Gysin, Knüsli, Swiss pat. **337,019** (1959 to Geigy), *C.A.* **57**, 14226c (1962). Crystal structure: A. J. Graham *et al., Cryst. Struct. Commun.* **7**, 227 (1978).

Crystals, mp 82-83°. LD_{50} orally in rats: 1830 mg/kg, *World Rev. Pest Contr.* **1**, 36 (1965).
USE: Herbicide.

8489. Simfibrate. *2-(4-Chlorophenoxy)-2-methylpropanoic acid 1,3-propanediyl ester; 2-(p-chlorophenoxy)-2-methylpropionic acid trimethylene ester;* 1,3-propanediol bis[α-(p-chlorophenoxy)isobutyrate]; 1,3-propanediol bis[2-(4-chlorophenoxy)-2-methylpropionate]; sinfibrate; CLY-503; Cholesolvin; Liposolvin. $C_{23}H_{26}Cl_2O_6$; mol wt 469.36. C 58.86%, H 5.58%, Cl 15.11%, O 20.45%. Prepn: Neth. Appl. **6,600,044** corresp 'to Nakanishi *et al.*, U.S. pat. **3,494,957** (1966, 1970 to Yoshitomi). Activity and toxicology studies: Nakanishi *et al., J. Pharm. Soc. Japan* **90**, 267, 921, 926 (1970). Activity on postheparin lipoprotein lipase: L. Cavallini *et al., Clin. Ter.* **92**, 25 (1980).

Crystals, mp 51-53°. $bp_{0.03}$ 197-200°, $bp_{0.15}$ 220-230°. LD_{50} in mice, rats (g/kg): 3.3-3.5; 7.3-8.0 orally (Nakanishi, *J. Pharm. Soc. Japan).*
THERAP CAT: Anticholesteremic.

8490. Simplesse. Protein-based fat substitute. Prepn from microparticulated, denatured dairy whey protein: N. S. Singer *et al.*, Eur. pat. Appl. **250,623**; *eidem*, U.S. pat. **4,734,287** (both 1988 to John LaBatt Ltd.). Brief description: D. D. Duxbury, *Food Process.* **49**, 21 (1988).
USE: Fat substitute for use in foods.

8491. Simvastatin. *[1S-[1α,3α,7β,8β(2S*,4S*),8aβ]]-2,2-Dimethylbutanoic acid 1,2,3,7,8,8a-hexahydro-3,7-dimethyl-8-[2-(tetrahydro-4-hydroxy-6-oxo-2H-pyran-2-yl)-ethyl]-1-naphthalenyl ester;* 2,2-dimethylbutyric acid 8-ester with (4R,6R)-6-[2-[(1S,2S,6R,8S,8aR)-1,2,6,7,8,8a-hexa-hydro-8-hydroxy-2,6-dimethyl-1-naphthyl]ethyl]tetrahy-dro-4-hydroxy-2H-pyran-2-one; synvinolin; MK-733; Zocor; Zocord. $C_{25}H_{38}O_5$; mol wt 418.57. C 71.74%, H 9.15%, O 19.11%. Competitive inhibitor of HMG-CoA reductase; synthetic analog of lovastatin, $q.v.$ Prepn: A. K. Willard *et al., Eur.* pat. **Appl. 33,538;** W. F. Hoffman *et al.,* **U.S.** pat. **4,444,784** (1981, 1984 both to Merck & Co.); W. F. Hoffman *et al., J. Med. Chem.* **29,** 849 (1986). Clinical evaluation in hypercholesterolemia: M. J. Mol *et al., Lancet* **2,** 936 (1986); L. A. Simons *et al., Med. J. Aust.* **147,** 65 (1987).

Crystals from *n*-butyl chloride + hexane, mp 135-138°.
THERAP CAT: Antihyperlipidemic.

8492. Sinalbin. Sinapine glucosinalbate. $C_{30}H_{42}N_2O_{15}S_2$; mol wt 734.82. C 49.03%, H 5.76%, N 3.81%, O 32.66%, S 8.73%. Mustard oil glucoside from white or yellow mustard *(Sinapis alba* L., *Cruciferae):* Gadamer, *Arch. Pharm.* **235,** 38, 83, 570 (1897); Viehoever, Nelson, *J. Assoc. Offic. Agr. Chem.* **21,** 488 (1938); *Am. J. Pharm.* **110,** 411 (1938). Structure: Schulz, Gmelin, *Z. Naturforsch.* **7b,** 500 (1952); Ettlinger, Lundeen, *J. Am. Chem. Soc.* **78,** 4172 (1956).

Needles from 95% ethanol, mp 100-102°. (Also reported as 83-84° for the hydrated, 138-140° for the anhydr.) $[\alpha]_D^{20}$ −8.76° (c = 0.29). Soluble in hot alcohol, cold water.

8493. Sinapine. 2-[[3-(4-Hydroxy-3,5-dimethoxyphen-yl)-1-oxo-2-propenyl]oxy]-N,N,N-trimethylethanaminium; sinapic acid choline ester. $[C_{16}H_{24}NO_5]^+$; mol wt 310.38. From black mustard seeds, the seeds of *Brassica nigra* Koch, *Cruciferae.* Extraction procedure: Remsen, Coale, *Am. Chem. J.* **6,** 52 (1884). Synthesis: Späth, *Monatsh.* **41,** 271 (1920).

Chloride, $(C_{16}H_{24}NO_5)Cl$, does not crystallize, very sol in water, alcohol.
Bromide trihydrate, $(C_{16}H_{24}NO_5)Br.3H_2O$, needles, mp 92°; mp 107-115° when anhydr, very sol in water.
Iodide trihydrate, $(C_{16}H_{24}NO_5)I.3H_2O$, crystals, becomes anhydr over H_2SO_4, mp 179° (anhydr). Sparingly sol in water.
Acid sulfate trihydrate, $(C_{16}H_{24}NO_5)HSO_4.3H_2O$, tablets, dec 127°; dec 187° when anhydr. Freely sol in water, hot alcohol; very sparingly sol in ether.
Neutral sulfate pentahydrate, $(C_{16}H_{24}NO_5)_2SO_4.5H_2O$, tablets from hot alcohol, mp 193° (anhydr). Not very stable; very sol in water, sparingly in alcohol.

8494. Sincalide. *1-De(5-oxo-*L*-proline)-2-de-*L*-gluta-mine-5-*L*-methioninecaerulein;* cholecystokinin C-terminal octapeptide; CCK C-terminal octapeptide; pancreozymin

C-terminal octapeptide; SQ 19844; Kinevac. $C_{49}H_{62}N_{10}$-$O_{16}S_3$; mol wt 1143.29. C 51.48%, H 5.47%, N 12.25%, O 22.39%, S 8.41%. The C-terminal octapeptide of cholecys-tokinin-pancreozymin (*see* cholecystokinin). Structurally similar to caerulein, *q.v.,* it evokes a variety of biological responses similar to CCK-PZ, including smooth-muscle contraction of the gall bladder and small intestine, relaxa-tion of the choledochoduodenal junction, protein secretion by the pancreas and acid secretion by the stomach. It is more active than cholecystokinin on a weight or molar basis. Initial isoln: V. Mutt, J. E. Jorpes, *Eur. J. Biochem.* **6,** 156 (1968). Prepn: M. A. Ondetti *et al., Ger.* pat. **1,922,185** corresp to **U.S.** pat. **3,723,406** (1969, 1970, both to Squibb); *eidem, J. Am. Chem. Soc.* **92,** 195 (1970). Mechanical and metabolic effects: K. E. Andersson *et al., Acta Physiol. Scand.* **96,** 495 (1976). *In vitro* study: M. Deschodt-Lanck-man *et al., Gastroenterology* **68,** 819 (1975). Use in clinical radiology: W. M. Thompson, J. R. Amberg, *Gastrointest. Radiol.* **3,** 191 (1978). Effect on food intake: M. A. Della-Fera, C. A. Baile, *Science* **206,** 471 (1979). Review of syn-thesis: L. Balaspiri, *Acta Physiol. Acad. Sci. Hung.* **47,** 299 (1976). Review of pharmacology: S. L. Engel *et al.,* in *Pharmacological and Biochemical Properties of Drug Sub-stances* vol. 2, M. E. Goldberg, Ed. (Am. Pharm. Assoc., Washington, DC, 1979) pp 516-526.

Solid. $[\alpha]_D^{23}$ −18.4° (c = 0.7 in 1N NH_3). uv max (0.1N NaOH): 280, 288 nm (ϵ 4850, 4230).
THERAP CAT: Choleretic.

8495. Sinigrin. *1-Thio-β-*D*-glucopyranose 1-[N-(sulfo-oxy)-3-butenimidate] monopotassium salt;* sinigroside; myro-nate potassium; potassium myronate; allyl glucosinolate. $C_{10}H_{16}KNO_9S_2$; mol wt 397.48. C 30.22%, H 4.06%, K 9.84%, N 3.52%, O 36.23%, S 16.13%. A β-glucopyranoside isolated from black mustard seeds and from the horseradish root *Alliaria officinalis* Andrz., *Cruciferae:* A. Bussy, *J. Pharm. Chim.* **26,** 39 (1840); Gadamer, *Arch. Pharm.* **235,** 44 (1897); Stoll, Seebeck, *Helv. Chim. Acta* **31,** 1432 (1948). Structure: Ettlinger, Lundeen, *J. Am. Chem. Soc.* **78,** 4173 (1956); **79,** 1764 (1957). Hydrolysis: Ettlinger *et al., Proc. Nat. Acad. Sci. USA* **47,** 1875 (1961). Crystal structure: Waser, Watson, *Nature* **198,** 1297 (1963). Synthesis: Benn, Ettlinger, *Chem. Commun.* **1965,** 445; A. Kjaer, *Acta Chem. Scand.* **22,** 3324 (1968).

Monohydrate, crystals, mp 127-129°. When anhydr mp 179°. $[\alpha]_D^{18}$ −16.4°. Freely sol in water, hot alcohol; insol in benzene, chloroform, ether.
Tetraacetate, $C_{18}H_{24}KNO_{13}S_2$, crystals, mp 193-195°. $[\alpha]_D^{26}$ −16° (c = 0.14).

8496. Sinomenine. *(9α,13α,14α)-7,8-Didehydro-4-hydroxy-3,7-dimethoxy-17-methylmorphinan-6-one;* cucol-ine; coculine; kukoline. $C_{19}H_{23}NO_4$; mol wt 329.38. C 69.28%, H 7.04%, N 4.25%, O 19.43%. An optical isomer of methoxythebainone. Configuration at the asymmetric cen-ters, C_9, C_{13}, and C_{14}, is the mirror image of those in mor-phine. From root of *Sinomenium acutum* (Thunb.) Rehd. & Wils. (*Cocculus diversifolius* DC.), *Menispermaceae:* Goto, *J. Chem. Soc. Japan* **44,** 795 (1923), *C.A.* **18,** 2710 (1924); Ohta, *Kitasato Arch. Exp. Med. (Tokyo)* **6,** 259 (1925), *C.A.* **21,** 3233^2 (1927). Structure: Kondo, Ochiai, *Ann.* **470,** 224 (1929); Sasaki, *Yakugaku Zasshi* **80,** 270 (1960), *C.A.* **54,** 11702g (1960). Biogenesis: Cohen, *Chem. & Ind. (London)* **1956,** 1391. Pharmacology: S. Fu *et al., Yao Hsueh Hsueh Pao* **10,** 673 (1963), *C.A.* **60,** 8525h (1964). Reviews: Holmes in R. H. F. Manske, H. L. Holmes, *The Alkaloids* vol. II

(Academic Press, New York, 1952) pp 219-260; K. W. Bentley, *The Chemistry of the Morphine Alkaloids* (Oxford, 1954).

Clusters of needles from benzene, mp 161°. After melting once, the mp is raised to 182°. $[\alpha]_D^{26}$ −71° (c = 2.1 in alc). Sol in alc, acetone, chloroform, dil alkali; slightly sol in water, ether, benzene. LD_{50} orally in mice: 580 mg/kg (Fu).

Hydrochloride dihydrate, $C_{19}H_{24}ClNO_4 \cdot 2H_2O$, long prisms from water, dec 231° when anhydr. $[\alpha]_D^{17}$ −82° (c = 4.4). Sol in about 1.5 parts water.

8497. Sisal. *Agave sisalana* Perrine, *Amaryllidaceae;* henequem; yaxci; potosina. One of the plants native to the American desert. Process of treating juice and extracting steroidal material and waxes: Spray, U.S. pat. 3,019,193 (1962). *Habit.* Principally the Mexican tableland. Abundant in the Mexican states Yucatan and Chiapas. *Constit.* Hemp, sapogenins such as hecogenin and tigogenin. *Review: Chem. Week* (April 20, 1957) pp 41-48.

USE: Manuf sisal ropes, steroid intermediates. *Caution:* The dust is irritating to respiratory tract.

8498. Sisomicin. *O-3-Deoxy-4-C-methyl-3-(methylamino)-β-*L*-arabinopyranosyl-(1 → 6)-O-[2,6-diamino-2,3,4,6-tetradeoxy-α-*D*-glycero-hex-4-enopyranosyl-(1 → 4)]-2-deoxy-*D*-streptamine; (2S-cis)-4-O-[3-amino-6-(aminomethyl)-3,4-dihydro-2H-pyran-2-yl]-2-deoxy-6-O-[3-deoxy-4-C-methyl-3-(methylamino)-β-*L*-arabinopyranosyl]-*D*-streptamine;* rickamicin; antibiotic 6640; Sch 13475; Siseptin; Sisolline. $C_{19}H_{37}N_5O_7$; mol wt 447.55. C 50.99%, H 8.33%, N 15.65%, O 25.03%. Gentamicin-like aminoglycoside antibiotic produced by *Micromonospora inyoesis* (NRRL 3292): M. J. Weinstein *et al.* Ger. pat. 1,932,309 corresp to U.S. pat. 3,832,286 (1970, 1974 both to Schering); *eidem, J. Antibiot.* **23,** 551 (1970). Isoln: G. H. Wagman *et al., ibid.* 555. Chemical and physical studies: Cooper *et al., Chem. Commun.* **1971,** 285, 924; Kugelman *et al., J. Antibiot.* **26,** 394 (1973). Structure consists of 2-deoxystreptamine, *q.v.,* linked to two saccharide units, *garosamine* and *sisosamine:* Keiman *et al., J. Org. Chem.* **39,** 1451 (1974). Synthesis: D. H. Davies *et al., J. Med. Chem.* **21,** 189 (1978). Synthesis of sisosamine derivs: Cleophax *et al., Chem. Commun.* **1975,** 11. Has broad spectrum antibiotic activity: Waitz *et al., J. Antibiot.* **23,** 559 (1970). Biotransformation of sisomicin to gentamicin: B. K. Lee *et al., Antimicrob. Ag. Chemother.* **12,** 335 (1977). Clinical studies: R. Lorber *et al., Clin. Ther.* **4,** 263 (1981); J. P. Sculier *et al., J. Antimicrob. Chemother.* **9,** 63 (1982). *Review:* P. Noone, *Drugs* **27,** 548-578 (1984).

Monohydrate, needles from ethanol, mp 198-201°. $[\alpha]_D^{26}$ +189° (c = 0.3).

Penta-*N*-acetyl deriv, $C_{29}H_{47}N_5O_{12}$, amorphous white powder from methanol-ethyl ether, mp 188-198° (dec). $[\alpha]_D^{26}$ +200° (c = 0.3).

Sulfate, $C_{38}H_{84}N_{10}O_{34}S_5$, *Baymicin, Extramycin, Mensiso,*

Pathomycin, Sisobiotic, Sisomin. LD_{50} in mice (mg/kg): 34 i.v., 221 i.p., 288 s.c. (Weinstein).

THERAP CAT: Antibacterial.

8499. α₁-Sitosterol. *(3β,4α,5α,24Z)-4-Methylstigmasta-7,24(28)-dien-3-ol;* citrostadienol. $C_{30}H_{50}O$; mol wt 426.70. C 84.44%, H 11.81%, O 3.75%. Occurs in plants. Isoln from wheat germ oil: Anderson *et al., J. Am. Chem. Soc.* **48,** 2987 (1926); Wallis, Fernholz, *ibid.* **58,** 2446 (1936). Structure: Bernstein, Wallis, *ibid.* **61,** 2308 (1939); Mazur, Sondheimer, *ibid.* **80,** 6293, 6296 (1958). Identity with citrostadienol: Schreiber, Osske, *Experientia* **19,** 69 (1963); *eidem, Tetrahedron* **20,** 2575 (1964). Stereochemistry: C. Brooks *et al., Steroids* **20,** 487 (1972).

Needles from alcohol, mp 164-166°. $[\alpha]_D^{28}$ −1.7° (c = 2 in chloroform). Is pptd by digitonin.

8500. β-Sitosterol. *(3β)-Stigmast-5-en-3-ol;* 22:23-dihydrostigmasterol; α-dihydrofucosterol; Δ^5-stigmasten-3β-ol; 24β-ethyl-Δ^5-cholesten-3β-ol; α-phytosterol; cinchol; cupreol; rhamnol; quebrachol; sitosterin; Flemun; Harzol; Prostasal; Sito-Lande. $C_{29}H_{50}O$; mol wt 414.69. C 83.99%, H 12.15%, O 3.86%. Common sterol in plants. Isoln from wheat germ oil, corn oil: Anderson *et al., J. Am. Chem. Soc.* **48,** 2987 (1926); from rye germ oil: Gloyer, Schuette, *ibid.* **61,** 1901 (1939); from cottonseed oil: Wallis, Chakravorty, *J. Org. Chem.* **2,** 335 (1937); from tall oil: Sandqvist, Bengtsson, *Ber.* **64,** 2167 (1931). Also occurs in soy and calabar beans, in rice embryos: Tanaka, *J. Biochem.* **17,** 483 (1933); in cascara, cinchona bark and cinchona wax: Dirscherl, *Z. Physiol. Chem.* **235,** 1 (1935). Prepn from tall oil: G. I. Fujimoto, A. E. Jacobson, *J. Org. Chem.* **29,** 3377 (1964). Identity with 22:23-dihydrostigmasterol: W. Dirscherl, H. Nahm, *Ann.* **555,** 57 (1944). Identity with cinchol: *eidem, Ann.* **558,** 231 (1947). Structure: Bernstein, Wallis, *J. Org. Chem.* **2,** 341 (1937); Bergmann, Low, *ibid.* **12,** 67 (1947); Shoppee, *J. Chem. Soc.* **1948,** 1032. Stereospecific synthesis: W. Sucrow, M. Slopianka, *Ber.* **108,** 3721 (1975). Clinical efficacy in treatment of type II hyperlipoproteinemia: R. S. Lees, A. M. Lees, in *Lipoprotein Metabolism,* H. Greten, Ed. (Springer-Verlag, New York, 1976) pp 119-124; P. Oster *et al., ibid.* pp 125-130. Studies on inhibition of cholesterol absorption: S. M. Grundy, H. Y. I. Mok, *ibid.* pp 112-118; I. Ikeda, M. Sugano, *Biochim. Biophys. Acta* **732,** 651 (1983). Inhibition of induced carcinogenesis: N. D. Nigro *et al., J. Natl. Cancer Inst.* **69,** 103 (1982). Clinical trial in treatment of prostatic adenoma: H.-P. Szutrely, *Med. Klin.* **77,** 520 (1982). Book: *Monographs on Atherosclerosis* Vol. 10, T. B. Clarkson *et al.,* Eds. entitled "Sitosterol" by O. J. Pollak, D. Kritchevsky (Karger, Basel, 1981) 219 pp.

Plates from alcohol, mp 140°. $[\alpha]_D^{25}$ −37° (c = 2 in chloroform).

Acetate, $C_{31}H_{52}O_2$, mp 127-128°. $[\alpha]_D^{25}$ −41° (c = 2 in chloroform).

THERAP CAT: Anticholesteremic. Treatment of prostatic adenoma.

8501. γ-Sitosterol. *(3β,24S)-Stigmast-5-en-3-ol;* clionasterol. $C_{29}H_{50}O$; mol wt 414.69. C 83.99%, H 12.15%, O 3.86%. Principal sterol of soybean oil: Bonstedt, *Z. Physiol. Chem.* **176**, 269 (1928). Isoln from corn oil: Anderson, Shriner, *J. Am. Chem. Soc.* **48**, 2976 (1926); from venom of Chinese toad, *Bufo gargarizans:* Rukstuhl, Meyer, *Helv. Chim. Acta* **40**, 1270 (1957). Differs from β-sitosterol in spatial configuration at C-24: Bergmann, Low, *J. Org. Chem.* **12**, 67 (1947). Structure: Pavanaram, *Helv. Chim. Acta* **46**, 1377 (1963). Stereospecific synthesis: W. Sucrow, M. Slopianka, *Ber.* **108**, 3721 (1975).

Monohydrate, plates from alcohol. When dry, mp 147-148°. $[\alpha]_D^{20}$ −43° (c = 1.9 in chloroform). Table of mp's and $[\alpha]_D$'s: Dirscherl, *Z. Physiol. Chem.* **257**, 242 (1939). Acetate, $C_{31}H_{52}O_2$, mp 143-144°. $[\alpha]_D^{20}$ −45.3° (c = 1.9 in chloroform).

8502. Sizofiran. Schizophyllan; SPG; Sonifilan. Immunostimulant polysaccharide produced by the fungus *Schizophyllum commune* Fries. Polymer of repeating tetrasaccharide units composed of three β-(1→3)-linked D-glucopyranose residues, to one of which is attached a single β-(1→6)-linked D-glucopyranosyl side chain. Forms triple-stranded helix of mol wt ~450,000 in neutral aqueous solution. Prepn by fermentation: S. Kikumoto et al., *Japan.* pat. **67 12,000** (1967 to Taito), *C.A.* **67**, 107386r (1967). Synthesis of the 3 possible tetrasaccharide units: K. Takeo, S. Tei, *Carbohydrate Res.* **145**, 293 (1986). Host-mediated antitumor activity: N. Komatsu et al., *Gann* **60**, 137 (1969). Pharmacology, toxicology and antitumor activity in mice: T. Matsuo et al., *Arzneimittel-Forsch.* **32**, 647 (1982). Macrophage stimulant activity *in vitro:* I. Sugawara et al., *Cancer Immunol. Immunother.* **16**, 137 (1984). Clinical trial in cervical cancer: K. Okamura et al., *Cancer* **58**, 865 (1986).

White, amorphous powder. Sol in water, forming highly viscous soln.

THERAP CAT: Antineoplastic.

8503. Skatole. *3-Methyl-1H-indole.* C_9H_9N; mol wt 131.17. C 82.40%, H 6.92%, N 10.68%. Constituent of feces, beetroot, nectandra wood, and coal tar. Obtained by fusing egg albumin with KOH.

White to brownish scales; fecal odor; mp 95°; bp 265-266°. Sol in hot water, alc, benzene, chloroform, ether. With potassium ferrocyanide and sulfuric acid it gives a violet color. MLD s.c. in frogs: 1.0 g/kg, *Handbook of Toxicology,* W. S. Spector, Ed. (Saunders, Philadelphia, 1956) p 268.

8504. Skellysolve. A brand of petroleum solvents.
Skellysolve—A: Essentially *n*-pentane, bp 28-30°.
Skellysolve—B: Essentially *n*-hexane, bp 60-68°.
Skellysolve—C: Essentially *n*-heptane, bp 90-100°.
Skellysolve—D: Mixed heptanes, bp 77-115°.

Skellysolve—E: Mixed octanes, bp 100-140°.
Skellysolve—F: Petr ether, bp 30-60°.
Skellysolve—G: Petr ether, bp 40-75°.
Skellysolve—H: Hexanes-heptanes, bp 69-96°.
Skellysolve—L: Essentially octanes, bp 95-127°.

8505. Skimmianine. *4,7,8-Trimethoxyfuro[2,3-b]quinoline;* β-fagarine; 7,8-dimethoxydictamnine. $C_{14}H_{13}NO_4$; mol wt 259.25. C 64.86%, H 5.05%, N 5.40%, O 24.69%. In *Skimmia japonica* Thunb., *Fagara* spp., *Glycosmis pentaphylla* Corr., *Ruta graveolens* L., *Rutaceae:* Honda, *Arch. Exp. Pathol. Pharmakol.* **52**, 83 (1904); Paris, Moyse-Mignon, *Ann. Pharm. Franc.* **5**, 410 (1947), *C.A.* **42**, 3909h (1948); Chatterjee, Majumdar, *J. Am. Chem. Soc.* **76**, 2459 (1954); Schneider, *Arzneimittel-Forsch.* **14**, 435 (1964). Identity with β-fagarine: Deulofeu et al., *J. Am. Chem. Soc.* **64**, 2326 (1942). Structure: Asahina, Inubuse, *Ber.* **63**, 2052 (1930).

Pyramids, octahedral rods from alcohol. mp 178°. uv max (ethanol): 212, 251, 321, 331 nm (log ε 4.10, 4.91, 3.89, 3.91). Neutral to litmus. Sol in alcohol, chloroform; slightly in ether, amyl alcohol and carbon disulfide; practically insol in water and petr ether.

Picrate, $C_{14}H_{13}NO_4 \cdot C_6H_3N_3O_7$, yellow prisms from alcohol, dec 195-197°.

8506. Skimmin. *7-(β-D-Glucopyranosyloxy)-2H-1-benzopyran-2-one;* *7-(glucosyloxy)coumarin;* 7-hydroxycoumarin-7-glucoside; umbelliferone glucoside. $C_{15}H_{16}O_8$; mol wt 324.28. C 55.55%, H 4.97%, O 39.47%. In bark and wood of *Skimmia japonica* Thunb., *Rutaceae.* Isoln, structure and synthesis: Späth, Neufeld, *Rec. Trav. Chim.* **57**, 535 (1938).

Crystals with one H_2O from water. After drying at 100° and 10 mm: mp 219-221° (evac tube). $[\alpha]_D^{18}$ −80° (c = 10 in pyridine). uv max (ethanol): 320 nm. Moderately sol in hot water, alcohol; practically insol in ether, chloroform.

8507. Skunk Oil. A fatty oil from the winter fat of the skunk [*Mephitis mephitis* Schreb. (North America) or *Putorius putorius* L. (Europe and Asia), Fam. *Mustelidae*]: Weil, *Am. Druggist* **59**, 8 (1912), *C.A.* **6**, 1655 (1912); Lobachev, *Sovet. Med.* **7**, no. 10, 21 (1943), *C.A.* **39**, 4193 (1945). Contains little, if any, of the odoriferous principles of the skunk (*n*-butyl mercaptan and dicrotyl sulfide): Stevens, *J. Am. Chem. Soc.* **67**, 407 (1945).

Neutral or faintly acid oil, of only slightly disagreeable odor. Solidifies around 21°. On standing, a deposit of stearic acid may separate which redissolves on heating. At 75° a pungent odor is evolved. d_{15}^{15} 0.931; iodine no. 83.48 (Hanus). Sapon no. 89.32 (Weil); 201 (Lobachev). Insol fatty acids mp 30°; their Hehner no. 93.45. The oil may be used directly or as an emulsion; the emulsion should be protected from the bacterial deterioration by an added antiseptic.

8508. Smilagenin. *(25R)-Spirostan-3β-ol;* isosarsasapogenin. $C_{27}H_{44}O_3$; mol wt 416.62. C 77.83%, H 10.65%, O 11.52%. Steroid sapogenin from *Smilax ornata* Hooker, *Liliaceae* (Sarsaparilla): Askew et al., *J. Chem. Soc.* **1936**, 1399; from *Agave lecheguilla* Tou., *Amaryllidaceae:* Nord, U.S. pat. **2,897,192** (1959). C_{25} isomer of sarsasapogenin: Scheer et al., *J. Am. Chem. Soc.* **77**, 641 (1955). Obtained by acid isomerization of sarsasapogenin: Marker, Rohrmann, *ibid.* **61**, 846 (1939); Wall et al., *ibid.* **77**, 3086 (1955). For structure and stereochemistry *see* Sarsasapogenin.

Silky needles from acetone, mp 185°. $[\alpha]_D^{25}$ −69° (c = 0.5

in chloroform). Absorption spectrum: Smith, Eddy, *Anal. Chem.* **31**, 1539 (1959).

Acetate, $C_{29}H_{46}O_4$, needles from methanol, mp 152°.

USE: In making compds of the pregnane series.

8509. Soap. Any salt of a fatty acid, usually made by saponification of a vegetable oil with caustic soda. Hard soap consists largely of sodium oleate (low titer) or of sodium palmitate or stearate (high titer); soft soap, of potassium and sodium salts of coconut and/or palm oil fatty acids. Rosin acid salts have detergent properties and are often incorporated in laundry soaps. Shaving soaps contain glycerol and gum to prevent rapid drying. Medicated soaps contain antiseptics such as phenols or mercury salts. Soaps also may be prepd from the fatty acid and an amine, *e.g.*, triethanolamine. Nonalkali-metal salts of the fatty acids are water insol and as such have certain uses.

Hard soap (olive-oil castile) is a white or yellowish white powder or bar, sol in water, alc. The aq soln is alkaline due to hydrolysis; the alc soln is only slightly alkaline. Transparent hard soap is obtained by adding additional glycerol during saponification. Floating soaps are made by adding air after saponification during the so-called "crutching" process. Soft soap (green soap) is a yellowish-white to brownish-yellow (or green) soft mass, more sol than hard soap.

USE: Hard soap—detergent; in soln as a vehicle for liniments; as ingredient of pills contg resinous drugs like aloe, etc. Soft soap—detergent; extern. as vehicle for active medicaments applied in ointment form or liniment. Insoluble soaps—fungicides (Cu), disinfectants (Pb, Cu, Hg), face powders and ointments (Zn), waxes and polishes (Al), greases (Al, Ca), and surface coatings (Ca).

THERAP CAT: Topical anti-infective. Antidote (mineral acid and heavy metal poisoning).

THERAP CAT (VET): Topically as detergent (cleansing agent) and mild antiseptic. In liniments as counterirritant. Has been used internally as laxative, antacid and as an antidote for mineral acids and heavy metals.

8510. Sobrerol. *5-Hydroxy-α,α,4-trimethyl-3-cyclohexene-1-methanol; p-menth-6-ene-2,8-diol;* 6,8-carvomenthenediol; 1-*p*-menthene-6,8-diol; pinol hydrate. $C_{10}H_{18}O_2$; mol wt 170.25. C 70.55%, H 10.65%, O 18.80%. An oxidation product of oil of turpentine formed by the autooxidation of α-pinene, *q.v.*, in the presence of water and especially in sunlight. First obtained by Margueron in 1797. Confused with terpin hydrate, *q.v.*, until fully characterized by Sobrero in 1851. The naturally occurring form is the *trans*-form. For a review of early structure and prepn work, *see* J. L. Simonsen, *The Terpenes* vol. II (University Press, Cambridge, 2nd ed., 1949) pp 137-138. Summary of general methods of prepn: Lombard, Heywang, *Bull. Soc. Chim. France* **1954**, 1210. Prepn of *dl-trans*-form: Ward, *J. Am. Chem. Soc.* **60**, 325 (1938); Schenck *et al.*, *Ann.* **584**, 177 (1953); Moore *et al.*, *J. Am. Chem. Soc.* **78**, 1173 (1956); Brook, Wright, *J. Org. Chem.* **22**, 1314 (1957); of *l-trans*-form: Klein, U.S. pat. **2,815,378** (1958 to Glidden). Prepn of *cis*- and *trans*-forms and analysis of stereochemistry: Blumann, Wood, *J. Chem. Soc.* **1952**, 4420; H. Schmidt, *Ber.* **86**, 1437 (1953). Pharmacology and chemistry of *dl-trans*-form: Brit. pat. **1,176,817**; C. Corvi-Mora, U.S. pat. **3,592,-908** (1970, 1971 to Camillo Corvi); V. Dalla Valle, *Boll. Chim. Farm.* **109**, 761 (1970). Clinical evaluation: C. F. Marchioni, G. Monzali, *Clin. Ter.* **60**, 135 (1972). Activity on mucociliary transport: P. Braga, *Clin. Trials J.* **18**, 30 (1981). Pharmacokinetics: P. C. Braga *et al.*, *Eur. J. Clin. Pharmacol.* **24**, 209 (1983). Metabolism: P. Ventura *et al.*, *Xenobiot.* **13**, 139 (1983); *eidem, ibid.* **15**, 317 (1985). GLC

determn in human plasma: U. A. Shulka, *J. Chromatog.* **308**, 189 (1984).

dl-trans-Form, **Lysmucol, Sobrepin.** Crystals, mp 130-131.5°. bp 270-271°. Soly in water: 3.3 g/100 ml (15°). LD_{50} in rats, mice (mg/kg): both 580 i.v. (Dalla Valle). *d-* or *l-trans*-Form, mp 150° from ethanol. $[\alpha]_D^{20}$ (+) or (−)150° resp. (c = 10 in ethanol) (Schmidt).

THERAP CAT: Mucolytic.

8511. Soda Lime. Sodium hydroxide with lime. A mixture of calcium oxide with 5-20% sodium hydroxide and contg 6-18% water. It absorbs 25-35% its wt of CO_2.

White or gray-white granules. Rapidly deteriorates on exposure to air through absorption of carbon dioxide. *Keep tightly closed.*

USE: To absorb carbon dioxide in basal metabolism tests, in rebreathing anesthesia systems, in submarines, and in carbon determinations. *Caution:* Strongly corrosive and irritating to skin, mucous membranes, eyes. Ingestion can cause severe damage to G.I. tract, death.

8512. Sodium. Natrium. Na; at. wt 22.98977; at. no. 11; valence 1. Alkali metal. Occurrence in earth's crust: 2.83% by wt; principal cation in hydrosphere. Natural isotopes: 23 (100%); radioactive isotopes (mass number): 20-22; 24-26. Prepd by Davy in 1807 by electrolysis of fused sodium hydroxide. Found in form of its compds, halides, silicates, carbonates; does not occur free. Industrial prepns: primarily in Downs cells; also in Castner cells: Batsford, *Chem. Met. Eng.* **26**, 888, 932 (1932); Regelsberger, *Chemische Technologie der Leichtmetalle*, Leipzig, 1926; Hardie, *Ind. Chemist* **30**, 161 (1954). *Reviews:* M. Sittig, *Sodium* A.C.S. Monograph Series, no. 133 (Reinhold, New York, 1956); Whaley, "Sodium, Potassium, Rubidium, Cesium and Francium" in *Comprehensive Inorganic Chemistry*, **Vol. 1**, J. C. Bailar, Jr. *et al.*, Eds. (Pergamon Press, Oxford, 1973) pp 369-529.

Light, silvery-white metal; body-centered cubic structure; lustrous when freshly cut; tarnishes on exposure to air, becoming dull and gray. Soft at ordinary temp, fairly hard at −20°. mp 97.82°; bp 881.4°; d^{20} 0.968. Heat capacity of solid: 0.292 cal/g deg; heat capacity of liquid at mp: 0.331 cal/g deg. Heat of fusion: 27.05 cal/g; thermal conductivity (cal/sec °C cm): 0.205 (97.82°) 0.170 (400°). E^0 (aqueous) Na/Na^+ 2.714 V. Violently decomposes water, forming sodium hydroxide and hydrogen which may ignite spontaneously. Decomposes alc. Reacts vigorously with oxygen, burning with a yellow flame. Combines directly with the halogens, with phosphorus. Reduces most oxides to the elemental state, reduces metallic chlorides. Dissolves in liq ammonia to give a blue soln; when heated in ammonia gas yields sodamide. Dissolves in mercury, forming sodium amalgam. *Keep under liquids contg no oxygen, such as kerosene, naphtha.*

USE: Manuf of sodium compds, such as the cyanide, azide, peroxide, etc.; in manuf of tetraethyllead; in org syntheses; for photoelectric cells; in sodium lamps. Alloyed with potassium in heat transfer media. *Human Toxicity:* Extremely caustic to all tissue.

8513. Sodium Acetate. $C_2H_3NaO_2$; mol wt 82.04. C 29.28%, H 3.69%, Na 28.03%, O 39.01%. CH_3COONa. Crystal structure of the trihydrate: K.-T. Wei, D. L. Ward, *Acta Crystallogr.* **B33**, 522 (1977).

Hygroscopic powder. Very sol in water; moderately sol in alcohol.

Trihydrate, transparent crystals or granules. Efflorescent in warm air. d 1.45. mp 58°; becomes anhydr at 120°; dec at higher temp. One gram dissolves in 0.8 ml water, 0.6 ml boiling water, 19 ml alcohol. pH of 0.1 molar aq soln at 25°: 8.9.

USE: Anhydrous form as auxiliary in acetylations. Trihydrate in photography; as reagent in anal. chemistry to eliminate effect of strong acids (buffer); for foot warmers and milk-bottle warmers. Technical grades are used as a mordant in dyeing. As acidulant in food. Pharmaceutic aid (in dialysis solutions).

THERAP CAT (VET): Has been used in bovine ketosis.

8514. Sodiumacetic Acid Sodium Salt. *(Carboxymethyl)- sodium sodium salt;* sodium acetate sodium derivative; sodium α-sodioacetate; α-sodioacetic acid sodium salt. C_2H_2-Na_2O_2; mol wt 104.02. C 23.09%, H 1.94%, Na 44.20%, O 30.76%. $NaCH_2COONa$. Prepd by fusing anhydr sodium acetate with sodamide in an atm of nitrogen at 200°: De-Pree, Closson, *J. Am. Chem. Soc.* **80**, 2311 (1958).

Light gray solid. Stable in dry air. Dec 280°. Insol in hydrocarbons and ethers. Dec by water into sodium acetate and sodium hydroxide.

8515. Sodium Acid Pyrophosphate. Disodium dihydro-gen pyrophosphate. $H_2Na_2O_7P_2$; mol wt 221.97. H 0.91%, Na 20.72%, O 50.46%, P 27.91%. $Na_2H_2P_2O_7$. Prepn: Bell, *Inorg. Syn.* **3**, 98 (1950).

White, fused masses or powder. Dec at 220°. d (hexahy-drate) 1.86. Sol in water, the soln having an acid reaction.

USE: Chiefly in baking powders.

8516. Sodium Alizarinesulfonate. *9,10-Dihydro-3,4-di-hydroxy-9,10-dioxo-2-anthracenesulfonic acid monosodium salt;* alizarine S; alizarine carmine; C.I. Mordant Red 3; C.I. 58005. $C_{14}H_7NaO_7S$; mol wt 342.25. C 49.13%, H 2.06%, Na 6.72%, O 32.72%, S 9.37%. Prepn: *Colour Index* vol. **4** (3rd ed., 1971) p 4514.

Monohydrate, orange-yellow powder. Freely sol in water with yellow color; sol in alcohol.

USE: As acid-base indicator and in the determination of fluorine, usually used as a 1% aq soln. pH: 3.7 yellow, 5.2 purple; as a reagent for aluminum and stain in microscopy.

8517. Sodium Aluminate. *Aluminum sodium oxide.* $AlNaO_2$; mol wt 81.97. Al 32.90%, Na 28.06%, O 39.04%. White granular mass. mp 1650°. Very sol in water. Insol in alcohol. The aq soln is strongly alkaline.

USE: Printing on fabrics; manuf lake colors; sizing paper; manuf milk-glass, soap; hardening building stones; water softener.

8518. Sodium Amalgam. An alloy of sodium and mercu-ry. Readily prepd by adding sodium metal in small portions at a time to mercury. The combination takes place with evolution of much heat. The sodium contents of the amal-gam can be varied, as desired, up to about 20% sodium, cor-responding to approx Na_2Hg. The consistency and mp vary with the sodium content; the amalgam corresponding to $NaHg_2$, 5.4% Na, has the highest mp, +360°.

Sodium amalgams are dec by water similarly to sodium metal, but more slowly. *Keep tightly closed.*

8519. Sodium Amide. Sodamide. H_2NNa; mol wt 39.02. Na 58.93%, N 35.90%, H 5.17%. $NaNH_2$. Prepd from sodi-um metal and gaseous ammonia (or liq ammonia with ferric nitrate catalyst): Dennis, Browne, *Inorg. Syn.* **1**, 74 (1939); Greenlee, Henne, *ibid.* **2**, 128 (1946); Schenk in *Handbook of Preparative Inorganic Chemistry* vol. **1**, G. Brauer, Ed. (Aca-demic Press, New York, 2nd ed., 1963) pp 465-468. Alter-nate method: Bergstrom, *Org. Syn.* **coll. vol. III**, 778 (1955).

Crystals, mp 210°. The commercial product may be a white to olive-green solid with sea-shell fracture. Begins to volatilize at 400° and dec into its elements between 500 and 600°. Heat of formation of solid (18°; 1 atm): -32.26 kcal/mole. Heat of soln (21°): -31.06 kcal/mole. Soly in liq ammonia at 20° = about 0.1%. Reacts violently with water forming NaOH and NH_3. The reaction with alcohol is considerably slower. Should be stored in sealed containers which prevent all contact with air during storage: Berg-strom, Fernelius, *Chem. Rev.* **12**, 63, 75, 78 (1933). When exposed to the atmosphere, sodium amide rapidly absorbs H_2O and CO_2. When only limited absorption takes place, as in poorly sealed containers, products are fomed which ren-der the resulting mixture highly explosive. The formation of oxidation products is accompanied by the deveopment of a yellow or brownish color. If such a change is noticed, the substance should be destroyed at once. This is conveniently

accomplished by covering with much benzene, toluene, or kerosene and slowly adding dil ethanol with stirring.

USE: Dehydrating agent. In the production of indigo and hydrazine. Intermediate in the prepn of sodium cyanide. In ammonolysis reactions, in Claisen condensations, alkylation of nitriles and ketones, synthesis of ethynyl compds, acetyl-enic carbinols. Fused $NaNH_2$ dissolves metallic Mg, Zn, Mo, W, quartz, glass, silicates and other substances. *Cau-tion:* Intensely irritating to skin, eyes, mucous membranes.

8520. Sodium Amylosulfate. Sulfated potato amylopec-tin sodium salt; sodium amylopectin sulfate; SN-263; De-pepsen. This product has substantially 1.6 sulfate groups per glucose unit and a mol wt of $2.0-8.0 \times 10^7$ and contains 15-16.5% of sulfur. Prepn: P. S. Camarata, S. Eich, **Belg.** pat. **645,762** corresp to **U.S.** pat. **3,271,388** (1964, 1966 to Searle).

THERAP CAT: Enzyme inhibitor.

8521. Sodium Arsanilate. *(4-Aminophenyl)arsonic acid sodium salt; arsanilic acid sodium salt;* sodium aminarsonate; sodium p-aminophenylarsonate; sodium anilarsonate; Arsa-min; Atoxyl; Nuarsol; Protoxyl; Soamin; Sonate; Piglet Pro-Gen V; Trypoxyl. $C_6H_7AsNNaO_3$; mol wt 311.08. C 30.15%, H 2.95%, As 31.34%, N 5.86%, Na 9.62%, O 20.08%. $NH_2C_6H_4AsO(OH)ONa$.

Tetrahydrate, white, odorless, cryst powder. *Poisonous!* Soluble in about 6 parts water, about 100 parts alcohol. The aq soln is moderately acid to litmus.

THERAP CAT: Formerly as antisyphilitic.

THERAP CAT (VET): Has been used as an anthelmintic in swine dysentery and other enteric conditions of pigs and poultry.

8522. Sodium Arsenate, Dibasic. Disodium arsenate. $AsHNa_2O_4$; mol wt 185.91. As 40.30%, H 0.54%, Na 24.73%, O 34.43%. Na_2HAsO_4. The salt of commerce is about 99% pure. Toxicity: Franke, Moxon, *J. Pharmacol. Exp. Ther.* **58**, 454 (1936).

Powder. Very sol in water; slightly sol in alcohol.

Heptahydrate, odorless crystals; effloresces in warm air. *Poisonous,* but less so than the arsenite. Loses $5H_2O$ (29%) at about 50°; becomes anhyd at 100°. At 150° or higher it is converted into pyroarsenate. d 1.87. mp 57° when rapidly heated. Sol in 1.3 parts water; sol in glycerol, slightly in alcohol. The aq soln is alkaline to litmus. LD_{75} i.p. in rats: 14-18 mg As/kg (Franke, Moxon).

USE: The technical grade, about 98% pure, is used in dye-ing with Turkey-red oil and in printing fabrics; manuf of other arsenates.

THERAP CAT: Formerly as antimalarial; dermatologic.

THERAP CAT (VET): Has been used in parasitism (internally and externally), also for nonparasitic skin and blood dis-eases, in rheumatism, asthma and heaves, and as an altera-tive.

8523. Sodium Arsenite. Sodium meta-arsenite. Approx $NaAsO_2$. The product of commerce is 95-98% pure.

White or grayish-white powder; somewhat hygroscopic. Absorbs CO_2 from air. *Very poisonous!* Freely sol in water, slightly in alcohol. *Keep well closed.* LD_{50} orally in rats: 0.041 g/kg, H. F. Smyth *et al., Am. Ind. Hyg. Assoc. J.* **30**, 470 (1969).

USE: The technical grade, 90-95% pure, is used in manuf of arsenical soap for use on skins, for treating vines against certain scale diseases; as insecticide especially for termites.

THERAP CAT (VET): Topical acaricide.

8524. Sodium Arsphenamine. *4,4'-Arsenobis[2-amino-phenol] disodium salt;* arsphenamine sodium. $C_{12}H_{10}As_2N_2$-

Na_2O_2; mol wt 410.75. C 35.09%, H 2.63%, As 36.47%, N 6.82%, Na 11.19%, O 7.79%. Usually contains some water and inert salts (NaCl) and hence only about 20% arsenic (As).

Bright yellow powder. *Poisonous!* Very unstable in air. Freely sol in water with alkaline reaction. *Keep as described for Neoarsphenamine.* Solution for injection must be prepared immediately after opening the container and must be administered promptly.

THERAP CAT: Formerly as antisyphilitic.

8525. Sodium Ascorbate. *Ascorbic acid sodium derivative;* vitamin C sodium; Ascorbin; Sodascorbate; Natrascorb; Cenolate; Ascorbicin; Cebitate. $C_6H_7NaO_6$; mol wt 198.12. C 36.38%, H 3.56%, Na 11.60%, O 48.46%. One mg of the sodium salt is equivalent to 0.8890 mg of ascorbic acid, or one mg of the acid is equivalent to 1.1248 mg of sodium ascorbate. Preparation: Holland, U.S. pat. **2,442,005** (1948); ascorbic acid is dissolved in water and an equiv amount of sodium bicarbonate is added. After cessation of effervescence the sodium ascobate is precipitated by the addn of isopropanol.

Minute crystals. Dec 218°. $[\alpha]_D^{20}$ +104.4°. Freely sol in water at 25°: 62 g/100 ml H_2O. Even more sol in warm water (78 g/100 ml H_2O at 75°). pH of aq solns 5.6-7.0 or even higher. A 10% soln, made from a commercial grade, may have a pH of 7.4-7.7. Aq solns are unstable and subject to quick oxidation by air at pH > 6.0. Solns can be buffered with ascorbic acid solns which have a pH of 2.3-2.5.

USE: In vitamin C prepns; antioxidant in chopped meat and other food, also in curing meat.

THERAP CAT: Vitamin C source.

8526. Sodium Azide. Smite. N_3Na; mol wt 65.02. N 64.64%, Na 35.36%. NaN_3. Cytochrome oxidase inhibitor. Prepd from $NaNH_2 + N_2O$: Dennis, Browne, *Z. Anorg. Allgem. Chem.* **40**, 95 (1904); Schenk in *Handbook of Preparative Inorganic Chemistry* Vol. 1, G. Brauer, Ed. (Academic Press, New York, 2nd ed., 1963) pp 474-475. Alternate procedures: *Inorg. Syn.* **1**, 79 (1939); **2**, 139 (1946). Large scale manuf processes: B. T. Fedoroff *et al., Encyclopedia of Explosives and Related Items,* **Vol. 1** (Picatinny Arsenal, Dover, N.J., 1960) pp A601-A619. *Review:* L. E. Audrieth, *Chem. Rev.* **15**, 169 (1934). Potent vasodilator; has been used therapeutically to control blood pressure. Review of toxicity, mutagenicity and carcinogenicity: K. A. Frederick, J. G. Babish, *Regul. Toxicol. Pharmacol.* **2**, 308-322 (1982).

Colorless hexagonal crystals, d 1.846. On heating dec into sodium and nitrogen. Highly sol in water. Rapidly converted to hydrazoic acid, *q.v.* Soly in water: 40.16% at 10°, 41.7% at 17°. pK = 4.8, aq solns contains HN_3 which escapes readily at 37°. Slightly sol in alcohol. Insol in ether. Sol in liquid ammonia. LD_{50} in rats (mg/kg): 45 orally (Frederick, Babish).

Caution: Highly toxic. May cause hypotension, tachycardia, tachypnea, hypothermia, convulsions and severe headache: *Clinical Toxicology of Commercial Products,* R. E. Gosselin *et al.,* Eds. (Williams & Wilkins, Baltimore, 5th ed., 1984), Section II, p 114; *Prudent Practices for Handling Hazardous Chemicals in Laboratories* (National Academy Press, Washington, D.C., 1981) pp 145-147.

USE: In organic syntheses; in the preparation of hydrazoic acid, lead azide, pure sodium. In the differential selection of bacteria; in automatic blood counters; as preservative for laboratory reagents. Propellant for inflating automotive safety bags. Agricultural nematocide; herbicide; in fruit rot control.

8527. Sodium Benzoate. $C_7H_5NaO_2$; mol wt 144.11. C 58.34%, H 3.50%, Na 15.96%, O 22.21%. Toxicity: Smyth, Carpenter, *J. Ind. Hyg. Toxicol.* **30**, 63 (1948).

White, odorless granules or crystalline powder; sweetish, astringent taste. One gram dissolves in 1.8 ml water, 1.4 ml boiling water, about 75 ml alcohol, in 50 ml of a mixture of 47.5 ml alcohol and 3.7 ml water. The aq soln is slightly alkaline to litmus. pH about 8. *Incompat:* Acids, ferric salts. LD_{50} orally in rats: 4.07 g/kg (Smyth, Carpenter).

USE: As preservative in pharmaceuticals and in food products, not more than 1 in 1000 being permitted. Its preservative effect is best exhibited in slightly acidic media; in alkaline media it is almost without effect. Clinical reagent (bilirubin assay).

THERAP CAT: Diagnostic aid (hepatic function).

8528. Sodium Bicarbonate. Sodium hydrogen carbonate; sodium acid carbonate; baking soda. $CHNaO_3$; mol wt 84.00. C 14.29%, H 1.20%, Na 27.37%, O 57.14%. $NaHCO_3$. The bicarbonate of commerce is about 99.8% pure. Prepd from sodium carbonate, water and carbon dioxide. Manuf: Faith, Keyes & Clark's *Industrial Chemicals,* F. A. Lowenheim, M. K. Moran, Eds. (Wiley-Interscience, New York, 4th ed., 1975) pp 702-705.

White cryst powder or granules. Begins to lose CO_2 at about 50° and at 100° it is converted into Na_2CO_3. Readily dec by weak acids. In aq soln it begins to break up into carbon dioxide and sodium carbonate at about 20° and completely on boiling. Sol in 10 parts water at 25°, in 12 parts water at about 18°; insol in alcohol. Its aq soln prepd with cold water and without agitation is only slightly alkaline to litmus or phenolphthalein; on standing or rise in temp the alkalinity increases. pH of freshly prepd 0.1 molar aq soln at 25°: 8.3.

USE: Manuf many sodium salts; source of CO_2; ingredient of baking powder, effervescent salts and beverages; in fire extinguishers, cleaning compds.

THERAP CAT: Antacid, urinary and systemic alkalizer.

THERAP CAT (VET): Antacid, systemic and urinary alkalizer. Locally in burns, erythema, to dissolve mucus, exudates, scabs.

8529. Sodium Bifluoride. F_2HNa; mol wt 62.01. F 61.29%, H 1.63%, Na 37.09%. NaF.HF.

White, cryst powder. Sol in water. The aq soln corrodes glass.

USE: As a "sour" in laundering.

8530. Sodium Bismuthate(V). $BiNaO_3$; mol wt 280.00. Bi 74.64%, Na 8.21%, O 17.14%. $NaBiO_3$. The bismuthate of commerce contains about 85% $NaBiO_3$; the balance is chiefly water and Bi_2O_3.

Yellow to yellowish-brown, somewhat hygroscopic. Slowly dec on keeping; decompn accelerated by moisture and higher temp. Insol in cold, dec by hot water forming Bi_2O_3, NaOH, and liberating oxygen; dec by acids; with HCl chlorine is formed; with oxy-acids oxygen is liberated. LD_{100} orally in rats: 720 mg/kg, Hanzlik *et al., J. Pharmacol. Exp. Ther.* **62**, 372 (1938).

USE: For the determination of manganese in iron and steel, etc., the manganese being oxidized by it in hot HNO_3 or H_2SO_4 soln to permanganate.

8531. Sodium Bisulfate. Sodium acid sulfate; sodium hydrogen sulfate; sodium pyrosulfate. $HNaO_4S$; mol wt 120.07. H 0.84%, Na 19.15%, O 53.30%, S 26.71%. $NaHSO_4$.

Fused $NaHSO_4$, hygroscopic pieces. d 2.435. mp about 315°. Sol in 2 parts water, 1 part boiling water; dec by alcohol into sodium sulfate and free H_2SO_4. *Keep well closed.*

Monohydrate, odorless crystals. When strongly heated it changes into pyrosulfate. Sol in about 0.8 part water; dec by alcohol into sodium sulfate and free H_2SO_4. The aq soln is strongly acid. pH of 0.1 molar soln: 1.4.

USE: Fusion of minerals to make them sol for analysis; for liberating CO_2 in carbonic acid baths. Technical grades are used for pickling metals, carbonizing wool, bleaching and swelling leather, manuf magnesia cements, etc.

8532. Sodium Bisulfide. Sodium sulfhydrate; sodium hydrosulfide; sodium hydrogen sulfide. HNaS; mol wt 56.07. H 1.80%, Na 41.02%, S 57.18%. NaSH. Prepd from sodium ethylate and hydrogen sulfide: Rule, *J. Chem. Soc.* **99**, 558 (1911); Teichert, Klemm, *Z. Anorg. Allgem. Chem.* **243**, 86 (1939); Eibeck, *Inorg. Syn.* **7**, 128 (1963). The technical grade may be obtained by reacting sodium bisulfate with calcium sulfide in the cold or by saturating NaOH solns with H_2S.

Rhombohedric-cubic crystals. White to colorless. Odor of hydrogen sulfide. Very hygroscopic. Readily hydrolyzed in moist air to NaOH and Na_2S. d 1.79. Turns yellow upon heating in dry air, changing to orange at higher temps. mp 350° forming a black liquid. Sol in water, alcohol, ether. Gives a blue-green soln in dimethylformamide.

Dihydrate, needles or flakes, mp 55°. Completely and rapidly sol in water, alcohol, ether. *Note:* The commercial product is usually the dihydrate. Can be shipped in lacquer-lined steel drums.

Trihydrate, shiny rhombs, mp 22°.

USE: Dehairing hides; desulfurizing viscose rayon; in the manuf of sulfur-contg dyes and other thio compds such as thioamides, thiourea, thioglycolic acid, thio- and dithiobenzoic acids, sodium thiosulfate.

8533. Sodium Bisulfite. Sodium acid sulfite. $HNaO_3S$; mol wt 104.07. H 0.97%, Na 22.10%, O 46.13%, S 30.81%. $NaHSO_3$. The bisulfite of commerce consists chiefly of sodium metabisulfite, $Na_2S_2O_5$, and for all practical purposes possesses the same properties as the true bisulfite. Toxicity: Hoppe, Goble, *J. Pharmacol. Exp. Ther.* **101**, 101 (1951).

White, crystalline powder; SO_2 odor; disagreeable taste; on exposure to air it loses some SO_2 and is gradually oxidized to sulfate. d 1.48. Sol in 3.5 parts cold water, 2 parts boiling water, in about 70 parts alcohol. Its aq soln is acid. *Keep well closed and in a cool place. Incompat:* Acids, oxidizers. LD_{50} i.v. in rats: 115 mg/kg (Hoppe, Goble).

USE: As disinfectant and bleach, particularly for wool; in dyeing for preparing hot and cold indigo vats; in paper-making in place of sodium hyposulfite to remove Cl from bleached fibers; as stripper (reducer) in laundering; to remove permanganate stains from skin and clothing; to render certain dyes sol; manuf sodium hydrosulfite; coagulating rubber latex; as preservative for deteriorative liqs or solns used for technical purposes; as antiseptic in fermentation industries. As preservative and bleach in food. Pharmaceutic aid (antioxidant). *Caution:* Concd solns are irritating to skin, mucous membranes.

8534. Sodium Bitartrate. Sodium acid tartrate. C_4H_5-NaO_6; mol wt 172.07. C 27.92%, H 2.93%, Na 13.36%, O 55.79%. $NaHC_4H_4O_6$.

Monohydrate, white crystals. Sol in about 9 parts water, 2 parts boiling water; almost insol in alcohol. The aq soln is acid.

USE: For detecting potassium; in nutrient media.

8535. Sodium Borate. Sodium biborate; sodium pyroborate; sodium tetraborate. $B_4Na_2O_7$; mol wt 201.27. B 21.50%, Na 22.84%, O 55.65%. $Na_2B_4O_7$. Toxicity: H. F. Smyth *et al., Am. Ind. Hyg. Assoc. J.* **30**, 470 (1969).

Anhydrous, *fused sodium borate, borax glass, fused borax.* Powder or glass-like plates becoming opaque on exposure to air. Slowly sol in water.

Decahydrate, *borax, Jaikin.* Hard odorless crystals, granules or cryst powder; efflorescent in dry air, the crystal often being coated with white powder. d 1.73. mp when rapidly heated at 75°; at 100° loses $5H_2O$; at 150° loses $9H_2O$; becomes anhydr at 320°. One gram dissolves in 16 ml water, 0.6 ml boiling water, about 1 ml glycerol. Insol in alcohol. The aq soln is alkaline to litmus and phenolphthalein. pH about 9.5. Borax dissolves many metallic oxides when fused with them. *Incompat:* Acids, alkaloidal and metallic salts. LD_{50} orally in rats: 5.66 g/kg (Smyth).

Human Toxicity: Ingestion of 5 to 10 g by young children can cause severe vomiting, diarrhea, shock, death, see *Clinical Toxicology of Commercial Products,* R. E. Gosselin *et al.,* Eds. (Williams & Wilkins, Baltimore, 4th ed., 1976) Section III, pp 63-66. *See also* Boric Acid.

USE: Soldering metals; manuf glazes and enamels; tanning; in cleaning compds; artificially aging wood; as preservative,

either alone or with other antiseptics against wood fungus; fireproofing fabrics and wood; curing and preserving skins; in cockroach control. Pharmaceutic aid (alkalizer).

THERAP CAT (VET): Has been used as antiseptic, detergent, astringent for mucous membranes.

8536. Sodium Borate Solution Compound. Dobell's soln. Made from 1.5 g sodium borate, 1.5 g sodium bicarbonate, 0.3 ml liquefied phenol, 3.5 ml glycerol and water to make 100 ml.

Yellowish, clear liquid.

THERAP CAT: Wash for mucous membranes.

THERAP CAT (VET): Has been used as a nonirritant wash for mucous membranes.

8537. Sodium Borohydride. *Sodium tetrahydroborate.* BH_4Na; mol wt 37.83. B 28.58%, H 10.65%, Na 60.77%. $NaBH_4$. Prepared from methyl borate and sodium hydride at elevated temps: Schlesinger *et al., J. Am. Chem. Soc.* **75**, 205 (1953). Review of sodium and other metal tetrahydroborates: James, Walbridge, *Prog. Inorg. Chem.* **11**, 99-231 (1970).

Hygroscopic, cubic crystals forming a dihydrate, mp 36-37°. The anhydr material (d 1.074) is stable in dry air to 300°; dec slowly at 400° and rapidly at 500°. Supports combustion. Soly (w/w) in water at 25°: 55%; at 60°: 88.5%; liq ammonia at 25°: 104%; ethylenediamine at 75°: 22%; morpholine at 25°: 1.4%; pyridine at 25°: 3.1%; methanol at 20°: 16.4% (reacts); ethanol at 20°: 4.0% (reacts slowly); tetrahydrofurfuryl alcohol at 20°: 14.0% (reacts slowly); tetrahydrofuran at 20°: 0.1%; diglyme at 25°: 5.5%; dimethylformamide at 20°: 18.0%. Aq solns are most stable in the presence of small amounts of NaOH (0.2% for a nearly satd soln contg 44% $NaBH_4$) and can be kept for several days. Solns are rapidly dec by boiling.

USE: Reducing agent for aldehydes, ketones and Schiff bases in nonaqueous solvents. Also reduces acids, esters, acid chlorides, disulfides, nitriles, inorganic anions. Further used to generate diborane, as foaming agent, as scavenger for traces of aldehyde, ketones and peroxides in organic chemicals.

8538. Sodium Bromate. $BrNaO_3$; mol wt 150.91. Br 52.96%, Na 15.24%, O 31.81%. $NaBrO_3$. The article of commerce contains about 99% $NaBrO_3$.

Colorless, odorless crystals, white granules or cryst powder. d 3.34. mp 381° with dec and liberation of oxygen. Sol in 2.5 parts water, 1.1 parts boiling water. The aq soln is neutral. *Keep from contact with organic matter.*

USE: As a mixture with sodium bromide for dissolving gold from its ores. *Caution: See* Potassium Bromate.

8539. Sodium Bromide. Sedoneural. BrNa; mol wt 102.91. Br 77.65%, Na 22.35%. NaBr. Prepd commercially by adding some excess bromine to a sodium hydroxide soln forming a mixture of bromide and bromate. The reaction products are evaporated to dryness and treated with carbon to reduce the bromate to bromide. *Ref:* van ter Meulen, U.S. pat. **1,775,598** (1930); Robertson, *Ind. Eng. Chem.* **34**, 133 (1942); T. O. Soine, C. O. Wilson, *Roger's Inorganic Pharmaceutical Chemistry* (Lea & Febiger, Philadelphia, 8th ed., 1967) pp 213-216.

White crystals, granules or powder; saline, feebly bitter taste. Absorbs moisture from air but is not deliquescent. d 3.21. mp 755°; volatilizes at somewhat higher temp. One gram dissolves in 1.1 ml water, about 16 ml alcohol, 6 ml methanol. The aq soln is practically neutral. pH 6.5-8.0. *Keep well closed.* From water of room temp, sodium bromide crystallizes with $2H_2O$ in the form of colorless crystals. *Incompat:* Acids, alkaloidal and heavy metal salts. LD_{50} orally in rats: 3.5 g/kg, Smith, Hambourger, *J. Pharmacol. Exp. Ther.* **55**, 200 (1935).

USE: In photography.

THERAP CAT: Sedative, hypnotic, anticonvulsant.

THERAP CAT (VET): Sedative. Has been used to control convulsions, chorea, hysteria.

8540. Sodium Cacodylate. *[(Dimethylarsino)oxy]sodium As-oxide;* sodium dimethylarsonate; Arsecodile; Arsicodile; Arsycodile; Rad-e-cate; Silvisar. $C_2H_6AsNaO_2$; mol wt 159.98. C 15.02%, H 3.78%, As 46.82%, Na 14.37%, O 20.00%. An organic compd of arsenic yielding inorganic, trivalent arsenic in the body (is excreted partly unchanged, also yields dimethylarsine oxide). Prepd by the distillation of a mixture of arsenic trioxide and potassium acetate which yields Cadet's liquid, contg mostly dimethylarsine oxide. This is oxidized with mercuric oxide yielding crystals of cacodylic acid which is neutralized with Na_2CO_3 or NaOH: Cadet de Gassicourt, *Mem. savants étrangers* **3**, 633 (1760); Valeur, Gaillot, *Compt. Rend.* **185**, 956 (1927). Use in treatment of psoriasis: S. I. Dawes, H. C. Jackson, *J. Am. Med. Assoc.* **48**, 2020 (1903). Metabolism and excretion: A. Heffter, *Arch. Exp. Pathol. Pharmakol.* **46**, 230 (1901). Use as herbicide: M. A. Sprague, U.S. pat. 3,056,663 (1962 to Ansul). Possible use in radiation dosimetry: H. S. Levinson, E.B. Garber, *Nature* **207**, 751 (1965).

$$\underset{\underset{CH_3}{|}}{\overset{\overset{NaO}{|}}{O=As-CH_3}}$$

Trihydrate, crystals, granules. Slight odor. Liquefies in its water of hydration at about 60°. Becomes anhyd at 120°. Burns with a bluish flame, emitting a garlic-like odor. One gram dissolves in 0.5 ml water, 2.5 ml alcohol. pH about 8-9. *Keep well closed.*

Human Toxicity: More toxic by mouth than by injection due to rapid release of inorganic arsenic by gastric acid. Large doses may cause nephritis, albuminuria, hematuria: *Martindale The Extra Pharmacopoeia*, N. W. Blacow, Ed. (Hazell, Watson & Viney, Aylesbury, Bucks, 1972) p 219.

USE: Herbicide.

THERAP CAT (VET): Has been used in chronic eczema, anemia, as a tonic.

8541. Sodium Carbonate. CNa_2O_3; mol wt 106.00. C 11.33%, Na 43.39%, O 45.29%. Na_2CO_3. Occurs in nature as the hydrate, *thermonatrite*, and the decahydrate, *natron* or *natrite*. Produced by the ammonia-soda or Solvay process, or from lake brines or sea water by electrolytic processes: Faith, Keyes & Clark's *Industrial Chemicals*, F. A. Lowenheim, M. K. Moran, Eds. (Wiley-Interscience, New York, 4th ed., 1975) pp 706-715. Toxicity: C. Norde *et al.*, *Compt. Rend.* **257**, 791 (1963). *Reviews:* Bailey in *Mellor's* vol II, suppl II, *The Alkali Metals* (part 1), 1058-1205 (1961).

Anhydrous, *Solvay soda.* The technical grade (about 99% pure) is known as *soda ash.* Odorless, hygroscopic powder; alkaline taste; d 2.53. mp 851° but begins to lose CO_2 even at 400°. On exposure to air it will gradually absorb one mol water—about 15%. Sol in glycerol; in 3.5 parts water at room temp, 2.2 parts water at 35°. Insol in alcohol. Dec by acids with effervescence. Combines with water with evolution of heat. Its aq soln is strongly alkaline. pH 11.6. *Keep well closed.* LD_{50} (30 day) i.p. in mice: 116.6 mg/kg (Norde).

Monohydrate, odorless, small crystals or cryst powder; alkaline taste. Stable at ordinary temps and atmospheric conditions; dries out somewhat in warm, dry air or above 50°; becomes anhyd at 100°. d 2.25; also reported as 1.55. Sol in 3 parts water, 1.8 parts boiling water, 7 parts glycerol. Insol in alcohol.

Decahydrate, *Nevite, Soda.* The technical product is known as *sal soda* or *washing soda.* Transparent crystals; readily effloresces on exposure to air. d 1.46. mp 34°. Sol in 2 parts cold, 0.25 part boiling water, in glycerol. Insol in alcohol. The aq soln is strongly alkaline to litmus. *Keep well closed and in a cool place.*

Human Toxicity: Skin irritant. Concd solns may produce local necrosis of mucous membranes: Patty's *Industrial Hygiene and Toxicology*, G. D. Clayton, F. E. Clayton, Eds. (Wiley-Interscience, New York, 3rd ed., 1981) pp 3059-3061.

USE: In manuf of Na salts, glass, soap; for washing wool; textiles, etc.; in bleaching linen, cotton; general cleanser; in water-softening; in photography; as reagent in analytical chemistry. Pharmaceutic aid (alkalizer).

THERAP CAT (VET): Has been used as an emetic. In solution to cleanse skin, in eczema, to soften scabs of ringworm.

8542. Sodium Cellulose Phosphate. SCP; Calcibind; Calcisorb. Non-absorbable ion exchange resin with high affinity for calcium ions. Prepn: G. P. Touey, U.S. pat. 2,759,924 (1956 to Eastman Kodak). Use in treatment of hypercalcemia: C. Y. C. Pak *et al.*, *J. Clin. Endocrinol. Metab.* **28**, 1828 (1968). Mechanism of action: C. Y. C. Pak, *J. Clin. Pharmacol.* **13**, 15 (1973). Clinical pharmacology: *idem, ibid.* **19**, 451 (1979). Use in treatment of calcium urolithiasis: U. Backman *et al.*, *J. Urol.* **123**, 9 (1980); C. Y. C. Pak, *Invest. Urol.* **19**, 187 (1981).

Powder; insol in water.

USE: As ion exchange resin.

THERAP CAT: Treatment of calcium urolithiasis.

8543. Sodium Chlorate. Atlacide; De-Fol-Ate. $ClNaO_3$; mol wt 106.45. Cl 33.31%, Na 21.60%, O 45.09%. $NaClO_3$. The chlorate of commerce is about 99% pure. Produced from sodium chloride by electrolysis: Faith, Keyes & Clark's *Industrial Chemicals*, F. A. Lowenheim, M. K. Moran, Eds. (Wiley-Interscience, New York, 4th ed., 1975) pp 716-721. Toxicity: Ulrich, *J. Pharmacol. Exp. Ther.* **35**, 1 (1929).

Colorless, odorless crystals or white granules. d 2.5. mp 248°; at about 300° liberates oxygen; entirely dec at higher temp. Sol in about 1 ml cold, 0.5 ml boiling water, about 130 ml alcohol, about 50 ml boiling alcohol, 4 ml glycerol. Sodium chloride diminishes its soly in water. The aq soln is neutral. *Keep out of contact with organic matter or other oxidizable substances.* LD orally in rats: 12,000 mg/kg (Ulrich).

USE: An oxidizer, like potassium chlorate, in manuf of dyes; explosives and matches; dyeing and printing fabrics; tanning and finishing leather; as herbicide. Pharmaceutic aid (oxidizing agent).

8544. Sodium Chloride. Salt; common salt. ClNa; mol wt 58.45. Cl 60.66%, Na 39.34%. NaCl. The article of commerce is also known as *table salt, rock salt* or *sea salt.* Occurs in nature as the mineral *halite.* Produced by mining (rock salt), by evaporation of brine from underground salt deposits and from sea water by solar evaporation: Faith, Keyes & Clark's *Industrial Chemicals*, F. A. Lowenheim, M. K. Moran, Eds. (Wiley-Interscience, New York, 4th ed., 1975) pp 722-730. Comprehensive monograph: D. W. Kaufmann, *Sodium Chloride*, ACS Monograph Series no. 145 (Reinhold, New York, 1960) 743 pp.

Cubic, white crystals, granules, or powder; colorless and transparent or translucent when in large crystals. d 2.17. The salt of commerce usually contains some calcium and magnesium chlorides which absorb moisture and make it cake. mp 804° and begins to volatilize at a little above this temp. One gram dissolves in 2.8 ml water at 25°, in 2.6 ml boiling water, in 10 ml glycerol; very slightly sol in alcohol. Its soly in water is decreased by HCl and it is almost insol in concd HCl. Its aq soln is neutral. pH: 6.7-7.3. d of satd aq soln at 25° is 1.202. A 23% aq soln of sodium chloride freezes at −20.5°C (5°F). LD_{50} orally in rats: 3.75 g/kg, Boyd, Shanas, *Arch. Int. Pharmacodyn. Ther.* **144**, 86 (1963).

Note: Blusalt, a brand of sodium chloride contg trace amounts of cobalt, iodine, iron, copper, manganese, zinc is used in farm animals.

Human Toxicity: Not generally considered poisonous. Accidental substitution of NaCl for lactose in baby formulas has caused fatal poisoning.

USE: Natural salt is the source of chlorine and of sodium as well as of all, or practically all, their compds, e.g., hydrochloric acid, chlorates, sodium carbonate, hydroxide, etc.; for preserving foods; manuf soap, dyes—to salt them out; in freezing mixtures; for dyeing and printing fabrics, glazing pottery, curing hides; metallurgy of tin and other metals.

THERAP CAT: Electrolyte replenisher, emetic; topical anti-inflammatory.

THERAP CAT (VET): Essential nutrient factor. May be given orally as emetic, stomachic, laxative or to stimulate thirst (prevention of calculi). Intravenously as isotonic solution to raise blood volume, to combat dehydration. Locally as wound irrigant, rectal douche.

8545. Sodium Chlorite. $ClNaO_2$; mol wt 90.45. Cl 39.20%, Na 25.42%, O 35.38%. $NaClO_2$. Prepd on a commercial scale by passing chlorine dioxide into a soln of sodium hydroxide contg carbonaceous matter and lime: Vincent, U.S. pats. **2,092,944/5** (1937 to Mathieson). Review of manuf and properties: Kesting, *Pulp Paper Mag. Can.* **53**, no. 8, 99-104 (1952).

Slightly hygroscopic crystals or flakes, does not cake. Dec 180-200°. Powerful oxidizer, but will not explode on percussion unless in contact with oxidizable material. Soly in water (g/100 g soln): at 5°: 34; at 17°: 39; at 30°: 46; at 45°: 53; at 60°: 55.

Trihydrate, triclinic leaflets, becomes anhydr at 38° or in desiccator over KOH at room temp.

USE: In the preparation of chlorine dioxide for immediate use; in water purification; as bleaching agent for textiles, paper pulp.

8546. Sodium 6-Chloro-5-nitrotoluene-3-sulfonate. *4-Chloro-5-nitro-m-toluenesulfonic acid sodium salt.* $C_7H_5Cl-NNaO_5S$; mol wt 273.64. C 30.72%, H 1.84%, Cl 12.96%, N 5.12%, Na 8.40%, O 29.24%, S 11.72%.

White or slightly yellow, cryst powder. Moderately sol in water, slightly in alcohol.

USE: As reagent for determination of potassium.

8547. Sodium Chromate(VI). Neutral sodium chromate. $CrNa_2O_4$; mol wt 161.97. Cr 32.10%, Na 28.38%, O 39.51%. Na_2CrO_4. Crystallizes as a tetra- or decahydrate. Manuf: Faith, Keyes & Clark's *Industrial Chemicals*, F. A. Lowenheim, M. K. Moran, Eds. (Wiley-Interscience, New York, 4th ed., 1975) pp 731-736.

Tetrahydrate, yellow, somewhat deliquesc crystals. Sol in about 1 part water, slightly in alcohol. The aq soln is alkaline. *Keep well closed.* The decahydrate is unstable with respect to water content. mp ~20°.

USE: Protection of iron against corrosion and rusting.

8548. Sodium Chromate(VI), Radioactive. Sodium radio-chromate(⁵¹Cr); sodium chromate-⁵¹Cr; Chromitope Sodium; Rachromate-51. $Na_2{}^{51}CrO_4$. Clinical use in platelet survival studies: L.-B. Olsson *et al.*, *Acta Haematol.* **58**, 3 (1977); in erythrocyte survival studies: F. G. Ebaugh, J. F. Ross, *Vox Sang.* **49**, 304 (1985).

Prepd from radioactive chromium (⁵¹Cr) which has a half-life of 26.5 days. The emission of gamma rays is applicable to biological tagging and tracing. Other properties identical with those of ordinary sodium chromate. Available as soln for intravenous injection or for mixing with blood. Unbound chromate in the plasma can be reduced with ascorbic acid or may be removed by separation and washing of cells.

THERAP CAT: Diagnostic aid (blood volume determination, blood cell survival).

8549. Sodium Citrate. Trisodium citrate; Citrosodine; Citnatin; Urisal. $C_6H_5Na_3O_7$; mol wt 258.07. C 27.92%, H 1.95%, Na 26.73%, O 43.40%. Toxicity data: Gruber, Halbeisen, *J. Pharmacol. Exp. Ther.* **94**, 65 (1948).

Dihydrate, white, odorless crystals, granules or powder; cool, saline taste. Stable in air; becomes anhydrous at 150°. Sol in 1.3 parts water, 0.6 part boiling water. Insol in alcohol. The aq soln is slightly alkaline to litmus. pH about 8. LD_{50} i.p. in rats: 6.0 mmoles/kg (Gruber, Halbeisen).

Pentahydrate, relatively large, colorless crystals or white granules. Not as stable as the dihydrate, drying out on exposure to air and also caking. *Keep well closed.*

USE: In photography; as sequestering agent to remove trace metals; *in vitro* anticoagulant; as emulsifier, acidulant and sequestrant in foods.

THERAP CAT: Systemic alkalizer; diuretic; expectorant; sudorific.

THERAP CAT (VET): Anticoagulant for collection of blood.

8550. Sodium Citrate, Acid. Disodium citrate; disodium hydrogen citrate; Alkacitron. $C_6H_6Na_2O_7$; mol wt 236.08. C 30.52%, H 2.56%, Na 19.48%, O 47.44%.

Sesquihydrate, white powder, saline taste. One gram dis-

solves in slightly less than 2 ml water; pH of a 3% w/v soln in water: 4.9 to 5.2.

USE: Anticoagulant, generally in soln with glucose, to prevent the clotting of blood intended for transfusion. Preferable to sodium citrate, since it prevents carmelization of glucose on sterilization because of its acidity. A suitable soln contains 1.7 to 2%, and 2.5% dextrose; 120 ml of this soln prevents the clotting of 420 ml blood.

8551. Sodium Cobaltinitrite. *Trisodium hexakis(nitrato-N)cobaltate(3—); sodium hexanitrocobaltate(III).* $CoN_6-Na_3O_{12}$; mol wt 403.98. Co 14.59%, N 20.81%, Na 17.08%, O 47.53%. $Na_3Co(NO_2)_6$.

Yellow to brownish-yellow, cryst powder. Very sol in water, slightly in alc. Dec by mineral acids, but unaffected by dil acetic or similar organic acids. The aq soln dec gradually but if a few drops of acetic acid are added it may be kept for about 3 months.

USE: For the detection of potassium with which it forms a slightly sol compd.

8552. Sodium Cyanate. *Cyanic acid sodium salt.* $CN-NaO$; mol wt 65.01. C 18.47%, N 21.55%, Na 35.36%, O 24.61%. NaOCN. Prepn and properties: *Gmelin's, Sodium* (8th ed.) **21**, 799-801 (1928) and supplement, part 4, 1382-1386 (1967). Used experimentally in treatment of sickle cell anemia. Effect of cyanate on sickling: May *et al.*, *Lancet* **1**, 658 (1972); P. N. Gillette *et al.*, *N. Engl. J. Med.* **290**, 654 (1974). Pharmacology and toxicology: Cerami *et al.* *J. Pharmacol. Exp. Ther.* **185**, 653 (1973). Clinical studies: Peterson *et al.*, *ibid.* **189**, 577 (1974).

Colorless needles from alcohol. d_4^{20} 1.893. mp 550°. Sol in water; decomposes to form Na_2CO_3 and urea. Soly in alc (0°): 0.22 g/100 g solvent. Insol in ether. LD_{50} i.p. in mice: 260 mg/kg, Cerami *et al.*, *loc. cit.*

8553. Sodium Cyanide. Cyanogran. $CNNa$; mol wt 49.02. C 24.50%, N 28.58%, Na 46.92%. NaCN. This cyanide of commerce is 95-98% pure. Mixtures of sodium cyanide with sodium chloride or carbonate for special uses are also marketed.

White granules or fused pieces. *Violent poison!* Odorless when perfectly dry; somewhat deliquesc in damp air and emits slight odor of HCN. mp 563°. Freely sol in water, slightly in alcohol. The aq soln is strongly alkaline and rapidly decomposes; the soln readily dissolves gold and silver in presence of air. *Keep well closed.* LD_{50} orally in rats: 15 mg/kg, Smyth *et al.*, *Am. Ind. Hyg. Assoc. J.* **30**, 470 (1969).

USE: Extracting gold and silver from ores; electroplating baths; fumigating citrus and other fruit trees, ships, railway cars, warehouses, etc.; manuf hydrocyanic acid and many other cyanides; case hardening of steel. *Caution:* See Hydrogen Cyanide.

8554. Sodium Cyanoborohydride. *Sodium (cyano-C)tri-hydroborate(1—); sodium borocyanohydride; sodium cyanohydridoborate.* CH_3BNNa; mol wt 62.84. C 19.11%, H 4.81%, B 17.20%, N 22.29%, Na 36.59%. $NaBH_3CN$. Reducing agent prepared from $NaBH_4$ and HCN: R. C. Wade *et al.*, *Inorg. Chem.* **9**, 2146 (1970); R. C.Wade, **Ger.** pat. **2,028,569** corresp to U.S. pat. **3,667,923** (1971, 1972 both to Ventron). Review of prepn, properties and use: C. F. Lane, *Synthesis* **1975**, 135-146.

White, hygroscopic powder, mp 240-242° (dec). d^{28} 1.199. Soly (g/100 g solvent) in water (29°): 212; in THF (28°): 37.2; in diglyme (25°): 17.6. Very sol in methanol; slightly sol in ethanol, isopropylamine; insol in ethyl ether, benzene, hexane. Stable in acid to pH 3; undergoes rapid hydrolysis in 12N HCl. Rate of hydrolysis 10^{-8} that of $NaBH_4$.

USE: Selective reducing agent for aldehydes, ketones, oximes, enamines; does not reduce amides, ethers, lactones, nitriles, nitro compds and epoxides. Also used for reductive amination of ketones and aldehydes, reductive alkylation of amines and hydrazines, reductive displacement of halides and tosylates, deoxygenation of aldehydes and ketones. See Lane, *loc. cit.*

8555. Sodium Diacetate. Sodium acid acetate; Dykon. $CH_3COONa.CH_3COOH$. Described as a "bound" compd of sodium acetate and acetic acid.

White powder, dec above 150°. Sol in water, liberating 42.25% available acetic acid.

USE: Acetic acid in solid form; as an inhibitor of molds and rope-forming bacteria in bread: Glabe, *Food Inds.* **14**, no. 2, 46 (1942); as sequestrant.

8556. Sodium Dichromate(VI). Sodium bichromate; bichromate of soda. $Cr_2Na_2O_7$; mol wt 261.96. Cr 39.70%, Na 17.55%, O 42.75%. $Na_2Cr_2O_7$. Usually prepd from Na_2-CrO_4 and H_2SO_4. Description of industrial processes: Müller, Glissmann in *Ullmann's Encyklopädie der Technischen Chemie* vol. 5 (Munich, 3rd ed., 1954) p 575; Faith, Keyes & Clark's *Industrial Chemicals*, F. A. Lowenheim, M. K. Moran, Eds. (Wiley-Interscience, New York, 4th ed., 1975) pp 731-736.

Dihydrate, reddish to bright orange, somewhat deliquescent crystals. Crystal system: monoclinic sphenoidal. Crystal habit: elongated prismatic. d_4^{25} 2.348. Bulk density: 96 lbs/cu ft. Becomes anhydr on prolonged heating at about 100°. The anhydr salt mp 356.7° and starts to dec at about 400°. Heat of soln −28.2 cal/g. Very sol in water. A satd aq soln contains at 0°: 70.6% $Na_2Cr_2O_7.2H_2O$; at 20°: 73.18%; at 40°: 77.09%; at 60°: 82.04%; at 80°: 88.39%; at 100°: 91.43%. A 20% soln freezes at −3.5°, a 30% soln at −6°, a 60% soln at −26°, a 69% soln at −48°. Specific heat of 20% soln at 25°: 0.85 cal/g/°C. Solns are acidic: pH of 1% soln: 4.0; pH of 10% soln: 3.5.

USE: Oxidizing agent in manuf of dyes, many other synthetic organic chemicals, inks, etc.; in chrome-tanning of hides; in electric batteries; bleaching fats, oils, sponges, resins; refining petroleum; manuf chromic acid, other chromates and chrome pigments; in corrosion-inhibitors, corrosion-inhibiting paints; in many metal treatments; electroengraving of copper; mordant in dyeing; for hardening gelatin; for the defoliation of cotton plants and other plants and shrubs, La Lande, U.S. pat. **2,760,854** (1956 to Pennsylvania Salt). *Caution:* Irritant and caustic to skin, mucous membranes.

THERAP CAT: Topical anti-infective.

8557. Sodium Dicyanoaurate(I). Gold sodium cyanide; sodium aurocyanide. C_2AuN_2Na; mol wt 272.03. C 8.82%, Au 72.44%, N 10.29%, Na 8.45%. $NaAu(CN)_2$.
White, cryst powder. Sol in water. *Poison!*
USE: Goldplating.

8558. Sodium Dithionate. *Dithionous acid disodium salt.* $Na_2O_6S_2$; mol wt 206.10. Na 22.31%, O 46.58%, S 31.11%. Prepd according to the equations $MnO_2 + 2SO_2 \rightarrow MnS_2O_6$ and $MnS_2O_6 + Na_2CO_3 \rightarrow MnCO_3 + Na_2S_2O_6$: de Baat, *Rec. Trav. Chim.* **45**, 237 (1926); Pfanstiel, *Inorg. Syn.* **2**, 170 (1946).

$$NaO-\overset{\overset{O}{\|}}{\underset{\underset{O}{\|}}{S}}-\overset{\overset{O}{\|}}{\underset{\underset{O}{\|}}{S}}-ONa$$

Dihydrate, colorless, water-clear, orthorhombic crystals. Very stable in air. d 2.189. Loses all of its water of crystn at 110°. When heated to 267° it is dissociated into Na_2SO_4 and SO_2. Soly in water at 0°: 6.05% (w/w); at 20°: 13.39%; at 30°: 17.32%. Insol in alc.

8559. Sodium Dodecylbenzenesulfonate. *Dodecylbenzenesulfonic acid sodium salt;* dodecylbenzene sodium sulfonate; Santomerse #1; Conoco C-50; Conoco SD 40; Conoco C-60. $C_{18}H_{29}NaO_3S$; mol wt 348.49. C 62.04%, H 8.39%, Na 6.60%, O 13.77%, S 9.20%. $C_{12}H_{25}C_6H_4SO_3Na$. Manuf: *Chem. Eng.* **61**, no. 6, 372 (1954); Huber *et al., J. Am. Oil Chem. Soc.* **33**, 57 (1956); Brit. pat. **761,095** (1956 to Monsanto); Seaton, U.S. pat. **2,782,230** (1957 to Monsanto); Brit. pat. **773,423** (1957 to Continental Oil); Gerhart, Karwacki, U.S. pat. **2,820,056** (1958 to Continental Oil).
LD_{50} in mice: 2 g/kg orally; 105 mg/kg i.v., Hopper *et al., J. Am. Pharm. Assoc. Sci. Ed.* **38**, 428 (1949).
USE: Anionic detergent. *Caution:* May cause skin irritation. If swallowed will cause vomiting.

8560. Sodium Ethoxide. Sodium ethylate; caustic alcohol. C_2H_5NaO; mol wt 68.06. C 35.29%, H 7.41%, Na 33.79%, O 23.51%.
White or yellowish, hygroscopic powder. Dec on exposure to air and becomes darker on keeping. Dec by water into NaOH and alcohol; sol without decompn in abs alc. *Keep tightly closed, protected from light and in a cool place.*

8561. Sodium Ethyl Sulfate. Sodium sulfovinate. C_2H_5-NaO_4S; mol wt 148.11. C 16.22%, H 3.40%, Na 15.52%, O 43.21%, S 21.65%. $NaC_2H_5SO_4$.
Monohydrate, white, very hygroscopic crystals. Sol in 0.7 part water, in alcohol. *Keep well closed.*
USE: In organic syntheses.

8562. Sodium Ferricyanide. *Trisodium hexakis(cyano-C)ferrate(3−); sodium hexacyanoferrate(III).* C_6Fe-N_6Na_3; mol wt 280.91. C 25.65%, Fe 19.88%, N 29.92%, Na 24.56%. $Na_3Fe(CN)_6$.
Monohydrate, ruby-red, deliquesc crystals. Sol in 5.5 parts cold water, 1.5 parts boiling water. *Keep well closed.*

8563. Sodium Ferrocyanide. *Tetrasodium hexakis(cyano-C)ferrate(4−); sodium hexacyanoferate(II);* yellow prussiate of soda; sodium prussiate yellow. $C_6FeN_6Na_4$; mol wt 303.91. C 23.71%, Fe 18.38%, N 27.65%, Na 30.26%. Na_4-$Fe(CN)_6$. Review of properties, chemistry and syntheses: *The Chemistry of Ferrocyanides*, American Cyanamid Co. (Beacon Press, New York, 1953) 112 pp.
Decahydrate, pale yellow, monoclinic, slightly efflorescent crystals. Steady dehydration occurs above 50°. Becomes anhydr at 81.5°. Dec 435°, forming sodium cyanide, iron, carbon, and nitrogen. Soly in water at 1°: 10.2% (calcd as the anhydr salt); at 17°: 14.7%; at 25°: 17.6%; at 53°: 28.1%; at 85°: 39%; at 96.6°: 39.7%. Practically insol in most organic solvents.
USE: Addition of sodium ferrocyanide solns to slightly acidic solns of iron salts causes precipitation of insol Prussian blue (ferric ferrocyanide), $Fe_4[Fe(CN)_6]_3$. Alkaline solns yield sol Prussian blue, $NaFe[Fe(CN)_6]$. Sodium ferrocyanide forms gels with heavy metals in general. Used in ore flotation. In photography for bleaching, toning, and fixing. To prevent caking of rock salt and table salt. Additive to pickling baths. Peptizing agent in rubber. Arc stabilizer in welding rod coatings. Emulsion polymerization catalyst. *Toxicity:* Because of strong chemical bondage between the cyanide groups and the iron, ferrocyanides have a low order of toxicity. *Caution:* Do not mix with hot or concd acids and do not expose solns to sunlight for any length of time to avoid generation of hydrogen cyanide. Waste ferrocyanides in streams and lakes should not exceed 2 ppm because irradiated solns become toxic to fish, G. E. Burdick, M. Lipscheutz, *C.A.* **44**, 10939f (1950).

8564. Sodium Fluoborate. *Sodium tetrafluoroborate;* sodium borofluoride. BF_4Na; mol wt 109.82. B 9.85%, F 69.21%, Na 20.94%. $NaBF_4$. Prepd according to the equation $2H_3BO_3 + 8HF + Na_2CO_3 \rightarrow 2NaBF_4 + 7H_2O + CO_2$: Balz, Wilke-Dörfurt, *Z. Anorg. Allgem. Chem.* **159**, 197 (1927); Kwasnik in *Handbook of Preparative Inorganic Chemistry*, vol. 1, G. Brauer, Ed. (Academic Press, New York, 2nd ed., 1963) p 222.
Orthorhombic, stout rectangular prisms d^{20} 2.47. mp 384° (slight decompn). Does not etch glass when absolutely dry. Soly in water (g/100 ml): 108 (26°); 210 (100°). Sparingly sol in alcohol. Aq solns have a bitter taste and are acid to litmus.
USE: Fluorinating agent, *see* Lawton, Levy, *J. Am. Chem. Soc.* **77**, 6083 (1955).

8565. Sodium Fluoride. Chemifluor; Duraphat; Fluoros; Luride-SF; Villiaumite; Florocid; Flura-Drops; Karidium; Lemoflur; Ossalin; Ossin; Osteofluor; Zymafluor. FNa; mol wt 42.00. F 45.24%, Na 54.75%. NaF. Prepd by fusing cryolite with NaOH; by adding equiv amounts of NaOH or Na_2CO_3 to 40% HF (precipitation is instantaneous and crystal size depends on pH, but too much HF yields sodium bifluoride, $NaHF_2$): Müller, *Chem.-Ztg.* **52**, 5 (1928); Kwasnik in *Handbook of Preparative Inorganic Chemistry*, vol. 1, G. Brauer, Ed. (Academic Press, New York, 2nd ed., 1963) pp 235-236. Technical grades are 90% and 95% NaF, light (37 cu in/lb) and dense (23 cu in/lb), and 98%. The impurities are mainly sodium and aluminum fluosilicates.

Pharmacology: Caruso *et al.*, *Handb. Exp. Pharmakol.* XX (Part 2), F. Smith, Ed. (Springer, Berlin, 1970) pp 144-165. Toxicity: H. F. Smyth *et al.*, *Am. Ind. Hyg. Assoc. J.* 30, 470 (1969).

Cubic or tetragonal crystals (NaCl lattice). d 2.78. mp 993°. bp 1704°. *Poisonous!* Soly in water (g/100 ml): 4.0 (15°); 4.3 (25°); 5.0 (100°). Insol in alc. Aq solns have an alkaline reaction caused by partial hydrolysis. pH of freshly prepd satd soln 7.4. Aq solns etch glass, but the dry crystals or powder may be kept in glass bottles. Sodium fluoride sold as household insecticide must be tinted Nile Blue. LD_{50} orally in rats: 0.18 g/kg (Smyth).

Human Toxicity: Severe symptoms from ingestion of less than one gram; death from 5 to 10 g. *Sublethal:* Nausea and vomiting, abdominal distress, diarrhea, stupor, weakness. *Lethal:* Muscular weakness, tremors, convulsions, collapse, dyspnea, respiratory and cardiac failure, death. *Chronic:* Mottling of tooth enamel, osteosclerosis. *See Clinical Toxicology of Commercial Products*, R. E. Gosselin *et al.*, Eds. (Williams & Wilkins, Baltimore, 4th ed., 1976) Section III, pp 159-163.

USE: As insecticide, particularly for roaches and ants; in other pesticide formulations; constituent of vitreous enamel and glass mixes; as a steel degassing agent; in electroplating; in fluxes; in heat-treating salt compositions; in the fluoridation of drinking water; for disinfecting fermentation apparatus in breweries and distilleries; preserving wood, pastes and mucilage; manuf of coated paper; frosting glass; in removal of HF from exhaust gases to reduce air pollution. Dental caries prophylactic.

THERAP CAT (VET): Anthelmintic, pediculicide, acaricide.

8566. Sodium Folate. *Folic acid sodium salt;* sodium pteroylglutamate; sodium Folvite. $C_{19}H_{18}N_7NaO_6$; mol wt 463.39. C 49.24%, H 3.92%, N 21.16%, Na 4.96%, O 20.72%.

Sold only as sterile soln in ampuls. Clear, mobile liquid. Yellow to orange-yellow color. pH between 8.5 and 11.0. For spectrophotometric data *see* Folic Acid.

THERAP CAT: Water-soluble hematopoietic vitamin.

8567. Sodium Formaldehydesulfoxylate. *Hydroxymethanesulfinic acid sodium salt;* formaldehyde sodium sulfoxylate; formaldehydesulfoxylic acid sodium salt; sodium hydroxymethanesulfinate; sodium methanalsulfoxylate; Aldanil; Rongalite; Rongalite C. CH_3NaO_3S; mol wt 118.09. C 10.17%, H 2.56%, Na 19.47%, O 40.65%, S 27.16%. Na-[HOCH_2SO_2]. Prepn: Heyl, Greer, *Am. J. Pharm.* 94, 80 (1922); Binns, U.S. pat. 2,013,125 (1935 to Virginia Smelting); Postnikov, Kunin, *J. Appl. Chem. USSR* 13, 185 (1940). Structure of dihydrate: Truter, *J. Chem. Soc.* 1955, 3064; 1962, 3400. Of limited value in treatment of mercuric chloride poisoning: Modell *et al.*, *J. Pharmacol. Exp. Ther.* 61, 66 (1937).

Dihydrate, crystals, mp 63-64°, dec at higher temp. Odorless when freshly prepd, but quickly develops a characteristic (garlic) odor. Freely sol in water; practically insol in abs alcohol, ether, benzene. Readily dec by dil acids. Aq soln is practically neutral. *Keep well closed in a cool place.* LD s.c. in mice, 4.0 g/kg: Rosenthal, *Public Health Rept. (U.S.)* 49, 908 (1934).

USE: In vat color printing pastes: Borstelmann, Fordemwalt, U.S. pat. 2,597,281 (1952 to Am. Cyanamid). In polymerization of ethylenic compds: Brit. pats. 816,252; 852,593 (1959 to Hercules Powder; 1960 to Air Reduction). In manuf of arsphenamines: Krumwiede, *J. Am. Pharm. Assoc.* 8, 795 (1919); Heyl, Miller, *ibid.* 11, 432 (1922).

THERAP CAT: Treatment of mercury poisoning.

8568. Sodium Formate. $CHNaO_2$; mol wt 68.02. C 17.66%, H 1.48%, Na 33.81%, O 47.05%. HCOONa.

White, deliquesc granules or cryst powder; slight odor of formic acid. d 1.92. mp 253°; at higher temp dec into sodium oxalate and hydrogen, then into sodium carbonate. Sol in about 1.3 parts water; sol in glycerol, slightly in alcohol. The aq soln is neutral. pH about 7. Has buffering action. *Keep well closed.*

USE: In dyeing and printing fabrics; also in anal. chemistry as a precipitant for the "noble" metals. Solubilizes trivalent metal ions in soln by forming complex ions. Buffering action adjusts the pH of strong mineral acids to higher values.

THERAP CAT: Caustic, astringent.

8569. Sodium Gluconate. *Gluconic acid sodium salt.* $C_6H_{11}NaO_7$; mol wt 218.13. C 33.04%, H 5.08%, Na 10.54%, O 51.34%. The normal sodium salt of gluconic acid.

Crystals. The technical grade may have a pleasant odor. Soly in water at 25°: 59 g/100 ml. Sparingly sol in alcohol. Insol in ether. Aq solns are stable to short boiling periods.

USE: As sequestering agent forming water-sol complexes with calcium in alkaline media and with iron in near neutral solns. Used in metal plating, mineral tanning of hides, mordanting fabrics, and in water-paste paints. Has been suggested as a photographic processing aid.

8570. Sodium Glycerophosphate. *1,2,3-Propanetriol mono(dihydrogen phosphate) disodium salt.* $C_3H_7Na_2O_6P$; mol wt 216.03. C 16.68%, H 3.26%, Na 21.28%, O 44.44%, P 14.34%. Three isomers exist: The β-glycerophosphoric acid disodium salt ((HOCH_2)_2CHOPO_3Na_2) and D(+)- and L(−)-α-glycerophosphoric acid disodium salt (HOCH_2CH(OH)CH_2OPO_3Na_2). Prepn: H. King, F. L. Pyman, *Pharm. J.* 92, 511 (1914). Exptl use in diagnosis of prostatic carcinoma: M. K. Schwartz *et al.*, *Ann. N.Y. Acad. Sci.* 166, 775 (1969). Chronic toxicity study of β-form: K. L. Raheja *et al.*, *Toxicology* 8, 115 (1977). Efficacy of β-form as cariostatic agent: T. H. Grenby, J. M. Bull. *Arch. Oral Biol.* 20, 717 (1975). GC determn: Y. Handa *et al.*, *J. Chromatog.* 206, 387 (1981).

β-Form hemiundecahydrate, white, odorless, scale-like crystals; dec above 130°. Sol in about 1.5 parts water; more sol in hot water. (pH of aq soln: about 9.5). Insol in alcohol.

THERAP CAT: Tonic.

THERAP CAT (VET): Has been used as a tonic.

8571. Sodium Hexachloroplatinate(IV). *Disodium hexachloroplatinate(2−);* sodium platinichloride; sodium chloroplatinate. Cl_6Na_2Pt; mol wt 453.77. Cl 46.87%, Na 10.13%, Pt 42.99%. $Na_2[PtCl_6]$. Prepn: Grube in *Handbook of Preparative Inorganic Chemistry* vol. 2, G. Brauer, Ed. (Academic Press, New York, 2nd ed., 1965) pp 1571-1572; Cox, Peters, *Inorg. Syn.* 13, 173 (1971).

Yellow, hygroscopic crystals. Easily forms hexahydrate at 25° and relative humidity > 50% (reconverted to anhyd salt by heat at 110° for one hour). uv max (1 formal HCl): 262 nm (ε 24,500). Sol in water, alcohol.

USE: Catalyst.

8572. Sodium Hexafluorosilicate. Sodium fluosilicate; sodium silicofluoride; Salufer. F_6Na_2Si; mol wt 188.05. F 60.62%, Na 24.46%, Si 14.92%. Na_2SiF_6. The average grade of commerce is about 98% pure.

White, granular powder. d 2.68. Melts at red heat with decompn. Sol in 150 parts cold, 40 parts boiling water; insol in alc. The soln in cold water is neutral. LD_{50} orally in rats: 125 mg/kg, *Handbook of Toxicology* vol. 1, W. S. Spector, Ed. (Saunders, Philadelphia, 1956) pp 278-279.

USE: In enamels for china and porcelain; manuf opal glass; as insect exterminator and poison for rodents; mothproofing of woolens.

THERAP CAT (VET): Pediculicide.

8573. Sodium Hydride. HNa; mol wt 24.00. H 4.21%, Na 95.79%. NaH. Prepd by passing hydrogen into molten sodium dispersed in oil or mixed with a catalyst such as anthracene above 250°: Hansley, Carlisle, *Chem. & Eng. News* 23, 1332 (1945). Laboratory procedure by hydrogenating sodium dispersions: Mattson, Whaley, *Inorg. Syn.* 5, 10 (1957). Book: J. Plesek, S. Hermanek, *Sodium Hydride, Its Use in the Laboratory and in Technology* (Iliffe Books, London, 1968) 185 pp.

Silvery needles; the commercial product is a gray-white powder. d 1.396. Dec 425°. Reacts explosively with water, violently with lower alcohols, ignites spontaneously on standing in moist air. Sol in molten sodium hydroxide, insol in liq ammonia but forms sodamide at moderate temps.

USE: At low temps where reducing properties of sodium are undesirable as in the condensation of ketones and aldehydes with acid esters; in soln with molten sodium hydroxide for the reduction of oxide scale on metals; at high temps as a reducing agent and reduction catalyst.

8574. Sodium Hydrosulfite. Sodium sulfoxylate; sodium dithionite. $Na_2O_4S_2$; mol wt 174.13. Na 26.41%, O 36.76%, S 36.83%. $Na_2S_2O_4$. The hydrosulfite of commerce contains 85-90% $Na_2S_2O_4$.

White or grayish-white, cryst powder; slight characteristic odor. Oxidizes in air (more readily so in presence of moisture or when in soln) to bisulfite and bisulfate and acquires an acid reaction. Very sol in water, slightly in alcohol.

USE: As reducing agent, particularly in dyeing with indigo and vat dyes; bleaching soaps, straw; removing dyes from dyed fabrics.

Note: The name sodium hydrosulfite is applied also to $NaHSO_2$, mol wt 88.06, sol in water, alcohol. Still more confusion results when "sodium hyposulfite" is applied to this compd ($Na_2S_2O_4$) see 1957 Subject Index to Chem. Abstracts, p 2218s under sodium dithionite.

8575. Sodium Hydroxide. Caustic soda; soda lye; sodium hydrate. HNaO; mol wt 40.01. H 2.52%, Na 57.48%, O 40.00%. NaOH. By reacting calcium hydroxide with sodium carbonate; from sodium chloride by electrolysis; from sodium metal and water vapor at low temp. Description of industrial processes: Faith, Keyes & Clark's Industrial Chemicals, F. A. Lowenheim, M. K. Moran, Eds. (Wiley-Interscience, New York, 4th ed., 1975) pp 737-745. Toxicity: Fazekas, Arch. Exp. Pathol. Pharmakol. **184,** 587 (1937).

Fused solid with crystalline fracture. Rapidly absorbs carbon dioxide and water from the air. Very corrosive (caustic) to animal and vegetable tissue and to aluminum metal in the presence of moisture. Sold as lumps, sticks, pellets, chips, etc. When kept in tight containers, the usual grades contain 97-98% NaOH. mp 318°. d^{25} 2.13. One gram dissolves in 0.9 ml water, 0.3 ml boiling water, 7.2 ml abs alcohol, 4.2 ml methanol, also sol in glycerol. Generates considerable heat while dissolving, or when the soln is mixed with an acid. Volumetric NaOH solns used in the laboratory must be protected from air to avoid formation of carbonate. Concentrated NaOH solns dissolve practically no sodium carbonate. The pH of a 0.05% w/w soln is about 12, of a 0.5% soln about 13, of a 5% soln about 14. Density, freezing and boiling point data of water solns: 5% soln (w/w) (d_4^{15}, fp, bp): 1.056, −4°, 102°; 10% soln: 1.111, −10°, 105°; 20% soln: 1.222, −26°, 110°; 30% soln: 1.333, 1°, 115°; 40% soln: 1.434, 15°, 125°; 50% soln: 1.530, 12°, 140°. LD orally in rabbits: 500 mg/kg (10% soln) (Fazekas).

Human Toxicity: Corrosive to all tissues. Ingestion: vomiting, prostration, collapse. Constrictive scarring may result. Inhalation of the dust or concd mist may cause damage to respiratory tract. Caution: Evacuation of stomach should not be attempted.

USE: NaOH solutions are used to neutralize acids and make sodium salts, e.g., in petroleum refining to remove sulfuric and organic acids; to treat cellulose in making viscose rayon and cellophane; in reclaiming rubber to dissolve out the fabric; in making plastics to dissolve casein. NaOH solns hydrolyze fats and form soaps; they precipitate alkaloids (bases) and most metals (as hydroxides) from water solns of their salts. Pharmaceutic aid (alkalizer).

THERAP CAT (VET): Caustic; dehorning of calves.

8576. Sodium Hypochlorite. ClNaO; mol wt 74.44. Cl 47.62%, Na 30.88%, O 21.49%. NaClO. Pentahydrate prepd from NaOH and Cl_2 in the presence of water: Sanfourche, Gardent, Bull. Soc. Chim. [4] **35,** 1089 (1924); Schmeisser in Handbook of Preparative Inorganic Chemistry vol. 1, G. Brauer, Ed. (Academic Press, New York, 2nd ed., 1963) pp 309-310.

Pentahydrate, highly unstable. Crystals, mp 18°. Dec by CO_2 from air. Anhyd NaClO may be obtained by freeze-drying in a vacuum (over concd H_2SO_4). Anhyd NaClO is very explosive. Soly of the pentahydrate at 0°: 29.3 g/100 ml H_2O. The hypochlorite ion in aq soln is remarkably sta-

ble. Names for aq solns are Eau de Labarraque; Clorox; Dazzle.

Human Toxicity: Ingestion may cause corrosion of mucous membranes, esophageal or gastric perforation, laryngeal edema. Inhalation may produce severe bronchial irritation, pulmonary edema. Prolonged skin contact may result in irritation, cf. Clinical Toxicology of Commercial Products, R. E. Gosselin et al., Eds. (Williams & Wilkins, Baltimore, 4th ed., 1976) Section III, pp 174-176.

USE: As bleach, disinfectant.

THERAP CAT: Disinfectant.

8577. Sodium Hypochlorite Solution, Alkaline. Antiformin. A strongly alk soln of sodium hypochloride; 100 ml contains 5.68 g active Cl, 7.8 g NaOH, 32 g Na_2CO_3.

Yellowish, clear, liq; odor of hypochlorites.

USE: Germicide, disinfectant, deodorizer. Also in bacteriology labs. Caution: See Sodium Hypochlorite.

8578. Sodium Hypochlorite Solution, Diluted. Modified Dakin's soln; surgical chlorinated soda soln; Carrel-Dakin soln. A soln of sodium hypochlorite contg 0.45-0.50 g of the salt in 100 ml. Prepd by diluting with distilled water a soln of sodium hypochlorite, adding a 5% soln of sodium bicarbonate, and adjusting to the proper strength and concn according to the procedure described in N.F. May be prepd also from 15.4 g chlorinated lime (30% available chlorine), 7.7 g anhyd sodium carbonate, and 6.4 g sodium bicarbonate per liter; this soln is then adjusted as in the previous method.

Colorless or slightly yellow liquid; faint chlorine-like odor. d of the soln should not exceed 1.025. Does not produce red color with phenolphthalein.

Human Toxicity: See Sodium Hypochlorite.

THERAP CAT: Topical anti-infective.

THERAP CAT (VET): Antiseptic for wound irrigation.

8579. Sodium Hypophosphate. $Na_4O_6P_2$; mol wt 249.91. Na 36.80%, O 38.41%, P 24.79%. $Na_4P_2O_6$.

Decahydrate, colorless or white crystals. Sol in water, the soln being alkaline.

8580. Sodium Hypophosphite. Phosphinic acid, sodium salt. H_2NaO_2P; mol wt 87.97. H 2.29%, Na 26.13%, O 36.37%, P 35.20%. $NaHPO_2$. Solubility data: Palit, J. Am. Chem. Soc. **69,** 3120 (1947).

Monohydrate, white, odorless, deliquesc granules; saline taste. When strongly heated, it dec with evolution of phosphine which ignites spontaneously in the air. It explodes when triturated with chlorates or other oxidizing agents. Sol in 1 part water, 0.15 part boiling water; freely sol in glycerol and in boiling alcohol; sol in cold alcohol, insol in abs alcohol. Insol in ether. Soly of anhyd NaH_2PO_2 at 25° in ethylene glycol: 33.0 g/100 g; in propylene glycol: 9.7 g/100 g. The aq soln is neutral. Keep well closed.

USE: As reagent for arsenic and iodates; prepn of hypophosphites syrup.

8581. Sodium Iodate. $INaO_3$; mol wt 197.90. I 64.13%, Na 11.62%, O 24.25%. $NaIO_3$. The article of commerce contains about 99% $NaIO_3$.

White, cryst powder. d 4.28. Sol in about 11 parts water, 3 parts boiling water; insol in alc. The aq soln is neutral. LD i.v. in dogs: 200 mg/kg, Handbook of Toxicology vol. 1, W. S. Spector, Ed. (Saunders, Philadelphia, 1956) pp 274-275.

THERAP CAT: Antiseptic (mucous membranes).

8582. Sodium Iodide. Ioduril; Anayodin. INa; mol wt 149.92. I 84.66%, Na 15.34%. NaI. U.S.P. NaI is at least 99% pure.

White, odorless, deliquesc crystals or granules. Gradually absorbs up to about 5% (½ mol) moisture on exposure to air. Slowly becomes brown in the air due to liberation of iodine. Its aq soln is similarly affected. d 3.67. mp 651°. One gram dissolves in 0.5 ml water, about 2 ml alc, 1 ml glycerol; sol in acetone. It is made slightly alkaline to render it more stable. pH: 8-9.5. Keep well closed and protected from light. At ordinary room temp crystallizes from water with $2H_2O$ in the form of colorless, prismatic crystals. Incompat. As of potassium iodide. MLD i.v. in rats: 1.3 g/kg, Loeser, Konwiser, J. Lab. Clin. Med. **15,** 35 (1929).

THERAP CAT: Iodine supplement; expectorant.

THERAP CAT (VET): Actinobacillosis, actinomycosis. Expectorant. Has been used for ringworm, hyperplastic fibrous lesions, paraplegia from pachymeningitis of dogs.

8583. Sodium Iodide, Radioactive. Sodium iodide—[131]I; sodium radio-iodide (^{131}I); Na[131]I; Iodotope; Oriodide; Radiocaps-131; Theriodide-131. Prepd from radioactive iodine (^{131}I) which has a half-life of 8 days and emits beta and gamma rays. Other properties identical with those of ordinary sodium iodide. Dispensed as carrier-free sodium radioiodide in capsules for oral use or in aq soln for oral or parenteral administration.

THERAP CAT: Diagnostic aid (thyroid function); antineoplastic.

8584. Sodium Iodomethamate. *1,4-Dihydro-3,5-diiodo-1-methyl-4-oxo-2,6-pyridinedicarboxylic acid disodium salt;* 3,5-diiodo-1-methylchelidamic acid sodium salt; disodium N-methyl-3,5-diiodo-4-pyridone-2,6-dicarboxylate; D40; Neo-Iopax R; Neo-Iopax Sodium; Iodoxyl; Pyelectan; Uropac; Uroselectan B; Urumbrin. $C_8H_3I_2NNa_2O_5$; mol wt 492.95. C 19.49%, H 0.61%, I 51.50%, N 2.84%, Na 9.33%, O 16.23%. Prepn: Chelidonic acid is converted to chelidamic acid by the action of NH_3, chelidamic acid is iodinated with iodine in a boiling aq alkaline soln, the iodinated product is methylated at the nitrogen with dimethyl sulfate in hot alkaline soln, cf. Dohrn, Diedrich, *Ann.* **494**, 284 (1932); **Ger.** pats. **545,916; 545,266; 556,142;** U.S. pat. **1,919,417** (1933).

Crystals, dec around 200° with effervescence. Freely sol in water (1:1); insol in chloroform, ether, acetone, petr ether.

THERAP CAT: Diagnostic aid (radiopaque medium).

8585. Sodium Isopropyl Xanthate. *Isopropylxanthic acid sodium salt;* Good-Rite Nix. $C_4H_7NaOS_2$; mol wt 158.22. C 30.36%, H 4.46%, Na 14.53%, O 10.11%, S 40.53%. $(CH_3)_2$-CHOCSSNa.

Deliquescent, white to yellowish powder or lumps, dec 150°. Slightly unpleasant odor. Soly in water: 30% at 4°; 46% at 24°; 54% at 35°.

USE: Control of annual weeds in bean and pea fields. *Caution:* Irritating to skin, eyes, mucous membranes, respiratory tract.

8586. Sodium Lactate. Lacolin. $C_3H_5NaO_3$; mol wt 112.07. C 32.15%, H 4.50%, Na 20.53%, O 42.83%. Commercially available as a mixture with water containing 70-80% sodium lactate. *Ref:* Shaw, U.S. pat. **2,856,326** (1958 to Nat. Dairy Prod. Corp.).

Colorless or almost colorless, thick, odorless liquid. Miscible with water, alcohol. The soln is neutral.

USE: Instead of glycerol in calico printing; as a plasticizer for casein; as a corrosion inhibitor in alc antifreeze mixture.

THERAP CAT: Electrolyte replenisher; systemic and urinary alkalizer.

THERAP CAT (VET): Has been used in bovine ketosis.

8587. Sodium Lauryl Sulfate. *Sulfuric acid monododecyl ester sodium salt;* sodium dodecyl sulfate; SDS; Irium. $C_{12}H_{25}NaO_4S$; mol wt 288.38. C 49.98%, H 8.74%, Na 7.97%, O 22.19%, S 11.12%. $CH_3(CH_2)_{10}CH_2OSO_3Na$. Anionic detergent prepd by sulfation of lauryl alcohol, followed by neutralization with sodium carbonate: A. Lottermoser, F. Stoll, *Kolloid-Z.* **63**, 50 (1933). Surfactant properties: J. Powney, C. C. Addison, *Trans. Faraday Soc.* **33**, 1244 (1937); E. E. Dreger *et al., Ind. Eng. Chem.* **36**, 610 (1944). Use in electrophoretic sepn and mol wt estimation of proteins: A. L. Shapiro *et al., Biochem. Biophys. Res. Commun.* **28**, 815 (1967); K. Weber, M. Osborn, *J. Biol. Chem.* **244**, 4406 (1969); of glycopolypeptides: B. S. Leach *et al., Biochemistry* **19**, 5734 (1980). Toxicity study: A. I. T. Walker

et al., Food Cosmet. Toxicol. **5**, 763 (1967). Review of toxicology: Ch. Gloxhuber, *Arch. Toxicol.* **32**, 245-270 (1974).

White or cream-colored crystals, flakes, or powder. Faint odor of fatty substances. Smooth feel. Neutral reaction. One gram dissolves in 10 ml water, giving an opalescent soln. Lowers the surface tension of aq solns. Emulsifies fats. LD_{50} orally in rats: 1288 mg/kg (Walker).

USE: Wetting agent, detergent, esp in the textile industry. Electrophoretic separation of proteins and lipids. Ingredient of toothpastes.

8588. Sodium-Lead Alloy. Lead-sodium alloy; sodium-lead; Drynap; Dri-Na. Usually contains a minimum of 9.5% active sodium. For tetraethyllead manufacture the sodium-lead alloy is produced in large quantities by making a melt of 90 parts of lead with 10.5 parts of sodium (w/w). The reaction is strongly exothermic and starts at 225°. Prepn of the alloy on a laboratory scale: Soroos, *Ind. Eng. Chem., Anal. Ed.* **11**, 657 (1939).

Obtained in brittle lumps which can be stored in an air-tight container. It reacts only slowly with air or water, yet dries ether as completely as sodium wire. Residues of ether or other liquids still containing some active alloy can be destroyed safely by the addition of water, as the reaction never reaches the violence observed with pure sodium metal.

USE: In manuf of tetraethyllead by reaction of alloy with ethyl chloride; for drying ether and for reductions: Tabei *et al., Bull. Chem. Soc. Japan* **40**, 1538 (1967); as a sodium substitute in many chemical reactions where the presence of lead is not objectionable. The alloy may be ground to a very fine powder *under* the surface of a non-polar solvent such as kerosine or ether. *Caution:* Finely ground powder, if not protected by a suitable liquid, may react with excess moisture from the air sufficiently to catch fire.

8589. Sodium Metabisulfite. *Sodium pyrosulfite.* $Na_2O_5S_2$; mol wt 190.13. Na 24.19%, O 42.08%, S 33.73%. $Na_2S_2O_5$. Crystallizes from cold water with $7H_2O$.

White crystals or powder; odor of SO_2. Freely sol in water, glycerol; slightly sol in alcohol. The aq soln is acid.

USE: Pharmaceutic aid (antioxidant).

8590. Sodium Metaborate. $BNaO_2$; mol wt 65.82. B 16.44%, Na 34.94%, O 48.62%. $NaBO_2$. Obtained by fusing equivalent mol wts of borax and sodium carbonate.

White pieces or powder. mp 966°. Sol in water, the soln being strongly alkaline.

8591. Sodium Metaperiodate. *Sodium periodate.* $INaO_4$; mol wt 213.91. I 59.33%, Na 10.75%, O 29.92%. $NaIO_4$. Synthesis starting with sodium iodide: *Inorg. Syn.* **2**, 212 (1946) and **1**, 170 (1939).

White, tetragonal crystals, d_4^{16} 3.865. Dec around 300°. Sol in cold water, sulfuric, nitric, acetic acids.

Trihydrate, white, efflorescent, trigonal crystals, d_4^{18} 3.219. Dec 175°. One gram dissolves in 8 ml water at 20°.

8592. Sodium Metaphosphate. $(NaPO_3)_n$ where"n" may be 2 or more. Prepd by dehydration of sodium phosphates (e.g. $Na_2H_2P_2O_7$, NaH_2PO_4, Na_2HPO_4). Several different forms are known. *Reviews:* J. R. Van Waser, *Phosphorus and Its Compounds* vol. **1** (Interscience, New York, 1958) pp 601-800; Pantony in *Mellor's* vol. **II** supplement II, *The Alkali Metals* (Part 1), 1316-1349 (1961); Thilo in *Advan. Inorg. Chem. Radiochem.* **4**, 1-75 (1962). *See also* Sodium Hexametaphosphate; Sodium Trimetaphosphate; Maddrell's Salt.

8593. Sodium Metasilicate. Na_2O_3Si; mol wt 122.08. Na 37.67%, O 39.32%, Si 23.01%. Na_2SiO_3. Usually prepd from sand (SiO_2) and soda ash (Na_2CO_3) by fusion: Schwarz, *Z. Anorg. Allgem. Chem.* **126**, 62 (1923); Faith, Keyes & Clark's *Industrial Chemicals*, F. A. Lowenheim, M. K. Moran, Eds. (Wiley-Interscience, New York, 4th ed., 1975) pp 755-761. *Review:* Gmelin's, *Sodium* (8th ed. supplement) **21**, pp 1474-1478 (1967).

Usually obtained as a glass; also orthorhombic crystals. d 2.614. mp 1089°. n_D^{25} (glass) 1.520. Specific heat (20°): 0.217. Heat of formation: −371.2 kcal/mol. Heat of soln (cryst) −7.45 kcal/mol. Heat of hydration for the nonahydrate: −24.15 kcal/mol. Heat of fusion: 10.3 kcal/mol.

Sol in cold water, hydrolyzed by hot water. Insol in alcohol, acids, salt solns.

Nonahydrate, efflorescent, orthorhombic bipyramidal platelets, mp 48° in water of crystn.

8594. Sodium Methoxide. Sodium methylate. CH_3NaO; mol wt 54.03. C 22.23%, H 5.60%, Na 42.56%, O 29.62%. CH_3ONa. Prepn: Burness, *Org. Syn.* **39**, 51 (1959).

White, free-flowing powder. Sensitive to air and moisture. Dec by water. Sol in methanol, ethanol. Also exists in solvated form, $CH_3ONa.2CH_3OH$, white powder. Apparent density of solvent-free material about 4.6 lb/gal.

USE: In organic syntheses.

8595. Sodium Methyl Sulfate. CH_3NaO_4S; mol wt 134.08. C 8.96%, H 2.26%, Na 17.15%, O 47.73%, S 23.91%. $NaCH_3SO_4$.

Monohydrate, white, hygrosopic crystals. Sol in water, alc, methanol. *Keep well closed.*

8596. Sodium Molybdate(VI). $MoNa_2O_4$; mol wt 205.92. Mo 46.59%, Na 22.33%, O 31.08%. Na_2MoO_4.

Dihydrate, $MoNaO_4.2H_2O$, *Molyhibit 100.* Cryst powder. Loses its water of crystn at 100°. Sol in 1.7 parts cold water, about 0.9 part boiling water. pH of 5% aq soln at 25° = 9.0-10.0. Has been reported to be less toxic than the other corresponding compds of group 6B in the periodic table: Fairhall *et al., Public Health Bull.* no. 293 (1945).

USE: Manuf of inorganic and organic pigments, corrosion inhibitor, bath additive for metals finishing, reagent for alkaloids, micronutrient for plants and animals.

8597. Sodium β-**Naphthoquinone-4-sulfonate.** *3,4-Dihydro-3,4-dioxo-1-naphthalenesulfonic acid sodium salt;* sodium 1,2-naphthoquinone-4-sulfonate; 1,2-naphthoquinone-4-sulfonic acid sodium salt; β-naphthoquinone-4-sulfonic acid sodium salt. $C_{10}H_5NaO_5S$; mol wt 260.20. C 46.16%, H 1.94%, Na 8.84%, O 30.74%, S 12.32%. Prepn: Folin, *J. Biol. Chem.* **51**, 377 (1922).

Yellow crystals from dil alc. Readily sol in water; slightly sol in 95% alcohol; moderately sol in acetone; practically insol in ether, chloroform, carbon disulfide, benzene, petr ether. Aq solns fade slowly on exposure to light but may be stabilized by HCl.

USE: Colorimetric determn of amino acids and amines.

8598. Sodium Nitrate. Chile saltpeter; cubic niter; soda niter. $NNaO_3$; mol wt 85.01. N 16.48%, Na 27.05%, O 56.47%. $NaNO_3$. The purified grade contains at least 99% $NaNO_3$. Occurs as a mineral in Chile.

Colorless, transparent crystals, white granules or powder; deliquesc in moist air. d 2.26. mp 308°. One gram dissolves in 1.1 ml water, 0.6 ml boiling water, 125 ml alcohol, 52 ml boiling alcohol, 3470 ml abs alcohol, 300 ml abs methanol. When dissolved in water the temp of the soln is lowered. The aq soln is neutral. *Keep well closed.* LD_{50} orally in rabbits: 1.955 g anion/kg, Dollahite, Rowe, *Southwest Vet.* **27**, 246 (1974).

USE: Manuf of nitric acid and as catalyst in the manuf of sulfuric acid. Manuf sodium nitrite, glass, enamels for pottery; in matches; for improving burning properties of tobacco; pickling meats; as color fixative in meats. The technical grade is used as fertilizer.

8599. Sodium Nitrite. *Nitrous acid sodium salt;* erinitrit. $NNaO_2$; mol wt 69.00. N 20.30%, Na 33.32%, O 46.38%. Contains 96-98% $NaNO_2$. Thermodynamic properties: Plekhotkun, *Zh. Prikl. Khim.* **40**, 1843 (1967), *C.A.* **68**, 33797x (1968). Toxicity studies: Faccini *et al., Ind. Aliment.* **8**, 77 (1969), *C.A.* **71**, 100583b (1969). Review of chemistry of $NaNO_2$ in a biological system as related to meat curing: Bard, Townsend, *Science of Meat & Meat Products,* J. F.

Price, B. S. Schweigert, Eds. (W. H. Freeman, 2nd ed. 1971) pp 452-470.

White or slightly yellow, hygroscopic granules, rods, or powder. Very slowly oxidizes to nitrate in air. d 2.17. mp 271°; dec above 320°. Sol in 1.5 parts cold water, 0.6 part boiling water, slightly in alc. Dec even by weak acids with evolution of brown fumes of N_2O_3. The aq soln is alkaline. pH about 9. *Keep well closed.* LD_{50} orally in rats: 180 mg/kg, Smyth *et al., Am. Ind. Hyg. Assoc. J.* **30**, 470 (1969).

Incompat: Acetanilide, antipyrine, chlorates, hypophosphites, iodides, mercury salts, permanganate, sulfites, tannic acid, vegetable astringent decoctions, infusions or tinctures.

USE: Manuf diazo dyes, nitroso compds, and in many other processes of manuf of organic chemicals; dyeing and printing textile fabrics; bleaching flax, silk, and linen; photography. In meat curing, coloring and preserving; in processing smoked chub. Also as reagent in anal. chemistry. *Caution:* Consult latest Government regulation to determine max amounts allowable in food.

THERAP CAT: Vasodilator; antidote (cyanide poisoning).

THERAP CAT (VET): In cyanide poisoning. Has been used as a vasodilator, as a circulatory (blood pressure) depressant and to relieve smooth muscle spasm.

8600. Sodium Nitroprusside. *Pentakis(cyano-C)nitrosylferrate(2−) disodium; sodium nitrosylpentacyanoferrate(III);* sodium nitroferricyanide; sodium nitroprussiate; Nipruss. $C_5FeN_6Na_2O$; mol wt 261.91. C 22.92%, Fe 21.32%, N 32.09%, Na 17.56%, O 6.11%. $Na_2[Fe(CN)_5NO]$. Prepn: L. Playfair, *Proc. Roy. Soc. London* **5**, 846 (1849). Pharmacology: Fernandez *et al., Arch. Inst. Farmacol. Exp. Madrid* **23**, 1-51 (1971). *Review:* I. H. Tuzel, *J. Clin. Pharmacol.* **14**, 494-503 (1974). Review of pharmacology, toxicology and therapeutic uses: J. H. Tinker, J. D. Michenfelder, *Anesthesiol.* **45**, 340 (1976). Comprehensive description: R. Rucki in *Analytical Profiles of Drug Substances* vol. 6, K. Florey, Ed. (Academic Press, New York, 1977) pp 487-513; A. Bult *et al., ibid.* vol. 15 (1986) pp 781-792.

Dihydrate, *Nipride, Nitropress.* Ruby-red, practically odorless, transparent crystals. Sol in about 2.3 parts water, slightly in alcohol. Slowly dec in aq soln.

USE: Reagent for the detection of many organic compds, e.g., acetone, aldehydes, also of alkali sulfides, zinc, SO_2.

THERAP CAT: Antihypertensive.

8601. Sodium Oxalate. *Ethanedioic acid disodium salt.* $C_2Na_2O_4$; mol wt 134.01. C 17.92%, Na 34.32%, O 47.76%. $Na_2C_2O_4$.

White, odorless, cryst powder. Sol in 27 parts water, 16 parts boiling water; insol in alcohol. The aq soln is practically neutral.

Caution: Ingestion of concd solns may cause severe G.I. irritation, hematemesis, CNS and cardiac depression, death. Dilute solns produce little G.I. distress but may cause weakness, muscular twitchings, rarely convulsions, coma, death. Chronic ingestion of small amounts may cause hypocalcemic tetany, urinary calculi, *cf. Clinical Toxicology of Commercial Products,* R. E. Gosselin *et al.,* Eds. (Williams & Wilkins, Baltimore, 4th ed., 1976) Section III, pp 260-263.

USE: Finishing textiles, tanning and finishing leather; for standardizing potassium permanganate soln.

8602. Sodium Oxide. Sodium monoxide. Na_2O; mol wt 61.98. Na 74.19%, O 25.81%.

White, amorphous pieces or powder. d 2.27. Melts at a dull red heat and begins to dec above 400° into sodium peroxide and metal. It is very reactive and combines violently with water, forming sodium hydroxide. *Handle with tongs and not with bare hands, and keep tightly closed.*

USE: As a dehydrating agent; in certain chemical reactions as a polymerizing or condensing agent.

8603. Sodium Oxybate. *4-Hydroxybutanoic acid monosodium salt;* sodium γ-hydroxybutyrate; Gamma OH; NSC-84223; Wy-3478; Anetamin; Somsanit. $C_4H_7NaO_3$; mol wt 126.09. C 38.10%, H 5.60%, Na 18.23%, O 38.07%. $HOCH_2CH_2CH_2COONa$. Prepd from γ-butyrolactone and NaOH: Marvel, Birkhimer, *J. Am. Chem. Soc.* **51**, 260 (1929).

Crystals from alcohol.

THERAP CAT: Hypnotic; adjunct to anesthesia.

8604. Sodium Perborate. Dexol. BNaO₃; mol wt 81.80. B 13.22%, Na 28.10%, O 58.68%. NaBO₃. Contains, when reasonably fresh, about 95% of the perborate corresp to 9.9% available oxygen. Prepn from sodium metaborate and hydrogen peroxide: Leblon, Lambert, U.S. pat. **3,109,706** (1963 to Solvay & Cie.).

Tetrahydrate, white, odorless, cryst powder; saline taste. Stable when kept cool and dry, but is dec with liberation of oxygen in warm or moist air. Dec above 60°. Sol in about 40 parts water, the soln being alkaline and dec with the liberation of H₂O₂, and then of oxygen. In the presence of acids, H₂O₂ is formed. *Keep well closed and in a cool place.*

USE: Bleaching straw and other fibers, ivory, sponges, bristles, waxes, textiles; in laundering, dentifrices, soaps. *Caution:* Prevent swallowing of soln.

THERAP CAT: Topical antiseptic.

THERAP CAT (VET): Mouthwash.

8605. Sodium Perchlorate. Irenat. ClNaO₄; mol wt 122.44. Cl 28.95%, Na 18.78%, O 52.27%. NaClO₄.

Monohydrate, white, deliquesc crystals. Dec at about 130°. d 2.02. Very sol in water. *Keep well closed.*

USE: In the explosives industry.

THERAP CAT: Thyroid inhibitor.

8606. Sodium Permanganate. MnNaO₄; mol wt 141.93. Mn 38.71%, Na 16.20%, O 45.09%. NaMnO₄.

Trihydrate, reddish-black, very hygroscopic granules. Very sol in water; dec by alcohol.

8607. Sodium Peroxide. Sodium dioxide; sodium super-oxide; Solozone. Na₂O₂; mol wt 77.99. Na 58.97%, O 41.04%. The product of commerce contains 90-95% Na₂O₂. Prepd by heating sodium metal to 300° in aluminum vessels with a current of air from which carbon dioxide has been removed. Prepn of the octahydrate: Penneman, *Inorg. Syn.* **3**, 1 (1950).

Yellowish-white, granular powder. Absorbs water and CO₂ from the air. Freely sol in water, forming sodium hydroxide and hydrogen peroxide, the latter quickly dec into oxygen and water. With dil acids H₂O₂ is formed which remains stable. In contact with organic matter or readily oxidizable substances ignition and explosion may take place. *Keep tightly closed and protected from contact with organic or oxidizable substances.*

USE: Bleaching animal and vegetable fibers, feathers, bones, ivory, wood, wax, sponges, coral; rendering air charged with CO₂ respirable as in torpedo boats, submarines, diving bells, etc.; purifying air in sick rooms; dyeing and printing textiles; chemical analysis. General oxidizing agent. *Caution:* Irritant and corrosive. *See* Sodium Hydroxide.

8608. Sodium Persulfate. *Sodium peroxydisulfate.* Na₂O₈S₂; mol wt 238.13. Na 19.31%, O 53.76%, S 26.93%. Na₂S₂O₈.

White, cryst powder. Gradually dec; decompn is promoted by moisture and higher temp. Initial soly in water at 20°: 549 g/l; dec by alcohol and silver ions. MLD i.v. in rabbits: 178 mg/kg, *Handbook of Toxicology* vol. **1**, W. S. Spector, Ed. (Saunders, Philadelphia, 1956) pp 278-279.

USE: Bleaching and oxidizing agent; promoter for emulsion polymerization reactions. *Caution:* Highly irritating to skin, mucous membranes.

8609. Sodium Pertechnetate ⁹⁹ᵐTc. Pertscan; Ultra-Technekow. NaO₄⁹⁹ᵐTc. Prepn: Keller, Kanellakopulos, *Radiochim. Acta* **1**, No. 2, 107 (1963), *C.A.* **59**, 1256a (1963); Kanellakopulos, **AEC Accession** No. **31424**, Rept. No. **KFK-197**, 73 pp (1964), *C.A.* **62**, 7350d (1965). Clinical application for labelling red blood cells: D. Ducassou *et al., Brit. J. Radiol.* **49**, 344 (1976). Diagnostic use in Meckel's diverticulum: D. R. Cooney *et al., J. Pediatr. Surg.* **17**, 611 (1982); in thyroid neoplasm: M. Vorne, K. Jarve, *Eur. J. Nucl. Med.* **13**, 362 (1987). Review of diagnostic use in brain scanning: J. G. McAfee *et al., J. Nucl. Med.* **5**, 811-827 (1964); in thyroid function: M. S. Sucupira *et al., Int. J. Nucl. Med. Biol.* **10**, 29-33 (1983).

THERAP CAT: Diagnostic aid (radioactive imaging agent).

8610. Sodium Phenolsulfonate. *Hydroxybenzenesulfonic acid sodium salt;* sodium sulfocarbolate. C₆H₅NaO₄S; mol wt 196.15. C 36.74%, H 2.57%, Na 11.72%, O 32.63%, S 16.34%. HOC₆H₄SO₃Na.

Dihydrate, white, odorless crystals; slightly bitter taste; somewhat efflorescent in dry air. One gram dissolves in 4.2 ml water, 0.8 ml boiling water, 140 ml alcohol, 13.5 ml boiling alcohol, 5 ml glycerol. The aq soln is neutral.

THERAP CAT: Intestinal antiseptic.

THERAP CAT (VET): Has been used as an intestinal antiseptic, in dusting powders for ulcers, slowly granulating wounds and in dilute solution in the eye.

8611. Sodium Phenoxide. Sodium phenate; sodium carbolate; sodium phenolate; phenol sodium. C₆H₅NaO; mol wt 116.10. C 62.07%, H 4.34%, Na 19.81%, O 13.78%. C₆H₅ONa. Prepn from phenol and NaOH in dil methanol: Kornblum, Lurie, *J. Am. Chem. Soc.* **81**, 2710 (1959).

White to reddish, deliquescent rods or granules. Decomposed by the CO₂ of the air. Very sol in water; sol in alcohol. The aq soln is caustic.

8612. Sodium Phosphate, Dibasic. Dibasic sodium phosphate; disodium hydrogen phosphate; disodium orthophosphate; disodium phosphate; DSP; phosphate of soda; secondary sodium phosphate. HNa₂O₄P; mol wt 141.98. H 0.71%, Na 32.39%, O 45.08%, P 21.82%. Na₂HPO₄. Industrial production: *Faith, Keyes, & Clark's Industrial Chemicals* (John Wiley, New York, 4th ed., 1975) pp 746-754. Toxicity of heptahydrate: H. F. Smyth *et al., Am. Ind. Hyg. Assoc. J.* **30**, 470 (1969).

Anhydr, *exsiccated sodium phosphate.* Hygroscopic powder. On exposure to air will absorb from 2 to 7 mols H₂O, depending on the humidity and temp. Sol in about 8 parts water, much more sol in hot water. Soly per 100 gal water increases from about 14 lbs at slightly > 0° to over 900 lbs at 95°. Insol in alc. pH of 1% aq soln at 25°: 9.1. *Keep well closed.*

Dihydrate, *Sorensen's phosphate, Sorensen's sodium phosphate.*

Heptahydrate, crystals or granular powder. Stable in the air. d about 1.7. Sol in 4 parts water, more sol in boiling water; practically insol in alcohol. The aq soln is alkaline, pH about 9.5. LD₅₀ orally in rats: 12.93 g/kg (Smyth).

Dodecahydrate, translucent crystals or granules; readily loses 5 mols of water on exposure to air at ordinary temp. mp 34-35° (when it contains the full 12 mols of H₂O). d about 1.5. Sol in 3 parts water; practically insol in alcohol. Aq soln is alkaline, pH about 9.5. *Keep well closed and in a cool place.*

Incompat: Alkaloids, antipyrine, chloral hydrate, lead acetate, pyrogallol, resorcinol.

Human Toxicity: Anhydr form may cause mild irritation to skin, mucous membranes; intern. causes purging.

USE: As sequestrant, emulsifier and buffer in foods. As mordant in dyeing; for weighting silk; in tanning; in manuf of enamels, ceramics, detergents, boiler compds; as fireproofing agent; in soldering and brazing instead of borax; as reagent and buffer in analytical chemistry.

THERAP CAT: Cathartic.

THERAP CAT (VET): Laxative.

8613. Sodium Phosphate, Monobasic. Sodium biphosphate; sodium dihydrogen phosphate; acid sodium phosphate; monosodium orthophosphate; primary sodium phosphate. H₂NaO₄P; mol wt 119.98. H 1.68%, Na 19.16%, O 53.34%, P 25.81%. NaH₂PO₄. It is about 99% pure.

Monohydrate, white, odorless, slightly deliquesc crystals or granules. At 100° loses all its water; when ignited it converts into metaphosphate. Freely sol in water; practically insol in alcohol. The aq soln is acid. pH of 0.1 molar aq soln at 25°: 4.5.

Dihydrate, orthorhombic bisphenoidal colorless crystals, mp 60°. d 1.915. At room temp crystallizes with 2H₂O. Directions for max yield: Beans, Kiehl, *J. Am. Chem. Soc.* **49**, 1878 (1927).

USE: In baking powders; in boiler water treatment; as dry acidulant and sequestrant for foods: Tidridge, Pals, U.S. pat. **3,030,213** (1962 to FMC).

THERAP CAT: Urinary acidifier.

THERAP CAT (VET): Urinary acidifier.

8614. Sodium Phosphate, Radioactive. Sodium phosphate—^{32}P; radioactive sodium phosphate; sodium radiophosphate (^{32}P); Phosphotope.

Aq soln of mixed radioactive phosphates with a pH range of 5.0-6.0. Contains radioactive monobasic sodium phosphate ($NaH_2^{32}PO_4$) and radioactive dibasic sodium phosphate ($Na_2H^{32}PO_4$). ^{32}P is a pure beta emitter with a half-life of 14.3 days.

THERAP CAT: Antineoplastic; antipolycythemic; diagnostic aid (neoplasm).

8615. Sodium Phosphate, Tribasic. Trisodium orthophosphate; trisodium phosphate; TSP; Oakite. Na_3O_4P; mol wt 163.94. Na 42.07%, O 39.04%, P 18.89%. Na_3PO_4. Crystallizes with 8 and 12 mols of H_2O.

Dodecahydrate, colorless or white crystals. When rapidly heated melts at about 75°. Does not lose the last mol of water even on moderate ignition. d 1.6. Sol in 3.5 parts water, 1 part boiling water; insol in alcohol. The aq soln is strongly alkaline. pH of 0.1% soln: 11.5; of 0.5% soln: 11.7; of 1.0% soln: 11.9. Technical crystals are sometimes made with excess alkali to prevent caking and give more alkaline solutions. LD_{50} orally in rats: 7.40 g/kg, H. F. Smyth et al., Am. Ind. Hyg. Assoc. J. **30**, 470 (1969).

USE: In photographic developers; clarifying sugar; removing boiler scale, softening water; manuf paper; laundering; tanning leather; in detergent mixture.

8616. Sodium Phosphite. HNa_2O_3P; mol wt 125.96. H 0.80%, Na 36.50%, O 38.11%, P 24.59%. Na_2HPO_3.

Pentahydrate, white, hygroscopic cryst powder. Heat of formation (25°): -684.2 kcal/mole. Freely sol in water. Keep well closed.

8617. Sodium Phosphomolybdate. Sodium molybdophosphate. $Mo_{12}Na_3O_{40}P$; mol wt 1891.37. Mo 60.88%, Na 3.65%, O 33.84%, P 1.64%. $Na_3PO_4.12MoO_3$.

White crystals. Freely sol in water.

USE: As reagent in chemical analysis.

8618. Sodium Phosphotungstate. Sodium tungstophosphate. Approx $2Na_2O.P_2O_5.12WO_3.18H_2O$.

White, granular powder. Sol in water.

USE: As reagent for alkaloids, uric acid, potassium.

8619. Sodium Plumbate(IV). Na_2O_3Pb; mol wt 301.17. Na 15.27%, O 15.94%, Pb 68.79%. Na_2PbO_3.

Trihydrate, light yellow, fused, hygroscopic lumps. Sol in water, but gradually dec with separation of PbO_2. Keep well closed.

8620. Sodium Polyanetholesulfonate. Polyanethesulfonic acid sodium salt; anetholesulfonic acid sodium salt polymer. A polymer of anetholesulfonic acid. Originally developed as an anticoagulant, it was soon found that it possesses anticomplement action and lowers the bactericidal action of blood. Ref: Demole, Reinert, Arch. Exp. Pathol. Pharmakol. **158**, 211 (1930); Friedmann, Klin. Wochenschr. **14**, 215 (1935); Stuart, J. Clin. Path. **1**, 311 (1948); Hoffmann-La Roche Biochemicals Catalog.

Light brown powder. Insol in alcohol. Swells in water and slowly goes in soln with neutral reaction. Aq solns are stable to heat, dil alkalies and dil acids.

USE: Under the trademark Liquoid to inhibit blood coagulation in vitro, and as diagnostic reagent to encourage the growth of pathogens in blood. Also to stabilize colloidal solns such as milk and gelatin. (Not to be confused with Liquoid, registered trademark of Johnson & Johnson for castor and olive oil emulsion.)

8621. Sodium Polymetaphosphate. Graham's salt; "sodium hexametaphosphate"; glassy sodium metaphosphate; Hy-Phos. $(NaPO_3)_x$. A mixture of polymeric metaphosphates; not a hexamer. Prepd by rapidly chilling molten sodium metaphosphate: Bell, Inorg. Syn. **3**, 103 (1950). Reviews: see Sodium Metaphosphate.

Clear, hygroscopic glass. mp 628°. Sol in water, but dissolves slowly. Depolymerizes in aqueous soln to form sodium trimetaphosphate and sodium orthophosphates.

Sodium hexametaphosphate detergents, Calgon, Giltex, Quadrafos, Hagan phosphate, Micromet. Mixtures contg Graham's salt as the principal agent. Supplied in the form of a powder, flakes, and as small, broken, glass-like particles. Sol in water (pH adjusted to 8-8.6). Insol in organic solvents. Possess dispersing and deflocculating properties, coagulate albumins, and inhibit the crystn of slightly sol compds such as calcium carbonate and calcium sulfate.

USE: Water softeners and detergents. For leather tanning, dyeing, laundry work, textile processing; for the "threshold treatment" of softening industrial water supplies.

8622. Sodium Polystyrene Sulfonate. Resonium A; Kayexalate. A cation exchange resin charged with sodium.

Marketed as a powder, insol in water; also as an emulsion with methyl cellulose.

THERAP CAT: Ion-exchange resin (potassium).

8623. Sodium Propionate. Propionic acid sodium salt; Mycoban. $C_3H_5NaO_2$; mol wt 96.07. C 37.50%, H 5.25%, Na 23.94%, O 33.31%. CH_3CH_2COONa.

Transparent crystals, granules. Deliquescent in moist air. Neutral or slightly alkaline reaction to litmus. One gram dissolves in about 1 ml water, in about 0.65 ml boiling water, in about 24 ml alcohol at 25°. Most active at acid pH: Wolford, Andersen, Food Ind. **17**, 622 (1945); Olsen, Macy, J. Dairy Sci. **29**, 173 (1946).

USE: Fungicide, mold preventative.

THERAP CAT: Topical antifungal.

THERAP CAT (VET): In ketosis of ruminants (glucose precursor). Antifungal agent. Has been used in dermatoses, wound infections, conjunctivitis.

8624. Sodium Pyroantimonate. Sodium antimonate. Approx $Na_2H_2Sb_2O_7$.

Monohydrate, white, granular powder. Slightly sol in water.

USE: Manuf opaque glass and opaque glazes.

8625. Sodium Rhodizonate. 5,6-Dihydroxy-5-cyclohexene-1,2,3,4-tetrone disodium salt; [(3,4,5,6-tetraoxo-1-cyclohexen-1,2-ylene)dioxy]disodium. $C_6Na_2O_6$; mol wt 214.05. C 33.66%, Na 21.49%, O 44.85%.

Violet crystals. Sol in water with an orange-yellow color; slightly sol in soda soln; insol in alc. Solns are unstable even in the refrigerator, and must be prepd fresh every other day.

USE: As a reagent for barium and strontium.

8626. Sodium Salicylate. 2-Hydroxybenzoic acid monosodium salt; Alysine; Ardall (from plant sources); Clin; Idocyl novum; Enterosalicyl; Enterosalil; Magsalyl; Parbocyl-Rev. $C_7H_5NaO_3$; mol wt 160.11. C 52.51%, H 3.15%, Na 14.36%, O 29.98%. It is at least 99.5% pure.

White, odorless crystals, scales or powder; becomes pinkish on long exposure to light. One gram dissolves in 0.9 ml water, 9.2 ml alcohol, 4 ml glycerol. The aq soln is slightly acid. pH 5-6. A satd aq soln contains 20.06% w/w at $-1.5°$. See Sidgwick, Ewbank, J. Chem. Soc. **121**, 1847, 1850 (1922). Protect from light. LD_{50} i.p. in rats: 780

mg/kg, E. I. Goldenthal, *Toxicol. Appl. Pharmacol.* **18**, 185 (1971).

Incompat: Ferric salts, lime water, spirit nitrous ether, mineral acids, iodine, lead acetate, silver nitrate, sodium phosphate in powder.

USE: As preservative for paste, mucilage, glue, hides.

THERAP CAT: Analgesic, antipyretic.

THERAP CAT (VET): Analgesic, antipyretic.

8627. Sodium Selenate. Na_2O_4Se; mol wt 188.94. Na 24.34%, O 33.87%, Se 41.79%. Na_2SeO_4.

Decahydrate, white crystals; very sol in water. LD_{100} (anhydrous) orally in rabbits: 4 mg/kg, *Handbook of Toxicology* vol. 1, W. S. Spector, Ed. (Saunders, Philadelphia, 1956) pp 278-279.

USE: Insecticide in some horticultural applications.

8628. Sodium Selenide. Na_2Se; mol wt 124.95. Na 36.81%, Se 63.19%. Prepd by adding selenium to a soln of sodium in liquid ammonia: Hugot, *Compt. Rend.* **129**, 299 (1899); *Ann. Chim. Phys.* [7] **21**, 34 (1900); Feher in *Handbook of Preparative Inorganic Chemistry* vol. 1, G. Brauer, Ed. (Academic Press, New York, 2nd ed., 1963) p 421.

Amorphous crystals. d^{10} 2.625. mp > 875°. Turns red on exposure to air and deliquesces. Dec in water. Insol in ammonia.

Hemienneahydrate, fine needles. Turns red on exposure to air and deliquesces.

Decahydrate, needles. Turns red and then brown on exposure to air.

Hexadecahydrate, prisms. mp 40°. Dec in air to sodium carbonate, selenium and a small amount of sodium selenide.

8629. Sodium Selenite. Na_2O_3Se; mol wt 172.95. Na 26.59%, O 27.75%, Se 45.65%. Na_2SeO_3. Prepared by evaporating an aqueous solution of sodium hydroxide and selenious acid between 60° and 100°: Krak, *J. Am. Ceram. Soc.* **12**, 530 (1929); by heating a mixture of sodium chloride and selenium oxide: Cameron, Macallan, *Proc. Roy. Soc.* **46**, 13 (1890). Metabolism: M. Sandholm, *Acta Pharmacol. Toxicol.* **33**, 6 (1973); H. W. Symonds *et al.*, *Brit. J. Nutr.* **45**, 117 (1981). Mutagenicity study: M. Nodo *et al.*, *Mutat. Res.* **66**, 175 (1979).

Tetragonal prisms. Stable in air. Freely sol in water; insol in alcohol. LD_{50} orally in rats: 7 mg/kg, Cummins, Kimura, *Toxicol. Appl. Pharmacol.* **20**, 89 (1971).

Pentahydrate, acicular crystals. Loses water of crystn in dry air.

USE: Removing green color from glass during its manuf; alkaloidal reagent.

8630. Sodium Sesquicarbonate. Urao; trona. $C_2HNa_3O_6$; mol wt 190.00. C 12.64%, H 0.53%, Na 36.30%, O 50.53%. $Na_2CO_3 \cdot NaHCO_3$. Found in nature as the dihydrate, e.g., Owens Lake, Searles Lake (U.S.A.); Lake Magadi (Kenya). Produced on a large scale from sodium carbonate and a slight excess of sodium bicarbonate: Schenk in *Winnacker-Weingaertner, Chemische Technologie* vol. I (München, 1950) p 427.

Dihydrate, monoclinic needles, d 2.112. Crystals are stable in air. Soly in water (g/100 ml) at 0°: 13; at 100°: 42. Aq solns are mildly alkaline. pH of 0.1*M* soln = 10.1.

USE: Chiefly in laundering in conjunction with soap. *Caution:* Irritating to skin, mucous membranes.

8631. Sodium Silicate. Water glass; soluble glass. Manuf: Faith, Keyes & Clark's *Industrial Chemicals*, F. A. Lowenheim, M. K. Moran, Eds. (Wiley-Interscience, New York, 4th ed., 1975) pp 755-761.

The compositions of the commonly available sodium silicates in dry form are: Na_2SiO_3, $Na_6Si_2O_7$, and $Na_2Si_3O_7$, with variable amounts of water, the first-named containing approx $5H_2O$. They are in the form of colorless to white or grayish-white, crystal-like pieces or lumps. These silicates are very slightly sol or almost insol in cold water. They are best brought into soln by heating with water under pressure. They are less readily sol in large amounts of water than in small amounts of water, and the anhydr dissolve with more difficulty than the hydrated silicates; also, the silicates containing more sodium dissolve more readily. The aq solns are

strongly alkaline. The dry sodium silicates are used relatively little.

See also Sodium Metasilicate.

USE: Lining Bessemer converters, acid concentrators. Manuf grindstones, abrasive wheels (as binder only). *Caution:* Irritating and caustic to skin, mucous membranes. If swallowed causes vomiting and diarrhea.

8632. Sodium Silicate Solution. Sodium silicate solns of varying composition with respect to the ratio between sodium and silica, as well as of various densities, are available. One of the most commonly used sodium silicate solns, also known as "egg preserver", contains about 40% $Na_2Si_3O_7$. Solns of sodium silicate are strongly alkaline and are readily dec by acids with separation of silicic acid. The greater the ratio of Na_2O to SiO_2 (the greater the alkalinity) the more tacky is the soln.

USE: For preserving eggs; fireproofing fabrics; as a detergent in soaps; as adhesive; waterproofing walls; in cements; in cold-water paints; manuf of abrasive wheels; weighting silk, etc.

8633. Sodium Stannate(IV). *Sodium tin oxide.* Na_2O_3Sn; mol wt 212.67. Na 21.62%, O 22.57%, Sn 55.81%. Na_2SnO_3.

Trihydrate, white or colorless crystals; gradually dec in the air; dec by weak acids. Sol in about 1.7 parts water; insol in alcohol. The aq soln is alkaline. *Keep well closed.*

USE: As mordant in dyeing and printing calico; for fireproofing of curtains, etc.

8634. Sodium Stearate. *Stearic acid sodium salt.* Approx $C_{18}H_{35}NaO_2$. Usually contains sodium palmitate.

White powder; soapy feel; slight, tallow-like odor. Slowly soluble in cold water or cold alcohol; freely sol in the hot solvents. The aq soln is strongly alkaline, due to hydrolysis; the alcohol soln is practically neutral.

USE: Pharmaceutic aid (emulsifying and stiffening agent). In glycerol suppositories; also in toothpaste; as waterproofing agent.

8635. Sodium Succinate. Succinic acid sodium salt; disodium succinate; Soduxin. $C_4H_4Na_2O_4$; mol wt 162.05. C 29.64%, H 2.49%, Na 28.38%, O 39.49%.

Hexahydrate, granules or crystalline powder; stable in air. Loses all its water at 120°. Sol in about 5 parts water; insol in alcohol. The aq soln is neutral or slightly alkaline. LD_{50} i.v. in mice: 4.5 g/kg, *Handbook of Toxicology* vol. 1, W. S. Spector, Ed. (Saunders, Philadelphia, 1956) pp 278-279.

THERAP CAT: Respiratory stimulant, analeptic; urinary alkalizer, diuretic; cathartic.

8636. Sodium Sulfate. Na_2O_4S; mol wt 142.06. Na 32.38%, O 45.06%, S 22.57%. Na_2SO_4. Occurs in nature as the minerals *mirabilite, thenardite.* Industrial production: Faith, Keyes & Clark's *Industrial Chemicals*, F. A. Lowenheim, M. K. Moran, Eds. (Wiley-Interscience, New York, 4th ed., 1975) p 762-768.

Anhydrous form, *salt cake* (technical grade). Powder or orthorhombic bipyramidal crystals, mp about 800°. d 2.7. Sol in about 3.6 parts water. Max soly at 33°: 1 in 2. Above this temp the soly gradually decreases and at 100° requires 2.4 parts water. Insol in alcohol.

Decahydrate, $Na_2SO_4 \cdot 10H_2O$, *Glauber's salt.* Odorless, efflorescent crystals or granules, mp 32.4°. d 1.46. Loses all its water at 100°. Sol in 1.5 parts water at 25°, in 3.3 parts water at 15°. Soly in water decreased by NaCl. Sol in glycerol; insol in alcohol. The aq soln is neutral. pH 6-7.5. *Keep well closed in a cool place.*

USE: For standardizing dyes; in freezing mixtures; in dyeing and printing textiles. The *anhydrous* form for drying organic liquids; in Kjeldahl nitrogen determination; in manuf of glass, ultramarine, paper pulp.

THERAP CAT: Cathartic.

THERAP CAT (VET): Purgative.

8637. Sodium Sulfide. Sodium monosulfide; sodium sulfuret. Na_2S; mol wt 78.05. Na 58.93%, S 41.07%. Best prepd from the elements in liq ammonia, also obtained by dehydration of the nonahydrate: Courtois, *Compt. Rend.* **207**, 1220 (1938); Klemm *et al.*, *Z. Anorg. Allgem. Chem.* **241**, 281 (1939); Feher in *Handbook of Preparative Inorganic*

Chemistry, vol 1, G. Brauer, Ed. (Academic Press, New York, 2nd ed., 1963) pp 358-360. *Review:* C. Drum in Kirk-Othmer *Encyclopedia of Chemical Techology* vol. 21 (Wiley-Interscience, New York, 3rd ed., 1983) pp 256-262.

Cubic crystals or granules. Extremely hygroscopic. Discolors upon exposure to air. d_4^{14} 1.856. mp 1180° *(in vacuo)*; also reported as 920°. Soly in water (g/100 g H_2O): 8.1 (−9.0°); 12.4 (0°); 18.6 (20°); 29.0 (40°); 35.7 (48°); 39.0 (50°). Slightly sol in alcohol. Insol in ether. Aq solns are strongly alkaline.

Pentahydrate, flat, shiny, four-sided, prismatic crystals. Loses 3 mols water at 100°. mp 120° (with loss of all water of crystn). Freely sol in water. Also sol in alcohol. Aq solns are strongly alkaline. Insol in ether. Dec by acids with evolution of H_2S.

Nonahydrate, tetragonal, deliquescent crystals. Odor of hydrogen sulfide. Discolors upon exposure to light and air (first yellow, then brownish-black). d_4^{16} 1.427. mp ~50°. One gram dissolves in 0.5 ml water, also reported as 18 g dissolve in 100 ml water at 25°. Slightly sol in alcohol, insol in ether. Dec by acids, even by carbonic acid. Aqueous solns are very alkaline and upon standing in contact with air are slowly converted according to the equation: $2Na_2S + 2O_2 + H_2O \rightarrow Na_2S_2O_3 + 2NaOH$. Exposure of the crystals to air produces hydrogen sulfide according to the equation $Na_2S + H_2O + CO_2 \rightarrow Na_2CO_3 + H_2S$. *Keep well closed and in a cool place. Do not handle with bare hands.*

USE: In dehairing hides and wool pulling; desulfurizing viscose rayon; in the manuf of rubber, sulfur dyes; in ore flotation, metal refining, engraving, cotton printing, as chemical intermediate, as laboratory reagent; in paper-pulping process.

8638. Sodium Sulfite. Na_2O_3S; mol wt 126.06. Na 36.49%, O 38.08%, S 25.44%. Na_2SO_3.

Small crystals or powder. It is fairly stable and does not oxidize as readily as the hydrated sulfite. Sol in 3.2 parts water; sol in glycerol; almost insol in alcohol. pH about 9. *Keep well closed.* LD_{50} i.v. in mice: 175 mg/kg, Hoppe, Goble, *J. Pharmacol. Exp. Ther.* **101**, 101 (1951).

Heptahydrate, efflorescent crystals. Unstable, oxidizing in the air to sulfate. Sol in 1.6 parts water, about 30 parts glycerol; sparingly sol in alc. The aq soln is alkaline and dissolves sulfur. pH ~9. *Keep well closed in a cool place.*

Note: The commercial heptahydrate (about 90% pure, the remainder consisting chiefly of the sulfate) has been largely replaced by the commercial anhyd salt (about 96-99% pure) which is more stable.

USE: Chiefly in photographic developers and instead of "hypo" for fixing prints; bleaching wool, straw, silk; generating SO_2; as reducer in manuf dyes; silvering glass; removing traces of Cl in bleached textiles and paper; preserving meat, egg yolks, etc.

8639. Sodium β-Sulfopropionitrile. *2-Cyanoethanesulfonic acid sodium salt;* sodium 2-cyanoethanesulfonate. $C_3H_4NNaO_3S$; mol wt 157.13. C 22.93%, H 2.56%, N 8.92%, Na 14.63%, O 30.55%, S 20.41%. $NaO_3SCH_2CH_2CN$.

Crystals, mp 243-244°. Sol in water, hot methanol, glacial acetic acid. Insol in most other organic solvents.

USE: In organic synthesis for introducing the sulfonic acid group into a variety of molecules, especially surface active agents.

8640. Sodium Tartrate. $C_4H_4Na_2O_6$; mol wt 194.05. C 24.76%, H 2.08%, Na 23.69%, O 49.47%. $Na_2C_4H_4O_6$.

Dihydrate, white crystals or granules. Loses its water at about 120°. d 1.82. Sol in about 3 parts cold water, 1.5 parts boiling water; insol in alcohol. The aq soln is slightly alkaline to litmus, pH 7 to 9.

USE: Dihydrate as standard for standardizing Karl Fischer reagent (determination of water).

THERAP CAT: Cathartic.

8641. Sodium Tellurate(IV). Sodium tellurite. Na_2O_3Te; mol wt 221.60. Na 20.75%, O 21.66%, Te 57.58%. Na_2TeO_3.

White powder. Soluble in water. LD_{50} orally in mice: 20 mg/kg, *Toxic Substances List,* H. E. Christensen, Ed. (1973) p 906.

8642. Sodium Tellurate(VI). Na_2O_4Te; mol wt 237.58. Na 19.35%, O 26.94%, Te 53.71%. Na_2TeO_4.

White powder. Sol in 130 parts cold water, 50 parts boiling water. MLD i.p. in rats: 37.2-55.8 mg/kg, *Handbook of Toxicology* vol. 1, W. S. Spector, Ed. (Saunders, Philadelphia, 1956) pp 280-281.

8643. Sodium Tetrachloroaluminate. Aluminum sodium chloride; sodium chloroaluminate. $AlCl_4Na$; mol wt 191.79. Al 14.06%, Cl 73.95%, Na 11.99%. $AlCl_3.NaCl$. Prepn: *Gmelin's, Aluminum* (8th ed.) **35B**, p 376 (1934); Good, Batha, U.S. pat. **2,867,499** (1959 to Olin Mathieson).

Yellowish powder. Sol in water.

USE: Catalyst for organic reactions.

8644. Sodium Tetrachloroaurate(III). Gold sodium chloride; sodium chloroaurate(III). $AuCl_4Na$; mol wt 362.03. Au 54.47%, Cl 39.18%, Na 6.35%. Prepd by evapn of a soln of $HAuCl_4$ and NaCl.

Dihydrate, orange-yellow, rhombic, bipyramidal crystals. Stable to 100°. Dec before all H_2O of hydration lost. Freely sol in water; sol in alcohol, ether.

USE: For paper-toning in photography, in electroplating baths.

8645. Sodium Tetradecyl Sulfate. *7-Ethyl-2-methyl-4-hendecanol sulfate sodium salt;* sodium 2-methyl-7-ethylundecyl sulfate-4; sodium 2-methyl-7-ethylundecanol-4-sulfate; STS; Sodium Sotradecol; Tergitol 4; Trombavar; Trombovar. $C_{14}H_{29}NaSO_4$; mol wt 316.44. C 53.14%, H 9.24%, Na 7.27%, O 20.22%, S 10.13%.

$$CH_3CHCH_2\overset{\overset{\displaystyle CH_3}{|}}{C}HCH_2CH_2\overset{\overset{\displaystyle OSO_3Na}{|}}{C}HCH_2\overset{\overset{\displaystyle CH_2CH_3}{|}}{C}H(CH_2)_2CH_3$$

White, waxy solid. Sol in water, alcohol, ether. The pH of a 5% soln is from 6.5 to 9.0. Surface tension (dynes/cm) of aq soln at 25°: 56.5 dynes/cm (0.05% w/w); 52 (0.10%); 47 (0.20%); 40 (0.50%); 35 (1.0%). LD_{50} orally in rats: 4.95 g/kg, H. F. Smyth, C. P. Carpenter, *J. Ind. Hyg. Toxicol.* **30**, 63 (1948).

Note: Tergitol 4 is the name applied to industrial grade; Sodium Sotradecol, applied to medicinal grade, not less than 85% pure.

USE: Wetting agent.

THERAP CAT: Sclerosing agent.

8646. Sodium Tetraphenylborate. Tetraphenylboron sodium; Kalignost. $C_{24}H_{20}BNa$; mol wt 342.24. C 84.23%, H 5.89%, B 3.16%, Na 6.72%. $Na[B(C_6H_5)_4]$. Prepn: Wittig, Raff, *Ann.* **573**, 195 (1950); U.S. pat. **2,853,525** (1958); Heyl, **Brit.** pat. **705,719** (1954); Kozlova, Pal'm, *Zh. Obshch. Khim.* **31**, 2922 (1961); Holtzapfel, Richter, *ibid.* **32**, 1358 (1962).

Snow-white crystals from chloroform. Freely sol in water, acetone. Less sol in ether, chloroform. Practically insol in petr ether. Aq solns should be adjusted to a pH of about 5 and can be stored at room temp or lower. Such solns have been stored at 45° for 5 days without deterioration. The soly in polar solvents increases as the temp decreases.

USE: As reagent for determination of potassium, ammonium, rubidium and cesium ions: Barnard, Buechl, *Chemist Analyst* **48**, 44, 49 (1959); Montequi, Serrano, *Anales Real Acad. Farm.* **26**, 107 (1960).

8647. Sodium Tetrathionate. Tetrathione. $Na_2O_6S_4$; mol wt 270.26. Na 17.01%, O 35.52%, S 47.46%. $Na_2S_4O_6$.

Colorless or white crystals. Usually the monohydrate. Freely sol in water.

8648. Sodium Thioantimonate(V). Sodium sulfantimonate; Schlippe's salt. Na_3S_4Sb; mol wt 318.96. Na 21.62%, S 40.21%, Sb 38.17%. Na_3SbS_4.

Nonahydrate, colorless or light yellow, large crystals; strong alk reaction. On exposure to air becomes covered with a reddish-brown coat of antimony sulfide. Sol in 3 parts cold water, 1 part boiling water; insol in alc; dec by weak acids.

8649. Sodium Thiocarbonate. Sodium sulfocarbonate.

CNa_2S_3; mol wt 154.20.   C 7.79%, Na 29.83%, S 62.39%. Na_2CS_3.

USE: Usually supplied as a 40°Bé soln for destroying insects injurious to plants, especially vines.

8650. Sodium Thioglycolate. *Mercaptoacetic acid sodium salt;* sodium mercaptoacetate. $C_2H_3NaO_2S$; mol wt 114.11. C 21.05%, H 2.65%, Na 20.15%, O 28.04%, S 28.10%. $HSCH_2COONa$.

Hygroscopic crystals. Slight characteristic odor. Discolors on exposure to air or iron. Freely sol in water; slightly sol in alc. LD_{50} i.p. in rats: 148 mg/kg, Freeman, Rosenthal, *Fed. Proc.* **11**, 347 (1952).

USE: In cold-waving of hair; as depilatory; in bacteriology for the prepn of thioglycolate media; as analytical reagent, *see* Thioglycolic Acid.

8651. Sodium Thiophosphate. Sodium phosphorothioate; sodium monothiophosphate. Na_3O_3PS; mol wt 180.00. Na 38.32%, O 26.67%, P 17.20%, S 17.81%. Na_3PO_3S. Conveniently prepd from $NaPO_3$ + Na_3S: Zintl, Bertram, *Z. Anorg. Allgem. Chem.* **245**, 16 (1940); from $PSCl_3$ + NaOH: Tridot, Tudo, *Bull. Soc. Chim. France* **1960**, 1231. Several alternate routes, e.g. from P_2S_5 and NaOH, *see* Klement in *Handbook of Preparative Inorganic Chemistry* vol. 1, G. Brauer, Ed. (Academic Press, New York, 2nd ed., 1963) pp 569-570.

Dodecahydrate, thin, six-sided leaflets from water. Effloresces in dry air. mp 60°. Freely sol in warm water. Aq solns are strongly alkaline.

8652. Sodium Thiosulfate. Sodium hyposulfite; "hypo"; antichlor; Sodothiol; Sulfothiorine; Ametox. $Na_2O_3S_2$; mol wt 158.13. Na 29.08%, O 30.36%, S 40.56%. $Na_2S_2O_3$. Review of mfg processes: Faith, Keyes & Clark's *Industrial Chemicals*, F. A. Lowenheim, M. K. Moran, Eds. (Wiley-Interscience, New York, 4th ed., 1975) pp 769-773. Crystal structure of the pentahydrate: A. A. Uraz, N. Armagan, *Acta Crystallogr.* **B33**, 1396 (1977). Toxicity of the pentahydrate: C. Voegtlin *et al., J. Pharmacol. Exp. Ther.* **25**, 297 (1925).

Powder. Sol in water; practically insol in alcohol.

Pentahydrate, odorless crystals or granules, mp 48° when rapidly heated. Effloresces in warm dry air; slightly deliquesces in moist air. d 1.69. Loses all its water at 100°; dec at higher temp. Sol in water; insol in alcohol. Slowly dec in aq soln at ordinary temp, more rapidly when heated. The aq soln is practically neutral. pH 6.5-8.0. It dissolves silver halides and many other salts of silver. *Incompat:* Iodine, acids; lead, mercury, and silver salts. LD i.v. in rats: > 2.5 g/kg (Voegtlin).

USE: To remove chlorine from solns according to the eq: $Na_2S_2O_3$ + $4Cl_2$ + $5H_2O$ → $2NaHSO_4$ + $8HCl$ and $Na_2S_2O_3$ + $2HCl$ → $2NaCl$ + H_2O + S + SO_2. As "antichlor" in bleaching of paper pulp; as fixer in photography; for extraction of silver from ores; as mordant in dyeing and printing textiles; reducer in chrome dyeing; manuf leather; bleaching bone, straw, ivory; also as reagent in analytical chemistry.

THERAP CAT: Antidote to cyanide.

THERAP CAT (VET): Antidote to cyanide. Has been used as a "general detoxifier", in bloat, and externally in ringworm, mange.

8653. Sodium Trimetaphosphate. $Na_3O_9P_3$; mol wt 305.92. Na 22.55%, O 47.07%, P 30.38%. $(NaPO_3)_3$. Prepd by tempering sodium hexametaphosphate at about 500° for 8-12 hrs; by heating Na_2HPO_4 and NH_4NO_3 to 320° or by heating NaH_2PO_4 at 530°: Fleitmann, Henneberg, *Ann.* **65**, 304 (1848); v. Knoore, *Z. Anorg. Allgem. Chem.* **24**, 381 (1900); Karbe, Jander, *Kolloid-Beihefte* **54**, 35 (1942); Bell, *Inorg. Syn.* **3**, 103 (1950); Lee, Bond, *J. Appl. Chem.* **18**, 345 (1968). Crystal structure of hexahydrate: Caglioti *et al., Atti Acad. Italia Rend.* **3** [7], 761 (1942), *C.A.* **40**, 6927b (1946); of anhydrate: Ondik, *Acta Cryst.* **18**, 226 (1965). *Reviews: see* Sodium Metaphosphate.

Hexahydrate, efflorescent, triclinic-rhombohedral prisms. d 1.786. mp 53°. Loses water on storage at 20°, becomes definitely anhyd at 100°. d (anhydr) 2.49. One gram dissolves in 4.5 ml water. Insol in alc. LD orally in mice:

> 100 mg/kg, Behrens, Seelkopf, *Arch. Exp. Pathol. Pharmakol.* **169**, 238 (1932).

USE: In detergent processing: Shen, *J. Am. Oil Chem. Soc.* **45**, 510 (1968).

8654. Sodium Tripolyphosphate. Sodium triphosphate; tripolyphosphate; STPP; poly; pentasodium triphosphate. $Na_5O_{10}P_3$; mol wt 367.91. Na 31.25%, O 43.49%, P 25.26%. $Na_5P_3O_{10}$. Prepared by molecular dehydration of mono- and disodium phosphates: Audrieth, Bell, *Inorg. Syn.* **3**, 85 (1950).

Slightly hygroscopic granules. Soly in water (g/100 ml) at 25°: 20; at 100°: 86.5. pH of 1% soln at 25° = 9.7-9.8. Its properties are between those of tetrasodium pyrophosphate and sodium metaphosphate. Stability: With prolonged heating of STPP solns, it tends to revert to the orthophosphate. More stable than the higher (i.e., meta-) phosphates, but less stable than tetrasodium pyrophosphate. LD_{50} orally in rats: 6.50 g/kg, H. F. Smyth *et al., Am. Ind. Hyg. Assoc. J.* **30**, 470 (1969).

USE: In water softening (calcium and magnesium hardness is sequestered from soln without precipitation); peptizing agent, emulsifier and dispersing agent, ingredient of cleansers in drilling fluids to control mud viscosity in oil fields; as preservative, sequestrant and texturizer in foods. *Caution:* Moderately irritating to skin, mucous membranes. Ingestion can cause violent purging.

8655. Sodium Tungstate(VI). Na_2O_4W; mol wt 293.83. Na 15.65%, O 21.78%, W 62.57%. Na_2WO_4.

Dihydrate, colorless crystals or white, cryst powder; effloresces in dry air. Loses its water at 100°. Sol in about 1.1 parts water; insol in alcohol. The aq soln is slightly alkaline, pH 8-9. LD orally in guinea pigs: 990 mg/kg, *Handbook of Toxicology* vol. 1, W. S. Spector, Ed. (Saunders, Philadelphia, 1956) pp 280-281.

USE: Fireproofing and waterproofing fabrics; preparing complex compds, such as phosphotungstate, silicotungstate, etc.; as a reagent for biological products; precipitant for alkaloids, etc.

8656. Sodium Uranate(VI). Sodium diuranate; uranium oxide yellow; uranium yellow. $Na_2O_7U_2$; mol wt 634.04. Na 7.25%, O 17.66%, U 75.08%. $Na_2U_2O_7$.

Monohydrate, yellow powder. Insol in water; sol in acids.

USE: Manuf yellowish-green fluorescent glass; painting on porcelain and enameling.

8657. Sodium Vanadate(V). Sodium metavanadate. NaO_3V; mol wt 121.93. Na 18.85%, O 39.37%, V 41.78%. $NaVO_3$.

Tetrahydrate, yellowish-white, cryst powder. Sol in hot water. LD_{75} i.p. in rats: 4-5 mg V/kg, Franke, Moxon, *J. Pharmacol. Exp. Ther.* **58**, 454 (1936).

USE: In photography and manuf of inks.

8658. Sofalcone. [5-[(3-Methyl-2-butenyl)oxy]-2-[3-[4-[(3-methyl-2-butenyl)oxy]phenyl]-1-oxo-2-propenyl]phenoxy]acetic acid; 2'-carboxymethoxy-4,4'-bis(3-methyl-2-butenyloxy)chalcone; SU-88; Solon. $C_{27}H_{30}O_6$; mol wt 450.54. C 71.98%, H 6.71%, O 21.31%. Substituted chalcone with anti-ulcer activity. Prepn: K. Kyogoku *et al.*, Ger. pat. **2,705,603**; *eidem*, U.S. pat. **4,085,135** (1977, 1978 both to Taisho); *eidem, Chem. Pharm. Bull.* **27**, 2943 (1979). Effects on exptl ulcer models: R. Saziki *et al., Oyo Yakuri* **18**, 579 (1979), *C.A.* **92**, 174433h (1980). Absorption, distribution, excretion: T. Nozu *et al., ibid.* 815, *C.A.* **92**, 191017f (1980). Reproduction studies: T. Yamada *et al., ibid.* **19**, 543 (1980), *C.A.* **93**, 230785j (1980). Toxicity study: Y. Tarumoto *et al., Yakuri To Chirgo* **10**, 61 (1982), *C.A.* **96**, 193205j (1982). Effect on gastric blood flow: Y. Matsuo *et al., Arzneimittel-Forsch.* **33**, 242 (1983). Gastric cytoprotective effect in rats: T. Suwa *et al., Japan. J. Pharmacol.* **35**, 47 (1984); S. Kishimoto *et al., Arzneimittel-Forsch.* **31**, 944 (1987).

Light yellow needles from ethanol, mp 143-144°. LD$_{50}$ in mice, rats: > 10 g/kg orally (Tarumoto).

THERAP CAT: Anti-ulcerative.

8659. Solan. *N-(3-Chloro-4-methylphenyl)-2-methyl-pentanamide; 3'-chloro-2-methyl-p-valerotoluidide;* Niagara 4512. C$_{13}$H$_{18}$ClNO; mol wt 239.75. C 65.12%, H 7.57%, Cl 14.79%, N 5.84%, O 6.67%. Prepn: Dorschner *et al.*, **Brit.** pat. **869,169;** U.S. pat. **3,020,142** (1961, 1962, both to FMC).

Crystals, mp 79-80°. Sol in pine oil, diisobutylketone, isophorone, xylene; practically insol in water. LD$_{50}$ orally in rats: 10,000 mg/kg, G. W. Bailey, J. L. White, *Residue Rev.* **10,** 97 (1965).

USE: Herbicide.

8660. Solanidine. *Solanid-5-en-3β-ol;* solatubine. C$_{27}$-H$_{43}$NO; mol wt 397.62. C 81.55%, H 10.90%, N 3.52%, O 4.02%. From *Solanum* spp., *Solanaceae.* Potato sprouts contain about 0.008%. Obtained by hydrolysis of solanine. Extraction procedure: Soltys, Wallenfels, *Ber.* **69,** 811 (1936). Review on structure: Reichstein, Reich, *Ann. Rev. Biochem.* **15,** 155 (1946). Stereochemistry: Rosen, Rosen, *Chem. & Ind. (London)* **1954,** 1581. Revised stereochemistry: Höhne *et al., Tetrahedron* **22,** 673 (1966). Synthesis: Schreiber, Roensch, *Ber.* **97,** 2362 (1964); *Tetrahedron* **21,** 645 (1965); Kessar *et al., ibid.* **27,** 2153 (1971).

Long needles from chloroform-methanol, mp 218-219°. Sublimes near mp with slight decompn. [α]$_D^{21}$ −29° (c = 0.5 in chloroform). Freely sol in benzene, chloroform; slightly in alcohol, methanol; almost insol in ether, water.

Hydrochloride, C$_{27}$H$_{43}$NO.HCl, prisms from 80% alcohol, dec 345°.

Methyliodide, C$_{27}$H$_{43}$NO.CH$_3$I, crystals from 50% alcohol, dec 286°.

Acetylsolanidine, C$_{29}$H$_{45}$NO$_2$, crystals from alc, mp 208°.

8661. Solanine. Solatunine. C$_{45}$H$_{73}$NO$_{15}$; mol wt 868.04. C 62.26%, H 8.48%, N 1.61%, O 27.65%. From *Solanum* species, especially from *S. tuberosum* L. (potato), *S. nigrum* L. (woody nightshade), and *Lycopersicon esculentum* Mill., *Solanaceae* (tomato). Fresh potato sprouts contain about 0.04%. Extraction procedure: Soltys, Wallenfels, *Ber.* **69,** 811 (1936). Separated by chromatography into six components: α-, β-, γ-solanine and α-, β-, γ-chaconine. Main constituent α-solanine: Kuhn, Löw, *Angew. Chem.* **66,** 639 (1954); Kuhn *et al., Ber.* **88,** 1492 (1955); Paseshnichenko, Guseva, *Biokhimiya* **21,** 585 (1956); *eidem, ibid.* **22,** 843 (1957). Pharmacology: Nishie *et al., Toxicol. Appl. Pharmacol.* **19,** 81 (1971).

Slender needles from 85% alc, browns and sinters about 190°, dec about 285°. [α]$_D^{20}$ −60° (pyridine). pK at 15° = 6.66; K = 2.2 × 10^{-7}. Readily sol in hot alc; practically insol in water (25 mg/l at pH 6.0), ether, chloroform.

Hydrochloride, C$_{45}$H$_{73}$NO$_{15}$.HCl, usually amorphous and gummy, but has been crystallized, dec about 212°. Sol in water. Has been used as an agricultural insecticide. LD$_{50}$ i.p. in mice: 42 mg/kg.

8662. Solanocapsine. *3-Amino-16,23-epoxy-16,28-seco-solanidan-23-ol;* solanocapsin; 3β-amino-22,26-imino-16β,-23-oxido-5α,22α,23β,25α-cholestan-23-ol; 3β-amino-22,-26-imino-16β,23-oxido-5α,22R,23S,25R-cholestan-23-ol. C$_{27}$H$_{46}$N$_2$O$_2$; mol wt 430.65. C 75.30%, H 10.77%, N 6.51%, O 7.43%. From *Solanum pseudocapsicum* L., *S. capsicastrum* Link, *S. hendersonii* hort., *Solanaceae.* Isolation: Barger, Fraenkel-Conrat, *J. Chem. Soc.* **1936,** 1537; Schreiber, Ripperger, *Z. Naturforsch.* **17B,** 217 (1962). Structure: *eidem, Experientia* **16,** 536 (1960). Stereochemistry: *eidem, Tetrahedron Letters* no. 27, 9 (1960); *Ann.* **655,** 114 (1962). Synthesis: Ripperger *et al., Tetrahedron Letters* **1970,** 5251; *eidem, Tetrahedron* **28,** 1629 (1972).

Monohydrate, needles from methanol, mp 213-215°. [α]$_D^{20}$ +26.3° (c = 1.80 in methanol); [α]$_D^{20}$ +24.9° (c = 2.13 in methanol).

N'-Acetylsolanocapsine, C$_{29}$H$_{48}$N$_2$O$_3$, needles from ethanol, mp 233-234°. [α]$_D^{23}$ −33.7° (c = 1.59 in pyridine).

N,N'-Diacetylsolanocapsine, C$_{31}$H$_{50}$N$_2$O$_4$, crystals from chloroform + acetone, mp 281-283°. [α]$_D^{22}$ −38.6° (c = 1.31 in pyridine).

N,N,N'-Trimethylsolanocapsine, C$_{30}$H$_{52}$N$_2$O$_2$, needles from abs methanol, mp 222-224°. [α]$_D^{20}$ +53.2° (c = 1.09 in pyridine).

N-Benzylidenesolanocapsine, C$_{34}$H$_{50}$N$_2$O$_2$, needles from pyridine + abs methanol, mp 249-251°. [α]$_D^{18}$ +31.4 (c = 0.78 in pyridine).

8663. Solanone. *[R-(E)]-8-Methyl-5-(1-methylethyl)-6,8-nonadien-2-one;* L-(+)-2-methyl-5-isopropyl-1,3-nonadien-8-one. C$_{13}$H$_{22}$O; mol wt 194.31. C 80.35%, H 11.41%, O 8.23%. Isoln from tobacco, structure and synthesis: Johnson, Nicholson *J. Org. Chem.* **30,** 2918 (1965).

Liquid. bp₁ 60°. n_D^{20} 1.4755. d_4^{20} 0.870. $[\alpha]_D^{23}$ +13.6° (neat). uv max (ethanal and hexane): 230 nm (log ε 4.07).

8664. Solanum. Bull nettle; radical weed; sand-brier; horse nettle. Air-dried ripe fruit of *Solanum carolinense* L., *Solanaceae*. *Habit.* South America, Florida and other sections of U.S.A. *Constit.* Solanine, solanidine.
THERAP CAT: Sedative, anticonvulsant.

8665. Solasodine. *Spirosol-5-en-3β-ol; solasod-5-en-3β-ol; Δ⁵-20β_F,22α_F,25α_F,27-azaspirosten-3β-ol;* solancarpidine; solanidine-s; purapuridine. $C_{27}H_{43}NO_2$; mol wt 413.62. C 78.40%, H 10.48%, N 3.39%, O 7.74%. Steroidal alkaloid isolated from various *Solanum* species. By hydrolysis of solasonine: Rochelmeyer, *Arch. Pharm.* **277**, 329 (1939). *See also* ref *under* Solasonine and Solanidine. Structure: Briggs *et al., J. Chem. Soc.* **1950**, 3013. Synthesis: Uhle, *J. Org. Chem.* **27**, 656 (1962); Schreiber, Rönsch, *Tetrahedron* **20**, 1939 (1964); Kessar *et al., ibid.* **27**, 2869 (1971).

Hexagonal plates from methanol or by sublimation in high vacuum, mp 200-202°. $[\alpha]_D^{25}$ −98° (c = 0.14 in methanol); $[\alpha]_D$ −113° (CHCl₃). Alkaline reaction to litmus in alcoholic soln. pKb 6.30. Absorption spectrum: Rochelmeyer, *loc. cit.* Freely sol in benzene, pyridine, and chloroform; moderately sol in alcohol, methanol, and acetone; slightly sol in water; practically insol in ether.
USE: Starting material for steroidal drugs.

8666. Solasonine. *(3β,22α,25R)-Spirosol-5-en-3-yl O-6-deoxy-α-L-mannopyranosyl-(1 → 2)-O-[β-D-glucopyranosyl-(1 → 3)]-β-D-galactopyranoside;* solanine-S; purapurine. $C_{45}H_{73}NO_{16}$; mol wt 884.04. C 61.13%, H 8.32%, N 1.58%, O 28.96%. From *Solanum aviculare* Forst F., *S. sodomeum* L., *S. xanthocarpum* Schrad & Wendl., *Solanaceae*: Bell, Briggs, *J. Chem. Soc.* **1942**, 1; Briggs *et al. ibid.* **1942**, 3; Kuhn, Löw, *Ber.* **88**, 289 (1955). Isoln from other *S.* spp: Briggs, Cambie, *J. Chem. Soc.* **1958**, 1422; Briggs *et al., ibid.* **1961**, 4645. Paper chromatography: Szendey, *Arch. Pharm.* **290**, 563 (1957). Structure: Briggs *et al., J. Chem. Soc.* **1963**, 2848. Enzymic decompn: Guseva, Pasechnichenko, *Biokhimiya* **24**, 563 (1959).

Needles from methanol, sinters 296°, mp 301-303°. $[\alpha]_D^{23}$ −88° (c = 1.01 in pyridine); $[\alpha]_D^{22}$ −74.5° (c = 0.51 in methanol). Sol in hot alc, hot dioxane; slightly sol in hot water, dil acetic acid. Practically insol in chloroform, ether.

8667. Solasulfone. *1,1'-[Sulfonylbis(4,1-phenyleneimino)]bis[3-phenyl-1,3-propanedisulfonic acid] tetrasodium salt;* di(γ-phenylpropylamino)-4,4'-diphenylsulfone α,α',-γ,γ'-sodium tetrasonate; tetrasodium 4,4'-bis(γ-phenyl-propylamino)diphenylsulfone α,γ,α',γ'-tetrasulfonate; phenopryldiasulfone sodium; solapsone; RP 3668; Cimedone; Sulphetrone. $C_{30}H_{28}N_2Na_4O_{14}S_5$; mol wt 892.89. C 40.35%, H 3.16%, N 3.14%, Na 10.30%, O 25.09%, S 17.96%. Prepared by reacting 4,4'-diaminodiphenyl sulfone with more than 2 mols cinnamic aldehyde in the cold and treating the resulting 4,4'-dicinnamylideneaminodiphenyl sulfone with NaHSO₃ soln: Gray, Henry, **Brit.** pat. 491,265 (1938 to Wellcome Foundation); Buttle *et al., Biochem. J.* **32**, 1101 (1938). Improved process: Henry, Gray, **Brit.** pat. 562,216 (1944 to Wellcome Foundation). Properties and estimation: Brownlee *et al., Brit. J. Pharmacol.* **3**, 15 (1948).

Crystals contg 5-7% H_2O from alc. Becomes anhydr at 110° *in vacuo.* Exceedingly sol in cold water, giving a neutral soln. 20% and 40% (w/v) aq solns are stable when neutral or slightly alkaline and may be autoclaved. Practically insol in cold alc. Solns are incompatible with acids. Absorption spectrum of dye obtained by diazotization and coupling to N-(1-naphthyl)ethylenediamine hydrochloride: Brownlee, *loc. cit.*
THERAP CAT: Antibacterial (leprostatic).

8668. Soman. *Methylphosphonofluoridic acid 1,2,2-trimethylpropyl ester;* pinacoloxymethylphosphoryl fluoride; pinacolyl methylphosphonofluoridate; GD. $C_7H_{16}FO_2P$; mol wt 182.19. C 46.15%, H 8.85%, F 10.43%, O 17.56%, P 17.00%. Nerve gas. Prepn: Holmstedt, *Acta Physiol. Scand.* **25**, suppl. 90, p 106 (1951); *Protar* **16**, 131 (1950); G. Schrader, *Die Entwicklung neuer Insektizide auf Grundlage organischer Fluor- und Phosphor-Verbindungen* [Monographie Nr. 62 zu Angewandte Chemie und Chemie-Ingenieur-Technik (Verlag Chemie 1951)]; B. C. Saunders, *Some Aspects of the Chemistry and Toxic Action of Organic Compounds Containing Phosphorus and Fluorine* (Cambridge, 1957) p 94.

LD₅₀ in mice (mg/100 g): 0.062 i.p.; 0.78 dermally, Loomis, Salafsky, *Toxicol. Appl. Pharmacol.* **5**, 685 (1963). Lethal dose for man may be as low as 0.01 mg/kg, *Chem. & Eng. News* **31**, 4676 (1953).
Caution: A potent cholinesterase inhibitor. Toxic effects similar to, but more intense than, parathion, *q.v. See also* Sarin and Tabun.
USE: Chemical warfare agent.

8669. Somatoliberin. *Growth hormone-releasing factor;* GH-RF; GH-RH; GRF; growth hormone-releasing hormone; hGRF; hpGRF; somatocrinin. Stimulatory growth-hormone releasing factor of the hypothalamus that mediates, together with somatostatin, *q.v.,* the neuroregulation of somatotropin secretion. The concept of hypothalamic regulation of growth hormone release was postulated on the basis of physiological and biochemical evidence, *cf.* A. V. Schally *et al., Recent. Progr. Horm. Res.* **24**, 497 (1968); J. B. Martin in *Frontiers in Neuroendocrinology,* L. Martini, W. F. Ganong, Eds. (Raven Press, New York, 1976) pp 129-168. Demonstration of the existence of a GH-RH: R. Deuben, J. Meites, *Endocrinology* **74**, 408 (1964); A. V. Schally *et al., ibid.* **82**, 271 (1968); E. Dickermann *et al., Neuroendocrinology* **4**, 75 (1969). Isoln of a decapeptide originally believed to be GRF: A. V. Schally *et al., Endocrinology* **84**, 1493 (1969). Review of early literature: *eidem, Science* **179**, 341 (1973). Isoln and characterization of hpGRF, a fully bioactive, 44 amino acid peptide from human pancreatic tumor: R. Guillemin *et al., Science* **218**, 585 (1982); J. Rivier *et al., Nature* **300**, 276 (1982). Cloning and sequence analysis of cDNA for hpGRF precursor: U. Gubler *et al., Proc. Natl. Acad. Sci. USA* **80**, 4311 (1983). Expression-cloning of cDNA for hpGRF: K. E. Mayo *et al., Nature* **306**, 86 (1983). Isoln, primary structure and synthesis of native human hypothalamic GRF (hGRF), identity with hpGRF and comparison with GRF from other species: N. Ling *et al., Proc. Natl. Acad. Sci. USA* **81**, 4302 (1984). Immunohistochemical detection of GRF in brain: B. Bloch *et al., ibid.* **301**, 607 (1983). Potent interaction of glucocorticoids with GRF *in vivo:* W. B. Wehrenberg *et al., Science* **221**, 556 (1983). Series of articles on clinical pharmacology: *Hormone Res.* **22**, 32-57 (1985). Clinical evaluation in treatment of hypopituitary dwarfism: M. O. Thorner *et al., N. Engl. J. Med.* **312**, 4 (1985). *Reviews:* R. Guillemin *et al., Recent Progr. Horm. Res.* **40**, 233-299 (1984); W. B. Wehrenberg *et al., Ann. Rev. Pharmacol. Toxicol.* **25**, 463-483 (1985).

Porcine GRF is destroyed by trypsin, chymotrypsin, pepsin. Biological activity remains after incubation at pH 8.5 for 4 hours at 37°, but is destroyed after similar incubation at pH 2.

8670. Somatomedins. A family of peptide hormones with mitogenic properties and insulin-like effects that mediates the action of growth hormone (somatotropin, *q.v.*) on skeletal tissue. They are synthesized in liver, pituitary, brain and in various fetal organs. Discovered and originally termed *"sulphation factors"* because of their ability to stimulate in cartilage the incorporation of sulfate into chondroitin sulfate, *cf.* W. D. Salmon, W. H. Daughaday, *J. Lab. Clin. Med.* **49**, 825 (1957). The designation *somatomedin* was later proposed to connote the hormonal and intermediary relationship to somatotropin: W. H. Daughaday *et al., Nature* **235**, 107 (1972). Within the group, several somatomedins have been described, all closely related chemically and biologically and having mol wts of about 7,500: *somatomedin A (SM-A), somatomedin C (SM-C), insulin-like growth factor I (IGF-I), insulin-like growth factor II (IGF-II),* and *multiplication stimulating activity (MSA).* The two IGF peptides have also been referred to as *NSILA-S,* or non-suppressible insulin-like acting substance. Biological activity and purification of human SM-A: K. Hall, *Acta Endocrinol.* **Suppl. 163**, 1 (1972); K. Uthne, *ibid.* **Suppl. 175**, 1 (1973); L. Fryklund *et al., Biochem. Biophys. Res. Commun.* **61**, 957 (1974). Purification of MSA from rat liver culture medium: N. C. Dulak, H. M. Temin, *J. Cell. Physiol.* **81**, 153 (1973). Purification of SM-C from human plasma: J. J. Van Wyk *et al., Recent Progr. Horm. Res.* **30**, 259 (1974); M. E. Svoboda *et al., Biochemistry* **19**, 790 (1980). Isoln, chemical characterization, biological properties of IGF-I and IGF-II: E. Rinderknecht, R. E. Humbel, *Proc. Nat. Acad. Sci. USA* **73**, 2365, 4379 (1976). Amino acid sequence of IGF-I: *eidem, J. Biol. Chem.* **253**, 2769 (1978); of IGF-II: *eidem, FEBS Letters* **89**, 283 (1978). Homology of rat MSA with human IGF-II: H. Marquardt *et al., J. Biol. Chem.* **256**, 6859 (1981). Identity of SM-C with IGF-I: D.G. Klapper *et al., Endocrinology* **112**, 2215 (1983). Total synthesis of human IGF-I: C. H. Li *et al., Proc. Natl. Acad. Sci. USA* **80**, 2216 (1983); of human IGF-II: *eidem, Biochem. Biophys. Res. Commun.* **127**, 420 (1985). All the somatomedins circulate in plasma in high molecular weight forms because of their binding to specific carrier proteins, *cf.* A. C. Moses *et al., Endocrinology* **104**, 536 (1979); D. J. Knauer *et al., J. Supramol. Struct.* **15**, 177 (1981). Although the serum concentrations of somatomedins are growth hormone-dependent, there is evidence that somatotropin levels may be influenced, in turn, by somatomedins through a "negative feedback loop" by stimulating release of somatostatin, *q.v.*, a GH-release-inhibiting polypeptide: M. Berelowitz *et al., Science* **212**, 1279 (1981); *see also* G. S. Tannenbaum *et al., ibid.* **220**, 77 (1983). Developmental patterns of IGF-I and IGF-II synthesis and regulation in rat fibroblasts: S. O. Adams *et al., Nature* **302**, 150 (1983). *Reviews:* M. Binoux, *Ann. Endocrinol.* **41**, 157-192 (1980); D. R. Clemmons, J. J. Van Wyk, *Handbook Exp. Pharmacol.* **57**, 161-208 (1981); C. D. Scher *et al., Progr. Cancer Res. Ther.* **23**, 57-71 (1982); K. A. Thomas, *Ann. Rep. Med. Chem.* **17**, 219-228 (1982); E. R. Froesch *et al., Ann. Rev. Physiol.* **47**, 443-467 (1985); W. E. Russell, *Semin. Liver Dis.* **5**, 46-58 (1985).

8671. Somatostatin. Growth hormone-release inhibiting factor; GH-RIF; somatotropin release inhibiting factor; SRIF; SRIF-14. $C_{76}H_{104}N_{18}O_{19}S_2$; mol wt 1637.92. C 55.73%, H 6.40%, N 15.39%, O 18.56%, S 3.92%. Widely occurring cyclic tetradecapeptide that mediates, together with somatoliberin, *q.v.*, the neuroregulation of somatotropin secretion. Inhibits release of growth hormone, insulin and glucagon. Also a potent inhibitor in a number of systems, including central and peripheral neural, gastrointestinal, and vascular smooth muscle. Isoln from ovine hypothalamic extracts: P. Brazeau *et al., Science* **179**, 77 (1973). Structure: Burgus *et al., Proc. Nat. Acad. Sci. USA* **70**, 684 (1973); Ling *et al., Biochem. Biophys. Res. Commun.* **50**, 127 (1973). Synthesis: Rivier *et al., Compt. Rend. Ser. D* **276**, 2737 (1973); Sarantakis, McKinley, *Biochem. Biophys. Res. Commun.* **54**, 234 (1973); Yamashiro, Li, *ibid.* 882; Coy *et al., ibid.* 1267; A. M. Felix *et al., Int. J. Peptide Protein Res.* **15**, 342 (1980); B. Hartrodt *et al., Pharmazie* **37**, 403 (1982). Synthesis and bacterial expression of somatostatin gene: K. Itakura *et al., Science* **198**, 1056 (1977); K. Itakura, A. D. Riggs, **Belg.** pat. **871,782**; K. Itakura, U.S. pat. **4,356,270**; A. D. Riggs, U.S. pat. **4,366,246** (1979, 1982, 1982 all to Genentech). Isoln of a 28-amino acid somatostatin, termed *SRIF-28,* from porcine gastrointestinal tract: L. Pradayrol *et al., Biochem. Biophys. Res. Commun.* **85**, 701 (1978). Primary structure: *eidem, FEBS Letters* **109**, 55 (1980). Synthesis: E. Wunsch, Z. *Naturforsch. B* **35**, 911 (1980). Isoln of a 25-amino acid somatostatin, *SRIF-25,* and structure: P. Brazeau *et al., Compt. Rend. Ser. D* **290**, 1369 (1980). Both of these peptides consist of an *N*-terminal extension of SRIF-14. They are more potent than SRIF-14 in inhibition of insulin release but less potent in inhibition of glucagon release: L. Mandarino *et al., Nature* **291**, 76 (1981).

Synthetic linear SRIF-14 exhibits the same bioactivity as the natural cyclic form: P. Brazeau *et al., Endocrinology* **94**, 184 (1974). Activity studies in humans: T. M. Siler *et al., J. Clin. Endocrinol. Metab.* **37**, 632 (1973). Inhibitory effects on the secretion of thyrotropin: Vale *et al., Fed. Proc.* **32**, 211 (1973); of insulin: K. G. Alberti *et al., Lancet* **2**, 1299 (1973); Koeker *et al., Science* **184**, 482 (1974). Role in diabetes: Maugh, *ibid.* **188**, 920 (1975). Clinical trial in controlling acute variceal hemorrhage: S. A. Jenkins *et al., Brit. Med. J.* **290**, 275 (1985). Review of distribution, secretion, physiology of gastrointestinal somatostatin: C. H. S. McIntosh, *Life Sci.* **37**, 2043-2058 (1985). Symposium of biosynthesis, bioactivity, and clinical applications: *Adv. Exp. Med. Biol.* **188**, 1-524 (1985). *Reviews:* Vale *et al., Recent Progr. Horm. Res.* **31**, 365-397 (1975); Guillemin, Gerich, *Ann. Rev. Med.* **27**, 379-388 (1976); R. L. Moss, *Ann. Rev. Physiol.* **41**, 617 (1979); A. Arimura, J. B. Fishback, *Neuroendocrinology* **33**, 246-256 (1981); several authors in *Gut Hormones,* S. R. Bloom, J. M. Polack, Eds. (Churchill, New York, 2nd ed., 1981) 605 pp; S. M. McCann, *Ann. Rev. Pharmacol. Toxicol.* **22**, 491-515 (1982); N. Bethge *et al., J. Clin. Chem. Clin. Biochem.* **20**, 603-613 (1982). Book: *Somatostatin* Vol. 2, M. T. McQuillan, Ed. (Eden Press, Quebec, 1980) 238 pp.

```
Ala-Gly-Cys-Lys-Asn-Phe-Phe-Trp
         |                      |
        Cys-Ser-Thr-Phe-Thr-Lys
```

Acetate, *SRIF-A, Aminopan, Modustatina, Somatofalk, Stilamin.*

THERAP CAT: Treatment of severe, acute hemorrhage of gastro-duodenal ulcers. Treatment of erosive or hemorrhagic gastritis. Experimental antidiabetic. Growth hormone inhibitor.

8672. Somatotropin. Adenohypophyseal growth hormone; GH; hypophyseal growth hormone; anterior pituitary growth hormone; growth hormone; phyone; pituitary growth hormone; somatotropic hormone; STH; Antuitrin-Growth; Phyol; Somacton. Species-specific anabolic protein that promotes somatic growth, stimulates protein synthesis, and regulates carbohydrate and lipid metabolism. Increases serum levels of somatomedins, *q.v.* Secreted by the anterior pituitary under the regulation of the hypothalamic hormones, somatoliberin and somatostatin, *q.q.v.* Growth hormones from various species differ in amino acid sequence, antigenicity, isoelectric point, and in the range of animals in which they can produce biological responses. Isoln from bovine anterior pituitary substance: C. H. Li *et al., J. Biol. Chem.* **159**, 353 (1945); A. E. Wilhelmi *et al., ibid.* **176**, 735

(1948); C. H. Li, **U.S.** pat. **3,118,815** (1964 to Upjohn). Isoln from human pituitaries: Lewis, Brink, **U.S.** pat. **2,974,088** (1961 to Merck & Co., Inc.); Reisfeld *et al., Endocrinology* **71**, 559 (1962). Amino acid sequence of HGH: C. H. Li *et al., J. Am. Chem. Soc.* **88**, 2050 (1966); *eidem, Arch. Biochem. Biophys.* **133**, 70 (1969). Total synthesis: C. H. Li, D. Yamashiro, *J. Am. Chem. Soc.* **92**, 7608 (1970). Revised sequence of HGH: H. D. Niall, *Nature New Biol.* **230**, 90 (1971). Homology of HGH with human placental lactogen: H. D. Niall *et al., Proc. Nat. Acad. Sci. USA* **68**, 866 (1971). Review of structural studies and comparison with prolactin, *q.v.: eidem, Recent Progr. Horm. Res.* **29**, 387-416 (1973). Amino acid sequence of ovine growth hormone: C. H. Li *et al., Int. J. Peptide Protein Res.* **4**, 151 (1972). Bovine and ovine growth hormones exist primarily as dimers in slightly alkaline soln. The molecular weight of these dimers is approx 41,000: T. A. Bewley, C. H. Li, *Biochemistry* **11**, 927 (1972). Cloning and expression of DNA for human growth hormone in bacteria: D. V. Goeddel *et al., Nature* **281**, 544 (1979); D. V. Goeddel, H. L. Heyneker, **Belg.** pat. **884,012**; *eidem,* **U.S.** pat. **4,342,832** (1980, 1982 both to Genentech); in mammalian cells: G. N. Pavakis *et al., Proc. Natl. Acad. Sci. USA* **78**, 7398 (1981); W. G. Roskam *et al.,* **Fr.** pat. **2,534,273** (1984 to Institut Pasteur), *C.A.* **101**, 84933w (1984). Purification and bioactivity of recombinant HGH: K. C. Olson *et al., Nature* **293**, 408 (1981). Review of GH gene studies: D. D. Moore *et al., Recent Progr. Horm. Res.* **38**, 197-225 (1982). Transfer of rat GH gene into mice: R. D. Palmiter *et al., Nature* **300**, 611 (1982).

Clinical efficacy in different types of growth failure: H. L. Lenko *et al., Eur. J. Pediatr.* **138**, 241 (1982). Clinical comparison of natural and recombinant HGH: R. L. Hintz *et al., Lancet* **1**, 1276 (1982). Veterinary use of bovine GH to increase milk production in cows: D. E. Bauman *et al., J. Dairy Sci.* **68**, 1352 (1985). Regulation of GH secretion: D. G. Johnston *et al., J. Roy. Soc. Med.* **78**, 319 (1985). Review of biological effects: O. G. P. Isaksson *et al., Ann. Rev. Physiol.* **47**, 483-499 (1985). *Reviews:* C. H. Li, H. M. Evans in *The Hormones,* G. Pincus, K. V. Thimann, Eds., **Vol. I** (Academic Press, New York, 1948) pp 674-685; P. D. Gluckman *et al., Endocrin. Rev.* **2**, 363-395 (1981); C. H. Li, *Mol. Cell Biochem.* **46**, 31-41 (1982). Books: *Growth and Growth Hormone,* A. Pecile, E. Muller, Eds. (Excerpta Medica, Amsterdam, 1972); *Human Growth Hormone and Gonadotrophins in Health and Disease,* P. Franchimont, H. Burger, Eds. (Elsevier, New York, 1975) 494 pp; *Growth Hormone and Other Biologically Active Peptides,* A. Pecile, E. Muller, Eds. (Elsevier, New York, 1980); *Evaluation of Growth Hormone Secretion,* Z. Laron, O. Butenandt, Eds. (S. Karger, New York, 1983).

Human growth hormone, $C_{990}H_{1529}N_{263}O_{299}S_7$, *HGH, somatropin, CB 311, Asellacrin, Bio-Tropin, Crescormon, Grorm, Nanormon, Norditropin, Saizen.* A single polypeptide chain of 191 amino acids having a molecular weight of 22,124. Isoelectric point 4.9. $[\alpha]^{25}$ $-38.7°$ (0.1M acetic acid).

Methionyl human growth hormone, somatrem, met-HGH, Genotonorm, Genotropin, Humatrope, Maxomat, Protropin, Somatonorm, Umatrope. Produced in bacteria from recombinant DNA. Contains the complete amino acid sequence of the natural hormone plus an additional N-terminal methionine.

THERAP CAT: Growth stimulant. HGH used to stimulate linear growth in hypopituitary dwarfism.

8673. Songorine. *(1α,15β)-21-Ethyl-1,15-dihydroxy-4-methyl-16-methylene-7,20-cycloveatchan-12-one; napellonine; zongorine.* $C_{22}H_{31}NO_3$; mol wt 357.48. C 73.91%, H 8.74%, N 3.92%, O 13.43%. From *Aconitum songoricum* Popov, *Ranunculaceae:* Yunusov, *J. Gen. Chem. USSR* **18**, 515 (1948); Kuzovkov, *ibid.* **23**, 504 (1953); **25**, 2006 (1955). Identity with napellonine: Kuzovkov, *Zh. Obshch. Khim.* **28**, 2283 (1958); **29**, 1728 (1959). Structure: Sugasawa, *Chem. Pharm. Bull.* **9**, 889, 897 (1961). Absolute configuration: Okamoto *et al., ibid.* **13**, 1270 (1965). Synthesis of the aromatic intermediate: Wiesner *et al., Can. J. Chem.* **51**, 3978 (1973). Pharmacology: Sadritdinov, *Farmakol. Alk.* No. 312 (1965), *C.A.* **66**, 93772d (1967). Mass spectra data: Yunusov *et al., Khim. Prir. Soedin.* **6**, 101 (1970), *C.A.* **73**, 131178u (1970).

Crystals, mp 201-202°. $[\alpha]_D^{20}$ $-135.4°$. uv max: 290 nm (log ϵ 2.6). LD_{50} in mice: 142.5 mg/kg.

Hydrochloride, crystals, mp 257-258°. $[\alpha]_D^{20}$ $-114°$ (c = 2 in water).

8674. Sophorabioside. *3-[4-[[2-O-(6-Deoxy-α-L-mannopyranosyl)-β-D-glucopyranosyl]oxy]phenyl]-5,7-dihydroxy-4H-1-benzopyran-4-one;* genistein-4'-glucosidorhamnoside. $C_{27}H_{30}O_{14}$; mol wt 578.54. C 56.05%, H 5.23%, O 38.72%. From fruits of *Sophora japonica* L., *Leguminosae.* Isoln and structure: Zemplén, Bognár, *Ber.* **75B**, 482 (1942). The biose is not identical with rutinose.

Trihydrate, needles from dil alcohol, mp 156-160°. The water of crystn can be removed by drying at 100° over P_2O_5 *in vacuo* for 12 hours. The anhydr substance mp 248° (slight decompn). $[\alpha]_D^{19}$ $-73°$ (0.27 g in 10 ml pyridine). Freely sol in pyridine; sol in hot alcohol, hot acetone; slightly sol in boiling water. The alcoholic soln gives a purple color with ferric chloride.

8675. Sophoricoside. *3-[4-(β-D-Glucopyranosyloxy)-phenyl]-5,7-dihydroxy-4H-1-benzopyran-4-one; 4',5,7-trihydroxyisoflavone-4'-D-glucoside; genistein-4'-glucoside.* $C_{21}H_{20}O_{10}$; mol wt 432.37. C 58.33%, H 4.66%, O 37.00%. From the green pods of *Sophora japonica* L., *Leguminosae:* Charaux, Rabate, *Bull. Soc. Chim. Biol.* **20**, 454 (1938). Structure: Zemplén *et al., Ber.* **76**, 267 (1943); Bognár, *C.A.* **46**, 8104h (1952). Synthesis: Bognár, Szabo, *Acta Chim. Acad. Sci. Hung.* **4**, 383 (1954); *Chem. & Ind. (London)* **1954**, 518. Hydrolyzed by emulsin or by dil acids.

Crystals from alcohol, mp 298°. $[\alpha]_D^{20}$ $-47°$ (pyridine). $[\alpha]_D^{20}$ $-32°$ (10% aq pyridine). uv max (abs ethanol): 262 nm. Sparingly sol in water, alc, acetic acid; more sol in hot alc, hot acetic acid; sol in pyridine, dil alkalies; practically insol in ethyl acetate, acetone. Ferric chloride gives the color of red wine to alcoholic solns which turn orange after addition of a few drops of Na-carbonate soln.

8676. Sophorose. *2-O-β-D-Glucopyranosyl-α-D-glucose.* $C_{12}H_{22}O_{11}$; mol wt 342.30. C 42.10%, H 6.48%, O 51.42%. From pods of *Sophora japonica* L., *Leguminosae:* Rebaté, *Bull. Soc. Chim. France* **7**, 565 (1940); Clancy, *J. Chem. Soc.* **1960**, 4213; Clancy in *Methods in Carbohydrate Chemistry* **vol. I,** R. L. Whistler, M. L. Wolfrom, Eds. (Academic Press, New York, 1962) pp 345-349. Structure and synthesis: Coxon, Fletcher, *J. Org. Chem.* **26**, 2892 (1961); Koeppen, *Carbohyd. Res.* **7**, 410 (1968). Crystal structure: J. Ohanessian *et al., Acta Crystallogr.* **B34**, 3666 (1978).

Monohydrate, needles from 80% aq methanol, mp 196-198°. $[\alpha]_D^{18} +19°$ (c = 1.2 in water).

Octa-O-acetyl-β-sophorose, $C_{28}H_{38}O_{19}$, needles from ethanol, mp 193-194°. $[\alpha]_D^{18} - 3.2°$ (c = 2.5 in chloroform).

8677. Sorbic Acid. *2,4-Hexadienoic acid;* 2-propenyl-acrylic acid. $C_6H_8O_2$; mol wt 112.12. C 64.27%, H 7.19%, O 28.54%. $CH_3CH=CHCH=CHCOOH$. May be obtained from berries of the mountain ash, *Sorbus aucuparia* L., *Rosaceae* where it occurs as the lactone, called parasorbic acid: Hofmann, *Ann.* **110**, 129 (1859). Synthesis by condensing crotonaldehyde and malonic acid in pyridine soln: Doebner, *Ber.* **33**, 2140 (1900); Allen, Van Allan, *Org. Syn.* **coll. vol. III**, 783 (1955); by condensing crotonaldehyde and ketene in the presence of boron trifluoride: Hagemeyer, Jr., *Ind. Eng. Chem.* **41**, 768 (1949). Prepn from 1,1,3,5-tetraalkoxyhex-ane: Parker, MacLean, U.S. pat. **2,921,090** (1960 to Celanese). Several other syntheses: Fernholz, Mundlos, Ger. pat. **1,049,852** (1959 to Hoechst); U.S. pat. **3,021,365** (1962 to Hoechst). Purification: Fernholz, **Ger.** pat. **1,044,803** (1958 to Hoechst). The *trans,trans*-isomer is usually obtained and is the commercial product.

Needles from water, mp 134.5°. Should be stored at temps below 40°. bp 228° (dec). Vapor pressure at 20° <0.01 mm, at 143° 50 mm. Flash pt 260°F (127°C). K at 25° = 1.73 × 10^{-5}. Soly in water at 30° 0.25%, at 100° 3.8%, in propylene glycol at 20° 5.5%, in abs ethanol or methanol 12.90%, in 20% ethanol 0.29%, in glacial acetic acid 11.5%, in acetone 9.2%, in benzene 2.3%, in carbon tetrachloride 1.3%, in cyclohexane 0.28%, in dioxane 11.0%, in glycerol 0.31%, in isopropanol 8.4%, in isopropyl ether 2.7%, in methyl acetate 6.1%, in toluene 1.9%. LD_{50} orally in rats: 7.36 g/kg, Smyth, Carpenter, *J. Ind. Hyg. Toxicol.* **30**, 63 (1948).

Sodium salt. Prepn: Horn, Fernholz, Ger. pat. **1,045,390** (1958 to Hoechst).

USE: Mold and yeast inhibitor. Fungistatic agent for foods, especially cheeses. To improve the characteristics of drying oils. In alkyd type coatings to improve gloss. To improve milling characteristics of cold rubber. Use of calcium sorbate: Gooding, U.S. pat. **3,139,378** (1964 to Corn Products Co.). *See also* Potassium Sorbate.

8678. Sorbic Alcohol. *2,4-Hexadien-1-ol;* 1-hydroxy-2,4-hexadiene; hexadenol; Hexene-Ol; Hexakose. $C_6H_{10}O$; mol wt 98.14. C 73.43%, H 10.27%, O 16.30%. $CH_3-CH=CHCH=CHCH_2OH$. Prepd from sorbic aldehyde by means of aluminum isopropylate in isopropanol: Reichstein *et al.*, *Helv. Chim. Acta* **15**, 264 (1932).

Long needles. Agreeable odor of new mown grass. mp 30.5-31.5°. bp_{12} about 80°. Volatile with steam. Insol in water. Sol in alcohol, ether, oils. Not stable to air; must be sealed in evacuated ampuls for storage.

Diphenylurethan, $C_{19}H_{19}NO_2$, crystals from petr ether, mp 78-79°.

3,5-Dinitrobenzoate, yellowish needles from petr ether, mp 85°.

8679. Sorbinil. *(S)-6-Fluoro-2,3-dihydrospiro[4H-1-benzopyran-4,4'-imidazolidine]-2',5'-dione;* (+)-(4S)-6-fluorospiro[chroman-4,4'-imidazolidine]-2',5'-dione; CP-45634. $C_{11}H_9FN_2O_3$; mol wt 236.20. C 55.94%, H 3.84%, F 8.04%, N 11.86%, O 20.32%. Spirohydantoin aldose reductase inhibitor. Prepn of racemate and resolution of isomers: R. Sarges, Ger. pat. **2,821,966**; *idem*, U.S. pat. **4,130,714** (both 1978 to Pfizer). Synthesis by asymmetric

induction: R. Sarges *et al.*, *J. Org. Chem.* **47**, 4081 (1982). Absolute configuration: R. Sarges *et al.*, *J. Med. Chem.* **28**, 1716 (1985). Effect on polyol accumulation in diabetic rats: M. J. Peterson *et al.*, *Metabolism* **28**, Suppl. 1, 456 (1979). Pharmacokinetics in humans: G. Foulds *et al.*, *Clin. Pharmacol. Ther.* **30**, 693 (1981). HPLC determn in human lens and plasma: P. Lloyd, M. J. C. Crabbe, *J. Chromatog.* **343**, 402 (1985). Preliminary clinical trials in diabetic neuropathy: R. J. Young *et al.*, *Diabetes* **32**, 938 (1983); J. Jaspan *et al.*, *Lancet* **2**, 758 (1983). Symposium on pharmacology and clinical efficacy: *Metabolism* **35**, Suppl 1, 1-121 (1986).

Crystals from ethanol, mp 241-243°. $[\alpha]_D^{25} +54.0°$ (c = 1 in methanol).

8680. Sorbitol. D-*Glucitol;* D-sorbitol; L-gulitol; sorbit; Sorbol; Sorbicolan; Sorbo; Sorbostyl; Nivitin; Cholaxine; Karion; Sionit; Sionon; Sorbilande; Diakarmon. $C_6H_{14}O_6$; mol wt 182.17. C 39.56%, H 7.75%, O 52.70%. First found in the ripe berries of the mountain ash *Pyrus aucuparia* Ehrh. (L.) (*Sorbus aucuparia* L.), *Rosaceae*. Occurs also in many other berries (except grapes) and in cherries, plums, pears, apples, seaweed and algae. Has been detected in blackstrap molasses. Isoln from berries: Embden, Griesbach, Z. *Physiol. Chem.* **91**, 268 (1914). Prepd industrially from glucose by high pressure hydrogenation or by electrolytic reduction: Boye, *Chem.-Ztg.* **82**, 657 (1958); Fedor *et al.*, *Ind. Eng. Chem.* **52**, 282 (1960); from dextrose by catalytic hydrogenation: Faith, Keyes & Clark's *Industrial Chemicals*, F. A. Lowenheim, M. K. Moran, Eds. (Wiley-Interscience, New York, 4th ed., 1975) pp 774-778. Review of uses: Kempf, *Die Stärke* **6**, 269-274 and 303-306 (1954); **9**, 234-237 (1955).

Needles with ½ or $1H_2O$. Sweet taste, ~60% as sweet as sugar (w/w). In the healthy human organism 1.0 g of sorbitol yields 3.994 calories which is comparable to 3.940 calories from 1.0 g of cane sugar. Seventy percent of orally ingested sorbitol is converted to CO_2 without appearing as glucose in the blood: Adcock, Gray, *Biochem. J.* **65**, 554 (1957). The hydrated crystals melt somewhat below 100°. When completely anhydr mp 110-112°. $[\alpha]_D^{20} - 2.0°$ (H_2O). In the presence of molybdate the rotation is reversed and increased to +56°. Freely sol in water (up to 83%). High % sorbitol solns are much more viscous than corresp glycerol solns. Quite sol in hot alcohol, sparingly sol in cold alcohol. Also sol in methanol, isopropanol, butanol, cyclohexanol, phenol, acetone, acetic acid, DMF, pyridine, acetamide solns. Practically insol in most other organic solvents. Not attacked in the cold when mixed with dil acids, alkalies or mild oxidizing substances. Ka at 17.5° = 2.5 × 10^{-14}. pH about 7.0. A commercial 70% aq soln may have the following characteristics: d_{20}^{20} 1.2879; n_D^{25} 1.45831; $[\alpha]_D^{20}$ -2.10°; bp_{760} 105°; pH between 6 and 7; viscosity (25°): 110 cp. d_4^{20} for various % solns: 5% 1.014; 10% 1.038; 25% 1.099; 50% 1.198; 60% 1.249; 70% 1.299; 83% 1.391. Viscosity in cp at 20°: 5% soln 1.230; 10% 1.429; 25% 2.689; 50% 11.09; 60% 35.73; 70% 185; 83% >10,000.

USE: In manuf of sorbose, ascorbic acid, propylene glycol, synthetic plasticizers and resins; as humectant (moisture conditioner) on printing rolls, in leather, tobacco. In writing inks to insure a smooth flow and to prevent crusting on the point of the pen. In antifreeze mixtures with glycerol or glycols. In candy manuf to increase shelf life by retarding the solidification of sugar; as humectant and softener in shredded coconut and peanut butter; as texturizer in foods; as sequestrant in soft drinks and wines. Used to reduce the undesirable aftertaste of saccharin in foodstuffs; as sugar substitute for diabetics. Pharmaceutic aid (flavor; tablet excipient); to increase absorption of vitamins and other nutrients in pharmaceutical preparations: *Chem. & Eng. News* **36**, 59 (Feb. 24, 1958).

THERAP CAT (VET): In ruminant ketosis, osmotic diuretic, laxative.

8681. Sorbose. L-*Sorbose;* sorbin; sorbinose. $C_6H_{12}O_6$; mol wt 180.16. C 40.00%, H 6.71%, O 53.29%. From sorbitol by fermentation with *Acetobacter suboxydans:* Perlman, *J. Chem. Ed.* **36**, 60 (1959); Lockwood, *Methods in Carbohydrate Chemistry* **1**, 151 (Academic Press, New York, 1962).

$$
\begin{array}{c}
CH_2OH \\
| \\
CO \\
| \\
HOCH \\
| \\
HCOH \\
| \\
HOCH \\
| \\
CH_2OH
\end{array}
$$

Orthorhombic, bisphenoidal crystals. About as sweet as sucrose. d^{15} 1.65. mp 165°. $[\alpha]_D^{30}$ —42.7° (c = 5). Ka at 17.5° = 2.8×10^{-12}. Freely sol in water; almost insol in alcohol. Reduces Fehling's soln.

USE: In the manuf of vitamin C (accounts for nearly 1000 tons of ascorbic acid produced every year). For conversion of L-sorbose to 2-keto-L-gulonic acid *see* Reichstein, Grüssner, *Helv. Chim. Acta* **17**, 311 (1934).

8682. Sotalol. *N-[4-[1-Hydroxy-2-[(1-methylethyl)amino]ethyl]phenyl]methanesulfonamide; 4'-[1-hydroxy-2-(isopropylamino)ethyl]methanesulfonanilide.* $C_{12}H_{20}N_2O_3S$; mol wt 272.36. C 52.92%, H 7.40%, N 10.28%, O 17.62%, S 11.77%. β-Adrenergic blocker. Prepn: Uloth *et al., J. Med. Chem.* **9**, 88 (1966). Pharmacology and toxicology: Larsen, Lish, *Nature* **203**, 1283 (1964); Lish *et al., J. Pharmacol. Exp. Ther.* **149**, 161 (1965); Stanton *et al., ibid.* 175; Kvam *et al., ibid.* 183. Activity of the *d*- and *l*-forms: Somani, Watson, *ibid.* **164**, 317 (1968); Somani, Bachand, *Eur. J. Pharmacol.* **7**, 239 (1969). HPLC determn in body fluids: W. P. Gluth *et al., Arzneimittel-Forsch.* **38**, 408 (1988).

Hydrochloride, $C_{12}H_{21}ClN_2O_3S$, *MJ-1999, Beta-Cardone, Betapace, Sotacor, Sotalex.* White crystalline solid, mp 206.5-207° (dec). Freely sol in water; slightly sol in chloroform. LD_{50} in mice, rats (mg/kg): 2600, 3450 orally; 670, 680 i.p.; LD_{50} orally in rabbits: 1000 mg/kg (Lish).

THERAP CAT: Antianginal; antiarrhythmic; antihypertensive.

8683. Soterenol. *N-[2-Hydroxy-5-[1-hydroxy-2-[(1-methylethyl)amino]ethyl]phenyl]methanesulfonamide; 2'-hydroxy-5'-[1-hydroxy-2-(isopropylamino)ethyl]methanesulfonanilide;* MJ-1992. $C_{12}H_{20}N_2O_4S$; mol wt 288.36. C 49.98%, H 6.99%, N 9.71%, O 22.19%, S 11.12%. β-Adrenergic agonist. Prepn: Larsen *et al., J. Med. Chem.* **10**, 462 (1967). Pharmacological study: K. W. Dungan *et al., J. Pharmacol. Exp. Ther.* **164**, 290 (1968).

Monohydrochloride, $C_{12}H_{21}ClN_2O_4S$, crystals from methanol-isopropyl ether, mp 195.5-196.5° (dec). LD_{50} (7 day) in mice (mg/kg): 41 i.v.; 315 i.p.; 660 orally (Dungan).

THERAP CAT: Bronchodilator.

8684. Soybean. Soya bean; soja bean; Lincoln bean; Manchurian bean; Chinese pea. Seed of *Glycine max* (L.) Merrill [*G. soja* Sieb. & Zucc., *G. hispida* (Moench) Maxim., *Soja hispida* Moench], *Leguminosae.* *Habit.* Eastern Asia, especially Manchuria. Cultivated in the midwestern U.S.A., Brazil, some European countries, such as Italy and Yugoslavia. *Constit.* Proteins 40%, carbohydrates 17%, oil 18%, ash 4.6%. The ash content equals K 1.67%, Na 0.34%, Ca 0.28%, Mg 0.22%, Cl 0.024%, I 0.000054%, Fe 0.0097%, Cu 0.0012%, Mn 0.0028%, Zn 0.0022%, P 0.66%, S 0.41%, Al 0.0007%. The chief proteins are glycinin, a globulin, phaseolin, another globulin, and 2 albumins: legumelin and soy legumelin. The carbohydrates are sucrose, raffinose, stachyose, and pentosans. Phosphorus compds (about 3%) are phospholipids, nucleic acid phosphates, phytin, and inorganic phosphates. The phospholipids contain lecithin, cephalin, and inositol. The vitamin content is moderate: Vitamin A 110 i.u./100g; thiamine 1.14 mg/100 g; riboflavin 0.31 mg/100 g; niacin 2.1 mg/100 g; ascorbic acid: trace. Food energy: 350 cal/100 g. Amino acid analysis of the protein fraction: Glutamic acid 18.4%, leucine 8.1%, arginine 7.5%, lysine 6.7%, valine 5.4%, isoleucine 5.3%, phenylalanine 5.2%, threonine 3.9%, histidine 2.3%, tryptophan 1.6%, methionine 1.4%. Ten kg of soybeans yield 15 g of the isoflavone glycoside genistin, the mother liquor contains daidzin. Soybeans contain a small amount of enzymes such as lipoxidase, urease, uricase, protease, antienzymes such as antitrypsin, and a growth-inhibiting factor. Bitter principles and saponins are also found. *Reviews:* K. S. Markley, *Soybeans and Soybean Products,* 2 vols. (Interscience, New York, 1951); Cowan in Kirk-Othmer *Encyclopedia of Chemical Technology* vol. 18 (Interscience, NewYork, 2nd ed., 1969) pp 599-614.

Whitish or yellowish-green to brownish-black, ovoid beans, about 8 mm long, 7 mm wide, and 6 mm thick. Most varieties have a wt of 20 g/100 seeds.

USE: In the production of soybean oil. As food and feedstuff. Debittered soybean flour contains practically no starch and is widely used in dietetic foods. Soybean meal obtained after expressing the oil is a preferred source of protein for feedstuffs. Other products are soybean lecithin, genistin, monosodium glutamate, soybean milk, soy sauce; *tofu,* a soybean curd; *miso,* a fermented mixture with barley or rice; *natto,* which is soybean cheese. Soybean proteins are used also in the adhesive and plastics industries.

THERAP CAT: Nutrient.

8685. Soybean Oil. Obtained from soybeans by solvent extraction using petroleum hydrocarbons or, to a lesser extent, by expression using continuous screw press operations. *Constit.* Triglycerides of oleic acid 26%, of linoleic acid 49%, of linolenic acid 11%, of saturated acids 14%. Free fatty acids are usually less than 1%. Phospholipids (lecithin) 1.5-4%. Another 0.8% consists of stigmasterol, sitosterols, and tocopherols. The phospholipids and sterols are removed by refining with alkali. Reviews and bibliographies: K. S. Markley, *Soybeans and Soybean Products,* 2 vols. (Interscience, New York, 1951); E. W. Eckey, *Vegetable Fats and Oils* (Reinhold, New York, 1954); W. J. Wolf in Kirk-Othmer, *Encyclopedia of Chemical Technology* vol. 21 (Wiley-Interscience, New York, 3rd ed., 1983) pp 417-442.

Pale yellow to brownish-yellow oil. Slight characteristic odor and taste. The flavor of crude soybean oil has been described as paint-like and grass-like, and is somewhat distasteful to the palate of occidentals. d_{25}^{25} 0.916-0.922. Flash pt 540°F (282°C). Ignition temp 833°F (445°C). n_D^{25} 1.471-1.475. Titer 22-27. Solidifies at —10 to —16°. Viscosity in centipoises: 172.9 at 0°; 99.7 at 10°; 50.09 at 25°; 28.86 at 40°. Acid value 0.3-3.0. Saponification value 189-195. Iodine value 127-138. Thiocyanogen value 77-85. Diene no. 0.7. Hydroxyl value 4-8. Reichert-Meissl value 0.2-0.7. Polenske value 0.2-1.0. Miscible with abs alcohol, ether, petr ether, chloroform, carbon disulfide.

USE: Directly as food in the Orient, also in blends with olive oil. In the manuf of margarine, shortenings, candy,

Consult the cross index before using this section.

soap. In drying oil industries (paints, varnishes, linoleum, printing ink). The foots or residue from refining has been used to obtain soybean fatty acids which can be used in the manuf of alkyd resins.

8686. Soy Sauce. Soy sauce is a hydrolysis product of soybeans. A combination of mold fermentaion and acid hydrolysis is used. The molds employed are *Aspergillus flavus, A. niger,* and *A. oryzae.* Soy sauce consists of a mixture of amino acids, peptides, polypeptides, peptones, simple proteins, purines, carbohydrates, and lesser organic compds suspended in an 18% sodium chloride soln. Adenine, arginine, choline, lysine, betaine, and glutamic acid have been isolated. In the manuf of soy sauce, during the fermentation process, 60-70% of the vitamins present in soybeans are destroyed. Review and bibliography: K. S. Markley, *Soybeans and Soybean Products,* 2 vols. (Interscience, N.Y., 1951).

8687. Sozoiodolic Acid. *4-Hydroxy-3,5-diiodo-benzenesulfonic acid;* 2,6-diiodophenol-4-sulfonic acid; Optojod. $C_6H_4I_2O_4S$; mol wt 425.95. C 16.92%, H 0.95%, I 59.59%, O 15.02%, S 7.53%. Made by iodizing potassium *p*-phenolsulfonic acid by means of ICl, or by KI and KIO$_3$. Prepn of salts: Rupp, Herman, *Arch. Pharm.* **254,** 488 (1916).

Trihydrate, white, almost odorless needles, becoming anhydr at 100°. mp 120° when anhydr; dec at 190°. Freely sol in water, alcohol, ether, glycerol.

Mercury deriv, $C_6H_2HgI_2O_4S$, *Sozoiodole-mercury.* Fine, orange-yellow powder. *Poisonous!* Practically insol in water, alcohol, ether, glycerol. Sol in NaCl or KI soln.

Sodium salt dihydrate, $C_6H_3I_2NaO_4S.2H_2O$, *Sozoiodole-sodium, Dijozol.* Crystals. Sol in water, alcohol, glycerol. Practically insol in ether.

Zinc salt hexahydrate, $C_{12}H_6I_4O_8S_2Zn.6H_2O$, *Sozoiodole-zinc.* Needles, sol in water, alc, glycerol. Practically insol in ether.

USE: In a test for albumin.

THERAP CAT: Diagnostic aid (radiopaque medium in retrograde pyelography). Hg deriv formerly in syphilis.

8688. Spandex. Lycra; Vyrene. An elastic, segmented polyurethane fiber obtained by the interaction of a diisocyanate with a glycol. The term "segmented" indicates that it is made from a block copolymer in which reasonably long flexible chains are joined to shorter stiff chains through the urethane linkages. *Ref: Modern Textiles Mag.* **40,** 38 (Dec. 1959); Hicks Jr., *Am. Dyestuff Reptr.* **52,** no. 1, 33 (1963). Book: Moncrieff, *Man-made Fibres* (Heywood, London, 4th ed., 1963) pp 396-403. Moncrieff cites U.S. pats. **2,692,873** (1954 to du Pont) and **2,751,363** (1956 to U.S. Rubber) in describing the historical development of spandex fibers. Compare also Smith; Frazer, Shivers, U.S. pats. **3,061,574** and **3,071,557** (1962, 1963 to du Pont). Historical review of industrial production with extensive patents list: L. Rose, *Repts. Progr. Appl. Chem. (London)* **51,** 609-612 (1966). Monograph: M. McDonald, *Spandex Manufacture* (Noyes, Park Ridge, N.J., 1970) 190 pp.

All linkages are hydrolytically stable to acids, alkalies. USE: Snap-back fiber, stronger and lighter than rubber. Used mainly for elastic garments such as belts, girdles, corsets, brassieres, garters, surgical stockings and sock tops. Can be used bare, giving lighter fabrics and eliminating the costly covering process needed for rubber. *Caution:* Can cause allergic contact dermatitis if any diisocyanate is present as impurity.

8689. Span® and Tween®. Nonionic surface active agents, complex esters, and ester-ethers, their chemical starting point being hexahydric alcohols, alkylene oxides, and fatty acids.

R = fatty acid residues

The hydrophilic character is supplied by free hydroxyl and oxyethylene groups, while the lipophile portion is found in the long chain fatty acids used.

The Span type materials are partial esters of the common fatty acids (lauric, palmitic, stearic, and oleic) and hexitol anhydrides (hexitans and hexides), derived from sorbitol.

The Tween type materials are derived from the Span products by adding polyoxyethylene chains to the nonesterified hydroxyls.

The Span products tend to be oil-soluble and dispersible or insol in water, while the Tween products are sol or well dispersible in water.

Tween 80 is official in pharmacy where it is called polysorbate 80, *see* separate entry. Use of Spans and Tweens in pharmaceutical formulations: Jowdy, *Carolina J. Pharm.* **33,** 465, 467 (1952).

8690. Sparassol. *2-Hydroxy-4-methoxy-6-methylbenzoic acid methyl ester; 4-methoxy-2,6-cresotic acid methyl ester;* everninic acid methyl ester; orsellinic acid methyl ester 4-methyl ether. $C_{10}H_{12}O_4$; mol wt 196.20. C 61.21%, H 6.17%, O 32.62%. Antibiotic substance produced by the fungus *Sparassis ramosa:* Falck, *Ber.* **56,** 2555 (1923). Also obtained in methanol extracts of the lichen *Evernia prunasti:* Stenhouse, *Ann.* **68,** 55 (1848); Späth, Jeschki, *Ber.* **57,** 471 (1924). Structure: Fischer, Hoesch, *Ann.* **391,** 347 (1912); Wedekind, Fleischer, *Ber.* **56,** 2556 (1923). Synthesis: G. Nicollier *et al., Helv. Chim. Acta* **61,** 2899 (1978).

Prisms from water, mp 67-68°. Slightly sol in hot water; freely sol in acetone, ether, chloroform; moderately sol in methanol, ethanol, petr ether. Alkaline hydrolysis yields free everninic acid, mp 170°.

8691. Sparsiflorine. *5,6,6a,7-Tetrahydro-2-methoxy-4H-dibenzo[de,g]quinoline-1,10-diol; 2-methoxy-6aβ-noraporphine-1,10-diol.* $C_{17}H_{17}NO_3$; mol wt 283.31. C 72.06%, H 6.05%, N 4.94%, O 16.94%. From the leaves of *Croton sparsiflorus* Morung, *Euphorbiaceae:* Saha, *Sci. Cult.* **24,** 572 (1959), *C.A.* **54,** 1580a (1960). Structure: Chatterjee *et al., Tetrahedron Letters* **1965,** 1539.

Needles from alc, dec 228°. Sol in methanol, ethanol, ethyl acetate, acetone; slightly sol in ether; practically insol in benzene, petr ether, hexane.

Hydrochloride, $C_{17}H_{17}NO_3 \cdot HCl$, crystals from water, dec 283-284°. $[\alpha]_D^{30}$ +43° (water). uv max (ethanol): 226, 266, 275, 310 nm (log ϵ 4.54, 4.09, 4.23, 3.95).

N-Methyl methiodide, $C_{18}H_{19}NO_3 \cdot CH_3I$, crystals, dec 236-238°. uv max (ethanol): 225, 267, 277, 310 nm (log ϵ 4.57, 4.08, 4.0, 3.85).

8692. Sparteine. *[7S-(7α,7aα,14α,14aβ)]-Dodecahydro-7,14-methano-2H,6H-dipyrido[1,2-a:1',2'-e][1,5]diazocine;* *l*-sparteine; lupinidine. $C_{15}H_{26}N_2$; mol wt 234.37. C 76.87%, H 11.18%, N 11.95%. In yellow and black lupin beans, *Lupinus luteus* L. and *L. niger* Hort.; also in *Cytisus scoparius* (L.) Link. and *Anagyris foetida* L., *Leguminosae.* Extraction procedure: Karrer *et al., Helv. Chim. Acta* **11**, 1062 (1928). Structure: Clemo *et al., J. Chem. Soc.* **1931**, 429; Ing, *ibid.* **1933**, 504; Schirm, Besendorf, *Arch. Pharm.* **280**, 64 (1942); Galinovsky, Stern, *Ber.* **77B**, 132 (1944); Clemo *et al., J. Chem. Soc.* **1949**, 663. Biosynthesis: Anet *et al., Nature* **165**, 35 (1950); Schöpf *et al., Angew. Chem.* **65**, 161 (1953); **69**, 69 (1957); van Tamelen, Foltz, *J. Am. Chem. Soc.* **82**, 2400 (1960). Absolute configuration: Okuda *et al., Chem. & Ind. (London)* **1961**, 1116. Conformation: Bohlmann *et al., Tetrahedron Letters* **1965**, 2705; Wiewiorowski *et al., Can. J. Chem.* **45**, 1447 (1967). Synthesis of racemate: Bohlmann *et al., Ber.* **106**, 3026 (1973); N. Takatsu *et al., Chem. Pharm. Bull.* **35** 4990 (1987). Chemistry: Binnig, *Arzneimittel-Forsch.* **24**, 752 (1974). Studies of cardiovascular effects: Raschak, *ibid.* 753.

Viscous, oily liquid. bp$_8$ 173°. Volatile with steam. $[\alpha]_D^{21}$ −16.4° (c = 10 in abs alc). n_D^{20} 1.5312. d_4^{20} 1.020. pK$_1$ at 20°: 2.24; pK$_2$: 9.46; pH of 0.01 molar soln 11.6. One gram dissolves in 325 ml water. Freely sol in alcohol, chloroform or ether.

Sulfate pentahydrate, $C_{15}H_{28}N_2O_4S \cdot 5H_2O$, *Actospar, Depasan, Spartocin, Spartein-Asal, Synastrin, Tocosamine Sulfate.* Columnar crystals, loses water of crystn at 100° turning brown, dec 136°. pH of 0.05 molar soln 3.3. One gram dissolves in 1.1 ml water, 3 ml alcohol. Practically insol in chloroform and ether.

Adenylate, *Spartopan.*

THERAP CAT: Oxytocic.

8693. Spasmolytol. *2-[(3,5-Dibromo-2-methoxyphenyl)-methoxy]-N,N-diethylethanamine; 2-(3,5-dibromo-2-methoxybenzyloxy)triethylamine;* β-diethylaminoethyl ether of 3,5-dibromo-2-methoxybenzyl alcohol; 3,5-dibromo-2-methoxybenzyl alcohol β-diethylaminoethyl ether; A-124. $C_{14}H_{21}Br_2NO_2$; mol wt 395.16. C 42.55%, H 5.36%, Br 40.45%, N 3.55%, O 8.10%. Preparation: Felton, U.S. pat. **2,766,238** (1956 to Hynson, Westcott & Dunning). Pharmacology and toxicology: Bryant *et al., J. Pharmacol. Exp. Ther.* **121**, 210 (1957).

Liquid, bp$_3$ 185-195°.

Hydrochloride, $C_{14}H_{21}Br_2NO_2 \cdot HCl$, crystals from methanol + ether, mp 124-127°.

THERAP CAT: Antispasmodic.

8694. Spearmint. Mint. Dried leaves and tops of *Mentha spicata* L. (*M. viridis* L.), *Labiatae.* Habit. Europe, Asia; widely cultivated in U.S. Principal active constit is a volatile oil which contains at least 50% carvone.

USE: As flavor.

8695. Spectinomycin. *Decahydro-4a,7,9-trihydroxy-2-methyl-6,8-bis(methylamino)-4H-pyrano[2,3-b][1,4]benzodioxin-4-one;* actinospectacin; espectinomicina; CHX-3101; M 141. $C_{14}H_{24}N_2O_7$; mol wt 332.35. C 50.59%, H 7.28%, N 8.43%, O 33.70%. Antibiotic isolated from fermentation broth of *Streptomyces spectabilis:* Mason *et al., Antibiot. & Chemother.* **11**, 118 (1961); Bergy *et al., ibid.* 661; Bergy, De Boer, U.S. pat. **3,234,092** (1966 to Upjohn). Purification and crystallization: Jahnke, U.S. pat. **3,206,360;** Peters, U.S. pat. **3,272,706** (1965 and 1966 to Upjohn). Isoln, characterization and identity with M 141: Sinclair, Winfield, *Antimicrob. Ag. Chemother.* **1961**, 503. Structure: Hoeksema *et al., J. Am. Chem. Soc.* **84**, 3212 (1962); Wiley *et al., ibid.* **85**, 2652 (1963). Stereochemistry and abs config: Cochran *et al., Chem. Commun.* **1972**, 494. Biosynthesis: Mitscher *et al., ibid.* **1971**, 1542; H. Otsuka *et al., J. Am. Chem. Soc.* **102**, 6817 (1980). Stereospecific synthesis of optically active spectinomycin: D. R. White *et al., Tetrahedron Letters* **1979**, 2737; S. Hanessian, R. Roy, *Can. J. Chem.* **63**, 163 (1985). Pharmacology: Wagner *et al., Int. Z. Klin. Pharmakol. Ther. Toxikol.* **1**, 261 (1968). Activity against penicillin resistant *Neisseria gonorrhoeae*: W. W. Karney *et al., N. Eng. J. Med.* **296**, 889 (1977); N. J. Fiumara, *J. Am. Med. Assoc.* **239**, 735 (1979). Review: R. N. Brogden, G. S. Avery, *Drugs* **3**, 314 (1972); W. M. McCormack, M. Finland, *Ann. Intern. Med.* **84**, 712-716 (1976).

Amorphous solid, mp 184-194°. pKa: 6.95, 8.70 (H$_2$O): $[\alpha]_D^{25}$ −20°. Sol in water, methanol, ethanol. Practically insol in acetone, hydrocarbon solvents.

Dihydrochloride pentahydrate, $C_{14}H_{26}Cl_2N_2O_7 \cdot 5H_2O$, *Spectam, Spectogard, Stanilo, Togamycin, Trobicin.* Exists as a ketone hydrate and not in the carbonyl form, Cochran *et al., loc. cit.* Colorless needles from aqueous acetone, mp 205-207° (dec). $[\alpha]_D$ +14.8° (c = 0.42 in water). Soly in water, methanol, propylene glycol, 0.1N NaOH, 0.1N HCl: > 20 mg/ml, Marsh, Weiss, *J. Assoc. Offic. Anal. Chem.* **50**, 457 (1967). Very slightly sol in benzene, chloroform, ethanol, acetone.

Sulfate dihydrate, $C_{14}H_{26}N_2O_{11}S \cdot 2H_2O$, crystals from water + acetone, dec about 185°. pKa: 7.00, 8.75 (H$_2$O). $[\alpha]_D^{25}$ +17°. LD$_{50}$ i.p. in mice: > 2 g/kg (Bergy, De Boer).

THERAP CAT: Antibacterial.

THERAP CAT (VET): Antibacterial.

8696. Spectrin. Tektin A. Major protein component of the membrane cytoskeleton of mammalian and avian erythroid cells. Responsible for maintaining the normal shape, strength and stability of the erythrocyte. A ubiquitous family of proteins, spectrin analogs have also been isolated from a variety of non-erythroid cells. Given the name spectrin because of its original isolation from hemoglobin-free red cell membranes known as erythrocyte "ghosts": V. T. Marchesi, E. Steers, *Science* **159**, 203 (1968). Comparison with actin, *q.v.*: E. Steers, V. Marchesi, *J. Gen. Physiol.* **54**, 65 S (1969). Isoln from human red cells and species comparison: T. W. Tillack *et al., Biochim. Biophys. Acta* **200**, 125 (1970). Spectrin is a heterodimer composed of 2 high mol wt polypeptide subunits referred to as band 1 or α-spectrin (mol wt ~240 kDa) and band 2 or β-spectrin (mol wt ~220 kDa). Structural studies: M. Clarke, *Biochem. Biophys. Res. Commun.* **45**, 1063 (1971); T. L. Steck, *J. Cell Biol.* **62**, 1 (1974); J. M. Anderson, *J. Biol. Chem.* **254**, 939 (1979); D. W. Speicher *et al.*, *ibid.* **257**, 9093 (1982). The β-subunit of spectrin is bound to the red cell membrane by proteins known as *ankyrin:* V. Bennett, P. J. Stenbuck, *ibid.* **254**, 2533 (1979); or *syndeins:* J. Yu, S. R. Goodman, *Proc. Nat. Acad. Sci. USA* **76**, 2340 (1979). The spectrin heterodimers aggregate on the membrane surface and, together with actin and other proteins, form a filamentous network that covers the surface of the cytoplasmic membrane: S. E. Lux, *Nature* **281**, 426 (1979); *idem, Sem. Hematol.* **16**, 21 (1979). Identification of spectrin analogs from nonerythroid cells: S. R. Goodman *et al.*, *Proc. Nat. Acad. Sci. USA* **78**, 7580 (1981). Isoln of *fodrin*, also known as brain or neural cell spectrin: J. Levine, M. Willard, *J. Cell Biol.* **90**, 631 (1981); V. Bennett *et al.*, *Nature* **299**, 126 (1982). Isoln of *TW 260/240*, a spectrin analog from chicken intestinal epithelial cells: J. R. Glenney *et al.*, *Cell* **28**, 843 (1982). Comparison of erythroid and non-erythroid spectrins: J. R. Glenney *et al.*, *Proc. Nat. Acad. Sci. USA* **79**, 4002 (1982); J. R. Glenney, P. Glenney, *Eur. J. Biochem.* **144**, 529 (1984). Review of erythrocyte spectrin: S. R. Goodman, K. Shiffer, *Am. J. Physiol.* **244**, C121-C141 (1983); of neural cell spectrin: S. R. Goodman, I. S. Zagon, *ibid.* **250**, C347-C360 (1986). Comprehensive review: S. R. Goodman *et al.*, *Crit. Rev. Biochem.* **23**, 171-234 (1988).

8697. Spermaceti. Cetaceum; Spermwax. A waxy substance from the head of the sperm whale. *Constit.* Chiefly cetyl palmitate; free cetyl alcohol present in appreciable amounts; esters of lauric, stearic, and myristic acids; esters of higher alcohols also present.

White, somewhat translucent, slightly unctuous masses with crystalline fracture and pearly luster; almost odorless and tasteless but becomes yellow and rancid on long exposure to air. d 0.938-0.944; mp 42-50°; n_D^{80} about 1.4330. Sapon no. 120-136. Iodine no. 3-4.4. Insol in water or cold alcohol. Sol in chloroform, ether, carbon disulfide, oils, boiling alcohol; slightly sol in petr ether.

USE: As a base for ointments, cerates, etc., and as emulsion with egg yolk or expressed almond oil. In manuf of candles, soaps, cosmetics, laundry wax; finishing and lustering linens. Emollient. A modified emulsifying form is marketed as *Cetina*.

8698. Spermidine. *N-(3-Aminopropyl)-1,4-butanediamine; N-(γ-aminopropyl)tetramethylenediamine.* $C_7H_{19}N_3$; mol wt 145.24. C 57.88%, H 13.19%, N 28.93%. NH_2-$(CH_2)_4NH(CH_2)_3NH_2$. Biogenic polyamine formed from putrescine; a precursor of spermine, *q.q.v.* First detected in human sperm, but occurs widely in nature. Essential for both normal and neoplastic tissue growth. Synthesis: H. W. Dudley *et al.*, *Biochem. J.* **20**, 1082 (1926); **21**, 97 (1927); J. V. Braun, W. Pinkernelle, *Ber.* **70**, 1234 (1937); M. Danzig, H. P. Schultz, *J. Am. Chem. Soc.* **74**, 1836 (1952). For reviews of the early literature, see M. Guggenheim, *Die biogenen Amine* (S. Karger, Basel, 4th ed., 1951) 619 pp; H. Tabor *et al.*, *Ann. Rev. Biochem.* **30**, 579-604 (1961). Role in cell growth processes: C. W. Tabor, H. Tabor, *ibid.* **45**, 285 (1976); J. Janne *et al.*, *Biochim. Biophys. Acta* **473**, 241 (1978); C. W. Porter, R. J. Bergeron, *Science* **219**, 1083 (1983). Formation and interconversion of spermidine and putrescine in mammalian cells: A. E. Pegg *et al.*, *Advan. Enzyme Regul.* **19**, 427 (1980). Effect on polypeptide chain elongation *in vitro:* A. K. Abraham, A. Pihl, *Eur. J. Biochem.* **106**, 257 (1980). Regulation of tRNA methyltransferase activity: M. Mach *et al.*, *Biochem. J.* **202**, 153 (1982). HPLC study: C. E. Prussak, D. H. Russell, *J. Chromatog.* **229**, 47 (1982). Interaction with actin, *q.v.:* C. Oriol-Audit, *Biochem. Biophys. Res. Commun.* **105**, 1096 (1982). Studies on use as a biochemical tool in cancer research: A. Thyss *et al.*, *Eur. J. Cancer Clin. Oncol.* **18**, 611 (1982); V. Quemener *et al.*, *J. Nat. Prod.* **45**, 608 (1982); C. W. Porter *et al.*, *Cancer Res.* **42**, 4072 (1982). Toxicity study: P. R. Langford *et al.*, *J. Antibiot.* **35**, 1387 (1982). *Reviews:* T. C. Theoharides, *Life Sci.* **27**, 703-713 (1980); O. Heby, *Differentiation* **19**, 1-20 (1981); L. Stevens, *Med. Biol.* **59**, 308-313 (1981). Book: *Polyamines in Biology and Medicine*, D. R. Morris, L. J. Marton, Eds. (Dekker, New York, 1981) 512 pp.

Liq, bp_{14} 128-130°. Sol in water, ethanol, ether.

Trihydrochloride, $C_7H_{22}Cl_3N_3$, cryst from ethanolic HCl (65:2). mp 256-258°.

USE: As a tool in biochemical research.

8699. Spermine. *N,N'-Bis(3-aminopropyl)-1,4-butanediamine; N,N'-bis(3-aminopropyl)tetramethylenediamine; gerontine; musculamine; neuridine.* $C_{10}H_{26}N_4$; mol wt 202.34. C 59.36%, H 12.95%, N 27.69%. $H_2N(CH_2)_3NH$-$(CH_2)_4NH(CH_2)_3NH_2$. Biogenic polyamine formed from spermidine, *q.v.*, and occurring in almost all tissues. Essential for both normal and neoplastic tissue growth. First observed in human semen and described as the cryst phosphate salt by A. von Leeuwenhoek, *Phil. Trans. Roy. Soc. London* **12**, 1040 (1678). For a description of the history, occurrence, formation, and early prepns of spermine, *see* Beilstein **vol. IV, Suppl. 1**, 704; M. Guggenheim, *Die biogenen Amine* (S. Karger, Basel, 4th ed., 1951) 619 pp; H. Tabor *et al.*, *Ann. Rev. Biochem.* **30**, 579-604 (1961). Prepn of spermine and its tetrahydrochloride: Israel *et al.*, *J. Med. Chem.* **7**, 710 (1964). Role in cell growth processes: C. W. Tabor, H. Tabor, *Ann. Rev. Biochem.* **45**, 285 (1976); J. Janne *et al.*, *Biochim. Biophys. Acta* **473**, 241 (1978). Modulation of calcium-dependent immune processes: T. C. Theoharides, *Life Sci.* **27**, 703 (1980). Biosynthesis in fungi: L. Stevens, *Med. Biol.* **59**, 308 (1981). HPLC study: C. E. Prussak, D. H. Russell, *J. Chromatog.* **229**, 47 (1982). Use as a biochemical marker for malignant tumors: Y. Horn *et al.*, *Cancer Res.* **42**, 3248 (1982). Metabolic study: A. E. Pegg *et al.*, *Biochemistry* **21**, 5082 (1982). Review of role in cell proliferation and differentiation: O. Heby, *Differentiation* **19**, 1-20 (1981). Book: *Polyamines in Biology and Medicine*, D. R. Morris, L. J. Marton, Eds. (Dekker, New York, 1981) 512 pp.

Needles; mp 55-60°. Liq, $bp_{0.5}$ 141-142°. Strong base, absorbs carbon dioxide from air. *Keep well closed.* Sol in water, lower alcohols, chloroform. Practically insol in ether, benzene, petr ether.

Diphosphate hexahydrate, $C_{10}H_{32}N_4O_8P_2.6H_2O$, *spermine phosphate.* Cryst from water, mp 230-234° (dec). Soly in water: 0.037% at 20°; 1% at 100°. Insol in alc, ether and other organic solvents. Sol in dilute acids and alkali. Known as *Charcot-Neumann crystals*, also found in spleen, blood, bone marrow in leukemia and secretions in asthma.

Tetrahydrochloride, $C_{10}H_{30}Cl_4N_4$, cryst from ethanol, mp 312-314.5°.

USE: As a tool in biochemical research.

8700. Sperm Oil. From sperm whale. *Constit.* Esters of fatty acids; small quantities of spermaceti.

Yellow, thin liquid; slight fishy odor if not of good quality. d 0.875-0.884. Sapon no. 123-147. Iodine no. 80-84. Insoluble in water, cold alcohol, or petr ether; sol in chloroform, ether.

USE: As lubricant, in lamps, hardening steel, manuf soap.

8701. Spherophysine. *N-(4-Aminobutyl)-N-(3-methyl-2-butenyl)guanidine; sphaerophysine.* $C_{10}H_{22}N_4$; mol wt 198.31. C 60.56%, H 11.18%, N 28.26%. Ganglion blocking agent. Isolated from the following Central Asiatic Leguminosae: *Swainsona salsula* (Pall.) Taub. *(Sphaerophysa salsula* (Pall.) DC.); *Smirnowia turkestana* Bunge; *Eremosparton flaccidum* Litw.: Rubinshtein, Menshikov, *J. Gen. Chem. USSR* **14**, 161 (1944); Rjabinin, *ibid.* **17**, 2265 (1947); Merlis, *ibid.* **22**, 347 (1952). Structure: Heesing, Eckard, *Ber.* **103**, 534 (1970).

Strong, diacidic base.

Dibenzoate, $C_{24}H_{34}N_4O_4$, crystals, mp 149-150°.

8702. Sphingomyelins. Rowamyelin. Sphingosine phosphatides occurring in the myelin sheaths of nerves. Prepn of beef heart sphingomyelin: Rapport, Lerner, *J. Biol. Chem.* **232**, 63 (1958). Isoln from human and rat brain: Hausheer et al., Helv. Chim. Acta **46**, 601 (1963). Structure: Fujino, J. Biochem. (Japan) **39**, 45 (1952); Rouser et al., J. Am. Chem. Soc. **75**, 310 (1953); Marinetti et al., ibid. 313. Structure of honey bee (Apis mellifera) sphingomyelin: Karlander et al., Biochim. Biophys. Acta **270**, 117 (1972). Synthesis of dihydrosphingomyelins: Shapiro et al., J. Am. Chem. Soc. **80**, 2339 (1958). Synthesis of sphingomyelins: Shapiro et al., ibid. **81**, 3743, 4360 (1959). Configuration: Shapiro, Flowers, ibid. **84**, 1047 (1962). Monograph: S. R. Korey, The Biology of Myelin (Hoeber-Harper, New York, 1959).

R = different fatty acids such as stearic, palmitic, lignoceric, nervonic acids.

Slightly hygroscopic powder. Occurs as a hydrate. Sol in hot abs alcohol, acetic acid, chloroform; slightly sol in pyridine; practically insol in ether, acetone, water.

8703. Sphingosine. 2-Amino-4-octadecene-1,3-diol; 1,3-dihydroxy-2-amino-4-octadecene; 4-sphingenine. $C_{18}H_{37}$-NO_2; mol wt 299.48. C 72.19%, H 12.45%, N 4.68%, O 10.69%. Important membrane component; long-chain amino-dialcohol; moiety of certain phosphatides, such as sphingomyelins, cerebrosides, and gangliosides. Natural sphingosine is D(+)-erythro-1,3-dihydroxy-2-amino-4-trans-octadecene. Does not exist in the free state in animals, plants or microorganisms. First obtained by hydrolysis of cerebrosides from brains: J. L. W. Thudichum, Die Konstitution des Gehirns des Menschen und der Tiere (Tübingen, 1901). Separation procedures: Carter et al., J. Biol. Chem. **170**, 269 (1947); Wittenberg, ibid. **216**, 379 (1955); Tipton, Biochem. Prepns. **9**, 127 (1962). Stereospecific synthesis of DL-erythro-trans-form and isomers: C. A. Grob, F. Gadient, Helv. Chim. Acta **40**, 1145 (1957). Total synthesis of DL- and D-erythro-trans-forms: D. Shapiro et al., J. Am. Chem. Soc. **80**, 1194 (1958); Y. Shoyama et al., J. Lipid Res. **19**, 250 (1978). Stereoselective synthesis of D-erythro-trans-form: H. Newman, J. Am. Chem. Soc. **95**, 4098 (1973); B. Bernet, A. Vasella, Tetrahedron Letters **24**, 5491 (1983). Diastereoselective synthesis of D,L-erythro-sphingosine: R. R. Schmidt, R. Kläger, Angew. Chem. Int. Ed. **21**, 210 (1982). Comprehensive review: C. A. Grob, Record Chem. Progr. (Kresge-Hooker Sci. Lib.) **18**, 55-66 (1957); D. Shapiro, Chemistry of Sphingolipids (Hermann, Paris, 1969) 111 pp. Bibliography: Rodd's Chemistry of Carbon Compounds, Vol. I, part E, S. Coffey, Ed. (Elsevier, New York, 2nd ed., 1976) pp 394-397.

DL-Sphingosine, waxy crystals from ether + pentane, mp 67°.

DL-Triacetylsphingosine, $C_{24}H_{43}NO_5$, crystals from pentane + ether, mp 91-92°.

Triacetyl derivative of natural sphingosine, $C_{24}H_{43}NO_5$, crystals, mp 101-102°. $[\alpha]_D^{25}$ -11.7° (chloroform).

8704. Spigelia. Pinkroot; Indian pink; Carolina pink; Maryland pink; wormgrass. Dried rhizome and roots of Spigelia marilandica L., Loganiaceae. Habit. North America (New Jersey to Florida and west to Wisconsin). Constit. Spigeline, resin, tannin, bitter principle, volatile oil.

THERAP CAT: Anthelmintic.

8705. α-Spinasterol. Stigmasta-7,22-dien-3-ol; α-spinasterin; bessisterol; hitodesterol. $C_{29}H_{48}O$; mol wt 412.67. C 84.40%, H 11.72%, O 3.88%. Stereoisomeric with chondrillasterol, q.v. Extracted from spinach leaves: Hart, Heyl, J.

Biol. Chem. **95**, 311 (1932); from Citrullus colocynthis Schrad., Cucurbitaceae: Hamilton, Kermack, J. Chem. Soc. **1952**, 5051. Structure: Fieser et al., J. Am. Chem. Soc. **71**, 2226 (1949). Prepn: Kircher, Rosenstein, J. Org. Chem. **38**, 2259 (1973); M. Anastasia et al., J. Chem. Soc. Perkin Trans. I **1981**, 2561.

Crystals from alcohol + light petr, mp 168-169°. $[\alpha]_D^{25}$ -3.6° (c = 2.8 in chloroform).

Benzoate, $C_{36}H_{52}O_2$, plates from ethyl acetate, mp 200°. $[\alpha]_D^{25}$ +1.8° (c = 1.7 in chloroform).

8706. Spinulosin. 2,5-Dihydroxy-3-methoxy-6-methyl-2,5-cyclohexadiene-1,4-dione; 2,5-dihydroxy-3-methoxy-6-methyl-p-benzoquinone; 3,6-dihydroxy-5-methoxy-p-toluquinone; 3,6-dihydroxy-4-methoxy-2,5-toluquinone; hydroxyfumigatin. $C_8H_8O_5$; mol wt 184.14. C 52.18%, H 4.38%, O 43.44%. Metabolic product of the fungi Penicillium spinulosum Thom. and Aspergillus fumigatus Fres.: Birckinshaw, Raistrick, Philos. Trans. **B220**, 245 (1931); Anslow, Raistrick, Biochem. J. **32**, 2288 (1938); Pettersson, Acta Chem. Scand. **17**, 1771 (1963). Structure: Aulin, Erdtman, Svensk Kem. Tidskr. **49**, 208 (1937). Synthesis: Anslow, Raistrick, Biochem. J. **32**, 803 (1938). Biosynthesis: Pettersson, Acta Chem. Scand. **19**, 1016 (1965).

Purple-black crystals, mp 202°. Slightly sol in cold water; more sol in hot water; sol in dil NaOH soln giving a purple soln with a bluish hue.

8707. Spiperone. 8-[4-(4-Fluorophenyl)-4-oxobutyl]-1-phenyl-1,3,8-triazaspiro[4.5]decan-4-one; 8-[3-(p-fluorobenzoyl)propyl]-1-phenyl-1,3,8-triazaspiro[4.5]decan-4-one; 4-phenyl-8-[3-(4-fluorobenzoyl)propyl]-1-oxo-2,4,8-triazaspiro[4.5]decane; Spiropitan. $C_{23}H_{26}FN_3O_2$; mol wt 395.48. C 69.85%, H 6.63%, F 4.80%, N 10.63%, O 8.09%. Prepn: Janssen, U.S. pats. **3,155,669**; **3,155,670** and **3,161,-644** (all 1964 to Janssen).

Crystals, mp 190-193.6°.

THERAP CAT: Antipsychotic.

8708. Spiramycin. RP 5337; Sequamycin; Selectomycin; Foromacidin; Rovamicina; Rovamycin; Provamycin. Antibiotic substance classified in the erythromycin-carbomycin group and produced by Streptomyces ambofaciens from soil of northern France: Cosar et al., Compt. Rend. Soc. Biol. **234**, 1498 (1952); Pinnert-Sindico et al., Antibiot. Ann. **1954-1955**, 724; Ninet, Verrier, U.S. pat. **2,943,023** (1960 to Rhône-Poulenc), see also U.S. pat. **3,000,785** (1961 to Rhône-Poulenc). Antibacterial activity and toxicity: H. Sous et al., Arzneimittel-Forsch. **8**, 386 (1958). Separation into 3 components named spiramycin I, II and III: Preud'homme, Charpentier, U.S. pats. **2,978,380** and **3,-011,947** (1961 to Rhône-Poulenc). Structure: Kuehne, Benson, *J. Am. Chem. Soc.* **87**, 4660 (1965). Revised structure: Omura *et al., ibid.* **91**, 3401 (1969); Mitscher *et al., J. Antibiot.* **26**, 55 (1973). Revised configuration at C-9: Freiberg *et al., J. Org. Chem.* **39**, 2474 (1974). Symposium on pharmacology, antibacterial spectrum, and clinical efficacy: *J. Antimicrob. Chemother.* **22**, Suppl. B, 1-213 (1988).

Amorphous base, slightly sol in water. $[\alpha]_D^{20}$ —80° (methanol). uv max (ethanol): 231 nm. Sol in most organic solvents. Active on gram-positive bacteria and rickettsiae. Cross resistance between microorganisms resistant to erythromycin and carbomycin. LD_{50} in rats (mg/kg): 9400 orally; 1000 s.c.; 170 i.v. (Sous).

Embonate, *Spira 200.*

Hexanedioate, *spiramycin adipate, Suanovil, Suanozil, Calactin vet.*

Spiramycin I, $C_{43}H_{74}N_2O_{14}$*, Foromacidin A.* R = H. Crystals, mp 134-137°. $[\alpha]_D^{20}$ —96°.

Spiramycin I triacetate, crystals, mp 140-142°. $[\alpha]_D^{20}$ —92.5°.

Spiramycin II, $C_{45}H_{76}N_2O_{15}$*, Foromacidin B.* R = $COCH_3$. Crystals, mp 130-133°. $[\alpha]_D^{20}$ —86°.

Spiramycin II diacetate, crystals from cyclohexane, mp 156-160°. $[\alpha]_D^{20}$ —98.4°.

Spiramycin III, $C_{46}H_{78}N_2O_{15}$*, Foromacidin C.* R = $COCH_2CH_3$. Crystals, mp 128-131°. $[\alpha]_D^{20}$ —83°.

Spiramycin III diacetate, crystals from cyclohexane, mp 140-142°. $[\alpha]_D^{20}$ —90.4°.

THERAP CAT: Antibacterial.

THERAP CAT (VET): Antibacterial; growth promotant.

8709. Spirilene. *8-[4-(4-Fluorophenyl)-3-pentenyl]-1-phenyl-1,3,8-triazaspiro[4.5]decan-4-one.* $C_{24}H_{28}FN_3O$; mol wt 393.50. C 73.26%, H 7.17%, F 4.83%, N 10.68%, O 4.07%. Prepn: Janssen, **Belg. pat. 633,914** (1963 to Janssen), *C.A.* **60**, 15882d (1964). Crystal structure: M. H. J. Koch, G. Evrard, *Acta Crystallog.* **B29**, 2971 (1973).

Solid, mp 171-172°.

THERAP CAT: Antipsychotic.

8710. Spirit of Ammonia, Aromatic. Spirit of hartshorn, aromatic. A soln contg 34 g ammonium carbonate, 90 ml 10% ammonia water, 10 ml lemon oil, 1 ml oil of lavender, 1 ml oil of myristica, 700 ml alcohol and a sufficient quantity of water to make 1 liter. Absolute alc by vol, about 66%.

Almost colorless to slightly yellow liquid with an aromatic, pungent odor and the taste of ammonia. d 0.90. Forms an opalescent mixture with water; miscible with alcohol.

THERAP CAT: Reflex respiratory stimulant.

THERAP CAT (VET): By inhalation: respiratory and circulatory stimulant. Internally: expectorant, diaphoretic, antacid, carminative. Externally: counterirritant, and in diluted form to relieve the irritation of insect stings and bites.

8711. Spirit of Camphor. A soln of camphor in alcohol contg 10 g camphor per 100 ml soln.

Colorless liquid; camphor odor.

THERAP CAT: Carminative; topical analgesic; antipruritic.

THERAP CAT (VET): Has been used as rubefacient, stimulant, expectorant, carminative.

8712. Spirit of Chloroform. An alcoholic soln of chloroform contg 6% by vol of chloroform, corresponding to 10.5% by wt, and about 89% abs alcohol by vol.

Colorless, clear liquid; chloroform odor. d about 0.85.

THERAP CAT: Carminative.

8713. Spirit of Ether. Hoffmann's drops. A mixture of alcohol and ether. Made by mixing 325 ml ether U.S.P. with sufficient U.S.P. alcohol (about 690 ml) to make 1000 ml. Absolute alc by vol, about 66%.

Colorless, volatile liquid. d 0.785.

THERAP CAT: Carminative.

8714. Spirit of Ether Compound. Hoffmann's anodyne. A mixture of 325 ml ether U.S.P., 25 ml ethereal oil and sufficient alcohol (about 665 ml) to make 1000 ml. Absolute alc by vol, about 64%.

Colorless, clear liquid. d about 0.785.

THERAP CAT: Carminative.

THERAP CAT (VET): Has been used as a stomachic, carminative, in colic.

8715. Spirit of Ethyl Nitrite. Spirit of nitrous ether; sweet spirit of niter. An alcoholic soln of ethyl nitrite, contg 3.5-4.5% ethyl nitrite. *Incompat:* Antipyrine, tannin, acetanilide, acetophenetidin, iodides, tincture guaiac, morphine salts, carbonates, acacia.

Pale yellow or faintly greenish-yellow, clear, mobile, volatile, flammable liquid; ethereal, pungent odor; sharp, burning taste. The ethyl nitrite evaporates rapidly; dec on exposure to air; dec in light. d not above 0.823 at 25°. Miscible with water and alcohol. *Keep tightly closed, protected from light, in a cool place, remote from fire.*

THERAP CAT: Diaphoretic, diuretic.

8716. Spirit of Formic Acid. Spirit of ants. Composed of 40 ml of 25% formic acid and 225 ml water in sufficient alcohol to make one liter; corresponds to about 1% formic acid and about 70% abs alcohol by vol.

Colorless liquid. Miscible with water, alcohol.

8717. Spirit of Glyceryl Trinitrate. Spirit of nitroglycerin; spirit of trinitroglycerin; spirit of glonoin; soln of glyceryl trinitrate. An alcoholic soln contg 1.0-1.1% glyceryl trinitrate. *Incompat:* Alkalies, carbonates, HCl, HI.

Colorless, clear liquid. d 0.814-0.820. Miscible with alcohol, chloroform, ether; 1 ml dissolves in 6 ml almond oil; very slightly sol in water; miscible with chloroform, ether.

Caution: Likely to produce violent headache when tasted or applied to the skin.

Note: When spilled, immediately pour NaOH soln over it or dry residue may explode by friction or shock.

THERAP CAT: Vasodilator.

8718. Spirit of Peppermint. An alcoholic soln contg per liter 100 ml oil of peppermint and the alcohol-soluble principles from 10 g of powdered peppermint previously macerated with water.

THERAP CAT: Carminative.

THERAP CAT (VET): Has been used as a carminative.

8719. Spirit of Spearmint. An alcoholic soln contg per liter 100 ml oil of spearmint and the alcohol-soluble principles from 10 g coarsely powdered spearmint leaves previously macerated with water.

THERAP CAT: Carminative.

THERAP CAT (VET): Has been used as a carminative.

8720. Spirogermanium. *8,8-Diethyl-N,N-dimethyl-2-aza-8-germaspiro[4.5]decane-2-propanamine;* 2-[3-(dimethylamino)propyl]-8,8-diethyl-2-aza-8-germaspiro[4.5]decane. $C_{17}H_{36}GeN_2$; mol wt 341.08. C 59.87%, H 10.64%, Ge 21.28%, N 8.21%. Cytostatic germanium deriv. Prepn: L. M. Rice, **Ger. pat. 2,243,550** corresp to **U.S. pat. 3,825,546** (1973, 1974 both to Geschickter Fund for Med. Res.); L. M. Rice *et al., J. Heterocycl. Chem.* **11**, 1041 (1974). Preclinical toxicological evaluation: M. C. Henry, C. D. Port, *U.S. NTIS Rep.* **PB-264117** (1977) 213 pp, *C.A.* **87**, 177620

(1977). Early clinical studies: P. S. Schein *et al.*, *Cancer Treat. Rep.* **64**, 1051 (1980); C. Tropie *et al.*, *ibid.* **65**, 119 (1981). Toxicity study: M. C. Henry *et al.*, *ibid.* **64**, 1207 (1980).

$$(C_2H_5)_2Ge \quad N-CH_2(CH_2)_2N(CH_3)_2$$

Oil, $bp_{0.03}$ 106-109°.

Dihydrochloride, $C_{17}H_{38}Cl_2GeN_2$, *NSC-192965, Spiro-32*. Cryst, mp 287-288°. LD_{50} in mice (mg/m²): 128.4 i.v.; 401.1 i.m., M. C. Henry, C. D. Port, *loc. cit.* LD_{50} in mice also reported as 324 mg/kg orally and 150 mg/kg i.v., L. M. Rice *et al.*, *loc. cit.*

THERAP CAT: Antineoplastic.

8721. Spironolactone. *7-(Acetylthio)-17-hydroxy-3-oxo-pregn-4-ene-21-carboxylic acid γ-lactone; 17-hydroxy-7-mercapto-3-oxo-17α-pregn-4-ene-21-carboxylic acid γ-lactone, 7-acetate;* 3-(3-oxo-7α-acetylthio-17β-hydroxy-4-androsten-17α-yl)propionic acid *γ-lactone; SC 9420; Aldace; Aldactone A; Aldopur; Alexan; Almatol; Altex; Aquareduct; Deverol; Diatensec; Dira; Duraspiron; Euteberol; Lacalmin; Lacdeen; Laractone; Nefurofan; Osiren; Osyrol; Sagisal; Sincomen; Spiretic; Spiroctan; Spirolone; Spiro-Tablinen; Supra-Puren; Suracton; Urusonin; Verospiron; Xenalon.* $C_{24}H_{32}O_4S$; mol wt 416.59. C 69.20%, H 7.74%, O 15.36%, S 7.70%. Aldosterone antagonist. Prepn: Cella, Tweit, *J. Org. Chem.* **24**, 1109 (1959); U.S. pat. **3,013,012** (1961 to Searle); Tweit *et al.*, *J. Org. Chem.* **27**, 3325 (1962); Kiprianov *et al.*, *Khim. Farm. Zh.* **3**, 10 (1969), *C.A.* **71**, 70779g (1969). Activity and metabolic studies: Gerhards, Engelhardt, *Arzneimittel-Forsch.* **13**, 972 (1963); Rosenthale *et al.*, *Proc. Soc. Exp. Biol. Med.* **118**, 806 (1965); Karim, Brown, *Steroids* **20**, 41 (1972). Crystal and molecular structure: Dideberg, Dupont, *Acta Crystallogr. Sect. B* **28**, 3014 (1972). Comprehensive description: J. L. Sutter, E. P. K. Lau, in *Analytical Profiles of Drug Substances* vol. **4**, K. Florey, Ed. (Academic Press, New York, 1975) pp 431-451.

Crystals from methanol, mp 134-135° (resolidifies and dec 201-202°). $[\alpha]_D^{20}$ −33.5° (chloroform). uv max: 238 nm (ε 20200). Practically insol in water. Sol in most organic solvents.

THERAP CAT: Diuretic.

THERAP CAT (VET): Has been used as diuretic.

8722. Spizofurone. *5-Acetylspiro[benzofuran-2(3H),1'-cyclopropan]-3-one; AG-629; Maon.* $C_{12}H_{10}O_3$; mol wt 202.21. C 71.28%, H 4.98%, O 23.74%. Prepn: H. Sugihara *et al.*, **Eur. pat. Appl. 3,084;** *eidem*, U.S. pat. **4,284,644** (1979, 1981 both to Takeda); M. Kawada *et al.*, *Chem. Pharm. Bull.* **32**, 3532 (1984). One step synthesis: M. Watanabe *et al.*, *ibid.* 3373. Anti-ulcerative effect in animals: N. Inatomi *et al.*, *Arzneimittel-Forsch.* **35**, 1553 (1985). Cytoprotective effects on rat gastric mucosa: *eidem*, *Eur. J. Pharmacol.* **112**, 81 (1985). Stimulation of endogenous prostaglandin synthesis *in vitro*: I. Inada *et al.*, *ibid.* **124**, 149 (1986).

Crystals from ethanol, mp 102-104° (Kawada); also reported as mp 106-107° (Watanabe).

THERAP CAT: Anti-ulcerative.

8723. Splenin. Splenin A and splenin B. Physiologically active substances which occur in extracts from the spleen of mammals. Splenin A decreases capillary permeability and inhibits inflammation. Splenin B increases capillary permeability and bleeding time. The effectiveness of splenin A is measured by the inhibition of inflammation produced in guinea pigs by the Arthus reaction. One Arthus unit is the amount of splenin A required to inhibit the inflammation by 50% and is roughly equivalent to 5000 bleeding time units. Isoln procedures: G. Ungar, *Endocrinology* **37**, 329 (1945); prepn of splenin A: Ungar, **Brit. pat. 637,309** (1950). S. A. Karjala, P. March, Y. Shigemura, *Chemical Studies on the Concentration of Splenin A* in A. F. Coburn *et al.*, *Splenin A in Rheumatic Fever* (Thomas, Springfield, Ill., 1955).

8724. Sporidesmins. Toxic fungal mixture from *Pithomyces chartarum* Ellis (*Sporidesmium bakeri* Sydow), composed of sporidesmins A-H and J, as causative agent of "facial eczema" in sheep. First isoln of A (major metabolite): R. L. M. Synge, E. P. White, *Chem. & Ind. (London)* **1959**, 1546. Isoln of A and B: J. W. Ronaldson *et al.*, *J. Chem. Soc. (C)* **1963**, 3172. Structure of A: J. Fridrichsons, A. M. Mathieson, *Tetrahedron Letters* **1962**, 1265. Abs config of A: A. F. Beecham *et al.*, *ibid.* **1966**, 3131. Isoln and structure of D and F: W. D. Jamieson *et al.*, *J. Chem. Soc. (C)* **1969**, 1564; of E: R. Rahman *et al.*, *ibid.* 1665; of G: E. Francis *et al.*, *J. Chem. Soc. Perkin Trans. I* **1972**, 470; of H and J: R. Rahman *et al.*, *ibid.* **1978**, 1476. Total synthesis of A: Y. Kishi *et al.*, *J. Am. Chem. Soc.* **95**, 6493 (1973).

sporidesmin A

Sporidesmin A, $C_{18}H_{20}ClN_3O_6S_2$, *9-chloro-2,3,5a,6,10b,11-hexahydro-10b,11-dihydroxy-7,8-dimethoxy-2,3,6-trimethyl-3,11a-epidithio-11aH-pyrazino[1',2':1,5]pyrrolo[2,3-b]indole-1,4-dione, sporidesmin.* Colorless needles with a faint green sheen, from aq methanol, mp 179° (dependent on rate of heating). $[\alpha]_D^{20}$ −45° (c = 0.98 in methanol). uv max: 219, 253, 305 nm ($E_{1cm}^{1\%}$ 700, 220, 40). Very slightly sol in water, light petroleum, CCl_4. Readily sol in most organic solvents.

8725. Sporidesmolides. Cyclodepsipeptides from the pasture fungus *Sporidesmium bakeri* Syd. (*Pithomyces chartarum* (Berk. & Curt.) Ellis): Russell, *Biochim. Biophys. Acta* **45**, 411 (1960); Bertaud *et al.*, *J. Gen. Microbiol.* **32**, 385 (1963); Bishop, Russell, *Biochem. J.* **92**, 19P (1964).

Sporidesmolide I. $C_{33}H_{58}N_4O_8$. x = 0, y = z = 1; $R^1 = R^3 = H$, $R^2 = CH_3$. Structure: Russell, *J. Chem. Soc.* **1962**, 753. Synthesis: Shemyakin *et al.*, *Tetrahedron* **19**, 995 (1963). Needles from 70% acetic acid, mp 261-263°. $[\alpha]_D^{17}$ −217° (c = 1.5 in chloroform). Practically insol in water;

very sol in chloroform; sparingly sol in other common organic solvents.

Sporidesmolide II. $C_{34}H_{60}N_4O_8$. $x = 0$, $y = z = 1$; $R^1 =$ H, $R^2 = R^3 = CH_3$. Structure and synthesis: *eidem, Tetrahedron Letters* **1963**, 1927. Crystals, mp 228-230°. $[\alpha]_D^{20}$ −195° (c = 0.6 in chloroform).

Sporidesmolide III. $C_{32}H_{56}N_4O_8$. $x = 0$, $y = z = 1$; $R^1 = R^2 = R^3 =$ H. Synthesis: Ovchinnikov *et al., ibid.* **1965**, 111. Crystals, mp 277-278°. $[\alpha]_D$ −80° (c = 1.6 in acetic acid).

Sporidesmolide IV. $C_{34}H_{60}N_4O_8$. $x = y = z = 1$; $R^1 = R^3 =$ H, $R^2 = CH_3$. Structure and synthesis: Kiryushkin *et al., ibid.* **1965**, 143; Bishop, Russell, *J. Chem. Soc. (C)* **1967**, 634. Crystals, mp 227-228°. $[\alpha]_D$ −195° (c = 1 in chloroform). Practically insol in water, sol in chloroform, moderately sol in common organic solvents.

8726. Squalane. *2,6,10,15,19,23-Hexamethyltetracosane;* perhydrosqualene; dodecahydrosqualene; spinacane; Cosbiol; Robane. $C_{30}H_{62}$; mol wt 422.80. C 85.22%, H 14.78%. $[(CH_3)_2CHCH_2CH_2CH_2CH(CH_3)CH_2CH_2CH_2CH(CH_3)-CH_2CH_2-]_2$. Prepn by complete hydrogenation of squalene, *q.v.*: Chapman, *J. Chem. Soc.* **123**, 770 (1923); Heilbron *et al., ibid.* **1926**, 3135. Commercial grades are obtained by direct hydrogenation of shark liver oil and may contain some batyl alcohol: Tsujimoto, *Chem. Umschau Fette* **34**, 256 (1927), *C.A.* **21**, 4081 (1927). Synthesis: J. W. Scott, D. Valentine, *Org. Prep. Proced. Int.* **12**, 7 (1980); T. Mandai *et al., Tetrahedron Letters* **22**, 763 (1981).

Oil. Stable to air and oxygen. d_4^{15} 0.8115. mp ~ −38°. bp_{760} ~350°; bp_{10} 263°; bp_5 248°. Flash pt 425°F (218°C). n_D^{15} 1.4530. Specific heat at 20° about 0.62. Viscosity (Engler) at 20° about 6.08. Readily sol in ether, gasoline, petr ether, benzene, chloroform, oils. Slightly sol in methanol, ethanol, acetone, glacial acetic acid. Concd H_2SO_4 at 70° is discolored, but the squalane remains unchanged.

USE: Lubricant, transformer oil. Ingredient of watch and chronometer oils. Perfume fixative. In pharmacy and cosmetics as skin lubricant, ingredient of suppositories, carrier of lipid-soluble drugs.

8727. Squalene. *(all-E)- 2,6,10,15,19,23-Hexamethyl-2,6,-10,14,18,22-tetracosahexaene;* Spinacene; Supraene. $C_{30}H_{50}$; mol wt 410.70. C 87.73%, H 12.27%. All *trans* isoprenoid contg six isoprene units. Found in large quantities in shark liver oil. Occurs in smaller amounts (0.1 to 0.7%) in olive oil, wheat germ oil, rice bran oil, and yeast. Intermediate in biosynthesis of cholesterol. Isoln: Heilbron *et al., J. Chem. Soc.* **1926**, 1630. Structure: *eidem, ibid.* **1929**, 873; Heilbron, Thompson, *ibid.* 883; Karrer *et al., Helv. Chim. Acta* **13**, 1084 (1930); Karrer, Helfenstein, *ibid.* **14**, 78 (1931); Karrer *et al., ibid.* 435. Crystal structure: J. Ernst *et al., Angew. Chem. Int. Ed.* **15**, 778 (1976). Synthesis: Trippett, *Chem. & Ind. (London)* **1956**, 80; Dicker, Whiting, *J. Chem. Soc.* **1958**, 1994; Cornforth *et al., ibid.* **1959**, 2539; Johnson *et al., J. Am. Chem. Soc.* **92**, 741 (1970); Hirai *et al., Tetrahedron Letters* **1971**, 4359; P. A. Grieco, Y. Masaki, *J. Org. Chem.* **39**, 2135 (1974). Synthesis of squalene and *trans-cis-trans-trans* isomer: Biellmann, Ducep, *Tetrahedron* **27**, 5861 (1971).

Oil. Faint, agreeable odor. Absorbs oxygen and becomes viscous like linseed oil. d_4^{20} 0.8584; bp_{25} 285°; bp_2 240°; $bp_{0.15}$ 203°. mp about −75°. n_D^{20} 1.4965. Viscosity at 25°: 12 cps. Iodine no. 360-380. Flash pt approx 200°. Practically insol in water. Freely sol in ether, petr ether, CCl_4, acetone, other fat solvents; sparingly sol in alc, glacial acetic acid.

USE: Bactericide; intermediate in manuf of pharmaceuti-

cals, organic coloring materials, rubber chemicals, aromatics and surface active agents.

8728. Squill. Sea onion; Bulbus Scillae; Meerzwiebel. The fleshy inner bulb scales of the white variety of *Urginea maritima* (L.) Baker (*Scilla maritima* L.), *Liliaceae. Habit.* Lands of the Mediterranean seacoast. *Constit.* Glucoscillaren A (scillarenin + rhamnose + glucose + glucose); scillaren A (scillarenin + rhamnose + glucose); proscillaridin A (scillarenin + rhamnose); scillaridin A; scilliglaucoside; scillipheoside; glucoscillipheoside; scillicyanoside; scillicoeloside; scilliazuroside; scillicryptoside. *Review:* G. Baumgarten, W. Förster, *Die Herzwirksamen Glykoside* (Thieme, Leipzig, 1963) pp 70-75 *et passim.*

USE: Rodenticide.

THERAP CAT: Diuretic, emetic, expectorant, cardiotonic.

THERAP CAT (VET): Has been used as expectorant, emetic.

8729. Stachydrine. *2-Carboxy-1,1-dimethylpyrrolidinium hydroxide inner salt;* methyl hygrate betaine; hygric acid methylbetaine. $C_7H_{13}NO_2$; mol wt 143.18. C 58.72%, H 9.15%, N 9.78%, O 22.35%. Occurs widely in nature, especially in alfalfa, chrysanthemum, citrus and stachys species. May be prepd by methylation of proline. Isoln: Planta, Schulze, *Ber.* **26**, 939 (1893); Jahns, *ibid.* **29**, 2065 (1896); Yoshimura, *Z. Physiol. Chem.* **88**, 334 (1913); Vickery, *J. Biol. Chem.* **61**, 117 (1924). Structure: Schulze, Trier, *Z. Physiol. Chem.* **67**, 59 (1910). Synthesis: Karrer, Widmer, *Helv. Chim. Acta* **8**, 364 (1925). Biosynthesis: Robertson, Marion, *Can. J. Chem.* **38**, 396 (1960). *Review:* Marion in *The Alkaloids* vol. 1, R. H. F. Manske, H. L. Holmes, Eds. (Academic Press, New York, 1950) pp 101-103.

Monohydrate, deliquescent crystals, sweetish taste, mp 235° when anhydr. Isomerizes at the mp to methyl hygrate. Sol in water, alcohol, dil acids; practically insol in ether, chloroform.

Hydrochloride, $C_7H_{13}NO_2$·HCl, large prisms from abs alcohol, dec 235°, very sol in water, sol in 13 parts alcohol.

Acid oxalate, $C_7H_{13}NO_2$·$C_2H_2O_4$, needles, mp 106°, practically insol in abs alcohol.

Picrate, $C_7H_{13}NO_2$·$C_6H_3N_3O_7$, yellow crystals, mp 196°, precipitates from concd solns only.

Aurichloride, $C_7H_{13}NO_2$·$HAuCl_4$, yellow leaflets, mp 225° (rapid heating), practically insol in cold water, quite sol in hot water.

Platinichloride tetrahydrate, $(C_7H_{13}NO_2)_2$·H_2PtCl_6·$4H_2O$, orange crystals, dec 210-220° (rapid heating), very sol in water and dil alc. Also obtained with 2 mols H_2O of crystn.

8730. Stallimycin. *N-[5-[[(3-Amino-3-iminopropyl)-amino]carbonyl]-1-methyl-1H-pyrrol-3-yl]-4-[[[4-(formylamino)-1-methyl-1H-pyrrol-2-yl]carbonyl]amino]-1-methyl-1H-pyrrole-2-carboxamide;* β-[1-methyl-4-[1-methyl-4-(1-methyl-4-formylaminopyrrole-2-carboxamido)pyrrole-2-carboxamido]pyrrole-2-carboxamido]propionamidine; distamycin A; F.I. 6426. $C_{22}H_{27}N_9O_4$; mol wt 481.56. C 54.88%, H 5.65%, N 26.18%, O 13.29%. Antibiotic substance with antiviral and oncolytic properties, obtained from *Streptomyces distallicus*: Arcamone *et al., Ger. pat.* **1,039,-198** (1958 to Farmaceutici Italia), *C.A.* **55**, 2012f (1961). Structure and synthesis: Arcamone *et al., Nature* **203**, 1064 (1964); *eidem, Gazz. Chim. Ital.* **97**, 1097, 1110 (1967). Total synthesis: M. Bialer *et al., Tetrahedron* **1978**, 2389; L. Grehn, U. Ragnarsson, *J. Org. Chem.* **46**, 3492 (1981). Toxicity and antitumor activity: A. DiMarco *et al., Cancer Chemother. Rep.* **18**, 15 (1962). *Review:* Hahn in *Antibiotics* vol. 3, J. W. Corcoran, F. E. Hahn, Eds. (Springer-Verlag, New York, 1975) pp 79-100.

mp 154-156°.

Hydrochloride, $C_{22}H_{28}ClN_9O_4$, *Herperal*. Crystals from dil HCl, mp 184-187° (Arcamone, 1967). Also reported as 189-193° from ethanol-ethyl acetate (Bialer). uv max (96% ethanol): 237, 303 nm (ϵ 30000, 37000). LD_{50} in mice (mg/kg): 75 i.v.; 500 i.p. (DiMarco).

THERAP CAT: Antiviral.

8731. Stannic Bromide. Tin tetrabromide. Br_4Sn; mol wt 438.36. Br 72.92%, Sn 27.08%. $SnBr_4$.

White, cryst mass; fumes strongly on exposure to air. d 3.34; mp 31°; bp 202°. Very sol in water with evolution of heat; also sol in alcohol. *Keep tightly closed.*

USE: In metallurgical separation of minerals.

8732. Stannic Chloride. Tin tetrachloride; fuming spirit of Libavius. Cl_4Sn; mol wt 260.53. Cl 54.44%, Sn 45.56%. $SnCl_4$. Improperly called "*tin bichloride*".

Fuming, caustic liquid. d 2.26; mp $-33°$; bp 114°. Sol in water and evolution of much heat; sol in alcohol, carbon tetrachloride, benzene, toluene, acetone, kerosene, gasoline. *Keep tightly closed.*

Pentahydrate, white or slightly yellow crystals or fused small lumps; slight HCl odor. Very sol in H_2O; sol in alc.

USE: As mordant; reviving colors; stabilizer for colors and perfumes in soap; in dyeing of fabrics, weighting silk, tinning vessels; dehydrating agent in organic syntheses; in ceramics to produce abrasion-resistant or light-reflecting coatings. *Caution:* May be highly irritating to eyes, mucous membranes.

8733. Stannic Chromate(VI). Cr_2O_8Sn; mol wt 350.72. Cr 29.66%, O 36.50%, Sn 33.84%. $Sn(CrO_4)_2$.

Brownish-yellow, cryst powder; dec when heated. Sol in water.

USE: Decorating porcelain and china in rose and violet colors.

8734. Stannic Fluoride. Tin tetrafluoride. F_4Sn; mol wt 194.70. F 39.03%, Sn 60.97%. SnF_4. Lewis acid. Prepd from $SnCl_4$ and HF: Ruff, Plato, *Ber.* 37, 673 (1904); or from stannous fluoride and chlorine or bromine: Forbes, Anderson, *J. Am. Chem. Soc.* 67, 1911 (1945); from stannous oxide or sulfide and fluorine: Haendler *et al., J. Am. Chem. Soc.* 76, 2179 (1954). *Review:* Kemmitt, Sharp, *Advan. Fluorine Chem.* 4, 185-186 (1965).

Snow-white, tetragonal crystals. Very hygroscopic. d_4^{19} 4.78. Sublimes at 705°. Hydrolyzes readily, but is more resistant to water than stannic chloride. Forms complexes with donor molecules.

USE: Friedel-Crafts catalyst.

8735. Stannic Iodide. Tin tetraiodide. I_4Sn; mol wt 626.38. I 81.05%, Sn 18.95%. SnI_4.

Yellow to reddish crystals. d 4.46. mp about 143°; sublimes at about 180°. bp 340°. Dec by water; sol in alcohol, benzene, chloroform, ether, carbon disulfide. MLD i.v. in rats: 200 mg/kg, *Handbook of Toxicology* vol. 1, W. S. Spector, Ed. (Saunders, Philadelphia, 1956) pp 282-283.

8736. Stannic Oxide. White tin oxide; tin dioxide; stannic anhydride; flowers of tin. O_2Sn; mol wt 150.70. O 21.23%, Sn 78.77%. SnO_2. Occurs in nature as the mineral *cassiterite*. The commercial grade is also known as *polishing powder, putty powder*, or *tin ash*.

White or slightly gray powder. d 6.95. Insol in water, alcohol, or cold acids; slowly sol in hot concd potassium or sodium hydroxide soln.

USE: Polishing glass and metals; manuf milk-colored, ruby and alabaster glass, enamels, pottery, putty; mordant in printing and dyeing fabrics; in fingernail polishes.

8737. Stannic Selenide. Tin diselenide. Se_2Sn; mol wt 276.62. Se 57.09%, Sn 42.91%. $SnSe_2$. Prepd by passing the vapor of selenium over heated tin: Little, *On Selenium and Some of the Metallic Selenides*, Göttingen (Thesis, 1859); by treating a soln of stannic chloride with hydrogen selenide: Berzelius, cited in *Mellor's* vol. X, 785 (1930); by treating a soln of an alkali selenostannate or sulfoselenostannate with hydrochloric acid: Ditte, *Compt. Rend.* 95, 641 (1882).

Red-brown crystals. d 5.133 (Little); d 4.85 [Schneider, *Pogg. Ann.* 127, 624 (1866)]. mp 650°. Sol in alkali, concd sulfuric acid, aqua regia, aq ammonia; insol in water, dilute acids; dec in nitric acid. Forms potassium selenostannate with potassium selenide; sodium selenostannate with sodium selenide.

8738. Stannic Selenite. O_6Se_2Sn; mol wt 372.62. O 25.76%, Se 42.38%, Sn 31.86%. $Sn(SeO_3)_2$. Prepd by treating stannic oxide with selenious acid: Berzelius, cited in *Mellor's* vol. X, 832 (1930).

Cryst powder. Insol in water; sol in excess warm HCl.

8739. Stannic Sulfide. Tin disulfide; mosaic gold; tin bronze. S_2Sn; mol wt 182.83. S 35.08%, Sn 64.92%. SnS_2.

Golden leaflets with metallic luster; fatty feel to the touch. d 4.5. Insol in water or dil acids; sol in aqua regia, in solns of alkali hydroxides or sulfides.

Note: The term "mosaic gold" is also used to designate an alloy consisting of 65.3% copper and 34.7% zinc.

USE: Gilding and bronzing metals; gypsum, wood and paper, usually by suspending in lacquer or varnish.

8740. Stannous Acetate. $C_4H_6O_4Sn$; mol wt 236.79. C 20.29%, H 2.55%, O 27.03%, Sn 50.13%. $Sn(C_2H_3O_2)_2$. Prepd by refluxing granulated tin with 98% acetic acid: Colonna, *Gazz. Chim. Ital.* 35 II, 224 (1905); by refluxing SnO with 50% (v/v) acetic acid under nitrogen: Donaldson *et al., J. Chem. Soc.* 1964, 5942.

White, orthorhombic crystals; dec by water. mp 182.5-183°. d 2.31. Sol in dil HCl. *Keep well closed.*

USE: Reducing agent.

8741. Stannous Bromide. Tin dibromide. Br_2Sn; mol wt 278.53. Br 57.38%, Sn 42.62%. $SnBr_2$.

Yellowish powder; oxidizes in air. d 5.12; mp 215°; bp 623°. Sol in little water, gradually dec by much water; sol in alcohol, ether, acetone. *Keep tightly closed and protected from light.*

8742. Stannous Chloride. Tin dichloride; tin protochloride; Stannochlor. Cl_2Sn; mol wt 189.61. Cl 37.40%, Sn 62.60%. $SnCl_2$. Prepn: Stephen, *J. Chem. Soc.* 1930, 2786; Williams, *Org. Syn. coll. vol.* III, 627 (1955).

Orthorhombic cryst mass or flakes; fatty appearance. bp 247°; d 3.95. Sol in water, ethanol, acetone, ether, methyl acetate, methyl ethyl ketone, isobutyl alcohol; practically insol in mineral spirits, petr naphtha, xylene. LD_{50} i.p. in mice: 66 mg/kg, *Toxic Substances List*, H. E. Christensen *et al.*, Eds. (1974) p 765.

Dihydrate, crystals; absorbs oxygen from air and forms insol oxychloride. d 2.71. mp 37-38° when rapidly heated; dec on strong heating. Sol in less than its own wt of water; with much water it forms an insol basic salt; very sol in dil or in concd hydrochloric acid; also sol in alcohol, ethyl acetate, glacial acetic acid, sodium hydroxide soln. *Keep tightly closed, in a cool place.*

USE: Powerful reducing agent, particularly in manuf of dyes; in tinning by galvanic methods; in liquor finishing of wire; in sensitizing of glass and plastics before metallizing; as soldering flux; as mordant in dyeing with cochineal; in manufacture of tin chemicals, color pigments, pharmaceuticals, sensitized paper, lubricating oil additives; as tanning agent; in removing ink stains; in yeast revivers; as reagent in analytical chemistry; as catalyst in organic reactions.

8743. Stannous Fluoride. Tin difluoride; Fluoristan. F_2Sn; mol wt 156.70. F 24.25%, Sn 75.75%. SnF_2. Prepd by evaporating a soln of stannous oxide in hydrofluoric acid in the absence of oxygen: Gay-Lussac, Thénard, *Mém. Phys.*

Chim. **2,** 317 (1809); Nebergall *et al., J. Am. Chem. Soc.* **74,** 1604 (1952); from tin and hydrogen fluoride: Muetterties, *Inorg. Chem.* **1,** 342 (1962). *Review:* Kemmitt, Sharp, *Advan. Fluorine Chem.* **4,** 186 (1965).

Monoclinic, lamellar plates. mp 213°. d^{25} 4.57. Sol in water (about 30%). Forms an oxyfluoride, $SnOF_2$, on exposure to air.

USE: Dental caries prophylactic.

8744. Stannous Hexafluorozirconate(IV). Stannous fluozirconate(IV). F_6SnZr; mol wt 323.92. F 35.19%, Sn 36.65%, Zr 28.16%. $SnZrF_6$. Prepd from ZrF_4 and SnF_2: Muhler, U.S. pat. **3,266,996** (1966 to Indiana University Foundation).

Crystals. d 4.21. Sol in water.

USE: In anticaries preparations.

8745. Stannous Iodide. Tin diiodide. I_2Sn; mol wt 372.54. I 68.14%, Sn 31.86%. SnI_2. Preparation and crystal structure: Moser, Trevena, *Chem. Commun.* **1969,** 25.

Red, cryst powder or needles. d 5.28; mp 320°; bp 720° with decompn. Slightly sol in and dec by water; sol in solns of alkali chlorides or iodides, in benzene, chloroform, carbon disulfide. *Keep tightly closed.*

8746. Stannous Oxalate. C_2O_4Sn; mol wt 206.72. C 11.62%, O 30.96%, Sn 57.42%. SnC_2O_4.

White, heavy powder. d 3.56. Insol in water; sol in dil hydrochloric acid.

USE: Dyeing and printing textiles.

8747. Stannous Oxide. Tin monoxide; tin protoxide. OSn; mol wt 134.70. O 11.88%, Sn 88.12%. SnO. Prepn of high purity SnO: Kwestroo, Vromans, *J. Inorg. Nucl. Chem.* **29,** 2187 (1967). Prepn of metastable, red, orthorhombic form: Donaldson *et al., J. Chem. Soc.* **1961,** 839.

Brownish-black powder; burns to SnO_2 on heating in air. d 6.45. Insol in water or alcohol; sol in acids, in concd sodium or potassium hydroxide solns.

USE: Reducing agent; prepn of stannous salts.

8748. Stannous Pyrophosphate. *Diphosphoric acid tin-(2+) salt (1:2);* ditin diphosphate; ditin pyrophosphate. $O_7P_2Sn_2$; mol wt 411.32. O 27.23%, P 15.06%, Sn 57.71%. $Sn_2P_2O_7$. Prepd by thermal dehydration of stannous hydrogen phosphate: Jablczynski, Wieckowski, *Z. Anorg. Allgem. Chem.* **152,** 207 (1926); from $Na_2H_2P_2O_7$ and $SnCl_2$: Klement, Haselbeck, *Ber.* **96,** 1022 (1958); from pyrophosphoric acid and stannous chloride: Nelson, U.S. pat. **3,401,012** (1968 to Monsanto). Clinical application for labelling red blood cells: D. Ducasson *et al., Brit. J. Radiol.* **49,** 344 (1976). Chemical characterization of technetium complexes: J. Kroesbergen *et al., Nucl. Med. Biol.* **15,** 209 (1988). Diagnostic application in bone scanning: N. S. Anderton *et al., Am. J. Roentgenol. Rad. Ther. Nucl. Med.* **124,** 625 (1975); in myocardial infarction: D. E. Jansen *et al., Circulation* **75,** 611 (1987).

Amorphous powder. $d^{16.4}$ 4.009. Insol in water. Sol in concd acid, excess alkali.

^{99m}Technetium complex, ^{99m}Tc-*PPi,* ^{99m}Tc-*PYP, Techne-Scan PYP.*

USE: Ingredient in caries-preventing toothpaste: Norris, Schweizer, U.S. pat. **2,946,725** (1960 to Proctor & Gamble).

THERAP CAT: Complex as diagnostic aid (radioactive imaging agent).

8749. Stannous Selenide. Tin monoselenide. SeSn; mol wt 197.66. Se 39.95%, Sn 60.05%. SnSe. Prepd by direct fusing of the elements: Ditte, *Compt. Rend.* **95,** 641 (1882); **96,** 1792 (1883); **97,** 44 (1883); Berzelius cited in *Mellor's* vol. X, 784 (1930); by adding powdered selenium to molten, anhydr stannous chloride: Schneider, *Pogg. Ann.* **127,** 624 (1866).

Steel-gray prisms. d^0 6.18; mp 861°. Insol in water; sol in aqua regia and alkali sulfides and selenides.

8750. Stannous Sulfate. O_4SSn; mol wt 214.77. O 29.80%, S 14.93%, Sn 55.27%. $SnSO_4$. Prepn: Donaldson, Moser, *J. Chem. Soc.* **1960,** 4000.

Snow-white, orthorhombic crystals; dec at 378° to SnO_2 and SO_2. Sol in water, the soln soon decomposing with pptn of a basic sulfate; sol in dil H_2SO_4. *Keep well closed.*

USE: In tin plating; prepn of stannous salts.

8751. Stannous Sulfide. Tin monosulfide; tin protosulfide. SSn; mol wt 150.77. S 21.27%, Sn 78.73%. SnS.

Dark-gray crystals or black, amorphous powder. d 5.08. Insol in water or alkali hydroxide or alkali sulfide soln; sol in concd HCl, hot concd H_2SO_4.

USE: Polymerization catalyst.

8752. Stannous Tartrate. $C_4H_4O_6Sn$; mol wt 266.77. C 18.01%, H 1.51%, O 35.99%, Sn 44.49%. $SnC_4H_4O_6$.

White powder. Sol in water, dil HCl. *Keep well closed.*

USE: Dyeing and printing textiles.

8753. Stanolone. *17-Hydroxyandrostan-3-one;* 17β-hydroxy-3-androstanone; 3-oxo-17β-hydroxyandrostane; androstan-17β-ol-3-one; 4-dihydrotestosterone; androstanolone; Anaboleen; Anabolex; Anaprotin; Andractim; Androlone; Cristerona MB; Neodrol; Proteina; Protona; Stanaprol. $C_{19}H_{30}O_2$; mol wt 290.43. C 78.57%, H 10.41%, O 11.02%. Prepd by hydrogenation of testosterone: Butenandt *et al., Ber.* **68,** 2097 (1935); from dehydroepiandrosterone: Ruzicka, Kagi, *Helv. Chim. Acta* **20,** 1557 (1937); Ruzicka *et al., ibid.* **24,** 1151 (1941); from 3,17-androstandione: Oliveto, Hershburg, U.S. pat. **2,927,921** (1960 to Schering).

Crystals from ethyl acetate + hexane. Sublimes$_{0.01}$ 135°. mp 181°. $[\alpha]_D^{20}$ +32.4° (alcohol). Infrared absorption data: *J. Am. Chem. Soc.* **75,** 903 (1953). Sol in acetone, ether, alcohol, ethyl acetate. Practically insol in water.

17-Valerate, $C_{24}H_{38}O_3$, *Apeton.*

THERAP CAT: Androgen.

8754. Stanozolol. *17-Methyl-2'H-androst-2-eno[3,2-c]-pyrazol-17-ol;* 1,2,3,3a,3b,4,5,5a,6,8,10,10a,10b,11,12,12a-hexadecahydro-1,10a,12a-trimethylcyclopenta[7,8]phenanthro[2,3-c]pyrazol-1-ol; 17β-hydroxy-17α-methylandrostano[3,2-c]pyrazole; androstanazole; stanazol; NSC-43193; WIN 14833; Stanozol; Stromba; Strombaject; Tevabolin; Winstrol. $C_{21}H_{32}N_2O$; mol wt 328.48. C 76.78%, H 9.82%, N 8.53%, O 4.87%. Prepn: Clinton *et al., J. Am. Chem. Soc.* **81,** 1513 (1959); **83,** 1478 (1961); Manson, U.S. pat. **3,030,-358** (1962 to Sterling Drug).

Crystals from alc, mp 229.8-242.0°. $[\alpha]_D$ +35.7° (chloroform); $[\alpha]_D$ +48.6° (methanol). uv max 223 nm (ϵ 4740).

THERAP CAT: Androgen.

THERAP CAT (VET): Anabolic steroid.

8755. Staphisagria. Stavesacre. Ripe seed of *Delphinium staphisagria* L., *Ranunculaceae.* *Habit.* Mediterranean basin; cultivated in France, Italy. *Constit.* Delphinine, delphinoidine, delphisine, staphisagrine, staphisagroine, malic acid, fixed oil.

THERAP CAT: Parasiticide (external).

8756. Star Anise. Chinese anise. Fruit of *Illicium verum* Hook. f., *Magnoliaceae.* *Habit.* Southeastern Asia and subtropical countries; commercial supply chiefly from China.

Constit. About 5% volatile and fixed oils; anisic acid, tannin, resin, pectin.

Note: Japanese star anise is *Illicium anisatum* L. (*I. religiosum* Sieb. & Zucc.; *I. japonicum* Sieb.) and contains a toxic lactone called anisatin. Chinese star anise does not contain this toxic principle. Shikimic acid has been found in both.

USE: Manufacture of liqueurs and the volatile oil. The fruit as source of oil of anise.

THERAP CAT: Hemostatic.

8757. Starch. Amylum. $(C_6H_{10}O_5)_n$. Stored by plants; analogous to storage of fats by animals. Occurs as discrete granules in the mature grain of corn, *Zea mays* Linné, *Gramineae* or of wheat, *Triticum aestivum* Linné, *Gramineae* or tubers of potato, *Solanum tuberosum* Linné, *Solanaceae* or rice, *Oryza sativa* Linné, *Gramineae*. Starches are mixtures of two polymers: *amylose*, a linear $(1 \to 4)$-α-D-glucan and *amylopectin*, a branched D-glucan with mostly α-D-$(1 \to 4)$ and approx 4% α-D-$(1 \to 6)$ linkages. The starch in corn contains approx 27% amylose and 73% amylopectin, with these two polymers so associated in the crystal lattice that they are practically insol in cold water or alcohol. *Refs:* J. N. BeMiller, "Starch Amylose" in *Industrial Gums*, R. L. Whistler, Ed. (Academic Press, New York, 2nd ed., 1973) pp 545-566; E. L. Powell, "Starch Amylopectin", *ibid.*, pp 567-576.

Although hydrolysis will not take place in cold water, and starch is comparatively resistant to naturally occurring enzymes, the reaction may be brought about by the use of acids or enzymes (α-amylase, β-amylase, amyloglucosidase). The hydrolysis reaction follows a different path depending on whether acids or enzymes are used. While acid hydrolysis produces a mixture of saccharides, the enzymes give more specific products. β-Amylase, for example, breaks off mostly maltose units, and amyloglucosidase yields mainly D-glucose. Chemistry and technology: R. L. Whistler, E. F. Paschall, Eds., *Starch Chemistry and Technology*, 2 vols. (Academic Press, New York, 1965); J. A. Radley, Ed., *Starch and Its Derivatives* (Chapman & Hall, London, 4th ed., 1968).

USE: Starching and sizing fabrics, etc.; paste; as indicator in iodometric analyses. In the food industry. Pharmaceutic aid (tablet disintegrant, filler, binder); dusting powder. Dietetic grades of corn starch are marketed as *Maizena; Mondamin.*

THERAP CAT: Antidote (iodine poisoning).

THERAP CAT (VET): Internally: demulcent, mild astringent, in diarrhea, as an antidote for iodine poisoning. Externally: absorbent, emollient, in dusting powders and in ointments.

8758. Starch, Soluble. Amylodextrin; amylogen. Prepd by treating potato or corn starch with dilute hydrochloric acid.

White, odorless, tasteless powder. Readily soluble in hot water; forms transparent mobile liquid.

USE: For determination of diastatic power of malt, etc.; as indicator in iodometric analyses.

8759. Statine. *[S-(R*,R*)]-4-Amino-3-hydroxy-6-methylheptanoic acid;* AHMHA. $C_8H_{17}NO_3$; mol wt 175.23. C 54.83%, H 9.78%, N 8.00%, O 27.39%. Amino acid present in pepstatin, *q.v.* Synthesis: H. Morishima *et al., J. Antibiot.* **26**, 115 (1973). Abs config and stereospecific synthesis of all four isomers: M. Kinoshita *et al., ibid.* 249. Crystal structure: H. Nakamura *et al., ibid.* 255. Biosynthesis: H. Morishima *et al., ibid.* **27**, 267 (1974). Alternate syntheses: M. Kinoshita *et al., Bull. Chem. Soc. Japan* **48**, 570 (1975); W.-S. Liu, G. I. Glover, *J. Org. Chem.* **43**, 754 (1978); D. H. Rich *et al., ibid.* 3624; K. E. Rittle *et al., ibid.* **47**, 3016 (1982). Distribution in rats: D. A. Grant *et al., Biochem. Pharmacol.* **31**, 2302 (1982).

$$CH_3CHCH_2CHCHCH_2COOH$$

mp 201-203° (dec). $[\alpha]_D^{15}$ $-20°$ (c = 0.64 in water).

8760. Statolon. An antiviral substance which appears to be a macromolecular polyanionic polysaccharide composed of galacturonic acid, galactose, galactosamine, glucose, arabinose, xylose, and rhamnose. Produced by submerged culture fermentation, using *Penicillium stoloniferum* var. ATCC 14586: Stark *et al.*, U.S. pat. **3,108,047** (1963 to Lilly).

Prophylactically active against a wide range of viruses, including those causing canine distemper, lymphomatosis in fowl, shipping fever in cattle, transmissible gastroenteritis in swine, and coryza and other upper respiratory illnesses, as well as against ECHO viruses, enteroviruses in monkeys, MM neurotropic virus, Semliki Forest virus, and NEF 1 poliomyelitis virus. Antitumor activity demonstrated in experimental leukemia and sarcoma.

THERAP CAT: Antiviral.

8761. Stearic Acid. *Octadecanoic acid;* Emersol 132; Promulsin; Proviscol Wax. $C_{18}H_{36}O_2$; mol wt 284.47. C 75.99%, H 12.76%, O 11.25%. $CH_3(CH_2)_{16}COOH$. Occurs as a glyceride in tallow and other animal fats and oils, as well as in some vegetable oils; also prepd synthetically by hydrogenation of cottonseed and other vegetable oils.

White leaflets. d^{70} 0.847; mp 69-70°; bp 383°; n_D^{80} 1.4299. Slowly volatilizes at 90-100°. Very slightly sol in water. One gram dissolves in 21 ml alcohol, 5 ml benzene, 2 ml chloroform, 26 ml acetone, 6 ml carbon tetrachloride, 3.4 ml carbon disulfide; also sol in amyl acetate, toluene. LD_{50} i.v. in mice, rats: 23 ± 0.7, 21.5 ± 1.8 mg/kg, L. Orö, A. Wretlind, *Acta Pharmacol. Toxicol.* **18**, 141 (1961).

U.S.P. stearic acid consists chiefly of a mixture of stearic and palmitic acids. It is in the form of white or slightly yellow, crystal masses, or a white to slightly yellow powder; slight tallow-like odor. Does not congeal below 54°.

Ethyl ester, $C_{20}H_{40}O_2$, *ethyl stearate.* White, cryst solid; odorless or practically so. mp 33-35°. bp 224°. Ethyl stearate of commerce solidifies at 20-24°; bp_4 180°. Insol in water; sol in alcohol or ether.

Methyl ester, $C_{19}H_{38}O_2$, *methyl stearate.* White crystals. mp 38-39°. bp_{15} 215°. Insol in water; sol in alcohol, ether.

USE: For suppositories, coating enteric pills, ointments, and for coating bitter remedies. Manuf stearates of aluminum, zinc, and other metals, stearin soap for opodeldoc, candles, phonograph records, insulators, modeling compds; impregnating plaster of Paris; in vanishing creams and other cosmetics.

8762. Stearyl Alcohol. *1-Octadecanol;* stenol. $C_{18}H_{38}O$; mol wt 270.48. C 79.92%, H 14.16%, O 5.91%. $CH_3(CH_2)_{16}CH_2OH$. The official substance is a mixture of solid alcohols consisting chiefly of stearyl alcohol. Preparation from ethyl stearate: Brown, Rao, *J. Am. Chem. Soc.* **78**, 2582 (1956); Hesse, Schrödel, *Ann.* **607**, 24 (1954). Prepn of technical grade from sperm whale oil: Maiorov *et al., Zh. Prikl. Khim.* **37**, 1344 (1964).

Unctuous white flakes or granules, mp 56-60° (the pure substance, mp 59.4-59.8°, bp_{15} 210°). Sol in alcohol, ether, benzene, acetone.

USE: Substitute for cetyl alc in pharmaceutical dispensing, in cosmetic creams, for emulsions, textile oils and finishes, as antifoam agent, lubricant, and chemical raw material.

8763. Stenbolone. *17β-Hydroxy-2-methyl-5α-androst-1-en-3-one;* 2-methyl-5α-androst-1-en-17β-ol-3-one; 2-methyl-17β-hydroxy-5α-androst-1-en-3-one; stenbolone (rescinded USAN). $C_{20}H_{30}O_2$; mol wt 302.44. C 79.42%, H 10.00%, O 10.58%. Preparation of free alcohol and acetate: Mauli, *J. Am. Chem. Soc.* **82**, 5494 (1960); Kaspar *et al., Ger.* pat. **1,096,356** (1961 to Schering AG), *C.A.* **55**, 27440b (1961); Counsell *et al., J. Org. Chem.* **27**, 248 (1962); *Brit.* pat. **925,849** (1963 to Syntex).

Crystals from acetone + hexane, mp 155-158°. $[\alpha]_D$ +52° (chloroform), $[\alpha]_D^{26}$ +47° (chloroform). uv max (95% ethanol): 241 nm (log ϵ 3.99).

Acetate, $C_{22}H_{32}O_3$, *Anatrofin*. Crystals from acetone + hexane, mp 146-149°. $[\alpha]_D$ +32° (chloroform), $[\alpha]_D^{26}$ +60° (chloroform): *see* Mauli *et al.*, and Counsell *et al.* uv max (95% ethanol): 241 nm (log ϵ 4.03).

THERAP CAT: Acetate as anabolic.

8764. Stepronin. *N-[1-Oxo-2-[(2-thienylcarbonyl)thio]-propyl]glycine;* 2-(α-thenoylthio)propionylglycine; *N*-(2-mercaptopropionyl)glycine 2-thiophenecarboxylate (ester); TTPG; prostenoglycine; Tiase. $C_{10}H_{11}NO_4S_2$; mol wt 273.33. C 43.94%, H 4.06%, N 5.13%, O 23.41%, S 23.46%. Prepn, properties and toxicology: F. Bolasco, **Belg.** pat. **875,186;** *eidem,* U.S. pat. **4,242,354** (1979, 1980 to Mediolanum Farmaceutici). Clinical trials in bronchitis: M. Lingetti *et al., Int. J. Clin. Pharm. Res.* **1**, 273 (1981); F. Dotta *et al., Clin. Ter.* **105**, 307 (1983); G. C. Morandini *et al., Curr. Ther. Res.* **35**, 783 (1984). Effect on human lung mucus: M. C. Morale *et al., Int. J. Tiss. Reac.* **5**, 231 (1983).

Colorless crystals from acetonitrile, mp 168-170°. uv max: 292 nm. LD_{50} in mice (mg/kg): > 2500 orally, > 1250 i.v.; in rats (mg/kg): > 2500 orally, 1801 i.m. (Bolasco).

Sodium salt, $C_{10}H_{10}NNaO_4S_2$, *Broncoplus, Tioten*.

THERAP CAT: Mucolytic.

8765. Steviol. *13-Hydroxykaur-16-en-18-oic acid;* hydroxydehydrostevic acid. $C_{20}H_{30}O_3$; mol wt 318.44. C 75.43%, H 9.50%, O 15.07%. Aglucon of stevioside, the sweet glucoside abundant in leaves and stems of *Stevia rebaudiana* Bert. Prepn by enzymatic hydrolysis of stevioside: Mosettig, Nes, *J. Org. Chem.* **20**, 884 (1955). Structure and stereochemistry: Dolder *et al., J. Am. Chem. Soc.* **82**, 246 (1960); Vorbrueggen, Djerassi, *ibid.* **84**, 2990 (1962). Absolute configuration: Mosettig *et al., ibid.* **85**, 2305 (1963). Biosynthesis: Ruddat *et al., Arch. Biochem. Biophys.* **110**, 496 (1965). Total synthesis of *dl*-form: Mori *et al., Tetrahedron Letters* **1970**, 2411; Cook, Knox, *ibid.* 4091; *eidem, Tetrahedron* **28**, 3217 (1972).

Needles from methanol or ethanol + water, mp 215°. $[\alpha]_D$ −94.7°.

8766. Stevioside. *13-[(2-O-β-D-Glucopyranosyl-α-D-glucopyranosyl)oxy]kaur-16-en-18-oic acid β-D-glucopyranosyl ester;* steviosin. $C_{38}H_{60}O_{18}$; mol wt 804.90. C 56.70%, H 7.51%, O 35.78%. Isolated from leaves of *Stevia rebaudiana* (Bert.) Hemsl. (*Eupatorium rebaudianum* Bert.), *Compositae,* also called *yerba dulce.* Habit. Paraguay. Review of botany: Samaniego, *Rev. farm. (Buenos Aires)* **88**, 199 (1946). Review of chemistry and use as sweetening agent: Bell, *Chem. & Ind. (London)* **1954**, 897; Fletcher, Jr., *Chemurgic Digest* **14**, 7, 18 (July-Aug. 1955); *Chem. & Eng. News* **34**, 124 (1956). Isoln and structure: Wood *et al., J. Org. Chem.* **20**, 875 (1955). Additional structure work: Vis, Fletcher, *J. Am. Chem. Soc.* **78**, 4709 (1956); Dolder *et al., ibid.* **82**, 246 (1960). Partial synthesis: T. Ogawa *et al., Carbohyd. Res.* **60**, C7 (1978). Total synthesis: *eidem, Tetrahedron* **36**, 2641 (1980). Synthesis and sensory evaluation of analogs: G. E. Dubois *et al., J. Med. Chem.* **24**, 1269 (1981).

Hygroscopic crystals, mp 198°. $[\alpha]_D^{25}$ −39.3° (c = 5.7 in H_2O). 300 times as sweet as cane sugar. One gram dissolves in 800 ml water. Sol in dioxane. Slightly sol in alc.

USE: Has been proposed as non-nutritive sweetening agent.

8767. Stibine. Antimony hydride. H_3Sb; mol wt 124.78. H 2.42%, Sb 97.57%. SbH_3. Conveniently prepd by dissolving zinc-antimony or magnesium-antimony alloy in dil HCl: Hurd, *Chemistry of the Hydrides* (Wiley, New York, 1952) p 132. Detailed directions (including prepn of the alloy from powdered Sb and Mg): Schenk in *Handbook of Preparative Inorganic Chemistry* vol. 1, G. Brauer, Ed. (Academic Press, New York, 2nd ed., 1963) pp 606-608. Review of preparative methods: Jolly, Norman, "Hydrides of Groups IV and V" in *Preparative Inorganic Reactions* vol. 4, W. L. Jolly, Ed. (Interscience, New York, 1968) pp 1-58.

Colorless gas. *Intensely poisonous.* Disagreeable odor. mp −88°. bp −18.4°. d 2.204 g/ml at bp. Heat of formation +34.68 kcal/mole: Gunn, Green, *J. Phys. Chem.* **65**, 779 (1961). Thermally less stable than arsine. Dec slowly on standing at room temp. Quickly destroyed at 200°. The decomposition products are hydrogen and metallic antimony, generally deposited in the form of a mirror. The gas is slightly sol in water. Freely sol in alcohol, carbon disulfide, other organic solvents. Lethal concn in air for mice: about 100 ppm.

USE: Has been used as fumigating agent. *Caution:* Toxic symptoms similar to arsine, *q.v.:* E. Browning, *Toxicity of Industrial Metals* (Appleton-Century-Crofts, New York, 2nd ed., 1969) pp 34-38.

8768. Stibocaptate. *2,2'-[(1,2-Dicarboxy-1,2-ethanediyl)bis(thio)]bis-1,3,2-dithiastibolane-4,5-dicarboxylic acid hexasodium salt;* 2,3-dimercaptosuccinic acid cyclic ester with antimonic acid, diester with 2,3-dimercaptosuccinic acid hexasodium salt; 2,3-dimercaptosuccinic acid cyclic thioantimonate (III) S,S-diester with 2,3-dimercaptosuccinate hexasodium salt; sodium antimony-2,3-*meso*-dimercaptosuccinate; "antimony dimercaptosuccinate"; Ro 4-1544/6; Sb-58; TWSb; Astiban. $C_{12}H_6Na_6O_{12}S_6Sb_2$; mol wt 787.74. C 18.30%, H 0.77%, Na 17.51%, O 24.37%, S 24.41%, Sb 30.91%. Prepn: Friedheim, U.S. pat. **2,880,222** (1959). Prepn and antischistosomal activity: J.-K. Hsu, M.-K. Ch'en, *K'o Hsueh Tung Pao* **1966**, 978, *C.A.* **65**, 17580e (1966). Toxicity data: Ercoli, *Proc. Soc. Exp. Biol. Med.* **129**, 284 (1968).

White or slightly yellowish-green powder. Hygroscopic,

unstable when moistened. Sol in water. LD_{50} s.c. in mice: 500 mg Sb/kg (Ercoli).

Note: Stibocaptate described as the tripotassium salt: Friedheim *et al., Am. J. Tropic. Med. Hyg.* **3**, 714 (1954).

THERAP CAT: Anthelmintic (Schistosoma).

8769. Stibophen. *Bis[4,5-dihydroxy-1,3-benzenedisulfonato(4−)-O⁴,O⁵]-antimonate(5−) pentasodium heptahydrate;* sodium antimony bis(pyrocatechol-2,4-disulfonate); antimony pyrocatechol sodium disulfonate; Sdt 91; Fuadin; Fouadin; Pyrostib; Corystibin; Trimon; Fantorin; Repodral; Neoantimosan; Sodium Antimosan. $C_{12}H_{18}Na_5O_{23}S_4Sb$; mol wt 895.21. C 16.10%, H 2.03%, Na 12.84%, O 41.11%, S 14.32%, Sb 13.60%. "Contains not less than 15.6 and not more than 16.0% of trivalent Sb, calculated on a moisture-free basis." Prepd from sodium pyrocatechol-3,5-disulfonate and antimony trioxide in alkaline soln: Schmidt, *Z. Angew. Chem.* **43**, 963 (1930); *Ger.* pats. 413,778; 414,854; 415,316; 448,800; 453,278; 453,279; 456,858; 458,089; 494,761; *Brit.* pat. 213,285; *U.S.* pats. 1,549,154; 1,873,668 (to Winthrop-Stearns). Toxicity: H. Eagle *et al., J. Pharmacol. Exp. Therap.* **89**, 196 (1947).

Fine crystals. Almost insol in abs alcohol, ether, chloroform, acetone, petr ether. Readily sol in cold water. Unless the colorless neutral water soln is acidified it soon acquires a yellowish tint and finally reaches lemon-yellow color. LD_{50} i.v. in rabbits: ~90 mg/kg (Eagle).

Incompat: Iron and its compds.

Potassium salt, *Antimosan, Heyden 611.*

THERAP CAT: Anthelmintic (Schistosoma).

THERAP CAT (VET): Antiprotozoan. Has been used as filaricide.

8770. Stigmastanol. *(3β,5α)-Stigmastan-3-ol;* dihydro-β-sitosterol; β-sitostanol; 24α-ethylcholestanol; fucostanol. $C_{29}H_{52}O$; mol wt 416.71. C 83.58%, H 12.58%, O 3.84%. Occurs along with β-sitosterol and stigmasterol. Formation from β-sitosterol: Bernstein, Wallis, *J. Org. Chem.* **2**, 341 (1937); from β-sitostenone: Marker, Wittle, *J. Am. Chem. Soc.* **59**, 2704 (1937). Antihypercholesteremic activity in rabbits: I. Ikeda *et al., J. Nutr. Sci. Vitaminol.* **27**, 243 (1981).

Monohydrate, crystals, mp 138-139°. When dry mp 144-145°. $[\alpha]_D^{20}$ +25° (c = 1.1 in chloroform).

Acetate, $C_{31}H_{54}O_2$, mp 137-138°. $[\alpha]_D^{20}$ +14° (c = 1.8 in chloroform).

8771. Stigmasterol. *(3β,22E)-Stigmasta-5,22-dien-3-ol;* 3β-hydroxy-24-ethyl-Δ⁵,²²-cholestadiene. $C_{29}H_{48}O$; mol wt 412.67. C 84.40%, H 11.72%, O 3.88%. Usually isolated from the phytosterol mixture from soy or calabar beans, *cf.* Thornton *et al., J. Am. Chem. Soc.* **62**, 2006 (1940); Byerrum, Ball, *Biochem. Prepns.* **7**, 86 (1959). Structure: Fernholz, *Ann.* **507**, 128 (1933); **508**, 215 (1933); Fernholz, Chakravorty, *Ber.* **67**, 2021 (1934). Identity with *guinea-pig-anti-stiffness factor:* H. Rosenkrantz *et al., Proc. Soc. Exp. Biol. Med.* **76**, 408 (1951). Synthesis: W. Sucrow, M. Slopianka, *Ber.* **108**, 3721 (1975). HPLC separation from plant oils: H. Colin *et al., Anal. Chem.* **51**, 1661 (1979).

Monohydrate, crystals from alcohol. When dry, mp 170°. $[\alpha]_D^{22}$ −51° (c = 2 in chloroform). Insol in water. Sol in the usual organic solvents.

Acetate, $C_{31}H_{50}O_2$, mp 144°. $[\alpha]_D^{20}$ −55.6° (c = 2 in chloroform).

p-Nitrobenzoate, $C_{36}H_{51}NO_4$, mp 203°. $[\alpha]_D^{17}$ −13° (c = 2 in alcohol).

8772. Stilbamidine. *4,4'-(1,2-Ethenediyl)bisbenzenecarboximidamide; 4,4'-stilbenedicarboxamidine; 4,4'-diamidinostilbene.* $C_{16}H_{16}N_4$; mol wt 264.33. C 72.70%, H 6.10%, N 21.20%. Prepn: A. J. Ewins, J. N. Ashley, *Brit.* pat. 510,097 (1939 to May & Baker); J. N. Ashley *et al., J. Chem. Soc.* **1942**, 103; of isethionate: G. Newbery, A. P. T. Easson, *U.S.* pat. 2,394,003 (1946 to May & Baker). Trypanocidal activity: E. M. Lourie, W. Yorke, *Ann. Trop. Med. Parasitol.* **33**, 289 (1939). Preliminary pharmacology: R. Wien, *ibid.* **37**, 1 (1943). Mechanism of action studies: M. J. Pine, *Biochem. Pharmacol.* **17**, 75 (1968); G. Weissmann *et al., ibid.* **19**, 1251 (1970). Crystal structure: C. Courseille *et al., C.R. Acad. Sci. Ser. C* **272**, 1115 (1971). DNA binding study: N. Gresh, B. Pullman, *Mol. Pharmacol.* **25**, 452 (1984). Early review of pharmacology, mode of action and clinical applications: E. B. Schoenbach, E. M. Greenspan, *Medicine* **27**, 327-377 (1948).

Dihydrochloride, $C_{16}H_{18}Cl_2N_4$, needles from water. LD_{50} in mice (mg/g): 0.031 i.v.; 0.18 s.c. (Wien).

Isethionate, $C_{20}H_{28}N_4O_8S_2$, M&B 744. Crystals, discolored by light, dec 290°. uv max: 330 nm ($E_{1cm}^{1\%}$ 750). Soly in water: approx 1 g in 2.5 to 3.0 ml H_2O; in methanol: 1.5 g/100 ml. pH of 0.5% aq soln 5.5-6.5. Neutral or alkaline solns are unstable. Stability of aq solns to heat is a function of pH. At pH 5.5 and 120°, decompn is 0.7% at 10 min, 6% at 30 min, 14% at 60 min. At pH 4.7 there is no decompn after 1 hr at 120°. Solns can be sterilized by brief boiling or autoclaving if pH is not too high.

THERAP CAT: Antiprotozoal (Leishmania, Trypanosoma).

8773. Stilbazium Iodide. *1-Ethyl-2,6-bis[2-[4-(1-pyrrolidinyl)phenyl]ethenyl]pyridinium iodide; 1-ethyl-2,6-bis[p-(1-pyrrolidinyl)styryl]pyridinium iodide; 2,6-bis(p-1-pyrrolidinylstyryl)-1-ethylpyridinium iodide; BW 61-32;* Monopar. $C_{31}H_{36}IN_3$; mol wt 577.56. C 64.47%, H 6.28%, I 21.97%, N 7.28%. Prepn: Phillips, Burrows, *U.S.* pat. 3,085,935 (1963 to Burroughs Wellcome). Toxicity data: R. B. Burrows, *Prog. Drug Res.* **17**, 110 (1973).

mp 282-283°. Nearly insol in hot methanol. LD_{50} in mice (mg/kg): 1360 orally; 7 i.p. (Burrows).

THERAP CAT: Anthelmintic (Nematodes).

8774. Stilbene. *1,1'-(1,2-Ethenediyl)bis[benzene];* α,β-diphenylethylene; bibenzal; bibenzylidene. $C_{14}H_{12}$; mol wt

180.24. C 93.29%, H 6.71%. *trans*-Form prepd by Clemmensen reduction of benzoin: Shriner, Berger, *Org. Syn. coll. vol. III*, 786 (1955); *cis*-form by copper-chromite decarboxylation of α-phenylcinnamic acid: Buckles, Wheeler, *ibid. coll. vol. IV*, 857 (1963). Synthesis of *cis*- and *trans*-forms by the Wittig reaction and by decarboxylation of phenylcinnamic acids: Wheeler, Batlle de Pabon, *J. Org. Chem.* **30**, 1473 (1965).

trans-Form, crystals from 95% ethanol, mp 124°. bp_{760} 306-307°. uv max (95% ethanol): 296, 305 nm (ε 28,100; 26,700). Volatile with steam. Practically insol in water; sol in 90 parts cold alc, 13 parts boiling alc, freely in benzene, ether.

cis-Form, liquid, solidifies at −5°. bp_{10} 135°; $bp_{1.0}$ 96°. n_D^{25} 1.6188. uv max (95% ethanol): 278 nm (ε 10,200). Completely sol in cold abs alcohol.

8775. Stillingia. Queen's root; yaw root; silver leaf. Dried roots of *Stillingia sylvatica* L., Euphorbiaceae. *Habit.* Southeastern U.S. *Constit.* Acrid resin (sylvacrol), fixed and volatile oils, a glucoside.

THERAP CAT: Emetic, cathartic.

8776. Stilonium Iodide. *N,N,N-Triethyl-2-[4-(2-phenylethenyl)phenoxy]ethanaminium iodide;* triethyl(4-stilbenehydroxyethyl)ammonium iodide; triethyl(2-*p*-styrylphenoxyethyl)ammonium iodide; MG 624; Elvetil. $C_{22}H_{30}INO$; mol wt 451.40. C 58.54%, H 6.70%, I 28.11%, N 3.10%, O 3.54%. Prepn: Cavallini *et al.*, *Farmaco, Ed. Sci.* **8**, 317 (1953).

Crystals, mp 211-212°. Sol in cold methanol, acetone, chloroform, hot water, ethanol; practically insol in ether, benzene, carbon tetrachloride.

THERAP CAT: Antispasmodic; ganglionic blocker.

8777. Stirofos. *Phosphoric acid 2-chloro-1-(2,4,5-trichlorophenyl)ethenyl dimethyl ester;* 2,4,5-trichloro-α-(chloromethylene)benzyl phosphate ester; 2-chloro-1-(2,4,5-trichlorophenyl)vinyl dimethyl phosphate; tetrachlorvinphos; ENT 25841; SD 8447; Dietreen; Gardona; Rabon. $C_{10}H_9$-Cl_4O_4P; mol wt 365.95. C 32.82%, H 2.48%, Cl 38.75%, O 17.49%, P 8.46%. Phosphate-type pesticide of low mammalian toxicity: Jennings, *Int. Pest Contr.* **12**, 28 (1970). Prepn of active Z-isomer: Phillips, Ward; Ramey, U.S. pats. **3,102,842** and **3,553,297** (1963, 1971 both to Shell). Properties: Whetstone *et al.*, *J. Agr. Food Chem.* **14**, 352 (1966).

mp 97-98°. Vapor pressure at 20° = 4.2 × 10^{-8}. Soly in water: 11 ppm; in xylene: <15%; 40-50% in chloroform at room temp. LD_{50} orally in male, female rats: 1100, 1125 mg/kg, T. B. Gaines, *Toxicol. Appl. Pharmacol.* **14**, 515 (1969).

USE: Insecticide. *Caution:* Cholinesterase inhibitor.

THERAP CAT (VET): Insecticide.

8778. Storax. Styrax; sweet oriental gum. Balsam obtained from the trunk of *Liquidambar orientalis* Mill., known as Levant Storax, or of *L. styraciflua* L., known as American Storax (family Hamamelidaceae). *Habit.* Levant storax is native to Asia Minor. American storax is produced chiefly

in Honduras; found along the Atlantic coast from Connecticut to Central America. *Constit.* 33-50% α- and β-storesin and its cinnamic ester; 5-10% styracin; 10% phenylpropyl cinnamate; small amounts of ethyl cinnamate; benzyl cinnamate; 5-15% free cinnamic acid; styrene; 0.4% levorotatory oil; $C_{10}H_{16}O$, and traces of vanillin.

Semiliquid, grayish, sticky, opaque mass (Levant storax) or a semisolid, sometimes solid mass softened by warming (American storax). Storax is transparent in thin layers, has characteristic taste and odor, and is denser than water. Insol in water; almost completely sol in 1 part warm alcohol, ether, acetone, carbon disulfide.

Ingredient of Compound Benzoin Tincture, *see* Benzoin.

USE: In fumigating pastilles and powders; in perfumery; as imbedding material in microscopy.

THERAP CAT: Topical protectant. Expectorant.

THERAP CAT (VET): As a component of Compound Benzoin Tincture. Has been used as a parasiticide.

8779. Stramonium. Thorn apple; Jamestown weed; Jimpson weed; Jimson weed; stinkweed; devil's apple; apple of Peru. Dried leaves and flowering tops of *Datura stramonium* L., Solanaceae. *Habit.* Europe, Asia, America. *Constit.* Leaves: 0.25-0.45% alkaloids consisting of atropine, hyoscyamine, and scopolamine; proteins, albumin. Seed: the same alkaloids but in lesser amounts; fixed oil, malic acid proteins. Comprehensive monograph: A. F. Blakeslee, *The Genus Datura* (Ronald Press, New York, 1959) 289 pp. *See also* Datura.

THERAP CAT: Anticholinergic.

THERAP CAT (VET): Same action as belladonna. Has been used to relieve signs of heaves in horses.

8780. Strepogenin. A name for the biologically active principle capable of stimulating growth of certain microorganisms, and originally found present in products of natural origin, such as liver extract, flour, yeast, and tomato juice. Its peptide nature was indicated by its presence in hydrolyzates of purified proteins. To compare the activities of various protein hydrolyzates, a strepogenin unit was introduced and defined as: 1 mg of a standard liver extract (Wilson's liver fraction L) has the strepogenin activity of one unit (using *Lactobacillus casei* as test organism). The different structures of strepogenin-active peptides subsequently obtained from natural sources or by synthesis, indicate that strepogenin is not a distinct, chemically defined, biologically active peptide, but rather represents a biological phenomenon of a multitude of peptides. From the large number of strepogenin-active peptides known, no conclusion can be drawn with regard to structure-function relationships; these peptides have been suspected to be nothing more than easily accessible amino acids, particularly since they (peptides) are not essential for the growth of the microorganisms. Early isolation from liver extract and from tryptic digests of pure proteins: Sprince, Woolley, *J. Exp. Med.* **80**, 213 (1944); *eidem, J. Am. Chem. Soc.* **67**, 1734 (1945). *Review:* E. Schröder, K. Lübke, *The Peptides* vol. II (Academic Press, New York, 1966) pp 267-270.

8781. Streptidine. *N,N′-Bis(aminoiminomethyl)streptamine; N,N′-diamidinostreptamine; N,N′′′-(2,4,5,6-tetrahydroxy-1,3-cyclohexanediyl)bisguanidine; 1,3-diguanido-2,4,5,6-cyclohexanetetrol.* $C_8H_{18}N_6O_4$; mol wt 262.27. C 36.63%, H 6.92%, N 32.05%, O 24.40%. The aglycone component of the streptomycin molecule joined with streptobiosamine through a glycosidic linkage. Obtained from streptomycin by acid hydrolysis: Peck *et al., J. Am. Chem. Soc.* **68**, 29 (1946); Fried *et al., J. Biol. Chem.* **162**, 391 (1946). Structure: Carter *et al., Science* **103**, 53, 540 (1946). Synthesis: Wolfrom *et al., J. Am. Chem. Soc.* **72**, 1724 (1950). Absolute configuration: Dyer, Todd, *ibid.* **85**, 3896 (1963);

Tatsuoka, Horii, *J. Antibiot.* **17A**, 88 (1964). Biosynthesis studies: Bruton *et al.*, *J. Biol. Chem.* **242**, 813 (1967).

Optically inactive, diacidic base. Isolated as the dipicrate dihydrate, $C_{20}H_{24}N_{12}O_{18}\cdot 2H_2O$, yellow needles from water, dec 283-284° (becomes anhydrous *in vacuo* at 56°, dec 284-285°).

Sulfate, $C_8H_{18}N_6O_4\cdot H_2SO_4$, crystals from dilute sulfuric acid + acetone, not melted at 300°. Forms solvated crystals from methanol.

Dihydrochloride, $C_8H_{18}N_6O_4\cdot 2HCl$, hygroscopic, amorphous powder, dec 170-210°. Forms solvated crystals from methanol.

8782. Streptobiosamine. *5-Deoxy-2-O-[2-deoxy-2-(methylamino)-α-L-glucopyranosyl]-3-C-formyl-L-lyxose.* $C_{13}H_{23}NO_9$; mol wt 337.33. C 46.29%, H 6.87%, N 4.15%, O 42.69%. A disaccharide representing two-thirds of the streptomycin molecule. It is built up from streptose and *N*-methyl-L-glucosamine. Structure and formation of derivatives by acid cleavage of streptomycin: Brink *et al.*, *J. Am. Chem. Soc.* **68**, 2096, 2405, 2557, 2679 (1946); Lemieux *et al.*, *ibid.* **69**, 1838 (1947); Kuehl *et al.*, *ibid.* **70**, 2613 (1948); **71**, 1445 (1948).

R = CH₃NH

8783. Streptogramin. Antibiotic complex produced by *Streptomyces graminofaciens* from Texas soil: Charney *et al.*, *Antibiot. & Chemother.* **3**, 1283 (1953); **Brit.** pat. **776,035** (1957 to Merck & Co.). Consists of an antibiotic factor, streptogramin A, capable of inhibiting protein synthesis; and a synergizing factor, streptogramin B, which is a cyclic polypeptide: Chabbert, Acar, *Ann. Inst. Pasteur* **107**, 777 (1964). The term streptogramin more broadly refers to a family of closely related antibiotics of which streptogramin itself is a member. For other members see Mikamycin, Pristinamycin, Vernamycin, Virginiamycin. Nomenclature: P. Crooy, R. De Neys, *J. Antibiot.* **25**, 371 (1972). *Review:* Vazquez, *Antibiotics vol. 1*, D. Gottlieb, P. D. Shaw, Eds. (Springer-Verlag, New York, 1967) pp 387-403; **vol. 3**, J. W. Corcoran, F. E. Hahn, Eds. (1975) pp 521-534.

Solid spherules, mp approx 155°. $[\alpha]_D$ −134°. Sol in methanol, ethanol, acetone, ethyl acetate; sparingly sol in water, dil acid; practically insol in petr ether. LD_{50} i.p. in mice: 450 mg/kg, Charney *et al.*, *Antibiot. Annual* **1953**, 171.

Streptogramin A, see Virginiamycin M₁.
Streptogramin B, see Mikamycin B.

8784. Streptokinase. Streptococcal fibrinolysin; plasminokinase; Awelysin; Kinalysin; Kabikinase; Streptase. An enzyme elaborated by hemolytic streptococci. Hydrolyzes —CONH— links. Activator of plasminogen, thus producing plasmin, which dissolves fibrin. Isoln from hemolytic streptococci: Christensen, *J. Gen. Physiol.* **28**, 363 (1945); *J. Clin. Invest.* **38**, 163 (1949); von Pölnitz *et al.*, *U.S.* pats. **3,063,913; 3,063,914; 3,138,542** (1962, 1962, 1964 all to Behringwerke); Siiteri, Mills, *U.S.* pat. **3,226,304** (1965 to Am. Cyanamid). Purification: Singher, Zuckerman, *U.S.* pat. **3,016,337** (1962 to Ortho); Siegel *et al.*, Baumgarten, Cole, *U.S.* pats. **3,042,586; 3,107,203** (1962, 1963 to Merck & Co.). Mechanism of action: Heimburger, *Thromb. Diath. Haemorrh.*, *Suppl.* **47**, 21 (1971); K. N. N. Reddy in *Fibrinolysis*, D. L. Kline, K. N. N. Reddy, Eds. (CRC, Boca Raton, 1980) pp 71-94. Use in thrombolytic therapy: G. Kolata, *Science* **214**, 1229 (1981). Clinical trial of intracoronary streptokinase in treatment of acute myocardial infarction: J. L. Anderson *et al.*, *N. Engl. J. Med.* **308**, 1312 (1983).

Comparison with nitroglycerin: K. P. Rentrop *et al.*, *ibid.* **311**, 1457 (1984). Experimental use in prevention of myocardial infarction: G. J. Davies *et al.*, *ibid.* 1488. Pharmacokinetics in humans: D. S. Grierson, T. D. Bjornsson, *Clin. Pharmacol. Ther.* **41**, 304 (1987). *Reviews:* Taylor, Tomar, *Methods Enzymol.* **19**, 807 (1970); Brogden *et al.*, *Drugs* **5**, 357 (1973).

Mixture with streptodornase, *Varidase.* See: *J. Am. Med. Assoc.* **145**, 1173 (1951); R. Pakula *et al.*, *C.A.* **49**, 568 (1955).

THERAP CAT: Thrombolytic.

THERAP CAT (VET): Mixture with streptodornase as debriding agent.

8785. Streptolydigin. Portamycin. $C_{32}H_{44}N_2O_9$; mol wt 600.69. C 63.98%, H 7.38%, N 4.66%, O 23.97%. Antibiotic isolated from culture filtrates of *Streptomyces lydicus* n. sp.: Eble *et al.*, *Antibiot. Annual* **1955-1956**, 893; **Brit.** pat. **779,-570** (1957 to Upjohn). Structure: Rinehart *et al.*, *J. Am. Chem. Soc.* **85**, 4083 (1963). Stereochemistry: Duchamp *et al.*, *ibid.* **95**, 4077 (1973). RNA-polymerase inhibitor: Cassani *et al.*, *Nature New Biol.* **230**, 197 (1971). NMR studies: V. J. Lee, K. L. Rinehart, *J. Antibiot.* **34**, 408 (1980).

Crystals from acetone + water, mp 147-148°. $[\alpha]_D^{25}$ −93° (c = 1.6 in chloroform). uv max (0.01N ethanolic KOH): 261, 291, 336 nm ($E_{1cm}^{1\%}$ 223.6, 270.7, 331.0); (0.01N ethanolic H_2SO_4): 234, 357.5, 370 nm ($E_{1cm}^{1\%}$ 139.1, 590.5, 560.3). Sol in ethanol, ethyl acetate, ether, chloroform. Practically insol in water, hydrocarbons. LD_{50} in mice: 533 mg/kg i.p., DeBoer *et al.*, *Antibiot. Annual* **1955-1956**, 886.

Sodium salt, $C_{32}H_{43}N_2NaO_9$, mp 225°. $[\alpha]_D^{25}$ +153° (c = 1.35 in chloroform). Sol in the usual organic solvents and in water. Practically insol in hydrocarbons.

THERAP CAT: Antibacterial.

8786. Streptomycin. *O-2-Deoxy-2-(methylamino)-α-L-glucopyranosyl-(1 → 2)-O-5-deoxy-3-C-formyl-α-L-lyxofuranosyl-(1 → 4)-N,N'-bis(aminoiminomethyl)-D-streptamine;* streptomycin A. $C_{21}H_{39}N_7O_{12}$; mol wt 581.58. C 43.37%, H 6.76%, N 16.86%, O 33.01%. Antibiotic substance produced by the soil Actinomycete *Streptomyces griseus* (Krainsky) Waksman et Henrici (Fam. *Actinomycetaceae*). Isolation: Schatz *et al.*, *Proc. Soc. Exp. Biol. Med.* **55**, 66 (1944). Production by aerobic fermentation and purification: Tishler in *Streptomycin*, Selman A. Waksman, Ed. (Williams & Wilkins, Baltimore, 1949) pp 32-54. Isoln and purification by ion exchange: Bartels *et al.*, *Chem. Eng. Progr.* **54** (8), 49-51 (Aug. 1958); Bartels *et al.*, *U.S.* pat. **2,868,779** (1959 to Olin Mathieson). Structure: Brink, Folkers, *J. Am. Chem. Soc.* **69**, 1234 (1947); Wolfrom *et al.*, *ibid.* **76**, 3675 (1950). Total synthesis: Umezawa *et al.*, *J. Antibiot.* **27**, 997 (1974). Mechanism of action: B. J. Wallace *et al.*, in *Antibiotics vol.* 5(pt. 1), F. E. Hahn, Ed. (Springer-Verlag, New York, 1979) pp 272-303. *Review:* Lemieux, Wolfrom, "The Chemistry of Streptomycin" in W. W. Pigman, M. L. Wolfrom, *Advan. Carbohyd. Chem.* **3**, 337-384 (1948). Comprehensive description: J. Mossa *et al.*, in *Analytical Profiles of Drug Substances vol. 16*, K. Florey, Ed. (Academic Press, New York, 1986) pp 507-609.

Streptomycin is usually available as the trihydrochloride, trihydrochloride-calcium chloride double salt, phosphate, or sesquisulfate, which occur as granules or powder. Odorless or nearly so, with a slightly bitter taste. Most salts are hygroscopic and deliquesce on exposure to air, but are not affected by air or light. The salts are very sol in water; but almost insol in alc, chloroform, ether. Solns are levorotatory.

Trihydrochloride, $C_{21}H_{42}Cl_3N_7O_{12}$, *streptomycin hydrochloride.* $[\alpha]_D^{25}$ −84°. Solubilities as determined by Weiss *et al., Antibiot. & Chemother.* **7**, 374 (1957) in mg/ml at about 28°: water > 20; methanol > 20; ethanol 0.90; isopropanol 0.12; isoamyl alcohol 0.117; petr ether 0.02; carbon tetrachloride 0.042; ether 0.01.

Trihydrochloride-calcium chloride double salt, $C_{42}H_{84}Ca$-$Cl_8N_{14}O_{24}$, *streptomycin hydrochloride-calcium chloride complex.* $(C_{21}H_{39}N_7O_{12}.3HCl)_2CaCl_2$. Prepn from the trihydrochloride: Peck, U.S. pat. **2,446,102** (1948 to Merck & Co.). Very hygroscopic, dec about 200°. $[\alpha]_D^{25}$ −76°.

Pantothenate, *Streptothenat.* Prepn: **Brit.** pat. **771,338** (1957 to Grünenthal). The commercial prepn may contain the sulfate.

Sesquisulfate, $C_{42}H_{84}N_{14}O_{36}S_3$, *streptomycin sulfate, Agri-Strep, Streptobrettin, Streptorex, Vetstrep.* White to light gray or pale buff powder with faint amine-like odor. Solubilities as determined by Weiss *et al., loc. cit.,* in mg/ml at about 28°: water > 20; methanol 0.85; ethanol 0.30; isopropanol 0.01; petr ether 0.015; carbon tetrachloride 0.035; ether 0.035.

THERAP CAT: Antibacterial (tuberculostatic).
THERAP CAT (VET): Antibacterial.

8787. Streptomycin B. Mannosidostreptomycin; man-

nosylstreptomycin. $C_{27}H_{49}N_7O_{17}$; mol wt 743.75. C 43.60%, H 6.64%, N 13.18%, O 36.57%. Isoln from streptomycin concentrates: Fried, Titus, *J. Biol. Chem.* **168**, 391 (1947); **174**, 57 (1948); Fried, Stavely, *J. Am. Chem. Soc.* **69**, 1549 (1947); Fried, Titus, *ibid.* **70**, 3615 (1948); O'Keeffe *et al., ibid.* **71**, 2452 (1949); *see also* Langlykke, Perlman, U.S pat. **2,493,489** (1950 to Squibb). Structure: Fried, Stavely, 113th Meeting *Am. Chem. Soc.,* Chicago, Ill. (April 20, 1948) *Abstracts of Papers* p 25C; Peck *et al., J. Am. Chem. Soc.* **70**, 3968 (1948); Stavely, Fried, *ibid.* **71**, 135 (1949); **74**, 5461 (1952).

On catalytic hydrogenation absorbs approx one mole of hydrogen and forms dihydrostreptomycin B. Yields maltol on alkaline hydrolysis. When treated with dil acids liberates streptidine (or an isomer of it) isolated as the sulfate monohydrate, $C_8H_{18}N_6O_4.H_2SO_4.H_2O$.

Trihydrochloride monohydrate, $C_{27}H_{49}N_7O_{17}.3HCl.H_2O$, white, amorphous powder; mp 190-200° (corr, decompn) after drying *in vacuo* at 100°. $[\alpha]_D^{25}$ −47° (c = 1.35 in water). Appears to have about one-fourth of the antibiotic activity of streptomycin trihydrochloride.

8788. Streptonicozid. *4-Pyridinecarboxylic acid hydrazide, hydrazone with O-2-deoxy-2-(methylamino)-α-L-glucopyranosyl-(1→2)-O-5-deoxy-3-C-formyl-α-L-lyxofuranosyl-(1→4)-N,N'-bis(aminoiminomethyl)-D-streptamine sulfate (2:3) (salt); isonicotinic acid hydrazide, hydrazone with streptomycin, sulfate (salt) (2:3); streptomyclidene isonicotinyl hydrazine sulfate; streptoniazide; Strazide; Streptohydrazid.* $C_{54}H_{94}N_{20}O_{36}S_3$; mol wt 1695.66. C 38.25%, H 5.59%, N 16.52%, O 33.97%, S 5.67%. Semi-synthetic antibiotic prepd by the condensation of streptomycin salts and isonicotinic acid hydrazide in methanol: Pennington *et al., J. Am. Chem. Soc.* **75**, 2261 (1953).

R' = $NHCH_3$

R'' = $CH=NHC$ (pyridine structure)

Crystals, dec ~230°. Freely sol in water with dissocn in dil solns. Solubilities determined by Weiss *et al., Antibiot. & Chemother.* **7**, 374 (1957) in mg/ml at about 28°: water > 20; ethanol 0.115; carbon tetrachloride 0.025; ether 0.31.
THERAP CAT: Antibacterial.

8789. Streptonigrin. *5-Amino-6-(7-amino-5,8-dihydro-6-methoxy-5,8-dioxo-2-quinolyl)-4-(2-hydroxy-3,4-dimethoxyphenyl)-3-methyl-2-pyridinecarboxylic acid; 5-amino-6-(7-amino-5,8-dihydro-6-methoxy-5,8-dioxo-2-quinolyl)-4-(2-hydroxy-3,4-dimethoxyphenyl)-3-methylpicolinic acid;* bruneomycin; NSC 45383; Nigrin. $C_{25}H_{22}N_4O_8$; mol wt 506.46. C 59.28%, H 4.38%, N 11.06%, O 25.27%. Antitumor antibiotic produced by *Streptomyces flocculus:* Rao,

Cullen, *Antibiot. Annual* **1959-1960,** 950. Structure: Rao *et al., J. Am. Chem. Soc.* **85,** 2532 (1963). Crystal and molecular structure: Y. Y. H. Chiu, W. N. Lipscomb, *ibid.* **97,** 2525 (1975). Identity with bruneomycin: Brazhnikova *et al., Antibiotiki* **13,** 99 (1968), *C.A.* **68,** 89890q (1968). Synthetic studies: Kametani *et al., C.A.* **76,** 14272w (1972); **79,** 31817g (1973). Total synthesis: F. Z. Basha *et al., J. Am. Chem. Soc.* **102,** 3962 (1980); A. S. Kende *et al., ibid.* **103,** 1271 (1981); S. M. Weinreb *et al., ibid.* **104,** 536 (1982). Biosynthesis: S. J. Gould, D. E. Cane, *ibid.* 343. *Reviews:* N. S. Mizuno in *Antibiotics* vol. 5, pt. 2, F. E. Hahn, Ed. (Springer-Verlag, New York, 1979) pp 372-384; S. J. Gould, S. M. Weinreb, *Fortschr. Chem. Org. Naturst.* **41,** 77-111 (1982).

Coffee-brown to almost black rectangular plates from acetone or dioxane, dec 275°; brown needles, mp 262-263° (Brazhnikova). uv max (methanol): 248, 375-380 nm (ϵ 38400, 17400). Weak acid. pKa 6.2-6.4 (1:1 aq dioxane). Sol in dioxane, pyridine, dimethylformamide, aq sodium bicarbonate with decompn; slightly sol in water, lower alcohols, ethyl acetate, chloroform.

Human Toxicity: Causes severe and prolonged bone marrow depression. Toxicity studies: Wilson *et al., Antibiot. Chemother.* **11,** 147 (1961) *sqq.*

THERAP CAT: Antineoplastic.

8790. L-Streptose. *5-Deoxy-3-C-formyl-L-lyxose;* 3-C-formyl-5-deoxy-L-lyxofuranose. $C_6H_{10}O_5$; mol wt 162.14. C 44.44%, H 6.22%, O 49.34%. A sugar which is part of the streptomycin molecule: Kuehl *et al., J. Am. Chem. Soc.* **68,** 2096 (1946). Structure: *eidem, ibid.* **68,** 2679 (1946). Configuration: Wolfrom, DeWalt, *ibid.* **70,** 3148 (1948); Kuehl *et al., ibid.* **71,** 1445 (1949). Biosynthesis: Hough, Jones, *Nature* **167,** 180 (1951). Synthesis of α- and β-L-streptose: Dyer *et al., J. Am. Chem. Soc.* **87,** 654 (1965); Paulsen *et al., Ber.* **105,** 1978 (1972).

Colorless, strongly hygroscopic glass. $[\alpha]_D^{20}$ $-18°$ (c = 0.65).

8791. Streptothricins. Racemomycins; yazumycins. A mixture of antibiotics, originally thought to be a single substance, differing only in the number of repeating residues in the peptide side chain. Streptothricins F, E, D, C, B, A, and X are known and have 1 to 7 β-lysine residues resp.: Khokhlov, Shutova, *J. Antibiot.* **25,** 501 (1972). Separation of isomers: Khokhlov, Reshetov, *J. Chromatog.* **14,** 495 (1964). Identity with racemomycins: H. Taniyama *et al., Chem. Pharm. Bull.* **19,** 1627 (1971). Identity with yazumycins: *eidem, J. Antibiot.* **24,** 390 (1971). Activity studies: Germanova, Goncharskaya, *Antibiotiki (Moscow)* **14,** 48, 137 (1969), *C.A.* **70,** 66547, 85906t (1969). Biosynthetic studies: Voronina *et al., ibid.* 1063, *C.A.* **72,** 77387a (1970); Y. Sawada *et al., Chem. Pharm. Bull.* **26,** 885 (1978). Partial syntheses: M. Kinoshita, Y. Suzuki, *Bull. Chem. Soc. Japan* **50,** 2375 (1977); S. Kusumoto *et al., Chem. Letters* **1981,** 1317. Prepn of semi-synthetic racemomycins: Y. Sawada, H. Taniyama, *Chem. Pharm. Bull.* **25,** 1302 (1977). *Streptolin*

was formerly used for the streptothricin components where n equals 2 or 3.

Streptothricin F, $C_{19}H_{34}N_8O_8$, *racemomycin A, streptothricin VI, yazumycin A.* n = 1. Originally known as streptothricin. Produced by variants of *Actinomyces lavendulae.* Isoln: Waksman, Woodruff, *Proc. Soc. Exp. Biol. Med.* **49,** 207 (1942). Purification: Peck *et al., J. Am. Chem. Soc.* **68,** 772 (1946); Kocholaty, Junowicz-Kocholaty, *Arch. Biochem.* **15,** 55 (1947). Identity with streptothricin VI: Hutchinson *et al., Arch. Biochem.* **22,** 16 (1949); Swart, *J. Am. Chem. Soc.* **71,** 2942 (1949). Prepn: Carter *et al., J. Am. Chem. Soc.* **76,** 566 (1954). Structure: van Tamelen *et al., ibid.* **83,** 4295 (1961); Johnson, Westley, *J. Chem. Soc.* **1962,** 1642. Revised structure: S. Kusumoto *et al., J. Antibiot.* **35,** 925 (1982). Biosynthesis: S. J. Gould *et al., J. Am. Chem. Soc.* **103,** 2871 (1981). Total synthesis: S. Kusumoto *et al., Tetrahedron Letters* **23,** 2961 (1982). Amorphous, white powder. Thermostable. Water sol; slightly sol in most organic solvents. Stable between pH 1 and 8.5. LD_{50} i.v. in mice: 300 mg/kg (Taniyama).

Hydrochloride, white powder. $[\alpha]_D^{25}$ $-51.3°$ (c = 1.4). Sol in water and dil mineral acids. Practically insol in ether, petr ether, chloroform.

8792. Streptovaricin. Streptovarycin. A complex of ansamycin antibiotics consisting of streptovaricins A, B, C, D, E, F, G, J and K of which streptovaricin C is the major component. Isolation from *Streptomyces spectabilis:* Siminoff *et al., Am. Rev. Tuberc. Pulm. Dis.* **75,** 576 (1957); Whitfield *et al., ibid.* 584; Dietz *et al.,* U.S. pat. **3,116,202** (1963 to Upjohn). Separation of components: Dietz *et al., loc. cit.;* Rinehart *et al., Chem. Commun.* **1974,** 861. Preliminary structural studies: Rinehart *et al., J. Am. Chem. Soc.* **88,** 3149, 3150 (1966); **90,** 6241 (1968). Revised structure of streptovaricin C: Sasaki *et al., J. Antibiot.* **25** 68 (1972); Rinehart, Antosz, *ibid.* 71. Streptovaricins A, B, D, E, F, G and J differ from streptovaricin C in their substituents at W, X, Y and Z: Rinehart *et al., J. Am. Chem. Soc.* **93,** 6273

Streptovaricin	W	X	Y	Z	R
A	OH	OH	Ac	OH	COOCH₃
B	H	OH	Ac	OH	COOCH₃
C	H	OH	H	OH	COOCH₃
D	H	OH	H	H	COOCH₃
E	H	O=	H	OH	COOCH₃
G	OH	OH	H	OH	COOCH₃
J	H	OAc	H	OH	COOCH₃
K	OH	OAc	H	OH	COOCH₃

(1971). Activity studies: Quintrell, McAuslan, *J. Virol.* **6**, 485 (1970); Tan, McAuslan, *Biochem. Biophys. Res. Commun.* **42**, 230 (1971); Carter *et al., Nature New Biol.* **232**, 212, 214 (1971); Rinehart *et al., loc. cit.* (1974). Biosynthetic study: P. V. Deshmukh *et al., J. Am. Chem. Soc.* **98**, 870 (1976). ^{13}C-NMR studies: K. B. Kakinuma *et al., J. Org. Chem.* **41**, 1358 (1976). Synthesis of a key intermediate for the total synthesis of streptovaricin A: P. A. McCarthy, *Tetrahedron Letters* **23**, 4199 (1982). Synthesis of the naphthalene core of streptovaricin D: B. M. Trost, W. H. Pearson, *ibid.* **24**, 269 (1983). Review of the chemistry of streptovaricins: K. L. Rinehart *et al., Fortschr. Chem. Org. Naturst.* **33**, 231-307 (1976).

The complex yields yellow crystals from ethyl acetate + hexane. Neutral or weakly acidic substance which hydrolyzes on treatment with dil alkali. Stable at room temp for 3-4 days at pH 2.0-6.0. Gradually loses its antibiotic activity while standing at room temp for 2-4 days at pH 7.8.

Streptovaricin A, $C_{42}H_{53}NO_{16}$, mol wt 827.89. mp 194-196°. $[\alpha]_D$ in chloroform +61°. uv max (95% ethanol): 245, 260, 320, 430 nm ($E_{1cm}^{1\%}$ 418.9, 352.1, 139.8, 136.2).

Streptovaricin B, $C_{42}H_{53}NO_{15}$, mol wt 811.89. mp 185-187°. $[\alpha]_D$ in chloroform +576° (Dietz, +454°). uv max (95% ethanol): 245, 266, 320, 432 nm ($E_{1cm}^{1\%}$ 408.3, 338.5, 140.1, 109.7).

Streptovaricin C, $C_{40}H_{51}NO_{14}$, mol wt 769.86. mp 189-191° (Dietz, 168-171°). $[\alpha]_D$ in chloroform +602° (Dietz, +317°). uv max (95% ethanol): 245.5, 260, 320, 430 nm ($E_{1cm}^{1\%}$ 443.5, 389.7, 153.1, 121.9). Solubility data: March, Weiss, *J. Assoc. Offic. Anal. Chem.* **50**, 457 (1967).

Streptovaricin D, $C_{40}H_{51}NO_{13}$, mol wt 753.86. mp 167-170° (Dietz, 115-118°). $[\alpha]_D$ in chloroform +436° (+102°). uv max (95% ethanol): 246, 264, 320, 433 nm ($E_{1cm}^{1\%}$ 431.4, 370.3, 147.7, 105.3).

Streptovaricin E, $C_{40}H_{49}NO_{14}$, mol wt 767.84. mp 102-105°. $[\alpha]_D^{24}$ in chloroform +164°, also reported as +613°. uv max (95% ethanol): 245, 273, 320, 437 nm ($E_{1cm}^{1\%}$ 382.2, 346.9, 190.4, 91.9).

Streptovaricin F, $C_{39}H_{47}NO_{14}$, mol wt 753.82.

Streptovaricin G, $C_{40}H_{51}NO_{15}$, mol wt 785.86. mp 190-192°. $[\alpha]_D$ in chloroform +473°.

Streptovaricin J, $C_{42}H_{53}NO_{15}$, mol wt 811.89. mp 177-180°. $[\alpha]_D$ +326° (c = 0.25 in ethanol); +436° (c = 0.094 in chloroform).

8793. Streptovirudin. Nucleoside antibiotic complex produced by *Streptomyces griseoflavus* subsp. *thuringiensis,* strain JA 10124, that has inhibitory activity vs gram-positive bacteria, mycobacteria, and several RNA and DNA viruses. Isoln: K. Eckardt *et al., Z. Allg. Mikrobiol.* **13**, 625 (1973); *eidem, J. Antibiot.* **28**, 274 (1975). Taxonomy, fermentation, production: H. Thrum *et al., ibid.* 514. Streptovirudin is related chemically to tunicamycin, *q.v.,* and contains at least 10 components, of which 8 have been isolated in pure form. The complex can be divided into 2 series of compounds; series I (streptovirudins A_1 through D_1) contains dihydrouracil and series II (streptovirudins A_2 through D_2) contains uracil. Differentiation of streptovirudin and tunicamycin by gel chromatography, HPLC, and hydrolysis: K. Eckardt *et al., ibid.* **33**, 908 (1980). Structures of streptovirudins and identity of components B_{2a} and C_2 with tunica-

mycin components: *eidem, ibid.* **34**, 1631 (1981); W. Ihn *et al., Tetrahedron* **38**, 1781 (1982). Biological activities of isolated streptovirudin and tunicamycin components, identity of streptovirudin C_2 with tunicamycin A: R. W. Keenan *et al., Biochemistry* **20**, 2968 (1981). Chemical and biological properties of streptovirudins: A. D. Elbein *et al., ibid.* 4210. Viral inhibition study: M. S. Kang *et al., Biochem. Biophys. Res. Commun.* **99**, 422 (1981).

White amorphous solid. uv max (methanol): 212 nm. Sol in methanol, pyridine. Slightly sol in water, ethanol. Insol in chloroform, acetone, benzene, ethyl acetate. Stable up to 5 hours at 90°. No loss of activity was observed after standing at 37° for 4 weeks. LD_{50} in mice (mg/kg) of the antibiotic complex from culture filtrate: 250 orally; 15 s.c.; 15-17.5 i.p.; 17.5 i.v.; of the complex from mycelium: 150 orally; 5-10 s.c.; 3.5-7.5 i.p., H. Thrum *et al., loc. cit.*

8794. Streptozocin. *2-Deoxy-2-[[(methylnitrosoamino)-carbonyl]amino]-D-glucopyranose; 2-deoxy-2-(3-methyl-3-nitrosoureido)-D-glucopyranose;* streptozotocin; NSC-85998; U-9889; Zanosar. $C_8H_{15}N_3O_7$; mol wt 265.22. C 36.23%, H 5.70%, N 15.84%, O 42.23%. Isoln from *Streptomyces achromogenes* fermentation broth: Herr *et al., Antibiot. Ann.* **1959-1960**, 236; Bergy *et al.,* Fr. pat. **1,434,920** (1966 to Upjohn), *C.A.* **65**, 17661h (1966). Structure and synthesis: Herr *et al., J. Am. Chem. Soc.* **89**, 4808 (1967); Hardegger *et al., Helv. Chim. Acta* **52**, 2555 (1969); Hessler, Jahnke, *J. Org. Chem.* **35**, 245 (1970); P. F. Wiley *et al., ibid.* **44**, 9 (1979). Diabetogenic effect: Rakieten *et al., Cancer Chemother. Rep. No.* **29**, 91 (May 1963). Comparison of streptozotocin- and alloxan-induced diabetes: C. Rerup, F. Tarding, *Eur. J. Clin. Pharmacol.* **7**, 89 (1969). Antileukemic activity: Bhuyan *et al., Cancer Chemother. Rep. Part 1* **56**, 709 (1972). Biosynthetic study: S. Singaram *et al., J. Antibiot.* **32**, 379 (1979). Review and evaluation of studies of carcinogenicity in laboratory animals: *IARC Monographs* **4**, 221-227 (1974). Toxicity data: M. Iwasaki *et al., J. Med. Chem.* **19**, 918 (1976). *Review:* B. Rudas, *Arzneimittel-Forsch.* **22**, 830-861 (1972); P. F. Wiley, *Anticancer Agents Based on Natural Product Models,* J. M. Cassidy, J. D. Douros, Eds. (Academic Press, New York, 1980) pp 167-200. *Book: Streptozotocin: Fundamentals and Therapy,* M. K. Agarwal, Ed. (Elsevier/North Holland Biomedical Press, New York, 1981) 309 pp.

Pointed platelets or prisms, from 95% ethanol, mp 115° (dec). Sol in H_2O, lower alcohols and ketones. uv max (ethanol): 228 nm (ε 6360). A mixture of α and β anomers; aq solns rapidly undergo mutarotation to an equilibrium value of $[\alpha]_D^{25}$ +39°. LD_{50} in female mice (mg/kg): 360 i.p.; 275 i.v.; in male dogs (mg/kg): 50 i.v. (Iwasaki).

Tetraacetate, $C_8H_{11}N_3O_7(COCH_3)_4$, crystals, mp 111-114° (dec). $[\alpha]_D^{25}$ +41° (c = 0.78 in 95% ethanol).

Note: This substance may reasonably be anticipated to be a carcinogen: *Fourth Annual Report on Carcinogens* (NTP 85-002, 1985) p 183.

USE: Production of experimental diabetes in laboratory animals.

THERAP CAT: Antineoplastic.

8795. Strigol. *3-[[(2,5-Dihydro-4-methyl-5-oxo-2-furanyl)oxy]methylene]-3,3a,4,5,6,7,8,8b-octahydro-5-hydroxy-8,8-dimethyl-2H-indeno[1,2-b]furan-2-one.* $C_{19}H_{22}O_6$; mol wt 346.39. C 65.88%, H 6.40%, O 27.71%. Potent seed germination stimulant for the root parasite, witchweed, *Striga lutea* Lour. isolated from root exudates of cotton, *Gossypium hirsutum* L.: C. E. Cook *et al., Science* **154**, 1189 (1966). Structure: *eidem, J. Am. Chem. Soc.* **94**, 6198 (1972). Crystal structure: P. Coggan *et al., J. Chem. Soc. Perkin Trans. 2* **1973**, 465. Synthesis: J. B. Heather *et al., J. Am. Chem. Soc.* **98**, 3661 (1976); of (±)-form: G. A. MacAlpine *et al., Chem. Commun.* **1974**, 834; *eidem, J. Chem.*

streptovirudins (series I)

A_1	: n = 6,	R = H
B_1	: n = 6,	R = CH_3
B_{1a}	: n = 7,	R = H
C_1	: n = 8,	R = H
D_1	: n = 8,	R = CH_3

Soc. Perkin Trans. 1 **1976,** 410. Synthesis of analogs: A. W. Johnson *et al., ibid.* **1981,** 1734. Witchweed seed germination stimulating activity: A. L. Hsiao *et al., Weed Sci.* **29,** 101 (1981); A. B. Pepperman *et al., ibid.* **30,** 561 (1982).

White needles from benzene-hexane, mp 200-202° (dec). $[\alpha]_D^{25}$ +293° (c = 0.15 in CHCl$_3$). uv max: 234 nm (ε 17,700). Sol in acetone, methylene chloride. Moderately sol in benzene; insol in hexane.

8796. Strobane®. Terpene polychlorinates; Dichloricide Aerosol; Dichloricide Mothproofer; Insecticide 3960-X14. A mixture of chlorinated terpene isomers. Preparation by chlorinating a mixture of camphene and pinenes contg dipentene: Schultz, Bloor, **U.S.** pat. **2,767,115** (1956 to B. F. Goodrich).

Viscous, amber liquid. d$_{25}$ 1.6267. Misc with aromatic and aliphatic hydrocarbons. Slightly sol in alcohol.

USE: Insecticide. *Compare* Toxaphene. *Caution:* May be mildly irritating to skin.

8797. Strontium. Sr; at. wt 87.62; at. no. 38; valence 2. An alkaline earth metal. Isotopes: 88 (82.56%); 86 (9.86%); 87 (7.02%); 84 (0.56%). Occurs as the sulfate, celestine; or the carbonate, strontianite; found in small quantities associated with calcium or barium minerals. First prepared in 1807 by Davy. Prepn: Glascock, *J. Am. Chem. Soc.* **32,** 1222 (1910); Matignon, *Compt. Rend.* **177,** 1116 (1923); *J. Chem. Soc.* **126**[ii], 44 (1924); Guntz *et al.,* cited in *Gmelin's, Strontium* (8th ed.) **29,** 35 (1931). Review of strontium and its compounds: Goodenough, Stenger, "Magnesium, Calcium, Strontium, Barium and Radium" in *Comprehensive Inorganic Chemistry,* vol. **1,** J. C. Bailar, Jr. *et al.,* Eds. (Pergamon Press, Oxford, 1973) pp 591-774.

Silvery-white metal; face-centered cubic structure; rapidly becomes yellow on exposure to air and assumes an oxide film. The finely divided metal ignites spontaneously in air. d 2.6; mp 757° ± 1°; bp 1366°. E⁰ (aq) Sr²⁺/Sr −2.89 V. For a description of reactions which are characteristic of alkaline earth metals *see* Calcium. *Keep under liquid containing no oxygen.* The heated metal combines with hydrogen to form *strontium hydride* and with nitrogen to form *strontium nitride.* Strontium salts impart brilliant red color to a flame.

USE: In fireworks, in red signal flares; on tracer bullets. The artificial isotope ⁹⁰Sr, with a half life of 28 years, is being considered as a source of electric power. Heat liberated by decay would be changed to electrical energy.

Note: The problems of internal radiation hazards from radiostrontium are discussed by A. Engström *et al., Bone and Radiostrontium* (Wiley, New York, 1958).

8798. Strontium Acetate. C$_4$H$_6$O$_4$Sr; mol wt 205.71. C 23.36%, H 2.94%, O 31.11%, Sr 42.59%. Sr(C$_2$H$_3$O$_2$)$_2$. Toxicity: V. V. Cole *et al., J. Pharmacol. Exp. Ther.* **71,** 1 (1941).

Hemihydrate, white crystalline powder. Loses its water at 150°; on ignition is converted into SrCO$_3$. Sol in 2.5 parts water; slightly sol in alcohol. The aq soln is practically neutral to litmus. LD i.v. in rats: 1.16 mmol/kg (Cole).

8799. Strontium Bromate. Br$_2$O$_6$Sr; mol wt 343.44. Br 46.54%, O 27.95%, Sr 25.51%. Sr(BrO$_3$)$_2$.

Monohydrate, white to slightly yellow, hygroscopic, cryst powder. Sol in 3 parts water. *Keep well closed.*

8800. Strontium Bromide. Br$_2$Sr; mol wt 247.44. Br 64.59%, Sr 35.41%. SrBr$_2$. Acute toxicity: K. W. Cochran *et al., Arch. Ind. Hyg. Occup. Med.* **1,** 637 (1950).

Hexahydrate, colorless, deliquesc crystals or white granules; bitter saline taste. mp 88° when rapidly heated; mp 643° when anhydr. Loses all its water at 180°. Sol in 0.35 part water; sol in alcohol. Insol in ether. The aq soln is neutral. *Incompat:* Soluble sulfates. LD$_{50}$ i.p. in rats: 1000 mg/kg (Cochran).

THERAP CAT: Has been used as anticonvulsant.

THERAP CAT (VET): Has been used as sedative, anticonvulsant.

8801. Strontium Carbonate. CO$_3$Sr; mol wt 147.64. C 8.14%, Sr 59.35%, O 32.51%. SrCO$_3$. Occurs in nature as the mineral *strontianite.*

White, odorless, tasteless powder. d 3.5. Dec at 1100° into SrO and CO$_2$. Sol in 100,000 parts water, in about 1000 parts of water saturated with CO$_2$; sol in dil acids.

USE: In pyrotechnics; manuf iridescent glass; refining sugar.

8802. Strontium Chlorate. Cl$_2$O$_6$Sr; mol wt 254.54. Cl 27.86%, O 37.72%, Sr 34.43%. Sr(ClO$_3$)$_2$.

Colorless or white crystals. d 3.15. mp 120° with decomposition and evolution of O$_2$. Sol in 0.6 part water, slightly in alcohol. *Handle with caution like potassium chlorate.*

USE: In pyrotechnics to produce red fire.

8803. Strontium Chloride. Cl$_2$Sr; mol wt 158.52. Cl 44.73%, Sr 55.27%. SrCl$_2$. Acute toxicity: I. B. Syed, F. Hosain, *Toxicol. Appl. Pharmacol.* **22,** 150 (1972).

Hexahydrate, colorless, odorless crystals or white granules. Effloresces in air; deliquesces in moist air. At 100° loses 5H$_2$O; at 150° all its H$_2$O; d 1.96; mp 61° when rapidly heated; the anhydr salt melts at 868°. Soluble in 0.8 part water, 0.5 part boiling water; sol in alcohol. The aq soln is neutral. *Keep well closed.* LD$_{50}$ i.v. in mice: 147.6 mg/kg (Syed, Hosain).

USE: Manuf other strontium salts; in pyrotechnics; as dental desensitizer under the name *Elecol.*

8804. Strontium Chromate(VI). CrO$_4$Sr; mol wt 203.64. Sr 43.03%, Cr 25.54%, O 31.43%. SrCrO$_4$.

Yellow powder. d 3.89. Soluble in 840 parts cold water, about 5 parts boiling water; freely sol in dil hydrochloric, nitric or acetic acids.

USE: Corrosion inhibitor in pigments; in electrochemical processes to control sulfate concn of solns.

8805. Strontium Fluoride. F$_2$Sr; mol wt 125.63. F 30.25%, Sr 69.75%. SrF$_2$. Prepd by dissolving SrCO$_3$ in an excess of 40% HF and evaporating to dryness: Berzelius, *Pogg. Ann.* **1,** 20 (1824); final drying should be done at 150° *in vacuo.*

White powder or cubic crystals. d 4.24. mp ~1400°. bp 2460°. Stable in air up to 1000°. Above this temp it is oxidized to strontium oxide. Heat of evaporation 78.2. Trouton const 28.6. Soly in water (18°) 11.7mg/100 ml. More sol in dil acids. Dec by strong acids. May be stored in glass bottles.

8806. Strontium Formate. C$_2$H$_2$O$_4$Sr; mol wt 177.66. C 13.52%, H 1.13%, O 36.02%, Sr 49.32%. Sr(HCOO)$_2$.

Dihydrate, white crystals or granules. d 2.69. Sol in water.

8807. Strontium Hydroxide. Strontium hydrate. H$_2$O$_2$Sr; mol wt 121.64. H 1.66%, O 26.31%, Sr 72.03%. Sr(OH)$_2$.

Octahydrate, colorless, deliquesc crystals or white powder; absorbs CO$_2$ from air forming carbonate. Loses part of its water at about 100°. Sol in 50 parts water, 2.1 parts boiling water. The soln is very alkaline. pH about 13.5. *Keep tightly closed.*

USE: Refining beet sugar; separating crystallizable sugar from molasses.

8808. Strontium Iodide. I$_2$Sr; mol wt 341.42. I 74.34%, Sr 25.66%. SrI$_2$. The article of commerce is hydrated. Toxicity data: K. W. Cochran *et al., Arch. Ind. Hyg. Occup. Med.* **1,** 637 (1950).

Hexahydrate, colorless to yellowish, deliquesc, fused masses or granules with bitterish saline taste; becomes yellow on exposure to air and light due to liberation of iodine. Melts when rapidly heated, at about 120°. mp 402° (anhydrous). d 4.42. Sol in 0.2 part water; sol in alcohol. The aq soln is practically neutral. *Keep well closed and protected from light.* LD$_{50}$ i.p. in rats: 800 mg/kg (Cochran).

THERAP CAT: Iodine source.
THERAP CAT (VET): Has been used in iodide therapy.

8809. Strontium Lactate. *2-Hydroxypropanoic acid strontium salt.* $C_6H_{10}O_6Sr$; mol wt 265.76. C 27.11%, H 3.79%, O 36.12%, Sr 32.97%. $(CH_3CHOHCOO)_2Sr$. Acute toxicity: K. W. Cochran *et al.*, *Arch. Ind. Hyg. Occup. Med.* **1**, 637 (1950).
Trihydrate, *Strontolac.* White, odorless, granular powder. Loses its water of crystn at 120°. Sol in 3 parts water, 0.5 part boiling water; slightly sol in alcohol. The aq soln is practically neutral. LD_{50} i.p. in rats: 900 mg/kg (Cochran).

8810. Strontium Nitrate. N_2O_6Sr; mol wt 211.65. N 13.24%, O 45.36%, Sr 41.40%. $Sr(NO_3)_2$.
White granules or powder. d 2.99. mp 570°. Sol in 1.5 parts water; slightly sol in alcohol or acetone. The aq soln is neutral. LD_{50} i.p. in rats: 540 mg/kg, Cochran *et al.*, *Arch. Ind. Hyg.* **1**, 637 (1950).
Note: At low temps strontium nitrate crystallizes with $4H_2O$ (25.4%).
USE: In pyrotechnics (red fire), signal lights, marine signals, railroad flares, matches.

8811. Strontium Oxalate. *Ethanedioic acid strontium salt.* C_2O_4Sr; mol wt 175.64. C 13.67%, O 36.44%, Sr 49.89%. SrC_2O_4.
Monohydrate, white, odorless, cryst powder. Soluble in 20,000 parts water; in 1900 parts of 3.5% acetic acid, in 1115 parts of the 23% acid, but less sol in the 35% acid; readily sol in dil HCl or HNO_3.

8812. Strontium Oxide. Strontia; strontium monoxide. OSr; mol wt 103.63. O 15.44%, Sr 84.56%. SrO.
White to grayish-white, porous, caustic mass. d 4.7. mp 2430°. When treated with water, forms the hydroxide with evolution of much heat. *Keep tightly closed.*
USE: Manuf of strontium salts.

8813. Strontium Peroxide. O_2Sr; mol wt 119.63. O 26.75%, Sr 73.25%. SrO_2. The peroxide of commerce contains about 85% SrO_2.
White, odorless, tasteless powder; gradually dec on exposure to air. Almost insol in water, but is gradually dec by it with evolution of oxygen; forms hydrogen peroxide with dil acids. *Keep well closed.*
THERAP CAT: Antiseptic.

8814. Strontium Phosphate, Tribasic. $O_8P_2Sr_3$; mol wt 452.85. O 28.27%, P 13.68%, Sr 58.05%. $Sr_3(PO_4)_2$.
White, odorless powder. Insol in water; sol in acids.

8815. Strontium Selenate. O_4SeSr; mol wt 230.59. Sr 38.00%, Se 34.24%, O 27.75%. $SrSeO_4$. Prepared by roasting strontium carbonate with selenium or selenium oxide: Lenher, Wechter, *J. Am. Chem. Soc.* **47**, 1522 (1925).
Orthorhombic crystals. d 4.25. Sol in hot hydrochloric acid; insol in water.

8816. Strontium Sulfate. O_4SSr; mol wt 183.70. O 34.84%, S 17.46%, Sr 47.70%. $SrSO_4$. Occurs in nature as the mineral *celestine* or *celestite.*
White, odorless, crystalline powder. d 3.96. One gram dissolves in about 8800 ml water, 800 ml 2% HCl, 700 ml 3% HNO_3; appreciably sol in alkali chloride solns.
USE: In ceramics, pyrotechnics.

8817. Strontium Sulfide. SSr; mol wt 119.70. S 26.79%, Sr 73.21%. SrS. The article of commerce contains 65-75% SrS. Strontium sulfide made by reduction of the sulfate is not luminous; the luminous sulfide is prepared by heating strontium hydroxide with sulfur.
Gray powder; has the odor of H_2S in moist air. d 3.70. Slightly sol in water; sol in acids with decompn. *Keep dry.*
USE: In luminous paints; as a depilatory.

8818. Strophanthidin. *3,5,14-Trihydroxy-19-oxocard-20(22)-enolide;* apocynamarin; convallatoxigenin; corchorin; cymarigenin; cynotoxin; corchorgenin; corchsularin. $C_{23}H_{32}O_6$; mol wt 404.49. C 68.29%, H 7.97%, O 23.73%. By acid or enzymic hydrolysis of glycosides present in several species of *Strophanthus, Apocynaceae.* Isoln: Jacobs, Heidelberger, *J. Biol. Chem.* **54**, 253 (1922). Structure: Kon, *Chem. & Ind. (London)* **53**, 593, 956 (1934); Jacobs, Elder-

field, *Science* **80**, 533 (1934). Identity with corchorin, corchorgenin and corchsularin: Sen *et al.*, *Helv. Chim. Acta* **40**, 588 (1957). Synthesis from pregnenolone acetate, *q.v.*: E. Yoshii *et al.*, *J. Org. Chem.* **43**, 3946 (1978). *Reviews:* Elderfield, *Chem. Rev.* **17**, 187 (1935); Tschesche, *Ergeb. Physiol.* **38**, 31 (1936); Stoll, *The Cardiac Glycosides* (Pharmaceutical Press, London, 1937); Strain in *Organic Chemistry*, vol. II, Gilman, Ed. (New York, 2nd ed., 1943); Fieser, Fieser, *Steroids* (Reinhold, New York, 1959) pp 729-730, 736-750.

Orthorhombic tablets from 5 parts methanol and 10 parts water. *Very poisonous!* Contains ½ mol H_2O which is given up at 110° over P_2O_5 at 20 mm Hg. mp 171-175° with effervescence. Occasionally a few isolated crystals are found which melt at about 230°. $[\alpha]_D^{25}$ +43.1° (c = 2.8 in methanol). Sol in alcohol, acetone, chloroform, benzene, and glacial acetic acid; practically insol in water, ether, petr ether. Color reaction with H_2SO_4 and Ac_2O like cholesterol. LD_{100} i.v. in cats: 0.337 mg/kg, Graebner, Geisel, *Arzneimittel-Forsch.* **22**, 1854 (1972).
3-Benzoylstrophanthidin, $C_{30}H_{36}O_7$, mp 230°. $[\alpha]_D^{26}$ +47.8° (c = 1.07 in acetone).
3-(p-Bromobenzoyl)strophanthidin, $C_{30}H_{35}BrO_7$, mp 222-224°. $[\alpha]_D^{20}$ +42.0° (c = 1.094 in acetone).
Dihydrostrophanthidin, $C_{23}H_{34}O_6$, mp 100-103°, resolidifies and again melts at 190-195°. $[\alpha]_D^{23}$ +34.9° (methanol).
16-Hydroxystrophanthidin, $C_{23}H_{32}O_7$, *strophadogenin.* Long prisms from ethanol, dec 238-241°. $[\alpha]_D^{25}$ +52.9° (c = 1.24 in 96% alc). uv max: 219 nm (log ε 4.3).

8819. Strophanthin. K-Strophanthin; K-strophanthoside; Combetin; Eustrophinum; Kombetin. A glycoside or a mixture of glycosides obtained from *Strophanthus kombé* Oliv., *Apocynaceae.* Properly speaking K-strophanthoside contains α-glucose 19%, β-glucose 19%, cymarose 15%, strophanthidin 47%. K-Strophanthin-β (Strophosid) contains β-glucose, cymarose, and strophanthidin: Geissberger, *Schweiz. Med. Wochenschr.* **91**, 241 (1961). Pharmacology and acute toxicity: H. Mehnert, *Arch. Pharmacol.* **184**, 181 (1937).
White or yellowish powder contg as much as 10% water, which it does not lose entirely without decompn. *Very poisonous!* Stable in air, but affected by light. Soluble in water and in dil alcohol; less sol in abs alcohol; nearly insol in chloroform, ether, benzene. Aq solns are neutral to litmus.
Strophanthin, when bioassayed, shall possess a potency per mg equivalent to 0.5 mg of U.S.P. Ouabain Reference Standard. A deviation of 20% is permitted. MLD in cats, rats (mg/kg): 0.11, 9.4 i.v. (Mehnert).
Note: For information on G-strophanthin *see* Ouabain.
THERAP CAT: Cardiotonic.
THERAP CAT (VET): Has been used as a cardiotonic.

8820. Strophanthobiose. *2,6-Dideoxy-4-O-β-D-glucopyranosyl-3-O-methyl-D-ribo-hexose;* periplobiose. $C_{13}H_{24}O_9$; mol wt 324.32. C 48.14%, H 7.46%, O 44.40%. Obtained by hydrolysis of K-strophanthin-β, isolated from seeds of *Strophanthus kombé* Oliv., *Apocynaceae:* Jacobs, Hoffmann, *J. Biol. Chem.* **69**, 153 (1926); Stoll *et al.*, *Helv. Chim. Acta* **20**, 1484 (1937). Identity with periplobiose: Barbier, Schindler, *ibid.* **42**, 1065 (1959). Structure: Lichti, von Wartburg, *ibid.* **44**, 238 (1961).

R = OCH₃

Needles from methanol, mp 144-146°. $[\alpha]_D^{20}$ + 33.8° (c = 2.068 in water).

Pentaacetate, $C_{23}H_{34}O_{14}$, needles, mp 174-175°. $[\alpha]_D^{20}$ +11.9° (c = 1.05 in chloroform).

8821. Strophanthus. Dried ripe seeds of *Strophanthus kombé* Oliv., or of *S. hispidus* DC., *Apocynaceae,* deprived of the awns. *Habit.* East and Central Africa. *Constit.* 2-5% Strophanthin, kombic acid, choline, trigonelline, about 30% oil. *Poisonous!*

USE: As arrow poison by African natives.

THERAP CAT: Cardiotonic.

THERAP CAT (VET): Has been used as a cardiotonic.

8822. Strychnine. *Strychnidin-10-one.* $C_{21}H_{22}N_2O_2$; mol wt 334.40. C 75.42%, H 6.63%, N 8.38%, O 9.57%. From seeds of *Strychnos nux-vomica* L. and other spp. of *Strychnos, Loganiaceae:* Hartwick, Geiger, *Arch. Pharm.* **239**, 491 (1901). Extraction procedure: Volck, U.S. pat. **1,548,566** (1925); Watson, Sen, *J. Indian Chem. Soc.* **3**, 397 (1926); Schwyzer, *Die Fabrikation der Alkaloide* (Berlin, 1927). Structure: Prelog, Szpilfogel, *Helv. Chim. Acta* **28**, 1669 (1945); Robinson, *Experientia* **2**, 28 (1946); Woodward *et al., J. Am. Chem. Soc.* **69**, 2250 (1947). Total synthesis: Woodward *et al., ibid.* **76**, 4749 (1954); *Tetrahedron* **19**, 247 (1963). Review of structure work: Woodward, *Experientia Suppl.* **II** (14th Int. Congr. Pure & Appl. Chem.) 213-228 (1955). Stereochemistry: Peerdeman, *Acta Cryst* **9**, 824 (1956); Edward, *Tetrahedron* **2**, 356 (1958). Abs config: Nagarajan *et al., Helv. Chim. Acta* **46**, 1212 (1963). ¹³C-NMR study: E. Wenkert *et al., J. Org. Chem.* **43**, 1099 (1978). Acute toxicity: I. Setnikar, M. J. Magistretti, *Arzneimittel-Forsch.* **14**, 996 (1964). *Review:* J. B. Hendrickson in *The Alkaloids* vol. VI, R. H. F. Manske, Ed. (Academic Press, New York, 1960) pp 179-195; G. F. Smith, *ibid.* **vol. VIII**(1965) pp 591-671. Comprehensive description: F. J. Muhtadi, M. S. Hifnawy in *Analytical Profiles of Drug Substances* vol. 15, K. Florey, Ed. (Academic Press, New York, 1986) pp 563-646.

Very bitter orthorhombic, sphenoidal prisms from alcohol, mp 268-290° (depending on the speed of heating). bp_5 270°; d_4^{20} 1.36. $[\alpha]_D^{18}$ −139.3° (CH₃Cl); $[\alpha]_D^{20}$ −104° (c = 0.5 in abs alcohol). pK_1 at 20°: 6.0; pK_2: 11.7. pH of satd soln 9.5. uv max (ethanol): 254, 278, 288 nm (log ε 4.10, 3.63, 3.53). One gram dissolves in 6400 ml water, 3100 ml boiling water, 150 ml alcohol, 35 ml boiling alcohol, 5 ml chloroform, 180 ml benzene, about 200 ml toluene, 260 ml methanol, 320 ml glycerol, 220 ml amyl alcohol; very slightly sol in ether, petr ether. Bitterness threshold: 1:130,000. LD_{50} i.v. (slow infusion) in rats: 0.96 mg/kg (Setnikar, Magistretti).

Acetate, $C_{23}H_{26}N_2O_4$, crystals, sol in 20 parts water, sol in alcohol.

Arsenate dihydrate, $C_{21}H_{25}AsN_2O_6.2H_2O$, efflorescent crystals, sol in about 20 parts water, slightly sol in alcohol. The aq soln is acid.

Dichromate, $C_{42}H_{46}Cr_2N_4O_{11}$, orange-yellow needles from water, one gram dissolves in 1800 ml water at 18°.

Formate, $C_{22}H_{24}N_2O_4$, crystals, sol in 2.5 parts water, sol in alcohol.

Gluconate pentahydrate, $C_{27}H_{34}N_2O_9.5H_2O$, crystals, darken above 80°, sol in 2 parts water, in about 40 parts alcohol; the aq soln is neutral.

Glycerophosphate hexahydrate, $C_{45}H_{53}N_4O_{10}P.6H_2O$, one gram dissolves in about 350 ml water, about 310 ml alcohol. Slightly sol in chloroform; very slightly in ether.

Hydrobromide monohydrate, $C_{21}H_{23}BrN_2O_2.H_2O$, efflorescent, crystals, sol in about 50 parts water, slightly sol in alcohol. pH of 0.01M soln 5.5.

Hydrochloride dihydrate, $C_{21}H_{23}ClN_2O_2.2H_2O$, efflorescent, trimetric prisms. One gram dissolves in about 40 ml water, about 80 ml alc. Insol in ether. pH of 0.01M soln 5.4.

Nitrate, see separate entry.

Phosphate, see separate entry.

Salicylate, $C_{28}H_{28}N_2O_5$, leaflets, sol in about 30 parts water; sparingly sol in alcohol.

Sulfate, see separate entry.

Caution: Extremely poisonous, see: G. D. Osweiler, *Curr. Vet. Ther.* **6**, 115 (1977).

USE: Chiefly in poison baits for rodents.

THERAP CAT (VET): Has been used as a tonic and central stimulant.

8823. Strychnine Nitrate. $C_{21}H_{23}N_3O_5$; mol wt 397.42. C 63.47%, H 5.83%, N 10.57%, O 20.13%.

Colorless, odorless needles or white, cryst powder. *Violent poison!* One gram dissolves in 42 ml water, 10 ml boiling water, 150 ml alcohol, 80 ml alcohol at 60°, 105 ml chloroform, 50 ml glycerol; insol in ether. pH about 5.7. *Protect from light.*

THERAP CAT (VET): Has been used as tonic, stimulant.

8824. Strychnine N^6-Oxide. Genostrychnine; Movellan (tabl.). $C_{21}H_{22}N_2O_3$; mol wt 350.40. C 71.98%, H 6.33%, N 8.00%, O 13.70%. Prepn: Pictet, Mattison, *Ber.* **38**, 2782 (1905); Bailey, Robinson, *J. Chem. Soc.* **1948**, 703; Zellner, U.S. pat. **2,758,113** (1956 to Donau-Pharmazie).

Monoclinic prisms from water, dec 207°. pK 5.17. Freely sol in alc, glacial acetic acid, chloroform; fairly sol in water; sparingly sol in benzene; practically insol in ether, petr ether.

Hydriodide, light yellow plates from hot water, dec 253°. Nitrate, small transparent prisms from water, dec 250°.

Note: Movellan (amps.) is strychnic acid sodium salt.

8825. Strychnine Phosphate. $C_{21}H_{25}N_2O_6P$; mol wt 432.41. C 58.32%, H 5.83%, N 6.48%, O 22.20%, P 7.16%. Usually as $C_{21}H_{22}N_2O_2.H_3PO_4.2H_2O$.

Dihydrate, colorless or white crystals or white powder. *Violent poison!* One gram dissolves slowly in about 30 ml water; more sol in hot water; slightly sol in alcohol. The aq soln is acid.

THERAP CAT (VET): Has been used as tonic, stimulant.

8826. Strychnine Sulfate. $C_{42}H_{46}N_4O_8S$; mol wt 766.92. C 65.78%, H 6.05%, N 7.31%, O 16.69%, S 4.18%. Usually crystallizes as $2C_{21}H_{22}N_2O_2.H_2SO_4.5H_2O$.

Pentahydrate, colorless, odorless, very bitter crystals, or white, cryst powder. *Violent poison!* Effloresces in dry air; loses all its water of crystn at 100°. mp when anhydr about 200° with decompn. One gram dissolves in 35 ml water, 7 ml boiling water, 81 ml alcohol, 26 ml alcohol at 60°, 220 ml chloroform, 6 ml glycerol; insol in ether. pH 5.5 in 1:100 soln. *Protect from light.* LD_{50} orally in rats: 5 mg/kg, E. W. Schafer, *Toxicol. Appl. Pharmacol.* **21**, 315 (1972).

Strychnine sulfate also crystallizes with $6H_2O$ (12.36% water). A mixture of the two hydrates contg 11.5% H_2O is also marketed.

Incompatibility of all strychnine salts: alkalies, alkali carbonates and bicarbonates, benzoates, dichromates, bromides, iodides, tannic and picric acids, salicylates, borax, gold chloride and other alkaloid precipitants, piperazine, potassium-mercuric iodide (not if acacia is present).

THERAP CAT (VET): Has been used as tonic, stimulant.

8827. Stylopine. *6,7,12b,13-Tetrahydro-4H-bis[1,3]benzodioxolo[5,6-a:4',5'-g]quinolizine; 2,3:9,10-bis(methylenedioxy)-13a-berbine;* tetrahydrocoptisine. $C_{19}H_{17}NO_4$; mol wt 323.33. C 70.57%, H 5.30%, N 4.33%, O 19.79%. From root of *Stylophorum diphyllum* (Michx.) Nutt, *Corydalis cava* Schwg. and *Chelidonium majus* L., *Papaveraceae*: Schlotterbeck, Watkins, *Ber.* **35**, 7 (1902); Späth, Julian, *Ber.* **64**, 1131 (1931); Manske, *Can. J. Res.* **B20**, 53 (1942); Slavik, *Coll. Czech. Chem. Commun.* **20**, 198 (1955); Bandelin, Malesh, *J. Am. Pharm. Assoc.* **45**, 702 (1956); Slavik, *Coll. Czech. Chem. Commun.* **26**, 2933 (1961).

dl-Form, crystals from chloroform + methanol, mp 222-224°.

Hydrochloride, $C_{19}H_{18}ClNO_4$, needles from methanol + HCl, mp 266-269°.

l-Form, needles from chloroform + ethanol, mp 203°. $[\alpha]_D^{25}$ −300° (c = 0.53 in chloroform). Practically insol in water; sol in alc, glacial acetic acid; slightly sol in dil acids.

d-Form, crystals from alcohol, mp 203-204°. $[\alpha]_D^{25}$ +310° (c = 0.88 in chloroform). Sol in chloroform; slightly sol in ether, ethanol, methanol.

8828. Styphnic Acid. *2,4,6-Trinitro-1,3-benzenediol; 2,4,6-trinitroresorcinol;* 2,4-dihydroxy-1,3,5-trinitrobenzene. $C_6H_3N_3O_8$; mol wt 245.11. C 29.40%, H 1.23%, N 17.14%, O 52.22%. May be prepd from natural sources, such as Pernambuco wood extract or quebracho extract by the action of nitric acid: Einbeck, Jablonski, *Ber.* **54**, 1086 (1921). Prepd industrially by first sulfonating, then nitrating resorcinol: Merz, Zetter, *Ber.* **12**, 681, 2037 (1879); Datta, Varma, *J. Am. Chem. Soc.* **41**, 2043 (1919). Prepd by nitrating 2,4-dinitroresorcinol: Kametani, Ogasawara, *Chem. Pharm. Bull.* **15**, 893 (1967).

Hexagonal yellow crystals from dil alc. Astringent, but not bitter taste. Solvated crystals from acetic acid. Sublimes in high vacuum. Becomes almost colorless upon vacuum sublimation, but turns deep yellow on contact with air. When dry, mp 175.5°; also reported as mp 179-180°. One gram dissolves in 156 ml water at 14°, in 88 ml water at 62°. Freely sol in alcohol, ether. Acid to litmus. Dibasic acid. Deflagrates on rapid heating.

USE: Forms salts and addn compds with organic compds. Used for the identification of organic compds through the formation of styphnates which, like picrates, have characteristic melting pts. Has been used in the manuf of explosives.

8829. Styramate. *1-Phenyl-1,2-ethanediol 2-carbamate;* carbamic acid β-hydroxyphenethyl ester; 2-phenyl-2-hydroxyethyl carbamate; 2-hydroxy-2-phenylethyl carbamate; Sinaxar; Linaxar. $C_9H_{11}NO_3$; mol wt 181.19. C 59.66%, H 6.12%, N 7.73%, O 26.49%. Prepn: **Brit.** pat. 841,626 corresp to C. D. Bossinger, K. G. Taylor, **U.S.** pat. 3,265,728 (1960, 1966 to Armour). Pharmacology: De Salva et al., *J. Pharmacol. Exp. Ther.* **126**, 318 (1959).

Crystals from chloroform, mp 111-112°. LD_{50} orally in mice: 1240 mg/kg, De Salva *et al.*, *loc. cit.*

Note: Ingredient of *Myospaz.*

THERAP CAT: Relaxant (skeletal muscle).

8830. Styrene. *Ethenylbenzene;* styrol; styrolene; cinnamene; cinnamol; phenylethylene; vinylbenzene. C_8H_8; mol wt 104.14. C 92.26%, H 7.74%. $C_6H_5CH=CH_2$. Isolated from storax by Bonastre in 1831. Manuf from benzene and ethylene: Faith, Keyes & Clark's *Industrial Chemicals*, F. A. Lowenheim, M. K. Moran, Eds. (Wiley-Interscience, New York, 4th ed., 1975) pp 779-785. Synthesis starting with 1-phenylethanol and leading to polystyrene: Wilen *et al.*, *J. Chem. Ed.* **38**, 304 (1961). When heated to 200° it is converted into the polymer, polystyrene, which is a clear plastic having excellent insulating properties even at ultra-high radio frequencies. Toxicity data: H. J. Meyer, R. Kretzschmar, *Arzneimittel-Forsch.* **19**, 617 (1969). Monograph: W. C. Teach, G. C. Kiessling, *Polystyrene* (Reinhold, New York, 1960). Reviews of styrene monomer and polymers: Boyer *et al.*, "Styrene Polymers" in *Encyclopedia of Polymer Science and Technology* **vol. 13** (Interscience, New York, 1970) pp 128-447; Coulter *et al.*, "Styrene and Related Monomers" in *Vinyl and Diene Monomers* (part 2), E. C. Leonard, Ed. (Wiley-Interscience, New York, 1971) pp 479-576.

Colorless to yellowish, very refractive, oily liq; penetrating odor. On exposure to light and air it slowly undergoes polymerization and oxidation with formation of peroxides, etc. d^{20} 0.9059. mp −30.6°. bp 145-146°. n_D^{20} 1.5463. Flash pt, closed cup: 87° F (31° C). Sparingly sol in water; sol in alcohol, ether, methanol, acetone, carbon disulfide. LD_{50} in mice (mg/kg): 660 ± 44.3 i.p.; 90 ± 5.2 i.v. (Meyer, Kretzschmar).

Polystyrene, Dylene, Trycite. Physical properties of unmodified polystyrene: d_4^{20} 1.04-1.065; n_D^{25} 1.60; water-clear solid plastic, begins to soften at about 85°. Dielectric constant at 100 megacycles: 2.4-2.65.

USE: Manuf plastics; synthetic rubber; resins; insulator. *Caution:* May be irritating to eyes, mucous membranes, and, in high concns, narcotic.

8831. Styrene Glycol. *1-Phenyl-1,2-ethanediol;* phenylglycol; phenylethyleneglycol; α,β-dihydroxyethylbenzene. $C_8H_{10}O_2$; mol wt 138.16. C 69.54%, H 7.30%, O 23.16%. Prepd by reduction of phenylglyoxylic acid with $LiAlH_4$: Nystrom, Brown, *J. Am. Chem. Soc.* **69**, 2548 (1947); by reduction of styrene peroxide: Russel, *J. Am. Chem. Soc.* **75**, 5011 (1953). Prepn of optically active styrene glycol: Wilhelm, Bright, *Helv. Chim. Acta* **37**, 221 (1954).

Needles from ligroin, mp 67-68°. bp_{755} 272-274°. Freely sol in water, alcohol, benzene, ether, chloroform, acetic acid; slightly sol in ligroin.

USE: Esters as plasticizers.

8832. Subathizone. *2-[[4-(Ethylsulfonyl)phenyl]methylene]hydrazinecarbothioamide; p-ethylsulfonylbenzaldehyde thiosemicarbazone;* 4-ethylsulfonylbenzaldehyde 3-thiosemicarbazone; Ethizone. $C_{10}H_{13}N_3O_2S_2$; mol wt 271.35. C 44.26%, H 4.83%, N 15.49%, O 11.79%, S 23.63%. Prepn from 4-ethylsulfonyltoluene: Bernstein *et al.*, *J. Am. Chem. Soc.* **73**, 906 (1951); other prepns are described in **Japan.** pats. 1119('52), 1768('52), 3977('52), 1872('53), 6032('54), 6427('54); in U.S. pat. 2,621,484 (1952 to Schenley).

Crystals from aq *n*-propanol, mp 234-235° (dec).

THERAP CAT: Antibacterial (tuberculostatic).

8833. Suberic Acid. *Octanedioic acid;* 1,6-hexanedicarboxylic acid. $C_8H_{14}O_4$; mol wt 174.19. C 55.16%, H 8.10%, O

36.74%. COOH(CH$_2$)$_6$COOH. Made by heating castor oil or ricinoleic acid with nitric acid: Dominguez, Lopez, *Ciencia* **20**, 73 (1959), *C.A.* **54**, 18990f (1960). Prepn from diethyl malonate + ω-bromocapronitrile: Cason *et al.*, *J. Org. Chem.* **14**, 37 (1949); by oxidation of oleic acid with nitric acid: Cavanaugh, Weir, U.S. pat. **2,560,156** (1951 to du Pont).

Crystals, mp 140-144°. bp$_{100}$ 279°. Sublimes at 300° without decompn. One gram dissolves in 625 ml water, 172 ml ether; sol in alcohol; almost insol in chloroform.

USE: In the plastics industry.

8834. Substance P. SP. C$_{63}$H$_{98}$N$_{18}$O$_{13}$S; mol wt 1347.66. C 56.15%, H 7.33%, N 18.71%, O 15.43%, S 2.38%. An undecapeptide belonging to a group of proteins named tachykinins characterized by contractile action on extravascular smooth muscle. Discovered by von Euler and Gaddum, *J. Physiol.* **72**, 74 (1931). Present in the brain of all vertebrates including man, in spinal ganglia, and in the intestines, esp the duodenum and jejunum. Isoln from cattle brain: Zuber, Jaques, *Angew. Chem. Int. Ed.* **1**, 160 (1962); from horse intestine: Studer *et al.*, *Helv. Chim. Acta* **56**, 860 (1973). Purification from bovine hypothalami and amino acid composition: N. N. Chang, S. E. Leeman, *J. Biol. Chem.* **245**, 4784 (1970). Amino acid sequence and solid-phase synthesis: Chang *et al.*, *Nature New Biol.* **232**, 86 (1971); Tregear *et al.*, *ibid.* 87; G. H. Fisher *et al.*, *J. Med. Chem.* **17**, 843 (1974). Additional syntheses: Yajima, Kitigawa, *Chem. Pharm. Bull.* **21**, 682 (1973); Bayer, Mutter, *Ber.* **107**, 1344 (1974); K. Neubert *et al.*, *Pharmazie* **36**, 10 (1981); A. Fournier *et al.*, *J. Med. Chem.* **25**, 64 (1981). Substance P acts as a vasodilator, a depressant, stimulates salivation and produces increased capillary permeability. It is also capable of producing both analgesia and hyperalgesia in animals, depending on dose and pain responsiveness of the animal, see R. C. A. Frederickson *et al.*, *Science* **199**, 1359 (1978); P. Oehme *et al.*, *ibid.* **208**, 305 (1980); role in sensory transmission and pain perception: T. M. Jessell, *Advan. Biochem. Psychopharmacol.* **28**, 189 (1981). Mechanism of action studies: Stern *et al.*, *Arch. Pharmacol.* **281**, 233 (1974); B. G. Livett *et al.*, *Nature* **278**, 256 (1979). Reviews: Haefeli, Huelimann, *Experientia* **18**, 297 (1962); Lembeck, Zetler, in *International Encyclopedia of Pharmacology and Therapeutics*, sect. **72**, vol. **1**, J. M. Walker, Ed. (Pergamon Press, New York, 1971) pp 29-71; J. L. Barker, *Physiol. Rev.* **56**, 435 (1976); D. R. Brown, R. J. Miller, *Ann. Rep. Med. Chem.* **17**, 271-280 (1982). Books: U.S. Von Euler, B. Pernow, Eds., *Substance P* (Raven Press, New York, 1977) 360 pp; *Substance P vol.* **2**, P. Skrabanek, D. Powell, Eds. (Eden Press, Quebec, 1980) 175 pp; *Substance P in The Nervous System: Ciba Foundation Symposium* **91**, R. Porter, M. O'Connor, Eds. (Pitman, London, 1982) 350 pp.

Arg-Pro-Lys-Pro-Gln-Gln-Phe-Phe-Gly-Leu-Met-NH$_2$

Behaves during isoln as a basic polypeptide (R$_f$ value = 0.5 in the thin layer chromatogram: silica gel—*n*-butanol/-pyridine/AcOH/H$_2$O = 30:20:6:24). Upon high voltage electrophoresis (pH 1.9; AcOH/formic acid/H$_2$O = 15:3:100) it migrates approx as far as glutamic acid or serine, and at pH 9.5 (0.25*M* triethylammonium carbonate buffer) it behaves in a manner similar to arginine. Biological activity is destroyed by pepsin and chymotrypsin. Aqueous solns of highly purified material lose biological activity in a few minutes. This loss is prevented by storage at low pH, under nitrogen, or with various antioxidants. Addition of Tween 80, gelatin, human plasma albumin, or bovine γ-globulin increases stability of aq SP solns. Crude SP solns are stable though rapidly destroyed above pH 8.

8835. Subtilin. Polypeptide antibiotic structurally similar to nisin, *q.v.*, but contains no methionine. Produced by *Bacillus subtilis* (NRRL no. B-543): E. F. Jansen, D. J. Hirshmann, *Arch. Biochem.* **4**, 297 (1944); Lewis *et al.*, *ibid.* **14**, 415 (1947); Dimick *et al.*, *ibid.* **15**, 1 (1947). Isoln by partition chromatography on silica gel: Alderton, Snell, *J. Am. Chem. Soc.* **81**, 701 (1959). Early structural studies: Stracher, Craig, *ibid.* **696**. Structure is a heterodetic pentacyclic peptide consisting of 32 amino acid residues and containing unusual amino acids such as lanthionine, β-methyl-

lanthionine, D-alanine, dehydroalanine and dehydrobutyrine: E. Gross *et al.*, *Z. Physiol. Chem.* **354**, 810 (1973). Biosynthetic studies: C. Nishio *et al.*, *Biochem. Biophys. Res. Commun.* **116**, 751 (1983); S. Banerjee, J. N. Hansen, *J. Biol. Chem.* **263**, 9508 (1988).

Amorphous powder. [α]$_D^{23}$ −29° to −35° (c = 1 in 1% acetic acid). Diffuses quickly through cellophane. Readily sol in dilute acids; sparingly sol (less than 5 mg/ml) in water at pH 6-9. Sol in methanol, 80% alc. Soly in butanol satd with water: about 5 mg/ml. Easily salted out of soln by NaCl.

USE: Food preservative.

8836. Succimer. (*R**,*S**)-*2,3-Dimercaptobutanedioic acid; meso-2,3-dimercaptosuccinic acid; DMS; DMSA; Ro 1-7977*. C$_4$H$_6$O$_4$S$_2$; mol wt 182.21. C 26.37%, H 3.32%, O 35.12%, S 35.19%. HOOCCH(SH)CH(SH)COOH. A water soluble chelating agent. Prepn: L. N. Owen, M. U. S. Sultanbawa, *J. Chem. Soc.* **1949**, 3109. Pharmacology, toxicity and activity in heavy metal poisoning in animals: E. Friedheim, C. Corvi, *J. Pharm. Pharmacol.* **27**, 625 (1975); J. H. Graziano *et al.*, *J. Pharmacol. Exp. Ther.* **207**, 1051 (1978). GC determn in urine: J. J. Knudsen, E. L. McGown, *J. Chromatog.* **424**, 231 (1988). Clinical evaluations in heavy metal poisoning: E. Friedheim *et al.*, *Lancet* **2**, 1234 (1978); L. Fournier *et al.*, *Med. Toxicol.* **3**, 499 (1988). Diagnostic use of complex with technetium in urinary tract imaging: I. G. Verber *et al.*, *Arch. Dis. Child.* **63**, 1320 (1988); in tumor imaging: H. Ohta *et al.*, *Nucl. Med. Commun.* **9**, 105 (1988). Review: H. V. Aposhian, "Biological Chelation: 2,3-Dimercaptopropanesulfonic Acid and Meso-Dimercaptosuccinic Acid" in *Advan. Enzyme Reg.* **20**, G. Weber, Ed. (Pergamon Press, Oxford, 1982) pp 301-319. Review of pharmacology and clinical use in heavy metal poisoning: *idem*, *Ann. Rev. Pharmacol. Toxicol.* **23**, 193-215 (1983); J. H. Graziano, *Med. Toxicol.* **1**, 155-162 (1986).

White crystals from aqueous methanol, mp 192-194°. LD$_{50}$ i.p. in mice: > 3000 mg/kg (Friedheim, Corvi).

USE: Chelating agent.

THERAP CAT: Antidote for heavy metal poisoning. ^{99m}Tc complex as diagnostic aid (radioactive imaging agent).

8837. Succinamide. *Butanediamide*. C$_4$H$_8$N$_2$O$_2$; mol wt 116.12. C 41.37%, H 6.94%, N 24.13%, O 27.56%. H$_2$NOC-CH$_2$CH$_2$CONH$_2$. Made by the action of ammonia water on the dimethyl or diethyl ester of succinic acid.

Needles. mp 260° with decompn; also stated as 242°. Sol in 220 parts cold, 9 parts boiling water; insol in alc, ether.

8838. Succinanil. *1-Phenyl-2,5-pyrrolidinedione; N-phenylsuccinimide; N-phenylbutanimide*. C$_{10}$H$_9$NO$_2$; mol wt 175.18. C 68.56%, H 5.18%, N 8.00%, O 18.27%. Prepd by treating succinanilic acid with acetyl chloride: Ruggli, *Ann.* **412**, 4 (1916); Warren, Briggs, *Ber.* **64**, 29 (1931); L. F. Fieser, *Experiments in Organic Chemistry* (Boston, 1955) p 106.

Monoclinic prisms from water or alcohol, mp 154.5°. d$_4^{20}$ 1.356. bp$_{760}$ about 400°. Soluble in ether, boiling alcohol. Slightly sol in boiling water.

8839. Succinanilic Acid. *4-Oxo-4-(phenylamino)butanoic acid; succinic acid monoanilide; N-phenylsuccinamic acid; N-phenylbutanedioic acid monoamide*. C$_{10}$H$_{11}$NO$_3$; mol wt 193.20. C 62.16%, H 5.74%, N 7.25%, O 24.84%. Prepd by boiling succinic anhydride with aniline in benzene: Menschutkin, *Ann.* **162**, 176 (1872); Warren, Briggs, *Ber.* **64**, 29 (1931); L. F. Fieser, *Experiments in Organic Chemistry* (Boston, 3rd ed., 1955) p 105.

Needles from benzene, mp 150°. K at 25° = 2.03 × 10^{-5}. Soluble in alcohol, ether, boiling water (some decompn).

8840. Succinic Acid. *Butanedioic acid;* amber acid; ethylenesuccinic acid; Bernsteinsäure (German); Asuccin. $C_4H_6O_4$; mol wt 118.09. C 40.68%, H 5.12%, O 54.20%. $HOOCCH_2CH_2COOH$. Observed by Agricola in 1546, in the distillate from amber. Occurs in fossils, fungi, lichens, etc. Prepn from acetic acid: Coffman *et al., J. Am. Chem. Soc.* **80,** 2864 (1958). Manuf: Crosby, Braunwarth, U.S. pat. **2,862,028** (1958 to Pure Oil); Chafetz, Patterson, U.S. pat. **2,978,473** (1961 to Texaco).

Odorless, monoclinic prisms; very acid taste; d 1.56; mp 185-187°; bp 235° with partial conversion into the anhydride. One gram dissolves in 13 ml cold water, 1 ml boiling water, 18.5 ml alcohol, 6.3 ml methanol, 36 ml acetone, 20 ml glycerol, 113 ml ether; practically insol in benzene, carbon disulfide, carbon tetrachloride, petr ether. pH of 0.1 molar aq soln 2.7.

Potassium salt trihydrate, $C_4H_4K_2O_4.3H_2O$, *potassium succinate*. Hygroscopic cryst powder. d 1.564. Freely sol in water; the aq soln is practically neutral. *Keep tightly closed.*

Diethyl ester, $C_8H_{14}O_4$, *ethyl succinate*. Colorless, clear liq; faint, pleasant odor. d_4^{20} 1.040. bp 217-218°. mp —21°. n_D^{20} 1.4201. Insol in water; miscible with alcohol, ether. LD_{50} orally in rats: 8.53 g/kg, H. F. Smyth *et al., Arch. Ind. Hyg. Occup. Med.* **4,** 119 (1951).

Dimethyl ester, $C_6H_{10}O_4$, *methyl succinate*. Colorless liquid. d_4^{18} 1.1202. mp 19.5°. bp_{760} 195.3°; bp_{25} 103.5°. Sol in 120 parts water, 35 parts alcohol.

USE: Manuf lacquers, dyes, esters for perfumes, succinates; in photography.

8841. Succinic Anhydride. *Dihydro-2,5-furandione;* succinic acid anhydride; succinyl oxide; 2,5-diketotetrahydrofuran; butanedioic anhydride; Bernsteinsäureanhydrid (German). $C_4H_4O_3$; mol wt 100.07. C 48.01%, H 4.03%, O 47.96%. Usually prepd by warming succinic acid with acetic anhydride, acetyl chloride, or phosphorus oxychloride: Shriner, Struck, *Org. Syn.* **12,** 66 (1932); L. F. Fieser, *Experiments in Organic Chemistry* (Boston, 3rd ed., 1955) p 105.

Orthorhombic prisms from abs alc. d 1.503 (also given as 1.104 and 1.234). Rate of sublimation: Kempf, *J. Prakt. Chem.* [2] **78,** 257 (1908). mp 119.6°; bp_{760} 261°; bp_{400} 237°; bp_{200} 212°; bp_{100} 189°; bp_{60} 174°; bp_{40} 163°; bp_{20} 145°; bp_{10} 128°. Sublimes at 115° and 5 mm pressure, at 92° and 1.0 mm pressure. Soluble in chloroform, carbon tetrachloride, alcohol; very slightly sol in ether and water.

8842. Succinimide. *2,5-Pyrrolidinedione;* butanimide; 2,5-diketopyrrolidine; 3,4-dihydropyrrole-2,5-dione; dihydro-3-pyrroline-2,5-dione; 2,5-dioxopyrrolidine; Succinimide-Sauba. $C_4H_5NO_2$; mol wt 99.09. C 48.48%, H 5.09%, N 14.14%, O 32.29%. Prepd by rapid distillation of ammonium succinate: Clarke, Behr, *Org. Syn.* **coll. vol. II** (1943) p 562; Fehling, *Ann.* **49,** 198 (1844); Bunge, *Ann. Suppl.* **7,** 118 (1870); Menschutkin, *Ann.* **162,** 166 (1872). Prepn by heating succinic acid with ammonia: Fehling, *loc. cit.;* Franchimont, Friedmann, *Rec. Trav. Chim.* **25,** 79 (1906); with urea: Ma, Sah, *C.A.* **28,** 6108 (1934); *see also* Tafel, Stern, *Ber.* **33,** 2232 (1900); Koller, *Ber.* **37,** 1598 (1904). Prevention of oxalic lithiasis in rats: J. M. Melon *et al., Therapie* **26,** 991 (1971).

Orthorhombic bipyramidal crystals from acetone or alc. mp 125-127°. d 1.41. bp 287-289° (slight decomp). pKa 9.5. One gram dissolves in 3 ml water, 0.7 ml boiling water, 24 ml alc, 5 ml alc at 60°. Insol in ether, chloroform. LD_{50} orally in rats: 14 g/kg (Melon).

THERAP CAT: Antiurolithic.

8843. Succinonitrile. *Butanedinitrile;* ethylene dicyanide; succinic acid dinitrile; *sym*-dicyanoethane; ethylene

cyanide; Dinile; Deprelin; Suxil. $C_4H_4N_2$; mol wt 80.09. C 59.98%, H 5.03%, N 34.98%. $NCCH_2CH_2CN$. Prepd from ethylene bromide and KCN in alcohol: Fauconnier, *Bull. Soc. Chim. France* **50,** 214 (1888).

Waxy isometric crystals. d_4^{45} 1.023. mp 57.15°. d_4^{60} (liq) 0.9868. n_D^{60} (liq) 1.41734. May be heated to 200° for 72 hrs without decompn. bp_{760} 265-267°; bp_{60} 185°; bp_{20} 158-160°. Sol in acetone, chloroform, dioxane. Slightly sol in water, ethanol, benzene, ether, carbon disulfide.

8844. Succinyl Chloride. *Butanedioyl chloride;* succinyl dichloride. $C_4H_4Cl_2O_2$; mol wt 154.99. C 31.00%, H 2.60%, Cl 45.76%, O 20.65%. $ClOCCH_2CH_2COCl$.

Fuming, highly refractive liq; d_4^{15} 1.395; n_D^{15} 1.473; mp 17°; also stated as 20°; bp 192-193°. Solidif at 0° to cryst leaflets.

8845. Succinylcholine Bromide. *2,2'-[(1,4-Dioxo-1,4-butanediyl)bis(oxy)]bis[N,N,N-trimethylethanaminium] dibromide;* bis[2-dimethylaminoethyl]succinate bis[methobromide]; 3,8-dioxadecane-4,7-dione-1,10-bis[trimethylammonium bromide]; suxamethonium bromide; IS 370; compd 48/268; LT 1; M & B 2207; Brevidil M. $C_{14}H_{30}Br_2N_2O_4$; mol wt 450.23. C 37.35%, H 6.72%, Br 35.50%, N 6.22%, O 14.21%. Prepd by reacting β-bromoethyl succinate with trimethylamine: Glick, *J. Biol. Chem.* **137,** 357 (1941); Fusco *et al., Gazz. Chim. Ital.* **79,** 129, 837 (1949); Walker, *J. Chem. Soc.* **1950,** 193. Alternate prepns: Phillips, *J. Am. Chem. Soc.* **71,** 3264 (1949); Tammelin, *Acta Chem. Scand.* **7,** 185 (1953).

Slightly hygroscopic crystals, mp 225°. Freely sol in water or normal saline, giving solns which are very slightly acidic. Aq solns undergo progressive hydrolysis with corresponding loss of activity and increase in acidity. *Prepare solns just before using.* The rate of this decompn increases with temp and autoclaving or prolonged exposure to warmth should be avoided.

THERAP CAT: Skeletal muscle relaxant.

8846. Succinylcholine Chloride. *2,2'-[(1,4-Dioxo-1,4-butanediyl)bis(oxy)]bis[N,N,N-trimethylethanaminium] dichloride;* bis[2-dimethylaminoethyl]succinate bis[methochloride]; 2-dimethylaminoethyl succinate dimethochloride; diacetylcholine dichloride; suxamethonium chloride; choline succinate dichloride; succinic acid bis[β-dimethylaminoethyl] ester dimethochloride; choline chloride succinate (2:1); listenon; Anectine chloride; Scoline; Lysthenon; Midarine; Quelicin chloride; Succinyl-Asta; Sucostrin chloride; Ultrapal chloride; Succicuran. $C_{14}H_{30}Cl_2N_2O_4$; mol wt 361.30. C 46.54%, H 8.37%, Cl 19.63%, N 7.75%, O 17.71%. Prepn: R. Fusco *et al., Gazz. Chim. Ital.* **79,** 129 (1949); Tammelin, *Acta Chem. Scand.* **7,** 185 (1953); O. Schmid, **Austrian** pat. **171,411** (1952 to OSSW), *C.A.* **47,** 4902f (1953); C. H. Wang *et al., Org. Prep. Proc. Int.* **11,** 93 (1979). Crystal structure: B. Jensen, *Acta Chem. Scand. B* **30,** 1002 (1976). Toxicity data: P. Anttila, P. Ertama, *Med. Biol.* **56,** 152 (1978). Comprehensive description: P. R. B. Foss, S. A. Benezra, in *Analytical Profiles of Drug Substances* vol. 10, K. Florey, Ed. (Academic Press, New York, 1981) pp 691-704.

Dihydrate, crystals, mp 156-163°. Anhydr form mp ~190°. Slightly bitter taste. Freely sol in water (about 1 g/1 ml water). Soly in 95% ethanol: 0.42 g/100 ml. Sparingly sol in benzene, chloroform. Practically insol in ether. The pH of a 2-5% aq soln may vary from 4.5 to 3.0. Solns without stabilizers (e.g., 0.1% methyl *p*-hydroxybenzoate) must be kept refrigerated, they are incompatible with alkaline agents, such as thiopental sodium. LD_{50} i.v. in mice: 0.45 mg/kg (Anttila, Ertama).

THERAP CAT: Skeletal muscle relaxant.

THERAP CAT (VET): Skeletal muscle relaxant (short duration).

8847. Succinylcholine Iodide.

O,O-Succinyldicholine iodide; diacetylcholine iodide; diacetylcholine diiodide; suxamethonium iodide; bis(β-dimethylaminoethyl)succinate bis-(methyl iodide); Celocurine; Ascuron; Curacit; Kuratsit. $C_{14}H_{30}I_2N_2O_4$; mol wt 544.22. C 30.89%, H 5.56%, I 46.64%, N 5.15%, O 11.76%. Prepd by reacting β-bromoethyl succinate with trimethylamine: Glick, *J. Biol. Chem.* **137**, 357 (1941); Fusco *et al., Gazz. Chim. Ital.* **79**, 129, 837 (1949); Walker, *J. Chem. Soc.* **1950**, 193. Succinic acid chloride may be coupled with choline chloride directly or with dimethylaminoethanol, followed by quaternization with methyl iodide: Tammelin, *Acta Chem. Scand.* **7**, 185 (1953). A third method is to start with the diethyl ester of succinic acid, then bring about the exchange reaction with dimethylaminoethanol, and quaternize with methyl iodide: Phillips, *J. Am. Chem. Soc.* **71**, 3264 (1949).

$$\left[\begin{array}{l} CH_2COOCH_2CH_2\overset{+}{N}(CH_3)_3 \\ CH_2COOCH_2CH_2\underset{+}{N}(CH_3)_3 \end{array} \right] 2I^-$$

Slightly hygroscopic crystals, mp 243-245°. Freely sol in water or normal saline, giving solns which are very slightly acidic. Most stable at pH 4-5. Aq solns undergo progressive hydrolysis with corresp loss of activity and increase in acidity. *Prepare solns just before using.* Incompatible with solns of alkaline salts.

THERAP CAT: Skeletal muscle relaxant.

THERAP CAT (VET): Skeletal muscle relaxant (short duration).

8848. Succinyl Peroxide.

4,4′-Dioxybis[4-oxobutanoic acid]; Alphozone. $C_8H_{10}O_8$; mol wt 234.16. C 41.03%, H 4.30%, O 54.66%. $(HOOCCH_2CH_2CO)_2O_2$. Obtained from succinic anhydride and H_2O_2.

White, odorless, cryst powder. mp about 127° with decompn. Dec gradually in light. Sol in 30 parts water; moderately sol in alc, acetone, slightly in ether. Insol in benzene, chloroform. *Keep well closed and protected from light.*

USE: Germicide, antiseptic.

THERAP CAT: Antiseptic.

8849. Succinylsalicylic Acid.

Bis(o-carboxyphenyl)succinate; succinyldisalicylic acid; Diaspirin. $C_{18}H_{14}O_8$; mol wt 358.29. C 60.34%, H 3.94%, O 35.72%. Prepd by the action of succinyl chloride on salicylic acid in the presence of benzene and dimethylaniline: Ger. pat. **196,634** (1906 to Bayer); Frdl. **IX**, 936.

$$\begin{array}{l} CH_2COOC_6H_4COOH \\ | \\ CH_2COOC_6H_4COOH \end{array}$$

Needles from glacial acetic acid or alcohol. mp 176-178°. Sparingly sol in water, cold alcohol and cold glacial acetic acid. Four grams dissolve in one liter ether at 20°.

8850. Succinylsulfathiazole.

4-Oxo-4-[[4-[(2-thiazolylamino)sulfonyl]phenyl]amino]butanoic acid; 4′-(2-thiazolylsulfamoyl)succinanilic acid; p-2-thiazolylsulfamylsuccinanilic acid; 2-(N⁴-succinylsulfanilamido)thiazole; Sulfenterone; Kaoxidin; Thiacyl; Sulfasuxidine. $C_{13}H_{13}N_3O_5S_2$; mol wt 355.38. C 43.93%, H 3.69%, N 11.82%, S 18.05%, O 22.51%. Prepd by refluxing sulfathiazole with a slight excess of succinic anhydride in alcohol: Moore, Miller, *J. Am. Chem. Soc.* **64**, 1572 (1942); U.S. pats. **2,324,013; 2,324,014** (both 1943); by refluxing *p*-succinimidobenzenesulfonyl chloride with 2-aminothiazole in benzene: **Brit.** pat. **578,004** (1946). Toxicology: A. D. Welch *et al., J. Pharmacol. Exp. Ther.* **75**, 231 (1942).

$$HOOCCH_2CH_2CONH-\!\!\!\bigcirc\!\!\!-SO_2NH\!\!-\!\!\langle \overset{S}{\underset{N}{\rangle}}$$

Monohydrate, crystals with unclear mp. Has been report-

ed as 184-186° and as 192-195°. Emits pungent fumes on heating to mp. One gram dissolves in about 4800 ml water. Soluble in solns of alkali hydroxides and in solns of sodium bicarbonate with the evolution of carbon dioxide; sparingly sol in alcohol and in acetone. Insol in chloroform and ether. LD_{50} i.p. in mice: approx 5.7 g/kg (Welch).

THERAP CAT: Antibacterial.

THERAP CAT (VET): Antibacterial, esp in enteric infections.

8851. Succisulfone.

4-[[4-[(4-Aminophenyl)sulfonyl]phenyl]amino]-4-oxobutanoic acid; 4′-sulfanilylsuccinanilic acid; 4-amino-4′-(β-carboxypropionylamino)phenylsulfonylbenzene; 4-amino-4′-β-carboxypropionylaminodiphenylsulfone; 4-(β-carboxypropionylamino)-4′-aminodiphenyl sulfone; 4-succinylamido-4′-aminodiphenylsulfone; F 1500; Fourneau 1500. $C_{16}H_{16}N_2O_5S$; mol wt 348.40. C 55.16%, H 4.63%, N 8.04%, O 22.96%, S 9.20%. Prepn: Fourneau, Tréfouel, **Fr.** pat. **866,619** (1941 to Rhône-Poulenc); Kharasch, Reinmuth, U.S. pat. **2,268,754** (1942 to Lilly); Bauer, *J. Am. Chem. Soc.* **70**, 2254 (1948); Rohls, Behnish, **Ger.** pat. **895,600** (1953 to Bayer); Liberman, **Fr.** pat. **1,020,713** (1953 to Chimie et Atomistique).

$$H_2N-\!\!\!\bigcirc\!\!\!-SO_2-\!\!\!\bigcirc\!\!\!-NHCOCH_2CH_2COOH$$

Crystals, mp 157°. Soluble in ammonia.

2,2′-Iminodiethanol salt, $C_{20}H_{27}N_3O_7S$, *Exosulfonyl.* Prepn: **Fr.** pat. **992,112** (1951 to Theraplix).

THERAP CAT: Antibacterial (leprostatic).

8852. Suclofenide.

3-Chloro-4-(2,5-dioxo-3-phenyl-1-pyrrolidinyl)benzenesulfonamide; 3-chloro-4-(phenylsuccinimido)benzenesulfonamide; 1-(α-phenylsuccinimido)-2-chloro-4-sulfamoylbenzene; 1-(2-chloro-4-sulfamoylphenyl)-3-phenylsuccinimide; CGP 8426; GS 385; Sulfalepsine. $C_{16}H_{13}ClN_2O_4S$; mol wt 364.81. C 52.68%, H 3.59%, Cl 9.72%, N 7.68%, O 17.54%, S 8.79%. Anti-epileptic agent. Prepn: R. W. Pfirrmann, **Belg.** pat. **752,109;** idem, **U.S.** pat. **3,789,056** (1970, 1974 both to Geistlich). Pharmacology: *Arzneimittel-Forsch.* **27**, 1942 (1977). HPLC determn in biological materials: F. Mikes *et al., ibid.* **29**, 1583 (1979).

$$\text{(structure)}$$

Crystals from methanol/ether-petroleum ether, mp 205-207°. LD_{50} orally in mice: >5000 mg/kg (Mikes).

THERAP CAT: Anticonvulsant.

8853. Sucralfate.

Hexadeca-μ-hydroxytetracosahydroxy-[μ₈-[1,3,4,6-tetra-O-sulfo-β-D-fructofuranosyl-α-D-glucopyranoside tetrakis(hydrogen sulfato)(8-)]]hexadecaaluminum; β-D-fructofuranosyl-α-D-glucopyranoside octakis(hydrogen sulfate) aluminum complex; sucrose octakis(hydrogen sulfate) aluminum complex; Antepsin; Carafate; Hexagastron; Keal; Succosa; Sucralfin; Sucrate; Sugast; Sulcrate; Ulcar; Ulcerban; Ulcerlmin; Ulcogant; Venter. $C_{12}H_{54}Al_{16}O_{75}S_8$; mol wt 2086.74. C 6.91%, H 2.61%, Al 20.69%, O 57.50%, S 12.29%. A basic aluminum sucrose sulfate complex which inhibits peptic hydrolysis and stomach acidity. Prepn: M. Nametaka *et al., Yakugaku Zasshi* **87**, 889 (1967), *C.A.* **68**, 20831d (1968); **Fr.** pat. **1,500,571** corresp to N. Yoshihiro *et al.,* U.S. pat. **3,432,489** (1967, 1969 both to Chugai). Structure and properties: R. Nagashima, N. Yoshida, *Arzneimittel-Forsch.* **29**, 1668 (1979). *In vitro* and *in vivo* study of antipeptic and antiulcer activity: L. E. Borella, *et al., ibid.* 793. Series of articles on mode of action: *ibid.* **30**, 73-88 (1980). Pharmacokinetics, metabolism and selective binding studies: K. Steiner *et al., ibid.* **32**, 512 (1982). Clinical studies: J. F. Mayberry *et al., Brit. J. Clin. Pract.* **32**, 291 (1978). Symposium on clinical studies: *J. Clin. Gastroenterol.* **3**, Suppl. 2, 103-184 (1981); *Scand. J. Gastro-*

enterol **18**, Suppl. 83, 1-82 (1983). Review of pharmacology and clinical studies: R. N. Brogden *et al., Drugs* **27**, 194 (1984). Series of articles on clinical efficacy in peptic ulcer disease and gastritis: *Am. J. Med.* **79**, Suppl. 2C, 1-64 (1985).

R = SO$_3$[Al$_2$(OH)$_5$]

White amorphous powder. Sol in dil HCL and NaOH solns. Practically insol in water, ethanol, CHCl$_3$. pKa = 0.43 to 1.19. Dissolution of aluminum occurs at pH < 3; of sucrose sulfate at pH > 4.

THERAP CAT: Anti-ulcerative.

8854. Sucralose. *1,6-Dichloro-1,6-dideoxy-β-D-fructo-furanosyl-4-chloro-4-deoxy-α-D-galactopyranoside;* 4,1',6'-trichloro-4,1',6'-trideoxy-*galacto*-sucrose; 1',4,6'-trichlorogalactosucrose; TGS. C$_{12}$H$_{19}$Cl$_3$O$_8$; mol wt 397.64. C 36.25%, H 4.81%, Cl 26.75%, O 32.19%. Chlorinated sucrose derivative with enhanced sweetness. Prepn: P. H. Fairclough *et al., Carbohyd. Res.* **40**, 285 (1975). Prepn of crystalline anhydrous and pentahydrate: M. R. Jenner, D. Waite, **Eur.** pat. **Appl.** 30,804 (1981 to Tate & Lyle; Talres Dev.); *eidem*, **U.S.** pat. **4,343,934** (1982 to Talres Dev.). Use as non-nutritive sweetener: **Belg.** pat. 850,180; L. Hough *et al.,* U.S. pat. **4,435,440** (1977, 1984 both to Tate & Lyle). *In vitro* activity vs cariogenic bacteria: J. Verran, D. B. Drucker, *Arch. Oral Biol.* **27**, 693 (1982). Structure-sweetness relationship: M. Mathlouthi *et al., Carbohyd. Res.* **152**, 47 (1986). *Review:* L. Hough, R. Khan, *Trends Biochem. Sci.* **3**, 61-63 (1978).

Syrup, [α]$_D$ +68.2° (c = 1.1 in ethanol). Anhydrous crystalline form: orthorhombic, needlelike crystals, mp 130°. Intensely sweet taste. LD$_{50}$ in mice: > 16 g/kg (Hough, Khan).

Pentahydrate, mp 36.5°.

8855. Sucrose. *β-D-Fructofuranosyl-α-D-glucopyranoside;* α-D-glucopyranosyl-β-D-fructofuranoside; sugar; saccharose; cane sugar; beet sugar. C$_{12}$H$_{22}$O$_{11}$; mol wt 342.30. C 42.10%, H 6.48%, O 51.42%. Obtained from sugar cane *(Saccharum officinarum* L., *Gramineae)* and sugar beet *(Beta valgaris* L., *Chenopodiaceae).* Sugar cane contains from 15-20% and sugar beet from 10-17% sucrose. Structure: Avery *et al., J. Chem. Soc.* **1927**, 2308; Beevers, Cochrane, *Proc. Roy. Soc.* **190A**, 257 (1947). Synthesis: Pictet, Vogel, *Helv. Chim. Acta* **11**, 436 (1928); Lemieux, Huber, *J. Am. Chem. Soc.* **78**, 4117 (1956). Ref. with extensive bibliography: Bates, *Polarimetry, Saccharimetry, and the Sugars, National Bureau of Standards Circular* C440, Washington, 1942; W. Pigman, *The Carbohydrates* (Academic Press, New York, 1957) pp 501-506. *Reviews:* M. R. Jenner, *Dev. Food Carbohyd.* **2**, 91-143 (1980); R. Khan, *Pure Appl. Chem.* **56**, 833 (1984).

Monoclinic sphenoidal crystals, cryst masses, blocks, or powder. Sweet taste. Stable in air. Finely divided sugar is hygroscopic and absorbs up to 1% moisture which is given up on heating to 90°. d$_4^{25}$ 1.587. Dec 160-186°. Chars and emits characteristic odor of caramel. [α]$_D^{20}$ not less than +65.9° (c = 26); usual value [α]$_D^{25}$ +66.47° to +66.49°. One gram dissolves in 0.5 ml water; in slightly more than 0.2 ml boiling water, in 170 ml alcohol; in about 100 ml methanol. Moderately sol in glycerol, pyridine. pKa 12.62. d$_4^{20}$ of water solns (g/100 g): 2% 1.0060; 6% 1.0219; 10% 1.0381; 20% 1.0810; 30% 1.1270; 40% 1.1764; 50% 1.2296; 60% 1.2865; 70% 1.3471; 76% 1.3854. n$_D^{20}$ of 10% soln 1.34783. Sucrose does not reduce Fehling's soln, form an osazone, or show mutarotation. It is hydrolyzed to glucose and fructose by dil acids and by invertase, a yeast enzyme. Upon hydrolysis the optical rotation falls and is negative when the hydrolysis is complete. The mixture of glucose and fructose is known as "invert sugar." Sucrose is fermentable, but resists bacterial decompn when in high concentrations.

USE: Sweetening agent and food. Starting material in the fermentative production of ethanol, butanol, glycerol, citric and levulinic acids. Used in pharmaceuticals as a flavor, as a preservative, as an antioxidant (in the form of invert sugar), as a demulcent, as substitute for glycerol, as granulation agent and excipient for tablets, as coating for tablets. In the plastics and cellulose industry, in rigid polyurethane foams, manuf of ink and of transparent soaps.

8856. Sucrose Octaacetate. C$_{28}$H$_{38}$O$_{19}$; mol wt 678.58. C 49.56%, H 5.64%, O 44.80%. Prepn from sucrose: Linstead *et al., J. Am. Chem. Soc.* **62**, 3260 (1940). Synthesis: Lemieux, Huber, *ibid.* **78**, 4117 (1956).

Hygroscopic, intensely bitter needles from alcohol, mp 89°; dec above 285°; bp$_1$ 260°. [α]$_D^{25.4}$ +58.5 (c = 2.56 in abs alcohol). n$_D$ 1.4660. Sol in 1100 parts water, 1.1 parts acetal, 0.7 part glacial acetic acid, 0.3 part acetone, 11 parts alcohol, 0.6 part benzene, 22 parts carbon tetrachloride, about 0.5 part methyl acetate, 7 parts paraldehyde, about 0.5 part toluene.

USE: Adhesive; impregnating and insulating papers; in lacquers and plastics; as a denaturant for alcohol.

8857. Sucrose Polyester. Olestra; SPE. Non-absorbable lipid; substitute for fat in foods to reduce cholesterol level. Mixture of hexa-, hepta- and octa-fatty acid esters of sucrose. Analysis by gel permeation chromatography: C. G. Birch, F. E. Crowe, *J. Am. Oil Chem. Soc.* **53**, 581 (1976). Effect on absorption of dietary cholesterol in rats: F. H. Mattson *et al., J. Nutr.* **106**, 747 (1976); L. Aust *et al., Ann. Nutr. Metab.* **25**, 255 (1981). Effects as a dietary agent for lowering plasma cholesterol in man: R. W. Fallat *et al., Am. J. Clin. Nutr.* **29**, 1024 (1976); C. J. Glueck *et al., ibid.* **32**, 1636 (1979). Effects on cholesterol metabolism in man: J. R. Crouse, S. M. Grundy, *Metab., Clin. Exp.* **28**, 994 (1979); R. J. Jandacek *et al., Am. J. Clin. Nutr.* **33**, 251 (1980). Caloric dilution study in obese patients: C. J. Glueck *et al., ibid.* **35**, 1352 (1982).

USE: As supplement in dietary foods.

8858. Sudan III. *1-[[4-(Phenylazo)phenyl]azo]-2-naph-thalenol;* 1-(p-phenylazophenylazo)-2-naphthol; tetrazobenzene-β-naphthol; D & C Red No. 17; C.I. Solvent Red 23; C.I. 26100; Oil Red; Oil Scarlet; Toney Red. C$_{22}$H$_{16}$N$_4$O; mol wt 352.40. C 74.98%, H 4.58%, N 15.90%, O 4.54%. Prepn from diazotized aminoazobenzene and 2-naphthol: *Colour Index* vol. 4 (3rd ed., 1971) p 4227. *Review:* H. J. Conn's *Biological Stains,* R. D. Lillie, Ed. (Williams & Wilkins, Baltimore, 9th ed., 1977) pp 168-169, 576.

Reddish-brown powder. Insol in water; sol in chloroform, glacial acetic acid; moderately sol in alcohol, ether, acetone, petr ether, fixed and volatile oils, hot glycerol.

USE: Coloring oils, spirit lacquers, etc.; also as stain for zoological, pathological, and vegetable objects like wax, cutin, resin, contents of lactiferous ducts, etc., which are colored red, while cellulose membranes remain uncolored. Approved by FDA for external use only.

8859. Suet, Prepared. Mutton suet. The internal fat of abdomen of sheep purified by melting and straining.

White, solid fat with slight odor and taste; bland if fresh, but rancid after long exposure to air. mp 45-50°; n_D^{60} 1.449-1.451. Sapon no. 192-195. Iodine no. 33-46. Insol in water or cold alcohol; one gram dissolves in 45 ml boiling alcohol, about 60 ml ether.

USE: Preparing ointments.

8860. Sufentanil. *N-[4-(Methoxymethyl)-1-[2-(2-thien-yl)ethyl]-4-piperidinyl]-N-phenylpropanamide;* N-[4-(methoxymethyl)-1-[2-(2-thienyl)ethyl]-4-piperidyl]propionanilide; sufentanyl; R 30730. $C_{22}H_{30}N_2O_2S$; mol wt 386.55. C 68.36%, H 7.82%, N 7.25%, O 8.28%, S 8.29%. Potent deriv of fentanyl, q.v. Prepn: P. A. J. Janssen, G. H. P. Van Daele, Ger. pat. 2,610,228; eidem, U.S. pat. 3,998,834 (both 1976 to Janssen); G. H. P. Van Daele et al., Arzneimittel-Forsch. 26, 1521 (1976). In vitro binding properties to mu-opiate receptor: J. E. Leysen, W. Gommeren, Arch. Int. Pharmacodyn. Ther. 260, 287 (1982); J. E. Leyson et al., Eur. J. Pharmacol. 87, 209 (1983). Analgesic activity and safety assessment: W. F. M. Van Bever et al., Arzneimittel-Forsch. 26, 1548 (1976); C. J. E. Niemegeers et al., ibid. 1551. GC determn in biological fluids: R. Woestenborghs et al., J. Chromatog. 224, 122 (1981); T. J. Gillespie et al., J. Anal. Toxicol. 5, 133 (1981). Radioimmunoassay: M. Michaels et al., J. Pharm. Pharmacol. 35, 86 (1983). Comparison to fentanyl for coronary artery surgery: S. de Lange et al., Anesthesiology 56, 112 (1982). EEG effects in man: J. G. Bovill et al., Brit. J. Anaesth. 54, 45 (1982). Review of pharmacokinetics: L. E. Mather, Clin. Pharmacokinet. 8, 422-426 (1983).

Crystals from petr ether, mp 96.6°. LD$_{50}$ i.v. in mice: 18.7 mg/kg (Van Bever).

Citrate, $C_{28}H_{38}N_2O_9S$, R 33800, Sufenta.

Note: This is a controlled substance (opiate) listed in the U.S. Code of Federal Regulations, Title 21 Part 1308.12 (1987).

THERAP CAT: Narcotic analgesic.

8861. Suint. Suint is that portion of the sheep's fleece which is sol in cold water after the wax has been removed. It is a complex mixture of metallic ions, organic acids, peptides, weak bases, neutral substances, and inorganic cations. The following acids have been identified: acetic, propionic, butyric, valeric, oxalic, succinic, and glutaric. Paper chromatography indicates the presence of adipic and pimelic acids: Deane, Truter, Biochim. Biophys. Acta 18, 435 (1955); J. Chem. Soc. 1959, 2746. Determination of ion content: Mohsin, Shah, Pak. J. Sci. Ind. Res. 12, 286 (1970), C.A. 73, 4778q (1970).

8862. Sulbactam. *(2S-cis)-3,3-Dimethyl-7-oxo-4-thia-1-azabicyclo[3.2.0]heptane-2-carboxylic acid 4,4-dioxide;* penicillanic acid sulfone; penicillanic acid 1,1-dioxide; CP

45899. $C_8H_{11}NO_5S$; mol wt 233.24. C 41.20%, H 4.75%, N 6.00%, O 34.30%, S 13.75%. Semi-synthetic β-lactamase inhibitor. Prepn and use with β-lactam antibiotics: **Belg. pat. 867,859;** W. E. Barth, U.S. pat. 4,234,579 (1978, 1980 both to Pfizer); R. A. Volkmann et al., J. Org. Chem. 47, 3344 (1982). β-Lactamase activity and antibacterial spectrum in vitro: A. R. English et al., Antimicrob. Ag. Chemother. 14, 414 (1978); R. N. Jones et al., Diagn. Microbiol. Infect. Dis. 3, 489 (1985). HPLC determn in human plasma and urine: J. Haginaka et al., J. Chromatog. 341, 115 (1985). Pharmacokinetics in humans: G. Foulds et al., Antimicrob. Ag. Chemother. 23, 692 (1983). Clinical study of synergistic effect with ampicillin: S. Mehtar et al., J. Antimicrob. Chemother. 17, 389 (1986); B. V. Stromberg et al., Surg. Gynecol. Obstet. 162, 575 (1986). Review of activity and therapeutic use of sulbactam/ampicillin: D. M. Campoli-Richards, R. N. Brogden, Drugs 33, 577-609 (1987).

White, crystalline solid, mp 148-151° (Barth); also reported as mp 154-155.5° (dec) (Barth); also reported as 170° (dec) (Volkmann). $[\alpha]_D^{20}$ +251° (c = 0.01 in pH 5.0 buffer). Sol in water.

Sodium salt, $C_8H_{10}NNaO_5S$, CP-45899-2. Combination with ampicillin, q.v., Loricin, Unasyn(a), Unacid. Combination with cefoperazone, q.v., Sulperazone.

THERAP CAT: Combination with β-lactam antibiotics as antibacterial.

8863. Sulbenicillin. *[2S-[2α,5α,6β(S*)]]-3,3-Dimethyl-7-oxo-6-[(phenylsulfoacetyl)amino]-4-thia-1-azabicyclo-[3.2.0]heptane-2-carboxylic acid;* α-sulfobenzylpenicillin. $C_{16}H_{18}N_2O_7S_2$; mol wt 414.45. C 46.37%, H 4.38%, N 6.76%, O 27.02%, S 15.47%. Semi-synthetic antibiotic related to penicillin. Prepn: S. Morimoto et al., Ger. pat. 1,948,-943 corresp to U.S. pat. 3,660,379 (1970, 1972 both to Takeda); eidem, J. Med. Chem. 15, 1105, 1108 (1972). In vitro activity studies: Tsuchiya et al., J. Antibiot. 24, 607 (1971); in vivo: Yamazaki, Tsuchiya, ibid. 620. Metabolism: Tsuchiya et al., ibid. 25, 336 (1972). Toxicity: Y. Murata et al., Takeda Kenkyusho Ho 30, 262 (1971), C.A. 76, 147x (1972). Pharmacokinetics: A. Montanari et al., Clin. Ter. 89, 163 (1979); eidem, Int. J. Clin. Pharmacol. Ther. Toxicol. 18, 225 (1980).

Disodium salt, $C_{16}H_{16}N_2Na_2O_7S_2$, Kedacillin, Sulfocillin, Sulpelin, Lilacillin (3:1 mixture of D(-) and L(+) isomers). Yellowish-white powder, mp 195-198° (dec). $[\alpha]_D^{22}$ +169-173°. uv max: 257, 262, 268 nm. Very sol in water; sol in methanol. Almost insol in n-propanol, acetone, chloroform, benzene, ethyl acetate. LD$_{50}$ in male, female mice, male, female rats (mg/kg): 7900, 8000, 6000, 6200 i.v.; 9600, 10000, 7200, 7500 i.p.; 11500, 13500, 11000, 11800 s.c.; 11000, 10500, 8300, 8600 i.m.; all >15000 orally (Murata).

THERAP CAT: Antibacterial.

8864. Sulbenox. *(4,5,6,7-Tetrahydro-7-oxobenzo[b]thien-4-yl)urea;* CL 206576; Vigazoo. $C_9H_{10}N_2O_2S$; mol wt 210.24. C 51.41%, H 4.79%, N 13.32%, O 15.22%, S 15.25%. Prepn: **Neth. pat. Appl. 7,610,270** (1977 to Am. Cyanamid), C.A. 88, 22607x (1978). Activity: G. Asato, R. D. Wilbur, Experientia 35, 1458 (1979).

Crystals, mp 245-246°. LD$_{50}$ orally in rats: > 5000 mg/kg, G. Asato, R. D. Wilbur, *loc. cit.*
THERAP CAT (VET): Growth stimulant.

8865. Sulbentine. *Tetrahydro-3,5-bis(phenylmethyl)-2H-1,3,5-thiadiazine-2-thione; 3,5-dibenzyltetrahydro-2H-1,3,5-thiadiazine-2-thione; 2-thioxo-3,5-dibenzyltetrahydro-1,3,5-thiadiazine; dibenzthione.* Afungin; Fungiplex; Refungine. C$_{17}$H$_{18}$N$_2$S$_2$; mol wt 314.48. C 64.93%, H 5.77%, N 8.91%, S 20.39%. Prepn: Rieche *et al., Arch. Pharm.* **293**, 957 (1960); **East Ger.** pat. **20,634** (1961), *C.A.* **56**, 487e (1962).

Crystals from acetone or methanol, mp 101-102°.
THERAP CAT: Antifungal.

8866. Sulconazole. *1-[2-[[(4-Chlorophenyl)methyl]thio]-2-(2,4-dichlorophenyl)ethyl]-1H-imidazole; (±)-1-[2,4-dichloro-β-[(p-chlorobenzyl)thio]phenethyl]imidazole.* C$_{18}$H$_{15}$Cl$_3$N$_2$S; mol wt 407.82. C 53.01%, H 6.18%, Cl 26.08%, N 6.87%, S 7.86%. Prepn: K. A. M. Walker, M. Mars, **Ger.** pat. **2,541,833** and **U.S.** pat. **4,038,409** (1976, 1977 both to Syntex). HPLC determn in plasma: M. Fass *et al., J. Pharm. Sci.* **70**, 1338 (1981). Percutaneous absorption in different species: B. A. Zaro *et al., Proc. West. Pharmacol. Soc.* **25**, 357 (1982), *C.A.* **97**, 55898 (1982).

Mononitrate, C$_{18}$H$_{16}$Cl$_3$N$_3$O$_3$S, *RS-44872, RS-44872-00-10-3, Exelderm, Myk, Sulcosyn.* Colorless cryst from acetone, mp 130.5-132°.
THERAP CAT: Antifungal.

8867. Sulfabenz. *4-Amino-N-phenylbenzenesulfonamide; sulfanilanilide; p-aminobenzenesulfonanilide; p-aminophenylsulfonamidobenzene; N¹-phenylsulfanilamide.* C$_{12}$H$_{12}$N$_2$O$_2$S; mol wt 248.32. C 58.04%, H 4.87%, N 11.28%, O 12.89%, S 12.91%. Prepn: Gelmo, *J. Prakt. Chem.* **77**, 369 (1908); Mastryukova *et al., Tetrahedron* **19**, 357 (1963).

Needles from dil alc, mp 200°. Practically insol in water; readily sol in warm alcohol, acetone. pKa: 10.94 (in 50% aq alc); 11.59 (in 80% aq alc).
THERAP CAT (VET): Antibacterial; coccidiostat (for poultry).

8868. Sulfabenzamide. *N-[(4-Aminophenyl)sulfonyl]benzamide; N-sulfanilylbenzamide; N¹-benzoylsulfanilamide; N-(p-aminobenzenesulfonyl)benzamide; Sulfabenzide.* C$_{13}$H$_{12}$N$_2$O$_3$S; mol wt 276.31. C 56.50%, H 4.38%, N 10.14%, O 17.37%, S 11.61%. Prepn: Crossley *et al., J. Am. Chem. Soc.* **61**, 2950 (1939); Dvornikoff, **U.S.** pat. **2,240,496** (1941 to Monsanto); **Brit.** pat. **541,958** (1942 to Schering AG); Siebenmann, Schnitzer, *J. Am. Chem. Soc.* **65**, 2126 (1943).

Long, hexagonal prisms from 60% alcohol, mp 181.2-182.3°. pKa 4.57; Ka at 25°: 2.7 × 10⁻⁵. Substantially insol in water. One gram dissolves in 3225 ml of water at 30°, in 33 ml of 95% ethanol, in 9 ml of acetone. Soly in 1N NaOH or KOH: ~16 g/100 ml; pH just above 7.0. Aq solns of the water-soluble sodium salt have the same pH.
THERAP CAT: Antibacterial.
THERAP CAT (VET): Antimicrobial.

8869. Sulfabromomethazine. *4-Amino-N-(5-bromo-4,6-dimethyl-2-pyrimidinyl)benzenesulfonamide; N¹-(5-bromo-4,6-dimethyl-2-pyrimidinyl)sulfanilamide; 5-bromo-4,6-dimethyl-2-sulfanilamidopyrimidine; 2-sulfanilamide-5-bromo-4,6-dimethyl pyrimidine; 5-bromosulfamethazine; SN 3517.* C$_{12}$H$_{13}$BrN$_4$O$_2$S; mol wt 357.22. C 40.34%, H 3.67%, Br 22.37%, N 15.68%, O 8.96%, S 8.98%. Prepn: English *et al., J. Am. Chem. Soc.* **68**, 453 (1946). Pharmacology: C. M. Stowe *et al., Am. J. Vet. Res.* **29**, 345 (1958).

Crystals, dec 250-252°. uv max (methanol): 238, 272 nm (A 428, 635); min: 250 nm (A 290). Sol in alkaline solns.
Sodium salt monohydrate, C$_{12}$H$_{12}$BrN$_4$NaO$_2$S.H$_2$O, *Sulfabrom.* Cream-colored powder. Sol in water.
THERAP CAT (VET): Antibacterial.

8870. Sulfacetamide. *N-[(4-Aminophenyl)sulfonyl]acetamide; N-sulfanilylacetamide; N¹-acetylsulfanilamide; p-aminobenzenesulfonoacetamide; Acetocid; Acetosulfamine; Albucid; Sulamyd; Sulfacet; Sulfacyl; Urosulfone.* C$_8$H$_{10}$N$_2$O$_3$S; mol wt 214.24. C 44.85%, H 4.71%, N 13.07%, O 22.40%, S 14.97%. Incorrectly called *Sulfacetimide.* Not to be confused with N⁴-acetylsulfanilamide, the substance formed in the animal body by acetylation of sulfanilamide. Prepn of base and sodium salt from sulfanilamide and acetic anhydride: Crossley *et al., J. Am. Chem. Soc.* **61**, 2950 (1939). *See also* Dohrn, Diedrich, *Münch. Med. Wochenschr.* **85**, 2017 (1938); **U.S.** pat. **2,411,495** (1946 to Schering). Toxicity: R. S. Fisher, H. B. Haag, *J. Urol.* **47**, 183 (1942). Stability study of ophthalmic solns: T. Ahmad, I. Ahmad, *Pharmazie* **36**, 619 (1981). Antimicrobial activity: R. D. Houlsby *et al., J. Pharm. Sci.* **72**, 1401 (1983).

Prisms, mp 182-184°. Soluble in 150 parts water at 20°, in 15 parts alcohol, in 7 parts acetone. Insol in ether. An aq soln is acid to litmus. LD$_{50}$ orally in dogs: 8000 mg/kg (Fisher).
Sodium salt monohydrate, C$_8$H$_9$N$_2$NaO$_3$S.H$_2$O, *soluble sulfacetamide, Ak-Sulf, Albucid Soluble, Antebor, Beocid-Isoptal, Bleph-10, Cetamide, Eye-Sul, Isopto-cetamide, Locula, Op-Sulfa, Prontamid, Sebizon, Sulamyd Sodium, Sulf-10, Sulf-30, Sulphacalyre, Sulten-10, Supracid.* Minute prisms from dil alcohol, mp 257°. Slightly bitter taste. Sol in 1.5 parts water. Sparingly sol in alc, acetone. pH of 10% (w/v) aq soln: 9.0. Commercially available 30% soln buffered to pH 7.4. Ampuled solns can be autoclaved with steam at 115° for 30 mins.
THERAP CAT: Antibacterial.
THERAP CAT (VET): Antibacterial, topically for eye and skin infections, orally for urinary tract.

8871. Sulfachlorpyridazine. *4-Amino-N-(6-chloro-3-pyridazinyl)benzenesulfonamide; N¹-(6-chloro-3-pyridazinyl)sulfanilamide; 3-chloro-6-sulfanilamidopyridazine; 3-sulfanilamido-6-chloropyridazine; 3-(p-aminophenylsulfon-*

amido)-6-chloropyridazine; Ciba 10370; Ba 10370; Cosulfa; Cosulid; Prinzone; Nefrosul; Sonilyn; Vetisulid. $C_{10}H_9ClN_4$-O_2S; mol wt 284.74. C 42.18%, H 3.19%, Cl 12.45%, N 19.68%, O 11.24%, S 11.26%. Prepn: Lester, English, U.S. pat. **2,790,798** (1957 to Am. Cyanamid).

THERAP CAT: Antibacterial.

THERAP CAT (VET): Antibacterial agent. In enteric infections.

8872. Sulfachrysoidine. *3,5-Diamino-2-[[4-(aminosulfonyl)phenyl]azo]benzoic acid;* 3,5-diamino-2-[p-(sulfamoylphenyl)azo]benzoic acid; carboxysulfamidochrysoidine; Rubiazol; Rubiazol C; Rubiazol II; Rubiazol IV; Girard's Rubiazol; Azo Compd No. 4; Prontosil III; Collubiazol. $C_{13}H_{13}N_5O_4S$; mol wt 335.35. C 46.56%, H 3.91%, N 20.89%, O 19.08%, S 9.56%. Prepd like Prontosil (4'-sulfamyl-2,4-diaminoazobenzene-HCl, *q.v.*) except that the coupling component is changed to 3,5-diaminobenzoic acid: Gley, Girard, *Compt. Rend. Soc. Biol.* **125,** 1027 (1937).

Deep red crystals. mp above 300°. More sol in water than Prontosil, but not sol enough for parenteral solns. Has been marketed also as the water-soluble sodium deriv (as a 5% aq solution).

THERAP CAT: Antibacterial.

8873. Sulfacytine. *4-Amino-N-(1-ethyl-1,2-dihydro-2-oxo-4-pyrimidinyl)benzenesulfonamide;* N'-(1-ethyl-1,2-dihydro-2-oxo-4-pyrimidinyl)sulfanilamide; 1-ethyl-N-sulfanilylcytosine; N-sulfanilyl-1-ethylcytosine; Cl-636; Renoquid. $C_{12}H_{14}N_4O_3S$; mol wt 294.34. C 48.97%, H 4.79%, N 19.04%, O 16.31%, S 10.90%. Prepn: **Neth.** pat. **Appl.** 6,610,815 corresp to Doub, Krolls, U.S. pat. **3,375,247** (1967, 1968 to Parke, Davis). Prepn and activity: Doub *et al., J. Med. Chem.* **13,** 242 (1970).

Crystals from butyl alcohol, methanol, mp 166.5-168°, (monohydrate, mp 104°). uv max (methanol): 263, 297 nm ($E^{1\%}_{1cm}$ 584, 762). Equilibrium soly in pH 5 buffer at 37°: approx 175 mg/100 ml. Soly increases with increasing pH. pK' 6.9.

THERAP CAT: Antibacterial.

8874. Sulfadiazine. *4-Amino-N-2-pyrimidinylbenzenesulfonamide;* 2-sulfanilamidopyrimidine; N'-2-pyrimidinylsulfanilamide; N'-2-pyrimidylsulfanilamide; 2-sulfanilylaminopyrimidine; sulfapyrimidine; Adiazine; Debenal; Delvoprim; Diazyl; Eskadiazine; Pyrimal; Sterazine; Sulfolex; Ultradiazin. $C_{10}H_{10}N_4O_2S$; mol wt 250.28. C 47.99%, H 4.03%, N 22.39%, O 12.79%, S 12.81%. Prepd by condensing 2-aminopyrimidine with acetylsulfanilyl chloride followed by hydrolysis of the acetyl group with NaOH: Roblin *et al., J. Am. Chem. Soc.* **62,** 2002 (1940); Barber, **Brit.** pat. **557,055** (1943); Sprague, U.S. pat. **2,407,966** (1946 to Sharp & Dohme); Winnek, Roblin, U.S. pat. **2,410,793** (1946 to Am. Cyanamid). Toxicological, chemotherapeutic and pharmacokinetic studies: Böhni *et al., Chemotherapy* **14,** 195-226 (1969). *In vitro* activity of silver salt: Carr *et al., Antimicrob. Ag. Chemother.* **4,** 585 (1973). Use of silver salt in burns: C. L. Fox, *Arch. Surg.* **96,** 184 (1968). Comprehensive description: H. Stober, W. DeWitte, in *Analytical*

Profiles of Drug Substances **vol. 11,** K. Florey, Ed. (Academic Press, New York, 1982) pp 523-551.

White or slightly yellow powder. mp 252-256°. Sparingly sol in water at 37°: 13 mg/100 ml at pH 5.5; 200 mg/100 ml at pH 7.5. Sparingly sol in alcohol, acetone. One gram dissolves in about 620 ml of human serum at 37°. Freely sol in dil mineral acids and in solns of potassium and sodium hydroxides and in ammonia water.

Mixture with trimethoprim (usually 5:1), *co-trimazine, Coptin, Geatrim-Boli, Norodine, Scorprin, Triglobe.*

Sodium salt, $C_{10}H_9N_4NaO_2S$; *sulfadiazine sodium, soluble sulfadiazine.* White powder. On prolonged exposure to humid air, it absorbs carbon dioxide with the liberation of sulfadiazine and becomes incompletely sol in water. One gram dissolves in about 2 ml of water. Slightly sol in alc. Solns are alkaline to phenolphthalein (pH 9-11).

Silver salt, *Flamazine, Flammazine, Silvadene.*

THERAP CAT: Antibacterial.

THERAP CAT (VET): Antibacterial.

8875. Sulfadicramide. *N-[(4-Aminophenyl)sulfonyl]-3-methyl-2-butenamide;* 3-methyl-N-sulfanilylcrotonamide; N'-senecioylsulfanilamide; N-sulfanilylseneciamide; N'-dimethylacroylsulfanilamide; N-sulfanilyl-β,β-dimethylacrylamide; Irgamide; Sulfirgamide. $C_{11}H_{14}N_2O_3S$; mol wt 254.30. C 51.95%, H 5.55%, N 11.02%, O 18.87%, S 12.61%. Prepd from β,β-dimethylacrylamide by treatment with sodamide and p-nitrobenzenesulfonyl chloride; the resulting 4-nitro-N-(β,β-dimethylacrylyl)benzenesulfonamide is reduced with Fe and dil AcOH: Martin, Häfliger, U.S. pat. **2,417,005** (1947), *cf.* U.S. pat. **2,383,874** (1945).

Crystals from alc. mp 184-185°. Slightly sol in water, ether. Freely sol in alcohol, acetone. For the prepn of solns the water-soluble sodium salt is marketed. A 20% aq soln of the sodium salt has a pH of 8.3.

THERAP CAT: Antibacterial.

8876. Sulfadimethoxine. *4-Amino-N-(2,6-dimethoxy-4-pyrimidinyl)benzenesulfonamide;* N'-(2,6-dimethoxy-4-pyrimidinyl)sulfanilamide; 2,6-dimethoxy-4-sulfanilamidopyrimidine; 2,4-dimethoxy-6-sulfanilamido-1,3-diazine; 6-sulfanilamido-2,4-dimethoxypyrimidine; 2,6-dimethoxy-4-(p-aminobenzenesulfonamido)pyrimidine; Agribon; Albon; Madribon; Sulxin; Bactrovet; Dinosol; Sudine; Radonin; Dimetazina; Memcozine; Metoxidon; Neostreptal; Sulfasol; Sulfabon; Theracanzan; Symbio; Suldixine; Diasulfa; Arnosulfan; Maxulvet; Diasulfyl; Roscosulf. $C_{12}H_{14}N_4O_4S$; mol wt 310.33. C 46.44%, H 4.55%, N 18.05%, O 20.62%, S 10.33%. Prepn: Bretschneider, Klötzer, U.S. pat. **2,703,800** (1955 to Oesterr. Stickstoffwerke); *eidem, Monatsh.* **87,** 136 (1956); Langley, **Brit.** pat. **866,843** (1961 to I.C.I.); Bretschneider *et al., Monatsh.* **92,** 128 (1961); U.S. pat. **3,127,398** (1964 to Hoffmann-La Roche); Shepherd *et al., J. Org. Chem.* **26,** 2764 (1961). Toxicity data: Seki *et al., Arzneimittel-Forsch.* **15,** 1441 (1965). Toxicological, chemotherapeutic and pharmacokinetic studies: Böhni *et al., Chemotherapy* **14,** 195-226 (1969); *see also* Aviado *et al., ibid.* 37.

Crystals from dil alc, mp 201-203°. Sol in dil HCl and in aq solns of sodium carbonate. Soly in water at 37° (mg/100 ml): 4.6 at pH 4.10; 29.5 at pH 6.7; 58.0 at pH 7.06; 5170 at pH 8.71. LD_{50} orally in mice: > 10 g/kg (Seki).

Sodium salt, $C_{12}H_{13}N_4NaO_4S$, crystals from methanol + ether. Freely sol in water. pH of 5% solution: 8.1; of 10% soln: 8.6.

THERAP CAT: Antibacterial.

THERAP CAT (VET): Antibacterial.

8877. Sulfadoxine. *4-Amino-N-(5,6-dimethoxy-4-pyrimidinyl)benzenesulfonamide; N'-(5,6-dimethoxy-4-pyrimidinyl)sulfanilamide;* 6-(4-aminobenzenesulfonamido)-4,5-dimethoxypyrimidine; 4-sulfanilamido-5,6-dimethoxypyrimidine; sulforthomidine; sulphormethoxine; Ro 4-4393; Fanasil; Fanzil. $C_{12}H_{14}N_4O_4S$; mol wt 310.34. C 46.44%, H 4.55%, N 18.05%, O 20.62%, S 10.33%. Prepn: **Belg.** pat. **618,639** corresp to Bretschneider *et al.,* **U.S.** pat. **3,132,139** (1962, 1964 both to Hoffmann-La Roche). Toxicological, chemotherapeutic and pharmacokinetic studies: E. Böhne *et al., Chemotherapy* **14**, 195-226 (1969); *see also* Aviado *et al., ibid.* 37. Comprehensive description: V. K. Kapoor in *Analytical Profiles of Drug Substances* **vol. 17**, K. Florey, Ed. (Academic Press, New York, 1988) pp 571-605.

Crystals from 50% aq alc, mp 190-194°. LD_{50} in mice (microcrystals, mg/kg): 5200 orally, 2900 s.c, 2900 i.p. (Böhne). Practically insol in ether. Very slightly sol in water, slightly sol in alcohol, methanol. Sol in dilute mineral acids, solutions of alkali hydroxides and carbonates.

THERAP CAT: Antibacterial.

8878. Sulfaethidole. *4-Amino-N-(5-ethyl-1,3,4-thiadiazol-2-yl)benzenesulfonamide; N¹-(5-ethyl-1,3,4-thiadiazol-2-yl)sulfanilamide;* 5-ethyl-2-sulfanilamido-1,3,4-thiadiazole; 2-(p-aminobenzenesulfonamido)-5-ethylthiadiazole; sulfaethylthiadiazole; VK 55; Globucid; Sulfa-Perlongit; Sethadil; Sul-Spantab. $C_{10}H_{12}N_4O_2S_2$; mol wt 284.36. C 42.24%, H 4.26%, N 19.70%, O 11.25%, S 22.55%. The commercial prepn (Schering AG) is described in P.B. reports, no. 245, p 5, and no. 1361, p 73. Prepn: Wojahn, Wuckel, *Arch. Pharm.* **284**, 53 (1951).

Crystals, mp 185.5-186.0°. One gram dissolves in 4000 ml water, in 40 g methanol, in 30 g ethanol, in 10 g acetone, in 1350 g ether, in 2800 g chloroform, in 20,000 g benzene. Faintly acid to litmus.

Sodium salt. pH of 20% soln: 7.5.

THERAP CAT: Antibacterial.

THERAP CAT (VET): Antimicrobial.

8879. Sulfaguanidine. *4-Amino-N-(aminoiminomethyl)benzenesulfonamide; 4-amino-N-(diaminomethylene)benzenesulfonamide; N¹-amidinosulfanilamide;* N¹-guanylsulfanilamide; p-aminobenzenesulfonylguanidine; sulfanilylguanidine; RP 2275; Guamide; Sulfaguine; Guanicil; Resulfon; Ruocid; Abiguanil; Diacta; Ganidan; Shigatox; Suganyl; Aterian; Sulfoguenil. $C_7H_{10}N_4O_2S$; mol wt 214.24. C 39.24%, H 4.70%, N 26.15%, O 14.94%, S 14.97%. The official product is the monohydrate. Prepd by condensing acetylsulfanilyl chloride with guanidine nitrate in the presence of much NaOH in water-acetone: Winnek, **U.S.** pats. **2,218,-490** (1940), **2,229,784** (1941); **2,233,569** (1941); or by fusing N⁴-acetylsulfanilamide and dicyanodiamide: Haworth, Rose, **Brit.** pat. **551,524** (1943). Also made from p-nitrobenzenesulfonyl chloride: Winnek, *loc. cit.* Several other syntheses, *cf.* E. H. Northey, *Sulfonamides.* A.C.S. monograph No. **106** (New York, 1948). Revised structure: G. R. Sullivan, J. D. Roberts, *J. Org. Chem.* **42**, 1095 (1977).

Monohydrate, needles. When anhydr, mp 190-193°. One gram dissolves in about 1000 ml water at 25° and in about 10 ml at 100°. Sparingly sol in alcohol or acetone. Freely sol in dil mineral acids. Insol in NaOH solns at room temp. LD_{100} i.p. in mice: 1.0 g/kg.

THERAP CAT: Antibacterial.

THERAP CAT (VET): Antimicrobial. In enteric infections.

8880. Sulfaguanole. *4-Amino-N-[[(4,5-dimethyl-2-oxazolyl)amino]iminomethyl]benzenesulfonamide; N¹-[(4,5-dimethyl-2-oxazolyl)amidino]sulfanilamide;* 1-(4,5-dimethyl-oxazol-2-yl)-3-sulfanilylguanidine; 1-(p-aminophenylsulfonyl)-3-(4,5-dimethyl-2-oxazolyl)guanidine; sulfadimethyloxazolylguanidine; Enterocura. $C_{12}H_{15}N_5O_3S$; mol wt 309.35. C 46.59%, H 4.89%, N 22.64%, O 15.52%, S 10.36%. Prepn: Loop *et al.,* **Brit.** pat. **1,185,139**, *C.A.* **73**, 3909w (1970); *eidem,* **U.S.** pat. **3,562,258** (1970, 1971 both to Nordmark-Werke); Loop, Kohlmann, *Arzneimittel-Forsch.* **23**, 171 (1973). Toxicology: J. Kuhne *et al., ibid.* 178. Series of articles on pharmacology: *ibid.* 172-192.

Almost colorless crystals from 9:1 acetone-water, mp 233-236°; from water-methanol, mp 228-230°. pKa 7.76. Practically insol in water. Sol in dil NaOH solns. No deaths in mice or rats for oral doses of 5 g/kg (Kuhne).

THERAP CAT: Antibacterial.

8881. Sulfalene. *4-Amino-N-(3-methoxypyrazinyl)benzenesulfonamide; N¹-(3-methoxy-2-pyrazinyl)sulfanilamide;* 2-(p-aminobenzenesulfonamido)-3-methoxypyrazine; 3-methoxy-2-sulfanilamidopyrazine; sulfamethopyrazine; 2-sulfanilamido-3-methoxypyrazine; sulfamethoxypyrazine; sulfapyrazinemethoxyine; Farmitalia 204/122; Kelfizina; Kelfizine W; Longum; Dalysep; Polycidal. $C_{11}H_{12}N_4O_3S$; mol wt 280.32. C 47.13%, H 4.32%, N 19.99%, O 17.12%, S 11.44%. Prepn: B. Camerino, G. Palamidessi, *Gazz. Chim. Ital.* **90**, 1815 (1960); **Brit.** pat. **928,151;** **U.S.** pat. **3,098,069** (both 1963 to Farmitalia). *Review:* Gasparini, *Veterinaria (Milan)* **20**, 302 (1971). Also used in combination with trimethoprim, *q.v.* Toxicology: V. De Pascale *et al., Boll. Chim. Farm.* **116**, 155 (1977). Pharmacodynamics: D. S. Reeves *et al., J. Antimicrob. Chemother.* **6**, 647 (1980). Clinical assessment: H. Harazim *et al., J. Int. Med. Res.* **11**, 197 (1983); M. L. Colombo *et al., Pharmatherapeutica* **3**, 556 (1984). Review of clinical efficacy: F. Celotti *et al., J. Int. Med. Res.* **14**, 236-241 (1986).

Crystals from alcohol, mp 176°. LD_{50} in mice (g/kg): 2.164 orally; 1.41 i.v. (**U.S.** pat.).

Mixture with trimethoprim (usually 5:4 w/w), *Kelfiprim.* LD_{50} in mice, rats (mg/kg): 3500; 3550 orally (De Pascale).

THERAP CAT: Antibacterial.

8882. Sulfallate. *Diethylcarbamodithioic acid 2-chloro-2-propenyl ester; diethyldithiocarbamic acid 2-chloroallyl ester;* 2-chloroallyl diethyldithiocarbamate; CDEC; CP 4742; Vegadex. $C_8H_{14}ClNS_2$; mol wt 223.79. C 42.93%, H 6.31%, Cl 15.85%, N 6.26%, S 28.66%. Selective pre-planting or pre-emergence herbicide. Prepn: Harman, D'Amico, *J. Am. Chem. Soc.* **75**, 4081 (1953); **Brit.** pat. **769,222** (1957 to Monsanto). Toxicity data: G. W. Bailey, J. L. White, *Residue Rev.* **10**, 97 (1965).

$$(C_2H_5)_2NCSSCH_2\overset{\displaystyle Cl}{\underset{\displaystyle |}{C}}=CH_2$$

Amber liquid, bp_1 128-130°. n_D^{25} 1.5822. d^{25} 1.088. Soly in water at 25°: 100 ppm. Sol in most organic solvents. LD_{50} orally in rats: 850 mg/kg (Bailey, White).
Caution: Prolonged contact with skin and eyes causes moderate irritation: *Clinical Toxicology of Commercial Products*, R. E. Gosselin *et al.*, Eds. (Williams & Wilkins, Baltimore, 5th ed., 1984) Section II, p 313. *Note:* This substance may reasonably be anticipated to be a carcinogen: *Fourth Annual Report on Carcinogens* (NTP 85-002, 1985) p 184.
USE: Herbicide.

8883. Sulfaloxic Acid. *2-[[[4-[[[[(Hydroxymethyl)amino]carbonyl]amino]sulfonyl]phenyl]amino]carbonyl]benzoic acid; 4'-(carbamoylsulfamoyl)phthalanilic acid hydroxymethyl derivative;* 4'-(carbamoylsulfamoyl)-N-(hydroxymethyl)phthalanilic acid; formophthaloylsulfanilyl urea; 2-[4-(hydroxymethylureidosulfonyl)phenylcarbamoyl]benzoic acid; sulphaloxic acid. $C_{16}H_{15}N_3O_7S$; mol wt 393.39. C 48.85%, H 3.84%, N 10.68%, O 28.47%, S 8.15%. Prepn: Wiedemann, Strassburger, Ger. pat. 960,190 (1957 to von Heyden), *C.A.* 53, 6547i (1959).

Crystals, mp 160-165°. Soluble in dil bases.
Calcium salt, *Enteromide, Intestin-Euvernil.*
THERAP CAT: Sulfonamide; antibacterial.

8884. Sulfamerazine. *4-Amino-N-(4-methyl-2-pyrimidinyl)benzenesulfonamide; N¹-(4-methyl-2-pyrimidyl)sulfanilamide; N¹-(4-methyl-2-pyrimidinyl)sulfanilamide; 2-sulfanilamido-4-methylpyrimidine;* sulfamethyldiazine; RP 2632; Methylpyrimal; Debenal M; Methyldebenal; Mesulfa; Pyrimal M; Percoccide; Veta-Merazine; Pyralcid; Sumédine. $C_{11}H_{12}N_4O_2S$; mol wt 264.30. C 49.98%, H 4.58%, N 21.20%, S 12.13%, O 12.11%. Prepd by condensing 2-amino-4-methylpyrimidine with acetylsulfanilyl chloride followed by hydrolysis of the acetyl group: Roblin *et al.*, *J. Am. Chem. Soc.* 62, 2002 (1940); Sprague *et al.*, *ibid.* 63, 3028 (1941); Sprague, U.S. pat. 2,407,966 (1946 to Sharp & Dohme). For prepn of 2-amino-4-methylpyrimidine *see* Benary, *Ber.* 63, 2601 (1930); Backer, Grevenstuk, *Rec. Trav. Chim.* 61, 291 (1942); *cf.* E. H. Northey, *Sulfonamides* (Reinhold, New York, 1948). Antimicrobial activity: Gill *et al., Indian J. Vet. Sci.* 32, 240 (1962); Vaichulis, Vedros, *Chemotherapia* 11, 315 (1966). Toxicity studies: Simunek *et al., Vet. Med. (Prague)* 13, 619 (1968). Kinetics of sulfamerazine decompn: Zajac, *Diss. Pharm. Pharmacol.* 22, 455 (1970). Comprehensive description: R. D. G. Woolfenden in *Analytical Profiles of Drug Substances* **vol.** 6, K. Florey, Ed. (Academic Press, New York, 1977) pp 515-577.

Crystals, mp 234-238°. uv max (water): 243, 257 nm ($E_{1cm}^{1\%}$ 875, 822); (0.1M HCl): 243, 307 nm ($E_{1cm}^{1\%}$ 625, 200); (ethanol): 271 nm ($E_{1cm}^{1\%}$ 835). Slowly darkens on exposure to light. Soly in water at 37°: 35 mg/100 ml at pH 5.5; 170 mg/100 ml at pH 7.5. Readily sol in dil mineral acids and in solns of potassium, ammonium and sodium hydroxides. Sparingly sol in acetone, slightly sol in alcohol, very slightly sol in ether, chloroform.
Monosodium salt, $C_{11}H_{11}N_4NaO_2S$, *soluble sulfamerazine, Solumédine.* Crystals. Bitter, caustic taste. Hygroscopic. On prolonged exposure to humid air, it absorbs CO_2 with the liberation of sulfamerazine and becomes incompletely sol in water. Its solns are alkaline to phenolphthalein (pH 10 or

more). One gram dissolves in 3.6 ml water. Slightly sol in alc. Insol in ether, chloroform.
THERAP CAT: Antibacterial.
THERAP CAT (VET): Antibacterial.

8885. Sulfameter. *4-Amino-N-(5-methoxy-2-pyrimidinyl)benzenesulfonamide; N¹-(5-methoxy-2-pyrimidinyl)-sulfanilamide;* sulfa-5-methoxypyrimidine; 2-sulfanilamido-5-methoxypyrimidine; 2-(p-aminobenzenesulfonamido)-5-methoxypyrimidine; 5-methoxysulfadiazine; sulfamethoxydiazine; sulfametin (rescinded USAN); sulfametorine; methoxypyrimal; AHR-857; I-2586; Bayrena; Berlicid; Dairena; Durenat; Juvoxin; Kinecid; Kiron; Kirocid; Longasulf; Sulla; Supramid; Ultrax. $C_{11}H_{12}N_4O_3S$; mol wt 280.32. C 47.13%, H 4.32%, N 19.99%, O 17.12%, S 11.44%. Prepn: Ger. pat. 1,101,428 corresp to P. Diedrich, U.S. pat. 3,214,-335 (1961, 1965 to Schering AG); Horstmann *et al., Arzneimittel-Forsch.* 11, 682 (1961); Budesinky *et al., Experientia* 17, 129 (1961); **Czech.** pat. 98,818; **Belg.** pat. 633,513 (1963 to SPOFA); *Coll. Czech. Chem. Commun.* 29, 2980 (1964). Toxicology, chemotherapeutics and pharmacokinetics: Böhne *et al., Chemotherapy* 14, 195-226 (1969).

Minute, bitter crystals, mp 214-216°. uv max: 230, 271 nm ($E_{1cm}^{1\%}$ 562, 726). Very sparingly sol in water, alcohol, ether. Sol in dil acids, alkalies.
Sodium salt, water soluble. pK about 6.8. pH of 1% aq soln: about 8.9. LD_{50} in mice, rats (g/kg): 1.1 ± 0.2, 1.2 i.v.; 1.5, 1.1 i.p.; 3.0, 1.0 orally, G. Hecht *et al., Arzneimittel-Forsch.* 11, 695 (1961).
THERAP CAT: Antibacterial.

8886. Sulfamethazine. *4-Amino-N-(4,6-dimethyl-2-pyrimidinyl)benzenesulfonamide; N¹-(4,6-dimethyl-2-pyrimidinyl)sulfanilamide; N¹-(4,6-dimethyl-2-pyrimidyl)sulfanilamide; 4,6-dimethyl-2-sulfanilamidopyrimidine;* sulfamezathine; sulfadimerazine; sulfadimidine; sulfamidine; sulfadimethylpyrimidine; Diazil; Dimezathine; Mefenal; Sulphix; S-Mez; Sulfadine; S-Dimidine; Dimidin-R; Vertolan; Neazina; Pirmazin; Sulmet; Azolmetazin. $C_{12}H_{14}N_4O_2S$; mol wt 278.32. C 51.78%, H 5.07%, N 20.13%, O 11.50%, S 11.52%. Prepn: Caldwell *et al., J. Am. Chem. Soc.* 63, 2188 (1941); Roblin *et al., ibid.* 64, 567 (1942); Weisner, Katscher, **Brit.** pat. 546,158 (1942 to Ward, Blenkinsop); Sprague, U.S. pat. 2,407,966 (1946 to Sharp & Dohme). Alternate prepn: Haworth, Rose, **Brit.** pat. 552,887 (1943 to I.C.I.); Garzia, U.S. pat. 3,119,818 (1964 to Ist. Chemioterap. Ital.). Anticoccidiosis activity: Zarin, Feodorova, *C.A.* 66, 114, 435p-436q (1967). Metabolic studies: Turco *et al., Clin. Pharmacol. Ther.* 7, 603 (1966); Parker *et al., Hum. Hered.* 19, 402 (1969). Solubility studies: Shkadova *et al., Farm. Zh. (Kiev)* 24, 39 (1969). Acute toxicity: B. Bobranski *et al., Arch. Immunol. Ther. Exp.* 16, 804 (1968). Comprehensive description: C. Papostephanou, M. Frantz in *Analytical Profiles of Drug Substances* **vol.** 7, K. Florey, Ed. (Academic Press, New York, 1978) pp 401-422.

Crystals from dioxane-water, mp 176° (also reported as 178-179°; 198-199°; 205-207°). uv max (water, pH 6.6): 241 nm ($E_{1cm}^{1\%}$ 670); (0.01N NaOH): 243, 257 nm ($E_{1cm}^{1\%}$ 765, 776); (0.01N HCl): 241, 297 nm ($E_{1cm}^{1\%}$ 561, 266). Soly in water at 29°: 150 mg/100 ml; at 37°: 192 mg/100 ml at pH 7.0. pK_1 7.4±0.2, pK_2 2.65±0.2. The soly increases rapidly with an increase in pH. LD_{50} i.p. in mice: 1.06 g/kg (Bobranski). Sodium salt, *Vesadin.*
THERAP CAT: Antibacterial.
THERAP CAT (VET): Antibacterial.

8887. Sulfamethizole. *4-Amino-N-(5-methyl-1,3,4-thia-diazol-2-yl)benzenesulfonamide; N¹-(5-methyl-1,3,4-thia-diazol-2-yl)sulfanilamide;* 2-sulfanilamido-5-methyl-1,3,4-thiadiazole; 5-methyl-2-sulfanilamido-1,3,4-thiadiazole; 2-(p-aminobenzenesulfonamido)-5-methylthiadiazole; sulfamethylthiadiazole; Famet; Urodiaton; Lucosil; Ayerlucil; Urolucosil; Tetracid; Thiosulfil; Sulfurine; Ultrasul; Uroz; Renasul; Thidicur; Rufol; Sulfa Gram; Sulfapyelon; Methazol; Salimol. $C_9H_{10}N_4O_2S_2$; mol wt 270.33. C 39.98%, H 3.73%, N 20.73%, O 11.84%, S 23.72%. Prepd by the reaction of acetaldehyde thiosemicarbazone with *p*-acetylamino-benzenesulfonyl chloride in pyridine: Hübner, **U.S.** pat. **2,447,702** (1948 to Lundbeck).

Crysts from water, mp 208°. pKa 5.45; Ka (25°) = 3.5 × 10^{-6}. One gram dissolves in 4000 ml water at pH 6.5, in 5 ml water at pH 7.5, in 40 g methanol, in 30 g ethanol, in 10 g acetone, in 1370 g ether, in 2800 g chloroform. Practically insol in benzene.
Sodium salt, pH of 20% soln: 7.5. Marketed as a 4% aq soln, pH 7.3-7.4.
THERAP CAT: Antibacterial.
THERAP CAT (VET): Antimicrobial.

8888. Sulfamethomidine. *4-Amino-N-(6-methoxy-2-methyl-4-pyrimidinyl)benzenesulfonamide; N¹-(6-methoxy-2-methyl-4-pyrimidinyl)sulfanilamide;* 2-methyl-4-methoxy-6-sulfanilamidopyrimidine; 4-sulfanilamido-2-methyl-6-methoxypyrimidine; 2-methyl-4-methoxy-6-sulfanilamido-1,3-diazine; 4-(p-aminobenzenesulfonyl)amino-2-methyl-6-methoxypyrimidine; Tanasul; Duroprocin; Methofadin; Telémid. $C_{12}H_{14}N_4O_3S$; mol wt 294.34. C 48.97%, H 4.79%, N 19.04%, O 16.31%, S 10.90%. Preparation: Loop, Lührs, *Ann.* **580**, 225 (1953); Loop, **Ger.** pat. **926,131** (1955 to Nordmark).

Crystals, mp 146°. Also obtained as the monohydrate.
THERAP CAT: Antibacterial.

8889. Sulfamethoxazole. *4-Amino-N-(5-methyl-3-isox-azolyl)benzenesulfonamide; N¹-(5-methyl-3-isoxazolyl)sulfanilamide;* 5-methyl-3-sulfanilamidoisoxazole; 3-sulfanilamido-5-methylisoxazole; 3-(p-aminophenylsulfonamido)-5-methylisoxazole; sulfisomezole; sulfamethylisoxazole; sulfamethoxizole; Gantanol; Sinomin. $C_{10}H_{11}N_3O_3S$; mol wt 253.31. C 47.42%, H 4.38%, N 16.60%, O 18.95%, S 12.66%. Prepn starting with ethyl 5-methylisoxazole-3-carbamate: Kano *et al.*, **U.S.** pat. **2,888,455** (1959 to Shionogi); *Ann. Rep. Shionogi Res. Lab.* **7**, 1 (1957), *C.A.* **51**, 17889 (1957). Toxicity data: Yamamoto *et al.*, *Chemotherapy (Tokyo)* **21**, 187 (1973), *C.A.* **79**, 73738n (1973). Clinical trial of mixture with trimethoprim in *Pneumocystis carinii* pneumonia: J. M. Wharton *et al.*, *Ann. Int. Med.* **105**, 37 (1986). Comprehensive description: B. C. Rudy, B. Z. Senkowski, in *Analytical Profiles of Drug Substances* vol. 2, K. Florey, Ed. (Academic Press, New York, 1973) pp 467-486. Review of antibacterial activity and clinical efficacy of mixture with trimethoprim: G. P. Wormser *et al.*, *Drugs* **24**, 459-518 (1982). Symposium on clinical intravenous therapy: *Rev. Infect. Dis.* **9**, Suppl. 2, S152-S229 (1987).

Bitter crystals from dil ethanol, mp 167°. LD_{50} orally in mice: 3662 mg/kg (Yamamoto).
Mixture with trimethoprim (usually 5:1), *co-trimoxazole, Abacin, Apo-Sulfatrim, Bactramin, Bactrim, Bactromin, Baktar, Chemotrim, Comox, Cotrim-Puren, Drylin, Duratri-met, Eltrianyl, Eusaprim, Fectrim, Gantaprim, Gantrim, Helveprim, Imexim, Kepinol, Laratrim, Linaris, Microtrim, Momentol, Nopil, Omsat, Septra, Septrim, Sigaprim, Sulfo-trim, Sulfotrimin, Sulprim, Sumetrolim, Supracombin, Suprim, Tacumil, Teleprim, Thiocuran, TMS 480, Trigonyl, Trimesulf, Trimforte, Uroplus, Uro-Septra.* LD_{50} orally in mice: 5513 mg/kg (Yamamoto).
N^4-Acetylsulfamethoxazole, crystals from alcohol, mp 209-210°.
THERAP CAT: Antibacterial. Antipneumocystis.

8890. Sulfamethoxypyridazine. *4-Amino-N-(6-meth-oxy-3-pyridazinyl)benzenesulfonamide; N¹-(6-methoxy-3-pyridazinyl)sulfanilamide;* 6-methoxy-3-sulfanilamidopyridazine; 3-(p-aminobenzenesulfamido)-6-methoxypyridazine; CL 13494; RP 7522; Davosin; Depovernil; Durox; Kynex; Lederkyn; Lentac; Midicel; Midikel; Myasul; Mylosul; Opinsul; Petrisul; Retasulfin; Spofadazine; Sulfalex; Sulfdurazin; Sultirene; Vinces. $C_{11}H_{12}N_4O_3S$; mol wt 280.32. C 47.13%, H 4.32%, N 19.99%, O 17.12%, S 11.44%. Prepn from 6-chloro-3-sulfanilamidopyridazine: Clark, **U.S.** pat. **2,712,012** (1955 to Am. Cyanamid). Toxicological, chemotherapeutic and pharmacokinetic studies: Böhni *et al.*, *Chemotherapy* **14**, 195-226 (1969).

Bitter crystals from water, mp 182-183°. pKa 6.7. Soly in water at 37° (mg/100 ml): 110 at pH 5; 120 at pH 6; 147 at pH 6.5. Slightly sol in methanol, ethanol (about 1:200); more sol in acetone (1:50), in dimethylformamide (1 g/1 ml). Freely sol in aq solns of alkali hydroxides. LD_{50} orally in mice: 1750 mg/kg, Seki *et al.*, *Arzneimittel-Forsch.* **15**, 1441 (1965).
N^1-Acetyl derivative, $C_{13}H_{14}N_4O_4S$, *Kynex Acetyl, Davosin Suspension.* The acetyl radical is attached to the nitrogen in the sulfonamido group. Prepn: Murphy, Shepherd, **U.S.** pat. **2,833,761** (1958 to Am. Cyanamid). Tasteless crystals, dec 186-187°. Has comparable antibacterial activity.
THERAP CAT: Antibacterial.
THERAP CAT (VET): Antibacterial.

8891. Sulfamethylthiazole. *4-Amino-N-(4-methyl-2-thiazole)benzenesulfonamide; 4-methyl-2-sulfanilamidothiazole; N¹-(4-methyl-2-thiazolyl)sulfanilamide;* 2-(p-amino-benzenesulfonamido)-4-methylthiazole; Sulfazole; Ultraseptyl; Toriseptin M; Novoseptale. $C_{10}H_{11}N_3O_2S_2$; mol wt 269.33. C 44.59%, H 4.12%, N 15.61%, O 11.88%, S 23.81%. Preparation: Fosbinder, Walter, *J. Am. Chem. Soc.* **61**, 2032 (1939). Alternate routes of synthesis: E. H. Northey, *The Sulfonamides and Allied Compounds* (Reinhold, New York, 1948) p 89.

Crystals from alcohol, mp 238-240°. Slightly sol in water (0.2 g dissolves in 100 ml H_2O). Insol in ether. Readily sol in solns of alkali hydroxides and carbonates and in dil mineral acids. Forms a water-sol sodium salt.

8892. Sulfametrole. *4-Amino-N-(4-methoxy-1,2,5-thia-diazol-3-yl)benzenesulfonamide; N¹-(4-methoxy-1,2,5-thia-diazol-3-yl)sulfanilamide;* 3-methoxy-4-(4'-aminobenzene-sulfonamido)-1,2,5-thiadiazole. $C_9H_{10}N_4O_3S_2$; mol wt 286.32. C 37.75%, H 3.52%, N 19.57%, O 16.76%, S 22.39%. Anti-infective sulfanilamide, related structurally to sulfadiazine, *q.v.* Prepn: **Belg.** pat. **629,551** corresp to K. Menzl, **U.S.** pat. **3,247,193** (1963, 1966 both to OSSW). Improved prepn: *idem*, **U.S.** pat. **4,151,164** (1979 to Chemie Linz). Bacteriological study (of mixture with trimethoprim): G. Nabert-Bock, H. Grims, *Arzneimittel-Forsch.* **27**, 1109 (1977). Tolerance, therapeutic effect, pharmacokinetics: S. Breyer *et al.*, *Wien Med. Wochenschr.* **130**, 448 (1980). Renal excretion mechanism, pharmacokinetics: T. B. Vree *et al.*, *Eur. J. Clin. Pharmacol.* **20**, 283 (1981). Pharmacokinetics of sulfametrole and its metabolite in man: Y. A. Hester *et al.*, *J. Antimicrob. Chemother.* **8**, 133 (1981). Crystal and molecular structures: C. H. Koo *et al.*, *Bull. Korean Chem. Soc.* **3**, 9 (1982), *C.A.* **96**, 208771 (1982). Single-dose therapy of chancroid with trimethoprim-sulfametrole: F. A. Plummer *et al.*, *N. Engl. J. Med.* **309**, 67 (1983).

Cryst from dilute acetic acid, mp 149-150°.

Mixture with trimethoprim (usually 5:1), *Lidaprim, Maderan.*

THERAP CAT: Antibacterial.

8893. Sulfamic Acid. Amidosulfonic acid. H_3NO_3S; mol wt 97.10. H 3.11%, N 14.43%, O 49.43%, S 33.02%. H_2NSO_3H. Obtained from chlorosulfonic acid and ammonia, or by heating urea with H_2SO_4. Purification: Sisler *et al.*, *Inorg. Syn.* **2**, 178 (1946). Toxicity data: Ambrose, *J. Ind. Hyg. Toxicol.* **25**, 26 (1943). *Reviews:* Audrieth *et al.*, *Chem. Rev.* **26**, 49 (1940); Burton, Nickless, "Amido- and Imido-Sulfonic Acids" in *Inorganic Sulphur Chemistry*, G. Nickless, Ed. (Elsevier, New York, 1968) pp 607-627, 661-667; E. B. Bell in Kirk-Othmer *Encyclopedia of Chemical Technology* **vol. 21** (Wiley-Interscience, New York, 3rd ed., 1983) pp 949-960.

Orthorhombic crystals. d 2.15. mp ~205° (dec). Stable when dry but in soln it slowly hydrolyzes forming ammonium bisulfate. Sol in 6.5 parts water at 0°, in about 2 parts water at 80°. Sulfuric acid decreases its soly in water. It is sparingly sol in alcohol, methanol; slightly sol in acetone; insol in ether. Freely sol in nitrogenous bases, e.g., liq ammonia, also in nitrogen contg organic solvents, e.g., pyridine, formamide, dimethylformamide. A strong acid; pH of a 1% soln at 25° 1.18. Can be titrated with bases by means of indicators showing color change between pH 4.5 to 9. MLD orally in rats: 1.6 g/kg (Ambrose).

USE: As standard in alkalimetry; in acid cleaning; in nitrite removal; in chlorine stabilization for use in swimming pools, cooling towers, paper mills. The acid or its ammonium salt has been recommended for flameproofing fabrics and wood. Metal salts are used in electroplating. Ammonium sulfamate, *q.v.*, is also widely used as a weed killer. *Caution:* Moderately irritating to skin, mucous membranes.

8894. Sulfamide. Sulfuryl amide. $H_4N_2O_2S$; mol wt 96.11. H 4.20%, N 29.15%, O 33.29%, S 33.36%. H_2NSO_2-NH_2. Prepd from sulfuryl chloride and gaseous ammonia: Traube, *Ber.* **25**, 2427 (1892); **26**, 610 (1893); Hantzsch, Holl, *Ber.* **34**, 3430 (1902); Schenk in *Handbook of Preparative Inorganic Chemistry* **vol. 1**, G. Brauer, Ed. (Academic Press, New York, 2nd ed., 1963) pp 482-483.

Orthorhombic, tasteless plates from abs alcohol, mp 93°; dec 250°. Dipole moment: 3.9. Freely sol in water, hot alcohol, acetone. Sparingly sol in cold alc. Solid and solns are stable.

Note: Not to be confused with ammonium sulfamate or with sulfanilamide.

USE: Similar to urea in many of its reactions, but is more acidic, and can act as a dibasic acid.

8895. Sulfamidochrysoidine. *p-[(2,4-Diaminophenyl)-azo]benzenesulfonamide;* 2,4-diaminoazobenzene-4'-sulfonamide; 4'-sulfamyl-2,4-diaminoazobenzene; Parazol; Prontosil; Prontosil flavum; Prontosil rubrum; Pronzin rubrum; Relbazon; Rubiazol I; Septosan; Streptocide [Evans]; Streptozon. $C_{12}H_{13}N_5O_2S$; mol wt 291.34. C 49.47%, H 4.50%, N 24.04%, O 10.98%, S 11.01%. Prepd by diazotizing sulfanilamide and coupling with *m*-phenylenediamine: Mietzsch,

Klarer, **Ger.** pat. **607,537** (1935 to I. G. Farbenind.), *C.A.* **29**, 4135 (1935); *eidem*, **U.S.** pat. **2,085,037** (1937 to Winthrop). Polymorphic forms: Kofler, Kofler, *Monatsh.* **81**, 321 (1950).

Three polymorphic forms found present in commercial preparations. Form A: red rectangular plates or large tablets from acetone, mp 232-235°. Form B: fine orange crystals from acetone, mp 218-220°. Form C: red crystals from water, changes to form A when heated above 160°.

Hydrochloride, $C_{12}H_{13}N_5O_2S.HCl$, orange-red crystals, mp 248-250°. One gram dissolves in 400 ml water; much more sol in hot water. Sol in alcohol, acetone, fats, oils.

THERAP CAT: Hydrochloride as antibacterial.

8896. Sulfamipyrine. *[(2,3-Dihydro-1,5-dimethyl-3-oxo-2-phenyl-1H-pyrazol-4-yl)amino]methanesulfonic acid monosodium salt; (antipyrinylamino)methanesulfonic acid sodium salt;* 2,3-dimethyl-1-phenyl-5-pyrazolone-4-aminomethanesulfonic acid sodium salt; 2,3-dimethyl-1-phenyl-3-pyrazolin-5-one-4-aminomethanesulfonic acid sodium salt; sodium 1-phenyl-2,3-dimethylpyrazolone-4-aminomethanesulfonate; melaminsulfone; sulphamipyrin; Melubrin; Metapyrin; Pyrazogin. $C_{12}H_{14}N_3NaO_4S$; mol wt 319.32. C 45.13%, H 4.42%, N 13.16%, Na 7.20%, O 20.04%, S 10.04%. Prepn from 4-aminoantipyrine: **Ger.** pats. **254,711** (1912) and **259,503** (1913), *Frdl.* **11**, 915, 917.

Crystals from methanol + ether, sinters at 231-233°. One gram dissolves in 1 ml water, in 10 ml methanol. Moderately sol in ethanol. Also sol in other alcohols, glycerol, and glycols. Practically insol in ether, acetone.

THERAP CAT: Formerly as antipyretic, analgesic.

THERAP CAT (VET): Has been used as antipyretic, analgesic.

8897. Sulfamoxole. *4-Amino-N-(4,5-dimethyl-2-oxazolyl)benzenesulfonamide; N^1-(4,5-dimethyl-2-oxazolyl)sulfanilamide;* 2-(p-aminobenzenesulfonamido)-4,5-dimethyloxazole; 4,5-dimethyl-2-sulfanilamidooxazole; *p*-aminobenzenesulfonyl-2-amino-4,5-dimethyloxazole; sulfadimethyloxazole; Depomide; Justamil; Sulfmidil; Sulfono; Sulfuno; Tardamide. $C_{11}H_{13}N_3O_3S$; mol wt 267.31. C 49.43%, H 4.90%, N 15.72%, O 17.96%, S 12.00%. Prepn: Loop *et al.*, **U.S.** pat. **2,809,966** (1957 to Nordmark-Werke). Series of articles on antimicrobial activity, pharmacology and clinical efficacy of mixture with trimethoprim: *Arzneimittel-Forsch.* **26**, 595-684 (1976). Toxicity data: F. Lagler *et al.*, *ibid.* 634.

Crystals, mp 193-194°. Soly in mg/100 ml at 20°: Water 85; 0.01N HCl 163; 0.01N NaOH 196; methanol 2315; chloroform 240. uv max (4.91 µg/ml methanol): 210, 250, 270 nm (ε 740, 546, 857). LD_{50} in mice, rats (g/kg): > 10.0, > 12.5 orally; 1.80, ~2.50 i.p. (Lagler).

Mixture with trimethoprim (5:1), *CN 3123, Co-Fram, Dibactil, Nevin, Supristol.* LD_{50} in mice, rats (g/kg): > 12.0, 14.0 orally; 1.87, 2.00 i.p. (Lagler).

THERAP CAT: Antibacterial.

8898. Sulfanilamide. *4-Aminobenzenesulfonamide; p*-anilinesulfonamide; *p*-sulfamidoaniline; 1162 F; Albexan;

Deseptyl; Prontalbin; Prontosil album; Prontylin; Septoplix; Streptocid album; Streptocide (Frosst). $C_6H_8N_2O_2S$; mol wt 172.21. C 41.84%, H 4.68%, N 16.27%, O 18.58%, S 18.62%. Prepd by the action of ammonia on acetylsulfanilyl chloride; the resulting N^4-acetylsulfanilamide is then hydrolyzed: Gelmo, *J. Prakt. Chem.* **77**, 369 (1908). Lab procedures: L. F. Fieser, *Experiments in Organic Chemistry* (Boston, 3rd ed., 1955) p 147; Hurdis, Yang, *J. Chem. Ed.* **46**, 697 (1969). Large-scale production and purification: Mietzsch, Klarer, **U.S.** pats. **2,132,178** (1938); **2,276,664** (1942). Alternate prepn: Simons, **U.S.** pat. **2,237,372** (1941).

Crystals, mp 164.5-166.5°. pKa 10.43; pKb 11.63. Neutral to litmus. pH (0.5% aq soln): 5.8-6.1. uv max: 257, 313 nm. Soly in water (g/liter): 2.6 (10°); 7.5 (25°); 17.0 (40°); 40 (60°); 477 (100°). One gram dissolves in about 37 ml alcohol; in about 5 ml acetone. Also sol in glycerol, propylene glycol, HCl, solns of K and Na hydroxides. Practically insol in chloroform, ether, benzene, petr ether. LD_{50} orally in dogs: 2.0 g/kg.

THERAP CAT: Antibacterial.
THERAP CAT (VET): Antimicrobial.

8899. Sulfanilamidomethanesulfonic Acid Triethanolamine Salt. *Sulfanilamidomethanesulfonic acid salt with 2,2',2''-nitrilotriethanol;* p-aminobenzenesulfonylaminomethanesulfonic acid triethanolamine salt; 4-aminophenylsulfonylaminomethanesulfonic acid triethanolamine salt; Salthion. $C_{13}H_{25}N_3O_8S_2$; mol wt 415.47. C 37.58%, H 6.06%, N 10.11%, O 30.81%, S 15.43%. Prepn: Zutavern, Kraft, **Ger.** pat. **894,560** (1953 to Knoll).

Crystals, mp 117-119°. Soly in water at 18°: 70% (w/w). pH of a satd aqueous soln 5.0. LD_{50} in mice: 11.0 g/kg s.c.; 3.5 g/kg i.v.

THERAP CAT: Topical antibacterial.
THERAP CAT (VET): Antimicrobial.

8900. 4-Sulfanilamidosalicylic Acid. p-Sulfanilamidosalicylic acid; 4-(p-aminobenzenesulfonamido)-2-hydroxybenzoic acid; Metrasil. $C_{13}H_{12}N_2O_5S$; mol wt 308.33. C 50.64%, H 3.92%, N 9.09%, O 25.95%, S 10.40%. Prepn: Benedikt, Eibl, *Monatsh.* **81**, 419 (1950); Goldberg, Salame, **Brit.** pat. **648,467** (1951 to Ward, Blenkinsop).

Crystals from dil alcohol, dec 220°. The water soly is described as high between pH 5 and 8.

THERAP CAT: Antibacterial.

8901. Sulfanilic Acid. *4-Aminobenzenesulfonic acid;* p-anilinesulfonic acid. $C_6H_7NO_3S$; mol wt 173.84. C 41.61%, H 4.07%, N 8.09%, O 27.72%, S 18.51%. Readily prepd from aniline and sulfuric acid: I. G. Farben, *FIAT Final Rept.* **1313** (I), 255 (1948); Fierz-David, Blangey, *Fundamental Processes of Dye Chemistry* (Interscience, New York, 1949) pp 126-128; A. I. Vogel, *Practical Organic Chemistry* (Longmans, London, 3rd ed., 1959) p 586; R. Q. Brewster *et al.*, *Unitized Experiments in Organic Chemistry* (Van Nostrand, Princeton, 2nd ed., 1964) pp 162-163. Improved procedure: Jacobs *et al.*, *Ind. Eng. Chem.* **35**, 321-323 (1943).

Monohydrate, orthorhombic plates from water (very slow crystn can yield the dihydrate), becomes anhydr at around 100°. Dec without melting at about 288° (also reported as >360°). K_2 at 25°: 5.93 × 10^{-4}. Slowly sol in water: about 1% at 20°, 1.45% at 30°, 1.94% at 40° (w/w anhydr). Almost insol in ethanol, benzene, ether. Slightly sol in hot methanol.

Sodium salt dihydrate, $C_6H_6NNaO_3S.2H_2O$, orthorhombic bipyramidal plates. Freely sol in water, sol in hot methanol.

Zinc salt tetrahydrate, $C_{12}H_{12}N_2O_6S_2Zn.4H_2O$, *zinc sulfanilate tetrahydrate, Nizin.*

Amide *see* Sulfanilamide.

USE: Manuf various dyes and organic chemicals; as reagent in anal. chemistry (Ehrlich's reagent, detn of nitrites).

THERAP CAT: Antibacterial.

8902. 2-p-Sulfanilylanilinoethanol. *2-[[4-(4-Aminophenyl)sulfonyl]phenyl]amino]ethanol;* 4-amino-4'-(β-hydroxyethylamino)diphenyl sulfone; p-aminophenyl p-(2-hydroxyethylamino)phenyl sulfone; Smith's sulfone. $C_{14}H_{16}N_2O_3S$; mol wt 292.37. C 57.51%, H 5.52%, N 9.58%, O 16.42%, S 10.97%. Preparation: Jackson, *J. Am. Chem. Soc.* **70**, 680 (1948); Weijlard, Swanezy, *ibid.* **71**, 4134 (1949); Jackson, *ibid.* **72**, 395 (1950); Anand *et al.*, *J. Sci. Ind. Res. (India)* **13B**, 260 (1954); Onishi, Fukagawa, **Japan.** pat. **2,975**('56) (to Yoshitomi Drug).

Exists in diamorphic forms, crystals, mp 130.5-131.5° and crystals, mp 144°. The lower melting crystals' soly in water: 74 mg/100 ml at 37°. The higher melting form is readily sol in acetone; sol in ethanol, dioxane, cold 1% HCl; fairly sol in hot water; slightly sol in cold water.

THERAP CAT: Antibacterial (Mycobacterium).

8903. p-Sulfanilylbenzylamine. *4-[(4-Aminophenyl)sulfonyl]benzenemethanamine;* p-sulfanilidobenzylamine; 4-aminomethyl-4'-aminodiphenyl sulfone; p-aminophenyl-p-aminomethylphenyl sulfone; Alphamide. $C_{13}H_{14}N_2O_2S$; mol wt 262.34. C 59.52%, H 5.39%, N 10.68%, O 12.20%, S 12.22%. Prepn: Dewing, *J. Chem. Soc.* **1946**, 466; Wenner, *J. Org. Chem.* **22**, 1508 (1957).

Needles from water, mp 159°.

Monohydrochloride monohydrate, $C_{13}H_{14}N_2O_2S.HCl.H_2O$, crystals, mp 195°. Very sol in water.

Dihydrochloride, $C_{13}H_{14}N_2O_2S.2HCl$, mp 285°.

THERAP CAT: Antibacterial.

8904. Sulfanilyl Fluoride. *p-Aminobenzenesulfonyl fluoride.* $C_6H_6FNO_2S$; mol wt 175.19. C 41.14%, H 3.45%, F 10.85%, N 8.00%, O 18.27%, S 18.30%. Prepd by heating potassium fluoride with 4-acetamidobenzenesulfonyl chloride: Parker, Hofmann, **U.S.** pat. **2,576,037** (1951 to Am. Cyanamid).

Crystals, mp 68-69°.

USE: In the prepn of dyes which pick up light readily.

8905. N^4-Sulfanilylsulfanilamide. *4'-Sulfamoylsulfanil-*

anilide; 4-aminobenzenesulfono-*p*-sulfamoylanilide; 4-(4′-aminobenzenesulfonamido)benzenesulfonamide; DB 32; Diseptal C; Uliron C; Disulon; Neosanamid II; Albasil C; Disulfan. $C_{12}H_{13}N_3O_4S_2$; mol wt 327.38. C 44.02%, H 4.00%, N 12.84%, O 19.55%, S 19.59%. Prepn: Fr. pat. **817,034** (to I. G. Farben.), *C.A.* **32**, 1714i (1938); Sakai, Yamamoto, *J. Pharm. Soc. Japan* **58**, 683 (1938); Golovchinskaya, *J. Appl. Chem. USSR* **18**, 647 (1945); Sasa, *J. Soc. Org. Syn. Chem. Japan* **12**, 211 (1954), *C.A.* **51**, 2780h (1957); Grigorovskii, Dykhanov, *J. Appl. Chem. USSR* **30**, 1284 (1957).

Needles from water, mp 133-134°. Absorption max: Maschka *et al., Monatsh.* **84**, 1071 (1953). Slightly sol in cold water; considerably more sol in hot water. Sol in methanol, ethanol, ether, dil NH_3, dil HCl; practically insol in petr ether, chloroform.

Note: The name Disulon is also used for disulfanilamide, $H_2NC_6H_4SO_2NHSO_2C_6H_4NH_2$, esp. in the French literature.

THERAP CAT: Topical antibacterial.

8906. Sulfanilylurea. *4-Amino-N-(aminocarbonyl)benzenesulfonamide; N*-sulfanilylcarbamide; sulfacarbamide; sulfaurea; Euvernil; Uractyl; Uramid; Urenil; Urosulfan. $C_7H_9N_3O_3S$; mol wt 215.23. C 39.06%, H 4.22%, N 19.53%, O 22.30%, S 14.90%. Prepd by treating N^4-acetylsulfanilamide with potassium cyanate or with carbamyl chloride or with urea (or with nitrourea and sodium carbonate) in 80% alcohol. The *p*-AcNHC_6H_4SO_2NHCONH_2 is saponified by slight warming with dil KOH and then acidified: Martin *et al.,* U.S. pat. **2,411,661** (1946 to Geigy). By boiling sulfanilamide with urea and sodium carbonate in 75% alcohol: Haack, *Alien Prop. Custodian, Serial* **369**, 118 (1943). By warming calcium acetylsulfanilylcyanamide with dil HCl: Winnek *et al., J. Am. Chem. Soc.* **64**, 1684 (1942); improved procedure: Leitch *et al., Can. J. Res.* **23B**, 139 (1945).

Crystals from water. mp 146-148° (slight dec). [Forms a monohydrate, mp 125-127°.] Solubility in water at 37°: 811 mg/100 ml. Soluble in alkalies. Forms a very soluble sodium salt.

THERAP CAT: Antibacterial.

8907. N-Sulfanilyl-3,4-xylamide. *N-[(-4-Aminophenyl)sulfonyl]-3,4-dimethylbenzamide; N*-(3,4-dimethylbenzoyl)sulfanilamide; Geigy 867; Irgafen. $C_{15}H_{16}N_2O_3S$; mol wt 304.36. C 59.19%, H 5.30%, N 9.20%, O 15.77%, S 10.54%. Prepd from *p*-nitrobenzenesulfonamide and 3,4-dimethylbenzoyl chloride; the resulting *p*-nitro-*N*-(3,4-dimethylbenzoyl)benzenesulfonamide is reduced with Fe and dil AcOH: Martin *et al.,* U.S. pat. **2,383,874** (1945).

Needles from alcohol, mp 222-223°. Sparingly sol in water. Marketed as the water-soluble sodium salt for making solns; a 5% aq soln having a pH of 8.2.

THERAP CAT: Antibacterial.

8908. Sulfanitran. *N-[4-[[(4-Nitrophenyl)amino]sulfonyl]phenyl]acetamide; 4′-[(p-nitrophenyl)sulfamoyl]acetanilide; N^4*-acetyl-*N*-(*p*-nitrophenyl)sulfanilamide; 4-acetaminobenzenesulfon-4′-nitroanilide; *N*-(*p*-acetylaminobenzene sulfonyl)-*p*-nitroaniline; APNPS. $C_{14}H_{13}N_3O_5S$; mol wt 335.34. C 50.14%, H 3.91%, N 12.53%, O 23.86%, S 9.55%. Preparation: Webster, Powers, *J. Am. Chem. Soc.* **60**, 1553 (1938); Kaufmann, Bückmann, *Arch. Pharm.* **279**, 194 (1941); Shepherd, *J. Org. Chem.* **12**, 275 (1947).

Crystals from dil alcohol, mp 239-240° (Kaufmann); mp 264° (Shepherd). Freely sol in acetone; sol in hot ethanol, methanol; sparingly sol in water, ether.

Note: Ingredient of *Polystat, Unistat, Novastat* (contains 25% Aklomide, 20% sulfanitran, 55% corn meal).

THERAP CAT: Antibacterial.

THERAP CAT (VET): Antibacterial; coccidiostat (poultry).

8909. Sulfaperine. *4-Amino-N-(5-methyl-2-pyrimidinyl)benzenesulfonamide; N*-(5-methyl-2-pyrimidinyl)sulfanilamide; 5-methyl-2-sulfanilamidopyrimidine; 2-sulfanilamido-5-methylpyrimidine; isosulfamerazine; 5-methylsulfadiazine; Demosulfan; Lentamid; Orosulfan; Pallidin; Pallimerck; Retardon; Rexulfa; Risulfasens; Sintosulfa; Sulfatreis; Sulfene; Ultrasulfon. $C_{11}H_{12}N_4O_2S$; mol wt 264.30. C 49.98%, H 4.58%, N 21.20%, O 12.11%, S 12.13%. Prepd by condensing 2-amino-5-methylpyrimidine with acetylsulfanilyl chloride followed by hydrolysis of the acetyl group with NaOH: Sprague, U.S. pat. **2,407,966** (1946 to Sharp & Dohme).

Minute, cream-colored crystals, 262-263°. Very sparingly sol in water, ethanol; about 40 mg/100 ml H_2O at pH 5.5. Sol in aq solns of acids and alkalies. Forms a water-sol sodium salt.

THERAP CAT: Antibacterial.

8910. Sulfaphenazole. *4-Amino-N-(1-phenyl-1H-pyrazol-5-yl)benzenesulfonamide; N′-(1-phenylpyrazol-5-yl)sulfanilamide;* 1-phenyl-5-sulfanilamidopyrazole; 3-(*p*-aminobenzenesulfonamido)-2-phenylpyrazole; 5-sulfanilamido-1-phenylpyrazole; Depocid; Isarol; Orisul; Orisulf; Sulfabid. $C_{15}H_{14}N_4O_2S$; mol wt 314.35. C 57.31%, H 4.49%, N 17.82%, O 10.18%, S 10.20%. Synthesis: Schmidt, Druey, *Helv. Chim. Acta* **41**, 309 (1958); Druey, Schmidt, U.S. pat. **2,858,309** (1958 to Ciba).

Crystals from alcohol, mp 179-183°. Very sparingly sol in water: 0.15 g/100 ml H_2O at pH 7.0 and 25°. More sol in methanol, ethanol, glacial acetic acid. LD_{50} orally in mice: 5800 mg/kg, Seki *et al., Arzneimittel-Forsch.* **15**, 1441 (1965).

Sodium salt monohydrate, $C_{15}H_{13}N_4NaO_2S.H_2O$, white powder, sol in water. A 10% soln (w/w) has a pH of ∼9.

Note: Commercial medicinal grades of sulfaphenazole also contain *5-sulfanilimido-1-phenyl-2H-pyrazoline:* Seydel, *Naturwiss.* **50**, 663 (1963).

THERAP CAT: Antibacterial.

THERAP CAT (VET): Antimicrobial.

8911. Sulfaproxyline. *p-Isopropoxy-N-sulfanilylbenzamide; N*-(*p*-isopropoxybenzoyl)sulfanilamide; *N*-(4-isopropoxybenzoyl)-*p*-aminobenzenesulfonamide; sulphaproxyline. $C_{16}H_{18}N_2O_4S$; mol wt 334.41. C 57.47%, H 5.42%, N 8.38%, O 19.14%, S 9.59%. Prepn: Gysin, U.S. pat. **2,503,820** (1950 to Geigy).

Crystals, mp 172-173°.

Ingredient of *Dosulfin* which also contains sulfamerazine in equal parts.

THERAP CAT: Antibacterial.

8912. Sulfapyrazine. *4-Amino-N-(pyrazinyl)benzenesulfonamide; N¹-2-pyrazinylsulfanilamide;* 2-sulfanilamidopyrazine. $C_{10}H_{10}N_4O_2S$; mol wt 250.28. C 47.99%, H 4.03%, N 22.39%, O 12.79%, S 12.81%. Prepd by reacting acetylsulfanilyl chloride with 2-aminopyrazine in pyridine, followed by hydrolysis and neutralization: Ellingson, U.S. pat. 2,420,703 (1947). Biological properties: Ghione *et al., Chemotherapia* **6,** 344 (1962).

Crystals, dec 250-254°. Very slightly sol in alc; slightly sol in acetone. Sol in aq solns of sodium, potassium, and barium hydroxides, in ammonia water, and in dil and concd mineral acid solns. Practically insol in water (5 mg/100 ml at 25°, and 5.2 mg/100 ml at 37°).
Sodium salt monohydrate, $C_{10}H_9N_4NaO_2S.H_2O$, bitter powder. Freely soluble in water (1 g/3.33 ml at 25°). Alkaline to litmus. pH of 10% aq soln 9.1. Very sol in acetone, slightly sol in alc. Insol in ether, chloroform. Aq solns of sulfapyrazine sodium may absorb carbon dioxide which causes precipitation of sulfapyrazine.
THERAP CAT: Antibacterial.
THERAP CAT (VET): Antibacterial.

8913. Sulfapyridine. *4-Amino-N-2-pyridinylbenzenesulfonamide; N¹-2-pyridylsulfanilamide;* 2-sulfanilamidopyridine; M & B 693; Dagenan; Eubasin; Pyriamid; Coccoclase; Piridazol; Septipulmon; Relbapiridina; Plurazol; Haptocil; Sulfidine. $C_{11}H_{11}N_3O_2S$; mol wt 249.29. C 52.99%, H 4.45%, N 16.86%, O 12.84%, S 12.86%. Prepd by reacting acetylsulfanilyl chloride with 2-aminopyridine followed by alkaline hydrolysis of the 2-(N^4-acetylsulfanilamido)pyridine and neutralization with SO₂: Ewins, Phillips, **Brit.** pat. 512,145 (1939); 530,187 (1940); U.S. pats. 2,259,222 (1941); 2,275,354 (1942); 2,312,032 (1943); 2,335,221 (1943). *Cf.* Winterbottom, *J. Am. Chem. Soc.* **62,** 160 (1940). Metabolism: R. Wien, J. Hampton, *J. Pharmacol. Exp. Ther.* **84,** 211 (1945). Toxicology: R. Wien *et al., ibid.* 203.

Crystals from alc, mp 191-193°. One gram dissolves in about 3500 ml water, 440 ml alcohol, 65 ml acetone. Freely sol in dil mineral acids and in aq solns of KOH and NaOH. More sol in warm sugar solns than in water alone. The aq soln is neutral. LD₅₀ orally in mice: 7.5 mg/g, R. Wien *et al., loc. cit.*
Calcium salt, *Orsulon* (obsolete).
Sodium salt monohydrate, $C_{11}H_{10}N_3NaO_2S.H_2O$, *soluble sulfapyridine, Soludagenan.* On prolonged exposure to humid air it absorbs CO₂, liberates sulfapyridine, becomes incompletely sol in water. One gram dissolves in about 1.5 ml water, 10 ml alcohol. pH of 5% aq soln: 11.4. LD₅₀ orally in mice: 2.7 mg/g, R. Wien *et al., loc. cit.*
THERAP CAT: Suppressant (dermatitis herpetiformis); antibacterial.
THERAP CAT (VET): Antibacterial.

8914. Sulfaquinoxaline. *4-Amino-N-2-quinoxalinylbenzenesulfonamide; N¹-(2-quinoxalinyl)sulfanilamide;* 2-sulfanilamidoquinoxaline; N^1-(2-quinoxalyl)sulfanilamide; sulfabenzpyrazine; Compd 3-120; S.Q.; Sulfa-Q 20; Sulquin. $C_{14}H_{12}N_4O_2S$; mol wt 300.33. C 55.99%, H 4.03%, N 18.66%, O 10.65%, S 10.68%. Prepd by treating 2-aminoquinoxaline with acetylsulfanilyl chloride in the presence of pyridine and hydrolyzing the resulting acetyl deriv: Weijlard *et al., J. Am. Chem. Soc.* **66,** 1957 (1944); Weijlard, Tishler, U.S. pat. 2,404,199 (1946 to Merck & Co.). Anticoccidial spectrum: G. F. Mathis *et al., Poultry Sci.* **63,** 1149 (1984). Evaluation of antibacterial and anticoccidial efficacy of mixture with trimethoprim: G. White, R. B. Williams, *Vet. Rec.* **113,** 608 (1983); D. W. T. Piercy *et al., ibid.* **114,** 60 (1984). Drug residue study in poultry muscle and liver: M. Patthy, *J. Chromatog.* **275,** 115 (1983). HPLC determn in

rabbit plasma and urine: J. G. Eppel, J. J. Thiessen, *J. Pharm. Sci.* **73,** 1635 (1984).

Minute crystals, mp 247-248°. uv max (pH 6.6 in H₂O): 252, 360 nm ($E^{1\%}_{1cm}$ 1110, 275). Solubility in water at pH 7: 0.75 mg/100 ml; in 95% alcohol: 73 mg/100 ml; in acetone: 430 mg/100 ml. Sol in aq Na₂CO₃ and NaOH solns.
Sodium salt, $C_{14}H_{11}N_4NaO_2S$, *Aviochina, Embazin.* Very sol in water; pH of 1% soln about 10. The amorphous salt is deliquescent and absorbs CO₂ which liberates the practically insol sulfaquinoxaline.
THERAP CAT (VET): Antimicrobial. Coccidiostat.

8915. Sulfarside. *[2-Amino-4-(aminosulfonyl)phenyl]arsinous acid; 4-sulfamoyl-o-arsanilic acid;* 2-amino-4-aminosulfonylphenylarsinic acid; RP 4482; Bemarside. C_6H_9-AsN_2O_5S; mol wt 296.14. C 24.34%, H 3.06%, As 25.30%, N 9.46%, O 27.01%, S 10.83%. Prepn: Trefouel, U.S. pat. 2,616,913 (1953 to Rhône-Poulenc).

THERAP CAT: Sodium salt as antiamebic.

8916. Sulfarsphenamine. *[1,2-Diarsenediylbis[(6-hydroxy-3,1-phenylene)imino]]bismethanesulfonic acid disodium salt;* disodium 3,3'-diamino-4,4'-dihydroxyarsenobenzene N-dimethylenesulfonate; sulfarsenobenzene; Karsulphan; Myosalvarsan; Myarsenol; Metarsenobillon; Thiosarmine. $C_{14}H_{14}As_2N_2Na_2O_8S_2$; mol wt 598.06. C 28.11%, H 2.36%, As 25.04%, N 4.68%, Na 7.69%, O 21.40%, S 10.72%. Prepn: C. Voegtlin, J. M. Johnson, *J. Am. Chem. Soc.* **44,** 2573 (1922); W. J. C. Dyke, H. King, *J. Chem. Soc.* **1935,** 1745.

Yellow, odorless, or almost odorless powder. Very sol in water, slightly in alcohol. The aq soln is usually slightly acid. *Keep in evacuated or in inert gas-filled ampuls.*
THERAP CAT: Formerly as antisyphilitic.
THERAP CAT (VET): Has been used as antibacterial.

8917. Sulfasalazine. *2-Hydroxy-5-[[4-[(2-pyridinylamino)sulfonyl]phenyl]azo]benzoic acid; 5-[p-(2-pyridylsulfamoyl)phenylazo]salicylic acid;* 5-[4-(2-pyridylsulfamoyl)phenylazo]-2-hydroxybenzoic acid; 4-(pyridyl-2-amidosulfonyl)-3'-carboxy-4'-hydroxyazobenzene; salicylazosulfapyridine; sulphasalazine; Colo-Pleon; Azopyrin; Azulfidine; Salazopyrin. $C_{18}H_{14}N_4O_5S$; mol wt 398.39. C 54.26%, H 3.54%, N 14.06%, O 20.08%, S 8.05%. Conjugate of 5-aminosalicylic acid and sulfapyridine, *q.v.* Prepn and bactericidal effect: E. E. A. Askelof *et al.,* U.S. pat. 2,396,145 (1946 to Aktiebolaget Pharmacia); *see also* Doraswamy, Guha, *J. Indian Chem. Soc.* **23,** 278 (1946). Physicochemical properties: B. Nygard *et al., Acta Pharm. Suecica* **3,** 313-342 (1966). Metabolism studies: H. Schroder, D. E. S. Campbell, *Clin. Pharmacol. Ther.* **13,** 539 (1972); M. A. Peppercorn, P. Goldman, *J. Pharmacol. Exp. Ther.* **181,** 555 (1972). HPLC determn in biological fluids: R. A. Hogezand *et al., J. Chromatog.* **305,** 470 (1984). Comparative clinical trial in rheumatoid arthritis: D. E. Bax, R. S. Amos, *Ann. Rheum. Dis.* **44,** 194 (1985). Review of pharmacology and clinical use in inflammatory bowel disease: M. A. Peppercorn, *Ann. Intern. Med.* **3**, 377 (1984). Comprehensive description: J. P. McDonnell in *Analytical Profiles of Drug Substances* vol. 5, K. Florey, Ed. (Academic Press, New York, 1976) pp 515-532.

Minute, brownish-yellow crystals, dec 240-245°. uv max: 237 ($E_{1cm}^{1\%}$ about 658) and 359 nm. Slightly sol in alc. Practically insol in water, benzene, chloroform, ether.

THERAP CAT: Treatment of ulcerative colitis and Crohn's disease.

THERAP CAT (VET): Has been used in granulomatous colitis.

8918. Sulfasomizole. *4-Amino-N-(3-methyl-5-isothiazolyl)benzenesulfonamide; 3-methyl-5-sulfanilamidoisothiazole;* 5-*p*-aminobenzenesulfonamido-3-methylisothiazole; 5-sulfanilamido-3-methylisothiazole; N^1-(3-methylisothiazol-5-yl)sulfanilamide; sulphasomizole; Bidizole; Amidozol. $C_{10}H_{11}N_3O_2S_2$; mol wt 269.35. C 44.59%, H 4.12%, N 15.60%, O 11.88%, S 23.81%. Preparation from N-acetylsulfanilyl chloride and 5-amino-3-methylisothiazole in pyridine followed by deacetylation in 2N NaOH: Adams, Slack, *J. Chem. Soc.* **1959**, 3070; Adams *et al.*, *Nature* **186**, 221 (1960); **Brit.** pat. **835,753** (1960 to May & Baker).

Cream-colored needles from H_2O, mp 192-192.5°. Soly in H_2O at pH 6.0: 248 mg/100 ml; at pH 7.0: 2.26 g/100 ml.

Sodium salt monohydrate, $C_{10}H_{10}N_3NaO_2S_2.H_2O$, crystals, when anhydr dec 345°. Freely sol in water. Aq solns of almost neutral pH contg more than 50% w/v of the sodium salt can be made.

THERAP CAT: Antibacterial.

8919. Sulfasymazine. *4-Amino-N-(4,6-diethyl-1,3,5-triazin-2-yl)benzenesulfonamide; N^1-(4,6-diethyl-s-triazin-2-yl)sulfanilamide;* 2-sulfanilamido-4,6-diethyl-1,3,5-triazine; 4,6-diethyl-2-sulfanilamido-1,3,5-triazine; sulphsymazine; Symasul; Prosymasul. $C_{13}H_{17}N_5O_2S$; mol wt 307.38. C 50.80%, H 5.57%, N 22.78%, O 10.41%, S 10.43%. Prepn by nucleophilic displacement of methoxy groups from 2-methoxy-4,6-disubstituted *s*-triazines with sulfanilamide anion: Taft *et al.*, *J. Med. Chem.* **8**, 784 (1965); Taft, **U.S.** pat. **3,344,137** (1967 to Am. Cyanamid). Pharmacological studies: Frisk, Hultman, *Antimicrob. Ag. Chemother.* **1965**, 672; Kruger-Thiemer *et al.*, *Chemotherapia* **10**, 325 (1966).

White precipitate from ethanol, mp 186.5-187.5° (Taft); 190-190.5° (Bader). Tasteless. pKa (30% acetone) = 5.4. Sol in water, acetate buffer (1 mg/ml at pH 5.9); also sol in human urine at 37° (2 mg/ml at pH 6; > 10 mg/ml at pH 7).

THERAP CAT: Antibacterial.

8920. Sulfathiazole. *4-Amino-N-2-thiazolylbenzenesulfonamide; N^1-2-thiazolylsulfanilamide;* 2-sulfanilamidothiazole; 2-(sulfanilylamino)thiazole; 2-(p-aminobenzenesulfonamido)thiazole; norsulfazole; RP 2090; M & B 760; Thiazamide; Cibazol; Eleudron; Thiacoccine; Neo-Strepsan; Duatok; Norsulfasol; Planomide; Sulfamul; Sulfavitina; Sulzol; Poliseptil. $C_9H_9N_3O_2S_2$; mol wt 255.32. C 42.34%, H 3.55%, N 16.46%, O 12.53%, S 25.12%. Prepd by reacting

acetylsulfanilyl chloride with 2-aminothiazole in dry pyridine. The resulting N^4-acetylsulfathiazole is deacetylated with NaOH. See Christiansen, **U.S.** pat. **2,242,237** (1941); Kyrides, **U.S.** pat. **2,330,223** (1943); Skrimshire, **Brit.** pat. **540,032** (1941); Leitch, Brickman, **U.S.** pats. **2,230,962** (1941); **2,339,083** (1944); Newbery, **U.S.** pat. **2,362,087** (1944). Sulfathiazole may be synthesized also by condensing halogenoketones, aldehydes, or esters with N^4-acetylsulfanilylthioureas: Földi *et al.*, **U.S.** pat. **2,332,906** (1943). Toxicity of the sodium salt: H. B. van Dyke *et al.*, *Proc. Soc. Exp. Biol. Med.* **42**, 410 (1939); P. H. Long *et al.*, *ibid.* **43**, 328 (1940). Review and bibliography: Northey, *Sulfonamides* (New York, 1948).

Dimorphous. Prismatic rods, mp 200-204°. The other phase forms six-sided plates and prisms which melt at about 175° or invert at a lower temp to the high-melting phase. Soly in water at pH 6 (mg/100 ml): 60 at 26°; 100 at 37°; at pH 7.5: 235 at 37°; in alcohol: 525 at 26°. One gram dissolves in about 1700 ml water, in about 200 ml alc. Soly in human blood serum: 330 mg/100 ml at 37°. Sol in acetone, dil mineral acids, KOH and NaOH solns, ammonia water.

Sodium salt sesquihydrate, $C_9H_8N_3NaO_2S_2.1\frac{1}{2}H_2O$, *soluble sulfathiazole*. Crystals or white powder or granules. On prolonged exposure to humid air, it absorbs carbon dioxide with the liberation of sulfathiazole and becomes incompletely sol in water. Also occurs as the monohydrate and pentahydrate. One gram dissolves in about 2.5 ml water, in about 15 ml alc. Alkaline to phenolphthalein. pH of 1% aq soln 9.35; of 10% soln 10.2. Aq solns are unstable, esp when heated. LD_{50} s.c. in mice: 1.45 g/kg (van Dyke); also reported as 1.95 g/kg (Long).

THERAP CAT: Antibacterial.

THERAP CAT (VET): Antimicrobial.

8921. Sulfathiourea. *4-Amino-N-(aminothioxomethyl)benzenesulfonamide; 1-sulfanilyl-2-thiourea;* sulfathiocarbamide; RP 2255; Badional; Fontamide. $C_7H_9N_3O_2S_2$; mol wt 231.28. C 36.35%, H 3.92%, N 18.17%, O 13.84%, S 27.72%. Prepd by reacting acetylsulfanilylcyanamide with hydrogen sulfide. Has also been prepd directly from sulfanilyl cyanamide and ammonium sulfide: Leitch *et al.*, *Can. J. Res.* **23B**, 139 (1945); alternate method: Földi *et al.*, **U.S.** pat. **2,332,906**.

Crystals, dec 171.5-172°. Soly in water at 37°: 1.1 g/100 ml.

Sodium salt, $C_7H_8N_3NaO_2S$, dec 245-245.5°. Very sol in water. pH of 1% aq soln: 7.

Water sol form, *Solufontamide*, marketed as a slightly viscous soln contg 33% of 1-sulfanilyl-2-thiourea. d_4^{15} 1.16. pH 6.8-7.2.

THERAP CAT: Antibacterial.

8922. Sulfatolamide. *4-Amino-N-(aminothioxomethyl)benzenesulfonamide compd with 4-(aminomethyl)benzenesulfonamide (1:1); 1-sulfanilyl-2-thiourea compd with α-amino-p-toluenesulfonamide;* α-amino-p-toluenesulfonamide salt with 1-sulfanilyl-2-thiourea; 4-homosulfanilamide salt with 1-sulfanilyl-2-thiourea; 1-sulfanilylthiocarbamide salt with 4-aminomethylbenzenesulfonamide; Aseptorid; Marbadal; Marfanil salt of Badional. $C_{14}H_{19}N_5O_4S_3$; mol wt 417.54. Prepn: Behnisch, Klarer, **U.S.** pat. **2,696,454** (1954 to Schenley Ind.).

Crystals, mp 179-181°. Slightly sol in water (to 0.78%)

with neutral reaction. Freely sol in dil hydrochloric acid; sol in dil sodium hydroxide and ammonia; slightly sol in alcohol and acetone. Practically insol in ether.

THERAP CAT: Antibacterial.

8923. Sulfazamet. *4-Amino-N-(3-methyl-1-phenyl-1H-pyrazol-5-yl)benzenesulfonamide; N¹-(3-methyl-1-phenyl-pyrazol-5-yl)sulfanilamide;* 5-(p-aminobenzenesulfonamido)-3-methyl-1-phenylpyrazole; 3-(p-aminobenzenesulfon-amido)-2-phenyl-5-methylpyrazole; 3-methyl-1-phenyl-5-(sulfanilamido)pyrazole; sulfamethylphenazole; sulfapyrazole; Vesulong. $C_{16}H_{16}N_4O_2S$; mol wt 328.41. C 58.52%, H 4.91%, N 17.06%, O 9.74%, S 9.77%. Prepn: Crippa, Guarneri, *Gazz. Chim. Ital.* **85**, 199 (1955); **Brit. pat. 848,-627** (1960 to Ciba); Seydel, Krueger-Theimer, *Arzneimittel-Forsch.* **14**, 1294 (1964). Amino-imino tautomerism: *eidem, ibid.*

Crystals from alc, mp 195° (Seydel, Krueger-Theimer, *loc. cit.*); mp 181-182° (**Brit. pat.** *loc. cit.*). pK'a 5.69.

THERAP CAT (VET): Antibacterial.

8924. Sulfazecin. D-γ-*Glutamyl-N-(3-methoxy-2-oxo-1-sulfo-3-azetidinyl)-D-alaninamide;* 3-γ-D-glutamyl-D-alanylamino-3-methoxyazetidin-2-one-1-sulfonic acid; (3R)-3-(γ-D-glutamyl-D-alanylamino)-3-methoxy-2-oxo-azetidine-1-sulfonic acid; antibiotic G-6302. $C_{12}H_{20}N_4O_9S$; mol wt 396.38. C 36.36%, H 5.09%, N 14.13%, O 36.33%, S 8.09%. Novel monocyclic β-lactam (*monobactam*) antibiotic, produced by *Pseudomonas acidophila* and *P. mesoacidophila.* Isoln, description of physico-chemical and bacterio-static properties: A. Imada *et al.,* **Ger. pat. 2,855,949** corresp to **U.S. pat 4,229,436** (1979, 1980 both to Takeda); *eidem, Nature* **289**, 590 (1981); M. Asai *et al., J. Antibiot.* **34**, 621 (1981). X-ray crystallographic structure determn: K. Kamiya *et al., Acta Crystallogr.* **B37**, 1626 (1981). Prepn of fluorinated sulfazecin analogs: K.Yoshioka *et al., J. Org. Chem.* **49**, 1427 (1984); P. F. Bevilacqua *et al., ibid.* 1430.

Crystallizes from methanol/water as the alcoholate-hemi-hydrate; colorless needles, mp 168-170°. $[\alpha]_D^{25}$ +82° (c = 1.0 in water). pKa: 3.4 (COO⁻); 9.2 (NH₃⁺). Sol in water, DMF, DMSO. Slightly sol in methanol, THF. Practically insol in ethanol, acetone, ethyl acetate, chloroform and other organic solvents. Relatively stable in neutral and weakly acidic solns; unstable in alkaline and strongly acidic solns.

Sodium salt monohydrate, $C_{12}H_{19}N_4NaO_9S.H_2O$. Color-less powder, browns at 170°, no sharp mp. $[\alpha]_D^{20}$ +85° (c = 0.37 in water).

8925. Sulfinalol. *4-Hydroxy-α-[[[3-(4-methoxyphenyl)-1-methylpropyl]amino]methyl]-3-(methylsulfinyl)benzene-methanol;* 4-hydroxy-α-[[[3-(p-methoxyphenyl)-1-methylpropyl]amino]methyl]-3-(methylsulfinyl)benzyl alcohol. $C_{20}H_{27}NO_4S$; mol wt 377.51. C 63.63%, H 7.21%, N 3.71%, O 16.95%, S 8.49%. β-Adrenergic blocker. Prepn: R. E. Philion, **Ger. pat. 2,728,641** (1978 to Sterling), *C.A.* **90**, 137468 (1979). Pharmacological activity profile: P. H. Hernández, *Fed. Proc.* **38**, 738 (1979). Chromatographic determn in human plasma and urine: G. B. Park *et al., Anal. Chem.* **53**, 604 (1981). Metabolism, disposition in animals: D. P. Benziger *et al., Drug. Metab. Dispos.* **9**, 493 (1981).

Hydrochloride, $C_{20}H_{28}ClNO_4S$, *Win 40808-7, Perifadil.* Crystals, mp 172-175°.

THERAP CAT: Antihypertensive.

8926. Sulfinpyrazone. *1,2-Diphenyl-4-[2-(phenylsulfinyl)ethyl]-3,5-pyrazolidinedione;* 1,2-diphenyl-3,5-dioxo-4-(2-phenylsulfinylethyl)pyrazolidine; 4-(phenylsulfoxyethyl)-1,2-diphenyl-3,5-pyrazolidinedione; 4-(2-benzenesulfinyl-ethyl)-1,2-diphenylpyrazolidine-3,5-dione; sulfoxyphenyl-pyrazolidine; G 28315; Anturan; Anturane; Anturano; Enturen. $C_{23}H_{20}N_2O_3S$; mol wt 404.48. C 68.28%, H 4.98%, N 6.93%, O 11.87%, S 7.93%. Synthesis: Pfister, Hafliger, *Helv. Chim. Acta* **44**, 232 (1961). Metabolism in man: W. Dieterle *et al., Arzneimittel-Forsch.* **30**, 989 (1980). Proposed use in prevention of sudden death from heart attacks: S. Sherry *et al., N. Engl. J. Med.* **298**, 289 (1978); **302**, 250 (1980). Symposium: *Cardiovascular Actions of Sulfinpyra-zone: Basic and Clinical Research,* M. M. McGregor, Ed. (Symposia Specialists, Hamilton, Bermuda, 1979) 342 pp. *Review:* E. H. Margulies, A. M. White, in *Pharmacological and Biochemical Properties of Drug Substances* vol. 2, M. E. Goldberg, Ed. (Am. Pharm. Assoc., Washington, DC, 1979) pp 255-278.

dl-Form, crystals from chloroform + heptane, mp 136-137°. Stable to light and air. uv max (1.0N NaOH): 255 nm. Sol in ethyl acetate, chloroform. Slightly sol in water, alcohol, ether, mineral oils, fats.

d-Form, crystals from ethanol, mp 130-133°. $[\alpha]_D^{25}$ +67.1° (c = 2.04 in ethanol); $[\alpha]_D^{10}$ +109.3° (c = 0.5 in CHCl₃). *l*-Form, crystals, mp 130-133°. $[\alpha]_D^{23}$ −64.2° (c = 2.14 in ethanol); $[\alpha]_D^{26}$ −104.5° (c = 0.5 in CHCl₃).

THERAP CAT: Uricosuric.

8927. 4,4'-Sulfinyldianiline. *4,4'-Sulfinylbisbenzene-amine;* di(p-aminophenyl) sulfoxide; 4,4'-diaminodiphenyl sulfoxide; bis(p-aminophenyl) sulfoxide; Medapsol. $C_{12}H_{12}N_2OS$; mol wt 232.30. C 62.04%, H 5.21%, N 12.06%, O 6.89%, S 13.81%. Analog of dapsone, *q.v.* Prepn: Gazdar, Smiles, *J. Chem. Soc.* **93**, 1835 (1908); **Brit. pat. 509,415** (1939 to I. G. Farbenind.), *C.A.* **34**, 3764 (1940); Berti *et al., Rev. Brasil. Leprol.* **20**, 104 (1952), *C.A.* **50**, 11520 (1956). *In vivo* activity vs *Mycobacterium leprae:* L. Levy, *Antimicrob. Ag. Chemother.* **14**, 791 (1978).

Prisms from alc or water, mp 175°. Slightly sol in water.

THERAP CAT: Antibacterial (leprostatic).

8928. Sulfiram. *Tetraethylthiodicarbonic diamide; bis-(diethylthiocarbamoyl)sulfide;* tetraethylthiuram monosulfide; monosulfiram; TTMS; Thiosan; Tetmosol. $C_{10}H_{20}N_2S_3$; mol wt 264.49. C 45.41%, H 7.62%, N 10.59%, S 36.37%. Prepn: Davies, Sexton, *Biochem. J.* **40**, 331 (1946); Ritter, **U.S. pat. 2,524,081** (1950 to Sharples Chem.); Sahasrabu-dhey, Radhakrishnan, *J. Indian Chem. Soc.* **31**, 853 (1954); Zbirovsky, Ettel, *Chem. Listy* **51**, 2094 (1957); **52**, 95 (1958).

Crystals from benzene + petr ether, mp 30-32°.
USE: Fungicide; as vulcanization agent.
THERAP CAT (VET): Ectoparasiticide.

8929. Sulfisomidine. *4-Amino-N-(2,6-dimethyl-4-pyrimidinyl)benzenesulfonamide; N¹-(2,6-dimethyl-4-pyrimidinyl)sulfanilamide;* 2,6-dimethyl-4-sulfanilamidopyrimidine; 6-sulfanilamido-2,4-dimethylpyrimidine; 6-(sulfanilamido)-2,4-dimethyl-1,3-diazine; 6-(p-aminophenylsulfonamido)-2,4-dimethylpyrimidine; 6-(p-aminobenzenesulfonyl)amino-2,4-dimethylpyrimidine; sulfadimetine; sulfaisodimidine; sulphasomidine; Elkosin; Elcosine; Elkosil; Domain; Aristamid. $C_{12}H_{14}N_4O_2S$; mol wt 278.34. C 51.78%, H 5.07%, N 20.13%, O 11.50%, S 11.52%. Prepn: **Fr. pat. 886,009** (1943 to Nordmark); Gysin, **U.S. pat. 2,351,333** (1944 to Geigy); Hartmann *et al.,* **U.S. pats. 2,386,852; 2,429,184** (1945, 1947, both to Ciba); Matsukawa *et al., J. Pharm. Soc. Japan* **70,** 283 (1950); Loop, Lührs, *Ann.* **580,** 225 (1953).

Needles from ethanol, mp 243°. Soly in water at 15°: 0.12 g/100 ml, at 30°: 0.30 g/100 ml. More sol in hot water (about 1:60). Aq solns are neutral to litmus. Soly in urine at 37°: 360 mg/100 ml at pH 5.5 to 1100 mg/100 ml at pH 7.5. Slightly sol in alcohol, acetone. Practically insol in benzene, ether, chloroform. Freely sol in dil HCl and NaOH. Compared with sulfanilamide it is but slightly acetylated in the body. *Do not confuse with sulfamethazine.*
THERAP CAT: Antibacterial.
THERAP CAT (VET): Antimicrobial.

8930. Sulfisoxazole. *4-Amino-N-(3,4-dimethyl-5-isoxazolyl)benzenesulfonamide; N¹-(3,4-dimethyl-5-isoxazolyl)sulfanilamide;* 3,4-dimethyl-5-sulfanilamidoisoxazole; 5-(4-aminophenylsulfonamido)-3,4-dimethylisoxazole; 5-(p-aminobenzenesulfonamido)-3,4-dimethylisooxazole; sulfafurazole; sulphafurazole; Entusil; Entusul; Sulfoxol; Gantrisin; Gantrosan; Renosulfan; Soxisol; Soxomide; Sulsoxin; Sosol; V-Sul; Sulfalar; Soxo; Sulfazin; Neazolin. $C_{11}H_{13}N_3O_3S$; mol wt 267.30. C 49.42%, H 4.90%, N 15.72%, O 17.96%, S 12.00%. Prepd by the action of p-acetaminobenzenesulfonyl chloride on 3,4-dimethyl-5-aminoisoxazole followed by deacetylation: Wuest, Hoffer, **U.S. pat. 2,430,094** (1947 to Hoffmann-La Roche); **Brit. pat. 595,775** (1947). Comprehensive description: B. C. Rudy, B. Z. Senkowski, in *Analytical Profiles of Drug Substances* vol. 2, K. Florey, Ed. (Academic Press, New York, 1973) pp 487-506.

Prisms, mp 194°. Acid to litmus. More sol in water than sulfanilamide. According to the patents cited, the soly at pH 5 to 7 is so high that neither the compd itself nor its N^4-acetyl derivative is deposited in the kidneys. Soly at pH 6.0: 350 mg/100 ml; acetyl derivative: 110 mg/100 ml. Sparingly sol in alcohol. Insol in chloroform. LD_{50} orally in mice: 6800 mg/kg, Seki *et al., Arzneimittel-Forsch.* **15,** 1441 (1965). Sodium salt, prisms contg 5 mols H_2O. Soly in water (calcd for the anhydr): 16% at 37°; 7.4% at 25°. If an aq soln of the sodium salt is shaken with an excess of sulfisoxazole, the pH changes from a little above 8 to 7.2. Such a soln can be sterilized at 100°.

Lithium salt, behaves similarly to sodium salt. Soly in water: over 20% at 5°.
Diethanolamine salt, $C_{15}H_{24}N_4O_5S$, *sulfisoxazole diolamine, Suladrin, Sulfium.* Crystals, very sol in water (>40%). A 4% soln is about isotonic with tears.
Note: See also Acetyl Sulfisoxazole.
THERAP CAT: Antibacterial.
THERAP CAT (VET): Antibacterial.

8931. Sulfitocobalamin. *Cobinamide sulfite phosphate 3'-ester with 5,6-dimethyl-1-α-D-ribofuranosylbenzimidazole;* sulphitocobalamin. $C_{62}H_{88}CoN_{13}O_{17}PS$; mol wt 1409.47. C 52.76%, H 6.43%, Co 4.18%, N 12.90%, O 19.27%, P 2.19%, S 2.27%. An analog of vitamin B_{12} where the CN group is replaced with $—SO_3$. Prepn from cyanocobalamin or hydroxocobalamin with excess water-soluble bisulfite: Fricke, **U.S. pat. 2,721,162** (1955 to Abbott). Structure: Dolphin *et al., Nature* **199,** 170 (1963).
Stable aq soln. Absorption max: 275, 365, 418, 516 nm ($E_{1cm}^{1\%}$ 328, 130, 49, 61).

8932. Sulfoacetic Acid. Sulfoethanoic acid. $C_2H_4O_5S$; mol wt 140.11. C 17.14%, H 2.88%, O 57.09%, S 22.88%. HO_3SCH_2COOH.
Monohydrate, hygroscopic crystals. mp 84-86° when anhydr. bp 245° with decompn. Sol in water, alcohol; insol in ether, chloroform. *Keep well closed.* LD_{50} orally in rats: 3.16 g/kg, Smyth *et al., J. Ind. Hyg. Toxicol.* **31,** 60 (1949).

8933. Sulfobromophthalein Sodium. *3,3'-(4,5,6,7-Tetrabromo-3-oxo-1(3H)-isobenzofuranylidene)bis[6-hydroxybenzenesulfonic acid] disodium salt;* disodium phenoltetrabromophthalein sulfonate; bromsulfophthalein; bromosulfophthalein; BSP; Hepartest; Bromthalein; Bromsulphalein Sodium; Brom-Tetragnost. $C_{20}H_8Br_4Na_2O_{10}S_2$; mol wt 838.05. C 28.66%, H 0.96%, Br 38.14%, Na 5.49%, O 19.09%, S 7.65%. Prepd like phenoltetrachlorophthalein by the condensation of phenol and tetrabromophthalic acid or its anhydride, followed by sulfonation and conversion to the disodium salt. Pharmacology and use as liver function test: S. M. Rosenthal, E. C. White, *J. Pharmacol.* **24,** 265 (1924); W. Häcki *et al., J. Lab. Clin. Med.* **88,** 1019 (1976).

Hygroscopic crystals. Bitter taste. Sol in water. Insol in alcohol, acetone. Alkaline solns have an intense bluish-purple color.
THERAP CAT: Diagnostic aid (hepatic function).
THERAP CAT (VET): Diagnostic aid (hepatic function).

8934. Sulfolane. *Tetrahydrothiophene 1,1-dioxide;* tetrahydrothiophene 1-dioxide; thiophan sulfone. $C_4H_8O_2S$; mol wt 120.16. C 39.98%, H 6.71%, O 26.63%, S 26.69%. Made by catalytic hydrogenation of sulfolene oxides: Mahan, Fauske, **U.S. pat. 2,578,565** (1951 to Phillips Petroleum); from butadiene and sulfur dioxide: Staaterman *et al., Chem. Eng. Progr.* **43,** 148 (1947); Evans, Morris, **U.S. pat. 2,360,-859** (1944 to Shell).

Liquid, d_4^{30} 1.2606. mp 27.4-27.8°. bp_{760} 285°. n_D^{30} 1.481. Viscosity at 30° = 10.34 cp. Flash pt, open cup: 350°F (176°C). At 30°, miscible with water, acetone, toluene; partially misc with octanes, olefins and naphthenes. LD_{50} orally in rats: 1.54 ml/kg, H. F. Smyth *et al., Am. Ind. Hyg. Assoc. J.* **30,** 470 (1962).

USE: Selective solvent for liquid-vapor extractions.

8935. 3-Sulfolene. *2,5-Dihydrothiophene 1,1-dioxide;* 1-thia-3-cyclopentene 1,1-dioxide; butadiene sulfone. C_4-H_6O_2S; mol wt 118.16. C 40.66%, H 5.12%, O 27.08%, S 27.14%. Preparation: Staudinger, Rizenthaler, *Ber.* **68,** 455 (1935); de Roy van Zuydewijn, *Rec. Trav. Chim.* **56,** 1047 (1937); Hooker *et al., U.S. pat.* **2,395,050** (1946 to Dow); Morris, Finch, **U.S. pat. 2,420,834** (1947 to Shell).

Crystals, mp 64-65.5°. Sol in water, organic solvents.

8936. Sulfometuron Methyl. *2-[[[[(4,6-Dimethyl-2-pyrimidinyl)amino]carbonyl]amino]sulfonyl]benzoic acid methyl ester; N-[(4,6-dimethylpyrimidin-2-yl)aminocarbonyl]-2-methoxycarbonylbenzenesulfonamide;* 2-[3-(4,6-dimethylpyrimidin-2-yl)ureidosulfonyl]benzoic acid methyl ester; Aa 5648; DPX 5648; Oust. $C_{15}H_{16}N_4O_5S$; mol wt 364.38. C 49.44%, H 4.43%, N 15.38%, O 21.95%, S 8.80%. Nonselective pre- and post-emergence sulfonylurea herbicide for noncropland use; inhibits branched chain amino acid biosynthesis. Prepn: G. Levitt, **Eur. pat. Appl. 7,687;** *idem,* **U.S. pat. 4,394,506** (1980, 1983 both to Du Pont); *idem, Pestic. Chem.: Hum. Welfare Environ., Proc. 5th Int. Congr. Pestic.* **1,** 243 (1983). Herbicidal activity and toxicity: R. H. Harding *et al., Proc. West. Soc. Weed Sci.* **34,** 120 (1981). Mechanism of action: T. B. Ray, *Proc. Brit. Crop Prot. Conf. - Weeds* **1985,** 131. HPLC determn in soil and water: E. W. Zahnow, *J. Agr. Food Chem.* **33,** 479 (1985). Degradation and bioaccumulation studies: J. Harvey *et al., ibid.* 590; J. J. Anderson, J. J. Dulka, *ibid.* 596. Field trials: R. L. Atkins *et al., Proc. South. Weed Sci. Soc.* **36,** 300 (1983); in afforested areas: A. Nir, Z. Arenstein, *Proc. Brit. Crop Prot. Conf. - Weeds* **1987,** 787.

White solid, mp 198-202°. Soly in water at 25° (ppm): 10 at pH 5, 300 at pH 7. pKa 5.7. LD_{50} in male, female rats (mg/kg): > 5000, > 5000 orally (Harding).

USE: Herbicide.

8937. Sulfonethylmethane. *2,2-Bis(ethylsulfonyl)butane;* diethylsulfonmethylethylmethane; methylsulfonal; Ethylsulfonal; Trional. $C_8H_{18}O_4S_2$; mol wt 242.36. C 39.64%, H 7.49%, O 26.41%, S 26.46%. Prepd by condensing ethylmercaptan with methyl ethyl ketone and oxidizing the resulting mercaptol with $KMnO_4$: Fromm, *Ann.* **253,** 135 (1889); *cf.* Baumann, *Ber.* **18,** 883 (1885); *Arch. Pharm.* **226,** 511 (1888); Bayer, *Pharm. Z.* **34,** 98 (1889); **Ger. pat. 46,333;** Moragas, Uthoff, *Chim. Ind. (Paris)* **17,** 284 (1927); Ramberg, Samén, *C.A.* **28,** 6103 (1934); Suter, *Organic Chemistry of Sulfur* (New York, 1944) pp 735, 742.

Lustrous leaflets, scales from water. Bitter taste. mp 74-76°. Upon further heating dec with evolution of SO_2. One gram dissolves in 200 ml cold, about 30 ml boiling water, in 8 ml alc. Sol in ether. A satd aq soln is neutral to litmus.

Incompat. A semiliquid paste results when sulfonethylmethane is triturated with chloral hydrate, salol or ethyl urethan.

Caution: May be habit forming. This is a controlled substance (depressant) listed in the U.S. Code of Federal Regulations, Title 21 Part 1308.13 (1985).

THERAP CAT: Hypnotic.

THERAP CAT (VET): Has been used as a hypnotic.

8938. Sulfoniazide. *4-Pyridinecarboxylic acid [(3-sulfophenyl)methylene]hydrazide; isonicotinic acid m-sulfobenzylidene hydrazide;* isonicotinyl hydrazonotoluene-*m*-sulfonic acid; isonicotinoyl hydrazone of *m*-sulfobenzaldehyde; G 605. $C_{13}H_{11}N_3O_4S$; mol wt 305.32. C 51.14%, H 3.63%, N 13.76%, O 20.96%, S 10.50%. Prepn: Girard, **U.S. pat. 2,727,041** (1955 to Lab. Français Chimiother.); Sugimoto, **Japan. pat. 1117('57)** (to Tanake Drug), *C.A.* **52,** 4696e (1958).

Needles, dec 250-253°. Slightly sol in water.

Sodium salt trihydrate, $C_{13}H_{10}N_3NaO_4S.3H_2O$, *Isorilone, Sulfon-Niazone.* Crystals from water. Soly in water at 18°: 1 g/33 cc.

THERAP CAT: Antibacterial (tuberculostatic).

8939. Sulfonmethane. *2,2-Bis(ethylsulfonyl)propane;* diethylsulfondimethylmethane; propane-diethyl sulfone; Sulfonal. $C_7H_{16}O_4S_2$; mol wt 228.33. C 36.82%, H 7.06%, O 28.03%, S 28.09%. Prepd by condensing ethyl mercaptan with acetone, then oxidizing with $KMnO_4$: Fromm, *Ann.* **253,** 140 (1889); *cf.* Baumann, *Ber.* **18,** 883 (1885); *Arch. Pharm.* **226,** 511 (1888); Bayer, *Pharm. Z.* **34,** 98 (1889); **Ger. pat. 46,333;** Moragas, Uthoff, *Chim. Ind. (Paris)* **17,** 284 (1927); Ramberg, Samén, *C.A.* **28,** 6103 (1934); Suter, *Organic Chemistry of Sulfur* (New York, 1944) pp 735, 742. Soly data: Falck, *J. Pharm. Chim.* [7] **21,** 279 (1920).

Crystals, almost tasteless, mp 124-126°. bp 300°. One gram dissolves in 365 ml water, 16 ml boiling water, 60 ml alcohol, 3 ml boiling alcohol, 64 ml ether, 11 ml chloroform. Soluble in benzene. Insol in glycerol.

Caution: May be habit forming. This is a controlled substance (depressant) listed in the U.S. Code of Federal Regulations, Title 21 Part 1308.13 (1985).

THERAP CAT: Hypnotic.

THERAP CAT (VET): Has been used as a hypnotic.

8940. Sulfonyldiacetic Acid. Sulfonediacetic acid; dimethylenesulfone-α,α'-dicarboxylic acid. $C_4H_6O_6S$; mol wt 182.15. C 26.37%, H 3.32%, O 52.70%, S 17.60%. HOOC-$CH_2SO_2CH_2$COOH. Prepd by oxidation of thiodiglycolic acid: Lovén, *Ber.* **17,** 2818 (1884).

Tabular crystals, mp 182°. Freely sol in water. Sol in alc. Slightly sol in ether.

USE: Detection of barium, lead, mercury and silver.

8941. *p,p'* -**Sulfonyldianiline -*N*,*N'* -digalactoside.** *N,N'-(Sulfonyldi-4,1-phenylene)bis[galactosylamine];* bis(4-aminophenyl)sulfone digalactoside; Eupatin II; Tibatin. $C_{24}H_{32}$-$N_2O_{12}S$; mol wt 572.58. C 50.34%, H 5.63%, N 4.89%, O 33.53%, S 5.60%. Prepd by refluxing a mixture of 4,4'-diaminodiphenylsulfone, galactose, methanol, and a small amount of salicylic acid, *cf. P.B. Report* **248,** p 14. Review of prepn and reactions of this type of compd: Suter, *Organic Chemistry of Sulfur* (New York, 1944) p 704. *See also* Glucosulfone Sodium.

White powder. Freely sol in water. Insol in ether. A 40% aq soln is neutral to litmus. Solns in ampuls darken on standing. This discoloration is said to have no effect on activity or toxicity.

THERAP CAT: Antibacterial.

8942. Sulforaphen. *4-Isothiocyanato-1-(methylsulfinyl)-1-butene; isothiocyanic acid 4-(methylsulfinyl)-3-butenyl ester;* raphanin. $C_6H_9NOS_2$; mol wt 175.28. C 41.12%, H 5.17%, N 7.99%, O 9.13%, S 36.59%. From seeds of radish, *Raphanus sativus* L., *Cruciferae:* Ivanovics, Horvath, *Nature* **160**, 297 (1947); *Proc. Soc. Exp. Biol. Med.* **66**, 625 (1947). Identity with raphanin: Koczka, Ivanovics, *C.A.* **44**, 5538c (1950). Structure: Schmid, Karrer, *Helv. Chim. Acta* **31**, 1017 (1948). Synthesis of ($\pm$)-form: Balenovic *et al., Tetrahedron* **22**, 2139 (1966).

$$CH_3\overset{\overset{O}{\|}}{S}CH=CHCH_2CH_2NCS$$

Liquid. $bp_{0.015}$ 125-130°. $[\alpha]_D^{19}$ $-107°$ (c = 1.37 in chloroform); $[\alpha]_D^{14}$ $-136°$ (c = 1.38 in alcohol). Sol in water, methanol, chloroform; moderately sol in ether; practically insol in petr ether.

8943. Sulforidazine. *10-[2-(1-Methyl-2-piperidinyl)ethyl]-2-(methylsulfonyl)-10H-phenothiazine;* 3-(methylsulfonyl)-10-[2-(1-methyl-2-piperidyl)ethyl]phenothiazine; thioridazine-2-sulfone; TPN-12; Imagotan; Inofal. $C_{21}H_{26}N_2O_2S_2$; mol wt 402.57. C 62.65%, H 6.51%, N 6.96%, O 7.95%, S 15.93%. Dopamine receptor blocker; metabolite of thioridazine and mesoridazine, q.q.v. Prepn: Renz *et al.,* **Fr. pats. 1,363,683** and **1,459,476** (1964, 1966, both to Sandoz). Pharmacology: Maruyama *et al., C.A.* **68**, 20799z (1968). Effects on dopaminergic function in comparison with mesoridazine and thioridazine: D. M. Niedzwiecki *et al., J. Pharmacol. Exp. Ther.* **228**, 636 (1984); C. D. Kilts *et al., ibid.* **231**, 334 (1984). Clinical evaluation as antipsychotic: R. Axelsson, *Curr. Ther. Res.* **21**, 587 (1977). HPLC determn in plasma: D. A. Ganes, K. K. Midha, *J. Chromatog.* **423**, 227 (1987).

Crystals from acetone, mp 121-123°.
THERAP CAT: Antipsychotic.

8944. Sulfosalicylic Acid. *2-hydroxy-5-sulfobenzoic acid; 3-carboxy-4-hydroxybenzenesulfonic acid;* 5-sulfosalicylic acid; 2-hydroxybenzoic-5-sulfonic acid; salicylsulfonic acid. $C_7H_6O_6S$; mol wt 218.18. C 38.53%, H 2.77%, O 44.00%, S 14.70%. Prepn: Pusca *et al., Rev. Chim. (Bucharest)* **13**, no. 1, 49-50 (1962), *C.A.* **57**, 14994b (1962).

Dihydrate, white crystals or cryst powder. Colored pink by traces of iron. mp about 120° when anhydr; at higher temp dec into phenol and salicylic acid. Very sol in water or alcohol; sol in ether. Generally sol in polar solvents. *Keep well closed and protected from light.*
Caution: Irritating to skin, mucous membranes.
USE: Clinical reagent for albumin in urine; colorimetric reagent for ferric ion with which it gives a violet color. It is a trifunctional aromatic compound, undergoing reactions typical of phenols, carboxylic and sulfonic acid groups. Suggested industrial uses: metal chelating agent, intermedi-

ate in the manuf of surface-active agents, organic catalysts and grease additives.

8945. Sulfotep. *Thiodiphosphoric acid tetraethyl ester; thiopyrophosphoric acid tetraethyl ester;* sulfotepp; thiotepp; dithio; dithione; dithiophos; TEDP; ASP-47; Bayer E 393; ENT 16273; Bladafum. $C_8H_{20}O_5P_2S_2$; mol wt 322.30. C 29.81%, H 6.26%, O 24.82%, P 19.22%, S 19.89%. Prepn: A. D. F. Toy, *J. Am. Chem. Soc.* **73**, 4670 (1951); G. Schrader, R. Mühlmann, **Ger. pat. 848,812** (1952 to Bayer), *C.A.* **47**, 5426a (1953); R. Appel, H. Einig, *Z. Anorg. Allgem. Chem.* **414**, 236 (1975). Activity: F. F. Smith *et al., J. Econ. Entomol.* **43**, 627 (1950). Toxicity data: B. Holmstedt, *Pharmacol. Rev.* **11**, 567 (1959).

Pale yellow liquid, bp_2 136-139°. *Caution: Poison!* Vapor press. at 20°: 1.7×10^{-4} mm Hg. d_4^{25} 1.196. n_D^{25} 1.4753. Soly in water: 25 mg/l. Miscible with most organic solvents. Corrosive to iron. LD_{50} s.c. in mice: 8 mg/kg (Holmstedt).
USE: Insecticide; miticide.

8946. Sulfoxide. *5-[2-(Octylsulfinyl)propyl]-1,3-benzodioxole; 1,2-(methylenedioxy)-4-[2-(octylsulfinyl)propyl]benzene;* n-octylsulfoxide of isosafrole; 1-methyl-2-(3,4-methylenedioxyphenyl)ethyl octyl sulfoxide; Sulfox-Cide. $C_{18}H_{28}O_3S$; mol wt 324.49. C 66.63%, H 8.70%, O 14.79%, S 9.88%. Prepn: M. E. Synerholm *et al., Contribs. Boyce Thompson Inst.* **15**, 35 (1947); M. E. Synerholm, **U.S. pat. 2,486,579** (1949 to Boyce Thompson Inst. for Plant Res.). Synergistic use with pyrethrum, allethrin, rotenone, ryania, etc.: *idem,* **U.S. pat. 2,486,445** (1949 to Boyce Thompson Inst. for Plant Res.). Acute toxicity: T. B. Gaines, R. E. Linder, *Fundam. Appl. Toxicol.* **7**, 299 (1986).

Pale yellow, sweet-smelling, viscous oil. Practically insol in water. Sol in most organic solvents except petr ether. LD_{50} in adult male, female rats (mg/kg): 3957, 3477 orally (Gaines, Linder).
USE: Insecticide.

8947. Sulfoxone Sodium. *Disodium[sulfonylbis(p-phenylenimino)]dimethanesulfinate;* 4,4'-diaminodiphenylsulfone disodium formaldehyde sulfoxylate; disodium formaldehydesulfoxylate-diaminodiphenylsulfone; aldesulfone sodium; Diazon; Novotrone; Diasone. $C_{14}H_{14}N_2Na_2O_6S_3$; mol wt 448.43. C 37.49%, H 3.15%, N 6.25%, Na 10.26%, O 21.41%, S 21.45%. Obtained as the di- or tetrahydrate. Prepd by combining p,p'-sulfonyldianiline with formaldehyde sulfoxylate in acetic acid or alcohol: Bauer, *J. Am. Chem. Soc.* **61**, 617 (1939); Rosenthal, Bauer, **U.S. pat. 2,234,981** (1941); Raiziss *et al., J. Am. Pharm. Assoc.* **33**, 43 (1944); **U.S. pat. 2,256,575** (1941).

Dihydrate, needles or amorphous powder. After drying at 100-110° it dec 263-265°. Easily sol in water. Slightly sol in alcohol; insol in the usual organic solvents. Must be stored in vacuum-sealed ampuls. The addition of 10% sodium bicarbonate helps to prevent oxidative changes.
THERAP CAT: Antibacterial (leprostatic).

8948. Sulfur. Brimstone; sulphur. S; at. wt 32.064 $\pm$ 0.003; at. no. 16; valences 2, 4, 6. Four naturally occurring isotopes: 32 (95.0%); 33 (0.76%); 34 (4.22%); 36 (0.014%); artificial, radioactive isotopes: 29-31; 35; 37; 38. Has been known from very early times. Occurs both in the free state and in combination, mainly as sulfides and sulfates; constitutes 0.05% of the crust of the earth. Industrial prepn by Frasch process: Faith, Keyes & Clark's *Industrial Chemicals*, F. A. Lowenheim, M. K. Moran, Eds. (Wiley-Interscience, New York, 4th ed., 1975) pp 786-794. Exists in several allotropic modifications; at S.T.P. only the orthorhombic, cyclooctasulfur (S_{α}) is thermodynamically stable: Meyer, *Chem. Rev.* **64**, 429 (1964). Prepn of cyclohexasulfur: Bartlett *et al.*, *J. Am. Chem. Soc.* **83**, 10 (1961); of cyclohepta- and cyclooctasulfur: Schmidt *et al.*, *Angew. Chem. Int. Ed.* **7**, 632 (1968); of cyclononasulfur: Schmidt, Wilhelm, *Chem. Commun.* **1970**, 1111; of cyclododecasulfur: *eidem, Angew. Chem. Int. Ed.* **5**, 964 (1966). Review of structures of elemental sulfur: Meyer in *Advan. Inorg. Chem. Radiochem.* **18**, 287-317 (1976). Reviews of sulfur and its compds: W. N. Tuller, *The Sulphur Data Book* (McGraw-Hill, New York, 1954); B. Meyer, *Elemental Sulfur* (Interscience-Wiley, 1965); *Inorganic Sulphur Chemistry*, G. Nickless, Ed. (Elsevier, New York, 1968); Schmidt, Siebert, in *Comprehensive Inorganic Chemistry* vol. 2, J. C. Bailar, Jr. *et al.*, Eds. (Pergamon, Oxford, 1973) pp 795-933.

Orthorhombic, cycloocta- or α-sulfur, amber-colored crystals; the stable form at ordinary temperature. d 2.06; when heated to 94.5° becomes opaque owing to the formation of monoclinic sulfur.

Monoclinic, cycloocta- or β-sulfur, light-yellow, opaque, brittle, needle-like crystals; stable between 94.5-120°. Passes slowly into the rhombic form on standing. d 1.96; mp 115.21°: West, *J. Am. Chem. Soc.* **81**, 29 (1959). bp about 444.6°. A second monoclinic form, γ-sulfur or mother-of-pearl sulfur is also known; mp 106.8°.

Sulfur is insol in water; sparingly sol in alcohol, in ether; sol in carbon disulfide (one gram/2 ml); sol in benzene (about 2.4% at 30°, much more at higher temps.), in toluene. Liquid ammonia (anhydr) dissolves 38.5% S at −78°; acetone dissolves 2.65% at 25°; methylene iodide dissolves 9.1% at 10°; chloroform dissolves about 1.5% at 18°. For other solvents *see* Tuller, *op. cit.* Ignites in air above 261°, in oxygen below 260°, burning to the dioxide; combines readily with hydrogen; combines in the cold with fluorine, chlorine, and bromine; combines with carbon at high temperatures; reacts with silicon, phosphorus, arsenic, antimony and bismuth at their melting points; combines with nearly all metals; with lithium, sodium, potassium, copper, mercury and silver in the cold on contact with the solid; with magnesium, zinc and cadmium very slightly in the cold, more readily on heating; with other metals at high temperatures. Does not react with iodine, nitrogen, tellurium, gold, platinum and iridium.

Polymeric sulfur, *Crystex*. Amorphous form; mol wt approx 200,000. Metastable, gradually reverts to the α-form. Insol in solvents used for orthorhombic form. *Reviews:* Meyer, *loc. cit.* (1964); A. V. Tobolsky, W. J. MacKnight, *Polymeric Sulfur and Related Polymers* (Interscience, New York, 1965) pp 87-97.

Note: See also Sulfur, Pharmaceutical.

USE: In mfg sulfuric acid, carbon disulfide, sulfites, insecticides, plastics, enamels, metal-glass cements; in vulcanizing rubber; in syntheses of dyes; in making gunpowder, matches; for bleaching wood pulp, straw, wool, silk, felt, linen. *Caution:* May cause irritation of skin, mucous membranes.

8949. Sulfur Chloride. Disulfur dichloride; sulfur monochloride; sulfur subchloride. Cl_2S_2; mol wt 135.03. Cl 52.52%, S 47.48%. S_2Cl_2. Prepared by passing chlorine into molten sulfur, can be purified by fractional distillation: Fehér in *Handbook of Preparative Inorganic Chemistry* vol. 1, G. Brauer, Ed. (Academic Press, New York, 2nd ed., 1963) pp 371-372. Hydrolysis: H. L. Olin, *J. Am. Chem. Soc.* **48**, 167 (1926).

Non-flammable, light amber to yellowish red, fuming, oily liquid. Penetrating odor. *Vapors are corrosive and irritate eyes, nose, and throat, cause tears and affect breathing.* $d_{15.5}^{15.5}$ 1.6885. n_D^{20} 1.670. mp −77°. bp_{760} 138°; bp_{100} 72.0°; bp_{10} 19.1°. Dielectric constant 4.9 at 22°. Dipole moment 1.60. Sol in alcohol, benzene, ether, carbon disulfide, carbon tetrachloride, oils. Readily dissolves sulfur (up to 67% at room

temp). Dec by water yielding sulfur, hydrogen chloride, sulfur dioxide, hydrogen sulfide, sulfite, thiosulfate. In acid solns pentathionic and other polythionic acids are formed. *Keep tightly closed and out of contact with water.*

USE: Intermediate and chlorinating agent in the manuf of organic chemicals, sulfur dyes, insecticides, synthetic rubbers; in cold vulcanization of rubber; as polymerization catalyst for vegetable oils; for hardening soft woods.

8950. Sulfur Dioxide. Sulfurous anhydride; sulfurous oxide. O_2S; mol wt 64.07. O 49.95%, S 50.05%. SO_2. Colorless, nonflammable gas; strong suffocating odor. Condenses at −10° and ordinary pressure to a colorless liquid. d (liq) 1.5. mp −72°. bp −10°. Mixed with oxygen and passed over red-hot Pt, it is converted into SO_3. With water forms sulfurous acid (H_2SO_3). Bleaches vegetable colors. Soly: in water 17.7% at 0°, 11.9% at 15°, 8.5% at 25°, 6.4% at 35°; in alcohol 25%; in methanol 32%; also sol in chloroform or ether. It is supplied compressed in cylinders.

USE: Preserving fruits, vegetables, etc.; disinfectant in breweries and food factories; bleaching textile fibers, straw, wicker ware, gelatin, glue, beet sugars. Liquid SO_2 used as solvent: Waddington, *Non-aqueous Solvent Systems*, T. C. Waddington, Ed. (Academic Press, New York, 1965) pp 253-284. *Caution:* Intensely irritating to eyes, respiratory tract.

8951. Sulfuretin. *2-[(3,4-Dihydroxyphenyl)methylene]-6-hydroxy-3(2H)-benzofuranone; 2-(3,4-dihydroxybenzylidene)-6-hydroxy-3(2H)-benzofuranone; 3′,4′,6-trihydroxyaurone; 3′,4′,6-trihydroxybenzalcoumaranone.* $C_{15}H_{10}O_5$; mol wt 270.23. C 66.67%, H 3.73%, O 29.60%. Flavonoid pigment responsible for the yellow color of certain species of *Compositae*. Isoln from *Cosmos sulphureus* Cav., *Compositae:* Shimokoriyama, Hattori, *J. Am. Chem. Soc.* **75**, 1900 (1953); from *Dahlia variabilis* Desf., *Compositae:* Nordström, Swain, *Arch. Biochem. Biophys.* **60**, 329 (1956). Prepn by condensation of 6-hydroxy-3-coumaranone and protocatechuic alcohol: Nordström, Swain, *ibid.* Prepn and structure studies: Geissman, Jurd, *J. Am. Chem. Soc.* **76**, 4475 (1954); Farkas *et al.*, *Ber.* **92**, 2847 (1959). Structure: Shimokoriyama, Geissman, *J. Org. Chem.* **25**, 1956 (1960).

Orange crystals from dil alc, mp 280-285° (Shimokoriyama), mp 315° (dec) (Nordström), mp 302-304° (Farkas). uv max: 398 nm.

Triacetate, $C_{21}H_{16}O_8$, pale yellow needles from methanol, mp 191-194° (Shimokoriyama), mp 167-168° (Nordström).

8952. Sulfur Hexafluoride. F_6S; mol wt 146.07. F 78.05%, S 21.95%. SF_6. Prepd by direct fluorination of sulfur or sulfur dioxide: Moissan, Lebeau, *Compt. Rend.* **130**, 865, 984 (1900); *eidem, Ann. Chim. Phys.* [7] **26**, 147 (1902); Schumb, *Inorg. Syn.* **3**, 119 (1950); Kwasnik in *Handbook of Preparative Inorganic Chemistry*, Vol. 1, G. Brauer, Ed. (Academic Press, New York, 2nd ed., 1963) pp 169-170. *Reviews:* Cady, "Fluorine-Containing Compounds of Sulfur" in *Advan. Inorg. Chem. Radiochem.* **2**, 105-157 (1960); Kemmitt, Sharp, *Advan. Fluorine Chem.* **4**, 218-219 (1965).

Colorless, odorless gas. One of the heaviest known gases; density approx 5 times that of air. mp −50.8°. Sublimes at −63.8°. Crit temp 45.6°. d (liq; −50.8°) 1.88. Sparingly sol in water, somewhat more in alcohol. At 25° and 1 atm 0.297 ml SF_6 dissolves in 1.0 ml of transformer oil. Thermodynamically unstable but kinetically stable gas. This stability explained by symmetrical, octahedral structure of the molecule. Inert to nucleophilic attack. Does not attack glass. No fluorine exchange with HF. Stable to silent electrical discharge. Unchanged at 500°.

USE: In electrical circuit interrupters. In electronic ultrahigh frequency piping.

8953. Sulfuric Acid. Oil of vitriol. H_2O_4S; mol wt

98.08. H 2.06%, O 65.25%, S 32.69%. H_2SO_4. Prepd by the Contact Process according to the reactions $2SO_2 + O_2 \rightarrow 2SO_3$, and $SO_3 + H_2O \rightarrow H_2SO_4$; by the Chamber Process according to the reactions $2NO + O_2 \rightarrow 2NO_2$, and $NO_2 + SO_2 + H_2O \rightarrow H_2SO_4 + NO$. Sulfuric acid of commerce contains 93-98% H_2SO_4; the remainder is water. Monograph: W. W. Duecker, J. R. West, *The Manufacture of Sulfuric Acid* (Reinhold, New York, 1959) 515 pp. Review of manuf: Pearce, "Sulphuric Acid: Physico-Chemical Aspects of Manufacture" in *Inorganic Sulphur Chemistry*, G. Nickless, Ed. (Elsevier, New York, 1968) pp 535-561; Faith, Keyes & Clark's *Industrial Chemicals*, F. A. Lowenheim, M. K. Moran, Eds. (Wiley-Interscience, New York,4th ed., 1975) pp 795-806. Toxicity data: H. F. Smyth *et al.*, *Am. Ind. Hyg. Assoc. J.* **30**, 470 (1969).

Clear, colorless, odorless, oily liquid. *Very corrosive!* Has a very great affinity for water, abstracting it from the air and also from many organic substances; hence it chars sugar, wood, etc. d ~1.84. bp ~290°; dec 340° into sulfur trioxide and water. mp 10° (anhydrous acid). 98% H_2SO_4 freezes at +3°; 93% at −32°; 78% at −38°; 74% at −44°; 65% at −64°. Misc with water and alcohol with the generation of much heat and with contraction in vol. *When diluting, the acid should be added to the diluent. Keep tightly closed. Handle with caution, avoiding contact with the skin as it produces severe burns.* LD_{50} orally in rats: 2.14 g/kg (Smyth).

Caution: Corrosive to all body tissues. Inhalation of concd vapor may cause serious lung damage. Contact with eyes may result in total loss of vision; skin contact may produce severe necrosis. Ingestion may cause severe injury and death. Frequent skin contact with dil soln has caused dermatitis, *cf. Clinical Toxicology of Commercial Products*, R. E. Gosselin *et al.*, Eds. (Williams & Wilkins, Baltimore, 5th ed., 1984) section III, pp 8-12.

Sulfuric acid, fuming. H_2SO_4 with free SO_3, designated in commerce as *oleum*. Available grades contain up to about 80% free SO_3. Colorless or slightly colored, viscous liquid, emitting choking fumes of sulfur trioxide. *Extremely corrosive. Handle with great care. Keep tightly closed in glass-stoppered bottles.*

USE: In manuf of fertilizers, explosives, dyestuffs, other acids, parchment paper, glue, purification of petroleum, pickling of metal.

THERAP CAT: Dil acid formerly in treatment of gastric hypoacidity. Concd acid formerly as a topical caustic.

8954. Sulfur Iodide. Iodosulfane. Approx SI. Obtained by fusing together 4 parts iodine with 1 part sulfur. It probably contains free iodine as well as free sulfur.

Grayish-black masses of metallic luster and iodine odor. Insol in water; sol in carbon disulfide, alcohol, ether, in about 50 parts glycerol. KI soln dissolves a portion of the iodine leaving free sulfur. *Keep well closed.*

THERAP CAT: Dermatologic.

THERAP CAT (VET): Has been used in chronic eczema, ringworm, mange.

8955. Sulfurous Acid. Sulfur dioxide soln. A soln of about 6% sulfur dioxide in water.

Colorless, clear, acid liquid; suffocating odor of sulfur dioxide. d about 1.03. Gradually oxidizes in the air to sulfuric acid. *Keep in nearly full, tightly closed containers, in a cool place.*

USE: Dental bleach.

THERAP CAT: Antiseptic.

8956. Sulfur, Pharmaceutical. Three forms of sulfur of 99.5% purity or better are recognized in pharmacy. *Precipitated sulfur*, also known as milk of sulfur, made by boiling sulfur with lime and pptg the filtered soln with hydrochloric acid. *Sublimed sulfur*, also known as flowers of sulfur. *Washed sulfur*, made by treating sublimed sulfur with ammonia to dissolve impurities, particularly arsenic (the U.S.A. sulfur contains practically no arsenic) and to remove traces of acid.

Sublimed and washed sulfur are in the form of a fine, yellow crystalline powder, with only a faint odor and taste. Both are slowly and usually incompletely sol in carbon disulfide. Precipitated sulfur is in the form of a very fine, pale yellow, amorphous or microcryst powder, odorless and tasteless. It dissolves much more quickly and usually more

completely in carbon disulfide than the other forms. Soly in anhydr lanolin at 45°: 0.38%; in olive oil at 15°: 2.2%, at 30°: 4.1%, at 100°: 20.0%.

THERAP CAT: Scabicide.

THERAP CAT (VET): Externally: antiseptic and parasiticidal; in lotions, ointments and dips, often with lime. Internally: has been used as a laxative, coccidiostat.

8957. Sulfur Tetrafluoride. F_4S; mol wt 108.07. F 70.33%, S 29.67%. SF_4. First prepd from cobalt trifluoride and elemental sulfur: Fischer, Jaenckner, *Angew. Chem.* **42**, 810 (1929); Kwasnik in *Handbook of Preparative Inorganic Chemistry* vol. 1, G. Brauer, Ed. (Academic Press, New York, 2nd ed., 1963) pp 168-169. Best prepared from SCl_2 and NaF suspended in acetonitrile; also from sulfur, chlorine and sodium fluoride: Tullock *et al.*, *J. Am. Chem. Soc.* **82**, 539 (1960); Arth, Fried, U.S. pat. **3,046,094** (1962 to Merck & Co.); Fawcett, Tullock, *Inorg. Syn.* **7**, 119 (1963). Reviews of prepn and chemistry: Smith, *Angew. Chem. Int. Ed.* **1**, 467-475 (1962); Kemmitt, Sharp, *Advan. Fluorine Chem.* **4**, 220-222 (1965); Martin, *Ann. N. Y. Acad. Sci.* **145**, 161-168 (1967).

Colorless gas. Thermostable to 600°. mp −121.0°. bp −38°. d (liq; −78°) 1.95. d (solid; −183°) 2.349. Freely sol in benzene. Violent reaction with water. Dec by concd H_2SO_4. Attacks glass but not quartz or mercury.

Human Toxicity: About as toxic as phosgene. Strong irritant, corrosive.

USE: Selective fluorinating agent.

8958. Sulfur Trioxide. Sulfuric anhydride; Sulfan. O_3S; mol wt 80.07. O 59.95%, S 40.05%. SO_3. Exists in 3 modifications. Prepd by the contact process, *i.e.*, by the action of oxygen on sulfur dioxide in the presence of catalysts such as platinized asbestos, platinized magnesium sulfate, ferric oxide, or vanadium compds. Extensive review and bibliography: Fairlie, *Sulfuric Acid Manufacture*, A.C.S. Monograph Series no. **69** (New York, 1936). May be prepd in the laboratory by heating fuming sulfuric acid and collecting the sublimate in a cooled receiver. If the vapor is condensed above 27°, the γ-form is obtained as a liquid. If the vapor is condensed below 27° and in the presence of a trace of moisture, a mixture of all 3 forms is obtained. The 3 forms can be separated by fractional distillation. The α-form is the stable modification, the β- and γ-forms are metastable.

α-Form, asbestos-like needles, mp 62.3°. Vapor pressure at 25° = 73 mm.

β-Form, asbestos-like neeedles, mp 32.5°. Vapor pressure at 25° = 344 mm.

γ-Form, ice-like mass, mp 16.8° or liquid, d 1.9224. bp_{760} 44.8°. Vapor pressure at 25° = 433 mm.

Melted SO_3 exists in the γ-form and on solidifying tends to the α-form. The difference in vapor pressures explains the so-called "alpha explosion" of sulfur trioxide. Heating of high-melting SO_3 in glass vessels should be avoided to prevent possible shattering of the container. Absolutely dry SO_3 is not corrosive to metals and shows no acid reaction. On exposure to air, it absorbs moisture rapidly, emitting dense white fumes. Combines with water with explosive violence (heat of dilution 504 cals/g) forming sulfuric acid. Due to this avidity for water, SO_3 chars many organic substances. On contact with wood shavings the heat produced by dehydration is sufficient to cause fire.

Under the trademark Sulfan, a product is marketed which contains 0.5% of a stabilizer to prevent polymerization to higher melting forms: Sulfan A consists largely of β-SO_3 and melts at 30-35°; Sulfan B consists largely of γ-SO_3 and melts at 17°. Sulfan C contains no stabilizer and will polymerize to α-SO_3.

USE: Intermediate in sulfuric acid manuf; in sulfonations for formation of addition compds with amines; in the manuf of explosives. *Caution:* Irritant and corrosive to mucous membranes. May cause coughing, choking and severe discomfort in a concn of 1 ppm.

8959. Sulfuryl Chloride. Cl_2O_2S; mol wt 134.98. Cl 52.54%, O 23.71%, S 23.76%. SO_2Cl_2. Prepd by passing a mixture of dry sulfur dioxide and chlorine through activated charcoal (other catalysts, such as camphor, can be used): Danneel, *Angew. Chem.* **39**, 1553 (1926); Durrans, *J. Soc. Chem. Ind.* **45**, 347 (1926); Meyer, *Angew. Chem.* **44**, 41

(1931); Danneel, Hesse, *Z. Anorg. Allgem. Chem.* **212**, 214 (1933); Allen, Maxson, *Inorg. Syn.* **1**, 114 (1939).

Colorless, mobile liquid, very pungent odor. *Even the vapors are corrosive to human skin and mucous membranes.* Turns yellow upon prolonged standing because of slight dissociation into SO_2 and Cl_2. d_4^{20} 1.6674. d_4^0 1.7045. mp $-54.1°$ (also given as $-46°$). bp 69.3°. n_D^{20} 1.4437. Dipole moment 1.86. Trouton constant 20.7. Slowly dec by water, forming H_2SO_4 and HCl. With ice-cold water it forms a hydrate, SO_2Cl_2.$15H_2O$, resembling camphor in appearance. Violent reaction on contact with alkalies. Miscible with benzene, toluene, ether, glacial acetic acid, other organic solvents.

USE: Chlorinating and sulfonating or chlorosulfonating agent in organic syntheses, e.g., in the manufacture of chlorophenol and chlorothymol. Has been used in war gas formulations.

8960. Sulfuryl Fluoride. Vikane. F_2O_2S; mol wt 102.07. F 37.23%, O 31.35%, S 31.42%. SO_2F_2. Prepd by heating $Ba(SO_3F)_2$: Trautz, Ehrmann, *J. Prakt. Chem.* [2] **142**, 91 (1935); Kwasnik in *Handbook of Preparative Inorganic Chemistry,* vol. 1, G. Brauer, Ed. (Academic Press, New York, 2nd ed., 1963) pp 173-174. Alternate prepn from AgF_2 and SO_2: Muetterties, *Inorg. Syn.* **6**, 158 (1960). *Review:* Kemmitt, Sharp, *Advan. Fluorine Chem.* **4**, 225-227 (1965).

Odorless, colorless gas. Not very reactive. Stable to 400°. mp $-135.82°$. bp $-55.38°$. Density and thermodynamic data: Bockhoff *et al., J. Chem. Phys.* **32**, 799 (1960). Soly (ml gaseous F_2SO_2/100 ml solvent): 4-5 (water); 24-27 (alcohol); 210-220 (toluene); 136-138 (CCl_4). Not hydrolyzed by water. Hydrolyzed by NaOH solns. May be stored in compressed form in steel cylinders or in a gasometer over H_2SO_4.

Caution: Highly irritating to the respiratory tract. For chronic effects *see* Sodium Fluoride.

USE: Fumigant.

8961. Sulindac. *(Z)-5-Fluoro-2-methyl-1-[[4-(methylsulfinyl)phenyl]methylene]-1H-indene-3-acetic acid; cis-5-fluoro-2-methyl-1-[p-(methylsulfinyl)benzylidene]indene-3-acetic acid;* MK-231; Aflodac; Algocetil; Arthrobid; Arthrocine; Artribid; Citireuma; Clinoril; Clisundac; Imbaral; Reumofil; Reumyl; Sudac; Sulinol; Sulreuma. $C_{20}H_{17}FO_3S$; mol wt 356.42. C 67.40%, H 4.81%, F 5.33%, O 13.47%, S 8.99%. Non-steroidal anti-inflammatory. Prepn: T.-Y. Shen *et al.,* **Ger. pat. 2,039,426;** *eidem,* **U.S. pat. 3,654,349;** D. F. Hinkley, J. B. Conn, U.S. pat. **3,647,858** (1971, 1972, 1972, all to Merck & Co.). Stereospecific synthesis: R. F. Shuman *et al., J. Org. Chem.* **42**, 1914 (1977). ^{13}C-NMR study: A. W. Douglas, *Can. J. Chem.* **56**, 2129 (1978). Pharmacological properties and metabolism: Van Arman *et al.,* and H. B. Hucker *et al, Fed. Proc.* **31**, 577 (1972); H. B. Hucker *et al., Drug Metab. Dispos.* **1**, 721 (1973). Renal metabolism: M. J. Miller *et al., Adv. Prostaglandin Thromboxane Leukotriene Res.* **11**, 487 (1983); D. C. Brater *et al., Kidney Int.* **27**, 66 (1985). HPLC determn in biological fluids: D. G. Musson *et al., J. Pharm. Sci.* **73**, 1270 (1984). Review of pharmacology and efficacy in rheumatic disease: R. N. Brogden *et al., Drugs* **16**, 97-114 (1978). Books: *Clinoril in the Treatment of Rheumatic Disorders,* E. C. Huskisson, P. Franchimont, Eds. (Raven Press, New York, 1976) 198 pp; *Current Concepts on Anti-inflammatory Drugs,* K. Miehlke, Ed. (Biomedical Information Corp., New York, 1980) 240 pp.

Yellow odorless crystals from ethyl acetate, mp 182-185° (dec). uv max (methanolic 0.1N HCl): 327, 285, 256, 226

nm ($E_{1cm}^{1\%}$ 375, 420, 410, 540). pKa (25°) 4.7. Sparingly sol in methanol, U.S.P. alcohol; slightly sol in ethyl acetate. Practically insol in water at pH below 4.5. Soly increases with rising pH to $\sim$3.0 mg/ml at pH 7. Stable in aq acid and base. Solid stable for at least three days in air at 100°.

THERAP CAT: Anti-inflammatory.

8962. Sulisatin. *1,3-Dihydro-7-methyl-3,3-bis[4-(sulfooxy)phenyl]-2H-indol-2-one; 3,3-bis(p-hydroxyphenyl)-7-methyl-2-indolinone bis(hydrogen sulfate) (ester).* $C_{21}H_{17}-NO_9S_2$; mol wt 491.50. C 51.32%, H 3.49%, N 2.85%, O 29.30%, S 13.04%. Deriv of isatin, *q.v.,* which stimulates motility of the large intestine. Prepn: A. Hosta Pujol, S. Brosa Rabassa, **Ger. pat. 2,521,966** corresp to U.S. pat. **4,053,483** (1975, 1977 both to Andreu); F. Garrido *et al., Eur. J. Med. Chem.* **10**, 143 (1975). Mechanism of action: J. Queralt *et al., Arch. Farmacol. Toxicol.* **1**, 137 (1975). Laxative properties: M. Moreto *et al., Eur. J. Pharmacol.* **36**, 221 (1976). Metabolism: *eidem, Arzneimittel-Forsch.* **29**, 1561 (1979).

Disodium salt, $C_{21}H_{15}NNa_2O_9S_2$, *DAN-603, Laxitex.* White crystals, mp >360°. Slightly hygroscopic.

THERAP CAT: Laxative.

8963. Sulisobenzone. *5-Benzoyl-4-hydroxy-2-methoxybenzenesulfonic acid;* 3-benzoyl-4-hydroxy-6-methoxybenzenesulfonic acid; 2-benzoyl-5-methoxy-1-phenol-4-sulfonic acid; 2-hydroxy-4-methoxybenzophenone-5-sulfonic acid; benzophenone-4; NSC-60584; Cyasorb UV 284 (obsolete); Spectra-Sorb UV 284; Sungard; Uval; Uvinul MS-40. $C_{14}-H_{12}O_6S$; mol wt 308.31. C 54.54%, H 3.92%, O 31.14%, S 10.40%. Prepn: A. J. Cofrancesco, **Brit. pat. 1,136,525** (1968 to GAF). Activity: J. M. Knox *et al., J. Invest. Dermatol.* **34**, 51 (1960).

Light-tan powder, mp 145°. 1 g dissolves in 2 ml methanol, 3.3 ml alc, 4 ml water, 100 ml ethyl acetate.

USE: Ultraviolet stabilizer in wool, cosmetics, pesticides, and lithographic plate coatings.

THERAP CAT: Ultraviolet screen.

8964. Sulmarin. *4-Methyl-6,7-bis(sulfooxy)-2H-1-benzopyran-2-one; 6,7-dihydroxy-4-methylcoumarin disulfate;* 4-methylesculetin bis(hydrogen sulfate); 4-methyl-6,7-dihydroxycoumarindisulfate; 4-methylesctletindisulfonic acid; 4-methylesculetin-6,7-disulfuric ester; MG 143; Idro P_2. $C_{10}H_8O_{10}S_2$; mol wt 352.30. C 34.09%, H 2.29%, O 45.42%, S 18.20%. Prepn: Cavallini, Mazzucchi, *Farm. Sci. e tec. (Pavia)* **3**, 297 (1948), *C.A.* **42**, 8900d (1948). Description: Maggiorelli, *Giorn. Med. Militare* **101**, 365 (1951), *C.A.* **47**, 6412d (1951); Banchetti, *Farmaco Ed. Sci.* **10**, 970 (1955). Polarographic method for determn: A. M. Contri, *Farmaco Ed. Prat.* **25**, 231 (1970).

Disodium salt trihydrate, $C_{10}H_6Na_2O_{10}S_2.3H_2O$, crystals from 60% alc, dec 252-253°. uv max (pH 11.85): 304 nm.
THERAP CAT: Hemostatic.

8965. Sulmazole. *2-[2-Methoxy-4-(methylsulfinyl)-phenyl]-1H-imidazo[4,5-b]pyridine;* AR-L 115BS; Vardax. $C_{14}H_{13}N_3O_2S$; mol wt 287.33. C 58.52%, H 4.56%, N 14.62%, O 11.14%, S 11.16%. Orally active non-glycoside, non-adrenergic inotropic agent. Prepn: **Neth. pat. Appl. 7,401,254** corresp to E. Kutler *et al.*, **U.S. pat. 3,985,891** (1974, 1976 both to Thomae). Series of articles on pharmacology, pharmacokinetics, metabolism, clinical studies: *Arzneimittel-Forsch.* **31**, 129-278 (1981). Brief review of pharmacology: H. Koch, *Pharm. Int.* **3**, 5 (1982).

Solid, mp 203-205°. LD_{50} in albino mice (mg/kg): 560 orally; 163 i.v., W. Diederen, R. Kadatz, *Arzneimittel-Forsch.* **31**, 141 (1981).
THERAP CAT: Cardiotonic.

8966. Sulmepride. *5-(Aminosulfonyl)-2-methoxy-N-[(1-methyl-2-pyrrolidinyl)methyl]benzamide;* N-[(1-methyl-2-pyrrolidinyl)methyl]-5-sulfamoyl-o-anisamide; TER-1546. $C_{14}H_{21}N_3O_4S$; mol wt 327.40. C 51.36%, H 6.47%, N 12.83%, O 19.55%, S 9.79%. Neuroleptic agent and analog of sulpiride, *q.v.*, which selectively blocks presynaptic dopaminergic receptors. Prepn: **Neth. pat. Appl. 6,500,326;** C. S. Miller *et al.*, **U.S. pat. 3,342,826** (1965, 1967 both to Soc. Etudes Sci. et Ind. de l'Ile de France); W. Liebenow, I. Grafe, **Ger. pat. 2,414,498** (1975 to Heumann), *C.A.* **84**, 4696 (1975). Comparison of activity vs apomorphine-induced effects: A. J. Puech *et al.*, *Eur. J. Pharmacol.* **50**, 291 (1978). Quantitative determn: F. Bressolle *et al.*, *J. Chromatog.* **174**, 421 (1979).

8967. Suloctidil. *4-[(1-Methylethyl)thio]-α-[1-(octylamino)ethyl]benzenemethanol;* erythro-p-(isopropylthio)-α-[1-(octylamino)ethyl]benzyl alcohol; 1-(4-isopropylthiophenyl)-2-octylaminopropanol; CP 556S; MJF 12637; Bemperil; Cerebro; Circleton; Dulasi; Duloctil; Euvasal; Fluversin; Fluvisco; Hemoantin; Iangene; Loctidon; Locton; Octamet; Polivasal; Sudil; Sulocton; Sulodene; Tamid. $C_{20}H_{35}$-NOS; mol wt 337.57. C 71.16%, H 10.45%, N 4.15%, O 4.74%, S 9.50%. Prepn: G. Lambelin *et al.*, **Ger. pat. 2,334,-404** (1974 to Continental Pharma), *C.A.* **82**, 97820w (1975). Pharmacology: R. Roncucci *et al.*, *Naturwiss.* **62**, 141 (1975); J. Roba *et al.*, *Eur. J. Pharmacol.* **37**, 265 (1976); *eidem*, *Arch. Int. Pharmacodyn.* **221**, 54 (1976); T. Godfraind, *ibid.* 342.

Crystals from *n*-pentane or methanol-water, mp 62-63°. LD_{50} orally in mice: 3700 mg/kg (Lambelin).
THERAP CAT: Peripheral vasodilator.

8968. Sulphan Blue. *N-[4-[[4-(Diethylamino)phenyl](2,4-disulfophenyl)methylene]-2,5-cyclohexadien-1-ylidene]-N-ethylethanaminium hydroxide, inner salt, sodium salt;* C.I. Acid Blue 1; [4-[α-[p-(diethylamino)phenyl]-2,4-disulfobenzylidene]-2,5-cyclohexadien-1-ylidene]diethylammonium hydroxide inner salt sodium salt; anhydro-4,4'-bis(diethylamino)triphenylmethanol-2'',4''-disulfonic acid monosodium salt; C.I. Food Blue 3; C.I. 42045; Disulphine Blue. $C_{27}H_{31}N_2NaO_6S_2$; mol wt 566.70. C 57.23%, H 5.51%, N 4.94%, Na 4.06%, O 16.94%, S 11.32%. Prepn from 4-formylbenzene-1,3-disulfonic acid and diethylaniline: *Colour Index* vol. **4** (3rd ed., 1971) p 4382.

Violet powder. One gram dissolves in 20 ml water at 20°. Dilute aq solns are blue and turn yellow upon the addition of concd hydrochloric acid. In the absence of strong acid or caustic the blue color is stable over a wide pH range. Partly sol in alcohol.
2,5-Disulfophenyl isomer, *isosulfan blue, Lymphazurin.*
USE: Coloring medicinal products; in dyeing and printing wool, silk.
THERAP CAT: Isosulfan blue as diagnostic aid (lymphangiography).

8969. Sulphenone®. *1-Chloro-4-(phenylsulfonyl)benzene;* p-chlorophenyl phenyl sulfone; 4-chlorodiphenyl sulfone; sulfenone; R-242. $C_{12}H_9ClO_2S$; mol wt 252.73. C 57.03%, H 3.59%, Cl 14.03%, O 12.66%, S 12.69%. The commercial product contains 80% sulfenone and 20% related diaryl sulfones. Prepn: Bender, Pitt, **U.S. pat. 2,593,001** (1952 to Stauffer Chem.). Toxicology: L. W. Hazleton *et al.*, *J. Agr. Food Chem.* **3**, 836 (1955).

Dimorphic crystals, mp 94°. Slight aromatic odor. Tasteless. Soly at 20° (g/100 ml): acetone 74.4; dioxane 65.6; isopropanol 21; hexane 0.4; benzene 44.4; toluene 29.4; xylene 18.2; carbon tetrachloride 4.9. Practically insol in water. Slightly sol in petr oils. Stable to oxidizing and reducing agents, acids and alkalies found in spray formulations. LD_{50} orally in mice: 2.7 g/kg, L. W. Hazleton *et al.*, *loc. cit.*
USE: Acaricide.

8970. Sulphurenic Acid. *3β,15α-Dihydroxy-24-methyl-enelanost-8-en-21-oic acid;* 15α-hydroxyeburicoic acid. $C_{31}H_{50}O_4$; mol wt 486.71. C 76.50%, H 10.36%, O 13.15%. From the mycelium of *Polyporus sulphureus* Bull. (Fr.), Polyporaceae. Isoln and structure: Fried *et al.*, *Tetrahedron* **20**, 2297 (1964).

Crystals from acetone, mp 252-254°. $[\alpha]_D^{23}$ +42° (c = 0.51 in pyridine).

Diacetate, $C_{35}H_{54}O_6$, crystals from methanol, mp 234-235°. $[\alpha]_D^{23}$ +58° (c = 1 in chloroform).

Methyl ester, $C_{32}H_{52}O_4$, crystals from acetone, mp 190-192°. $[\alpha]_D^{23}$ +66° (c = 0.42 in chloroform).

8971. Sulpiride. *5-(Aminosulfonyl)-N-[(1-ethyl-2-pyrrolidinyl)methyl]-2-methoxybenzamide; N-[(1-ethyl-2-pyrrolidinyl)methyl]-5-sulfamoyl-o-anisamide; N-[(1-ethyl-2-pyrrolidinyl)methyl]-2-methoxy-5-sulfamoylbenzamide;* Abilit; Aiglonyl; Coolspan; Dobren; Dogmatil; Dogmatyl; Dolmatil; Guastil; Meresa; Miradol; Mirbanil; Misulvan; Neogama; Omperan; Pyrikappl; Sernevin; Splotin; Sulpitil; Sursumid; Synedil; Trilan. $C_{15}H_{23}N_3O_4S$; mol wt 341.43. C 52.77%, H 6.79%, N 12.31%, O 18.74%, S 9.39%. Dopamine D_2-receptor antagonist. Prepn: C. S. Miller *et al.*, U.S. pat. **3,342,826** (1964 to Soc. Etudes Sci. Ind. l'Ile-de-France). Pharmacology: Justin-Besançon *et al.*, *C.R. Acad. Sci. Ser. D* **265**, 1253 (1967). Synthesis of *l*-form: F. Mauri, Ger. pat. **2,903,891** (1979 to Ravizza), *C.A.* **91**, 211259h (1979). Structure of racemate and *l*-form: L. Y. Y. Ma *et al.*, *Acta Crystallogr.* **B38**, 2861 (1982). Physical properties: D. Pitrè, E. Valoti, *Arch. Pharmacol.* **320**, 859 (1987). Dopamine D_2-receptor binding activity: P. Jenner *et al.*, *Adv. Biochem. Psychopharmacol.* **35**, 109 (1982); comparative activity of enantiomers: P. Jenner *et al.*, *J. Pharm. Pharmacol.* **32**, 39 (1980). Series of articles on clinical studies: *Sem. Hop.* **46**(29B), 1-132 (1970). Review of clinical trials in psychiatry: E. D. Peselow, M. Stanley, *Adv. Biochem. Psychopharmacol.* **35**, 163-194 (1982). Toxicity data: P. Dostert *et al.*, *Eur. J. Med. Chem. - Chim. Ther.* **17**, 437 (1982). Comprehensive description: D. Pitrè *et al.*, in *Analytical Profiles of Drug Substances* **vol. 17**, K. Florey, Ed. (Academic Press, New York, 1988) pp 607-641.

White, odorless crystalline powder, mp 175-182° (dec). Sparingly sol in methanol. Practically insol in water, ether, chloroform, benzene. pka$_1$ 9.00, pKa$_2$ 10.19. $[\alpha]_D^{25}$ −66.8° (c = 0.5 in DMF). LD$_{50}$ in mice (mg/kg): 170 i.p.; 2250 orally (Dostert).

l-Form, *levosulpiride, S-(−)-sulpiride, Levopraid.* Monomorphic crystalline solid, mp 185-187°.

THERAP CAT: Antidepressant. Antipsychotic. Digestive aid.

8972. Sulprofos. *O-Ethyl O-[4-(methylthio)phenyl]phosphorodithioic acid S-propyl ester; O-ethyl O-[4-(methylmercapto)phenyl]-S-n-propylphosphorothionothiolate; O-ethyl O-[4-(methylthio)phenyl] S-propylphosphorodithioate;* BAY NTN 9306; Bolstar. $C_{12}H_{19}O_2PS_3$; mol wt 322.43. C 44.70%, H 5.94%, O 9.92%, P 9.61%, S 29.83%. Prepn: S. Kishino *et al.*, Ger. pat. **2,111,414** corresp to U.S. pat. **3,947,529** (1971, 1976, both to Bayer). Metabolism: D. L. Bull, G. W. Ivie, *J. Agr. Food Chem.* **24**, 143 (1976); G. W. Ivie *et al.*, *ibid.* 147; D. L. Bull *et al.*, *ibid.* 601. Photodegradation: G. W. Ivie, D. L. Bull, *ibid.* 1053.

Tan liquid, phosphorus odor, bp$_{0.1}$ 155-158°. d$_{20}^{20}$ 1.20, n$_D^{20}$ 1.5859. Low sol in water, sol in organic solvents. Toxic to fish and wildlife. LD$_{50}$ orally in rats: 227 mg/kg, D. L. Bull, G. W. Ivie, *loc. cit.*

USE: Insecticide.

8973. Sulprostone. *7-[3-Hydroxy-2-(3-hydroxy-4-phenoxy-1-butenyl)-5-oxocyclopentyl]-N-(methylsulfonyl)-5-heptenamide;* CP-34089; SHB-286; ZK-57671; Nalador. $C_{23}H_{31}NO_7S$; mol wt 465.57. C 59.33%, H 6.71%, N 3.01%, O 24.06%, S 6.89%. Analog of prostaglandin E_2, *q.v.*, with uterine stimulant activity. Prepn: J. Bindra, M. R. Johnson, Ger. pat. **2,355,540** (1974 to Pfizer), *C.A.* **81**, 49330u (1974); *eidem*, U.S. pat. **4,024,179** (1977 to Pfizer). Antifertility effects: H. Hess *et al.*, *Experientia* **33**, 1076 (1977). Influence on platelet function: R. C. Briel, T. H. Lippert, *Adv. Prostaglandin Thromboxane Res.* **6**, 351 (1980). Biological action and half-life: *eidem*, *Prostaglandins Med.* **6**, 1 (1981). Induction of abortion: K. Schmidt-Gollwitzer, *Int. J. Fertil.* **26**, 86 (1981).

Colorless oil.
THERAP CAT: Abortifacient.

8974. Sulthiame. *4-(Tetrahydro-2H-1,2-thiazin-2-yl)-benzenesulfonamide S,S-dioxide; N-(4'-sulfamylphenyl)-1,4-butanesultam; tetrahydro-2-(p-sulfamoylphenyl)-1,2-thiazine 1,1-dioxide; 2-(p-aminosulfonylphenyl)tetrahydro-1,2-thiazine dioxide; 1-(p-amidosulfonylphenyl)-2-thiapiperidine 2,2-dioxide; N-(p-aminosulfonylphenyl)-1,4-butanesultam;* Elisal; Ospolot; Trolone; Contravul; Conadil. $C_{10}H_{14}N_2O_4S_2$; mol wt 290.37. C 41.36%, H 4.86%, N 9.65%, O 22.04%, S 22.09%. Prepn of this type of compd: Helferich, Behnisch, U.S. pat. **2,916,489** (1959 to Schenley).

Crystalline powder, mp 180-182°. Practically insol in cold water. Partly sol in boiling water; slightly sol in alcohol, acids; readily sol in alkalies.
THERAP CAT: Anticonvulsant.

8975. Sultopride. *N-[(1-Ethyl-2-pyrrolidinyl)methyl]-5-(ethylsulfonyl)-2-methoxybenzamide; N-[(1-ethyl-2-pyrrolidinyl)methyl]-5-(ethylsulfonyl)-o-anisamide.* $C_{17}H_{26}N_2O_4S$; mol wt 354.47. C 57.60%, H 7.39%, N 7.90%, O 18.06%, S 9.04%. Substituted benzamide with neuroleptic activity related to sulpiride, *q.v.* Prepn: C. S. Miller, Fr. pat. M **5,916** (1968 to Soc. Etudes Sci. Ind. l'Ile-de-France), *C.A.* **71**, 70484a (1969). Chemical and biological properties: L. Justin-Besançon *et al.*, *C.R. Acad. Sci. D* **279**, 375 (1974). Pharmacology: C. Lavelle, J. Margarit, *J. Pharmacol.* **5**, Suppl. 2, 58 (1974); B. Bruguerolle *et al.*, *ibid.* **12**, 27 (1981). Pharmacokinetics, metabolism: L. Jung, *Sem. Hop.* **54**, 1347 (1978). Assessment of neuroleptic potential: B. Costall *et al.*, *J. Pharm. Pharmacol.* **30**, 771 (1978). Clinical study: M. Cornely, *Med. Klin.* **73**, 1281 (1978). Review of clinical trials in psychiatry: E. D. Peselow, M. Stanley: *Adv. Biochem. Psychopharmacol.* **35**, 163-194 (1982).

Hydrochloride, $C_{17}H_{27}ClN_2O_4S$, *LIN-1418, Barnetil, Barnotil, Topral.* Cryst from methyl ethyl ketone, mp 181-182°.

THERAP CAT: Antidepressant.

8976. Sultosilic Acid, Piperazine Salt. *2-Hydroxy-5-[[(4-methylphenyl)sulfonyl]oxy]benzenesulfonic acid compd with piperazine (1:1); mono-[2-hydroxy-5-[[(4-methylphenyl)sulfonyl]oxy]benzenesulfonate]piperazine;* 2-hydroxy-5-tosyloxybenzenesulfonic acid, piperazine salt; diethylenediamine sultosylate; A 585; Mimedran. $C_{17}H_{22}N_2O_7S_2$; mol wt 430.49. C 47.43%, H 5.15%, N 6.51%, O 26.02%, S 14.89%. Prepn: A. Esteve Subirana, **Belg. pat.** 819,330, *C.A.* **83**, 84869q (1975); idem, **Belg. pat.** 825,369; idem, **U.S. pat.** 3,954,767 (1974, 1975, 1976 all to Lab. del Dr. Esteve, Geneva). Hypolipemic activity and acute toxicity: J. Esteve *et al., Eur. J. Med. Chem.* **11**, 43 (1976). Metabolism in rat, dog, man: S. G. Wood *et al., Xenobiotica* **12**, 165 (1982). Clinical comparison with bezafibrate, *q.v.*: H. Vinazzer, J. C. Farine, *Atherosclerosis* **49**, 109 (1983). Toxicology: L. Rodriguez *et al., Arch. Farmacol. Toxicol.* **5**, 281 (1979).

Crystals from ethanol, mp 171-174°. LD_{50} orally in male mice, male rats (mg/kg): > 10750, > 11000 (Esteve); i.p. in male rats, dogs (mg/kg): 833.6, 605 (Rodriguez).

THERAP CAT: Antihyperlipoproteinemic.

8977. Sultroponium. *endo-(±)-3-(3-Hydroxy-1-oxo-2-phenylpropoxy)-8-methyl-8-(3-sulfopropyl)-8-azoniabicyclo-[3.2.1]octane hydroxide, inner salt; 3α-hydroxy-8-(3-sulfopropyl)-1αH,5αH-tropanium hydroxide, (−)-tropate, inner salt;* sulfo-betaine of atropine; tropyl tropate sulfo-betaine; A 118; Sultropan. $C_{20}H_{29}NO_6S$; mol wt 411.52. C 58.37%, H 7.10%, N 3.40%, O 23.33%, S 7.79%. Prepn: Savini *et al., Therapie* **22**, 843 (1967); Wahl, *Prod. Probl. Pharm.* **25**, 260 (1970); Raudnitz *et al.*, **Belg. pat.** 672,926 (1966) and **Brit. pat.** 1,082,445 (1967).

Crystals from alc, mp 220° (dec). Sol in water and warm ethanol; insol in ether, acetone, benzene, chloroform. uv max (water): 251.5, 257.5, 263.5 nm.

THERAP CAT: Anticholinergic, antispasmodic.

8978. Sumach. Name used for several plants of the *Rhus* species. Leaves of *Rhus glabra* L. and *Rhus typhina* L., *Anacardiaceae* are known commercially as *North American Sumach.* They contain fisetin, dihydrofisetin, 27% tannin and gallic acid esters. Constituents of the *Rhus typhina* fruit: J. Tischer, *Pharmazie* **15**, 83 (1960).

USE: North American sumach as black dye; as scenting agent for tobacco.

8979. Sumatriptan. *3-[2-(Dimethylamino)ethyl]-N-methyl-1H-indole-5-methanesulfonamide;* GR 43175. $C_{14}H_{21}N_3O_2S$; mol wt 295.40. C 56.92%, H 7.17%, N 14.22%, O 10.83%, S 10.85%. Selective serotonin 5HT$_1$-like receptor agonist. Prepn: M. D. Dowle, I. H. Coates, **Ger. pat.** 3,320,521, *C.A.* **100**, 103175y (1984); A. W. Oxford, **Brit. pat. Appl.** 2,162,522 (1983, 1986 both to Glaxo). Effects on 5HT-receptor subtypes: P. P. A. Humphrey *et al., Brit. J. Pharmacol.* **94**, 1123 (1988). Clinical evaluation in migraine: A. Doenicke *et al., Lancet* **1**, 1309 (1988).

mp 169-171°.
Succinate, $C_{18}H_{27}N_3O_6S$, GR 43175C. mp 165-166°.
THERAP CAT: Antimigraine.

8980. Sumatrol. *1,2,12,12a-Tetrahydro-5-hydroxy-8,9-dimethoxy-2-(1-methylethenyl)-[1]benzopyrano[3,4-b]furo-[2,3-h][1]benzopyran-6(6aH)-one.* $C_{23}H_{22}O_7$; mol wt 410.41. C 67.31%, H 5.40%, O 27.29%. Present in resin from *Derris* sp., *Leguminosae:* Robertson, Rusby, *J. Chem. Soc.* **1937**, 497. Isolated also from *Tephrosia toxicaria* Pers., *Papilionaceae:* Harper, *ibid.* **1940**, 1178. Structure and stereochemistry: Harper, *loc. cit.;* Crombie, Peace, *ibid.* **1961**, 5445.

Needles from alcohol, mp 195-196°. Needles from acetone, mp 183°; melts again at 194° after keeping for several days. $[\alpha]_D^{20} - 182°$ (c = 2.84 in benzene). Absorption max: 670 nm (ε 4660). Practically insol in water; sparingly sol in methanol, cold acetic acid, moderately sol in benzene, ethyl acetate; sol in chloroform.

8981. Sumbul. Musk root. Dried rhizome and roots of *Ferula sumbul* (Kauffm.) Hook. f. *(Euryangium sumbul* Kauffm.), or of other closely related species of *Ferula* having a musk-like odor *(Umbelliferae). Habit.* Central Asia, East Indies. *Constit.* Angelic (sumbulic) acid, valeric acid, methylcrotonic acid, resin, 0.2-0.4% volatile oil.

THERAP CAT: Sedative.

8982. Sunflower Seed Oil. The oil obtained by milling the seeds of *Helianthus annuus* L., *Compositae.* Classified as a semidrying oil. Alternately classified as an oleic-linoleic acid oil. Produced on a large scale in U.S.S.R., the Baltic region, India, Egypt, Canada, Argentina. Composition: palmitic acid 6.4%, stearic acid 1.3%, arachidic acid 4.0%, behenic acid 0.8%, oleic acid 21.3%, linoleic acid 66.2% (the glycerides in sunflower oil consist mainly of mixed triglycerides, each contg one or two linoleic acid radicals), linolenic acid < 0.1%. Vitamin E (mixed tocopherols) 75 mg/100 g. This approaches the tocopherol content of wheat germ oil which is 103 mg/100 g. *Reviews:* E. W. Eckey, *Vegetable Fats and Oils* (Reinhold, New York, 1954) pp 772-777; L. H. Bailey, *Standard Cyclopedia of Horticulture* vol. II (Macmillan, New York, 1953) pp 1445-1449.

Pale yellow oil. Bland, agreeable taste. d_{25}^{15} 0.922-0.926. d_{25}^{25} 0.915-0.919. mp −18°. Titer: 16-20°. Acid value: 0.6. Saponif value: 188-194. Iodine value: 125-136. Thiocyanogen value: 78. Hydroxyl value: 14-16. Reichert-Meissl value below 0.5. Polenske value below 0.5. Unsaponifiable below 1.5%. n_D^{25} 1.472-1.474. n_D^{40} 1.466-1.468. Slightly sol in alc. Miscible with benzene, chloroform, carbon tetrachloride. Forms a "skin" after exposure to air for 2-3 weeks.

USE: Food and salad oil, in candy manuf, in oleomargarine. Like wheat germ oil in dietary supplements. Industrially in oil-modified alkyd resins and soap manuf.

8983. Sunset Yellow FCF. *6-Hydroxy-5-[(4-sulfophenyl)azo]-2-naphthalenesulfonic acid disodium salt;* 1-*p*-sulfophenylazo-2-naphthol-6-sulfonic acid disodium salt; FD & C Yellow No. 6; C.I. Food Yellow 3; C.I. 15985. $C_{16}H_{10}N_2Na_2O_7S_2$; mol wt 452.37. C 42.48%, H 2.23%, N 6.19%, Na 10.16%, O 24.76%, S 14.17%. Prepn: P. Griess, *Ber.* **11**, 2191 (1878). Metabolism: J. L. Radomski, T. J. Mellinger,

J. Pharmacol. Exp. Ther. **136**, 259 (1962). Toxicology: I. F. Gaunt *et al., Food Cosmet. Toxicol.* **5**, 747 (1967); **7**, 9 (1969); **12**, 1 (1974). *Review: IARC Monographs* **8**, 257-266 (1967). *See also: Colour Index* vol. 4 (3rd ed., 1971) p 4087.

Orange-red crystals. Absorption max (0.02 N CH$_3$-COONH$_4$): 480 nm. Sol in water; slightly sol in ethanol. Reddish-orange soln in conc H$_2$SO$_4$, changing to yellow on dilution. LD$_{50}$ orally in rats, mice: > 10, > 6 g/kg, I. F. Gaunt *et al., loc. cit.* (1967).

USE: Provisionally listed for use in food, drugs and cosmetics.

8984. Suprasterol II. *(2α,7α,8R,19α,22E)-7,19:8,19-Dicyclo-9,10-secoergosta-5(10),22-dien-2-ol.* C$_{28}$H$_{44}$O; mol wt 396.63. C 84.78%, H 11.18%, O 4.03%. Ultraviolet irradiation product of vitamin D$_2$. Isoln: Windaus *et al., Ann.* **483**, 17 (1930). Structure and stereochemistry: Dauben, Baumann, *Tetrahedron Letters* **1961**, 565. Crystal structure: C. P. Saunderson *et al., ibid.* 573. Improved isoln and spectral data: T. Kobayashi *et al., J. Nutr. Sci. Vitaminol.* **23**, 291 (1977).

Rough prisms from acetone or methanol, mp 110°. bp$_{0.005}$ 190°. $[\alpha]_D^{19}$ +62.9° (methanol); $[\alpha]_D^{22}$ +47.8° (methanol). Sol in all organic solvents.

8985. Suprofen. *α-Methyl-4-(2-thienylcarbonyl)benzeneacetic acid; p-(2-thenoyl)hydratropic acid;* sutoprofen; R-25061; Maldocil; Masterfen; Supranol; Suprocil; Suprol. C$_{14}$H$_{12}$O$_3$S; mol wt 260.31. C 64.59%, H 4.65%, O 18.44%, S 12.32%. Inhibitor of prostaglandin biosynthesis with analgesic, antipyretic, and anti-inflammatory properties. Prepn: P. A. Janssen *et al.,* **Ger.** pat. **2,353,357;** *eidem,* **U.S.** pat. **4,035,376** (1974, 1977 both to Janssen); P. G. H. Van Daele *et al., Arzneimittel-Forsch.* **25**, 1495 (1975). Pharmacology: C. J. E. Niemegeers *et al., ibid.,* 1537. Synthesis of ³H and ¹⁴C-labeled suprofen: Y. Mori *et al., Radioisotopes* **30**, 584 (1981). HPLC determn in plasma and urine: H. Muller *et al., Arzneimittel-Forsch.* **32**, 257 (1982). Series of articles on pharmacology, antipyretic and anti-inflammatory activity, toxicity studies, clinical trials: *ibid.* **25**, 1501-1542 (1975); *ibid.* **33**, 1322-1338 (1983); on pharmacology, efficacy and safety: *Pharmacology* **27**, Suppl. 1, 1-96 (1983); on pharmacokinetics, local and systemic tolerability, and comparison with pentazocine: *Arzneimittel-Forsch.* **35**, 738-759 (1985); on sustained release kinetics, pediatric efficacy and comparison with indomethacin: *ibid.* **36**, 941-971 (1986).

White to slightly yellow microcryst powder, mp 124.3°. uv max (0.01N HCl-90% 2-propanol): 266, 292 nm (ϵ 15700, 15600). pKa 3.91. Freely sol in methanol, ethanol, chloroform, acetone, polyethylene glycol, 1.0N NaOH. Sol in ether. Slightly sol in 0.1N NaOH. Practically insol in *n*-hexane. Very slightly sol in water. LD$_{50}$ (based on mortality 7 days post-drug) in mice, rats, guinea pigs, dogs (mg/kg): 590, 353, 280, 160 orally (Niemegeers).

THERAP CAT: Anti-inflammatory. Analgesic.

8986. Suramin Sodium. *8,8'-[Carbonylbis[imino-3,1-phenylenecarbonylimino(4-methyl-3,1-phenylene)carbonylimino]]bis-1,3,5-naphthalenetrisulfonic acid hexasodium salt;* hexasodium *sym*-bis(*m*-aminobenzoyl-*m*-amino-*p*-methylbenzoyl-1-naphthylamino-4,6,8-trisulfonate) carbamide; Bayer 205; 309F; Antrypol; Germanin; Moranyl; Naganol; Naganin; Naphuride Sodium. C$_{51}$H$_{34}$N$_6$Na$_6$O$_{23}$S$_6$; mol wt 1429.21. C 42.86%, H 2.40%, N 5.88%, Na 9.65%, O 25.75%, S 13.46%. Discovered in 1917 by O. Dressel and R. Kothe. Prepn: I. G. Farben (Bayer) in O.P.B. repts; E. Fourneau *et al., Compt. Rend.* **178**, 675 (1924); J. Trefouel, E. Fourneau, **Brit.** pat. **224,849** (1923); B. Heymann, *Angew. Chem.* **37**, 585 (1924). History of discovery: J. Dressel, *J. Chem. Ed.* **38**, 620 (1961); *see also ibid.* **39**, 320 (1962). Inhibition of reverse transcriptase activity in certain avian and murine retroviruses: E. De Clercq, *Cancer Letters* **8**, 9 (1979). *In vitro* activity vs HIV-1 (HTLV-III/LAV) virus: H. Mitsuya *et al., Science* **226**, 172 (1984). Clinical pharmacokinetics: J. M. Collins *et al., J. Clin. Pharmacol.* **26**, 22 (1986). Clinical pharmacology and virustatic effect in acquired immunodeficiency syndrome (AIDS): S. Broder *et al., Lancet* **2**, 627 (1985); A. M. Levine *et al., Ann. Int. Med.* **105**, 32 (1986). Clinical use in onchocerciasis: H. Schultz-Key *et al., Trop. Med Parasit.* **36**, 244 (1985). Review of pharmacology, toxicology and clinical antiparasitic activity: F. Hawking, *Adv. Pharmacol. Chemother.* **15**, 289-322 (1978). *Review:* Olenick in *Antibiotics* vol. 3, J. W. Corcoran, F. E. Hahn, Eds. (Springer-Verlag, New York, 1975) pp 699-703.

White or slightly pink or cream-colored powder. Slightly bitter taste. Hygroscopic. Freely sol in water, in physiological saline; sparingly sol in 95% alcohol. Insol in benzene, ether, petr ether, chloroform. Aq solns are neutral to litmus. LD$_{50}$ in mice (mg/kg): $\sim$620 i.v. (Hawking).

THERAP CAT: Anthelmintic (Onchocerca). Antiprotozoal (Trypanosoma).

THERAP CAT (VET): Antiprotozoal (Trypanosoma).

8987. Suriclone. *4-Methyl-1-piperazinecarboxylic acid 6-(7-chloro-1,8-naphthyridin-2-yl)-2,3,6,7-tetrahydro-7-oxo-5H-1,4-dithiino[2,3-c]pyrrol-5-yl ester;* 6-(7-chloro-1,8-naphthyridin-2-yl)-5-(4-methylpiperazin-1-yl)carbonyloxy-7-oxo-2,3,6,7-tetrahydro-1,4-dithiino[2,3-c]pyrrole; RP-31264; Celexane; Clexane. C$_{20}$H$_{20}$ClN$_5$O$_3$S$_2$; mol wt 477.98. C 50.26%, H 4.22%, Cl 7.42%, N 14.65%, O 10.04%, S 13.41%. Cyclopyrrolone derivative; structural analog of zopiclone, *q.v.* Prepn: C. Jeanmart *et al.,* **Ger.** pat. **2,360,362;** *eidem,* **U.S.** pat. **3,948,917** (1974, 1976 both to Rhone-Poulenc). Benzodiazepine receptor binding studies: J.-C. Blanchard, L. Julou, *J. Neurochem.* **40**, 601 (1983); R. R. Trifiletti, S. H. Snyder, *Mol. Pharmacol.* **26**, 458 (1984); J. L. Zundel *et al., Life Sci.* **36**, 2247 (1985). Pharmacology: L.

Julou *et al.*, *Pharmacol. Biochem. Behav.* **23**, 653 (1985). Clinical evaluation as an anxiolytic: Y. D. Lapierre, K. L. Oyewumi, *Prog. Neuro-Psychopharmacol. Biol. Psychiat.* **7**, 805 (1983); W. E. Falk *et al.*, *Psychopharmacol. Bull.* **23**, 134 (1987).

Crystals from acetonitrile, mp 280°.
THERAP CAT: Anxiolytic.

8988. Surinamine. *N-Methyltyrosine;* andirine; angeline; geoffroyine; ratanhine. $C_{10}H_{13}NO_3$; mol wt 195.21. C 61.52%, H 6.71%, N 7.18%, O 24.59%. Isoln from bark of *Andira retusa* Kunth. *(Geoffroya retusa* Lam.), *Leguminosae:* Hiller-Bombein, *Arch. Pharm.* **230**, 513 (1892). Prepn of inactive compd: Kanevskaya, *J. Prakt. Chem.* **124**, 48 (1929). Prepn of L-form: Kanao, *J. Pharm. Soc. Japan* **66**, 4 (1946); Huguenin, Boissonnas, *Helv. Chim. Acta* **44**, 213 (1961). Prepn of D-form: Izumiya, Nagamatsu, *Bull. Chem. Soc. Japan* **25**, 265 (1952).

L-surinamine

L-Form, crystals, mp 292-295°. $[\alpha]_D^{21}$ +19.6° (c = 3.8 in 10% HCl). Almost insol in water; slightly sol in alcohol; sol in dil acids.

D-Form, crystals, mp 273-274°. $[\alpha]_D^{15}$ -18.9° (c = 1.74 in 3N HCl).

DL-Form, crystals from water.

8989. Suxethonium Bromide. *2,2'-[(1,4-Dioxo-1,4-butanediyl)bis(oxy)]bis[N-ethyl-N,N-dimethylethanaminium] dibromide;* bis[2-dimethylaminoethyl]succinate bis[ethobromide]; 3,8-dioxadecane-4,7-dione-1,10-bis[ethyldimethyl-ammonium bromide]; IS 362; M 115; M & B 2210; Brevidil E. $C_{16}H_{34}Br_2N_2O_4$; mol wt 478.28. C 40.18%, H 7.17%, Br 33.42%, O 13.38%, N 5.86%.

Slightly hygroscopic crystals, mp 158°. Freely sol in water or normal saline, giving solns which are slightly acidic. Aq solns undergo progressive hydrolysis with corresp loss of activity and increase in acidity. The rate of this decompn increases with temp, and autoclaving or prolonged exposure to heat should be avoided. When solns are mixed with alkaline solns of thiopental sodium or other intravenous barbiturates, a precipitate of barbituric acid forms which redissolves when more thiopental sodium or other alkali is added. Such solns rapidly lose activity to the extent of 20% within 5 min. If mixed solns of this type are required, they should, therefore, be used immediately.
THERAP CAT: Neuromuscular blocker.

8990. Suxibuzone. *Butanedioic acid mono[(4-butyl-3,5-dioxo-1,2-diphenyl-4-pyrazolidinyl]methyl] ester; succinic acid monoester with 4-butyl-4-(hydroxymethyl)-1,2-diphenyl-3,5-pyrazolidinedione;* 4-butyl-4-(hydroxymethyl)-1,2-diphenyl-3,5-pyrazolidinedione hydrogen succinate (ester); 1,2-diphenyl-4-n-butyl-4-hydroxymethyl-3,5-dioxopyrazolidine hemisuccinate; 4-hydroxymethylbutazolidine hemisuccinate; AE-17; Alfide; Calibène; Danilon; Flogos; Solurol. $C_{24}H_{26}N_2O_6$; mol wt 438.48. C 65.74%, H 5.98%, N 6.39%, O 21.89%. Prodrug of phenylbutazone, *q.v.* Prepd

(not claimed) and use as anti-inflammatory: A. Esteve, **Ger.** pat. 1,936,747; *idem*, U.S. pat. 3,752,894 (1970, 1973 both to Lab. Esteve). Pharmacology: A. Esteve *et al.*, *Quim. Ind. (Madrid)* **17**, 107 (1971). HPLC determn in plasma and urine: T. Marunaka *et al.*, *J. Pharm. Sci.* **69**, 1258 (1980). Metabolism in humans: Y. Yasuda *et al.*, *ibid.* **71**, 565 (1982). Comparative clinical study with phenylbutazone in rheumatoid arthritis: Y. Mizushima *et al.*, *Int. J. Tissue React.* **5**, 35 (1983).

White, bitter crystalline powder from alcohol, mp 126-127°. Sol in most organic solvents. Insol in water. LD_{50} orally in mice: 5.683 micromol/kg (Esteve, 1973).
THERAP CAT: Anti-inflammatory.

8991. Swertiamarin. *[4aR-(4aα,5β,6α)]-5-Ethenyl-6-(β-D-glucopyranosyloxy)-4,4a,5,6-tetrahydro-4a-hydroxy-1H,3H-pyrano[3,4-c]pyran-1-one;* 4,4a,5,6-tetrahydro-4aα-hydroxy-1-oxo-5β-vinyl-1H,3H-pyrano[3,4-c]pyran-6-yl β-D-glucopyranoside. $C_{16}H_{22}O_{10}$; mol wt 374.34. C 51.33%, H 5.92%, O 42.74%. Bitter principle of *Swertia japonica* (Maxim.) Makino, *Gentianaceae.* Yields erythrocentaurin on hydrolysis with emulsin. Isoln: Kubota, Tomita, *Chem. & Ind. (London)* **1958**, 229; Inouye *et al.*, *Chem. Pharm. Bull.* **18**, 1856 (1970). Occurrence in gentianaceous plants: Inouye, Nakamura, *J. Pharm. Sci. Japan* **91**, 755 (1971), *C.A.* **75**, 95431b (1971). Structure: Kubota, Tomita, *Tetrahedron Letters* **1961**, 176; *Bull. Chem. Soc. Japan* **34**, 1345 (1961); Koch *et al.*, *Bull. Soc. Chim. France* **1964**, 405. Stereochemistry: Inouye *et al.*, *Tetrahedron Letters* **1968**, 4429. Biosynthesis studies: Inouye *et al.*, *Chem. Pharm. Bull.* **18**, 2043 (1970).

Plates from ethanol + chloroform + ether, mp 113-114°. $[\alpha]_D^{20}$ -127° (c = 1 in 96% ethanol). uv max (methanol): 238 nm (log ε 3.93).

Tetraacetate, $C_{24}H_{30}O_{14}$, prisms, mp 190-191°. $[\alpha]_D$ -100.3° (CHCl₃). uv max: 206, 234 nm (log ε 3.20, 4.00).

8992. Sydnones. A term originally applied to the compd obtained by action of acetic anhydride on *N*-nitrosophenylglycine: Eade, Earl, *J. Chem. Soc.* **1946**, 591; Earl, *Chem. & Ind. (London)* **1953**, 746; *Rec. Trav. Chim.* **75**, 346 (1956). Now used to describe 5-membered mesoionic compounds that are best represented by resonance structures: Y. Noel, *Bull. Soc. Chim. France* **1964**, 173. Photochromic properties: S. Nespurek, M. Sorm, *Coll. Czech. Chem. Commun.* **42**, 811 (1977). ^{14}N-NMR study: L. Stefaniak, *Tetrahedron* **33**, 2571 (1977). ^{15}N-NMR study: L. Stefaniak *et al.*, *Org. Magn. Res.* **13**, 274 (1980). Review of chemistry: F. H. C. Stewart, *Chem. Rev.* **64**, 129-147 (1964).

8993. Symclosene. *1,3,5-Trichloro-1,3,5-triazine-2,4,6-*
(1H,3H,5H)-trione; trichloroiminocyanuric acid; trichloro-
isocyanuric acid; ACL-85; Chloreal. $C_3Cl_3N_3O_3$; mol wt
232.42. C 15.50%, Cl 45.77%, N 18.08%, O 20.65%. Prepd
by chlorinating cyanuric acid in NaOH soln: Hands, Whitt,
J. Soc. Chem. Ind. **67**, 66 (1948); Hardy, U.S. pat. **2,607,738**
(1952 to Monsanto); Christian, U.S. pat. **2,956,056** (1960 to
W. R. Grace). Purification by dissolving in concd H_2SO_4
and diluting with ice water: Lorenz, U.S. pat. **2,828,308**
(1958 to Purex). Structure studies and prepn: Petterson *et*
al., J. Org. Chem. **25**, 1595 (1960).

Needles from ethylene chloride, mp 246-247° (dec). May
be stored in the dry state for at least a year. Releases hypochlorous acid on contact with water. Available chlorine
about 90%. pH of aq solns about 4.4. Soly in water at 25°
about 0.2%. Sol in chlorinated and highly polar solvents.
USE: Chlorinating agent, disinfectant, industrial deodorant. In household cleansers, such as **Bab-O.** *Caution:*
Moderately irritating to eyes, skin, mucous membranes.
THERAP CAT: Topical anti-infective.

8994. Synephrine. *4-Hydroxy-α-[(methylamino)methyl]-*
benzenemethanol; p-hydroxy-α-[(methylamino)methyl]benzyl
alcohol; 1-(4-hydroxyphenyl)-2-methylaminoethanol;
p-methylaminoethanolphenol; *β*-methylamino-*α*-(4-hydr-
oxyphenyl)ethyl alcohol; methylaminomethyl 4-hydroxy-
phenyl carbinol; Analeptin; Ethaphene; Oxedrine; Parasympatol; Simpalon; Synephrin; Synthenate. $C_9H_{13}NO_2$;
mol wt 167.20. C 64.65%, H 7.84%, N 8.38%, O 19.14%.
Prepd by hydrogenating *ω*-methylamino-4-hydroxyaceto-
phenone in water in the presence of Pt or Pd: **Ger.** pat.
566,578 (1931 to Boehringer, Ing.), *Frdl.* **18**, 3025.

Crystals, mp 184-185°. Stable to air and light.
Hydrochloride, $C_9H_{13}NO_2·HCl$, crystals, mp 151-152°.
Freely sol in water.
Tartrate, $C_{22}H_{32}N_2O_{10}$, *Corvasymton, Simpadren, Sympa-*
thol, Sympatol. Crystals, mp 188-190° (some decompn).
Freely sol in water; sol in alcohol.
Tartaric acid monoester, $C_{13}H_{17}NO_7$, *p-methylaminoetha-*
nolphenol tartrate, Neupentedrin, Pentedrin.
THERAP CAT: Adrenergic; vasopressor.

8995. Synhexyl. *3-Hexyl-7,8,9,10-tetrahydro-6,6,9-tri-*
methyl-6H-dibenzo[b,d]pyran-1-ol; 1-hydroxy-3-hexyl-6,6,-
9-trimethyl-7,8,9,10-tetrahydro-6H-dibenzo[b,d]pyran;
parahexyl; pyrahexyl. $C_{22}H_{32}O_2$; mol wt 328.50. C 80.43%,
H 9.82%, O 9.74%. Psychotomimetic synthetic analog of the
tetrahydrocannabinols, *q.v.* Prepn: Adams *et al., J. Am.*
Chem. Soc. **63**, 1971 (1941); Adams, U.S. pat. **2,419,935**
(1947); Hughes *et al.,* U.S. pat. **3,576,887** (1971 to Am.
Home Products). Comparison of physiological effects with
those of tetrahydrocannabinols: Hollister *et al., Clin. Phar-*
macol. Ther. **9**, 783 (1968). Physical constants: Farmilo *et*
al., Bull. Narcotics **6**, 7 (1954). Toxicity data: Loewe, *J.*
Pharmacol. Exp. Ther. **88**, 154 (1946).

Liquid, $bp_{1.0}$ 190-192°; n_D^{20} 1.5504. uv and ir spectra: *Bull.*
Narcotics **6**, 27 (1954); **7**, 42 (1955). LD_{50} in mice, dogs,
rabbits (mg/kg): 170, 223, 143 i.v. (Loewe).
Caution: This is a controlled substance (hallucinogen)
listed in the U.S. Code of Federal Regulations, Title 21 Part
1308.11 (1985).

8996. Syringaldehyde. *4-Hydroxy-3,5-dimethoxybenzal-*
dehyde; syringic aldehyde; 3,5-dimethoxy-4-hydroxybenzene
carbonal; gallaldehyde 3,5-dimethyl ether. $C_9H_{10}O_4$; mol wt
182.17. C 59.33%, H 5.53%, O 35.13%. Widely distributed
in plants. Hydrolysis product of the naturally occurring
glycosyringic aldehyde: Körner, *Gazz. Chim. Ital.* **18**, 215
(1888). Prepn from heat-treated beechwood and lignin:
Kratzl, Silbernagel, *C.A.* **50**, 6040 (1956). Synthesis from
pyrogallol 1,3-dimethyl ether: Pearl, *J. Am. Chem. Soc.* **70**,
1746 (1948), *cf.* Graebe, Martz, *Ber.* **36**, 1032 (1903); Allen,
Leubner, *Org. Syn.* **31**, 92 (1951); **coll. vol. IV,** 866 (1963).

Very pale yellow needles from petr ether, mp 113°. bp_{14}
192-193°. uv max (dioxane): 305 nm. Very sparingly sol in
water, petr ether. Sol in alcohol, ether, chloroform, hot benzene, glacial acetic acid. Forms yellow sodium and potassium salts.

8997. Syringin. *4-(3-Hydroxy-1-propenyl)-2,6-dimeth-*
oxy-β-D-glucopyranoside; 4-(3-hydroxypropenyl)-2,6-di-
methoxyphenyl-D-glucoside; syringoside; ligustrin; lilacin;
methoxyconiferin. $C_{17}H_{24}O_9$; mol wt 372.36. C 54.83%, H
6.50%, O 38.67%. First isolated by Meillet in 1841 from
bark of *Syringa vulgaris* L. (lilac): *Ann.* **40**, 319 (1841).
Prepn: Pauly, Strassberger, *Ber.* **62**, 2277 (1929); Freudenberg, Schraube, *Ber.* **88**, 16 (1955). Isoln from lilac bark:
Freudenberg *et al., Ber.* **84**, 472 (1951); from various plants:
Plouvier, *Compt. Rend.* **254**, 4196 (1962); from cambial sap of
spruce: Freudenberg, Harkin, *Phytochemistry* **2**, 189 (1963).

Monohydrate, crystals from water, mp 192°. $[α]_D^{20}$ −8.2°
(c = 2.43 in chloroform), −17.25° (water). Slightly sol in
cold water; sol in hot water, alc; practically insol in ether.

8998. Syrosingopine. *18-[[4-[(Ethoxycarbonyl)oxy]-3,5-*
dimethoxybenzoyl]oxy]-11,17-dimethoxyyohimban-16-carb-
oxylic acid methyl ester; methyl carbethoxysyringoyl reser-

pate; carbethoxysyringoyl methyl reserpate; methyl *O*-(*O'*-carbethoxysyringoyl)reserpate; syringopine; Su 3118; Iso-tense; Londomin; Seniramin; Singoserp; Siringina; Raunova. $C_{35}H_{42}N_2O_{11}$; mol wt 666.70. C 63.06%, H 6.35%, N 4.20%, O 26.40%. Prepn: Lucas *et al., J. Am. Chem. Soc.* **81,** 1928 (1959); U.S. pat. **2,813,871** (1957 to Ciba).

Crystals from acetone, mp 175-179°.

THERAP CAT: Antihypertensive.

T

8999. 2,4,5-T. *(2,4,5-Trichlorophenoxy)acetic acid;* Esterone 245; Trioxone; Weedone. $C_8H_5Cl_3O_3$; mol wt 255.49. C 37.61%, H 1.97%, Cl 41.63%, O 18.79%. Post-emergence herbicide. Prepd from 2,4,5-trichlorophenol: Pokorny, *J. Am. Chem. Soc.* **63**, 1768 (1941); from benzenehexachloride: Galat, *ibid.* **74**, 3890 (1952). Activity: C. L. Hamner, T. B. Tukey, *Science* **100**, 154 (1944). Contains trace levels of TCDD, *q.v.*, as a contaminant: J. Smith, *Science* **203**, 1090 (1979); *Chem. & Eng. News* **59**, 6 (Jan. 5, 1981). Toxicity: V. A. Rowe, T. A. Hymas, *Am. J. Vet. Res.* **15**, 622 (1954). *See also* 2,4-D.

Crystals from benzene, mp 153°. d_{20}^{20} 1.80. Soly in water at 30°: 238 mg/kg. Sol in alcohol. Forms water-soluble sodium and alkanolamine salts. Commercial products are usually in the form of amines or esters, often in mixture with 2,4-D. LD_{50} orally in mice, rats: 389, 500 mg/kg (Rowe, Hymas).
USE: Formerly as herbicide. *Caution:* Irritant.
Note: In March 1985 the E.P.A. terminated all registrations for the use of this herbicide on rice fields, orchards, sugarcane, rangeland and other noncrop sites. This follows the 1970 action of the Department of Agriculture halting the use of the pesticide on all food crops except rice: *Chem. & Eng.* News **63**, 6 (Mar. 25, 1985).

9000. Tabernanthine. *13-Methoxyibogamine.* $C_{20}H_{26}$-N_2O; mol wt 310.42. C 77.38%, H 8.44%, N 9.03%, O 5.15%. Indole alkaloid isolated from root of *Tabernanthe iboga* Baill., *Apocynaceae:* Delourme-Houdé, *Ann. Pharm. Franc.* **4**, 30 (1946); Dickel *et al., J. Am. Chem. Soc.* **80**, 123 (1958). Also in *Tabernaemontana* and *Stemmadenia* spp.; usually found in ibogaine mother liquors: Walls *et al., Tetrahedron* **2**, 173 (1958). Isoln from genus *Conopharingia, Apocynaceae:* Renner, Prins, U.S. pat. **3,008,954** (1961 to Geigy). Structure: Bartlett *et al., J. Am. Chem. Soc.* **80**, 126 (1958). Mass spectrum: Biemann, Friedmann-Spiteller, *ibid.* **83**, 4805 (1961). Derivs: Taylor, U.S. pat. **2,877,229** (1959 to Ciba). Interaction with benzodiazepine receptors: J.-H. Trouvin *et al., Eur. J. Pharmacol.* **140**, 303 (1987).

Needles or shiny leaflets from ethanol, mp 213.5-215°. Sublimes at 160° (0.005 mm pressure). $[\alpha]_D^{20}$ −40° (acetone). pKa 6.04 in 80% methylcellosolve. uv max (ethanol): 228, 271, 299 nm (log ε 4.53, 3.64, 3.77). Sol in alcohol, benzene, ether, chloroform. Practically insol in water.
Hydrochloride, $C_{20}H_{27}ClN_2O$, crystals from water, dec 275-277°. $[\alpha]_D^{25}$ −66° (methanol, Dickel, *loc. cit.*); mp 210°, $[\alpha]_D^{20}$ −76.5° (methanol, Delourme-Houdé). Sol in water. More sol in chloroform than ibogaine hydrochloride.

9001. Tabun. *Dimethylphosphoramidocyanidic acid, ethyl ester;* ethyl *N*-dimethylphosphoramidocyanidate; dimethylamidoethoxyphosphoryl cyanide; GA. $C_5H_{11}N_2O_2P$; mol wt 162.12. C 37.04%, H 6.84%, N 17.28%, O 19.74%, P 19.10%. Military nerve gas; prepd from dimethylamidophosphoryl dichloride and sodium cyanide in the presence of ethanol: Holmstedt, *Acta Physiol. Scand.* **25**, Suppl. 90, 26 (1951). The synthesis of dimethylamidophosphoryl dichloride is also described by Michaelis, *Ann.* **326**, 129 (1903). Alternate route: B. C. Saunders, *Some Aspects of the Chemistry and Toxic Action of Organic Compounds Containing*

Phosphorus and Fluorine (Cambridge, 1957) p 91. Brief review: Schrader, *Die Entwicklung neuer insektizider Phosphorsaüre-Ester* (Verlag Chemie, Weinheim, 1963) p 3.

Liquid. Fruity odor reminiscent of bitter almonds. d 1.077. mp −50°. bp_{760} 240°; bp_{10} 120°; bp_9 100-108°. n_D^{20} 1.4250. IR absorption: *Acta Chem. Scand.* **5**, 1179 (1951). Readily sol in organic solvents. Miscible with water, but quickly hydrolyzed. Bleaching powder (chlorinated lime) destroys Tabun, but gives rise to cyanogen chloride. Extremely poisonous! LD_{50} i.p. in mice: 0.6 mg/kg, B. Holmstedt, *Pharmacol. Rev.* **11**, 567 (1959). The lethal dose for man may be as low as 0.01 mg/kg, *Chem. & Eng. News* **31**, 4676 (1953).
Caution: Potent cholinesterase inhibitor. Toxic not only by inhalation but by absorption through skin and eyes. Inhalation produces constriction of pupils of the eye, difficulty in breathing followed by bronchial constriction, convulsions, death.
USE: Chemical warfare agent.

9002. Tachysterol. *(6E,22E)-9,10-Secoergosta-5(10),6,-8,22-tetraen-3-ol.* $C_{28}H_{44}O$; mol wt 396.63. C 84.78%, H 11.18%, O 4.03%. From ergosterol or lumisterol by ultraviolet irradiation: Windaus *et al., Ann.* **492**, 226 (1932); *Ann.* **499**, 188 (1932); Dimroth, *Ber.* **70**, 1631 (1937). From calciferol by adsorption on acid clay: Thibaudet, *Compt. Rend.* **220**, 751 (1945). From precalciferol: Velluz, Goffinet, U.S. pat. **2,847,426** (1958 to UCLAF). Structure: Grundmann, *Z. Physiol. Chem.* **252**, 151 (1938); Thibaudet, *loc. cit.* Stereochemistry of the tachysterol system: Inhoffen, *Ber.* **88**, 1424 (1955); Verloop, *Rec. Trav. Chim.* **76**, 689 (1957); Delaroff *et al., Bull. Soc. Chim. France* **1963**, 1739.

Oil. $[\alpha]_D^{18}$ −70° (24.6 mg in 2 ml petr ether); $[\alpha]_{546}^{18}$ −86.3° (petr ether). uv max: 280 nm. Not pptd by digitonin. Insol in water; sol in organic solvents, but not in methanol. Very easily oxidized by air.
4-Methyl-3,5-dinitrobenzoate, $C_{36}H_{48}N_2O_6$, pale yellow crystals, mp 155°.
Dihydro derivative, *see* dihydrotachysterol.

9003. Tacrine. *1,2,3,4-Tetrahydro-9-acridinamine;* *9-amino-1,2,3,4-tetrahydroacridine;* 5-amino-1,2,3,4-tetra-hydroacridine; 1,2,3,4-tetrahydro-5-aminoacridine. $C_{13}H_{14}$-N_2; mol wt 198.26. C 78.75%, H 7.12%, N 14.13%. Centrally active anticholinesterase. Prepn: Braun *et al., Ber.* **64**, 227 (1931); Albert, Gledhill, *J. Soc. Chem. Ind.* **64**, 169T (1945); Petrow, *J. Chem. Soc.* **1947**, 634; Moore, Kornreich, *Tetrahedron Letters* **1963**, 1277. Pharmacology and toxicity: S. Gershon, F. H. Shaw, *J. Pharm. Pharmacol.* **10**, 638 (1958). Enzyme inhibiting activity: P. N. Kaul, *ibid.* **14**, 243 (1962). Neuropharmacology: B. Drukarch *et al., Life Sci.* **42**, 1011 (1988). Clinical trial in severe Alzheimer's disease: W. K. Summers *et al., N. Engl. J. Med.* **315**, 1241 (1986). *See also: ibid.* **316**, 1603-1606. Review of pharmacology and clinical uses: W. K. Summers *et al., Clin. Toxi-col.* **16**, 269-281 (1980).

Octahedra from very dil alc, mp 183-184°.
Hydrochloride, $C_{13}H_{15}ClN_2$, *Romotal, THA.* Yellow needles from concd hydrochloric acid, mp 283-284°. Bitter taste. Sol in water. pH of 1.5% soln: 4.5-6.
THERAP CAT: Cognition activator. Antidote to curare. Respiratory stimulant.

9004. Tacryl®. Acrylic fiber of a specific multichain type. The structure could be described as a spider molecule with up to six long straight linear legs which can orient independently and build up a fibrous structure. The molecular structure is built up by a controlled cross-linking process, so that Tacryl is like wool as it has a specific chain molecular interlinking which contributes to its mechanical and elastic properties. Synthesis of multichain polymers: Schaefgen, Flory, *J. Am. Chem. Soc.* **70**, 2709 (1948). Tacryl has a higher shear modulus and a higher strength than linear acrylics with the same degrees of orientation. It undergoes very slow hydrolysis under hot acid aq conditions, and is very resistant to dry and wet heat (neutral conditions): Sundén, *Tappi* **41**, 173A (1958).

9005. D-Tagatose. D-*lyxo*-Hexulose. $C_6H_{12}O_6$; mol wt 180.16. C 40.00%, H 6.71%, O 53.29%. Occurs in *Sterculia setigera* gum, a partially acetylated acidic polysaccharide: Hirst *et al., J. Chem. Soc.* **1949**, 3145. Prepn from D-galactose: Reichstein, Bosshard, *Helv. Chim. Acta* **17**, 753 (1934); Jones, Wall, *Can. J. Chem.* **38**, 2290 (1960); by biochemical oxidation of D-talitol: Totton, Lardy, *J. Am. Chem. Soc.* **71**, 3076 (1949); from D-galacturonic acid: Gorin *et al., Can. J. Chem.* **33**, 1116 (1955); from lactose in heated milk: Adachi, *Nature* **181**, 840 (1958). Synthesis: Wolfrom, Bennett, *J. Org. Chem.* **30**, 1284 (1965); Al-Gobore *et al., Carbohyd. Res.* **16**, 466 (1971). Structure: Takagi, Rosenstein, *ibid.* **11**, 156 (1969). Stereoselective synthesis of L-*tagatose*, antipode of the naturally occurring D-tagatose: T. Mukaiyama *et al., Chem. Letters* **1982**, 1169.

Crystals from dil alc, mp 134-135°. $[\alpha]_D^{20}$ −2.3° (c = 2.19 in water); $[\alpha]_D^{25}$ −5° (c = 1 in water).
Phenylosazone, $C_{18}H_{22}N_4O_4$, crystals from dil ethanol, dec 187-188°. $[\alpha]_D^{23}$ +47° (c = 0.82 in 2-methoxyethanol).

9006. Taglutimide. *2-(2,6-Dioxo-3-piperidinyl)hexahy-dro-4,7-methano-1H-isoindole-1,3(2H)-dione; (cis-endo)-N-(2,6-dioxo-3-piperidyl)-2,3-norbornanedicarboximide; 3-(1,4-endomethylenecyclohexane-2,3-endo-cis-dicarboximi-do)piperidine-2,6-dione; biglumide; K 2004; Synval.* $C_{14}H_{16}N_2O_4$; mol wt 276.29. C 60.86%, H 5.84%, N 10.14%, O 23.16%. Sedative-hypnotic agent, related structurally to thalidomide, *q.v.* Prepn: H. Koch, J. Kotlan, *Monatsch.* **97**, 1648 (1966); *eidem*, Fr. pat. **1,570,456** corresp to U.S. pat. **3,625,946** (1969, 1971 both to Kwizda). Synthesis of metabolites: H. Koch *et al., Ann.* **755**, 51 (1972). Teratological studies: H. Koch, F. Köhler, *Arch. Toxicol.* **35**, 63 (1976); W. Jurecka *et al., ibid.* **37**, 165 (1977). Pharmacokinetics, metabolism in man: H. Koch *et al., Arch. Pharm.* **309**, 609 (1976); in rats: F. Fiebrich *et al., Arzneimittel Forsch.* **29**, 1036 (1979). Pharmacological properties: W. G. Schützenberger *et al., ibid.* 1146. Clinical and electroencephalographic study: G. Argyropoulos *et al., Wien Med. Wochenschr.* **126**, 75 (1976).

Colorless or off-white cryst from DMF/water, mp 235-236°. Slightly sol in water and organic solvents. Unstable in alkaline soln. Taglutimide did not produce any symptoms of toxicity after oral administration of 5 g/kg in mice or 400 mg/kg in dogs, W. G. Schützenberger *et al., loc. cit.*
THERAP CAT: Sedative, hypnotic.

9007. Taka-Diastase. *α-Amylase (Aspergillus oryzae);* Koji; Aspergillus diastase; Sanzyme. A purified multienzyme produced by the microorganism *Aspergillus oryzae* (Ahl.) Cohn, *Aspergillaceae,* grown on sterilized wheat bran or on rice hulls. Represents more than 30 different enzymatic functions. Unlike diastase of malt, it is not only amylolytic but digests proteins and fats also.
Whitish-yellow, very hygroscopic powder. Converts 450 times its wt of starch into maltose. *Keep tightly closed.*
USE: Preparing the Japanese national drink "sake"; in converting maizes into sugar in manuf of whiskey.
THERAP CAT: Amylolytic.

9008. Talampicillin. *6-[(Aminophenylacetyl)amino]-3,3-dimethyl-7-oxo-4-thia-1-azabicyclo[3.2.0]heptane-2-carboxylic acid 1,3-dihydro-3-oxo-1-isobenzofuranyl ester; (2S,5R,-6R)-6-[(R)-2-amino-2-phenylacetamido]-3,3-dimethyl-7-oxo-4-thia-1-azabicyclo[3.2.0]heptane-2-carboxylic acid ester with 3-hydroxyphthalide; 6-[D(−)-α-aminophenyl-acetamido]penicillanic acid phthalide ester; ampicillin 1-oxo-1,3-dihydroisobenzofuran-3-yl ester; phthalidyl D-α-aminobenzylpenicillanate; yamacillin; BRL-8988.* $C_{24}H_{23}N_3O_6S$; mol wt 481.52. C 59.86%, H 4.81%, N 8.73%, O 19.94%, S 6.66%. Semi-synthetic antibiotic related to penicillin. Prepn: Murakami *et al.,* Ger. pat. **2,225,149** corresp to U.S. pat. **3,951,954** (1972, 1976, both to Yamanouchi Pharm.); Ferres, Clayton, Ger. pats. **2,228,012** and **2,228,-255** corresp to U.S. pat. **3,860,579** (1972, 1972, 1975, all to Beecham Group). Pharmacology: Clayton *et al., Antimicrob. Ag. Chemother.* **5**, 670 (1974).

Hydrochloride, $C_{24}H_{24}ClN_3O_6S$, *BRL 8988, Talat, Talpen.* White powder, mp 154-157° (dec).
THERAP CAT: Antibacterial.

9009. Talastine. *2-[2-(Dimethylamino)ethyl]-4-(phenyl-methyl)-1(2H)-phthalazinone; 4-benzyl-2-[2-(dimethylami-no)ethyl]-1(2H)-phthalazinone.* $C_{19}H_{21}N_3O$; mol wt 307.38. C 74.24%, H 6.89%, N 13.67%, O 5.21%. Prepn: Engelbrecht *et al.,* Ger. pat. **1,046,625**; *eidem,* U.S. pat. **3,017,411** (1958, 1962, both to VEB Deutsches Hydrierwerk Rodleben). Pharmacology: Lenke, *Arzneimittel-Forsch.* **7**, 678 (1957).

Liquid, $bp_{0.3}$ 215-222°.
Hydrochloride, $C_{19}H_{22}ClN_3O$, *HL 2186, Ahanon.* Crystals, mp 178°. LD_{50} i.p. in mice: 116 mg/kg.
THERAP CAT: Antihistaminic.

9010. Talbutal. *5-(1-Methylpropyl)-5-(2-propenyl)-2,4,-6(1H,3H,5H)-pyrimidinetrione; 5-allyl-5-sec-butylbarbituric* *acid;* 5-allyl-5-(1-methylpropyl)barbituric acid; Lotusate; Profundol. $C_{11}H_{16}N_2O_3$; mol wt 224.25. C 58.91%, H 7.19%, N 12.49%, O 21.40%. Prepn: Volwiler, *J. Am. Chem. Soc.* **47**, 2236 (1925). Acute toxicity: E. W. Schafer, *Toxicol. Appl. Pharmacol.* **21**, 315 (1972).

Crystals from water or dil alcohol, mp 108-110°. Slightly bitter taste. Practically insol in water and petr ether. Sol in alcohol, chloroform, ether, acetone, glacial acetic acid, also in solns of fixed alkali hydroxides. A satd aq soln is acid to litmus. LD_{50} orally in rats: 57.5 mg/kg (Schafer).

Caution: May be habit forming. This is a controlled substance (depressant) listed in the U.S. Code of Federal Regulations, Title 21 Parts 329.1 and 1308.13 (1987).

THERAP CAT: Sedative, hypnotic.

9011. Talc. Talcum; French chalk. The lumps are also known as *soapstone* or *steatite.* Finely powdered native hydrous magnesium silicate.

White to grayish-white, very fine odorless, crystalline powder; unctuous, and adheres readily to the skin. Insol in water, cold acids or in alkalies.

USE: Dusting powder, either alone or with starch or boric acid, for medicinal and toilet prepns; excipient and filler for pills, tablets and for dusting tablet molds; clarifying liquids by filtration. As pigment in paints, varnishes, rubber; filler for paper, rubber, soap; in fireproof and cold-water paints for wood, metal and stone; lubricating molds and machinery; glove and shoe powder; electric and heat insulator.

9012. Talinolol. (±)-*N-Cyclohexyl-N'-[4-[3-[(1,1-dimethylethyl)amino]-2-hydroxypropoxy]phenyl]urea;* (±)-1-[*p*-[3-(*tert*-butylamino)-2-hydroxypropoxy]phenyl]-3-cyclohexylurea; 02-115; Cordanum. $C_{20}H_{33}N_3O_3$; mol wt 363.50. C 66.09%, H 9.15%, N 11.56%, O 13.20%. β-Adrenergic blocker structurally related to practolol, *q.v.* Prepn: R. Eckardt *et al.*, Ger. pat. **2,153,024** (1972 to Arzneimittelwerke VEB), *C.A.* **77**, 61639b (1972); M. Wilhelm, Swiss pat. **566,302** corresp to U.S. pats. **3,935,259** and **4,-038,313** (1975, 1976, 1977, all to Ciba-Geigy); R. Eckardt *et al.*, *Pharmazie* **30**, 633 (1975). Series of articles on pharmacology, toxicology: *ibid.* 638-683. Acute toxicity: K. Femmer *et al.*, *ibid.* 642. Clinical pharmacology: K. O. Haustein *et al.*, *Int. J. Clin. Pharmacol. Biopharm.* **17**, 465 (1979). Cardiovascular, β-blocking effects: S. Heer, K. Femmer, *Pharmazie* **34**, 317 (1980).

Cryst from isopropanol, mp 142-144°. LD_{50} in rats, mice (mg/kg): 1180, 593 orally; 54.3, 74.7 i.p.; 29.7, 25.0 i.v. (Femmer).

THERAP CAT: Antihypertensive, antiarrhythmic.

9013. Tall Oil. Liquid rosin; Acintol C; tallol; talleol. ["Tall" is Swedish for "pine".] A by-product of the wood pulp industry. Usually recovered from pine wood "black liquor" of the sulfate or kraft paper process. Contains rosin acids, oleic and linoleic acids. Long chain alcohols and small amounts of sterols, especially phytosterol, have also been found. Comprehensive collection of 1660 abstracts: J. Weiner, *Tall Oil* (The Institute of Paper Chemistry, Appleton, Wisconsin, 3rd ed., 1959) 450 pp; J. Weiner, J. Byrne, 1st supplement (1965).

Dark brown liquid. Acrid odor similar to that of burnt rosin. d 0.95 to 1.0. n_D^{20} approx 1.5. Acid no. 170-180°. Sapon no. 172-185. Iodine no. 120-188. Fatty acids 50-60%. Rosin acids 34-40%. Unsaponifiable matter 5-10%.

USE: Mfg soap pastes, flotation agents, greases, paint, alkyd resins, linoleum, soaps, fungicides, asphalt emulsions,

rubber formulations, cutting oils, sulfonated oils. Review of possible uses: Cannon, *Chem. Eng.* **61**, 142 (June 1954).

9014. Tallow. In North America designates the fat from the fatty tissue of bovine cattle and sheep only. It may be offered separately as beef tallow and as sheep or mutton tallow. The term horse tallow is generally no longer admitted. *Oleo stock* is the highest grade of beef tallow. Contains (as glycerides): Oleic acid (37-43%), palmitic (24-32%), stearic (20-25%), myristic (3-6%), linoleic (2-3%). Minor constituents are cholesterol, arachidonic, elaidic, and vaccenic acids. Perhaps the most observed characteristic of tallow is its titer (solidif pt) which ranges from 40° to 46°.

9015. Tallow Alcohol. A name for commercial mixtures of *n*-octadecanol and *n*-hexadecanol.

Fatty crystalline mass, mp 46-47°.

USE: Defoaming agent, emollient, intermediate for surface active agents.

9016. Tallysomycin. Talisomycin; BU-2231. Antitumor antibiotic complex and third generation analog of bleomycins, *q.v.*, produced by *Streptoalloteichus hindustanus* E 465-94. Prodn, isoln, properties of the major components, tallysomycins A and B: H. Kawaguchi *et al.*, Belg. pat. **845,513** corresp to U.S. pat. **4,051,237** (both 1977 to Bristol-Myers); *eidem, J. Antibiot.* **30**, 779 (1977). Structure of A and B, based on originally proposed bleomycin structure: M. Konishi *et al.*, *ibid.* 789; *cf.* bleomycin for revised structure. Antitumor activity: H. Imanishi *et al.*, *ibid.* **31**, 667 (1978). Radioimmunoassay: A. Broughton *et al.*, *Cancer Treat. Rep.* **63**, 1829 (1979). Pharmacokinetics: J. E. Strong *et al.*, *ibid.* 1821. ^{13}C-NMR spectra of tallysomycin and its zinc complex: F. T. Greenaway *et al.*, *Org. Magn. Reson.* **13**, 270 (1980). Biosynthetic derivs: T. Miyaki *et al.*, *J. Antibiot.* **34**, 658, 665 (1981). Relative pulmonary toxicity: A. Broughton *et al.*, *Cancer Treat. Rep.* **64**, 659 (1980). *Review:* S. T. Crooke *et al.*, *Recent Results Cancer Res.* **76**, 83-90 (1981).

tallysomycin A

tallysomycin B

Tallysomycin A, $C_{68}H_{110}N_{22}O_{27}S_2$, *$N^1$-[4-amino-6-[[3-[(4-aminobutyl)amino]propyl]amino]-6-oxohexyl]-13-[(4-amino-4,6-dideoxy-α-L-talopyranosyl)oxy]-19-demethyl-12-hydroxybleomycinamide; talisomycin; BU-2231A.* White amorphous solid. No definite mp; gradually dec >210°. Sol in water, methanol, DMF; slightly sol in ethanol. Practically insol in other organic solvents. $[\alpha]_D^{23}$ $-21°$ (c = 0.5 in water). uv max (water): 290 nm ($E_{1cm}^{1\%}$ 67). LD_{50} in mice: 28 mg/kg s.c., H. Kawaguchi et al., U.S. pat. 4,051,237.

Tallysomycin B, $C_{62}H_{98}N_{20}O_{26}S_2$, N^1-[3-[(4-aminobutyl)-amino]propyl]-13-[(4-amino-4,6-dideoxy-α-L-talopyranosyl)oxy]-19-demethyl-12-hydroxybleomycinamide; talisomycin B; BU-2231B. Mp, soly similar to tallysomycin A. $[\alpha]_D^{23}$ $-19°$ (c = 0.5 in water). uv max (water): 289.5 ($E_{1cm}^{1\%}$ 77). LD_{50} in mice: > 50 mg/kg s.c., H. Kawaguchi et al., U.S. pat. 4,051,237.

9017. Talniflumate. 2-[[3-(Trifluoromethyl)phenyl]amino]-3-pyridinecarboxylic acid 1,3-dihydro-3-oxo-1-isobenzofuranyl ester; phthalidyl 2-(3-trifluoromethylanilino)nicotinate; phthalidyl 2-(α,α,α-trifluoro-m-toluidino)nicotinate; BA 7602-06; Somalgen. $C_{21}H_{13}F_3N_2O_4$; mol wt 414.34. C 60.87%, H 3.16%, F 13.76%, N 6.76%, O 15.45%. Deriv of niflumic acid, q.v. Prepn: Belg. pat. 858,864 (1978 to Bago), C.A. 89, 109104 (1978); S. Bago, U.S. pat. 4,168,313 (1979). Synthesis and pharmacologic study: M. Los et al., Farmaco Ed. Sci. 36, 372 (1981).

White or pale yellow cryst powder, mp 165-166°. uv max (chloroform): 287, 357 nm (ϵ 25600, 7800). LD_{50} in rats: 12,000 mg/kg orally, M. Los et al., loc. cit.

THERAP CAT: Anti-inflammatory; analgesic.

9018. Tamarind. Partially dried ripe fruit of Tamarindus indica L., Leguminosae; preserved in sugar or syrup. Habit. East Indies, India, Africa; naturalized in West Indies. Constit. The pulp contains about 10% tartaric acid, also some citric and malic acids; 25-40% invert sugar, pectin. Review: Rao, Srivastava, in Industrial Gums, R. L. Whistler, Ed. (Academic Press, New York, 2nd ed., 1973) pp 369-411.

USE: The pulp as souring agent in Indian curries. The seed kernel powder with water as sizing agent.

9019. Tamoxifen. [Z]-2-[4-(1,2-Diphenyl-1-butenyl)-phenoxy]-N,N-dimethylethanamine; 1-p-β-dimethylaminoethoxyphenyl-trans-1,2-diphenylbut-1-ene. $C_{26}H_{29}NO$; mol wt 371.53. C 84.05%, H 7.87%, N 3.77%, O 4.31%. Nonsteroidal estrogen antagonist. Prepn: Belg. pat. 637,389 (1964 to ICI). Identification and separation of isomers: G. R. Bedford, D. N. Richardson, Nature 212, 733 (1966); Belg. pat. 678,807; M. J. K. Harper et al., Brit. pat. 1,064,629; eidem, U.S. pat. 4,536,516 (1966, 1967, 1985 all to ICI). Synthesis of isomers: D. W. Robertson, J. A. Katzenellenbogen, J. Org. Chem. 47, 2387 (1982). Stereospecific synthesis of isomers: R. B. Miller, M. S. Al-Hassan, J. Org. Chem. 50, 2121 (1985). Comparative activity of isomers: M. J. K. Harper, A. L. Walpole, Nature 212, 87 (1966). Pharmacology: eidem, J. Endocrinol. 37, 83 (1967). Review of biochemistry: R. I. Nicholson et al., Adv. Sex Horm. Res. 4, 119-152 (1980); of pharmacology and use in breast cancer: V. C. Jordan et al., Cancer Treat. Rep. 64, 745-759 (1980); of pharmacokinetics and metabolism: V. C. Jordan, Breast Cancer Treat. Res. 2, 123-138 (1982); of chemistry, pharmacology and clinical uses: B. J. A. Furr, V. C. Jordan, Pharmacol. Ther. 25, 127-205 (1984). Brief review of use in mastalgia: I. S. Fentiman, Drugs 32, 477-480 (1986).

Crystals from petr ether, mp 96-98°.

Citrate, $C_{32}H_{37}NO_8$, ICI-46474, Kessar, Noltam, Nolvadex, Tamofen, Tamoxasta, Terimon, Zynoplex. Fine, white, odorless crystalline powder, mp 140-142°. Slightly sol in water; sol in ethanol, methanol, acetone. Hygroscopic at high relative humidities. Sensitive to uv light. LD_{50} in mice, rats (mg/kg): 200, 600 i.p.; 62.5, 62.5 i.v.; 3000-6000, 1200-2500 orally (Furr, Jordan).

cis-Form base, mp 72-74° from methanol.

cis-Form citrate, $C_{32}H_{37}NO_8$, ICI-47699. mp 126-128°.

THERAP CAT: Anti-estrogen. Palliative treatment of breast cancer.

9020. Tanacetin. Decahydro-6β,9aα-dihydroxy-5aβ-methyl-3,9-bis(methylene)naphtho[1,2-b]furan-2(3H)-one; 1β,5α-dihydroxy-6β,7αH-selina-4(15),11(13)-dien-6,12-olide. $C_{15}H_{20}O_4$; mol wt 264.31. C 68.16%, H 7.63%, O 24.21%. Isoln from seed, herb, and flowers of Tanacetum vulgare L., Compositae: Homolle, J. Pharm. Chim. 7, 57 (1845); Jaretzky, Kühne, Arch. Pharm. 271, 353 (1933); Suchy, Coll. Czech. Chem. Commun. 27, 1058 (1962). Structure and absolute config: Samek et al., ibid. 38, 1971 (1973).

Crystals, mp 205°. $[\alpha]_D^{22}$ +179.5° (c = 2.3 in ethanol).

9021. Tanghinigenin. 7β,8-Epoxy-3β,14-dihydroxy-5β-card-20(22)-enolide. $C_{23}H_{32}O_5$; mol wt 388.49. C 71.10%, H 8.30%, O 20.59%. Isoln from glucosides: Sigg et al., Helv. Chim. Acta 38, 166 (1955). Structure: Flury, Reichstein, Ann. Chim. (Rome) 53, 23 (1963); Flury et al., Helv. Chim. Acta 48, 1113 (1965).

Prisms from acetone + petr ether, mp 187-188°. $[\alpha]_D^{25}$ +14.1° (c = 1.138 in chloroform). uv max: 217 nm (log ϵ 4.22). LD_{50} in cats: 1 mg/kg i.v., Chen, Henderson, J. Pharmacol. Exp. Ther. 111, 365 (1954).

Acetate, $C_{25}H_{34}O_6$, acetyltanghinigenin. Prisms from acetone + petr ether, mp 241-243°. $[\alpha]_D^{24}$ +14.9° (c = 1.075 in chloroform).

9022. Tanghinin. 3β-[(2-O-Acetyl-6-deoxy-3-O-methyl-α-L-glucopyranosyl)oxy]-7β,8-epoxy-14-hydroxy-5β-card-20(22)-enolide. $C_{32}H_{46}O_{10}$; mol wt 590.69. C 65.06%, H 7.85%, O 27.09%. From the seed of Tanghinia madagascariensis Pet., Apocynaceae and Tanghinia venenifera Poir., Apocynaceae. Isoln: Arnaud, Compt. Rend. 108, 1255 (1889); Frèrejacque et al., ibid. 226, 268 (1948). Tentative structure: Sigg et al., Helv. Chim. Acta 38, 166 (1955). Revised structure: Flury, Reichstein, Ann. Chim. (Rome) 53, 23 (1963).

Leaflets from methanol + ether, mp 128-131°. $[\alpha]_D^{21}$ −81.5° (c = 1.092 in methanol). uv spectra: Frèrejacque *et al., Helv. Chim. Acta* **39**, 1900 (1956). LD_{50} in cats: 0.4 mg/kg i.v., Chen, Henderson, *J. Pharmacol. Exp. Ther.* **111**, 365 (1954).

9023. Tannic Acid. Tannin; gallotannin; gallotannic acid. Incorrectly *"digallic acid"*. Tannic acid of commerce usually contains about 10% H_2O. Occurs in the bark and fruit of many plants, notably in the bark of the oak species, in sumac and myrobalan. It is produced from Turkish or Chinese nutgall, the former contg 50-60%, the latter about 70%. The chemistry of the tannins is most complex and non-uniform. Tannins may be divided into 2 groups: *(a)* derivatives of flavanols, so-called condensed tannins and *(b)* hydrolyzable tannins (the more important group) which are esters of a sugar, usually glucose, with one or more trihydroxybenzenecarboxylic acids. The structure given here is that of a tannin named *corilagin:* Schmidt *et al., Ann.* **587**, 67 (1954). The empirical formula of corilagin is $C_{27}H_{24}O_{18}$. For the commercial tannic acid, whose specifications follow, the empirical formula is usually given as $C_{76}H_{52}O_{46}$. *Comprehensive reviews:* M. Nierenstein, *The Natural Organic Tannins* (London, 1934); O. Th. Schmidt, "Gallotannine" in *Fortschr. Chem. Org. Naturst.* **13**, 70-136 (1956); *Symposium on the Chemistry of Vegetable Tannins* (Soc. Leather Trades Chemists, Croydon 1956).

corilagin

Yellowish-white to light brown, amorphous, bulky powder or flakes, or spongy masses; faint characteristic odor; astringent taste. Gradually darkens on exposure to air and light; at 210-215° dec mostly into pyrogallol and CO_2. Gives insol ppts with albumin, starch, gelatin, most alkaloidal and metallic salts; produces a bluish-black color or precip with ferric salts. One gram dissolves in 0.35 ml water, 1 ml warm glycerol; very sol in alc, acetone; practically insol in benzene, chloroform, ether, petr ether, carbon disulfide, carbon tetrachloride. *Keep well closed and protected from light.* LD_{100} orally in mice: 6.0 g/kg, Robinson, Graessle, *J. Pharmacol. Exp. Ther.* **77**, 63 (1943).

Incompat: Salts of heavy metals, alkaloids, gelatin, albu-

min, starch, oxidizing substances—e.g., permanganates, chlorates; spirit nitrous ether; lime water.

USE: Mordant in dyeing; manuf ink; sizing paper and silk; printing fabrics; with gelatin and albumin for manuf of imitation horn and tortoise shell; tanning; clarifying beer or wine; in photography; as coagulant in rubber manuf; manuf gallic acid and pyrogallol; as reagent in analytical chemistry.

THERAP CAT: Astringent.

THERAP CAT (VET): Astringent, hemostatic, in solutions for burns. Has been used internally as an astringent and as a heavy metal antidote.

9024. Tannoform. *Methyleneditannin;* tannin-formaldehyde; Helgotan. Prepd by condensing one mole formaldehyde with two moles tannin: Chemnitius, *Pharm. Zentralh.* **68**, 273 (1927); Schwyzer, *Pharm. Ztg.* **74**, 1334 (1929).

Reddish, odorless, tasteless, bulky powder, mp about 230° with decompn. Practically insol in water; sol in alcohol, alkaline fluids.

THERAP CAT: Astringent.

THERAP CAT (VET): Externally as astringent, antiseptic (skin lesions and otorrhea). Has been used internally for diarrhea.

9025. Tantalum. Ta; at. wt 180.9479; at. no. 73; valence 5, also 4, 3, 2. Two naturally occurring isotopes: 181 (99.9877%); 180 (0.0123%; $T_{1/2} > 10^{12}$ years); artificial radioactive isotopes: 172-179; 182-186. Occurs almost invariably with niobium; less abundant than niobium. Found in the minerals columbite, *q.v.*, tantalite $[(Fe,Mn)(Ta,Nb)_2O_6]$ and *microlite* $[(Na,Ca)_2Ta_2O_6(O,OH,F)]$. Discovered by Ekeberg in 1802; first obtained pure by Bolton: *Z. Elektrochem.* **11**, 45 (1905). Prepn: Schoeller, Powell, *J. Chem. Soc.* **119**, 1927 (1921). Reviews of tantalum and its compounds: G. L. Miller, *Tantalum and Niobium* (Academic Press, New York, 1959) 767 pp; Brown, "The Chemistry of Niobium and Tantalum" in *Comprehensive Inorganic Chemistry* vol. 3, J. C. Bailar, Jr. *et al.*, Eds. (Pergamon Press, Oxford, 1973) pp 553-622.

Gray, very hard, malleable, ductile metal; can readily be drawn in fine wires. mp 2996°. bp 5429°. d 16.69. Spec heat (0°): 0.036 cal/g/°C. Electrical resistivity (18°): 12.4 μohm-cm. Insol in water. Very resistant to chemical attack; not attacked by acids other than hydrofluoric; not attacked by aq alkalies; slowly attacked by fused alkalies. Reacts with fluorine, chlorine, and oxygen only on heating. At high temps absorbs several hundred times its volume of hydrogen; combines with nitrogen, with carbon.

USE: In pen points; analytical weights; apparatus and instruments for chemical, surgical, and dental use instead of platinum, in tantalum capacitors (a type of electrolytic condenser, trademarked "Tantalytic").

9026. Tantalum Pentachloride. Cl_5Ta; mol wt 358.24. Cl 49.50%, Ta 50.50%. $TaCl_5$. Prepn: Rolsten, *J. Am. Chem. Soc.* **80**, 2952 (1958). Review of tantalum halides: Fairbrother in *Halogen Chemistry* vol. 3, V. Gutmann, Ed. (Academic Press, New York, 1967) pp 123-178. Acute toxicity: K. W. Cochran *et al., Arch. Ind. Hyg. Occup. Med.* **1**, 637 (1950).

White or light yellow, cryst powder; monoclinic; dec in moist air. d 3.68; mp 216.5-220°. Begins to volatilize at 144°, bp 239.3°. Dec by water; sol in abs alcohol. LD_{50} in rats (mg/kg): 75 i.p.; 1900 orally (Cochran).

9027. Tantalum Pentafluoride. F_5Ta; mol wt 275.95. F 34.43%, Ta 65.57%. TaF_5. Prepd from tantalum pentachloride by the halide exchange method according to the equation $TaCl_5 + 5HF \rightarrow TaF_5 + 5HCl$: Ruff, Zedner, *Ber.* **42**, 492 (1909); Ruff, Schiller, *Z. Anorg. Allgem. Chem.* **72**, 329 (1911); Kwasnik in *Handbook of Preparative Inorganic Chemistry* vol. 1, G. Brauer, Ed. (Academic Press, New York, 2nd ed, 1963) pp 255-256. Prepn from the elements: Fairbrother, Frith, *J. Chem. Soc.* **1951**, 3051. Review of transition metal pentafluorides: Peacock, *Advan. Fluoride Chem.* **7**, 113-145 (1973).

Deliquescent, strongly refractive prisms. d^{20} 4.74. mp 96.8°. Also reported as 95.1°: Fairbrother, Frith, *loc. cit.* bp 229.5°. Sol in water and ether with formation of oxyfluoro complexes. Also sol in concd nitric acid, more sol in fuming nitric acid. Sparingly sol in hot carbon disulfide and hot carbon tetrachloride. Etches glass slowly.

USE: Friedel-Crafts catalyst.

9028. Tantalum Pentoxide. Tantalic acid anhydride. O_5Ta_2; mol wt 441.90. O 18.10%, Ta 81.89%. Ta_2O_5. Acute toxicity: K. W. Cochran *et al., Arch. Ind. Hyg. Occup. Med.* **1**, 637 (1950).

White, microcrystalline, infusible powder. Insol in water, alcohol, mineral acids. Sol in HF. Dec by fusing with $KHSO_4$ or KOH, forming potassium tantalate with the latter. LD_{50} orally in rats: 8000 mg/kg (Cochran).

9029. Tar Acids. Tar acids are the phenols obtained from coal tar distillates or synthesized from coal tar hydrocarbons. Examples: Phenol, cresol, cresylic acid, xylenol.

9030. Taraxacum. Dandelion; lion's tooth. Dried rhizome and roots of *Taraxacum palustre* (Lyons) Lam. & DC. (*T. officinale* Weber, *Leontodon taraxacum* L.), *Compositae. Habit.* Europe; naturalized in North America. *Constit.* Taraxerol, choline, levulin, inulin, pectin.

9031. Taraxasterol. *18α,19α-Urs-20(30)-en-3β-ol;* taraxast-20(30)-en-3β-ol; anthesterin; α-lactucerol; taraxasterin. $C_{30}H_{50}O$; mol wt 426.70. C 84.44%, H 11.81%, O 3.75%. A monohydroxy triterpene. Isoln from *Taraxacum officinale*, Wiggers, *Compositae:* Power, Browning, *J. Chem. Soc.* **101**, 2411 (1912). Structure and configuration: Ames *et al., ibid.* **1954**, 1905. Identity with anthesterin: Power, Browning *ibid.* **105**, 1829 (1914); with α-lactucerol: Zellner, *Monatsh.* **47**, 681 (1926).

Needles from alcohol, mp 221-222°. $[\alpha]_D$ +96.3° $(CHCl_3)$. Very sol in alcohol, ether, petr ether; slightly sol in chloroform, benzene, carbon disulfide, acetone.

Acetate, $C_{32}H_{52}O_2$, *lactucerin, lactucon.* Hexagonal plates. mp 251-252° (from ethyl acetate + alcohol). $[\alpha]_D$ +100.5°.

9032. Taraxein. A protein complex isolated from the blood serum of schizophrenics by chromatography on diethylaminoethyl cellulose. The taraxein fraction precedes the ceruloplasmin fraction. Contains copper bound to protein. Method of isolation: Heath *et al., Am. J. Psychiat.* **114**, 14 (1957); *(Lippincott's) Medical Science* **6**, 401 (1959). Processing and identification: Heath *et al., Proc. 3rd World Congr. Psychiat., Montreal, 1961* **1**, 619 (1962). Studies of taraxein in schizophrenics: Heath *et al., Arch. Gen. Psychiat.* **16**, 1, 10, 24 (1967).

9033. Taraxerol. *D-Friedoolean-14-en-3β-ol;* isoolean-14-en-3β-ol; skimmiol; alnulin; tiliadin. $C_{30}H_{50}O$; mol wt 426.70. C 84.44%, H 11.81%, O 3.75%. Found in *Tilia cordata* Mill., *Tiliaceae:* Bräutigam, *Arch. Pharm.* **238**, 555 (1900); in *Alnus glutinosa* (L.) Gaertn., *Betulaceae:* Zellner, Weiss, *Sitzber. Akad. Wiss Wien* **132**, 258 (1923); in *Taraxacum officinale* Weber, *Compositae:* Burrows, Simpson, *J. Chem. Soc.* **1938**, 2042; in *Litsea dealbata* Nees, *Lauraceae:* Dunstan *et al., Aust. J. Chem.* **6**, 321 (1953); from *Befaria racemosa* (Vent.), *Ericaceae:* Euda *et al., J. Org. Chem.* **26**, 271 (1961). Structure: Beaton *et al., J. Chem. Soc.* **1955**, 2131. Partial synthesis from β-amyrin: *eidem, Chem. & Ind. (London)* **1955**, 35.

Plates from chloroform + methanol, needles from benzene, mp 282-285°. uv max (ethanol): 210, 215, 220, 223 nm (ε 3900, 2400, 700, 250). Sol in benzene, chloroform, ether, ethyl acetate, acetic anhydride, acetic acid, phenol, pyridine, xylene; less sol in alcohol.

Acetate, $C_{32}H_{52}O_2$, plates from chloroform + methanol, mp 303-305°. $[\alpha]_D$ +10.5° (c = 1.8 in chloroform).

Benzoate, $C_{37}H_{54}O_2$, needles from benzene or chloroform + alc, mp 292-293°. $[\alpha]_D^{23}$ +35.7° (c = 0.7 in chloroform).

9034. Tar Oil. Volatile oil distilled from wood tar. *Principal constit.* Phenolic substances and hydrocarbons.

Almost colorless liquid when fresh, but soon becomes dark brownish-red. d 0.860-0.900. Insol in water; sol in alcohol, ether.

9035. Tar Oil, Rectified. Pine tar oil. The volatile oil from pine tar rectified by steam distillation. Chief active constituents are phenolic substances.

Dark reddish-brown, thin liquid; strong empyreumatic odor and taste. d_{25}^{25} 0.960-0.990. Insol in water; miscible with alcohol.

THERAP CAT: Topical antiseptic; dermatologic.

THERAP CAT (VET): Antiseptic, antipruritic. For chronic skin conditions and in hoof dressings. Has been used internally as an expectorant.

9036. Tarragon. Estragon. The dried leaves and flowering tops of the perennial herb, *Artemisia dracunculus* L., *Compositae. Habit.* Siberia, Caspian Sea region; cultivated in Western Europe. Yields up to 0.8% oil of tarragon. *Constit.* p-Allylanisole (estragole; methyl chavicol); ocimene; myrcene; phellandrene(?); p-methoxycinnamaldehyde.

USE: The herb for culinary purposes, the oil as flavoring agent in liqueurs, soups, sauces and salad dressings. In perfumery to improve the note of chypre type perfumes.

9037. D-Tartaric Acid. *[S-(R*,R*)]-2,3-Dihydroxybutanedioic acid;* unusual tartaric acid; unnatural tartaric acid; *l*-tartaric acid; (−)-tartaric acid; levotartaric acid; D-*threo*-2,3-dihydroxysuccinic acid. $C_4H_6O_6$; mol wt 150.09. C 32.01%, H 4.03%, O 63.96%. Levorotatory tartaric acid having a dextro configuration. Although termed "unnatural," its occurrence in nature has been demonstrated. Obtained in small amounts from racemic tartaric acid through biochemical cleavage using *Penicillium notatum, Aspergillus griseus, A. niger* or other microorganisms: Pasteur, *Compt. Rend.* **51**, 298 (1860). Alternate route using salt formation with *d*-methylamphetamine: Walton, *J. Soc. Chem. Ind.* **64**, 219 (1945). Monograph: K. Freudenberg, *Stereochemie* I, (1933), reprinted by J. W. Edwards (Ann Arbor, 1945). Crystallographic data: A. N. Winchell, *The Optical Properties of Organic Compounds* (Academic Press, New York, 2nd ed., 1954) p 47.

Monoclinic sphenoidal prisms. d_4^{20} 1.7598. mp 168-170°. $[\alpha]_D^{20}$ −12.0° (c = 20 in H_2O). pKa_1 2.93; pKa_2 4.23. One gram dissolves in 0.75 ml water at room temp, in 0.5 ml boiling water, 1.7 ml methanol, 3 ml ethanol, 0.5 ml propanol, 250 ml ether. Also sol in glycerol. Insol in chloroform. Maximum soly in water at 20°: 139 g/100 ml.

9038. DL-Tartaric Acid. *2,3-Dihydroxybutanedioic acid;* racemic tartaric acid; racemic acid; *dl*-tartaric acid; resolvable tartaric acid; uvic acid; paratartaric acid; *dl*-Weinsäure (German); Vogesensäure (German); Traubensäure (German). $C_4H_6O_6$; mol wt 150.09. C 32.01%, H 4.03%, O 63.96%. Probably never a natural product, although sometimes found in small amounts during wine-making. Prepn from L-tartaric acid by boiling with aq NaOH (*meso*-tartaric acid is obtained as a byproduct): Holleman, *Org. Syn.* **6**, 82 (1926); **coll. vol. I** (2nd ed., 1941) p 497. Synthesis by oxidation of fumaric acid: Milas, Terry, *J. Am. Chem. Soc.* **47**, 1412 (1925); Milas, Sussman, *ibid.* **58**, 1302 (1936); **U.S. pat. 2,000,213** (1935 to Standard Brands). From maleic acid: Church, Blumberg, *Ind. Eng. Chem.* **43**, 1780 (1951).

$$
\begin{array}{c}
\text{COOH} \\
|\\
\text{CHOH} \\
|\\
\text{CHOH} \\
|\\
\text{COOH}
\end{array}
$$

Anhydr acid, triclinic pinacoidal crystals from abs alc, from water above 73°, or by drying the monohydrate at 100°. mp 206°. pKa_1 2.96; pKa_2 4.24. Less soluble in water than L-tartaric acid. pH of 0.1M aq soln: 2.0. Soly in alcohol (g/100 g): 2.006 at 0°; 3.153 at 15°; 5.01 at 25°; 6.299 at 40°. Soly in ether about 1%.

Monohydrate, triclinic pinacoidal crystals from water. d_4^{20} 1.697. One hundred parts (w/w) of water dissolve 14.00 parts at 10°; 20.60 at 20°; 29.10 at 30°; 43.32 at 40°; 99.88 at 70°; 184.91 at 100°.

9039. L-Tartaric Acid. *[R-(R*,R*)]-2,3-Dihydroxybutanedioic acid;* ordinary tartaric acid; natural tartaric acid; *d*-tartaric acid; (+)-tartaric acid; L-2,3-dihydroxybutanedioic acid; *d*-α,β-dihydroxysuccinic acid; Weinsäure (German); Weinsteinsäure (German). $C_4H_6O_6$; mol wt 150.09. C 32.01%, H 4.03%, O 63.96%. Dextrorotatory tartaric acid having a levo configuration. Widely distributed in nature, classified as a fruit acid. Occurs in many fruits, free and combined with potassium, calcium or magnesium. Observed in antiquity as the acid potassium salt found deposited as a fine crystalline crust during fermentation of grape juice or tamarind juice and termed *faecula* (little yeast) by the Romans. The derivation from *Tartarus* is of medieval, alchemical origin. In modern processes the acid potassium tartrate obtained during wine-making is first converted to calcium tartrate which is then hydrolyzed to tartaric acid and calcium sulfate: Metzner, *Chem. Eng. Progress* **43**, 160 (1947); several modifications, e.g., **Ital. pat. 490,221** (1954 to Procedimenti Chimici), *C.A.* **50**, 11607c (1956). Extraction from tamarind pulp in about 10% yield: **Indian pat. 52,167** (1955), *C.A.* **50**, 5249g (1956). Synthesis by hydroxylation of maleic acid: Church, Blumberg, *Ind. Eng. Chem.* **43**, 1780 (1951). Practically all of the L-tartaric acid sold today is a byproduct of the wine industry. Monograph: U. Roux, *La Grande Industrie des Acides Organiques* (Dounod, Paris, 1939). Example of a modern process: Dabul, **U.S. pat. 3,114,770** (1963 to Orandi & Massera).

$$
\begin{array}{c}
\text{COOH} \\
|\\
\text{HCOH} \\
|\\
\text{HOCH} \\
|\\
\text{COOH}
\end{array}
$$

Monoclinic sphenoidal prisms, mp 168-170°. Stable to air and light. Strong acid taste. Refreshing when in dil aq soln. d_4^{20} 1.7598. Odor of burnt sugar when heated to mp. $[\alpha]_D^{20}$ +12.0° (c = 20 in H_2O). Strong organic acid. At 25° pKa_1 2.93; pKa_2 4.23. pH of 0.1N soln: 2.2. Heat of combustion: −275.1 kcal/mol. Specific heat: 0.288 cal/g/°C at 21 to 51°; 0.296 at 0 to 99.6°. Dielectric constant 36.0 for 1200 cm waves. Freely sol in water. d_4^{15} of aq solns (w/w at 15°): 1% 1.0045; 10% 1.0469; 20% 1.0969; 30% 1.1505; 40% 1.2078; 50% 1.2696. Max soly in water in g/100 ml at various temps: 0° = 115; 10° = 126; 20° = 139; 30° = 156; 40° = 176; 50° = 195; 60° = 217; 70° = 244; 80° = 273; 90° =

307; 100° = 343. One gram dissolves in 0.75 ml water at room temp, in 0.5 ml boiling water, 1.7 ml methanol, 3 ml ethanol, 10.5 ml propanol, 250 ml ether. Also sol in glycerol. Insol in chloroform.

Human Toxicity: Strong solns are mildly irritating.

USE: In the soft drink industry, confectionery products, bakery products, gelatin desserts, as an acidulant. In photography, tanning, ceramics, manuf tartrates. The common commercial esters are the diethyl and dibutyl derivs used for lacquers and in textile printing. Pharmaceutic aid (buffering agent).

9040. *meso*-Tartaric Acid. Mesotartaric acid; internally compensated tartaric acid; unresolvable tartaric acid; Antiweinsäure (German). $C_4H_6O_6$; mol wt 150.09. C 32.01%, H 4.03%, O 63.96%. Prepd by boiling L-tartaric acid with alkali; as byproduct of racemization: Winther, *Z. Physik. Chem.* **56**, 507 (1906); Holleman, *Org. Syn.* **coll. vol. I** (2nd ed., 1941) p 497; Milas, **U.S. pat. 2,414,385** (1947). Microbial prepn: Martin, Foster, *J. Bacteriol.* **70**, 405 (1955); Foster, **U.S. pat. 2,947,665** (1960). Explanation of optical inactivity: Noller, *Science* **102**, 508 (1945); C. R. Noller, *Chemistry of Organic Compounds* (Philadelphia, 2nd ed., 1957) p 339.

$$
\begin{array}{c}
\text{COOH} \\
|\\
\text{HOCH} \\
|\\
\text{HOCH} \\
|\\
\text{COOH}
\end{array}
$$

Monohydrate, rectangular plates, d_4^{20} 1.666 (also reported as 1.737). mp 140° (also reported as 159-160°). pKa_1 3.11; pKa_2 4.80. Maximum soly in water at 20°: 125 g/100 ml.

9041. Tartrazine. *4,5-Dihydro-5-oxo-1-(4-sulfophenyl)-4-[(4-sulfophenyl)azo]-1H-pyrazole-3-carboxylic acid trisodium salt; C.I. acid yellow 23;* 3-carboxy-5-hydroxy-1-p-sulfophenyl-4-p-sulfophenylazopyrazole trisodium salt; 5-hydroxy-1-(p-sulfophenyl)-4-[(p-sulfophenyl)azo]pyrazole-3-carboxylic acid trisodium salt; hydrazine yellow; C.I. 19140; FD & C Yellow No. 5; C.I. Food Yellow 4. $C_{16}H_9N_4Na_3O_9S_2$; mol wt 534.39. C 35.96%, H 1.70%, Na 12.91%, O 26.95%, N 10.49%, S 12.00%. Prepn: **U.S. pat. 2,457,823** (1949 to Ilford); Freeman *et al.*, *J. Assoc. Offic. Agr. Chem.* **33**, 937 (1950). *See also: Colour Index* vol. **4** (3rd ed., 1971) p 4132.

Bright orange-yellow powder. Freely sol in water. The aq soln is not changed by HCl but becomes redder with sodium hydroxide.

USE: As a dye for wool and silk. In biochemistry as an adsorption-elution indicator for chloride estimations. Approved by FDA for use in food, ingested drugs, externally applied drugs and cosmetics. FDA requires labelling of FD & C No. 5 in food and drugs: *Fed. Regist.* **44**, 37212 (1979); **50**, 35774 (1985).

9042. Tartronic Acid. *Hydroxypropanedioic acid;* hydroxymalonic acid. $C_3H_4O_5$; mol wt 120.06. C 30.01%, H 3.36%, O 66.63%. $HOCH(COOH)_2$. Prepd by ozonization of malonic acid in aq soln: Dobinson, *Chem. & Ind. (London)* **1959**, 853.

Colorless, odorless crystals. Crystallizes from water with ½ and 1H_2O. Becomes anhyd at 60°. Sublimes 110-120°. mp 158-160° with decompn (CO_2 evolved). pK_1 2.42; pK_2 4.54. Grandjean, *Bull. Soc. Roy. Sci. Liege* **38**, 288 (1969), *C.A.* **72**, 42684t (1970). Very sol in water, alcohol; the anhydrous acid is sol in ether.

9043. Taurine. *2-Aminoethanesulfonic acid.* C_2H_7-

NO$_3$S; mol wt 125.14. C 19.19%, H 5.64%, N 11.19%, O 38.35%, S 25.62%. NH$_2$CH$_2$CH$_2$SO$_3$H. Conditionally essential nutrient, important during mammalian development. Conjugates bile acids, present in most milk (but only minimally in that of dairy cows); occurs also in lungs and flesh extract of oxen, in shark blood, in mussels, in oysters. Generally isolated from ox bile: Hammarsten, *Z. Physiol. Chem.* **32**, 456 (1901), or from the large muscle of abalone *(Haliotis):* Schmidt, Watson, *J. Biol. Chem.* **33**, 499 (1918). Synthesis starting with 2-bromoethanesulfonate: C. S. Marvel, C. F. Bailey, *Org. Syn. coll. vol.* **II**, 563 (1943). Synthesis by sodium sulfite sulfonation of ethylene chloride followed by ammonolysis with anhydr NH$_3$ or with aq NH$_3$ and ammonium carbonate: Schick, Degering, *Ind. Eng. Chem.* **39**, 906 (1947). Importance to retinal function in cats: A. R. Rabin *et al., Invest. Ophthalmol.* **12**, 694 (1973); K. C. Ha *et al., Science* **188**, 949 (1975); E. L. Berson *et al., Invest. Ophthalmol.* **15**, 52 (1976); K. C. Hayes, *Can. Vet. J.* **23**, 2 (1982). Comparison of taurine depletion in humans and other mammals: K. C. Hayes, J. A. Sturman, *Adv. Exp. Med. Biol.* **139**, 79 (1982). Presence in human milk: G. E. Gaull, *J. Pediatr. Gastroenterol. Nutr.* **2**, Suppl. 1, 266 (1983); idem, *Acta Paediatr. Scand., Suppl.* **296**, 38 (1982). Concn in plasma and urine of breast-fed and formula-fed infants: A.-L. Järvenpää *et al., Pediatrics* **70**, 221 (1982). Association of retinal dysfunction with low plasma taurine levels in adults: H. S. Geggel *et al., N. Engl. J. Med.* **312**, 142 (1985). Book: *Taurine,* R. Huxtable, A. Barbeau, Eds. (Raven Press, New York, 1975) 480 pp.

Large monoclinic prismatic rods, dec about 300°. pK$_1$' 1.5; pK$_2$' 8.74. Sol in 15.5 parts of water at 12°. 100 parts of 95% alc dissolve 0.004 parts at 17°. Insol in abs alc.

Note: Some infant formulas are being reformulated to increase the concn of taurine to that found in human milk, *Chem. & Eng. News* **62**, 22 (July 30, 1984).

9044. Taurocholic Acid. 2-[[(3α,5β,7α,12α)-3,7,12-Trihydroxy-24-oxocholan-24-yl]amino]ethanesulfonic acid; *N-choloyltaurine;* cholaic acid; cholyltaurine. C$_{26}$H$_{45}$NO$_7$S; mol wt 515.69. C 60.55%, H 8.78%, N 2.72%, O 21.72%, S 6.22%. The product of conjugation of cholic acid with taurine. Its sodium salt is the chief ingredient of the bile of carnivorous animals. Prepn from dog bile: Hammarsten in Abderhalden, *Handbuch der Biol. Arbeitsmethoden,* Abt. I, Teil 6, p 219 (1925). Separation from bile acids: Ahrens, Craig, *J. Biol. Chem.* **195**, 763 (1952). Enzymic synthesis: Siperstein, Murray, *Science* **123**, 377 (1956). Prepn of the sodium salt: Cortese, *J. Am. Chem. Soc.* **59**, 2532 (1937); Norman, *Arkiv. Kemi* **8**, 331 (1955); of the barium salt: Kazuno, Yanazaki, *Z. Physiol. Chem.* **224** 160 (1934). Binding of taurocholic acid to serum proteins: Burke *et al., Clin. Chim. Acta* **32**, 207 (1971). Toxicity data: Klaassen, *Toxicol. Appl. Pharmacol.* **24**, 37 (1973).

Clusters of slender, four-sided prisms from alcohol + ether, stable to air. (Commercial prepns are usually amorphous, very hygroscopic, and of yellow color.) Dec about 125°. [α]$_D^{18}$ +38.8° (c = 2 in alcohol). pK 1.4. Freely sol in water, sol in alcohol. Almost insol in ether, ethyl acetate. Is hydrolyzed to cholic acid and taurine by acids and alkalies. LD$_{50}$ in newborn rats: 380 mg/kg (Klaassen).

Sodium salt, C$_{26}$H$_{44}$NNaO$_7$S, *sodium taurocholate.* Crystals with 1.5 and 2 mols H$_2$O. Sweet taste with bitter aftertaste. Dec about 230°. [α]$_D^{20}$ +24° (c = 3). Very freely sol in water or alcohol. Solvent action on cholesterol: Rosin, *Z. Physiol. Chem.* **124**, 282 (1923).

Barium salt, C$_{52}$H$_{88}$BaN$_2$O$_{14}$S$_2$, *barium taurocholate.* Crystals with 5 mols H$_2$O, dec 225-227°. [α]$_D^{20}$ +25.6°. Converted to taurocholic acid by treatment with sulfuric acid.

USE: Sodium salt is a lipase accelerator. *See also* Ox Bile Extract.

THERAP CAT: Choleretic.

9045. Taxicatin. 3,5-Dimethoxyphenol glucoside. C$_{14}$H$_{20}$O$_8$; mol wt 316.30. C 53.16%, H 6.37%, O 40.47%. Isoln from leaves of *Taxus baccata* L., *Taxaceae:* Lefebvre, *Arch. Pharm.* **245**, 486 (1907); Merz, Preuss, *ibid.* **279**, 134 (1941). Structure and synthesis from 3,5-dimethoxyphenol and α-acetobromoglucose: Merz, Preuss, *loc. cit.*

Crystals from ethyl acetate, mp 170-170.5°. [α]$_D^{20}$ −72° (0.12 g/10 ml of aq soln). Sol in water, alcohol, ethyl acetate, methanol, acetic acid, acetone, dioxane, pyridine.

9046. Taxicins. Acyl-free polyols from which the taxines are prepared by partial esterfication: Baxter *et al., J. Chem. Soc.* **1962**, 2964. Structure of taxicin-I and -II: Eyre *et al, Proc. Chem. Soc.* **1963**, 271. Structure and stereochemical studies of taxicin-I and -II: Dukes *et al., Tetrahedron Letters* **1965**, 4765; *J. Chem. Soc. (C)* **1967**, 448; Eyre *et al., ibid.* 452. Structure of *O*-cinnamoyltaxicin-II triacetate (taxinine): Kurono *et al., Tetrahedron Letters* **1963**, 2153; Nakanishi, Kurono, *ibid.* **1963**, 2161; Ueda *et al., ibid.* **1963**, 2167; Uyeo *et al., J. Pharm Soc. Japan* **84**, 762 (1964). Stereochemistry of taxinine: Kurono *et al., Tetrahedron Letters* **1965**, 1917; Shiro *et al., Chem. Commun.* **1966**, 97.

Taxicin-I. C$_{20}$H$_{30}$O$_6$. R = OH; R' = R'' = H.
O-Cinnamoyltaxicin-I, C$_{29}$H$_{36}$O$_7$. R = OH; R' = H; R'' = COCH=CHC$_6$H$_5$. mp 233-234°. [α]$_D^{21}$ +285° (chloroform). uv max (alc): 282 nm (ε 28,200).
O-Cinnamoyltaxicin-I triacetate, C$_{35}$H$_{42}$O$_{10}$. R = OH; R' = COCH$_3$; R'' = COCH=CHC$_6$H$_5$. Prisms from alc. mp 237-239°. [α]$_D^{18}$ +218° (chloroform).
Taxicin-II. C$_{20}$H$_{30}$O$_5$. R = R' = R'' = H.
O-Cinnamoyltaxicin-II triacetate, C$_{35}$H$_{42}$O$_9$, *taxinine.* R = H; R' = COCH$_3$; R'' = COCH=CHC$_6$H$_5$. Prisms from alc, mp 265-267°. [α]$_D^{18}$ +137° (chloroform). uv max: 217, 223, 278 nm (ε 20,500; 16,000; 28,600).

9047. Taxine(s). A mixture of alkaloids from needles and berries of the yew tree, *Taxus baccata* L., *Taxaceae:* Lucas, *Arch. Pharm.* **85**, 145 (1856); Winterstein, Iatrides, *Z. Physiol. Chem.* **117**, 240 (1921); Winterstein, Guyer, *ibid.* **128**, 175 (1923); Callow *et al., J. Chem. Soc.* **1931**, 2138. Structural studies of *taxine-I:* Baxter *et al., ibid.* **1962**, 2964e; of *taxine-II:* Dukes *et al., ibid.* **1967**, 448; Eyre *et al., ibid.* 452. Sepn of cryst *taxine A* and amorph *taxine B* from the amorph mixture: E. Graf, H. Bertholdt, *Pharm. Zentral.* **96**, 385 (1957). Structure of taxine A: E. Graf *et al., Ann.* **1982**, 376. Review: B. Lythgoe, *The Alkaloids* vol. X, R. H. F. Manske, Ed. (Academic Press, New York, 1968) pp 597-626.

taxine A

Granular amorph powder, mp 121-124°. $[\alpha]_D^{17}$ +95.7° (c = 4.59 in ethanol). Sol in ether, chloroform, alcohol; practically insol in water, petr ether. Undoubtedly responsible for the poisonous properties of the yew. Fatalities among domestic animals due to yew poisoning are not uncommon today. Human fatal symptoms are those of gastrointestinal irritation, cardiac and respiratory failure.

Taxine A, $C_{35}H_{47}NO_{10}$, mp 204-206°. $[\alpha]_D$ −140° (CHCl₃).

9048. Taxodione. (4bS-trans)-4b,5,6,7,8,8a-Hexahydro-4-hydroxy-4b,8,8-trimethyl-2-(1-methylethyl)-3,9-phenanthrenedione; 11-hydroxy-13-isopropylpodocarpa-7,9(11),13-triene-6,12-dione. $C_{20}H_{26}O_3$; mol wt 314.43. C 76.40%, H 8.34%, O 15.26%. Diterpenoid quinone methide with tumor-inhibitory properties. Isoln of naturally occurring (+)-form from Taxodium distichum Rich, Taxodiaceae: S. M. Kupchan et al., J. Am. Chem. Soc. 90, 5923 (1968). Structure: eidem, J. Org. Chem. 34, 3912 (1969). Total synthesis of the racemate: K. Mori, M. Matsui, Tetrahedron 26, 3467 (1970); T. Matsumoto et al., Bull. Chem. Soc. Japan 44, 2766 (1971); 50, 1575 (1977); D. L. Snitman et al., Tetrahedron Letters 1979, 2477; R. V. Stevens, G. S. Bisacchi, J. Org. Chem. 47, 2396 (1982). Total synthesis of the (+)-form: T. Matsumoto et al., Bull. Chem. Soc. Japan 50, 266 (1977); R. H. Burnell et al., Can. J. Chem. 65, 775 (1987). Antitumor activity studies: Hanson et al., Science 168, 378 (1970).

Golden plates from methanol, mp 115-116°. $[\alpha]_D^{28}$ +56° (c = 1 in CHCl₃). uv max (methanol): 320, 332, 400 nm (ε 25000, 26000, 2000).

9049. Taxol. [2aR-[2aα,4β,4aβ,6β,9α(αR*,βS*),11α,-12α,12aα,12bα]]-β-(Benzoylamino)-α-hydroxybenzenepropanoic acid 6,12b-bis(acetyloxy)-12-(benzoyloxy)-2a,3,4,4a,5,-6,9,10,11,12,12a,12b-dodecahydro-4,11-dihydroxy-4a,8,13,-13-tetramethyl-5-oxo-7,11-methano-1H-cyclodeca[3,4]benz-[1,2-b]oxet-9-yl ester; 5β,20-epoxy-1,2α,4,7β,10β,13α-hexahydroxytax-11-en-9-one 4,10-diacetate 2-benzoate 13-ester with (2R,3S)-N-benzoyl-3-phenylisoserine; taxol A; NSC 125973. $C_{47}H_{51}NO_{14}$; mol wt 853.92. C 66.11%, H 6.02%, N 1.64%, O 26.23%. Antileukemic and antitumor agent first isolated from the bark of the Pacific yew tree, Taxus breviofolia, Taxaceae; promotes the assembly of microtubules and inhibits the tubulin disassembly process. Isoln and structure: M. C. Wani et al., J. Am. Chem. Soc. 93, 2325 (1971). In vitro promotion of microtubule assembly: P. B. Schiff et al., Nature 277, 665 (1979). Isoln from Taxus baccala L. and in vitro inhibition of depolymerization of microtubules into tubulin: G. Chauviere et al., C.R. Acad. Sci. Paris Ser. II 293, 501 (1981). Enantioselective synthesis of taxol side chain: J.-N. Denis et al., J. Org. Chem. 51, 46 (1986); of a taxol A-ring building unit: L. Pettersson et al., Tetrahedron Letters 28, 2753 (1987). Total synthesis of taxusin, which

contains the entire ring skeleton of taxol: R. A. Holton et al., J. Am. Chem. Soc. 110, 6558 (1988). Biogenic approach to synthesis: F. Gueritte-Voegelein et al., J. Nat. Prod. 50, 9 (1987). Partial synthesis of taxol from 10-deacetyl baccatin III: M. Colin et al., Eur. pat. Appl. 253,739 (1988 to Rhone-Poulenc); J.-N. Denis et al., J. Am. Chem. Soc. 110, 5917 (1988). Synthesis and anticancer activity of taxol derivs: D. G. I. Kingston et al., Studies in Organic Chemistry vol. 26, entitled "New Trends in Natural Products Chemistry 1986", Atta-ur-Rahman, P. W. Le Quesne, Eds. (Elsevier, Amsterdam, 1986) pp 219-235. Use in study of structure and function of microtubules: S. B. Horwitz et al., Cold Spring Harbor Symp. Quant. Biol. 46, 219 (1982). Review of mechanism of action: J. J. Manfredi, S. B. Horwitz, Pharmacol. Ther. 25, 83-125 (1984); S. B. Horwitz et al., Ann. N.Y. Acad. Sci. 466, 733-744 (1986).

Needles from aq methanol, mp 213-216° (dec). $[\alpha]_D^{20}$ −49° (methanol). uv max (methanol): 227, 273 nm (ε 29800, 1700).

USE: Tool in study of structure and function of microtubules.

THERAP CAT: Antineoplastic.

9050. Tazettine. Sekisanine; sekisanoline; ungernine. $C_{18}H_{21}NO_5$; mol wt 331.26. C 65.24%, H 6.39%, N 4.23%, O 24.14%. From Narcissus tazetta L., Lycoris radiata Herb., Ungernia sewerzowi (Rgl.) Fedtsch., and other Amaryllidaceae: Späth, Kahovec, Ber. 67, 1501 (1934). Structure and stereochemistry: Ikeda et al., J. Chem. Soc. 1956, 4749. Abs config: Highet, Highet, Tetrahedron Letters 1966, 4099. Synthesis: Hendrickson et al., J. Am. Chem. Soc. 92, 5538 (1970); Tsuda et al., Tetrahedron Letters 1972, 3153. Biosynthesis: Fales, Wildman, J. Am. Chem. Soc. 86, 294 (1964). Identity with sekisanine and sekisanoline: Ikeda et al., loc. cit. Stereospecific total synthesis: Hendrickson et al., J. Am. Chem. Soc. 96, 7781 (1974); S. Danishefsky et al., ibid. 102, 2838 (1980); 104, 7591 (1982).

Crystals, mp 210-211° (evac tube); racemate reported as mp 237-238° (Tsuda) and mp 175-176° (Danishefsky). $[\alpha]_D^{25}$ +150.3° (82 mg in 2 ml chloroform). Sol in methanol, ethanol, choroform. Sparingly sol in ether.

Hydrochloride, crystals, mp 206°, water soluble.

Methiodide, crystals, dec 220° (evacuated tube).

9051. Taziprinone Hydrochloride. (4α,4aβ,9bβ)-(±)-N-(1,2,3,4,4a,9b-Hexahydro-8,9b-dimethyl-3-oxo-4-dibenzofuranyl)-4-methyl-1-piperazinepropanamide dihydrochloride; 1,2,3,4,4a,9b-hexahydro-8,9b-dimethyl-4-[3-(4-methylpiperazin-1-yl)propionamido]dibenzofuran-3-one dihydrochloride; RU-20201; Azipranone; Drazifon. $C_{22}H_{33}Cl_2N_3O_3$; mol wt 458.43. C 57.64%, H 7.25%, Cl 15.47%, N 9.17%, O 10.47%. Prepn and pharmacology: S. S. Matharu et al., Ger. pat. 2,518,289; eidem, U.S. pat. 4,010,268 (1975, 1977 both to Roussel-UCLAF). Metabolism, tissue distribution in man and animals: J. W. Daniel *et al., Xenobiotica* **8,** 321 (1978). Antitussive activity in animals: R. W. Pick-ering, G. W. James, *Arzneimittel-Forsch.* **29,** 287 (1979). General pharmacology: *eidem, ibid.* 642. Preliminary mode of action study in animals: S. Yanaura *et al., Japan. J. Pharmacol.* **34,** 289 (1984).

Crystals from methanol, mp 215-218°. LD_{50} orally in gui-nea pigs: 1475 mg/kg (Matharu).

THERAP CAT: Antitussive.

9052. TCDD. *2,3,7,8-Tetrachlorodibenzo[b,e][1,4]di-oxin; 2,3,7,8-tetrachlorodibenzo-p-dioxin; 2,3,6,7-tetrachlo-rodibenzodioxin; dioxin; TCDBD.* $C_{12}H_4Cl_4O_2$; mol wt 321.96. C 44.77%, H 1.25%, Cl 44.04%, O 9.94%. Highly toxic and teratogenic contaminant of 2,4,5-trichlorophenol and 2,4,5-T, *q.q.v.,* can be formed during the manufacture of trichlorophenol. Prepn by chlorination of dibenzo-*p*-dioxin: W. Sandermann, *Ber.* **90,** 690 (1957); M. Tomita *et al., Yakugaku Zasshi* **79,** 186 (1959), *C.A.* **53,** 13152d (1959); by condensation of potassium 2,4,5-trichlorophenate: O. Ani-line in *Chlorodioxins—Origin and Fate,* E. H. Blair, Ed., *Advances in Chemistry Series* **120** (A.C.S., Washington, D.C., 1973) pp 126-135. Crystal structure: F. P. Boer *et al., Acta Crystallogr.* **28B,** 1023 (1972). Toxicity and metabolism studies: R. J. Kociba *et al., Toxicol. Appl. Pharmacol.* **35,** 553 (1976); J. Q. Rose *et al., ibid.* **36,** 209 (1976); A. Poland, A. Kende, *Fed. Proc.* **35,** 2404 (1976). Acute toxicity data: B. A. Schwetz *et al.,* in *Chlorodioxins-Origin and Fate, loc. cit.* pp 55-69. Environmental degradation: D. G. Crosby, A. S. Wong, *Science* **195,** 1337 (1976). Review of carcino-genicity studies: *IARC Monographs* **15,** 41-102 (1977). Comprehensive reviews of formation, chemistry, and toxic and environmental effects: *Chlorodioxins—Origin and Fate,* E. H. Blair, Ed., *loc. cit.* 141 pp; *Environ. Health Perspect.* **5,** 313 pp (1973); R. D. Kimbrough, *Crit. Rev. Toxicol.* **2,** 445-498 (1974); A. Poland, J. C. Knutson, *Ann. Rev. Pharmacol. Toxicol.* **22,** 517-554 (1982). *See also: Dioxin—Toxicological and Chemical Aspects,* F. Cattabeni *et al.,* Eds. (Wiley, New York, 1978) 222 pp; special issue, *Chem. & Eng. News* **61** (June 6, 1983).

Needles, mp 295° (Tomita); crystals from anisole, mp 320-325° (Sandermann). LD_{50} in male, female rats (mg/kg): 0.022, 0.045 orally (Schwetz).

Note: An industrial accident during the manufacture of 2,4,5-trichlorophenol in Seveso, Italy on July 10, 1976 caused the release of an estimated two to ten pounds of TCDD into the environment. Concentrations as high as 51.3 ppm TCDD were found in some samples: R. Rawls, D. A. O'Sullivan, *Chem. & Eng. News* **54,** 27 (Aug. 23, 1976); A. Hay, *Nature* **262,** 636 (1976).

TCDD, as a contaminant created in the manufacture of *Agent Orange,* a widely used defoliant in Vietnam during the 1960's, has also been implicated as the causative agent of various symptoms described by veterans exposed to the defoliant, *see* C. Holden, *Science* **205,** 770 (1979).

Caution: Extremely potent, low molecular weight toxin. Toxic effects in animals include anorexia, severe weight loss, hepatotoxicity, hepatoporphyria, vascular lesions, chloracne, gastric ulcers, teratogenicity and delayed death. Industrial workers exposed to TCDD have developed chloracne, por-phyrinuria and porphyria cutanea tarda. *See* Poland, Kende, *loc. cit.;* C. D. Carter *et al., Science* **188,** 738 (1975). This substance may reasonably be anticipated to be a carci-

nogen: *Fourth Annual Report on Carcinogens* (NTP 85-002, 1985) p 185.

9053. Tebuthiuron. *N-[5-(1,1-Dimethylethyl)-1,3,4-thiadiazol-2-yl)-N,N'-dimethylurea; 1-(5-tert-butyl-1,3,4-thiadiazol-2-yl)-1,3-dimethylurea;* EL-103; Graslan; Spike; Perflan. $C_9H_{16}N_4OS$; mol wt 228.31. C 47.35%, H 7.06%, N 24.54%, O 7.01%, S 14.04%. Broad spectrum pre- and post-emergent herbicide. Prepn: **Brit.** pat. **1,266,172** (1972 to Air Prod. Chem.); E. V. P. Tao, **Belg.** pat. **799,575;** *idem,* **U.S.** pat. **3,803,164** (1973, 1974 both to Lilly); *idem, Synth. Commun.* **4,** 249 (1974). Herbicidal effects, phytotoxicity: J. R. Baur, R. W. Bovey, *Agron. J.* **67,** 547 (1975). Metabo-lism in animals: D. M. Morton, D. G. Hoffman, *J. Toxicol. Environ. Health* **1,** 757 (1976). Determn by HPLC: J. H. Kennedy, *J. Chromatog. Sci.* **15,** 79 (1977); by gas chroma-tography: A. Loh *et al., J. Agr. Food Chem.* **26,** 410 (1978). Toxicity in animals: G. C. Todd *et al., Food Cosmet. Toxi-col.* **12,** 461 (1974). Use in noxious shrub control: C. H. Herbel *et al., J. Range Management* **38,** 391 (1985).

mp 160-163°. Moderately sol in water (2500 ppm); sol in methanol (170,000 ppm) (company literature). LD_{50} in mice, rats, rabbits (mg/kg): 579, 644, 286 orally (Todd).

USE: Herbicide.

9054. Technetium. Tc; at. wt (longest-lived isotope) 98; at. no. 43. Usual valences 4 and 7. Trivalent Tc less com-mon. Radioactive element. Discovery claimed by Noddack, Tacke, and Berg who called it "masurium"; the existence of masurium has never been confirmed by isoln of the element. Element no. 43 is the first artificially produced element. Named from the Greek word for "artificial"; separated from a molybdenum plate that had been bombarded for a few months with a strong beam of deuterons in the Berkeley cyclotron: Perrier, Segré, *Nature* **140,** 193 (1937); *eidem, J. Chem. Phys.* **5,** 712 (1937); Cacciapuoti, Segré, *Phys. Rev.* **52,** 1252 (1937). The most commonly available isotope, ^{99}Tc, has an at. wt of 98.9062; $T_{1/2}$ 2.12 $\times$ 10^5 years. Other long-lived isotopes are ^{97}Tc, $T_{1/2}$ 2.6 $\times$ 10^6 years; ^{98}Tc, $T_{1/2}$ 1.5 $\times$ 10^6 years. Isotopes have been prepared with mass numbers 92-107. Prepn of metal: Cobble *et al., J. Am. Chem. Soc.* **74,** 1852 (1952). Comprehensive reviews: Boyd, *J. Chem. Ed.* **36,** 3-14 (1959); Schwochau, *Angew. Chem.* **76,** 9-19 (1964); Kotegov *et al.,* in *Advan. Inorg. Chem. Radiochem.* **11,** 1-90 (1968); Peacock in *Comprehensive Inorganic Chemistry,* **vol. 3,** J. C. Bailar Jr. *et al.,* Eds. (Pergamon Press, Oxford, 1973) pp 877-903. Pharmacokinetics and organ distribution of radiopharmaceuticals: O. P. D. Noronha, K. S. Venkateswarlu, *Eur. J. Nucl. Med.* **6,** 121 (1980).

Close-packed hexagonal structure; isomorphous with rho-dium, ruthenium, and osmium. The element obtained by hydrogen reduction of ammonium pertechnate is a silver-gray spongy mass which tarnishes slowly in moist air. mp 2250 $\pm$ 50°. In its chemical behavior resembles rhenium. The element is precipitated by hydrogen sulfide from hydro-chloric acid (diluted up to 5*N*); for higher acid concns the precipitation is not complete. Is oxidized by hydrogen per-oxide in alkaline soln to soln anions: Perrier, Segré, *Atti R. accad. Lincei* [6] **27,** 579 (1938); Segré, *Nature* **143,** 460 (1939). Reacts with dil or concd nitric acid, in aqua regia, in concd sulfuric acid; not with HCl. Burns in fluorine to form penta- and hexafluorides. Combines with sulfur at high temp to form disulfide; with carbon to form TcC.

USE: Minute quantities of TcO_4^- ion exert remarkable inhibition of the corrosion of soft iron in neutral aq soln: Cartledge, *J. Am. Chem. Soc.* **77,** 2658 (1955).

THERAP CAT: ^{99m}Tc as diagnostic aid (radioactive imaging agent).

9055. Teclozan. *N,N'-[1,4-Phenylenebis(methylene)]-bis[2,2-dichloro-N-(2-ethoxyethyl)acetamide]; N,N'-bis(eth-oxyethyl)-N,N'-bis(dichloroacetyl)-1,4-xylylenediamine; N,N'-bis(dichloroacetyl)-N,N'-bis(2-ethoxyethyl)-1,4-bis-(aminomethyl)benzene; teclosan; teclosine; teclozine;* NSC-

107433; Win -13146; Win AM 13146; Falmonox. $C_{20}H_{28}Cl_4$-N_2O_4; mol wt 502.29. C 47.82%, H 5.62%, Cl 28.24%, N 5.58%, O 12.74%. Prepn: Surrey, Mayer, *J. Med. Pharm. Chem.* **3**, 409 (1961). Amebicidal activity and toxicity: Berberian *et al.*, *Am. J. Trop. Med. Hyg.* **10**, 503 (1961).

CH_3CH_2OCH_2CH_2 — NCH_2 — CH_2N — CH_2CH_2OCH_2CH_3
Cl_2CHCO   COCHCl_2

Crystals, mp 137.6-143.9°. LD_{50} orally in mice: > 8000 mg/kg (Berberian).

THERAP CAT: Antiamebic.

9056. Tecomanine. *[4R-(4α,7β,7aβ)]-1,2,3,4,7,7a-Hexahydro-2,4,7-trimethyl-6H-2-pyrindin-6-one;* tecomine. $C_{11}H_{17}NO$; mol wt 179.25. C 73.70%, H 9.56%, N 7.81%, O 8.93%. Principal alkaloid isolated from *Tecoma stans* (L.) H.B.K. (*Bignonia stans* L.), Bignoniaceae: Y. Hammouda, M. M. Motawi, *Egypt. Pharm. Bull.* **41**, 73 (1959), *C.A.* **54**, 21646c (1960). Structure: G. Jones *et al.*, *Tetrahedron Letters* **1963**, 397. Stereochemistry, abs config and crystal structure: G. Jones *et al.*, *Chem. Commun.* **1971**, 994; G. Ferguson, W. C. Marsh, *J. Chem. Soc. Perkin Trans. II* **1975**, 1124. Hypoglycemic activity of salts: Y. Hammouda *et al.*, *J. Pharm. Pharmacol.* **16**, 833 (1964); Y. Hammouda, M. S. Amer, *J. Pharm. Sci.* **55**, 1452 (1966). Stereoselective total synthesis of ±-form: T. Imanishi *et al.*, *Tetrahedron Letters* **22**, 667 (1981). Facile synthesis of (+)-form: T. Kametani *et al.*, *Heterocycles* **26**, 1491 (1987).

Liquid. $bp_{0.1}$ 125°. $[\alpha]_D^{24}$ −175° (c = 1.17 in $CHCl_3$). uv max (alc): 226 nm (log ε 4.10). Aver. lethal dose in mice: 300 mg/kg (Hammouda).

Methiodide, $C_{12}H_{20}INO$, crystals, dec 240-242°.

Methoperchlorate, $C_{12}H_{20}ClNO_5$, crystals from methanol, mp 242°.

9057. Tectorigenin. *5,7-Dihydroxy-3-(4-hydroxyphenyl)-6-methoxy-4H-1-benzopyran-4-one; 4',5,7-trihydroxy-6-methoxyisoflavone.* $C_{16}H_{12}O_6$; mol wt 300.26. C 64.00%, H 4.03%, O 31.97%. The aglucone of tectoridin: Shibata, *J. Pharm. Soc. Japan* No. **543**, 380 (1927), *C.A.* **21**, 3050[8] (1927). Structure: Asahina *et al.*, *J. Pharm. Soc. Japan* **48**, 1087 (1928), *C.A.* **23**, 2718[1] (1929); Shriner *et al.*, *J. Am. Chem. Soc.* **61**, 2322 (1939). Synthesis: Farkas, Várady, *Ber.* **93**, 1269 (1960); Baker *et al.*, *Tetrahedron Letters* no. 5, 6 (1960); Várady, *ibid.* **1965**, 4273.

Yellow needles from 30% alc, dec 230°.
Triacetate, $C_{22}H_{18}O_9$, prisms from acetone, mp 190°.
7-Glucoside, $C_{22}H_{22}O_{11}$, *tectoridin, shekanin.* From rhizomes of *Iris tectorum* Maxim., Iridaceae: Shibata, *loc. cit.* Identity with shekanin and structure: Mannich *et al.*, *Arch. Pharm.* **275**, 317 (1937). Synthesis: Várady, *Acta Chim. Acad. Sci. Hung.* **48**, 181 (1966). Needles from alc, dec 257-258°. $[\alpha]_D^{20}$ −29.4° (pyridine). Sol in pyridine, alcohol, water; practically insol in organic solvents.

9058. Tedelparin. Low molecular weight fragment of heparin, *q.v.*, prepd by nitrous acid depolymerization of porcine mucosal heparin. Mean mol wt 4000-6000 daltons.

Prepn: U. Lindahl *et al.*, **PCT Int. pat. Appl. 80 1383;** *eidem*, **U.S. pat. 4,303,651** (1980, 1981 both to Kabi AB). *See also:* U. Lindahl *et al.*, *Proc. Nat. Acad. Sci. USA* **76**, 3198 (1979). Exhibits antithrombotic activity comparable to standard heparin but with diminished hemorrhagic effects: E. Holmer *et al.*, *Thromb. Res.* **18**, 861 (1980). Pharmacokinetics in humans: G. Bratt *et al.*, *Thromb. Haemostasis* **53**, 208 (1985); G. Bratt *et al.*, *Thromb. Res.* **42**, 613 (1986). Clinical comparison with heparin in deep venous thrombosis: G. Bratt *et al.*, *Thromb. Haemostasis* **54**, 813 (1985); in perioperative thromboprophylaxis: M. Koller *et al.*, *ibid.* **56**, 243 (1986). Symposium on pharmacology and clinical efficacy: *Haemostasis* **16**, Suppl. 2, 1-71 (1986).

Sodium salt, *Kabi 2165, Fragmin.*

THERAP CAT: Antithrombotic.

9059. Teflurane. *2-Bromo-1,1,1,2-tetrafluoroethane; 1,1,2-tetrafluoro-2-bromoethane;* Abbott-16900; DA 708; Terflurane. C_2HBrF_4; mol wt 180.95. C 13.27%, H 0.56%, Br 44.17%, F 42.00%. $FBrCHCF_3$. Prepn: Larsen, U.S. pat. **2,971,990** (1961 to Dow). Pharmacology: Black *et al.*, *Brit. J. Anaesth.* **41**, 288 (1969).

Gas, bp 8°. Nonexplosive; nonflammable.

THERAP CAT: Anesthetic (inhalation).

9060. Tegafur. *5-Fluoro-1-(tetrahydro-2-furanyl)-2,4-(1H,3H)-pyrimidinedione; 5-fluoro-1-(tetrahydro-2-furyl)-uracil; N_1-(2'-furanidyl)-5-fluorouracil;* FT 207; MJF-12264; NSC-148958; Carzonal; Citofur; Coparogin; Exonal; Fental; Franroze; Ftorafur; Fulaid; Fulfeel; Furafluor; Furofutran; Futraful; Lamar; Lifril; Neberk; Nitobanil; Riol; Sinoflurol; Sunfural; Tefsiel C. $C_8H_9FN_2O_3$; mol wt 200.16. C 48.00%, H 4.53%, F 9.49%, N 14.00%, O 23.98%. Prepn: S. A. Hillers *et al.*, *Doklady Akad. Nauk SSSR* **176**, 332 (1967), *C.A.* **68**, 29664j (1968); *eidem*, **Brit. pat. 1,168,391** (1969), *C.A.* **72**, 43715r (1970) and **Fr. pat. 1,574,684** (1969), *C.A.* **73**, 77281g (1970). Alternate synthesis: T. Kametani *et al.*, *J. Heterocycl. Chem.* **14**, 473 (1977). Synthesis, antitumor activity, toxicity study: M. Yasumoto *et al.*, *J. Med. Chem.* **21**, 738 (1978). Crystal structure: Y. Nakai, *Chem. Pharm. Bull.* **30**, 2629 (1982). Pharmacokinetics, metabolism in man: J. L. Au *et al.*, *Cancer Treat. Rep.* **63**, 343 (1979). Evaluation of efficacy and toxicity: C. R. Smart *et al.*, *Cancer* **36**, 103 (1975). Evaluation in colorectal cancer: T. Buroker *et al.*, *Cancer* **44**, 48 (1979). *In vivo* and *in vitro* studies on Walker carcinoma in rats: J. Mattern *et al.*, *Arzneimittel-Forsch.* **30**, 981 (1980).

Crystals from ethanol, mp 164-165°. uv max: 270 nm [ε 8460 (pH 2); ε 8050 (pH 7); ε 6700 (pH 12)]. Easily sol in hot water, alcohol, DMF. Practically insol in ether. LD_{50} in mice (mg/kg): 900 orally (3 days) (Yasumoto); 750 i.p. (Hillers, Fr. pat.), also reported as 1150 i.p. (Smart).

THERAP CAT: Antineoplastic.

9061. Teichoic Acids. Major components of walls and membranes of a number of bacteria, accounting for 20 to 60% of the dry weight of cell walls. Teichoic acids vary considerably in structure but are all rich in phosphodiester linkages. Depending on their location, they can be divided into two classes: membrane and cell-wall teichoic acids. *Reviews:* Baddiley, *Endeavour* **23**, 33 (1964); *idem*, *Proc. Roy. Soc. London* **170B**, 331 (1968); *idem*, *Accounts Chem. Res.* **3**, 98 (1970); M. Duckworth in *Surface Carbohydrates of the Procaryotic Cell*, I. W. Sutherland, Ed. (Academic Press, London, 1977) pp 177-208.

R = glycosyl

membrane teichoic acid

R = glycosyl

wall teichoic acid

Membrane teichoic acids contain polyglycerol phosphate chains linking positions 1 and 3 on adjacent glycerol units through the phosphodiesters, with glycosyl substituents and alanine residues on some or all of the 2 positions. Structural studies: Kelemen, Baddiley, *Biochem. J.* **80**, 246 (1961). Biosynthetic studies: Burger, Glaser, *J. Biol. Chem.* **239**, 3168, 3187 (1964); **241**, 494 (1966). Synthesis of a membrane teichoic acid fragment of *Staphylococcus aureus:* J. Oltvoort *et al., Rec. Trav. Chim.* **101**, 87 (1982).

Wall teichoic acids have greater structural diversity and include also polyribitol phosphate chains linking positions 1 and 5 on adjacent ribitol residues. Polymers may contain 6 to 20 repeating units. Structural studies: Armstrong *et al., Biochem. J.* **76**, 610 (1960); Baddiley, *ibid.* **85**, 49 (1962). Biosynthetic studies: Glaser, *J. Biol. Chem.* **239**, 3178 (1964). Molecular arrangement in cell wall of *Staphylococcus lactis:* Archibald *et al., Nature New Biol.* **241**, 29 (1973).

9062. Teicoplanin. Teicoplanin A₂; teichomycin A₂; MDL-507; Targocid; Targosid. Glycopeptide antibiotic complex produced by *Actinoplanes teichomyceticus* nov. sp.; structurally related to vancomycin, *q.v.* Comprised of 5 major components differentiated by a specific fatty acid moiety. Inhibits peptidoglycan synthesis in the cell wall of gram-positive bacteria. Isoln: **Belg.** pat. **839,259;** C. Coronelli *et al.,* **U.S.** pat. **4,239,751** (1976, 1980 both to Lepetit); F. Parenti *et al., J. Antibiot.* **31**, 276 (1978); M. R. Bardone *et al., ibid.* 170. Separation of components: A. Borghi *et al.,* **Ger.** pat. **3,320,342;** *eidem,* **U.S.** pat. **4,542,018** (1983, 1985 both to Lepetit); *eidem, J. Antibiot.* **37**, 615 (1984). Struc-

Teicoplanin A₂-1: R = (Z)-4-decanoic acid
         A₂-2:   R = 8-methylnonanoic acid
         A₂-3:   R = n-decanoic acid
         A₂-4:   R = 8-methyldecanoic acid
         A₂-5:   R = 9-methyldecanoic acid

tural studies: C. Coronelli *et al., ibid.* 621; J. C. J. Barna *et al., J. Am. Chem. Soc.* **106**, 4895 (1984). Comparative *in vitro* activity: M. H. Cynamon, P. A. Granato, *Antimicrob. Ag. Chemother.* **21**, 504 (1982); H. C. Neu, P. Labthavikul, *idem.* **24**, 425 (1983). Human pharmacokinetics: L. Verbist *et al., ibid.* **26**, 881 (1984); P. L. Carver *et al., ibid.* **33**, 82 (1989). Mechanism of action: S. Somma *et al., ibid.* **26**, 917 (1984). Synergism with rifampin, *q.v.:* C. Watanakunakorn, *J. Antimicrob. Chemother.* **19**, 439 (1987). HPLC determn in human serum: J. Levy *et al., ibid.* 533. Clinical trial in gram-positive bacterial infections: Y. Glupczynski *et al., Antimicrob. Ag. Chemother.* **29**, 52 (1986). Brief review: A. H. Williams, R. N. Grüneberg, *J. Antimicrob. Chemother.* **14**, 441 (1984). Symposium on chemistry, pharmacology, activity, and clinical trials: *J. Hosp. Infect.* **7**, Suppl. A, 47-112 (1986).

Amorphous powder, mp 260° (dec). uv max in 0.1*N* HCl: 278 (E$_{1cm}^{1\%}$ 53); in 0.1*N* NaOH: 297 (E$_{1cm}^{1\%}$ 74). Sol in aq soln at pH 7.0; partially sol in methanol, ethanol. Insol in dil mineral acids, in non-polar organic solvents.

Teicoplanin A₂-1, $C_{88}H_{95}Cl_2N_9O_{33}$, white amorphous powder, darkens at 220°, dec 255°.

Teicoplanin A₂-2, $C_{88}H_{97}Cl_2N_9O_{33}$, white amorphous powder, darkens at 210°, dec 250°.

Teicoplanin A₂-3, $C_{88}H_{97}Cl_2N_9O_{33}$, white amorphous powder, darkens at 210°, dec 250°.

Teicoplanin A₂-4, $C_{89}H_{99}Cl_2N_9O_{33}$, white amorphous powder, darkens at 210°, dec 250°.

Teicoplanin A₂-5, $C_{89}H_{99}Cl_2N_9O_{33}$, white amorphous powder, darkens at 210°, dec 250°.

THERAP CAT: Antibacterial.

9063. Telluric(VI) Acid. Orthotelluric acid. H_6O_6Te; mol wt 229.66. H 2.63%, O 41.80%, Te 55.57%. $Te(OH)_6$. Most stable form. Prepd by oxidizing tellurium or its dioxide with chromic acid or potassium permanganate in nitric acid: Staudenmaier, *Z. Anorg. Chem.* **10**, 189 (1895); Mathers *et al., Inorg. Syn.* **3**, 145 (1950); from tellurium by oxidation with chloric acid: Meyer, Franke, *Z. Anorg. Allgem. Chem.* **193**, 191 (1930); Meyer, Holowatyi, *Ber.* **81**, 119 (1948); Fehér in *Handbook of Preparative Inorganic Chemistry* vol. **1**, G. Brauer, Ed. (Academic Press, New York, 2nd ed., 1963) pp 451-453°. Forms *polymetatelluric acid* $(H_2TeO_4)_n$ when heated in air at 100-200°; *n* approx 11. Concd soln of polymerized form called *allotelluric acid;* formed when orthotelluric acid is heated in a sealed tube and dissolves in its water of constitution. *Review:* Datton, Cooper, *Chem. Rev.* **66**, 657-675 (1966).

White solid; dimorphic: monoclinic, d 3.068; cubic, d 3.163: Avinens, Petit, *Compt. Rend. Ser. C* **266**, 981 (1968). Crystallizes as tetrahydrate at temp below 10°. Very weak acid: $K_1 = 2 \times 10^{-8}$; $K_2 = 1 \times 10^{-11}$. Soly in water about 33% at 30°: Mylius, *Ber.* **34**, 2208 (1901). Strong tendency to polymerize (like stannic acid) as the mol wt increases the soly becomes less and less and aq solns become truly colloidal. Sparingly sol in concd nitric acid. Soly in dil nitric acid: *Inorg. Syn. (loc. cit.).*

9064. Tellurium. Aurum paradoxum; metallum problematum. Te; at. wt 127.60; at. no. 52; valence 2, 4, 6. Diatomic (Te₂) in the vapor state. Eight stable isotopes: 120 (0.089%); 122 (2.46%); 123 (0.87%); 124 (4.61%); 125 (6.99%); 126 (18.71%); 128 (31.79%); 130 (34.48%); artificial radioactive isotopes: 114-119; 121; 127; 129; 131-134. Present in the earth's crust to the extent of 0.002 ppm. Discovered by von Reichenstein in 1782; named by Klaproth in 1798. Occurs as tellurides in combination with metals in the minerals tetradymite, altaite, coloradolite; found as the dioxide, tellurite; found also native, associated with silver and gold. Prepn: Kracek, *J. Am. Chem. Soc.* **63**, 1989 (1941); Fehér in *Handbook of Preparative Inorganic Chemistry* vol. **1**, G. Brauer, Ed. (Academic Press, New York, 2nd ed., 1963) pp 437-438. Prepn of spectrally pure Te for semiconductor devices: Weidel, *Z. Naturforsch.* **9a**, 697 (1954). Symposium on organic selenium and tellurium compds: Y. Okamoto, W. H. H. Gunther, Eds., *Ann. N.Y. Acad. Sci.* **192**, 1-225 (1972). *Reviews:* Stone, Caron in *Rare Metals Handbook,* C. A. Hampel, Ed. (Reinhold, New York, 1954) pp 405-415; Bagnall in *Comprehensive Inorganic Chemistry* vol. **2**, J. C. Bailar, Jr. *et al.,* Eds. (Pergamon Press, Oxford, 1973) pp

935-1008; E. M. Elkin in Kirk-Othmer *Encyclopedia of Chemical Technology* vol. 22 (Wiley-Interscience, New York, 3rd ed., 1983) pp 658-679.

Grayish-white, lustrous, brittle, crystalline solid, hexagonal, rhombohedral structure, or dark-gray to brown, amorphous powder with metal characteristics. d (cryst) 6.11-6.27. mp 449.8°. bp 989.9°. Electrical resistivity (19.6°): 200,000 μ-ohms-cm. Latent heat of fusion: 4.27 kcal/mole. Linear coefficient of thermal expansion: $16.8 \times 10^{-6}/°C$. Magnetic susceptibility (18°): -0.31×10^{-6} cgs. Hardness (Mohs): 2.3. Modulus of elasticity: 6,000,000 psi. Specific heat (solid): 0.047 cal/g/°C. Thermal conductivity: 0.014 at 20°. Burns in air with a greenish-blue flame, forming the dioxide. Insol in water, in benzene, in carbon disulfide. Not attacked by hydrochloric acid; reacts with nitric acid; with concd or fuming sulfuric acids, forming a red soln; in presence of air dissolves in potassium hydroxide with formation of a deep-red soln. Combines with the halogens; does not react with sulfur or selenium.

USE: As coloring agent in chinaware, porcelains, enamels, glass; reagent in producing black finish on silverware; in manuf special alloys of marked electrical resistance; in semiconductor research. *Caution:* Causes nausea, vomiting, CNS depression. Garlic odor to breath: E. Browning, *Toxicity of Industrial Metals* (Appleton-Century-Crofts, New York, 1969) pp 310-316.

9065. Tellurium Dibromide. Tellurous bromide. Br_2Te; mol wt 287.44. Br 55.60%, Te 44.40%. $TeBr_2$. Prepd from $CBrF_3$ and Te: Aynsley, Watson, *J. Chem. Soc.* **1955**, 2603. Chocolate-brown. Sol in ether; less sol in chloroform. Decomposed by water.

9066. Tellurium Dichloride. Tellurous chloride. Cl_2Te; mol wt 198.52. Cl 35.72%, Te 64.28%. $TeCl_2$. Prepn from CCl_2F_2 and Te: Aynsley, *J. Chem. Soc.* **1953**, 3016.

Black, amorphous solid; black liquid; purple vapor. mp 208°. bp 328°. Disproportionates in ether, dioxane, dibutyl ether. Insol in CCl_4.

9067. Tellurium Dioxide. O_2Te; mol wt 159.61. O 20.05%, Te 79.95%. TeO_2. Prepd by oxidation of Te by HNO_3: Norris, *J. Am. Chem. Soc.* **28**, 1675 (1906); Marshall, *Inorg. Syn.* **3**, 143 (1950); Fehér in *Handbook of Preparative Inorganic Chemistry* vol. 1, G. Brauer, Ed. (Academic Press, New York, 2nd ed., 1963) pp 447-449. The orthorhombic form occurs in the mineral *tellurite*. *Review:* Dutton, Cooper, *Chem. Rev.* **66**, 657-675 (1966).

White crystals, dimorphic: tetragonal, d 5.75; orthorhombic, d 6.04. Turns yellow on heating, mp 733°, forming a deep yellow liquid. Soly in water about 1:150,000. Sol in sodium hydroxide solns and in hydrochloric acid. Practically insol in ammonia water.

9068. Tellurium Hexafluoride. F_6Te; mol wt 241.61. F 47.18%, Te 52.82%. TeF_6. Prepd by direct fluorination of tellurium metal: Prideaux, *J. Chem. Soc.* **89**, 322 (1906); Klemm, Henkel, *Z. Anorg. Allgem. Chem.* **207**, 74 (1932); Yost, Claussen, *J. Am. Chem. Soc.* **55**, 885 (1933); Yost, *Inorg. Syn.* **1**, 121 (1939).

Colorless gas. Repulsive odor. mp $-37.6°$. Sublimes at $-38.9°$. Critical temperature 83°. d (solid; $-191°$) 4.006; d (liq; $-10°$) 2.499. Not as inert chemically as SeF_6 and SF_6 because the covalence maximum of tellurium is higher than 6. Slowly absorbed by water with hydrolysis to telluric acid, H_6TeO_6; more quickly hydrolyzed by aq KOH. Does not attack glass when pure. Corrodes mercury.

9069. Tellurium Tetrabromide. Telluric bromide. Br_4Te; mol wt 447.27. Br 71.47%, Te 28.53%. $TeBr_4$. Prepd from the elements: Fehér in *Handbook of Preparative Inorganic Chemistry* vol. 1, G. Brauer, Ed. (Academic Press, New York, 2nd ed., 1963) pp 445-446.

Orange crystals when cold, red when hot. d 4.3. mp about 380°. bp 414-420°, dec into dibromide and bromine; can be sublimed without dec at 300° in vacuum. Soluble in a little water, but hydrolyzed by much water. Sol in HBr, ether, glacial acetic acid. *Keep well closed.*

9070. Tellurium Tetrachloride. Telluric chloride. Cl_4-Te; mol wt 269.44. Cl 52.64%, Te 47.36%. $TeCl_4$. Prepd from the elements: Suttle, Smith, *Inorg. Syn.* **3**, 140 (1950).

White, very hygroscopic, cryst solid. d 3.01. mp 225°. Melts to a yellow liquid, becoming dark red at higher temp. bp 380° without decompn. Dec by water into TeO_2 and HCl. Sol in abs alcohol and toluene. *Keep tightly closed.*

9071. Tellurium Tetraiodide. I_4Te; mol wt 635.29. I 79.91%, Te 20.09%. TeI_4. Prepd from $Te(OH)_6$ and HI, Gutbier, Flury, *Z. Anorg. Allgem. Chem.* **32**,108 (1902); Damiens, *Ann. Chimie* [9] **19**, 44 (1923); Fehér in *Handbook of Preparative Inorganic Chemistry* vol. 1, G. Brauer, Ed. (Academic Press, New York, 2nd ed., 1963) p 447.

Gunmetal-gray crystals. Stable in moist air. d_4^{15} 5.05. mp 280°. Gives off I_2 when heated. Hydrolyzed slowly by cold, quickly by hot water, forming TeO_2 and HI. Sol in HI; somewhat sol in acetone.

9072. Tellurous Acid. *Telluric(IV) acid.* H_2O_3Te; mol wt 177.63. H 1.13%, O 27.02%, Te 71.84%. H_2TeO_3.

White crystals or cryst powder. d 3.0. Slightly sol in water; sol in dil acids or alkalies. Its potassium salt is reduced by many microorganisms, producing dark-colored solns.

9073. Telomycin. $C_{59}H_{77}N_{13}O_{19}$; mol wt 1272.36. C 55.70%, H 6.10%, N 14.31%, O 23.89%. Polypeptide antibiotic produced by *Streptomyces* spp. from Florida soil: Misiek *et al., Antibiot. Annual* **1957-1958**, 852. Structure: Sheehan *et al., J. Am. Chem. Soc.* **85**, 2867 (1963); **90**, 462 (1968). NMR spectrum and conformation: Kumar, Urry, *Biochemistry* **12**, 3811, 4392 (1973). Antibacterial spectrum: A. Gourevitch *et al., Antibiot. Ann.* **1957-1958**, 856. Pharmacology: D. E. Tisch *et al., ibid.* 863.

```
HO₂CCHCH₂CO—Ser—Thr—allo—Thr—Ala—Gly—trans—3-HOPro
      |                |
     NH₂               0
                       |
                       C— cis-3-HOPro-Δ-Try-β-Me-Try— β-HOLeu
                       ‖
                       0
```

Amorphous, gray solid. $[\alpha]_D^{28}$ $-133°$ (c = 1 in 2:1 methanol:water). uv max (ethanol:water, 2:1): 222.5, 277, 290, 339 nm (ϵ 63732; 13746; 11890; 22058). Minimum soly in water at pH 3.0 to 3.3 = 4 mg/ml. Maximum soly is above pH 8.5 = > 150 mg/ml. Soly in 10% sodium chloride soln less than 1 mg/ml. Moderately sol in methanol, ethanol. Very slightly sol in acetone, ethyl acetate. Insol in ether, chloroform, hydrocarbons. Aq solns are stable to heat. LD_{50} in mice (mg/kg): > 1000 orally, i.v., i.p., i.m. (Tisch).

9074. Temazepam. *7-Chloro-1,3-dihydro-3-hydroxy-1-methyl-5-phenyl-2H-1,4-benzodiazepin-2-one*; 3-hydroxydiazepam; N-methyloxazepam; oxydiazepam; ER 115; K 3917; Ro 5-5345; Wy 3917; Cerepax; Euhypnos; Levanxene; Levanxol; Mabertin; Normison; Planum; Remestan; Restoril; Signopam. $C_{16}H_{13}ClN_2O_2$; mol wt 300.74. C 63.90%, H 4.36%, Cl 11.79%, N 9.31%, O 10.64%. Pharmacologically active metabolite of diazepam, *q.v.* Prepn: S. C. Bell, S. J. Childress, *J. Org. Chem.* **27**, 1691 (1962); S. C. Bell, U.S. pat. 3,197,467 (1965 to Am. Home. Prod.). *See also:* E. Reeder *et al.,* U.S. pats. 3,340,253 and 3,374,225 (1967, 1968, both to Hoffmann-La Roche). Metabolism: H. J. Schwandt *et al., Xenobiotica* **4**, 733 (1974); S. H. Curry *et al., Brit. J. Pharmacol.* **57**, 427P (1976). Pharmacology: L. O. Randall *et al., Arch. Int. Pharmacodyn.* **185**, 135 (1970); S. Garattini *et al.,* "Metabolic Studies on Benzodiazepines in Various Animal Species" in *Benzodiazepines,* S. Garattini, Ed. (Raven Press, New York, 1973) pp 73-97. Pharmacology and toxicity study: L. O. Randall *et al., Curr. Ther. Res.* **7**, 590 (1965). Clinical study: P. Sarteschi *et al., Arzneimittel-Forsch.* **22**, 93 (1972). Review of pharmacology and therapeutic efficacy: R. C. Heel *et al., Drugs* **21**, 321-340 (1981).

Crystals from cyclohexane, mp 119-121°.

Caution: May be habit forming. This is a controlled substance (depressant) listed in the U.S. Code of Federal Regulations, Title 21 Part 1308.14 (1987).

THERAP CAT: Sedative, hypnotic.

9075. Temephos. *O,O'-(Thiodi-4,1-phenylene)phosphorothioic acid O,O,O',O'-tetramethyl ester; O,O'-(thiodi-4,1-phenylene)bis(O,O'-dimethylphosphorothioate); O,O,-O',O'-tetramethyl O,O'-thiodi-p-phenylene phosphorothioate; phosphorothioic acid O,O-dimethyl ester O,O-di-ester with 4,4'-thiodiphenol;* ENT 27165; AC 52160; Abate; Biothion. $C_{16}H_{20}O_6P_2S_3$; mol wt 466.46. C 41.20%, H 4.32%, O 20.58%, P 13.28%, S 20.62%. Prepn: J. B. Lovell, R. W. Baer, Belg. pat. **648,531,** *C.A.* **63,** 11433e (1965) and *eidem,* U.S. pat. **3,317,636** (1964 and 1967 to Am. Cyanamid). Metabolism: R. C. Blinn, *J. Agr. Food Chem.* **17,** 118 (1969).

Cryst solid, mp 30.0-30.5°. Optimum stability at pH 5-7. Sol in acetonitrile, carbon tetrachloride, ether, dichloroethane, toluene. Almost insol in water, hexane. LD_{50} orally in male, female rats: 8600, 13000 mg/kg, T. B. Gaines, *Toxicol. Appl. Pharmacol.* **14,** 515 (1969).

USE: Insecticide.

9076. Temocillin. *6-[(Carboxy-3-thienylacetyl)amino]-6-methoxy-3,3-dimethyl-7-oxo-4-thia-1-azabicyclo[3.2.0]heptane-2-carboxylic acid; (6S)-6-[2-carboxy-2-(3-thienyl)-acetamido]-6-methoxypenicillanic acid.* $C_{16}H_{18}N_2O_7S_2$; mol wt 414.45. C 46.37%, H 4.38%, N 6.76%, O 27.02%, S 15.47%. Semi-synthetic injectable penicillin deriv with high activity vs a large number of gram-negative bacteria, but with little activity vs gram-positive organisms. Prepn: J. P. Clayton, P. H. Bentley, Ger. pat. **2,600,866** (1976 to Beecham), *C.A.* **86,** 55420t (1977); P. H. Bentley *et al., J. Chem. Soc. Perkin Trans. I* **1979,** 2455. *In vitro* antibacterial activity and β-lactamase susceptibility: I. Phillips *et al., J. Antimicrob. Chemother.* **10,** 271 (1982). Comparison to other penicillins vs *H. influenzae* and intestinal gram-negative rods: H. Y. Chen, J. D. Williams, *ibid.* 279. Pharmacokinetics and tissue penetration in healthy volunteers: R. M. Brown *et al., ibid.* 295. *In vivo* and *in vitro* comparison to ampicillin: R. Yogev *et al., Antimicrob. Ag. Chemother.* **23,** 182 (1983). Series of articles on microbiology, pharmacology and clinical studies: *Drugs* **29,** Suppl. 5, 1-243 (1985).

Disodium salt, $C_{16}H_{16}N_2Na_2O_7S_2$, *BRL 17421, Temopen.* Amorphous solid.

THERAP CAT: Antibacterial.

9077. Teniloxazine. *2-[[2-(2-Thienylmethyl)phenoxy]-methyl]morpholine; 2-[[(α-2-thienyl-o-tolyl)oxy]methyl]-morpholine.* $C_{16}H_{19}NO_2S$; mol wt 289.39. C 66.40%, H 6.62%, N 4.84%, O 11.06%, S 11.08%. Analog of viloxazine, *q.v.* Prepn: T. Muro *et al.,* **Japan. Kokai 75 89,380;** *eidem,* U.S. pat. **4,005,084** (1975, 1977 both to Yoshitomi); and pharmacology: T. Muro *et al., Yakugaku Zasshi* **106,** 964 (1986), *C.A.* **107,** 39720n (1987). Effects on cerebral glucose metabolism in mice: M. Nakanishi, *Kanazawa Ika Daigaku Zasshi* **9,** 8 (1984), *C.A.* **101,** 163540u (1984). Pharmacology and pharmacokinetics in humans: C. Ogura *et al., Brit. J. Clin. Pharmacol.* **23,** 537 (1987). Toxicity studies: Y. Kamimura *et al., Oyo Yakuri* **33,** 873 (1987); K. Okumura *et al., ibid.* 879; K. Okuda *et al., ibid.* 907, *C.A.* **107,** 190379d, 168276w, 168277x (1987).

Maleate, $C_{20}H_{23}NO_6S$, *Y-8894, sufoxazine, Metatone.* Crystals from methanol and isopropyl ether (1:1), mp 155-156°. LD_{50} in male, female mice, male, female rats (mg/kg): 719, 684, 1674, 1576 orally; 169, 161, 124, 122 i.p.; 213, 238, 310, 276 s.c. (Kamimura).

THERAP CAT: Antidepressant.

9078. Teniposide. *[5R-[5α,5aβ,8aα,9β(R*)]]-5,8,8a,9-Tetrahydro-5-(4-hydroxy-3,5-dimethoxyphenyl)-9-[[4,6-O-(2-thienylmethylene)-β-D-glucopyranosyl]oxy]furo[3',4':6,7]naphtho[2,3-d]-1,3-dioxol-6(5aH)-one; 4'-demethylepipodophyllotoxin 9-(4,6-O-2-thenylidene-β-D-glucopyranoside); 4'-demethylepipodophyllotoxin-β-D-thenylidine glucoside;* ETP; NSC 122819; VM-26; Vehem; Vumon. $C_{32}H_{32}O_{13}S$; mol wt 656.67. C 58.53%, H 4.91%, O 31.67%, S 4.88%. Semi-synthetic derivative of podophyllotoxin, *q.v.* Prepn: A. Von Wartburg, **S. Afr. pat. 66 07,585;** C. Keeler-Juslen *et al.,* U.S. pat. **3,524,844** (1968, 1970 both to Sandoz). Mechanism of action: H. Stählin, *Eur. J. Cancer* **6,** 303 (1970). Pharmacology: M. Hacker, D. Roberts, *Cancer Res.* **37,** 3287 (1977); S. M. Sieber *et al., Teratology* **18,** 31 (1978); T. J. Vietti *et al., Cancer Treat. Rep.* **62,** 1313 (1978). Metabolism: L. Allen, *Drug Metab. Rev.* **8,** 119 (1978); *Cancer Res.* **38,** 2549 (1978). Clinical studies: N. M. Gad-el-Mawla *et al., Cancer Treat. Rep.* **62,** 993 (1978); R. E. Bellet *et al., ibid.* 445. Studies on delayed toxicity in mice after i.p. injections: M. Hacker, D. Roberts, *Cancer Res.* **35,** 1756 (1975); H. Stählin, *Eur. J. Cancer* **12,** 925 (1976). Review of pharmacology, pharmacokinetics and assay methods: P. I. Clark, M. L. Slevin, *Clin. Pharmacokinet.* **12,** 223-252 (1987).

Crystals from abs ethanol, mp 242-246°. $[α]_D^{20}$ −107° (9:1 chloroform/methanol).

THERAP CAT: Antineoplastic.

9079. Tenonitrozole. *N-(5-Nitro-2-thiazolyl)-2-thiophenecarboxamide; 2-(α-thenoylamino)-5-nitrothiazole;* thenitrazole; TC 109; Atrican; Moniflagon. $C_8H_5N_3O_3S_2$; mol wt 255.28. C 37.64%, H 1.97%, N 16.46%, O 18.80%, S 25.12%. Prepd from 2-thenoyl chloride and 2-amino-5-nitrothiazole: Fr. pat. **M715** (1961 to Chantereau), *C.A.* **59,** 7533g (1963).

Crystals from dioxane or DMF, mp 255-256°.

THERAP CAT: Antiprotozoal (Trichomonas); antifungal.

9080. Tenoxicam. *4-Hydroxy-2-methyl-N-2-pyridinyl-2H-thieno[2,3-e]-1,2-thiazine-3-carboxamide 1,1-dioxide;* Ro 12-0068; Liman; Mobiflex; Tilatil; Tilcotil. $C_{13}H_{11}N_3$-O_4S_2; mol wt 337.37. C 46.28%, H 3.29%, N 12.45%, O 18.97%, S 19.01%. Non-steroidal anti-inflammatory agent. Prepn: O. Hromatka *et al.,* **Ger. pat.** 2,537,070 (1976 to Hoffmann-La Roche), *C.A.* **85,** 63077 (1976). Pharmacology: Y. Tanaka *et al., Nippon Yakurigaku Zasshi* **77,** 531 (1981), *C.A.* **95,** 35473 (1981). HPLC determn in plasma: M. E. Pickup *et al., J. Chromatog.* **225,** 493 (1981). Comparative study vs aspirin in normal volunteers: H. A. Bird *et al., Curr. Med. Res. Opin.* **8,** 9 (1982). Preliminary review of pharmacology, therapeutic efficacy and mechanism of action: J. P. Gonzalez, P. A. Todd, *Drugs* **34,** 289-310 (1987).

Crystals from xylene, mp 209-213° (dec).
THERAP CAT: Anti-inflammatory; analgesic.

9081. Tenuazonic Acid. *3-Acetyl-1,5-dihydro-4-hydroxy-5-(1-methylpropyl)-2H-pyrrol-2-one; 3-acetyl-5-sec-butyl-4-hydroxy-3-pyrrolin-2-one;* 3-acetyl-5-*sec*-butyltetramic acid. $C_{10}H_{15}NO_3$; mol wt 197.23. C 60.89%, H 7.67%, N 7.10%, O 24.34%. A metabolite found in a strain of the fungus *Alternaria tenuis* Auct.: Rosett *et al., Biochem. J.* **67,** 390 (1957); Kaczka *et al., Biochem. Biophys. Res. Commun.* **14,** 54 (1964). Gitterman *et al., Cancer Res.* **24,** 440 (1964). Structure: Stickings, *Biochem. J.* **72,** 332 (1959); Biosynthesis: Stickings, Townsend, *ibid.* **78,** 412 (1961). Chemical synthesis from diketene and L-isoleucine: Lacey, *J. Chem. Soc.* **1954,** 850; Harris *et al., J. Med. Chem.* **8,** 478 (1965).

Pale brown, viscous, gummy substance, $[\alpha]_{5461}^{20}$ −136 ±5°; −132 ± 2° (c = 0.2, 0.5 in $CHCl_3$). $bp_{0.035}$ 117°. Readily sol in organic solvents incl petr ether; sparingly sol in water. On long standing, changes into the crystalline *iso-*form.
Copper salt monohydrate, $C_{20}H_{28}CuN_2O_6 \cdot H_2O$, green needles, mp above 175°. $[\alpha]_{5461}^{19}$ −124 ± 5° (c = 0.2 in methanol). Sol in chloroform, methanol, ethanol, acetone; sparingly sol in water; practically insol in benzene.
Sodium salt, $C_{10}H_{14}NNaO_3$, *sodium* L-*tenuazonate.* $[\alpha]_{546}$ −96.7° (c = 2 in methanol). uv max (pH 7): 280, 241 nm ($E_{1 cm}^{1\%}$573, 400).
THERAP CAT: Antineoplastic.

9082. Tephrosin. *13,13a-Dihydro-7a-hydroxy-9,10-dimethoxy-3,3-dimethyl-3H-bis[1]benzopyrano[3,4-b:6',5'-e]pyran-7(7aH)-one;* hydroxydeguelin; toxicarol. $C_{23}H_{22}O_7$; mol wt 410.41. C 67.31%, H 5.40%, O 27.29%. Occurs in leaves of *Tephrosia vogelii* Hook. f., *Leguminosae,* in derris root, cubé root: Hanriot, *Compt. Rend.* **144,** 150 (1907); Clark, *J. Am. Chem. Soc.* **53,** 729 (1931). Structure: Butenandt, Hilgetag, *Ann.* **495,** 172 (1932).

Prisms. mp 198° (218-220°). Practically insol in water; sol in chloroform, ether, acetone, sparingly in methanol. It is toxic to fish, crustacea, insects, but apparently only slightly, or not at all, to humans.

9083. Teprenone. *6,10,14,18-Tetramethyl-5,9,13,17-nonadecatetraen-2-one;* geranylgeranylacetone; GGA; E 0671; E36U31; Selbex. $C_{23}H_{38}O$; mol wt 330.55. C 83.57%, H 11.59%, O 4.84%. Acyclic polyisoprenoid 2:3 mixture of *(5Z,9E,13E)* and all *trans (5E,9E,13E)* isomers. Prepn from geranylgeranyl bromide: L. Ruzicka, L. Castro, *Helv. Chim. Acta* **28,** 590 (1945); from substituted acetylenes: K. Sato *et al., J. Org. Chem.* **35,** 565 (1970). Anti-ulcer and toxicity studies: M. Murakami *et al., Arzneimittel-Forsch.* **31,** 799 (1981). *In vivo* effects on aspirin induced gastric ulcer: *eidem, Japan. J. Pharmacol.* **32,** 299 (1982); on stress induced gastric ulcer: *eidem, ibid.* **33,** 549 (1983). Mechanism of action studies: Y. Nishizawa *et al., Biochem. Biophys. Res. Commun.* **103,** 706 (1981); K. Oketani *et al., Japan. J. Pharmacol.* **33,** 593 (1983). Metabolism in rats: Y. Nishizawa *et al., Xenobiotica* **17,** 575 (1987). Determn in serum by GC-MS: M. Tanaka *et al., J. Chromatog.* **231,** 301 (1982); by HPLC: T. Seki *et al., ibid.* **424,** 410 (1988).

Yellow oil, $bp_{0.01}$ 155-160°. $d_4^{20.5}$ 0.9081. n_D^{20} 1.4947.
THERAP CAT: Anti-ulcerative.

9084. Terazosin. *1-(4-Amino-6,7-dimethoxy-2-quinazolinyl)-4-[(tetrahydro-2-furanyl)carbonyl]piperazine;* 2-[(4-tetrahydro-2-furoyl)-1-piperazinyl]-4-amino-6,7-dimethoxyquinazoline. $C_{19}H_{25}N_5O_4$; mol wt 387.44. C 58.90%, H 6.50%, N 18.08%, O 16.52%. α_1-Adrenergic blocker related to prazosin, *q.v.* Prepn of anhydrous hydrochloride: M. Winn *et al.,* **Ger. pat.** 2,646,186; *eidem,* **U.S. pat.** 4,026,894 (both 1977 to Abbott). Prepn of the more stable hydrochloride dihydrate: R. Roteman, **Ger. pat.** 2,831,112; *idem,* **U.S. pat.** 4,251,532 (1979, 1981 both to Abbott). Comparison with prazosin: J. J. Kyncl *et al., The Pharmacologist* **22,** 272 (1980). Selective α-adrenergic inhibition: *eidem, Fed. Proc.* **41,** 1648 (1982). Hypotensive activity: S. Mizogami, M. Hanazuka, *Japan. J. Pharmacol.* **32,** Suppl., 174P (1982). Toxicity in rats: F. L. Fort *et al., Drug Chem. Toxicol.* **7,** 435 (1984). Clinical study in essential hypertension: P. A. Abraham *et al., Pharmacotherapy* **5,** 285 (1985). Symposium on pharmacology, pharmacokinetics, clinical efficacy and safety, and comparison with prazosin: *Am. J. Med.* **80,** Suppl. 5B, 1-105 (1986). Review of pharmacodynamics and therapeutic efficacy: S. Titmarsh, J. P. Monk, *Drugs* **33,** 461-477 (1987).

Hydrochloride, $C_{19}H_{26}ClN_5O_4$, hygroscopic crystals from isopropyl alc, mp 278-279°. Soly in water: 761.2 mg/ml. LD_{50} in mice (mg/kg): 259.3 i.v. (Winn).

Hydrochloride dihydrate, $C_{19}H_{26}ClN_5O_4.2H_2O$, *Abbott 45975, Heitrin, Hytracin, Hytrin, Hytrinex, Vasocard, Vasomet.* mp 271-274°. Soly in water: 24.2 mg/ml.

THERAP CAT: Antihypertensive.

9085. Terbacil. *5-Chloro-3-(1,1-dimethylethyl)-6-methyl-2,4(1H,3H)-pyrimidinedione; 3-tert-butyl-5-chloro-6-methyluracil;* Sinbar; Du Pont Herbicide 732. $C_9H_{13}ClN_2O_2$; mol wt 216.65. C 49.89%, H 6.04%, Cl 16.36%, N 12.93%, O 14.77%. Prepn: Loux, U.S. pat. 3,235,357 (1966 to du Pont). Herbicidal activity and physical properties: G. D. Hill, Jr. *et al., Symp. N. Herbic., Eur. Weed Res. Counc., 2nd, Paris* **1965**, 313, *C.A.* **66**, 18234b (1967). Metabolism and mode of action: Herholdt, *Diss. Abstr. Int. B* **30**, 1978 (1969).

Crystals, mp 175-177°. Soly in water at 25°: 710 ppm. Sol in DMF, dimethylacetamide, cyclohexanone; moderately sol in methyl isobutyl ketone, butyl acetate, xylene. Approx. lethal dose orally in rats: > 5000 mg/kg (Hill).

USE: Herbicide.

9086. Terbinafine. *(E)-N-(6,6-Dimethyl-2-hepten-4-ynyl)-N-methyl-1-naphthalene methanamine; trans-N-methyl-N-(1-naphthylmethyl)-6,6-dimethylhept-2-en-4-ynyl-1-amine;* SF 86-327; Lamisil. $C_{21}H_{25}N$; mol wt 291.44. C 86.55%, H 8.65%, N 4.80%. Antimycotic allylamine related to naftifine, *q.v.* Prepn: A. Stütz, *Eur. pat. Appl.* **24,587** (1981 to Sandoz); A. Stütz, G. Petranyi, *J. Med. Chem.* **27**, 1539 (1984). Specific inhibitor of squalene epoxidase, a key enzyme in fungal ergosterol biosynthesis: G. Petranyi *et al., Science* **224**, 1239 (1984); N. S. Ryder, *Antimicrob. Ag. Chemother.* **27**, 252 (1985). *In vitro* antifungal activity: S. Shadomy *et al., Sabouraudia* **23**, 125 (1985). Series of articles on activity, pharmacokinetics and metabolism: *Proc. 13th Int. Congr. Chemother., Vienna 1983*, **6**, 116/1-116/59. Toxicology: U. Ganzinger *et al., ibid.* 116/52.

Hydrochloride, $C_{21}H_{26}ClN$, crystals from 2-propanol + diethyl ether, mp 195-198° (change in crystal structure begins ~150°). LD_{50} in mice, rats (mg/kg): 4000, 4000 orally; 393, 213 i.v. (Ganzinger).

THERAP CAT: Topical antifungal.

9087. Terbium. Tb; at. wt 158.9254; at. no. 65; valences 3, 4. A lanthanide; belongs to the yttrium group of rare earths. One naturally occurring isotope: [159]Tb; artificial radioactive isotopes: 147-158; 160-164. Abundance in earth's crust: approx 1 ppm; occurs in small quantities in monazite, cerite, gadolinite and other rare earth minerals. Discovered by Mosander, *Skand. Naturför. Förh.* **3**, 387 (1842); *Phil. Mag.* [3] **23**, 241 (1843). Separation of fractional crystn and precipitation of the double nickel nitrates: Urbain, *Compt. Rend.* **139**, 736 (1904); **141**, 521 (1905); by fractional precipitation of its bromates: James, Bissel, *J. Am. Chem. Soc.* **36**, 2060 (1914); by ion exchange: Spedding *et al., ibid.* **76**, 2557 (1954). Prepn of metal by electrodeposition: *eidem, J. Electrochem. Soc.* **100**, 442 (1953). Absorption spectrum: Urbain, *loc. cit.* Reviews of prepn, properties and compds of terbium and other lanthanides: *The Rare Earths*, F. H. Spedding, A. H. Daane, Eds. (Krieger, Huntington, N.Y., 1971, reprint of 1961 ed.) 641 pp; Hulet,

Bode, "Separation Chemistry of the Lanthanides and Transplutonium Actinides" in *MTP Int. Rev. Sci.: Inorg. Chem., Ser. One*, **vol. 7**, K. W. Bagnall, Ed. (University Park Press, Baltimore, 1972) pp 1-45; Moeller, "The Lanthanides" in *Comprehensive Inorganic Chemistry*, **vol. 4**, J. C. Bailar, Jr. *et al.*, Eds. (Pergamon Press, Oxford, 1973) pp 1-101.

Silver-gray metal, easily oxidized in air. Hexagonal close-packed structure at room temp; d 8.27: Spedding, Daane, *loc. cit.* p. 183. mp 1356°.

Oxide, O_3Tb_2, *terbia*, a white solid.

Oxide, non-stoichiometric, approx compn, Tb_4O_7, for this formula Tb^{3+} and Tb^{4+} are present in equal amounts. Dark brown or black solid; obtained by igniting the oxalate or the sulfate. Dissolves in hot concd acids with formation of salts; loses oxygen on heating; forms the oxide when heated in hydrogen.

Nitrate, $Tb(NO_3)_3$, hexahydrate, monoclinic crystals, mp 89.3°. Urbain, cited by Mellor, *A Comprehensive Treatise on Inorganic and Theoretical Chemistry* **5**, 695 (1929). LD_{50} in rats: 260 mg/kg i.p.; > 5000 mg/kg orally, Haley, *J. Pharm. Sci.* **54**, 663 (1965).

Chloride hexahydrate, $TbCl_3.6H_2O$, prismatic deliquesc crystals. Very sol in water; forms supersaturated solns. Dehydrated on heating in hydrogen chloride at 180-200°. The anhyd chloride, crystals, d^0 4.35; mp 588°; dissolves in water without hydrolysis. LD_{50} in mice: 550 mg/kg i.p.; 5100 mg/kg orally, Haley, *loc. cit.*

9088. Terbufos. *Phosphorodithioic acid S-[[(1,1-dimethylethyl)thio]methyl] O,O-diethyl ester; phosphorodithioic acid S-[(tert-butylthio)methyl] O,O-diethyl ester;* AC 92100; Counter. $C_9H_{21}O_2PS_3$; mol wt 288.41. C 37.48%, H 7.34%, O 11.09%, P 10.74%, S 33.35%. Organophosphate insecticide. Prepn: Ikeda *et al.,* **Japan. pats. 66 16,799; 67 1743; 67 4455** (all to Hokko), *C.A.* **66**, 18510p, 85467p, **67**, 11205e (all 1967). As insecticidal soil treatment: F. M. Gordon, *Ger. pat.* **2,258,528** (1973 to Am. Cyanamid), *C.A.* **79**, 62601m (1973). Toxicity to birds: E. F. Hill, M. B. Camardese, *Echotoxicol. Environ. Saf.* **8**, 551 (1984).

Technical product (85 to 88% purity): clear, colorless to pale yellow liq, d^{24} 1.105. $bp_{0.01}$ 69°. mp −29.2°. Flash pt 88° (tag open cup). Sol in acetone, alcs, aromatic and chlorinated hydrocarbons. Soly in water: 10-15 ppm. LD_{50} orally in quail: 15 mg/kg (Hill, Camardese). *Caution:* Cholinesterase inhibitor.

USE: Soil insecticide.

9089. Terbutaline. *5-[2-[(1,1-Dimethylethyl)amino]-1-hydroxyethyl]-1,3-benzenediol; α-[(tert-butylamino)methyl]-3,5-dihydroxybenzyl alcohol;* 1-(3,5-dihydroxyphenyl)-2-(tert-butylamino)ethanol. $C_{12}H_{19}NO_3$; mol wt 225.29. C 63.97%, H 8.50%, N 6.22%, O 21.30%. β-Adrenergic agonist. Prepn: K. Wetterlin, L. A. Svensson, **Belg. pat. 704,-932**; *eidem,* U.S. pat. **3,937,838** (1968, 1976 both to Draco). Pharmacology: Bergman *et al., Experientia* **25**, 899 (1969). Resolution of isomers and activity studies: K. Wetterlin, *J. Med. Chem.* **15**, 1182 (1972). Clinical study in treatment of preterm labor: S. N. Caritis *et al., Am. J. Obstet. Gynecol.* **150**, 7 (1984). *Review:* J. J. McPhillips in *Pharmacological and Biochemical Properties of Drug Substances* **vol. 1**, M. E. Goldberg, Ed. (Am. Pharm. Assoc., Washington, DC, 1977) pp 311-328.

Crystals from abs ether, mp 119-122°.

Sulfate, $C_{24}H_{40}N_2O_{10}S$, *Bricanyl, Brethine* (tabl), *Dracanyl, Filair, Monovent, Solutin, Terbasmin.* mp 246-248°.

THERAP CAT: Bronchodilator; tocolytic.

9090. Terconazole. *cis-1-[4-[[2-(2,4-Dichlorophenyl)-2-(1H-1,2,4-triazol-1-ylmethyl)-1,3-dioxolan-4-yl]methoxy]phenyl]-4-(1-methylethyl)piperazine;* triaconazole; R 42470; Fungistat; Gyno-Terazol; Panlomyc; Terazol; Tercospor. $C_{26}H_{31}Cl_2N_5O_3$; mol wt 532.48. C 58.65%, H 5.87%, Cl 13.32%, N 13.15%, O 9.01%. Topical triazole antifungal. Prepn: J. Heeres *et al.,* **Ger.** pat. **2,804,096;** *eidem,* **U.S.** pats. **4,144,346; 4,223,036** (1978, 1979, 1980 all to Janssen); *eidem, J. Med. Chem.* **26,** 611 (1983). Pharmacology: J. Van Cutsem *et al., Chemotherapy* **29,** 322 (1983). Clinical comparison with clotrimazole, *q.v.,* in vaginal candidiasis: A. Kjaeldgaard, *Pharmacotherapeutica* **4,** 525 (1986).

Crystals from isopropyl ether, mp 126.3°.
THERAP CAT: Antifungal.

9091. Terebene. A mixture of dipentene and other hydrocarbons obtained by shaking oil of turpentine with successive quantities of sulfuric acid: Howard, *Pharm. J.* **103,** 76 (1919).
Colorless liquid; thyme-like odor, bp 160-172°. Resinifies on exposure to air and light. d_2^{25} 0.860-0.865. Practically optically inactive. Almost insol in water. Miscible with chloroform, ether, abs alcohol; 1 ml dissolves in 3 ml 95% alcohol. *Keep well closed and protected from light.*
USE: Treatment of cellulosic matter with terebene to render it water and oil resistant.
THERAP CAT: Expectorant; antiseptic.
THERAP CAT (VET): Orally or by inhalation: antiseptic and expectorant.

9092. Terebic Acid. *Tetrahydro-2,2-dimethyl-5-oxo-3-furancarboxylic acid; tetrahydro-2,2-dimethyl-5-oxo-3-furoic acid;* terebinic acid; (1-hydroxy-1-methylethyl)succinic acid γ-lactone. $C_7H_{10}O_4$; mol wt 158.15. C 53.16%, H 6.37%, O 40.47%. Prepared from fumaric or maleic acid: Schenck, Steinmetz, *Tetrahedron Letters* no. 21, 1 (1960); Lipp *et al., Ann.* **644,** 37 (1961). Prepn of optical isomers: Fredga, *C.A.* **42,** 123g (1948); Delépine, Badoche, *Compt. Rend.* **235,** 1069 (1952).

Crystals, mp 174-175°, but begins to volatilize at 100°. d 0.815. Slightly sol in cold water, freely in boiling water or warm alcohol.
(+)-Form, $[\alpha]_D^{25}$ +13.2° (c = 0.03 in acetone).
(−)-Form, mp 201-205° (dec). $[\alpha]_D^{25}$ −13.2° (c = 0.03 in acetone).

9093. Terephthalic Acid. *1,4-benzenedicarboxylic acid;* p-phthalic acid; Tephthol. $C_8H_6O_4$; mol wt 166.13. C 57.83%, H 3.64%, O 38.52%. Prepd by oxidation of p-methylacetophenone: Koelsch, *Org. Syn.* **coll. vol. III,** 791 (1955). Manuf processes: U.S. pat. **3,014,961** (1959 to VEB Chemie Werke Buna); Sherwood, *Chem. & Ind. (London)* **1960,** 1096. *Review:* Faith, Keyes & Clark's *Industrial Chemicals,* F. A. Lowenheim, M. K. Moran, Eds. (Wiley-Interscience, New York, 4th ed., 1975) pp 807-813; A. G. Bemis *et al.,* "Phthalic Acids" in Kirk-Othmer *Encyclopedia of Chemical Technology* **vol. 17** (Wiley-Interscience, New York, 3rd ed., 1982) pp 732-777.

Crystals. Sublimes at 402°. Practically insol in water, chloroform, ether, acetic acid; slightly sol in cold alcohol, more in hot alcohol; sol in alkalies.
Dimethyl ester, $C_{10}H_{10}O_4$, *dimethyl terephthalate, DMT.* White crystals, mp 140.6°. bp 288°.
USE: Forms polyesters with glycols which are made into plastic films and sheets; in analytical chemistry. *Caution:* Mild irritant.

9094. Terfenadine. *α-[4-(1,1-Dimethylethyl)phenyl]-4-(hydroxydiphenylmethyl)-1-piperidinebutanol;* 1-(p-tert-butylphenyl)-4-[4'-(α-hydroxydiphenylmethyl)-1'-piperidyl]-butanol; α-(p-tert-butylphenyl)-4-(α-hydroxy-α-phenylbenzyl)-1-piperidinebutanol; MDL 9918; Aldaban; Allerplus; Nebralin; Seldane; Teldane; Teldanex; Terdin; Terfen; Ternadin; Triludan. $C_{32}H_{41}NO_2$; mol wt 471.69. C 81.48%, H 8.76%, N 2.97%, O 6.78%. Nonsedating-type histamine H_1-receptor antagonist. Prepn: A. A. Carr, C. R. Kinsolving, **Ger.** pat. **2,303,306;** *eidem,* **U.S.** pat. **3,878,217** (1973, 1975 both to Richardson-Merrell); A. A. Carr, D. R. Meyer, *Arzneimittel-Forsch.* **32,** 1157 (1982). Antihistamine activity: A. A. Carr *et al., Pharmacologist* **15,** 221 (1973); C. R. Kinsolving *et al., ibid.* 221. Metabolism in rats: G. A. Leeson, *Fed. Proc.* **34,** 2911 (1975). Bioavailability: C. R. Kinsolving, N. L. Munro, *Drug. Metab. Rev.* **4,** 285 (1975). Clinical study: M. L. Brandon, M. Weiner, *Ann. Allergy* **44,** 71 (1980). Acute toxicity: C. R. Kinsolving *et al., Pharmacologist* **15,** 221 (1973). Series of articles on chemistry, pharmacology, clinical studies, toxicity studies: *Arzneimittel-Forsch.* **32,** 1153-1218 (1982). Review of pharmacology, clinical efficacy: J. T. Connell, *Pharmacotherapy* **5,** 201-208 (1985).

Crystals from acetone, mp 146.5-148.5°. LD_{50} in rats, mice, guinea pigs (mg/kg): > 2000 orally (Kinsolving).
THERAP CAT: Antihistaminic.

9095. Terguride. *N,N-Diethyl-N'-[(8α)-6-methylergolin-8-yl]urea;* N-(D-6-methyl-8-isoergolin-1-yl)-N',N'-diethylurea; 6-methyl-8α-(diethylcarbamoylamino)ergoline; 9,10α-dihydrolisuride; 9,10-transdihydrolisuride; TDHL; $C_{20}H_{28}N_4O$; mol wt 340.47. C 70.55%, H 8.29%, N 16.46%, O 4.70%. Ergot derivative; dihydrogenated analog of lisuride, *q.v.* Exhibits dopamine agonist and antagonist activity. Prepn: V. Zikán *et al., Coll. Czech. Chem. Commun.* **37,** 2600 (1972); *eidem,* **Ger.** pat. **2,238,540;** *eidem,* **U.S.** pat. **3,953,454** (1973, 1976 both to Spofa). Physical properties: A. Cerny *et al., Coll. Czech. Chem. Commun.* **52,** 1331 (1987). Pharmacology in animals and humans: H. Wachtel, R. Dorow, *Life Sci.* **32,** 421 (1983). Receptor binding studies in rat brain: W. Kehr *et al., Acta Pharm. Suec.* **1983,** Suppl. 2, 98; M. W. Valchár *et al., Eur. J. Pharmacol.* **136,** 97 (1987). Evaluation in animal models of Parkinson's disease: W. C. Koller, G. Herbster, *Neurology* **37,** 723 (1987); T. Brücke *et al., Eur. J. Pharmacol.* **148,** 445 (1988). Radioreceptor assay in biological fluids: R. Lapka *et al., J. Pharmacol. Methods* **11,** 263 (1984). Pharmacokinetics in humans: W. Krause *et al., Eur. J. Clin. Pharmacol.* **27,** 335 (1984). Clinical evaluation in Huntington's disease: S. Bassi *et al., Neurology* **36,** 984 (1986); in Parkinson's disease: T. Brücke *et al., Advan. Neurol.* **45,** 573 (1986); I. Suchy *et al., ibid.* 577; in hyperprolactinemia and acromegaly: D. Dallabonzana *et al., J. Clin. Endocrinol. Metab.* **63,** 1002 (1986).

Crystals from ethanol, mp 203-204° (dec). $[\alpha]_D^{20}$ +30° (c = 1 in pyridine). Also reported as crystals from ethanol, mp 205-207° (dec) (Cerny). $[\alpha]_D^{20}$ +29.0 (c = 0.2 in pyridine). uv max: 292, 281, 224 nm (log ϵ 3.72, 3.81, 4.42). Practically insol in water.

Hydrogen maleate, $C_{24}H_{32}N_4O_5$, *SH-406, VUFB-6638, ZK-31224, Dironyl, Mysalfon.* Crystals from ethanol, mp 190-191°. (Hydrate: crystals from ethanol, mp 150-153°. $[\alpha]_D^{20}$ −15.0° (c = 0.1 in H_2O). Soly in water: 1.26 mg/ml.)

THERAP CAT: Antiparkinsonian; antihyperprolactinemic.

9096. Teriparatide Acetate. hPTH 1-34 acetate; MN-10T; Parathar. $C_{181}H_{292}N_{55}O_{51}S_2 \cdot yC_2H_4O_2 \cdot xH_2O$. Synthetic, biologically active polypeptide consisting of the 1-34 amino-terminal fragment of human parathyroid hormone (hPTH), *q.v.* Amino acid sequence: H. B. Brewer *et al., Proc. Nat. Acad. Sci. USA* **69**, 3585 (1972). Prepn: R. H. Andreatta *et al., Helv. Chim. Acta* **56**, 470 (1973). Revised structure: H. D. Niall *et al., Proc. Nat. Acad. Sci. USA* **71**, 384 (1974). Prepn: R. L. Colescott, **Ger.** pat. **2,649,727;** idem, **Ger.** pat. **2,649,848;** G. W. Tregear, **U.S.** pat. **4,086,196** (1977, 1977, 1978 all to Armour Pharm.). Solid phase synthesis and biological activity: G. W. Tregear *et al., Z. Physiol. Chem.* **355**, 415 (1974). Solution synthesis and biological activity: M. Takai *et al., Peptide Chem.* **17**, 187 (1980). Clinical pharmacology: R. M. Neer *et al., J. Clin. Endocrinol. Metab.* **38**, 420 (1977). RIA determn in plasma: J. M. Zanelli *et al., J. Immunoassay* **1**, 289 (1980). Pharmacokinetics in humans: G. N. Kent *et al., Clin. Sci.* **68**, 171 (1985). Clinical evaluation in osteoporosis: J. Reeve *et al., Brit. Med. J.* **280**, 1340 (1980); D. M. Slovik *et al., J. Clin. Invest.* **68**, 1261 (1981). Diagnostic use: T. Igarashi *et al., Pharmatherapeutica* **3**, 79 (1982); L. E. Mallette *et al., J. Clin. Endocrinol. Metab.* **67**, 964 (1988). Review of diagnostic use in modified Ellsworth-Howard test: L. E. Mallette, *Ann. Int. Med.* **109**, 800-804 (1988).

THERAP CAT: Treatment of osteoporosis. Diagnostic aid (hypocalcemia).

9097. Terlipressin. *N-[N-(N-Glycylglycyl)glycyl]-8-*L-*lysinevasopressin; N^α-glycylglycylglycylvasopressin; 1-trigly-cyl-8-lysinevasopressin;* Glypressin. $C_{52}H_{74}N_{16}O_{15}S_2$; mol wt 1227.39. C 50.89%, H 6.08%, N 18.26%, O 19.55%, S 5.22%. Analog of lypressin, *q.v.* Prepn: E. Kasafirek *et al., Coll. Czech. Chem. Commun.* **31**, 4581 (1966); Z. Procházka *et al., ibid.* **43**, 1285 (1978). Pharmacological studies: J. Kyncl *et al., Eur. J. Pharmacol.* **28**, 294 (1974); P. O. B. Sjöquist *et al., Acta Pharmacol. Toxicol.* **40**, 369 (1977). Use in treatment of uterine bleeding: V. Pavlin *et al., Brit. J. Obstet. Gynaecol.* **85**, 801 (1978). Hemostatic effects in cirrhosis patients: C. V. Prowse *et al., Eur. J. Clin. Invest.* **10**, 49 (1980). Conversion to lysine vasopressin in man: M. L. Forsling *et al., J. Endocrinol.* **85**, 237 (1980). Uterine effects in first trimester of pregnancy: T. Laudanski, M. Aakerlund, *Contraception* **22**, 199 (1980). Single-dose effects on renal potassium excretion in humans: H. Nadvornikova, O. Schueck, *Int. J. Clin. Pharmacol. Ther. Toxicol.* **20**, 155 (1982).

```
Gly-Gly-Gly-Cys-Tyr-Phe-Gln-Asn-Cys-Pro-Lys-GlyNH₂
```

Diacetate pentahydrate, $C_{56}H_{82}N_{16}O_{19}S_2 \cdot 5H_2O$, *Glycylpressin.* $[\alpha]_D^{25}$ −82° (c = 0.2 in 1M acetic acid).

THERAP CAT: Vasopressor (treatment of uterine and esophageal bleeding).

9098. Terodiline. *N-(1,1-Dimethylethyl)-α-methyl-γ-phenylbenzenepropanamine; N-tert-butyl-1-methyl-3,3-di-phenylpropylamine; N-(1-methyl-3,3-diphenylpropyl)-tert-butylamine; 4,4-diphenyl-2-(tert-butylamino)butane.* $C_{20}H_{27}N$; mol wt 281.44. C 85.35%, H 9.67%, N 4.98%. Calcium antagonist with anticholinergic and vasodilatory activity. Prepn and use in treatment of angina: **Brit.** pat. **923,942;** S. Carlsson, **U.S.** pat. **3,371,014** (1963, 1968 both to AB Recip). Pharmacokinetics: B. Karlen *et al., Eur. J. Clin. Pharmacol.* **23**, 267 (1982). Series of articles on pharmacology and clinical efficacy in urinary frequency and motor urge incontinence: *Scand. J. Urol. Nephrol.* **Suppl. 87**, 1-61 (1984). Clinical trial in treatment of urinary incontinence: U. Ulmsten *et al., Am. J. Obstet. Gynecol.* **153**, 619 (1985).

Yellow liquid, $bp_{1.0}$ 130-132°.

Hydrochloride, $C_{20}H_{28}ClN$, *Bicor, Mictrol, Micturin, Micturol.* Crystals from water, mp 178-180°. Sol in ethanol. Slightly sol in ether.

THERAP CAT: Anti-anginal. Treatment of urinary incontinence.

9099. Terofenamate. *2-[(2,6-Dichloro-3-methylphenyl)-amino]benzoic acid ethoxymethyl ester; N-(2,6-dichloro-m-tolyl)anthranilic acid ethoxymethyl ester;* A₃; etoclofene; Etofen. $C_{17}H_{17}Cl_2NO_3$; mol wt 354.24. C 57.64%, H 4.84%, Cl 20.01%, N 3.95%, O 13.55%. Anti-inflammatory deriv of anthranilic acid, *q.v.* Prepn: **Brit.** pat. **1,199,386** corresp to E. Manghisi, **U.S.** pat. **3,642,864** (1970, 1972 both to Luso-farmaco); A. Salembeni *et al., Farmaco Ed. Sci.* **30**, 276 (1975). Biological properties: G. B. Fregnan *et al., ibid.* **353**. Absorption, distribution, excretion: *eidem, Boll. Chim. Farm.* **114**, 85 (1975). Pharmacological and toxicity studies: T. Chieli *et al., ibid.* **115**, 41 (1976).

Crystals from ethanol, mp 73-74°. LD_{50} in mice, rats (mg/kg): 918, 307 orally; 300, 274 i.p., T. Chieli *et al., loc. cit.*

THERAP CAT: Anti-inflammatory; analgesic.

9100. Terpenylic Acid. *Tetrahydro-2,2-dimethyl-5-oxo-3-furanacetic acid; 3-(1-hydroxy-1-methylethyl)glutaric acid γ-lactone;* tetrahydro-5-keto-2,2-dimethyl-3-furanacetic acid; terpenolic acid. $C_8H_{12}O_4$; mol wt 172.19. C 55.80%, H 7.03%, O 37.17%. Prepd from α-terpineol, terebic acid, or terpin hydrate: Suga, Sakoda, *J. Sci. Hiroshima Univ. Ser. A* **22**, 69 (1958), *C.A.* **53**, 10273f; Lipp *et al., Ann.* **644**, 37 (1961). Resolution of optical isomers: Fredga, Sandberg, *C.A.* **52**, 11747g (1958).

Monohydrate, prisms, mp 57°. When anhydrous, melts at 90°. Sublimes 130-140°. Moderately sol in cold water; very sol in hot water.

(+)-Form, mp 92-94°, $[\alpha]_D^{25}$ +56.3°.

(−)-Form, mp 92-94°, $[\alpha]_D^{25}$ −56.5°.

9101. Terpin. *4-Hydroxy-α,α,4-trimethylcyclohexane-methanol; p-menthane-1,8-diol;* dipenteneglycol. $C_{10}H_{20}O_2$;

mol wt 172.27. C 69.72%, H 11.70%, O 18.58%. Both *cis*- and *trans*-modifications are known. The *cis*-compd is obtained most readily in the hydrated form, *cis*-terpin hydrate. Prepn of *cis*-form from oil of turpentine: Hempel, *Ann.* **180**, 71 (1876); Wallach, *Ann.* **230**, 225 (1885); Schmitt, *Mfg. Chemist* **26**, 350 (1955). From *d*-limonene: Sword, *J. Chem. Soc.* **127**, 1632 (1925). Prepn of *trans*-form from 1,8-cineole, α-terpineol or *cis*-terpin hydrate: Matsuura *et al.*, *Bull. Chem. Soc. Japan* **31**, 990 (1958); Lombard, Ambroise, *Bull. Soc. Chim. France* **1961**, 230. Structure of *cis*- and *trans*-forms: Baeyer, *Ber.* **26**, 2861 (1893); Wagner, *ibid.* **27**, 1636 (1894).

cis-Form hydrate, **terpin hydrate, terpinol**. Rhombic pyramids from water, mp 116-117°; sublimes at about 100° when heated slowly; slight characteristic odor and slightly bitter taste; efflorescent in dry air. Anhydr *cis*-form: mp 104-105°; bp 258°; rapidly re-forms hydrate on exposure to air. One gram dissolves in 34 ml boiling water, 13 ml alcohol, 3 ml boiling alcohol, 135 ml chloroform, 140 ml ether, about 1 ml boiling glacial acetic acid. At 20°, one gram dissolves in 13 ml methanol, 13 ml ethyl acetate, 250 ml water, 77 ml benzene, 290 ml carbon tetrachloride, 250 ml carbon disulfide. Practically insol in petr ether.

trans-Form, monoclinic prisms, mp 158-159°. One gram dissolves at 20° in 11 ml methanol, 20 ml ethyl acetate, 100 ml water, 250 ml benzene, 250 ml carbon tetrachloride, 500 ml carbon disulfide.

THERAP CAT: *cis*-Form hydrate as expectorant.

THERAP CAT (VET): Expectorant.

9102. Terpinene. $C_{10}H_{16}$; mol wt 136.23. C 88.16%, H 11.84%. Mixture of three isomeric hydrocarbons: α-terpinene and γ-terpinene which occur naturally and β-terpinene which has been prepd synthetically. Isoln of α-terpinene from cardamom and marjoram oils: Weber, *Ann.* **238**, 101 (1887); Beltz, *Ber.* **32**, 996 (1899). Isoln of α-terpinene from oils of *Mosla japonica*, *M. grosserrata* and *Cupressus macrocarpa*: Richter, Wolff, *Ber.* **60**, 477 (1927); Briggs, Sutherland, *J. Org. Chem.* **7**, 397 (1942). Prepn of β-terpinene from sabinine: Wallach, *Ann.* **357**, 64 (1907); **362**, 285 (1908). Structure: Wallach, *ibid.* **362**, 293 (1908). *Review:* J. L. Simonsen, *The Terpenes* vol. I (University Press, Cambridge, 2nd ed., 1947) pp 172-193.

α β γ

α-Terpinene, oil, pleasant odor of lemons. bp 173.5-174.8°; $bp_{13.5}$ 65.4-66°; $d_4^{19.6}$ 0.8375. n_D^{20} 1.4784. Practically insol in water. Miscible with alcohol, ether.

β-Terpinene, oil, bp 173-174°. d^{22} 0.838. n_D^{22} 1.4754.
γ-Terpinene, oil, bp 183°. d_4^{15} 0.853. $n_D^{15.6}$ 1.4754.
Dihydrochloride, $C_{10}H_{18}Cl_2$, crystals, mp 51-52°.

9103. α-Terpineol. *α,α,4-Trimethyl-3-cyclohexene-1-methanol; p-menth-1-en-8-ol.* $C_{10}H_{18}O$; mol wt 154.24. C 77.86%, H 11.76%, O 10.37%. Terpineol exists as three isomers, α-, β-, and γ-terpineol: J. L. Simonsen, *The Terpenes* vol. I (University Press, Cambridge, 2nd ed., 1947) pp 256-274. Isoln of *d*-α-terpineol from petitgrain oil: Walbaum, Hüthig, *J. Prakt. Chem.* **67**, 322 (1903). Isoln from *l*-α-terpineol from long leaf pine oil: Teeple, *J. Am. Chem. Soc.* **30**, 412 (1908). Isoln of *dl*-α-terpineol from cajeput oil: Voiry, *Compt. Rend.* **106**, 1540 (1888). Synthesis of *d*-α-terpineol:

Cologne, Crabalona, *Bull. Soc. Chim. France* **1960**, 102. Synthesis of *l*-α-terpineol: *eidem, ibid.* **1959**, 1505. Stereochemistry: Henbest, McElkinney, *J. Chem. Soc.* **1959**, 1834. *Review:* Wagner, *Mfg. Chemist* **22**, 98, 153 (1951).

d-Form, liquid. $bp_{4.5}$ 81-82°; bp_{731} 206-207°. d_4^{20} 0.9338. n_D^{20} 1.4818. $[α]_D^{20}$ +92.45°. Solidifies at 31°.
Phenylurethan, $C_{17}H_{23}NO_2$, crystals from petr ether, mp 111°. $[α]_D^{20}$ +30.50° (benzene).
l-Form, liquid. bp_5 80-81.5°. d_4^{20} 0.935. n_D^{20} 1.4820. $[α]_D^{20}$ −100° (c = 20 in alc). Solidifies at 36.4°.
Dinitrobenzoate, $C_{17}H_{20}N_2O_6$, crystals from alcohol, mp 101.5°. $[α]_D^{20}$ −31° (c = 9.5 in carbon tetrachloride).
dl-Form, liquid. bp_3 85°; bp_{752} 218.8-219.4°, d^{15} 0.9386. n_D^{20} 1.4831.
Phenylurethan, $C_{17}H_{23}NO_2$, crystals, mp 113°.
USE: Perfumes; denaturing fats for soap manufacture.
THERAP CAT: Antiseptic.

9104. Terreic Acid. *(1R)-3-Hydroxy-4-methyl-7-oxa-bicyclo[4.1.0]hept-3-ene-2,5-dione;* 2-hydroxy-3-methyl-1,4-benzoquinone 5,6-epoxide; 5,6-epoxy-3-hydroxy-*p*-toluquinone. $C_7H_6O_4$; mol wt 154.12. C 54.55%, H 3.92%, O 41.53%. Antibiotic metabolite produced by the mold *Aspergillus terreus*: Wilkins, Harris, *Brit. J. Exp. Pathol.* **23**, 166 (1942); Abraham, Florey, in H. W. Florey *et al.*, *Antibiotics* vol. I (Oxford Univ. Press, New York, 1949) p 337; Kaplan *et al.*, *Antibiot. & Chemother.* **4**, 746 (1954). Structure: Sheehan *et al.*, *J. Am. Chem. Soc.* **80**, 5536 (1958). Synthesis of the racemate: Rashid, Read, *J. Chem. Soc. (C)* **1967**, 1323. Alternate synthesis and resolution of isomers: Sheehan, Lo, *J. Med. Chem.* **17**, 371 (1974).

(−)-Form (natural), pale yellow plates from benzene or hexane. Easily sublimed *in vacuo*. mp 127-127.5°. Rotation varies considerably with the solvent: $[α]_D^{22}$ −16.6° (chloroform); $[α]_D^{22}$ −28.6° (methanol-benzene 1:1); $[α]_D^{22}$ +74.3° (pH 7 phosphate buffer). uv max (ethanol): 214, 316 nm (log ε 4.03, 3.88). Enol-type acid, pKa 4.5. Slightly sol in water. Soluble in ether, lower alcohols, acetone, hot cyclohexane. Moderately stable to mineral acid, but dec rapidly in alkaline soln.

9105. Tertatolol. (±)-*1-[(3,4-Dihydro-2H-1-benzo-thiopyran-8-yl)oxy]-3-[(1,1-dimethylethyl)amino]-2-pro-panol;* (±)-1-*tert*-butylamino-3-(1-thiachroman-8-yloxy)-2-propanol; *dl*-8-[2-hydroxy-3-[(*tert*-butylamino)propyl]-oxy]thiochromane. $C_{16}H_{25}NO_2S$; mol wt 295.44. C 65.05%, H 8.53%, N 4.74%, O 10.83%, S 10.85%. Nonselective β-adrenergic blocker with no intrinsic sympathomimetic activity. Prepn: C. Malen, M. Laubie, *Ger. pat.* **2,115,201**; *eidem*, *U.S. pat.* **3,960,891** (1971, 1976 both to Sci. Union et Cie). Pharmacology in animals: M. Laubie *et al.*, *Arch. Int. Pharmacodyn. Ther.* **201**, 323, 334 (1973). Cardiovascular and renal effects in dogs: B. R. Walker *et al.*, *J. Cardiovasc. Pharmacol.* **7**, 1193 (1985). Synergistic antihypertensive effect with indapamide, *q.v.*: E. Marmo *et al.*, *Drugs Exptl. Clin. Res.* **11**, 709 (1985). GC-MS determn in plasma and urine: S. Staveris *et al.*, *J. Chromatog.* **339**, 97 (1985). Pharmacology in humans and mechanism of action study: A. De Blasi *et al.*, *Clin. Pharmacol. Ther.* **39**, 245 (1986).

OCH$_2$CHCH$_2$NHC(CH$_3$)$_3$
OH

Crystals from hexane, mp 70-72°.

Hydrochloride, C$_{16}$H$_{26}$ClNO$_2$S, S 2395, SE 2395, Artex. Crystals from acetonitrile, mp 180-183°. pKa 9.8. LD$_{50}$ in rats, mice (mg/kg): 40, 37 i.v.; 90, 120 i.p. (Laubie).

THERAP CAT: Antihypertensive.

9106. α-Terthienyl. *2,2':5'2''-Terthiophene; 5-(2-thienyl)-2,2'-bithiophene.* C$_{12}$H$_8$S$_3$; mol wt 248.37. C 58.03%, H 3.25%, S 38.72%. Biocidal constituent of various species of marigolds. Isoln from *Tagetes erecta* L., *Compositae:* L. Zechmeister, J. W. Sease, *J. Am. Chem. Soc.* **69**, 273 (1947). Isoln and distribution in *Tageteae:* K. R. Downum, G. H. N. Towers, *J. Nat. Prod.* **46**, 98 (1983). Synthesis: W. Steinkopf *et al.*, *Ann.* **546**, 180 (1941); H. J. Kooreman, H. Wynberg, *Rec. Trav. Chim.* **86**, 37 (1967); J.-P. Beny *et al.*, *J. Org. Chem.* **47**, 2201 (1982). Biological activity as nematocide: J. H. Uhlenbroek, J. D. Bijloo, *Rec. Trav. Chim.* **77**, 1004 (1958); J. D. Bijloo *et al.*, Ger. pat. **1,075,891**; *eidem,* U.S. pat. **3,050,442** (1960, 1962, both to North American Philips); J. Bakker *et al.*, *J. Biol. Chem.* **254**, 1841 (1979); as herbicide: J. Harvey, Jr., U.S. pat. **3,086,854** (1963 to Du Pont); G. Campbell *et al.*, *J. Chem. Ecol.* **8**, 961 (1982); as antimicrobial: J. R. Kagan *et al.*, *Photochem. Photobiol.* **31**, 465 (1980); T. Arnason *et al.*, *ibid.* **33**, 821 (1981); F. DiCosmo *et al.*, *Pestic. Sci.* **13**, 589 (1982).

Yellow-orange plates from methanol, mp 93-94°. uv max (methanol): 254, 350 nm (ε 7100, 21300). Sol in carbon bisulfide, ether, benzene, acetone, petr ether; slightly sol in methanol, ethanol. Insol in water.

9107. Tertiomycins. Antibiotic substances produced by *Streptomyces eurocidicus.* Isoln of *Tertiomycin A* and *Tertiomycin B:* Osato *et al.*, *J. Antibiot.* **8A**, 105, 161 (1955). Tertiomycin A also isolated from *S. albireticuli:* Miyake *et al.*, *ibid.* **12A**, 59 (1959). Manuf by culture of *S. eurocidicus:* Umezawa *et al.*, *Japan.* pat. **5200('67)** (to Sanraku Ocean Co.), *C.A.* **67**, 10386c (1967). Belongs to the erythromycin-carbomycin group of antibiotics.

9108. Testolactone. *D-Homo-17a-oxaandrosta-1,4-diene-3,17-dione; 13-hydroxy-3-oxo-13,17-secoandrosta-1,4-dien-17-oic acid δ-lactone;* 1,2,3,4,4a,4b,7,9,10,10a-decahydro-2-hydroxy-2,4b-dimethyl-7-oxo-1-phenanthrenepropionic acid δ-lactone; delta-1-testololactone; 1-dehydrotestololactone; 17α-oxo-D-homo-1,4-androstadiene-3,17-dione; Δ^1-testolactone; NSC-23759; SQ 9538; Fludestrin; Teslac. C$_{19}$H$_{24}$O$_3$; mol wt 300.38. C 75.97%, H 8.05%, O 15.98%. Obtained by microbial transformation of progesterone, Reichstein's substance S, or testosterone: Fried *et al.*, *J. Am. Chem. Soc.* **75**, 5764 (1953); *eidem,* U.S. pat. **2,744,120** (1956 to Olin Mathieson); Brannon *et al.*, *J. Org. Chem.* **30**, 760 (1965). Comprehensive description: K. Florey, Ed. in *Analytical Profiles of Drug Substances* vol. **5** (Academic Press, New York, 1976) pp 533-553.

Crystals from acetone, mp 218-219°. [α]$_D^{23}$ -45.6° (c = 1.24 in chloroform). uv max (ethanol): 242 nm (ε 15,800).

THERAP CAT: Antineoplastic.

9109. Testosterone. *17β-Hydroxyandrost-4-en-3-one;* Δ^4-androsten-17β-ol-3-one; *trans*-testosterone; Géno-cristaux Gremy; Malestrone (amps); Orquisteron; Percutacrine Androgénique; Primotest; Sustanon; Mertestate; Testobase; Virosterone; Testryl; Testrone; Homosteron; Oreton-F; Teslen. C$_{19}$H$_{28}$O$_2$; mol wt 288.41. C 79.12%, H 9.79%, O 11.09%. Principal hormone of the testes, produced by the interstitial cells. Isoln in minute amounts from testes, esp bull testes: David *et al.*, *Z. Physiol. Chem.* **233**, 281 (1935). Prepd by conversion of other steroids such as cholesterol. The important intermediate dehydroandrosterone is efficiently transformed into testosterone by a microbial process: Mamoli, Vercellone, *Ber.* **70**, 470 (1937), and later papers; U.S. pat. **2,236,574**. Structure: Butenandt, Hanisch, *Ber.* **68**, 1859 (1935); *Z. Physiol. Chem.* **237**, 89 (1935); Ruzicka, Wettstein, *Helv. Chim. Acta* **18**, 1264 (1935); and Kägi, *ibid.* **18**, 1478 (1935); Fieser, Fieser, *Steroids* (New York, 1959) *passim*. Structure determn by x-ray crystallography: P. J. Roberts *et al.*, *J. Chem. Soc., Perkin Trans. II* **1973**, 1978. Physical data of β-maltoside: Ch. Meystre, K. Miescher, *Helv. Chim. Acta* **27**, 1153 (1944). Toxicity data: H. Selye, *Proc. Soc. Exp. Biol. Med.* **46**, 116 (1941).

Needles from dil acetone, mp 155°. [α]$_D^{24}$ +109° (c = 4 in alc). uv max: 238 nm. Insol in water; sol in alcohol, ether, and other organic solvents. 0.015 mg = 1 international unit.

Acetate, C$_{21}$H$_{30}$O$_3$, *Aceto-Sterandryl, Aceto-Testoviron, Perandrone A.* mp 140-141°.

Isobutyrate, C$_{23}$H$_{34}$O$_3$, *Perandren M.*

3-Oxododecanoate, C$_{31}$H$_{48}$O$_4$, *testosterone ketolaurate, Androdurin.*

β-Maltoside, C$_{31}$H$_{48}$O$_{12}$, crystals from methanolacetone, mp 250-255° (dec). [α]$_D^{19}$ +73° (c = 0.992 in methanol). Freely sol in water.

Propionate, see separate entry. See also 17-Methyltestosterone.

Undecanoate, C$_{29}$H$_{48}$O$_3$, *Andriol, Pantestone, Restandol.*

THERAP CAT: Androgen.

9110. Testosterone 17-Chloral Hemiacetal. *17β-(2,2,2-Trichloro-1-hydroxyethoxy)androst-4-en-3-one;* 17β-(1-hydroxy-2,2,2-trichloroethoxy)androst-4-en-3-one; testosterone 17-hemiacetal with chloral; Caprosem. C$_{21}$H$_{29}$Cl$_3$O$_3$; mol wt 435.83. C 57.87%, H 6.71%, Cl 24.41%, O 11.01%. Prepn: P. Borrevang, U.S. pat. **2,933,514** (1960 to Leo Pharm.); *idem, Acta Chem. Scand.* **16**, 883 (1962).

Crystals from ethyl acetate, mp 200-201°. uv max (ethanol): 241 nm (ε 16,300).

Acetate, C$_{23}$H$_{31}$Cl$_3$O$_4$, crystals from ethyl acetate, mp 192-193°. uv max (ethanol): 241 nm (ε 16,400).

THERAP CAT: Androgen.

9111. Testosterone 17β-Cypionate. *17β-(3-Cyclopentyl-1-oxopropoxy)androst-4-en-3-one;* depo-testosterone; testosterone 17β-cyclopentanepropionate; testosterone cyclopentylpropionate; Depovirin; Pertestis; Testergon; Testodrin prolongatum. $C_{27}H_{40}O_3$; mol wt 412.59. C 78.59%, H 9.77%, O 11.63%. *Ref.* Ott *et al.*, XIIth *International Congress of Pure and Applied Chemistry, Abstracts of Papers*, p 294 (New York, 1951); *J. Clin. Endocrinol. Metabol.* **12**, 15 (1952); *Am. Prof. Pharmacist* **18**, 555 (1952).

Crystals, mp 101-102°. $[\alpha]_D^{25}$ +87° (CHCl$_3$). Sol in oils.
THERAP CAT: Androgen.

9112. Testosterone Enanthate. *17-[(1-Oxoheptyl)oxy]-androst-4-en-3-one;* testosterone oenanthate; testosterone heptoate; Androtardyl; Delatestryl; Malogen L.A.; Orquisteron-E; Primoteston; Testate; Testinon; Testoenant; Testroval. $C_{26}H_{40}O_3$; mol wt 400.58. C 77.95%, H 10.07%, O 11.98%. Prepn: Junkmann *et al.*, U.S. pat. **2,840,508** (1958 to Schering, AG); Span. pat. **241,206** (1958 to Alter, SA), *C.A.* **54**, 3532b (1960). Comprehensive description: K. Florey, Ed. in *Analytical Profiles of Drug Substances* vol. 4 (Academic Press, New York, 1975) pp 452-465.

Crystals, mp 36-37.5°.
THERAP CAT: Androgen.

9113. Testosterone Nicotinate. *17-[(3-Pyridinylcarbonyl)oxy]androst-4-en-3-one;* testosterone 17-nicotinic acid ester; Bolfortan. $C_{25}H_{31}NO_3$; mol wt 393.51. C 76.30%, H 7.94%, N 3.56%, O 12.20%. Prepared from testosterone and nicotinic acid anhydride: Zirm, Weichsel, U.S. pat. **3,057,-856** (1962 to Lannacher Heilmittel).

Crystals from acetone, mp 187-188°. Sol in dil HCl and other dil mineral acids.
THERAP CAT: Androgen.

9114. Testosterone Phenylacetate. *17-[(Phenylacetyl)-oxy]androst-4-en-3-one;* Perandren Phenylacetate. $C_{27}H_{34}O_3$; mol wt 406.54. C 79.76%, H 8.43%, O 11.81%. Prepd from testosterone and phenylacetyl chloride in pyridine: Miescher *et al.*, *Biochem. Z.* **294**, 39 (1937); Gould *et al.*, *J. Am. Chem. Soc.* **79**, 4472 (1957).

Crystals from hexane, mp 129-131° (Miescher *et al.*), mp 126.5-127.5° (Gould *et al.*). $[\alpha]_D^{25}$ +80.0° (dioxane).
THERAP CAT: Androgen.

9115. Testosterone Propionate. Δ^4-Androstene-17β-propionate-3-one; Oreton; Neo-Hombreol; Perandren (Injectable); Sterandryl (amps.); Testodet; Testoviron (amps.); Androsan (amps.); Androtest P; Bio-Testiculina; Orchisterone-P; Orquisteron-P; Testrex; Testodrin; Testosid; Testoxyl; Uniteston; Homandren (amps.); Hormoteston; Nasdol; Propiokan; Recthormone Testosterone; Testaform; Testogen; Testonique; Testormol; Tostrin; Enarmon; Anertan; Virormone. $C_{22}H_{32}O_3$; mol wt 344.48. C 76.70%, H 9.36%, O 13.93%. Prepn from Δ^4-androsten-17-ol propionate and CrO$_3$: U.S. pats. **2,311,067**; **2,374,369**; **2,374,370**. From testosterone: *C.A.* **35**, 3040^4. From androstenediol propionate by bacteria: U.S. pat. **2,236,574**. Prepn of oxime: Mazur, *J. Org. Chem.* **28**, 248 (1963).

Stout prisms from alcohol + water, mp 118-122°. $[\alpha]_D^{25}$ +83° to +90° (100 mg in 10 ml dioxane). Insol in water. Freely sol in alcohol, ether, pyridine, in other organic solvents. Sol in vegetable oils.
Oxime, crystals from methanol, mp 177-183°. Prepn: Mazur, *J. Org. Chem.* **28**, 248 (1963).
THERAP CAT: Androgen.
THERAP CAT (VET): Androgenic hormonal activity. In deficiencies of male sex hormone, cryptorchidism, mammary tumors in females.

9116. Tetraamminecopper Sulfate. Cuprammonium sulfate; ammonium cupric sulfate; cupric sulfate, ammoniated; Eau Celeste. $CuH_{12}N_4O_4S$; mol wt 227.73. Cu 27.90%, H 5.31%, N 24.60%, O 28.10%, S 14.08%. $[Cu(NH_3)_4]SO_4 \cdot$ Prepd by dissolving copper sulfate in ammonia water and pptg with alcohol: Mazzi, *Acta Cryst.* **8**, 137 (1955). Crystal structure of monohydrate: Morosin, *ibid* **25B**, 19 (1969).
Monohydrate, large, dark blue crystals. d$_4^{20}$ 1.81. Ammonia odor, dec in air. Loses H_2O and $2NH_3$ on heating to 120°, remaining $2NH_3$ at 160°. Soly in water at 21.5°: 18.5 g/100 ml. Practically insol in the lower alcohols.
USE: In textile printing, especially in calico finishing. As fungicide.

9117. Tetrabarbital. *5-Ethyl-5-(1-ethylbutyl)-2,4,6-(1H,3H,5H)-pyrimidinetrione; 5-ethyl-5-(1-ethylbutyl)barbituric acid;* 5-ethyl-5-(1-ethylbutyl)malonylurea; JL 991; Butysal; Butysedal. $C_{12}H_{20}N_2O_3$; mol wt 240.31. C 59.98%, H 8.39%, N 11.66%, O 19.97%. Prepd from ethyl 2-ethyl-2-(1-ethylbutyl)malonate + urea: Kopp, Tchoubar, *Bull. Soc. Chim. France* **1951**, 30; Tchoubar, Fr. pat. **1,020,357** (1953), *C.A.* **53**, 1389a.

Crystals, mp 122°.
Note: This is a controlled substance (depressant) listed in the U.S. Code of Federal Regulations, Title 21 Part 1308.13 (1987).
THERAP CAT: Sedative, hypnotic.

9118. Tetrabenazine. *1,3,4,6,7,11b-Hexahydro-9,10-dimethoxy-3-(2-methylpropyl)-2H-benzo[a]quinolizin-2-one;* 2-oxo-3-isobutyl-9,10-dimethoxy-1,2,3,4,6,7-hexahydro-11bH-benzo[a]quinolizine; Ro 1-9569. $C_{19}H_{27}NO_3$; mol wt 317.41. C 71.89%, H 8.57%, N 4.41%, O 15.12%. Dopamine depleting agent. Prepn: Brossi *et al., Helv. Chim. Acta* **41**, 119 (1958); **U.S.** pat. **2,830,993** (1958 to Hoffmann-La Roche); Osbond, *J. Chem. Soc.* **1961**, 4711. Pharmacology: Leusen *et al., Arch. Int. Pharmacodyn.* **119**, 225 (1959). Metabolism: Schwartz *et al., Biochem. Pharmacol.* **15**, 645 (1966). Clinical trials in hyperkinetic movement disorders: J. Janovic, *Ann. Neurol.* **11**, 41 (1982); J. Janovic, J. Orman, *Neurology* **38**, 391 (1988).

Prisms from methanol, mp 125-126°. Oxime, crystals from dil alcohol, mp 158°.

Hydrochloride, $C_{19}H_{28}ClNO_3$, crystals, mp 208-210°. Soluble in hot water. Practically insol in acetone. uv max (alcohol): 230, 284 nm (ϵ 7780, 3820).

Methanesulfonate, $C_{20}H_{31}NO_6S$, *Nitoman.* Bitter crystals, sensitive to light, mp 126-130°. Sparingly sol in water, sol in alcohol. Practically insol in ether.

THERAP CAT: Antidyskinetic. Antipsychotic.

9119. Tetraborane(10). Tetraboron decahydride; borobutane. B_4H_{10}; mol wt 53.36. B 81.11%, H 18.89%. Prepd by the reaction of magnesium boride with hydrochloric or phosphoric acid: Stock, Kuss, *Ber.* **56B**, 789 (1923); from dihydropentaborane: Burg, Schlesinger, *J. Am. Chem. Soc.* **55**, 4009 (1933); from diborane: Stock, Mathing, *Ber.* **69B**, 1456 (1936); Klein *et al., J. Am. Chem. Soc.* **80**, 4149 (1958). *Review:* Greenwood in *Comprehensive Inorganic Chemistry*, vol. 1, J. C. Bailar, Jr. *et al.,* Eds. (Pergamon Press, Oxford, 1973) pp 785-791.

Gas; disagreeable odor; mp −120°; bp 18°. Vapor pressure 580 mm Hg at 6°; 388 mm at 0°. Dec at room temp in a few hours. Dec rapidly at 100°. Ignites spontaneously in air or oxygen. Hydrolyzes in water to boric acid and hydrogen. Reacts with ammonia to form a tetraammoniate.

9120. 3,4,5,6-Tetrabromo-*o*-cresol. $C_7H_4Br_4O$; mol wt 423.76. C 19.84%, H 0.95%, Br 75.43%, O 3.78%. Prepd by bromination of *o*-cresol or its methyl or ethyl ether: Zincke, Hedenström, *Ann.* **350**, 269 (1906); Bonneaud, *Bull. Soc. Chim. France* **7**, 776 (1910); Treacy, **U.S.** pat. **2,319,960** (1943 to Merck & Co.).

White to buff, cryst powder, mp 205-208° with dec. Practically insol in water; sol in alcohol, ether, alkali hydroxides.

Acetate, $C_9H_6Br_4O_2$, needles from dil alcohol, mp 154°. Slightly sol in alcohol and glacial acetic acid.

USE: Fungicide. *Caution:* Irritating to skin, mucous membranes.

9121. *sym*-Tetrabromoethane. *1,1,2,2-Tetrabromoethane;* acetylene tetrabromide; Muthmann's liquid. $C_2H_2Br_4$; mol wt 345.70. C 6.95%, H 0.58%, Br 92.47%. $Br_2CH-CHBr_2$. Manuf by bromination of acetylene: **Brit.** pat. **889,649** (1962 to Associated Ethyl Co.). Toxicity data: D. L. Wolff, *Acta Biol. Med. Ger.* **41**, 945 (1982).

Yellowish, heavy, very refractive liquid; odor of camphor and iodoform. d 2.964; bp_{54} 151°; mp 0°; n_D^{20} 1.638. Insol in water. Miscible with alc, chloroform, ether, aniline, glacial acetic acid. LD_{50} i.p. in mice: 443.3 mg/kg (Wolff).

Caution: Irritant, narcotic. Has caused liver, kidney injury in exptl animals: Gray, *Arch. Ind. Hyg. Occup. Med.* **2**, 407 (1950).

USE: In microscopy, as solvent, separating minerals by density.

9122. 3′,3′′,5′,5′′-Tetrabromophenolphthalein. *3,3-Bis(3,5-dibromo-4-hydroxyphenyl)-1(3H)-isobenzofuranone.* $C_{20}H_{10}Br_4O_4$; mol wt 633.94. C 37.89%, H 1.59%, Br 50.42%, O 10.10%. Prepd by direct bromination of phenolphthalein: Blicke *et al., J. Am. Chem. Soc.* **54**, 1465 (1932). Diagnostic use: E. A. Graham, W. H. Cole, *J. Am. Med. Assoc.* **250**, 2975 (1984), reprint of *J. Am. Med. Assoc.* **82**, 613 (1924).

Crystals from alcohol, glacial acetic acid, or ether, mp 295-297° (dec). Practically insol in water. Slightly sol in alc, glacial acetic acid; sol in ether, in alkali with a violet color.

Disodium salt, $C_{20}H_8Br_4Na_2O_4$, *tetrabromophenolphthalein sodium, Brom-Tetragnost, Cholegnostyl, Tetrabrom.* Blue crystals, dec on exposure to air becoming partially insol. Freely sol in water; slightly sol in alcohol. Insol in ether. pH of aq soln is 9-10. *Keep well closed and protected from light.*

THERAP CAT: Disodium salt as diagnostic aid (radiopaque medium).

9123. Tetracaine Hydrochloride. *4-(Butylamino)benzoic acid 2-(dimethylamino)ethyl ester hydrochloride;* *p*-butylaminobenzoyl-2-dimethylaminoethanol hydrochloride; 2-dimethylaminoethyl 4-*n*-butylaminobenzoate hydrochloride; dicain; Amethocaine Hydrochloride; Anethaine; Butethanol; Curtacain; Decicain; Gingicain M; Menonasal; Pantocaine; Pontocaine Hydrochloride;Tonexol. $C_{15}H_{25}ClN_2O_2$; mol wt 300.83. C 59.89%, H 8.38%, Cl 11.78%, N 9.31%, O 10.64%. Prepn: **U.S.** pat. **1,889,645** (1932); **Brit.** pat. **815,144** (1959 to Abbott). Prepn of pharmaceutical dosage forms: Shupe, **U.S.** pat. **3,272,700** (1966 to Sterling Drug). Mechanism of action studies: Y.-W. Leung *et al., J. Infect. Dis.* **136**, 679 (1977).

Faintly bitter crystals producing transient numbness of the tongue. mp 147-150°. Sol in 7 parts water, in alcohol; insol in ether, benzene. The aq soln is neutral to litmus. Aq solns are stable and may be sterilized by brief boiling. LD_{50} i.p. in mice: 70 mg/kg, Dawes, *Brit. J. Pharmacol. Chemother.* **1**, 90 (1946).

THERAP CAT: Local anesthetic.

THERAP CAT (VET): Topical anesthetic.

9124. Tetrachlormethiazide. *6-Chloro-3,4-dihydro-3-trichloromethyl-2H-1,2,4-benzothiadiazine-7-sulfonamide 1,1-dioxide;* 6-chloro-3,4-dihydro-7-sulfamoyl-3-trichloromethyl-2H-1,2,4-benzothiadiazine 1,1-dioxide; 3-trichloromethylhydrochlorothiazide; teclothiazide; PS 207; K 33; Depleil. $C_8H_7Cl_4N_3O_4S_2$; mol wt 415.13. C 23.15%, H 1.70%, Cl 34.17%, N 10.12%, O 15.42%, S 15.45%. Prepn: Close *et al., J. Am. Chem. Soc.* **82**, 1132 (1960); Novello *et al., J. Org. Chem.* **25**, 970 (1960).

Consult the cross index before using this section.

Crystals, mp 300-303° (Close); mp 287° (Novello).
THERAP CAT: Diuretic.

9125. Tetrachloroethane. *1,1,2,2-Tetrachloroethane;*
sym-tetrachloroethane; acetylene tetrachloride; Cellon;
Bonoform. $C_2H_2Cl_4$; mol wt 167.86. C 14.31%, H 1.20%, Cl
84.49%. $Cl_2CHCHCl_2$. Manuf by catalytic addition of chlorine to acetylene: Peters, Neumann, *Angew. Chem.* **45**, 261
(1932); by chlorination of ethylene: Pye, U.S. pat. **2,752,402**
(1956 to Dow); by catalytic chlorination of ethane: Joseph,
U.S. pat. **2,752,401** (1956 to Dow); by chlorination of 1,2-
dichloroethane: Conrad, U.S. pat. **2,725,412** (1955 to Ethyl
Corp.); Fox, U.S. pat. **2,846,484** (1958 to Monsanto). Toxicity: E. Browning, *Toxicity and Metabolism of Industrial
Solvents* (Elsevier, New York, 1965) pp 220-229.
  Nonflammable, heavy, mobile liquid. Sweetish, suffocating, chloroform-like odor. d_4^{25} 1.58658. mp $-44°$. bp_{760}
146.5°. n_D^{20} 1.49419. Very sparingly sol in water. At 25° one
gram dissolves in 350 ml H_2O. Misc with methanol, ethanol, benzene, ether, petr ether, carbon tetrachloride, chloroform, carbon disulfide, dimethylformamide, oils. Has the
highest solvent power of the chlorinated hydrocarbons.
LD_{50} orally in rats: 0.20 ml/kg, H. F. Smyth *et al., Am. Ind.
Hyg. Assoc. J.* **30**, 470 (1969).
  USE: Nonflammable solvent for fats, oils, waxes, resins,
cellulose acetate, rubber, copal, phosphorus, sulfur. As solvent in certain types of Friedel-Crafts reactions or phthalic
anhydride condensations. In the manuf of paint, varnish,
and rust removers. In soil sterilization and weed killer and
insecticide formulations. In the determination of theobromine in cacao. As immersion fluid in crystallography. In
the biological laboratory to produce pathological changes in
gastrointestinal tract, liver, and kidneys. Intermediate in the
manuf of trichloroethylene and other chlorinated hydrocarbons having two carbon atoms. *Caution:* Powerful narcotic;
liver poison. For symptoms *see* Carbon Tetrachloride.

9126. Tetrachloroethylene. *Tetrachloroethene;* perchloroethylene; ethylene tetrachloride; tetrachlorethylene; Nema;
Tetracap; Tetropil; Perclene; Ankilostin; Didakene; C_2Cl_4;
mol wt 165.85. C 14.48%, Cl 85.52%. $Cl_2C{=}CCl_2$. Prepd
by Faraday in 1821. Manuf by catalytic oxidation of 1,1,-
2,2-tetrachloroethane: Ellsworth, Vancamp, U.S. pat.
2,951,103 (1960 to Columbia-Southern Chem.); Feathers,
Rogerson, U.S. pat. **3,040,109** (1962 to Pittsburgh Plate
Glass); by catalytic chlorination of acetylene: Thermet,
Parvi, U.S. pat. **2,938,931** (1960 to Société d'électrochimie,
d'électrométallurgie et des aciéries électriques d'Ugine).
Review of mfg processes: Faith, Keyes & Clark's *Industrial
Chemicals*, F. A. Lowenheim, M. K. Moran, Eds. (Wiley-
Interscience, New York, 4th ed., 1975) pp 604-611. Physical properties: Mumford, Phillips, *J. Chem. Soc.* **1950**, 75.
Toxicity data: Dybing, *Acta Pharmacol. Toxicol.* **2**, 223
(1946); Lazarew, *Arch. Exp. Pathol. Pharmacol.* **141**, 19
(1929).
  Colorless, nonflammable liq; ethereal odor; d_4^{15} 1.6311; d_4^{20}
1.6230. bp 121°. mp $\sim{-}22°$. n_D^{20} 1.5055. Sol in about
10,000 vol water; misc with alcohol, ether, chloroform,
benzene. LD_{50} orally in mice: 8.85 g/kg (Dybing). LC for
mice in air: 5925 ppm (Lazarew).
  Human Toxicity: Narcotic in high concns. Defatting action on skin can lead to dermatitis.
  USE: Dry cleaning; degreasing metals; solvent.
  THERAP CAT: Anthelmintic (Nematodes, Trematodes).
  THERAP CAT (VET): Anthelmintic.

9127. 3,3′,4′,5-Tetrachlorosalicylanilide. *3,5-Dichloro-
N-(3,4-dichlorophenyl)-2-hydroxybenzamide;* Irgasan BS200.
$C_{13}H_7Cl_4NO_2$; mol wt 351.03. C 44.48%, H 2.01%, Cl
40.40%, N 3.99%, O 9.12%. Prepn of polyhalosalicylanil-
ides: Bindler, Model, U.S. pat. **2,703,332** (1955 to Geigy).
Bacteriostat in the manuf of thermoplastic articles: Teller,
U.S. pat. **3,005,720** (1961 to Weco Products).

Crystals, mp 161°. Fluoresces under ultraviolet light.
Practically insol in water. Sol in alkaline aq solns and in
solns of wetting agents. Sol in many organic solvents.
  USE: Bacteriostat in formulations of surgical soaps, laundry soaps, rinses, polishes, shampoos, deodorants. Also as
preservative in textile finishes, certain petroleum products,
cellulose esters, cutting oils, coolants. Banned by FDA from
use in cosmetics.

9128. Tetracosamethylhendecasiloxane. *Tetracosameth-
ylundecasiloxane.* $C_{24}H_{72}O_{10}Si_{11}$; mol wt 829.83. C 34.75%,
H 8.75%, O 19.28%, Si 37.21%. Prepd by reaction of hexamethyldisiloxane with octamethylcyclotetrasiloxane and sulfuric acid: Patnode, Wilcock, *J. Am. Chem. Soc.* **68**, 362
(1946); by cohydrolysis of ethoxytrimethylsilane and diethoxydimethylsilane: Hyde, U.S. pat. **2,457,677** (1948 to
Corning Glass).

$$(CH_3)_3Si{-}\left[O{-}\underset{\underset{CH_3}{|}}{\overset{\overset{CH_3}{|}}{Si}}{-}O\right]_9{-}Si(CH_3)_3$$

Liquid; $bp_{4.7}$ 201°; $bp_{0.5}$ 152°; d_4^{25} 0.9247; n_D^{20} 1.3994. Flash
pt 188.33°. Stable. Inert to most chemical reagents and
rubber. Maintains about the same viscosity over a wide
temperature range. Sol in benzene and the lighter hydrocarbons; slightly sol in alcohol and the heavy hydrocarbons.
  USE: As a basis for silicone oils or fluids designed to withstand extremes of temperature; as a foam suppressant in
petroleum lubricating oil.

9129. Tetracyanoethylene. *Ethenetetracarbonitrile;*
TCNE. C_6N_4; mol wt 128.10. C 56.26%, N 43.74%. (NC)$_2$-
C{=}C(CN)$_2$. Prepd by debromination of dibromomalononi-
trile with copper powder in boiling benzene: Cairns *et al., J.
Am. Chem. Soc.* **80**, 2775 (1958).
  Crystals, mp 200°. Begins to sublime at 120°.
  USE: In the synthesis of spiro compds, in modified Diels-
Alder reactions, as aromatizing agent: Longone, Smith,
Tetrahedron Letters **1962**, 205.

9130. Tetracycline. *4-(Dimethylamino)-1,4,4a,5,5a,6-
11,12a-octahydro-3,6,10,12,12a-pentahydroxy-6-methyl-
1,11-dioxo-2-naphthacenecarboxamide;* deschlorobiomycin;
tsiklomitsin; Abricycline; Achromycin; Agromicina; Ambramicina; Ambramycin; Bio-Tetra; Bristaciclina; Cefracycline suspension; Criseociclina; Cyclomycin; Democracin;
Hostacyclin; Omegamycin; Panmycin; Polycycline; Purocyclina; Sanclomycine; Steclin; Tetrabon; Tetracyn; Tetradecin. $C_{22}H_{24}N_2O_8$; mol wt 444.43. C 59.45%, H 5.44%, N
6.30%, O 28.80%. Antibiotic substance produced by *Streptomyces spp.* Prepn: Boothe *et al., J. Am. Chem. Soc.* **75**,
4621 (1953); Conover *et al.,* ibid. 4622; Conover, U.S. pat.
2,699,054 (1955). Production by *Streptomyces viridifaciens:*
Gourevitch, Lein, Heinemann *et al.,* U.S. pats. **2,712,517;
2,886,595** (1955, 1959 both to Bristol Labs.); by *S. aureofaciens:* Miller, Arishima, Sekizwa, U.S. pats. **3,005,023;
3,019,173** (1961, 1962 both to Am. Cyanamid). Purification:
Kaplan, Granatek, U.S. pat. **3,301,899** (1967 to Bristol-
Myers). Prepn of phosphate complex: Seiger, Weiden-
heimer, U.S. pat. **3,053,892** (1962 to Am. Cyanamid). Total
synthesis of tetracyclines: Boothe *et al., J. Am. Chem. Soc.*
81, 1006 (1959); Conover *et al.,* ibid. **84**, 3222 (1962).
Graphic outline of Woodward synthesis: *Chem. & Eng.
News* **40**, 36 (Oct. 8, 1962). Abs config: Dobrynin *et al.,
Tetrahedron Letters* **1962**, 901. Solubility studies: Weiss *et
al., Antibiot. & Chemother.* **7**, 374 (1957). Toxicity: E. I.
Goldenthal, *Toxicol. Appl. Pharmacol.* **18**, 185 (1971). Review: "Tetracycline" in *The Technology of the Tetracyclines,*
vol. I, R. C. Evans, Ed. (Quadrangle Press, New York,
1968) pp 209-426. Mechanism of action: A. Kaji, M. Ryoji
in *Antibiotics* vol. 5(pt. 1), F. E. Hahn, Ed. (Springer-Verlag,
New York, 1979) pp 304-328. Review of biosynthesis: C.
R. Hutchinson in *Antibiotics* vol. 4, J. W. Corcoran, Ed.
(Springer-Verlag, New York, 1981) pp 1-12.

Trihydrate, crystals. Swells at 165°. Dec 170-175°. Becomes anhydr by drying *in vacuo* at 60° for 8 hrs. $[\alpha]_D^{25}$ −257.9° (0.1N HCl); $[\alpha]_D^{25}$ −239° (methanol). uv max (0.1N HCl): 220, 268, 355 nm (ϵ 13000, 18040, 13320). pKa (50% aq DMF): 8.3, 10.2. Stable in neutral and in alkaline soln. Soly at about 28°: 1.7 mg/ml water; >20 mg/ml methanol. LD_{50} in rats, mice (mg/kg): 807, 808 orally (Goldenthal).

Hydrochloride, $C_{22}H_{25}ClN_2O_2$, *Achro, Achromycin V, Ala Tet, Ambracyn, Artomycin, Cefracycline tablets, Cyclopar, Diacycline, Dumocyclin, Helvecyclin, Imex, Mephacyclin, Partrex, Quadracycline, Quatrex, Remicyclin, Ricycline, Ro-cycline, Stilciclina, Subamycin, Supramycin, Sustamycin, Tefilin, Teline, Telotrex, Tetrabakat, Tetrabid, Tetrablet, Tetrachel, Tetracompren, Tetra-D, Tetrakap, Tetralution, Tetramavan, Tetramycin, Tetrosol, Tetra-Wedel, Topicycline, Totomycin, Triphacyclin, Unicin, Unimycin, Vetquamycin-324.* Crystals from butanol +HCl, dec 214°. $[\alpha]_D^{25}$ −257.9° (c = 0.5 in 0.1N HCl). Freely sol in water, sol in methanol, ethanol. Insol in ether, hydrocarbons. pH (2% aq soln): 2.1-2.3. LD_{50} orally in rats: 6443 mg/kg (Goldenthal).

Phosphate complex, *Panmycin Phosphate, Sumycin, Tetradecin Novum, Tetrex, Upcyclin.* (Elemental analysis shows about 45% C; 6-8% P; 4.8% N; 3.9% H; 0.7-1.4% NA). Yellow, odorless powder. Sparingly sol in water; slightly sol in ethanol.

Lauryl sulfate, *Lauracycline.*

THERAP CAT: Anti-amebic; antibacterial; antirickettsial.

THERAP CAT (VET): Antibacterial.

9131. Tetradecamethylhexasiloxane. $C_{14}H_{42}O_5Si_6$; mol wt 459.03. C 36.63%, H 9.22%, O 17.43%, Si 36.72%. Prepd by hydrolysis of dimethyldichlorosilane and trimethylchlorosilane: Patnode, Wilcock, *J. Am. Chem. Soc.* **68**, 358 (1946); by reaction of hexamethyldisiloxane with octamethylcyclotetrasiloxane and sulfuric acid: *eidem, ibid.* 362; Patnode, U.S. pat. **2,469,888** (1949 to General Electric); by cohydrolysis of ethoxytrimethylsilane and diethoxydimethylsilane: Hunter *et al., J. Am. Chem. Soc.* **68**, 2284 (1946); Hyde, U.S. pat. **2,457,677** (1948 to Corning Glass).

Liquid; bp_{20} 142°; d_4^{20} 0.8910; n_D^{20} 1.3948. Flash pt 118.33°. Freezes below −100°. Stable. Inert to most chemical reagents and rubber. Maintains about the same viscosity over a wide temp range. Sol in benzene and the lighter hydrocarbons; slightly sol in alcohol and the heavy hydrocarbons.

USE: As a basis for silicone oils or fluids designed to withstand extremes of temp; as a foam suppressant in petroleum lubricating oil.

9132. Tetradifon. *1,2,4-Trichloro-5-[(4-chlorophenyl)sulfonyl]benzene; p-chlorophenyl 2,4,5-trichlorophenyl sulfone;* 2,4,5,4'-tetrachlorodiphenyl sulfone; Duphar; Tedion; Tedion-V_{18}. $C_{12}H_6Cl_4O_2S$; mol wt 356.06. C 40.48%, H 1.70%, Cl 39.83%, O 8.99%, S 9.01%. Prepd by modified Friedel-Crafts reaction between 2,4,5-trichlorobenzene sulfonyl chloride and monochlorobenzene; also by Sandmeyer diazotization of 2,4,5-trichloro-4'-aminodiphenyl sulfone in the presence of CuCl: Meltzer, Huisman, U.S. pat. **2,812,281** (1957 to Phillips); Huisman *et al., Rec. Trav. Chim.* **77**, 103 (1958).

Crystals from benzene; mp 146.5-147.5°. Stable to concd and dil alkalies, mineral acids, high temp, and u.v. light. Soly data: Huisman *et al., loc. cit.* LD_{50} in rats: 556 mg/kg orally, Ben-dyke *et al., World Rev. Pest Contr.* **9**, 119 (1970).

USE: Acaricide. Ovicide on deciduous fruits, citrus, cotton and other crops.

9133. Tetraethylammonium Bromide. *N,N,N-Triethylethanaminium bromide;* TEAB; TMD-10; Etylon; Etambro; Sympatektoman; Tetranium. $C_8H_{20}BrN$; mol wt 210.16. C 45.72%, H 9.59%, Br 38.03%, N 6.67%. $(C_2H_5)_4NBr$. Ganglion blocking agent. Prepd from triethylamine and ethyl bromide: Hofmann, *Ann.* **78**, 263 (1851). Review of the pharmacology of the tetraethylammonium ion: Moe, Freyburger, *Pharmacol. Rev.* **2**, 61-95 (1950).

Deliquesc crystals. Freely sol in water, alc, chloroform, acetone. Slightly sol in benzene. pH of a 10% aq soln 6.5. The pH is not changed by heating for 28 hrs at 95°.

9134. Tetraethylammonium Chloride. Etamon chloride; T.E.A. chloride. $C_8H_{20}ClN$; mol wt 165.71. C 57.98%, H 12.17%, Cl 21.40%, N 8.45%. $(C_2H_5)_4NCl$. Ganglion blocking agent. *See* Tetraethylammonium Bromide.

Deliquescent crystals. d_4^{21} 1.0801. Freely sol in water, alcohol, chloroform, acetone; slightly sol in benzene. pH of 10% aq soln 6.48. The pH is not changed by heating for 28 hrs at 95°.

Tetrahydrate, monoclinic prismatic crystals. mp 37.5°. d 1.084.

9135. Tetraethylammonium Hydroxide. $C_8H_{21}NO$; mol wt 147.26. C 65.25%, H 14.37%, N 9.51%, O 10.86%. $(C_2H_5)_4NOH$. Made from the corresp halide by treating with silver oxide or with a soln of potassium hydroxide in methanol.

Marketed as an aq soln. A 10% soln has a d_4^{25} of about 1.01. The free base is known only in soln or as hydrates; tetrahydrate, mp 49-50°; hexahydrate, mp 55°. Dec on boiling. It is a very strong base readily absorbing CO_2 from the air. The aq soln is colorless, odorless, bitter, caustic, strongly alkaline, and imparts a soapy feel to the skin. *Keep well closed.*

9136. Tetraethyllead. *Tetraethylplumbane;* lead tetraethyl; TEL. $C_8H_{20}Pb$; mol wt 323.45. C 29.70%, H 6.23%, Pb 64.06%. $Pb(C_2H_5)_4$. Prepd by the action of $PbCl_2$ on zinc ethyl or on a Grignard reagent; by heating C_2H_5Cl and sodium-lead alloy in an autoclave. The production from lead, ethylene, and hydrogen using triethylaluminum as intermediate was first described by K. Ziegler at the 14th International Congress of Pure and Applied Chemistry (July 1955): *Chem. & Eng. News* **33**, 3486 (1955). Alternate synthesis using nonhalide compds: Pearson *et al., Advances in Chemistry Series* **23**, 299-305 (1959).

Colorless liq; burns with an orange-colored flame with green margin. *Extremely poisonous!* d^{20} 1.653. bp about 200° also stated as 227.7° with decompn. n_D^{20} 1.5198. Practically insoluble in water; soluble in benzene, petr ether, gasoline, slightly in alcohol. LD_{50} orally in rats: 12.3 mg/kg, Schroeder *et al., Experientia* **28**, 923 (1972).

USE: As a gasoline additive to prevent "knocking" in motors. *See also* Milde, Beatty, "Chemical Reactions of Tetraethyllead" in *Advances in Chemistry Series* **23**, 306-318 (1959). *Caution:* Acute or chronic poisoning may occur if inhaled or absorbed through skin. See E. Browning, *Toxicity of Industrial Metals* (Appleton-Century-Crofts, London, 2nd ed., 1969) pp 192-199.

9137. N,N,N',N'-Tetraethylphthalamide. *N,N,N',N'-Tetraethyl-1,2-benzenedicarboxamide;* orthophthalic acid didiethylamide; *o*-phthalic acid bis[diethylamide]; tetraethylbis(phthalamide); Analetil; Neo-Cardiamine; Neospiran; Unispiran. $C_{16}H_{24}N_2O_2$; mol wt 276.37. C 69.53%, H 8.75%, N 10.14%, O 11.58%. Prepd by treating phthalyl chloride with diethylamine: **Fr. pat. 785,428; Brit. pat.**

443,396; U.S. pat. 2,057,145 (1936 to Chem. Fabrik Grünau); by heating sodium phthalate with diethylamine phosphate: Fr. pat. 866,229 (1941 to Corbière).

Crystals, mp 39°. bp 175-180°. Soluble in water, physiol saline.

THERAP CAT: Analeptic.

9138. Tetraethyl Pyrophosphate. *Diphosphoric acid tetraethyl ester; pyrophosphoric acid tetraethyl ester;* bis-*O,O*-diethylphosphoric anhydride; TEPP; Bladan; Nifos T; Kilmite 40; Vapotone; Tetron; Killax; Mortopal. $C_8H_{20}O_7P_2$; mol wt 290.20. C 33.11%, H 6.95%, P 21.35%, O 38.59%. Prepd commercially by controlled hydrolysis of *O,O*-diethylphosphoric acid chloride: Kosolapoff, U.S. pat. 2,479,939 (1947 to Monsanto); Toy, *J. Am. Chem. Soc.* **70**, 3882 (1948). Chemical history and comparison of various syntheses: G. Schrader, *Die Entwicklung Neuer Insektizider Phosphorsäure-Ester* (Verlag Chemie, Weinheim, 3rd ed., 1963) pp 68-79.

Mobile liquid. Agreeable odor. Hygroscopic. d_4^{20} 1.185. Thermal decompn range 170-213° with copious formn of ethylene. $bp_{0.05}$ 82°; $bp_{1.0}$ 124°; $bp_{2.3}$ 138°. Vapor pressure at 30° = 4.7 × 10⁻⁴ mm Hg. n_D^{20} 1.4196. Misc with water, but quickly hydrolyzed by it (half life at 25° about 7 hrs in a 50 v/v mixt.). Also misc with acetone, methanol, ethanol, benzene, chloroform, carbon tetrachloride, glycerol, ethylene glycol, propylene glycol, toluene, xylene. Not misc with petr ether, kerosene, other petr oils. LD_{50} orally in male rats: 1.1 mg/kg, T. B. Gaines, *Toxicol. Appl. Pharmacol.* **14**, 515 (1969).

USE: Insecticide. *Caution:* Cholinesterase inhibitor.

9139. 2,2,3,3-Tetrafluoro-1-propanol. C_3-Fluoroalcohol. $C_3H_4F_4O$; mol wt 132.06. C 27.28%, H 3.05%, F 57.55%, O 12.12%. $HCF_2CF_2CH_2OH$. Prepn: Bestian, Rehn, Ger. pat. 1,007,771 (1957 to Hoechst).

Liquid. d_4^{20} 1.4853. mp −15°. bp_{760} 109-110°. n_D^{20} 1.3197. Surface tension at 20° = 27.6 dyn/cm.

p-Nitrobenzoate, mp 47°.

USE: To introduce fluoroalkyl groups into an organic molecule. Proposed intermediate for plastics, surface active agents, lubricants, elastomers.

9140. Tetraglycine Hydroperiodide. Globaline; Potable Aqua. $C_{16}H_{42}I_7N_8O_{16}$; mol wt 1490.95. C 12.89%, H 2.84%, I 59.58%, N 7.52%, O 17.17%. $2[(NH_2CH_2COOH)_4HI]·2½I_2$. Prepn: Frost, Eddy, *J. Am. Chem. Soc.* **74**, 1346 (1952); Morris et al., *Ind. Eng. Chem.* **45**, 1013 (1953).

Flat needles with brassy-bronze metallic luster in reflected light, dec 162-167°. Soly in water at 25° = 380 g/l.

USE: Decontamination of drinking water in emergencies. Used in amounts sufficient to yield 8 ppm of active iodine. A tablet contg 20 mg plus 96 mg $Na_2H_2PO_7$ plus 4 mg talc will decontaminate one quart of water. Such tablets after 7 days' storage at 60° retained 60% of their original active iodine. Less stable than aluminum hexaurea sulfate triiodide.

9141. Tetraglyme. *2,5,8,11,14-Pentaoxapentadecane;* tetraethylene glycol dimethyl ether; dimethoxytetraethylene glycol. $C_{10}H_{22}O_5$; mol wt 222.28. C 54.03%, H 9.98%, O 35.99%. $CH_3O(CH_2CH_2O)_4CH_3$. Prepd from ethylene glycol methyl ether and 2,2'-dichlorodiethyl ether: Zellhoefer, U.S. pat. 2,111,234 (1935); *Ind. Eng. Chem.* **29**, 550 (1937). Purification: Vogel, *J. Chem. Soc.* **1948**, 618.

Liquid. d_4^{20} 1.0087; d_4^{86} 0.9514. mp −27°. bp_{760} 275.3°; bp_2 118°. n_D^{20} 1.4325. Sol in water. Misc with hydrocarbon solvents. LD_{50} orally in rats: 5.14 g/kg, H. F. Smyth et al., *J. Ind. Hyg. Toxicol.* **23**, 259 (1941).

USE: Solvent.

9142. Tetrahydrocannabinols. *Tetrahydro-6,6,9-trimethyl-3-pentyl-6H-dibenzo[b,d]pyran-1-ol.* $C_{21}H_{30}O_2$; mol wt 314.45. C 80.21%, H 9.62%, O 10.18%. Active constituents of marihuana (hashish). The Δ^1-3,4-*trans* isomer, also referred to as Δ^9-THC, is the only major active constituent in hashish; the Δ^6-3,4-*trans* isomer, although physiologically active, represents no more than 1%: R. Mechoulam et al., *Science* **169**, 611 (1970). Isoln of Δ^1-3,4-*trans*-form from marihuana: Gaoni, Mechoulam, *J. Am. Chem. Soc.* **86**, 1646 (1964). Isoln of Δ^6-3,4-*trans*-form: Hively et al., ibid. **88**, 1832 (1966). Synthesis of dl-Δ^1-3,4-*trans*-form: Fahrenholtz et al., ibid. **89**, 5934 (1967); Razden et al., ibid. **96**, 5860 (1974); eidem, *Experientia* **31**, 16 (1975); of dl-Δ^6-3,4-*trans*-form: Taylor et al., *J. Am. Chem. Soc.* **88**, 367 (1966). Stereospecific synthesis of (−)-Δ^1-3,4-*trans*-form and (−)-Δ^6-3,4-*trans*-form: Mechoulam et al., ibid. **89**, 4552 (1967). Abs config of naturally occurring (−) Δ^1-3,4-*trans* form: Mechoulam, Gaoni, *Tetrahedron Letters* **1967**, 1109. IR, NMR, mass spec data for Δ^9THC: Petrzilka, Sikemeier, *Helv. Chim. Acta* **50**, 2111 (1967); for Δ^6 THC: eidem, ibid. 1416. Metabolism in mice: H. D. Christensen et al., *Science* **172**, 165 (1971); in man: M. M. Halldin et al., *Arzneimittel-Forsch.* **32**, 764 (1982). Clinical studies of Δ^9-THC as an anti-emetic in cancer patients: S. E. Sallan et al., *N. Engl. J. Med.* **302**, 135 (1980); A. E. Chang et al., *Cancer* **47**, 1746 (1981); D. S. Poster et al., *J. Am. Med. Assoc.* **245**, 2047 (1981). Topical use in hypertensive glaucomas: J. C. Merrit et al., *J. Pharm. Pharmacol.* **33**, 40 (1981). Effects of long-term THC treatment on the menstrual cycle of rhesus monkeys: C. G. Smith et al., *Science* **219**, 1453 (1983). Toxicity studies: R. N. Phillips et al., *Proc. Soc. Exp. Biol. Med.* **136**, 260 (1971); H. Rosenkranz et al., *Toxicol. Appl. Pharmacol.* **28**, 18 (1974); H. Yoshimura et al., *J. Med. Chem.* **21**, 1079 (1978). Review of analytical methods: L. Vollner et al., *Reg. Toxicol. Pharmacol.* **6**, 348-358 (1986). *See also* Cannabis and Hashish.

alternate numbering systems

$$\Delta^1\text{-THC} = \Delta^9\text{-THC}$$
$$\Delta^6\text{-THC} = \Delta^{1(6)}\text{-THC} = \Delta^8\text{-THC}$$

(−)-Δ^1-3,4-*trans*-Form, Δ^1-THC, Δ^9-THC, QCD 84924, *dronabinol, Marinol.* $bp_{0.02}$ 200°. $[\alpha]_D^{20}$ −150.5° (c = 0.53 in $CHCl_3$). uv max (ethanol): 283, 276 nm (log ε 3.21, 3.20). LD_{50} in Fischer rats (mg/kg): 1270 (males), 730 (females) orally, sesame oil vehicle; 800 (males) orally, sesame oil, 1% polysorbate 80, saline emulsion; 40 (males, females) i.v.; 105.7 (males, females) inhalation, corrected for particulate losses and pulmonary absorption to 42 mg/kg (Rosenkranz).

(−)-Δ^6-3,4-*trans*-Form, Δ^6-THC, Δ^8-THC. $bp_{0.001}$ 200°. $[\alpha]_D^{18}$ −264° (c = 0.11 in ethanol). uv max (ethanol): 282, 275 nm (log ε 3.22, 3.22); shoulder at 230 nm (log ε 4.07). LD_{50} i.v. in mice: 27.5 mg/kg (Yoshimura).

Caution: This is a controlled substance (hallucinogen) listed in the U.S. Code of Federal Regulations, Title 21 Part 1308.11 (1985).

THERAP CAT: Anti-emetic.

9143. Tetrahydrocortisone. *3α,17α,21-Trihydroxy-5β-pregnane-11,20-dione;* 3α,17α,21-pregnanetriol-11,20-dione; 11,20-pregnanedione-3α,17α,21-triol; THE. $C_{21}H_{32}O_5$; mol wt 364.47. C 69.20%, H 8.85%, O 21.95%. A normal mammalian metabolite of cortisone: Schneider, *J. Biol. Chem.* **183**, 365 (1950). Prepn by microbial reduction using a *Streptomyces* sp.: Barkemeyer et al., *Appl. Microbiol.* **8**, 237

(1960). Prepn of the triacetate: **Brit.** pat. **737,291** (1955 to Merck & Co.). Prepn of the 21-acetate: Julian *et al.*, **U.S.** pat. **2,752,339** (1956 to Glidden).

Crystals from ethyl acetate, mp 190°. $[\alpha]_D^{25}$ +85.5° (abs ethanol).

21-Acetate, $C_{23}H_{34}O_6$, crystals from acetone, mp 227°.

Triacetate, $C_{27}H_{38}O_8$, crystals from methanol, mp 150-152°; solvated crystals from ethyl acetate, mp 112-118°.

9144. Tetrahydrofuran. Diethylene oxide; tetramethylene oxide. C_4H_8O; mol wt 72.10. C 66.63%, H 11.18%, O 22.19%. Prepn from 1,4-butanediol: Schmoyer, Case, *Nature* **187**, 592 (1960). Manuf by catalytic hydrogenation of maleic anhydride: Gilbert, Howk, U.S. pat. **2,772,293** (1956 to du Pont); of furan: Banford, Manes, U.S. pat. **2,846,449** (1958 to du Pont); Manly, U.S. pat. **3,021,342** (1962 to Quaker Oats). Stabilization to prevent excessive peroxide formation on storage with 0.05-1.0% *p*-cresol, 0.05-0.1% hydroquinone, or less than 0.01-0.1% 4,4'-thiobis(6-*tert*-butyl-*m*-cresol): Bordner, Hinegardner, and Campbell, **U.S.** pats. **2,489,260, 2,525,410,** and **3,029,257** (1949, 1950, and 1962, all to du Pont).

Liquid. Ether-like odor. mp −108.5°. d_4^{20} 0.8892. bp_{760} 66°; bp_{176} 25°. Flash pt 1°F. n_D^{20} 1.4070. Dipole moment: 1.70. uv cut-off for spectro grade: 220 nm. Miscible with water, alcohols, ketones, esters, ethers, and hydrocarbons.

Caution: Distil only in presence of a reducing agent, such as ferrous sulfate; peroxide explosions have occurred: *Angew. Chem.* **68**, 182 (1956).

USE: Solvent for high polymers, esp polyvinyl chloride. As reaction medium for Grignard and metal hydride reactions. In the synthesis of butyrolactone, succinic acid, 1,4-butanediol diacetate. Solvent in histological techniques. May be used under Federal Food, Drug & Cosmetic Act for fabrication of articles for packaging, transporting, or storing of foods if residual amount does not exceed 1.5% of the film: *Fed. Reg.* **27**, 3919 (Apr. 25, 1962). *Caution:* Irritating to skin, eyes, mucous membranes. Narcotic in high concns.

9145. 2,5-Tetrahydrofurandimethanol. THF glycol; 2,5-bis(hydroxymethyl)tetrahydrofuran. $C_6H_{12}O_3$; mol wt 132.16. C 54.53%, H 9.15%, O 36.32%. Prepd from diallyl by oxidation with perbenzoic acid in chloroform, boiling with dil sulfuric acid and hydrolyzing the reaction product with KOH soln: Böeseken, *Rec. Trav. Chim.* **45**, 838 (1926); by Raney nickel reduction of 5-hydroxymethylfurfural or of dimethyl furan-2,5-dicarboxylate: Cope, Baxter, *J. Am. Chem. Soc.* **77**, 393 (1955). The usual form obtained is the *cis* form, described here.

Hygroscopic liquid. Faint odor. *Avoid contact with eyes.* d_0^0 1.1719; d_4^{25} 1.1542; d_4^{50} 1.1359. mp below −50°. bp_{760} 265°; bp_{96} 200°; bp_{11} 155°; $bp_{0.25}$ 105°. n_D^{25} 1.4766. Viscosity: 1926 cp at 0°; 225 at 25°; 51.9 at 50°. Coefficient of expansion: 0.00063 per °C at 25°. Specific heat 0.5 cal/g/°C. Heat of vaporization 115 cal/g. Miscible with water, methanol, ethanol, acetone, benzene, methyl acetate, methyl ethyl

ketone, chloroform. Moderately sol in ether, toluene. Almost insol in heptane, methylcyclohexane.

USE: Solvent, softener, humectant. In the synthesis of plasticizers, resins, surfactants, agricultural chemicals. *Caution:* Highly irritating to eyes, skin, mucous membranes.

9146. Tetrahydrofurfuryl Alcohol. *Tetrahydro-2-furanmethanol;* tetrahydro-2-furancarbinol; tetrahydro-2-furylmethanol; THFA. $C_5H_{10}O_2$; mol wt 102.13. C 58.80%, H 9.87%, O 31.33%. Prepn by catalytic hydrogenation of furfuryl alcohol: Lukes, Nelson, *J. Org. Chem.* **21**, 1096 (1956). Manuf by catalytic hydrogenation of furfural or furfuryl alcohol: Dunlop, Schegulla, U.S. pat. **2,838,523** (1958 to Quaker Oats). Occurs in two isomeric forms: D-isomer (levorotatory), L-isomer (dextrorotatory). Abs config: Gagnaire, Butt, *Bull. Soc. Chim. France* **1961**, 312; Hartman, Barker, *J. Org. Chem.* **29**, 873 (1964). *Review:* A. P. Dunlop, F. N. Peters, *The Furans* (Reinhold, New York, 1953).

Liquid. Hygroscopic. d_{20}^{20} 1.0543; d_{24}^{24} 1.0511; d_{31}^{31} 1.0450. Melts below −80°. bp_{760} 178°. n_D^{20} 1.4520; n_D^{25} 1.4499. Flash pt, open cup: 183°F (84°C). Flammability in air: Upper limit 9.7% by vol, lower limit 1.5% by vol. Heat capacity at 30-37°: 0.432 cal/g/°C. Heat of combustion at constant vol: 708.6 cal/g mole. Viscosity at 20°: 6.24 cp. Surface tension at 25°: 37 dyn/cm. Octane no. 82.5. Evaporation rate: 7 (*n*-butyl acetate = 100). Kauributanol value 71.5. Dilution ratio (lacquer ingredients): 4.5. Dielectric constant at 23°: 13.6. Miscible with water, alcohol, ether, acetone, chloroform, benzene.

L-Isomer, prepn see Hartman, Barker, *loc. cit.* $[\alpha]_D^{24}$ +14.9° (c = 5.0 in nitromethane).

USE: Solvent for fats, waxes, resins. In organic synthesis: Undergoes the reactions of a primary alcohol, while the ring exhibits characteristics of a saturated cyclic ether. *Caution:* Moderately irritating to skin, mucous membranes.

9147. Tetrahydropalmatine. *5,8,13,13a-Tetrahydro-2,3,-9,10-tetramethoxy-6H-dibenzo[a,g]quinolizine;* 2,3,9,10-tetramethoxyberbine; 2,3,9,10-tetramethoxydibenzo[a,g]quinolizidine; hyndarin. $C_{21}H_{25}NO_4$; mol wt 355.42. C 70.96%, H 7.09%, N 3.94%, O 18.01%. Synthesis of *dl*-form: Haworth *et al.*, *J. Chem. Soc.* **1927**, 548; Bradsher, Dutta, *J. Org. Chem.* **26**, 2231 (1961); T. Kametani, M. Ihara, *J. Chem. Soc. (C)* **1967**, 530; G. D. Pandey, K. P.Tiwari, *Indian J. Chem.* **18B**, 545 (1979); Z. Kiparissides *et al.*, *Can. J. Chem.* **58**, 2770 (1980); N. S. Narasimhan *et al.*, *Tetrahedron Letters* **22**, 2797 (1981). Biosynthesis: D. S. Bhakuni *et al.*, *Tetrahedron* **36**, 2491 (1980). *See also* Palmatine. Both optically active forms are found in plants.

l-Form, *caseanine, gindarine, rotundine.* Crystals from dil methanol, mp 147°. $[\alpha]_D^{20}$ −291° (c = 0.8 in 95% alcohol). Also a hydrate, mp 115° (effervescence).

Hydrochloride, $C_{21}H_{25}NO_4 \cdot HCl$, crystals.

d-Form, crystals from methanol, mp 141-142° (evac tube). $[\alpha]_D^{14}$ +292° (c = 0.8 in 95% alcohol).

Hydrochloride, crystals, mp 266°.

dl-Form, crystals from methanol upon addition of water. mp 148-149°.

Hydrochloride, needles from methanol, mp 215-216°.

9148. Tetrahydropapaveroline. *1-[(3,4-Dihydroxyphenyl)methyl]-1,2,3,4-tetrahydro-6,7-isoquinolinediol;* 1-(3,4-dihydroxybenzyl)-1,2,3,4-tetrahydro-6,7-isoquinolinediol; 6,7-

dihydroxy-1-(3,4-dihydroxybenzyl)-1,2,3,4-tetrahydroiso-quinoline; norlaudanosoline; THP. $C_{16}H_{17}NO_4$; mol wt 287.33. C 66.88%, H 5.97%, N 4.88%, O 22.27%. Alkaloid deriv of dopamine and a biosynthetic precursor of morphine, *q.v.* Prepn of the racemic hydrochloride: F. L. Pyman, *J. Chem. Soc.* **95**, 1610 (1909); of the (+)- and (−)-form hydrochlorides: S. Teitel *et al., J. Med. Chem.* **15**, 845 (1972). Chromatographic study: K. D. McMurtrey *et al., J. Liq. Chromatog.* **3**, 663 (1980). Possible role of THP in the biochemical mediation of alcohol addiction: V. E. Davis, M. J. Walsh, *Science* **167**, 1005 (1970); G. Cohen, M. Collins, *ibid.* 1749; R. D. Meyers, C. L. Melchior, *ibid.* **196**, 554 (1977); M. Sandler *et al., Prog. Clin. Biol. Res.* **90**, 215 (1982). Biosynthetic study: D. K. Choudhary, B. L. Kaul, *Indian Drugs* **19**, 229 (1982). *In vivo* and *in vitro* effects on rat pituitary function: D. R. Britton *et al., Biochem. Pharmacol.* **31**, 1205 (1982). Effect on dopaminergic neurons: I. S. Hoffman, L. X. Cubeddu, *J. Pharmacol. Exp. Ther.* **220**, 16 (1982). Comparison to opioid effects in brain regions: G. R. Siggins *et al., Prog. Clin. Biol. Res.* **90**, 275 (1982).

(±)-Form hydrochloride, $C_{16}H_{18}ClNO_4$, colorless microscopic prisms, mp 291-293° (dec). Practically insol in water, alc.

(−)-Form hydrochloride, cryst from 6*N* HCl, mp 285-286°. $[\alpha]_D^{25}$ −32.4° (c = 1 in water). uv max (ethanol): 230, 286 nm (ε 11100, 6700).

(+)-Form hydrochloride, cryst, mp 285-286°. $[\alpha]_D^{25}$ +32.1° (c = 1 in water).

USE: As a research tool in neurological biochemistry.

9149. Tetrahydropyran. Pentamethylene oxide. $C_5H_{10}O$; mol wt 86.13. C 69.72%, H 11.70%, O 18.58%. Prepd by hydrogenation of dihydropyran (from furfural): Paul, *Bull. Soc. Chim. France* [4] **53**, 1489 (1933); Andrus, Johnson, *Org. Syn.* **23**, 90 (1943); Cass, *Ind. Eng. Chem.* **40**, 219 (1948).

Mobile, flammable liquid. Pungent, sweetish odor. d_4^{20} 0.8814. mp −49.2°. bp 88°. n_D^{20} 1.4211. Dipole moment 1.87. Flash pt −4.0°F. Azeotrope with water, bp 71°, contains 8.5% H_2O. Sol in water [relative solubilities: Bennett, Philip, *J. Chem. Soc.* **1928**, (1939)]. Miscible with alcohol, ether, many other organic solvents. Forms peroxides on exposure to air. All technical tetrahydropyran is stabilized against peroxide formation.

9150. Tetrahydrozoline. *4,5-Dihydro-2-(1,2,3,4-tetra-hydro-1-naphthalenyl)-1H-imidazole; 2-(1,2,3,4-tetrahydro-1-naphthyl)-2-imidazoline; tetryzoline.* $C_{13}H_{16}N_2$; mol wt 200.27. C 77.96%, H 8.05%, N 13.99%. Prepn from ethylenediamine, ethylenediamine hydrochloride, and methyl 1,2,3,4-tetrahydro-1-naphthoate: Synerholm *et al., U.S. pat.* **2,731,471** (1956 to Sahyun Labs.). Use as potentiator for veterinary depressants: Gardocki *et al., U.S. pat.* **2,842,-478** (1958 to Pfizer).

Hydrochloride, $C_{13}H_{17}ClN_2$, *Nasan, Rhinopront, Tina-rhinin, Tyzanol, Tyzine, Visine, Yxin.* Crystals from alcohol, dec 256-257°. uv max: 264.5, 271.5 nm ($A_{1cm}^{1\%}$ 17.5, 15.5). Freely sol in water, alcohol. Very slightly sol in chloroform. Practically insol in ether. pH of a 1% aq soln 5.0 to 6.5.

THERAP CAT: Adrenergic (vasoconstrictor); nasal decongestant.

9151. Tetraiodoethylene. *Tetraiodoethene;* diiodoform; ethylene periodide; ethylene tetraiodide. C_2I_4; mol wt 531.70. C 4.52%, I 95.48%. $I_2C=CI_2$. Prepd by the action of iodine on diiodoacetylene obtained from calcium carbide and iodine. Crystal and molecular structure: B. C. Haywood, R. Shirley, *Acta Crystallogr.* **B33**, 1765 (1977).

Light yellow, heavy, small, practically odorless crystals; characteristic odor. On exposure to light turns brown. d 2.98. mp 187°. Insol in water; sol in benzene, chloroform, toluene, CS_2, slightly in ether. *Protect from light.*

9152. Tetralin®. *1,2,3,4-Tetrahydronaphthalene;* Tetra-nap. $C_{10}H_{12}$; mol wt 132.20. C 90.85%, H 9.15%. Prepd by catalytic hydrogenation of purified naphthalene. *See* ref *under* Decalin.

Liquid. Odor resembling that of a mixture of benzene and menthol. d_4^{20} 0.9702; d_4^{25} 0.9662. Volatile with steam; mp −31.0°; bp_{760} 207.2°; bp_{400} 181.8°; bp_{200} 157.2°; bp_{100} 135.3°; bp_{60} 121.3°; bp_{40} 110.4°; bp_{20} 93.8°; bp_{10} 79.0°; bp_5 65.3°; $bp_{1.0}$ 38.0°. n_D^{20} 1.54135; n_D^{25} 1.53919. Flash pt, open cup 171°F (77°C), closed cup 180°F (82°C). Insol in water; miscible with ethanol, butanol, acetone, benzene, ether, chloroform, petr ether, Decalin; soluble in methanol: 50.6% w/w. Prolonged, intimate contact with air may cause the formn of tetralin peroxide which may cause explosion of tetralin distn residues. Peroxide formn is prevented by the addn of an antioxidant, such as hydroquinone. LD_{50} orally in rats: 2.86 g/kg, Smyth *et al., Arch. Ind. Hyg. Occup. Med.* **4**, 119 (1951).

USE: Degreasing agent. Solvent for naphthalene, fats, resins, oils, waxes, used instead of turpentine in lacquers, shoe polishes, floor waxes. *Caution:* Irritating to skin, eyes, mucous membranes, and, in high concns, narcotic. In exptl animals has produced cataracts: E. Browning, *Toxicity and Metabolism of Industrial Solvents* (Elsevier, New York, 1965) pp 119-124.

9153. Tetralol. *1,2,3,4-Tetrahydro-2-naphthol; ac-tetra-hydro-β-naphthol; ac-β-tetralol.* $C_{10}H_{12}O$; mol wt 148.20. C 81.04%, H 8.16%, O 10.80%. Prepd by hydrogenation of 2-naphthol in the presence of a palladium catalyst: Foreman, Stork, *U.S. pat.* **2,526,859** (1950 to Lakeside Labs.); by reduction of 2-naphthol with sodium: Hueckel *et al., Ann.* **645**, 162 (1961).

Liquid, bp_{12} 140°; mp 15.5°. Crystallizes on prolonged storage at −50° and then at −15°. LD_{50} orally in rats: 1.0 ml/kg, Draize *et al., J. Pharmacol. Exp. Ther.* **93**, 26 (1948).

p-Tosylate, $C_{10}H_{11}O_3SC_6H_4CH_3$, crystals from petr ether + benzene, mp 86°.

Methyl ether, $C_{10}H_{11}OCH_3$, bp_{11} 114.5°, d_4^{20} 1.0239, n_D^{20} 1.5326.

9154. Tetramethrin. *2,2-Dimethyl-3-(2-methyl-1-pro-penyl)cyclopropanecarboxylic acid (1,3,4,5,6,7-hexahydro-1,3-dioxo-2H-isoindol-2-yl)methyl ester; 2,2-dimethyl-3-(2-methylpropenyl)cyclopropanecarboxylic acid ester with N-(hy-droxymethyl)-1-cyclohexene-1,2-dicarboximide; N-(3,4,5,6-tetrahydrophthalimide)methyl-cis,trans-chrysanthemate; N-(chrysanthemoxymethyl)-1-cyclohexene-1,2-dicarbox-imide; phthalthrin; FMC-9260; SP 1103; Neo-Pynamin.* $C_{19}H_{25}NO_4$; mol wt 331.42. C 68.86%, H 7.60%, N 4.23%, O 19.31%. Potent synthetic pyrethroid insecticide. Prepn of racemic mixture: T. Kato *et al.*, **Japan. pat. 65 8535** corresp to **U.S. pat. 3,268,398** (1965, 1966 to Sumitomo). Activity: *eidem, Agr. Biol. Chem.* **28**, 914 (1965). Comparative activity of isomers: Y. Okuno *et al.*, **Ger. pat. 2,348,930** corresp to **U.S. pat. 3,934,023** (1973, 1976 to Sumitomo). Metabolism: J. Miyamoto *et al.*, *Agr. Biol. Chem.* **32**, 628 (1968). Photodecompn: Y.-L. Chen, J. E. Casida, *J. Agr. Food Chem.* **17**, 208 (1969).

The commercial product is a mixture of isomers. White cryst solid, mp 65-80°. d_{20}^{20} 1.108; $n_D^{21.5}$ 1.5175. LD_{50} orally in mice: 1000 mg/kg, T. Kato *et al.*, *Agr. Biol. Chem.* **28**, 914 (1965).

USE: Insecticide.

9155. Tetramethylammonium Hydroxide. *N,N,N-Trimethylmethanaminium hydroxide.* $C_4H_{13}NO$; mol wt 91.15. C 52.70%, H 14.38%, N 15.37%, O 17.55%. $(CH_3)_4NOH$.
Usually marketed in 10% aq soln. Strong ammonia-like odor. d_4^{25} about 1.00. The free base is known only in soln or as a solid pentahydrate, forming colorless, deliquesc needles, mp 63°. On distillation, it dec to trimethylamine and CH_3OH. It is a very strong base, readily absorbing CO_2 from the air. *Keep well closed.*

9156. Tetramethylammonium Iodide. $C_4H_{12}IN$; mol wt 201.06. C 23.89%, H 6.02%, I 63.14%, N 6.97%. $(CH_3)_4NI$.
Pale yellow crystals. Begins to dec at about 230°. d 1.84. Sparingly sol in water, freely in abs alcohol; insol in chloroform, ether.
USE: Emergency disinfection of drinking water. Required dosage: 8 ppm of iodine.

9157. Tetramethyldiaminobutane. *N,N,N',N'-Tetramethyl-1,4-butanediamine; N,N,N',N'*-tetramethylputrescine. $C_8H_{20}N_2$; mol wt 144.26. C 66.60%, H 13.98%, N 19.42%. $(CH_3)_2N(CH_2)_4N(CH_3)_2$. From root and herb of *Hyoscyamus reticulatus* L. and *H. muticus* L., *Solanaceae*: Willstätter, Heuber, *Ber.* **40**, 3869 (1907); Konowalowa, Magidson, *Arch. Pharm.* **266**, 449 (1928). Synthesis: Lunsford *et al.*, *J. Org. Chem.* **22**, 1225 (1957); Solov'ev, Skoldinov, *Zh. Obshch. Khim.* **33**, 1821 (1963), *C.A.* **59**, 7360f (1963).
White crystals; penetrating odor and sharp, scratching taste. bp 169°; bp_{28} 78-80°; bp_7 43°. n_D^{20} 1.4280. d^{20} 0.7861. Volatile with steam. Sol in water, alc, ether.
Dihydrochloride, $C_8H_{20}N_2 \cdot 2HCl$, crystals from alc, mp 273°.
Dipicrate, $C_8H_{20}N_2 \cdot 2C_6H_3N_3O_7$, needles from hot water, mp 203-205°.

9158. Tetramethylenedisulfotetramine. *2,6-Dithia-1,3,-5,7-tetraazatricyclo[3.3.1.1³,⁷]decane 2,2,6,6-tetraoxide; 2,6-dithia-1,3,5,7-tetraazaadamantane 2,2,6,6-tetraoxide.* C_4H_8-$N_4O_4S_2$; mol wt 240.27. C 20.00%, H 3.35%, N 23.32%, O 26.64%, S 26.69%. Prepd from sulfamide, $H_2NSO_2NH_2$ and formaldehyde in 60% H_2SO_4: Hecht, Henecka, *Angew. Chem.* **61**, 365 (1949). Toxicity: Hagen, *Deut. Med. Wochenschr.* **75**, 183 (1950).

Cubic crystals from acetone, dec 255-260°. *Violent convulsive poison!* Stable to acids and alkalies in dilutions up to 0.1N. Dec upon prolonged boiling of aq solns. Soly in

water about 0.25 mg/ml. Slightly sol in acetone. Insol in methanol, ethanol. LD in mice (mg/kg): 0.20 orally or s.c. (Hagen).

9159. Tetramethyl-*p*-phenylenediamine. *N,N,N',N'-Tetramethyl-1,4-benzenediamine;* Wurster's reagent; Wurster's blue. $C_{10}H_{16}N_2$; mol wt 164.24. C 73.12%, H 9.82%, N 17.06%. Prepn: Meyer, *Ber.* **36**, 2979 (1903); Cox, Smith, *J. Org. Chem.* **29**, 488 (1964).

Crystals from petr ether, mp 51-52°. bp 260°. Slightly sol in cold water; more sol in hot water; freely sol in alc, chloroform, ether, petr ether.
USE: In the form of the hydrochloride as a reagent in analytical chemistry.

9160. Tetramethylurea. TMU; Temur. $C_5H_{12}N_2O$; mol wt 116.16. C 51.70%, H 10.41%, N 24.12%, O 13.77%. $[(CH_3)_2N]_2CO$. Prepn: Lawson, Croom, *J. Org. Chem.* **28**, 232 (1963). Review of prepn, manuf, properties and use as solvent and as reagent: Lüttringhaus, Dirksen, *Angew. Chem. Int. Ed.* **3**, 260 (1964).
Liquid with faint, pleasant odor, bp 176.5°, bp_{740} 174.5°, bp_{12} 63-64°. mp −1.2°. Flash pt about 75°. d_4^{20} 0.9687. n_D^{25} 1.4493. Dipole moment: 3.47 D in benzene. pKb 2. uv max: 217.5 nm (ε 1940). Miscible with water, and with all common organic solvents including petr ether. LD_{50} i.v. in rats, 1.1 g/kg: Lüttringhaus, Dirksen, *loc. cit.*
USE: As solvent and reagent.

9161. Tetramisole. *2,3,5,6-Tetrahydro-6-phenylimidazo[2,1-b]thiazole;* DL-6-phenyl-2,3,5,6-tetrahydroimidazo-[2,1-*b*]thiazole; tetramizole. $C_{11}H_{12}N_2S$; mol wt 204.31. C 64.67%, H 5.92%, N 13.71%, S 15.70%. Prepn: Raeymaekers *et al.*, **U.S. pat. 3,274,209** (1966 to Janssen); *eidem, J. Med. Chem.* **9**, 545 (1966); **Fr. pat. 1,544,972** (1968 to ICI), *C.A.* **71**, 124417b (1969). Absolute configuration: Raeymaekers *et al.*, *Tetrahedron Letters* **1967**, 1467. Resolution of isomers: Bullock *et al.*, *J. Med. Chem.* **11**, 169 (1968); **U.S. pat. 3,565,907** (1971 to Am. Cyanamid); Dewar *et al.*, **U.S. pat. 3,579,530** (1971 to ICI). Mechanism of action studies: Van den Bossche, Janssen, *Life Sci.* **6**, 1781 (1967); *Biochem. Pharmacol.* **18**, 35 (1969). Immunopotentiating action of levamisole: J. W. Hadden *et al.*, *Ann. N. Y. Acad. Sci.* **284**, 139 (1977); H. Schneiden, *Int. J. Immunopharmacol.* **3**, 9 (1981). Toxicity data: Thienpont *et al.*, *Nature* **209**, 1084 (1966). Review of pharmacology of levamisole: J. Symoens *et al.*, in *Pharmacological and Biochemical Properties of Drug Substances* vol. **2**, M. E. Goldberg, Ed. (Am. Pharm. Assoc., Washington, DC, 1979) pp 407-464. Series of articles on immunopharmacology and clinical use of levamisole: *Drugs* **20**, 89-136 (1980).

Crystals, mp 87-89°.
Hydrochloride, $C_{11}H_{13}ClN_2S$, *Bayer 9051, McN-JR-8299, R 8299, Anthelvet, Citarin, Immunol, Meglum, Nilverm, Orovermol, Ripercol, Spartakon.* Crystals, mp 264-265°. Sol in water (21 g/100 ml at 20°), methanol, propylene glycol; sparingly sol in ethanol. Slightly sol in chloroform, hexane, acetone. LD_{50} in mice, rats (mg/kg): 22, 24 i.v.; 84, 130 s.c.; 210, 480 orally (Thienpoint).
D-(+)-Form, *dexamisole.* mp 60-61.5°. $[\alpha]_D^{25}$ +85.1° (c = 10 in chloroform).
D-(+)-Form hydrochloride, *R 12563.* mp 227-227.5°. $[\alpha]_D^{20}$ +125° (c = 0.7 in water).
L-(−)-Form, *levamisole, Totalon.* mp 60-61.5°. $[\alpha]_D^{25}$ −85.1° (c = 10 in chloroform).
L-(−)-Form hydrochloride, *R-12564, Ergamisol, Levasole, Nemicide, Solaskil, Stimamizol, Tramisol, Worm-Chek.* mp 227-229°. $[\alpha]_D^{20}$ −124° ± 2° (c = 0.9, water). Activity

studies: Ciordia, McCampbell, *Am. J. Vet. Res.* **32**, 545 (1971).

THERAP CAT: Anthelmintic (Nematodes). Immunostimulant.

THERAP CAT (VET): Anthelmintic (Nematodes).

9162. Tetrandrine. *6,6',7,12-Tetramethoxy-2,2'-dimethylberbaman.* $C_{38}H_{42}N_2O_6$; mol wt 622.73. C 73.29%, H 6.80%, N 4.50%, O 15.42%. From root of *Stephania tetrandra* S. Moore, *Menispermaceae.* Isoln: Kondo, Yano, *Ann.* **497**, 90 (1932). Structure: Fujita, Murai, *J. Pharm. Soc. Japan* **71**, 1039 (1951). Synthesis: Kataoka, *C.A.* **51**, 16501i (1957). Total synthesis: Inubushi *et al., Tetrahedron Letters* **1968**, 3399; *eidem, J. Chem. Soc. (C)* **1969**, 1547. Present in the Chinese drug han-fang-chi.

Needles. mp 217-218°. $[\alpha]_D^{26}$ +252.4° (chloroform). Practically insol in water, petr ether; sol in ether and some other organic solvents.

l-Tetrandrine, **Phaenthine.** mp 210°. $[\alpha]_D^{20}$ −278° (chloroform).

THERAP CAT: Analgesic; antipyretic.

9163. Tetranectin. Tetrameric protein isolated from human plasma. Enhances plasminogen activation catalyzed by tissue plasminogen activator, *q.v.* Composed of four identical, non-covalently bound, polypeptide chains each containing 181 amino acids, with mol wt 20,100 daltons. Isoln from human serum: I. Duhl Clemmensen, C. Kluft, **Eur. pat. Appl. 206,400** (1986 to Ned. Cent. Org. Toegepast-Natuurwetenschappelijk Onderzoek). Purification, characterization and plasminogen binding activity: I. Clemmensen *et al., Eur. J. Biochem.* **156**, 327 (1986). Primary structure: J. Fuhlendorff *et al., Biochemistry* **26**, 6757 (1987). Enzyme-linked immunoassay in human plasma: B. A. Jensen *et al., J. Lab. Clin. Med.* **110**, 612 (1987). Distribution in human endocrine tissue: L. Christensen *et al., Histochemistry* **87**, 195 (1987). Possible role in cancer metastasis: B. A. Jensen, I. Clemmensen, *Cancer* **62**, 869 (1988). Isoelectric point 5.8. $E_{280 nm}^{1\%}$ 12.5.

9164. Tetranitromethane. CN_4O_8; mol wt 196.04. C 6.13%, N 28.58%, O 65.29%. $C(NO_2)_4$. Prepd by nitration of acetic anhydride with anhydrous nitric acid: Liang, *Org. Syn. coll. vol.* III, p 803 (1955).

Pale yellow liquid. d_4^{25} 1.6229; d_4^{25} 1.638 (tech). mp +13.8°. bp_{760} 126°; $bp_{25.8}$ 40°; $bp_{14.9}$ 30°; $bp_{8.4}$ 20°; $bp_{5.7}$ 13.8°; $bp_{1.9}$ 0°. n_D^{20} 1.4384; n_D^{25} 1.4358. Viscosity at 20° = 1.76 cp. Energy of decompn: Tschinkel, *Ind. Eng. Chem.* **48**, 732 (1956). Insol in water. Freely sol in alcohol, ether, alcoholic KOH. Attacks iron, copper, brass, zinc, rubber.

USE: Oxidizer in rocket propellants. As explosive in admixture with toluene. To increase cetane number of diesel fuels. Reagent for detecting the presence of double bonds in organic compds. Has been proposed as irritant war gas. *Caution:* Skin and lung irritant. Highly explosive in the presence of impurities.

9165. Tetrantoin. *3',4'-Dihydrospiro[imidazolidine-4,-2'(1'H)-naphthalene]-2,5-dione;* 7,8-benzo-1,3-diazaspiro-[4.5]decane-2,4-dione; S 2-676; Spirodon. $C_{12}H_{12}N_2O_2$; mol wt 216.23. C 66.65%, H 5.59%, N 12.96%, O 14.80%. Prepn: Novelli, *Anales Farm. Bioquim.* **21**, 81 (1954), *C.A.* **50**, 4922 (1956); Faust *et al., J. Am. Pharm. Assoc.* **46**, 118 (1957); Jules *et al., U.S. pat.* **2,716,648** (1955 to Cutter Labs.).

Crystals from ethanol or glacial acetic acid, mp 267-268°.

THERAP CAT: Anticonvulsant.

9166. Tetraphenylarsonium Bromide. $C_{24}H_{20}AsBr$; mol wt 463.25. C 62.22%, H 4.35%, As 16.17%, Br 17.25%. $(C_6H_5)_4AsBr$.

Dihydrate, crystals. *Poisonous!* Loses all its water over sulfuric acid or at 100°. mp 281-284°. Sol in about 60 parts water; sol in alcohol or methanol, sparingly in acetone.

9167. Tetraphenylarsonium Chloride. $C_{24}H_{20}AsCl$; mol wt 418.79. C 68.83%, H 4.81%, As 17.89%, Cl 8.47%. $(C_6H_5)_4AsCl$. Willard, Smith, *Ind. Eng. Chem. Anal. Ed.* **11**, 186 (1939).

Dihydrate, crystals. *Poisonous!* Loses all its water at 100°. mp 258-260°. Freely sol in water; sol in alcohol or methanol, sparingly in acetone.

USE: As a reagent for Cd, Hg, Zn, perchlorate, periodate, and other ions.

9168. Tetraphosphorus Trisulfide. Phosphorus sesquisulfide; trisulfurated phosphorus. P_4S_3; mol wt 220.08. P 56.30%, S 43.70%. Conveniently prepd by fusing red phosphorus with sulfur: Stock, *Ber.* **43**, 150 (1910); also from white phosphorus and sulfur in a high-boiling solvent such as α-chloronaphthalene: Frary, *Ger. pat.* **309,618** (1918); *Chem. Zentr.* **1919**, II, 55.

Yellowish-green, long, rhombic needles from benzene. Stable to air. d_4^{20} 2.03. mp 172.5°. bp 407.5°. Insol in cold water. Dec by hot water, yielding H_2S. Soly in carbon disulfide (20°): about 60% (w/w). Sol in benzene, similar hydrocarbons, phosphorus trichloride. LD orally in rabbits: 100 mg/kg, *Handbook of Toxicology,* vol. 1, W. S. Spector, Ed. (Saunders, Philadelphia, 1956) pp 236-237.

USE: In match tips.

9169. Tetrasilane. Tetrasilicon decahydride; silicobutane; tetrasilicane; tetrasilicobutane. $H_{10}Si_4$; mol wt 122.44. H 8.24%, Si 91.76%. Si_4H_{10}. Prepn by the action of hydrochloric acid on magnesium silicide: Stock, Somiesky, *Ber.* **49**, 111 (1916); **54B**, 524 (1921); **56B**, 247 (1923); Stock *et al., Ber.* **56**, 1695 (1923); Emeleus, Maddock, *J. Chem. Soc.* **1946**, 1131.

Liquid. mp approx −90°; bp 109°; vapor pressure 7.8 mm Hg at 0° (Stock *et al., loc. cit.*). mp −84.3°; bp (calc) 107.4°; vapor pressure 9.1 mm Hg at 0° (Emeleus, Maddock, *loc. cit.*). d^0 0.825. Dec at room temp; explodes in air. Reacts vigorously with CCl_4 and $CHCl_3$. Dec in water.

9170. Tetrasodium Pyrophosphate. TSPP; pyro; sodium pyrophosphate. $Na_4O_7P_2$; mol wt 265.94. Na 34.59%, O 42.11%, P 23.29%. $Na_4P_2O_7$. Available alkalinity as Na_2O 4.4%, total alkalinity 22.7%. Produced by molecular dehydration of dibasic sodium phosphate at 500°: Bell, *Inorg. Syn.* **3**, 98 (1950).

Crystals, d 2.534. mp 988°. Soly in water (g/100 ml) at 0°: 2.61; at 25°: 6.70; at 100°: 42.2. pH of a 1% soln = 10.2. Hydrolyzes to orthophosphate in aq soln, but the rate of hydrolysis is much slower than for the more acid pyrophosphate. No noticeable hydrolysis within 60 hrs at 70°: Bell, *Ind. Eng. Chem.* **39**, 136 (1947).

Decahydrate, crystals, d 1.82. mp 79.5°. Slight efflorescence in dry air. Soly in water (g/100 ml) at 0°: 3.16; at 20°: 6.23; at 25°: 8.14; at 60°: 21.83; at 80°: 30.04. pH of 1% soln at 25° = 10.2. Insol in alc.

USE: In cleansing compds, oil-well drilling, water treatment, cheese emulsification, as general sequestering agent, to remove rust stains, as ingredient of one-fluid ink eradicators, in electrodeposition of metals.

9171. Tetrasulfur Tetranitride. Schwefelstickstoff. N_4S_4; mol wt 184.27. N 30.41%, S 69.59%. S_4N_4. Prepd by the interaction of disulfur dichloride and ammonia: Becke-Goehring, *Inorg. Syn.* **6**, 124 (1960).

Orange-red, monoclinic needles from benzene, mp 178°. Additional purification by sublimation in high vacuum, (bath temp 100°), mp 180°. bp$_{760}$ ca. 185°. Further heating results in deflagration and explosion. Practically insol in cold water, hydrolyzed by boiling water. Slightly sol in benzene, abs ethanol, carbon disulfide. *Handle with caution:* May dec explosively on striking or at temps much above 100°.

9172. Tetrazepam. *7-Chloro-5-(1-cyclohexen-1-yl)-1,3-dihydro-1-methyl-2H-1,4-benzodiazepin-2-one;* 7-chloro-5-(1-cyclohexenyl)-1-methyl-2-oxo-2,3-dihydro-1H-[1,4]-benzo[f]diazepine; CB 4261; Clinoxan; Musaril; Myolastan. $C_{16}H_{17}ClN_2O$; mol wt 288.78. C 66.55%, H 5.93%, Cl 12.28%, N 9.70%, O 5.54%. Prepn: Schmitt, **Neth.** pat. **Appl. 6,600,095** and **U.S.** pats. **3,426,014; 3,551,412** (1966, 1969, 1970 to Clin-Byla). Synthesis and pharmacology: Schmitt *et al., Chim. Ther.* **2,** 254 (1967). Spectroscopic and chromatographic studies of tetrazepam, its metabolites, and its acid hydrolysis products: Lafargue *et al., Ann. Pharm. Franc.* **28,** 343, 477 (1970).

Yellow-brown crystals from ethyl acetate, mp 144°. uv max (ethanol): 227 nm (ε 28500). LD$_{50}$ in mice (mg/kg): 415 i.p.; 2000 orally (Schmitt).

Note: This is a controlled substance (depressant) listed in the U.S. Code of Federal Regulations, Title 21 Part 1308.14 (1987).

THERAP CAT: Muscle relaxant (skeletal).

9173. Tetrazolium Blue. *3,3'-(3,3'-Dimethoxy[1,1'-biphenyl]-4,4'-diyl)bis[2,5-diphenyl-2H-tetrazolium] dichloride;* 3,3'-dianisolebis[4,4'-(3,5-diphenyl)tetrazolium chloride]; blue tetrazolium; dimethoxy neotetrazolium; ditetrazolium chloride; BT. $C_{40}H_{32}Cl_2N_8O_2$; mol wt 727.67. C 66.02%, H 4.43%, Cl 9.75%, N 15.40%, O 4.40%. Prepn: A. M. Rutenburg *et al., Cancer Res.* **10,** 113 (1950); L. J. Pannone, J. B. Rust, **U.S.** pat. **2,713,581** (1955 to Montclair Res. Corp. and Ellis-Foster Co.).

Lemon-yellow crystals, dec 242-245°. Freely sol in methanol, ethanol, chloroform. Slightly sol in water. Insol in ethyl acetate, acetone, ether. Reduction potential about −0.08 volt. Yields a dark blue diformazan pigment in the presence of a reducing agent.

USE: For research in seed germination, as stain for bacteria and molds, in histochemical studies, to demonstrate oxidation-reduction enzymes in normal and cancerous tissues. *See also* Triphenyltetrazolium Chloride.

9174. Tetrin. Polyene antifungal antibiotic produced by Streptomyces *Illinois* #155-2: Pote, *Diss. Abstr.* **19,** 2778 (1959); Gottlieb, Pote, *Phytopathology* **50,** 817 (1960). Isoln of the two tetraenes, tetrin A and B: Rinehart *et al., Ann.* **668,** 77 (1963); German, *Diss. Abstr.* **25,** 97 (1964). Structure of tetrin A: Pandey *et al., J. Am. Chem. Soc.* **93,** 3738 (1971); of tetrin B: Rinehart *et al., ibid.* 3747. Revised structure: R. C. Pandey, K. L. Rinehart, *J. Antibiot.* **29,**

1035 (1976). Mode of action of tetrin A: van Etten, Gottlieb, *J. Gen. Microbiol.* **46,** 377 (1967).

Tetrin A. $C_{34}H_{51}NO_{13}$. R = H. Fine, colorless needles from methanol or aqueous n-butanol, mp > 350° (dec). [α]$_D^{21}$ +8.3° (c = 0.72 in pyridine). [α]$_D^{28}$ +27.5° (c = 1.0 in pyridine). uv max: 214, 278, 290, 303, 318 nm (E$_{1cm}^{0.1\%}$ 19.4, 44.2, 81.2, 115.0, 110.9). Monobasic, pKa' 8.30 in 60% ethanol. Sol in pyridine, dil alkalies, dil mineral acids; moderately sol in lower alcohols; practically insol in acetone, ether, water.
Tetrin B. $C_{34}H_{51}NO_{14}$. R = OH. Brown, amorphous powder, mp > 360° (darkens at 160-165°, blackens at 250-295°). [α]$_D^{24}$ +43.5° (c = 0.14 in methanol); [α]$_D^{28}$ +45° (c = 0.3 in pyridine). uv max: 214, 278, 290, 303, 318 nm (E$_{1cm}^{0.1\%}$ 18.6, 51.4, 80.1, 112.8, 108.9). Readily sol in ethanol + water, dioxane + water; fairly sol in water, lower alcohols, dioxane, pyridine, dimethyl sulfoxide; slightly sol in acetone. Practically insol in ethyl acetate, chloroform, ether, ethylene dichloride.

9175. Tetrodotoxin. *Octahydro-12-(hydroxymethyl)-2-imino-5,9:7,10a-dimethano-10aH-[1,3]dioxocino[6,5-d]pyrimidine-4,7,10,11,12-pentol;* maculotoxin; spheroidine; tarichatoxin; tetrodontoxin; fugu poison; TTX. $C_{11}H_{17}N_3O_8$; mol wt 319.28. C 41.38%, H 5.37%, N 13.16%, O 40.09%. Toxin from the ovaries and liver of many species of *Tetraodontidae,* esp the globe fish *(Spheroides rubripes):* Yokoo, *J. Chem. Soc. Japan* **71,** 590 (1950), *C.A.* **45,** 6759c (1951). Identity with tarichatoxin: Buchwald *et al., Science* **143,** 474 (1963); with maculotoxin: D. D. Sheumack *et al., ibid.* **199,** 188 (1978). Structure studies: Goto *et al., Tetrahedron Letters* **1963,** 2105, 2115; **1964,** 779, 1831. Structure: Woodward, *Pure Appl. Chem.* **9,** 49 (1964); Tsuda *et al., Chem. Pharm. Bull.* **12,** 1357 (1964); Goto *et al., Tetrahedron* **21,** 2059 (1965). Synthetic studies: Kishi *et al., Tetrahedron Letters* **1970,** 5127, 5129. Total synthesis: Kishi *et al., J. Am. Chem. Soc.* **94,** 9219 (1972). Pharmacology: Evans, *Brit. Med. Bull.* **25,** 263 (1969); Kao, *Fed. Proc.* **31,** 1117 (1972). Mechanism of action: Narahashi, *ibid.* 1124. Review: Scheuer, *Fortschr. Chem. Org. Naturst.* **22,** 265 (1964); Mosher *et al., Science* **144,** 1100 (1964); Kao, *Pharmacol. Rev.* **18,** 997 (1966); Evans, *Int. Rev. Neurobiol.* **15,** 83 (1972).

Darkens above 220° without dec. [α]$_D^{25}$ −8.64° (c = 8.55 in dil acetic acid). pKa: 8.76 (water); 9.4 (50% alc). Sol in dil acetic acid; slightly sol in water, dry alc, ether. Practically insol in other organic solvents. Toxin destroyed in strong acids and in alkaline solns. LD$_{50}$ i.p. in mice: 10 μg/kg, C. Y. Kao, F. A. Fuhrman, *J. Pharmacol. Exp. Ther.* **140,** 31 (1963).

9176. Tetronasin. *(6S,7S,8R,12S,15R,16R,19S,22S,23R,-24S,26R,27S)-4-Demethylene-22,24-dimethyltetronomycin;* antibiotic M139603; ICI 139603; M139603. $C_{35}H_{54}O_8$; mol wt 602.81. C 69.74%, H 9.03%, O 21.23%. Polyether antibiotic produced by *Streptomyces longisporoflavus* NCIB 11426. Possesses a biosynthetically rare acid grouping in the form of an acyl tetronic acid moiety. Isoln and use in ruminants: D. H. Davies, G. L. F. Norris, **Brit. pat. Appl. 2,-027,013;** *eidem,* **U.S. pat. 4,279,894** (1980, 1981 both to ICI). Physical data and crystal structure: D. H. Davies *et al., Chem. Commun.* **1981,** 1073. Solution conformation and cation-binding properties: J. Grandjean, P. Laszlo, *Tetrahedron Letters* **24,** 3319 (1983). Synthetic studies: A. M. Doherty, S. V. Ley, *ibid.* **27,** 105 (1986). Biosynthetic studies: J. M. Bulsing *et al., Chem. Commun.* **1984,** 1301; D. M. Doddrell *et al., ibid.* 1302; A. K. Demetriadou *et al., ibid.* **1985,** 408. Antimicrobial activity: C. J. Newbold *et al., Appl. Environ. Microbiol.* **54,** 544 (1988). Effect on gain efficiency in cattle: S. J. Bartle *et al., J. Anim. Sci.* **66,** 1502 (1988).

Sodium salt, $C_{35}H_{53}NaO_8$. mp 176-178°. $[\alpha]_D^{23}$ −82° (c = 0.2 in methanol). uv max (ethanol): 234, 270 nm (ε 13000, 11000). pKa 1.8 ±0.3 (methanol/H_2O 1:9). Sol in most organic solvents. Insol in water.

THERAP CAT (VET): Ruminant performance enhancer.

9177. Tetroquinone. *2,3,5,6-Tetrahydroxy-2,5-cyclo-hexadiene-1,4-dione; tetrahydroxy-p-benzoquinone;* tetrahydroxyquinone; THQ; HPEK-1; NSC-112931; Kelox. $C_6H_4O_6$; mol wt 172.09. C 41.87%, H 2.34%, O 55.78%. Prepn from glyoxal: A. J. Fatiadi, W. F. Sager, *Org. Syn. coll. vol.* **V,** 1011 (1973).

Blue-black crystals appearing yellow in transmitted light. Slightly sol in cold water; freely sol in hot water or in alcohol; slightly sol in ether. It acts like a strong dibasic acid. Disodium salt, $Na_2C_6H_2O_6$, almost black crystals with a green metallic luster, sparingly sol in water.

USE: Disodium salt is used as an indicator in the volumetric determination of sulfate by means of barium chloride solution.

THERAP CAT: Keratolytic (systemic).

9178. Tetroxoprim. *5-[[3,5-Dimethoxy-4-(2-methoxyethoxy)phenyl]methyl]-2,4-pyrimidinediamine; 2,4-diamino-5-[3,5-dimethoxy-4-(2-methoxyethoxy)benzyl]pyrimidine;* He 781; Sterinor. $C_{16}H_{22}N_4O_4$; mol wt 334.39. C 57.47%, H 6.63%, N 16.76%, O 19.14%. Analog of trimethoprim, *q.v.* Prepn: W. Liebenow, J. Prikryl, **Fr. pat. 2,221,147** corresp to **U.S. pat. 3,992,379** (1974, 1976 both to Heumann). Series of articles on tetroxoprim and other antibacterial folate inhibitors: *J. Antimicrob. Chemother.* **5,** Suppl. B, 1-239 (1979); on synthesis of radioactive tetroxoprim, tissue distribution, kinetics, HPLC determn: *Arzneimittel-Forsch.* **30,** 307-319 (1980). Effect on bacterial growth kinetics: J. K. Seydel, E. Wempe, *Chemotherapy* **26,** 361 (1980). *In vitro* activity of the tetroxoprim-sulfadiazine combination: H. Hahn, A. Kirov, *Arzneimittel-Forsch.* **30,** 1047 (1980).

Cryst from water, mp 153-156°, W. Liebenow, J. Prikryl, *loc. cit.,* also reported as mp 160.1°, *eidem, J. Antimicrob. Chemother.* **5,** Suppl B, 15 (1979). Soly at 30° (mg/ml): water, 2.65; chloroform 69; *n*-octanol 1.61. pKb 8.25. LD_{50} in rats: 1357 mg/kg orally, *eidem,* **U.S. pat. 3,992,379.**

Mixture with sulfadiazine (usually 1:2.5), *co-tetroxazine, Raslogin, Tibirox.*

THERAP CAT: Antibacterial.

9179. Tevenel®. *N-[2-[4-(Aminosulfonyl)phenyl]-2-hydroxy-1-(hydroxymethyl)ethyl]-2,2-dichloroacetamide;* D-*threo-*(−)*-2,2-dichloro-N-[β-hydroxy-α-(hydroxymethyl)-p-sulfamoylphenethyl]acetamide;* D-*threo*-1-(*p*-sulfamoylphenyl)-2-(2',2'-dichloroacetamido)-1,3-propanediol; AMP-3; D-AMP-3. $C_{11}H_{14}Cl_2N_2O_5S$; mol wt 357.22. C 36.98%, H 3.95%, Cl 19.85%, N 7.84%, O 22.39%, S 8.98%. Sulfamoyl analog of chloramphenicol, *q.v.* Prepn: Gregory, **U.S. pat. 2,680,135** (1954 to Du Pont). Toxicology: R. M. Jiji *et al., Arch. Int. Med.* **111,** 70 (1963).

Crystals from acetonitrile, mp 155-158°. Sol in ethanol, propylene glycol. Slightly sol in water, dil aq solns of sodium bicarbonate, acetic acid. Aq solns are neutral.

THERAP CAT: Anti-infective.

9180. TFM. *4-Nitro-3-(trifluoromethyl)phenol; α,α,α-trifluoro-4-nitro-m-cresol;* 3-trifluoromethyl-4-nitrophenol; lamprecid 2770; Lamprecid. $C_7H_4F_3NO_3$; mol wt 207.11. C 40.59%, H 1.95%, F 27.52%, N 6.76%, O 23.18%. A substance toxic to the parasitic sea lamprey, *Petromyzon marinus,* which preys upon commercial fish species of the Great Lakes: Scherer *et al.,* **Ger. pat. 1,068,505,** *C.A.* **55,** 9774d (1961) corresp to **U.S. pat. 3,157,571** (1959, 1964, both to Hoechst). Toxicological studies in fish, and bibliography: Kawatski, McDonald, *Comp. Gen. Pharmacol.* **5,** 67 (1974). Physical properties: Smith, *J. Chem. Eng. Data* **6,** 607 (1961). *Review:* Schnick, *Investigations in Fish Control,* no. 44 (Sport Fisheries and Wildlife Bureau) 31 pp.

mp 76°. pK 6.07. uv max (acid): 280 nm (ε 1930); in 1% NaOH: 300, 395 nm (ε 4650, 13,130); in 95% ethanol: 290 nm (ε 14,700).

USE: Lamprey killer.

9181. Thalicarpine. *9-[4,5-Dimethoxy-2-[(1,2,3,4-tetra-hydro-6,7-dimethoxy-2-methyl-1-isoquinolinyl)methyl]phenoxy]-5,6,6a,7-tetrahydro-1,2,10-trimethoxy-6-methyl-4H-dibenzo[de,g]quinoline;* mol wt 696.81. C 70.67%, H 6.94%, N 4.02%, O 18.37%. Tumor-inhibitory alkaloid from *Thalictrum dasycarpum* Fisch. & Lall., *Ranunculaceae:* Kupchan *et al., J. Pharm. Sci.* **52,** 985 (1963). Structure: Kupchan, Yokoyama, *J. Am. Chem. Soc.* **86,** 2177 (1964); Tomita *et al., Tetrahedron Letters* **1965,** 4309. Total synthesis: Kupchan, Liepa, *Chem. Commun.* **1971,**

599; Kupchan *et al., J. Am. Chem. Soc.* **95**, 2995 (1973). Pharmacology: Herman, Chadwick, *Toxicol. Appl. Pharmacol.* **26**, 137 (1973); S. M. Sieber *et al., Cancer Treat. Rep.* **60**, 1127 (1976). Antimicrobial and hypotensive activity: W. N. Wu *et al., Lloydia* **40**, 508 (1977). Biosynthesis: D. S. Bhakuni, S. Jain, *Tetrahedron* **38**, 729 (1982).

Needles from ethyl acetate, mp 160-161°. $[\alpha]_D^{25}$ +133° (c = 0.83 in methanol); $[\alpha]_D^{25}$ +89° (c = 0.88 in chloroform). uv max (methanol): 282, 302 nm (ε 17,000, 13,000).

9182. Thalidomide. 2-(2,6-Dioxo-3-piperidinyl)-1H-isoindole-1,3(2H)-dione; N-(2,6-dioxo-3-piperidyl)phthalimide; α-phthalimidoglutarimide; 3-phthalimidoglutarimide; 2,6-dioxo-3-phthalimidopiperidine; N-phthalylglutamic acid imide; N-phthaloylglutamimide; K-17; Distaval; Softenon; Sedalis; Talimol; Pantosediv; Neurosedyn; Kevadon; Contergan. $C_{13}H_{10}N_2O_4$; mol wt 258.23. C 60.46%, H 3.90%, N 10.85%, O 24.78%. Prepn: **Brit.** pat. **768,821** (1957 to Chemie Grünenthal). Teratogenicity studies: I. D. Fratta *et al., Toxicol. Appl. Pharmacol.* **7**, 268 (1965); H. Schumacher *et al., J. Pharmacol. Exp. Ther.* **160**, 189 (1968). Thalidomide has been used in the treatment of leprosy, *cf.* J. Sheskin, *Int. J. Dermatol.* **19**, 318 (1980); E. J. Shannon *et al., Scand. J. Immunol.* **13**, 553 (1981). Evidence for a toxic arene oxide metabolite as the basis of thalidomide teratogenesis: G. B. Gordon *et al., Proc. Nat. Acad. Sci. USA* **78**, 2545 (1981).

Needles, mp 269-271°. uv max (neutral soln): 220, 300 nm. Sparingly sol in water, methanol, ethanol, acetone, ethyl acetate, butyl acetate, glacial acetic acid. Very sol in dioxane, DMF, pyridine. Practically insol in ether, chloroform, benzene.

Note: Withdrawn from market because of association with fetal abnormalities.

THERAP CAT: Formerly as sedative, hypnotic.

9183. Thallium. Tl; at. wt 204.383; at. no. 81; valence 1,3. Natural isotopes: 203 (29.50%), 205 (70.50%); artificial, radioactive isotopes: 191-202; 204; 206-210. Occurs in crookesite, $(Cu,Tl,Ag)_2Se$, found in Sweden; in lorandite, $TlAgS_2$, found in Greece; in hutchinsonite, $(Tl,Cu,Ag)_2S.$-$PbS.2As_2S_3$, found in Switzerland. Occurrence in the earth's crust: 0.7 ppm. Discovered by Crookes in 1861. Prepn: Sanderson, *Can. Mining J.* **65**, 624 (1944). Use in organic syntheses: McKillop *et al., Tetrahedron Letters* **1970**, 5281; Taylor *et al., ibid.* 5285. *Review:* Wade, Banister in *Comprehensive Inorganic Chemistry* vol. 1, J. C. Bailar, Jr. *et al.,* Eds. (Pergamon Press, Oxford, 1973) pp 997-1000, 1119-1172.

Bluish-white, very soft, inelastic, easily fusible, heavy metal; leaves a streak on paper. Oxidizes superficially in air forming a coating of Tl_2O. Forms alloys with other metals and readily amalgamates with mercury. d 11.85. Begins to volatilize at 174°. mp 303.5°. bp 1457°. Specific heat at 20° 0.031 cal/g/°C. Latent heat of fusion 5.04 cal/g. Brinell

hardness: 2. May be distilled in a stream of hydrogen. Insol in water; reacts with nitric or sulfuric acid; difficultly with hydrochloric acid.

Caution: Symptoms of *acute* toxicity include nausea, vomiting, diarrhea, tingling, pain in extremities, weakness, coma, convulsions, death. *Chronic:* weakness and pain in extremities (polyneuritis), loss of hair: E. Browning, *Toxicity of Industrial Metals* (Appleton-Century-Crofts, New York, 2nd ed., 1969) pp 317-322.

USE: In admixture with 97-98% of inert substances the salts are used as poison for rats and other rodents. In semiconductor research. Alloyed with mercury for switches and closures which operate at subzero temps.

9184. Thallium Acetate. Thallous acetate. $C_2H_3O_2Tl$; mol wt 263.43. C 9.12%, H 1.15%, O 12.15%, Tl 77.59%. White, deliquesc crystals. *Poisonous!* Sol in water, alcohol. *Keep well closed.* MLD orally in dogs: 18.5 mg/kg, *Handbook of Toxicology* vol. 1, W. S. Spector, Ed. (Saunders, Philadelphia, 1956) pp 294-295.

9185. Thallium Bromide. Thallous bromide. BrTl; mol wt 284.31. Br 28.11%, Tl 71.89%. TlBr. Pale yellow, cryst powder. *Poisonous!* d 7.5. mp about 460°. Sol in 2360 parts water.

9186. Thallium Carbonate. Thallous carbonate. CO_3Tl_2; mol wt 468.78. C 2.56%, O 10.24%, Tl 87.20%. Tl_2CO_3. White crystals. *Poisonous!* d 7.1. mp 272°. Soluble in 24 parts water, 3.7 parts boiling water; insol in alc. USE: Manuf imitation diamonds.

9187. Thallium Chloride. Thallous chloride. ClTl; mol wt 239.85. Cl 14.78%, Tl 85.22%. TlCl. White, cryst powder. *Poisonous!* d 7.0. mp 430°. Sol in 260 parts cold water, 70 parts boiling water; insol in alcohol; HCl decreases its soly in water. USE: As catalyst in chlorinations.

9188. Thallium Cyanide. Thallous cyanide. CNTl; mol. wt. 230.39. C 5.21%, N 6.08%, Tl 88.71%. TlCN. Prepn: *Gmelin's, Thallium (8th ed.)* **38**, 390 (1940); E. C. Taylor *et al., J. Org. Chem.* **43**, 2280 (1978). White, hexagonal platelets. d 6.523. Soly in water: 16.8 g/100 ml. Aqueous soln is alkaline. USE: In organic synthesis.

9189. Thallium Fluoride. Thallous fluoride. FTl; mol wt 223.39. F 8.51%, Tl 91.49%. TlF. Prepared from Tl_2CO_3 and HF: Ketelaar, *Z. Kristallogr.* **92**, 30 (1935); Hayek, *Z. Anorg. Allgem. Chem.* **225**, 47 (1935); Barrow *et al., Trans. Faraday Soc.* **51**, 1650 (1955); from Tl and HF: Keneshea, Cubicciotti, *J. Phys. Chem.* **69**, 3910 (1965); Tranquard, *Bull. Soc. Chim. France* **1967**, 2578.

Hard, shiny, crystals which deliquesce when breathed upon, but which resolidify immediately in dry air. Not hygroscopic in the usual sense. Orthorhombic (deformed NaCl lattice). d_4^{25} 8.36. mp 322°. Begins to sublime at 300°. Very freely sol in water. Concd solns show strong alkalinity. USE: In the prepn of fluoro esters.

9190. Thallium Hydroxide. Thallous hydroxide. HOTl; mol wt 221.38. H 0.45%, O 7.23%, Tl 92.32%. TlOH. Yellow needles. *Poisonous!* Very sol in water; the soln is strongly alkaline and turns turmeric paper brown.

9191. Thallium Iodide. Thallous iodide. ITl; mol wt 331.31. I 38.31%, Tl 61.69%. TlI. Yellow, cryst powder. *Poisonous!* d 7.1; mp 440°; bp 824°. Almost insol in water; insol in alcohol; sol in KI soln.

9192. Thallium Nitrate. Thallous nitrate. NO_3Tl; mol wt 266.40. N 5.26%, O 18.02%, Tl 76.72%. $TlNO_3$. White crystals. *Poisonous!* d 5.55. mp 206°; dec at 450°. Sol in 10 parts cold water, 0.3 part boiling water; insol in alcohol. LD orally in dogs: 45 mg/kg, *Handbook of Toxicology* vol. 1, W. S. Spector, Ed. (Saunders, Philadelphia, 1956) pp 294-295.

USE: As a reagent in analytical chemistry, esp for the determination of iodine in presence of Br and Cl; also with $KClO_3$, HgCl and resin for green fire for signalling at sea.

9193. Thallium Oxide. Thallous oxide. OTl_2; mol wt 424.78. O 3.77%, Tl 96.23%. Tl_2O.

Black powder. *Poisonous!* mp about 300°. Sol in water, forming the hydroxide; also sol in alcohol. On exposure to air it gradually oxidizes to thallic oxide and becomes insol.

USE: In manuf of glass of a high coefficient of refraction for optical purposes (thallium flint glass) and for artificial gems.

9194. Thallium Selenate. Thallous selenate. O_4SeTl_2; mol wt 551.74. O 11.60%, Se 14.31%, Tl 74.09%. Tl_2SeO_4. Prepd from a soln of thallous carbonate in selenic acid: Kuhlmann, *Bull. Soc. Chim.* [2] **1**, 330 (1864).

Orthorhombic crystals; d 6.875; mp >400°; 2.13 g dissolve in 100 g water at 9.3°, 2.4 g at 12°, 10.86 g at 100°. Insol in alcohol and ethyl ether.

9195. Thallium Selenide. Thallous selenide. $SeTl_2$; mol wt 487.74. Tl 83.81%, Se 16.19%. Tl_2Se. Prepared by the action of hydrogen selenide on a soln of thallous carbonate: Kuhlmann, *Bull. Soc. Chim.* [2] **1**, 330 (1864).

Dark gray plates with a metallic luster. mp 340° (Kuhlmann); mp 338° [Palabon, *Compt. Rend.* **145**, 118 (1907); **173**, 142 (1921)]. Insol in water and acids.

9196. Thallium Sesquioxide. Thallic oxide; thallium peroxide. O_3Tl_2; mol wt 456.78. O 10.51%, Tl 89.49%. Tl_2O_3. Brown powder. d 9.65. mp 717°. Insol in water; dec by HCl with evolution of chlorine and by H_2SO_4 with evolution of oxygen.

9197. Thallium Sulfate. Thallous sulfate; Eccothal. O_4STl_2; mol wt 504.85. O 12.68%, S 6.35%, Tl 80.97%. Tl_2SO_4.

White, rhomboid prisms. *Poisonous!* d 6.77; mp 632°. Soly in 100 ml water at 0°: 2.70 g; at 20°: 4.87 g; at 100°: 18.45 g. LD_{50} orally in rats: 25 mg/kg, E. W. Schafer, *Toxicol. Appl. Pharmacol.* **21**, 315 (1972).

USE: As rat poison, as an ant bait and as a reagent in analytical chemistry. *Caution:* Ingestion causes G.I. colic, vomiting trembling, convulsions, paralysis, dyspnea, collapse, death, cf. *Clinical Toxicology of Commercial Products*, R. E. Gosselin *et al.*, Eds. (Williams & Wilkins, Baltimore, 4th ed., 1976) section III, pp 307-311. *See also* Thallium.

9198. Thallium Sulfide. Thallous sulfide. STl_2; mol wt 440.85. S 7.27%, Tl 92.73%. Tl_2S.

Bluish-black, cryst powder. d 8.39. mp 448.5°. Almost insol in water, alkali hydoxides, sulfides or cyanides; sol in mineral acids.

9199. Thallium Trifluoride. Thallic fluoride. F_3Tl; mol wt 261.39. F 21.81%, Tl 78.19%. TlF_3. Prepd from Tl_2O_3 and F_2: Hannebohn, Klemm, *Z. Anorg. Allgem. Chem.* **229**, 343 (1936); Kwasnik in *Handbook of Preparative Inorganic Chemistry*, Vol. 1, G. Brauer, Ed. (Academic Press, New York, 2nd ed., 1963) pp 230-231.

Orthothombic crystals. Very sensitive to moisture. d_4^{25} 8.36. d 8.65 also reported: Hebecker, Hoppe, *Naturwiss.* **53**, 104 (1966). mp 550°. Dec on heating in air, but may be melted in an atm of fluorine. Instantly dec by water, forming thallium hydroxide and a dark brown sediment. May be stored in sealed quartz tubes.

9200. Thaumatin. Talin. Sweet-tasting basic protein extracted from the fruit of the tropical plant, *Thaumatococcus danielli* Benth., *Marantaceae*, found in Western Africa from Sierra Leone to Zaire, in Sudan and Uganda. Composed of five different forms, thaumatins I, II, III, b and c; thaumatins I and II predominate. All are nearly 100,000 times sweeter than sucrose and have molecular weights of about 22,000. Isoln and characterization: H. van der Wel, K. Loeve, *Eur. J. Biochem.* **31**, 221 (1972). Extraction process: J. Higgenbotham, *U.S. pat.* **4,011,206** (1976 to Tate and Lyle Ltd.). Electrophysical study of effects on taste receptors: Brouwer *et al.*, *Acta Physiol. Scand.* **89**, 550 (1973). Spectrometric investigation: O. Korver *et al.*, *Eur. J. Biochem.* **35**, 554 (1973). Studies on the primary structure of thaumatin I: R. B. Iyengar *et al.*, *ibid.* **96**, 193 (1979). Thaumatins I and II consist of almost identical sequences of 207 amino acids. Thaumatin I crystallizes in two different forms: H. van der Wel *et al.*, *FEBS Letters* **56**, 316 (1975). The crystal structure shows two distinct regions with the amino acids either in sheets or in complex loops. Crystal

structure: A. M. de Vos *et al.*, *Proc. Nat. Acad. Sci. USA* **82**, 1406 (1985). Cloning and expression in *E. coli* of the structural gene of thaumatin II: L. Edens *et al.*, *Gene* **18**, 1 (1982); C. T. Verrips, *Eur. pat.* **Appls. 54,330** and **54,331** (both 1982 to Unilever). Cloning of the natural gene: A. M. Ledeboer *et al.*, *Gene* **30**, 23 (1984). *Reviews:* R. Cagan, *Science* **181**, 32 (1973); H. van der Wel, *Chem. & Ind. (London)* **1983**, 19.

Intensely sweet taste, licorice aftertaste. Strongly cationic, isoelectric pt greater than or equal to 11.7. uv max: 278 nm (pH 5.6); 283, 290 nm (pH 13.0). About 750-1600 times sweeter than sucrose on a wt basis; 30,000-100,000 times on a molar basis. Threshold values are near 10^{-4}%. The proteins lose sweetness on heating, on splitting of disulfide bridges and also at pHs <2.5 which points to the importance of the tertiary structure for the sweetness. *See* Korver *et al., loc. cit.*

USE: Potential low-calorie sweetener.

9201. Theaflavine. *1,8-Bis(3,4-dihydro-3,5,7-trihydroxy-2H-1-benzopyran-2-yl)-3,4,6-trihydroxy-5H-benzocyclohepten-5-one.* $C_{29}H_{24}O_{12}$; mol wt 564.51. C 61.70%, H 4.29%, O 34.01%. From black tea extracts: Roberts, Myers, *J. Sci. Food Agr.* **10**, 176 (1959). Structure: Takino *et al.*, *Tetrahedron Letters* **1965**, 4019. Configuration: Brown *et al., ibid.* **1966**, 1193.

Crystals from water, dec 237-240°. Absorption max (ethanol): 216, 271, 384, 470 nm (ε 35,500, 19,500, 8700, 3600). Nonaacetate, $C_{47}H_{42}O_{21}$, crystals, mp 167-168°. uv max (alc): 211, 250, 314, 353 nm (ε 25,100, 13,200, 7100, 4700).

9202. Thebaine. *6,7,8,14-Tetradehydro-4,5-epoxy-3,6-dimethoxy-17-methylmorphinan;* paramorphine. $C_{19}H_{21}NO_3$; mol wt 311.37. C 73.29%, H 6.80%, N 4.50%, O 15.42%. From opium, which contains from 0.3-1.5% depending on its origin. Discussion of structure and bibliography: Small, Lutz, *Chemistry of the Opium Alkaloids, U.S. Public Health Reports,* **Suppl. No. 103**, Washington (1932); Small, Browning, *J. Org. Chem.* **3**, 618 (1938); Cherbuliez, Araqui, *Helv. Chim. Acta* **26**, 2251 (1943); Ghosh, Robinson, *J. Chem. Soc.* **1944**, 506; K. W. Bentley, *The Chemistry of the Morphine Alkaloids* (Oxford, 1954) p 184 sqq. Config: Kalvoda *et al., Helv. Chim. Acta* **38**, 1847 (1955). Syntheses: Rapoport *et al., J. Am. Chem. Soc.* **89**, 1942 (1967); Schwartz, Mami, *ibid.* **97**, 1239 (1975); Barber, Rapoport, *J. Med. Chem.* **18**, 1074 (1975). Absorption spectrum: Csokán, *Z. Anal. Chem.* **124**, 344 (1942). Toxicity data: Eddy, *J. Pharmacol. Exp. Ther.* **66**, 182 (1939).

Orthorhombic, rectangular plates by sublimation at 170-180° under atmospheric pressure and a 1 mm distance. mp 193° (rapid heating). $[\alpha]_D^{15}$ −219° (p = 2 in alc); $[\alpha]_D^{23}$ −230° (p = 5 in chloroform). pK at 15° = 6.05. pH of satd water soln 7.6. One gram dissolves in 1460 ml water at 15° [Koltthoff, *Biochem. Z.* **162**, 336 (1925)], in about 15 ml hot alcohol or 13 ml chloroform, about 200 ml ether, 25 ml benzene, 12 ml pyridine; not very sol in petr ether.

Hydrochloride monohydrate, $C_{19}H_{21}NO_3 \cdot HCl \cdot H_2O$, orthorhombic prisms from alcohol, $[\alpha]_D^{23}$ −164° (p = 2). Sol in about 12 parts water, in alcohol; pH of 0.05 molar soln 4.95. LD_{50} s.c. in rabbits: 14 mg/kg (Eddy).

Oxalate hexahydrate, $2C_{19}H_{21}NO_3 \cdot C_2H_2O_4 \cdot 6H_2O$, prisms, sol in about 10 parts water, in alcohol. Almost insol in ether.

Binoxalate monohydrate, $C_{19}H_{21}NO_3 \cdot C_2H_2O_4 \cdot H_2O$, prisms, sol in 45 parts water.

Bitartrate monohydrate, $C_{19}H_{21}NO_3 \cdot C_4H_6O_6 \cdot H_2O$, prisms, sol in 130 parts water, quite sol in hot water, hot alcohol.

Salicylate, $C_{19}H_{21}NO_3 \cdot C_7H_6O_3$, crystals, sol in 750 parts water.

Caution: Produces strychnine-like convulsions rather than narcosis. May be habit forming. This is a controlled substance (opiate) listed in the U.S. Code of Federal Regulations, Title 21 Part 1308.12 (1985).

9203. Thebainone. *7,8-Didehydro-4-hydroxy-3-methoxy-17-methylmorphinan-6-one;* thebainone-A. $C_{18}H_{21}NO_3$; mol wt 299.36. C 72.21%, H 7.07%, N 4.68%, O 16.03%. Prepn from thebaine, codeinone or β-ethylthiocodide: Morris, Small, *J. Am. Chem. Soc.* **56**, 2159 (1934). Earlier references and discussion of structure: Small, Lutz, *Chemistry of the Opium Alkaloids, U.S. Public Health Reports* **Suppl. No. 103**, Washington (1932). About anomalies in nomenclature and difference from metathebainone *see* Henry, *Plant Alkaloids* (London, 1939) p 249. Description of all thebainones: K. W. Bentley, *The Chemistry of the Morphine Alkaloids* (Oxford, 1954) p 219.

Crystals from ethyl acetate, mp 146°. $[\alpha]_D^{28}$ −47° (c = 1.16 in 95% alc). One gram dissolves in 250 ml water, about 120 ml boiling water. Sol in chloroform, benzene, acetone; sparingly sol in ether, alcohol, methanol.

Sesquihydrate, crystals from water, mp 90°.

Methanolate, crystals from methanol, mp 118°.

Hydrochloride, $C_{18}H_{21}NO_3 \cdot HCl$, crystals from alcohol, mp 256° (turns red at mp). $[\alpha]_D^{30}$ −25° (c = 1.63).

Hydriodide, $C_{18}H_{21}NO_3 \cdot HI$, crystals, mp 165°, solidifies and remelts 260°.

Methiodide, mp 251°.

9204. Thenaldine. *1-Methyl-N-phenyl-N-(2-thienylmethyl)-4-piperidinamine; 1-methyl-4-N-2-thenylanilinopiperidine;* 1-methyl-4-amino-*N*-phenyl-*N*-(2-thenyl)piperidine; thenophenopiperidine; 1-methyl-4-[phenyl-(2-thenyl)amino]piperidine; thenalidine; Sandostene. $C_{17}H_{22}N_2S$; mol wt 286.46. C 71.28%, H 7.74%, N 9.78%, S 11.20%. Prepn: Stoll, Bourquin, U.S. pats. **2,717,251** and **2,757,175** (1955 and 1956 to Sandoz).

bp$_{0.02}$ 158-160°, mp 95-97°.

Tartrate, crystals, mp 170-172°.

THERAP CAT: Tartrate as antihistaminic; antipruritic.

9205. Thenium Closylate. *N,N-Dimethyl-N-(2-phenoxyethyl)-2-thiophenemethanaminium salt with 4-chlorobenzenesulfonic acid (1:1); dimethyl(2-phenoxyethyl)-2-thenylammonium p-chlorobenzenesulfonate;* 611C55; Bancaris;

Canopar. $C_{21}H_{24}ClNO_4S_2$; mol wt 454.02. C 55.55%, H 5.33%, Cl 7.81%, N 3.09%, O 14.10%, S 14.12%. Prepn: Copp, Brit. pat. **864,885** (1961 to Wellcome Found.). Anthelmintic activity: Burrows, Lillis, *Am. J. Vet. Res.* **23**, 77 (1962).

Crystals from isopropanol + ether, mp 159-160°. Soly in water at 20°: 0.6% w/v.

THERAP CAT (VET): Anthelmintic.

9206. 3-Thenoic Acid. *3-Thiophenecarboxylic acid; β-*thiophenic acid. $C_5H_4O_2S$; mol wt 128.15. C 46.86%, H 3.14%, O 24.97%, S 25.03%. Prepd by the oxidation of 3-thenaldehyde with silver oxide: Campaigne, LeSuer, *Org. Syn.* **33**, 94 (1953).

Crystals from water, mp 137-138°. Soly in water at 25°: 0.43 g/100 g. pKa 6.23. Volatile with steam.

9207. Thenyldiamine. *N,N-Dimethyl-N′-2-pyridinyl-N′-(3-thienylmethyl)-1,2-ethanediamine; 2-[(2-dimethylaminoethyl)-3-thenylamino]pyridine; N,N-dimethyl-N′-(α-pyridyl)-N′-(3-methylthienyl)ethylenediamine; N-(α-pyridyl)-N-(β-thenyl)-N′,N′-dimethylethylenediamine; N-(2-dimethylaminoethyl)-N-2-pyridyl-3-thenylamine; dethylandiamine;* WIN-2848; Thenfadil; Tenfidil. $C_{14}H_{19}N_3S$; mol wt 261.36. C 64.33%, H 7.33%, N 16.08%, S 12.27%. Prepd by condensing *N,N*-dimethylaminoethyl-α-aminopyridine with 3-thenyl bromide: Campaigne, LeSuer, *J. Am. Chem. Soc.* **71**, 333 (1949).

Free base, liquid. bp$_{1.0}$ 169-172°. n_D^{20} 1.5915.

Hydrochloride, $C_{14}H_{20}ClN_3S$, crystals from methanol, mp 169.5-170°. Bitter taste. Sol in water up to 20%. Slightly sol in alc. pH of 1% aq soln 6.5. LD_{50} orally in rats: 525 mg/kg, Hoppe, Lands, *J. Pharmacol. Exp. Ther.* **97**, 371 (1949).

THERAP CAT: Hydrochloride as antihistaminic.

9208. Theobroma Oil. Cacao butter; cocoa butter. From roasted seeds of *Theobroma cacao* L., *Sterculiaceae. Constit.* Chiefly glycerides of stearic, palmitic, oleic, arachidic, and linoleic acids.

Yellowish-white solid; brittle below 25°; chocolate odor and taste; d$_{25}^{100}$ 0.858-0.864; mp 30-35°; n_D^{40} 1.4537-1.4578. Sapon. no. 188-195. Iodine no. 35-40. Insoluble in water; slightly sol in alcohol; sol in boiling abs alc; very sol in chloroform, ether, benzene, petr ether.

USE: Lubricant in massage; base for suppositories and ointments. Manuf chocolate, toilet soaps, creams, etc.

9209. Theobromine. *3,7-Dihydro-3,7-dimethyl-1H-purine-2,6-dione; 3,7-dimethylxanthine.* $C_7H_8N_4O_2$; mol wt 180.17. C 46.66%, H 4.48%, N 31.10%, O 17.76%. The principal alkaloid of the cacao bean which contains 1.5-3% of the base. Also present in cola nuts and in tea. Usually extracted from the hull of cacao beans which contains 0.7-1.2%. Extraction process: Schwyzer, *Die Fabrikation Phar-*

mazeutischer und Chemisch-Technischer Produkte (Berlin, 1931). Synthesis starting with 3-methyluric acid: Fischer, Ach, *Ber.* **31**, 1980 (1898); *cf.* Gebner, Krebs, *J. Gen. Chem. USSR* **16**, 179-186 (1946). Comparison with theophylline, *q.v.*, of metabolism in humans: D. J. Birkett *et al., Drug Metabol. Dispos.* **13**, 725 (1985). Pharmacokinetics: A. Lelo *et al., Brit. J. Clin. Pharmacol.* **22**, 177 (1986). Bronchodilator effect in asthma: F. E. R. Simons *et al., J. Allerg. Clin. Immunol.* **76**, 703 (1985).

Monoclinic needles (lamellar twinning on 001) from water, mp 357°. Sublimes 290-295°. Kb at 18°: 1.3×10^{-14}; Ka 0.9×10^{-10}. Absorption spectrum: Hartley, *J. Chem. Soc.* **87**, 1803, 1810 (1905). One gram dissolves in about 2000 ml water, 150 ml boiling water, 2220 ml 95% alcohol; sol in the fixed alkali hydroxides, concd acids, in about 22 parts of 20% aq tribasic sodium phosphate soln; moderately sol in ammonia. Almost insol in benzene, ether, chloroform, carbon tetrachloride. Forms salts which are dec by water, and compds with bases which are more stable.

Calcium salicylate, *Theocalcin, Calcium Diuretin, Theosol, Theosalin, Theocal, Calcotheobromine.* Double salt or mixture of calcium theobromine and calcium salicylate. Contains no less than 44% theobromine. White, amorphous powder. Slightly saline taste. Partly sol in water. Aq solns are alkaline to phenolphthalein.

Sodium acetate, *Agurin, Thesodate.* Equimolar mixture of sodium theobromine and sodium acetate, containing $1H_2O$. Theobromine 59.6%, anhyd sodium acetate 27.1%. White, odorless or almost odorless, hygroscopic powder. Absorbs CO_2 from the air becoming incompletely sol. Very sol in water, sparingly sol in cold alcohol; the solns are strongly alkaline. pH about 10. *Keep tightly closed.*

Sodium salicylate, *Diuretin.* Equimolar mixture of sodium theobromine and sodium salicylate, containing $1H_2O$. Theobromine 47.3%, sodium salicylate 42.1%. White, odorless or almost odorless, hygroscopic powder. Absorbs CO_2 from the air and becomes incompletely soluble. Soluble in 1 part water; slightly sol in alcohol; the solns are strongly alkaline. pH about 10.

THERAP CAT: Diuretic, bronchodilator, cardiotonic.

THERAP CAT (VET): Diuretic, myocardial stimulant, vasodilator.

9210. 1-Theobromineacetic Acid. *2,3,6,7-Tetrahydro-3,7-dimethyl-2,6-dioxo-1H-purine-1-acetic acid; 3,6-dihydro-3,7-dimethyl-2,6-dioxo-1(2H)-purineacetic acid;* 3,6-dihydro-2,6-diketo-3,7-dimethyl-1(2)-purineacetic acid. $C_9H_{10}N_4O_4$; mol wt 238.21. C 45.38%, H 4.23%, N 23.52%, O 26.87%. Prepd from sodium theobromine + chloroacetic acid: E. Merck *et al.,* Ger. pat. 352,980, *C.A.* **17**, 1307[1] (1923).

Crystals, mp 260°.

Sodium salt, $C_9H_9N_4NaO_4$, *sodium theobromine acetate, Técarine.* Crystalline powder, freely sol in water, forming a neutral soln.

THERAP CAT: Bronchodilator.

9211. Theofibrate. *2-(4-Chlorophenoxy)-2-methylpropanoic acid 2-(1,2,3,6-tetrahydro-1,3-dimethyl-2,6-dioxo-7H-purin-7-yl)ethyl ester; 2-(p-chlorophenoxy)-2-methylpropionic acid ester with 7-(2-hydroxyethyl)theophylline; 1-(the-*

ophyllin-7-yl)ethyl 2-(p-chlorophenoxy)isobutyrate; etofylline clofibrate; ML-1024; Duolip. $C_{19}H_{21}ClN_4O_5$; mol wt 420.86. C 54.22%, H 5.03%, Cl 8.42%, N 13.31%, O 19.01%. Deriv of clofibric acid, *q.v.*, with antilipemic, antithrombotic, and platelet-aggregation inhibitory activity: Prepn: G. Metz, M. Specker, Ger. pat. **2,308,826** corresp to U.S. pat. **3,984,413** (1974, 1976 both to Merckle); *eidem, Arzneimittel-Forsch.* **25**, 1686 (1975). Hypolipemic activity: G. Metz *et al., ibid.* **27**, 1173 (1977). Series of articles on chemistry, biopharmaceutic evaluation, pharmacology, toxicology, clinical studies: *ibid.* **30**, 2013-2074 (1980).

Colorless cryst from ethanol, mp 133-135°. Practically insol in water at pH 2-7.4 and in cold alcohols. Sol in acetone, chloroform, hot alcohols. LD_{50} in mice, rats, dogs (g/kg): 11.7, 17.0, > 10.0 orally, G. Metz *et al., loc. cit.*

THERAP CAT: Antihyperlipoproteinemic.

9212. Theophylline. *3,7-Dihydro-1,3-dimethyl-1H-purine-2,6-dione;* 1,3-dimethylxanthine; theocin; Accurbron; Aerobin; Afonilum; Armophylline; Bilordyl; Bronchoretard; Bronkodyl; Cetraphylline; Duraphyl; Duraphyllin; Diffumal; Elixicon; Euphylline L.A.; Euphylong; LaBID; Labophylline; Lasma; Physpan; Pro-Vent; PulmiDur; Pulmo-Timelets; Respbid; Rona-Phyllin; Sabidal; Slo-Phyllin; Solosin; Sustaire; Teobid; Tesona; Theal tabl.; Theocap; Theoclear; TheoChron; Theocontin; Theo-Dur; Theolair; Theolan; Theophyl-SR; Theolix; Theograd; Theoplus; Theosol; Theostat; Theovent; Uni-Dur; Unifyl; Uniphyl; Uniphyllin; Xanthium. $C_7H_8N_4O_2$; mol wt 180.17. C 46.66%, H 4.48%, N 31.10%, O 17.76%. Xanthine derivative with diuretic, cardiac stimulant and smooth muscle relaxant activities; isomeric with theobromine, *q.v.* Small amounts occur in tea. Synthesis starting with dimethylurea and ethyl cyanoacetate: Traube, *Ber.* **33**, 3035 (1900); Grinberg, *J. Appl. Chem. USSR* **13**, 1461 (1940); Gebner, Krebs, *J. Gen. Chem. USSR* **16**, 179 (1946). Bioavailability: K. Svedmyr *et al., Allergy* **37**, 111 (1982). Comprehensive description: J. L. Cohen in *Analytical Profiles of Drug Substances* vol. 4, K. Florey, Ed. (Academic Press, New York, 1975) pp 466-493. Symposium on clinical experience in asthma and allergy: *Am. J. Med.* **79**(6A), 1-78 (1985); on pharmacology, bioavailability, pharmacokinetics and efficacy in obstructive pulmonary disease: *J. Allerg. Clin. Immunol.* **78**(4 Part 2), 669-824 (1986).

Monohydrate, thin monoclinic tablets from water, mp 270-274°. Bitter taste. pKa at 25°: 8.77; pKb: 13.5, 11.5. One gram dissolves in 120 ml water, 80 ml alcohol, about 110 ml chloroform. Sol in hot water, in alkali hydroxides, ammonia, dil HCl or HNO_3; sparingly sol in ether. uv max (0.1N NaOH): 274 nm.

Diethanolamine, *Adisné.* A compd of one mol of theophylline with one or two mols of diethanolamine (and $1H_2O$) corresponding to 40-57% hydrated theophylline. White, odorless powder. The product with one mol diethanolamine is sol in 20 parts water, with 2 mols in 5 parts water.

Ethanolamine, $C_9H_{15}N_5O_3$, *Monotheamin, Theamin.* A white, crystalline, nonhygroscopic powder contg 75% anhydr theophylline and 25% ethanolamine. Sol in water.

Isopropanolamine, $C_{10}H_{17}N_5O_3$, *Oxyphyllin, Theopropanol.* Prisms from alcohol. Very sol in water. pH of aq solns 9.20. Prepn: Greenbaum, *Am. J. Pharm.* **109**, 550 (1937).

Sodium acetate, *Theocin Soluble, Acet-Theocin Sodium.* Equimolar mixture of theophylline sodium and sodium acetate, containing $1H_2O$. White, odorless, crystalline powder; bitter taste. Sol in about 25 parts water. Insol in alcohol, chloroform or ether.

Sodium glycinate, *Biophylline, Englate, Glytheonate, Nuelin, Panophylline.* Equimolar mixture of theophylline sodium and glycine. Prepn: Krantz *et al., J. Am. Pharm. Assoc.* **36**, 248 (1947). Darkens at 180°. Soly in water 18% w/v. Saturated soln has a pH of 8.7-9.1, d^{20} 1.05, and is stable to carbon dioxide.

THERAP CAT: Bronchodilator.

THERAP CAT (VET): Bronchodilator.

9213. Thermolysin. Bacillus thermoproteolyticus neutral proteinase; E.C. 3.4.24.4. Proteolytic enzyme of mol wt 37,500 that hydrolyzes protein bonds on the *N*-terminal side of hydrophobic amino acid residues. Contains a zinc atom essential for activity and four Ca^{2+} ions essential for thermal and conformational stability. Isoln from *Bacillus thermoproteolyticus:* S. Endo, *J. Ferment. Technol.* **40**, 346 (1962). Properties and amino acid composition: Y. Ohta *et al., J. Biol. Chem.* **241**, 5919 (1966). Site of enzymatic hydrolysis: Y. Ohta, Y. Ogura, *J. Biochem.* **58**, 607 (1965). Substrate specificity studies: H. Matsubara *et al., Biochem. Biophys. Res. Commun.* **21**, 242 (1965); **24**, 427 (1966); K. Morihara, H. Tsuzuki, *Biochim. Biophys. Acta* **118**, 215 (1966). Stability studies: Y. Ohta, *J. Biol. Chem.* **242**, 509 (1967). Inhibition studies: H. Matsubara *et al., Biochem. Biophys. Res. Commun.* **34**, 719 (1969); J. Murphy *et al., Arch. Biochem. Biophys.* **202**, 405 (1980). Purification: H. Matsubara, *Methods Enzymol.* **19**, 642 (1970). Structure studies: K. Titani *et al., Nature* **238**, 35 (1972); B. W. Matthews *et al., ibid.* **37**, 41; P. M. Colman *et al., J. Mol. Biol.* **70**, 701 (1972). Function of the metal ions: J. Feder *et al., Biochemistry* **10**, 4552 (1971); G. Voordouw, R. S. Roche, *ibid.* **14**, 4667 (1975); R. S. Roche, G. Voordouw, *CRC Crit. Rev. Biochem.* **5**, 1 (1978). Effect of the histidyl residue on the mechanism of action: S. Blumberg *et al., Isr. J. Chem.* **12**, 643 (1974); M. K. Pangburn, K. A. Walsch, *Biochemistry* **14**, 4050 (1975). Review: *Experientia* **26**, Suppl., 31-59 (1976).

Crystals. uv max: 280 nm (ϵ 66,300). Optimum pH 7.0-8.5; stable at pH 6.0-9.0. The refrigerated lyophilized enzyme is stable for months; frozen enzyme soln can be kept for weeks without significant loss of activity. Not deactivated at 65°, but loses half of its activity upon heating at 80° for 1 hr.

USE: In studies of protein sequences.

9214. Thermorubin. Antibiotic substances isolated from *Thermoactinomyces antibioticus:* Craveri *et al., Clin. Med.* **71**, 511 (1964); *eidem,* **Ger. pat.** **1,180,891;** *eidem,* **U.S. pat.** **3,300,379** (1964, 1967 both to Lepetit). Structure of thermorubin A, the major component: Moppett *et al., J. Am. Chem. Soc.* **94**, 3269 (1972); revised structure: F. Johnson *et al., ibid.* **102**, 5580 (1980). Inhibition of bacterial protein synthesis: G. Pirali *et al., Biochim. Biophys. Acta* **366**, 310 (1974). Studies on mechanism of action: F. Lin, A. Wishnia, *Biochemistry* **21**, 477, 484 (1982).

Orange-red rosettes and needles from chloroform, darkens 190°, chars 300° without melting. $[\alpha]_D^{25}$ -14° (c = 0.4 in dioxane). Absorption max [dioxane-cyclohexane (1:1)]: 300, 328, 414, 430 nm ($E_{1cm}^{1\%}$ 1041, 1066, 313, 332). Sol in dioxane, pyridine, tetrahydrofuran, DMF, DMSO, concd alkalies, concd H_2SO_4, glacial acetic acid; slightly sol in

methanol, ethanol, butanol, ethyl acetate, acetone, chloroform, benzene, cyclohexane. Practically insol in water, ether, petr ether, hexane. LD_{50} i.p. in mice: 300 mg/kg (Craveri, 1967).

Diacetate, crystals from ethyl acetate. Absorption max [0.1*N* HCl-dioxane (1:1)]: 337, 436, 462 nm ($E_{1cm}^{1\%}$ 1500, 113, 162).

Thermorubin A, $C_{32}H_{24}O_{12}$, dihydrate, orange-red rosettes from ethyl acetate. Absorption max (ethanol): 250, 300, 328, 435 nm (ϵ 33700, 54640, 49980, 16300). pKa values: 4.7, 7.0, 9.0. Easily tautomerizes in soln. Thermally labile, decomposes rapidly in soln at temps > 60°.

9215. Thevetin A. *3-[(O-β-D-Glucopyranosyl-(1 → 6)-O-D-glucopyranosyl-(1 → 4)-6-deoxy-3-O-methyl-α-L-glucopyranosyl)oxy]-14-hydroxy-19-oxocard-20(22)-enolide.* $C_{42}H_{64}O_{19}$; mol wt 872.93. C 57.78%, H 7.39%, O 34.83%. Glycoside isolated from *Thevetia neriifolia* Juss., *Apocynaceae:* Bloch *et al., Helv. Chim. Acta* **43**, 652 (1960). Separation from thevetin B *(see under* Cerberoside*):* Delalande, Baisse, **U.S. pats.** **3,030,355; 3,043,829** (1962). Structure: K. Tori *et al., Tetrahedron Letters* **1977**, 717.

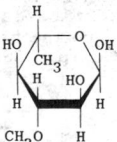

O-thevetose-gentiobiose

Crystals from water. mp 208-210°. $[\alpha]_D^{24}$ -72.0° ±1.5° (c = 1.48 in methanol).

Acetyl deriv, $C_{58}H_{80}O_{27}$, crystals from methanol + ether, mp 143-149°. $[\alpha]_D^{26}$ -54.2° ±1° (c = 1.86 in chloroform).

9216. Thevetose. *6-Deoxy-3-O-methylglucose;* 3-O-methylglucomethylose; quinovose 3-methyl ether. $C_7H_{14}O_5$; mol wt 178.18. C 47.18%, H 7.92%, O 44.90%. Isoln of L-thevetose from *Thevetia nereifolia* Juss. and *Tanghinia venenifera* Poir., *Apocynaceae:* Frèrejacque, *Compt. Rend.* **221**, 645 (1945); Frèrejacque, Hasenfratz, *ibid.* **222**, 815 (1946); Sigg *et al., Helv. Chim. Acta.* **38**, 166 (1955); from thevetin: Helfenberger, Reichstein, *ibid.* **31**, 1470 (1948); from *Bowiea volubilis* Harv., *Liliaceae:* Tschesche, Dölberg, *Ber.* **90**, 2378 (1957). Synthesis: Blindenbacher, Reichstein, *Helv. Chim. Acta* **31**, 1669 (1948). Isoln of D-thevetose from *Adenium honghel* D.C., *Apocynaceae:* Frèrejacque, *Compt. Rend.* **230**, 127 (1950). Rare instance of two optical isomers of the same sugar being present in the vegetable kingdom.

α-L-thevetose

L-Form, needles from acetone + ether. mp 128-130°. $[\alpha]_D^{19}$ -66° → -36.9° (c = 1.46 in water).

β-L-Triacetate, $C_{13}H_{20}O_8$, crystals, mp 118-119°. $[\alpha]_D^{18}$ -7.5° (c = 1.12 in acetone).

L-Thevetosazone, crystals, mp 136-139°. $[\alpha]_D^{24}$ $+110.8$° → $+40.3$°.

D-Form, crystals, mp 116°. $[\alpha]_D^0$ $+84$° → $+33$°.

β-D-Triacetate, $C_{13}H_{20}O_8$, crystals, mp 121°. $[\alpha]_D^{20}$ $+6$° (acetone).

α-D-Triacetate, crystals, mp 105°. $[\alpha]_D^{20}$ $+122$° (acetone).

D-Thevetosazone, crystals, mp 123-128°. $[\alpha]_D^{16}$ -117.8° → -26.4°.

9217. Thiabendazole. *2-(4-Thiazolyl)-1H-benzimida-*

zole; 4-(2-benzimidazolyl)thiazole; MK-360; Omnizole; Thiaben; Thibenzole; Bovizole; Eprofil; Equizole; Mintezol; Top Form Wormer; Mertect; Lombristop; Minzolum; Nemapan; Polival; TBZ; Tecto. $C_{10}H_7N_3S$; mol wt 201.26. C 59.68%, H 3.51%, N 20.88%, S 15.93%. Prepd by the reaction of 4-thiazolecarboxamide with *o*-phenylenediamine in polyphosphoric acid: H. D. Brown *et al., J. Am. Chem. Soc.* **83**, 1764 (1961); L. H. Sarett, H. D. Brown, U.S. pat. **3,017,415** (1962 to Merck & Co.). Synthesis of labeled thiabendazole: D. J. Tocco *et al., J. Med. Chem.* **7**, 399 (1964). Alternate route of synthesis: V. J. Grenda *et al., J. Org. Chem.* **30**, 259 (1965). Anthelmintic props: H. D. Brown *et al., loc. cit.;* K. C. Kates *et al., J. Parasitol.* **57**, 356 (1971). Fungicidal props: H. J. Robinson *et al., J. Invest. Dermatol.* **42**, 479 (1966). Systemic props in plants: D. C. Erwin *et al., Phytopathology* **58**, 860 (1968). Toxicity: H. J. Robinson *et al., Toxicol. Appl. Pharmacol.* **7**, 53 (1965). Residue analysis: IUPAC Appl. Chem. Div., *Pure Appl. Chem.* **52**, 2567 (1980). Comprehensive description: V. K. Kapoor in *Analytical Profiles of Drug Substances* **vol. 16**, K. Florey, Ed. (Academic Press, New York, 1986) pp 611-639.

Colorless crystals, mp 304-305°. uv max (methanol): 298 nm (ϵ 23330). Fluorescence max in acid soln: 370 nm (310 nm excitation). Max soly in water at pH 2.2: 3.84%. Soluble in DMF, DMSO. Slightly soluble in alcohols, esters, chlorinated hydrocarbons. LD_{50} in mice, rats, rabbits (g/kg): 3.6, 3.1, >3.8 orally (Robinson).

Hypophosphite, *Arbotect.* Amber liquid. d^{25} 1.103.

USE: Systemic fungicide for spoilage control of citrus fruit; for treatment and prevention of Dutch elm disease in trees; for control of fungal diseases of seed potatoes.

THERAP CAT: Anthelmintic (Nematodes).

THERAP CAT (VET): Anthelmintic, fungicide.

9218. Thiacetazone. *N-[4-[[(Aminothioxomethyl)hydrazono]methyl]phenyl]acetamide;* 2-[[4-(acetylamino)phenyl]-methylene]hydrazinecarbothioamide; 4'-formylacetanilide thiosemicarbazone; *p*-acetamidobenzaldehyde thiosemicarbazone; *p*-acetylaminobenzaldehyde thiosemicarbazone; *p*-acetaminobenzylidenethiosemicarbazone; amithiozone; thibone; thioacetazone; Tb I/698; Sdt 1041; RP 4207; Conteben; Thizone; Novakol; Thionicid; Domäkol; Aktivan; Ambathizon; Seroden; Benthiozone; Berkazon; Diasan; Ilbion; Tebalon; Siocarbazone; Tibicur; Thiomicid; Tibione; Tebethion; Panrone; Thioparamizone; Myvizone; Tiobicina; Tubercazon; Berculon A; Antib; Neustab; Nuclon Argentinian; Livazone; Benzothiozon; Parazone; Tebezon; Thiotebezin; Thiocarbazil; Tiocarone; Tiosecolo. $C_{10}H_{12}N_4OS$; mol wt 236.29. C 50.83%, H 5.12%, N 23.71%, O 6.77%, S 13.57%. Prepd by treating *p*-acetamidobenzaldehyde with thiosemicarbazide in alcohol: Domagk *et al., Naturwiss.* **33**, 315 (1946); Behnisch *et al., Angew. Chem.* **60A**, 113 (1948); G. Domagk, *Chemotherapie mit den Thiosemikarbazonen* (Thieme, Stuttgart, 1950); Chabrier, Cattelain, *Bull. Soc. Chim. France* **1950M**, 52. Many alternate synthetic routes have been described, *e.g.,* Das, Mukherjee, *J. Am. Chem. Soc.* **75**, 1241 (1953); Carvajal, Espinosa, *Ciencia (Mexico)* **12**, 231 (1953). Toxicity data: Bavin *et al., J. Pharm. Pharmacol.* **2**, 764 (1950).

Minute, pale yellow crystals from abs alc. Bitter taste. Darkens on exposure to light. Dec 225-230°. uv max (ethanol): 328 nm, extinction 0.580. Sol in hot alc; very sparingly sol in cold alc. Insol in water, acetone, benzene, carbon tetrachloride, chloroform, carbon disulfide, petr ether. Practically insol in the other common organic solvents except glycols. Soly in propylene glycol about 1%. It is pos-

sible to prepare a 4% soln in 90% methyl acetamide (methyl acetamide contg 10% H_2O). LD_{50} s.c. in mice: 1-2 g/kg (Bavin).

THERAP CAT: Antibacterial (tuberculostatic).

9219. Thialbarbital. *5-(2-Cyclohexen-1-yl)dihydro-5-(2-propenyl)-2-thioxo-4,6(1H,5H)-pyrimidinedione;* 5-allyl-5-(2-cyclohexen-1-yl)-2-thiobarbituric acid; 5-(2-cyclohexen-1-yl)-5-allyl-2-thiobarbituric acid; thialbarbitone; Intranarcon; Kemithal. $C_{13}H_{16}N_2O_2S$; mol wt 264.36. C 59.06%, H 6.10%, N 10.60%, O 12.10%, S 12.13%. Prepn: Volwiler, Tabern, U.S. pat. **2,153,730** (1939 to Abbott).

Crystals, mp 148-150°.

Sodium salt, $C_{13}H_{15}N_2NaO_2S$, pale yellow crystals from methanol, mp 130-132°. Very sol in water, alc; practically insol in benzene, ether. pH of 2.5% soln in water: 10.5.

Note: This is a controlled substance (depressant) listed in the U.S. Code of Federal Regulations, Title 21 Part 1308.13 (1987).

THERAP CAT: Anesthetic (intravenous).

THERAP CAT (VET): Anesthetic.

9220. Thiambutene. *N,N-Diethyl-1-methyl-3,3-di-2-thienylallylamine;* 3-diethylamino-1,1-di(2'-thienyl)but-1-ene; diethylthiambutene; 191C49; NIH-4185; Diethibutin; Themalon; Kemithal. $C_{16}H_{21}NS_2$; mol wt 291.49. C 65.93%, H 7.26%, N 4.81%, S 22.00%. Prepn: Adamson, *J. Chem. Soc.* **1950**, 885; U.S. pat. **2,561,899** (1951 to Burroughs Wellcome). Use: Hayes, *Vet. Rec.* **83**, 528 (1968). Toxicity data: J. S. McKenzie, N. R. Beechey, *Arch. Int. Pharmacodyn. Ther.* **135**, 376 (1962).

$bp_{0.03}$ 122-128°. LD_{50} in mice: 90 mg/kg i.p. (McKenzie, Beechey).

Hydrochloride, $C_{16}H_{21}NS_2$·HCl, crystals, mp 152-153°.

Caution: May be habit forming. This is a controlled substance (opiate) listed in the U.S. Code of Federal Regulations, Title 21 Part 1308.11 (1985).

THERAP CAT (VET): Narcotic analgesic.

9221. Thiamine Disulfide. *N,N'-[Dithiobis[2-(2-hydroxyethyl)-1-methyl-2,1-ethenediyl]]bis[N-[(4-amino-2-methyl-5-pyrimidinyl)methyl]formamide];* aneurine disulfide; vitamin B_1 disulfide; Neolamin. $C_{24}H_{34}N_8O_4S_2$; mol wt 562.72. C 51.23%, H 6.09%, N 19.91%, O 11.37%, S 11.40%. Isoln: Zima, Williams, *Ber.* **73**, 941 (1940). Prepn: Warnat, U.S. pat. **2,458,453** (1949 to Hoffmann-La Roche); Matsukawa, Hirano, *J. Pharm. Soc. Japan* **73**, 379 (1953); Hirano, Iwatsu, *ibid.* **73**, 1115 (1953); Kawasaki, *ibid.* **76**, 706 (1956); Yurugi, *C.A.* **51**, 16486b (1957); Kawasaki, Yonemoto, *C.A.* **51**, 16487c (1957); Kawasaki, Tomita, *C.A.* **53**, 5273a (1958).

When anhydr mp 177° with intense yellow coloration. Very sparingly sol in water, benzene, acetone, ether, ethanol; freely sol if the crystals contain solvated acetone or water.

Hydrochloride, crystals from methanol + water, dec 231°.

Phosphate, *thiamine monophosphate disulfide, Biotinin, Vitamogen.*

O,O-Diisobutyrate, $C_{32}H_{46}N_8O_6S_2$, *O-isobutyrylthiamine disulfide, bisibutiamine, sulbutiamine, Arcalion, Neodaian, Vitaverin.*

THERAP CAT: Enzyme co-factor vitamin. *O,O*-Diisobutyrate as a psychotropic agent.

9222. Thiamine Hydrochloride. *3-[(4-Amino-2-methyl-5-pyrimidinyl)methyl]-5-(2-hydroxyethyl)-4-methylthiazolium chloride monohydrochloride;* vitamin B_1 hydrochloride; thiamine chloride hydrochloride; aneurine hydrochloride; thiaminium chloride hydrochloride; Aneurin-AS; Bedome; Begiolan; Benerva; Bequin; Berin; Betabion hydrochloride; Betalin S; Betaxin; Bethiazine; Bevitex; Bewon; Biuno; Bivatin; Bivita; Clotiamina; Metabolin; Thiadoxine; Thiavit; Timidon; Tiaminal; Vitaneuron. $C_{12}H_{18}Cl_2N_4OS$; mol wt 337.28. C 42.73%, H 5.38%, Cl 21.03%, N 16.61%, O 4.74%, S 9.51%. Occurs in plants and in animal tissues, notably in rice husk, cereal grains, yeast, liver, eggs, milk, green leaves, roots and tubers. Isoln from rice bran: Jansen, Donath, *Chem. Weekblad* **23**, 201 (1926). Structure: Williams, *J. Am. Chem. Soc.* **58**, 1063 (1936); Williams, Cline, *ibid.* **58**, 1504 (1936); Williams *et al.*, *ibid.* **59**, 526 (1937); Gravin, *J. Appl. Chem. USSR* **16**, 105 (1943). Review of syntheses: Léon Velluz, *Substances Naturelles de Synthese* vol. III (Paris, 1951) pp 59-80; Knobloch in H. Vogel, *Chemie und Technik der Vitamine* vol. II (Stuttgart, 1953) pp 1-128. Practically all vitamin B_1 sold is synthetic. Toxicity data: D. Winter *et al.*, *Int. Z. Vitaminforsch.* **37**, 82 (1967). Comprehensive review: several authors in *The Vitamins*, vol. 5, W. H. Sebrell, R. S. Harris, Eds. (Academic Press, New York, 2nd ed., 1972) pp 97-164; *Ann. N.Y. Acad. Sci.* **378**, entitled "Thiamin—Twenty Years of Progress", H. Z. Sable, C. J. Grubier, Eds. (1982) 470 pp.

Monoclinic plates in rosette-like clusters. Slight thiazole odor. Bitter taste. dec 248°. One gram dissolves in about 1 ml water, 18 ml glycerol, 100 ml 95% alcohol, 315 ml abs alcohol; more sol in methanol. Sol in propylene glycol. Practically insol in ether, benzene, hexane, chloroform. pH of a 1% w/v soln in water 3.13; pH of a 0.1% w/v soln in water 3.58. On exposure to air of average humidity, the vitamin absorbs an amount of water corresponding to nearly one mol, forming a hydrate. The article of commerce contains about 4% water which is removable by drying at 100° or in a vacuum over H_2SO_4. In the dry form the vitamin is stable, and heating at 100° for 24 hrs does not diminish its potency. In aq soln it can be sterilized at 110°, but if the pH of the soln is above 5.5, it is destroyed rapidly. One gram of cryst vitamin B_1 hydrochloride is equivalent to 333,00 I.U. Absorption curves and analytical thiochrome procedures depending upon the oxidation of thiamine to thiochrome which fluoresces in ultraviolet light, are given in *Methods of Vitamin Assay*, edited by the Assoc. of Vitamin Chemists, Inc. (Interscience, New York, 2nd ed., 1951). LD_{50} in mice (mg/kg): 89.2 i.v.; 8224 orally (Winter).

Incompat: The vitamin is destroyed by alkalies and alkaline drugs such as phenobarbital sodium and by oxidizing and reducing agents. It is precipitated by tannins (which occur in wine) and by reagents which precipitate alkaloids, *e.g.* Mayer's reagent, mercuric chloride, picric acid, iodine. For stability in dry prepns over several years *see* Partington, Waterhouse, *J. Pharm. Pharmacol.* **5**, 715, 721 (1953). Chemical fate in alkaline solns: Maier, Metzler, *J. Am. Chem. Soc.* **79**, 4386 (1957).

THERAP CAT: Enzyme co-factor vitamin.
THERAP CAT (VET): Nutritional factor (vitamin B_1).

9223. Thiamine Mononitrate. Vitamin B_1 mononitrate; aneurine mononitrate; 3-(4-amino-2-methylpyrimidyl-5-methyl)-4-methyl-5-(β-hydroxyethyl)thiazolium nitrate; Betabion mononitrate. $C_{12}H_{17}N_5O_4S$; mol wt 327.36. C 44.02%, H 5.23%, N 21.40%, O 19.55%, S 9.80%. Prepd by removing all chloride ions from thiamine chloride hydro-

chloride by treatment with alkali and reconstituting the half salt by treatment of the free base with a stoichiometric amount of nitric acid. Alternate route: Turner, Schmitt, U.S. pat. **2,844,579** (1958 to Am. Cyanamid).

Crystals, dec 196-200°. Practically nonhygroscopic. pKa 4.8. Soly in water 2.7 g/100 ml at 25° and approx 30 g/100 ml at 100°. pH of 2% aq soln 6.5 to 7.1. Solns of pH 4.0 show greater stability than neutral solns and can be prepd in concns as high as 18.5 g/100 ml at room temp. For preparing solns of pH 4.0, approx 2.6 ml of 1.0*N* HCl is required for each gram of thiamine mononitrate when no other acidic or basic substances are present. One gram of thiamine mononitrate is equal to 343,000 international units. More stable than the chloride hydrochloride; especially recommended for enrichment of flour mixes and prepn of multivitamin capsules and tablets.

THERAP CAT: Enzyme co-factor vitamin.
THERAP CAT (VET): *See* Thiamine Hydrochloride.

9224. Thiamine Phosphoric Acid Ester Chloride. *3-[(4-Amino-2-methyl-5-pyrimidinyl)methyl]-4-methyl-5-[2-(phosphonooxy)ethyl]thiazolium chloride;* thiamine monophosphate ester chloride; thiamine orthophosphate ester chloride. $C_{12}H_{18}ClN_4O_4PS$; mol wt 380.78. C 37.85%, H 4.76%, Cl 9.31%, N 14.71%, O 16.81%, P 8.13%, S 8.42%. Prepn by sulfuric acid hydrolysis of cocarboxylase: Lohmann, Schuster, *Biochem. Z.* **294**, 188 (1937); Karrer, Viscontini, *Helv. Chim. Acta* **29**, 711 (1946); Zima *et al.*, *E. Merck's Jahresber.* **67**, 13 (1953).

Monohydrate, plates, tablets, or stout prisms from dil acetone or water, mp about 200°. Freely sol in water.
Hydrochloride, $C_{12}H_{18}ClN_4O_4PS$·HCl, crystals. Freely sol in water.

9225. Thiamine Phosphoric Acid Ester Phosphate Salt. Thiamine monophosphate ester phosphoric acid salt; Umbeon. $C_{12}H_{20}N_4O_8P_2S$; mol wt 442.35. C 32.58%, H 4.56%, N 12.67%, O 28.94%, P 14.00%, S 7.25%. Obtained by boiling an aq soln of thiamine triphosphoric acid ester for 2 hrs: Velluz *et al.*, *Bull. Soc. Chim. France* **15**, 871 (1948); from thiamine phosphoric acid ester chloride hydrochloride by ion exchange: Zima *et al.*, *E. Merck's Jahresber.* **67**, 10 (1953).

Monohydrate, needles from dil acetone, mp 195-197°. When anhydr, dec 228-230°. Freely sol in water. Insol in the usual organic solvents.

9226. Thiamine 1,5-Salt. Aneurin-1,5-salt; thiamine chloride naphthalene-1,5-disulfonic acid salt. $C_{22}H_{24}N_4O_7S_3$. Prepd as $[C_{12}H_{18}N_4OS]^{2+}·C_{10}H_6(SO_3^-)_2·H_2O$, by mixing a 10% aq soln of thiamine chloride hydrochloride with a 10% aq soln of 1,5-naphthalenedisulfonic acid, cooling the mixture and recovering the cryst product: Westphal *et al.*,

U.S. pat. **2,694,642** (1954 to Schenley Ind.); *see also* **Ger. pat. 831,247** (1952 to Bayer).

Said to be more stable than the hydrochloride, comparable to thiamine mononitrate. Soly in water: 1 g/150 ml.

9227. Thiamine Triphosphoric Acid Ester. Thiamine triphosphate ester; vitamin B_1 triphosphate ester. $C_{12}H_{19}N_4$-$O_{10}P_3S$; mol wt 504.28. C 28.58%, H 3.80%, N 11.11%, O 31.73%, P 18.42%, S 6.36%. Prepn: Velluz *et al., Bull. Soc. Chim. France* **15**, 871 (1948); Roux *et al., Bull. Soc. Chim. Biol.* **30**, 592 (1948); Viscontini *et al., Helv. Chim. Acta* **32**, 1478 (1949); Velluz, Bartos, *Bull. Soc. Chim. France* **18**, 118 (1951). According to Roux the third phosphoric acid group is attached to the NH_2 of the pyrimidine moiety. This opinion is not shared by Velluz and Karrer's group: *Experientia* **6**, 386 (1950). Comparison with other thiamine phosphate esters and salts: Zima *et al., E. Merck's Jahresber.* **67**, 1-58 (1953). Determn of thiamine and its phosphate esters by electrophoresis and fluorometry: H. K. Penttinen, *Acta Chem. Scand. B* **32**, 609 (1978).

Hemihydrate, very hygroscopic, rhomb-shaped microcrystals from water-alcohol-acetone, dec 228-232°. Freely sol in water. Insol in the usual organic solvents. Easily hydrolyzes to cocarboxylase monophosphate salt especially when dissolved in acidulated water. Restores the cocarboxylase activity of washed yeast.

Orthophosphoric acid salt monohydrate, $C_{12}H_{22}N_4O_{14}$-$P_4S.H_2O$, *Trifosfaneurina.* Very hygroscopic crystals. Sol in water.

9228. Thiamiprine. *6-[(1-Methyl-4-nitro-1H-imidazol-5-yl)thio]-1H-purin-2-amine; 2-amino-6-[(1-methyl-4-nitroimidazol-5-yl)thio]purine;* 2-amino-6-(1'-methyl-4'-nitro-5'-imidazolyl)mercaptopurine; Guaneran. $C_9H_8N_8O_2S$; mol wt 292.29. C 36.98%, H 2.76%, N 38.34%, O 10.95%, S 10.97%. Prepd from thioguanine and 1-methyl-4-nitro-5-chloroimidazole: Hitchings, Elion, U.S. pat. **3,056,785** (1962 to Burroughs Wellcome).

Crystals, dec slowly > 200°. uv max: 320 nm at pH 1; 315 nm at pH 11.
THERAP CAT: Antineoplastic.

9229. Thiamorpholine. *Thiomorpholine;* tetrahydro-1,4-thiazine; parathiazan. C_4H_9NS; mol wt 103.18. C 46.56%, H 8.79%, N 13.58%, S 31.07%. Prepd by heating β,β'-dichlorodiethyl sulfide with alcoholic ammonia under pressure: Davies, *J. Chem. Soc.* **117**, 298, 306 (1920).

Mobile liquid. Strong odor resembling that of piperidine. Absorbs CO_2 from air. bp_{758} 169°; bp_{743} 166-167°. Volatile with steam. Miscible with water and many organic liquids.

Hydrochloride, $C_4H_9NS.HCl$, hygroscopic needles from alcohol + ether, dec 163°. Freely sol in water.

Picrate, $C_4H_9NS.C_6H_3N_3O_7$, orange needles, dec 198°. Sol in acetone, slightly sol in alcohol.

Picrolonate, $C_4H_9NS.C_{10}H_8N_4O_5$, deep orange prisms from alcohol, dec 242° (darkens at 210°).

9230. Thiamphenicol. D-*threo-2,2-Dichloro-N-[β-hydroxy-α-(hydroxymethyl)-p-(methylsulfonyl)phenethyl]acetamide;* D-*d-threo*-2-dichloroacetamido-1-(4-methylsulfonyl)-1,3-propanediol; dextrosulphenidol; thiophenicol; Win 5063-2; 8053CB; Hyrazin; Igralin; Macphenicol; Neomyson; Thiamcol; Thiocymetin; Urfamycine; Urophenyl; Descocin; Rigelon; Thionicol; Vicemycetin; Propacin. $C_{12}H_{15}Cl_2NO_5S$; mol wt 356.23. C 40.46%, H 4.24%, Cl 19.91%, N 3.93%, O 22.46%, S 9.00%. Synthesis: Cutler *et al., J. Am. Chem. Soc.* **74**, 5475 (1952); Suter *et al., ibid.* **75**, 4330 (1953); Suter, U.S. pats. **2,759,927; 2,759,970-1, 2, 6** (1956 to Sterling Drugs); **Brit.** pat. **770,277** (1957 to Parke, Davis).

Crystals, mp 164.3-166.3°. $[\alpha]_D^{25}$ +12.9° (ethanol). uv max (95% ethanol): 224, 266, 274 nm (ε 13,700, 800, 700). Appreciably sol in water. Soluble in alcohol.

DL-Form, *racephenicol, Dexawin, Raceophenidol.*
THERAP CAT: Antibacterial.
THERAP CAT (VET): D-Form as antimicrobial; DL-form in control of fowl cholera.

9231. Thiamylal. *Dihydro-5-(1-methylbutyl)-5-(2-propenyl)-2-thioxo-4,6(1H,5H)-pyrimidinedione; 5-allyl-5-(1-methylbutyl)-2-thiobarbituric acid;* thioseconal. $C_{12}H_{18}N_2$-O_2S; mol wt 254.36. C 56.66%, H 7.13%, N 11.02%, O 12.58%, S 12.61%. Prepn: **Brit.** pat. **613,704** (1948 to Lilly); Abe *et al., J. Pharm. Soc. Japan* **75**, 891 (1955); Izumi, Nakanishi, **Japan.** pat. **8785('56)** (to Yoshitomi); Donnison, U.S. pat. **2,876,225** (1959 to Abbott).

Crystals from dil ethanol, mp 132-133°.
Sodium salt, $C_{12}H_{17}N_2NaO_2S$, *Surital.*
Caution: May be habit forming. This is a controlled substance (depressant) listed in the U.S. Code of Federal Regulations, Title 21 Parts 329.1 and 1308.13 (1987).
THERAP CAT: Anesthetic (intravenous).
THERAP CAT (VET): Anesthetic (intravenous).

9232. Thianaphthene. *Benzo[b]thiophene;* benzothiofuran. C_8H_6S; mol wt 134.19. C 71.60%, H 4.51%, S 23.90%. Occurs in lignite tar. Catalytic synthesis from styrene and hydrogen sulfide: Moore, Greensfelder, *J. Am. Chem. Soc.* **69**, 2008 (1947); from ethyl benzene and hydrogen sulfide: Hansch, Hawthorne, *ibid.* **70**, 2495 (1948).

Leaflets. Odor similar to that of naphthalene. mp 32°; bp_{760} 221°; bp_{20} 103-105°. Volatile with steam. Sol in the usual organic solvents.
Picrate, yellow needles from alcohol, mp 149°.
USE: Manuf of pharmaceuticals, thioindigo.

9233. Thiazesim. *5-[2-(Dimethylamino)ethyl]-2,3-dihy-dro-2-phenyl-1,5-benzothiazepin-4(5H)-one;* thiazenone; tiazesim. $C_{19}H_{22}N_2OS$; mol wt 326.48. C 69.90%, H 6.79%, N 8.58%, O 4.90%, S 9.82%. Prepn of the hydrochloride: Krapcho *et al., J. Med. Chem.* **6**(5), 544 (1963); Krapcho, U.S. pat. **3,075,967** (1963 to Olin Mathieson). Metabolism: J. Dreyfuss *et al., J. Pharm. Sci.* **57**, 1497, 1505 (1968).

Hydrochloride, $C_{19}H_{23}ClN_2OS$, *SQ-10496, Altinil.* Crystals from acetonitrile, mp 222-224°.

THERAP CAT: Antidepressant.

9234. Thiazinamium Methylsulfate. *N,N,N,α-Tetra-methyl-10H-phenothiazine-10-ethanaminium methyl sulfate; trimethyl(1-methyl-2-phenothiazin-10-ylethyl)ammonium methyl sulfate;* trimethyl[1-methyl-2-(10-phenothiazinyl)-ethyl]ammonium methyl sulfate; N-[β-(10-phenothiazinyl)-propyl]trimethylammonium methyl sulfate; RP 3554; Multergan; Padisal; Multezin; Valan. $C_{19}H_{26}N_2O_4S_2$; mol wt 410.55. C 55.58%, H 6.38%, N 6.82%, O 15.59%, S 15.62%. Prepd by treating promethazine with dimethyl sulfate: **Brit. pat. 641,452** (1950 to Rhône-Poulenc). Crystal structure: P. Marsau, Y. Cam, *Acta Crystallogr.* **B29**, 980 (1973). Determn in body fluids: J. H. G. Jonkman *et al., J. Pharm. Pharmacol.* **27**, 849 (1975). Bioavailability: J. H. G. Jonkman *et al., Clin. Pharmacol. Ther.* **21**, 457 (1977). Pharmacokinetics: J. H. G. Jonkman *et al., Arzneimittel-Forsch.* **23**, 223 (1983); J. H. G. Jonkman *et al., Int. J. Clin. Pharmacol. Ther. Toxicol.* **21**, 454 (1983).

Crystals, mp 206-210° (some decompn). Discolors on exposure to light. Soly in water at 25° about 10%. Freely sol in abs ethanol. Sparingly sol in acetone. Practically insol in ether, benzene.

THERAP CAT: Antihistaminic.

9235. Thiazole. C_3H_3NS; mol wt 85.13. C 42.32%, H 3.55%, N 16.46%, S 37.67%. First described as the pyridine of the thiophene series: A. Hantzsch, J. Weber, *Ber.* **20**, 3118 (1887); A. Hantzsch, *Ann.* **249**, 1 (1888). Book: *The Chemistry of Heterocyclic Compounds,* A. Weissberger, E. C. Taylor, Eds., **vol. 34** (New York, Wiley, 1979).

Colorless or pale yellow liquid. Characteristic foul odor; d^{17} about 1.20; bp 115-118°. Slightly sol in water; sol in many organic solvents. Forms compds with auric, mercuric, and platinic chlorides.

9236. Thiazolinobutazone. *4-Butyl-1,2-diphenyl-3,5-pyrazolidinedione compd with 4,5-dihydro-2-thiazolamine (1:1);* 2-thiazoline-2-ammonium 4-n-butyl-1,2-diphenyl-3,5-pyrazolidinedionate; phenylbutazone 2-amino-2-thiazoline salt; TZB; LAS 11871; Fordonal. $C_{22}H_{26}N_4O_2S$; mol wt 410.54. C 64.36%, H 6.38%, N 13.65%, O 7.80%, S 7.81%. A heterocyclic derivative of phenylbutazone, *q.v.* Prepn: J.

Moragues *et al., Arzneimittel-Forsch.* **24**, 1785 (1974). Pharmacology: M. Márquez, D. J. Roberts, *ibid.* 1786, 1790; M. Colombo *et al., ibid.* **26**, 1347 (1976).

White crystals, mp 164-166°. LD_{50} orally in rats, mice: 1425, 1650 mg/kg (Colombo).

THERAP CAT: Anti-inflammatory.

9237. Thiazolsulfone. *5-[(4-Aminophenyl)sulfonyl]-2-thiazolamine; 2-amino-5-sulfanilylthiazole;* 4-aminophenyl-2'-aminothiazolyl-5'-sulfone; thiazosulfone; thiazolesul-fone; Promizole. $C_9H_9N_3O_2S_2$; mol wt 255.32. C 42.34%, H 3.55%, N 16.46%, O 12.53%, S 25.12%. Prepd from the re-action products of *p*-nitrobenzenesulfonyl chloride and 2 mols of 2-aminothiazole: Bambas, *J. Am. Chem. Soc.* **67**, 671 (1945); Bambas, U.S. pat. **2,389,126** (1945 to Parke, Davis).

Fine needles from alc, mp 219-221° (some decompn). Sol in water at pH 6.5 to the extent of 30-40 mg per 100 ml at 28-30°. Freely sol in acetone, dioxane, 70% alc, dil acids. Moderately sol in abs alcohol, ethyl acetate and ether. It is a very weak acid, dissolves in 10% alkali (decompn) and forms an alkali metal salt at pH 10.

THERAP CAT: Antibacterial.

9238. Thiazol Yellow G. *2,2'-(1-Triazene-1,3-diyldi-4,1-phenylene)bis[6-methyl-7-benzothiazolesulfonic acid] di-sodium salt;* C.I. *Direct Yellow 9;* 2,2'-[(diazoamino)di-*p*-phenylene]bis[6-methyl-7-benzothiazolesulfonic acid] di-sodium salt; C.I. 19540; Titan Yellow; Chlorazol Yellow 2G; Diazamine Golden Yellow T; Clayton Yellow. $C_{28}H_{19}N_5$-$Na_2O_6S_4$; mol wt 695.73. C 48.33%, H 2.75%, N 10.07%, Na 6.61%, O 13.80%, S 18.44%. Prepn from 2-(*p*-aminophen-yl)-6-methyl-7-benzothiazolesulfonic acid ("dehydrothio-*p*-toluidinesulfonic acid"): *Beilstein* **vol. 27**, 2nd suppl, 509; F. J. Welcher, *Organic Analytical Reagents* **vol. 4** (Van Nos-trand, New York, 1948) p 391; H. King *et al., Analyst* **92**, 695 (1967). *See also: Colour Index* **vol. 2** (3rd ed., 1971) p 2010; H. J. Conn's *Biological Stains,* R. D. Lillie, Ed. (Willi-ams & Wilkins, Baltimore, 9th ed., 1977) p 371.

Yellowish-brown powder. Sol in water, or alc (yellow solns), NaOH (reddish-yellow soln), H_2SO_4 (brownish-yellow soln). *Protect from light.*

USE: Dyeing cotton, viscose rayon, natural silk; as biologi-cal stain; as an analytical reagent, most often for the deter-mination of Mg; as pH indicator, yellow 11.0 to red 13.0; as fluorescent dye for microscopy.

9239. Thibenzazoline. *1,3-Dihydro-1,3-bis(hydroxy-methyl)-2H-benzimidazole-2-thione;* 1,3-bis(hydroxymeth-yl)-2-benzimidazolinethione; 2-mercaptobenzimidazole-1,3-dimethylol; Thyreocordon. $C_9H_{10}N_2O_2S$; mol wt 210.26. C 51.41%, H 4.79%, N 13.33%, O 15.22%, S 15.25%. Prepn: Monti, Venturi, *Gazz. Chim. Ital.* **76**, 365 (1946).

Very bitter. mp 160-162°. Soluble dil in alkalies.
THERAP CAT: Antihyperthyroid.

9240. Thienamycin. *[5R-[5α,6α(R*)]]-3-[(2-Aminoethyl)thio]-6-(1-hydroxyethyl)-7-oxo-1-azabicyclo[3.2.0]hept-2-ene-2-carboxylic acid.* $C_{11}H_{16}N_2O_4S$; mol wt 272.32. C 48.51%, H 5.93%, N 10.29%, O 23.50%, S 11.77%. The first member of a family of des-thia-carbapenem nucleus antibiotics having a thioethylamine side-chain on the enamine portion of the fused 5-membered ring. Produced by *Streptomyces cattleya:* J. S. Kahan *et al.,* U.S. pat. **3,950,357** (1976 to Merck & Co.). Discovery, isoln, taxonomy, physical properties: *eidem, J. Antibiot.* **32,** 1 (1979). Structure, absolute configuration: G. Albers-Schönberg *et al., J. Am. Chem. Soc.* **100,** 6491 (1978). Synthesis of the (±)-form: D. B. R. Johnston *et al., ibid.* 313; F. A. Bouffard *et al., J. Org. Chem.* **45,** 1130 (1980); S. M. Schmitt *et al., ibid.* 1135, 1142; D. G. Melillo *et al., Tetrahedron Letters* **21,** 2783 (1980); *eidem, ibid.* **22,** 913 (1981); M. Shiozaki, T. Hiraoka, *Tetrahedron* **38,** 3457 (1982). Alternate prepn of key intermediate: J. D. Buynak *et al., Chem. Commun.* **1986,** 941. Stereocontrolled synthesis of the naturally occurring (+)-form: T. Salzmann *et al., J. Am. Chem. Soc.* **102,** 6161 (1980); S. Karady *et al., ibid.* **103,** 6765 (1981). Alternate synthetic routes: S. T. Hodgson *et al., Tetrahedron* **41,** 5871 (1985); T. Iimori, M. Shibasaki, *Tetrahedron Letters* **26,** 1523 (1985); T. Chiba, T. Nakai, *ibid.* 4647; D. J. Hart, D. C. Ha, *ibid.* 5493; T. Kametani *et al., J. Org. Chem.* **50,** 2327 (1985); H. Maruyama, T. Hiraoka, *ibid.* **51,** 399 (1986); D. G. Melillo *et al., ibid.* 1498. Continuous production in immobilized cell systems: E. J. Arcuri *et al., Biotechnol. Bioeng.* **28,** 842 (1986). Biosynthesis: J. M. Williamson *et al., J. Biol. Chem.* **260,** 4637 (1985). *In vitro* antibacterial activity: F. P. Tally *et al., Antimicrob. Ag. Chemother.* **14,** 436 (1978); S. S. Weaver *et al., ibid.* **15,** 518 (1979). Evaluation of *in vitro* and *in vivo* activities: H. Kropp *et al., Antimicrob. Ag. Chemother* **17,** 993 (1980). Comparative study vs gram-positive and gram-negative aerobic and anaerobic species and β-lactamase stability: H. C. Neu, P. Labthavikul, *ibid.* **21,** 180 (1982). *N*-Acyl derivatives as anti-inflammatory agents: J. B. Doherty *et al.,* U.S. pat. **4,465,687** (1984 to Merck & Co). Pharmacokinetics, bacteriological efficacy: P. Patamasucon, G. H. McCracken, *Antimicrob. Ag. Chemother.* **21,** 390 (1982). Review of early syntheses of the carbapenems: T. Kametani, *Heterocycles* **17,** 463-506 (1982).

White hygroscopic solid. $[α]_D^{27}$ +82.7° (c = 1.0 in water). uv max (water, pH 4-8): 296.5 nm (ε 7900); (pH 2): 309 nm; (pH 12): 300.5 nm. Freely sol in water, sparingly sol in methanol. In dilute soln, stability is optimal between pH 6-7, declining with unusual rapidity above that range. Susceptible to inactivation by dilute solns of hydroxylamine and cysteine.

N-Formimidoylthienamycin monohydrate, see imipenem.

9241. Thiethylperazine. *2-(Ethylthio)-10-[3-(4-methyl-1-piperazinyl)propyl]phenothiazine;* 3-ethylmercapto-10-(1'-methylpiperazinyl-4'-propyl)phenothiazine. $C_{22}H_{29}N_3S_2$; mol wt 399.62. C 66.12%, H 7.31%, N 10.51%, S 16.05%. Prepn: Bourquin *et al., Helv. Chim. Acta* **41,** 1072 (1958).

Crystals from acetone, mp 62-64°. bp$_{0.01}$ 227°.
Dimaleate, $C_{30}H_{37}N_3O_8S_2$, *Torecan Maleate, Toresten, Tresten.* Crystals from methanol, dec 188-190°.
Dihydrochloride, $C_{22}H_{29}N_3S_2$.2HCl, crystals from ethanol, mp 214-216°.
Dimalate, $C_{22}H_{29}N_3S_2$.2C$_4$H$_6$O$_5$, crystals from ethanol, mp 139°.
THERAP CAT: Antiemetic.

9242. Thihexinol. *α-[4-(Diethylamino)cyclohexyl]-α-2-thienyl-2-thiophenemethanol; (4-dimethylaminocyclohexyl)-di-2-thienylmethanol; α,α'-dithienyl-4-dimethylaminocyclohexyl carbinol.* $C_{17}H_{23}NOS_2$; mol wt 321.51. C 63.51%, H 7.21%, N 4.36%, O 4.98%, S 19.95%. Prepn: Villani, U.S. pat. **2,764,519** (1956 to Schering).

Crystals from benzene, mp 156-157°.
Methyl bromide, $C_{18}H_{26}BrNOS_2$, *Entoquel.* Crystals from ethanol + ether, mp 231-232°.
THERAP CAT: Anticholinergic.

9243. Thimerfonate Sodium. *Ethyl(4-mercaptobenzene-sulfonato-S⁴)mercury sodium salt;* sodium *p*-ethylmercurithiophenylsulfonate; ethyl[(*p*-sulfophenyl)thio]mercury sodium salt; Sulfo-Merthiolate. $C_8H_9HgNaO_3S_2$; mol wt 440.89. C 21.79%, H 2.06%, Hg 45.50%, Na 5.21%, O 10.89%, S 14.55%. Prepn: Waldo, *J. Am. Chem. Soc.* **53,** 992 (1931); Kharasch, U.S. pat. **1,672,615.**

Powder, very sol in water. Forms a stable solution.
THERAP CAT: Topical anti-infective.

9244. Thimerosal. *Ethyl[2-mercaptobenzoato(2—)-O,S]-mercurate(1—) sodium; [(o-carboxyphenyl)thio]ethylmercury sodium salt;* sodium ethylmercurithiosalicylate; thiomersalate; mercurothiolate; Merthiolate; Mercurin; Mertorgan; Merfamin. $C_9H_9HgNaO_2S$; mol wt 404.84. C 26.70%, H 2.24%, Hg 49.55%, Na 5.68%, O 7.90%, S 7.92%. Prepd by reacting ethylmercuric chloride (or ethylmercuric hydroxide) with thiosalicylic acid: Kharasch, U.S. pat. **1,672,615** (1928); Trikojus, *Nature* **158,** 472 (1946); Swirska *et al., Przemysl Chem.* **39,** 371 (1960), *C.A.* **55,** 3507a (1961). Toxicity: Mason *et al., Clin. Toxicol.* **4,** 185 (1971).

Cream-colored, crystalline powder. Stable in air, but not in sunlight. One gram dissolves in about 1 ml water, in about 8 ml alcohol. Practically insol in ether and benzene.

Stabilization of solns with EDTA: Davisson, U.S. pat. **2,864,844** (1958 to Lilly). pH of 1% aq soln: 6.7. LD_{50} s.c. in rats: 98 mg/kg (Mason).

USE: Pharmaceutic aid (preservative).

THERAP CAT: Anti-infective.

THERAP CAT (VET): Antibacterial, antifungal (topical).

9245. Thioacetaldehyde. *2,4,6-Trimethyl-s-trithiane;* trithioacetaldehyde. $C_6H_{12}S_3$; mol wt 180.35. C 39.95%, H 6.71%, S 53.34%. Occurs in α- and β-forms. Prepn: Baumann, Fromm, *Ber.* **22**, 2600 (1889); Fromm, Engler, *Ber.* **58**, 1916 (1925). Molecular structure: Hassel, Viervoll, *Acta Chem. Scand.* **1**, 164 (1947). Stereochemistry: Schönberg, Barakat, *J. Chem. Soc.* **1947**, 693.

α-Form, monoclinic plates, mp 101°.
β-Form, needles from acetone, mp 126°.

9246. Thioacetamide. *Ethanethioamide.* C_2H_5NS; mol wt 75.14. C 31.97%, H 6.71%, N 18.64%, S 42.68%. CH_3-$CSNH_2$. Prepd by heating ammonium acetate and aluminum sulfide: Kindler, Finndorf, *Ber.* **54**, 1080 (1921); from acetonitrile + H_2S: Kindler, *Ann.* **431**, 203 (1923); from acetamide + K_3PS_4: Schultz, Ranke, *Arch. Pharm.* **294**, 82 (1961). Toxicity data: Ambrose *et al., J. Ind. Hyg. Toxicol.* **31**, 158 (1949). Review of uses: H. F. Walton, *Thioacetamide as Analytical Reagent,* Arapahoe Chemicals, Inc., Boulder, Colorado.

Crystals from benzene, mp 113-114°. Slight odor of mercaptans. uv max (water): 210, 261, 318 nm (log ε 3.66, 4.08, 1.8). Soly in water at 25° 16.3 g/100 ml; in ethanol 26.4 g/100 g. Sparingly sol in ether. MLD orally in rats: 200 mg/kg (Ambrose).

Note: This substance may reasonably be anticipated to be a carcinogen: *Fourth Annual Report on Carcinogens* (NTP 85-002, 1985) p 186.

USE: Substitute for H_2S in the laboratory.

9247. Thioacetic Acid. *Ethanethioic acid; thiolacetic acid;* thiacetic acid. C_2H_4OS; mol wt 76.12. C 31.56%, H 5.30%, O 21.02%, S 42.13%. CH_3COSH. Prepn by distilling glacial acetic acid with phosphorus pentasulfide: Kekulé, *Ann.* **90**, 309 (1854); from acetic anhydride and hydrogen sulfide: Ellingboe, *Org. Syn.* **31**, 105 (1951).

Yellow, fuming liquid; pungent odor. d_4^{10} 1.075; bp 93°. Not solidified at −17°. Sol in water, particularly hot; very sol in alcohol.

9248. Thiobarbital. *5,5-Diethyldihydro-2-thioxo-4,6-(1H,5H)-pyrimidinedione; 5,5-diethyl-2-thiobarbituric acid;* Ibition; Sedothyron; Thiothyr. $C_8H_{12}N_2O_2S$; mol wt 200.26. C 47.98%, H 6.04%, N 13.99%, O 15.98%, S 16.01%. Prepn: Fischer, Dilthey, *Ann.* **335**, 350 (1904); Carrington, *J. Chem. Soc.* **1944**, 124. Metabolism in man: Bush *et al., J. Pharmacol. Exp. Ther.* **134**, 110 (1961).

Pale yellow needles from water, mp 180°. Sol in about 88 parts hot water, in ethanol, chloroform, ether, acetone, ammonia, alkalies. Sparingly sol in toluene. Practically insol in benzene. *Note:* This is a controlled substance (depressant) listed in the U.S. Code of Federal Regulations, Title 21 Part 1308.13 (1987).

THERAP CAT: Thyroid inhibitor.

9249. Thiobenzyl Alcohol. *Benzenemethanethiol;* benzyl mercaptan; α-toluenethiol. C_7H_8S; mol wt 124.20. C 67.69%, H 6.49%, S 25.82%. $C_6H_5CH_2SH$.

Colorless liquid; odor of leek; d^{20} 1.058; bp 194-195°. Oxidizes in air to dibenzyl disulfide.

Caution: Can cause mild irritation to mucous membranes.

9250. Thiobutabarbital. *5-Ethyldihydro-5-(1-methylpropyl)-2-thioxo-4,6(1H,5H)-pyrimidinedione;* 5-sec-butyl-5-ethyl-2-thiobarbituric acid; 5-ethyl-5-(1-methylpropyl)-2-thiobarbituric acid; thibutabarbital. $C_{10}H_{16}N_2O_2S$; mol wt 228.32. C 52.61%, H 7.06%, N 12.27%, O 14.02%, S 14.05%. Prepn: Volwiler, Tabern, U.S. pats. **2,153,729; 2,153,731** (both 1939 to Abbott).

Crystals, mp 163-165°.

Sodium salt, $C_{10}H_{15}N_2NaO_2S$, Inactin, Inaktin, Brevinarcon, Narkothion. Crystalline, slightly hygroscopic solid. Readily sol in water.

Note: This is a controlled substance (depressant) listed in the U.S. Code of Federal Regulations, Title 21 Part 1308.13 (1987).

THERAP CAT: Anesthetic (intravenous).

9251. Thiocarbamizine. *2,2'-[[[4-[(Aminocarbonyl)amino]phenyl]arsinidene]bis(thio)]bis[benzoic acid]; (p-ureidophenylarsylenedithio)di-o-benzoic acid; (p-ureidobenzenearsylenedithio)di-o-benzoic acid; 4-carbamidophenyl bis[o-carboxyphenylthio]arsenite; bis[o-carboxyphenylmercapto]-(p-ureidophenyl)arsine; thiocarbamisin.* $C_{21}H_{17}AsN_2O_5S_2$; mol wt 516.40. C 48.84%, H 3.32%, As 14.51%, N 5.43%, O 15.49%, S 12.42%. Prepn by heating *o*-HSC_6H_4COOH and Carbarsone oxide in aq alkali, purified by repeated precipitation with acid and dissolving in alkali: Rohrmann, U.S. pat. **2,516,831** (1950 to Lilly). *Compare* Arsenamide, Thiocarbarsone, Carbarsone.

White, cryst powder. Sparingly sol in water, alcohols. Insol in acids. Freely sol in dil alkali [*see* Arsenamide].

THERAP CAT: Antiamebic.

9252. Thiocarbarsone. *2,2'-[[[4-[(Aminocarbonyl)amino]phenyl]arsinidene]bis(thio)]bis[acetic acid]; (p-ureidophenylarsylenedithio)diacetic acid; 4-carbamidophenyl bis[carboxymethylthio]arsenite; bis[carboxymethylmercapto]-(p-ureidophenyl)arsine.* $C_{11}H_{13}AsN_2O_5S_2$; mol wt 392.27. C 33.68%, H 3.34%, As 19.10%, N 7.14%, O 20.39%, S 16.35%. Prepd by heating $HSCH_2COOH$ and Carbarsone oxide in aq alkali, purified by repeated precipitation with acid and dissolving in alkali: Rohrmann, U.S. pat. **2,516,831** (1950 to Lilly). *Compare* Arsenamide.

White, cryst powder. Sparingly sol in water, alcohols. Insol in acids. Freely sol in dil alkali [*see* Arsenamide].

THERAP CAT: Antiamebic.

9253. Thiocolchicine. (S)-N-[5,6,7,9-Tetrahydro-1,2,3-trimethoxy-10-(methylthio)-9-oxobenzo[a]heptalen-7-yl]-acetamide. $C_{22}H_{25}NO_5S$; mol wt 415.50. C 63.59%, H 6.06%, N 3.37%, O 19.25%, S 7.72%. Prepn: Velluz, Muller, Bull. Soc. Chim. France **1954**, 755; Muller, Velluz, U.S. pat. **2,820,029** (1958 to UCLAF). Prepn of 2-glucoside analog: Velluz, Muller, Bull. Soc. Chim. France **1955**, 194; Muller, Velluz, loc. cit.

Yellow cubic crystals from ethyl acetate, mp 192-194°. $[\alpha]_D^{20}$ −221. Sol in ethanol, acetone, chloroform. Practically insol in water, ether.

2-Glucoside analog, $C_{27}H_{33}NO_{10}S$, **thiocolchicoside, 2-demethoxy-2-glucosidoxythiocolchicine, Colcamyl, Coltramyl, Coltromyl, Coltrax, Musco-Ril.** Crystals from ethanol + 1N NaOH, dec 220°. $[\alpha]_D$ −609° (water), −240° (ethanol).

THERAP CAT: Thiocolchicoside as muscle relaxant (skeletal).

9254. Thiocresol. Methylbenzenethiol; tolylmercaptan; toluenethiol. C_7H_8S; mol wt 124.20. C 67.69%, H 6.49%, S 25.82%. $CH_3C_6H_4SH$.

o-Thiocresol melts at 15°; volatile with steam. m-Thiocresol, liquid, mp below −20°. p-Thiocresol, leaflets, mp 43-44°. All three isomers boil at about 195°, are insol in water; sol in alcohol or ether.

THERAP CAT: Antiseptic; dermatologic.

9255. Thioctic Acid. 1,2-Dithiolane-3-pentanoic acid; 1,2-dithiolane-3-valeric acid; 6,8-thioctic acid; α-lipoic acid; 5-(1,2-dithiolan-3-yl)valeric acid; 5-[3-(1,2-dithiolanyl)]-pentanoic acid; δ-[3-(1,2-dithiacyclopentyl)]pentanoic acid; protogen A; acetate replacing factor; pyruvate oxidation factor; Biletan; Lipoicin; Thioctacid; Thioctan; Tioctan. $C_8H_{14}O_2S_2$; mol wt 206.32. C 46.57%, H 6.84%, O 15.51%, S 31.08%. Growth factor for many bacteria and protozoa; prosthetic group, coenzyme, or substrate in plants, microorganisms, and animal tissues. Isoln of naturally occurring d-form: L. J. Reed et al., Science **114**, 93 (1951); eidem, J. Am. Chem. Soc. **75**, 1267 (1953); Patterson et al., ibid. **76**, 1823 (1954). Syntheses of dl-form: Bullock et al., ibid. **74**, 1868, 3455 (1952); Hornberger et al., ibid. 2382; Reed, U.S. pats. **2,980,716** and **3,049,549** (1961, 1962 to Res. Corp.); Lewis, Raphael, J. Chem. Soc. **1962**, 4263; Dee et al., U.S. pat. **3,223,712** (1965 to Yamanouchi); J. Tsuji et al., J. Org. Chem. **43**, 3606 (1978). Biosynthesis via linoleic acid: J. P. Carreau in Methods in Enzymology vol. **62**, D. McCormick, L. Wright, Eds. (Academic Press, New York, 1974) pp 152-158. Enantioselective synthesis of d-form: P. C. Bulman-page et al., Chem. Commun. **1986**, 1408. Clinical study in treatment of Wilson's disease: S. F. Gomes da Costa, Arzneimittel-Forsch. **20**, 1210 (1970). Use in treatment of mushroom poisoning: R. Plotzker et al., Am. J. Med. Sci. **283**, 79 (1982); J. P. Hanrahan, M. A. Gordon, J. Am. Med. Assoc. **251**, 1057 (1984). Reviews: Wagner, Folkers, Vitamins and Coenzymes (Interscience, New York, 1964) pp 244-263; Schmidt et al., Angew. Chem. Int. Ed. **4**, 846 (1965); Schmidt et al., Advan. Enzymol. Relat. Areas Mol. Biol. **32**, 423 (1969).

d-Form, crystals by vacuum sublimation (at 85-90° and 25 microns). mp 46-48° (microblock). $[\alpha]_D^{23}$ +104° (c =

0.88 in benzene). uv max (methanol): 333 nm (ε 150). pKa 5.4. Practically insol in water. Sol in fat solvents.

Sodium salt, $C_8H_{13}NaO_2S_2$, white powder, sol in water. pH of aq solns about 7.4.

dl-Form, yellow needles from cyclohexane, mp 60-61°. bp 160-165°. uv spectrum: Calvin, Fed. Proc. **13**, 703 (1954). Practically insol in water. Sol in fat solvents. Forms a water-soluble sodium salt.

Ethylenediamine, **Tioctidasi.**

l-Form, crystals from cyclohexane, mp 45-47.5° (microblock). $[\alpha]_D^{23}$ −113° (c = 1.88 in benzene). uv max (methanol): 330 nm (ε 140).

THERAP CAT: Treatment of liver disease; antidote to poisonous mushrooms (Amanita species).

9256. Thiocyanate Sodium. Sodium sulfocyanate; sodium rhodanide; sodium thiocyanate. CNNaS; mol wt 81.08. C 14.81%, N 17.28%, Na 28.36%, S 39.55%. NaSCN. Contains at least 98% NaSCN.

Colorless or white, deliquesc crystals. mp about 300°. Sol in about 0.6 part water; freely sol in alcohol, acetone. When dissolved in water the temp is considerably lowered. The solns are neutral. Keep well closed. LD_{50} orally in rats: 764 mg/kg; i.v. in mice: 484 mg/kg, Anderson, Chen, J. Am. Pharm. Assoc. **29**, 152 (1940).

USE: Manuf other thiocyanates, especially organic.

9257. Thiocyanic Acid. Hydrogen thiocyanate; Rhodanwasserstoffsäure (German). CHNS; mol wt 59.09. C 20.33%, H 1.71%, N 23.71%, S 54.25%. HSCN. Thiocyanic acid is believed to be a tautomeric mixture of HSCN and HNCS (isothiocyanic acid): Beard, Dailey, J. Chem. Phys. **18**, 1437 (1950). Conveniently prepd in the laboratory from KNCS + $KHSO_4$: Birckenbach, Bucher, Ber. **73**, 1153 (1940). Dil aq solns may be prepd from ammonium thiocyanate by the action of ion exchange resins: Klement, Z. Anorg. Allgem. Chem. **260**, 268 (1949). Toxicology of organic thiocyanates: Y. Yokoi, Japan. J. Pharmacol. **3**, 99 (1954); C.A. **48**, 13965f (1954).

Colorless gas or white solid depending upon degree of polymerization. Freely sol in water. Very strong acid. A 5% aq soln may be kept refrigerated for several weeks. Also sol in some organic solvents.

Ethyl ester, C_3H_5NS, **ethyl thiocyanate, ethyl sulfocyanate.** Liquid. d_4^4 1.007. bp 146°. n_D^{15} 1.4684. Insol in water; misc with alcohol, ether. MLD in mice (mg/kg): 52 orally; 39.1 s.c.; 18.3 i.p.; 6 i.v. (Yokoi).

9258. Thiodicarb. N,N'-[Thiobis[(methylimino)carbonyloxy]]bisethanimidothioic acid dimethyl ester; bis-[O-(1-methylthioethylimino)-N-methylcarbamic acid]-N,N'-sulfide; N,N'-bis[1-methylthioacetaldehyde O-(N-methylcarbamoyl)oxime]sulfide; O-[[N-[N'-(1-methylthioethylidene-iminooxycarbonyl)-N'-methylaminosulfenyl]-N-methylcarbamoyl]]-S-methylacetohydroximate; dicarbosulf; bis-methomyl thioether; UC 51762; CGA 45156; Larvin; Lepicron. $C_{10}H_{18}N_4O_4S_3$; mol wt 354.46. C 33.88%, H 5.12%, N 15.81%, O 18.06%, S 27.13%. Prepn: **Neth.** pat. **Appl. 7,-508,197** (1976 to Ciba-Geigy); T. D. J. D'Silva, **Ger.** pat. **2,654,331**; eidem, **U.S.** pat. **4,382,957** (1977, 1983 both to Union Carbide). Activity: A. A. Sousa et al., J. Econ. Entomol. **70**, 803 (1977). Field trials: E. P. Pieters, D. L. Pitts, J. Ga. Entomol. Soc. **15**, 207 (1980); J. R. Bradley, Jr., A. M. Agnello, J. Econ. Entomol. **81**, 706 (1988).

Crystals, mp 173-174°. LD_{50} in rats (mg/kg): 160 orally, > 1600 dermally (Sousa).

USE: Insecticide.

9259. 2,2'-Thiodiethanol. Thiodiglycol; thiodiethylene glycol; bis(hydroxyethyl)sulfide. $C_4H_{10}O_2S$; mol wt 122.18. C 39.32%, H 8.25%, O 26.19%, S 26.24%. $HOCH_2CH_2S-CH_2CH_2OH$. Prepd from ethylene oxide and hydrogen sulfide: Chichibabin, Bestuzher, Compt. Rend. **200**, 242 (1935); Nenitzescu, Scarlatescu, Ber. **68**, 587 (1935); Headlee, cited in G. O. Curme, F. Johnston, *Glycols* (Reinhold, New York, 1952) p 103.

Liquid. d_4^{20} 1.1824. mp $-16°$. bp_{14} 168°. n_D^{20} 1.519. Flash pt, open cup: 320°F (160°C). Misc with water, alc. Slightly sol in ether.

9260. Thiodiglycolic Acid. *2,2'-Thiobis[acetic acid];* dimethylsulfide-α,α'-dicarboxylic acid; mercaptodiacetic acid. $C_4H_6O_4S$; mol wt 150.15. C 31.99%, H 4.03%, O 42.62%, S 21.36%. HOOCCH$_2$SCH$_2$COOH. Prepn from sodium chloroacetate and hydrogen sulfide: Loven, *Ber.* **27**, 3059 (1894).

Crystals from water. mp 129°. Sol in water, alcohol.
USE: Detection of copper, lead, mercury, silver: Dubsky *et al.*, *C.A.* **34**, 6185 (1940).

9261. 3,3'-Thiodipropionic Acid. *3,3'-Thiobis[propanoic acid];* β,β-thiodipropionic acid; thiodihydracrylic acid; diethyl sulfide 2,2'-dicarboxylic acid. $C_6H_{10}O_4S$; mol wt 178.20. C 40.44%, H 5.66%, O 35.91%, S 17.99%. Prepn: **Brit.** pat. **571,628** (1945 to Am. Cyanamid); Gresham, Shaver, **U.S.** pat. **2,449,992** (1948 to B. F. Goodrich).

$$CH_2CH_2COOH$$
$$|$$
$$S$$
$$|$$
$$CH_2CH_2COOH$$

Nacreous leaflets from hot water, mp 134°. K at 25° = 7.8 × 10^{-5}. One gram dissolves in 26.9 ml water at 26°. Freely sol in hot water, alcohol, acetone.
USE: Antioxidant for soap products and polymers of ethylene. In plasticizers and lubricants. Proposed for edible fats, oils, other foods.

9262. Thioformamide. CH_3NS; mol wt 61.11. C 19.65%, H 4.95%, N 22.92%, S 52.47%. HCSNH$_2$. Prepd by the action of phosphorus pentasulfide on formamide: Willstätter, Wirth, *Ber.* **42**, 1911 (1909). Improved procedures: Erlenmeyer, Menzi, *Helv. Chim. Acta* **31**, 2071 (1948); Schmitz, U.S. pat. **2,682,558** (1954 to du Pont).

Prisms from ethyl acetate or from ether + petr ether. Turns yellow and dec to a dark-colored mass. May be stored as soln in abs ether, preferably over P$_2$O$_5$. mp 29°. Sol in water (with considerable cooling). Freely sol in tetrahydrofuran, alcohol, ether, acetone, ethyl acetate; sparingly sol in cold chloroform, benzene, petr ether, carbon disulfide. The decompn products are hydrogen cyanide, hydrogen sulfide, ammonia, and some solid, amorphous, sulfur-contg compounds.

Monohydrate, yellow oil.
USE: In the synthesis of the thiazole part of vitamin B$_1$.

9263. 5-Thio-D-glucose. *5-Thio-α-D-glucopyranose;* α-D-glucothiopyranose. $C_6H_{12}O_5S$; mol wt 196.22. C 36.72%, H 6.17%, O 40.77%, S 16.34%. The nearest analog of normal D-glucose available to date. Thought to interfere with cellular transport systems utilizing D-glucose due to this structural similarity. The first chemical other than a hormone or alkylating agent which can interfere reversibly with spermatogenesis and which also has been effective against malignant cultured cells. Prepn: Feather, Whistler, *Tetrahedron Letters* **1962**, 667; Rowell, Whistler, *J. Org. Chem.* **31**, 1514 (1966); Nayak, Whistler, *ibid.* **34**, 97 (1969); Abd El-Rahman, Whistler, *Org. Prep. Proced. Int.* **5**, 245 (1973); H. Driguez, B. Henrissat, *Tetrahedron Letters* **22**, 5061 (1981). Pharmacology: Hoffman, Whistler, *Biochemistry* **7**, 4479 (1968); Whistler, Lake, *Biochem. J.* **130**, 919 (1972). Brief review: *Science* **186**, 431 (1974).

Crystals from methanol, mp 135-136°. $[\alpha]_D^{20}$ +188° (c =

1.56 in water). LD$_{50}$ in mice: 14 g/kg, Whistler, Lake, *loc. cit.*
USE: Tool for examination of D-glucose biochemistry.

9264. Thioglycerol. *3-Mercapto-1,2-propanediol;* α-monothioglycerol; thioglycerin. $C_3H_8O_2S$; mol wt 108.16. C 33.31%, H 7.46%, O 29.59%, S 29.65%. HOCH$_2$CH(OH)-CH$_2$SH. Prepd by heating glycerol-α-monochlorohydrin with KHS in alcoholic solution. *Ref:* Sutton, *J. Am. Med. Assoc.* **104**, 2168 (1935).

Yellowish, very viscous liq; slight sulfidic odor. d 1.295. Slightly sol in water; miscible with alcohol; insol in ether.
THERAP CAT: In wound healing.
THERAP CAT (VET): Has been used in 1:5000 concn to promote wound healing.

9265. Thioglycolic Acid. *Mercaptoacetic acid;* thioglycollic acid. $C_2H_4O_2S$; mol wt 92.12. C 26.07%, H 4.38%, O 34.74%, S 34.81%. HSCH$_2$COOH. Prepd by the action of sodium sulfhydrate on sodium chloroacetate; by electrolysis of dithioglycollic acid (from sodium sulfide and sodium chloroacetate).

Liquid, strong, unpleasant odor. Readily oxidized by air. d 1.325. mp $-16.5°$. bp_{29} 123°. bp_{15} 108°. bp_5 96°. K_1 at 25° = 2.1 × 10^{-4}; K_2 = 2.1 × 10^{-11}. Miscible with water, alcohol, ether, chloroform, benzene, and many other organic solvents. LD$_{50}$ orally in rats: 0.15 ml/kg, Deichmann, Mergard, *J. Ind. Hyg. Toxicol.* **30**, 373 (1948).
USE: Sensitive reagent for iron, molybdenum, silver, tin. With ferric iron a blue color appears, and when an alkali hydroxide is added to a soln contg ferrous salts and thioglycolic acid, a yellow precipitate forms. Used in the manuf of thioglycolates. The ammonium and sodium salts are commonly used for cold waving and the calcium salt is a depilatory. The sodium salt also is used in bacteriology in the prepn of thioglycolate media. *Caution:* Can cause severe burns and blistering of skin.

9266. Thioguanine. *2-Amino-1,7-dihydro-6H-purine-6-thione; 2-aminopurine-6-thiol;* Lanvis. $C_5H_5N_5S$; mol wt 167.21. C 35.92%, H 3.01%, N 41.89%, S 19.18%. Prepn: Elion, Hitchings, *J. Am. Chem. Soc.* **77**, 1676 (1955); Hitchings, Elion, U.S. pats. **2,697,709; 2,800,473; 2,884,667;** Hitchings *et al.*, U.S. pats. **3,019,224; 3,132,144** (1954, 1957, 1959, 1962, and 1964, all to Burroughs Welcome).

Needles from water, mp > 360°.
THERAP CAT: Antineoplastic.

9267. Thioguanosine. *2-Amino-6-mercapto-9-β-D-ribofuranosylpurine.* $C_{10}H_{13}N_5O_4S$; mol wt 299.31. C 40.13%, H 4.38%, N 23.40%, O 21.38%, S 10.71%. Prepn: Fox *et al.*, *J. Am. Chem. Soc.* **80**, 1669 (1948).

Hemihydrate, tiny tapered prisms from water, dec 224-227°. $[\alpha]_D^{22}$ $-64°$ (c = 1.3 in 0.1N NaOH). uv max (pH 4-6): 257, 342 nm (ε 8820, 24800). pKa 8.33. Used in cancer research, *compare* 6-Mercaptopurine and Nebularine.

9268. Thiokol®. *Polysulfide rubber;* thiorubber. Polysulfide polymers prepd from dihaloalkanes and sodium polysulfide. Reviews of prepn, chemistry and applications: Berenbaum in *High Polymers*, H. Mark *et al.*, Eds., vol. 13 entitled *Polyethers* part 3, N. G. Gaylord, Ed. (Interscience, New York, 1962) pp 43-114; Panek, *ibid.* pp 115-224; Berenbaum in *Encyclopedia of Polymer Science and Technology* vol. 11 (Interscience, New York, 1969) pp 425-447.

$$-(CH_2)_2S-S-$$

Thiokol A

Thiokol A, Ethanite, Perduren. The first commercial polysulfide polymer, prepd from ethylene dichloride and sodium polysulfide. Sulfur content 84%; d about 1.6. Mixes with natural rubber. Cured polymer retains unpleasant odor; irritating fumes evolve during manuf. Stable to the usual organic solvents and dil mineral acids. Unstable to alkalies and oxidizing substances. Of low tensile strength and abrasion resistance. Not recommended where tropic or arctic climates prevail.

Thiokol FA. Prepd from ethylene dichloride, dichlorodiethyl formal and sodium polysulfide. Sulfur content 47%; d 1.34. No odor. Excellent solvent resistant characteristics but not as good as Thiokol A. Low temperature flexibility to −50°F.

USE: In rubber and resin manuf. As lining for flexible oil pipes and self-sealing gasoline tanks.

9269. Thiolactic Acid. *2-Mercaptopropanoic acid;* 2-mercaptopropionic acid; 2-thiolpropionic acid; α-mercaptopropanoic acid. $C_3H_6O_2S$; mol wt 106.14. C 33.95%, H 5.70%, O 30.14%, S 30.21%. CH₃CH(SH)COOH. First prepd from α-chloropropionic acid and potassium hydrosulfide: Schacht, *Ann.* 129, 3 (1864); prepn from α-chloropropionic acid and sodium thiosulfate: Martin, U.S. pat. 2,413,361 (1946 to Martin Labs). From sodium sulfide, sulfur and α-bromopropionic acid: Dumesnil, U.S. pat. 2,985,557 (1961 to Frank E. Jonas). *Review:* Reid, *Organic Chemistry of Bivalent Sulfur* vol. I (Chemical Publ. Co., New York, 1958) p 449.

Oil, disagreeable odor. Solidif on cooling, crystals mp about 10°. d_4^{15} 1.220. bp_{16} 117°. n_D^{16} 1.4823. Miscible with water, alcohol, ether, acetone.

Copper salt, $C_3H_5CuO_2S$, yellow precipitate, practically insol in water and in dil HNO₃.

Barium salt, $Ba(C_3H_5O_2S)_2$, rubbery compd, very sol in water, practically insol in alcohol.

Mercury salt, $Hg(C_3H_5O_2S)_2$, small shiny platelets, sparingly sol in water, very sol in alcohol.

Sodium salt pentahydrate, $C_3H_5NaO_2S.5H_2O$, crystals. Dec about 250°. Sol in water, methanol, ethanol. Less sol in propanol.

Calcium salt decahydrate, $Ca(C_3H_5O_2S)_2.10H_2O$, crystals. Dec 270°. Sol in water, methanol, ethanol. Less sol in propanol.

Platinum salt, $Pt(C_3H_5O_2S)_2$, greenish-yellow precipitate, practically insol in water and dil acids, sol in sodium hydroxide and sodium carbonate.

Silver salt, $C_3H_5AgO_2S$, yellow precipitate, practically insol in water and dil HNO₃.

USE: In depilatory and hair waving prepns.

9270. 2-Thiolhistidine. *α-Amino-2,3-dihydro-2-thioxo-1H-imidazole-4-propanoic acid;* 2-mercaptohistidine; α-amino-2-mercapto-4-imidazolepropionic acid. $C_6H_9N_3O_2S$; mol wt 187.23. C 38.49%, H 4.84%, N 22.44%, O 17.09%, S 17.13%. Prepn from histidine: Ashley, Harington, *J. Chem. Soc.* **1930**, 2586; from aspartic acid: Harington, Overhoff, *Biochem. J.* **27**, 338 (1933). Alternate procedures: Hegedüs, *Helv. Chim. Acta* **38**, 22 (1955); Marei, Raphael, *J. Chem. Soc.* **1958**, 2624. *Review:* J. P. Greenstein, M. Winitz, *Chemistry of the Amino Acids*, vol. 3 (John Wiley, New York, 1961) pp 2671-2675.

L-2-thiolhistidine

L-Form, plates from water, darkens at 290°, not melted at 310°. $[\alpha]_D^{20}$ −10° (c = 2 in N HCl). pK_1^1 1.84; pK_2^1 8.47; pK_3^1 11.4. Readily sol in hot water, sparingly sol in water at room temp. Tends to form supersaturated solns. Practically insol in alcohol, other organic solvents.

Dihydrochloride, $C_6H_9N_3O_2.2HCl$, large prisms, dec 197-199°. Freely sol in water, sparingly sol in alcohol.

9271. Thiolutin. *6-(Acetamido)-4-methyl-1,2-dithiolo-[4,3-b]pyrrol-5(4H)-one;* N-(4,5-dihydro-4-methyl-5-oxo-1,2-dithiolo[4,3-b]pyrrol-6-yl)acetamide; 3-acetamido-5-methylpyrrolin-4-one[4,3-d]-1,2-dithiole; acetopyrrothine. $C_8H_8N_2O_2S_2$; mol wt 228.29. C 42.09%, H 3.53%, N 12.28%, O 14.02%, S 28.07%. Antibiotic isolated from several strains of *Streptomyces albus:* F. W. Tanner, Jr. *et al.*, 118th Am. Chem. Soc. Meet. (Chicago, Sept. 1950), *Abstracts of Papers*, p 18A; F. W. Tanner, Jr. *et al.*, U.S. pat. **2,689,854** (1954 to Pfizer). Characterization, similarity to aureothricin, *q.v.:* W. D. Celmer *et al.*, *J. Am. Chem. Soc.* **74**, 6304 (1952). Structure: W. D. Celmer, I. A. Solomons, *ibid.* **77**, 2861 (1955). Total synthesis: U. Schmidt, F. Geiger, *Ann.* **664**, 168 (1963); K. Hagio, N. Yoneda, *Bull. Chem. Soc. Japan* **47**, 1484 (1974). Bactericidal, protozoicidal, fungicidal properties: H. Seneca *et al.*, *Antibiot. & Chemother.* **2**, 357 (1952). Mode of action: A. Jimenez *et al.*, *Antimicrob. Ag. Chemother.* **3**, 729 (1973). Inhibition of microbiological growth in beer: J. B. Bockelmann, F. B. Strandskov, U.S. pat. **2,798,-811** (1957 to Schaefer Brewing Co.). Activity against soil borne pathogens: P. R. Deb, B. K. Dutta, *Curr. Sci. (India)* **53**, 659 (1984). As allergy inhibitor: P. Stahl *et al.*, *Ger. pat.* **3,434,562** (1986 to Boehringer, Mannheim), *C.A.* **105**, 54609k (1986).

Brilliant yellow needles from *n*-butanol, dec 273-276°. Sublimes 200°/0.1 mm. uv max (methanol): 250, 311, 388 nm (ε 6300, 5700, 11,000) (Celmer, Solomons). Sparingly sol in water (210 mg/l); more sol in methanol, ethanol, chloroform, acetone (1% solns in acetone have been prepd), glacial acetic acid, methyl isobutyl ketone. Less sol in ether, benzene, hexane. Stable in acid and neutral solns, dec in alkaline soln. Effective vs gram-positive and gram-negative bacteria, fungi and ameboid parasites. LD_0 in mice (mg/kg): 5-10 s.c.; 10 orally; LD_{50} in mice (mg/kg): 25 s.c.; 25 orally (Seneca).

9272. Thiomalic Acid. *Mercaptobutanedioic acid;* mercaptosuccinic acid. $C_4H_6O_4S$; mol wt 150.15. C 31.99%, H 4.03%, O 42.62%, S 21.35%. HOOCCH(SH)CH₂COOH.

White crystals; sulfidic odor. mp 149-150°. Soly in water at 40° about 50 g/100 ml; in ethanol at 25° about 50 g/100 ml. Also sol in acetone. Moderately sol in ether. Practically insol in benzene. The aq soln gives a transitory blue color with ferric chloride.

9273. Thionalide. *2-Mercapto-N-2-naphthalenylacetamide;* thioglycollic-β-aminonaphthalide. $C_{12}H_{11}NOS$; mol wt 217.29. C 66.33%, H 5.10%, N 6.45%, O 7.36%, S 14.75%.

White to ivory-colored needles. mp 111-112°. Insol in water. Freely sol in most organic solvents.

USE: As a reagent for copper, mercury, silver, thallium, and bismuth.

9274. Thionaphthene-2-carboxylic Acid. *Benzo[b]thiophene-2-carboxylic acid;* 2-benzothiophenecarboxylic acid; TNCA; BL-5583. $C_9H_6O_2S$; mol wt 178.21. C 60.66%, H 3.39%, O 17.96%, S 17.99%. Prepn: R. Weissgerber, O. Kruber, *Ber.* **53**, 1551 (1920); eidem, *Ger. pat.* 341,837 (1921), *Frdl.* **13**, 279 (1921); D. A. Shirley, M. D. Cameron, *J. Am. Chem. Soc.* **72**, 2788 (1950); T. Higa, A. J. Krubsack, *J. Org. Chem.* **41**, 3399 (1976). Use as anti-osteoporotic agent: C. M. Samour, J. A. Vida, *U.S. pat.* **4,101,668** (1978 to Bristol-Myers). Effect on total skeletal calcium in mice: J. C. Robin et al., *J. Med.* **11**, 15 (1980). Mechanism of action study: J. C. Robin et al., *Calcif. Tissue Int.* **36**, 194 (1984). Hypocalcemic effect: A. J. Johannesson et al., *Endocrinology* **117**, 1508 (1985).

Prisms from alcohol or needles from water, mp 236° (Weissgerber). Also reported as crystals from benzene-ethyl acetate, mp 236-238° (Higa).

9275. Thionazin. *Phosphorothioic acid O,O-diethyl O-pyrazinyl ester; O,O-diethyl-O-(2-pyrazinyl) phosphorothioate;* ethyl pyrazinyl phosphorothioate; EN-18133; ENT 25580; American Cyanamid 18133; Cynem; Nemafos; Zinophos. $C_8H_{13}N_2O_3PS$; mol wt 248.26. C 38.71%, H 5.28%, N 11.29%, O 19.33%, P 12.49%, S 12.92%. Prepn: Dixon et al.; Gordon; Miller, Forbes; Gagliardi; *U.S. pats.* **2,918,468; 2,938,831; 3,091,614; 3,340,262** (1959, 1960, 1963, 1967, all to Am. Cyanamid).

Amber liquid, mp −1.7°. bp 80°. n_D^{25} 1.5131. Vapor press at 30°: 3 × 10⁻³ mm Hg. Slightly sol in water; misc with most organic solvents. LD_{50} in female, male rats: 3.5, 6.4 mg/kg orally; 11, 17 mg/kg dermally, T. B. Gaines, *Toxicol. Appl. Pharmacol.* **14**, 515 (1969).

USE: Nematocide; insecticide. *Caution:* Cholinesterase inhibitor. *See* Parathion.

9276. Thionine. *3,7-Diaminophenothiazin-5-ium chloride;* Lauth's violet; C.I. 52000. $C_{12}H_{10}ClN_3S$; mol wt 263.74. C 54.65%, H 3.82%, Cl 13.44%, N 15.93%, S 12.16%. Prepn from *p*-phenylenediamine: Lauth, *Compt. Rend.* **82**, 1441 (1876); *Bull. Soc. Chim. France* [2] **26**, 422 (1876); Bernthsen, *Ger. pat.* **25,150** (1883 to BASF), *Frdl.* **1**, 253; Loiseleur, *Ann. Inst. Pasteur* **86**, 262 (1954); Loiseleur, Petit, *Compt. Rend.* **250**, 2573 (1960); Balestic, Magat, *C.A.* **55**, 23542g (1961). *See also: Colour Index* vol. **4** (3rd ed., 1971) p 4469.

Blackish-green glistening needles. Absorption max (water): 602.5 nm. Difficultly sol in cold, easily sol in hot water, giving first a blue, then a violet soln. The soln becomes somewhat bluer when HCl is added. Brownish-red precipitate with NaOH. Yellowish-green soln with concd H_2SO_4 which upon dilution turns blue, then violet on further dilution.

USE: For general nuclear staining; for counting bacteria in milk; as antioxidant for linseed oil. Has been used as catalyst in condensation of 1,3-butadienyl ethyl ether with maleic anhydride.

9277. Thionyl Bromide. Br_2OS; mol wt 207.90. Br 76.88%, O 7.70%, S 15.42%. $SOBr_2$. Prepd from thionyl chloride and hydrogen bromide: Besson, *Compt. Rend.* **122**, 320 (1896); **123**, 884 (1896); Hibbert, Pullman, *Inorg. Syn.* **1**, 113 (1939).

Orange-yellow liquid. d_4^{20} 2.688. mp −52°. bp_{773} 138°; bp_{20} 48°. Somewhat less stable than thionyl chloride. Dec slowly on standing in glass-stoppered bottles. Hydrolyzed by water. Miscible with benzene, chloroform, carbon tetrachloride.

Caution: Highly irritating to skin, eyes, mucous membranes, respiratory tract.

9278. Thionyl Chloride. Sulfurous oxychloride. Cl_2OS; mol wt 118.98. Cl 59.60%, O 13.45%, S 26.95%. $SOCl_2$. Usually prepd by the oxidation of sulfur dichloride with sulfur trioxide: Michaelis, *Ann.* **274**, 173 (1893); Edwards, *U.S. pat.* **2,362,057** (1944 to Hooker Electrochemical); Fehér in *Handbook of Preparative Inorganic Chemistry,* vol. **1**, G. Brauer, Ed. (Academic Press, New York, 2nd ed., 1963) pp 382-383; Macaluso, "Sulfur Compounds" in Kirk-Othmer *Encyclopedia of Chemical Technology,* vol. **19** (Interscience, New York, 2nd ed., 1969) pp 398-401. Purification by distilling from quinoline and boiled linseed oil: Martin, Fieser, *Org. Syn.* coll. vol. **II**, 570 (1943). Alternate methods of purification: Cottle, *J. Am. Chem. Soc.* **68**, 1380 (1946); Kunkel; Rosenberg, Flaxman, *U.S. pats.* **3,155,457; 3,156,529** (both 1964 to Hooker Chem.); Friedman, Wetter, *J. Chem. Soc. (A)* **1967**, 36.

Colorless to pale yellow or reddish, fuming, refractive liquid. *Suffocating odor.* d_4^0 1.676; d_4^{10} 1.655; d_4^{20} 1.638. mp −104.5°. bp_{760} 76°; $bp_{96.6}$ 20°. n_D^{20} 1.517. Dec when heated above 140° forming Cl_2, SO_2, and S_2Cl_2. Hydrolyzed by water forming SO_2 and HCl. Miscible with benzene, chloroform, carbon tetrachloride.

Caution: Vapors and liquid are strongly irritating and corrosive to skin, mucous membranes and eyes.

USE: For making acyl chlorides, to replace OH or SH groups with chlorine atoms; reacts with Grignard reagents to form the corresp sulfoxides. Review of use in organic synthesis: J. S. Pizey, *Synthetic Reagents,* vol. **1** (John Wiley, New York, 1974) pp 321-357.

9279. Thionyl Fluoride. Thionyl difluoride. F_2OS; mol wt 86.07. F 44.15%, O 18.59%, S 37.26%. SOF_2. Usually prepd by the action of antimony trifluoride on thionyl chloride in the presence of antimony pentafluoride: Booth, Mericola, *J. Am. Chem. Soc.* **62**, 640 (1940); Smith, Muetterties, *Inorg. Syn.* **6**, 162 (1960). May also be prepared by the action of other fluorides on thionyl chloride: Kemmitt, Sharp, *Advan. Fluorine Chem.* **4**, 228-229 (1965).

Colorless gas. Suffocating odor. Does not attack glass. Stored in compressed form in steel cylinders. mp −129.5°. bp −43.8°. Crit temp +88°. d (liq; −100°) 1.780. d (solid; −183°) 2.095. Trouton constant 22.6. Very slowly hydrolyzed by cold water. Sol in ether, benzene.

Caution: Highly irritating to eyes, respiratory tract.

9280. Thiopental Sodium. *5-Ethyldihydro-5-(1-methylbutyl)-2-thioxo-4,6(1H,5H)-pyrimidinedione monosodium salt; 5-ethyl-5-(1-methylbutyl)-2-thiobarbituric acid sodium salt;* thiomebumal sodium; penthiobarbital sodium; thiopentone sodium; thionembutal; Pentothal Sodium; Nesdonal Sodium; Intraval Sodium; Trapanal; Thiothal Sodium. $C_{11}H_{17}N_2NaO_2S$; mol wt 264.33. C 49.98%, H 6.48%, N 10.60%, Na 8.70%, O 12.11%, S 12.13%. Prepn: *U.S. pats.* **2,153,729** (1939); **2,876,225** (1959 to Abbott). Prepn of nonhygroscopic crystals: Hartop, *U.S. pat.* **3,109,001** (1963 to Abbott). Acute toxicity: Christensen, Lee, *Toxicol. Appl. Pharmacol.* **26**, 495 (1973).

Yellowish-white, hygroscopic powder. Alliaceous, garlic-like odor. Sol in water, alcohol. Insol in ether, benzene, petr ether. Aq solns are alkaline to litmus. Solns dec on standing; on boiling precipitation occurs. LD_{50} in mice (mg/kg): 149 i.p.; 78 i.v. (Christensen, Lee).

Note: Sterile Thiopental Sodium is a mixture of sterilized thiopental sodium (91.7%) with anhydr sodium carbonate as a buffer. A 2.5% w/v soln in water is strongly alkaline, having a pH of about 10.5.

Caution: May be habit forming. This is a controlled substance (depressant) listed in the U.S. Code of Federal Regulations, Title 21 Parts 329.1 and 1308.13 (1987).

THERAP CAT: Anesthetic (intravenous).

THERAP CAT (VET): Short-acting anesthetic.

9281. Thiopeptin. *Thiofeed.* Sulfur containing peptide antibiotic complex produced by *Streptomyces tateyamensis* no. 7906: Miyairi *et al.,* **Ger. pat. 1,929,355** (1969 to Fujisawa), *C.A.* **72,** 88921w (1970). Consists of the a and b series of thiopeptins A_1, A_2, A_3, A_4, and B (major component). Structurally similar to thiostrepton, *q.v.* Biological and chemical studies: Miyairi *et al., Antimicrob. Ag. Chemother.* **1,** 192 (1972). Characterization of thiopeptins B: Miyairi *et al., J. Antibiot.* **23,** 113 (1970); structural studies: Muramatsu *et al., ibid.* **25,** 537 (1972); **30,** 383 (1977). Total structures of components: O. D. Hensens, G. Albers-Schönberg, *Tetrahedron Letters* **1978,** 3649.

Thiopeptins B, faint yellow crystals, mp 219-222° (dec). $[\alpha]_D^{23} -80°$ (c = 1 in chloroform). uv spectrum (methanol): shoulders at 230-250, 295, 305 nm. Sol in dioxane, DMSO, DMF, pyridine, chloroform. Insol in ether, benzene, *n*-hexane, petr ether, water. Fairly sol in methanol, acetone, ethyl acetate.

THERAP CAT (VET): Antibiotic feed additive.

9282. Thiophanate. *[1,2-Phenylenebis(iminocarbonothioyl)]biscarbamic acid diethyl ester; 4,4'-o-phenylenebis[3-thioallophanic acid]diethyl ester;* 1,2-bis(3-ethoxycarbonyl-2-thioureido)benzene; Cercobin; Topsin; Nemafax. $C_{14}H_{18}N_4O_4S_2$; mol wt 370.44. C 45.39%, H 4.90%, N 15.12%, O 17.28%, S 17.31%. Prepn: Noguchi *et al.,* **Ger. pat. 1,806,123** (1969 to Nippon Soda Co.), *C.A.* **71,** 70347h (1969), corresp to U.S. pats. **3,745,187** and **3,769,308** (both 1973). Photodegradation: H. Buchenauer *et al., Pestic. Sci.* **5,** 343 (1973). Activity against nematode infection in ruminants: Eichler, *Brit. Vet. J.* **129,** 533 (1973). Toxicological studies in animals: *idem, ibid.* **130,** 570 (1974).

Colorless plates, mp 195°. LD_{50} orally in mice and rats: > 15 g/kg (Eichler).

O,O-Dimethyl analog, $C_{12}H_{14}N_4O_4S_2$, *thiophanate-methyl, Cercobin-M, Topsin-M.* See above for prepn. Persistence in soil: J. R. Fleeker *et al., J. Agr. Food Chem.* **22,** 592 (1974). Pharmacological properties: Hashimoto *et al., Toxicol. Appl. Pharmacol.* **23,** 616 (1972). Toxicity data: *eidem, ibid.* 606. Colorless prisms, mp 181.5-182.5°. Sol in acetone, methanol, chloroform, acetonitrile. Slightly sol in other organics; insol in water. LD_{50} in rats, mice, guinea pigs, rabbits (g/kg): 3.40, 6.64, 3.64, 2.27 orally (Hashimoto).

USE: Systemic fungicide.

THERAP CAT (VET): Anthelmintic.

9283. Thiophene. Thiofuran; thiofurfuran; thiole; thiotetrole; divinylene sulfide. C_4H_4S; mol wt 84.14. C 57.10%, H 4.79%, S 38.11%. Found in coal tar, in coal gas, and in technical benzene: V. Meyer, *Ber.* **16,** 1471 (1883); **17,** 2642 (1884). Made available in commercial quantities by a process utilizing the dehydrogenation of butane with sulfur as the dehydrogenating agent, followed by cyclization with sulfur to form the thiophene ring: Rasmussen, Ray, *Chem. Inds.* **60,** 593, 620 (1947). Laboratory prepn by heating sodium succinate with phosphorus trisulfide: R. Phillips, *Org. Syn.* **coll. vol. II,** 578 (1943). Also prepd by passing

ethylene or acetylene into boiling sulfur; or by passing acetylene and hydrogen sulfide over hot bauxite or nickel hydroxide: U.S. pat. **1,421,743,** *C.A.* **16,** 3093 (1922). *Review:* B. Buchholz in Kirk-Othmer *Encyclopedia of Chemical Technology* vol. **22** (Wiley-Interscience, New York, 3rd ed., 1983) pp 965-973.

Liquid. Slight aromatic odor resembling that of benzene. d_4^0 1.0873; d_4^{25} 1.0573; d_4^{50} 1.0285. mp -38.3°. bp_{760} 84.4°; bp_{400} 64.7°; bp_{200} 46.5°; bp_{100} 30.5°; bp_{60} 20.1°; bp_{40} +12.5°; bp_{20} 0.0°; bp_{10} -10.9°; bp_5 -20.8°; n_D^{25} 1.52684. Absorption spectrum: Purvis, *J. Chem. Soc.* **97,** 1653, 1656 (1910). Insol in water; miscible with most organic solvents. May be heated to 850° without decompn.

USE: Solvent similar to benzene, but suitable for lower and higher temps; manuf of resins from thiophene-phenol mixtures and formaldehyde; manuf of dyes and pharmaceuticals.

9284. 2-Thiophenecarboxylic Acid. 2-Thenoic acid. $C_5H_4O_2S$; mol wt 128.16. C 46.86%, H 3.14%, O 24.97%, S 25.02%. Prepn: Voerman, *Rec. Trav. Chim.* **26,** 293 (1907); Sy, de Malleray, *Bull. Soc. Chim. France* **1963,** 1276; Gross *et al., Ber.* **96,** 1382 (1963).

Needles from water, mp 128.5°. Very sol in ether, alc, hot water; moderately sol in $CHCl_3$; slightly sol in petr ether. Sodium salt, $C_5H_3NaO_2S$, *sodium 2-thiophenecarboxylate, sodium 2-thenoate, Trophires.*

USE: Sodium salt as lubricating-grease thickener, Morway, Kolfenbach, U.S. pat. **2,576,031** (1951 to Standard Oil).

9285. Thiophenol. *Benzenethiol;* phenylmercaptan. C_6H_6S; mol wt 110.17. C 65.41%, H 5.49%, S 29.12%. C_6H_5SH. Prepd by the reduction of benzenesulfonyl chloride with zinc dust in sulfuric acid: Adams, Marvel, *Org. Syn.* **1,** 71 (1921).

Liquid. Repulsive, penetrating, garlic-like odor, esp when impure. d_4^{25} 1.0728; bp_{760} 168.3°; bp_{100} 103.6°; bp_{50} 86.2°; bp_{20} 69.7°; $bp_{1.0}$ 18.6°; n_D^{25} 1.58603. Heat of fusion 24.90 cal/g; spec heat at 25°: 0.3829; entropy at 25°: 52.6. Insol in water. Very sol in alc; miscible with ether, benzene, CS_2. Feebly acidic. Oxidizes in air, esp when dissolved in alcoholic ammonia, forming diphenyl disulfide, $C_6H_5SSC_6H_5$. The hydrogen of the SH group is easily replaced by metals.

9286. Thiopropazate. *4-[3-(2-Chlorophenothiazin-10-yl)propyl]-1-piperazineethanol acetate;* 2-chloro-10-[3-[1-(2-acetoxyethyl)-4-piperazinyl]propyl]phenothiazine; 10-[3-[1-(2-acetoxyethyl)-4-piperazinyl]propyl]-2-chlorophenothiazine; *N*-(β-acetoxyethyl)-*N'*-[γ-(2'-chloro-10'-phenothiazinyl)propyl]piperazine; 1-(2-acetoxyethyl)-4-[3-(2-chloro-10-phenothiazinyl)propyl]piperazine. $C_{23}H_{28}ClN_3O_2S$; mol wt 446.00. C 61.93%, H 6.33%, Cl 7.95%, N 9.42%, O 7.17%, S 7.19%. Prepd from 2-chloro-10-(γ-chloropropyl)phenothiazine and piperazine in butanone followed by treatment with β-bromoethyl acetate in toluene: Cusic, U.S. pat. **2,766,235** (1956); Anderson *et al., Arzneimittel-Forsch.* **12,** 937 (1962).

Free base, $bp_{0.1}$ 214-218°. Sol in ether.

Dihydrochloride, $C_{23}H_{30}Cl_3N_3O_2S$, *Dartal, Dartalan.* Crystals from 95% ethanol, dec 223-229°. Freely sol in water; much less sol in alc, chloroform; almost insol in ether.

Dimaleate, $C_{23}H_{28}ClN_3O_2S.2C_4H_4O_4$, crystals from methanol + ethanol, mp 167-169°.

THERAP CAT: Antipsychotic.

9287. Thioproperazine. *N,N-Dimethyl-10-[3-(4-methyl-1-piperazinyl)propyl]phenothiazine-2-sulfonamide;* 3-dimethylsulfamoyl-10-[3-(4-methylpiperazino)propyl]phenothiazine; 2-dimethylsulfamoyl-10-[3'-(4''-piperazino)propyl]phenothiazine; thioperazine; RP 7843; SKF 5883; Vontil; Sulfenazin. $C_{22}H_{30}N_4O_2S_2$; mol wt 446.64. C 59.16%, H 6.77%, N 12.54%, O 7.16%, S 14.36%. Prepn: **Brit. pat. 814,512** (1959 to Rhône-Poulenc).

Crystals, mp 140°.
Fumarate, $C_{22}H_{30}N_4O_2S_2.C_4H_4O_4$, crystals, mp 182°.
Dimethanesulfonate, $C_{24}H_{38}N_4O_8S_4$, *Mayeptil, Majeptil.*
THERAP CAT: Neuroleptic; anti-emetic.

9288. Thioquinox. *1,3-Dithiolo[4,5-b]quinoxaline-2-thione; trithiocarbonic acid cyclic 2,3-quinoxalinediyl ester;* 2,3-quinoxalinedithiol cyclic trithiocarbonate; trithiocarbonic acid cyclic ester with 2,3-quinoxalinedithiol; 2-thio-1,3-dithiolo[4,5-b]quinoxaline; chinothionat; quinothionate; Eradex; Readex. $C_9H_4N_2S_3$; mol wt 236.31. C 45.74%, H 1.70%, N 11.85%, S 40.70%. Prepn starting with 2,3-dimercaptoquinoxaline and thiophosgene: Sasse *et al.*, **U.S. pat. 3,141,886** (1964 to Bayer); **Ger. pat. 1,088,965** (1960) and **Belg. pat. 580,478** (1958).

Brownish to yellow powder, mp 180°. The commercial product begins to melt at 165° but does not decompose until the temp exceeds 200°. Vapor press. 1×10^{-7} mm Hg at 20°. Practically insol in water. Slightly sol in acetone, methanol, anhyd ethanol, petr ether, kerosene. LD_{50} orally in rats: 3.4 g/kg, Sasse, *Hoefchen-Briefe* **13**, 197 (1960), *C.A.* **56**, 1801h (1962).

USE: Acaricide; fungicide.

9289. Thioredoxin. Ubiquitous, small, hydrogen carrier protein that participates in a wide variety of biochemical reactions, e.g., ribonucleotide reduction, methionine sulfoxide, sulfate and disulfide reduction, phosphate transfer reactions. Thioredoxin is oxidized from a dithiol to a disulfide during ribonucleotide reduction, and regenerated by reduced triphosphopyridine nucleotide (TPNH) and the flavoprotein thioredoxin reductase: Laurent *et al.*, *J. Biol. Chem.* **239**, 3436 (1964). Prepn from *Escherichia coli:* Williams *et al., ibid.* **242**, 5226 (1967). Complete amino acid sequence: A. Holmgren, *Eur. J. Biochem.* **6**, 475 (1968). Purification and characterization: A. Holmgren *et al., J. Biol. Chem.* **256**, 3118 (1981). Prepn of bacteriophage T4-induced thioredoxin: Berglund, Sjoberg, *ibid.* **245**, 6030 (1970). Amino acid sequence: Sjoberg, Holmgren, *ibid.* **247**, 8063 (1972). The T4 thioredoxin shows no structural homology with thioredoxin from *E. coli.* Purification, characterization and amino acid sequence of thioredoxin from *Corynebacterium nephridii:* M. Meng, H. P. C. Hogenkamp, *ibid.* **256**, 9174 (1981).

9290. Thioridazine. *10-[2-(1-Methyl-2-piperidinyl)ethyl]-2-(methylthio)-10H-phenothiazine;* 2-methylmercapto-10-[2-(1-methyl-2-piperidinyl)ethyl]phenothiazine; 3-methylmercapto-N-[2'-(N'-methyl-2-piperidyl)ethyl]phenothiazine; Melleretten. $C_{21}H_{26}N_2S_2$; mol wt 370.56. C 68.06%, H 7.07%, N 7.56%, S 17.31%. Dopamine receptor blocker; parent compound of sulforidazine and mesoridazine, *q.q.v.* Prepn: Bourquin *et al., Helv. Chim. Acta* **41**, 1072 (1958). Toxicity: E. I. Goldenthal, *Toxicol. Appl. Pharmacol.* **18**,

185 (1971). Effect on dopaminergic function: D. M. Niedzwiecki *et al., J. Pharmacol. Exp. Ther.* **228**, 636 (1984); C. D. Kilts *et al., ibid* **231**, 334 (1984). GLC determn of thioridazine and metabolites in plasma: E. C. Dinovo *et al., J. Pharm. Sci.* **65**, 667 (1976). HPLC determn: R. Whelpton *et al., J. Chromatog.* **426**, 223 (1988). Clinical evaluation as antipsychotic: R. Axelsson *et al., Curr. Ther. Res.* **21**, 587 (1977).

Crystals from acetone, mp 72-74°. $bp_{0.02}$ 230°. LD_{50} orally in rats: 995 ±39 mg/kg (Goldenthal).

Hydrochloride, $C_{21}H_{27}ClN_2S_2$, *TP 21, Aldazine, Mellaril, Melleril, Mallorol, Novoridazine, Orsanil, Ridazin, Sonapax, Stalleril.* Crystals from acetone, mp 158-160°. Sol in water and ethanol.

Fumarate, $C_{25}H_{30}N_2O_4S_2$, crystals from ethanol, mp 158-161°.

Tartrate monohydrate, $C_{25}H_{32}N_2O_6S_2.H_2O$, dec 130°.

THERAP CAT: Antipsychotic.

9291. Thiosalicylic Acid. *2-Mercaptobenzoic acid;* o-sulfhydrylbenzoic acid. $C_7H_6O_2S$; mol wt 154.18. C 54.53%, H 3.92%, O 20.75%, S 20.80%. Prepd by heating o-halogenated benzoic acids with an alkaline hydrosulfide in the presence of copper: **Ger. pat. 189,200** (1906); by reduction of dithiosalicylic acid: Claasz, *Ber.* **45**, 2427 (1912); **Ger. pat. 205,450;** C. F. H. Allen, D. D. MacKay, *Org. Syn. coll. vol. II,* 580 (1943).

Sulfur-yellow flakes, plates, needles from glacial acetic acid or alcohol. Softens at 158°. mp 164-165°. Sublimes. Slightly sol in hot water; freely in glacial acetic acid, alc. Exposure to air yields dithiosalicylic acid, which is produced also when an alcoholic soln of thiosalicylic acid comes in contact with $FeCl_3$ (resulting in a transitory blue color).

Sodium salt, $C_7H_5NaO_2S$, *Jecto-Sal, Thiocyl.* Crystals, sol in water.

USE: Manuf thioindigo dyes, reagent for determn of iron.

9292. Thiosemicarbazide. *Hydrazinecarbothioamide.* CH_5N_3S; mol wt 91.14. C 13.18%, H 5.53%, N 46.12%, S 35.18%. $NH_2CSNHNH_2$.

White, cryst powder. mp 182-184°. Sol in water or alc. LD_{50} orally in adult Norway rats: 13 mg/kg, Dieke, *Proc. Soc. Exp. Biol. Med.* **70**, 688 (1949).

USE: As a reagent for detection of metals.

9293. Thiosinamine. *(2-Propenyl)thiourea;* allyl thiourea; allylthiocarbamide; Aminosin; Rhodalline. $C_4H_8N_2S$; mol wt 116.19. C 41.35%, H 6.94%, N 24.11%, S 27.60%. $CH_2=CHCH_2NHCSNH_2$. Made by warming a mixture of equal parts of allyl mustard oil and abs alcohol with an equal amount of 30% ammonia. Growth regulating activity: Karanov, Vasilev, *Izv. Inst. Fiziol. Rast. Bulg. Akad. Nauk* **16**, 167 (1970), *C.A.* **73**, 119454y (1970). Mechanism of action studies as an egg suppressive agent in schistosomiasis, A. B. Machado *et al., J. Parasitol.* **56**, 392 (1970). *In vivo* and *in vitro* effects: I. Popiel, D. A. Erasmus, *Trans. Roy. Soc. Trop. Med. Hyg.* **75**, 287 (1981).

White crystals; bitter taste; slight garlic odor; d 1.22; mp 78°. Sol in about 30 parts water; sol in alcohol, slightly sol in ether; insol in benzene. LD s.c. in rats: 850 mg/kg.

THERAP CAT (VET): Has been used to minimize scar tissue.

9294. 1-Thiosorbitol. *1-Thio-*D*-glucitol.* $C_6H_{14}O_5S$; mol wt 198.24. C 36.35%, H 7.12%, O 40.36%, S 16.18%. CH_2-$OH(CHOH)_4CH_2SH$. Prepd by heating an aq soln of glucose with a catalyst, sulfur, and hydrogen at 150° and 1500 psi for 3 hrs: Farlow *et al., J. Am. Chem. Soc.* **70**, 1392 (1948).

Crystals from abs alcohol, mp 92-93°. Nonhygroscopic. $[\alpha]_D^{27}$ −1.9° (c = 2). Readily sol in water, pyridine, ethylene glycol, formamide; practically insol in benzene, petr ether, carbon tetrachloride; carbon disulfide. Solubilities at 20° in g/100 ml solvent: 1.7 in abs alcohol; 1.2 in dioxane; 0.016 in ethyl ether; 0.016 in trichloroethylene; 0.010 in acetone. Strong reducing agent.

USE: Suggested as a polymer stabilizer; anticorrosion agent for pickling baths; in plating baths.

9295. Thiostrepton. Gargon; Thiactin; Bryamycin. C_{72}-$H_{85}N_{19}O_{18}S_5$; mol wt 1664.83. C 51.94%, H 5.15%, N 15.98%, O 17.30%, S 9.63%. Polypeptide antibiotic contg sulfur. Produced by *Streptomyces azureus* isolated from New Mexican soil: Pagano *et al., Antibiot. Ann.* **1955-56**, 554; Vandeputte, Dutcher, *ibid.* 560; **Brit.** pat. **795,570** (1958); Donovick *et al.;* Platt, **U.S.** pats. **2,982,689; 2,982,698** (1961 all to Olin Mathieson). Structure studies: Bodanszky *et al., J. Am. Chem. Soc.* **84**, 2003 (1962); **86**, 2478 (1964); Anderson *et al., Nature* **225**, 233 (1970). Total structure determn: K. Tori *et al., Tetrahedron Letters* **1976**, 185; *eidem, J. Antibiot.* **32**, 1072 (1979). ^{13}C-NMR study: *eidem, ibid.* **34**, 124 (1981). Mode of action as an inhibitor of protein synthesis: Cannon, Burns, *FEBS Letters* **18**, 1 (1971); Cundliffe, *Biochem. Biophys. Res. Commun.* **44**, 912 (1971). Identity with thiactin: Bodanszky *et al., J. Antibiot.* **16A**, 76 (1963). Comprehensive description: K. Florey, Ed. in *Analytical Profiles of Drug Substances* **vol.** 7 (Academic Press, New York, 1978) pp 423-444. *Review:* Pestka, Bodley, in *Antibiotics* **vol.** 3, J. W. Corcoran, F. E. Hahn, Eds. (Springer-Verlag, New York, 1975) pp 551-573; E. Cundliffe, *ibid.* **vol.** 5(pt. 1), F. E. Hahn, Ed. (1979) pp 329-343.

Crystals from chloroform + methanol, dec 246-256°. $[\alpha]_D^{23}$ −98.5° (glacial acetic acid); −61° (dioxane); −20° (pyridine). Sol in chloroform, dioxane, pyridine, glacial acetic acid, DMF. Practically insol in water, the lower alcohols, nonpolar organic solvents, dil aq acid or base. Dissolved by methanolic acid or base with decompn. Stable in the presence of gastric and intestinal juices and urine. No uv maxima but shows characteristic shoulders at 225, 250, 280 nm ($E_{1cm}^{1\%}$ 520, 380, 255).

Hemisuccinate, mp 200-220°, forms a water-soluble potassium salt. Prepn: Bodanszky, Fried, **U.S.** pat. **3,181,995** (1965 to Olin Mathieson).

THERAP CAT: Antibacterial.

THERAP CAT (VET): Has been used as an antibacterial.

9296. Thiothiamine. *3-[(4-Amino-2-methyl-5-pyrimidinyl)methyl]-5-(2-hydroxyethyl)-4-methyl-2(3H)-thiazolethione;* 3-[(2-methyl-4-amino-5-pyrimidyl)methyl]-4-methyl-5-(β-hydroxyethyl)thiothiazol-2-one; Sulbone. $C_{12}H_{16}$-N_4OS_2; mol wt 296.42. C 48.62%, H 5.44%, N 18.90%, O 5.40%, S 21.63%. Prepn: Matsukawa, Iwatsu, *J. Pharm. Soc. Japan* **71**, 455 (1951); *eidem,* **U.S.** pats. **2,592,930/1** (both 1952 to Takeda); Maxion, **U.S.** pat. **2,799,676** (1957 to Chase Chem.).

Crystals from aq ethanol, mp 239°.

Hydrochloride, $C_{12}H_{16}N_4OS_2.HCl$, dec 245°.

USE: Intermediate in the prepn of thiamine and related compds: Maxion, *loc. cit.;* Turner, Schmitt, **U.S.** pat. **2,844,-579** (1958 to Am. Cyanamid).

9297. Thiothixene. *N,N-Dimethyl-9-[3-(4-methyl-1-piperazinyl)propylidene]thioxanthene-2-sulfonamide; cis-9-*[3-(4-methyl-1-piperazinyl)propylidene]-2-(dimethylsulfonamido)thioxanthene; tiotixene; Navane; Navaron (obsolete); Orbinamon. $C_{23}H_{29}N_3O_2S_2$; mol wt 443.63. C 62.27%, H 6.59%, N 9.47%, O 7.21%, S 14.46%. Prepn of *cis/trans*-isomer mixture and sepn of isomers: Bloom, Muren, **Belg.** pat. **647,066** corresp to **U.S.** pat. **3,310,553** (1964, 1967 both to Pfizer); Muren, Bloom, *J. Med. Chem.* **13**, 17 (1970). Only *cis*-isomer exhibits therapeutic activity. Structure studies; J. P. Schaefer, *Chem. Commun.* **1967**, 743. Pharmacological studies: A. Weissman, *Psychopharmacologia* **12**, 142 (1968). Book: T. A. Ban, *Psychopharmacology of Thiothixene* (Raven Press, New York, 1978) 485 pp.

cis-Isomer, crystals, mp 147.5-149°. uv max (methanol): 228, 260, 310 nm (log ∈ 4.6, 4.3, 3.9). Has greater pharmacologic activity than *trans*-isomer. LD_{50} in mice, rats (mg/kg): 100, 55 i.p. (Weissman).

trans-Isomer, mp 123-124.5°. uv max (methanol): 229, 252, 301 nm (log ∈ 4.5, 4.2, 3.9). LD_{50} i.p. in mice: 235 mg/kg (Weissman).

cis-trans-Isomer mixture, crystals, mp 114-118°.

Dimaleate, mp 158-160.5°.

Dioxalate, mp 229°.

THERAP CAT: Antipsychotic.

9298. 2-Thiouracil. *2,3-Dihydro-2-thioxo-4(1H)-pyrimidinone;* 2-mercapto-4-hydroxypyrimidine; 4-hydroxy-2(1H)-pyrimidinethione; 2-mercapto-4(1H)-pyrimidinone; 6-hydroxy-2-mercaptopyrimidine; 2-mercapto-4-pyrimidinone; Deracil. $C_4H_4N_2OS$; mol wt 128.15. C 37.49%, H 3.15%, O 12.48%, N 21.86%, S 25.02%. Occurs in seeds of *Brassica, Cruciferae:* Purves, *Brit. J. Exp. Pathol.* **22**, 241 (1941). Prepd by condensing ethyl formylacetate with thiourea: Wheeler, Liddle, *Am. Chem. J.* **40**, 550 (1908). Antithyroid activity results from its interference with the iodination of thyroxine precursors: *see* Maloof, Soodak, *Pharmacol. Rev.* **15**, 72-79 (1963). Inhibition of nucleic acid metabolism: Cardeilhac, *Proc. Soc. Exp. Biol. Med.* **125**, 692 (1967). Toxicology: K. K. Carroll, R. L. Noble, *J. Pharmacol. Exp. Ther.* **97**, 478 (1949).

Minute, bitter crystals, no definite mp. Very slightly soluble in water (1:2000); practically insol in alcohol, ether, acids. Readily sol in alkaline solns. LD_{100} i.p. in rats: 1500 mg/kg (Carroll, Noble).

THERAP CAT: Treatment of hyperthyroidism; angina pectoris; congestive heart failure.

THERAP CAT (VET): Thyroid depressant. In hyperthyroidism and to promote fattening.

9299. Thiourea. Thiocarbamide. CH_4N_2S; mol wt 76.12. C 15.78%, H 5.30%, N 36.80%, S 42.12%. H_2NCS-NH_2. Made by fusing ammonium thiocyanate: Powers and Powers, Mitchell, U.S. pats. **2,552,584**; **2,560,596** (both 1951 to Koppers); by treating cyanamide with hydrogen sulfide: Robin, Jr., U.S. pat. **2,173,067** (1940 to Am. Cyanamid); Lewis, U.S. pat. **2,393,917** (1946 to Monsanto); Van de Kamp, U.S. pat. **2,357,149** (1944 to Merck & Co.). Toxicity data: Dieke et al., J. Pharmacol. Exp. Ther. **90**, 262 (1947).

Crystals, mp 176-178°. d 1.405. Soluble in 11 parts water, in alcohol; sparingly sol in ether. Neutral reaction. Forms addition compds with metallic salts. LD_{50} orally in wild Norway rats: 1830 mg/kg (Dieke).

Note: Chronic administration in rats has resulted in hepatic tumors, bone marrow depression and goiters: Clinical Toxicology of Commercial Products, R. E. Gosselin et al., Eds. (Williams & Wilkins, Baltimore, 5th ed., 1984) Section II, p 350. This substance may reasonably be anticipated to be a carcinogen: Fourth Annual Report on Carcinogens (NTP 85-002, 1985) p 187.

USE: Photographic fixing agent and to remove stains from negatives; manuf resins; vulcanization accelerator; a reagent for bismuth, selenite ions.

9300. Thioxanthene. Thiaxanthene; dibenzothiopyran; dibenzopenthiophene; diphenylenemethane sulfide. $C_{13}H_{10}S$; mol wt 198.27. C 78.75%, H 5.08%, S 16.17%. Conveniently obtained from thioxanthone in 74% yield by reduction wth lithium aluminum hydride: Mustafa, Hilmy, J. Chem. Soc. **1952**, 1345.

Needles or rods from alcohol + chloroform. Sublimes easily. mp 128°. bp_{730} 340°. Dipole moment: 1.44 D. Freely sol in chloroform. Moderately sol in alcohol, ether.

9301. Thioxanthone. Thioxanthen-9-one; thiaxanthone; 9-oxothioxanthene. $C_{13}H_8OS$; mol wt 212.25. C 73.56%, H 3.80%, O 7.54%, S 15.10%. Prepd by the interaction of diphenyl sulfide and phosgene with aluminum chloride as catalyst: Szmant et al., J. Org. Chem. **18**, 745 (1953).

Yellow needles from chloroform, mp 211°. bp_{715} 273°. Freely sol in benzene, chloroform, carbon disulfide, hot glacial acetic acid. Slightly sol in alc. Practically insol in water, alkaline solns. Sol in concd sulfuric acid, giving a yellow soln with strong green fluorescence in visible light.

9302. THIP. 4,5,6,7-Tetrahydroisoxazolo[5,4-c]pyridin-3(2H)-one; 4,5,6,7-tetrahydroisoxazolo[5,4-c]pyridin-3-ol; gaboxadol. $C_6H_8N_2O_2$; mol wt 140.13. C 51.42%, H 5.75%, N 19.99%, O 22.84%. Structural analog of muscimol, q.v., with potent GABA agonist activity. Prepn: P. Krogsgaard-Larsen, Acta Chem. Scand. B **31**, 584 (1977); idem, Japan. Kokai 79 36290 corresp to U.S. pat. **4,278,676** (1979, 1981 both to H. Lundbeck & Co.). GABA agonist effects: P.

Krogsgaard-Larsen et al., Nature **268**, 53 (1977). Antinociceptive effects: D. A. Kendall et al., J. Pharmacol. Exp. Ther. **220**, 482 (1982). HPLC determn: S. M. Madsen, J. Chromatog. **238**, 509 (1982). Muscle relaxant activity: R. C. Hill, **Belg.** pat. **890,136** (1982 to Sandoz), C.A. **97**, 11847k (1982). ^{3}H-THIP binding study: E. Falch, P. Krogsgaard-Larsen, J. Neurochem. **38**, 1123 (1982). Pharmacodynamics and potential therapeutic uses: A. V. Christensen et al., Pharm. Weekbl., Sci. Ed. **4**, 145 (1982). Effects in tardive dyskinesia: S. Korsgaard et al., Arch. Gen. Psychiatry **39**, 1017 (1982).

Zwitterionic. Colorless cryst, mp 242-244° (dec). uv max (methanol): 212 nm (log ϵ 3.64). pKa (water, 25°): 4.44 ±0.03; 8.48 ±0.04.

Hydrobromide, $C_6H_9BrN_2O_2$. Faintly reddish cryst from methanol-ether, mp 162-163° (dec). LD_{50} in mice (mg/kg): 80 i.v., 145 i.p.; >320 orally (U.S. patent).

USE: As a molecular probe to study GABA receptors.

9303. Thiphenamil. α-Phenylbenzeneethanethioic acid S-[2-(diethylamino)ethyl]ester; diphenylthioacetic acid S-(2-diethylaminoethyl) ester; diphenylthiolacetic acid 2-diethylaminoethyl ester; S-[2-(diethylamino)ethyl]diphenylthioacetate; β-diethylaminoethyl diphenylthioacetate. $C_{20}H_{25}NOS$; mol wt 327.50. C 73.35%, H 7.69%, N 4.28%, O 4.89%, S 9.79%. Prepn: Richardson, U.S. pat. **2,390,555** (1945 to Wm. P. Poythress); Clinton, U.S. pat. **2,510,773** (1950 to Sterling Drug); Richardson, J. Pharm. Sci. **55**, 1316 (1966).

$$(C_6H_5)_2CHCSCH_2CH_2N(C_2H_5)_2$$

Hydrochloride, $C_{20}H_{26}ClNOS$, **Thiphen, Trocinate**. Rosettes of tiny needles from benzene + petr ether, large prisms from abs ethanol + ethyl acetate, mp 129-130°. Sol in water. Aq solns are about neutral to litmus.

THERAP CAT: Anticholinergic; relaxant (smooth muscle).

9304. Thiram. Tetramethylthioperoxydicarbonic diamide; bis(dimethylthiocarbamoyl) disulfide; bis(dimethylthiocarbamyl) disulfide; tetramethylthiuram disulfide; TMTD; ENT 987; SQ 1489; NSC 1771; Thiurad; Thiosan; Thylate; Tiuramyl; Thiuramyl; Puralin; Fernasan; Nomersan; Rezifilm; Pomarsol; Tersan; Tuads; Tulisan; Arasan. $C_6H_{12}N_2S_4$; mol wt 240.44. C 29.97%, H 5.03%, N 11.65%, S 53.35%. Prepn: v. Braun, Stechele, Ber. **35**, 820 (1902); **36**, 2280 (1903); Romani, C.A. **16**, 854 (1922); Cummings, Simmons, Ind. Eng. Chem. **20**, 1173 (1928). Acute toxicity: Gaines, Toxicol. Appl. Pharmacol. **14**, 515 (1969). Review: IARC Monographs **12**, 225 (1976).

$$(CH_3)_2NC\text{—}SS\text{—}CN(CH_3)_2$$

Crystals from chloroform + alcohol, mp 155-156° (commercial grades, mp 146°). d 1.29. Insol in water, dil caustic, gasoline. Soly in alcohol and in ether less than 0.2%; soly in acetone 1.2%; in benzene 2.5%. Sol in chloroform, acetone. LD_{50} orally in rats: 640 mg/kg (Gaines).

USE: Rubber accelerator; vulcanizer; seed disinfectant; fungicide; bacteriostat in soap; animal repellent. Main ingredient of antiseptic spray Nobecutan. Caution: Irritant of mucous membranes and skin; a skin sensitizer. Toxicity greater in presence of fats, oils, fat solvents.

THERAP CAT: Antiseptic.

9305. Thomas Phosphate. Thomas flour; calcium phosphate, tetrabasic. $Ca_3(PO_4)_2 \cdot CaO$ also represented by $4CaO \cdot P_2O_5$. A byproduct of the Thomas steel process used in Europe: Iron ore of the "Minette" type from Alsace and the Rhineland contains about 2% phosphates which are combined with calcium and separated during the refining process.

USE: Agricultural fertilizer.

9306. Thonzonium Bromide. *N-[2-[[(4-Methoxyphenyl)-methyl]-2-pyrimidinylamino]ethyl]-N,N-dimethyl-1-hexa-decanaminium bromide; hexadecyl[2-[(p-methoxybenzyl)-2-pyrimidinylamino]ethyl]dimethylammonium bromide;* cetyl-[2-[(p-methoxybenzyl)-2-pyrimidinylamino]ethyl]dimethyl-ammonium bromide; *N-(2-pyrimidyl)-N-(p-methoxybenz-yl)-N',N'-dimethyl-N'-cetyl-N'-bromoethylenediamine;* Thonzide. $C_{32}H_{55}BrN_4O$; mol wt 591.74. C 64.95%, H 9.37%, Br 13.51%, N 9.47%, O 2.70%. Prepn: Fand, U.S. pat. **2,742,470** (1956 to Nepera Chem.).

Crystals, mp 91-92°.

USE: Detergent.

9307. Thonzylamine Hydrochloride. *N-[(4-Methoxy-phenyl)methyl]-N',N'-dimethyl-N-2-pyrimidinyl-1,2-eth-anediamine monohydrochloride; 2-[(2-dimethylaminoethyl)-(p-methoxybenzyl)amino]pyrimidine hydrochloride; N,N-di-methyl-N'-(p-methoxybenzyl)-N'-(2-pyrimidyl)ethylenedi-amine hydrochloride;* Anahist; Neohetramine hydrochloride; Novohetramin; Resistab. $C_{16}H_{23}ClN_4O$; mol wt 322.83. C 59.52%, H 7.18%, Cl 10.98%, O 4.96%, N 17.36%. Prepd by treating the sodium salt of 2-(p-methoxybenzyl)aminopy-rimidine with N,N-dimethyl-2-chloroethylamine: Fried-man, Tolstoouhov, U.S. pat. **2,465,865** (1949).

Crystals, mp 173-176° (free base, oily liq, bp$_{2.2}$ 185-187°). Freely sol in water; sol in alcohol, chloroform; practically insol in ether. pH 5.1-5.7 (2% aq solution). LD$_{50}$ orally in guinea pigs: 493 mg (base)/kg, Reinhard, Seudi, *Proc. Soc. Exp. Biol. Med.* **66**, 512 (1947).

Dipicrate, yellow needles, mp 141-145°.

Other names used for this substance (alone or in combina-tion with aspirin, caffeine, etc.) are: *NH-188; Anohist.*

THERAP CAT: Antihistaminic.

9308. Thorium. Th; at. wt 232.0381; at. no. 90; valence 4. Long-lived, natural isotope: 232; other isotopes: 224-231; 233-235. Discovered by Berzelius in 1828. Occurs in the minerals thorite, thorianite, orangite, yttrocrasite; in monazite sand; present to the extent of about 15 ppm in the crust of the earth. ^{232}Th usually found in nature eventually will decay to an isotope of lead by the emission of α-, β-, and γ-rays. Prepn: van Arkel, de Boer, *Z. Anorg. Chem.* **148**, 345 (1925); Marden, Rentschler, *Ind. Eng. Chem.* **19**, 97 (1927); Marden, *Trans. Electrochem. Soc.* **66**, 39 (1934). Monographs: F. L. Cuthbert, *Thorium Production Technol-ogy* (Addison-Wesley, Reading, Mass., 1958); L. Grainger, *Uranium and Thorium* (Pitman, London, 1958).

Grayish-white, lustrous, radioactive metal; somewhat ductile and malleable. mp 1842 ±30° (Marden, Rentschler); 1690° (Cuthbert). d 11.3-11.7. $T_{1/2}$ (^{232}Th) 1.41 × 10^{10} yrs. On heating burns in air with formation of the oxide. At-tacked by concd acids. Reacts with the halogens at a red heat. Forms the sulfide when heated with sulfur, the nitride when heated with nitrogen.

USE: In nuclear reactors usually in conjunction with ^{235}U. When excess neutrons strike ^{232}Th nuclei, they can produce ^{233}U which is potential fissionable fuel. With more nuclei

^{238}U can be produced from ^{232}Th. In manuf incandescent mantles; as a reducing agent in metallurgy; for filament coatings in incandescent lamps and vacuum tubes; as cata-lyst in organic syntheses.

9309. Thorium Chloride. Thorium tetrachloride. Cl_4Th; mol wt 373.88. Cl 37.93%, Th 62.07%. ThCl$_4$. It may con-tain from 7 to 9 mols H$_2$O. Prepn of the octahydrate: Kremer, *J. Am. Chem. Soc.* **64**, 1009 (1942).

White, odorless crystals; d 4.59; mp 770°. bp 921°. Sol in water, alcohol. Sol in ethylenediamine.

9310. Thorium Iodide. Thorium tetraiodide. I_4Th; mol wt 739.69. I 68.63%, Th 31.37%. ThI$_4$. Prepd by passing iodine vapors in an inert gas over the metal at elevated temp: Allen, Yost, *J. Chem. Phys.* **22**, 855 (1954). *See also* Cuth-bert, *Thorium Production Technology* (Addison-Wesley, Reading, Mass., 1958).

Pale yellow crystals, mp 566°. bp 837°. Dec by light or heat. When heated with Th gives ThI$_3$.

9311. Thorium Nitrate. $N_4O_{12}Th$; mol wt 480.06. N 11.67%, O 40.00%, Th 48.33%. Th(NO$_3$)$_4$. Commercial form obtained as tetrahydrate; also crystallizes with 6 and 12 mols H$_2$O. Obtained from monazite: Pearce *et al.*, *Inorg. Syn.* **II**, 38 (1946).

Tetrahydrate, white, slightly deliquesc crystals. Very sol in water, alcohol, the soln having an acid reaction. *Keep well closed.*

USE: Thorium nitrate with 1% cerium nitrate constitutes the usual impregnating liq for incandescent mantles. As a reagent for determination of fluorine.

9312. Thorium Oxide. Thorium dioxide; thoria. O_2Th; mol wt 264.05. O 12.12%, Th 87.88%. ThO$_2$. Review: Kel-ler in *Comprehensive Inorganic Chemistry* vol. 5, J. C. Bailar, Jr. *et al.*, Eds. (Pergamon Press, Oxford, 1973) pp 221-223.

White, heavy, infusible, cryst powder; when heated is incandescent. d 10.0. mp 3390°. Insol in water or alkalies. Sol in acids with difficulty.

Suspension contg 20-25% ThO$_2$, *Thorotrast.* Reviews of tumor induction by Thorotrast: Looney, *Am. J. Roentgenol., Radium Ther. Nucl. Med.* **83**, 163-185; Smoron, Battifora, *Cancer* **30**, 1252-1259 (1972).

Note: This substance has been listed as a known carcino-gen: *Fourth Annual Report on Carcinogens* (NTP 85-002, 1985) p 189.

THERAP CAT: Has been used as diagnostic aid (radiopaque medium).

9313. Thorium Sulfate. O_8S_2Th; mol wt 424.15. O 30.18%, S 15.12%, Th 54.70%. Th(SO$_4$)$_2$. Crystallizes with 4, 6, 8 and 9 mols H$_2$O.

Nonahydrate, colorless or white, monoclinic crystals; dec when strongly heated; d 2.8. Sol in about 70 parts cold, in about 15 parts hot water.

9314. Thorium Tetracyanoplatinate(II). *Thorium(4+) tetrakis(cyano-C)platinate(2−) (1:2);* platinous thorium cya-nide; thorium platinocyanide. $C_8N_8Pt_2Th$; mol wt 830.35. C 11.57%, N 13.49%, Pt 47.00%, Th 27.94%. Th[Pt(CN)$_4$]$_2$.

Hexadecahydrate, yellow crystals. Sparingly sol in water.

USE: For fluorescent screens, *see* Barium Platinous Cya-nide.

9315. Thozalinone. *2-(Dimethylamino)-5-phenyl-4(5H)-oxazolone; 2-(dimethylamino)-5-phenyl-2-oxazolin-4-one;* 5-phenyl-2-(dimethylamino)-2-oxazolin-4-one; tozalinone; Stimsen. $C_{11}H_{12}N_2O_2$; mol wt 204.22. C 64.69%, H 5.92%, N 13.72%, O 15.67%. Prepn: Hardy, Jr. *et al.*, U.S. pat. **3,037,990** (1962 to Am. Cyanamid). Pharmacological stud-ies: B. M. Bernstein, C. N. Latimer, *Psychopharmacologia* **12**, 338 (1968); U. H. Lindberg, J. Pedersen, *Acta Pharm. Suecica* **5**, 15 (1968). In treatment of Parkinsonism: W. D. Gray, C. E. Edward, U.S. pat. **3,665,075** (1972 to Am. Cyanamid).

Crystals from water, mp 133-136°.
THERAP CAT: Antidepressant.

9316. Threonine. Thr (IUPAC abbrev.); 2-amino-3-hydroxybutyric acid; α-amino-β-hydroxybutyric acid; 2-amino-3-hydroxybutanoic acid. $C_4H_9NO_3$; mol wt 119.12. C 40.33%, H 7.62%, N 11.76%, O 40.29%. $CH_3CH(OH)CH(NH_2)COOH$. An amino acid classified as essential in maintaining the growth of rats. "Of the four stereoisomers of α-amino-β-hydroxybutyric acid, only one, L-threonine, is capable of supporting growth of the rat", Schmidt, *Amino Acids*, p 1265 (1944). Essential component in human nutrition, not synthesized by the human body. Occurs in whole egg (5.3%), skim milk (4.6%), casein (4%), gelatin (1.4%), also in other proteins. Isoln from oat protein: Schryver, Buston, *Proc. Roy. Soc. London* **99B**, 476 (1925/6). From casein: Rimington, *Biochem. J.* **21**, 1187 (1927); Czarnetzky, Schmidt, *Z. Physiol. Chem.* **204**, 129 (1932); Rose *et al.*, *J. Biol. Chem.* **92**, LXVII (1931); **94**, 173 (1931/2); **109**, LXXVII (1935); **112**, 275, 283 (1935). From *E. coli*: Huang, U.S. pat. **2,937,121** (1960 to Pfizer). Synthesis: Abderhalden, Heyns, *Ber.* **67**, 530 (1934); Carter, *J. Biol. Chem.* **112**, 769 (1935); Wood *et al., ibid.* **117**, 1 (1937); Adkins, Reeve, *J. Am. Chem. Soc.* **60**, 1328 (1938); Attenburrow *et al.*, *J. Chem. Soc.* **1948**, 310; Pfister *et al.*, *J. Am. Chem. Soc.* **71**, 1101 (1949); Reeve, U.S. pat **3,038,007** (1962). Synthesis and physico-chemical study: Maldonado *et al., Bull. Soc. Chim. France* **1971**, 2933. Resolution of *dl*-threonine: **Brit.** pat. **1,197,809** (1970 to Noguchi Res. Found.), *C.A.* **73**, 99204m (1970). Review on microbial production: Daoust, *Develop. Ind. Microbiol.* **7**, 41 (1966).

L-Threonine, crystals, dec 255-257°. $[\alpha]_D^{26}$ −28.3° (c = 1.1). pK_1' 2.15; pK_2' 9.12. Freely sol in water; insol in abs alcohol, ether, chloroform.

Monobenzoyl deriv, $C_{11}H_{13}NO_4$, rectangular plates from abs alc, mp 151°.

DL-Threonine hemihydrate, orthorhombic crystals, dec 229-230°.

THERAP CAT: Nutrient.

9317. D-Threose. *[S-(R*,S*)]-2,3,4-Trihydroxybutanal.* $C_4H_8O_4$; mol wt 120.10. C 40.00%, H 6.71%, O 53.29%. From calcium D-xylonate by oxidation with H_2O_2: Ruff, *Ber.* **34**, 1370 (1901). Improved procedure using strontium D-xylonate and ferric acetate catalyst: Hockett, *J. Am. Chem. Soc.* **57**, 2260, 2265 (1935). From tetraacetyl-D-xylonitrile: Maquenne, *Compt. Rend.* **130**, 1403 (1900); *Ann. Chim.* [7], **24**, 404 (1901); Bonner, Roth, *J. Am. Chem. Soc.* **81**, 5454 (1959); from monobenzylidene-D-arabitol: Steiger, Reichstein, *Helv. Chim. Acta* **19**, 1016 (1939); from 1,1-diethylsulfonyl-D-*threo*-3,4,5-trihydroxypent-1-ene: Hough, Taylor, *J. Chem. Soc.* **1955**, 1212; from D-galactose: Perlin, Brice, *Can. J. Chem.* **34**, 541 (1956). Synthesis of DL-threose: Lake, Glattfeld, *J. Am. Chem. Soc.* **66**, 1091 (1944); Schmid, Grob, *Helv. Chim. Acta* **32**, 77 (1949); Sonogashira, Nakagawa, *Bull. Chem. Soc. Japan* **45**, 2616 (1972).

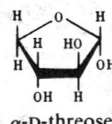

α-D-threose

Syrup. Shows mutarotation. Final $[\alpha]_D^{20}$ − 12.3° (20 min, c = 4). Very sol in water; slightly in alcohol. Practically insol in ether, petr ether.

Phenylosazone $C_{16}H_{18}N_4O_2$, dec 164-165°. Identical with D-erythrose phenylosazone.

Triacetate, $C_{10}H_{14}O_7$, prisms from abs ethanol, mp 117-118°. $[\alpha]_D^{25}$ +34.4° (c = 2 in chloroform). Soluble in hot water, chloroform, acetone, ethyl acetate; sparingly sol in abs alcohol, methanol, ether.

9318. L-Threose. $C_4H_8O_4$; mol wt 120.10. C 40.00%, H 6.71%, O 53.29%. By degradation of L-xylose through the oxime and tetraacetyl-L-xylononitrile to L-threose diacetamide: Hockett *et al., J. Am. Chem. Soc.* **60**, 278 (1938); from D-glucitol: Hutson, Weigel, *J. Chem. Soc.* **1961**, 1546.

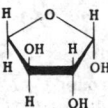

Not isolated. Shows mutarotation; final $[\alpha]_D^{20}$ +13.2° (c = 4.5). Sol in water.
Phenylosazone, $C_{16}H_{18}N_4O_2$, dec 162°.

9319. Thrombin. Serine proteinase present in blood plasma in the form of a precursor, prothrombin *q.v.* Mol wt about 33,580. Plays a central part in the mechanism of blood coagulation by converting the sol plasma protein, fibrinogen, *q.v.*, into insol fibrin, *q.v.* Bovine thrombin consists of two polypeptide chains, designated as A and B. The A chain has 49 amino acid residues of determined sequence: Magnussen, *Biochem. J.* **110**, 25P (1968). It is bound by a disulfide bridge to the B chain which contains about 265 residues and carbohydrate. Further structural studies: Seegers *et al., Thromb. Diath. Haemorrh.*, **Suppl. 47**, 325 (1971); **Suppl. 51**, 265 (1972); Magnussen, *ibid.* **Suppl. 54**, 31 (1973); **Suppl. 57**, 153 (1974). *Reviews:* Waugh *et al.*, "Thrombin" in *The Enzymes*, **vol. 4**, P. D. Boyer *et al.*, Eds. (Academic Press, New York, 2nd ed., 1960) pp 215-232; Magnussen, *ibid.* **vol. 3** (3rd ed., 1971) pp 277-321; W. H. Seegers, *Prothrombin* (Harvard Univ. Press, 1962) 728 pp; Laki, *Fed. Proc.* **24**, no. 4, part 1, 794 (1965); R. L. Lundblad *et al., Methods Enzymol.* **45B**, 156-176 (1976).
THERAP CAT: Local hemostatic.

9320. Thrombin, Topical. Thrombofort; Topostasin. A prepn of thrombin isolated from the prothrombin concentrate of bovine plasma, *see* W. H. Seegers, *Prothrombin* (Harvard Univ. Press, 1962) p 582. Prepn: Seegers, *Arch. Biochem.* **3**, 363 (1943-44); U.S. pat. **2,433,299** (1947 to Parke, Davis).

White powder, usually contg added sucrose. Completely sol in water or in isotonic soln of NaCl. Preserve dry product at a temperature between 2° and 8°, preferably at the lower limit. It may be applied as a dry powder or dissolved in sterile isotonic saline soln.
Human thrombin, *Fibrindex.*
THERAP CAT: Local hemostatic.

9321. Thromboplastin. *Blood-coagulation factor III;* factor III; cytozyme; thrombokinase; thrombokinin; tissue factor; tissue thromboplastin; zymoplastic substance; thrombostop; Cytocym; Tachostyptan; Thrombol. An integral membrane glycoprotein which, in the presence of Ca^{2+} ions, initiates coagulation by augmenting the proteolytic attack of factor VII on factors IX and X, *q.q.v.*: Nemerson, *Biochemistry* **5**, 601 (1966). Prepn from cattle brains: Hess, *J. Am. Med. Assoc.* **66**, 558 (1916). From rabbit brains or lung tissue: Singher, Swart, U.S. pat. **2,842,480** (1958 to Ortho); *see also* U.S. pats. **2,847,347** and **2,847,350** (1958 to Ortho). Purification of bovine tissue factor: R. Bach *et al., J. Biol. Chem.* **256**, 8324 (1981). The tissue thromboplastin activity seems to be brought about by a lipoprotein comprising a combination of phospholipids: phosphatidyl ethanolamine (PE), phosphatidyl serine (PS), and phosphatidyl choline (PC): R. B. Hunter, *et al., Fibrinogen and Fibrin Turnover of Clotting Factors* (Schattauer, Stuttgart, 1963) pp 429-454. Removal of phospholipids from tissue factor results in loss of coagulant activity as well as solubilizing the protein component. Purification and characterization of the protein component: Nemerson, Pitlick, *Biochemistry* **9**, 5100 (1970); of tissue factor apoprotein: F. A. Pitlick, Y. Nemerson, *Methods Enzymol.* **45B**, 37-48 (1976). Located on the plasma membrane of endothelial cells readily available for complexing with clotting factors: Zeldis *et al., Science* **175**, 766 (1972). Synthesis by cultured cells: Zacharski, McIntyre, *J. Med.* **4**, 118 (1973). Possibly a vitamin K dependent clotting factor: L. R. Zacharski *et al., Ann. N.Y. Acad. Sci.* **370**, 311 (1981). Use as clinical reagent: F. D. Ziegler *et al., Ann. Clin. Lab. Sci.* **11**, 202 (1981); A. L. Suchman, P. F. Griner, *Ann. Int. Med.* **104**, 810 (1986).
Note: Clauden is a "partial thrombokinase" obtained from lung tissue of animals.
THERAP CAT: Local hemostatic.

9322. Thrombospondin. Glycoprotein G; thrombin-sensitive protein; TSP. A major glycoprotein constituent of human platelet α-granules that is released in response to platelet activation by α-thrombin, and plays an important role in mediating platelet aggregation. Discovery and initial isoln and properties: N. L. Baenziger et al., Proc. Nat. Acad. Sci. USA **68**, 240 (1971); eidem, J. Biol. Chem. **247**, 2723 (1972). Thrombospondin was initially thought to be a single polypeptide of mol wt about 190,000, but has been shown to be a disulfide-linked trimer of mol wt 450,000: I. Hagen, Biochim. Biophys. Acta **392**, 242 (1975); D. R. Phillips, P. P. Agin, J. Biol. Chem. **252**, 2121 (1977); J. W. Lawler et al., Thromb. Haemostasis **37**, 355 (1977); eidem, J. Biol. Chem. **253**, 8609 (1978); S. S. Margossian et al., ibid. **256**, 7495 (1981). Synthesis and secretion by cells in culture: G. J. Raugi et al., J. Cell Biol. **95**, 351 (1982). Quantitative analysis: J. N. George et al., J. Lab. Clin. Med. **92**, 430 (1978). The binding of secreted thrombospondin to platelet membranes is dependent on Ca^{2+}: D. R. Phillips et al., J. Biol. Chem. **255**, 11629 (1980). Evidence for calcium-sensitive structure: J. Lawler et al., ibid. **257**, 12257 (1982). Radioimmunoassay: S. D. Saglio, H. S. Slayter, Blood **59**, 162 (1982). Identity of thrombospondin with the endogenous lectin secreted by activated platelets: E. A. Jaffee et al., Nature **295**, 246 (1982). See fibrinogen for additional ref.
Partial specific volume: 0.714 ml/g. Intrinsic viscosity: 40 ml/g in buffered saline at pH 7.6, 20°.

9323. Thromboxanes. Compounds derived from prostaglandin endoperoxides that cause platelet aggregation, contraction of arteries and other biological effects. Found in platelets, leucocytes, lung tissue, spleen, kidney, and umbilical artery. They are important mediators of the actions of polyunsaturated fatty acids transformed by cyclooxygenase. Discovery and structure of thromboxane B$_2$ (originally referred to as PHD): M. Hamberg, B. Samuelsson, Proc. Nat. Acad. Sci. USA **71**, 3400 (1974). Discovery and structure of thromboxane A$_2$ and identity with the unstable component of RCS (rabbit aorta contracting substance) see P. J. Piper, J. R. Vane, Nature **223**, 29 (1969); M. Hamberg et al., Proc. Nat. Acad. Sci. USA **72**, 2994 (1975). Biosynthesis and biological properties: P. Needleman et al., Science **193**, 163 (1976). Physiological review: J. R. Vane et al., Int. Rev. Exp. Pathol. **23**, 161-207 (1982). Reviews: B. Samuelsson in Organic Chemistry, A. T. Blomquist, H. H. Wasserman, Eds. vol. **36**, entitled "Prostaglandin Research", P. Crabbe, Ed. (Academic Press, New York, 1977) pp 17-46; E. Granström et al., Advan. Prostaglandin Thromboxane Leukotriene Res. **10**, 15-58 (1982); L. J. Roberts et al., ibid. 211-225. Books: Advances in Prostaglandin and Thromboxane Research vols. **1-8**, B. Samuelsson, R. Paoletti, Eds. (Raven Press, New York, 1976-1980); New Synthetic Routes to Prostaglandins and Thromboxanes, S. M. Roberts, F. Scheinmann, Eds. (Academic Press, New York, 1982).

thromboxane A$_2$

thromboxane B$_2$

Thromboxane A$_2$, $C_{20}H_{32}O_5$, [1S-[1α,3α(1E,3R*)4β(Z),-5α]]-7-[3-(3-hydroxy-1-octenyl)-2,6-dioxabicyclo[3.1.1]-hept-4-yl]-5-heptenoic acid, TXA$_2$. Highly unstable, biologically active bicyclic oxitane-oxane compound derived from the endoperoxide PGG$_2$ and rapidly converted to thromboxane B$_2$ by addition of water. Formed by incubation of

arachidonic acid or PGG$_2$ with washed platelets. It induces irreversible platelet aggregation and causes contraction of the isolated rabbit aorta and release of serotonin and ADP from platelets in platelet-rich plasma. Synthesis of stable analogs: E. J. Corey et al., Tetrahedron Letters **1980**, 137; K. M. Massey, G. M. Bundy, ibid. 445; S. Ohuchida et al., J. Am. Chem. Soc. **103**, 4597 (1981); V. N. Kale, D. L. J. Clive, J. Org. Chem. **49**, 1554 (1984). Synthesis of biologically active unstable analogs: S. S. Bhagwat et al., Tetrahedron Letters **1985**, 1955. Total synthesis and structure of thromboxane A2: eidem, Nature **315**, 511 (1985). Formation and effects in human platelets: J. Svensson et al., Acta Physiol. Scand. **98**, 285 (1970). Biological half-life: 32 $\pm$2 sec at 37°.
Thromboxane B$_2$, $C_{20}H_{34}O_6$, [2R-[2α(1E,3S*),3β(Z),4β,-6α]]-7-[tetrahydro-4,6-dihydroxy-2-(3-hydroxy-1-octenyl)-2H-pyran-3-yl]-5-heptenoic acid, TXB$_2$, PHD. A stable metabolite of thromboxane A$_2$ in platelets, initially considered biologically inactive. It is released during anaphylaxis in isolated guinea pig lungs and has been isolated from guinea pig brain homogenates and carrageenin-induced granuloma. TXB$_2$ has also been reported as possessing chemotactic properties. Total synthesis: N. A. Nelson, R. W. Jackson, Tetrahedron Letters **1976**, 3275; R. C. Kelly et al., ibid. 3279; from a prostaglandin F$_{2\alpha}$ derivative: W. P. Schneider, R. A. Morge, ibid. 3283; stereospecific synthesis from D-glucose: S. Hanessian, P. Lavallee, Can. J. Chem. **55**, 562 (1977); E. J. Corey et al., Tetrahedron Letters **1977**, 1625; S. Hanessian, P. Lavallee, Can. J. Chem. **59**, 870 (1981). Metabolism: L. J. Roberts et al., J. Biol. Chem. **252**, 7415 (1966). Biological properties: J. R. Boot et al., J. Physiol. **257**, 47P (1976); L. S. Wolfe et al., Biochem. Biophys. Res. Commun. **70**, 907 (1976); W.-C. Chang et al., Prostaglandins **13**, 3 (1977). Plates from ethyl acetate/ether/petr ether, mp 95-96°. $[\alpha]_D^{25}$ +57.4° (c = 0.26 in ethyl acetate).

9324. Thuja. Arbor vitae; yellow cedar; false white cedar; tree of life. Dried, leafy, young twigs of Thuja occidentalis L., Cupressaceae. Habit. North America; cultivated in Europe. Constit. Thujone, fenchone, thujetic acid, tannin, pinipicrin.

9325. Thujic Acid. 5,5-Dimethyl-1,3,6-cycloheptatriene-1-carboxylic acid; 4,4-dimethylcyclohepta-2,5,7-trienecarboxylic acid; dehydroperillic acid. $C_{10}H_{12}O_2$; mol wt 164.20. C 73.14%, H 7.37%, O 19.49%. Antibiotic substance from heartwood of Thuja plicata D. Don, Cupressaceae (Western red cedar). Isoln: Anderson, Sherrard, J. Am. Chem. Soc. **55**, 3813 (1933); Erdtman, Gripenberg, Acta Chem. Scand. **2**, 625 (1948). Structure: Gripenberg, ibid. **3**, 1137 (1949); **5**, 995 (1951); **10**, 487 (1956); Davis, Tulinsky, Tetrahdron Letters **1962**, 839.

Crystals from petr ether, mp 88-89°. uv max: 220, 280 nm (log ϵ 4.3, 3.7); min: 240 nm (log ϵ 3.0).
Methyl ester, $C_{11}H_{14}O_2$, crystals, mp 34.5-35° or liq with pleasant odor, bp$_{14}$ 112-113°. d_4^{22} 1.0225. n_D^{22} 1.5130.
Hexahydrothujic acid, $C_{10}H_{18}O_2$, liq, bp$_{14}$ 150-152°. n_D^{20} 1.4671.

9326. Thujone. 4-Methyl-1-(1-methylethyl)bicyclo[3.1.0]-hexan-3-one; 3-thujanone. $C_{10}H_{16}O$; mol wt 152.23. C 78.89%, H 10.59%, O 10.51%. A constituent of many essential oils; present in thuja, etc. Equilibrium mixture contains 33% α-thujone and 67% β-thujone: Eastman, Winn, J. Am. Chem. Soc. **82**, 5908 (1960). α- and β-Thujones differ only in the stereochemistry of the 4-methyl group. Conformation: Hach et al., Tetrahedron Letters **1970**, 3175. Chemistry: J. P. Kutney et al., Bioorg. Chem. **7**, 289 (1978); eidem, Can. J. Chem. **57**, 3145 (1979); **58**, 2641 (1980). Review: J. L. Simonsen, The Terpenes vol. II (University Press, Cambridge, 1949) pp 32-52.

β-thujone

Colorless or almost colorless liquid. uv max (isooctane): 300 nm (ε 23). Practically insol in water; sol in alc and many other organic solvents. LD_{50} s.c. in mice: 134.2 mg/kg, K. C. Rice, R. S. Wilson, *J. Med. Chem.* **19**, 1054 (1976).

l-Form, α-*thujone.* bp_{17} 83.8-84.1°. d_4^{25} 0.9109. n_D^{15} 1.4490. $[\alpha]_D^{20}$ −19.2°. LD_{50} s.c. in mice: 87.5 mg/kg, *eidem, ibid.*

d-Form, *d-isothujone,* β*-thujone.* bp_{17} 85.7-86.2°. d_4^{25} 0.9135. n_D^{25} 1.4500. $[\alpha]_D^{15}$ +72.5°. LD_{50} s.c. in mice: 442.2 mg/kg, *eidem, ibid.*

Caution: Ingestion may cause convulsions.

9327. Thujopsene. *1,1a,4,4a,5,6,7,8-Octahydro-2,4a,8,8-tetramethylcyclopropa[d]naphthalene;* widdrene. $C_{15}H_{24}$; mol wt 204.34. C 89.04%, H 10.96%. From wood oil of the Japanese Hiba tree, *Thujopsis dolobrata* Sieb. and Zucc., *Cupressaceae:* Yano, *J. Soc. Chem. Ind. Japan* **16**, 443 (1913); Uchida, *ibid* **31**, 501 (1928). Identity with widdrene: Erdtman, Thomas, *Acta Chem. Scand.* **12**, 267 (1958). Structure: Norin, *ibid.* **15**, 1676 (1961). Stereochemistry: Sisido *et al., J. Org. Chem.* **26**, 1964 (1961); Norin *Acta Chem. Scand.* **17**, 738 (1963). Synthesis: Dauben, Ashcraft, *J. Am. Chem. Soc.* **85**, 3673 (1963); Büchi, White, *ibid.* **86**, 2884 (1964).

Liquid. bp_{10} 120°. $[\alpha]_D$ −110° (c = 2 in chloroform). n_D^{25} 1.5031. d^{24} 0.932. uv max (alc): 212 nm (ε 4680).

9328. Thulium. Tm; at. wt 168.9342; at. no. 69; valence 3. A rare earth element of the yttrium group. One naturally occurring isotope: [169]Tm; artificial, radioactive isotopes: 153; 154; 161-168; 170-176. Estimated abundance in earth's crust: 0.2-1 ppm. Found in small quantities in euxenite, ytterspar, sipylite, gadolinite, and other rare earth minerals. Discovered by Cleve in crude erbium oxide: Cleve, *Compt. Rend.* **89**, 478, 521, 708 (1879). Obtained in a state of high purity by fractional crystn of its bromide: James, *J. Am. Chem. Soc.* **32**, 517 (1910); **33**, 1332 (1911). Sepn from other rare earths by ion exchange: Spedding *et al., ibid.* **76**, 2557 (1954). Review of prepn, properties and compounds of thulium and other lanthanides: *The Rare Earths,* F. H. Spedding, A. H. Daane, Eds. (Krieger, Huntington, N.Y., 1971, reprint of 1961 ed.) 641 pp; Hulet, Bode, "Separation Chemistry of the Lanthanides and Transplutonium Actinides" in *MTP Int. Rev. Sci.: Inorg. Chem., Ser. One,* vol. 7, K. W. Bagnall, Ed. (University Park Press, Baltimore, 1972) pp 1-45; Moeller, "The Lanthanides" in *Comprehensive Inorganic Chemistry,* vol. 4, J. C. Bailar, Jr. *et al.,* Eds. (Pergamon Press, Oxford, 1973) pp 1-101.

Silvery-white, easily worked metal. Hexagonal close-packed structure, d 9.332. mp 1545°. The solns of thulium salts show a characteristic absorption spectrum: Exner, Haschek, cited in *Mellor's* vol. V, 698 (1929).

Oxide, O_3Tm_2, *thulia.* Dense powder of greenish-white color. Prepd by igniting the oxalate; dissolves slowly in strong acids; exhibits a reddish glow on gentle heating.

Hydroxide, $Tm(OH)_3$, white precipitate.

Chloride heptahydrate, $TmCl_3.7H_2O$, deliquesc crystals. Sol in water, in alcohol. LD_{50} in mice: 485 mg/kg i.p.; 6.25 g/kg orally, Haley, *J. Pharm. Sci.* **54**, 663 (1965).

Sulfate octahydrate, $Tm_2(SO_4)_3.8H_2O$, obtained by pptg an aq soln of thulium chloride and sulfuric acid with alc.

Oxalate hexahydrate, $Tm_2(C_2O_4)_3.6H_2O$, greenish-white precipitate. Sol in aq alkali oxalates with formation of double oxalates.

9329. Thurfyl Nicotinate. *Nicotinic acid tetrahydrofurfuryl ester;* tetrahydrofurfuryl nicotinate; nicotafuryl; Trafuril; Trafuryl. $C_{11}H_{13}NO_3$; mol wt 207.22. C 63.75%, H 6.32%, N 6.76%, O 23.16%. Prepn: Hartmann, Merz, **U.S.** pat. **2,485,152** (1949 to Ciba).

Oil, $bp_{0.25}$ 114-116°. Sol in water, oil.
THERAP CAT: Topical vasodilator.

9330. Thyme. Dried leaves and flowering tops of *Thymus vulgaris* L., *Labiatae. Habit.* Southern Europe; cultivated in gardens. *Constit.* Volatile oil; tannin, gum.
THERAP CAT: Carminative.

9331. Thymidine. 1-(2-Deoxy-β-D-ribofuranosyl)-5-methyluracil; thymine-2-desoxyriboside. $C_{10}H_{14}N_2O_5$; mol wt 242.23. C 49.58%, H 5.83%, N 11.57%, O 33.03%. Constituent of deoxyribonucleic acid, *q.v.* Isoln from thymonucleic acid: Levene, London, *J. Biol. Chem.* **83**, 793 (1929). Structure: Levene, Tipson, *ibid.* **109**, 623 (1935). Conformation: Lemieux, *Can. J. Chem.* **39**, 116 (1961); Tollin *et al., Nature* **217**, 1148 (1968). Prepn of thymidine-3'-phosphate and of thymidine-5'-phosphate: Tener, *J. Am. Chem. Soc.* **83**, 165 (1961). *Review in Basic Principles in Nucleic Acid Chemistry,* vol. 1, P. O. P. Ts'o, Ed. (Academic Press, New York, 1974) *passim. See also* Nucleic Acids.

Rosettes of needles from ethyl acetate. mp 185°. Yields a sublimate of thymine when heated. $[\alpha]_D^{25}$ +30.6° (c = 1.029). uv max (pH 7.2): 206.5, 267 nm (ε × 10^{-3} 9.8, 9.7), Voet *et al., Biopolymers* **1**, 193 (1963). Sol in water, methanol, hot alcohol, hot acetone, hot ethyl acetate, pyridine, glacial acetic acid; sparingly sol in hot chloroform.

Monotrityl thymidine, $C_{29}H_{28}N_2O_5$, prepd by the action of triphenylmethyl chloride on thymidine in pyridine. mp 125°. $[\alpha]_D^{24}$ +11.4° (c = 1.01 in acetone).

9332. Thymine. *5-Methyl-2,4(1H,3H)-pyrimidinedione;* 5-methyluracil; 2,4-dihydroxy-5-methylpyrimidine. $C_5H_6N_2O_2$; mol wt 126.11. C 47.62%, H 4.80%, N 22.22%, O 25.37%. A pyrimidine derivative; constituent of nucleic acids. Originally isolated from thymus nucleic acid: Levene, *Z. Physiol. Chem.* **39**, 4 (1903). Prepn by heating 2-ethylmercapto-4-hydroxy-5-methylpyrimidine: Wheeler, Merriam, *Am. Chem. J.* **29**, 478 (1903); **43**, 29 (1910). From methylcyanacetylurea by catalytic reduction: Bergmann, Johnson, *J. Am. Chem. Soc.* **55**, 1733 (1933). From β-methylmalic acid: Scherp, *J. Am. Chem. Soc.* **68**, 912 (1946). Crystal structure of monohydrate: Gerdil, *Acta Cryst.* **14**, 333 (1961). *Review:* Ts'o, "Bases, Nucleosides and Nucleotides" in *Basic Principles in Nucleic Acid Chemistry* vol. 1, P. O. P. Ts'o, Ed. (Academic Press, New York, 1974) pp 453-584. *See also* Nucleic Acids.

Dendritic or star-shaped plates from water, sometimes short needles. Sublimes in platelets. Dec 335-337° (Kofler stage). Weak acid, pK at 25° = 9.94. uv max (pH 7.0): 205, 264.5 nm ($\epsilon \times 10^{-3}$ 9.5, 7.9). Sol in cold water (4 g/l at 25°). Somewhat sol in alc; sparingly sol in ether; readily sol in alkalies with formation of salts. Oxidation yields urea, ethanal, pyruvic acid, formic acid. Hydrazine reacts with thymine forming urea and 4-methylpyrazolone. Thymine forms a silver salt which is sol in excess ammonia. Its mercuric and lead salts are insol.
Thymine-2-desoxyriboside, see Thymidine.
USE: In biochemical research.

9333. Thymol. 5-*Methyl-2-(1-methylethyl)phenol;* 5-methyl-2-isopropyl-1-phenol; 1-methyl-3-hydroxy-4-isopropylbenzene; 3-p-cymenol; 3-hydroxy-p-cymene; thyme camphor; m-thymol. $C_{10}H_{14}O$; mol wt 150.21. C 79.95%, H 9.39%, O 10.65%. Isolated by Neumann in 1719. Obtained from the essential oil of *Thymus vulgaris* L. and *Monarda punctata* L., *Labiatae:* Arppe, *Ann.* **58**, 41 (1846); Meyer, *Pharm. Ztg.* **81**, 192, 205 (1936). Also occurs in other volatile oils. Produced synthetically from p-cymene, piperitone, or m-cresol: Austerweil, **Brit. pat. 221,227** (1923); Jennen, Verdroncken, *Compt. Rend.* **245**, 183 (1957); Bottoms, **U.S. pat. 2,840,616** (1958 to Natl. Cylinder Gas). Bactericidal activity: J. M. Schaffer, F. W. Tilley, *J. Bacteriol.* **14**, 259 (1927). Mold elimination on surfaces: O. W. Richards, K. J. Hawley, *J. Chem. Ed.* **16**, 6 (1939). In vitro antifungal activity: H. B. Myers, *J. Am. Med. Assoc.* **89**, 1834 (1927). Effectiveness as antifungal preservative: M. Dersarkissian, M. Goodberry: *Studies Conserv.* **25**, 28 (1980). Use as clinical preservative: T. Z. Liu, *Clin. Chem.* **25**, 336 (1979); T. Z. Liu *et al., ibid.* **27**, 1144 (1981). Toxicity: P. M. Jenner *et al., Food Cosmet. Toxicol.* **2**, 327 (1964).

Crystals, mp 51.5°. bp about 233°. Appreciably volatile at 100°; volatilizes in water vapors. Characteristic odor; pungent, somewhat caustic taste. d_4^{25} 0.9699. n_D^{20} 1.5227; n_D^{25} 1.5204. One gram dissolves in about 1000 ml water, 1 ml alcohol, 0.7 ml chloroform, 1.5 ml ether, 1.7 ml olive oil at 25°. Sol in glacial acetic acid, oils, fixed alkali hydroxides. LD_{50} orally in rats: 980 mg/kg (Jenner).
Incompat: Acetanilide, antipyrine, camphor, monobromated camphor, chloral hydrate, menthol, quinine sulfate, salol, urethane, spirit nitrous ether; in triturations because of liquefaction.
Acetate, $C_{12}H_{16}O_2$, *acetylthymol, thymyl acetate.* Yellowish, oily liq; thymol odor. d⁰ 1.009. bp 243.5-245.5°. Practically insol in water. Miscible with alcohol, benzene chloroform, ether. *Caution:* Mild irritant.
Carbonate, $C_{21}H_{26}O_3$, white crystals; thymol odor; volatilizes with steam. mp 49°. Insol in water, acids, alkalies. Sol in hot alcohol, chloroform, ether, carbon tetrachloride.
USE: For destroying mold; preserving documents, art objects and urine. Stabilizer (antioxidant) for trichloroethylene, halothane.
THERAP CAT: Topical antiseptic; anthelmintic (Nematodes).
THERAP CAT (VET): Has been used as anthelmintic, and as an antiseptic, external and internal.

9334. Thymol Blue. 4,4'-(3H-2,1-Benzoxathiol-3-ylidene)bis[5-methyl-2-(1-methylethyl)phenol] S,S-dioxide; α-hydroxy-α,α-bis(5-hydroxycarvacryl)-o-toluenesulfonic

acid γ-sultone; thymolsulfonephthalein. $C_{27}H_{30}O_5S$; mol wt 466.58. C 69.50%, H 6.48%, O 17.15%, S 6.87%.
Brownish-green, cryst powder; characteristic odor. Insol in water. Sol in alcohol, dil alkali solns.
USE: As as acid-base indicator; pH: red 1.2 to yellow 2.8; also yellow 8.0 to blue 9.6.

9335. Thymol Iodide. 4,4'-*Bis(iodooxy)-2,2'-dimethyl-5,5'-bis(1-methylethyl)-1,1'-biphenyl; bithymol diiodide;* dithymol diiodide; diiododithymol; Aristol; Annidalin; Iodistol; Iodothymol; Iodosol; Iosol; Iothymol; Thymiode; Thymiodol; Thymodin. $C_{20}H_{24}I_2O_2$; mol wt 550.23. C 43.65%, H 4.40%, I 46.13%, O 5.82%. A mixture of iodine derivs of thymol, principally dithymol diiodide. The official prepn contains not less than 43% iodine, when dried over H_2SO_4 for 18 hours. Made by treating a soln of thymol in NaOH soln with an iodine-potassium iodide soln: Messinger, Vortmann, **Ger. pats. 49,739; 52,828; 52,833;** *Ber.* **22**, 2312 (1889).

Reddish-brown or reddish-yellow, bulky powder. Gives off purple iodine vapors when heated above 100°. Loses iodine on prolonged exposure to light. Readily sol in chloroform, ether, collodion, fixed and volatile oils, usually leaving a slight residue. Slightly sol in alcohol. Insol in water, glycerol, alkaline solns.
THERAP CAT: Antifungal; anti-infective.
THERAP CAT (VET): Has been used externally as an antiseptic, internally as a source of iodine.

9336. Thymolphthalein. 3,3-*Bis[4-hydroxy-2-methyl-5-(1-methylethyl)phenyl]-1(3H)-isobenzofuranone; 5',5''-di-isopropyl-2',2''-dimethylphenolphthalein.* $C_{28}H_{30}O_4$; mol wt 430.52. C 78.11%, H 7.02%, O 14.86%. Obtained by heating phthalic anhydride with thymol at 110° in the presence of stannic chloride.

Needles, mp about 253°. Insol in water; sol in alcohol, acetone; also sol in dil alkalies with a blue color, in H_2SO_4 with a carmine-red color.
USE: As pH indicator: colorless 9.3 to blue 10.5. Also as reagent for blood after decolorizing the alkaline soln by boiling with zinc dust.

9337. Thymomodulin. Leucotrofina. Cell-free thymic hormone preparation extracted from calf thymus by acid hydrolysis. Composed of a mixture of biologically active acidic peptides of mol wt < 10,000. Modulates the maturation of T-cells. Prepn of crude extract from calf thymus: B. Brunetti, E. Pini, **U.S. pat. 3,657,417** (1972 to Ellem). Electrophoretic characterization: C. Secchi *et al., Riv. Eur. Sci. Med. Farmacol.* **4**, 499 (1982), *C.A.* **101**, 21878m (1984). Biological activity and comparison with other thymic hormones: J. J. Twomey, N. M. Kouttab, *Cell. Immunol.* **72**, 186 (1982). Purification and pharmacodynamics: *eidem, Drugs Exptl. Clin. Res.* **10**, 921 (1984). Antileukopenic activity in myelodepressed cancer patients: Z. Uray *et al., Aggressologie* **21**, 215 (1980); C. Gallo Curcio *et al., Int. J. Immunother.* **2**, 189 (1986). Use in prophylaxis of infantile asthma: R. Genova, A. Guerra, *Ped. Med. Chir.* **5**, 395

(1983). Preliminary study in AIDS: G. Valesini *et al.*, *Eur. J. Cancer Clin. Oncol.* **22**, 531 (1986).

THERAP CAT: Immunoregulator.

9338. Thymopentin. *N*-(*N*-(*N*-(*N*²-L-*arginyl*-L-*lysyl*)-L-α-*aspartyl*)-L-*valyl*)-L-*tyrosine;* thymopoietin pentapeptide; TP-5; ORF 15244; Immunox; Sintomodulina; Timunox. $C_{30}H_{49}N_9O_9$; mol wt 679.77. C 53.01%, H 7.27%, N 18.54%, O 21.18%. Thymic hormone analog corresponding to residues 32-36 of thymopoietin, *q.v.*, which exhibits the full biological activity of the natural hormone. Synthesis: G. Goldstein *et al.*, *Science* **204**, 1309 (1979); *eidem*, U.S. pat. **4,190,646** (1980 to Sloan-Kettering). Bioavailability: T. Audhya, G. Goldstein, *Int. J. Pept. Protein Res.* **22**, 187 (1983). Pharmacology: K. Bolla *et al.*, *Int. J. Clin. Pharmacol. Res.* **4**, 431 (1984). Comparison of biological activity with thymopoietin and splenin: T. Audhya *et al.*, *Proc. Nat. Acad. Sci. USA* **81**, 2847 (1984). Clinical study in treatment of primary immunodeficiencies: F. Aitui *et al.*, *Lancet* **1**, 551 (1983); in AIDS: N. Clumeck *et al.*, *Int. J. Clin. Pharmacol. Res.* **4**, 459 (1984); in rheumatoid arthritis: M. G. Malaise *et al.*, *Lancet* **1**, 832 (1985). *Review:* E. A. Boyse, *Surv. Immunol. Res.* **4**, 6-10 (1985).

Arg-Lys-Asp-Val-Tyr

THERAP CAT: Immunoregulator.

9339. Thymopoietin. Thymin (formerly); TP. mol wt ~5000. A polypeptide thymic hormone which selectively induces T-cell maturation from prothymocytes and inhibits B-cell differentiation. First detected as factor causing impaired neuromuscular transmission in myasthenia gravis, a neuromuscular disease associated with thymic abnormalities: G. Goldstein, S. Whittingham, *Lancet* **2**, 315 (1966); G. Goldstein, *ibid.* **2**, 119 (1968). Thymopoietin is a single chain 49 amino acid protein which exists in 2 closely homologous forms (thymopoietins I and II). Both exhibit the same biological activity; the active site corresponds to residues 32-36. Isoln from bovine thymus: G. Goldstein, *Nature* **247**, 11 (1974); *Ann. N.Y. Acad. Sci.* **249**, 177 (1975); U.S. pat. **4,077,949** (1978 to Sloan-Kettering). Amino acid sequence of bovine thymopoietin II: D. H. Schlesinger, G. Goldstein, *Cell* **5**, 361 (1975). Revised sequence of II, sequence of bovine I, isoln and sequence of *splenin*, a thymopoietin-like protein found in the spleen: T. Audhya *et al.*, *Biochemistry* **20**, 6195 (1981). Comparison of biological activity of thymopoietin and splenin: T. Audhya *et al.*, *Proc. Natl. Acad. Sci. USA* **81**, 2847 (1984). First synthesis of a biologically active fragment: D. H. Schlesinger *et al.*, *Cell* **5**, 367 (1975). Complete synthesis of originally proposed amino acid sequence of II: M. Fujino *et al.*, *Chem. Pharm. Bull.* **25**, 1486 (1977); E. G. Bliznakov *et al.*, *Biochem. Biophys. Res. Commun.* **80**, 631 (1978). Solid phase synthesis: R. Colombo, *Experientia* **37**, 798 (1981). Synthesis of thymopentin, *q.v.*, the biologically active pentapeptide corresp to residues 32-36: G. Goldstein *et al.*, *Science* **204**, 1309 (1979). Synthesis of the octadecapeptide corresp to residues 32-49 of the revised amino acid sequence of II: T. Abiko, H. Sekino, *Chem. Pharm. Bull.* **30**, 3271 (1982). Effect on lymphocyte differentiation: R. S. Basch, G. Goldstein, *Proc. Nat. Acad. Sci. USA* **71**, 1474 (1974); M. P. Scheid *et al.*, *Ann. N.Y. Acad. Sci.* **249**, 531 (1975). Clinical studies: M. I. Joffe *et al.*, *J. Clin. Lab. Immunol.* **8**, 69 (1982); E. H. Veys *et al.*, *Ann. Rheum. Dis.* **41**, 441 (1982). *Reviews:* J. F. Bach, *J. Immunopharmacol.* **1**, 277-310 (1979); G. Goldstein, C. Lau in *Polypeptide Hormones*, R. F. Beers, E. G. Bassett, Eds. (Raven Press, New York, 1980) pp 459-467; E. A. Boyse, *Surv. Immunol. Res.* **4**, 6-10 (1985).

9340. Thymosin. Cell free, lymphocytopoietic hormone and biologically active thymic factor, comprising a family of acidic, heat-stable polypeptides, that plays an important part in the development of T-cells (thymus-dependent lymphocytes) and immunological functions in animals and man. Extraction from calf thymus and partial purification: A. L. Goldstein *et al.*, *Proc. Nat. Acad. Sci. USA* **56**, 1010 (1966). Purification and characterization of thymosins and isoln of one of the active components, *thymosin fraction 5: eidem, ibid.* **69**, 1800 (1972). Improved isoln and properties: J. A. Hooper *et al.*, *Ann. N.Y. Acad. Sci.* **249**, 125 (1975).

Thymosin fraction 5 is a mixture of small polypeptides ranging in mol wt from approx 1000 to 15,000, it is carbohydrate and lipid-free and contains no nucleotides or unusual amino acids. It is an immunopotentiating agent that can act in place of the thymus gland in immuno-deprived or thymus-deprived patients to restore some immune functions. Nomenclature of the polypeptide fragments of fraction 5 is based on isoelectric points: α-region below 5.0, β-region 5.0-7.0, and γ-region above 7.0. Isoln and preliminary structural determination of *thymosin* α_1, a biologically active, 28-residue polypeptide component in the highly acidic region of fraction 5: A. L. Goldstein *et al.*, *Proc. Nat. Acad. Sci. USA* **74**, 725 (1977). Confirmation of primary structure: T. Michalewsky *et al.*, *Int. J. Peptide Protein Res.* **21**, 93 (1983). Isoln, characterization, biological activities, amino acid sequence analysis of α_1 and *polypeptide* β_1, a biologically inactive fragment: T. L. K. Low *et al.*, *J. Biol. Chem.* **254**, 981 (1979); T. L. K. Low, A. L. Goldstein, *ibid.* 987. Synthesis of α_1 by solution methods: C. Birr, U. Stollenwerk, *Angew. Chem. Int. Ed.* **18**, 394 (1979); S. S. Wang *et al.*, *J. Am. Chem. Soc.* **101**, 253 (1979). Automated solid phase synthesis: *eidem, Int. J. Peptide Protein Res.* **15**, 1 (1980). *Thymosin* β_4 is a biologically active fragment that appears to act on stem cells to form prothymocytes. Extrathymic occurrence of β_4: E. Hannappel *et al.*, *Proc. Nat. Acad. Sci. USA* **79**, 2172 (1982). Amino acid sequence of β_4: T. L. K. Low *et al.*, *ibid.* **78**, 1162 (1981); T. L. K. Low, A. L. Goldstein, *J. Biol. Chem.* **257**, 1000 (1982). Automated solid phase synthesis of β_4: S. S. Wang *et al.*, *Int. J. Peptide Protein Res.* **18**, 413 (1981); T. L. K. Low *et al.*, *Biochemistry* **22**, 733 (1983). In vitro synthesis of β_4 by rat spleen mRNA: A. W. Filipowicz, B. L. Horecker, *Proc. Nat. Acad. Sci. USA* **80**, 1811 (1983). Biosynthesis of β_4 by spleen cells: G. Xu *et al.*, *ibid.* **79**, 4006 (1982). Thymosin β_4 (and fraction 5) have been shown to stimulate secretion of luteinizing hormone-releasing factor (LH-RH, *q.v.*): R. W. Rebar *et al.*, *Science* **214**, 669 (1981). Isoln of *thymosin* β_8 and *thymosin* β_9: E. Hannappel *et al.*, *Proc. Nat. Acad. Sci. USA* **79**, 1708 (1982). Synthesis of β_8: T. Abiko, H. Sekino, *Chem. Pharm. Bull.* **31**, 1320 (1983). Synthesis of β_9: *eidem, J. Appl. Biochem.* **4**, 449 (1982). For specific immunological activities of thymosin see Zisblatt *et al.*, *Proc. Nat. Acad. Sci. USA* **66**, 1170 (1970); Goldstein *et al.*, *Radiat. Res.* **41**, 579 (1970); Goldstein *et al.*, *J. Immunol.* **104**, 359 (1970); Bach *et al.*, *Proc. Nat. Acad. Sci. USA* **68**, 2734 (1971); A. L. Goldstein *et al.*, in *Progress in Cancer Research and Therapy*, **vol. 2**, M. A. Chirigos, Ed. (Raven Press, New York, 1977) pp 241-254. Evidence for mediation of thymosin fraction 5 effects on lymphocytes by prostaglandin E_2, *q.v.*: G. R. Garaci *et al.*, *Science* **220**, 1163 (1983). Clinical studies: D. W. Wara *et al.*, *N. Engl. J. Med.* **292**, 70 (1975); L. A. Schafer *et al.*, *Ann. N.Y. Acad. Sci.* **277**, 609 (1976); J. Costanzi *et al.*, *ibid.* **332**, 148 (1979). Review of clinical studies: G. D. Marshall *et al.*, *Recent Results Cancer Res.* **75**, 100-105 (1980). *Reviews:* A. L. Goldstein *et al.*, *Recent Progr. Horm. Res.* **26**, 505-538 (1970); Robey, "Thymosin Reviewed" in *Thymic Hormones*, T. D. Luckey, Ed. (University Park Press, Baltimore, 1973) pp 159-16; A. L. Goldstein, *Trans. Am. Clin. Climatol. Assoc.* **88**, 79-94 (1976); A. L. Goldstein *et al.*, *Recent Progr. Horm. Res.* **37**, 369-416 (1981).

Soluble in 5% perchloric acid. Activity remains relatively stable at 100°.

9341. Thymostatin. Calf thymic extract which markedly inhibits *in vitro* and *in vivo* the incorporation of labelled nucleosides into the DNA and RNA of lymphocytes. Also inhibited incorporation of labelled nucleosides *in vitro* into other cell types both lymphoid and non-lymphoid, but did not affect incorporation *in vitro* of labelled amino acids into cellular protein. A carbohydrate containing peptide with a particle size < 2000. Prepn and properties: Goldstein *et al.*, *Proc. Nat. Acad. Sci. USA* **57**, 821 (1967).

Relatively heat stable; insol in chloroform but sol in chloroform-methanol (2:1).

9342. o-Thymotic Acid. *2-Hydroxy-6-methyl-3-(1-methylethyl)benzoic acid; 3-hydroxy-2-p-cymenecarboxylic acid; o-thymotinic acid;* 6-methyl-3-isopropylsalicylic acid. $C_{11}H_{14}O_3$; mol wt 194.22. C 68.02%, H 7.27%, O 24.71%. Prepn from thymol: Kolbe, Lautemann, *Ann.* **115**, 205

(1861); Spallino, Provenzal, *Gazz. Chim. Ital.* **39** (II), 326 (1909); Royer *et al.*, *Bull. Soc. Chim. France* **1955**, 1421.

Monoclinic prismatic needles from water, mp 127°. Volatile with steam. One gram dissolves in 10 liters of water at 20°. Sol in alc, ether, chloroform, benzene, petr ether.

Sodium salt, $C_{11}H_{13}NaO_3$, white cryst mass, freely sol in water.

Amide, $C_{11}H_{15}NO_2$, needles from alc, mp 137°, dec 205°. Sol in org solvents and in aq solns of sodium carbonate.

Acetonyl ester, $C_{14}H_{18}O_4$, needles from alcohol, mp 75°. Sol in alcohol.

The methyl and ethyl esters appear to be liquids at room temperature.

USE: As salicylic acid and salicylates.

9343. Thymyl N-Isoamylcarbamate. *Isoamylcarbamic acid thymyl ester;* isopropyl-*m*-cresyl ester of isoamylcarbamic acid; Egressin. $C_{16}H_{25}NO_2$; mol wt 263.37. C 72.96%, H 9.57%, N 5.32%, O 12.15%. Prepd by interaction of isoamylamine and thymyl chloroformate: Zima, v. Werder, U.S. pat. 2,524,185 (1949 to E. Merck).

Needles from petr ether, mp 57°. Practically insol in water (< 1:50,000). Upon alkaline saponification, it is split into isoamylamine and thymol (along with CO_2).

THERAP CAT: Anthelmintic (Nematodes).

9344. Thyroid. Tiroidina; Thyradin; Thyrocrine. Thyroid gland of domesticated animals that are used as food by man, freed from connective tissue and fat, dried and powdered. Contains not less than 0.17% and not over 0.23% iodine in thyroid combination. 1 part = approx 5 parts fresh gland. Chemistry and physiology: Rawson *et al.*, *The Hormones* vol. III (Academic Press, New York, 1955) pp 433-519.

Yellowish powder; slight meat-like odor; saline taste.

THERAP CAT: Thyroid hormone.

THERAP CAT (VET): In myxedema. Has been used in obesity, renal insufficiency, chronic skin conditions and to increase spermatogenesis, libido, lactation.

9345. Thyroidin. Iodothyrin. A dried extract of thyroid diluted with milk sugar or other suitable diluent, equal in potency to official thyroid.

THERAP CAT: Thyroid hormone.

9346. Thyropropic Acid. *4-(4-Hydroxy-3-iodophenoxy)-3,5-diiodohydrocinnamic acid;* 3,3',5-triiodothyropropionic acid; β-[4-(3'-iodo-4'-hydroxyphenoxy)-3,5-diiodophenyl]propionic acid; Birodan; Triopron. $C_{15}H_{11}I_3O_4$; mol wt 635.99. C 28.33%, H 1.74%, I 59.86%, O 10.06%. Prepd by iodination of 3,5-diiodothyropropionic acid: Tomita, Lardy, *J. Biol. Chem.* **219**, 595 (1956).

Crystals from abs ethanol, mp 200°.

THERAP CAT: Anticholesteremic.

9347. Thyroprotein. *Thyroactive protein;* Protamone. Iodinated casein. Use in feed supplements: Kohler, *The Use of Thyroactive Protein in Dairy Cattle Feeding,* The Texas Nutrition Conference Oct. **1953**, Abstracts of papers, pp 5-7. Prepn: Reineke *et al.*, *J. Biol. Chem.* **143**, 285 (1942); **147**, 115 (1943); Turner, Reineke, U.S. pats. 2,329,445, 2,379,842, 2,385,117, 2,478,065 (1944, 1945, 1945, 1949 to Am. Dairies & Quaker Oats); Whitmoyer, Moore, U.S. pat. 2,382,193 (1945 to Whitmoyer Labs.); West, Van Bruggen, U.S. pats. 2,642,426, 2,709,671 (1953, 1955 to Feed Prods).

USE: Feed supplement, *Fed. Regist.* **45**, 41360 (1980).

9348. Thyroxine. *O-(4-Hydroxy-3,5-diiodophenyl)-3,5-diiodotyrosine; 3-[4-(4-hydroxy-3,5-diiodophenoxy)-3,5-diiodophenyl]alanine; β-[(3,5-diiodo-4-hydroxyphenoxy)-3,5-diiodophenyl]alanine; 3,5,3',5'-tetraiodothyronine.* $C_{15}H_{11}I_4NO_4$; mol wt 776.93. C 23.19%, H 1.43%, I 65.34%, N 1.80%, O 8.24%. An amino acid of the thyroid gland, exerts a stimulating effect on metabolism. Its natural occurrence in the free form has not been demonstrated. It may be a cleavage product of thyroglobulin. The L-form is twice as active physiologically as the racemic product, the D-form has very little activity. Obtained from thyroid glands of animals: Kendall, *J. Am. Med. Assoc.* **64**, 2042 (1915); *J. Biol. Chem.* **39**, 125 (1919). Structure: Harington, *Biochem. J.* **20**, 293, 300 (1926). Synthesis: Harington, Barger, *ibid.* **21**, 169 (1927); Canzanelli, *et al.*, *ibid.* **29**, 1617 (1935); Ginger, Anthony, U.S. pats. 2,889,363; Anthony, Ginger, U.S. pat. 2,889,364 (both 1959 to Baxter Labs.). *See also* R. Pitt-Rivers, J. R. Tata, *The Thyroid Hormones* (Pergamon Press, 1959). Commercial syntheses: Nahm, Siedel in *Forschung in Hoechst* (Jubiläumsjahr 1963) p 663; Nahm, Siedel, *Ber.* **96**, 1 (1963).

DL-Thyroxine, needle-like crystals. Dec 231-233°. Insol in water, in alcohol, and in the other usual organic solvents, but in the presence of mineral acids or alkalies it dissolves in alcohol; sol in solns of the alkali hydroxides and in hot solns of the alkali carbonates. When alkali hydroxide solns of thyroxine are satd with sodium chloride, the sodium salt of thyroxine separates.

L-Thyroxine, crystals, dec 235-236°. $[\alpha]_{546}^{25}$ −3.2° (0.66 g in 6.07 g of 0.5N NaOH and 13.03 g alc); $[\alpha]_D^{20}$ −4.4° (3% in 0.13N NaOH in 70% EtOH).

L-Thyroxine Sodium Salt. *See* separate entry *under* Sodium Levothyroxine.

D-Thyroxine, *Debetrol.* Crystals, dec 237°. $[\alpha]_{546}^{21}$ +2.97° (0.74 g in 6 g of 0.5N NaOH and 14 g of alcohol).

D-Thyroxine Sodium Salt, $C_{15}H_{10}I_4NNaO_4$, *dextrothyroxine sodium, Biotirmone, Choloxin, Detyroxin, Dethyrona, Dextroid, Dynothel, Eulipos.*

THERAP CAT: L-Form as thyroid hormone. D-Form as antihyperlipoproteinemic.

9349. Tiadenol. *2,2'-[1,10-Decanediylbis(thio)]bisethanol; 2,2'-(decamethylenedithio)diethanol;* 1,10-bis(2-hydroxyethylthio)decane; LL-1558; Delipid; Eulip; Fonlipol; Tiaden; Tiaterol. $C_{14}H_{30}O_2S_2$; mol wt 294.51. C 57.09%, H 10.27%, O 10.87%, S 21.77%. $HOCH_2CH_2S(CH_2)_{10}SCH_2-CH_2OH$. Prepn: Williams, Cossar, U.S. pat. 3,021,215 (1962 to Eastman Kodak); Lafon, Ger. pat. 2,038,836 (1971 to Orsymonde), *C.A.* **75**, 35528d (1971). Series of articles on pharmacology, metabolism, and toxicology: *Therapie* **27**, 395-444 (1972).

Crystals, mp 69.5°. uv max (ethanol): 212 nm. Soluble in ethanol, chloroform. Practically insol in water.

THERAP CAT: Antilipidemic.

9350. Tiamenidine. *N-(2-Chloro-4-methyl-3-thienyl)-4,5-dihydro-1H-imidazol-2-amine;* 2-[(2-chloro-4-methyl-3-thienyl)amino]-2-imidazoline; 2-chloro-4-methyl-3-(2'-imidazolin-2'-ylamino)thiophene. $C_8H_{10}ClN_3S$; mol wt 215.70. C 44.55%, H 4.67%, Cl 16.43%, N 19.48%, S 14.86%. A thiophene analog of clonidine, *q.v.* Prepn: R.

Rippel *et al.*, **Ger.** pat. **1,941,761;** *eidem*, **U.S.** pat. **3,758,476** (1971, 1973, both to Hoechst). Structural studies: J. M. Leger, *Compt. Rend. Ser. C* **289**, 93 (1979); A. Carpy *et al.*, *Mol. Pharmacol.* **21**, 400 (1981). GC mass spectrometry determn in human plasma: T. A. Bryce, J. L. Burrows, *Biomed. Mass Spectrom.* **6**, 27 (1979). Pharmacology: E. Lindner, J. Kaiser, *Arch. Int. Pharmacol.* **211**, 305 (1974). Activity as α-adrenoceptor agonist: A. G. Roach *et al.*, *J. Pharmacol. Exp. Ther.* **227**, 421 (1983).

Crystals from isopropanol + petr ether, mp 152°.
Hydrochloride, $C_8H_{11}Cl_2N_3S$, *HOE 440, Sundralen, Symcor, Thiamenidine.* Crystals from isopropanol + petr ether, mp 228-229°. LD_{50} in rats, mice (mg/kg): 40, 45 i.v.; in mice: 170 s.c., 400 orally (Lindner, Kaiser).

THERAP CAT: Antihypertensive.

9351. Tiamulin. *[(2-(Diethylamino)ethyl)thio]acetic acid 6-ethenyldecahydro-5-hydroxy-4,6,9,10-tetramethyl-1-oxo-3a,9-propano-3aH-cyclopentacyclooocten-8-yl ester;* 14-desoxy-14-[(2-diethylaminoethyl)mercaptoacetoxy]mutilin; thiamutilin; tiamutin; SQ 14055; Dynalin Injectable. C_{28}-$H_{47}NO_4S$; mol wt 493.76. C 68.11%, H 9.60%, N 2.84%, O 12.96%, S 6.49%. Deriv of pleuromutilin, *q.v.* Prepn: H. Egger, **Ger.** pat. **2,248,237** corresp to **U.S.** pat. **3,919,290** (1973, 1975 both to Sandoz); H. Egger, H. Reinshagen, *J. Antibiot.* **29**, 915 (1976). Biosynthesis: F. Knauseder, E. Brandl, *ibid.* 125. *In vitro* activity: H. Werner *et al.*, *ibid.* **31**, 756 (1978). Metabolism: J. Dreyfuss *et al.*, *ibid.* **32**, 496 (1979). Effect vs *Mycoplasma:* C. O. Baughn *et al.*, *Avian Dis.* **22**, 620 (1978); R. F. Goodwin, *Vet. Rec.* **104**, 194 (1979). Mechanism of action: G. Högenauer in *Antibiotics* vol. **5**(pt. 1), F. E. Hahn, Ed. (Springer-Verlag, New York, 1979) pp 344-360. Treatment of swine dysentery: M. D. Anderson, *Vet. Med. Small Anim. Clin.* **78**, 98 (1983).

Fumarate, $C_{32}H_{51}NO_8S$, *81723 hfu, SQ 22947, Denagard, Dynalin Soluble Powder, Dynalin Feed Premix, Dynamutilin.* Crystals from acetone, mp 147-148° (after stirring in ethyl acetate and drying at 60° and 80° overnight).

THERAP CAT (VET): Antibacterial.

9352. Tianeptine. *7-[(3-Chloro-6,11-dihydro-6-methyl-dibenzo[c,f][1,2]thiazepin-11-yl)amino]heptanoic acid S,S-dioxide.* $C_{21}H_{25}ClN_2O_4S$; mol wt 436.95. C 57.72%, H 5.77%, Cl 8.11%, N 6.41%, O 14.65%, S 7.34%. Tricyclic compound with psychostimulant, anti-ulcer and anti-emetic properties. Prepn: C. Malen *et al.*, **Ger.** pat. **2,011,806** corresp to **U.S.** pat. **3,758,528** (1970, 1973 both to Sci. Union et Cie-Soc. Franc. Rech. Med.). Neuropharmacology study: C. Malen, J. Poignant, *Experientia* **28**, 811 (1972).

Sodium salt, $C_{21}H_{24}ClN_2NaO_4S$, *S-1574, Stablon.* Solid, mp 180°.

THERAP CAT: Antidepressant.

9353. Tiapride. *N-[2-(Diethylamino)ethyl]-2-methoxy-5-(methylsulfonyl)benzamide;* N-[2-(diethylamino)ethyl]-5-(methylsulfonyl)-o-anisamide; thiapride; FLC 1374; Gramalil; Tiapridal. $C_{15}H_{24}N_2O_4S$; mol wt 328.43. C 54.86%, H 7.36%, N 8.53%, O 19.49%, S 9.76%. Dopamine receptor antagonist structurally related to sulpiride, *q.v.* Prepn: G. Bulteau *et al.*, **Ger.** pat. **2,327,192** corresp to **Brit.** pat. **1,394,563;** Belg. pat. 800,232 corresp to **Brit.** pat. **1,394,559; Ger.** pat. **2,327,414** corresp to **Brit.** pat. **1,422,221** (1973, 1975, 1973, 1975, 1973, 1976 all to Soc. Etudes Sci. Ind. L'Ile de France). Crystal structure: C. Houttemane *et al.*, *Acta Cryst.* **C39**, 585 (1983). Dopamine receptor binding studies: T. Arima *et al.*, *Japan. J. Pharmacol.* **41**, 419 (1986). Pharmacokinetics: E. Rey *et al.*, *Int. J. Clin. Pharmacol. Ther. Toxicol.* **20**, 62 (1982). Evaluation in levodopa induced dyskinesia: P. Price *et al.*, *Lancet* **2**, 1106 (1978); in neuroleptic-induced dyskinesia: W. Greil *et al.*, *Neuropsycobiology* **14**, 17 (1985). Clinical evaluation in alcohol withdrawal syndrome: R. Agricola *et al.*, *J. Int. Med. Res.* **10**, 160 (1982); G. K. Shaw *et al.*, *Brit. J. Psychiatry* **150**, 164 (1987). Review of clinical trials in psychiatry: E. D. Peselow, M. Stanley, *Adv. Biochem. Psychopharmacol.* **35**, 163-194 (1982).

Crystals, mp 123-125°.
Hydrochloride, $C_{15}H_{25}ClN_2O_4S$, *Tiapridex, Luxoben, Sereprile, Italprid.*

THERAP CAT: Antidyskinetic.

9354. Tiaprofenic Acid. *5-Benzoyl-α-methyl-2-thiopheneacetic acid;* α-methyl-5-benzoyl-2-thienylacetic acid; FC 3001; RU 15060; Suralgan; Surgam. $C_{14}H_{12}O_3S$; mol wt 260.31. C 64.59%, H 4.65%, O 18.44%, S 12.32%. Nonsteroidal anti-inflammatory. Prepn: F. Clémence, O. Le Martret, **Ger.** pat. **2,055,264** and **Fr.** pat. **2,112,111** (1971 and 1972 to Roussel-UCLAF), *C.A.* **75**, 63597u (1971) and **78**, 97473c (1973); F. Clémence *et al.*, *Eur. J. Med. Chem. - Chem. Ther.* **9**, 390 (1974). Pharmacology: *eidem, ibid.;* H. Fujimura *et al.*, *Oyo Yakuri* **9**, 715 (1975), *C.A.* **83**, 188321w (1975). Review of pharmacology and efficacy in rheumatic diseases and pain control: E. M. Sorkin, R. N. Brogden, *Drugs* **29**, 208-235 (1985).

mp 96° (isopropyl ether).

THERAP CAT: Anti-inflammatory.

9355. Tiaprost. *7-[3,5-Dihydroxy-2-[3-hydroxy-4-(3-thienyloxy)-1-butenyl]cyclopentyl]-5-heptenoic acid;* (15-R,S)-16-(3-thienyloxy)-ω-tetranor-PGF$_{2α}$. $C_{20}H_{28}O_6S$; mol wt 396.50. C 60.58%, H 7.12%, O 24.21%, S 8.09%. Analog of prostaglandin F$_{2α}$, *q.v.* Prepn: W. Bartmann *et al.*, **Ger.** pat. **2,524,955** corresp to **U.S.** pat. **4,258,053** (1977, 1981 both to Hoechst); *eidem, Prostaglandins* **17**, 301 (1979). Efficacy in bovine endometritis: W. Bentele *et al.*, *Tierarztliche Umschau* **35**, 676, 678, 683 (1980). Pharmacological effects: W. v. Rechenberg *et al.*, *Blue Book for the Vet. Profession* (Hoechst AG) **30**, 417 (1981). Administration in cattle: R. Humble, *ibid.* 425.

Tromethamine salt, $C_{24}H_{39}NO_9S$, *tiaprost trometamol, Iliren.*

THERAP CAT (VET): Luteolytic.

9356. Tiaramide. *4-[(5-Chloro-2-oxo-3(2H)-benzothia-zolyl)acetyl]-1-piperazineethanol;* 5-chloro-3-[4-(2-hydroxyethyl)-1-piperazinyl]carbonylmethyl-2-benzothiazolinone; tialamide. $C_{15}H_{18}ClN_3O_3S$; mol wt 355.84. C 50.63%, H 5.10%, Cl 9.96%, N 11.81%, O 13.49%, S 9.01%. Prepn: Umio et al., **Japan.** pats. **15,302('71); 18,752('71),** *C.A.* **75,** 36127j, 63824r (1971); Umio, **U.S. pat. 3,661,921** (1972) (all three to Fujisawa). Pharmacology: Takashima et al., *Arzneimittel-Forsch.* **22,** 711 (1972); Tsurumi et al., *ibid.* 716, 724. Metabolism: Noda et al., *ibid.* 732. Toxicity studies: Watanabe et al., *ibid.* **23,** 504 (1973). Mode of antiasthmatic action: G. C. Folco et al., *Pharmacol. Res. Commun.* **11,** 703 (1979).

pKa 6.2.

Hydrochloride, $C_{15}H_{19}Cl_2N_3O_3S$, *NTA-194, FK 1160, Solantal.* White, odorless, bitter tasting crystalline powder, mp 159-161°. pH (10% aqueous soln): 3.4-3.7. Very soluble in water; slightly sol in organic solvents. LD_{50} in male mice, rats: 178, 203 mg/kg i.v.; 298, 540 mg/kg i.p.; 564, 3600 mg/kg orally, Watanabe, *loc. cit.*

THERAP CAT: Antiasthmatic; anti-inflammatory.

9357. Tibezonium Iodide. *N,N-Diethyl-N-methyl-2-[[4-[4-(phenylthio)phenyl]-3H-1,5-benzodiazepin-2-yl]thio]-ethanaminium iodide;* diethylethyl[2-[[4-[p-(phenylthio)-phenyl]-3H-1,5-benzodiazepin-2-yl]thio]ethyl]ammonium iodide; 2-[β-(N-diethylamino)ethylthio]-4-(p-phenylthio)-phenyl-1,5-benzodiazepine methiodide; thiabenzazonium iodide; Rec-15/0691; Antoral. $C_{28}H_{32}IN_3S_2$; mol wt 601.61. C 55.90%, H 5.36%, I 21.09%, N 6.98%, S 10.66%. 1,5-Benzodiazepine deriv with bactericidal activity. Prepn: D. Nardi et al., **Swiss pat. 555,347** (1974 to Recordati), *C.A.* **82,** 43480 (1975). Synthesis, physical characteristics, antibacterial activity: eidem, *Experientia* **31,** 440 (1975); eidem, *Farmaco Ed. Sci.* **30,** 248 (1975). Structure activity study: C. Greico et al., *ibid.* **32,** 909 (1977). Antimicrobial activity: M. Veronese et al., *Chemotherapy* **23,** 90 (1977).

Cryst from isopropanol, mp 162°. LD_{50} in mice, rats (mg/kg): 9000, > 10,000 orally; 42, 35 i.p., D. Nardi et al., *Experientia* **31,** 440 (1975).

THERAP CAT: Antibacterial.

9358. Tibolone. *(7α,17α)-17-Hydroxy-7-methyl-19-nor-pregn-5(10)-en-20-yn-3-one;* 7α-methyl-17α-ethynyl-17β-hydroxy-19-norandrost-5(10)-en-3-one; 7α-methyl-17α-ethynyl-17β-hydroxyestr-5(10)-en-3-one; Org OD 14; Livial. $C_{21}H_{28}O_2$; mol wt 312.45. C 80.73%, H 9.03%, O 10.24%. Synthetic anabolic steroid. Prepn: **Neth. pat. Appl. 6,406,797** corresp to H. P. de Jongh, N. P. van Vliet, **U.S. pat. 3,340,279** (1965, 1967 to Organon). Improved process: M. S. de Winter, E. A. Harryvan, **U.S. pat. 3,475,-465** (1969 to Organon). Clinical trial in treatment of postmenopausal osteoporosis: R. Lindsey et al., *Brit. Med. J.* **280,** 1207 (1980). Clinical effects on lipid and carbohydrate

metabolism: N. Crona et al., *Acta Endocrinol.* **102,** 451 (1983). Endocrinological profile: J. de Visser et al., *Arzneimittel-Forsch.* **34,** 1010 (1984).

Crystals, mp 165-169°.

THERAP CAT: Treatment of menopausal syndrome.

9359. Ticarbodine. *2,6-Dimethyl-N-[3-(trifluoromethyl)phenyl]-1-piperidinecarbothioamide;* α,α,α-trifluoro-2,6-dimethylthio-1-piperidinecarboxy-m-toluidide; EL-974; Tribodine. $C_{15}H_{19}F_3N_2S$; mol wt 316.39. C 56.95%, H 6.05%, F 18.02%, N 8.85%, S 10.13%. Deriv of thiourea, q.v. Prepn: H. D. Porter, H. M. Taylor, **U.S. pat. 3,659,012** (1972 to Lilly). Anthelmintic efficacy in dogs: R. J. Boisvenue et al., *Am. J. Vet. Res.* **33,** 709 (1972); G. F. Slonka et al., *ibid.* 1075; D. K. Hass, J. A. Collins, *Proc. Helminthol. Soc. Wash.* **43,** 135 (1976).

Light yellow cryst, mp 123-126°.

THERAP CAT (VET): Anthelmintic.

9360. Ticarcillin. *6-[(Carboxy-3-thienylacetyl)amino]-3,3-dimethyl-7-oxo-4-thia-1-azabicyclo[3.2.0]heptane-2-carboxylic acid;* N-(2-carboxy-3,3-dimethyl-7-oxo-4-thia-1-azabicyclo[3.2.0]hept-6-yl)-3-thiophenemalonamic acid; 6-[D(−)-α-carboxy-3-thienylacetamido]penicillanic acid; α-carboxy-3-thienylmethylpenicillin. $C_{15}H_{16}N_2O_6S_2$; mol wt 384.43. C 46.87%, H 4.19%, N 7.29%, O 24.97%, S 16.68%. Broad spectrum semi-synthetic antibiotic related to penicillin. Prepn: **Belg. pat. 646,991** corresp to E. G. Brain, J. H. Nayler, **U.S. pat. 3,282,926** (1964, 1966 to Beecham Group Ltd.). *In vitro* studies: H. C. Neu, E. B. Winshell, *Antimicrob. Ag. Chemother.* **1970,** 385; R. Sutherland et al., *ibid.* 390; N. J. Legakis, J. Papavassiliou, *J. Antibiot.* **28,** 912 (1975). *In vivo* studies: P. Acred et al., *Antimicrob. Ag. Chemother.* **1970,** 396. Absorption and excretion: R. Sutherland, P. J. Wise, *ibid.* 402. Clinical pharmacology: V. Rodriguez et al., *Antimicrob. Ag. Chemother.* **4,** 31 (1973); R. D. Libke et al., *Clin. Pharmacol. Ther.* **17,** 441 (1975). Review of pharmacology and therapeutic efficacy: R. N. Brogden et al., *Drugs* **20,** 325-352 (1980).

Disodium salt, $C_{15}H_{14}N_2Na_2O_6S_2$, *BRL 2288, Aerugipen, Monapen, Ticar, Ticarpenin.* Creamy-white hygroscopic non-crystalline powder. Readily sol in water (> 100 g/100 ml water) giving a clear soln with pH between 6.0 and 8.0. Aq solns are relatively stable; acid solns relatively unstable, R. Sutherland et al., *loc. cit.*

THERAP CAT: Antibacterial.

9361. Ticlopidine. *5-[(2-Chlorophenyl)methyl]-4,5,6,7-tetrahydrothieno[3,2-c]pyridine;* 5-(o-chlorobenzyl)-4,5,6,7-tetrahydrothieno[3,2-c]pyridine. $C_{14}H_{14}ClNS$; mol wt 263.78. C 63.75%, H 5.35%, Cl 13.44%, N 5.31%, S 12.15%. Prepn: **Ger. pat. 2,404,308** corresp to A. R. J. Castaigne, **U.S. pat. 4,051,141** (1974, 1977 both to Cent. Etudes Ind.

Pharm.) and E. Braye, **U.S. pat. 4,127,580** (1978 to Parcor). Metabolism: P. Godard *et al., Eur. J. Drug Metab. Pharmacokinet.* **3**, 67 (1978); *eidem, ibid.* **4**, 133 (1979); A. Tuong *et al., ibid.* **6**, 91 (1981). Mode of action: G. Leblondel, P. Allain, *Biochem. Pharmacol.* **27**, 2099 (1978); J. R. O'Brien *et al., Thromb. Res.* **13**, 245 (1978); J. J. Bruno, *ibid.* **1983**, Suppl. 4, 59. Pharmacology: A. Akashi *et al., Arzneimittel-Forsch.* **30**, 409, 415 (1980). Clinical studies: J. J. Thebault *et al., J. Int. Med. Res.* **5**, 405 (1977); C. Lecrubier *et al., Therapie* **32**, 189 (1977); T. Katsumura *et al., Angiology* **33**, 357 (1982).

Hydrochloride, $C_{14}H_{15}Cl_2NS$, **4-C-32, 53-32 C, Caudaline, Panaldine, Ticlid, Ticlodix, Ticlodone, Tiklid.** Cryst from ethanol, mp 1890°. LD_{50} in mice (mg/kg/24 hrs): 55 i.v.; > 300 orally (Castaigne).
THERAP CAT: Platelet aggregation inhibitor.

9362. Ticrynafen. *[2,3-Dichloro-4-(2-thienylcarbonyl)phenoxy]acetic acid; [2,3-dichloro-4-(2-thenoyl)phenoxy]acetic acid; [2,3-dichloro-4-(2-thiophenecarbonyl)phenoxy]-acetic acid; tienilic acid; thienylic acid;* ANP 3624; CE 3624; SKF 62698; Diflurex; Selacryn; Ticrex. $C_{13}H_8Cl_2O_4S$; mol wt 331.16. C 47.15%, H 2.43%, Cl 21.41%, O 19.33%, S 9.68%. A heterocyclic derivative of phenoxyacetic acid. Prepn: J. Godfroid, J. Thuillier, **Ger. pat. 2,048,372** corresp to **U.S. pat. 3,758,506** (1971, 1973, both to C.E.R.P.H.A.) and **Fr. pat. 2,115,042** (1972 to C.E.R.P.H.A.). Synthesis and pharmacology: G. Thuillier *et al., Eur. J. Med. Chem.* **9**, 625 (1974). Pharmacokinetics in healthy volunteers: A. L. Kerremans *et al., Eur. J. Clin. Pharmacol.* **22**, 515 (1982). Comparative study in hypertensive patients: B. T. Emmerson *et al., ibid.* 203. Hepatotoxicity study: J. W. Manier *et al., Am. J. Gastroenterol.* **77**, 401 (1982).

Crystals from 50% ethanol, mp 148-149°; also reported as mp 157°. LD_{50} i.v., orally in mice: 225, 1275 mg/kg, **U.S. pat. 3,758,506.**
THERAP CAT: Diuretic; uricosuric; antihypertensive.

9363. Tiemonium Iodide. *4-[3-Hydroxy-3-phenyl-3-(2-thienyl)propyl]-4-methylmorpholinium iodide; N-methyl-N-[3-hydroxy-3-phenyl-3-(α-thienyl)propyl]morpholinium iodide; Visceralgina.* $C_{18}H_{24}INO_2S$; mol wt 445.38. C 48.54%, H 5.43%, I 28.49%, N 3.15%, O 7.18%, S 7.20%. Prepn: **Brit. pat. 953,386** (1964 to C.E.R.M.).

Solid, mp 189-191°.
Marketed also in form of the methyl sulfate.
THERAP CAT: Anticholinergic, antispasmodic.

9364. Tigemonam. *[S-(Z)]-[[[1-(2-Amino-4-thiazolyl)-2-[[2,2-dimethyl-4-oxo-1-(sulfooxy)-3-azetidinyl]amino]-2-oxoethylidene]amino]oxy]acetic acid; [[[(Z)-(2-amino-4-thiazolyl)[[(3S)-1-hydroxy-2,2-dimethyl-4-oxo-3-azetidinyl]-carbamoyl]methylene]amino]oxy]acetic acid hydrogen sulfate (ester).* $C_{12}H_{15}N_5O_9S_2$; mol wt 437.40. C 32.95%, H 3.46%, N 16.01%, O 32.92%, S 14.66%. Orally active synthetic monosulfactam, structurally similar to the monobactam aztreonam, *q.v.* Prepn: C. Yoshida *et al., J. Antibiot.* **38**,

1536 (1985); W. A. Slusarchyk *et al.,* **Belg. pat. 904,121;** W. A. Slusarchyk, W. H. Koster, **U.S. pat. 4,638,061** (1986, 1987 both to Squibb). Antibacterial spectrum and β-lactamase stability *in vitro:* S. K. Tanaka *et al., Antimicrob. Ag. Chemother.* **31**, 219 (1987); N.-X. Chin, H. C. Neu, *ibid.* **32**, 84 (1988). Antibacterial activity and pharmacokinetics in animals: J. M. Clark *et al., ibid.* **31**, 226 (1987).

Disodium salt, $C_{12}H_{13}N_5Na_2O_9S_2$, *SQ 30213.* mp 140-145° (dec).
Dicholine salt, $C_{22}H_{41}N_2O_{11}S_2$, *SQ 30836, Tigemen.*
THERAP CAT: Antibacterial.

9365. Tiglic Acid. *(E)-2-Methyl-2-butenoic acid; (E)-2-methylcrotonic acid; trans-2,3-dimethylacrylic acid.* $C_5H_8O_2$; mol wt 100.11. C 59.98%, H 8.06%, O 31.96%. The stable isomer of angelic acid. Found as glyceride in croton oil, as butyl ester in the oil of the Roman camomile, *Anthemis nobilis* L., *Compositae,* and as geranyl tiglate in oil of geranium. Is formed during the charcoaling of maple wood. Formation by the intestinal roundworm, *Ascaris lumbricoides:* Bueding, *J. Biol. Chem.* **202**, 505 (1953). Has been found in crude sodium penicillin: Cram, Tishler, *J. Am. Chem. Soc.* **70**, 4238 (1948). Sepn by partition chromatography: Bueding, *loc. cit.* Synthesis from 2-hydroxy-2-methylbutyronitrile: Crawford, *J. Soc. Chem. Ind. (London)* **64**, 231 (1945). Review and bibliography: Buckles *et al., Chem. Rev.* **55**, 659-677 (1955).

Triclinic plates, rods from water. Spicy odor. *Vesicant.* d 0.972. mp 63.5-64°. bp_{760} 198.5°; $bp_{11.5}$ 95.0-96°. Volatile with steam. n_D^{81} 1.4342. K at 25° = 9.6×10^{-6}. uv max (H_2O): 216-217 nm (ε 10,700). Molar heat of combustion 635.1 kcal. Sparingly sol in cold water; freely sol in hot water. Sol in alcohol, ether.
Calcium salt trihydrate, $Ca(C_5H_7O_2)_2 \cdot 3H_2O$, leaflets. Much less sol in water than calcium angelate: 100 parts of aq soln satd at 17° contains 6.05 parts of anhydr calcium tiglate.
Amide, C_5H_9NO, crystals, mp 75-76°.
Methyl ester, $C_6H_{10}O_2$, liquid; d_4^{20} 0.9498; bp_{766} 139.6°; n_D^{20} 1.4370.
Ethyl ester, $C_7H_{12}O_2$, liquid; $d_4^{19.5}$ 0.9247; bp_{752} 156°; bp_{11} 55.5°. n_D^{20} 1.4350. Heat of formn at constant vol: 953.2 kcal, at constant pressure: 954.4 kcal.
Geranyl ester, $C_{15}H_{24}O_2$, liquid; pleasant odor; d_{15}^{15} 0.9279; bp_7 149-151°.
USE: The esters in perfumes and flavoring agents. The free acid as a breaker of emulsions.

9366. Tigloidine. *2-Methyl-2-butenoic acid [1α,3α(E),-5α]-8-methyl-8-azabicyclo[3.2.1]oct-3-yl ester; (E)-1αH,5αH-tropan-3β-ol 2-methylcrotonate; tiglylpseudotropeine; 3β-tigloyloxytropane; tiglic acid ester with pseudotropine; Tiglyssin.* $C_{13}H_{21}NO_2$; mol wt 223.31. C 69.92%, H 9.48%, N 6.27%, O 14.33%. Isoln from *Duboisia myoporoides* R. Br., *Solanaceae* and prepn from tropine and tigloyl chloride: Barger *et al., J. Chem. Soc.* **1937**, 1820; isoln from *Datura innoxia* Miller, *Solanaceae:* Evans, Wellendorf, *ibid.* **1959**, 1406. Pharmacology: Sanghvi *et al., Eur. J. Pharmacol.* **4**, 246 (1968).

Hydrobromide, $C_{13}H_{22}BrNO_2$, mp 234-235°. Soluble in chloroform.

THERAP CAT: CNS depressant, antiparkinsonian.

9367. Tigogenin. *(25R)-5α-Spirostan-3β-ol.* $C_{27}H_{44}O_3$; mol wt 416.62. C 77.83%, H 10.65%, O 11.52%. The aglycon of tigonin. Usually separated from crude gitogenin obtained from leaves of *Digitalis lanata* Ehrh., *Scrophulariaceae*: Windhaus, *Z. Physiol. Chem.* **150**, 205 (1925); Jacobs, Fleck, *J. Biol. Chem.* **88**, 545 (1930). From the sisal plant *Agave sisalana* L., *Amaryllidaceae*: Rubin, U.S. pats. **2,991,-282** (1961); **3,303,187** (1967). Differs from smilagenin *(q.v.)* in the configuration at C-5. Structure: Marker, Rohrmann, *J. Am. Chem. Soc.* **62**, 898 (1940). Structure and synthesis: Mazur, Sondheimer, *ibid.* **81**, 3161 (1959); Caglioti, Magi, *Tetrahedron* **19**, 1127 (1963).

Crystals from dil methanol, mp 203°. $[α]_D^{20}$ −62°. More sol in acetone, in ether, and esp in petr ether than gitogenin. Pptd by digitonin.

Acetyltigogenin, $C_{29}H_{46}O_4$, crystals from methanol, mp 206°. $[α]_D^{18}$ −74.4°.

9368. Tigonin. $C_{56}H_{92}O_{27}$; mol wt 1197.30. C 56.17%, H 7.75%, O 36.08%. Isoln from *Digitalis lanata* Ehrh., *Scrophulariaceae*: Tschesche, *Ber.* **69**, 1665 (1936). Also occurs in *D. purpurea* L., *Scrophulariaceae*, in *Chlorogalum pomeridianum* (DC.) Kunth, *Liliaceae*: Liang, Noller, *J. Am. Chem. Soc.* **57**, 525 (1935). Built from 2 glucose, 2 galactose, 1 xylose, and 1 tigogenin unit.

Hygroscopic, amorphous flakes from 95% alc. After drying at 118° over P_2O_5 *in vacuo* sinters at 220°. mp ~260° (*in vacuo*). Sol in water. Not pptd by ether from water (difference from digitonin); amorphous ppt with amyl alcohol.

Cholesterol tigonide, an equimolar addition compd, long needles from methanol, dec above 200°.

9369. Tiliacorine. *6',7-Didemethoxy-6',7-epoxyrodiasine.* $C_{36}H_{36}N_2O_5$; mol wt 576.70. C 74.98%, H 6.29%, N 4.86%, O 13.87%. From bark of *Tiliarcora acuminata* Miers, and *T. racemosa* Colebr., *Menispermaceae*. Isoln: Van Itallie, Steenhauer, *Pharm. Weekbl.* **59**, 1381 (1922). Structure: Anjaneyulu *et al.*, *Chem. & Ind. (London)* **1959**, 1119; Rao, Row, *J. Org. Chem.* **25**, 981 (1960). Revised structure: Anjaneyula *et al.*, *J. Sci. Ind. Res. (India)* **21B**, 602 (1962); *eidem*, *Tetrahedron* **25**, 3091 (1969); M. Shamma, J. E. Foy, *J. Org. Chem.* **41**, 1293 (1976). Abs config and biosynthesis: D. S. Bhakuni *et al.*, *Chem. Commun.* **1978**, 226.

Bitter crystals, mp 271-272°. $[α]_D$ +105.3°. uv max: 295, 265 nm (log ε 3.91, 3.48). Sol in alcohol, benzene, chloroform, ether.

9370. Tilidine. *2-(Dimethylamino)-1-phenyl-3-cyclohexene-1-carboxylic acid ethyl ester;* ethyl 2-(dimethylamino)-1-phenyl-3-cyclohexene-1-carboxylate; 3-dimethylamino-4-phenyl-4-carbethoxy-Δ¹-cyclohexene. $C_{17}H_{23}NO_2$; mol wt 273.38. C 74.69%, H 8.48%, N 5.12%, O 11.70%. Prepn: **S. Afr.** pat. **66 06,476; Brit.** pat. **1,120,186** corresp to G. Satzinger, **U.S.** pat. **3,557,127** (1967, 1968, 1971, all to Warner-Lambert). Synthesis of 6-C^{14} hydrochloride: K.-O. Vollmer, F. W. Koss, *Arzneimittel-Forsch.* **20**, 990 (1970). Pharmacology and toxicity data: M. Hermann *et al.*, *ibid.* 977, 983. Metabolism: K.-O. Vollmer, H. Achenbach, *ibid.* **24**, 1237 (1974).

DL-*trans,trans*-Hydrochloride hemihydrate, $C_{17}H_{24}Cl-NO_2.½H_2O$, *Go 1261 C, W 5759 A, Lucayan, Perdolat, Valoron.* Crystals, mp 125°. Easily sol in water. LD_{50} (7 day) in mice, rats (mg/kg): 437.0, 417.7 i.g.; 490.0, 400.0 s.c.; 52.0, 74.1 i.v. (Herrmann). Anhydrous: crystals from ethyl acetate-methyl ethyl ketone, mp 159°.

Free base, $bp_{0.01}$ 95.5-96°.

DL-*cis,cis*-Hydrochloride, $C_{17}H_{24}ClNO_2$, crystals from ethyl acetate-methyl ethyl ketone, mp 84°.

Free base, $bp_{0.01}$ 97.5-98°.

Caution: May be habit forming. This is a controlled substance (opiate) listed in the U.S. Code of Federal Regulations, Title 21 Part 1308.11 (1987).

THERAP CAT: Narcotic analgesic.

9371. Tilorone. *2,7-Bis[2-(diethylamino)ethoxy]-9H-fluoren-9-one;* bis-DEAE-fluorenone. $C_{25}H_{34}N_2O_3$; mol wt 410.56. C 73.14%, H 8.34%, N 6.82%, O 11.69%. The first recognized synthetic, small mol wt compound that is an orally active interferon inducer: Krueger, Mayer, *Science* **169**, 1213 sqq. (1970). Prepn: Fleming *et al.*, **Ger.** pat. **1,964,761** corresp to **U.S.** pat. **3,592,819** (1970, 1971 to Richardson Merrell); E. R. Andrews *et al.*, *J. Med. Chem.* **17**, 882 (1974); H. M. Burke, M. M. Joullie, *ibid.* **21**, 1084 (1978). Mechanism of action: De Clercq, Merigan, *J. Infec. Dis.* **123**, 190 (1971). Toxicity studies: Kaufman *et al.*, *Proc. Soc. Exp. Biol. Med.* **137**, 357 (1971). Review: P. Chandra *et al.*, "Tilorone Hydrochloride" in *Antibiotics* vol. 5(pt. 2), F. E. Hahn, Ed. (Springer-Verlag, New York, 1979) pp 385-413.

Hydrochloride, $C_{25}H_{36}Cl_2N_2O_3$, crystals from butanone-methanol, mp 235-237°. uv max (water): 269 nm ($E_{1cm}^{1\%}$

1600). LD_{50} in mice, rats (single dose): 959, 852 mg/kg orally; 145, 244 mg/kg i.p. (Krueger, Mayer).

9372. Timepidium Bromide. *3-(Di-2-thienylmethylene)-5-methoxy-1,1-dimethylpiperidinium bromide;* SA 504; Mepidium; Sesden. $C_{17}H_{22}BrNOS_2$; mol wt 400.41. C 51.00%, H 5.54%, Br 19.96%, N 3.50% O 3.99%, S 16.01%. Prepn: M. Kawazu *et al.*, **Ger. pat. 2,128,808** corresp to U.S. pat. **3,764,607** (1971, 1973 both to Tanabe). Pharmacological properties: H. Tamaki *et al.*, *Japan. J. Pharmacol.* **22**, 685 (1972), *C.A.* **78**, 52812 (1973). Metabolism: J. Sugihara, N. Taga, *Radioisotopes* **26**, 238 (1977). Absorption, distribution, excretion: M. Yoshikawa, *Oyo Yakuri* **14**, 179 (1977), *C.A.* **88**, 163772 (1978). Teratological study: Y. Fujisawa *et al., ibid.* **7**, 1293 (1973), *C.A.* **81**, 99413 (1974). Chronic toxicity study: K. Doi *et al., ibid.* **13**, 851 (1977), *C.A.* **88**, 32014 (1978).

Colorless cryst from acetone/ether, mp 198-200°.
THERAP CAT: Anticholinergic.

9373. Timiperone. *4-[4-(2,3-Dihydro-2-thioxo-1H-benzimidazol-1-yl)-1-piperidinyl]-1-(4-fluorophenyl)-1-butanone;* 1-[1-[3-(4-fluorobenzoyl)propyl]-4-piperidyl]-2-mercaptobenzimidazole; 4-fluoro-4-[4-(2-thioxo-1-benzimidazolinyl)piperidino]butyrophenone; 1-[1-[3-(4-fluorobenzoyl)propyl]-4-piperidyl]-2,3-dihydrobenzimidazole-2-thione; DD 3480; Tolopelon. $C_{22}H_{24}FN_3OS$; mol wt 397.51. C 66.47%, H 6.09%, F 4.78%, N 10.57% O 4.02%, S 8.07%. Butyrophenone derivative with neuroleptic activity. Prepn: M. Sato *et al.*, **Japan. Kokai 75 84,578**; K. Ueno *et al.*, U.S. pat. **3,963,727** (1975, 1976 both to Daiichi); M. Sato *et al., J. Med. Chem.* **21**, 1116 (1978). Improved synthesis: M. Sato, M. Arimoto, *Chem. Pharm. Bull.* **30**, 719 (1982). Pharmacology: T. Yamasaki *et al., Japan. J. Pharmacol.* **27** (Suppl.), 124P (1977); *eidem, Arzneimittel-Forsch.* **31**, 701, 707 (1981); T. Shibuya *et al., Int. J. Clin. Pharmacol. Ther. Toxicol.* **20**, 251 (1982). Pharmacokinetics and metabolism: H. Tachizawa *et al., Drug. Metabol. Dispos.* **9**, 442 (1981); K. Sudo *et al., Xenobiotica* **11**, 685 (1981). Multicenter controlled clinical trials in schizophrenia: R. Takahashi *et al., J. Int. Med. Res.* **10**, 257 (1982); T. Kariya *et al., ibid.* **11**, 66 (1983).

Crystals from acetone, mp 201-203°. uv max (ethanol): 226.5, 246, 309 nm. Slightly sol in water. LD_{50} in male rats, mice (mg/kg): 232, 478 orally (Yamasaki).
THERAP CAT: Antipsychotic.

9374. Timolol. *(S)-1-[(1,1-Dimethylethyl)amino]-3-[[4-(4-morpholinyl)-1,2,5-thiadiazol-3-yl]oxy]-2-propanol;* S-(−)-3-(3-tert-butylamino-2-hydroxypropoxy)-4-morpholino-1,2,5-thiadiazole; (−)-3-morpholino-4-(3-tert-butylamino-2-hydroxypropoxy)-1,2,5-thiadiazole. $C_{13}H_{24}N_4O_3S$; mol wt 316.42. C 49.35%, H 7.64%, N 17.71%, O 15.17%, S 10.13%. β-Adrenergic blocker. Prepn: B. K. Wasson, **Ger. pat. 1,925,956;** *idem,* U.S. pat. **3,655,663** (1969, 1972 both to Frosst). Manufacturing process: L. M. Weinstock *et al.,* **Ger. pat. 1,925,955;** *eidem,* U.S. pat. **3,-657,237** (1970, 1972 both to Frosst). Synthesis and activity data: Wasson *et al., J. Med. Chem.* **15**, 651 (1972). Pharmacology: Franciosa *et al., Clin. Pharmacol. Ther.* **13**, 138 (1972); Ulrych *et al., ibid.* 232. Review of efficacy in glau-

coma: R. C. Heel *et al., Drugs* **17**, 38-55 (1979). Clinical evaluation in hypertension: B. A. Rofman *et al., Hypertension* **2**, 643 (1980). Multicenter study of effect in myocardial infarction: N. Engl. J. Med. **304**, 801 (1981); of efficacy in limiting infarct size: M. Sederholm *et al., ibid.* **310**, 9 (1984). Comprehensive description: D. J. Mazzo, A. E. Loper in *Analytical Profiles of Drug Substances* **vol. 16**, K. Florey, Ed. (Academic Press New York, 1987) pp 641-692.

Hydrogen maleate salt, $C_{17}H_{28}N_4O_7S$, *MK-950, Betim, Blocadren, Proflax, Temserin, Tenopt, Timacar, Timacor, Timolate, Timoptic, Timoptol.* White crystals from ethanol, mp 201.5-202.5°. $[\alpha]_{405}^{24}$ −12.0° (c = 5 in 1N HCl), $[\alpha]_D^{25}$ −4.2°. uv max (0.1N HCl): 294 nm ($A_{1cm}^{1\%}$ 200). Sol in water, ethanol, methanol; sparingly sol in chloroform; very slightly sol in cyclohexane. Practically insol in isooctane, ether. Stable in soln up to pH 12.

(±)-Free base, crystalline solid from isopropyl ether, mp 71.5-72.5°.

THERAP CAT: Antihypertensive, antiarrhythmic, antianginal, antiglaucoma agent.

9375. Timonacic. *4-Thiazolidinecarboxylic acid;* ATC; norgamen; thioproline; NSC 25855; Detoxepa; Hépalidine; Héparégène; Thiobiline; Tiazolidin. $C_4H_7NO_2S$; mol wt 133.18. C 36.07%, H 5.30%, N 10.52%, O 24.03%, S 24.08%. Prepd from DL- or L-cysteine and formaldehyde. Prepn of *l*-form: Ratner, Clarke, *J. Am. Chem. Soc.* **59**, 200 (1937); of *dl*- and *l*-forms: Werner *et al., Helv. Chim. Acta* **30**, 432 (1947); **Fr. pat. M3184** (1965 to Sogespar S.A.), *C.A.* **63**, 12980h (1965). Proton magnetic resonance and conformation: Martin, Mathur, *J. Am. Chem. Soc.* **87**, 1065 (1965). Series of articles on pharmacology and toxicology: *Gazz. Med. Ital.* **131**, 251-286 (1972). Treatment of cancer through induced reverse transformation: A. Brugarolas, M. Gosalvez, *Lancet* **1**, 68 (1980).

dl-Form, crystals, mp 195°. LD_{50} orally in mice: 400 mg/kg, Bertrand, Piton, *Gazz. Med. Ital.* **131**, 265 (1972). *l*-Form, crystals from water, dec 196-197°. Readily sol in hot water, acid, alkali; sparingly sol in cold water; practically insol in alcohol.
THERAP CAT: Choleretic.

9376. Tin. Sn; at. wt 118.69; at. no. 50; valence 2, 4. Ten naturally occurring isotopes: 112 (0.95%); 114 (0.65%); 115 (0.34%); 116 (14.24%); 117 (7.57%); 118 (24.01%); 119 (8.59%); 120 (32.97%); 122 (4.71%); 124 (5.98%); artificial, radioactive isotopes: 108-111; 113; 121; 123; 125-132. Found in cassiterite, stannite, and tealite. Occurrence in earth's crust: 6×10^{-4}%. The metal of commerce is about 99.8% pure. Prepn of high purity tin: Baralis, Marone, *Met. Ital.* **59**, 494 (1967), *C.A.* **67**, 119613a (1967). Physical properties: Kirshenbaum, Cahill, *J. Inorg. Nucl. Chem.* **25**, 232 (1963). *Monograph:* C. L. Mantell, *Tin: Its Mining, Production, Technology and Applications* (Reinhold, New York, 1949). *Reviews:* Abel in *Comprehensive Inorganic Chemistry* **vol. 2**, J. C. Bailar, Jr. *et al.*, Eds. (Pergamon Press, Oxford, 1973) pp 43-104; W. Germain *et al.*, in *Kirk-Othmer Encyclopedia of Chemical Technology* **vol. 23** (Wiley-Interscience, New York, 3rd ed., 1983) pp 18-42.

Almost silver-white, lustrous, soft, very malleable and ductile metal; only slightly tenacious; easily powdered. When being bent, emits the crackling "tin cry". Brittle at 200°. At −40° crumbles to gray amorphous powder ("gray tin"), slowly changing back above 20° to white tin. Available in the form of bars, foil, powder, shot, etc. Stable in air, but when in powder form it oxidizes, esp in presence of moisture. d 7.31. mp 231.9°. bp 2507° (2780°K). Specific heat (25°) 0.053 cal/g/°C. Brinell hardness 2.9. Reacts slowly with cold dil HCl or dil HNO_3, hot dil H_2SO_4; readily with concd HCl, aqua regia; very slowly attacked by acetic acid; slowly attacked by cold, more readily by hot caustic alkali; concd HNO_3 converts it into insol metastannic acid.

USE: Chiefly for tin-plating, soldering alloys, babbitt and type metals, manuf tin salts, collapsible tubes.

9377. Tinidazole. *1-[2-(Ethylsulfonyl)ethyl]-2-methyl-5-nitro-1H-imidazole;* ethyl[2-(2-methyl-5-nitro-1-imidazolyl)ethyl]sulfone; CP 12574; Fasigin; Fasigyn; Pletil; Simplotan; Sorquetan; Tricolam; Trimonase. $C_8H_{13}N_3O_4S$; mol wt 247.26. C 38.86%, H 5.30%, N 16.99%, O 25.88%, S 12.97%. Prepn: Butler, **S. Afr.** pat. **66 07,466,** *C.A.* **71,** 3384e (1969) and **U.S.** pat. **3,376,311** (both 1968 to Pfizer); Miller *et al., Antimicrob. Ag. Chemother.* **1969,** 257; *eidem, J. Med. Chem.* **13,** 849 (1970). Activity studies: Howes *et al., Antimicrob. Ag. Chemother.* **1969,** 261. Pharmacokinetics: Taylor *et al., ibid.* 267. Clinical results: Diwald, *Wien. Med. Wochenschr.* **121,** 492 (1971). *Review: J. Antimicrob. Chemother.* **10,** Suppl. 8, 1-184 (1982). Review of antibacterial activity, pharmacology and therapeutic efficacy: A. A. Carmine *et al., Drugs* **24,** 85-117 (1982).

$$CH_2CH_2SO_2CH_2CH_3$$
$$O_2N-\underset{N}{\overset{N}{|}}-CH_3$$

Colorless crystals from benzene, mp 127-128°. LD_{50} in mice (mg/kg): > 3600 orally; > 2300 i.p. (Miller).

THERAP CAT: Antiprotozoal (Trichomonas, Giardia); antiamebic.

9378. Tinofedrine. *α-[1-[(3,3-Di-3-thienyl-2-propenyl)amino]ethyl]benzenemethanol;* α-[1-[(3,3-di-3-thienylallyl)amino]ethyl]benzyl alcohol. $C_{20}H_{21}NOS_2$; mol wt 355.52. C 67.57%, H 5.95%, N 3.94%, O 4.50%, S 18.04%. Prepn: K. Thiele, K. Posselt, **Ger.** pat. **1,921,453** (1970 to Degussa), *C.A.* **74,** 76320c (1971); K. Posselt, H. Offermanns, **Ger.** pat. **2,150,977** corresp to **U.S.** pat. **3,767,675** (1972, 1973 both to Degussa); K. Thiele *et al., Arzneimittel-Forsch.* **28,** 2047 (1978). Absolute configuration determined by circular dichroism: J. Engle *et al., Chem. Ztg.* **105,** 85 (1981). Series of articles on pharmacology, hemodynamics, pharmacokinetics, metabolism: *Arzneimittel-Forsch.* **28,** 1335-1367 (1978). Synthesis of radioactive tinofedrine: A. Saus, K. Posselt, *ibid.* **30,** 917 (1980). Spectroscopic study: K. Thiele *et al., ibid.* 1057. EEG study in geriatric patients: B. Saletu, P. Anderer, *ibid.* 1218. General effects on heart and blood circulation: F. Stroman, K. Thiemer, *ibid.* **31,** 1892 (1981).

Hydrochloride, $C_{20}H_{22}ClNOS_2$, D-8955, *Novocebrin.* Cryst from isopropanol, mp 226-229° (dec). LD_{50} in mice, rats (mg/kg): 1890, 6600 orally, 20.15, 14.00 i.v., P. St. Janiak, *Drugs of the Future* **4,** 286 (1979).

THERAP CAT: Cerebral vasodilator.

9379. Tinoridine. *2-Amino-4,5,6,7-tetrahydro-6-(phenylmethyl)thieno[2,3-c]pyridine-3-carboxylic acid ethyl ester;* 2-amino-6-benzyl-4,5,6,7-tetrahydrothieno[2,3-c]pyridine-3-carboxylic acid ethyl ester; 2-amino-3-ethoxycarbonyl-6-benzyl-4,5,6,7-tetrahydrothieno[2,3-c]pyridine; ethyl 2-amino-6-benzyl-4,5,6,7-tetrahydrothieno[2,3-c]pyridine-3-carboxylate; Y-3642. $C_{17}H_{20}N_2O_2S$; mol wt 316.42. C 64.53%, H 6.37%, N 8.85%, O 10.11%, S 10.13%. Prepn: Nakanishi *et al., Ger.* pat. **1,812,404;** *eidem,* **U.S.** pat. **3,-563,997** (1969, 1971 both to Yoshitomi). Pharmacological

studies: Nakanishi *et al., J. Pharm. Soc. Japan* **90,** 272-334 (1970), *C.A.* **73,** 54504t-54510s (1970); *eidem, Arzneimittel-Forsch.* **20,** 998, 1004 (1970); *Yakuri to Chiro* **2,** 1028 (1974). Protective effect vs. carbon tetrachloride hepatotoxicity: H. Yasuda *et al., Toxicol. Appl. Pharmacol.* **52,** 407 (1980). Brief review: *Japan Med. Gaz.* **8**(8), 7 (1971).

Slightly sol in H_2O, mp 112-113°. LD_{50} in mice, rats (mg/kg): 5400, > 10200 orally; 1600, 1250 i.p. (Nakanishi, *Arzneimittel-Forsch.,* 1970).

Hydrochloride, $C_{17}H_{21}ClN_2O_2S$, *Dimaten, Nonflamin.* Yellowish white to yellow powder, mp 234-235° (dec). Practically odorless with a bitter taste. Slightly sol in methanol; very slightly sol in water, ether, acetone, benzene. LD_{50} orally in mice: 1601 mg/kg (*Yakuri to Chiro*).

THERAP CAT: Analgesic; antipyretic; anti-inflammatory.

9380. Tin Phosphides. Several stoichiometries reported: *Mellor's* **vol. VIII,** 847-849 (1931). Crystal structure of Sn_4P_3 and discussion of other phases: Olofsson, *Acta Chem. Scand.* **21,** 1659 (1967); **24,** 1153 (1970).

USE: Tin phosphide with a phosphorus content > 10% used in manuf of phosphor bronze.

9381. Tinuvin® P. *2-(2H-Benzotriazol-2-yl)-4-methylphenol;* 2-(2H-benzotriazol-2-yl)-p-cresol; 2-(2'-hydroxy-5'-methylphenyl)benzotriazole. $C_{13}H_{11}N_3O$; mol wt 225.25. C 69.32%, H 4.92%, N 18.65%, O 7.10%. Prepn: Heller *et al.,* **U.S.** pats. **3,004,896; 3,189,615** (1961, 1965 to Geigy).

Minute crystals, mp 131-133°. Sol in ethyl acetate, acetone, caprolactam solns, dioctylphthalate, oleyl alcohol, hot petrolatum. Stable to conditions and chemicals used in polymerization or compounding of plastics.

USE: An ultraviolet light absorber for stabilizing plastics and other organic materials against discoloration and deterioration. Effective in protecting polyesters, chlorinated polyesters, polystyrene, polyvinyls, polypropylene, alkyds, cellulose acetate, ethyl cellulose, acrylates, dyes, synthetic and natural fibers, waxes, detergent solns, cosmetic formulations. *Ref:* Dunn, Fogg, *J. Appl. Polymer Sci.* **2,** 367 (1959).

9382. Tiocarlide. *N,N'-[4-(3-Methylbutoxy)phenyl]thiourea;* 4,4'-bis(isopentyloxy)thiocarbanilide; 4,4'-di(isoamyloxy)thiocarbanilide; 1,3-bis(p-isoamyloxyphenyl)-2-thiourea; thiocarlide; Amixyl; Datanil; Disocarban; Isoxyl; DATC; Sarbamyl. $C_{23}H_{32}N_2O_2S$; mol wt 400.58. C 68.96%, H 8.05%, N 6.99%, O 7.99%, S 8.00%. Prepn: Buu-Hoi, Xuong, *Compt. Rend.* **237,** 498 (1953); Huebner, Scholz, **U.S.** pat. **2,703,815** (1955 to Ciba).

$$(CH_3)_2CHCH_2CH_2O-\bigcirc-HNCSNH-\bigcirc-OCH_2CH_2CH(CH_3)_2$$

Crystals from ethanol, mp 134-145°.

THERAP CAT: Antibacterial (tuberculostatic).

9383. Tioclomarol. *3-[3-(4-Chlorophenyl)-1-(5-chloro-2-thienyl)-3-hydroxypropyl]-4-hydroxy-2H-1-benzopyran-2-one;* 3-[5-chloro-α-(p-chloro-β-hydroxyphenethyl)-2-thenyl]-4-hydroxycoumarin; Apegmone. $C_{22}H_{16}Cl_2O_4S$; mol wt 447.33. C 59.07%, H 3.60%, Cl 15.85%, O 14.31%, S 7.17%. Slow-acting anticoagulant, related structurally to warfarin, *q.v.* Prepn: E. Boschetti *et al.,* **S. Afr.** pat. **67 07,267** corresp to **U.S.** pat. **3,574,234** (1968, 1971 both to Lipha). Synthesis and anticoagulant activity: *eidem, Chim. Ther.* **7,** 20 (1972).

White cryst from methanol, mp 104°.
THERAP CAT: Anticoagulant.

9384. Tioconazole. *1-[2-[(2-Chloro-3-thienyl)methoxy]-2-(2,4-dichlorophenyl)ethyl]-1H-imidazole;* 1-[2,4-dichloro-β-[(2-chloro-3-thienyl)oxy]phenethyl]imidazole; UK-20349; Fungibacid; Gyno-Trosyd; Trosyd; Trosyl; Zoniden. $C_{16}H_{13}Cl_3N_2OS$; mol wt 387.70. C 49.57%, H 3.38%, Cl 27.43%, N 7.22%, O 4.13%, S 8.27%. Antimycotic deriv of imidazole, *q.v.* Prepn: G. E. Gymer, **Belg.** pat. 841,309; *idem,* **U.S.** pat. **4,062,966** (1976, 1977 both to Pfizer). Antifungal activity: S. Jevons, *Antimicrob. Ag. Chemother.* **15,** 597 (1979). Laboratory evaluation vs other imidazole derivs: F. C. Odds, *J. Antimicrob. Chemother.* **6,** 749 (1980); E. Lefler, D. A. Stevens, *Antimicrob. Ag. Chemother.* **25,** 450 (1984). Pharmacology: M. S. Marriott *et al., Dermatologica* **166,** Suppl. 1, 1 (1983). Use in dermatomycosis: K. Kuokkanen, *Mykosen* **25,** 274 (1982). Series of articles on clinical efficacy in gynecological use: *Gynak. Rundsch.* **23,** Suppl. 1, 1-60 (1983).

Hydrochloride, $C_{16}H_{14}Cl_4N_2OS$, crystals, mp 168-170°.
THERAP CAT: Topical antifungal.

9385. Tiomesterone. *1,7-Bis(acetylthio)-17-hydroxy-17-methylandrost-4-en-3-one; 17β-hydroxy-1α,7α-dimercapto-17-methylandrost-4-en-3-one 1,7-diacetate;* 1α,7α-bis(acetylthio)-17α-methyltestosterone; 1α,7α-bis(acetylmercapto)-17α-methyltestosterone; thiomesterone; STA 307; Emdabol; Emdabolin; Protabol. $C_{24}H_{34}O_4S_2$; mol wt 450.66. C 63.96%, H 7.60%, O 14.20%, S 14.23%. Prepn: Krämer, *Ber.* **96,** 2803 (1963); Brueckner *et al.,* **U.S.** pat. **3,087,942** (1963 to E. Merck).

Crystals, mp 205-206°. $[\alpha]_D$ —66° (dioxane). uv max: 237.5 nm (ε 19,800).
THERAP CAT: Androgen.

9386. Tiopronin. *N-(2-Mercapto-1-oxopropyl)glycine;* N-(2-mercaptopropionyl)glycine; α-mercaptopropionylglycine; Acadione; Capen; Epatiol; Mucolysin; Thiola; Thiosol. $C_5H_9NO_3S$; mol wt 163.20. C 36.79%, H 5.56%, N 8.58%, O 29.41%, S 19.65%. Prepn: Mita *et al.,* **U.S.** pat. **3,246,025** (1966 to Santen). Antidotal effects in mice: H. Fujimura *et al., Nippon Yakurigaku Zasshi* **60,** 278 (1964), *C.A.* **62,** 972h (1965). Anti-inflammatory pharmacology in animals: F. Capasso, *Agents Actions* **11,** 741 (1981). Clinical trial in chronic hepatitis: F. Ichida *et al., J. Int. Med. Res.* **10,** 325 (1982). Clinical comparison with D-penicillamine in rheu-

matoid arthritis: B. Amor *et al., Arthritis Rheum.* **25,** 698 (1982); G. Pasero *et al., ibid.* 923.

$$CH_3CHCONHCH_2COOH$$
$$|$$
$$SH$$

Crystals from ethyl acetate, mp 95-97°. LD_{50} i.v. in mice: 2.1 g/kg (Fujimura).
THERAP CAT: Hepatoprotectant. Mucolytic. Antidote to heavy metal poisoning.

9387. Tioxidazole. *(6-Propoxy-2-benzothiazolyl)carbamic acid methyl ester;* methyl 6-propoxy-2-benzothiazolylcarbamate; Sch 21480; Tiox. $C_{12}H_{14}N_2O_3S$; mol wt 266.31. C 54.12%, H 5.30%, N 10.52%, O 18.02%, S 12.04%. Structurally related to benzimidazole anthelmintics. Prepn and anthelmintic activity: M. M. Nafissi-Varchei, **Belg.** pat. **840,945;** **U.S.** pat. **4,006,242** (1976, 1977 both to Schering). Physical data, activity against roundworms: E. Panitz *et al., Experientia* **34,** 733 (1978). Anthelmintic effect against gastrointestinal parasites in naturally infected horses: J. H. Drudge *et al., Am. J. Vet. Res.* **41,** 1383 (1980); E. T. Lyons *et al., ibid.* **42,** 1048 (1981).

White odorless solid, mp 178-180°. Insol in water. Slightly sol in organic solvents.
THERAP CAT (VET): Anthelmintic for horses.

9388. Tioxolone. *6-Hydroxy-1,3-benzoxathiol-2-one;* 6-hydroxy-2-oxo-1,3-benzoxathiole; thioxolone; Acnosan; Camyna; Stepin. $C_7H_4O_3S$; mol wt 168.18. C 49.99%, H 2.40%, O 28.54%, S 19.07%. Prepd from resorcinol and ammonium thiocyanate in presence of copper sulfate: Werner, **U.S.** pat. **2,332,418** (1943 to Winthrop); Urushibara, Koga, *Bull. Chem. Soc. Japan* **29,** 419 (1956); Berg, Fiedler, **U.S.** pat. **2,886,488** (1959 to Thomae). Formerly believed to be the 4-hydroxy isomer: Fiedler, *Ber.* **95,** 1771 (1962).

Crystals from water, mp 160°. Practically insol in water. Sol in ethanol, isopropanol, propylene glycol, ether, benzene, toluene. Hydrolyzed by alkali.
Combination with hydrocortisone, *Psoil.*
THERAP CAT: Antiseborrheic.

9389. Tipepidine. *3-(Di-2-thienylmethylene)-1-methylpiperidine;* 1-methyl-3-piperidylidenedi(2-thienyl)methane; tipedine; AT 327; CR/662. $C_{15}H_{17}NS_2$; mol wt 275.45. C 65.41%, H 6.22%, N 5.09%, S 23.28%. Prepn: Okumura *et al., Tanabe Seiyaku Kenkyu Nempo* **3,** 30 (1958), *C.A.* **53,** 10214 (1959); Ponomarev, Martem'yanova, **USSR** pat. **176,903** (1965), *C.A.* **64,** 12648b (1966). Synthesis: *eidem, Khim. Geterosikl. Soedin.* **1967,** 174, *C.A.* **67,** 73501g (1967). Prepn of deriv: **Brit.** pat. **924,544** (1963); Yamamoto, Yoshikawa, **Japan.** pat. **17,988('62)** (both to Tanabe Seiyaku), *C.A.* **59,** 11443f, 11446b (1963). Pharmacology and clinical efficacy: Higaki *et al., Tanabe Seiyaku Kenkyu Nempo* **4,** 35 (1959), *C.A.* **54,** 3725h (1960); Kase *et al., Chem. Pharm. Bull.* **7,** 372 (1959). Metabolism: Watanabe *et al., Yakugaku Zasshi* **89,** 29 (1969); Sasaki *et al., ibid.* 345.

Yellow crystals from petr ether, mp 64-65°. $bp_{4.5}$ 178-

184°. LD$_{50}$ in mice (mg/kg): 294 i.p.; 308 i.m.; 867 orally (Higaki).

Citrate monohydrate, C$_{21}$H$_{25}$NO$_7$S$_2$.H$_2$O, *bithiodine*. Yellow crystals, mp 138-139°. Sol in water, ethanol, propylene glycol.

Hibenzate, C$_{29}$H$_{27}$NO$_4$S$_2$, *Asverin, Sotal*. mp 187-190°.

THERAP CAT: Antitussive.

9390. Tiquizium Bromide. *trans-3-(Di-2-thienylmethylene)octahydro-5-methyl-2H-quinolizinium bromide;* 3-(di-2-thienylmethylene)-5-methyl-*trans*-quinolizidinium bromide; HSR-902; HS-902; Thiaton. C$_{19}$H$_{24}$BrNS$_2$; mol wt 410.43. C 55.60%, H 5.89%, Br 19.47%, N 3.41%, S 15.62%. Quaternary ammonium salt with anticholinergic activity. Prepn: H. Kato *et al.*, Belg. pat. 866,988 (1978 to Hokuriku*)*, *C.A.* 90, 151996p (1979); H. Kato, E. Koshinaka, U.S. pat. 4,205,074 (1980 to Hokuriku). Synthesis and anticholinergic activity: E. Koshinaka *et al.*, *Chem. Pharm. Bull.* 27, 1454 (1979). Pharmacological comparison with other antispasmodics: S. Kubo *et al.*, *Japan. J. Pharmacol.* 30, 103P (1980). Pharmacology: M. Yamazaki *et al.*, *Oyo Yakuri* 23, 417, 423 (1981), *C.A.* 97, 66016u, 66017v (1982). HPLC determn in biological fluids: T. Yamada *et al.*, *Yakugaku Zasshi* 103, 1319 (1983), *C.A.* 100, 131957y (1984).

Needles from methanol-acetone, mp 278-281° (dec).

THERAP CAT: Antispasmodic.

9391. Tiratricol. *[4-(4-Hydroxy-3-iodophenoxy)-3,5-diiodophenyl]acetic acid;* 3,3',5-triiodo-4-(4-hydroxyphenoxy)phenylacetic acid; 3,5-diiodo-4-(3-iodo-4-hydroxyphenoxy)phenylacetic acid; 3,3',5-triiodothyroacetic acid; Triac; Triacana. C$_{14}$H$_9$I$_3$O$_4$; mol wt 621.95. C 27.04%, H 1.46%, I 61.22%, O 10.29%. Prepd by iodination of 4-(4'-hydroxyphenoxy)-3,5-diiodophenylacetic acid: Wilkinson, *Biochem. J.* 63, 601 (1956); Hems, Brit. pat. 803,149 (1958 to Glaxo); Wilkinson, **Brit.** pat. 805,761 (1958 to Nat. Res. Dev. Corp.); Meltzer *et al.*, *J. Org. Chem.* 26, 1418 (1961).

Needles from methanol + water, mp 65°, resolidifies 110°, mp 180-183°.

THERAP CAT: Thyroid replacement therapy.

9392. Tiron®. *4,5-Dihydroxy-1,3-benzenedisulfonic acid disodium salt;* 1,2-dihydroxybenzene-3,5-disulfonic acid disodium salt; disodium-1,2-dihydroxybenzene-3,5-disulfonate; sodium catechol disulfonate; sodium pyrocatechol-2,4-disulfonate; disodium pyrocatechol-3,5-disulfonate; Tiferron. C$_6$H$_4$Na$_2$O$_8$S$_2$; mol wt 314.22. C 22.94%, H 1.28%, Na 14.63%, O 40.74%, S 20.41%. Prepd by heating 5-bromo-4-hydroxy-*m*-benzenedisulfonic acid with sodium hydroxide in water and acidifying product with hydrochloric acid: Fukayama *et al.*, Japan. pat. 4327('52) (to Sanwa Pure Chem.), *C.A.* 48, 5215c (1954).

Crystals, nonhygroscopic. Very freely sol in water. Produces water-sol, colored compds with metal salts. With ferric chloride the resulting complex yields a deep blue soln at pH below 5. Titanium salts give an orange color, copper salts produce a green-yellow, and hexavalent molybdenum a canary-yellow color.

USE: Colorimetric reagent for iron, manganese, titanium, molybdenum: Yoe, Jones, *Ind. Eng. Chem., Anal. Ed.* 16, 111 (1944). Zirconium complexes: Intorre, Martell, *J. Am. Chem. Soc.* 82, 358 (1960).

9393. Tiropramide. *α-(Benzoylamino)-4-[2-(diethylamino)ethoxy]-N,N-dipropylbenzenepropanamide;* DL-α-benzamido-*p*-[2-(diethylamino)ethoxy]-*N,N*-dipropylhydrocinnamamide; *O*-(2-diethylaminoethyl)-*N*-benzoyl-DL-tyrosyl-di-*n*-propylamide; CR 605; Maiorad. C$_{28}$H$_{41}$N$_3$O$_3$; mol wt 467.66. C 71.91%, H 8.84%, N 8.99%, O 10.26%. Antispasmodic deriv of tyrosine, q.v. Smooth muscle relaxant. Prepn: F. Makovec *et al.*, **Ger.** pat. 2,503,992; *eidem*, U.S. pat. 4,004,008 (1975, 1977 both to Rotta). Activity on gastrointestinal motility: P. Senin *et al.*, *Gastrointestinal Motility in Health and Disease*, H. L. Duthie, Ed. (University Park Press, Baltimore, 1978) pp 417-427. Mechanism of action: R. R. Vidal y Plana *et al.*, *J. Pharm. Pharmacol.* 33, 19 (1981).

Cryst from petr ether, mp 65-67°. LD$_{50}$ in rats: 33.9 mg/kg i.v. (Makovec).

Hydrochloride, C$_{28}$H$_{42}$ClN$_3$O$_3$, *Alfospas*.

THERAP CAT: Antispasmodic.

9394. Tissue Plasminogen Activator. Fibrinokinase; extrinsic plasminogen activator; t-PA; TPA; Actilyse. Mol wt ~70,000. Serine protease catalyzing the enzymatic conversion of plasminogen to plasmin, *q.q.v.*, through the hydrolysis of a single Arg-Val bond. Component of the mammalian fibrinolytic system responsible for the specific activation of plasminogen associated with fibrin clots. Distribution of t-PA in various organs differs widely and appears to be related to the degree of tissue vascularization. Identification of plasminogen-activating ability of tissue fragments: T. Astrup, P. M. Permin, *Nature* 159, 681 (1947); H. J. Tagnon, M. L. Petermann, *Proc. Soc. Exp. Biol. Med.* 70, 359 (1949). Extraction of soluble activator with potassium thiocyanate: T. Astrup, A. Stage, *Nature* 170, 929 (1952). Localization of plasminogen activator in vascular endothelial cells: A. J. Todd, *J. Pathol. Bacteriol.* 78, 281 (1959). Tissue distribution and physiological significance: T. Astrup, *Fed. Proc.* 25, 42 (1966). Purification: F. Bachmann *et al.*, *Biochemistry* 3, 1578 (1964); P. Kok, T. Astrup, *ibid.* 8, 79 (1969); D. C. Rijken *et al.*, *Biochim. Biophys. Acta* 580, 140 (1979). Immunological identity with plasminogen activators from vascular wall and circulating blood: *eidem*, *Thromb. Res.* 18, 815 (1980). Differentiation from urokinase: O. Matsuo *et al.*, *Thromb. Haemostasis* 45, 225 (1981). Purification and characterization of human melanoma t-PA: D. C. Rijken, D. Collen, *J. Biol. Chem.* 256, 7035 (1981). Cloning and expression of cDNA for human t-PA: D. Pennica *et al.*, *Nature* 301, 214 (1983). t-PA exists in 2 forms. The native, single chain form (t-PA I) can be broken by proteolytic cleavage into 2 polypeptides connected by a disulfide bond (t-PA II): P. Wallen *et al.*, *Biochim. Biophys. Acta* 719, 318 (1982). Conversion of t-PA I to t-PA II during fibrinolysis and comparison of biological activity: D. C. Rijken *et al.*, *J. Biol. Chem.* 257, 2920 (1982). Activity is markedly enhanced by fibrin, primarily as a result of specific binding of the activator to the fibrin clot: S. M. Camiolo *et al.*, *Proc. Soc. Exp. Biol. Med.* 138, 277 (1971); B. Wiman, D. Collen, *Nature* 272, 549 (1978). Fibrin-bound t-PA exhibits greater affinity for plasminogen than does non-bound activator, localizing fibrinolysis at the clot site: M. Hoylaerts et al., J. Biol. Chem. **257**, 2912 (1982). Experimental thrombolysis in rabbits: O. Matsuo et al., Nature **291**, 590 (1981); in humans: W. Weimar et al., Lancet **2**, 1018 (1981). Clinical study in coronary occlusion of myocardial infarction: The Thrombolysis in Myocardial Infarction (TIMI) Study Group, N. Engl. J. Med. **312**, 932 (1985); M. Verstraete et al., Lancet **1**, 842 (1985). Comparative clinical study with urokinase, q.v., in acute pulmonary embolism: S. Z. Goldhaber et al., Lancet **2**, 293 (1988). Review of the fibrinolytic system: D. Collen, Thromb. Haemostasis **43**, 77-89 (1980). Review of pharmacology and bioactivity: D. Collen, R. Lijnen, Arteriosclerosis **4**, 579-585 (1984); of pharmacology and clinical efficacy: S. D. Rogers et al., Pharmacotherapy **7**, 111-121 (1987).

Practically insol in solns of low ionic strength at neutral pH. Sol in 1.6M KSCN at neutral pH; 0.3M K-acetate at pH 4.2.

THERAP CAT: Thrombolytic.

9395. Titanic(IV) Acid. Orthotitanic acid; titanic hydroxide. $Ti(OH)_4 \cdot xH_2O$. The H_2O content depends upon the method and time of drying.

White, amorphous powder. Insol in water; when freshly pptd it is sol in dil HCl or H_2SO_4 and in boiling caustic alkali solns; after drying it is almost insol.

9396. Titanium. Ti; at. wt 47.88; at. no. 22; valence 2, 3, 4 (mostly tetravalent). Five natural isotopes (mass numbers): 48 (73.94%); 46 (7.93%); 47 (7.28%); 49 (5.51%); 50 (5.34%). Artificial isotopes: 43-45; 51. Ninth most abundant element in earth's crust; 0.63% by wt. Occurs as the oxide in the minerals rutile, ilmenite, perovskite, anatase, or octahedrite and brookite; other minerals include sphene or titanite ($CaTiSiO_5$) and benitoite ($BaTiSi_3O_9$). Discovered by Gregor in 1789; investigated and named by Klaproth in 1795; isolated by Berzelius in 1825. Prepn: de Boer, Ind. Eng. Chem. **19**, 1256 (1927); Fast, Z. Anorg. Chem. **241**, 42 (1939); Ehrlich in Handbook of Preparative Inorganic Chemistry vol. 2, G. Brauer, Ed. (Academic Press, New York, 2nd ed., 1965) pp 1161-1172. Reviews: Gmelin's, Titanium (8th ed.) **41** (1951); Everhart, Titanium and Titanium Alloys (Reinhold, New York, 1954); Brophy et al., Titanium Bibliography 1900-1951 + suppl (Washington, 1954); McQuillan & McQuillan, Titanium (Butterworth's, London, 1956); Barksdale, Titanium, Its Occurrence, Chemistry and Technology (Ronald Press, New York, 2nd ed, 1966); Clark, "Titanium" in Comprehensive Inorganic Chemistry vol. 3, J. C. Bailar, Jr. et al., Eds. (Pergamon Press, Oxford, 1973) pp 355-417.

Dark gray, lustrous metal. Dimorphic; α-form: hexagonal structure below 882.5°; β-form: body-centered cubic crystals above 882.5°. Brittle when cold; malleable when hot. Ductile only when free of oxygen; traces of oxygen or nitrogen increase strength. mp 1677°. bp 3277°. Calculated d (α-form): 4.506 (25°); (β-form): 4.400 (900°). Specific heat (25°): 5.98 cal/g-atom/°C. Decomposes steam at 700-800°. Combines with oxygen at a red heat. Can burn in an atm of oxygen under certain conditions: Chem. & Eng. News **36**, 36 (Aug. 4, 1958). Attacked by acids only on heating; oxidized by nitric acid to the dioxide. Reacts with fluorine at 150°, with chlorine at 300°, with bromine at 360°, with iodine above 360°. Combines with nitrogen at 800°. Forms alloys with aluminum, chromium, cobalt, copper, iron, lead, nickel, tin.

USE: As alloy with copper and iron in titanium bronze; as addition to steel to impart great tensile strength; to aluminum to impart resistance to attack by salt solns and by organic acids; to remove traces of oxygen and nitrogen from incandescent lamps. Surgical aid (fracture fixation).

9397. Titanium Dichloride. Cl_2Ti; mol wt 118.81. Ti 40.32%, Cl 59.68%. $TiCl_2$. Prepn: Clifton, McWood, J. Phys. Chem. **60**, 311 (1956); Ehrlich et al., Z. Anorg. Allgem. Chem. **292**, 139 (1957).

Black crystals. mp 1035°. d 3.13. Burns like tinder when heated in air; dec by water. Sol in alcohol; practically insol in chloroform, ether, carbon disulfide. Keep dry and well closed.

9398. Titanium Dioxide. Unitane. O_2Ti; mol wt 79.90. O 40.05%, Ti 59.95%. TiO_2. Found in nature as the minerals rutile (tetragonal), anatase or octahedrite (tetragonal), brookite (orthorhombic), ilmenite ($FeTiO_3$), and perovskite ($CaTiO_3$). May be prepd by direct combination of titanium and oxygen; by treatment of titanium salts in aq soln; by the reaction of volatile, inorganic titanium compds with oxygen; by oxidation or hydrolysis of organic compds of titanium. Industrial prepn from ilmenite or rutile: Faith, Keyes & Clark's Industrial Chemicals, F. A. Lowenheim, M. K. Moran, Eds. (Wiley-Interscience, New York, 4th ed., 1975) pp 814-821. Prepn of synthetic rutile: Merker, U.S. pat. **2,760,874** (1956 to National Lead). Prepn of spectroscopically pure material by dissolving titanium in an ammoniacal soln of 90% H_2O_2: Czanderna et al., J. Am. Chem. Soc. **79**, 5407 (1957).

White powder, mp 1855°. d (rutile): 4.23; (anatase): 3.90; (brookite): 4.13. Insol in water, HCl, HNO_3, or dil H_2SO_4; sol in hot concd H_2SO_4, in HF. The reactivity depends on a previous heat treatment; prolonged heating produces a less sol material. Also made sol by fusion with potassium bisulfate or with alkali hydroxides or carbonates to form alkali titanates. Possesses perhaps the greatest hiding power of all inorganic white pigments. Titania is a name applied to large TiO_2 crystals (translucent water-white or with yellowish cast) suitable for use in jewelry. These crystals have a refractive index (2.7) higher than diamonds (2.4), but lack the hardness of diamonds. When substantially pure, a massive single crystal (boule) of rutile has the properties of a precious gem with a very light straw color and with reflectance, refraction and brilliance measuring greater than those of a diamond.

Note: The rutile structure is common among metal fluorides and oxides of the type MF_2 and MO_2.

USE: Airfloated ilmenite is used for titanium pigment manuf. Rutile sand is suitable for welding-rod-coating materials, as ceramic colorant, as source of titanium metal. As color in the food industry. Anatase titanium dioxide is used for welding-rod-coatings, acid resistant vitreous enamels, in specification paints, exterior white house paints, acetate rayon, white interior air-dry and baked enamels and lacquers, inks and plastics, for paper filling and coating, in water paints, tanners' leather finishes, shoe whiteners, and ceramics. High opacity and tinting values are claimed for rutile-like pigments.

THERAP CAT: Protectant (topical).

9399. Titanium Hydride. H_2Ti; mol wt 49.92. Ti 95.95%, H 4.04%. TiH_2. Prepd by the reduction of titanium oxide with calcium hydride in the presence of hydrogen above 600°: Alexander, U.S. pat. **2,427,338** (1947).

Gray-black metallic powder. Stable in air. Dissociation appreciable at 450°.

USE: Additive in powder metallurgy; getter for oxygen and nitrogen in electronic tubes; wetting agent for ceramic to metal seals; source of pure hydrogen.

9400. Titanium Isopropylate. Isopropyl titanate(IV). $C_{12}H_{28}O_4Ti$; mol wt 284.26. C 50.70%, H 9.93%, O 22.51%, Ti 16.85%. $Ti[OCH(CH_3)_2]_4$. Prepn: Bradley et al., J. Chem. Soc. **1952**, 5020.

Liquid. Fumes in air. mp approx 20°: R. Feld, P. L. Cowe, The Organic Chemistry of Titanium (Butterworths, Washington, 1965) p 36. bp_{760} 220°; bp_{10} 104°. d_4^{20} 0.9711. Apparent viscosity 2.11 cp at 25°. Dielectric constant: 3.64 at 62 kilocycles. Dec rapidly in water. Sol in anhydr ethanol, ether, benzene, chloroform. Precautions must be taken in handling to exclude water from solvents and diluents.

USE: Polymerization catalyst.

9401. Titanium Sesquisulfate. Titanous sulfate. $O_{12}S_3Ti_2$; mol wt 384.00. O 50.00%, S 25.05%, Ti 24.95%. $Ti_2(SO_4)_3$. (Crystallizes also as a hydrate with $8H_2O$).

Green, cryst powder. Insol in water, alc, concd H_2SO_4; sol in dil HCl or dil H_2SO_4, giving violet-colored solns.

USE: Same as titanium trichloride.

9402. Titanium Sulfate. Titanyl sulfate; titanoxy sulfate. O_5STi; mol wt 159.97. O 50.01%, S 20.05%, Ti 29.94%. $TiOSO_4$. Prepn: Hayek, Engelbrecht, Monatsh. **80**, 640 (1949).

White or slightly yellow powder; dec by water.
Note: The name titanyl sulfate is sometimes applied to Ti(SO$_4$)$_2$.
USE: As mordant in dyeing.
THERAP CAT: Dermatologic.

9403. Titanium Tetrabromide. Br$_4$Ti; mol wt 367.56. Br 86.97%, Ti 13.03%. TiBr$_4$. Prepn: Olsen, Ryan, *J. Am. Chem. Soc.* **54**, 2215 (1932).
Amber-yellow or orange, very hygroscopic crystals. mp 28.25°. bp 233.45°. d^{20} 3.25. Also reported as 2.6: Duppa, *Proc. Roy. Soc. London* **8**, 42 (1857). Dec by water; sol in abs ether, abs alcohol. *Keep tightly closed.*

9404. Titanium Tetrachloride. Cl$_4$Ti; mol wt 189.73. Cl 74.75%, Ti 25.25%. TiCl$_4$. Purification: Baxter *et al., J. Am. Chem. Soc.* **45**, 1228 (1923); **48**, 3117 (1926). Survey of preparative methods: Ehrlich in *Handbook of Preparative Inorganic Chemistry*, **Vol. 2**, G. Brauer, Ed. (Academic Press, New York, 2nd ed., 1965) pp 195-199.
Colorless liquid; penetrating acid odor; absorbs moisture from the air and evolves dense white fumes. d 1.726. mp −24.1°. bp 136.4°. Sol in cold water, alcohol; dec by hot water. *Keep tightly closed.*
USE: Formerly used with potassium bitartrate as a mordant in textile industry, and with dyewoods in dyeing leather; also as smoke-producing screen with ammonia; manuf iridescent glass and artificial pearls. *Caution:* Irritant to eyes, respiratory tract.

9405. Titanium Tetrafluoride. F$_4$Ti; mol wt 123.90. F 61.34%, Ti 38.66%. TiF$_4$. Prepd by the reaction of hydrogen fluoride with titanium tetrachloride: Ruff, Ipsen, *Ber.* **36**, 1777 (1903); Ruff, Plato, *ibid.* **37**, 673 (1904); by the action of fluorine on the metal or the dioxide: Haendler *et al., J. Am. Chem. Soc.* **76**, 2177 (1954).
Powdery white mass. d$_4^{20}$ 2.798. Sublimes at 284°. mp >400°. Hisses on contact with water. Very hygroscopic; the dihydrate may be crystallized from aqueous soln (also obtained from solns of TiO$_2$ in HF). Aq solns hydrolyze slowly, and the existence of an oxyfluoride, TiOF$_2$, has been established. Also sol in alcohol and pyridine from which the compds TiF$_4$.2EtOH and TiF$_4$.C$_5$H$_5$N have been isolated. Insol in ether. Dry ammonia is absorbed at room temp to form TiF$_4$.4NH$_3$, but at 120° TiF$_4$.2NH$_3$ is the stable phase. This is sol in water and is sufficiently stable to sublime. TiF$_4$ is also sol in phosphorus oxychloride, but at 30° a reaction occurs and POF$_3$ is evolved. *See also* Eméleus in *Fluorine Chemistry* vol. I, J. H. Simons, Ed. (Academic Press, New York, 1950) p 47.

9406. Titanium Trichloride. Titanous chloride. Cl$_3$Ti; mol wt 154.27. Cl 68.95%, Ti 31.05%. TiCl$_3$. Prepn by reduction of the tetrachloride: Ingraham *et al., Inorg. Syn.* **6**, 52 (1960); Sherfey, *ibid.* **57**.
Dark red-violet, unstable, deliquesc crystals; dec on heating above 500°. The dry powder is pyrophoric in air. Very reactive and readily dissociated by moisture in air. Sol in water, exothermic process; sol in alcohol. Practically insol in ether. *Keep tightly closed.*
USE: A powerful reducing agent. Reduces nitrate to ammonia; when boiled with aq SO$_2$, sulfur is separated; hence is used as an aq soln for estimation of nitro groups, ferric ions, per-salts, etc. Removes stains, etc. (stripper) in laundering.

9407. Titanocene Dichloride. *Dichlorobis(η^5-2,4-cyclopentadien-1-yl)titanium; dichlorodi-π-cyclopentadienyltitanium;* biscyclopentadienyltitanium(IV) dichloride. C$_{10}$H$_{10}$Cl$_2$Ti; mol wt 248.99. C 48.24%, H 4.05%, Cl 28.47%, Ti 19.24%. (C$_5$H$_5$)$_2$TiCl$_2$. Also Cp$_2$TiCl$_2$. Prepn: Wilkinson, Birmingham, *J. Am. Chem. Soc.* **76**, 4281 (1954). Structure: Clearfield *et al., Can. J. Chem.* **53**, 1622 (1975).
Bright red acicular crystals from toluene. mp 289 ± 2°. d 1.60. Moderately sol in toluene, chloroform, alcohol, other hydroxylic solvents; sparingly sol in water, petr ether, benzene, ether, carbon disulfide, carbon tetrachloride.
USE: Catalyst; with aluminum alkyls as Ziegler-Natta polymerization catalyst.

9408. Tixocortol. *(11β)-11,17-Dihydroxy-21-mercaptopregn-4-ene-3,20-dione;* 11β,17α-dihydroxy-21-thio-3,20-dioxo-4-pregnene. C$_{21}$H$_{30}$O$_4$S; mol wt 378.53. C 66.63%, H

7.99%, O 16.91%, S 8.47%. Synthesis and biological activity: S. S. Simons *et al., J. Steroid Biochem.* **13**, 311 (1980). Use in fluorescent chemoaffinity labeling: *eidem, Biochemistry* **18**, 4915 (1979). Prepn of the 21-pivalate: D. R. Torossian *et al.,* Ger. pat. **2,357,778**; *eidem,* U.S. pat. **4,014,909** (1974, 1977 both to Jouveinal). Series of articles on pharmacology, *in vitro* and *in vivo* activity of the pivalate: *Arzneimittel-Forsch.* **31**, 453-469 (1981).

Fine white solid, dec 220-221°. uv max (95% ethanol): 241 nm (ε 1.65 × 10^4).
21-Pivalate, C$_{26}$H$_{38}$O$_5$S, *(11β)-21-[(2,2-dimethyl-1-oxopropyl)thio]-11,17-dihydroxypregn-4-ene-3,20-dione, JO 1016, Pivalone, Procolon, Rectovalone, Tivalon-NT, Tiprederm.* Crystals from ethanol, mp 195-200°. [α]$_D^{20}$ +145° (c = 1 in dioxane). uv max (methanol): 229 nm (log ε 4.259).
USE: Free thiol in chemoaffinity labeling.
THERAP CAT: Anti-inflammatory.

9409. Tizanidine. *5-Chloro-N-(4,5-dihydro-1H-imidazol-2-yl)-2,1,3-benzothiadiazol-4-amine;* 5-chloro-4-(2-imidazolin-2-ylamino)-2,1,3-benzothiadiazole; DS 103-282; Sirdalud; Ternelin. C$_9$H$_8$ClN$_5$S; mol wt 253.70. C 42.61%, H 3.18%, Cl 13.97%, N 27.60%, S 12.64%. α$_2$-Adrenergic agonist; centrally active myotonolytic. Prepn and antitremor activity: P. Neumann, Neth. pat. **Appl. 7,306,228**; *eidem,* U.S. pat. **3,843,668** (1973, 1974 both to Wander-Sandoz). Pharmacology: A. C. Sayers *et al., Arzneimittel-Forsch.* **30**, 793 (1980); and mechanism of action studies: L. Turski *et al., Brain Res.* **379**, 367 (1986); H. Ono *et al., Gen. Pharmacol.* **17**, 137 (1986). Pharmacokinetics: V. Heazlewood *et al., Eur. J. Clin. Pharmacol.* **25**, 65 (1983). Clinical evaluation in spasticity of multiple sclerosis: P. M. Newman *et al., Eur. J. Clin. Pharmacol.* **23**, 31 (1982); M. C. Hoogstraten *et al., Acta Neurol. Scand.* **77**, 224 (1988).

Crystals from methanol, mp 221-223°. LD$_{50}$ orally in mice: 235 mg/kg (Sayers).
THERAP CAT: Muscle relaxant (skeletal).

9410. TMD. *Decahydro-1,1,4a-trimethyl-2-naphthalenol;* 1,1,10-trimethyl-*trans*-2-decalol; 4,4,10β-trimethyl-*trans*-decal-3β-ol. C$_{13}$H$_{24}$O; mol wt 196.33. C 79.53%, H 12.32%:, O 8.15%. Prepn: B. Gaspert *et al., J. Chem. Soc.* **1958**, 624; C. Djerassi, D. Marshall, *J. Am. Chem. Soc.* **80**, 3986 (1958); T. G. Halsall *et al., J. Chem. Soc.* **1959**, 2798; S. L. Mukherjee, P. C. Dutta, *ibid.* **1960**, 67. Inhibitor of cholesterol biosynthesis: J. A. Nelson *et al., J. Am. Chem. Soc.* **100**, 4900 (1978); T.-Y. Chang *et al., J. Biol. Chem.* **254**, 11258 (1979).

Crystals, mp 60-65° (subl). Sol in alcohol; slightly sol in water.

d-Form, crystals, mp 87-88° (subl). $[\alpha]_D^{26}$ +15.9° (CHCl$_3$).
l-Form, crystals, mp 87-88° (subl). $[\alpha]_D^{26}$ −16.9° (CHCl$_3$).

9411. TNF. Tumor necrosis factor. Cytokine produced by activated macrophages as part of the cellular immune response. Originally characterized by its selective hemorrhagic necrosis of tumor cells. Identification in the sera of endotoxin treated mice previously sensitized with bacillus Calmette-Guerin (BCG): E. A. Carswell *et al., Proc. Nat. Acad. Sci. USA* **72**, 3666 (1975). Partial purification: S. Green *et al., ibid.* **73**, 381 (1976). Preliminary characterization of murine TNF: T. Haranaka, N. Satomi, *Japan. J. Exp. Med.* **51**, 191 (1981); F. C. Kull, P. Cuatrecasas, *J. Immunol.* **126**, 1279 (1981); of rabbit TNF: M. R. Ruff, G. E. Gifford, *ibid.* **125**, 1671 (1980); N. Matthews *et al., Brit. J. Cancer* **42**, 416 (1980). Identification of macrophages as cellular source of TNF: D. N. Männel *et al., Infect. Immun.* **30**, 523 (1980); N. Santomi *et al., Japan. J. Exp. Med.* **51**, 317 (1981). Differentiation from interferon: N. Bloksma *et al., Cancer Immunol. Immunother.* **14**, 41 (1982). Cytotoxic effect on the malaria parasite *Plasmodium falciparum:* C. G. Haidaris *et al., Infect. Immun.* **42**, 385 (1983); A. O. Wozencraft *et al., ibid.* **43**, 664 (1984). Activity against transplanted human and murine tumors in mice: T. Haranaka *et al., Int. J. Cancer* **34**, 263 (1984). Human TNF is a trimer of 3 identical subunits with 157 amino acid residues and mol wt 17,350 Da. Cloning and expression of cDNA for human TNF in *E. coli:* D. Pennica *et al., Nature* **312**, 724 (1984); T. Shirai *et al., ibid.* **313**, 803 (1985); A. M. Wang *et al., Science* **228**, 149 (1985). Identity with *cachectin:* B. Buetler, A. Cerami, *Nature* **320**, 584 (1986). Structure: E. Y. Jones *et al., ibid.* **338**, 225 (1989). Preliminary evaluation with interferon-γ in metastatic melanoma: S. Retsas *et al., Brit. Med. J.* **298**, 1290 (1989). *Reviews:* M. R. Ruff, G. E. Gifford, *Lymphokines* **2**, 235 (1981); L. J. Old, *Sci. Am.* **258**(5), 59-60, 70-75 (1988).

Note: **Lymphotoxin**, a TNF-like factor produced by lymphocytes has been referred to as *TNF-β.*

9412. Tobacco Mosaic Virus. TMV. Most thoroughly investigated and best known virus. A ribonucleoprotein contg 95% protein and 5% ribonucleic acid. Molecular weight is 39.0×10^6. TMV is a rod-shaped particle, about 3000 Å long, consisting of 2130 polypeptide chains arranged in some regular order to form a tubular molecule containing a core of RNA. The polypeptide chains consist of 158 amino acid residues each; their mol wt is 17,533. The RNA is responsible for the infectivity of the virus. The protein functions to protect the RNA and determine the macroscopic properties of the virus such as morphology and serological specificity. Isoln from diseased tobacco plants: Stanley, *Science* **81**, 644 (1935). Prepn and characterization of essentially uniform TMV particles: Boedtker, Simmons, *J. Am. Chem. Soc.* **80**, 2550 (1958); Knight, *Biochem. Prepns.* **9**, 132 (1962). Sequential arrangement of the 158 amino acid residues of the protein subunit: Tsugita *et al., Proc. Nat. Acad. Sci. USA* **46**, 1463 (1960). Synthesis of TMV-RNA by cell free extracts from infected tobacco leaves: Kim, Wildman, *Biochem. Biophys. Res. Commun.* **8**, 394 (1962); Cochran, *Chem. & Eng. News* **40**, 64 (Sept. 17, 1962). *Reviews:* Anderer, *Advan. Protein Chem.* **18**, 1 (1963); Caspar, *ibid.* 37; Klug, Caspar, *Advan. Virus Res.* **7**, 233-277 (1960); Lauffer, Stevens, *ibid.* **13**, 1 (1968); Reddi, *ibid.* **17**, 51 (1972); L. Hirth, K. E. Richards, *ibid.* **26**, 145-199 (1981).

USE: Popular tool for studying the correlation between chemical structure and biological function.

9413. Tobramycin. *O-3-Amino-3-deoxy-α-D-glucopyranosyl-(1→6)-O-[2,6-diamino-2,3,6-trideoxy-α-D-ribohexopyranosyl-(1→4)]-2-deoxy-D-streptamine;* 4-[2,6-diami-

no-2,3,6-trideoxy-α-D-glycopyranosyl]-6-[3-amino-3-deoxy-α-D-glycopyranosyl]-2-deoxystreptamine; nebramycin factor 6; NF 6; Distobram; Gernebcin; Obramycin; Tobradistin; Tobralex; Tobramaxin; Tobrex. C$_{18}$H$_{37}$N$_5$O$_9$; mol wt 467.54. C 46.24%, H 7.98%, N 14.98%, O 30.80%. Single factor antibiotic comprising about 10% of *nebramycin,* (formerly called *tenebrimycin, tenemycin*), the aminoglycosidic antibiotic complex produced by *Streptomyces tenebrarius. See also* apramycin. Series of articles on isolation, separation and evaluation of the nebramycin complex: Stark *et al.,* Higgens, Kastner; Thompson, Presti; Wick, Welles, *Antimicrob. Ag. Chemother.* **1967**, 314-348. *Review:* Stark *et al., Folia Microbiol. (Prague)* **16**, 205-217 (1971). Tobramycin is structurally similar to the kanamycins and the gentamicins. Elucidation of structure: Koch, Rhoades, *Antimicrob. Ag. Chemother.* **1970**, 309. Synthesis: Y. Takagi *et al., Bull. Chem. Soc. Japan* **49**, 3649 (1976); M. Tanabe *et al., Tetrahedron Letters* **1977**, 3607. Activity studies: Black, Griffith; Preston, Wick, *ibid.* **314**, 322; Meyers, Hirschman, *J. Clin. Pharmacol.* **12**, 313, 321 (1972). Toxicology: Welles *et al., Toxicol. Appl. Pharmacol.* **22**, 332 (1972); De Rosa *et al., J. Int. Med. Res.* **2**, 100 (1974).

Basic, water-sol substance. $[\alpha]_D$ 128°. LD$_{50}$ in mice, rats (mg/kg): 441, 969 s.c. (Welles).

Sulfate, *Nebcin, Nebicina, Obracine, Tobra, Tobracin.*

THERAP CAT: Antibacterial.

9414. Tocainide. *2-Amino-N-(2,6-dimethylphenyl)propanamide;* 2-aminopropiono-2',6'-xylidide. C$_{11}$H$_{16}$N$_2$O; mol wt 192.26. C 68.72%, H 8.39%, N 14.57%, O 8.32%. Anti-arrhythmic agent related to lidocaine, *q.v.* Prepn: R. N. Boys *et al.,* **Ger.** pat. **2,235,745** (1973 to Astra), *C.A.* **78**, 140411 (1973); and resolution of isomers: E. W. Byrnes *et al., J. Med. Chem.* **22**, 1171 (1979). *In vitro* effects on muscle contractions: R. Dengler, R. Rüdel, *Arzneimittel-Forsch.* **29**, 270 (1979). Inhibition of leucocyte locomotion in dogs: G. J. Stewart *et al., Lab. Invest.* **42**, 302 (1980). Biotransformation in humans: A. T. Elvin *et al., J. Pharm. Sci.* **69**, 47 (1980). Kinetics: C. Groffner, *Clin. Pharmacol. Ther.* **27**, 64 (1980). Use in refractory ventricular arrhythmias: D. M. Roden *et al., Am. Heart J.* **100**, 15 (1980). HPLC resolution of enantiomers: K. M. McErlane, G. K. Pillai, *J. Chromatog.* **274**, 129 (1983); determn in plasma: A. J. Sedman, J. Gal, *J. Chromatog.* **306**, 155 (1984). Pharmacokinetics of enantiomers: A. H. Thomson *et al., Brit. J. Pharmacol.* **21**, 149 (1986). *In vitro* and *in vivo* pharmacodynamics of enantiomers: A. J. Block *et al., J. Cardiovasc. Pharmacol.* **11**, 216 (1988). Review of pharmacology and therapeutic efficacy: B. Holmes *et al., Drugs* **26**, 93-123 (1983); D. M. Roden, R. L. Woosley, *N. Engl. J. Med.* **315**, 41-45 (1986).

DL-Form hydrochloride, C$_{11}$H$_{17}$ClN$_2$O, *W-36095, Taquidil, Tonocard, Xylotocan.* Crystals from ethanol/ether, mp 246-247°.

D-Form hydrochloride, crystals from ethanol-diethyl ether, mp 265-266°. $[\alpha]_D$ −42.16° (c = 2.63 in methanol).

L-Form hydrochloride, crystals from ethanol-diethyl ether, mp 264.5°. $[\alpha]_D$ +42.35° (c = 2.63 in methanol).

THERAP CAT: Cardiac depressant (anti-arrhythmic).

9415. Tocamphyl. *1,2,2-Trimethyl-1,3-cyclopentanedicarboxylic acid 1-[1-(4-methylphenyl)ethyl] ester, compd with 2,2'-iminobis[ethanol] (1:1); camphoric acid 1-(p,α-dimethylbenzyl) ester, compd with 2,2'-imidodiethanol (1:1); p,α-dimethylbenzyl camphorate diethanolamine salt; p-toluoylmethylcarbinolmono-d-camphoric acid ester, diethanolamine salt; methyl p-tolylcarbinol camphorate, diethanolamine salt; p-tolylmethylcarbinol camphoric acid ester diethanolamine salt; diethanolamine p-tolylmethylcarbinol camphorate; diethanolamine d-methyltoluylcarbinol camphorate; Licarbin; Bilagen; Hepasynthyl; Hepatoxane; Synthobilin; Biliphorine; Syncuma; Lymethol.* $C_{23}H_{37}NO_6$; mol wt 423.53. C 65.22%, H 8.81%, N 3.31%, O 22.67%. Prepn: *Swiss.* pat. 211,203 (1940 to Chemiewerk Homburg); *Chem. Zentr.* **1941,** I, 2972.

Crystals, sol in water.
THERAP CAT: Choleretic.

9416. Tocol. *3,4-Dihydro-2-methyl-2-(4,8,12-trimethyltridecyl)-2H-1-benzopyran-6-ol; 2-methyl-2-(4,8,12-trimethyltridecyl)-6-chromanol; 2-methyl-2-phytyl-6-chromanol; 6-hydroxy-2-methyl-2-phytylchroman; 2-methyl-2-phytyl-6-hydroxychroman.* $C_{26}H_{44}O_2$; mol wt 388.61. C 80.35%, H 11.41%, O 8.23%. Synthesis by the condensation of hydroquinone and phytol in the presence of anhydr formic acid: Pendse, Karrer, *Helv. Chim. Acta* **40,** 1837 (1957). Antioxidant activity of tocol and its methyl derivs: Olcott, van der Veen, *Lipids* **3,** 331 (1968).

Colorless, viscous oil. bp$_{0.001}$ 165-175°.
Acetate, $C_{28}H_{46}O_3$, viscous oil. bp$_{0.001}$ 180-185°.
USE: Antioxidant.

9417. β-Tocopherol. *3,4-Dihydro-2,5,8-trimethyl-2-(4,8,12-trimethyltridecyl)-2H-1-benzopyran-6-ol; 2,5,8-trimethyl-2-(4,8,12-trimethyltridecyl)-6-chromanol; 5,8-dimethyltocol; cumotocopherol; neotocopherol; p-xylotocopherol.* $C_{28}H_{48}O_2$; mol wt 416.66. C 80.71%, H 11.61%, O 7.68%. Accompanies vitamin E (α-tocopherol) and γ-tocopherol from natural sources. Is biologically less active than α-tocopherol. May be separated by fractional crystn of the allophanates: Emerson *et al., Science* **83,** 421 (1936); *J. Biol. Chem.* **113,** 319 (1936); Baxter *et al., J. Am. Chem. Soc.* **65,** 918 (1943).

Pale yellow, viscous oil. bp$_{0.1}$ 200-210°. $[\alpha]_D^{20}$ +6.37°. uv max: 297 nm (E$_{1cm}^{1\%}$ 86.4). Insol in water. Freely sol in oils, fats, acetone, alcohol, chloroform, ether, other fat solvents. Very stable to heat and alkalies. Slowly oxidized by atmospheric oxygen, rapidly by ferric and silver salts. Gradually darkens on exposure to light. Not pptd by digitonin.
Allophanate, $C_{30}H_{50}N_2O_4$, crystals, mp 138-139°. $[\alpha]_D^{18}$ +6.7° (chloroform).
Acetate, $C_{30}H_{50}O_3$, pale yellow, oily liquid. bp$_{0.3}$ 215-220°. Insol in water. Freely sol in acetone, chloroform, ether. Less readily sol in alcohol.
THERAP CAT: *See* Vitamin E.

THERAP CAT (VET): In "stiff lamb disease".

9418. γ-Tocopherol. *3,4-Dihydro-2,7,8-trimethyl-2-(4,8,12-trimethyltridecyl)-2H-1-benzopyran-6-ol; 2,7,8-trimethyl-2-(4,8,12-trimethyltridecyl)-6-chromanol; 7,8-dimethyltocol; o-xylotocopherol.* $C_{28}H_{48}O_2$; mol wt 416.66. C 80.71%, H 11.61%, O 7.68%. Accompanies vitamin E (α-tocopherol) and β-tocopherol from natural sources. Is biologically less active than α-tocopherol. May be separated by fractional crystn of the allophanates: Emerson *et al., Science* **83,** 421 (1936); *J. Biol. Chem.* **113,** 319 (1936); Baxter *et al., J. Am. Chem. Soc.* **65,** 918 (1943).

Pale yellow, viscous oil. Has been crystallized at −30°. bp$_{0.1}$ 200-210°. $[\alpha]_D^{20}$ −2.4° (alc). uv max: 298 nm (E$_{1cm}^{1\%}$ 92.8). Insol in water. Freely sol in oils, fats, acetone, alcohol, chloroform, ether, other fat solvents. Very stable to heat and alkalies. Slowly oxidized by atmospheric oxygen, rapidly by ferric and silver salts. Gradually darkens on exposure to light. Not pptd by digitonin.
Allophanate, $C_{30}H_{50}N_2O_4$, crystals, mp 136-138°. $[\alpha]_D^{18}$ +3.4° (chloroform).
Acetate, $C_{30}H_{50}O_3$, pale yellow, oily liquid. bp$_{0.3}$ 215-220°. Insol in water. Freely sol in acetone, chloroform, ether. Less readily sol in alcohol.
THERAP CAT: *See* Vitamin E.
THERAP CAT (VET): *See* Vitamin E.

9419. δ-Tocopherol. *3,4-Dihydro-2,8-dimethyl-2-(4,8,12-trimethyltridecyl)-2H-1-benzopyran-6-ol; 8-methyltocol.* $C_{27}H_{46}O_2$; mol wt 402.64. C 80.54%, H 11.52%, O 7.95%. Appears to be a rather common member of the vitamin E complex. Ingredient of Mixed Tocopherols Concentrate, N.F. It was found to constitute approx 30% of the mixed tocopherols in soybean oil, 5% of those in wheat germ oil, and there is evidence of its occurrence in cottonseed and peanut oils. Claimed to be the most potent antioxidant of the tocopherols. It has only one hundredth of the activity of natural α-tocopherol in the Evans resorption sterility test for vitamin E. Isoln from soybean oil: Stern *et al., J. Am. Chem. Soc.* **69,** 869 (1947). Synthesis: Green *et al., J. Chem. Soc.* **1959,** 3374; Brit. pat. 900,085 (1961 to Hoffmann-La Roche).

Pale yellow, viscous oil. $[\alpha]_{546}^{25}$ +3.4° (c = 15.5 in alc); $[\alpha]_{546}^{25}$ +1.1° (c = 10.9 in benzene). uv max: 298 nm (E$_{1cm}^{1\%}$ 91.2). Solubilities similar to γ-tocopherol, q.v.
THERAP CAT: *See* Vitamin E.

9420. ε-Tocopherol. *3,4-Dihydro-2,5,8-trimethyl-2-(4,8,12-trimethyl-3,7,11-tridecatrienyl)-2H-1-benzopyran-6-ol; 2,5,8-trimethyl-2-(4,8,12-trimethyltrideca-3,7,11-trienyl)-chroman-6-ol; 5-methyltocol.* $C_{28}H_{42}O_2$; mol wt 410.62. C 81.90%, H 10.31%, O 7.79%. Isoln from wheat germ oil and from bran: Eggitt, Ward, *J. Sci. Food Agr.* **4,** 569 (1953); Eggitt, Norris, *ibid.* **6,** 689 (1955); **7,** 496 (1956). Green *et al., J. Chem. Soc.* **1959,** 3362; *Chem. & Ind. (London)* **1960,** 73; McHale *et al., J. Chem. Soc.* **1963,** 784. Synthesis of all-trans-ε-tocopherol: Schudel *et al., Helv. Chim. Acta* **46,** 2517 (1963).

Pale yellow oil. uv max (ethanol): 296 nm ($E_{1cm}^{1\%}$ 87).
4-Phenylazobenzoate, $C_{41}H_{50}N_2O_3$, orange crystals from isopropanol, mp 70-71°.

9421. ζ₁-Tocopherol. *3,4-Dihydro-2,5,7,8-tetramethyl-2-(4,8,12-trimethyl-3,7,11-tridecatrienyl)-2H-1-benzopyran-6-ol; 2,5,7,8-tetramethyl-2-(4,8,12-trimethyl-3,7,11-tridecatrienyl)-6-chromanol; 5,7,8-trimethyltocotrien-3',7',11'-ol.* $C_{29}H_{44}O_2$; mol wt 424.64. C 82.02%, H 10.44%, O 7.54%. Isoln from wheat bran: Green et al., *J. Sci. Food Agr.* **6**, 274 (1955); Green et al., *Chem. & Ind. (London)* **1960**, 73. Structure: Green et al., *J. Chem. Soc.* **1959**, 3362. Attempt at synthesis: McHale et al., *ibid.* **1963**, 784. *Review:* M. Kofler et al., "Physiochemical Properties and Assay of the Tocopherols" in R. S. Harris, I. G. Wood, *Vitam. Horm. (New York)* **20**, 407-439 (1962). Synthesis of all-*trans*-ζ₁-tocopherol: Schudel et al., *Helv. Chim. Acta* **46**, 2517 (1963).

uv max (ethanol): 292.5 nm ($E_{1cm}^{1\%}$ 91).

9422. ζ₂-Tocopherol. *3,4-Dihydro-2,5,7-trimethyl-2-(4,8,12-trimethyltridecyl)-2H-1-benzopyran-6-ol; 2,5,7-trimethyl-2-(4,8,12-trimethyltridecyl)-6-chromanol; 5,7-dimethyltocol.* $C_{28}H_{48}O_2$; mol wt 416.66. C 80.71%, H 11.61%, O 7.68%. Isoln from rice: Green, Marcinkiewicz, *Nature* **177**, 86 (1956). Structure: Green et al., *J. Chem. Soc.* **1959**, 3362. Synthesis: Karrer, Fritsche, *Helv. Chim. Acta* **21**, 1234 (1938); Bergel et al., *J. Chem. Soc.* **1938**, 1382; McHale et al., *ibid.* **1958**, 1600.

Pale yellow oil. Crystallizes in prisms at −4°. uv max (ethanol): 292 nm ($E_{1cm}^{1\%}$ 83.0).
4-Phenylazobenzoate, $C_{41}H_{56}N_2O_3$, crystals from 2-propanol, mp 61°.
3',5'-Dinitrophenylurethane, $C_{35}H_{51}N_3O_7$, plates from 2-propanol + water, mp 65°.

9423. η-Tocopherol. *3,4-Dihydro-2,7-dimethyl-2-(4,8,12-trimethyltridecyl)-2H-1-benzopyran-6-ol; 2,7-dimethyl-2-(4,8,12-trimethyltridecyl)-6-chromanol; 7-methyltocol.* $C_{27}H_{46}O_2$; mol wt 402.64. C 80.54%, H 11.52%, O 7.95%. Isolation from rice: Green, Marcinkiewicz, *Nature* **177**, 86 (1956). Synthesis: McHale et al., *J. Chem. Soc.* **1958**, 1600; Green et al., *ibid.* **1959**, 3374; Marcinkiewicz et al., *ibid.* 3377.

Pale yellow oil. uv max (ethanol): 298 nm ($E_{1cm}^{1\%}$ 104.1).
4-Phenylazobenzoate, $C_{40}H_{54}N_2O_3$, crystals from 2-propanol, mp 55-56°.
3',5'-Dinitrophenylurethane, $C_{34}H_{49}N_3O_7$, crystals from ethanol + water, mp 115-117°.
p-Nitrophenylurethane, $C_{34}H_{50}N_2O_5$, crystals from methanol, mp 112°.

9424. α-Tocopherol Acid Succinate. Vitamin E acid succinate. $C_{33}H_{54}O_5$; mol wt 530.76. C 74.67%, H 10.26%, O 15.07%. Prepd by treating α-tocopherol with succinic anhydride in pyridine: Demole et al., *Helv. Chim. Acta* **22**, 65

(1939); McArthur, Watson, *Can. Chem. Process Inds.* **23**, 350 (1939); Baxter et al., *J. Am. Chem. Soc.* **65**, 918 (1943); Cawley, Stern, U.S. pat. **2,680,749** (1954 to Eastman Kodak).

Needles from petr ether, mp 76-77°. uv max (ethanol): 286 nm ($E_{1cm}^{1\%}$ 38.5). Practically insol in water.
Sodium salt, $C_{33}H_{53}NaO_5$, forms an 0.2% aq soln, pH 3.5.

9425. Todralazine. *2-(1-Phthalazinyl)hydrazinecarboxylic acid ethyl ester; 3-(1-phthalazinyl)carbazic acid ethyl ester; N^1-carbethoxy-N^2-phthalazinehydrazine; carboethoxyphthalazinohydrazine; ecarazine.* $C_{11}H_{12}N_4O_2$; mol wt 232.25. C 56.89%, H 5.21%, N 24.12%, O 13.78%. Prepn: S. Biniecki et al., *Bull. Acad. Polon. Sci. Ser. Sci.* **6**, 227 (1958), *C.A.* **52**, 18424g (1958); *eidem,* Belg. pat. **647,722** corresp to U.S. pat. **3,591,588** (1964, 1971 both to Polfa). Spectrofluorometric determn in plasma: A. Ishii, T. Deguchi, *Chem. Pharm. Bull.* **26**, 2241 (1978). Pharmacological study: M. Filczewski, E. Boguka, *Pol. J. Pharmacol. Pharm.* **31**, 127 (1979), *C.A.* **91**, 204508 (1979). Absorption, distribution, excretion in rats and humans: A. Ishii et al., *Oyo Yakuri* **18**, 61 (1979), *C.A.* **92**, 104012 (1980). Clinical study: W. Reiterer, H. Czitober, *Arzneimittel-Forsch.* **27**, 2163 (1977).

Hydrochloride, $C_{11}H_{13}ClN_4O_2$, **BT 621**, **CEPH**, **Apiracohl**, **Aperdor**, **Apride**, **Atapren**, **Binazine**, **Illcut**, **Propat**. LD_{50} in mice: 500 mg/kg i.p., F. Parravincini et al., *Farmaco Ed. Sci.* **34**, 299 (1979).
THERAP CAT: Antihypertensive.

9426. Tofenacin. *N-Methyl-2-[(2-methylphenyl)phenylmethoxy]ethanamine; N-methyl-2-[(o-methyl-α-phenylbenzyl)oxy]ethanamine; N-demethylorphenadrine; N-methylaminoethyl 2-methylbenzhydryl ether; phenyl-(o-tolylmethyl) methylaminoethyl ether; [2-(phenyl-o-tolylmethoxy)ethyl]methylamine; N-methyl-2-[α-(2-tolylbenzyl)oxy]ethylamine.* $C_{17}H_{21}NO$; mol wt 255.37. C 79.96%, H 8.29%, N 5.49%, O 6.26%. Tofenacin is a metabolite of orphenadrine, *q.v.,* formed by a process of *N*-demethylation. Prepn: Belg. pat. **628,167** corresp to Harms, U.S. pat. **3,407,258** (1963, 1968, both to Brocades-Stheeman & Pharmacia). Resolution of isomers: van der Stelt et al., *Arzneimittel-Forsch.* **19**, 2010 (1969). Physical and chemical properties: Doorenbos et al., *Pharm. Weekblad* **101**, 525 (1966). Pharmacology: Funcke et al., *Arch. Int. Pharmacodyn.* **177**, 28 (1969); Den Besten et al., *Arzneimittel-Forsch.* **20**, 538 (1970). Metabolic studies: Hespe, Prins, *Eur. J. Pharmacol.* **8**, 119 (1969); Hespe, Kafoe, *ibid.* **13**, 113 (1970). Clinical studies: Bram, Shanmuganathan, *Curr. Ther. Res.* **13**, 625 (1971).

Liquid, $bp_{0.7}$ 139-143°.
Hydrochloride, $C_{17}H_{22}ClNO$, **BS 7331**, **Elamol**, **Tofacine**. White to off-white crystals, mp 143-147° (Doorenbos); 147-148° (Harms). Freely sol in water, methanol, ethanol,

chloroform. Slightly sol in acetone; practically insol in ethyl ether. LD$_{50}$ in mice (mg/kg): 182 orally, 82 s.c., 58 i.p., 36.5 i.v.; in rats: 72 i.p.; in guinea pigs: 92 s.c., Funcke *et al., loc. cit.*

THERAP CAT: Antidepressant.

9427. Tofisopam. *1-(3,4-Dimethoxyphenyl)-5-ethyl-7,8-dimethoxy-4-methyl-5H-2,3-benzodiazepine;* EGYT 341; Grandaxin; Seriel. C$_{22}$H$_{26}$N$_2$O$_4$; mol wt 382.46. C 69.09%, H 6.85%, N 7.32%, O 16.73%. The first 5H-2,3-benzodiazepine. Prepn: J. Korosi *et al., Hung.* pat. 155,572 (1969 to Pharm. Res. Inst.), *C.A.* 70, 115026a (1969); **Brit.** pat. 1,202,579 corresp to U.S. pat. 3,736,315 (1970, 1973 both to EGYT); J. Korosi, T. Lang, *Ber.* 107, 3883 (1974). Synthesis and conformation: *eidem, Ther. Hung.* 23, 132 (1975). FT ^{13}C NMR study: A. Neszmelyi *et al., Ber.* 107, 3894 (1974). Pharmacology: L. Petocz, I. Kosoczky, *Ther. Hung.* 23, 134 (1975). Human pharmacokinetics: S. Ronai *et al., ibid.* 139. Comparative efficacy: H. L. Goldberg, R. J. Finnerty, *Am. J. Psychiatry* 136, 196 (1979).

Colorless to light cream cryst powder from isopropyl alcohol, mp 156-157°. uv max (methanol): 310, 272, 239 nm (ε 16100, 11200, 26300).

THERAP CAT: Anxiolytic.

9428. Tolan. *1,1'-(1,2-Ethynediyl)bisbenzene; diphenylacetylene;* diphenylethyne. C$_{14}$H$_{10}$; mol wt 178.22. C 94.34%, H 5.66%. C$_6$H$_5$C≡CC$_6$H$_5$. Prepd by the oxidation of benzil dihydrazone with mercuric oxide: Schlenk, Bergmann, *Ann.* 463, 76 (1928); by dehydrohalogenation of stilbene dibromide: Söderbäck, *Ann.* 443, 161 (1925); Smith, Hoehn, *J. Am. Chem. Soc.* 63, 1180 (1941); Smith, Falkof, *Org. Syn. coll. vol. III,* 350 (1955). Improved procedure: L. F. Fieser, *Experiments in Organic Chemistry* (Boston, 3rd ed., 1955) p 181.

Monoclinic, pseudo-rhombic rods or large spears from 95% ethanol. mp 60-61° (also reported as 62.5°). bp$_{760}$ 300°; bp$_{19}$ 170°. Dipole moment 0.3. Specific heat at 20°: 0.297. uv max: 216, 221, 269, 272, 279, 288, 297 nm (ε 20,600, 20,300, 23,450, 25,200, 33,000, 23,250, 29,400). Insol in water. Freely sol in ether, hot alcohol.

9429. Tolazamide. *N-[[(Hexahydro-1H-azepin-1-yl)amino]carbonyl]-4-methylbenzenesulfonamide; 1-(hexahydro-1H-azepin-1-yl)-3-(p-tolylsulfonyl)urea; N-(p-toluenesulfonyl)-N'-hexamethyleniminourea;* tolazolamide; U 17835; Diabewas; Norglycin; Tolanase; Tolinase; Tolonase. C$_{14}$H$_{21}$N$_3$O$_3$S; mol wt 311.41. C 54.00%, H 6.80%, N 13.49%, O 15.41%, S 10.30%. Prepn: Wright, Willette, *J. Med. Pharm. Chem.* 5, 815 (1962); Wright, **Brit.** pat. 887,-886 (1962 to Upjohn). Mode of action study: Marshall *et al., Metab. Clin. Exp.* 19, 1046 (1970). Clinical experience and review of literature: Balodimos, Marble, *Curr. Ther. Res.* 13, 6-12 (1971).

Crystals, mp 170-173°.
THERAP CAT: Antidiabetic.

9430. Tolazoline. *4,5-Dihydro-2-(phenylmethyl)-1H-imidazole; 2-benzyl-2-imidazoline;* 2-benzyl-1-imidazole; benzazoline; 2-benzyl-4,5-imidazoline; phenylmethylimidazoline; Priscoline; Priscol; Vasodil; Prefaxil; Benzidazol;

Vasimid; Kasimid; Artonil; Vasodilatan; Lambril; Olitensol. C$_{10}$H$_{12}$N$_2$; mol wt 160.21. C 74.96%, H 7.55%, N 17.49%. Prepd from α-phenylthioacetamide and ethylenediamine: Sonn, **U.S.** pat. 2,161,938 (1939 to Ciba). Review and bibliography: Scholz, *Ind. Eng. Chem.* 37, 120 (1945).

Hydrochloride, C$_{10}$H$_{13}$ClN$_2$, bitter crystals, mp 174°. Freely sol in water, alcohol. Sol in chloroform. Very slightly sol in ether, ethyl acetate. Aq solns are slightly acid to litmus. pH of 2.5% soln 4.9-5.3.

Picrate, yellow crystals, mp 144-149°.
THERAP CAT: Anti-adrenergic.

9431. Tolboxane. *5-Methyl-2-(4-methylphenyl)-5-propyl-1,3,2-dioxaborinane; p-tolueneboronic acid cyclic 2-methyl-2-propyltrimethylene ester;* 5-methyl-5-n-propyl-2-p-tolyl-1,3,2-dioxaborinane; 2-methyl-2-propyl-1,3-propanediol *p*-methylbenzeneboronate; IS 813; Clarmil; Clarphoril. C$_{14}$H$_{21}$BO$_2$; mol w 232.13. C 72.44%, H 9.12%, B 4.66%, O 13.79%. Prepn: Farthouat, **U.S.** pat. 3,038,926 (1962 to Roussel-UCLAF). Structure-activity studies: Caujolle, Pham-Huu-Chanh, *Arch. Int. Pharmacodyn. Ther.* 172, 467 (1968). Toxicity studies: Caujolle *et al., Med. Pharmacol. Exp.* 15, 130 (1966).

Crystals from methanol, mp 54°.
THERAP CAT: Tranquilizer.

9432. Tolbutamide. *N-[(Butylamino)carbonyl]-4-methylbenzenesulfonamide; 1-butyl-3-(p-tolylsulfonyl)urea;* tolylsulfonylbutylurea; 3-(p-tolyl-4-sulfonyl)-1-butylurea; *N-n-*butyl-*N'*-tosylurea; *N'*-4-methylbenzenesulfonyl-*N''*-butylurea; *N*-(sulfonyl-*p*-methylbenzene)-*N'*-*n*-butylurea; D 860; U 2043; Diabetamid; Diasulfon; Glyconon; Orinase; Rastinon; Diabuton; Mobenol; Oterben; Toluina; Diaben; Diabesan; Ipoglicone; Orabet; Oralin; Artosin; Dolipol; Tolbet; Tarasina; Tolbusal; Pramidex; Tolbutone; Willbutamide. C$_{12}$H$_{18}$N$_2$O$_3$S; mol wt 270.34. C 53.31%, H 6.71%, N 10.36%, O 17.75%, S 11.86%. Description: Ehrhart, *Naturwiss.* 43, 93 (1956). Prepn from butyl isocyanate and 4-toluenesulfonamide sodium and use as antidiabetic: **Brit.** pat. 808,071 and Aumüller, Herr, **Ger.** pat. 1,066,575 (both 1959 to Hoechst); Ruschig *et al.,* **U.S.** pat. 2,968,158 (1961 to Upjohn). Comprehensive description: W. F. Beyer, E. H. Jensen, in *Analytical Profiles of Drug Substances* vol. 3, K. Florey, Ed. (Academic Press, New York, 1974) pp 513-543.

Crystals, mp 128.5-129.5°. (The corresponding isobutyl compd, mp 172-173°.)

Sodium salt tetrahydrate, C$_{12}$H$_{17}$N$_2$NaO$_3$S.4H$_2$O, crystals, mp 41-43°. When anhyd, mp 130-133°.
THERAP CAT: Antidiabetic.
THERAP CAT (VET): Hypoglycemic agent.

9433. Tolciclate. *Methyl(3-methylphenyl)carbamothioic acid O-(1,2,3,4-tetrahydro-1,4-methanonaphthalen-6-yl) ester;* O-(1,4-methano-1,2,3,4-tetrahydro-6-naphthyl)-*N*-methyl-*N*-(*m*-tolyl)thiocarbamate; KC 9147; Fungifos; Kilmicen; Tolmicen. C$_{20}$H$_{21}$NOS; mol wt 323.46. C 74.27%, H 6.54%, N 4.33%, O 4.95%, S 9.91%. Topical antimycotic agent with high liposolubility. Prepn: P. Melloni *et al.,* **Ger.** pat. 2,313,845 corresp to U.S. pat. 3,855,263 (1973, 1974 both to Carlo Erba). *In vitro* and *in vivo* study: I. deCarneri *et al., Arzneimittel-Forsch.* 26, 769 (1976). Antimycotic studies: A. Bianchi *et al., Antimicrob. Ag. Chemother.* **12**, 429 (1977). Clinical studies in dermatomycosis: L. C. Cucé *et al., J. Int. Med. Res.* **8**, 144 (1980); C. Intini *et al., Pharmatherapeutica* **2**, 439 (1980).

White cryst powder from isopropanol, mp 92-94°. Practically insol in water. Soly (mg/ml): 14.9 in *n*-hexane; 23.9 in *n*-octanol. LD$_{50}$ in mice, rats, dogs (mg/kg): 4000, 6000, 5000 orally (deCarneri).
THERAP CAT: Antifungal.

9434. Tolcyclamide. *N-[(Cyclohexylamino)carbonyl]-4-methylbenzenesulfonamide;* 1-cyclohexyl-3-*p*-tolylsulfonylurea; tolhexamide; glycyclamide; cyclamide; K 386; Diaboral. C$_{14}$H$_{20}$N$_2$O$_3$S; mol wt 296.39. C 56.73%, H 6.80%, N 9.45%, O 16.19%, S 10.82%. Prepn: Logemann, Artini, *Ber.* **90**, 2527 (1957).

Crystals from trichloroethylene, mp 174-176°.
THERAP CAT: Antidiabetic.

9435. Toldimfos Sodium. *(4-Dimethylamino-o-tolyl)-phosphonous acid sodium salt;* sodium (4-dimethylamino-*o*-tolyl)phosphonate; *p*-dimethylamino-*o*-toluenephosphonous acid sodium salt. C$_9$H$_{13}$NNaO$_2$P; mol wt 221.16. C 48.87%, H 5.92%, N 6.33%, Na 10.40%, O 14.47%, P 14.01%. Prepd from *N,N*-dimethyl-*m*-toluidine and phosphorus trichloride: Benda, Schmidt, Ger. pat. **397,813** (1924 to Cassella), *Frdl.* **14**, 1409.

Trihydrate, *Novofosfan, Phiniphos, Phosodyl, Tonofosfan, Tonophosphan.* Scales, needles, or prisms from alc. Freely sol in cold water, hot alcohol.
THERAP CAT: Tonic.

9436. Tolfenamic Acid. *2-[(3-Chloro-2-methylphenyl)-amino]benzoic acid;* N-(3-chloro-o-tolyl)anthranilic acid; N-(2-methyl-3-chlorophenyl)anthranilic acid; GEA 6414; Clotam; Tolfedine. C$_{14}$H$_{12}$ClNO$_2$; mol wt 261.71. C 64.25%, H 4.62%, Cl 13.55%, N 5.35%, O 12.23%. Deriv of anthranilic acid, related structurally to mefenamic and flufenamic acids, *q.q.v.* Prepn: **Neth.** pat. **Appl. 6,600,251** (1966 to Gea A/S), *C.A.* **66**, 2377 (1967); R. A. Scherrer, F. W. Short, U.S. pat. **3,313,848** (1967 to Parke, Davis). Inhibition of prostaglandin biosynthesis: I. B. Linden *et al., Scand. J. Rheumatol.* **5**, 129 (1976). HPLC determn: F. Nielsen-Kudsk, *Acta Pharmacol. Toxicol.* **47**, 267 (1980). Metabolism: T. Kuninaka *et al., Yakugaku Zasshi* **101**, 232 (1981), *C.A.* **95**, 168 (1981). Human pharmacokinetics: P. Pentikaeinen *et al., Eur. J. Clin. Pharmacol.* **19**, 359 (1981). Pharmacological studies: S. Yamashita *et al., Toho Igakkai Zasshi* **28**, 76-105 (1981), *C.A.* **95**, 16183, 180846 (1981). Clinical study: V. Rejholec *et al., Scand. J. Rheumatol.* **Suppl. 33**, 50 (1980); **Suppl. 36**, 1 (1980).

Crystals from abs ethanol, mp 207-207.5°.
THERAP CAT: Anti-inflammatory; analgesic.

9437. o-Tolidine. *3,3'-Dimethyl-[1,1'-biphenyl]-4,4'-diamine;* 3,3'-dimethylbenzidine; 4,4'-diamino-3,3'-dimethylbiphenyl. C$_{14}$H$_{16}$N$_2$; mol wt 212.28. C 79.21%, H 7.60%, N 13.20%. Made by alkaline reduction of *o*-nitrotoluene with zinc, and subsequent rearrangement of the *o*-hydrazotoluene formed, by boiling with HCl: Van Loon, *Chem. Weekbl.* **5**, 689 (1907). *See also* Schultz *et al., Ann.* **352**, 111 (1907). Crystal and molecular structure: Chawdhury *et al., Acta Crystallogr. Sect. B,* **24**, 1222 (1968). Metabolism: Dieteren, *Arch. Environ. Health* **12**, 30 (1966). Carcinogenic activity: Pliss, Zebenzhinskii, *J. Nat. Cancer Inst.* **45**, 283 (1970).

White to reddish crystals or cryst powder. mp 129-131°. Slightly sol in water; sol in alcohol, ether, dil acids. *Keep well closed and protected from light.*
Sulfate, C$_{14}$H$_{16}$N$_2$.H$_2$SO$_4$, white to gray mass. Slightly sol in water, alcohol; sol in dil acids.
Note: This substance may reasonably be anticipated to be a carcinogen: *Fourth Annual Report on Carcinogens* (NTP 85-002, 1985) p 90.
USE: Manuf dyes; also as very sensitive reagent for gold (1:10 million detectable), and for free chlorine in water.

9438. Tolindate. *Methyl (3-methylphenyl)carbamothioic acid O-(2,3-dihydro-1H-inden-5-yl) ester;* m,N-dimethylthiocarbanilic acid O-5-indanyl ester; O-(5-indanyl) m,N-dimethylthiocarbanilate; Dalnate. C$_{18}$H$_{19}$NOS; mol wt 297.42. C 72.69%, H 6.44%, N 4.71%, O 5.38%, S 10.78%. Prepared by treating 5-indanyl thionochloroformate with *N*-methyl-*m*-toluidine: Elpern, Youlus, U.S. pat. **3,509,200** (1970 to USV).

Crystals, mp 94-95°.
THERAP CAT: Antifungal.

9439. Toliprolol. *1-[(1-Methylethyl)amino]-3-(3-methylphenoxy)-2-propanol;* 1-(isopropylamino)-3-(m-tolyloxy)-2-propanol; 1-(3-methylphenoxy)-3-(isopropylamino)-2-propanol; 1-(3-methylphenoxy)-2-hydroxy-3-isopropylaminopropane; MHIP. C$_{13}$H$_{21}$NO$_2$; mol wt 223.32. C 69.92%, H 9.48%, N 6.27%, O 14.33%. β-Adrenergic blocker. Prepn: **Neth.** pat. **Appl. 6,409,883**; H. Koppe *et al.,* **U.S.** pat. **3,459,782** (1965, 1969 both to Boehringer, Ing.); **Neth.** pat. **Appl. 6,410,522**; R. Howe, U.S. pat. **3,432,545** (1965, 1969 both to I.C.I.). The (−)-isomer is the more potent adrenergic β-receptor antagonist. Resolution of isomers: Howe, Rao, *J. Med. Chem.* **11**, 1118 (1968). Structure-activity studies: Crowther *et al., ibid.* **12**, 638 (1969); Somani, Laddu, *Eur. J. Pharmacol.* **14**, 209 (1971). Metabolism: Stock, Westermann, *Biochem. Pharmacol.* **14**, 227 (1965). Review of pharmacology and clinical data: Marmo *et al., Clin. Ter.* **62**, 11-51, 117-163 (1972).

Crystals from ethyl acetate-petr ether, mp 75-76°. Also reported as mp 79°.
Hydrochloride, C$_{13}$H$_{22}$ClNO$_2$, *ICI 45763, Ko 592, Doberol,*

Consult the cross index before using this section.

Sinorytmal. Crystals from ethanol-ether, mp 120-121°. uv max (water): 270 nm ($E_{1cm}^{1\%}$ 49.8).

THERAP CAT: Antianginal, antihypertensive.

9440. Tollens Reagent. A solution prepared from equal amounts of 10% silver nitrate and 10% sodium hydroxide solutions to which enough dilute ammonia solution has been added to dissolve the precipitated silver oxide. Tollens reagent oxidizes aldehydes to the corresponding acids; during the reaction the silver, bound in form of a complex, is reduced to metallic silver and forms a characteristic silver mirror. *Refs:* B. Tollens, *Ber.* **15,** 1635 (1882); W. Ponndorf, *ibid.* **64B,** 1913 (1937); S. Siggia, E. Segel, *Anal. Chem.* **25,** 640 (1953); J. M. Kolthoff, P. J. Elving, *Treatise on Analytical Chemistry* vol. **13** (New York, 1966) p 183. *Caution:* Tollens reagent should always be prepared freshly; old, opaque or "dried out" solutions are explosive, H. Waldmann, *Chimia* **13,** 297 (1959).

USE: Reagent in characterization of sugars, aldehydes, hydrazides. As oxidizing agent.

9441. Tolmetin. *1-Methyl-5-(4-methylbenzoyl)-1H-pyrrole-2-acetic acid; 1-methyl-5-p-toluoylpyrrole-2-acetic acid; 5-(p-toluoyl)-1-methylpyrrole-2-acetic acid;* McN-2559. $C_{15}H_{15}NO_3$; mol wt 257.30. C 70.02%, H 5.88%, N 5.44%, O 18.66%. Prepn: Carson, **Fr.** pat. **1,574,570** (1969 to McNeil Labs.), *C.A.* **72,** 100498y (1969). Pharmacology: Carson *et al., J. Med. Chem.* **14,** 646 (1971); S. Wong *et al., J. Pharmacol. Exp. Ther.* **185,** 127 (1973). *Review:* S. Wong in *Pharmacological and Biochemical Properties of Drug Substances* vol. **1,** M. E. Goldberg, Ed. (Am. Pharm. Assoc., Washington, DC, 1977) pp 233-255. Review of pharmacology and therapeutic efficacy: R. N. Brogden *et al., Drugs* **15,** 429-450 (1978).

Crystals from acetonitrile, mp 155-157° (dec). Sodium salt dihydrate, $C_{15}H_{14}NNaO_3 \cdot 2H_2O$, *McN-2559-21-98, Reutol, Tolectin, Tolmene.*

THERAP CAT: Anti-inflammatory.

9442. Tolnaftate. *Methyl(3-methylphenyl)carbamothioic acid O-2-naphthalenyl ester; m,N-dimethylthiocarbanilic acid O-2-naphthyl ester; O-2-naphthyl m,N-dimethylthiocarbanilate; 2-naphthyl N-methyl-N-(3-tolyl)thionocarbamate; naphthiomate T; Sch 10144; Aftate; Chinofungin; Focusan; Fungistop; Hi-Alazin; Sorgoa; Sporiline; Timoped; Tinactin; Tinaderm; Tonoftal.* $C_{19}H_{17}NOS$; mol wt 307.43. C 74.23%, H 5.58%, N 4.56%, O 5.20%, S 10.43%. Prepn: **Fr.** pat. **1,337,797** corresp to Miyazaki *et al.,* **U.S.** pat. **3,334,126** (1963, 1967 to Japan Soda); Noguchi *et al., J. Pharm. Soc. Japan* **88,** 335 (1968). Pharmacology and toxicology: Noguchi *et al., Antimicrob. Ag. Chemother.* **1962,** 259; Hashimoto *et al., Toxicol. Appl. Pharmacol.* **8,** 380 (1966); Noguchi *et al., ibid.* 368.

Crystals from alc, mp 110.5-111.5°. Insol in water. Sparingly sol in methanol, ethanol. Sol in chloroform (1:1.5), acetone (1:7), CCl_4 (1:9). LD_{50} in mice, rats (g/kg): > 10, > 6 orally; > 6, > 4 s.c. (Hashimoto).

THERAP CAT: Antifungal.

THERAP CAT (VET): Antifungal.

9443. Tolonidine. *N-(2-Chloro-4-methylphenyl)-4,5-dihydro-1H-imidazol-2-amine; 2-(2-chloro-p-toluidino)-2-imidazoline; 2-(2-chloro-4-methylphenyl)amino-1,3-diazacyclopentene-2.* $C_{10}H_{12}ClN_3$; mol wt 209.68. C 57.28%, H 5.77%, Cl 16.91%, N 20.04%. Orally active antihypertensive agent, related structurally to clonidine, *q.v.* Prepn: **Neth.**

pat. **Appl. 6,411,516** corresp to **U.S.** pats. **3,236,857** and **3,454,701** (1965, 1966, 1969 all to Boehringer, Ing.). Synthesis and hypotensive activity: P. B. M. Timmermans *et al., Rec. Trav. Chim.* **97,** 51 (1978). Pharmacological properties: C. Cosnier *et al., Arzneimittel-Forsch.* **25,** 1557, 1802, 1926 (1975). Structure-activity study: P. B. M. Timmermans, P.A. Van Zwieten, *J. Med. Chem.* **20,** 1636 (1977). Chromatographic study: P. B. M. Timmermans *et al., J. Chromatog.* **144,** 215 (1977). Quantum chemical studies: *eidem, Arzneimittel-Forsch.* **27,** 2266 (1977). Radioimmunoassay disposition study: B. Jarrott, S. Spector, *J. Pharmacol. Exp. Ther.* **207,** 195 (1978).

Cryst from benzene/petr ether, mp 148-150°. pKa 9.41. Nitrate, $C_{10}H_{13}ClN_4O_3$, *CERM 10,137, Euctan.* Cryst, mp 162-164°. LD_{50} in male mice, rats (mg/kg): 160, 420 orally; 21.25, 42 i.v., D. Cosnier *et al., Arzneimittel-Forsch.* **25,** 1926 (1975).

THERAP CAT: Antihypertensive.

9444. Tolonium Chloride. *3-Amino-7-(dimethylamino)-2-methylphenothiazin-5-ium chloride; 3-amino-7-dimethylamino-2-methylphenazathionium chloride;* blutene chloride; toluidine blue O; dimethyltoluthionine chloride; C.I. Basic Blue 17; C.I. 52040; Klot; Tolazul. $C_{15}H_{16}ClN_3S$; mol wt 305.83. C 58.91%, H 5.27%, Cl 11.59%, N 13.74%, S 10.48%. Prepd from dimethyl-*p*-phenylenediamine, sodium thiosulfate, and *o*-toluidine: Dändliker, Bernthsen, **U.S.** pat. **416,055** (1888 to BASF). Prepn of hemostatic compositions contg tolonium chloride: D. A. Hoff, **U.S.** pat. **2,809,913** (1957 to Warren-Teed). Clinical studies in bleeding disorders: J. Allen *et al., Surg. Gynecol. Obstet.* **89,** 692 (1949). Acute and chronic toxicity study: T. J. Haley, F. Stolarsky, *Stanford Med. Bull.* **9,** 96 (1951). Clinical use for parathyroid identification during thyroidectomy: R. M. Yeager, E. T. Krementz, *Ann. Surg.* **169,** 829 (1969). Review of therapeutic and diagnostic use: A. Mashberg, *J. Am. Dent. Assoc.* **106,** 319-323 (1983). *See also Colour Index* vol. **4** (3rd ed., 1971) p 4471; H. J. Conn's *Biological Stains,* R. D. Lillie, Ed. (Williams & Wilkins, Baltimore, 9th ed., 1977) p 428.

Dark green powder. Sol in water (3.82 g/100 ml), giving a blue to violet soln; sol in alc (0.57 g/100 ml), giving a blue soln. Absorption max (water): 640.4 nm. LD_{50} in mice, rats, rabbits (mg/kg): 27.56, 28.93, 13.44 i.v. (Haley, Stolarsky).

Clear, colorless, stable, isotonic solns of purified leuco toluidine blue O are prepd by reducing toluidine blue O with sodium hydrosulfite at pH 2.5-3.5, see B. March, E. E. Moore U.S. pat. **2,571,593** (1951 to Abbott).

USE: Direct dyeing, printing of wool, silk. Biological stain.

THERAP CAT: Hemostatic. Diagnostic aid (oral carcinoma).

9445. Toloxatone. *5-(Hydroxymethyl)-3-(3-methylphenyl)-2-oxazolidinone; 5-(hydroxymethyl)-3-m-tolyl-2-oxazolidinone;* 69276 MD; MD 690276; Humoryl; Perenum. $C_{11}H_{13}NO_3$; mol wt 207.23. C 63.76%, H 6.32%, N 6.76%, O 23.16%. Reversible monoamine oxidase inhibitor. Prepn: C. Fauran *et al.,* **Ger.** pat. **2,012,120;** *eidem,* **U.S.** pats. **3,-641,036; 3,655,687** (1970, 1972, 1972 all to Delalande); *eidem, Chim. Ther.* **8,** 324 (1973). Pharmacology: G. Raynaud *et al., ibid.* 328. Psychopharmacological profile: J.-P. Kan *et al., Eur. J. Med. Chem.* **12,** 13 (1977); H. Giono-Barber *et al., Arzneimittel-Forsch.* **27,** 1188 (1977). Pharmacokinetics: M. S. Benedetti *et al.*, *ibid.* **32**, 276 (1982). Metabolism: A. Malnoe, M. S. Benedetti, *Xenobiotica* **9**, 281 (1979). GLC determn in plasma: S. Vajta *et al.*, *J. Chromatog.* **274**, 139 (1983).

Crystals from isopropyl alcohol, mp 76°. LD$_{50}$ orally in mice (mg/kg): 1850 (Fauran); also reported as 1500 (Raynaud).

THERAP CAT: Antidepressant.

9446. Toloxychlorinol. 1,1'-[(*o*-Tolyloxymethyl)ethyl-enedioxy]bis[2,2,2-trichloroethanol]; 1,1'-(3-*o*-tolyloxypropylenedioxy)bis[2,2,2-trichloroethanol]; 1-(*o*-toloxy)-2,3-bis(2,2,2-trichloro-1-hydroxyethoxy)propane; 1-(*o*-methylphenoxy)-2,3-bis(2,2,2-trichloro-1-hydroxy)propane; toloxychloral; NC 1318; Myavan. C$_{14}$H$_{16}$Cl$_6$O$_5$; mol wt 477.02. C 35.25%, H 3.38%, Cl 44.60%, O 16.77%. Prepn: Scudi, Tenenbaum, U.S. pat. **2,666,082** (1954 to Nepera Chem.). Pharmacology and toxicity: J. F. Reinhard *et al.*, *J. Pharmacol. Exp. Ther.* **106**, 444 (1952).

Crystals, mp 103-106°. Practically insol in water. Sol in alc, acetone, chloroform, benzene, propylene glycol; limited soly in toluene and hexane. LD$_{50}$ in rats (mg/kg): 880 orally, 480 i.p. (Reinhard).

9447. Tolperisone. *2-Methyl-1-(4-methylphenyl)-3-(1-piperidinyl)-1-propanone; 2,4'-dimethyl-3-piperidinopropiophenone;* 1-piperidino-2-methyl-3-(*p*-tolyl)-3-propanone; 2-methyl-3-piperidino-1-*p*-tolylpropan-1-one; mydetone. C$_{16}$H$_{23}$NO; mol wt 245.35. C 78.32%, H 9.45%, N 5.71%, O 6.52%. Prepn: Nádor *et al.*, **Hung.** pat. **144,997** (1956); Yokoyama *et al.*, **Japan.** pat. **20,390('65)** (to Eisai). Pharmacology: J. Porszasz *et al.*, *Acta Physiol. Acad. Sci. Hung.* **18**, 149 (1960); *eidem, Arzneimittel-Forsch.* **11**, 257 (1961).

Hydrochloride, C$_{16}$H$_{24}$ClNO, *N-553, Abbsa, Atmosgen, Arantoick, Besnoline, Isocalm, Kineorl, Menopatol, Metosomin, Minacalm, Muscalm, Mydocalm, Naismeritin, Tolisartine.* Crystals from methyl ethyl ketone, mp 176-177°. LD$_{50}$ in mice: 620 mg/kg s.c., J. Porszasz *et al., loc. cit.* (1961).

THERAP CAT: Relaxant (skeletal muscle).

9448. Tolpovidone ^{131}**I.** 1-(ω-*p*-Iodo-^{131}I-benzylpolyethylene)-2-pyrrolidinone; iodinated *p*-toluidine polyvinylpyrrolidone; radio-iodinated (^{131}I) tolpovidone; Raovin.

USE: Differential diagnosis of source of hypoalbuminemia.

9449. Tolpronine. *3,6-Dihydro-α-[(2-methylphenoxy)-methyl]-1(2H)-pyridineethanol;* 1-(2-methylphenoxy)-3-(3-piperidino)-2-propanol; 1-(1,2,3,6-tetrahydropyridino)-3-

o-tolyloxypropan-2-ol; 1-Δ^3-piperidino-3-*o*-toloxypropan-2-ol; α-Δ^3-piperidino-β-hydroxy-γ-*o*-toloxypropane; Proponesin. C$_{15}$H$_{21}$NO$_2$; mol wt 247.33. C 72.84%, H 8.56%, N 5.66%, O 12.94%. Prepn: Beasley *et al., J. Pharm. Pharmacol.* **10**, 47 (1958). Prepd (not claimed): **Brit.** pat. **889,075;** F. M. Berger, B. J. Ludwig, U.S. pat. **3,085,938** (1962, 1963, both to Carter Prods.). Pharmacology: Monacelli, *Pathologica* **60**, 11 (1968).

Crystals, mp 56-58°.
Hydrochloride, C$_{15}$H$_{22}$ClNO$_2$, crystals, dec 136-137.5°. Bitter taste. Freely sol in water. pH of a 10% aq soln about 5.8. Aq solns are stable to heat, but unstable to acids and alkalies. LD$_{50}$ i.p. in mice: 146 ±5.4 mg/kg (Berger, Ludwig).

THERAP CAT: Analgesic.

9450. Tolpropamine. *N,N,4-Trimethyl-γ-phenylbenzenepropanamine; N,N-dimethyl-3-phenyl-3-p-tolylpropylamine;* 3-dimethylamino-1-phenyl-1-*p*-tolylpropane; Pragman; Tylagel. C$_{18}$H$_{23}$N; mol wt 253.37. C 85.32%, H 9.15%, N 5.53%. Prepn: Bockmühl, Stein, **Ger.** pat. **925,468** (1955 to Hoechst); Klosa, *J. Prakt. Chem.* **34**, 312 (1966). Pharmacology: Sendrail, Gleizes, *Therapie* **15**, 119 (1960).

Hydrochloride, C$_{18}$H$_{23}$N.HCl, mp 182-184°.
THERAP CAT: Topical antihistaminic, antipruritic.

9451. Tolrestat. *N-[[6-Methoxy-5-(trifluoromethyl)-1-naphthalenyl]thioxomethyl]-N-methylglycine;* tolrestatin; AY 27773; Alredase. C$_{16}$H$_{14}$F$_3$NO$_3$S; mol wt 357.35. C 53.78%, H 3.95%, F 15.95%, N 3.92%, O 13.43%, S 8.97%. Orally active aldose reductase inhibitor. Prepn and pharmacology: K. Sestanj *et al.,* **Eur.** pat. **Appl. 59,596** (1982 to Ayerst); *ibid., J. Med. Chem.* **27**, 255 (1984). Prevention of cataracts in galactosemic rats: N. Simard-Duquesne *et al., Proc. Exp. Biol. Med.* **178**, 599 (1985). Kinetics and metabolism in man: D. R. Hicks *et al., Clin. Pharmacol. Ther.* **36**, 493 (1984). Ultraviolet and HPLC determn in serum: D. R. Hicks, M. Kraml, *Ther. Drug Monit.* **6**, 328 (1984). Effect on erythrocyte sorbitol levels in diabetic patients: P. Raskin *et al., Clin. Pharmacol. Ther.* **38**, 625 (1985).

mp 164-165°.
Methyl ester, C$_{17}$H$_{16}$F$_3$NO$_3$S, mp 109-110°.
THERAP CAT: Treatment of diabetic neuropathy.

9452. Toltrazuril. *1-Methyl-3-[3-methyl-4-[4-[(trifluoromethyl)thio]phenoxy]phenyl]-1,3,5-triazine-2,4,6(1H,3H,-5H)-trione;* 1-methyl-3-[4-[*p*-[(trifluoromethyl)thio]phenoxy]-*m*-tolyl]-*s*-triazine-2,4,6(1H,3H,5H)-trione; Bay Vi 9142; Baycox. C$_{18}$H$_{14}$F$_3$N$_3$O$_4$S; mol wt 425.38. C 50.82%, H 3.32%, F 13.40%, N 9.88%, O 15.04%, S 7.54%. Triazinetrione anticoccidial. General prepn: **Belg.** pat. **826,900;** J. H. Reisdorff *et al.,* U.S. pat. **3,966,725** (1975, 1976 both to Bayer). Prepn and use as animal growth promotant: **Belg.** pat. **866,389;** A. Haberkorn *et al.,* U.S. pat. **4,219,552** (1978, 1980 both to Bayer). Series of articles on efficacy vs coccidia in chickens: E. Kutzer *et al., Wien. Tierärztl. Monatsschr.* **72**,

321-340 (1985). Field trial in sheep: B. Gjerde, O. Helle, *Acta Vet. Scand.* **27,** 124 (1986); in chickens: H. D. Chapman, *J. Comp. Path.* **97,** 21 (1987).

mp 194°.
THERAP CAT (VET): Coccidiostat.

9453. *o*-Tolualdehyde. *2-Methylbenzaldehyde; o-*toluyl-aldehyde. C_8H_8O; mol wt 120.14. C 79.97%, H 6.71%, O 13.32%. Prepd by reacting nitropropane with *o*-xylyl bromide in the presence of sodium ethanoate: Hass, Bender, *Org. Syn.* **30,** 99 (1950).

Liquid. d_4^{19} 1.0386. bp_{760} 200-202°; bp_{15} 94-96°; bp_6 68-72°. n_D^{25} 1.5423; n_D^{19} 1.549; n_α^{19} 1.5423; n_β^{19} 1.5650; n_γ^{19} 1.5798.

9454. *o*-Toluamide. *2-Methylbenzamide.* C_8H_9NO; mol wt 135.16. C 71.09%, H 6.71%, N 10.36%, O 11.84%. Prepd by reacting *o*-tolunitrile, hydrogen peroxide, 95% alcohol, and sodium hydroxide at 40-50°: Noller, *Org. Syn.* coll. **vol. II,** 586 (1943). Also prepared by reacting the nitrile with boron fluoride in dil acetic acid: Hauser, Hoffenberg, *J. Org. Chem.* **20,** 1448 (1955).

Crystals from water, mp 144-145°. Very sol in alcohol, hot water, concd HCl, less sol in ether. Sparingly sol in benzene. Practically insol in cold water.
Caution: Explosive. Keep away from open flame.

9455. Toluene. *Methylbenzene;* toluol; phenylmethane; Methacide. C_7H_8; mol wt 92.13. C 91.25%, H 8.75%. Obtained mainly from tar oil. Review of mfg processes: Faith, Keyes & Clark's *Industrial Chemicals,* F. A. Lowenheim, M. K. Moran, Eds. (Wiley-Interscience, New York, 4th ed., 1975) pp 822-830. Solubility: F. P. Schwarz, *Anal. Chem.* **52,** 10 (1980). Myelotoxic potential: L. Greenburg *et al., J. Am. Med. Assoc.* **118,** 573 (1942). Comparison with benzene of effects on hematopoiesis and bone marrow metabolism: H. W. Gerard, *AMA Arch. Ind. Health* **13,** 468 (1956). Acute toxicity: H. F. Smyth *et al., Am. Ind. Hyg. Assoc. J.* **30,** 470 (1969). Evaluation of chronic occupational exposure: H. Tahti *et al., Int. Arch. Occup. Environ. Health* **48,** 61 (1981). *Review:* M. C. Hoff in Kirk-Othmer *Encyclopedia of Chemical Technology* vol. **23** (Wiley-Interscience, New York, 3rd ed., 1983) pp 246-273.

Flammable, refractive liq; benzene-like odor. d_4^{20} 0.866. mp −95°. bp 110.6°. n_D^{20} 1.4967. Flash pt, closed cup: 40°F (4.4°C). Soly in water at 23.5°C (w/w): 0.067%. Very slightly sol in water; misc with alc, chloroform, ether, acetone, glacial acetic acid, carbon disulfide. LD_{50} orally in rats: 7.53 g/kg (Smyth).
Caution: Narcotic in high concns: E. Browning, *Toxicity*

and Metabolism of Industrial Solvents (Elsevier, New York, 1965) pp 66-76; Patty's *Industrial Hygiene and Toxicology* vol. **2B,** G. D. Clayton, F. E. Clayton, Eds. (Wiley-Interscience, New York, 3rd ed., 1981) pp 3283-3291.
USE: In manuf benzoic acid, benzaldehyde, explosives, dyes, and many other organic compds; as a solvent for paints, lacquers, gums, resins, in the extraction of various principles from plants; as gasoline additive.

9456. Toluene 2,4-Diisocyanate. 2,4-Diisocyanatotoluene; 2,4-tolylene diisocyanate; TDI; Nacconate 100. $C_9H_6N_2O_2$; mol wt 174.15. C 62.07%, H 3.47%, N 16.09%, O 18.37%. Usually prepd from toluene-2,4-diamine and phosgene. *Review:* Astle, *Industrial Organic Nitrogen Compounds* (New York, 1961) pp 284-313; Faith, Keyes & Clark's *Industrial Chemicals,* F. A. Lowenheim, M. K. Moran, Eds. (Wiley-Interscience, New York, 4th ed., 1975) pp 831-835.

Liquid at room temperature. Sharp, pungent odor. mp 19.5-21.5°. d_4^{20} liq 1.2244. bp_{760} 251°; bp_{11} 126°. Darkens on exposure to sunlight. Reacts with water with evolution of carbon dioxide. Flash pt, open cup: 132° (270°F). Misc with alcohol (decompn), diglycol monomethyl ether, ether, acetone, carbon tetrachloride, benzene, chlorobenzene, kerosene, olive oil. Concd alkaline compds such as NaOH or *tert*-amines may cause run-away polymerization.
Caution: Vapor is irritating to eyes, skin and respiratory tract: *Clinical Toxicology of Commercial Products,* R. E. Gosselin *et al.,* Eds. (Williams & Wilkins, Baltimore, 5th ed., 1984) Section II, p 414. This substance may reasonably be anticipated to be a carcinogen: *Fourth Annual Report on Carcinogens* (NTP 85-002, 1985) p 190.
USE: In the manuf of polyurethane foams and other elastomers.

9457. Toluene-3,4-dithiol. 1,2-Dimercapto-4-methylbenzene; "dithiol". $C_7H_8S_2$; mol wt 156.27. C 53.80%, H 5.16%, S 41.04%. Prepd from toluene-3,4-disulfonyl chloride with tin and hydrochloric acid: Mills, Clark, *J. Chem. Soc.* **1936,** 178.

Crystals, mp 31°. bp_{84} 185-187°. Sol in benzene, in aq alkali hydroxide solns.
USE: For the detection of bismuth, molybdenum, rhenium, tin, tungsten, *see* Bickford *et al., J. Am. Pharm. Assoc., Sci. Ed.* **37,** 255 (1948).

9458. *p*-Toluenesulfinic Acid. $C_7H_8O_2S$; mol wt 156.20. C 53.82%, H 5.16%, O 20.49%, S 20.53%. $CH_3C_6H_4SO_2H$. Prepd by reduction of *p*-toluenesulfonyl chloride with zinc dust: Whitmore, Hamilton, *Org. Syn.* **2,** 89 (1922). Because the sulfinic acid is difficult to dry without partial conversion to the sulfonic acid, the sodium salt, $CH_3C_6H_4SO_2Na.2H_2O$, is usually prepd. The free sulfinic acid is then obtained as needed by dissolving the sodium salt in cold water and carefully acidifying the soln with the exact amt of HCl needed.
Long, rhombic plates or needles from water. mp 85°. Freely sol in alc, ether; sparingly sol in water, hot benzene.

9459. *p*-Toluenesulfonic Acid. *4-Methylbenzenesulfonic acid;* tosic acid. $C_7H_8O_3S$; mol wt 172.20. C 48.82%, H 4.68%, O 27.87%, S 18.62%. $CH_3C_6H_4SO_3H$. Prepd by sulfonation of toluene with 96-100% H_2SO_4; when carried out at 75° the compn of the reaction product is 75% *para-*, 19% *ortho-* and 6% *meta-*toluenesulfonic acid. Convenient lab prepn: L. F. Fieser, *Experiments in Organic Chemistry,*

(Boston, 3rd ed., 1955) p 144. The separation of toluene from petroleum fractions can be accomplished by sulfonation with H_2SO_4 at 60°.

Monoclinic leaflets or prisms. Also reported as crystallizing with $1H_2O$ or $4H_2O$. When anhydrous, mp 106-107°. Metastable form, mp 38°. bp_{20} 140°. $bp_{0.1}$ 185-187°. Freely sol in water, about 67 g/100 ml. Sol in alc and ether.

Sodium salt, $C_7H_7NaO_3S$, orthorhombic plates, very sol in water.

USE: In dye chemistry; in manuf of oral antidiabetic drugs. *Caution:* Highly irritating to skin, mucous membranes.

9460. *p*-Toluenesulfonyl Chloride. Tosyl chloride. C_7-H_7ClO_2S; mol wt 190.65. C 44.10%, H 3.70%, Cl 18.60%, O 16.78%, S 16.82%. $CH_3C_6H_4SO_2Cl$. Made by treating toluene with chlorosulfonic acid.

Crystals, mp 69-71°. bp_{15} 146°. Insol in water; freely sol in alcohol, benzene, ether.

9461. Toluic Acid. *Methylbenzoic acid.* $C_8H_8O_2$; mol wt 136.14. C 70.57%, H 5.92%, O 23.50%. Prepn of *m*- and *o*-forms by oxidation of corresponding xylene: Toland, U.S. pat. **2,903,480** (1959 to California Res. Corp.); Hay *et al.*, *J. Org. Chem.* **25**, 616 (1960). Prepn of *p*-form by reaction of *p*-tolyldiazonium tetrafluoroborate with nickel carbonyl and acetic acid: Clark, Cookson, *J. Chem. Soc.* **1962**, 686; by oxidation of *p*-xylene: Taves, U.S. pat. **3,030,413** (1962 to Hercules Powder). Manuf of *p*-form from toluene: Braunworth, U.S. pat. **3,046,305** (1962 to Pure Oil).

m-Toluic acid, prisms from water, mp 111-113°, bp 263°. Sublimes. Sol in 1170 parts water at 15°, 60 parts boiling water; very sol in alcohol, ether.

o-Toluic acid, crystals, mp 107-108°; bp 258-260°; volatile with steam. Slightly sol in cold water; sol in 35 parts boiling water; very sol in alcohol.

p-Toluic acid, crystals, mp 180-181°; bp 274-275°. Sparingly sol in hot water; very sol in alcohol, ether, methanol.

9462. Toluidine. C_7H_9N; mol wt 107.15. C 78.46%, H 8.47%, N 13.07%. Prepn: J. S. Muspratt, A. W. Hofmann, *Ann.* **54**, 1 (1845); of each isomer: F. Beilstein, A. Kuhlberg, *ibid.* **156**, 66 (1870); P. Kovacic, J. L. Foote, *J. Am. Chem. Soc.* **83**, 743 (1961); P. Kovacic *et al.*, *ibid.* **84**, 759 (1962). Toxicity data: H. F. Smyth, *Am. Ind. Hyg. Assoc. J.* **23**, 95 (1962). GC determn in urine: K. El-Bayoumy *et al.*, *Cancer Res.* **46**, 6064 (1986).

m-Toluidine, *3-methylbenzamine, 3-aminotoluene, 3-methylaniline.* Liquid, mp ~ − 50°. bp 203-204°. d_{25}^{25} 0.990. n_D^{22} 1.5711. Slightly sol in water; sol in alcohol, ether, dil acids.

o-Toluidine, *2-methylbenzamine, 2-aminotoluene, 2-methylaniline.* Light yellow liquid becoming reddish brown on exposure to air and light. bp 200-202°. d_{20}^{20} 1.008. n_D^{20} 1.5688. Flash pt, closed cup: 185°F (85°C). Slightly sol in water; sol in alcohol, ether, dil acids. Keep well closed and protected from light. LD_{50} orally in rats: 0.94 g/kg (Smyth).

p-Toluidine, *4-methylbenzamine, 4-aminotoluene, 4-methylaniline.* Lustrous plates or leaflets, mp 44-45°. bp 200-201°. d_{20}^{20} 1.046. n_D^{59} 1.5532. Flash pt, closed cup: 188°F (86°C). Sol in about 135 parts water; freely sol in alcohol, ether, acetone, methanol, carbon disulfide, oils, dil acids.

Note: *o*-Toluidine and its hydrochloride may reasonably be anticipated to be carcinogens: *Fourth Annual Report on Carcinogens* (NTP 85-002, 1985) p 191.

USE: Manufacture of various dyes and other organic chemicals. *o*-Isomer also in printing textiles blue black; making colors fast to acids. *p*-Isomer also as a reagent for lignin, nitrite, phloroglucinol.

9463. *o*-Tolunitrile. *2-Methylbenzonitrile; 2-methylbenzenecarbonitrile;* *o*-cyanotoluene; *o*-toluenecarbonitrile. C_8-H_7N; mol wt 117.14. C 82.02%, H 6.02%, N 11.96%. CH_3-C_6H_4CN. Prepd from *o*-toluidine by diazotization in HCl soln and treatment of the diazonium chloride with potassium cuprocyanide: Herb, *Ann.* **258**, 9 (1890); Clarke, Read, *Org. Syn.* coll. vol. I (2nd ed., 1941) p 514.

Liquid; d_4^{20} 0.9955; d_4^{45} 0.9737; d_4^{75} 0.9481; mp −13°; bp_{760} 205.2°; bp_{100} 135°; bp_{40} 110°; bp_{20} 93°; bp_{10} 77.9°; bp_5 64°; $bp_{1.0}$ 36.7°; n_D^{23} 1.52720. Absorption spectrum: Baly, Ewbank, *J. Chem. Soc.* **87**, 1357 (1905); Purvis, *ibid.* **107**, 503 (1915). Insol in water. Miscible with alc, ether.

9464. *p*-Tolunitrile. Prepd from *p*-toluidine in the manner described for *o*-tolunitrile.

Needles from alc. d_4^{30} 0.9785; d_4^{45} 0.9640; d_4^{60} 0.9512; d_4^{75} 0.9390; mp 29.5°; bp_{760} 217.6°; bp_{100} 145.2°; bp_{60} 130°; bp_{40} 109.5°; bp_{20} 101.7°; bp_{10} 85.8°; bp_5 71.3°; $bp_{1.0}$ 42.5°. Absorption spectrum: Baly, Ewbank, *J. Chem. Soc.* **87**, 1357 (1905); Purvis, *ibid.* **107**, 503 (1915). Insol in water. Very sol in alcohol, ether.

9465. 2-(*p*-Toluyl)benzoic Acid. *p*-Toluyl-*o*-benzoic acid; 4'-methylbenzophenone-2-carboxylic acid. $C_{15}H_{12}O_3$; mol wt 240.25. C 74.98%, H 5.03%, O 19.98%. Prepd from phthalic anhydride and toluene in the presence of aluminum chloride: Friedel, Crafts, *Ann. Chim. Phys.* [6] **14**, 447 (1888); Fieser, *Org. Syn.* **4**, 73 (1925).

Monohydrate, triclinic pinacoidal crystals from alcohol. Sweet taste. Becomes anhyd at 100°, then melts at 146°. Slightly sol in boiling water; freely sol in alcohol, benzene, ether, acetone, boiling toluene.

Methyl ester, $C_{16}H_{14}O_3$, plates from methanol, mp 53°. Soluble in alcohol, benzene.

9466. Toluylene Blue. *N-[4-[(2,4-Diamino-5-methylphenyl)imino]-2,5-cyclohexadien-1-ylidene]-N-methylmethanaminium chloride; [4-(4,6-diamino-m-tolyl)imino-2,5-cyclohexadien-1-ylidene]dimethylammonium chloride;* C.I. 49410. $C_{15}H_{19}ClN_4$; mol wt 290.79. C 61.95%, H 6.59%, N 19.27%. Prepd by condensation of *p*-nitrosodimethylaniline with 2,4-diaminotoluylene: Witt, *Ber.* **12**, 933 (1879); Ger. pat. **15,272**; *Frdl.* **1**, 274. Also formed by irradiating a mixture of dimethyl-*p*-phenylenediamine and *m*-toluylenediamine: Loiseleur, *Compt. Rend.* **237**, 461 (1953). *See also Colour Index* vol. **4** (3rd ed., 1971) p 4443.

Monohydrate, prismatic, copper-brown, shiny crystals. Gives blue soln with cold water, alc, acetic acid.

USE: Biological stain.

9467. Tolycaine. *2-(2-Diethylaminoacetamido)-m-toluic acid methyl ester; 2-methyl-6-carbomethoxy-N-diethylaminoacetanilide; methyl 2-diethylaminoacetamido-m-toluate; 3-methyl-2-diethylaminoacetylaminobenzoic acid methyl ester.* $C_{15}H_{22}N_2O_3$; mol wt 278.34. C 64.72%, H 7.97%, N 10.07%, O 17.24%. Prepn: Hiltmann *et al.*, Ger. pat. **1,018,**-070 (1957 to Bayer); U.S. pat. **2,921,077** (1960 to Schenley).

Oil, bp$_5$ 190-192°.

Hydrochloride, $C_{15}H_{23}ClN_2O_3$, *Baycain*. Crystals, mp 139-140.5°.

THERAP CAT: Local anesthetic.

9468. Tolylhydrazine. $C_7H_{10}N_2$; mol wt 122.17. C 68.82%, H 8.25%, N 22.93%. Prepd by stannous chloride reduction of the diazonium salt of the corresponding toluidine: Hunsberger *et al.*, *J. Org. Chem.* **21**, 394 (1956).

m-Tolylhydrazine. Oily liquid. bp$_{760}$ 243°, bp$_{16}$ 132-134°. d$_{15}^{15}$ 1.061-1.062, d$_4^{20}$ 1.057-1.058. Insol in water; sol in alc, chloroform, ether.

Nitrate, $C_7H_{10}N_2$.HNO$_3$, needles, mp 145-147°.

o-Tolylhydrazine. Needles, mp 56-59°. Sparingly sol in water; sol in alcohol, chloroform, ether, slightly in cold petr ether.

Nitrate, leaflets, mp 98-100°. Very sol in water, alcohol; insol in ether.

p-Tolylhydrazine. Rhombic bipyramids, mp 61° or 65-66°. bp 240-244° with slight decompn. Sparingly sol in water; sol in alcohol, benzene, ether.

Nitrate, leaflets, mp 152-153°.

USE: *o*-Tolylhydrazine as reagent for galactose.

9469. p-Tolylsulfonylmethylnitrosamide. *N,4-Dimethyl-N-nitrosobenzenesulfonamide; N-methyl-N-nitroso-p-toluenesulfonamide; Diazald.* $C_8H_{10}N_2O_3S$; mol wt 214.25. C 44.85%, H 4.70%, N 13.08%, O 22.40%, S 14.97%. Prepd by the action of nitrous acid on *p*-tolylsulfonylmethylamide: **Ger. pat. 224,388** (1910 to Bayer); *Frdl.* **10**, 1216; *Chem. Zentr.* **1910, II**, 609; Takizawa, *J. Pharm. Soc. Japan* **70**, 490 (1950); de Boer, Backer, *Rec. Trav. Chim.* **73**, 229 (1954); *Org. Syn.* **34**, 96 (1954).

Yellow crystals from benzene + petr ether, mp 62°. Stable in an ordinary brown bottle for several years. A white coating (formed of *p*-tolylsulfonylmethylamide) does no harm. Insol in water. Sol in ether, petr ether, benzene, chloroform, carbon tetrachloride. Yields diazomethane on treatment with alkali.

USE: In the laboratory prepn of diazomethane. Directions for use: de Boer, Backer, *Rec. Trav. Chim.* **73**, 232 (1954).

9470. Tomatidine. (3β,5α,22β,25S)-*Spirosolan-3-ol*; 5α-*tomatidan-3β-ol*; 5α,20β$_F$,22α$_F$,25β$_F$,27-*azaspirostan-3β-ol*. $C_{27}H_{45}NO_2$; mol wt 415.64. C 78.02%, H 10.91%, N 3.37%, O 7.70%. By hydrolysis of tomatine: Kuhn *et al.*, *Ber.* **83**, 448 (1950). Isoln from the roots of Rutgers tomato plant [*Lycopersicon esculentum* Mill., cultivar. "Rutgers"]: Brink, Folkers, *J. Am. Chem. Soc.* **73**, 4018 (1951); Fontaine *et al.*, *ibid.* 878; Sato *et al.*, *ibid.* 880; Kuhn, Low, **U.S. pat. 2,770,618** (1956 to Amer. Home Prod.). Structure: Sato *et al.*, *J. Org. Chem.* **25**, 783 (1960); Schreiber, Adams, *Experientia* **17**, 13 (1961). Synthesis: Uhle, Moore, *J. Am. Chem. Soc.* **76**, 6412 (1954); Uhle, *ibid.* **83**, 1460 (1961); Kessar *et al.*, *Tetrahedron* **27**, 2869 (1971).

Plates from ethyl acetate, mp 202-206°. [α]$_D^{25}$ +8° (chloroform).

Hydrochloride, $C_{27}H_{45}NO_2$.HCl, crystals from abs ethanol, mp 265-270°. [α]$_D^{25}$ −5° (methanol).

9471. Tomatine. *Lycopersicin.* $C_{50}H_{83}NO_{21}$; mol wt 1034.22. C 58.07%, H 8.09%, N 1.35%, O 32.49%. Occurs in the extract of leaves of wild tomato plants: Fontaine *et al.*, *Arch. Biochem.* **18**, 467 (1948); Kuhn, Low, *Ber.* **81**, 552 (1948); Kuhn *et al.*, *ibid.* **83**, 448 (1950); Bognar, Makleit, *Pharmazie* **11**, 376 (1956). Yields on partial hydrolysis, besides α-tomatine, the main constituent, β$_1$-, β$_2$-, γ- and δ-tomatine: Kuhn *et al.*, *Ber.* **90**, 203 (1957). α-Tomatine consists of one mol tomatidine linked to a tetrasaccharide composed of 2 mols D-glucose, 1 mol D-xylose and 1 mol D-galactose: Kuhn *et al.*, *Angew. Chem.* **68**, 212 (1956). Structure: Reichstein, *ibid.* **74**, 887 (1962).

Needles from methanol, mp 263-268°. [α]$_D^{20}$ −18° (c = 0.55 in pyridine). Sol in ethanol, methanol, dioxane, propylene glycol; practically insol in water, ether, petr ether. Stable to strong alkali but hydrolyzed by acids to produce cryst tomatidine and a soln rich in reducing sugars. Has been found to inhibit the growth of various fungi and bacteria. LD orally in rats: 900-1000 mg/kg, Wilson *et al.*, *Toxicol. Appl. Pharmacol.* **3**, 39 (1961).

USE: Precipitating agent for steroids. Proposed as an alternate precipitant to digitonin: Schultz, Sander, *Z. Physiol. Chem.* **308**, 122 (1957).

9472. Tonin. β-Angiotensin-I converting enzyme (formerly). A converting enzyme that differs from renin, *q.v.*, in its ability to form angiotensin II directly from angiotensinogen by cleaving the Phe-His bond; it can also convert angiotensin I to angiotensin II. Its presence was discovered in rat submaxillary glands: R. Boucher *et al.*, *Hypertension* 72, J. Genest, E. Koiw, Eds. (Springer-Verlag, New York, 1972) p 512; *eidem, Circ. Res.* **Suppl. I**, 203 (1974). Purification and characterization: S. Demassieux *et al.*, *Can. J. Biochem.* **54**, 788 (1976). Crystal data: K. Hayakawa *et al.*, *J. Mol. Biol.* **123**, 107 (1978). Radioimmunoassay: J. Gutkowska *et al.*, *Can. J. Biochem.* **56**, 769 (1978). Purification by affinity chromatography: M. Ikeda *et al.*, *Hypertension* **3**, 81 (1981); by gel permeation and HPLC: C. Lazure *et al.*, *Anal. Biochem.* **125**, 406 (1982). Isoln using chromatofocusing: E. S. P. Cheng, B. J. Morris, *ibid.* **126**, 295 (1982). N-Terminal amino acid sequence of rat tonin: N. G. Seidah *et al.*, *Can. J. Biochem.* **56**, 920 (1978). Substrate specificity studies: *eidem, Proc. Am. Peptide Symp. 6th*, E. Gross, J. Meienhofer, Eds. (Pierce Chem. Co., Rockford, Ill., 1979) p 921; M. Chretien *et al.*, *FEBS Letters* **113**, 173 (1980). Formation of angiotensin II by tonin from partially purified human angiotensinogen: C. Grise et al., Can. J. Biochem. **59**, 250 (1981). Pressor effect in anephric animals: E. L. Schiffrin et al., Can. J. Physiol. Pharmacol. **59**, 864 (1981). Sequence homologies between tonin and other peptides: C. Lazure et al., Nature **292**, 383 (1981). Immunohistochemical study: T. B. Oerstavik et al., J. Histochem. Cytochem. **30**, 1123 (1982). Role as renin activator: J. Gutkowska et al., Can. J. Biochem. **60**, 843 (1982). Role in exptl hypertension: R. Garcia et al., Hypertension, Int. Symp., 3rd, H. Villarreal, Ed. (Wiley, New York, 1981) p 79. Reviews: R. Boucher et al., Circ. Res. Suppl. II, 26-29 (1977); R. Boucher, J. Genest, Endocrine Functions of the Brain, M. Motta, Ed. (Raven Press, New York, 1980) pp 373-384; J. Genest, Heterogeneity of Renin and Renin Substrate, M. P. Sambhi, Ed. (Elsevier, New York, 1981) pp 11-24.

Mol wt determn is 31,400 by gel filtration and 28,700 by sedimentation equilibrium. Activity is not affected by pepstatin. Can be incubated at 20° for 150 min, between pH 3.4-8 without significant loss of enzymatic activity; loses 15% of its original activity at pH 2.8. After 5 min of incubation at 100°, 60-65% activity remains.

9473. Torasemide. N-[[(1-Methylethyl)amino]carbonyl]-4-[(3-methylphenyl)amino]-3-pyridinesulfonamide; 1-isopropyl-3-[(4-m-toluidino-3-pyridyl)sulfonyl]urea; 3-isopropylcarbamylsulfonamido-4-(3'-methylphenyl)aminopyridine; torsemide; AC 4464; BM 02015; JDL-464. $C_{16}H_{20}N_4O_3S$; mol wt 348.42. C 55.16%, H 5.78%, N 16.08%, O 13.78%, S 9.20%. Sulfonylurea loop diuretic. Prepn: J. E. DeLarge et al., **Ger. pat. 2,516,025**; eidem, **U.S. pat. 4,018,929** (1975, 1977 both to A. Christiaens, S.A.); J. DeLarge, C. L. Lapiere, Ann. Pharm. Fr. **36**, 369 (1978). Pharmacokinetics in humans: M. Lesne et al., Int. J. Clin. Pharmacol. Ther. Toxicol. **20**, 382 (1982). Preliminary evaluation in acute heart failure: R. Stroobandt et al., Arch. Int. Pharmacodyn. **260**, 151 (1982). Clinical pharmacology: D. C. Brater et al., Clin. Pharmacol. Ther. **42**, 187 (1987). Series of articles on pharmacology, mode of action and renal effects in animals: Arzneimittel-Forsch. **35**, 1520-1541 (1985); on pharmacology, pharmacokinetics and clinical studies: Eur. J. Clin. Pharmacol. **31**, Suppl., 1-55 (1986); Arzneimittel-Forsch. **38**, 143-214 (1988).

mp 163-164°. pKa 6.44.
THERAP CAT: Diuretic.

9474. Toremifene. (Z)-2-[4-(4-Chloro-1,2-diphenyl-1-butenyl)phenoxy]-N,N-dimethylethanamine; (Z)-4-chloro-1,2-diphenyl-1-[4-[2-(N,N-dimethylamino)ethoxy]phenyl]-1-butene. $C_{26}H_{28}ClNO$; mol wt 405.97. C 76.92%, H 6.95%, Cl 8.73%, N 3.45%, O 3.94%. Nonsteroidal antiestrogen structurally similar to tamoxifen, q.v. Prepn: R. J. Toivola et al., **Eur. pat. Appl. 95,875**; **U.S. pat. 4,696,949** (1983, 1987 both to Farmos). Pharmacology: S. Kallio et al., Cancer Chemother. Pharmacol. **17**, 103 (1986). Antitumor effects in vitro and in vivo: L. Kangas et al., ibid. 109. HPLC determn in plasma: W. M. Holleran et al., Anal. Letters **20**, 871 (1987). Clinical evaluation in advanced breast cancer: R. Valavaara et al., Eur. J. Cancer Clin. Oncol. **24**, 785 (1988). Review: L. Kangas, Prog. Cancer Res. Ther. **35**, Proc. 3rd Int. Congr. Hormones Cancer, Hamburg, 1987, 374-377 (1988).

mp 108-110°.
Citrate, $C_{32}H_{36}ClNO_8$, FC-1157a, Fareston. mp 160-162°.

THERAP CAT: Antiestrogen; antineoplastic.

9475. Toril Oil. From the fruit of Torilis anthriscus (L.) Gmel., Umbelliferae.
A Japanese folk remedy for ascaris. It is relatively nontoxic for higher animals and very toxic for Lumbricus, leech and ascaris.

9476. Torularhodin. 3',4'-Didehydro-β,ψ-caroten-16'-oic acid. $C_{40}H_{52}O_2$; mol wt 564.82. C 85.05%, H 9.28%, O 5.67%. Carotenoid pigment found in Torula rubra and Rhodotorula mucilaginosa yeasts. Isoln: Karrer, Rutschmann, Helv. Chim. Acta **26**, 2109 (1943). Structure and synthesis: Isler et al., ibid. **42**, 864 (1959).

Fine dark purple needles from methanol + ether or toluene, mp 210-212° (vac, some decompn). Absorption max in CS_2: 582, 541, 502 nm; in methanol: 529, 493, 460 nm. Freely sol in carbon disulfide, chloroform, pyridine; less sol in ether, benzene, hot ethanol; sparingly sol in methanol; practically insol in petr ether.
Methyl ester, $C_{41}H_{54}O_2$, dark red needles from benzene + methanol, mp 172-173°.

9477. Tosufloxacin. (±)-7-(3-Amino-1-pyrrolidinyl)-1-(2,4-difluorophenyl)-6-fluoro-1,4-dihydro-4-oxo-1,8-naphthyridine-3-carboxylic acid; A-61827. $C_{19}H_{15}F_3N_4O_3$; mol wt 404.35. C 56.44%, H 3.74%, F 14.09%, N 13.86%, O 11.87%. Trifluorinated quinolone antibacterial. Prepn: H. Narita et al., **Ger. pat. 3,514,076**, C.A. **104**, 129888r (1986); **Belg. pat. 904,086**, C.A. **105**, 208850w (1986); Y. Todo et al., **U.S. pat. 4,704,459** (1985, 1985, 1987 all to Toyama); and activity: D. T. W. Chu et al., J. Med. Chem. **29**, 2363 (1986); H. Narita et al., Yakugaku Zasshi **106**, 802 (1986), C.A. **106**, 196291v (1987). In vitro activity studies of the base: P. B. Fernandes et al., Antimicrob. Ag. Chemother. **32**, 27 (1988); and in vivo animal studies of the toluenesulfonate: M. Takahata et al., J. Antimicrob. Chemother. **22**, 143 (1988). Series of articles on antibacterial activity and clinical evaluation of the toluenesulfonate: Chemotherapy (Tokyo) **36**, Suppl. 9, 1-1538 (1988).

Hydrochloride, $C_{19}H_{16}ClF_3N_4O_3$, A-60969. Crystals from conc HCl-ethanol (1:3), mp 247-250° (dec).
Toluenesulfonic acid salt monohydrate, $C_{26}H_{23}F_3N_4O_6S \cdot H_2O$, A-64730, T-3262, tosufloxacin tosilate. mp 258-260°.
THERAP CAT: Antibacterial.

9478. Toxaphene. Chlorinated camphene; camphechlor; polychlorocamphene; Hercules 3956; Alltox; Geniphene; Motox; Penphene; Phenacide; Phenatox; Strobane-T; Toxakil. A very complex, but reproducible mixture of at least 177 C_{10} polychloro derivs., having an approx overall empirical formula of $C_{10}H_{10}Cl_8$. Produced by the chlorination of camphene to 67-69% chlorine by weight and made up of compds of $C_{10}H_8Cl_{10}$, $C_{10}H_{18-n}Cl_n$ (mostly polychlorobornanes) and $C_{10}H_{16-n}Cl_n$ (polychlorobornenes and/or polychlorotricyclenes) with n = 6 to 9. Prepn: Buntin, **U.S. pat. 2,565,471** (1951 to Hercules Powder). Isoln of compo-

Consult the cross index before using this section.

nents in crystalline form: Casida *et al.*, *Science* **183**, 520 (1974); *eidem*, *J. Agr. Food Chem.* **22**, 939 (1974). Acute toxicity data: T. B. Gaines, *Toxicol. Appl. Pharmacol.* **14**, 515 (1969). Mutagenicity studies: N. K. Hooper *et al.*, *Science* **205**, 591 (1979). Livestock toxicity and tissue residues: L. Penumarthy *et al.*, *Vet. Toxicol.* **18**, 60 (1976). *Review:* Liebmann *et al.*, *Arch. Pflanzenschutz* **7**, 131-150 (1971); F. Korte *et al.*, *Pure Appl. Chem.* **51**, 1583-1601 (1979).

Yellow waxy solid, mp 65-90°. Pleasant piney odor. Dehydrochlorinates in the presence of alkali, prolonged exposure to sunlight, and at temps about 155°. Practically insol in water. Freely sol in aromatic hydrocarbons. Corrosive to iron. LD$_{50}$ in male, female rats (mg/kg): 90, 80 orally (Gaines).

Caution: Can cause mild irritation of, and be absorbed through, skin. Causes CNS stimulation with tremors, convulsions, death. Liver injury has been reported in exptl animals. *See: Clinical Toxicology of Commercial Products*, R. E. Gosselin *et al.*, Eds. (Williams & Wilkins, Baltimore, 4th ed., 1976) Section III, pp 313-315. This substance may reasonably be anticipated to be a carcinogen: *Fourth Annual Report on Carcinogens* (NTP 85-002, 1985) p 191.

USE: Insecticide. *Compare* Strobane. Not recommended for use in dairy barns or on milking animals, L. Penumarthy *et al.*, *loc. cit.*

9479. Toxiferine I. *C*-Toxiferine I. $[C_{40}H_{46}N_4O_2]^{2+}$; mol wt 614.80. From calabash curare: Schmid, Karrer, *Helv. Chim. Acta* **30**, 1162 (1947); from *Strychnos toxifera* Schomb., *Loganiaceae:* Wieland *et al.*, *Ann.* **547**, 156 (1941); King, *J. Chem. Soc.* **1949**, 3263. Identity with *toxiferine V* and *toxiferine XI:* Battersby *et al.*, *ibid.* **1960**, 1848. Structure: Arnold *et al.*, *Helv. Chim. Acta* **44**, 620 (1961). Synthesis: Berlage *et al.*, *ibid.* **42**, 394 (1959); Grdinic *et al.*, *J. Am. Chem. Soc.* **86**, 3357 (1964). Pharmacokinetics: P. G. Waser, J. Reller, *Agents Actions* **2**, 170 (1972). ^{13}C-NMR study: E. Wenkert *et al.*, *J. Org. Chem.* **43**, 1099 (1978).

Dichloride, $C_{40}H_{46}Cl_2N_4O_2$, crystals. $[\alpha]_D^{22}$ −546° (c = 0.30). uv max (ethanol): 292 nm (log ε 4.62). Sol in water.

9480. Toxoflavin. *1,6-Dimethylpyrimido[5,4-e]-1,2,4-triazine-5,7(1H,6H)-dione;* 1,6-dimethyl-5,7-dioxo-1,5,6,7-tetrahydropyrimido[5,4-e]-*as*-triazine; xanthothricin. C_7H_7-N_5O_2; mol wt 193.17. C 43.52%, H 3.65%, N 36.26%, O 16.57%. Highly toxic antibiotic from cultures of *Pseudomonas cocovenenans* which also produces bongkrekic acid, *q.v.* Isoln: van Veen, Mertens, *Rec. Trav. Chim.* **53**, 257, 398 (1934); Machlowitz *et al.*, *Antibiot. & Chemother.* **4**, 259 (1954). Structure: Van Damm *et al.*, *Rec. Trav. Chim.* **79**, 255 (1960). Synthesis: Daves *et al.*, *J. Am. Chem. Soc.* **83**, 3904 (1961); **84**, 1724 (1962); Yoneda *et al.*, *Tetrahedron Letters* **1971**, 851. Mode of action: Latuasan, Berends, *Biochim. Biophys. Acta* **52**, 502 (1961). Biosynthesis: Levenberg, Linton, *J. Biol. Chem.* **241**, 846 (1966).

Bright yellow platelets from propanol, dec 172-173°. uv max: 257.5, 394 nm (ε 16400, 2500). Acts as a pH indicator with a sharp loss of color at pH 10.5, but is destroyed by alkali. Sol in water, chloroform, ethyl acetate, ethanol. LD$_{50}$ in mice: 1.7 mg/kg i.v.; 8.4 mg/kg orally.

9481. Toxohormone. Name given in 1948 by Nakahara and Fukuoka to a factor produced by living cancer cells and released into the circulation to produce decreases in liver catalase activity, tryptophan pyrrolase activity, liver ferritin and plasma iron. *See* Nakahara, Fukuoka, *Japan. J. Cancer Res.* **40**, 45 (1949). This tumor-specific concept has been questioned: Greenfield, Meister, *J. Nat. Cancer Inst.* **11**, 997 (1951); Olivares *et al.*, *Science* **157**, 327 (1967); Kampschmidt, Upchurch, *Proc. Soc. Exp. Med. Biol.* **127**, 632 (1968). Described as a polypeptide of low mol wt having 30-40 amino acid residues of which 12-13 are different, and with a high content of glycine, glutamic acid, aspartic acid, alanine and leucine. Amino acid and lipid composition of a highly purified toxohormone prepd from human malignant tissue: Yunoki, Griffin, *Cancer Res.* **21**, 537 (1961); from cell-free fluid of ascites sarcoma 180: H. Masuno *et al.*, *ibid.* **41**, 284 (1981). *Reviews:* W. Nakahara, F. Fukuoka, *Chemistry of Cancer Toxin Toxohormone* (C. C. Thomas, Springfield, Illinois, 1961) 75 pp; Nakahara, *Methods Cancer Res.* **2**, 203-237 (1967); Olivares, Kampschmidt in *Oncology, Proc. Int. Cancer Congr., 10th, Houston 1970* **3**, 158-170 (1971); S. Fujii, *Gann Monogr. Cancer Res.* **24**, 215-222 (1979).

Thermostable, non-heat-coagulable, water-sol and alcohol precipitable.

9482. Toxopyrimidine. *4-Amino-2-methyl-5-pyrimidinemethanol;* 6-amino-5-hydroxymethyl-2-methylpyrimidine; 4-amino-5-hydroxymethyl-2-methylpyrimidine; pyramin; pyramine. $C_6H_9N_3O$; mol wt 139.16. C 51.78%, H 6.52%, N 30.20%, O 11.50%. A metabolite of thiamine. Prepd from 4-amino-5-aminomethyl-2-methylpyrimidine dihydrochloride or ethyl 4-amino-2-methyl-5-pyrimidine-carboxylate: Dornow, Petsch, *Ber.* **86**, 1404 (1953); DiBella, Hennessy, *J. Org. Chem.* **26**, 2017 (1961).

Needles from water, mp 193-198°; crystals from methanol + ether, mp 198-200°. Sublimes at 0.01 mm between 155-170°.

9483. Toyocamycin. *4-Amino-7-β-D-ribofuranosyl-7H-pyrrolo[2,3-d]pyrimidine-5-carbonitrile;* 4-amino-5-cyano-7-(D-ribofuranosyl)-7H-pyrrolo[2,3-d]pyrimidine; uramycin B; vengicide; E-212; antibiotic 1037. $C_{12}H_{13}N_5O_4$; mol wt 291.26. C 49.48%, H 4.50%, N 24.05%, O 21.97%. Antibiotic substance extracted from the culture filtrate and mycelium of *Streptomyces toyocaensis.* Isoln: Nishimura *et al.*, *J. Antibiot.* **9A**, 60 (1956). Structure: Ohkuma, *ibid.* **14A**, 343 (1961). Synthesis of aglycone: Taylor, Hendess, *J. Am. Chem. Soc.* **86**, 951 (1964). Total synthesis: Tolman *et al.*, *ibid.* **90**, 524 (1968); **91**, 2102 (1969). Biosynthesis: Uematsu, Suhadolnik, *Biochemistry* **9**, 1260 (1970); *eidem*, *J. Biol. Chem.* **245**, 4365 (1970). Crystal and molecular structure: P. Prusiner, M. Sundaralingam, *Acta Crystallogr.* **B34**, 517 (1978).

Fine needles from methanol or acetone, mp 243°. Recrystallization from water yields the hydrate, $C_{12}H_{13}N_5O_4 \cdot H_2O$, mp 239-243°. $[\alpha]_D^{16}$ -45.7° (c = 1.05 in 0.1N HCl). uv max (H_2O): 230, 277 nm ($E_{1cm}^{1\%}$ 400, 548). Soluble in acetic acid, acidic solns. Moderately sol in methanol, ethanol, acetone, dioxane, butanol, water, ether. Practically insol in chloroform, ethyl acetate, petr ether. LD_{100} in mice: 10-20 mg/kg s.c., Nishimura *et al., loc. cit.*

9484. Tralkoxydim. *2-[1-(Ethoxyimino)propyl]-3-hydroxy-5-(2,4,6-trimethylphenyl)-2-cyclohexen-1-one;* 2-[1-(ethoxyimino)propyl]-3-hydroxy-5-mesitylcyclohex-2-en-1-one; PP604; Grasp; Splendor. $C_{20}H_{27}NO_3$; mol wt 329.44. C 72.92%, H 8.26%, N 4.25%, O 14.57%. Cereal selective post-emergent herbicide. Prepn: R. B. Warner *et al.,* **Eur. pat. Appl. 80,301;** *eidem,* **U.S. pat. 4,717,418** (1983, 1988 both to ICI). Physical properties and herbicidal activity: R. B. Warner *et al., Proc. Brit. Crop Prot. Conf. - Weeds* **1987,** 19. Field trials on grass weeds: J. Rola, *ibid.* 363; P. B. Sutton *et al., ibid.* 389. Mechanism of action study: J. Secor, C. Cseke, *Plant Physiol.* **86,** 10 (1988).

White crystalline solid, mp 106°. Vapor pressure at 20°: 4×10^{-10} kPa. Soly at 20° (mg/l): water 6 at pH 6.5, 5 at pH 5.0; at 24° (g/l): hexane 18; toluene 213; dichloromethane > 500; methanol 25; acetone 89; ethyl acetate 110. LD_{50} orally in male and female rats, male and female mice, male rabbit (mg/kg): 1324, 934, 1231, 1100, 519 mg/kg; dermally in male and female rats: > 2000, > 2000 mg/kg (Warner 1987).

USE: Post-emergent herbicide.

9485. Tramadol. *2-[(Dimethylamino)methyl]-1-(3-methoxyphenyl)cyclohexanol;* E-265; CG 315E; U-26,225A. $C_{16}H_{25}NO_2$; mol wt 263.39. C 72.96%, H 9.57%, N 5.32%, O 12.15%. Prepn: **Brit. pat. 997,399** (1965 to Grünenthal), *C.A.* **63,** 9871g (1965); K. Flick, E. Frankus, **U.S. pat. 3,652,589** (1972 to Grünenthal); *Arzneimittel-Forsch.* **28,** 107 (1978). Series of articles pertaining to pharmacology, toxicology and clinical studies: *ibid.* 114-219.

Hydrochloride, $C_{16}H_{26}ClNO_2$, *Crispin, Tramal.* White crystals, mp 180-181°. Sol in water. LD_{50} in mice, rats (mg/kg): 350, 228 orally, 200, 286 s.c., F. Lagler *et al., Arzneimittel-Forsch.* **28,** 164 (1978).

THERAP CAT: Analgesic.

9486. Tramazoline. *4,5-Dihydro-N-(5,6,7,8-tetrahydro-1-naphthalenyl)-1H-imidazol-2-amine;* 2-[(5,6,7,8-tetrahydro-1-naphthyl)amino]-2-imidazoline. $C_{13}H_{17}N_3$; mol wt 215.29. C 72.52%, H 7.96%, N 19.52%. Prepn: Berg, **Ger. pats. 1,191,381; 1,195,323** (both 1965 to Thomae), *C.A.* **63,** 8373c; 13274d (1965). Pharmacology and toxicity data: R. Engelhorn, H. Klupp, *Arzneimittel-Forsch.* **12,** 971 (1962). Activity studies: Sachsenröder *et al., ibid.* **22,** 392 (1972).

Crystals from isopropanol, mp 142-143°.

Hydrochloride monohydrate, $C_{13}H_{18}ClN_3 \cdot H_2O$, *KB 227, Biciron, Ellatun, Rhinaspray, Rhinogutt, Rhinospray, Rinogutt, Towk.* Crystals from alc + ether or acetone + ether, mp 172-174°. Sol in water. LD_{50} orally in mice: 195 mg/kg (Engelhorn, Klupp).

THERAP CAT: Adrenergic.

9487. Tranexamic Acid. *4-(Aminomethyl)cyclohexanecarboxylic acid;* RP 18429; AMCHA; Amikapron; Amstat; Anvitoff; Cyklokapron; Emorhalt; Exacyl; Frenolyse; Hexapromin; Hexatron; Rikavarin; Spiramin; Tranex; Tranexan; Transamin; Trasamlon; Ugurol. $C_8H_{15}NO_2$; mol wt 157.21. C 61.12%, H 9.62%, N 8.91%, O 20.35%. Prepn: Einhorn, Ladisch, *Ann.* **310,** 194 (1900); Levine, Sedlecky, *J. Org. Chem.* **24,** 115 (1959); **Neth. pat. Appl. 6,503,605;** *eidem,* Naito *et al.,* **U.S. pat. 3,499,925** (1965, 1970 to both Daiichi Seiyaku and Mitsubishi Chem.). Pharmacology: Andersson *et al., Scand. J. Haematol.* **2,** 230 (1965), *C.A.* **64,** 2608c (1966). Controlled clinical study in treatment of acute upper gastrointestinal tract bleeding: D. Barer *et al., N. Engl. J. Med.* **308,** 1571 (1983). Toxicity data: B. Melander *et al., Acta Pharmacol. Toxicol.* **22,** 340 (1965). *Review:* Kjellman, Schannong, *Nord. Med.* **83,** 166 (1970); *Japan. Med. Gaz.* **9**(6), 10 (1972).

Crystals, softening at 270°, but still not melted at 280°: Einhorn, Ladisch, *loc. cit.* Also reported as mp 386-392° (dec). Soly in water: about 1 g/6 ml. Very slightly sol in alcohol, ether. Practically insol in most other organic solvents. Chemically stable; not hygroscopic. LD_{50} in mice, rats (mg/kg): 1500, 1200 i.v. (Melander).

THERAP CAT: Hemostatic.

9488. Tranid®. *5-Chloro-6-[[[(methylamino)carbonyl]oxy]imino]bicyclo[2.2.1]heptane-2-carbonitrile; exo-5-chloro-6-oxo-endo-6-norbornanecarbonitrile syn-O-(methylcarbamoyl)oxime; exo-3-chloro-endo-6-cyanobicyclo[2.2.1]heptan-2-one syn-(N-methylcarbamyl)oxime; syn-exo-3-chloro-endo-6-cyano-2-norbornanone O-(methylcarbamoyl)oxime;* UC-20047. $C_{10}H_{12}ClN_3O_2$; mol wt 241.68. C 49.70%, H 5.01%, Cl 14.67%, N 17.39%, O 13.24%. Prepn: Kilsheimer, Manning; Payne, **U.S. pats. 3,231,599** and **3,328,457** (1966, 1967, both to Union Carbide). Stereoisomers: Payne *et al., J. Med. Chem.* **8,** 525 (1965).

mp 159-160°. LD_{50} orally in female, male rats: 19, 30 mg/kg, T. B. Gaines, *Toxicol. Appl. Pharmacol.* **14,** 515 (1969).

USE: Insecticide, miticide.

9489. Tranilast. *2-[[3-(3,4-Dimethoxyphenyl)-1-oxo-2-propenyl]amino]benzoic acid; N-(3',4'-dimethoxycinnamoyl)anthranilic acid; N-5';* Rizaben. $C_{18}H_{17}NO_5$; mol wt 327.35. C 66.04%, H 5.24%, N 4.28%, O 24.44%. Orally active anti-allergic agent. Prepn: K. Harita *et al.,* **Ger. pat. 2,402,398** corresp to **U.S. pat. 3,940,422** (1974, 1976 to Kissei). Pharmacological properties: H. Azuma *et al., Brit. J. Pharmacol.* **58,** 483 (1976). Mechanism of action study: Y. Iijima *et al., Biochem. Biophys. Res. Commun.* **93,** 912 (1980). Clinical study in pediatric bronchial asthma: H. Shioda, *Allergy* **34,** 213 (1979). Toxicity studies: M. Nakazawa *et al., Oyo Yakuri* **12,** 385, 407 (1976), *C.A.* **88,** 115327-8 (1978). Series of articles on teratogenicity tests: *Iyakuhin Kenkyu* **9,** 148-193 (1978), *C.A.* **88,** 130930f, 146300m-303q (1978).

Consult the cross index before using this section.

Cryst from chloroform, mp 211-213°. LD_{50} in male mice, rats (mg/kg): 780, 1600 orally, 410, 405 i.p.; 2630, 3630 s.c., M. Nakazawa *et al.*, *Oyo Yakuri* **12**, 305 (1976), *C.A.* **88**, 115327 (1978).

THERAP CAT: Antiallergic.

9490. Transferrins. A group of homologous non-heme, iron-binding glycoproteins of approx mol wts of 76,000-81,000. They are widely distributed in a variety of physiological fluids and cells, esp in the sera of most vertebrates, in egg whites and in mammalian milk, tears and leukocytes. They are involved in iron transport to developing red cells for hemoglobin synthesis. Each protein molecule specifically binds with two Fe^{3+} ions to form salmon-pink complexes; bicarbonate or carbonate ions are involved in the formation of these colored complexes. Reviews of isoln, properties and biological functions: Feeney, Komatsu, *Struct. Bonding* **1**, 149-206 (1966); Aisen, "The Transferrins" in *Inorganic Biochemistry*, vol. 1, G. L. Eichhorn, Ed. (Elsevier, New York, 1973) pp 280-305; Bezkorovainy, Zschocke, *Arzneimittel-Forsch.* **24**, 476-485, 726-737 (1974); P. Aisen, A. Leibman, *Bioinorganic Chemistry* **II**, K. N. Raymond, Ed. (A.C.S., Washington, 1977) pp 104-126; P. Aisen, I. Listowsky, *Ann. Rev. Biochem.* **49**, 357-393 (1980). Review of transferrin receptors: R. Newman *et al.*, *Trends Biochem. Sci.* **7**, 397-399 (1982).

Serum transferrin, β₁-metal-combining protein, siderophilin. Commonly called transferrin. Structure of carbohydrate moiety: Jamieson *et al.*, *J. Biol. Chem.* **246**, 3686 (1971). Composed of two homologous domains, each containing a binding site for metal ions. The sites are similar, but not identical, in their metal-binding properties. X-ray studies: L. J. DeLucas *et al.*, *J. Mol. Biol.* **123**, 285 (1978). Resolution of the two sites by Eu(III) excitation spectroscopy: P. B. O'Hara, R. Bersohn, *Biochemistry* **21**, 5269 (1982). N-Terminal amino acid sequence of human serum transferrin: M.-H. Metz-Boutigue *et al.*, *Biochim. Biophys. Acta* **670**, 243 (1981). Absorption max of human serum Fe^{3+}-transferrin: about 465 nm ($E_{1cm}^{1\%}$ 0.57); uv max: 280 nm ($E_{1cm}^{1\%}$ 14.3).

Conalbumin, ovotransferrin, Diarconal. Isolated from egg white; distinguished from ovalbumin by its lower thermal coagulation point: Osborne, Campbell, *J. Am. Chem. Soc.* **22**, 422 (1900). Sepn from other egg-white proteins: Longworth *et al.*, *ibid.* **62**, 2580 (1940). Primary structure of hen ovotransferrin: J. Williams *et al.*, *Eur. J. Biochem.* **122**, 297 (1982). Purification, characterization and function of the iron-binding fragments: W.-M. Keung *et al.*, *J. Biol. Chem.* **257**, 1177, 1184 (1982). Antibacterial activity: P. Valenti *et al.*, *Antimicrob. Ag. Chemother.* **21**, 840 (1982). Absorption max of Fe^{3+}-complex: 470 nm ($E_{1cm}^{1\%}$ 0.62).

Lactoferrin, lactotransferrin. Important component of the human milk bacteriostatic system; also found in human and bovine tear proteins. Isoln from human whey by a single chromatographic step: L. Bläckberg, O. Hernell, *FEBS Letters* **109**, 180 (1980). Sequential purification of lactoferrin, lysozyme, and secretory IgA from human milk: M. Boesman-Finkelstein, R. A. Finkelstein, *ibid.* **144**, 1 (1982). Partial C-terminal amino acid sequence of human lactoferrin: M.-H. Metz-Boutigue *et al.*, *ibid.* **142**, 107 (1982).

9491. Tranylcypromine. *trans-(±)-2-Phenylcyclopropanamine*; SKF *trans-385*; Parnitene. $C_9H_{11}N$; mol wt 133.19. C 81.16%, H 8.33%, N 10.52%. Monoamine oxidase inhibitor. Prepn: Burger, Yost, *J. Am. Chem. Soc.* **70**, 2198 (1948); Tedeschi, U.S. pat. **2,997,422** (1961 to SK & F).

Liquid, $bp_{1.5-1.6}$ 79-80°.

Hydrochloride, $C_9H_{12}ClN$, crystals from ethyl acetate + ether, mp 164-166°.

Sulfate, $C_{18}H_{24}N_2O_4S$, *Parnate, Tylciprine.* Crystals, sol in water; very slightly sol in alcohol, ether. Practically insol in chloroform.

THERAP CAT: Antidepressant.

9492. Trapidil. *N,N-Diethyl-5-methyl-[1,2,4]triazolo-[1,5-a]pyrimidin-7-amine; 7-diethylamino-5-methyl-s-triazolo[1,5-a]pyrimidine;* trapymin; AR 12008; Rocornal. $C_{10}H_{15}N_5$; mol wt 205.27. C 58.51%, H 7.37%, N 34.12%. First triazolopyrimidine registered as a drug. Prepn: E. Tenor *et al.*, *East Ger. pat.* **55,956** (1967), *C.A.* **67**, 90830f (1967); E. Tenor, R. Ludwig, *Pharmazie* **26**, 534 (1971). Physical and chemical properties: S. Pfeifer *et al.*, *ibid.* 539. Pharmacology and toxicology: H. Fuller *et al.*, *ibid.* 554. Metabolic studies: S. Pfeifer *et al.*, *ibid.* 549; *ibid.* **27**, 752 (1972); I. Bornschein *et al.*, *ibid.* **33**, 51 (1978). Mechanism of action: K. Satoh *et al.*, *Arzneimittel-Forsch.* **30**, 1264 (1980). Antiarrhythmic activities in rabbits: M. Sakanashi *et al.*, *ibid.* **33**, 215 (1983). Clinical hemodynamic effects: M. Di Donato *et al.*, *ibid.* **35**, 1295 (1985). GC determn in biological fluids: A. Marzo *et al.*, *ibid.* **37**, 947 (1987).

White to yellowish, odorless and bitter crystalline powder, mp 98-99.4° (Pfeifer); 102-104° from heptane (Tenor). Eutectic temp of mixture with azobenzene: 48°. Very sol in water, 1N sulfuric acid, 10% ammonium hydroxide; easily sol in methanol, isopropanol, *n*-butanol, chloroform and benzene; sol in ether. Practically insol in hexane, heptane. $pK_s = 2.79$. uv max (methanol): 222, 270, 307 nm (log ε 4.28, 3.83, 4.28). Very stable except under extremely alk conditions. LD_{50} in mice, rats (mg/kg): 115, 76 i.v.; 380, 235 orally; 155, 100 i.p.; 132, 100 s.c. (Fuller).

Hydrochloride, $C_{10}H_{16}ClN_5$, mp 212°.

THERAP CAT: Coronary vasodilator.

9493. Traumatic Acid. *2-Dodecenedioic acid;* 1-decene-1,10-dicarboxylic acid. $C_{12}H_{20}O_4$; mol wt 228.28. C 63.13%, H 8.83%, O 28.03%. A wound hormone of plants. Isoln from pods of green beans: English *et al.*, *Proc. Nat. Acad. Sci. USA* **25**, 323 (1939). The naturally occurring traumatic acid is the *trans*-form. Synthesis of the *trans*-form: *eidem, J. Am. Chem. Soc.* **61**, 3434 (1939); U.S. pats. **2,339,259** (1944); **2,391,824** (1945); Truscheit, Eiter, *Ann.* **658**, 86 (1962); Dolezal, *Coll. Czech. Chem. Commun.* **35**, 1932 (1970); Schreurs *et al.*, *Rec. Trav. Chim. Pays-Bas Belg.* **90**, 1331 (1971); Prakasa Rao, Nayak, *Synthesis* **1975**, 608; J. H. Babler, R. K. Moy, *Syn. Commun.* **9**, 669 (1979). Synthesis of the *cis*-form: Lauer, Gensler, *J. Am. Chem. Soc.* **67**, 1171 (1945).

trans-Form, crystals from alc, acetone, or 1,2-dimethoxyethane, mp 166-167°. $bp_{0.001}$ 150-160°. Very sparingly sol in water; sol in alc, ether, benzene, chloroform.

cis-Form, crystals from ethyl acetate and petr ether, mp 67-68°.

9494. Traxanox. *9-Chloro-7-(1H-tetrazol-5-yl)-5H-[1]benzopyrano[2,3-b]pyridin-5-one;* 9-chloro-7-(5-1H-tetrazolyl)-5-oxo-5H-[1]benzopyrano[2,3-b]pyridine. $C_{13}H_6$-ClN_5O_2; mol wt 299.68. C 52.10%, H 2.02%, Cl 11.83%, N 23.37%, O 10.68%. Antiallergic agent which selectively inhibits release of allergic mediators from mast cells. Prepn: T. Oe, M. Tsuruda, Ger. pat. **2,521,980**; *eidem*, U.S. pat. **4,085,111** (1975, 1978 both to Yoshitomi). Antianaphylactic activity in animals: K. Goto *et al.*, *Japan. J. Pharmacol.* **30**, 537 (1980). Mechanism of action study: K. Goto *et al.*, *Int. Arch. Allergy Appl. Immunol.* **68**, 332 (1982). Clinical pharmacology and pharmacokinetics: A. Ebihara *et al.*, *Arzneimittel-Forsch.* **37**, 1388 (1987). Metabolism in humans: M.

Tateno *et al., Yakuri to Chiryo* **16**, 3251 (1988), *C.A.* **110**, 18040n (1989).

mp > 300°.
Sodium salt pentahydrate, $C_{13}H_5ClN_5NaO_2 \cdot 5H_2O$, *Y-12141; Clearnal.*
THERAP CAT: Antiallergic; antiasthmatic.

9495. Trazodone. *2-[3-[4-(3-Chlorophenyl)-1-piperazinyl]propyl]-1,2,4-triazolo[4,3-a]pyridin-3(2H)-one.* C_{19}-$H_{22}ClN_5O$; mol wt 371.88. C 61.37%, H 5.96%, Cl 9.53%, N 18.83%, O 4.30%. Prepn: Palazzo, Silvestrini, U.S. pat. **3,381,009** (1968 to Angelini Francesco). Pharmacology: Catanese, Lisciani, *Boll. Chim. Farm.* **109**, 369 (1970); B. Silvestrini, E. Quadri, *Eur. J. Pharmacol.* **12**, 231 (1970). Analytical data: Baiocchi *et al., Arzneimittel-Forsch.* **23**, 400 (1973). Crystal structure: J. P. Fillers, S. W. Hawkinson, *Acta Crystallogr.* **B35**, 498 (1979). Review of pharmacological properties and therapeutic use: R. N. Brogden *et al., Drugs* **21**, 401-429 (1981); B. Silvestrini *et al.*, in *Pharmacological and Biochemical Properties of Drug Substances* **vol. 3**, M. E. Goldberg, Ed. (Am. Pharm. Assoc., Washington, DC, 1981) pp 94-119. Comprehensive description: D. Gorecki, R. Verbeeck in *Analytical Profiles of Drug Substances* **vol. 16**, K. Florey, Ed. (Academic Press, New York, 1986) pp 693-729. Symposium on clinical efficacy: *Psychopharmacology* **95**, Suppl., 1-56 (1988).

Crystals, mp 86-87°. Also reported as mp 96° (Baiocchi). pKa (50% ethanol): 6.14.
Hydrochloride, $C_{19}H_{23}Cl_2N_5O$, *AF 1161, Desyrel, Bimaran, Molipaxin, Pragmazone, Tombran, Thombran, Trazolan, Trialodine, Trittico.* White, odorless plates from ethanol, mp 223°. Soluble in chloroform; sparingly sol in water, ethanol, methanol. Practically insol in common organic solvents. uv max (water): 211, 246, 274, 312 nm (ε 50100, 11730, 3840, 3840). LD$_{50}$ i.v. in mice: 96 mg/kg (Silvestrini, Quadri).
THERAP CAT: Antidepressant.

9496. Trehalose. *α-D-Glucopyranosyl-α-D-glucopyranoside; α,α-trehalose; natural trehalose; mycose; (α-D-glucosido)-α-D-glucoside.* $C_{12}H_{22}O_{11}$; mol wt 342.31. C 42.10%, H 6.48%, O 51.42%. About 23-30% is found in trehala manna, the cocoons of a parasitic beetle (*Larinus* species), found on *Echinops pesicus.* Trehalose also occurs in fungi, e.g., *Amanita muscaria.* Extraction procedure: Harding, *Chem. Zentr.* 1924, I, 2017. Isoln from compressed baker's yeast: Stewart *et al., J. Am. Chem. Soc.* **72**, 2059 (1950). Structure: Schlubach, Maurer, *Ber.* **58**, 1179 (1925); Bredereck, *ibid.* **63**, 959 (1930); Hassid, Ballou in *The Carbohydrates*, W. Pigman, Ed. (Academic Press, New York, 1957) p 507. Synthesis: Haworth, Hickenbottom, *J. Chem. Soc.* **1931**, 2847; Lemieux, Bauer, *Can. J. Chem.* **32**, 340 (1954); A. A. Pavia *et al., Carbohyd. Res.* **79**, 79 (1980). *Review:* C. K. Lee, *Dev. Food Carbohyd.* **2**, 1-89 (1980).

Dihydrate, orthorhombic, bisphenoidal crystals from dil alcohol. Sweet taste. mp 96.5-97.5°. The water of crystn escapes around 130°. Anhydrous trehalose melts at 203°. $[\alpha]_D^{20} + 178°$ (c = 7 of the dihydrate). Sol in water, hot alcohol; insol in ether. Does not reduce Fehling's soln. Is fermented by yeast. Is not split by α-glucosidase. Acid hydrolysis gives 2 mols D-glucose.

9497. Tremetone. *1-[2,3-Dihydro-2-(1-methylethenyl)-5-benzofuranyl]ethanone; 2,3-dihydro-2-isopropenyl-5-benzofuranyl methyl ketone; 2-isopropenyl-2,3-dihydro-5-acetylbenzofuran.* $C_{13}H_{14}O_2$; mol wt 202.24. C 77.20%, H 6.98%, O 15.82%. Principal ketone suspected of being the active toxin of *Eupatorium urticaefolium* Reichard, *Compositae* (white snakeroot). Isoln and structure: Bonner, De-Graw, *Tetrahedron* **18**, 1295 (1962). Synthesis of dihydrotremetone: DeGraw, Bonner, *ibid.* 1311. Synthesis of racemic tremetone: DeGraw *et al., ibid.* **19**, 19 (1963); Bohlmann, Buehmann, *Ber.* **105**, 863 (1972). Abs config: Bonner *et al., Tetrahedron* **20**, 1419 (1964). *Review:* Christensen, *Econ. Bot.* **19**, 293-300 (1965).

Liquid. $[\alpha]_D^{28} - 59.6°$ (c = 5.52 in absolute ethanol). n_D^{25} 1.5658. d_4^{28} 1.080. uv max (ethanol): 227, 280, 285 nm (ε 11,950, 12,600, 12,300).
Dihydrotremetone, $C_{13}H_{16}O_2$, *(R)-1-[2,3-dihydro-2-(1-methylethyl)-5-benzofuranyl]ethanone.* Liquid. bp 216-221°. $[\alpha]_D^{25} - 47.0°$ (c = 1.78 in abs ethanol). uv max (ethanol): 231, 279 nm (ε 39,500; 18,800).

9498. Tremorine. *1,1'-(2-Butyne-1,4-diyl)bispyrrolidine; 1,1'-(2-butynylene)dipyrrolidine;* 1,4-dipyrrolidino-2-butyne. $C_{12}H_{20}N_2$; mol wt 192.30. C 74.95%, H 10.48%, N 14.57%. Prepn from pyrrolidine + 1,4-dichloro-2-butyne: Maier, Ger. pat 896,810 (1953 to BASF); Reppe *et al., Ann.* **596**, 79 (1955); Biel, DiPierro, *J. Am. Chem. Soc.* **80**, 4609 (1958). Review of chemical and biological studies: Karlen, *Acta Pharm. Suecica* **1970**, 169-200; *see also* Kolla, Obvintseva, *Farmakol. Toksikol.* **36**, 736-745 (1973).

Liquid. bp$_{2.5}$ 116-116.5°; bp$_{0.1}$ 93-95°.
Methiodide, $C_{12}H_{20}N_2 \cdot 2CH_3I$, crystals, mp 239-240°.
USE: To produce exptl parkinsonism: Everett *et al., Science* **124**, 79 (1956).

9499. Trenbolone. *17-Hydroxyestra-4,9,11-trien-3-one; 4,9,11-estratrien-17β-ol-3-one; 17β-hydroxy-19-norandrosta-4,9,11-trien-3-one; 19-norandrosta-4,9,11-trien-17β-ol-3-one; trienbolone; trienolone.* $C_{18}H_{22}O_2$; mol wt 270.38. C 79.96%, H 8.20%, O 11.84%. Prepn of base: Velluz *et al., C. R. Acad. Sci.* **257**, 569 (1963); Heller *et al., Steroids* **10**, 211 (1967). Base and 17-acetate: **Fr. pat. M1958** and **Brit.**

pat. **1,035,683** (1963 and 1966 to Roussel-UCLAF), *C.A.* **60**, 3039h (1964); **65**, 17027c (1966). 17-Cyclohexylmethyl-carbonate: Nedelec, Costerousse, **Fr.** pat. **M5979** (1968 to Roussel-UCLAF), *C.A.* **71**, 50356g (1969). Pharmacology of the acetate: Krüskemper *et al., Arzneimittel-Forsch.* **17**, 449 (1967). Animal studies: Beranger, Malterre, *C. R. Soc. Biol.* **162**, 1157 (1968); Best, *Vet. Rec.* **91**, 624 (1972).

Crystals, mp 186°. $[\alpha]_D^{20}$ +19° (c = 0.45 in ethanol). Also reported as mp 183-186° from acetone-water. uv max: 239, 340.5 nm (ε 5260, 28,000) (Heller *et al.*).

Acetate, $C_{20}H_{24}O_3$, *17β-acetoxy-3-oxoestra-4,9,11-triene, 17β-acetoxyestra-4,9,11-trien-3-one, Finaplix.* Crystals, mp 96-97°. $[\alpha]_D^{20}$ +36.8° (c = 0.37 in methanol).

Cyclohexylmethylcarbonate, $C_{26}H_{35}O_4$, *trenbolone hexahydrobenzylcarbonate, Hexabolan, Parabolan.* Crystals from cyclohexane-petr ether, mp 90-95°. $[\alpha]_D^{20}$ +41.6° (c = 0.5 in ethanol).

THERAP CAT: Anabolic.
THERAP CAT (VET): Anabolic.

9500. Trengestone. *6-Chloropregna-1,4,6-triene-3,20-dione;* 6-chloro-1,6-didehydroretroprogesterone; 6-chloro-1,6-bisdehydroretroprogesterone; Ro 4-8347; Retroid. $C_{21}H_{25}ClO_2$; mol wt 344.88. C 73.13%, H 7.31%, Cl 10.28%, O 9.28%. A member of a class of hormonally active steroids termed "retrosteroids", which are characterized by a (9β, 10α)-configuration in contrast to the usual (9α,10β)-configuration of steroids. Prepn: Threadgold, **Belg.** pat. **652,597** corresp to Reerink *et al.,* **U.S.** pat. **3,422,122** (1965, 1969 to Phillips). Activity studies: Sadovsky *et al., Gynecol. Invest.* **1**, 319 (1970); Kalra *et al., J. Endocrinol.* **51**, 675 (1971). Metabolism: Breuer *et al., Acta Endocrinol. (Copenhagen)* **74**, 127 (1973); Dixon *et al., Steroids* **22**, 35 (1973).

Crystals from acetone, mp 208-209° (dec). uv max: 229, 253, 302 nm (ε 11,500, 10,520, 10,650).

THERAP CAT: Progestogen.

9501. Trepibutone. *2,4,5-Triethoxy-γ-oxobenzenebutanoic acid;* 3-(2,4,5-triethoxybenzoyl)propionic acid; AA 149; Cholibil; Supacal. $C_{16}H_{22}O_6$; mol wt 310.36. C 61.92%, H 7.15%, O 30.93%. Prepn: T. Murata *et al.,* **Ger.** pat. **2,244,-324** corresp to **U.S.** pat. **3,943,169** (1973, 1976 both to Takeda). Properties and stabilities: M. Mitani *et al., Takeda Kenkyusho Ho* **36**, 206 (1977), *C.A.* **88**, 126284f (1978). Pharmacokinetics: *eidem, ibid.* 215, *C.A.* **88**, 115067m (1978). Biotransformation: T. Kobayashi *et al., Xenobiotica* **8**, 535 (1978). Metabolism studies: S. Tanayama *et al., ibid.* 365, 377. Spasmolytic action in dogs: H. Satoh *et al., Eur. J. Pharmacol.* **48**, 309 (1978). Mechanism of choleretic action: *eidem, ibid.* 125. Toxicity studies: S. Sato *et al., Takeda Kenkyusho Ho* **36**, 263 (1977), *C.A.* **88**, 11535x (1978). Toxicity data: *Japan. Med. Gaz.* **18**(6), 13 (1981).

Colorless needles from aq ethanol or plates from aq acetone, mp 150-151°. Stable to heat, humidity, indoor diffused sunlight. Aq solns heated to 100° for 10 hr showed no degradation. LD_{50} in male mice, rats (mg/kg): 1340, 2450 orally, 530, 410 i.p. *(Japan. Med. Gaz.).*

THERAP CAT: Choleretic; antispasmodic.

9502. Tretoquinol. *1,2,3,4-Tetrahydro-1-(3,4,5-trimethoxybenzyl)-6,7-isoquinolinediol;* 1-(3',4',5'-trimethoxybenzyl)-6,7-dihydroxy-1,2,3,4-tetrahydroisoquinoline; trimethoquinol. $C_{19}H_{23}NO_5$; mol wt 345.38. C 66.07%, H 6.71%, N 4.06%, O 23.16%. Synthesis of *dl*-form: Yamato *et al., Tetrahedron, Suppl.* **8**, 129 (1966). Pharmacology: Fogelman, Grundy, *Brit. J. Pharmacol.* **38**, 416 (1970). Metabolic studies: Meshi *et al., Biochem. Pharmacol.* **19**, 2937 (1970); Satoh *et al., Chem. Pharm. Bull.* **19**, 667 (1971).

dl-Form hydrochloride, pale yellow crystals from methanol + ether, decomp 224.5-226°.

l-Form hydrochloride, $C_{19}H_{24}ClNO_5$, *AQ 110, Inolin, Vems.* Pale yellow crystals, freely sol in water; sol in alcohol.

THERAP CAT: The *l*-form hydrochloride as bronchodilator.

9503. TRH. *5-Oxo-L-prolyl-L-histidyl-L-prolinamide; thyrotropin-releasing factor;* thyrotropin releasing hormone; TRF; protirelin; lopremone (rescinded USAN); TSH-releasing factor; pyroglutamylhistidylprolinamide; thyroliberin; TRH-Roche; Antepan; Relefact TRH; Stimu-T.S.H.; Thymone; Thypinone; Thyrefact; Xantium. $C_{16}H_{22}N_6O_4$; mol wt 362.40. C 53.03%, H 6.12%, N 23.19%, O 17.66%. A hypothalamic neurohormone which stimulates the release and synthesis of TSH, *q.v.,* from the anterior pituitary via the hypophyseal portal system; the first of the hypothalamic regulatory hormones to be isolated, characterized, and synthesized. Activity does not appear to be species specific in mammals, porcine and ovine TRH being identical. Isoln from bovine hypothalami: Schreiber *et al., Experientia* **18**, 338 (1962); Schally *et al., Endocrinology* **78**, 726 (1966). Purification from ovine hypothalami: Guillemin *et al., C.R. Acad. Sci., Ser. D* **262**, 2278 (1966). Isoln from porcine hypothalami and properties: Schally *et al., Biochem. Biophys. Res. Commun.* **25**, 165 (1966); Schally *et al., J. Biol. Chem.* **244**, 4077 (1969). Structural studies: Folkers *et al., Biochem. Biophys. Res. Commun.* **37**, 123 (1969); Burgus *et al., C.R. Acad. Sci., Ser. D* **269**, 226 (1969). Identity of isolated TRH with synthetic tripeptide: Burgus *et al., ibid.* 1870; Bowers *et al., Endocrinology* **86**, 573, 1143 (1970). Solubility: Burgus *et al., Experientia* **23**, 417 (1967). Synthesis: Flouret, *J. Med. Chem.* **13**, 843 (1970); Chang *et al., ibid.* **14**, 481 (1971); E. Gross *et al., Angew. Chem., Int. Ed.* **12**, 664 (1972); P. G. Pietta *et al., J. Org. Chem.* **39**, 44 (1974). Review, bibliography of synthetic methods and additional synthesis: Rivier, *Methods Enzymol.* **37**, 408 (1975). TRH is also believed to induce the secretion of the pituitary lactogenic hormone, prolactin, *q.v.:* Tasjian *et al., Biochem. Biophys. Res. Commun.* **43**, 516 (1971); Bowers *et al., ibid.* **51**, 512 (1973). It has been shown to block and reverse leukotriene-induced hypotension in the unanesthetized guinea pig: W. E. Lux *et al., Nature* **302**, 822 (1983). Clinical studies: Hershman, *N. Engl. J. Med.* **290**, 886 (1974). Reviews of TRH and other hypothalamic releasing hormones: Schally *et al., Recent Progr. Horm. Res.* **24**, 497

(1968); Burgus, Guillemin, *Ann. Rev. Biochem.* **39**, 499 (1970); *Polypeptide Hormones*, R. F. Beers, E. G. Bassett, Eds. (Raven Press, New York, 1980) pp 165-278.

pyroGlu-His-ProNH₂

Purified TRH is partially sol in chloroform, highly sol in absolute methanol. Completely insol in pyridine. Inactivated by diazotized sulfanilic acid (Pauly reagent) and by plasma, serum, or whole blood *in vitro*. Resists inactivation by proteolytic enzymes.

Tartrate, $C_{16}H_{22}N_6O_4 \cdot xC_4H_6O_6$, *Irtonin*.

THERAP CAT: Prohormone.

9504. Triacetin. *1,2,3-Propanetriol triacetate;* glyceryl triacetate; triacetyl glycerine; Enzactin; Fungacetin; Vanay. $C_9H_{14}O_6$; mol wt 218.20. C 49.54%, H 6.47%, O 43.99%. $C_3H_5(OCOCH_3)_3$. Prepd by acetylation of glycerol: Dunbar, Bolstad, *J. Org. Chem.* **21**, 1041 (1956); by reaction of oxygen with a liquid phase mixture of allyl acetate and acetic acid using a bromide as catalyst: Keith, U.S. pat. **2,911,437** (1959 to Sinclair Refining). As an antifungal: **Brit.** pat. **845,029** (1960 to Wisc. Alumni Res. Found.). Acute toxicity: A. Wretlind, *Acta Physiol. Scand.* **40**, 338 (1957).

Colorless, somewhat oily liquid having a slight, fatty odor and a bitter taste. d_4^{25} 1.1562, d_4^{20} 1.1596, d_{20}^{20} 1.163. mp −78°. bp 258-260°. bp_{40} 172°. n_D^{20} 1.4307. Sol in 14 parts water. Miscible with alcohol, ether, chloroform. Slightly sol in carbon disulfide. LD_{50} i.v. in mice: 1600 ±81 mg/kg (Wretlind).

USE: As fixative in perfumery; solvent in manuf celluloid, photographic films. Technical triacetin (a mixture of mono-, di-, and small quantities of triacetin) as a solvent for basic dyes, particularly indulines, and tannin in dyeing.

THERAP CAT: Topical antifungal.

9505. Triacetyldiphenolisatin. *1-Acetyl-3,3-bis[4-(acetyloxy)phenyl]-1,3-dihydro-2H-indol-2-one;* 1-acetyl-3,3-bis(p-hydroxyphenyl)oxindole diacetate; TDI; Isatex; Laxagen; Phenisatin; Trisatin; Unilax. $C_{26}H_{21}NO_6$; mol wt 443.44. C 70.42%, H 4.77%, N 3.16%, O 21.65%. Prepn: A. Baeyer, M. J. Lazarus, *Ber.* **18**, 2641 (1885). Toxicology: Woislawski *et al.*, *J. Am. Pharm. Assoc.*, *Sci. Ed.* **42**, 468 (1953); E. G. Jung, *Arch. Klin. Exp. Dermatol.* **231**, 39 (1967).

Crystals from alc, mp 201-202°. Freely sol in chloroform. LD_{50} in rats (mg/kg): 500 orally; 350 i.p., Woislawski *et al.*, *loc. cit.*

THERAP CAT: Cathartic.

9506. 1-Triacontanol. Melissyl alcohol; myricyl alcohol; 1-hydroxytriacontane. $C_{30}H_{62}O$; mol wt 438.80. C 82.11%, H 14.24%, O 3.65%. $CH_3(CH_2)_{28}CH_2OH$. Present in plant cuticle waxes and in beeswax as the palmitate. Prepn: Robinson, *J. Chem. Soc.* **1934**, 1545; K. Maruyama *et al.*, *J. Org. Chem.* **45**, 737 (1980). Use as a plant growth regulator: S. K. Ries *et al.*, *Science* **195**, 1339 (1977); S. K. Ries, **Belg.** pat. **854,587** corresp to U.S. pat. **4,150,970** (1977, 1979 to

Michigan State Univ.). Use of colloidally dispersed triacontanol in plant growth enhancement studies: R. G. Laughlin *et al.*, *Science* **219**, 1219 (1983).

Crystals. d_{95} 0.777. mp 87°. Practically insol in water; freely sol in benzene, ether; very slightly sol in cold, more sol in hot alcohol.

USE: Plant growth regulator.

9507. Triadimefon. *1-(4-Chlorophenoxy)-3,3-dimethyl-1-(1H-1,2,4-triazol-1-yl)-2-butanone;* MEB 6447; Bay 6681 F; Bayleton. $C_{14}H_{16}ClN_3O_2$; mol wt 293.75. C 57.24%, H 5.49%, Cl 12.07%, N 14.30%, O 10.89%. Systemic fungicide active against mildews and rusts of grains, fruits, vegetables and ornamentals. Prepn and activity: W. Meiser *et al.*, **Ger.** pat. **2,201,063** corresp to U.S. pat. **3,912,752** (1973, 1975 to Bayer). Fungal metabolism: M. Gasztonyi, *Pestic. Sci.* **12**, 433 (1981). *Reviews:* F. Michel, P. Pourcharesse, *Def. Veg.* **31**, 97-109 (1977); T. J. Martin, D. B. Morris, *Pflanzenschutz-Nachr.* **32**, 31-79 (1979).

Crystals, mp 82°. Soly in water at 20°: 260 mg/l. Moderately sol in most organic solvents except aliphatics. LD_{50} in male, female rats (mg/kg): 568, 363 orally (Michel, Pourcharesse).

USE: Systemic agricultural fungicide.

9508. Triadimenol. *β-(4-Chlorophenoxy)-α-(1,1-dimethylethyl)-1H-1,2,4-triazole-1-ethanol;* 1-(4-chlorophenoxy)-3,3-dimethyl-1-(1H-1,2,4-triazol-1-yl)butan-2-ol; BAY KWG 0519; Baytan; Spinnaker; Summit. $C_{14}H_{18}ClN_3O_2$; mol wt 295.77. C 56.85%, H 6.13%, Cl 11.99%, N 14.21%, O 10.82%. Agricultural fungicide systemically active against powdery mildews and rusts of grains. Commercial product is mixture of 2 diastereoisomers. Prepn: **Belg.** pat. **814,831**; W. Kramer *et al.*, U.S. pat. **3,952,002** (1974, 1976 both to Bayer AG). Comprehensive description: P. E. Frohberger, *Pflanzenschutz-Nachr.* **31**, 11 (1978). Field trials in cereal diseases: J. Trägner-Born, T. van den Boom, *ibid.* 25. Inhibits fungal ergosterol biosynthesis: H. Buchenauer, *Pestic. Sci.* **9**, 507 (1978). Active fungal metabolite of triadimefon, *q.v.*: M. Gasztonyi, *ibid.* **12**, 433 (1981). GC separation of enantiomers: T. Clark, A. H. B. Deas, *J. Chromatog.* **329**, 181 (1985).

Crystals, mp 112-117°. Slight, non-specific odor. Soly in water at 20°: 0.012 g/100 g. Soluble in alcohol, ketones. LD_{50} in male, female rats (mg/kg): 1161, 1105 orally; > 5000 dermally, 24-hr; LD_{50} in quail: > 10,000 mg/kg (Frohberger).

USE: Systemic agricultural fungicide; cereal seed protectant.

9509. Triafur. *2-Amino-5-(5-nitro-2-furyl)-1,3,4-thiadiazole;* 2-(5-nitro-2-furyl)-5-amino-1,3,4-thiadiazole; Furidiazina. $C_6H_4N_4O_3S$; mol wt 212.20. C 33.96%, H 1.90%, N 26.40%, O 22.62%, S 15.11%. Prepn: Skagius *et al.*, *Acta Chem. Scand.* **14**, 1054 (1960); Skagius, **Brit.** pat. **852,795** (1960 to Pharmacia); Sherman, *J. Org. Chem.* **26**, 88 (1961). Bacteriostatic and fungistatic activity: Jeney, Zsolnai, *Zentralbl. Bakteriol., Parasitenk., Infektionskr., Hyg., Abt. I Orig.* **204**, 430 (1967).

Yellow crystals from dimethylformamide + water or acetic acid, mp 280°.

9510. Triallate. *Bis(1-methylethyl)carbamothioic acid S-(2,3,3-trichloro-2-propenyl) ester; diisopropylthiocarbamic acid S-(2,3,3-trichloroallyl) ester;* 2,3,3-trichloro-2-propene-1-thiol diisopropylcarbamate; *S*-2,3,3-trichloroallyl diisopropylthiocarbamate; *S*-(2,3,3-trichloro-2-propenyl) bis(1-methylethyl)carbamothioate; CP 23426; Avadex BW; Far-Go. $C_{10}H_{16}Cl_3NOS$; mol wt 304.66. C 39.42%, H 5.29%, Cl 34.91%, N 4.60%, O 5.25%, S 10.52%. Prepn: M. W. Harman, J. J. D'Amico, U.S. pat. **3,330,821** (1967 to Monsanto). Herbicidal activity: R. Grover *et al., Weed Res.* **19**, 363 (1979). Mutagenic evaluation *in vitro* (Ames Test): F. De Lorenzo *et al., Cancer Res.* **38**, 13 (1978); G. R. Douglas *et al., Mutat. Res.* **85**, 45 (1981). Soil persistence: A. E. Smith, B. J. Hayden, *Bull. Environm. Contam. Toxicol.* **29**, 240 (1982). Field studies for pre-emergent use: T. G. Reeves, C. L. Touhey, *Aust. J. Exp. Agr. Anim. Husb.* **12**, 55 (1972); E. M. Randall, R. H. Jarvis, *Exp. Husb.* **38**, 32 (1982); post-emergent use: R. P. Garnett, *Aspects Appl. Biol.* **13**, 73 (1980). GC determn in soils: A. E. Smith, *J. Chromatog.* **97**, 103 (1974). HPLC determn in soils: A. Pena Heras, F. Sanchez-Rasero, *ibid.* **358**, 302 (1986).

USE: Herbicide.

9511. Triamcinolone. *9-Fluoro-11,16,17,21-tetrahydroxypregna-1,4-diene-3,20-dione;* Δ^1-9α-fluoro-16α-hydroxyhydrocortisone; 9α-fluoro-16α-hydroxyprednisolone; Δ^1-16α-hydroxy-9α-fluorohydrocortisone; 16α-hydroxy-9α-fluoroprednisolone; CL 19823; Adcortyl; Aristocort; Celeste; Cinolone; Delphicort; Extracort; Kenacort; Ledercort; Omcilon; Orion; Triamcet; Tricortale; Volon. $C_{21}H_{27}FO_6$; mol wt 394.45. C 63.94%, H 6.90%, F 4.82%, O 24.34%. Prepn: Bernstein *et al., J. Am. Chem. Soc.* **78**, 5693 (1956); **81**, 1689 (1959); Thoma *et al., ibid.* **79**, 4818 (1957); Bernstein *et al.,* Allen *et al.,* U.S. pats. **2,789,118; 3,021,347** (1957, 1962, both to Am. Cyanamid). Comprehensive description: K. Florey, Ed. in *Analytical Profiles of Drug Substances* vol. 1 (Academic Press, New York, 1972) pp 367-396, 423-442; D. H. Sieh, *ibid.* vol. **11**(1982) pp 593-614, 651-661.

Crystals, mp 269-271°. mp also reported as 260-262.5°. $[\alpha]_D^{25}$ +75° (acetone). uv max: 238 nm (ε 15800).

16α,21-Diacetate, $C_{25}H_{31}FO_8$, *16α,21-diacetoxy-9α-fluoro-11β,17-dihydroxy-1,4-pregnadiene-3,20-dione,* Polcortolon, CINO-40, Cenocort, Tracilon, Triamcin. Solvated crystals, mp 186-188° (with effervescence, mp 235° after drying). $[\alpha]_D^{25}$ +22° (chloroform). uv max: 239 nm (ε 15200).

THERAP CAT: Glucocorticoid.

THERAP CAT (VET): Adrenocortical steroid. Anti-inflammatory glucocorticoid.

9512. Triamcinolone Acetonide. *9-Fluoro-11,21-dihydroxy-16,17-[1-methylethylidenebis(oxy)]pregna-1,4-diene-3,20-dione;* 9α-fluoro-11β,16α,17,21-tetrahydroxypregna-1,4-diene-3,20-dione cyclic 16,17-acetal with acetone; 9α-fluoro-16α-hydroxyprednisolone acetonide; triamcinolone 16α,17-acetonide; 9α-fluoro-11β,21-dihydroxy-16α,17α-isopropylidenedioxy-1,4-pregnadiene-3,20-dione; 9α-fluoro-16α,17-isopropylidenedioxyprednisolone; Adcortyl-A; Aristoderm; Cutinolone Simple; Flutex; Ftorocort; Kena-

cort-A; Kenalog; Kenaquart; Ledercort D; Omcilon-A; Respicort; Rineton; Solodelf; Tramacin; Tricinolon; Vetalog; Volon A; Volonimat. $C_{24}H_{31}FO_6$; mol wt 434.49. C 66.34%, H 7.19%, F 4.37%, O 22.09%. Prepd by stirring a suspension of triamcinolone in acetone in the presence of a trace of perchloric acid: Fried *et al., J. Am. Chem. Soc.* **80**, 2338 (1958); Bernstein *et al., ibid.* **81**, 1689 (1959); Bernstein, Allen, U.S. pat. **2,990,401** (1961 to Am. Cyanamid). Alternate synthesis using 2,3-dibromo-5,6-dicyanoquinone: Hydorn, U.S. pat. **3,035,050** (1962 to Olin Mathieson). Clinical trial in chronic asthma: I. L. Bernstein *et al., Chest* **81**, 20 (1982). Comprehensive description: K. Florey, Ed. in *Analytical Profiles of Drug Substances* vol. 1 (Academic Press, New York, 1972) pp 397-421; D. H. Sieh, *ibid.* vol. **11** (1982) pp 615-649.

Crystals, mp 292-294°. $[\alpha]_D^{23}$ +109° (c = 0.75 in chloroform). uv max (abs alc.): 238 nm (ε 14,600). Sparingly sol in methanol, acetone, ethyl acetate.

21-Acetate, crystals, mp 268-270°. $[\alpha]_D^{23}$ +92° (c = 0.59 in chloroform).

21-Disodium phosphate, $C_{24}H_{30}FNa_2O_9P$, Aristosol.

21-Hemisuccinate, $C_{28}H_{35}FO_9$, Soluredarol.

THERAP CAT: Glucocorticoid; anti-inflammatory; antiasthmatic (inhalant).

THERAP CAT (VET): Glucocorticoid, anti-inflammatory.

9513. Triamcinolone Benetonide. *(11β,16α)-21-[3-(Benzoylamino)-2-methyl-1-oxopropoxy]-9-fluoro-11-hydroxy-16,17-[(1-methylethylidene)bis(oxy)]pregna-1,4-diene-3,20-dione;* 9-fluoro-11β,16α,17,21-tetrahydroxypregna-1,4-diene-3,20-dione cyclic 16,17-acetal with acetone 21-ester with N-benzoyl-2-methyl-β-alanine; 9α-fluoro-16α-hydroxyprednisolone 16α,17α-acetonide 21-(β-benzoylamino)-isobutyrate; triamcinolone acetonide β-benzoylaminoisobutyrate; TBI; Tibicorten. $C_{35}H_{42}FNO_8$; mol wt 623.73. C 67.40%, H 6.79%, F 3.05%, N 2.24%, O 20.52%. Prepn: C. Cavazza *et al.,* Ger. pat. **2,047,218;** eidem, U.S. pat. **3,749,712** (1971, 1973 both to Sigma-Tau). Pharmacology: E. T. Ordonez, *Arzneimittel-Forsch.* **21**, 248 (1971). Percutaneous absorption by rats and rabbits: W. H. Down *et al., Toxicol. Letters* **1**, 95 (1977). Clinical study: D. J. Tazelaar, *J. Int. Med. Res.* **5**, 338 (1977). HPLC analysis: S. Muck *et al., Boll. Chim. Farm.* **120**, 240 (1981).

Crystalline powder, mp 203-207°. $[\alpha]_D^{20}$ +96° ±3° (c = 1 in ethanol). Sol in methanol, acetone, ethanol, dioxane, pyridine, DMF, chloroform. Insol in water.

THERAP CAT: Glucocorticoid; topical anti-inflammatory.

9514. Triamcinolone Hexacetonide. *21-(3,3-dimethyl-1-oxobutoxy)-9-fluoro-11-hydroxy-16,17-[(1-methylethylidene)bis(oxy)]pregna-1,4-diene-3,20-dione;* 9-fluoro-11β,16α,17,21-tetrahydroxypregna-1,4-diene-3,20-dione cyclic 16,17-acetal with acetone, 21-(3,3-dimethylbutyrate); 21-tert-butylacetate-9α-fluoro-11β-hydroxy-16α,17α-(isopropylidenedioxy)pregna-1,4-diene-3,20-dione; 21-(3,3-dimethylbutyryloxy)-9α-fluoro-11β-hydroxy-16α,17α-(isopropylidene-dioxy)pregna-1,4-diene-3,20-dione; triamcinolone acetonide *tert*-butyl acetate; TATBA; CL-34433; Aristospan; Azmacort; Hexatrione; Lederlon; Lederspan. $C_{30}H_{41}FO_7$; mol wt 532.66. C 67.65%, H 7.76%, F 3.57%, O 21.02%. The hexacetonide ester of the potent glucocorticoid, triamcinolone, *q.v.* Prepn of syringeable suspension: Nash, Naeger, U.S. pat. **3,457,348** (1969 to Am. Cyanamid). Anti-inflammatory activity in rabbits: I. M. Hunneyball, *Agents Actions* **11**, 490 (1981). Early clinical studies: Bilka, *Minnesota Med.* **50**, 483 (1967); Layman, Peterson, *ibid.* 669. Clinical studies of intra-articular therapy in arthritis: R. C. Allen *et al., Arthritis Rheum.* **29**, 997 (1986); M. Talke, *Fortschr. Med.* **104**, 742 (1986). Toxicity study: Tonelli, *Steroids* **8**, 857 (1966). Comprehensive description: V. Zbinovsky, G. P. Chrekian in *Analytical Profiles of Drug Substances* **vol. 6**, K. Florey, Ed. (Academic Press, New York, 1977) pp 579-595.

Fine, white, needle-like crystals, mp 295-296° (dec), also reported as mp 271-272° (dec). uv max (ethanol): 238 nm (ε 15500). $[\alpha]_D^{25}$ +90°±2 (c = 1.13% in chloroform). Soly in g/100 ml at 25°: chloroform and dimethylacetamide > 5; ethyl acetate 0.77, methanol 0.59, diethyl carbonate 0.50, glycerin 0.42, propylene glycol 0.13; absolute alcohol 0.03; water 0.0004.

THERAP CAT: Anti-inflammatory.

9515. Triamterene. *6-Phenyl-2,4,7-pteridinetriamine; 2,4,7-triamino-6-phenylpteridine;* 6-phenyl-2,4,7-triaminopteridine; ademin(e); pterofen; pterophene; NSC-77625; SKF 8542; Triteren; Jatropur; Dytac; Dyren; Dyrenium; Noridyl; Teriam; Urocaudal. $C_{12}H_{11}N_7$; mol wt 253.26. C 56.91%, H 4.38%, N 38.72%. Prepn: Spickett, Timmis, *J. Chem. Soc.* **1954**, 2887; Pachter, *J. Org. Chem.* **28**, 1191 (1963); Weinstock, Wiebelhaus, U.S. pat. **3,081,230** (1963 to SK & F); Osdene *et al., J. Med. Chem.* **10**, 431 (1967). *Review:* Ther. Triamteren, *Wiener Symp.* **1966**, K. Fellinger, Ed. (Georg Thieme, Vienna, 1967).

Yellow plates from butanol, mp 316° (Spickett, Timmis); also reported as crystals from DMF, mp 327° (Osdene *et al.*). uv max (4.5% formic acid): 356 nm (ε 21,000).

THERAP CAT: Diuretic.

9516. s-Triazaborane. *Borazine;* hexahydro-s-triazaborine; borazane; borazole; triborine triamine; triboron nitride. $B_3H_6N_3$; mol wt 80.53. H 7.51%, B 40.31%, N 52.18%. Prepd by heating an equimolar mixture of ammonia and a boron hydride at 250-300° for 30 min: Stock, Pohland, *Ber.* **59**, 2215 (1926); Wiberg, Bolz, *Ber.* **73**, 209 (1940). Structure and dipole moment: Watanabe, Kubo, *J. Am. Chem. Soc.* **82**, 2428 (1960). *Review:* E. M. Smolin, L. Rapoport, *s-Triazines and Derivatives* (Interscience, New York, 1959) pp 597-626.

Mobile liquid; mp −58°; bp +53°. Stable up to 500° when pure and totally anhydrous. d_4^0 0.824; d_4^{57} 0.898. n_D^{20} 1.3821. Critical temp 252°. Heat of vaporization 7.0 kcal. Mol vol at bp = 99.7 ml. Surface tension at mp = 31.09 dynes/cm. Parachor 207.9. Trouton's constant 21.4. Eötvös' constant 2.0. Dipole moment in benzene at 25° = 0.50. Dissolves in water, giving a soln with strong reducing properties. Hydrolyzes slowly in aq soln yielding hydrogen, boric acid, and ammonia.

9517. Triaziquone. *2,3,5-Tris(1-aziridinyl)-2,5-cyclohexadiene-1,4-dione; 2,3,5-tris(1-aziridinyl)-p-benzoquinone;* 2,3,5-tris(aziridino)-1,4-benzoquinone; 2,3,5-tris(ethyleneimino)benzoquinone; Bayer 3231; Trenimon; Tris (Trenimon). $C_{12}H_{13}N_3O_2$; mol wt 231.25. C 62.32%, H 5.67%, N 18.17%, O 13.84%. Prepn: Gauss, Domagk, U.S. pat. **2,976,279** (1961 to Schenley).

Purple acicular crystals from ethyl acetate, mp 162.5-163°. Sparingly sol in cold water; sol in acetone, benzene, chloroform, ethyl acetate, methanol and warm acetic acid. Active against *Entamoeba histolytica.*

THERAP CAT: Antineoplastic.

9518. Triazolam. *8-Chloro-6-(2-chlorophenyl)-1-methyl-4H-[1,2,4]triazolo[4,3-a][1,4]benzodiazepine; 8-chloro-6-(o-chlorophenyl)-1-methyl-4H-s-triazolo[4,3-a][1,4]benzodiazepine;* clorazolam; U-33,030; Halcion; Novidorm; Songar. $C_{17}H_{12}Cl_2N_4$; mol wt 343.22. C 59.49%, H 3.53%, Cl 20.66%, N 16.32%. Prepn: J. B. Hester, Ger. pat. **2,012,190** corresp to U.S. pat. **3,701,782** (1970, 1972 both to Upjohn); J. B. Hester *et al., J. Med. Chem.* **14**, 1078 (1971). Pharmacology: T. Furukawa *et al., Igaku Kenkyu* **45**, 285 (1975), *C.A.* **84**, 130354p. Metabolism: F. S. Eberts, *Drug Metab. Dispos.* **5**, 547 (1977). Clinical studies: K. K. Okawa, G. S. Allens, *J. Int. Med. Res.* **6**, 343 (1978); A. J. Bowen, *ibid.* 337; A. J. Puech *et al., Therapie* **33**, 287 (1978). Toxicity: *Pharm. Weekblad.* **113**, 725 (1978). Review of pharmacology and therapeutic efficacy: G. E. Pakes *et al., Drugs* **22**, 81-110 (1981).

Tan crystals from 2-propanol, mp 233-235°. LD_{50} in mice, rats (mg/kg): > 100, > 5000 orally (*Pharm. Weekblad.*).

Note: This is a controlled substance (depressant) listed in the U.S. Code of Federal Regulations, Title 21 Part 1308.14 (1985).

THERAP CAT: Sedative. Hypnotic.

9519. 1H-1,2,4-Triazole. Pyrrodiazole. $C_2H_3N_3$; mol wt 69.07. C 34.78%, H 4.38%, N 60.84%. Prepn starting with thiosemicarbazide and formic acid: Ainsworth, *Org. Syn.* **40**, 99 (1960).

Needles from ethanol + benzene, mp 120-121°. bp_{760} 260° (decompn). Appreciably sol in water, alcohol.
Hydrochloride, $C_2H_3N_3\cdot HCl$, platelets, dec 169°.

9520. Tribenoside. *Ethyl 3,5,6-tris-O-(phenylmethyl)-D-glucofuranoside;* Ba 21401; Alven; Flebosan; Glyvenol; Hemocuron; Venex. $C_{29}H_{34}O_6$; mol wt 478.56. C 72.78%, H 7.16%, O 20.06%. Prepn: Druey, Huber, U.S. pat. **3,157,-634** (1964 to Ciba). Pharmacology: Lecomte, *C.R. Soc. Biol.* **163**, 1469 (1969); Helfer, Jaques, *Pharmacology* **5**, 23 (1971). Increase in capillary resistance in patients with rheumatoid arthritis: W. C. Dick *et al., Ann. Rheum. Dis.* **28**, 187 (1969). GLC determn in plasma: A. Sioufi, F. Pommier, *J. Pharm. Sci.* **69**, 167 (1980). Excretion as hippuric acid: *eidem, Eur. J. Drug Metab. Pharmacokinet.* **7**, 223 (1982). Review of pharmacology: R. Jaques, *ibid.* **15**, 445-460 (1977).

R = $C_6H_5CH_2$

$bp_{1.2}$ 270-280°. $[\alpha]_D^{26}$ +8° (chloroform).
THERAP CAT: Sclerosing agent.

9521. Tribromoacetic Acid. $C_2HBr_3O_2$; mol wt 296.78. C 8.09%, H 0.34%, Br 80.78%, O 10.78%. CBr_3COOH. Prepn: Müller, U.S. pat. **2,057,964** (1936 to Winthrop Chem.); A. M. Kovalevskaya, S. A. Shkylar, *Zh. Org. Khim.* **1**, 1540 (1965). Study of far IR spectrum: G. Statz, E. Lippert, *Ber. Bunsenges. Phys. Chem.* **71**, 673 (1967).
Monoclinic prisms, mp 129-135°. bp 245° (dec). Sol in water, alcohol, ether; slightly sol in petr ether. Dec in boiling water to bromoform.
USE: Catalyst for polymerization; as brominating agent: W. J. Szczepek, *Polish J. Chem.* **55**, 709 (1981).

9522. 2,4,6-Tribromoaniline. Aniline tribromide. $C_6H_4Br_3N$; mol wt 329.85. C 21.85%, H 1.22%, Br 72.68%, N 4.25%. Prepd by controlled bromination of aniline: Suthers *et al., J. Org. Chem.* **27**, 447 (1962).

Needles; d 2.35; mp 120-122°; bp 300°. Insol in water. Sol in hot alc, chloroform, ether; slightly sol in cold alcohol.

9523. Tribromo-*tert*-butyl Alcohol. *1,1,1-Tribromo-2-methyl-2-propanol;* acetone-bromoform; Brometone. $C_4H_7Br_3O$; mol wt 310.85. C 15.45%, H 2.27%, Br 77.13%, O 5.15%. $(CH_3)_2C(OH)CBr_3$. Prepd from acetone and bromoform in the presence of sodium or potassium hydroxide: Aldrich, *J. Am. Chem. Soc.* **33**, 386 (1911); Viehe *et al., Ber.* **96**, 426 (1963).
Crystals from dil alcohol, mp 167-176° (Aldrich), from ether, mp 169° (Viehe). Camphor-like odor and taste. Volatilizes in air. Slightly sol in water; sol in alcohol, ether.
Caution: This substance may be habit forming and is listed in the U.S. Code of Federal Regulations, Title 21 Part 329.1 (1987).
USE: As modifier in polymerization of vinyl chloride, Seymour, U.S. pat. **2,716,112** (1955 to U.S. Rubber).

9524. 2,4,6-Tribromo-*m*-cresol. *2,4,6-Tribromo-3-methylphenol;* 2,4,6-tribromo-3-hydroxytoluene; Micatex. $C_7H_5Br_3O$; mol wt 344.87. C 24.38%, H 1.46%, Br 69.52%, O 4.64%. Prepn by bromination of *m*-cresol: Biilmann, Rimbert, *Bull. Soc. Chim. France* [4] **33**, 1473 (1923).

Crystals from 50% aq alcohol, mp 84°.
THERAP CAT: Topical antifungal.

9525. Tribromoethanol. Tribromoethyl alcohol; Avertin; Bromethol; Ethobrom; Narcolan; Narkolan. $C_2H_3Br_3O$; mol wt 282.79. C 8.49%, H 1.07%, Br 84.78%, O 5.66%. Br_3CCH_2OH. Prepd by the reduction of bromal with aluminum isopropylate. *See* U.S. pats. **1,572,742** (1926); **1,725,-054** (1929); **1,882,944** (1932). Acute toxicity: R. R. Burtner, G. Lehmann, *J. Pharmacol. Exp. Ther.* **63**, 183 (1938).
Crystals, mp 79-82°, decompn starting at 70°. bp_{10} 92-93°. Slight ethereal or aromatic odor and taste. Sol in 40 parts water at 40°; sol in alcohol, ether, benzene, warm petr ether; very sol in amylene hydrate (*tert*-amyl alc). An aq soln (1 in 50), prepd at 35-40° and tested immediately, is not acid to methyl red. The aq and alcoholic solns dec on exposure to light; in aq soln dibromoacetaldehyde and HBr are formed, both of which are strongly irritant. *Protect from light.* LD_{50} orally in rats: 1.09 g/kg (Burtner, Lehmann).
Note: To avoid danger of decompn it is marketed under the name Avertin with Amylene Hydrate (Solution), as Tribromoethanol Solution. Each ml contains 1 g tribromoethanol.
THERAP CAT: Anesthetic (inhalation).
THERAP CAT (VET): Has been used as a general anesthetic and narcotic.

9526. 2,4,6-Tribromophenol. Bromol. $C_6H_3Br_3O$; mol wt 330.83. C 21.78%, H 0.91%, Br 72.47%, O 4.84%. Prepd by controlled bromination of phenol: Konecny, *J. Am. Chem. Soc.* **76**, 4993 (1954).

Long crystals; d 2.55; mp 94-96°; bp 244°. Sublimable. Sol in 14,000 parts water at 15°; sol in alcohol, chloroform, ether, glycerol oils. LD_{50} orally in rats: < 2000 mg/kg, E. F. Stohlman, *Pub. Health Reports* **66**, 1303 (1951).

9527. 1,2,3-Tribromopropane. *sym*-Tribromopropane; glycerol tribromohydrin; tribromohydrin; allyl tribromide. $C_3H_5Br_3$; mol wt 280.82. C 12.83%, H 1.79%, Br 85.37%. $BrCH_2CHBrCH_2Br$. Prepd by the addn of bromine to allyl bromide in carbon tetrachloride: Perkin, Simonsen, *J. Chem. Soc.* **87**, 859 (1905); Johnson, McEwen, *Org. Syn.* **5**, 99 (1925). By gentle heating of propylene bromide with bromine in the presence of iron wire: Kronstein, *Ber.* **54**, 7 (1921); Tapley, Giesy, *J. Am. Pharm. Assoc.* **15**, 173 (1926); by reaction of bromotrichloromethane with allyl bromide initiated by γ-rays: Heiba, Anderson, *J. Am. Chem. Soc.* **79**, 4940 (1957).
Liquid; d^{23} 2.436; mp 16.5°; bp_{760} 220°; bp_{200} 170°; bp_{100} 148°; bp_{60} 134°; bp_{40} 123°; bp_{20} 106°; bp_{10} 90°; bp_5 76°; $bp_{1.0}$ 47.5°; n_D^{18} 1.58436. Insol in water; sol in alcohol, ether, chloroform.
USE: As a nematocide: Youngson, Goring, U.S. pat. **3,003,914** (1959 to Dow).

9528. Tribromosilane. Tribromomonosilane; silicobromoform. Br_3HSi; mol wt 268.85. Br 89.18%, H 0.37%, Si 10.45%. $SiHBr_3$. Prepd by passing hydrogen bromide over heated silicon or a silicide such as copper silicide: Schumb, Young, *Inorg. Syn.* **1**, 38 (1939); Schenk in *Handbook of Preparative Inorganic Chemistry* vol. **1**, G. Brauer, Ed. (Academic Press, New York, 2nd ed., 1963) pp 692-694.
Mobile liquid. Spontaneously flammable when the contact surface with air is large. d_4^{17} 2.7. mp −73.5°. bp_{760}

111.8°. The vapor pressure at 0° is given as 8.8 mm Hg. Dipole moment 0.79. Hydrolyzed by water with the formation of silicoformic anhydride, $H_2Si_2O_3$ and HBr. Sol in chlorinated hydrocarbons.

9529. Tribromsalan. *3,5-Dibromo-N-(4-bromophenyl)-2-hydroxybenzamide; 3,4',5-tribromosalicylanilide;* TBS; Temasept IV; Tuasol 100. $C_{13}H_8Br_3NO_2$; mol wt 449.96. C 34.70%, H 1.79%, Br 53.28%, N 3.11%, O 7.11%. Preparation: **Brit.** pat. **840,366** (1960 to Unilever); Lemaire *et al., J. Pharm. Sci.* **50,** 831 (1961); Lamberti; Schramm, Lemaire; Schramm, U.S. pats. **2,967,885; 3,064,048; 3,057,920** (1961, 1962 and 1962, all to Lever Bros.); Majewski, U.S. pat. **3,254,121** (1966 to Dow).

Crystals, mp 227-228°. Practically insol in water. Sol in hot acetone; very sol in DMF.

Note: Tribromsalan is a component of *Diaphene* which also contains 4',5-dibromosalicylanilide, *dibromsalan*.

USE: Bacteriostat in detergents. Banned by FDA from use in cosmetics.

9530. Tributylamine. *N,N-Dibutyl-1-butanamine.* $C_{12}H_{27}N$; mol wt 185.34. C 77.76%, H 14.68%, N 7.56%. $(CH_3CH_2CH_2CH_2)_3N$. Prepd by vapor phase alkylation of ammonia with butanol: Lemon, Myerly, U.S. pat. **3,022,349** (1962 to Union Carbide Corp.).

Hygroscopic liquid; characteristic odor. d_{20}^{20} 0.7782; bp 216-217°. Sparingly sol in water; very sol in alcohol, ether. *Keep well closed.*

Caution: Causes CNS stimulation, skin irritation, sensitization.

9531. Tributyl Phosphate. $C_{12}H_{27}O_4P$; mol wt 266.32. C 54.12%, H 10.22%, O 24.03%, P 11.64%. $(C_4H_9)_3PO_4$. Prepd by reaction of $POCl_3$ with butyl alcohol: Pianfetti, Janey, U.S. pat. **3,020,303** (1962 to FMC).

Colorless, odorless liq; d_{25}^{25} 0.976; bp 289° with decompn; bp_{27} 177-178°; mp below −80°. Flash pt 146°. n_D^{25} 1.4215. One ml dissolves in about 165 ml water; miscible with usual organic solvents. LD_{50} orally in rats: 3.0 g/kg, Smyth, Carpenter, *J. Ind. Hyg. Toxicol.* **26,** 269 (1944).

USE: Plasticizer for cellulose esters, lacquers, plastics, and vinyl resins. *Caution:* Irritating to mucous membranes.

9532. Tributyrin. *Butanoic acid 1,2,3-propanetriyl ester;* glyceryl tributyrate. $C_{15}H_{26}O_6$; mol wt 302.36. C 59.58%, H 8.67%, O 31.75%. $(C_3H_7COO)_3C_3H_5$. Prepd by esterification of glycerol with excess butyric acid: Weatherby *et al., J. Am. Chem. Soc.* **47,** 2249 (1925).

Oily liq; bitter taste; d_4^{20} 1.032; mp −75°; bp_{760} 305-310°. bp_{15} about 190°; n_D^{20} 1.4358. Insol in water. Very sol in alc, ether.

9533. Tricaine. *3-Aminobenzoic acid ethyl ester methanesulfonate;* ethyl *m*-aminobenzoate methanesulfonate; MS-222; Finquel; Metacaine. $C_{10}H_{15}NO_5S$; mol wt 261.31. C 45.96%, H 5.79%, N 5.36%, O 30.61%, S 12.27%. Prepn: Billeter *et al.,* U.S. pat. **1,678,317** (1928 to Sandoz); **Ger.** pat. **454,698** (1927); Frdl. **16,** 2427 (1931). Pharmacology in fish: Jolly *et al., Vet. Rec.* **91,** 424 (1972).

Fine needles from alcohol + ethyl acetate, mp 149-150°. Soly in water at 20°: 1 g/0.8 ml. Also given as 1:9. Slightly acid reaction. Aq solns stable to boiling.

THERAP CAT (VET): Anesthetic for fish.

9534. Tricarballylic Acid. *1,2,3-Propanetricarboxylic acid;* β-carboxyglutaric acid. $C_6H_8O_6$; mol wt 176.12. C 40.92%, H 4.58%, O 54.51%. $HOOCCH_2CH(COOH)CH_2-COOH$. The calcium salt is found in maple-sugar sand, and on the walls of evaporators in beet sugar manuf. Prepared by hydrolysis of ethyl propane-1,1,2,3-tetracarboxylate: Clarke, Murray, *Org. Syn.* **coll. vol. I** (2nd ed., 1941) p 523; from aconitic acid by reduction with sodium amalgam: Fittig, *Ann.* **314,** 15 (1901); by hydrolysis of 1-cyano-2,3-dicarbomethoxypropane, obtained by reaction of dimethyl itaconate with HCN in the presence of KCN: Bavley, Tate, U.S. pat. **2,992,268** (1958 to Pfizer).

Large, orthorhombic prisms from water or ether, mp 166°. K_1 at 30° = 3.25×10^{-4}; K_2 = 2.65×10^{-5}; K_3 = 1.48×10^{-6}. At 18° 50 g dissolve in 100 ml water and 0.9 g dissolve in 100 ml ether. Quite sol in alc. Absorption spectrum: Bielecki, Henri, *Ber.* **46,** 2596 (1913). The trisodium salt is neutral to litmus.

Trimethyl ester, $C_9H_{14}O_6$, d_4^{20} 1.1822; bp_{13} 150°; n_D^{20} 1.4398.

9535. Tricetamide. *N-[2-(Diethylamino)-2-oxoethyl]-3,4,5-trimethoxybenzamide; N-(diethylcarbamoylmethyl)-3,4,5-trimethoxybenzamide;* 3,4,5-trimethoxybenzamido-*N,N*-diethylacetamide; *N*-(3,4,5-trimethoxybenzoyl)glycine diethylamide; Riker 548; Trimeglamide. $C_{16}H_{24}N_2O_5$; mol wt 324.37. C 59.24%, H 7.46%, N 8.64%, O 24.66%. Prepd from *N*-(3,4,5-trimethoxybenzoyl)glycyl chloride and diethylamine: Kuehne, Lambert, *J. Am. Chem. Soc.* **81,** 4278 (1959); from *N*-(3,4,5-trimethoxybenzoyl)glycine hydrazide, nitrous acid and diethylamine: Kusserow, Draper, U.S. pat. **2,956,081** (1960 to Riker).

Crystals from water, mp 133-134°.
THERAP CAT: Sedative.

9536. Trichlorfon. *(2,2,2-Trichloro-1-hydroxyethyl)-phosphonic acid dimethyl ester; O,O-*dimethyl-1-hydroxy-2,2,2-trichloroethylphosphonate; *O,O-*dimethyl 2,2,2-trichloro-1-hydroxyethylphosphonate; chlorofos; metrifonate; trichlorphene; Bayer L 13/59; Vermicide Bayer 2349; Combot Equine; Danex; Dipterex; Neguvon; Dyrex; Anthon; Dylox; Bilarcil; Tugon; Proxol; Foschlor. $C_4H_8Cl_3O_4P$; mol wt 257.45. C 18.66%, H 3.13%, Cl 41.32%, O 24.86%, P 12.03%. Prepd by reaction of chloral with dimethyl phosphite: Lorenz, U.S. pat. **2,701,225** (1955 to Bayer); Barthel *et al., J. Am. Chem. Soc.* **76,** 4186 (1954); Lorenz *et al., ibid.* **77,** 2554 (1955). Activity: R. L. Metcalf *et al., J. Econ. Entomol.* **52,** 44 (1959). Identification of metabolites: C. Lange, *Z. Chem.* **20,** 446 (1980).

White crystals, mp 83-84°. d_4^{20} 1.73. n_D^{20} 1.3439. Vapor press at 20°: 7.8×10^{-6} mm Hg. Soly at 25°: water 15.4 g/100 ml; chloroform 75 g/100 mg; ether 17 g/100 ml; benzene 15.2 g/100 ml. Very slightly sol in pentane and hexane. Dec by alkali. LD_{50} orally in male, female rats: 630, 560 mg/kg, T. B. Gaines, *Toxicol. Appl. Pharmacol.* **14,** 515 (1969).

Note: Neguvon A contains trichlorfon and, as a minor ingredient, coumaphos, *q.v.*

USE: Insecticide for the control of flies and roaches, also in anthelmintic compositions for animals: Trace, Webster, U.S. pat. **3,111,457** (1963 to Am. Home Prods.). *Caution:* A cholinesterase inhibitor.

THERAP CAT (VET): Anthelmintic. For grubs, bots, screwworms.

9537. Trichlormethiazide. *6-Chloro-3-(dichloromethyl)-3,4-dihydro-2H-1,2,4-benzothiadiazine-7-sulfonamide 1,1-dioxide;* 6-chloro-3-dichloromethyl-7-sulfamyl-3,4-dihydro-1,2,4-benzothiadiazine 1,1-dioxide; 3-dichloromethyl-6-chloro-7-sulfamyl-3,4-dihydro-1,2,4-benzothiadiazine 1,1-dioxide; 3-dichloromethylhydrochlorothiazide; hydrotrichlorothiazide; trichloromethiazide; Achletin; Anatran; Anistadin; Aponorin; Carvacron; Diurese; Triclordiuride; Esmarin; Eurinol; Flutra; Fluitran; Intromene; Kubacron; Metahydrin; Tachionin; Tolcasone; Triflumen; Gangesol; Salurin; Naqua. $C_8H_8Cl_3N_3O_4S_2$; mol wt 380.67. C 25.24%, H 2.12%, Cl 27.94%, N 11.04%, O 16.81%, S 16.85%. Prepn: deStevens *et al., Experientia* **16**, 113 (1960); Sherlock *et al., ibid.* 184. Toxicity: E. I. Goldenthal, *Toxicol. Appl. Pharmacol.* **18**, 185 (1971).

Crystals from methanol + acetone + water, dec 266-273°, also reported as 248-250°. Soly (mg/ml) at 25°: water 0.8; ethanol 21; methanol 60. LD_{50} orally in rats: > 20,000 mg/kg (Goldenthal).
THERAP CAT: Diuretic, antihypertensive.

9538. Trichloroacetaldehyde. Chloral; anhydr chloral. C_2HCl_3O; mol wt 147.40. C 16.30%, H 0.68%, Cl 72.16%, O 10.85%. Cl_3CCHO. Made by chlorinating alcohol, treating with H_2SO_4 and then distilling: Liebig, *Ann.* **1**, 189 (1832); Personne, *J. Pharm. Chim.* [4] **10**, 350 (1869); **11**, 205 (1870); Brochet, *Bull. Soc. Chim. France* [3] **17**, 228 (1897); *Ann. Chim.* [7] **10**, 332 (1897); Trillat, *Bull. Soc. Chim. France* [3] **17**, 230 (1897); Besson, **Ger.** pat. **133,021** (1902); *Chem. Zentr.* **1902, II**, 553; Ohse, **Ger.** pat. **734,723** (1943), *C.A.* **38**, 3671 (1944). Prepn by chlorination of a mixture of alcohol and acetaldehyde: Société d'Electrochimie, **Fr.** pat. **612,396**; *Chimie et Industrie* **21**, 567 (1929). Prepn from chloral hydrate by azeotropic distn: Mahoney, Pierson, **U.S.** pat. **2,584,036** (1952 to Merck & Co.). From hypochlorous acid and trichloroethylene: Stevens *et al.,* **U.S.** pat. **2,759,978** (1956 to Columbia-Southern). In the laboratory anhydr chloral may be quickly obtained by shaking pharmaceutical grade chloral hydrate with concd H_2SO_4, separating the two layers, and distilling: *cf.* Gattermann, Wieland, *Praxis des Organischen Chemikers* (de Gruyter, Berlin, 40th ed., 1961) p 334. *See also* Chloral Hydrate.
Oily liquid. Pungent, irritating odor; d_4^{20} 1.510; d_4^{25} 1.5050. mp −57.5°. bp_{760} 97.8°. n_D^{20} 1.45572; $n_{He}^{21.4}$ 1.45412. Freely sol in water forming chloral hydrate. Sol in alcohol forming chloral alcoholate; sol in ether. Polymerizes under the influence of light and in presence of sulfuric acid forming a white solid trimer called *metachloral.*
Caution: This substance may be habit forming and is listed in the U.S. Code of Federal Regulations, Title 21 Part 329.1 (1987).
USE: Manuf chloral hydrate, DDT.

9539. Trichloroacetic Acid. TCA. $C_2HCl_3O_2$; mol wt 163.40. C 14.70%, H 0.62%, Cl 65.10%, O 19.58%. CCl_3-COOH. Prepd by oxidation of chloral hydrate with nitric acid: Parkes, Hollingshead, *Chem. & Ind. (London)* **1954**, 222. Manuf by chlorination of acetic acid: Eaker, **U.S.** pat. **2,832,803** (1958 to Monsanto).
Very deliquesc crystals; slight characteristic odor. d_4^{61} 1.629; mp 57-58°; bp 196-197°. Sol in 0.1 part water; very sol in alcohol, ether; dec by heating with caustic alkalies into chloroform and alkali carbonate. Its aq soln is very acid. pH of 0.1 molar aq soln 1.2. *Keep tightly closed in a cool place.* Storage of trichloroacetic acid solns in water of less than 30% strength is not recommended. Decompn products are chloroform, hydrochloric acid, carbon dioxide and carbon monoxide. LD_{50} orally in rats: 5000 mg/kg, G. W. Bailey, J. L. White, *Residue Rev.* **10**, 97 (1965).
Ethyl ester, $C_4H_5Cl_3O_2$, *ethyl trichloroacetate.* Clear liq; odor resembling menthol. d_4^{20} 1.383. bp 168°. n_D^{20} 1.4507. Insol in water; misc with alcohol, ether.

Sodium salt, $C_2Cl_3NaO_2$, *sodium trichloroacetate, Konesta, Varitox.* Yellow deliquesc powder, mp > 300°. Soly in water at 25°: 1.2 kg/l. Sol in ethanol.
Caution: Very corrosive! When sufficient penetration has occurred, the parts should be irrigated with sodium carbonate soln.
USE: As a decalcifier and fixative in microscopy; also as a precipitant of protein. As herbicide.
THERAP CAT: Caustic.
THERAP CAT (VET): Caustic, vesicant.

9540. Trichloroacetonitrile. Trichloromethylnitrile; Tritox. C_2Cl_3N; mol wt 144.40. C 16.63%, Cl 73.66%, N 9.70%. CCl_3CN. Prepared from ethyl trichloroacetate and aq ammonia: Davies, Jenkin, *J. Chem. Soc.* **1954**, 2374; by action of phosphorus pentoxide on trichloroacetamide: Carpenter, *J. Org. Chem.* **27**, 2085 (1962). Manuf by reaction of methylnitrile, HCl and chlorine gas: Käbisch, **U.S.** pat. **2,745,868** (1956 to Degussa). Physical and thermodynamic properties: Davies, Jenkin, *loc. cit.*
Liquid, bp_{760} 85.7°; d_4^{25} 1.4403, d_4^{35} 1.4223; $n_D^{20.0}$ 1.4409, $n_D^{27.0}$ 1.4375. LD_{50} orally in rats: 0.25 g/kg, H. F. Smyth *et al., Am. Ind. Hyg. Assoc. J.* **23**, 95 (1962).
USE: Insecticide. *Caution:* A strong irritant to eyes, skin.

9541. 2,4,6-Trichloroanisole. Tyrene. $C_7H_5Cl_3O$; mol wt 211.49. C 39.75%, H 2.38%, Cl 50.30%, O 7.57%. Prepd by reaction of 2,4,6-trichlorophenol with dimethyl sulfate: Kohn, Heller, *Monatsh.* **46**, 91 (1925).

Monoclinic needles from alc, mp 60°, $bp_{738.2}$ 240°, bp_{28} 132°. Faint odor similar to that of acetophenone. Sublimes slowly at room temp. Volatile with steam. Practically insol in water. Sol in methanol, dioxane, benzene, cyclohexanone.
USE: Formerly as a dye assistant for polyester fibers.

9542. 1,2,3-Trichlorobenzene. *vic*-Trichlorobenzene. $C_6H_3Cl_3$; mol 181.46. C 39.71%, H 1.67%, Cl 58.62%. Prepared from 3,4,5-trichloroaniline by diazotization: Cohen, Hartley, *J. Chem. Soc.* **87**, 1365 (1905); Holleman, *Rec. Trav. Chim.* **37**, 196 (1918); from 2,3,4-trichloroaniline by diazotization: Dadien *et al., Monatsh.* **61**, 431 (1932); from 2,3,4-trichloroaniline by treatment with ethyl nitrite: Beilstein, Kurbatow, *Ann.* **192**, 234 (1878).

Platelets from alc, d 1.69. mp 52.6°. bp 221°. n_D^{19} 1.5776. Flash pt 113° (235.4°F). Volatile with steam. Insol in water. Sparingly sol in alc. Freely sol in benzene, carbon disulfide.
USE: A commercial grade (mixture of isomeric trichlorobenzenes) is used to combat termites. *Caution:* Irritating to eyes, mucous membranes.

9543. 1,2,4-Trichlorobenzene. *unsym*-Trichlorobenzene. $C_6H_3Cl_3$; mol wt 181.46. C 39.71%, H 1.67%, Cl 58.62%. Prepd from 2,4-dichloroaniline or 2,5-dichloroaniline or 3,4-dichloroaniline by diazotization and treatment with Cu_2-Cl_2: Beilstein, Kurbatow, *Ann.* **192**, 230 (1878); van der Lande, *Rec. Trav. Chim.* **51**, 104, 110 (1932); from 1,3-diaminobenzene by tetrazotization and treatment with Cu_2Cl_2: Cohn, Fischer, *Monatsh.* **21**, 278 (1900).

Liquid; mp 17°; d_{25}^{25} 1.4634; bp 213°; n_D^{25} 1.5524. Flash pt 110° (230°F). Volatile with steam. Insol in water. Sparingly sol in alc. Miscible with ether, benzene, petr ether, carbon disulfide.

9544. 1,3,5-Trichlorobenzene. *sym*-Trichlorobenzene. $C_6H_3Cl_3$; mol wt 181.46. C 39.71%, H 1.67%, Cl 58.62%. Prepd from 2,4,6-trichloraniline by diazotization and treatment with alcohol: Jackson, Lamar, *Am. Chem. J.* **18**, 667 (1896); Backer, van der Baan, *Rec. Trav. Chim.* **56**, 1177 (1937).

Crystals; mp 63.4°; bp 208.4°. n_D^{19} 1.5662. Flash pt 107° (224.6°F). Volatile with steam. Insol in water. Sparingly sol in alc. Freely sol in ether, benzene, petr ether, carbon disulfide, glacial acetic acid.

9545. α,α,β-Trichloro-n-butyraldehyde. *2,2,3-Trichlorobutanal;* butylchloral; anhydr butylchloral; butyrchloral; crotonchloral. $C_4H_5Cl_3O$; mol wt 175.45. C 27.38%, H 2.87%, Cl 60.63%, O 9.12%. $CH_3CHClCCl_2CHO$. Prepd by the action of chlorine on acetaldehyde: Krämer, Pinner, *Ber.* **3**, 883 (1870); Pinner, *Ann.* **179**, 24 (1875); on paraldehyde: *idem;* on crotonaldehyde after saturation with gaseous HCl: High, **U.S.** pat. **2,280,290** (1942); Brown, Plump, **U.S.** pat. **2,351,000** (1944). From crotonaldehyde and chlorine: Ropp *et al., Org. Syn.* **coll. vol. IV**, 130 (1963). *See also* Butylchloral Hydrate.

Oily liquid. Pungent, disagreeable odor; d_4^{20} 1.3956; bp_{760} 164.5-165.5°; bp_{25} 57-60°. n_D^{20} 1.47554. Freely sol in water forming butylchloral hydrate; sol in alcohol forming an alcoholate; sol in ether. Polymerizes.

9546. 2,3,6-Trichloro-p-cresol. *2,3,6-Trichloro-4-methylphenol;* 2,3-trichloro-4-hydroxytoluene; 2,3,5-trichloro-4-hydroxy-1-methylbenzene. Prepared by chlorination of *p*-cresol: Zincke *et al., Ann.* **328**, 268 (1903).

Long needles from glacial acetic acid or petr ether. mp 66-67°. Slightly sol in water; freely sol in the usual organic solvents, and in dil alkali and soda solns.

9547. 2,4,6-Trichloro-m-cresol. *2,4,6-Trichloro-3-methylphenol;* 2,4,6-trichloro-3-hydroxytoluene. $C_7H_5Cl_3O$; mol wt 211.48. C 39.75%, H 2.38%, Cl 50.30%, O 7.57%. Prepd by chlorination of *m*-cresol, chlorination of thymol in the presence of iron, or by action of concd sulfuric acid on 2,4,4-trichloro-3-methyl-6-isopropyl-$\Delta^{2,5}$-cyclohexadienone: Crowther, McCombie, *J. Chem. Soc.* **103**, 536 (1913).

Needles or plates from water or petr ether. mp 45-47°; bp 265°, bp_{14} 142-144°. Volatile with steam. Slightly sol in water, benzene, petr ether, glacial acetic acid. Freely sol in methanol, alc, chloroform, xylene, ether and in alkali solns.

9548. 4,5,6-Trichloro-o-cresol. *2,3,4-Trichloro-6-methylphenol;* 3,4,5-trichloro-2-hydroxytoluene; 1-methyl-2-hydroxy-3,4,5-trichlorobenzene. Prepd by chlorination of 1-methyl-2-hydroxy-4-chlorobenzene: Kohn, Syreia, *J. Am. Chem. Soc.* **70**, 3950 (1948).

Needles from petr ether, mp 77°. Slightly soluble in water; freely sol in the customary organic solvents and in caustic alkali solns.

9549. 1,1,1-Trichloroethane. Methylchloroform; Chlorothene. $C_2H_3Cl_3$; mol wt 133.42. C 18.00%, H 2.27%, Cl 79.72%. CH_3CCl_3. Prepd by the action of chlorine on 1,1-dichloroethane: Sutton, *Proc. Roy. Soc.* **A133**, 673 (1931); by the catalytic addition of HCl to 1,1-dichloroethylene: **Ger.**

pat. **523,436** (1931); **U.S.** pat. **2,209,000** (Nutting, Huscher, 1940). Review of mfg processes: Faith, Keyes & Clark's *Industrial Chemicals,* F. A. Lowenheim, M. K. Moran, Eds. (Wiley-Interscience, New York, 4th ed., 1975) pp 836-843.

Liquid. Nonflammable. mp −32.5°. d_4^{20} 1.3376. bp_{760} 74.1°. n_D^{20} 1.43838. Insoluble in water. Absorbs some water. Sol in acetone, benzene, carbon tetrachloride, methanol, ether.

USE: In cold type metal cleaning, also in cleaning plastic molds. *Caution:* Irritating to eyes, mucous membranes, and, in high concns, narcotic.

9550. 1,1,2-Trichloroethane. Vinyl trichloride. $C_2H_3Cl_3$; mol wt 133.42. C 18.00%, H 2.27%, Cl 79.73%. CH_2-$ClCHCl_2$. Prepd by catalytic chlorination of ethane or ethylene: Joseph, **U.S.** pat. **2,752,401** and Pye, **U.S.** pat. **2,752,402** (both 1956 to Dow); Reynolds, **U.S.** pat. **2,783,286** (1957 to Olin Mathieson). Toxicity data: H. F. Smyth *et al., Am. Ind. Hyg. Assoc. J.* **30**, 470 (1969).

Nonflammable liquid; pleasant odor; d_4^{20} 1.4416. mp −35°. bp 113-114°. n_D^{20} 1.4711. Insol in water; misc with alcohol, ether, and many other organic liquids. LD_{50} orally in rats: 0.58 ml/kg (Smyth).

USE: Solvent for fats, waxes, natural resins, alkaloids. *Caution:* Irritating to eyes, mucous membranes, and, in high concns, narcotic.

9551. 2,2,2-Trichloroethanol. Trichloroethyl alcohol. $C_2H_3Cl_3O$; mol wt 149.42. C 16.08%, H 2.02%, Cl 71.19%, O 10.71%. CCl_3CH_2OH. Prepd by reduction of the corresponding ester, acid chloride, or acid with lithium aluminum hydride: Sroog *et al., J. Am. Chem. Soc.* **71**, 1710 (1949). Manufacture by reduction of chloral hydrate with an amine borane: Chamberlain, Schechter, **U.S.** pat. **2,898,379** (1959 to Callery Chem.).

Hygroscopic liquid, ethereal odor. At low temps it crystallizes in rhombic tablets. mp at 18°; bp 151-153°; d_D^{20} 1.55. Sol in about 12 parts water; miscible with alcohol or ether. pH of aq soln is 5-6, but on prolonged contact with water some free acid is formed. *Keep well closed and protected from light.* LD_{50} orally in rats: 600 mg/kg, *Handbook of Toxicology vol. 1,* W. S. Spector, Ed. (Saunders, Philadelphia, 1955) pp 302-303.

THERAP CAT: Sedative, hypnotic.

9552. Trichloroethylene. *Trichloroethene;* ethinyl trichloride; Tri-Clene; Trielene; Trilene; Trichloran; Trichloren; Algylen; Trimar; Triline; Tri; Trethylene; Westrosol; Chlorylen; Gemalgene; Germalgene. C_2HCl_3; mol wt 131.40. C 18.28%, H 0.77%, Cl 80.95%. $CCl_2=CHCl$. Usually prepd from *sym*-tetrachloroethane by elimination of HCl (by boiling with lime): **Ger.** pat. **171,900.** By passing tetrachloroethane vapor over $CaCl_2$ catalyst at 300°: **Ger.** pat. **263,457;** without catalyst at 450-470°: **Brit.** pat. **575,-530** (1946 to du Pont). Review of mfg processes: S. A. Miller, *Chem. Process Eng.* **47**, 268 (1966); Faith, Keyes & Clark's *Industrial Chemicals,* F. A. Lowenheim, M. K. Moran, Eds. (Wiley-Interscience, New York, 4th ed., 1975) pp 844-848. Toxicity and metabolism: E. Browning, *Toxicity and Metabolism of Industrial Solvents* (Elsevier, New York, 1965) pp 189-212. Toxicity data: Smyth *et al., Am. Ind. Hyg. Assoc. J.* **30**, 470 (1969).

Nonflammable, mobile liquid. Characteristic odor resembling that of chloroform. d_4^4 1.4904; d_4^{15} 1.4695; d_4^{20} 1.4649. Vapor density: 4.53 (air = 1.00). mp −84.8°. bp_{760} 86.7°; bp_{400} 67.0°; bp_{200} 48.0°; bp_{100} 31.4°; bp_{60} 20.0°; bp_{20} −1.0°; bp_{10} −12.4°; bp_5 −22.8°; $bp_{1.0}$ −43.8°; n_D^{17} 1.47914; n_D^{25} 1.45560. Practically insol in water; misc with ether, alcohol, chloroform. Dissolves most fixed and volatile oils. Slowly dec (with formn of HCl) by light in the presence of moisture. Trichloroethylene for medicinal purposes may contain some thymol or ammonium carbonate (not more than 20 mg/100 ml) as a stabilizer. Industrial grades of trichloroethylene may contain other stabilizers such as triethanolamine stearate and cresol. LD_{50} orally in rats: 4.92 ml/kg; LC (4 hrs) in rats: 8000 ppm (Smyth).

Caution: Use with adequate ventilation. Preserve trichloroethylene in sealed, light-resistant ampuls or in frangible, light-resistant glass tubes. Avoid prolonged exposure of the product to excessive heat. It must be dispensed in the unopened glass container in which it was placed by the manu-facturer.

Human Toxicity: Moderate exposures can cause symptoms similar to alcohol inebriation. Higher concns can have narcotic effect. Deaths occurring after heavy exposure have been attributed to ventricular fibrillation. Liver injury is not definitely established in occupational exposures. Found to induce hepatocellular carcinomas in National Cancer Institute tests on mice: *Chem. & Eng. News* **54**, 4 (Apr. 5, 1976). USE: Solvent for fats, waxes, resins, oils, rubber, paints, and varnishes. Solvent for cellulose esters and ethers. Used for solvent extraction in many industries. In degreasing, in dry cleaning. In the manuf of organic chemicals, pharmaceuticals, such as chloroacetic acid.

THERAP CAT: Inhalation analgesic.

THERAP CAT (VET): Inhalation anesthetic.

9553. Trichlorofluoromethane. Trichloromonofluoromethane; fluorotrichloromethane; Freon 11; Frigen 11; Arcton 11. CCl_3F; mol wt 137.38. C 8.74%, Cl 77.43%, F 13.83%. Prepn: Henne, *Organic Reactions* **2**, 64 (1944). Manuf: Faith, Keyes & Clark's *Industrial Chemicals*, F. A. Lowenheim, M. K. Moran, Eds. (Wiley-Interscience, New York, 4th ed., 1975) pp 325-330.

Liquid at temps below 23.7°. Faint ethereal odor. Nonflammable. $d_4^{17.2}$ 1.494; d_{gas}^{25} 5.04 (air = 1). mp −111°. bp_{760} 23.7°; bp_{400} +6.8°; bp_{200} −9.1°; bp_{100} −23.0°; bp_{60} −32.3°; bp_{40} −39.0°; bp_{20} −49.7°; bp_{10} −59.0°; bp_5 −67.6°; $bp_{1.0}$ −84.3°. Crit temp 198°; crit press. 43.2 atm (635 lb/sq inch, abs). $n_D^{18.5}$ 1.3865. Dipole moment 0.45. Practically insol in water. Sol in alcohol, ether, other, organic solvents. Less toxic than carbon dioxide, but decomposes into harmful materials by flames or high heat.

Note: Consult latest Government regulations on use as aerosol propellant.

USE: In refrigeration machinery requiring a refrigerant effective at negative pressures. As aerosol propellant. *Caution:* May be narcotic in high concentrations.

9554. 3,4,6-Trichloro-2-nitrophenol. 2-Nitro-3,4,6-trichlorophenol; 2,4,5-trichloro-6-nitrophenol; Dowlap. $C_6H_2Cl_3NO_3$; mol wt 242.44. C 29.72%, H 0.83%, Cl 43.87%, N 5.78%, O 19.80%. Prepd by dissolving 2,4,5-trichlorophenol in glacial acetic acid and treating with concd nitric acid: Kohn, Fink, *Monatsh.* **58**, 73 (1931); Harrison *et al.*, *J. Chem. Soc.* **1943**, 235.

Pale yellow crystals from petr ether, mp 92-93°.

USE: To combat the sea lamprey, an eel-like fish which attacks trout, especially in the Great Lakes region.

9555. 2,4,5-Trichlorophenol. Collunosol; Dowicide 2. $C_6H_3Cl_3O$; mol wt 197.46. C 36.49%, H 1.53%, O 8.10%, Cl 53.87%. Prepd by treating 1,2,4,5-tetrachlorobenzene with methanolic NaOH in autoclave at 160° for several hrs: Harrison *et al.*, *J. Chem. Soc.* **1943**, 235; Agfa, **Ger.** pat. 411,052 (1925); *Chem. Zentr.* **1925**, I, 2411. Prepn of complex with triisobutylphosphate: Bouillenne-Wallrand *et al.*, **Fr.** pat. M149 (1961 to Pechiney). Toxicity data: Deichmann, *Fed. Proc.* **2**, 76 (1943).

Needles from alcohol or ligroin. Strong phenolic odor. mp 67°. Sublimes. bp_{746} 248°. bp_{760} 253°. Weak monobasic acid. K at 25° = 4.3 × 10⁻⁸. Soly (g/100 g of solvent at 25°): acetone 615; benzene 163; carbon tetrachloride 51;

ether 525; denatured alcohol formula 30, 525; methanol 615; liquid petrolatum (at 50°) 56; soybean oil 79; toluene 122; water < 0.2. LD_{50} orally in rats: 0.82 g/kg (Deichmann).

Sodium salt sesquihydrate, *Dowicide B.* Flakes [prepd according to U.S. pat. **1,991,329** (1935 to Dow)]. Solubility (g/100 g solvent at 25°): acetone 163; denatured alcohol formula 30, 186; ethylene glycol 33; methanol 241; water 113. pH of satd aq soln 11.0-13.0.

Complex with triisobutyl phosphate, $C_{18}H_{30}ClO_5P$, *Trichlorex.* Liquid. $bp_{0.01}$ 94-103°.

USE: Fungicide, bactericide.

9556. 2,4,6-Trichlorophenol. Dowicide 2S; Omal. $C_6H_3Cl_3O$; mol wt 197.46. C 36.49%, O 8.10%, H 1.53%, Cl 53.87%. Prepd by direct chlorination of phenol: Tiessens, *Rec. Trav. Chim.* **50**, 115 (1931); Chulkov *et al.*, *Org. Chem. Ind. USSR* **3**, 97 (1937); *Chem. Zentr.* **1938**, I, 1419; *C.A.* **31**, 4967 (1937). Prepn of sodium salt monohydrate: Hunter, Seyfried, *J. Am. Chem. Soc.* **43**, 154 (1921).

Crystals from ligroin. Strong phenolic odor. Volatile with steam, but not from alkaline soln. d 1.4901. mp 69°. bp_{760} 246°. Soly (g/100 g of solvent): Acetone 525; benzene 113; carbon tetrachloride 37; diacetone alcohol 335; ether 354; denatured alcohol formula 30, 400; methanol 525; pine oil 163; Stoddard solvent 16; toluene 100; turpentine 37; water < 0.1.

Sodium salt monohydrate, flaky crystals. Freely sol in water, alcohol, ether, acetone. pH of satd aq soln 11.0-13.0.

Note: This substance may reasonably be anticipated to be a carcinogen: *Fourth Annual Report on Carcinogens* (NTP 85-002, 1985) p 194.

USE: Fungicide, bactericide, preservative.

9557. 1,1,1-Trichloro-2-propanol. 1,1,1-Trichloroisopropyl alcohol; trichloroisopropanol; Isopral. $C_3H_5Cl_3O$; mol wt 163.44. C 22.04%, H 3.08%, Cl 65.08%, O 9.79%. $CCl_3CH(OH)CH_3$. Prepd by reaction of chloral and methylmagnesium bromide: Kharasch *et al.*, *J. Am. Chem. Soc.* **63**, 2305 (1941).

Monoclinic crystals; camphor-like odor; pungent taste; mp 50°; bp 161-162°. Sublimes at ordinary temp. Sol in about 35 parts water; freely sol in alcohol or ether. *Keep well closed in a cool place.* LD_{50} orally in rats: 1 g/kg, Burtner, Lehmann, *J. Pharmacol. Exp. Ther.* **63**, 183 (1938).

9558. 3′,4′,5-Trichlorosalicylanilide. 3,4-Dichloroanilide of 5-chlorosalicylic acid; 5-chlorosalicylic acid 3′,4′-dichloroanilide; Anobial. $C_{13}H_8Cl_3NO_2$; mol wt 316.58. C 49.32%, H 2.55%, Cl 33.60%, N 4.42%, O 10.11%. Prepn: Bindler, Model, **U.S.** pat. **2,703,332** (1955 to Geigy).

Crystals from chlorobenzene mp 246-248°. Forms a water sol sodium salt.

USE: As skin antiseptic and deodorant in soaps and cosmetics.

THERAP CAT: Topical antiseptic; antifungal.

9559. Trichlorosilane. Trichloromonosilane; silicochloroform. Cl_3HSi; mol wt 135.47. Cl 78.52%, H 0.74%, Si 20.74%. $SiHCl_3$. Prepd from Si and HCl or from SiH_4 and HCl in presence of $AlCl_3$: Gattermann, *Ber.* **22**, 190 (1889); Ruff, Albert, *Ber.* **38**, 2226 (1905); Stock, Zeidler, *Ber.* **56**, 986 (1923); Schenk in *Handbook of Preparative Inorganic Chemistry* **vol. 1**, G. Brauer, Ed. (Academic Press, New York, 2nd ed., 1963) pp 691-692.

Volatile, mobile liquid. Fumes in air. Supports combustion. d_4^0 1.3830; d_4^{20} 1.3417; d_4^{25} 1.3313. mp $-126.5°$. bp$_{760}$ 31.8°, also reported as 36.5°. Viscosity in cp: 0.397 at 0°; 0.332 at 20°; 0.316 at 25°. n_D^{20} 1.4020; n_D^{25} 1.3983. Dec by water. Sol in benzene, carbon disulfide, chloroform, carbon tetrachloride. Dipole moment 0.97. LD$_{50}$ orally in rats: 1.03 g/kg, H. F. Smyth *et al.*, *J. Ind. Hyg. Toxicol.* **31**, 60 (1949).

USE: In organic synthesis.

9560. 2,2',2''-Trichlorotriethylamine. *2-Chloro-N,N-bis(2-chloroethyl)ethanamine;* tris(β-chloroethyl)amine; RA$_2$; Stickstoff Lost; Nitrogen Lost; HN3. $C_6H_{12}Cl_3N$; mol wt 204.53. C 35.23%, H 5.91%, Cl 52.01%, N 6.85%. N(CH$_2$CH$_2$Cl)$_3$. Prepd by the action of thionyl chloride on triethanolamine: Ward, *J. Am. Chem. Soc.* **57**, 914 (1935); U.S. pat. **2,072,348** (1937 to Hercules Powder). Improved procedure: Contardi, Dumontel, *Chim. Ind. (Milan)* **29**, 169 (1947), *C.A.* **42**, 5849 (1948); Witten in Kirk-Othmer, *Encyclopedia of Chemical Technology* **vol. 7** (Interscience, New York, 1951) p 130; Wilson, Tishler, *J. Am. Chem. Soc.* **73**, 3635 (1951).

Mobile liquid. Faint odor of fish + soap. *Vesicant, necrotizing irritant. Never use without appropriate gas mask.* Volatility at 25° = 0.120 mg/l. d_4^{25} 1.2347, mp $-4°$. bp$_{15}$ 144°. n_D^{25} 1.4925. Very slightly sol in water. Miscible with dimethylformamide, carbon disulfide, carbon tetrachloride, many other organic solvents and oils. The undiluted liq dec on standing and forms polymeric quaternary ammonium salts which are insol in the free base.

Hydrochloride, $C_6H_{13}Cl_4N$, *trimustine, trichlormethine, Sinalost, Trillekamin.* Crystals, mp 130-131°. Freely sol in water, sol in alcohol.

THERAP CAT: Hydrochloride as antineoplastic.

9561. Trichlorourethan. *2,2,2-Trichloroethanol carbamate;* trichloroethyl urethan; 2,2,2-trichloroethyl carbamate; carbamic acid trichloroethyl ester; Voluntal. $C_3H_4Cl_3NO_2$; mol wt 192.44. C 18.72%, H 2.09%, Cl 55.27%, N 7.28%, O 16.63%. NH$_2$COOCH$_2$CCl$_3$. Prepd from 2,2,2-trichloroethyl alcohol and carbamoyl chloride in ether: Willstätter, Duisberg, *Ber.* **56**, 2283 (1923).

Needles from petr ether, mp 64-65°. Soluble in about 100 parts of water. Very sol in chloroform, alcohol, ether. Sol in benzene. Sparingly sol in petr ether.

Complex with aminopyrine, *Compral, Compralgyl.*

THERAP CAT: Sedative; hypnotic.

9562. Trichodermin. *12,13-Epoxytrichothec-9-en-4-ol acetate;* WG 696. $C_{17}H_{24}O_4$; mol wt 292.36. C 69.83%, H 8.27%, O 21.89%. Antifungal metabolite from *Trichoderma viride* ND8: Neth. pat. **Appl. 302,527** (1964 to Loevens); *C.A.* **62**, 1050f (1965); also isolated from *Myrothecium roridum.* Structure: Godtfredsen, Vangedal, *Proc. Chem. Soc.* **1964**, 188; Gutzwiller *et al.*, *Helv. Chim. Acta* **47**, 2234 (1964); Godtfredsen, Vangedal, *Acta Chem. Scand.* **19**, 1088 (1965); Abrahamsson, Nilsson, *ibid.* **20**, 1044 (1966). Total synthesis: Colvin *et al.*, *Chem. Commun.* **1971**, 858; *eidem, J. Chem. Soc. Perkin Trans. I* **1973**, 1989. Activity studies: Yamamoto *et al.*, *Takeda Kenkyusho Nempo* **28**, 69 (1969), *C.A.* **72**, 76058g (1970).

Crystals from pentane at $-70°$. mp 46°, 58-60° (Colvin). bp$_{0.05}$ 110-112°. $[\alpha]_D^{20}$ $-11°$ (chloroform). uv max (ethanol): 205 nm (ϵ 2400). Sparingly soluble in water; soluble in all common organic solvents. Alkaline hydrolysis yields *trichodermol (Roridin C)*, crystals, mp 124-125°, *see also* verrucarins. LD$_{50}$ in mice (mg/kg): 500-1000 s.c.; >1000 orally (Neth. pat.).

9563. Trichostatin(s). Antibiotic A-300; A-300-I. Antifungal antibiotic isolated from metabolites of *Streptomyces hygroscopicus.* It is composed of trichostatins A (ma-

jor), B, the ferric chelate of trichostatin A, and C, the first glycosyl hydroxamate from a natural source. Isoln of A and B: N. Tsuji *et al.*, **Japan. Kokai 74 14,691**, *C.A.* **81**, 48547h (1974). Structural elucidation of A and B: *eidem, J. Antibiot.* **29**, 1 (1976). Isoln and structure of C: N. Tsuji, M. Kobayashi, *ibid.* **31**, 939 (1978).

Trichostatin A, $C_{17}H_{22}N_2O_3$, *7-[4-(dimethylamino)phenyl]-N-hydroxy-4,6-dimethyl-7-oxo-2,4-heptadienamide.* Crystals from ethyl acetate, mp 150-151°. $[\alpha]_D^{20.5}$ +62.8° ±1.1° (c = 1.007 in ethanol). uv max (ethanol): 252, 265, 341 nm (E$_{1cm}^{1\%}$ 531, 582, 648). Sol in lower alcohols, sparingly sol in chloroform, ethyl acetate, acetone, benzene.

Trichostatin B, $C_{51}H_{63}FeN_6O_9$, *tris[7-[4-(dimethylamino)phenyl]-N-hydroxy-4,6-dimethyl-7-oxo-2,4-heptadienamidato-O^NO^1]iron.* Trihydrate, dark reddish purple prisms from methanol, mp 192° (dec). uv max (ethanol): 253, 277, 341, 450 nm (E$_{1cm}^{1\%}$ 624, 651, 918, 50).

Trichostatin C, $C_{23}H_{32}N_2O_8$, *7-[4-(dimethylamino)phenyl]-N-(β-D-glucopyranosyloxy)-4,6-dimethyl-7-oxo-2,4-heptadienamide.* Colorless prisms from methanol, mp 171-173°. $[\alpha]_D^{24}$ +50.5° (±0.9°) (c = 0.987 in methanol). uv max (methanol): 268, 344 nm (ϵ 14,600; 14,300).

9564. Trichothecin. *12,13-Epoxy-4-[(1-oxo-2-butenyl)oxy]trichothec-9-en-8-one; 12,13-epoxy-4-hydroxytrichothec-9-en-8-one crotonate.* $C_{19}H_{24}O_5$; mol wt 332.38. C 68.65%, H 7.28%, O 24.07%. Mycotoxin with antibiotic activity. Produced by *Trichothecium roseum:* Freeman, Morrison, *Nature* **162**, 30 (1948). Activity: G. G. Freeman, *J. Gen. Microbiol.* **12**, 213 (1955). Structure: Godtfredsen, Vangedal, *Proc. Chem. Soc.* **1964**, 188; Gutzwiller *et al.*, *Helv. Chim. Acta* **47**, 2234 (1964). Biogenesis: Jones, Lowe, *J. Chem. Soc.* **1960**, 3959. Biosynthetic studies: Achilladelis *et al.*, *J. Chem. Soc. Perkin Trans. I* **1972**, 1425; Machida, Nozoe, *Tetrahedron Letters* **1972**, 1969; *eidem, Tetrahedron* **28**, 5113 (1972).

Slender needles from petrol ether, mp 118°. $[\alpha]_D^{18}$ +44° (c = 1 in CHCl$_3$). uv max (hexane): 217 nm (ϵ 18,000); (methanol): 215 nm (ϵ 19,000). Slightly sol in water (400 mg/l at 25°). Freely sol in most organic solvents. Aq solns are stable at pH 1-10 for at least 48 hrs at 20°. At pH 12 the antifungal activity is destroyed rapidly. Aq solns at pH 7 can be maintained at 100° for at least one hour without loss of activity. LD$_{50}$ i.v. in mice: $\sim$300 mg/kg, G. G. Freeman, *loc. cit.*

9565. Tricine. *N-[2-Hydroxy-1,1-bis(hydroxymethyl)ethyl]glycine; N-[tris(hydroxymethyl)methyl]glycine.* $C_6H_{13}NO_5$; mol wt 179.17. C 40.22%, H 7.31%, N 7.82%, O 44.65%. One of the zwitterionic amino acids known as "Good" buffers; active in the pH range 6-8.5. Prepn: N. E. Good, *Arch. Biochem. Biophys.* **96**, 653 (1962); and characterization: N. E. Good *et al.*, *Biochemistry* **5**, 467 (1966). Temperature effects on pK: R. N. Roy *et al.*, *J. Am. Chem. Soc.* **95**, 8231 (1973); M. L. Soni, R. C. Kapoor, *Int. J. Quant. Chem.* **20**, 385 (1981); on pH: R. N. Roy *et al.*, *Cryobiology* **22**, 589 (1985). Scavenger of hydroxyl radicals: M. Hicks, J. M. Gebicki, *FEBS* **199**, 92 (1986); B. Halliwell *et al.*, *Anal. Biochem.* **165**, 215 (1987). Use as buffer: R. S. Gardner, *J. Cell. Biol.* **42**, 320 (1969); R. S. Spendlove *et al.*,

Proc. Soc. Exp. Biol. Med. **137,** 258 (1971); H. Schaegger, G. Von Jagow, *Anal. Biochem.* **166,** 368 (1987).

$$HOCH_2 - \underset{\underset{CH_2OH}{|}}{\overset{\overset{CH_2OH}{|}}{C}} - \overset{+}{N}H_2CH_2COO^-$$

Crystals from alcohol/water, mp 187°. $pKa_1 \sim 2.3$; pKa_2 (0.1M): 0°, 8.6; 20°, 8.15; 37°, 7.8. pKa_2 (20°): 8.15 (0.2M); 8.15 (0.01M). $\Delta pKa/°C$ −0.021. Saturated aqueous soln is 0.8M at 0°.

USE: Biological buffer.

9566. Triclabendazole. *5-Chloro-6-(2,3-dichlorophenoxy)-2-methylthio-1H-benzimidazole;* CGA-89317; Fasinex. $C_{14}H_9Cl_3N_2OS$; mol wt 359.65. C 46.75%, H 2.52%, Cl 29.57%, N 7.79%, O 4.45%, S 8.91%. Benzimidazole derivative effective against liver fluke. Prepn: J.-J. Gallay *et al.,* **Belg.** pat. **865,870** corresp to **U.S.** pat. **4,197,307** (1978, 1980 both to Ciba-Geigy). Efficacy against immature and mature *Fasciola hepatica* in sheep, goats: K. Wolff *et al., Vet. Parasitol.* **13,** 145 (1983); in sheep: J. C. Boray *et al., Vet. Rec.* **113,** 315 (1983). Effective also against *F. gigantica* in sheep: N. Güralp, R. Tinar, *J. Helminthol.* **58,** 113 (1984). Brief review: J. Eckert *et al., Biol. Muench. Tieraerztl. Wochenschr.* **97,** 349 (1984).

Crystals, mp 175-176°.

THERAP CAT (VET): Anthelmintic (fasciola).

9567. Triclobisonium Chloride. *N,N,N',N'-Tetramethyl-N,N'-bis[1-methyl-3-(2,2,6-trimethylcyclohexyl)propyl]-1,6-hexanediaminium dichloride;* hexamethylenebis[dimethyl-[1-methyl-3-(2,2,6-trimethylcyclohexyl)propyl]ammonium chloride]; *N,N'*-bis[1-methyl-3-(2,2,6-trimethylcyclohexyl)propyl]-*N,N'*-dimethyl-1,6-hexanediamine bis(methochloride); Ro 5-0810/i; Triburon. $C_{36}H_{74}Cl_2N_2$; mol wt 605.92. C 71.36%, H 12.31%, Cl 11.70%, N 4.62%. Obtained as the hemihydrate. Prepn: Goldberg, Teitel, **U.S.** pat. **3,064,052** (1962 to Hoffmann-La Roche). Clinical application: Edelson *et al., Antibiot. Ann.* **1958-1959,** 110; Robinson, Harmon, *ibid.* 113. Comprehensive description: B. C. Rudy, B. Z. Senkowski in *Analytical Profiles of Drug Substances,* **vol. 2,** K. Florey, Ed. (Academic Press, New York, 1973) pp 507-521.

White cryst powder, mp 243-253° (dec). Sol in water, chloroform, alcohol.

THERAP CAT: Anti-infective, topical.

9568. Triclocarban. *N-(4-Chlorophenyl)-N'-(3,4-dichlorophenyl)urea; 3,4,4'-trichlorocarbanilide;* 1-(3',4'-dichlorophenyl)-3-(4'-chlorophenyl)urea; TCC; Solubacter. $C_{13}H_9Cl_3N_2O$; mol wt 315.59. C 49.47%, H 2.87%, Cl 33.71%, N 8.88%, O 5.07%. Prepn from 3,4-dichloroaniline and 4-chlorophenyl isocyanate: Beaver *et al., J. Am. Chem. Soc.* **79,** 1236 (1957); Beaver, Stoffel, **U.S.** pat. **2,818,390** (1957 to Monsanto); **Brit.** pat. **769,273** (1957).

Fine white plates, mp 255.2-256°. Constituent of *Anafung, Cutisan, Nobacter, Septivon-Lavril.*

USE: Bacteriostat and antiseptic in soaps and other cleansing compositions.

THERAP CAT: Disinfectant.

9569. Triclodazol. *5,5-Diphenyl-3-(2,2,2-trichloro-1-hydroxyethyl)-4-imidazolidinone; N-(α-hydroxy-β-trichloroethyl)-5,5-diphenyltetrahydroglyoxalin-4-one; 3-(2,2,2-trichloro-1-hydroxyethyl)-5,5-diphenyltetrahydroglyoxalin-4-one;* Hypnofon. $C_{17}H_{15}Cl_3N_2O_2$; mol wt 385.70. C 52.94%, H 3.92%, Cl 27.58%, N 7.26%, O 8.30%. Prepn: Lafon, **U.S.** pat. **3,140,290** (1964 to Orsymonde).

Yellowish-white powder, mp 138°, also reported as mp 148° (Maquenne block), *see* Lafon, *loc. cit.* Sol in hot acetone, in alcohol, in benzene, in hot 50% alcohol; practically insol in water.

THERAP CAT: Tranquilizer.

9570. Triclofenol Piperazine. *2,4,5-Trichlorophenol compd with piperazine (2:1);* bis(2,4,5-trichlorophenol) piperazine; piperazine salt of bis(2,4,5-trichlorophenol); CI 416; Ranestol (obsolete). $C_{16}H_{16}Cl_6N_2O_2$; mol wt 481.07. C 39.95%, H 3.35%, Cl 44.22%, N 5.82%, O 6.65%. Prepn: Short, Elslager, **U.S.** pat. **2,980,681** (1961 to Parke, Davis); *J. Med. Pharm. Chem.* **5,** 642 (1962).

Crystals, mp 109-110°.

THERAP CAT: Anthelmintic (Nematodes).

9571. Triclofos. *2,2,2-Trichloroethanol dihydrogen phosphate;* trichloroethyl phosphate. $C_2H_4Cl_3O_4P$; mol wt 229.39. C 10.47%, H 1.76%, Cl 46.37%, O 27.90%, P 13.50%. Prepn: Hems *et al., Brit. Med. J.* **1,** 1834 (1962).

Monosodium salt, $C_2H_3Cl_3NaO_4P$, *Sch 10159,* Trichloryl, Tricloryl, Triclos. Sol in water. LD_{50} orally in mice: 1.4 g/kg, Hems *et al., loc. cit.*

THERAP CAT: Hypnotic; sedative.

9572. Triclopyr. *[(3,5,6-Trichloro-2-pyridinyl)oxy]acetic acid;* Dowco 233; Garlon; Timbrel. $C_7H_4Cl_3NO_3$; mol wt 256.46. C 32.78%, H 1.57%, Cl 41.47%, N 5.46%, O 18.72%. Selective post-emergence herbicide. Prepn: L. D. Markley, **U.S.** pat. **3,862,952** (1975 to Dow). Activity: B. C. Byrd *et al., Proc. South. Weed Sci. Soc.* **28,** 251 (1975). Acute toxicity: E. E. Kenaga, *Down Earth* **35,** 25 (1979).

Fluffy solid, mp 148-150°. Vapor pressure at 25°: 1.26 × 10^{-6} mm Hg. pKa 2.68. Subject to photolysis. Soly in water at 25°: 440 mg/l. Soly at 25° (g/kg): acetone 989; 1-octanol 307. LD_{50} orally in rats: 713 mg/kg (Kenaga).
USE: Herbicide.

9573. Triclosan. *5-Chloro-2-(2,4-dichlorophenoxy)phenol;* 2,4,4′-trichloro-2′-hydroxydiphenyl ether; CH 3635 (formerly); Irgasan CH 3635 (formerly); Irgasan DP 300; Ster-Zac. $C_{12}H_7Cl_3O_2$; mol wt 289.53. C 49.78%, H 2.44%, Cl 36.73%, O 11.05%. Prepn: **Neth.** pat. **Appl. 6,401,526** corresp to E. Model, J. Bindler, **U.S.** pat. **3,506,720** (1964, 1970 to Geigy). Physical and bacteriostatic properties: C. A. Savage, *Drug Cosmet. Ind.* **109**(3), 36, 161 (1971). Metabolism: J. G. Black *et al., Toxicology* **3**, 33 (1975). Toxicology: F. L. Lyman, T. Furia, *Ind. Med. Surg.* **38**(2), 64 (1969), *C.A.* **71**, 89601h (1969). Use as disinfectant and textile preservative: E. Model, J. Bindler, **U.S.** pat. **3,629,477** (1971 to Geigy).

White to off-white crystalline powder, slight, faintly aromatic odor. mp 54-57.3°. Vapor pressure (20°C) 4 × 10^{-6} mm Hg. pKa 7.9. Insol in water. Readily sol in alkaline solns and many organic solvents.
USE: Bacteriostat and preservative for cosmetic and detergent prepns.
THERAP CAT: Disinfectant.

9574. Tricromyl. *3-Methyl-4H-1-benzopyran-4-one;* 3-methylchromone; 3-methyl-γ-benzopyrone; Crodimyl; Cromonalgina; Spasmocromona. $C_{10}H_8O_2$; mol wt 160.16. C 74.99%, H 5.03%, O 19.98%. Based on an ancient Egyptian drug now termed *bezr el khelda.* Prepn from o-hydroxypropiophenone: Clerc-Bory *et al., Bull. Soc. Chim. France* **1955**, 1083; Mentzer, Meunier, **Fr.** pat. **980,785** (1951 to Lab. Franc. Chimiother.); Mentzer, **U.S.** pat. **2,769,015** (1956 one-half to Laroche-Navarron).

Crystals from ethanol, mp 68°. uv max (alc): 304 nm.
THERAP CAT: Antispasmodic; coronary vasodilator.

9575. Tridecylbenzene. *1-Phenyltridecane;* Detergent Alkylate #5; Tridane. $C_{19}H_{32}$; mol wt 260.45. C 87.62%, H 12.38%. $C_6H_5(CH_2)_{12}CH_3$. Prepn: Ziegler *et al., Ann.* **511**, 13 (1934). Manufacture: Williamson, Bieneman, **U.S.** pat. **3,207,800** (1965 to Continental Oil).

Liquid, bp 346°, bp_{10} 188-189.5°. mp 10°. d_4^{20} 0.8550, d_4^{25} 1.8515. n_D^{20} 1.4821, n_D^{25} 1.4800. Forms stable foams in the presence of fat.
USE: In manuf of detergents and surface-active agents. Can be sulfonated.

9576. Tridemorph. *2,6-Dimethyl-4-tridecylmorpholine;* N-tridecyl-2,6-dimethylmorpholine; 2,6-dimethyl-4-tridecyltetrahydro-1,4-oxazine. $C_{19}H_{39}NO$; mol wt 297.52. C 76.70%, H 13.21%, N 4.71%, O 5.38%. Prepn: W. Sanne *et al.,* **Belg.** pat. **614,214**; *eidem,* **U.S.** pat. **3,468,885** (1962, 1969 both to BASF); K.-H. König *et al., Angew. Chem. Int. Ed. Engl.* **4**, 336 (1965). Structure determn of two diastereomers: D. Kost, E. Gurfinkel, *J. Chromatog.* **108**, 207 (1975). Fungicidal activity: J. Kradel *et al., Proc. 5th Brit. Insectic. Fungic. Conf.,* 16 (1969); *in vitro* activity comparison of tridemorph and commercial formulation Calixin: A. Kerke-

naar, A. A. Sijpesteijn, *Pest. Biochem. Physiol.* **12**, 124 (1979). Uptake and distribution in barley: E.-H. Pommer *et al., Proc. 5th Brit. Insectic. Fungic. Conf.,* 347 (1969); R. H. Waring, M. S. Wolfe, *Pestic. Sci.* **6**, 169 (1975). Behavior in soil: S. Otto, N. Drescher, *Proc. 7th Brit. Insectic. Fungic. Conf.,* 57 (1973). Metabolic fate in rats: D. R. Hawkins *et al., Pestic. Sci.* **5**, 535 (1974); of Calixin: R. H. Waring, *Xenobiotica* **4**, 717 (1974). Mode of action: A. Kerkenaar *et al., Pest. Biochem. Physiol.* **12**, 195 (1979). Comparison of toxicity of tridemorph and Calixin: J. Merkle *et al., Teratology* **29**, 259 (1984). Field trials: P. Lakshmanan, S. Mohan, *Pesticides* **22**, 27 (1988).

Oil, $bp_{0.7}$ 130-133°; $bp_{1.3}$ 139-142°. n_D^{25} 1.4568.
Note: The commercial product, *Calixin,* is a reaction mixture of C_{11}-C_{14} 4-alkyl-2,6-dimethylmorpholine homologs containing 60-70% of tridemorph.
USE: Systemic agricultural fungicide.

9577. Tridihexethyl Iodide. *γ-Cyclohexyl-N,N,N-triethyl-γ-hydroxybenzenepropanaminium iodide;* (3-cyclohexyl-3-hydroxy-3-phenylpropyl)triethylammonium iodide; 3-diethylamino-1-cyclohexyl-1-phenyl-1-propanol ethiodide; 3-diethylamino-1-phenyl-1-cyclohexyl-1-propanol ethiodide; α-(2-diethylaminoethyl)-α-phenylcyclohexanemethanol ethiodide; propethonum iodide; tridihexethide; 921 C; Claviton. $C_{21}H_{36}INO$; mol wt 445.44. C 56.63%, H 8.15%, I 28.49%, N 3.14%, O 3.59%. Prepn: Lobby, **U.S.** pat. **2,913,494** (1959 to Am. Cyanamid).

Bitter crystals, mp 179-184°. Solubility in water at 25°: 1.1 g/100 ml. Freely sol in alc, chloroform. Very slightly sol in ether. pH of a 1% aq soln 5.5-7.
Note: The name *Pathilon* previously used for the iodide is now used for *tridihexethyl chloride.*
THERAP CAT: Anticholinergic.

9578. Tridiphane. *2-(3,5-Dichlorophenyl)-2-(2,2,2-trichloroethyl)oxirane;* Dowco 356; Tandem. $C_{10}H_7Cl_5O$; mol wt 320.43. C 37.48%, H 2.20%, Cl 55.32%, O 4.99%. Prepn: L. D. Markley, E. J. Norton, **Ger.** pat. **2,519,073**; *eidem,* **U.S.** pat. **4,211,549** (1975, 1980 both to Dow). Herbicidal activity: J. A. Jagschitz, *Proc. Ann. Meet. Northeast Weed Sci. Soc.* **39**, 274 (1985); D. B. Vitolo *et al., ibid.* **40**, 272 (1986). Field trial in combination with atrazine, *q.v.*: P. J. Dryden, M. J. Watson, *Proc. 38th N.Z. Weed and Pest Control Conf.* **1985**, 191. Mechanism of action: G. L. Lamoureux, D. G. Rusness, *Pestic. Biochem. Physiol.* **26**, 323 (1986). Metabolism in mouse: J. Magdalou, B. D. Hammock, *Toxicol. Appl. Pharmacol.* **91**, 439 (1987). Toxicological studies: T. R. Hanley *et al., Fundam. Appl. Toxicol.* **8**, 179 (1987); J. A. John-Greene *et al., Toxicology* **43**, 325 (1987).

Yellow oil, n_D^{25} 1.5720. LD_{50} for technical grade (89-91% pure) in male, female mice, male, female rats (mg/kg): ~1200, ~740, 1700-2300, 1500-1900 orally (Hanley).
USE: Herbicide.

9579. Trientine. *N,N′-Bis(2-aminoethyl)-1,2-ethanediamine; triethylenetetramine;* 1,8-diamino-3,6-diazaoctane; 3,6-diazaoctane-1,8-diamine; 1,4,7,10-tetraazadecane; trien; TETA; TECZA. $C_6H_{18}N_4$; mol wt 146.23. C 49.28%, H 12.40%, N 38.32%. $(NH_2CH_2CH_2NHCH_2—)_2$. Prepn: A. W. von Hoffmann, *Ber.* **23,** 3711 (1890); J. van Alphen, *Rec. Trav. Chim.* **55,** 412 (1936); G. D. Jones *et al., J. Org. Chem.* **9,** 125 (1944); of the dihydrochloride: H. B. Dixon *et al., Lancet* **1,** 853 (1972). Effects on the metabolism of copper in rats: F. W. Sunderman *et al., Toxicol. Appl. Pharmacol.* **38,** 177 (1976); H. Harders *et al., Arzneimittel-Forsch.* **30,** 254 (1980). Use of the dihydrochloride in the treatment of Wilson's disease: J. M. Walshe, *Prog. Clin. Biol. Res.* **34,** 271 (1979); R. H. Haslam *et al., Dev. Pharmacol. Ther.* **1,** 318 (1980); J. M. Walshe, *Lancet* **1,** 643 (1982). Review of chemistry, physical properties and industrial uses: R. D. Spitz, "Diamines and Higher Amines, Aliphatic" in *Kirk-Othmer Encyclopedia of Chemical Technology* vol. 7 (Wiley-Interscience, New York, 3rd ed., 1979) pp 580-602. Orphan drug developed by Merck. *Review:* J. M. Walshe, *Orphan Drugs,* F. E. Karch, Ed. (Marcel Dekker, New York, 1982) pp 57-71.

Oily liquid, mp 12°. bp 266-267°. n_D^{20} 1.4971; d^{15} 0.9817. Flash pt 290°F (143°C). pH 14. Sol in water, alc. *Corrosive!* LD_{50} orally in rats: 2.5 g/kg (Spitz).

Dihydrochloride, $C_6H_{20}Cl_2N_4$, *Cuprid, Syprine.* Hygroscopic crystals, mp 115-118°.

USE: As thermosetting resin; epoxy curing agent; lubricating oil additive; analytical reagent for Cu, Ni.

THERAP CAT: Chelating agent; in treatment of Wilson's disease.

9580. Trietazine. *6-Chloro-N,N,N′-triethyl-1,3,5-triazine-2,4-diamine;* 2-chloro-4-diethylamino-6-ethylamino-s-triazine; 2-ethylamino-4-diethylamino-6-chloro-s-triazine. G 27901; NC 1667; Gesafloc. $C_9H_{16}ClN_5$; mol wt 229.73. C 47.06%, H 7.02%, Cl 15.44%, N 30.49%. Prepn: Pearlman, Banks, *J. Am. Chem. Soc.* **70,** 3726 (1948).

Crystals from propanol, mp 100-102°. Soly at 25°: 20 ppm in water; 17% in acetone; 20% in benzene; > 50% in chloroform; 10% in dioxane; 3% in ethanol. LD_{50} orally in rats: 1750 mg/kg, G. W. Bailey, J. L. White, *Residue Rev.* **10,** 97 (1965).

USE: Herbicide.

9581. Triethanolamine. *2,2′,2″-Nitrilotrisethanol;* trihydroxytriethylamine; tris(hydroxyethyl)amine; triethylolamine. $C_6H_{15}NO_3$; mol wt 149.19. C 48.30%, H 10.13%, N 9.39%, O 32.17%. $N(CH_2CH_2OH)_3$. Produced along with mono- and diethanolamine by ammonolysis of ethylene oxide. *See the refs under* Ethanolamine. Monograph: E. J. Fischer, *Triäthanolamin und andere Alkanolamine* (Heidelberg, 4th ed., 1954).

Very hygroscopic, viscous liq. Slight ammoniacal odor. Turns brown on exposure to air and light. d_4^{20} 1.1242; d_4^{60} 1.0985. One gallon weighs 9.37 lbs. mp 21.57°. bp_{760} 335-.4°. *See:* McDonald *et al., J. Chem. Eng. Data* **4,** 311 (1959). Viscosity (centipoise) at 25°: 590.5; viscosity at 60°: 65.7. Strong base. K at 25° = 3.15 × 10^{-10}. pH of 0.1N aq soln 10.5. n_D^{20} 1.4852. Flash pt 365°F. Miscible with water, methanol, acetone. Soly at 25° in benzene, 4.2%; in ether, 1.6%; in carbon tetrachloride, 0.4%; in *n*-heptane, <0.1%.

Hydrochloride, crystals from ethanol, mp 177°.

Salicylate, *Mobisyl, Myoflex.*

USE: Intermediate in the manuf of surface active agents, textile specialties, waxes, polishes, herbicides, petroleum demulsifiers, toilet goods, cement additives, cutting oils. In making emulsions with mineral and vegetable oils, paraffin and waxes. Solvent for casein, shellac, dyes; manuf synthetic resins; increasing the penetration of organic liquids into wood and paper. In the production of lubricants for the textile industry. Pharmaceutic aid (alkalizer).

9582. Triethylamine. *N,N-Diethylethanamine.* $C_6H_{15}N$; mol wt 101.19. C 71.21%, H 14.94%, N 13.84%. $(C_2H_5)_3N$. Prepd by reaction of *N,N*-diethylacetamide with lithium aluminum hydride: Uffer, Schlittler, *Helv. Chim. Acta* **31,** 1397 (1948). Manuf by vapor phase alkylation of ammonia with ethanol: Lemon, Myerly, U.S. pat. **3,022,349** (1962 to Union Carbide).

Liquid; strong ammoniacal odor; d^{25} 0.7255; mp −115°; bp 89-90°; n_D^{20} 1.4003. Flash pt, closed cup: 20°F (−6°C). Slightly sol in water above 18.7°; misc with alcohol, ether, also with water below 18.7°. *Keep well closed.* LD_{50} orally in rats: 0.46 g/kg, H. F. Smyth *et al., Arch. Ind. Hyg. Occup. Med.* **4,** 119 (1951).

Hydrochloride, $C_6H_{15}N.HCl$, crystals from alcohol; mp 253-254°, sublimes at 245°; d 1.069. Sol in 0.7 parts water; sol in alcohol, chloroform; very slightly sol in benzene; practically insol in ether.

USE: In the prepn of quaternary ammonium compds. *Caution:* May be irritating to skin, mucous membranes.

9583. Triethylammonium Formate. $C_7H_{17}NO_2$; mol wt 147.21. C 57.11%, H 11.64%, N 9.52%, O 21.74%. $(C_2H_5)_3N.HOOCH$. Prepd by neutralizing 50% formic acid with triethylamine and evaporating the resulting soln on a steambath for twelve hours at 20 mm: Alexander, Wildman, *J. Am. Chem. Soc.* **70,** 1187 (1948).

Light brown syrup. Very sol in water. Sol in alc.

USE: In syntheses employing the Leuckart reaction.

9584. Triethylenediamine. *1,4-Diazabicyclo[2.2.2]octane;* Dabco. $C_6H_{12}N_2$; mol wt 112.17. C 64.24%, H 10.78%, N 24.98%. Prepn: Krause *et al., Brit. pat.* **871,754** (1958 to Houdry Process).

Crystals, extremely hygroscopic. Sublimes readily at room temp. mp 158°. bp 174°. pKa_1 3.0; pka_2 8.7. Soly (g/100 g at 25°) in water: 45; acetone: 13; benzene: 51; ethanol: 77; methyl ethyl ketone: 26.1.

USE: Catalyst in making urethane foams.

9585. Triethylene Glycol. *2,2′-[1,2-Ethanediylbis(oxy)]-bisethanol;* 2,2′-ethylenedioxybis(ethanol). $C_6H_{14}O_4$; mol wt 150.17. C 47.99%, H 9.40%, O 42.62%. Prepd from ethylene oxide and ethylene glycol in the presence of sulfuric acid: Matignon *et al., Bull. Soc. Chim.* [5] **1,** 1308 (1934). Manuf by forming ether-ester of $HOCH_2COOH$ with glycol and then hydrogenating: Gresham, U.S. pat. **2,654,786** (1953 to du Pont). Toxicity data: Stenger *et al., Arzneimittel-Forsch.* **18,** 1536 (1968).

$$CH_2OCH_2CH_2OH$$
$$|$$
$$CH_2OCH_2CH_2OH$$

Colorless, hygroscopic, practically odorless liquid. d_4^{15} 1.1274; bp 285°; bp_{14} 165°. mp −7.2°. Viscosity at 20°: 47.8 cp; n_D^{15} 1.4578. Misc with water, alcohol, benzene, toluene; sparingly sol in ether; practically insol in petr ether. LD_{50} in mice, rats (g/kg): 21, 15-22 orally; 7.3-9.5, 11.7 i.v. (Stenger).

USE: In various plastics to increase pliability; in air disinfection.

9586. Triethylenemelamine. *2,4,6-Tris(1-aziridinyl)-s-triazine;* 2,4,6-tris(ethylenimino)-s-triazine; 2,4,6-triethylenimino-1,3,5-triazine; TEM; tretamine; NSC-9706; Triamelin; Triethanomelamine; Persistol Hö 1/193. $C_9H_{12}N_6$; mol wt 204.23. C 52.92%, H 5.92%, N 41.15%. Prepd from ethylenimine and cyanuric chloride: Bestian, *Ann.* **566,** 210 (1950); Wystrach, Kaiser, U.S. pat. **2,520,619** (1950 to Am. Cyanamid); Kaiser, Schaefer, U.S. pat. **2,653,934** (1953); Wystrach *et al., J. Am. Chem. Soc.* **77,** 5915 (1955).

Minute crystals from chloroform, dec 139°. Soly (w/w) at 26° in water 40%, chloroform 28.1%, methylene chloride 19.7%, methanol 12.5%, acetone 10.6%, dioxane 9.6%, ethanol 7.7%, benzene 5.6%, dimethyl Cellosolve 4.8%, methyl ethyl ketone 4.7%, ethyl acetate 4.5%, carbon tetrachloride 3.6%. Ampuled aq solns stored at 4° are stable for about 3 months. At room temp polymerization sets in, forming a physiologically inactive, apparently nontoxic substance. LD_{50} i.p. in mice: 4 mg/kg, Kraus *et al.*, *Proc. Soc. Exp. Biol. Med.* **76**, 489 (1951).

USE: Manuf of resinous products, textile finishing agents. As insect sterilant.

THERAP CAT: Antineoplastic.

9587. Triethylenephosphoramide. *1,1',1''-Phosphinylidynetrisaziridine; tris(1-aziridinyl)phosphine oxide;* phosphoric acid triethyleneimide; aphoxide; APO; TEPA. $C_6H_{12}N_3OP$; mol wt 173.15. C 41.62%, H 6.99%, N 24.27%, O 9.24%, P 17.88%. Prepn: Bestian, *Ann.* **566**, 231 (1950).

Crystals, mp 41°. bp_{23} 90-91°. Extremely sol in water. Very sol in alcohol, ether, acetone. LD_{50} orally in male rats: 37 mg/kg, Gaines, *Toxicol. Appl. Pharmacol.* **14**, 515 (1969).

USE: Insect chemosterilant; in dyeing, creaseproofing and flameproofing textiles; stabilizer for polymers; in photographic emulsion hardening.

THERAP CAT: Antineoplastic.

9588. Triethylenethiophosphoramide. *1,1',1''-Phosphinothioylidynetrisaziridine; tris(1-aziridinyl)phosphine sulfide;* thio-TEPA; Tifosyl; Tespamin. $C_6H_{12}N_3PS$; mol wt 189.23. C 38.09%, H 6.39%, N 22.21%, P 16.37%, S 16.95%. Preparation: Kuh, Seeger, U.S. pat. **2,670,347** (1954 to Am. Cyanamid); Saijo, Endo, **Japan.** pat. **55 218** (1955 to Semimoto Chem.). Toxicity data: Scherf *et al.*, *Arzneimittel-Forsch.* **20**, 1467 (1970). Review of carcinogenicity studies: *IARC Monographs* **9**, 85-94 (1975).

Crystals from pentane or ether, mp 51.5°. Soly in water at 25°: 19 g/100 ml. Freely sol in alc; sol in benzene, ether, chloroform. LD_{50} i.v. in rats: 15 mg/kg (Scherf).

Note: This substance may reasonably be anticipated to be a carcinogen: *Fourth Annual Report on Carcinogens* (NTP 85-002, 1985) p 196.

USE: Insect sterilant.

THERAP CAT: Antineoplastic.

9589. Triethyl Phosphate. Ethyl phosphate. $C_6H_{15}O_4P$; mol wt 182.16. C 39.56%, H 8.30%, O 35.13%, P 17.01%. $(C_2H_5)_3PO_4$. Prepd from tetraethyl hypophosphate with ethanol in the presence of aluminum ethoxide: Mukaiyama *et al., J. Org. Chem.* **27**, 1815 (1962). Manuf by treating triethyl phosphite with diethyl hydrogen phosphate: McCall, Coover, U.S. pat. **2,960,529** (1960 to Kodak).

Liquid; d^{19} 1.0725; bp 215-216°; bp_{10} 90-95°; n_D^{17} 1.4067. Sol in water with some decompn; sol in alcohol, ether.

USE: As ethylating agent; formation of polyesters which are used as insecticides.

9590. Triethyl Phosphine. $C_6H_{15}P$; mol wt 118.16. C

60.99%, H 12.80%, P 26.22%. $(C_2H_5)_3P$. Prepd from ethyllithium and phosphorus trichloride: Screttas, Isbell, *J. Org. Chem.* **27**, 2573 (1962). Manuf from white phosphorus, ethylene and hydrogen at elevated pressures: Oppegard, U.S. pat. **2,687,437** (1954 to du Pont).

Colorless liquid; odor of hyacinths; d_4^{15} 0.800; bp_{744} 127-128°. Practically insol in water; miscible with alcohol, ether.

USE: In organic syntheses.

9591. Trifenmorph. *4-(Triphenylmethyl)morpholine; 4-tritylmorpholine;* WL 8008; Frescon. $C_{23}H_{23}NO$; mol wt 329.44. C 83.85%, H 7.04%, N 4.25%, O 4.86%. Prepn: Adams *et al.,* U.S. pat. **3,577,413** (1971 to Shell). Lab tests as molluscicide: Webbe, Sturrock, *Ann. Trop. Med. Parasitol.* **58**, 234 (1964); *C.A.* **61**, 15286g (1964). Field trials: Boyce *et al., Nature* **210**, 1140 (1966). Degradn: K. I. Beynon *et al., Pestic. Sci.* **3**, 689 (1972). Neurophysiological action: R. B. Morton, D. R. Gardner, *Experientia* **32**, 611 (1976).

Crystalline solid, mp 174-176°, resolidifies and melts again at 185-187°. Vapor press at 20°: 1.4×10^{-7} mm Hg. Soly in water at 20°: 0.02 mg/l. Soly at 20° (g/l): carbon tetrachloride 300; chloroform 450; tetrachloroethylene 255. Stable to heat and alkali but hydrolyzed by mild acidic conditions. LD_{50} orally in rats: 1200-2000 mg/kg, Boyce *et al., loc. cit.*

USE: As molluscicide, to kill snails which carry bilharzia flukes.

9592. Triflumuron. *2-Chloro-N-[[[4-(trifluoromethoxy)phenyl]amino]carbonyl]benzamide; N-(2-chlorobenzoyl)-N'-[4-(trifluoromethoxy)phenyl]urea;* trifluron; SIR-8514; BAY-SIR 8514; Alsystin. $C_{15}H_{10}ClF_3N_2O_3$; mol wt 358.70. C 50.23%, H 2.81%, Cl 9.88%, F 15.89%, N 7.81%, O 13.38%. Arthropod insecticide which inhibits chitin biosynthesis: Prepn: W. Sirrenberg *et al.,* **Ger.** pat. **2,601,780** corresp to U.S. pat. **4,139,636** (1977, 1979 both to Bayer AG); **Belg.** pat. **867,046** corresp to R. H. Rigterink, U.S. pat. **4,170,657** (1978, 1979 both to Dow Chemical). Insecticidal activity: C. H. Schaefer *et al., J. Econ. Entomol.* **71**, 427 (1978); S. C. Chang, *ibid.* **72**, 479 (1979). Stability under field conditions: C. H. Schaefer, E. F. Dupras, *J. Agr. Food Chem.* **27**, 1031 (1979). Comprehensive description: G. Zoebelein *et al., Z. Angew. Entomol.* **89**, 289-297 (1980).

Crystals, mp 198° (Sirrenberg). Also reported as mp 188-190° (Rigterink). Slightly toxic to fish. Safe to bees. LD_{50} in rats, mice (mg/kg): > 5000, > 5000 ip., s.c., orally (Zoebelein).

USE: Insecticide (larvicide).

9593. Trifluomeprazine. *N,N,β-Trimethyl-2-(trifluoromethyl)-10H-phenothiazine-10-propanamine; 10-[3-(dimethylamino)-2-methylpropyl]-2-(trifluoromethyl)phenothiazine;* RP 7746. $C_{19}H_{21}F_3N_2S$; mol wt 366.45. C 62.28%, H 5.78%, F 15.55%, N 7.64%, S 8.75%. Prepn: **Brit.** pat. **813,-861** corresp to G. E. Ullyot, U. S. pat. **2,921,069** (1959, 1962 both to SK&F). Metabolism study: T. L. Flanagan *et al., J. Pharm. Sci.* **51**, 996 (1962). Chromatographic studies: M. W. Anders, G. J. Mannering, *J. Chromatog.* **7**, 258 (1962); R. J. Warren *et al., J. Pharm Sci.* **55**, 144 (1966); B. B. Wheals, *J. Chromatog.* **187**, 65 (1980). Pharmacological studies: R. C. Kelsey, G. J. Frishmuth, *Arch Int. Pharmacodyn. Ther.* **173**, 44 (1968); E. W. Baur, *J. Pharmacol. Exp. Ther.* **177**, 219 (1971).

uv max (95% ethanol): 308, 258 nm (log ϵ 3.60, 4.55). Maleate, $C_{23}H_{25}F_3N_2O_4S$, **Nortran.**
THERAP CAT (VET): Tranquilizer.

9594. Trifluoperazine. *10-[3-(4-Methyl-1-piperazinyl)-propyl]-2-(trifluoromethyl)-10H-phenothiazine; 2*-trifluoromethyl-10-[3'-(1-methyl-4-piperazinyl)propyl]phenothiazine. $C_{21}H_{24}F_3N_3S$; mol wt 407.49. C 61.89%, H 5.94%, F 13.99%, N 10.31%, S 7.87%. Prepn: Craig *et al., J. Org. Chem.* **22,** 709 (1957); **Brit. pat. 813,861;** G. E. Ullyot, **U.S. pat. 2,921,069** (1959 and 1960 to SK & F). Metabolism: T. L. Flanagan *et al., J. Pharm. Sci.* **51,** 996 (1962); C. L. Huang, K. G. Bhansali, *ibid.* **57,** 1511 (1968). Toxicity: P. J. Fowler *et al., Arzneimittel-Forsch.* **27,** 866 (1977). Comprehensive description: A. Post *et al.,* in *Analytical Profiles of Drug Substances* vol. 9, K. Florey, Ed. (Academic Press, New York, 1980) pp 543-581.

$bp_{0.6}$ 202-210°. uv max (ethanol): 258, 307.5 nm (log ϵ 4.50, 3.50). LD_{50} orally in rats, mice: 542.7, 424.0 mg/ml (Fowler).
Dihydrochloride, $C_{21}H_{26}Cl_2F_3N_3S$, **triftazin, triphthasine, Iatroneural, Jatroneural, Eskazinyl, Eskazine, Modalina, Stelazine, Terfluzine, Triflurin.** Cream colored fine powder from abs alc, mp 242-243°. Hygroscopic. Freely sol in water. Insol in dil base, ether, benzene. pK_1 3.9, pK_2 8.1. pH of 5% aq soln 2.2.
THERAP CAT: Antipsychotic.

9595. Trifluoracetic Acid. Perfluoroacetic acid. C_2HF_3-O_2; mol wt 114.03. C 21.06%, H 0.89%, F 49.99%, O 28.06%. CF_3COOH. Prepn: F. Swarts, *Bull. Sci. Acad. Roy. Belg.* **8,** 343 (1922); by oxidation of fluorine olefins: A. L. Henne, **U.S. pat. 2,371,757** (1945 to du Pont); A. L. Henne *et al., J. Am. Chem. Soc.* **67,** 918 (1945). Improved prepn: A. L. Henne, P. Trott, *ibid.* **69,** 1820 (1947). Photochemical prepn: R. N. Hazeldine, F. Nyman, *J. Chem. Soc.* **1959,** 387; *eidem, ibid.* 420. Toxicity and metabolism: M. M. Airaksinen, T. Tammisto, *Ann. Med. Exp. Biol. Fenn.* **46,** 242 (1968). *Review:* J. B. Milne in *Chem. Non-Aqueous Solvents* vol. 5B, J. J. Lagowski, Ed. (Academic Press, New York, 1978) pp 1-52.
Liquid, sharp biting odor. bp 72.4°. mp −15.4°. d^{20} 1.5351. Miscible with ether, acetone, ethanol, benzene, CCl_4, hexane. Strong, non-oxidizing acid. pka 0.3. LD_{50} i.v. in mice: 1200 mg/kg, M. M. Airaksinen, T. Tammisto, *loc. cit.*
USE: In organic synthesis; dissolves protein when mixed with liquid SO_2.

9596. Trifluperidol. *4'-Fluoro-4-[4-hydroxy-4-(α,α,α-trifluoro-m-tolyl)piperidino]butyrophenone; p*-fluoro-4-[4'-hydroxy-4'-(3''-trifluoromethyl)phenyl]piperidinobutyrophenone; 1-(3'-p-fluorobenzoylpropyl)-4-hydroxy-4-(3''-trifluoromethylphenyl)piperidine; ω-[4-hydroxy-4-(m-trifluoromethylphenyl)piperidino]-p-fluorobutyrophenone; flumoperone; Psicoperidol; Psychoperidol. $C_{22}H_{23}F_4NO_2$; mol wt 409.43. C 64.54%, H 5.66%, F 18.56%, N 3.42%, O 7.82%. Prepn: P. A. J. Janssen, **Brit. pat. 895,309** corresp to **U.S. pat. 3,438,991** (1962, 1969 to Janssen). Pharmacology: *idem, Arzneimittel-Forsch.* **11,** 932 (1961).

Hydrochloride, $C_{22}H_{24}ClF_4NO_2$, **R2498, Triperidol.** Crystals from acetone, mp 200.5-201.3°. Sol in water. LD_{50} in rats: 14 mg/kg i.v.; 70 mg/kg s.c., P. A. J. Janssen, *loc. cit.*
THERAP CAT: Antipsychotic.

9597. Triflupromazine. *N,N-Dimethyl-2-(trifluoromethyl)-10H-phenothiazine-10-propanamine; 10-[3-(dimethylamino)propyl]-2-trifluoromethylphenothiazine; 2*-trifluoromethyl-10-(γ-dimethylaminopropyl)phenothiazine; **Vetame.** $C_{18}H_{19}F_3N_2S$; mol wt 352.44. C 61.35%, H 5.43%, F 16.17%, N 7.95%, S 9.10%. Prepn: Yale *et al., J. Am. Chem. Soc.* **79,** 4375 (1957); **Brit. pat. 813,861** (1959 to SK & F). Comprehensive description of the hydrochloride: K. Florey, Ed. in *Analytical Profiles of Drug Substances* vol. 2 (Academic Press, New York, 1973) pp 523-550.

Liquid, $bp_{0.7}$ 176°; $bp_{0.4}$ 162-164°. n_D^{23} 1.5780.
Hydrochloride, $C_{18}H_{20}ClF_3N_2S$, **Adazine, Fluomazina, Fluorofen, Nivoman, Psyquil, Siquil, Vespral, Vesprin.** Crystals from xylene, dec 173-174°. uv max: 255, 305 nm ($E_{1cm}^{1\%}$ 700, 90). Sol in water, ethanol, acetone. pH of 2% aq soln 4.1, if the pH is raised to 6.4, pptn of the free base results.
THERAP CAT: Antipsychotic.
THERAP CAT (VET): Tranquilizer.

9598. Trifluralin. *2,6-Dinitro-N,N-dipropyl-4-(trifluoromethyl)benzenamine; α,α,α-trifluoro-2,6-dinitro-N,N-dipropyl-p-toluidine;* 2,6-dinitro-N,N-dipropyl-α,α,α-trifluoro-p-toluidine; 2,6-dinitro-N,N-dipropyl-4-trifluoromethylaniline; N,N-dipropyl-2,6-dinitro-4-trifluoromethylaniline; L-36352; Lilly 36352; **Treflan; Triflurex.** $C_{13}H_{16}F_3N_3O_4$; mol wt 335.29. C 46.57%, H 4.81%, F 17.00%, N 12.53%, O 19.09%. Pre-emergence herbicide. Prepn: **Brit. pat. 917,253,** *C.A.* **59,** 9889b (1963); Soper, **U.S. pat. 3,403,180** (1963; 1968, both to Lilly). Photodecomposition: E. Leitis, D. G. Crosby, *J. Agr. Food Chem.* **22,** 842 (1974). Soil degradation: T. Golab *et al., ibid.* **27,** 163 (1979). Toxicity data: Goldenthal, *Toxicol. Appl. Pharmacol.* **18,** 185 (1971). Carcinogenicity study: J. F. Robens, *Vet. Human Toxicol.* **22,** 328 (1980).

Yellow crystals, mp 46-47°. $bp_{4.2}$ 139-140°. Slightly sol in water (0.0024 g/100 ml); freely sol in acetone, Stoddard solvent, xylene. LD_{50} orally in rats: 500 mg/kg (Goldenthal).
USE: Herbicide.

9599. Trifluridine. *α,α,α-Trifluorothymidine; 2'-deoxy-5-(trifluoromethyl)uridine; 5-(trifluoromethyl)-2'-deoxyuridine;* F3TDR; NSC 75,520; **TFT Thilo; Virophta; Viroptic.** $C_{10}H_{11}F_3N_2O_5$; mol wt 296.21. C 40.55%, H 3.74%, F 19.24%, N 9.46%, O 27.01%. Prepn: C. Heidelberger *et al., J. Am. Chem. Soc.* **84,** 3597 (1962); *eidem, J. Med. Chem.* **7,** 1 (1964); C. Heidelberger, **U.S. pat. 3,201,387** (1965 to U.S. Dept. HEW). Crystal structure: A. H. Tench, *Diss. Abstr. Int. B* **33,** 3587 (1973). NMR study: R. J. Cushley *et al., J. Am. Chem. Soc.* **90,** 709 (1968). Metabolism: D. L. Dexter *et al., Cancer Res.* **32,** 247 (1972); W. J. O'Brien, H. F.

Edelhauser, *Invest. Ophthalmol. Vis. Sci.* **16**, 1093 (1977). Pharmacodynamics: B. L. Wigdahl, J. R. Parkhurst, *Antimicrob. Ag. Chemother.* **14**, 470 (1978); G. J. Smith *et al.*, *Biochem. Biophys. Res. Commun.* **83**, 1538 (1978). Teratogenicity study: M. Itoi *et al.*, *Arch. Ophthalmol.* **93**, 46 (1975). Cytotoxicity and mutagenicity study: E. Huberman, C. Heidelberger, *Mutat. Res.* **14**, 130 (1972). Clinical studies: H. E. Kaufman, *Invest. Ophthalmol. Vis. Sci.* **17**, 941 (1978); R. A. Hyndiuk *et al.*, *Arch. Ophthalmol.* **96**, 1839 (1978). Review of mechanism of antiviral activity: C. Heidelberger, *Ann. N.Y. Acad. Sci.* **255**, 317 (1975). Review of pharmacology and therapeutic use: A. A. Carmine *et al.*, *Drugs* **23**, 329-353 (1982).

Cryst from ethyl acetate, mp 186-189°. uv max (0.1N HCl): 260 nm (ε 9960); (0.1N NaOH): 260 nm (ε 6590).
THERAP CAT: Ophthalmic antiviral.

9600. Triflusal. *2-(Acetyloxy)-4-(trifluoromethyl)benzoic acid; α,α,α-trifluoro-2,4-cresotic acid acetate;* acetyl-4-trifluoromethylsalicylic acid; UR 1501; Disgren. $C_{10}H_7F_3O_4$; mol wt 248.16. C 48.40%, H 2.84%, F 22.97%, O 25.79%. Analog of aspirin, *q.v.*, with an inhibitory effect on platelet aggregation. Prepn: M. Hauptschein, U.S. pat. **3,019,253** (1962 to Pennsalt); E. F. Barra, A. C. M. Boga, **Ger.** pat. **2,641,556** corresp to U.S. pat. **4,096,252** (1977, 1978 both to Uriach). Inhibition of platelet aggregation in man and rat: J. Garcia-Rafanell, J. Morell, *Therapie* **32**, 337 (1977). Effect on blood coagulation: M. Rutllant *et al.*, *Curr. Ther. Res.* **22**, 510 (1977). Pharmacokinetics: V. Rimbau *et al.*, *Arch. Farmacol. Toxicol.* **7**, 11 (1981). Clinical studies: R. M. Masso *et al.*, *Curr. Ther. Res.* **25**, 791 (1979); E. Sala-Planell *et al.*, *Angiologia* **33**, 71 (1981).

White cryst solid from petr ether/ether, mp 120-122° (upon slow heating); 110-112° (upon quick heating). Misc with ethanol; practically insol in water. LD_{50} in mice, rats (mg/kg): 437, 402 orally; 380, 217 i.p., *RTECS* Vol. I, R. J. Lewis, R. L. Tatken, Eds. (1980) p 546.
THERAP CAT: Antithrombotic.

9601. Trifolium. Meadow clover; red clover; purple clover; cow clover. Dried inflorescence of *Trifolium pratense* L., *Leguminosae*. *Habit*. Europe, Asia, Northern Africa; naturalized in U.S. *Constit*. Tannin, resins, fat, trifolianol, trifoliin.
THERAP CAT: Antispasmodic; expectorant.

9602. Triforine. *N,N'-[1,4-Piperazinediylbis(2,2,2-trichloroethylidene)]bisformamide;* 1,4-di(2,2,2-trichloro-1-formamidoethyl)piperazine; 1,4-bis(1-formamido-2,2,2-trichloroethyl)piperazine; Cela W-524; CME 74770; Basforin; Funginex; Saprol. $C_{10}H_{14}Cl_6N_4O_2$; mol wt 434.95. C 27.61%, H 3.24%, Cl 48.90%, N 12.88%, O 7.36%. Prepn: W. Ost *et al.*, **Ger.** pat. **1,901,421** corresp to U.S. pat. **3,595,916** (1969, 1971, both to Boehringer, Ing.).

White crystals, mp 155°. Vapor pressure at 25° = 2×10^{-7} mm Hg. Soly in water at 20°: 27-29 ppm. Sol in CMF, DMSO, N-methylpyrrolidone; moderately sol in tetrahydrofuran. Insol in acetone, benzene, carbon tetrachloride, chloroform, methylene chloride, petroleum ether. Dec to chloral and piperazine salts by conc H_2SO_4 and HCl; slowly dec to chloroform and piperazine by conc alk. Half life in the soil ~3 weeks. Low toxicity to fish and bees.
USE: Fungicide.

9603. Trigentisic Acid. *3,3',3''-Methylidynetrigentisic acid;* polycondensed hexahydroxytricarboxytriphenylmethane; 21P; Rehibin. Prepn: **Brit.** pat. **723,525** (1955 to Ferrosan). As hyaluronidase inhibitor: *Brit. J. Pharmacol.* **8**, 30 (1953).

LD_{50} orally in mice: > 20 mg/g.

9604. Triglyme. *2,5,8,11-Tetraoxadodecane;* triethylene glycol dimethyl ether. $C_8H_{18}O_4$; mol wt 178.22. C 53.91%, H 10.18%, O 35.91%. $CH_3O(CH_2)_2O(CH_2)_2O(CH_2)_2OCH_3$. Prepd by high pressure hydrogenation of 1,1,2,2-tetrakis(2-methoxyethoxy)ethane: McNamee, MacDowell, U.S. pat. **2,425,042** (1947 to Carbide & Carbon Chem.).
Liquid. d_4^{20} 0.990. Flash pt 111°. mp −45°. bp_{760} 216°; bp_{10} 103.5°; $bp_{0.9}$ 20°. n_D^{20} 1.4233. Miscible with water, hydrocarbon solvents.
USE: Solvent.

9605. Trigonellamide Chloride. *3-(Aminocarbonyl)-1-methylpyridinium chloride;* N^1-methylnicotinamide chloride; 1-methylpyridine-3-carboxylic acid amide chloride; nicotinamide chloromethylate; nicotinamide methyl chloride. $C_7H_9ClN_2O$; mol wt 172.62. C 48.71%, H 5.25%, Cl 20.54%, N 16.23%, O 9.27%. One of the principal excretion products of the metabolism of nicotinic acid in man, dog, and rat. Coenzyme action: Warburg, Christian, *Biochem. Z.* **287**, 291 (1936). Synthesis by refluxing nicotinamide with methyl iodide in methanol, then shaking the nicotinamide methiodide with AgCl: Karrer *et al.*, *Helv. Chim. Acta* **19**, 826 (1936); and isolation from urine: Huff, Perizweig, *J. Biol. Chem.* **150**, 395 (1943). Differentiation from NAD: Carpenter, Kodicek, *Biochem. J.* **46**, 421 (1950). *In vitro* metabolism: G. S. Johnson, *Eur. J. Biochem.* **112**, 635 (1980); H. Hoshino *et al.*, *Biochim. Biophys. Acta* **801**, 250 (1984). HPLC determn in urine: M. A. Kutnink *et al.*, *J. Liq. Chromatog.* **7**, 969 (1984).

Crystals from methanol. Dec 240°. Moderately sol in water. More sol in alcohol, butanol, isobutanol. Insol in amyl alcohol, octyl alcohol, benzene, chlorobenzene, chloroform. Destroyed upon boiling the aq soln, more rapidly in the presence of alkali. At room temp it is destroyed in alkaline soln. Reacts with ketones in aq alkaline soln to produce a greenish-blue fluorescence; on acidification the fluorescence changes to blue, and is intensified by heating.

9606. Trigonelline. *3-Carboxy-1-methylpyridinium hydroxide inner salt;* nicotinic acid *N*-methylbetaine; coffearine; caffearine; gynesine; trigenolline. $C_7H_7NO_2$; mol wt 137.13. C 61.31%, H 5.15%, N 10.22%, O 23.33%. In seeds of *Trigonella foenumgraecum* L., *Leguminosae*, in coffee beans, in seeds of *Strophanthus* spp, *Apocynaceae* and of *Cannabis sativa* L., *Moraceae*, in seeds of many other plants; also in sea urchin, *Arabacia pustulosa*, and in jellyfish, *Velella spirans*. Excreted in urine after taking nicotinic acid: Ackermann, Z. *Biol.* **59**, 17 (1912). Isoln from normal urine: Linnewah, Renwein, Z. *Physiol. Chem.* **207**, 48 (1932); **209**, 110 (1932). Synthesis: Turnau, *Monatsh.* **26**, 551 (1905); Sarett *et al.*, J. *Biol. Chem.* **135**, 483 (1940); Green, Tong, J. *Am. Chem. Soc.* **78**, 4896 (1956); Kosower, Patton, J. *Org. Chem.* **26**, 1318 (1961).

Monohydrate, crystals from ethanol, mp 230-233°. Salty taste. Very sol in water; sol in alcohol; practically insol in ether, chloroform. LD_{50} s.c. in rats: 5.0 g/kg, Brazda, Coulson, *Proc. Soc. Exp. Biol. Med.* **62**, 19 (1946).

Hydrochloride, $C_7H_7NO_2$·HCl, crystals from 90% alcohol, mp 258-259°. Very sol in water; slightly in alcohol; practically insol in ether, benzene.

9607. Trihexyphenidyl Hydrochloride. *α-Cyclohexyl-α-phenyl-1-piperidinepropanol hydrochloride;* 3-(1-piperidyl)-1-cyclohexyl-1-phenyl-1-propanol hydrochloride; 1-phenyl-1-cyclohexyl-3-piperidyl-1-propanol hydrochloride; benzhexol chloride; Aparkan; Artane; Artilan; Bentex; Broflex; Cyclodol; Tsiklodol; Parkinsan; Parkinane retard; Romparkin; Paralest; Triphenidyl; Triesifenidile; Triexifenidila; Sedrena; Triphedinon; Pacitane; Parkan; Tremin; Peragit; Pargitan; Parkopan; Pipanol. $C_{20}H_{32}ClNO$; mol wt 337.92. C 71.08%, H 9.55%, Cl 10.49%, N 4.15%, O 4.73%. Synthesis: Denton *et al.*, J. *Am. Chem. Soc.* **71**, 2053 (1949); Adamson, Wilkinson, U.S. pat. **2,682,543** (1954 to Burroughs Wellcome); Denton, U.S. pat. **2,716,121** (1955 to Am. Cyanamid). Resolution into isomers: Adamson, Duffin, **Brit.** pat. **750,156** (1956 to Wellcome Found.).

Crystals, dec 258.5°. Free base, mp 114.3-115.0°. Soly (g/100 ml): water at 25°, 1.0; alcohol 6; chloroform 5. More sol in methanol; very slightly sol in ether, benzene. pH of a 1% aq soln 5.5-6.0.

l-Form, crystals from isopropyl alcohol, mp 264°. $[\alpha]_D^{20}$ −30° (c = 0.4 in chloroform). Free base mp 112-113°. $[\alpha]_D^{20}$ −25° (c = 0.4 in ethanol).

THERAP CAT: Anticholinergic; antiparkinsonian.

9608. Trihydrazine Dihydriodide. Hydrazine dihydriodide. $H_{14}I_2N_6$; mol wt 351.98. H 4.01%, I 72.11%, N 23.88%. $(N_2H_5^+I^-)_2\cdot N_2H_4$. Prepared from hydrazine (85-100%) and hydriodic acid in ethanol at pH 8-9: Curtius, Schulz, J. *Prakt. Chem.* **42**, 540 (1890); Gilbert, Decius, J. *Am. Chem. Soc.* **80**, 3871 (1958).

Large, white, biaxial needles from abs ethanol, mp 90-92°.

Crystals stored in a glass-stoppered bottle at room temp showed no deterioration over a six month period. Freely sol in water with formation of hydrazine monoiodide and hydriodic acid. The hydrazine is very strongly held in the dry crystals. Exposure in a vacuum still at 25 microns and room temperature for 5½ hours showed no loss in weight. The neutral hydrazine molecule is probably hydrogen-bonded to the $N_2H_5^+$ ions.

9609. Trillium. Beth root; Indian balm; ground lily; birthroot. Dried rhizome of *Trillium erectum* L. and other spp of *Trillium*, *Liliaceae*. *Habit.* Canada, south to Tennessee and Missouri; also Japan. *Constit.* Trilline, fixed oil, tannin.

THERAP CAT: Formerly in metrorrhagia, menorrhagia, various types of hemorrhage; astringent in diarrhea.

9610. Trilobine. *6′,7-Epoxy-6,12′-dimethoxy-2′-methyl-1′α-oxyacanthan.* $C_{35}H_{34}N_2O_5$; mol wt 562.64. C 74.71%, H 6.09%, N 4.98%, O 14.22%. R = H. From root of *Cocculus trilobus* DC. and *C. sarmentosus* Diels, *Menispermaceae*: Tomita *et al.*, J. *Pharm. Soc. Japan* **48**, 83 (1928); **50**, 127 (1930); **62**, 468, 481 (1942). Structure: Tomita, Inubushi, *Pharm. Bull. (Japan)* **3**, 7 (1955), *C.A.* **50**, 1854c (1956); Inubushi, Nomura, *Tetrahedron Letters* **1962**, 1133; Inubushi *et al.*, J. *Pharm. Soc. Japan* **83**, 282, 288 (1963); Tomita, Furukawa, *ibid.* 760, *C.A.* **59**, 5212b,h; 15336c (1963). Total synthesis of trilobine and isotrilobine: Y. Inubushi *et al.*, *Chem. Pharm. Bull.* **25**, 1636 (1977).

Crystals from benzene, mp 237°. $[\alpha]_D$ +307° (chloroform). Practically insol in water. Sparingly sol in alcohol, acetone, ether. Sol in chloroform.

N-Methyl deriv., $C_{36}H_{36}N_2O_5$, *isotrilobine, homotrilobine, N-methyltrilobine.* R = CH_3. Crystals from acetone, mp 215°. $[\alpha]_D$ +317° (chloroform). Structure: Kondo, Tomita, *Ann.* **497**, 104 (1932); Inubushi, Nomura, *loc. cit.;* Inubushi *et al.*, *loc. cit.*, 1963.

9611. Trilostane. *(4α,5α,17β)-4,5-Epoxy-3,17-dihydroxyandrost-2-ene-2-carbonitrile; 4,5-epoxy-17-hydroxy-3-oxoandrostane-2-carbonitrile;* 2α-cyano-4α,5α-epoxyandrostan-17β-ol-3-one; Win 24540; Desopam; Modrastane; Modrenal. $C_{20}H_{27}NO_3$; mol wt 329.45. C 72.92%, H 8.26%, N 4.25%, O 14.57%. Prepn: R. O. Clinton, A. J. Manson, U.S. pat. **3,296,255** (1967 to Sterling); H. C. Neumann *et al.*, J. *Med. Chem.* **13**, 948 (1970). Inhibition of steroid biosynthesis: G. O. Potts *et al.*, *Steroids* **32**, 257 (1978). Disposition in animals: J. F. Baker *et al.*, *Arch. Int. Pharmacodyn. Ther.* **243**, 4 (1980). Metabolism in rats: Y. Mori *et al.*, *Chem. Pharm. Bull.* **29**, 2646 (1981); in humans: D. T. Robinson *et al.*, J. *Steroid Biochem.* **21**, 601 (1984). HPLC determn in plasma: P. Powles *et al.*, J. *Chromatog.* **311**, 434 (1984); and of major metabolite, *17-ketotrilostane*: R. R. Brown *et al.*, *ibid.* **339**, 440 (1985). Clinical study in Cushing's syndrome: P. Komanicky *et al.*, J. *Clin. Endocrinol. Metab.* **47**, 1042 (1978). Clinical trials in breast cancer: C. G. Beardwell *et al.*, *Cancer Chemother. Pharmacol.* **10**, 158 (1983); C. J. Williams *et al.*, *Cancer Treat. Rep.* **71**, 1197 (1987).

Tan crystals from pyridine/dioxane, mp 257.8-270° (dec). $[\alpha]_D^{25}$ +137.4° (c = 1 in pyridine). uv max (ethanol): 252 nm (ϵ 8300).

THERAP CAT: Adrenocortical suppressant; in treatment of breast cancer.

9612. Trimazosin. *4-(4-Amino-6,7,8-trimethoxy-2-quinazolinyl)-1-piperazinecarboxylic acid 2-hydroxy-2-methylpropyl ester.* $C_{20}H_{29}N_5O_6$; mol wt 435.49. C 55.16%, H 6.71%, N 16.08%, O 22.04%. Prepn: H. J. Hess, **Ger. pat.** **2,120,495** corresp to U.S. pat. **3,669,968** (1971, 1972 both to Pfizer). Pharmacological study in rats: J. P. Buyniski *et al., Clin. Exp. Hypertens.* **2**, 1039 (1980), *C.A.* **94**, 76779x (1981). Cardiocirculatory effects in congestive heart failure: N. A. Awan *et al., Am. J. Cardiol.* **44**, 126 (1979). Long-term therapeutic use: K. T. Weber *et al., N. Engl. J. Med.* **303**, 242 (1980).

White cryst from chloroform/diisopropyl ether, mp 158-159°.

Hydrochloride monohydrate, $C_{20}H_{30}ClN_5O_6 \cdot H_2O$, *CP-19,106-1, Cardovar, Supres.* White cryst, mp 166-169° (dec).

THERAP CAT: Antihypertensive.

9613. Trimebutine. *3,4,5-Trimethoxybenzoic acid 2-(dimethylamino)-2-phenylbutyl ester.* $C_{22}H_{29}NO_5$; mol wt 387.48. C 68.19%, H 7.54%, N 3.62%, O 20.65%. A gastrointestinal antispasmodic. Prepn by esterification: C. P. J. Roux, D. R. Torossian, **Fr. pat.** **1,344,455**, *C.A.* **60**, 15777g (1964); by transesterification: D. R. Torossian, G. G. Aubard, **Ger. pat.** **1,151,716** corresp to **Brit. pat.** **1,342,547** (1964, 1972, 1974 all to Jouveinal). Clinical trials in treatment of spastic colon: K. Lüttecke, *J. Int. Med. Res.* **6**, 86 (1978); M. G. Moshal, M. Herron, *ibid.* **7**, 231 (1979). Effect on gastrointestinal tract contractions in dogs: K. Yamada *et al., Japan. J. Pharmacol.* **33**, 301 (1983); on spontaneous contraction of guinea pig colon: H. Takenaga *et al., ibid.* **34**, 177 (1984). HPLC determn in plasma: A. Astier, A. M. Deutsch, *J. Chromatog.* **224**, 149 (1981).

Crystals from ethanol, mp 78-80°C. Soluble in methylene chloride.

Maleate, $C_{26}H_{33}NO_9$, *TM 906, Cerekinon, Debridat, Digerent Polifarma, Foldox, Miopropan, Polibutin, Spabucol, Trimedat.* Crystals from water, mp 105-106°.

THERAP CAT: Antispasmodic.

9614. Trimecaine. *2-Diethylamino-2',4',6'-trimethylacetanilide; N-sym-trimethylphenyldiethylaminoacetamide; 2-diethylaminoacetyl-2',4',6'-trimethylanilide; Mesocaine; Mesidicaine; Mesokain.* $C_{15}H_{24}N_2O$; mol wt 248.36. C 72.54%, H 9.74%, N 11.28%, O 6.44%. Prepn: Löfgren, Lundqvist, U.S. pat. **2,441,498** (1948 to Astra); Borovansky *et al., J. Am. Pharm. Assoc.* **48**, 402 (1959).

Crystals, mp 44°. bp$_6$ 187°; bp$_{0.6}$ 154-155°.

Hydrochloride, $C_{15}H_{24}N_2O \cdot HCl$, crystals from acetone, mp 140°. LD$_{50}$ s.c. in mice: 295 mg/kg.

THERAP CAT: Hydrochloride as local anesthetic.

9615. Trimedlure. *4(or 5)-Chloro-2-methylcyclohexanecarboxylic acid, 1,1-dimethylethyl ester; tert*-butyl 4(or 5)-chloro-2-methylcyclohexanecarboxylate. $C_{12}H_{21}ClO_2$; mol wt 232.75. C 61.92%, H 9.09%, Cl 15.23%, O 13.75%. Substituted cyclohexane developed as attractant for the Mediterranean fruit fly, or medfly, *Ceratitis capitata* (Weidemann). Prepn and physical props: M. Beroza *et al., J. Agr. Food Chem.* **9**, 361 (1961). Commercial product consists mainly of the four isomers having the methyl and ester substituents on the ring in a *trans* configuration. Separation and identification of isomers: T. P. McGovern, M. Beroza, *J. Org. Chem.* **31**, 1472 (1966). Extension of activity by fixatives: *eidem, J. Econ. Entomol.* **60**, 379 (1967). Controlled release: S. Nakagawa *et al., ibid.* **72**, 625 (1979). Repellent effect of high concns: *eidem, ibid.* **64**, 762 (1971). Rate of loss from wicks: J. R. King, P. J. Landolt, *ibid.* **77**, 221 (1984). Toxicity studies: M. Beroza *et al., Toxicol. Appl. Pharmacol.* **31**, 421 (1975).

Oil, bp 107-113°. n_D^{20} 1.460. LD$_{50}$ in rats (mg/kg): 4556 (±1136) orally; in rabbits (mg/kg): >2025 dermally. LC$_{50}$ in rainbow trout, bluegill sunfish (ppm): 11.5, 14.7 (24 hr) (Beroza).

USE: In traps for monitoring medfly infestations.

9616. Trimellitic Acid. *1,2,4-Benzenetricarboxylic acid;* 1,2,4-tricarboxybenzene. $C_9H_6O_6$; mol wt 210.14. C 51.44%, H 2.88%, O45.68%. Obtained by oxidation of coal with nitric acid: Grosskinsky, *Glückauf* **88**, 376 (1952), *C.A.* **46**, 7731 (1952); by oxidation of β-indancarboxylic acid with nitric acid: Braun *et al., Ber.* **53**, 1160 (1920). Manuf by oxidation of pseudocumene: Backlund, U.S. pat. **3,009,953** (1961 to Union Oil of Calif.).

Crystals from acetic acid or dilute alcohol. mp 218-220° (also reported 229-234° dec). Solubilities at 25° in g/100 g solvent: carbon tetrachloride 0.004; ligroin 0.03; mixed xylenes 0.006; dimethylformamide 31.3; ethyl acetate 1.7; acetone 7.9; water 2.1; ethanol 25.3. Practically insol in chloroform, benzene, carbon disulfide.

USE: Intermediate in the prepn of resins, plasticizers, dyes, inks, adhesives.

9617. Trimellitic Anhydride. *1,3-Dihydro-1,3-dioxo-5-isobenzofurancarboxylic acid;* trimellitic acid 1,2-anhydride; anhydrotrimellitic acid; 1,3-dioxo-5-phthalancarboxylic acid. $C_9H_4O_5$; mol wt 192.12. C 56.26%, H 2.10%, O 41.64%. Prepd by subliming trimellitic acid above its mp: Alder, Dortmann, *Ber.* **85**, 556 (1952); by heating crude trimellitic acid with V_2O_5: McKinnis, U.S. pat. **2,998,431** (1959 to Union Oil of Calif.).

Crystals, mp 161-163.5°. bp_{14} 240-245°. Solubilities at 25° in g/100 g solvent: carbon tetrachloride 0.002; ligroin 0.06; mixed xylenes 0.4; dimethylformamide 15.5; acetone 49.6; ethyl acetate 21.6.

USE: In the preparation of resins, adhesives, polymers, dyes, printing inks.

9618. Trimeprazine. *N,N,β-Trimethyl-10H-phenothia-zine-10-propanamine; 10-[3-(dimethylamino)-2-methylprop-yl]phenothiazine;* alimemazine; methylpromazine; Bayer 1219; RP 6549. $C_{18}H_{22}N_2S$; mol wt 298.44. C 72.44%, H 7.43%, N 9.39%, S 10.75%. Prepn: Jacob, Robert, U.S. pat. **2,837,-518** (1958 to Rhône-Poulenc). Metabolism studies: Robinson, *J. Pharm. Pharmacol.* **18**, 19 (1966).

Crystals, mp 68°. $bp_{0.3}$ 150-175°.
Tartrate, $C_{40}H_{50}N_4O_6S_2$, *Panectyl, Vanectyl, Repeltin, Temaril, Theralene, Vallergan.* Crystals. Soluble in water; slightly sol in alcohol.

THERAP CAT: Antipruritic.

9619. Trimetazidine. *1-[(2,3,4-Trimethoxyphenyl)meth-yl]piperazine.* $C_{14}H_{22}N_2O_3$; mol wt 266.33. C 63.13%, H 8.33%, N 10.52%, O 18.02%. Prepn: Regnier, Canevari, Fr. pat. **1,302,958** (1962 to Sci. Union et Cie, Soc. Franç. Rech. Med.); Fr. pat. **M805**; J. Servier, U.S. pat. **3,262,852** (1961, 1966 to Biofarma). Pharmacology: Fujita, *Japan. J. Phar-macol.* **17**, 19 (1967); Nagata et al., ibid. **19**, 628 (1969); **21**, 337 (1971).

bp_2 200-205°.
Dihydrochloride, $C_{14}H_{24}Cl_2N_2O_3$, *Kyurinett, Vastarel, Yosimilon.* Crystals, mp 225-228°. LD_{50} in mice (mg/kg): 125-135 i.v.; 305-315 i.p. (Servier, 1966).

THERAP CAT: Coronary vasodilator.

9620. Trimethadione. *3,5,5-Trimethyl-2,4-oxazolidine-dione; 3,5,5-trimethyl-2,4-dioxooxazolidine;* troxidone; Absentol;Tridione; Trimedal; Petidon; Epidione; Ptimal. C_6H_9-NO_3; mol wt 143.14. C 50.34%, H 6.34%, N 9.79%, O 33.53%. Synthesis: Spielman, *J. Am. Chem. Soc.* **66**, 1244 (1944); U.S. pat. **2,575,692** (1951 to Abbott); Davies, Hook, U.S. pat. **2,559,011** (1951 to Brit. Schering). Metabolism to dimethadione, *q.v.*: T. C. Butler, *J. Pharmacol. Exp. Ther.* **108**, 11 (1953); T. C. Butler, W. J. Waddell, ibid. **110**, 241 (1954). Anticonvulsant activity: H. Ferngren, *Acta Phar-macol. Toxicol.* **26**, 177 (1968); H. H. Frey, ibid. **27**, 295 (1969). Pharmacodynamics and metabolism in rats: D. O. Thueson et al., *Epilepsia* **15**, 563 (1974). Evaluation in tera-togenesis: J. German et al., *Teratology* **3**, 349 (1970); A. B. Rifkind, *Toxicol. Appl. Pharmacol.* **30**, 452 (1974). GC de-termn in serum: E. Tanaka, S. Misawa, *J. Chromatog.* **413**, 376 (1987); LC determn in serum: M. Okamoto et al., *Chromatographia* **23**, 325 (1987). Use in treatment of petit mal epilepsy: S. Livingston et al., *J. Am. Med. Assoc.* **194**, 227 (1965). Use in dissolution of pancreatic stones in

humans and dogs: A. Noda et al., *Lancet* **2**, 351 (1984); eidem, *Gastroenterology* **93**, 1002 (1987).

Granules, crystals, mp 46-46.5°. bp_5 78-80°. Slight cam-phor-like odor. Burning, faintly bitter taste. Soly in water about 5%; increased by the addition of urethan. Freely sol in alcohol, benzene, chloroform, ether. Practically insol in petr ether. The pH of a 5% soln is about 6.0.

THERAP CAT: Anticonvulsant.
THERAP CAT (VET): Anticonvulsant.

9621. Trimethaphan Camsylate. *Decahydro-2-oxo-1,3-bis(phenylmethyl)thieno[1',2':1,2]thieno[3,4]imidazol-5-ium, salt with (+)-7,7-dimethyl-2-oxobicyclo[2.2.1]heptane-1-methanesulfonic acid (1:1);* trimetaphan camphorsulfo-nate; methioplegium; trimethaphan camphorsulfonate; tri-methaphen camphorsulfonate; 4,6-dibenzyl-5-oxo-1-thia-4,6-diazatricyclo[6.3.0.0$^{3.7}$]undecanium (+)-β-camphorsul-fonate; d-3,4-(1',3'-dibenzyl-2'-ketoimidazolido)-1,2-tri-methylenethiophanium d-camphorsulfonate; Nu-2222; Ar-fonad. $C_{32}H_{40}N_2O_5S_2$; mol wt 596.80. C 64.40%, H 6.76%, N 4.69%, O 13.40%, S 10.75%. Available from the Roche synthesis of biotin: Randall et al., *J. Pharmacol. Exp. Ther.* **97**, 48 (1949); Scurr, Wyman, *Lancet* **266**, 338 (Feb. 13, 1954). May be prepd by treating (+)-3,4-(1,3-dibenzyl-ureylene)tetrahydro-2-oxothiophene with 3-ethoxypropyl-magnesium bromide, followed by reduction, ring closure, and conversion into the (+)-β-camphorsulfonate. Compre-hensive description: K. W. Blessel et al., in *Analytical Pro-files of Drug Substances* vol. 3, K. Florey, Ed. (Academic Press, New York, 1974) pp 545-564.

Bitter crystals, dec about 245°. $[\alpha]_D^{20}$ +22.0° (c = 4 in water). One gram dissolves in less than 5 ml water and in less than 2 ml alcohol. Slightly sol in acetone, ether; pH of a 1% aq soln is 5.0-6.0.

THERAP CAT: Antihypertensive.

9622. Trimethidinium Methosulfate. *1,3,8,8-Tetrameth-yl-3-[3-(trimethylammonio)propyl]-3-azoniabicyclo[3.2.1]-octane bis(methyl sulfate); 3-[3-(dimethylamino)propyl]-1,3,-8,8-tetramethyl-3-azoniabicyclo[3.2.1]octane methyl sulfate methosulfate;* 1-(3-dimethylaminopropyl)-3,4,4-trimethyl-3,5-ethylenepiperidinium methyl sulfate methosulfate; N-(γ-trimethylammoniumpropyl)-N-methylcamphidinium methyl sulfate; camphidonium; Wy 1395; HA 106; Ostensin. C_{19}-$H_{42}N_2O_8S_2$; mol wt 490.67. C 46.51%, H 8.63%, N 5.71%, O 26.09%, S 13.07%. Prepd from 1-(3-dimethylaminopropyl)-3,4,4-trimethyl-3,5-ethylenepiperidine and methyl sulfate: Schmidt, Ger. pat. **1,086,703** (1960 to Thomae), *C.A.* **55**, 27379i (1961). Ganglion blocking agent.

Crystals, mp 192-193°.
THERAP CAT: Antihypertensive.

9623. Trimethobenzamide. *N-[(2-Dimethylaminoeth-oxy)benzyl]-3,4,5-trimethoxybenzamide; 4-(2-dimethylami-*

noethoxy)-*N*-(3,4,5-trimethoxybenzoyl)benzylamine. C_{21}-$H_{28}N_2O_5$; mol wt 388.45. C 64.93%, H 7.27%, N 7.21%, O 20.59%. Prepn: Goldberg, Teitel, **U.S. pat. 2,879,293** (1959 to Hoffmann-La Roche). Pharmacology: K. W. Blessel *et al.*, *Analytical Profiles of Drug Substances* **vol. 2**, K. Florey, Ed. (Academic Press, New York, 1973) pp 551-570.

Hydrochloride, $C_{21}H_{29}ClN_2O_5$, *Ro 2-9578, Anaus, Emamin, Nauseton, Tigan, Xametina.* Crystals, mp 187.5-190°. Freely sol in water (approx soly at 25° > 50%). A 5% aq soln is stable to autoclaving at 120° for 20 min at pH 3-7.
THERAP CAT: Anti-emetic.

9624. Trimethoprim. *5-[(3,4,5-Trimethoxyphenyl)methyl]-2,4-pyrimidinediamine; 2,4-diamino-5-(3,4,5-trimethoxybenzyl)pyrimidine;* Monotrim; Proloprim; Syraprim; Tiempe; Trimanyl; Trimogal; Trimopan; Trimpex; Uretrim; Wellcoprim. $C_{14}H_{18}N_4O_3$; mol wt 290.32. C 57.92%, H 6.25%, N 19.30%, O 16.52%. Prepn from guanidine and β-ethoxy-3,4,5-trimethoxybenzylbenzalnitrile: Stenbuck, Hood, **U.S. pat. 3,049,544** (1962 to Burroughs Wellcome); Hoffer, **U.S. pat. 3,341,541** (1967 to Hoffmann-La Roche). Improved synthesis: B. Roth *et al., J. Med. Chem.* **23**, 379, 535 (1980). Toxicity data: Yamamoto *et al., Chemotherapy* (Tokyo) **21**, 187 (1973). *Review:* Burchall in *Antibiotics* **vol. 3**, J. W. Corcoran, F. E. Hahn, Eds. (Springer-Verlag, New York, 1975) pp 304-320. Comprehensive description: G. J. Manius in *Analytical Profiles of Drug Substances* **vol. 7**, K. Florey, Ed. (Academic Press, New York, 1978) pp 445-475. Review of antibacterial activity, pharmacokinetics and therapeutic use: R. N. Brogden *et al., Drugs* **23**, 405-430 (1982).

White to cream, bitter crystalline powder, mp 199-203°. Soly in g/100 ml at 25°: DMAC 13.86; benzyl alcohol 7.29; propylene glycol 2.57; chloroform 1.82; methanol 1.21; water 0.04; ether 0.003; benzene 0.002. pKa 6.6. LD$_{50}$ orally in mice: 7000 mg/kg (Yamamoto).
Note: See Sulfamethoxazole, Sulfadiazine, Sulfametrole, Sulfamoxole, and Sulfalene for lists of trade names of mixtures with Trimethoprim.
THERAP CAT: Antibacterial.

9625. Trimethylamine. *N,N-Dimethylmethanamine.* C_3H_9N; mol wt 59.11. C 60.95%, H 15.35%, N 23.70%. $(CH_3)_3N$. Together with other amines it is a degradation product of nitrogenous plant and animal substances. It is formed during the distillation of sugar beet residues which contain betaine. In conjugated form it is widely distributed in animal tissue and especially in fish. It is converted to the free tertiary amine during putrefaction. It has been detected in menstrual blood, and in urine which was stored at room temp. Prepn from paraformaldehyde and ammonium chloride: Adams, Brown, *Org. Syn.* **1**, 75 (1921); **coll. vol. I**, 2nd ed., p 528; by the action of formaldehyde and formic acid on ammonia: Sommelet, Ferrand, *Bull. Soc. Chim.* [4] **35**, 446 (1924). Physical properties: J. G. Aston *et al., J. Am. Chem. Soc.* **66**, 1171 (1944).
Gas. Pungent, fishy, ammoniacal odor, saline taste. Liquefiable by pressure at ordinary temp or by condensation. mp −117.08°. bp$_{760}$ 2.87°; bp$_{747}$ 3.2-3.8°. d$_4^0$ 0.6709. Strong base. pK$_b$ (25°): 4.13. Readily absorbed by water, alcohol with which it is miscible; also sol in ether, benzene, toluene, xylene, ethylbenzene, chloroform. Sold as 25% water soln or as liquefied gas.

Hydrochloride, $C_3H_{10}ClN$, *trimethylammonium chloride.* Prepn: *Org. Syn.* **coll. vol. I**, 2nd ed., p 531. Monoclinic deliquesc crystals from alc. Odor less intense than that of base. Dec 277-278°. Sinters and sublimes at 200°. Sol in water, alcohol; moderately sol in chloroform. Insol in ether. *Keep well closed.*
Oxide, C_3H_9NO, *trimethyloxamine.* Needles with $2H_2O$ from water. Anhydr at 96°. mp 257°. Sol in water, methanol. Strong alkaline reaction.
USE: In the manuf of quaternary ammonium compds; as insect attractant; as warning agent for natural gas.

9626. Trimethyl Borate. *Boric acid trimethyl ester;* methyl borate. $C_3H_9BO_3$; mol wt 103.92. C 34.67%, H 8.73%, B 10.41%, O 46.19%. $B(OCH_3)_3$. Prepn from pyridine-boron trichloride complex: Gerrard, Lappert, *Chem. & Ind. (London)* **1952**, 53; from methanol and boric oxide, borax or boric acid: Schlesinger, *J. Am. Chem. Soc.* **75**, 213 (1953); from methyl orthosilicates and boron halide: Wiberg, Krüerke, *Z. Naturforsch.* **8b**, 608 (1953); from boric acid and methanol: Steinberg, Hunter, *Ind. Eng. Chem.* **49**, 174 (1957). Several mfg processes: **U.S. pats. 2,689,259; 2,884,439; 2,937,195** (to Callery Chem.); **U.S. pats. 2,880,227** and **2,884,440** (to Olin Mathieson); **U.S. pat. 2,855,427** (to Am. Potash & Chem.); **U.S. pat. 2,739,979** (to USAEC). Acute toxicity: H. F. Smyth *et al., Am. Ind. Hyg. Assoc. J.* **23**, 95 (1962).
Liquid, d 0.915. bp 67-68°. Flash pt 29° (84.2°F). Miscible with tetrahydrofuran, ether, isopropylamine, hexane, methanol, Nujol and other organic liquids. Stable in the absence of moisture, but hydrolyzes in the presence of water to methanol and boric acid. Forms an azeotrope with methanol: 70% $B(OCH_3)_3$ +30% methanol, d 0.87; bp 52-54°; flash pt 34° (93.2°F). LD$_{50}$ orally in rats: 6.14 ml/kg (Smyth).
USE: As solvent for waxes, resins, oils; catalyst in the manuf of ketones; analysis of paint and varnish ingredients; as neutron detector gas in the presence of a scintillation counter; as a promoter of diborane reactions.

9627. N,N-1-Trimethyl-3,3-diphenylpropylamine. *N,N,α-Trimethyl-γ-phenylbenzenepropanamine;* 4,4-diphenyl-2-dimethylaminobutane; 3-dimethylamino-1,1-diphenylbutane; Recipavrin. $C_{18}H_{23}N$; mol wt 253.37. C 85.32%, H 9.15%, N 5.53%. Prepn: Perrine, *J. Org. Chem.* **18**, 898 (1953).

l-Form hydrochloride, $C_{18}H_{23}N.HCl$, rods from acetone, mp 180-182°. $[\alpha]_D^{20}$ −43.3° (c = 1.04 in water).
d-Form hydrochloride, $C_{18}H_{23}N.HCl$, mp 179-181°. $[\alpha]_D^{20}$ +43.1° (c = 0.53 in water).
Nitrate, crystals from water, mp 118-120°.
THERAP CAT: Antispasmodic.

9628. Trimethylene Bromide. *1,3-Dibromopropane;* α,γ-dibromopropane; ω,ω'-dibromopropane; trimethylene dibromide. $C_3H_6Br_2$; mol wt 201.91. C 17.84%, H 3.00%, Br 79.16%. $BrCH_2CH_2CH_2Br$. Prepd by the action of hydrobromic acid on trimethylene glycol in the presence of sulfuric acid: Kamm, Marvel, *Org. Syn.* **coll. vol. I**, p 30 (1941); *cf.* Derick, Hess, *J. Am. Chem. Soc.* **40**, 545 (1918); Norris, Mulliken, *ibid.* **42**, 2096 (1920): Kamm, Newcomb, *ibid.* **43**, 2229 (1921). From trimethylene glycol and PBr$_3$: Bogert, Slocum, *ibid.* **46**, 765 (1924).
Colorless liquid; d$_{25}^{25}$ 1.9712; bp$_{760}$ 167° (mp −36°); n$_D^{15}$ 1.5249. Slightly sol in water (1.68 g/l at 30°); sol in alc, ether. Upon prolonged heating trimethylene bromide dec and part of it is converted to propylene bromide (1,2-dibromopropane). Boiling with water yields trimethylene glycol.

9629. Trimethylene Glycol. *1,3-Propanediol.* $C_3H_8O_2$; mol wt 76.09. C 47.35%, H 10.60%, O 42.05%. $HOCH_2CH_2CH_2OH$. Prepd by reduction of ethyl glycidate with lithium aluminum hydride: Walborsky, Colombini, *J. Org. Chem.* **27**, 2387 (1962).
Colorless to pale yellow, very viscid, sweet liquid. d$_4^{20}$ 1.0597; bp 210-212°; n$_D^{20}$ 1.4398. Miscible with water, alc.

9630. Trimethylene Oxide. *Oxetane;* 1,3-epoxypropane. C_3H_6O; mol wt 58.08. C 62.04%, H 10.41%, O 27.55%. Prepd by dropwise addition of 3-chloropropyl acetate to hot potassium hydroxide soln: Noller, *Org. Syn.* **29**, 92 (1949); modified procedure: Searles, *J. Am. Chem. Soc.* **73**, 124 (1951).

Oil, agreeable aromatic odor. d_0^0 0.8975; d_4^{25} 0.8930. bp_{750} 48°; bp_{736} 45-46°. n_D^{25} 1.3895; n_D^{23} 1.3905. Reacts with Grignard reagents and organolithium compds to give, after hydrolysis, 3-substituted propanols.

9631. Trimethylolmelamine. *(s-Triazine-2,4,6-triyltriimino)trimethanol;* 2,4,6-tris(methylamino)-1,3,5-triazine; N^2,N^4,N^6-tris(hydroxymethyl)melamine; Cilag 61; C 61; Cealysin. $C_6H_{12}N_6O_3$; mol wt 216.20. C 33.33%, H 5.59%, N 38.88%, O 22.20%. Prepn: **Brit.** pat. **801,404** (1958 to Svenska Oljeslargeri).

Note: Cealysin, Cilag 61, C 61, formerly known as *hexamethylolmelamine*: Eichler, Staib, *Arzneimittel-Forsch.* **6**, 119 (footnote) (1956).

THERAP CAT: Antineoplastic.

9632. 2,4,6-Trimethylpyridine. γ-Collidine; *sym*-collidine; 2,4,6-collidine; α,γ,α'-collidine. $C_8H_{11}N$; mol wt 121.18. C 79.29%, H 9.15%, N 11.56%. Found in small amounts in coal tar, in shale oil. Produced commercially from coal tar to some extent; also produced by the Hantzsch pyridine synthesis: Hantzsch, *Ann.* **215**, 1 (1882); Mosher in *Heterocyclic Compounds* vol. 1, R. C. Elderfield, Ed., (John Wiley, New York, 1950) pp 462-472; from acetone and ammonia: Mosher in Kirk-Othmer *Encyclopedia of Chemical Technology* vol. **11** (New York, 1953) p 287, cf. Dürkopf, *Ber.* **21**, 2713 (1888); from paraldehyde, acetone, ammonium acetate, and ammonia water: **Ger.** pat. **349,267**; *Frdl.* **14**, 539; from 3,5-dimethyl-2-cyclohexen-1-one, ammonium acetate, and ammonia water: Frank, Meikle, *J. Am. Chem. Soc.* **72**, 4184 (1950).

Liquid. Aromatic odor. $d_4^{16.4}$ 0.9191; $d_4^{22.1}$ 0.9166 (commercial grade: $d_{15.5}^{15.5}$ 0.920-0.935, approx 7.74 lbs/gal). mp −46°. bp_{762} 170.5°; bp_{760} 171°; bp_{71} 65°; $bp_{2.7}$ 10°. n_D^{25} 1.4959; $n_D^{22.1}$ 1.49770. pKa (25°): 6.69. Flash pt 136°F. Dielectric constant at 22° = 6.6 (λ = 70 cm). More sol in cold water than in hot water: 20.8 g/100 ml at 6°; 3.5 g/100 ml at 20°; 1.8 g/100 ml at 100°. Miscible with ether. Sol in methanol, ethanol, chloroform, benzene, toluene, dil acids.

9633. Trimethylsilyl Triflate. *Trifluoromethanesulfonic acid trimethylsilyl ester;* trimethylsilyltrifluoromethane sulfonate. $C_4H_9F_3O_3SSi$; mol wt 222.26. C 21.61%, H 4.08%, F 25.65%, O 21.60%, S 14.42%, Si 12.64%. $CF_3SO_2OSi(CH_3)_3$. Prepn: M. Schmeisser *et al.*, *Ber.* **103**, 868 (1970); H. W. Roesky, H. H. Giere, *Z. Naturforsch.* **25B**, 773 (1970); D. Haebich, F. Effenberger, *Synthesis* **1978**, 755; T. Morita *et al., ibid.* **1981**, 745. As catalytic reagent: H. Vorbrueggen, K. Krolikiewicz, *Angew. Chem. Int. Ed.* **14**, 421 (1975); as silylating agent: G. A. Olah *et al., J. Org. Chem.* **46**, 5212 (1981).

Liquid, bp_{760} 140°, bp_{10} 36.5°. n_D^{20} 1.3630. Fumes in air; sensitive to atm moisture.

USE: Catalyst and silylating agent for organic syntheses.

9634. Trimetozine. *4-(3,4,5-Trimethoxybenzoyl)morpholine;* N-(3,4,5-trimethoxybenzoyl)tetrahydro-1,4-oxazine; V-7; Opalene; Trioxazine. $C_{14}H_{19}NO_5$; mol wt 281.30. C 59.77%, H 6.81%, N 4.98%, O 28.44%. Prepd from morpholine and 3,4,5-trimethoxybenzoyl chloride: **Brit.** pat. **872,-350** (1961 to Egyesült Gyogyszer és Tápszergyár); Pettit *et al., J. Med. Pharm. Chem.* **5**, 800 (1962).

Crystals, mp 120-122°. Slightly sol in water, alcohol.

THERAP CAT: Sedative.

9635. Trimetrexate. *5-Methyl-6-[[(3,4,5-trimethoxyphenyl)amino]methyl]-2,4-quinazolinediamine;* 2,4-diamino-5-methyl-6-[(3,4,5-trimethoxyanilino)methyl]quinazoline; TMQ; NSC-249008; JB-11; CI-898. $C_{19}H_{23}N_5O_3$; mol wt 369.42. C 61.77%, H 6.28%, N 18.96%, O 12.99%. Lipophilic dihydrofolate reductase inhibitor structurally related to methotrexate, *q.v.*, with antimicrobial and antitumor activity. Prepn: E. F. Elslager, L. M. Werbel, **Brit.** pat. **1,345,502** (1974 to Parke, Davis); E. F. Elslager *et al., J. Med. Chem.* **26**, 1753 (1983); of water soluble salts: N. L. Colbry, **Eur.** pat. **Appl.** **51,415**; *idem*, **U.S.** pat. **4,376,858** (1982, 1983 both to Warner-Lambert). *In vitro* antifolate activity and *in vivo* antitumor effect: J. R. Bertino *et al., Biochem. Pharmacol.* **28**, 1983 (1979). Pharmacology: E. C. Weir *et al., Cancer Res.* **42**, 1696 (1982); R. C. Jackson *et al., Advan. Enzyme Regul.* **22**, 187 (1984). GC-MS determn in human plasma: P. L. Stetson, W. D. Ensminger, *J. Chromatog.* **383**, 69 (1986). Clinical evaluation with leucovorin vs *Pneumocystis carinii* in patients with AIDS: C. J. Allegra *et al., N. Engl. J. Med.* **317**, 978 (1987). Review of pharmacology and clinical efficacy: J. T. Lin, J. R. Bertino, *J. Clin. Oncol.* **5**, 2032-2040 (1987); P. J. O'Dwyer *et al., NCI Monographs* **5**, 105-109 (1987).

Monoacetate monohydrate, $C_{21}H_{27}N_5O_5 \cdot H_2O$, crystals from aqueous acetic acid, mp 215-217°. Poorly sol in water. LD_{50} i.p. in mice: 175 mg/kg (Jackson).

D-Glucuronate, $C_{25}H_{33}N_5O_{10}$, NSC-352122, Oncotrex. Tan colored solid. Soly in water: > 50 mg/ml.

THERAP CAT: Antineoplastic.

9636. Trimipramine. *10,11-Dihydro-N,N,β-trimethyl-5H-dibenz[b,f]azepine-5-propanamine;* 5-[3-(dimethylamino)-2-methylpropyl]-10,11-dihydro-5H-dibenz[b,f]azepine; 5-(3-dimethylamino-2-methylpropyl)iminodibenzyl; trimeprimine; trimeproprimine; RP 7162; Sapilent. $C_{20}H_{26}N_2$; mol wt 294.42. C 81.58%, H 8.90%, N 9.52%. Prepn: Jacob, Messer, *Compt. Rend.* **252**, 2117 (1961). Toxicity studies: Okamoto, *C.A.* **72**, 77299y (1970). Chemistry: Bever, Bredenstein, *Deut. Apoth.-Ztg.* **113**, 1562 (1973). Comprehensive description of the maleate: A. A. Al-Badr in *Analytical Profiles of Drug Substances* vol. 12, K. Florey, Ed. (Academic Press, New York, 1983) pp 683-712.

Crystals, mp 45°.

Maleate, $C_{24}H_{30}N_2O_4$, *Stangyl* (*tabl*), *Surmontil* (*tabl*). Crystals, mp 142°. Sol in chloroform; slightly sol in water, ethanol; practically insol in ether.

Methanesulfonate, *Stangyl* (*amp*), *Surmontil* (*amp*).

THERAP CAT: Antidepressant.

9637. Trimoprostil. (5Z,11α,13E,15R)-15-Hydroxy-11,16,16-trimethyl-9-oxoprosta-5,13-dien-1-oic acid; (Z)-7-[(1R,2R,3R)-2-[(E)-(3R)-3-hydroxy-4,4-dimethyl-1-octenyl]-3-methyl-5-oxocyclopentyl]-5-heptenoic acid; Nat-11R,16,16-trimethyl-15R-hydroxy-9-oxoprosta-cis-5-trans-13-dienoic acid; 11R,16,16-trimethyl-(11-desoxyprostaglandin E₂); 11-deoxy-11α,16,16-trimethyl-PGE₂; TM-PGE₂; Ro 21-6937; Ulstar. $C_{23}H_{38}O_4$; mol wt 378.55. C 72.98%, H 10.12%, O 16.90%. Synthetic prostaglandin E₂ analog with antisecretory activity. Prepn: G. W. Holland *et al.*, *Ger.* pat. 2,437,622; *eidem*, *U.S.* pats. 4,052,446; 4,190,-587 (1975, 1977, 1980 all to Hoffmann-La Roche). Inhibition of gastric acid secretion: D. E. Wilson, S. L. Winter, *Prostaglandins* **16**, 127 (1978); S. P. Lee *et al.*, *Eur. J. Clin. Invest.* **17**, 1 (1987). Effect on bicarbonate secretion: M. Feldman, *J. Clin. Invest.* **72**, 295 (1983). Effect of food on bioavailability: R. J. Wills, *J. Clin. Pharmacol.* **24**, 194 (1984); on inhibition of gastric acid secretion: R. J. Wills *et al.*, *Clin. Pharmacol. Ther.* **37**, 113 (1985). Clinical pharmacokinetics: R. J. Wills *et al.*, *J. Clin. Pharmacol.* **26**, 48 (1986). Metabolism in rats: S. J. Kolis *et al.*, *Drug Metab. Dispos.* **14**, 465 (1986). Toxicity data: M. Shimizu *et al.*, *Shin'yaku to Rinsho* **35**, 2199 (1986), *C.A.* **106**, 150180e (1987). Clinical evaluation: H. G. Dammann *et al.*, *Arzneimittel-Forsch.* **36**, 500 (1986). Multicenter clinical comparison with cimetidine, *q.v.*: K. D. Bardhan *et al.*, *Scand. J. Gastroenterol.* **23**, 134 (1988).

Colorless oil. $[\alpha]_D$ −51.54° (c = 1 in $CHCl_3$). LD_{50} in mice, rats (mg/kg): 41, 23 orally; 70, 21 i.p.; 68, 29 s.c. (Shimizu).

THERAP CAT: Antiulcerative.

9638. Trimyristin. Myristin; glyceryl trimyristate. $C_{45}H_{86}O_6$; mol wt 723.14. C 74.73%, H 11.99%, O 13.27%. Occurs in many vegetable fats and oils, notably in coconut oil and nutmeg butter.

White to yellowish-gray solid; d_4^{60} 0.885; mp 56-57°; n_D^{60} 1.4429. Insol in H_2O; sol in alc, benzene, chloroform, ether.

9639. sym-Trinitrobenzene. 1,3,5-Trinitrobenzene; benzite. $C_6H_3N_3O_6$; mol wt 213.11. C 33.81%, H 1.42%, N 19.72%, O 45.05%. Prepared by decarboxylation of trinitrobenzoic acid, obtained by oxidation of TNT: Clarke, Hartman, *Org. Syn.* **2**, 93 (1922); by the action of alkali on 2,4,6-trinitrobenzaldehyde: Secareanu, *Bull. Soc. Chim.* **51**, 591 (1932).

Orthorhombic bipyramidal plates from glacial acetic acid. d_4^{20} 1.76; mp 122.5°; d_4^{152} 1.4775. Can be sublimed by careful

heating, explodes when heated rapidly. Absorption spectrum: Hatzsch, Picton, *Ber.* **42**, 2121 (1909). Soly (g/100 g solvent): Water 0.035; benzene 6.2; methanol 4.9; alcohol 1.9; ether 1.5; carbon disulfide 0.25; petr ether 0.05. Freely sol in dil Na_2SO_3 soln. Trinitrobenzene is dimorphous, the other (rare) form melts at 61°.

USE: Explosive, less sensitive to impact than TNT but more powerful and brisant: Robertson, *J. Chem. Soc.* **119**, 8 (1921); van Duin, *Rec. Trav. Chim.* **39**, 687 (1920).

9640. 2,4,6-Trinitrobenzoic Acid. *sym*-Trinitrobenzoic acid. $C_7H_3N_3O_8$; mol wt 257.12. C 32.70%, H 1.18%, N 16.34%, O 49.78%. Prepd by chromic acid oxidation of 2,4,-6-trinitrotoluene: Clarke, Hartman, *Org. Syn.* **2**, 95 (1922); Kastens, Kaplan, *Ind. Eng. Chem.* **42**, 402 (1950).

Orthorhombic crystals from water, mp 228.7°. Sublimes with decompn forming CO_2 and trinitrobenzene. Soly at 25°: 2.05% (w/w) in water, 26.6% in alcohol, 14.7% in ether. Also sol in acetone, methanol; slightly sol in benzene.

9641. 2,4,7-Trinitrofluorenone. 2,4,7-Trinitro-9H-fluoren-9-one; TNF. $C_{13}H_5N_3O_7$; mol wt 315.19. C 49.54%, H 1.60%, N 13.33%, O 35.53%. Prepn by nitration of fluorenone: Schmidt, Bauer, *Ber.* **38**, 3758 (1905); Orchin *et al.*, *J. Am. Chem. Soc.* **69**, 1225 (1947); by nitration of 2,5-dinitrofluorenone: Ray, Francis, *J. Org. Chem.* **8**, 58 (1943). Structure: Bell, *J. Chem. Soc.* **1928**, 1990. Crystal structure: D. L. Dorset *et al.*, *Acta Crystallogr.* **B28**, 3122 (1972); H. L. Ammon, *ibid.* **B29**, 2314 (1973). Carcinogenicity study: C. Huggins, N. C. Yang, *Science* **137**, 257 (1962). Evaluation of employee exposure to TNF in workplace using HPLC: M. J. Seymour, *J. Chromatog.* **236**, 530 (1982); R. E. McCullen, A. N. Sanghvi, *ASTM Spec. Tech. Publ.* **786**, 26 (1982).

Yellow needles from acetic acid, mp 175.2-176°.

USE: In photocopiers; forms charge-transfer complexes with aromatic hydrocarbons and amines.

9642. Trinitromethane. Nitroform. CHN_3O_6; mol wt 151.04. C 7.95%, H 0.67%, N 27.82%, O 63.56%. CH-$(NO_2)_3$. Prepn from tetranitromethane and $K_4[Fe(CN)_6]$ in aq soln: Chattaway, Harrison, *J. Chem. Soc.* **109**, 171 (1916); by nitration of acetylene with nitric acid: Hager, *Ind. Eng. Chem.* **41**, 2168 (1949).

Crystals, mp 15° (the unstable *aci*-form, mp 50°). d_4^{25} (liq) 1.469. Heat of combustion 746 cal/g. Dipole moment 2.61 (benzene). Dec above 25°. Explodes when heated rapidly. Sol in water, giving an intensely yellow soln, although the dry crystals are pure white.

Potassium salt, $CK(NO_2)_3$, moderately stable crystals. Soly in water at 0°: 16.7 g/100 ml; at 60°: 193.8 g/100 ml. Soly in ethanol: 5.29 g/l.

USE: In the manuf of explosives and propellants. *Caution*: Slightly irritating to eyes, mucous membranes.

9643. 2,4,6-Trinitrotoluene. TNT; α-trinitrotoluol; *sym*-trinitrotoluene; 1-methyl-2,4,6-trinitrobenzene; trotyl; Tolit; Trilit. $C_7H_5N_3O_6$; mol wt 227.13. C 37.01%, H 2.22%, N 18.50%, O 42.27%. Prepn by nitration of toluene with mixed acid ($HNO_3 + H_2SO_4$) in three steps or by continuous flow according to the Schmid-Meissner and Biazi processes: Swift, Tittensor, *J. Soc. Chem. Ind.* **59**, 92 (1940);

Johnston in *McGraw-Hill Encyclopedia of Science and Technology* **9**, 104 (1960).

CH₃ structure (2,4,6-trinitrotoluene):

O_2N — CH₃ ring with NO_2 positions

Monoclinic rhombohedra from alcohol. The commercial crystals (needles) are yellow. mp 80.1°. d_4^{20} 1.654. Burns at 295° when not confined. Can be distilled under reduced pressure. Vapors are toxic. Dipole moment 1.37. Very sparingly sol in water: About 0.01% at 25°, one gram dissolves in 700 ml of boiling water. Sol in acetone, benzene. Less sol than 2,4,6-trinitrophenol in alcohol, ether, carbon disulfide. Reacts vigorously with reducing agents.

USE: High explosive. Must be detonated by a high velocity initiator such as nitramine or by efficient concussion. For physical constants and applications *see* Lothrop, Handrick, *Chem. Rev.* **44**, 419-445 (1949) and M. A. Cook, *The Science of High Explosives*, ACS monograph no. 139 (Reinhold, New York, 1958). *Caution:* Can cause headache, weakness, anemia, liver injury. May be absorbed through skin.

9644. Triolein. *9-Octadecenoic acid 1,2,3-propanetriyl ester;* olein; glyceryl trioleate. $C_{57}H_{104}O_6$; mol wt 885.40. C 77.32%, H 11.84%, O 10.84%. One of the chief constituents of nondrying oils and fats. From Palestine olive oil: Hilditch, Madison, *J. Soc. Chem. Ind.* **60**, 258 (1941); from cacao butter: Meara, *J. Chem. Soc.* **1949**, 2154. Prepn by esterification of oleic acid: Wheeler *et al., J. Biol. Chem.* **132**, 687 (1940); Swicklik *et al., J. Am. Oil Chem. Soc.* **32**, 69 (1955). Synthesis: Serebrennikova *et al., Dokl. Akad. Nauk SSSR* **140**, 1083 (1961).

$CH_2OCO(CH_2)_7CH{=}CH(CH_2)_7CH_3$
$CHOCO(CH_2)_7CH{=}CH(CH_2)_7CH_3$
$CH_2OCO(CH_2)_7CH{=}CH(CH_2)_7CH_3$

Colorless to yellowish, oily liquid; tasteless, odorless. d_4^{15} 0.915. mp −4° to −5°. bp_{15} 235-240°. n_D^{20} 1.4676; n_D^{60} 1.4561. Practically insol in water; sol in chloroform, ether, carbon tetrachloride; slightly sol in alcohol.

9645. Triostins. A *"quinoxaline antibiotic complex"* similar to echinomycin, *q.v.* Powerful, selective inhibitor of nucleic acid synthesis *in vitro*. Isoln of triostin C from *Streptomyces* S-2-210 resembling *S. aureus*: J. Shoji, K. Katagiri, *J. Antibiot.* **14A**, 335 (1961). Isoln of triostins A and B: H. Otsuka, J. Shoji, *ibid.* **19A**, 128 (1966). Structure of C:

(quinoxaline depsipeptide structure)

eidem, Tetrahedron **21**, 2931 (1965); of minor components: *eidem, ibid.* **23**, 1535 (1967); of A: H. Otsuka *et al., J. Antibiot.* **29**, 107 (1976). Biosynthesis: T. Yoshida, K. Katagiri, *Biochemistry* **8**, 2645 (1969). Synthesis of A: P. K. Chakravarty, R. K. Olsen, *Tetrahedron Letters* **1978**, 1613; M. Shin *et al., Pept. Chem., 18th* **1980**, 207, *C.A.* **95**, 98304 (1981). Conformation of A in soln: J. R. Kalman *et al., J. Chem. Soc. Perkin Trans. I* **1979**, 1313. Review of chemistry and biochemistry: M. J. Waring in *Antibiotics* vol. **5**(pt. 2), F. E. Hahn, Ed. (Springer-Verlag, New York, 1979) pp 173-194. Needles from chloroform + methanol.

Triostin A, $C_{50}H_{62}N_{12}O_{12}S_2$, R = CH_3. mp 245-248° (dec). $[\alpha]_D^{25}$ −157° (c = 0.97 in chloroform). uv max (methanol): 243, 320 nm (log ϵ 4.75, 4.11).

Triostin C, $C_{54}H_{70}N_{12}O_{12}S_2$, R = $CH(CH_3)_2$. mp > 260° (dec). $[\alpha]_D^{24}$ −143.9° (c = 1.2 in chloroform). uv max (methanol): 243, 320 nm (log ϵ 4.87, 4.13).

9646. s-Trioxane. *1,3,5-Trioxane;* 1,3,5-trioxacyclohexane; metaformaldehyde; trioxymethylene; Triformol. $C_3H_6O_3$; mol wt 90.08. C 40.00%, H 6.71%, O 53.28%. *Review:* Walker, Carlisle, *Chem. & Eng. News.* **21**, 1250 (1943).

(1,3,5-trioxane ring structure)

Stable, cyclic trimer of formaldehyde possessing characteristic chloroform-like odor. Crystalline solid, mp 64°, bp_{759} 114.5° without dec. Sublimes readily. Easily sol in water (17.2 g/100 ml at 18°, 21.1 g/100 ml at 25°), alcohols, ketones, ether, acetone, chlorinated and aromatic hydrocarbons, and other organic solvents. Only slightly sol in pentane, petroleum ether, and lower paraffins. d^{65} 1.17. Flash pt 45°. On distillation with water, forms an azeotrope boiling at 91.4° containing approx 70% trioxane by wt. Slowly depolymerized by strong acids in aq soln, but inert to alkalies. In non-aqueous systems, readily converted to monomeric formaldehyde by small concentrations of strong acids at a rate determined by the acid concentration.

9647. Trioxsalen. *2,5,9-Trimethyl-7H-furo[3,2-g][1]-benzopyran-7-one;* 6-hydroxy-β,2,7-trimethyl-5-benzofuranacrylic acid δ-lactone; 4,5',8-trimethylpsoralen; NSC-71047; Trisoralen. $C_{14}H_{12}O_3$; mol wt 228.24. C 73.67%, H 5.30%, O 21.03%. Synthetic trimethyl psoralen deriv. Prepn: K. D. Kaufman, *J. Org. Chem.* **26**, 117 (1961); U.S. pat. **3,201,421** (1965). Protective effect vs UV-B erythema: P. G. Agache, L. Coupez, *Arch. Dermatol. Res.* **268**, 85 (1980). Clinical study: N. Vaatainen *et al., Clin. Exp. Dermatol.* **6**, 133 (1981). Comprehensive description: M. M. A. Hassan, M. A. Loutfy, in *Analytical Profiles of Drug Substances* vol. 10, K. Florey, Ed. (Academic Press, New York, 1981) pp 705-727. *See also* Psoralen, Methoxsalen.

(trioxsalen structure with CH₃ groups)

Prisms from chloroform, mp 234.5-235°. uv max (methanol): 250, 295, 335 nm (log ϵ 4.35, 3.99, 3.80). Sol in methylene chloride. Slightly sol in alc, chloroform. Practically insol in water.

THERAP CAT: Pigmentation agent (photosensitizer).

9648. Tripalmitin. *Hexadecanoic acid 1,2,3-propanetriyl ester;* palmitin; glyceryl tripalmitate. $C_{51}H_{98}O_6$; mol wt 807.29. C 75.87%, H 12.24%, O 11.89%. $(C_{15}H_{31}COO)_3C_3H_5$. Occurs in fats. Prepd from glycerol and palmitic acid in the presence of Twitchell reagent: Ozaki, *Biochem. Z.* **177**, 159 (1926); or in the presence of trifluoroacetic anhydride: Bourne *et al., J. Chem. Soc.* **1949**, 2976.

Needles from ether, mp 66° (occurs also in lower-melting, unstable forms). bp 310-320°; d_4^{70} 0.8730; d_4^{80} 0.8663. n_D^{80} 1.43807. Saponification value 208.5. Insol in water. Practically insol in alcohol (0.0043 parts/100 parts of abs alcohol at 21°). Freely sol in ether, benzene, chloroform.

9649. Tripamide. *3-(Aminosulfonyl)-4-chloro-N-(octahydro-4,7-methano-2H-isoindol-2-yl)benzamide;* 4-chloro-N-(endo-hexahydro-4,7-methanoisoindolin-2-yl)-3-sulfamoylbenzamide: toripamide; ADR-033; E-614; Normonal. $C_{16}H_{20}ClN_3O_3S$; mol wt 369.87. C 51.96%, H 5.45%, Cl 9.58%, N 11.36%, O 12.98%, S 8.67%. Sulfonamide with diuretic and peripheral vasodilator activity. Prepn: H. Hamano *et al.*, **Japan.** pat. **73 05,585** (1973 to Eisai), *C.A.* **78**, 136070r (1973). Synthesis of ^{14}C-tripamide: T. Nakamura *et al.*, *J. Label. Comp. Radiopharm.* **14**, 191 (1978). Pharmacological study: T. Satoh *et al.*, *Oyo Yakuri* **21**, 607 (1981), *C.A.* **96**, 28414u (1982). Metabolism: T. Horie *et al.*, *Xenobiotica* **11**, 197, 693 (1981).

Colorless needles.
THERAP CAT: Antihypertensive; diuretic.

9650. Triparanol. *4-Chloro-α-[4-[2-(diethylamino)ethoxy]phenyl]-α-(4-methylphenyl)benzeneethanol;* 1-[p-(2-diethylaminoethoxy)phenyl]-1-(p-tolyl)-2-(p-chlorophenyl)-ethanol; 2-(p-chlorophenyl)-1-[p-(2-diethylaminoethoxy)-phenyl]-1-(p-tolyl)ethanol; MER-29; Trianel; Hipocolestina; Triparin; Acosterina; Metasclene; Diticyl; Drenaren; Clotrox; Tropalin; Trikosterol; Valip; Verdiana; Metasqualene; Sclane. $C_{27}H_{32}ClNO_2$; mol wt 438.00. C 74.03%, H 7.37%, Cl 8.10%, N 3.20%, O 7.31%. Prepn: Allen, Palopoli *et al.*, U.S. pat. **2,914,562** (1959 to Wm. S. Merrell).

Crystals, mp 102-104°. Sol in alcohol; slightly sol in olive oil; practically insol in water.
Human Toxicity: Withdrawn from market because of association with formation of irreversible cataracts; *see* Laughlin, Carey, *J. Am. Med. Assoc.* **181**, 339 (1962).
THERAP CAT: Antilipemic.

9651. Tripelennamine. *N,N-Dimethyl-N'-(phenylmethyl)-N'-2-pyridinyl-1,2-ethanediamine;* 2-[benzyl(2-dimethylaminoethyl)amino]pyridine; N-benzyl-N',N'-dimethyl-N-(2-pyridyl)ethylenediamine; N,N-dimethyl-N'-benzyl-N'-(α-pyridyl)ethylenediamine; β-dimethylaminoethyl-2-pyridylbenzylamine; β-dimethylaminoethyl 2-pyridylaminotoluene; Dehistin; Azaron; Pyribenzamine; PBZ; Tonaril; Vetibenzamina. $C_{16}H_{21}N_3$; mol wt 255.35. C 75.25%, H 8.29%, N 16.46%. Prepn: C. P. Huttrer *et al.*, *J. Am. Chem. Soc.* **68**, 1999 (1946); C. Djerassi *et al.*, U.S. pat. **2,406,594** (1946 to Ciba); R. J. Horclois, U.S. pat. **2,502,151** (1950 to Rhône-Poulenc). Crystal structure: M. Parvez, *Acta Crystallogr.* **C43**, 1408 (1987). Toxicity studies: D. P. Waller *et al.*, *Clin. Toxicol.* **16**, 17 (1980). Comprehensive description: H. G. Piskorik in *Analytical Profiles of Drug Substances* vol. **14**, K. Florey, Ed. (Academic Press, New York, 1985) pp 108-133.

Yellow oil. Amine odor. $bp_{0.1}$ 138-142°, $bp_{1.7}$ 185-190°, bp_{20} 193-205°. n_D^{25} 1.5759-1.5765. Miscible with water.
Hydrochloride, $C_{16}H_{22}ClN_3$, *Pyrinamine, Resistamine.* Crystals from ethyl acetate + methanol, mp 192-193°. Bitter taste, produces temporary numbness of the tongue. uv max (water): 244, 305 (ε 14470, 4780). One gram dissolves in 0.77 ml water, in 6 ml alcohol, in 6 ml chloroform, in about 350 ml acetone. Practically insol in benzene, ether, ethyl acetate. About neutral to litmus. pH of aq soln contg 25 mg/ml: 6.71; 50 mg/ml: 6.67; 100 mg/ml: 5.56. LD_{50} in mice (mg/kg): 47 i.p. (Walker).
Citrate, $C_{22}H_{29}N_3O_7$, crystals, mp 106-110°. Less bitter than the hydrochloride. Freely sol in water, alcohol. Very slightly sol in ether. Practically insol in benzene, chloroform. 1% aq soln has a pH of 4.25.
THERAP CAT: Antihistaminic.
THERAP CAT (VET): Antihistaminic.

9652. Triphal®. *[(4-Carboxy-2-benzimidazolyl)thio]-gold;* sodium 2-aurothiobenzimidazole-4-carboxylate; Aurothiol. $C_8H_4AuN_2NaO_2S$; mol wt 412.13. C 23.30%, H 0.98%, Au 47.82%, N 6.79%, Na 5.58%, O 7.76%, S 7.78%. Prepn: **Brit.** pat. **225,875** (1923 to Hoechst), *C.A.* **19**, 1615 (1925).

Dihydrate, light yellow powder. Freely sol in water, hot glycerol. Insol in alcohol, ether.

9653. Triphenylcarbinol. *Triphenylmethanol;* tritanol. $C_{19}H_{16}O$; mol wt 260.32. C 87.66%, H 6.20%, O 6.15%. $(C_6H_5)_3COH$. Prepd by the action of phenylmagnesium bromide on benzophenone: Acree, *Ber.* **37**, 2755 (1904); Peters *et al.*, *J. Am. Chem. Soc.* **47**, 452 (1925); Dubsky, Jacot-Guillarmod, *Helv. Chim. Acta* **53**, 1965 (1970); by the action of potassium permanganate on triphenylfluoro- or triphenylchloromethane: Blicke, *J. Am. Chem. Soc.* **46**, 1518 (1924). Convenient lab prepn from methyl benzoate and phenylmagnesium bromide: L. F. Fieser, *Experiments in Organic Chemistry* (Boston, 3rd ed., 1955) p 77.
Trigonal crystals from benzene; d_4^0 1.199; mp 164.2°. Distills between 360 and 380° without decompn. Absorption spectrum in alc: Orndorff *et al.*, *J. Am. Chem. Soc.* **49**, 1543 (1927). Insol in water, petr ether. Easily sol in alc, ether, benzene. Sol in concd H_2SO_4 with an intensely yellow color, in glacial acetic acid without color.

9654. Triphenylene. *9,10-Benzphenanthrene;* 1,2,3,4-dibenznaphthalene; isochrysene. $C_{18}H_{12}$; mol wt 228.28. C 94.70%, H 5.30%. Occurs in coal tar: Kaffer, *Ber.* **68**, 1812 (1935). Synthesis from cyclohexanone: Mannich, *Ber.* **40**, 163 (1907); Nenitzescu, Curcaneanu, *Ber.* **70**, 346 (1937); from 9-phenanthrylmagnesium bromide and succinic anhydride: Bergmann, *J. Am. Chem. Soc.* **59**, 1441 (1937); from 1,2,3,4-tetrahydrophenanthrene: Bachmann, Struve, *J. Org. Chem.* **4**, 472 (1937); by reaction of 3 mols o-bromoiodobenzene in the presence of lithium: Heaney, Millar, *Org. Syn.* **40**, 105 (1960).

Long needles from alc or chloroform. Sublimes; d 1.302; mp 199°; bp 425°. Absorption spectrum: Clar, Lombardi, *Ber.* **65**, 1414 (1932). Solns have blue fluorescence. Does not react with maleic anhydride.

9655. Triphenylmethane. *1,1',1''-Methylidenetris[benzene];* Tritan. $C_{19}H_{16}$; mol wt 244.32. C 93.40%, H 6.60%.

$(C_6H_5)_3CH$. Prepd by ether reduction of triphenylchloromethane obtained by reaction of carbon tetrachloride and benzene in the presence of aluminum chloride: Norris, *Org. Syn.* **4**, 81 (1925); from diphenylchloromethane and phenylmagnesium bromide: Sayles, Kharasch, *J. Org. Chem.* **26**, 4210 (1961).

Orthorhombic pyramidal, solvated crystals containing one mol benzene, mp 78.2°, dries on exposure to air. When dry, mp 93.4° (stable form; there are 2 metastable forms). d_4^{100} 1.0134; bp_{760} 360°; bp_{200} 239.7°; b_{100} 228°; bp_{60} 221°; bp_{40} 215.5°; bp_{20} 206.8°; bp_{10} 197°; bp_5 188°; $bp_{1.0}$ 170°. n_D^{100} 1.59546. Absorption spectrum: Orndorff, *J. Am. Chem. Soc.* **49**, 1543 (1927); Anderson, *ibid.* **50**, 209 (1928); **51**, 1890 (1929). Very sol in ether, hot alcohol, chloroform. Sol in petr ether, benzene, CS_2. Slightly sol in glacial acetic acid. Soly (parts/100 parts satd soln (w/w) at 30°) 48.6 in chloroform; 12.5 in hexane; 53 in CS_2; 7.24 parts in benzene (19°).

9656. Triphenyl Phosphate. $C_{18}H_{15}O_4P$; mol wt 326.28. C 66.26%, H 4.63%, O 19.61%, P 9.49%. $(C_6H_5O)_3PO$. Prepd from P_2O_5 and phenol: Prahl, **U.S.** pat. **2,805,240** (1957); by reaction of triethyl phosphite with chloramine-T: Cadogan, Moulden, *J. Chem. Soc.* **1961**, 3079. Toxicology: Sutton *et al., Arch. Environ. Health* **1**, 33 (1960).

Nonflammable needles; mp 49-50°; bp_{10} 245°. Insol in water; sol in benzene, chloroform, ether, acetone, moderately sol in alcohol.

USE: Noncombustible substitute for camphor in celluloid; rendering acetylcellulose, nitrocellulose, airplane "dope", etc., stable and fireproof; impregnating roofing paper; plasticizer in lacquers and varnishes.

9657. Triphenylphosphine. $C_{18}H_{15}P$; mol wt 262.28. C 82.42%, H 5.76%, P 11.81%. $(C_6H_5)_3P$. Prepd from phenylmagnesium bromide and phosphorus trichloride: Pfeiffer, Pietsch, *Ber.* **37**, 4621 (1904); Sauvage, *Compt. Rend.* **139**, 675 (1904); Dodonon, Medox, *Ber.* **61**, 910 (1928); Denney *et al., J. Am. Chem. Soc.* **83**, 1729 (1961).

Odorless monoclinic platelets or prisms from ether, mp 80.5°. bp > 360° (in inert gas). d_4^{25} 1.194; d_4^{80}(liq) 1.075. For more physical constants *see* Forward *et al., J. Chem. Soc.* **1949**, 5121. Is triboluminescent. Freely sol in ether; sol in benzene, chloroform, glacial acetic acid; less sol in alcohol; practically insol in water.

USE: In organic synthesis; polymerization initiator.

9658. Triphenyltetrazolium Chloride. *2,3,5-Triphenyl-2H-tetrazolium chloride*; red tetrazolium; TPTZ; TTC; RT; VitaStain; Uroscreen. $C_{19}H_{15}ClN_4$; mol wt 334.80. C 68.16%, H 4.52%, Cl 10.59%, N 16.74%. Prepn: H. v. Pechmann, P. Runge, *Ber.* **27**, 2920 (1894); Atkinson *et al., Science* **111**, 385 (1950); R. Kuhn, D. Jerchel, *Ber.* **74**, 945 (1941); W. Ried, *Angew. Chem.* **14**, 391 (1952); R. Price, *J. Chem. Soc. (A)* **1971**, 3379. Review on the significance of tetrazolium salts as indicators in biochemical and biological oxidation-reduction processes: W. Ried, *loc. cit.*

Solvated nearly colorless needles from alcohol or chloroform, dry at 105°. Turns yellow on exposure to light. Dec 243°. Sol in water, alcohol, acetone; insol in ether. Oxidizes aldoses and ketoses, as well as other α-ketols, and is thereby reduced to a water-insoluble, deep red pigment, a triphenylformazan. LD_{50} i.v. in mice: 5600 μg/kg, *RTECS* Vol. **II**, R. J. Lewis, R. L. Tatken, Eds. (1980) p 690.

USE: In analytical chemistry as a sensitive reagent for reducing sugars, and to distinguish between α-ketols and simple aldehydes. For staining plant and animal tissue. Germination indicator in testing the ability of seeds to germinate. If the seed is alive, its embryonic tissue hydrogenates the colorless tetrazolium salt to a deep red, insol triphenylformazan: G. Lakon, *Ber. Dtsch. Bot. Ges.* **60**, 299, 434 (1942). In histochemical studies. In determination of antibiotics; of dehydrogenases.

9659. Triphenyltin Hydroxide. *Hydroxytriphenylstannane; hydroxytriphenyltin;* fentin hydroxide; fenolovo; Duter. $C_{18}H_{16}OSn$; mol wt 367.03. C 58.91%, H 4.39%, O 4.36%, Sn 32.34%. $(C_6H_5)_3SnOH$. Prepn by alkaline hydrolysis of $(C_6H_5)_3SnCl$: Kushlefsky *et al., Inorg. Chem.* **2**, 187 (1963); of $(C_6H_5)_3SnI$: Poller, *J. Inorg. Nucl. Chem.* **24**, 593 (1963). GC determn in vegetable crops: H. H. Van den Broek *et al., Analyst (London)* **113**, 1237 (1988). Review of analytical methods: A. Van Rossum *et al., Anal. Methods Pestic. Plant Growth Regul.* **11**, 227-246 (1980). *Reviews:* Ingham, *Chem. Rev.* **60**, 459-539 (1960); R. Bock, *Residue Rev.* **79**, 1-270 (1981).

Crystals. mp 122-123.5°. Thermally dec to $(C_6H_5)_4Sn$, $(C_6H_5)_2SnO$ and H_2O. Slightly sol in alcohol, toluene. Practically insol in water.

Note: The EPA has determined that triphenyltin hydroxide is teratogenic in rats: *Fed. Reg.* **50**, 1107 (1985).

Acetate, $C_{20}H_{18}O_2Sn$, *triphenyltin acetate, acetoxytriphenyltin, fentin acetate, Brestan.* Prepn from acetic acid and $(C_6H_5)_3SnOH$: van der Kirk, Luijten, *J. Appl. Chem. (London)* **6**, 49 (1956); and $(C_6H_5)_3SnH$: Weber, Becker, *J. Org. Chem.* **27**, 1258 (1962). Small needles, mp 122-124°. Sol in ether, slightly sol in alcohol, benzene.

USE: Antifeeding compounds for insect pest control; non-systemic fungicide.

9660. Tripoli. Finely granulated, white or gray, porous siliceous rock. Used as abrasive and paint filler. The original tripoli, mined near Tripoli in Northern Africa was a diatomaceous earth, but this definition is no longer correct. Tripoli grains are an amorphous form of SiO_2, soft, porous, and free from sharp cutting edges, making it especially suitable as an ingredient of metal polishes. Also used in oil well drilling muds.

9661. Triprolidine. *2-[1-(4-Methylphenyl)-3-(1-pyrrolidinyl)-1-propenyl]pyridine; trans-2-[3-(1-pyrrolidinyl)-1-p-tolylpropenyl]pyridine; trans-1-(2-pyridyl)-3-pyrrolidino-1-p-tolylprop-1-ene; trans-1-(4-methylphenyl)-1-(2-pyridyl)-3-pyrrolidinoprop-1-ene.* $C_{19}H_{22}N_2$; mol wt 278.38. C 81.97%, H 7.97%, N 10.06%. H_1 receptor antagonist. Prepn: Adamson, **U.S.** pats. **2,712,020; 2,712,023** (both 1955 to Burroughs Wellcome); Adamson *et al., J. Chem. Soc.* **1958**, 312. Structure-activity studies: Ison, Casy, *J. Pharm. Pharmacol.* **23**, 848 (1971). Crystal and molecular structure: James, Williams, *Can. J. Chem.* **52**, 1880 (1974). Pharmacokinetics and antihistaminic effects in humans: K. J. Simons *et al., J. Allergy Clin. Immunol.* **77**, 326 (1986). Comprehensive description: S. A. Benezra, C.-H. Yang, in *Analytical Profiles of Drug Substances* vol. **8**, K. Florey, Ed. (Academic Press, New York, 1979) pp 509-528.

Crystals from light petr, mp 59-61°. uv max (ethanol): 236, 285 nm (ϵ 15300, 6800).

Hydrochloride monohydrate, $C_{19}H_{23}ClN_2 \cdot H_2O$, *Actidil, Actidilon, Alleract, Pro-Actidil, Pro-Entra, Venen,* 295.C.51. Crystals from water, mp 116-118°. uv max (ethanol): 235, 283 nm (ϵ 15000, 7400). Moderately sol in water, ethanol, methanol.

Oxalate, $C_{21}H_{24}N_2O_4$, crystals from methanol, dec 173-174°. uv max (ethanol): 233, 283 nm (ϵ 16200, 8200).

THERAP CAT: Antihistaminic.

9662. Triptorelin. *6-D-Tryptophanluteinizing hormone-releasing factor(pig)*; 6-D-tryptophan-LH-RH; D-trp^6-LHRH; D-Trp^{6}LRH; D-trp^6-gonadorelin; détryptoréline; AY 25650; Wy 42462; Wy 42422. $C_{64}H_{82}N_{18}O_{13}$; mol wt 1311.47. C 58.61%, H 6.30%, N 19.22%, O 15.86%. Synthetic peptide agonist analog of LH-RH, *q.v.* Prepn: A. V. Schally, D. H. Coy, **Ger.** pat. **2,625,843**; *eidem,* **U.S.** pat. **4,010,125** (1976, 1977); D. H. Coy *et al., J. Med. Chem.* **19**,

423 (1976). Comparison with LH-RH of *in vitro* activity: D. H. Coy *et al.*, *Biochem. Biophys. Res. Commun.* **67**, 576 (1975). Pharmacokinetics and metabolism in humans: J. L. Barron *et al.*, *J. Clin. Endocrinol. Metab.* **54**, 1169 (1982). HPLC analysis: D. C. Serti *et al.*, *J. Liq. Chromatog.* **4**, 1135 (1981). Radioimmunoassay in human serum: M. Mason-Garcia *et al.*, *Proc. Nat. Acad. Sci. USA* **82**, 1547 (1985). Clinical trial in prostatic carcinoma: H. Parmar *et al.*, *Lancet* **2**, 1201 (1985). Evaluation in precocious puberty: M. Roger *et al.*, *J. Clin. Endocrinol. Metab.* **62**, 670 (1986).

5-oxoPro-His-Trp-Ser-Tyr-D-Trp-Leu-Arg-Pro-Gly-NH$_2$

Fluffy, white solid. $[\alpha]_D^{23}$ −58.8° (c = 0.33 in acetic acid). Acetate, C$_{64}$H$_{82}$N$_{18}$O$_{13}$.xC$_2$H$_4$O$_2$, (x = 1-1.5), *Decapeptyl*. THERAP CAT: Treatment of prostatic carcinoma.

9663. Triptycene. *9,10-Dihydro-9,10-o-benzenoanthracene.* C$_{20}$H$_{14}$; mol wt 254.31. C 94.45%, H 5.55%. Synthesis by three different methods: Bartlett *et al.*, *J. Am. Chem. Soc.* **64**, 2649 (1942); Wittig, *Org. Syn.* **39**, 75 (1959); Friedman, Logullo, *J. Am. Chem. Soc.* **85**, 1549 (1963).

Crystals from cyclohexane or methylcyclohexane, mp 253-254°.

9664. 2,4,6-Tripyridyl-*s*-triazine. *2,4,6-Tri-2-pyridyl-s-triazine;* 2,4,6-tripyridyl-1,3,5-triazine; tripyridyltriazine; TPTZ. C$_{18}$H$_{12}$N$_6$; mol wt 312.34. C 69.22%, H 3.87%, N 26.91%. Synthesis: Case, Kroft, *J. Am. Chem. Soc.* **81**, 905 (1959). Preparation: Schaefer, U.S. pat. **3,294,798** (1966 to Am. Cyanamid). Thermal stability data: Johns *et al.*, *J. Chem. Eng. Data* **7**, 227 (1962). *Review:* "2,4,6-Tripyridyl-s-triazine" in Diehl *et al.*, *The Iron Reagents* (The G. Frederick Smith Chem. Co., Columbus, Ohio, 1965) pp 41-56.

Crystals, mp 210-220° (Schaefer); trihydrate from aqueous ethanol, mp 244-245° (Case, Kroft). Reacts with ferrous ions to yield intense violet color over pH range 3.4-5.8. Absorption max Fe(TPTZ)$_2$$^{2+}$ (water): 593 nm (ϵ 22,600), Collins *et al.*, *Anal. Chem.* **31**, 1862 (1959). USE: Reagent for the spectrophotometric determn of iron.

9665. Tris-BP. *2,3-Dibromo-1-propanol phosphate(3:1);* phosphoric acid tris(2,3-dibromopropyl) ester; tris(2,3-dibromopropyl) phosphate; Apex 462-5; Flammex AP; Flammex T 23P; Firemaster LV-T 23P; Firemaster T 23P; T 23P; Fyrol HB 32. C$_9$H$_{15}$Br$_6$O$_4$P; mol wt 697.93. C 15.49%, H 2.17%, Br 68.73%, O 9.17%, P 4.44%. Prepn: G. E. Walter, I. Hornstein, U.S. pat. **2,574,515** (1951 to Glenn. L. Martin Co.); D. E. Overbeek, R. C. Nametz, U.S. pat. **3,046,297** (1962 to Michigan Chem. Co.); R. W. Rimmer, U.S. pat. **3,223,755** (1965 to duPont). Use in flameproofing: W. D. Paist, N. Van Gorder, U.S. pat. **2,662,834** (1953 to Celanese). Mutagenicity studies: M. J. Prival *et al.*, *Science* **195**, 76 (1977); A. Nakamura *et al.*, *Mutat. Res.* **66**, 373 (1979). Carcinogenicity studies: B. L. Van Duuren *et al.*, *Cancer Res.* **38**, 3236 (1978); G. Reznik *et al.*, *Nat. Cancer Inst.* **63**, 205 (1979). Review of toxicology: F. A. Daniher, *Proc. Symp. Text. Flammability* **4**, 126-143 (1976). *Review:* A. Blum, B. N. Ames, *Science* **195**, 17 (1977).

Viscous liquid. LD$_{50}$ orally in rats: > 5.0 g/kg (Daniher). *Note:* This substance may reasonably be anticipated to be a carcinogen: *Fourth Annual Report on Carcinogens* (NTP 85-002, 1985) p 196. USE: Flame retardant. Formerly used in children's sleepwear.

9666. Tris(ethylenediamine)cadmium Dihydroxide. Tris(ethylenediamine)cadmium hydroxide; tri(en)cadmium hydroxide; Cadoxen. C$_6$H$_{26}$CdN$_6$O$_2$; mol wt 326.74. C 22.06%, H 8.02%, Cd 34.40%, N 25.72%, O 9.79%. [Cd(H$_2$NCH$_2$-CH$_2$NH$_2$)$_3$](OH)$_2$. Prepared by shaking a given amount of cadmium oxide in 10 times its wt of 30% aq ethylenediamine soln for 15 minutes and centrifuging; the supernatant liquor is the product: Jayme, Neuschäfer, *Naturwiss.* **44**, 62 (1957). The soln contains about 4.5% Cd (w/w) and dissolves about 3% (w/w) cellulose, giving a clear, highly viscous soln. USE: Solvent for cellulose: Jayme, **Ger.** pat. **1,079,318** (1960 to E. Merck), *C.A.* **55**, 18107d (1961); solvent for sulfite pulps.

9667. Tris(hydroxymethyl)nitromethane. *2-(Hydroxymethyl)-2-nitro-1,3-propanediol;* 2-nitro-2-(hydroxymethyl)-1,3-propanediol; trimethylolnitromethane. C$_4$H$_9$NO$_5$; mol wt 151.12. C 31.79%, H 6.00%, N 9.27%, O 52.94%. Prepn from trioxymethylene and nitromethane: Boileau, *Mém. Poudres* **35**, Annexe 7-76 (1953).

Crystals from ethyl acetate + benzene, mp 214° (pure); mp 180° (usual laboratory product); mp 175-176° (tech). Soly in water at 20°: 220 g/100 ml. Freely sol in alcohols. Sparingly sol in benzene, other hydrocarbons. pH of 0.1M aq soln 4.5. USE: Bactericide for inanimate objects. To inhibit bacterial growth in circulating industrial water systems, cutting oils, nonprotein glues and sizings. *Caution:* Irritating to skin, mucous membranes.

9668. Trisilane. Trisilicopropane; trisilicon octahydride; silicopropane; trisilicane. H$_8$Si$_3$; mol wt 92.33. H 8.73%, Si 91.27%. Si$_3$H$_8$. Obtained by separation of mixed silanes prepared from magnesium silicide and hydrochloric acid: Stock, Somiesky, *Ber.* **49**, 111 (1916); **54B**, 524 (1921); **56B**, 247 (1923); Culbertson, U.S. pat. **2,551,571** (1951 to Union Carbide); prepared by conversion of silane to higher silanes in an ozonizer type of electric discharge: Spanier, MacDiarmid, *Inorg. Chem.* **1**, 432 (1962). Liquid. mp −117.4°; bp 52.9°; d^0 0.743; vapor pressure 95.5 mm Hg at 0°. Much less stable than silane or disilane. Detonates in air. Dec in water. Reacts vigorously with CCl$_4$ and CHCl$_3$.

9669. Tristearin. *Octadecanoic acid 1,2,3-propanetriyl ester;* stearin; glyceryl tristearate. C$_{57}$H$_{110}$O$_6$; mol wt 891.45. C 76.79%, H 12.44%, O 10.77%. Present in many animal and vegetable fats, especially the hard ones like cacao butter and tallow. Prepd from stearic acid and glycerol in the presence of Al$_2$O$_3$: Ingram, **Brit.** pat. **663,566** (1951 to I.C.I.); by catalytic hydrogenation of many oils: *Bailey's Industrial Oil and Fat Products* (Wiley, New York, 3rd ed., 1964) pp 881-882. White powder; d$_4^{80}$ 0.862; mp about 55°; on further heating solidifies and melts again at 72°. n$_D^{80}$ 1.4385. Insol in water. Sol in benzene, chloroform, hot alcohol; almost insol in cold alcohol, ether, petr ether. USE: In textile sizes. Formerly in making candles.

9670. Tristriphenylphosphine Rhodium Carbonyl Hydride. *Carbonylhydrotris(triphenylphosphine)rhodium;* hydridocarbonyltris(triphenylphosphine)rhodium(I). C$_{55}$H$_{46}$-

OP$_3$Rh; mol wt 918.79. C 71.90%, H 5.05%, O 1.74%, P 10.11%, Rh 11.20%. HRh(CO)[P(C$_6$H$_5$)$_3$]$_3$. Prepn: Bath, Vaska, *J. Am. Chem. Soc.* **85**, 3500 (1963); Levison, Robinson, *J. Chem. Soc. (A)* **1970**, 2947; Ahmad *et al.*, *J. Chem. Soc. Dalton Trans.* **1972**, 843; Ahmad *et al.*, *Inorg. Syn.* **15**, 45 (1974). Structure: LaPlaca, Ibers, *J. Am. Chem. Soc.* **85**, 3501 (1963); *eidem*, *Acta Crystallogr.* **18**, 511 (1965). Chemistry: Evans *et al.*, *J. Chem. Soc. (A)* **1968**, 2660; O'Connor, Wilkinson, *ibid.* 2665.

Yellow microcrystals, d 1.33. mp 120-122° in air; 172-174° under nitrogen. Moderately sol in benzene, chloroform, dichloromethane; sparingly sol in cyclohexane; insol in light petroleum. Dissociates extensively in organic solvents.

USE: Catalyst.

9671. Trithiocarbonic Acid. *Carbonotrithioic acid;* dihydrogen thiocarbonate. CH$_2$S$_3$; mol wt 110.21. C 10.90%, H 1.83%, S 87.27%. Prepared by the treatment of BaCS$_3$ with ice-cold 10% HCl: v. Halban *et al.*, *Z. Elektrochem.* **29**, 445 (1923); Mills, Robinson, *J. Chem. Soc.* **1928**, 2326; Gattow, Krebs, *Angew. Chem.* **74**, 29 (1962).

Highly refractive red oil. mp −26.9°. bp +57.8°. d$_4^{20}$ 1.483; d$_4^{25}$ 1.476. n$_D^{20}$ 1.8225. Dec by water, alcohol. Addition of sulfur produces tetrathiocarbonic acid CH$_2$S$_4$.

9672. Trithiozine. *4-[Thioxo(3,4,5-trimethoxyphenyl)methyl]morpholine;* 4-(3,4,5-trimethoxythiobenzoyl)morpholine; sulmetozine (rescinded INN); tritiozine; ISF 2001; Tresanil. C$_{14}$H$_{19}$NO$_4$S; mol wt 297.37. C 56.55%, H 6.44%, N 4.71%, O 21.52%, S 10.78%. Non-anticholinergic gastric secretion inhibitor. Prepn: G. Pifferi, *Ger. pat.* **2,102,246** corresp to U.S. pat. **3,862,138** (1972, 1975 both to ISF). Synthesis and pharmacology: G. Pifferi *et al.*, *Chim. Ther.* **8**, 462 (1973). HPLC determn in plasma and urine: T. Crolla *et al.*, *J. Chromatog.* **222**, 257 (1981). Clinical studies: M. Elakovic *et al.*, *J. Int. Med. Res.* **8**, 347 (1980); U. Marini *et al.*, *Clin. Ter.* **92**, 399 (1980). Evaluation of gastric acid suppression: K. Gibinski, *Curr. Med. Res. Opin.* **7**, 516 (1981).

Pale yellow solid from ethanol, mp 141-143°. LD$_{50}$ in mice: 2000 mg/kg i.p., G. Pifferi *et al.*, *loc. cit.*

THERAP CAT: Antisecretory (gastric).

9673. Triticum. Agropyrum; couch grass; dog grass; graminis; quick grass. Dried rhizome and roots of *Agropyron repens* L., Beauv., *Gramineae*, gathered in spring. *Habit.* Europe, Northern Asia; naturalized in the U.S. *Constit.* Triticin, glucose, mannite, inosite.

9674. Tritium. Triterium. T or $^3_1$H. at. wt 3.016. Exists in the diatomic state; mol wt 6.032. Naturally occurring radioactive isotope of hydrogen. Under normal conditions the total atmospheric content of molecular T$_2$ gas is only 11 g. Half-life 12.26 years, low β-emitter. First prepd by the bombardment of deuterophosphoric acid with fast deuterons: M. L. E. Oliphant *et al.*, *Proc. Roy. Soc.* **A144**, 692 (1934); T. W. Bonner, *Phys. Rev.* **53**, 711 (1938). Produced commercially from ^{6}Li by slow neutron bombardment: $^6_3$Li + $^1_0$n → $^3_1$H + $^4_2$He. Details of process producing tritium by neutron irradiation of lithium fluoride: Jenks *et al.*, U.S. pat. **3,079,317** (1963). *Reviews:* E. A. Evans, *Tritium and Its Compounds* (Butterworth, London, 1966) 441 pp; Mackay, Dove, "Deuterium and Tritium" in *Comprehensive Inorganic Chemistry* vol. 1, J. C. Bailar Jr. *et al.*, Eds. (Pergamon Press, Oxford, 1973) pp 77-116; J. J. Katz in Kirk-Othmer *Encyclopedia of Chemical Technology* vol. 7 (Wiley-Interscience, New York, 3rd ed., 1979) pp 554-564.

Gas, having the properties of hydrogen. mp −254.54° (20.62°K) at 162 mm (triple point). bp −248.12° (25.04°K). Crit temp −232.56°. Crit press. 18.317 atm. Molar density of liquid: 45.35 moles/l (20.62°K).

USE: In fusion-based thermonuclear weapons (hydrogen bombs): energy is released by deuteron bombardment ac-

cording to the reaction: $^3_1$H + $^2_1$H → $^4_2$He + $^1_0$n + 18 Mev. Widely used as a radioactive tracer in chemical, biochemical and biological research.

9675. Tritolyl Phosphate. *Phosphoric acid tris(methylphenyl) ester;* tricresyl phosphate; TCP; PX-917; Celluflex; Kronitex; Lindol. C$_{21}$H$_{21}$O$_4$P; mol wt 368.36. C 68.47%, H 5.75%, O 17.37%, P 8.41%. A mixture of isomeric tritolyl phosphates, usually excluding the very toxic *ortho*-isomer as much as possible. Prepd from cresol and phosphoric oxychloride, phosphoric acid or pentachloride: Prahl, U.S. pat. **2,805,240** (1957); Bondy, Gumb, **Brit.** pat. **890,642** (1962 to Coalite and Chem. Prod.); Faith, Keyes & Clark's *Industrial Chemicals*, F. A. Lowenheim, M. K. Moran, Eds. (Wiley-Interscience, New York, 4th ed., 1975) pp 849-853.

Oily, flame resistant liquid. d$_{25}^{25}$ 1.16; bp$_{10}$ about 265°. Pour point −28°. n$_D^{25}$ 1.55. Flash pt (closed cup) 410°. Insol in water (<0.002% at 85°). Sp heat 0.38. Misc with all the common organic solvents and thinners, linseed oil, china wood oil, castor oil.

USE: As plasticizer in vinyl plastics manuf, as flame-retardant, solvent for nitrocellulose, in cellulosic molding compositions, as additive to extreme pressure lubricants, as a nonflammable fluid in hydraulic systems, as lead scavenger in gasoline: Yust, Bame, U.S. pat. **2,889,212** (1959 to Shell); to sterilize certain surgical instruments.

9676. Tri-*o*-tolyl Phosphate. Tri-*o*-cresyl phosphate.
Colorless or pale yellow liquid. *Poisonous!* bp about 410° with slight decompn. Sparingly sol in water; freely in alcohol, benzene, ether.

USE: As plasticizer in lacquers and varnishes. *Caution:* Ingestion may cause nausea, vomiting, diarrhea. Polyneuritis progressing to paralysis of extremities has been seen in severe poisoning: Fassett in *Industrial Hygiene and Toxicology* vol. 2, F. A. Patty, Ed. (Interscience, New York, 2nd ed., 1962) pp 1914-1923.

9677. Tritoqualine. *7-Amino-4,5,6-triethoxy-3-(5,6,7,8-tetrahydro-4-methoxy-6-methyl-1,3-dioxolo[4,5-g]isoquinolin-5-yl)-1(3H)-isobenzofuranone;* 7-amino-4,5,6-triethoxy-3-(5,6,7,8-tetrahydro-4-methoxy-6-methyl-1,3-dioxolo-[4,5-g]isoquinolin-5-yl)phthalide; tritocaline; 554L; Hypostamine; Inhibostamin; Livalfa. C$_{26}$H$_{32}$N$_2$O$_8$; mol wt 500.57. C 62.39%, H 6.44%, N 5.60%, O 25.57%. Prepn: Fr. pat. **1,295,309** (1962 to Lab. de Recherches Biol. Laborec). Activity: Hahn *et al.*, *Arzneimittel-Forsch.* **20**, 1490 (1970).

Crystals, mp 183°.
THERAP CAT: Antihistaminic.

9678. Triuret. *Diimidotricarbonic diamide;* 1,3-dicarbamylurea; carbonyldiurea. C$_3$H$_6$N$_4$O$_3$; mol wt 146.11. C 24.66%, H 4.14%, N 38.35%, O 32.85%. H$_2$NCONHCO-NHCONH$_2$. Prepared by the action of phosgene on urea: Schiff, *Ann.* **291**, 374 (1896); Blair, *J. Am. Chem. Soc.* **48**, 101 (1926); Haworth, Mann, *J. Chem. Soc.* **1943**, 603; Werner, Gray, *Sci. Proc. Roy. Dublin Soc.* **24**, 111 (1946).
Crystals from ammonia water, dec 233°. Freely sol in liq

ammonia. Forms a mono- and dipotassium salt, *see* Blair, *loc. cit.*

9679. Troclosene Potassium. *1,3-Dichloro-1,3,5-triazine-2,4,6(1H,3H,5H)-trione potassium salt;* 3,5-dichlorotetrahydro-2,4,6-trioxo-*s*-triazin-1(2H)-yl potassium; potassium troclosene; potassium dichloroisocyanurate; ACL-59. $C_3Cl_2KN_3O_3$; mol wt 236.06. C 15.26%, Cl 30.04%, K 16.56%, N 17.80%, O 20.33%. Structure studies and prepn: Petterson *et al., J. Org. Chem.* **25,** 1595 (1960). Prepn from trisodium isocyanurate and gaseous Cl: Symes *et al.,* U.S. pats. **3,035,056/7** (1962 to Monsanto).

USE: Useful source of available Cl in solid bleach and detergent formulations.

THERAP CAT: Topical anti-infective.

9680. Trofosfamide. *N,N,3-Tris(2-chloroethyl)tetrahydro-2H-1,3,2-oxazaphosphorin-2-amine 2-oxide;* 2-[bis(2-chloroethyl)amino]-3-(2-chloroethyl)tetrahydro-2H-1,3,2-oxaphosphorine 2-oxide; N,N,N'-tris(2-chloroethyl)-N',O-propylene phosphoric acid ester diamide; trilophosphamide; trophosphamide; NSC 109723; Z 4828; Ixoten. $C_9H_{18}Cl_3N_2O_2P$; mol wt 323.56. C 33.41%, H 5.60%, Cl 32.87%, N 8.66%, O 9.89%, P 9.57%. The 3-(2-chloroethyl) deriv. of cyclophosphamide, *q.v.* Prepn: **Brit.** pat. **1,188,159,** *C.A.* **73,** 44892d (1970); Arnold *et al.,* **Ger.** pat. **2,107,936** [*C.A.* **77,** 152238m (1972)] (1970, 1972, both to Asta-Werke); K. Pankiewicz *et al., J. Am. Chem. Soc.* **101,** 7712 (1979). Pharmacology: N. Brock, *Int. Congr. Chemother., Proc. 5th,* K. H. Spitzy, H. Haschek, Eds. (Verlag Wiener Med. Akad., Vienna, 1967) **II**(1), pp 155-161; J. Potel, N. Brock, *Arzneimittel-Forsch.* **21,** 1250 (1971); N. Brock, J. Potel, *ibid.* **24,** 1149 (1974); Harrison, Fuquay, *Proc. Soc. Exp. Biol. Med.* **139,** 957 (1972). Crystal and molecular structure: A. Perales, S. Garcia-Blanco, *Acta Crystallogr.* **33B,** 1939 (1977).

Crystals from ether, mp 50-51°. $[\alpha]_D^{25}$ −28.6° (c = 2 in CH_3OH). LD_{50} i.p. in mice: 212 mg/kg (Brock, Potel).

THERAP CAT: Antineoplastic.

9681. Troleandomycin. *Oleandomycin triacetate ester;* triacetyloleandomycin (obsolete); TAO; NSC-108166; Cyclamycin; Evramicina; Oleandocetine; Wytrion; Evramycin; Cyclamin; Spectrobact; Treolmicina; Matromicina; Olicin. $C_{41}H_{67}NO_{15}$; mol wt 813.96. C 60.50%, H 8.30%, N 1.72%, O 29.49%. For structure *see* Oleandomycin. Semi-synthetic macrolide antibiotic. Prepd from oleandomycin: Celmer *et al., Antibiot. Ann.* **1957-1958,** p 476; **Brit.** pat. **877,730** (1958 to Pfizer); M. Khristov, N. Petkov, *Farmatsiya* **27,** 1 (1977), *C.A.* **90,** 23458c (1979). Review: S. Ross, *Antimicrobial Therapy,* B. M. Kagan, Ed. (Saunders, Philadelphia, 1970) pp 134-144.

Crystals from isopropanol. Practically tasteless. Dec 176°. $[\alpha]_D^{25}$ −23° (methanol). pKa 6.6. Soly in water < 0.1 g/100 ml.

THERAP CAT: Antibacterial.

9682. Trolnitrate Phosphate. *2,2′2″-Nitrilotrisethanol trinitrate (ester) phosphate (1:2) (salt);* triethanolamine trinitrate biphosphate; trinitrotriethanolamine diphosphate; tri-(2-nitroxyethyl)amine biphosphate; tris(2-nitroxyethyl)-amine biphosphate; Angitrit; Bentonyl; Duronitrin; Kardin; Metamine; Nitranol; Nitretamin; Nitroduran; Nitro-tabl.; Ortin; Praenitron; Thibetine; Tricoryl; Trisustan; Vasomed. $C_6H_{18}N_4O_{17}P_2$; mol wt 480.18. C 15.01%, H 3.78%, N

11.67%, O 56.65%, P 12.90%. Prepd by the nitration of triethanolamine followed by precipitation with phosphoric acid: Metadiet, **Fr.** pat. **984,523** (1951); Junkmann *et al.,* **Ger.** pat. **830,955** (1952 to Schering).

$$N \underset{CH_2CH_2ONO_2}{\overset{CH_2CH_2ONO_2}{\underset{\textstyle|}{\overset{\textstyle|}{-CH_2CH_2ONO_2}}}} \quad 2H_3PO_4$$

Crystals, mp 107-109°.

Note: **Aminotrate phosphate** formerly indicated this compound. Aminotrate® now designates a dietary food supplement composed of proteins and vitamins.

THERAP CAT: Vasodilator.

9683. Tromantadine. *2-[2-(Dimethylamino)ethoxy]-N-tricyclo[3.3.1.1³,⁷]dec-1-ylacetamide;* N-1-adamantyl-N-[2-(dimethylamino)ethoxy]acetamide; 1-(dimethylaminoethoxyacetamido)adamantane. $C_{16}H_{28}N_2O_2$; mol wt 280.41. C 68.54%, H 10.06%, N 9.99%, O 11.41%. Prepn: Scherm, Peteri, **Ger.** pat. **1,941,218** (1971 to Merz), *C.A.* **74,** 99516k (1971); May, Peteri, *Arzneimittel-Forsch.* **23,** 718 (1973). Chemistry and toxicology: Peteri, Sterner, *ibid.* 577.

$$HNCOCH_2OCH_2CH_2N(CH_3)_2$$

Hydrochloride, $C_{16}H_{29}ClN_2O_2$, *D 41, Viru-Merz, Viru-serol.* Crystals, mp 157-158°. LD_{50} orally in rats: 630 mg/kg; i.v. in mice: 71.0 mg/kg (Peteri, Sterner).

THERAP CAT: Antiviral.

9684. Tromethamine. *2-Amino-2-hydroxymethyl-1,3-propanediol;* trimethylol aminomethane; tris(hydroxymethyl)aminomethane; trisamine; tris buffer; trometamol; Tromethane; THAM; TRIS; Talatrol; Tris Amino; Tris-steril; Trizma. $C_4H_{11}NO_3$; mol wt 121.14. C 39.66%, H 9.15%, N 11.56%, O 39.62%. May be prepd by reduction or catalytic hydrogenation of the corresp nitro compd. Prepn of similar compds: Hass, Vanderbilt, U.S. pat. **2,174,242** (1940); Johnson, Degering, *J. Org. Chem.* **8,** 7 (1943); Boileau, *Mém. Poudres* **35,** Annexe 7-76 (1953). Prepn by electrolytic reduction: McMillan, U.S. pat. **2,485,982** (1949 to Comm. Solvents Corp.). Titrimetric standard: Whitehead, *J. Chem. Ed.* **36,** 297 (1959). Monograph: *Ann. New York Acad. Sci.* **92,** Art. 2, pp 333-812 (June 17, 1961). Crystal structure: R. Rudman *et al., Science* **200,** 531 (1978). GC determn in plasma: H. Hulshoff, H. B. Kostenbauder, *J. Chromatog.* **145,** 155 (1978). Pharmacokinetics in rabbits: H. Brasch, H. Iven, *Arch. Int. Pharmacodyn.* **254,** 4 (1981). Use as a biological buffer: R. A. Durst, B. R. Staples, *Clin. Chem.* **18** 206 (1972); S. P. Fling, D. S. Gregerson, *Anal. Biochem.* **155,** 83 (1986); T. Higa, D. M. Desiderio, *ibid.* **173,** 463 (1988). Interaction with hydroxyl radicals: M. Hicks, J. M. Gebicki, *FEBS Letters* **199,** 92 (1986). Review: Nahas, *Pharmacol. Rev.* **14,** 447-472 (1962).

$$HOCH_2\underset{CH_2OH}{\overset{NH_2}{\underset{\textstyle|}{\overset{\textstyle|}{CCH_2OH}}}}$$

Crystalline mass, mp 171-172°. bp_{10} 219-220°. Weak, monoacidic base: pKb (25°): 5.91. pKa (20°): 8.3. pKa (37°): 7.82. pH of 0.1 molar aq soln 10.4. Aq solns do not absorb CO_2 from the air. Soly at 25° (mg/ml) in water: 550; ethylene glycol: 79.1; methanol: 26; anhydr ethanol: 14.6; 95% ethanol: 22.0; DMF: 14; acetone: 2.0; ethyl acetate: 0.5; olive oil: 0.4; cyclohexane: 0.1; chloroform: 0.05; carbon tetrachloride: < 0.05.

USE: In the synthesis of surface-active agents, vulcanization accelerators, pharmaceuticals. As emulsifying agent for cosmetic creams and lotions, mineral oil and paraffin wax emulsions, leather dressings, textile specialties, polishes, cleaning compds, so-called soluble oils. Absorbent for acidic gases. Biological buffer.

THERAP CAT: Alkalizer.

9685. Tropacine. *endo-α-Phenylbenzeneacetic acid 8-methyl-8-azabicyclo[3.2.1]oct-3-yl ester; 1αH,5αH-tropan-3α-ol diphenylacetate;* diphenylacetic acid 3α-tropanyl ester; 3α-tropanyl diphenylacetate; tropine diphenylacetate. C_{22}-$H_{25}NO_2$; mol wt 335.43. C 78.77%, H 7.51%, N 4.18%, O 9.54%. Prepd from tropine and diphenylacetyl chloride: **Swiss** pat. **202,181** (1939 to Ciba), *C.A.* **33,** 8922[8] (1939); Friess *et al., Toxicol. Appl. Pharmacol.* **2,** 574 (1960).

Hydrochloride, crystals from chloroform + ether, mp 217-218°.

THERAP CAT: Anticholinergic.

9686. Tropacocaine. *exo-8-Methyl-8-azabicyclo[3.2.1]-octan-3-ol benzoate; 1αH,5αH-tropan-3β-ol benzoate;* benzoylpseudotropeine; benzoyl-ψ-tropeine; pseudotropine benzoate; ψ-tropine benzoate; tropacaine. $C_{15}H_{19}NO_2$; mol wt 245.31. C 73.44%, H 7.81%, N 5.71%, O 13.04%. From Javanese coca leaves. Prepared by heating pseudotropine with water and benzoic anhydride: Wilstätter, *Ber.* **29,** 943 (1896). Stereochemistry: Beyerman *et al., Rec. Trav. Chim.* **75,** 1445 (1956).

Plates, tablets, mp 49°. Distills *in vacuo* without decompn. pK at 15° = 4.32; K = 1.9 × 10^{-5}; pH of 0.06M soln 8.4. Freely sol in alc, ether, chloroform, benzene, petr ether, dil acids; slightly sol in water. MLD i.v. in rats: 15-20 mg/kg, Hirschfelder, *Physiol. Rev.* **12,** 262 (1932).

Hydrochloride, $C_{15}H_{19}NO_2 \cdot HCl$, strongly refractive prisms from alcohol, dec 283°. Sol in water, slightly in abs alc, practically insol in ether. Aq solns are stable to boiling water for about 20 minutes. pH of 0.1M soln 5.8.

9687. Tropaeolin O. *4-[(2,4-Dihydroxyphenyl)azo]benzenesulfonic acid monosodium salt; C.I. Acid Orange 6;* sodium *p*-(2,4-dihydroxyphenylazo)benzenesulfonate; sodium azoresorcinolsulfanilate; C.I. Food Yellow 8; C.I. 14270; Resorcinol Yellow; Yellow T; Tropeolin O; Tropaeolin R; Chrysoine; Gold Yellow. $C_{12}H_9N_2NaO_5S$; mol wt 316.27. C 45.57%, H 2.87%, N 8.86%, Na 7.27%, O 25.29%, S 10.14%. Prepd from resorcinol and a *p*-sulfobenzenediazonium salt, followed by reaction with an inorganic sodium salt: Sisley, *Bull. Soc. Chim. France* [3] **25,** 869 (1901). See also: *Colour Index* vol. **4** (3rd ed., 1971) p 4064. Brief review: H. J. Conn's *Biological Stains,* R. D. Lillie, Ed. (Williams & Wilkins, Baltimore, 9th ed., 1977) p 106.

Brown powder. Sol in water with reddish-yellow color; sol in alcohol.

USE: As indicator. pH: yellow 11, orange-brown 12.7. Occasionally used as plasma stain.

9688. Tropaeolin OO. *4-[[4-(Phenylamino)phenyl]azo]-*

benzenesulfonic acid monosodium salt; C.I. Acid Orange 5; p-[(*p*-anilinophenyl)azo]benzenesulfonic acid sodium salt; sodium *p*-[(*p*-anilinophenyl)azo]benzenesulfonate; Diphenylamine Orange; Orange GS; Orange N; Orange IV; Fast Yellow; Acid Yellow D; C.I. 13080. $C_{18}H_{14}N_3NaO_3S$; mol wt 375.38. C 57.59%, H 3.76%, N 11.20%, Na 6.13%, O 12.79%, S 8.54%. Prepd from diphenylamine and *p*-sulfobenzenediazonium chloride, followed by reaction with NaOH: Witt, *Ber.* **12,** 258 (1879); *Colour Index* vol. **4** (3rd ed., 1971) p 4045.

Orange-yellow scales or yellow powder. Sol in water. USE: As indicator. pH: red 1.4 to yellow 2.6.

9689. Tropane. *8-Methyl-8-azabicyclo[3.2.1]octane; 1αH,5αH-tropane;* 2,3-dihydro-8-methylnortropidine. C_8-$H_{15}N$; mol wt 125.21. C 76.74%, H 12.08%, N 11.19%. Structure and synthesis: Ladenburg, *Ber.* **16,** 1408 (1883); Merling, *Ber.* **25,** 3124 (1892); Willstätter, *Ber.* **30,** 721 (1897); **33,** 1170 (1900); *eidem, Ann.* **317,** 315, 350 (1901); Robinson, *J. Chem. Soc.* **111,** 762, 876 (1917); Hess, *Ber.* **51,** 1007 (1918); Schöpf, *Angew. Chem.* **50,** 779, 797 (1937); Ruggli, Maeder, *Helv. Chim. Acta* **27,** 436 (1944); Keagle, Hartung, *J. Am. Chem. Soc.* **68,** 1608 (1946).

Liquid; bp 163-169°; d_{15}^{15} 0.9259. Sparingly sol in water, the soly decreasing with an increase in temp. Tropane and water will mix provided the volume of tropane is in excess.

9690. Tropenzile. *α-Hydroxy-α-phenylbenzeneacetic acid 6-methoxy-8-methyl-8-azabicyclo[3.2.1]oct-3-yl ester; 6-methoxy-1αH,5αH-tropan-3α-ol benzilate;* 6-methoxytropine benzilate; benzilic acid 6-methoxytropine ester. $C_{23}H_{27}$-NO_4; mol wt 381.45. C 72.42%, H 7.13%, N 3.67%, O 16.78%. Prepn: Stoll *et al., Helv. Chim. Acta* **38,** 571 (1955); **Swiss** pat. **325,296** (1957 to Sandoz).

Needles from benzene + petr ether, mp 99-101°. Hydrochloride, $C_{23}H_{27}NO_4 \cdot HCl$, mp 146-148°. Methyl bromide, $C_{24}H_{30}BrNO_4$, *Tropenzilium bromide.* Ingredient of *Palerol (Pelerol).*

THERAP CAT: Antispasmodic.

9691. Tropic Acid. *α-(Hydroxymethyl)benzeneacetic acid;* 2-phenylhydracrylic acid; α-phenyl-β-hydroxypropionic acid; tropaic acid; tropeic acid. $C_9H_{10}O_3$; mol wt 166.17. C 65.05%, H 6.07%, O 28.88%. Degradation product of tropane alkaloids, esp atropine: Lossen, *Ann.* **133,** 351, 370 (1865). Resolution of isomers: McKenzie, Wood, *J. Chem. Soc.* **115,** 828 (1919). Absolute config of isomers: Fodor, Csepreghy, *ibid.* **1961,** 3222; Watson, Youngson, *J. Chem. Soc. Perkin Trans. I* **1972,** 1597. Prepn: Sletzinger, Paulsen, **U.S.** pat. **2,390,278** (1945 to Merck & Co.); Blicke, **U.S.** pat. **2,716,650** (1955 to U. of Michigan); **Ger.** pat. **923,426** (1955 to Sterling Drug). Biosynthetic studies: Louden, Leete, *J. Am. Chem. Soc.* **84,** 1510, 4507 (1962).

(±)-Form, needles or plates from water or benzene, mp 118°. K at 25° = 7.5 × 10⁻⁵. Absorption spectrum: Dobbie, Fox, *J. Chem. Soc.* **103**, 1194 (1913). 1 gram dissolves in 50 ml water; freely sol in boiling water; sol in alcohol, ether, slightly in benzene; practically insol in petr ether.

(+)-Form, mp 107°. [α]$_D^{20}$ +72° (c = 0.5 in water).

(−)-Form, mp 126-128°. [α]$_D^{20}$ −72° (c = 0.5 in water).

9692. Tropicamide. *N-Ethyl-α-(hydroxymethyl)-N-(4-pyridinylmethyl)benzeneacetamide; N-ethyl-2-phenyl-N-(4-pyridylmethyl)hydracrylamide; N-ethyl-N-(γ-picolyl)tropamide; Mydriacyl.* $C_{17}H_{20}N_2O_2$; mol wt 284.35. C 71.80%, H 7.09%, N 9.85%, O 11.25%. Prepn: Rey-Bellet, Spiegelberg, **U.S. pat. 2,726,245** (1955 to Hoffmann-La Roche). Comprehensive description: K. W. Blessel *et al.*, in *Analytical Profiles of Drug Substances* **vol. 3**, K. Florey, Ed. (Academic Press, New York, 1974) pp 565-580.

Crystals, mp 96-97°. uv max (0.025 mg/ml in 0.1N HCl): 254 nm (ε 5.1 × 10³).

THERAP CAT: Anticholinergic (ophthalmic).

9693. Tropine. *endo-8-Methyl-8-azabicyclo[3.2.1]octan-3-ol; 1αH,5αH-tropan-3α-ol; 2,3-dihydro-3α-hydroxy-8-methylnortropidine; 2,3-dihydro-3α-hydroxytropidine.* $C_8H_{15}NO$; mol wt 141.21. C 68.04%, H 10.71%, N 9.92%, O 11.33%. Prepd by reaction of tropidine and HBr, followed by hydrolysis and separation of isomers, tropine and pseudotropine: Ladenburg, *Ber.* **35**, 1159 (1902); by hydrogenation of tropinone in the presence of Raney nickel: Van de Kamp, Sletzinger, **U.S. pat. 2,366,760** (1945 to Merck & Co.); Stoll *et al.*, **U.S. pat. 2,746,976** (1956 to Sandoz). Yield of tropine and pseudotropine under varying conditions of reduction: Beckett *et al.*, *Tetrahedron* **6**, 319 (1959). Separation of tropine and pseudotropine by gold salt formation: Ladenburg, *loc. cit.;* by fractional distillation under reduced pressure: Friess *et al.*, *Toxicol. Appl. Pharmacol.* **2**, 574 (1960). Stereochemistry: Beyerman *et al.*, *Rec. Trav. Chim.* **75**, 1445 (1956).

Hygroscopic plates from ether, mp 63°. bp 233°. pK at 15° = 3.80. pH of 0.05 molar soln 11.5. Freely sol in water and alcohol; sol in ether and chloroform.

Note: Esters of tropine are known as *tropeines.*

9694. Tropine Benzylate. *endo-α-Hydroxy-α-phenylbenzeneacetic acid 8-methyl-8-azabicyclo[3.2.1]oct-3-yl ester; 1αH,5αH-tropan-3α-ol benzilate; benzilic acid 3α-tropanyl ester; glykin; BAT; BTE; BETE.* $C_{22}H_{25}NO_3$; mol wt 351.43. C 75.18%, H 7.17%, N 3.99%, O 13.66%. Prepn: Hromatka *et al.*, *Monatsh.* **83**, 1321 (1952).

Prisms from ether or benzene, mp 152-153°. Hydrochloride, $C_{22}H_{25}NO_3$·HCl, mp 239-240°.

9695. Tropomyosins. Fibrous proteins involved in the regulation of muscle relaxation. They are present in all forms of striated and smooth muscles and probably in nonmuscle cells as well. Native tropomyosin consisting of two proteins, tropomyosin and *troponin*, is the Ca²⁺-sensitive regulatory protein that controls the interaction between actin and myosin necessary for the production of force in muscle. In all skeletal tissues, there are two forms of tropomyosin chains, designated α and β. Their ratio depends on the muscle source. Troponin consists of three subunits, troponin T (the tropomyosin binding subunit), troponin I (the actomyosin ATPase inhibitory subunit) and troponin C (the Ca²⁺-binding subunit). All three subunits, in addition to tropomyosin, are responsible for the native tropomyosin activity. Isoln of tropomyosin from skeletal muscle and cardiac muscle: K. Bailey, *Nature* **157**, 368 (1946); idem, *Biochem. J.* **43**, 271 (1948). Structure studies: R. S. Hodges, L. B. Smillie, *Biochem. Biophys. Res. Commun.* **41**, 987 (1970). Amino acid sequence studies: J. Sodek *et al.*, *Proc. Nat. Acad. Sci. USA* **69**, 3800 (1972). X-ray crystal structure of troponin C: M. Sundaralingam *et al.*, *Science* **227**, 945 (1985). *Reviews:* C. E. Bodwell, K. Laki, in *Contractile Proteins and Muscle*, K. Laki, Ed. (Dekker, New York, 1971); W. F. Harrington in *The Proteins*, vol. 4, H. Neurath, R. L. Hill, Eds. (Academic Press, New York, 1979) pp 317-327. Comprehensive review of isolation, preparation, identification and role in the contractile process: *Methods Enzymol.* vol. 85, Part B, entitled "Structural and Contractile Proteins", D. W. Frederiksen, L. W. Cunningham, Eds. (Academic Press, New York, 1982) 774 pp.

9696. Tropylium Bromide. *Cycloheptatrienylium bromide;* cycloheptatrienocarbonium bromide. C_7H_7Br; mol wt 171.04. C 49.15%, H 4.13%, Br 46.72%. Prepd by bromination of 1,3,5-cycloheptatriene in carbon tetrachloride, followed by removal of the CCl₄ and heating the residue *in vacuo* for several days: von Doering, Knox, *J. Am. Chem. Soc.* **79**, 352 (1957); King, Stone, *Inorg. Syn.* **7**, 99 (1963).

Yellow prisms from ethanol, mp 203°. Freely sol in water. Practically insol in ether.

9697. Trospium Chloride. *endo-3-[(Hydroxydiphenylacetyl)oxy]spiro[8-azoniabicyclo[3.2.1]octane-8,1'-pyrrolidinium] chloride; 3α-hydroxyspiro[1αH,5αH-nortropane-8,1'-pyrrolidinium] chloride benzilate; azoniaspiro(3α-benziloyloxynortropane-8,1'-pyrrolidine) chloride; azoniaspiro(3α-diphenylglycoloyloxynortropan-8,1'-pyrrolidine) chloride; 3α-benziloyloxyspiro(nortropane-8,1'-pyrrolidinium) chloride; Relaspium; Spasmex.* $C_{25}H_{30}ClNO_3$; mol wt 427.97. C 70.16%, H 7.07%, Cl 8.28%, N 3.27%, O 11.22%. Tropine derivative with anticholinergic activity. Prepn: **Neth. pat. Appl. 6,402,155;** R. Pfleger *et al.*, **U.S. pat. 3,480,626** (1964, 1969 both to Pfleger); H. Bertholdt *et al.*, *Arzneimittel-Forsch.* **17**, 719 (1967). Pharmacology and toxicology in animals: H. Antweiler *et al., ibid.* **16**, 1581 (1966). Inhibition of gastric motility and acid secretion in humans: G. Lux, P. Frühmorgen, *Fortschr. Med.* **96**, 2113 (1978). Fluorimetric determn in plasma and urine: G. Schladitz-Keil *et al., J. Chromatog.* **345**, 99 (1985). Bioavailability: *eidem, Arzneimittel-Forsch.* **36**, 984 (1986).

Crystals from ethanol-ether, mp 255-257° (dec). LD₅₀ in mice (mg/kg): 12.3 i.v. (Antweiler).

THERAP CAT: Antispasmodic.

Consult the cross index before using this section.

9698. Troxerutin. *2-[3,4-Bis(2-hydroxyethoxy)phenyl]-3-[[6-O-(6-deoxy-α-L-mannopyranosyl)-β-D-glucopyranos-yl]oxy]-5-hydroxy-7-(2-hydroxyethoxy)-4H-1-benzopyran-4-one;* 7,3',4'-tris[O-(2-hydroxyethyl)]rutin; trioxyethylrutin; tri(hydroxyethyl)rutoside; Posorutin; Ruven; Vastribil; Veinamitol. $C_{33}H_{42}O_{19}$; mol wt 742.70. C 53.37%, H 5.70%, O 40.93%. The principal component of a mixture, the *O-(β-hydroxyethyl)rutosides,* which also contains mono-, di-, tetra- and other trihydroxyethyl derivs of rutin, *q.v.* The mixture is prepd by the hydroxyethylation of the phenolic groups of rutin with glycochlorohydrin in alk medium: J. Favre, **Swiss** pat. **349,614** (1957); *see also* **Brit.** pat. **833,174** (1960 to Zyma), *C.A.* **54,** 21135i (1960). Isolation and identification of major components of the mixture: P. Courbat *et al., Helv. Chim. Acta* **49,** 1203, 1420 (1966). Prepn of troxerutin: P. J. Courbat, U.S. pat. **3,420,815** (1969 to Zyma). Metabolism in man: A. M. Hackett *et al., Arzneimittel-Forsch.* **26,** 925 (1976).

Yellow powder, mp 181°. Sol in water, glycerol, propylene glycol. Practically insol in cold ethanol, methanol (forms alcoholate), ether, benzene, chloroform.

*O-(β-*Hydroxyethyl)rutinosides (mixture), *HR, Paroven, Relvene, Varemoid, Venoruton.* Yellow powder, mp 156°. Sol in water, methanol, glycerol, propylene glycol. Practically insol in cold ethanol (forms alcoholate), ether, benzene, chloroform.

THERAP CAT: Treatment of venous disorders.

9699. Troxipide. (±)-*3,4,5-Trimethoxy-N-3-piperidin-ylbenzamide;* 3-(3,4,5-trimethoxybenzamido)piperidine; KU 54; Aplace; Lefron. $C_{15}H_{22}N_2O_4$; mol wt 294.35. C 61.21%, H 7.53%, N 9.52%, O 21.74%. Prepn: **Belg.** pat. **736,840;** T. Irikura *et al.,* **U.S.** pat. **3,647,805** (1969, 1972 both to Kyorin). Anti-ulcer activity: T. Irikura, K. Kasuga, *J. Med. Chem.* **14,** 357 (1971). Pharmacokinetics in man: T. Irikura *et al., Iyakuhin Kenkyu* **12,** 971 (1981), *C.A.* **96,** 79377s (1982). Effect on gastric mucosa in rats: Y. Abe *et al., Nippon Yakurigaku Zasshi* **83,** 317 (1984), *C.A.* **101,** 456g (1984); on glucosamine synthetase: Y. Abe *et al., Oyo Yakuri* **27,** 521 (1984), *C.A.* **101,** 17001c (1984). Metabolism in rats: K. Tagaki *et al., ibid.* 1151, *C.A.* **101,** 103555t (1984); K. Tagaki, K. Endo, *ibid.* 1167, *C.A.* **101,** 103556u (1984). Toxicity: T. Irikura *et al., Kiso to Rinsho* **12,** 3422 (1978).

Needles from acetonitrile, mp 179-181.5°. Sol in ethanol. LD_{50} in male, female rats, male, female mice (mg/kg): 2500, 2100, 2200, 2000 orally; >4150, >4150, 1600, 1550 s.c.; 340, 340, 300, 305 i.p. (Irikura, 1978).

Hydrochloride hemihydrate, $C_{15}H_{23}ClN_2O_4 \cdot \frac{1}{2}H_2O$. Needles from acetonitrile, mp 206-209°. Sol in ethanol.

THERAP CAT: Anti-ulcerative.

9700. Truxillic Acid. *2,4-Diphenyl-1,3-cyclobutanedi-carboxylic acid.* $C_{18}H_{16}O_4$; mol wt 296.31. C 72.96%, H 5.44%, O 21.60%. Cinnamic acid polymers obtained from the minor alkaloids of cocaine: Liebermann, *Ber.* **21,** 2342 (1888). Five stereoisomers have been obtained: α-, γ-, ε-, *peri-* and *epi-*isomers. Stereochemical configurations: Stoermer, Bacher, *Ber.* **57B,** 15-23 (1924).

α-Isomer (d = C_6H_5; b,c,e = H; a,f = COOH), *γ-isatropaic acid; cocaic acid.* Prepd by irradiation of cinnamic acid in water: White, Dunathan, *J. Am. Chem. Soc.* **78,** 6055 (1956). Crystals from acetic acid, mp 284-285°. Sparingly sol in boiling water, in ether, benzene, carbon disulfide; sol in hot glacial acetic acid, hot alc; sparingly sol in acetone.

γ-Isomer (d = C_6H_5; b,c,f = H; a,e = COOH), *ε-isatropaic acid.* γ-Truxillic anhydride, obtained by heating α-truxillic acid with acetic anhydride and sodium acetate, was heated with alkali and the free acid pptd with HCl: Liebermann, *Ber.* **22,** 124 (1889). Needles from dil alc, mp 228°. Very slightly sol in hot water; sol in ether.

ε-Isomer (c = C_6H_5; a,d,e = H; b,f = COOH), *β-cocaic acid.* Prepd by fusion of α-truxillic acid with KOH: Hesse, *Ann.* **271,** 180 (1892). Needles from ether, mp 192°. Freely sol in glacial acetic acid, abs alcohol, chloroform; less sol in benzene; practically insol in ligroin.

*peri-*Isomer (c = C_6H_5; b,d,f = H; a,e = COOH), *η-truxillic acid.* Prepn: γ-Truxillic anhydride heated at low pressure forms the *peri-*anhydride, converted to the acid by warming with alcoholic KOH: Stoermer, Bacher, *loc. cit.* Crystals from benzene + ligroin, mp 266° (effervescence). Soluble in alcohol; practically insol in ether, benzene.

*epi-*Isomer (c = C_6H_5; b,d,e = H; a,f = COOH). Prepd by boiling *peri-*truxillic acid with excess 10% NaOH: Stoermer, Bacher, *loc. cit.* Crystals from dil alc, mp 285-287°. By melting or heating with acetic anhydride in a tube, ε-isomer forms. Practically insol in ether, benzene.

9701. Trypan Blue. *3,3'-[(3,3'-Dimethyl[1,1'-biphen-yl]-4,4'-diyl)bis(azo)]bis[5-amino-4-hydroxy-2,7-naphtha-lenedisulfonic acid] tetrasodium salt;* C.I. Direct Blue 14; C.I. 23850; 3,3'-[(3,3'-dimethyl-4,4'-biphenylene)bis(azo)]bis-(5-amino-4-hydroxy-2,7-naphthalenedisulfonic acid) tetra-sodium salt; tetrasodium 3,3'-[(3,3'-dimethyl-4,4'-biphen-ylene)bis(azo)]bis(5-amino-4-hydroxy-2,7-naphthalenedi-sulfonate); sodium ditolyl-diazobis-8-amino-1-naphthol-3,6-disulfonate; Benzamine Blue; Diamine Blue; Benzo Blue; Congo Blue; Dianil Blue; Naphthylamine Blue; Niagara Blue. $C_{34}H_{24}N_6Na_4O_{14}S_4$; mol wt 960.83. C 42.50%, H 2.52%, N 8.75%, Na 9.57%, O 23.31%, S 13.35%. Prepared by coupling diazotized *o-*tolidine with 5-amino-4-hydroxy-2,7-naphthalenedisulfonic acid in sodium carbonate soln: Lewers, Lowy, *Ind. Eng. Chem.* **17,** 1289-1290 (1925); *Colour Index* vol. 4 (3rd ed., 1971) p 4198. Review of teratogenicity studies: R. L. Cahen, *Clin. Pharmacol. Ther.* **5,** 480 (1964).

Bluish-gray powder. Sol in water forming a deep blue soln with violet tinge; almost insol in alcohol. LD_{100} i.v. in rats: 300 mg/kg, Anderson *et al., Proc. Soc. Exp. Biol. Med.* **31,** 825 (1934).

USE: Biological stain.

9702. Trypan Red. *4,4'-[(3-Sulfo[1,1'-biphenyl]-4,4'-diyl)bis(azo)]bis[3-amino-2,7-naphthalenedisulfonic acid] pentasodium salt;* 4,4'-[(3-sulfo-4,4'-biphenylene)bis(azo)]-bis(3-amino-2,7-naphthalenedisulfonic acid) pentasodium salt; pentasodium 4,4'-[(3-sulfo-4,4'-biphenylene)bis(azo)]-bis(3-amino-2,7-naphthalenedisulfonate); C.I. 22850. $C_{32}H_{19}N_6Na_5O_{15}S_5$; mol wt 1002.84. C 38.32%, H 1.91%, N

8.38%, Na 11.47%, O 23.93%, S 16.00%. Prepd by coupling diazotized 4,4'-diaminobiphenyl-3-sulfonic acid with sodium 2-amino-3,6-naphthalenedisulfonate: Krauss, *J. Am. Chem. Soc.* **36**, 961 (1914); *Colour Index* **vol. 4** (3rd ed., 1971) p 4181.

Reddish-brown powder. Sol in water. Practically insol in alcohol.

USE: Biological stain.

THERAP CAT: Antiprotozoal (Trypanosoma).

THERAP CAT (VET): Has been used as a trypanocide.

9703. Tryparsamide. *[4-[(2-Amino-2-oxoethyl)amino]-phenyl]arsonic acid monosodium salt; N-(carbamoylmethyl)-arsanilic acid monosodium salt;* monosodium *N-phenylgly-cinamide-p-arsonate; Glyphenarsine; Tryparsone; Trypon-arsyl; Trypothane.* $C_8H_{10}AsN_2NaO_4$; mol wt 296.08. C 32.45%, H 3.40%, As 25.30%, N 9.46%, Na 7.77%, O 21.62%. Prepd by heating a soln of arsanilic acid in aq NaOH, Na_2CO_3 and chloroacetamide: Jacobs, Heidelberger, *J. Am. Chem. Soc.* **41**, 1590 (1919); *Org. Syn.* **8**, 100 (1928). U.S. pats. **1,280,119; 1,280,124; 1,280,126** (Sept. 24, 1918); **2,465,308** (1949).

Hemihydrate, platelets, slowly affected by light, stable to air. One gram dissolves in about 2 ml water. Slightly sol in alcohol. Insol in ether, chloroform. pH of 1:20 aq soln 6.5. *Keep tryparsamide in tight containers, preferably at a temp not above 20° and protected from light.*

THERAP CAT: Antiprotozoal (Trypanosoma).

9704. Trypsin. Parenzyme; Parenzymol; Tryptar; Try-pure. Mol wt 24,000. Proteolytic enzyme formed in the small intestine by the action of a peptidase, enterokinase, on the pancreatic cell product, trypsinogen. Acts on lysyl and arginyl bonds of peptide chains and hydrolyzes even esters and amides. *Reviews:* Desnuelle, "Trypsin" in *The Enzymes* **vol. 4**, P. D. Boyer *et al.*, Eds. (Academic Press, New York, 2nd ed., 1960) pp 119-132; Keil, *ibid.* **vol. 3** (3rd ed., 1971) pp 250-275; Inagami, "Trypsin" in *Proteins, Structure and Function* **vol. 1**, M. Funatsu *et al.*, Eds. (Kodansha, Tokyo, Wiley, New York, 1972) pp 1-83.

Yellow to grayish-yellow powder or crystals. Stable indefinitely in dry form at room temp. Sol in water; practically insol in alcohol or glycerol. Readily sol in Sorensen's sodium phosphate buffer soln. Acts optimally at pH values between 7 and 9. Solns lose 75% of their potency within 3 hrs at room temp. Prepn of stabilized trypsin compositions contg partially hydrolyzed gelatin: Sullivan, Martin, U.S. pat. **2,930,736** (1960 to National Drug).

THERAP CAT: Enzyme (proteolytic).

THERAP CAT (VET): Enzyme (proteolytic).

9705. Tryptamine. *1H-Indole-3-ethanamine; 3-(2-ami-noethyl)indole;* 2-(3-indolyl)ethylamine. $C_{10}H_{12}N_2$; mol wt 160.21. C 74.96%, H 7.55%, N 17.49%. Occurs in plants. Synthesis starting with nitroethylene and indole: Noland, Hartman, *J. Am. Chem. Soc.* **76**, 3227 (1954). Alternate routes: Thesing, Schulde, *Ber.* **85**, 324 (1952); Jackson, Smith, *J. Chem. Soc.* **1965**, 3498; Tacconi, *Farmaco Ed. Sci.* **20**, 902 (1965); S. Takano *et al.*, *Heterocycles* **6**, 1167 (1977); I. Fleming, M. Woolias, *J. Chem. Soc. Perkin Trans. I* **1979**, 829. X-ray structure determn: Wakahara *et al.*, *Tetrahedron Letters* **1970**, 4999. *Review* of tryptamine syntheses: J.

E. Saxton in R. H. F. Manske, *The Alkaloids,* **vol. VIII** (1965) pp 8-10.

Needles from petr ether, mp 118°. uv max (ethanol): 222, 282, 290 nm (log ε 4.56, 3.78, 3.71). Sol in ethanol, acetone. Practically insol in water, ether, benzene, chloroform.

Hydrochloride, $C_{10}H_{12}N_2 \cdot HCl$, needles from ethanol + ethyl acetate, mp 248°. uv max (95% ethanol): 221, 275, 281, 290 nm (log ε 4.52, 3.73, 3.75, 3.69).

9706. Tryptazan. *α-Amino-1H-indazole-3-propanoic acid; α-amino-1H-indazole-3-propionic acid;* 3-indazoleala-nine. $C_{10}H_{11}N_3O_2$; mol wt 205.21. C 58.53%, H 5.40%, N 20.48%, O 15.59%. Synthesis from isatin: Snyder *et al.*, *J. Am. Chem. Soc.* **74**, 2009 (1952).

Hemihydrate, crystals from water, darkens at 240°, dec 267-268°. pKa_1 2.20; pKa_2 8.95. Soly in water (g/100 ml): 0.8 at 29°; 12 at 100°; in 95% ethanol: 0.1 at 29°.

9707. L-Tryptophan. L-Trp (IUPAC abbrev.); *l-α-ami-noindole-3-propionic acid; l-α-amino-3-indolepropionic acid;* 2-amino-3-indolylpropanoic acid; *l-β-3-indolylala-nine; l(−)-tryptophan;* Ardeydorm; Kalma; Neurocalm; Optimax; Pacitron; Sedanoct; Trofan; Tryptan. $C_{11}H_{12}N_2O_2$; mol wt 204.22. C 64.69%, H 5.92%, N 13.72%, O 15.67%. An amino acid classified as essential with respect to its growth effect in rats. Essential component in human nutrition, not synthesized by the human body. Isoln from casein, which contains about 1.2%: Hopkins, Cole, *J. Physiol.* **27**, 418 (1902); Dakin, *Biochem. J.* **12**, 302 (1918); Onslow, *ibid.* **15**, 392 (1921); Cox, King, *Org. Syn. coll. vol. II*, 613 (1943). Synthesis starting with β-indolylaldehyde and hippuric acid: Ellinger, *Ber.* **39**, 2515 (1906); Ellinger, Flamand, *Ber.* **40**, 3029 (1907); from hydantoin: Boyd, Robson, *Biochem. J.* **29**, 2256 (1935). Alternate route starting with 3-indoleace-tonitrile: J. N. Coker *et al.*, *J. Org. Chem.* **27**, 850 (1962). Process starting with α-ketoglutaric acid phenylhydrazone: Sakurai *et al.*, U.S. pat. **3,019,232** (1962 to Ajinomoto).

Leaflets or plates from dil alc, dec 289° (rapid heating). $[\alpha]_D^{23}$ −31.5° (c = 1); $[\alpha]_D^{20}$ +2.4° (0.5N HCl); $[\alpha]_D^{20}$ +0.15° (c = 2.43 in 0.5N NaOH). pK_1 2.38; pK_2 9.39. Soly in water (g/l): 0.23 at 0°; 11.4 at 25°; 17.1 at 50°; 27.95 at 75°; 49.9 at 100°. Sol in hot alcohol, in alkali hydroxides. Insol in chloroform.

Hydrochloride, $C_{11}H_{13}ClN_2O_2$, needles from methanol, dec 251°.

THERAP CAT: Antidepressant. Nutrient.

9708. Tryptophol. *3-Indoleethanol;* 3-ω-hydroxyethyl-indole; 2-(3-indolyl)ethyl alcohol; 2-indolyl(3)-ethanol; *β-indolylethyl alcohol;* 3-β-hydroxyethylindole. $C_{10}H_{11}NO$; mol wt 161.20. C 74.51%, H 6.88%, N 8.69%, O 9.93%. Prepd by treatment of indolemagnesium bromide with ethylene oxide: Oddo, Cambieri, *Gazz. Chim. Ital.* **60**, 19 (1939); Snyder, Pilgrim, *J. Am. Chem. Soc.* **70**, 1962 (1948); by Bouveault-Blanc reduction of 3-indoleacetic ester: Jackson, *J. Biol. Chem.* **88**, 659 (1930); by treatment of indolemagne-sium iodide with ethylene chlorohydrin: Majima, Hoshino, *Ber.* **58**, 2042 (1925); from gramine: Snyder, Pilgrim, *J. Am.*

Chem. Soc. **70,** 3770 (1948). By lithium aluminum hydride reduction of methyl 3-indoleglycolate: Speeter, Anthony, U.S. pat. **3,076,814** (1963 to Upjohn).

Platelets from ether + petr ether, mp 59°. $bp_{2.0}$ 174°. Slightly sol in water. Sol in methanol, ethanol, ether, acetone, chloroform, ethyl acetate, glacial acetic acid. Moderately sol in benzene, amyl alcohol, hot carbon disulfide. Sparingly sol in petr ether.

Picrate, red needles from water, mp 101°.

9709. TSH. *Thyrotropin;* thyrotropic hormone; thyreotrophic hormone; thyroid-stimulating hormone; TTH; Pretiron; Thytropar. Member of the glycoprotein family of hormones, produced by the anterior lobe of the pituitary gland. Appears to be essential for the correct functioning of the thyroid gland. Stimulates the production of thyroxine, thereby raising the basal metabolic rate. Massive doses of vitamin A inhibit the secretion of thyrotropic hormone. Like HCG, LH, and FH, *q.q.v.,* TSH is composed of α and β subunits, *cf.* T. H. Liao, J. G. Pierce, *J. Biol. Chem.* **245,** 3275 (1970); J. G. Pierce, T. H. Liao, *Fed. Proc.* **29,** 600 (1970). Amino acid sequence of bovine TSH: Shome *et al., J. Biol. Chem.* **246,** 833 (1971); Liao, Pierce, ibid. 850; of human α and β subunits: J. G. Pierce *et al., Recent Progr. Horm. Res.* **27,** 165 (1971). Diagnostic use: C. W. H. Havard, M. Boss, *Brit. Med. J.* **3,** 678 (1974). *Reviews:* Salter, *Hormones* **2,** 301-349 (1950); Sonnenberg, *Vitam. Horm. (New York)* **16,** 205-261 (1958); J. G. Pierce, T. F. Parsons, *Ann. Rev. Biochem.* **50,** 465-495 (1981). Monograph: S. C. Werner, *Thyrotropin* (C. C. Thomas, Springfield, Ill., 1963) 392 pp.

Behaves like a globulin with a broad isoelectric zone. Biological activity is destroyed by heating and by proteolysis with pepsin, trypsin, and chymotrypsin. Also inactivated by oxidizing agents, such as potassium permanganate and elemental iodine.

THERAP CAT: Thyrotropic hormone; diagnostic aid (thyroid function).

9710. Tsuduranine. *(R)-5,6,6a,7-Tetrahydro-1,2-dimethoxy-4H-dibenzo[de,g]quinolin-10-ol; 1,2-dimethoxy-6aβ-noraporphin-10-ol;* tuduranine. $C_{18}H_{19}NO_3$; mol wt 297.34. C 72.70%, H 6.44%, N 4.71%, O 16.14%. In sinomenine mother liquors. Isoln: Gotô, *Ann.* **521,** 175 (1935). Structure: Gotô, Shishido, *Ann.* **539,** 262 (1939). Synthesis of dl-form: Narayanaswami *et al., Indian J. Chem.* **7,** 945 (1969).

Difficult to crystallize. Minute needles from slowly evaporating ether, mp about 125° (softens at 105°), or 204° depending on cryst form. $[\alpha]_D^{20}$ −127.5° (c = 0.855 in ethanol). Freely sol in the usual organic solvents.

Hydrochloride, shiny scales from water, dec 286°. $[\alpha]_D^{15}$ −148° (c = 0.88 in water + methanol). Sparingly sol in water.

9711. T-2 Toxin. *12,13-Epoxytrichothec-9-ene-3,4,8,15-tetrol 4,15-diacetate 8-(3-methylbutanoate);* 3α-hydroxy-4β,-15-diacetoxy-8α-(3-methylbutyryloxy)-12,13-epoxy-Δ⁹-tri-cothecene; 8α-(3-methylbutyryloxy)-4β,15-diacetoxyscirp-9-en-3α-ol; fusariotoxin T-2; insariotoxin; mycotoxin T-2; NSC 138780. $C_{24}H_{30}O_9$; mol wt 462.50. C 62.33%, H 6.54%, O 31.13%. *Tricothecene mycotoxin* isolated from *Fusarium tricinctum:* J. R. Bamburg *et al., Tetrahedron* **24,**

3329 (1968). Metabolism: T. S. Robison *et al., J. Agr. Food Chem.* **27,** 1141 (1979); T. Yoshizawa *et al., Appl. Environ. Microbiol.* **40,** 901 (1980). Toxicology: W. F. O. Marasas *et al., Toxicol. Appl. Pharmacol.* **15,** 471 (1969). Implicated as a chemical warfare agent in Southeast Asia with nivalenol, *q.v.:* N. Wade, *Science* **214,** 34 (1981); R. T. Rosen, J. D. Rosen, *Biomed. Mass Sectrom.* **9,** 443 (1982).

Crystals, mp 151-152°. $[\alpha]_D^{26}$ + 15° (c = 2.58 in ethanol). LD_{50} orally in female rats: 4.0 mg/kg (Marasas).

Toxicity: Caustic skin irritant! Causes blisters, necrosis of tissues, dizziness, nausea, vomiting, diarrhea, hemorrhaging; may result in death.

9712. Tuaminoheptane. *2-Heptanamine; 1-methylhexylamine;* 2-aminoheptane; Heptamine; Heptin; Heptedrine; Tuamine. $C_7H_{17}N$; mol wt 115.21. C 72.97%, H 14.87%, N 12.16%. $CH_3(CH_2)_4CH(CH_3)NH_2$. Prepd by heating 2-bromoheptane on steam bath with alcoholic ammonia in pressure tube: Clarke, *J. Am. Chem. Soc.* **21,** 1027 (1899); by hydrogenation of methyl amyl ketone and ammonia in the presence of Raney nickel: Norton *et al., J. Org. Chem.* **19,** 1054 (1954).

Volatile liq; d_4^{25} 0.7600-0.7660; bp_{760} 142-144°; n_D^{25} 1.4150-1.4200. Slightly soluble in water. pH of 1% aq soln 11.45. Freely sol in alcohol, ether, petr ether, chloroform, benzene.

Hydrochloride, $C_7H_{17}N.HCl$, needles, mp 133°. Soluble in water. pH about 5.4.

Sulfate, $2C_7H_{17}N.H_2SO_4$, crystals, readily sol in water. The pH of a 1% soln is about 5.4.

THERAP CAT: Sulfate as adrenergic.

9713. Tuberactinomycin. Tuberactin. Polypeptide antibiotic mixture produced by *Streptomyces griseoverticillatus* var *tuberacticus:* Nagata *et al., J. Antibiot.* **21,** 681 (1968); eidem, U.S. pat. **3,639,580** (1972). Composed of tuberactinomycins A, B, N, and O. Structures: Yoshioka *et al., Tetrahedron Letters* **1971,** 2043; Wakamiya *et al., Bull. Chem. Soc. Japan* **46,** 949 (1973); Wakamiya, Shiba, *J. Antibiot.* **27,** 900 (1974); eidem, **28,** 292 (1975). Total synthesis of tuberactinomycin O: T. Teshima *et al., Tetrahedron Letters* **1976,** 2343; eidem, *J. Antibiot.* **30,** 1073 (1977). Chemical studies on tuberactinomycin: T. Wakamiya *et al., Heterocycles* **15,** 999 (1981).

Hydrochloride, white solid, mp 244-264° (dec). $[\alpha]_D^{25}$

$-31.5°$ (c = 1 in water). uv max in water: 268 nm ($E_{1cm}^{1\%}$ 330); in 1N HCl: 268.5 nm ($E_{1cm}^{1\%}$ 313); in 0.1N NaOH: 285 nm ($E_{1cm}^{1\%}$ 206.5). Sol in water; weakly sol in methanol. Practically insol in ethanol, pyridine, ether, chloroform, dioxane; insol in acetone, benzene. pKa_1 7.2; pKa_2 10.3.

Tuberactinomycin A, $C_{25}H_{43}N_{13}O_{11}$, *1-(L-threo-3,6-di-amino-4-hydroxyhexanoic acid)viomycin*. R_1 = OH, R_2 = OH.

Tuberactinomycin B, R_1 = H, R_2 = OH, see Viomycin.
Tuberactinomycin N, R_1 = OH, R_2 = H, see Enviomycin.

Tuberactinomycin O, $C_{25}H_{43}N_{13}O_9$, *(R)-6-[L-2-(2-amino-1,4,5,6-tetrahydro-4-pyrimidinyl)glycine]viomycin*. R_1 = H, R_2 = H.

THERAP CAT: Antibacterial (tuberculostatic).

9714. Tubercidin. *7-β-D-Ribofuranosyl-7H-pyrrolo[2,3-d]pyrimidin-4-amine; 4-amino-7-β-D-ribofuranosyl-7H-pyrrolo[2,3-d]pyrimidine;* 7-deazaadenosine; sparsamycin A; U 10071. $C_{11}H_{14}N_4O_4$; mol wt 266.25. C 49.62%, H 5.30%, N 21.04%, O 24.04%. Antibiotic substance produced in the culture broth of *Streptomyces tubericidus*. Isoln: Anzai *et al., J. Antibiot.* **10A**, 201 (1957). Structure: Susuki, Maru-mo, *ibid.* **14A**, 34 (1961). Total synthesis: Tolman *et al., J. Am. Chem. Soc.* **91**, 2102 (1969). Crystal structure: Stroud, *Acta Crystallogr.* **29B**, 690 (1973); Abola, Sundaralingham, *ibid.* 697.

Needles from water, dec 247-248°. $[\alpha]_D^{17}$ $-67°$ (50% acetic acid). uv max (0.01N NaOH): 270 nm (ε 12,100). Sol in acidic and alkaline soln. One gram dissolves in 330 ml water, 200 ml methanol, 2000 ml ethanol. Practically insol in acetone, ethyl acetate, chloroform, benzene, petr ether. LD_{50} i.v. in mice: 45 mg/kg (Anzai).

THERAP CAT: Antifungal, antibacterial (tuberculostatic); antineoplastic.

9715. Tuberculin. A filtrate from triturated *Mycobacteria tuberculosis*. *Old Tuberculin* is the culture filtrate prepd by boiling and then filtering the *M. tuberculosis* culture. Purified protein derivative is a more refined tuberculin prepd by precipitating the culture filtrate with ammonium sulfate. Purification and dermal reactivity: F. B. Seibert, J. T. Glenn, *Am. Rev. Tuberc.* **44**, 9 (1941). Review of tuberculin skin test: G. W. Comstock *et al., Am. Rev. Resp. Dis.* **124**, 356 (1981).

Purified Protein Derivative, PPD, Aplisol, Tubersol.

THERAP CAT: Diagnostic aid (tuberculosis).

9716. Tuberin. *N-[2-(4-Methoxyphenyl)ethenyl]form-amide; N-trans-(p-Methoxystyryl)formamide; N-formyl trans-p-methoxystyrylamine.* $C_{10}H_{11}NO_2$; mol wt 177.20. C 67.78%, H 6.26%, N 7.91%, O 18.06%. Antitubercular antibiotic isolated from the broth filtrate of *Streptomyces amakusaensis*: Ohkuma *et al., J. Antibiot.* **15A**, 115 (1962); Sumiki *et al.,* **Japan.** pat. **64 7399** (1964 to Inst. Phys. & Chem. Res.), *C.A.* **62**, 8355g (1965). Structure and synthesis: Anzai *et al., J. Antibiot.* **15A**, 110, 117, 123 (1962). Alternate synthesis: I. J. Massey, I. T. Harrison, *Chem. Ind. (London)* **1977**, 920. Biosynthetic studies: K. M. Cable *et al., J. Chem. Soc., Perkin Trans. I* **1987**, 1593.

Prisms from benzene, mp 132-133°. Stable in weakly acidic or weakly alkaline solns. uv max (methanol): 219, 285 nm ($E_{1cm}^{1\%}$ 870, 1710). Sol in the lower alcohols, ethyl acetate, acetone; moderately sol in carbon tetrachloride, chloroform; sparingly sol in water, benzene. Practically insol in petr ether.

Dihydrotuberin, $C_{10}H_{13}NO_2$, liquid. n_D^{19} 1.5349.

9717. Tubocurarine Chloride. *7',12'-Dihydroxy-6,6'-dimethoxy-2,2',2'-trimethyltubocuraranium chloride hydrochloride.* $C_{37}H_{42}Cl_2N_2O_6$; mol wt 681.66. C 65.20%, H 6.21%, Cl 10.40%, N 4.11%, O 14.08%. Isoln of d-form from *Chondodendron tomentosum* R. & P., *Menispermaceae:* Dutcher, *J. Am. Chem. Soc.* **68**, 419 (1946). Purification: *ibid.* **74**, 2221 (1952). Isoln of l-form from *Ch. tomentosum:* King, *Nature* **158**, 515 (1946); *J. Chem. Soc.* **1947**, 936. Structure: King, *ibid.* **1948**, 265. Stereochemistry: Hultin, *Acta Chem. Scand.* **17**, 753 (1963). Revised structure: Everett *et al., Chem. Commun.* **1970**, 1020; Codding, James, *ibid.* **1972**, 1174. Synthesis of d-form: J. Naghaway, T. O. Soine, *J. Pharm. Sci.* **68**, 655 (1979). Synthesis of dl-form: Veronin *et al., Doklady Akad. Nauk SSSR* **122**, 77 (1958), *C.A.* **53**, 1345f (1959). Pharmacology and toxicology: R. D. Sofia, L. C. Knobloch, *Toxicol. Appl. Pharmacol.* **28**, 227 (1974); C. A. Winter, J. T. Lehman, *J. Pharmacol. Exp. Ther.* **100**, 489 (1950). Toxicity data: H. Rosen *et al., Proc. Soc. Exp. Biol. Med.* **120**, 511 (1965). Use as diagnostic aid for myasthenia gravis: F. F. Foldes *et al., J. Am. Med. Assoc.* **203**, 133 (1968). Comprehensive description: C. Papastephanou in *Analytical Profiles of Drug Substances* **vol. 7**, K. Florey, Ed. (Academic Press, New York, 1978) pp 477-500.

(+)tubocurarine chloride

*d-*Form, *Jexin, Tubadil, Delacurarine, Curarin-HAF, Tubarine, Intocostrin.* Hexagonal and pentagonal micro-platelets from water. Can exist in the form of various hydrates. The anhydrous material (dec 274-275°) takes up water in moist atm until it reaches the pentahydrate stage, dec ~270°. $[\alpha]_D^{22}$ $+190°$ (c = 0.5); $[\alpha]_D^{23}$ $+219°$ (c = 0.785 in methanol). Corresp calcd values for the anhydr salt: $[\alpha]_D^{22}$ $+190°$; $[\alpha]_D^{23}$ $+245°$ (methanol). uv max (H_2O): 280 nm ($E_{1cm}^{1\%}$ 118). Soly (25°): ~50 mg/ml water; but supersatd solns are formed readily (Dutcher). Soly also reported as ~1 g/40 ml water; ~1 g/75 ml ethanol. Presence of 1.0N HCl diminishes soly by about one-third. Also sol in methanol. Insol in pyridine, chloroform, benzene, acetone, ether. LD_{50} in mice, rats (mg/kg): 33.2, 27.8 orally (Rosen).

*l-*Form pentahydrate, needles from water, mp 268° (effervescence). $[\alpha]_D^{20}$ $-258°$ (c = 0.38) for the anhydr salt.

Dimethyl ether, $C_{39}H_{46}Cl_2N_2O_6$, *d-tubocurarine chloride dimethyl ether, Mecostrin chloride.* Crystals, dec about 236° (effervescence). $[\alpha]_D^{25}$ $+185°$; to $+195°$ (c = 0.5 in H_2O). uv max (acidified 0.005% aq soln): 280 nm ($A_{1cm}^{1\%}$ ~89). Soluble in water, dil sodium hydroxide. Sparingly sol in alc, in dil HCl, chloroform. Practically insol in benzene, ether.

Human Toxicity: Large doses and overdoses may cause respiratory paralysis and hypotension.

THERAP CAT: The *d*-form and dimethyl ether as skeletal muscle relaxants. *d-*Form as diagnostic aid (myasthenia gravis).

THERAP CAT (VET): Muscle relaxant.

9718. Tubulin. Colchicine-binding protein. The subunit protein of *microtubules*, which are large protein assemblies that play an important role in eukaryotic cell form determination and dynamics. Microtubules have the general

structure of long hollow cylinders within which 13 protofilaments of tubulin are arranged in a parallel manner to the
cylinder axis. The axial arrangement of the protofilaments
with respect to each other results in the appearance of a helical structure, and the *in vitro* microtubule assembly process
generally follows the laws of helical protein polymerization,
cf. Thermodynamics of the Polymerization of Proteins, F.
Oosawa, S. Asakura, Eds. (Academic Press, New York,
1975). Tubulin is an asymmetric dimer consisting of two
nearly identical molecules, *α-tubulin* and *β-tubulin*, each
having mol wts of about 55,000. The two molecules can be
separated due to differences in electrophoretic mobilities.
Isoln from mammalian brain using colchicine binding: R. C.
Weisenberg *et al., Biochemistry* **7,** 4466 (1968). Discovery of
conditions for microtubule assembly *in vitro:* R. C. Weisenberg, *Science* **177,** 1104 (1972); G. G. Borisy, J. B. Olmstead,
ibid. 1196. Prepn of large quantities of brain tubulin
through successive assembly-disassembly cycles: M. L.
Shelanski *et al., Proc. Nat. Acad. Sci. USA* **70,** 765 (1973).
Purification of tubulin from rat pancreas: J. F. Launay *et
al., Biochem. Biophys. Res. Commun.* **111,** 253 (1983). Structure of two human α-tubulin genes: C. Wilde *et al., Proc.
Nat. Acad. Sci. USA* **79,** 96 (1982). Structure and arrangement of protofilaments in microtubules and tubulin sheets:
B. F. McEwen, *Diss. Abstr. B* **43,** 942 (1982). Series of articles on prepn, isoln, and purification of tubulin from various
sources: *Methods Enzymol.* **85,** Pt. B, 376-417 (1982). *Reviews:* J. A. Snyder, J. R. McIntosh, *Ann. Rev. Biochem.* **45,**
699-720 (1976); S. N. Timasheff, L. M. Grisham, *ibid.* **49,**
565-591 (1980); M. F. Carlier, *Mol. Cell. Biochem.* **47,** 97-
113 (1982).
    Purified calf brain tubulin retains many of its *in vivo* bio-
chemical characteristics, such as the ability to self-assemble
into microtubules and the response of the assembly reaction
to inhibitory effects of cold temperature, Ca^{2+}, and anti-
microtubule agents, *e.g.* vinblastine and colchicine, *q.q.v.* uv
max (PG buffer): 278 nm (ϵ 1.33 ml mg^{-1} cm^{-1}).

9719. Tuftsin. N^2-[1-(N^2-L-*Threonyl*-L-*lysyl*)-L-*prolyl*]-
L-*arginine.* $C_{21}H_{40}N_8O_6$; mol wt 500.61. C 50.38%, H
8.05%, N 22.38%, O 19.18%. A naturally occurring tetrapeptide having a variety of immunopotentiating properties,
especially stimulation and enhancement of phagocytosis. It
also exhibits antitumor and antibacterial activity and has
been shown to possess chemotactic, migration-enhancing,
and mitogenic properties for leukocytes. Discovered during
research on the physiological role of cytophilic gamma-
globulin: V. A. Najjar, K. Nishioka, *Nature* **228,** 672
(1970). Produced in the spleen; present in mammalian blood
in the gamma globulin fraction as part of the larger molecule
leukokinin. Isoln and characterization: K. Nishioka *et al.,
Biochim. Biophys. Acta* **310,** 217 (1973); V. A. Najjar, U.S.
pat. **3,778,426** (1973 to Research Corp.). Solid phase synthesis: K. Nishioka *et al., Biochem. Biophys. Res. Commun.*
47, 172 (1972); *eidem, Biochim. Biophys. Acta* **310,** 230
(1973). Synthesis by fragment condensation: J. Vicar *et al.,
Coll. Czech. Chem. Commun.* **41,** 3467 (1976); by liquid
phase method: S. Nozaki *et al., Bull. Chem. Soc. Japan* **50,**
422 (1977). ^{13}C-NMR and circular dichroism studies: I. Z.
Siemion *et al., Eur. J. Biochem.* **112,** 339 (1980). Conformational studies have provided conflicting evidence on the
structure, *cf.* M. Blumenstein *et al., Biochemistry* **18,** 4247
(1979). Specific receptors on macrophages, monocytes, and
granulocytes are thought to mediate the biological activity of
tuftsin: A. Constantopoulos, V. A. Najjar, *J. Biol. Chem.*
248, 3819 (1973); R. M. G. Nair *et al., Immunochemistry* **15,**
901 (1978); Z. Bar-Shavit *et al., Biochem. Biophys. Res.
Commun.* **94,** 1445 (1980). Its physiological significance has
been shown in patients in whom tuftsin deficiency has resulted in a human syndrome with increased incidence of
severe infections: *Macrophages and Lymphocytes, Part A,* M.
R. Escobar, H. Friedman, Eds. (Plenum Press, New York,
1980) pp 131-147; *Lymphokine Reports,* E. Pick, Ed. (Academic Press, New York, 1980) pp 157-159; V. A. Najjar,
Med. Biol. **59,** 134 (1981). General biological proprties: V.
A. Najjar, *Mol. Cell. Biochem.* **41,** 1 (1981). Antitumor
activity: K. Nishioka *et al., ibid.* 13. Bactericidal activity:
J. Martinez, F. Winternitz, *ibid.* 123. Analogs: F. Z. Siemion, *ibid.* 99. *Reviews:* V. A. Najjar, *Exp. Cell. Biol.* **46,** 114-
126 (1978); *eidem, Adv. Exp. Med. Biol.* **121A,** 131-147

(1980); K. Nishioka *et al., Life Sci.* **28,** 1081-1090 (1981); V.
A. Najjar, *Mol. Cell. Biochem.* **41,** 73-98 (1981). Conference
proceedings: *Ann. N.Y. Acad. Sci.* **419,** entitled "Antineo-
plastic, Immunogenic and Other Effects of the Tetrapeptide
Tuftsin: a Natural Macrophage Activator", V. A. Najjar,
M. Fridkin, Eds. (1983) pp 1-273.

9720. Tulobuterol. 2-*Chloro-α-[[(1,1-dimethylethyl)-
amino]methyl]benzenemethanol;* α-[(*tert*-butylamino)meth-
yl]-*o*-chlorobenzyl alcohol. $C_{12}H_{18}ClNO$; mol wt 227.73. C
63.29%, H 7.96%, Cl 15.57%, N 6.15%, O 7.03%. A β-adrenergic receptor agonist, related structurally to terbutaline,
q.v. Prepn: H. Kato, S. Kurata, Ger. pat. **2,244,737** (1973
to Hokuriku), *C.A.* **78,** 147538a (1973). Pharmacology: S.
Kubo *et al., Arzneimittel-Forsch.* **25,** 1028 (1975); *eidem,
ibid.* **27,** 1433 (1977); I. Uesaka *et al., ibid.* 1439. Metabolism: T. Fujiihashi *et al., Oyo Yakuri* **18,** 347 (1979), *C.A.*
92, 121781 (1980); K. Matsumura *et al., Yakugaku Zasshi*
101, 198 (1981), *C.A.* **94,** 167404 (1981). Determn of tulobuterol and its metabolites in human urine: *eidem, J. Chromatog.* **222,** 53 (1981). Toxicological studies: S. Kubo *et al.,
Oyo Yakuri* **13,** 197, 317 (1977), *C.A.* **88,** 83591-2 (1978).

    Crystals, mp 89-91°. LD_{50} in male mice, rats, rabbits
(mg/kg): 305, 850, 563 orally; 170, 417, 164 s.c. (Kubo,
1975).
    Hydrochloride, $C_{12}H_{19}Cl_2NO$, *C-78, Atenos, Berachin,
Brelomax, Bremax, Hokunalin, Respacal.* White crystalline
powder, mp 161-163°.
    THERAP CAT: Bronchodilator.

9721. Tung Oil. China wood oil. A drying oil from
seeds of *Aleurites cordata* Steud., *Euphorbiaceae,* indigenous
to China and Japan, but now grown also in Florida. Chief
fatty acid component is eleostearic acid. Unlike linseed and
soybean oils, it need not be refined.
    Pale yellow liquid; characteristic disagreeable odor. On
long keeping, or on heating for a short time at 300°, polymerizes to a stiff jelly. d 0.936-0.943. Iodine no. 163-171.
Sapon no. 190-197. Sol in chloroform, ether, carbon disulfide, oils; the polymerized product is practically insol in the
usual organic solvents.
    USE: Manuf quick-drying wood varnishes, linoleum, and
floor cloth; in India rubber substitutes, insulating masses;
for waterproofing paper and other tissues.

9722. Tungsten. Wolfram. W; at. wt 183.85; at. no. 74;
valences 6, 5, 4, 3, 2. Naturally occurring isotopes: 180
(0.135%); 182 (26.4%); 183 (14.4%); 184 (30.6%); 186
(28.4%); artificial radioactive isotopes: 173-179; 181; 185;
187-189. Discovered by C. W. Scheele in 1781, isolated in
1783 by J. J. and F. de Elhuyar. One of the rarer metals,
comprises about 1.5 ppm of the earth's crust. Chief ores are
wolframite [(Fe,Mn)WO_4] and scheelite (CaWO_4). Found
chiefly in China, Malaya, Mexico, Alaska, South America
and Portugal. Scheelite ores mined in the U.S. carry from
0.4-1.0% WO_3. Description of isoln processes: K. C. Li, C.
Y. Wang, *Tungsten,* A.C.S. Monograph Series no. **94** (Reinhold, New York, 3rd ed., 1955) pp 113-269; G. D. Rieck,
Tungsten and Its Compounds (Pergamon Press, New York,
1967) 154 pp. *Reviews:* Parish, *Advan. Inorg. Chem. Radiochem.* **9,** 315-354 (1966); Rollinson, "Chromium, Molybdenum and Tungsten" in *Comprehensive Inorganic Chemistry*
vol. 3, J. C. Bailar, Jr. *et al.,* Eds. (Pergamon Press, Oxford,
1973) pp 623-624, 742-769.
    Steel-gray to tin-white metal; body centered cubic struc-
ture. d_4^{20} 18.7-19.3; depends on extent of working. Hardness 6.5-7.5. mp 3410°. bp$_{760}$ 5900°. Spec heat (20°): 0.032
cal/g/°C. Heat of fusion 44 cal/g. Heat of vaporization
1150 cal/g. Electrical resistivity (20°): 5.5 μohm-cm. Stable
in dry air at ordinary temps, forms the trioxide at red heat.
Not attacked by water, but oxidized to the dioxide by steam.
Very stable to acids, attacked only superficially by concd
nitric acid or aqua regia. Powdered tungsten can be pyrophoric under the right conditions. Slowly sol in fused potas-

sium hydroxide or sodium carbonate in presence of air; sol in a fused mixture of NaOH and nitrate. Attacked by fluorine at room temp; by chlorine at 250-300° giving the hexachloride in absence of air, and the trioxide and oxychloride in the presence of air. LD_{50} i.p. in rats: 5 g/kg, *Toxic Substances List*, H. E. Christensen, Ed. (1973) p 965.

USE: To increase hardness, toughness, elasticity, and tensile strength of steel; manuf alloys; manuf filaments for incandescent lamps and in electron tubes; in contact points for automotive, telegraph, radio and television apparatus; in phonograph needles. *Tungsten carbides* (W_2C, WC) used in rock drills, metal-cutting tools, wire-drawing dies. WC used as catalyst instead of platinum: Bennett *et al., Science* **184,** 563 (1974).

9723. Tungsten Hexafluoride. F_6W; mol wt 297.86. F 38.27%, W 61.73%. WF_6. Prepd by direct fluorination of powdered tungsten. Can be purified by distillation, preferably under pressure: Ruff, Ascher, *Z. Anorg. Allgem. Chem.* **196,** 413 (1931); Henkel, Klemm, *ibid.* **222,** 68 (1935); Marchi, *Inorg. Syn.* **3,** 181 (1950); from ClF and tungsten: Pitts, Jache, *Inorg. Chem.* **7,** 1661 (1968).

Colorless gas or pale yellow liquid. Orthorhombic, deliquescent crystals when solid. d_{liq}^{15} 3.441. mp +2.3°. bp 17.5°. Soly in anhydr HF: 3.14 moles/1000 g HF, Frlec, Hyman, *Inorg. Chem.* **6,** 1596 (1967). May be stored in glass pressure ampuls.

9724. Tungsten Trioxide. Tungstic anhydride. O_3W; mol wt 231.86. O 20.70%, W 79.30%. WO_3. Prepn from sodium tungstate: Hein, Herzog, in *Handbook of Preparative Inorganic Chemistry* vol. 2, G. Brauer, Ed. (Academic Press, New York, 2nd ed., 1965) pp 1423-1424.

Canary yellow, heavy powder; dark orange when heated, regaining the original color on cooling. Insol in water; sol in caustic alkalies; very slightly sol in acids.

USE: Manuf tungstates which are used for x-ray screens and for fireproofing fabrics.

9725. Tungstic(VI) Acid. H_2O_4W; mol wt 249.88. H 0.81%, O 25.61%, W 73.59%. H_2WO_4. Prepn: Morley, *J. Chem. Soc.* **1930,** 1990.

Yellow or greenish-yellow powder. Insol in water and acid except hydrofluoric acid; slowly sol in solns of caustic alkalies.

Note: When freshly pptd from a soluble tungstate it contains 1 mol H_2O, and is appreciably sol in water.

9726. Tunicamycin. A family of nucleoside antibiotics produced by *Streptomyces lysosuperificus.* Isoln, characterization: A. Takatsuki *et al., J. Antibiot.* **24,** 215 (1971). Biological properties: A. Takatsuki, G. Tamura, *ibid.* 224. Effect on microorganisms: A. Takatsuki *et al., ibid.* **25,** 75 (1972). Tunicamycin is produced as a mixture of at least 10 homologous antibiotics, the main components being tunicamycins V, VII, II and X, also referred to respectively as A, B, C, D. They contain uracil, *N*-acetylglycosamine, an 11-carbon aminodialdose called *tunicamine*, and a fatty acid linked to the amino group of tunicamine. The homologs differ in their fatty acid components, which vary in degree of saturation and chain length and branching. Structural elucidation: A. Takatsuki *et al., Agr. Biol. Chem.* **41,** 2307 (1977); T. Ito *et al., ibid.* **44,** 695 (1980). HPLC sepn of components: W. C. Mahoney, D. Duskin, *J. Biol. Chem.* **254,** 6572 (1979). Approaches to synthesis: Y. Fukuda *et al., Bull. Chem. Soc. Japan* **55,** 880 (1982). Tunicamycin has been shown to interfere with glycoprotein synthesis in yeast and mammalian systems: A. Takatsuki *et al., Agr. Biol. Chem.* **39,** 2089 (1975); S. Kuo, J. O. Lampen, *Arch. Biochem. Biophys.* **172,** 574 (1976). Effect on epidermal glycoprotein and glycosaminoglycan synthesis *in vitro:* I. A. King, A. Tabiowo, *Biochem. J.* **198,** 331 (1981). Enhancement of antiviral and anticellular activity of interferon, *q.v.:* R. K. Maheshwari *et al., Science* **219,** 1339 (1983). *Review:* A. D. Elbein, *Trends Biochem. Sci.* **6,** 219-221 (1981). *Book: Tunicamycin,* G. Tamura, Ed. (Japan Sci. Soc., Tokyo, 1982) 220 pp. *See also* Streptovirudin.

tunicamycins II, V, VII, X
(n = 8, 9, 10, 11)

White cryst powder, mp 234-235° (dec). $[\alpha]_D^{20}$ +52° (c = 0.5 in pyridine). uv max (methanol): 205, 260 nm ($E_{1cm}^{1\%}$ 230, 110). Sol in alk water, pyridine, hot methanol. Slightly sol in ethanol, butanol. Practically insol in acetone, ethyl acetate, chloroform, benzene, acidic water. When dissolved in water at 100° and held for 30 min, stable at neutral and alk pH, unstable at acidic pH.

USE: As a tool in studying glycoproteins in a wide variety of biological systems.

9727. Tunichrome B-1. *(E,Z)-3,5-Dihydroxy-L-tyrosyl-α,β-didehydro-3,5-dihydroxy-N-[2-(3,4,5-trihydroxyphenyl)-ethenyl]tyrosinamide;* TB-1. $C_{26}H_{25}N_3O_{11}$; mol wt 550.50. C 56.22%, H 4.54%, N 7.56%, O 31.68%. Major component of a group of polyphenolic yellow blood pigments isolated from the sea squirt (tunicate) *Ascidia nigra* (Linnaeus). The *tunichromes* are strong biological reducing agents involved in the selective accumulation of vanadium by *A. nigra.* Isoln of crude tunichrome mixture from the blood cells of *A. nigra,* action in converting vanadium(V) to vanadium(III): I. G. Macara *et al., Biochem. J.* **181,** 457 (1979). Isoln of tunichrome B-1 by HPLC under anaerobic conditions, characterization and structure determn by mass spec, nmr: R. C. Bruening *et al., J. Am. Chem. Soc.* **107,** 5298 (1985). Brief account of the role of tunichromes in studies into the biochemical nature of vanadium and other trace metals: R. J. Seltzer, *Chem. & Eng. News* **63,** 67 (Sept. 16, 1985). *Review:* R. C. Bruening *et al., J. Nat. Prod.* **49,** 193 (1986).

Yellow solid, sol in methanol. uv max (methanol): 210, 340 nm (ϵ 68000, 19600).

9728. Turanose. *3-O-α-D-Glucopyranosyl-D-fructose;* 3-(α-D-glucosido)-D-fructose. $C_{12}H_{22}O_{11}$; mol wt 342.30. C 42.10%, H 6.48%, O 51.42%. Prepd from melezitose: Tanret, *Compt. Rend.* **142,** 1424 (1906); Bridel, Aagard, *Bull. Soc. Chim. Biol.* **9,** 884 (1927); Hudson, Pacsu, *J. Am. Chem. Soc.* **52,** 2522 (1930); Pacsu in *Methods in Carbohydrate Chemistry* vol. I (Academic Press, New York, 1962) p 353. Structure: Isbell, Pigman, *J. Res. N.B.S.* **20,** 787 (1938); Isbell, *ibid.* **26,** 35 (1941); Hudson, *J. Org. Chem.* **9,** 117, 470 (1944); Hassid, Ballou in *The Carbohydrates,* W. Pigman, Ed. (Academic Press, New York, 1957) p 508. Crystal structure: A. Neuman *et al., Acta Crystallogr.* **B34,** 242 (1978).

Nonhygroscopic prisms from water + alc. Very sweet taste. Dec 157°. Shows mutarotation. $[\alpha]_D^{20}$ +27.3° → +75.8° (c = 4 in water). Freely sol in water, methanol. One gram dissolves in 19 ml of 95% alc. Acid hydrolysis yields 1 mol D-glucose and 1 mol D-fructose. Reducing power about ½ of D-glucose.

9729. Turicine. *cis-2-Carboxy-4-hydroxy-1,1-dimethyl-pyrrolidinium hydroxide inner salt; d-N,N-dimethyl-4-hydroxypyrrolidine-2-carboxylic acid betaine; d-4-hydroxyproline betaine; d-4-hydroxystachydrine; allohydroxy-D-proline betaine.* $C_7H_{13}NO_3$; mol wt 159.18. C 52.81%, H 8.23%, N 8.80%, O 30.15%. Isolated from *Stachys (betonica) officinalis* L., Trev., and *Stachys sylvatica* L., Wald-Ziest, *Labiatae*: Küng, Trier, *Z. Physiol. Chem.* **85**, 209 (1913); Guggenheim, *Die Biogenen Amine* (S. Karger, New York, 1951) p 247. Stereoisomeric with betonicine. Prepd by the reaction of allohydroxy-D-proline with methyl iodide and silver oxide: Patchett, Witkop, *J. Am. Chem. Soc.* **79**, 185 (1957).

Crystals from water, ethanol and ethyl acetate, mp 259-260°. $[\alpha]_D^{20}$ +37.8° (c = 1 in water).

Monohydrate, efflorescent prisms from water. Sweet taste. Less sol in water than betonicine; slightly sol in alcohol. Neutral to litmus.

Hydrochloride, $C_7H_{13}NO_3 \cdot HCl$, needles from abs alcohol, dec 222°. $[\alpha]_D^{24}$ +25° (c = 7).

Aurichloride, $C_7H_{13}NO_3 \cdot HAuCl_4$, scaly clusters from water, mp 232°.

Platinichloride monohydrate, $(C_7H_{13}NO_3)_2 \cdot H_2PtCl_6 \cdot H_2O$, dec 223°.

L-*Turicine, allohydroxy-L-proline betaine*, crystals, mp 252-254°. $[\alpha]_D^{20}$ −39.0° (c = 1 in water). *Ref:* Friess *et al.*, *J. Am. Chem. Soc.* **79**, 459 (1957).

9730. Turkey-Red Oil. Sulfated castor oil; red oil. Prepd by treating castor oil with 15-30% H_2SO_4 at 25-30° for several hrs, followed by washing and neutralizing with NaOH soln. The product is an anion-active wetting agent. The commercial mfg process actually esterifies the 12-hydroxyl group in ricinoleic acid (the main constit of castor oil) thus producing a true sulfate ester, and not a sulfonate.

USE: In finishing cotton and linen; to obtain bright, clear colors in dyeing fabrics.

9731. Turks Island Salt. A salt mixture used in biological laboratories to make artificial sea water. When the natural product is not available, the following recipe may be used: sodium chloride (NaCl) 2815 g; potassium chloride (KCl) 67 g; magnesium chloride ($MgCl_2 \cdot 6H_2O$) 551 g; magnesium sulfate ($MgSO_4 \cdot 7H_2O$) 692 g. Dissolve each salt (technical grade) separately and dilute the combined solns to 95 l; then add calcium chloride ($CaCl_2$) 145 g dissolved in 5 l H_2O. *Ref:* Abderhalden, *Handbuch der Biologischen Arbeitsmethoden*, Abt. **IX**, Teil 1, 2. Hälfte, Band 2, p 1101.

9732. Turmeric. *Curcuma longa*; tumeric; saffron Indian. From rhizome of *Curcuma longa* Linn. (*C. domestica*

Valeton), *Zingiberaceae*. *Habit.* India, China, East Indies. *Constit.* Yellow coloring matter (curcumin), p,p-dihydroxy-dicinnamoylmethane, p-hydroxycinnamoylferuloylmethane, p,α-dimethylbenzyl alcohol, 1-methyl-4-acetyl-1-cyclohexene, turmerone, α-phellandrene, sabinene, zingiberene, cineol, borneol, caprylic acid. Isoln of curcumin from turmeric: Janaki, Bose, *J. Indian Chem. Soc.* **44**, 985 (1967).

Turmeric has an aromatic pepper-like but somewhat bitter taste and gives curry dishes their characteristic yellowish color.

USE: Condiment (as curry powder), color for ointments.

9733. ar-Turmerone. *2-Methyl-6-(4-methylphenyl)-2-hepten-4-one; 2-methyl-6-p-tolyl-2-hepten-4-one.* $C_{15}H_{20}O$; mol wt 216.31. C 83.28%, H 9.32%, O 7.40%. Isoln of naturally occurring (+)-form from *Curcuma longa* Linn., *Zingiberaceae*): H. Rupe *et al.*, *Helv. Chim. Acta* **17**, 372 (1934). Structure: H. Rupe, A. Gassmann, *ibid.* **19**, 569 (1936). Abs config: V. K. Honwad, A. S. Rao, *Tetrahedron* **20**, 2921 (1964). Synthesis of racemic form: J. Colonge, J. Chambion, *C.R. Acad. Sci.* **222**, 557 (1946); R. P. Gandhi *et al.*, *Tetrahedron* **7**, 236 (1959); P. A. Grieco, R. S. Finkelhor, *J. Org. Chem.* **38**, 2909 (1973); Y. Masaki *et al.*, *Chem. Commun.* **1979**, 855; T.-L. Ho, *Syn. Commun.* **11**, 579 (1981). Total synthesis of (+)-form: A. I. Meyers, R. K. Smith, *Tetrahedron Letters* **1979**, 2749; T. Sato *et al.*, *ibid.* **1980**, 3377.

Oil. bp_{10} 159-160°. $[\alpha]_D^{20}$ +82.21° (Rupe, Gassman); also reported as $[\alpha]_D^{24}$ +55.0° (c = 4.69 in benzene) (T. Sato *et al.*). d_4^{20} 0.9634. n_D^{21} 1.5218.

9734. Turpentine. Gum thus. Oleoresin from *Pinus palustris* Mill. and from other species of *Pinus, Pinaceae*.

Yellowish, opaque, sticky masses; characteristic odor and taste. Insol in water. Sol in alc, chloroform, ether, glacial acetic acid.

Note: "Turpentine" also is used to designate "oil of turpentine".

Human Toxicity: Absorbed through skin, lungs, intestine. Irritates skin, mucous membranes. Causes skin eruption, G.I. irritation, delirium, ataxia, kidney damage, coma. Inhalation causes palpitation, dizziness, nervous disturbances, chest pain, bronchitis, nephritis. Benign skin tumors from chronic contact, *cf. Clinical Toxicology of Commercial Products*, R. E. Gosselin *et al.*, Eds. (Williams & Wilkins, Baltimore, 4th ed., 1976) section III, pp 315-317.

USE: As source of turpentine oil; also as constituent of stimulating ointments. Industrially, as an insecticide, solvent for waxes, in production of synthetic camphor; in shoe, stove, furniture, etc., polishes.

THERAP CAT: Rubefacient; counterirritant.

THERAP CAT (VET): Internally: antiseptic, carminative, expectorant. Externally: counterirritant, rubefacient.

9735. Tutin. *[1aS-(1aα,1bβ,2β,5β,6α,6aβ,7β,7aα,8S*)]-Hexahydro-1b,6-dihydroxy-6a-methyl-8-(1-methylethenyl)-spiro(2,5-methano-7H-oxireno[3,4]cyclopent[1,2-d]oxepin-7,2'-oxiran)-3(2H)-one.* $C_{15}H_{18}O_6$; mol wt 294.29. C 61.21%, H 6.17%, O 32.62%. Poisonous constituent of *Coriaria ruscifolia* L. or *C. japonica* A. Gray, (tutu) *Coriariaceae*. Naturally occurring form is (+)-tutin. Isoln: Easterfield, Aston, *J. Chem. Soc.* **79**, 125 (1901). Structure: Johns, Markham, *ibid.* **1961**, 3006; Craven, *Nature* **197**, 1193 (1963); Mackay, Mathieson, *Tetrahedron Letters* **1963**, 1399. Absolute configuration: Okuda, Yoshida, *ibid.* **1965**, 2137. Biosynthesis: A. Corbella *et al.*, *Chem. Commun.* **1969**, 634. NMR analysis: J. W. Blunt *et al.*, *Aust. J. Chem.* **32**, 1339 (1979). Stereospecific synthesis: K. Watamatsu *et al.*, *Tetrahedron* **42**, 5551 (1986). Determn in toxic honey: J. L. Love *et al.*, *N.Z. J. Technol.* **2**, 179 (1986). Effect on central nervous system: D. R. Curtis *et al.*, *Brain Res.* **63**, 419

(1973); A. Nistri *et al., Lancet* **1**, 996 (1974). Toxicity data: C. H. Jarboe *et al., J. Med. Chem.* **11**, 729 (1968).

White, odorless crystals. mp 209-212°; also reported as mp 204-205° (Wakamatsu). $[\alpha]_D^{20}$ +9.25° (alc); $[\alpha]_D^{17}$ +13.9° (c = 0.75, methanol). Sol in water, alcohol, ether. LD_{50} i.p. in female mice: 3.0 mg/kg (Jarboe).

Diacetate, $C_{19}H_{22}O_8$, crystals from ethanol-water, mp 197-201°.

Caution: Poisoning from ingestion of *C. ruscifolia* has been reported and is due to presence of tutin. Symptoms include giddiness, stupor, coma, delirium, convulsions.

9736. Twistane. *Tricyclo[4.4.0.0^{3,8}]decane.* $C_{10}H_{16}$; mol wt 136.23. C 88.16%, H 11.84%. Synthesis: Whitlock, *J. Am. Chem. Soc.* **84**, 3412 (1962); Gautier, Deslongchamps, *Can. J. Chem.* **45**, 297 (1967); Whitlock, Siefken, *J. Am. Chem. Soc.* **90**, 4929 (1968). Absolute configuration: Keiichi *et al., Tetrahedron Letters* **1968**, 5467.

Crystals, mp 163-164.8°.

9737. Tybamate. *Butylcarbamic acid 2-[[(aminocarbonyl)oxy]methyl]-2-methylpentyl ester; carbamate of 2-(hydroxymethyl)-2-methylpentyl ester of butylcarbamic acid; N-butyl-2-methyl-2-propyl-1,3-propanediol dicarbamate; 2-methyl-2-propyltrimethylene butylcarbamate carbamate;* Nospan; Solacen; Solacin; Tybatran. $C_{13}H_{26}N_2O_4$; mol wt 274.35. C 56.91%, H 9.55%, N 10.21%, O 23.33%. Prepd from 2-methyl-2-propyl-3-hydroxypropyl carbamate + butyl isocyanate: Berger, Ludwig, U.S. pat. **2,937,119** (1960 to Carter Prod.). Comprehensive description: P. Reisberg *et al.,* in *Analytical Profiles of Drug Substances* vol. 4, K. Florey, Ed. (Academic Press, New York, 1975) pp 494-515.

Crystals from 1,1,2-trichloroethane + hexane (1:2), mp 49-51°, bp$_{0.06}$ 150-152°.
THERAP CAT: Anxiolytic.

9738. Tylocrebrine. *9,11,12,13,13a,14-Hexahydro-2,3,-5,6-tetramethoxydibenzo[f,h]pyrrolo[1,2-b]isoquinoline; 2,3,-5,6-tetramethoxyphenanthro[9,10:6',7']indolizidine.* $C_{24}H_{27}NO_4$; mol wt 393.46. C 73.26%, H 6.92%, N 3.56%, O 16.27%. Phenanthroindolizidine alkaloid; isomeric with tylophorine *(q.v.).* Isoln of *l*-isomer from *Tylophora crebriflora* S. T. Blake, *Asclepiadaceae:* E. Gellert *et al., J. Chem. Soc.* **1962**, 1008; K. V. Rao *et al., J. Pharm. Sci.* **59**, 1501 (1970). Antileukemic activity: E. Gellert, R. Rudzats, *J. Med. Chem.* **7**, 361 (1964). *d*-Isomer isolated from *Ficus septica, Moraceae.* Synthesis of the *dl*-form and structure: E. Gellert *et al., loc. cit.;* B. Chauncy, E. Gellert, *Austr. J. Chem.* **23**, 2503 (1970). Inhibits protein synthesis: G. R. Donaldson *et al., Biochem. Biophys. Res. Commun.* **31**, 104 (1968); E. Battaner, D. Vasquez, *Biochim. Biophys. Acta* **254**, 316 (1971). Mode of action studies: M. T. Huang, A. P.

Grollman, *Mol. Pharmacol.* **8**, 538 (1972). *Review:* R. S. Gupta, *Antibiotics (N.Y.)* **6**, 47 (1983).

dl-Form, needles from chloroform + methanol, mp 219-221°.

l-Form, crystals from methanol, dec 218-220°. uv max: 263, 342, 360 nm (log ε 4.81, 3.25, 3.09). $[\alpha]_D^{24}$ −45° (c = 0.74 in chloroform). pKa (50% aq ethanol): 6.7.

d-Form, crystals from methanol, mp 220-222°. $[\alpha]_D^{22}$ +20.5°.

9739. Tylophorine. *(S)-9,11,12,13,13a,14-Hexahydro-2,3,6,7-tetramethoxydibenzo[f,h]pyrrolo[1,2-b]isoquinoline; 2,3,6,7-tetramethoxyphenanthro[9,10:6',7']indolizidine.* $C_{24}H_{27}NO_4$; mol wt 393.46. C 73.26%, H 6.92%, N 3.56%, O 16.27%. Major alkaloid from *Tylophora asthmatica* Wight et Arn., *Asclepiadaceae.* Also found in other *Asclepiadaceae, Moraceae, Urticaceae* and *Lauraceae.* Naturally occurring form originally isolated and reported as levorotatory: A. N. Ratnagiriswaran, K. Venkatachalam, *Indian J. Med. Res.* **22**, 433 (1935); R. N. Chopra *et al., Arch. Pharm.* **275**, 236 (1937); T. R. Govindachari *et al., J. Chem. Soc.* **1954**, 2801. Structure: *eidem, Tetrahedron* **9**, 53 (1960). Absolute configuration: *eidem, J. Chem. Soc. Perkin Trans. I* **1974**, 1161. Synthesis of the *dl*-form: *eidem, Chem. & Ind. (London)* **1960**, 664; *eidem, Tetrahedron* **14**, 284 (1961); S. M. Weinreb *et al., J. Am. Chem. Soc.* **101**, 5073 (1979); V. K. Mangla, D. S. Bhakuni, *Tetrahedron* **36**, 2489 (1980); N. A. Khatri *et al., J. Am. Chem. Soc.* **103**, 6387 (1981). Synthesis and verification of dextrorotation of naturally occurring form: T. F. Buckley 3rd, H. Rapoport, *J. Org. Chem.* **48**, 4222 (1983); J. E. Nordlander, F. G. Njoroge, *ibid.* **52**, 1627 (1987). Stereoselective synthesis of *d*-form: M. Ihara *et al., Tetrahedron Letters* **29**, 4135 (1988). Biosynthesis: Mulchandani *et al., Phytochemistry* **8**, 1931 (1969); *ibid* **10**, 1047 (1971); D. S. Bhakuni, V. K. Mangla, *Tetrahedron* **37**, 401 (1981). Pharmacology: C. Gopalakrishnan *et al., Indian J. Med. Res.* **69**, 513 (1979); **71**, 940 (1980).

l-Form, colorless crystals, dec 286-287°. $[\alpha]_D^{27}$ −11.6° (c = 1.07 in chloroform). uv max (255, 290, 340, 352 nm) (log ε 4.74, 4.49, 3.30, 2.93). Sol in chloroform; slightly sol in abs alc, ether, cold benzene. Practically insol in water.

d-Form, crystals, dec 282-284°. $[\alpha]_D^{23}$ +15° (c = 0.7 in chloroform); $[\alpha]_D^{21}$ +73° (c = 0.7 in chloroform). uv max in ethanol: 257, 286, 339, 356 nm (log ε 4.7, 4.42, 3.28, 3.19). Unstable in solutions, decomposition with yellowing sets in promptly accompanied by decreasing rotatory strength.

dl-Form, crystals from chloroform + ethanol, mp 292°.

9740. Tylosin. Tylan; Tylon. $C_{46}H_{77}NO_{17}$; mol wt 916.14. C 60.30%, H 8.47%, N 1.53%, O 29.69%. Macrolide antibiotic isolated from a strain of *Streptomycetes fradiae* found in soil from Thailand: Hamill *et al., Antibiot. & Chemother.* **11**, 328 (1961); *eidem,* U.S. pat. **3,178,341** (1965 to

Lilly). Prodn in batch and chemostat cultures: P. P. Gray, S. Bhuwapathanapun, *Biotech. Bioeng.* **22**, 1785 (1980). Partial structure: Morin, Gorman, *Tetrahedron Letters* **1964**, 2339. Structure: Morin *et al., ibid.* **1970**, 4737; Achenbach *et al., Ber.* **108**, 2481 (1975). Configurational study: S. Omura *et al., J. Antibiot.* **33**, 915 (1980); N. D. Jones *et al., ibid.* **35**, 420 (1982). Synthesis of *tylonolide*, the aglycone: S. Masamune *et al., J. Am. Chem. Soc.* **98**, 7874 (1976); K. Tatsuta *et al., Tetrahedron Letters* **22**, 3997 (1981). Relationship of ribosomal binding and antibacterial properties: J. W. Corcoran *et al., J. Antibiot.* **30**, 1012 (1977). Biosynthesis studies: E. T. Seno *et al., Antimicrob. Ag. Chemother.* **11**, 455 (1977); S. Omura *et al., J. Antibiot.* **31**, 254 (1978).

Crystals from water, mp 128-132°. $[\alpha]_D^{25}$ −46° (c = 2 in methanol). uv max: 282 nm (E_{1cm}^1 245). Soly in water at 25°: 5 mg/ml. Sol in lower alcohols, esters and ketones, in chlorinated hydrocarbons, benzene, ether. Solns are stable at pH 4-9; at pH < 4 another active compd, *desmycosin* is formed.

Hydrochloride, $C_{46}H_{77}NO_{17}\cdot HCl$, crystals from ethanol + ether, mp 141-145°.

THERAP CAT (VET): Antibacterial.

9741. Tyloxapol. *4-(1,1,3,3-Tetramethylbutyl)phenol polymer with formaldehyde and oxirane;* oxyethylated tertiary octylphenol formaldehyde polymer; oxyethylated tertiary octylphenol-polymethylene polymer; *p-isooctylpolyoxyeth-ylenephenol formaldehyde polymer;* tyloxypal; Alevaire; Superinone; Triton A-20; Triton WR-1339. Nonionic detergent with surface-tension-reducing properties. Prepn: Bock, Rainey, U.S. pat. **2,454,541** (1948 to Rohm & Haas); Cornforth *et al., Nature* **168**, 150 (1951).

Freely sol in water; sol in benzene, toluene, chloroform, carbon tetrachloride, carbon disulfide, acetic acid. Alkaline pH. Oxidized by metals. Aq solns grow molds in contact with air.

THERAP CAT: Mucolytic.

9742. Tymazoline. *4,5-Dihydro-2-[[5-methyl-2-(1-methylethyl)phenoxy]methyl]-1H-imidazole; 2-[(thymyloxy)methyl]-2-imidazoline;* 2-(thymyloxymethyl)glyoxalidine; 2-[(p-mentha-1,3,5-trien-2-yloxy)methyl]-2-imidazoline; Pernazene. $C_{14}H_{20}N_2O$; mol wt 232.32. C 72.38%, H 8.68%, N 12.06%, O 6.89%. Prepd from an alkyl thymyloxyacet-imidate and ethylenediamine: Sonn, U.S. pat. **2,149,473** (1939 to Ciba); Djerassi, Scholz, *J. Am. Chem. Soc.* **69**, 1688 (1947). Pharmacology: Pham-Huu-Chanh *et al., Therapie* **24**, 797 (1969).

Hydrochloride, $C_{14}H_{21}ClN_2O$, crystals, mp 215-217° (Sonn); from water or methyl ethyl ketone + ethanol, mp 223.5-225° (Djerassi, Scholz).

THERAP CAT: Nasal decongestant.

9743. Tyramine. *4-(2-Aminoethyl)phenol;* 4-hydroxy-phenethylamine; tyrosamine; 2-p-hydroxyphenylethylamine; *p-β-aminoethylphenol;* α-(4-hydroxyphenyl)-β-aminoeth-ane. $C_8H_{11}NO$; mol wt 137.18. C 70.04%, H 8.08%, N 10.21%, O 11.66%. Decarboxylation product of tyrosine. Found in mistletoes, putrefied animal tissue, ripe cheese, ergot. Synthesis: Barger, *J. Chem. Soc.* **95**, 1127 (1909); Waser, *Helv. Chim. Acta* **8**, 766 (1925); Buck, *J. Am. Chem. Soc.* **55**, 3389 (1933). Crystal and molecular structure: A. Podder *et al., Acta Crystallogr.* **B35**, 649 (1979).

Crystals from benzene or alcohol, mp 164-165°. bp_{25} 205-207°; bp_2 166°. Alkaline reaction. One gram dissolves in 95 ml water at 15°; in 10 ml boiling alcohol. Sparingly sol in benzene, xylene.

Hydrochloride, $C_8H_{12}ClNO$, **Mydrial, Uteramin.** Crystals from alcohol + ether, mp 269°. Sol in water with neutral reaction.

THERAP CAT: Adrenergic.

9744. Tyrocidine. Peptide antibiotic mixture produced by *Bacillus brevis:* major constituent of tyrothricin, *q.v.* Isoln: Hotchkiss *et al., J. Biol. Chem.* **141**, 155, 163 (1941); Moses, U.S. pat. **3,265,572** (1966 to Penick). Separated into the three components, tyrocidines A, B, and C. Review of chemistry and biosynthesis: E. Katz, A. L. Demain, *Bacteriol. Rev.* **41**, 449-474 (1977). Regulatory role in bacterial sporogenesis: H. Ristow *et al., Nature* **280**, 165 (1979); eidem, *Eur. J. Biochem.* **129**, 395 (1982); W. Pschorn *et al., ibid.* 403. Structure-activity relationship: W. Danders *et al., Antimicrob. Ag. Chemother.* **22**, 785 (1982).

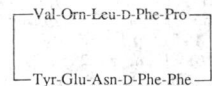

Tyrocidine A

Hydrochloride mixture, **Brevicidin, Rapicidin,** rods or needles from methanol, dec 240°. $[\alpha]_D^{20}$ −101° (c = 1.2 in 95% alc). Soluble in 95% alc, acetic acid, pyridine; slightly sol in water, acetone, abs alcohol. Practically insol in ether, chloroform, hydrocarbons.

Tyrocidine A, $C_{66}H_{87}N_{13}O_{13}$. Separation from the tyrocidine mixture: Battersby, Craig, *J. Am. Chem. Soc.* **74**, 4019, 4023 (1952). Structure: Paladine, Craig, *ibid.* **76**, 584 (1954). Synthesis: Ohno *et al., Bull. Chem. Soc. Japan* **39**, 1738 (1966); K. Okamoto *et al., ibid.* **50**, 231 (1977).

Tyrocidine A hydrochloride, crystals from methanol + water, mp 240-242°. $[\alpha]_D^{25}$ −111° (c = 1.37 in 50% alc). Freely sol in aq methanol or alc; slightly sol in methanol, ethanol. Practically insol in chloroform, acetone, ether.

Tyrocidine B, $C_{68}H_{88}N_{14}O_{13}$. Purification and amino acid sequence determination: King, Craig, *J. Am. Chem. Soc.* **77**, 6624, 6627 (1955). Synthesis: Kuromizu, Izumiya, *Experientia* **26**, 587 (1970). Possesses the same structure as tyrocidine A except that L-tryptophan replaces the L-phenylalanine. Crystals from methanol + isopropyl ether.

Tyrocidine B hydrochloride pentahydrate, crystals, mp 236-237°. $[\alpha]_D$ −93.0° (c = 0.5 in methanol).

Tyrocidine C, $C_{70}H_{89}N_{15}O_{13}$. Separation from the tyrocidine mixture: Ruttenberg *et al., Biochemistry* **4**, 11 (1965). Possesses same structure as tyrocidine B except that D-tryptophan replaces D-phenylalanine attached to L-asparagine. Synthesis: Kuromizu, Izumiya, *Tetrahedron Letters* **1970**, 1471.

THERAP CAT: Antibacterial.

9745. Tyropanoate Sodium. α-*Ethyl-2,4,6-triiodo-3-[(1-oxobutyl)amino]benzenepropanoic acid monosodium salt; 3-butyramido-α-ethyl-2,4,6-triiodohydrocinnamic acid sodium salt;* α-ethyl-β-(2,4,6-triiodo-3-butyramidophenyl)propionic acid sodium salt; sodium tyropanoate; Win 8851-2; Bilopac; Bilopaque; Lumopaque; Tyropaque. $C_{15}H_{17}I_3$-$NNaO_3$; mol wt 663.04. C 27.17%, H 2.59%, I 57.42%, N 2.11%, Na 3.47%, O 7.24%. Prepn: S. Archer, J. O. Hoppe, U.S. pat. **2,895,988** (1959 to Sterling Drug); and toxicity: J. O. Hoppe *et al., J. Med. Chem.* **13**, 997 (1970).

Colorless solid from ethyl ether, mp 208-210°. Soluble in water. LD_{50} i.v. in mice: 720 mg/kg (Hoppe).

Free acid, $C_{15}H_{18}I_3NO_3$, crystals from ethyl oxyacetate, mp 182-184°.

THERAP CAT: Diagnostic aid (radiopaque medium—cholecystographic).

9746. Tyrosinase. A copper-containing enzyme widely distributed in plants, animals, and man. Catalyzes the hydroxylation of tyrosine in the liver and in melanin-forming cells to 3,4-dihydroxyphenylalanine (dopa). Causes the cut surface of many fruits and plants to darken. The enzyme, as isolated from the common edible mushroom or potato, is characterized by its ability to catalyze aerobic oxidation of both monohydric and *o*-dihydric phenols. These activities are commonly referred to as *cresolase* (*monophenolase, monophenoloxidase*) and *catecholase* (*o-dihydric phenolase*) activities. Since monophenolase activity is frequently lost during purification of the enzyme, the name *polyphenolase* (*polyphenoloxidase*) is preferred by some workers for prepns which have mainly *o*-dihydric phenolase activity. *Ref:* C. R. Dawson, W. B. Tarpley "Copper Oxidases" in J. B. Sumner, K. Myrbäck, *The Enzymes* vol. **II**, (Academic Press, New York, 1951) pp 456-483. Prepn from potato peels: Kubowitz, *Biochem.* **Z. 299**, 32 (1938). Extraction from mushrooms: Cohen, Lerner, U.S. pat. **2,956,929** (1960 to Gillette). Separation of α-, β-, γ-, and δ-tyrosinases of mushroom tyrosinase: Bouchilloux *et al., J. Biol. Chem.* **238**, 1699 (1963). Isoln and properties of crystalline tyrosinase from *Neurospora:* Fling *et al., ibid.*, p 2045. Isoln and properties of β-tyrosinase: Kumagai *et al., J. Biol. Chem.* **245**, 1767 (1970).

THERAP CAT: Antihypertensive.

9747. Tyrosine. Tyr (IUPAC abbrev.); β-(*p*-hydroxyphenyl)alanine; α-amino-*p*-hydroxyhydrocinnamic acid. $C_9H_{11}NO_3$; mol wt 181.19. C 59.66%, H 6.12%, N 7.73%, O 26.49%. A widely distributed amino acid, classified as nonessential in respect to the growth effect in rats. Isoln from silk waste: Abderhalen, Teruuchi, *Z. Physiol. Chem.* **48**, 528 (1906), *see also* Morrow in *Biochemical Lab. Methods* (New York, 1935); Stein, Moore, *Biochem. Prepn.* vol. **I** (New York, 1949) p 11. From casein: Marshall, *J. Biol. Chem.* **15**, 85 (1913); Cox, King, *Org. Syn.* **coll. vol. II**, 612 (1943). From corn: U.S. pat. **2,178,210** (1940).

L-tyrosine

L-Tyrosine, natural product. Fine silky needles, dec 342-344° (closed capillary, bath preheated to 280°, rapid heating). d 1.456. pK_1' 2.20; pK_2' 9.11; pK_3' 10.07. $[\alpha]_D^{22}$ −10.6° (c = 4 in *N* HCl); $[\alpha]_D^{18}$ −13.2° (c = 4 in 3*N* NaOH). R_f

value 0.62. Soly in water (g/l): 0.196 at 0°; 0.453 at 25°; 1.052 at 50°; 2.438 at 75°; 5.65 at 100°. Insol in abs alcohol, ether, acetone. Sol in alkaline solns.

DL-Tyrosine, synthetic product. Stout needles. Dec 316°. Soly in water (g/l): 0.147 at 0°; 0.351 at 25°; 0.836 at 50°. D-Tyrosine. Prepn: *Biochem. Prepn.* **vol. I** (New York, 1949) p 71. Crystals. Decomp 310-314°. $[\alpha]_D^{25}$ +10.3° (c = 4 in *N* HCl). Soly in water (g/l): 0.196 at 0°; 0.453 at 25°; 1.052 at 50°.

9748. m-Tyrosine. 3-*Hydroxy*-L-*phenylalanine;* α-amino-3-hydroxyhydrocinnamic acid; metatyrosine. C_9H_{11}-NO_3; mol wt 181.19. C 59.66%, H 6.12%, N 7.73%, O 26.49%. A possible precursor of catecholamines: Sourkes *et al., Nature* **189**, 576 (1961). An intermediate in an alternate pathway for the biosynthesis of catecholamines, where with the existing hydroxylating enzymes *m*-hydroxylation of phenylalanine to *m*-tyrosine occurs before *p*-hydroxylation (forming dopa) and is followed by subsequent decarboxylation to dopamine. Formation *in vitro* of dopa from L-*m*-tyrosine: Tong *et al., Biochem. Biophys. Res. Commun.* **43**, 819 (1971); *in vivo:* Hollunger, Persson, *Acta Pharmacol. Toxicol.* **34**, 391 (1974). Biosynthesis and metabolism studies: D'Iorio *et al., Advan. Neurol.* **5**, 265 (1974). Has also been isolated from a plant source, *Euphorbia myrsinites* L. *Euphorbiaceae:* Mothes *et al., Z. Naturforsch.* **19b**, 1161 (1964). *m*-Tyrosine has the ability to cross the blood-brain barrier and is decarboxylated to *m*-tyramine which stimulates dopamine receptors, presumably accounting for the demonstrated pharmacological effects of *m*-tyrosine. Pharmacological studies: Carlsson, Lindqvist, *Eur. J. Pharmacol.* **2**, 187 (1967); Rubenson, *J. Pharm. Pharmacol* **23**, 228, 412 (1971); Sandler *et al., Nature* **229**, 414 (1971); Ungerstedt *et al., Eur. J. Pharmacol.* **21**, 230 (1973). Crystal and molecular structure: Byrkjedal *et al., Acta Chem. Scand.* **28B**, 750 (1974).

mp 267-270° (dec). $[\alpha]_D^{22}$ −14.5° (70% ethanol); $[\alpha]_D^{22}$ +8.9° (70% ethanol, 2*N* HCl).

9749. Tyrothricin. "Dubos crude crystals"; Coltirot; Martricin; Hydrotricine; Dermotricine; Tyroderm; Solutricine; Tyri 10. Polypeptide antibiotic mixture extracted from cultures of soil bacilli belonging to the *Tyrothrix* group of bacteria. Usually obtained by extracting acidified cultures of *Bacillus brevis* with alc and precipitating with NaCl soln. Surface, submerged, and aerated cultures are used. Tyrothricin contains from 10 to 20% gramicidin and from 40 to 60% tyrocidine. Solubility data: Weiss *et al., Antibiot. & Chemother.* **7**, 374 (1957). Function in the sporulation process: Sarkar, Paulus, *Nature New Biol.* **239**, 228 (1972). Toxicity data: H. Sous *et al., Arzneimittel-Forsch.* **7**, 98 (1957).

Gray to brown powder, decomp 215-220°. Has 2-3% absorbed or combined moisture. Stable to light, air and temps up to 50°. Practically insol in water. Sol in alcohol (about 28 mg/ml), in methanol and in propylene glycol. One part of a 2% alcoholic soln contg 10% formaldehyde soln U.S.P. dissolves in 49 parts of water without cloudiness. Tyrothricin should not be dissolved in ether or acetone as only the gramicidin component is soluble. Soly in mg/ml at about 28°: water 2.1; isopropanol 5.6; benzene 0.30; isooctane 0.042; carbon tetrachloride 0.455; ethyl acetate 2.65; acetone 6.8; ether 3.25; dioxane 11.1; chloroform 1.6. Aqueous suspensions may be attacked by microorganisms. Solutions and emulsions should not be sterilized by heat. Incompatible with alkalies and strong acids. Tyrothricin gives a positive biuret and Hopkins-Cole test. LD_{50} in mice (mg/kg): > 1500 s.c.; 100 i.p.; > 3000 orally (Sous).

THERAP CAT: Topical antibacterial.

THERAP CAT (VET): Topical antibacterial.

U

9750. Ubenimex. *[S-(R*,S*)]-N-(3-Amino-2-hydroxy-1-oxo-4-phenylbutyl)-L-leucine;* [(2S,3R)-3-amino-2-hydroxy-4-phenylbutanoyl]-L-leucine; NK-421; Bestatin. C_{16}-$H_{24}N_2O_4$; mol wt 308.38. C 62.32%, H 7.84%, O 20.75%. Dipeptide antitumor antibiotic produced by *Streptomyces olivoreticuli.* Competitive inhibitor of aminopeptidase B and leucine aminopeptidase. Prepn from fermentation broth: H. Umezawa *et al.,* **Ger. pat. 2,528,984;** *eidem,* U.S. pats. **4,029,547; 4,052,449** (1976, 1977, 1977 all to Microbiochem. Res. Found., Japan); *eidem, J. Antibiot.* **29,** 97 (1976). Structure: H. Suda *et al., ibid.* 100. Syntheses: *eidem, ibid.* 600; R. Nishizawa *et al., ibid.* **36,** 695 (1983). Synthesis of stereoisomers and structure-activity study: *eidem, J. Med. Chem.* **20,** 510 (1977). Stereocontrolled synthesis: S. Kobayashi *et al., Tetrahedron Letters* **25,** 5079 (1984). Crystal structure: J. S. Ricci *et al., J. Org. Chem.* **47,** 3063 (1982). Cell surface binding studies: H. Umezawa *et al., J. Antibiot.* **29,** 857 (1976); W. E. Müller *et al., Int. J. Immunopharmacol.* **4,** 393 (1982). Mitogenic effect on lymphocytes *in vitro:* M. Ishizuka *et al., J. Antibiot.* **33,** 653 (1980). Immunopotentiating and antitumor activities in mice: M. Ishizuka *et al., ibid.* 642. Acute toxicity: T. Sakakibara *et al., Japan. J. Antibiot.* **36,** 2971 (1983), *C.A.* **100,** 132296u (1984). Clinical evaluation in bladder cancer: H. Blomgren *et al., Biomed. Pharmacother.* **38,** 143 (1984); in acute nonlymphocytic leukemia: K. Ota *et al., Cancer Immunol. Immunother.* **23,** 5 (1986). Review of bioactivity and clinical studies: H. Umezawa, *Drugs Exptl. Clin. Res.* **10,** 519-531 (1984).

Colorless needles, mp 233-236°. $[\alpha]_D^{20}$ −15.5° (c = 1.0 in 1N HCl). pKa 8.1, 3.1. uv max: 241.5, 248, 253, 258, 264.5, 268 nm ($E_{1cm}^{1\%}$ 3.8, 4.0, 5.0, 6.0, 4.6, 2.7). Sol in acetic acid, DMSO, methanol. Less sol in water. Insol in ethyl acetate, benzene, hexane, chloroform. LD_{50} (14 day) in mice, rats (g/kg): 1.3-1.9, 1.9-2.1 s.c.; 0.19, 0.78-0.90 i.p.; > 4.0, > 2.0 orally (Sakakibara).

THERAP CAT: Immunomodulator; antineoplastic.

9751. Ubiquinones. Coenzymes Q; mitoquinones; SA; Q-275; 272-substance. A group of lipid-soluble benzoquinones involved in electron transport in mitochondrial preparations, i.e., in the oxidation of succinate or reduced nicotine adenine dinucleotide (NADH) via the cytochrome system. Occurs in the majority of aerobic organisms, from bacteria to higher plants and animals. Ubiquinone structures, analogous to the vitamins K_2, *q.v.,* are based on the 2,3-dimethoxy-5-methylbenzoquinone nucleus with a variable terpenoid side chain contg one to twelve mono-unsaturated *trans*-isoprenoid units with 10 units being the most common in animals. According to the existing dual system of nomenclature the compds can be described as: coenzyme Q_n in which $n = 1$-12, or ubiquinone(x) in which x designates the total number of carbon atoms in the side chain and can be any multiple of 5. Differences in properties are due to the difference in length of the side chain. Naturally occurring members are the coenzymes Q_6-Q_{10}. The entire series has been prepd synthetically. Recent syntheses of ubiquinone (50): S. Terao *et al., J. Org. Chem.* **44,** 868 (1979); Y. Naruta, *ibid.* **45,** 4097 (1980); K. Sato *et al., Chem. Commun.* **1982,** 153. Reviews: *Ciba Foundation Symposium on Quinones in Electron Transport,* Wolstenholme, O'Connor, Eds. (Churchill, London, 1961) 453 pp; Wagner, Folkers, *Vitamins and Coenzymes* (Interscience, New York, 1964) pp 435-468; Crane in *Progr. Chem. Fats Lipids,* vol. 7(2), Holman, Ed. (Pergamon Press, 1964) pp 267-289; *Methods Enzymol.* **18C,** 135-237 (1971); R. A. Morton in *The Vitamins* vol. V, W. H. Sebrell, R. S. Harris, Eds. (Academic Press, New York, 2nd ed., 1972) pp 355-391. Series of

books: *Biomedical and Clinical Aspects of Coenzyme Q* vols. 1-3, K. Folkers, Y. Yamamura, Eds. (Elsevier, New York, 1977, 1980, 1981).

Ubiquinone (50), $C_{59}H_{90}O_4$, *coenzyme Q_{10}, coenzyme Q-199, ubidecarenone,* NSC 140865, *Adelir, Caomet, Decorenone, Dymion, Heartcin, Inokiton, Iuvacor, Mitocor, Neuquinone, Quasar, Taidecanone, Ubiquasar, Ubiten, Udekinon.*
Ubichromenol (50), cyclic isomer of ubiquinone (50).
THERAP CAT: Cardiovascular agent.

9752. Ubiquitin. ATP-dependent proteolytic factor; APF-1; Ub; ubiquitous immunopoietic polypeptide; UBIP. Ubiquitous polypeptide, mol wt 8500 daltons, so named due to its presence in all eukaryotes including plants; widely distributed in the organism. Highly conserved sequence of 76 amino acids is identical in a wide variety of sources including humans, fish, and insects. Participates in diverse cellular functions, such as protein degradation, chromatin structure, and heat shock, by conjugation to other proteins through its carboxyl terminus. Isoln: G. Goldstein *et al., Proc. Nat. Acad. Sci. USA* **72,** 11 (1975); as APF-1 and recognition of role as marker in protein degradation: A. Ciechanover *et al., Biochem. Biophys. Res. Commun.* **81,** 1100 (1978); A. Ciechanover *et al., Proc. Nat. Acad. Sci. USA* **77,** 1365 (1980); A. Hershko *et al., ibid.* 1783. Amino acid sequence: D. H. Schlesinger *et al., Biochemistry* **14,** 2214 (1975). Identity of APF-1 with ubiquitin and revised amino acid sequence: A. Ciechanover *et al., J. Biol. Chem.* **255,** 7525 (1980); K. D. Wilkinson *et al., ibid.* 7529; K. D. Wilkinson, T. K. Audhya, *ibid.* **256,** 9235 (1981). Structural studies: W. J. Cook *et al., J. Mol. Biol.* **130,** 353 (1979); S. Vijay-Kumar *et al., J. Biol. Chem.* **262,** 6396 (1987); P. L. Weber *et al., Biochemistry* **26,** 7282 (1987). Role of chromatin structure: D. C. Watson *et al., Nature* **276,** 196 (1978); R. D. Mueller *et al., J. Biol. Chem.* **260,** 5147 (1985); in receptor structure: M. Siegelman *et al., Science* **231,** 823 (1986). Identity as heat shock protein: U. Bond, M. J. Schlesinger, *Mol. Cell. Biol.* **5,** 949 (1985). Role as marker for protein degradation in polymeric form: A. Hershko, H. Heller, *Biochem. Biophys. Res. Commun.* **128,** 1079 (1985); V. Chau *et al., Science* **243,** 1576 (1989). Enzymic determn: I. A. Rose, J. V. B. Warms, *Proc. Nat. Acad. Sci. USA* **84,** 1477 (1987). Brief review of functions and mechanisms: D. Finley, A. Varshavsky, *Trends Biochem. Sci.* **10,** 343 (1985). Review of protein degradation: M. Rechsteiner, *Ann. Rev. Cell. Biol.* **3,** 1-30 (1987); A. Hershko, *J. Biol. Chem.* **263,** 15327-15240 (1988). *Review:* K. D. Wilkinson, *Anti-Cancer Drug Des.* **2,** 211-229 (1987). Book: *Ubiquitin,* M. Rechsteiner, Ed. (Plenum Press, New York, 1988).

9753. Uglow Black Silver. A sand coated with silver and manganese. Used in the sterilization of large amounts of water: Uglow, Gan, *Z. Hyg. Infektionskrankh.* **117,** 488 (1935); Granata, *Ann. Igiene* **56,** 284 (1946); *Chimie & Industrie* **58,** 148 (1947).

9754. Uintahite. Gilsonite; uintaite. A variety of asphalt from the Uinta (or Uintah) valley, near Fort Duchesne, Utah: E. S. Dana, *A System of Mineralogy* (John Wiley, New York, 6th ed., 1901) p 1020.

Black, lustrous masses; streak and powder a rich brown. Nonconductor of electricity. Burns easily with a brilliant flame. Sol in heavy lubricating petroleum, warm oil of turpentine, and alc.

USE: Potential raw material for gasoline and coke. In the making of asphalt floor tile, storage battery boxes, inks, paints, and insulation for hot underground piping.

9755. Ujothion. *5-Benzyldihydro-6-thioxo-2H-1,3,5-thiadiazine-3(4H)-acetic acid.* $C_{12}H_{14}N_2O_2S_2$; mol wt 282.39. C 51.04%, H 5.00%, N 9.92%, O 11.33%, S 22.71%.

Prepn: Rieche *et al.*, *Arch. Pharm.* **296**, 770 (1963); Schade *et al.*, Ger. (East) pat. **31,793** (1964), *C.A.* **63**, 16371a (1965).

Crystals, mp 152°.
THERAP CAT: Antifungal.

9756. Uliginosins. Antibiotics isolated from *Hypericum uliginosum* HBK, a woody herb found in Mexico and Central America. Isoln: Taylor, Brooker, *Lloydia* **32**, 217 (1969). Structure: Parker, Johnson, *J. Am. Chem. Soc.* **90**, 4716 (1968); Parker *et al.*, *ibid.* 4723. Synthesis of uliginosin A and dihydrouliginosin B: Meikle, Stevens, *Chem. Commun.* **1972**, 123; *eidem, J. Chem. Soc., Perkin Trans. I* **1978**, 1303. Synthesis of uliginosin B: *eidem, Tetrahedron Letters* **1972**, 4787; *eidem, J. Chem. Soc., Perkin Trans. I* **1979**, 2563.

uliginosin A

uliginosin B

Uliginosin A. $C_{28}H_{36}O_8$. *3'-[(2,4-Dihydroxy-5-isobutyryl-3,3-dimethyl-6-oxo-1,4-cyclohexadien-1-yl)methyl]-2',4',6'-trihydroxy-2-methyl-5'-(3-methyl-2-butenyl)propiophenone.* Pale yellow crystals, mp 160.5-161.5° from acetonitrile-chloroform (4:1). uv max (cyclohexane): 229, 293 nm (ε 31,500, 25,000).

Uliginosin B. $C_{28}H_{34}O_8$. *2-[(5,7-Dihydroxy-8-isobutyryl-2,2-dimethyl-2H-1-benzopyran-6-yl)methyl]-3,5-dihydroxy-6-isobutyryl-4,4-dimethyl-2,5-cyclohexadien-1-one.* Pale yellow crystals, mp 139.5-142.0° from nitromethane. uv max (cyclohexane): 230, 270 nm (ε 34,000, 37,000).

9757. Ultramarine. C.I. Pigment Blue 29; C.I. 77007. A blue pigment occurring naturally as the mineral lapis lazuli. Made by igniting a mixture of kaolin, Na_2CO_3 (or Na_2SO_4), S and carbon. The resulting aluminosulfosilicates resemble zeolites structurally and have the approx formula $Na_7Al_6Si_6$-$O_{24}S_3$. *See: Colour Index* vol. **4** (3rd ed., 1971) p 4653.
Blue lumps or powder. Insol in water; readily dec by acids, even carbonic acid, with liberation of H_2S.
USE: As a pigment in calico printing, wall paper, mottled soap; bluing in laundry use; for coloring tiles, cements, rubber, but is now largely replaced by coal tar dyes; as a color in food.

9758. Umbelliferone. *7-Hydroxy-2H-1-benzopyran-2-one; 7-hydroxycoumarin;* hydrangin; skimmetin. $C_9H_6O_3$; mol wt 162.14. C 66.66%, H 3.73%, O 29.60%. The aglucon of skimmin. Present in many plants. Obtained by distillation of resins from umbelliferae: Zwenger, *Ann.* **115**, 1, 15 (1860). Prepn: Bert, *Compt. Rend.* **214**, 230 (1942); Austerweil, *ibid.* **248**, 1810 (1959); Dressler, Reabe, U.S. pat. **3,503,996** (1970 to Koppers). Main product of metabolism of coumarin in man: Schilling *et al.*, *Nature* **221**, 664 (1969). Metabolism studies of umbelliferone: Indahl, Scheline, *Xenobiotica* **1**, 13 (1971). Use in brain intracellular pH measurements: T. M. Sundt, R.E. Anderson, *Brain Res.* **186**, 355 (1980); *eidem, J. Neurophysiol.* **44**, 60 (1980). Use

in fluorescent immunoassays: S. G. Thompson, J. F. Burd, *Antimicrob. Ag. Chemother.* **18**, 264 (1980); T. M. Li *et al.*, *Anal. Biochem.* **118**, 102 (1981).

Needles from water, mp 225-228°. Develops odor of coumarin on heating. Sublimes. Absorption spectrum: Sen, Bagchi, *J. Org. Chem.* **24**, 316 (1959). One gram dissolves in about 100 ml boiling water; freely sol in alcohol, chloroform, acetic acid; sol in dil alkalies; sparingly sol in ether. Solns show blue fluorescence.
USE: In sunscreen lotions and creams; as intracellular and pH sensitive fluorescent indicator and blood-brain barrier probe.

9759. Undecoylium Chloride. Acylcolaminoformylmethylpyridinium chloride. The acyl group is a mixture containing 8-14 carbon atoms, but principally a 10-11 carbon atom chain. $[C_5H_5N^+ \cdot CONHCH_2CH_2OOC(CH_2)_n$-$CH_3]Cl^-$, where $n = 6$-12.
Complex with iodine, $[C_5H_5N^+ \cdot CONHCH_2CH_2OOC$-$(CH_2)nCH_3]Cl^- \cdot I_2$, *Virac.* Freely sol in water.
THERAP CAT: Topical antiseptic.

9760. Undecylenic Acid. *10-Undecenoic acid; 10-hendecenoic acid;* 9-undecylenic acid; Declid; Renselin; Sevinon. $C_{11}H_{20}O_2$; mol wt 184.27. C 71.69%, H 10.94%, O 17.36%. $CH_2=CH(CH_2)_8CO_2H$. Occurs in sweat. Obtained by pyrolysis of ricinoleic acid. Prepn by vacuum distn of castor oil: Krafft, *Ber.* **10**, 2035 (1877); Perkins, Cruz, *J. Am. Chem. Soc.* **49**, 1073 (1927), found that distn at 400° under a pressure of 50 mm produced a distillate composed of about 40% heptaldehyde and 20% undecylenic acid. Toxicology: G. W. Newell *et al.*, *J. Invest. Dermatol.* **13**, 145 (1949).
Liquid or crystals. Odor suggestive of perspiration. d_4^{24} (vac) 0.9072; d_{25}^{25} 0.9102; d_{45}^{45} 0.8993; $d_{79.9}^{79.9}$ (vac) 0.8653. mp 24.5°; bp_{760} 275° (dec); bp_{182} 232-235°; bp_{130} 230-235°; bp_{100} 213.5°; bp_{90} 198-200°; bp_{15} 168.3°; $bp_{1.0}$ 131°. n_D^{25} 1.4486. Neutralization value 304.5; iodine value 137.8. Insol in water. Sol in alcohol, chloroform, ether. LD_{50} in mice (g/kg): 8.15 orally; 0.960 i.p. (Newell).
Zinc salt, $C_{22}H_{38}O_4Zn$, *zinc undecylenate.* Amorphous white powder, mp 115-116°. Resembles zinc stearate in appearance and physical properties. Can be prepd by dissolving zinc oxide in dil undecylenic acid and concentrating the solution.
Methyl ester, $C_{12}H_{22}O_2$, liq; d_4^{15} 0.889; bp_{760} 248°; bp_{10} 124°; mp −27.5°; n_D^{20} 1.43928; n_D^{25} 1.43727. Sol in alcohol, chloroform, ether, petr ether, oils.
THERAP CAT: Topical antifungal.
THERAP CAT (VET): Topical antifungal.

9761. Uracil. *2,4(1H,3H)-Pyrimidinedione;* 2,4-dioxopyrimidine; 2,4-pyrimidinediol; 4-hydroxy-2(1H)-pyrimidinone; 2-hydroxy-4(1H)-pyrimidinone; 2-hydroxy-4(3H)-pyrimidinone. $C_4H_4N_2O_2$; mol wt 112.09. C 42.86%, H 3.60%, N 24.99%, O 28.55%. Obtained by hydrolysis of nucleic acids, *cf.* Levene, Bass, *Nucleic Acids* (New York, 1931). Alternate routes of formation: Johnson, *J. Am. Chem. Soc.* **63**, 263 (1941); Fox, Harada, *Science* **133**, 1923 (1961); Takemoto, Yamamoto, *Synthesis* **1971**, 154. Several desmotropic forms. Crystal structure: Stewart, *Acta Cryst.* **23**, 1102 (1967). *Review:* Ts'o "Bases, Nucleosides and Nucleotides" in *Basic Principles in Nucleic Acid Chemistry* vol. **1**, P. O. P Ts'o, Ed. (Academic Press, New York, 1974) pp 453-584. *See also* Nucleic Acids.

Needles from water, mp 335° with effervescence. uv max (pH 7.0): 202.5, 259.5 nm (ε × 10^{-3} 9.2, 8.2). Freely sol in hot water; sparingly in cold water (100 parts of water at 25° dissolves 0.358 part of uracil). Almost insol in alc, ether; sol in ammonia water and in other alkalies. pK = 9.45.

USE: In biochemical research.

9762. Uracil Mustard. 5-[Bis(2-chloroethyl)amino]-2,4-(1H,3H)-pyrimidinedione; 5-[bis(2-chloroethyl)amino]uracil; 2,6-dihydroxy-5-bis[2-chloroethyl]aminopyramidine; 5-[di-(β-chloroethyl)amino]uracil; uramustine; demethyldopan; desmethyldopan; NSC-34462; U-8344. $C_8H_{11}Cl_2N_3O_2$; mol wt 252.10. C 38.11%, H 4.40%, Cl 28.13%, N 16.67%, O 12.69%. Synthesis from 5-aminouracil: Lyttle, Petering, J. Am. Chem. Soc. **80**, 6459 (1958); Lyttle, U.S. pat. **2,969,364** (1961 to Upjohn). Teratology study: S. Chaube, M. L. Murphy, "The Teratogenic Effects of the Recent Drugs Active in Cancer Chemotherapy" in Advances in Teratology **3**, D. H. M. Woollam, Ed. (Academic Press, New York, 1968) pp 181-237.

Crystals from methanol + water, dec 206°. Sparingly sol in water. uv max (0.01N H_2SO_4 in 95% ethanol): 257 nm (ε 5675). LD_{50} i.p. in rats: approx. 1.25-2.5 mg/kg (Chaube, Murphy).

THERAP CAT: Antineoplastic.

9763. Uramil. 5-Amino-2,4,6(1H,3H,5H)-pyrimidinetrione; 5-aminobarbituric acid; dialuramide; 5-amino-2,4,6-pyrimidinetriol; Murexan. $C_4H_5N_3O_3$; mol wt 143.10. C 33.57%, H 3.52%, N 29.37%, O 33.54%. Prepd by boiling alloxantin with ammonium chloride: Wöhler, Liebig, Ann. **26**, 241 (1838); by reduction of 5-nitrobarbituric acid with tin and HCl: Hartman, Sheppard, Org. Syn. coll. vol. II, 617 (1943); from barbituric acid, sodium nitrite and sodium hydrosulfite: Koppel, Robins, J. Am. Chem. Soc. **80**, 2751 (1958).

Crystals. Discolors in the air. Does not melt below 400°. Insol in cold water, in ether, chloroform; sparingly sol in hot water; sol in solns of alkali hydroxides, in ammonia and in sulfuric acid. Forms a slightly sol lead salt.

9764. Uramite®. A polymerized urea-formaldehyde prepn, contg 38% nitrogen. Prepn of a fertilizer contg 40% nitrogen from urea and formaldehyde: Darden, U.S. pat. **2,766,283** (1956 to du Pont).

Yellow, granular material of small particle size. Essentially insol in water.

USE: Nitrogen fertilizer for turfs and ornamentals.

9765. Uranediol. 17-Methyl-D-homoandrostane-3,17a-diol. $C_{21}H_{36}O_2$; mol wt 320.50. C 78.69%, H 11.32%, O 9.98%. From pregnant mares' urine: Marker et al., J. Am. Chem. Soc. **60**, 210, 1061, 1561, 2719 (1938); Klyne, Biochem. J. **43**, 611 (1948); Brooks et al., ibid. **51**, 694 (1952). Structure: Klyne, Nature **166**, 559 (1950). Formation by acid hydrolysis of 5α-pregnane-3β,20β-diol: Hirschmann, Williams, J. Biol. Chem. **238**, 2305 (1963). Configuration: Hirschmann et al., J. Org. Chem. **31**, 375 (1966).

Needles from aq ethanol, mp 216-219°. Sublimes at 180° and 0.06-0.1 mm. $[\alpha]_D^{15}$ +3.7° (c = 1.8 in chloroform). Diacetate, $C_{25}H_{40}O_4$, plates from aq methanol, mp 159.5-160.5°. Sublimes at 190° and 0.1 mm. $[\alpha]_D^{20}$ −30.4° (c = 1.4 in chloroform); $[\alpha]_D^{22}$ −29.6° (ethanol). Dibenzoate, $C_{35}H_{44}O_4$, leaflets from chloroform + methanol, mp 209-210°. $[\alpha]_D^{29}$ +18.6° (c = 1.05 in chloroform). uv max: 230, 272 nm (log ε 4.43, 3.26).

9766. Uranium. U; at. wt 238.0289; at. no. 92; valence 6, 5, 4, 3. Three naturally occurring isotopes: 238 (99.276%), 235 (0.718%), 234 (0.0056%); artificial isotopes: 227-233; 236; 237; 239; 240. Occurrence in the earth's crust 2×10^{-5}%. Discovered by Klaproth in 1789; isolated by Peligot in 1841: Peligot, cited in Mellor's vol. XII, 10 (1932). Occurs in pitchblende, which is chiefly U_3O_8, found in Colorado, Utah, Bear Lake in Canada, Zaire, Joachimstahl in Czechoslovakia, Cornwall. The uranium ore is digested with nitric acid to form uranyl nitrate, this is extracted with tributylphosphate + kerosene. The nitrate is calcined to UO_3. Uranium trioxide is reduced to UO_2 in furnaces supplied with hydrogen. The uranium dioxide is converted to UF_4 with anhyd HF. Pure uranium metal is obtained by reduction of UF_4 in a Thermit type of reaction carried out in metal bombs. Flowsheet and details: Chem. Eng. **62**, no. 10, 113 (1955); Spedding et al., U.S. pat. **2,852,-364** (1958 to U.S.A.E.C.). Reviews: C. D. Harrington, A. R. Ruehle: Uranium Production Technology (Van Nostrand, Princeton, 1959); E. H. P. Cordfunke, The Chemistry of Uranium (Elsevier, New York, 1969) 250 pp; several authors in Handb. Exp. Pharmakol. **36**, 3-306 (1973); F. Weigel in Kirk-Othmer Encyclopedia of Chemical Technology vol. **23** (Wiley-Interscience, New York, 3rd ed., 1983) pp 502-547.

Silver-white, lustrous, radioactive metal; malleable and ductile. On vigorous shaking the metallic particles exhibit luminescence. A black powder when obtained by reduction. Three allotropic modifications: orthorhombic α-form to 667.7°; tetragonal, β-form from 667.7° to 774.8°; body-centered cubic, γ-form from 774.8° to mp. mp 1132.3°. d about 19.05. $T_{1/2}$ (^{238}U) 4.51×10^9 years. Spec heat (25°): 6.65 cal/g-atom/°C; heat of fusion: 2.9 kcal/g-atom. Ignites in oxygen at about 170°; the ignition temp depends on the state of subdivision of the metal. Burns in air at 150-175° with formation of U_3O_8. When finely powdered, slowly decomposes cold water, more quickly boiling water. Burns in fluorine to produce mainly a green volatile tetrafluoride; in chlorine at 180°, in bromine at 240°; forms an iodide at 260°. Attacked by dry hydrogen chloride at a dull red heat with formation of a stable chloride; combines with sulfur at 500°, with nitrogen at 1000°. Reacts with acids with liberation of hydrogen and formation of salts of tetravalent uranium. Not attacked by alkalies.

Caution: Uranium and its salts are highly toxic. Dermatitis, renal damage, acute necrotic arterial lesions, death may occur. Radiation hazard from inhalation of fine particles of approx 1μ. Insol particles in lung may be long-term carcinogenic hazard. See L. T. Fairhall, Industrial Toxicology (Hafner, New York, 2nd ed., 1969) pp 129-131.

USE: ^{235}U is used in atom and hydrogen bombs. ^{234}U and ^{235}U are used as nuclear fuel in power reactors.

9767. Uranium Dioxide. Uranous oxide; black uranium oxide. O_2U; mol wt 270.07. O 11.85%, U 88.15%. UO_2. Occurs in nature as the minerals uraninite or pitchblende. Review of uranium oxides: Keller in Comprehensive Inorganic Chemistry vol. 5, J. C. Bailar, Jr. et al., Eds. (Pergamon Press, Oxford, 1973) pp 224-233.

Brown to black powder, or cubic crystals. d 10.97. mp 2865°. When obtained by heating the urano-uranic oxide or the oxalate in hydrogen, it is brown or copper-red in color and is pyrophoric. Insol in water, dil acids; sol in concd acids.

USE: Nuclear fuel.

9768. Uranium Hexafluoride. F_6U; mol wt 352.07. F 32.38%, U 67.62%. UF_6. Prepd by the action of fluorine on uranium metal or carbide; on uranium pentachloride; on uranium tetrafluoride; or on triuranium octaoxide in the presence of carbon. It appears that all uranium compds when heated with fluorine to a sufficiently high temp give UF_6: National Nuclear Energy Series **VIII-5**, J. J. Katz, E.

Rabinowitch, *The Chemistry of Uranium,* part **I** (New York, 1951) pp 396-449.

Volatile, white monoclinic cryst solid. $d_2^{20.7}$ 5.09. d^{70} (liq) 3.595. mp 64.8°. Sublimes at 56.5°. Critical temp 230.2°, crit pressure 45.5 atm: Oliver *et al., J. Am. Chem. Soc.* **75**, 2827 (1953). Reacts vigorously with water, forming mainly UO_2F_2 and HF. Sol in liq chlorine and bromine. Dissolves in nitrobenzene to give a dark red soln fuming in air. Sol in carbon tetrachloride, chloroform, and *sym*-tetrachloroethane, $Cl_2CHCHCl_2$, which forms the most stable soln, extensive reaction occurring only after several days at room temp. Sol in fluorocarbons (C_6F_{12} or C_7F_{16}) without reaction. Best handled in copper apparatus.

9769. Uranium Peroxide. O_4U; mol wt 302.03. O 21.19%, U 78.81%. UO_4. Exists as a di- or tetrahydrate. Prepd from $UO_2(NO_3)_2.6H_2O$: Watt *et al., J. Am. Chem. Soc.* **72**, 3341 (1950); Boggs, El-Chehabi, *ibid.* **79**, 4258 (1957); Sato, *J. Appl. Chem.* **13**, 361 (1963).

Dihydrate, hygroscopic, pale yellow crystals. Dec slowly in the temp range 90-195° to form a different peroxide U_2O_7, an orange, hygroscopic solid which dec to UO_3 and oxygen on contact with water, and to uranyl salts and oxygen on contact with acids. At temps of 200° and above, the dec of $UO_4.2H_2O$ leads to UO_3 without the formation of U_2O_7.

9770. Uranium Tetrachloride. Cl_4U; mol wt 379.90. Cl 37.33%, U 62.67%. UCl_4. Prepd from uranium(VI) oxide dihydrate and hexachloropropene: Hermann, Suttle, *Inorg. Syn.* **5**, 143 (1957).

Dark green octahedral crystals (tetragonal symmetry). Oxidizes in air and dec on contact with water. Should be stored in sealed ampuls. d_4^{25} 4.725. mp 590°, bp_{760} 791°. Heat of formation (solid) 250.9 kcal/mol at 0°. Freely sol in water (dec). Also sol in polar organic solvents. Insol in nonpolar solvents such as hydrocarbons and ethyl ether.

9771. Uranium Tetrafluoride. F_4U; mol wt 314.07. F 24.20%, U 75.80%. UF_4. Prepd from uranyl oxide and dichlorodifluoromethane: Booth *et al., J. Am. Chem. Soc.* **68**, 1969 (1946); also prepd by the action of hydrogen fluoride on uranium dioxide or by addition of hydrofluoric acid to a U(IV) soln: Eméleus in *Fluorine Chemistry,* vol. I, J. H. Simons, Ed. (Academic Press, New York, 1950) p 59.

Monoclinic green crystals, mp >1100°. Changes to U_3O_8 when heated in air. Insol in water. Sol in concd acids and alkalies with decompn.

Hemipentahydrate, orthorhombic green crystals, insol in water.

9772. Uranium Trichloride. Cl_3U; mol wt 344.44. Cl 30.88%, U 69.12%. UCl_3. Prepd by hydrogen reduction of uranium(IV) chloride: Suttle, *Inorg. Syn.* **5**, 145 (1957).

Dark purple crystals. Less hygroscopic than uranium(IV) chloride. d 5.51. mp 842°. Heat of formation at 25° 212.0 kcal/mol. Freely sol in water, giving a purple soln, which evolves hydrogen and turns green because of oxidation to uranium(IV). Less sol in polar organic solvents than uranium(IV) chloride and insol in nonpolar solvents.

9773. Uranium Trioxide. Uranic oxide; red uranium oxide. O_3U; mol wt 286.07. O 16.78%, U 83.22%. UO_3. Review of uranium oxides: Keller in *Comprehensive Inorganic Chemistry* vol. 5, J. C. Bailar, Jr. *et al.,* Eds. (Pergamon Press, New York, 1973) pp 224-233.

Red or brownish-yellow powder. d 7.29. Insol in water; sol in acids.

9774. Uranyl Acetate. $C_4H_6O_6U$; mol wt 388.15. C 12.38%, H 1.56%, O 24.73%, U 61.33%. $UO_2(C_2H_3O_2)_2$.

Dihydrate, yellow, cryst powder; slight acetic odor. d 2.89. Sol in 10 parts of water, usually incompletely, due to the presence of basic salt. Freely sol in water acidulated with acetic acid, slightly in alcohol.

USE: As a reagent for pptn of sodium; in dry copying inks, and as activator in bacterial oxidation processes.

9775. Uranyl Chloride. Uranium dioxydichloride. Cl_2O_2U; mol wt 340.98. U 69.82%, O 9.38%, Cl 20.80%. UO_2Cl_2. Prepd by the reaction of O_2 with uranium(IV) chloride at 300-350°: Leary, Suttle, *Inorg. Syn.* **5**, 148 (1957).

Bright yellow crystals. Orthorhombic system. Very hy-groscopic. Appreciably volatile above 775°. Dec *in vacuo* above 450° yielding chlorine and UO_2 plus U_3O_8. Very freely sol in water. Forms cryst hydrates. Sol in polar organic solvents such as acetone and alcohol, but insol in less polar solvents such as benzene. Aq solns are unstable.

9776. Uranyl Nitrate. N_2O_8U; mol wt 394.04. N 7.11%, O 32.48%, U 60.41%. $UO_2(NO_3)_2$.

Hexahydrate, yellow crystals; greenish luster by reflected light. When shaken, rubbed, or crushed, the cystals show remarkable tribolum inescence with occasional detonations. d 2.807; mp 60°. Sol in about 1.5 parts of water, freely in alcohol, ether. The aq soln is acid. Solns nf uranium nitrate in ether should not be allowed to stand in sunlight as explosion may occur.

USE: As intensifier in photography; manuf uranium glaze, decorating porcelain; also as reagent in anal. chemistry.

9777. Uranyl Phosphate. HO_6PU; mol wt 366.01. H 0.28%, O 26.23%, P 8.46%, U 65.03%. UO_2HPO_4.

Tetrahydrate, yellow, microcryst powder. Insol in water; sol in acids.

9778. Uranyl Sulfate. O_6SU; mol wt 366.09. O 26.22%, S 8.76%, U 65.02%. UO_2SO_4.

Trihydrate, lemon-yellow, cryst mass. d 3.28. Sol in about 5 parts water, 25 parts alcohol.

9779. Urapidil. *6-[[3-[4-(2-Methoxyphenyl)-1-piperazinyl]propyl]amino]-1,3-dimethyl-2,4(1H,3H)-pyrimidinedione;* 6-[[3-[4-(o-methoxyphenyl)-1-piperazinyl]propyl]-amino]-1,3-dimethyluracil; B-66256; Ebrantil; Eupressyl; Uraprene. $C_{20}H_{29}N_5O_3$; mol wt 387.49. C 61.99%, H 7.54%, N 18.08%, O 12.39%. α_1-Adrenergic antagonist; deriv of uracil, *q.v.* Prepn: W. Pruesse *et al.,* **Ger. pat.** 1,942,405; *eidem,* **U.S. pat.** 3,957,786 (1971, 1976 both to Byk Gulden); K. Klemm *et al., Arzneimittel-Forsch.* **27**, 1895 (1977). Series of articles on pharmacology, pharmacokinetics, metabolism: *ibid.* 1898-1932. Toxicity data: J. Koenig *et al., ibid.* 1919. Mode of action: M. Eltze, *Eur. J. Pharmacol.* **59**, 1 (1979); H. R. Kaplan, R. D. Smith, *Fed. Proc.* **40**, 2268 (1981). Hemodynamic responses in man: G. G. Belz *et al., Clin. Pharmacol. Ther.* **37**, 48 (1985). Clinical trials: A. Barankay *et al., Arzneimittel-Forsch.* **31**, 849 (1981); H. Liebau *et al., J. Hypertens.* **4**, Suppl. 6, S141 (1986). Series of articles on clinical efficacy in hypertension: *Drugs* **35**, Suppl. 6, 147-192 (1988).

Crystals from water, mp 156-158°. uv max (methanol): 237, 268 nm (ϵ 1.10 × 10^4, 2.67 × 10^4). pKa 7.10. LD$_{50}$ in male mice, rats (mg/kg): 750, 550 orally; 260, 145 i.v. (Koenig).

THERAP CAT: Antihypertensive.

9780. Urazole. *1,2,4-Triazolidine-3,5-dione; bicarbamimide;* 1H-1,2,4-triazole-3,5(2H,4H)-dione; hydrazodicarbonimide; 3,5-diketotriazolidine. $C_2H_3N_3O_2$; mol wt 101.07. C 23.77%, H 2.99%, N 41.58%, O 31.66%. Prepd by heating biuret with hydrazine hydrate at 108°: Stolle, Krauch, *J. Prakt. Chem.* [2] **88**, 313 (1913). From methyl or ethyl allophanate: Gordon, Audrieth, *Inorg. Syn.* **5**, 52 (1957).

Leaflets from water, dec 249-250°. Weak acid, Ka: 1.6 × 10^{-6}. pH of satd aqueous soln at 25° = 3.15. Soly in water (g/100 ml): 2.83 (0°); 23.7 (65°). Sparingly sol in alc, practically insol in ether. Sol in concd HCl from which it crystallizes without change. Forms ammonium and sodium salts.

9781. Urea. Carbamide; carbonyldiamide; Aquacare; Aquadrate; Basodexan; Hyanit; Keratinamin; Nutraplus; Onychomal; Pastaron; Ureaphil; Ureophil; Urepearl. CH_4-N_2O; mol wt 60.06. C 20.00%, H 6.71%, N 46.65%, O 26.64%. H_2NCONH_2. Product of protein metabolism. From cyanamide by hydrolysis, from CO_2 by ammonolysis: One mol of carbon dioxide reacts with two mols of ammonia to yield ammonium carbamate. The carbamate then dec to form urea and water: Wöhler, *Ann. Physik* **12**, 253 (1828). Continuous process: **Brit. pat. 844,110** (1960 to Inventa, AG), *C.A.* **55**, 7298b (1961). Review of commercial processes: Tonn, Jr., *Chem. Eng.* **62**, no. 10, 186-190 (1955); Moore, Parks, *Chem. & Eng. News* **37**, 84-90 (Nov. 23, 1959); Faith, Keyes & Clark's *Industrial Chemicals*, F. A. Lowenheim, M. K. Moran, Eds. (Wiley-Interscience, New York, 4th ed., 1975) pp 854-861. Prepn from ammonia, carbon monoxide and sulfur in methanol: Applegath, **U.S. pat. 2,857,430** (1958 to Monsanto); Franz, Applegath, *J. Org. Chem.* **26**, 3304, 3306, 3309 (1961). Comprehensive review: G. T. Y. Chao, *Urea, Its Properties and Manufacture* (Chao's Institute, W. Covina, Calif., 1967) 508 pp.

Tetragonal prisms. Develops odor of NH_3. Cooling, saline taste. mp 132.7°. On further heating it decomposes to biuret, NH_3, and cyanuric acid. d_4^{18} 1.32; d_4^{18} of water solns (w/w): 10% 1.027; 20% 1.054; 50% 1.145. pH of a 10% water soln: 7.2. One gram dissolves in 1 ml water, 10 ml 95% alc, 1 ml boiling 95% alc, 20 ml abs alc, 6 ml methanol, 2 ml glycerol. Almost insol in chloroform, ether. Sol in concd HCl. Water solns dec on heating, giving off some NH_3. Pure urea should not give the biuret reaction unless heated above the melting point. In practice all reagent grade urea gives a positive biuret reaction. *See* Biuret.

USE: Fertilizer, because of easily available high nitrogen content. In animal feeds. Is reacted with aldehydes to make resins and plastics. Is condensed with malonic ester to form barbituric acid. Used extensively in the paper industry to soften cellulose. In ammoniated dentifrices.

THERAP CAT: Diuretic.

THERAP CAT (VET): Nutritional factor (partial source of dietary nitrogen in ruminants). To enhance action of sulfonamides. Has beeen used to promote healing in infected wounds, septic metritis. Diuretic. Antiseptic.

9782. Urea Hydrochloride. CH_5ClN_2O; mol wt 96.52. C 12.44%, H 5.22%, O 16.58%, Cl 36.74%, N 29.03%. $CO(NH_2)_2 \cdot HCl$.

White to faintly yellow, deliquesc crystals. Dec at 145°. Sol in water. The aq soln is acid. *Keep well closed.*

9783. Urea Hydrogen Peroxide. Hydrogen peroxide carbamide; Exterol; Ortizon; Hyperol; Perhydrit; Perhydrol-Urea. $CH_6N_2O_3$; mol wt 94.07. C 12.77%, H 6.43%, O 51.02%, N 29.78%. $CO(NH_2)_2 \cdot H_2O_2$. Usually contains 34-35% H_2O_2.

White crystals or cryst powder. Dec in air into urea, oxygen, and water. Sol in 2.5 parts water; partly dec by alcohol and ether into H_2O_2 and urea.

USE: For extemporaneous prepn of H_2O_2.

THERAP CAT: Disinfectant.

9784. Urea Nitrate. Acidogen nitrate. $CH_5N_3O_4$; mol wt 123.07. C 9.76%, H 4.10%, N 34.15%, O 52.00%. $CO(NH_2)_2 \cdot HNO_3$.

White, odorless leaflets. mp 152° with decompn. Has tendency to explode, especially when rubbed while heating. Sparingly sol in cold water or in HNO_3; sol in hot water or alcohol. The aq soln is acid.

Caution: Mildly irritating to skin, mucous membranes.

9785. Urease. Urea amidohydrolase; Urastrat. An enzyme which hydrolyzes urea to ammonium carbonate. The molecule (mol wt 489,000) contains many amino acids, among them there are about 77 methionyl, 29 cystinyl, and 47 cysteinyl sidues. Isolated from jack bean, *Canavalia ensiformis* (L.) D.C., *Leguminosae:* Hanabuse, *Nature* **193**, 1078 (1962); Gorin *et al., Biochemistry* **1**, 911 (1962). Structural studies: Bailey, Boulter, *Biochem. J.* **113**, 669 (1969). *Review:* Varner, "Urease" in *The Enzymes* vol. 4, P. D. Boyer *et al.,* Eds. (Academic Press, New York, 1960) pp 247-256; Reithel, *ibid.* vol. 4 (3rd ed., 1971) pp 1-21.

Microscopic, octahedral cystals. Soluble in water. uv max: 278.5 nm. Isoelectric point pH 5.0-5.1 (also reported as pH 4.8, *see* Reithel, *loc. cit.*). Urease is inhibited by heavy metal ions.

USE: In determination of urea in body fluid.

9786. Urea Stibamine. Carbostibamide. No definitely known and agreed structure. Originally reported to consist of urea and *p*-aminophenylstibinic acid with the formula $C_7H_{12}N_3O_4Sb$: Brahmachari, *Ind. J. Med. Res.* **10**, 492 (1922). Later suggested that the substance was identical with ammonium *p*-carbamidophenylstibinate, NH_2CONH-$C_6H_4SbO(OH)ONH_4$: Niyogi, *J. Ind. Chem. Soc.* **5**, 285 (1928); Ghosh *et al., Ind. J. Med. Res.* **16**, 461 (1928), and Gray *et al., Proc. Roy. Soc.* **B108**, 54 (1931), produced evidence that urea stibamine was not a single substance and attempted to prove that the active principle was a disubstituted urea: *sym*-diphenylcarbamido-4,4-distibinic acid. The antimony content of commercial samples seems to vary between 39 and 42%: Guha *et al., Nature* **151**, 108 (1943). Prepd by the addition of urea to a warm aq suspension of stibanilic acid: Datta *et al., Sci. Cult.* **11**, 385 (1946), *C.A.* **41**, 105 (1947); Christiansen, Green, **U.S. pat. 2,488,268** (1949 to Squibb).

White powder. Sol in water. Partially sol in alc, ether.

THERAP CAT: Anthelmintic (Nematodes, Schistosoma); antiprotozoal (Leishmania).

9787. Uredepa. *[Bis(1-aziridinyl)phosphinyl]carbamic acid ethyl ester;* ethyl [bis(1-aziridinyl)phosphinyl]carbamate; ethyl *N*-[bis(ethyleneimido)phosphoro]carbamate; bis-(ethylenimido)phosphorylurethane; urethimine; AB-100; NSC-37095; Avinar. $C_7H_{14}N_3O_3P$; mol wt 219.17. C 38.36%, H 6.44%, N 19.17%, O 21.90%, P 14.13%. Outline of prepn: Bardos *et al., Nature* **183**, 399 (1959); Bardos, Papanastassiou, **U.S. pat. 3,201,313** (1965 to Armour Pharm.).

Crystals from benzene + cyclohexane, mp 88-90°. Readily sol in water but dec in aq soln.

THERAP CAT: Antineoplastic.

THERAP CAT (VET): Has been used experimentally as an insect chemosterilant.

9788. Urena. A bast fiber from the plant *Urena lobata* Dill. ex L., *Malvaceae.* The plant is tropical and subtropical and grows wild, sometimes becoming a noxious weed in South and Middle America, Asia, Indonesia, the Philippines, Madagascar, and Africa. It is cultivated to a small extent in Brazil, India, and Madagascar, and to a much larger extent in the former Belgian Congo and French Equatorial Africa. Harvested when in full bloom (pink flowers). The fiber is creamy white, lustrous, soft and flexible, comparable to jute. The presence of steroids in the rest of the plant has been demonstrated. *Ref:* E. E. Stout, *Introduction to Textiles* (John Wiley, New York, 1960) pp 56-57.

9789. Urethan. *Carbamic acid ethyl ester; ethyl carbamate;* urethane; ethyl urethan. $C_3H_7NO_2$; mol wt 89.09. C 40.44%, H 7.92%, N 15.72%, O 35.92%. $NH_2COOC_2H_5$. Prepd by heating urea with alcohol under pressure; by warming urea nitrate with alcohol and sodium nitrite. Toxicity data: K. J. Franklin, *J. Pharmacol. Exp. Ther.* **42**, 1 (1931). Review of carcinogenic action and metabolism: Mirvish in *Advan. Cancer Res.* **11**, 1-42 (1968).

Crystals, mp 48-50°. Cooling saline taste, d 1.1. bp 182-184°. Sublimes readily at 103° and 54 mm pressure. One gram dissolves in 0.5 ml water, 0.8 ml alcohol, 0.9 ml chloroform, 1.5 ml ether, 2.5 ml glycerol, 32 ml olive oil. The aq soln is neutral. MLD i.p. in mice: 2.1-2.2 g/kg (Franklin).

Incompat: Alkalies, acids, antipyrine, chloral hydrate, camphor, menthol, salol, or thymol.

Note: This substance may reasonably be anticipated to be a carcinogen: *Fourth Annual Report on Carcinogens* (NTP 85-002, 1985) p 198.

USE: Molten urethan is a good solvent for various organic materials. Intermediate in organic synthesis. In the prepn and modification of amino resins. As solubilizer and cosolvent for pesticides, fumigants.

THERAP CAT: Antineoplastic.

THERAP CAT (VET): Anesthetic.

9790. Uric Acid. *7,9-Dihydro-1H-purine-2,6,8(3H)-trione;* 8-hydroxyxanthine; purine-2,6,8-triol; purine-2,6,8-(1*H*,3*H*,9*H*)-trione; 2,6,8-trioxypurine. $C_5H_4N_4O_3$; mol wt 168.11. C 35.72%, H 2.40%, N 33.33%, O 28.55%. Discovered by Scheele and independently by Bergman in 1776. It forms the chief end-product of the nitrogenous metabolism of birds and of scaly reptiles and is found in their excrement; present in the urine of all carnivorous animals. Prepn from urea: Bills *et al., J. Org. Chem.* **27,** 4633 (1962). Role in biological processes: Bishop, Talbott, *Pharmacol. Revs.* **5,** 231 (1953).

White, odorless, tasteless crystals; dec by heat without melting and with evolution of HCN. d 1.89. One gram dissolves in about 15,000 parts cold water, about 2000 parts boiling water; sol in glycerol, in solns of alkali hydroxides, their carbonates, sodium acetate and sodium phosphate; insol in alcohol, ether. Gives murexide reaction.

9791. Uricase. *Urate oxidase; uric oxidase;* urico-oxidase; uriKoxidase. An enzyme responsible for the oxidative scission of the purine skeleton, and therefore of key importance to the catabolism of nitrogenous compds in general in organisms that do not eliminate intact uric acid or some other purine. All mammals except the primates (including man) are uricolytic organisms. Porcine liver or bovine kidney is the starting material for various purified prepns of uricase. Isoln from porcine liver: Holmberg, *Biochem. J.* **33,** 1901 (1939); Miller *et al., J. Biol. Chem.* **216,** 625 (1955); Robbins, Grant, U.S. pat. **2,878,161** (1959 to Armour). *Review:* Mahler, "Uricase" in *The Enzymes,* vol. **8,** P. D. Boyer *et al.,* Eds. (Academic Press, New York, 1963) pp 285-296.

Pale, brownish-green crystals or shiny, transparent, striated plates. Practically insol in water. Slightly sol in buffered alkali solns. Solns at pH 7.5-10.5 are relatively stable. Shows unusually high absorption in the region of 330-350 nm (for highly purified uricase: $A_{276}^{1\%} = 11.3$, $A_{330}^{1\%} = 2.0$, both in 1% Na_2CO_3; $A_{280}/A_{330} = 5.6$). Isoelec. pt. pH 6.3. Copper content of the enzyme (specific activity 120-125) equals ~0.56 mg/g of enzyme protein. There appears to be a satisfactory correlation between copper content and activity of the purified enzyme. The enzyme is sensitive to cyanide ion; the presence of $10^{-4}M$ KCN inhibits its activity.

USE: In the determination of serum and urine uric acid.

9792. Uridine. 1-β-D-Ribofuranosyluracil; uracil riboside. $C_9H_{12}N_2O_6$; mol wt 244.20. C 44.26%, H 4.95%, N 11.47%, O 39.31%. Nucleoside; widely distributed in nature. Prepd by hydrolysis of yeast nucleic acid with weak alkali, *cf.* Levene, Bass, *Nucleic Acids* (New York, 1931). Improved isolns: Harris, Thomas, *J. Chem. Soc.* **1948,** 1936; Elmore, *ibid.* **1950,** 2084; Lorine, Ploeser, *J. Biol. Chem.* **178,** 439 (1949). Crystal structure: Green *et al., Chem. Commun.* **1971,** 53. Review: *Basic Principles in Nucleic Acid Chemistry*

vol. 1, P. O. P. Ts'o, Ed. (Academic Press, New York, 1974) *passim. See also* Nucleic Acids.

Needles from dil alc, mp 165°. $[\alpha]_D^{20}$ +4° (c = 2). uv max (pH 7.3): 261, 205 nm ($\epsilon \times 10^{-3}$ 10.1, 9.8), Voet *et al., Biopolymers* **1,** 193 (1963). Sol in water. Upon prolonged refluxing with HCl, furfural is formed.

9793. Uridine 5′-Diphosphate. *Uridine 5′-(trihydrogen diphosphate);* UDP; uridine 5′-pyrophosphate; uridine-5-pyrophosphoric acid. $C_9H_{14}N_2O_{12}P_2$; mol wt 404.18. C 26.75%, H 3.49%, N 6.93%, O 47.50%, P 15.33%. Can be isolated from calf's liver, thymus, and yeast. The commercial product is derived from yeast. Pentose nucleic acids (isolated from yeast) are digested with rattlesnake venom (freed of 5′-monoesterase) and the nucleotides are separated by chromatography: Cohn, Volkin, *Arch. Biochem. Biophys.* **35,** 465 (1952); *J. Biol. Chem.* **203,** 319 (1953). For alternate procedures *see* the refs under Uridine Diphosphate Glucose. Syntheses: Chambers, *J. Am. Chem. Soc.* **81,** 3032 (1959); Moffatt, Khorana, *ibid.* **83,** 649 (1961).

Lithium salt monohydrate, $C_9H_{12}N_2O_{12}P_2Li_2.H_2O$, crystals, sol in water.

Sodium salt trihydrate, $C_9H_{12}N_2O_{12}P_2Na_2.3H_3O$, crystals. pKa'_1 6.5; pKa'_2 9.4. uv max (pH 7): 262 nm; (pH 11): 261 nm. Freely sol in water.

9794. Uridine Diphosphate Glucose. *Uridine 5′-(trihydrogen diphosphate) mono-α-D-glucopyranosyl ester;* UDPG; uridine 5′-pyrophosphate glucose ester; uridine-5′-diphosphoglucose; co-waldenase; co-galactoisomerase. $C_{15}H_{24}N_2O_{17}P_2$; mol wt 566.33. C 31.81%, H 4.27%, N 4.95%, O 48.03%, P 10.94%. The coenzyme of the galactowaldenase system which catalyzes the conversion of galactose-1-phosphate into glucose-1-phosphate. Isoln from baker's yeast: Caputto *et al., J. Biol. Chem.* **184,** 333 (1950). Also present in animal tissue. Synthesis: Michelson, Todd, *J. Chem. Soc.* **1956,** 3459; Moffatt, Khorana, *J. Am. Chem. Soc.* **80,** 3756 (1958). *Reviews:* Leloir, Cardini in *The Enzymes* vol. **2A,** P. D. Boyer *et al.,* Eds. (Academic Press, New York, 2nd ed., 1960) pp 39-61; A. M. Michelson, *The Chemistry of Nucleosides and Nucleotides* (Academic Press, New York, 1963) pp 153-250; D. W. Hutchison, *Nucleotides and Coenzymes* (John Wiley, New York, 1964) pp 36-82.

Isolated as the barium or calcium salt, white powder, sol in water.

Sodium salt dihydrate, $C_{15}H_{23}N_2O_{17}P_2Na.2H_2O$, white powder. uv max (pH 2.0): 262 nm. Freely sol in water.

9795. Uridine 5′-Triphosphate. *Uridine 5′-(tetrahydrogen triphosphate);* UTP; Uteplex. $C_9H_{15}N_2O_{15}P_3$; mol wt 484.17. C 22.33%, H 3.12%, N 5.79%, O 49.57%, P 19.19%. Pyrimidine analog of ATP. Isoln from rabbit muscle: Lipton *et al., J. Am. Chem. Soc.* **75,** 5450 (1953). Synthesis: Kenner *et al., J. Chem. Soc.* **1954,** 2288; Hall, Khorana, *J.*

Am. Chem. Soc. **76,** 5056 (1954). Prepd enzymatically from uridine diphosphate or from ribonucleic acid.

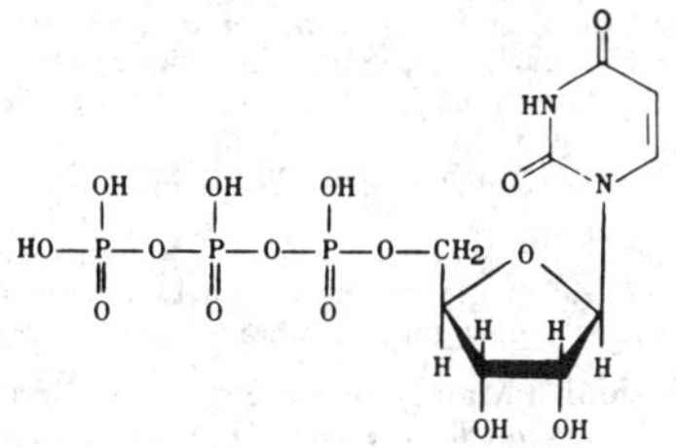

Barium salt tetrahydrate,$(C_9H_{12}N_2O_{15}P_3)_2Ba_3.4H_2O$, white powder.

Trisodium salt dihydrate, $C_9H_{12}N_2Na_3O_{15}P_3.2H_2O$, *Utipina.* White powder, sol in water, very sparingly sol in alcohol. pKa_1 6.6; pKa_2 9.5. uv max (pH 7): 262 nm (a_M 10,000); (pH 11): 261 nm (a_M 8100).

9796. 5′-Uridylic Acid. Uridine 5′-phosphoric acid; uridine 5′-monophosphate; UMP. $C_9H_{13}N_2O_9P$; mol wt 324.19. C 33.34%, H 4.04%, N 8.64%, O 44.42%, P 9.56%. Nucleotide; widely distributed in nature. Synthesis by phosphorylation of 2′,3′-O-benzylidene uridine with diphenyl phosphorochloridate: Brown *et al., J. Chem. Soc.* **1950,** 408; Smith, *Biochem. Prepn.* **8,** 130 (1961). Monograph on the synthesis of nucleotides: G. R. Pettit, *Synthetic Nucleotides* vol. **1** (Van Nostrand Reinhold, New York, 1972) 252 pp. Crystal structure of hydrated barium salt: Shefter, Trueblood, *Acta Cryst.* **18,** 1067 (1965). *Reviews:* see Uridine; Nucleic Acids.

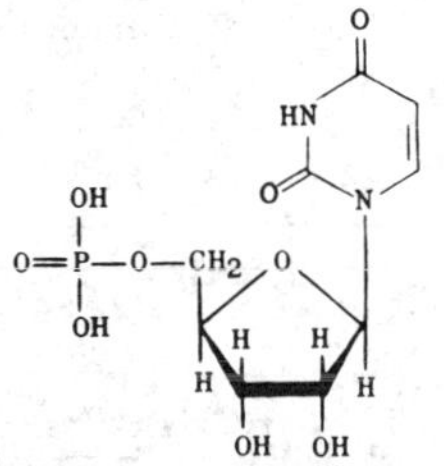

Free acid: pKa_1' 6.4, pKa_2' 9.5. uv max (pH 7.0): 262 nm (a_M 10,000).

Disodium salt dihydrate, $C_9H_{11}N_2Na_2O_9P.2H_2O$, white powder, characteristic meaty taste. uv max (0.1M HCl): 262 nm (ϵ 10,000). Soly in water at 20°: about 41 g/100 ml H_2O.

9797. Urobilins. Bile pigments which occur in feces; and in urine. More specifically described as dipyrrylmethene compds joined at the two α-pyrrolyl positions by methylene bridges to either pyrrolenone or pyrrolidone rings. Reviews of occurrence, proposed structures, stereoisomerism, and interrelationships of *i-urobilin, d-urobilin, stercobilin, urobilin IXα,* and *d-urobilin IXα:* Gray, Nicholson, *Nature* **209,** 581 (1966); *eidem, Medicine* **46,** 83-90 (1967); Rüdiger, *Fortschr. Chem. Org. Naturst.* **29,** 60-139 (1971). Structure of stercobilin and *d*-urobilin: Gray, Nicholson, *J. Chem. Soc.* **1958,** 3085. Revised structure of *d*-urobilin: Killilea, O'Carra, *Biochem. J.* **129,** 1179 (1972). Stereochemistry of urobilins: Cole *et al., ibid.* **1965,** 4085. Spectral absorption of *d*-urobilin hydrochloride, *d*-urobilin IXα, stercobilin hydrochloride, and stercobilin: Cole *et al., J. Chem. Soc. (C)* **1966,** 1321. Synthesis of optically active stercobilin: Plieninger, Ruppert, *Ann.* **736,** 43 (1970); of optically active urobilins: Plieninger *et al., ibid.* 62. Stercobilin, proposed structure (M = CH_3, E = C_2H_5, P = CH_2CH_2COOH):

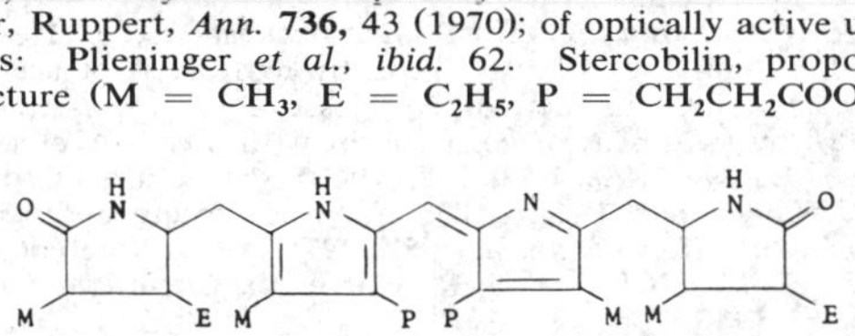

USE: Spot-test reagent for detection of copper, mercury, zinc: F. J. Welcher, *Organic Analytical Reagents,* **vol. IV** (Van Nostrand, New York, 1949).

9798. Urochloralic Acid. *2,2,2-Trichloroethyl β-D-glucopyranosiduronic acid; 2,2,2-trichloroethyl β-D-glucosiduronic acid; β,β,β-trichloroethyl*-D-glucuronide. C_8H_{11}-Cl_3O_7; mol wt 325.54. C 29.51%, H 3.41%, Cl 32.68%, O 34.40%. Isolated from urine of man and dog after ingestion of chloral hydrate: von Mering, Musculus, *Ber.* **8,** 663 (1875); from urine of calves fed trichloroethylene: Seto, Schultze, *J. Am. Chem. Soc.* **78,** 1616 (1956).

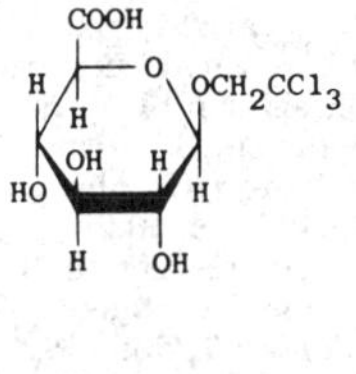

Needles, mp 142°. $[\alpha]_D^{27}$ $-50.0°$ (c = 0.7). Freely soluble in water, alcohol. One gram dissolves in 234 ml of anhydr ether at 20°.

9799. Urokinase. WIN 22005; Abbokinase; Actosolv; Breokinase; Persolv; Purochin; Ukidan; Uronase; Win-Kinase. Serine protease which activates plasminogen to plasmin; present in mammalian blood and urine. Produced as prourokinase, *q.v.,* and converted to active form by plasmin or kallikrein, *q.q.v.* Description of fibrinolytic activity: J. R. B. Williams, *Brit. J. Exp. Pathol.* **32,** 530 (1951). Isolation from human urine and activity: T. Astrup, I. Sterndorff, *Proc. Soc. Exp. Biol. Med.* **81,** 675 (1952); G. W. Sobel *et al., Am. J. Physiol.* **171,** 768 (1952). Species specificity and distribution in mammals: S. R. Mohler *et al., Am. J. Physiol.* **192,** 186 (1958). Isoln from human male urine: H. O. Singher, L. Zuckerman, U.S. pats. **2,961,382** and **2,989,440** (1960, 1961 to Ortho); N. O. Kjeldgaard, J. Ploug, U.S. pat. **2,983,647** (1961 to Leo Pharm.); J. Doczi, U.S. pat. **3,081,-236** (1963 to Warner-Lambert). Prepn of crystalline form: A. Lesuk *et al., Science* **147,** 880 (1965). Two variants of bioactive urokinase, high molecular weight (HMW-UK, $\sim$50000 Da) and low molecular weight (LMW-UK, $\sim$30000 Da) have been identified: W. F. White *et al., Biochemistry* **5,** 2160 (1966). Both are disulfide-linked dimers consisting of a heavy chain (B) and a light chain (A). HMW-UK is converted to LMW-UK by proteolytic cleavage of the A-chain to form the A_1-chain. Structural characterization: M. Nobuhara *et al., J. Biochem.* **90,** 225 (1981). Structure and amino acid sequences: W. A. Günzler *et al., Z. Physiol. Chem.* **363,** 133 (1982); G. J. Steffens *et al., ibid.* 1043; W. A. Günzler *et al., ibid.* 1155. Expression of gene coding for human urokinase in *E. coli:* B. Ratzkin *et al., Proc. Nat. Acad. Sci. USA* **78,** 3313 (1981). Clinical evaluation in pulmonary embolism: P. Petitpretz *et al., Circulation* **70,** 861 (1984). Double-blind multicenter clinical trial in coronary thrombosis following myocardial infarction: H. Kambara *et al., Japan. Circ. J.* **51,** 1072 (1987). Clinical applications of urokinase-treated tubing: T. Ohshiro *et al., Methods Enzymol.* **137,** 529 (1988). Series of articles on plasminogen activation, fibrinolysis and clinical efficacy: *Proc. Serono Symp., Thrombosis and Urokinase* **9,** 1-257 (1977). *Review:* F. Duckert, *Handb. Exp. Pharmacol.* **46,** 209-237 (1978).

THERAP CAT: Thrombolytic.

9800. Urothion. *2-Amino-7-(1,2-dihydroxyethyl)-6-(methylthio)thieno[3,2-g]pteridin-4(3H)-one.* $C_{11}H_{11}N_5O_3S_2$; mol wt 325.37. C 40.60%, H 3.41%, N 21.53%, O 14.75%, S 19.71%. Constituent of normal human urine: Koschara, *Z. Physiol. Chem.* **277,** 284 (1943). Structure: Tschesche *et al., Ber.* **88,** 1251 (1955). Revised structure: Goto *et al., Tetrahedron Letters* **1967,** 4507; *eidem, J. Biochem. (Tokyo)* **65,** 611 (1969). Synthesis: Sakurai, Goto, *Tetrahedron Letters* **1968,** 2941; *eidem, J. Biochem. (Tokyo)* **65,** 755 (1969).

Orange-colored clusters of crystals, not melted at 360°. Shows mutarotation: $[\alpha]_D^{20} -20°$ (after 15 hrs in $0.05N$ NaOH). Soly in water at pH 6.6 about 1:10,000. More sol in alkaline soln. Also sol in acids. Practically insol in the usual organic solvents.

9801. Ursodiol. *(3α,5β,7β)-3,7-Dihydroxycholan-24-oic acid;* 17β-(1-methyl-3-carboxypropyl)etiocholane-3α,7β-diol; 3α,7β-dioxycholanic acid; ursodeoxycholic acid; Actigall; Arsacol; Cholit-Ursan; Delursan; Destolit; Deursil; Litursol; Lyeton; Peptarom; Solutrat; Ursacol; Urso; Ursobilin; Ursochol; Ursodamor; Ursofalk; Ursolvan. $C_{24}H_{40}O_4$; mol wt 392.56. C 73.43%, H 10.27%, O 16.30%. Epimeric with chenodiol, *q.v.,* with respect to the hydroxyl group at C_7. Found in bear bile (combined with taurine). Isoln: Shoda, *J. Biochem. (Japan)* **7**, 505 (1927). Structure: Kaziro, *Z. Physiol. Chem.* **185**, 151 (1929); **197**, 206 (1931); Iwasaki, *ibid.* **244**, 181 (1936). Toxicity data: M. Ardenne, P. G. Reitnauer, *Arzneimittel-Forsch.* **20**, 323 (1970). Review of pharmacology, toxicology, efficacy: A. Ward *et al., Drugs* **27**, 95-131 (1984). Brief review of clinical effects and comparison with chenodiol: H. Fromm, *Gastroenterology* **87**, 229-233 (1984). Effect on cholesterol and bile acid metabolism: G. S. Tint *et al., ibid.* **91**, 1007 (1986).

Bitter plates from alc. mp 203°. $[\alpha]_D^{20} +57°$ (c = 2 in abs ethanol). Freely sol in ethanol, glacial acetic acid; slightly sol in chloroform, sparingly sol in ether. Practically insol in water. LD_{50} in mice (g/kg): 0.1 i.v. (Ardenne, Reitnauer); in rats, mice (mg/kg): 2000, 6000 s.c.; 1000, 1200 i.p., 310, 260 i.v. (Ward).

Diformate, $C_{26}H_{40}O_6$, crystals, mp 170°.
Diacetate, $C_{28}H_{44}O_6$, crystals, mp 98-102°.
THERAP CAT: Anticholelithogenic.

9802. Ursolic Acid. *3β-Hydroxyurs-12-en-28-oic acid;* urson; prunol; micromerol; malol. $C_{30}H_{48}O_3$; mol wt 456.68. C 78.90%, H 10.59%, O 10.51%. In leaves and berries of *Arctostaphylos uva-ursi* (L.) Spreng (bearberry), of *Vaccinium macrocarpon* Ait. (cranberry), *Rhododendron hymenanthes* Makino, *Ericaceae.* In the protective wax-like coating of apples, pears, prunes, and other fruits. Isoln from apple peelings: Sando, *J. Biol. Chem.* **56**, 457 (1923). Structure: Ruzicka *et al., Helv. Chim. Acta* **28**, 199 (1945); Zurcher *et al., ibid.* **37**, 2145 (1954). Conversion from α-amyrin: Boar *et al., J. Chem. Soc. (C)* **1970**, 678. Chemistry: Mezzetti *et al., Planta Med.* **20**, 244 (1971).

Large, lustrous prisms from abs alcohol, fine hair-like needles from dil alcohol, mp 285-288°. $[\alpha]_D^{21} +67.5°$ (c = 1 in *N* alc KOH). Soly at 15°: One part dissolves in 88 parts methanol, 178 alcohol (35 boiling alcohol), 140 ether, 388 chloroform, 1675 carbon disulfide. Moderately sol in acetone. Sol in hot glacial acetic acid and in 2% alcoholic NaOH. Insol in water and petr ether.

Acetate, $C_{32}H_{50}O_4$, mp 289-290°. $[\alpha]_D +62.3°$ (c = 1.15 in chloroform).

Methyl ester, $C_{31}H_{50}O_3$, mp 171°. $[\alpha]_D^{20} +58°$ (c = 1.2 in pyridine). Acetate of methyl ester, $C_{33}H_{52}O_4$, mp 246-247°.

USE: As emulsifying agent in pharmaceuticals, foods.

9803. Urushiol. Main constituent of the irritant oil of poison ivy, *Toxicodendron radicans* (L.) Kuntze, poison oak, *T. diversilobum,* the Asiatic lacquer tree, *T. verniciferum* D.C. and other plants of the genus *Toxicodendron, Anacardiaceae.* Yoshida, *J. Chem. Soc.* **43**, 472 (1883); Hill *et al., J. Am. Chem. Soc.* **56**, 2736 (1934). It is a mixture of several compds which are derivatives of catechol with unsaturated C_{15} or C_{17} side chains and which, upon hydrogenation, yield the same 3-pentadecylcatechol, *q.v.* or 3-heptadecylcatechol: Majima *et al., Ber.* **55**, 172 (1922); Corbett, Billets, *J. Pharm. Sci.* **64**, 1715 (1975). Urushiol from *T. verniciferum* is a mixture of I-IV and urushiol from *T. radicans* is a mixture of I, II, III, and V: Symes, Dawson, *J. Am. Chem. Soc.* **76**, 2959 (1954); Sunthankar, Dawson, *ibid.* **76**, 5070 (1954). Separation of components: Markiewitz, Dawson, *J. Org. Chem.* **30**, 1610 (1965). Structural characterization of urushiol from *T. diversilobum:* Corbett, Billets, *loc. cit.* Separation of urushiol congeners: C. Y. Ma *et al., J. Chromatog.* **200**, 163 (1980). Isoln of pure urushiol from poison ivy or oak extracts and separation of congeners by reversed phase chromatography: M. A. El Sohly *et al., J. Nat. Prod.* **45**, 532 (1982). Allergenic properties of congeners: E. S. Watson *et al., J. Pharm. Sci.* **70**, 785 (1981). Review of extraction and compositional analysis: J. H. P. Tyman, *Chem. Soc. Rev.* **8**, 499 (1979).

I R = $(CH_2)_{14}CH_3$
II R = $(CH_2)_7CH=CH(CH_2)_5CH_3$
III R = $(CH_2)_7CH=CHCH_2CH=CH(CH_2)_2CH_3$
IV R = $(CH_2)_7CH=CHCH_2CH=CHCH=CHCH_3$
V R = $(CH_2)_7CH=CHCH_2CH=CHCH_2CH=CH_2$

Pale yellow liq; $d_4^{21.5}$ 0.9687; bp 200-210°. Soluble in alcohol, ether, benzene. Moderately sol in petr ether.

Human Toxicity: An extremely active allergen causing skin reactions similar to poison ivy.

THERAP CAT: Antiallergic (hyposensitization therapy).

9804. Urylon. A synthetic fiber, condensation product of nonamethylenediamine and urea. Prepn of such polyurea polymers: **Brit. pat. 900,787** (1962 to Toyo Koatsu Ind.).

Fiber softens at 205-210°, mp at about 237°. Sp gr 1.07. Loses not more than 10% of its strength after 10 hrs of exposure to 40% sulfuric acid or 40% caustic soda at room temp. Also loses strength on prolonged exposure to light or to a temp of 150°. Hydrolyzes at 130° in the presence of steam.

USE: For fishing nets and lines, knitted fabric, and blends of staple fiber with wool and cotton.

9805. Uscharidin. *2,14-Dihydroxy-19-oxo-3-[(tetrahydro-6-methyl-2,3-dioxo-2H-pyran-2-yl)oxy]card-20(22)-enolide.* $C_{29}H_{38}O_9$; mol wt 530.59. C 65.64%, H 7.22%, O 27.14%. African arrow poison and cardiac glycoside produced by *Calotropis procera* R. Br., *Asclepiadaceae:* Hesse *et al., Ann.* **566**, 130 (1950). Prepn by hydrolysis of uscharin with $2N$ H_2SO_4 in methanol: Hesse *et al., Ann.* **537**, 67 (1938); by treatment of voruscharin with mercuric chloride: Hesse, Ludwig, *Ann.* **632**, 158 (1960). Structure: Crout *et al., J. Chem. Soc.* **1964**, 2187. Revised structure: Brüschweiler *et al., Helv. Chim. Acta* **52**, 2276 (1969). Stereochemistry: H. T. A. Cheung, T. R. Watson, *J. Chem. Soc. Perkin Trans. I* **1980**, 2162.

Consult the cross index before using this section.

Rhombic plates, decomp 290°. $[\alpha]_D^{20}$ +38° (c = 0.9 in methanol).

9806. Usnic Acid. *2,6-Diacetyl-7,9-dihydroxy-8,9b-dimethyl-1,3(2H,9bH)-dibenzofurandione;* usninic acid; usnein; Usniacin. $C_{18}H_{16}O_7$; mol wt 344.31. C 62.79%, H 4.68%, O 32.53%. Antibacterial substance found in lichens. Isolation from varieties of *Usnea barbata* (L.) Wigg., *Usneaceae:* Rochleder, Heldt, *Ann.* **48,** 11 (1843); Widman, *Ann.* **310,** 230 (1900); **324,** 139 (1902). Isoln from *Ramalina reticulata:* Marshak, *Public Health Reports* **62,** 3 (1947); Stark *et al., J. Am. Chem. Soc.* **72,** 1819 (1950). Occurs in nature in both the *d*- and *l*-forms as well as a racemic mixture. Structure: Curd, Robertson, *J. Chem. Soc.* **1937,** 894; Schöpf, Ross, *Ann.* **546,** (1941); Barton, Brunn, *J. Chem. Soc.* **1953,** 603. Resolution of (±)-usnic acid: Dean *et al., ibid.* 1250. Synthesis: Barton *et al., ibid.* **1956,** 530. Biosynthesis *in vitro:* Penttila, Fales, *Chem. Commun.* **1966,** 656. Abs config of (+)-form: S. Huneck *et al., Tetrahedron Letters* **22,** 351 (1981).

Yellow orthorhombic prisms from acetone, mp 204°. $[\alpha]_D^{16}$ +509.4° (c = 0.697 in chloroform). Monobasic acid. Soly at 25° (g/100 ml): water < 0.01; acetone 0.77; ethyl acetate 0.88; ethanol 0.02; methyl Cellosolve 0.22; ethyl Cellosolve 0.32; furfural 7.32; furfuryl alcohol 1.21. LD_{50} i.v. in mice: 25 mg/kg, *Antibiotics* **I,** D. Gottlieb, P. Shaw, Eds. (Springer Verlag, New York, 1967) p 611.

Sodium salt dihydrate, $C_{18}H_{15}NaO_7 \cdot 2H_2O$, pale yellow, silky needles, slightly sol in water.

9807. Ustilagic Acid. Ustizeain B. $C_{37}H_{62-66}O_{17}$; mol wt about 780. A mixture of partially acylated derivatives of a di-D-glucosyldihydroxyhexadecanoic acid. Antibiotic substance produced by the corn smut fungus *Ustilago zeae:* Haskins, *Can. J. Res.* **28C,** 213 (1950); Lemieux *et al., Can. J. Chem.* **29,** 409, 415 (1951); Reed, Holder, *Can. J. Med. Sci.* **31,** 505 (1953); Haskins, **U.S.** pat. **2,698,843** (1955 to National Research Council, Canada). Biosynthesis: Boothroyd *et al., Can. J. Biochem. Physiol.* **33,** 289 (1955). Production by submerged culture: Lemieux, **Can.** pat. **600,121** (1960 to National Research Council, Canada).

Long crystals from ether, mp 146-147°. $[\alpha]_D^{23}$ +7° (pyridine). Freely sol in methanol, pyridine, 2,3-butanediol, 1,2-propanediol. Sparingly sol in ethanol, butanol, acetone. Insol in water, glycerol, ethyl acetate, ether, benzene, petr ether. Shows *in vitro* activity against *Cryptococcus neoformans, Candida albicans,* and some saprophytic fungi. Practically ineffective in rabbits and mice suffering from fungus diseases.

9808. Uva Ursi. Bearberry. Dried leaves of *Arctostaphylos uva-ursi* (L.), Spreng., *Ericaceae. Habit.* Northern Europe, North America, Asia. *Constit.* Volatile oil, arbutin, quercetin, ericolin, methylarbutin, myricetin, tannin (6-7%); ursolic, gallic and malic acids.

THERAP CAT: Antiseptic (urinary).

THERAP CAT (VET): Has been used as a diuretic.

9809. Uzarin. *3β-[(6-O-β-D-Glucopyranosyl-β-D-glucopyranosyl)oxy]-14-hydroxy-5α-card-20(22)-enolide.* $C_{35}H_{54}O_{14}$; mol wt 698.78. C 60.15%, H 7.79%, O 32.05%. Isoln from the African drug uzara, the dried root of a *Gomphocarpus* sp, *Asclepiadaceae:* Windaus, Haack, *Ber.* **63,** 1377 (1930); Tschesche, *ibid.* **85,** 1042 (1952); Schmid *et al., Helv. Chim. Acta* **42,** 72 (1959). Yields uzarigenin by enzymic cleavage. Structure: Tschesche, Bohle, *Ber.* **68,** 2252 (1935). Structure of uzarigenin: Rangaswami, Reichstein, *Helv. Chim. Acta* **32,** 939 (1949); Russel *et al., ibid.* **44,** 1320 (1961). Unlike all other known cardiac glycosides, has the A/B-*trans* configuration: L. F. Fieser, M. Fieser, *Steroids* (Reinhold, New York, 1959) pp 762-763. *Review:* Heusser, *Fortschr. Chem. Org. Naturst.* **7,** 101 (1950).

uzarigenin

Prisms from pyridine + water, mp 266-270°; stout needles from methanol + ether, mp 206-208°. $[\alpha]_D^{20}$ −27° (c = 1.075 in pyridine); $[\alpha]_D^{19}$ −1.4° (c = 0.85 in methanol). uv max: 217 nm (log ε 4.23). Sol in pyridine, hot methyl Cellosolve; sparingly sol in water. Practically insol in ether, chloroform, acetone.

Uzarigenin, $C_{23}H_{34}O_4$, *3β,14-dihydroxy-5α-card-20(22)-enolide, odorigenin.* Synthesis: Stache *et al., Ann.* **726,** 136 (1969); Kamano *et al., J. Org. Chem.* **39,** 2319 (1974); eidem, *J. Chem. Soc. Perkin Trans. I* **1975,** 1972. Crystals from methanol, mp 240-256°. $[\alpha]_D^{20}$ +10.5° (c = 1.056 in alc).

3-O-Acetyluzarigenin, $C_{25}H_{36}O_5$, hexagonal plates from methanol, mp 262-266°. $[\alpha]_D^{22}$ +4.6° (c = 1.09 in chloroform).

THERAP CAT: Antidiarrheal.

V

9810. Vaccenic Acid. *trans-11-Octadecenoic acid.* $C_{18}H_{34}O_2$; mol wt 282.45. C 76.54%, H 12.13%, O 11.33%. $CH_3(CH_2)_5CH=CH(CH_2)_9COOH$. Neutralization equivalent 282.5; iodine no. 89.9. Found in butterfat and in other animal fats. Growth-promoting factor for rats. Isoln: Bertram, *Biochem. Z.* **197**, 433 (1928). Synthesis: Böeseken, Hoagland, *Rec. Trav. Chim.* **46**, 632 (1927); Ahmad *et al.*, *J. Am. Chem. Soc.* **70**, 3391 (1948). Configuration: Rao, Daubert, *ibid.* 1102.

Platelets from acetone, mp 43-44°; n_D^{60} 1.4439; n_D^{70} 1.4402. Infrared absorption: Rao, Daubert, *loc. cit.*

Methyl ester, $C_{19}H_{36}O_2$, bp_3 172-173°.

9811. Vacciniin. *β-D-Glucopyranose 6-benzoate;* 6-benzoyl-D-glucose. $C_{13}H_{16}O_7$; mol wt 284.26. C 54.93%, H 5.67%, O 39.40%. From the berries of *Vaccinium vitisidaea* L., *V. macrocarpum* Ait., and *V. oxycoccus* L., Ericaceae: Griebel, *Z. Untersuch. Nahr.-u. Genussm.* **19**, 241 (1910); Brigl, Zerrweck, *Z. Physiol. Chem.* **229**, 117 (1934). Prepn and chromatography: Bock *et al.*, *Nahrung* **3**, 1036 (1960).

Crystalline powder from water, mp 120-123°. Bitter taste. $[\alpha]_D^{20}$ +47.4° (c = 3.13 in methanol). Sol in water, alcohol, acetone, methanol; sparingly sol in chloroform, benzene; practically insol in ether, petr ether.

Phenylhydrazone, yellow cryst powder from ethanol + water, mp 134-135°.

9812. Valacidin. $C_{26}H_{24}N_4O_8$. Antibiotic substance produced by a *Streptomyces* (NRRL 2675) similar to *S. lavendulae*: Bromer, McGuire, U.S. pat. **2,970,943** (1961 to Lilly).

Reddish-brown solid. pKa' 7.0. uv max (methanol): 375, 294, 246 nm (ϵ 16,800, 27,100, 40,400). Sol in alkaline aq solns and in most polar solvents. Slightly sol in lower alcohols. Substantially insol in most nonpolar organic solvents and in aq acid solns. Stable in aq solns up to pH 9.0.

USE: Preserving agent. In embalming fluids (compatible with formaldehyde). To preserve biological specimens.

9813. n-Valeraldehyde. *Pentanal;* valeral; valeric aldehyde. $C_5H_{10}O$; mol wt 86.13. C 69.72%, H 11.70%, O 18.58%. $CH_3(CH_2)_3CHO$. Prepn: Lieben, Rossi, *Ann.* **159**, 70 (1871); Olsen, U.S. pat. **2,548,171** (1951 to GAF); Sisti *et al.*, *J. Org. Chem.* **27**, 279 (1962).

Liquid, bp 102-103°. d_4^{20} 0.8095. n_D^{20} 1.3944. Very slightly sol in water; miscible with many organic solvents. LD_{50} orally in rats: 5.66 ml/kg, H. F. Smyth *et al.*, *Am. Ind. Hyg. Assoc. J.* **30**, 470 (1969).

USE: In flavoring compds, resin chemistry, rubber accelerators. *Caution:* A mild irritant.

9814. Valerian. Dried rhizome and roots of *Valeriana officinalis* L., Valerianaceae. *Habit.* Europe, Northern Asia; naturalized in eastern U.S. *Constit.* Volatile oil (~1%); valerine (valerianin) and chatinine (alkaloids); valeric, formic and malic acids; tannin, resin.

Component of *Valerbé.*

THERAP CAT: Sedative.

9815. Valeric Acid, Normal. *Pentanoic acid;* valerianic acid; propylacetic acid. $C_5H_{10}O_2$; mol wt 102.13. C 58.80%, H 9.87%, O 31.33%. $CH_3(CH_2)_3COOH$. Obtained by decompn of *n*-propylmalonic acid: Fürth, *Monatsh.* **9**, 308 (1888); from *n*-butyl chloride: Gilman, Kirby, *Org. Syn. coll. vol.* I, 363 (2nd ed., 1941). Industrially by oxidation of amyl alcohol or by fermentation processes.

Colorless liquid; unpleasant odor; d_4^{20} 0.939; mp −34.5°; bp 186-187°; bp_{23} 96°; n_D^{20} 1.4086. Sol in 30 parts water; free-

ly sol in alcohol, ether. LD_{50} i.v. in mice: 1290±53 mg/kg, L. Orö, A. Wretlind, *Acta Pharmacol. Toxicol.* **18**, 141 (1961).

Ethyl ester, $C_7H_{14}O_2$, *ethyl n-valerate.* Liquid. d_4^{20} 0.877. bp 145-146°. n_D^{20} 1.3732. Insol in water; misc with alcohol, ether.

USE: Intermediate in perfumery.

9816. Valethamate Bromide. *N,N-Diethyl-N-methyl-2-[(3-methyl-1-oxo-2-phenylpentyl)oxy]ethanaminium bromide; 3-methyl-2-phenylvaleric acid diethyl(3-hydroxyethyl)-methylammonium bromide ester;* 2-phenyl-3-methylvaleric acid β-(diethylamino)ethyl ester bromomethylate; 3-methyl-2-phenylvaleric acid 2-diethylaminoethyl ester methyl bromide; 2-diethylaminoethyl 2-phenyl-3-methylvalerate methyl bromide; diethyl(2-hydroxyethyl)methylammonium 3-methyl-2-phenylvalerate bromide; Resitan; Epidosin; Murel. $C_{19}H_{32}BrNO_2$; mol wt 386.38. C 59.06%, H 8.35%, Br 20.68%, N 3.63%, O 8.28%. Prepn: Stühmer, Funke, **Ger.** pats. **969,245; 971,136; 1,112,989** (all 1958 to Kali-Chemie); Martin, Habicht, **Ger.** pat. **1,091,124** and U.S. pat. **2,987,517** (1960, 1961 to Cilag-Chemie).

Crystals from ethanol + ether or acetone, mp 100-101°. Freely sol in water, very sol in alcohol. Practically insol in ether. Aq solns are stable to storage; 0.6% ampuled solns showed no loss after one year at room temp.

THERAP CAT: Anticholinergic.

9817. Validamycins. Antibiotic complex produced by *Streptomyces hygroscopicus* var *limoneus.* Consists of validamycins A (major component), B, C, D, E, and F. Isoln, characterization, and biological properties of validamycins A and B: Iwasa *et al.*, *J. Antibiot.* **24**, 107, 119 (1971). Isoln and characterization of validamycins C, D, E, F: Horii *et al.*, *ibid.* **25**, 48 (1972). Manuf: Horii *et al.*, **Japan.** pat. **39,697('72)** (to Takeda), *C.A.* **78**, 122657a (1973). Structure of validamycin A: Horii, Kameda, *Chem. Commun.* **1972**, 747; revised structure: T. Suami *et al.*, *J. Antibiot.* **33**, 98 (1980). Bioassay methods: Iwasa *et al.*, *ibid.* **24**, 114 (1971). Total synthesis of (±)-validoxylamines A and B, constituents of validamycins: S. Ogawa *et al.*, *J. Org. Chem.* **49**, 2594 (1984). Total synthesis of validamycin B: S. Ogawa, Y. Miyamoto, *Chem. Commun.* **1987**, 1843; of validamycin A: *eidem, Chem. Letters* **1988**, 889.

validamycin A

Validamycin A, $C_{20}H_{35}NO_{13}$. *1,5,6-Trideoxy-3-o-β-D-glucopyranosyl-5-(hydroxymethyl)-1-[[4,5,6-trihydroxy-3-(hydroxymethyl)-2-cyclohexen-1-yl]amino]-D-chiro-inositol, Validacin, Valimon.* Colorless hydrophilic powder. Does not show sharp mp; softens at 100°, dec at 135°. $[\alpha]_D^{24}$ +110° (c = 1 in water or pyridine), 92° (c = 1 in DMF). pKa 6.0. Sol in water, methanol, DMF, DMSO; sparingly sol in ethanol, acetone. Insol in ethyl acetate, diethyl ether.

Hydrochloride, $C_{20}H_{36}ClNO_{13}$, crystalline powder, mp 95° (dec). $[\alpha]_D^{22}$ +49° (c = 1 in water). Sol in water, alcohols. Insol in acetone, diethyl ether.

USE: Fungicide.

9818. Valine. Val (IUPAC abbrev.); 2-aminoisovaleric

acid; 2-amino-3-methylbutyric acid; α-aminoisovaleric acid; 2-amino-3-methylbutanoic acid. $C_5H_{11}NO_2$; mol wt 117.15. C 51.26%, H 9.46%, N 11.96%, O 27.32%. An amino acid classified as "essential" for maintaining growth of rats. Essential component in human nutrition, not synthesized by the human body. Occurs esp in fibrous proteins. Isoln by hydrolysis of fish proteins: Abderhalden, Landan, Z. Physiol. Chem. **71**, 458 (1911). Structure: Fischer, Ber. **39**, 2320 (1906). Prepared by the action of ammonia and hydrogen cyanide on isobutyraldehyde (Strecker synthesis) followed by hydrolysis: Lipp, Ann. **205**, 18 (1880); by the action of ammonia on α-bromoisovaleric acid: Clark, Fittig, Ann. **139**, 202 (1866); Schmidt, Sachtleben, Ann. **193**, 105 (1878); C. S. Marvel, Org. Syn. coll. vol. III, 848 (1955). Production through a hydantoin intermediate: Goldsmith, Tishler, U.S. pat. **2,480,644** (1949 to Merck & Co.); White, U.S. pats. **2,557,920**; **2,700,054** (1951, 1955 to Dow).

$$(CH_3)_2CH\text{-}\overset{\overset{NH_2}{|}}{\underset{\underset{H}{|}}{C}}\text{-}COOH$$

L-valine

L-Form: leaflets from water + alcohol, mp 315° (closed capillary). d 1.230. Sublimes. $[M]_D$ +33.1° (5N HCl); +72.6° (glacial acetic acid); $[\alpha]_D^{26}$ +13.9° (p = 0.9); $[\alpha]_D^{23}$ +22.9° (c = 0.8 in 20% HCl). Soly in water at 0°: 83.4 g/l; at 25°: 88.5 g/l; at 50°: 96.2 g/l; at 65°: 102.4 g/l. Almost insol in cold alcohol, ether, acetone.

DL-Form, sublimes without melting at ordinary speed of heating. Dec 298° (closed capillary, very rapid heating). pK_1 2.29; pK_2 9.72. One part dissolves in 11.7 parts of water at 15°, in 14.1 parts of water at 25°. Almost insol in cold alcohol and ether.

9819. Valinomycin. $C_{54}H_{90}N_6O_{18}$; mol wt 1111.36. C 58.36%, H 8.16%, N 7.56%, O 25.92%. Cyclododecadepsi-peptide ionophore antibiotic produced by Streptomyces fulvissimus and related to the enniatins, q.v. Composed of 3 moles each of L-valine, D-α-hydroxyisovaleric acid, D-valine, and L-lactic acid linked alternately to form a 36-membered ring: Brockmann et al., Ber. **88**, 57 (1955); Ann. **603**, 216 (1957). Structural studies: Shemyakin et al., Tetrahedron Letters **1963**, 351; Tetrahedron **19**, 995 (1963). Proposed structure: Brockmann et al., Naturwiss. **50**, 689 (1963). Structure and synthesis: Shemyakin et al., Tetrahedron Letters **1963**, 1921. Solid phase synthesis: Gisin et al., J. Am. Chem. Soc. **91**, 2691 (1969); Losse, Klengel, Tetrahedron **27**, 1423 (1971). Biosynthesis: Smirnova et al., C.A. **73**, 97347m (1970); Ristow et al., FEBS Letters **42**, 127 (1974). Conformation: Ivanov et al., Biochem. Biophys. Res. Commun. **34**, 803 (1969); Onishi, Urry, ibid. **36**, 194 (1969); Duax et al.,

Science **176**, 911 (1972). Review: Y. A. Ovchinnokov, V. T. Ivanov, "The Cyclic Peptides: Structure, Conformation, and Function", in The Proteins, vol. V, H. Neurath, R. L. Hill, Eds. (Academic Press, New York, 3rd ed., 1982) pp 563-573.

Shiny rectangular platelets from dibutyl ether, mp 190° (hot stage). $[\alpha]_D^{20}$ +31.0° (c = 1.6 in benzene). Neutral reaction. Practically insol in water. Freely sol in petr ether, ether, benzene, chloroform, glacial acetic, butyl acetate, acetone. Active in vitro against Mycobacterium tuberculosis.
USE: Insecticide, nematocide, Patterson, Wright, U.S. pat. **3,520,973** (1970 to Am. Cyanamid).

9820. Valnoctamide. 2-Ethyl-3-methylpentanamide; 2-ethyl-3-methylvaleramide; α-ethyl-β-methylvaleramide; valmethamide; McN-X-181; Axiquel; Nirvanil. $C_8H_{17}NO$; mol wt 143.22. C 67.09%, H 11.96%, N 9.78%, O 11.17%. Prepn by hydrolysis of 5-ethyl-5-(1-methylpropyl)barbituric acid: Freifelder et al., J. Org. Chem. **26**, 203 (1961).

$$CH_3CH_2CH_2\overset{\overset{CH_3}{|}}{\underset{\underset{C_2H_5}{|}}{CH}}CHCONH_2$$

Crystals, mp 113.5-114°.
THERAP CAT: Tranquilizer.

9821. Valproic Acid. 2-Propylpentanoic acid; 2-propylvaleric acid; di-n-propylacetic acid; Mylproin. $C_8H_{16}O_2$; mol wt 144.21. C 66.62%, H 11.19%, O 22.19%. $(CH_3CH_2CH_2)_2CHCOOH$. Prepn: Oberreit, Ber. **29**, 1998 (1896); Keil, Z. Physiol. Chem. **282**, 137 (1947); Wiemann, Thuan, Bull. Soc. Chim. France **1958**, 199. Pharmacology: Meunier et al., Therapie **18**, 435 (1963); Lebreton et al., ibid. **19**, 451 (1964); Carraz et al., ibid. **20**, 419 (1965). Activity of deriv: Meunier, U.S. pat. **3,325,361** (1967 to Chemetron); idem, Fr. pat. CAM **244** (1969 to Lab. Berthier), C.A. **74**, 146380y (1971). Mechanism of action: B. Meldrum, Brain Res. Bull. **5**, Suppl. 2, 579 (1980). Review of early clinical studies: Carraz et al., Encephale **54**, 458 (1965). Use in infantile spasms: D. S. Bachman, Arch. Neurol. **39**, 49 (1982). Toxicity data: Jenner et al., Food Cosmet. Toxicol. **2**, 327 (1964). Discussion of potential teratogenicity: R. W. Hurd et al., Lancet **1**, 181 (1983). Review of pharmacology, therapeutic indications, and adverse reactions: E. M. Rimmer, A. Richens, Pharmacother. **5**, 171-184 (1985). Comprehensive description: Z. L. Chang in Analytical Profiles of Drug Substances vol. **8**, K. Florey, Ed. (Academic Press, New York, 1979) pp 529-556. Reviews: R. M. Pinder et al., Drugs **13**, 81-123 (1977); A. T. Dren et al., in Pharmacological and Biochemical Properties of Drug Substances vol. **2**, M. E. Goldberg, Ed. (Am. Pharm. Assoc., Washington, DC, 1979) pp 58-97.

Colorless liquid, bp_{14} 120-121°; bp_{20} 128-130°. $n_D^{24.5}$ 1.425. d_{25} 0.904. pKa 4.6. Very slightly sol in water. LD_{50} orally in rats: 670 mg/kg (Jenner).

Sodium salt (1:1), $C_8H_{15}NaO_2$, sodium valproate, DPA sodium, Convulex, Depakene, Depakin, Dépakine, Epilim, Ergenyl, Eurekene, Labazene, Leptilan, Orfiril, Valcote. White crystals. pKa 4.8. Hygroscopic. LD_{50} orally in mice: 1700 mg/kg (Meunier).

Sodium salt (2:1), $C_{16}H_{31}NaO_4$, sodium hydrogen bis(2-propylpentanoate), divalproex sodium, Abbott-50711, Depakote.

Magnesium salt, $C_{16}H_{30}MgO_4$, Logical.
THERAP CAT: Anticonvulsant. Anti-epileptic.

9822. Valpromide. 2-Propylpentanamide; 2-propylvaleramide; 2-propylpentamide; dipropylacetamide; Dépamide. $C_8H_{17}NO$; mol wt 143.23. C 67.08%, H 11.97%, N 9.78%, O 11.17%. $(CH_3CH_2CH_2)_2CHCONH_2$. Method of prepn: Fischer, Dilthey, Ber. **35**, 853 (1902); Tiffeneau, Deux, Compt. Rend. **212**, 105 (1941); Wiemann, Thuan, Bull. Soc. Chim. France **1958**, 199; Benoit-Guyod et al., ibid. **1965**, 1660. Patents: Meunier, U.S. pats. **3,301,754** and **3,325,361** (both 1967 to Chemetron). Pharmacology: Carraz et al., Therapie **19**, 468 (1964); Eymard, Werbenac, ibid. **27**, 11 (1972).

White, odorless and bitter crystalline powder, mp 125-126°. Practically insol in water.

THERAP CAT: Anticonvulsant.

9823. Vanadium. V; at. wt 50.9415; at. no. 23; valences 2, 3, 4, 5. Two naturally occurring isotopes: ^{51}V (99.75%); ^{50}V (0.25%); the latter is radioactive: $T_{1/2}$ 6 × 10^{15} years. Artificial isotopes: 46-49; 52-54. Abundance in earth's crust: 0.01% by wt. Widespread in nature; over 65 minerals known including *patronite* (polysulfide), *vanadinite* (9PbO.-3V$_2$O$_5$.PbCl$_2$), *roscoelite* [2K$_2$O.2Al$_2$O$_3$.(Mg,Fe)O.3V$_2$O$_5$.-10SiO$_2$.4H$_2$O] and *carnotite* (K$_2$O.2U$_2$O$_3$.V$_2$O$_5$.3H$_2$O). [Carnotite is also an imporant source of uranium]. Discovered by Selström in 1830; prepd by Roscoe in 1869. Prepn: Prandtl, Manz, *Z. Anorg. Allgem. Chem.* **79**, 209 (1912); Marden, Rich, *J. Ind. Eng. Chem.* **19**, 786 (1927); McKechnie, Seybolt, *J. Electrochem. Soc.* **97**, 311 (1950); Gregory, Lilliendahl, *ibid.* **98**, 395 (1951); *Handbook of Preparative Inorganic Chemistry* vol. 2, G. Brauer, Ed. (Academic Press, New York, 2nd ed., 1965) pp 1252-1255. *Review:* Clark, "Vanadium" in *Comprehensive Inorganic Chemistry* vol. 3, J. C. Bailar, Jr. *et al.*, Eds. (Pergamon Press, Oxford, 1973) pp 491-551.

Light gray or white lustrous powder, fused hard lumps or body-centered cubic crystals; not tarnished in air and not appreciably affected by moisture at ordinary temp. mp 1917°. d$^{18.7}$ 6.11. Sp heat (20-100°) 0.12 cal/g/°C. Electrical resistivity 24.8 microhms/cm. Insol in water. Not attacked by hot or cold hydrochloric acid, by cold sulfuric acid. Reacts with hot sulfuric acid, hydrofluoric acid, nitric acid, aqua regia. Not attacked by bromine water, or by aq alkalies. The metal precipitates gold, silver and platinum from their salts; reduces mercuric salts to mercurous, ferric salts to ferrous.

USE: Manuf rust-resistant vanadium steel.

9824. Vanadium Carbonyl. Vanadium hexacarbonyl. C$_6$O$_6$V; mol wt 219.00. C 32.90%, O 43.84%, V 23.26%. V(CO)$_6$. Prepn: Natta *et al.*, *C.A.* **54**, 16252 (1960); Ercoli *et al.*, *J. Am. Chem. Soc.* **82**, 2966 (1960); Hileman in *Preparative Inorganic Reactions* **1**, 107 (1964).

Blue-green pyrophoric crystals. Sensitive to air. Should be stored under nitrogen. Dec 60-70° in a nitrogen-filled tube.

9825. Vanadium Pentafluoride. F$_5$V; mol wt 145.95. V 34.91%, F 65.09%. VF$_5$. Prepd from vanadium tetrafluoride by thermal disproportionation at 650° in N$_2$ current according to the eq 2VF$_4$ → VF$_5$ + VF$_3$: Ruff, Lickfett, *Ber.* **44**, 2548 (1911); Kwasnik in *Handbook of Preparative Inorganic Chemistry*, vol. 1, G. Brauer, Ed. (Academic Press, New York, 2nd ed., 1963) p 253; from the elements: Trevorrow, *et al.*, *J. Am. Chem. Soc.* **79**, 5167 (1957); Clark, Emeleus, *J. Chem. Soc.* **1957**, 2119. Review of transition metal pentafluorides: Peacock, *Advan. Fluorine Chem.* **7**, 113-145 (1973).

Liquid. mp 19.5°; bp 47.9°. d$_4^{19.5}$ 2.502. Heat of vaporization 10.60 kcal/mole. Appreciable vapor pressure at room temp. Hydrolyzed by water, dil alkali. Soly in anhydrous HF: 3.3 moles/l. Freely sol in alcohol, chloroform, acetone, ligroin. Insol in carbon disulfide. Dec toluene and ether. Etches glass slowly at room temp; more rapidly in presence of moisture with formation of yellow color. May be stored in sealed vessels made from iron, nickel, copper, platinum.

9826. Vanadium Pentoxide. Vanadic anhydride. O$_5$V$_2$; mol wt 181.90. O 43.98%, V 56.02%. V$_2$O$_5$. Prepd by heating ammonium metavanadate, NH$_4$VO$_3$.

Yellow to rust-brown orthorhombic crystals; d 3.35; mp 690°. Loses oxygen reversibly in the region 700-1125°. One gram dissolves in about 125 ml water; sol in concd acids, forming red to yellow solns; sol in alkalies, forming vanadates; insol in alcohol. Its acid solutions are reduced by SO$_2$, Zn + HCl, and by evaporation with HCl.

Human Toxicity: Reported to be a respiratory irritant and to cause skin pallor, greenish-black tongue, chest pain, cough, dyspnea, palpitation, lung changes. When ingested, causes G.I. disturbances. May also cause a papular skin rash: Zenz *et al.*, *Arch. Environ. Health* **5**, 542 (1962); E. Browning, *Toxicity of Industrial Metals* (Appleton-Century-Crofts, New York, 2nd ed., 1969) pp 340-347.

USE: As catalyst in the oxidation of SO$_2$ to SO$_3$, alcohol to

acetaldehyde, etc.; for the manuf of yellow glass; inhibiting ultraviolet light transmission in glass; depolarizer; as developer in photography; in form of ammonium vanadate as mordant in dyeing and printing fabrics and in manuf of aniline black.

9827. Vanadium Tetrafluoride. F$_4$V; mol wt 126.95. V 40.13%, F 59.87%. VF$_4$. Prepd from VCl$_4$ and HF or from the elements: Ruff, Lickfett, *Ber.* **44**, 2539 (1911); Cavell, Clark, *J. Chem. Soc.* **1962**, 2692; Kwasnik in *Handbook of Preparative Inorganic Chemistry*, vol. 1, G. Brauer, Ed. (Academic Press, New York, 2nd ed., 1963) pp 252-253.

Bright lime-green powder; very hygroscopic; hydrolyzes to form a brown powder and eventually a blue paste. d 3.15. Sublimes at 100-120° in a vacuum. Also disproportionates to VF$_3$ and VF$_5$ at 100-120° in a vacuum. Freely sol in water with blue color, in acetone deep green, in glacial acetic acid blue green. Sparingly sol in alc, chloroform, SO$_2$Cl$_2$. May be stored in sealed iron or copper vessels.

9828. Vanadium Trifluoride. F$_3$V; mol wt 107.95. V 47.20%, F 52.80%. VF$_3$. Prepd from VCl$_3$ and HF: Ruff, Lickfett, *Ber.* **44**, 2539 (1911); from VCl$_2$ and HF: Emeleus, Gutmann, *J. Chem. Soc.* **1949**, 2979; by thermal decompn in an inert atm of (NH$_4$)$_3$VF$_6$: Sturm, Sheridan, *Inorg. Syn.* **7**, 87 (1963).

Greenish-yellow powder. d 3.363. mp approx 1406°. Sublimes at bright red heat. Almost insol in water, alcohol, acetone, ethyl acetate, acetic anhydride, glacial acetic acid, toluene, carbon tetrachloride, chloroform, carbon disulfide. Black color with NaOH soln. Aq solns have strong reducing properties.

Trihydrate, dark green rhombohedral crystals, obtained by crystallizing vanadium trioxide dissolved in hydrofluoric acid, or by electrolytic reduction of a soln of vanadium pentoxide in aq HF. Loses one molecule of water at 100°. Dissolves in water to some extent, forming autocomplexes.

9829. Vanadium Trioxide. Vanadium sesquioxide; vanadic oxide. O$_3$V$_2$; mol wt 149.90. V 67.98%, O 32.02%. V$_2$O$_3$. Prepd by reduction of V$_2$O$_5$ with hydrogen or carbon monoxide.

Black powder. On exposure to air it is gradually converted into indigo-blue crystals of V$_2$O$_4$. d 4.87; mp 1940°. Insol in water; difficultly sol in acids.

USE: Catalyst, e.g., when making ethanol from ethylene.

9830. Vanadium Trisulfate. O$_{12}$S$_3$V$_2$; mol wt 390.10. O 49.22%, S 24.66%, V 26.12%. V$_2$(SO$_4$)$_3$. Prepd by the reduction of V$_2$O$_5$ in sulfuric acid with elemental sulfur: Auger, *Compt. Rend.* **173**, 306 (1921); Riveng, *Bull. Soc. Chim. France* **4**, 1697 (1937); Claunch, Jones, *Inorg. Syn.* **7**, 92 (1963).

Lemon-yellow powder. When heated in vacuum at or slightly below 410°, it dec into VOSO$_4$ and SO$_2$. Stable in dry air. Upon exposure to moist air for several weeks, a green hydrate forms. Very slowly sol in water at room temp, somewhat faster in boiling water; practically insol in concd sulfuric acid; sol in dil and concd nitric acid. Powerful reducing agent.

9831. Vanadium Trisulfide. Vanadium sesquisulfide. S$_3$V$_2$; mol wt 198.10. V 51.44%, S 48.56%. V$_2$S$_3$.

Greenish-black powder; d 4.7. Dec when heated. Insol in water, cold HCl, dil H$_2$SO$_4$; sol in hot HCl, hot dil H$_2$SO$_4$, HNO$_3$.

9832. Vanadyl Dichloride. Vanadium oxydichloride. Cl$_2$OV; mol wt 137.86. V 36.96%, O 11.61%, Cl 51.44%. VOCl$_2$. Usually contains some water. Prepd according to the eq V$_2$O$_5$ + 3VCl$_3$ + VOCl$_3$ → 6VOCl$_2$: Funk, Weiss, *Z. Anorg. Allgem. Chem.* **295**, 327 (1958); Oppermann, *ibid.* **351**, 113 (1967).

Green, very deliquesc tabular crystals. d 2.88. Disproportionates at 384° to VOCl and VOCl$_3$. Slowly dec by water. Sol in abs alcohol, glacial acetic acid. *Keep tightly closed.*

USE: Has been used as mordant in printing fabrics.

9833. Vanadyl Sulfate. Vanadium oxysulfate. O$_5$SV; mol wt 163.00. O 49.08%, S 19.67%, V 31.25%. VOSO$_4$. Dihydrate, blue, cryst powder. Sol in water.

USE: Dihydrate as mordant in dyeing and printing textiles; manuf colored glass; for blue and green glazes on pottery.

9834. Vanadyl Trichloride. *Trichlorooxovanadium;* vanadium oxytrichloride. Cl_3OV; mol wt 173.32. Cl 61.37%, O 9.23%, V 29.40%. $VOCl_3$. Prepn by the action of dry Cl_2 on V_2O_3 or V_2O_5: Brown, Griffitts, *Inorg. Syn.* **1**, 106 (1939); **4**, 80 (1953); Oppermann, *Z. Anorg. Allgem. Chem.* **351**, 113 (1967). Toxicity data: Smyth *et al., Am. Ind. Hyg. Assoc. J.* **30**, 470 (1969).

Yellow liquid, emitting red fumes on exposure to moist air. d 1.84. bp 126-127°. mp −77°. Dec in presence of moisture into vanadic acid and HCl. When a small quantity of water is added, it becomes thick and almost blood-red because of formation of vanadic acid. *Keep tightly closed.* LD_{50} orally in rats: 0.14 g/kg (Smyth).

9835. Vanaspati. Manuf by hydrogenation of vegetable oils, peanut, cottonseed and sesame in the presence of activated-Ni catalyst: Gupta, *Indian Chem. Engr.* **3**, 163 (1961), *C.A.* **56**, 15895h (1962). Product mp 33-37°; max free fatty acid (as oleic) 0.25%.

USE: Hydrogenated shortening. Used in the Orient.

9836. Vancomycin. Vancocin. $C_{66}H_{75}Cl_2N_9O_{24}$; mol wt 1449.22. C 54.69%, H 5.22%, Cl 4.89%, N 8.70%, O 26.50%. Amphoteric glycopeptide antibiotic substance produced by *Streptomyces orientalis* from Indonesian and Indian soil which inhibits bacterial mucopeptide biosynthesis by formation of complexes. Isoln: M. H. McCormick *et al., Antibiot. Ann.* **1955-56**, 606; U.S. pat. 3,067,099 (1962 to Lilly). Purif: H. M. Higgins *et al., Antibiot. Ann.* **1957-58**, 906. Structure studies: F. J. Marshall, *J. Med. Chem.* **8**, 18 (1965); W. D. Weringa *et al., J. Chem. Soc. Perkin Trans. I* **1972**, 443; K. A. Smith *et al., ibid.* **1974**, 2369. Total structure determination by x-ray analysis: G. M. Sheldrick *et al., Nature* **271**, 223 (1978). Revised configuration: M. P. Williamson, D. H. Williams, *J. Am. Chem. Soc.* **103**, 6580 (1981). Further revision of vancomycin structure: C. M. Harris *et al., ibid.* **105**, 6915 (1983). Complex formation: M. Nieto, H. P. Perkins, *Biochem. J.* **123**, 789 (1971). Mode of binding: D. H. Williams, J. R. Kalman, *J. Am. Chem. Soc.* **99**, 2768 (1977). Toxicology: R. C. Anderson *et al., Antibiot. Ann.* **1956-57**, 75. Series of articles on antimicrobial activity, pharmacokinetics and clinical usage: *J. Antimicrob. Chemother.* **14**, Suppl. D, 1-109 (1984). Review of clinical studies: J. W. Lightbown, *Experimental Chemotherapy* vol. III, R. J. Schitzer, F. Hawking, Eds. (Assoc. Press, London, 1964) pp 278-289. *General review:* Jordan, Reynolds, in *Antibiotics* vol. 3, J. W. Corcoran, F. E. Hahn, Eds. (Springer-Verlag, New York, 1975) pp 704-718.

Monohydrochloride, $C_{66}H_{76}Cl_3N_9O_{24}$, *Lyphocin, Vancor.* White solid; uv max (H_2O): 282 nm ($E_{1cm}^{1\%}$ 40). Soly in water: > 100 mg/ml. Moderately sol in dil methanol. Insol in the higher alcohols, acetone, ether. Low concns of urea increase the soly in neutral aq solns. Ammonium sulfate and sodium chloride ppt the antibiotic from acidic solns. LD_{50} in mice (mg/kg): 489 i.v.; 1734 i.p.; 5000 s.c. and orally (Anderson).

THERAP CAT: Antibacterial.

9837. Vanilla. Cured, full-grown, unripe fruit of *Vanilla planifolia* Andr., *Orchidaceae. Habit.* Mexico, West Indies, Reunion, Mauritius, Seychelles. *Constit.* 2-2.75% vanillin, about 4% resin; vanillic acid, about 10% sugar.

USE: In manuf of confectionery and in various bakery products; perfumery; flavor for beverages; pharmaceutic aid (flavor).

9838. Vanillic Acid. *4-Hydroxy-3-methoxybenzoic acid.* $C_8H_8O_4$; mol wt 168.14. C 57.14%, H 4.80%, O 38.06%. Obtained from vanillin by oxidation with silver oxide or by controlled caustic fusion: Pearl, *Org. Syn.* **30**, 101 (1950).

White, odorless needles. mp 210°. Sublimes undecomposed. Sol in 860 parts water; very sol in alc; sol in ether. Not colored by $FeCl_3$. Its salts are freely sol in water.

9839. Vanillin. *4-Hydroxy-3-methoxybenzaldehyde;* methylprotocatechuic aldehyde; vanillic aldehyde; 3-methoxy-4-hydroxybenzaldehyde. $C_8H_8O_3$; mol wt 152.14. C 63.15%, H 5.30%, O 31.55%. Occurs naturally in vanilla, in potato parings, in Siam benzoin, etc.; made synthetically from eugenol or guaiacol; also from the waste (lignin) of the wood pulp industry: Sörensen, Mehlum, U.S. pat. 2,752,394 (1956). Production from lignosulfonic acid compds: Craig, Logan, U.S. pats. 3,054,659 and 3,054,825 (both 1962 to Ontario Paper Co.). Toxicity: P. M. Jenner *et al., Food Cosmet. Toxicol.* **2**, 327 (1964).

White or very slightly yellow needles; pleasant aromatic vanilla odor and taste. Slowly oxidizes somewhat on exposure to moist air. Affected by light. d 1.056; mp 80-81° (81-83°); bp 285°; bp_{15} 170°. One gram dissolves in 100 ml water, 16 ml water at 80°, about 20 ml glycerol; freely sol in alc, chloroform, ether, carbon disulfide, glacial acetic acid, pyridine; also sol in oils and aq solns of alkali hydroxides. Solns are acid to litmus. *Preserve in tight, light-resistant containers.* LD_{50} orally in rats: 1580 mg/kg (Jenner).

USE: Pharmaceutic aid (flavor). As a flavoring agent in confectionery, beverages, foods; in perfumery. One part vanillin equals 400 parts vanilla pods; in manuf liqueurs, 2.5-3 parts vanillin replace 500 parts tincture vanilla. Also as reagent in analytical chemistry.

9840. Vanilmandelic Acid. *α,4-Dihydroxy-3-methoxybenzeneacetic acid;* 3-methoxy-4-hydroxymandelic acid; 4-hydroxy-3-methoxymandelic acid; VMA. $C_9H_{10}O_5$; mol wt 198.17. C 54.54%, H 5.09%, O 40.37%. Catecholamine metabolite. Urine levels elevated in various pathologies. Misnamed *vanillinemandelic acid* and *vanillylmandelic acid.* Prepn from vanillin cyanohydrin: Gardner, Hibbert, *J. Am. Chem. Soc.* **66**, 608 (1944). Improved procedure: Shaw *et al., J. Org. Chem.* **23**, 30 (1958); E. F. Recondo, H. Rinderknecht, *ibid.* **25**, 2248 (1960); I. Goodman *et al., Biochem. Prep.* **13**, 75 (1971). Resolution: Armstrong *et al., Biochim. Biophys. Acta* **25**, 422 (1957). Determination in urine: T. C. Stewart, J. A. Freeman, *Vanilmandelic Acid & Catecholamine Determinations* (Am. Soc. Clin. Pathol., Chicago, 1976) pp 1-81.

DL-Form, scales from ether + benzene, dec 131-133°; also reported as mp 134-135° (Goodman, *loc. cit.*). uv max (0.1*N* HCl): 230, 279 nm (ε 6320, 2810); (0.1*N* NaOH): 247, 285, 345 nm (ε 6860, 3960, 630). Readily resinifies on heating or on prolonged exposure to air. Freely sol in water, acetone. Mod sol in ether, acetonitrile. Sparingly sol in benzene. Mass spectral data: T. R. Sharp, *Org. Mass Spectrom.* **15**, 381 (1980).

L-Form, crystals, dec 152°. $[\alpha]_D^{22}$ +128° (c = 0.7).
D-Form, crystals, dec 152°. $[\alpha]_D^{23}$ −131°.

9841. Vanitiolide. *4-[(4-Hydroxy-3-methoxyphenyl)thiomethyl]morpholine; 4-(thiovanilloyl)morpholine; 4-hydroxy-3-methoxybenzoic acid morpholide thione; 2-hydroxy-5-(morpholinothiocarbonyl)anisole; 4-[4-hydroxy(thio-*m*-anisoyl)]morpholine; 2-methoxy-4-(morpholinothiocarbonyl)phenol; 4-(morpholinothiocarbonyl)guaiacol;* Bildux. $C_{12}H_{15}NO_3S$; mol wt 253.33. C 56.89%, H 5.97%, N 5.53%, O 18.95%, S 12.66%. Prepn of similar compds, 4-[*o*- and *p*-methoxy(thiobenzoyl)]morpholine and 4-(thiosalicyloyl)morpholine: Chabrier *et al., Bull. Soc. Chim. France* **1950**, 1167.

THERAP CAT: Choleretic.

9842. Vasicine. *1,2,3,9-Tetrahydropyrrolo[2,1-b]quinazolin-3-ol;* peganine. $C_{11}H_{12}N_2O$; mol wt 188.22. C 70.19%, H 6.43%, N 14.88%, O 8.50%. Isoln from *Adhatoda vasica* Nees, *Acanthaceae:* Hooper, *Pharm. J.* **18**, 841 (1888); Sen, Ghose, *J. Indian Chem. Soc.* **1**, 315 (1924); Mehta *et al., J. Org. Chem.* **28**, 445 (1963). Isoln from *Peganum harmala* L., *Zygophyllaceae:* Späth, Nikawitz, *Ber.* **67**, 45 (1934); Späth, Kuffner, *ibid.* **67**, 868 (1934). Structure and synthesis: Späth *et al., ibid.* **68**, 699 (1935); Späth, Platzer, *ibid.* **69**, 255 (1936). Synthesis of *dl*-vasicine: Southwick, Casanova, *J. Am. Chem. Soc.* **80**, 1168 (1958). *Reviews:* Späth, *Monatsh.* **72**, 115 (1938); H. T. Openshaw, "The Quinazoline Alkaloids" in *The Alkaloids* vol. III, R. H. F. Manske, H. L. Holmes, Eds. (Academic Press, New York, 1953) pp 101-118; Ray, *J. Indian Chem. Soc.* **35**, 697 (1958).

dl-Form, needles from alc. mp 210°. Sublimes in high vacuum. Sol in acetone, alcohol, chloroform; slightly sol in water, ether, benzene.

l-Form, needles from alc, mp 212°. $[\alpha]_D^{14}$ −254° (c = 2.4 in CHCl$_3$); $[\alpha]_D^{14}$ −62° (c = 2.4 in alc). In dil HCl this alkaloid is dextrorotatory.

Hydrochloride dihydrate, needles, mp 208° (dry).
Hydriodide dihydrate, needles, mp 195° (dry).
Methiodide, needles from methanol, mp 187°.
Acetylvasicine, $C_{11}H_{11}N_2OCOCH_3$, crystals, mp 123°, bp$_{0.01}$ 230-240°.

9843. Vasopressin. Antidiuretic hormone; beta-hypophamine; β-hypophamine; Leiormone; Pitressin; Tonephin; Vasophysin. The water-soluble pressor principle prepd by synthesis or obtained from the posterior lobe of the pituitary of healthy domestic animals used for food by man: *U.S.P.* XVIII, 770. Separation from oxytocin: Kamm *et al., J. Am.*

Chem. Soc. **50**, 573 (1928). Purification: Turner *et al., J. Biol. Chem.* **191**, 21 (1951). Two vasopressins, differing only in the amino acid at position 8, have been isolated: arginine vasopressin, from human, beef, chicken, horse and sheep pituitaries, and lysine vasopressin (lypressin, *q.v.*), from porcine pituitaries. *Review:* E. Schröder, K. Lübke, *The Peptides* vol. II (Academic Press, New York, 1966) pp 336-350; C. R. W. Edwards, "Vasopressin" in *Hormones in Blood* vol. 2, C. H. Gray, V. H. T. James, Eds. (Academic Press, London, 3rd ed., 1979) pp 423-450. Book: *Antidiuretic Hormone* vol. 4, M. L. Forsling, Ed. (Eden Press, Quebec, 1980) 165 pp.

arginine vasopressin

Arginine vasopressin, $C_{46}H_{65}N_{15}O_{12}S_2$, *argipressin, rinder-vasopressin.* Structure: du Vigneaud *et al., J. Am. Chem. Soc.* **75**, 4880 (1953). Synthesis: du Vigneaud *et al., ibid.* **80**, 3355 (1958); Bodanszky *et al., ibid.* **86**, 4452 (1964); Meienhofer *et al., ibid.* **92**, 7199 (1970); Jones *et al., J. Org. Chem.* **38**, 2865 (1973). Fever suppressing activity: K. E. Cooper *et al., J. Physiol. (London)* **295**, 33 (1979). Effect on female sexual behavior in rats: P. Södersten *et al., Nature* **301**, 608 (1983).

Ornithine vasopressin see ornipressin.

THERAP CAT: Antidiuretic and vasopressor hormone; hemostatic.

THERAP CAT (VET): Antidiuretic hormone; in deficiency conditions.

9844. Veatchine. $C_{22}H_{33}NO_2$; mol wt 343.49. C 76.92%, H 9.68%, N 4.08%, O 9.32%. From bark of *Garrya veatchii* Kellogg, *Garryaceae,* where it occurs together with garryine and other alkaloids: Oneto, *J. Am. Pharm. Assoc.* **35**, 204 (1946); Wiesner *et al., Can. J. Chem.* **30**, 608 (1952). Structure: Wiesner *et al., J. Am. Chem. Soc.* **76**, 6068 (1954); Pelletier, *ibid.* **82**, 2398 (1960). Stereochemistry: Vorbrüggen, Djerassi, *Tetrahedron Letters* **1961**, 119; *J. Am. Chem. Soc.* **84**, 2990 (1962). Coexistence of epimers found in crystal and molecular structure: S. W. Pelletier *et al., ibid.* **100**, 7976 (1978). ^{13}C-NMR study of epimers: N. V. Mody, S. W. Pelletier, *Tetrahedron* **34**, 2421 (1978). Racemic syntheses and resolution: Nagata *et al., ibid.* **86**, 929 (1964), **89**, 1499 (1967); Guthrie *et al., Coll. Czech. Chem. Commun.* **31**, 602 (1966). Total synthesis of optically active form: Wiesner *et al., Experientia* **26**, 471 (1970).

Crystals from dil acetone, mp 119-120°. Bitter taste. pH 11.5. $[\alpha]_D^{27.5}$ −69.01° (c = 1.06 in ethanol). Readily sol in water, ethanol.

Hydrochloride, $C_{22}H_{33}NO_2 \cdot HCl$, crystals from abs ethanol + ether, dec 267-271°. Sol in water.

9845. Vecuronium Bromide. *1-[3,17-Bis(acetyloxy)-2-(1-piperidinyl)androstan-16-yl]-1-methylpiperidinium bromide;* NC-45; Org-NC-45; Musculax; Norcuron. $C_{34}H_{57}BrN_2O_4$; mol wt 637.75. C 64.03%, H 9.01%, Br 12.53%, N 4.39%, O 10.04%. A non-depolarizing neuromuscular blocking agent and homolog of pancuronium bromide, *q.v.* Prepn: W. R. Buckett *et al., J. Med. Chem.* **16**, 1116 (1973); I. C. Carlyle *et al., Eur.* pat. Appl. 8824 (1980 to Akzo), *C.A.* **93**, 155850 (1980). Series of articles on pharmacology, pharmacokinetics, pharmacodynamics, clinical studies: *Brit. J. Anaesth.* **52**, Suppl. 1, 1S-72S (1980). Clinical pharmacology: M. R. Fahey *et al., Anesthesiology* **55**, 6 (1981). Comparison with pancuronium bromide: S. L. Son *et al., ibid.* 12.

Cryst, mp 227-229°. LD_{50} in mice: 0.061 mg/kg i.v. (Buckett).

THERAP CAT: Skeletal muscle relaxant.

9846. Vellosimine. *Sarpagan-17-al;* velosimine. $C_{19}H_{20}$-N_2O; mol wt 292.39. C 78.05%, H 6.89%, N 9.58%, O 5.47%. From bark of *Geissosperum vellosii* Allem., *Apocynaceae:* Rapoport *et al., J. Am. Chem. Soc.* **80,** 1601 (1958). Structure: Rapoport, Moore, *J. Org. Chem.* **27,** 2981 (1962); Ohashi *et al., Tetrahedron* **19,** 2241 (1963).

Crystals from methanol, mp 305-306°. Can be sublimed at 180-100° (0.01 mm). $[\alpha]_D^{26}$ +48°. uv max (ethanol): 280, 289 nm (ϵ 8000, 6430).

9847. Venice Turpentine. Larch turpentine. Oleoresin from *Larix decidua* Mill. (*L. europaea* Lam. & DC.), *Pinaceae. Habit.* Middle and Southern Europe. *Constit.* Volatile oil, resin. Use in light microscopy: D. A. Johansen, *Plant Microtechnique* (McGraw Hill, New York, 1940) pp 115-116.

Yellow, sometimes greenish, limpid, tenacious, thick liquid; pleasant, aromatic odor; hot, pungent, somewhat bitter taste. Becomes hard and brittle on prolonged exposure to air. Insol in water; sol in glacial acetic acid, amyl alcohol, acetone, caustic alkalies; slowly but freely sol in alc.

USE: As clearing agent and mounting medium for light microscopy.

9848. Venturicidins. Antifungal antibiotics isolated from *Streptomyces aureofaciens* strains. Preliminary isolation work: Rhodes *et al., Nature* **192,** 952 (1962). Isoln of venturicidins A and B and activity studies: Brufani *et al., Helv. Chim. Acta* **51,** 1293 (1968); *see also* Langcake *et al., Biochem. Soc. Trans.* **2,** 202 (1974). Final structures: Brufani *et al., Experientia* **27,** 604 (1971); *eidem, Helv. Chim. Acta* **55,** 2329 (1972).

Venturicidin A. $C_{41}H_{67}NO_{11}$. R = NH_2CO. Needles from chloroform-petr ether, mp 145-147°. Also reported as mp 140-142°. $[\alpha]_D$ +119° (c = 0.5 in chloroform). uv max

(alcohol): 206, 247 (shoulder), ~300 nm (shoulder) (log ϵ 3.80, 2.23, 2.08).

Venturicidin B. $C_{40}H_{66}O_{10}$, *(3-decarbamoyloxy)-3-hydroxyventuricidin A.* R = H. Amorphous white powder from chloroform-ether, mp 168-170°. Also reported as mp 145-149° (ethyl acetate-petr ether). $[\alpha]_D$ +100° (c = 0.847 in chloroform).

9849. Veralipride. *5-(Aminosulfonyl)-2,3-dimethoxy-N-[[1-(2-propenyl)-2-pyrrolidinyl]methyl]benzamide; N-[(1-allyl-2-pyrrolidinyl)methyl]-5-sulfamoyl-o-veratramide;* LIR 1660; Agréal; Agradil; Veralipril. $C_{17}H_{25}N_3O_5S$; mol wt 383.47. C 53.25%, H 6.57%, N 10.96%, O 20.86%, S 8.36%. Prepn: **Neth. pat. Appl. 7,707,982** (1978 to Soc. Etud.-Sci. et Ind. de l'Ile-de-France), *C.A.* **89,** 30768 (1978). Pharmacological studies: P. Bouyard *et al., Sem. Hop.* **56,** 1475 (1980); J. C. Czyba, *ibid.* 1483. Clinical studies: R. Renaud, J. Macler, *ibid.* **57,** 353 (1981); S. Angeli, P. Fougère, *ibid.* **58,** 111 (1982).

THERAP CAT: Treatment of menopausal disorders.

9850. Veralkamine. *17-Methyl-18-nor-16,28-secosolanida-5,12-diene-3β,16β-diol; 17-methyl-20α-((2S,5S)-5-methyl-2-piperidyl)-18-nor-17α-pregna-5,12-diene-3β,16β-diol; (22S:25S)-12,26-epimino-17β-methyl-18-norcholesta-5,12-diene-3β,16β-diol; (17S:12S:25S)-22,26-epimino-18-(13 → 17)-abeo-cholesta-5,12-diene-3β,16β-diol;* veralcamine. $C_{27}H_{43}NO_2$; mol wt 413.62. C 78.40%, H 10.48%, N 3.39%, O 7.74%. Steroidal alkaloid isolated from *Veratrum album* sp. *lobelianium* (Bernh.) Suessenguth, *Liliaceae:* Tomko *et al., Pharm. Zentralhalle* **99,** 373 (1960), *C.A.* **55,** 2013e. Structure studies: Tomko *et al., Coll. Czech. Chem. Commun.* **27,** 1404 (1962). Complete structure: Tomko *et al., Tetrahedron Letters* **1967,** 3907; *eidem, Tetrahedron* **24,** 4865 (1968); Hoehne *et al., ibid.* 4875.

Crystals from ethanol, mp 119-123° and 165-169°; $[\alpha]_D^{24}$ −84.1° ±3 (c = 0.533 in $CHCl_3$).
N,O,O-Triacetate, mp 152-154°. $[\alpha]_D^{27}$ −8.0° ($CHCl_3$).
N-Monoacetate, mp 191-193°. $[\alpha]_D^{23}$ −79.1° ($CHCl_3$).

9851. Verapamil. *α-[3-[[2-(3,4-Dimethoxyphenyl)ethyl]methylamino]propyl]-3,4-dimethoxy-α-(1-methylethyl)benzeneacetonitrile; 5-[(3,4-dimethoxyphenethyl)methylamino]-2-(3,4-dimethoxyphenyl)-2-isopropylvaleronitrile; α-isopropyl-α-[(N-methyl-N-homoveratryl)-γ-aminopropyl]-3,4-dimethoxyphenylacetonitrile; iproveratril;* D 365. C_{27}-$H_{38}N_2O_4$; mol wt 454.59. C 71.33%, H 8.43%, N 6.16%, O 14.08%. Coronary vasodilator with calcium blocking activity. Prepn: **Belg. pat. 615,861;** Dengel, **U.S. pat. 3,261,859** (1962, 1966 both to Knoll). Physical and chemical data: Appel, *Arzneimittel-Forsch.* **12,** 562 (1962). Synthesis and absolute configuration of enantiomers: Ramuz, *Helv. Chim. Acta* **58,** 2050 (1975). Pharmacology: Haas, Hartfelder, *ibid.* 549; Schlepper, Witzleb, *ibid.* 559. Metabolism: McIlhenny, *J. Med. Chem.* **14,** 1178 (1971). HPLC determn in plasma: C. Horne *et al., Arzneimittel-Forsch.* **37,** 956 (1987). Series of articles on clinical studies: *ibid.* **20,** 1277-1336

(1970). Comparative study in variant angina: D. D. Waters et al., *Am. J. Cardiol.* **47**, 179 (1981); in arrhythmias: B. N. Singh et al., *Drugs* **25**, 125 (1983). Symposium on pharmacology and clinical efficacy in hypertension: *Am. J. Cardiol.* **57**, 1D-107D (1986). Comprehensive description: Z. L. Chang in *Analytical Profiles of Drug Substances* vol. 17, K. Florey, Ed. (Academic Press, New York, 1988) pp 643-674. *Reviews:* S. H. Baky, E. B. Kirsten, in *Pharmacological and Biochemical Properties of Drug Substances* vol. 3, M. E. Goldberg, Ed. (Am. Pharm. Assoc., Washington, DC, 1981) pp 226-261; D. J. Triggle, V. C. Swamy, *Circ. Res.* **52**(2), Pt. 2, 117-128 (1983).

Viscous, pale yellow oil, $bp_{0.01}$ 243-246°. n_D^{25} 1.5448. Practically insol in water. Sparingly sol in hexane; sol in benzene, ether. Freely sol in the lower alcohols, acetone, ethyl acetate, chloroform.

Hydrochloride, $C_{27}H_{39}ClN_2O_4$, *Berkatens, Calan, Cardiagutt, Cardibeltin, Cordilox, Dignover, Drosteakard, Geangin, Isoptin, Securon, Univer, Vasolan, Veramex, Veraptin, Verelan, Verexamil.* Crystals, dec 138.5-140.5°. Soly (21°): 7 g/100 g water (pH = 4.24). pH of 0.1% aq soln: 5.25. uv max: 232, 278 nm. Sparingly sol in chloroform. Sol in ethanol, isopropanol, acetone, ethyl acetate; freely sol in methanol, DMF. Soly (mg/ml): water 83, ethanol (200 proof) 26, propylene glycol 93, ethanol (190 proof) > 100, methanol > 100, 2-propanol 4.6, ethyl acetate 1.0, DMF > 100, methylene chloride > 100, hexane 0.001. LD_{50} in rats, mice (mg/kg): 16, 8 i.v. (Baky, Kirsten).

THERAP CAT: Anti-anginal; anti-arrhythmic.

9852. Veratraldehyde. *3,4-Dimethoxybenzaldehyde;* 3,4-dimethoxybenzenecarbonal; veratric aldehyde; protocatechualdehyde dimethyl ether. $C_9H_{10}O_3$; mol wt 166.17. C 65.05%, H 6.07%, O 28.88%. Prepd by methylation of vanillin: Kostanecki, Tambor, *Ber.* **39**, 4022 (1906); Buck, *Org. Syn.* **13**, 102 (1933); Alt, U.S. pat. **3,007,968** (1957 to Monsanto); by oxidation of veratryl alcohol with chromium(VI) oxide-pyridine complex: Holum, *J. Org. Chem.* **26**, 4814 (1961).

Needles from ether, petr ether, toluene, or carbon tetrachloride. Odor of vanilla beans; mp 42-43°; bp_{760} 281°; bp_{53} 201°; bp_{10} 155°. Slightly sol in hot water; freely sol in alcohol and ether. Solns are oxidized to veratric acid under the influence of light.

9853. Veratramine. $C_{27}H_{39}NO_2$; mol wt 409.59. C 79.17%, H 9.60%, N 3.42%, O 7.81%. Secondary base from *Veratrum grandiflorum* (Maxim.) Loes. f., and from *V. viride* Ait., *Liliaceae.* Isoln and structure: Saito, *Bull. Chem. Soc. Japan* **15**, 22 (1940); Jacobs, Craig, *J. Biol. Chem.* **160**, 555 (1945); Jacobs, Sato, *ibid.* **181**, 55 (1949); **191**, 71 (1951); Tamm, Wintersteiner, *J. Am. Chem. Soc.* **74**, 3842 (1952); Wintersteiner, *Festschrift Arthur Stoll* (Birkhäuser-Verlag, Basel) pp 166-176. Total synthesis: Masamune et al., *J. Am. Chem. Soc.* **89**, 4521 (1967); Johnson et al., *ibid.* 4523; Masamune et al., *Tetrahedron* **27**, 3369 (1971); Kutney et al., *Can. J. Chem.* **53**, 1796 (1975). Stereochemistry: Sicher, Tichy, *Tetrahedron Letters* **1959**(12), 6 (1959); Kataoka, *Chem. & Ind. (London)* **1961**, 512; Bailey et al., *Tetrahedron Letters* **1963**, 555. Revised stereochemistry: Scott et al., *ibid.* **1967**, 2381; Kupchan, Suffness, *J. Am. Chem. Soc.* **90**, 2730 (1968); Sprague et al., *Tetrahedron* **27**, 4857 (1971).

Crystals, mp 206-107°. Slightly sol in water. Sol in methanol, alcohol. Precipitated by digitonin. $[\alpha]_D^{25}$ − 71.8° (c = 1.21); $[\alpha]_D^{25}$ −70° (c = 1.56 in methanol). uv max: 268 nm. Dihydroveratramine, crystals, mp 192.5-194°. $[\alpha]_D^{25}$ +26° (c = 1.26 in acetic acid).

THERAP CAT: *See* Veratrum Viride.

9854. Veratric Acid. *3,4-Dimethoxybenzoic acid;* dimethylprotocatechuic acid. $C_9H_{10}O_4$; mol wt 182.17. C 59.33%, H 5.53%, O 35.13%. Isolated from seed of *Schoenocaulon officinale* (Schlecht. & Cham.) A. Gray (*Sabadilla officinarum* Brandt). Prepn: Arthur, Ng, *J. Chem. Soc.* **1959**, 3094.

Monohydrate, odorless crystals. At 100° becomes anhydr; mp 180-181° when anhydr. Sublimes in rhombic crystals. Sol in 2150 parts cold, 165 parts boiling water; very sol in alcohol or ether. Its barium salt is but slightly sol in water.

9855. Veratridine. *4,9-Epoxycevane-3,4,12,14,16,17,20-heptol 3-(3,4-dimethoxybenzoate);* 3-veratroylveracevine. $C_{36}H_{51}NO_{11}$; mol wt 673.81. C 64.17%, H 7.63%, N 2.08%, O 26.12%. From seed of *Schoenocaulon officinale* (Schlecht. & Cham.) A. Gray and also from the rhizome of *Veratrum album* L. *Liliaceae.* Can be isolated from commercial veratrine as the sparingly sol nitrate: Blount, *J. Chem. Soc.* **1935**, 122; Vejdelek et al., *Chem. Listy* **50**, 603 (1956); *Coll. Czech. Chem. Commun.* **22**, 98 (1957). Review of chemistry and structure of veratridine and other *Veratrum* alkaloids: Kupchan, By, in *The Alkaloids* vol. 10, R. H. F. Manske, Ed. (Academic Press, New York, 1968) pp 193-285.

Yellowish-white, amorphous powder. Retains water tenaciously. mp 180° (after drying at 130°). $[\alpha]_D^{20}$ +8.0° (ethanol). Insol in water. Slightly sol in ether. LD_{50} i.p. in mice: 1.35 mg/kg, Swiss, Bauer, *Proc. Soc. Exp. Biol. Med.* **76**, 847 (1951).

Nitrate, amorphous powder, sparingly sol in water.

Sulfate, slender needles, very hygroscopic.

Perchlorate, long thin needles from water, mp 259-260° (after drying at 120° *in vacuo*).

9856. Veratrine (Mixture). This veratrine is a mixture of the alkaloids cevadine, veratridine, cevadilline, sabadine, cevine from seeds of *Schoenocaulon officinale* (Schlecht. & Cham.) A. Gray, *Liliaceae.*

White or grayish-white powder. mp 145-155°. One gram dissolves in about 1800 ml water, 1000 ml boiling water, 2.8 ml alcohol, 0.7 ml chloroform, 4.2 ml ether, 80 ml olive oil; freely sol in dil acids, benzene, amyl alcohol, slightly in glycerol; insol in petr ether.

Toxicity: Poisonous! Exceedingly irritating to mucous membranes, causing violent sneezing when inhaled.

THERAP CAT: Topical counterirritant.

THERAP CAT (VET): Has been used as emetic, ruminatoric.

9857. Veratrole. *1,2-Dimethoxybenzene; o-dimethoxy-benzene;* pyrocatechol dimethyl ether. $C_8H_{10}O_2$; mol wt 138.16. C 69.54%, H 7.30%, O 23.16%. Prepd by methylation of pyrocatechol: Ullmann, *Ann.* **327**, 104 (1903); Drahowzal, Klamann, *Monatsh.* **82**, 588 (1951).

Crystals or liquid; d_{25}^{25} 1.084; mp 22-23°; bp 206-207°. Slightly sol in water; sol in alcohol, ether, fatty oils. LD_{50} orally in rats, mice: 1360, 2020 mg/kg, P. M. Jenner *et al., Food Cosmet. Toxicol.* **2**, 327 (1964).

9858. Veratrum viride. American hellebore; green hellebore; American veratrum; Indian poke. Dried rhizome and roots of *Veratrum viride* Ait., *Liliaceae. Habit.* North America. *Constit.* The alkaloids jervine, pseudojervine, rubijervine, cevadine, germitrine, germidine, veratralbine, veratroidine. *See* Seiferle *et al., J. Econ. Entomol.* **35**, 35 (1942); Fried *et al., J. Am. Chem. Soc.* **72**, 4621 (1950). *Review:* Kupchan, *J. Pharm. Sci.* **50**, 273-287 (1961).

Mixture of the alkaloids of *Veratrum viride, alkavervir, Veriloid.* Partially purified prepn of the ester alkaloids, *Vergitryl.*

THERAP CAT: Antihypertensive.

THERAP CAT (VET): Has been used as a circulatory depressant, stomachic, emetic, parasiticide.

9859. Verazide. *4-Pyridinecarboxylic acid [(3,4-dimeth-oxyphenyl)methylene]hydrazide; isonicotinic acid veratrylidenehydrazide;* 1-isonicotinoyl-2-veratrylidenehydrazine; 3,4-dimethoxybenzal isonicotinoylhydrazone. $C_{15}H_{15}N_3O_3$; mol wt 285.29. C 63.15%, H 5.30%, N 14.73%, O 16.82%. Prepd from isonicotinylhydrazine and veratraldehyde: Fox, Gibas, *J. Org. Chem.* **18**, 983 (1953).

Crystals from 2-propanol, mp 189-190°.

THERAP CAT: Antibacterial (tuberculostatic).

9860. Verbascose. *β-D-Fructofuranosyl O-α-D-galactopyranosyl-(1 → 6)-O-α-D-galactopyranosyl(1 → 6)-O-α-D-galactopyranosyl-(1 → 6)-α-D-glucopyranoside; O-α-D-galactopyranosyl-(1 → 6)-[O-α-D-galactopyranosyl-(1 → 6)-]₂-O-α-D-glucopyranosyl-(1 → 2)-β-D-fructofuranoside.* $C_{30}H_{52}O_{26}$; mol wt 828.75. C 43.48%, H 6.33%, O 50.20%. Oligosaccharide isolated from roots of mullein *Verbascum thapsus* L., *Scrophulariaceae;* Bourquelot, Bridel, *Compt. Rend.* **151**, 760 (1910); from birch wood, *Betula verrucosa* Ehrh., *Betulaceae;* Lindberg, Selleby, *Acta Chem. Scand.* **12**, 1512 (1958). Structure: Wickström *et al., Bull. Soc. Chim. France* **1956**, 827. *Review:* D. French "The Raffinose Family of Oligosaccharides" in *Advan. Carbohyd. Chem.* **9**, 180-181 (1954).

Needles, mp 219-220°, mp 253°. $[\alpha]_D^{20}$ +170° (water); $[\alpha]_D^{25}$ +146° (c = 2.1 in water).

9861. Verbenalin. *1-(β-D-Glucopyranosyloxy)-1,4a,5,6,-7,7a-hexahydro-7-methyl-5-oxocyclopenta[c]pyran-4-carboxylic acid methyl ester;* cornin. $C_{17}H_{24}O_{10}$; mol wt 388.36. C 52.57%, H 6.23%, O 41.20%. A glycoside isolated from *Verbena officinalis* L., *Verbenaceae, Cornus florida* L., *Cornaceae. Refs:* Bourdier, *J. Pharm. Chim.* **27**(6), 49, 101 (1908); Holste, *Arch. Exp. Path. Pharmakol.* **101**, 46 (1924); Miller, *J. Am. Pharm. Assoc.* **17**, 744 (1928). Identity with cornin: Reichert, *Arch. Pharm.* **273**, 357 (1935). Structure: Büchi, Manning, *Tetrahedron Letters* **1960**, 5 (1960); *eidem, Tetrahedron* **18**, 1049 (1962). Synthesis of *verbenalol,* the aglycone: P. Callant *et al., ibid.* **37**, 2085 (1981).

Bitter needles, mp 182-183°. $[\alpha]_D^{25}$ −173° (water). uv max: 238, 290 nm (ϵ 9600, 105) in ethanol. Freely sol in water, slightly in alcohol, ethyl acetate, acetone; practically insol in chloroform, ether.

9862. *d*-Verbenone. *(1R-cis)-4,6,6-Trimethylbicyclo-[3.1.1]hept-3-en-2-one; 2-pinen-4-one.* $C_{10}H_{14}O$; mol wt 150.21. C 79.95%, H 9.39%, O 10.65%. A constituent of Spanish verbena oil (from *Verbena triphylla* L., *Verbenaceae*): Kerschbaum, *Ber.* **33**, 885 (1900). Prepn from α-pinene: Bain, Gary, U.S. pat. **2,911,442** (1959 to Glidden). Structure: Blumann, Zeitschel, *Ber.* **46**, 1178 (1913); Weinhaus, Schumm, *Ann.* **439**, 20 (1924). Absolute configuration: Hurst, Whitham, *J. Chem. Soc.* **1960**, 2864. *Review:* J. L. Simonsen, *The Terpenes* vol. II (University Press, Cambridge, 2nd ed., 1949) p 232-239.

Oil. Characteristic odor. mp 6.5°. bp_{12} 102-105°; bp_{760} 227-228°. d^{20} 0.9780. n_D^{18} 1.4993. $[\alpha]_D^{18}$ +249.62°. uv max (ethanol): 253 nm (ϵ 6730). Practically insol in water; miscible in all proportions with the usual organic solvents. LD in mice: 250 mg/kg i.p. (*Toxic Substances List,* 1972 Edition, H. E. Christensen, Ed.).

9863. Vermiculite. Hydrated magnesium-aluminum-iron silicate. Typical analysis of Montana vermiculite ore: SiO_2 38.64%, MgO 22.68%, Al_2O_3 14.94%, Fe_2O_3 9.29%, K_2O 7.84%, CaO 1.23%, Cr_2O_3 0.29%, Mn_3O_4 0.11%, Cl 0.28%.

Monoclinic crystals, pseudo-hexagonal characteristics. Soft, resilient, glabrous feel. Dissolves in hot concd H_2SO_4. Has base exchange capacities. Sold as the ore or in "expanded" form. The expanded grades can absorb large quantities of liquids and remain free flowing. The ore is not sol in weak acids or bases, but expanded grades may be attacked. Vermiculites should be classed as montmorillonoids: Rustum Roy, *Chem. & Eng. News* **32**, 4842 (1954).

Note: A food grade vermiculite is known as *Verxite.*

USE: As catalyst; loose fill insulation, filler and packing material.

9864. Vernadigin. $C_{30}H_{44}O_{10}$; mol wt 564.65. C 63.81%, H 7.85%, O 28.34%. Isoln from *Adonis vernalis* L., *Ranunculaceae* and structure: Poláková, Cekan, *Chem. & Ind. (London)* **1963**, 1766.

Crystals, mp 228.5-230°. $[\alpha]_D^{20}$ +39.2° (c = 0.76 in pyridine); $[\alpha]_D^{20}$ +35.5° (c = 0.19 in methanol). uv max: 218 nm (log ϵ 4.2).

9865. Vernamycin B. Group of peptide lactones produced by *Streptomyces loidensis:* Donovick *et al.,* U.S. pat. **2,990,325** (1961 to Olin Mathieson). Antimicrobial antibiotic structurally similar to virginiamycin S_1, *q.v.:* Bodanszky, Ondetti, *Antimicrob. Ag. Chemother.* **1963**, 360. Identity with pristinamycins: Preud'homme *et al.,* *Bull. Soc. Chim. France* **1968**, 585. Nomenclature: Crooy, De Neys, *J. Antibiot.* **25**, 371 (1972).

Crystals, dec 130-135°. $[\alpha]_D^{25}$ −72°. uv max (methanol): 230 nm ($E_{1cm}^{1\%}$ 590). Sol in methanol, ethanol, isopropanol, butanol, acetone, methyl ethyl ketone, ethyl acetate, amyl acetate, ether, dioxane, dimethylformamide, benzene, acetic acid; slightly sol in water, hexane.
Vernamycin B_α. R_1 = CH_2CH_3; R_2 = CH_3. *See* Mikamycin B.
Vernamycin B_β, *4-[N-methyl-4-(methylamino)-L-phenylalanine]virginiamycin S_1, ostreogrycin B_2, pristinamycin I_{B},* R_1 = CH_2CH_3; R_2 = H.
Vernamycin B_γ, *2-D-alanine-4-[4-(dimethylamino)-N-methyl-L-phenylalanine]virginiamycin S_1, ostreogrycin B_1, pristinamycin I_C,* R_1 = CH_3, R_2 = CH_3.
Vernamycin B_δ, *2-D-alanine-4-[N-methyl-4-(methylamino)-L-phenylalanine]virginiamycin S_1,* R_1 = CH_3, R_2 = H.

9866. Vernolate. *Dipropylcarbamothioic acid S-propyl ester; dipropyl thiocarbamic acid S-propyl ester; S-propyl di-*

propylthiocarbamate; R 1607; Vernam. $C_{10}H_{21}NOS$; mol wt 203.35. C 59.06%, H 10.41%, N 6.89%, O 7.87%, S 15.77%. Selective soil incorporated herbicide. Prepn: **Brit. pat. 808,-753**; H. Tilles, J. Antognini, U.S. pat. **2,913,327** (both 1959 to Stauffer); H. Tilles, *J. Am. Chem. Soc.* **81**, 714 (1959).

Clear liquid, bp_{30} 149-150°. Vapor pressure at 25°: 1.04 × 10^{-2} mm Hg. d_4^{30} 0.9440; n_D^{30} 1.4736. Soly in water at 25°: 107 mg/l. Misc with most organic solvents. LD_{50} orally in rats: 1780 mg/kg, G. W. Bailey, J. L. White, *Residue Rev.* **10**, 97 (1965).
USE: Herbicide.

9867. Vernolepin. *5a-Ethenyloctahydro-4-hydroxy-3,9-bis(methylene)-2H-furo[2,3-f][2]benzopyran-2,8(3H)-dione.* $C_{15}H_{16}O_5$; mol wt 276.30. C 65.21%, H 5.84%, O 28.95%. The first recognized naturally occurring elemanolide dilactone; a sesquiterpene lactone with tumor inhibiting properties. Isoln from *Vernonia hymenolepis* A. Rich, *Compositae* and structure: S. M. Kupchan *et al., J. Am. Chem. Soc.* **90**, 3596 (1968); improved isoln and x-ray crystallographic structural elucidation: *eidem, J. Org. Chem.* **34**, 3903 (1969). Total synthesis of (±)-form: P. A. Grieco *et al., J. Am. Chem. Soc.* **98**, 1612 (1976); **99**, 5773 (1977); S. Danishefsky *et al., ibid.* 6066; G. R. Kieczykowski, R. H. Schlessinger, *ibid.* **100**, 1938 (1978); M. Isobe *et al., ibid.* 1940; F. Zutterman *et al., Tetrahedron* **35**, 2389 (1979). Biological properties: S. M. Kupchan *et al., J. Med. Chem.* **14**, 1147 (1971).

(+)-Form, crystals, mp 181-182°. $[\alpha]_D^{28}$ +72° (c = 1.04 in acetone). uv max (methanol): 208 nm (ϵ 20,300).
(±)-Form, crystals from chloroform, mp 210-211°.

9868. Vernolic Acid. *11-(3-Pentyloxiranyl)-9-undecenoic acid; 12,13-epoxyoleic acid; cis-12,13-epoxyoctadec-cis-9-enoic acid; octadec-cis-12,13-epoxy-cis-9-enoic acid.* $C_{18}H_{32}O_3$; mol wt 296.44. C 72.92%, H 10.88%, O 16.19%. Principal seed-oil acid from *Vernonia anthelmintica* Willd. (*Serratula anthelmintica* Roxb., *Conyza* Linn.) *Compositae.* Isoln and structure: Gunstone, *J. Chem. Soc.* **1954**, 1611. Synthesis and absorption spectrum: Osbond, *ibid.* **1961**, 5270; *idem,* **Brit.** pat **909,354** (1962 to Roche).

Crystals from acetone at −25°, mp 30-31°. Liq at room temp. $[\alpha]_D$ −8°.

9869. Verrucarins. Macrocyclic tricothecane derivs which are secondary metabolites of the soil fungi *Myrothecium verrucaria* (Albertini et Schweinitz) Ditmar ex Fries; they are characterized by antibiotic, antifungal, and cytostatic activity. Verrucarins are triesters of the sesquiterpene alcohol *verrucarol* and closely related to the *roridins,* which are diesters of the same alcohol. Isoln of verrucarins A (major), B, C, D, E, F, G: E. Haerri *et al., Helv. Chim. Acta* **45**, 840 (1962). Isoln of A: Symth, Kraskin, U.S. pat. **3,087,859** (1963 to Rohm & Haas); Vittimberga, *J. Org.Chem.* **28**, 1786 (1963). Identity with muconomycin A: Vittimberga, Vittimberga, *ibid.* **30**, 746 (1965). Structure of A: J. Gutzwiller, C. Tamm, *Helv. Chim. Acta* **48**, 157 (1965). Stereochemistry: W. Zürcher *et al., ibid.* 840; A. T. McPhail, G. T. Sim, *J. Chem. Soc. (C)* **1966**, 1394. Synthesis of A: W. C. Still,

H. Ohmizu, *J. Org. Chem.* **46**, 5242 (1981). Isoln of B: W. Loeffler *et al.*, **Belg.** pat. **627,002** (1963 to Sandoz), *C.A.* **60**, 10485b (1964). Structure of B: J. Gutzwiller, C. Tamm, *Helv. Chim. Acta* **48**, 177 (1965). Toxicological studies of A and B: Guarino *et al.*, *Biotechnol. Bioeng.* **10**, 457 (1968); of A: Mortimer *et al.*, *Res. Vet. Sci.* **12**, 508 (1971). Biosynthetic studies of A and B: B. Müller, C. Tamm, *Helv. Chim. Acta* **58**, 483 (1975); G. A. Cordell, *Chem. Rev.* **76**, 425 (1976). Isoln of J (muconomycin B): Vittimberga, Vittimberga, *loc. cit.*; of H and J: B. Boehner *et al.*, *Helv. Chim. Acta* **48**, 1079 (1965). Structure of J: E. Fetz *et al.*, *ibid.* 1669. Partial synthesis of tetrahydroverrucarin J and revised structure of J: W. Breitenstein, C. Tamm, *ibid.* **61**, 1975 (1978). Isoln and structure of K: *eidem, ibid.* **60**, 1522 (1977). Isoln of L: B. B. Jarvis *et al.*, *J. Antibiot.* **34**, 121 (1981). Total synthesis of verrucarol: R. H. Schlessinger, R. A. Nugent, *J. Am. Chem. Soc.* **104**, 1116 (1982). Comprehensive review of the verrucarins and roridins: C. Tamm, *Fortschr. Chem. Org. Naturst.* **31**, 64-117 (1974). For a description of the stereochemistry of the roridins *see* B. R. Jarvis *et al.*, *J. Nat. Prod.* **45**, 440 (1982).

verrucarin A

Verrucarin A, $C_{27}H_{34}O_9$, *muconomycin A*. Colorless rectangular plates from ether/acetone, mp > 360° (dec). $[\alpha]_D^{23}$ +206° (c = 1.012 in chloroform); uv max (ethanol): 260 nm (log ε 4.25). LD_{50} in mice, rats, rabbits (mg/kg): 1.5, 0.87, 0.54 i.v., E. M. Rüsch, H. Stählin, *Arzneimittel-Forsch.* **15**, 893 (1965).
Verrucarin B, $C_{27}H_{32}O_9$. Colorless needles from acetone/-ether, decomp > 330°. $[\alpha]_D^{23}$+147° (c = 1.066 in benzene). $[\alpha]_D^{22}$ +101° ±1.5° (c = 1.416 in dioxane). uv max (methanol): 258.5 nm (log ε 4.37).
Verrucarin J, $C_{27}H_{32}O_8$, *muconomycin B*, *2',3'-didehydro-2'-deoxyverrucarin A*. Crystals from ether, decomp > 235°. $[\alpha]_D^{19}$ +54° (benzene). $[\alpha]_D^{23}$ +20° ± 2° (c = 1.011 in chloroform). uv max: 261, 220.5 nm (ε 22,000, 21,600).
Verrucarin K, $C_{27}H_{34}O_8$. Crystals from methylene chloride/ether, decomp > 320°. $[\alpha]_D^{23}$ +218° ±2° (c = 0.58 in chloroform). uv max (ethanol): 259 nm (log ε 4.19).
Caution: Verrucarins are extremely toxic; can cause severe local irritation and inflammation of the skin.

9870. Versalide®. *1-(3-Ethyl-5,6,7,8-tetrahydro-5,5,-8,8-tetramethyl-2-naphthalenyl)ethanone; 3'-ethyl-5',6',7',8'-tetrahydro-5',5',8',8'-tetramethyl-2'-acetonaphthone.* $C_{18}H_{26}O$; mol wt 258.41. C 83.67%, H 10.14%, O 6.19%. Prepn: Davidson, Lusskin, U.S. pat. **3,045,047** (1962 to Trubek Lab.). GC determn in fragrances: H. H. Wisneski *et al.*, *J. Assoc. Offic. Anal. Chem.* **65**, 598 (1982). Toxicity: K. R. Butterworth, P. L. Mason, *Food Cosmet. Toxicol.* **19**, 753 (1981).

Crystals, mp 46.5°. bp₂ 130°. Soluble in alc.
USE: As musk for perfumes, cosmetics, soaps.

9871. Versen-Ol®. *N-[2-[Bis(carboxymethyl)amino]-ethyl]-N-(2-hydroxyethyl)glycine trisodium salt; N-(carboxymethyl)-N'-(2-hydroxyethyl)-N,N'-ethylenediglycine trisodi-*

um salt; N-hydroxyethylethylenediaminetriacetic acid trisodium salt; trisodium N-hydroxyethylethylenediaminetriacetate. $C_{10}H_{15}N_2Na_3O_7$; mol wt 344.22. C 34.89%, H 4.39%, N 8.14%, Na 20.04%, O 32.54%. Prepn: Young, U.S. pat. **2,811,550** (1957 to Refined Prod.); Kroll, Dexter, U.S. pat. **2,845,457** (1958 to Geigy).

USE: As chelating agent for trivalent iron from pH 7 to pH 10. Forms strong 1:1 ferric chelates. The iron chelates formed will then chelate other heavy metal or alkaline earth ions.

9872. Verticillins. Three antibiotics, verticillins A, B and C, produced by a species of *Verticillium* (strain TM-759), an imperfect fungus isolated from a basidiocarp of *Coltricia cinnamomea (Polystictus cinnamomeus)*. Isoln, IR, NMR spectra of verticillin A: Katagiri *et al.*, *J. Antibiot.* **23**, 420 (1970); *see also* Chepenko *et al.*, *C. A.* **78**, 156654n (1973). Structure: Minato *et al.*, *Chem. Commun.* **1971**, 44. Structure of verticillins A, B and C and absolute configuration of verticillins A and B: *eidem, J. Chem. Soc. Perkin Trans. I* **1973**, 1819. Synthetic studies: Häusler, Schmidt, *Ber.* **107**, 2804 (1974); Schmidt *et al.*, *ibid.* 2816.

verticillin A

Verticillin A, $C_{30}H_{28}N_6O_6S_4$. Pale yellow plates from chloroform, $C_{30}H_{28}N_6O_6S_4$·CHCl₃, mp 199-213° (dec); pale yellow needles from pyridine, $C_{30}H_{28}N_6O_6S_4$·2/3C₅H₅N, mp 202-217° (dec); pale yellow amorphous powder from tetrahydrofuran, mp 203-214° (dec). $[\alpha]_D$ +703.7° (c = 0.422 in dioxane). uv max (dioxane): 306 nm (ε 5960). LD_{50} i.p. in mice: 7.6 mg/kg, Katagiri *et al.*, *loc. cit.* Showed delayed toxicity.
Verticillin B, $C_{30}H_{28}N_6O_7S_4$. The mono-3-hydroxymethyl analog of verticillin A. Pale yellow prisms from chloroform, mp 230-233° (dec). $[\alpha]_D^{21}$ +704.7° (c = 0.493 in dioxane). uv max (dioxane): 306 nm (ε 5600).
Verticillin C, $C_{30}H_{28}N_6O_7S_5$. Differs from verticillin B by having a trisulfide rather than disulfide bridge in one of the two dioxopiperazine rings. Pale yellow amorphous powder from methanol-water, mp 230-235° (dec). $[\alpha]_D^{21}$ +765.0° (c = 0.506 in dioxane). uv max (dioxane): 303 nm (ε 5500).

9873. Verticine. *Cevane-3,6,20-triol; peimine.* $C_{27}H_{45}$-NO_3; mol wt 431.64. C 75.13%, H 10.51%, N 3.25%, O 11.12%. A member of the hexacyclic *Ceveratrum* group of alkaloids. Isoln from *Fritillaria verticillata* Willd. var. *Thunbergii* Baker, *Liliaceae:* Fukuda: *Sci. Repts. Tohoku Univ.*, Ser. A, **18**, 323 (1929); Morimoto, Kimata, *Chem. Pharm. Bull.* **8**, 302 (1960). Also isolated from *F. roylei* Hook: Chou, Chen, *Chinese J. Physiol.* **6**, 265 (1932), *C.A.* **26**, 5703 (1932). Structure: Ito *et al.*, *Chem. Pharm. Bull.* **11**, 1337 (1963). Abs config: S. Ito *et al.*, *Tetrahedron Letters* **1968**, 5373. Total synthesis: J. P. Kutney *et al.*, *J. Am. Chem. Soc.* **99**, 964 (1977).

Needles from ethanol, mp 223-224°. $[\alpha]_D^{16}$ −19.4° (ethanol); $[\alpha]_D^{17}$ −20° (CHCl$_3$). pK'a 9.5. uv max (ethanol + HCl): 215 nm (ϵ 10).

Hydrochloride, $C_{27}H_{45}NO_3 \cdot HCl$, prisms from water, dec 291-294°. $[\alpha]_D^{18}$ −18.5°.

Perchlorate, $C_{27}H_{45}NO_3 \cdot HClO_4$, prisms from water, dec 273°. $[\alpha]_D^{23}$ −15° (c = 0.05).

Methiodide, $C_{27}H_{45}NO_3 \cdot CH_3I$, crystals from butanol + ether, dec 205-210°. $[\alpha]_D^{18}$ −14.0° (water). Sol in water, methanol, ethanol; practically insol in chloroform, benzene, acetone.

3-Glucoside, $C_{33}H_{55}NO_8$, *peiminoside*. Pale brown powder. Sol in water, methanol, alc; sparingly sol in benzene.

9874. Vetivones. $C_{15}H_{22}O$; mol wt 218.33. C 82.51%, H 10.16%, O 7.33%. Ketonic components of vetiver oil; from roots of *Vetiveria zizanioides (Andropogon muricatus* Retz., *Gramineae).* Two isomers, the α-form having the stronger odor. Isoln and structure: Pfau, Plattner, *Helv. Chim. Acta* **22**, 640 (1939); Naves, Perrottet, *ibid.* **24**, 3 (1941).

α-vetivone   β-vetivone

α-Vetivone. *(4R-cis)-4,4a,5,6,7,8-Hexahydro-4,4a-dimethyl-6-(1-methylethylidene)-2(3H)-naphthalenone; 4βH,-5α-eremophila-1(10),7(11)-dien-2-one; isonootketone.* Structure: Endo, DeMayo, *Chem. Commun.* **1967,** 89; Marshall, Andersen, *Tetrahedron Letters* **1967,** 1611. Total syntheses of the racemate: Marshall *et al., Chem. Commun.* **1967,** 753; Marshall, Warne, *J. Org. Chem.* **36,** 178 (1971); Vandergen *et al., Rec. Trav. Chim.* **90,** 1034, 1045 (1971); Dastur, *J. Am. Chem. Soc.* **96,** 2605 (1974).

Crystals from pentane, strong agreeable odor. mp 51.5° (Naves, Perrottet); mp of racemate 30-35° (Marshall, Andersen). n_D^{20} 1.5384. $[\alpha]_D^{20}$ +248° (chloroform). d_4^{20} 1.003 (liquid, supercooled). uv max (ethanol): 233 nm (ϵ 13,200).

β-Vetivone. *(5R-cis)-6,10-Dimethyl-2-(1-methylethylidene)spiro[4.5]dec-6-en-8-one; 2-isopropylidene-6,10-dimethylspiro[4.5]dec-6-en-8-one.* Structure: Marshall *et al., J. Am. Chem. Soc.* **89,** 2748, 2750 (1967). Syntheses of racemate: Marshall, Johnson, *Chem. Commun.* **1968,** 391; *eidem, J. Org. Chem.* **35,** 192 (1970); K. Uneyama *et al., Chem. Letters* **1977,** 493; S. Torii *et al., Bull. Chem. Soc. Japan* **51,** 3590 (1978); E. Wenkert *et al., J. Am. Chem. Soc.* **100,** 1267 (1978). Stereospecific synthesis of (−)-β-vetivone: Deighton *et al., Chem. Commun.* **1975,** 662; M. Asaoka *et al., Chem. Letters* **1988,** 1225.

Crystals from pentane, mp 44-46°. d_4^{20} 1.000 (liquid, supercooled). d_4^{45} 0.9804. $[\alpha]_D^{20}$ −38.9° (c = 10 in alcohol). n_D^{20} 1.5309 (liquid, supercooled), n_D^{45} 1.5216.

USE: Of potential interest in the perfume industry.

9875. Vetrabutine. *α-(3,4-Dimethoxyphenyl)-N,N-dimethylbenzenebutanamine; N,N-dimethyl-α-(3-phenylpropyl)veratrylamine; 1-(3,4-dimethoxyphenyl)-1-dimethylamino-4-phenylbutane; 3,4-dimethoxy-N,N-dimethyl-α-(3-*

phenylpropyl)benzylamine; dimophebumine. $C_{20}H_{27}NO_2$; mol wt 313.45. C 76.64%, H 8.68%, N 4.47%, O 10.21%. Prepn: **Brit. pat. 802,723** (1958 to Thomae).

bp$_{0.1}$ 166-168°.
Hydrochloride, $C_{20}H_{28}ClNO_2$, *Sp 281, Monzal, Monzaldon.* Solid, mp 146-148°.
THERAP CAT: Uterine relaxant.
THERAP CAT (VET): Uterine relaxant.

9876. Viburnum opulus. Cramp bark; high bush cranberry; cranberry tree; water elder; squaw bush; snowball bush. Dried bark of *Viburnum opulus* L., var. *americanum* Ait., *Caprifoliaceae. Habit.* Europe, Asia, northern North America, south to Pennsylvania. *Constit.* Viburnin, bitter resin, tannin, sugar; citric, malic, oxalic, and valeric acids.
THERAP CAT: Antispasmodic (uterine).

9877. Viburnum prunifolium. Black haw; sweet viburnum; sheep berry; stag bush; sloe-leaved viburnum. Dried root bark of *Viburnum prunifolium* L., *Caprifoliaceae. Habit.* U.S. *Constit.* Viburnin, bitter resin, tannin; citric, malic, oxalic, and valeric acids.
THERAP CAT: Antispasmodic (uterine).

9878. Vicianin. Mandelonitrile vicianoside. $C_{19}H_{25}NO_{10}$; mol wt 427.40. C 53.39%, H 5.90%, N 3.28%, O 37.43%. In seeds of *Vicia angustifolia* Clos, and *V. sativa* L., *Leguminosae.* Isoln: Bertrand, *Compt. Rend.* **143,** 832 (1906); *Bull. Soc. Chim.* [4] **1,** 151 (1907); Bertrand, Weisweiller, *Compt. Rend.* **147,** 252 (1908); **151,** 325 (1910); *Bull. Soc. Chim.* [4] **9,** 147 (1911). Preparation of hexaacetyl-β-vicianoside of (−)-mandelonitrile, believed to be vicianin hexaacetate: Chaudhury, Robertson, *J. Chem. Soc.* **1949,** 2054.

Monohydrate, needles from water. When anhydr mp 160° (copper block). $[\alpha]_D^{18}$ −21° (satd aq soln). Sol in hot water, slightly in cold water, in alcohol; practically insol in chloroform, benzene, petr ether.

9879. Vicianose. *6-O-α-L-Arabinopyranosyl-D-glucose; 6-(α-L-arabinosido)-D-glucose.* $C_{11}H_{20}O_{10}$; mol wt 312.28. C 42.31%, H 6.46%, O 51.24%. By enzymatic hydrolysis of vicianin, gein, and violutin: Bertrand, Weisweiller, *Compt. Rend.* **150,** 181 (1910); *Bull. Soc. Chim.* [4] **9,** 38 (1911). Synthesis: Helferich, Bredereck, *Ann.* **465,** 166 (1928); Wallenfels, Beck, *Ann.* **630,** 46 (1960).

Needles from 75% alc, dec 210°. Weak sweetish taste. Shows mutarotation. $[\alpha]_D^{20}$ +63° →40° (22 hrs, c = 8). Very sol in water, practically insol in abs alc. Reduces Fehling's soln somewhat more than maltose.

9880. Vicine. *2,6-Diamino-5-(β-D-glucopyranosyloxy)-4(1H)-pyrimidinone; 2,4-diamino-6-oxypyrimidine-5-(β-D-glucopyranoside); divicine 5-glucoside; vicioside; divicine-β-glucoside.* $C_{10}H_{16}N_4O_7$; mol wt 304.26. C 39.47%, H 5.30%, N 18.42%, O 36.81%. Isoln from vetch seeds (*Vicia sativa* L., *Leguminosae):* Ritthausen, *Ber.* **9,** 301 (1876); Levene, *J. Biol. Chem.* **18,** 306 (1914); Levene, Senior, *ibid.* **25,** 611 (1916); Herissey, Cheymol, *Bull. Soc. Chim. Biol.* **13,** 29 (1931); Gmelin, Hasenmaier, *Arzneimittel-Forsch.* **7,** 755

(1957). Structure: Bendich, *Trans. N.Y. Acad. Sci.* **15**, 58 (1952); Bendich, Clements, *Biochim. Biophys. Acta* **12**, 462 (1953).

Needles from water, dec 243-244°. $[\alpha]_D^{26}$ −11.7° (c = 3.9 in 0.2N NaOH). uv max (0.1N HCl): 274 nm (ε 16,400); (0.1N NaOH): 269, 248, 235, 230 nm (ε 9520, 4360, 5180, 5090). One gram dissolves in about 100 ml water; slightly sol in alcohol; readily sol in dil acids and alkalies.

Note: Another *vicin*, a rhamnosidoglucoside from vetch flowers, is described by Karrer, Widmer, *Helv. Chim. Acta* **10**, 67 (1927). It has the formula: $C_{21}H_{21}O_{12} \cdot 2H_2O$.

9881. Vidarabine. *9-β-D-Arabinofuranosyl-9H-purine-6-amine monohydrate; 9-β-D-arabinofuranosyladenine monohydrate;* arabinosyladenine; adenine arabinoside; spongoadenosine; ara-A; CI 673; Arasena-A; Vira-A. $C_{10}H_{13}$-$N_5O_4 \cdot H_2O$; mol wt 285.27. C 42.10%, H 5.30%, N 24.55%, O 28.04%. Purine nucleoside first synthesized as a potential anticancer agent: Lee *et al., J. Am. Chem. Soc.* **82**, 2648 (1960); Reist *et al., J. Org. Chem.* **27**, 3274 (1962); Glaudemans, Fletcher, *ibid.* **28**, 3004 (1963); Reist *et al., ibid.* **29**, 3725 (1964). Production by culturing a strain of *Streptomyces antibioticus:* **Brit. pat.** **1,159,290** (1969 to Parke, Davis), *C.A.* **71**, 79757z (1969). Crystal and molecular structure: Bunick, Voet, *Acta Crystallogr.* **30B**, 1641 (1974). Metabolism and distribution studies: J. J. Brink, G. A. LePage, *Cancer Res.* **24**, 1042 (1964). *In vitro* antiviral activity: J. de Rudder, M. Privat de Garilhe, *Antimicrob. Ag. Chemother.* **1965**, 578. Series of articles on *in vitro* and *in vivo* antiviral activity: *ibid.* **1968**, 136-179. Comparative clinical evaluation with acyclovir, *q.v.,* in varicella-zoster infection: J. L. Vildé *et al., J. Med. Virol.* **20**, 127 (1986); in herpes simplex encephalitis: R. J. Whitley *et al., Infection* **15**, Suppl. 1, S 3 (1987). Pharmacokinetics in children: T. C. Shope *et al., J. Infect. Dis.* **148**, 721 (1983). HPLC determn in plasma and urine: W. P. McCann *et al., Antimicrob. Ag. Chemother.* **28**, 265 (1985). Toxicity data: S. M. Kurtz *et al., ibid.* **1968**, 180. Book: *Adenine Arabinoside: An Antiviral Agent,* D. Paven-Langston *et al.,* Eds. (Raven Press, New York, 1975) xviii + 425 pp. *Reviews:* C. E. Cass in *Antibiotics* vol. 5(pt. 2), F. E. Hahn, Ed. (Springer-Verlag, New York, 1979) pp 85-109; R. A. Buchanan, F. Hess, *Pharmacol. Ther.* **8**, 143-171 (1980). Comprehensive description: W. Hong *et al.,* in *Analytical Profiles of Drug Substances* vol. 15, K. Florey, Ed. (Academic Press, New York, 1986) pp 647-672.

Crystals from water, mp 257.0-257.5° (0.4 H_2O). $[\alpha]_D^{27}$ −5° (c = 0.25). uv max (pH 1): 257.5 nm (ε 12700); pH 7: 259 nm (ε 13400); pH 13: 259 nm (ε 14000). LD$_{50}$ in mice (mg/kg): 4677 i.p.; > 7950 orally (Kurtz).

THERAP CAT: Antiviral.

9882. Vigabatrin. *4-Amino-5-hexenoic acid;* γ-vinyl-γ-aminobutyric acid; gamma-vinyl GABA; γ-vinyl GABA; GVG; MDL 71754; RMI-71754; Sabril. $C_6H_{11}NO_2$; mol wt

129.16. C 55.80%, H 8.58%, N 10.84%, O 24.77%. Irreversible inhibitor of γ-aminobutyric acid transaminase, the enzyme responsible for the degradation of the neurotransmitter, γ-aminobutyric acid (GABA). Prepn: B. W. Metcalf, M. Jung, **U.S. pat.** **3,960,927** (1976 to Richardson-Merrell); and *in vitro* enzyme inactivation: B. Lippert *et al., Eur. J. Biochem.* **74**, 441 (1977). Mechanism of action study: P. J. Schechter *et al., Eur. J. Pharmacol.* **45**, 319 (1977). Anticonvulsant activity and toxicity studies: W. Löscher, *Neuropharmacol.* **21**, 803 (1982). HPLC determn in plasma and urine: J. A. Smithers *et al., J. Chromatog.* **341**, 232 (1985). The $S(+)$-enantiomer is the pharmacologically active form. Pharmacokinetics of enantiomers in humans: K. D. Haegele, P. J. Schechter, *Clin. Pharmacol. Ther.* **40**, 581 (1986). Clinical studies in treatment resistant epilepsy: C. A. Tassinari *et al., Arch. Neurol.* **44**, 907 (1987); T. R. Browne *et al., Neurology* **37**, 184 (1987). Reviews of early literature and mechanism of action: M. J. Iadarola, K. Gale, *Mol. Cell. Biochem.* **39**, 305-330 (1981); of pharmacology and toxicology: E. J. Hammond, B. J. Wilder, *Clin. Neuropharmacol.* **8**, 1-12 (1985).

Crystals from acetone/water, mp 209°. Freely sol in water. LD$_{50}$ i.p. in mice: > 2500 mg/kg (Löscher).

THERAP CAT: Anticonvulsant.

9883. Villikinin. An intestinal villi-stimulating material extracted from canine intestinal mucosa and human plasma: Kokas *et al., Fed. Proc.* **22**, no. 2, part I, 225 (Mar.-Apr. 1963). *Review:* Grossman, *Physiol. Rev.* **30**, 33 (1950).

Heat stable, sol in water, dil acids, alcohol, acetone.

9884. Viloxazine. *2-[(2-Ethoxyphenoxy)methyl]morpholine;* 2-(2-ethoxyphenoxymethyl)tetrahydro-1,4-oxazine; ICI-58834. $C_{13}H_{19}NO_3$; mol wt 237.30. C 65.80%, H 8.07%, N 5.90%, O 20.23%. Prepn: K. B. Mallion *et al.,* **Brit. pat.** **1,138,405**; *idem,* **U.S. pat.** **3,714,161** (1969, 1973 both to I.C.I.). Manufacturing process: S. A. Lee, **Brit. pat.** **1,260,886**; *idem,* **U.S. pat.** **3,712,890** (1972, 1973 both to I.C.I.). Pharmacology: Mallion *et al., Nature* **238**, 157 (1972); *see also* Bereen: *Lancet* **1**, 379 (1973). Toxicity: R. D. Brosnan *et al., J. Int. Med. Res.* **4**, 83 (1976). Clinical evaluation in narcolepsy: C. Guilleminault *et al., Sleep* **9**, 275 (1986). Review of pharmacology and therapeutic efficacy: R. M. Pinder *et al., Drugs* **13**, 401 (1977).

Hydrochloride, $C_{13}H_{20}ClNO_3$, *Catatrol, Vivalan, Vicilan, Vivarint.* mp 185-186°. LD$_{50}$ in mice (mg/kg): 1000 orally, 60 i.v. (Brosnan).

THERAP CAT: Antidepressant.

9885. Viminol. *α-[[Bis(1-methylpropyl)amino]methyl]-1-[(2-chlorophenyl)methyl]-1H-pyrrole-2-methanol; 1-(o-chlorobenzyl)-α-[(di-sec-butylamino)methyl]pyrrole-2-methanol; 1-[α-(N-o-chlorobenzyl)pyrryl]-2-di-sec-butylaminoethanol; diviminol; Z 424.* $C_{21}H_{31}ClN_2O$; mol wt 362.94. C 69.49%, H 8.61%, Cl 9.77%, N 7.72%, O 4.41%. Prepn: U. M. Teotino, D. Della Bella, **S. Afr. pat.** **67 02,732**, *idem,* **U.S. pat.** **3,539,589** (1970 to Whitefin Holding). Synthesis and metabolic fate of ^{14}C-viminol: Cameron *et al., Arzneimittel-Forsch.* **23**, 708 (1973). Viminol possesses 3 asymmetric carbon atoms and thus is composed of 6 stereoisomers. Role of stereoisomers in determining specific pharmacological effects: D. Della Bella *et al., Nature New Biol.* **241**, 282 (1973). Additional pharmacology data: M. Babbini *et al., Eur. J. Pharmacol.* **23**, 137 (1973). *Review:* D. Della Bella, *Boll. Chim. Farm.* **111**, 5-19 (1972).

$bp_{0.1 \text{ mm}}$ 160-165°.

p-Hydroxybenzoate, $C_{28}H_{37}ClN_2O_4$, *Dividol.* mp 128-130° (dec). LD_{50} i.p. in rats: 205.9 mg/kg (Babbini).
THERAP CAT: Analgesic.

9886. Vinbarbital Sodium. *5-Ethyl-5-(1-methyl-1-but-enyl)-2,4,6(1H,3H,5H)-pyrimidinetrione sodium salt; 5-ethyl-5-(1-methyl-1-butenyl)barbituric acid sodium salt;* sodium 5-ethyl-5-(1-methyl-1-butenyl)barbiturate; *Delvinal Sodium.* $C_{11}H_{15}N_2NaO_3$; mol wt 246.24. C 53.65%, H 6.14%, N 11.38%, Na 9.34%, O 19.49%. Prepn: Cope, Hancock, *J. Am. Chem. Soc.* **61**, 776 (1939); Cope, U.S. pat. 2,187,703 (1940 to Sharp & Dohme). Pharmacology and toxicity: J. P. Hendrix, *J. Pharmacol. Exp. Ther.* **68**, 22 (1940).

Hygroscopic crystals, bitter taste. Slightly sol in ether and chloroform. A 1% aq soln is alkaline to phenolphthalein and has a pH between 8.5 and 9.5. "Unbuffered aq solns of vinbarbital sodium are unstable. The powder is hygroscopic and if capsules contg it are broken or exposed to high humidity the contents are affected by both moisture and carbon dioxide." LD_{50} orally in rats: 130 mg/kg (Hendrix).
Free acid, $C_{11}H_{15}N_2O_3$, mp 161-163°.
Caution: May be habit forming. This is a controlled substance (depressant) listed in the U.S. Code of Federal Regulations, Title 21 Parts 329.1 and 1308.13 (1987).
THERAP CAT: Sedative, hypnotic.
THERAP CAT (VET): Sedative, hypnotic.

9887. Vinblastine. *Vincaleukoblastine;* VLB; 29060-LE. $C_{46}H_{58}N_4O_9$; mol wt 811.00. C 68.13%, H 7.21%, N 6.91%, O 17.75%. Antitumor alkaloid isolated from *Vinca rosea* Linn., *Apocynaceae:* Noble et al., *Ann. N.Y. Acad. Sci.* **76**, Art. 3, 882-894 (1958); Gorman et al., *J. Am. Chem. Soc.* **81**, 4745, 4754 (1959); Beer et al., U.S. pat. 3,097,137 (1963 to Can. Pats. Dev.); Svoboda, U.S. pat. 3,225,030 (1965 to Lilly). Structural studies: Neuss et al., *J. Am. Chem. Soc.* **86**, 1440 (1964). Structure and stereochemistry: Moncrief, Lipscomb, *ibid.* **87**, 4963 (1965). Partial synthesis of vinblastine-type alkaloids: Potier et al., *Chem. Commun.* **1975**, 670. Studies on total synthesis: J. P. Kutney, *Lloydia* **40**, 107 (1977). Biosynthesis from catharanthine and vindoline, *q.v.:* P. Mangeney et al., *J. Am. Chem. Soc.* **101**, 2243 (1979). ¹³C-NMR spectral analysis: Wenkert et al., *Helv. Chim. Acta* **58**, 1560 (1975). Pharmacology: S. M. Sieber et al., *Cancer Treat. Rep.* **60**, 1127 (1976). Toxicology: C. Lu, M. Meistrich, *Cancer Res.* **39**, 3575 (1979). Comprehensive description: J. H. Burns in *Analytical Profiles of Drug Substances* vol. 1, K. Florey, Ed. (Academic Press, New York, 1972) pp 443-462.

Solvated needles from methanol, mp 211-216°. $[\alpha]_D^{26}$ +42° (chloroform). uv max (ethanol): 214, 259 nm (log ϵ 4.73, 4.21). Practically insol in water, petr ether. Sol in alcohols, acetone, ethyl acetate, $CHCl_3$.
Sulfate, $C_{46}H_{60}N_4O_{13}S$, *Exal, Velban, Velbe.* Crystals, mp 284-285°. $[\alpha]_D^{26} -28°$ (c = 1.01 in methanol). LD_{50} i.v. in mice: 9.5 mg/kg (Lu, Meistrich).
Dihydrochloride dihydrate, $C_{46}H_{60}Cl_2N_4O_9\cdot 2H_2O$, crystals, dec 244-246°.
THERAP CAT: Antineoplastic.

9888. Vincamine. *(3α,14β,16α)-14,15-Dihydro-14-hydroxyeburnamenine-14-carboxylic acid methyl ester;* 13a-ethyl-2,3,5,6,12,13,13a,13b-octahydro-12-hydroxy-1H-indolo[3,2,1-de]pyrido[3,2,1-ij][1,5]naphthyridine-12-carboxylic acid methyl ester; Angiopac; Anasclerol; Arteriovinca; Cerebroxine; Cincuental; Devincan; Equipur; Novicet; Ocu-Vinc; Oxygeron; Perval; Pervincamine; Pervone; Sostenil; Tripervan; Vincadar; Vincafarm; Vincafolina; Vincafor; Vincagil; Vincalen; Vincamidol; Vincapan; Vincapront; Vincasaunier; Vincimax; Vinodrel Retard; Vraap. $C_{21}H_{26}$-N_2O_3; mol wt 354.43. C 71.16%, H 7.39%, N 7.90%, O 13.54%. Major alkaloid of *Vinca minor* L., *Apocynaceae* occurring naturally in the *d*-form: Schlittler, Furlenmeier, *Helv. Chim. Acta* **36**, 2017 (1953); Pailer, Belohlav, *Monatsh.* **85**, 1055 (1954); King et al., *J. Chem. Soc.* **1955**, 4206; Trojanek et al., *Coll. Czech. Chem. Commun.* **26**, 867 (1961). Isoln from *Tabernaemontana rigida* Miers, *Apocynaceae:* Cava et al., *J. Org. Chem.* **33**, 1055 (1968). Structure: Trojanek et al., *Tetrahedron Letters* **1961**, 702; Mokry et al., *ibid.* **1962**, 433; Clauder, *ibid.* 1147; Plat et al., *Bull. Soc. Chim. France* **1962**, 1082. Abs config: Trojanek et al., *Chem. & Ind. (London)* **1965**, 1261; Blaha et al., *Coll. Czech. Chem. Commun.* **33**, 3833 (1968). X-ray determn of molecular structure: H. P. Weber, T. J. Petcher, *J. Chem. Soc. Perkin Trans. II* **1973**, 2001. Total synthesis: C. Szantay et al., *Tetrahedron Letters* **1973**, 191; eidem, *Tetrahedron* **33**, 1803 (1977); Pfaffli et al., *Helv. Chim. Acta* **58**, 1131 (1975); W. Oppolzer et al., *ibid.* **60**, 1801 (1977); B. Danieli et al., *Chem. Commun.* **1981**, 908; of *dl*-form: Kuehne, *J. Am. Chem. Soc.* **86**, 2946 (1964); Herrmann et al., *ibid.* **96**, 3702 (1974); K. H. Gibson, J. E. Saxton, *Tetrahedron* **33**, 833 (1977); K. Irie, Y. Ban, *Heterocycles* **18**, 255 (1982); T. R. Govindachari, Rajeswari, *Indian J. Chem.* **22B**, 531 (1983). Synthesis of stereoisomers: J. Warnant et al., Ger. pat. 2,115,718; eidem, U.S. pat. 3,770,724 (1971, 1973 both to Roussel-UCLAF). HPLC determn: A. Amato et al., *J. Chromatog.* **270**, 387 (1983). Toxicity data: L. Szporny, K. Szász, *Arch. Exp. Pathol. Pharmakol.* **236**, 296 (1959). Hypotensive activity in dogs and man: Z. Szabo, Z. Nagy, *Arzneimittel-Forsch.* **10**, 811 (1960). Clinical pharmacology: C. C. Lim et al., *Brit. J. Clin. Pharmacol.* **9**, 100 (1980). Clinical pharmacokinetics: H. Millart, D. Lamiable et al., *Int. J. Clin. Pharmacol. Ther.*

Toxicol. **21,** 581 (1983). Brief review: P. Cook, I. James, *N. Engl. J. Med.* **305,** 1562 (1981).

Yellow crystals from acetone or methanol, mp 232-233°. $[\alpha]_D^{23}$ +41° (in pyridine). uv max: 225, 278 nm (log ϵ 4.14, 3.61). LD_{50} in mice (mg/kg): 75 i.v.; > 1000 s.c. (Szporny, Szász); 1000 orally (Szabo, Nagy).

Hydrochloride, $C_{21}H_{27}ClN_2O_3$, *Cetal, Esberidin.*

Note: Other alkaloids found in the "vincamine fraction" of *Vinca minor* L. are: *vincine, vincaminine* and *vincinine:* Holubek *et al., Tetrahedron Letters* **1963,** 897.

THERAP CAT: Vasodilator.

9889. Vincetoxin. Cynanchin. $C_{50}H_{82}O_{20}$; mol wt 1003.16. C 59.86%, H 8.24%, O 31.90%. $C_{46}H_{70}O_{16}(OCH_3)_4$. From roots of *Vincetoxicum officinale* Moench., *Asclepiadaceae:* Tanret, *Compt. Rend.* **100,** 277 (1885); Kubler, *Arch. Pharm.* **246,** 660 (1908); Gager, Zechner, *ibid.* **276,** 431 (1938). Constitution: Korte, *Ber.* **88,** 1527 (1955); Korte, Ripphahn, *Ann.* **621,** 58 (1959). Known in two forms: water-soluble and water-insoluble.

The water-soluble form is a light yellow, amorphous powder, mp 132° (dec). $[\alpha]_D$ − 50°. Sol in alcohol, chloroform; insol in ether.

The water-insoluble form is an amorphous powder mp 59°. Sol in alcohol, chloroform, ether.

9890. Vinclozolin. *3-(3,5-Dichlorophenyl)-5-ethenyl-5-methyl-2,4-oxazolidinedione;* 3-(3,5-dichlorophenyl)-5-methyl-5-vinyloxazolidine-2,4-dione; BAS 352F; Ronilan. $C_{12}H_9Cl_2NO_3$; mol wt 286.11. C 50.37%, H 3.17%, Cl 24.78%, N 4.90%, O 16.78%. Prepn: D. Mangold *et al.,* Ger. pat. 2,207,576 (1973 to BASF), *C.A.* **79,** 137120q (1973). Activity: C. Hess, F. Locher, *Proc. Brit. Insectic. Fungic. Conf.* **2,** 693 (1975). Comparative mechanism of action: A. C. Pappas, D. J. Fisher, *Pestic. Sci.* **10,** 239 (1979).

Cryst solid, mp 108°. Soly in water at 20°: 1 g/l. Soly in (g/kg): acetone 435; benzene 146; chloroform 319; ethyl acetate 253. Slowly hydrolyzed in alkaline soln. LD_{50} orally in rats: 10 g/kg, C. Hess, F. Locher, *loc. cit.*

USE: Fungicide.

9891. Vincristine. *22-Oxovincaleukoblastine; leurocristine;* VCR; LCR. $C_{46}H_{56}N_4O_{10}$; mol wt 824.94. C 66.97%, H 6.84%, N 6.79%, O 19.40%. Antitumor alkaloid isolated from *Vinca rosea* Linn. (*Catharanthus roseus* G. Don), *Apocynaceae:* Svoboda, *Lloydia* **24,** 173 (1961). Structure: Neuss *et al., J. Am. Chem. Soc.* **86,** 1440 (1964); Moncrief, Lipscomb, *ibid.* **87,** 4963 (1965). Pharmacology: R. H. Adamson *et al., Arch. Int. Pharmacodyn. Ther.* **157,** 299 (1965); S. M. Sieber *et al., Cancer Treat. Rep.* **60,** 127 (1976). Prepn and pharmacology of [³H]vincristine: Owellen, Donigian, *J. Med. Chem.* **15,** 894 (1972). Symposium on vincristine: *Cancer Chemother. Rep.* **52,** 453-535 (1968). Biosynthesis from catharanthine, *q.v.* and vindoline, *q.v.*: A. Rahman *et al., Tetrahedron Letters* **1976,** 2351; P. Mangeney *et al., J. Am. Chem. Soc.* **101,** 2243 (1979). Comprehensive description of the sulfate: J. H. Burns in *Analytical Profiles of Drug Substances* vol. 1, K. Florey, Ed. (Academic Press, New York, 1972) pp 463-480. For structure *see* Vinblastine, R = CHO.

Blades from methanol, mp 218-220°. $[\alpha]_D^{25}$ +17°; $[\alpha]_D^{25}$ +26.2° (ethylene chloride). pKa: 5.0, 7.4 in 33% DMF. uv max (ethanol): 220, 255, 296 nm (log a_m 4.65, 4.21, 4.18). LD_{50} i.p. in mice: 5.2 mg/kg (Adamson).

Sulfate, $C_{46}H_{58}N_4O_{14}S$, *Kyocristine, Oncovin, Vincosid, Vincrex.* Crystals from ethanol.

THERAP CAT: Antineoplastic.

9892. Vindesine. *3-(Aminocarbonyl)-O⁴-deacetyl-3-de(methoxycarbonyl)vincaleukoblastine;* desacetylvinblastine amide; VDS; Compound 112531; NSC 245467. $C_{43}H_{55}N_5O_7$;

mol wt 753.95. C 68.50%, H 7.35%, N 9.29%, O 14.86%. Synthetic deriv of the dimeric *Catharanthus* alkaloid vinblastine, *q.v.* Prepn: G. J. Cullinan, K. Gerzon, **Ger.** pat. **2,415,980** (1974 to Lilly), *C.A.* **82,** 72191r (1975); *see also* C. J. Burnett *et al., J. Med. Chem.* **21,** 88 (1978). ¹³C-NMR spectroscopy: D. E. Dorman, J. W. Paschal, *Org. Magn. Reson.* **8,** 413 (1976). Antitumor activity in rodents: M. J. Sweeney *et al., Cancer Res.* **38,** 2886 (1978). Pharmacokinetics: R. J. Owellen *et al., ibid.* **37,** 2603 (1977). Use in acute lymphocytic leukemia of childhood: W. Krivit *et al., Cancer Chemother. Pharmacol.* **2,** 267 (1979). Brief review of preclinical and early clinical data: R. W. Dyke *et al., ibid.* 229. Toxicology: G. C. Todd *et al., Toxicol. Environ. Health* **1,** 843 (1976); R. J. Owellen *et al., Biochem. Pharmacol.* **26,** 1213 (1977).

Crystals from ethanol-methanol, mp 230-232°. $[\alpha]_D^{25}$ +39.4° (c = 1.0 in methanol). uv max (methanol): 214, 266, 288, 296 nm (ϵ 53400, 17450, 13950, 12500). pKa' (DMF 66%) 5.39, 7.36; (H₂O) 6.04, 7.67.

Sulfate salt, $C_{43}H_{57}N_5O_{11}S$, *LY 099094, Eldisine, Fildesin.* Amorphous solid from ethanol-isopropyl alcohol, mp > 250° (solvated form). LD_{50} i.v. in mice, rats (mg/kg): 6.3 ± 0.6, 2.0 ± 0.2 (Todd); i.p. in mice: 8.8 ± 2.5 (Owellen, *Biochem. Pharmacol.*).

THERAP CAT: Antineoplastic.

9893. Vindoline. *4-(Acetyloxy)-6,7-didehydro-3-hydroxy-16-methoxy-1-methylaspidospermidine-3-carboxylic acid methyl ester.* $C_{25}H_{32}N_2O_6$; mol wt 456.52. C 65.77%, H 7.07%, N 6.14%, O 21.03%. Major alkaloid from the leaves of *Vinca rosea* Linn. (*Catharanthus roseus* G. Don.), *Apocynaceae:* Gorman *et al., J. Am. Pharm. Assoc.* **48,** 256 (1959); Svoboda *et al., ibid.* 659; Moza, Trojánek, *Coll. Czech. Chem. Commun.* **28,** 1419 (1963). Structure: Gorman *et al., J. Am. Chem. Soc.* **84,** 1058 (1962); Neuss, *Bull. Soc. Chim. France* **1963,** 1509. Stereochemistry: Moncrief, Lipscomb, *J. Am. Chem. Soc.* **87,** 4963 (1965). Review of chemistry: Neuss *et al., Advan. Chemother.* **1,** 133 (1964). Total synthesis of ($\pm$)-vindoline: Ando *et al., J. Am. Chem. Soc.* **97,** 6880 (1975); Y. Ban *et al., Tetrahedron Letters* **1978,** 151; J. P. Kutney *et al., J. Am. Chem. Soc.* **100,** 4220 (1978).

R = OCH_3. Needles from acetone + petr ether, mp 164-165°; prisms, mp 174-175°; $[\alpha]_D^{26}$ − 18° (chloroform); pKa 5.5 in 66% DMF [Moza, Trojánek, *loc. cit.*]. Also reported as crystals, mp 154-155°; $[\alpha]_D^{27}$ +42° (chloroform) [Gorman *et al., loc. cit.*]. uv max (ethanol): 212, 250, 304 nm (log ϵ 4.49, 3.74, 3.57).

Hydrochloride, $C_{25}H_{32}N_2O_6$·HCl, crystals from acetone, mp 161-164°.

Demethoxyvindoline, $C_{24}H_{30}N_2O_5$, *vindorosine, vindolidine.* R = H. Isoln: Moza, Trojánek, *loc. cit.* Structure: *eidem, Coll. Czech. Chem. Commun.* **28,** 1427 (1963). Needles from benzene + petr ether, mp 167°. $[\alpha]_D^{16}$ −31° (chloroform). uv max (methanol): 250, 302 nm (log ϵ 3.98, 3.52).

Note: Lacks physiological activity alone but is contained as the pentacyclic moiety in the antineoplastic agents vinblastine, *q.v.*, and vincristine, *q.v.*

9894. Vinpocetine. *Eburnamenine-14-carboxylic acid ethyl ester;* $3\alpha,16\alpha$*-apovincaminic acid ethyl ester; ethyl apovincamin-22-oate;* RGH-4405; Cavinton; Ceractin. $C_{22}H_{26}$-N_2O_2; mol wt 350.46. C 75.40%, H 7.48%, N 7.99%, O 9.13%. Deriv of vincamine, *q.v.*, with vasodilating activity. Prepn: C. Lörincz *et al.*, **Ger.** pat. **2,253,750** corresp to **U.S.** pat. **4,035,370** (1973, 1977 both to Gedeon Richter); *eidem, Arzneimittel-Forsch.* **26,** 1907 (1976). Series of articles on pharmacology, biochemistry, metabolism, pharmacokinetics, clinical studies: *ibid.* 1908-1989. Toxicity studies: E. Pálosi, L. Szporny, *ibid.* 1926; E. Cholnoky, L. I. Dömök, *ibid.* 1939. HPLC studies: G. Szepesi, M. Gazdag, *J. Chromatog.* **205,** 57, 341 (1981). Evaluation of effectiveness as antimotion drug: E. I. Matsnev, D. Bodo, *Aviat. Space Environ. Med.* **55,** 281 (1984).

Crystals from benzene, mp 147-153° (dec). $[\alpha]_D^{20}$ +114° (c = 1 in pyridine). uv max (96% ethanol): 229, 275, 315 nm (log ϵ 4.45, 4.08, 3.85). LD_{50} in mice, rats (mg/kg): 534, 503 orally; 240, 133.8 i.p.; 58.7, 42.6 i.v. (Pálosi, Szporny), also reported as 161.2 mg/kg i.p. in mice (Cholnoky, Dömök).

THERAP CAT: Cerebral vasodilator.

9895. Vintiamol. *N-[(4-Amino-2-methyl-5-pyrimidinyl)methyl]-N-[4-hydroxy-1-methyl-2-[(3-oxo-3-phenyl-1-propenyl)thio]-1-butenyl]formamide; N-[(4-amino-2-methyl-5-pyrimidinyl)methyl]-N-[2-[(2-benzoylvinyl)thio]-4-hydroxy-1-methyl-1-butenyl]formamide; S-(benzoylvinyl)thiamine.* $C_{21}H_{24}N_4O_3S$; mol wt 412.52. C 61.14%, H 5.86%, N 13.58%, O 11.64%, S 7.77%. Prepn and sepn of isomers: Fusco, Tenconi, *Farmaco Ed. Sci.* **20,** 866 (1965).

Higher melting isomer: crystals from ethanol, mp 202-204°, very sparingly sol in chloroform. Lower melting isomer: separated from mixture of both isomers by its greater soly in chloroform; crystals from ethanol contg 1 molecule of ethanol, mp 100-103°.

THERAP CAT: Vitamin, enzyme co-factor.

9896. Vinyl Acetate. *Acetic acid ethenyl ester; acetic acid vinyl ester.* $C_4H_6O_2$; mol wt 86.09. C 55.80%, H 7.03%, O 37.17%. $CH_3COOCH=CH_2$. Prepn: Schnizer, **U.S.** pat. **2,859,241** (1958 to Celanese); Sharp, Steitz, **U.S.** pat. **2,860,-159** (1958 to Pan Am. Petroleum); Foster, Tobler, *J. Am. Chem. Soc.* **83,** 851 (1961). Toxicity data: Smyth, Carpenter, *J. Ind. Hyg. Toxicol.* **30,** 63 (1948). Study of chronic human exposure to vinyl acetate: Deese, Joyner, *Am. Ind. Hyg. Assoc. J.* **30,** 449 (1969). *Reviews:* Leonard "Vinyl Acetate" in *Vinyl and Diene Monomers* (part 1), E. C. Leonard, Ed. (Wiley-Interscience, New York, 1970) pp 263-328; Faith, Keyes & Clark's *Industrial Chemicals*, F. A. Lowenheim, M. K. Moran, Eds (Wiley-Interscience, New York, 4th ed., 1975) pp 862-867; W. Daniels, "Poly(Vinyl Acetate)" in Kirk-Othmer *Encyclopedia of Chemical Technology* vol. 23 (Wiley-Interscience, New York, 3rd ed., 1983) pp 817-847.

Liquid. Polymerizes in light to a colorless, transparent mass. bp 72.7°. Two reported melting pts: −100°; −93° (Daniels). d_4^{20} 0.932. Flash pt, closed cup: 18°F (−8°C). Soly in water (20°): 1 g/50 ml. Misc with alc, ether. LD_{50} orally in rats: 2.92 g/kg (Smyth, Carpenter).

USE: In polymerized form for plastic masses, films and lacquers.

9897. Vinylbital. *5-Ethenyl-5-(1-methylbutyl)-2,4,6-(1H,3H,5H)-pyrimidinetrione; 5-(1-methylbutyl)-5-vinylbarbituric acid; 5-vinyl-5-(1-methylbutyl)barbituric acid; butyvinal; Speda; Optanox.* $C_{11}H_{16}N_2O_3$; mol wt 224.25. C 58.91%, H 7.19%, N 12.49%, O 21.40%. Prepn: Seefelder, *Festschr. Carl Wurster* 1960, 71, *C.A.* **56,** 9928d (1962).

Crystals, mp 90-91.5°.

Caution: May be habit forming. This is a controlled substance (depressant) listed in the U.S. Code of Federal Regulations, Title 21 Part 1308.13 (1987).

THERAP CAT: Sedative, hypnotic.

9898. Vinyl Chloride. *Chloroethylene.* C_2H_3Cl; mol wt 62.50. C 38.43%, H 4.84%, Cl 56.73%. $CH_2=CHCl$. Prepd from ethylene dichloride and alcoholic potassium: Regnault, *Ann.* **14,** 22 (1835); by halogenation of ethylene: Miller, Jenks, **U.S.** pat. **2,896,000** (1959 to National Distillers and Chemical Corp.). Review of mfg processes: Faith, Keyes & Clark's *Industrial Chemicals*, F. A. Lowenheim, M. K. Moran, Eds. (Wiley-Interscience, New York, 4th ed., 1975) pp 868-873. Comprehensive monograph on toxicity of vinyl chloride: *Ann. N.Y. Acad. Sci.* **246,** 1-337 (1975). Review of carcinogenicity studies: *IARC Monographs* **19,** 377-437 (1979).

Colorless gas; liquefies in a freezing mixture. Polymerizes in light or in presence of catalyst. mp −153.8°. bp −13.37°. d_4^{20} 0.9106. n_D^{10} 1.3700. Vapor pressure at 20°: 2530 mm Hg. *Flammable!* Flash pt, closed cup: −78° (−112°F). Sol in alc, ether, carbon tetrachloride, benzene. Slightly sol in water.

Caution: Causes "vinyl chloride disease". May be narcotic in high concns. Consult latest Government regulations on allowable concentrations in air. If spilled on skin rapid evaporation can cause local frostbite. This substance has been listed as a known carcinogen: *Fourth Annual Report on Carcinogens* (NTP 85-002, 1985) p 199.

USE: In the plastics industry; as refrigerant; in organic syntheses.

9899. Vinyl Ether. *1,1'-Oxybisethene;* divinyl ether; divinyl oxide; ethenyloxyethene; Vinethene; Vinesthene. C_4H_6O; mol wt 70.09. C 68.54%, H 8.63%, O 22.83%. $CH_2=CHOCH=CH_2$. Prepn: Major, Ruigh, **U.S.** pat. **2,021,872** (1935 to Merck & Co.); Mittag, Smidt, **U.S.** pat. **2,832,807** (1958 to Consortium für electrochem. Ind. GmbH).

Very volatile, flammable liq; characteristic odor. Dec on exposure to light or to acid fumes, forming acetaldehyde, and polymerizes to a solid glass-like mass. d_{20}^{20} 0.774; d_4^{20} 0.773. bp 28.4°. n_D^{20} 1.3989. Crit temp 183° (calc). Critical press. about 30 atm. Specific heat about 0.53 cal/g. Flash point (closed cup) < −30° (< −22°F). Explosives limits (% by vol in air), lower: 1.7, upper: 27.0. Autoignition temp 360° (680°F). 0.53 g dissolves in 100 g water at 37°. Oil-water soly coefficient at 37°: 41. Miscible with alcohol, ether, oils and other organic solvents. *Keep protected from light and acid fumes, in a cool place, and away from an open flame.* Lethal concn for mice in air: 51,233 ppm.

Note: Pharmaceutical product contains not less than 96.0% and not more than 97.0% vinyl ether, the remainder consisting of dehydrated alcohol. It may contain not more than 0.025% of suitable preservative. *U.S.P.* **XXI** (1985).

THERAP CAT: Anesthetic (inhalation).

9900. Vinylidene Chloride. *1,1-Dichloroethene; 1,1-di-*

Consult the cross index before using this section.

chloroethylene; *asym-*dichloroethylene. $C_2H_2Cl_2$; mol wt 96.95. C 24.78%, H 2.08%, Cl 73.14%. $CH_2=CCl_2$. Prepn from ethylene chloride: Reilly, U.S. pat. **2,140,548** (1938 to Dow). By dehydrochlorination of 1,1,2-trichloroethane: Conrad, Gould, U.S. pat. **2,989,570** (1961 to Ethyl Corp.).

Liquid. Mild, sweet odor resembling that of chloroform. d_4^{20} 1.2129. mp $-122.5°$. bp_{760} 31.7°. n_D^{20} 1.4249. Flash pt $-15°$. Practically insol in water. Sol in organic solvents. At temps above 0° and especially in the presence of oxygen or other suitable catalysts polymerizes to a plastic. Several inhibitors to preserve the monomer have been invented. Uncontrolled polymerization may lead to explosive reaction products with oxygen or ozone: Reinhardt, *Chem. Eng. News* **25**, 2136 (1947).

USE: Intermediate in the production of "vinylidene polymer plastics" such as *Saran* and *Velon*. *Caution:* Irritant to skin, mucous membranes; narcotic in high concentrations; has caused liver, kidney injury in exptl animals: Irish in *Industrial Hygiene and Toxicology* vol. 2, F. A. Patty, Ed. (Interscience, New York, 2nd ed., 1962) pp 1305-1307; Prendergast *et al., Toxicol. Appl. Pharmacol.* **10**, 270 (1967).

9901. Violacein. 3-[1,2-Dihydro-5-(5-hydroxy-1H-indol-3-yl)-2-oxo-3H-pyrrol-3-ylidene]-1,3-dihydro-2H-indol-2-one; 3-[2-(5-hydroxyindol-3-yl)-5-oxo-2-pyrrolin-4-ylidene]-2-indolinone; 3-[2-(5-hydroxyindol-3-yl)-5-oxo-pyrrolin-4-ylidene]oxindole; 5-(5-hydroxy-3-indolyl)-3-(3-oxindolylidene)-2-oxopyrroline. $C_{20}H_{13}N_3O_3$; mol wt 343.33. C 69.96%, H 3.82%, N 12.24%, O 13.98%. Pigment isolated from *Chromobacterium violaceum (Bacillus violaceus):* Tobie, *J. Bact.* **29**, 223 (1935); Strong, *Science* **100**, 287 (1944); Ballantine *et al., J. Chem. Soc.* **1958**, 755. Structure and synthesis: Ballantine *et al., ibid.* **1960**, 2292. *Review:* DeMoss in *Antibiotics* vol. 2, D. Gottlieb, P. D. Shaw, Eds. (Springer, New York, 1967).

Purplish-black needles, prisms. Practically insol in water; slightly sol in alcohol; moderately sol in dioxane.

9902. Violaxanthin. 5,6:5',6'-Diepoxy-5,5',6,6'-tetrahydro-β,β-carotene-3,3'-diol; zeaxanthin diepoxide. $C_{40}H_{56}O_4$; mol wt 600.85. C 79.95%, H 9.39%, O 10.65%. Widely distributed carotenoid pigment. Formed in plants from zeaxanthin. Isoln from yellow pansies *(Viola tricolor):* Kuhn, Winterstein, *Ber.* **64**, 326 (1931). Structure: Karrer *et al., Helv. Chim. Acta* **14**, 1044 (1931); **16**, 977 (1933); **19**, 1024 (1936); **27**, 1684 (1944). Partial synthesis: Karrer, Jucker, *ibid.* **28**, 300 (1945). Abs config: Bartlett *et al., J. Chem. Soc. (C)* **1969**, 2527. Isoln from *Viola tricolor* and configuration of the 15-*cis-*isomer: P. Molnar, J. Szabolcs, *Phytochemistry* **19**, 623 (1980).

Orange prisms from methanol, reddish-brown acicular crystals from CS_2. mp 200°. $[\alpha]_{Cd}^{20}$ $+35°$ (c = 0.08 in chloroform). Absorption max (in alc): 471.5, 442.5, 417.5 nm. Sol in alcohol, methanol, carbon disulfide, ether; almost insol in petr ether.

15-*cis-*Isomer, irregular yellow plates from benzene/petrol, mp 109°. uv max (benzene): 479, 448, 423, 337 nm (log ε 4.91, 4.98, 4.83, 4.77). Friction causes crystals to form orange prisms.

9903. Viologen. A term coined by Michaelis to designate the chlorides of certain quaternary bases derived from γ,γ'-

dipyridyl. Presently used as dichlorides, dibromides and diiodides. Prepn: Michaelis, *Biochem. Z.* **250**, 564 (1932); Michaelis, Hill, *J. Am. Chem. Soc.* **55**, 1481 (1933); *J. Gen. Physiol.* **16**, 859-873 (1933). Viologens are useful as oxidation-reduction indicators because their potential range is very negative. In contrast to other redox indicators, the color is exhibited by the reduced form, whereas usually the oxidized form is the colored one and secondly, the redox potential of these substances is independent of pH. Review of electrochemistry: C. L. Bird, A. T. Kuhn, *Chem. Soc. Rev.* **10**, 49-82 (1981).

Ethyl viologen, 1,1-diethyl-4,4'-bipyridinium, *N,N'-diethyl-γ,γ'-dipyridylium.* Normal potential at 30°: -0.449 volts.

Benzyl viologen, 1,1'-bis(phenylmethyl)-4,4'-bipyridinium, *N,N'-dibenzyl-γ,γ'-dipyridylium.* Normal potential at 30°: -0.359 volts.

Betaine viologen, N,N'-dibetaine-γ,γ'-dipyridylium. Normal potential at 30°: -0.444 volts.

Dimethyl analog, *see* Paraquat.

USE: As biological oxidation-reduction indicators.

9904. Violuric Acid. 2,4,5,6(1H,3H)-Pyrimidinetetrone 5-oxime; alloxan 5-oxime; 5-isonitrosobarbituric acid; 5-(hydroxyimino)barbituric acid. $C_4H_3N_3O_4$; mol wt 157.09. C 30.58%, H 1.93%, N 26.76%, O 40.74%. Known to exist as a mixture of keto-enol tautomers. Prepd from alloxan and hydroxylamine: Cresole, *Ber.* **16**, 1133 (1883); Guinchard, *Ber.* **32**, 1723 (1899).

Orthorhombic crystals, dec 240-241°. pK 4.7. Sparingly sol in water to a violet-colored soln; sol in alcohol. With $FeCl_3$ produces blue color.

USE: Analytical reagent for chromatographic separation of cations. Forms chelates: Leermakers, Hoffman, *J. Am. Chem. Soc.* **80**, 5663 (1958).

9905. Viomycin. Celiomycin; florimycin; tuberactinomycin B; Vinactane; Viocin; Vionactane. $C_{25}H_{43}N_{13}O_{10}$; mol wt 685.71. C 43.79%, H 6.32%, N 26.55%, O 23.33%. Antibiotic substance produced by various *Streptomyces* species including *S. puniceus, S. floridae, S. vinaceus.* Isoln: A. C. Finlay *et al., Am. Rev. Tuberc.* **63**, 1 (1951); Bartz *et al., ibid.* 4. Production: Marsh *et al.,* U.S. pat. **2,633,445** (1953 to Ciba); Freaney, U.S. pat. **2,828,245** (1958 to C.S.C.). Purification and chemical characterization: T. Kitagawa *et al., Chem. Pharm. Bull.* **20**, 2176 (1972). Proposed structure: Yoshioka *et al., Tetrahedron Letters* **1971**, 2043. Alternate structure: B. W. Bycroft *et al., ibid.* **1968**, 5901; *eidem, Experientia* **27**, 501 (1971); *eidem, J. Chem. Soc., Perkin Trans. I* **1972**, 820, 827. Confirmed structure: Noda *et al., J. Antibiot.* **25**, 427 (1972); T. Kitagawa *et al., Chem. Pharm. Bull.* **20**, 2215 (1972). X-ray crystallography: B. W. Bycroft, *Chem. Commun.* **1972**, 660. Biosynthesis: J. H. Carter II *et al., Biochemistry* **13**, 1221, 1227 (1974). Toxicological study: H. Keller *et al., Arzneimittel-Forsch.* **6**, 61 (1956). Enzyme immunoassay for determn in body fluids: T. Kitagawa *et al., Chem. Pharm. Bull.* **30**, 2487 (1982). For structure *see* Tuberactinomycin.

Sulfate, hygroscopic plates, mp 266° (dec). $[\alpha]_D^{18}$ $-29.5°$ (c = 1 in H_2O). uv max (water or 0.1N HCl; 0.1N NaOH): 268 nm (log ε 4.4); 285 nm (log ε 4.2). LD_{50} in mice (mg/kg): 240 i.v. (Finlay), 1750 s.c. (Keller). Sol in water; relatively insol in most organic solvents. Solns adjusted to pH 5-6 are quite stable. Soly data: Weiss *et al., Antibiot. & Chemother.* **7**, 374 (1957).

Hydrochloride, hygroscopic plates, mp 270° (dec). $[\alpha]_D^{18}$ $-16.6°$ (c = 1 in H_2O). uv max (water; 0.1N HCl; 0.1N NaOH): 268 nm (log ε 4.5); 268 nm (log ε 4.4); 285 nm (log ε 4.3).

THERAP CAT: Antibacterial (tuberculostatic).

9906. Viomycin Pantothenate. Semi-synthetic polypeptide antibiotic. Prepd from viomycin sulfate and calcium pantothenate, pantothenic acid, or an anion-exchange resin loaded with pantothenic acid: Keller, Mückter, **Ger.** pats. **954,874** and **1,011,800** (1956 and 1957 to Grünenthal); *C.A.* **53**, 2110b and 13519c (1959); **Brit.** pat. **852,334** (1960 to Lab. Atral).

Crystals, mp 156°. dec 242°.

A component of *Viothenat, Vionactan.*

THERAP CAT: Antibacterial (tuberculostatic).

9907. VIP. *Vasoactive intestinal polypeptide;* vasoactive intestinal peptide. Neuroactive polypeptide gastrointestinal hormone, related to glucagon and secretin, *q.q.v.* It exhibits a wide variety of biological activities, including relaxation of systemic and vascular smooth muscle and stimulation of the exocrine pancreas, of secretion of insulin and of cyclic-AMP formation in the small intestine. Initially isolated from porcine upper intestinal wall: S. I. Said, V. Mutt, *Science* **169**, 1217 (1970); *eidem, Eur. J. Biochem.* **28**, 199 (1972). Porcine VIP is a 28-residue peptide having an *N*-terminal seryl sequence: M. Bodanszky *et al., Proc. Nat. Acad. Sci. USA* **70**, 382 (1973); V. Mutt, S. I. Said, *Eur. J. Biochem.* **42**, 581 (1974). Synthesis: M. Bodanszky *et al., J. Am. Chem. Soc.* **96**, 4973 (1974); *see also* S. I. Said *et al.,* **U.S.** pat. **3,862,927** (1975). Solid-phase synthesis: D. H. Coy, J. Gardner, *Int. J. Peptide Protein Res.* **15**, 73 (1980); R. Colombo, *Experientia* **38**, 773 (1982). Isoln of chicken VIP: A. Nilsson, *FEBS Letters* **47**, 284 (1974). Amino acid sequence: *idem, ibid.* **60**, 322 (1975). Synthesis of C-terminal fragment: M. Bodanszky *et al., Bioorg. Chem.* **5**, 339 (1976). Solid-phase synthesis: R. Colombo, *Int. J. Peptide Protein Res.* **19**, 71 (1982). Multiple immunoreactive forms of VIP have also been found in human colonic mucosa: R. Dimaline, G. J. Dockray, *Gastroenterology* **75**, 387 (1978). The presence of a common form of VIP in brain and gastrointestinal tract has been demonstrated by radioimmunoassay and radioreceptor assay: J. Besson *et al., Acta Endocrinol.* **87**, 799 (1978). Although the precise effect of VIP on neuronal function is not completely known, it is believed to play an important role in neurotransmission, *cf.* M. G. Bryant *et al., Lancet* **1**, 991 (1976); Y. Matsuzaki *et al., Science* **210**, 1252 (1980). Review of possible functions as a neural peptide: S. I. Said *et al., Advan. Biochem. Psychopharmacol.* **22**, 75-82 (1980). Book: *Vasoactive Intestinal Peptide,* S. I. Said, Ed. (Raven Press, New York, 1982) 528 pp.

```
1                                         10
His-Ser-Asp-Ala-Val-Phe-Thr-Asp-Asn-Tyr-

11                                        20
Thr-Arg-Leu-Arg-Lys-Gln-Met-Ala-Val-Lys-

21                           28
Lys-Tyr-Leu-Asn-Ser-Ile-Leu-Asn
```

porcine VIP

9908. Viquidil. *3-(3-Ethenyl-4-piperidinyl)-1-(6-methoxy-4-quinolinyl)-1-propanone;* **quinicine;** 1-(6-methoxy-4-quinolyl)-3-(3-vinyl-4-piperidyl)-1-propanone; chinicine; mequiverine; quinotoxine; quinotoxol; LM-192. $C_{20}H_{24}N_2O_2$; mol wt 324.41. C 74.04%, H 7.47%, N 8.64%, O 9.86%. An isomer of quinine. Present in small quantities in cinchona barks; formed by heating quinine with glycerol at 180°: Howard, *J. Chem. Soc.* **24**, 61 (1871); **25**, 101 (1872); Miller, Rohde, *Ber.* **33**, 3214 (1900); Howard, Chick, *Pharm. J.* **99**, 143 (1917). Conversion to quinine: Rabe, Kindler, *Ber.* **51**, 466 (1918). Partial synthesis: Prostenik, Prelog, *Helv. Chim. Acta* **26**, 1965 (1943). Total synthesis: Woodward, Doering, *J. Am. Chem. Soc.* **66**, 849 (1944); **67**, 860 (1945); **U.S.** pat. **2,500,444** (1950 to Polaroid); Grethe *et al., Helv. Chim. Acta* **56**, 1485 (1973). Review of synthesis, chemistry and pharmacology: Quevauviller *et al., Ann. Pharm. Fr.* **24**, 39 (1966). Metabolic studies: Guérémy *et al., Arzneimittel-Forsch.* **22**, 1336 (1972); Uzan *et al., ibid.* 1341.

d-Form, yellow viscous oil, $[\alpha]_D$ +43°. Slightly sol in water; freely sol in alcohol, chloroform, ether.

Hydrochloride, $C_{20}H_{15}ClN_2O_2$, *Desclidium, Permiran.* Yellow, odorless and bitter tasting powder, mp 184 ± 4°. uv max (chloroform): 246, 355 nm. Sol in alc; sparingly sol in water; practically insol in acetone.

THERAP CAT: Vasodilator; antiarrhythmic.

9909. Viractin. Active principle in the mycelium of *Streptomyces griseus* cultures which produce cycloheximide: B. E. Leach *et al., Nature* **204**, 788 (1964). Production: B. E. Leach, **U.S.** pat. **3,421,981** (1969). It is postulated that viractin acts prophylactically to prevent upper respiratory infections by combining competitively with the sites where the virus attaches itself to the host cell. Polemic about efficacy: Tyrell, Walker, *Nature* **210**, 386 (1966); B. E. Leach, *ibid.* 387.

Liquid. bp_{200} 105-135°. LD_{50} in mice (mg/kg): 200 intranasally; 300 i.v.; > 300 s.c. (Leach, 1964).

9910. Virginiamycin. Staphylomycin; virgimycin; antibiotic no. 899; SKF 7988; Eskalin V; Stafac; Staphylomycine. Antibiotic complex produced by a *Streptomyces* related to *S. virginiae:* Somer, Van Dijck, *Antibiot. & Chemother.* **5**, 632 (1955). A mixture of two principal components, virginiamycin M_1 (factor M_1) and virginiamycin S_1 (factor S). Separation of factors M_1 and S: Vanderhaeghe *et al., ibid.* **7**, 606 (1957); of the complex into six components: Gosselinckx, Parmentier, *Chomatog. Sym., 2nd Brussels* **1962**, 181; of factor S into three components: Vanderhaeghe *et al., Tetrahedron Letters* **1971**, 2687. The product of commerce contains about 75% of fraction M_1, 5% of fraction S. Structure of virginiamycin S_1: Vanderhaeghe, Parmentier, *J. Am. Chem. Soc.* **82**, 4414 (1960); of virginiamycin M_1: Delpierre *et al., Tetrahedron Letters* **1966**, 369; *eidem, J. Chem. Soc.*

virginiamycin S_1

virginiamycin M_1

(C) **1966,** 1653; Kingston *et al., ibid.* 1669. Nomenclature: Crooy, De Neys, *J. Antibiot.* **25,** 371 (1972). Prepn of derivatives of virginiamycin S_1: G. Janssen *et al., J. Antibiot.* **30,** 141 (1977). Biosynthesis of virginiamycin M_1: D. G. I. Kingston, M. X. Kolpak, *J. Am. Chem. Soc.* **102,** 5964 (1980).

Antibiotic complex, amorphous, white powder. Dec 138-140°. $[\alpha]_D^{20}$ −124° (c = 0.5 in ethanol). Sparingly sol in water and dil acid. Dissolves in aq alkali (above pH 9.5) with rapid inactivation. Sol in methanol, ethanol, acetone, ethyl acetate, chloroform, benzene. Practically insol in ligroin.

Virginiamycin S_1, $C_{43}H_{49}N_7O_{10}$, *staphylomycin S.* Crystals from methanol, mp 240-242°. $[\alpha]_D^{20}$ −28° (c = 1 in ethanol). uv max (ethanol): 305 nm (log ε 3.85). Approx solys (percent): ether 0.1; methanol 0.5; ethanol 1.5; benzene 2.5; acetone and ethyl acetate 3; dioxane 4. Very sol in chloroform, DMF. Practically insol in water, petr ether.

Virginiamycin M_1, $C_{28}H_{35}N_3O_7$, *mikamycin A, ostreogrycin A, pristinamycin II_A, staphylomycin M_1, streptogramin A, vernamycin A.* Tan powder from acetone + petr ether, dec 165-167°. $[\alpha]_D$ −190° ± 2° (c = 0.5 in ethanol). uv max (methanol): 216 nm ($E_{1cm}^{1\%}$ 582). Approx solys (percent): ether, 0.1; benzene 0.3; ethyl acetate 0.5; acetone 2; methanol and ethanol 4; dioxane and tetrahydrofuran 5. Very sol in chloroform, DMF. Practically insol in water, petr ether.

THERAP CAT: Antibacterial.

THERAP CAT (VET): Antibacterial; food additive.

9911. Viridicatin. *3-Hydroxy-4-phenyl-2(1H)-quinolinone; 3-hydroxy-4-phenylcarbostyril;* 2,3-dihydroxy-4-phenylquinoline. $C_{15}H_{11}NO_2$; mol wt 237.25. C 75.93%, H 4.67%, N 5.90%, O 13.49%. Antibiotic substance from the mycelium of *Penicillium viridicatum* Westling, the chief mold on stored corn: Cunningham, Freeman, *Biochem. J.* **53,** 328 (1953); from various strains of *P. cyclopium* Westling: Bracken *et al., ibid.* **57,** 587 (1954); from *P. puberulum* Bainier: Austin, Meyers, *J. Chem. Soc.* **1964,** 1197. Biosynthesis: Luckner, *Tetrahedron Letters* **1962,** 1035; Luckner, Mothes, *Arch. Pharm.* **296,** 18 (1963); Framm *et al., Eur. J. Biochem.* **37,** 78 (1973). Synthesis: Eistert, Selzer, *Z. Naturforsch.* **17b,** 202 (1962). Mass spectra: Luckner *et al., Tetrahedron* **25,** 2575 (1969). Antibacterial activity and inhibitory effect on plant growth: Taniguchi, Satomura, *Agr. Biol. Chem.* **34,** 506 (1970).

Lustrous needles from methanol or ethanol, mp 268°. Sublimes unchanged in high vacuum at 160-170°. Very weak acid. uv and ir curves: Cunningham, Freeman, *loc. cit.* Practically insol in water, aq $NaHCO_3$. Sparingly sol in cold organic solvents, cold concd HCl, dil mineral acids; sol in cold, aq *2N* KOH, glacial acetic acid.

Sodium salt, $C_{15}H_{10}NNaO_2$, prismatic needles from aq NaOH, dec 260-265°. Sol in water.

Monoacetylviridicatin, $C_{17}H_{13}NO_3$, crystals from aq ethanol, mp 200-201°. Dec slowly on standing.

O,O-Dimethylviridicatin, $C_{17}H_{15}NO_2$, plates from ethanol, mp 86-87°.

O,N-Dimethylviridicatin, $C_{17}H_{15}NO_2$, plates from ethanol, mp 197-198°.

9912. Viridin. *1β-Hydroxy-2β-methoxy-18-norandrosta-5,8,11,13-tetraeno[6,5,4-bc]furan-3,7,17-trione.* $C_{20}H_{16}O_6$; moi wt 352.35. C 68.18%, H 4.58%, O 27.24%. Antibiotic substance from *Gliocladium virens:* Grove *et al., J. Chem. Soc.* **1965,** 3803. Previously reported isolation from *Trichoderma viride:* Brian, McGowan, *Nature* **156,** 144 (1945); Brian *et al., Ann. Appl. Biol.* **33,** 190 (1946). Epimerizes to β-viridin where methoxyl group is equatorial. Separation of the two isomers: Vischer *et al., Nature* **165,** 528 (1950). Structure: Grove *et al., Chem. Commun.* **1965,** 343. Configuration: *eidem, J. Chem. Soc. (C)* **1966,** 743; Neidle, Hursthouse, *J. Chem. Soc. Perkin Trans. II* **1972,** 760. Biosynthe-

sis: Blight *et al., Chem. Commun.* **1968,** 1117; Grove, *J. Chem. Soc. (C)* **1969,** 549.

Prisms from benzene, mp 245° (dec). $[\alpha]_D^{19}$ −224°. uv max: 242, 300 nm (log ε 4.49, 4.22). Needles from acetone, mp 222-224° (dec). Prisms from glacial acetic acid, mp 200-205° (dec). Sol in water, chloroform; sparingly sol in carbon disulfide, carbon tetrachloride. Practically insol in ether, camphor. Aq solns lose activity rapidly unless acidified to pH 3.

β-Isomer, prisms from benzene, mp 240-245° (dec). $[\alpha]_D^{16}$ −23°. uv max: 243, 300 nm (log ε 4.45, 4.25).

THERAP CAT: Antifungal.

9913. Viscose. A viscous orange-red aq soln of sodium cellulose xanthogenate obtained by dissolving wood pulp cellulose in sodium hydroxide soln and treating with carbon disulfide. Manuf: Bachlott, U.S. pat. **2,855,321** (1958 to du Pont); Von Kohorn, U.S. pat. **2,985,647** (1961).

USE: Intermediate in the manuf of rayon and cellophane.

9914. Viscosin. $C_{54}H_{95}N_9O_{16}$; mol wt 1126.41. C 57.58%, H 8.50%, N 11.19%, O 22.73%. Antibiotic substance produced by *Pseudomonas viscosa.* Isoln: Kochi *et al., Bact. Proc.* **1951,** 29; Groupé *et al., Proc. Soc. Exptl. Biol. Med.* **78,** 354 (1951); Ohno *et al., J. Agr. Chem. Soc. Japan* **27,** 665 (1953); *C.A.* **49,** 3012 (1955); Toki, Ohno, *C.A.* **53,** 243b (1959). Synthesis studies: Fukuda, Bani, *C.A.* **56,** 10265i (1962). Structure: Toki, Ohno, *loc. cit.* Revised structure: Hiramoto *et al., Tetrahedron Letters* **1970,** 1087; *eidem, Chem. Pharm. Bull.* **19,** 1308, 1315 (1971). Shows antitubercular activity.

$$\text{D-CH}_3(\text{CH}_2)_6\text{CH(OH)CH}_2\text{CO—L-Leu—D-Glu—D-allo-Thr—D-Val—L-Leu—D-Ser}$$
$$\text{L-Ile←D-Ser←L-Leu}$$

Powder, dec 270-273°. $[\alpha]_D^{29}$ −168.3° (ethanol). Stable to heat. Practically insol in water. Sol in methanol, ethanol, ether, acetone, alkaline phosphate buffer solns.

9915. Visnadine. *2-Methylbutyric acid 9-ester with 9,10-dihydro-9,10-dihydroxy-8,8-dimethyl-2H,8H-benzo[1,2-b: 3,4-b']dipyran-2-one acetate; 8,8-dimethyl-2-oxo-9,10-dihydro-2H,8H*-benzo[1,2-b:3,4-b']dipyran-9,10-diyl-10-acetate-9-(α-methylbutyrate); 3,4,5-trihydroxy-2,2-dimethyl-6-chromanacrylic acid δ-lactone 4-acetate 3-(2-methylbutyrate); 4'-acetoxy-3'-(α-methylbutyryloxy)-2',2'-dimethyldihydropyrano(7,8:6',5')coumarin; 3-(α-methylbutyryloxy)-4-acetoxy-3,4-dihydroseseline; Cardine; Carduben; Vibeline; Visnamine. $C_{21}H_{24}O_7$; mol wt 388.40. C 64.93%, H 6.23%, O 28.84%. Isoln from seeds of *Ammi visnaga* L., *Umbelliferae* (bishop's weed): Smith *et al., Science* **115,** 520 (1952); *eidem, J. Am. Chem. Soc.* **79,** 3534 (1957); Smith, Haber; Smith, Hosansky, U.S. pats. **2,816,118; 2,980,699** (1957, 1961, both to Penick); Le Men, U.S. pat. **2,995,574** (1961 to Lab. Roger Bellon); Baddar *et al., J. Am. Chem. Soc.*

1963, 4522. Synthesis of ($\pm$)-form: Shanbhag *et al., Tetrahedron* **21,** 3591 (1965). Pharmacology and toxicology: Nkondi *et al., Therapie* **21,** 1267 (1966); Erbring *et al., Arzneimittel-Forsch.* **17,** 283 (1967); Eyraud, Aurousseau, *ibid.* **23,** 201 (1973).

Needles from light petr, or ether + hexane, mp 85-88°. $[\alpha]_D^{20}$ +9.2° (alc), $[\alpha]_D^{30}$ +42.5° (c = 2 in dioxane). Slightly sol in water; quite sol in ethanol, methanol. Very sol in chloroform, acetone, ether, benzene, DMF. LD_{50} in mice: 2240 mg/kg orally; > 370 mg/kg s.c., Erbring *et al., loc. cit.* (±)-Form, crystals from petr ether, mp 150-152° (Shanbhag).

THERAP CAT: Vasodilator (coronary).

9916. Visnagin. *4-Methoxy-7-methyl-5H-furo[3,2-g][1]-benzopyran-5-one;* 5-methoxy-2-methylfuranochromone; Visnacorin. $C_{13}H_{10}O_4$; mol wt 230.21. C 67.82%, H 4.38%, O 27.80%. A constituent of *Ammi visnaga* Lam., *Umbelliferae.* Isoln and structure: Späth, Gruber, *Ber.* **74,** 1492 (1941). Synthesis: Aneja *et al., Tetrahedron* **3,** 230 (1958); Badawi, Fayez, *Tetrahedron Letters* **1967,** 1029.

Thread-like needles from water. mp 142-145°. Very slightly sol in water. Sparingly sol in alc. Freely sol in chloroform.

9917. Vital Red. *3-Amino-4-[[4'-[(2-amino-6-sulfo-1-naphthalenyl)azo]-3,3'-dimethyl[1,1'-biphenyl]-4-yl]azo]-2,7-naphthalenedisulfonic acid trisodium salt; C.I. Direct Red 34;* C.I. 23570; brilliant vital red; Vital Red Evans; ditolyldiazo-3,6-disulfo-β-naphthylamine-β-naphthylamine-6-sulfonic acid sodium salt. $C_{34}H_{25}N_6Na_3$-O_9S_3; mol wt 826.77. C 49.39%, H 3.05%, N 10.17%, Na 8.34%, O 17.42%, S 11.64%. Prepd by diazotization of *o*-tolidine and coupling the resulting tetrazotolidine first with 2-naphthylamine-3,6-disulfonic acid (amino R salt) and next with 2-naphthylamine-6-sulfonic acid (Brönner acid): Palkin, Evans, *J. Am. Chem. Soc.* **47,** 429 (1925); **Ger. pat. 41,095;** *Frdl.* **1,** 476; *Colour Index* vol. 4 (3rd ed., 1971) p 4191.

Cherry-red crystals from water. Absorption max: 500 nm (aq soln). Sol in water giving a brownish-red soln. Gives blue solution with concd HCl. Slightly sol in alc. USE: For blood volume determinations.

9918. Vitamin A. *Retinol;* 3,7-dimethyl-9-(2,6,6-trimethyl-1-cyclohexen-1-yl)-2,4,6,8-nonatetraen-1-ol; antiinfective vitamin; lard-factor; antixerophthalmic vitamin; axerophthol; biosterol; oleovitamin A; ophthalamin (obsolete); vitamin A_1; vitamin A alcohol; Acon; Afaxin; Agiolan; Alphalin; Anatola; Aoral; Apexol; Apostavit; Atav; Avibon; Avita; Avitol; Axerol; "Dohyfral" A; Epiteliol; Nio-A-Let; Prepalin; Testavol; Vaflol; Vi-Alpha; Vitpex; Vogan; Vo-

gan-Neu. $C_{20}H_{30}O$; mol wt 286.44. C 83.86%, H 10.56%, O 5.59%. Occurs in the animal organism (not in plants); carotenoids are converted into vitamin A by the liver. Extracted from fish liver oils where it occurs mostly in esterified form. Structure: Karrer *et al., Helv. Chim. Acta* **14,** 1036, 1431; Karrer, Morf, *ibid.* **16,** 625 (1933); Heilbron *et al., Biochem. J.* **26,** 1178, 1194 (1932). Stereochemistry: *see* L. Zechmeister, *Chem. Rev.* **34,** 267 (1944). Identity of lard-factor and vitamin A: Ames, Harris, *Science* **120,** 391 (1954). Total synthesis of vitamin A from β-ionone and a propargyl halide: Eiter, Truscheit, U.S. pat. **3,060,229** (1962 to Bayer); by conversion of retinal: Wendler *et al., J. Am. Chem. Soc.* **72,** 234 (1950); Klein, Kapp, U.S. pat. **2,972,634** (1961 to Nopco); T. Mukaiyama, A. Ishida, *Chem. Letters* **1975,** 1201. Stereospecific synthesis of all *(E)*-form: P. S. Manchand *et al., Helv. Chim. Acta* **59,** 567 (1976); A. Fischli *et al., ibid.* 397; G. Cardillo *et al., J. Chem. Soc., Perkin Trans. I* **1979,** 1729. Toxicology: J. J. Kamm, *J. Am. Acad. Dermatol.* **6,** 652 (1982). Comprehensive monograph of the chemistry, physics, physiology of vitamin A and its provitamins: W. H. Sebrell, R. S. Harris, *The Vitamins,* Vol. I (Academic Press, New York, 2nd ed., 1967) 570 pp. Book: *The Retinoids,* Vol. 1-2, M. B. Sporn *et al.,* Eds. (Academic Press, New York, 1984). *See also:* Neovitamin A.

Yellow prisms from propylene oxide or petr ether. Solvated crystals from more polar solvents, such as methanol or ethyl formate. mp 62-64°. Distills at 120-125° at 5 × 10⁻³ mm pressure. n_D^{22} 1.6410 (calculated from refractive indexes of 20-70% solns in mineral oil). uv max: 324-325 nm ($E_{1cm}^{1\%}$ 1835): Baxter, Robeson, *J. Am. Chem. Soc.* **64,** 2407 (1942). Practically insol in water or glycerol; sol in abs alcohol, methanol, chloroform, ether, fats and oils. Ultraviolet light inactivates vitamin A and its solns which exhibit a characteristic green fluorescence. The free alcohol is sensitive to air-oxidation, but oil solns of it are quite stable. Esters of vitamin A are more stable to oxidation. LD_{50} (10 day) in mice (mg/kg): 1510 i.p.; 2570 orally (Kamm).

Acetate, $C_{22}H_{32}O_2$, pale yellow prismatic crystals from methanol, mp 57-58°. uv max (ethanol): 326 nm ($E_{1cm}^{1\%}$ 1550). Biopotency 2.904 × 10⁶ I.U./g. LD_{50} (10 day) orally in mice: 4100 mg/kg (Kamm).

Palmitate, $C_{36}H_{60}O_2$, *Arovit, Optovit-A,* is the ester preponderant in fish liver oils. Amorphous or cryst. mp 28-29°. uv max (ethanol): 325-328 nm ($E_{1cm}^{1\%}$ 975). Biopotency 1.817 × 10⁶ I.U./g. LD_{50} (10 day) in mice, rats (mg/kg): 6060, 7910 orally (Kamm).

Note: The U.S.P. unit of vitamin A is equal to 0.30 microgram of vitamin A equivalent to 0.344 microgram of vitamin A acetate. Observed biopotency 3.33 × 10⁶ I.U./g.

THERAP CAT: Antixerophthalmic vitamin.

THERAP CAT (VET): Nutritional factor.

9919. Vitamin A_2. *3,4-Didehydroretinol;* retinol₂; dehydroretinol. $C_{20}H_{28}O$; mol wt 284.42. C 84.45%, H 9.92%, O 5.63%. Combines with opsin to form the visual pigment, porphyropsin, *q.v.,* in fresh-water fish. A mixture of stereoisomers. Isoln from pike liver oils: Shantz, *Science* **108,** 417 (1948). *See also* Shantz, Brinkman, *J. Biol. Chem.* **183,** 467 (1950); Farrar *et al., J. Chem. Soc.* **1952,** 503. Synthesis and characteristics of stereoisomers: Schwieter, *Chimia* **14,** 362 (1960); *Helv. Chim. Acta* **45,** 517, 528, 541, 548 (1962). *Review: The Vitamins,* vol. 1, W. H. Sebrell, R. S. Harris, Eds. (Academic Press, New York, 2nd ed., 1967) *passim.*

Stereoisomeric mixture, golden yellow oil. Readily affected by oxygen. uv max (ethanol): 351, 287 nm ($E_{1cm}^{1\%}$ 1460, 820). Has about 40% of the bioactivity of cryst vitamin A_1.

Stereoisomers: All-*trans*, mp 17-19°; 13-*cis*, mp 73-74°; 9-*cis*, mp 77-79°; 9,13-di-*cis*, mp $< -30°$.

9920. Vitamin B₁₀ and Vitamin B₁₁. Two factors reported to be necessary for growth and feathering of chicks: Briggs *et al., J. Biol. Chem.* **148**, 163 (1943). Prepn from culture filtrates of *Mycobacterium tuberculosis:* Mills *et al., Proc. Soc. Exp. Biol. Med.* **56**, 240 (1944).

9921. Vitamin B₁₂. Cyanocobalamin; 5,6-dimethylbenz-imidazolyl cyanocobamide; cobinamide cyanide phosphate 3'-ester with 5,6-dimethyl-1-α-D-ribofuranosylbenzimida-zole inner salt; LLD factor; Lactobacillus lactis Dorner factor; extrinsic factor; antipernicious anemia principle; Anacobin; Antipernicin; Bedoce; Bedodeka; Bedoz; Behepan; Berubi; Berubigen; Betalin-12; Betolvex; Bevatine-12; Bevidox; Bexii; Bexil; Biocobalamine; Biocres; Bitevan; B-Telve; B-Twelv; Byladoce; Claretin-12; Cobalin; Cobamin; Cobamine; Cobione; Covit; Crystamin; Cycobemin; Cycolamin; Cykobeminet; Cytacon; Cytamen; Cytobion; Distivit (B₁₂ peptide); Dobetin; Docemine; Docibin; Docigram; Docivit; Dodecabee; Dodecavite; Dodex; Ducobee; Duodecibin; Embiol; Emociclina; Eritrone; Erycytol; Erythrotin; Euhaemon; Fresmin; Hemo-B-Doze; Hemomin; Hepagon; Hepavis; Hepcovite; Hydoxamin; Hydroxobase; Macrabin; Megabion (Indian); Megalovel; Milbedoce; Millevit; Nagravon; Normocytin; Peraemon; Pernaevit; Pernipur; Plecyamin; Poyamin; Redamina; Redisol; Rhodacryst; Rubesol; Rubivitan; Rubramin; Rubripca; Rubrocitol; Sytobex; Vibalt; Vibisone; Virubra; Vitarubin; Vita-Rubra; Vitral. $C_{63}H_{88}CoN_{14}O_{14}P$; mol wt 1355.38. C 55.83%, H 6.54%, Co 4.35%, N 14.47%, O 16.53%, P 2.28%. A cobalt containing coordination compd produced by intestinal microorganisms. Found also in soil and water, the richest sources being activated sewage sludge (*see* Milorganite), manure, and dried estuarine mud. Higher plants do not concentrate vitamin B₁₂ from the soil and so are a poor source as compared with animal tissues. First isolated from liver: Rickes *et al., Science* **107**, 396 (1948). Also from cultures of *Streptomyces griseus:* Rickes *et al., ibid.* **108**, 634 (1948); Rickes, Wood, U.S. pats. **2,563,-794** and **2,703,302-3** (1951 and 1955 to Merck & Co.); Speedie, Hull, U.S. pat. **2,951,017** (1960 to Distillers Co.); McDaniel; Long, U.S. pats. **3,000,793** and **3,018,225** (1961, 1962, both to Merck & Co.). From sewage sludge: Van Melle, U.S. pat. **3,057,851** (1962 to Armour); Bernhauer *et al.,* U.S. pat. **3,120,509** (1964 to Hoffmann-La Roche). Absorption through the intestinal wall is dependent on the presence of Castle's Intrinsic Factor, *q.v.* Although requirement for the vitamin is minute, deficiency states have been observed in individuals who abstain from all animal products.

Nomenclature: IUPAC rules, *Pure Appl. Chem.* **48**, 497 (1976). *Cobalamin* refers to all of the molecule except the cyano group. The fundamental ring system without cobalt or side chains is called *corrin* and the octadehydrocorrin is called *corrole*. The Co-contg heptacarboxylic acid resulting from hydrolysis of all the amide groups without the CN and the nucleotide, is designated *cobyrinic acid.* The corresp hexacarboxylic acid which still retains the 2-hydroxypropionamide group on a side chain is called *cobinic acid* and the hexacarboxylic acid having the ribofuranosidophosphoryl-propionamide side chain is called *cobamic acid.* Thus *cobamide* is the hexaamide of cobamic acid, *cobyric acid* is the hexaamide of cobyrinic acid and *cobinamide* is the hexaamide of cobinic acid.

Structure announced by A. Todd and team (Cambridge), D. Hodgkin and team (Oxford), and E. L. Smith (Glaxo), *see Nature* **176**, 325, 328, 551 (1955); **178**, 64 (1956). X-ray structure analysis: D. C. Hodgkin, *Fortschr. Chem. Org. Naturst.* **15**, 167-227 (1958). Stereochemistry: Stora, *Bull. Soc. Chim. France* **1959**, 1421. Total synthesis: Woodward, *Pure Appl. Chem.* **33**, 145 (1973). Monograph: Smith, *Vitamin B₁₂* (Methuen & Co., London, 3rd ed., 1965); B. Zagalak, W. Friedrich, Eds. *Vitamin B₁₂* (de Gruyter, New York, 1979). *Review:* several authors in *The Vitamins,* vol. **2**, W. H. Sebrell, R. S. Harris, Eds. (Academic Press, New York, 2nd ed., 1968) pp 119-259; B. T. Golding in *Compre-*

hensive Organic Chemistry vol. **5**, E. Haslam, Ed. (Pergamon, New York, 1979) pp 549-584. Comprehensive description: J. Kirschbaum in *Analytical Profiles of Drug Substances* vol. **10**, K. Florey, Ed. (Academic Press, New York, 1981) pp 183-288. Book: *B₁₂* vols. 1 and 2, D. Dolphin, Ed. (Wiley-Interscience, New York, 1982) 672 and 506 pp.

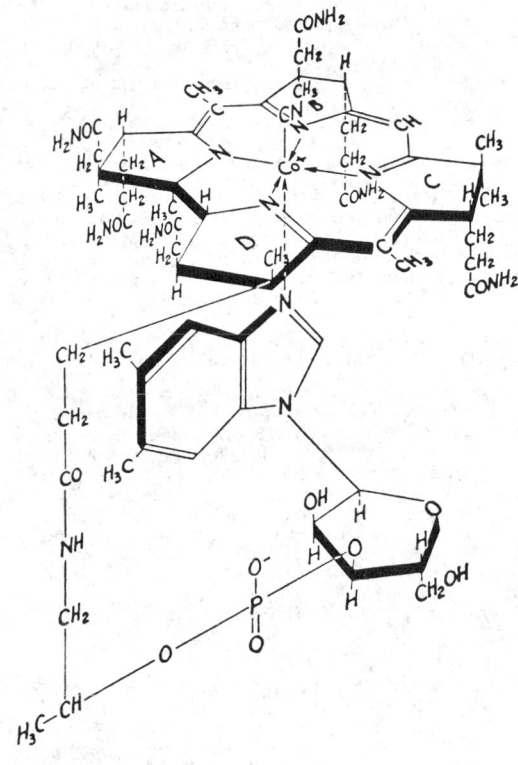

Note: Structure reproduced through the courtesy of John Wiley & Sons, Inc. from E. Lester Smith, *Vitamin B₁₂* (3rd ed., 1965) p 32.

Hygroscopic, dark red crystals. When exposed to air, may absorb about 12% water. The hydrated crystals are stable to air. Darkens at 210-220°. Not melted at 300°. $[\alpha]_{656}^{23}$ −59 ± 9° (dil aq soln). Absorption max (water): 278, 361, 550 nm ($A_{1cm}^{1\%}$ 115, 204, 64). Odorless and tasteless. One gram dissolves in about 80 ml water. Aq solns are neutral, maximum stability in the pH range 4.5-5. Solns in this pH range can be autoclaved for 20 min at 120°. Soluble in alc. Insol in acetone, $CHCl_3$, ether. Aq solns decomp in the presence of acacia, aldehydes, ascorbic acid, ferrous gluconate, ferrous sulfate, vanillin; are stabilized by the addn of ammonium sulfate. Talc has a tenacious affinity for vitamin B₁₂; although this is not an incompatibility, it precludes the use of talc as a filter aid or lubricant for tablets, particularly in view of possible assay difficulties.

Co-methylcobalamin, $C_{63}H_{91}CoN_{13}O_{14}P$, *mecobalamin, cobinamide Co-methyl deriv. hydroxide dihydrogen phosphate (ester) inner salt 3'-ester with 5,6-dimethyl-1-α-D-ribofura-nosyl-1H-benzimidazole, cobaltmethyl-5,6-dimethylbenz-imidazolecobalamin, methyl vitamin B₁₂ Algobaz, Hitocoba-min-M, Lyomethyl, Methycobal, Methylcobaz.* Prepn: O. Müller, G. Müller, *Biochem. Z.* **336**, 299 (1962); D. Dolphin, *Methods Enzymol.* **18**, Pt. C, 34 (1971); M. Tohda *et al., et al.,* **Ger.** pat. **2,019,176** (1971 to Eisai), *C.A.* **76**, 46473a (1972); D. Autissier *et al., Bull. Soc. Chim. France* **1980**, part 2, 192. Bright red crystals from water/acetone. uv max (pH 7): 522, 342, 266 nm (ε 9357, 14416, 19897); (0.1N HCl): 462, 304, 264 nm (ε 9599, 22855, 24737).

THERAP CAT: Hematopoietic vitamin.

THERAP CAT (VET): Nutritional factor (growth and anti-anemic factor).

9922. Vitamin B₁₂, Radioactive. Radioactive cyano-cobalamin; radiocyanocobalamin. Prepn of ⁶⁰Co labelled compound: Chaiet *et al.*, *Science* **111**, 601 (1950). Properties identical with unlabelled vitamin B₁₂, except for the presence of radioactive cobalt. Evaluation of dual-isotope Schilling test for pernicious anemia: L. S. Zuckier, L. R. Chervu, *J. Nucl. Med.* **25**, 1032 (1984).

⁶⁰Co Vitamin B₁₂, *Racobalamin-60, Rubratope-60*, half life 5.27 years, emitting beta and gamma rays.

⁵⁸Co Vitamin B₁₂, half-life 70.8 days, emitting beta, gamma and x-rays.

⁵⁷Co Vitamin B₁₂, *Rubratope-57*, half-life 271 days, emitting gamma and x-rays.

Combination of ⁵⁷Co and ⁵⁸Co labelled Vitamin B₁₂, *Dicopac.*

THERAP CAT: Diagnostic aid (pernicious anemia).

9923. Vitamin B₁₂—Zinc Tannate Complex. Cyanocobalaminzinc tannate complex; zinc-vitamin B₁₂-tannate. Prepn: Thompson, U.S. pat. **2,920,015** (1960 to Armour).

Practically insol in water. On reconstitution forms a colloidal suspension, incompatible with iron salts. After intramuscular injection of 500γ a high cyanocobalamin serum level persists for 28 days.

Ingredient of *Depinar; Bevitam.*

THERAP CAT: Long acting injectable vitamin B₁₂ prepn.

9924. Vitamin B₁₂ᶜ. *Cobinamide hydroxide nitrite salt, dihydrogen phosphate ester, inner salt, 3′-ester with 5,6-dimethyl-1-α-D-ribofuranosyl-1H-benzimidazole;* nitrosocobalamin; nitrocobalamin; nitritocobalamin. Isolated from *Streptomyces griseus* fermentation liquors and prepared by treating vitamin B₁₂ᵦ with nitrous acid: Smith *et al.*, **Brit. pat. 688,132** (1953 to Glaxo); Smith *et al.*, *Biochem. J.* **52**, 389 (1952). Prepared from hydroxocobalamin, sodium nitrite and acetic acid: Kaczka *et al.*, *J. Am. Chem. Soc.* **73**, 3569 (1951).

Red crystalline solid. Absorption max in water: 352, 527.5 nm ($E^{1\%}_{1cm}$ 153.2, 59.5); in 0.01N NaOH: 357, 535 nm ($E^{1\%}_{1cm}$ 139, 62.5).

9925. Vitamin B₁₂ᵣ. *Vitamin B₁₂-Co(II).* A brown reduction product of vitamin B₁₂: Diehl, Murie, *Iowa State Coll. J. Sci.* **26**, 555 (1952); also obtained by photolysis of coenzyme B₁₂: Brady, Barker, *Biochem. Biophys. Res. Commun.* **4**, 373 (1961); and by reduction of vitamin B₁₂ₐ: Hill *et al.*, *Chem. & Ind. (London)* **1964**, 197. Prepn of a stable soln by catalytic hydrogenation of vitamin B₁₂ in methanol: Thesing, Koppe, **Ger. pat. 1,178,073** (1964 to E. Merck), *C.A.* **61**, 14479f (1964). *Review:* Bonnett, *Chem. Rev.* **63**, 573 (1963).

Obtained as a brown material in soln. Oxidized by air to a red material, vitamin B₁₂ₐ. Absorption max: 473, 405, and 312 nm ($E^{1\%}_{1cm}$ 54, 43, and 153).

9926. Vitamin B₁₂ₛ. *Cobinamide hydride hydroxide, dihydrogen phosphate ester, inner salt 3′-ester with 5,6-dimethyl-1-α-D-ribofuranosyl-1H-benzimidazole;* hydridocobalamin. $C_{62}H_{89}CoN_{13}O_{14}P$; mol wt 1330.40. C 55.97%, H 6.74%, Co 4.43%, N 13.69%, O 16.84%, P 2.33%. Occurrence in marine algae: Ericson, Lewis, *Arkiv. Kemi* **6**, 427 (1953), *C.A.* **48**, 7684f (1954). Prepd by reduction of vitamin B₁₂ₐ, but not isolated: Hill *et al.*, *Chem. & Ind. (London)* **1964**, 197. Postulated hydride form of vitamin B₁₂ₛ questioned: Collat, Abbott, *J. Am. Chem. Soc.* **86**, 2308 (1964). *Review:* Bonnett, *Chem. Rev.* **63**, 573 (1963).

Obtained as a gray-green product in solution.

9927. Vitamin D₁. $C_{56}H_{88}O_2$; mol wt 793.32. C 84.79%, H 11.18%, O 4.03%. A 1:1 molecular compd of lumisterol and vitamin D₂. Prepn: Windaus *et al.*, *Ann.* **489**, 252 (1931); **493**, 259 (1932).

Crystals from acetone, mp 124-125°. Sublimes in high vacuum at 135°. Dec 180°. Soly at 16° (g/ml): 0.037 of acetone; 0.024 of petr ether (bp 35-40°); 0.020 of methanol. uv max (0.02% sol in petr ether): 265 nm (ε 1.56). $[\alpha]^{20}_D$ +140.5° (8.9 mg in 2 ml acetone), +140.5° (ethanol), +127° (chloroform).

9928. Vitamin D₂. *9,10-Secoergosta-5,7,10(19),22-tetraen-3-ol;* calciferol; ergocalciferol; oleovitamin D₂; activated ergosterol; viosterol; Drisdol; D-Tracetten; Divit Urto; Os-

telin; Condol; Ergorone; Davitin; Metadee; Mina D₂; Mulsiferol; Mykostin; Radsterin; Shock-Ferol; Dee-Ron; Decaps; Deltalin; De-Rat Concentrate; Deratol; Hi-Deratol; Detalup; Diactol; Doral; Vio-D; Ertron; Infron; Radiostol; Sterogyl; Fortodyl. $C_{28}H_{44}O$; mol wt 396.63. C 84.78%, H 11.18%, O 4.03%. The synthetic form of vitamin D. Prepd from ergosterol by uv irradiation in a suitable solvent: Askew *et al.*, *Proc. Roy. Soc. London* **B109**, 488 (1931/32). The best wave lengths for production of vitamin D₂ seem to be from 275 to 300 nm: Knudson, Benford, *J. Biol. Chem.* **124**, 287 (1938); Bunker *et al.*, *J. Am. Chem. Soc.* **62**, 508 (1940). Prepn by electron bombardment of ergosterol using longer waves: **Austrian pat. 119,210.** Some U.S. pats. covering the manuf of vitamin D₂ are **1,902,785** (1933); **2,030,-792** (1936); also **1,680,818** (1928) and **1,871,136** (1932). Direct total synthesis: B. Lythgoe *et al.*, *Tetrahedron Letters* **1977**, 3685. Discussion of structural problems: Papers by Askew and Windaus *et al.*, beginning in 1930. Stereochemistry: Crowfoot, Dunitz, *Nature* **162**, 608 (1948); *Chem. & Ind. (London)* **1957**, 1149. General review: Inhoffen, *Angew. Chem.* **72**, 875 (1960). A polar, biologically active metabolite of vitamin D₂, *25-hydroxyergocalciferol,* which is about 1.5 times more active in curing rickets in rats, has been isolated from pig plasma. See Suda *et al.*, *Biochem. Biophys. Res. Commun.* **35**, 182 (1969); *eidem, Biochemistry* **8**, 3515 (1969).

Prisms from acetone, mp 115-118°. Sublimes in very high vacuum (0.0006 mm) without dec. $[\alpha]^{25}_D$ +82.6° (c = 3 in acetone); $[\alpha]^{20}_D$ +102.5° (alcohol); $[\alpha]^{20}_D$ +52° (CHCl₃). uv max (hexane): 264.5 nm ($E^{1\%}_{1cm}$ 458.9 ± 7.5). Not precipitated by digitonin (diff from ergosterol). Insol in water; sol in the usual organic solvents. 1 ml acetone dissolves 0.0695 g at 7°. Slightly sol in vegetable oils. Commercial solns are usually made with propylene glycol or sesame oil. Vitamin D₂ crystals have a potency of 40 units of vitamin D (U.S.P.) per μg. The behavior in storage is described under vitamin D₃. See also Huber, Barlow, *J. Biol. Chem.* **149**, 125 (1943) for stability information.

THERAP CAT: Antirachitic vitamin.

THERAP CAT (VET): Nutritional factor (antirachitic). Low activity in poultry. Also as rodenticide.

9929. Vitamin D₃. *9,10-Secocholesta-5,7,10(19)-trien-3-ol;* activated 7-dehydrocholesterol; oleovitamin D₃; cholecalciferol; colecalciferol; CC; Duphafral D₃ 1000; Delsterol; Deparal; Ebivit; Micro-Dee; Neo Dohyfral D₃; Provitina; Ricketon; Trivitan; D₃-Vicotrat; Vi-De-3-hydrosol; Vigantol; Vigorsan. $C_{27}H_{44}O$; mol wt 384.62. C 84.31%, H 11.53%, O 4.16%. The vitamin that mediates intestinal calcium absorption, bone calcium metabolism and probably, muscle activity. It usually acts as a hormone precursor as it requires two stages of metabolism, first to 25-hydroxycholecalciferol, *q.v.*, and then to 1α,25-dihydroxycholecalciferol, *q.v.*, before reaching actual hormonal form. Occurs in and is isolated from fish liver oils. Methods of sepn include chromatography, molecular distillation, esterification and fractionation of the esters. Prepd by irradiation of its provitamin 7-dehydrocholesterol, *q.v.*: Windaus *et al.*, *Z. Physiol. Chem.* **241**, 100 (1936); Windaus *et al.*, *Ann.* **533**, 118 (1938); Akhtar, Gibbons, *Tetrahedron Letters* **1965**, 509. Direct total synthesis: B. Lythgoe *et al.*, *Tetrahedron Letters* **1977**, 3685. Laser photochemical production: V. Malatesta *et al.*,

J. Am. Chem. Soc. **103**, 6781 (1981). General review: Inhoffen, Angew. Chem. **72**, 875-881 (1960). Review of metabolism: Haussler, Rasmussen, J. Biol. Chem. **247**, 2328-2335 (1972); Nature **245**, 180-182 (1973).

Fine needles from dilute acetone, mp 84-85°. $[\alpha]_D^{20}$ +84.8° (c = 1.6 in acetone); $[\alpha]_D^{20}$ +51.9° (c = 1.6 in chloroform). uv max (alcohol or hexane): 264.5 nm. $(E_{1cm}^{1\%}$ 450-490): Huber et al., J. Am. Chem. Soc. **67**, 609 (1945). Not precipitated by digitonin (diff from 7-dehydrocholesterol). Practically insol in water; sol in the usual organic solvents; slightly sol in vegetable oils. Oxidized and inactivated by moist air within a few days. Deterioration of pure cryst vitamin D_3 is negligible after storage of 1 year in amber evacuated ampuls at refrigerator temps; vitamin D_2 may be kept for 9 months under the same conditions. Additional stability information: Huber, Barlow, J. Biol. Chem. **149**, 125 (1942). Generally vitamin D_3 is considered more stable than vitamin D_2.

Note: Vitamin D_3 is approx as effective as vitamin D_2 in the human and in the rat. It is also fully active in chicks. Vitamin D_2 is only 1-2 percent as potent for the chick as vitamin D_3. Because of this difference it is important that poultry feeds are supplemented with vitamin D_3 rather than D_2. One unit (U.S.P. or international) is defined as the activity of 0.025 γ of vitamin D_3 contained in the U.S.P. vitamin D reference standard.

THERAP CAT: Antirachitic vitamin.

THERAP CAT (VET): Nutritional factor (antirachitic).

9930. Vitamin D₄. *9,10-Secoergosta-5,7,10(19)-trien-3-ol; 22:23-dihydrovitamin D_2; 22,23-dihydroergocalciferol.* $C_{28}H_{46}O$; mol wt 398.65. C 84.35%, H 11.63%, O 4.01%. Prepd from 22:23-dihydroergosterol by irradiation with light of the magnesium arc: Windaus, Trautmann, Z. Physiol. Chem. **247**, 185 (1937). Synthesis: P. J. Kocienski et al., J. Chem. Soc., Perkin Trans. I **1979**, 1290. Its biological activity seems doubtful.

Platelets from dil acetone, mp 96-98°. Originally given as 107-108°, see Windaus, Guntzel, Ann. **538**, 122 (1939). $[\alpha]_D^{18}$ +89.3° (c = 0.47 in acetone). uv max: 265 nm. Not precipitated by digitonin. Practically insol in water. Sol in the usual organic solvents except petr ether; slightly sol in vegetable oils.

9931. Vitamin E. *3,4-Dihydro-2,5,7,8-tetramethyl-2-(4,8,12-trimethyltridecyl)-2H-1-benzopyran-6-ol; 2,5,7,8-tetramethyl-2-(4',8',12'-trimethyltridecyl)-6-chromanol; α-tocopherol; 5,7,8-trimethyltocol; antisterility vitamin; Ep-*

rolin-S; Epsilan; Ephynal; Syntopherol; E-Vimin; Evipherol; Etavit; Phytogermine; Profecundin; Tokopharm; Viprimol; Viteolin; Esorb; Vascuals; Covitol; Evion. $C_{29}H_{50}O_2$; mol wt 430.69. C 80.87%, H 11.70%, O 7.43%. Found largely in plant materials. Present in highest concns (0.1-0.3%) in wheat germ, corn, sunflower seed, rapeseed, soybean oils, alfalfa and lettuce. Natural α-tocopherol is usually found with β- and γ-tocopherols, q.q.v. Isoln from wheat germ: Evans et al., J. Biol. Chem. **113**, 319 (1936). Structure: Fernholz, J. Am. Chem. Soc. **59**, 1154 (1937); **60**, 700 (1938). Synthesis: Karrer et al., Helv. Chim. Acta **21**, 520, 820 (1938); Bergel et al., J. Chem. Soc. **1938**, 1382; Smith et al., Science **88**, 37 (1938); Smith, Sprung, J. Am. Chem. Soc. **65**, 1276 (1943). Recent syntheses: N. Cohen et al., Helv. Chim. Acta **61**, 837 (1978); eidem, J. Am. Chem. Soc. **101**, 6710 (1979); R. Barner, M. Schmid, Helv. Chim. Acta **62**, 2384 (1979). Abs config of natural α-tocopherol: Mayer et al., ibid. **46**, 963 (1963). Stereoselective synthesis of the side chain: C. H. Heathcock, E. T. Jarvi, Tetrahedron Letters **23**, 2825 (1982). Review of industrial processes: Rubel, Vitamin E Manufacture (Noyes Dev. Corp., Park Ridge, N.J., 1969). Reviews: The Vitamins Vol. 5, W. H. Sebrell, R. S. Harris, Eds. (Academic Press, New York, 1972) pp 165-317; J. M. Bieri, P. M. Farrell, Vitam. Horm. (New York) **34**, 31-75 (1976). Comprehensive description: B. C. Rudy, B. Z. Senkowski, in Analytical Profiles of Drug Substances vol. 4, K. Florey, Ed. (Academic Press, New York, 1975) pp 111-126. Book: Ann. N.Y. Acad. Sci. **393**, entitled "Vitamin E: Biochemical, Hematological and Clinical Aspects", B. Lubin, L. J. Machlin, Eds. (1982) 506 pp. Review of medical uses: J. G. Bieri et al., N. Engl. J. Med. **308**, 1063-1071 (1983).

dl-Form, slightly viscous, pale yellow oil. Natural α-tocopherol has been crystallized, mp 2.5°-3.5°. d_4^{25} 0.950; $bp_{0.1}$ 200-220°; n_D^{25} 1.5045. uv max: 294 nm $(E_{1cm}^{1\%}$ 71). Practically insol in water. Freely sol in oils, fats, acetone, alcohol, chloroform, ether, other fat solvents. Stable to heat and alkalies in the absence of oxygen. Not affected by acids up to 100°. Slowly oxidized by atm oxygen, rapidly by ferric and silver salts. Gradually darkens on exposure to light. The I.U. of vitamin E is equal to one mg of standard dl-α-tocopheryl acetate.

d-Form, $[\alpha]_{5461}^{25}$ −3.0° (benzene); $[\alpha]_{5461}^{25}$ +0.32° (alc).

d-α-Tocopheryl succinate, white powder, mp 76-77°.

α-Tocopheryl nicotinate, $C_{35}H_{53}NO_3$, *Hijuven, Juvela Nicotinate, Renascin.*

USE: As an antioxidant in vegetable oils and shortening.

THERAP CAT: Treatment of vitamin E deficiency.

THERAP CAT (VET): Nutritional factor. Interrelationship with selenium. (Prevents muscle degeneration, also encephalomalaria and exudative diathesis.) Has been used to promote fertility.

9932. Vitamin E Acetate. *3,4-Dihydro-2,5,7,8-tetramethyl-2-(4,8,12-trimethyltridecyl)-2H-1-benzopyran-6-ol acetate; 2,5,7,8-tetramethyl-2-(4,8,12-trimethyltridecyl)-6-chromanol acetate; α-tocopherol acetate; α-tocopheryl acetate;* Alfacol; Contopheron; Detulin; Ecofrol; Econ; E-Ferol; Endo E Dompé; Ephynal Acetate; Epsilan-M; E-Toplex; Eusovit; Evipherol; Fertilvit; Gevex; Juvela; Optovit-E; Taxofit; Tocopherex; Tocophrin; Tofaxin; Vitagutt. $C_{31}H_{52}O_3$; mol wt 472.73. C 78.76%, H 11.09%, O 10.15%. Prepn from dl-α-tocopherol and acetic anhydride: Surmatis, Weber, U.S. pat. 2,723,278 (1955 to Hoffmann-La Roche). Prepn of d- and l-forms: Robeson, Nelan, J. Am. Chem. Soc. **84**, 3196 (1962). Stereoselective synthesis: K.-K. Chan et al., J. Org. Chem. **43**, 3435 (1978). Total synthesis of all eight stereoisomers: N. Cohen et al., Helv. Chim. Acta **64**, 1158 (1981).

dl-Form, pale yellow, viscous liquid. mp $-27.5°$. $d_4^{21.3}$ 0.9533. $bp_{0.01}$ 184°; $bp_{0.025}$ 194°; $bp_{0.3}$ 224°. n_D^{20} 1.4950-1.4972. uv max (cyclohexane): 285.5 nm. Practically insol in water. Freely sol in acetone, chloroform, ether. Less readily sol in alc. Unlike the free vitamins, the acetate is practically unaffected by the oxidizing influence of air, light, and ultraviolet light.

d-Form, *Spondyvit 500*. Crystals, mp 28°. $[\alpha]_D^{25}$ +0.25° (c = 10 in chloroform); $[\alpha]_D^{25}$ +3.2° (in ethanol).

l-Form, crystals, mp 23°. $[\alpha]_D^{25}$ −2.0° (in ethanol).

Note: The international unit of vitamin E is equal to one mg of standard *dl*-α-tocopheryl acetate. The *d*-form is more active: 1 mg = 1.36 I.U. *l*-α-Tocopheryl acetate has 42% of the activity of *d*-α-tocopheryl acetate in the rat hemolysis test. Based on this activity a potency ratio of 1.4:1.0 for *d*-α-tocopheryl acetate compared to *dl*-α-tocopheryl acetate has been established.

THERAP CAT: Vitamin E supplement.

THERAP CAT (VET): *See* Vitamin E.

9933. Vitamin K₁. *2-Methyl-3-(3,7,11,15-tetramethyl-2-hexadecenyl)-1,4-naphthalenedione; 2-methyl-3-phytyl-1,4-naphthoquinone;* 3-phytylmenadione; phytomenadione; phytonadione; phylloquinone; antihemorrhagic vitamin; K-Ject; Konakion; Mephyton; Mono-Kay. $C_{31}H_{46}O_2$; mol wt 450.68. C 82.61%, H 10.29%, O 7.10%. A fat-soluble vitamin occurring naturally as the *trans* isomer. The designation K is derived from the German "Koagulationsvitamin": H. Dam, *Biochem. Z.* **215**, 475 (1929); **220**, 158 (1930); *Nature* **135**, 652 (1935). First isolated from alfalfa; also shows widespread distribution in higher green plants: H. Dam *et al., Helv. Chim. Acta* **22**, 310 (1939). D. W. MacCorquodale, L. C. Cheney *et al., J. Biol. Chem.* **131**, 357 (1939); L. F. Fieser, *J. Am. Chem. Soc.* **61**, 3467 (1939). Early syntheses: H. J. Almquist, A. A. Klose, *J. Am. Chem. Soc.* **61**, 2557 (1939); S. B. Binkley *et al., ibid.* 2558; L. F. Fieser, *ibid.* 2559. Synthesis yields a mixture of the *cis* and *trans* isomers of which the *trans* form predominates. Commercial prepns may contain up to 20% of the *cis* isomer. Stereochemistry and total synthesis: H. Mayer *et al., Helv. Chim. Acta* **47**, 221 (1964); L. M. Jackman *et al., ibid.* **48**, 1332 (1965). Synthesis using a π-allylic nickel(I) complex: Sato *et al., Chem. Commun.* **1972**, 953; *J. Chem. Soc. Perkin Trans. I* **1973**, 2289. Alternate synthesis: Y. Tachibana, *Chem. Letters* **1977**, 901. The commercial product is prepared by synthesis. Metabolic studies: M. J. Shearer *et al., Brit. J. Haematol.* **18**, 297 (1970); **22**, 579 (1972). The *cis* isomer has little, if any, vitamin K activity: J. T. Matschiner *et al., J. Nutr.* **102**, 625 (1972). Comprehensive description: M. M. A. Hassan *et al.,* in *Analytical Profiles of Drug Substances* **vol. 17**, K. Florey, Ed. (Academic Press, New York, 1988) pp 449-531.

Trans-form, yellow viscous oil. $[\alpha]_D^{25}$ −0.28° (dioxane). n_D^{20} = 1.5263. uv max (petr ether): 242, 248, 260, 269, 325 nm ($E_{1cm}^{1\%}$ 396, 419, 383, 387, 68). Insol in water. Sparingly sol in methanol; sol in ethanol, acetone, benzene, petr ether, hexane, dioxane, chloroform, ether, other fat solvents and in vegetable oils. A soln of one part vitamin K₁ and 20 parts alcohol is neutral to litmus. Vitamin K₁ is stable to air and moisture, but dec in sunlight. Unaffected by dil acids, but is destroyed by solns of alkali hydroxides and by reducing agents. *Keep well closed and protected from light.*

Diacetyl-dihydro derivative, crystals, mp 62-63°.

Note: A colloidal soln is marketed under the name *Aqua-MEPHYTON.*

THERAP CAT: Prothrombogenic vitamin.

THERAP CAT (VET): In hypoprothrombinemias. Antidote for dicoumarol poisoning.

9934. Vitamin K₁ Oxide. *1a,7a-Dihydro-7a-methyl-1a-(3,7,11,15-tetramethyl-2-hexadecenyl)naphth[2,3-b]oxirene-2,7-dione;* Vitamin K₁ 2,3-oxide; 2-methyl-3-phytyl-1,4-naphthoquinone 2,3-oxide. $C_{31}H_{46}O_3$; mol wt 466.68. C 79.78%, H 9.94%, O 10.28%. Biologically equivalent to vitamin K₁. Occurs in nature: Fernholz *et al., J. Am. Chem. Soc.* **61**, 1613 (1939); *Proc. Soc. Exp. Biol. Med.* **42**, 655 (1939). Prepn by treatment of vitamin K₁ with hydrogen peroxide: Fieser *et al., J. Am. Chem. Soc.* **61**, 3216 (1939); Tishler *et al., ibid.* **62**, 2866 (1940); Tishler, U.S. pat. **2,312,-504** (1943 to Res. Corp.).

Colorless oil. Much more stable to light than the vitamin itself. (A soln of 1 g of the oxide in 15 ml dioxane + 1.5 ml water was heated in a sealed tube at 140-150° for six hours and the product was recovered by ether extraction and found to be of unchanged composition.) uv max (95% alc): 259, 305 nm (log E_M 3.79, 3.31). Insol in water. Sol in fat solvents. A stable emulsion for intravenous use may be prepd in 10% glucose soln.

Note: The stability advantages stressed in earlier reports are not confirmed by later, more extensive investigations.

THERAP CAT: Prothrombogenic vitamin.

9935. Vitamin(s) K₂. Menaquinones; 2-methyl-3-*all-trans*-polyprenyl-1,4-naphthoquinones. Antihemorrhagic vitamins possessing side chains varying in length from C_5 (n = 1) to C_{65} (n = 13) which have been isolated from gram-positive bacteria, anaerobic and facultative nonphotosynthetic gram-negative bacteria and photosynthetic bacteria. It is suggested that most menaquinone-containing organisms contain a series of menaquinones, the major homolog (usually n = 7, 8 or 9) constituting some 85-95% of the total. *Reviews:* Isler, *Angew. Chem.* **71**, 7 (1959); Isler, Wiss, *Vitam. Horm.* (New York) **17**, 53-90 (1959); Threlfall, *ibid.* **29**, 153-200 (1971).

Vitamin K₂(35); n = 7. $C_{46}H_{64}O_2$. *(all-E)-2-(3,7,11,15,-19,23,27-Heptamethyl-2,6,10,14,18,22,26-octacosaheptaenyl)-3-methyl-1,4-naphthalenedione; 2-(3,7,11,15,19,23,27-heptamethyl-2,6,10,14,18,22,26-octacosaheptaenyl)-3-methyl-1,4-naphthoquinone; menaquinone 7.* Isoln from putrefied fish meal: McKee *et al., J. Biol. Chem.* **131**, 327 (1939); from cultures of spore forming soil bacillus, *B. brevis*: Tishler, Sampson, *Proc. Soc. Exp. Biol. Med.* **68**, 136 (1948). Structure and synthesis: Isler *et al., Helv. Chim. Acta* **41**, 786 (1958). Light yellow microcrystalline plates from petr ether, mp 54°. $bp_{0.0002}$ 200° (some dec). uv max (petr ether): 243, 248, 261, 270, 325-328 nm ($E_{1cm}^{1\%}$ 278, 195, 266, 267, 48). Slightly less sol than vitamin K₁ in the same organic solvents.

Dihydrodiacetate, $C_{50}H_{70}O_4$, crystals from alcohol, mp 57°. uv max (alcohol): 230, 278-282 nm ($E_{1cm}^{1\%}$ 1157, 71).

Vitamin K₂(30); n = 6. $C_{41}H_{56}O_2$. *(all-E)-2-(3,7,11,15,-19,23-Hexamethyl-2,6,10,14,18,22-tetracosahexaenyl)-3-methyl-1,4-naphthalenedione; 2-(3,7,11,15,19,23-hexamethyl-2,6,10,14,18,22-tetracosahexaenyl)-3-methyl-1,4-naphthoquinone; 2-difarnesyl-3-methyl-1,4-naphthoquinone; farnoquinone; menaquinone 6.* Isoln from putrefied fish meal, structure and synthesis: Isler *et al.*, *loc. cit.* Yellow crystals from acetone + alcohol or petr ether, mp 50°. uv max (petr ether): 243, 248, 261, 270, 325-328 nm ($E_{1cm}^{1\%}$ 304, 320, 290, 292, 53).

Dihydrodiacetate, $C_{45}H_{62}O_4$, crystals from alc, mp 49-50°. uv max (alc): 230, 278-280 nm ($E_{1cm}^{1\%}$ 1298, 192).

THERAP CAT: Antihemorrhagic vitamin.

THERAP CAT (VET): *See* Vitamin K_1.

9936. Vitamin K_5. *4-Amino-2-methyl-1-naphthalenol; 4-amino-2-methyl-1-naphthol;* 1-hydroxy-2-methyl-4-aminonaphthalene; 2-methyl-4-amino-1-hydroxynaphthalene; 2-methyl-4-amino-1-naphthol; 3-methyl-4-hydroxy-1-naphthylamine. $C_{11}H_{11}NO$; mol wt 173.21. C 76.27%, H 6.40%, N 8.09%, O 9.24%. Prepd from 2-methylnaphthalene or from menadione: Sah, *Rec. Trav. Chim.* **59**, 458 (1940); **60**, 373 (1941); Veldstra, Wiardi, *Rec. Trav. Chim.* **61**, 547 (1942); **62**, 75 (1943). Use as food preservative: Yang, Steele, *Food Technol.* **12**, 501 (1958).

Hydrochloride, $C_{11}H_{12}ClNO$, *Kayvisyn, Synkamin.* Needles from dil HCl, darkens at 262°, dec 280-282°. Turns pink to dark violet on exposure to air and light. Freely sol in water. Slightly sol in alc. Insol in ether.

N-Acetyl analog, $C_{13}H_{13}NO_2$, *4-acetamido-2-methyl-1-naphthol, K-Vitrat.* Needles, mp 208-210°. Prepn: Veldstra, Wiardi, *Rec. Trav. Chim.* **62**, 75 (1943).

USE: Food preservative.

THERAP CAT: Antihemorrhagic vitamin.

9937. Vitamin K_6. *2-Methyl-1,4-naphthalenediamine;* 1,4-diamino-2-methylnaphthalene; 2-methyl-1,4-diaminonaphthalene. $C_{11}H_{12}N_2$; mol wt 172.22. C 76.71%, H 7.02%, N 16.27%. Prepn from 1,4-dihydroxy-2-methylnaphthalene: Baker, Carlson, *J. Am. Chem. Soc.* **64**, 2657, 2661 (1942); from 2-methyl-4-nitro-1-naphthylamine: Vesely, Kapp, *Rec. Trav. Chim.* **44**, 360, 369, 373 (1925); from 1-amino-2-methyl-4-*p*-nitrophenylazonaphthalene: Veldstra, Wiardi, *ibid.* **62**, 84 (1943). Has high antihemorrhagic activity combined with objectionable toxicity.

Dihydrochloride, $C_{11}H_{14}Cl_2N_2$, crystals from dil hydrochloric acid. Sinters around 300°. Freely sol in water.

9938. Vitamin K_7. *4-Amino-3-methyl-1-naphthalenol; 4-amino-3-methyl-1-naphthol;* 1-hydroxy-3-methyl-4-aminonaphthalene; 3-methyl-4-amino-1-hydroxynaphthalene; 3-methyl-4-amino-1-naphthol; 2-methyl-4-hydroxy-1-naphthylamine; 1-amino-2-methyl-4-naphthol. $C_{11}H_{11}NO$; mol wt 173.21. C 76.27%, H 6.40%, N 8.09%, O 9.24%. Prepn analogous to that of vitamin K_5: Baker, Carlson, *J. Am. Chem. Soc.* **64**, 2659 (1942); Sah, *Z. Vitamin-, Hormon-Fermentforsch.* **3**, 324 (1950), *C.A.* **45**, 6179 (1951); Bean, U.S. pat. 2,525,515 (1950 to Kodak).

Hydrochloride, $C_{11}H_{11}NO.HCl$, crystals from HCl contg $SnCl_2$, dec 270°. Turns pink to dark violet on exposure to air and light. Soluble in water. At least one gram dissolves in 25 ml H_2O at 75°.

9939. Vitamin K-S(II). *3-[(1,4-Dihydro-3-methyl-1,4-dioxo-2-naphthalenyl)thio]propanoic acid; 3-(1,4-dihydro-3-methyl-1,4-dioxo-2-naphthylthio)propionic acid; S-*(2-methyl-1,4-naphthoquinonyl-3)-β-mercaptopropionic acid. $C_{14}H_{12}O_4S$; mol wt 276.31. C 60.86%, H 4.38%, O 23.16%, S 11.60%. Prepn: Hanna, *J. Am. Chem. Soc.* **74**, 2120 (1952). Clinical studies: Carter, Warner, *J. Clin. Invest.* **37**, 70 (1958); Hoak, Carter, *Arch. Intern. Med.* **107**, 715 (1961).

Bright orange needles from benzene, mp 161°. Sol in chloroform, ether; less sol in benzene, ligroin, ethanol (5 g/l in 95% ethanol); practically insol in water.

THERAP CAT: Antihemorrhagic vitamin.

9940. Vitamins L. Factors presumably necessary for lactation. Vitamin L_1: *See o*-Aminobenzoic Acid. Vitamin L_2 was shown to be 7-[tetrahydro-3,4-dihydroxy-5-(methyl-mercaptomethyl)-2-furyl]adenine: Nakahara *et al.*, *Sci. Papers Inst. Phys. Chem. Research (Tokyo)* **40**, 433 (1943), *C.A.* **41**, 6317c (1947). Manuf: Sato *et al.*, **Japan.** pats. 7244('62) and 16,015 ('69), *C.A.* **59**, 1746f (1963) and *C.A.* **77**, 102146j (1972).

9941. Vitamin T. Tegotin; termitin; torutulin; factor T; vitamin T Goetsch; Goetsch's vitamin; Temina. A complex of growth-promoting substances, originally obtained from termites: Goetsch, *Oesterr. Zool. Z.* **1**, 58-85 (1946). Also obtainable from roaches, yeasts and fungi. The isoln procedure has never been described. Isoln of "vitamin T complexes" from yeast: Koch *et al.*, **Ger. pat.** 1,000,962 (1957 to Aschaffenburger Zellstoffwerke), *C.A.* **54**, 18898d (1960). Vitamin T may be a mixture of known vitamins and growth-promoting factors: Koch *et al.*, *Naturwiss.* **38**, 339 (1951). Polemic by Barkow, Goetsch, *ibid.* **42**, 346 (1955).

9942. Vitamin U. *(3-Amino-3-carboxypropyl)dimethyl sulfonium chloride;* methylmethioninesulfonium chloride; MMSC; Cabagin-U; Epadyn-U; Vitas-U. $C_6H_{14}ClNO_2S$; mol wt 199.69. C 36.09%, H 7.07%, Cl 17.75%, N 7.02%, O 16.02%, S 16.05%. [$(CH_3)_2S^+CH_2CH_2CH(NH_2)COOH$]$Cl^-$. The anti-ulcer vitamin reported in cabbage leaves and other green vegetables: G. Cheney, *California Med.* **77**, 248 (1952). Activity: V. Z. Szabo, G. Vargha, *Arzneimittel-Forsch.* **10**, 23 (1960); K. Seri *et al.*, *ibid.* **29**, 1517 (1979). Prepn: Y. Kanai, Y. Kawamura, **Japan. pat.** 4757('62) (to Nippon Kayaku), *C.A.* **58**, 12672e (1963); H. Wagner, **Ger. pat.** 1,239,697 (1967 to Degussa), *C.A.* **67**, 73866e (1967). Needles, mp 134° (dec). LD_{50} i.v. in mice: 2760 mg/kg, I. V. Zaikonnikova, L. G. Urazaeva, *C.A.* **80**, 10388y (1974).

Bromide analog, $C_6H_{14}BrNO_2S$, *methylmethioninesulfonium bromide, Ardésyl.* mp 139° (dec).

THERAP CAT: Treatment of gastric disorders.

9943. Vitride®. *Dihydrobis(2-methoxyethanolato-O,O')aluminate(1−) sodium;* sodium bis(2-methoxyethoxy)aluminum hydride. $C_6H_{16}AlNaO_4$; mol wt 202.16. C 35.64%, H 7.98%, Al 13.34%, Na 11.37%, O 31.66%.

NaAlH$_2$(OCH$_2$CH$_2$OCH$_3$)$_2$. Prepn: Casensky *et al.*, U.S. pat. **3,507,895** (1970). Properties: Vit, *Eastman Organic Chemical Bulletin* **42**(3), 1 (1970).

Highly viscous liquid at room temp. Miscible in a great variety of inert solvents (e.g., aromatic hydrocarbons and ethers). Immiscible in paraffinic hydrocarbons. Thermally stable at 200°C; thermal decomposition starts at 205°C and is vigorous.

USE: Reducing agent.

9944. Voacamine. *12-Methoxy-13-[(3α)-17-methoxy-17-oxovobasan-3-yl]ibogamine-18-carboxylic acid methyl ester;* voacanginine. C$_{43}$H$_{52}$N$_4$O$_5$; mol wt 704.88. C 73.26%, H 7.44%, N 7.95%, O 11.35%. Bisindole alkaloid found in *Voacanga africana* Stapf., *V. thouarsii* R. & Sch., var *obtusa* (K. Sch.) Pichon and *V. schweinfurthii* Staph., *Apocynaceae:* Janot, Goutarel, *Compt. Rend.* **240**, 1719 (1955); Rao, *J. Org. Chem.* **23**, 1455 (1958); Janot, Goutarel, U.S. pat. **2,823,204** (1958 to Lab. Gobey). Identity with voacangi-nine: LeBarre, Gillo, *Bull. Acad. Roy. Med. Belg.* **20**, 194 (1956). Cleaved by acid catalysis to voacangine: Winkler, *Naturwiss.* **48**, 694 (1961). Structure: Goutarel *et al.*, *Compt. Rend.* **243**, 1670 (1956); Percheron, *Ann. Chim. (Paris)* **4**, 303 (1959); Büchi *et al.*, *J. Am. Chem. Soc.* **85**, 1893 (1963). *Review:* Gorman *et al.* in *The Alkaloids* **1**, J. E. Saxton, Ed. (The Chem. Soc., London, 1971) pp 242-249.

voacangine

Prisms from acetone + methanol, dec 223°. [α]$_D^{20}$ −52° (chloroform). uv max: 225, 295 nm (log ε 4.72, 4.28). Sol in chloroform, acetone; slightly sol in methanol, ethanol. *Voacangine*, C$_{22}$H$_{28}$N$_2$O$_3$, *carbomethoxyibogaine, 12-methoxyibogamine-18-carboxylic acid methyl ester.* Structure: Bartlett *et al.*, *J. Am. Chem. Soc.* **80**, 126 (1958). Prismatic needles from ethanol, mp 136-137°. Sublimes$_{0.01}$ 135°. [α]$_D^{20}$ −42° (c = 1.26 in chloroform). pKa 7.4 (40% aq methanol); pKa 5.73 (33% DMF). uv max (methanol): 225, 287 300 nm (log ε 4.43, 3.97, 3.93). Freely sol in acetone, chloroform; slightly sol in methanol, ethanol.

9945. Vodka. Little water. Defined as ethyl alcohol obtained from grain or potatoes, filtered through charcoal, and diluted with distilled water. Manuf by "spirit flow" process: Jacobs, U.S. pat. **2,946,687** (1960 to Heublein).

9946. Vomicine. *4-Hydroxy-19-methyl-16,19-secostrychnidine-10,16-dione;* 12-hydroxy-*N*-methylpseudostrychnine. C$_{22}$H$_{24}$N$_2$O$_4$; mol wt 380.43. C 69.45%, H 6.36%, N 7.36%,

O 16.82%. From seed of *Strychnos nux vomica* L., *Loganiaceae:* Wieland, Oertel, *Ann.* **469**, 193 (1929). Structure: Huisgen *et al.*, *ibid.* **573**, 121 (1951). Synthesis: Rosenmund, *Angew. Chem.* **75**, 1127 (1963). *Reviews:* R. Robinson in *Progress in Organic Chemistry* vol. I (Butterworths, London, 1952) pp 2-21; J. B. Hendrickson in *The Alkaloids* vol. VI, R. H. F. Manske, H. L. Holmes, Eds. (Academic Press, New York, 1960) pp 195-204.

Hexagonal prisms from acetone, mp 284°. [α]$_D^{22}$ +80° (c = 0.5 in alc). Weak, mono-acidic base forming salts with an acid reaction. Freely sol in chloroform, sol in hot alcohol, acetone, slightly in ether, ethyl acetate.

Hydrochloride, C$_{22}$H$_{24}$N$_2$O$_4$.HCl, prisms from water, dec 245°. uv max (ethanol): 222, 263, 297 nm (log ε 4.27, 3.77, 3.52). Sparingly sol in water.

Methyl ether, C$_{23}$H$_{26}$N$_2$O$_4$, needles from alcohol, dec 290°. [α]$_D^{20}$ +16° (c = 0.5 in alcohol).

9947. Vomitoxin. *12,13-Epoxy-3,7,15-trihydroxytrichothec-9-en-8-one;* deoxynivalenol; dehydronivalenol. C$_{15}$H$_{20}$O$_6$; mol wt 296.33. C 60.80%, H 6.81%, O 32.40%. Isoln of the trichothecene mycotoxin from *Fusarium roseum* and structure: N. Morooka *et al.*, *J. Food Hyg. Soc. Japan* **13**, 368 (1972); T. Yoshizawa, N. Morooka, *Agr. Biol. Chem.* **37**, 2933 (1973). Isoln from *F. graminearum:* R. F. Vesonder *et al.*, *Appl. Microbiol.* **26**, 1008 (1973); *eidem, Appl. Environ. Microbiol.* **31**, 280 (1976). Emetic and refusal activity in swine: D. M. Forsyth *et al.*, *ibid.* **34**, 547 (1977). HPLC analysis: G. A. Bennett *et al.*, *J. Am. Oil Chem. Soc.* **58**, 1002A (1981). Implicated as a chemical warfare agent with nivalenol, *q.v.* in Southeast Asia: N. Wade, *Science* **214**, 34 (1981).

Fine needles from ethyl acetate + petr ether, mp 151-153°. [a]$_D^{25}$ +6.35° (c = 0.07 in ethanol). uv max (ethanol): 218 nm (ε 4500). LD$_{50}$ i.p. in male, female mice (mg/kg): 70.0, 76.7 i.p. (Yoshizawa, Morooka).

9948. VX. *Methylphosphonothioic acid S-[2-[bis(1-methylethyl)amino]ethyl] O-ethyl ester;* *O*-ethyl *S*-[2-(diisopropylamino)ethyl]methylphosphonothioate; Tx 60. C$_{11}$H$_{26}$NO$_2$PS; mol wt 267.37. C 49.42%, H 9.80%, N 5.24%, O 11.97%, P 11.58%, S 11.99%. Nerve gas. Prepn: R. V. Ley, G. L. Sainsbury, **Brit. pat. 1,346,409** and A. W. Wardrop, C. Stratford, **Brit. pat. 1,346,410** (both 1974 to U.K. Sec. of State for Defence), *C.A.* **81**, 4068y, 4069z (1974); S. R. Eckhaus *et al.*, U.S. pat. **3,911,059** (1975 to U.S.A. Sec. for Army). Activity studies in humans: F. R. Sidell, W. A. Groff, *Toxicol. Appl. Pharmacol.* **27**, 241 (1974); F. N. Craig *et al.*, *J. Invest. Dermatol.* **68**, 357 (1977).

Nonvolatile, odorless liquid. pKa′ 7.9. LD$_{50}$ s.c. in rabbits: 15.4 μg/kg, J. J. Gordon, L. Leadbeater, *Toxicol. Appl. Pharmacol.* **40**, 109 (1977). *Caution:* Cholinesterase inhibitor; more potent than sarin, *q.v.*

USE: Chemical warfare agent.

W

9949. Warburganal. *[1S-(1α,4aα,8aβ)]-1,4,4a,5,6,7,8,-8a-Octahydro-1-hydroxy-5,5,8a-trimethyl-1,2-naphthalenedicarboxaldehyde.* $C_{15}H_{22}O_3$; mol wt 250.34. C 71.97%, H 8.86%, O 19.17%. Drimane sesquiterpene with antifeedant activity against the African army worm. Biological activity includes plant growth regulation, cytotoxic, antimicrobial and molluscicidal properties. Isoln from *Warburgia ugandensis, Canellaceae* and structure: I. Kubo *et al., Chem. Commun.* **1976**, 1013. Relationship between structure and antifeedant activity: K. Nakanishi, I. Kubo, *Isr. J. Chem.* **16**, 28 (1977). Synthesis of (±)-warburganal: S. P. Tanis, K. Nakanishi, *J. Am. Chem. Soc.* **101**, 4398 (1979); T. Nakata *et al., ibid.* 4400; A. S. Kende, T. J. Blacklock, *Tetrahedron Letters* **1980**, 3119; P. A. Wender, S. L. Eck, *ibid.* **1982**, 1871; D. M. Hollinshead *et al., J. Chem. Soc. Perkin Trans. I* **1983**, 1579. *See also:* **Japan. Kokai 80 136,238**, and **80 136,240** (both 1980 to Inst. Phys. Chem. Res.); **Japan. Kokai 81 43,236** (1981 to Suntory Ltd.); **Japan. Kokai 83 38,232** (1983 to Teikoku Zoki). Synthesis of (−)-warburganal: H. Okawara *et al., Tetrahedron Letters* **1982**, 1087.

mp 98-99°. uv max (methanol): 224 nm (ε 6300). $[\alpha]_D^{21}$ −260° (c = 0.350 in CHCl$_3$).

9950. Warfarin. *4-Hydroxy-3-(3-oxo-1-phenylbutyl)-2H-1-benzopyran-2-one; 3-(α-acetonylbenzyl)-4-hydroxycoumarin;* 1-(4'-hydroxy-3'-coumarinyl)-1-phenyl-3-butanone; 3-α-phenyl-β-acetylethyl-4-hydroxycoumarin; compound 42; WARF compound 42; Co-Rax; Rodex. $C_{19}H_{16}O_4$; mol wt 308.32. C 74.01%, H 5.23%, O 20.76%. The commercial product is the racemic mixture; the *S*(−)-form is more active than the *R*-isomer. Prepd by the Michael condensation of benzylidene-acetone with 4-hydroxycoumarin: Stahmann *et al.,* U.S. pat. **2,427,578** (1947); Schroeder, Link, U.S. pat. **2,765,321** (1956 to Wisconsin Alumni Res. Found.); Link, U.S. pat. **2,777,859** (1957). Resolution and abs configuration: West *et al., J. Am. Chem. Soc.* **83**, 2676 (1961); Preis, *Dissertation Abstr.* **18**, 793 (1958); Preis *et al.,* U.S. pat. **3,239,529** (1966 to Wisconsin Alumni Res. Found.). Mechanism of action: Bell *et al., Biochemistry* **11**, 1959 (1972). Conformation in soln: E. J. Valente *et al., J. Med. Chem.* **20**, 1849 (1977); **21**, 141, 231 (1978). Human metabolism: R. J. Lewis, W. F. Trager, *Ann. N.Y. Acad. Sci.* **179**, 205 (1971). Stereospecific HPLC determn in plasma: C. Banfield, M. Rowland, *J. Pharm. Sci.* **72**, 921 (1983). Antimetastatic effect in lung cancer: L. R. Zacharski *et al., Cancer* **53**, 2046 (1984); in rat adenocarcinoma: B. L. Neubauer *et al., J. Urol.* **135**, 163 (1986). Animal toxicity studies: E. C. Hagan, J. L. Radomski, *J. Am. Pharm. Assoc.-Sci. Ed.* **42**, 379 (1953); N. Back *et al., Pharmacol. Res. Commun.* **10**, 445 (1978). Review of clinical uses: J. V. Lloyd, *Med. J. Aust.* **142**, 197-201 (1985). Comprehensive description: S. A. Babhair *et al.,* in *Analytical Profiles of Drug Substances* vol. **14**, K. Florey, Ed. (Academic Press, New York, 1985) pp 423-452.

Crystals from alc, mp 161°. uv max (water, pH 10): 308 nm (ε 13610). Soluble in acetone, dioxane. Moderately sol in methanol, ethanol, isopropanol, some oils. Freely sol in alkaline aq solns (forms a water-soluble sodium salt). Practically insol in water, benzene, cyclohexane, Skellysolves A and B. Warfarin has an acidic enol which forms metallic salts and an acetate, mp 117-118°, and a ketone which forms an oxime, mp 182-183° and a 2,4-dinitrophenylhydrazone, mp 215-216°.

Sodium salt, $C_{19}H_{15}NaO_4$, *Coumadin, Marevan, Panwarfin, Prothromadin, Tintorane, Warfilone, Waran.* Slightly bitter, cryst powder. Freely sol in water, alcohol; very slightly sol in chloroform, ether. LD_{50} in male rats, female rats, mice, rabbits (mg/kg): 323, 58, 374, ~800 orally (Hagen); also reported as LD_{50} in male, female rats (mg/kg): 100.3, 8.7 orally (Back).

Potassium salt, $C_{19}H_{15}KO_4$, *Athrombin-K.*

Compound with 2-(dimethylamino)ethanol, $C_{23}H_{27}NO_5$, *warfarin-deanol, MD 6134, Adoisine.*

Human Toxicity: Depresses formation of prothrombin and increases capillary fragility, leading to hemorrhages, *Clinical Toxicology of Commercial Products,* R. E. Gosselin *et al.,* Eds. (Williams & Wilkins, Baltimore, 5th ed., 1984) Section III, pp 395-397.

USE: Rodenticide.

THERAP CAT: Anticoagulant.

9951. Water. Hydrogen oxide. H_2O; mol wt 18.016. H 11.19%, O 88.81%. *Reviews:* N. E. Dorsey, *Properties of Ordinary Water-Substance,* A.C.S. Monograph Series no. **81**, (Reinhold, New York, 1940) 673 pp; D. Eisenberg, W. Kauzmann, *The Structure and Properties of Water* (Oxford University Press, New York, 1969) 296 pp; Ebsworth *et al.,* in *Comprehensive Inorganic Chemistry* vol. **2**, J. C. Bailar, Jr. *et al.,* Eds. (Pergamon Press, Oxford, 1973) pp 741-747.

Liquid. Temp of max density 3.98°. $d^{3.98}$ 1.000000 g/ml (0.999972 g/cc). d_4^{25} 0.997. d^0 (ice) 0.917 g/cc; d_4^0 (liq) 0.999868. Density tables: Bigg, *Brit. J. Appl. Phys.* **18**, 521 (1967); Kell, *J. Chem. Eng. Data* **12**, 66 (1967). Expands on freezing. mp 0°. bp 100°. One liter satd vapor weighs 0.5974 g at 100° and 760 mm. Crit temp 374.2°; crit pressure 218 atm. Sp. heat (liq; 14°) 1.000 cal/g/°C. Latent heat of fusion: 1.436 kcal/mole. Latent heat of vaporization: 9.717 kcal/mole. n_D^{20} 1.3330. Dielectric const (0°) 87.740. Dipole moment (25°) in benzene 1.76; in dioxane 1.86. Ionization const for pure water only: K (25°) 1.008 × 10^{-14}; at moderate concn of solutes (e.g. 1.0M KOH): K (25°) 0.971 × 10^{-14}. The most universal solvent known.

Pyrogen-free water (water for injection) is distilled water rendered free of fever-producing proteins (bacteria and their metabolic products). Method of prepn: Ishizuka *et al., C.A.* **49**, 15177 (1955). *See also* Pyrogens.

9952. Water Gas. Blue gas. Obtained by blowing steam through incandescent coke. *Composition:* 6% CO_2; 42% CO; 51% H_2; 1% N_2.

Caution: Asphyxiant.

USE: In the manuf of ammonia as source of hydrogen. *Cf.* Producer Gas.

9953. Watermelon. Arbuse. *Citrullus vulgaris* Schrad., *Cucurbitaceae,* cultivated in hot and temperate zones the world over. Contains diuretic principles: Bliss *et al., Am. J. Pharm.* **105**, 53 (1933); Roby *et al., ibid.* **111**, 68 (1939).

9954. Wheat Germ Oil. Cav-Ecol; Myopone; Denamone. Obtained by hydraulic expression or solvent extraction of wheat germ which constitutes about 2% of a wheat grain, the seed of *Triticum aestivum* L. (*T. sativum* Lam., *T. vulgare* Vill.), *Gramineae.* *Constit.* (of the oil): linoleic acid 44.1%, oleic acid 30.0%, satd acids 15.1%, linolenic acid 10.8%, unsaponifiable matter 4.7%. The unsaponifiable matter contains vitamin E-active tocopherols (reported as 0.5% of the oil and as 2 international vitamin E units per gram of the oil), sitosterols, dihydrositosterol and other cryst alcohols and phospholipids. *Review:* E. W. Eckey, *Vegetable Fats and Oils* (Reinhold, New York, 1954) pp 291-293.

Bland yellow oil resembling corn oil. d_{25}^{25} 0.925-0.933. n_D^{40} 1.469-1.478. Acid value 6-20. Saponif value 179-190. Iodine value 115-129. Thiocyanogen value 80-85. Hydroxyl value 10-48. Reichert-Meissl value 0.3-1.4. Polenske value 0.4-2.1. Hehner value 76-95. Miscible with chloroform, ether, benzene, petr ether. Slightly sol in alc.

USE: Nutritional supplement. Source of natural vitamin E and unsatd fatty acids (vitamin F).

THERAP CAT (VET): Dietary source of vitamin E.

9955. Whisky. Whiskey. A liquid produced by distillation of the fermented mash of malted cereal grains, which has been stored in wood containers for not less than 4 years. Straight whisky contains 47-53% abs alcohol by vol, 0.05-0.16% acid calculated as acetic acid, and 0.038-0.15% esters as ethyl acetate. Whisky marked 100 proof contains 50% ethanol (v/v). It also contains small quantities of other natural constituents (congeners), which vary according to the grain used, method of fermentation, etc., and which are largely responsible for the characteristic aroma and flavor.

Light to deep amber liquid; characteristic odor and taste. d 0.935-0.923 at 25°.

Blended whisky contains at least 40% ethanol (v/v) and is made from at least 20% of 100 proof straight whisky mixed with neutral spirits, q.v. The most popular blends are made from 65 vols neutral spirits and 35 vols straight whisky.

THERAP CAT: Sedative, peripheral vasodilator.

9956. White Pine. Deal pine; Northern pine; Weymouth pine. The dried inner bark of *Pinus strobus* L., *Pinaceae*. *Constit.* Coniferin glycoside, coniferyl alcohol, tannin, oleoresin, volatile oils.

Weak yellowish-orange to light yellowish-brown when powdered, with a slightly terebinthinate odor and a slightly mucilaginous, sweet, then bitter and astringent taste.

THERAP CAT: The bark as expectorant.

9957. Wieland-Gumlich Aldehyde. *19,20-Didehydro-17,18-epoxycuran-17-ol; 1-deacetyldiaboline;* caracurine VII. $C_{19}H_{22}N_2O_2$; mol wt 310.38. C 73.52%, H 7.14%, N 9.03%, O 10.31%. Decompn product of isonitrosostrychnine: Wieland, Kaziro, *Ann.* **506**, 60 (1933). Isoln from *Strychnos toxifera* Benth., and *S. subcordata* Spruce, *Loganiaceae*: Asmis *et al., Helv. Chim. Acta* **37**, 1983 (1954); Penna *et al., Gazz. Chim. Ital.* **87**, 1163 (1957). Identity with caracurine VII: Bernauer *et al., Helv. Chim. Acta* **41**, 1405 (1958). Structure: Deyrup *et al., ibid.* **45**, 2266 (1962). Chemistry of the degradation from strychnine, *q.v.*: Hymon *et al., ibid.* **52**, 1564-1602 (1969). Synthesis: L. Szabo, O. Clauder, *Acta Chim. Acad. Sci. Hung.* **95**, 85 (1977). ¹³C-NMR study: E. Wenkert *et al., J. Org. Chem.* **43**, 1099 (1978).

Crystals from acetone + methanol, decomp 213-214°. $[\alpha]_D^{22}$ −133.8° (c = 0.52 in methanol). Sol in methanol, ethanol, chloroform; slightly sol in acetone, ethyl acetate.

Hydrochloride hemihydrate, crystals from ethanol, mp over 300°. uv max (water): 240, 290 nm (log ε 3.80, 3.40). Sol in water, ethanol, methanol, warm chloroform; slightly sol in acetone, chloroform.

9958. Wild Cherry. Wild black cherry bark. Dried stem bark of *Prunus serotina* Ehrh., *Rosaceae,* collected in autumn. *Habit.* North America. *Constit.* Prunasin; the enzyme emulsin capable of hydrolyzing prunasin to benzaldehyde, glucose, and hydrocyanic acid; benzoic, trimethylgallic and p-coumaric acids; tannin, volatile oil.

USE: Flavoring in foods. Pharmaceutic aid (flavor).

9959. Wildfire Toxin. *N-[2-Amino-4-(3-hydroxy-2-oxo-3-azetidinyl)-1-oxobutyl]-ʟ-threonine.* $C_{11}H_{19}N_3O_6$; mol wt 289.29. C 45.67%, H 6.62%, N 14.53%, O 33.18%. Isoln from *Pseudomonas tabaci,* the bacterium responsible for wildfire disease of tobacco: Woolley *et al., J. Biol. Chem.* **197**, 409 (1952). Highly toxic to a variety of organisms including bacteria, algae, higher plants and animals: Sinden *et al., Toxicol. Appl. Pharmacol.* **14**, 82 (1969). Mode of action: Lovrekovich *et al., Nature* **197**, 917; **198**, 710 (1963). Struc-

tural studies: Woolley *et al., J. Biol. Chem.* **215**, 485 (1955); Stewart, *J. Am. Chem. Soc.* **83**, 435 (1961). Revised structure: *idem, Nature* **229**, 174 (1971).

Colorless, very hygroscopic substance. Very soluble in water; readily sol in methanol. Moderately sol in ethanol; sparingly sol in *n*-butanol and higher alcohols. Practically insol in acetone, ethyl acetate, and less polar organic liquids. Shows marked instability. Slowly loses potency in aq soln at pH 6. Complete loss of activity at these conditions and 0° in one month. Inactivation more rapid in methanol or when pH is shifted to either side of 6, esp in alkaline range.

9960. Withaferin A. *5,6-Epoxy-4,22,27-trihydroxy-1-oxoergosta-2,24-dien-26-oic acid δ-lactone; 4β,27-dihydroxy-1-oxo-5β,6β-epoxywitha-2,24-dienolide.* $C_{28}H_{38}O_6$; mol wt 470.58. C 71.46%, H 8.14%, O 20.40%. Isolated from the leaves of *Withania somnifera* Dun., *Solanaceae:* Lavie, Yarden, *J. Chem. Soc.* **1962**, 2925; Kirson *et al., Tetrahedron* **26**, 2209 (1970); Subramanian, Sethi, *Indian J. Pharm.* **32**, 16 (1970); from roots of *Withania coagulans:* eidem, *Curr. Sci.* **38**, 267 (1969); from *Acnistus arborescens* (L.): Kupchan *et al., J. Org. Chem.* **34**, 3858 (1969). Structure: Lavie *et al., J. Chem. Soc.* **1965**, 7517; Kupchan, *loc. cit.* Crystal structure: McPhail, Sim, *J. Chem. Soc. (C)* **1968**, 962. Antitumor activity studies: B. Shohat *et al., Cancer Chemother. Rep.* **51**, 271 (1967); B. Shohat, H. Joshua, *Int. J. Cancer* **8**, 487 (1971). Synthetic studies: M. Hirayama *et al., J. Chem. Soc. Perkin Trans. I* **1981**, 88. Stereoselective synthesis: *eidem, Tetrahedron Letters* **23**, 4725 (1982).

White prisms from acetone-petroleum ether, mp 252-253° (Kupchan). Also reported as mp 243-245° (ethyl acetate). $[\alpha]_D^{28}$ +125° (c = 1.30 in CHCl₃). uv max (ethanol): 214, 335 nm (ε 17,300, 165).

9961. Woodruff. Woodward herb; petit muguet; Waldmeister (German). Leaves of *Asperula odorata* L., *Rubiaceae. Habit.* Europe, Siberia, North Africa, Australia. *Constit.* Coumarin, tannin, asperuloside, fatty oil, essential oil, bitter principle. Discussion of its fragrance, prepn of *asperule absolute,* use of *Asperula* in various fragrant formulations: *Perfumery Essent. Oil Record,* **1963**, (June) p 382.

USE: Flavoring agent; the fresh leaves in flavoring May wine, the dried leaves in sachets.

9962. Woodward's Reagent K. *2-Ethyl-5-m-sulfophenylisoxazolium hydroxide, inner salt; N-ethyl-5-phenylisoxazolium-3′-sulfonate.* $C_{11}H_{11}NO_4S$; mol wt 253.29. C 52.16%, H 4.38%, N 5.53%, O 25.27%, S 12.66%. Prepd by reacting 5-phenylisoxazole with chlorosulfonic acid, followed by alkylation with triethyloxonium fluoroborate, and acid hydrolysis: Woodward *et al., J. Am. Chem. Soc.* **83**, 1010 (1961).

Consult the cross index before using this section.

Crystals, dec 207-208°. Stable. Non-hygroscopic.
USE: Peptide-bond former.

9963. Worenine. *5,6-Dihydro-14-methylbis[1,3]benzo-dioxolo[5,6-a:5',6'-g]quinolizinium;* 13-methyl-ψ-coptisine. [C$_{20}$H$_{16}$NO$_4$]$^+$; mol wt 334.36. Protoberberine alkaloid isolated from root of *Coptis japonica* Makino, *Ranunculaceae:* Kitasato, *J. Pharm. Soc. Japan* **No. 542,** 48, 315 (1927); *Acta Phytochim.* **3,** 175 (1927), *C.A.* **22,** 1779-1780 (1928); in *Coptis chinensis* Franch., *Ranunculaceae:* Schramm, Wej-ds, *Pharmazie* **14,** 405 (1959). Synthesis of 13-methyl-ψ-coptisine and its probable identity with worenine: T. R. Govindachari *et al., Indian J. Chem.* **9,** 1313 (1971).

Chloride, C$_{20}$H$_{16}$ClNO$_4$, yellow needles from water, mp 288-289°.
Iodide, C$_{20}$H$_{16}$INO$_4$, crystals from ethanol, mp 297-299°.

9964. Wortmannin. *[1S-(1α,6bα,9aβ,11α,11bβ)]-11-(Acetyloxy)-1,6b,7,8,9a,10,11,11b-octahydro-1-(methoxymethyl)-9a,11b-dimethyl-3H-furo[4,3,2-de]indeno[4,5-h]-2-benzopyran-3,6,9-trione.* C$_{23}$H$_{24}$O$_8$; mol wt 428.44. C 64.48%, H 5.65%, O 29.87%. Antifungal antibiotic from *Penicillium wortmanni* Klocker: Brian *et al., Trans. Brit. Mycol. Soc.* **40,** 365 (1957). Similar to viridin, *q.v.* Structure: MacMillan *et al., Chem. Commun.* **1968,** 613; *eidem, J. Chem. Soc. Perkin Trans. I* **1972,** 2898. Absolute stereochemistry: Petcher *et al., Chem. Commun.* **1972,** 1061; MacMillan *et al., ibid.* 1063.

Neutral solid, mp 240°. Unstable in aq solns pH 3-8.

X

9965. Xamoterol. $(\pm)$-*N*-[2-[[2-*Hydroxy*-3-(4-*hydroxyphenoxy*)*propyl]amino]ethyl]*-4-*morpholinecarboxamide*; 1-(4-hydroxyphenoxy)-3-[2-(4-morpholinocarboxamido)-ethylamino]-2-propanol. $C_{16}H_{25}N_3O_5$; mol wt 339.39. C 56.62%, H 7.42%, N 12.38%, O 23.57%. β_1-Adrenoceptor partial agonist. Prepn: **Belg.** pat. **867,376**; B. G. Main, J. J. Barlow, **U.S.** pat. **4,143,140** (1978, 1979 both to ICI); prepn and adrenoceptor stimulant activity: J. J. Barlow *et al., J. Med. Chem.* **24**, 315 (1981). Cardiovascular activity: A. Nuttall, H. M. Snow, *Brit. J. Pharmacol.* **77**, 381 (1982). Hemodynamics, cardioselectivity and kinetics: G. Jennings *et al., Clin. Pharmacol. Ther.* **35**, 594 (1984). Hemodynamic effects: J.-M. R. Detry *et al., Eur. Heart J.* **4**, 584 (1983); H. Sato *et al., Circulation* **75**, 213 (1987). Metabolism: T. R. Marten *et al., Drug Metab. Dispos.* **12**, 652 (1984). HPLC determn in biological fluids: C. J. Oddie *et al., J. Chromatog.* **308**, 370 (1984). Clinical evaluations: F. L. Tseu *et al., Brit. Heart J.* **56**, 469 (1986); L. Barrios *et al., Eur. J. Clin. Pharmacol.* **29**, 667 (1986). Clinical comparison with atenolol, *q.v.*, in asthmatic patients: J. W. J. Lammers *et al., Brit. J. Clin. Pharmacol.* **22**, 595 (1986).

Hemifumarate, $C_{36}H_{54}N_6O_{14}$, *ICI 118587*, *Carwin, Corwin, Xamtol.* Crystals from ethanol, mp 168-169° (dec).
THERAP CAT: Cardiotonic.

9966. Xanthan Gum. Polysaccharide B-1459; Keltrol F; Kelzan. Mol wt $>10^6$. Polysaccharide gum produced by the bacterium *Xathomonas campestris*: Jeanes *et al., J. Polymer Sci.* **5**, 519 (1961). Composed of D-glucosyl, D-mannosyl and D-glucosyluronic acid residues and differing proportions of *O*-acetyl and pyruvic acid acetal. The primary structure consists of a cellulose backbone with trisaccharide side chains, the repeating unit being a pentasaccharide. Structural studies: Sloneker *et al., Can. J. Chem.* **40**, 2066, 2188 (1962); **42**, 1261 (1964); P. E. Jansson *et al., Carbohyd. Res.* **45**, 275 (1975); L. D. Melton *et al., ibid.* **46**, 245 (1976); E. R. Morris *et al., J. Mol. Biol.* **110**, 1 (1977); G. Holzwarth, E. B. Prestridge, *Science* **197**, 757 (1977). Secondary and tertiary structure in solutions and gels: D. A. Rees, E. J. Welsh, *Angew. Chem. Int. Ed.* **16**, 214 (1977). *Review:* McNeely, Kang in *Industrial Gums*, R. L. Whistler, Ed. (Academic Press, New York, 2nd ed., 1973) pp 486-497.

Cream-colored, odorless, free-flowing powder. Dissolves readily in water with stirring to give highly viscous solns at low concns. Forms strong films on evaporation of aq solns. Resistant to heat degradation. Aq solns are highly pseudo-plastic.

USE: In foods, non-foods, and cosmetics as stabilizer and emulsifying agent. Potential use in chemically enhanced oil recovery.

9967. Xanthatin. 3,3a,4,7,8,8a-*Hexahydro*-7-*methyl*-3-*methylene*-6-(3-*oxo*-1-*butenyl*)-2H-*cyclohepta*[b]*furan*-2-*one*. $C_{15}H_{18}O_3$; mol wt 246.29. C 73.14%, H 7.37%, O 19.49%. Antimicrobial agent from *Xanthium strumarium* L., sens. lat.; *X. pennsylvanicum* Wallr., *Compositae*; *X. riparium*. Isoln: Little *et al., Arch. Biochem.* **27**, 247 (1950); Geissman *et al., J. Am. Chem. Soc.* **76**, 685 (1954). Structure: Deuel, Geissman, *ibid.* **79**, 3778 (1957); Pashchenko *et al., Farm. Zh. (Kiev)* **24**, 70 (1969); *C.A.* **71**, 67924p (1969). Prepn of isomer and deriv: Minato, **Japan.** pat. **04,259('68)** (to Shionogi), *C.A.* **70**, 37944g (1969).

Flat needles from methanol or ethanol, mp 114.5-115°. uv max: 275, 213 nm (ϵ 22,800, 7300). $[\alpha]_D^{30}$ $-20°$ (ethanol). Sol in ether, acetone, alcohol; slightly sol in hot water (neutral pH). Practically insol in petr ether, 5% NaOH, 5% HCl.

9968. Xanthine. 3,7-*Dihydro*-1H-*purine*-2,6-*dione*; 2,6-(1H,3H)-*purinedione*; 2,6-*dioxopurine*. $C_5H_4N_4O_2$; mol wt 152.11. C 39.48%, H 2.65%, N 36.84%, O 21.04%. Occurs in animal organs, yeast, potatoes, coffee beans, tea. First isolated from urinary bladder stones: *Beilstein* **26**, 447 (1937). Prepd by treating a sulfuric acid soln of guanine with sodium nitrite: Fischer, *Ann.* **215**, 309 (1882). Several other syntheses, *cf.* Levene, Bass, *Nucleic Acids* (New York, 1931).

Scales, plates from water. Dec on heating without melting and with partial sublimation. Kb at 40° $= 6.09 \times 10^{-14}$; Ka at 40° $= 1.19 \times 10^{-10}$. Absorption spectrum: Kalckar, *J. Biol. Chem.* **167**, 429 (1947). One gram dissolves in 14.5 liters of water at 16°, in 1.4 liters boiling water; less sol in alcohol. Sol in mineral acids, freely sol in NH_4OH and in NaOH solns.

1,3-Dimethylxanthine, *see* Theophylline.
3,7-Dimethylxanthine, *see* Theobromine.
1,3,7-Trimethylxanthine, *see* Caffeine.

9969. Xanthinol Niacinate. 3-*Pyridinecarboxylic acid compd with* 3,7-*dihydro*-7-[2-*hydroxy*-3-[(2-*hydroxyethyl*)-*methylamino*]*propyl*]-1,3-*dimethyl*-1H-*purine*-2,6-*dione* (1:1); 7-[2-*hydroxy*-3-[(2-*hydroxyethyl*)*methylamino*]*propyl*]*theophylline, compd with nicotinic acid;* 7-[3-[N-(2-hydroxyethyl)amino]-2-hydroxypropyl]theophylline nicotinate; xanthinol nicotinate; SK 331 A; Angiomin; Complamex; Complamin; Sadamin; Xavin. $C_{19}H_{26}N_6O_6$; mol wt 434.45. C 52.52%, H 6.03%, N 19.35%, O 22.10%. Prepn: Bestian, **Ger.** pat. **1,102,750** (1961 to Wuelfing), *C.A.* **56**, 11602i (1962). Alternate synthesis: Korbonits *et al., Acta Pharm. Hung.* **38**, 98 (1968). Pharmacology: Hemmer, Diezemann, *Arzneimittel-Forsch.* **12**, 672 (1962); Lennartz, *ibid.* 675; Stamm, *ibid.* 679.

Crystals, mp 180°. Freely sol in water.
THERAP CAT: Vasodilator (peripheral).

9970. Xanthocillin. Xantocillin; Brevicid. Antibiotic complex produced by *Penicillium notatum* Westling: Rothe, *Pharmazie* **5**, 190 (1950); Barwald, **Brit.** pat. **898,498** (1962 to Arzneimittelwerk VEB), *C.A.* **57**, 9013a (1962). Consists of at least three antibiotics, xanthocillins X, Y_1 and Y_2; the first being the major component (about 70%).

Xanthocillin X. $C_{18}H_{12}N_2O_2$. *Bis(p-hydroxybenzylidene)-ethylene isocyanide; 1,4-bis(p-hydroxyphenyl)-2,3-diisonitrilo-1,3-butadiene.* $R_1 = R_2 = H$. Structure: Hagedorn, Tönjes, *Pharmazie* **11**, 409 (1956); **12**, 567 (1957); Hagedorn *et al.*, *Ber.* **93**, 1584 (1960). Synthesis: Hagedorn, Eholzer, *Angew. Chem.* **74**, 215 (1962). Biosynthetic studies: Achenbach, Grisebach, *Z. Naturforsch.* **20b**, 137 (1965); Achenbach, König, *Ber.* **105**, 784 (1972). Crystal structure: D. Britton *et al.*, *Cryst. Struct. Commun.* **10**, 1497 (1981). Studies on antiviral activity: Takatsuki *et al.*, *J. Antibiot.* **21**, 671 (1968); *eidem, ibid.* **22**, 151 (1969); Kitahara, *ibid.* **34**, 1556 (1981). Clusters of yellow needles from alc; yellow rhombs from ethyl acetate. Chars at about 210°. Sol (up to 1%) in alc, ether, acetone, dioxane. Freely sol in aqueous alkaline solns. Practically insol in water, petr ether, benzene, chloroform. Forms a water-sol dipotassium salt, the concd solns of which have a pH of 10.5.

Dimethyl ether, $C_{20}H_{16}N_2O_2$, yellow needles from dioxane + alcohol, dec 181°.

Xanthocillin Y_1, $C_{18}H_{12}N_2O_3$. $R_1 = OH$, $R_2 = H$.
Xanthocillin Y_2, $C_{18}H_{12}N_2O_4$. $R_1 = R_2 = OH$. Structure of xanthocillins Y_1 and Y_2: Achenbach *et al.*, *Ber.* **105**, 3061 (1972).

USE: In feed supplements.

THERAP CAT: Antibacterial.

9971. Xanthone. *9H-Xanthen-9-one;* 9-oxoxanthene; diphenylene ketone oxide; dibenzo-γ-pyrone; Genicide. $C_{13}H_8O_2$; mol wt 196.19. C 79.58%, H 4.11%, O 16.31%. Prepd by heating phenyl salicylate: Graebe, *Ann.* **254**, 279 (1889); Holleman, *Org. Syn.* **coll. vol. I** (2nd ed., 1941) p 552.

Polymorphic needles from alcohol, mp 174°. bp$_{730}$ 351°. Sparingly volatile with steam. Dipole moment 3.0. Solubility: 0.55 g/100 ml of cold alcohol; 6.71 g/100 ml of boiling alcohol. Freely sol in chloroform. Slightly sol in hot water, ether, petr ether, benzene, toluene, xylene. Dissolves in concd H_2SO_4 to a yellow soln with pale blue fluorescence.

Oxime, *xanthoxime*, mp 161°.

USE: In the prepn of xanthydrol. Ovicide for codling moth eggs; less effective as larvicide.

9972. Xanthophyll. *β,ε-Carotene-3,3'-diol;* lutein; vegetable lutein; vegetable luteol; Bo-Xan. $C_{40}H_{56}O_2$; mol wt 568.85. C 84.45%, H 9.92%, O 5.63%. One of the most widespread carotenoid alcohols in nature. Originally isolated from egg yolk, now usually isolated (by chromatography) from nettles, algae, and petals of many yellow flowers. Occurs also in colored feathers of birds: Volker, *Z. Physiol. Chem.* **288**, 20 (1951). Extraction from petals of *Tagetes patula* L., *Compositae:* Karrer *et al.*, *Helv. Chim. Acta* **30**, 531 (1947). Occurs together with zeaxanthin, *q.v.* Dipalmitate occurs in *Helenium autumnale* L., *Compositae* and other flowers: Kuhn, Winterstein, *Naturwiss.* **18**, 754 (1930). Conversion to zeaxanthin with sodium alcoholate: Karrer, Jucker, *ibid.* 266. Does not possess vitamin A potency: Schumacher *et al.*, *Poultry Sci.* **23**, 529 (1944). Stereochemistry: Zechmeister, *Chem. Rev.* **34**, 267 (1944). Structure: Karrer, *Helv. Chim Acta* **34**, 2160 (1951). Abs config: Goodfellow *et al.*, *Chem. Commun.* **1970**, 1578; Buchecker *et al.*, *Chimia* **25**, 192 (1971); *eidem, Helv. Chim. Acta* **57**, 631 (1974). Synthesis: H. Mayer, A. Rüttimann, *ibid.* **63**, 1451 (1980). Sepn and determn of configurational isomers: A. Rüttiman *et al.*, *J. High Resolut. Chromatog. Chromatog. Commun.* **6**, 612 (1983). *Reviews:* Zechmeister, *Carotinoide* (Berlin, 1934); Mayer, *The Chemistry of Natural Coloring*

Matters (New York, 1943); Karrer, Jucker, *Carotenoids* (New York, 1950).

Yellow prisms with metallic luster from ether + methanol, mp 190° (corr), (a higher mp indicates impure material). Also reported as mp 183° [Buchecker (1974)]. $[\alpha]_{Cd}^{18}$ +165° (c = 0.7 in benzene). Absorption max (dioxane): 481, 453, 429, 333, 268 nm (ε 142,000, 152,000, 100,000, 15,500, 35,000). Insol in water, sol in fats and in fat solvents. More sol in boiling methanol (1:700) than zeaxanthin.

Dipalmitate, $C_{72}H_{116}O_4$, *helenien, Adaptinol*. Red needles from alcohol, mp 92°.

9973. Xanthopterin. *2-Amino-1,5-dihydro-4,6-pteridinedione; 2-amino-4,6-pteridinediol;* 2-amino-4,6-dihydroxypteridine; 2-amino-4,6-dihydroxypyrimido[4,5-b]pyrazine. $C_6H_5N_5O_2$; mol wt 179.14. C 40.23%, H 2.81%, N 39.10%, O 17.86%. Pigment first found in wings of butterflies. Widely distributed in insects and in animals. Has been separated from urine and from the crab. Isoln and structure: Schöpf, Becker, *Ann.* **507**, 266 (1933); **524**, 55, 126 (1936); Schöpf, Kottler, *Ann.* **539**, 128 (1939); Wieland, Purrmann, *Ann.* **544**, 163 (1940). See also Fukushima, Shiota, *J. Biol. Chem.* **247**, 4549 (1972). Synthesis: Purrmann, *Ann.* **546**, 98 (1940); **548**, 284 (1941); Koschara, *Z. Physiol. Chem.* **277**, 159 (1943); Totter, *J. Biol. Chem.* **154**, 105 (1944); Elion *et al.*, *J. Am. Chem. Soc.* **71**, 741 (1949); Koschara, **Ger.** pat. 859,471 (1952 to Bayer), *C.A.* **52**, 10222f (1958); Stuart, Wood, *J. Chem. Soc.* **1963**, 4186. Facile synthesis: Taylor, Jacobi, *J. Am. Chem. Soc.* **95**, 4455 (1973); Taylor *et al.*, *J. Org. Chem.* **40**, 2341 (1975). Exhibits tumor inhibitory properties.

Monohydrate, orange-yellow crystals, sinters around 360°, decomp above 410°. Practically insol in water. Freely sol in dil NH_4OH and NaOH giving yellow solns and in 2N HCl giving colorless solns. uv max at pH 11: 255, 390 nm ($E_{1cm}^{1\%}$ 0.92, 0.355). Can be converted by yeast into folic acid.

9974. Xanthosine. Xanthine riboside; 9-β-D-ribofuranosyl xanthine; 9-β-D-ribofuranosyl-9H-purine-2,6-diol; 9-β-D-ribofuranosyl-9H-purine-2,6-(1H,3H)-dione. $C_{10}H_{12}N_4O_6$; mol wt 284.23. C 42.25%, H 4.26%, N 19.71%, O 33.77%. Prepd from guanosine by treatment with sodium nitrite and acetic acid: Levene, Jacobs, *Ber.* **43**, 3163 (1910); Gulland, Macrae, *J. Chem. Soc.* **1933**, 662. Synthesis: Howard *et al.*, *ibid.* **1949**, 232. Biosynthesis: Magasanik, Brooke, *J. Biol. Chem.* **206**, 83 (1954); Korn, Buchanan, *ibid.* **217**, 183 (1955); Bolis *et al.*, *ibid.* **219**, 917 (1956).

Dihydrate, long prisms from water. Anhydr felted clusters (warts) from alc. Dec on heating, no distinct melting range. $[\alpha]_D^{30}$ −51.2° (p = 8 in 0.3N NaOH). uv max: 253 nm (ε 8.790). a_M (molar absorbancy): 11.4 × 10^3 at 248.5

nm in water at pH 8.0. Sparingly sol in cold water; freely sol in hot water; sol in hot dil alcohol. Easily hydrolyzed by mineral acids.

9975. Xanthoxyletin. *5-Methoxy-8,8-dimethyl-2H,8H-benzo[1,2-b:5,4-b']dipyran-2-one;* 7-hydroxy-5-methoxy-2,2-dimethyl-2H-1-benzopyran-6-acrylic acid δ-lactone; xanthoxylin N; xanthoxyloin. $C_{15}H_{14}O_4$; mol wt 258.26. C 69.75%, H 5.46%, O 24.78%. Found in *Xanthoxylum americanum* Mill., *Melicope ternata* Forst., *M. mantelli* Buch., *Halifordia scleroxyla* F. Muell., *Rutaceae; Chloroxylon swietenia* DC., *Meliaceae.* Isoln: Staples, *Am. J. Pharm.* **1829**, 123; Witte, *Arch. Pharm.* **212**, 283 (1878); Lloyd, *Am. J. Pharm.* **1890**, 229; Gordin, *J. Am. Chem. Soc.* **28**, 1649 (1906); Bell *et al., J. Chem. Soc.* **1936**, 627; Robertson, Subramaniam, *ibid.* **1937**, 286; Dieterle, Kruta, *Arch. Pharm.* **275**, 45 (1937); Briggs, Locker, *J. Chem. Soc.* **1951**, 3131; King *et al., ibid.* **1954**, 1392; Hegarty, Lahey, *Aust. J. Chem.* **9**, 120 (1956); Cambie, *J. Chem. Soc.* **1960**, 2376. Synthesis: Joshi, Kamat, *Tetrahedron Letters* **1966**, 5767.

Elongated prisms from methanol, mp 132-133°. uv max: 277, 269, 322, 347 nm (log ε 4.28, 4.32, 3.99, 4.05 in ethanol). Very easily sol in benzene, chloroform, hot alcohol; sol in acetone; sparingly sol in ether. Sol in 49 parts cold acetone and 25,000 parts cold water.

9976. Xanthoxylin. *1-(2-Hydroxy-4,6-dimethoxyphenyl)ethanone;* 2'-hydroxy-4',6'-dimethoxyacetophenone; phloracetophenone 4,6-dimethyl ether; 2,4-dimethoxy-6-hydroxyacetophenone; brevifolin. $C_{10}H_{12}O_4$; mol wt 196.20. C 61.21%, H 6.17%, O 32.62%. Found in *Xanthoxylum piperitum* DC., *X. alatum* Roxb., *Rutaceae, Artemisia brevifolia* Wallich, *Compositae, Hippomane mancinella* L., *Sapium sebiferum* Roxb., *Euphorbiaceae.* Isoln: Stenhouse, *Ann.* **89**, 251 (1854); **104**, 236 (1857); Semmler, Schossberger, *Ber.* **44**, 2885 (1911); Smith, Smith, *Pharm. J.* **119**, 688 (1927); **123**, 604, 611 (1929); Schaeffer *et al., J. Am. Pharm. Assoc.* **43**, 43 (1954); Chu *et al., C.A.* **53**, 10532g (1959). Synthesis: Kostanecki, Lambor, *Ber.* **32**, 2260 (1899); Canter *et al., J. Chem. Soc.* **1931**, 1245; Belton *et al., Sci. Proc. Roy. Dublin Soc.* **25**, 19 (1949); Schmid, Bolleter, *Helv. Chim. Acta* **33**, 917 (1950); MacKenzie *et al., J. Chem. Soc.* **1950**, 2965; Dean, Robertson, *ibid.* **1953**, 1244; Kawano, *Chem. & Ind. (London)* **1959**, 368.

Crystals from ethanol, mp 81-83°. Practically insol in water; sol in alcohol, ether.

9977. Xanthurenic Acid. *4,8-Dihydroxy-2-quinolinecarboxylic acid;* 4,8-dihydroxyquinaldic acid. $C_{10}H_7NO_4$; mol wt 205.16. C 58.54%, H 3.44%, N 6.83%, O 31.19%. Excreted by pyridoxine-deficient animals after the ingestion of tryptophan. Isoln from the urine of albino rats fed almost exclusively on fibrin: Musajo, *Gazz. Chim. Ital.* **67**, 165 (1937). Synthesis from ethyl oxalacetate and *o*-anisidine: Musajo, Minchilli, *Ber.* **74**, 1842 (1941); Mebane, Oroshnik, *J. Am. Chem. Soc.* **73**, 3520 (1951); Furst, Olsen, *J. Org. Chem.* **16**, 412 (1951).

Sulfur-yellow crystals, mp 286°. uv max (water): 243, 342 nm (ε 30,000, 6500). Insol in water. Sol in aq alkali hydroxides and carbonates (yellow solns) and in hot dil HCl. It gives a red color with Millon reagent, intense ruby-red with alkali diazobenzenesulfonates, and intense green with $FeSO_4$. When the acid is dissolved in very dil $NaHCO_3$, the green color with $FeSO_4$ is still visible at a diln of 1:200,000.
Sodium salt, $C_{10}H_6NNaO_4 \cdot nH_2O$, yellow crystals, anhydr at 170°. Slightly sol in water.
Methyl ester, $C_{11}H_9NO_4$, yellow crystals, dec 262°. Sol in NH_4OH, alkali hydroxides and carbonates.

9978. Xanthyletin. *8,8-Dimethyl-2H,8H-benzo[1,2-b:-5,4-b']dipyran-2-one;* 7-hydroxy-2,2-dimethyl-2H-1-benzopyran-6-acrylic acid δ-lactone; 2,2-dimethylchromenocoumarin. $C_{14}H_{12}O_3$; mol wt 228.24. C 73.67%, H 5.30%, O 21.03%. Linear pyranocoumarin found in *Xanthoxylum americanum* Mill., *Luvanga scandens* Ham., *Citrus acida* Roxb., *Rutaceae; Chloroxylon swietenia* DC., *Meliaceae.* Isoln: Bell, Robertson, *J. Chem. Soc.* **1936**, 1828; Bell *et al., ibid.* **1937**, 1542; Späth *et al., Ber.* **72**, 1450 (1939); Bose, Mookerjee, *J. Indian Chem. Soc.* **21**, 181 (1944); Mookerjee, *ibid.* **23**, 41 (1946); King *et al., J. Chem. Soc.* **1954**, 1392. Synthesis: Späth, Hillel, *Ber.* **72**, 2093 (1939); Steck, *Can. J. Chem.* **49**, 2297 (1971); P. Waykole *et al., Indian J. Chem.* **19B**, 238 (1980); V. K. Ahluwalia *et al., Monatsh.* **112**, 119 (1981); J. Banerji *et al., Indian J. Chem.* **21B**, 496 (1982).

Elongated flat prisms from methanol, mp 130-131°. $bp_{0.1}$ 140-145°. uv max: 266, 348 nm (log ε 4.34, 4.15).

9979. Xenazoic Acid. *4-[(2-[1,1'-Biphenyl]-4-yl-1-ethoxy-2-oxoethyl)amino]benzoic acid;* p-[(α-ethoxy-p-phenylphenacyl)amino]benzoic acid; p-[N-(α-ethoxy-β-keto-β-para-biphenyl)ethylamino]benzoic acid; p-(α-ethoxy-p-phenylphenacylamido)benzoic acid; xenalamine; CV 58903; SKF 8318; Xenovis. $C_{23}H_{21}NO_4$; mol wt 375.41. C 73.58%, H 5.64%, N 3.73%, O 17.05%. Prepn: Cavallini *et al., J. Med. Pharm. Chem.* **2**, 99 (1960).

Crystals from ethanol, mp 192° (dec).
THERAP CAT: Antiviral.

9980. Xenbucin. *α-Ethyl-[1,1'-biphenyl]-4-acetic acid;* 2-(4-biphenylyl)butyric acid; α-(4-biphenylyl)butyric acid; α-(4-diphenylyl)butyric acid; α-(p-xenyl)butyric acid; 4-diphenylylethylacetic acid; 4-biphenylylethylacetic acid; ethyl-p-xenylacetic acid; α-ethyl-p-phenyl-α-toluic acid; MG 1559; Maggioni 1559; Liosol; Liposana. $C_{16}H_{16}O_2$; mol wt 240.29. C 79.97%, H 6.71%, O 13.32%. Prepn: Blicke, Grier, *J. Am. Chem. Soc.* **65**, 1725 (1943); Cavallini, Massarani, *Farmaco Ed. Sci.* **11**, 167 (1956). Fibrinolytic activity: R. J. Gryglewski, M. Eckstein, *Nature* **214**, 626 (1967).

Crystals from glacial acetic acid, mp 123-125°. Soluble in ethanol, methanol, acetone, carbon tetrachloride, benzene, ether, dimethylformamide. Practically insol in water.
trans-4-Phenylcyclohexylamine salt (1:1), $C_{28}H_{33}NO_2$, *MG*

5771, butixirate, Flectar. Prepn, analgesic and anti-inflammatory properties: G. Cantarelli *et al., Farmaco Ed. Sci.* **24**, 140 (1969); **Brit. pat. 1,168,542** (1969 to Maggioni), *C.A.* **72**, 43197 (1970). Therapeutic effectiveness: P. Croce, *Clin. Ter.* **50**, 241 (1969). Crystals, mp 224-227°. LD$_{50}$ i.p. in mice: 183 μM/kg (Cantarelli).

THERAP CAT: Free acid as antihyperlipoproteinemic; *trans*-4-phenylcyclohexylamine salt as analgesic, anti-inflammatory.

9981. Xenon. Xe; at. wt 131.29; at. no. 54. Nine naturally occurring isotopes: 124 (0.096%); 126 (0.090%); 128 (1.919%); 129 (26.44%); 130 (4.08%); 131 (21.18%); 132 (26.89%); 134 (10.44%); 136 (8.87%); artificial, radioactive isotopes: 118-123; 125; 127; 133; 135; 137-144. Discovered in the final residues obtained after evaporating liq air: Ramsay, Travers, *Proc. Roy. Soc.* **63** [A], 405 (1898). Occurs frequently in gases evolved from thermal springs; present in the air to the extent of about 1×10^{-5}%. Extraction from liq air residues: Allen, Moore, *J. Am. Chem. Soc.* **53**, 2512 (1931). *Xenon platinum hexafluoride* was the first reported xenon compound: N. Bartlett, *Proc. Chem. Soc.* **1962**, 218. Teratogenicity study: G. A. Lane *et al., Science* **210**, 899 (1980). Review of biology, chemistry and anesthetic properties: R. M. Featherstone, C. A. Muelbaecher, *Pharmacol. Rev.* **15**, 97 (1963). Review of diagnostic use of radioactive compounds for pulmonary studies: F. Fazio, P. Wollman, *Clin. Physiol.* **1**, 323 (1981); for cerebral blood flow: H. Yonas *et al., Adv. Tech. Stand. Neurosurg.* **15**, 3 (1987). Reviews of chemistry: *Noble-Gas Compounds*, H. H. Hyman, Ed. (Univ. Chicago Press, Chicago, 1963) 404 pp; J. H. Holloway, *Noble-Gas Chemistry* (Methuen, London, 1968) 213 pp; Sladky, "Noble Gases" in *MTP Int. Rev. Sci.: Inorg. Chem., Ser. One* vol. **3**, V. Gutman, Ed. (Butterworths, London, 1972) pp 1-52; Bartlett, Sladky in *Comprehensive Inorganic Chemistry* vol. **1**, J. C. Bailar, Jr. *et al.*, Eds. (Pergamon Press, Oxford, 1973) pp 213-330. Comprehensive review: Cockett, Smith, *ibid.*, pp. 139-211.

Gas. d 5.8971 g/l. bp $-108.10°$; triple pt temp $-111.8°$; triple pt pressure 612 mm. Crit temp 16.59°; crit press. 57.64 atm. Spectrum: Collie, *Proc. Roy. Soc.* **97** [A], 349 (1920).

Xenon-133, *Xeneisol*® *Xe 133.*

A hydrate, Xe.xH$_2$O, mp 24°, and a deuterate, Xe.6D$_2$O, have been prepd: R. de Forcrand, *Compt. Rend.* **176**, 355 (1923); **181**, 15 (1925); Tamman, Krige, *Z. Anorg. Chem.* **146**, 179 (1925).

Xenon difluoride, F$_2$Xe. Prepd from the elements: Weeks *et al., J. Am. Chem. Soc.* **84**, 4612 (1962); Hoppe *et al., Z. Anorg. Allgem. Chem.* **324**, 214 (1963). Colorless crystals; d 4.32. Triple pt. 129.03°: Schreiner *et al., J. Phys. Chem.* **72**, 1162 (1968). Sublimes without decompn. Soly in water at 0°: 25 g/l.

Xenon tetrafluoride, F$_4$Xe. First prepd by direct combination of the elements at 6 atm and 400°: Claassen *et al., J. Am. Chem. Soc.* **84**, 3593 (1962). Colorless crystals; d 4.04. Triple pt 117.10°: Schreiner *et al., loc. cit.* Sublimes without decompn. Hydrolyzes to form Xe, O$_2$, HF and XeO$_3$.

Xenon hexafluoride, F$_6$Xe. Laboratory prepn: Chernick *et al., Inorg. Syn.* **8**, 258 (1966). Colorless solid; greenish-yellow vapor; vapor press. about 30 mm at 25°. mp 49.48°; bp 75.57°; d$^{24.4}$(solid) 3.411; d$^{55.2}$(liq) 3.173: Schreiner *et al., J. Chem. Phys.* **51**, 4838 (1969). Hydrolyzed by water to form XeOF$_4$ and XeO$_3$. More powerful oxidizing and fluorinating agent than XeF$_2$ and XeF$_4$. Cannot be stored in glass or quartz containers.

Xenon trioxide, O$_3$Xe. *Caution: Powerful explosive*, formed when XeF$_4$ and XeF$_6$ are hydrolyzed. Prepn: Williamson, Koch, *Science* **139**, 1046 (1963); Jaselskis *et al., J. Am. Chem. Soc.* **88**, 2149 (1966). Colorless, hygroscopic solid; d 4.55. Aqueous solns, "*xenic acid*", may be prepd; concns > 2*M* may be obtained.

USE: In lamps designed to resemble natural daylight; in lamps of extremely high brilliance.

THERAP CAT: Anesthetic (inhalation). Xenon[133] as diagnostic aid (radioactive imaging agent—regional blood flow, pulmonary function).

9982. *p*-Xenylcarbimide. *4-Isocyanato-1,1'-biphenyl; isocyanic acid 4-biphenylyl ester; p*-diphenyl isocyanate.

C$_{13}$H$_9$NO; mol wt 195.21. C 79.98%, H 4.65%, N 7.18%, O 8.20%. C$_6$H$_5$C$_6$H$_4$NCO. Prepd from the corresponding amine and phosgene: Kaplan, *J. Chem. Eng. Data* **6**, 272 (1961).

Needles, mp 56°; bp 283° (decomp), bp$_{2.3}$ 137-140°. Very sol in ether.

USE: As a reagent for the identification of hydroxy compounds, with which it gives crystalline derivatives of definite melting point.

9983. Xenytropium Bromide. *endo-(±)-8-([1,1'-Biphenyl]-4-ylmethyl)-3-(3-hydroxy-2-phenyl-1-oxopropoxy)-8-methyl-8-azoniabicyclo[3.2.1]octane bromide; 3α-hydroxy-8-(p-phenylbenzyl)-1αH,5αH-tropanium bromide (±)-tropate;* 8-*p*-phenylbenzylatropinium bromide; *p*-biphenylmethyl-(*dl*-tropyl-*α*-tropinium)bromide; 4-diphenylmethyl-*dl*-tropeyltropinium bromide; N-399; Gastropin; Gastripon. C$_{30}$H$_{34}$BrNO$_3$; mol wt 536.53. C 67.16%, H 6.39%, Br 14.90%, N 2.61%, O 8.95%. Anticholinergic. Prepn: Nador, Gyermek, **U.S. pat. 2,833,773** (1958 to Licencia-Budapest).

Crystals from ethanol, dec 220-222°.
THERAP CAT: Antispasmodic.

9984. Xibenolol. (±)-*1-[(Dimethylethyl)amino]-3-(2,3-dimethylphenoxy)-2-propanol;* 3-(*tert*-butylamino)-3-(2',3'-dimethylphenoxy)-2-propanol; (±)-1-(*tert*-butylamino)-3-(2,3-xylyloxy)-2-propanol. C$_{15}$H$_{25}$NO$_2$; mol wt 251.37. C 71.67%, H 10.02%, N 5.57%, O 12.73%. Non-selective β-adrenergic blocker. Prepn: W. Kunz *et al.*, **Ger. pat. 1,236,523** (1967 to Sanol-Arzneimittel), *C.A.* **67**, 64046k (1967); K. Tsukamoto *et al.*, **U.S. pat. 4,018,824** (1977 to Teikoku Hormone). Pharmacology in dogs: N. Himori *et al., Arch. Int. Pharmacodyn.* **220**, 4 (1976); *eidem, Arch. Pharmacol.* **316**, 19 (1986). Metabolism in animals: S. Honma, A. Kambekawa, *Chem. Pharm. Bull.* **23**, 1045 (1975). Stereoselective metabolism and pharmacokinetics in man: S. Honma *et al., ibid.* **33**, 760 (1985). Toxicology: T. Usui *et al., Yakuri to Chiryo* **12**, Suppl. 6, 969, 987 (1984); *C.A.* **102**, 125280x, 125281y (1985).

Crystals from ether, mp 57°. bp$_{0.7mm}$ 134-136°. uv max (ethanol): 271.2, 274, 279.3 nm (ϵ 1.08 $\times$ 10^3, 1.07 $\times$ 10^3, 1.11 $\times$ 10^3).

Hydrochloride, C$_{15}$H$_{25}$ClNO$_2$, *D-32, Selapin, Rhythminal, Rythminal.* mp 135-137°.
THERAP CAT: Antiarrhythmic.

9985. Xibornol. *6-Isobornyl-3,4-xylenol;* 6-isobornyl-3,4-dimethylphenol; 6-(2-isobornyl)-3,4-xylen-1-ol; 3,4-dimethyl-6-isobornylphenol; Nanbacine. C$_{18}$H$_{26}$O; mol wt 258.41. C 83.67% H 10.14% O 6.19%. Syntheses: Moldovanskaya *et al., C.A.* **66**, 76163p (1967); Starkov, Glushkova, *J. Appl. Chem. USSR* **40**, 209 (1967), *C.A.* **67**, 72960u (1967), *C.A.* **71**, 70225s (1969). Prepn of cryst substance: **Brit. pat. 1,206,774** and Mardiguian, Fournier, **Ger. Offen. pat. 2,032,170** (1970 and 1971 to MARPHA Soc. d'Etudes et d'Exploitation de Marques), *C.A.* **74**, 13299g, 88171t (1971). Activity study: Capponi, *Bull. Soc. Pathol. Exot.* **62**, 658 (1969), *C.A.* **73**, 118883a (1970).

Crystals from petr ether, mp 94-96°, bp$_3$ 165-168°. Also reported as very viscous, pale yellow liquid; bp$_9$ 185-189°. d$_4^{20}$ 1.0240. n$_D^{20}$ 1.5382 (Starkov, Glushkova).

USE: Rubber antioxidant.

THERAP CAT: Antibacterial.

9986. Xipamide. *4-Chloro-5-sulfamoyl-2',6'-salicyloxylidide;* 4-chloro-2',6'-dimethyl-5-sulfamoylsalicylanilide; 4-chloro-5-sulfamylsalicyloyl-2',6'-dimethylanilide; Bei 1293; Aquaphor; Diurexan; Lumitens. C$_{15}$H$_{15}$ClN$_2$O$_4$S; mol wt 354.81. C 50.78%, H 4.26%, Cl 9.99%, N 7.89%, O 18.04%, S 9.04%. Salicylic acid derivative. Prepn: **Neth. pat. Appl. 6,607,680** corresp to Liebenow, **U.S. pat. 3,567,-777** (1966 and 1970 to P. Beiersdorf). Clinical studies: O. Hammer, U. Dembowski, *Med. Klin.* **64,** 1862 (1969); R. Fischer, A. Lenhartz, *Med. Welt* **1970,** 270. Review of pharmacology and therapeutic efficacy: B. N. C. Prichard, R. N. Brogden, *Drugs* **30,** 313-332 (1985).

Crystals from methanol-water, mp 256°.

THERAP CAT: Diuretic. Antihypertensive.

9987. Xylazine. *N-(2,6-Dimethylphenyl)-5,6-dihydro-4H-1,3-thiazin-2-amine;* 5,6-dihydro-2-(2,6-xylidino)-4H-1,3-thiazine; 2-(2,6-dimethylphenylamino)-4H-5,6-dihydro-1,3-thiazine; BAY 1470; BAY VA 1470; Wh7286; Rompun (also the hydrochloride). C$_{12}$H$_{16}$N$_2$S; mol wt 220.33. C 65.42%, H 7.32%, N 12.71%, S 14.55%. Prepn: Behner *et al.,* **Ger. pat. 1,173,475** (1964), *C.A.* **61,** 13323c (1964); *eidem,* **Belg.** pat. **634,552** (1964) corresp to **U.S. pat. 3,235,550** (1966) (all to Bayer). Pharmacology: G. Sagner *et al., Deut. Tieraerztl. Wochenschr.* **75,** 565 (1968); Kroneberg, Schlossman, *Arch. Pharmakol. Exp. Pathol.* **268,** 348 (1971). Metabolic studies: Duhm *et al., Berlin. München. Tieraerztl. Wochenschr.* **82,** 104 (1969). Veterinary clinical studies: Clarke, Hall, *Vet. Rec.* **85,** 512 (1969); Burns, McMullan, *Vet. Med.* **67,** 77 (1972).

Colorless, almost tasteless crystals from benzene + ligroin, mp 140-142°. Also reported as mp 136-139°. Soluble in dilute acids, benzene, acetone, chloroform; sparingly sol in petr ether. Practically insol in water, alkalies. LD$_{50}$ in mice (mg/kg): 43 i.v.; 121 s.c.; 240 orally (Sagner).

Hydrochloride, C$_{12}$H$_{17}$ClN$_2$S, crystals from ethanol + ethyl ether, mp 236-239°.

THERAP CAT (VET): Sedative, analgesic, muscle relaxant.

9988. Xylene. *Dimethylbenzene;* xylol. C$_8$H$_{10}$; mol wt 106.16. C 90.50%, H 9.50%. C$_6$H$_4$(CH$_3$)$_2$. First isolated from a crude wood distillate: Cahours, *Compt. Rend.* **30,** 319 (1850). Obtained from coal tar: Fittig, *Ann.* **153,** 265 (1870). The xylene of commerce is a mixture of the three isomers o-, m- and p-xylene, the m-isomer predominating. Manuf from pseudocumene: Seubold, **U.S. pat. 2,960,545** (1960 to Union Oil); by catalytic isomerization of a hydrocarbon fraction: Berger, **U.S. pat. 3,078,318** (1963 to Universal Oil Prod.). Separation of isomers by clathration: Schaeffer, **U.S. pat. 3,029,300** (1962 to Union Oil). Toxicol-

ogy: E. Browning, *Toxicity and Metabolism of Industrial Solvents* (Elsevier, New York, 1965) pp 77-89. Use as clearing agent: K. Kubota, *J. Polym. Sci.* **5,** 1179 (1967); J. B. Matthews, *J. Clin. Pathol.* **34,** 103 (1981). Review of mfg processes: Faith, Keyes & Clark's *Industrial Chemicals,* F. A. Lowenheim, M. K. Moran, Eds. (Wiley-Interscience, New York, 4th ed., 1975) pp 874-881. Toxicity: H. F. Smyth *et al., Am. Ind. Hyg. Assoc. J.* **23,** 95 (1962).

Mobile, flammable liquid; d about 0.86; bp 137-140°. Flash pt 29°. Practically insol in water. Miscible with abs alcohol, ether, and many other organic liquids.

m-Xylene, colorless liquid; d$_4^{15}$ 0.8684; mp −47.4°; bp 139.3°; n$_D^{20}$ 1.4973. Flash pt, closed cup: 77°F (25°C). Insol in water. Miscible with alcohol, ether, and many other organic solvents. LD$_{50}$ orally in rats: 7.71 ml/kg (Smyth).

o-Xylene, colorless liquid; d$_4^{20}$ 0.8801; mp −25°; bp 144°; n$_D^{20}$ 1.5058. Flash pt, closed cup: 63°F (17°C). Insol in water. Miscible with alc, ether.

p-Xylene, colorless plates or prisms at low temp; d$_4^{20}$ 0.86104; mp 13-14°; bp 137-138°; n$_D^{20}$ 1.49575: Thorne *et al., Ind. Eng. Chem. Anal. Ed.* **17,** 481 (1945). Flash pt, closed cup: 77°F (25°C). Insol in water. Sol in alcohol, ether, and many other organic solvents.

Caution: May be narcotic in high concns. Chronic toxicity not well defined, but is less toxic than benzene.

USE: As solvent; raw material for production of benzoic acid, phthalic anhydride, isophthalic and terephthalic acids as well as their dimethyl esters used in the manufacture of polyester fibers; manuf dyes and other organics; sterilizing catgut; with Canada balsam as oil-immersion in microscopy; clearing agent in microscope technique.

9989. Xylenol. *Dimethylphenol.* C$_8$H$_{10}$O; mol wt 122.16. C 78.65%, H 8.25%, O 13.10%. (CH$_3$)$_2$C$_6$H$_3$OH. Constituent of "cresylic acid." There are 6 isomers of xylenol. They are only slightly sol in water but freely sol in alcohol, chloroform, ether, benzene, etc.; also sol in NaOH soln. Prepn and physical properties: R. J. L. Andon *et al., J. Chem. Soc.* **1960,** 5246.

2,3-Dimethylphenol, *vic-o-xylenol.* Needles from water or dil alc, mp 75°, bp 218°: Thöl, *Ber.* **18,** 2561 (1885); also reported as mp 72.57° ±0.02°. bp$_{760}$ 216.87° ±0.001° (Andon).

2,4-Dimethylphenol, *as-m-xylenol.* Crystals, bp$_{766}$ 211.5°; mp 25.4-26°: Jacobsen, *Ber.* **11,** 17 (1887); **18,** 3463 (1885); also reported as mp 24.54° ±0.01°. bp$_{760}$ 210.931° ±0.001° (Andon).

2,5-Dimethylphenol, *p-xylenol.* Crystals from alcohol + ether, mp 74.5°, bp$_{762}$ 211.5°: Jacobsen, *loc. cit.* bp 213.5°: Würtz, *Ann.* **147,** 372 (1868); also reported as mp 74.85° ±0.02°. bp$_{760}$ 211.132° ±0.002° (Andon).

2,6-Dimethylphenol, *vic-m-xylenol.* Needles, mp 49°, bp 203°: Gattermann, *Ann.* **357,** 313 (1907); also reported as mp 45.62° ±0.01°. bp$_{760}$ 201.030° ±0.001° (Andon).

3,4-Dimethylphenol, *as-o-xylenol.* Needles from water, mp 62.5°, bp 225°: Jacobsen, *Ber.* **17,** 159 (1884); also reported as mp 65.11° ±0.01°. bp$_{760}$ 226.947° ±0.001° (Andon).

3,5-Dimethylphenol, *sym-m-xylenol.* Needles from water, mp 64°, bp 219.5°: Thöl, *Ber.* **18,** 359 (1885); also reported as mp 63.27° ±0.02°. bp$_{760}$ 221.962° ±0.003° (Andon).

USE: For the prepn of coal tar disinfectants; manuf of artificial resins.

9990. Xylenol Blue. *4,4'-(3H-2,1-Benzoxathiol-3-ylidene)bis[2,5-dimethylphenol] S,S-dioxide;* p-xylenolsulfonephthalein; 1,4-dimethyl-5-hydroxybenzenesulfonephthalein; α,4,4'-trihydroxy-2,5,2',5'-tetramethyltriphenylmethane-

2''-sulfonic acid γ-sultone. $C_{23}H_{22}O_5S$; mol wt 410.50. C 67.30%, H 5.40%, O 19.49%, S 7.81%. Prepd from o-sulfobenzoic acid dichloride or o-sulfobenzoic acid anhydride and p-xylenol in the presence of zinc chloride: Cohen, *Biochem. J.* **16**, 31 (1922).

Brown crystals from alcohol.

USE: Indicator, used in 0.02% soln: pH 1.2 red, 2.8 yellow, 9.6 blue.

9991. Xylidine. *ar,ar-Dimethylbenzenamine;* dimethylaniline; aminodimethylbenzene. $C_8H_{11}N$; mol wt 121.18. C 79.29%, H 9.15%, N 11.56%. $(CH_3)_2C_6H_3NH_2$. There are six isomeric xylidines. Prepd by reduction of corresponding nitro-compounds: Allchin, U.S. pat. **1,867,962** (to I.C.I.); physical properties of all six isomers also given: Birch *et al., J. Am. Chem. Soc.* **71**, 1362 (1949); van Loon *et al., Rec. Trav. Chim.* **79**, 977 (1960); Bergmann, Berkovic, *J. Org. Chem.* **26**, 919 (1961).

All except o-4-xylidine are liquids above 20°. d 0.97-0.99, and bp between 213° and 226°. They are sparingly sol in water, sol in alcohol and form more or less sol salts with the strong mineral acids.

USE: Chiefly in the manuf of dyes.

9992. Xylitol. *xylo-Pentane-1,2,3,4,5-pentol;* xylite; Eutrit; Kannit; Klinit; Kylit; Newtol; Torch; Xyliton. $C_5H_{12}O_5$; mol wt 152.15. C 39.47%, H 7.95%, O 52.58%. Intermediate in metabolism of D-glucose through glucuronate cycle in livers. Prepd by reduction of xylose: G. Bertrand, *Bull. Soc. Chim. France* [3] **5**, 555 (1891); E. Fischer, R. Stahel, *Ber.* **24**, 538 (1891). Prepn of metastable crystals: M. L. Wolfrom, E. J. Kohn, *J. Am. Chem. Soc.* **64**, 1739 (1942); of stable form: J. F. Carson *et al., ibid.* **65**, 1777 (1943). Crystal structure: H. S. Kim, G. A. Jeffrey, *Acta Crystallogr.* **25B**, 2607 (1969). Use in prevention of dental caries: E. Grunberg *et al., Int. J. Vit. Nutr. Res.* **43**, 227 (1973); A. Scheinin, K. K. Makinen, Ger. pat. **2,606,533** (1976 to Hoffmann-La Roche), *C.A.* **85**, 149140h (1976). Acute toxicity: S. Salminen *et al., Toxicol. Lett.* **18**, Suppl. 1, 37 (1983). Reviews of toxicity, metabolism and use as dietary additive: *International Symposium on Metabolism, Physiology and Clinical Uses of Pentoses and Pentitols,* B. L. Horecker *et al.,* Eds. (Springer-Verlag, New York, 1969) 408 pp; *Sugars in Nutrition,* H. L. Sipple, K. W. McNutt, Eds. (Academic Press, New York, 1974) *passim;* G. E. Demetrakopoulos, H. Amos, *World Rev. Nutr. Diet* **32**, 96-122 (1978); R. Ylikahri, *Adv. Food Res.* **25**, 159-180 (1979). Book: *Xylitol,* J. N. Counsel, Ed. (Applied Science, London, 1978) 191 pp.

$$
\begin{array}{c}
CH_2OH \\
| \\
HCOH \\
| \\
HOCH \\
| \\
HCOH \\
| \\
CH_2OH
\end{array}
$$

Stable form: orthorhombic needles from THF, prisms from ethanol; mp 93-94.5°; d 1.52. Metastable form: colorless, monoclinic, lath-shaped crystals from anhydrous methanol; hygroscopic; mp 61-61.5°. Soly of stable form (g/100 g soln): abs methanol 6.0; abs ethanol 1.2; water 64.2. Relative sweetness equal to sucrose. LD_{50} orally in mice: approx. 22 g/kg (Salminen).

USE: As oral and intravenous nutrient; in anticaries preparations.

9993. Xylometazoline. *2-[[4-(1,1-Dimethylethyl)-2,6-dimethylphenyl]methyl]-4,5-dihydro-1H-imidazole; 2-(4-tert-butyl-2,6-dimethylbenzyl)-2-imidazoline.* $C_{16}H_{24}N_2$; mol wt 244.37. C 78.63%, H 9.90%, N 11.46%. Prepn: Hüni, U.S. pat. **2,868,802** (1959 to Ciba). Pharmacological studies: Morimoto, Tanaka, *C.A.* **72**, 20437n (1970).

mp 131-133°.

Hydrochloride, $C_{16}H_{25}ClN_2$, *Novorin, Olynth, Otriven, Otrivin, Otrix, Xymelin.* Soly in water: up to 3%; also sol in methanol, ethanol. Practically insol in ether, benzene.

THERAP CAT: Adrenergic (vasoconstrictor); nasal decongestant.

9994. Xylopropamine. *α,3,4-Trimethylbenzeneethanamine; α,3,4-trimethylphenethylamine;* 1-(3,4-dimethylphenyl)-2-aminopropane. $C_{11}H_{17}N$; mol wt 163.25. C 80.92%, H 10.50%, N 8.58%. Prepd from 1-(3,4-dimethylphenyl)-2-propanone and ammonia in methanol followed by catalytic reduction: **Swiss** pat. **230,368** (1944 to Hoffmann-La Roche); *C.A.* **43**, 3454i (1949).

Oil, bp_{12} 116-118°. Slightly sol in water.

Hydrobromide, $C_{11}H_{18}BrN$, crystals, mp 132-133°. Ingredient of *Esanin.*

THERAP CAT: Adrenergic.

9995. Xylose. D-Xylose; wood sugar; Xylomed; Xylo-Pfan. $C_5H_{10}O_5$; mol wt 150.13. C 40.00%, H 6.71%, O 53.29%. Widely distributed in plant materials, especially in wood (maple, cherry), in straw, in hulls. Not found in free state, but in form of xylan, a polysaccharide built from D-xylose units and occurring in association with cellulose. Xylose occurs also as part of glycosides. Isoln from corn cobs by boiling with 8% H_2SO_4: Monroe, *J. Am. Chem. Soc.* **41**, 1002 (1919). Peanut shells and cottonseed hulls also are practical sources of xylose: Ling, Nanji, *J. Chem. Soc.* **1923**, 620. Configuration: Hudson, Yanovsky, *J. Am. Chem. Soc.* **39**, 1029 (1917); Haworth, *Nature* **116**, 430 (1925). Review on history, constitution and prepn: Harding, *Sugar* **24**, 14 (1922).

α-D-xylose

Monoclinic needles or prisms. Very sweet taste. mp 144-145° (Wheeler, Tollens, *Ann.* **254**, 309); mp 153-154° (Hébert, *Compt. Rend.* **110**, 970). d_4^{20} 1.525. Shows mutarotation. $[\alpha]_D^{20}$ +92° → +18.6° (16 hrs c = 10). One gram dissolves in 0.8 ml water. Sol in pyridine, hot alcohol. pKa (18°) = 12.14. Reduces warm Fehling's soln. Upon heating with water in closed tube to 140° or by boiling with dil H_2SO_4 furfurol is formed.

USE: Xylose is used in tanning, dyeing, and as a diabetic food.

THERAP CAT: Diagnostic aid (intestinal function).

9996. Xylulose. *threo-Pentulose.* $C_5H_{10}O_5$; mol wt 150.13. C 40.00%, H 6.71%, O 53.29%. L-Form has been found in the urine of humans with pentosuria. Prepn of DL-form: Gascoigne, *Chem. & Ind. (London)* **1959**, 402; of D-form: Mendicino, *J. Am. Chem. Soc.* **82**, 4975 (1960); of L-form: Wolfrom, Bennett, *J. Org. Chem.* **30**, 458 (1965).

Isoln of DL-form from the acid hydrolysate of bagasse hemicellulose: Banerjee *et al., Sci. Cult. (Calcutta)* **27,** 498 (1961), *C.A.* **56,** 11682d (1962). Enzymic prepn of L-form: Hough, Jones, *Chem. & Ind. (London)* **1952,** 907; *eidem, J. Chem. Soc.* **1952,** 4047. Formation of L-form in normal humans and guinea pigs, and its utilization by guinea-pig liver prepns: Touster *et al., J. Am. Chem. Soc.* **76,** 5005 (1954). *Reviews: The Carbohydrates,* W. Pigman, Ed. (Academic Press, New York, 1957) pp 80, 86-87, 759, 795; *Methods in Carbohydrate Chemistry* **vol. 1,** R. L. Whistler, M. L. Wolfrom, Eds. (Academic Press, New York, 1962) pp 94-101.

$$
\begin{array}{c}
CH_2OH \\
| \\
CO \\
| \\
HCOH \\
| \\
HOCH \\
| \\
CH_2OH \\
\\
\underline{L\text{-isomer}}
\end{array}
$$

D-Isomer, syrup. $[\alpha]_D^{18}$ −33° (c = 2.5).

D-Isomer *p*-bromophenylhydrazone, $C_{11}H_{15}BrN_2O_4$, pale yellow crystals from abs ethanol + water, mp 128-129°. $[\alpha]_D^{20}$ +24° (15 min) → −31° (7 days, in pyridine). *Ref:* Whistler, Wolfrom, *loc. cit.*

L-Isomer, syrup. $[\alpha]_D^{21}$ +31°.

L-Isomer *p*-bromophenylhydrazone, yellow plates from dil alc, mp 128°. $[\alpha]_D^{20}$ −20° (10 min) → +22° (5 hrs, c = 0.5 in ethanol).

9997. 1-Xylylazo-2-naphthol. *1-[(2,4-Dimethylphenyl)-azo]-2-naphthalenol; C.I. Solvent Orange 7;* 1-(2,4-xylyl-azo)-2-naphthol; FD & C Red no. 32; Oil Red XO; Ext. D & C Red no. 14; C.I. 12140. $C_{18}H_{16}N_2O$; mol wt 276.32. C 78.24%, H 5.84%, N 10.14%, O 5.79%. Once reported as 2,5-xylylazo deriv. Prepd by coupling diazotized *m*-xylidene with 2-naphthol: J. M. Tedder, *J. Chem. Soc.* **1957,** 4003; R. B. Smyth, G. G. McKeown, *J. Chromatog.* **5,** 395 (1961). Metabolism: J. L. Radomski, *J. Pharm. Exp. Ther.* **136,** 378 (1962).

Red needles, mp 166°. Insol in water; sol in ethanol, acetone, benzene.

Caution: Delisted for use in foods, drugs, and cosmetics by the FDA.

9998. Xylyl Bromide. C_8H_9Br; mol wt 185.07. C 51.92%, H 4.90%, Br 43.18%. Prepn of *m*-isomer: Wenner, *J. Org. Chem.* **17,** 523 (1952); of *o*-isomer: Dev, *J. Indian Chem. Soc.* **32,** 403 (1955); of *p*-isomer: Cockburn *et al., J. Chem. Soc.* **1960,** 3340.

m-Xylyl bromide, 1-(bromomethyl)-3-methylbenzene, α-bromo-m-xylene, m-methylbenzyl bromide, ω-bromo-m-xylene. Liquid, bp 212-215° with slight dec. d^{23} 1.371. Practically insol in water; sol in alcohol, ether.

o-Xylyl bromide. Prisms, mp 21°. bp 223-234°, bp_{742} 216-217°, bp_{15} 102°. n_D^{27} 1.5730. d^{23} 1.381. Practically insol in water; sol in alcohol, ether.

p-Xylyl bromide. Needles from alcohol, mp 38°, bp_{740} 218-220°, bp_{15} 120°. d 1.324. Practically insol in water; very sol in chloroform, hot ether.

USE: In organic syntheses; in war-gas formulations. *Caution:* Powerful lacrimator.

9999. Xylyl Chloride. C_8H_9Cl; mol wt 140.61. C 68.33%, H 6.45%, Cl 25.22%. Prepn of *o*-isomer: Rabjohn, *J. Am. Chem. Soc.* **76,** 5479 (1954); of *m*-isomer: van Zanten, Nauta, *Rec. Trav. Chim.* **79,** 1211 (1960); of *p*-isomer: Newman, George, *J. Org. Chem.* **26,** 4306 (1961).

m-Xylyl chloride, 1-(chloromethyl)-3-methylbenzene, α-chloro-m-xylene, m-methylbenzyl chloride, ω-chloro-m-xylene. Liquid, bp 195-196°. bp_{10}90-92°. d^{20} 1.064. Practically insol in water; miscible with abs alcohol, ether.

o-Xylyl chloride. Liquid, bp 195-203°, bp_{25} 95-96°. n_D^{20} 1.5391. Practically insol in water; miscible with abs alcohol, ether.

p-Xylyl chloride. Fuming liquid, characteristic odor. bp 200-202°, bp_1 48-50°. n_D^{17} 1.5360. Practically insol in water; miscible with abs alcohol, ether.

Caution: Powerful lacrimator.

Y

10000. Yam, Mexican. Giant yam; cabeza de negro; nigerhead. *Dioscorea macrostachya* Benth., *Dioscoreaceae* or one of approx 600 other *Dioscoreae*, native to subtropical countries. The underground tubers weigh as much as 90 lbs. and contain steroidal sapogenins. Account of nature, origins, cultivation and utilization of the useful members of the *Dioscoreaceae*: D. G. Coursey, *Yams* (Longmans, London, 1967) 230 pp. *See also* Dioscorea and Barbasco.

USE: In the partial synthesis of hormones having a steroid structure.

10001. Yangonin. *4-Methoxy-6-[2-(4-methoxyphenyl)ethenyl]-2H-pyran-2-one; 4-methoxy-6-(p-methoxystyryl)-2H-pyran-2-one;* 5-hydroxy-3-methoxy-7-(p-methoxyphenyl)-2,4,6-heptatrienoic acid δ-lactone; 4-methoxy-6-[β-(p-anisyl)vinyl]-α-pyrone; 6-(p-methoxystyryl)-4-methoxy-α-pyrone. $C_{15}H_{14}O_4$; mol wt 258.26. C 69.76%, H 5.46%, O 24.78%. From root of *Piper methysticum* Forst., *Piperaceae* (kava): Winzhermer, *Arch. Pharm.* **246**, 338 (1908); Borsche, Gerhardt, *Ber.* **47**, 2902 (1914). From *Ranunculus quelpaertensis* Nakai, *Ranunculaceae:* Shibata *et al., Bull. Chem. Soc. Japan* **45**, 930 (1972). Structure: Chmielewska *et al., Tetrahedron* **4**, 36 (1958); Bu'Lock, Smith, *J. Chem. Soc.* **1960**, 502. Synthesis: Harris, Combs, *J. Org. Chem.* **33**, 2399 (1968); R. Bacardit, *J. Heterocycl. Chem.* **19**, 157 (1982). Molecular and crystal structure: Engel, Nowacki, *Z. Kristallogr., Kristallgeometrie, Kristallphys., Kristallchem.* **134**, 180 (1971). Activity studies: Kretzschmar *et al., Arch. Int. Pharmacodyn. Ther.* **180**, 471 (1969).

Crystals from methanol, mp 155-157°. uv max (ethanol): 360 nm (log ε 4.33). Practically insol in water; sol in hot alcohol, glacial acetic acid, ethyl acetate, acetone; slightly sol in benzene, ether.

Dihydroyangonin, $C_{15}H_{16}O_4$, needles from alc, mp 106-107°. uv max (methanol): 274, 228 nm (log ε 3.82, 3.71).

10002. Yeast. The moist, living cells of a fungus or fungi whose usual and dominant growth form is unicellular. The term yeast is not one with an exact botanical meaning, *see* Henrici's *Molds, Yeasts and Actinomycetes* (New York, 1947) p 264 sqq. Normally produces alcoholic fermentation in fluids contg sugar. Moist yeast is usually combined with a starchy or absorbent base and comes on the market in the form of white or yellowish-white, soft, easily broken masses of a characteristic, slightly sour odor. Description of a modern process of manuf: Schultz, Swift, **U.S. pat. 2,717,837** (1955 to Standard Brands).

The dried yeast of pharmacopoeias consists of the dry cells of any suitable strain of *Saccharomyces cerevisiae* Meyen, *Saccharomytaceae* or *Candida utilis* (Hanneberg) Lodder and Kreger-Van Rij, *Cryptococcaceae* (torula yeast), usually obtained as a by-product from the brewing of beer made from an extract of cereal grains and hops. The yeast cells are washed free of beer and dried, and may be debittered. These yeasts are commonly known as "Brewer's Dried Yeast" and "Debittered Brewer's Dried Yeast." Dried yeast may be obtained also by growing suitable strains of yeast, using media other than those required for the production of beer, and under appropriate environmental conditions. The yeast thus obtained is commonly known as "Primary Dried Yeast."

Dried yeast that fulfills pharmacopoeal requirements con-

tains not less than 40% protein, and, in each gram, the equivalent of not less than 0.12 mg of thiamine hydrochloride, 0.04 mg of riboflavin, and 0.25 mg of nicotinic acid. It occurs as yellowish-white to weak yellowish-orange flakes, granules or powder, having an odor indicative of the type. It is inactive in fermenting power.

USE: Moist yeast in baking bread, brewing; producing alcohol by fermentation of sugar, molasses, and cereals; as a source of vitamins. Dried yeast as a source of vitamins; in baking.

THERAP CAT: Source of protein and vitamin B complex.

THERAP CAT (VET): Dietary source of B vitamins.

10003. Yellow AB. *1-Phenylazo-2-naphthalenamine; C.I. Solvent Yellow 5;* FD & C Yellow no. 3; Ext. D & C Yellow no. 9; C.I. 11380. $C_{16}H_{13}N_3$; mol wt 247.29. C 77.71%, H 5.30%, N 16.99%. Prepd from diazotized aniline and β-naphthylamine: Lawson, *Ber.* **18**, 796 (1885); *Colour Index* **vol. 4** (3rd ed., 1971) p 4021. Toxicity studies: Allmark *et al., J. Pharm. Pharmacol.* **7**, 591 (1955); Hansen *et al., Toxicol. Appl. Pharmacol.* **5**, 16 (1963). Formerly used in the U.S. for coloring oleomargarine.

Red platelets from abs alc, mp 102-104°. Practically insol in water. Sol in alcohol, carbon tetrachloride, acetic acid, vegetable oils. An alcoholic soln becomes redder with HCl addition, and is unaltered by NaOH addition.

Caution: Delisted for internal use in 1959: *Fed. Reg.* **24**, 883 (1959), *C.A.* **53**, 12505a. Delisted also for external use in 1960. The FDA has declared this color unsafe due to the impurity β-naphthylamine, a proven carcinogen.

10004. Yellow OB. *1-[(2-Methylphenyl)azo]-2-naphthalenamine; C.I. Solvent Yellow 6;* 1-o-tolylazo-2-naphthylamine; FD & C Yellow no. 4; Ext. D & C Yellow no. 10; C.I. 11390. $C_{17}H_{15}N_3$; mol wt 261.31. C 78.13%, H 5.79%, N 16.08%. Prepd from diazotized o-toluidine and β-naphthylamine: Krüss, *Z. Physik. Chem.* **51**, 270 (1905); Norman, *J. Chem. Soc.* **101**, 1913 (1912); Fischer, *J. Prakt. Chem.* **104**, 102 (1922); Hodgson, Foster, *J. Chem. Soc.* **1942**, 30. Toxicity studies: Allmark *et al., J. Pharm. Pharmacol.* **7**, 591 (1955); Hansen *et al., Toxicol. Appl. Pharmacol.* **5**, 16 (1963). Formerly used in the U.S. for coloring oleomargarine.

Deep red crystals from alc. Also reported as bright yellow needles (Krüss, *loc. cit.*). mp 125-126° (Fischer, *loc. cit.*). Practically insol in water; sol in alc, ether, benzene, carbon tetrachloride, vegetable oils, glacial acetic acid. An alcoholic soln becomes redder with HCl addition, but is unaltered by NaOH addition.

Caution: Delisted for internal use in 1959: *Fed. Reg.* **24**, 883 (1959), *C.A.* **53**, 12505a (1959). Delisted also for external use in 1960. The FDA has declared this color unsafe due to the impurity β-naphthylamine, a proven carcinogen.

10005. Yellow Phenolphthalein. In the mfg process of phenolphthalein, *q.v.*, a stage is reached where certain byproducts formed in the synthesis have not yet been removed: Ebert, *et al.*, **U.S. pat. 1,711,048,** *C.A.* **23**, 2990 (1929); Hubacher, **U.S. at. 1,940,495,** *C.A.* **28**, 1366 (1934); **U.S. pat. 2,192,485,** *C.A.* **34**, 4396 (1940). These are yellow to brownish-yellow bodies mixed with the white phenolphthalein and imparting a characteristic yellow color to the product. Compds isolated from a sample prepd according to U.S. pat. **2,168,346** are: white phenolphthalein 93.16%, fluoran 0.32%, isophenolphthalein 0.08%, white compd mp 250° 0.04%, 2-(4-hydroxybenzoyl)benzoic acid 0.10%. Total compds isolated: 93.70%. Review of chemistry and pharmacology: Hubacher, Doernberg, *J. Am. Pharm. Assoc., Sci. Ed.* **37**, 261 (1948).

The commercially available yellow phenolphthalein is a powder of moderate yellow color. d_4^{20} 1.290-1.296. mp 255-260°. One gram dissolves in 12 ml alcohol, in 102 ml ether. Solns show a slight greenish fluorescence. Found to be 2.5 times more active as a laxative in rhesus monkeys than phenolphthalein.

THERAP CAT: Cathartic.

10006. Yig. *Iron yttrium oxide;* yttrium iron garnet. $Fe_5O_{12}Y_3$; mol wt 738.01. Fe 37.84%, O 26.01%, Y 36.15%. $Y_3Fe_5O_{12}$. Man-made mineral. Prepn by molten salt technique: Nielsen, *Electronics*, Jan. 31, **1964**, pp 44-45. Advances in manufacture and applications: Gundlach, *ibid.*, Oct. 14, **1968**, pp 104-118.

Hard, brittle, garnet-type crystals. Hardness about equal to that of quartz. Melts above 1040°. The melt is easily amenable to doping and crystal nucleation by epitaxy. Crystals may be oriented along the simple axes (100), (110), and (111) using the natural (110) face of the crystal.

USE: In microwave tunable filters and limiters.

10007. Ylangene. *1,3-Dimethyl-8-(1-methylethyl)tricyclo[4.4.0.0^{2,7}]dec-3-ene; (1S,2R,6R,7R,8S)-(+)-8-isopropyl-1,3-dimethyltricyclo[4.4.0.0^{2,7}]dec-3-ene;* α-ylangene. $C_{15}H_{24}$; mol wt 204.34. C 88.16%, H 11.84%. From ylang-ylang oil: Heraut, Dimitrov, *Chem. Listy* **46**, 432 (1952); from oil of birch buds: Holub *et al.*, *Coll. Czech. Chem. Commun.* **24**, 3730 (1959); from oil of *Juniperus oxycedrus* Linn., *Cupressaceae:* Motl *et al.*, *ibid.* **25**, 1656 (1960). Structure: Motl *et al.*, *Chem. & Ind. (London)* **1963**, 1759; Hunter, Brogden, *J. Org. Chem.* **29**, 982 (1964); Motl *et al.*, *Tetrahedron Letters* **1965**, 451. Stereoisomer of copaene, *q.v.* Total synthesis of (±)-form: Heathcock *et al.*, *J. Am. Chem. Soc.* **89**, 4133 (1967); Corey, Watt, *ibid.* **95**, 2303 (1973).

Oil. d_4^{20} 0.9091. n_D^{20} 1.4934. $[\alpha]_D^{20}$ +15.4°.

10008. Ylang-Ylang Oil. Cananga oil. Volatile oil distilled in Madagascar, Reunion Island, Comoro Islands and the Philippine Islands from freshly picked flowers of *Cananga odorata* Hook & Thoms., *Anonaceae.* The first distillate yields the so-called "ylang-ylang extra" and is the finest oil. The first, second and third fractions follow, the first two having insignificant commercial value. *Constit.* Geraniol and linalool esters of acetic and benzoic acids; *p*-cresol methyl ether, cadinene, a sesquiterpene, a phenol.

Light yellow, very fragrant liquid. d_{20}^{20} 0.930-0.950. Rotation: −27° to −50° in 100-mm tube.

USE: In delicate perfumes.

10009. Ylides. Compounds in which an atom from group V or VI of the periodic table, bearing a positive charge, is connected to a carbon atom carrying an unshared pair of electrons. *Ref:* A. William Johnson, *Ylid Chemistry* (Academic Press, New York, 1966); B. M. Trost, L. S. Melvin, *Sulfur Ylides* (Academic Press, New York, 1975); J. March, *Advanced Organic Chemistry* (McGraw-Hill, New York, 2nd ed., 1977) p 40.

10010. Yogurt. Yoghurt; joghurt; kisselo-mleko; mazun; leben raib; dahi. Fermented whole milk. Produced by evaporating milk to about 50% of its volume and maintaining at 50°C for 12 hrs after adding "maya" a mixture of *Lactobacillus bulgaricus,* L. *acidophilus,* and *Streptococcus lactis.* Used as food. Contains antibacterial principles. Differentiation of yogurt and *clabber* as produced in Kentucky: *J. Am. Med. Assoc.* **209**, 778 (1969).

10011. Yohimbine. *(16α,17α)-17-Hydroxyyohimban-16-carboxylic acid methyl ester;* quebrachine; corynine; aphrodine. $C_{21}H_{26}N_2O_3$; mol wt 354.43. C 71.16%, H 7.39%, N 7.90%, O 13.54%. Indole alkaloid with α_2-adrenergic blocking activity. Found in *Corynanthe johimbe* K. Schum., *Rubiaceae* and related trees, also in *Rauwolfia serpentina* (L.) Benth., *Apocynaceae:* Raymond-Hamet, *J. Pharm. Chim.* **19**, 209 (1934); Hofmann, *Helv. Chim. Acta* **37**, 849 (1954); Stoll, Jucker, *Ullmanns Encyklopädie der Technischen Chemie,* vol. 3 (Munich, 3rd ed., 1953) p 266; Bader *et al.*, *J. Am. Chem. Soc.* **76**, 1695 (1954). Structure: Witkop, *Ann.* **554**, 83 (1943); Clemo, Swan, *J. Chem. Soc.* **1946**, 617. Stereochemistry: Janot *et al.*, *Bull. Soc. Chim. France* **1952**, 1085; Godfredsen, Vandegal, *Acta Chim. Scand.* **10**, 1414 (1956); Van Tamelen *et al.*, *J. Am. Chem. Soc.* **78**, 4628 (1956); Ban, Yonemitsu, *Tetrahedron* **20**, 2877 (1964). Synthesis: Van Tamelen *et al.*, *J. Am. Chem. Soc.* **80**, 5006 (1958); Liljegren, Potts, *J. Org. Chem.* **27**, 377 (1962). Total synthesis: Van Tamelen *et al.*, *J. Am. Chem. Soc.* **91**, 7315 (1969); Stork, Guthikonda, *ibid.* **94**, 5109 (1972); T. Kametani *et al.*, *Chem. Pharm. Bull.* **24**, 2500 (1976); E. Wenkert *et al.*, *J. Am. Chem. Soc.* **100**, 4894 (1978); **101**, 5370 (1979); **104**, 2244 (1982); I. Ninomiya *et al.*, *Heterocycles* **14**, 631 (1980). Pharmacokinetics in humans: J. A. Owen *et al.*, *Eur. J. Clin. Pharmacol.* **32**, 577 (1987). Clinical studies in impotence: K. Reid *et al.*, *Lancet* **2**, 421 (1987); A. Morales *et al.*, *J. Urol.* **137**, 1168 (1987). Review of pharmacology and use in molecular studies of α_2-adrenoreceptor: M. R. Goldberg, D. Robertson, *Pharmacol. Rev.* **35**, 143-180 (1987). Comprehensive description: A. G. Mekkawi, A. A. Al-Badr in *Analytical Profiles of Drug Substances* vol. 16, K. Florey, Ed. (Academic Press, New York, 1986) pp 731-768.

Orthorhombic needles from dil alc, mp 234°. Also mp 235-237°. $[\alpha]_D^{20}$ +50.9° to +62.2° (ethanol); $[\alpha]_D^{20}$ +108° (pyridine); $[\alpha]_{546}^{20}$ +129° (c = 0.5 in pyridine). uv max (methanol): 226, 280, 291 nm (log ϵ 4.56, 3.88, 3.80). Sparingly sol in water. Sol in alcohol, chloroform, hot benzene; moderately sol in ether.

Hydrochloride, $C_{21}H_{27}ClN_2O_3$, *Aphrodyne, Yocon, Yohimex, Yohydrol.* Orthorhombic plates, prisms from alc; dec 302°. $[\alpha]_D^{22}$ +105° (H_2O). Sol in about 120 ml water, 400 ml alc. The aq soln is about neutral.

USE: Pharmacological probe for the study of α_2-adrenoceptor.

THERAP CAT: α-Adrenergic blocker; mydriatic.

THERAP CAT (VET): Adrenergic blocking agent. Has been used as an aphrodisiac.

10012. allo-Yohimbine. *17-Hydroxyyohimban-16-carboxylic acid methyl ester;* alloyohimbine; dihydroyohimbine.

$C_{21}H_{26}N_2O_3$; mol wt 354.43. C 71.16%, H 7.39%, N 7.90%, O 13.54%. From bark of *Corynanthe johimbe* K. Schum., *Rubiaceae:* Warnat, *Ber.* **59**, 2388 (1926); Hahn, Brandenberg, *ibid.* **60**, 699 (1927). Identity with dihydroyohimbine: Heinemann, *ibid.* **67**, 15 (1934). Structure and stereochemistry: LeHir *et al., Bull. Soc. Chim. France* **1953**, 1027; Janot *et al., ibid.* **1961**, 637. Revised structure and total synthesis: Töke *et al., J. Org. Chem.* **38**, 2496 (1973). Total synthesis: E. Wenkert *et al., J. Am. Chem. Soc.* **101**, 5370 (1979). *Review:* A. Chatterjee *et al.,* "Rauwolfia Alkaloids" in Zechmeister, *Progress in the Chemistry of Organic Natural Products* **vol. XIII** (Springer Verlag, Vienna, 1956) pp 354-6.

Trihydrate, needles from 50% alcohol, mp 98-99°. After drying at 100° and 0.01 mm Hg for 24 hrs, the anhydrous form melts at 135-140°. $[\alpha]_D^{19}$ +84° (c = 0.40 in pyridine). uv max (methanol): 225, 280, 290 nm (log ϵ 4.52, 3.91, 3.74). Sol in ethanol, methanol, pyridine, dil acids; slightly sol in benzene; practically insol in water.

10013. α-**Yohimbine.** *17α-Hydroxy-20α-yohimban-16β-carboxylic acid methyl ester;* corynanthidine; isoyohimbine; mesoyohimbine; rauwolscine. $C_{21}H_{26}N_2O_3$; mol wt 354.43. C 71.16%, H 7.39%, N 7.90%, O 13.54%. From bark of *Corynanthe johimbe* K. Schum., *Rubiaceae:* Hahn, Brandenburg, *Ber.* **60**, 669 (1927); Wilbaut, van Gastel, *Rec. Trav. Chim.* **54**, 88 (1935); from *Rauwolfia canescens* L., *Apocynaceae:* Mookerjee, *J. Indian Chem. Soc.* **18**, 33 (1941), *C.A.* **35**, 7967 (1941); Stoll *et al., Helv. Chim. Acta* **38**, 270 (1955). Identity with corynanthidine: Janot, Goutarel, *Bull. Soc. Chim. France* **1946**, 535. Identity with isoyohimbine: Heinemann, *Ber.* **60**, 15 (1934). Identity with rauwolscine: Chatterjee *et al., Chem. & Ind. (London)* **1954**, 491. Structure and stereochemistry: Le Hir *et al., Bull. Soc. Chim. France* **1953**, 1027; Wenkert, Liu, *Experientia* **11**, 302 (1955); Aldrich *et al., J. Am. Chem. Soc.* **81**, 2481 (1959); Janot *et al., Bull. Soc. Chim France* **1961**, 637. Total synthesis: Töke *et al., J. Org. Chem.* **38**, 2496 (1973).

Crystals, mp 243-244°. $[\alpha]_D^{19}$ −18° (pyridine); $[\alpha]_D^{19}$ −27° (abs alcohol). pKa 6.34. uv max (methanol): 227, 281 nm (log ϵ 4.50, 3.93). Sol in warm methanol, ethanol; slightly sol in ether, benzene; practically insol in petr ether, water. Hydrochloride, $C_{21}H_{26}N_2O_3 \cdot HCl$, crystals, mp 288°. $[\alpha]_D^{20}$ +55.5° (water).

10014. Ytterbium. Yb; at. wt 173.04; at. no. 70; valences 2, 3. A rare earth metal of the yttrium group. Seven naturally occurring isotopes: 168 (0.140%); 170 (3.03%); 171 (14.31%); 172 (21.82%); 173 (16.13%); 174 (31.84%); 176 (12.73%); artificial radioactive isotopes: 154; 155; 162; 164-167; 169; 175; 177. Estimated abundance in earth's crust: 2.7-8 ppm. Occurs in xenotime, ytterbite (gadolinite), monazite. Discovered in 1907 by Urbain and independently in 1908 by von Welsbach who called it *aldebaranium:* Urbain, *Compt. Rend.* **145**, 759 (1907); von Welsbach, *Monatsh.* **29**, 181 (1908); **34**, 1713 (1913). Sepn and puri-

fication: Urbain, *Congress of Applied Chemistry* [X] **94** (1909); Prandtl, *Z. Anorg. Chem.* **238**, 321 (1938); Spedding *et al., J. Am. Chem. Soc.* **74**, 2783 (1952); **76**, 2557 (1954). Spectrum: Exner, Haschek *et al.,* cited by Mellor, *A Comprehensive Treatise on Inorganic and Theoretical Chemistry* **5**, 706 (1929). Reviews of prepn, properties, and compds of ytterbium and other lanthanides: *The Rare Earths,* F. H. Spedding, A. H. Daane, Eds. (Krieger, Huntington, N.Y., 1971, reprint of 1961 ed.) 641 pp; Hulet, Bode, "Separation Chemistry of the Lanthanides and Transplutonium Actinides", in *MTP Int. Rev. Sci.: Inorg. Chem.,* Ser. One, vol. 7, K. W. Bagnall, Ed. (University Park Press, Baltimore, 1972) pp 1-45; Moeller, "The Lanthanides", in *Comprehensive Inorganic Chemistry,* **vol. 4**, J. C. Bailar Jr. *et al.,* Eds. (Pergamon Press, Oxford, 1973) pp 1-101.

Silvery, ductile metal. Face-centered cubic structure at room temp, d 6.977; body-centered cubic above 798°, d 6.54. mp 824°. Forms both di- and trivalent salts.

Oxide, O_3Yb_2, *ytterbia.* Colorless mass, sol in dil acids.

Chloride, $YbCl_3$, hexahydrate, deliquescent crystals, d 2.575; mp 150-155°. LD_{50} in mice: 395 mg/kg i.p.; 6.7 g/kg orally, Haley, *J. Pharm. Sci.* **54**, 663 (1965).

Nitrate, $Yb(NO_3)_3$, tetrahydrate, transparent hygroscopic prisms from concd nitric acid. LD_{50} (hexahydrate) in rats: 255 mg/kg i.p.; 3.1 g/kg orally, Haley, *loc. cit.*

Sulfate, $Yb_2(SO_4)_3$, octahydrate, lustrous colorless crystals, soly decreases with rise in temp.

10015. Yttrium. Y; at. wt 88.9059; at. no. 39; valence 3. A rare earth metal. Naturally occurring isotope: ^{89}Y; artificial isotopes (mass numbers): 82-88; 90-96. Estimated abundance in earth's crust: 28-70 ppm. Occurs in nature as a phosphate, *xenotime,* as mixed oxides, *fergusonite* and *samarskite,* as silicates in *yttrialite* and *gadolinite;* also found in other rare-earth minerals. Discovered in 1794 by Gadolin; separated as yttria in 1843 by Mosander; named by Ekeberg. Sepn by fractional precipitation: Bonardi, James, *J. Am. Chem. Soc.* **37**, 2642 (1915); Willand, James, *ibid.* **38**, 1198 (1916); Wichers *et al., ibid.* **40**, 1615 (1918). Sepn of yttrium and other rare-earths by ion exchange: Spedding *et al., ibid.* **69**, 2812 (1947); Mayer, Freiling, *ibid.* **75**, 5647 (1953). Reviews of prepn, properties and compds of yttrium and other rare earths: *The Rare Earths,* F. H. Spedding, A. H. Daane, Eds. (Krieger, Huntington, N.Y., 1971, reprint of 1961 ed.) 641 pp; Vickery, "Scandium, Yttrium and Lanthanum", in *Comprehensive Inorganic Chemistry,* **vol. 3**, J. C. Bailar Jr. *et al.,* Eds. (Pergamon Press, Oxford, 1973) pp 329-353; Moeller, "The Lanthanides", *ibid.* **vol. 4**, pp 1-101.

Iron-gray, lustrous powder; darkens on exposure to light. Forms crystals with hexagonal, close-packed structure. d 4.472; mp 1509°. bp about 3000°. $E°(aq) Y^{3+}/Y$ −2.37 V (calc). Oxidizes on heating in air or oxygen; dec cold water slowly, boiling water rapidly.

Oxide, O_3Y_2, *yttria.* White powder, body-centered cubic structure, d 5.03. Obtained by igniting yttrium or its salts. Sol in dil acids; readily absorbs ammonia from the air; displaces ammonia from ammonium salts. LD_{50} in rats: 500 mg/kg i.p., Cochran *et al., Arch. Ind. Hyg. Occup. Med.* **1**, 637 (1950).

Hydroxide, $Y(OH)_3$, white gelatinous ppt, dries to a white powder which absorbs CO_2 from the air, obtained by the action of ammonium or alkali hydroxides on a soln of an yttrium salt.

Chloride, YCl_3, hexahydrate, colorless, deliquesc crystals. Sol in water, alc. Anhydr chloride obtained by heating in a stream of HCl.

Carbonate, $Y_2(CO_3)_3$, trihydrate, white to reddish-white powder, prepd by hydrolysis of yttrium trichloroacetate. Insol in water, sol in dil mineral acids.

Sulfate, $Y_2(SO_4)_3$, octahydrate, monoclinic crystals. Sol in concd sulfuric acid with formation of $Y(HSO_4)_3$. Soly in water decreases with increase in temp. Forms double salts with alkali sulfates.

Nitrate, $Y(NO_3)_3$, hexahydrate, deliquesc crystals, sol in water, produces basic nitrates on partial decompn. LD_{50} in rats: 350 mg/kg i.p., Cochran *et al., loc. cit.*

USE: Yttrium doped with rare earths as phosphors for color television receivers. Oxide for mantles in gas and acetylene lights. Chloride in prepn of pure metal. Yttrium aluminum garnets (YAGS) in lasers.

Z

10016. Zea. Corn silk. Fresh styles and stigmas of *Zea mays* L., *Gramineae*. *Habit.* Found on all continents except Antarctica. *Constit.* Maizenic acid, fixed oil, resin, mucilage.

10017. Zearalenone. *3,4,5,6,9,10-Hexahydro-14,16-dihydroxy-3-methyl-1H-2-benzoxacyclotetradecin-1,7(8H)-dione;* 6-(10-hydroxy-6-oxo-*trans*-1-undecenyl)-β-resorcylic acid lactone; Compound F-2; FES. $C_{18}H_{22}O_5$; mol wt 318.36. C 67.91%, H 6.97%, O 25.13%. Isoln of *l*-form from the mycelia of the fungus *Gibberella zeae (Fusarium graminearum):* Stob *et al., Nature* **196**, 1318 (1962); Andrews, Stob, U.S. pat. **3,196,019** (1965 to Purdue Res. Found.); Hodge *et al.,* U.S. pats. **3,239,341; 3,239,342** (both 1966 to Commercial Solvents). Structure: Urry *et al., Tetrahedron Letters* **1966**, 3109. Synthesis of *dl*-form: Taub *et al., Chem. Commun.* **1967**, 225; **Neth.** pat. **6,812,148** (1968 to Merck & Co.); Vlattas *et al., J. Org. Chem.* **33**, 4176 (1968); T. Takahashi *et al., J. Am. Chem. Soc.* **101**, 5072 (1979); *eidem, Tetrahedron Letters* **22**, 1363 (1981). Total synthesis and abs config: Taub *et al., Tetrahedron* **24**, 2443 (1968); Girotra, Wendler, U.S. pat. **3,551,455** (1970 to Merck & Co.).

Crystals, mp 164-165°. $[\alpha]_{546}^{25}$ −170.5° (c = 1.0 in CH_3-OH). uv max (CH_3OH): 236, 274, 316 nm (ε 29,700; 13,900; 6020). Sol in aq alkali, ether, benzene, alcohols. Practically insol in water.

Diacetate, $C_{22}H_{26}O_7$, crystals, mp 123-125°.

dl-Form, crystals, mp 187-189°.

Note: Zearalenone is one of a group of compounds known under the more general name of *RALs* or *resorcylic acid lactones.*

THERAP CAT (VET): Anabolic.

10018. Zeatin. (E)-2-Methyl-4-(1H-purin-6-ylamino)-2-buten-1-ol; *trans*-zeatin. $C_{10}H_{13}N_5O$; mol wt 219.24. C 54.78%, H 5.98%, N 31.94%, O 7.30%. Naturally occurring plant growth hormone; cytokinin originally isolated from sweet corn kernels, *Zea mays* L. *Gramineae.* Isoln and structure determn: D. S. Letham *et al., Proc. Chem. Soc.* **1964**, 230. Synthesis: G. Shaw, D. V. Wilson, *ibid.* 231; G. Shaw *et al., J. Chem. Soc. (C)* **1966**, 921; J. Corse, J. Kuhnle, *Synthesis* **1972**, 618; G. M. Gray, *ibid.* **1983**, 488; *idem,* Eur. pat. **Appl. 86,454** (1983 to J. T. Baker). Inhibition of mitochondrial function: C. O. Miller, *Plant Physiol.* **69**, 1274 (1982); translocation in soybean explants: L. Noodén, D. S. Letham, *J. Plant Growth Regul.* **2**, 265 (1984). *Reviews:* D. S. Letham, *Ann. Rev. Plant Physiol.* **18**, 349-363 (1967); D. S. Letham, L. M. S. Palni, *ibid.* **34**, 163-197 (1983).

Crystals from water, mp 207-208°. uv max in 0.1*M* HCl: 207, 275 nm (ε 14500, 14650); at pH 7.2: 212, 270 nm (ε 17050, 16150); in 0.1*M* NaOH: 220, 276 nm (ε 15900, 14650).

10019. Zeaxanthin. β,β-*Carotene-3,3'-diol; all trans-β-carotene-3,3'-diol;* (3R,3'R)-dihydroxy-β-carotene; zeaxanthol; anchovyxanthin. $C_{40}H_{56}O_2$; mol wt 568.85. C 84.45%, H 9.92%, O 5.63%. One of the most widespread carotenoid alcohols in nature. Occurs together and is isomeric with

xanthophyll, *q.v.* It is the pigment of yellow corn *Zea mays* L., *Gramineae.* Isoln from yellow corn grits: Karrer *et al., Helv. Chim. Acta* **13**, 268 (1930). Isoln by chromatography: Kuhn, Grundmann, *Ber.* **67**, 596 (1934). Isoln from algae, bacteria: A. J. Aasen *et al., Acta Chem. Scand.* **26**, 404 (1972). Isoln of dipalmitate from *Physalis* petal: Kuhn *et al., Ber.* **63**, 1489 (1930). Does not possess vitamin A potency: Schumacher *et al., Poultry Sci.* **23**, 529 (1944). Stereochemistry: Zechmeister, *Chem. Rev.* **34**, 267 (1944). Abs config of natural zeaxanthin: T. E. de Ville *et al., Chem. Commun.* **1969**, 1311; J. R. Hlubucek *et al., J. Chem. Soc. Perkin Trans. I* **1974**, 848. Total synthesis: Isler *et al., Helv. Chim. Acta* **39**, 2041 (1956); **40**, 456 (1957); Loeber *et al., J. Chem. Soc. (C)* **1971**, 404; A. Rüttimann, H. Mayer, *Helv. Chim. Acta* **63**, 1456 (1980); P. R. Ellis *et al., ibid.* **64**, 1092 (1980); E. Widmer *et al., ibid.* **65**, 944, 958 (1982). Sepn of configurational isomers by HPLC: A. Rüttimann, *J. High Resolut. Chromatog. Chromatog. Commun.* **6**, 612 (1983). First isoln of enantiomeric and *meso*-zeaxanthin in nature: T. Maoka *et al., Comp. Biochem. Physiol.* **83B**, 121 (1986). Mfg procedure: Isler *et al.,* U.S. pat. **2,917,539** (1959 to Hoffmann-La Roche). *Reviews:* Zechmeister, *Carotinoide* (Berlin, 1934); Mayer, *The Chemistry of Natural Coloring Matters* (New York, 1943); Karrer, Jucker, *Carotenoids* (New York, 1950).

Yellow rhombic plates with steel-blue metallic luster from ethanol, mp 207° (Zechmeister); mp 215.5° (Kuhn). Optically inactive. Absorption bands (ethanol) 483, 451.5 nm. Practically insol in water. Slightly sol in petr ether, methanol; sol in carbon disulfide, benzene, chloroform, carbon tetrachloride, pyridine, ethyl acetate; sol in glacial acetic acid upon addition of hexane. Less sol in boiling methanol (1:1550) than xanthophyll.

Dipalmitate, $C_{72}H_{116}O_4$, *physalien, physalin.* Fine yellow or red needles from benzene, mp 97°.

10020. Zein. A prolamine; an alcohol-soluble protein present in amounts of 2.5-10% (dry basis) in corn *(Zea mays* L., *Gramineae).* The greater part of zein has a mol wt of 38,000. Does not contain lysine or tryptophan. Extracted commercially from gluten meal with dil isopropanol: Swallen, *Ind. Eng. Chem.* **33**, 394 (1941). Improved method: Carter, Reck, Ger. pat. **2,002,337** (1971 to Nutrilite Prod.), *C.A.* **75**, 117341b (1971). *Review:* Mossé, *Ann. Physiol. Végétale* **3**, 105 (1961).

White to slightly yellow powder. d 1.226. When completely dry may be heated to 200° without visible signs of decompn. Soluble in aq alcohols, the glycols, the ethyl ether of ethylene glycol, furfuryl alcohol, tetrahydrofurfuryl alcohol, aq alkaline solns of pH 11.5 (or greater). Tends to become denatured when in soln and becomes insol. Insol in water, acetone; but readily sol in acetone-water mixtures between the limits of 60-80% acetone by volume. Insol in anhydr alcohols except methanol.

USE: Manuf plastics, paper coatings, adhesives, substitutes for shellac, laminated board, in solid color printing, films, edible coatings for foodstuffs.

10021. Zeolites. Crystalline, hydrated alkali-aluminum silicates of the general formula $M_{2/n}O.Al_2O_3.ySiO_2.wH_2O$ where M represents a group IA or IIA element, *n* is the cation valence, *y* is 2 or greater and *w* is the number of water molecules contained in the channels or interconnected voids within the zeolite. The cations are mobile and capable of undergoing ion exchange. Zeolites occur naturally in sedimentary and volcanic rocks, altered basalts, ores, clay deposits. Some 40 known zeolite minerals and a great number of synthetic zeolites are available commercially. Ref: Milton, U.S. pat. **2,882,243** (1959 to Union Carbide). *Reviews:* D. W. Breck, R. A. Anderson, "Molecular Sieves" in Kirk-Othmer *Encyclopedia of Chemical Technology* vol. 15 (Wiley-Interscience, New York, 3rd ed., 1981) pp 638-669; P. L. Layman, *Chem. & Eng. News* **60**, 10 (Sept. 27, 1982); J. Haggin, *ibid.* **60**, 9 (Dec. 13, 1982). *Books: Molecular Sieve*

Zeolites - I, II, Advances in Chemistry Series **101, 102** (A.C.S., Washington, 1971) 526 pp, 459 pp; *Molecular Sieves,* Advances in Chemistry Series **121** (A.C.S., Washington, 1973) 634 pp; D. W. Breck, *Zeolite Molecular Sieves* (John Wiley, New York, 1974) 771 pp; *Natural Zeolites: Occurrence, Properties, Use,* L. B. Sand, F. A. Mumpton, Eds. (Pergamon, Oxford, 1978). Studies on fibrous zeolites as possible environmental hazards: A. N. Rohl *et al., Science* **216,** 518 (1982); Y. I. Baris, *Arch. Environ. Health* **37,** 177 (1982).

USE: As molecular sieves, filters, adsorbents, catalysts, drying agents, cation exchangers, dispersing agents, detergent builders.

10022. Zeranol. *3,4,5,6,7,8,9,10,11,12-Decahydro-7,14,16-trihydroxy-3-methyl-1H-2-benzoxacyclotetradecin-1-one;* 6-(6,10-dihydroxyundecyl)-β-resorcylic acid μ-lactone; zearalanol; MK-188; P-1496; Ralgro; Ralabol; Ralone; Zerano. $C_{18}H_{26}O_5$; mol wt 322.41. C 67.05%, H 8.13%, O 24.81%. Prepn from zearalenone, *q.v.*: Hodge *et al.,* U.S. pat. **3,239,345** (1956 to Commercial Solvents). Metabolism in steers: Sharp, Dyer, *J. Anim. Sci.* **34,** 176 (1972).

Isopropyl alcohol-water mixtures yield two crystalline diastereoisomers. The more soluble has mp 146-148°; $[\alpha]_D^{25}$ ~+39° (methanol). The less soluble has mp 178-180°; $[\alpha]_D^{25}$ ~+46° (methanol) with uv max (methanol): 218, 265 and 304 nm.

THERAP CAT (VET): Anabolic.

10023. Zidovudine. *3'-Azido-3'-deoxythymidine;* azidothymidine; AZT; BW A509U; Retrovir. $C_{10}H_{13}N_5O_4$; mol wt 267.24. C 44.94%, H 4.90%, N 26.21%, O 23.95%. Pyrimidine nucleoside analog; reverse transcriptase inhibitor. Prepn: J. P. Horwitz *et al., J. Org. Chem.* **29,** 2076 (1964); R. P. Glinski *et al., ibid.* **38,** 4299 (1973). Total synthesis: C. K. Chu *et al., Tetrahedron Letters* **29,** 5349 (1988). *In vitro* antiviral, antimetabolite and antineoplastic activity: E. De Clercq *et al., Biochem. Pharmacol.* **29,** 1849 (1980). *In vitro* activity against HIV-1 (HTLV-III/LAV) virus: H. Mitsuya *et al., Proc. Nat. Acad. Sci. USA* **82,** 7096 (1985). Preliminary clinical trial in AIDS: R. Yarchoan *et al., Lancet* **1,** 575 (1986). Clinical pharmacokinetics: R. W. Klecker *et al., Clin. Pharmacol. Ther.* **41,** 407 (1987). Use in treatment of AIDS and AIDS-related complex: J. L. Rideout *et al.,* **Ger.** pat. **3,608,606** (1986 to Wellcome Found.); *eidem,* U.S. pat. **4,724,232** (1988 to Burroughs Wellcome). Symposium on clinical experience: *J. Infect.* **18,** Suppl. 1, 1-101 (1989).

Needles from petr ether, mp 106-112° (Horwitz). Also reported as crystals from water, mp 120-122° (Glinski). $[\alpha]_D^{25}$ +99° (c = 0.5 in water). uv max (water): 266.5 nm (ϵ 11650).

THERAP CAT: Antiviral.

10024. Zimeldine. *(Z)-3-(4-Bromophenyl)-N,N-dimethyl-3-(3-pyridinyl)-2-propen-1-amine; (Z)-3-(4'-bromophenyl)-3-(3''-pyridyl)dimethylallylamine;* zimelidine; *cis-*H 102/09. $C_{16}H_{17}BrN_2$; mol wt 317.24. C 60.58%, H 5.40%, Br 25.19%, N 8.83%. Antidepressant that inhibits membranal 5-hydroxytryptamine uptake. Prepn: P. B. Berntsson *et al.,* **S. Afr.** pat. **72 01,503;** *eidem,* U.S. pat. **3,928,369** (1972, 1975 both to AB Hassle). X-ray crystallographic study: S. Abrahamsson *et al., Acta Chem. Scand.* **A30,** 609 (1976). *In vivo* study in rats and humans: S. B. Ross *et al., Life Sci.* **19,** 205 (1976). Pharmacokinetics: D. Brown, *Eur. J. Clin. Pharmacol.* **17,** 111 (1980). Mode of action: K. Fuxe *et al., Neurosci. Letter* **13,** 307 (1979). Metabolism: J. Lundström *et al., Arzneimittel-Forsch.* **31,** 486 (1981). Clinical studies: A. Georgotas *et al., Am. J. Psychiatry* **139,** 1057 (1982); E. Syvalahti *et al., J. Int. Med. Res.* **10,** 250 (1982). Profile of antidepressant action: S. A. Montgomery *et al., Advan. Biochem. Psychopharmacol.* **32,** 35 (1982). Reversal of ethanol-induced memory impairment in human subjects: H. Weingartner *et al., Science* **221,** 472 (1983). Review of pharmacology and therapeutic efficacy: R. C. Heel *et al., Drugs* **24,** 169-206 (1982). Symposium on pharmacology, pharmacokinetics, metabolism, clinical studies: *Brit. J. Clin. Pract.* **1982,** Suppl. 19, 1-122.

Dihydrochloride monohydrate, $C_{16}H_{19}BrCl_2N_2 \cdot H_2O$, *Normud, Zelmid.* Crystals, mp 193°.

THERAP CAT: Antidepressant.

10025. Zinc. Zn; at. wt 65.38; at. no. 30; valence 2. Group 2b element. Abundance in earth's crust: 0.02% by wt. Natural isotopes: 64 (48.89%); 66 (27.81%); 68 (18.57%); 67 (4.11%); 70 (0.62%); eight radioactive isotopes and two isomers. Occurs in smithsonite or zinc spar, sphalerite or zinc blende, zincite, willemite, *franklinite,* [(Zn,Mn,-Fe)O.(Fe.Mn₂)O₃] or *gahnite* (ZnAl₂O₄). Has been known since very early times. Commercial forms: ingots; lumps; sheets; wire; shot; strips; sticks; granules; mossy; powder (dust). Prepn: Gowland, Bannister, *Metallurgy of Non-Ferrous Metals* (Griffin, London, 1930); *Zinc Production, Properties and Uses* (Zinc Development Association, London, 1968). *Reviews:* Zinc, C. H. Mathewson, Ed., A.C.S. Monograph Series no. **142** (Reinhold, New York, 1959) 721 pp; Schlechter, Thompson, "Zinc and Zinc Alloys" in Kirk-Othmer, *Encyclopedia of Chemical Technology,* **vol. 22** (Interscience, New York, 2nd ed., 1970) pp 555-603; Aylett, "Group IIB" in *Comprehensive Inorganic Chemistry,* **vol. 3,** J. C. Bailar, Jr. *et al.,* Eds. (Pergamon Press, Oxford, 1973) pp 187-328.

Bluish-white, lustrous metal; distorted hexagonal close-packed structure; stable in dry air; becomes covered with a white coating of basic carbonate on exposure to moist air. mp 419.5°. bp 908°. d^{25} 7.14. Heat capacity at constant pressure (25°): 6.07 cal/mole deg. Mohs' hardness 2.5. When heated to 100-150° becomes malleable, at 210° becomes brittle and pulverizable. Burns in air with a bluish-green flame. Loses electrons in aqueous systems to form Zn^{2+} E° (aq) Zn/Zn^{2+} 0.763 V. Slowly attacked by H_2SO_4 or HCl; oxidizing agents or metal ions, e.g. Cu^{2+}, Ni^{2+}, Co^{2+}, accelerate the process. Reacts slowly with ammonia water and acetic acid; rapidly with HNO_3. Reacts with alkali hydroxides to form "zincates", ZnO_2^{2-}, which are actually hydroxo complexes such as $Zn(OH)_3^-$; $Zn(OH)_4^-$, $[Zn(OH)_4(H_2O)_2]^{2-}$.

Human Toxicity: Inhalation of fumes may result in sweet taste, throat dryness, cough, weakness, generalized aching, chills, fever, nausea, vomiting. Zinc chloride fumes have caused injury to mucous membranes and skin irritation. Ingestion of sol salts may cause nausea, vomiting, purging.

See E. Browning, *Toxicity of Industrial Metals* (Appleton-Century-Crofts, New York, 2nd ed., 1969) pp 348-355.

USE: Galvanizing sheet iron; as ingredient of alloys such as bronze, brass, Babbitt metal, German silver, and special alloys for die-casting; as a protective coating for other metals to prevent corrosion; for electrical apparatus, especially dry cell batteries, household utensils, castings, printing plates, building materials, railroad car linings, automotive equipment; as reducing agent in organic chemistry; for deoxidizing bronze; extracting gold by the cyanide process, purifying fats for soaps; bleaching bone glue; manuf sodium hydrosulfite; insulin zinc salts; as reagent in analytical chemistry, e.g., in the Marsh and Gutzeit test for arsenic; as a reducer in the determination of iron. It is a nutritional trace element.

10026. Zinc Acetate. $C_4H_6O_4Zn$; mol wt 183.46. C 26.18%, H 3.30%, O 34.88%, Zn 35.64%. $Zn(C_2H_3O_2)_2$. Prepn of anhydr salt from zinc nitrate and acetic anhydride: Späth, *Monatsh.* **33**, 235 (1912). Clinical evaluations in Wilson's disease: G. M. Hill *et al., Hepatology* **7**, 522 (1987); G. J. Brewer *et al., J. Lab. Clin. Med.* **109**, 526 (1987). Toxicity: H. F. Smyth *et al., Am. Ind. Hyg. Assoc. J.* **30**, 470 (1969).

Dihydrate, crystallizes from dil acetic acid; faint, acetous odor; astringent taste; slightly efflorescent. d 1.735; mp 237°. One gram dissolves in 2.3 ml water, 1.6 ml boiling water, 30 ml alcohol, about 1 ml boiling alcohol. The aq soln is neutral or slightly acid to litmus; pH about 5-6. *Keep in well-closed containers. Incompat.* with zinc salts in general: Acacia, alkalies and their carbonates, oxalates, phosphates, sulfides, lime water, vegetable astringent decoctions and infusions. LD_{50} orally in rats: 2.46 g/kg (Smyth).

USE: Preserving wood; as mordant in dyeing; manuf glazes for painting on porcelain. As a reagent in testing for albumin, tannin, urobilin, phosphate, blood.

THERAP CAT: Styptic, astringent. Formerly used as an emetic.

THERAP CAT (VET): Antiseptic, astringent, protective (topical). Has been used as an emetic.

10027. Zinc Bacitracin. *Bacitracin zinc complex*; bacitracin zinc salt; Baciferm. Contains about 7% Zn. Prepd by the action of zinc salts on bacitracin broth: Anker *et al., J. Bacteriol.* **55**, 249 (1948); Hodge, Lafferty as cited in U.S. pat. 2,803,584 (1957 to CSC). The potency is usually between 50 and 60 units/mg of bacitracin activity: Gross, *Drug Cosmet. Ind.* **75**, 612 (1954); Gross *et al., J. Am. Pharm. Assoc., Sci. Ed.* **45**, 447 (1956).

Creamy powder contg 1-4% H_2O. Less bitter than bacitracin. Soly in water at 25° = 0.23-0.45% (w/w). Solubilities determined by Weiss *et al., Antibiot. & Chemother.* **7**, 374 (1957) in mg/ml at about 28°: water 5.1; methanol 6.55; ethanol 2.0; isopropanol 0.16; ethyl acetate 1.3; chloroform 0.01; petr ether 0.025. More stable than bacitracin at room and elevated temps. Zinc bacitracin can be used in formulations requiring heat processing.

USE: In ointments, tablets, implantation pellets, suppositories, and troches, either alone or in combination with other antibiotics or therapeutic agents: Hodge, Lafferty, *loc. cit.*

THERAP CAT: Antibacterial.

THERAP CAT (VET): Antimicrobial.

10028. Zinc Bromide. Br_2Zn; mol wt 225.21. Br 70.97%, Zn 29.03%. $ZnBr_2$. Usually contains at least 97% $ZnBr_2$, the remainder being chiefly water.

Very hygroscopic, granular powder; sharp, metallic taste. d 4.22; mp 394°; bp 697° with partial decompn. One gram dissolves in 0.25 ml water, 0.5 ml 90% alcohol; sol in ether, solns of alkali hydroxides. The aq soln is acid to litmus; pH about 4. *Keep tightly closed.*

USE: Making silver bromide collodion emulsions for photography; in the shielding of viewing windows for nuclear reactions.

10029. Zinc Caprylate. *Octanoic acid zinc salt.* $C_{16}H_{30}$-O_4Zn; mol wt 351.79. C 54.63%, H 8.60%, O 18.19%, Zn 18.58%. $Zn(C_8H_{15}O_2)_2$. Prepd from ammonium caprylate and zinc sulfate: van Renesse, *Ann.* **171**, 380 (1874).

Lustrous scales from alc, mp 136°. Sparingly sol in boiling water; moderately sol in boiling alcohol. *Keep well closed.* Dec in moist atm giving off caprylic acid.

USE: As fungicide like zinc propionate.

10030. Zinc Carbonate. CO_3Zn; mol wt 125.38. C 9.58%, O 38.28%, Zn 52.14%. $ZnCO_3$. Occurs in nature as the minerals *smithsonite, zincspar.* Prepn: Hüttig *et al., Monatsh.* **72**, 31 (1939).

Rhombohedral structure. Solubility in water at 15° 0.001 g/100 g; sol in dil acids, alkalies, solns of NH_4^+ salts. Basic carbonate, *zinc carbonate hydroxide, zinc subcarbonate.* Variable composition, usually characterized as $3Zn(OH)_2 \cdot 2ZnCO_3$. Occurs as the mineral *hydrozincite.* Reagent specification: 70% ZnO minimum.

USE: As pigment; manuf of porcelains, pottery, rubber.

THERAP CAT: Astringent, topical antiseptic.

THERAP CAT (VET): Astringent, antiseptic, protective (topical). Also used in rations to prevent Zn deficiency diseases.

10031. Zinc Chloride. Butter of zinc. Cl_2Zn; mol wt 136.29. Zn 47.97%, Cl 52.03%. $ZnCl_2$. Usually contains at least 95% $ZnCl_2$; remainder is chiefly water and oxychloride. Toxicity: Bruner, *Fed. Proc.* **9**, 260 (1950).

White, odorless, very deliquesc granules, or fused pieces or rods. d^{25} 2.907; mp about 290°; bp 732°. Solubility in H_2O: 432 g/100 g (25°); 614 g/100 g (100°). One gram dissolves in 0.25 ml of 2% HCl, in 1.3 ml alcohol, 2 ml glycerol; freely sol in acetone. With much water some zinc oxychloride is formed. The aq soln is acid to litmus; pH about 4. *Keep tightly closed.* LD i.v. in rats: 60-90 mg/kg (Bruner).

USE: Deodorant, disinfecting and embalming material; alone or with phenol and other antiseptics for preserving railway ties; fireproofing lumber; with ammonium chloride as flux for soldering; etching metals; manuf parchment paper, artificial silk, dyes, activated carbon, cold-water glues, vulcanized fiber; browning steel, galvanizing iron, copperplating iron; in magnesia cements; petroleum oil refining; cement for metals and for facing stone; mordant in printing and dyeing textiles; carbonizing woolen goods; producing crepe and crimping fabrics; mercerizing cotton; sizing and weighting fabrics; vulcanizing rubber; solvent for cellulose; preserving anatomical specimens; in microscopy for separating silk, wool, and plant fibers; as dehydrating agent in chemical syntheses. Dentin desensitizer. *Caution:* Moderately irritating to skin, mucous membranes.

THERAP CAT: Astringent.

THERAP CAT (VET): Antiseptic, astringent. Has been used in ulcers, fistulas, pododermatitis.

10032. Zinc Chromate(VI) Hydroxide. Zinc yellow; buttercup yellow; C.I. Pigment Yellow 36. A basic salt of somewhat variable composition. Approx $Zn_2CrO_4(OH)_2$.

Hydrate, yellow, odorless, fine powder. Slightly sol in water; sol in dil acids, including acetic acid.

USE: As pigment in paints, varnishes, oil colors, linoleum, rubber, etc.

10033. Zinc Citrate. *2-Hydroxy-1,2,3-propanetricarboxylic acid zinc salt.* $C_{12}H_{10}O_{14}Zn_3$; mol wt 574.32. C 25.09%, H 1.76%, O 39.00%, Zn 34.15%. $Zn_3(C_6H_5O_7)_2$. Prepd from zinc carbonate and citric acid: Heldt, *Ann.* **47**, 157 (1843).

Dihydrate, odorless powder. Slightly sol in water; sol in dil mineral acids, in alkali hydroxides.

USE: In toothpaste and mouthwash.

10034. Zinc Cyanide. C_2N_2Zn; mol wt 117.42. C 20.46%, N 23.86%, Zn 55.68%. $Zn(CN)_2$. Usually contains about 85% zinc cyanide, some water and oxide.

White powder. *Poison!* Insol in water; sol in solns of alkali cyanides or hydroxides; not appreciably attacked by organic acids, but readily dec by dil mineral acid with evolution of hydrogen cyanide.

USE: Electroplating; removing NH_3 from producer gas.

10035. Zinc Fluoride. F_2Zn; mol wt 103.38. F 36.76%, Zn 63.24%. ZnF_2. Prepd from $ZnCO_3$ and HF: Ruff, *Die Chemie des Fluors* (Berlin, 1920) p 36; Emeleus in *Fluorine Chemistry* vol. 1, J. H. Simons, Ed. (Academic Press, New York, 1950) p 38; Kwasnik in *Handbook of Preparative Inorganic Chemistry* vol. 1, G. Brauer, Ed. (Academic Press, New York, 2nd ed., 1963) p 242.

Tetragonal needles (rutile lattice) or white cryst mass. d^{25} 5.00: Haendler *et al., J. Am. Chem. Soc.* **74**, 3167 (1952). mp 872°. bp 1500°. Soly in water: 5×10^{-5} moles/l,

Kwasnik, *loc. cit.* Slightly sol in aq HF, more sol in HCl, HNO_3, NH_4OH. Zinc fluoride used for fluorinations should be slightly hydrated.

Tetrahydrate, rhombohedral crystals, becomes anhydr at 100°. Soly in water: 1.516 g/100 ml. May be stored in glass bottles.

USE: In the fluorination of organic compds, manuf phosphors for fluorescent electric lights, glazes and enamels for porcelain, preserving wood, in electroplating baths.

10036. Zinc Formate. $C_2H_2O_4Zn$; mol wt 155.41. C 15.45%, H 1.30%, O 41.18%, Zn 42.07%. $Zn(HCOO)_2$. Forms dihydrate readily. Prepd from zinc carbonate and formic acid: Kendall, Adler, *J. Am. Chem. Soc.* **43**, 1470 (1921).

Dihydrate, crystals. d 2.207. Solubility in H_2O (20°): 5.2 g/100 g. Practically insol in alc.

10037. Zinc Hexafluorosilicate. Zinc fluosilicate; zinc silicofluoride. F_6SiZn; mol wt 207.46. F 54.95%, Si 13.54%, Zn 31.51%. $ZnSiF_6$.

Hexahydrate, white crystals. Soluble in water. pH of 1% aqueous solution 3.2.

USE: Mothproofing agent; laundry sour; hardener for concrete.

10038. Zinc Insulin Crystals. Crystalline preparation of the active antidiabetic principle of the internal secretion of Langerhans' islands of the pancreas. The crystals contain a small amount of zinc (not less than 0.45% nor more than 0.9%), which is chemically combined with the active principle. Each milligram of the crystals is equivalent to not less than 22 units of insulin. The product is marketed in the form of cryst zinc-insulin injection. Study of the binding of zinc by insulin: Tanford, Epstein, *J. Am. Chem. Soc.* **76**, 2170 (1954); Cunningham, *et al., ibid.* **77**, 5703 (1955). Proposed crystallographic structure: Marcker, Graae, *Acta Chem. Scand.* **16**, 41 (1962).

Flat rhombohedral crystals, usually in pairs. Sparingly sol in water; sol in dil acid and dil alkali. Insol in alcohol, chloroform, ether. Isoelectric point is about 5.3. Stable at low temp. The crystals brown rapidly when heated above 220°. mp 230-240° (dec).

THERAP CAT: Antidiabetic.

10039. Zinc Iodate. I_2O_6Zn; mol wt 415.22. I 61.13%, O 23.12%, Zn 15.75%. $Zn(IO_3)_2$. May contain 1-2 mols of water.

White, crystalline powder. Soluble in 115 parts cold, 77 parts hot water.

THERAP CAT: Topical antiseptic.

10040. Zinc Iodide. I_2Zn; mol wt 319.22. I 79.52%, Zn 20.48%. ZnI_2. Usually contains at least 98% ZnI_2.

White or almost white, odorless, hygroscopic, granular powder; sharp, saline taste; becomes brown on exposure to air and light, due to liberation of iodine. d^{25} 4.74; mp about 446°; bp about 625° with decompn. One gram dissolves in 0.3 ml water, 0.2 ml boiling water, 2 ml glycerol; freely sol in alcohol, ether. The aq soln is acid to litmus; pH about 5. *Keep well closed and protected from light.*

THERAP CAT: Topical antiseptic, astringent.

10041. Zinc Iodide-Starch. Trommsdorff's starch. A soln is prepd by heating 4 parts starch, 20 $ZnCl_2$, and 2 ZnI_2 with 1 liter water. Deteriorates with age and acquires a blue color.

USE: For detecting nitrites, free Cl, and other oxidizing agents.

10042. Zinc Lactate. $C_6H_{10}O_6Zn$; mol wt 243.51. C 29.59%, H 4.14%, O 39.42%, Zn 26.85%. $Zn(C_3H_5O_3)_2$. Prepd from zinc carbonate and lactic acid: Pederson *et al., J. Biol. Chem.* **68**, 151 (1926).

Trihydrate, crystals. Soluble in 60 parts cold, 6 parts boiling water.

10043. Zinc Meta-arsenite. ZMA. As_2O_4Zn; mol wt 279.20. As 53.66%, O 22.92%, Zn 23.42%. $Zn(AsO_2)_2$. White powder. Sol in acids.

USE: Wood preservative, insecticide.

10044. Zinc Nitrate. N_2O_6Zn; mol wt 189.38. N 14.79%, O 50.69%, Zn 34.52%.

Hexahydrate, colorless, odorless crystals; d 2.065; mp about 36°. Sol in about 0.5 part water, freely in alcohol. The aq soln is acid to litmus; pH of 5% aq soln 5.1. *Keep well closed and in a cool place.* Also available in the form of fused pieces. In this form it contains only about 20% water. Technical flake usually contains 25.6% H_2O.

USE: As a mordant in dyeing.

10045. Zinc Nitride. N_2Zn_3; mol wt 224.16. N 12.50%, Zn 87.50%. Zn_3N_2. Prepd by heating zinc amide, $Zn(NH_2)_2$, to 330° or by heating zinc dust to 600° in a stream of ammonia: Juza *et al., Z. Anorg. Chem.* **239**, 273 (1938).

Blackish-gray, cryst solid.

10046. Zinc Nitrite. N_2O_4Zn; mol wt 157.40. N 17.80%, O 40.66%, Zn 41.54%. $Zn(NO_2)_2$. Prepd by treating sodium nitrite with zinc sulfate in alcohol: Ephraim, Bolle, *Ber.* **48**, 643 (1915).

Hydrolyzes so quickly that it cannnot be prepd from water, from which only basic salts will separate.

10047. Zinc Oleate. *9-Octadecenoic acid zinc salt.* $C_{36}H_{66}O_4Zn$; mol wt 628.30. C 68.82%, H 10.59%, O 10.19%, Zn 10.40%. $(C_{17}H_{33}CO_2)_2Zn$. Usually contains small amounts of the zinc salts of other fatty acids. Prepn: Grabner, *Monatsh.* **42**, 287 (1921).

White, dry, greasy powder. Insol in water; sol in alcohol, ether, carbon disulfide, benzene, petr ether.

10048. Zinc Ortho-arsenate. $As_2O_8Zn_3$; mol wt 473.95. As 31.61%, O 27.01%, Zn 41.38%. $Zn_3(AsO_4)_2$. Occurs in nature as an octahydrate *koettigite,* $Zn_3(AsO_4)_2.8H_2O$, and as a basic arsenate *adamite,* $Zn_3(AsO_4)_2.Zn(OH)_2$.

White, odorless powder. *Poison!* Insol in water; sol in acids, alkali hydroxides, ammonia.

10049. Zinc Oxalate. C_2O_4Zn; mol wt 153.39. C 15.66% O 41.72%, Zn 42.62%. ZnC_2O_4. Prepn: Chatterjee, Dhar, *J. Phys. Chem.* **28**, 1020 (1924).

Dihydrate, powder. Very slightly sol in water; sol in dil mineral acids, ammonia.

10050. Zinc Oxide. Flowers of zinc; philosopher's wool; zinc white; C.I. Pigment White 4; C.I. 77947. OZn; mol wt 81.38. Zn 80.34%, O 19.66%. ZnO. Occurs as the mineral *zincite.* Prepd by vaporization of metallic zinc and oxidation of the vapors with preheated air (French process); also from franklinite, (American process) or from zinc sulfide: Faith, Keyes & Clark's *Industrial Chemicals,* F. A. Lowenheim, M. K. Moran, Eds. (Wiley-Interscience, New York, 4th ed., 1975) pp 882-888. Purification: Depew, U.S. pat. **2,372,367** (1945 to American Zinc, Lead & Smelting). The medicinal grade contains 99.5% or more ZnO; technical grades contain 90-99% ZnO and a few tenths of 1% of lead. *See also: Colour Index* vol. 4 (3rd ed., 1971) p 4687.

White or yellowish-white, odorless powder. Hexagonal crystals: d 5.67. Also reported as d_4^{20} 5.607. Sublimes at normal pressure. n_D 2.0041, 2.0203. American process zinc oxide pH 6.95. French process zinc oxide pH 7.37. Practically insol in water; sol in dil acetic or mineral acids, ammonia, ammonium carbonate, fixed alkali hydroxide solns.

Human Toxicity: Freshly formed fumes, as from welding, may cause metal fume fever with chills, fever, tightness in chest, cough and leukocytes.

USE: As pigment in white paints instead of lead carbonate; in cosmetics, driers, quick-setting cements; with syrupy phosphoric acid or $ZnCl_2$ in dental cements; manuf opaque glass and certain types of transparent glass; manuf enamels, automobile tires, white glue, matches, white printing inks, porcelains, zinc green; as a reagent in analytical chemistry; in electrostatic copying paper; as flame retardant; in electronics as semiconductor.

THERAP CAT: Astringent, topical protectant.

THERAP CAT (VET): Antiseptic, astringent, protective (topical).

10051. Zinc Perchlorate. Cl_2O_8Zn; mol wt 264.27. Cl 26.83%, O 48.43%, Zn 24.74%. $Zn(ClO_4)_2$.

Hexahydrate, deliquesc crystals, mp 106°. Freely sol in water. Sol in alcohol. Also forms a tetrahydrate.

10052. Zinc Permanganate. Mn_2O_8Zn; mol wt 303.25. Mn 36.23%, O 42.21%, Zn 21.56%. $Zn(MnO_4)_2$. The article of commerce is about 95% pure.

Hexahydrate, violet-brown or almost black, deliquesc crystals; similar to potassium permanganate in appearance. Deteriorates on exposure to light and air. Sol in 3 parts water; dec by alc. *Keep well closed and protected from light.*

THERAP CAT: Antiseptic; astringent.

10053. Zinc Peroxide. ZPO; zinc superoxide. O_2Zn; mol wt 97.38. O 32.86%, Zn 67.14%. ZnO_2. The peroxide of commerce contains 50-60% ZnO_2, the remainder is ZnO. White to yellowish-white, odorless powder. Dec above 150°. Insol in, but gradually dec by, water. Sol in dil acids, liberating hydrogen peroxide.

USE: Accelerator in rubber compounding; curing agent for synthetic elastomers. Deodorant for wounds and skin diseases.

THERAP CAT: Topical antiseptic, astringent.

10054. Zinc *p*-Phenolsulfonate. *p-Hydroxybenzenesulfonic acid zinc salt;* 1-phenol-4-sulfonic acid zinc salt; zinc *p*-hydroxybenzenesulfonate; zinc sulfocarbolate; zinc sulfophenate; Phenozin. $C_{12}H_{10}O_8S_2Zn$; mol wt 411.70. C 35.01%, H 2.45%, O 31.09%, S 15.58%, Zn 15.88%. $Zn(HO\text{-}C_6H_4SO_3)_2$. It is at least 99.5% pure. Prepn: Rojahn, *Deut. Apoth.-Ztg.* **50**, 1095 (1935), *C.A.* **31**, 6816[8] (1937).

Octahydrate, crystals or cryst powder; odorless. Effloresces in dry air; loses all its H_2O at about 120°. One gram dissolves in 1.6 ml water, 0.4 ml boiling water, 1.8 ml alcohol. The aq soln is acid to litmus; pH about 4.

USE: In insecticide formulations.

THERAP CAT: Astringent.

THERAP CAT (VET): Has been used as an intestinal antiseptic, and externally to promote healing of ulcers, slowly granulating wounds.

10055. Zinc Phosphate. $O_8P_2Zn_3$; mol wt 386.05. O 33.16%, P 16.04%, Zn 50.80%. $Zn_3(PO_4)_2$. The tetrahydrate occurs in nature as the mineral *hopeite;* the article of commerce is about 98% pure.

Tetrahydrate, white, odorless powder. Insol in water or alcohol; sol in dil mineral acids, acetic acid, ammonia, and in alkali hydroxide solns.

USE: In dental cements.

10056. Zinc Phosphide. P_2Zn_3; mol wt 258.09. Zn 75.99%, P 24.01%. Zn_3P_2. Usually prepd from the elements. Dark gray tetragonal crystals, lustrous or dull powder; faint phosphorus odor. d 4.55. When strongly heated with the exclusion of air it melts and finally sublimes. The mp has been given as 420° and the bp as 1100°. When kept dry it is quite stable. Insol in water, alcohol; sol in benzene, carbon disulfide; reacts with HCl, H_2SO_4 with evolution of spontaneously flammable phosphine. Reacts violently with concd H_2SO_4, HNO_3 and other oxidizing agents. *Keep dry.* LD_{50} orally in rats: 40.5-46.7 mg/kg, *Handbook of Toxicology*, vol. 1, W. S. Spector, Ed. (Saunders, Philadelphia, 1956) pp 316-317.

USE: In rat and field mice poison preparations.

10057. Zinc Propionate. $C_6H_{10}O_4Zn$; mol wt 211.52. C 34.07%, H 4.77%, O 30.26%, Zn 30.91%. $Zn(C_3H_5O_2)_2$. Prepd by dissolving zinc oxide in dil propionic acid and concentrating the soln: Gaze, *Arch. Pharm.* **229**, 488 (1891).

Clusters of plates, tablets; crystallizes also in needles as the monohydrate. The soly of the anhydr salt is 32% (w/w) in water at 15°; 2.8% in alc at 15°; 17.2% in boiling alc. *Keep well closed.* Dec in moist atm, giving off propionic acid.

USE: As fungicide on adhesive tape to reduce plaster irritation caused by molds, fungi, and bacterial action.

THERAP CAT: Topical antifungal.

10058. Zinc Pyrophosphate. *Diphosphoric acid, zinc salt* (1:2). $O_7P_2Zn_2$; mol wt 304.72. O 36.76%, P 20.33%, Zn 42.91%. $Zn_2P_2O_7$. White, crystalline powder. d^{23} 3.75. Insoluble in water; soluble in dilute mineral acids.

10059. Zinc Salicylate. *2-Hydroxybenzoic acid zinc salt.* $C_{14}H_{10}O_6Zn$; mol wt 339.60. C 49.51%, H 2.97%, O 28.27%, Zn 19.25%. $Zn[C_6H_4(OH)COO]_2$. Prepd from sodium salic-

ylate and zinc sulfate: Clark, Kao, *J. Am. Chem. Soc.* **70**, 2151 (1948).

Trihydrate, needles or cryst powder. Sol in water, alc. The aq soln is practically neutral to litmus.

THERAP CAT: Antiseptic; astringent.

10060. Zinc Selenate. O_4SeZn; mol wt 208.34. O 30.72%, Se 37.90%, Zn 31.38%. $ZnSeO_4$. Heptahydrate prepd by the action of selenic acid on zinc carbonate: Banks, *J. Chem. Soc.* **1934**, 1010.

Pentahydrate, triclinic crystals. Dec above 50°. d_4^{20} 2.591. Sol in water.

Hexahydrate, tetragonal crystals. d 2.325.

10061. Zinc Selenide. SeZn; mol wt 144.34. Se 54.70%, Zn 45.30%. ZnSe. Prepd by mixing a soln of a zinc salt and potassium selenide: Berzelius, cited in *Mellor's Comprehensive Treatise on Inorganic and Theoretical Chemistry* **10**, 776 (1930); by passing selenium vapor over zinc heated *in vacuo:* Moser, Doctor, *Z. Anorg. Chem.* **118**, 284 (1921); by passing H_2Se into a methanol soln of zinc acetate: Nitsche, U.S. pat. **2,805,917** (1957 to du Pont); from $ZnSe.N_2H_4$ (zinc selenide hydrazinate): Conn *et al.,* U.S. pat. **3,014,779** (1961 to Merck & Co.).

Yellow, cubic crystals. d_4^{15} 5.42. mp >1100°. Dec in air. Insol in water. Dec in dil nitric acid.

USE: Commercial phosphor.

10062. Zinc Silicate. Zinc orthosilicate. O_4SiZn_2; mol wt 222.85. O 28.72%, Si 12.60%, Zn 58.68%. Zn_2SiO_4. Occurs in nature as the mineral *willemite.* Prepd by heating the proper amounts of ZnO and SiO_2 at about 1200°.

White powder, insol in water or dil acids.

USE: In television screens.

10063. Zinc Stearate. *Octadecanoic acid zinc salt.* $C_{36}H_{70}O_4Zn$; mol wt 632.33. C 68.38%, H 11.16%, O 10.12%, Zn 10.34%. $Zn(C_{18}H_{35}O_2)_2$. Usually occurs as a mixture of the zinc salts of stearic and palmitic acids, and usually with some excess of zinc oxide. Contains 13.5-15% ZnO. Prepd from stearic acid and zinc chloride: Vold, Hattiangdi, *Ind. Eng. Chem.* **41**, 2311 (1949).

Fine, soft, bulky powder; slight characteristic odor; neutral reaction. Repels water. mp about 120°. Insol in water, alcohol, ether. Sol in benzene; dec by dil acids.

USE: In tablet manuf; in cosmetic and pharmaceutical powders and ointments; as a flatting and sanding agent in lacquers; as a drying lubricant and dusting agent for rubber; as a plastic mold releasing agent; as a waterproofing agent for concrete, rock wool, paper, textiles.

THERAP CAT (VET): Antiseptic, astringent, protective (topical).

10064. Zinc Sulfate. White vitriol; zinc vitriol; Medizinc; Optraex; Solvezink; Solvazinc; Zincaps; Zincate; Zincomed; Z Span. O_4SZn; mol wt 161.44. O 39.64%, S 19.86%, Zn 40.50%. $ZnSO_4$. Prepn and physical properties: *Gmelin's, Zink* (8th ed.) **32**, 936-960 (1956). Effects of oral zinc sulfate in acne treatment: G. Michaelsson *et al., Arch. Dermatol.* **113**, 31 (1977). Clinical evaluations in Wilson's Disease: T. U. Hoogenraad, C. J. A. Van den Hamer, *Acta Neurol. Scand.* **67**, 356 (1983); T. U. Hoogenraad *et al., J. Neurolog. Sci.* **77**, 137 (1987).

Monohydrate, *dried zinc sulfate.* Powder or granules. Loses H_2O above 238°. Sol in water. Practically insol in alcohol.

Heptahydrate, *Op-Thal-Zin, Verazinc.* Odorless crystals or granules or powder; astringent taste. Efflorescent in dry air. d 1.97. mp 100°. At 280° loses all H_2O; dec above 500°. One gram dissolves in 0.6 ml water, 2.5 ml glycerol. Insol in alcohol. The aq soln is acid to litmus; pH about 4.5. *Keep well closed.*

Note: The monohydrate does not cake as the heptahydrate does, and hence is more convenient for use during the warm season and in warm climates.

Caution: Irritating to skin, mucous membranes: H. E. Stokinger, *Patty's Industrial Hygiene and Toxicology* **2A**, G. D. Clayton, F. E. Clayton, Eds. (John Wiley & Sons, New York, 1981) pp 2033-2049.

USE: As mordant in calico-printing; preserving wood and skins; with hypochlorite for bleaching paper; manuf lithopone and other zinc salts; clarifying glue; electrodepositing Zn; also as reagent in analytical chemistry.

THERAP CAT: Ophthalmic astringent. Zinc supplement.

THERAP CAT (VET): Astringent. Has been used as an emetic.

10065. Zinc Sulfide. Zinc Blende. SZn; mol wt 97.45. S 32.91%, Zn 67.09%. ZnS. Occurs in nature as the minerals *wurtzite* (hexagonal, d 4.087) and *sphalerite* (cubic, d 4.102). Precipitated zinc sulfide of commerce usually contains 15-20% water of hydration. The dried precipitate may have been heated to 725° in the absence of air to obtain substantial conversion to wurtzite, the form preferred by the pigment industry.

White to grayish-white or yellowish powder. When contg water, it slowly oxidizes in air to sulfate. Insol in water, alkalies; sol in dil mineral acids.

USE: Pigment for paints, oilcloths, linoleum, leather, dental rubber, etc., especially in the form of lithopone; mixed with ZnO as "mineral white." Anhydr zinc sulfide is used in x-ray screens and with a trace of a radium or mesothorium salt in luminous dials of watches, etc.; also television screens.

10066. Zinc Tannate. Sal barnit. Compd of ZnO and tannin in variable proportions. Prepd from an aq tannin soln and zinc hydroxide in excess ammonium hydroxide plus an ammonium salt: Ger. pat. **479,229** (1926 to Lab. Reumella Adolf Boas), *C.A.* **23**, 4778[1] (1929).

Yellowish, odorless powder. Practically insol in water, alcohol; sol in dil acids.

THERAP CAT: Astringent; antiseptic.

10067. Zinc Tartrate. $C_4H_4O_6Zn$; mol wt 213.44. C 22.51%, H 1.89%, O 44.97%, Zn 30.63%. $ZnC_4H_4O_6$. Prepd from potassium tartrate and zinc chloride: Cantoni, Zachoder, *Bull. Soc. Chim.* [3] **33**, 747 (1905).

Dihydrate, cryst powder. Soly in water: Cantoni, Zachoder, *loc. cit.*

10068. Zinc Telluride. TeZn; mol wt 192.99. Te 66.12%, Zn 33.88%. ZnTe. Prepd by fusing Zn and Te: Braithwaite, *Proc. Phys. Soc. (London)* **64B**, 274 (1951); Bube, *Proc. IRE* **43**, 1836 (1955), *see also* Dennis, Anderson, *J. Am. Chem. Soc.* **36**, 882 (1914). From zinc oxide and powdered tellurium in aq alkaline medium: Montignie, *Bull. Soc. Chim. (France)* **1947**, 750.

Gray or brownish-red powder. Ruby-red crystals (cubic system) by sublimation. Stable in dry air. d_4^{15} 6.34. mp 1239°. Forms a monohydrate. Prolonged contact with water or dil HCl yields H_2 and H_2Te.

USE: In semiconductor research, as photoconductor.

10069. Zinc Thiocyanate. Zinc sulfocyanate. $C_2N_2S_2Zn$; mol wt 181.55. C 13.23%, N 15.43%, S 35.32%, Zn 36.02%. $Zn(SCN)_2$.

White, deliquesc crystals. Sol in water, alcohol. The aq soln is only slightly acid to litmus. *Keep well closed and protected from light.*

USE: To assist in textile dyeing.

10070. Zinc Valerate. *Pentanoic acid zinc salt.* $C_{10}H_{18}O_4Zn$; mol wt 267.62. C 44.88%, H 6.78%, Zn 24.43%, O 23.91%. $Zn(C_5H_9O_2)_2$. The article of commerce is usually basic. Prepd from valeric acid and zinc hydrate: Lieben, Rossi, *Ann.* **159**, 58 (1871).

Dihydrate, lustrous scales or powder; valerian odor; sweetish taste. Gradually dec on exposure to air. One grain dissolves in 70 ml water or in 22 ml alcohol when free from basic salt; dec by acid. *Keep tightly closed and in cool place.*

10071. Zineb. *[[1,2-Ethanediylbis[carbamodithioato]]-(2−)]zinc; [ethylenebis(dithiocarbamato)]zinc;* zinc ethylenebis(dithiocarbamate); ethylenebis(dithiocarbamic acid) zinc salt; ENT 14874; Parzate; Lonacol; Dithane Z-78; Polyram Z; Tiezene. $C_4H_6N_2S_4Zn$. Polymeric salt of *ethylenebisdithiocarbamic acid;* related to maneb and mancozeb, *q.q.v.* Prepd from a zinc salt and nabam, *q.v.*: C. B. Luginbuhl, U.S. pat. **2,690,448** (1954 to du Pont). Fungicidal activity is due to degradation products, principally *ethylenethiuram monosulfide:* R. A. Ludwig *et al., Can. J. Bot.* **33**, 42 (1955). Activity against potato blight: B. K. De, S. B. Chattopadhyay, *Pesticide* **18**, 52 (1984). Decomposition of zineb to ethylenebisdicarbamate and ethylene thiourea, *q.v.*, in toma-

toes: B. D. Ripley, D. F. Cox, *J. Agr. Food Chem.* **26**, 1137 (1978). HPLC determn: K. Gustafsson *et al., ibid.* **29**, 729 (1981). Toxicity data: T. B. Gaines, *Toxicol. Appl. Pharmacol.* **14**, 515 (1969). Aquatic toxicology: C. J. Van Leeuwen *et al., Aquat. Toxicol.* **7**, 145 (1985).

$$\left[-ZnSCNHCH_2CH_2NHCS- \right]_n$$

Powder or crystals from chloroform + alcohol. Practically insol in water. The powder spreads easily on water, also forms aq suspensions. Sol in carbon disulfide, pyridine. LD_{50} orally in rats: > 5000 mg/kg (Gaines).

Caution: Irritation of skin and mucous membranes has been reported: *Clinical Toxicology of Commercial Products,* R. E. Gosselin *et al.,* Eds. (Williams & Wilkins, Baltimore, 5th Ed., 1984) Section II, p 313.

USE: Agricultural fungicide.

10072. Zingerone. *4-(4-Hydroxy-3-methoxyphenyl)-2-butanone;* (4-hydroxy-3-methoxyphenyl)ethyl methyl ketone; vanillylacetone; zingerone; zingiberone. $C_{11}H_{14}O_3$; mol wt 194.22. C 68.02%, H 7.27%, O 24.71%. Isolated from ginger root or prepd from vanillin and acetone followed by catalytic hydrogenation: Nomura, *J. Chem. Soc.* **111**, 769 (1917); *idem,* U.S. pats. **1,263,796** (1918); **1,306,710** (1919); Cotton, U.S. pat. **2,381,210** (1945 to Penn. Coal Prod.); K. Banno, T. Mukaiyama, *Bull. Chem. Soc. Japan* **49**, 1453 (1976).

Crystals from acetone, petr ether, ether + petr ether, mp 40-41°. bp_{14} 187-188°. Sparingly sol in water, petr ether; sol in ether, dil alkalies.

10073. Zinostatin. *Neocarzinostatin;* neocarcinostatin; NCS; NSC 69856; Neocarzinostatin K. An antitumor acidic antibiotic consisting of 2 components: a protein of mol wt 10,700 containing 109 amino acid residues of 18 kinds and a labile non-protein chromophore which possesses the full biological activity of NCS. Isoln from the culture filtrate of *Streptomyces carcinostaticus* var. F-41: N. Ishida *et al., J. Antibiot.* **18A**, 68 (1965). Characterization: H. Maeda *et al., ibid.* **19A**, 253 (1966). Prodn: N. Ishida *et al.,* U.S. pat. **3,334,022** (1967 to Empire Corp.). Amino acid sequence: J. Meienhofer *et al., Science* **178**, 875 (1972). Review of protein structural studies: *eidem, Progress in Peptide Research* vol. **II**, S. Lande, Ed. (Gordon and Breach, New York, 1972) pp 295-306. Spectral characterization and separation of the chromophore: M. A. Napier *et al., Biochem. Biophys. Res. Commun.* **89**, 635 (1979). Partial structure of the chromophore: G. Albers-Schönberg *et al., ibid.* **95**, 1351 (1980). Roles of chromophore and apo-protein in biological activity: L. S. Kappen *et al., Proc. Nat. Acad. Sci. USA* **77**, 1970 (1980). Molecular basis of action: I. H. Goldberg *et al., Mol. Biol. Biochem. Biophys.* **32**, 308 (1980). Absorption, distribution, excretion: K. Toriyama *et al., Japan. J. Antibiot.* **28**, 24 (1975). Review of pharmacology: T. A. Beerman *et al.,* in *Advances in Enzyme Regulation* vol. **14**, G. Weber, Ed. (Pergamon Press, New York, 1976) pp 207-228.

White powder, mp 260° (dec). uv max (water): 278 nm $(E_{1cm}^{1\%}$ 15). Isoelectric pt: pH 3.26 [*Japan. Med. Gaz.* **14**(3), 11 (1977)]. Sol in water. Insol in most common org solvents. *See* M. Kohno *et al., Japan. J. Antibiot.* **27**, 707 (1974) for stability of aq solns. LD_{50} in mice (mg/kg): 1050 orally, 0.96 i.v. (Toriyama).

THERAP CAT: Antineoplastic.

10074. Zipeprol. *4-(2-Methoxy-2-phenylethyl)-α-(methoxyphenylmethyl)-1-piperazineethanol;* α-(α-methoxybenzyl)-4-(β-methoxyphenethyl)-1-piperazineethanol; 1-(2-hydroxy-3-methoxy-3-phenylpropyl)-4-(2-methoxy-2-phen-ylethyl)piperazine; 1-(2-methoxy-2-phenylethyl)-4-(2-hy-droxy-3-methoxy-3-phenylpropyl)piperazine. $C_{23}H_{32}N_2O_3$; mol wt 384.52. C 71.84%, H 8.39%, N 7.28%, O 12.48%. Prepn: Mauvernay et al., Ger. pat. 2,109,366 corresp to U.S. pat. 3,718,650 (1971, 1973 both to Mauvernay). Pharmacology: G. Rispat et al., Arzneimittel-Forsch. **26**, 523 (1976); D. Cosnier et al., ibid. 848.

Crystals from abs ethanol, mp 83°.

Dihydrochloride, $C_{23}H_{34}Cl_2N_2O_3$, *3024 CERM, Antituxil-Z, Citizeta, Mirsol, Respilene, Respirase, Zitoxil.* Crystals from abs ethanol, mp 231°. Very stable in aq soln. LD_{50} orally in mice: 301 mg/kg (Rispat).

THERAP CAT: Antitussive.

10075. Ziram. *Bis(dimethylcarbamodithioato-S,S')zinc; bis(dimethyldithiocarbamato)zinc;* zinc dimethyldithiocarbamate; dimethyldithiocarbamic acid zinc salt; zinc bis(dimethylthiocarbamoyl) disulfide; methyl cymate; Methasan; Zimate; Zirberk; Karbam White; Corozate; Fuclasin; Fuklasin; Zerlate. $C_6H_{12}N_2S_4Zn$; mol wt 305.82. C 23.56%, H 3.96%, N 9.16%, S 41.94%, Zn 21.38%. Prepd from zinc oxide, dimethylamine, and carbon disulfide: Olin, Deger, U.S. pat. 2,492,314 (1949 to Sharples Chemicals). Crystal structure: Klug, Acta Cryst. **21**, 536 (1966).

Crystals from hot chloroform + alcohol, mp 250°. Can form a flammable dust. d_4^{25} 1.66. Practically insol in water. Soly per 100 ml of solvent at 25°: <0.2 g, alcohol; <0.5 g, acetone; <0.5 g, benzene; <0.2 g, carbon tetrachloride, more sol in chloroform; <0.2 g, ether; 0.5 g, naphtha. Sol in dil caustic solns. LD_{50} orally in rats: 1.4 g/kg, Hodge et al., J. Pharmacol. Exp. Ther. **118**, 174 (1956).

USE: Rubber vulcanization accelerator; agricultural fungicide. Caution: Irritant to skin and mucous membranes.

10076. Zirconium. Zr; at. wt 91.22; at. no. 40; valence 4; also 3. Five naturally occurring isotopes: 90 (51.46%); 91 (11.23%); 92 (17.11%); 94 (17.40%); 96 (2.80%); artificial radioactive isotopes: 81-89, 93, 95, 97-99. Occurrence in earth's crust: 0.023%. Occurs in the minerals zircon, malacon, baddeleyite, zirkelite, eudialyte; frequently found in the rare-earth minerals; in monazite sand. Discovered by Klaproth in 1789; prepd by Berzelius in 1824. Prepn: Fast, Z. Anorg. Chem. **239**, 145 (1938); purification of zirconium by ion exchange columns: Ayres, J. Am. Chem. Soc. **69**, 1879 (1947). Sepn of zirconium and hafnium: Fischer et al., Angew. Chem. Int. Ed. **5**, 15 (1966). Reviews of zirconium and its compds: W. B. Blumenthal, The Chemical Behavior of Zirconium (Van Nostrand, Princeton, 1958); Gmelin's Handb. Anorg. Chem., Zirconium (8th ed.) **42**, (1958) 448 pp; Larsen, "Zirconium and Hafnium Chemistry" in Advan. Inorg. Chem. Radiochem. **13**, 1-333 (1970); Bradley, Thornton, "Zirconium and Hafnium" in Comprehensive Inorganic Chemistry, vol. 3, J. C. Bailar, Jr. et al., Eds. (Pergamon Press, Oxford, 1973) pp 419-490.

Bluish-black, amorphous powder or grayish-white lustrous metal (platelets or flakes) of hexagonal lattice below 865°, body-centered cubic above 865°, mp 1857°; bp 3577°. d 6.5. Brinnell hardness: 85. Can absorb up to 10 atoms per cent of oxygen or nitrogen. Reacts with hydrofluoric acid, aqua regia, hot phosphoric acid. Not attacked by cold, very slightly attacked by hot, concd sulfuric or hydrochloric acid; not attacked by nitric acid. Attacked by fused potassium hydroxide or nitrate. On prolonged heating the compact form combines with oxygen, nitrogen, carbon, and the halo-

gens. The powder form has a very low ignition temp and is very explosive when mixed with oxidizing agents.

Human Toxicity: Zirconium and its salts generally have low systemic toxicity. A granulomatous disease of the skin, particularly in the axilla, has been reported in users of a deodorant contg sodium zirconium lactate: see E. Browning, Toxicity of Industrial Metals (Appleton-Century-Crofts, New York, 2nd ed., 1969) pp 356-360. Consult latest Government regulations on use in aerosol antiperspirants.

USE: Pure zirconium (hafnium-free) is a valuable structural material for atomic reactors because of its low nuclear cross-section and high corrosion and heat resistance. Because of hafnium's high neutron absorption characteristics, it must be removed from zirconium which is to be used in nuclear reactors; removal unnecessary for other commercial purposes. As an ingredient of priming or explosive mixtures; flashlight powders; as deoxidizer in metallurgy; as "getter" in vacuum tubes; in constructing rayon spinnerets in lamp filaments, flash bulbs.

10077. Zirconium Chloride. Zirconium tetrachloride. Cl_4Zr; mol wt 233.05. Cl 60.86%, Zr 39.14%. $ZrCl_4$. Lewis acid. In large-scale prepns zirconium oxide is converted to the carbide, which is chlorinated to yield the tetrachloride: Kroll et al., J. Electrochem. Soc. **94**, 1 (1948). Process not easily adaptable to lab use. Lab prepn based on the equation $ZrO_2 + 2CCl_4 \rightarrow ZrCl_4 + 2COCl_2$: Hummers et al., Inorg. Syn. **4**, 121 (1953).

Lustrous monoclinic crystals: Krebs, Angew. Chem. Int. Ed. **8**, 146 (1969); idem, Z. Anorg. Allgem. Chem. **378**, 263 (1970). Tetrahedral symmetry in gas phase with the Zr-Cl distance of 2.33 Å. Extremely hygroscopic, forms HCl vapor and gives off fumes in moist air. d 2.803. Sublimes at 331°. mp 437° under its own pressure which is about 25 atm at this temp. Decomposed by water to form $ZrOCl_2$ and HCl; sol in alcohol, ether. LD_{50} in mice, rats: 655, 1688 mg/kg orally, Toxic Substances List, H. E. Christensen, Ed. (1972) p. 547.

USE: Friedel-Crafts catalyst. Component of Ziegler-type catalysts in the condensation of ethylene. Starting material in the synthesis of a number of organic derivs of zirconium, such as alkoxides and zirconocene. The alkoxides have been shown to be of value in the curing of silicone plastic films. The alkoxyzirconium carboxylates are said to be useful in the water-repellent treatment of textiles and other fibrous materials. Review: Blumenthal, J. Chem. Ed. **39**, 604-610 (1962).

10078. Zirconium Fluoride. Zirconium tetrafluoride. F_4Zr; mol wt 167.22. Zr 54.55%, F 45.45%. ZrF_4. Prepd by thermal decompn of $(NH_4)_2ZrF_6$: v. Hevesy, Dullenkamp, Z. Anorg. Allgem. Chem. **221**, 161 (1934); according to the equation $ZrCl_4 + 4HF \rightarrow ZrF_4 + 4HCl$: Wolter, Chemiker-Ztg. **51**, 607 (1908); from zirconium oxide and fluorine: Haendler et al., J. Am. Chem. Soc. **76**, 2177 (1954).

Strongly refractive, crystalline mass (monoclinic system). d^{16} 4.6. Sublimes above 600°. Solubility in water (20°): 1.32 g/100 ml. Does not react with water; forms stable trihydrate. Freely soluble in hydrofluoric acid. LD_{50} in mice: 98 mg/kg orally: Toxic Substances List, H. E. Christensen, Ed. (1972) p 547.

10079. Zirconium Hydride. Ideal composition: ZrH_2. Prepd by the reduction of zirconium oxide with calcium hydride in the presence of hydrogen above 600°: Alexander, U.S. pat. 2,427,339 (1947 to Metal Hydride).

Gray-black metallic powder. Stable, no reaction with water.

USE: Powerful reducing agent in acid solution or at high temps; hydrogenation catalyst; in the vacuum tube industry and powder metallurgy.

10080. Zirconium Hydroxide. H_4O_4Zr; mol wt 159.25. H 2.53%, O 40.19%, Zr 57.28%. $Zr(OH)_4$. Proposed structures of freshly pptd and aged compds, cyclic tetramers: Zaitsev, Russ. J. Inorg. Chem. **11**, 900 (1966).

White, bulky, amorphous powder; d 3.25. Insol in water; sol in mineral acids when freshly pptd; less sol when aged. It colors turmeric paper brown.

USE: In the pigment, dye, and glass industries.

10081. Zirconium Iodide. Zirconium tetraiodide. I_4Zr;

mol wt 598.86. I 84.77%, Zr 15.23%. ZrI_4. Prepd from the elements at a furnace temp of 450°: Eberly, *Inorg. Syn.* **7**, 52 (1963).

Orange-colored, crystalline solid. Sublimes at 431°. mp 499° (elevated pressure). Fumes heavily in air. Dissolves in water with the liberation of steam. Heat of formation: 90 kcal/mol at 25°. Magnetic susceptibility -0.238×10^{-6} c.g.s. electromagnetic units.

10082. Zirconium Nitrate. $N_4O_{12}Zr$; mol wt 339.25. N 16.52%, O 56.59%, Zr 26.89%. $Zr(NO_3)_4$. Prepn according to the equation $ZrCl_4 + 4N_2O_5 \rightarrow Zr(NO_3)_4 + 4NO_2Cl$: Field, Hardy, *Proc. Chem. Soc.* **1962**, 76. Pentahydrate obtained from strong nitric acid. The zirconium nitrate of commerce is usually somewhat basic.

Pentahydrate, white, very hygroscopic crystals or white pieces or scales. Very sol in water; sol in alcohol. The aq soln is acid to litmus. *Keep tightly closed.*

10083. Zirconium Oxide. Zirconia; zirconium dioxide; zirconic anhydride. O_2Zr; mol wt 123.22. Zr 74.03%, O 25.97%. ZrO_2. Occurs in nature as the mineral *baddeleyite.* Prepn: Clark, Reynolds, *Ind. Eng. Chem.* **29**, 711 (1937); Henderson, Higbie, *J. Am. Chem. Soc.* **76**, 5878 (1954).

White, heavy, amorphous, odorless, tasteless powder or monoclinic crystals. Also forms tetragonal crystals above ~1100°; cubic above ~1900°. d 5.85. mp 2680°. bp 4300°. Practically insol in water; slightly sol in HCl, HNO_3; slowly sol in HF; dissolves on heating with a mixt of 2 parts H_2SO_4 and 1 part H_2O. Fusion with sodium carbonate results in the formation of sodium zirconate which dec hydrolytically with water to form sodium hydroxide and practically insol zirconium hydroxide.

USE: Instead of lime for the oxyhydrogen light; with earths of the yttrium group in incandescent lighting (Nernst lamps); as pigment, abrasive; manuf enamels, white glass, refractory crucibles, and furnace linings.

THERAP CAT: Dermatologic.

10084. Zirconium Silicate. Zirconium orthosilicate. O_4SiZr; mol wt 183.31. O 34.91%, Si, 15.32%, Zr 49.76%. $ZrSiO_4$. Occurs in nature as the mineral *zircon.* Widely scattered in all kinds of rocks and in beach sand (South Carolina, Northeast Florida). Usually separated from sand by electrostatic and electromagnetic elutriation, yielding 99% pure $ZrSiO_4$. Prepn from SiO_2 and ZrO_2 in an arc furnace: Curtis, Sowman, *J. Am. Ceram. Soc.* **36**, 195 (1953); can also be made from solns of zirconium salts and sodium silicate.

Tetragonal, bipyramidal crystals. Should be colorless. Colors are from impurities and radioactive bombardment, often removable by calcination. Dissociates to ZrO_2 and SiO_2 when heated above 1540°. Recombines when cooled slowly, but forms a mixture of monoclinic ZrO_2 and vitreous SiO_2 when rapidly quenched. Undissociated (α-form) $ZrSiO_4$ has a high sp gr, 4.7, and a high birefringence; the dissociated, γ-form has a lower sp gr, 3.9-4.0, and a low birefringence. Hardness of α-form about the same as that of quartz (6-7.5). Very inert chemically. Unaffected by aq reagents.

USE: In refractories, ceramics, glazes, cements, coatings for casting molds, polishing materials, gem stones, catalyst in alkyl and alkenyl hydrocarbon manuf, in fritted glass filters, as stabilizer in silicone rubbers. In Europe in cosmetic creams and powders.

10085. Zirconium Sulfate. Disulfatozirconic acid. O_8S_2Zr; mol wt 283.34. O 45.18%, S 22.63%, Zr 32.19%. $Zr(SO_4)_2$. Tetrahydrate usually prepd by treating zirconium oxychloride with hot concd H_2SO_4: Blumenthal, *J. Chem. Ed.* **39**, 604 (1962).

Tetrahydrate, crystalline solid. Converted to monohydrate at 100°; to anhydr form at 380°. Soly in water (18°): 52.5 g/100 g of soln. A soln at room temp deposits a solid on standing. The more dilute the soln, the more rapid the deposition. The composition of the solid is $4ZrO_2 \cdot 3SO_3 \cdot 15H_2O$, known as *Hauser's salt:* Hauser, Herzfeld, *Z. Anorg. Allgem. Chem.* **67**, 369 (1910). This pptn does not take place if the soln has been heated above 64°. LD_{50} in rats: 3.5 g/kg orally; 175 mg/kg i.p.: Cochran *et al.*, *Arch. Ind. Hyg. Occup. Med.* **1**, 637 (1950).

USE: Catalyst support; precipitation of amino acids and proteins; in the tanning industry.

10086. Zirconyl Acetate. Diacetatozirconic acid. Approx formula: $Zr(OH)_2(C_2H_3O_2)_2$. Obtained as aq soln by adding acetic acid to an aq slurry of carbonated hydrous zirconia: Blumenthal, *J. Chem. Ed.* **39**, 604 (1962). Toxicity data: Cochran *et al.*, *Arch. Ind. Hyg. Occup. Med.* **1**, 637 (1950).

Available as 22% ZrO_2 soln. d 1.46. mp −7°. pH 3.8-4.2. Stable at room temp. Also available as 13% ZrO_2 soln. d 1.20. pH 3.3-4.0. Undergoes exchange with anion exchange resins, but not with cation exchangers. Evaporation of the solns under reduced pressure yields the solid compd. It is resinous or glue-like in appearance and amorphous under x-rays. Its properties indicate it to be a highly polymerized product. The powdered solid readily dissolves when added slowly to rapidly swirling water. On heating, the solns yield a solid hydrolyzate, which is a more basic acetate with the compn $ZrOOHC_2H_3O_2$. When the mixt is allowed to cool and stand at room temp, this ppt slowly redissolves. LD_{50} in rats: 4.1 g/kg orally; 300 mg/kg i.p. (Cochran).

USE: Precipitating agent for gelatin and starch on textiles and paper; water-repellent for textiles (especially in combination with silicones).

10087. Zirconyl Chloride. *Dichlorooxozirconium;* zirconium oxychloride; basic zirconium chloride. Cl_2OZr; mol wt 178.13. Cl 39.81%, O 8.98%, Zr 51.21%. $ZrOCl_2$. Prepn from $ZrCl_4$ and Cl_2O: Dehnicke, Meyer, *Z. Anorg. Allgem. Chem.* **331**, 121 (1964). Octahydrate conveniently prepd by crystn of an aq soln of zirconium chloride: v. Siemens, Zander, *Wissenschaftl. Veröffentl. Siemens* **2**, 484 (1922); Blumenthal, *J. Chem. Ed.* **39**, 607 (1962).

Octahydrate, tetragonal crystals from water. Crystals contain tetramers of the form $[Zr_4(OH)_8(H_2O)_{16}]^{8+}$; d 1.91: Mak, *Can. J. Chem.* **46**, 3491 (1968). Freely sol in water, alcohol. The pH is about equal to that of HCl of the same molarity. LD_{50} in rats: 400 mg/kg i.p.; 3.5 g/kg orally, Cochran *et al.*, *Arch. Ind. Hyg. Occup. Med.* **1**, 637 (1950).

USE: To make other zirconium compds; to precipitate acid dyes; to prepare high quality pigment toners; to improve the properties of color lakes. Aq $ZrOCl_2$ solns have considerable solvent action on sparingly sol sulfates, such as calcium sulfate. The free acid in $ZrOCl_2$ solns may be neutralized, and a sol compd $ZrOOHCl \cdot xH_2O$ can be recovered. It is highly polymerized in soln and amorphous in the solid state; has been used in the prepn of body deodorants and antiperspirant preparations.

10088. Zoapatanol. *9-[3-Hydroxy-6-(2-hydroxyethylidene)-2-methyl-2-oxepanyl]-2,6-dimethyl-2-nonen-5-one.* $C_{20}H_{34}O_4$; mol wt 338.49. C 70.97%, H 10.12%, O 18.91%. Oxepane diterpenoid isolated from the leaves of the zoapatle plant, *Montanoa tomentosa, Compositeae,* which has been used by Mexican women to prepare "tea" to induce menses and labor. Isoln and structure: M. P. Wachter, R. M. Kanojia, U.S. pat. **4,086,358** (1978 to Ortho); S. D. Levine *et al.*, *J. Am. Chem. Soc.* **101**, 3404 (1979); R. M. Kanojia *et al.*, *J. Org. Chem.* **47**, 1310 (1982). Spasmogenic effects: J. B. Smith *et al.*, *Life Sci.* **23**, 2743 (1981). Total syntheses of (±)-form: R. Chen, D. A. Rowand, *J. Am. Chem. Soc.* **102**, 6609 (1980); K. C. Nicolau *et al.*, *ibid.* 6611; V. V. Kane, D. L. Doyle, *Tetrahedron Letters* **22**, 3027, 3031 (1981). Mass spec: C. J. Shaw, *Org. Mass Spectrom.* **16**, 281 (1981). ^{13}C-NMR study: M. L. Cotter, *Org. Magn. Res.* **17**, 14 (1981).

Pale yellow oil.

10089. Zolamine. *N-[(4-Methoxyphenyl)methyl]-N',N'-dimethyl-N-2-thiazolyl-1,2-ethanediamine; 2-[(2-dimethylaminoethyl)(p-methoxybenzyl)amino]thiazole; N',N'-dimethyl-N-(2-thiazolyl)-N-(p-methoxybenzyl)ethylenedi-*

amine. $C_{15}H_{21}N_3OS$; mol wt 291.42. C 61.82%, H 7.26%, N 14.42%, O 5.49%, S 11.00%. This is an isostere of pyrilamine, *q.v.*, and is prepd accordingly: Shigeya Saijo, *J. Pharm. Soc. Japan* **72**, 1009 (1952).

Oily liquid. Putrid odor. bp_7 217-219°.

Hydrochloride, $C_{15}H_{22}ClN_3OS$, *194-B*, *Wl 291*. Odorless crystals from alc. Slightly bitter taste. mp 167.5-167.8°. Sol in water. Ingredient of *Otodyne* (obsolete).

THERAP CAT: Antihistaminic; topical anesthetic.

10090. Zolimidine. *2-[4-(Methylsulfonyl)phenyl]imidazo[1,2-a]pyridine;* zoliridine; Solimidin. $C_{14}H_{12}N_2O_2S$; mol wt 272.32. C 61.75%, H 4.44%, N 10.29%, O 11.75%, S 11.77%. Non-anticholinergic gastroprotective agent. Prepn: **Brit. pat. 991,589** corresp to L. Almirante *et al.*, **U.S. pat. 3,318,880** (1965, 1967 to Selvi); *eidem, J. Med. Chem.* **8**, 305 (1965). Metabolism: *eidem, Farmaco Ed. Sci.* **29**, 941 (1974). Series of articles on pharmacology: *Panminerva Med.* **16**, 301-359 (1974). Pharmacokinetics: E. Schraven, D. Trottnow, *Arzneimittel-Forsch.* **26**, 213 (1976). Comparative study of effects on rat gastric mucosa: A. P. Green *et al.*, *J. Pharm. Pharmacol.* **33**, 348 (1981). Clinical study: A. Materia *et al.*, *Clin. Ter.* **97**, 183 (1981).

Cryst, mp 242-244°. LD_{50} in rats: 3710 mg/kg orally, U.S. pat. **3,318,880**. LD_{50} of the hydrochloride monohydrate in mice: 800 mg/kg i.p., L. Almirante *et al.*, *J. Med. Chem.* **8**, 305 (1965).

THERAP CAT: Anti-ulcerative.

10091. Zolpidem. *N,N,6-Trimethyl-2-(4-methylphenyl)-imidazo[1,2-a]pyridine-3-acetamide;* N,N,6-trimethyl-2-*p*-tolylimidazo[1,2-a]pyridine-3-acetamide. SL 80.0750. $C_{19}H_{21}N_3O$; mol wt 307.40. C 74.24%, H 6.89%, N 13.67%, O 5.20%. Selective benzodiazepine receptor agonist not related chemically to benzodiazepines. Prepn: J. P. Kaplan, P. George, **Eur. pat. Appl. 50,563**; *eidem*, **U.S. pat. 4,382,938** (1982, 1983 both to Synthelabo). Neuropharmacology: S. Arbilla *et al.*, *Arch. Pharmacol.* **330**, 248 (1985); H. Depoortere *et al.*, *J. Pharmacol. Exp. Ther.* **237**, 649 (1986). Neurochemical profile: B. Scatton *et al., ibid.* 659. Binding study in rat brain: S. Arbilla *et al.*, *Eur. J. Pharmacol.* **130**, 257 (1986). HPLC determn in plasma: P. Guinebault *et al., J. Chromatog.* **383**, 206 (1986). Clinical evaluation of hypnotic activity: A. N. Nicholson, P. A. Pascoe, *Brit. J. Clin. Pharmacol.* **21**, 205 (1986). Evaluation as pre-anesthetic medication: J. N. Cashman *et al., ibid.* **24**, 85 (1987).

mp 196°. pKa 6.2.

L-(+)-Hemitartrate, $C_{42}H_{48}N_6O_8$, *SL 80.0750-23N, Stilnox*. Soly in water (20°): 23 mg/ml.

THERAP CAT: Hypnotic.

10092. Zomepirac. *5-(4-Chlorobenzoyl)-1,4-dimethyl-1H-pyrrole-2-acetic acid;* 1,4-dimethyl-5-(p-chlorobenzoyl)-pyrrole-2-acetic acid. $C_{15}H_{14}ClNO_3$; mol wt 291.74. C 61.76%, H 4.84%, Cl 12.15%, N 4.80%, O 16.45%. Prepn: J.

R. Carson, **Ger. pat. 2,102,746**; *idem*, **U.S. pat. 3,752,826** (1971, 1973 both to McNeil); J. R. Carson, S. Wong, *J. Med. Chem.* **16**, 172 (1973). Pharmacology: R. Sofia *et al.*, *Pharmacol. Res. Commun.* **11**, 179 (1979); P. O'Neill *et al.*, *J. Pharmacol. Exp. Ther.* **209**, 366 (1979). Determn in plasma by HPLC: K.-T. Ng, T. Snyderman, *J. Chromatog.* **178**, 241 (1979). Metabolism: J. M. Grindel *et al.*, *Drug Metab. Dispos.* **8**, 343 (1980); W. N. Wu *et al., ibid.* 349. Pharmacokinetics: R. K. Nayak *et al.*, *Clin. Pharmacol. Ther.* **27**, 395 (1980). Series of articles on pharmacology, kinetics, clinical studies: *J. Clin. Pharmacol.* **20**, 213-424 (1980). Preclinical narcotic abuse liability evaluation: J. H. Woods *et al.*, *Arzneimittel-Forsch.* **33**, 218 (1983). Multicenter clinical study in painful conditions: C. E. Steele, W. L. Jefferson, *Curr. Med. Res. Opin.* **8**, 382 (1983). Review of pharmacology and therapeutic efficacy: P. A. Morley *et al.*, *Drugs* **23**, 250-275 (1982). Comprehensive description: M. Zinic *et al.* in *Analytical Profiles of Drug Substances* vol. 15, K. Florey, Ed. (Academic Press, New York, 1986) pp 673-698.

White cryst from 2-propanol, mp 178-179°.

Sodium salt dihydrate, $C_{15}H_{13}ClNNaO_3 \cdot 2H_2O$, *Zomax, Zomaxin, Zopirac*. Crystals from isopropanol + water, mp 295-296°.

THERAP CAT: Analgesic; anti-inflammatory.

10093. Zometapine. *4-(3-Chlorophenyl)-1,6,7,8-tetrahydro-1,3-dimethylpyrazolo[3,4-e][1,4]diazepine;* CI-781. $C_{14}H_{15}ClN_4$; mol wt 274.25. C 61.20%, H 5.50%, Cl 12.90%, N 20.39%. A pyrazolodiazepine, related structurally to the benzodiazepines, but having primarily antidepressant properties. Prepn: **Belg. pat. 808,599** (1974 to Parke-Davis) corresp to H. A. De Wald, S. J. Lobbestael, **U.S. pat. 3,823,157** (1974); H. A. De Wald *et al.*, *J. Med. Chem.* **24**, 982 (1981). Pharmacology: B. P. H. Poschel *et al.*, *Drugs Exp. Clin. Res.* **7**, 139 (1981). Initial clinical studies: V. B. Tuason *et al.*, *Curr. Ther. Res.* **27**, 94 (1980); N. M. James *et al., ibid.* 100.

Cryst from acetonitrile, mp 180-185°.

Hydrochloride, $C_{14}H_{16}Cl_2N_4$. Cryst from THF, mp 279-281°.

THERAP CAT: Antidepressant.

10094. Zonisamide. *1,2-Benzisoxazole-3-methanesulfonamide;* 3-(sulfamoylmethyl)-1,2-benzisoxazole; AD-810; CI-912; Aleviatin; Exceglan; Excegram. $C_8H_8N_2O_3S$; mol wt 212.22. C 45.28%, H 3.80%, N 13.20%, O 22.62%, S 15.10%. Prepn and anticonvulsant activity: H. Uno *et al.*, **Japan Kokai 78 77,057**; *eidem*, **U.S. pat. 4,172,896** (1978, 1979 both to Dainippon); *eidem, J. Med. Chem.* **22**, 180 (1979). Anticonvulsant activity and neurotoxic effects: Y. Masuda *et al.*, *Epilepsia* **20**, 623 (1979). Mechanism of action study: T. Ito *et al.*, *Arzneimittel-Forsch.* **30**, 603 (1980). Pharmacology and toxicity: Y. Masuda *et al., ibid.* 477. Pharmacological effects on CNS: M. Hori *et al., ibid.* **37**, 1124 (1987). Pharmacology: K. Nakatsuji *et al., ibid.* 1131. Pharmacokinetics: T. Ito *et al., ibid.* **32**, 1581 (1982); J. G. Wagner *et al.*, *Ther. Drug Monit.* **6**, 277 (1984). Absorption and distribution: K. Matsumoto *et al.*, *Arzneimittel-Forsch.* **33**, 961 (1983); E. M. Cornford, K. P. Landon, *Ther. Drug*

Monit. **7**, 247 (1985). HPLC determn in serum: U. Juergens, *J. Chromatog.* **385** 233 (1987). Clinical evaluations in partial seizures: J. C. Sackellares *et al., Epilepsia* **26**, 206 (1985); comparison with carbamazepine, *q.v.* in epilepsy: A. J. Wilensky *et al., ibid.* 212. Brief review: E. J. Hammond *et al., Gen. Pharmacol.* **18**, 303 (1987).

Tasteless, odorless, white needles from ethyl acetate, mp 160-163° (Uno); also reported as mp 162-166° (Masuda, 1980). Sparingly sol in water, chloroform, *n*-hexane. Sol in methanol, ethanol, ethyl acetate, and acetic acid. LD_{50} in mice, rats (mg/kg): 1892, 2001 orally; 1273, 2569 s.c.; 699, 733 i.p.; 604, 748 i.v. (Masuda, 1980).

THERAP CAT: Anti-epileptic.

10095. Zopiclone. *4-Methyl-1-piperazinecarboxylic acid 6-(5-chloro-2-pyridinyl)-6,7-dihydro-7-oxo-5H-pyrrolo-[3,4-b]pyrazin-5-yl ester;* 6-(5-chloropyrid-2-yl)-5-(4-methylpiperazin-1-yl)carbonyloxy-7-oxo-6,7-dihydro-5*H*-pyrrolo[3,4-*b*]pyrazine; RP-27267; Amoban; Amovane; Imovance; Imovane; Zimovane. $C_{17}H_{17}ClN_6O_3$; mol wt 388.82. C 52.52%, H 4.41%, Cl 9.12%, N 21.61%, O 12.34%. The first of a family of non-benzodiazepines showing a pharmacological profile similar to chlordiazepoxide, *q.v.,* and nitrazepam, *q.v.* Prepn: C. Cotrel *et al.,* **Ger. pat. 2,300,491;** *eidem,* **U.S. pat. 3,862,149** (1973, 1975 both to Rhone-Poulenc); C. Jeanmart, C. Cotrel, *Compt. Rend. Ser. C* **287**, 377 (1978). *In vitro* and *in vivo* inhibition of benzodiazepine binding: J. C. Blanchard *et al., Life Sci.* **24**, 2417 (1979); *see also* P. H. Wu *et al., ibid.* **28**, 1023 (1981). Clinical study in sleep disorders: R. Duriez *et al., Therapie* **34**, 317 (1979). Comparative study with benzodiazepines: E. Wickström, K. E. Giercksky, *Eur. J. Clin. Pharmacol.* **17**, 93 (1980). Experimental and clinical pharmacology, clinical trials including sleep laboratory studies and investigations of dependence inducing potential: *Pharmacology* **27**, Suppl. 2, 1-250 (1983). Series of articles on pharmacokinetics, pharmacology and efficacy in insomnia: *Sleep* **10**, Suppl 1, 1-79 (1987). Review of pharmacology and therapeutic efficacy: K. L. Goa, R. C. Heel, *Drugs* **32**, 48-65 (1986).

Crystals from acetonitrile/diisopropyl ether (1:1), mp 178°.

THERAP CAT: Sedative; hypnotic.

10096. Zorubicin. *Benzoic acid [1-[4-[(3-amino-2,3,6-trideoxy-α-L-lyxo-hexopyranosyl)oxy]-1,2,3,4,6,11-hexahydro-2,5,12-trihydroxy-7-methoxy-6,11-dioxo-2-napthacenyl]ethylidene]hydrazide;* benzoic acid hydrazide 3-hydrazone with daunorubicin; RP 22,050. $C_{34}H_{35}N_3O_{10}$; mol wt 645.67. C 63.25%, H 5.46%, N 6.51%, O 24.78%. Semi-synthetic antibiotic related to daunorubicin, *q.v.* Prepn: G. Jolles, **Ger. pat. 2,327,211** (1974 to Rhone-Poulenc), *C.A.* **82**, 171381x (1975). Biological activity: R. Maral *et al., Compt. Rend. Ser. D* **275**, 301 (1972); R. Maral, *Cancer Chemother. Pharmacol.* **2**, 31 (1979). Distribution and metabolism in mice: R. Baurain *et al., ibid.* 37. Mechanism of action: G. P. Sartiano *et al., J. Antibiot.* **32**, 1038 (1979). Acute cardiovascular effects in dogs: E. H. Herman, R. S. Young, *Cancer Treat. Rep.* **63**, 1771 (1979). Clinical study in breast cancer: J. N. Ingle, *ibid.* 1701.

Hydrochloride, $C_{24}H_{36}ClN_3O_{10}$, NSC 164011, *Rubidazone*. Red-orange cryst powder from ethanol. $[\alpha]_D^{20}$ − 50° (c = 0.2 in water). uv max (methanol): 232.5, 253, 480, 495 nm (ϵ 40225, 35300, 10480, 10300). LD_{50} in mice (mg/kg): 13.66 s.c., 4.42 i.p., 8.50 i.v., R. Maral *et al., loc. cit.*

THERAP CAT: Antineoplastic.

10097. Zotepine. *2-[(8-Chlorodibenzo[b,f]thiepin-10-yl)-oxy]-N,N-dimethylethanamine;* 2-chloro-11-(2-dimethyl-aminoethoxy)dibenzo[*b,f*]thiepine; Lodopin. $C_{18}H_{18}ClNOS$; mol wt 331.86. C 65.15%, H 5.47%, Cl 10.68%, N 4.22%, O 4.82%, S 9.66%. A tricyclic enol-ether compound with psychotropic and neurotropic activity. Prepn: S. Umio *et al.,* **Ger. pat. 1,907,670;** *eidem,* **U.S. pat. 3,704,245** (1969, 1972 both to Fujisawa); I. Ueda *et al., Chem. Pharm. Bull.* **26**, 3058 (1978). Pharmacological study: S. Uchida *et al., Arzneimittel-Forsch.* **29**, 1588 (1979). Pharmacokinetics and metabolism: K. Noda *et al., ibid.* 1595. Toxicological and teratological studies: K. Fukuhara *et al., ibid.* 1600.

Crystals from cyclohexane, mp 90-91°. uv max (95% ethanol): 266 nm. LD_{50} in male mice, rats (mg/kg): 108, 458 orally; 43.3, 39.7 i.v.; 40.0, 97.0 i.p.; 84.9, 2080 s.c. (Fukuhara).

THERAP CAT: Antipsychotic.

10098. Zoxazolamine. *5-Chloro-2-benzoxazolamine; 2-amino-5-chlorobenzoxazole;* McN-485; Deflexol; Flexilon; Flexin; Zoxamin; Zoxine. $C_7H_5ClN_2O$; mol wt 168.59. C 49.87%, H 2.99%, Cl 21.03%, N 16.62%, O 9.49%. Prepd by reacting 2-amino-5-chlorophenol with ethanolic cyanogen bromide: Nagana *et al., J. Am. Chem. Soc.* **75**, 2770 (1953); Marsh, Sam, **U.S. pats. 2,780,633** (1957); **2,890,985** (1959). From 4-chloro-2-hydroxyphenylthiourea: Sam, **U.S. pat. 2,978,458** (1961 to McNeil Labs.); from 2,5-dichlorobenzoxazole: Button *et al.,* **U.S. pat. 2,969,370** (1961 to Dow). *Review:* C. K. Cain, A. P. Roszkowski, "Benzoxazoles, Benzothiazoles and Benzimidazoles" in *Medicinal Chemistry: A Series of Monographs* **4**, G. DeStevens, Ed. (Academic Press, New York, 1964) pp 325-357.

Plates from benzene, mp 184-185°. uv max (methanol): 244, 285 nm. Slightly sol in water; sol in alcohol, propylene glycol. LD_{50} in mice, rats (mg/kg): 376, 102 i.p.; 678, 730 orally (Cain, Roszkowski).

Hydrochloride, $C_7H_5ClN_2O \cdot HCl$, needles, mp 229° (dec). Hydrobromide, $C_7H_5ClN_2O \cdot HBr$, plates, mp 240° (dec). 2-Acetylzoxazolamine, $C_9H_7ClN_2O_2$, crystals, mp 212-214°. LD_{50} i.p. in mice: 1500 mg/kg (Cain, Roszkowski).

THERAP CAT: Skeletal muscle relaxant; uricosuric.

10099. Zygadenine. *4α,9-Epoxycevane-3β,4,14,15α,16β,-20-hexol.* $C_{27}H_{43}NO_7$; mol wt 493.62. C 65.69%, H 8.78%,

N 2.84%, O 22.69%. Occurs together with germine in various species of *Zygadenus* (death camas) and *Veratrum (Liliaceae)*: Heyl *et al., J. Am. Chem. Soc.* **35,** 258 (1913); Heyl, Herr, *ibid.* **71,** 1751 (1949); Kupchan, Deliwala, *ibid.* **75,** 1025 (1953); Klohs *et al., ibid.* 4925; Stoll, Seebeck, *Helv. Chim. Acta* **36,** 1570 (1953); Kupchan *et al., J. Am. Chem. Soc.* **77,** 689, 755 (1955). Structure: Kupchan, *ibid.* **78,** 3546 (1956); **81,** 1925 (1959).

Fine needles from benzene. mp 214-216°. $[\alpha]_D^{20}$ −48.4° (c = 1.26 in chloroform). Not precipitated by digitonin. Soluble in alcohol, benzene, chloroform.

Hydrochloride, $C_{27}H_{44}ClNO_7$, crystals, dec 231-234°.

Sulfate, $C_{54}H_{88}N_2O_{18}S$, crystals, dec 237-242°.

10100. Zymosan. Anticomplementary factor. Crude yeast cell wall prepns consisting chiefly of protein-carbohydrate complexes. Prepn from whole yeast cells: Pillemer, Ecker, *J. Biol. Chem.* **137,** 139 (1941); Pillemer *et al., J. Exp. Med.* **103,** 1 (1956); from yeast cell walls: Z. Holan, *Folia Microbiol.* **25,** 501 (1980). Composition of zymosan: F. J. DiCarlo, J. V. Fiore, *Science* **127,** 756 (1958). Enhances the specific defenses of an organism through the activation of the properdin system, *q.v.* Activity studies: Brade *et al., J. Immunol.* **111,** 1389 (1973); **112,** 1115 (1974). Induces aggregation and a release reaction in human platelet-rich plasma: Zucker, Grant, *ibid.* 1219. *Review:* Fitzpatrick, DiCarlo, *Ann. N.Y. Acad. Sci.* **118,** art. 4, 233 (1964).

Light gray powder. Practically insol in water, but readily disperses to give a homogeneous suspension.

USE: In the assay of properdin.

MISCELLANEOUS TABLES

ALPHABETICAL LIST OF TABLES

Cancer Chemotherapy Drug Regimens

Listed below are selected acronyms for combination cancer chemotherapy regimens comprising substances in *The Merck Index*.

Acronym	Drug regimens
ABP	doxorubicin + bleomycin + prednisone
ABVD	doxorubicin + bleomycin + vinblastine + dacarbazine
AC	doxorubicin + cyclophosphamide
ADIC	doxorubicin + dacarbazine
APO	doxorubicin + prednisone + vincristine + 6-mercaptopurine + asparaginase + methotrexate
AV	doxorubicin + vincristine
BACOP	bleomycin + doxorubicin + cyclophosphamide + vincristine + prednisone
BAPP	bleomycin + doxorubicin + cisplatin + prednisone
BCVPP	carmustine + cyclophosphamide + vinblastine + procarbazine + prednisone
BEP	bleomycin + etoposide + cisplatin
BMP	bleomycin + methotrexate + cisplatin
BOLD	bleomycin + vincristine + lomustine + dacarbazine
CA	cyclophosphamide + doxorubicin
CAF	cyclophosphamide + doxorubicin + fluorouracil
CAP	cyclophosphamide + doxorubicin + cisplatin
CAV	cyclophosphamide + doxorubicin + vincristine
CBV	cyclophosphamide + carmustine + etoposide
CFP	cyclophosphamide + fluorouracil + prednisone
CFPMV	cyclophosphamide + fluorouracil + prednisone + methotrexate + vincristine
CFPT	cyclophosphamide + fluorouracil + prednisone + tamoxifen
CHAD	cyclophosphamide + hexamethylmelamine + doxorubicin + cisplatin
CHAMOCA	cyclophosphamide + hydroxyurea + dactinomycin + methotrexate + vincristine + doxorubicin
CHAP-5	cyclophosphamide + hexamethylmelamine + doxorubicin + cisplatin
CHF	cyclophosphamide + hexamethylmelamine + fluorouracil
ChlVPP	chlorambucil + vinblastine + procarbazine + prednisone
CHO	cyclophosphamide + doxorubicin + vincristine
CHOP	cyclophosphamide + doxorubicin + vincristine + prednisone
CHOP-B	cyclophosphamide + doxorubicin + vincristine + prednisone + bleomycin
CMF	cyclophosphamide + methotrexate + fluorouracil
CMFP	cyclophosphamide + methotrexate + fluorouracil + prednisone
CMFVP	cyclophosphamide + methotrexate + fluorouracil + vincristine + prednisone
C-MOPP	cyclophosphamide + mechlorethamine + vincristine + procarbazine + prednisone
COAP	cyclophosphamide + vincristine + cytarabine + prednisolone
COMLA	cyclophosphamide + vincristine + methotrexate* + cytarabine
COMP	cyclophosphamide + vincristine + methotrexate + prednisone
COP	cyclophosphamide + vincristine + prednisone
COP-BLAM	cyclophosphamide + vincristine + prednisone + bleomycin + doxorubicin + procarbazine
COPP	cyclophosphamide + vincristine + prednisone + procarbazine
CVF	cyclophosphamide + vincristine + fluorouracil
CVP	cyclophosphamide + vincristine + prednisone
CYVADIC	cyclophosphamide + vincristine + doxorubicin + dacarbazine
EMA-CO	etoposide + methotrexate + dactinomycin + cyclophosphamide + vincristine
FAC	fluorouracil + doxorubicin + cyclophosphamide
FAM	fluorouracil + doxorubicin + mitomycin C
FAP	fluorouracil + doxorubicin + cisplatin
FUVAC	fluorouracil + vinblastine + doxorubicin + cyclophosphamide
HAD	hexamethylmelamine + doxorubicin + cisplatin
Hexa-CAF	hexamethylmelamine + cyclophosphamide + methotrexate + fluorouracil
LOPP	chlorambucil + vincristine + procarbazine + prednisone
LSA$_2$-L$_2$	cyclophosphamide + vincristine + prednisone + daunorubicin + methotrexate + cytarabine + thioguanine + colaspase + hydroxyurea + carmustine
MAC	methotrexate + dactinomycin + chlorambucil
MACOP-B	methotrexate* + doxorubicin + cyclophosphamide + vincristine + prednisone + bleomycin + co-trimoxazole
M-BACOD	methotrexate + bleomycin + doxorubicin + cyclophosphamide + vincristine + dexamethasone
MBD	methotrexate + bleomycin + cisplatin
MCF	mitoxantrone + cyclophosphamide + fluorouracil
MOP	mechlorethamine + vincristine + procarbazine
MOPP	mechlorethamine + vincristine + procarbazine + prednisone
MVPP	mechlorethamine + vinblastine + procarbazine + prednisone
PAC	cisplatin + doxorubicin + cyclophosphamide
PE	cisplatin + etoposide
PMF	cisplatin + mitomycin C + fluorouracil
ProMACE	prednisone + methotrexate + doxorubicin + cyclophosphamide + etoposide
ProMACE-CytaBOM	prednisone + methotrexate + doxorubicin + cyclophosphamide + etoposide + cytarabine + bleomycin + vincristine + methotrexate*
ProMACE-MOPP	prednisone + methotrexate* + doxorubicin + cyclophosphamide + etoposide + mechlorethamine + vincristine + procarbazine + prednisone
PVB	cisplatin + vinblastine + bleomycin
VAB-6	vinblastine + dactinomycin + bleomycin + cisplatin + cyclophosphamide
VAC	vincristine + dactinomycin + cyclophosphamide
VAMP	vincristine + prednisone + methotrexate + 6-mercaptopurine
VMF	etoposide + methotrexate + fluorouracil
VP	vindesine + cisplatin

*with folinic acid rescue

Abbreviated Terms Used by the U.S. Adopted Names Council (USAN) for Pharmacologically Inactive Radicals

Abbreviation	Chemical name
aceturate	N-acetylglycinate
acistrate	acetate(ester) and octadecanoate(salt) *or* acetate(ester) and stearate(salt)
axetil	1-acetoxyethyl
besylate	benzenesulfonate
buteprate	butyrate(ester) and propionate(ester)
camsylate	camphorsulfonate
caproate	hexanoate
closylate	p-chlorobenzenesulfonate
cyclotate	4-methylbicyclo[2.2.2]oct-2-ene-1-carboxylate
cypionate	cyclopentanepropionate
diolamine	diethanolamine
edetate*	ethylenediaminetetraacetate
edisylate	1,2-ethanedisulfonate
enanthate	heptanoate
erbumine	*tert*-butylamine
estolate	propionate lauryl sulfate
esylate	ethanesulfonate
gluceptate	glucoheptonate
hybenzate	o-(4-hydroxybenzoyl)benzoate
hyclate	monohydrochloride hemiethanolate hemihydrate
isethionate	2-hydroxyethanesulfonate
meglumine	N-methylglucamine
mesylate	methanesulfonate
napsylate	2-naphthalenesulfonate
olamine	ethanolamine
pamoate	4,4'-methylenebis(3-hydroxy-2-naphthoate)
phenpropionate	3-phenylpropionate
pivalate	trimethylacetate
pivoxetil	1-(2-methoxy-2-methyl-1-oxopropoxy)ethyl *or* 1-[(2-methoxy-2-methylpropionyl)-oxy]ethyl
pivoxil	(2,2-dimethyl-1-oxopropoxy)methyl *or* (pivaloyloxy)methyl
tebutate	tertiary butyl acetate
tosylate	p-toluenesulfonate
trolamine	triethanolamine

* The sodium salts are named as follows: monosodium edetate (one Na ion); disodium edetate (two Na ions); trisodium edetate (three Na ions); sodium edetate (four Na ions).

International Non-proprietary Names for
Radicals and Groups Proposed by the
World Health Organization (WHO)

Proposed name	Chemical name
acetofenide	methylphenylmethylene
aceturate	N-acetylglycinate
amsonate	4,4'-diaminostilbene-2,2'-disulfonate
axetil	1-acetoxyethyl
besilate	benzenesulfonate
bunapsilate	3,7-di-*tert*-butyl-1,5-naphthalenedisulfonate
camsilate	camphorsulfonate
carbesilate	p-carboxybenzenesulfonate
ciclotate	4-methylbicyclo[2.2.2]oct-2-ene-1-carboxylate
cipionate	cyclopentanepropionate
closilate	p-chlorbenzenesulfonate
cromacate	[(6-hydroxy-4-methyl-2-oxo-2H-1-benzopyran-7-yl)oxy]acetate
cromesilate	6,7-dihydroxycoumarin-4-methanesulfonate
deanil	2-(dimethylamino)ethyl
decil	decyl
dibudinate	2,6-di-*tert*-butyl-1,5-naphthalenedisulfonate
dibunate	2,6-di-*tert*-butyl-1-naphthalenesulfonate
digolil	2-(2-hydroxyethoxy)ethyl
diolamine	diethanolamine
edisilate	1,2-ethanedisulfonate
embonate	4,4'-methylenebis(3-hydroxy-2-naphthoate)
enantate	heptanoate
esilate	ethanesulfonate
estolate	propionate lauryl sulfate
fendizoate	o-[(2'-hydroxy-4-biphenylyl)carbonyl]benzoate
gluceptate	glucoheptonate
hibenzate	o-(4-hydroxybenzoyl)benzoate
isetionate	2-hydroxyethanesulfonate
lauril	n-dodecyl
laurilsulfate	n-dodecylsulfate
megallate	3,4,5-trimethoxybenzoate
mesilate	methanesulfonate
metembonate	3-methoxy-2-naphthoate
metilsulfate	methylsulfate
napadisilate	1,5-naphthalenedisulfonate
napsilate	2-naphthalenesulfonate
olamine	ethanolamine
oxoglurate	2-oxoglutarate
pivalate	trimethylacetate
pivoxetil	1-(2-methoxy-2-methyl-1-oxopropoxy)ethyl
pivoxil	(pivaloyloxy)methyl
proxetil	1-[(isopropoxycarbonyl)oxy]ethyl
steaglate	stearoyl-glycolate
tebutate	tertiary butyl acetate
tenoate	2-thiophenecarboxylate
teoclate	8-chlorotheophyllinate
teprosilate	1,2,3,6-tetrahydro-1,3-dimethyl-2,6-dioxopurine-7-propanesulfonate
tofesilate	1,2,3,6-tetrahydro-1,3-dimethyl-2,6-dioxopurine-7-ethanesulfonate
tosilate	p-toluenesulfonate
triclofenate	2,4,5-trichlorophenolate
trolamine	triethanolamine
troxundate	[2-(2-ethoxyethoxy)ethoxy]acetate

Code Letters Used by Companies for Experimental Substances

A	Abbott Laboratories, USA
	Ajinomoto Co., Inc., Japan
	Amchem Products, Inc., USA
	Asta Werke AG, W. Germany
	AB Astra, Sweden
	Eli Lilly & Co., USA
	Warner-Lambert Pharm. Co., USA
AB	Takeda Chem. Ind. Ltd., Japan
	Armour Pharmaceutical Co., USA
	Dainippon Pharmaceutical Co., Japan
	Roswell Park Mem. Inst., USA
AC	Ajinomoto Co., Inc., Japan
	Am. Cyanamid Co., USA
	A. Christiaens, Belgium
	Hercules Powder Co., USA
ACL	Monsanto Chemical Co., USA
ACP	Amchem Products Inc., USA
Ad	C. H. Boehringer Sohn, Ingelheim, W. Germany
AD	Dainippon Pharm. Co., Ltd., Japan
	Lab. Miquel, Spain
ADL	A. D. Little, Inc., USA
ADR	Adria Labs., Inc., USA
A-E	Dow Chemical Co., USA
AF	Angelini Francesco, Italy
AG	Abbott Laboratories, USA
AH	Allen & Hanburys, England
	Glaxo Labs., Ltd., England
AHR	A. H. Robins Co., USA
AI	Archifar S.p.A., Italy
AK	E. Merck AG, W. Germany
AM	Kyorin Pharm. Co., Ltd., Japan
AMA	Takeda Chem. Ind. Ltd., Japan
AMR	Affiliated Medical Research Inc., USA
AN	Rhone-Poulenc, France
ANP	Albert Rolland, France
	Lab. Anphar, France
AP	Ciba-Geigy, France
AR	Stanlabs, Inc., USA
AS	Dr. Madaus & Co., W. Germany
ASL	Du Pont Critical Care, USA
AT	Atlas Chemical, USA
AW	Wander AG, Switzerland
AY	Ayerst Labs., USA
B	Bayer AG, W. Germany
	Biosedra, France
	Kelco Co., USA
	Laake Oy, Finland
	Takeda Pharm. Industries, Japan
Ba	Ciba-Geigy AG, Switzerland
	Siegfried AG, Switzerland
BAS	Badische Anilin- und Soda-Fabrik, W. Germany
BAX	Baxter Travenol Labs., Inc., USA
BAY, Bay	Bayer AG, W. Germany
BB	Bristol-Banyu Research Institute (BBRI), Japan
BC	Bristol Labs. of Canada (BLOC)
	Sandoz AG, Switzerland
BDH	British Drug Houses, England
BL	Bristol Labs., Inc., USA
BM	C. F. Boehringer & Soehne, Mannheim, W. Germany
BMY	Bristol Myers Co., USA
BN	Ipsen Beaufour, England
BO	Bottu, France
BRL	Beecham Res. Labs., Ltd., England
BS	Brocades-Stheeman, Netherlands
BTC	Organon Labs., Ltd., England
BTS	Boots Pure Drug Co., Ltd., England
BW	Burroughs-Wellcome Res. Inst., England

C	Byk-Gulden Lomberg, W. Germany
	Cassenne, France
	Ciba-Geigy AG, Switzerland
	Cilag-Chemie AG, Switzerland
	Institut Pasteur, France
	Lab. Sopharga, France
	E. R. Squibb & Sons, USA
CB	Allen & Hanburys, England
	Chester Beatty Res. Inst., England
	Clin-Byla, France (Labs. Byla)
C-C	Cilag-Chemie AG, Switzerland
CERM	Centre Europeen de Recherches Mauvernay (RIOM Labs.), France
CFT	Chem. Fabr. Tempelhof, W. Germany
CG	Glaxo Labs., Ltd., England
	Chemie Grunenthal GmbH, W. Germany
CGA	Ciba-Geigy AG, Switzerland
CGP	Ciba-Geigy AG, Switzerland
CGS	Ciba-Geigy AG, Switzerland
CH	Chinoin, Hungary
	A. Nattermann & Cie, W. Germany
CI	Parke, Davis & Co., USA
CJ	Pharmacia AB, Sweden
CK	Schering AG, W. Germany
CL	Am. Cyanamid Co., USA
	Chemie Linz AG, Austria
	Cutter Labs., USA
	Lederle Labs., USA
CLY	Yoshitomi Pharmaceutical Ind., Co., Ltd., Japan
CM	Sanofi S.A., France
CN	Parke, Davis & Co., USA
CP	Continental Pharma, Belgium
	Monsanto Chemical Co., USA
	Chas. Pfizer & Co., Inc., USA
CRA	Sandoz Pharmaceuticals, USA
CRD	Monsanto Chemical Co., USA
	Tanabe Seiyaku Co., Ltd., Japan
CRL	Labs. Lafon, France
CS	Cassenne, France
	Sankyo Co., Ltd., Japan
	Sumitomo Chemical Co., Japan
CSAG	Philips Roxane Labs., USA
CV	Takeda Chem. Ind. Ltd., Japan
CY	Choay, France
D	Labs. Dr. J. Auclair, France
	Chemiewerke Homburg, W. Germany
	Troponwerke Dinklage & Co., W. Germany
	Dominion Rubber Co., Canada
	Knoll AG, W. Germany
	Ortho Pharm. Corp., USA
	Siegfried AG, Switzerland
DA	Istituto De Angeli, Italy
DAN	Labs. Delagrange, France
DD	Daiichi Seiyaku Co. Ltd., Japan
DE	Santen Pharm. Co. Ltd., Japan
DH	Dow Chemical Co., USA
Diu	Sanol Arzneimittel Dr. Schwartz, W. Germany
DJ	Daiichi Seiyaku Co., Ltd., Japan
DL	Dow-Lepetit, USA
DN	Dow Chemical Co., USA
Do	Byk-Gulden Lomberg, W. Germany
DO	Uniroyal, Inc., USA
DPX	E. I. du Pont de Nemours & Co., USA
DRA	Sandoz AG, W. Germany
DS	Sandoz AG, Switzerland; Inc., USA
DU	Philips-Duphar N.V., Netherlands
DW	Sandoz AG, Switzerland
Dup	Du Pont Pharm., USA
E	Bayer AG, W. Germany
	Eisai Co. Ltd., Japan

EA	Parke, Davis & Co., USA		HSR	Hokuriku Seiyaku Co. Ltd., Japan
EB	L'Equilibre Biologique, France			
EC	Worthington Biochemical Corp., USA		I	Carlo Erba, Italy
EGIS	EGIS Pharmaceutical Works, Hungary		IBD	Sandoz, USA
EGYT	EGIS Pharmaceutical Works, Hungary		IBI	Istituto Biochimico Italiano S.p.A., Italy
EI	Am. Cyanamid Co., USA		ICI	ICI Pharmaceuticals, England
EL	Eli Lilly & Co., USA			Istituto Chemioterapico Italiano, Italy
EM	Evans Medical Ltd., England		ICN	ICN Pharmaceuticals, USA
EMD	E. Merck AG, W. Germany		ID	Sumitomo Chemical Co., Japan
Emfac	Emery Industries Inc., USA		IDB	Iverni & Della Beffa, Italy
EN	Endo Labs., Inc., USA		IL	Ives Labs., USA
ES	Abbott Labs. Ltd., England		IMI	Farmitalia Carlo Erba, Italy
	Eprova Ltd., Switzerland		IN	Neisler Labs., Inc., USA
Eu	Norwich Eaton Pharmaceuticals, USA		INF	Parke, Davis & Co., England
EX	Lakeside Labs., Inc., USA		IS	Istituto Superiore di Sanita, Italy
EXP	Du Pont Pharm, USA		ISF	ISF S.p.A., Italy
			ITF	Italfarmaco, Italy
F	Dominion Rubber Co., Canada		IZ	Spofa Ltd., Czechoslovakia
	Eaton Labs., USA			
	AB Ferrosan, Sweden		J	VEB Fahlberg-List, E. Germany
	Lab. Funai, Japan		JB	Lakeside Labs., Inc., USA
FB	Bayer AG, W. Germany		JDL	Christiaens S.A., Belgium
FBA	Bayer AG, W. Germany		JL	Lab. Jacques Logeais, France
	FBA Pharmaceuticals, Ltd., England		JM	Bristol Myers Co., USA
FC	Farmos-Yhtyma Oy, Finland			
FG	Ferrosan, Sweden		K	Carlo Erba, Italy
FH	C. H. Boehringer Sohn, Ingelheim, W. Germany			Kaken Chemical Co., Japan
	Clin-Byla, France			Klinge Pharma, W. Germany
FI	Farmitalia Carlo Erba, Italy		KA	Kali-Chemie Pharma GmbH, W. Germany
FK	Fujisawa Pharmaceutical Co., Ltd., Japan		KB	Kanebo Pharm. Ltd., Japan
FL	Chinoin, Hungary		KBT	Kanebo Pharm. Ltd., Japan
FLC	Labs. Delagrange, France		KD	Klinge Pharma, W. Germany
FMC	FMC Corp., USA		KF	Dr. Karl Thomae, W. Germany
FPL	Fisons Pharm. Ltd., England		Ko	C. H. Boehringer Sohn, Ingelheim, W. Germany
FR	Fujisawa Pharm. Co. Ltd., Japan		KSD	C. H. Boehringer Sohn, Ingelheim, W. Germany
FUT	Torii & Co. Ltd., Japan			
FW	Rohm & Haas Co., USA		KSW	C. H. Boehringer Sohn, Ingelheim, W. Germany
FWH	Frank W. Horner Ltd., Canada			
			KU	Kyorin Pharm. Co. Ltd., Japan
G	Ciba-Geigy AG, Switzerland		KWD	Draco, Sweden
	Givaudan Corp., USA			
GC	Allied Chemical Corp., Agric. Div., USA		L	Ives Labs. Inc., USA
GEA	Bayer AG, W. Germany			Labaz, Belgium, France
Go	Godecke AG, W. Germany			Lederle Labs., USA
GP	Ciba-Geigy AG, Switzerland			Lepetit, Italy
	Du Pont Pharm, USA			Merck & Co., Inc., USA
GPA	Ciba-Geigy, USA			Sandoz, USA
	GPC Gulf Oil Chemicals Co., USA			Alfa Wassermann S.p.a., Italy
GR	Glaxo, England		LA	Lab Aron, France
GS	Ciba-Geigy AG, Switzerland			Hoffmann-La Roche, Inc., USA
	Pfizer Ltd., England		LAC	AB Bofors, Sweden
			LAS	Almirall, Spain
H	Bracco Industria Chimica, Italy		LB	Hoechst AG, W. Germany
	AB Hassle, Sweden		LBC	Bruneau & Cie, France
HA	Chem. Werke Albert, W. Germany		LD	Lab. Dausse, France
HB	Hoechst AG, W. Germany		LG	Guidotti & C., Italy
HC	Pechiney-Progil, France		LJ	Joullie, France
	Sandoz AG, Switzerland		LJC	Lederle Labs., USA
	Toyo Jozo Co. Ltd., Japan		LL	Lab. Lafon, France
HF	Dorsey Labs., USA (Div. Wander) Wander AG, Switzerland			Lederle Labs., USA
HH	Hommel AG, Switzerland		LM	Lipha S.A., France
HL	VEB Deutsche Hydrierwerk, E. Germany			Maestretti, Italy
HM	Sandoz, Italy			Pharmuka S.F., France
HOE, Hoe	Hoechst-Roussel Pharmaceuticals, Inc., USA		LS	AB Leo, Sweden
				Rhone-Poulenc, France
HP	Smith & Nephew Pharmaceuticals, Ltd., England		LT	Sandoz AG, Switzerland; Inc., USA
HPEK	Paul B. Elder Co., USA		Lu, LU	H. Lundbeck, Denmark
HQ	Tanabe Seiyaku Co. Ltd., Japan		LV	Valda, France
HR	Hoechst-Roussel Pharmaceuticals, Inc., USA		LW	Lab. Wander, France
HS	Sandoz, USA		LY	Eli Lilly & Co., USA

M	Farmoplant, Italy	P	Armour Pharm. Co., USA
	Imperial Chemical Industries, England		Asta Werke AG, W. Germany
	J. F. Macfarlan & Co., Scotland		Bayer AG, W. Germany
	Montavit Co., Austria		Dainippon Pharmaceutical Co., Japan
	Reckitt & Sons Ltd., England		Farmitalia Carlo Erba, Italy
MA	Miles Labs., Inc., USA		Leo Pharm, Denmark
MB	May & Baker Ltd., England		Lepetit, Italy
MBR	3M Co., USA		Chas. Pfizer & Co., USA
MC	Mobil Chemical Co., USA		Pitman-Moore Co., USA
MCI	Mitsubishi Chem. Ind., Ltd., Japan		UCB (Union Chimique Belge), Belgium
MD	Delalande, France		Warner-Lambert Pharm. Co., USA
MDL	Merrell Dow Pharm. Inc., USA	PA	Chas. Pfizer & Co., USA
McN	McNeil Labs., Inc., USA	PAA	Parke, Davis & Co., USA
Me	C. H. Boehringer Sohn, Ingelheim, W.	PB	Dr. Karl Thomae GmbH, W. Germany
	Germany	PC	Sumitomo Chemical Co., Japan
MER	Merrell Dow Pharmaceuticals, Inc., USA	PD	Parke, Davis & Co., USA
MF	C. F. Boehringer & Soehne, Mannheim,	PGA	The Upjohn Co., USA
	W. Germany	PH	Philips-Duphar N.V., Netherlands
	Spofa Ltd., Czechoslovakia	PN	Sandoz Inc., USA
MG	Maggioni & C., Italy	R	Denver Labs., England
MI	Lab. Miquel, Spain	PP	Eisai Co., Ltd., Japan
MJ	Mead Johnson & Co., USA		ICI, Plant Protection Div., England
MK	Merck & Co., Inc., USA	PR	Abbott Labs., USA
MO	Abbott Laboratories, USA	PR-G	Pharma Research, Canada
MON	Monsanto Chemical Co., USA	PS	Abbott Labs., USA
MP	Mallinckrodt Chem. Works, USA	PSC	Sandoz AG, Switzerland
MPV	Farmos-Yhtyma Oy, Finland	Pe	Warner-Lambert Pharm. Co., USA
MR	Medea Researches, Italy	Ph	Pharmacia, Sweden
MRL	Merrell Dow Pharmaceuticals, Inc., USA		
MS	Mitsui Pharm. Inc., Japan	R	Cilag-Chemie, Switzerland
MT	Meiji Seika Kaisha, Japan		Janssen, Belgium
MY	Mitsubishi Chem. Ind. Ltd., Japan		Riker Labs., Inc., USA
MZ	Chinoin, Hungary		Roussel, France
			Stauffer Chemical Co., USA
		RB	Lab. Roger Bellon, France
		RC	Robert & Carriere, France
		RD	Boots Pure Drug Co., England
N	Cementex, USA	REV	William H. Rorer, Inc., USA
	Lundbeck, Denmark	Rd	Lab. Albert Rolland, France
	Nelson Res. & Dev., USA	Rec	Recordati, Italy
	Stauffer Chemical Co., USA	RG	VEB Fahlberg-List, E. Germany
	Dr. Karl Thomae GmbH, W. Germany		William H. Rorer, Inc., USA
NA	Organon Inc., USA	RGH	Gedeon Richter, Hungary
NAT	National Drug Co., USA	RH	Rohm & Haas Co., USA
NC	Fisons Ltd., England	RM	Lab. Roland-Marie, France
	Nissan Kagaku Kogyo K.K., Japan	RMI	Merrell Dow Pharmaceuticals, Inc., USA
	Sandoz, USA	RN	Bristol-Myers Co., USA
	Warner-Lambert Pharm. Co., USA	Ro	Hoffmann-La Roche, Switzerland, USA
NCI	Nissan Kagaku Kogyo K.K., Japan	RP	Rhone-Poulenc, France
NDR	National Drug Co., USA		Specia, France
NE	Norwich Eaton Pharm. Inc., USA	RS	Syntex Labs., Inc., USA
NF	Eaton Labs., Inc., USA	RU	Roussel-UCLAF, France
	Fujisawa Pharmaceutical Co., Japan	RV	Knoll Chemische Fabriken AG, W.
NIH	Natl. Inst. Health, USA		Germany
NK	Nippon Kayaku Co., Ltd., Japan		Ravizza S.p.A., Italy
NKK	Nihon Nohyaku Co. Ltd., Japan	RW	Rowa-Wagner KG, W. Germany
NP	Nippon Soda Co., Ltd., Japan	RX	Reckitt & Colman Ltd., England
NSC	Natl. Cancer Inst., USA		
NSD	Ferrosan, Denmark	S	C. H. Boehringer Sohn, Ingelheim,
NTA	Fujisawa Pharmaceutical Co., Japan		W. Germany
			Hoechst-Roussel, USA
			VEB Isis-Chemie, E. Germany
			Pitman-Moore Co., USA
OA	Sanraku Inc., Japan		Selvi, Italy
OHB	L'Equilibre Biologique, France		Servier, France
OK	Otsuka Pharm. Co. Ltd., Japan		Shionogi & Co., Ltd., Japan
OM	Olin Chemicals, USA		Simes, Italy
OMP	Espe Fabrik, W. Germany		Solco Basel AG, Switzerland
ONO	Ono Pharm. Co. Ltd., Japan		Sumitomo Chemical Co., Ltd., Japan
OP	Ono Pharm. Co. Ltd., Japan		Wyeth Labs., USA
OPC	Otsuka Pharm. Co. Ltd., Japan	SA	Siegfried AG, Switzerland
ORF	Ortho Pharm. Co., USA		VEB Farbenfabrik Wolfen, E. Germany
ORG	Organon, Netherlands	SaH	Sandoz Pharmaceutical, USA
OS	Shell Dev. Corp., USA	SAN	Sandoz AG, Switzerland

SAS	Schiapparelli, Italy		TH	Theraplix, France
SB	Simes S.p.A., Italy		TI	Takeda Chemical Ind. Co., Ltd., Japan
SC	Manetti Roberts, Italy		TMS	Toyo Jozo Co. Ltd., Japan
	G. D. Searle & Co., USA		TN	Taiyo Pharm. Ind. Co., Ltd., Japan
SCH	Sch, Schering Corp., USA		TPN	Sandoz Pharmaceuticals, USA
SCO	Manetti Roberts, Italy		TPO	Sandoz Pharmaceuticals, USA
SD	Diamant, France		TR	Miles Labs. Inc., USA
	Shell Development Co., USA		TRK	Roche Products Ltd., England
	Siegfried AG, Switzerland		TSAA	Takeda Chem. Ind. Ltd., Japan
SDZ	Sandoz AG, Switzerland		TVX	Troponwerke Dinklage & Co., W.
SE	Lab. Servier, France			Germany
SF	Meiji Seika Kaisha Ltd., Japan		TZU	Teikoku Hormone Mfg. Co. Ltd., Japan
	Proter S.p.A., Italy			
	Sandoz, Inc., USA		U	Benzon, Denmark
	E. R. Squibb & Sons, USA			Eaton Labs., Inc., USA
	SG, Chugai Pharmaceutical Co., Japan			The Upjohn Co., USA
SGD	Siegfried AG, Switzerland		UC	Union Carbide Corp., USA
SH	Schering AG, W. Germany		UCB	U.C.B., Belgium
	Shell Dev. Corp., USA			U.C.B. Chemie, W. Germany
SI	Sankyo Co., Ltd., Japan		UK	Pfizer Ltd., England
SKF	Smith, Kline & French, USA		UP	Lab. UPSA, France
SL	Synthelabo, France		UR	Lab. J. Uriach & Cia S.A., Spain
SM	Greenwich Pharm., USA			
	Dr. Schwarz Arzneimittelfabrik GmbH, W.		V	Philips-Duphar N.V., Netherlands
	Germany			Sanol-Arzneimittel Dr. Schwarz, W.
	Sumitomo Pharm. Co., Ltd., Japan			Germany
SMS	Sandoz, Inc., USA		VC	Mobil Chemical Co., USA
SN	Parke, Davis & Co., USA		VC-K	Eli Lilly & Co., Canada
	Sandoz AG, W. Germany; Inc., USA		VCS	Velsicol Chemical Corp., USA
	Schering AG, Switzerland		VD	Leo Pharm. Products, Denmark
	Winthrop Sterling Res. Labs.,			
	Rensselaer, USA		W	Wallace Labs., USA
SPA-S	Societa Prodotti Antibiotici, Italy			Dr. A. Wander AG, Switzerland
SPE	Unilabo, France			Warner-Lambert Pharm. Co., USA
SQ	E. R. Squibb & Sons, USA		WAC	Wassermann S.A., Spain
SR	Chevron Chemical Corp., USA		WG	Riker Laboratories Ltd., England
	Sanofi S.A., France		WHR	William H. Rorer, USA
SRG	Schering Corp., USA		WIN, Win	Winthrop Labs., USA
SW	Sankyo Chemical Co., Ltd., Japan		WL	Warner-Lambert Pharm. Co., USA
St	C. H. Boehringer Sohn, Ingelheim, W.			White Labs. Inc., USA
	Germany			Wyeth Labs., USA
	Chemie Linz AG, Austria		WR	Warner-Lambert Pharm. Co., USA
	E. Merck AG, W. Germany		WS	Washine Chemical Corp., USA
Su	Ciba-Geigy Pharm. Prod., Inc., USA		WV	Diwag, W. Germany
SW	Sankyo Chemical Co., Ltd., Japan		Wy	Wyeth Labs., USA
T	Takeda Chem. Ind. Ltd., Japan		X	Central Drug Res. Inst., India
	Toyama Chem. Co. Ltd., Japan			Simes, Italy
TA	VEB Fahlberg-List, E. Germany		XZ	Chas. Pfizer & Co., Inc., USA
	Tanabe Seiyaku Co., Japan			
TAN	Takeda Chemical Ind. Co., Ltd., Japan		Y	Yoshitomi Pharmaceutical Ind., Japan
TAP	Takeda-Abbott Pharmaceuticals, USA		YM	Yamanouchi Pharm. Co. Ltd., Japan
TC	Chinoin, Hungary			
Tc	Hoechst-Roussel, USA		Z	Zambon, Italy
TEM	Roswell Park Mem. Inst., USA			Zyma AG, Switzerland
TEPA	Roswell Park Mem. Inst., USA		ZK	Schering AG, W. Germany
Th	C. H. Boehringer Sohn, Ingelheim, W.		ZP	Revlon Inc., USA
	Germany		ZY	Zyma GmbH, W. Germany

Company Register

The following table lists the full company names and addresses which correspond to the abbreviations used in the Cross Index to designate company ownership of trademarks.

[Aaciphar]	Aaciphar (subsidiary of Akzo), Brussels, Belgium
[Aandersen]	Aandersen Farmaceutisk Institut S.p.A., Rome, Italy
[Abbott]	Abbott Labs., North Chicago, Illinois
[ABC]	Istituto Biologico Chemioterapico ABC S.p.A., Torino, Spain
[ABIC]	ABIC, Ramat Gan, Israel
[AB Leo]	AB Leo (subsidiary of Pharmacia), Helsingborg, Sweden
[Adria]	Adria Labs., Inc. (subsidiary of Erbamont), Columbus, Ohio
[Adrian-Marinier]	Labs. Adrian-Marinier, Paris, France
[Adroka]	Adroka AG, Basel, Switzerland
[AFI]	AFI, Oslo, Norway
[AGIPS]	AGIPS Pharmaceutici s.r.l., Rapallo, Italy
[Agpharm]	Agpharm AG, Lucerne, Switzerland
[Ajinomoto]	Ajinomoto Co., Inc., Tokyo, Japan
[Akorn]	Akorn, Inc., Abita Springs, Louisiana
[Akzo]	Akzo Zout Chemie Nederland N.V., Amsterdam, The Netherlands
[Albert-Farma]	Albert-Farma S.p.A., Scoppito, Italy
[Albert-Roussel]	Albert-Roussel Pharma, GmbH (subsidiary of Hoechst), Wiesbaden, W. Germany
[Alcolac]	Alcolac Inc., Baltimore, Maryland
[Alcon]	Alcon Labs., Inc. (subsidiary of Nestle), Fort Worth, Texas
[Alconox]	Alconox Inc., New York, New York
[Aldepha]	Aldepha AG, Zurich, Switzerland
[Aldrich]	Aldrich Chemical Co., Inc., Milwaukee, Wisconsin
[Alfa]	Alfa Wassermann S.p.A., Bologna, Italy
[Allard]	Lab. Allard (subsidiary of Bristol-Myers Co.), Paris, France
[Allen & Hanburys]	Allen & Hanburys Ltd. (subsidiary of Glaxo), London, England
[Allergan]	Allergan Pharmaceuticals, Irvine, California
[Allergopharma]	Allergopharma GmbH, Hamburg, W. Germany
[Allied]	Allied Chemical Corp., New York, New York
[Almirall]	Almirall, Barcelona, Spain
[Alpinapharm]	Alpinapharm AG., Zurich, Switzerland
[Alza]	Alza Corp., Palo Alto, California
[Amchem]	Amchem Products Inc. (division of Henkel Corp.), Ambler, Pennsylvania
[Am. Crit. Care]	American Critical Care (subsidiary of Du Pont), McGraw Park, Illinois
[Am. Cyanamid]	American Cyanamid Co., Wayne, New Jersey
[Ames]	Miles Diagnostics Division [formerly Ames Co. (division of Miles Labs., Inc.)], Elkhart, Indiana
[Amfre-Grant]	Amfre-Grant, Inc. (division of Ormont Drug & Chem. Co.), Englewood, New Jersey
[Am. Home]	American Home Products, New York, New York
[Amid]	Amid Labs., Marion, Alabama
[Amino]	Amino AG, Neuenhof-Wettingen, Switzerland
[Am. McGaw]	American McGaw, Honolulu, Hawaii
[Amsa]	A.M.S.A., Berberino di Mugello, Italy
[Am. Urologicals]	American Urologicals, Inc. (subsidiary of Key Pharmaceuticals, Inc.), Miami, Florida
[ANA]	Labs. Nicholas-Ana, Gaillard, France
[Anaquest]	Anaquest (subsidiary of British Oxygen Co.), Madison, Wisconsin
[Anasco]	Anasco GmbH (subsidiary of Boehringer, Ing.), Wiesbaden, W. Germany
[Anca]	Anca Labs., Whitby, Ontario, Canada
[Andreu]	Labs. Dr. Andreu, S.A., Barcelona, Spain
[Andrómaco]	Labs. Andrómaco, S.A., Madrid, Spain
[Angelini Francesco]	Angelini Francesco S.p.A., Rome, Italy
[Ankermann]	Ankermann & Co., GmbH, Friesoythe, W. Germany
[Anphar]	Lab. Anphar, Arcueil, France
[Ansul]	The Ansul Company, Marinette, Wisconsin
[Antibioticos]	Antibioticos A.A. (subsidiary of Montedison), Madrid, Spain
[Apia]	Lab. Apia, Lahr, W. Germany
[Apogepha]	Apogepha, Dresden, E. Germany
[Apotex]	Apotex, Inc., Weston, Ontario, Canada
[Apothekernes]	Apothekernes Labs., Oslo, Norway
[APS]	Approved Prescription Services Ltd., West Yorkshire, England
[Archifar]	Archifar Laboratori S.p.A., Milan, Italy
[Arco]	Arco S.A., Lugano, Switzerland
[Arcum]	Arcum Pharmaceutical Corp., Vienna, Virginia
[Ardeypharm]	Ardeypharm Heilmittel GmbH, Herdecke/Ruhr, W. Germany
[Aristochimica]	Aristochimica S.p.A., Milan, Italy
[Ariston]	Labs. Ariston, Buenos Aires, Argentina
[Armak]	Armak Industrial Chemicals, Div., Akzona Inc., Chicago, Illinois
[Armour Montagu]	Lab Armour Montagu, Paris, France
[Armour Pharm.]	Armour Pharmaceutical Co. (subsidiary of Rorer), Blue Bell, Pennsylvania
[Armstrong]	Labs. Armstrong S.A.C.I.F., Buenos Aires, Argentina
[Arnar-Stone]	Arnar-Stone Labs., Inc., Mt. Prospect, Illinois
[Arnolds]	Arnolds Veterinary Products Ltd., Reading, England
[Aron]	S.N.E. des Lab. Aron, Suresnes, France

[Arsac]	Lab. Arsac (subsidiary of Zambon), Antibes, France
[Artesan]	Artesan GmbH, Lüchow, W. Germany
[Arun]	Arun Products Ltd., Bognor Regis, W. Sussex, England
[Arzneimittelwerk VEB]	Kombinat VEB, Arzneimittelwerk, Dresden, E. Germany
[Asahi]	Asahi Chemical Industry Co., Ltd., Tokyo, Japan
[Asche]	Asche AG (subsidiary of Schering AG), Hamburg, W. Germany
[Ascher]	B. F. Ascher & Co., Inc., Lenexa, Kansas
[Ashland]	Ashland Chemical Co. (division of Ashland Oil Inc.), Dublin, Ohio
[Asla]	Labs. Asla, S.A., Madrid, Spain
[Aspro-Nicholas]	Aspro-Nicholas, Ltd., Slough, Bucks, England
[Assia]	Assia Chemical Labs., Ltd., Tel Aviv, Israel
[Asta]	Asta Pharma AG (subsidiary of Degussa), Frankfurt, W. Germany
[Astier]	Lab du Dr. P. Astier, Paris, France
[Astra]	AB Astra, Södertälje, Sweden
[Astra Pharm.]	Astra Pharmaceutical Products, Inc., Westboro, Massachusetts
[Astra-Syntex]	Astra-Syntex, Södertälje, Sweden
[Athenstaedt]	Athenstaedt & Redeker KG, Bremen, W. Germany
[Atlas]	Atlas Powder Co., Wilmington, Delaware
[Atmos]	Pharma Atmos GmbH & Co., Arzneimittel, Viernheim, W. Germany
[Auclair]	Labs. Auclair, Montrouge, France
[Ausonia]	Ausonia Farmaceutici, Pomezia (Rome), Italy
[Avlon]	Labs. Avlon (subsidiary of ICI Pharma), Cergy, France
[A.V.P.]	A.V.P. Pharmaceuticals, Inc., Clarence, New York
[Ayerst]	Wyeth-Ayerst (division of American Home Products), New York, New York
[Azuchemie]	Azuchemie, Dr. med. R. Müller GmbH & Co., Gerlingen, W. Germany
[Azusa]	Azusa Pharm. Co., Ltd., Tokyo, Japan
[Badarznei]	Badische Arzneimittel GmbH, Baden-Baden, W. Germany
[Badrial]	Labs. Badrial, Reims, France
[Baeschlin]	Dr. E. Baeschlin GmbH, Germering, W. Germany
[Bago]	Labs. Bago s.r.l., Buenos Aires, Argentina
[Bailly]	Lab. A. Bailly S.P.E.A.B., Paris, France
[Baldacci]	Labs. Baldacci, S.p.A., Pisa, Italy
[Banyu]	Banyu Pharmaceutical Co., Ltd., Tokyo, Japan
[Barnes-Hind]	Barnes-Hind Pharmaceuticals Inc., Sunnyvale, California
[BASF]	Badische Anilin- und Soda-Fabrik, Ludwigshafen/Rhein, W. Germany
[BASF-Wyandotte]	BASF-Wyandotte Corp., Wyandotte, Michigan
[Basotherm]	Basotherm GmbH (subsidiary of Boehringer, Ing.), Biberach an der Riss, W. Germany
[Bastian]	Bastian-Werk, GmbH, Munich, W. Germany
[Bausch & Lomb]	Bausch & Lomb Pharmaceuticals, Clearwater, Florida
[Baxter Travenol]	Baxter Travenol (formerly Baxter Labs.), Morton Grove, Illinois
[Bayer]	Bayer AG, Leverkusen, W. Germany
[Baylor]	Baylor Labs., Hurst, Texas
[Bayropharm]	Bayropharm GmbH (subsidiary of Bayer AG), Cologne, W. Germany
[Bayvet]	Bayvet Division of Bayer AG, Shawnee, Kansas
[BDH]	BDH Pharmaceuticals Ltd., London, England
[Beard]	Beard, Glynn A., Inc., Denver, Colorado
[Beecham]	Beecham Research Labs., Ltd., Brentford, Middlesex, England. Beecham-Massengill Pharmaceuticals, Bristol, Tennessee
[Behringwerke]	Behringwerke AG (subsidiary of Hoechst), Marburg/Lahn, W. Germany
[Beiersdorf]	Beiersdorf AG, Hamburg, W. Germany
[Belupo]	Belupo, Zagreb, Yugoslavia
[Bencard]	Bencard, Middlesex, England
[Bene-Arzneimittel]	Bene-Arzneimittel GmbH, Munich, W. Germany
[Bengue]	Bengue & Co., Ltd., Wembley, Middlesex, England
[Bentex]	Bentex Pharmaceuticals Co. (division of International Chemical & Nuclear Corp.), Houston, Texas
[Benvegna Neoterapici]	Benvegna s.r.l., Palermo, Italy
[Benzon]	Alfred Benzon A/S, Copenhagen, Denmark
[Bergamon]	Bergamon S.p.A., Rome, Italy
[Berk]	Berk Pharmaceuticals Ltd., Gadalming, Surrey, England
[Berlex]	Berlex Labs. (subsidiary of Schering AG), Cedar Knolls, New Jersey
[Berlin-Chemie]	VEB Berlin-Chemie, Berlin, E. Germany
[Berlin Labs.]	Berlin Labs., Inc., New York, New York
[Bernabo]	Labs. Bernabo y Cia, S.A., Buenos Aires, Argentina
[Besins-Iscovesco]	Labs. Besins-Iscovesco, Paris, France
[Beta]	Lab. Biologico Chemioterapico Beta s.r.l., Brescia, Italy
[Beta, Argentina]	Labs. Beta, Buenos Aires, Argentina
[Beytout]	Labs. Beytout, Saint-Mandé, France
[Bika]	Bika Arzneimittel Fabrik, Stuttgart, W. Germany
[Biochemie]	Biochemie GmbH, Vienna, Austria
[Biocodex]	Lab. Biocodex, Montrouge, France
[Bio-Derivatives]	Bio-Derivatives Corp., Deer Park, New York
[Bioindustria]	Bioindustria Farmaceutici S.p.A., Novi Ligure, Italy
[Biologici]	Biologici Italia, Lucca, Italy
[Biopharma]	Biopharma (Labs. Biopharmaceutiques de France) (division of Servier), Neuilly, France

[Biorex]   Biorex Labs, London, England
[Biosedra]   Lab. Biosedra (subsidiary of Knoll AG), Malakoff, France
[Biosintetica]   Biosintetica, São Paulo, Brazil
[Biothérax]   Lab. Biothérax (subsidiary of Boehringer, Ing.), Plaine-St-Denis, France
[Bitterfeld]   VEB Chemiekombinat Bitterfeld, Wolfen, E. Germany
[Blue Line]   The Blue Line Chemical Co., St. Louis, Missouri
[Bock]   Bock Pharmacal Co., St. Louis, Missouri
[Boehringer, Ing.]   C. H. Boehringer Sohn, Ingelheim, W. Germany
[Boehringer, Mann.]   C. F. Boehringer & Soehne GmbH, Mannheim, W. Germany
[Bofors]   AB Bofors, Nobel-Pharma, Mölndal, Sweden
[B.O.I.]   Labs. B.O.I., Barcelona, Spain
[Boizot]   Boizot S.A. Laboratorios, Madrid, Spain
[Bolar]   Bolar Pharmaceutical Co., Inc., Copiague, New York
[Bonomelli]   Bonomelli S.p.A. (subsidiary of Glaxo), Dolzato, Italy
[Boots]   Boots Pure Drug Co., Ltd., Nottingham, England
[Boots-Dacour]   Labs. Boots-Dacour, Courbevoie, France
[Borden]   The Borden Chemical Co. (division of Borden, Inc.), New York, New York
[Bottu]   Lab. Bottu, Nanterre, France
[Bouchara]   Lab. du Dr. E. Bouchara, Paris, France
[Bowman]   Bowman Pharmaceuticals, Canton, Ohio
[Boyle]   Boyle & Co. Pharmaceuticals, Pasadena, California
[Bracco]   Bracco Industria Chimica S.p.A., Milan, Italy
[Braun]   Braun Melsungen, Melsungen, W. Germany
[Brayten]   Brayten Pharmaceutical Co. (division of Chattem Drug & Chemical Co.), Chattanooga, Tennessee
[Brenner]   Georg A. Brenner Arzneimittel Fabrik GmbH, Alpirsbach, Sulzburg, W. Germany
[Breon]   Sterling Drug Inc. [formerly Breon Labs., Inc. (subsidiary of Sterling Drug Inc.)], New York, New York
[Bristol]   Bristol Labs. (division of Bristol-Myers Co.), Syracuse, New York
[Bristol-Banyu]   Bristol-Banyu, Tokyo, Japan
[Bristol-Myers]   Bristol-Myers Co., New York, New York
[Britannia]   Britannia Pharmaceuticals Ltd., Reigate, Surrey, England
[Brocades]   Brocades Ltd., West Byfleet, Surrey, England
[Brocades-Stheeman]   Brocades-Stheeman & Pharmacia, Amsterdam, Netherlands
[Brocchieri]   Stabilimento Chimico Farmaceutico dr. L. Brocchieri s.r.l., Rome, Italy
[Bruneau]   Lab. Bruneau, Boulogne, France
[Brunnengräber]   Dr. Christian Brunnengräber GmbH (subsidiary of Ciba-Geigy), Lübeck, W. Germany
[Brunner]   Ludwig Brunner GmbH, Nürnberg, W. Germany
[Brunton]   Brunton Chemists Ltd., Sunderland, Durham, England
[Bruschettini]   Bruschettini s.r.l., Genoa, Italy
[Burns-Biotec]   Burns-Biotec Laboratories (subsidiary of Schering Corp.), Omaha, Nebraska
[Burroughs Wellcome]   Burroughs Wellcome Co., (subsidiary of The Wellcome Foundation Ltd.), Research Triangle Park, North Carolina
[Byk-Essex]   Essex Pharm. GmbH (formerly Byk-Essex Pharm.), Munich, W. Germany
[Byk-Gulden]   Byk-Gulden Lomberg Chem. Fabrik, GmbH, Konstanz, W. Germany
[Caber]   Farmaceutici Caber s.r.l., Pisa, Italy
[Caffaro]   Caffaro S.p.A., Milan, Italy
[Calbiochem]   Calbiochem, San Diego, California
[Calgon]   Calgon Corp. (subsidiary of Merck & Co., Inc.), Pittsburgh, Pennsylvania
[Calipe]   Calipe, S.A., Reus, Spain
[Calmic]   Calmic (division of The Wellcome Foundation), Berkhamsted, England
[Camden]   Camden Chemical Co., Ltd., London, England
[Care Labs]   Care Laboratories Ltd., Wilmslow, Cheshire, England
[Carlton]   The Carlton Corporation, Tenafly, New Jersey
[Carnegie]   Carnegie Medical, Loughborough, Leicestershire, England
[Carnrick]   G. W. Carnrick, Co., Cedar Knolls, New Jersey
[Carrion]   Labs. Carrion, Courbevoie, France
[Casasco]   Labs. Casasco, Buenos Aires, Argentina
[Cascan]   Cascan GmbH (associated with E. Merck AG), Wiesbaden, W. Germany
[Cassella-Riedel]   Cassella-Riedel Pharma, Frankfurt, W. Germany
[Cassenne]   Lab. Cassenne, Paris, France
[Causyth]   Causyth, S.p.A., Milan, Italy
[CCP]   Comptoir Chimico Pharmaceutique S.A., Brussels, Belgium
[Cehasol]   Cehasol, Vienna, Austria
[Celamerck]   Celamerck GmbH & Co., KG, Ingelheim/Rhein, W. Germany
[Cenci]   H. R. Cenci Labs., Inc., Fresno, California
[Centerchem]   Centerchem Pharmaceuticals, Inc., Tarrytown, New York
[Central Pharm.]   Central Pharmaceuticals, Inc., Seymour, Indiana
[Century]   Century Pharmaceuticals, Inc., Indianapolis, Indiana
[C.E.P.A.]   Compañía Española de Penicilina y Antibioticos, S.A., Madrid, Spain
[Cétrane]   Lab. Cétrane (subsidiary of Schering-Plough), Levallois, France
[C.F.T.S.]   Centres Francais de Transfusion Sanguine, Paris, France
[Chantereau]   Labs. Chantereau (Prod. Innothera), Arcueil, France
[Chassot]   Chassot & Cie AG, Koniz, Switzerland
[Chauvin-Blache]   Chauvin-Blache, S.A., Montpellier, France

[Chemagro]   Chemagro Agricultural Chemicals Div. (division of Mobay Chemical Corp.), Kansas City, Missouri
[Chemiefarma]   ACF Chemiefarma N.V., Amsterdam, Netherlands
[Chemie Linz]   Chemie Linz AG, Linz/Donau, Austria
[Chemil]   Chemil Farmaceutici s.r.l., Milan, Italy
[Chemioterapico]   Istituto Chemioterapico Italiano, Milan, Italy
[Chemipharm]   Chemipharm GmbH & Co., Chemisch-pharmazeutische Fabrik KG, Cologne, W. Germany
[Chemofux]   Chemofux GmbH, Vienna, Austria
[Chephasaar]   Chephasaar, Chemische-Pharmazeutische Fabrik KG, Cologne, W. Germany
[Chesebrough]   Chesebrough Manufacturing Co., Consolidated, New York, New York
[Chevron]   Chevron Chemical Co. (subsidiary of Standard Oil of California), San Francisco, California
[Chibret]   Labs. Merck Sharp & Dohme-Chibret S.A. (subsidiary of Merck & Co., Inc.), Paris, France
[Chiesi]   Chiesi Farmaceutici S.p.A., Parma, Italy
[Chinoin]   Chinoin Gyógyszer és Vegyészeti Termékek Gyára, Budapest, Hungary
[Chinosolfabrik]   Chinosolfabrik (subsidiary of Riedel-de Haen AG), Seelze, W. Germany
[Choay]   Lab. Choay (subsidiary of Sanofi), Paris, France
[Christiaens]   Christiaens S.A., Brussels, Belgium
[Chugai]   Chugai Pharmaceutical Co., Ltd., Tokyo, Japan
[Ciba]   Ciba-Geigy, Basel, Switzerland (formerly Ciba AG, Basel, Switzerland; Ciba Pharmaceuticals, Summit, New Jersey)
[Ciba-Geigy]   Ciba-Geigy, Basel, Switzerland
[Cidán]   Labs. Cidán, Benicarló (Castellón), Spain
[Cifa]   Cifa Farmaceutici S.p.A., Torino, Italy
[Cilag-Chemie]   Cilag-Chemie, AG (subsidiary of Johnson & Johnson), Schaffhausen, Switzerland
[Circle]   Circle Pharmaceuticals, Inc., Indianapolis, Indiana
[Cleary]   Cleary Corp., New Brunswick, New Jersey
[Clin-Comar-Byla]   Lab. Clin-Comar-Byla, Paris, France
[Clin Midy]   Lab. Clin Midy (subsidiary of Sanofi), Paris, France
[Cochard]   G. A. Cochard S.A., Brussels, Belgium
[Cole]   Cole Pharmacal Co., Inc., St. Louis, Missouri
[Coli]   Farmaceutici Coli S.p.A., Rome, Italy
[Collins]   L. D. Collins & Co., Ltd., Hertfordshire, England
[Combe]   Combe, White Plains, New York
[Comm. Solvents]   Commercial Solvents Corp., Terre Haute, Indiana
[Conal]   Conal Pharmaceuticals (subsidiary of Alcon Labs., Inc.), Fort Worth, Texas
[Conoco]   Conoco Chemicals, Continental Oil Co., Houston, Texas
[Consolidated Chem.] Consolidated Chemicals Ltd., Wrexham, Denbighshire, England
[Consolidated Midland] Consolidated Midland Corp., Brewster, New York
[Cont. Pharma]   Continental Pharma S.A. (subsidiary of Monsanto), Brussels, Belgium
[Cooper]   Cooper Labs., Inc. (division of Burroughs Wellcome), Parsippany, New Jersey
[CooperVision]   CooperVision Pharmaceuticals Inc., San German, Puerto Rico
[Coop. Farm.]   Cooperative Farmaceutica Societa Coop., a.r.l., Milan, Italy
[Corbiere]   Corbiere Pharma Lab., Paris, France
[Corvi]   Camillo Corvi S.p.A., Piacenzi, Italy
[COSMA]   Compagnia Specialita Medicinali, Brescia, Italy
[Cosmopharma]   Cosmopharma N.V., Zwanenburg, Netherlands
[Cox Pharm]   A.H. Cox & Co. Ltd., Devon, England
[C.R.E.A.T.]   C.R.E.A.T. (Centre de Recherches d'Études et d'Applications Thérapeutiques), Neuilly, France
[Crinos]   Industria Farmacobiologica Crinos S.p.A., Como Villaguardia, Italy
[Cristalfarma]   Cristalfarma s.r.l., Milan, Italy
[Critikon]   Critikon (subsidiary of Johnson & Johnson), Irvine, California
[Croda]   Croda, Inc., New York, New York
[Crookes]   Crookes Labs, Basingstoke, England
[Crookes Vet.]   Crookes Veterinary, Basingstoke, Hantshire, England
[Crosara]   Lab. Farmaco Biologico Crosara S.p.A., Rome, Italy
[CT]   C.T.-Lab. Farmaceutico s.r.l., San Remo, Italy
[Cusi]   Laboratorios Cusi SA, El Masnou (Barcelona), Spain
[Cutter]   Cutter Labs., Inc. (division of Miles Labs.), Berkeley, California
[C-Vet]   C-Vet Ltd., Bury St. Edmonds, Suffolk, England
[Dagra]   Dagra N.V., Diemen, Netherlands
[Daiichi]   Daiichi Seiyaku Co., Ltd., Tokyo, Japan
[Dainippon]   Dainippon Pharmaceutical Co., Ltd., Osaka, Japan
[Daisan]   Daisan Koeki Co., Ltd., Tokyo, Japan
[Damancy]   Damancy & Co., Ltd., Slough, Bucks., England
[Damor]   Farmaceutici Damor S.p.A., Naples, Italy
[Danal]   Danal Labs., Inc., St. Louis, Missouri
[Dandoy]   Dandoy S.A., Brussels, Belgium
[Dauelsberg]   Dauelsberg & Co., Penicillin-Gesellschaft, Gottingen, W. Germany
[Dausse]   Labs. Dausse, Paris, France
[Davis & Geck]   Davis & Geck (subsidiary of American Cyanamid Co.), Pearl River, New York
[DDSA]   DDSA Pharmaceuticals, London, England

[De Angeli]	Istituto de Angeli S.p.A.(subsidiary of Boehringer, Ing.), Milan, Italy
[Debat]	Lab. Debat, Paris, France
[Decenta]	Decenta Werk VEB, Döbeln, E. Germany
[Declimed]	Declimed (subsidiary of Desitin-Werk Carl Klinke GmbH), Hamburg, W. Germany
[Dedieu]	Labs. Dedieu S.A., Bordeaux, France
[Deglaude]	Deglaude, Merignac, France
[Degussa]	Deutsche Gold- und Silber-Schneideanstalt vormals Roessler, Frankfurt, W. Germany
[Deiglmayr]	Dr. Ivo Deiglmayr, Munich, W. Germany
[Delagrange]	Labs. Delagrange, Paris, France
[Delalande]	Lab. Delalande, Courbevoie, France
[Delandale]	Delandale Labs., (subsidiary of Delalande), Canterbury, Kent, England
[Delbay]	Delbay Pharmaceuticals Inc. (division of Schering Corp.), Bloomfield, New Jersey
[Delmar]	Delmar Chemicals Ltd., Lazalle, Quebec, Canada
[Delta]	Delta Drug Corp., Jacksonville, Florida
[Delta-Chemie]	Delta-Chemie GmbH u. Co., Neu-Isenburg, W. Germany
[Denver]	Denver Labs., Ltd., Wembley, Middlesex, England
[DePree]	The DePree Co., Holland, Michigan
[Dermik]	Dermik Labs., Inc. (subsidiary of William H. Rorer, Inc.), Fort Washington, Pennsylvania
[Dermohr]	Dermohr, Bryan, Ohio
[Desitin]	Desitin Arzneimittel GmbH, Hamburg, W. Germany
[Dessy]	Istituto Biologico Dessy S.p.A., Florence, Italy
[Deut. Hydrierwerk]	VEB Deutsches Hydrierwerk, Rodleben, E. Germany
[Dexo]	Labs. Dexo, Nanterre, France
[Dexter]	Dexter, Buenos Aires, Argentina
[DHA]	Drug Houses of Australia Ltd., Melbourne, Victoria, Australia
[Diagnostic Data]	Diagnostic Data, Inc., Mountain View, California
[Diamant]	Labs. Diamant (subsidiary of Roussel-UCLAF), Pateaux, France
[Diamond]	Diamond Shamrock Chemical Co., Cleveland, Ohio
[Dieckmann]	Dieckmann Arzneimittel GmbH (subsidiary of Boehringer, Ing.), Bielefeld, W. Germany
[DIMA]	DIMA, Lab. Biofarmaceutici s.r.l., Rome, Italy
[Dista]	Dista Products, Ltd. (subsidiary of Lilly), Liverpool, England
[Diwag]	Diwag Chemische Fabrieken GmbH, Amsterdam, Netherlands
[Doak]	Doak Pharmacal Co., Inc., Westbury, Long Island, New York
[Doetsch, Grether]	Doetsch, Grether & Cie, Basel, Switzerland
[Dolder]	Dolder AG, Basel, Switzerland
[Dolorgiet]	Dolorgiet GmbH u. Co. KG, Bonn, W. Germany
[Dome]	Miles Pharmaceutical Division [formerly Dome Labs. (division of Miles Labs., Inc.)], West Haven, Connecticut
[Dompé]	Dompé Farmaceutici S.p.A., Milan, Italy
[Donau-Pharmazie]	Donau-Pharmazie GmbH (subsidiary of Rhone-Poulenc), Linz, Austria
[Dooner]	Dooner Labs., Inc., Haverhill, Massachusetts
[Dorsch]	Dorsch GmbH & Co., KG, Munich, W. Germany
[Dorsey]	Dorsey Labs. (division of Sandoz, Inc.), Lincoln, Nebraska
[Dow]	Dow Chemical Co., Midland, Michigan
[Doyle]	Doyle Pharmaceutical Co. (division of The Delmark Company, Inc.), Minneapolis, Minnesota
[Draco]	AB Draco (subsidiary of AB Astra), Lund, Sweden
[Dukron]	Dukron Italiana S.p.A., Campoverde (Latina), Italy
[Dulcis]	Laboratoire Dulcis, Monaco, France
[Dumex]	Dumex Ltd., Copenhagen, Denmark
[Duncan Farm]	Duncan Farmaceutici S.p.A., Verona, Italy
[Duncan, Flockhart]	Duncan, Flockhart & Co., Ltd. (subsidiary of Glaxo), London, England
[Duphar]	Duphar Labs., Ltd. (subsidiary of Solvay Corp.), Southampton, England
[Du Pont]	E. I. du Pont de Nemours & Co., Inc., Wilmington, Delaware
[Du Pont Pharm.]	Du Pont Pharmaceuticals (division of E. I. du Pont de Nemours & Co., Inc.), Garden City, New York
[Durachemie]	Durachemie GmbH & Co., Rottach-Egern, W. Germany
[Durst]	S. F. Durst & Co., Inc., Philadelphia, Pennsylvania
[Eastman-Kodak]	Eastman-Kodak Co., Rochester, New York
[Eaton]	Norwich-Eaton Pharmaceuticals, formerly Eaton Labs. (subsidiary of Procter & Gamble), Norwich, New York
[Ebewe]	Ebewe Arzneimittelfabrik (subsidiary of BASF), Unterach am Attersee, Austria
[Edwin Burgess]	Edwin Burgess Ltd., Aylesbury, Bucks, England
[Efeka]	Efeka Friedrich & Kaufman GmbH & Co., KG, Hannover, W. Germany
[Egic]	Laboratoires Egic-Joullie, Montargis, France
[EGIS]	Egis Pharmaceutical Works, Budapest Hungary
[EGYT]	Egis Pharmaceutical Works (formerly Egyesült Gyógyszer és Tápszer Gyár), Budapest, Hungary
[Eisai]	Eisai Co., Ltd., Tokyo, Japan
[Elan]	Elan Corp., Westmeath, Ireland
[Elanco]	Elanco Products Co. (subsidiary of Eli Lilly & Co.), Indianapolis, Indiana and Omaha, Nebraska
[Elder]	Paul B. Elder Co., Inc., Bryan, Ohio
[E.L.E.A.]	Labs. E.L.E.A., Buenos Aires, Argentina

[Elmu]	Especialidades Latinas Medicamentos Universales S.A., Madrid, Spain
[E. Merck]	E. Merck AG, Darmstadt, W. Germany
[E. Merck-Clévenot]	Labs. Merck-Clévenot, Nogent-sur-Marne, France
[Emery]	Emery Industries, Inc., Mauldin, South Carolina
[Emser]	Emser Werke AG, Zurich, Switzerland
[Endo]	Du Pont Pharmaceuticals, formerly Endo Labs., Inc. (subsidiary of E. I. du Pont de Nemours & Co., Inc.), Garden City, New York
[Endopharm]	Endopharm Arzneimittelfabrik GmbH, Sprendlinger, W. Germany
[Engelhard]	Engelhard Corp., Edison, New Jersey
[Engelhard KG]	Karl Engelhard, Fabrik pharmazeutischer Präparate, Frankfurt/Main, W. Germany
[Enzomedic]	Enzomedic Labs., Inc., Seattle, Washington
[Erba]	Carlo Erba S.p.A. (subsidiary of Montedison), Milan, Italy
[Erco]	Ercopharm A/S, Vedbaeck, Denmark
[Errekappa]	Errekappa Euroterapici, Milan, Italy
[Espe]	Espe Fabrik pharmazeutischer Präparate GmbH, Seefeld/Oberbayern, W. Germany
[Esseti]	Esseti s.a.s. Laboratorio Chimico Farmaco Biologico, S. Giorgi-Cremano, Italy
[Essex]	Essex Italia, Milan, Italy
[Esta]	Esta Medical Labs., Inc., Evanston, Illinois
[Ester]	Ester Farmaceutico Lab., Madrid, Spain
[Esteve]	Labs. Dr. Esteve, S.A., Barcelona, Spain
[Ethicon]	Ethicon, Inc. (subsidiary of Johnson & Johnson), Somerville, New Jersey
[Europa]	Labs. Europa, S.L., Vigo (Pontevedra), Spain
[Europharma]	Labs. Europharma, S.A., Madrid, Spain
[Eutherapie]	Lab. Eutherapie (division of Servier), Neuilly-sur-Seine, France
[Evans]	Evans Healthcare Ltd. (subsidiary of Glaxo), Liverpool, England
[Exa]	Exa Impex Ltda., Buenos Aires, Argentina
[Fabre]	Lab. Pierre Fabre, Departement medical de Fimex S.A., Paris, France
[Fahlberg-List]	VEB Fahlberg-List, Magdeburg, E. Germany
[F.A.I.R.]	F.A.I.R. Labs., Twickenham, Middlesex, England
[Fairfield]	Fairfield American Corp., Medina, New York
[Falk]	Dr. Falk GmbH & Co., Freiburg, W. Germany
[Falorni]	Istituto Farmochimico Falorni S.p.A., Florence, Italy
[Fardeco]	Fardeco S.p.A., Piacenza, Italy
[Fargal]	Fargal-Pharmasint s.r.l., Rome, Italy
[Farillon]	Farillon Ltd., Romford, Essex, England
[Farmacologico]	Lab. Farmacologico Milanese s.r.l., Caronno Pertusella, Italy
[Farmacosmici]	Farmacosmici s.r.l., Rome, Italy
[Farmades]	Farmades S.p.A., Rome, Italy
[Farmaroma]	Farmaroma s.r.l., Lab. Farmacobiologici, Rome, Italy
[Farmaryn]	Farmaryn-Arzneimittel, Berlin, W. Germany
[Farmasa]	Farmasa, São Paulo, Brazil
[Farmetrusca]	Farmetrusca s.a.s. di C. Pini e C., Tavarnuzze (Firenze), Italy
[Farmex]	Farmex, Courbevoie, France
[Farmigea]	Farmigea S.p.A.-Industria Chimico Farmaceutica, Pisa, Italy
[Farmila]	Farmila-Farmaceutici Milano S.p.A., Milan, Italy
[Farmitalia]	Farmitalia Carlo Erba (subsidiary of Erbamont), Milan, Italy
[Farmochim. Ital.]	La Farmochimica Italiana S.p.A., Milan, Italy
[Farmoplant]	Farmoplant (subsidiary of Montedison), Milan, Italy
[Farmos]	Farmos-Yhtyma Oy, Turku, Finland
[Farnex]	Laboratori Farnex S.p.A., Codagno, Italy
[Fawns & McAllen]	Fawns & McAllen Pty, Ltd., Croydon, Victoria, Australia
[FBA Pharmaceuticals]	FBA Pharmaceuticals, Inc., New York, New York
[Federal]	Federal Pharmacal Corp. (subsidiary of Ormont Drug & Chemical Co., Inc.), Fort Lauderdale, Florida
[Fellows-Testagar]	Fellows-Testagar, Oak Park, Michigan
[Ferndale]	Ferndale Labs., Inc., Ferndale, Michigan
[Ferrer]	Labs. Ferrer, S.L., Barcelona, Spain
[Ferring]	Ferring AB, Malmö, Sweden
[Ferrosan]	AB Ferrosan, Malmö, Sweden
[Fides]	Lab. Fides, Barcelona, Spain
[Fidia]	Farmaceutici Italiani Derivati Industriali e Affini S.p.A., Albano Terme, Italy
[Fine Organics]	Fine Organics Inc. (subsidiary of Hexcel Corp.), Lodi and Sayreville, New Jersey
[FIRMA]	FIRMA, Fabbr. Ital. Ritrov. Medic. Aff. S.p.A., Florence, Italy
[Firmenich]	Firmenich et Cie, Geneva, Switzerland
[Fischer]	Fischer Arzneimittelwerk, Buhl (Baden), W. Germany
[Fisons]	Fisons Pharmaceuticals Ltd., Loughborough, England
[Flint]	Flint Labs. (subsidiary of Boots Pure Drug Co.), Deerfield, Illinois
[Fluxine]	Lab. Fluxine Synthemedica S.A., Villefranche-sur-Saone, France
[FMC]	FMC Corp. (American Viscose Div.), New York, New York
[Fontoura-Wyeth]	Industrias Farmaceuticos Fontoura-Wyeth S.A., Sao Paulo, Brazil
[Formenti]	Formenti Dott., S.p.A., Milan, Italy
[Fort Dodge]	Fort Dodge Labs. Co., Inc. (subsidiary of American Home Products Corp.), Fort Dodge, Iowa
[Foscama]	Biomedica Foscama Industria Chimico-Farmaceutica S.p.A., Rome, Italy
[Fougera]	E. Fougera & Co., Inc., Hicksville, Long Island, New York

MISC-14

[Fournier]	Laboratoires Fournier S.A., Dijon, France
[Fournier Frères]	Lab. Fournier Frères, Clichy, France
[Foy]	Foy Labs. Div., Wernersville, Pennsylvania
[Francia]	Francia Farmaceutici s.r.l., Industria Farmaco Biologica, Milan, Italy
[Franç. Thérap.]	Lab. Français de Thérapeutique S.A., Bordeaux, France
[Fraysse]	Labs. Fraysse et Cie, Nanterre, France
[Fresenius]	Dr. E. Fresenius Chemische-Pharmazeutische Industrie, K.G., Bad Homburg, W. Germany
[Frosst]	Frosst Division, Merck Frosst Canada, Inc., Montreal, Canada
[Fujinaga]	Fujinaga Pharmaceutical Co., Ltd., Tokyo, Japan
[Fujisawa]	Fujisawa Pharmaceutical Co., Ltd., Osaka, Japan
[Fumouze]	Labs. Fumouze, S.A., Ile-Saint-Denis, France
[Funai]	Funai Pharmaceutical Co., Ltd. (subsidiary of Merrell-Dow), Osaka, Japan
[Fuso]	Fuso Pharmaceutical Industries, Ltd., Osaka, Japan
[Gador]	Lab. Dr. Gador y Cia. S.A.C.I., Buenos Aires, Argentina
[GAF]	General Aniline & Film Corp., New York, New York
[Galactina]	Galactina Arzneimittel-Fabrik, GmbH, Neu-Isenburg bei Frankfurt/Main, W. Germany
[Galen]	Galen Ltd., Portadown, Co., Armagh, Northern Ireland
[Galenica]	Galenica Vertretungen AG, Bern, Switzerland
[Galenika]	Galenika, Zemun, Yugoslavia
[Galenus]	Galenus Mannheim GmbH (subsidiary of Boehringer, Mann.), W. Germany
[Gallier]	Labs. Gallier S.A., Aubervilliers, France
[Galma]	Galma Pharmaceutical Chemical Factory, Tel Aviv, Israel
[Gea]	Gea A/S, Copenhagen, Denmark
[Gebro]	G. Broschek Gebro KG, Chem.-Pharm. Fabrik, Tirol, Austria
[Gedeon Richter]	Chemical Works of Gedeon Richter Ltd., Budapest, Hungary
[Geigy]	Ciba-Geigy, Basel, Switzerland (formerly J. R. Geigy AG, Basel, Switzerland; Geigy Pharmaceuticals, Ardsley, New York)
[Geistlich]	Ed. Geistlich Söhne AG, Wolhusen, Switzerland
[Genentech]	Genentech, San Francisco, California
[Geneva]	Geneva Generics, Broomfield, Colorado
[Génévrier]	Labs. Génévrier, Neuilly-sur-Seine, France
[Gentili]	Istituto Gentili S.p.A., Pisa, Italy
[Gerot]	Gerot-Pharmazeutika, GmbH (subsidiary of Rhone-Poulenc), Vienna, Austria
[Geymonat]	Farmaceutici Geymonat S.p.A., Turin, Italy
[Ghimas]	Ghimas S.p.A., Reno, Italy
[Gibipharma]	Gibipharma S.p.A., Pero (Milano), Italy
[Giorgio Zoja]	Lab. Chimico Farmaceutico Giorgio Zoja, Milan, Italy
[Gist-Brocades]	Gist-Brocades N.V., Delft, Netherlands
[Giuliani]	Giuliani S.p.A., Milan, Italy
[Giulini]	Giulini Pharma GmbH (subsidiary of Kali-Chemie), Hannover, W. Germany
[Givaudan]	Givaudan Corp., New York, New York and Clifton, New Jersey
[Glaxo]	Glaxo Labs., Ltd., Greenford, Middlesex, England
[Glenbrook]	Glenbrook Labs. (subsidiary of Sterling Drug Inc.), New York, New York
[Glenn]	Glenn Chemical Co., Inc., Chicago, Illinois
	Glenwood Glenwood Labs., Inc., Tenafly, New Jersey
[Globopharm]	Globopharm AG, Kusnacht/Zurich, Switzerland
[Glovers]	Glovers Chemicals Ltd., Leeds, England
[Glyco]	Glyco-Chemicals, Inc., Greenwich, Connecticut
[Gobbi-Novag]	Gobbi-Novag, Buenos Aires, Argentina
[Gödecke]	Gödecke AG (subsidiary of Warner-Lambert), Berlin, W. Germany
[GP]	GP, Alexandria, N.S.W., Australia
[Grant]	Grant Labs., Oakland, California
[Gray]	Gray Pharmaceutical Co., Norwalk, Connecticut
[Green Cross]	The Green Cross Corp., Osaka, Japan
[Greenwich]	Greenwich Pharmaceuticals, Greenwich, Connecticut
[Grelan]	Grelan Pharmaceutical Co., Ltd., Tokyo, Japan
[Grémy-Longuet]	Labs. Grémy-Longuet, Clichy, France
[Grossman]	Labs. Grossman S.A., Mexico City, Mexico
[Grünenthal]	Chemie Grünenthal, GmbH, Stolberg/Rheinland, W. Germany
[Guardian]	Guardian Chemical Corp., Happauge, New York
[Guerbet]	Lab. André Guerbet, Aulnay-sous-Bois, France
[Guidi]	Guidi, Milan, Italy
[Guidotti]	Lab. Guidotti & C., S.p.A., Pisa, Italy
[Halsey]	Halsey Drug Co., Brooklyn, New York
[Hamilton]	Hamilton Labs., Pty. Ltd., Adelaide, Australia
[Hartz]	Hartz Standard Ltd., Scarborough, Ontario, Canada
[Hässle]	AB Hässle (subsidiary of AB Astra), Molndal, Sweden
[Hauck]	W. E. Hauck, Inc., Roswell, Georgia
[Haury]	Heinz Haury, Chemische Fabrik, Munich, W. Germany
[Hausmann]	Hausmann AG, St. Gallen, Switzerland
[Haver-Lockhart]	Haver-Lockhart Labs. (division of Bayvet), Shawnee Mission, Kansas
[Hefa-Frenon]	Hefa-Frenon Arzneimittel GmbH & Co. KG, Werne, W. Germany
[Heilit]	Heilit Arzneimittel GmbH, Reinbek, W. Germany
[Heilmittelwerke]	Heilmittelwerke Wien GmbH, Vienna, Austria

[Hek]	Hek Pharma GmbH, Lübeck, W. Germany
[Helfenberg]	Helfenberg AG, Wewelinghofen/Rheinland, W. Germany
[Hellwig]	Hellwig Pharmaceutical, Chicago, Illinois
[Helopharm]	Helopharm W. Petrik GmbH & Co. KG, Berlin, W. Germany
[Henkel]	Henkel, Inc., Teaneck, New Jersey
[Hennig]	Hennig Arzneimittel GmbH & Co., KG, Florsheim-am-Main, W. Germany
[Henning]	Henning Berlin GmbH Chemie und Pharmawerk, Berlin, W. Germany
[Hépatrol]	Lab. de l'Hépatrol, Paris, France
[Herbert]	Apotheker A. Herbert GmbH, Wiesbaden-Bierstadt, W. Germany
[Herbrand]	Dr. Herbrand KG, Gengenback, W. Germany
[Hercules]	Hercules Inc., Wilmington, Delaware
[Hermal]	Hermal Kurt Herrmann, Reinbek/Hamburg, W. Germany
[Hermes]	Hermes Arzneimittel GmbH, Grosshesselohe, W. Germany
[Hermes SA]	Hermes SA, Barcelona, Spain
[Heterochemical]	Heterochemical Corp., Valley Stream, New York
[Heumann]	Heumann Pharma GmbH & Co., Nürenberg, W. Germany
[Heyden]	von Heyden GmbH (subsidiary of Squibb), Munich, W. Germany
[Heyl]	Heyl Chemisch-pharmazeutische Fabrik GmbH, Berlin, W. Germany
[Hirsch]	Hirsch Industries, Richmond, Virginia
[Hishiyama]	Hishiyama Pharmaceutical Co., Ltd., Osaka, Japan
[Hobein]	Dr. Hobein & Co., Nachf. GmbH & Co., Arzneimittel, Meckenheim-Merl, W. Germany
[Hodag]	Hodag Chemical Corp., Skokie, Illinois
[Hoechst]	Farbwerke Hoechst AG, Frankfurt, W. Germany
[Hoei]	Hoei Pharmaceutical Co., Ltd., Osaka, Japan
[Hokuriku]	Hokuriku Seiyaku Co., Ltd., Tokyo, Japan
[Hollister-Stier]	Hollister-Stier Labs., Spokane, Washington
[Holphar]	Holphar, Arzneimittel GmbH, Sulzbach-Neuweiler, W. Germany
[Holzinger]	Holzinger, Dr. et Mr. L., pharm. Erzeugungs- u. Vertriebsgesellschaft mbH, Vienna, Austria
[Homburg]	Asta Pharma AG (formerly Chemiewerk Homburg) (subsidiary of Degussa), Frankfurt, W. Germany
[Hommel]	Chemische Werke Hommel GmbH (subsidiary of Zyma), Mullheim, W. Germany
[Hooker]	Hooker Chemicals and Plastics Corp., Niagara Falls, New York
[Hopkins]	Hopkins Agricultural Chemical Co., Madison, Wisconsin
[Hor-Fer-Vit]	Hor-Fer-Vit Pharmazeutische Fabrik, Oldenburg, W. Germany
[Horii]	Horii Pharmaceutical Ind., Ltd., Osaka, Japan
[Horita]	Horita Yakuhin Gosei K.K., Nagoya, Japan
[Horlicks]	Horlicks Ltd., Slough, Bucks, England
[Hormon-Chemie]	Hormon-Chemie GmbH, Munich, W. Germany
[Hormosan]	Hormosan-Kwizda GmbH, Frankfurt/Main, W. Germany
[Horner]	Frank W. Horner, Ltd. (subsidiary of Carter-Wallace), Montreal, Quebec, Canada
[Hosbon]	Labs. Hosbon, S.A., São Paulo, Brazil
[Houdé]	Lab. Houdé, Paris, France
[Hough, Hoseason]	Hough, Hoseason & Co., Ltd., Levenshulme, Manchester, England
[Hoyer]	Hoyer GmbH & Co., Neuss, W. Germany
[Hoyt]	Hoyt Labs., Needham, Massachusetts
[HW & D]	Hynson, Westcott & Dunning, Inc., Baltimore, Maryland
[Hyland]	Hyland Labs. (division of Travenol Labs.), Costa Mesa, California
[Hyrex-Key]	Hyrex-Key Pharmaceuticals, Memphis, Tennessee
[IBI]	Istituto Biochimico Italiano S.p.A. G. Lorenzini, Milan, Italy
[Ibirn]	Istituto Bioterapico Nazionale s.r.l., Rome, Italy
[IBP]	Istituto Biochimico Pavese s.p.a., Pavia, Italy
[Icar]	Icar S.p.A., Rome, Italy
[ICB]	Industria Chimica Biologica S.p.A., Genoa, Italy
[Ichthyol]	Ichthyol-Gesellschaft, Hamburg, W. Germany
[ICI]	Imperial Chemical Industries, Ltd., Macclesfield, Cheshire, England
[ICI-Pharma]	ICI-Pharma Arzneimittelwerk, Plankstadt, W. Germany
[ICN]	ICN Canada Ltd., Montreal, Quebec, Canada
[ICN-Pharm.]	ICN Pharmaceuticals, Covina, California
[I.F.C.I.]	I.F.C.I. (Industria Farmaceutica Cosmetica Italiana S.p.A.), Casalecchio di Reno, Italy
[Ika]	Ikapharm, Ramat-Gan, Israel
[IMC]	International Mineral & Chemical Corp., Des Plaines, Illinois
[Importex]	Importex Chimici Farmaceutici S.p.A., Trieste, Italy
[Inava]	Labs. Inava, Departement Medical de Fimex S.A., Paris, France
[Inca]	Inca, Buenos Aires, Argentina
[Indian Health]	Indian Health Institute and Lab., Ltd., Calcutta, India
[Innothera]	Innothera (Lab. Chantereau), Arcueil, France
[Inpharzam]	Inpharzam (subsidiary of Zambon), Milan, Italy
[Inter-Alia]	Inter-Alia Pharmceutical Services, Ltd., Thetford, Norfolk, England
[Intermuti]	Intermuti Pharma, GmbH, Eschwege, W. Germany
[Intersint]	Intersint Italiana s.r.l., Rome, Italy
[Intervet]	Intervet Labs., Ltd. (subsidiary of Akzo), Cambridge, Craftshill, England
[INTERx]	INTERx Research Corp. (subsidiary of Merck & Co., Inc.), Lawrence, Kansas
[INTES]	INTES (Industria Terapeutica Splendore), Casoria, Italy
[Invenex]	Invenex Pharmaceuticals, Grand Island, New York

[Inverni]	Inverni Della Beffa S.p.A., Milan, Italy
[ION]	ION (Istituto Opoterapico Nazionale Pisa S.p.A.), Pisa, Italy
[IPG]	IPG Internationale Pharma, GmbH, Cuxhaven, W. Germany
[I.P.S.E.N.]	Institut de Produits de Synthese et d'Extraction Naturelle, Paris, France
[Ipsen-Beaufour]	Ipsen-Beaufour, London, England
[IRBI]	IRBI (Istituto Ricerche Biochimische Italiane A.A. Neri s.c.), Rome, Italy
[Iromedica]	Iromedica AG, St. Gallen, Switzerland
[Isei]	Isei Co., Ltd., Yamagata, Japan
[ISF]	ISF S.p.A., Milan, Italy
[I.S.H.]	Labs. I.S.H., Paris, France
[ISI]	Istituto Sierovaccinogeno Italiano S.p.A., Antimo, Italy
[Isis-Chemie]	Isis-Chemie KG., Zwickau, E. Germany
[ISM]	Istituto Sieroterapico Milanese, Milan, Italy
[Isnardi]	Pietro Isnardi & C., S.p.A., Imperia Oneglia, Italy
[Isola-Ibi]	Isola-Ibi, Genova-Quarto, Italy
[I.S.O.M.]	I.S.O.M.-Medicinali S.p.A., Milan, Italy
[Ist. Chim. Inter]	Istituto Chimico Internazionale di Rendre dr. Giuseppe s.n.c., Rome, Italy
[Italchemi]	Italchemi Pharma S.p.A., S. Polo di Torrile, Italy
[Italchimici]	Italchimici S.p.A., Rome, Italy
[Italfarmaco]	Italfarmaco S.p.A., Milan, Italy
[Ital Research]	Ital Research, Pomezia (Rome), Italy
[Ital Suisse]	Ital Suisse Co. s.a.s. di Giancarlo Ceroni & C., Casarile, Italy
[Itting]	Franz Itting KG, Ludwigsstadt, W. Germany
[Ives]	Ives Labs. (subsidiary of Am. Home Products Corp.), New York, New York
[Iwaki]	Iwaki Seiyaku Co., Ltd., Tokyo, Japan
[J & J]	Johnson & Johnson, New Brunswick, New Jersey
[Jackson-Mitchell]	Jackson-Mitchell Pharmaceuticals, Inc., Santa Barbara, California
[Janssen]	Janssen Pharmaceutica (subsidiary of Johnson & Johnson), Beerse, Belgium
[Jenapharm]	Kombinat VEB Jenapharm, Jena, E. Germany
[Jenkins]	Jenkins Labs., Inc., Auburn, New York
[Joullié]	Labs. Joullié (S.A.), Puteaux, France
[Jouveinal]	Jouveinal Labs., Fresnes, France
[Kabi]	AB Kabi, Stockholm, Sweden
[KabiVitrum]	KabiVitrum Ltd., Uxbridge, Middlesex, England
[Kade]	Dr. Kade Pharmazeutische Fabrik GmbH, Berlin, W. Germany
[Kaigai]	Kaigai Seiyaku Co., Ltd., Tokyo, Japan
[Kaken]	Kaken Pharmaceutical Co., Ltd., Tokyo, Japan
[Kakenyaku]	Kakenyaku Kako Co., Ltd., Tokyo, Japan
[Kali-Chemie]	Kali-Chemie Pharma GmbH (subsidiary of Solvay), Hannover, W. Germany
[Kanebo]	Kanebo Pharmaceuticals, Ltd., Tokyo, Japan
[Kanegafuchi]	Kanegafuchi Chemical Industry Co., Ltd., Osaka, Japan
[Kanoldt]	Kanoldt Arzneimittel GmbH, Hochstadt, W. Germany
[Kanto]	Kantoisi Pharmaceutical Co., Ltd., Tokyo, Japan
[Karlspharma]	Karlspharma pharmazeutische Produkte GmbH, Karlsruhe, W. Germany
[Katwijk]	Katwijk N.V., Amsterdam, Netherlands
[Kayaku]	Kayaku Antibiotics Research Co., Ltd., Tokyo, Japan
[Kelco]	Kelco Division of Merck & Co., Inc., San Diego, California
[Kettelhack]	Kettelhack Riker Pharma GmbH, Borken/Westfalen, W. Germany
[Key]	Key Pharmaceuticals, Inc., Miami, Florida
[Kinney]	Kinney & Co., Inc., Columbus, Indiana
[Kirby-Warrick]	Kirby-Warrick Pharm. Ltd. (subsidiary of Schering-Plough), Bury-St.-Edmunds, Suffolk, England
[Kissei]	Kissei Pharmaceutical Co., Ltd., Nagano, Japan
[Klinge]	Klinge Pharma GmbH, Munich, W. Germany
[Klinge-Nattermann]	Klinge-Nattermann Puren, GmbH (subsidiary of Rhone-Poulenc), Munich, W. Germany
[Knoll]	Knoll Chemische Fabriken AG (subsidiary of BASF), Ludwigshafen, W. Germany
[Köbányai]	Köbányai Gyógyszergyár, Budapest, Hungary
[Kodama]	Kodama Ltd., Tokyo, Japan
[Kodipharm]	Kodipharm GmbH, Mainz, W. Germany
[Köhler]	Dr. Franz Köhler Chemie KG, Alsbach-Hahnlein, W. Germany
[Kotani]	Kotani Seiyaku Co., Ltd., Osaka, Japan
[Kotobuki]	Kotobuki Pharmaceutical Co., Ltd., Nagano, Japan
[Kowa]	Kowa Company Ltd., Nagoya, Japan
[Kowa Yakuhin]	Kowa Yakuhin Ind. Co., Ltd., Tokyo, Japan
[Kowa Yakuko]	Kowa Yakuko, Itabishi, Japan
[Kremers-Urban]	Kremers-Urban Co., Milwaukee, Wisconsin
[Kreussler]	Kreussler & Co., Chemische Fabrik, GmbH, Wiesbaden-Biebrich, W. Germany
[Krewel]	Krewel-Werke, GmbH, Koln, W. Germany
[Kureha]	Kureha Chemical Industry Co., Ltd., Tokyo, Japan
[Kwizda]	F. J. Kwizda Chemische Fabrik, Vienna, Austria
[Kyorin]	Kyorin Pharmaceutical Co., Ltd., Tokyo, Japan
[Kyoritsu]	Kyoritsu Pharmaceutical Co., Ltd., Yamagata, Japan
[Kyowa]	Kyowa Hakko Kogyo Co., Ltd., Tokyo, Japan
[Kyowa Yakuhin]	Kyowa Yakuhin, Osaka, Japan
[L.A.B.]	Laboratories for Applied Biology, Ltd., London, England

[Labatec]	Labatec Pharma S.A., Geneva, Switzerland
[Labaz]	Labaz AG (subsidiary of Sanofi), Basel, Switzerland
[Labcatal]	Lab. Labcatal, Montrouge, France
[Labinca]	Laboratorios Labinca S.A., Cramer, Argentina
[Labopharma]	Labopharma, Chemisch-Pharmazeutische Fabrik, GmbH, Berlin, W. Germany
[Lafare]	Lab. Farmaceutico Reggiano s.n.c., Ercolana Resina, Italy
[Lafayette]	Lafayette Pharmacal, Inc., Lafayette, Indiana
[Lafi]	Lab. Lafi Ltda., São Paulo, Brazil
[Lafon]	Labs. Lafon, Maisons-Alfort, France
[Lagap]	Lagap Pharmaceuticals Ltd., Bordon, Hants., England
[Lakeside]	Lakeside Labs. (subsidiary of Merrell Dow), Cincinnati, Ohio
[Lampugnani]	Lampugnani Farmaceutici S.p.A., Milan, Italy
[Lancet]	Lancet s.r.l. Industria Farmaceutica, Aprilia, Italy
[Landerlán]	Labs. Landerlán, Madrid, Spain
[Lannett]	The Lannett Co., Inc., Philadelphia, Pennsylvania
[Lappe]	Produpharm Lappe, Bensberg-Frankenforst, W. Germany
[Larma]	Labs. Larma, S.A., Madrid, Spain
[Laroche Navarron]	Laroche Navarron (subsidiary of Syntex), Puteaux, France
[Lasa]	Lasa Laboratorios, Barcelona, Spain
[Laser]	Laser, Inc., Crown Point, Indiana
[Laserson]	Laserson & Sabetay, Etampes, France
[Latéma]	Latéma, Labs. de Therapeutique Moderne, Paris, France
[Lazar]	Dr. Lazar y cia S.A., Velez Sarfield, Argentina
[Leciva]	Leciva, Prague, Czechoslovakia
[Lederle]	Lederle Labs. (subsidiary of American Cyanamid Co.), Pearl River, New York
[Leeming]	Leeming/Pacquin (division of Chas. Pfizer & Co.), New York, New York
[Lefrancq]	Lab. Lefrancq, Romainville, France
[Lefranq]	Lefranq, Romainville, France
[Lek]	Lek, Ljubljana, Yugoslavia
[Lematte et Boinot]	Lab. Lematte et Boinot, Paris, France
[Lemmon]	Lemmon Pharmacal Co., Sellersville, Pennsylvania
[Lemoine]	Labs. Lemoine, Lille, France
[Lemonier]	Lemonier, Buenos Aires, Argentina
[Len-Tag]	Len-Tag Co., Detroit, Michigan
[Lentia]	Lentia GmbH, Munich, W. Germany
[Lenza]	Farmaceutici Lenza s.r.l., Naples, Italy
[Leo Pharm]	Leo Pharmaceutical Products, Ballerup, Denmark
[Lepetit]	Gruppo Lepetit S.p.A. (subsidiary of Merrell Dow), Milan, Italy
[L.I.B.S.]	Lab. International Biologique et Scientifique, Mareuil-sur-Lay, France
[Lifasa]	Lab. Lifasa, Valencia, Spain
[Lifepharma]	Lifepharma S.p.A., Milan, Italy
[Lilly]	Eli Lilly & Co., Indianapolis, Indiana
[Lincoln]	Lincoln Labs., Inc., Decatur, Illinois
[Lindopharm]	Lindopharm GmbH, Hilden, W. Germany
[Lipha]	Lipha, S.A., Lyon, France
[Lipo]	Lipo Chemicals, Inc., New York, New York
[Lisapharma]	Lisapharma S.p.A., Como, Italy
[Locatelli]	Farmaceutici Locatelli s.r.l., Rome, Italy
[Logeais]	Labs. Jacques Logeais, Issy-les-Moulineaux, France
[Lucien]	Lab. Lucien, Colombes, France
[Luitpold]	Luitpold-Werk GmbH & Co., Chemisch-pharmazeutische Fabrik, Munich, W. Germany
[Lumière]	Labs. Lumière, Paris, France
[Lundbeck]	H. Lundbeck & Co., Copenhagen, Denmark
[Lusofarmaco]	Istituto Luso Farmaco d'Italia S.p.A., Milan, Italy
[Lutetia]	Lab. Lutetia S.A., Rio de Janeiro, Brazil
[Lyocentre]	Lab. Lyocentre (subsidiary of Akzo), Aurillac, France
[Lysoform]	Lysoform Dr. Hans Rosemann GmbH, Berlin, W. Germany
[Lyssia]	Lyssia Chemisch-pharmazeutisch Fabrik (subsidiary of Kali-Chemie), Wiesbaden, W. Germany
[3M]	Minnesota Mining & Manufacturing Co., St. Paul, Minnesota
[M & B]	May & Baker, Ltd. (subsidiary of Rhone-Poulenc), Dagenham, Essex, England
[M & M]	Mowatt & Moore Ltd., Pointe Claire, Quebec, Canada
[Macfarlan Smith]	Macfarlan Smith Ltd. (subsidiary of Glaxo), Edinburgh, Scotland
[Mack, Illert.]	Heinrich Mack Nachf., Chemisch-Pharmazeutische Fabrik (subsidiary of Pfizer Corp.), Illertissen, W. Germany
[Madaus]	Dr. Madaus & Co., Cologne, W. Germany
[Maestretti]	Maestretti Lab. Farmaceutici S.p.A., Milan, Italy
[Maggioni]	Maggioni Farmaceutici S.p.A., Milan, Italy
[Magis]	Lab. Magis Farmaceutici, Brescia, Italy
[Makara]	Makara GmbH, Dusseldorf, W. Germany
[Makhteshim-Agan]	Makhteshim-Agan, Beer-Sheva, Israel
[Malesci]	Malesci Istituto Farmacobiologico, S.p.A., Florence, Italy
[Mallard]	Mallard, Inc., Detroit, Michigan
[Mallinckrodt]	Mallinckrodt, Inc., St. Louis, Missouri
[Manetti Roberts]	Manetti L. & Roberts H. & Co., Florence, Italy

[Maney]	Paul Maney Labs. (division of Canapharm Industries Inc.), Toronto, Canada
[Mann]	Dr. Gerhard Mann, Chem.-pharm. Fabrik GmbH, Berlin, W. Germany
[Marion]	Marion Labs., Inc., Kansas City, Missouri
[Martinet]	Lab. Martinet, Paris, France
[Maruishi]	Maruishi Pharmaceutical Co., Ltd., Osaka, Japan
[Maruko]	Maruko Seiyaku Co., Ltd., Nagoya, Japan
[Master]	Master Pharma s.r.l., Parma, Italy
[Mauchant]	Lab. Mauchant, Clichy-La Garenne, France
[Maurry Biol.]	H. E. Maurry Biological Co., Inc., Los Angeles, California
[Max Ritter]	Max Ritter Pharma AG, Zurich, Switzerland
[Mayrand]	Mayrand Inc., Greensboro, North Carolina
[Mazer]	Mazer Chemicals, Inc., Gurnee, Illinois
[McGaw]	McGaw Labs., Glendale, California
[McKesson]	McKesson Lab., Inc. (division of Foremost McKesson Inc.), Fairfield, Connecticut
[McNeil]	McNeil Lab., Inc. (subsidiary of Johnson & Johnson), Fort Washington, Pennsylvania
[MCP]	MCP Pharmaceuticals Ltd., Livingston, West Lothian, England
[Mead Johnson]	Mead Johnson & Co. (subsidiary of Bristol-Myers Co.), Evansville, Indiana
[Mect]	Mect Corp, Tokyo, Japan
[Medea]	Medea Researches, Milan, Italy
[Medial]	Labs. Medial-Riker, Pithiviers, France
[Medica]	Medica Oy, Helsinki, Finland
[Medice]	Medice, Chem.-Pharm. Fabrik Putter GmbH & Co. KG, Iserlohn, W. Germany
[Medichemie]	Medichemie AG, Basel, Switzerland
[Medici]	Lab. Farm. Dr. Medici s.r.l., Rome, Italy
[Medicia]	Lab. Medicia, Lyon, France
[Medici Domus]	Medici Domus s.r.l., Vittore-Olona, Italy
[Medics]	Medics Pharmaceutical Corp., Decatur, Georgia
[Medimpex]	Medimpex, Budapest, Hungary
[Medinova]	Medinova AG, Zurich, Switzerland
[Mediolanum]	Mediolanum Farmaceutici s.r.l., Milan, Italy
[Medix]	Medix S.p.A., Milan, Italy
[Medo]	Medo-Chemicals Ltd., London, England
[Medosan]	Medosan S.p.A. Industrie Biochimiche Riunite, Cecchina, Italy
[Med. Prod. Quim.]	Medicamentos y Productos Quimicos, S.A., Barcelona, Spain
[Meiji]	Meiji Seika Kaisha, Ltd., Tokyo, Japan
[Meiji Yakuhin]	Meiji Yakuhin, K.K., Tokyo, Japan
[Mekos]	AB Mekos, Helsingborg, Sweden
[Melusin]	Melusin Schwarz GmbH, Monheim, W. Germany
[Menarini]	Menarini A. s.a.s., Florence, Italy
[Mendelejeff]	Mendelejeff s.r.l., Rome, Italy
[Mentholatum]	Mentholatum Co., Buffalo, New York
[Mepha]	Mepha AG, Neu-Aesch/Basel, Switzerland
[Mepros]	Mepros, Bladel, Netherlands
[Merck & Co.]	Merck & Co., Inc., Rahway, New Jersey
[Merck-Clévenot]	Labs. Merck-Clévenot (subsidiary of E. Merck AG), Nogent-sur-Marne, France
[Merckle]	L. Merckle Chemisch-pharmazeutische Fabrik, Blaubeuren/Württ., W. Germany
[Merieux]	Institut Merieux (subsidiary of Rhone-Poulenc), Lyon, France
[Merminod]	Lab. Merminod (S.A.R.L.), Paris, France
[Merrell]	Merrell Dow Pharmaceuticals Inc. (subsidiary of The Dow Chemical Co.), formerly Merrell-National Labs., Cincinnati, Ohio
[Merz]	Merz & Co., GmbH & Co., Frankfurt/Main, W. Germany
[Metadier]	Labs. Paul Metadier, Tours, France
[Meyer]	Meyer Labs., Inc., Fort Lauderdale, Florida
[Midy]	Lab. Midy, Paris, France
[Mikasa]	Mikasa Seiyaku Co., Ltd., Tokyo, Japan
[Miles]	Miles Labs., Inc. (subsidiary of Bayer AG), Elkhart, Indiana
[Millet Roux]	Millet Roux & Co., Quebec, Canada
[Millot-Solac]	Labs. Millot-Solac (subsidiary of Sanofi), Paris, France
[Minden]	Minden Pharma GmbH (subsidiary of Knoll AG), Minden, W. Germany
[Minsa]	Minsa, Basel, Switzerland
[Miquel]	Laboratorios Miquel S.A., Barcelona, Spain
[Misemer]	Misemer Pharmaceuticals, Inc., Springfield, Missouri
[Mission Pharmacal]	Mission Pharmacal Co., San Antonio, Texas
[Mitsubishi]	Mitsubishi Chemical Industries, Ltd., Tokyo, Japan
[Mitsui]	Mitsui Pharmaceuticals, Inc., Tokyo, Japan
[Mobil]	Mobil Oil Corp., New York, New York
[Mochida]	Mochida Pharmaceutical Co., Ltd., Tokyo, Japan
[Mohan]	Mohan Medicine Research Inst., Tokyo, Japan
[Molimin]	Molimin Arzneimittel GmbH, Hallstadt/Ofr., W. Germany
[Molteni]	Molteni L. & C. dei F. Ili Alitti Societa Esercizio S.p.A., Scandicci, Italy
[Montecatini]	Montecatini Edison, Milan, Italy
[Morishita]	Morishita Pharmaceutical Co., Ltd., Osaka, Japan
[Morton-Norwich]	Morton-Norwich Products, Inc., Chicago, Illinois
[Much]	Prof. Dr. med. Much AG (subsidiary of American Home Products Corp.), Bad Soden, W. Germany

[Mulford]	Mulford Colloid Labs., Philadelphia, Pennsylvania
[Mulli]	Dr. Kurt Mulli Nachf., GmbH & Co., AG, Neuenburg, W. Germany
[Mundipharma]	Mundipharma GmbH, Limburg/Lahn, W. Germany
[Murphy]	Murphy Chemical Company, Ltd., Wheathampstead, St. Albans, Herts., England
[Nadrol]	Nadrol-Chemie-Pharma Keizer KG, Osnabruck, W. Germany
[Nagase]	Nagase & Co., Ltd., Osaka, Japan
[Nakano]	Nakano Yakuhin Ind. Co., Ltd., Gifu, Japan
[Nakataki]	Nakataki Pharmaceutical Ind. Co., Ltd., Tokyo, Japan
[Napp]	Napp Labs. Ltd., Cambridge, England
[Nativelle]	Lab. Nativelle S.A., Paris, France
[Nattermann]	A. Nattermann & Cie, GmbH (subsidiary of Rhone-Poulenc), Cologne, W. Germany
[NEGMA]	Labs. NEGMA, Meudon, France
[Negroni]	Negroni Pietro S.p.A., Cremona, Italy
[Nelson]	Nelson Research & Development (subsidiary of Ethyl Corp.), Irvine, California
[Nemi]	Laboratorios Nemi SACIFI, Avda, Argentina
[Neos]	Neos Donner KG, Berlin, W. Germany
[Newport]	Newport Pharmaceuticals, Inc., Newport Beach, California
[Nichiiko]	Nihon Iyakuhin Kogyo Co., Ltd., Toyama, Japan
[Nicholas]	Nicholas Labs., Ltd., Slough, Bucks, England
[Nigy]	Lab. Nigy, Arcueil, France
[Nikken]	Nikken Chemicals Co., Ltd., Tokyo, Japan
[Nion]	Nion Corp., El Monte, California
[Nippon Chemiphar]	Nippon Chemiphar Co., Ltd., Tokyo, Japan
[Nippon Kayaku]	Nippon Kayaku Co., Ltd., Tokyo, Japan
[Nippon Roche]	Nippon Roche K.K., Tokyo, Japan
[Nippon Shinyaku]	Nippon Shinyaku Co., Ltd., Kyoto, Japan
[Nippon Shoji]	Nippon Shoji Kaisha, Ltd., Osaka, Japan
[Nippon Soda]	Nippon Soda Co., Ltd., Tokyo, Japan
[Nippon Zoki]	Nippon Zoki Pharmaceutical Co., Ltd., Osaka, Japan
[Nissan]	Nissan Kagaku Kogyo K.K., Tokyo, Japan
[Nisshin]	Nisshin Pharmaceutical Co., Ltd., Osaka, Japan
[Nissui]	Nissui Seiyaku Co., Ltd., Tokyo, Japan
[Nobel]	Société Nobel Française, Paris, France
[Nor-Am]	Nor-Am Agricultural Products, Inc., Woodstock, Illinois
[Norden]	Norden Labs. Inc. (subsidiary of SmithKline Beckman Corp.), Lincoln, Nebraska
[Nordic]	Nordic Pharmaceuticals, Ltd., Quebec, Canada
[Nordmark]	Nordmark Arzneimittel GmbH (subsidiary of Knoll AG), Uetersen/Holstein, W. Germany
[Norgine]	Norgine GmbH, Marburg/Lahn, W. Germany
[Norma]	Norma Chem. Fabrik, GmbH, Köln, W. Germany
[North Am. Pharm.]	North American Pharmacal, Inc., Dearborn, Michigan
[Norton]	H. N. Norton & Co., Ltd., London, England
[Norwich]	Norwich-Eaton Pharmaceuticals (subsidiary of Procter & Gamble), Norwich, New York
[Nourypharma]	Nourypharma GmbH (subsidiary of Akzo), Oberschleissheim, W. Germany
[Novo]	Novo Industri A/S, Copenhagen, Denmark
[Novocol]	Novocol Chemical Mfg. Co., Inc., Brooklyn, New York
[Nuovo Cons.]	Nuovo Consorzio Sanitario Nazionale di Malizia Dr. Paolo, Rome, Italy
[Nyegaard]	Nyegaard & Co., A/S, Oslo, Norway
[Oberval]	Lab. Oberval, Lyon, France
[Ohio Med.]	Ohio Medical Products, Madison, Wisconsin
[Ohio Med. Anaesth.]	Ohio Medical Anaesthetics Ltd., Swindon, Wiltshire, England
[Ohta]	Ohta Pharmaceutical Co., Ltd., Tokyo, Japan
[OJ & F]	O'Neal, Jones & Feldman, Inc., St. Louis, Missouri
[Olin]	Olin Mathieson Chemical Corp., New York, New York
[OM]	Lab. OM, S.A., Meyrin, Switzerland
[Omegin]	Omegin Dr. Schmidgall & Co., Köngen, W. Germany
[Ono]	Ono Pharmaceutical Co., Ltd., Osaka, Japan
[O.P.G.]	Onderlinge Pharmaceutische Groothandel, Cooperatieve Apothekers vereniging, Utrecht, Netherlands
[Organon]	Organon, Oss, Netherlands; also Organon Co. (subsidiary of Akzo), West Orange, New Jersey
[Orion]	Orion Pharmaceutica, Espoo, Finland
[Ormont]	Ormont Drug & Chem. Co., Inc., Englewood, New Jersey
[Ortega]	Ortega Pharm. Co., Jacksonville, Florida
[Ortho]	Ortho Pharmaceutical Corp. (subsidiary of Johnson & Johnson), Raritan, New Jersey
[Oryx]	Oryx Pharmazeutica AG, Zurich, Switzerland
[Oscar]	Oscar s.r.l., Milan, Italy
[OSSW]	Oesterreichische Stickstoffwerke AG., Linz, Austria
[Otsuka]	Otsuka Pharmaceutical Co., Ltd., Osaka, Japan
[OTW]	Organotherapeutische Werke GmbH, Karlsruhe, W. Germany
[Owen]	Owen Labs. (division of Alcon Labs., Inc.), Dallas, Texas
[Ozothine]	Labs. de l'Ozothine S.A., Nanterre Cedex, France
[Paidostene]	Paidostene s.a.s., Milan, Italy
[Paines & Byrne]	Paines & Byrne Ltd., Greenford, Middlesex, England
[Palmedico]	Palmedico, Inc., Columbia, South Carolina

[Panray]	Panray (division of Ormont Drug & Chemical Co., Inc.), Englewood, New Jersey
[Panthox & Burck]	Panthox & Burck S.p.A., Milan, Italy
[Parke, Davis]	Parke, Davis & Co. (division of Warner-Lambert Pharmaceutical Co.), Detroit, Michigan
[Parmed]	Parmed Pharmaceuticals, Niagara Falls, New York
[Pearson]	Pearson & Co., Cologne, W. Germany
[Penick]	S. B. Penick & Co., New York, New York
[Penn]	Penn Pharmaceuticals Ltd., Penn, Bucks, England
[Pennwalt]	Pennwalt Prescription Products, Rochester, New York
[Penreco]	Pennsylvania Refining Co., Butler, Pennsylvania
[Perga]	Lab. Perga S.A., Barcelona, Spain
[Permamed]	Permamed AG, Basel, Switzerland
[Permicutan]	Permicutan KG, Dr. Euler, Paderborn, W. Germany
[Perrier]	Lab. Perrier, Toulouse, France
[Person & Covey]	Person & Covey, Inc., Glendale, California
[Petroleum Spec.]	Petroleum Specialties, Inc., New York, New York
[Pfizer]	Chas. Pfizer & Co., Inc., New York, New York
[Pfleger]	Dr. R. Pfleger Chemische Fabrik GmbH, Bamberg, W. Germany
[Pharmachim]	Pharmachim, Sofia, Bulgaria
[Pharmacia]	Pharmacia Labs., Inc., Piscataway, New Jersey Pharmacia AB, Uppsala, Sweden
[Pharma Investi]	Pharma Investi S.A., Madrid, Spain
[Pharm-Allergan]	Pharm-Allergan GmbH, Karlsruhe, W. Germany
[Pharma Schwarz]	Pharma Schwarz GmbH, Monheim, W. Germany
[Pharmascience]	Laboratoires Pharmascience, Courbevoie, France
[Pharma-Selz]	Pharma-Selz Worrstadt GmbH & Co., KG, Worrstadt, W. Germany
[Pharma-Stern]	Pharma-Stern GmbH, Wedel/Holstein, W. Germany
[Pharma Stroschein]	Pharma Stroschein GmbH, Hamburg, W. Germany
[Pharmax]	Pharmax Ltd., Dartforth, Kent, England
[Pharmex]	Pharmex, Inc., Hollywood, Florida
[Pharm. Mfg.]	Pharmaceutical Manufacturing Company, Bolton, Lanc., England
[Pharmuka]	Pharmuka S.F. (subsidiary of Rhone-Poulenc), Genevilliers, France
[Philadelphia Labs.]	Philadelphia Labs., Inc., Philadelphia, Pennsylvania
[Piam]	Vecchi & C. Piam di G. Assereto, Genoa, Italy
[Pickles]	J. Pickles & Sons, Knaresborough, N. Yorks, England
[Pierrel]	Pierrel, S.p.A. (subsidiary of Fermenta), Milan, Italy
[Pirri]	Pirri Ist Biochimico, Milan, Italy
[Pitman-Moore]	Pitman-Moore, Inc. (subsidiary of Johnson & Johnson), Washington Crossing, New Jersey
[Plantorgan]	Plantorgan Werk KG, Bad Zwischenahn, W. Germany
[Plant Protection]	Plant Protection Ltd. (division of ICI), Yalding, Kent, England
[Plessner]	The Paul Plessner Co., St. Petersburg, Florida
[Pliva]	Pliva, Zagreb, Yugoslavia
[Plough]	Plough Inc. (subsidiary of Schering Plough), Memphis, Tennessee
[Pohl]	G. Pohl-Boskamp, Hohenlockstedt/Holstein, W. Germany
[Polfa]	Polfa, Krakow, Poland
[Poli]	Poli Industria Chimica, S.p.A., Milan, Italy
[Polifarma]	Polifarma S.p.A., Rome, Italy
[P.O.S.]	Labs. P.O.S., Kayserberg, France
[Poythress]	Wm. P. Poythress & Co., Inc., Richmond, Virginia
[PPG]	PPG Industries, Inc., Pittsburgh, Pennsylvania
[Pradel]	Lab. Pradel-Ibero, Madrid, Spain
[Premo]	Premo Pharmaceutical Labs., Inc. South Hackensack, New Jersey
[Primedics]	Primedics Labs., Los Angeles, California
[Procter & Gamble]	Procter & Gamble, Cincinnati, Ohio
[Prodes]	Lab. Quim. Farm. Prodes, S.A., Barcelona, Spain
[Profarmi]	Profarmi s.r.l., Milan, Italy
[Promedica]	Lab. Promedica, Levallois-Perret, France
[Promonta]	Chemische Fabrik Promonta, GmbH, Hamburg, W. Germany
[Protea]	Protea, Czechoslovakia
[Proter]	Proter S.p.A., Milan, Italy
[Protina]	Protina Chemische Gesellschaft mbH, Ismaning, W. Germany
[Provet]	Provet, Lyssach, Switzerland
[Pulitzer]	Pulitzer Italiana S.p.A., Rome, Italy
[Purdue Frederick]	The Purdue Frederick Co., Norwalk, Connecticut
[Puropharma]	Puropharma s.r.l., Milan, Italy
[PVO Intl]	PVO International, Inc., Chem. Spec. Div., Boonton, New Jersey
[Quinoderm]	Quinoderm Ltd., Oldham, Lanc., England
[Rachelle]	Rachelle Labs., Inc. (subsidiary of International Rectifier Corp.), Long Beach, California
[Radiumpharma]	Radiumpharma s.r.l., Milan, Italy
[RAFA]	RAFA Labs., Ltd., Jerusalem, Israel
[Ralston-Purina]	Ralston-Purina Co., St. Louis, Missouri
[Ratiopharm]	Ratiopharm GmbH, Arzneimittel, Blaubeuren, W. Germany
[Ravasini Organon]	Ravasini Dr. R. & C.ia S.p.A., Rome, Italy
[Ravensberg]	Ravensberg Chemische Fabrik, GmbH, Konstanz, W. Germany

[Ravizza]	Ravizza S.p.A. (subsidiary of Knoll AG), Muggiò (Milan), Italy
[Recip]	AB Recip, Stockholm, Sweden
[Reckitt & Colman]	Reckitt & Colman Ltd., Hull, Yorkshire, England
[Recordati]	Recordati Industria Chimica e Farmaceutica S.p.A., Milan, Italy
[Reed & Carnrick]	Reed & Carnrick, Piscataway, New Jersey
[Reid-Provident]	Reid-Provident Labs., Inc., Atlanta, Georgia
[Reiss]	Dr. Rudolf Reiss Chemische Werke GmbH & Co., KG, Cologne, W. Germany
[Rentschler]	Dr. Rentschler Arzneimittel GmbH & Co., Laupheim, W. Germany
[Resistoflex]	Resistoflex, Belleville, New Jersey
[Rexall]	Rexall Drug Co., St. Louis, Missouri
[Rheingold]	Rheingold Arzneimittel GmbH, Neuss, W. Germany
[Rhein-Pharma]	Rhein-Pharma Arzneimittelwerk GmbH, Plankstadt, W. Germany
[Rhodia]	Rhodia Pharma GmbH (subsidiary of Rhône-Poulenc), Hamburg, W. Germany
[Rhône-Poulenc]	Rhône-Poulenc, Paris, France
[Ricar]	Ricar Labs., Moran, Argentina
[Richard Daniel]	Richard Daniel & Son Ltd., Derby, England
[Richardson-Merrell]	Merrell Dow Pharmaceuticals Inc. (subsidiary of The Dow Chemical Co.), formerly Richardson-Merrell, Inc., Cincinnati, Ohio
[Richter]	Richter & Cie, GmbH, Chemische Fabrik, Osnabruck, W. Germany
[Riker]	Riker Labs. (subsidiary of the 3M Company), Northridge, California
[Riom]	Riom Labs. (subsidiary of Akzo), Riom, France
[Ripari-Gero]	Istituto Farmaco Biologico Ripari-Gero, S.p.A., Monteriggioni, Italy
[RIT]	Recherche et Industrie Therapeutiques, Paris, France
[Ritsert]	Dr. E. Ritsert, GmbH & Co., KG, Eberbach/Neckar, W. Germany
[Robeco]	Robeco Chemicals, Inc., New York, New York
[Robert et Carrière]	Lab. Robert et Carrière, Paris, France
[Roberts]	Roberts Labortories Inc., Eatontown, New Jersey
[Robin]	Boehringer Biochemia Robin S.p.A., Milan, Italy
[Robins]	A. H. Robins Co., Inc. (subsidiary of American Home Products Corp.), Richmond, Virginia
[Robinson]	Robinson Labs., Inc., San Francisco, California
[Robisch]	Chemische Fabrik G. Robisch GmbH, Munich, W. Germany
[Robugen]	Robugen GmbH, Neckar, W. Germany
[Roche]	Hoffmann-La Roche & Co., AG, Basel, Switzerland Roche Labs. (division of Hoffmann-La Roche, Inc.), Nutley, New Jersey
[Roemmers]	Roemmers, Buenos Aires, Argentina
[Roerig]	J. B. Roerig (division of Pfizer Corp.), New York, New York
[Roger]	Labs. Roger S.A., Barcelona, Spain
[Roger Bellon]	Lab. Roger Bellon (subsidiary of Rhone-Poulenc), Neuilly-sur-Seine, France
[Röhm]	Röhm Pharma GmbH, Darmstadt, W. Germany
[Rohm & Haas]	Rohm & Haas Co., Philadelphia, Pennsylvania
[Roland]	Roland Arzneimittel GmbH, Hamburg, W. Germany
[Roleca]	Roleca, Rolf H. J. Lechner GmbH & Co., KG, Ratingen, W. Germany
[Rolland]	Albert Rolland S.A. (Lab. de l'Hepatrol), Paris, France
[Rona]	Rona Labs., Ltd., Hitchin, Herts., England
[Rorer]	William H. Rorer, Inc., Fort Washington, Pennsylvania
[Rosa-Phytopharma]	Lab. Rosa-Phytopharma S.A., Pantin, France
[Ross]	Ross Labs. (division of Abbott Labs.), Columbus, Ohio
[Roter]	Roter (Pharmazeutische Fabrik), Hilversum, Netherlands
[Rotta]	Rotta Farmaceutici S.p.A., Milan, Italy
[Roussel]	Lab. Roussel, Paris, France
[Roussel-Amor Gil]	Roussel-Amor Gil S.A., Madrid, Spain
[Roussel-Maestretti]	Roussel-Maestretti S.p.A., Milan, Italy
[Roussel-UCLAF]	Roussel-UCLAF (subsidiary of Hoechst), Paris, France
[Roux-Ocefa]	Roux-Ocefa, Buenos Aires, Argentina
[Rowa]	Rowa Ltd., Newtown, Co. Cork, Ireland
[Rowell]	Rowell Labs., Inc., Baudette, Minnesota
[Roxane]	Roxane Labs., Inc. (subsidiary of Boehringer, Ing.), Columbus, Ohio
[R. P. Drugs]	R. P. Drugs Ltd., Leeds, Yorkshire, England
[Rybar]	Rybar Labs., Ltd., Tankerton, Kent, England
[Rystan]	Rystan Co., Mt. Vernon, New York
[Saarstickstoff-Fatol]	Saarstickstoff-Fatol, GmbH, Schiffweiler/Saar, W. Germany
[Sabona]	Sabona GmbH, Feldkirchen, W. Germany
[Sagitta]	Sagitta Arzneimittel GmbH, Feldkirchen, W. Germany
[Saita]	Saita s.r.l. Biofarmaceutici, Milan, Italy
[Salsbury]	Salsbury Labs., Charles City, Iowa
[Salus]	Salus Researches S.p.A., Rome, Italy
[Salvoxyl-Wander]	Lab. Salvoxyl-Wander, Paris, France
[Samil]	Samil S.p.A., Rome, Italy
[Sana]	Sana Yakuhin Kogyo Co., Ltd., Tokyo, Japan
[San Carlo]	San Carlo Farmaceutici S.p.A., Rome, Italy
[Sandoz]	Sandoz AG, Basel, Switzerland; also Sandoz, Inc., East Hanover, New Jersey
[Sanko]	Sanko Seiyaku Kogyo Co., Ltd., Tokyo, Japan
[Sankyo]	Sankyo Co., Ltd., Tokyo, Japan
[Sankyo Zoki]	Sankyo Zoki Co., Tokyo, Japan

[Sanofi]	Sanofi S.A., Paris, France
[Sanol]	Sanol Schwarz GmbH, Monheim, W. Germany
[Sanorania]	Sanorania, Dr. Gerhard Strohscheer, Berlin, W. Germany
[Sanraku]	Sanraku Inc., Tokyo, Japan
[Santen]	Santen Pharmaceutical Co., Ltd., Osaka, Japan
[Sanwa]	Sanwa Kagaku Kenkyusho Co., Ltd., Nagoya, Japan
[Sanzen]	Sanzen Pharmaceutical Co., Ltd., Tokyo, Japan
[Sapos]	Sapos S.A., Geneva, Switzerland
[S.A.R.E.P.]	S.A.R.E.P.-Pharmeurop, La Plaine-Saint-Denis, France
[Sarm]	Sarm s.r.l., Rome Italy
[Sarva]	Sarva, Brussels, Belgium
[Sasse]	Dr. Friedrich Sasse (division of Gödecke AG), Berlin, W. Germany
[Sato]	Sato Pharmaceutical Co., Ltd., Tokyo, Japan
[Sato Yakuhin]	Sato Yakuhin Kogyo Co., Ltd., Nara, Japan
[Sauba]	Lab. Sauba, Montreuil, France
[Saunier-Daguin]	Labs. Saunier-Daguin, Paris, France
[Sauter]	Labs. Sauter S.A., Geneva, Switzerland
[Savage]	Savage Labs. (division of Altana), Melville, New York
[Savoma]	Savoma Medicinali S.p.A., Parma, Italy
[Sawai]	Sawai Pharmaceutical Co., Ltd., Osaka, Japan
[Scarium]	Scarium AG, Lucerne, Switzerland
[Schaper & Brummer]	Schaper & Brummer, Salzgitter/Ringelheim, W. Germany
[Scharper]	Scharper S.p.A. per l'industria Farmaceutica, Milan, Italy
[Schering]	Schering-Plough Corp., Kenilworth, New Jersey
[Schering AG]	Schering AG, Berlin, W. Germany
[Schering Corp.]	Schering Corp. (subsidiary of Schering-Plough Corp.), Kenilworth, New Jersey
[Scheurich]	E. Scheurich, Pharmwerk GmbH, Appenweier, W. Germany
[Schiapparelli]	Schiapparelli S.p.A., Turin, Italy
[Schieffelin]	Schieffelin & Co., New York, New York
[SCHI-WA]	SCHI-WA GmbH, Glandorf, W. Germany
[Schmid]	Schmid Labs., Inc., Little Falls, New Jersey
[Schmiden]	Pharmawerk Schmiden, Schmiden/Fellbach, W. Germany
[Schur]	Werner Schur O.W.G. Chemie, Kiel, W. Germany
[Schurholz]	Schurholz Arzneimittel GmbH, Munich, W. Germany
[Schwarzhaupt]	Kommandit gesellschaft Schwarzhaupt GmbH, Cologne, W. Germany
[Sclavo]	Sclavo S.p.A., Siena, Italy
[Scott]	O. M. Scott and Sons Co., Marysville, Ohio
[Scott-Cord]	Scott-Cord Labs., Inc., Pharm. Div., Glen Head, New York
[Searle]	G. D. Searle & Co. (subsidiary of Monsanto), Skokie, Illinois
[SEDAPH]	Sté d'Applications Pharmaceutiques, Clichy, France
[Seiko]	Seiko Eiyo Yakuhin Co., Ltd., Osaka, Japan
[Selvi]	Selvi & C., S.p.A., Milan, Italy
[Senju]	Senju Pharmaceutical Co., Ltd., Osaka, Japan
[Septa]	Lab. Septa, Madrid, Spain
[Serono]	Industria Farmaceutica Serono S.p.A., Rome Italy; also Serono Labs., Inc., Randolph, Massachusetts
[Serpero]	Serpero, Industria Galenica Milanese S.p.A., Milan, Italy
[Servier]	Lab. Servier, Orleans, France
[Sharp & Dohme]	Sharp & Dohme GmbH, Munich, W. Germany
[Shawinigan]	Shawinigan, New York, New York
[Shell]	Shell Oil Co., Houston, Texas
[Sheraton]	Sheraton Labs., Inc., Santa Clara, California
[Sherwood]	Sherwood Refining Co., Inc., Kansas City, Missouri
[Shionogi]	Shionogi & Co., Ltd., Osaka, Japan
[Showa Shinyaku]	Showa Shinyaku Co., Ltd., Nagoya, Japan
[Showa Yakuhin]	Showa Yakuhin Kako Co., Ltd., Tokyo, Japan
[Sidus]	Instituto Sidus S.A., Buenos Aires, Argentina
[Siegfried]	Siegfried Fabrik fur Chemisch-pharmazeutische Produkte, GmbH, Sackingen, W. Germany
[SIFI]	Societa Industria Farmaceutica Italiana S.p.A., Catania, Italy
[Sigma]	Sigma Chemical Co., St. Louis, Missouri
[Sigma-Tau]	Sigma-Tau S.p.A., Rome, Italy
[Sigurta]	Sigurta s.r.l., Milan, Italy
[Silva Araújo Roussel]	Silva Araújo Roussel S.A., Rio de Janeiro, Brazil
[SIMES]	Simes S.p.A. (subsidiary of Zambon), Milan, Italy
[Sinclair]	Sinclair Pharmaceuticals Ltd., Godalming, Surrey, England
[Sintopharma]	Labs. Sintopharma S.A., São Paulo, Brazil
[Sintyal]	Sintyal, Buenos Aires, Argentina
[Siphar]	Siphar GmbH, Köln, W. Germany
[Sis-Ter]	Sistemi Terapeutici S.p.A., Palazzo Pignano, Italy
[SIT]	Specialita Igienico Terapeutiche S.p.A., Mede (Pavia), Italy
[SK & F]	Smith Kline & French Labs. (subsidiary of SmithKline Beecham), Philadelphia, Pennsylvania
[SKB]	SmithKline Beecham, Philadelphia, Pennsylvania
[SM & P]	Smith, Miller & Patch (division of CooperVision Inc. which is a subsidiary of Cooper Labs.), San German, Puerto Rico

[Smith & Nephew]	Smith & Nephew Pharmaceuticals Ltd., Welwyn Garden City, Herts., England
[Sobio]	Lab. Sobio, Paris, France
[Soc. d'Etudes Prod. Chim.]	Société d'Études de Produits Chimiques, Issy-les-Moulineaux, France
[Solac]	Lab. Solac (subsidiary of Sanofi), Toulouse, France
[Solco]	Solco GmbH Pharm., Grenzach/Whylen, W. Germany
[Sonneborn]	L. Sonneborn Sons, Inc., New York, New York
[SPA]	Societa Prodotti Antibiotici S.p.A., Milan, Italy
[Specia]	"Specia" (Société Parisienne d'Expansion Chimique), (subsidiary of Rhone-Poulenc), Paris, France
[Speywood]	Speywood Labs., Ltd., Nottingham, England
[Spitzner]	W. Spitzner, Arzneimittelfabrik GmbH, Ettlingen, W. Germany
[Spofa]	Spofa, Prague, Czechoslovakia
[S.P.R.E.T.]	Knoll (S.P.R.E.T.), Clichy-La Garenne, France
[Spret-Mauchant]	Spret-Mauchant, Gennevilliers, France
[Squibb]	E. R. Squibb & Sons, Princeton, New Jersey
[SS Pharm.]	SS Pharmaceutical Co., Ltd., Tokyo, Japan
[Stada]	Stada-Arzneimittel AG & Stada-Chemie GmbH, Bad Vilbel, W. Germany
[Stafford-Miller]	Stafford-Miller Ltd., Hatfield, Herts., England
[Stago]	Laboratoire Stago, Sannois, France
[Standard Oil]	Standard Oil Co., Chicago, Illinois
[Stauffer]	Stauffer Chemical Co., New York, New York
[Stecker]	Stecker Chemicals, Inc., Hohokus, New Jersey
[Steiner]	Steiner & Co., Deutsche Arzneimittel Gesellschaft, Berlin, W. Germany
[Steinhard]	M. A. Steinhard Ltd., London, England
[Stepan]	Stepan Chemical Co., Northfield, Illinois
[Sterling]	Sterling Drug Inc. (subsidiary of Eastman Kodak), New York, New York
[Sterwin]	Sterwin Labs., Inc. (subsidiary of Sterling Drug, Inc.), New York, New York
[Stiefel]	Stiefel Labs., Inc., Oak Hill, New York
[Stockli]	Dr. J. Stockli, Basel, Switzerland
[Streuli]	G. Streuli & Co., Uznach, Switzerland
[Stroder]	Ist. Farmaco Biologico Stroder s.r.l., Florence, Italy
[Stuart]	Stuart Pharmaceuticals (division of ICI Americas), Wilmington, Delaware
[Substantia]	Lab. Substantia, Courbevoie, France
[Südmedica]	Südmedica Chemisch-pharmazeutische Fabrik, GmbH, Munich, W. Germany
[Sumitomo]	Sumitomo Pharmaceuticals Co., Ltd., Osaka, Japan
[Summit]	Summit Hill Laboratories, Avalon, New Jersey
[Sutliff & Case]	Sutliff & Case Co., Inc., Peoria, Illinois
[Syncro]	Syncro Argentina S.A., Buenos Aires, Argentina
[Synochem]	Synochem-Präparate, Hamburg, W. Germany
[Syntex]	Syntex Labs., Inc., Palo Alto, California
[Synthelabo]	Synthelabo, Paris, France
[Synthetic]	A/S Synthetic, Grinsted, Denmark
[Szabo]	Laboratorios Szabo Kessler, Humahuaca, Argentina
[Table Rock]	Table Rock Labs., Inc., Greenville, South Carolina
[TAD]	TAD Pharmaceutisches Werk GmbH, Cuxhaven, W. Germany
[Taiho]	Taiho Pharmaceutical Co., Ltd., Tokyo, Japan
[Taisho]	Taisho Pharmaceutical Co., Ltd., Tokyo, Japan
[Taiyo]	Taiyo Pharmaceutical Industry Co., Ltd., Gifu, Japan
[Taiyo Yakuko]	Taiyo Yakuko, Takayama, Japan
[Takata]	Takata Pharmaceutical Co., Ltd., Tokyo, Japan
[Takeda]	Takeda Chem. Ind. Ltd., Osaka, Japan
[Tanabe Seiyaku]	Tanabe Seiyaku Co., Ltd., Osaka, Japan
[TAP]	TAP (Takeda-Abbott Pharmaceuticals), N. Chicago, Illinois
[Taro]	Taro Pharm Ind. Ltd., Haifa Bay, Isreal
[Tatsumi]	Tatsumi Kagaku Co., Ltd., Ishikawa, Japan
[Teikoku Kagaku]	Teikoku Chemical Ind. Co., Ltd., Osaka, Japan
[Teikoku Zoki]	Teikoku Hormone Mfg. Co., Ltd., Osaka, Japan
[Teisan]	Teisan Chemical Ind. Co., Ltd., Osaka, Japan
[Temmler]	Temmler Pharma GmbH, Marburg/Lahn, W. Germany
[Tempelhof]	Chemische Fabrik Tempelhof, Preuss & Temmler, Berlin, W. Germany
[Tenneco]	Tenneco Chemicals, Inc., New York, New York
[Tennessee Pharm.]	Tennessee Pharmaceutical Co., Inc., Memphis, Tennessee
[Teva]	Teva Pharmaceutical Industries Ltd., Jerusalem, Israel
[Texas Pharmacal]	Texas Pharmacal Co., San Antonio, Texas
[Thames]	Thames Labs., Ltd., London, England
[Théranol]	Lab. Théranol, Paris, France
[Théraplix]	Théraplix S.A. (subsidiary of Rhone-Poulenc), Paris, France
[Thiemann]	Dr. Thiemann Arzneimittel GmbH (subsidiary of Akzo), Waltrop, W. Germany
[Thilo]	Dr. Thilo & Co., GmbH, Sauerlach, W. Germany
[Thomae]	Dr. Karl Thomae GmbH (subsidiary of C. H. Boehringer Sohn), Biberach/Riss, W. Germany
[Thompson Hayward]	Thompson Hayward Chemical Co. (subsidiary of North American Philips Corp.), Kansas City, Kansas
[Thuron]	Thuron Industries, Inc., Dallas, Texas

[Tiber]	Tiber Prodotti Chimico Biologici s.r.l., Rome, Italy
[Tilden-Yates]	Tilden-Yates Labs., Inc., Worcester, Massachusetts
[Tillotts]	Tillotts Labs., Henlow, Bedfordshire, England
[Toa Eiyo]	Toa Eiyo Co., Ltd., Tokyo, Japan
[Tobishi]	Tobishi Pharmaceutical Co., Ltd., Tokyo, Japan
[Toho]	Toho Pharmaceutical Ind. Co., Ltd., Osaka, Japan
[Toho Iyaku]	Toho Iyaku Kenkyusho, Tokyo, Japan
[Tokyo Hosei]	Tokyo Hosei Seiyaku Co., Ltd., Tokyo, Japan
[Tokyo Tanabe]	Tokyo Tanabe Seiyaku Co., Ltd., Tokyo, Japan
[Torii]	Torii & Co., Ltd., Tokyo, Japan
[Tosi]	Istituto Franco Tosi S.p.A., Milan, Italy
[Tosse]	E. Tosse & Co. mbH, Hamburg, W. Germany
[Towa Yakuhin]	Towa Yakuhin Co., Ltd., Osaka, Japan
[Toyama]	Toyama Chemical Co., Ltd., Tokyo, Japan
[Toyo Jozo]	Toyo Jozo Co., Ltd., Tokyo, Japan
[Toyo Pharmar]	Toyo Pharmar Co., Ltd., Toyama, Japan
[Travenol]	Travenol Labs. (subsidiary of Baxter Travenol), Deerfield, Illinois
[Trent]	Trent Pharmaceuticals, Elmsford, New York
[Trommsdorff]	H. Trommsdorff GmbH & Co., Arzneimittel, Alsdorf, W. Germany
[Troponwerke]	Tropenwerke Dinklage & Co. (subsidiary of Bayer AG), Koln, W. Germany
[Tsuruhara]	Tsuruhara Pharmaceutical Co., Ltd., Osaka, Japan
[Tubi Lux]	Tubi Lux Farma, Rome, Italy
[Tutag]	Tutag Pharmaceuticals, Inc., Broomfield, Colorado
[TVL]	Tasman Vaccine Laboratory, Suffolk, England
[UCB]	Union Chimique Belge, Brussels, Belgium
[UCB-Smit]	UCB-Smit, Turin, Italy
[UCEPHA]	UCEPHA, Nanterre, France
[UCM-Difme]	Unione Chimica Medicamenti Difme S.p.A., Grugliasco, Italy
[Uhlhorn]	Dr. E. Uhlhorn & Co., GmbH, Wiesbaden/Biebrich, W. Germany
[Ulmer]	The Ulmer Pharmacal Co., Minneapolis, Minnesota
[Unicet]	Unicet (subsidiary of Schering-Plough Corp.), Levallois-Perret, France
[Unifa]	Unifa S.A., Buenos Aires, Argentina
[Unilabo]	"Unilabo", Levallois, France
[Unimed]	Unimed, Inc., Somerville, New Jersey
[Union Carbide]	Union Carbide Corp., New York, New York
[Uniroyal]	Uniroyal, Inc., New York, New York
[UOP]	Universal Oil Products Chemical Div., East Rutherford, New Jersey
[Upjohn]	The Upjohn Co., Kalamazoo, Michigan
[UPSA]	UPSA, Rueil-Malmaison, France
[Upsher-Smith]	Upsher-Smith Labs., Inc., Minneapolis, Minnesota
[Uriach]	J. Uriach y Cia, Barcelona, Spain
[Ursapharm]	Ursapharm, Bernhard Buxmann & Co., GmbH, Bubingen, W. Germany
[Usafarma]	Usafarma S.A., São Paulo, Brazil
[U.S. Ethicals]	U.S. Ethicals, Inc., Long Island City, New York
[Usines Fournier-Cimag]	Usines Fournier-Cimag, Marseille, France
[USV]	USV Pharmaceuticals Corp. (subsidiary of Rorer), New York, New York
[Vaillant-Defresne]	Lab. Vaillant-Defresne, Paris, France
[Valderrama]	Valderrama Labs., Bilbao, Spain
[Vale]	The Vale Chem. Co., Inc., Allentown, Pennsylvania
[Valeas]	Valeas S.p.A., Milan, Italy
[Valpan]	Lab. Valpan, Le Mee-sur-Seine, France
[Vanderbilt]	R. T. Vanderbilt Co., Inc., Norwalk, Connecticut
[Van Dyk]	Van Dyk & Company, Inc., Belleville, New Jersey
[Vangard]	Vangard, Glasgow, Kentucky
[Van Pelt & Brown]	Van Pelt & Brown, Inc. (division of Mallinckrodt, Inc.), Richmond, Virginia
[Varialab]	Varialab, Lausanne, Switzerland
[Velsicol]	Velsicol Chemical Corp., Chicago, Illinois
[Verla-Pharm]	Verla-Pharm. Arzneimittelfabrik, Tutzing, W. Germany
[Veyron-Froment]	Lab. Veyron-Froment, Marseille, France
[Vifor]	Vifor S.A., Geneva, Switzerland
[Vikwood]	Vikwood Ltd., Sheboygan, Wisconsin
[Villela]	Lab. Mauricio Villela S.A., Rio de Janeiro, Brazil
[Villette]	Labs. H. Villette, Paris, France
[Vineland]	Vineland Chemical Co., Vineland, New Jersey
[Violani]	Violani-Farmavigor S.p.A., Milan, Italy
[Virax]	Virax Ethicals Ltd., Melbourne, Victoria, Australia
[Vister]	Vister Vismara Terapeutici S.p.A., Casatenovo Brianza (Como), Italy
[Vita]	Vita Farmaceutici S.p.A., Turin, Italy
[Vitacain]	Vitacain Pharmaceutical Co., Ltd., Osaka, Japan
[Viti]	Marco Viti Industria Farmaceutica s.r.l., Milan, Italy
[Vitrum]	Apoteksvarucentralen Vitrum Apotekare AB, Stockholm, Sweden
[Voigt]	Dr. Hans Voigt, Pharmazeutische Fabrik GmbH, Limburg/Lahn, W. Germany
[Von Boch]	Von Boch Arzneimittel s.r.l. Istituto Farmacobiologico, Rome, Italy
[Wacker]	Wacker Chemie, Munich, W. Germany
[Wakamoto]	Wakamoto Pharmaceutical Co., Ltd., Tokyo, Japan

[Walker]	Walker Corp. & Co., Inc., Syracuse, New York
[Wallace]	Wallace Mfg. Chemists, Ltd., London, England
[Wallace Labs.]	Wallace Labs. (division of Carter-Wallace, Inc.) Cranbury, New Jersey
[Wampole]	Wampole Labs. (division of Carter-Wallace, Inc.), Cranbury, New Jersey
[Wander]	Sandoz-Wander Pharma (subsidiary of Sandoz), Bern, Switzerland
[Warner-Chilcott]	Warner-Chilcott Labs. (division of Warner-Lambert Co.), Morris Plains, New Jersey
[Warner-Lambert]	Warner-Lambert Pharmaceutical Co., Morris Plains, New Jersey
[Warren-Teed]	Warren-Teed Pharmaceuticals, Inc. (subsidiary of Montedison), Columbus, Ohio
[Warrick]	Warrick Pharmaceuticals Ltd., Bracknell, Berkshire, England
[Wassermann]	Sociedad Española de Espec. Fármaco-Terapéuticas Wassermann, S.A., Barcelona, Spain
[W. B. Pharm.]	W. B. Pharmaceuticals, Ltd. (subsidiary of C. H. Boehringer Sohn), Wembley, Middlesex, England
[Webcon]	Webcon Pharmaceuticals (division of Alcon Labs., Inc.), Fort Worth, Texas
[Weddel]	Weddel Pharmaceuticals, London, England
[Weimer]	Waldemar Weimer Chem.-pharm. Fabrik GmbH, Rastatt, W. Germany
[Weisskopf]	Weisskopf KG Pharmazeutisches Werk, Berlin, W. Germany
[Welcker-Lyster]	Welcker-Lyster Ltd., Montreal, Canada
[Wellcome]	The Wellcome Foundation Ltd., London, England
[Westerfield]	Westerfield Labs. (division of O'Neal, Jones & Feldman, Inc.), Cincinnati, Ohio
[West-Ward]	West-Ward, Inc., Bronx, New York
[Westwood]	Westwood Pharmaceuticals, Inc. (subsidiary of Bristol-Myers Co.), Buffalo, New York
[Wharton]	Wharton s.r.l., Bologna, Italy
[White]	White Labs., Inc. (subsidiary of Schering-Plough Corp.), Kenilworth, New Jersey
[Whitehall]	Whitehall Labs. (subsidiary of American Home Products Corp.), New York, New York
[Willows]	Willows Francis Ltd., Bolton, Lancs., England
[Will-Pharma]	Will-Pharma, Amsterdam, Netherlands; Brussels, Belgium
[Winston]	Winston Pharmaceuticals, Inc., Winston-Salem, North Carolina
[Winthrop]	Winthrop Labs. (subsidiary of Sterling Drug Inc.), New York, New York
[Winzer]	Dr. Winzer, Chem.-Pharm. Fabrik, Konstanz, W. Germany
[Wisconsin Pharmacal]	Wisconsin Pharmacal Co., Jackson, Wisconsin
[Wm. R. Warner]	Wm. R. Warner & Co., Ltd., Eastleigh, Hants., England
[Woelfer]	Otto A. H. Woelfer GmbH, Hamburg, W. Germany
[Woelm]	Woelm Pharma GmbH, Eschwege, W. Germany
[Wolff]	Dr. August Wolff Chem.-pharm. Fabrik GmbH & Co., KG, Bielefeld, W. Germany
[Wülfing]	Johann A. Wülfing-Bauer & Cie., Neuss, W. Germany
[Wyandotte]	Wyandotte Chemicals Corp. (subsidiary of BASF), Wyandotte, Michigan
[Wyeth]	Wyeth-Ayerst Labs. (division of American Home Products Corp.), Radnor, Pennsylvania
[Xttrium]	Xttrium Labs., Inc., Chicago, Illinois
[Yamagata]	Yamagata Doctors Pharmaceutical Co., Ltd., Yamagata-shi, Japan
[Yamanouchi]	Yamanouchi Pharmaceutical Co., Ltd., Tokyo, Japan
[Yoshitomi]	Yoshitomi Pharmaceutical Ind. Co., Ltd., Osaka, Japan
[Zambeletti]	Dr. L. Zambeletti S.p.A. (subsidiary of Beecham), Milan, Italy
[Zambon]	Zambon S.p.A., Bresso (Milan), Italy
[Zdravlje]	"Zdravlje" Factory for Pharmaceutical Chemical Products, Leskovac, Yugoslavia
[Zeria]	Zeria Pharmaceutical Co., Ltd., Tokyo, Japan
[Zirkulin]	Zirkulin Werke, GmbH, Herdecke, W. Germany
[Zoecon]	Zoecon Corp., Palo Alto, California
[Zyma]	Zyma GmbH (subsidiary of Ciba-Geigy), Munich, W. Germany

Greek Alphabet

Name of letter	Capital	Lower case	Transliteration	Name of letter	Capital	Lower case	Transliteration
alpha	A	α	a	nu	N	ν	n
beta	B	ß or β	b	xi	Ξ	ξ	x
gamma	Γ	γ	g	omicron	O	o	o short
delta	Δ	δ	d	pi	Π	π	p
epsilon	E	ϵ	e short	rho	P	ρ	r
zeta	Z	ζ	z	sigma	Σ	σ or ς	s
eta	H	η	e long	tau	T	τ	t
theta	Θ	θ	th	upsilon	Υ	υ	y
iota	I	ι	i	phi	Φ	ϕ or φ	f
kappa	K	κ	k, c	chi	X	χ	ch as in German echt
lambda	Λ	λ	l	psi	Ψ	ψ	ps
mu	M	μ	m	omega	Ω	ω	o long

Russian Alphabet

Cyrillic print		Transliteration	Pronunciation	Cyrillic print		Transliteration	Pronunciation
А	а	a	*a* in far	С	с	s	*s* in say
Б	б	b	*b*	Т	т	t	*t*
В	в	v	*v*	У	у	u	*oo* in boot
Г	г	g (h)	*g* in gay	Ф	ф	f	*f*
Д	д	d	*d*	Х	х	kh	like German *ch*
Е	е	e	*e* in fell; also *ye* in yell	Ц	ц	t͡s	*ts* in hoots
Ж	ж	zh	*z* in azure	Ч	ч	ch	*ch* in church
З	з	z	*z* in zeal	Ш	ш	sh	*sh*
И	и	i	*i* in meet	Щ	щ	shch	*shch*, as in fre*sh ch*eese
Й	й	ĭ	*y* in boy	Ъ	ъ	ʺ	mute*
К	к	k	*k*	Ы	ы	y	*y* in rhythm (hard)
Л	л	l	*l*	Ь	ь	ʹ	mute (softens preceding consonant)
М	м	m	*m*				
Н	н	n	*n*	Э	э	ė	*e* in met
О	о	o	*o* in or	Ю	ю	i͡u	*u* in union
П	п	p	*p*	Я	я	i͡a	*ya* in yard
Р	р	r	*r*				

* Hard sign; used to separate a consonant from a soft vowel especially in foreign words; frequently replaced by an apostrophe.

Roman Numerals

I	II	III	IV	V	VI	VII	VIII	IX	X
1	2	3	4	5	6	7	8	9	10
XX	XXX	XL	L	LX	LXX	LXXX	XC	IC	C
20	30	40	50	60	70	80	90	99	100
CC	CCC	CD	D	DC	DCC	DCCC	CM	XM	M
200	300	400	500	600	700	800	900	990	1000

Numerical Prefixes Commonly Used in Forming Chemical Names

Numeral	Prefix	Numeral	Prefix	Numeral	Prefix
½	hemi-	13	trideca-	28	octacosa-
1	mono-	14	tetradeca-	29	nonacosa-
1½	sesqui-	15	pentadeca-	30	triaconta-
2	di-, bi-	16	hexadeca-	40	tetraconta-
2½	hemipenta-	17	heptadeca-	50	pentaconta-
3	tri-	18	octadeca-	60	hexaconta-
4	tetra-	19	nonadeca-	70	heptaconta-
5	penta-	20	eicosa-	80	octaconta-
6	hexa-	21	heneicosa-	90	nonaconta-
7	hepta-	22	docosa-	100	hecta-
8	octa-	23	tricosa-	101	henhecta-
9	ennea-, nona-	24	tetracosa-	102	dohecta-
10	deca-	25	pentacosa-	110	decahecta-
11	hendeca-, undeca-	26	hexacosa-	120	eicosahecta-
12	dodeca-	27	heptacosa-	200	dicta-

Alchemical Symbols Used in Biology and Botany

☉	Sun; gold; annual plant.
☉ ☉	Biennial plant.
○ or ☽	Moon; silver; *sometimes* female sex.
☿	Mercury.
♀	Venus; copper; ***female sex.***
♁	Earth; terra; *sometimes* male sex.
♁ ♀	Having male and female flowers separate.
♁ — ♀	Having male and female flowers on the same plant.
♁ : ♀	Having male and female flowers on different plants.
♂	Mars; iron; ***male sex.***
☌	Conjunction; mating; mated.
♃	Jupiter; tin; perennial herb.
♄	Saturn; lead.
🜍	Sulfur.
☊	Arsenic.

Prescription Notation

The units of quantities written on prescriptions in the Apothecaries System are usually designated by symbols. Occasionally abbreviations are used. The following list provides these symbols and abbreviations.

Apothecaries' Weight

Symbol	*Abbreviation*	*Unit*
—	gr.	grain
℈	sc.	scruple
ʒ	dr.	dram
℥	oz.	ounce
℔	lb.	pound

Apothecaries' Volume

Symbol	*Abbreviation*	*Unit*
♍	min.	minim
f ʒ	fl. dr.	fluid dram
f ℥	fl. oz.	fluid ounce
O.	pt.	pint
Cong.	gal.	gallon

Roman numerals (lower case) are always used following the symbol to designate the number of units required, but, if the abbreviation is used, arabic numerals are used and precede the abbreviation. For example:

ʒiv but 4 dr.

For less than one unit, one-half may be designated by ss following the symbol, but other fractions must be designated by arabic numeral fractions. For example:

ʒss but gr. $\frac{1}{8}$

Roman numerals are also usually used to designate the number of dosage forms required. For example:

Caps. No. xlv

The following list contains other common abbreviations and notations used in prescription writing.

Latin Terms

Abbreviation	Expanded form	Meaning	Abbreviation	Expanded form	Meaning
a.	ante	before	d.	dexter	right
a., aur.	auris	ear	d.	dies	a day
aa.	ana	of each	d.	dosis	a dose
a.c.	ante cibos	before meals	da	da	give
ad	ad	to, up to	dent., d.	dentur	let be given, give
add.	adde, addantur, addendus, addendo	add, let them be added, to be added, by adding	det.	detur	let it be given
			dieb. alt.	diebus alternis	on alternate days
			dieb. secund.	diebus secundis	every second day
ad lib.	ad libitum	at pleasure	dil.	dilue	dilute
adm.	admove	apply	disp.	dispensa, dispensetur	dispense
ad man. med.	ad manus medici	(to be delivered) into the hands of the (pre-scribing) physician	div.	divide	divide
			div. in par. aeq.	dividatur in partes aequales	divide in equal parts
ad satur.	ad saturandum	to saturation			
agit.	agita	shake			
alb.	albus	white	d.t.d.	dentur tales doses	give such doses
alt. hor.	alternis horis	every other hour	dulc.	dulcis	sweet
A.M.	ante meridiem	before noon	dur.	durus	hard
ampl.	amplus	large	e.m.p.	ex modo prescripto	after the manner prescribed
ampul.	ampulla	ampul, ampule			
applicand.	applicandus	to be applied	et	et	and
aq.	aqua	water	ex	ex	out of
aq. bull.	aqua bulliens	boiling water	ext.	extractum	an extract
aq. dest.	aqua destillata	distilled water	f., ft.	fac, fiat, fiant	make, let be made
aq. ferv.	aqua fervens	hot water	ferv.	fervens	hot
aq. frig.	aqua frigida	cold water	filt.	filtra	filter (imperative form)
b.	bis	twice	fl.	fluidus	fluid
ben.	bene	well	flav.	flavus	yellow
bib.	bibe	drink	fldext.	fluidextractum	fluidextract
b.i.d.	bis in die	twice a day	fort.	fortis	strong
bol.	bolus	a large pill	frig.	frigidus	cold
brevis	brevis	short	garg.	gargarisma	a gargle
bull.	bulliens, bulliat, bulliant	boiling, let boil	Gm.	gramma	gram
			gr.	granum	grain
c̄	cum	with	gran.	granulatus	granulated
C	centum	a hundred	gtt.	gutta	a drop
cap.	capiat	let the patient take	h.s.	hora somni	at the hour of sleep, at bedtime
caps.	capsula	a capsule			
cerat.	ceratum	wax ointment	hydrarg.	hydrargyrum	mercury
chart.	charta	paper, a powder in paper	i.c.	inter cibos	between meals
			juxt.	juxta	near
chart. cerat.	charta cerata	waxed paper, parch-ment paper	Kal.	Kalium	potassium
			l.a.	lege artis	according to the art
chirurg.	chirurgicalis	surgical	laev.	laevus	left
chord. chirurg.	chorda chirurgicalis	surgical "catgut"	lb.	libra	pound
			lev.	levis	light
cito disp!	cito dispensetur!	let it be dispensed quickly	liq.	liquor	a liquor, a solution
			lot.	lotio	lotion
coch. amp.	cochleare amplum	a tablespoonful	m.	mane	in the morning
			M.	misce	mix
coch. mag.	cochleare magnum	a tablespoonful	m.	mitte	send
			mag.	magnus	large
coch. med.	cochleare medium	a dessertspoonful	m. dict.	more dicto	as directed
			min.	minimum	minim
coch. mod.	cochleare modicum	a dessertspoonful	mist.	mistura	mixture
			mixt.	mixtura	mixture
coch. parv.	cochleare parvum	a teaspoonful	m.p.	modo praescripto	in the manner prescribed
col.	cola	strain (imperative form)	m.t.d.	mitte tales doses	send such doses
			n.	naris	nostril
colet.	coletur	let it be strained	Natr.	Natrium	sodium
collun.	collunarium	a nose wash	nebul.	nebula	a spray
collut.	collutorium	a mouth wash	n. et m.	nocte maneque	night and morning
collyr.	collyrium	an eye wash	nig.	niger	black
comp.	compositus	compound, compounded	no.	numero	number
			noct.	nocte	at night
cong.	congius	a gallon	non rep.	non repetatur	do not repeat
consperg.	consperge, conspergetur	dust, sprinkle	O.	Octarius	a pint
			o.d.	oculus dexter	right eye
cont.	contra	against	o.l.	oculus laevus	left eye
contus.	contusus	bruised	omn. hor.	omni hora	at every hour
coq.	coque, coquatur	boil, let it boil	omn. man.	omni mane	on every morning

Abbreviation	Expanded form	Meaning	Abbreviation	Expanded form	Meaning
opt.	optimus	best	s.a.	secundum artem	according to art
o.s.	oculus sinister	left eye	sat.	saturatus	saturated
o.u.	oculi uterque	both eyes	scat.	scatula	box
p. ae.	partes aequales	equal parts	scat. orig.	scatula originalis	original package, manufacturer's package (and label)
parv.	parvus	small			
p.c.	post cibos	after meals			
phial.	phiala	bottle	sic.	siccus	dried
Plumb.	Plumbum	lead	s.n.	secundum naturam	according to nature
P.M.	post meridiem	after noon			
p.o.	per os	by mouth	sol.	solubilis	soluble
pond.	ponderosus	heavy	sol.	solutio	solution
ppt.	praecipitatus	precipitated	solv.	solve	dissolve
p.r.n.	pro re nata	as occasion arises, as needed	s.o.s.	si opus sit	if there is need
			spir.	spiritus	spirit
pro rect.	pro recto	rectal	spiss.	spissus	dried
pt.	perstetur	let it be continued	ss.	semis	one half
pulv.	pulvis, pulveres	powder, powders	stat.	statim	immediately
q., qq.	quodque, quaeque	each, every	suc.	succus	juice
			sum.	sume, sumendus	take, to be taken
q.i.d.	quater in die	four times a day	S.V.R.	spiritus vini rectificatus	alcohol
qq. hor.	quaque hora	every hour			
Q.R.	quantum rectum	the quantity is correct	syr.	syrupus	syrup
			tab.	tabella	tablet
q.s.	quantum sufficiat, quantum satis	a sufficient quantity	tal.	talis, tales, talia	such
			ter.	tere	rub
			t.i.d.	ter in die	three times a day
quot. op. sit	quoties opus sit	as often as necessary	tinct., tr.	tinctura	tincture
q.v.	quantum voleris	as much as you wish	trit.	tritura	triturate
℞	recipe	you take	ult.	ultime	lastly
recen.	recens	fresh	unct.	unctus	smeared
rect.	rectificatus	rectified	ung., ungt.	unguentum	ointment
ren. sem.	renovetur semel	shall be renewed (only) once	ust.	ustus	burnt
			ut dict.	ut dictum	as directed
rept.	repetatur	let it be repeated	v.	vel	or
rub.	ruber	red	vesp.	vesper	evening
S., Sig.	signa, signetur	write, it shall be written (as instruction to the patient)	vir.	viridis, viride	green
			vol.	volatilis, volatile	volatile

Table of Radioactive Isotopes

The following table is based principally on the *Table of Isotopes*, C. M. Lederer, V. S. Shirley, Eds. (Wiley-Interscience, New York, 7th ed., 1978) and the *Table of Radioactive Isotopes*, V. S. Shirley, Ed. (Wiley-Interscience, New York, 8th ed., 1986). Additional references, particularly for the transcurarium elements, are cited in the monographs for each element in the monograph section of *The Merck Index*.

Columns 1, 2 and 3 list, respectively, the atomic number (Z), the name of the element and the mass number (A) of the isotope. An asterisk preceding the mass number indicates that the radioisotope occurs in nature. An m following the mass number indicates a metastable isotope. Customary designations for naturally occurring radioisotopes are given in parentheses after the mass number.

The total half-life of the isotope is listed in column 4. The half-life reported represents the most accurate value as published in the *Table of Radioactive Isotopes*. Except in the case of the transuranium elements, isotopic modifications for which the mass number assignment is doubtful are omitted. Generally, only those isotopes with a half-life of 60 minutes or longer are included in the table. Exceptions are made in the following instances:

1. Radioisotopes already used as biological tracers
2. Short-lived products, including nuclear isomers, of radioactive parents with half-lives of 60 minutes or greater
3. Naturally occurring radioisotopes
4. Isotopes of the transuranium elements

The following abbreviations for time units are employed: y = years, d = days, h = hours, min = minutes, s = seconds, ms = milliseconds, μs = microseconds, ns = nanoseconds.

In the last column are the principal modes of disintegration and energies of the radiations in million electronvolts (MeV). Symbols used to represent the various modes of decay are:

α	alpha particle emission	K	electron capture
β^-	beta particle	IT	isomeric transition
β^+	positron	e^-	internal conversion electron
γ	gamma ray	x	x-ray
		S.F.	spontaneous fission

When x-radiation can be associated with a specific element, the symbol of this element precedes the x designation, as As-x. The energies of these radiations are enclosed in parentheses following the appropriate symbol for a given type of radiation. For β^- and β^+, values of E_{max} are listed. An attempt was made to arrange the modes of disintegration and energies in order of decreasing importance. Radiation types and energies of minor importance are omitted unless useful for identification purposes. Annihilation radiation associated with β^+ emission is not explicitly indicated. The reader should consult the literature cited for detailed decay schemes.

Z	Element	A	Half-life	Radiation (MeV)
1	Hydrogen	3	12.33y	β^-(0.0186)
4	Beryllium	7	53.29d	K, γ(0.478)
		10	1.6×10^6y	β^-(0.555), no γ
6	Carbon	11	20.39min	β^+(0.961), no γ
		14	5730y	β^-(0.156), no γ
7	Nitrogen	13	9.965min	β^+(1.190), no γ
9	Fluorine	18	1.8295h	β^+(0.635), K, O-x
11	Sodium	22	2.602y	β^+(0.545, 1.83), K, Ne-x, γ(1.275)
		24	14.659h	β^-(1.39), γ(2.75, 1.37)
12	Magnesium	28	20.90h	β^-(0.46), γ(1.34, 0.94, 0.40, 0.031)
13	Aluminum	26	7.2×10^5y	β^+(1.16), K, Mg-x, γ(1.81, 1.12)
14	Silicon	31	2.622h	β^-(1.48), γ(1.27)
		32	104y	β^-(0.213), no γ
15	Phosphorus	32	14.282d	β^-(1.71), no γ
		33	25.34d	β^-(0.25), no γ
16	Sulfur	35	87.51d	β^-(0.167)
		38	2.84h	β^-(1.0, 3.0), γ(1.94)
17	Chlorine	36	3.01×10^5y	β^-(0.71), K, S-x, no γ
		38	37.24min	β^-(4.81, 1.11, 2.77), γ(2.17, 1.60)
		39	55.6min	β^-(1.91, 2.18, 3.45), γ(1.27, 0.25, 1.52)
18	Argon	37	35.04d	K, Cl-x, no γ
		41	1.827h	β^-(1.20, 2.49), γ(1.29)
		42	32.9y	β^-

Z	Element	A	Half-life	Radiation (MeV)
19	Potassium	*40	1.277×10^9y	β^-(1.34), K, Ar-x, γ(1.46)
		42	12.360h	β^-(3.52, 1.97), γ(1.52, 0.31)
		43	22.3h	β^-(0.83, 0.46, 1.22, 1.82), γ(0.618, 0.373, 0.39, 0.59, 0.22)
20	Calcium	41	1.03×10^5y	K, K-x
		45	163.8d	β^-(0.255)
		47	4.536d	β^-(1.98, 0.67), γ(1.30, 0.81, 0.49)
21	Scandium	43	3.891h	β^+(1.2, 0.82, 0.39), K, γ(0.373)
		44m	2.442d	IT, Sc-x, γ(0.271), e$^-$
		44	3.927h	β^+(1.47), K, γ(1.16)
		46	83.83d	β^-(0.357), Ti-x, γ(1.12, 0.889)
		47	3.341d	β^-(0.439, 0.60), γ(0.159)
		48	1.821d	β^-(0.65), γ(1.31, 1.04, 0.984)
22	Titanium	44	47.3y	K, γ(0.068, 0.078)
		45	3.080h	β^+(1.044), K, Sc-x,
23	Vanadium	48	15.976d	β^+(0.696), K, Ti-x, γ(1.31, 0.983)
		49	330d	K, Ti-x, no γ
24	Chromium	48	21.56h	K, V-x, γ(0.308, 0.116), no β^+
		51	27.704d	K, V-x, γ(0.32)
25	Manganese	52	5.591d	K, β^+(0.574), Cr-x, γ(0.744, 0.935, 1.434)
		53	3.74×10^6y	K, Cr-x, no γ
		54	312.20d	K, Cr-x, γ(0.835), no β^+, no β^-
		56	2.5785h	β^-(2.84, 1.03, 0.72), γ(0.847, 1.81, 2.11)
26	Iron	52	8.275h	β^+(0.804), K, Mn-x, γ(0.169)
		55	2.73y	K, Mn-x, no γ
		59	44.496d	β^-(0.273, 0.475), γ(1.10, 1.29, 0.192)
		60	$\sim 1 \times 10^5$y	β^-
27	Cobalt	55	17.53h	β^+(1.04, 1.50), K, Fe-x, γ(0.932, 0.480, 1.41)
		56	77.7d	K, β^+(1.46), Fe-x, γ(0.847, 1.04, 1.24, 1.77, 2.60, 3.26, 2.02)
		57	271.77d	K, Fe-x, γ(0.136, 0.122, 0.014)
		58m	9.15h	IT, Co-x, γ(0.025), no β^+
		58	70.916d	K, β^+(0.474), Fe-x, γ(0.811)
		60	5.271y	β^-(0.318), γ(1.173, 1.332)
		61	1.650h	β^-(1.22), γ(0.067)
28	Nickel	56	6.10d	K, Co-x, γ(0.158, 0.270, 0.480, 0.75, 0.812, 1.56), no β^+
		57	1.503d	K, β^+(0.85, 0.72), Co-x, γ(1.378, 0.127, 1.76)
		59	7.5×10^4y	K, Co-x, no γ
		63	100.1y	β^-(0.066), no γ
		65	2.520h	β^-(2.14, 0.65, 1.02), γ(1.48, 0.367, 1.12)
		66	2.275d	β^-(0.20), no γ
29	Copper	61	3.408h	β^+(1.21), K, Ni-x, γ(0.28, 0.66)
		64	12.701h	K, β^-(0.571), β^+(0.657), Ni-x, γ(1.34)
		67	2.580d	β^-(0.395, 0.484, 0.577), γ(0.185, 0.092)
30	Zinc	62	9.26h	K, β^+(0.66), Cu-x, γ(0.041, 0.25, 0.51, 0.60)
		65	244.1d	K, β^+(0.325), γ(1.12), Cu-x
		69m	13.76h	IT, Zn-x, γ(0.439), e$^-$
		69	55.6min	β^-(0.90)
		72	1.938d	β^-(0.3), γ(0.016, 0.145, 0.192)
31	Gallium	66	9.49h	β^+(4.15), K, Zn-x, γ(1.04, 2.75)
		67	3.261d	K, Zn-x, γ(0.093, 0.184, 0.300, 0.393), e$^-$, no β^+
		68	1.135h	β^+(1.9), K, Zn-x, γ(1.08)
		72	14.10h	β^-(0.64, 0.96, 1.51, 2.53, 3.17), γ(0.835, 0.63, 2.20, 2.50)
		73	4.87h	β^-(1.20), γ(0.297, 0.534, 0.326)
32	Germanium	66	2.26h	K, β^+(1.02), Ga-x, γ(0.382, 0.044, 0.065, 0.109, 0.190, 0.246, 0.273, 0.338, 0.471)

Table of Radioactive Isotopes (Continued)

Z	Element	A	Half-life	Radiation (MeV)
		68	270.8d	K, Ga-x
		69	1.627d	K, β^+(1.22, 0.61), γ(1.11, 0.573, 0.872)
		71	11.15d	K, Ga-x
		73m	499ms	IT, γ(0.053, 0.0133), e$^-$
		75	1.3797h	β^-(1.19, 0.98, 0.92, 0.72), γ(0.265, 0.199)
		77	11.30h	β^-(2.20, 1.38, 0.71), γ(0.264, 0.211, 0.216, 0.416, 0.714)
		78	1.467h	β^-(0.71), γ(0.277)
33	Arsenic	71	2.70d	K, β^+(0.81), Ge-x, γ(0.175, 0.023), e$^-$
		72	1.083d	β^+(3.323, 2.490), K, Ge-x, γ(0.834)
		73	80.30d	K, γ(0.054, 0.0133), Ge-x, e$^-$
		74	17.78d	K, β^-(1.35, 0.72), β^+(1.53, 0.94), Ge-x, γ(0.596, 0.635)
		76	1.097d	β^-(2.97, 2.41, 1.79), γ(0.559, 0.657)
		77	1.6179d	β^-(0.68), γ(0.239, 0.250, 0.088)
		78	1.512h	β^-(4.42, 3.70, 3.00, 2.50), γ(0.62, 0.70, 0.83, 1.31)
34	Selenium	72	8.40d	K, As-x, γ(0.046), e$^-$
		73	7.15h	β^+(1.32), K, As-x, γ(0.361, 0.067)
		75	119.77d	K, γ(0.265, 0.136, 0.280, 0.121, 0.401), As-x
		79	$<6.5 \times 10^4$y	β^-(0.16), no γ
35	Bromine	75	1.62h	β^+(1.74, 1.59, 1.45, 1.34, 1.13), K, γ(0.286)
		76	16.2h	β^+(3.98, 3.44), K, Se-x, γ(0.559, 0.657, 1.216)
		77	2.37650d	K, β^+(0.34), Se-x, γ(0.24, 0.30, 0.52, 0.58, 0.82)
		80m	4.42h	IT, Br-x, γ(0.037, 0.049), e$^-$
		80	17.68min	β^-(1.997, 1.38), K, β^+(0.85), Se-x, γ(0.616, 0.667)
		82	1.4708d	β^-(0.444), γ(0.554, 0.619, 0.698, 0.777, 0.828, 1.04, 1.32, 1.48), no K, no β^+
		83	2.39h	β^-(0.925), γ(0.530)
36	Krypton	76	14.8h	K, γ(0.252), no β^+
		77	1.24h	β^+(1.875, 1.700, 1.550), K, Br-x, γ(0.130)
		79	1.460d	K, β^+(0.613), Br-x, γ(0.044–1.34)
		81m	13s	IT, Kr-x, γ(0.19)
		81	2.10×10^5y	K, Br-x, γ(0.276)
		83m	1.86h	IT, Kr-x, γ(0.032, 0.009)
		85m	4.480h	β^-(0.84), IT, Kr-x, γ(0.151, 0.305)
		85	10.72y	β^-(0.67), γ(0.517)
		87	1.272h	β^-(3.8, 1.3), γ(0.403, 2.55, 0.85)
		88	2.84h	β^-(0.52, 2.8, 0.9), γ(2.392, 0.028, 0.166, 0.196, 0.362)
37	Rubidium	81	4.58h	K, β^+(1.05), Kr-x, γ(0.446)
		82m	6.472h	K, β^+(0.800), γ(0.776, 0.554)
		82	1.273min	β^+(3.15), K, Kr-x, γ(0.776)
		83	86.2d	K, Kr-x, γ(0.521, 0.530, 0.553, 0.009, 0.79), no β^+
		84	32.87d	K, β^+(1.66, 0.78), β^-(0.89), Kr-x, γ(0.882)
		86	18.66d	β^-(1.78, 0.71), γ(1.08)
		*87	4.80×10^{10}y	β^-(0.272), no γ
		88	17.8min	β^-(5.080, 3.240, 2.350), γ(1.836, 0.898, 2.678)
38	Strontium	82	25.55d	K, Rb-x, no γ
		83	1.3504d	K, β^+(1.227), Rb-x, γ(0.763)
		85m	1.1258h	IT, K, Rb-x, Sr-x, γ(0.150, 0.231), e$^-$, no β^+
		85	64.84d	K, Rb-x, γ(0.514)
		87m	2.795h	IT, γ(0.388), e$^-$
		89	50.55d	β^-(1.488), γ(0.91 with Y^{89m})
		90	28.5y	β^-(0.546), no γ
		91	9.52h	β^-(2.67, 1.36, 1.09), γ(1.024, 0.556, 0.750)
		92	2.71h	β^-(0.55, 1.5), γ(1.384)
39	Yttrium	85m	4.86h	β^+(2.24), K, Sr-x, γ(0.767, 0.232, 2.124), no IT
		85	2.68h	β^+(1.58, 1.15), K, Sr-x, γ(0.504, 0.232), no IT
		86	14.74h	K, β^+(1.04, 1.25, 1.60, 2.02, 2.34, 3.15), Sr-x, γ(1.077)
		87m	12.9h	IT, β^+?(1.150), Y-x, γ(0.381)

Z	Element	A	Half-life	Radiation (MeV)
		87	3.346d	K, β^+?(0.7), Sr x, γ(0.485, 0.39)
		88	106.61d	K, β^+(0.76), Sr-x, γ(0.898, 1.84)
		90	2.671d	β^-(2.288), γ(2.186)
		91m	49.71min	IT, Y-x, γ(0.556)
		91	58.51d	β^-(1.543), γ(1.21)
		92	3.54h	β^-(3.64), γ(0.448, 0.560, 0.934, 1.40, 1.85)
		93	10.25h	β^-(2.89), γ(0.267, 0.95, 1.92)
40	Zirconium	86	16.5h	K, Y-x, γ(0.243), no β^+
		87	1.733h	β^+(2.260), K, Y-x, γ(1.228)
		88	83.4d	K, Y-x, γ(0.394)
		89	3.268d	K, β^+(0.897), Y-x, γ(0.909)
		90m	809.2ms	IT, Zr-x, γ(2.319)
		93	1.53×10^6y	β^-(0.060)
		95	64.02d	β^-(0.360, 0.396), γ(0.724, 0.757)
		97	16.90h	β^-(1.91), γ(0.743)
41	Niobium	89	2.03h	β^+(3.320), γ(1.627)
		90m	18.82s	IT, Nb-x, γ(0.122)
		90	14.60h	β^+(1.50), K, Zr-x, γ(0.141, 2.319)
		91m	62d	IT, Nb-x, γ(0.1045, 1.2), K
		91	680y	K, Zr-x
		92m	10.15d	K, γ(0.934)
		93m	13.6y	IT, Nb-x, γ(0.0304)
		94	2.03×10^4y	β^-(0.473), γ(0.703, 0.871)
		95m	3.61d	IT, β^-(1.160), Nb-x, γ(0.204, 0.236), e$^-$
		95	34.97d	β^-(0.160), γ(0.766)
		96	23.35h	β^-(0.748, 0.500), γ(0.778, 1.091)
		97m	1.00min	IT, γ(0.743)
		97	1.202h	β^-(1.267), γ(0.658)
42	Molybdenum	90	5.67h	K, β^+(1.085), Nb-x, γ(0.122, 0.257)
		93m	6.85h	IT, Mo-x, γ(0.264, 0.685, 1.477), e$^-$, no β^+
		93	3500y	K, Nb-x
		99	2.7477d	β^-(1.214), Tc-x, γ(0.041, 0.181, 0.740, 0.780)
43	Technetium	93	2.75h	K, β^+(0.807), Mo-x, γ(1.363, 1.520)
		95m	61d	K, IT, β^+(0.71), Mo-x, γ(0.204, 0.582, 0.835)
		95	20.0h	K, Mo-x, γ(0.766), no β^+
		96	4.28d	K, Mo-x, γ(0.778, 0.813, 0.850), no β^+
		97m	90.5d	IT, Tc-x, γ(0.096)
		97	2.6×10^6y	K, Mo-x
		98	4.2×10^6y	β^-(0.397), γ(0.652, 0.745)
		99m	6.006h	IT, Tc-x, γ(0.141, 0.002), e$^-$
		99	2.13×10^5y	β^-(0.292),
44	Ruthenium	95	1.64h	K, β^+(1.200, 0.910), Tc-x, γ(0.340, 0.625, 1.09)
		97	2.88d	K, Tc-x, γ(0.216, 0.324), no β^+
		103	39.254d	β^-(0.12, 0.22), γ(0.497)
		105	4.44h	β^-(1.187, 1.109), γ(0.316, 0.469, 0.676, 0.724)
		106	1.020y	β^-(0.0392), no γ
45	Rhodium	99	16d	K, β^+(1.100), Ru-x, γ(0.090, 0.353, 0.528)
		99m	4.7h	K, β^+(0.74), Ru-x, γ(0.090, 0.277, 0.341, 0.618, 1.261)
		100	20.8h	K, β^+(2.62, 2.07, 1.26), Ru-x, γ(0.446, 0.540, 1.107, 2.376)
		101m	4.34d	K, IT, Ru-x, Rh-x, γ(0.307)
		101	3.3y	K, γ(0.127, 0.198, 0.325)
		102m	207d	K, β^-(1.15), β^+(1.29, 0.82), Ru-x, γ(0.475)
		102	~2.9y	K, Ru-x, γ(0.418, 0.475, 0.632, 0.698, 0.768, 1.05, 1.11), no β^+
		103m	56.12min	IT, Rh-x, γ(0.040), e$^-$
		105m	45s	IT, Rh-x, γ(0.129), e$^-$
		105	1.4733d	β^-(0.568, 0.249), γ(0.306, 0.319)
		106m	2.17h	β^-(0.925, 0.730), γ(0.451, 0.512, 0.717, 0.749)
		106	29.80s	β^-(3.53, 3.1, 2.4), γ(0.512, 0.622, 1.128)
46	Palladium	100	3.63d	K, Rh-x, γ(0.0748, 0.0840, 0.126)
		101	8.47h	K, β^+(0.776, 0.488), Rh-x, γ(0.270, 0.296)
		103	16.97d	K, Rh-x, γ(0.040, 0.357, 0.497)

Z	Element	A	Half-life	Radiation (MeV)
		107	6.5×10^6y	$\beta^-(0.04)$
		109	13.7h	$\beta^-(1.028)$, Ag-x, $\gamma(0.088, 0.311, 0.636)$, e$^-$
		111m	5.5h	IT, β^-, Pd-x, $\gamma(0.391)$
		111	23.4min	$\beta^-(2.10)$, $\gamma(0.060, 0.071, 0.377, 0.547, 0.580,$ $0.623, 0.651)$
		112	21.04h	$\beta^-(0.28)$, $\gamma(0.0185)$
47	Silver	103	1.095h	K, $\beta^+(1.680, 1.3, 1.15)$ Pd-x, $\gamma(0.12, 0.15, 0.24,$ $0.27, 1.01)$
		104	1.153h	$\beta^+(0.99)$, K, Pd-x, $\gamma(0.556, 0.768)$
		105	41.29d	K, $\beta^+(0.325)$, Pd-x, $\gamma(0.064, 0.280, 0.344, 0.443)$
		106m	8.46d	K, Pd-x, $\gamma(0.451, 0.512, 0.717, 1.046)$, no β^+
		107m	44.3s	IT, Ag-x, $\gamma(0.93)$, e$^-$
		108m	127y	K, IT, Pd-x, Ag-x, $\gamma(0.434, 0.614, 0.723)$
		109m	39.6s	IT, Ag-x, $\gamma(0.088)$, e$^-$
		110m	249.76d	$\beta^-(0.087, 0.53)$, IT, $\gamma(0.658, 0.885, 0.937)$
		110	24.6s	$\beta^-(2.891, 2.22)$, K, $\gamma(0.658)$
		111m	1.080min	IT, $\gamma(0.060)$
		111	7.45d	$\beta^-(1.04, 0.69)$, $\gamma(0.342)$
		112	3.14h	$\beta^-(3.94, 3.35, 1.96)$, $\gamma(0.617)$
		113	5.37h	$\beta^-(2.03)$, $\gamma(0.298)$
48	Cadmium	107	6.50h	K, $\beta^+(0.302)$, Ag-x, $\gamma(0.093, 0.796, 0.829)$
		109	1.2665y	K, Ag-x, $\gamma(0.088)$
		113m	13.7y	$\beta^-(0.590)$, IT, $\gamma(0.264)$
		115m	44.6d	$\beta^-(1.62)$, $\gamma(0.934, 1.29, 0.485)$, no IT
		115	2.228d	$\beta^-(1.11, 0.58)$, In-x, $\gamma(0.336, 0.528)$
		117m	3.36h	$\beta^-(0.715, 0.415)$, In-x, $\gamma(0.564, 0.860, 1.029,$ $1.066, 1.235, 1.433, 1.997)$
		117	2.49h	$\beta^-(2.220, 1.80, 1.29, 0.67)$, In-x, $\gamma(0.273, 0.344,$ $0.434)$
49	Indium	109	4.2h	K, $\beta^+(0.79)$, Cd-x, $\gamma(0.203)$, e$^-$
		110	4.9h	K, Cd-x, $\gamma(0.937, 0.884, 0.658)$, no β^+, no IT
		110	1.152h	$\beta^+(2.25)$, K, Cd-x, $\gamma(0.658)$
		111	2.807d	K, Cd-x, $\gamma(0.172, 0.247)$
		113m	1.658h	IT, In-x, $\gamma(0.393)$, e$^-$, no K
		114m	49.51d	IT, K, In-x, $\gamma(0.192, 0.558, 0.725)$
		114	1.1983min	$\beta^-(1.99, 0.67)$, K, $\beta^+(0.40)$, Cd-x, $\gamma(1.30)$
		115m	4.486h	IT, $\beta^-(0.83)$, In-x, $\gamma(0.336)$, e$^-$
		*115	4.41×10^{14}y	$\beta^-(0.495)$, no γ
		117m	1.942h	$\beta^-(1.77, 1.62)$, IT, In-x, $\gamma(0.159)$
		117	43.8min	$\beta^-(0.74)$, Sn-x, $\gamma(0.159, 0.553)$
50	Tin	110	4.11h	K, In-x, $\gamma(0.283)$
		113	115.09d	K, In-x, $\gamma(0.392)$
		117m	13.61d	IT, Sn-x, $\gamma(0.159)$, e$^-$
		119m	293.0d	IT, Sn-x, $\gamma(0.065, 0.024)$, e$^-$
		121m	55y	$\beta^-(0.354)$, Sb-x, $\gamma(0.037)$
		121	1.1275d	$\beta^-(0.383)$, no γ
		123	129.2d	$\beta^-(1.42)$, $\gamma(1.030, 1.089)$
		125	9.64d	$\beta^-(2.35, 0.40)$, $\gamma(0.823, 0.916, 1.067, 1.089)$
		127	2.10h	$\beta^-(1.45)$, $\gamma(1.114)$
51	Antimony	115	32.1min	K, $\beta^+(1.51)$, Sn-x, $\gamma(0.497)$
		116m	1.005h	K, $\beta^+(1.16)$, Sn-x, $\gamma(0.100, 0.136, 0.407, 0.543,$ $0.973, 1.072, 1.294)$
		117	2.80h	K, $\beta^+(0.57)$, Sn-x, $\gamma(0.159, 0.862)$
		118m	5.00h	K, Sn-x, $\gamma(0.041, 0.254, 1.05, 1.23)$, $\beta^+(?)$
		118	3.6min	K, $\beta^+(2.7)$, Sn-x, $\gamma(1.23)$
		119	1.587d	K, Sn-x, $\gamma(0.024)$, e$^-$
		120	5.76d	K, Sn-x, $\gamma(0.090, 0.200, 1.03, 1.17)$, no β^+, no IT
		122	2.70d	$\beta^-(1.414, 1.980, 0.723)$, K, Sn-x, $\gamma(0.564)$
		124	60.20d	$\beta^-(2.301)$, $\gamma(0.603, 1.69, 0.722)$, no K
		125	2.73y	$\beta^-(0.621, 0.444, 0.302)$, $\gamma(0.428, 0.464, 0.601,$ $0.636)$
		126	12.4d	$\beta^-(1.9)$, $\gamma(0.666, 0.695, 0.721)$
		127	3.85d	$\beta^-(1.493, 1.244)$, $\gamma(0.473, 0.686, 0.784)$
		128	9.01h	$\beta^-(2.30)$, $\gamma(0.314, 0.53, 0.64, 0.75)$
		129	4.40h	$\beta^-(1.80, 1.53, 1.38, 0.62, 0.52)$, $\gamma(0.545, 0.813,$ $0.915)$

Table of Radioactive Isotopes (Continued)

Z	Element	A	Half-life	Radiation (MeV)
52	Tellurium	116	2.49h	K, Sb-x, $\gamma(0.094)$, β^+(?)
		117	1.03h	K, $\beta^+(1.75)$, Sb-x, $\gamma(0.720)$
		118	6.00d	K, Sb-x, no γ
		119m	4.69d	K, Sb-x, $\gamma(0.153, 0.270, 1.22)$, no β^+, no IT
		119	16.05h	K, $\beta^+(0.627)$, Sb-x, $\gamma(0.644)$
		121m	154d	IT, K, Te-x, Sb-x, $\gamma(0.212)$
		121	16.8d	K, Sb-x, $\gamma(0.573, 0.508)$, no β^+
		123m	119.7d	IT, Te-x, $\gamma(0.0885, 0.159)$, e^-
		125m	58d	IT, Te-x, $\gamma(0.110, 0.035)$, e^-
		127m	109d	IT, Te-x, $\gamma(0.058, 0.089)$, β^-
		127	9.35h	$\beta^-(0.70)$, I-x, $\gamma(0.360, 0.418)$
		129m	33.6d	IT, $\beta^-(1.60, 0.91)$, Te-x, $\gamma(0.696, 0.730)$
		129	1.160h	$\beta^-(1.453, 0.989, 0.69, 0.29)$, I-x, $\gamma(0.460, 0.487)$
		131m	1.25d	$\beta^-(2.46, 0.57, 0.42)$, IT, Te-x, I-x, $\gamma(0.150, 0.774, 0.794)$
		131	25.0min	$\beta^-(2.14, 1.69, 1.35)$, I-x, $\gamma(0.150, 0.453)$
		132	3.26d	$\beta^-(0.22)$, I-x, $\gamma(0.228)$
53	Iodine	120	1.35h	K, $\beta^+(4.595, 4.030, 3.400, 2.490, 1.54)$, Te-x, $\gamma(0.560, 1.523)$
		121	2.12h	K, $\beta^+(1.2)$, Te-x, $\gamma(0.212, 0.532)$
		122	3.62min	$\beta^+(3.1)$, K, Te-x, $\gamma(0.564)$
		123	13.2h	K, Te-x, $\gamma(0.159)$, no β^+
		124	4.18d	K, $\beta^+(2.14, 1.53)$, Te-x, $\gamma(0.603)$
		125	60.14d	K, Te-x, $\gamma(0.035)$
		126	13.02d	K, $\beta^-(1.25, 0.865, 0.385)$, $\beta^+(1.13)$, Te-x, (0.389)
		129	1.57×10^7y	$\beta^-(0.150)$, Xe-x, $\gamma(0.038)$
		130	12.36h	$\beta^-(0.62, 1.04, 1.7)$, $\gamma(0.418, 0.539, 0.669, 0.739)$
		131	8.040d	$\beta^-(0.607, 0.336)$, Xe-x, $\gamma(0.080, 0.284, 0.364, 0.637, 0.723)$
		132	2.284h	$\beta^-(2.16, 1.61, 1.22, 1.04, 0.80)$, $\gamma(0.667, 0.773)$
		133	20.8h	$\beta^-(1.27, 0.94, 0.7)$, $\gamma(0.53, 0.87)$
		135	6.55h	$\beta^-(1.32, 0.87)$, $\gamma(0.527, 1.132, 1.260, 1.458, 1.678)$
54	Xenon	122	20.1h	K, I-x, $\gamma(0.149, 0.350)$
		123	2.08h	K, $\beta^+(1.51)$, I-x, $\gamma(0.149, 0.178)$
		125	16.9h	K, I-x, $\gamma(0.055, 0.188, 0.242)$
		127m	1.153min	IT, Xe-x, $\gamma(0.125, 0.173)$
		127	36.41d	K, I-x, $\gamma(0.058, 0.145, 0.172, 0.203, 0.375)$
		129m	8.89d	IT, Xe-x, $\gamma(0.040, 0.197)$, e^-
		131m	11.9d	IT, Xe-x, $\gamma(0.164)$, e^-
		133m	2.19d	IT, Xe-x, $\gamma(0.233)$, e^-
		133	5.245d	$\beta^-(0.346)$, Cs-x, $\gamma(0.081)$
		135m	15.65min	IT, Xe-x, $\gamma(0.527)$
		135	9.104h	$\beta^-(0.905)$, $\gamma(0.250, 0.61)$
55	Cesium	126	1.64min	$\beta^+(3.4)$, K, Xe-x, $\gamma(0.389)$
		127	6.25h	K, $\beta^+(1.063, 0.685)$, Xe-x, $\gamma(0.125, 0.287, 0.411, 0.462, 0.587)$
		128	3.62min	$\beta^+(2.88, 2.44)$, K, Xe-x, $\gamma(0.443)$
		129	1.3358d	K, Xe-x, $\gamma(0.372, 0.411, 0.549)$
		131	9.69d	K, Xe-x
		132	6.475d	K, $\beta^+(0.40)$, β^-, Xe-x, $\gamma(0.668)$
		134m	2.91h	IT, $\beta^-(0.55)$, Cs-x, $\gamma(0.127)$
		134	2.062y	$\beta^-(0.658, 0.089)$, $\gamma(0.605, 0.796)$
		135	3.0×10^6y	$\beta^-(0.21)$, no γ
		136	13.16d	$\beta^-(0.341, 0.657)$, Ba-x, $\gamma(0.818, 1.048)$
		137	30.0y	$\beta^-(0.514, 1.18)$, Ba-x, $\gamma(0.662)$
56	Barium	126	1.67h	K, $\gamma(0.218, 0.234, 0.241, 0.258)$
		128	2.43d	K, Cs-x, $\gamma(0.273)$
		129(129m)	2.23(2.17)h	K, $\beta^+(1.43, 1.24, 1.0)$, Cs-x, $\gamma(0.129, 0.182, 0.21, 0.22, 1.45)$
		131	11.8d	K, Xe-x, $\gamma(0.124, 0.216, 0.373, 0.496)$, no β^+
		133m	1.621d	IT, Ba-x, $\gamma(0.276, 0.012)$, e^-
		133	10.54y	K, Cs-x, $\gamma(0.080, 0.276, 0.302, 0.356, 0.382)$, e^-
		135m	1.196d	IT, Ba-x, $\gamma(0.268)$
		137m	2.552min	IT, Ba-x, $\gamma(0.662)$

Z	Element	A	Half-life	Radiation (MeV)
		139	1.41h	β^-(2.38, 2.23, 0.95), La-x, γ(0.166, 1.421)
		140	12.75d	β^-(1.02, 0.83, 0.59, 0.46), La-x, γ(0.537)
57	Lanthanum	131	59min	K, β^+(1.94, 1.42, 0.70), Ba-x, γ(0.418)
		132	4.8h	β^+(3.66, 3.20, 2.62), K, Ba-x, γ(0.465)
		133	3.91h	K, β^+(1.2), Ba-x, γ(0.279, 0.290, 0.302, 0.618, 0.633)
		134	6.45min	β^+(2.67), K, Ba-x, γ(0.563, 0.605, 1.555)
		135	19.48h	K, Ba-x, γ(0.481, 0.588, 0.87), no β^+
		136	9.87min	K, β^+(1.85), Ba-x, γ(0.819)
		137	6×10^4y	K, Ba-x, no γ
		*138	1.06×10^{11}y	K, β^-(0.205), Ba-x, γ(0.788, 1.436)
		140	1.68d	β^-(2.164, 1.680, 1.365, 1.150, 0.857, 0.510), γ(0.487, 1.596)
		141	3.93h	β^-(2.43, 0.9), γ(1.354)
		142	1.54h	β^-(4.52, 3.85, 2.98, 2.31, 2.11, 1.98, 1.79, 1.23, 0.87), γ(0.641, 2.398)
58	Cerium	132	3.51h	K, γ(0.182)
		133	5.40h	K, β^+(1.3), La-x, γ(0.058, 0.131, 0.477, 0.510)
		134	3.16d	K, La-x
		135	17.8h	K, β^+(0.80), La-x, γ(0.266, 0.300, 0.518, 0.607)
		137m	1.43d	IT, K, Ce-x, γ(0.169, 0.254, 0.762, 0.825), e$^-$
		137	9.0h	K, La-x, γ(0.477), e$^-$
		139	137.66d	K, La-x, γ(0.166)
		141	32.50d	β^-(0.444, 0.582), Pr-x, γ(0.145)
		143	1.38d	β^-(1.40, 1.125, 0.74), Pr-x, γ(0.293), e$^-$
		144	284.9d	β^-(0.316), Pr-x, γ(0.080, 0.134)
59	Praseodymium	136	13.1min	K, β^+(3.00, 2.56), Ce-x, γ(0.540, 0.552)
		137	1.28h	K, β^+(1.7), Ce-x, γ(0.837)
		138m	2.1h	K, β^+(1.65), Ce-x, γ(0.302, 0.789, 1.038)
		139	4.41h	K, β^+(1.09), Ce-x, γ(0.255, 1.347, 1.376, 1.631)
		140	3.39min	K, β^+(2.366), Ce-x, γ(0.307, 1.596)
		142	19.13h	β^-(2.164), γ(1.576)
		143	13.58d	β^-(0.932)
		144	17.28min	β^-(2.99, 2.3, 0.81), γ(0.697, 1.49, 2.19)
		145	5.98h	β^-(1.8), γ(0.072, 0.676, 0.748, 0.921, 0.979, 1.051, 1.150)
		146	24.15min	β^-(4.1, 3.6, 2.8, 2.1), γ(0.454)
60	Neodymium	139m	5.5h	K, IT, β^+(1.170), Nd-x, Pr-x, γ(0.114, 0.231, 0.708, 0.738, 0.982)
		140	3.37d	K, Pr-x, no γ
		141	2.49h	K, β^+(0.79), Pr-x, γ(0.145, 1.127, 1.147, 1.293)
		*144	2.1×10^{15}y	α(1.83)
		147	10.98d	β^-(0.810, 0.369), γ(0.091, 0.531), e$^-$
		149	1.73h	β^-(1.555, 1.425, 1.130, 1.025), Pm-x, γ(0.030, 0.114, 0.655)
61	Promethium	142	40.5s	β^+(3.80), K, Nd-x, γ(1.576)
		143	265d	K, Nd-x, γ(0.742), no β^+
		144	363d	K, Nd-x, γ(0.477, 0.618, 0.696), no β^+
		145	17.7y	K, Nd-x, γ(0.067, 0.072), e$^-$
		146	5.53y	K, β^-(0.795), Nd-x, γ(0.453, 0.75)
		147	2.6234y	β^-(0.224), γ(0.122)
		148m	41.29d	β^-(0.69, 0.50, 0.40), IT, Pm-x, Sm-x, γ(0.550, 0.630)
		148	5.37d	β^-(2.48, 1.93, 1.02), γ(0.551, 1.47)
		149	2.2117d	β^-(1.064, 0.784), γ(0.286)
		150	2.68h	β^-(3.260, 2.930, 2.100, 1.600), γ(0.344)
		151	1.1833d	β^-(1.19, 1.13, 1.05, 0.98, 0.84, 0.73, 0.50, 0.35), Sm-x, γ(0.168, 0.275, 0.340)
62	Samarium	142	1.2082h	K, β^+(1.03), Pm-x
		145	340d	K, Pm-x, γ(0.061), e$^-$
		146	1.03×10^8y	α(2.55)
		*147	1.06×10^{11}y	α(2.23)
		151	90y	β^-(0.076), γ(0.022), e$^-$
		153	1.946d	β^-(0.81, 0.71, 0.64), Eu-x, γ(0.103, 0.070), e$^-$
		156	9.4h	β^-(0.715, 0.43), Eu-x, γ(0.0226–0.291)

Z	Element	A	Half-life	Radiation (MeV)
63	Europium	145	5.93d	K, β^+(1.67, 0.78), Sm-x, γ(0.656, 0.894, 1.66)
		146	4.59d	K, β^+(2.11, 1.47, 0.80), Sm-x, γ(0.633, 0.634, 0.747)
		147	24d	K, β^+, α(2.91; 0.002%), Sm-x
		148	54.5d	K, β^+(0.92), α(2.63; 9 × 10^{-7}%), Sm-x, γ(0.550, 0.630)
		149	93.1d	K, Sm-x, γ(0.277, 0.328)
		150	12.62h	β^-(1.01), K, β^+(1.24), Sm-x, γ(0.334, 0.407)
		150	35.8y	K, Sm-x, γ(0.334, 0.439, 0.584)
		152m$_1$	9.32h	β^-(1.89, 1.55, 0.56), K, β^+(0.89, 0.77), γ(0.122, 0.344, 0.841, 0.963)
		152m$_2$	1.600h	IT, Eu-x, γ(0.090), e$^-$, no β
		152	13.33y	K, β^-(1.492, 0.690, 0.360), β^+(0.727, 0.479), Gd-x, Sm-x, γ(0.122, 0.344, 0.964, 1.086, 1.112, 1.408)
		154	8.8y	β^-(0.843, 0.579, 0.274), Gd-x, γ(0.123, 1.005, 1.274)
		155	4.96y	β^-(0.152, 0.252), Gd-x, γ(0.087, 0.105)
		156	15.19d	β^-(2.460, 1.215, 0.495, 0.300), GD-x, γ(0.089, 0.812, 1.230, 1.242)
		157	15.15h	β^-(1.28), Gd-x, γ(0.064, 0.373)
		158	46.0min	β^-(3.40, 2.43), γ(0.079, 0.977)
64	Gadolinium	146	48.27d	K, β^+(0.350), Eu-x, γ(0.115, 0.155)
		147	1.588d	K, β^+(1.250, 0.970), Eu-x, γ(0.229, 0.370, 0.396)
		148	75y	α(3.18)
		149	9.25d	K, α(3.018; 0.00046%), Eu-x, γ(0.150)
		150	1.79 × 10^{6}y	α(2.73)
		151	120d	K, α(2.60; 8 × 10^{-7}%), Eu-x, γ(0.154, 0.175, 0.243)
		*152	1.08 × 10^{14}y	α(2.4)
		153	241.6d	K, Eu-x, γ(0.070, 0.097, 0.103)
		159	18.56h	β^-(0.95, 0.89, 0.60), Tb-x, γ(0.363)
65	Terbium	148	1.00h	K, β^+(4.6), Gd-x, γ(0.489, 0.783)
		149	4.15h	K, α(3.967), β^+(1.78), Gd-x, γ(0.165, 0.352, 0.388, 0.652, 0.817, 0.853)
		150	3.27h	K, β^+(3.70, 3.04), α(3.492), Gd-x, γ(0.496, 0.638)
		151	17.6h	K, α(3.41; 0.0095%), Gd-x, γ(0.108, 0.252, 0.287, 0.479)
		152	17.5h	K, β^+(2.800, 2.456, 1.890), Gd-x, γ(0.344)
		153	2.34d	K, β^+(0.570, 0.520, 0.460, 0.339) Gd-x, γ(0.212)
		154	21.4d	K, β^+, Gd-x, γ(0.123, 1.274, 1.291, 1.997, 2.064, 2.187)
		154m$_1$	9.0h	K, IT, Gd-x, γ(0.123, 0.248)
		154m$_2$	22.6h	K, IT, γ(0.226, 0.347, 0.427, 1.420)
		155	5.32d	K, Gd-x, γ(0.105)
		156m	5.0h	IT, K, β^+(2.640), γ(0.0884)
		156	5.34d	K, Gd-x, γ(0.089, 0.199, 0.534, 1.222)
		157	150y	K, Gd-x, γ(0.054)
		158	150y	K, β^-(0.853), Gd-x, γ(0.944, 0.962)
		160	72.3d	β^-(1.76, 0.87, 0.57, 0.45), Dy-x, γ(0.299, 0.879, 0.966)
		161	6.91d	β^-(0.590, 0.520, 0.460), Dy-x, γ(0.075)
66	Dysprosium	152	2.38h	K, α(3.63), Tb-x, γ(0.257)
		153	6.4h	K, β^+(0.980, 0.670), α(3.464), Tb-x, γ(0.081, 0.099, 0.214, 0.254)
		154m	13h	α(3.37)
		154	2.9 × 10^{6}y	α(2.872)
		155	10.0h	K, β^+(1.08, 0.85), Tb-x, γ(0.227)
		157	8.1h	K, Tb-x, γ(0.326), no β^+
		159	144.4d	K, Tb-x, γ(0.058)
		165	2.334d	β^-(1.305, 1.215), Ho-x, γ(0.095, 0.362)
		166	3.400d	β^-(0.402, 0.48), Ho-x, γ(0.082)
67	Holmium	160m	5.02h	IT, K, β^+(1.9, 1.01, 0.57, 0.31), Dy-x, γ(0.060, 0.087, 0.197, 0.645, 0.728, 0.879, 0.966), includes daughter radiation from Ho^{160}
		160	25.6min	K, γ(see Ho^{160m})
		161	2.48h	K, Dy-x, γ(0.026)
		162m	1.133h	IT, K, Ho-x, Dy-x, γ(0.081, 0.185, 1.220)

Z	Element	A	Half-life	Radiation (MeV)
		162	15min	K, β^+(1.200, 1.120), Dy-x, γ(0.080, 1.319)
		166m	1200y	β^-($\lesssim$0.065), Er-x, γ(0.184, 0.712, 0.810)
		166	1.117d	β^-(1.85, 1.78), Er-x, γ(0.081)
		167	3.1h	β^-(0.970, 0.610, 0.315), Er-x, γ(0.347)
68	Erbium	160	1.191d	K, Ho-x, no γ
		161	3.24h	K, β^+(0.820), Ho-x, γ(0.211, 0.826)
		163	1.250h	K, β^+(0.188), Ho-x, γ(0.298, 0.436, 1.113)
		165	10.36h	K, Ho-x, no γ
		167m	2.28s	IT, Er-x, γ(0.208)
		169	9.40d	β^-(0.34), γ(0.008)
		171	7.52h	β^-(1.49, 1.06, 0.58), Tm-x, γ(0.112, 0.296, 0.308)
		172	2.054d	β^-(0.356), Tm-x, γ(0.407, 0.610)
69	Thulium	162	21.7min	K, β^+(3.82, 2.3, 0.90), Er-x, γ(0.102, 0.227, 0.799)
		163	1.81h	K, β^+(1.05, 0.71), Er-x, γ(0.104)
		165	1.2525d	K, β^+(0.30), Er-x, γ(0.054, 0.243, 0.807)
		166	7.70h	K, β^+(1.928, 1.219), Er-x, γ(0.081, 0.084, 0.184, 0.705, 2.052)
		167	9.24d	K, Er-x, γ(0.208, 0.532)
		168	93.1d	K, β^-(?), Er-x, γ(0.184, 0.198, 0.447, 0.816)
		170	128.6d	β^-(0.968, 0.884), Yb-x, γ(0.084)
		171	1.92y	β^-(0.097), Yb-x, γ(0.067)
		172	2.650d	β^-(1.870, 1.790), Yb-x, γ(1.094, 1.387, 1.466)
		173	8.24d	β^-(0.89, 1.3), Yb-x, γ(0.399)
70	Ytterbium	164	1.26h	K, Tm-x, γ(0.097)
		166	2.362d	K, Tm-x, γ(0.082)
		169	32.022d	K, Tm-x, γ(0.063, 0.110, 0.131, 0.177, 0.198, 0.308)
		175	4.19d	β^-(0.466, 0.073), Lu-x, γ(0.114, 0.283, 0.396)
		177	1.9h	β^-(1.40), Lu-x, γ(1.080)
71	Lutetium	169	1.4192d	K, β^+(1.271), Yb-x, γ(0.191, 0.960)
		170	2.00d	K, β^+(2.39), Yb-x, γ(0.985, 1.054, 1.226, 1.280, 1.365, 2.042, 2.126)
		171	8.24d	K, Yb-x, γ(0.740)
		172	6.70d	K, Yb-x, γ(0.181, 0.901, 1.094), no β^+
		173	1.37y	K, Yb-x, γ(0.0787, 0.272)
		174m	142d	IT, K, Lu-x, γ(0.045, 0.067), no β^-
		174	3.31y	K, β^+(0.380), Yb-x, γ(0.0765, 1.24)
		176m	3.635h	β^-(1.31, 1.22), Hf-x, γ(0.0884), no IT
		*176	3.59×10^{10}y	β^-(0.565), Hf-x, γ(0.089, 0.203, 0.306), no K
		177m	160.9d	β^-, IT, Lu-x, Hf-x, γ(0.105, 0.208, 0.229, 0.379)
		177	6.71d	β^-(0.497, 0.385, 0.175), Hf-x, γ(0.113, 0.208)
		179	4.59h	β^-(1.35), γ(0.214)
72	Hafnium	169	1.87h	K, β^+(1.3), Lu-x, γ(0.115)
		170	16.01h	K, Lu-x, γ(0.165, 0.573)
		171	12.1h	K, Lu-x, γ(0.122, 1.072)
		172	1.87y	K, Lu-x, γ(0.024, 0.082, 0.125), no α
		173	23.6h	K, Lu-x, γ(0.124, 0.297)
		*174	2.0×10^{15}y	α(2.50)
		175	70d	K, Lu-x, γ(0.343)
		180m	5.519h	IT, Hf-x, γ(0.058, 0.215, 0.333, 0.444)
		181	42.39d	β^-(0.408), Ta-x, γ(0.133, 0.346, 0.482)
		182	9.0×10^6y	β^-, γ(0.271)
73	Tantalum	173	3.65h	K, β^+(2.48), Hf-x, γ(0.070, 0.090, 0.172)
		174	1.18h	K, β^+(2.525), Hf-x, γ(0.206)
		175	10.5h	K, β^+, Hf-x, γ(0.082, 0.207, 0.349)
		176	8.08h	K, β^+, Hf-x, γ(0.088, 1.159)
		177	2.358d	K, Hf-x, γ(0.113, 0.208)
		178	9.31min	K, β^+(0.89, 0.80), Hf-x, γ(1.106, 1.341, 1.351, 1.403)
		178	2.45h	K, Hf-x, γ(0.089, 0.213, 0.326, 0.426)
		179	1.79y	K, Hf-x, no γ
		180m	8.152h	K, β^-(0.71, 0.61), Hf-x, γ(0.093, 0.102)
		182	115.0d	β^-(1.713, 1.470), W-x, γ(1.121, 1.221)
		183	5.1d	β^-(0.615, 0.470), W-x, γ(0.108, 0.246, 0.354)
		184	8.7h	β^-(1.19, 1.45, 2.64), W-x, γ(0.111, 0.253, 0.318, 0.414, 0.921)

Z	Element	A	Half-life	Radiation (MeV)
74	Tungsten	176	2.5h	K, Ta-x, γ(0.061, 0.095)
		177	2.21h	K, Ta-x, γ(0.030, 0.427, 1.036)
		178	21.5d	K, Ta-x
		181	121.2d	K, Ta-x, γ(0.0062, 0.136, 0.152)
		185	75.1d	β^-(0.433), γ(0.125)
		187	23.9h	β^-(1.312, 0.622), Re-x, γ(0.072, 0.134, 0.480)
		188	69.4d	β^-(0.349), Re-x, γ(0.227, 0.290)
75	Rhenium	181	19.9h	K, W-x, γ(0.361, 0.366)
		182	12.7h	K, β^+(1.74), W-x, γ(0.100, 1.121, 1.189, 1.222)
		182	2.667d	K, W-x, γ(0.100, 0.131, 0.169, 0.229, 1.121, 1.222, 1.231), no β^+
		183	70.0d	K, W-x, γ(0.046, 0.162, 0.292)
		184m	165d	IT, K, Re-x, W-x, γ(0.111, 0.161, 0.217, 0.253, 0.921)
		184	38.0d	K, W-x, γ(0.111, 0.792, 0.895, 0.903)
		186	3.777d	β^-(1.07, 0.933), K, W-x, Os-x, γ(0.137, 0.632, 0.768)
		*187	4.6×10^{10}y	β^-(0.0026)
		188	16.98h	β^-(2.12, 1.96), Os-x, γ(0.155)
		189	1.013d	β^-(1.00), Os-x, γ(0.031, 0.696, 0.217, 0.219, 0.245)
76	Osmium	182	21.6h	K, Re-x, γ(0.180, 0.263, 0.510)
		183m	9.9h	K, IT, Os-x, γ(1.102, 1.108)
		183	13.0h	K, Re-x, γ(0.382)
		185	93.6d	K, Re-x, γ(0.646, 0.875), no β^+
		189m	5.8h	IT, Os-x, γ(0.0308), e$^-$
		190m	9.9min	IT, Os-x, γ(0.187, 0.361, 0.616) e$^-$
		191m	13.10h	IT, Os-x, γ(0.074)
		191	15.4d	β^-(0.143), Ir-x, γ(0.129), e$^-$
		193	1.271d	β^-(1.13, 1.06, 0.99, 0.85, 0.67), Ir-x, γ(0.073, 0.139, 0.460)
		194	6.0y	β^-(0.053), Ir-x, γ(0.043)
77	Iridium	183	57min	K, Os-x, γ(0.088, 0.229)
		184	3.02h	K, β^+(2.9, 2.5), Os-x, γ(0.120, 0.264, 0.390)
		185	14.0h	K, Os-x, γ(0.037, 0.060, 0.254)
		186	15.8h	K, β^+(1.94), Os-x, γ(0.137, 0.297, 0.434)
		187	10.5h	K, Os-x, γ(0.026, 0.065, 0.913)
		188	1.729d	K, β^+(1.66), Os-x, γ(0.155, 0.478)
		189	13.2d	K, Os-x, γ(0.031, 0.070, 0.245)
		190m$_1$	1.2h	IT, Ir-x, γ(0.0263)
		190m$_2$	3.2h	K, IT, Os-x, Ir-x, γ(0.187, 0.361, 0.503, 0.617)
		190	11.78d	K, Os-x, γ(0.187, 0.371, 0.519, 0.558, 0.569, 0.605), no β^+
		192	73.831d	β^-(0.672, 0.536), K, Os-x, Pt-x, γ(0.296, 0.308, 0.316, 0.468)
		193m	10.60d	IT, Ir-x, γ(0.0802)
		194	19.15h	β^-(2.24, 1.92, 1.62, 0.98), γ(0.328)
		195	2.8h	β^-(1.15, 0.98), Pt-x, γ(0.10, 0.13)
		196m	1.40h	β^-(1.16), γ(0.356, 0.393, 0.447, 0.521, 0.647), no IT
78	Platinum	186	2.0h	K, α(4.23; 1.4×10^{-4}%), Ir-x, γ(0.690)
		187	2.35h	K, Ir-x, γ(0.110, 0.285)
		188	10.2d	K, α(3.93; 3×10^{-5}%), Ir-x, γ(0.188, 0.195)
		189	10.89h	K, β^+(0.885, 0.479), Ir-x, γ(0.094, 0.244, 0.545, 0.569, 0.608, 0.721)
		*190	6.0×10^{11}y	α(3.18)
		191	2.9d	K, Ir-x, γ(0.082, 0.360, 0.409)
		*192	$>6.0 \times 10^{16}$y	α(2.6)
		193m	4.33d	IT, Pt-x, γ(0.0126, 0.136)
		193	50y	K, Ir-x, no γ
		195m	4.02d	IT, Pt-x, γ(0.0311, 0.0991, 0.130)
		197m	1.573h	IT, β^-, Pt-x, γ(0.279, 0.346)
		197	18.3h	β^-(0.719, 0.642, 0.451), Au-x, γ(0.077, 0.191)
		200	12.5h	β^-, γ(0.076)

Z	Element	A	Half-life	Radiation (MeV)
79	Gold	191	3.18h	K, Pt-x, γ(0.278, 0.284, 0.400, 0.586, 0.674)
		192	4.94h	K, β^+(2.492, 2.192), Pt-x, γ(0.316)
		193	17.65h	K, Pt-x, γ(0.256), no α
		194	1.646d	K, β^+(1.49), Pt-x, γ(0.328)
		195	186.09d	K, Pt-x, γ(0.099)
		196m$_2$	9.7h	IT, Au-x, γ(0.020, 0.085, 0.148, 0.175, 0.188)
		196	6.183d	K, β^-(0.259), Pt-x, γ(0.333, 0.356, 0.426)
		197m	7.8s	IT, Au-x, γ(0.279)
		198	2.6935d	β^-(1.371), γ(0.412, 0.676, 1.09), no K
		199	3.139d	β^-(0.296, 0.250, 0.462), Hg-x, γ(0.158, 0.208)
80	Mercury	191	49min	K, Au-x, γ(0.252)
		192	4.85h	K, Au-x, γ(0.157, 0.275)
		193m	11.8h	K, IT, β^+(1.17), Hg-x, Au-x, γ(0.258, 0.408)
		193	3.80h	K, Au-x, γ(0.038)
		194	520y	K, Au-x, no γ
		195m	1.73d	K, IT, Hg-x, Au-x, γ(0.262, 0.560)
		195	9.5h	K, Au-x, γ(0.061, 0.780)
		197m	23.8h	IT, K, Hg-x, γ(0.134), e$^-$
		197	2.6725d	K, Au-x, γ(0.077)
		203	46.60d	β^-(0.21), γ(0.279)
81	Thallium	195	1.16h	K, β^+(1.8), Hg-x, γ(0.242, 0.279, 0.564, 0.885, 1.363)
		196m	1.41h	K, IT, Hg-x, γ(0.426, 0.635, 0.695)
		196	1.84h	K, Hg-x, γ(0.426, 0.611, 0.635, 1.496)
		197	2.84h	K, β^+(1.2), Hg-x, γ(0.152, 0.426)
		198m	1.87h	K, IT, Hg-x, Tl-x, γ(0.412, 0.587, 0.637)
		198	5.3h	K, β^+(2.4), Hg-x, γ(0.412)
		199	7.42h	K, Hg-x, γ(0.158, 0.208, 0.247, 0.455), no β^+
		200	1.087d	K, β^+(1.44, 1.07), Hg-x, γ(0.579, 0.828)
		201	3.046d	K, Hg-x, γ(0.135, 0.167)
		202	12.23d	K, Hg-x, γ(0.439, 0.52, 0.96)
		204	3.78y	β^-(0.763), K, Hg-x
		*206(RaE')	4.20min	β^-(1.53)
		*207(AcC″)	4.77min	β^-(1.43), γ(0.90)
		*208(ThC″)	3.053min	β^-(1.80, 1.52, 1.29), γ(2.61, 0.583)
		209	2.20min	β^-(1.8), Pb-x, γ(0.12, 0.45, 1.56)
		*210(RaC″)	1.30min	β^-(2.3, 1.9, 1.3), γ(0.296, 0.795)
82	Lead	198	2.4h	K, Tl-x, γ(0.117, 0.173, 0.290), no α
		199	1.50h	K, β^+(2.8), Tl-x, γ(0.353, 0.367, 0.721, 1.658)
		200	21.5h	K, Tl-x, γ(0.148)
		201m	1.02min	IT, Pb-x, γ(0.629)
		201	9.33h	K, β^+(0.55), Tl-x, γ(0.331, 0.361)
		202m	3.62h	IT, K, Tl-x, Pb-x, γ(0.422, 0.787, 0.961)
		202	5.3×10^4y	K, Tl-x
		203	2.169d	K, Tl-x, γ(0.279)
		204m	1.120h	IT, Pb-x, γ(0.375, 0.899, 0.912)
		205	1.9×10^7y	K, Tl-x
		207m	796ms	IT, γ(0.570, 1.06)
		209	3.253h	β^-(0.645), no γ
		*210(RaD)	22.3y	β^-(0.017), α(3.72; 1.7×10^{-6}%), γ(0.0465)
		*211(AcB)	36.1min	β^-(1.36, 0.53), γ(0.405, 0.427, 0.832)
		*212(ThB)	10.64h	β^-(0.569, 0.331), Bi-x, γ(0.239)
		*214(RaB)	26.8min	β^-(0.59, 0.65, 1.03), γ(0.352)
83	Bismuth	201m	59.1min	K, α(5.240), Pb-x
		201	1.80h	K, Pb-x, γ(0.629 with Pb^{201m})
		202	1.72h	K, β^+, Pb-x, γ(0.422, 0.961)
		203	11.76h	K, β^+(1.35, 0.74), Pb-x, γ(0.820)
		204	11.22h	K, Pb-x, γ(0.899, 0.984), no β^+
		205	15.31d	K, β^+(0.990), Pb-x, γ(0.703, 0.988, 1.764)
		206	6.243d	K, β^+(0.98), Pb-x, γ(0.516, 0.537, 0.803, 0.881, 1.719)
		207	32.2y	K, Pb-x, γ(0.570, 1.06, 1.77)
		208	3.68×10^5y	K, Pb-x, γ(2.61), no β^+
		210m	3.00×10^6y	α(4.96, 4.92, 4.57), γ(0.266, 0.305), no β^-
		*210(RaE)	5.013d	β^-(1.16), α(4.69, 4.65)
		*211(AcC)	2.14min	α(6.62, 6.28), β^-, γ(0.351)

Table of Radioactive Isotopes (Continued)

Z	Element	A	Half-life	Radiation (MeV)
		*212(ThC)	1.0092h	β^-(2.25), α(6.05, 6.09), Tl-x, γ(0.727)
		213	45.59min	β^-(1.420, 1.018), α(5.87, 5.55), γ(0.439)
		*214(RaC)	19.9min	β^-(3.26, 1.88, 1.51, 1.0, 0.4), α(5.512, 5.448), γ(0.609)
		215	7.6min	β^-
84	Polonium	204	3.53h	K, α(5.38), γ(0.137, 0.270, 0.534, 0.884, 1.016)
		205	1.80h	K, α(5.220), γ(0.849, 0.872, 1.001)
		206	8.8d	K, α(5.22), Bi-x, γ(0.011, 0.286, 0.511, 0.807, 1.032)
		207	5.83h	K, α(5.11), β^+(1.14, 0.89), Pb-x, γ(0.743, 0.912, 0.992)
		208	2.898y	α(5.11), K, Bi-x, γ(0.292, 0.571, 0.603, 0.862)
		209	102y	α(4.88), K, Bi-x, γ(0.26)
		*210(RaF)	138.376d	α(5.30), γ(0.803)
		*211(AcC′)	516ms	α(7.45, 6.89), γ(0.56, 0.88)
		*212(ThC′)	298ns	α(8.78)
		213	4.2μs	α(8.38)
		*214(RaC′)	163.69μs	α(7.69), γ(0.799)
		*215(AcA)	1.780ms	α(7.38), β^-
		*216(ThA)	150ms	α(6.78)
		*218(RaA)	3.11min	α(6.00)
85	Astatine	207	1.80h	K, α(5.76), γ(0.588, 0.814)
		208	1.63h	K, α(5.641), Po-x, γ(0.177, 0.660, 0.685)
		209	5.41h	K, α(5.65), Po-x, γ(0.545, 0.782, 0.790)
		210	8.1h	K, α(5.52, 5.44, 5.36), Po-x, γ(0.245, 1.181, 1.483)
		211	7.214h	K, α(5.87), Po-x, γ(0.67)
		215	100μs	α(8.03)
		216	300μs	α(7.80)
		217	32.3ms	α(7.07)
		218	1.6s	α(6.70, 6.65)
		219	54s	α(6.28), β^-
86	Radon	210	2.4h	α(6.04), K, γ(0.073, 0.233, 0.458, 0.649)
		211	14.6h	K, α(5.85, 5.78, 5.62), At-x, γ(0.442, 0.674, 0.678, 0.947)
		218	35ms	α(7.13), γ(0.609)
		*219(An)	3.96s	α(6.82, 6.55, 6.42), Po-x, γ(0.272, 0.401)
		*220(Tn)	55.6s	α(6.29), γ(0.550)
		*222(Rn)	2.825d	α(5.49), γ(0.510)
87	Francium	212	20.0min	K, α(6.41, 6.38, 6.26)
		220	27.4s	α(6.68, 6.64, 6.58), β^-, γ(0.045, 0.106, 0.154, 0.162)
		221	4.9min	α(6.34, 6.12), At-x, γ(0.218)
		222	14.4min	β^-(1.78), α
		*223(AcK)	21.8min	β^-(1.15), α(5.34), Ra-x, γ(0.050, 0.080, 0.234)
88	Radium	222	38.0s	α(6.56), γ(0.324)
		*223(AcX)	11.43d	α(5.74, 5.71, 5.60, 5.54), Rn-x, γ(0.154, 0.270, 0.324)
		*224(ThX)	3.66d	α(5.68, 5.45), Rn-x, γ(0.241)
		225	14.8d	β^-(0.32), Ac-x, γ(0.040), no α
		*226(Ra)	1600y	α(4.78, 4.60), Rn-x, γ(0.187)
		*228(MsTh$_1$)	5.75y	β^-(0.039, 0.015), γ(0.135), no α
89	Actinium	224	2.9h	K, α(6.20, 6.14, 6.04), Ra-x, γ(0.133, 0.217)
		225	10.0d	α(5.83, 5.79, 5.64), Fr-x, γ(0.099, 0.150, 0.187)
		226	1.2d	β^-(1.105, 0.885), K, α(5.399), Th-x, Ra-x, γ(0.158, 0.185, 0.230, 0.253)
		*227(Ac)	21.77y	β^-(0.046), α(4.95, 4.94), Th-x, γ(0.086, 0.100, 0.160)
		*228(MsTh$_2$)	6.13h	β^-(2.18, 1.85, 1.7, 1.11, 0.64, 0.45), Th-x, γ(0.339, 0.911, 0.969)
		229	1.045h	β^-(1.09), γ(0.135, 0.146, 0.165, 0.262)
90	Thorium	226	31min	α(6.34, 6.23), Ra-x, γ(0.111, 0.242)
		*227(RaAc)	18.718d	α(6.04, 5.98, 5.76), Ra-x, γ(0.050, 0.236, 0.256)
		*228(RaTh)	1.913y	α(5.43, 5.34), Ra-x, γ(0.084, 0.132, 0.167, 0.214)

Z	Element	A	Half-life	Radiation (MeV)
		229	7340y	α(4.90, 4.84, 4.81), Ra-x, γ(0.137, 0.20)
		*230(Io)	7.54×10^4y	α(4.68, 4.62), Ra-x, γ(0.068)
		*231(UY)	1.0633d	β^-(0.302, 0.218, 0.138), Pa-x, γ(0.026, 0.084)
		*232(Th)	1.405×10^{10}y	α(4.01, 3.95), γ(0.059)
		*234(UX₁)	24.10d	β^-(0.199, 0.104), Pa-x, γ(0.029, 0.063, 0.092, 0.093, 0.113)
91	Protactinium	228	22h	K, α(6.11, 6.08, 6.03, 5.80), Th-x, γ(0.410, 0.463, 0.965, 0.969)
		229	1.4d	K, α(5.67, 5.62, 5.58, 5.54), Th-x
		230	17.4d	K, β^-(0.509), α(5.34, 5.33, 5.31, 5.30), Th-x, γ(0.444, 0.455, 0.899, 0.919, 0.952)
		*231(Pa)	3.276×10^4y	α(5.06, 5.02, 5.01, 4.95, 4.73), Ac-x, γ(0.027, 0.284, 0.300, 0.303)
		232	1.31d	β^-(0.32, 1.19, 1.30), U-x, γ(0.150, 0.894, 0.969)
		233	27.0d	β^-(0.257, 0.15, 0.568), U-x, γ(0.312)
		*234m(UX₂)	1.17min	β^-(2.29), IT, U-x, γ(0.765, 1.00)
		*234(UZ)	6.70h	β^-(0.68, 0.51, 0.28), U-x, γ(0.131, 0.570)
92	Uranium	230	20.8d	α(5.89, 5.82), Th-x, γ(0.072, 0.154, 0.231)
		231	4.2d	K, α(5.46), Pa-x, γ(0.026, 0.084)
		232	68.9y	α(5.32, 5.27), Pa-x, γ(0.058, 0.129)
		233	1.592×10^5y	α(4.82, 4.78), Th-x, γ(0.029, 0.042, 0.055, 0.097, 0.317)
		*234(U$_{II}$)	2.454×10^5y	α(4.77, 4.72), Th-x, γ(0.053, 0.121), e$^-$
		235m	26min	IT
		*235(AcU)	7.037×10^8y	α(4.40, 4.37, 4.22), Th-x, γ(0.186)
		236	2.342×10^7y	α(4.49, 4.44), γ(0.049), e$^-$
		237	6.75d	β^-(0.248, 0.09), γ(0.060, 0.208)
		*238(U$_I$)	4.468×10^9y	α(4.20, 4.15), γ(0.050), e$^-$
		239	23.47min	β^-(1.21, 1.29), Np-x, γ(0.044, 0.075)
		240	14.1h	β^-(0.36), Np-x, γ(0.044)
93	Neptunium	231	48.8min	K, α(6.28), γ(0.264, 0.348, 0.371), no β^+
		232	14.7min	K, U-x, γ(0.282, 0.327, 0.864), no β^+
		233	36.2min	K, α(5.54), U-x, γ(0.299, 0.312)
		234	4.4d	K, β^+(0.8), U-x, γ(1.528, 1.559, 1.602)
		235	1.085y	K, α(5.02), U-x, γ(0.026, 0.084)
		236	22.5h	K, β^-(0.537), U-x, γ(0.045, 0.643, 0.688)
		236	1.550×10^5y	K, β^-, γ(0.160)
		237	2.140×10^6y	α(4.79, 4.77), Pa-x, γ(0.029, 0.086)
		238	2.117d	β^-(1.24, 0.28, 0.25), γ(0.984, 1.026, 1.029)
		239	2.355d	β^-(0.437, 0.332, 0.393, 0.713), Pu-x, γ(0.106, 0.228, 0.278)
		240m	7.22min	β^-(2.18, 1.60, 1.30), γ(0.555, 0.597)
		240	1.032h	β^-(0.89), γ(0.448, 0.566, 0.974)
		241	13.9min	β^-(1.4), γ(0.135, 0.175), no α
		241	3.4h	β^-
94	Plutonium	232	34.1min	K, α(6.60, 6.54)
		233	20.9min	K, α(6.30), γ(0.235, 0.500, 0.535, 0.688, 1.004)
		234	8.8h	K, α(6.20, 6.15), Np-x, no γ
		235	25.3min	K, α(5.86), Np-x, γ(0.049)
		236	2.851y	α(5.77, 5.72), U-x, γ(0.047, 0.110, 0.165), e$^-$
		237m	180ms	IT, Pu-x, γ(0.145), e$^-$
		237	45.17d	K, α(5.37, 5.66), Np-x, γ(0.033, 0.044, 0.060)
		238	87.74y	α(5.50, 5.46), U-x, γ(0.0435, 0.0998, 0.153)
		239	2.411×10^4y	α(5.16, 5.14, 5.11), U-x, γ(0.052, 0.129)
		240	6563y	α(5.17, 5.12), U-x, γ(0.045)
		241	14.4y	β^-(0.021), α(4.90, 4.85), U-x, γ(0.149)
		242	3.763×10^5y	α(4.90, 4.86), γ(0.045, 0.103)
		243	4.956h	β^-(0.579, 0.490), Am-x, γ(0.084)
		244	8.26×10^7y	α(4.59, 4.55), S.F.
		245	10.5h	β^-(1.210, 0.930), γ(0.327)
		246	10.85d	β^-(0.15, 0.33), Am-x, γ(0.044, 0.180, 0.224)
95	Americium	237	1.217h	K, α(6.04), γ(0.280)
		238	1.63h	K, α(5.94), Pu-x, γ(0.561, 0.919)
		239	11.9h	K, α(5.78), Pu-x, γ(0.228, 0.278)
		240	2.117d	K, α(5.38, 5.34), Pu-x, γ(0.889, 0.988)

Z	Element	A	Half-life	Radiation (MeV)
		241	432.7y	α(5.49, 5.44), Np-x, γ(0.060), e$^-$
		242m	141y	IT, α(5.21), Am-x, Np-x, γ(0.0486), e$^-$
		242	16.01h	β^-(0.625, 0.667), K, Pu-x, Cm-x, γ(0.042, 0.045)
		243	7380y	α(5.28, 5.23), Np-x, γ(0.075)
		244m	~26min	β^-(1.50), K, Cm-x
		244	10.1h	β^-(0.387), Cm-x, γ(0.744, 0.898)
		245	2.05h	β^-(0.91), Cm-x, γ(0.253), e$^-$
		246	25.0min	β^-(1.31, 1.60, 2.10), Cm-x, γ(0.799, 1.062, 1.079)
96	Curium	238	2.4h	K, α(6.52)
		239	~3h	K, Am-x, γ(0.188), no α
		240	27d	α(6.29, 6.25)
		241	32.8d	K, α(5.94), Am-x, γ(0.015, 0.472)
		242	162.94d	α(6.12, 6.07), Pu-x, γ(0.561, 0.605)
		243	28.5y	α(5.79, 5.74), K, Pu-x, γ(0.210, 0.228, 0.278), e$^-$
		244	18.11y	α(5.81, 5.77), γ(0.043)
		245	8500y	α(5.36, 5.31), Pu-x, γ(0.133, 0.174)
		246	4730y	α(5.39, 5.34), γ(0.045)
		247	1.56×10^7y	α(4.87, 5.27), γ(0.402)
		248	3.40×10^5y	α(5.08, 5.04), S.F.
		249	1.0692h	β^-(0.86), γ(0.560, 0.634)
		250	<1.3×10^4y	S.F.
97	Berkelium	243	4.5h	K, α(6.76, 6.72, 6.57, 6.54, 6.21), Cm-x, γ(0.755, 0.84, 0.946)
		244	4.35h	K, α(6.67, 6.62), Cm-x, γ(0.218, 0.892)
		245	4.94h	K, α(6.35, 6.31, 6.15, 6.12), Cm-x, γ(0.253, 0.39)
		246	1.80d	K, Cm-x, γ(0.799)
		247	1380y	α(5.53, 5.69, 5.71), Am-x, γ(0.084, 0.27)
		248	23.7h	β^-(0.65), K, Cm-x
		249	320d	β^-(0.125), α(5.42), γ(0.327)
		250	3.217h	β^-(0.73, 1.76), Cf-x, γ(0.990, 1.03)
		251	56min	β^-
98	Californium	240	1.06min	α(7.59)
		241	3.8min	α(7.34)
		242	3.5min	α(7.39, 7.35)
		243	10.7min	K, α(7.06, 7.17)
		244	19.4min	α(7.22, 7.18)
		245	43.6min	K, α(7.14)
		246	1.487d	α(6.76, 6.72), Cm-x
		247	3.11h	K, α(6.30), Bk-x, γ(0.294, 0.418, 0.448)
		248	334d	α(6.26), S.F.
		249	350.6y	α(5.81), Cm-x, γ(0.388, 0.333, 0.253)
		250	13.08y	α(6.03, 5.99), γ(0.0429), S.F.
		251	898y	α(5.85, 5.67), Cm-x, γ(0.18)
		252	2.645y	α(6.12, 6.08), Cm-x, γ(0.043), S.F.
		253	17.81d	β^-(0.27), α(5.98)
		254	60.5d	S.F., α(5.84)
99	Einsteinium	243	21s	α(7.89)
		244	37s	K, α(7.57)
		245	1.33min	K, α(7.73)
		246	7.7min	K, α(7.36)
		247	4.7min	K, α(7.31)
		248	27min	K, α(6.87)
		249	1.703h	K, α(6.77), γ(0.379, 0.813)
		250	8.6h	K, γ(0.303, 0.829), no α
		251	1.38d	K, α(6.49), γ(0.153, 0.178)
		252	1.291y	α(6.63, 6.56), K, Bk-x, γ(0.139, 0.785)
		253	20.4d	α(6.64), Bk-x, γ(0.387, 0.389, 0.429)
		254m	1.638d	β^-(0.48, 1.13), α(6.38), K, Fm-x, γ(0.648, 0.692)
		254	275.7d	α(6.44), Bk-x, γ(0.034, 0.036, 0.043), no β^-
		255	38.3d	β^-, α(6.30), S.F.
100	Fermium	244	3.7ms	S.F.
		245	4.2s	α(8.15)
		246	1.1s	α(8.24), S.F.
		248	36s	α(7.87, 7.83), S.F.

Z	Element	A	Half-life	Radiation (MeV)
		249	2.6min	α(7.53)
		250	30min	α(7.44)
		251	5.30h	K, α(6.83), γ(0.406, 0.453, 0.881)
		252	1.0579d	α(7.04)
		253	3.00d	K, α(6.94, 6.67)
		254	3.240h	α(7.20, 7.16), γ(0.042, 0.094), S.F.
		255	20.07h	α(7.02), γ(0.204)
		256	2.627h	S.F., α(6.92)
		257	100.5d	α(6.52), Cf-x, γ(0.115, 0.241), S.F.
101	Mendelevium	248	7s	K, α(8.36, 8.32)
		249	24s	K, α(8.03)
		250	52s	K, α(7.82, 7.75)
		251	4.0min	K, α(7.55)
		252	2.3min	K
		254	10min	K, no α
		255	27min	K, α(7.33)
		256	1.27h	K, α(7.22)
		257	5.2h	K, α(7.07)
		258	55d	α(6.79, 6.73)
102	Nobelium	251	800ms	α(8.68, 8.60)
		252	2.30s	α(8.42, 8.37), S.F.
		253	1.7min	α(8.01)
		254	55s	α(8.10)
		255	3.1min	α(8.11), K
		256	3.3s	α(8.43), S.F.
		257	25s	α(8.32, 8.27, 8.22)
		258	1.2ms	S.F.
		259	1.00h	α(7.49, 7.52), K
103	Lawrencium	255	22s	α(8.43, 8.37), no K
		256	28s	α(8.52, 8.47, 8.43, 8.39)
		257	646ms	α(8.86, 8.80)
		258	4.3s	α(8.62, 8.59)
		259	5.4s	α(8.46)
		260	3.0min	α(8.04)
104	Unnilquadium	257	3.8s	α(9.00, 8.95, 8.78, 8.70)
		258	13ms	S.F.
		259	3.4s	α(8.77, 8.86)
		260	21ms	S.F.
		261	1.08min	α(8.30)
105	Unnilpentium	260	1.52s	α(9.12, 9.07, 9.04), S.F.
		261	1.8s	α(8.93), S.F.
		262	34s	α(8.45, 8.66)
106	Unnilhexium	263	800ms	α(9.25, 9.06)
107	Unnilseptium	262	~115ms	α(9.704)
		262	4.7ms	α(10.376)

Radioactive Isotopes Used in Medical Diagnosis and Therapy

Name and Symbol	Principal Nuclear Properties	Form	Use	Mode of Administration
Americium ^{241}Am	Half-life 432.7y α(5.49, 5.44) γ(0.060)	Encapsulated source	Diagnostic: Radiation source for bone mineral analyzer Therapeutic: Intracavitary radiation source in treatment of malignancies	External irradiation Intracavitary irradiation
Calcium ^{47}Ca	Half-life 4.536d β^-(0.67, 1.98) γ(1.297)	Calcium chloride	Diagnostic: Calcium metabolism studies	i.v.
Cesium ^{137}Cs Daughter ^{137m}Ba	Half-life 30.0y β^-(1.176, 0.514) Half-life 2.552min γ(0.662)	Cesium chloride or cesium sulfate (encased in needles or applicator cells)	Therapeutic: Teletherapy source or intracavitary or interstitial radiation source in treatment of malignancies	External, intracavitary or interstitial irradiation
Chromium ^{51}Cr	Half-life 27.704d K; γ(0.32)	Chromic chloride	Diagnostic: Determn of serum protein loss into the gastrointestinal tract	i.v.
		Chromium disodium edetate	Diagnostic: Determn of glomerular filtration rate	i.v.
		Labeled human serum albumin	Diagnostic: Placenta localization; gastrointestinal protein loss	i.v.
		Sodium chromate-labeled red blood cells	Diagnostic: Determn of red cell volume or mass; red cell survival time; evaluation of blood loss; spleen imaging; placenta localization	i.v.
Cobalt ^{60}Co	Half-life 5.271y β^-(0.318, 1.48) γ(1.173, 1.332)	Metallic cobalt	Therapeutic: Teletherapy source or intracavitary or interstitial radiation source in treatment of malignancies	External, intracavitary or interstitial irradiation
		Radioactive vitamin B$_{12}$	Diagnostic: In Schilling test–for absence of intrinsic factor (pernicious anemia) or other defects of intestinal vitamin B$_{12}$ absorption	Orally
^{57}Co	Half-life 271.77d K; γ(0.122)	Radioactive vitamin B$_{12}$	Diagnostic: In Schilling test–for absence of intrinsic factor (pernicious anemia) or other defects of intestinal vitamin B$_{12}$ absorption	Orally
^{58}Co	Half-life 71.916d K; β^+(0.48) γ(0.811)	Radioactive vitamin B$_{12}$	Diagnostic: In Schilling test–for absence of intrinsic factor (pernicious anemia) or other defects of intestinal vitamin B$_{12}$ absorption	Orally
Copper ^{64}Cu	Half-life 12.701h β^-(0.571) β^+(0.657) γ(1.34)	Copper versenate Copper acetate	Diagnostic: Brain scan Diagnostic: Study of Wilson's disease	i.v. Orally or i.v.
Fluorine ^{18}F	Half-life 1.8295h β^+(0.635)	Fluorodeoxy glucose	Diagnostic: Functional brain imaging	i.v.
		Sodium fluoride (reactor produced)	Diagnostic: Bone scan	Orally or i.v.
Gadolinium ^{153}Gd	Half-life 241.6d K; γ(0.70, 0.097, 0.103)	Sealed source	Diagnostic: Radiation source for bone mineral analyzer	External irradiation
Gallium ^{67}Ga	Half-life 3.261d K; γ(0.093, 0.184 0.300, 0.393)	Gallium citrate	Diagnostic: Detection of neoplastic and inflammatory lesions; tumor seeking agent	i.v.

Name and Symbol	Principal Nuclear Properties	Form	Use	Mode of Administration
Gold ^{198}Au	Half-life 2.6935d β^-(1.371, 0.962) γ(0.412)	Colloidal gold	Diagnostic: Liver imaging	i.v.
		Colloidal gold or seeds	Therapeutic: Treatment of widespread abdominal carcinomatosis with ascites; carcinomatosis of pleura with effusion; lymphomas; interstitially in metastatic tumors	Intraperitoneal or intrapleural irradiation
Indium ^{113m}In	Half-life 1.658h γ(0.393)	Indium-colloid	Diagnostic: Liver and spleen imaging	i.v.
		Indium-DTPA (pentetate)	Diagnostic: Brain scan; renal function studies	i.v.
		Indium-Fe(OH)$_3$	Diagnostic: Pulmonary perfusion imaging; cardiac output	i.v.
		Indium-labeled red blood cells	Diagnostic: Determn of blood volume	i.v.
		Indium-transferrin	Diagnostic: Static cardiovascular blood pool imaging; hepatic and placenta blood pool imaging; placenta localization	i.v.
^{111}In	Half-life 2.807d K; γ(0.172, 0.247)	Indium bleomycin	Diagnostic: Tumor detection	i.v.
		Indium chloride	Diagnostic: Hematopoietic bone marrow imaging; tumor detection	i.v.
		Indium-DTPA (pentetate)	Diagnostic: Cisternography	Intrathecal, intracisternal or intraventricular
			Diagnostic: Gastric emptying studies Cardiac output; renal scintigraphy	Orally i.v
		Indium oxoquinoline (oxine) - labeled leukocytes	Diagnostic: Detection of abscesses, infections and inflammation	i.v.
		Indium oxoquinoline (oxine) - labeled platelets	Diagnostic: Detection of deep vein thrombosis; cardiac thrombosis; renal transplant rejection	i.v.
		Indium oxoquinoline (oxine) - labeled red blood cells	Diagnostic: Detection of gastrointestinal bleeding	i.v.
Iodine ^{131}I	Half-life 8.040d β^-(0.607, 0.81, 0.336) γ(0.080, 0.284, 0.364, 0.637, 0.723)	Diiodofluorescein	Diagnostic: Brain scan	i.v.
		Iodinated fats and fatty acids e.g. oleic acid, triolein	Diagnostic: Pancreatic function; intestinal fat absorption	Orally
		Iodinated fibrinogen	Diagnostic: In vitro determn of fibrinolytic enzymes	In vitro
		Iodinated human serum albumin (IHSA)	Diagnostic: Plasma volume determn; peripheral vascular flow; cardiac output; circulation time; cerebral vascular flow	i.v.
			Diagnostic: Brain scan; placenta localization	i.v.
			Diagnostic: Cisternography	Intrathecal
		Iodinated human serum albumin (macroaggregated)	Diagnostic: Pulmonary perfusion imaging	i.v.

Name and Symbol	Principal Nuclear Properties	Form	Use	Mode of Administration
Iodine (Cont'd)		Iodinated human serum albumin (microaggregated)	Diagnostic: Hepatic blood pool imaging	i.v.
		Iodinated levothyroxine	Diagnostic: Metabolic study of endogenous thyroxine	i.v.
			Diagnostic: *In vitro* determn of thyroid function	*In vitro*
		Iodinated liothyronine	Diagnostic: *In vitro* determn of thyroid function	*In vitro*
		Iodinated povidine	Diagnostic: Protein-loss enteropathy	i.v.
		Iodinated rose bengal	Diagnostic: Liver function–hepatic excretion studies	i.v.
		Iodohippurate sodium	Diagnostic: Determn of renal function, renal blood flow, urinary tract obstruction; renal imaging	i.v.
		Sodium iodide	Diagnostic: Thyroid function studies; thyroid imaging	Orally or i.v.
			Therapeutic: Hyperthyroidism; cancer of the thyroid	Orally or i.v.
[125]I	Half-life 60.14d K; γ(0.035)	Iodinated fats or fatty acids	Diagnostic: Pancreatic function; intestinal fat absorption	Orally
		Iodinated fibrinogen	Diagnostic: Localization of deep vein thrombosis; study of fibrinogen metabolism	i.v.
			Diagnostic: *In vitro* determn of fibrinolytic enzymes	*In vitro*
		Iodinated human serum albumin (IHSA)	Diagnostic: Determn of blood or plasma volume; circulation time; cardiac output	i.v.
		Iodinated levothyroxine	Diagnostic: Metabolic study of endogenous thyroxine	i.v.
			Diagnostic: *In vitro* determn of thyroid function	*In vitro*
		Iodinated liothyronine	Diagnostic: *In vitro* determn of thyroid function	*In vitro*
		Iodinated povidone	Diagnostic: Protein-loss enteropathy	i.v.
		Iodinated rose bengal	Diagnostic: Liver function–hepatic excretion studies	i.v.
		Sealed source	Diagnostic: Radiation source for bone mineral analyzer	External irradiation
		Sodium iodide	Diagnostic: Thyroid function studies; thyroid imaging	Orally or i.v.
[123]I	Half-life 13.2h K; γ(0.159)	Iodohippurate sodium	Diagnostic: Determn of renal function, renal blood flow, urinary obstruction; renal imaging	i.v.
		Iofetamine hydrochloride	Diagnostic: Cerebral imaging	i.v.
		Sodium iodide	Diagnostic: Thyroid function studies; thyroid imaging	Orally or i.v.

Name and Symbol	Principal Nuclear Properties	Form	Use	Mode of Administration
Iridium ^{192}Ir	Half-life 73.831d $\beta^-(0.67)$ $\gamma(0.296, 0.308, 0.317, 0.468, 0.589, 0.604, 0.612)$	Seed encased in nylon ribbon	Therapeutic: Interstitial treatment of tumors	Interstitial irradiation
Iron ^{59}Fe	Half-life 44.496d $\beta^-(0.273, 0.475)$ $\gamma(1.095, 1.292)$	Ferric chloride Ferrous citrate Ferrous sulfate	Diagnostic: Ferrokinetics	i.v.
		Labeled red blood cells	Diagnostic: Red cell maturation studies	i.v.
^{55}Fe	Half-life 2.73y K	Labeled red blood cells	Diagnostic: Red cell maturation studies	i.v.
Krypton ^{85}Kr	Half-life 10.72y $\beta^-(0.67)$ $\gamma(0.517)$	Gas	Diagnostic: Cardiac abnormalities; skeletal muscle, coronary or cerebral blood flow	Intramuscular or intra-arterial injection
^{81m}Kr	Half-life 13s $\gamma(0.19)$	Gas	Diagnostic: Lung ventilation studies	Inhalation
Lead RaD(^{210}Pb)	Half-life 22.3y $\beta^-(0.017)$ $\gamma(0.047)$	Beta ray applicator	Therapeutic: See Strontium(^{90}Sr)	External irradiation
Daughter RaE(^{210}Bi)	Half-life 5.013d $\beta^-(1.16)$			
Mercury ^{197}Hg	Half-life 2.6725d K; $\gamma(0.077)$	Chlormerodrin	Diagnostic; Brain scan; renal imaging	i.v.
		Merisoprol	Diagnostic: Determn of renal function	i.v.
^{203}Hg	Half-life 46.60d $\beta^-(0.214)$ $\gamma(0.279)$	Chlormerodrin	Diagnostic: Brain scan; renal imaging	i.v.
Phosphorous ^{32}P	Half-life 14.282d $\beta^-(1.71)$	Chromic phosphate	Therapeutic: Neoplastic suppressant; treatment for peritoneal or pleural effusions caused by metastatic disease	Intrapleural or intraperitoneal
		Labeled red blood cells	Diagnostic: Blood volume determn	i.v.
		Sodium phosphate	Diagnostic: Study of peripheral vascular disease; localization of ocular, brain and skin tumors; study of breast carcinomas	Orally or i.v.
			Therapeutic: Polycythemia vera; chronic myelocytic leukemia; chronic lymphocytic leukemia; skeletal metastases; neoplastic suppressant	Orally or i.v.
Potassium ^{43}K	Half-life 22.3h $\beta^-(0.83, 0.46, 1.22, 1.82)$ $\gamma(0.618, 0.373, 0.39, 0.59, 0.22)$	Potassium chloride	Diagnostic: Myocardial scan; determn of total exchangeable potassium	i.v.
^{42}K	Half-life 12.360h $\beta^-(3.52)$ $\gamma(1.524)$	Potassium carbonate	Diagnostic: Localization of brain tumors; determn of intracellular fluid space	Orally or i.v.
		Potassium chloride	Diagnostic: Tumor detection; renal blood flow; determn of total exchangeable potassium	i.v.

Name and Symbol	Principal Nuclear Properties	Form	Use	Mode of Administration
Radium ^{226}Ra	Half-life 1600y α(4.78, 4.60) γ(0.187)	Radium bromide; α and β particles filtered by platinum	Therapeutic: Treatment of malignancies such as cancer of uterine cervix and fundus, oral pharynx, urinary bladder, skin and metastatic cancer of lymph nodes	Interstitial irradiation
Radon (Radium Emanation) ^{222}Rn Daughter of ^{226}Ra)	Half-life 2.825d α(5.49) γ(0.510)	Gaseous radon; α and β particles filtered by 0.3 mm of gold	Therapeutic: See Radium (^{226}Ra)	Interstitial irradiation
Ruthenium ^{106}Ru Daughter (^{106}Rh)	Half-life 1.020y β^-(0.039) Half-life 29.80s β^-(3.53, 3.1, 2.4) γ(0.512), 0.622, 1.128)	Beta ray applicator	Therapeutic: See Strontium(^{90}Sr)	External irradiation
Selenium ^{75}Se	Half-life 119.77d K; γ(0.265, 0.136, 0.121, 0.280, 0.401)	Selenomethionine	Diagnostic: Imaging of pancreas and parathyroid glands	i.v.
Sodium ^{24}Na	Half-life 14.659h β^-(1.389, 4.17) γ(1.369, 2.754)	Sodium chloride	Diagnostic: Determn of circulation times, sodium space, total exchangeable sodium	i.v.
^{22}Na	Half-life 2.602y β^+(0.545, 1.83) K; γ(1.275)	Sodium chloride	Diagnostic: Determn of sodium space and total exchangeable sodium	i.v.
Strontium ^{85}Sr	Half-life 64.84d K; γ(0.514)	Strontium chloride Strontium nitrate	Diagnostic: Bone imaging	i.v.
^{87m}Sr	Half-life 2.795h γ(0.388)			
^{90}Sr Daughter ^{90}Y	Half-life 28.5y β^-(0.546) Half-life 2.671d β^-(2.288) γ(2.186)	Beta ray applictor	Therapeutic: Treatment of benign conditons of eye such as pterygia, traumatic corneal ulceration, corneal scars, vernal conjunctivitis, hemangioma of eyelid, vascularization of cornea and in preparation for a corneal transplant	External irradiation
Sulfur ^{35}S	Half-life 87.51d β^-(0.167)	Sodium sulfate	Diagnostic: Determn of extracellular fluid volume	i.v.
Technetium ^{99m}Tc	Half-life 6.006h IT, γ(0.141)	Sodium pertechnetate	Diagnostic: Brain imaging. Cerebral angiography; thyroid imaging; salivary gland imaging; placenta localization; blood pool imaging; gastric mucosa imaging; cardiac function sutides; renal blood flow studies Urinary bladder imaging Nasolcrimal drainage system imaging	Orally or i.v. i.v. Urethral catheterization Direct instillation
		Sodium pertechnetate labeled red blood cells	Diagnostic: Determn of red blood cell volume, short-term survival studies In vitro compatibility studies	i.v. In vitro

Name and Symbol	Principal Nuclear Properties	Form	Use	Mode of Administration
Technetium ^{99m}Tc (Cont'd)		Tc-albumin	Diagnostic: Blood pool imaging; cardiovascular studies; placenta localization; determn of blood or plasma volumes	i.v.
		Tc-albumin (aggregated)	Diagnostic: Pulmonary perfusion imaging	i.v.
		Tc-albumin (microaggregated)	Diagnostic: Liver imaging	i.v.
		Tc-disofenin	Diagnostic: Hepatobiliary imaging	i.v.
		Tc-DTPA (pentetate)	Diagnostic: Brain imaging; renal imaging; assess renal and brain perfusion; estimate glomerular filtration rate Lung ventilation studies	i.v. Inhalation
		Tc-DTPA iron ascorbate	Diagnostic: Renal imaging	i.v.
		Tc-glucoheptonate	Diagnostic: Brain imaging; renal imaging; assess renal and brain perfusion	i.v.
		Tc-HDP (oxidronate)	Diagnostic: Bone imaging	i.v.
		Tc-HIDA (lidofenin)	Diagnostic: Hepatobiliary imaging	i.v.
		Tc-HM-PAO (exametazime)	Diagnostic: Cerebral perfusion imaging	i.v.
		Tc-MDP (medronate)	Diagnostic: Bone imaging	i.v.
		Tc-mebrofenin	Diagnostic: Hepatobiliary imaging	i.v.
		Tc-polyphosphates	Diagnostic: Bone imaging; myocardial imaging; blood pool imaging; detection of gastrointestinal bleeding	i.v.
		Tc-pyrophosphate	Diagnostic: Bone imaging; cardiac imaging; blood pool imaging; detection of gastrointestinal bleeding	i.v.
		Tc-succimer	Diagnostic: Renal imaging	i.v.
		Tc-sulfur colloid	Diagnostic: Liver, spleen and bone marrow imaging Esophageal transit studies; gastroesophageal reflux scintigraphy; determn of pulmonary aspiration of gastric contents; detection of intrapulmonary and lower gastrointestinal bleeding Lung ventilation imaging	i.v. Orally Inhalation
		Tc-stannous etidronate	Diagnostic: Bone imaging	i.v.
Thallium ^{201}Tl	Half-life 3.046d K; γ(0.135, 0.167)	Thallous chloride	Diagnostic: Myocardial perfusion imaging; localization of sites of parathyroid hyperactivity	i.v.
Xenon ^{133}Xe	Half-life 5.245d β^-(0.346) γ(0.081)	Gas	Diagnostic: Pulmonary perfusion imaging; cerebral blood flow; lung ventilation studies	Inhalation
		Gas in saline solution	Diagnostic: Pulmonary perfusion imaging; regional blood flow; lung ventilation studies	Intra-arterial or intramuscular injection

Radioactive Isotopes Used in Medical Diagnosis and Therapy (Continued)

Name and Symbol	Principal Nuclear Properties	Form	Use	Mode of Administration
[127]Xe	Half-life 36.41d K; γ(0.172, 0.203)	Gas	Diagnostic: Pulmonary perfusion imaging; lung ventilation studies	Inhalation
Ytterbium [169]Yb	Half-life 32.022d K; γ(0.063, 0.100, 0.131, 0.177, 0.198, 0.308)	Yb-DTPA (pentetate)	Diagnostic: Cisternography; brain scan	Intrathecal instillation or i.v.

References

1. S. Baum, R. Bramlet, *Basic Nuclear Medicine* (Appleton-Century-Crofts, New York, 1975).
2. C. Behrns *et al., Atomic Medicine* (The Williams & Wilkins Co., Baltimore, 5th ed., 1969).
3. H. M. Chilton, R. L. Witcofski, *Nuclear Pharmacy* (Lea & Febiger, Philadelphia, 1986).
4. *CRC Handbook of Radioactive Nuclides*, Y. Wang, Ed. (The Chemical Rubber Co., Cleveland, Ohio, 1969).
5. *Diagnostic and Investigational Uses of Radiolabeled Blood Elements*, R. J. Davey, M. E. Wallace, Eds. (Am. Assoc. of Blood Banks, Arlington, Va., 1987).
6. *Martindale*, J. E. F. Reynolds, Ed. (Pharmaceutical Press, London, 29th ed., 1989) pp. 1377-1382.
7. *Physician's Desk Reference for Radiology and Nuclear Medicine*, L. M. Freeman, M. D. Blaufox, Editorial Consultants (Medical Economics Co., Oradell, N.J., 4th ed., 1974/75, 5th ed., 1975/76).
8. *Remington's Pharmaceutical Sciences*, A. R. Gennaro, Ed. (Mack Publ. Co., Easton, Penn., 17th ed., 1985) pp. 471-499.

Maximum Allowable Concentrations of Air Contaminants

The tables have been extracted from 29 CFR (Code of Federal Regulations) Section 1910.1000 (revised January, 1989). The substances were selected to correspond to those found in *The Merck Index*.

Table 1

Permissible Exposure Limit (PEL) and Time Weighted Average (TWA) - the concentration for a normal 8-hour workday of a 40-hour workweek to which nearly all workers may be repeatedly exposed without adverse effect.

Short Term Exposure Limit (STEL) - a 15-minute, unless otherwise noted, time weighted average exposure which should not be exceeded at any time during a work day.

Ceiling Value - the concentration from instantaneous monitoring that should not be exceeded during any part of the working exposure. The ceiling values, in bold italics, are found in the PEL and STEL columns.

Skin designation - a substance which is followed by a skin designation in italics may be absorbed through the skin. The use of skin designation does not indicate that the substance may irritate the skin. Similarly, lack of a skin designation does not mean that the substance will not irritate the skin. The letters following the skin designation are T for transitional limits and F for final limits which are explained below.

Transitional limits - limits effective from March 1, 1989 through December 31, 1992.

Final limits - limit compliance by any method effective September 1, 1989.

   - limit compliance by engineering controls effective December 31, 1992.

ppm - parts of vapor or gas per million parts of contaminated air by volume at 25 deg C and 760 torr.

mg/m³ - approximate milligrams of substance per cubic meter of air.

Substance	Transitional Limits PEL ppm	mg/m³	Final Limits TWA ppm	mg/m³	STEL ppm	mg/m³
Acetaldehyde	200	360	100	180	150	270
Acetic acid glacial	10	25	10	25	-	-
Acetic anhydride	5	20	-	-	*5*	*20*
Acetone	1,000	2,400	750	1,800	1,000	2,400
Acetonitrile	40	70	40	70	60	105
Acetylene dichloride	200	790	200	790	-	-
Acrolein	0.1	0.25	0.1	0.25	0.3	0.8
Acrylamide, *skin*(T,F)	-	0.3	-	0.03	-	-
Acrylic acid, *skin*(F)	-	-	10	30	-	-
Acrylonitrile[a]						
Aldrin, *skin*(T,F)	-	0.25	-	0.25	-	-
Allyl alcohol, *skin*(T,F)	2	5	2	5	4	10
Allyl chloride	1	3	1	3	2	6
Aluminum(as Al)						
Metal						
Total dust	-	15	-	15	-	-
Respirable fraction	-	5	-	5	-	-
Pyro powders	-	-	-	5	-	-
Welding fumes**	-	-	-	5	-	-
Soluble salts	-	-	-	2	-	-
Alkyls	-	-	-	2	-	-
Aluminum oxide						
Total dust	-	15	-	10	-	-
Respirable fraction	-	5	-	5	-	-
α-Aminopyridine	0.5	2	0.5	2	-	-
Amitrole	-	-	-	0.2	-	-
Ammonia	50	35	-	-	35	27
Ammonium chloride fume	-	-	-	10	-	20
Ammonium sulfamate						
Total dust	-	15	-	10	-	-
Respirable fraction	-	5	-	5	-	-
Aniline and homologs, *skin*(T,F)	5	19	2	8	-	-
Anisidine (*o-*,*p*-isomers)	-	0.5	-	0.5	-	-
Antimony and compounds (as Sb)	-	0.5	-	0.5	-	-
ANTU	-	0.3	-	0.3	-	-
Arsenic, organic compounds (as As)	-	0.5	-	0.5	-	-
Arsenic, inorganic compounds (as As)[a]						
Arsine	0.05	0.2	0.05	0.2	-	-
Asbestos[a]						
Aspirin	-	-	-	5	-	-
Atrazine	-	-	-	5	-	-
Azinphos-methyl, *skin*(T,F)	-	0.2	-	0.2	-	-
Barium, soluble compounds (as Ba)	-	0.5	-	0.5	-	-
Barium sulfate						
Total dust	-	15	-	10	-	-
Respirable fraction	-	5	-	5	-	-
Benomyl						
Total dust	-	15	-	10	-	-
Respirable fraction	-	5	-	5	-	-
Benzene[a]						
Benzidine[a]						

Table 1 (Continued)

Substance	Transitional Limits PEL ppm	mg/m³	Final Limits TWA ppm	mg/m³	STEL ppm	mg/m³
Benzo[*a*]pyrene	-	0.2	-	0.2	-	-
Benzoyl peroxide	-	5	-	5	-	-
Benzyl chloride	1	5	1	5	-	-
Beryllium and beryllium compounds (as Be)[b]	-	-	0.002	-	*0.025#*	-
p-Biphenylamine[a]						
Bis(2-ethylhexyl) phthalate	-	5	-	5	-	10
Bismuth telluride, Un-doped						
Total dust	-	15	-	15	-	-
Respirable fraction	-	5	-	5	-	-
Bismuth telluride, Se-doped	-	-	-	5	-	-
Boric anhydride						
Total dust	-	15	-	10	-	-
Respirable fraction	-	5	-	5	-	-
Boron tribromide	-	-	-	-	1	10
Boron trifluoride	*1*	*3*	-	-	1	3
Bromacil	-	-	1	10	-	-
Bromine	0.1	0.7	0.1	0.7	0.3	2
Bromine pentafluoride	-	-	0.1	0.7	-	-
Bromoform, *skin*(T,F)	0.5	5	0.5	5	-	-
1,3-Butadiene[c]	1,000	2,200	-	-	-	-
Butane	-	-	800	1,900	-	-
n-Butyl acetate	150	710	150	710	200	950
sec-Butyl acetate	200	950	200	950	-	-
tert-Butyl acetate	200	950	200	950	-	-
n-Butyl acrylate	-	-	10	55	-	-
n-Butyl alcohol, *skin*(F)	100	300	-	-	*50*	*150*
sec-Butyl alcohol	150	450	100	305	-	-
tert-Butyl alcohol	100	300	100	300	150	450
n-Butylamine, *skin*(T,F)	*5*	*15*	-	-	*5*	*15*
Butylated hydroxytoluene	-	-	-	10	-	-
Butyl Cellosolve®, *skin*(T,F)	50	240	25	120	-	-
n-Butyl mercaptan	10	35	0.5	1.5	-	-
n-Butyl phthalate	-	5	-	5	-	-
Cadmium fume (as Cd)[b,c]						
Cadmium dust (as Cd)[b,c]						
Calcium carbonate						
Total dust	-	15	-	15	-	-
Respirable fraction	-	5	-	5	-	-
Calcium cyanamide	-	-	-	0.5	-	-
Calcium hydroxide	-	-	-	5	-	-
Calcium oxide	-	5	-	5	-	-
Calcium silicate						
Total dust	-	15	-	15	-	-
Respirable fraction	-	5	-	5	-	-
Calcium sulfate						
Total dust	-	15	-	15	-	-
Respirable fraction	-	5	-	5	-	-
Camphor, synthetic	-	2	-	2	-	-
Caprolactam						
Dust	-	-	-	1	-	3
Vapor	-	-	5	20	10	40
Captafol	-	-	-	0.1	-	-
Captan	-	-	-	5	-	-
Carbaryl	-	5	-	5	-	-
Carbofuran	-	-	-	0.1	-	-
Carbon, amorphous	-	3.5	-	3.5	-	-
Carbon dioxide	5,000	9,000	10,000	18,000	30,000	54,000
Carbon disulfide[b], *skin*(F)	-	-	4	12	12	36
Carbon monoxide	50	55	35	40	*200*	*229*
Carbon tetrachloride[b]	-	-	2	12.6	-	-
Carbonyl fluoride	-	-	2	5	5	15
Cellulose						
Total dust	-	15	-	15	-	-
Respirable fraction	-	5	-	5	-	-
Cesium hydroxide	-	-	-	2	-	-
Chlordan(e), *skin*(T,F)	-	0.5	-	0.5	-	-
Chlorine	*1*	*3*	0.5	1.5	1	3
Chlorine dioxide	0.1	0.3	0.1	0.3	0.3	0.9
Chlorine trifluoride	*0.1*	*0.4*	-	-	*0.1*	*0.4*

Table 1 (Continued)

Substance	Transitional Limits PEL ppm	mg/m³	Final Limits TWA ppm	mg/m³	STEL ppm	mg/m³
Chloroacetaldehyde	*1*	*3*	-	-	*1*	*3*
ω-Chloroacetophenone	0.05	0.3	0.05	0.3	-	-
Chlorobenzene	75	350	75	350	-	-
o-Chlorobenzylidene						
malononitrile, *skin*(F)	0.05	0.4	-	-	*0.05*	*0.4*
Chloroform	*50*	*240*	2	9.78	-	-
Chloromethyl methyl ether[a]						
Chloronitrobenzene, (*p*-isomer),						
skin(T,F)	-	1	-	1	-	-
Chloropicrin	0.1	0.7	0.1	0.7	-	-
o-Chlorotoluene	-	-	50	250	-	-
Chlorpyrifos, *skin*(F)	-	-	-	0.2	-	-
Chromium (II) compounds (as Cr)	-	0.5	-	0.5	-	-
Chromium (III) compounds (as Cr)	-	0.5	-	0.5	-	-
Chromium metal (as Cr)	-	1	-	1	-	-
Chromium trioxide and chromates (as CrO₃)[b]	-	-	-	-	*0.1*	
Chrysene	-	0.2	-	0.2	-	-
Clopidol						
Total dust	-	15	-	15	-	-
Respirable fraction	-	5	-	5	-	-
Coal tar pitch volatiles (benzene soluble						
fraction), anthracene, BaP,						
phenanthrene, acridine, chrysene,						
pyrene	-	0.2	-	0.2	-	-
Cobalt metal dust and fume (as Co)	-	0.1	-	0.05	-	-
Cobalt hydrocarbonyl* (as Co)	-	-	-	0.1	-	-
Copper						
Fume (as Cu)	-	0.1	-	0.1	-	-
Dusts and mists (as Cu)	-	1	-	1	-	-
Cresol(s), all isomers, *skin*(T,F)	5	22	5	22	-	-
Crotonaldehyde	2	6	2	6	-	-
Crufomate	-	-	-	5	-	-
Cryofluorane	1,000	7,000	1,000	7,000	-	-
Cumene, *skin*(T,F)	50	245	50	245	-	-
Cyanamide	-	-	-	2	-	-
Cyanides (as CN)	-	5	-	5	-	-
Cyanogen	-	-	10	20	-	-
Cyanogen chloride	-	-	-	-	0.3	0.6
Cyclohexane	300	1,050	300	1,050	-	-
Cyclohexanol, *skin*(F)	50	200	50	200	-	-
Cyclohexanone, *skin*(F)	50	200	25	100	-	-
Cyclohexene	300	1,015	300	1,015	-	-
Cyclohexylamine	-	-	10	40	-	-
Cyclonite, *skin*(F)	-	-	-	1.5	-	-
Cyclopentadiene	75	200	75	200	-	-
Cyclopentane	-	-	600	1,720	-	-
Cyhexatin	-	-	-	5	-	-
2,4-D	-	10	-	10	-	-
Dalapon	-	-	1	6	-	-
DDT, *skin*(T,F)	-	1	-	1	-	-
Decaborane(14), *skin*(T,F)	0.05	0.3	0.05	0.3	0.15	0.9
Demeton, *skin*(T,F)	-	0.1	-	0.1	-	-
Diacetone alcohol	50	240	50	240	-	-
Diazinon®, *skin*(F)	-	-	-	0.1	-	-
Diazomethane	0.2	0.4	0.2	0.4	-	-
Diborane(6)	0.1	0.1	0.1	0.1	-	-
1,2-Dibromo-3-chloropropane[a]						
o-Dichlorobenzene	*50*	*300*	-	-	*50*	*300*
p-Dichlorobenzene	75	450	75	450	110	675
3,3'-Dichlorobenzidine[a]						
Dichlorodifluoromethane	1,000	4,950	1,000	4,950	-	-
1,3-Dichloro-5,5-dimethyl hydantoin	-	0.2	-	0.2	-	0.4
sym-Dichloroethyl ether, *skin*(T,F)	*15*	*90*	5	30	10	60
1,3-Dichloropropene, *skin*(F)	-	-	1	5	-	-
Dichlorvos, *skin*(T,F)	-	1	-	1	-	-
Dicrotophos, *skin*(F)	-	-	-	0.25	-	-
Dieldrin, *skin*(T,F)	-	0.25	-	0.25	-	-
Diethanolamine	-	-	3	15	-	-
Diethylamine	25	75	10	30	25	75

Table 1 (Continued)

Substance	Transitional Limits PEL ppm	mg/m³	Final Limits TWA ppm	mg/m³	STEL ppm	mg/m³
2-Diethylaminoethanol, *skin*(T,F)	10	50	10	50	-	-
Diethyl ketone	-	-	200	705	-	-
Diethyl phthalate*	-	-	-	5	-	-
Diisopropylamine, *skin*(T,F)	5	20	5	20	-	-
N,N-Dimethylacetamide, *skin*(T,F)	10	35	10	35	-	-
Dimethylamine	10	18	10	18	-	-
p-Dimethylaminoazobenzene[a]						
N,N-Dimethylaniline, *skin*(T,F)	5	25	5	25	10	50
N,N-Dimethylformamide, *skin*(T,F)	10	30	10	30	-	-
1,1-Dimethylhydrazine, *skin*(T,F)	0.5	1	0.5	1	-	-
Dimethylphthalate	-	5	-	5	-	-
Dimethyl sulfate, *skin*(T,F)	1	5	0.1	0.5	-	-
Dinitolmide	-	-	-	5	-	-
Dinitrobenzene (all isomers), *skin*(T,F)	-	1	-	1	-	-
Dinitrocresol, *skin*(T,F)	-	0.2	-	0.2	-	-
Dioxane, *skin*(T,F)	100	360	25	90	-	-
Dioxathion, *skin*(F)	-	-	-	0.2	-	-
Diphenyl	0.2	1	0.2	1	-	-
Diphenylamine	-	-	-	10	-	-
Dipropyl ketone	-	-	50	235	-	-
Diquat dibromide	-	-	-	0.5	-	-
Disulfiram	-	-	-	2	-	-
Disulfoton, *skin*(F)	-	-	-	0.1	-	-
Disul-sodium						
Total dust	-	15	-	10	-	-
Respirable fraction	-	5	-	5	-	-
Diuron	-	-	-	10	-	-
Endosulfan, *skin*(F)	-	-	-	0.1	-	-
Endrin, *skin*(T,F)	-	0.1	-	0.1	-	-
Epichlorohydrin, *skin*(T,F)	5	19	2	8	-	-
EPN, *skin*(T,F)	-	0.5	-	0.5	-	-
Ethanethiol	*10*	*25*	0.5	1	-	-
Ethanolamine	3	6	3	8	6	15
Ethion, *skin*(F)	-	-	-	0.4	-	-
2-Ethoxyethanol[c], *skin*(T)	200	740	-	-	-	-
2-Ethoxyethyl acetate[c], *skin*(T)	100	540	-	-	-	-
Ethyl acetate	400	1,400	400	1,400	-	-
Ethyl acrylate, *skin*(T,F)	25	100	5	20	25	100
Ethyl alcohol	1,000	1,900	1,000	1,900	-	-
Ethylamine	10	18	10	18	-	-
Ethyl amyl ketone	25	130	25	130	-	-
Ethyl benzene	100	435	100	435	125	545
Ethyl bromide	200	890	200	890	250	1,110
Ethyl chloride	1,000	2,600	1,000	2,600	-	-
Ethylene chlorohydrin, *skin*(T,F)	5	16	-	-	*1*	*3*
Ethylenediamine	10	25	10	25	-	-
Ethylene dibromide[b,c]						
Ethylene dichloride[b]	-	-	1	4	2	8
Ethylene glycol	-	-	-	-	*50*	*125*
Ethylene oxide[a]						
Ethylenimine[a]						
Ethyl ether	400	1,200	400	1,200	500	1,500
Ethyl formate	100	300	100	300	-	-
Ethylidene chloride	100	400	100	400	-	-
Ethyl silicate	100	850	10	85	-	-
Fensulfothion	-	-	-	0.1	-	-
Fenthion, *skin*(F)	-	-	-	0.2	-	-
Ferbam						
Total dust	-	15	-	10	-	-
Respirable fraction	-	5	-	5	-	-
Ferric oxide dust and fume (as Fe)						
Total particulate	-	10	-	10	-	-
Ferrocene						
Total dust	-	15	-	10	-	-
Respirable fraction	-	5	-	5	-	-
N-2-Fluorenylacetamide[a]						
Fluorides[b] (as F)	-	-	2.5	-	-	-
Fluorine	0.1	0.2	0.1	0.2	-	-
Fluorine monoxide	0.05	0.1	-	-	*0.05*	*0.1*
Fonofos, *skin*(F)	-	-	-	0.1	-	-

Table 1 (Continued)

Substance	Transitional Limits PEL ppm	mg/m³	Final Limits TWA ppm	mg/m³	STEL ppm	mg/m³
Formaldehyde[a,b]						
Formamide	-	-	20	30	30	45
Formic acid	5	9	5	9	-	-
Furfural, *skin*(T,F)	5	20	2	8	-	-
Furfuryl alcohol, *skin*(F)	50	200	10	40	15	60
Gasoline	-	-	300	900	500	1,500
Glutaraldehyde	-	-	-	-	*0.2*	*0.8*
Glycerol (mist)						
Total dust	-	15	-	10	-	-
Respirable fraction	-	5	-	5	-	-
Glycidol	50	150	25	75	-	-
Graphite, synthetic						
Total dust	-	15	-	10	-	-
Respirable fraction	-	5	-	5	-	-
Gypsum*						
Total dust	-	15	-	15	-	-
Respirable fraction	-	5	-	5	-	-
Hafnium	-	0.5	-	0.5	-	-
Heptachlor, *skin*(T,F)	-	0.5	-	0.5	-	-
n-Heptane	500	2,000	400	1,600	500	2,000
2-Heptanone	100	465	100	465	-	-
Hexachloroethane, *skin*(T,F)	1	10	1	10	-	-
n-Hexane	500	1,800	50	180	-	-
Hexylene glycol	-	-	-	-	*25*	*125*
Hydrazine, *skin*(T,F)	1	1.3	0.1	0.1	-	-
Hydrogen bromide	3	10	-	-	*3*	*10*
Hydrogen chloride	*5*	*7*	-	-	*5*	*7*
Hydrogen cyanide, *skin*(T,F)	10	11	-	-	4.7	5
Hydrogen fluoride[b] (as F)	-	-	3	-	6	-
Hydrogen peroxide	1	1.4	1	1.4	-	-
Hydrogen selenide (as Se)	0.05	0.2	0.05	0.2	-	-
Hydrogen sulfide[b]	-	-	10	14	15	21
Hydroquinone	-	2	-	2	-	-
Indene	-	-	10	45	-	-
Indium and compounds (as In)	-	-	-	0.1	-	-
Iodine	*0.1*	*1*	-	-	*0.1*	*1*
Iodoform	-	-	0.6	10	-	-
Iron soluble salts (as Fe)	-	-	-	1	-	-
Iron pentacarbonyl (as Fe)	-	-	0.1	0.8	0.2	1.6
Isoamyl acetate	100	525	100	525	-	-
Isobutyl acetate	150	700	150	700	-	-
Isobutyl alcohol	100	300	50	150	-	-
Isooctyl alcohol, *skin*(F)	-	-	50	270	-	-
Isopentyl alcohol						
Primary and secondary	100	360	100	360	125	450
Isopropyl acetate	250	950	250	950	310	1,185
Isopropylacetone	100	410	50	205	75	300
Isopropyl alcohol	400	980	400	980	500	1,225
Isopropylamine	5	12	5	12	10	24
Isopropyl ether	500	2,100	500	2,100	-	-
Kaolin						
Total dust	-	15	-	10	-	-
Respirable fraction	-	5	-	5	-	-
Ketene	0.5	0.9	0.5	0.9	1.5	3
Lead inorganic[a] (as Pb)						
Ligroin	-	-	300	1,350	400	1,800
Limestone						
Total dust	-	15	-	15	-	-
Respirable fraction	-	5	-	5	-	-
Lindane, *skin*(T,F)	-	0.5	-	0.5	-	-
Lithium hydride	-	0.025	-	0.025	-	-
Magnesite*						
Total dust	-	15	-	15	-	-
Respirable fraction	-	5	-	5	-	-
Magnesium oxide, fume						
Total dust	-	15	-	10	-	-
Respirable fraction	-	5	-	5	-	-
Malathion, *skin*(T,F)						
Total dust	-	15	-	10	-	-
Respirable fraction	-	5	-	5	-	-

Table 1 (Continued)

Substance	Transitional Limits PEL		Final Limits TWA		STEL	
	ppm	mg/m³	ppm	mg/m³	ppm	mg/m³
Maleic anhydride	0.25	1	0.25	1	-	-
Manganese compounds (as Mn)	-	5	-	-	-	5
Manganese, fume (as Mn)	-	5	-	1	-	3
Manganese oxide (as Mn)	-	-	-	1	-	-
Mecrylate	-	-	2	8	4	16
Mercury, (as Hg) skin(F)						
Aryl and inorganic[b]	-	-	-	-	-	*0.1*
Organo alkyl compounds[b]	-	-	-	0.01	-	0.03
Vapor[b]	-	-	-	0.05	-	-
Mesityl oxide	25	100	15	60	25	100
Methacrylic acid, skin(F)	-	-	20	70	-	-
Methacrylonitrile, skin(F)	-	-	1	3	-	-
Methanethiol	*10*	*20*	0.5	1	-	-
Methanol, skin(F)	200	260	200	260	250	310
Methomyl	-	-	-	2.5	-	-
Methoxychlor						
Total dust	-	15	-	10	-	-
Respirable fraction	-	5	-	5	-	-
Methyl acetate	200	610	200	610	250	760
Methyl acrylate, skin(T,F)	10	35	10	35	-	-
Methylal	1,000	3,100	1,000	3,100	-	-
Methylamine	10	12	10	12	-	-
Methylaniline, skin(T,F)	2	9	0.5	2	-	-
Methyl bromide, skin(T,F)	*20*	*80*	5	20	-	-
Methyl butyl ketone	100	410	5	20	-	-
Methyl Cellosolve®,c, skin(T)	25	80	-	-	-	-
Methyl Cellosolve® acetate[c], skin(T)	25	120	-	-	-	-
Methyl chloride[b]	-	-	50	105	100	205
Methyl demeton, skin(F)	-	-	-	0.5	-	-
Methylene chloride[b,c]						
Methyl ethyl ketone	200	590	200	590	300	885
Methyl formate	100	250	100	250	150	375
Methyl hydrazine, skin(T,F)	*0.2*	*0.35*	-	-	*0.2*	*0.35*
Methyl iodide, skin(T,F)	5	28	2	10	-	-
Methyl methacrylate*	100	410	100	410	-	-
Methyl parathion, skin(F)	-	-	-	0.2	-	-
Methyl propyl ketone	200	700	200	700	250	875
Metribuzin	-	-	-	5	-	-
Mevinphos®, skin(T,F)	-	0.1	0.01	0.1	0.03	0.3
Molybdenum (as Mo)						
Soluble compounds	-	5	-	5	-	-
Insoluble compounds						
Total dust	-	15	-	10	-	-
Respirable fraction	-	5	-	5	-	-
Monocrotophos	-	-	-	0.25	-	-
Morpholine, skin(T,F)	20	70	20	70	30	105
Naled, skin(F)	-	3	-	3	-	-
Naphthalene	10	50	10	50	15	75
1-Naphthylamine[a]						
2-Naphthylamine[a]						
Nickel carbonyl (as Ni)	0.001	0.007	0.001	0.007	-	-
Nickel (as Ni)						
Metal and insoluble compounds	-	1	-	1	-	-
Soluble compounds	-	1	-	0.1	-	-
Nicotine, skin(T,F)	-	0.5	-	0.5	-	-
Nitramine, skin(T,F)	-	1.5	-	0.1	-	-
Nitrapyrin						
Total dust	-	15	-	15	-	-
Respirable fraction	-	5	-	5	-	-
Nitric acid	2	5	2	5	4	10
Nitric oxide	25	30	25	30	-	-
p-Nitroaniline, skin(T,F)	1	6	-	3	-	-
Nitrobenzene, skin(T,F)	1	5	1	5	-	-
p-Nitrobiphenyl[a]						
Nitroethane	100	310	100	310	-	-
Nitrogen dioxide	*5*	*9*	-	-	1	1.8
Nitrogen fluoride	10	29	10	29	-	-
Nitroglycerin, skin(T,F)	*0.2*	*1*	-	-	-	0.1
Nitromethane	100	250	100	250	-	-
1-Nitropropane	25	90	25	90	-	-

MISC-58

Table 1 (Continued)

Substance	Transitional Limits PEL ppm	mg/m³	Final Limits TWA ppm	mg/m³	STEL ppm	mg/m³
2-Nitropropane	25	90	10	35	-	-
N-Nitrosodimethylamine[a]						
Nitrotoluene, all isomers, skin(T,F)	5	30	2	11	-	-
Octane	500	2,350	300	1,450	375	1,800
Osmium tetroxide (as Os)	-	0.002	0.0002	0.002	0.0006	0.006
Oxalic acid	-	1	-	1	-	2
Ozone	0.1	0.2	0.1	0.2	0.3	0.6
Paraffin, fume	-	-	-	2	-	-
Paraquat, skin(T,F)						
Respirable dust	-	0.5	-	0.1	-	-
Parathion, skin(T,F)	-	0.1	-	0.1	-	-
Pentaborane(9)	0.005	0.01	0.005	0.01	0.015	0.03
Pentachlorophenol, skin(T,F)	-	0.5	-	0.5	-	-
Pentaerythritol						
Total dust	-	15	-	10	-	-
Respirable fraction	-	5	-	5	-	-
Pentane	1,000	2,950	600	1,800	750	2,250
Perchloryl fluoride	3	13.5	3	14	6	28
Petrolatum, liquid mist	-	5	-	5	-	-
Petroleum	500	2,000	400	1,600	-	-
Petroleum benzin	100	400	100	400	-	-
Phenol, skin(T,F)	5	19	5	19	-	-
Phenothiazine, skin(F)	-	-	-	5	-	-
p-Phenylenediamine, skin(T,F)	-	0.1	-	0.1	-	-
Phenyl ether, vapor	1	7	1	7	-	-
Phenyl ether-biphenyl mixture, vapor	1	7	1	7	-	-
Phenylhydrazine, skin(T,F)	5	22	5	20	10	45
Phorate, skin(F)	-	-	-	0.05	-	0.2
Phosgene	0.1	0.4	0.1	0.4	-	-
Phosphine	0.3	0.4	0.3	0.4	1	1
Phosphoric acid	-	1	-	1	-	3
Phosphorous (yellow)	-	0.1	-	0.1	-	-
Phosphorous oxychloride	-	-	0.1	0.6	-	-
Phosphorous pentachloride	-	1	-	1	-	-
Phosphorous pentasulfide	-	1	-	1	-	3
Phosphorous trichloride	0.5	3	0.2	1.5	0.5	3
Phthalic anhydride	2	12	1	6	-	-
Picloram						
Total dust	-	15	-	10	-	-
Respirable fraction	-	5	-	5	-	-
Picric acid, skin(T,F)	-	0.1	-	0.1	-	-
Pindone	-	0.1	-	0.1	-	-
Plaster of Paris*						
Total dust	-	15	-	15	-	-
Respirable fraction	-	5	-	5	-	-
Platinum (as Pt)						
Metal	-	-	-	1	-	-
Soluble salts	-	0.002	-	0.002	-	-
Potassium hydroxide	-	-	-	-	-	2
Propane	1,000	1,800	1,000	1,800	-	-
Propargyl alcohol, skin(F)	-	-	1	2	-	-
β-Propiolactone[a]						
Propionic acid	-	-	10	30	-	-
Propoxur	-	-	-	0.5	-	-
Propyl acetate	200	840	200	840	250	1,050
n-Propyl alcohol	200	500	200	500	250	625
Propylene dichloride	75	350	75	350	110	510
Propylene oxide	100	240	20	50	-	-
n-Propyl nitrate	25	110	25	105	40	170
Pyridine	5	15	5	15	-	-
Pyrocatechol, skin(F)	-	-	5	20	-	-
Quinone	0.1	0.4	0.1	0.4	-	-
Resorcinol	-	-	10	45	20	90
Rhodium (as Rh)						
Metal fume and insoluble compounds	-	0.1	-	0.1	-	-
Soluble compounds	-	0.001	-	0.001	-	-
Ronnel	-	15	-	10	-	-
Rotenone	-	5	-	5	-	-
Selenium compounds (as Se)	-	0.2	-	0.2	-	-
Selenium hexafluoride (as Se)	0.05	0.4	0.05	0.4	-	-

Table 1 (Continued)

Substance	Transitional Limits PEL		Final Limits TWA		STEL	
	ppm	mg/m³	ppm	mg/m³	ppm	mg/m³
Silane	-	-	5	7	-	-
Silicon						
Total dust	-	15	-	10	-	-
Respirable fraction	-	5	-	5	-	-
Silicon carbide						
Total dust	-	15	-	10	-	-
Respirable fraction	-	5	-	5	-	-
Silver, metal dust and fume (as Ag)	-	0.01	-	0.01	-	-
Sodium azide, *skin*(F)						
(as HN₃)	-	-	-	-	*0.1*	-
(as NaN₃)	-	-	-	-	-	*0.3*
Sodium bisulfite	-	-	-	5	-	-
Sodium borate						
Anhydrous	-	-	-	10	-	-
Decahydrate	-	-	-	10	-	-
Sodium hydroxide	-	2	-	-	-	2
Sodium metabisulfite	-	-	-	5	-	-
Stannous oxide	-	-	-	2	-	-
Starch						
Total dust	-	15	-	15	-	-
Respirable fraction	-	5	-	5	-	-
Stibine	0.1	0.5	0.1	0.5	-	-
Stoddard solvent	500	2,900	100	525	-	-
Strychnine	-	0.15	-	0.15	-	-
Styrene[b]	-	-	50	215	100	425
Sucrose						
Total dust	-	15	-	15	-	-
Respirable fraction	-	5	-	5	-	-
Sulfotep, *skin*(T,F)	-	0.2	-	0.2	-	-
Sulfur chloride	1	6	-	-	*1*	*6*
Sulfur dioxide	5	13	2	5	5	10
Sulfur hexafluoride	1,000	6,000	1,000	6,000	-	-
Sulfuric acid	-	1	-	1	-	-
Sulfur tetrafluoride	-	-	-	-	*0.1*	*0.4*
Sulfuryl fluoride	5	20	5	20	10	40
Sulprofos	-	-	-	1	-	-
2,4,5-T	-	10	-	10	-	-
Tantalum, metal and oxide dust	-	5	-	5	-	-
Tellurium and compounds (as Te)	-	0.1	-	0.1	-	-
Tellurium hexafluoride (as Te)	0.02	0.2	0.02	0.2	-	-
Temephos						
Total dust	-	15	-	10	-	-
Respirable fraction	-	5	-	5	-	-
sym-Tetrabromoethane	1	14	1	14	-	-
Tetrachloroethane, *skin*(T,F)	5	35	1	7	-	-
Tetrachloroethylene[b]	-	-	25	170	-	-
Tetraethyl lead (as Pb), *skin*(T,F)	-	0.075	-	0.075	-	-
Tetraethyl pyrophosphate, *skin*(T,F)	-	0.05	-	0.05	-	-
Tetrahydrofuran	200	590	200	590	250	735
Tetranitromethane	1	8	1	8	-	-
Tetrasodium pyrophosphate	-	-	-	5	-	-
Thallium, *skin*(T,F)						
Soluble compounds (as Ti)	-	0.1	-	0.1	-	-
Thioglycolic acid, *skin*(F)	-	-	1	4	-	-
Thionyl chloride	-	-	-	-	*1*	*5*
Thiophenol	-	-	0.5	2	-	-
Thiram	-	5	-	5	-	-
Tin, inorganic compounds						
(except oxides) (as Sn)	-	2	-	2	-	-
Tin, organic compounds (as Sn), *skin*(F)	-	0.1	-	0.1	-	-
Titanium dioxide						
Total dust	-	15	-	10	-	-
Respirable fraction	-	5	-	5	-	-
Toluene[b]	-	-	100	375	150	560
Toluene 2,4-diisocyanate	*0.02*	*0.14*	0.005	0.04	0.02	0.15
m-Toluidine, *skin*(F)	-	-	2	9	-	-
o-Toluidine, *skin*(T,F)	5	22	5	22	-	-
p-Toluidine, *skin*(F)	-	-	2	9	-	-
Toxaphene, *skin*(T,F)	-	0.5	-	0.5	-	1
Tributyl phosphate	-	5	0.2	2.5	-	-

Table 1 (Continued)

Substance	Transitional Limits PEL		Final Limits TWA		STEL	
	ppm	mg/m³	ppm	mg/m³	ppm	mg/m³
Trichloroacetic acid	-	-	1	7	-	-
1,2,4-Trichlorobenzene	-	-	-	-	5	*40*
1,1,1-Trichloroethane	350	1,900	350	1,900	450	2,450
1,1,2-Trichloroethane, *skin*(T,F)	10	45	10	45	-	-
Trichloroethylene[b]	-	-	50	270	200	1,080
Trichlorofluoromethane	1,000	5,600	-	-	*1,000*	*5,600*
Triethylamine	25	100	10	40	15	60
Trimellitic anhydride	-	-	0.005	0.04	-	-
Trimethylamine	-	-	10	24	15	36
2,4,6-Trinitrotoluene, *skin*(T,F)	-	1.5	-	0.5	-	-
Triphenyl phosphate	-	3	-	3	-	-
Tri-*o*-tolyl phosphate, *skin*(F)	-	0.1	-	0.1	-	-
Tungsten (as W)						
Insoluble compounds	-	-	-	5	-	10
Soluble compounds	-	-	-	1	-	3
Turpentine	100	560	100	560	-	-
Uranium (as U)						
Soluble compounds	-	0.05	-	0.05	-	-
Insoluble compounds	-	0.25	-	0.2	-	0.6
n-Valeraldehyde	-	-	50	175	-	-
Vanadium (as V₂O₅)						
Respirable dust	-	*0.5*	-	0.05	-	-
Fume	-	*0.1*	-	0.05	-	-
Vinyl acetate	-	-	10	30	20	60
Vinyl chloride[a]						
Vinylidene chloride	-	-	1	4	-	-
Warfarin	-	0.1	-	0.1	-	-
Xylene (*o*-,*m*-,*p*-isomers)	100	435	100	435	150	655
Xylidine, *skin*(T,F)	5	25	2	10	-	-
Yttrium	-	1	-	1	-	-
Zinc chloride, fume	-	1	-	1	-	2
Zinc oxide, fume	-	5	-	5	-	10
Zinc oxide						
Total dust	-	15	-	10	-	-
Respirable fraction	-	5	-	5	-	-
Zinc stearate						
Total dust	-	15	-	10	-	-
Respirable fraction	-	5	-	5	-	-
Zirconium compounds (as Zr)	-	5	-	5	-	10

*Derivative of title compound.
**As determined from breathing-zone air samples.
#The STEL value for 30 minutes is 0.005 ppm.
[a]See Table 3.
[b]See Table 2.
[c]Substances undergoing separate final limit rulemaking.

Table 2

(1) **8-hour time weighted averages**—An employee's exposure to any material listed in Table 2, in any 8-hour work shift of a 40-hour work week, shall not exceed the 8-hour time weighted average limit given for that material in the table.

(2) **Acceptable ceiling concentrations**—An employee's exposure to a material listed in Table 2 shall not exceed at any time during an 8-hour shift the acceptable ceiling concentration limit given for the material in the table, except for a time period, and up to a concentration not exceeding the maximum duration and concentration allowed in the column under "acceptable maximum peak above the acceptable ceiling concentration for an 8-hour shift".

Material	8-hour time weighted average	Acceptable ceiling concentration	Acceptable maximum peak above the acceptable ceiling concentration for an 8-hour shift.	
			concentration	maximum duration
Benzene[a]	10 ppm	25 ppm	50 ppm	10 minutes
Beryllium and Beryllium compounds	2 ug/m^3	5 ug/m^3	25 ug/m^3	30 minutes
Cadmium fume	0.1 mg/m^3	0.3 mg/m^3		
Cadmium dust	0.2 mg/m^3	0.6 mg/m^3		
Carbon disulfide	20 ppm	30 ppm	100 ppm	30 minutes
Carbon tetrachloride	10 ppm	25 ppm	200 ppm	5 minutes in any 4 hours
Chromic acid and chromates		1 mg/10 m^3		
Ethylene dibromide	20 ppm	30 ppm	50 ppm	5 minutes
Ethylene dichloride	50 ppm	100 ppm	200 ppm	5 minutes in any 3 hours
Formaldehyde[b]	3 ppm	5 ppm	10 ppm	30 minutes
Hydrogen fluoride	3 ppm			
Hydrogen sulfide		20 ppm	50 ppm	10 minutes once only if no other measurable exposure occurs
Fluoride as dust	2.5 mg/m^3			
Mercury		1 mg/10 m^3		
Methyl chloride	100 ppm	200 ppm	300 ppm	5 minutes in any 3 hours
Methylene chloride	500 ppm	1,000 ppm	2,000 ppm	5 minutes in any 2 hours
Organo (alkyl) mercury	0.01 mg/m^3	0.04 mg/m^3		
Styrene	100 ppm	200 ppm	600 ppm	5 minutes in any 3 hours
Tetrachloroethylene	100 ppm	200 ppm	300 ppm	5 minutes in any 3 hours
Toluene	200 ppm	300 ppm	500 ppm	10 minutes
Trichloroethylene	100 ppm	200 ppm	300 ppm	5 minutes in any 2 hours

[a]This standard applies to the industry segments exempt from the 1 ppm 8-hour TWA and 5 ppm STEL of the benzene standard at 1910.1028. This standard also applies to any industry for which 1910.1028 is stayed or otherwise not in effect.
[b]This standard applies to any industry for which 1910.1048 is stayed or otherwise not in effect.

Table 3

Substances subjected to particular regulations (29 CFR Sections) with limits dependent upon the method and equipment employed.

Substance	29 CFR Section	Substance	29 CFR Section
Acrylonitrile	1910.1045	Ethylene oxide	1910.1047
Arsenic, inorganic compounds (as As)	1910.1018	Ethylenimine	1910.1012
Asbestos	1910.1001	N-2-Fluorenylacetamide	1910.1014
Benzene	1910.1028[a]	Formaldehyde	1910.1048
Benzidine	1910.1010	1-Naphthylamine	1910.1004
p-Biphenylamine	1910.1011	2-Naphthylamine	1910.1009
Chloromethyl methyl ether	1910.1006	p-Nitrobiphenyl	1910.1003
1,2-Dibromo-3-chloropropane	1910.1044	N-Nitrosodimethylamine	1910.1016
3,3'-Dichlorobenzidine	1910.1007	β-Propiolactone	1910.1013
p-Dimethylaminoazobenzene	1910.1015	Vinyl chloride	1910.1017

[a]The final benzene standard in 1910.1028 applies to all occupational exposures to benzene except some subsegments of industry where exposures are consistently under the action level (i.e., distribution and sale of fuels, sealed containers and pipelines, coke production, oil and gas drilling and production, natural gas processing, and the percentage exclusion for liquid mixtures); for the excepted subsequents, the benzene limits in Table 2 apply.

Boiling Points of Solvents

The following tables list boiling points which have been extracted from pertinent monographs of *The Merck Index*. Only those solvents whose boiling points are reported at 1 atmospheric pressure (760 mm Hg) have been included. Table 1 lists compounds arranged in ascending order by boiling point (°C). Table 2 lists the solvents alphabetically.

Table 1

Solvent	Boiling Point	Solvent	Boiling Point
Trimethylamine	2.87	Ethyl acetate	77
Cryofluorane	4.1	Acrylonitrile	77.3
Dimethylamine	7	n-Butylamine	78
Neopentane	9.5	n-Butyl chloride	78.5
Ethylene oxide	10.7	Ethyl alcohol	78.5
Ethyl chloride	12.3	Methyl ethyl ketone	79.6
Acetaldehyde	21	Benzene	80.1
Trichlorofluoromethane	23.7	Cyclohexane	80.7
1-Pentene	30.1	Propyl formate	81-82
Furan	31.36	Acetonitrile	81.6
Methyl formate	31.5	1,2-Dimethoxyethane	82-83
Vinylidene chloride	31.7	tert-Butyl alcohol	82.41
Isopropylamine	33-34	Isopropyl alcohol	82.5
Isoprene	34.067	Cyclohexene	83
Propylene oxide	34.23	Ethylene dichloride	83-84
Ethyl ether	34.6	Diisopropylamine	84
Ethanethiol	34.7-35.04	Thiophene	84.4
Isopropyl chloride	35-36	Fluorobenzene	84.73
2-Pentene (*trans*-form)	35.85	Trichloroethylene	86.7
Pentane	36.1	Propylidene chloride	87
Methyl sulfide	36.2	Tetrahydropyran	88
2-Pentene (*cis*-form)	37.0	Pyrrolidine	88.5-89
Amylene	37.5-38.5	Isopropyl acetate	89
Ethyl bromide	38.2	Isopropyl iodide	89-90
Methylene chloride	39.75	Triethylamine	89-90
Methylal	41.6	Propyl ether	89-91
Methyl iodide	42.5	Methacrylonitrile	90.3
Allyl chloride	44-45	sec-Butyl bromide (*dl*-form)	91.2
tert-Butylamine	44-46	Isobutyl bromide	91.5
Propyl chloride	46-47	Ethyl sulfide	92
Carbon disulfide	46.5	Ethyl chloroformate	95
Propylamine	48-49	Propylene dichloride	95-96
Isoflurane	48.5	Allyl alcohol	96-97
Propionaldehyde	49	Methylene bromide	97
Cyclopentane	49.3	Propionitrile	97.2
tert-Butyl chloride	51.0	n-Propyl alcohol	97.2
Acrolein	52.5	Isobutyl formate	98
Ethyl formate	53-54	n-Butyl mercaptan	98.4
Allylamine	55-58	n-Heptane	98.4
Diethylamine	55.5	Ethyl propionate	99
Ethylenimine	56-57	Isooctane	99.3
Acetone	56.5	Ethyl acrylate	99.4
Methyl acetate	56.9	sec-Butyl alcohol (*dl*-form)	99.5
Ethylidene chloride	57.3	Formic acid	100.5
Isopropyl bromide	59-60	Dioxane	101.1
Acetylene dichloride (*cis*-form)	60	Nitromethane	101.2
Chloroform	61-62	n-Butyl bromide	101.3
sec-Butylamine (all isomers)	63	Diethyl ketone	101.5
3-Chloro-1-butene (*dl*-form)	63.9-64.2	Propyl acetate	101.6
Isobutyraldehyde	64	Methyl propyl ketone	102
Methanol	64.7	Propyl iodide	102-103
Tetrahydrofuran	66	Acetal	102.7
Trimethyl borate	67-68	Benzotrifluoride	103.46
sec-Butyl chloride (*dl*-form)	68	n-Amylamine	104
Isobutylamine	68-69	Crotonaldehyde	104.0
Isobutyl chloride	68-69	Phosphorus oxychloride	105.8
Isopropyl ether	68-69	Piperidine	106
1-Chloro-2-methylpropene	68.1	Amyl chloride	107.8
n-Hexane	69	Isobutyl alcohol	108
Propyl bromide	71	n-Dipropylamine	110
Ethyl iodide	72	Toluene	110.6
Vinyl acetate	72-73	sec-Butyl acetate (*dl*-form)	112-113
Trifluoroacetic acid	72.4	1,1,2-Trichloroethane	113-114
tert-Butyl bromide	73.3	o-Fluorotoluene	114
1,1,1-Trichloroethane	74.1	Neopentyl alcohol	114
Butyraldehyde	74.8	Nitroethane	114-115
Carbon tetrachloride	76.7	Propargyl alcohol	114-115

Table 1 (Continued)

Solvent	Boiling Point	Solvent	Boiling Point
Pyridine	115-116	2,2,2-Trichloroethanol	151-153
N-Methylmorpholine	115.4	2-Heptanone	151.5
3-Pentanol	115.6	Cumene	152-153
m-Fluorotoluene	116	Isobutyric acid	152-155
p-Fluorotoluene	116	N,N-Dimethylformamide	153
sec-Butyl acetate (d-form)	116-117	Ethyl lactate	154
Ethylenediamine	116-117	α-Pinene (all isomers)	155-156
n-Butyl alcohol	117-118	Anisole	155.5
Isopropylacetone	117-118	Cyclohexanone	155.6
Butyronitrile	117.5	2-Ethoxyethyl acetate	156
Epichlorohydrin	117.9	Bromobenzene	156.2
Acetic acid glacial	118	1-Hexanol	157
Isobutyl acetate	118	Ethyl amyl ketone	157-162
Allyl cyanide	119	2-Heptanol (dl-form)	158-160
Methyl isothiocyanate	119	o-Chlorotoluene	158.97
Propylenediamine	119-120	n-Dibutylamine	159-160
2-Pentanol	119.3	n-Propylbenzene	159.2
Chloroacetone	119.7	Cyclohexanol	161
Isobutyl iodide	120	Pentachloroethane	161-162
Ethyl butyrate	120-121	m-Chlorotoluene	161.75
Isoamyl bromide	120-121	Furfural	161.8
2-Nitropropane	120.3	p-Chlorotoluene	161.99
Caprylene	121	Diglyme	162
Tetrachloroethylene	121	β-Pinene (l-form)	162-163
Crotyl alcohol (trans-form)	121.2	Methacrylic acid	163
Crotyl alcohol (cis-form)	123.6	Methyl oxalate	163-164
n-Butyl acetate	125-126	N,N-Dimethylacetamide	163-165
Octane	125.6	Butyric acid	163.5
Bromine trifluoride	125.75	β-Pinene (d-form)	164-166
Ethyl carbonate	126	Mesitylene	164.7
sec-Propylene chlorohydrin	126-127	2-Amino-2-methyl-1-propanol	165
Methyl butyl ketone	127	n-Butyl n-butyrate	165
2-Methyl-1-butanol	128	Pyruvic acid	165 (dec)
α-Picoline	128-129	β-Pinene (dl-form)	165-166
Ethylene chlorohydrin	128-130	Propionic anhydride	167.0
Morpholine	128.9	Diacetone alcohol	167.9
1H-Pyrrole	129.8	Isocrotonic acid	168-169
Mesityl oxide	130	Thiophenol	168.3
Ethyl isothiocyanate	130-132	tert-Butylbenzene	168.5
Methyl chloroacetate	130-132	Methyl acetoacetate	169-171 (dec)
Methyl thiocyanate	130-133	Furfuryl alcohol	170
n-Butyl iodide	130.4	Isobutylbenzene	170.5
Cyclopentanone	130.6	Ethanolamine	170.8
Chlorobenzene	131-132	2,4,6-Trimethylpyridine	171
Ethylene dibromide	131-132	Butyl Cellosolve®	171-172
1-Nitropropane	131.6	Phenetole	171-173
Isopentyl alcohol	132.0	Isoamyl ether	172
Cyclohexylamine	134.5	Hexyl methyl ketone	172-173
2-Ethoxyethanol	135	m-Dichlorobenzene	173
Ethyl isovalerate	135	sec-Butylbenzene	173.5
Ethylbenzene	136.25	p-Dichlorobenzene	174.12
p-Xylene	137-138	1,3-Dichloro-2-propanol	174.3
1-Pentanol	137.5	Isovaleric acid	175-177
Acetic anhydride	139	Limonene (d-form)	175.5-176
m-Xylene	139.3	Limonene (dl,l-forms)	175.5-176.5
Propylene dibromide	140-142	Eucalyptol	176-177
2,3-Dibromopropene	140-143	Tetramethylurea	176.5
Acetylacetone	140.5	Methyl carbamate	177
Acrylic acid	141.0	p-Cymene	177.10
Propionic acid	141.1	sym-Dichloroethyl ether	178
Isoamyl acetate	142	Tetrahydrofurfuryl alcohol	178
n-Butyl ether	142-143	2-Octanol (dl-form)	178.5
Methyl Cellosolve® acetate	143	Benzaldehyde	179
2,6-Lutidine	144	Benzyl chloride	179
o-Xylene	144	Isoamyl butyrate	179
Methyl lactate	144-145	2,3-Butylene glycol (l-form)	179-182
Ethyl chloroacetate	144-146	o-Dichlorobenzene	180.5
γ-Picoline	145	Ethyl acetoacetate	180.8
Styrene	145-146	Methylene iodide	181
Tetrachloroethane	146.5	n-Butyl sulfide	182
Isobutyl isobutyrate	147	Phenol	182
Allyl isothiocyanate	148-154	Urethan	182-184
Bromoform	149-150	n-Butylbenzene	183.1

Table 1 (Continued)

Solvent	Boiling Point	Solvent	Boiling Point
2-Ethyl-1-hexanol	184-185	Ethyl benzoate	211-213
Aniline	184-186	2,5-Xylenol	211.132
Benzylamine	185	Benzyl acetate	213
Crotonic acid	185.0	1,2,4-Trichlorobenzene	213
Diethyl oxalate	185.7	Triethyl phosphate	215-216
Benzyl ethyl ether	186	Triglyme	216
Valeric acid, normal	186-187	Tributylamine	216-217
n-Amyl ether	186.75	2,3-Xylenol	216.87
Hexachloroethane	186.8	Naphthalene	217.9
Propylene glycol (l-form)	187-189	Carbitol® acetate	218.5
Decalin® (trans-form)	187.25	1,2,3-Tribromopropane	220
Acetonylacetone	188	Methyl salicylate	220-224
Diethyl Carbitol®	188	Benzotrichloride	220.8
Iodobenzene	188-189	3,5-Xylenol	221.692
Propylene glycol (dl-form)	188.2	Acetamide	222
Dimethyl sulfoxide	189	Ethyl maleate	223
Ethylene glycol diacetate	190-191	3,4-Xylenol	226.947
Benzonitrile	190.7	Ethylene cyanohydrin	228 (dec)
o-Cresol	191-192	Butyl Carbitol®	230.4
Isoamyl isovalerate	191-194	m-Nitrotoluene	231.9
Glycerol formal	191-195	Benzyl cyanide	233.5
Succinyl chloride	192-193	Quinoline	237.7
N,N-Dimethylaniline	192-194	Dimethyl sulfone	238
Methyl Carbitol®	193	1,5-Pentanediol	239
1-Octanol	194-195	Caprylic acid	239.7
Phenyl acetate	195-196	Isoquinoline	243.25
Decalin® (cis-form)	195.7	Diethylene glycol	244-245
Carbitol®	196	Ethohexadiol	244.2
Salicylaldehyde	196-197	2-Pyrrolidone	245
Trichloroacetic acid	196-197	2-Phenoxyethanol	245.2
Ethylene glycol	197.6	Cinnamaldehyde	246.0 (dec)
Hexylene glycol	198	n-Butyl benzoate	250
Ethyl malonate	198-199	Eugenol	255
Methyl benzoate	198-200	Triacetin	258-260
Octyl acetate	199	Diacetin	259
Butyric anhydride	199.4-201.4	Phenyl ether	259
p-Toluidine	200-201	1-Chloronaphthalene	259.3
o-Toluidine	200-202	Isoamyl benzoate	260-262
Pine oil	200-220	2,5-Tetrahydrofurandimethanol	265
2,6-Xylenol	201.03	Succinonitrile	265-267
p-Cresol	201.8	Triethylenetetramine	266-267
Acetophenone	202	Diethanolamine	268.8
m-Cresol	202	Ethyl cinnamate	271
Benzyl formate	203	Isoamyl salicylate	274-278
m-Toluidine	203-204	Tetraglyme	275.3
Butyrolactone	204	o-Nitroanisole	277
Camphor	204	2,4-Dimethylsulfolane	280-281 (dec)
Benzyl alcohol	204.7	1-Bromonaphthalene	281.1
Benzal chloride	205	Dimethyl phthalate	283.7
n-Caproic acid	205	Sulfolane	285
Veratrole	206-207	Triethylene glycol	285
Ethyl cyanoacetate	206.0	Glycerol	290.0 (dec)
Tetralin®	207.2	Diethyl phthalate	295
1,3-Butylene glycol	207.5	Benzyl ether	295-298 (dec)
Bromoacetic acid	208	Phenyl carbonate	302-306
o-Chloroaniline	208.84	Benzyl benzoate	323-324
Nitrobenzene	210-211	Triethanolamine	335.4
Trimethylene glycol	210-212	n-Butyl phthalate	340
Formamide	210.5 (dec)	Butyl stearate	343
2,4-Xylenol	210.931	Methyl abietate	360-365 (dec)

Table 2

Solvent	Boiling Point	Solvent	Boiling Point
Acetal	102.7	*n*-Butyl phthalate	340
Acetaldehyde	21	Butyl stearate	343
Acetamide	222	*n*-Butyl sulfide	182
Acetic acid glacial	118	Butyraldehyde	74.8
Acetic anhydride	139	Butyric acid	163.5
Acetone	56.5	Butyric anhydride	199.4-201.4
Acetonitrile	81.6	Butyrolactone	204
Acetonylacetone	188	Butyronitrile	117.5
Acetophenone	202	Camphor	204
Acetylacetone	140.5	*n*-Caproic acid	205
Acetylene dichloride (*cis*-form)	60	Caprylene	121
Acrolein	52.5	Caprylic acid	239.7
Acrylic acid	141.0	Carbitol®	196
Acrylonitrile	77.3	Carbitol® acetate	218.5
Allyl alcohol	96-97	Carbon disulfide	46.5
Allylamine	55-58	Carbon tetrachloride	76.7
Allyl chloride	44-45	Chloroacetone	119.7
Allyl cyanide	119	*o*-Chloroaniline	208.84
Allyl isothiocyanate	148-154	Chlorobenzene	131-132
2-Amino-2-methyl-1-propanol	165	3-Chloro-1-butene (*dl*-form)	63.9-64.2
n-Amylamine	104	Chloroform	61-62
Amyl chloride	107.8	1-Chloro-2-methylpropene	68.1
Amylene	37.5-38.5	1-Chloronaphthalene	259.3
n-Amyl ether	186.75	*m*-Chlorotoluene	161.75
Aniline	184-186	*o*-Chlorotoluene	158.97
Anisole	155.5	*p*-Chlorotoluene	161.99
Benzal chloride	205	Cinnamaldehyde	246.0 (dec)
Benzaldehyde	179	*m*-Cresol	202
Benzene	80.1	*o*-Cresol	191-192
Benzonitrile	190.7	*p*-Cresol	201.8
Benzotrichloride	220.8	Crotonaldehyde	104.0
Benzotrifluoride	103.46	Crotonic acid	185.0
Benzyl acetate	213	Crotyl alcohol (*cis*-form)	123.6
Benzyl alcohol	204.7	Crotyl alcohol (*trans*-form)	121.2
Benzylamine	185	Cryofluorane	4.1
Benzyl benzoate	323-324	Cumene	152-153
Benzyl chloride	179	Cyclohexane	80.7
Benzyl cyanide	233.5	Cyclohexanol	161
Benzyl ether	295-298 (dec)	Cyclohexanone	155.6
Benzyl ethyl ether	186	Cyclohexene	83
Benzyl formate	203	Cyclohexylamine	134.5
Bromine trifluoride	125.75	Cyclopentane	49.3
Bromoacetic acid	208	Cyclopentanone	130.6
Bromobenzene	156.2	*p*-Cymene	177.10
Bromoform	149-150	Decalin® (*cis*-form)	195.7
1-Bromonaphthalene	281.1	Decalin® (*trans*-form)	187.25
n-Butyl acetate	125-126	Diacetin	259
sec-Butyl acetate (*d*-form)	116-117	Diacetone alcohol	167.9
sec-Butyl acetate (*dl*-form)	112-113	2,3-Dibromopropene	140-143
n-Butyl alcohol	117-118	*n*-Dibutylamine	159-160
sec-Butyl alcohol (*dl*-form)	99.5	*m*-Dichlorobenzene	173
tert-Butyl alcohol	82.41	*o*-Dichlorobenzene	180.5
n-Butylamine	78	*p*-Dichlorobenzene	174.12
sec-Butylamine (all isomers)	63	*sym*-Dichloroethyl ether	178
tert-Butylamine	44-46	1,3-Dichloro-2-propanol	174.3
n-Butylbenzene	183.1	Diethanolamine	268.8
sec-Butylbenzene	173.5	Diethylamine	55.5
tert-Butylbenzene	168.5	Diethyl Carbitol®	188
n-Butyl benzoate	250	Diethylene glycol	244-245
n-Butyl bromide	101.3	Diethyl ketone	101.5
sec-Butyl bromide (*dl*-form)	91.2	Diethyl oxalate	185.7
tert-Butyl bromide	73.3	Diethyl phthalate	295
n-Butyl *n*-butyrate	165	Diglyme	162
Butyl Carbitol®	230.4	Diisopropylamine	84
Butyl Cellosolve®	171-172	1,2-Dimethoxyethane	82-83
n-Butyl chloride	78.5	*N,N*-Dimethylacetamide	163-165
sec-Butyl chloride (*dl*-form)	68	Dimethylamine	7
tert-Butyl chloride	51.0	*N,N*-Dimethylaniline	192-194
1,3-Butylene glycol	207.5	*N,N*-Dimethylformamide	153
2,3-Butylene glycol (*l*-form)	179-182	Dimethyl phthalate	283.7
n-Butyl ether	142-143	2,4-Dimethylsulfolane	280-281 (dec)
n-Butyl iodide	130.4	Dimethyl sulfone	238
n-Butyl mercaptan	98.4	Dimethyl sulfoxide	189

Table 2 (Continued)

Solvent	Boiling Point	Solvent	Boiling Point
Dioxane	101.1	Isobutyl acetate	118
n-Dipropylamine	110	Isobutyl alcohol	108
Epichlorohydrin	117.9	Isobutylamine	68-69
Ethanethiol	34.7-35.04	Isobutylbenzene	170.5
Ethanolamine	170.8	Isobutyl bromide	91.5
Ethohexadiol	244.2	Isobutyl chloride	68-69
2-Ethoxyethanol	135	Isobutyl formate	98
2-Ethoxyethyl acetate	156	Isobutyl iodide	120
Ethyl acetate	77	Isobutyl isobutyrate	147
Ethyl acetoacetate	180.8	Isobutyraldehyde	64
Ethyl acrylate	99.4	Isobutyric acid	152-155
Ethyl alcohol	78.5	Isocrotonic acid	168-169
Ethyl amyl ketone	157-162	Isoflurane	48.5
Ethylbenzene	136.25	Isooctane	99.3
Ethyl benzoate	211-213	Isopentyl alcohol	132.0
Ethyl bromide	38.2	Isoprene	34.067
Ethyl butyrate	120-121	Isopropyl acetate	89
Ethyl carbonate	126	Isopropylacetone	117-118
Ethyl chloride	12.3	Isopropyl alcohol	82.5
Ethyl chloroacetate	144-146	Isopropylamine	33-34
Ethyl chloroformate	95	Isopropyl bromide	59-60
Ethyl cinnamate	271	Isopropyl chloride	35-36
Ethyl cyanoacetate	206.0	Isopropyl ether	68-69
Ethylene chlorohydrin	128-130	Isopropyl iodide	89-90
Ethylene cyanohydrin	228 (dec)	Isoquinoline	243.25
Ethylenediamine	116-117	Isovaleric acid	175-177
Ethylene dibromide	131-132	Limonene (d-form)	175.5-176
Ethylene dichloride	83-84	Limonene (dl,l-forms)	175.5-176.5
Ethylene glycol	197.6	2,6-Lutidine	144
Ethylene glycol diacetate	190-191	Mesitylene	164.7
Ethylene oxide	10.7	Mesityl oxide	130
Ethylenimine	56-57	Methacrylic acid	163
Ethyl ether	34.6	Methacrylonitrile	90.3
Ethyl formate	53-54	Methanol	64.7
2-Ethyl-1-hexanol	184-185	Methyl abietate	360-365 (dec)
Ethylidene chloride	57.3	Methyl acetate	56.9
Ethyl iodide	72	Methyl acetoacetate	169-171 (dec)
Ethyl isothiocyanate	130-132	Methylal	41.6
Ethyl isovalerate	135	Methyl benzoate	198-200
Ethyl lactate	154	2-Methyl-1-butanol	128
Ethyl maleate	223	Methyl butyl ketone	127
Ethyl malonate	198-199	Methyl carbamate	177
Ethyl propionate	99	Methyl Carbitol®	193
Ethyl sulfide	92	Methyl Cellosolve® acetate	143
Eucalyptol	176-177	Methyl chloroacetate	130-132
Eugenol	255	Methylene bromide	97
Fluorobenzene	84.73	Methylene chloride	39.75
o-Fluorotoluene	114	Methylene iodide	181
m-Fluorotoluene	116	Methyl ethyl ketone	79.6
p-Fluorotoluene	116	Methyl formate	31.5
Formamide	210.5 (dec)	Methyl iodide	42.5
Formic acid	100.5	Methyl isothiocyanate	119
Furan	31.36	Methyl lactate	144-145
Furfural	161.8	N-Methylmorpholine	115.4
Furfuryl alcohol	170	Methyl oxalate	163-164
Glycerol	290.0 (dec)	Methyl propyl ketone	102
Glycerol formal	191-195	Methyl salicylate	220-224
n-Heptane	98.4	Methyl sulfide	36.2
2-Heptanol (dl-form)	158-160	Methyl thiocyanate	130-133
2-Heptanone	151.5	Morpholine	128.9
Hexachloroethane	186.8	Naphthalene	217.9
n-Hexane	69	Neopentane	9.5
1-Hexanol	157	Neopentyl alcohol	114
Hexylene glycol	198	o-Nitroanisole	277
Hexyl methyl ketone	172-173	Nitrobenzene	210-211
Iodobenzene	188-189	Nitroethane	114-115
Isoamyl acetate	142	Nitromethane	101.2
Isoamyl benzoate	260-262	1-Nitropropane	131.6
Isoamyl bromide	120-121	2-Nitropropane	120.3
Isoamyl butyrate	179	m-Nitrotoluene	231.9
Isoamyl ether	172	Octane	125.6
Isoamyl isovalerate	191-194	1-Octanol	194-195
Isoamyl salicylate	274-278	2-Octanol (dl-form)	178.5

Table 2 (Continued)

Solvent	Boiling Point	Solvent	Boiling Point
Octyl acetate	199	Styrene	145-146
Pentachloroethane	161-162	Succinonitrile	265-267
Pentane	36.1	Succinyl chloride	192-193
1,5-Pentanediol	239	Sulfolane	285
1-Pentanol	137.5	Tetrachloroethane	146.5
2-Pentanol	119.3	Tetrachloroethylene	121
3-Pentanol	115.6	Tetraglyme	275.3
1-Pentene	30.1	Tetrahydrofuran	66
2-Pentene (cis-form)	37.0	2,5-Tetrahydrofurandimethanol	265
2-Pentene (trans-form)	35.85	Tetrahydrofurfuryl alcohol	178
Phenetole	171-173	Tetrahydropyran	88
Phenol	182	Tetralin®	207.2
2-Phenoxyethanol	245.2	Tetramethylurea	176.5
Phenyl acetate	195-196	Thiophene	84.4
Phenyl carbonate	302-306	Thiophenol	168.3
Phenyl ether	259	Toluene	110.6
Phosphorus oxychloride	105.8	m-Toluidine	203-204
α-Picoline	128-129	o-Toluidine	200-202
γ-Picoline	145	p-Toluidine	200-201
α-Pinene (all isomers)	155-156	Triacetin	258-260
β-Pinene (d-form)	164-166	1,2,3-Tribromopropane	220
β-Pinene (dl-form)	165-166	Tributylamine	216-217
β-Pinene (l-form)	162-163	Trichloroacetic acid	196-197
Pine oil	200-220	1,2,4-Trichlorobenzene	213
Piperidine	106	1,1,1-Trichloroethane	74.1
Propargyl alcohol	114-115	1,1,2-Trichloroethane	113-114
Propionaldehyde	49	2,2,2-Trichloroethanol	151-153
Propionic acid	141.1	Trichloroethylene	86.7
Propionic anhydride	167.0	Trichlorofluoromethane	23.7
Propionitrile	97.2	Triethanolamine	335.4
Propyl acetate	101.6	Triethylamine	89-90
n-Propyl alcohol	97.2	Triethylene glycol	285
Propylamine	48-49	Triethylenetetramine	266-267
n-Propylbenzene	159.2	Triethyl phosphate	215-216
Propyl bromide	71	Trifluoracetic acid	72.4
Propyl chloride	46-47	Triglyme	216
sec-Propylene chlorohydrin	126-127	Trimethylamine	2.87
Propylenediamine	119-120	Trimethyl borate	67-68
Propylene dibromide	140-142	Trimethylene glycol	210-212
Propylene dichloride	95-96	2,4,6-Trimethylpyridine	171
Propylene glycol (dl-form)	188.2	Urethan	182-184
Propylene glycol (l-form)	187-189	Valeric acid, normal	186-187
Propylene oxide	34.23	Veratrole	206-207
Propyl ether	89-91	Vinyl acetate	72-73
Propyl formate	81-82	Vinylidene chloride	31.7
Propylidene chloride	87	p-Xylene	137-138
Propyl iodide	102-103	m-Xylene	139.3
Pyridine	115-116	o-Xylene	144
1H-Pyrrole	129.8	2,5-Xylenol	211.132
Pyrrolidine	88.5-89	3,5-Xylenol	221.692
2-Pyrrolidone	245	3,4-Xylenol	226.947
Pyruvic acid	165 (dec)	2,6-Xylenol	201.03
Quinoline	237.7	2,3-Xylenol	216.87
Salicylaldehyde	196-197	2,4-Xylenol	210.931

Molar Volumes (at 25°) of Frequently Used Laboratory Chemicals

Chemical	Molar Volume mol. wt./density, ml	Chemical	Molar Volume mol. wt./density, ml
Acetal	143.82	Caprylic acid	159.07
Acetaldehyde	56.62 (20°)	Carbitol	136.34
Acetic acid	57.54	Carbitol acetate	174.54 (20°)
Acetic anhydride	95.49 (30°)	Carbon disulfide	61.00
Acetone	74.05	Carbon tetrachloride	97.09
Acetonitrile	52.86	o-Chloroaniline	105.63
Acetonylacetone	117.23 (20°)	Chlorobenzene	101.74 (20°)
Acetophenone	117.36	Chloroform	80.67
Acetylacetone	103.00	l-Chloronaphthalene	136.22 (20°)
cis-Acetylene dichloride	75.52 (20°)	β-Chloropropionitrile	78.79
trans-Acetylene dichloride	77.26 (20°)	Cineole	167.81
Acrolein	66.83 (20°)	Citraconic anhydride	89.89 (16°)
Acrylic acid	68.56 (20°)	m-Cresol	104.97
Acrylonitrile	66.30	trans-Crotonaldehyde	82.31 (20°)
Alcohol, anhydrous	58.69	cis-Crotyl alcohol	84.44 (20°)
Allyl acetate	108.94 (28°)	trans-Crotyl alcohol	85.29
Allyl alcohol	68.97 (30°)	Cumene	140.17
Allylamine	75.37	Cyclohexane	108.74
Allyl chloride	82.78 (30°)	Cyclohexanone	104.68 (30°)
Amyl chloride	121.56	Cyclohexene	101.91
n-Amyl ether	203.19	Cyclohexylamine	115.62 (30°)
Aniline	91.53	Cyclohexylbenzene	170.73
Anisole	109.31	Cyclopentane	94.71
Benzaldehyde	101.71	Cyclopentanone	88.67 (20°)
Benzene	89.41	p-Cymene	157.30
Benzonitrile	103.06	Decalin	157.30
Benzyl acetate	142.82	cis-Decalin	154.84
Benzyl alcohol	103.85	trans-Decalin	159.67
Benzyl benzoate	190.86	Decane	195.91
Benzyl cyanide	115.71	1-Decene	190.35
Benzyl ether	190.13 (20°)	Diacetone alcohol	124.34
Benzyl ethyl ether	144.18	Diacetyl	87.78 (18.5°)
Bicyclohexyl	188.45	1,2-Dibromotetrafluoroethane	120.13
Bis(2-ethylhexyl) phthalate	398.95	n-Dibutylamine	170.58
Bis(2-methoxyethyl) ether	142.14	m-Dichlorobenzene	114.60
Bromobenzene	105.51	o-Dichlorobenzene	112.82
Bromoform	87.89	sym-Dichloroethyl ether	117.90
1-Bromonaphthalene	140.06	Diethanolamine	96.47 (30°)
cis-2-Butene-1,4-diol	82.04 (20°)	Diethylamine	104.25
n-Butyl acetate	132.54	Diethylene glycol	95.06 (20°)
sec-Butyl acetate	133.21 (20°)	Diethyl ketone	106.41
n-Butyl alcohol	91.97	Diethyl oxalate	136.98 (30°)
sec-Butyl alcohol	92.35	Diisopropylamine	142.53
n-Butylamine	99.56	1,2-Dimethoxyethane	104.54
sec-Butylamine	101.57	N,N-Dimethylacetamide	93.02
tert-Butylamine	105.88	2,2-Dimethylbutane	133.71
n-Butylbenzene	156.78	2,3-Dimethylbutane	131.17
sec-Butylbenzene	156.44	N,N-Dimethylformamide	77.43
tert-Butylbenzene	155.64	2,3-Dimethylpentane	145.04
Butyl borate	269.66	2,4-Dimethylpentane	149.94
Butyl Cellosolve	132.10 (27°)	Dimethyl sulfoxide	71.30
n-Butyl chloride	105.08	p-Dioxane	85.70
(±)-sec-Butyl chloride	106.01 (20°)	n-Dipropylamine	138.07
tert-Butyl chloride	110.72	Dodecane	228.58
1,3-Butylene glycol	89.65 (20°)	Epichlorohydrin	78.77
n-Butyl ether	170.44	1,2-Epoxybutane	86.91 (20°)
Butyl ethyl ether	137.19	Ethanolamine	60.38
Butyl formate	114.54 (20°)	2-Ethoxyethanol	97.41
Butyl maleate	230.43	2-Ethoxyethyl acetate	135.83 (20°)
n-Butyl mercaptan	107.79	Ethyl acetate	98.49
Butyl phosphate	272.87	Ethyl acetoacetate	127.43
n-Butyl phthalate	266.98	Ethyl acrylate	108.42 (20°)
Butyl sebacate	337.27	Ethylbenzene	123.08
Butyl vinyl ether	129.63	Ethyl benzoate	144.79 (30°)
Butyraldehyde	90.54	Ethyl bromide	75.13
Butyric acid	92.43	2-Ethyl-l-butanol	123.18
Butyric anhydride	164.45	Ethyl butyrate	132.92
γ-Butyrolactone	76.50	Ethyl carbonate	121.88
Butyronitrile	87.87	Ethyl cinnamate	168.32
n-Caproic acid	125.85	Ethyl cyanoacetate	107.08

Chemical	Molar Volume mol. wt./density, ml	Chemical	Molar Volume mol. wt./density, ml
Ethylcyclohexane	143.15	Methyl acetate	79.84
Ethylene chlorohydrin	66.99 (20°)	Methyl acetoacetate	108.28
Ethylenediamine	67.83 (30°)	Methyl acrylate	90.29 (20°)
Ethylene dibromide	86.63	Methylal	89.79 (30°)
Ethylene dichloride	79.43	Methyl benzoate	126.18 (30°)
Ethylene glycol	55.92	2-Methylbutane	117.40
Ethylene glycol diacetate	132.34 (20°)	2-Methyl-1-butanol	108.16
Ethylenimine	51.77	3-Methyl-2-butanol	108.32
Ethyl ether	104.75	Methyl Carbitol	118.18
Ethyl formate	81.44 (30°)	Methyl carbonate	84.19 (20°)
2-Ethyl-l-hexanol	157.08	Methyl Cellosolve	79.25
Ethylidene chloride	84.73	Methyl Cellosolve acetate	117.56 (20°)
Ethyl iodide	81.05	Methyl cyanoacetate	88.28
Ethyl isovalerate	150.47 (20°)	Methylcyclohexane	128.34
Ethyl lactate	115.01	Methylcyclohexanol	123.92 (24.7°)
Ethyl maleate	161.87	cis-2-Methylcyclohexanol	122.55
Ethyl malonate	153.33 (30°)	trans-2-Methylcyclohexanol	124.00
Ethyl propionate	116.19 (30°)	3-Methylcyclohexanol	124.55 (20°)
Ethyl sulfide	108.50	cis-3-Methylcyclohexanol	124.73 (20°)
Ethyl vinyl ether	95.75 (20°)	trans-3-Methylcyclohexanol	123.93 (20°)
Fluorobenzene	94.86 (30°)	4-Methylcyclohexanol	124.35 (22.5°)
m-Fluorotoluene	110.42 (20°)	cis-4-Methylcyclohexanol	125.08 (21.5°)
o-Fluorotoluene	110.00 (17.3°)	trans-4-Methylcyclohexanol	125.76
p-Fluorotoluene	111.12 (26.9°)	Methylcyclopentane	113.14
Formamide	39.89	Methylene chloride	64.50
Formic acid	37.91	Methylene iodide	80.97
Furan	72.59 (20°)	Methyl ethyl ketone	90.17
Furfural	83.23	N-Methylformamide	59.14
Furfuryl alcohol	86.93 (20°)	Methyl formate	61.64 (20°)
Glycerol	73.02 (20°)	2-Methylhexane	148.58
n-Heptane	147.47	3-Methylhexane	146.71
(±)2-Heptanol	142.86	Methyl iodide	62.67
1-Heptene	141.75	Methyl maleate	125.74
n-Hexadecane	294.09	Methyl methacrylate	106.14 (20°)
Hexafluorobenzene	115.79	Methyl oleate	340.72
Hexametapol	174.29 (20°)	2-Methylpentane	132.89
n-Hexane	131.61	3-Methylpentane	130.61
Hexanenitrile	121.27	2-Methyl-1-pentanol	124.52
1-Hexanol	125.23	4-Methyl-2-pentanol	127.15
1-Hexene	125.90	N-Methylpropionamide	93.63
Hydracrylonitrile	68.32	1-Methyl-2-pyrrolidinone	96.44
Isoamyl cyanide	121.56	Methyl salicylate	129.14
Isoamyl ether	205.22 (28°)	Methyl sulfide	73.77
Isoamyl isovalerate	201.70	Morpholine	87.52
Isobutyl acetate	133.60	Neopentane	123.31
Isobutyl alcohol	92.91	o-Nitroanisole	123.42
Isobutylamine	100.23	Nitrobenzene	102.73
Isobutyl chloride	106.19	Nitroethane	71.86
Isobutyl formate	115.35 (20°)	Nitromethane	53.96
Isobutyl isobutyrate	168.83 (20°)	1-Nitropropane	89.44
Isobutyraldehyde	92.02	2-Nitropropane	90.65
Isobutyric acid	93.50	Nonane	179.69
Isobutyronitrile	90.27	1-Nonene	174.06
Isooctane	166.08	Octane	163.54
Isopentyl acetate	150.26	Octanenitrile	154.64
Isopentyl alcohol	109.22	1-Octanol	158.41
Isopropyl acetate	117.83	1-Octene	157.85
Isopropylacetone	125.82	Octyl acetate	197.60 (20°)
Isopropyl alcohol	76.92	Oleic acid	318.46
Isopropylamine	86.66	Pentachloroethane	121.48 (30°)
Isopropyl bromide	94.18	Pentane	116.11
Isopropyl chloride	91.72	1-Pentanol	108.63
Isopropyl ether	142.28	2-Pentanol	109.45
Isopropyl iodide	100.32	3-Pentanol	108.03
Isovaleric acid	111.37 (30°)	1-Pentene	110.40
Mesitylene	139.58	2-Pentene	107.16 (20°)
Methacrylic acid	84.79 (20°)	cis-2-Pentene	107.83
Methacrylonitrile	83.85 (20°)	trans-2-Pentene	109.06
Methanol	40.73	Pentyl acetate	149.52
Methoxyacetonitrile	74.88 (20°)	tert-Pentyl alcohol	109.50

Chemical	Molar Volume mol. wt./density, ml	Chemical	Molar Volume mol. wt./density, ml
Phenetole	127.19	Tetrahydrofuran	81.09 (20°)
Phenyl ether	159.66 (30°)	Tetrahydrofurfuryl alcohol	97.43 (24°)
2-Pinene	159.55	Tetrahydropyran	98.19
2(10)-Pinene	157.19	Tetrahydrothiophene	88.72
Piperidine	99.92 (30°)	Tetralin	136.83
Propargyl acetate	98.28 (20°)	Tetramethylurea	119.92 (20°)
Propargyl alcohol	59.33	2,2'-Thiodiethanol	103.61
Propionaldehyde	73.41	Thiophene	79.47
Propionic acid	74.98	Thiophenol	102.71
Propionic anhydride	129.41	Toluene	106.86
Propionitrile	70.91	m-Toluidine	107.91 (15°)
Propyl acetate	115.67	o-Toluidine	107.77
n-Propyl alcohol	75.14	1,1,1-Trichloroethane	99.73 (20°)
n-Propylamine	83.01	Trichloroethylene	90.53 (30°)
Propyl benzoate	160.48 (20°)	1,1,2-Trichloro-1,2,2-trifluoroethane	119.84
Propyl bromide	91.43	Triethanolamine	133.25
Propyl chloride	88.75	Triethylamine	139.96
Propylene glycol	73.68	Triethylene glycol	133.67 (20°)
Propylene oxide	70.09 (20°)	Trifluoroacetic acid	77.12
Propyl ether	137.72	Trimethylene glycol	72.47
Propyl formate	98.54 (30°)	2,2,5-Trimethylhexane	182.39
Propyl iodide	97.73	2,2,3-Trimethylpentane	160.42
Pyridine	80.86	Trimethyl phosphate	117.03 (19.5°)
Pyrrole	69.47	Valeric acid	109.29
2-Pyrrolidone	76.88	Valeronitrile	104.57
Quinoline	118.96 (30°)	Veratrole	127.71
Salicylaldehyde	105.97	Vinyl acetate	92.45 (20°)
Styrene	115.57	Vinylidene chloride	82.53
Sulfolane	95.27 (30°)	Water	18.07
sym-Tetrabromoethane	117.06	m-Xylene	123.47
1,1,2,2-Tetrachlorodifluoroethane	123.93	o-Xylene	121.20
sym-Tetrachloroethane	105.79	p-Xylene	123.93
Tetrachloroethylene	103.23 (30°)		

Weight of Liquids

(Applicable to the usual purity of the compounds listed)

Liquid	lbs/gal	kg/l	Liquid	lbs/gal	kg/l
acetal	6.9	0.83	butyl butyrate	7.2	0.86
acetic acid, glacial	8.7	1.04	butyl Carbitol	8.0	0.96
acetic anhydride	9.0	1.08	butyl Cellosolve	7.5	0.90
acetone	6.6	0.79	n-butyl chloride	7.4	0.89
acetonitrile	6.5	0.78	butyl citrate	8.7	1.04
acetophenone	8.6	1.03	1,3-butyleneglycol	8.3	0.99
acetyl chloride	9.2	1.10	n-butyl ether	6.4	0.77
acetylene dichloride	10.7	1.28	butyl ethyl ether	6.3	0.75
acetylene tetrachloride	13.3	1.59	butyl lactate	8.1	0.97
acrylonitrile	6.6	0.79	butyl oxalate	8.3	0.99
alcohol	6.8	0.81	n-butyl phthalate	8.7	1.04
alcohol, anhydrous	6.6	0.79	butyl propionate	7.2	0.86
aldehol	6.9	0.83	butyl stearate	7.2	0.86
aldol	9.2	1.10	n-butyraldehyde	6.8	0.81
allyl alcohol	7.1	0.85	butyric acid	8.0	0.96
allyl bromide	11.6	1.39	butyric anhydride	8.1	0.97
allyl chloride	7.7	0.92	n-caproic acid	7.7	0.92
aluminum subacetate soln	8.7	1.04	caprylic acid	7.6	0.91
ammonia solution, diluted (10%)	8.0	0.96	caraway oil	7.5	0.90
ammonia solution, strong (28%)	7.5	0.90	Carbitol	8.2	0.98
amyl acetate	7.3	0.87	Carbitol acetate	8.5	1.02
n-amyl alcohol	6.8	0.81	carbon disulfide	10.5	1.26
amyl alcohol, secondary	6.8	0.81	carbon tetrachloride	13.3	1.59
amyl alcohol, tertiary	6.7	0.80	cardamom oil	7.8	0.93
amylamine	6.3	0.75	carvacrol	8.1	0.97
amylene hydrate	6.7	0.80	carvone	8.0	0.96
amyl mercaptan	7.1	0.85	castor oil	8.0	0.96
amyl nitrite	7.1	0.85	Cellosolve	7.8	0.93
amyl oleate	7.2	0.86	Cellosolve acetate	8.1	0.97
p-tert-amylphenol	7.7	0.92	chenopodium oil	8.0	0.96
amyl phthalate	8.5	1.02	chlordane	13.4	1.61
amyl propionate	7.3	0.87	o-chloroaniline	10.1	1.21
amyl salicylate	8.8	1.05	chlorobenzene	9.2	1.10
anethole	8.2	0.98	β-chloroethyl acetate	9.6	1.15
aniline	8.5	1.02	chloroethylbenzene	8.7	1.04
anise oil	8.2	0.98	chloroform	12.3	1.47
o-anisidine	9.1	1.09	β-chlorophenetole	9.6	1.15
anisole	8.3	0.99	o-chlorophenol	10.3	1.23
aromatic ammonia spirit	7.5	0.90	chloropicrin	13.8	1.65
balsam canada	8.3	0.99	chlorosulfonic acid	14.7	1.76
bay oil	8.1	0.97	chlorotoluene	9.0	1.08
benzal chloride	10.5	1.26	cinnamon oil	8.8	1.05
benzaldehyde	8.7	1.04	citronellal	7.1	0.85
benzene	7.3	0.87	citronella oil	7.5	0.90
benzotrichloride	11.5	1.38	clove oil	8.7	1.04
benzoyl chloride	10.2	1.22	coconut oil	7.7	0.92
benzyl acetate	8.8	1.05	cod liver oil	7.7	0.92
benzyl alcohol	8.7	1.04	collodion	6.4	0.77
benzyl benzoate	9.3	1.11	coriander oil	7.2	0.86
benzyl Cellosolve	8.9	1.07	corn oil	7.6	0.91
benzyl chloride	9.2	1.10	cottonseed oil	7.6	0.91
benzyl ether	8.7	1.04	creosote	9.0	1.08
bergamot oil	7.3	0.87	cresol	8.6	1.03
bitter almond oil	8.7	1.04	m-cresol	8.6	1.03
bitter orange oil	7.1	0.85	o-cresol	8.7	1.04
bornyl acetate	8.2	0.98	p-cresol	8.6	1.03
bromine	26.0	3.11	crotonaldehyde	7.2	0.86
bromobenzene	12.5	1.50	crotonyl alcohol	7.3	0.87
bromoethylbenzene	11.6	1.39	cubeb oil	7.6	0.91
bromoform	24.0	2.88	cumene	7.2	0.86
o-bromotoluene	11.9	1.43	cyclohexane	6.5	0.78
p-bromotoluene	11.7	1.40	cyclohexanol	8.0	0.96
n-butyl acetate	7.3	0.87	cyclohexanone	7.9	0.95
butyl acetate, secondary	7.2	0.86	cyclohexene	6.7	0.80
butyl acetyl ricinoleate	7.8	0.93	cyclohexyl glycolate	8.6	1.03
n-butyl alcohol	6.8	0.81	cyclopentane	6.2	0.74
butyl alcohol, secondary	6.7	0.80	cyclopentanol	7.9	0.95
butyl alcohol, tertiary	6.6	0.79	cyclopentanone	7.9	0.95
butylamine, tertiary	5.8	0.69	cyclopentene	6.4	0.77
n-butyl bromide	10.6	1.27	m-cymene	7.2	0.86

Liquid	lbs/gal	kg/l	Liquid	lbs/gal	kg/l
o-cymene	7.3	0.87	ethylene glycol	9.3	1.11
p-cymene	7.1	0.85	ethylene glycol diacetate	9.2	1.10
decalin	7.5	0.90	ethylene glycol monacetate	9.2	1.10
n-decyl alcohol	6.9	0.83	ethyl formate	7.7	0.92
diacetone alcohol	7.8	0.93	ethyl iodide	16.1	1.93
diallyl ether	6.7	0.80	ethyl isobutyrate	7.2	0.86
dibutylamine	6.4	0.77	ethyl isopropyl ether	6.2	0.74
dibutyl phthalate	8.7	1.04	ethyl isovalerate	7.2	0.86
dichloroacetic acid	13.1	1.57	ethyl lactate	8.6	1.03
o-dichlorobenzene	10.9	1.31	ethyl maleate	8.9	1.07
sym-dichloroethyl ether	10.2	1.22	ethyl malonate	8.8	1.05
dichloroisopropyl ether	9.3	1.11	ethyl methyl ether	6.0	0.72
diethanolamine	9.1	1.09	ethyl nitrite	7.5	0.90
diethylamine	5.9	0.71	ethyl nitrite spirit	6.9	0.83
2-diethylaminoethanol	7.2	0.86	ethyl oxalate	9.0	1.08
diethylaniline	7.8	0.93	ethylphenylethanolamine	8.7	1.04
diethyl carbinol	6.8	0.81	ethyl phenyl ketone	8.5	1.02
diethyl Carbitol	7.6	0.91	ethyl phthalate	9.4	1.13
diethylene glycol	9.3	1.11	ethyl propionate	7.4	0.89
diethylene glycol monolaurate	8.0	0.96	2-ethyl-3-propylacrolein	7.1	0.85
diethylformamide	7.5	0.90	ethyl n-propyl ether	6.1	0.73
diethyl ketone	6.8	0.81	ethyl propyl ketone	6.8	0.81
diethyl phthalate	9.4	1.13	ethyl salicylate	9.4	1.13
diethyl sulfate	9.8	1.17	ethyl silicate	7.8	0.93
diglycol chlorohydrin	9.7	1.16	ethylsulfuric acid	11.0	1.32
diglycol laurate	8.1	0.97	ethyl trichloroacetate	11.5	1.38
diglycol oleate	7.7	0.92	ethyl urethan	8.2	0.98
diisobutyl ketone	6.7	0.80	eucalyptol	7.7	0.92
diisopropanolamine	8.3	0.90	eucalyptus oil	7.6	0.91
diisopropylamine	6.0	0.72	eugenol	8.9	1.07
dimercaprol	10.3	1.23	expressed almond oil	7.6	0.91
dimethoxytetraglycol	8.4	1.01	fennel oil	8.0	0.96
dimethylamine	5.7	0.68	ferric chloride soln	11.0	1.32
dimethylaniline	8.0	0.96	ferric subsulfate soln	12.9	1.55
dimethylformamide	7.9	0.95	ferrous iodide syrup	11.4	1.37
dimethyl furan	7.4	0.89	formaldehyde soln	9.0	1.08
dimethyl glyoxal	8.3	0.99	formamide	9.5	1.14
dimethyl phthalate	9.9	1.19	formic acid, 90%	10.0	1.20
dimethyl sulfate	11.1	1.33	formic acid, 25%	8.8	1.05
dioctylamine	6.7	0.80	furan	7.8	0.93
1,4-dioxane	8.6	1.03	furfural	9.7	1.16
dioxolane	8.9	1.07	furfuryl alcohol	9.4	1.13
diphenylamine	9.7	1.16	geranial	7.4	0.89
dipropylamine	6.1	0.73	geraniol	7.4	0.89
dipropylene glycol	8.6	1.03	glycerin	10.4	1.25
dipropyl ketone	6.8	0.81	glycerophosphoric acid	13.2	1.58
dwarf pine needle oil	7.2	0.66	glyceryl laurate	8.2	0.98
epichlorohydrin	9.9	1.19	halibut liver oil	7.7	0.92
ethanolamine	8.5	1.02	heptadecanol	7.1	0.85
ether (at 20°)	6.0	0.72	2,4-heptadiene	6.1	0.73
ethohexadiol	7.8	0.93	n-heptane	5.7	0.68
ethyl acetate	7.5	0.90	heptyl ether	6.8	0.81
ethyl acetoacetate	8.6	1.03	1,5-hexadiene	5.7	0.68
ethylamine	5.9	0.71	2,4-hexadiene	5.9	0.71
m-ethylaniline	8.0	0.96	hexaethyltetraphosphate	10.8	1.29
ethyl benzoate	8.8	1.05	n-hexaldehyde	6.9	0.83
ethyl bromide	11.9	1.43	n-hexyl alcohol	6.8	0.81
2-ethylbutyl alcohol	6.9	0.83	n-hexyl methyl ketone	6.8	0.81
ethyl butyl ketone	6.8	0.81	hydrazine	8.4	1.01
2-ethylbutyraldehyde	6.8	0.81	hydriodic acid, 47%	12.5	1.50
ethyl n-butyrate	7.3	0.87	hydrobromic acid, 40%	11.5	1.38
2-ethylbutyric acid	7.7	0.92	hydrochloric acid, 36%	9.8	1.17
ethyl caproate	7.5	0.90	hydrocyanic acid, 97%	5.8	0.69
ethyl carbonate	8.1	0.97	hydrofluoric acid, 55%	9.8	1.17
ethyl chloroacetate	9.7	1.16	hydrofluosilicic acid, 30%	10.6	1.27
ethyl chloroformate	9.5	1.14	hydrogen peroxide, anhyd	12.2	1.46
ethyl cinnamate	8.8	1.05	hydrogen peroxide, 30%	9.2	1.10
ethyl citrate	9.5	1.14	hydroxybutyraldehyde	9.2	1.10
ethylene chlorohydrin	10.1	1.21	hypophosphorous acid, 30%	9.4	1.13
ethylenediamine, anhyd	7.5	0.90	indalone	9.0	1.08
ethylene dibromide	18.2	2.18	indan	8.0	0.96
ethylene dichloride	10.5	1.26	β-ionone	7.9	0.95

Liquid	lbs/gal	kg/l	Liquid	lbs/gal	kg/l
isoamyl acetate	7.3	0.87	methyl isobutyrate	7.4	0.89
isoamyl alcohol	6.8	0.81	methyl lactate	9.0	1.08
isoamyl benzoate	8.3	0.99	methyl phthalate	9.9	1.19
isoamyl bromide	10.2	1.22	methyl propyl ether	6.1	0.73
isoamyl butyrate	7.1	0.85	methyl propyl ketone	6.7	0.80
isoamyl chloride	7.4	0.89	methyl salicylate	9.9	1.19
isoamyl ether	6.5	0.78	monacetin	10.1	1.21
isoamyl isovalerate	7.1	0.85	α-monochlorohydrin	11.1	1.33
isoamyl nitrite	7.3	0.87	monoisopropanolamine	8.2	0.98
isoamyl phthalate	8.6	1.03	morpholine	8.3	0.99
isoamyl salicylate	8.9	1.07	mustard oil	7.6	0.91
isobutenyl chloride	7.7	0.92	myristica oil	7.5	0.90
isobutyl acetate	7.3	0.87	neatsfoot oil	7.6	0.91
isobutyl alcohol	6.8	0.81	nitric acid (68%)	11.7	1.40
isobutylamine	6.0	0.72	nitric acid, fuming	12.5	1.50
isobutyl benzoate	8.3	0.99	nitrobenzene	10.0	1.20
isobutyl bromide	10.5	1.26	nitroglycerin	13.3	1.59
isobutyl n-butyrate	7.2	0.86	nitromethane	9.4	1.13
isobutyl chloride	7.4	0.89	m-nitrotoluene	9.6	1.15
isobutyl ether	6.3	0.75	o-nitrotoluene	9.7	1.16
isobutyl formate	7.3	0.87	n-nonane	5.9	0.71
isobutyl nitrite	7.2	0.86	octanal	6.8	0.81
isobutyl propionate	7.4	0.89	octane	5.9	0.71
isobutyric acid	7.9	0.95	octyl acetate	7.3	0.87
isoeugenol	9.1	1.09	octyl alcohol	6.9	0.83
isoprene	5.7	0.68	octyl amine	6.6	0.79
isopropyl acetate	7.3	0.87	oleic acid	7.5	0.90
isopropyl alcohol	6.5	0.78	olive oil	7.6	0.91
isopropylamine	5.8	0.69	orange flower oil	7.2	0.86
isopropylbenzene	7.2	0.86	orange oil	7.0	0.84
isopropyl bromide	10.9	1.31	paraldehyde	8.3	0.99
isopropyl chloride	7.2	0.86	peanut oil	7.6	0.91
isopropyl ether	6.0	0.72	pennyroyal oil, American	7.8	0.93
isovaleric acid	7.8	0.93	pentachloroethane	14.0	1.68
isovaleronitrile	6.5	0.78	pentane	5.2	0.62
lactic acid, 85–90%	10.1	1.21	peppermint oil	7.5	0.90
lavender oil	7.4	0.89	perchloric acid, 70%	14.0	1.68
lead subacetate soln	10.4	1.25	persic oil	7.6	0.91
lemon oil	7.1	0.85	peruvian balsam	9.7	1.16
lemon grass oil	7.5	0.90	petroleum benzin	5.5	0.66
d-limonene	7.0	0.84	phenetole	8.0	0.96
linoleic acid	7.5	0.90	phenylacetaldehyde	8.6	1.03
linseed oil	7.7	0.92	phenyl acetate	8.9	1.07
liquefied phenol	8.9	1.07	phenyl Cellosolve	9.2	1.10
liquid petrolatum	7.4	0.89	phenylethanolamine	9.0	1.08
liquid petrolatum, light	7.1	0.85	phenyl ethyl acetate	8.6	1.03
l-menthol	7.4	0.89	phenylethyl alcohol	8.5	1.02
mercury	112.9	13.53	phosphoric acid, 85%	14.2	1.70
mesityl oxide	7.1	0.85	o-phthalyl chloride	11.7	1.90
methyl acetate	7.7	0.92	3-picoline	8.0	0.96
methyl acetoacetate	8.9	1.07	pimenta oil	8.6	1.03
methylal	7.1	0.85	dl-pinene	7.1	0.85
methyl alcohol	6.6	0.79	pine needle oil	7.4	0.89
methyl amyl acetate	7.1	0.85	pine oil	7.8	0.93
methyl amyl alcohol	6.7	0.80	piperidine	7.2	0.86
methyl amyl carbinol	6.8	0.81	polyethylene glycol 300	9.4	1.13
methyl-n amyl ketone	6.8	0.81	polyethylene glycol 400	9.4	1.13
N-methylaniline	8.2	0.98	polysorbate 80	9.0	1.08
methyl butyl ketone	6.9	0.83	poppy oil	7.7	0.92
methyl Carbitol	8.6	1.03	potassium hydroxide soln, 50%	12.7	1.52
methyl Cellosolve	8.0	0.96	propionaldehyde	6.7	0.80
methyl Cellosolve acetate	8.4	1.01	propionic acid	8.3	0.99
methyl cyclohexanone	7.7	0.92	propionic anhydride	8.3	0.99
methylene bromide	20.8	2.49	propionitrile	6.5	0.78
methylene chloride	11.1	1.33	propyl acetate	7.4	0.89
methylene iodide	27.7	3.32	n-propyl alcohol	6.7	0.80
methyl ethyl ketone	6.7	0.80	propyl chloride	7.4	0.89
methyl formate	8.1	0.97	propylene chlorohydrin	9.2	1.10
methyl hexyl ketone	6.8	0.81	propylenediamine	7.3	0.87
methyl-3-hydroxybutyrate	8.8	1.05	propylene dichloride	9.7	1.16
methyl iodide	18.8	2.25	propylene glycol	8.6	1.03
methyl isobutyl ketone	6.7	0.80	propylene oxide	6.9	0.83

Liquid	lbs/gal	kg/l	Liquid	lbs/gal	kg/l
propyl ether	6.1	0.73	thyme oil	7.7	0.92
pyridine	8.2	0.98	toluene	7.2	0.86
quinoline	9.1	1.09	m-toluidine	8.2	0.98
rapeseed oil	7.6	0.91	o-toluidine	8.4	1.01
rosemary oil	7.5	0.90	triacetin	9.7	1.16
rose oil	7.1	0.85	tributylamine	6.5	0.78
safrol	9.2	1.10	tributyl citrate	8.7	1.04
salicylaldehyde	9.7	1.16	tributyl phosphate	8.1	0.97
santal oil	8.1	0.97	1,1,2-trichloroethane	12.0	1.44
sassafras oil	8.9	1.07	trichloroethylene	12.2	1.46
sesame oil	7.6	0.91	triethanolamine	9.4	1.13
sodium carbonate soln, 10%	9.1	1.09	triethylamine	6.0	0.72
sodium hydroxide soln, 50%	12.8	1.53	triethyleneglycol	9.4	1.13
sorbitol	11.0	1.32	triethylphosphate	8.9	1.07
sorbitol soln	10.4	1.25	trimethylene glycol	8.8	1.05
soya oil	7.7	0.92	trimethyl phosphate	10.1	1.21
spearmint oil	7.7	0.92	triolein	7.6	0.91
sperm oil	7.3	0.87	tung oil	7.8	0.93
styrene	7.6	0.91	turpentine oil	7.2	0.86
succinaldehyde	8.9	1.07	turpentine oil. rectified	7.2	0.86
sulfuric acid, 96%	15.3	1.83	undecane	6.2	0.74
sulfurous acid, 6%	8.5	1.02	l-undecanol	6.9	0.83
syrup	11.0	1.32	undecylenic acid	7.6	0.91
terebene	7.2	0.86	valeric acid	7.9	0.95
terpineol	7.8	0.93	valeronitrile	6.7	0.80
tetrachloroethylene	13.6	1.63	vinyl acetate	7.8	0.93
tetradecanol	7.0	0.84	vinyl ether	6.4	0.77
tetraethylene glycol	9.4	1.13	vitamin K_1	8.1	0.97
tetrahydrofuran	7.4	0.89	water	8.3	0.99
tetrahydrofurfuryl alcohol	8.8	1.05	whiskey	7.7	0.92
tetralin	8.1	0.97	m-xylene	7.2	0.86
thioglycolic acid	11.0	1.32	o-xylene	7.5	0.90
thionyl chloride	13.6	1.63	ylang-ylang oil	7.8	0.93
thiophene	8.9	1.07			

Specific Gravity Comparisons

Degrees Baumé (U.S. Standard) corresponding to specific gravities at $\dfrac{60°}{60°}$ F $\left(\dfrac{15.56°}{15.56°} C \right)$ for liquids lighter than water.

(Calculated from the formula: $°Bé = \dfrac{140}{\text{sp gr}} - 130$.)

Sp gr	°Bé	Sp gr	°Bé	Sp gr	°Bé	Sp gr	°Bé
0.600	103.33	0.700	70.00	0.800	45.00	0.900	25.56
.605	101.40	.705	68.58	.805	43.91	.905	24.70
.610	99.51	.710	67.18	.810	42.84	.910	23.85
.615	97.64	.715	65.80	.815	41.78	.915	23.01
.620	95.81	.720	64.44	.820	40.73	.920	22.17
.625	94.00	.725	63.10	.825	39.70	.925	21.35
.630	92.22	.730	61.78	.830	38.67	.930	20.54
.635	90.47	.735	60.48	.835	37.66	.935	19.73
.640	88.75	.740	59.19	.840	36.67	.940	18.94
.645	87.05	.745	57.92	.845	35.68	.945	18.15
.650	85.38	.750	56.67	.850	34.71	.950	17.37
.655	83.74	.755	55.43	.855	33.74	.955	16.60
.660	82.12	.760	54.21	.860	32.79	.960	15.83
.665	80.53	.765	53.01	.865	31.85	.965	15.08
.670	78.96	.770	51.82	.870	30.92	.970	14.33
.675	77.41	.775	50.65	.875	30.00	.975	13.59
.680	75.88	.780	49.49	.880	29.09	.980	12.86
.685	74.38	.785	48.34	.885	28.19	.985	12.13
.690	72.90	.790	47.22	.890	27.30	.990	11.41
.695	71.44	.795	46.10	.895	26.42	.995	10.70

Specific Gravity Comparisons

Degrees Baumé (U.S. Standard) corresponding to specific gravities at $\frac{60°}{60°}$F $\left(\frac{15.56°}{15.56°}C\right)$ for liquids heavier than water.

(Calculated from the formula: $°Bé = 145 - \dfrac{145}{sp\ gr}$.)

Sp gr	°Bé	Sp gr	°Bé	Sp gr	°Bé	Sp gr	°Bé	Sp gr	°Bé
1.005	0.72	1.205	24.67	1.405	41.80	1.605	54.66	1.805	64.67
1.010	1.44	1.210	25.17	1.410	42.16	1.610	59.94	1.810	64.89
1.015	2.14	1.215	25.66	1.415	42.53	1.615	55.22	1.815	65.11
1.020	2.84	1.220	26.15	1.420	42.89	1.620	55.49	1.820	65.33
1.025	3.54	1.225	26.63	1.425	43.25	1.625	55.77	1.825	65.55
1.030	4.22	1.230	27.11	1.430	43.60	1.630	56.04	1.830	65.77
1.035	4.90	1.235	27.59	1.435	43.95	1.635	56.32	1.835	65.98
1.040	5.58	1.240	28.06	1.440	44.31	1.640	56.59	1.840	66.20
1.045	6.24	1.245	28.53	1.445	44.65	1.645	56.85	1.845	66.41
1.050	6.91	1.250	29.00	1.450	45.00	1.650	57.12	1.850	66.62
1.055	7.56	1.255	29.46	1.455	45.34	1.655	57.39	1.855	66.83
1.060	8.21	1.260	29.92	1.460	45.68	1.660	57.65	1.860	67.04
1.065	8.85	1.265	30.38	1.465	46.02	1.665	57.91	1.865	67.25
1.070	9.49	1.270	30.83	1.470	46.36	1.670	58.17	1.870	67.46
1.075	10.12	1.275	31.27	1.475	46.69	1.675	58.43	1.875	67.67
1.080	10.74	1.280	31.72	1.480	47.03	1.680	58.69	1.880	67.87
1.085	11.36	1.285	32.16	1.485	47.36	1.685	58.95	1.885	68.08
1.090	11.97	1.290	32.60	1.490	47.68	1.690	59.20	1.890	68.28
1.095	12.58	1.295	33.03	1.495	48.01	1.695	59.45	1.895	68.48
1.100	13.18	1.300	33.46	1.500	48.33	1.700	59.71	1.900	68.68
1.105	13.78	1.305	33.89	1.505	48.65	1.705	59.96	1.905	68.88
1.110	14.37	1.310	34.31	1.510	48.97	1.710	60.20	1.910	69.08
1.115	14.96	1.315	34.73	1.515	49.29	1.715	60.45	1.915	69.28
1.120	15.54	1.320	35.15	1.520	49.61	1.720	60.70	1.920	69.48
1.125	16.11	1.325	35.57	1.525	49.92	1.725	60.94	1.925	69.68
1.130	16.68	1.330	35.98	1.530	50.23	1.730	61.18	1.930	69.87
1.135	17.25	1.335	36.39	1.535	50.54	1.735	61.43	1.935	70.06
1.140	17.81	1.340	36.79	1.540	50.84	1.740	61.67	1.940	70.26
1.145	18.36	1.345	37.19	1.545	51.15	1.745	61.91	1.945	70.45
1.150	18.91	1.350	37.59	1.550	51.45	1.750	62.14	1.950	70.64
1.155	19.46	1.355	37.99	1.555	51.75	1.755	62.38	1.955	70.83
1.160	20.00	1.360	38.38	1.560	52.05	1.760	62.61	1.960	71.02
1.165	20.54	1.365	38.77	1.565	52.35	1.765	62.85	1.965	71.21
1.170	21.07	1.370	39.16	1.570	52.64	1.770	63.08	1.970	71.40
1.175	21.60	1.375	39.55	1.575	52.94	1.775	63.31	1.975	71.58
1.180	22.12	1.380	39.93	1.580	53.23	1.780	63.54	1.980	71.77
1.185	22.64	1.385	40.31	1.585	53.52	1.785	63.77	1.985	71.95
1.190	23.15	1.390	40.68	1.590	53.81	1.790	63.99	1.990	72.14
1.195	23.66	1.395	41.06	1.595	54.09	1.795	64.22	1.995	72.32
1.200	24.17	1.400	41.43	1.600	54.38	1.800	64.44	2.000	72.50

Percentage Solution Table

Although the making of a solution of given percentage strength is a simple matter when both the substance and the solvent are *weighed*, the preparation of a *definite volume* of solution of given percentage strength is not so easy, because the displacement volume of the substance is not known. Hence, it is practically impossible to determine the ultimate volume readily produced by a given *weight* of substance and a definite *volume* of solvent. For weight-in-volume solutions of low concentration, the margin of error is relatively low but as the concentration increases the error increases also. Accurate strength solutions of higher concentration are best made by weighing.

However, weight-in-volume solutions are prescribed by the U.S.P. to be used in compounding prescriptions whenever gases or solids are dissolved in liquids, since most physicians have in mind a certain weight of substance in a definite volume of solution. Such solutions are defined by the U.S.P. as "the number of grams of an active constituent in 100 milliliters of solution" regardless of whether water or some other liquid is the vehicle.

For such weight-in-volume solutions, the following tables will afford a ready means of ascertaining the quantities (by weight) of substance required to prepare varying volumes of definite w/v percentage strength. The calculations are based on 1 fl oz (480 minims) of water = 455 grains (round number) in the Apothecaries table and 100 ml of water = 100 g in the Metric table.

APOTHECARIES

Strength of solution (% w/v)	Grains of substance required per given volume of solution										
	¼ fl oz	½ fl oz	1 fl oz	2 fl oz	3 fl oz	4 fl oz	6 fl oz	8 fl oz	10 fl oz	12 fl oz	16 fl oz
0.01	0.0114	0.023	0.046	0.091	0.137	0.182	0.273	0.364	0.455	0.546	0.728
0.02	0.023	0.046	0.091	0.182	0.273	0.364	0.546	0.728	0.910	1.092	1.456
0.04	0.046	0.091	0.182	0.364	0.546	0.728	1.092	1.456	1.820	2.184	2.912
0.05	0.057	0.114	0.2275	0.455	0.683	0.910	1.365	1.820	2.275	2.730	3.640
0.1	0.114	0.228	0.455	0.910	1.365	1.820	2.730	3.640	4.55	5.46	7.28
0.2	0.2275	0.455	0.910	1.820	2.730	3.640	5.460	7.28	9.10	10.92	14.56
0.25	0.284	0.569	1.138	2.275	3.413	4.55	6.83	9.10	11.38	13.65	18.20
0.5	0.569	1.138	2.275	4.55	6.83	9.10	13.65	18.20	22.75	27.3	36.4
1	1.138	2.275	4.55	9.10	13.65	18.2	27.3	36.4	45.5	54.6	72.8
2	2.275	4.55	9.10	18.20	27.30	36.4	54.6	72.8	91.0	109.2	145.6
3	3.413	6.83	13.65	27.30	40.95	54.6	81.9	109.2	136.5	163.8	218.4
4	4.55	9.10	18.20	36.40	54.60	72.8	109.2	145.6	182.0	218.4	291.2
5	5.69	11.37	22.75	45.50	68.25	91.0	136.5	182.0	227.5	273	364
10	11.38	22.75	45.50	91.0	136.50	182	273	364	455	546	728
15	17.06	34.13	68.25	136.5	204.75	273	409.5	546	682.5	819	1092
20	22.75	45.50	91.0	182	273	364	546	728	910	1092	1456
25	28.44	56.90	113.75	227.5	341.25	455	682.5	910	1137.5	1365	1820
30	34.13	68.25	136.5	273	409.5	546	819	1092	1365	1638	2184
40	45.5	91.0	182	364	546	728	1092	1456	1820	2184	2912

METRIC

Strength of solution (% w/v)	Grams of substance required per given volume of solution										
	10 ml	15 ml	25 ml	30 ml	60 ml	90 ml	100 ml	120 ml	150 ml	200 ml	500 ml	1000 ml
0.01	0.001	0.0015	0.0025	0.003	0.006	0.009	0.01	0.012	0.015	0.02	0.05	0.1
0.02	0.002	0.003	0.005	0.006	0.012	0.018	0.02	0.024	0.03	0.04	0.1	0.2
0.05	0.005	0.008	0.013	0.015	0.03	0.045	0.05	0.06	0.075	0.1	0.25	0.5
0.1	0.01	0.015	0.025	0.03	0.06	0.09	0.1	0.12	0.15	0.2	0.5	1.0
0.2	0.02	0.03	0 05	0.06	0.12	0.18	0.2	0.24	0.30	0.4	1.0	2.0
0.25	0.025	0.038	0.063	0.075	0.15	0.225	0.25	0.3	0.375	0.5	1.25	2.5
0.5	0.05	0.075	0.125	0.15	0.3	0.45	0.5	0.6	0.75	1.0	2.5	5.0
1	0.1	0.15	0.25	0.3	0.6	0.9	1.0	1.2	1.5	2.0	5.0	10.0
1.5	0.15	0.225	0.375	0.45	0.9	1.35	1.5	1.8	2.25	3.0	7.5	15.0
2	0.2	0.3	0.5	0.6	1.2	1.8	2.0	2.4	3.0	4.0	10.0	20.0
3	0.3	0.45	0.75	0.9	1.8	2.7	3.0	3.6	4.5	6.0	15.0	30.0
4	0.4	0.6	1.0	1.2	2.4	3.6	4.0	4.8	6.0	8.0	20.0	40.0
5	0.5	0.75	1.25	1.5	3.0	4.5	5.0	6.0	7.5	10.0	25.0	50.0
10	1.0	1.5	2.5	3.0	6.0	9.0	10.0	12.0	15.0	20.0	50.0	100.0
15	1.5	2.25	3.75	4.5	9.0	13.5	15.0	18.0	22.5	30.0	75.0	150.0
20	2.0	3.0	5.0	6.0	12.0	18.0	20.0	24.0	30.0	40.0	100.0	200.0
25	2.5	3.75	6.25	7.5	15.0	22.5	25.0	30.0	37.5	50.0	125.0	250.0
40	4.0	6.0	10.0	12.0	24.0	36.0	40.0	48.0	60.0	80.0	200.0	400.0

Concentration of Acids and Bases

Common Commercial Strengths

	Molecular weight	Moles per liter	Grams per liter	Percent by weight	Specific gravity
acetic acid, glacial	60.05	17.4	1045	99.5	1.05
acetic acid	60.05	6.27	376	36	1.045
butyric acid	88.1	10.3	912	95	0.96
formic acid	46.02	23.4	1080	90	1.20
		5.75	264	25	1.06
hydriodic acid	127.9	7.57	969	57	1.70
		5.51	705	47	1.50
		0.86	110	10	1.1
hydrobromic acid	80.92	8.89	720	48	1.50
		6.82	552	40	1.38
hydrochloric acid	36.5	11.6	424	36	1.18
		2.9	105	10	1.05
hydrocyanic acid	27.03	25	676	97	0.697
		0.74	19.9	2	0.996
hydrofluoric acid	20.01	32.1	642	55	1.167
		28.8	578	50	1.155
hydrofluosilicic acid	144.1	2.65	382	30	1.27
hypophosphorous acid	66.0	9.47	625	50	1.25
		5.14	339	30	1.13
		1.57	104	10	1.04
lactic acid	90.1	11.3	1020	85	1.2
nitric acid	63.02	15.99	1008	71	1.42
		14.9	938	67	1.40
		13.3	837	61	1.37
perchloric acid	100.5	11.65	1172	70	1.67
		9.2	923	60	1.54
phosphoric acid	98	14.7	1445	85	1.70
sulfuric acid	98.1	18.0	1766	96	1.84
sulfurous acid	82.1	0.74	61.2	6	1.02
ammonia water	17.0	14.8	252	28	0.898
potassium hydroxide	56.1	13.5	757	50	1.52
		1.94	109	10	1.09
sodium carbonate	106.0	1.04	110	10	1.10
sodium hydroxide	40.0	19.1	763	50	1.53
		2.75	111	10	1.11

Isotonic Solutions

The following table provides data for adjusting the osmolality of aqueous solutions, so that they will be isosmotic with normal saline solution (0.9% NaCl) and therefore isotonic with blood and tears. This can be achieved by using either the freezing point depression method or the sodium chloride equivalent method. The freezing point depression values have been determined experimentally and published. (Refs. 1-6) NaCl equivalent values have been calculated from these data.

The top values listed for each chemical are NaCl equivalents. The second values, in italics, are freezing point depression values in degrees Celsius. The percentage concentration (w/v) at isotonicity is given in bold face in the last column.

Using the freezing point depression method an isotonic solution is prepared by matching the freezing point depression value of 0.9% NaCl which is accepted as 0.52. Freezing point depression values are additive in this concentration range; therefore, to make a 2% Hepes buffer solution isotonic with tears, the calculation is as follows:

Freezing point depression of isotonic solution	0.52
Freezing point depression of 2% Hepes from table	0.163
Difference to be supplied by NaCl	0.357

0.9% NaCl has a freezing point depression of 0.52 ,
x% NaCl has a freezing point depression of 0.357.
Therefore $x/0.9 = 0.357/0.52$,
$$x = 0.618$$

Thus 0.618 g/100ml NaCl combined with 2 g/100ml Hepes produces an isotonic solution. Note: If more than one component is in solution the freezing point depressions of each must be added together before the difference value is obtained.

Using the NaCl equivalents method to prepare a solution involves: Multiplying the number of grams of each component corrected for volume by its NaCl equivalent from the table at the nearest listed concentration. For example to make 60 ml of 1% boric acid isotonic: Add the NaCl equivalents of each solution component.

60 ml of 1% boric acid soln (0.6g boric acid x 0.5g NaCl equiv.) = 0.3g

Calculate the number of grams of NaCl required for the volume being used. (60 ml need 0.54 g) Subtract the NaCl equivalents of the components from the total NaCl required. (0.54 - 0.3 = 0.24g) The difference is the amount of NaCl which must be added to that specific volume. Note: If the difference is less than zero then the solution is already hypertonic and cannot be adjusted without altering the concentration of the components.

Sodium Chloride Equivalents and Freezing Point Depressions (°C) for Certain Concentrations (w/v) of Solution

Chemical	Concentration of Solution, NaCl Equivalents					At Isosmotic Concentration	
	0.5%	1%	2%	3%	5%		
Acetazolamide sodium	0.24	0.23	0.23	0.23	—	0.23	**3.85%**
	0.068°	*0.13°*	*0.27°*	*0.40°*	—	*0.52°*	**3.85%**
Acetrizoate methylglucamine	0.09	0.08	0.08	0.08	0.08	0.07	**12.12%**
	0.024°	*0.04°*	*0.09°*	*0.13°*	*0.22°*	*0.52°*	**12.12%**
Acetrizoate sodium	0.10	0.10	0.10	0.10	0.10	0.09	**9.64%**
	0.027°	*0.05°*	*0.10°*	*0.16°*	*0.27°*	*0.52°*	**9.64%**
Acetylcysteine	0.20	0.20	0.20	0.20	—	0.20	**4.58%**
	0.055°	*0.11°*	*0.22°*	*0.34°*	—	*0.52°*	**4.58%**
Acetylsulfanilamide sodium	0.24	0.23	0.23	0.23	—	0.23	**3.85%**
	0.066°	*0.13°*	*0.26°*	*0.40°*	—	*0.52°*	**3.85%**
Acriflavine	0.10	0.10	0.09	0.09	—	—	
	0.025°	*0.05°*	*0.10°*	*0.15°*	—	—	
Acyclovir sodium	0.26	0.24	0.22	0.21	—	0.20	**4.50%**
	0.074°	*0.13°*	*0.25°*	*0.36°*	—	*0.52°*	**4.50%**
Adenosine phosphate	0.50	0.41	—	—	—	—	
	0.140°	*0.23°*	—	—	—	—	
Adiphenine hydrochloride	0.28	0.22	0.17	0.15	0.12	—	
	0.083°	*0.12°*	*0.19°*	*0.25°*	*0.34°*	—	
Adrenalone hydrochloride	0.30	0.27	0.24	0.22	—	0.21	**4.24%**
	0.086°	*0.15°*	*0.27°*	*0.38°*	—	*0.52°*	**4.24%**
Alcohol	0.65	0.65	—	—	—	0.65	**1.39%**
	0.188°	*0.37°*	—	—	—	*0.52°*	**1.39%**

Chemical	Concentration of Solution, NaCl Equivalents						
	0.5%	1%	2%	3%	5%	At Isosmotic Concentration	
Alcohol, dehydrate	0.70	0.70	—	—	—	0.70	**1.28%**
	0.203°	*0.40°*	—	—	—	*0.52°*	**1.28%**
Alphaprodine hydrochloride	0.19	0.19	0.18	0.18	—	0.18	**4.98%**
	0.053°	*0.10°*	*0.21°*	*0.31°*	—	*0.52°*	**4.98%**
Alum, potassium	0.20	0.18	0.16	0.15	0.15	0.14	**6.35%**
	0.054°	*0.10°*	*0.18°*	*0.26°*	*0.41°*	*0.52°*	**6.35%**
Amantadine hydrochloride	0.31	0.31	0.31	—	—	0.31	**2.95%**
	0.090°	*0.18°*	*0.35°*	—	—	*0.52°*	**2.95%**
Amdinocillin	0.11	0.10	0.10	0.10	0.10	—	
	0.032°	*0.06°*	*0.11°*	*0.17°*	*0.29°*	—	
Amidoxyl benzoate	0.20	0.20	0.20	0.20	—	0.20	**4.42%**
	0.059°	*0.11°*	*0.23°*	*0.35°*	—	*0.52°*	**4.42%**
Amikacin	0.06	0.05	0.05	0.05	0.05	—	
	0.016°	*0.03°*	*0.06°*	*0.09°*	*0.15°*	—	
Aminacrine hydrochloride	0.20	0.17	—	—	—	—	
	0.052°	*0.09°*	—	—	—	—	
Aminoacetic acid	0.42	0.41	0.41	—	—	0.41	**2.20%**
	0.119°	*0.23°*	*0.47°*	—	—	*0.52°*	**2.20%**
Aminocaproic acid	0.26	0.26	0.26	0.26	—	0.26	**3.52%**
	0.075°	*0.14°*	*0.29°*	*0.44°*	—	*0.52°*	**3.52%**
p-Aminohippuric acid	0.13	0.13	—	—	—	—	
	0.035°	*0.07°*	—	—	—	—	
Aminophylline	0.18	0.17	—	—	—	—	
	0.056°	*0.10°*	—	—	—	—	
p-Aminosalicylate sodium	0.30	0.29	0.29	0.28	—	0.27	**3.27%**
	0.086°	*0.16°*	*0.32°*	*0.47°*	—	*0.52°*	**3.27%**
Amitriptyline hydrochloride	0.24	0.18	0.11	0.08	0.06	—	
	0.070°	*0.10°*	*0.12°*	*0.14°*	*0.17°*	—	
Ammonium carbonate	0.70	0.70	—	—	—	0.70	**1.29%**
	0.202°	*0.40°*	—	—	—	*0.52°*	**1.29%**
Ammonium chloride	1.10	—	—	—	—	1.07	**0.84%**
	0.315°	—	—	—	—	*0.52°*	**0.84%**
Ammonium lactate	0.33	0.33	0.33	—	—	0.33	**2.76%**
	0.093°	*0.18°*	*0.37°*	—	—	*0.52°*	**2.76%**
Ammonium nitrate	0.69	0.69	—	—	—	0.69	**1.30%**
	0.200°	*0.40°*	—	—	—	*0.52°*	**1.30%**
Ammonium phosphate, dibasic	0.58	0.55	—	—	—	0.51	**1.76%**
	0.165°	*0.31°*	—	—	—	*0.52°*	**1.76%**
Ammonium sulfate	0.55	0.55	—	—	—	0.54	**1.68%**
	0.158°	*0.31°*	—	—	—	*0.52°*	**1.68%**
Amobarbital sodium	0.26	0.25	0.25	0.25	—	0.25	**3.6%**
	0.074°	*0.14°*	*0.29°*	*0.44°*	—	*0.52°*	**3.6%**
Amphetamine phosphate	0.38	0.34	0.30	0.27	—	0.26	**3.47%**
	0.114°	*0.19°*	*0.33°*	*0.46°*	—	*0.52°*	**3.47%**
Amphetamine sulfate	0.22	0.22	0.22	0.21	—	0.21	**4.23%**
	0.066°	*0.12°*	*0.25°*	*0.37°*	—	*0.52°*	**4.23%**

Isotonic Solutions (Continued)

Chemical	Concentration of Solution, NaCl Equivalents						
	0.5%	1%	2%	3%	5%	At Isosmotic Concentration	
Ampicillin sodium	0.16 0.045°	0.16 0.09°	0.16 0.18°	0.16 0.27°	0.16 0.45°	0.16 0.52°	5.78% 5.78%
Amprotropine phosphate	0.19 0.058°	0.18 0.10°	0.17 0.19°	0.16 0.28°	0.15 0.44°	— —	
Amydricaine hydrochloride	0.28 0.080°	0.24 0.13°	0.20 0.23°	0.18 0.31°	0.16 0.46°	0.16 0.52°	5.74% 5.74%
Amydricaine nitrate	0.20 0.058°	0.19 0.10°	0.18 0.19°	0.17 0.28°	0.16 0.46°	0.16 0.52°	5.68% 5.68%
Anileridine hydrochloride	0.19 0.052°	0.19 0.10°	0.19 0.21°	0.18 0.31°	0.18 0.50°	0.18 0.52°	5.13% 5.13%
Antazoline hydrochloride	0.25 0.073°	0.23 0.13°	0.21 0.24°	— —	— —	— —	
Antazoline phosphate	0.20 0.062°	0.20 0.11°	0.18 0.20°	0.17 0.29°	0.15 0.44°	— —	
Antimony potassium tartrate	0.22 0.065°	0.18 0.10°	0.15 0.17°	0.13 0.23°	0.10 0.33°	— —	
Antipyrine	0.18 0.050°	0.17 0.09°	0.16 0.17°	0.14 0.25°	0.14 0.39°	0.13 0.52°	6.81% 6.81%
Apomorphine hydrochloride	0.14 0.041°	0.14 0.08°	0.14 0.15°	— —	— —	— —	
Arecoline hydrobromide	0.030 0.084°	0.27 0.15°	0.25 0.28°	0.24 0.41°	— —	0.23 0.52°	3.88% 3.88%
Arginine glutamate	0.17 0.048°	0.17 0.09°	0.17 0.19°	0.17 0.29°	0.17 0.48°	0.17 0.52°	5.37% 5.37%
L-Arginine hydrochloride	0.31 0.087°	0.30 0.17°	0.28 0.32°	0.27 0.46°	— —	0.26 0.52°	3.43% 3.43%
Arsenic trioxide	0.30 0.085°	0.30 0.16°	— —	— —	— —	— —	
Ascorbic acid	0.20 0.053°	0.18 0.10°	0.18 0.20°	0.18 0.31°	0.18 0.51°	0.18 0.52°	5.94% 5.94%
Atracurium besylate	0.08 0.019°	0.06 0.03°	0.05 0.05°	0.04 0.07°	0.04 0.10°	— —	
Atropine methylnitrate	0.20 0.055°	0.18 0.10°	0.16 0.18°	0.15 0.26°	0.14 0.41°	0.14 0.52°	6.52% 6.52%
Atropine sulfate	0.14 0.039°	0.13 0.07°	0.12 0.13°	0.11 0.19°	0.11 0.31°	0.10 0.52°	8.85% 8.85%
Aurothioglucose	0.03 0.007°	0.03 0.01°	0.03 0.02°	0.03 0.04°	0.03 0.07°	— —	
Azlocillin sodium	0.15 0.043°	0.13 0.07°	0.12 0.13°	0.11 0.19°	0.11 0.30°	— —	
Bacitracin	0.06 0.016°	0.05 0.02°	0.05 0.05°	0.04 0.07°	0.04 0.12°	— —	
Barbital sodium	0.32 0.087°	0.30 0.17°	0.29 0.33°	0.29 0.50°	— —	0.29 0.52°	3.12% 3.12%
Benoxinate hydrochloride	0.20 0.061°	0.18 0.10°	0.15 0.17°	0.14 0.23°	— —	— —	

Chemical	Concentration of Solution, NaCl Equivalents						
	0.5%	1%	2%	3%	5%	At Isosmotic Concentration	
Benzalkonium chloride	0.18 *0.048°*	0.16 *0.09°*	0.15 *0.17°*	0.14 *0.24°*	0.13 *0.38°*	— —	
Benzethonium chloride	0.08 *0.022°*	0.05 *0.02°*	0.03 *0.03°*	0.02 *0.04°*	0.02 *0.05°*	— —	
Benzpyrinium bromide	0.02 *0.061°*	0.20 *0.11°*	0.19 *0.21°*	0.18 *0.30°*	0.17 *0.48°*	— —	
Benzquinamide hydrochloride	0.14 *0.041°*	0.14 *0.07°*	0.13 *0.15°*	0.12 *0.21°*	— —	— —	
Benztropine mesylate	0.26 *0.073°*	0.21 *0.11°*	0.15 *0.17°*	0.12 *0.20°*	0.09 *0.24°*	— —	
Benzyl alcohol	0.18 *0.049°*	0.17 *0.09°*	0.16 *0.18°*	0.15 *0.26°*	— —	— —	
Benzylpenicillin potassium	0.18 *0.052°*	0.18 *0.10°*	0.17 *0.19°*	0.17 *0.29°*	0.16 *0.47°*	0.16 *0.52°*	**5.48%** **5.48%**
Benzylpenicillin sodium	0.18 *0.052°*	0.18 *0.10°*	0.17 *0.19°*	0.16 *0.28°*	0.16 *0.45°*	— —	
Betaxolol	0.18 *0.050°*	0.17 *0.09°*	0.16 *0.18°*	0.16 *0.27°*	0.16 *0.46°*	— —	
Betazole hydrochloride	0.54 *0.158°*	0.51 *0.29°*	— —	— —	— —	0.47 *0.52°*	**1.91%** **1.91%**
Bethanechol chloride	0.50 *0.140°*	0.39 *0.22°*	0.32 *0.36°*	0.30 *0.51°*	— —	0.30 *0.52°*	**3.05%** **3.05%**
Bismuth potassium tartrate	0.10 *0.033°*	0.09 *0.05°*	0.07 *0.08°*	0.06 *0.10°*	0.05 *0.14°*	— —	
Bismuth sodium tartrate	0.14 *0.041°*	0.13 *0.07°*	0.13 *0.13°*	0.12 *0.19°*	0.11 *0.31°*	0.10 *0.52°*	**8.91%** **8.91%**
Boric acid	0.52 *0.146°*	0.50 *0.28°*	— —	— —	— —	0.47 *0.52°*	**1.9%** **1.9%**
Bretylium tosylate	0.16 *0.043°*	0.14 *0.08°*	0.13 *0.14°*	0.12 *0.20°*	0.11 *0.32°*	— —	
Bromodiphenhydramine hydrochloride	0.20 *0.067°*	0.17 *0.10°*	0.14 *0.16°*	0.10 *0.18°*	0.07 *0.20°*	— —	
Brompheniramine maleate	0.10 *0.026°*	0.09 *0.05°*	0.08 *0.08°*	— —	— —	— —	
Bupivacaine hydrochloride	0.17 *0.048°*	0.17 *0.09°*	0.17 *0.19°*	0.17 *0.29°*	0.17 *0.48°*	0.17 *0.52°*	**5.38%** **5.38%**
Butabarbital sodium	0.27 *0.078°*	0.27 *0.15°*	0.27 *0.31°*	0.27 *0.47°*	— —	0.27 *0.52°*	**3.33%** **3.33%**
Butacaine sulfate	0.26 *0.073°*	0.20 *0.11°*	0.16 *0.17°*	0.13 *0.22°*	0.10 *0.30°*	— —	
Butethamine formate	0.28 *0.077°*	0.26 *0.14°*	0.24 *0.26°*	0.21 *0.37°*	— —	0.20 *0.52°*	**4.56%** **4.56%**
Butethamine hydrochloride	0.28 *0.079°*	0.25 *0.14°*	0.22 *0.25°*	— —	— —	— —	
Caffeine	0.08 *0.025°*	0.08 *0.04°*	— —	— —	— —	— —	

Isotonic Solutions (Continued)

Chemical	Concentration of Solution, NaCl Equivalents					At Isosmotic Concentration	
	0.5%	1%	2%	3%	5%		
Calcium aminosalicylate	0.30	0.27	0.23	0.21	—	—	
	0.091°	*0.15°*	*0.26°*	*0.36°*	—	—	
Calcium chloride dihydrate	0.50	0.51	—	—	—	0.53	**1.70%**
	0.145°	*0.29°*	—	—	—	*0.52°*	**1.70%**
Calcium chloride hexahydrate	0.34	0.35	0.36	—	—	0.36	**2.5%**
	0.097°	*0.20°*	*0.41°*	—	—	*0.52°*	**2.5%**
Calcium chloride, anhydrous	0.70	0.70	—	—	—	0.70	**1.29%**
	0.206°	*0.40°*	—	—	—	*0.52*	**1.29%**
Calcium disodium edetate	0.21	0.21	0.21	0.20	—	0.20	**4.50%**
	0.061°	*0.12°*	*0.24°*	*0.35°*	—	*0.52°*	**4.50%**
Calcium gluconate	0.18	0.16	0.15	0.14	—	—	
	0.050°	*0.09°*	*0.16°*	*0.23°*	—	—	
Calcium lactate	0.26	0.23	0.22	0.21	—	0.20	**4.5%**
	0.073°	*0.13°*	*0.25°*	*0.37°*	—	*0.52°*	**4.5%**
Calcium lactobionate	0.08	0.08	0.08	0.07	0.07	—	
	0.022°	*0.04°*	*0.08°*	*0.12°*	*0.19°*	—	
Calcium levulinate	0.30	0.27	0.26	0.25	—	—	
	0.080°	*0.15°*	*0.30°*	*0.44°*	—	—	
Calcium pantothenate	0.20	0.19	0.18	0.17	0.16	0.16	**5.6%**
	0.055°	*0.10°*	*0.20°*	*0.29°*	*0.47°*	*0.52°*	**5.6%**
Capreomycin sulfate	0.04	0.04	0.04	0.04	0.04	—	
	0.011°	*0.02°*	*0.04°*	*0.06°*	*0.10°*	—	
Carbachol	0.04	0.36	0.34	—	—	0.32	**2.82%**
	0.108°	*0.20°*	*0.38°*	—	—	*0.52°*	**2.82%**
Carbazochrome salicylate	0.38	0.36	0.36	—	—	0.35	**2.57%**
	0.106°	*0.21°*	*0.41°*	—	—	*0.52°*	**2.57%**
Carbenicillin disodium	0.20	0.20	0.20	0.20	—	0.20	**4.40%**
	0.059°	*0.11°*	*0.23°*	*0.35°*	—	*0.52°*	**4.40%**
Cefamandole nafate	0.16	0.14	0.12	0.11	0.10	—	
	0.045°	*0.07°*	*0.13°*	*0.18°*	*0.29°*	—	
Cefazolin sodium	0.14	0.13	0.12	0.11	0.11	—	
	0.042°	*0.07°*	*0.13°*	*0.19°*	*0.30°*	—	
Ceforanide	0.14	0.12	—	—	—	—	
	0.040°	*0.06°*	—	—	—	—	
Cefotaxime sodium	0.16	0.15	0.14	0.13	0.12	—	
	0.046°	*0.08°*	*0.15°*	*0.22°*	*0.35°*	—	
Cefoxitin sodium	0.18	0.16	0.15	0.14	0.13	—	
	0.050°	*0.09°*	*0.16°*	*0.23°*	*0.38°*	—	
Ceftazidime pentahydrate	0.09	—	—	—	—	—	
	0.022°	—	—	—	—	—	
Ceftizoxime sodium	0.16	0.15	0.14	0.13	0.12	—	
	0.045°	*0.08°*	*0.15°*	*0.22°*	*0.35°*	—	
Ceftriaxone sodium	0.14	0.13	0.13	0.12	0.12	—	
	0.040°	*0.07°*	*0.14°*	*0.21°*	*0.33°*	—	
Cefuroxime sodium	0.13	0.13	0.13	0.13	0.13	—	
	0.037°	*0.07°*	*0.15°*	*0.22°*	*0.36°*		

Chemical	Concentration of Solution, NaCl Equivalents						
	0.5%	1%	2%	3%	5%	At Isosmotic	Concentration
Cephaloridine	0.09	0.07	0.06	0.06	0.05	—	
	0.023°	0.04°	0.07°	0.10°	0.14°	—	
Cephalothin sodium	0.18	0.17	0.16	0.15	0.14	0.13	**6.80%**
	0.050°	0.09°	0.17°	0.25°	0.40°	0.52°	**6.80%**
Cephapirin sodium	0.14	0.13	0.13	0.13	0.12	0.11	**7.80%**
	0.038°	0.07°	0.14°	0.22°	0.36°	0.52°	**7.80%**
Cetrimonium bromide	0.10	0.09	0.09	0.09	0.08	—	
	0.030°	0.05°	0.10°	0.14°	0.23°	—	
Chiniofon	0.14	0.13	0.12	0.11	—	—	
	0.039°	0.07°	0.13°	0.20°	—	—	
Chloramine-T	0.24	0.23	0.22	0.22	—	0.22	**4.1%**
	0.064°	0.12°	0.25°	0.38°	—	0.52°	**4.1%**
Chloramphenicol sodium succinate	0.14	0.14	0.14	0.13	0.13	0.13	**6.38%**
	0.038°	0.07°	0.15°	0.23°	0.38°	0.52°	**6.38%**
Chlorcyclizine hydrochloride	0.24	0.17	0.12	0.09	0.07	—	
	0.068°	0.09°	0.13°	0.16°	0.20°	—	
Chlordiazepoxide hydrochloride	0.24	0.22	0.19	0.18	0.17	0.16	**5.50%**
	0.068°	0.12°	0.22°	0.31°	0.48°	0.52°	**5.50%**
Chlorobutanol, hydrated	0.24	—	—	—	—	—	
	0.071°	—	—	—	—	—	
Chlorophyll	0.14	0.10	0.08	0.06	0.05	—	
	0.037°	0.05°	0.08°	0.11°	0.15°	—	
2-Chloroprocaine hydrochloride	0.20	0.20	0.18	—	—	—	
	0.054°	0.10°	0.21°	—	—	—	
Chloroquine phosphate	0.14	0.14	0.14	0.14	0.13	0.13	**7.15%**
	0.039°	0.08°	0.16°	0.24°	0.37°	0.52°	**7.15%**
Chloroquine sulfate	0.10	0.09	0.08	0.07	0.07	—	
	0.028°	0.05°	0.09°	0.12°	0.19°	—	
Chlorpheniramine maleate	0.18	0.17	0.14	0.12	0.09	—	
	0.049°	0.08°	0.16°	0.22°	0.26°	—	
Chlorpromazine hydrochloride	0.18	0.10	0.06	0.05	0.03	—	
	0.052°	0.05°	0.06°	0.07°	0.10°	—	
Chlortetracycline hydrochloride	0.10	0.10	0.10	—	—	—	
	0.030°	0.06°	0.12°	—	—	—	
Chlortetracycline sulfate	0.16	0.13	0.11	0.10	—	—	
	0.047°	0.07°	0.12°	0.17°	—	—	
Citric acid	0.18	0.18	0.17	0.17	0.16	0.16	**5.52%**
	0.050°	0.09°	0.19°	0.28°	0.47°	0.52°	**5.52%**
Clindamycin phosphate	0.08	0.08	0.08	0.08	0.08	0.08	**10.73%**
	0.022°	0.04°	0.09°	0.14°	0.24°	0.52°	**10.73%**
Cocaine hydrochloride	0.16	0.16	0.16	0.15	0.14	0.14	**6.33%**
	0.047°	0.09°	0.17°	0.25°	0.41°	0.52°	**6.33%**
Codeine hydrochloride	0.16	0.15	0.15	0.15	—	—	
	0.045°	0.08°	0.17°	0.25°	—	—	
Codeine phosphate	0.14	0.14	0.13	0.13	0.13	0.12	**7.29%**
	0.040°	0.07°	0.15°	0.22°	0.36°	0.52°	**7.29%**

Chemical	Concentration of Solution, NaCl Equivalents						
	0.5%	1%	2%	3%	5%		At Isosmotic Concentration
Colistimethate sodium	0.15 *0.045°*	0.15 *0.08°*	0.15 *0.17°*	0.15 *0.25°*	0.14 *0.41°*	0.13 *0.52°*	**6.73%** **6.73%**
Congo red	0.05 *0.015°*	0.05 *0.03°*	0.05 *0.05°*	0.05 *0.09°*	0.05 *0.15°*	— —	
Cromolyn sodium	0.16 *0.046°*	0.14 *0.08°*	0.11 *0.12°*	0.09 *0.14°*	0.05 *0.15°*		
Cupric sulfate	0.20 *0.054°*	0.18 *0.09°*	0.16 *0.17°*	0.15 *0.25°*	0.14 *0.39°*	0.13 *0.52°*	**6.85%** **6.85%**
Cupric sulfate, anhydrous	0.30 *0.084°*	0.27 *0.15°*	0.25 *0.28°*	0.23 *0.39°*	— —	0.22 *0.52°*	**4.09%** **4.09%**
Cyclizine hydrochloride	0.20 *0.060°*	— —	— —	— —	— —	— —	
Cyclomethycaine sulfate	0.16 *0.046°*	0.13 *0.07°*	0.11 *0.12°*	0.10 *0.16°*	0.09 *0.24°*	— —	
Cyclopentamine hydrochloride	0.36 *0.104°*	0.36 *0.20°*	0.35 *0.39°*	— —	— —	0.34 *0.52°*	**2.68%** **2.68%**
Cyclopentolate hydrochloride	0.22 *0.061°*	0.20 *0.11°*	0.19 *0.21°*	0.18 *0.31°*	0.17 *0.49°*	0.17 *0.52°*	**5.30%** **5.30%**
Cyclophosphamide	0.10 *0.031°*	0.10 *0.06°*	0.10 *0.12°*	— —	— —	— —	
Cytarabine	0.11 *0.034°*	0.11 *0.06°*	0.11 *0.13°*	0.11 *0.19°*	0.11 *0.31°*	0.10 *0.52°*	**8.92%** **8.92%**
Decamethonium bromide	0.29 *0.084°*	0.25 *0.14°*	0.22 *0.25°*	0.20 *0.35°*	0.18 *0.52°*	0.18 *0.52°*	**5.0%** **5.0%**
Deferoxamine mesylate	0.09 *0.023°*	0.09 *0.04°*	0.09 *0.09°*	0.09 *0.14°*	0.09 *0.24°*		
Demecarium bromide	0.14 *0.038°*	0.12 *0.06°*	0.10 *0.10°*	0.08 *0.13°*	0.07 *0.19°*	— —	
Dexamethasone sodium phosphate	0.18 *0.050°*	0.17 *0.09°*	0.16 *0.18°*	0.15 *0.26°*	0.14 *0.41°*	0.13 *0.52°*	**6.75%** **6.75%**
Dexchlorpheniramine maleate	0.17 *0.048°*	0.15 *0.08°*	0.14 *0.16°*	0.13 *0.22°*	0.09 *0.26°*	— —	
Dexpanthenol	0.20 *0.053°*	0.18 *0.10°*	0.17 *0.19°*	0.17 *0.28°*	0.16 *0.46°*	0.16 *0.52°*	**5.60%** **5.60%**
Dextroamphetamine hydrochloride	0.34 *0.097°*	0.34 *0.19°*	0.34 *0.39°*	— —	— —	0.34 *0.52°*	**2.64%** **2.64%**
Dextroamphetamine phosphate	0.25 *0.072°*	0.25 *0.14°*	0.25 *0.28°*	0.25 *0.43°*	— —	0.25 *0.52°*	**3.62%** **3.62%**
Dextroamphetamine sulfate	0.24 *0.069°*	0.23 *0.13°*	0.22 *0.25°*	0.22 *0.38°*	— —	0.22 *0.52°*	**4.16%** **4.16%**
Dextrose	0.16 *0.045°*	0.16 *0.09°*	0.16 *0.18°*	0.16 *0.27°*	0.16 *0.47°*	0.16 *0.52°*	**5.51%** **5.51%**
Dextrose, anhydrous	0.18 *0.050°*	0.18 *0.10°*	0.18 *0.20°*	0.18 *0.31°*	0.18 *0.51°*	0.52 *0.52°*	**5.05%** **5.05%**
Diatrizoate sodium	0.10 *0.025°*	0.09 *0.04°*	0.09 *0.09°*	0.09 *0.14°*	0.09 *0.24°*	0.09 *0.52°*	**10.55%** **10.55%**

Chemical	Concentration of Solution, NaCl Equivalents					At Isosmotic Concentration	
	0.5%	1%	2%	3%	5%		
Dibucaine hydrochloride	0.14 *0.040°*	0.13 *0.07°*	0.12 *0.13°*	0.11 *0.18°*	0.08 *0.22°*	— —	
Dibutoline sulfate	0.18 *0.049°*	0.16 *0.09°*	0.15 *0.17°*	0.15 *0.25°*	0.14 *0.41°*	— —	
Dichlorophenarsine hydrochloride	0.55 *0.150°*	0.55 *0.31°*	— —	— —	— —	0.55 *0.52°*	**1.64%** **1.64%**
Dicloxacillin sodium monohydrate	0.10 *0.030°*	0.10 *0.06°*	0.10 *0.12°*	0.10 *0.18°*	— —	— —	
Dicyclomine hydrochloride	0.18 *0.052°*	0.18 *0.10°*	0.17 *0.20°*	0.17 *0.29°*	— —	— —	
Diethanolamine	0.31 *0.089°*	0.31 *0.17°*	0.31 *0.35°*	— —	— —	0.31 *0.52°*	**2.90%** **2.90%**
Diethylcarbamazine citrate	0.14 *0.042°*	0.14 *0.08°*	0.14 *0.16°*	0.14 *0.24°*	0.14 *0.41°*	0.14 *0.52°*	**6.29%** **6.29%**
Dihydrocodeinone enol acetate hydrochloride	0.15 *0.042°*	0.14 *0.08°*	0.13 *0.15°*	0.13 *0.21°*	0.12 *0.34°*	0.12 *0.52°*	**7.76%** **7.76%**
Dihydrostreptomycin sulfate	0.08 *0.017°*	0.06 *0.03°*	0.06 *0.05°*	0.05 *0.08°*	0.05 *0.13°*	0.04 *0.52°*	**21.4%** **21.4%**
Dimethindene maleate	0.13 *0.039°*	0.12 *0.07°*	0.11 *0.12°*	— —	— —	— —	
Dimethyl sulfoxide	0.42 *0.122°*	0.42 *0.24°*	0.42 *0.48°*	— —	— —	0.42 *0.52°*	**2.16%** **2.16%**
Diperodon hydrochloride	0.15 *0.045°*	0.14 *0.07°*	0.13 *0.14°*	— —	— —	— —	
Diphemanil methylsulfate	0.16 *0.047°*	0.15 *0.08°*	— —	— —	— —	— —	
Diphenhydramine hydrochloride	0.34 *0.099°*	0.27 *0.15°*	0.22 *0.25°*	0.20 *0.33°*	0.17 *0.47°*	— —	
Diphenidol hydrochloride	0.16 *0.045°*	0.16 *0.09°*	0.16 *0.18°*	— —	— —	— —	
Dipivefrin hydrochloride	0.19 *0.052°*	0.17 *0.09°*	0.15 *0.17°*	0.14 *0.24°*	0.12 *0.32°*	— —	
Dipyrone	0.20 *0.057°*	0.19 *0.11°*	0.19 *0.22°*	0.19 *0.33°*	— —	0.19 *0.52°*	**4.65%** **4.65%**
Disodium edetate	0.24 *0.070°*	0.23 *0.13°*	0.22 *0.24°*	0.21 *0.36°*	— —	0.20 *0.52°*	**4.44%** **4.44%**
Dobutamine hydrochloride	0.20 *0.053°*	0.18 *0.10°*	0.16 *0.18°*	— —	— —	— —	—
Dopamine hydrochloride	0.30 *0.085°*	0.30 *0.17°*	0.29 *0.33°*	0.29 *0.50°*	— —	0.29 *0.52°*	**3.11%** **3.11%**
Doxapram hydrochloride	0.12 *0.035°*	0.12 *0.07°*	0.12 *0.14°*	0.12 *0.21°*	— —	— —	
Doxycycline hyclate	0.12 *0.035°*	0.12 *0.07°*	0.12 *0.13°*	0.11 *0.18°*	0.09 *0.26°*	— —	
Dyclonine hydrochloride	0.26 *0.073°*	0.24 *0.13°*	0.17 *0.19°*	— —	— —	— —	

Chemical	Concentration of Solution, NaCl Equivalents						At Isosmotic Concentration
	0.5%	1%	2%	3%	5%		
Dyphylline	0.10 *0.025°*	0.10 *0.05°*	0.09 *0.10°*	0.09 *0.15°*	0.08 *0.24°*	— —	
Echothiopate iodide	0.16 *0.045°*	0.16 *0.09°*	0.16 *0.17°*	— —	— —	— —	
Edrophonium chloride	0.32 *0.093°*	0.31 *0.17°*	0.29 *0.32°*	0.27 *0.47°*	— —	0.27 *0.52°*	**3.36%** **3.36%**
Emetine hydrochloride	0.12 *0.033°*	0.10 *0.06°*	0.10 *0.11°*	0.10 *0.17°*	0.10 *0.27°*	— —	
Encainide hydrochloride	0.16 *0.045°*	0.15 *0.08°*	0.14 *0.15°*	0.11 *0.22°*	0.10 *0.35°*	— —	
Ephedrine hydrochloride	0.32 *0.087°*	0.30 *0.16°*	0.29 *0.33°*	0.28 *0.48°*	— —	0.28 *0.52°*	**3.2%** **3.2%**
Ephedrine lactate	0.28 *0.075°*	0.26 *0.14°*	0.25 *0.28°*	0.24 *0.42°*	— —	0.24 *0.52°*	**3.72%** **3.72%**
Ephedrine sulfate	0.24 *0.070°*	0.23 *0.13°*	0.22 *0.24°*	0.20 *0.35°*	— —	0.20 *0.52°*	**4.54%** **4.54%**
Epinephrine bitartrate	0.18 *0.050°*	0.18 *0.09°*	0.17 *0.19°*	0.16 *0.28°*	0.16 *0.45°*	0.16 *0.52°*	**5.7%** **5.7%**
Epinephrine hydrochloride	0.30 *0.088°*	0.29 *0.16°*	0.27 *0.31°*	0.26 *0.45°*	— —	0.26 *0.52°*	**3.47%** **3.47%**
Ergonovine maleate	0.20 *0.055°*	0.16 *0.08°*	0.13 *0.14°*	— —	— —	— —	
Erythromycin glucoheptonate	0.08 *0.021°*	0.07 *0.04°*	0.07 *0.08°*	0.07 *0.12°*	0.07 *0.19°*	— —	
Erythromycin lactobionate	0.08 *0.020°*	0.07 *0.04°*	0.07 *0.07°*	0.07 *0.11°*	0.06 *0.18°*	— —	
Ethaverine hydrochloride	0.14 *0.037°*	0.12 *0.07°*	— —	— —	— —	— —	
Ethylenediamine	0.46 *0.130°*	0.44 *0.25°*	0.43 *0.50°*	— —	— —	— —	
Ethylhydrocupreine hydrochloride	0.22 *0.063°*	0.17 *0.09°*	0.13 *0.15°*	0.11 *0.19°*	0.09 *0.27°*	— —	
Ethylmorphine hydrochloride	0.16 *0.045°*	0.16 *0.08°*	0.15 *0.17°*	0.15 *0.25°*	0.15 *0.42°*	0.15 *0.52°*	**6.18%** **6.18%**
Ethylnorepinephrine hydrochloride	0.36 *0.104°*	0.32 *0.14°*	0.29 *0.33°*	0.28 *0.47°*	— —	0.27 *0.52°*	**3.32%** **3.32%**
Etidocaine hydrochloride	0.18 *0.051°*	0.18 *0.10°*	0.18 *0.20°*	0.18 *0.30°*	0.18 *0.51°*	0.18 *0.52°*	**5.08%** **5.08%**
Evans blue	0.06 *0.017°*	0.06 *0.03°*	0.06 *0.06°*	0.05 *0.09°*	0.05 *0.14°*	— —	
Ferric ammonium citrate, green	0.18 *0.054°*	0.17 *0.09°*	0.16 *0.17°*	0.15 *0.25°*	0.14 *0.39°*	— —	
Ferric cacodylate	0.10 *0.023°*	0.09 *0.04°*	0.08 *0.09°*	— —	— —	— —	
Ferrous gluconate	0.16 *0.048°*	0.15 *0.08°*	0.14 *0.15°*	0.12 *0.21°*	0.11 *0.33°*	— —	

Chemical	Concentration of Solution, NaCl Equivalents						
	0.5%	1%	2%	3%	5%	At Isosmotic Concentration	
Ferrous lactate	0.22 *0.062°*	0.21 *0.12°*	0.21 *0.23°*	— —	— —	— —	
Floxuridine	0.14 *0.040°*	0.13 *0.07°*	0.13 *0.14°*	0.12 *0.21°*	0.12 *0.33°*	0.12 *0.52°*	**8.47%** **8.47%**
Fluorescein sodium	0.36 *0.099°*	0.31 *0.18°*	0.29 *0.33°*	0.27 *0.47°*	— —	0.27 *0.52°*	**3.34%** **3.34%**
Fluorouracil	0.16 *0.045°*	0.13 *0.07°*	— —	— —	— —	— —	
Fluphenazine dihydrochloride	0.14 *0.041°*	0.14 *0.08°*	0.12 *0.14°*	0.09 *0.15°*	— —	— —	
Folinic acid-SF calcium	0.06 *0.013°*	0.05 *0.02°*	0.05 *0.05°*	0.04 *0.07°*	0.04 *0.12°*	— —	
D-Fructose	0.18 *0.050°*	0.18 *0.10°*	0.18 *0.20°*	0.18 *0.31°*	0.18 *0.51°*	0.18 *0.52°*	**5.05%** **5.05%**
Furtrethonium iodide	0.24 *0.070°*	0.24 *0.13°*	0.22 *0.25°*	0.21 *0.36°*	— —	0.20 *0.52°*	**4.44%** **4.44%**
Galactose, anhydrous	0.18 *0.053°*	0.18 *0.10°*	0.18 *0.21°*	0.18 *0.31°*	— —	0.18 *0.52°*	**4.92%** **4.92%**
Gallamine triethiodide	0.08 *0.022°*	0.08 *0.04°*	0.08 *0.09°*	0.08 *0.13°*	0.08 *0.22°*	— —	
Gentamicin sulfate	0.05 *0.015°*	0.05 *0.03°*	0.05 *0.06°*	0.05 *0.09°*	0.05 *0.15°*	— —	
Glucoheptonate calcium	0.12 *0.037°*	0.12 *0.06°*	0.11 *0.12°*	0.10 *0.17°*	0.10 *0.27°*	— —	
Glucosulfone sodium	0.18 *0.049°*	0.16 *0.08°*	0.14 *0.16°*	0.13 *0.23°*	0.13 *0.36°*	— —	
D-Glucuronic acid	0.20 *0.061°*	0.20 *0.11°*	0.19 *0.22°*	0.19 *0.32°*	0.18 *0.51°*	0.18 *0.52°*	**5.02%** **5.02%**
L-Glutamic acid	0.25 *0.070°*	0.25 *0.14°*	0.25 *0.29°*	— —	— —	— —	
Glycerin	0.36 *0.104°*	0.35 *0.20°*	0.35 *0.40°*	— —	— —	0.35 *0.52°*	**2.6%** **2.6%**
Glycine	0.41 *0.118°*	0.41 *0.23°*	0.41 *0.47°*	— —	— —	0.41 *0.52°*	**2.19%** **2.19%**
Glycopyrrolate	0.15 *0.042°*	0.15 *0.08°*	0.15 *0.16°*	0.14 *0.24°*	0.13 *0.38°*	0.12 *0.52°*	**7.22%** **7.22%**
Gnoscopine hydrochloride	0.11 *0.032°*	0.10 *0.05°*	0.09 *0.10°*	0.08 *0.14°*	0.08 *0.22°*	— —	
Gold sodium thiomalate	0.10 *0.032°*	0.10 *0.06°*	0.10 *0.11°*	0.09 *0.15°*	0.09 *0.25°*	— —	
Guanidine hydrochloride	0.72 *0.208°*	0.65 *0.37°*	— —	— —	— —	0.61 *0.52°*	**1.47%** **1.47%**
Heparin sodium	0.07 *0.021°*	0.07 *0.04°*	0.07 *0.08°*	0.07 *0.12°*	0.07 *0.21°*	0.07 *0.52°*	**12.2%** **12.2%**
Hepes	0.17 *0.048°*	0.16 *0.08°*	0.14 *0.16°*	0.14 *0.23°*	0.13 *0.38°*	0.13 *0.52°*	**6.80%** **6.80%**

Chemical	Concentration of Solution, NaCl Equivalents						
	0.5%	1%	2%	3%	5%	At Isosmotic Concentration	
Hetacillin potassium	0.17 *0.048°*	0.17 *0.09°*	0.17 *0.19°*	0.17 *0.28°*	0.17 *0.47°*	0.17 *0.52°*	**5.50%** **5.50%**
Hexafluorenium bromide	0.12 *0.033°*	0.11 *0.06°*	— —	— —	— —	— —	
Hexamethonium bromide	0.24 *0.069°*	0.22 *0.12°*	0.20 *0.23°*	0.19 *0.33°*	— —	0.18 *0.52°*	**4.99%** **4.99%**
Hexamethonium chloride	0.27 *0.078°*	0.27 *0.15°*	0.27 *0.31°*	0.27 *0.47°*	— —	0.27 *0.52°*	**3.3%** **3.3%**
Hexamethonium tartrate	0.16 *0.045°*	0.16 *0.08°*	0.16 *0.18°*	0.16 *0.27°*	0.16 *0.45°*	0.16 *0.52°*	**5.68%** **5.68%**
Hexamethylenamine sodium acetaminosalicylate	0.18 *0.049°*	0.18 *0.09°*	0.17 *0.19°*	0.17 *0.29°*	0.16 *0.48°*	0.16 *0.52°*	**5.48%** **5.48%**
Hexobarbital sodium	0.28 *0.078°*	0.26 *0.14°*	0.25 *0.28°*	0.24 *0.40°*	— —	0.23 *0.52°*	**3.88%** **3.88%**
Hexylcaine hydrochloride	0.28 *0.084°*	0.26 *0.15°*	0.24 *0.27°*	0.22 *0.38°*	— —	— —	
Histamine dihydrochloride	0.40 *0.115°*	0.40 *0.23°*	0.40 *0.46°*	— —	— —	0.40 *0.52°*	**2.24%** **2.24%**
Histamine phosphate	0.28 *0.080°*	0.25 *0.14°*	0.24 *0.27°*	0.23 *0.39°*	— —	0.22 *0.52°*	**4.1%** **4.1%**
Histidine monohydrochloride	0.30 *0.082°*	0.29 *0.16°*	0.28 *0.31°*	0.26 *0.46°*	— —	— —	
Homatropine hydrobromide	0.18 *0.049°*	0.17 *0.09°*	0.17 *0.18°*	0.16 *0.28°*	0.16 *0.46°*	0.16 *0.52°*	**5.67%** **5.67%**
Homatropine methylbromide	0.20 *0.060°*	0.19 *0.10°*	0.17 *0.18°*	0.15 *0.25°*	0.13 *0.39°*	—	
Hyaluronidase	0.01 *0.004°*	0.01 *0.00°*	0.01 *0.01°*	0.01 *0.02°*	0.01 *0.03°*	—	
Hydralazine hydrochloride	0.44 *0.126°*	0.37 *0.21°*	— —	— —	— —	—	
Hydrastine hydrochloride	0.18 *0.052°*	0.15 *0.08°*	0.14 *0.15°*	0.12 *0.20°*	0.11 *0.31°*	—	
Hydromorphone hydrochloride	0.26 *0.073°*	0.22 *0.12°*	0.19 *0.21°*	0.17 *0.28°*	0.15 *0.42°*	0.14 *0.52°*	**6.39%** **6.39%**
Hydroxyamphetamine hydrobromide	0.28 *0.083°*	0.26 *0.15°*	0.26 *0.29°*	0.25 *0.43°*	— —	0.24 *0.52°*	**3.71%** **3.71%**
Hydroxychloroquine phosphate	0.20 *0.059°*	0.18 *0.10°*	0.16 *0.18°*	0.15 *0.25°*	0.13 *0.38°*	—	
8-Hydroxyquinoline sulfate	0.26 *0.071°*	0.21 *0.11°*	0.16 *0.18°*	0.14 *0.23°*	0.12 *0.33°*	0.11 *0.52°*	**9.75%** **9.75%**
Hydroxystilbamidine isethionate	0.20 *0.060°*	0.16 *0.09°*	0.12 *0.13°*	0.10 *0.17°*	0.07 *0.21°*	—	
Hydroxyzine hydrochloride	0.26 *0.075°*	0.25 *0.13°*	0.22 *0.25°*	0.20 *0.34°*	0.16 *0.45°*	0.14 *0.52°*	**6.32%** **6.32%**
Hyoscyamine hydrobromide	0.20 *0.059°*	0.19 *0.10°*	0.17 *0.19°*	0.16 *0.27°*	0.14 *0.41°*	—	

Chemical	Concentration of Solution, NaCl Equivalents						At Isosmotic Concentration
	0.5%	1%	2%	3%	5%		
Hyoscyamine sulfate	0.17 *0.048°*	0.15 *0.08°*	0.13 *0.14°*	0.12 *0.20°*	0.11 *0.31°*	— —	
Imipramine hydrochloride	0.20 *0.058°*	0.20 *0.11°*	— —	— —	— —	— —	
Indigotindisulfonate sodium	0.30 *0.085°*	0.30 *0.17°*	— —	— —	— —	— —	
o-Iodohippurate sodium	0.16 *0.047°*	0.16 *0.09°*	0.16 *0.18°*	0.15 *0.26°*	0.15 *0.44°*	0.15 *0.52°*	**5.92%** **5.92%**
Iodophthalein sodium	0.20 *0.055°*	0.17 *0.09°*	0.14 *0.15°*	0.12 *0.21°*	0.11 *0.31°*	0.09 *0.52°*	**9.58%** **9.58%**
Iodopyracet	0.12 *0.036°*	0.11 *0.06°*	0.11 *0.12°*	0.11 *0.18°*	0.10 *0.29°*	0.10 *0.52°*	**9.21%** **9.21%**
Iodopyracet diethylamine	0.14 *0.035°*	0.12 *0.06°*	0.12 *0.13°*	0.11 *0.19°*	0.11 *0.30°*	0.10 *0.52°*	**8.73%** **8.73%**
Iopamidol	0.03 *0.008°*	0.03 *0.01°*	0.03 *0.03°*	0.03 *0.05°*	0.03 *0.08°*	— —	
Isoetharine hydrochloride	0.24 *0.068°*	0.23 *0.13°*	0.22 *0.25°*	0.21 *0.36°*	— —	0.21 *0.52°*	**4.27%** **4.27%**
Isometheptene mucate	0.18 *0.048°*	0.18 *0.09°*	0.18 *0.19°*	0.18 *0.30°*	— —	0.18 *0.52°*	**4.95%** **4.95%**
Isoniazid	0.28 *0.079°*	0.25 *0.14°*	0.23 *0.26°*	0.22 *0.37°*	— —	0.21 *0.52°*	**4.35%** **4.35%**
Isopropyl Alcohol	0.53 *0.153°*	0.53 *0.30°*	— —	— —	— —	0.53 *0.52°*	**1.71%** **1.71%**
Isoproterenol sulfate	0.14 *0.039°*	0.14 *0.07°*	0.14 *0.15°*	0.14 *0.23°*	0.14 *0.38°*	0.14 *0.52°*	**6.65%** **6.65%**
Kanamycin sulfate	0.08 *0.021°*	0.07 *0.04°*	0.07 *0.08°*	0.07 *0.12°*	0.07 *0.21°*	— —	
Ketamine hydrochloride	0.21 *0.061°*	0.21 *0.12°*	0.21 *0.24°*	0.21 *0.36°*	— —	0.21 *0.52°*	**4.29%** **4.29%**
Labetalol hydrochloride	0.20 *0.059°*	0.19 *0.10°*	— —	— —	— —	— —	
Lactic acid	0.44 *0.124°*	0.41 *0.23°*	0.39 *0.45°*	— —	— —	0.39 *0.52°*	**2.3%** **2.3%**
Lactose	0.06 *0.019°*	0.07 *0.04°*	0.08 *0.08°*	0.08 *0.13°*	0.09 *0.24°*	0.09 *0.52°*	**9.75%** **9.75%**
Levallorphan tartrate	0.13 *0.036°*	0.13 *0.07°*	0.13 *0.14°*	0.12 *0.21°*	0.12 *0.32°*	0.10 *0.52°*	**9.40%** **9.40%**
Levobunolol hydrochloride	0.12 *0.035°*	0.12 *0.07°*	0.12 *0.01°*	0.12 *0.21°*	0.12 *0.36°*	— —	
Levorphanol tartrate	0.12 *0.033°*	0.12 *0.06°*	0.12 *0.13°*	0.12 *0.20°*			
Lidocaine hydrochloride	0.22 *0.065°*	0.22 *0.12°*	0.21 *0.24°*	0.21 *0.35°*	— —	0.20 *0.52°*	**4.42%** **4.42%**
Lincomycin hydrochloride	0.16 *0.045°*	0.16 *0.09°*	0.15 *0.17°*	0.14 *0.24°*	0.14 *0.40°*	0.14 *0.52°*	**6.60%** **6.60%**

	Concentration of Solution, NaCl Equivalents						
Chemical	0.5%	1%	2%	3%	5%	At Isosmotic Concentration	
Lithium carbonate	1.06	—	—	—	—	0.98	**0.92%**
	0.303°	—	—	—	—	*0.52°*	**0.92%**
Lithium chloride	1.40	—	—	—	—	1.40	**0.65%**
	0.405°	—	—	—	—	*0.52°*	**0.65%**
Lobeline hydrochloride	0.16	0.16	0.16	—	—	—	
	0.047°	*0.09°*	*0.17°*	—	—	—	
Lyapolate sodium	0.10	0.09	0.09	0.09	0.09	0.09	**9.96%**
	0.025°	*0.05°*	*0.10°*	*0.15°*	*0.26°*	*0.52°*	**9.96%**
Mafenide hydrochloride	0.27	0.27	0.27	0.26	—	0.25	**3.55%**
	0.075°	*0.15°*	*0.30°*	*0.44°*	—	*0.52°*	**3.55%**
Magnesium chloride	0.48	0.45	0.45	—	—	0.45	**2.02%**
	0.136°	*0.26°*	*0.51°*	—	—	*0.52°*	**2.02%**
Magnesium sulfate	0.18	0.17	0.16	0.15	0.15	0.14	**6.3%**
	0.049°	*0.09°*	*0.17°*	*0.26°*	*0.41°*	*0.52°*	**6.3%**
Magnesium sulfate, anhydrous	0.34	0.32	0.30	0.29	—	0.28	**3.18%**
	0.093°	*0.18°*	*0.34°*	*0.49°*	—	*0.52°*	**3.18%**
Mannitol	0.16	0.17	0.17	0.17	0.18	0.18	**5.07%**
	0.047°	*0.09°*	*0.20°*	*0.30°*	*0.51°*	*0.52°*	**5.07%**
Menadiol sodium diphosphate	0.27	0.25	0.23	0.21	—	—	
	0.078°	*0.14°*	*0.26°*	*0.37°*	—	—	
Menadione sodium bisulfite	0.20	0.20	0.19	0.18	0.18	0.18	**5.07%**
	0.057°	*0.11°*	*0.21°*	*0.31°*	*0.51°*	*0.52°*	**5.07%**
Meperidine hydrochloride	0.24	0.22	0.21	0.20	—	0.19	**4.8%**
	0.066°	*0.12°*	*0.23°*	*0.34°*	—	*0.52°*	**4.8%**
Mephenesin	0.19	0.19	—	—	—	—	
	0.055°	*0.10°*	—	—	—	—	
Mephentermine sulfate	0.24	0.22	0.21	0.20	—	0.19	**4.74%**
	0.069°	*0.13°*	*0.24°*	*0.34°*	—	*0.52°*	**4.74%**
Mepivacaine hydrochloride	0.21	0.21	0.20	0.20	—	0.20	**4.6%**
	0.060°	*0.11°*	*0.23°*	*0.34°*	—	*0.52°*	**4.6%**
Merbromin	0.16	0.14	0.12	0.11	0.09	—	
	0.044°	*0.08°*	*0.13°*	*0.18°*	*0.27°*	—	
Mercaptomerin sodium	0.19	0.18	0.18	0.18	0.17	—	
	0.056°	*0.10°*	*0.20°*	*0.30°*	*0.49°*	—	
Mercuric cyanide	0.16	0.15	0.15	0.14	0.13	—	
	0.047°	*0.08°*	*0.16°*	*0.23°*	*0.38°*	—	
Mercurophylline	0.14	0.13	0.11	0.10	0.09	—	
	0.042°	*0.07°*	*0.12°*	*0.17°*	*0.26°*	—	
Mercury bichloride	0.14	0.13	0.12	0.12	0.10	—	
	0.038°	*0.07°*	*0.14°*	*0.20°*	*0.33°*	—	
Mersalyl	0.14	0.12	0.11	0.11	0.10	0.10	**9.06%**
	0.041°	*0.06°*	*0.12°*	*0.18°*	*0.29°*	*0.52°*	**9.06%**
Mesoridazine besylate	0.10	0.07	0.05	0.04	0.03	—	
	0.024°	*0.04°*	*0.05°*	*0.07°*	*0.08°*	—	
Metaraminol bitartrate	0.20	0.20	0.19	0.18	0.17	0.17	**5.17%**
	0.060°	*0.11°*	*0.21°*	*0.30°*	*0.50°*	*0.52°*	**5.17%**

Chemical	Concentration of Solution, NaCl Equivalents					At Isosmotic Concentration	
	0.5%	1%	2%	3%	5%		
Methacholine bromide	0.29	0.28	0.26	0.24	—	0.24	3.77%
	0.087°	0.16°	0.29°	0.42°	—	0.52°	3.77%
Methacholine chloride	0.34	0.32	0.30	0.28	—	0.28	3.21%
	0.099°	0.18°	0.38°	0.49°	—	0.52°	3.21%
Methadone hydrochloride	0.22	0.18	0.15	0.14	0.12	0.10	8.59%
	0.060°	0.10°	0.17°	0.23°	0.34°	0.52°	8.59%
Methamphetamine hydrochloride	0.38	0.37	0.34	—	—	0.33	2.75%
	0.112°	0.20°	0.38°	—	—	0.52°	2.75%
Methantheline bromide	0.22	0.15	0.11	0.09	0.07	—	
	0.063°	0.08°	0.12°	0.15°	0.19°	—	
Methapyrilene hydrochloride	0.20	0.19	0.18	0.18	0.17	0.17	5.35%
	0.060°	0.11°	0.21°	0.30°	0.48°	0.52°	5.35%
Methdilazine hydrochloride	0.12	0.10	0.08	0.06	0.04	—	
	0.035°	0.05°	0.08°	0.09°	0.11°	—	
Methenamine	0.22	0.23	0.24	0.24	—	0.24	3.68%
	0.061°	0.12°	0.27°	0.41°	—	0.52°	3.68%
Methicillin sodium	0.18	0.18	0.17	0.16	0.15	0.15	6.00%
	0.050°	0.09°	0.19°	0.28°	0.44°	0.52°	6.00%
Methiodal sodium	0.24	0.24	0.24	0.24	—	0.24	3.81%
	0.068°	0.13°	0.27°	0.41°	—	0.52°	3.81%
Methionine	0.32	0.28	0.25	—	—	—	
	0.091°	0.16°	0.28°	—	—	—	
Methitural sodium	0.26	0.25	0.24	0.23	—	0.23	3.85%
	0.074°	0.14°	0.27°	0.40°	—	0.52°	3.85%
Methocarbamol	0.10	0.10	—	—	—	—	
	0.030°	0.06°	—	—	—	—	
Methotrimeprazine hydrochloride	0.23	0.10	0.07	0.06	0.04	—	
	0.034°	0.06°	0.07°	0.09°	0.12°	—	
Methoxamine hydrochloride	0.28	0.26	0.25	0.24	—	0.24	3.82%
	0.078°	0.14°	0.28°	0.41°	—	0.52°	3.82%
Methoxyphenamine hydrochloride	0.26	0.26	0.26	0.26	—	0.52	3.47%
	0.075°	0.15°	0.30°	0.45°	—	0.52°	3.47%
Methylatropine bromide	0.15	0.15	0.14	0.14	0.13	0.13	7.03%
	0.045°	0.08°	0.16°	0.23°	0.38°	0.52°	7.03%
Methyldopa ethyl ester hydrochloride	0.21	0.21	0.21	0.21	—	0.21	4.28%
	0.063°	0.12°	0.24°	0.36°	—	0.52°	4.28%
Methylergonovine maleate	0.10	0.10	—	—	—	—	
	0.028°	0.05°	—	—	—	—	
N-Methylglucamine	0.20	0.20	0.18	0.18	0.18	0.18	5.02%
	0.057°	0.11°	0.21°	0.31°	0.51°	0.52°	5.02%
Methylphenidate hydrochloride	0.22	0.22	0.22	0.22	—	0.20	4.07%
	0.065°	0.12°	0.25°	0.38°	—	0.52°	4.07%
Methylprednisolone sodium succinate	0.10	0.09	0.09	0.08	0.07	—	
	0.025°	0.05°	0.10°	0.14°	0.20°	—	
Metoclopramide hydrochloride	0.16	0.15	0.13	0.12	0.11	—	
	0.045°	0.08°	0.15°	0.21°	0.31°	—	

Chemical	Concentration of Solution, NaCl Equivalents						
	0.5%	1%	2%	3%	5%	At Isosmotic Concentration	
Metrizamide	0.04 *0.010°*	0.04 *0.02°*	0.03 *0.04°*	0.03 *0.06°*	— —	— —	
Mezlocillin sodium	0.13 *0.037°*	0.11 *0.06°*	0.11 *0.12°*	0.10 *0.16°*	0.09 *0.25°*	— —	
Minocycline hydrochloride	0.10 *0.030°*	0.10 *0.05°*	0.09 *0.10°*	0.08 *0.14°*	— —	— —	
Monoethanolamine	0.53 *0.154°*	0.53 *0.30°*	— —	— —	— —	0.53 *0.52°*	**1.70%** **1.70%**
Morphine hydrochloride	0.16 *0.044°*	0.15 *0.08°*	0.15 *0.16°*	0.14 *0.24°*	— —	— —	
Morphine nitrate	0.22 *0.061°*	0.19 *0.10°*	0.16 *0.18°*	0.15 *0.25°*	— —	— —	
Morphine sulfate	0.16 *0.046°*	0.14 *0.07°*	0.12 *0.13°*	0.11 *0.17°*	0.09 *0.25°*	— —	
Moxalactam disodium	0.20 *0.054°*	0.17 *0.09°*	0.16 *0.17°*	0.15 *0.25°*	0.14 *0.41°*	— —	
Naepaine hydrochloride	0.24 *0.067°*	0.22 *0.12°*	0.20 *0.23°*	0.19 *0.33°*	— —	0.18 *0.52°*	**4.98%** **4.98%**
Nafcillin sodium	0.14 *0.039°*	0.14 *0.07°*	0.14 *0.15°*	0.13 *0.21°*	0.10 *0.28°*	— —	
Nalbuphine hydrochloride	0.16 *0.045°*	0.15 *0.08°*	0.14 *0.15°*	— —	— —	— —	
Nalorphine hydrochloride	0.24 *0.070°*	0.21 *0.12°*	0.18 *0.21°*	0.17 *0.28°*	0.15 *0.43°*	0.24 *0.52°*	**6.36%** **6.36%**
Naloxone hydrochloride	0.14 *0.042°*	0.14 *0.08°*	0.14 *0.15°*	0.13 *0.23°*	0.13 *0.36°*	0.11 *0.52°*	**8.07%** **8.07%**
Naltrexone hydrochloride	0.17 *0.047°*	0.16 *0.09°*	0.15 *0.17°*	0.14 *0.25°*	— —	— —	
Naphazoline hydrochloride	0.30 *0.084°*	0.27 *0.15°*	0.25 *0.28°*	0.24 *0.41°*	— —	0.22 *0.52°*	**3.99%** **3.99%**
Neoarsphenamine	0.42 *0.116°*	0.40 *0.22°*	0.39 *0.44°*	— —	— —	0.39 *0.52°*	**2.32%** **2.32%**
Neomycin sulfate	0.14 *0.041°*	0.12 *0.06°*	0.10 *0.11°*	0.09 *0.15°*	0.08 *0.22°*	— —	
Neostigmine bromide	0.23 *0.065°*	0.22 *0.12°*	0.20 *0.23°*	0.19 *0.33°*	— —	— —	
Neostigmine methyl sulfate	0.22 *0.056°*	0.20 *0.10°*	0.18 *0.20°*	0.18 *0.30°*	0.17 *0.50°*	0.17 *0.52°*	**5.22%** **5.22%**
Netilmicin sulfate	0.09 *0.023°*	0.07 *0.04°*	0.06 *0.07°*	0.06 *0.10°*	0.06 *0.15°*	— —	
Nicotinamide	0.30 *0.083°*	0.26 *0.14°*	0.23 *0.26°*	0.21 *0.37°*	— —	0.20 *0.52°*	**4.49%** **4.49%**
Nicotinic acid	0.26 *0.074°*	0.25 *0.14°*	— —	— —	— —	— —	
Nikethamide	0.20 *0.053°*	0.18 *0.10°*	0.17 *0.19°*	0.16 *0.27°*	0.15 *0.44°*	0.15 *0.52°*	**5.94%** **5.94%**

Chemical	Concentration of Solution, NaCl Equivalents					At Isosmotic Concentration	
	0.5%	1%	2%	3%	5%		
Novobiocin sodium	0.10	0.08	0.08	0.07	0.07	—	
	0.25°	0.04°	0.08°	0.12°	0.19°	—	
Oleandomycin phosphate	0.08	0.08	0.08	0.08	0.08	0.08	10.82%
	0.017°	0.03°	0.08°	0.12°	0.25°	0.52°	10.82%
Orphenadrine citrate	0.13	0.13	0.13	0.12	0.10	—	
	0.037°	0.07°	0.14°	0.20°	0.28°	—	
Oxacillin sodium	0.18	0.17	0.16	0.15	0.14	0.14	6.64%
	0.050°	0.09°	0.17°	0.25°	0.40°	0.52°	6.64%
Oxophenarsine hydrochloride	0.24	0.24	0.24	0.24	—	0.24	3.67%
	0.067°	0.13°	0.28°	0.42°	—	0.52°	3.67%
Oxycodone	0.16	0.14	0.14	0.13	0.13	0.12	7.4%
	0.043°	0.08°	0.15°	0.22°	0.36°	0.52°	7.4%
Oxymetazoline hydrochloride	0.22	0.22	0.29	0.19	—	0.18	4.92%
	0.063°	0.12°	0.23°	0.33°	—	0.52°	4.92%
Oxymorphone hydrochloride	0.16	0.16	0.15	0.14	0.13	—	
	0.044°	0.08°	0.16°	0.24°	0.38°	—	
Oxytetracycline hydrochloride	0.17	0.14	0.11	0.08	—	—	
	0.052°	0.08°	0.11°	0.14°	—	—	
Pancuronium bromide	0.16	0.13	0.11	0.10	0.10	—	
	0.046°	0.07°	0.12°	0.17°	0.28°	—	
Papaverine hydrochloride	0.10	0.10	0.10	—	—	—	
	0.028°	0.06°	0.12°	—	—	—	
Paraldehyde	0.25	0.25	0.25	0.25	—	0.25	3.65%
	0.071°	0.14°	0.28°	0.43°	—	0.52°	3.65%
Parethoxycaine hydrochloride	0.20	0.20	—	—	—	—	
	0.058°	0.11°	—	—	—	—	
Pargyline hydrochloride	0.30	0.29	0.29	0.28	—	0.28	3.18%
	0.093°	0.16°	0.32°	0.49°	—	0.52°	3.18%
Pentazocine lactate	0.15	0.15	0.15	0.15	0.15	—	
	0.042°	0.08°	0.16°	0.25°	0.42°	—	
Pentobarbital sodium	0.26	0.25	0.24	0.23	—	—	
	0.076°	0.14°	0.27°	0.39°	—	—	
Pentolinium tartrate	0.18	0.17	0.16	0.15	0.15	—	
	0.050°	0.09°	0.18°	0.26°	0.44°	—	
Pentylenetetrazole	0.24	0.22	0.21	0.19	—	0.18	4.91%
	0.069°	0.12°	0.23°	0.33°	—	0.52°	4.91%
Phenacaine hydrochloride	0.22	0.20	—	—	—	—	
	0.061°	0.10°	—	—	—	—	
Phenarsone sulfoxylate	0.36	0.33	0.31	0.29	—	0.29	3.07%
	0.104°	0.19°	0.35°	0.50°	—	0.52°	3.07%
Phenindamine tartrate	0.22	0.17	0.14	0.12	0.10	—	
	0.064°	0.10°	0.15°	0.20°	0.28°	—	
Pheniramine maleate	0.18	0.16	0.15	0.14	0.13	—	
	0.052°	0.09°	0.17°	0.24°	0.38°	—	
Phenobarbital sodium	0.24	0.24	0.23	0.23	—	0.23	3.95%
	0.069°	0.13°	0.26°	0.39°	—	0.52°	3.95%

Chemical	Concentration of Solution, NaCl Equivalents						
	0.5%	1%	2%	3%	5%	At Isosmotic Concentration	
Phenol	0.38 *0.104°*	0.35 *0.19°*	0.33 *0.38°*	— —	— —	0.32 *0.52°*	**2.8%** **2.8%**
Phentolamine mesylate	0.18 *0.052°*	0.17 *0.09°*	0.16 *0.17°*	0.14 *0.24°*	0.13 *0.36°*	0.11 *0.52°*	**8.23%** **8.23%**
Phenylbutazone sodium	0.19 *0.054°*	0.18 *0.10°*	0.17 *0.20°*	0.17 *0.29°*	0.17 *0.48°*	0.17 *0.52°*	**5.34%** **5.34%**
Phenylephrine hydrochloride	0.34 *0.096°*	0.32 *0.18°*	0.31 *0.35°*	0.30 *0.52°*	— —	0.30 *0.52°*	**3.0%** **3.0%**
Phenylephrine tartrate	0.20 *0.055°*	0.19 *0.10°*	0.17 *0.19°*	0.16 *0.28°*	0.16 *0.44°*	0.15 *0.52°*	**5.9%** **5.9%**
Phenylethyl alcohol	0.25 *0.070°*	0.25 *0.14°*	0.25 *0.28°*	— —	— —	— —	
Phenylpropanolamine hydrochloride	0.40 *0.117°*	0.38 *0.21°*	0.35 *0.40°*	— —	— —	0.35 *0.52*	**2.6%** **2.6%**
Phenylpropylmethylamine hydrochloride	0.42 *0.123°*	0.38 *0.22°*	0.34 *0.39°*	— —	— —	0.33 *0.52°*	**2.7%** **2.7%**
Physostigmine salicylate	0.16 *0.045°*	0.16 *0.09°*	— —	— —	— —	— —	
Physostigmine sulfate	0.14 *0.040°*	0.13 *0.07°*	0.13 *0.14°*	0.12 *0.21°*	0.12 *0.34°*	0.12 *0.52°*	**7.74%** **7.74%**
Pilocarpine hydrochloride	0.24 *0.069°*	0.24 *0.13°*	0.23 *0.26°*	0.22 *0.38°*	— —	0.22 *0.52°*	**4.08%** **4.08%**
Pilocarpine nitrate	0.24 *0.070°*	0.23 *0.13°*	0.21 *0.24°*	0.20 *0.35°*	— —	— —	
Piperacillin sodium	0.11 *0.032°*	0.11 *0.06°*	0.11 *0.12°*	0.10 *0.17°*	— —	— —	
Piperocaine hydrochloride	0.22 *0.066°*	0.21 *0.12°*	0.19 *0.22°*	0.19 *0.31°*	0.17 *0.49°*	— —	
Piridocaine hydrochloride	0.24 *0.072°*	0.24 *0.13°*	— —	— —	— —	— —	
Polyethylene glycol 300	0.12 *0.034°*	0.12 *0.06°*	0.12 *0.14°*	0.12 *0.21°*	0.13 *0.37°*	0.13 *0.52°*	**6.73%** **6.73%**
Polyethylene glycol 400	0.08 *0.022°*	0.08 *0.04°*	0.09 *0.09°*	0.09 *0.15°*	0.09 *0.27°*	0.11 *0.52°*	**8.50%** **8.50%**
Polyethlene glycol 1500	0.06 *0.015°*	0.06 *0.03°*	0.07 *0.07°*	0.07 *0.12°*	0.07 *0.21°*	0.09 *0.52°*	**10.00%** **10.00%**
Polyethylene glycol 1540	0.02 *0.005°*	0.02 *0.01°*	0.02 *0.02°*	0.03 *0.04°*	0.03 *0.09°*	—	
Polyethylene glycol 4000	0.02 *0.004°*	0.02 *0.00°*	0.02 *0.02°*	0.02 *0.03°*	0.02 *0.06°*	—	
Polymyxin B sulfate	0.10 *0.033°*	0.09 *0.04°*	0.07 *0.07°*	0.06 *0.09°*	0.04 *0.13°*	—	
Polysorbate 80	0.02 *0.005°*	0.02 *0.01°*	0.02 *0.02°*	0.02 *0.03°*	0.02 *0.05°*	—	
Polyvinyl alcohol (99% hydrolyzed)	0.02 *0.004°*	0.02 *0.00°*	0.02 *0.02°*	0.02 *0.03°*	0.03 *0.07°*	—	

Chemical	Concentration of Solution, NaCl Equivalents					At Isosmotic Concentration	
	0.5%	1%	2%	3%	5%		
Potassium acetate	0.59	0.59	—	—	—	0.59	**1.53%**
	0.172°	*0.34°*	—	—	—	*0.52°*	**1.53%**
Potassium chlorate	0.50	0.49	—	—	—	0.48	**1.88%**
	0.140°	*0.27°*	—	—	—	*0.52°*	**1.88%**
Potassium chloride	0.76	0.76	—	—	—	0.76	**1.19%**
	0.219°	*0.43°*	—	—	—	*0.52°*	**1.19%**
Potassium iodide	0.34	0.34	0.34	—	—	0.34	**2.59%**
	0.104°	*0.20°*	*0.40°*	—	—	*0.52°*	**2.59%**
Potassium nitrate	0.58	0.56	—	—	—	0.56	**1.62%**
	0.163°	*0.32°*	—	—	—	*0.52°*	**1.62%**
Potassium permanganate	0.39	0.39	0.39	—	—	—	
	0.112°	*0.22°*	*0.44°*	—	—	—	
Potassium phosphate, anhydrous	0.50	0.46	0.43	—	—	0.43	**2.11%**
	0.140°	*0.26°*	*0.49°*	—	—	*0.52°*	**2.11%**
Potassium phosphate, monobasic	0.48	0.44	0.42	—	—	0.41	**2.18%**
	0.133°	*0.25°*	*0.48°*	—	—	*0.52°*	**2.18%**
Potassium sorbate	0.44	0.41	0.40	—	—	0.40	**2.23%**
	0.125°	*0.23°*	*0.46°*	—	—	*0.52°*	**2.23%**
Potassium sulfate	0.46	0.44	0.43	—	—	0.43	**2.11%**
	0.132°	*0.25°*	*0.49°*	—	—	*0.52°*	**2.11%**
Potassium thiocyanate	0.61	0.59	—	—	—	0.59	**1.52%**
	0.180°	*0.34°*	—	—	—	*0.52°*	**1.52%**
Povidone	0.01	0.01	0.01	0.01	0.01	—	
	0.004°	*0.00°*	*0.01°*	*0.01°*	*0.03°*	—	
Pralidoxime chloride	0.32	0.32	0.32	—	—	0.32	**2.87%**
	0.092°	*0.18°*	*0.36°*	—	—	*0.52°*	**2.87%**
Pramoxine hydrochloride	0.18	0.18	0.17	0.15	0.10	—	
	0.056°	*0.10°*	*0.19°*	*0.25°*	*0.28°*	—	
Prilocaine hydrochloride	0.22	0.22	0.22	0.22	—	0.22	**4.18%**
	0.062°	*0.12°*	*0.25°*	*0.37°*	—	*0.52°*	**4.18%**
Probarbital calcium	0.28	0.25	—	—	—	—	
	0.079°	*0.14°*	—	—	—	—	
Probarbital sodium	0.38	0.32	0.30	0.29	—	0.29	**3.1%**
	0.110°	*0.18°*	*0.35°*	*0.50°*	—	*0.52°*	**3.1%**
Procainamide hydrochloride	0.24	0.22	0.20	0.19	0.17	—	
	0.071°	*0.12°*	*0.23°*	*0.33°*	*0.50°*	—	
Procaine hydrochloride	0.24	0.21	0.20	0.19	0.18	0.18	**5.05%**
	0.065°	*0.12°*	*0.22°*	*0.32°*	*0.51°*	*0.52°*	**5.05%**
Prochlorperazine edisylate	0.08	0.06	0.05	0.30	0.02	—	
	0.020°	*0.03°*	*0.04°*	*0.05°*	*0.06°*	—	
Promazine hydrochloride	0.18	0.13	0.09	0.07	0.05	—	
	0.050°	*0.07°*	*0.10°*	*0.11°*	*0.13°*	—	
Promethazine hydrochloride	0.28	0.18	0.12	0.10	0.07	—	
	0.084°	*0.11°*	*0.15°*	*0.18°*	*0.22°*	—	
Propantheline bromide	0.11	0.11	—	—	—	—	
	0.032°	*0.06°*	—	—	—	—	

Chemical	Concentration of Solution, NaCl Equivalents					At Isosmotic Concentration	
	0.5%	1%	2%	3%	5%		
Proparacaine hydrochloride	0.16 *0.044°*	0.15 *0.08°*	0.15 *0.16°*	0.14 *0.24°*	0.13 *0.38°*	0.12 *0.52°*	**7.46%** **7.46%**
Propiomazine hydrochloride	0.18 *0.050°*	0.15 *0.08°*	0.12 *0.13°*	0.10 *0.16°*	0.08 *0.21°*	— —	
Propoxycaine hydrochloride	0.22 *0.063°*	0.19 *0.11°*	0.17 *0.19°*	0.16 *0.28°*	0.15 *0.42°*	— —	
Propranolol hydrochloride	0.20 *0.060°*	0.20 *0.12°*	0.20 *0.23°*	— —	— —	— —	
Propylene glycol	0.44 *0.126°*	0.43 *0.25°*	0.43 *0.49°*	— —	— —	0.43 *0.52°*	**2.10%** **2.10%**
Pyrathiazine hydrochloride	0.22 *0.065°*	0.17 *0.09°*	0.11 *0.12°*	0.08 *0.14°*	0.06 *0.17°*	— —	
Pyridostigmine bromide	0.22 *0.062°*	0.22 *0.12°*	0.22 *0.25°*	0.22 *0.37°*	— —	0.22 *0.52°*	**4.13%** **4.13%**
Pyridoxine hydrochloride	0.41 *0.118°*	0.36 *0.20°*	0.32 *0.36°*	0.29 *0.52°*	— —	— —	
Pyrilamine maleate	0.24 *0.072°*	0.18 *0.10°*	0.14 *0.15°*	0.11 *0.19°*	0.09 *0.25°*	— —	
Quinacrine hydrochloride	0.20 *0.056°*	0.18 *0.10°*	0.16 *0.17°*	— —	— —	— —	
Quinacrine mesylate	0.12 *0.034°*	0.11 *0.06°*	0.11 *0.12°*	0.10 *0.17°*	0.10 *0.28°*	— —	
Quinidine gluconate	0.14 *0.037°*	0.12 *0.06°*	0.11 *0.12°*	0.10 *0.17°*	— —	— —	
Quinidine sulfate	0.14 *0.041°*	0.10 *0.06°*	0.08 *0.08°*	— —	— —	— —	
Quinine bisulfate	0.09 *0.029°*	0.09 *0.05°*	0.09 *0.10°*	0.09 *0.15°*	— —	— —	
Quinine dihydrochloride	0.26 *0.072°*	0.23 *0.12°*	0.20 *0.23°*	0.19 *0.33°*	0.18 *0.51°*	0.18 *0.52°*	**5.07%** **5.07%**
Quinine hydrochloride	0.16 *0.043°*	0.14 *0.07°*	0.13 *0.14°*	0.11 *0.19°*	— —	— —	
Quinine urea hydrochloride	0.26 *0.073°*	0.23 *0.13°*	0.22 *0.25°*	0.21 *0.37°*	— —	0.20 *0.52°*	**4.5%** **4.5%**
Racephedrine hydrochloride	0.32 *0.093°*	0.31 *0.17°*	0.30 *0.34°*	0.30 *0.51°*	— —	0.29 *0.52°*	**3.07%** **3.07%**
Ranitidine hydrochloride	0.20 *0.057°*	0.18 *0.10°*	0.17 *0.19°*	0.16 *0.27°*	0.15 *0.42°*	— —	
Resorcinol	0.28 *0.082°*	0.28 *0.16°*	0.28 *0.31°*	0.27 *0.47°*	— —	0.27 *0.52°*	**3.3%** **3.3%**
Riboflavin phosphate (sodium)	0.08 *0.022°*	0.08 *0.04°*	0.08 *0.09°*	0.08 *0.15°*	— —	— —	
Ritodrine hydrochloride	0.21 *0.062°*	0.20 *0.11°*	0.18 *0.21°*	0.18 *0.30°*	0.16 *0.46°*	— —	
Rolitetracycline	0.11 *0.032°*	0.11 *0.06°*	0.10 *0.11°*	0.09 *0.15°*	0.07 *0.20°*	— —	

Chemical	Concentration of Solution, NaCl Equivalents					At Isosmotic Concentration	
	0.5%	1%	2%	3%	5%		
Rose bengal	0.08	0.07	0.07	0.07	0.07	0.06	**14.9%**
	0.020°	*0.04°*	*0.08°*	*0.12°*	*0.19°*	*0.52°*	**14.9%**
Rose bengal B	0.08	0.08	0.08	0.08	0.08	—	
	0.022°	*0.04°*	*0.08°*	*0.15°*	*0.21°*	—	
Saralasin acetate	0.10	0.09	0.08	0.08	0.08	—	
	0.028°	*0.05°*	*0.09°*	*0.13°*	*0.18°*	—	
Scopolamine hydrobromide	0.12	0.12	0.12	0.12	0.12	0.11	**7.85%**
	0.034°	*0.06°*	*0.13°*	*0.20°*	*0.33°*	*0.52°*	**7.85%**
Scopolamine methyl nitrate	0.18	0.16	0.15	0.14	0.13	0.13	**6.95%**
	0.049°	*0.09°*	*0.17°*	*0.24°*	*0.38°*	*0.52°*	**6.95%**
Secobarbital sodium	0.25	0.24	0.23	0.23	—	0.23	**3.9%**
	0.071°	*0.13°*	*0.27°*	*0.40°*	—	*0.52°*	**3.9%**
Silver nitrate	0.33	0.33	0.33	—	—	0.33	**2.74%**
	0.095°	*0.19°*	*0.38°*	—	—	*0.52°*	**2.74%**
Silver protein, mild	0.17	0.17	0.17	0.17	0.16	0.16	**5.51%**
	0.047°	*0.09°*	*0.18°*	*0.28°*	*0.47°*	*0.52°*	**5.51%**
Silver protein, strong	0.12	0.08	0.06	0.05	0.04	—	
	0.033°	*0.04°*	*0.06°*	*0.08°*	*0.10°*	—	
Sodium acetate	0.47	0.46	0.45	—	—	0.45	**2.03%**
	0.136°	*0.26°*	*0.51°*	—	—	*0.52°*	**2.03%**
Sodium acetate, anhydrous	0.08	0.77	—	—	—	0.76	**1.18%**
	0.226°	*0.44°*	—	—	—	*0.52°*	**1.18%**
Sodium antimonyl tartrate	0.14	0.13	0.13	0.12	0.12	0.11	**7.9%**
	0.039°	*0.07°*	*0.14°*	*0.20°*	*0.33°*	*0.52°*	**7.9%**
Sodium arsenate, dibasic	0.26	0.25	0.25	0.24	—	0.24	**3.83%**
	0.074°	*0.14°*	*0.27°*	*0.41°*	—	*0.52°*	**3.83%**
Sodium ascorbate	0.34	0.32	0.30	—	—	0.30	**2.99%**
	0.097°	*0.18°*	*0.35°*	—	—	*0.52°*	**2.99%**
Sodium benzoate	0.40	0.40	0.40	—	—	0.40	**2.25%**
	0.116°	*0.23°*	*0.46°*	—	—	*0.52°*	**2.25%**
Sodium bicarbonate	0.68	0.65	—	—	—	0.65	**1.39%**
	0.197°	*0.38°*	—	—	—	*0.52°*	**1.39%**
Sodium biphosphate monohydrate	0.45	0.43	0.41	—	—	0.41	**2.21%**
	0.128°	*0.24°*	*0.47°*	—	—	*0.52°*	**2.21%**
Sodium biphosphate dihydrate	0.40	0.36	0.34	—	—	0.32	**2.77%**
	0.109°	*0.20°*	*0.38°*	—	—	*0.52°*	**2.77%**
Sodium biphosphate, anhydrous	0.50	0.46	0.43	—	—	0.43	**2.1%**
	0.142°	*0.26°*	*0.49°*	—	—	*0.52°*	**2.1%**
Sodium bismuth thioglycolate	0.20	0.19	0.18	0.18	0.17	0.17	**5.29%**
	0.055°	*0.10°*	*0.20°*	*0.30°*	*0.49°*	*0.52°*	**5.29%**
Sodium bisulfite	0.64	0.61	—	—	—	0.60	**1.5%**
	0.186°	*0.35°*	—	—	—	*0.52°*	**1.5%**
Sodium borate	0.48	0.42	0.37	—	—	0.35	**2.6%**
	0.137°	*0.24°*	*0.42°*	—	—	*0.52°*	**2.6%**
Sodium bromide	0.58	0.58	—	—	—	0.57	**1.6%**
	0.166°	*0.32°*	—	—	—	*0.52°*	**1.6%**

Chemical	Concentration of Solution, NaCl Equivalents						
	0.5%	1%	2%	3%	5%	At Isosmotic Concentration	
Sodium cacodylate	0.38	0.32	0.30	0.28	—	0.27	3.3%
	0.104°	*0.18°*	*0.33°*	*0.48°*		*0.52°*	3.3%
Sodium carbonate, anhydrous	0.74	0.70	—	—	—	0.68	1.32%
	0.214°	*0.40°*	—	—	—	*0.52°*	1.32%
Sodium carbonate, monohydrated	0.64	0.60	—	—	—	0.58	1.56%
	0.183°	*0.34°*	—	—	—	*0.52°*	1.56%
Sodium carboxymethyl cellulose	0.03	0.03	—	—	—	—	
	0.007°	*0.01°*	—	—	—	—	
Sodium chloride	1.00	—	—	—	—	1.00	0.9%
	0.289°	—	—	—	—	*0.52°*	0.9%
Sodium citrate	0.32	0.31	0.30	0.30	—	0.30	3.02%
	0.091°	*0.17°*	*0.34°*	*0.51°*	—	*0.52°*	3.02%
Sodium colistimethate	0.16	0.15	0.14	0.14	0.13	0.13	6.85%
	0.045°	*0.08°*	*0.16°*	*0.23°*	*0.38°*	*0.52°*	6.85%
Sodium folate	0.14	0.12	0.11	0.10	—	—	
	0.040°	*0.06°*	*0.12°*	*0.16°*	—	—	
Sodium hypophosphite	0.68	0.61	—	—	—	—	
	0.190°	*0.35°*	—	—	—	—	
Sodium iodide	0.41	0.39	0.39	—	—	0.38	2.37%
	0.113°	*0.22°*	*0.44°*	—	—	*0.52°*	2.37%
Sodium lactate	0.58	0.55	—	—	—	0.52	1.72%
	0.164°	*0.31°*	—	—	—	*0.52°*	1.72%
Sodium lauryl sulfate	0.10	0.08	0.07	0.05	—	—	—
	0.029°	*0.04°*	*0.06°*	*0.08°*	—	—	
Sodium metabisulfite	0.70	0.67	—	—	—	0.65	1.38%
	0.206°	*0.38°*	—	—	—	*0.52°*	1.38%
Sodium nitrate	0.74	0.68	—	—	—	0.66	1.36%
	0.214°	*0.39°*	—	—	—	*0.52°*	1.36%
Sodium nitrite	0.86	0.84	—	—	—	0.83	1.08%
	0.248°	*0.48°*	—	—	—	*0.52°*	1.08%
Sodium nitroferricyanide	0.30	0.29	0.28	0.28	—	0.27	3.30%
	0.086°	*0.16°*	*0.32°*	*0.47°*	—	*0.52°*	3.30%
Sodium phosphate	0.30	0.29	0.28	0.27	—	0.27	3.33%
	0.086°	*0.16°*	*0.32°*	*0.47°*	—	*0.52°*	3.33%
Sodium phosphate, dibasic dihydrate	0.44	0.42	0.41	—	—	0.40	2.23%
	0.127°	*0.24°*	*0.47°*	—	—	*0.52°*	2.23%
Sodium phosphate, dibasic dodecahydrate	0.24	0.22	0.21	0.21	—	0.20	4.45%
	0.064°	*0.12°*	*0.24°*	*0.35°*	—	*0.52°*	4.45%
Sodium phosphate, exsiccated	0.56	0.53	—	—	—	0.51	1.75%
	0.159°	*0.30°*	—	—	—	*0.52°*	1.75%
Sodium propionate	0.62	0.61	—	—	—	0.61	1.47%
	0.177°	*0.35°*	—	—	—	*0.52°*	1.47%
Sodium ricinoleate	0.10	0.10	0.10	0.09	0.09	—	
	0.033°	*0.06°*	*0.11°*	*0.16°*	*0.25°*	—	
Sodium salicylate	0.38	0.36	0.36	—	—	0.36	2.53%
	0.106°	*0.20°*	*0.41°*	—	—	*0.52°*	2.53%

Chemical	Concentration of Solution, NaCl Equivalents						
	0.5%	1%	2%	3%	5%	At Isosmotic Concentration	
Sodium succinate	0.32 *0.092°*	0.32 *0.18°*	0.31 *0.36°*	— —	— —	0.31 *0.52°*	**2.90%** **2.90%**
Sodium sulfate	0.28 *0.079°*	0.26 *0.14°*	0.25 *0.28°*	0.23 *0.40°*	— —	0.23 *0.52°*	**3.95%** **3.95%**
Sodium sulfate, anhydrous	0.58 *0.165°*	0.54 *0.30°*	— —	— —	— —	0.52 *0.52°*	**1.78%** **1.78%**
Sodium sulfite, exsiccated	0.72 *0.204°*	0.65 *0.37°*	— —	— —	— —	— —	
Sodium tartrate	0.33 *0.098°*	0.33 *0.19°*	0.33 *0.38°*	— —	— —	0.33 *0.52°*	**2.72%** **2.72%**
Sodium thiosulfate	0.32 *0.092°*	0.31 *0.18°*	0.31 *0.35°*	— —	— —	0.30 *0.52°*	**2.98%** **2.98%**
Sorbitol hemihydrate	0.16 *0.045°*	0.16 *0.09°*	0.16 *0.19°*	0.16 *0.28°*	0.16 *0.48°*	0.16 *0.52°*	**5.48%** **5.48%**
Sparteine sulfate	0.10 *0.030°*	0.10 *0.05°*	0.10 *0.11°*	0.10 *0.16°*	0.10 *0.27°*	0.10 *0.52°*	**9.46%** **9.46%**
Spectinomycin hydrochloride	0.16 *0.045°*	0.16 *0.09°*	0.16 *0.18°*	0.16 *0.28°*	0.16 *0.46°*	0.16 *0.52°*	**5.66%** **5.66%**
Streptomycin sulfate	0.08 *0.020°*	0.07 *0.03°*	0.07 *0.07°*	0.06 *0.10°*	0.06 *0.17°*	— —	
Stibamine glucoside	0.16 *0.046°*	0.14 *0.07°*	0.12 *0.14°*	0.11 *0.19°*	— —	— —	
Stibophen	0.20 *0.059°*	0.18 *0.10°*	0.17 *0.19°*	0.16 *0.28°*	0.15 *0.43°*	— —	
Streptomycin calcium chloride complex	0.20 *0.057°*	0.20 *0.11°*	0.19 *0.21°*	0.19 *0.32°*	0.18 *0.52°*	0.18 *0.52°*	**5.0%** **5.0%**
Streptomycin hydrochloride	0.18 *0.050°*	0.17 *0.09°*	0.17 *0.19°*	0.16 *0.28°*	0.16 *0.45°*	— —	
Streptozocin	0.14 *0.041°*	0.13 *0.07°*	0.12 *0.14°*	0.12 *0.21°*	0.12 *0.35°*	— —	
Strychnine hydrochloride	0.20 *0.060°*	0.18 *0.09°*	0.14 *0.16°*	— —	— —	— —	
Strychnine nitrate	0.12 *0.035°*	0.12 *0.06°*	— —	— —	— —	— —	
Succinylcholine chloride	0.20 *0.059°*	0.20 *0.11°*	0.20 *0.23°*	0.20 *0.35°*	— —	0.20 *0.52°*	**4.48%** **4.48%**
Sucrose	0.08 *0.023°*	0.08 *0.04°*	0.09 *0.09°*	0.09 *0.15°*	0.09 *0.26°*	0.10 *0.52°*	**9.25%** **9.25%**
Sulbactam sodium	0.24 *0.070°*	0.24 *0.14°*	0.24 *0.27°*	0.24 *0.41°*	— —	0.24 *0.52°*	**3.75%** **3.75%**
Sulfadiazine sodium	0.26 *0.073°*	0.24 *0.13°*	0.23 *0.26°*	0.22 *0.38°*	— —	0.21 *0.52°*	**4.24%** **4.24%**
Sulfamerazine sodium	0.24 *0.069°*	0.23 *0.13°*	0.22 *0.24°*	0.21 *0.36°*	— —	0.20 *0.52°*	**4.53%** **4.53%**
Sulfamethazine sodium	0.22 *0.066°*	0.21 *0.12°*	0.20 *0.22°*	0.19 *0.32°*	0.18 *0.51°*	— —	

Chemical	Concentration of Solution, NaCl Equivalents						
	0.5%	1%	2%	3%	5%	At Isosmotic Concentration	
Sulfapyridine sodium	0.26 *0.073°*	0.23 *0.13°*	0.22 *0.24°*	0.21 *0.35°*	— —	0.20 *0.52°*	**4.55%** **4.55%**
Sulfathiazole sodium	0.23 *0.067°*	0.22 *0.12°*	0.21 *0.23°*	0.20 *0.34°*	— —	0.19 *0.52°*	**4.82%** **4.82%**
Sulfisoxazole diethanolamine	0.20 *0.059°*	0.18 *0.10°*	0.16 *0.18°*	0.15 *0.26°*	— —	— —	
Sulfobromophthalein sodium	0.07 *0.019°*	0.06 *0.03°*	0.05 *0.06°*	0.05 *0.08°*	0.04 *0.12°*	— —	
Suramin sodium	0.10 *0.030°*	0.10 *0.05°*	0.10 *0.11°*	0.10 *0.16°*	0.10 *0.27°*	— —	
Synephrine tartrate	0.18 *0.048°*	0.17 *0.09°*	0.16 *0.19°*	0.16 *0.28°*	0.16 *0.45°*	0.16 *0.52°*	**5.83%** **5.83%**
Tannic acid	0.03 *0.009°*	0.03 *0.01°*	0.03 *0.03°*	0.03 *0.05°*	0.03 *0.08°*	— —	
Tartaric acid	0.26 *0.075°*	0.25 *0.14°*	0.24 *0.27°*	0.23 *0.40°*	— —	0.23 *0.52°*	**3.9%** **3.9%**
Terbutaline sulfate	0.14 *0.042°*	0.14 *0.08°*	0.14 *0.16°*	0.14 *0.23°*	0.13 *0.39°*	0.33 *0.52°*	**6.75%** **6.75%**
Tetracaine hydrochloride	0.20 *0.062°*	0.18 *0.10°*	0.17 *0.18°*	0.15 *0.26°*	0.12 *0.35°*	— —	
Tetracycline hydrochloride	0.16 *0.046°*	0.14 *0.07°*	0.12 *0.12°*	0.10 *0.17°*	— —	— —	
Tetraethylammonium bromide	0.36 *0.098°*	0.33 *0.18°*	0.30 *0.34°*	0.28 *0.49°*	— —	0.28 *0.52°*	**3.17%** **3.17%**
Tetraethylammonium chloride	0.36 *0.100°*	0.34 *0.19°*	0.33 *0.38°*	— —	— —	0.33 *0.52°*	**2.67%** **2.67%**
Tetrahydrozoline hydrochloride	0.30 *0.090°*	0.28 *0.16°*	0.25 *0.28°*	0.23 *0.40°*	— —	— —	
Theophylline	0.10 *0.028°*	— —	— —	— —	— —	— —	
Theophylline sodium glycinate	0.32 *0.090°*	0.31 *0.18°*	0.31 *0.35°*	— *0.35°*	— —	0.31 *0.52°*	**2.94%** **2.94%**
Thiamine hydrochloride	0.26 *0.074°*	0.25 *0.13°*	0.23 *0.26°*	0.22 *0.37°*	— —	0.21 *0.52°*	**4.24%** **4.24%**
Thiethylperazine maleate	0.10 *0.030°*	0.09 *0.05°*	0.08 *0.08°*	0.07 *0.11°*	0.05 *0.15°*	— —	
Thiocyanate sodium	0.71 *0.205°*	0.17 *0.41°*	— —	— —	— —	0.17 *0.52°*	**1.27%** **1.27%**
Thiopental sodium	0.28 *0.079°*	0.27 *0.15°*	0.27 *0.30°*	0.26 *0.44°*	— —	0.26 *0.52°*	**3.5%** **3.5%**
Thiopropazate dihydrochloride	0.20 *0.053°*	0.16 *0.09°*	0.12 *0.13°*	0.10 *0.17°*	0.08 *0.22°*	— —	
Thioridazine hydrochloride	0.06 *0.015°*	0.05 *0.02°*	0.04 *0.04°*	0.03 *0.05°*	0.03 *0.07°*	— —	
Thiotepa	0.16 *0.045°*	0.16 *0.09°*	0.16 *0.18°*	0.16 *0.27°*	0.16 *0.46°*	0.16 *0.52°*	**5.67%** **5.67%**

Chemical	Concentration of Solution, NaCl Equivalents					At Isosmotic Concentration	
	0.5%	1%	2%	3%	5%		
Ticarcillin disodium	0.20 0.056°	0.20 0.11°	0.20 0.22°	0.19 0.33°	— —	0.19 0.52°	**4.62%** **4.62%**
Timolol maleate	0.14 0.038°	0.13 0.07°	0.12 0.14°	— —	— —	— —	
Tobramycin	0.08 0.019°	0.07 0.03°	0.07 0.07°	0.07 0.11°	0.06 0.18°	— —	
Tolazoline hydrochloride	0.36 0.107°	0.34 0.19°	0.31 0.35°	0.30 0.51°	— —	0.30 0.52°	**3.05%** **3.05%**
Tribromoethanol	0.06 0.015°	0.05 0.03°	0.05 0.05°	— —	— —	— —	
Tridihexethyl chloride	0.16 0.047°	0.16 0.09°	0.16 0.19°	0.16 0.28°	0.16 0.46°	0.16 0.52°	**5.62%** **5.62%**
Triethanolamine	0.20 0.058°	0.21 0.12°	0.22 0.25°	0.22 0.38°	— —	0.22 0.52°	**4.05%** **4.05%**
Trifluoperazine dihydrochloride	0.18 0.052°	0.18 0.10°	— —	— —	— —	— —	
Triflupromazine hydrochloride	0.10 0.031°	0.09 0.05°	0.05 0.06°	0.04 0.07°	0.03 0.09°	— —	
Trimeprazine tartrate	0.10 0.023°	0.06 0.03°	0.04 0.04°	0.03 0.05°	0.02 0.06°	— —	
Trimethadione	0.23 0.069°	0.23 0.13°	0.22 0.25°	0.22 0.37°	— —	0.21 0.52°	**4.22%** **4.22%**
Trimethaphan camsylate	0.12 0.033°	0.10 0.06°	0.10 0.11°	0.09 0.15°	0.09 0.24°	— —	
Trimethobenzamide hydrochloride	0.12 0.033°	0.10 0.06°	0.10 0.10°	0.09 0.15°	0.08 0.23°	— —	
Tripelennamine hydrochloride	0.38 0.110°	0.30 0.17°	0.24 0.26°	0.20 0.35°	— —	— —	
Trisodium edetate (monohydrate)	0.29 0.079°	0.29 0.15°	0.28 0.31°	0.27 0.47°	— —	0.27 0.52°	**3.31%** **3.31%**
Tromethamine	0.26 0.075°	0.26 0.15°	0.26 0.30°	0.26 0.45°	— —	0.26 0.52°	**3.41%** **3.41%**
Tropacocaine hydrochloride	0.30 0.085°	0.25 0.14°	0.22 0.25°	0.20 0.34°	— —	0.18 0.52°	**4.92%** **4.92%**
Tropicamide	0.10 0.030°	0.09 0.05°	— —	— —	— —	— —	
Trypan blue	0.26 0.075°	0.26 0.15°	— —	— —	— —	— —	
Tryparsamide	0.20 0.057°	0.20 0.11°	0.20 0.22°	0.20 0.33°	— —	0.19 0.52°	**4.62%** **4.62%**
Tuaminoheptane sulfate	0.28 0.078°	0.27 0.15°	0.27 0.30°	0.27 0.46°	— —	0.26 0.52°	**3.4%** **3.4%**
Tubocurarine chloride	0.14 0.042°	0.13 0.07°	0.11 0.12°	0.10 0.17°	0.09 0.26°	— —	
Urea	0.55 0.158°	0.52 0.30°	— —	— —	— —	0.52 0.52°	**1.73%** **1.73%**

Chemical	Concentration of Solution, NaCl Equivalents						
	0.5%	1%	2%	3%	5%	At Isosmotic Concentration	
Urethan	0.31	0.31	0.31	—	—	0.31	**2.93%**
	0.089°	*0.17°*	*0.35°*	—	—	*0.52°*	**2.93%**
Uridine	0.12	0.12	0.12	0.12	0.12	0.11	**8.18%**
	0.035°	*0.06°*	*0.13°*	*0.20°*	*0.33°*	*0.52°*	**8.18%**
Valethamate bromide	0.16	0.15	0.15	0.14	0.11	—	
	0.044°	*0.08°*	*0.16°*	*0.23°*	*0.32°*	—	
Vancomycin hydrochloride	0.06	0.05	0.04	0.04	0.04	—	
	0.015°	*0.02°*	*0.04°*	*0.06°*	*0.09°*	—	
Verapamil hydrochloride	0.16	0.13	0.10	0.09	0.07	—	
	0.044°	*0.07°*	*0.12°*	*0.15°*	*0.20°*	—	
Vinbarbital sodium	0.26	0.26	0.26	0.25	—	0.25	**3.55%**
	0.074°	*0.14°*	*0.29°*	*0.44°*	—	*0.52°*	**3.55%**
Vindesine sulfate	0.10	0.08	0.08	0.07	0.07	—	
	0.024°	*0.04°*	*0.08°*	*0.12°*	*0.19°*	—	
Viomycin sulfate	0.08	0.08	0.07	0.07	0.07	—	
	0.025°	*0.04°*	*0.08°*	*0.12°*	*0.19°*	—	
Warfarin sodium	0.18	0.17	0.16	0.15	0.15	0.15	**6.10%**
	0.049°	*0.09°*	*0.18°*	*0.26°*	*0.43°*	*0.52°*	**6.10%**
Xylometazoline hydrochloride	0.22	0.21	0.20	0.20	—	0.19	**4.68%**
	0.065°	*0.12°*	*0.23°*	*0.31°*	—	*0.52°*	**4.68%**
Zinc chloride	0.66	0.61	—	—	—	—	
	0.190°	*0.35°*	—	—	—	—	
Zinc p-phenosulfonate	0.18	0.18	0.18	0.17	0.17	—	
	0.053°	*0.10°*	*0.19°*	*0.29°*	*0.48°*	—	
Zinc sulfanilate	0.22	0.21	0.20	0.19	0.18	—	
	0.066°	*0.12°*	*0.22°*	*0.32°*	*0.51°*	—	
Zinc sulfate	0.16	0.15	0.14	0.13	0.12	0.12	**7.65%**
	0.045°	*0.08°*	*0.15°*	*0.22°*	*0.35°*	*0.52°*	**7.65%**
Zinc sulfate, dried	0.24	0.23	0.22	0.21	—	0.20	**4.52%**
	0.072°	*0.13°*	*0.25°*	*0.36°*	—	*0.52°*	**4.52%**

References
1. E. R. Hammarlund, K. Pedersen-Bjergaard, *J. Am. Pharm. Assoc., Sci. Ed.* **47**, 107 (1958).
2. E. R. Hammarlund *et al., J. Pharm. Sci.* **54**, 160 (1965).
3. E. R. Hammarlund, G. L. Van Pevenage, *ibid.* 55, 1448 (1966).
4. W. E. Fassett *et al., ibid.* **58**, 1540 (1969).
5. C. Sapp *et al., ibid.* **64**, 1884 (1975).
6. E. R. Hammarlund, *ibid.* **70**, 1161 (1981).
7. A. R. Gennaro (Ed)., *Remington's Pharmaceutical Sciences* (Mack Publ. Co., Easton, PA., 17th ed., 1985) pp 1455-1472.
8. E. R. Hammarlund, *J. Pharm. Sci.* **78**, 519 (1989).

Saturated Solutions

The following table provides the data for making saturated solutions of the substances listed at the temperature designated. Data are provided for making saturated solutions by weight (g of substance per 100 g of saturated solution) and by volume (g of substance per 100 ml of saturated solution and the ml of water required to make such a solution).

To make one *fluid ounce* of a saturated solution: multiply the grams of substance per 100 ml of saturated solution by 4.55 to obtain the number of grains required, by 0.01039 to obtain the number of avoirdupois ounces, by 0.00947 to obtain the number of apothecaries (Troy) ounces; also multiply the ml of water by 16.23 to obtain the number of minims, or divide by 100 to obtain the number of fluid ounces.

To make one *fluid dram:* multiply the grams of substance per 100 ml of saturated solution by 0.5682 to obtain the number of grains required; also multiply the ml of water by 0.60 to obtain the number of minims required.

Substance	Formula	Temp, °C	g/100 g satd soln	g/100 ml satd soln	ml water/ 100 ml satd soln	Specific gravity
acetanilide	$C_6H_5NHCOCH_3$	25	0.54	0.54	99.2	0.997
p-acetophenetidin	$C_6H_4(OC_2H_5)NHCH_3CO$	25	0.0766	0.0766	99.92	1.00
p-acetotoluide	$CH_3CONHC_6H_4CH_3$	25	0.12	0.12	99.7	0.9979
alanine	$CH_3CH(NH_2)COOH$	25	14.1	14.7	89.5	1.042
aluminum ammonium sulfate	$Al_2(SO_4)_3(NH_4)_2SO_4.24H_2O$	25	12.4	13	92	1.05
aluminum chloride hydrated	$AlCl_3.6H_2O$	25	55.5	75	60	1.35
aluminum fluoride	$Al_2F_6.5H_2O$	20	0.499	0.5015	100.0	1.0051
aluminum potassium sulfate	$AlK(SO_4)_2$	25	6.62	7.02	99.1	1.061
aluminum sulfate	$Al_2(SO_4)_3.18H_2O$	25	48.8	63	66	1.29
o-aminobenzoic acid	$C_6H_4NH_2COOH$	25	0.52	0.519	99.4	0.999
DL-α-amino-*n*-butyric acid	$CH_3CH_2CH(NH_2)COOH$	25	17.8	18.6	86.2	1.046
DL-α-aminoisobutyric acid	$(CH_3)_2C(NH_2)COOH$	25	13.3	13.7	89.5	1.031
ammonium arsenate	$NH_4H_2AsO_4$	20	32.7	40.2	83.0	1.228
ammonium benzoate	$NH_4C_7H_5O_2$	25	18.6	19.4	84.7	1.040
ammonium bromide	NH_4Br	15	41.7	53.8	75.2	1.290
ammonium carbonate		25	20	22	88	1.10
ammonium chloride	NH_4Cl	15	26.3	28.3	79.3	1.075
ammonium citrate, dibasic	$(NH_4)_2HC_6H_5O_7$	25	48.7	60.5	61.5	1.22
ammonium dichromate	$(NH_4)_2Cr_2O_7$	25	27.9	33	85	1.18
ammonium iodide	NH_4I	25	64.5	106.2	58.3	1.646
ammonium molybdate	$(NH_4)_6Mo_7O_{24}.4H_2O$	25	30.6	39	88	1.27
ammonium nitrate	NH_4NO_3	25	68.3	90.2	41.8	1.320
ammonium oxalate	$(NH_4)_2C_2O_4.H_2O$	25	4.95	5.06	97.0	1.019
ammonium perchlorate	NH_4ClO_4	25	21.1	23.7	88.7	1.123
ammonium periodate	NH_4IO_4	16	2.63	2.68	99.2	1.018
ammonium persulfate	$(NH_4)_2S_2O_8$	25	42.7	53	71	1.24
ammonium phosphate, dibasic	$(NH_4)_2HPO_4$	14.5	56.2	75.5	58.8	1.343
ammonium phosphate, monobasic	$NH_4H_2PO_4$	25	28.4	33	83	1.16
ammonium salicylate	$NH_4C_7H_5O_3$	25	50.8	58.2	56.4	1.145
ammonium silicofluoride	$(NH_4)_2SiF_6$	17.5	15.7	17.2	92.3	1.095
ammonium sulfate	$(NH_4)_2SO_4$	20	42.6	53.1	71.7	1.248
ammonium sulfite	$(NH_4)_2SO_3.H_2O$	25	39.3	47.3	73.2	1.204
ammonium thiocyanate	NH_4CNS	25	62.2	71	43	1.14
amyl alcohol	$C_5H_{11}OH$	25	2.61	2.60	96.9	0.995
aniline	$C_6H_5NH_2$	22	3.61	3.61	96.2	0.998
aniline hydrochloride	$C_6H_5NH_2.HCl$	25	49	54	56	1.10
aniline sulfate	$(C_6H_5NH_2)_2.H_2SO_4$	25	5.88	6	96	1.02
L-asparagine	$NH_2COCH_2CH(NH_2)COOH$	25	2.44	2.46	98.2	1.007
barium bromide	$BaBr_2$	20	51	87.2	83.8	1.710
barium chlorate	$Ba(ClO_3)_2$	25	28.5	36.8	92.6	1.294
barium chloride	$BaCl_2$	20	26.3	33.4	93.8	1.27
barium iodide	$BaI_2.7\frac{1}{2}H_2O$	25	68.8	157.0	71.1	2.277
barium nitrate	$Ba(NO_3)_2$	25	9.4	10.2	97.9	1.080
barium nitrite	$Ba(NO_2)_2$	17	40	59.6	89.4	1.490
barium perchlorate	$Ba(ClO_4)_2$	25	75.3	145.8	47.8	1.936
benzamide	$C_6H_5CONH_2$	25	1.33	1.33	98.6	0.999
benzoic acid	$C_7H_6O_2$	25	0.367	0.367	99.63	1.00
beryllium sulfate	$BeSO_4.4H_2O$	25	28.7	37.3	93.0	1.301
boric acid	H_3BO_3	25	4.99	5.1	97	1.02
n-butyl alcohol	$CH_3(CH_2)_2CH_2OH$	25	79.7	67.3	17.1	0.845
cadmium bromide	$CdBr_2.4H_2O$	25	52.9	94.0	83.9	1.775
cadmium chlorate	$Cd(ClO_3)_2.12H_2O$	18	76.4	174.5	54.0	2.284

Substance	Formula	Temp, °C	g/100 g satd soln	g/100 ml satd soln	ml water/ 100 ml satd soln	Specific gravity
cadmium chloride	$CdCl_2.2\frac{1}{2}H_2O$	25	54.7	97.2	80.8	1.778
cadmium iodide	CdI_2	20	45.9	73.0	86.3	1.590
cadmium sulfate	$3(CdSO_4).8H_2O$	25	43.4	70.3	91.8	1.619
calcium bromide	$CaBr_2$	20	58.8	107.2	75.0	1.82
calcium chlorate	$Ca(ClO_3)_2.2H_2O$	18	64.0	110.7	62.3	1.729
calcium chloride	$CaCl_2.6H_2O$	25	46.1	67.8	79.2	1.47
calcium chromate	$CaCrO_4.2H_2O$	18	14.3	16.4	98.7	1.149
calcium ferrocyanide	$Ca_2Fe(CN)_6$	25	36.5	49.6	86.2	1.357
calcium iodide	CaI_2	20	67.6	143.8	69.0	2.125
calcium lactate	$Ca(C_3H_5O_3)_2.5H_2O$	25	4.95	5	96	1.01
calcium nitrite	$Ca(NO_2)_2.4H_2O$	18	45.8	65.7	77.8	1.427
calcium sulfate	$CaSO_4.2H_2O$	25	0.208	0.208	99.70	0.999
camphoric acid	$C_8H_{14}(COOH)_2$	25	0.754	0.754	99.246	1.00
carbon disulfide	CS_2	22	0.173	0.173	99.63	0.998
cerium nitrate	$Ce(NO_3)_3.6H_2O$	25	63.7	119.9	68.2	1.880
cesium bromide	$CsBr$	21.4	53.1	89.8	79.5	1.693
cesium chloride	$CsCl$	25	65.7	126.3	65.9	1.923
cesium iodide	CsI	22.8	48.0	74.1	80.5	1.545
cesium nitrate	$CsNO_3$	25	21.9	26.1	92.9	1.187
cesium perchlorate	$CsClO_4$	25	2.01	2.03	99.0	1.010
cesium periodate	$CsIO_4$	15	2.10	2.13	99.5	1.017
cesium sulfate	Cs_2SO_4	25	64.5	129.8	71.7	2.013
chloral hydrate	$CCl_3CHO.H_2O$	25	79.4	120	31	1.51
chloroform	$CHCl_3$	29.4	0.703	0.705	99.57	1.0028
chromic oxide	CrO_3	18	62.5	106.3	64.0	1.703
chromium potassium sulfate	$Cr_2K_2(SO_4)_4.24H_2O$	25	19.6	22	90	1.12
citric acid	$(CH_2)_2COH(COOH)_3.H_2O$	25	67.5	88.6	42.7	1.311
cobalt chlorate	$Co(ClO_3)_2$	18	64.2	119.3	66.5	1.857
cobalt nitrate	$Co(NO_3)_2$	18	49.7	78.2	79.1	1.572
cobalt perchlorate	$Co(ClO_4)_2$	26	71.8	113.5	44.7	1.581
cupric ammonium chloride	$CuCl_2.2NH_4Cl.2H_2O$	25	30.3	35.5	82	1.17
cupric ammonium sulfate	$CuSO_4.(NH_4)_2SO_4$	19	15.3	17.3	96.0	1.131
cupric bromide	$CuBr_2$	25	55.8	102.5	81.2	1.84
cupric chlorate	$Cu(ClO_3)_2$	18	62.2	105.2	64.1	1.692
cupric chloride	$CuCl_2.2H_2O$	25	53.3	80	70	1.50
cupric nitrate	$Cu(NO_3)_2.6H_2O$	20	56.0	94.5	74.3	1.688
cupric selenate	$CuSeO_4$	21.2	14.7	17.2	99.4	1.165
cupric sulfate	$CuSO_4.5H_2O$	25	18.5	22.3	98.7	1.211
dextrose	$C_6H_{12}O_6.H_2O$	25	49.5	59	60	1.19
ether	$(C_2H_5)_2O$	22	5.45	5.34	93.0	0.985
ethyl acetate	$CH_3COOC_2H_5$	25	7.47	7.44	92.1	0.996
ferric ammonium citrate		25	67.7	97	46	1.43
ferric ammonium oxalate	$Fe(NH_4)_3(C_2O_4)_3.3H_2O$	25	51.5	65	61	1.26
ferric ammonium sulfate	$FeSO_4.(NH_4)_2SO_4$	16.5	19.1	22.4	94.3	1.165
ferric chloride	$FeCl_3$	25	73.1	131.1	48.3	1.793
ferric nitrate	$Fe(NO_3)_3$	25	46.8	70.2	79.8	1.50
ferric perchlorate	$Fe(ClO_4)_3.10H_2O$	25	79.9	132.1	33.2	1.656
ferrous sulfate	$FeSO_4.7H_2O$	25	42.1	52.8	72.7	1.255
gallic acid	$C_6H_2(OH)_3COOH.H_2O$	25	1.15	1.15	99.05	1.002
D-glutamic acid	$C_5H_9O_4N$	25	0.86	0.86	99.15	1.0002
glycine	NH_2CH_2COOH	25	20.0	21.7	86.8	1.083
hydroquinone	$C_6H_4(OH)_2$	20	6.7	6.78	94.4	1.012
m-hydroxybenzoic acid	$C_6H_4OHCOOH$	25	0.975	0.975	99.03	1.000
lactose	$C_{12}H_{22}O_{11}.H_2O$	25	15.9	17	90	1.07
lead acetate	$Pb(C_2H_3O_2)_2$	25	36.5	49.0	85.1	1.340
lead bromide	$PbBr_2$	25	0.97	0.98	99.6	1.006
lead chlorate	$Pb(ClO_3)_2$	18	60.2	117.0	77.3	1.944
lead chloride	$PbCl_2$	25	1.07	1.08	99.6	1.007
lead iodide	PbI_2	25	0.08	0.08	99.7	0.998
lead nitrate	$Pb(NO_3)_2$	25	37.1	53.6	91.0	1.445
DL-leucine	$C_6H_{13}O_2N$	25	0.976	0.975	98.9	0.999
L-leucine	$C_6H_{13}O_2N$	25	2.24	2.24	97.85	1.0012
lithium benzoate	$LiC_7H_5O_2$	25	27.7	30.4	79.6	1.100
lithium bromate	$LiBrO_3$	18	60.4	110.5	72.5	1.830

Substance	Formula	Temp, °C	g/100 g satd soln	g/100 ml satd soln	ml water/ 100 ml satd soln	Specific gravity
lithium carbonate	Li_2CO_3	15	1.36	1.38	100.0	1.014
lithium chloride	$LiCl.H_2O$	25	45.9	59.5	70.2	1.296
lithium citrate	$Li_3C_6H_5O_7$	25	31.8	38.6	82.8	1.213
lithium dichromate	$Li_2Cr_2O_7.H_2O$	18	52.6	82.9	74.8	1.574
lithium fluoride	LiF	18	0.27	0.27	99.9	1.002
lithium formate	$LiCHO_2$	18	27.9	31.8	80.4	1.140
lithium iodate	$LiIO_3$	18	44.6	69.9	86.8	1.566
lithium nitrate	$LiNO_3$	19	48.9	64.5	67.5	1.318
lithium perchlorate	$LiClO_4.3H_2O$	25	37.5	47.6	79.5	1.269
lithium salicylate	$LiC_7H_5O_3$	25	52.7	63.6	57.1	1.206
lithium sulfate	$Li_2SO_4.H_2O$	25	27.2	33	88.5	1.21
magnesium bromide	$MgBr_2.6H_2O$	18	50.1	83.1	82.8	1.655
magnesium chlorate	$Mg(ClO_3)_2$	18	56.3	90.0	69.7	1.594
magnesium chloride	$MgCl_2.6H_2O$	25	62.5	79	47.5	1.26
magnesium chromate	$MgCr_2O_4.7H_2O$	18	42.0	59.7	82.5	1.422
magnesium dichromate	$MgCrO_7.5H_2O$	25	81.0	138.8	32.6	1.712
magnesium iodate	$Mg(IO_3)_2.4H_2O$	18	6.44	6.95	100.8	1.078
magnesium iodide	$MgI_2.8H_2O$	18	59.7	114.0	77.1	1.909
magnesium molybdate	$MgMoO_4$	25	15.9	18.4	97.4	1.159
magnesium nitrate	$Mg(NO_3)_2.6H_2O$	25	42.1	58.6	80.5	1.388
magnesium perchlorate	$Mg(ClO_4)_2.6H_2O$	25	49.9	73.6	73.9	1.472
magnesium selenate	$MgSeO_4$	20	35.3	50.8	93.0	1.440
magnesium sulfate	$MgSO_4.7H_2O$	25	55.3	72	58.5	1.30
manganese chloride	$MnCl_2$	25	43.6	63.2	82.0	1.449
manganese nitrate	$Mn(NO_3)_2.6H_2O$	18	57.3	93.2	69.2	1.624
manganese silicofluoride	$MnSiF_6$	17.5	37.7	54.5	90.1	1.446
manganese sulfate	$MnSO_4$	25	39.4	59.1	90.8	1.499
mercuric acetate	$Hg(C_2H_3O_2)_2$	25	30.2	38	88	1.26
mercuric bromide	$HgBr_2$	25	0.609	0.610	99.6	1.0023
mercury bichloride	$HgCl_2$	25	6.6	6.96	98.5	1.054
methylene blue	$C_{16}H_{18}N_3ClS.3H_2O$	25	4.25	4.3	97	1.01
methyl salicylate	$C_6H_4OHCOOCH_3$	25	0.12	0.12	99.88	1.00
monochloracetic acid	$CH_2ClCOOH$	25	78.8	105	28	1.33
β-naphthalenesulfonic acid	$C_{10}H_7SO_3H$	30	56.9	67.9	51.4	1.193
nickel ammonium sulfate	$NiSO_4(NH_4)_2SO_4.6H_2O$	25	9.0	9.5	96	1.05
nickel chlorate	$Ni(ClO_3)_2$	18	56.7	94.2	72.0	1.658
nickel chlorate	$Ni(ClO_3)_2.6H_2O$	18	64.5	107.2	59.1	1.661
nickel nitrate	$Ni(NO_3)_2.6H_2O$	25	77	122	36	1.58
nickel perchlorate	$Ni(ClO_4)_2$	26	70.8	112.2	46.4	1.584
nickel perchlorate	$Ni(ClO_4)_2.9H_2O$	18	52.4	82.7	75.1	1.576
nickel sulfate	$NiSO_4.6H_2O$	25	47.3	64	71	1.35
DL-norleucine	$C_6H_{13}NO_2$	25	1.13	1.13	98.97	0.999
oxalic acid	$H_2C_2O_4.2H_2O$	25	9.81	10.3	94.2	1.044
phenol	C_6H_5OH	20	6.1	6.14	94.5	1.0057
β-phenylalanine	$C_6H_5CH_2CH(NH_2)COOH$	25	2.88	2.89	97.5	1.0035
m-phenylenediamine	$C_6H_8N_2$	20	23.1	23.8	79.3	1.032
p-phenylenediamine	$C_6H_8N_2$	20	3.69	3.70	96.67	1.0038
phenyl salicylate	$C_6H_4OHCOOC_6H_5$	25	0.015	0.015	99.84	0.999
phenyl thiourea	$CS(NH_2)NHC_6H_5$	25	0.24	0.24	99.6	0.998
phosphomolybdic acid	$20MoO_3.2H_3PO_4.48H_2O$	25	74.3	135	46	1.81
phosphotungstic acid	Approx. $20WO_3.2H_3PO_4.25H_2O$	25	71.4	160	64	2.24
potassium acetate	$KC_2H_3O_2$	25	68.7	97.1	44.3	1.413
potassium antimony tartrate	$KSbOC_4H_4O_6$	25	7.64	8.02	96.9	1.049
potassium bicarbonate	$KHCO_3$	25	26.6	31.6	87.5	1.188
potassium bitartrate	$KC_4H_5O_6$	25	0.65	0.65	99.3	0.999
potassium bromate	$KBrO_3$	25	7.53	7.89	97.5	1.054
potassium bromide	KBr	25	40.6	56.0	82.0	1.380
potassium carbonate	$K_2CO_3.1\frac{1}{2}H_2O$	25	52.9	82.2	73.5	1.559
potassium chlorate	$KClO_3$	25	8.0	8.41	96.6	1.051
potassium chloride	KCl	25	26.5	31.2	86.8	1.178
potassium chromate	K_2CrO_4	25	39.4	54.1	83.7	1.381
potassium citrate	$K_3C_6H_5O_7$	25	60.91	92.1	59.2	1.514
potassium dichromate	$K_2Cr_2O_7$	25	13.0	14.2	95.0	1.092
potassium ferricyanide	$K_3Fe(CN)_6$	22	32.1	38.1	80.8	1.187

Substance	Formula	Temp, °C	g/100 g satd soln	g/100 ml satd soln	ml water/ 100 ml satd soln	Specific gravity
potassium ferrocyanide	$K_4Fe(CN)_6$	25	24.0	28.2	89.2	1.173
potassium fluoride	$KF.2H_2O$	18	48.0	72.0	78.0	1.500
potassium formate	$KCHO_2$	18	76.8	120.6	36.4	1.571
potassium hydroxide	KOH	15	51.7	79.2	74.2	1.536
potassium iodate	KIO_3	25	8.40	8.99	98.0	1.071
potassium iodide	KI	25	59.8	103.2	69.1	1.721
potassium meta-antimonate	$KSbO_3$	18	2.73	2.81	99.7	1.025
potassium nitrate	KNO_3	25	28.0	33.4	86.0	1.193
potassium nitrite	KNO_2	20	74.3	121.5	42.3	1.649
potassium oxalate	$K_2C_2O_4.H_2O$	25	28.3	34	86	1.20
potassium perchlorate	$KClO_4$	25	2.68	2.72	99.0	1.014
potassium periodate	KIO_4	13	0.658	0.661	99.83	1.005
potassium permanganate	$KMnO_4$	25	7.10	7.43	97.3	1.046
potassium sodium tartrate	$KNaC_4H_4O_6.4H_2O$	25	39.71	51.9	78.8	1.308
potassium stannate	K_2SnO_3	15.5	42.7	69.2	92.9	1.620
potassium sulfate	K_2SO_4	25	10.83	11.8	96.9	1.086
quinine salicylate	$C_{20}H_{24}N_2O_2.C_6H_4(OH)COOH.2H_2O$	25	0.065	0.065	99.84	0.999
resorcinol	$C_6H_4(OH)_2$	25	58.8	67.2	47.2	1.142
rubidium bromate	$RbBrO_3$	16	2.15	2.18	99.4	1.016
rubidium bromide	$RbBr$	25	52.7	85.6	76.9	1.625
rubidium chloride	$RbCl$	25	48.6	72.8	77.1	1.50
rubidium iodate	$RbIO_3$	15.6	2.72	2.78	99.5	1.022
rubidium iodide	RbI	24.3	63.6	117.7	67.3	1.850
rubidium nitrate	$RbNO_3$	25	40.1	55.0	82.4	1.375
rubidium perchlorate	$RbClO_4$	25	1.88	1.90	99.3	1.012
rubidium periodate	$RbIO_4$	16	0.645	0.648	99.85	1.0052
rubidium sulfate	Rb_2SO_4	25	33.8	45.6	89.7	1.354
silicotungstic acid	$H_4SiW_{12}O_{40}$	18	90.6	258	26.8	2.843
silver acetate	$Ag(C_2H_3O_2)$	25	1.10	1.11	99.40	1.0047
silver bromate	$AgBrO_3$	25	0.204	0.2037	99.65	0.9985
silver fluoride	$AgF.2H_2O$	15.8	64.5	168.4	92.7	2.61
silver nitrate	$AgNO_3$	25	71.5	164	65.5	2.29
silver perchlorate	$AgClO_4.H_2O$	25	84.5	237.1	43.5	2.806
sodium acetate	$NaC_2H_3O_2$	25	33.6	40.5	80.0	1.205
sodium ammonium sulfate	$NaNH_4SO_4$	15	25.2	29.6	87.9	1.174
sodium arsenate	$Na_3AsO_4.12H_2O$	17	21.1	23.5	88.0	1.119
sodium benzenesulfonate	$NaC_6H_5SO_3$	25	16.4	17.6	90.1	1.076
sodium benzoate	$NaC_7H_5O_2$	25	36.0	41.5	73.9	1.152
sodium bicarbonate	$NaHCO_3$	15	8.28	8.80	97.6	1.061
sodium bisulfate	$NaHSO_4.H_2O$	25	59	87	60	1.47
sodium bromide	$NaBr.2H_2O$	25	48.6	75.0	79.4	1.542
sodium carbonate	$Na_2CO_3.10H_2O$	25	22.6	28.1	96.5	1.242
sodium chlorate	$NaClO_3$	25	51.7	74.3	69.6	1.440
sodium chloride	$NaCl$	25	26.5	31.7	88.1	1.198
sodium chromate	Na_2CrO_4	18	40.1	57.4	85.7	1.430
sodium citrate	$Na_3C_6H_5O_7.5H_2O$	25	48.1	61.2	66.0	1.272
sodium dichromate	$Na_2Cr_2O_7$	18	63.9	111.4	63.0	1.743
sodium ferrocyanide	$Na_4Fe(CN)_6$	25	17.1	19.4	93.9	1.131
sodium fluoride	NaF	25	3.98	4.14	99.7	1.038
sodium formate	$NaCHO_2$	18	44.7	58.9	73.0	1.316
sodium hydroxide	$NaOH$	25	50.8	77	74	1.51
sodium hypophosphite	NaH_2PO_2	16	52.1	72.4	66.6	1.386
sodium iodate	$NaIO_3.H_2O$	25	8.57	9.21	98.5	1.075
sodium iodide	NaI	25	64.8	124.3	67.7	1.919
sodium molybdate	Na_2MoO_4	18	39.4	56.6	87.0	1.435
sodium nitrate	$NaNO_3$	25	47.9	66.7	72.5	1.391
sodium nitrite	$NaNO_2$	20	45.8	62.3	73.8	1.359
sodium oxalate	$Na_2(CO_2)_2$	25	3.48	3.58	99.1	1.025
sodium paratungstate	$(Na_2O)_3(WO_3)_7.16H_2O$	0	26.7	35.2	96.5	1.316
sodium perchlorate	$NaClO_4$	25	67.8	114.1	54.1	1.683
sodium periodate	$NaIO_4.3H_2O$	25	12.6	13.9	96.2	1.103
sodium phenolsulfonate	$C_6H_4(OH)SO_3Na$	25	16.1	17.4	90.5	1.079
sodium phosphate dibasic	Na_2HPO_4	17	4.2	4.4	99.9	1.043
sodium phosphate tribasic	Na_3PO_4	14	9.5	10.5	99.8	1.103

Substance	Formula	Temp, °C	g/100 g satd soln	g/100 ml satd soln	ml water/ 100 ml satd soln	Specific gravity
sodium pyrophosphate	$Na_2H_2P_2O_7.6H_2O$	25	13.0	14.4	95.8	1.104
sodium salicylate	$NaC_7H_5O_3$	25	53.6	67.0	58.0	1.248
sodium selenate	Na_2SeO_4	18	29.0	38.1	93.4	1.313
sodium silicofluoride	$NaSiF_6$	20	0.773	0.737	99.76	1.0054
sodium sulfate	Na_2SO_4	25	21.8	26.4	94.5	1.208
sodium sulfate	$Na_2SO_4.10H_2O$	25	27.7	33.3	87.0	1.207
sodium sulfide	$Na_2S.9H_2O$	25	52.3	63	57	1.20
sodium sulfite, anhydrous	Na_2SO_3	25	23	28.5	95.5	1.24
sodium thiocyanate	$NaCNS$	25	62.9	87	51	1.38
sodium thiosulfate	$Na_2S_2O_3.5H_2O$	25	66.8	93	46	1.39
sodium tungstate	$Na_2WO_4.10H_2O$	18	42.0	66.1	91.3	1.573
stannous chloride	$SnCl_2$	15	72.9	133.1	49.5	1.827
strontium chlorate	$Sr(ClO_3)_2$	18	63.6	117.0	67.0	1.839
strontium chloride	$SrCl_2.6H_2O$	15	33.4	45.5	90.7	1.36
strontium iodide	$SrI_2.6H_2O$	20	64.0	137.8	77.5	2.15
strontium nitrate	$Sr(NO_3)_2$	25	44.2	65.3	82.5	1.477
strontium nitrite	$Sr(NO_2)_2$	19	39.3	56.8	87.8	1.445
strontium perchlorate	$Sr(ClO_4)_2$	25	75.6	158.5	50.8	2.084
strontium salicylate	$Sr(C_7H_5O_3)_2$	25	4.58	4.68	97.5	1.019
succinic acid	$(CH_2)_2(COOH)_2$	25	7.67	7.82	94.5	1.021
succinimide	$(CH_2CO)_2NH.H_2O$	25	30.6	32.7	74.2	1.067
sucrose	$C_{12}H_{22}O_{11}$	25	67.89	90.9	43.0	1.340
tartaric acid	$C_2H_2(OH)_2(COOH)_2$	15	58.5	76.9	54.7	1.31
tetraethyl ammonium iodide	$N(C_2H_5)_4I$	25	32.9	36.2	74.0	1.102
tetramethyl ammonium iodide	$N(CH_3)_4I$	25	5.51	5.60	96.1	1.016
thallium chloride	$TlCl$	25	0.40	0.40	99.6	1.0005
thallium nitrate	$TlNO_3$	25	10.4	11.4	98.0	1.093
thallium nitrite	$TlNO_2$	25	32.1	43.7	92.5	1.360
thallium perchlorate	$TlClO_4$	25	13.5	15.2	97.1	1.122
thallium sulfate	Tl_2SO_4	25	5.48	5.74	99.0	1.047
trichloroacetic acid	CCl_3COOH	25	92.3	149.6	12.41	1.615
uranyl chloride	UO_2Cl_2	18	76.2	208.5	65.2	2.736
uranyl nitrate	$UO_2(NO_3)_2.6H_2O$	25	68.9	120	54.5	1.74
urea	$(NH_2)_2CO$	25	53.8	62	53.5	1.15
urea phosphate	$CO(NH_2)_2.H_3PO_4$	24.5	52.4	66.1	60.1	1.26
urethan	$NH_2CO_2C_2H_5$	25	82.8	88.8	18.5	1.073
D-valine	$(CH_3)_2CHCH(NH_2)COOH$	25	8.14	8.26	93.3	1.015
DL-valine	$(CH_3)_2CHCH(NH_2)COOH$	25	6.61	6.68	94.5	1.012
zinc acetate	$Zn(C_2H_3O_2)_2$	25	25.7	30.0	86.5	1.165
zinc benzenesulfonate	$Zn(C_6H_5SO_3)_2$	25	29.5	34.9	83.4	1.182
zinc chlorate	$Zn(ClO_3)_2$	18	65.0	124.4	67.0	1.914
zinc chloride	$ZnCl_2$	25	67.5	128	61	1.89
zinc iodide	ZnI_2	18	81.2	221.3	51.2	2.725
zinc phenolsulfonate	$(C_6H_5OSO_3)_2Zn.8H_2O$	25	39.8	47.3	71.5	1.185
zinc selenate	$ZnSeO_4$	22	37.8	58.9	97.0	1.559
zinc silicofluoride	$ZnSiF_6.6H_2O$	20	32.9	47.2	96.3	1.434
zinc sulfate	$ZnSO_4.7H_2O$	25	36.7	54.6	94.7	1.492
zinc valerate	$Zn(C_5H_9O_2)_2$	25	1.27	1.27	98.8	1.001

Cooling Mixtures

Substance	Quantity of substance	Quantity of water	Temp of resulting mixtures in °C	Approx temp, °F
ammonium nitrate	100	94	−4.0	25
sodium acetate	85	100	−4.7	23
sodium nitrate	75	100	−5.3	22.5
sodium thiosulfate cryst	110	100	−8.0	18
calcium chloride, $6H_2O$	100	246 (ice)	−9.0	16
sodium chloride	36	100	−10.0	14
ammonium nitrate	45	100 (ice)	−16.8	1.5
sodium nitrate	50	100 (ice)	−17.8	0
ammonium thiocyanate	133	100	−18.0	0
sodium chloride	33	100 (ice)	−21.3	−6
calcium chloride, $6H_2O$	100	123 (ice)	−21.5	−6.5
sodium bromide	66	100 (ice)	−28	−18
magnesium chloride cryst	85	100 (ice)	−34	−29
sulfuric acid (66.1% H_2SO_4)	100	109.7 (snow)	−37.0	−34.6
calcium chloride, $6H_2O$	100	81 (ice)	−40.3	−40
calcium chloride, $6H_2O$	100	70 (ice)	−55	−67
alcohol at 4°C with solid carbon dioxide			−72	−98
chloroform with solid carbon dioxide			−77	−106
acetone with solid carbon dioxide			−86	−123
ether with solid carbon dioxide			−100	−148
solid carbon dioxide (dry ice) sublimes at			−78.5	−109

Constant Humidity Solutions

A saturated aqueous solution in contact with an excess of the solute when kept in an enclosed space will maintain a constant humidity at a given temperature.

Substance dissolved and solid phase	Temp, °C	% Humidity
lead nitrate, $Pb(NO_3)_2$	20	98
dibasic sodium phosphate, $Na_2HPO_4.12H_2O$	20	95
monobasic ammonium phosphate, $NH_4H_2PO_4$	20–25	93
zinc sulfate, $ZnSO_4.7H_2O$	20	90
potassium chromate, K_2CrO_4	20	88
potassium bisulfate, $KHSO_4$	20	86
potassium bromide, KBr	20	84
ammonium sulfate, $(NH_4)_2SO_4$	20	81
ammonium chloride, NH_4Cl	20–25	79
sodium acetate, $NaC_2H_3O_2.3H_2O$	20	76
sodium chlorate, $NaClO_3$	20	75
sodium nitrite, $NaNO_2$	20	66
sodium bromide, $NaBr.2H_2O$	20	58
magnesium nitrate, $Mg(NO_3)_2.6H_2O$	18.5	56
sodium dichromate, $Na_2Cr_2O_7.2H_2O$	20	52
potassium thiocyanate, $KSCN$	20	47
zinc nitrate, $Zn(NO_3)_2.6H_2O$	20	42
chromium trioxide, CrO_3	20	35
calcium chloride, $CaCl_2.6H_2O$	24.5	31
potassium acetate, $KC_2H_3O_2$	20	20
lithium chloride, $LiCl.H_2O$	20	15

Indicators
For Volumetric Work and pH Determinations

Indicator	Chemical name	Acid color	pH range	Basic color	Preparation
methyl violet	tetra- penta- and hexamethyl-*p*-rosaniline HCl	Y	0.15–3.2	V	pH: 0.25% water
metanil yellow	4-phenylamino-azobenzene-3'-sulfonic acid	R	1.2–2.3	Y	pH: 0.25% in ethanol
metacresol purple (acid range)	*m*-cresolsulfonphthalein	R	1.2–2.8	Y	pH: 0.10 g in 13.6 ml 0.02*N* NaOH, diluted to 250 ml with water
p-xylenol blue (acid range)	1,4-dimethyl-5-hydroxybenzenesulfonphthalein	R	1.2–2.8	Y	pH: 0.04% in ethanol
thymol blue (acid range)	thymolsulfonphthalein	R	1.2–2.8	Y	pH: 0.1 g in 10.75 ml 0.02*N* NaOH, diluted to 250 ml with water
tropaeolin OO	sodium *p*-diphenylamino-azobenzenesulfonate	R	1.3–3.2	Y	pH: 0.1% in water vol: 1% in water
benzopurpurine 4B	ditolyl-diazo-bis-α-naphthyl-amine-4-sulfonic acid	B-V	1.3–4.0	R	pH, vol: 0.1% in water
quinaldine red	2-(*p*-dimethylaminostyryl)-quinoline ethiodide	C	1.4–3.2	R	vol: 0.1% in ethanol
2,4-dinitrophenol		C	2.4–4.0	Y	pH, vol: 0.1 g in 5 ml ethanol, diluted to 100 ml with water
methyl yellow	*p*-dimethylaminoazobenzene	R	2.9–4.0	Y	pH, vol: 0.05% in ethanol
bromphenol blue	tertabromophenolsulfon-phthalein	Y	3.0–4.6	B	pH: 0.1 g in 7.45 ml 0.02*N* NaOH, diluted to 250 ml with water
tetrabromophenol blue	tetrabromophenol-tetrabro-mosulfonphthalein	Y	3.0–4.6	B	pH: 0.1 g in 5.00 ml 0.02*N* NaOH, diluted to 250 ml with water
Congo red	diphenyl-diazo-bis-l-naphthyl-amine-4-sodium sulfonate	B	3.0–5.0	R	pH: 0.1% in water
methyl orange	4'dimethylaminoazobenzene-4-sodium sulfonate	R	3.1–4.4	Y	vol: 0.1% in water
brom-chlorphenol blue	dibromodichlorophenol-sulfonphthalein	Y	3.2–4.8	Pu	pH: 0.1 g in 8.6 ml 0.02*N* NaOH, diluted to 250 ml with water vol: 0.04% in ethanol
p-ethoxychrysoidine	4'-ethoxy-2,4-diaminoazo-benzene	R	3.5–5.5	Y	vol: 0.1% in ethanol
α-naphthyl red	α-naphthylaminoazobenzene	R	3.7–5.0	Y	vol: 0.1% in ethanol
sodium alizarin-sulfonate	dihydroxyanthraquinone sodium sulfonate	Y	3.7–5.2	V	pH, vol: 1% in water
bromcresol green	tetrabromo-*m*-cresolsulfon-phthalein	Y	3.8–5.4	B	pH: 0.10 g in 7.15 ml 0.02*N* NaOH, diluted to 250 ml with water
2,5-dinitrophenol		C	4.0–5.8	Y	pH, vol: 0.10 g in 20 ml ethanol, then dilute to 100 ml with water
methyl red	4'dimethylaminoazobenzene-2-carboxylic acid	R	4.4–6.2	Y	pH: 0.10 g in 18.6 ml 0.02*N* NaOH, diluted to 250 ml with water vol: 0.1% in ethanol
lacmoid		R	4.4–6.4	B	vol: 0.5% in ethanol
litmus		R	4.5–8.3	B	vol: 0.5% in water
chlorphenol red	dichlorophenolsulfon-phthalein	Y	4.8–6.4	R	pH: 0.1 g in 11.8 ml 0.02*N* NaOH, diluted to 250 ml with water vol: 0.04% in ethanol
benzoyl auramine G		V	5–5.6	Y	vol: 0.04% in ethanol pH: 0.25% in methanol
azolitmin		R	5.0–8.0	B	vol: 0.5% in water
bromcresol purple	dibromo-*o*-cresolsulfon-phthalein	Y	5.2–6.8	Pu	pH: 0.1 g in 9.25 ml 0.02*N* NaOH, diluted to 250 ml with water vol: 0.02% in ethanol
bromphenol red	dibromophenolsulfon-phthalein	Y	5.2–6.8	R	pH: 0.1 g in 9.75 ml 0.02*N* NaOH, diluted to 250 ml with water vol: 0.04% in ethanol

Indicator	Chemical name	Acid color	pH range	Basic color	Preparation
dibromophenol-tetrabromophenol-sulfonphthalein		Y	5.6–7.2	Pu	pH: 0.1 g in 1.21 ml 0.1N NaOH, diluted to 250 ml with water
p-nitrophenol		C	5.6–7.6	Y	pH, vol: 0.25% in water
bromothymol blue	dibromothymolsulfon-phthalein	Y	6.0–7.6	B	pH: 0.1 g in 8 ml 0.02N NaOH, diluted to 250 ml with water vol: 0.1% in 50% ethanol
phenol red	phenolsulfonphthalein	Y	6.4–8.2	R	pH: 0.1 g in 14.20 ml 0.02N NaOH, diluted to 250 ml with water vol: 0.1% in ethanol
neutral red	2-methyl-3-amino-6-di-methylaminophenazine	R	6.8–8.0	Y	pH: 0.1 g in 70 ml ethanol, diluted to 100 ml with water
rosolic acid aurin; corallin		Y	6.8–8.2	R	pH, vol: 1% in 50% ethanol
quinoline blue	cyanine	C	7.0–8.0	V	vol: 1% in ethanol
cresol red	o-cresolsulfonphthalein	Y	7.2–8.8	R	pH: 0.1 g in 13.1 ml 0.02N NaOH, diluted to 250 ml with water vol: 0.1% in ethanol
α-naphtholphthalein		P	7.3–8.7	G	pH, vol: 0.1% in 50% ethanol
metacresol purple (alkaline range)	m-cresolsulfonphthalein	Y	7.4–9.0	Pu	pH: 0.1 g in 13.1 ml 0.02N NaOH, diluted to 250 ml with water vol: 0.1% in ethanol
ethyl bis[2,4-dinitro-phenyl]acetate		C	7.5–9.1	B	vol: saturated soln in equal volumes of acetone and ethanol
tropaeolin OOO no. 1	sodium α-naphtholazoben-zenesulfonate	Y	7.6–8.9	R	vol: 0.1% in water
thymol blue (alka-line range)	thymolsulfonphthalein	Y	8.0–9.6	B	pH: 0.1 g in 10.75 ml 0.02N NaOH, diluted to 250 ml with water vol: 0.1% in ethanol
o-cresolphthalein		C	8.2–9.8	R	pH, vol: 0.04% in ethanol
phenolphthalein	3,3-bis(p-hydroxyphenyl)-phthalide	C	8.2–10	Pu	vol: 1% in ethanol
thymolphthalein		C	9.3–10.5	B	pH, vol: 0.1% in ethanol
Nile blue A	aminonaphthodiethylamino-phenoxazine sulfate	B	10–11	P	vol: 0.1% in water
α-naphtholbenzein	dimethylphenolphthalein	Y	9.8–11.0	B	pH, vol: 1% in ethanol
alizarin yellow GG	3-carboxy-4-hydroxy-3′-nitroazobenzene	Y	10.0–12.1	O	pH, vol: 0.1% in 50% ethanol
alizarin yellow R	3-carboxy-4-hydroxy-4′-nitro-azobenzene sodium salt	Y	10.0–12.1	R	pH, vol: 0.1% in water
nitramine	2,4,6-trinitrophenylmethyl-nitramine	C	10.8–13	Br	pH: 0.1% in 70% ethanol
Poirrier blue C4B	Na or K triphenylrosaniline-sulfonate	B	11–13	Pu	pH: 0.2% in water
tropaeolin O	p-benzenesulfonic acid-azoresorcinol	Y	11–13	O	pH: 0.1% in water
1,3,5-trinitrobenzene		C	11.5–14	O	pH: 0.1% in ethanol
indigo carmine	sodium indigodisulfonate	B	11.6–14	Y	pH: 0.25% in 50% ethanol

The indicator colors are abbreviated as follows: B, blue; Br, brown; C, colorless; G, green; O, orange; P, pink; Pu, purple; R, red; V, violet; and Y, yellow.

References

I.M. Kolthoff and V.A. Stenger, *Volumetric Analysis* **Vol. I** (Interscience, New York, 2nd ed., 1942).
R.G. Bates, *Determination of pH* (Wiley & Sons, New York, 1954).
E. Bishop, *Indicators* (Pergamon Press, Oxford, 1972).

Mixed Indicators

Composition	Solvent	Transition pH	Acid color	Transition color	Basic color
dimethyl yellow, 0.05% + methylene blue, 0.05%	alc	3.25	blue-violet	—	green
methyl orange, 0.02% + xylene cyanole FF, 0.28%	50% alc	3.8	violet	gray	green
methyl yellow, 0.08% + methylene blue, 0.004%	alc		pink	straw pink	yellow-green
methyl orange, 0.1% + indigocarmine, 0.25%	aq	4.1	violet	gray	green
bromcresol green, 0.1% + methyl orange, 0.02%	aq	4.3	orange	light green	green
bromcresol green, 0.075% + methyl red, 0.05%	alc	5.1	wine-red	—	green
methyl red, 0.1% + methylene blue, 0.05%	alc	5.4	red-violet	dirty blue	green
bromcresol green, 0.05% + chlorphenol red, 0.05%	aq	6.1	yellow-green	—	blue-violet
bromcresol purple, 0.05% + bromthymol blue, 0.05%	aq	6.7	yellow	violet	violet-blue
neutral red, 0.05% + methylene blue, 0.05%	alc	7.0	violet-blue	violet-blue	green
bromthymol blue, 0.05% + phenol red, 0.05%	aq	7.5	yellow	violet	violet
cresol red, 0.025% + thymol blue, 0.075%	aq	8.3	yellow	rose	violet
phenolphthalein, 0.033% + methyl green, 0.067%	alc	8.9	green	gray-blue	violet
phenolphthalein, 0.075% + thymol blue, 0.025%	50% alc	9.0	yellow	green	violet
phenolphthalein, 0.067% + naphtholphthalein, 0.033%	50% alc	9.6	pale rose	—	violet
phenolphthalein, 0.033% + Nile blue, 0.133%	alc	10.0	blue	violet	red
alizarin yellow, 0.033% + Nile blue, 0.133%	alc	10.8	green	—	red-brown

Standard Buffers for Calibrations of pH Meters

Temperature °C	pH values					
	0.1 M Hydrochloric acid	Saturated potassium hydrogen tartrate	0.05 M Potassium hydrogen phthalate	0.05 M Phosphate	0.01 M Borax	Saturated calcium hydroxide
0	1.10		4.003	6.984	9.464	13.423
5	1.10		3.999	6.951	9.395	13.207
10	1.10		3.998	6.923	9.332	13.003
15	1.10		3.999	6.900	9.276	12.810
20	1.10		4.002	6.881	9.225	12.627
25	1.10	3.557	4.008	6.865	9.180	12.454
30	1.10	3.552	4.015	6.853	9.139	12.289
35	1.10	3.549	4.024	6.844	9.102	12.133
38	1.10	3.548	4.030	6.840	9.081	12.043
40	1.10	3.547	4.035	6.838	9.068	11.984
45	1.10	3.547	4.047	6.834	9.038	11.841
50	1.10	3.549	4.060	6.833	9.011	11.705
55	1.11	3.554	4.075	6.834	8.985	11.574
60	1.11	3.560	4.091	6.836	8.962	11.449
70	1.11	3.580	4.126	6.845	8.921	
80	1.11	3.609	4.164	6.859	8.885	
90	1.12	3.650	4.205	6.877	8.850	
95	1.12	3.674	4.227	6.886	8.833	

Preparation of Solutions

0.1M Hydrochloric Acid. Prepare 0.1N hydrochloric acid by diluting reagent-grade hydrochloric acid. Determine the molarity by titration with standard alkali and adjust to 0.1000 ± 0.005M.

Saturated Potassium Hydrogen Tartrate. Shake an excess of reagent-grade potassium hydrogen tartrate with water at 25 ± 5°C for about 3 minutes.

0.05M Potassium Hydrogen Phthalate. Dissolve 10.21 ± 0.05 g of reagent-grade potassium hydrogen phthalate in sufficient water to make 1000 ml of solution.

0.05M Phosphate. Dissolve 3.40 ± 0.01 g of reagent-grade potassium dihydrogen phosphate and 3.55 ± 0.01 g of reagent-grade anhydrous disodium hydrogen phosphate in sufficient water to make 1000 ml of solution.

0.01M Borax. Dissolve 3.81 ± 0.01 g of reagent-grade sodium tetraborate decahydrate in sufficient water to make 1000 ml of solution.

Saturated Calcium Hydroxide. Shake large excess of finely granulated calcium hydroxide with water in polyethylene bottle at 25° and filter prior to use.

pH Values of Standard Solutions

Normality	pH values			
	HCl	CH₃COOH	NaOH	NH₃
1	0.10	2.37	14.05	11.77
0.1	1.07	2.87	13.07	11.27
0.01	2.02	3.37	12.12	10.77
0.001	3.01	3.87	11.13	10.27
0.0001	4.01			

THERMOMETRIC EQUIVALENTS

Temperature Scales

Symbol	Designation	Zero point	Freezing point of water	Boiling point of water at standard atm press.
°C	degree Celsius or Centigrade	freezing point of water	0°C	100°C
°K	degree Kelvin or absolute temperature in degrees Centigrades	absolute zero	273.15°K	373.15°K
°F	degree Fahrenheit	−17.8°C	+32°F	212°F
°Rank	degree Rankine or absolute temperature in degrees Fahrenheit	absolute zero	491.4°Rank	671.4°Rank

The following formulas may be used to convert temperatures from one scale to another:

Temperature given in	Temperature wanted in			
	°C	°K	°F	°Rank
°C	C	C + 273.15	1.8C + 32	1.8C + 491.4
°K	K − 273.15	K	1.8K − 459.4	1.8K
°F	0.556F − 17.8	0.556F + 255.3	F	F + 459.4
°Rank	0.556Rank − 273.1	0.556Rank	Rank − 459.4	Rank

Conversion Table (Thermometric Equivalents)

To convert a temperature from one scale to the other, find the desired temperature (Fahrenheit or Celsius) in one of the columns of bold face figures. For a Fahrenheit value, the Celsius equivalent will appear in the column to the left. For a Celsius value, the Fahrenheit equivalent will be listed in the column to the right. For example, the Celsius equivalent of 100° F is 37.78° C; the Farenheit equivalent of 100° C is 212° F.

To Convert Degrees			To Convert Degrees			To Convert Degrees		
To C	←F or C→	To F	To C	←F or C→	To F	To C	←F or C→	To F
−40	−40	−40	−12.22	10	50	15.56	60	140
−39.44	−39	−38.2	−11.67	11	51.8	16.11	61	141.8
−38.89	−38	−36.4	−11.11	12	53.6	16.67	62	143.6
−38.33	−37	−34.6	−10.56	13	55.4	17.22	63	145.4
−37.78	−36	−32.8	−10	14	57.2	17.78	64	147.2
−37.22	−35	−31	−9.44	15	59	18.33	65	149
−36.67	−34	−29.2	−8.89	16	60.8	18.89	66	150.8
−36.11	−33	−27.4	−8.33	17	62.6	19.44	67	152.6
−35.56	−32	−25.6	−7.78	18	64.4	20	68	154.4
−35	−31	−23.8	−7.22	19	66.2	20.56	69	156.2
−34.44	−30	−22	−6.67	20	68	21.11	70	158
−33.89	−29	−20.2	−6.11	21	69.8	21.67	71	159.8
−33.33	−28	−18.4	−5.56	22	71.6	22.22	72	161.6
−32.78	−27	−16.6	−5	23	73.4	22.78	73	163.4
−32.22	−26	−14.8	−4.44	24	75.2	23.33	74	165.2
−31.67	−25	−13	−3.89	25	77	23.89	75	167
−31.11	−24	−11.2	−3.33	26	78.8	24.44	76	168.8
−30.56	−23	−9.4	−2.78	27	80.6	25	77	170.6
−30	−22	−7.6	−2.22	28	82.4	25.56	78	172.4
−29.44	−21	−5.8	−1.67	29	84.2	26.11	79	174.2
−28.89	−20	−4	−1.11	30	86	26.67	80	176
−28.33	−19	−2.2	−0.56	31	87.8	27.22	81	177.8
−27.78	−18	−0.4	0	32	89.6	27.78	82	179.6
−27.22	−17	1.4	.56	33	91.4	28.33	83	181.4
−26.67	−16	3.2	1.11	34	93.2	28.89	84	183.2
−26.11	−15	5	1.67	35	95	29.44	85	185
−25.56	−14	6.8	2.22	36	96.8	30	86	186.8
−25	−13	8.6	2.78	37	98.6	30.56	87	188.6
−24.44	−12	10.4	3.33	38	100.4	31.11	88	190.4
−23.89	−11	12.2	3.89	39	102.2	31.67	89	192.2
−23.33	−10	14	4.44	40	104	32.22	90	194
−22.78	−9	15.8	5	41	105.8	32.78	91	195.8
−22.22	−8	17.6	5.56	42	107.6	33.33	92	197.6
−21.67	−7	19.4	6.11	43	109.4	33.89	93	199.4
−21.11	−6	21.2	6.67	44	111.2	34.44	94	201.2
−20.56	−5	23	7.22	45	113	35	95	203
−20	−4	24.8	7.78	46	114.8	35.56	96	204.8
−19.44	−3	26.6	8.33	47	116.6	36.11	97	206.6
−18.89	−2	28.4	8.89	48	118.4	36.67	98	208.4
−18.33	−1	30.2	9.44	49	120.2	37.22	99	210.2
−17.78	0	32	10	50	122	37.78	100	212
−17.22	1	33.8	10.56	51	123.8	38.33	101	213.8
−16.67	2	35.6	11.11	52	125.6	38.89	102	215.6
−16.11	3	37.4	11.67	53	127.4	39.44	103	217.4
−15.56	4	39.2	12.22	54	129.2	40	104	219.2
−15	5	41	12.78	55	131	40.56	105	221
−14.44	6	42.8	13.33	56	132.8	41.11	106	222.8
−13.89	7	44.6	13.89	57	134.6	41.67	107	224.6
−13.33	8	46.4	14.44	58	136.4	42.22	108	226.4
−12.78	9	48.2	15	59	138.2	42.78	109	228.2

To Convert Degrees			To Convert Degrees			To Convert Degrees		
To C	←F or C→	To F	To C	←F or C→	To F	To C	←F or C→	To F
43.33	**110**	230	73.89	**165**	329	104.44	**220**	428
43.89	**111**	231.8	74.44	**166**	330.8	105	**221**	429.8
44.44	**112**	233.6	75	**167**	332.6	105.56	**222**	431.6
45	**113**	235.4	75.56	**168**	334.4	106.11	**223**	433.4
45.56	**114**	237.2	76.11	**169**	336.2	106.67	**224**	435.2
46.11	**115**	239	76.67	**170**	338	107.22	**225**	437
46.67	**116**	240.8	77.22	**171**	339.8	107.78	**226**	438.8
47.22	**117**	242.6	77.78	**172**	341.6	108.33	**227**	440.6
47.78	**118**	244.4	78.33	**173**	343.4	108.89	**228**	442.4
48.33	**119**	246.2	78.89	**174**	345.2	109.44	**229**	444.2
48.89	**120**	248	79.44	**175**	347	110	**230**	446
49.44	**121**	249.8	80	**176**	348.8	110.56	**231**	447.8
50	**122**	251.6	80.56	**177**	350.6	111.11	**232**	449.6
50.56	**123**	253.4	81.11	**178**	352.4	111.67	**233**	451.4
51.11	**124**	255.2	81.67	**179**	354.2	112.22	**234**	453.2
51.67	**125**	257	82.22	**180**	356	112.78	**235**	455
52.22	**126**	258.8	82.78	**181**	357.8	113.33	**236**	456.8
52.78	**127**	260.6	83.33	**182**	359.6	113.89	**237**	458.6
53.33	**128**	262.4	83.89	**183**	361.4	114.44	**238**	460.4
53.89	**129**	264.2	84.44	**184**	363.2	115	**239**	462.2
54.44	**130**	266	85	**185**	365	115.56	**240**	464
55	**131**	267.8	85.56	**186**	366.8	116.11	**241**	465.8
55.56	**132**	269.6	86.11	**187**	368.6	116.67	**242**	467.6
56.11	**133**	271.4	86.67	**188**	370.4	117.22	**243**	469.4
56.67	**134**	273.2	87.22	**189**	372.2	117.78	**244**	471.2
57.22	**135**	275	87.78	**190**	374	118.33	**245**	473
57.78	**136**	276.8	88.33	**191**	375.8	118.89	**246**	474.8
58.33	**137**	278.6	88.89	**192**	377.6	119.44	**247**	476.6
58.89	**138**	280.4	89.44	**193**	379.4	120	**248**	478.4
59.44	**139**	282.2	90	**194**	381.2	120.56	**249**	480.2
60	**140**	284	90.56	**195**	383	121.11	**250**	482
60.56	**141**	285.8	91.11	**196**	384.8	121.67	**251**	483.8
61.11	**142**	287.6	91.67	**197**	386.6	122.22	**252**	485.6
61.67	**143**	289.4	92.22	**198**	388.4	122.78	**253**	487.4
62.22	**144**	291.2	92.78	**199**	390.2	123.33	**254**	489.2
62.78	**145**	293	93.33	**200**	392	123.89	**255**	491
63.33	**146**	294.8	93.89	**201**	393.8	124.44	**256**	492.8
63.89	**147**	296.6	94.44	**202**	395.6	125	**257**	494.6
64.44	**148**	298.4	95	**203**	397.4	125.56	**258**	496.4
65	**149**	300.2	95.56	**204**	399.2	126.11	**259**	498.2
65.56	**150**	302	96.11	**205**	401	126.67	**260**	500
66.11	**151**	303.8	96.67	**206**	402.8	127.22	**261**	501.8
66.67	**152**	305.6	97.22	**207**	404.6	127.78	**262**	503.6
67.22	**153**	307.4	97.78	**208**	406.4	128.33	**263**	505.4
67.78	**154**	309.2	98.33	**209**	408.2	128.89	**264**	507.2
68.33	**155**	311	98.89	**210**	410	129.44	**265**	509
68.89	**156**	312.8	99.44	**211**	411.8	130	**266**	510.8
69.44	**157**	314.6	100	**212**	413.6	130.56	**267**	512.6
70	**158**	316.4	100.56	**213**	415.4	131.11	**268**	514.4
70.56	**159**	318.2	101.11	**214**	417.2	131.67	**269**	516.2
71.11	**160**	320	101.67	**215**	419	132.22	**270**	518
71.67	**161**	321.8	102.22	**216**	420.8	132.78	**271**	519.8
72.22	**162**	323.6	102.78	**217**	422.6	133.33	**272**	521.6
72.78	**163**	325.4	103.33	**218**	424.4	133.89	**273**	523.4
73.33	**164**	327.2	103.89	**219**	426.2	134.44	**274**	525.2

To Convert Degrees			To Convert Degrees			To Convert Degrees		
To C	←F or C→	To F	To C	←F or C→	To F	To C	←F or C→	To F
135	275	527	165.56	330	626	196.11	385	725
135.56	276	528.8	166.11	331	627.8	196.67	386	726.8
136.11	277	530.6	166.67	332	629.6	197.22	387	728.6
136.67	278	532.4	167.22	333	631.4	197.78	388	730.4
137.22	279	534.2	167.78	334	633.2	198.33	389	732.2
137.78	280	536	168.33	335	635	198.89	390	734
138.33	281	537.8	168.89	336	636.8	199.44	391	735.8
138.89	282	539.6	169.44	337	638.6	200	392	737.6
139.44	283	541.4	170	338	640.4	200.56	393	739.4
140	284	543.2	170.56	339	642.2	201.11	394	741.2
140.56	285	545	171.11	340	644	201.67	395	743
141.11	286	546.8	171.67	341	645.8	202.22	396	744.8
141.67	287	548.6	172.22	342	647.6	202.78	397	746.6
142.22	288	550.4	172.78	343	649.4	203.33	398	748.4
142.78	289	552.2	173.33	344	651.2	203.89	399	750.2
143.33	290	554	173.89	345	653	204.44	400	752
143.89	291	555.8	174.44	346	654.8	205	401	753.8
144.44	292	557.6	175	347	656.6	205.56	402	755.6
145	293	559.4	175.56	348	658.4	206.11	403	757.4
145.56	294	561.2	176.11	349	660.2	206.67	404	759.2
146.11	295	563	176.67	350	662	207.22	405	761
146.67	296	564.8	177.22	351	663.8	207.78	406	762.8
147.22	297	566.6	177.78	352	665.6	208.33	407	764.6
147.78	298	568.4	178.33	353	667.4	208.89	408	766.4
148.33	299	570.2	178.89	354	669.2	209.44	409	768.2
148.89	300	572	179.44	355	671	210	410	770
149.44	301	573.8	180	356	672.8	210.56	411	771.8
150	302	575.6	180.56	357	674.6	211.11	412	773.6
150.56	303	577.4	181.11	358	676.4	211.67	413	775.4
151.11	304	579.2	181.67	359	678.2	212.22	414	777.2
151.67	305	581	182.22	360	680	212.78	415	779
152.22	306	582.8	182.78	361	681.8	213.33	416	780.8
152.78	307	584.6	183.33	362	683.6	213.89	417	782.6
153.33	308	586.4	183.89	363	685.4	214.44	418	784.4
153.89	309	588.2	184.44	364	687.2	215	419	786.2
154.44	310	590	185	365	689	215.56	420	788
155	311	591.8	185.56	366	690.8	216.11	421	789.8
155.56	312	593.6	186.11	367	692.6	216.67	422	791.6
156.11	313	595.4	186.67	368	694.4	217.22	423	793.4
156.67	314	597.2	187.22	369	696.2	217.78	424	795.2
157.22	315	599	187.78	370	698	218.33	425	797
157.78	316	600.8	188.33	371	699.8	218.89	426	798.8
158.33	317	602.6	188.89	372	701.6	219.44	427	800.6
158.89	318	604.4	189.44	373	703.4	220	428	802.4
159.44	319	606.2	190	374	705.2	220.56	429	804.2
160	320	608	190.56	375	707	221.11	430	806
160.56	321	609.8	191.11	376	708.8	221.67	431	807.8
161.11	322	611.6	191.67	377	710.6	222.22	432	809.6
161.67	323	613.4	192.22	378	712.4	222.78	433	811.4
162.22	324	615.2	192.78	379	714.2	223.33	434	813.2
162.78	325	617	193.33	380	716	223.89	435	815
163.33	326	618.8	193.89	381	717.8	224.44	436	816.8
163.89	327	620.6	194.44	382	719.6	225	437	818.6
164.44	328	622.4	195	383	721.4	225.56	438	820.4
165	329	624.2	195.56	384	723.2	226.11	439	822.2

To Convert Degrees			To Convert Degrees			To Convert Degrees		
To C	←F or C→	To F	To C	←F or C→	To F	To C	←F or C→	To F
226.67	440	824	248.89	480	896	271.11	520	968
227.22	441	825.8	249.44	481	897.8	271.67	521	969.8
227.78	442	827.6	250	482	899.6	272.22	522	971.6
228.33	443	829.4	250.56	483	901.4	272.78	523	973.4
228.89	444	831.2	251.11	484	903.2	273.33	524	975.2
229.44	445	833	251.67	485	905	273.89	525	977
230	446	834.8	252.22	486	906.8	274.44	526	978.8
230.56	447	836.6	252.78	487	908.6	275	527	980.6
231.11	448	838.4	253.33	488	910.4	275.56	528	982.4
231.67	449	840.2	253.89	489	912.2	276.11	529	984.2
232.22	450	842	254.44	490	914	276.67	530	986
232.78	451	843.8	255	491	915.8	277.22	531	987.8
233.33	452	845.6	255.56	492	917.6	277.78	532	989.6
233.89	453	847.4	256.11	493	919.4	278.33	533	991.2
234.44	454	849.2	256.67	494	921.2	278.89	534	993.2
235	455	851	257.22	495	923	279.44	535	995
235.56	456	852.8	257.78	496	924.8	280	536	996.8
236.11	457	854.6	258.33	497	926.6	280.56	537	998.6
236.67	458	856.4	258.89	498	928.4	281.11	538	1000.4
237.22	459	858.2	259.44	499	930.2	281.67	539	1002.2
237.78	460	860	260	500	932	282.22	540	1004
238.33	461	861.8	260.56	501	933.8	282.78	541	1005.8
238.89	462	863.6	261.11	502	935.6	283.33	542	1007.6
239.44	463	865.4	261.67	503	937.4	283.89	543	1009.4
240	464	867.2	262.22	504	939.2	284.44	544	1011.2
240.56	465	869	262.78	505	941	285	545	1013
241.11	466	870.8	263.33	506	942.8	285.56	546	1014.8
241.67	467	872.6	263.89	507	944.6	286.11	547	1016.6
242.22	468	874.4	264.44	508	946.4	286.67	548	1018.4
242.78	469	876.2	265	509	948.2	287.22	549	1020.2
243.33	470	878	265.56	510	950	287.78	550	1022
243.89	471	879.8	266.11	511	951.8	288.33	551	1023.8
244.44	472	881.6	266.67	512	953.6	288.89	552	1025.6
245	473	883.4	267.22	513	955.4	289.44	553	1027.4
245.56	474	885.2	267.78	514	957.2	290	554	1029.2
246.11	475	887	268.33	515	959			
246.67	476	888.8	268.89	516	960.8			
247.22	477	890.6	269.44	517	962.6			
247.78	478	892.4	270	518	964.4			
248.33	479	894.2	270.56	519	966.2			

International System of Units (SI)

SI Basic and Supplementary Units

Name	Symbol	Physical Quantity
SI BASE UNITS		
meter	m	length
kilogram	kg	mass
second	s	time
ampere	A	electric current
kelvin	K	thermodynamic temperature
mole	mol	amount of substance
candela	cd	luminous intensity
SI SUPPLEMENTARY UNITS		
radian	rad	plane angle
steradian	sr	solid angle

SI Derived Units with Special Names

Formula	Symbol	Special Name	Physical Quantity
$1 \text{ kg} \cdot \text{m/s}^2$	$= 1 \text{ N}$	1 newton	force
1 N/m^2	$= 1 \text{ Pa}$	1 pascal	pressure or stress
$1 \text{ N} \cdot \text{m}$	$= 1 \text{ J}$	1 joule	work, energy, or quantity of heat
1 J/s	$= 1 \text{ W}$	1 watt	power or radiant energy flux
1 W/A	$= 1 \text{ V}$	1 volt	electric potential, potential difference, or electromotive force
1 A/V	$= 1 \text{ S}$	1 siemens	electric conductance
1 V/A	$= 1 \Omega$	1 ohm	electric resistance
$1 \text{ A} \cdot \text{s}$	$= 1 \text{ C}$	1 coulomb	quantity of electricity or electric charge
1 C/V	$= 1 \text{ F}$	1 farad	electric capacitance
$1 \text{ V} \cdot \text{s}$	$= 1 \text{ Wb}$	1 weber	magnetic flux
1 Wb/A	$= 1 \text{ H}$	1 henry	inductance
1 Wb/m^2	$= 1 \text{ T}$	1 tesla	magnetic flux density or magnetic induction
$1 \text{ cd} \cdot \text{sr}$	$= 1 \text{ lm}$	1 lumen	luminous flux
1 lm/m^2	$= 1 \text{ lx}$	1 lux	illuminance
1 J/kg	$= 1 \text{ Gy}$	1 gray	absorbed dose (of ionizing radiation)
1 (disintegration)/s	$= 1 \text{ Bq}$	1 becquerel	activity (of a radionuclide)
1 (cycle)/s	$= 1 \text{ Hz}$	1 hertz	frequency (of a periodic phenomenon)

SI Prefixes

Factor	Prefix	Symbol	Factor	Prefix	Symbol
10^{-18}	atto	a	10	deca	da
10^{-15}	femto	f	10^2	hecto	h
10^{-12}	pico	p	10^3	kilo	k
10^{-9}	nano	n	10^6	mega	M
10^{-6}	micro	μ	10^9	giga	G
10^{-3}	milli	m	10^{12}	tera	T
10^{-2}	centi	c	10^{15}	peta	P
10^{-1}	deci	d	10^{18}	exa	E

References
G. G. Stoner, *Textile Research Journal* **46**, 623, 850 (1976).
Federal Register **47**, 8399 (1982).

Universal Conversion Factors

To convert units of one system into another, find the given units in the table and multiply by the appropriate conversion factor. For example,
To convert 26.5 centimeters into inches, find centimeters in the table, read across to inches and multiply by 0.3937.

$$26.5 \text{ cm} \times \frac{0.3937 \text{ inch}}{\text{cm}} = 10.4 \text{ inches}$$

To convert 8.75 inches into centimeters, find inches in the table, read across to centimeters and multiply by 2.540.

$$8.75 \text{ inches} \times \frac{2.540 \text{ cm}}{\text{inch}} = 22.2 \text{ cm}$$

TO CONVERT	INTO	MULTIPLY BY	TO CONVERT	INTO	MULTIPLY BY
A			baryes	dynes/sq cm	1.000
abcoulombs	statcoulombs	2.998×10^{10}	bolts (US cloth)	meters	36.576
acres	sq chains (Gunter's)	10.	Btu	liter – atmosphere	10.409
acres	sq rods	160.	Btu	ergs	1.0550×10^{10}
acres	square links		Btu	foot-lb	778.3
	(Gunter's)	1×10^5	Btu	gram-calories	252.0
acres	hectares or		Btu	horsepower-hr	3.931×10^{-4}
	sq hectometers	.4047	Btu	joules	1,054.8
acres	sq ft	43,560.0	Btu	kilogram-calories	0.2520
acres	sq meters	4,047.	Btu	kilogram-meters	107.5
acres	sq miles	1.562×10^{-3}	Btu	kilowatt-hr	2.928×10^{-4}
acres	sq yd	4,840.	Btu/hr	foot-lb/sec	0.2162
acre-feet	cu ft	43,560.0	Btu/hr	gram-cal/sec	0.0700
acre-feet	gallons	3.259×10^5	Btu/hr	horsepower-hr	3.929×10^{-4}
amperes/sq cm	amps/sq in	6.452	Btu/hr	watts	0.2931
amperes/sq cm	amps/sq meter	10^4	Btu/min	foot-lb/sec	12.96
amperes/sq in	amps/sq cm	0.1550	Btu/min	horsepower	0.02356
amperes/sq in	amps/sq meter	1,550.0	Btu/min	kilowatts	0.01757
amperes/sq meter	amps/sq cm	10^{-4}	Btu/min	watts	17.57
amperes/sq meter	amps/sq in	6.452×10^{-4}	Btu/sq ft/min	watts/sq in	0.1221
ampere-hours	coulombs	3,600.0	bucket (Br dry)	cubic cm	1.818×10^4
ampere-hours	faradays	0.03731	bushels	cu ft	1.2445
ampere-turns	gilberts	1.257	bushels	cu in	2,150.42
ampere-turns/cm	amp-turns/in	2.540	bushels	cu meters	0.03524
ampere-turns/cm	amp-turns/meter	100.0	bushels	liters	35.24
ampere-turns/cm	gilberts/cm	1.257	bushels	pecks	4.0
ampere-turns/in	amp-turns/cm	0.3937	bushels	pints (dry)	64.0
ampere-turns/in	amp-turns/meter	39.37	bushels	quarts (dry)	32.0
ampere-turns/in	gilberts/cm	0.4950			
ampere-turns/meter	amp-turns/cm	0.01			
ampere-turns/meter	amp-turns/in	0.0254		**C**	
ampere-turns/meter	gilberts/cm	0.01257	calories, gram		
angstrom unit	inches	3937×10^{-9}	(mean)	Btu (mean)	3.9685×10^{-3}
angstrom unit	meters	1×10^{-10}	candle/sq cm	lamberts	3.142
angstrom unit	microns or (mu)	1×10^{-4}	candle/sq inch	lamberts	.4870
ares	acres (US)	.02471	centares (centiares)	sq meters	1.0
ares	sq yd	119.60	centigrade	Fahrenheit	$1.8C° + 32$
ares	acres	0.02471	centigrams	grams	0.01
ares	sq meters	100.0	centiliters	ounces (fl)	.3382
astronomical units	kilometers	1.495×10^8	centiliters	cu in	.6103
atmospheres	ton/sq in	.007348	centiliters	drams	2.705
atmospheres	cm of mercury	76.0	centiliters	liters	0.01
atmospheres	mm of mercury	760.0	centimeters	feet	3.281×10^{-2}
atmospheres	torrs	760.0	centimeters	inches	0.3937
atmospheres	ft of water (at 4°C)	33.90	centimeters	kilometers	10^{-5}
atmospheres	in of mercury (at 0°C)	29.92	centimeters	meters	0.01
atmospheres	kg/sq cm	1.0333	centimeters	miles	6.214×10^{-6}
atmospheres	kg/sq meter	10,332.	centimeters	millimeters	10.0
atmospheres	lb/sq in	14.70	centimeters	mils	393.7
atmospheres	tons/sq ft	1.058	centimeters	yards	1.094×10^{-2}
			centimeter-dynes	cm-grams	1.020×10^{-3}
	B		centimeter-dynes	meter-kg	1.020×10^{-8}
barrels (US, dry)	cu in	7056.	centimeter-dynes	lb-ft	7.376×10^{-8}
barrels (US, dry)	quarts (dry)	105.0	centimeter-grams	cm-dynes	980.7
barrels (US, liq)	gallons	31.5	centimeter-grams	meter-kg	10^{-5}
barrels (oil)	gallons (oil)	42.0	centimeter-grams	lb-ft	7.233×10^{-5}
bars	atmospheres	0.9869	centimeters of		
bars	dynes/sq cm	10^6	mercury	atmospheres	0.01316
bars	kg/sq meter	1.020×10^4	centimeters of		
bars	lb/sq ft	2,089.	mercury	ft of water	0.4461
bars	lb/sq in	14.50			

TO CONVERT	INTO	MULTIPLY BY	TO CONVERT	INTO	MULTIPLY BY
centimeters of mercury	kg/sq meter	136.0	cubic inches	bushels	4.650×10^{-4}
centimeters of mercury	lb/sq ft	27.85	cubic inches	cu cm	16.39
centimeters of mercury	lb/sq in	0.1934	cubic inches	cu ft	5.787×10^{-4}
centimeters/sec	ft/min	1.1969	cubic inches	cu meters	1.639×10^{-5}
centimeters/sec	ft/sec	0.03281	cubic inches	cu yards	2.143×10^{-5}
centimeters/sec	kilometers/hr	0.036	cubic inches	drams (fl)	4.4329
centimeters/sec	knots	0.1943	cubic inches	gallons (US)	4.329×10^{-3}
centimeters/sec	meters/min	0.6	cubic inches	gallons (UK)	3.605×10^{-3}
centimeters/sec	miles/hr	0.02237	cubic inches	gills	0.1385
centimeters/sec	miles/min	3.728×10^{-4}	cubic inches	liters	0.01639
centimeters/sec/sec	ft/sec/sec	0.03281	cubic inches	milliliters	16.39
centimeters/sec/sec	km/hr/sec	0.036	cubic inches	mil-ft	1.061×10^{5}
centimeters/sec/sec	meters/sec/sec	0.01	cubic inches	minims	265.974
centimeters/sec/sec	miles/hr/sec	0.02237	cubic inches	ounces (fl)	0.5541
chain	inches	792.00	cubic inches	pecks	1.860×10^{-3}
chain	meters	20.12	cubic inches	pints (dry)	0.0298
chains (surveyors' or Gunter's)	yards	22.00	cubic inches	pints (liq)	0.03463
circular mils	sq cm	5.067×10^{-6}	cubic inches	quarts (dry)	0.0149
circular mils	sq mils	0.7854	cubic inches	quarts (liq)	0.01732
circular mils	sq in	7.854×10^{-7}	cubic meters	bushels	28.38
circumference	radians	6.283	cubic meters	cu cm	10^{6}
cords	cord ft	8	cubic meters	cu ft	35.31
cord feet	cu ft	16	cubic meters	cu in	61,023.74
coulomb	statcoulombs	2.998×10^{9}	cubic meters	cu yards	1.308
coulombs	faradays	1.036×10^{-5}	cubic meters	gallons (US)	264.2
coulombs/sq cm	coulombs/sq in	64.52	cubic meters	gallons (UK)	220.0
coulombs/sq cm	coulombs/sq meter	10^{4}	cubic meters	liters	1,000.0
coulombs/sq in	coulombs/sq cm	0.1550	cubic meters	milliliters	10^{6}
coulombs/sq in	coulombs/sq meter	1,550.	cubic meters	pecks	113.51
coulombs/sq meter	coulombs/sq cm	10^{-4}	cubic meters	pints (dry)	1,816.166
coulombs/sq meter	coulombs/sq in	6.452×10^{-4}	cubic meters	pints (liq)	2,113.0
cubic centimeters	cu ft	3.531×10^{-5}	cubic meters	quarts (dry)	908.083
cubic centimeters	cu in	0.06102	cubic meters	quarts (liq)	1,057.
cubic centimeters	cu meters	10^{-6}	cubic yards	cu cm	7.646×10^{5}
cubic centimeters	cu yards	1.308×10^{-6}	cubic yards	cu ft	27.0
cubic centimeters	drams (fl)	0.2705	cubic yards	cu in	46,656.0
cubic centimeters	gallons (US)	2.642×10^{-4}	cubic yards	cu meters	0.7646
cubic centimeters	gills	8.454×10^{-3}	cubic yards	gallons (US)	202.0
cubic centimeters	liters	0.001	cubic yards	gallons (UK)	168.2
cubic centimeters	milliliters	1.0	cubic yards	liters	764.6
cubic centimeters	minims	16.231	cubic yards	milliliters	7.646×10^{5}
cubic centimeters	ounces (fl)	0.0338	cubic yards	pints (liq)	1,615.9
cubic centimeters	pints (liq)	2.113×10^{-3}	cubic yards	quarts (liq)	807.9
cubic centimeters	quarts (liq)	1.057×10^{-3}	cubic yards/min	cubic ft/sec	0.45
cubic feet	bushels	0.8036	cubic yards/min	gallons/sec	3.367
cubic feet	cu cm	28,316.85	cubic yards/min	liters/sec	12.74
cubic feet	cu in	1,728.0			
cubic feet	cu meters	0.02832			
cubic feet	cu yards	0.03704		**D**	
cubic feet	drams (fl)	7660.05	Dalton	gram	1.650×10^{-24}
cubic feet	gallons (US)	7.48052	days	seconds	86,400.0
cubic feet	gallons (UK)	6.229	decigrams	grams	0.1
cubic feet	gills	239.38	deciliters	liters	0.1
cubic feet	liters	28.32	decimeters	meters	0.1
cubic feet	milliliters	28,316.85	degrees (angle)	quadrants	0.01111
cubic feet	minims	459,603.1	degrees (angle)	radians	0.01745
cubic feet	ounces (fl)	957.51	degrees (angle)	seconds	3,600.0
cubic feet	pecks	3.2143	degrees/sec	radians/sec	0.01745
cubic feet	pints (dry)	51.428	degrees/sec	revolutions/min	0.1667
cubic feet	pints (liq)	59.84	degrees/sec	revolutions/sec	2.778×10^{-3}
cubic feet	quarts (dry)	25.714	dekagrams	grams	10.0
cubic feet	quarts (liq)	29.92	dekaliters	liters	10.0
cubic feet/min	cu cm/sec	472.0	dekameters	meters	10.0
cubic feet/min	gallons/sec	0.1247	drams (avdp)	drams (apoth)	0.4557
cubic feet/min	liters/sec	0.4720	drams (avdp)	grains	27.3437
cubic feet/min	lb of water/min	62.43	drams (avdp)	grams	1.7718
cubic feet/sec	million gals/day	0.646317	drams (avdp)	kilograms	1.7718×10^{-3}
cubic feet/sec	gallons/min	448.831	drams (avdp)	milligrams	1771.85
			drams (avdp)	ounces (apoth or troy)	0.0570
			drams (avdp)	ounces (avdp)	0.0625
			drams (avdp)	pennyweights	1.139

TO CONVERT	INTO	MULTIPLY BY	TO CONVERT	INTO	MULTIPLY BY
drams (avdp)	pounds (apoth or troy)	4.747×10^{-3}	faradays	coulombs	9.649×10^4
drams (avdp)	pounds (avdp)	3.906×10^{-3}	fathoms	meters	1.828804
drams (avdp)	scruples	1.367	fathoms	feet	6.0
drams (apoth)	drams (avdp)	2.1943	feet	centimeters	30.48
drams (apoth)	grains	60.0	feet	inches	12.0
drams (apoth)	grams	3.8879	feet	kilometers	3.048×10^{-4}
drams (apoth)	kilograms	3.888×10^{-3}	feet	meters	0.3048
drams (apoth)	milligrams	3887.93	feet	miles (naut)	1.645×10^{-4}
drams (apoth)	ounces (apoth or troy)	0.125	feet	miles (stat)	1.894×10^{-4}
			feet	millimeters	304.8
drams (apoth)	ounces (avdp)	0.1371429	feet	mils	1.2×10^4
drams (apoth)	pennyweights	2.5	feet	yards	0.333
drams (apoth)	pounds (apoth or troy)	0.0104	feet of water	atmospheres	0.02950
			feet of water	in of mercury	0.8826
drams (apoth)	pounds (avdp)	8.571×10^{-3}	feet of water	kg/sq cm	0.03048
drams (apoth)	scruples	3.0	feet of water	kg/sq meter	304.8
drams (fl)	cu cm	3.6967	feet of water	lb/sq ft	62.43
drams (fl)	cu ft	1.3055×10^{-4}	feet of water	lb/sq in	0.4335
drams (fl)	cu in	0.2256	feet/min	cm/sec	0.5080
drams (fl)	gallons (US)	9.7656×10^{-4}	feet/min	feet/sec	0.01667
drams (fl)	gills	0.03125	feet/min	km/hr	0.01829
drams (fl)	liters	3.6967×10^{-3}	feet/min	meters/min	0.3048
drams (fl)	milliliters	3.6967	feet/min	miles/hr	0.01136
drams (fl)	minims	60.0	feet/sec	cm/sec	30.48
drams (fl)	ounces (fl)	0.125	feet/sec	km/hr	1.097
drams (fl)	pints (liq)	7.8125×10^{-3}	feet/sec	knots	0.5921
drams (fl)	quarts (liq)	3.9063×10^{-3}	feet/sec	meters/min	18.29
dyne/cm	erg/sq mm	.01	feet/sec	miles/hr	0.6818
dyne/sq cm	atmospheres	9.869×10^{-7}	feet/sec	miles/min	0.01136
dyne/sq cm	inch of mercury at 0°C	2.953×10^{-5}	feet/sec/sec	cm/sec/sec	30.48
			feet/sec/sec	km/hr/sec	1.097
dyne/sq cm	inch of water at 4°C	4.015×10^{-4}	feet/sec/sec	meters/sec/sec	0.3048
dynes	grams	1.020×10^{-3}	feet/sec/sec	miles/hr/sec	0.6818
dynes	joules/cm	10^{-7}	feet/100 feet	percent grade	1.0
dynes	joules/meter (newtons)	10^{-5}	foot-candles	lumens/sq meter	10.764
			foot-pounds	Btu	1.286×10^{-3}
dynes	kilograms	1.020×10^{-6}	foot-pounds	ergs	1.356×10^7
dynes	poundals	7.233×10^{-5}	foot-pounds	gram-calories	0.3238
dynes	pounds	2.248×10^{-6}	foot-pounds	hp-hr	5.050×10^{-7}
dynes/sq cm	bars	10^{-6}	foot-pounds	joules	1.356
			foot-pounds	kg-calories	3.24×10^{-4}
E			foot-pounds	kg-meters	0.1383
ell	cm	114.30	foot-pounds	kilowatt-hr	3.766×10^{-7}
ell	inches	45.	foot-pounds/min	Btu/min	1.286×10^{-3}
em, pica	inches	.167	foot-pounds/min	foot-lb/sec	0.01667
em, pica	cm	.4233	foot-pounds/min	horsepower	3.030×10^{-5}
erg/sec	dyne−cm/sec	1.000	foot-pounds/min	kg-calories/min	3.24×10^{-4}
ergs	Btu	9.480×10^{-11}	foot-pounds/min	kilowatts	2.260×10^{-5}
ergs	dyne-centimeters	1.0	foot-pounds/sec	Btu/hr	4.6263
ergs	foot-lb	7.367×10^{-8}	foot-pounds/sec	Btu/min	0.07717
ergs	gram-calories	0.2389×10^{-7}	foot-pounds/sec	horsepower	1.818×10^{-3}
ergs	gram-cm	1.020×10^{-3}	foot-pounds/sec	kg-calories/min	0.01945
ergs	horsepower-hr	3.7250×10^{-14}	foot-pounds/sec	kilowatts	1.356×10^{-3}
ergs	joules	10^{-7}	furlongs	miles (US)	0.125
ergs	kg-calories	2.389×10^{-11}	furlongs	rods	40.0
ergs	kg-meters	1.020×10^{-8}	furlongs	feet	660.0
ergs	kilowatt-hr	0.2778×10^{-13}			
ergs	watt-hr	0.2778×10^{-10}	**G**		
ergs/sec	Btu/min	5.688×10^{-6}	gallons (US)	cu cm	3,785.0
ergs/sec	ft-lb/min	4.427×10^{-6}	gallons (US)	cu ft	0.1337
ergs/sec	ft-lb/sec	7.3756×10^{-8}	gallons (US)	cu in	231.0
ergs/sec	horsepower	1.341×10^{-10}	gallons (US)	cu meters	3.785×10^{-3}
ergs/sec	kg-calories/min	1.433×10^{-9}	gallons (US)	cu yards	4.951×10^{-3}
ergs/sec	kilowatts	10^{-10}	gallons (US)	drams (fl)	1024.0
			gallons (US)	gallons (UK)	0.83267
F			gallons (US)	gills	32.0
Fahrenheit	centigrade	$0.556 F° - 17.8$	gallons (US)	liters	3.785
farads	microfarads	10^6	gallons (US)	milliliters	3,785.0
faradays/sec	ampere (abs)	9.6500×10^4	gallons (US)	minims	61,440.0
faradays	ampere-hr	26.80	gallons (US)	ounces (fl)	128.0
			gallons (US)	pints (liq)	8.0

TO CONVERT	INTO	MULTIPLY BY	TO CONVERT	INTO	MULTIPLY BY
gallons (US)	quarts (liq)	4.0	grams	tons (short)	1.102×10^{-6}
gallons (UK)	cu ft	0.1605	grams/cm	lb/in	5.600×10^{-3}
gallons (UK)	cu in	277.4	grams/cu cm	lb/cu ft	62.43
gallons (UK)	cu meters	4.546×10^{-3}	grams/cu cm	lb/cu in	0.03613
gallons (UK)	cu yards	5.946×10^{-3}	grams/cu cm	lb/mil-ft	3.405×10^{-7}
gallons (UK)	gallons (US)	1.20095	grams/liter	grains/gal	58.417
gallons (UK)	liters	4.546	grams/liter	lb/1,000 gal	8.345
gallons of water	lb of water	8.3453	grams/liter	lb/cu ft	0.062427
gallons/min	cu ft/sec	2.228×10^{-3}	grams/liter	parts/million	1,000.0
gallons/min	liters/sec	0.06308	grams/sq cm	lb/sq ft	2.0481
gallons/min	cu ft/hr	8.0208	gram-calories	Btu	3.9683×10^{-3}
gausses	lines/sq in	6.452	gram-calories	ergs	4.1868×10^{7}
gausses	webers/sq cm	10^{-8}	gram-calories	ft-lb	3.0880
gausses	webers/sq in	6.452×10^{-8}	gram-calories	horsepower-hr	1.5596×10^{-6}
gausses	webers/sq meter	10^{-4}	gram-calories	kilowatt-hr	1.1630×10^{-6}
gilberts	ampere-turns	0.7958	gram-calories	watt-hr	1.1630×10^{-3}
gilberts/cm	amp-turns/cm	0.7958	gram-calories/sec	Btu/hr	14.286
gilberts/cm	amp-turns/in	2.021	gram-centimeters	Btu	9.297×10^{-8}
gilberts/cm	amp-turns/meter	79.58	gram-centimeters	ergs	980.7
gills	cu cm	118.2941	gram-centimeters	joules	9.807×10^{-5}
gills	cu ft	4.1775×10^{-3}	gram-centimeters	kg-cal	2.343×10^{-8}
gills	cu in	7.21875	gram-centimeters	kg-meters	10^{-5}
gills	drams (fl)	32.0			
gills	gallons (US)	0.03125			
gills	liters	0.1183	**H**		
gills	milliliters	118.2941	hand	cm	10.16
gills	minims	1920.0	hectares	acres	2.471
gills	ounces (fl)	4.0	hectares	sq ft	1.076×10^{5}
gills	pints (liq)	0.25	hectograms	grams	100.0
gills	quarts (liq)	0.125	hectoliters	liters	100.0
gills (UK)	cu cm	142.07	hectometers	meters	100.0
grade	radian	0.01571	hectowatts	watts	100.0
grains	drams (apoth)	0.0167	henries	millihenries	1,000.0
grains	drams (avdp)	0.03657143	hogsheads (Br)	cu ft	10.114
grains	grams	0.0648	hogsheads (US)	cu ft	8.42184
grains	kilograms	6.479×10^{-5}	hogsheads (US)	gallons (US)	63
grains	milligrams	64.799	horsepower	Btu/min	42.44
grains	ounces (apoth or		horsepower	ft-lb/min	33,000.
	troy)	2.083×10^{-3}	horsepower	ft-lb/sec	550.0
grains	ounces (avdp)	2.286×10^{-3}	horsepower	horsepower (metric)	1.014
grains	pennyweights	0.04167	horsepower	kg-calories/min	10.68
grains	pounds (apoth or		horsepower	kilowatts	0.7457
	troy)	1.736×10^{-4}	horsepower	watts	745.7
grains	pounds (avdp)	1.423×10^{-4}	horsepower (boiler)	Btu/hr	33,479.
grains	scruples	0.05	horsepower (boiler)	kilowatts	9.803
grains (troy)	grains (avdp)	1.0	horsepower (metric)	horsepower	0.9863
grains (troy)	grams	0.06480	horsepower (metric)	ft-lb/sec	542.5
grains (troy)	ounces (avdp)	2.286×10^{-3}	horsepower-hours	Btu	2,547.
grains (troy)	pennyweight (troy)	0.04167	horsepower-hours	ergs	2.6845×10^{13}
grains/US gal	parts/million	17.118	horsepower-hours	ft-lb	1.98×10^{6}
grains/US gal	lb/million gal	142.86	horsepower-hours	gm-cal	641,190.
grains/UK gal	parts/million	14.286	horsepower-hours	joules	2.684×10^{6}
grams	drams (apoth)	0.2572	horsepower-hours	kg-calories	641.1
grams	drams (avdp)	0.5644	horsepower-hours	kg-meters	2.737×10^{5}
grams	dynes	980.7	horsepower-hours	kilowatt-hr	0.7457
grams	grains	15.43	hours	days	4.167×10^{-2}
grams	joules/cm	9.807×10^{-5}	hours	weeks	5.952×10^{-3}
grams	joules/meter		hundredweights		
	(newtons)	9.807×10^{-3}	(long)	cwt (short)	1.12
grams	kilograms	0.001	hundredweights		
grams	milligrams	1,000.0	(long)	kilograms	50.802
grams	ounces (apoth or		hundredweights		
	troy)	0.03215	(long)	ounces	1792.0
grams	ounces (avdp)	0.03527	hundredweights		
grams	pennyweights	0.643	(long)	pounds	112.
grams	pounds (apoth or		hundredweights		
	troy)	2.679×10^{-3}	(long)	slugs	3.4811
grams	pounds (avdp)	2.205×10^{-3}	hundredweights		
grams	poundals	0.07093	(long)	tons (long)	0.05
grams	scruples	0.7716	hundredweights		
grams	slugs	6.852×10^{-5}	(long)	tons (short)	0.056

TO CONVERT	INTO	MULTIPLY BY	TO CONVERT	INTO	MULTIPLY BY
hundredweights (short)	cwt (short)	0.8929	kilograms	joules/meter (newtons)	9.807
hundredweights (short)	kilograms	45.359	kilograms	milligrams	10^6
hundredweights (short)	ounces	1600.0	kilograms	ounces (apoth or troy)	32.151
hundredweights (short)	pounds	100.0	kilograms	ounces (avdp)	35.274
hundredweights (short)	slugs	3.1081	kilograms	pennyweights	643.015
hundredweights (short)	tons (metric)	0.0453592	kilograms	pounds (apoth or troy)	2.679
hundredweights (short)	tons (long)	0.0446429	kilograms	pounds (avdp)	2.205
hundredweights (short)	tons (short)	0.05	kilograms	poundals	70.93
			kilograms	scruples	771.62
			kilograms	slugs	0.0685
	I		kilograms	tons (metric)	0.001
inches	centimeters	2.540	kilograms	tons (long)	9.842×10^{-4}
inches	feet	0.0833	kilograms	tons (short)	1.102×10^{-3}
inches	meters	2.540×10^{-2}	kilograms/cu meter	gm/cu cm	0.001
inches	miles	1.578×10^{-5}	kilograms/cu meter	lb/cu ft	0.06243
inches	miles (naut)	1.3715×10^{-5}	kilograms/cu meter	lb/cu in	3.613×10^{-5}
inches	millimeters	25.40	kilograms/cu meter	lb/mil-foot	3.405×10^{-10}
inches	mils	1,000.0	kilograms/meter	lb/ft	0.6720
inches	yards	2.778×10^{-2}	kilograms/sq cm	dynes	980,665.
inches of mercury	atmospheres	0.03342	kilograms/sq cm	atmospheres	0.9678
inches of mercury	ft of water	1.133	kilograms/sq cm	ft of water	32.81
inches of mercury	kg/sq cm	0.03453	kilograms/sq cm	in of mercury	28.96
inches of mercury	kg/sq meter	345.3	kilograms/sq cm	lb/sq ft	2,048.
inches of mercury	lb/sq ft	70.73	kilograms/sq cm	lb/sq in	14.22
inches of mercury	lb/sq in	0.4912	kilograms/sq meter	atmospheres	9.678×10^{-5}
inches of water (at 4°C)	atmospheres	2.458×10^{-3}	kilograms/sq meter	bars	9.807×10^{-5}
inches of water (at 4°C)	in of mercury	0.07355	kilograms/sq meter	ft of water	3.281×10^{-3}
inches of water (at 4°C)	kg/sq cm	2.540×10^{-3}	kilograms/sq meter	in of mercury	2.896×10^{-3}
inches of water (at 4°C)	oz/sq in	0.5781	kilograms/sq meter	lb/sq ft	0.2048
inches of water (at 4°C)	lb/sq ft	5.204	kilograms/sq meter	lb/sq in	1.422×10^{-3}
inches of water (at 4°C)	lb/sq in	0.03613	kilograms/sq mm	kg/sq meter	10^6
international ampere	ampere (abs)	.9998	kilogram-calories	Btu	3.968
international volt	volts (abs)	1.0003	kilogram-calories	ft-lb	3,088.
international volt	joules	9.654×10^4	kilogram-calories	hp-hr	1.560×10^{-3}
			kilogram-calories	joules	4,186.
	J		kilogram-calories	kg-meters	426.9
joules	Btu	9.480×10^{-4}	kilogram-calories	kilojoules	4.186
joules	ergs	10^7	kilogram-calories	kilowatt-hr	1.163×10^{-3}
joules	ft-lb	0.7376	kilogram meters	Btu	9.294×10^{-3}
joules	kg-calories	2.389×10^{-4}	kilogram meters	ergs	9.804×10^7
joules	kg-meters	0.1020	kilogram meters	ft-lb	7.233
joules	watt-hr	2.778×10^{-4}	kilogram meters	joules	9.804
joules/cm	grams	1.020×10^4	kilogram meters	kg-calories	2.342×10^{-3}
joules/cm	dynes	10^7	kilogram meters	kilowatt-hr	2.723×10^{-6}
joules/cm	joules/meter (newtons)	100.0	kilolines	maxwells	1,000.0
joules/cm	poundals	723.3	kiloliters	liters	1,000.0
joules/cm	pounds	22.48	kilometers	centimeters	10^5
			kilometers	feet	3,281.
	K		kilometers	inches	3.937×10^4
kilograms	cwt (long)	0.0197	kilometers	meters	1,000.0
kilograms	cwt (short)	0.022	kilometers	miles	0.6214
kilograms	drams (apoth)	257.21	kilometers	millimeters	10^6
kilograms	drams (avdp)	564.38	kilometers	yards	1,094.
kilograms	dynes	980,665.	kilometers/hr	cm/sec	27.78
kilograms	grains	15,432.36	kilometers/hr	ft/min	54.68
kilograms	grams	1000.0	kilometers/hr	ft/sec	0.9113
kilograms	joules/cm	0.09807	kilometers/hr	knots	0.5396
			kilometers/hr	meters/min	16.67
			kilometers/hr	miles/hr	0.6214
			kilometers/hr/sec	cm/sec/sec	27.78
			kilometers/hr/sec	ft/sec/sec	0.9113
			kilometers/hr/sec	meters/sec/sec	0.2778
			kilometers/hr/sec	miles/hr/sec	0.6214
			kilowatts	Btu/min	56.92
			kilowatts	foot-lb/min	4.426×10^4
			kilowatts	foot-lb/sec	737.6
			kilowatts	horsepower	1.341
			kilowatts	kg-calories/min	14.34

TO CONVERT	INTO	MULTIPLY BY	TO CONVERT	INTO	MULTIPLY BY
kilowatts	watts	1,000.0	meters	millimeters	1,000.0
kilowatt-hours	Btu	3,413.	meters	rods	0.1988
kilowatt-hours	ergs	3.600×10^{13}	meters	yards	1.094
kilowatt-hours	ft-lb	2.655×10^6	meters	varas	1.179
kilowatt-hours	gram-calories	859,850.	meters/min	cm/sec	1.667
kilowatt-hours	horsepower-hr	1.341	meters/min	ft/min	3.281
kilowatt-hours	joules	3.6×10^6	meters/min	ft/sec	0.05468
kilowatt-hours	kg-calories	860.5	meters/min	km/hr	0.06
kilowatt-hours	kg-meters	3.671×10^5	meters/min	knots	0.03238
kilowatt-hours	lb of water evap-		meters/min	miles/hr	0.03728
	orated from and at		meters/sec	feet/min	196.8
	212°F.	3.53	meters/sec	feet/sec	3.281
kilowatt-hours	lb of water raised		meters/sec	kilometers/hr	3.6
	from 62° to 212°F.	22.75	meters/sec	kilometers/min	0.06
knots	ft/hr	6,080.	meters/sec	miles/hr	2.237
knots	kilometers/hr	1.8532	meters/sec	miles/min	0.03728
knots	nautical miles/hr	1.0	meters/sec/sec	cm/sec/sec	100.0
knots	statute miles/hr	1.151	meters/sec/sec	ft/sec/sec	3.281
knots	yd/hr	2,027.	meters/sec/sec	km/hr/sec	3.6
knots	ft/sec	1.689	meters/sec/sec	miles/hr/sec	2.237
			meter-kilograms	cm-dynes	9.807×10^7
L			meter-kilograms	cm-gm	10^5
league	miles (approx.)	3.0	meter-kilograms	lb-ft	7.233
light year	miles	5.9×10^{12}	microfarad	farads	10^{-6}
light year	kilometers	9.46091×10^{12}	micrograms	grams	10^{-6}
lines/sq cm	gausses	1.0	microhms	megohms	10^{-12}
lines/sq in	gausses	0.1550	microhms	ohms	10^{-6}
lines/sq in	webers/sq cm	1.550×10^{-9}	microliters	liters	10^{-6}
lines/sq in	webers/sq in	10^{-8}	microns	meters	1×10^{-6}
lines/sq in	webers/sq meter	1.550×10^{-5}	miles (naut)	feet	6,080.27
links (engineer's)	inches	12.0	miles (naut)	inches	7.2913×10^4
links (surveyor's)	inches	7.92	miles (naut)	kilometers	1.853
liters	bushels	0.02838	miles (naut)	meters	1,853.
liters	cu cm	1,000.0	miles (naut)	miles (stat)	1.1516
liters	cu ft	0.03531	miles (naut)	yards	2,027.
liters	cu in	61.024	miles (statute)	centimeters	1.609×10^5
liters	cu meters	0.001	miles (statute)	feet	5,280.
liters	cu yards	1.308×10^{-3}	miles (statute)	inches	6.336×10^4
liters	drams (fl)	270.512	miles (statute)	kilometers	1.609
liters	gallons (US)	0.2642	miles (statute)	meters	1,609.
liters	gallons (UK)	0.220	miles (statute)	miles (naut)	0.8684
liters	gills	8.454	miles (statute)	rods	320.
liters	milliliters	1,000.0	miles (statute)	yards	1,760.
liters	minims	16,230.73	miles/hr	cm/sec	44.70
liters	ounces (fl)	33.814	miles/hr	ft/min	88.
liters	pecks	0.1135	miles/hr	ft/sec	1.467
liters	pints (dry)	1.8162	miles/hr	km/hr	1.609
liters	pints (liq)	2.113	miles/hr	km/min	0.02682
liters	quarts (dry)	0.9081	miles/hr	knots	0.8684
liters	quarts (liq)	1.057	miles/hr	meters/min	26.82
liters/min	cu ft/sec	5.886×10^{-4}	miles/hr	miles/min	0.1667
liters/min	gal/sec	4.403×10^{-3}	miles/hr/sec	cm/sec/sec	44.70
lumens/sq ft	foot-candles	1.0	miles/hr/sec	ft/sec/sec	1.467
lumen	spherical candle		miles/hr/sec	km/hr/sec	1.609
	power	.07958	miles/hr/sec	meters/sec/sec	0.4470
lumen	watt	.001496	miles/min	cm/sec	2,682.
lumen/sq ft	lumen/sq meter	10.76	miles/min	ft/sec	88.
lux	foot-candles	0.0929	miles/min	km/min	1.609
			miles/min	knots/min	0.8684
M			miles/min	miles/hr	60.0
maxwells	kilolines	0.001	mil-feet	cu in	9.425×10^{-6}
maxwells	webers	10^{-8}	milliers	kilograms	1,000.
megalines	maxwells	10^6	milligrams	drams (apoth)	2.572×10^{-4}
megohms	microhms	10^{12}	milligrams	drams (avdp)	5.644×10^{-4}
megohms	ohms	10^6	milligrams	grains	0.01543236
meters	centimeters	100.0	milligrams	grams	0.001
meters	feet	3.281	milligrams	kilograms	10^{-6}
meters	inches	39.37	milligrams	ounces (apoth or	
meters	kilometers	0.001		troy)	3.215×10^{-5}
meters	miles (naut)	5.396×10^{-4}	milligrams	ounces (avdp)	3.527×10^{-5}
meters	miles (stat)	6.214×10^{-4}	milligrams	pennyweights	6.43×10^{-4}

TO CONVERT	INTO	MULTIPLY BY	TO CONVERT	INTO	MULTIPLY BY
milligrams	pounds (apoth or troy)	2.679×10^{-6}	ounces (avdp)	milligrams	28,349.5
milligrams	pounds (avdp)	2.2046×10^{-6}	ounces (avdp)	ounces (apoth or troy)	0.9115
milligrams	scruples	7.7162×10^{-4}	ounces (avdp)	pennyweights	18.23
milligrams/liter	parts/million	1.0	ounces (avdp)	pounds (apoth or troy)	0.0759
millihenries	henries	0.001	ounces (avdp)	pounds (avdp)	0.0625
milliliters	cu cm	1.0	ounces (avdp)	scruples	21.875
milliliters	cu ft	3.531×10^{-5}	ounces (avdp)	slugs	1.9426×10^{-3}
milliliters	cu in	0.06102	ounces (avdp)	tons (long)	2.790×10^{-5}
milliliters	drams (fl)	0.2705	ounces (avdp)	tons (metric)	2.835×10^{-5}
milliliters	gallons (US)	2.642×10^{-4}	ounces (avdp)	tons (short)	3.125×10^{-5}
milliliters	gills	8.454×10^{-3}	ounces (apoth or troy)	drams (apoth)	8.0
milliliters	liters	0.001	ounces (apoth or troy)	drams (avdp)	17.554
milliliters	minims	16.231	ounces (apoth or troy)	grains	480.0
milliliters	ounces (fl)	0.0338	ounces (apoth or troy)	grams	31.103481
milliliters	pints (liq)	2.113×10^{-3}	ounces (apoth or troy)	kilograms	0.0311
milliliters	quarts (liq)	1.057×10^{-3}	ounces (apoth or troy)	milligrams	31,103.48
millimeters	centimeters	0.1	ounces (apoth or troy)	ounces (avdp)	1.09714
millimeters	feet	3.281×10^{-3}	ounces (apoth or troy)	pennyweights	20.0
millimeters	inches	0.03937	ounces (apoth or troy)	pounds (apoth or troy)	0.08333
millimeters	kilometers	10^{-6}	ounces (apoth or troy)	pounds (avdp)	0.0686
millimeters	meters	0.001	ounces (apoth or troy)	scruples	24.0
millimeters	miles	6.214×10^{-7}	ounces (fl)	cu cm	29.5735
millimeters	mils	39.37	ounces (fl)	cu ft	1.0444×10^{-3}
millimeters	yards	1.094×10^{-3}	ounces (fl)	cu in	1.805
millimeters of Hg	atmospheres	1.316×10^{-3}	ounces (fl)	drams (fl)	8.0
millimeters of Hg	torrs	1.0	ounces (fl)	gallons (US)	7.8125×10^{-3}
millimicrons	meters	1×10^{-9}	ounces (fl)	gills	0.25
million gals/day	cu ft/sec	1.54723	ounces (fl)	liters	0.02957
mils	centimeters	2.540×10^{-3}	ounces (fl)	milliliters	29.5735
mils	feet	8.333×10^{-5}	ounces (fl)	minims	480.0
mils	inches	0.001	ounces (fl)	pints (liq)	0.0625
mils	kilometers	2.540×10^{-8}	ounces (fl)	quarts (liq)	0.03125
mils	yards	2.778×10^{-5}	ounces/sq in	dynes/sq cm	4309.
miner's inches	cu ft/min	1.5	ounces/sq in	pounds/sq in	0.0625
minims (Br)	cu cm	0.059192			
minims	cu cm	0.061612			
minims	cu ft	2.176×10^{-6}	**P**		
minims	cu in	3.7598×10^{-3}	parsecs	miles	19×10^{12}
minims	drams (fl)	0.0167	parsecs	kilometers	3.084×10^{13}
minims	gallons (US)	1.628×10^{-5}	parts/million	gr/US gal	0.0584
minims	gills	5.208×10^{-4}	parts/million	gr/UK gal	0.07016
minims	liters	6.161×10^{-5}	parts/million	lb/million gal	8.345
minims	milliliters	0.061612	pecks (Br)	cu in	554.6
minims	ounces (fl)	0.0021	pecks (Br)	liters	9.091901
minims	pints (liq)	1.302×10^{-4}	pecks	bushels	0.25
minims	quarts (liq)	6.51×10^{-5}	pecks	cu ft	0.3111
minutes(angles)	degrees	0.01667	pecks	cu in	537.605
minutes (angles)	quadrants	1.852×10^{-4}	pecks	cu meters	8.8098×10^{-3}
minutes (angles)	radians	2.909×10^{-4}	pecks	liters	8.8098
minutes (angles)	seconds	60.0	pecks	pints (dry)	16.0
myriagrams	kilograms	10.0	pecks	quarts (dry)	8.0
myriameters	kilometers	10.0	pennyweights	drams (apoth)	0.4
myriawatts	kilowatts	10.0	pennyweights	drams (avdp)	0.8777
			pennyweights	grains	24.0
			pennyweights	grams	1.55517
N			pennyweights	kilograms	1.555×10^{-3}
nepers	decibels	8.686	pennyweights	milligrams	1,555.17
newtons	dynes	1×10^5	pennyweights	ounces (apoth or troy)	0.05
O					
ohm (Int)	ohm (abs)	1.0005			
ohms	megohms	10^{-6}			
ohms	microhms	10^6			
ounces (avdp)	cwt (long)	5.5804×10^{-4}			
ounces (avdp)	cwt (short)	6.25×10^{-4}			
ounces (avdp)	drams (apoth)	7.292			
ounces (avdp)	drams (avdp)	16.0			
ounces (avdp)	grains	437.5			
ounces (avdp)	grams	28.349527			
ounces (avdp)	kilograms	0.0283			

TO CONVERT	INTO	MULTIPLY BY	TO CONVERT	INTO	MULTIPLY BY
pennyweights	ounces (avdp)	0.0549	pounds (apoth or troy)	ounces (apoth or troy)	12.0
pennyweights	pounds (apoth or troy)	4.1667×10^{-3}	pounds (apoth or troy)	ounces (avdp)	13.1657
pennyweights	pounds (avdp)	3.428×10^{-3}	pounds (apoth or troy)	pennyweights	240.0
pennyweights	scruples	1.2	pounds (apoth or troy)	pounds (avdp)	0.822857
pints (dry)	bushels	0.0156	pounds (apoth or troy)	scruples	288.0
pints (dry)	cu ft	0.0194	pounds (apoth or troy)	tons (long)	3.6753×10^{-4}
pints (dry)	cu in	33.60	pounds (apoth or troy)	tons (metric)	3.7324×10^{-4}
pints (dry)	cu meters	5.506×10^{-4}	pounds (apoth or troy)	tons (short)	4.1143×10^{-4}
pints (dry)	liters	0.5506			
pints (dry)	pecks	0.0625	pounds of water	cu ft	0.01602
pints (dry)	quarts (dry)	0.5	pounds of water	cu in	27.68
pints (liq)	cu cm	473.2	pounds of water	gallons	0.1198
pints (liq)	cu ft	0.01671	pounds of water/min	cu ft/sec	2.670×10^{-4}
pints (liq)	cu in	28.875	pound-feet	cm-dynes	1.356×10^{7}
pints (liq)	cu meters	4.732×10^{-4}	pound-feet	cm-gm	13,825.
pints (liq)	cu yards	6.189×10^{-4}	pound-feet	meter-kg	0.1383
pints (liq)	drams (fl)	128.0	pounds/cu ft	gm/cu cm	0.01602
pints (liq)	gallons (US)	0.125	pounds/cu ft	kg/cu meter	16.02
pints (liq)	gills	4.0	pounds/cu ft	lb/cu in	5.787×10^{-4}
pints (liq)	liters	0.4732	pounds/cu ft	lb/mil-foot	5.456×10^{-9}
pints (liq)	milliliters	473.2	pounds/cu in	gm/cu cm	27.68
pints (liq)	minims	7680.0	pounds/cu in	kg/cu meter	2.768×10^{4}
pints (liq)	ounces (fl)	16.0	pounds/cu in	lb/cu ft	1,728.
pints (liq)	quarts (liq)	0.5	pounds/cu in	lb/mil-foot	9.425×10^{-6}
Planck's quantum	erg-sec	6.624×10^{-27}	pounds/ft	kg/meter	1.488
poise	gm/cm sec	1.00	pounds/in	gm/cm	178.6
poundals	dynes	13,826.	pounds/mil-foot	gm/cu cm	2.306×10^{6}
poundals	grams	14.10	pounds/sq ft	atmospheres	4.725×10^{-4}
poundals	joules/cm	1.383×10^{-3}	pounds/sq ft	ft of water	0.01602
poundals	joules/meter (newtons)	0.1383	pounds/sq ft	in of mercury	0.01414
poundals	kilograms	0.01410	pounds/sq ft	kg/sq meter	4.882
poundals	pounds	0.03108	pounds/sq ft	lb/sq in	6.944×10^{-3}
pounds (avdp)	cwt (long)	8.929×10^{-3}	pounds/sq in	atmospheres	0.06804
pounds (avdp)	cwt (short)	0.01	pounds/sq in	ft of water	2.307
pounds (avdp)	drams (apoth)	116.67	pounds/sq in	in of mercury	2.036
pounds (avdp)	drams (avdp)	256.0	pounds/sq in	kg/sq meter	703.1
pounds (avdp)	dynes	44.4823×10^{4}	pounds/sq in	lb/sq ft	144.0
pounds (avdp)	grains	7,000.0			
pounds (avdp)	grams	453.5924		**Q**	
pounds (avdp)	joules/cm	0.04448	quadrants (angle)	degrees	90.0
pounds (avdp)	joules/meter (newtons)	4.448	quadrants (angle)	minutes	5,400.0
pounds (avdp)	kilograms	0.4536	quadrants (angle)	radians	1.571
pounds (avdp)	milligrams	453,592.37	quadrants (angle)	seconds	3.24×10^{5}
pounds (avdp)	ounces (apoth or troy)	14.5833	quarts (dry)	bushels	0.0313
pounds (avdp)	ounces (avdp)	16.0	quarts (dry)	cu ft	0.0389
pounds (avdp)	pennyweights	291.667	quarts (dry)	cu in	67.20
pounds (avdp)	poundals	32.17	quarts (dry)	cu meters	1.101×10^{-3}
pounds (avdp)	pounds (apoth or troy)	1.21528	quarts (dry)	liters	1.1012
pounds (avdp)	scruples	350.0	quarts (dry)	pecks	0.125
pounds (avdp)	slugs	3.108×10^{-2}	quarts (dry)	pints (dry)	2.0
pounds (avdp)	tons (long)	4.464×10^{-4}	quarts (liq)	cu cm	946.4
pounds (avdp)	tons (metric)	4.536×10^{-4}	quarts (liq)	cu ft	0.03342
pounds (avdp)	tons (short)	5.0×10^{-4}	quarts (liq)	cu in	57.75
pounds (apoth or troy)	drams (apoth)	96.0	quarts (liq)	cu meters	9.464×10^{-4}
pounds (apoth or troy)	drams (avdp)	210.65	quarts (liq)	cu yards	1.238×10^{-3}
pounds (apoth or troy)	grains	5,760.0	quarts (liq)	drams (fl)	256.0
pounds (apoth or troy)	grams	373.2417	quarts (liq)	gallons (US)	0.25
pounds (apoth or troy)	kilograms	0.3732	quarts (liq)	gills	8.0
pounds (apoth or troy)	milligrams	373,241.72	quarts (liq)	liters	0.9464
			quarts (liq)	milliliters	946.4
			quarts (liq)	minims	15,360.0
			quarts (liq)	ounces (fl)	32.0
			quarts (liq)	pints (liq)	2.0

TO CONVERT	INTO	MULTIPLY BY
R		
radians	degrees	57.30
radians	minutes	3,438.
radians	quadrants	0.6366
radians	seconds	2.063×10^5
radians/sec	degrees/sec	57.30
radians/sec	rev/min	9.549
radians/sec	rev/sec	0.1592
radians/sec/sec	rev/min/min	573.0
radians/sec/sec	rev/min/sec	9.549
radians/sec/sec	rev/sec/sec	0.1592
revolutions	degrees	360.0
revolutions	quadrants	4.0
revolutions	radians	6.283
revolutions/min	degrees/sec	6.0
revolutions/min	radians/sec	0.1047
revolutions/min	rev/sec	0.01667
revolutions/min/min	radians/sec/sec	1.745×10^{-3}
revolutions/min/min	rev/min/sec	0.01667
revolutions/min/min	rev/sec/sec	2.778×10^{-4}
revolutions/sec	degrees/sec	360.0
revolutions/sec	radians/sec	6.283
revolutions/sec	rev/min	60.0
revolutions/sec/sec	radians/sec/sec	6.283
revolutions/sec/sec	rev/min/min	3,600.0
revolutions/sec/sec	rev/min/sec	60.0
rods	chain (Gunter's)	0.25
rods	meters	5.029
rods (surveyors' measure)	yards	5.5
rods	feet	16.5
S		
scruples	drams (apoth)	0.3333
scruples	drams (avdp)	0.7314
scruples	grains	20.0
scruples	grams	1.296
scruples	kilograms	1.296×10^{-3}
scruples	milligrams	1295.97
scruples	ounces (apoth or troy)	0.0417
scruples	ounces (avdp)	0.0457
scruples	pennyweights	0.8333
scruples	pounds (apoth or troy)	3.472×10^{-3}
scruples	pounds (avdp)	2.857×10^{-3}
seconds (angle)	degrees	2.778×10^{-4}
seconds (angle)	minutes	0.01667
seconds (angle)	quadrants	3.087×10^{-6}
seconds (angle)	radians	4.848×10^{-6}
slugs	cwt (long)	0.2873
slugs	cwt (short)	0.3217
slugs	grams	1.459×10^4
slugs	kilograms	14.59
slugs	ounces (avdp)	514.79
slugs	pounds (avdp)	32.17
slugs	tons (long)	1.436×10^{-2}
slugs	tons (short)	1.609×10^{-2}
sphere	steradians	12.57
square centimeters	circular mils	1.973×10^5
square centimeters	sq ft	1.076×10^{-3}
square centimeters	sq in	0.1550
square centimeters	sq meters	0.0001
square centimeters	sq miles	3.861×10^{-11}
square centimeters	sq mm	100.0
square centimeters	sq yd	1.196×10^{-4}
square feet	acres	2.296×10^{-5}
square feet	circular mills	1.833×10^8
square feet	sq cm	929.0
square feet	sq in	144.0
square feet	sq meters	0.09290
square feet	sq miles	3.587×10^{-8}
square feet	sq mm	9.290×10^4
square feet	sq yd	0.1111
square inches	circular mils	1.273×10^6
square inches	sq cm	6.452
square inches	sq ft	6.944×10^{-3}
square inches	sq mm	645.2
square inches	sq mils	10^6
square inches	sq yd	7.716×10^{-4}
square kilometers	acres	247.1
square kilometers	sq cm	10^{10}
square kilometers	sq ft	1.076×10^7
square kilometers	sq in	1.550×10^9
square kilometers	sq meters	10^6
square kilometers	sq miles	0.3861
square kilometers	sq yd	1.196×10^6
square meters	acres	2.471×10^{-4}
square meters	sq cm	10^4
square meters	sq ft	10.76
square meters	sq in	1,550.
square meters	sq miles	3.861×10^{-7}
square meters	sq mm	10^6
square meters	sq yd	1.196
square miles	acres	640.0
square miles	sq ft	2.788×10^7
square miles	sq km	2.590
square miles	sq meters	2.590×10^6
square miles	sq yd	3.098×10^6
square millimeters	circular mils	1,973.
square millimeters	sq cm	0.01
square millimeters	sq ft	1.076×10^{-5}
square millimeters	sq in	1.550×10^{-3}
square mils	circular mils	1.273
square mils	sq cm	6.452×10^{-6}
square mils	sq in	10^{-6}
square yards	acres	2.066×10^{-4}
square yards	sq cm	8,361.
square yards	sq ft	9.0
square yards	sq in	1,296.
square yards	sq meters	0.8361
square yards	sq miles	3.228×10^{-7}
square yards	sq mm	8.361×10^5
T		
temperature (°C) +273	absolute temperature (°C)	1.0
temperature (°C) +17.78	temperature (°F)	1.8
temperature (°F) +460	absolute temperature (°F)	1.0
temperature (°F) −32	temperature (°C)	5/9
tons (long)	cwt (long)	20.
tons (long)	cwt (short)	22.4
tons (long)	kilograms	1,016.
tons (long)	ounces (avdp)	35,840.0
tons (long)	pounds (avdp)	2,240.0
tons (long)	slugs	69.621
tons (long)	tons (metric)	1.0160
tons (long)	tons (short)	1.120
tons (metric)	cwt (short)	22.046
tons (metric)	kilograms	1,000.0
tons (metric)	ounces (avdp)	35,273.96
tons (metric)	pounds (avdp)	2,205.
tons (metric)	tons (long)	0.9842
tons (metric)	tons (short)	1.1023
tons (short)	cwt (long)	17.857
tons (short)	cwt (short)	20.0
tons (short)	grams	9.072×10^5
tons (short)	kilograms	907.1847

TO CONVERT	INTO	MULTIPLY BY	TO CONVERT	INTO	MULTIPLY BY
tons (short)	ounces (apoth or troy)	29,166.66	watts (abs)	Btu (mean)/min	0.056884
tons (short)	ounces (avdp)	32,000.0	watts (abs)	joules/sec	1.
tons (short)	pounds (apoth or troy)	2,430.56	watt-hours	Btu	3.413
tons (short)	pounds (avdp)	2,000.0	watt-hours	ergs	3.60×10^{10}
tons (short)	slugs	62.16	watt-hours	ft-lb	2,656.
tons (short)	tons (long)	0.89286	watt-hours	gm-cal	859.85
tons (short)	tons (metric)	0.9072	watt-hours	horsepower-hr	1.341×10^{-3}
tons (short)/sq ft	kg/sq meter	9,765.	watt-hours	kg-cal	0.8605
tons (short)/sq ft	lb/sq in	2,000.	watt-hours	kg-meters	367.2
tons of water/24 hr	lb of water/hr	83.333	watt-hours	kilowatt-hr	0.001
tons of water/24 hr	gal/min	0.16643	watt (int)	watt (abs)	1.0002
tons of water/24 hr	cu ft/hr	1.3349	webers	maxwells	10^8
torrs	mm of mercury	1.0	webers	kilolines	10^5
torrs	atmospheres	1.316×10^{-3}	webers/sq in	gausses	1.550×10^7
	V		webers/sq in	lines/sq in	10^8
volt/inch	volt/cm	0.39370	webers/sq in	webers/sq cm	0.1550
volt (abs)	statvolts	0.003336	webers/sq in	webers/sq meter	1,550.
	W		webers/sq meter	gausses	10^4
watts	Btu/hr	3.4129	webers/sq meter	lines/sq in	6.452×10^4
watts	Btu/min	0.05688	webers/sq meter	webers/sq cm	10^{-4}
watts	ergs/sec	107.	webers/sq meter	webers/sq in	6.452×10^{-4}
watts	ft-lb/min	44.27		**Y**	
watts	ft-lb/sec	0.7378	yards	centimeters	91.44
watts	horsepower	1.341×10^{-3}	yards	feet	3.0
watts	horsepower (metric)	1.360×10^{-3}	yards	inches	36.0
watts	kg-calories/min	0.01433	yards	kilometers	9.144×10^{-4}
watts	kilowatts	0.001	yards	meters	0.9144
			yards	miles (naut)	4.934×10^{-4}
			yards	miles (stat)	5.682×10^{-4}
			yards	millimeters	914.4

CHEMICAL ABSTRACTS
REGISTRY NUMBERS

CHEMICAL ABSTRACTS SERVICE REGISTRY NUMBERS

The following tables provide the Chemical Abstracts Service (CAS) registry numbers for the monograph title compounds and selected derivatives. The latter are indicated by an asterisk following the name. Two formats are employed: the first table is ordered alphabetically by compound name; the second table is ordered numerically by CAS registry number.

CHEMICAL ABSTRACTS REGISTRY NUMBERS-I

Abamectin [71751-41-2]
Abietic Acid [514-10-3]
Abikoviromycin [31774-33-1]
Abrin [1393-62-0]
Abrine [526-31-8]
Abscisic Acid [21293-29-8]
Absinthin [1362-42-1]
Acacetin [480-44-4]
Acacia [9000-01-5]
Acacic Acid [1962-14-7]
Acarbose [56180-94-0]
Acebutolol [37517-30-9]
Acecainide [32795-44-1]
Acecainide Hydrochloride*
 [34118-92-8]
Acecarbromal [77-66-7]
Acedapsone [77-46-3]
Acediasulfone [80-03-5]
Acediasulfone Sodium* [127-60-6]
Acefylline [652-37-9]
Aceglatone [642-83-1]
Aceglutamide [2490-97-3]
Aceglutamide Aluminum*
 [12607-92-0]
Acemetacin [53164-05-9]
Acenaphthene [83-32-9]
Acenocoumarol [152-72-7]
Aceperone [807-31-8]
Acephate [30560-19-1]
Acepromazine [61-00-7]
Acepromazine Maleate* [3598-37-6]
Acerin [8001-08-9]
Acesulfame [33665-90-6]
Acetal [105-57-7]
Acetaldehyde [75-07-0]
Acetaldehyde Ammonia [75-39-8]
Acetaldehyde Sodium Bisulfite
 [918-04-7]
Acetaldoxime [107-29-9]
Acetamide [60-35-5]
Acetamidine Hydrochloride
 [124-42-5]
ε-Acetamidocaproic Acid [57-08-9]
Acetamidoeugenol [305-13-5]
Acetaminophen [103-90-2]
Acetaminosalol [118-57-0]
Acetanilide [103-84-4]
p-Acetanisidine [51-66-1]
Acetarsone [97-44-9]
Acetarsone Diethylamine Salt*
 [534-33-8]
Acetazolamide [59-66-5]
Acetazolamide Sodium* [1424-27-7]
Acetiamine [299-89-8]
Acetic Acid Glacial [64-19-7]
Acetic Anhydride [108-24-7]
Acetimidoquinone [50700-49-7]
Acetoacetanilide [102-01-2]
Acetoacetic Acid [541-50-4]
Acetobromglucose [572-09-8]
Acetohexamide [968-81-0]
Acetohydroxamic Acid [546-88-3]
Acetoin [513-86-0]
Acetomeroctol [584-18-9]
Acetone [67-64-1]
Acetone Cyanohydrin [75-86-5]
Acetonedicarboxylic Acid [542-05-2]
Acetone Sodium Bisulfite [540-92-1]
Acetonitrile [75-05-8]

Acetonylacetone [110-13-4]
Acetophenazine [2751-68-0]
Acetophenazine Dimaleate*
 [5714-00-1]
Acetophenone [98-86-2]
Acetosulfone Sodium [128-12-1]
Acetosulfone* [80-80-8]
Acetoxime [127-06-0]
Acetoxolone [6277-14-1]
21-Acetoxypregnenolone [566-78-9]
Acetozone [644-31-5]
Acetrizoate Sodium [129-63-5]
Acetulan® [8028-98-6]
Aceturic Acid [543-24-8]
Acetylacetone [123-54-6]
Acetyl Bromide [506-96-7]
α-Acetylbutyrolactone [517-23-7]
Acetylcarnitine [14992-62-2]
Acetylcarnitine (L-Form) Hydrochlo-
 ride* [5080-50-2]
Acetyl Chloride [75-36-5]
Acetylcholine Bromide [66-23-9]
Acetylcholine Chloride [60-31-1]
Acetylcysteine [616-91-1]
α-Acetyldigitoxin* [1111-39-3]
Acetylene [74-86-2]
Acetylene Dibromide [540-49-8]
Acetylene Dichloride [540-59-0]
Acetyleneurea [496-46-8]
Acetyl Iodide [507-02-8]
Acetylleucine Monoethanolamine
 [149-90-6]
N-Acetylmethionine [65-82-7]
5-Acetyl-2-methoxybenzaldehyde
 [531-99-7]
Acetyl Nitrate [591-09-3]
N-Acetylpenicillamine [15537-71-0]
Acetylpheneturide [13402-08-9]
Acetylsalicylsalicylic Acid [530-75-6]
Acetyl Sulfamethoxypyrazine
 [3590-05-4]
N⁴-Acetylsulfanilamide [121-61-9]
N'-Acetylsulfanilamide Sodium Salt
 Monohydrate* [6209-17-2]
N-Acetylsulfanilic Acid [121-62-0]
N-Acetylsulfanilyl Chloride
 [121-60-8]
Acetyl Sulfisoxazole [80-74-0]
Acetyltannic Acid [1397-74-6]
Acid Fuchsin [3244-88-0]
Acid Violet 7B [5905-34-0]
Acifluorfen [50594-66-6]
Acifran [72420-38-3]
Acifran (−)-Form* [77103-92-5]
Acifran (+)-Form* [77103-91-4]
Acipimox [51037-30-0]
Acitretin [55079-83-9]
Aclatonium Napadisilate
 [55077-30-0]
Aconine [509-20-6]
Aconitic Acid [499-12-7]
Aconitine [302-27-2]
Acranil [1684-42-0]
Acridine [260-94-6]
Acriflavine [8048-52-0]
Acrisorcin [7527-91-5]
Acrivastine [87848-99-5]
Acrolein [107-02-8]
Acrylamide [79-06-1]

Acrylic Acid [79-10-7]
Acrylonitrile [107-13-1]
ACTH [9002-60-2]
Actinium [7440-34-8]
Actinobolin [24397-89-5]
Actinodaphnine [517-69-1]
Actinomycetin [1402-37-5]
Actinomycin F₁ [1402-44-4]
Actinoquinol [15301-40-3]
Actinoquinol Sodium* [7246-07-3]
Actinorhodine [1397-77-9]
Actiphenol [526-02-3]
ACV [32467-88-2]
Acyclovir [59277-89-3]
Adamantane [281-23-2]
Adenine [73-24-5]
A-Denopterin [25663-23-4]
Adenosine [58-61-7]
Adenosine Diphosphate [58-64-0]
Adenosine Triphosphate [56-65-5]
S-Adenosylmethionine [29908-03-0]
3'-Adenylic Acid [84-21-9]
5'-Adenylic Acid [61-19-8]
Adinazolam [37115-32-5]
Adinazolam Mesylate* [57938-82-6]
Adiphenine Hydrochloride [50-42-0]
Adipic Acid [124-04-9]
Adipsin [104118-48-1]
Adlumidine [550-49-2]
Adlumine [524-46-9]
Adonitol [488-81-3]
Adonitoxin [17651-61-5]
Adrafinil [63547-13-7]
Adrenalone [99-45-6]
Adrenochrome [54-06-8]
Adrenoglomerulotropin [1210-56-6]
Adrenolutin [642-75-1]
Adrenosterone [382-45-6]
AET [56-10-0]
Affinin [25394-57-4]
Afloqualone [56287-74-2]
Agar [9002-18-0]
Agaricic Acid [666-99-9]
Agaritine [2757-90-6]
Agmatine [306-60-5]
Agroclavine [548-42-5]
Agrocybin [544-44-5]
Ahistan [518-61-6]
AICAR [3031-94-5]
AIR [25635-88-5]
Ajacine [509-17-1]
Ajaconine [545-61-9]
Ajmaline [4360-12-7]
Ajoene [92285-01-3]
Ajugarin I* [62640-05-5]
Ajugarin II* [62640-06-6]
Ajugarin III* [62640-07-7]
Ajugarin IV* [82225-47-6]
Ajugarin V* [82231-14-9]
Aklomide [3011-89-0]
Akuammicine [639-43-0]
Akuammine [3512-87-6]
AL 721 [99751-63-0]
Alacepril [74258-86-9]
Alachlor [15972-60-8]
Alafosfalin [60668-24-8]
L-Alanine [56-41-7]
Alanine (DL-Form)* [302-72-7]
β-Alanine [107-95-9]

L-Alanosine [5854-93-3]
Alantolactone [546-43-0]
Alazopeptin [1397-84-8]
Albaspidin [58409-52-2]
Albendazole [54965-21-8]
Albizziin [1483-07-4]
Albofungin [37895-35-5]
Albomycin [1414-39-7]
Alborixin [57760-36-8]
Albuterol [18559-94-9]
Albuterol Sulfate* [51022-70-9]
Albutoin [830-89-7]
Alcian Blue [12040-44-7]
Alclofenac [22131-79-9]
Alclometasone [67452-97-5]
Alclometasone Dipropionate*
   [66734-13-2]
Alcuronium [23214-96-2]
Alcuronium Chloride* [15180-03-7]
C_{14}-Aldehyde [32791-31-4]
Aldicarb [116-06-3]
Aldol [107-89-1]
Aldosterone [52-39-1]
Aldrin [309-00-2]
Aleuritic Acid [533-87-9]
Alexidine [22573-93-9]
Alexitol Sodium [66813-51-2]
Alfadolone Acetate [23930-37-2]
Alfaprostol [74176-31-1]
Alfaxalone [23930-19-0]
Alfentanil [71195-58-9]
Alfentanil Hydrochloride*
   [70879-28-6]
Alfuzosin [81403-80-7]
Alfuzosin Hydrochloride*
   [81403-68-1]
Algestone [595-77-7]
Algestone Acetonide* [4968-09-6]
Algestone Acetophenide [24356-94-3]
Algin [9005-38-3]
Alginic Acid [9005-32-7]
Alibendol [26750-81-2]
Alinidine [33178-86-8]
Alitame [80863-62-3]
Alizapride [59338-93-1]
Alizarin [72-48-0]
Alizarin Cyanine Green F [4403-90-1]
Alizarine Blue [568-02-5]
Alizarine Orange [568-93-4]
Alizarine Yellow R [2243-76-7]
Alkannin [23444-65-7]
Alkofanone [7527-94-8]
Allantoin [97-59-6]
Allenolic Acid [553-39-9]
Allicin [539-86-6]
Allidochlor [93-71-0]
Alliin [556-27-4]
Allobarbital [52-43-7]
Allocholesterol [517-10-2]
Alloclamide [5486-77-1]
Allocryptopine [485-91-6]
Allocupreide Sodium [5965-40-2]
Allopregnane [641-85-0]
Allopregnane-3α,20α-diol [566-58-5]
Allopregnane-3α,20β-diol [566-57-4]
Allopregnane-3β,20α-diol [566-56-3]
Allopregnane-3β,20β-diol [516-53-0]
Allopregnane-3β,21-diol-11,20-dione
   [566-02-9]
Allopregnane-3β,17α-diol-20-one
   [570-54-7]
3,20-Allopregnanedione [566-65-4]
Allopregnane-3β,11β,17α,20β,21-
   pentol [516-39-2]
Allopregnane-3β,17α,20β,21-tetrol
   [566-41-6]
Allopregnane-3α,11β,17α,21-tetrol-
   20- one [302-91-0]
Allopregnane-3β,11β,17α,21-tetrol-
   20- one [651-43-4]

Allopregnane-3β,17α,20α-triol
   [570-50-3]
Allopregnane-3β,17α,20β-triol
   [520-86-5]
Allopregnane-3β,17α,21-triol-11,20-
   dione [516-45-0]
Allopregnane-3β,11β,21-triol-20-one
   [516-16-5]
Allopregnane-3β,17α,21-triol-20-one
   [516-47-2]
Allopregnan-3α-ol-20-one [516-54-1]
Allopregnan-3β-ol-20-one [516-55-2]
Allopregnan-20α-ol-3-one [516-59-6]
Allopregnan-20β-ol-3-one [516-58-5]
Allopurinol [315-30-0]
D-Allose [2595-97-3]
Allotetrahydrocortisone [547-77-3]
Alloxan [50-71-5]
Alloxantin [76-24-4]
Allura® Red AC [25956-17-6]
Allyl Alcohol [107-18-6]
Allylamine [107-11-9]
Allyl Bromide [106-95-6]
Allyl Chloride [107-05-1]
Allyl Cyanide [109-75-1]
Allylestrenol [432-60-0]
Allyl Ether [557-40-4]
Allyl Ethyl Ether [557-31-3]
Allyl Iodide [556-56-9]
Allyl Isothiocyanate [57-06-7]
Allylprodine [25384-17-2]
Allyl Sulfide [592-88-1]
Allylurea [557-11-9]
Almagate [66827-12-1]
Aminoprofen [39718-89-3]
Almitrine [27469-53-0]
Aloe-Emodin [481-72-1]
Aloin [5133-19-7]
Aloxidone [526-35-2]
Aloxiprin [9014-67-9]
Alphaprodine [77-20-3]
Alphaprodine Hydrochloride*
   [561-78-4]
Alpidem [82626-01-5]
Alpiropride [81982-32-3]
Alprazolam [28981-97-7]
Alprenolol [13655-52-2]
Alprenolol Hydrochloride*
   [13707-88-5]
Alsactide [34765-96-3]
Alstonidine [25394-75-6]
Alstonine [642-18-2]
Althiazide [5588-16-9]
Altrenogest [850-52-2]
Altretamine [645-05-6]
D-Altrose [1990-29-0]
Aluminon [569-58-4]
Aluminum [7429-90-5]
Aluminum Acetate Solution
   [8006-13-1]
Aluminum Ammonium Sulfate
   [7784-25-0]
Aluminum Antimonide [25152-52-7]
Aluminum Benzoate [555-32-8]
Aluminum Bis(acetylsalicylate)
   [23413-80-1]
Aluminum Borate [11121-16-7]
Aluminum Boroformate [8011-44-7]
Aluminum Borohydride [16962-07-5]
Aluminum Bromide [7727-15-3]
Aluminum tert-Butoxide [556-91-2]
Aluminum Calcium Hydride
   [16941-10-9]
Aluminum Carbide [1299-86-1]
Aluminum Cesium Sulfate
   [14284-36-7]
Aluminum Chlorate [15477-33-5]
Aluminum Chloride [7446-70-0]
Aluminum Diacetate [142-03-0]
Aluminum Ethoxide [555-75-9]
Aluminum Fluoride [7784-18-1]

Aluminum Hexafluorosilicate
   [17099-70-6]
Aluminum Hexaurea Sulfate
   Triiodide
   [15304-14-0]
Aluminum Hydride [7784-21-6]
Aluminum Hydroxide [21645-51-2]
Aluminum Hydroxychloride
   [1327-41-9]
Aluminum Hypophosphite
   [7784-22-7]
Aluminum Iodide [7784-23-8]
Aluminum Isopropoxide [555-31-7]
Aluminum Lactate [18917-91-4]
Aluminum Lithium Hydride
   [16853-85-3]
Aluminum Magnesium Silicate
   [12511-31-8]
Aluminum β-Naphtholdisulfonate
   [1300-81-8]
Aluminum Nicotinate [1976-28-9]
Aluminum Nitrate [13473-90-0]
Aluminum Nitride [24304-00-5]
Aluminum Oleate [688-37-9]
Aluminum Oxalate [814-87-9]
Aluminum Oxide [1344-28-1]
Aluminum Oxide (Brockmann)
   [1344-28-1]
Aluminum Palmitate [555-35-1]
Aluminum Phosphate [7784-30-7]
Aluminum Phosphide [20859-73-8]
Aluminum Potassium Sulfate
   [10043-67-1]
Aluminum Rubidium Sulfate
   [13530-57-9]
Aluminum Selenide [1302-82-5]
Aluminum Silicate [12141-46-7]
Aluminum Sodium Sulfate
   [10102-71-3]
Aluminum Stearate [637-12-7]
Aluminum Subacetate Solution
   [8000-61-1]
Aluminum Sulfate [10043-01-3]
Aluminum Sulfide [1302-81-4]
Aluminum Tartrate [815-78-1]
Aluminum Thiocyanate [538-17-0]
Aluminum Zinc Sulfate [22992-10-5]
Alverine [150-59-4]
Alverine Citrate* [5560-59-8]
Amanitin [11030-71-0]
Amanozine [537-17-7]
Amantadine [768-94-5]
Amantadine Hydrochloride*
   [665-66-7]
Amantanium Bromide [58158-77-3]
Amaranth (Dye) [915-67-3]
Amarogentin [21018-84-8]
Amarolide [29913-86-8]
Ambazone [539-21-9]
Ambenonium Chloride [115-79-7]
Amber [9000-02-6]
Ambergris [8038-65-1]
Amberlite® [9079-25-8]
Ambrosin [509-93-3]
Ambroxol [18683-91-5]
Ambucaine [119-29-9]
Ambucetamide [519-88-0]
Ambuphylline [5634-34-4]
Ambuside [3754-19-6]
Ambutonium Bromide [115-51-5]
Amcinonide [51022-69-6]
Amdinocillin [32887-01-7]
Amdinocillin Pivoxil [32886-97-8]
Americium [7440-35-9]
Ametryn [834-12-8]
Amezinium Methyl Sulfate
   [30578-37-1]
Amfenac [51579-82-9]
Amfenac Sodium Monohydrate*
   [61618-27-7]
Amicarbalide [3459-96-9]

Denotes derivative of title compound

Amicetin [17650-86-1]
Amicibone [23271-63-8]
Amicoumacin A [78654-44-1]
Amidephrine [3354-67-4]
Amidephrine Mesylate* [1421-68-7]
Amidinomycin [3572-60-9]
Amidochlor [40164-67-8]
Amido-G-Acid [86-65-7]
Amidomycin [552-33-0]
Amido-R-Acid [92-28-4]
Amifloxacin [86393-37-5]
Amifloxacin Mesylate* [88036-80-0]
Amikacin [37517-28-5]
Amikacin Sulfate* [39831-55-5]
Amiloride [2609-46-3]
Amiloride Hydrochloride Dihydrate*
  [17440-83-4]
Aminacrine [90-45-9]
Aminacrine Hydrochloride*
  [134-50-9]
Amine 220® [95-38-5]
Amineptine [57574-09-1]
Aminitrozole [140-40-9]
p-Aminoacetanilide [122-80-5]
Aminoacetonitrile [540-61-4]
D-Amino Acid Oxidase [9000-88-8]
L-Amino Acid Oxidase [9000-89-9]
α-Aminoadipic Acid [542-32-5]
1-Aminoanthraquinone [82-45-1]
1-Aminoanthraquinone-2-carboxylic
  Acid [82-24-6]
p-Aminoazobenzene [60-09-3]
m-Aminobenzoic Acid [99-05-8]
o-Aminobenzoic Acid [118-92-3]
p-Aminobenzoic Acid [150-13-0]
2-Aminobenzothiazole [136-95-8]
6-Aminobenzothiazole [533-30-2]
N-(p-Aminobenzoyl)glutamic Acid
  [4271-30-1]
2-Amino-1-butanol [96-20-8]
α-Aminobutyric Acid [80-60-4]
β-Aminobutyric Acid [541-48-0]
γ-Aminobutyric Acid [56-12-2]
ε-Aminocaproic Acid [60-32-2]
Aminocarb [2032-59-9]
7-Aminocephalosporanic Acid
  [957-68-6]
4-Amino-4′-chlorodiphenyl
  [135-68-2]
Aminochlorthenoxazin [3567-76-8]
2-Amino-4,6-dichlorophenol
  [527-62-8]
2-Amino-1,2-diphenylethanol
  [530-36-9]
1-[(2-Aminoethyl)amino]-2-propanol
  [123-84-2]
2-Amino-2-ethyl-1,3-propanediol
  [115-70-8]
Aminoglutethimide [125-84-8]
Aminoguanidine [79-17-4]
p-Aminohippuric Acid [61-78-9]
Aminohippurate Sodium* [94-16-6]
3-Amino-4-hydroxybutyric Acid
  [589-44-6]
4-Amino-3-hydroxybutyric Acid
  [352-21-6]
α-Aminoisobutyric Acid [62-57-7]
δ-Aminolevulinic Acid [106-60-5]
β-Amino-α-methylphenethyl Alcohol
  [52500-61-5]
2-Amino-2-methyl-1,3-propanediol
  [115-69-5]
2-Amino-2-methyl-1-propanol
  [124-68-5]
2-Amino-4-methylthiazole
  [1603-91-4]
Aminometradine [642-44-4]
3-Amino-2-naphthoic Acid
  [5959-52-4]
4-Amino-1-naphthol [2834-90-4]

1-Amino-2-naphthol-4-sulfonic Acid
  [116-63-2]
1-Amino-2-naphthol-6-sulfonic Acid
  [5639-34-9]
6-Aminonicotinic Acid [3167-49-5]
2-Amino-5-nitrothiazole [121-66-4]
6-Aminopenicillanic Acid [551-16-6]
Aminopentamide [60-46-8]
m-Aminophenol [591-27-5]
o-Aminophenol [95-55-6]
p-Aminophenol [123-30-8]
p-Aminophenylacetic Acid
  [1197-55-3]
4-Amino-3-phenylbutyric Acid
  [1078-21-3]
Aminophylline [317-34-0]
2-Amino-4-picoline [695-34-1]
Aminopromazine [58-37-7]
2-Aminopropanol [78-91-1]
3-Aminopropionitrile [151-18-8]
p-Aminopropiophenone [70-69-9]
α-(α-Aminopropyl)benzyl Alcohol
  [5897-76-7]
Aminopropylon [3690-04-8]
Aminopterin [54-62-6]
α-Aminopyridine [504-29-0]
β-Aminopyridine [462-08-8]
Aminopyrine [58-15-1]
Aminoquinuride [3811-56-1]
Aminorex [2207-50-3]
p-Aminosalicylic Acid [65-49-6]
Aminosalicylate Calcium* [133-15-3]
Aminosalicylate Potassium*
  [133-09-5]
p-Aminosalicylic Acid Hydrazide
  [6946-29-8]
Aminosalicylate Sodium Dihydrate*
  [6018-19-5]
4-Amino-2-sulfobenzoic Acid
  [527-76-4]
2-Aminothiazole [96-50-4]
2-Amino-1,1,3-tricyanopropene
  [868-54-2]
Amioca [9037-22-3]
Amiodarone [1951-25-3]
Amiphenazole [490-55-1]
Amiprilose [56824-20-5]
Amiprilose Hydrochloride*
  [60414-06-4]
Amisometradine [550-28-7]
Amisulpride [71675-85-9]
Amiton [78-53-5]
Amitraz [33089-61-1]
Amitriptyline [50-48-6]
Amitriptyline Hydrochloride*
  [549-18-8]
Amitriptylinoxide [4317-14-0]
Amitrole [61-82-5]
Amixetrine [24622-72-8]
Amixetrine Hydrochloride*
  [24622-52-4]
Amlexanox [68302-57-8]
Amlodipine [88150-42-9]
Amlodipine Maleate* [88150-47-4]
Ammonia [7664-41-7]
Ammoniacum [9000-03-7]
Ammonium Acetate [631-61-8]
Ammonium Acetate Solution*
  [8013-61-4]
Ammonium Benzoate [1863-63-4]
Ammonium Bicarbonate [1066-33-7]
Ammonium Bifluoride [1341-49-7]
Ammonium Bimalate [5972-71-4]
Ammonium Binoxalate [5972-72-5]
Ammonium Bisulfate [7803-63-6]
Ammonium Bisulfide [12124-99-1]
Ammonium Bisulfite [10192-30-0]
Ammonium Bitartrate [3095-65-6]
Ammonium Borate [12007-58-8]
Ammonium Bromide [12124-97-9]
Ammonium Caprylate [5972-76-9]

Ammonium Carbamate [1111-78-0]
Ammonium Carbonate [10361-29-2]
Ammonium Ceric Nitrate
  [16774-21-3]
Ammonium Cerous Sulfate
  [21995-38-0]
Ammonium Chloride [12125-02-9]
Ammonium Chromate(VI)
  [7788-98-9]
Ammonium Chromic Sulfate
  [13548-43-1]
Ammonium Citrate, Dibasic
  [3012-65-5]
Ammonium Cobaltous Phosphate
  [14590-13-7]
Ammonium Cobaltous Sulfate
  [13596-46-8]
Ammonium Cupric Chloride
  [15610-76-1]
Ammonium Dichromate(VI)
  [7789-09-5]
Ammonium Dithiocarbamate
  [513-74-6]
Ammonium Ferric Chromate
  [7789-08-4]
Ammonium Ferric Citrate
  [1185-57-5]
Ammonium Ferric Oxalate
  [14221-47-7]
Ammonium Ferric Sulfate
  [10138-04-2]
Ammonium Ferricyanide
  [14221-48-8]
Ammonium Ferrocyanide
  [14481-29-9]
Ammonium Ferrous Sulfate
  [10045-89-3]
Ammonium Fluoride [12125-01-8]
Ammonium Formate [540-69-2]
Ammonium Hexafluoroaluminate
  [7784-19-2]
Ammonium Hexafluorogallate
  [14639-94-2]
Ammonium Hexafluorophosphate
  [16941-11-0]
Ammonium Hexafluorosilicate
  [16919-19-0]
Ammonium Hypophosphite
  [7803-65-8]
Ammonium Iodide [12027-06-4]
Ammonium Lactate [52003-58-4]
Ammonium Magnesium Chloride
  [39733-35-2]
Ammonium Mandelate [530-31-4]
Ammonium Mercuric Chloride
  [33445-15-7]
Ammonium Molybdate(VI)
  [12027-67-7]
Ammonium Nickel Sulfate
  [15699-18-0]
Ammonium Nitrate [6484-52-2]
Ammonium Nitroferricyanide
  [14402-70-1]
Ammonium Oleate [544-60-5]
Ammonium Osmium Chloride
  [12125-08-5]
Ammonium Oxalate [1113-38-8]
Ammonium Palmitate [593-26-0]
Ammonium Pentachlorozincate
  [14639-98-6]
Ammonium Perchlorate [7790-98-9]
Ammonium Peroxydisulfate
  [7727-54-0]
Ammonium Phosphate, Dibasic
  [7783-28-0]
Ammonium Phosphate, Monobasic
  [7722-76-1]
Ammonium Phosphite [51503-61-8]
Ammonium Phosphomolybdate
  [12026-66-3]

Denotes derivative of title compound

Ammonium Phosphotungstate [1311-90-6]
Ammonium Picrate [131-74-8]
Ammonium Platinic Chloride [16919-58-7]
Ammonium Platinous Chloride [13820-41-2]
Ammonium Salicylate [528-94-9]
Ammonium Selenate [7783-21-3]
Ammonium Selenite [7783-19-9]
Ammonium Sodium Phosphate [13011-54-6]
Ammonium Stearate [1002-89-7]
Ammonium Sulfamate [7773-06-0]
Ammonium Sulfate [7783-20-2]
Ammonium Sulfide [12135-76-1]
Ammonium Sulfite [10196-04-0]
Ammonium Tetrachloroaluminate [7784-14-7]
Ammonium Tetrachlorozincate [14639-97-5]
Ammonium Thiocyanate [1762-95-4]
Ammonium Thiosulfate [7783-18-8]
Ammonium Titanium Oxalate [10580-02-6]
Ammonium Tungstate(VI) [11120-25-5]
Ammonium Uranate(VI) [7783-22-4]
Ammonium Uranium Carbonate [18077-77-5]
Ammonium Uranium Fluoride [18433-40-4]
Ammonium Valerate [42739-38-8]
Ammonium Vanadate(V) [7803-55-6]
Ammonium Zirconyl Carbonate [12616-24-9]
Amobarbital [57-43-2]
Amobarbital Sodium* [64-43-7]
Amodiaquin [86-42-0]
Amodiaquine Dihydrochloride Dihydrate* [6398-98-7]
Amolanone [76-65-3]
Amolanone Hydrochloride* [6009-67-2]
Amoproxan [22661-76-3]
Amorolfine [78613-35-1]
Amorolfine Hydrochloride* [78613-38-4]
Amoscanate [26328-53-0]
Amosulalol [85320-68-9]
Amotriphene [5585-64-8]
Amoxapine [14028-44-5]
Amoxicillin [26787-78-0]
Amoxicillin Trihydrate* [61336-70-7]
Ampelopsin [27200-12-0]
Amperozide [75558-90-6]
Amphecloral [5581-35-1]
Amphenidone [134-37-2]
Amphenone B [2686-47-7]
Amphetamine [300-62-9]
Amphetamine Phosphate* [139-10-6]
Amphetamine Sulfate* [60-13-9]
Amphetamine d-Form Tannate* [1407-85-8]
Amphetaminil [17590-01-1]
Amphomycin [1402-82-0]
Amphotalide [1673-06-9]
Amphotericin B [1397-89-3]
Ampicillin [69-53-4]
Ampicillin Sodium* [69-52-3]
Ampicillin Trihydrate* [7177-48-2]
Amprolium [121-25-5]
Amprotropine Phosphate [134-53-2]
Ampyrone [83-07-8]
Amrinone [60719-84-8]
Amsacrine [51264-14-3]
Amsonic Acid [81-11-8]
Amygdalin [29883-15-6]
n-Amylamine [110-58-7]
Amylase [9000-92-4]
α-Amylase (Bacterial) [9000-85-5]

α-Amylase (Swine Pancreas) [9000-90-2]
β-Amylase (Sweet Potato) [9000-91-3]
Amylbenzene [538-68-1]
d-Amyl Bromide [534-00-9]
n-Amyl Bromide [110-53-2]
tert-Amyl Bromide [507-36-8]
n-Amyl Butyrate [540-18-1]
n-Amyl Caproate [540-07-8]
Amyl Carbamate, Tertiary [590-60-3]
Amyl Chloride [543-59-9]
6-n-Amyl-m-cresol [53043-14-4]
Amylene [513-35-9]
Amylene Dichloride [507-45-9]
n-Amyl Ether [693-65-2]
tert-Amyl Isovalerate [542-37-0]
n-Amyl Mercaptan [110-66-7]
Amylocaine Hydrochloride [532-59-2]
Amylpenicillin Sodium [575-47-3]
α-Amyrin [638-95-9]
β-Amyrin [559-70-6]
Anabasine [494-52-0]
Anabsinthin [6903-12-4]
Anagestone [2740-52-5]
Anagestone Acetate* [3137-73-3]
Anagrelide [68475-42-3]
Anagrelide Hydrochloride* [58579-51-4]
Anagyrine [486-89-5]
Anatabine [581-49-7]
Anazolene Sodium [3861-73-2]
Ancitabine [31698-14-3]
Ancrod [9046-56-4]
Ancymidol [12771-68-5]
Andrographolide [5508-58-7]
Androisoxazole [360-66-7]
Androstane [438-22-2]
Androstane-3β,11β-diol-17-one [514-17-0]
Androstenediol [521-17-5]
4-Androstene-3,17-dione [63-05-8]
Androst-16-en-3-ol [1153-51-1]
Androsterone [53-41-8]
Anemonin [508-44-1]
Anethole [104-46-1]
Anethole Trithione [532-11-6]
Angelic Acid [565-63-9]
Angelica Lactone [1333-38-6]
Angiotensin [1407-47-2]
Angiotensinamide* [53-73-6]
Anhalamine [643-60-7]
Anhalonidine [17627-77-9]
Anhalonine [519-04-0]
Anilazine [101-05-3]
Anileridine [144-14-9]
Anileridine Dihydrochloride* [126-12-5]
Aniline [62-53-3]
Aniline Mustard [553-27-5]
Anilinephthalein [509-77-3]
1-Anilino-8-naphthalenesulfonate [82-76-8]
A-Ninopterin [751-19-9]
Aniracetam [72432-10-1]
p-Anisaldehyde [123-11-5]
Anise Alcohol [105-13-5]
p-Anisic Acid [100-09-4]
Anisindione [117-37-3]
Anisole [100-66-3]
Anisomycin [22862-76-6]
Anisotropine Methylbromide [80-50-2]
o-(p-Anisoyl)benzoic Acid [1151-15-1]
p-Anisoyl Chloride [100-07-2]
Annatto [1393-63-1]
Annotinine [559-49-9]
p-Anol [539-12-8]
Anot [3572-44-9]

Anserine [584-85-0]
Antazoline [91-75-8]
Antazoline Hydrochloride* [2508-72-7]
Antazoline Phosphate* [154-68-7]
Antheridiol [22263-79-2]
Anthiolimine [305-97-5]
Anthracene [120-12-7]
Anthragallol [602-64-2]
Anthralin [1143-38-0]
Anthramycin [4803-27-4]
Anthranol [529-86-2]
Anthraquinone [84-65-1]
Anthrarobin [577-33-3]
Anthrarufin [117-12-4]
Anthrimide [82-22-4]
Anthrone [90-44-8]
α-Antiarin [23605-05-2]
Antimony [7440-36-0]
Antimony Chloride Oxide [7791-08-4]
Antimony Dichlorotrifluoride [7791-16-4]
Antimony Pentachloride [7647-18-9]
Antimony Pentafluoride [7783-70-2]
Antimony Pentasulfide [1315-04-4]
Antimony Pentoxide [1314-60-9]
Antimony Potassium Oxalate [5965-33-3]
Antimony Potassium Tartrate [28300-74-5]
Antimony Sodium Gluconate [16037-91-5]
Antimony Sodium Tartrate [34521-09-0]
Antimony Sodium Thioglycollate [539-54-8]
Antimony Sulfate [7446-32-4]
Antimony Thioglycollamide [6533-78-4]
Antimony Tribromide [7789-61-9]
Antimony Trichloride [10025-91-9]
Antimony Trifluoride [7783-56-4]
Antimony Triiodide [7790-44-5]
Antimony Trioxide [1309-64-4]
Antimony Triselenide [1315-05-5]
Antimony Trisulfide [1345-04-6]
Antimycin A₁ [642-15-9]
Antimycin A₃ [522-70-3]
Antipyrine [60-80-0]
Antipyrine Salicylate [520-07-0]
α₁-Antitrypsin [9041-92-3]
Antrafenine [55300-29-3]
ANTU [86-88-4]
Apalcillin [63469-19-2]
Apalcillin Sodium* [58795-03-2]
Apamin [24345-16-2]
Apazone [13539-59-8]
Aphidicolin [38966-21-1]
Apholate [52-46-0]
Aphylline [577-37-7]
Apicycline [15599-51-6]
Apigenin [520-36-5]
Apigetrin [578-74-5]
Apiin [26544-34-3]
Apiole (Dill) [484-31-1]
Apiole (Parsley) [523-80-8]
Apiose [639-97-4]
Aplasmomycin [61230-25-9]
Apoatropine [500-55-0]
Apocodeine [641-36-1]
Apocynin [498-02-2]
Apo-β-erythroidine [478-85-3]
Apomorphine [58-00-4]
Apomorphine Hydrochloride* [41372-20-7]
Apomorphine Methylbromide* [602-81-3]
Apoquinine [5985-94-4]
Aporeine [2030-53-7]
Apraclonidine [66711-21-5]

Denotes derivative of title compound

Apramycin [37321-09-8]
Aprindine [37640-71-4]
Aprindine Hydrochloride*
  [33237-74-0]
Aprobarbital [77-02-1]
Apronalide [528-92-7]
Aprotinin [9087-70-1]
APSAC [81660-57-0]
Apyrase [9000-95-7]
Aquocobalamin [13422-52-1]
Arabinose [147-81-9]
Arabitol [2152-56-9]
D-Araboflavin [5978-87-0]
Arachidic Acid [506-30-9]
Arachidonic Acid [506-32-1]
Aramite® [140-57-8]
Araroba [1393-64-2]
Arbaprostil [55028-70-1]
Arbekacin [51025-84-5]
Arborescin [6831-14-7]
Arbutin [497-76-7]
Arecaidine [499-04-7]
Arecoline [63-75-2]
Arecoline Hydrobromide* [300-08-3]
Arecoline p-Stibonobenzoic Acid
  [17162-36-6]
Argatroban [74863-84-6]
Arginine [74-79-3]
Arginine Hydrochloride* [1119-34-2]
Arginine Glutamate [4320-30-3]
Argol [8007-14-5]
Argon [7440-37-1]
Aricine [482-91-7]
Aristolochic Acid [313-67-7]
Armepavine [524-20-9]
Armstrong's Acid [81-04-9]
Arogenic Acid [53078-86-7]
Arotinolol [68377-92-4]
Arprinocid [55779-18-5]
Arsacetin [618-22-4]
Arsanilic Acid [98-50-0]
Arsenamide [531-72-6]
Arsenic [7440-38-2]
Arsenic Acid [7778-39-4]
Arsenic Disulfide [1303-32-8]
Arsenic Hemiselenide [1303-35-1]
Arsenic Pentafluoride [7784-36-3]
Arsenic Pentaselenide [1303-37-3]
Arsenic Pentasulfide [1303-34-0]
Arsenic Pentoxide [1303-28-2]
Arsenic Tribromide [7784-33-0]
Arsenic Trichloride [7784-34-1]
Arsenic Trifluoride [7784-35-2]
Arsenic Triiodide [7784-45-4]
Arsenic Trioxide [1327-53-3]
Arsenic Triselenide [1303-36-2]
Arsenic Trisulfide [1303-33-9]
Arsenoacetic Acid [544-27-4]
Arsine [7784-42-1]
Arsonoacetic Acid [107-38-0]
Arsphenamine [139-93-5]
Arsthinol [119-96-0]
Arteether [75887-54-6]
Artemether [71963-77-4]
Artemisin [481-05-0]
Artemisinin [63968-64-9]
Artesunate [88495-63-0]
Asafetida [9000-04-8]
Asaprol [516-18-7]
Asarinin [133-04-0]
Ascaridole [512-85-6]
Ascorbic Acid [50-81-7]
Ascorbigen [8075-98-7]
Asiaticoside [16830-15-2]
L-Asparaginase [9015-68-3]
Asparagine [70-47-3]
Aspartame [22389-47-0]
Aspartic Acid [56-84-8]
Aspergillic Acid [490-02-8]
Asperlicin [93413-04-8]
Asperlicin B* [93413-08-2]

Asperlicin C* [93413-06-0]
Asperlicin D* [93413-07-1]
Asperlicin E* [93413-05-9]
Asperuloside [14259-45-1]
Aspidin [584-28-1]
Aspidinol [519-40-4]
Aspidospermine [466-49-9]
Aspirin [50-78-2]
Aspoxicillin [63358-49-6]
Astacin [514-76-1]
Astatine [7440-68-8]
Astaxanthin [472-61-7]
Astemizole [68844-77-9]
Atenolol [29122-68-7]
Athamantin [1892-56-4]
Atisine [466-43-3]
Atractyloside [17754-44-8]
Atracurium Besylate [64228-81-5]
Atranorin [479-20-9]
Atrazine [1912-24-9]
Atrial Natriuretic Factor
  [85637-73-6]
Atrolactamide [2019-68-3]
Atrolactic Acid [515-30-0]
Atropic Acid [492-38-6]
Atropine [51-55-8]
Atropine Methylnitrate* [52-88-0]
Atropine N-Oxide [4438-22-6]
Atropine Oxide Hydrochloride*
  [4574-60-1]
Aucubin [479-98-1]
Auranofin [34031-32-8]
Aurantiogliocladin [483-54-5]
Aureothin [2825-00-5]
Aureothricin [574-95-8]
Aurin [603-45-2]
Aurodox [12704-90-4]
Aurothioglucose [12192-57-3]
Aurothioglycanide [16925-51-2]
Avilamycin [11051-71-1]
Avoparcin [37332-99-3]
Azacitidine [320-67-2]
Azacosterol [313-05-3]
Azacosterol Dihydrochloride*
  [1249-84-9]
Azacyclonol [115-46-8]
Azacyclonol Hydrochloride*
  [1798-50-1]
Azadirachtin [11141-17-6]
Azafrin [507-61-9]
8-Azaguanine [134-58-7]
Azamethonium Bromide [306-53-6]
Azanidazole [62973-76-6]
Azaperone [1649-18-9]
Azaserine [115-02-6]
Azatadine [3964-81-6]
Azatadine Maleate* [3978-86-7]
Azathioprine [446-86-6]
6-Azathymine [932-53-6]
6-Azauridine [54-25-1]
Azelaic Acid [123-99-9]
Azelastine [58581-89-8]
Azelastine Hydrochloride*
  [79307-93-0]
2-Azetidinecarboxylic Acid
  [2517-04-6]
Azidamfenicol [13838-08-9]
Azidocillin [17243-38-8]
Azinphos-methyl [86-50-0]
Azintamide [1830-32-6]
Azithromycin [83905-01-5]
Azlocillin [37091-66-0]
Azlocillin Sodium* [37091-65-9]
Azobenzene [103-33-3]
2,2'-Azobisisobutyronitrile [78-67-1]
Azodicarbonamide [123-77-3]
Azolitmin [1395-18-2]
Azomycin [527-73-1]
Azosemide [27589-33-9]
Azosulfamide [133-60-8]
Azoxybenzene [495-48-7]

Aztreonam [78110-38-0]
Azulene [275-51-4]
Azure A [531-53-3]
Azure B [531-55-5]
Azure C [531-57-7]
Bacampicillin [50972-17-3]
Bacampicillin Hydrochloride*
  [37661-08-8]
Bacilysin [29393-20-2]
Bacimethrin [3690-12-8]
Bacitracin [1405-87-4]
Bacitracin Methylenedisalicylic Acid
  [1405-88-5]
Baclofen [1134-47-0]
Badische Acid [86-60-2]
Baicalein [491-67-8]
Bakankosin [1398-17-0]
Bambermycins [11015-37-5]
Bambuterol [81732-65-2]
Bambuterol Hydrochloride*
  [81732-46-9]
Bamethan [3703-79-5]
Bamethan Sulfate* [5716-20-1]
Bamifylline [2016-63-9]
Bamifylline Hydrochloride*
  [20684-06-4]
Bamipine [4945-47-5]
Baptigenin [5908-63-4]
Barban [101-27-9]
Barbital [57-44-3]
Barbital Sodium* [144-02-5]
Barbituric Acid [67-52-7]
Barium [7440-39-3]
Barium Acetate [543-80-6]
Barium Benzenesulfonate [515-72-0]
Barium Bromate [13967-90-3]
Barium Bromide [10553-31-8]
Barium Carbonate [513-77-9]
Barium Chlorate [13477-00-4]
Barium Chloride [10361-37-2]
Barium Chromate(VI) [10294-40-3]
Barium Cyanide [542-62-1]
Barium Dithionate [13845-17-5]
Barium Ferrocyanide [13821-06-2]
Barium Fluoride [7787-32-8]
Barium Formate [541-43-5]
Barium Hexafluorosilicate
  [17125-80-3]
Barium Hydroxide [17194-00-2]
Barium Hypophosphite [14871-79-5]
Barium Iodate [10567-69-8]
Barium Iodide [13718-50-8]
Barium Manganate(VI) [7787-35-1]
Barium Mercuric Iodide [10048-99-4]
Barium Nitrate [10022-31-8]
Barium Nitrite [13465-94-6]
Barium Oxalate [516-02-9]
Barium Oxide [1304-28-5]
Barium Perchlorate [13465-95-7]
Barium Permanganate [7787-36-2]
Barium Peroxide [1304-29-6]
Barium Phosphate, Dibasic
  [10048-98-3]
Barium Platinous Cyanide [562-81-2]
Barium Selenide [1304-39-8]
Barium Silicide [1304-40-1]
Barium Sulfate [7727-43-7]
Barium Sulfide [21109-95-5]
Barium Sulfide, Black [8011-62-9]
Barium Sulfite [7787-39-5]
Barium Thiocyanate [2092-17-3]
Barium Thiosulfate [35112-53-9]
Barium Titanate(IV) [12047-27-7]
Barium Uranium Oxide [10380-31-1]
Barthrin [70-43-9]
Basic Aluminum Carbonate Gel
  [1339-92-0]
Basic Lead Carbonate [1344-36-1]
Batrachotoxin [23509-16-2]
Batrachotoxinin A [19457-37-5]
Batroxobin [9039-61-6]

Batyl Alcohol [544-62-7]
Bebeerine [477-60-1]
Becanthone [15351-04-9]
Becanthone Hydrochloride*
[5591-22-0]
Beclamide [501-68-8]
Beclobrate [55937-99-0]
Beclomethasone [4419-39-0]
Beclomethasone Dipropionate*
[5534-09-8]
Beclotiamine [13471-78-8]
Beeswax [8012-89-3]
Befunolol [39552-01-7]
Behenic Acid [112-85-6]
Belladonnine [510-25-8]
Bemegride [64-65-3]
Benactyzine [302-40-9]
Benalaxyl [71626-11-4]
Benapryzine [22487-42-9]
Benapryzine Hydrochloride*
[3202-55-9]
Bencyclane [2179-37-5]
Bencyclane Fumarate* [14286-84-1]
Bendazac [20187-55-7]
Bendazol [621-72-7]
Bendiocarb [22781-23-3]
Bendroflumethiazide [73-48-3]
Benexate Hydrochloride [78718-25-9]
Benfluorex [23602-78-0]
Benfluralin [1861-40-1]
Benfotiamine [22457-89-2]
Benfuracarb [82560-54-1]
Benfurodil Hemisuccinate
[3447-95-8]
Benmoxine [7654-03-7]
Benomyl [17804-35-2]
Benorylate [5003-48-5]
Benoxaprofen [67434-14-4]
Benoxinate [99-43-4]
Benoxinate Hydrochloride*
[5987-82-6]
Benperidol [2062-84-2]
Benproperine [2156-27-6]
Benserazide [322-35-0]
Bentazon [25057-89-0]
Bentiromide [37106-97-1]
Bentonite [1302-78-9]
β-Benzalbutyramide [7236-47-7]
Benzal Chloride [98-87-3]
Benzaldehyde [100-52-7]
Benzamide [55-21-0]
Benzanilide [93-98-1]
1,2-Benzanthracene [56-55-3]
Benzanthrone [82-05-3]
Benzarone [1477-19-6]
Benzathine [140-28-3]
Benzbromarone [3562-84-3]
Benzene [71-43-2]
Benzenearsonic Acid [98-05-5]
Benzeneboronic Acid [98-80-6]
Benzenestibonic Acid [535-46-6]
Benzenesulfonic Acid [98-11-3]
Benzenesulfonic Anhydride
[512-35-6]
Benzenesulfonyl Chloride [98-09-9]
1,2,4-Benzenetriol [533-73-3]
Benzestrol [85-95-0]
Benzethonium Chloride [121-54-0]
Benzetimide [14051-33-3]
Benzetimide Hydrochloride*
[5633-14-7]
Benzhydrylamine [91-00-9]
Benzidine [92-87-5]
Benzil [134-81-6]
Benzil Dioxime [23873-81-6]
Benzilic Acid [76-93-7]
Benzilonium Bromide [1050-48-2]
Benzimidazole [51-17-2]
2-Benzimidazolethiol [583-39-1]
Benziodarone [68-90-6]
Benznidazole [22994-85-0]

Benzo Azurine G [2429-71-2]
Benzoctamine [17243-39-9]
Benzoctamine Hydrochloride*
[10085-81-1]
Benzodepa [1980-45-6]
Benzofuran [271-89-6]
Benzoguanamine [91-76-9]
Benzohydrol [91-01-0]
Benzoic Acid [65-85-0]
Benzoic Anhydride [93-97-0]
Benzoin [119-53-9]
Benzoin Oxime [441-38-3]
6,7-Benzomorphan [575-19-9]
Benzonatate [104-31-4]
Benzonitrile [100-47-0]
Benzophenone [119-61-9]
Benzophenone-6 [131-54-4]
Benzopinacol [464-72-2]
Benzopurpurine 4B [992-59-6]
1,2-Benzopyran [254-04-6]
Benzo[a]pyrene [50-32-8]
Benzo[e]pyrene [192-97-2]
Benzo[f]quinoline [85-02-9]
Benzoquinonium Chloride [311-09-1]
Benzoresorcinol [131-56-6]
Benzothiazole [95-16-9]
1H-Benzotriazole [95-14-7]
Benzotrichloride [98-07-7]
Benzotrifluoride [98-08-8]
Benzoxiquine [86-75-9]
Benzoxonium Chloride [19379-90-9]
Benzoyl Chloride [98-88-4]
Benzoylecgonine [519-09-5]
Benzoyl Isothiocyanate [532-55-8]
Benzoylpas [13898-58-3]
Benzoyl Peroxide [94-36-0]
3,4-Benzphenanthrene [195-19-7]
Benzphetamine [156-08-1]
Benzphetamine Hydrochloride*
[5411-22-3]
Benzpiperylon [53-89-4]
Benzpyrinium Bromide [587-46-2]
Benzquinamide [63-12-7]
Benzthiazide [91-33-8]
Benztropine Mesylate [132-17-2]
Benzydamine [642-72-8]
Benzydamine Hydrochloride*
[132-69-4]
Benzyl Acetate [140-11-4]
Benzyl Alcohol [100-51-6]
Benzylamine [100-46-9]
Benzylaniline [103-32-2]
Benzyl Benzoate [120-51-4]
Benzyl Bromide [100-39-0]
Benzyl Chloride [100-44-7]
Benzyl Cinnamate [103-41-3]
Benzyl Cyanide [140-29-4]
Benzyl Ether [103-50-4]
Benzyl Ethyl Ether [539-30-0]
Benzyl Formate [104-57-4]
Benzyl Fumarate [538-64-7]
Benzylhydrochlorothiazide
[1824-50-6]
Benzylideneacetone [122-57-6]
Benzylideneaniline [538-51-2]
Benzylimidobis(p-methoxyphenyl)-
methane [524-96-9]
Benzyl Methyl Ether [538-86-3]
Benzylmorphine [14297-87-1]
Benzylpenicillinic Acid [61-33-6]
Benzylpenicillin Sodium [69-57-8]
o-Benzylphenol [28994-41-4]
p-Benzylphenol [101-53-1]
Benzyl Salicylate [118-58-1]
Benzylsulfamide [104-22-3]
Benzyl Sulfide [538-74-9]
p-(Benzylsulfonamido)benzoic Acid
[536-95-8]
S-Benzylthiuronium Chloride
[538-28-3]
Benzylurea [538-32-9]

Bephenium [7181-73-9]
Bephenium Hydroxynaphthoate*
[3818-50-6]
Bepridil [64706-54-3]
Bepridil Hydrochloride Monohydrate*
[74764-40-2]
Berbamine [478-61-5]
Berberine [2086-83-1]
Berbine [483-49-8]
Bergapten(e) [484-20-8]
Bergenin [477-90-7]
Berkelium [7440-40-6]
Berninamycin [58798-97-3]
Beryllium [7440-41-7]
Beryllium Acetate [543-81-7]
Beryllium Acetate, Basic [1332-52-1]
Beryllium Acetylacetonate
[10210-64-7]
Beryllium Borohydride [17440-85-6]
Beryllium Bromide [7787-46-4]
Beryllium Carbide [506-66-1]
Beryllium Chloride [7787-47-5]
Beryllium Fluoride [7787-49-7]
Beryllium Formate [1111-71-3]
Beryllium Hydride [7787-52-2]
Beryllium Hydroxide [13327-32-7]
Beryllium Iodide [7787-53-3]
Beryllium Nitrate [13597-99-4]
Beryllium Nitride [1304-54-7]
Beryllium Oxide [1304-56-9]
Beryllium Perchlorate [13597-95-0]
Beryllium Potassium Fluoride
[7787-50-0]
Beryllium Potassium Sulfate
[53684-48-3]
Beryllium Selenate [10039-31-3]
Beryllium Sodium Fluoride
[13871-27-7]
Beryllium Sulfate [13510-49-1]
Bestrabucil [75219-46-4]
Betahistine [5638-76-6]
Betahistine Dihydrochloride*
[5579-84-0]
Betaine [107-43-7]
Betaine Aspartate* [52921-08-1]
Betaine Hydrochloride* [1477-10-7]
Betamethasone [378-44-9]
Betamethasone Acetate* [987-24-6]
Betamethasone Benzoate*
[22298-29-9]
Betamethasone Dipropionate*
[5593-20-4]
Betamethasone Sodium Phosphate*
[151-73-5]
Betamethasone 17-Valerate*
[2152-44-5]
Betasine [3734-24-5]
Betaxolol [63659-18-7]
Betaxolol Hydrochloride*
[63659-19-8]
Betazole [105-20-4]
Betazole Dihydrochloride* [138-92-1]
Bethanechol Chloride [590-63-6]
Bethanidine [55-73-2]
Bethanidine Sulfate* [114-85-2]
Betonicine [515-25-3]
Betoxycaine [3818-62-0]
Betoxycaine Hydrochloride*
[5003-47-4]
Betulin [473-98-3]
Bevantolol [59170-23-9]
Bevantolol Hydrochloride*
[42864-78-8]
Bevonium Methyl Sulfate [5205-82-3]
Bezafibrate [41859-67-0]
Bezitramide [15301-48-1]
Bialamicol [493-75-4]
Bialamicol Hydrochloride*
[3624-96-2]
Bibenzonium Bromide [15585-70-3]
Bibenzyl [103-29-7]

Denotes derivative of title compound

Bibrocathol [6915-57-7]
Bicine [150-25-4]
Bicozamycin [38129-37-2]
Bicuculline [485-49-4]
Bietamiverine [479-81-2]
Bietamiverine Hydrochloride*
  [479-81-2]
Bietanautine [6888-11-5]
Bietaserpine [53-18-9]
Bifemelane [90293-01-9]
Bifenox [42576-02-3]
Bifenthrin [82657-04-3]
Bifidus Factor [9007-03-8]
Bifluranol [34633-34-6]
Bifonazole [60628-96-8]
Biguanide [56-03-1]
Bikhaconitine [6078-26-8]
Bilirubin [635-65-4]
Biliverdine [114-25-0]
Binapacryl [485-31-4]
Binedaline [60662-16-0]
Binifibrate [69047-39-8]
Biocytin [576-19-2]
Biopterin [22150-76-1]
Bioresmethrin [28434-01-7]
Biotin [58-85-5]
Biotin l-Sulfoxide [3376-83-8]
Biperiden [514-65-8]
Biperiden Hydrochloride*
  [1235-82-1]
Biphenamine [3572-52-9]
Biphenamine Hydrochloride*
  [5560-62-3]
p-Biphenylamine [92-67-1]
2,4'-Biphenyldiamine [492-17-1]
Bipiperidyl Mustard [6802-93-3]
Bisacodyl [603-50-9]
Bisacodyl Tannex* [1336-29-4]
Bis(4-amino-1-anthraquinonyl)amine
  [128-87-0]
Bisantrene [78186-34-2]
Bisantrene Dihydrochloride*
  [71439-68-4]
2,5-Bis(1-aziridinyl)-3,6-bis(2-meth-
  oxyethoxy)-1,4-benzoquinone
  [800-24-8]
Bisbentiamine [2667-89-2]
Bis(p-chlorophenoxy)methane
  [555-89-5]
Bisdequalinium Chloride
  [52951-36-7]
Bis(p-dimethylaminobenzylidene)-
  benzidine [6001-51-0]
Bis(1,2-dimethylpropyl)borane
  [1069-54-1]
Bis(2-ethylhexyl) Phthalate
  [117-81-7]
Bis(2-ethylhexyl) Sebacate
  [122-62-3]
Bismark Brown R [8005-78-5]
Bismark Brown Y [10114-58-6]
Bis(1-methylamyl) Sodium Sulfosuc-
  cinate [6001-97-4]
Bis[methylthio]methane [1618-26-4]
Bismuth [7440-69-9]
Bismuth Aluminate [12284-76-3]
Bismuth Bromide [7787-58-8]
Bismuth Bromide Oxide [7787-57-7]
Bismuth Butylthiolaurate
  [53897-25-9]
Bismuth Chloride [7787-60-2]
Bismuth Chloride Oxide [7787-59-9]
Bismuth Ethyl Camphorate
  [52951-37-8]
Bismuth Fluoride [7787-61-3]
Bismuth Hydroxide [10361-43-0]
Bismuthine [18288-22-7]
Bismuth Iodide [7787-64-6]
Bismuth Iodide Oxide [7787-63-5]
Bismuth Iodosubgallate [138-58-9]
Bismuth Nitrate [10361-44-1]

Bismuth Oleate [52951-38-9]
Bismuth Oxalate [6591-55-5]
Bismuth Oxide [1304-76-3]
Bismuth Pentafluoride [7787-62-4]
Bismuth Phosphate [10049-01-1]
Bismuth Potassium Iodide
  [41944-01-8]
Bismuth Potassium Tartrate
  [5798-41-4]
Bismuth Selenide [12068-69-8]
Bismuth Sodium Iodide [53778-50-0]
Bismuth Sodium Tartrate
  [31586-77-3]
Bismuth Sodium Triglycollamate
  [5798-43-6]
Bismuth Subacetate [5142-76-7]
Bismuth Subcarbonate [5892-10-4]
Bismuth Subcitrate Sol (Dried)
  [57644-54-9]
Bismuth Subgallate [22650-86-8]
Bismuth Subnitrate [1304-85-4]
Bismuth Subsalicylate [14882-18-9]
Bismuth Sulfate [7787-68-0]
Bismuth Sulfide [1345-07-9]
Bismuth Telluride [1304-82-1]
Bismuth Tetroxide [12048-50-9]
Bismuth Tribromophenate
  [5175-83-7]
Bismuth Valerate, Basic [5798-47-0]
Bis(1-naphthylmethyl)amine
  [5798-49-2]
Bisobrin [22407-74-5]
Bisoprolol [66722-44-9]
Bisoprolol Fumarate* [104344-23-2]
Bisoxatin Acetate [14008-48-1]
Bisphenol A [80-05-7]
Bisphenol B [77-40-7]
1,4-Bis(trichloromethyl)benzene
  [68-36-0]
Bis(triphenylphosphine)dicarbonyl-
  nickel [13007-90-4]
Bitertanol [55179-31-2]
Bithionol [97-18-7]
Bithionolate Sodium* [6385-58-6]
Bitolterol [30392-40-6]
Bitolterol Mesylate* [30392-41-7]
Bitoscanate [4044-65-9]
Biuret [108-19-0]
Bixin [6983-79-5]
Blancophor® R [2606-93-1]
Blasticidin S [2079-00-7]
Bleomycin Sulfate* [9041-93-4]
Bolandiol [19793-20-5]
Bolandiol Dipropionate* [1986-53-4]
Bolasterone [1605-89-6]
Boldenone [846-48-0]
Boldenone Undecylenate*
  [13103-34-9]
Boldine [476-70-0]
Bombesin [31362-50-2]
Bomyl® [122-10-1]
Bongkrekic Acid [11076-19-0]
Bopindolol [62658-63-3]
Boric Acid [10043-35-3]
Boric Anhydride [1303-86-2]
Borneol [507-70-0]
Bornyl Acetate [76-49-3]
d-Bornyl α-Bromoisovalerate
  [52964-40-6]
Bornyl Chloride [464-41-5]
d-Bornyl Isovalerate [53022-14-3]
Bornyl Salicylate [560-88-3]
Boromycin [34524-20-4]
Boron [7440-42-8]
Boron Carbide [12069-32-8]
Boron Monoxide [12505-77-0]
Boron Nitride [10043-11-5]
Boron Tribromide [10294-33-4]
Boron Trichloride [10294-34-5]
Boron Trifluoride [7637-07-2]

Boron Trifluoride Etherate
  [109-63-7]
Bostrycoidin [4589-33-7]
β-Boswellic Acid [631-69-6]
Bottromycin [1393-68-6]
Bradykinin [58-82-2]
Brallobarbital [561-86-4]
Brassidic Acid [506-33-2]
Brassinolide [72962-43-7]
Brazilin [474-07-7]
Brefeldin A [20350-15-6]
Bretylium Tosylate [61-75-6]
Brilliant Blue FCF [3844-45-9]
Brilliant Green [633-03-4]
Brodifacoum [56073-10-0]
Brodimoprim [56518-41-3]
Bromacil [314-40-9]
Bromadiolone [28772-56-7]
Bromal [115-17-3]
Bromal Hydrate [507-42-6]
Bromazepam [1812-30-2]
Bromcresol Green [76-60-8]
Bromcresol Purple [115-40-2]
Bromelain [9001-00-7]
Bromethalin [63333-35-7]
Bromhexine [3572-43-8]
Bromhexine Hydrochloride*
  [611-75-6]
Bromic Acid [7789-31-3]
Bromindione [1146-98-1]
Bromine [7726-95-6]
Bromine Pentafluoride [7789-30-2]
Bromine Trifluoride [7787-71-5]
Bromisovalum [496-67-3]
N-Bromoacetamide [79-15-2]
p-Bromoacetanilide [103-88-8]
Bromoacetic Acid [79-08-3]
Bromoacetone [598-31-2]
p-Bromoacetophenone [99-90-1]
ω-Bromoacetophenone [70-11-1]
p-Bromoaniline [106-40-1]
5-Bromoanthranilic Acid [5794-88-7]
Bromobenzene [108-86-1]
p-Bromobenzenesulfonyl Chloride
  [98-58-8]
p-Bromobenzoic Acid [586-76-5]
p-Bromobenzyl Bromide [589-15-1]
p-Bromobenzyl Chloride [589-17-3]
p-Bromobenzyl Chloroformate
  [5798-78-7]
α-Bromobenzyl Cyanide [5798-79-8]
α-Bromobutyric Acid [80-58-0]
3-Bromo-d-camphor [76-29-9]
α-Bromo-n-caproic Acid [616-05-7]
Bromocriptine [25614-03-3]
Bromocriptine Mesylate*
  [22260-51-1]
Bromodiphenhydramine [118-23-0]
Bromodiphenhydramine Hydrochloride* [1808-12-4]
Bromofenofos [21466-07-9]
Bromoform [75-25-2]
α-Bromoisobutyric Acid [2052-01-9]
α-Bromoisovaleric Acid [565-74-2]
β-Bromoisovaleric Acid [5798-88-9]
Bromolysergide [478-84-2]
p-Bromomandelic Acid [6940-50-7]
1-Bromonaphthalene [90-11-9]
2-Bromonaphthalene [580-13-2]
p-Bromophenacyl Bromide [99-73-0]
m-Bromophenol* [591-20-8]
o-Bromophenol [95-56-7]
p-Bromophenol* [106-41-2]
p-Bromophenylhydrazine [589-21-9]
p-Bromophenyl Isocyanate
  [2493-02-9]
Bromophos [2104-96-3]
Bromopride [4093-35-0]
β-Bromopropionic Acid [590-92-1]
Bromopropylate [18181-80-1]
Bromosalicylchloranilide [3679-64-9]

5-Bromosalicylhydroxamic Acid [5798-94-7]
5-Bromosalicylic Acid Acetate [1503-53-3]
Bromosaligenin [2316-64-5]
Bromosuccinic Acid [923-06-8]
N-Bromosuccinimide [128-08-5]
5-Bromouracil [51-20-7]
Bromoxynil [1689-84-5]
Bromperidol [10457-90-6]
Brompheniramine [86-22-6]
Brompheniramine Maleate* [980-71-2]
Brompheniramine (d-Form)* [132-21-8]
Brompheniramine d-Form Maleate* [2391-03-9]
Bromphenol Blue [115-39-9]
Bromthymol Blue [76-59-5]
p-Bromtripelennamine [531-09-9]
Bronopol [52-51-7]
Broparoestrol [479-68-5]
Brotizolam [57801-81-7]
Brovincamine [57475-17-9]
Broxyquinoline [521-74-4]
Bruceantin [41451-75-6]
Brucine [357-57-3]
Bucetin [1083-57-4]
Bucillamine [65002-17-7]
Bucladesine [362-74-3]
Buclizine [82-95-1]
Buclizine Dihydrochloride* [129-74-8]
Buclosamide [575-74-6]
Bucloxic Acid [32808-51-8]
Bucolome [841-73-6]
Bucrylate [1069-55-2]
Bucumolol [58409-59-9]
Budesonide [51333-22-3]
Budipine [57982-78-2]
Budralazine [36798-79-5]
Bufalin [465-21-4]
Bufencarb [8065-36-9]
Bufeniode [22103-14-6]
Bufetolol [53684-49-4]
Bufexamac [2438-72-4]
Buflomedil [55837-25-7]
Bufogenin B [465-19-0]
Buformin [692-13-7]
Bufotalin [471-95-4]
Bufotenine [487-93-4]
Bufotoxin [464-81-3]
Bufuralol [54340-62-4]
Bulan® [117-26-0]
Bulbocapnine [298-45-3]
Bumadizon [3583-64-0]
Bumetanide [28395-03-1]
Bunaftine [32421-46-8]
Bunamidine [3748-77-4]
Bunamidine Hydrochloride* [1055-55-6]
Bunamiodyl Sodium [1923-76-8]
Bunazosin [80755-51-7]
Bunitrolol [34915-68-9]
Buparvaquone [88426-33-9]
Buphanamine [6793-24-4]
Buphanitine [4673-18-1]
Bupirimate [41483-43-6]
Bupivacaine [2180-92-9]
Bupivacaine Hydrochloride* [14252-80-3]
Bupranolol [14556-46-8]
Buprenorphine [52485-79-7]
Buprenorphine Hydrochloride* [53152-21-9]
Bupropion [34911-55-2]
Bupropion Hydrochloride* [31677-93-7]
Buquinolate [5486-03-3]
Buramate [4663-83-6]
Burgundy Mixture [11125-96-5]

Buserelin [57982-77-1]
Buserelin Acetate* [68630-75-1]
Buspirone [36505-84-7]
Buspirone Hydrochloride* [33386-08-2]
Busulfan [55-98-1]
Butabarbital Sodium [143-81-7]
Butabarbital* [125-40-6]
Butacaine [149-16-6]
Butacaine Sulfate* [149-15-5]
Butacetin [2109-73-1]
Butachlor [23184-66-9]
Butaclamol [36504-93-5]
1,3-Butadiene [106-99-0]
Butalamine [22131-35-7]
Butalbital [77-26-9]
Butallylonal [1142-70-7]
Butamben [94-25-7]
Butamben Picrate* [577-48-0]
Butamirate [18109-80-3]
Butamirate Citrate* [18109-81-4]
Butamisole [54400-59-8]
Butamisole Hydrochloride* [54400-62-3]
Butane [106-97-8]
Butanilicaine [3785-21-5]
Butaperazine [653-03-2]
Butaverine [55837-14-4]
Butazolamide [16790-49-1]
Butedronic Acid [51395-42-7]
Butedronic Acid Technetium Salt* [108544-30-5]
Butedronic Acid Tetrasodium Salt* [97772-42-7]
1-Butene [106-98-9]
2-Butene [107-01-7]
Butethal [77-28-1]
Butethamate [14007-64-8]
Butethamine [2090-89-3]
Butethamine Hydrochloride* [553-68-4]
Metabutethamine Hydrochloride* [553-58-2]
Buthalital Sodium [510-90-7]
Buthiazide [2043-38-1]
Buthiobate [51308-54-4]
Buthionine Sulfoximine [71765-30-5]
Butibufen [55837-18-8]
Butidrine Hydrochloride [1506-12-3]
Butirosin [12772-35-9]
Butirosin Sulfate* [51022-98-1]
Butobendine [55769-65-8]
Butobendine Hydrochloride* [55769-64-7]
Butoconazole [64872-76-0]
Butoconazole Nitrate* [64872-77-1]
Butoctamide [32838-26-9]
Butofilolol [64552-17-6]
Butonate [126-22-7]
Butopyronoxyl [532-34-3]
Butorphanol [42408-82-2]
Butorphanol Tartrate* [58786-99-5]
Butoxycaine [2350-32-5]
Butralin [33629-47-9]
Butriptyline [35941-65-2]
Butriptyline Hydrochloride* [5585-73-9]
Butropium Bromide [29025-14-7]
n-Butyl Acetate [123-86-4]
sec-Butyl Acetate [105-46-4]
tert-Butyl Acetate [540-88-5]
tert-Butylacetic Acid [1070-83-3]
n-Butyl Acrylate [141-32-2]
n-Butyl Alcohol [71-36-3]
sec-Butyl Alcohol [78-92-2]
tert-Butyl Alcohol [75-65-0]
n-Butylamine [109-73-9]
sec-Butylamine [13952-84-6]
tert-Butylamine [75-64-9]
Butylate [2008-41-5]

Butylated Hydroxyanisole [25013-16-5]
Butylated Hydroxytoluene [128-37-0]
n-Butylbenzene [104-51-8]
sec-Butylbenzene [135-98-8]
tert-Butylbenzene [98-06-6]
n-Butyl Benzoate [136-60-7]
n-Butyl Bromide [109-65-9]
sec-Butyl Bromide [78-76-2]
tert-Butyl Bromide [507-19-7]
n-Butyl n-Butyrate [109-21-7]
Butyl Carbitol® [112-34-5]
n-Butyl Carbonate [542-52-9]
Butyl Cellosolve® [111-76-2]
n-Butyl Chloride [109-69-3]
sec-Butyl Chloride [78-86-4]
tert-Butyl Chloride [507-20-0]
tert-Butyl Chloroacetate [107-59-5]
Butyl Citrate [77-94-1]
α-Butylene Dibromide [533-98-2]
1,3-Butylene Glycol [107-88-0]
2,3-Butylene Glycol [513-85-9]
n-Butyl Ether [142-96-1]
tert-Butyl Hydroperoxide [75-91-2]
tert-Butyl Hypochlorite [507-40-4]
Butylidene Chloride [541-33-3]
n-Butyl Iodide [542-69-8]
sec-Butyl Iodide [513-48-4]
n-Butylmalonic Acid [534-59-8]
n-Butyl Mercaptan [109-79-5]
sec-Butyl Mercaptan [513-53-1]
tert-Butyl Mercaptan [75-66-1]
n-Butylmercuric Chloride [543-63-5]
1-Butyl-3-metanilylurea [4618-41-1]
Butyl Methoxydibenzoylmethane [70356-09-1]
n-Butyl Nitrite [544-16-1]
tert-Butyl Nitrite [540-80-7]
Butylparaben [94-26-8]
p-tert-Butylphenol [98-54-4]
4-tert-Butylphenyl Salicylate [87-18-3]
n-Butyl Phthalate [84-74-2]
n-Butyl Propionate [590-01-2]
N-Butylscopolammonium Bromide [149-64-4]
Butyl Stearate [123-95-5]
n-Butyl Sulfide [544-40-1]
Butyraldehyde [123-72-8]
n-Butyramide [541-35-5]
Butyric Acid [107-92-6]
Butyric Anhydride [106-31-0]
Butyroin [496-77-5]
Butyrolactone [96-48-0]
Butyronitrile [109-74-0]
n-Butyryl Chloride [141-75-3]
Buzepide [3691-21-2]
Cacodyl [471-35-2]
Cacodylic Acid [75-60-5]
Cacotheline [561-20-6]
Cactinomycin [8052-16-2]
Cadalene [483-78-3]
Cadaverine [462-94-2]
Cadmium [7440-43-9]
Cadmium Acetate [543-90-8]
Cadmium Bromide [7789-42-6]
Cadmium Carbonate [513-78-0]
Cadmium Chloride [10108-64-2]
Cadmium Cyanide [542-83-6]
Cadmium Fluoride [7790-79-6]
Cadmium Hydroxide [21041-95-2]
Cadmium Iodide [7790-80-9]
Cadmium Nitrate [10325-94-7]
Cadmium Oxide [1306-19-0]
Cadmium Potassium Cyanide [14402-75-6]
Cadmium Salicylate [19010-79-8]
Cadmium Selenate [13814-62-5]
Cadmium Selenide [1306-24-7]
Cadmium Succinate [141-00-4]
Cadmium Sulfate [10124-36-4]

Denotes derivative of title compound

Cadmium Sulfide [1306-23-6]
Cadmium Telluride [1306-25-8]
Cadmium Tungstate(VI) [7790-85-4]
Cadralazine [64241-34-5]
Cafaminol [30924-31-3]
Cafestol [469-83-0]
Caffeic Acid [331-39-5]
Caffeine [58-08-2]
Calcifediol [19356-17-3]
Calcimycin [52665-69-7]
Calcitonin [9007-12-9]
Calcitriol [32222-06-3]
Calcium [7440-70-2]
Calcium Acetate [62-54-4]
Calcium Acetylsalicylate [69-46-5]
Calcium Aluminosilicate [1327-39-5]
Calcium Arsenate [7778-44-1]
Calcium Arsenite [52740-16-6]
Calcium Ascorbate [5743-27-1]
Calcium 3-Aurothio-2-propanol-1-
  sulfonate [5743-29-3]
Calcium Bisulfite, Solution
  [13780-03-5]
Calcium Borate [12007-56-6]
Calcium Borogluconate [5743-34-0]
Calcium Bromide [7789-41-5]
Calcium Bromolactobionate
  [33659-28-8]
Calcium N-Carbamoylaspartate
  [16649-79-9]
Calcium Carbide [75-20-7]
Calcium Carbonate [471-34-1]
Calcium Chlorate [10137-74-3]
Calcium Chloride [10043-52-4]
Calcium Chromate(VI) [13765-19-0]
Calcium Citrate [813-94-5]
Calcium Cyanamide [156-62-7]
Calcium Cyanamide Citrated
  [8013-88-5]
Calcium Cyanide [592-01-8]
Calcium Cyclamate [139-06-0]
Calcium Dichromate(VI)
  [14307-33-6]
Calcium 2-Ethylbutanoate [136-91-4]
Calcium Ferrous Citrate [53684-61-0]
Calcium Fluoride [7789-75-5]
Calcium Fluorophosphate [7789-74-4]
Calcium Formate [544-17-2]
Calcium Gluconate [299-28-5]
Calcium Glycerophosphate
  [27214-00-2]
Calcium Hexafluorosilicate
  [16925-39-6]
Calcium Hydride [7789-78-8]
Calcium Hydroxide [1305-62-0]
Calcium Hypochlorite [7778-54-3]
Calcium Hypophosphite [7789-79-9]
Calcium Iodate [7789-80-2]
Calcium Iodide [10102-68-8]
Calcium Iodobehenate [1319-91-1]
Calcium Iodostearate [1301-16-2]
Calcium Lactate [814-80-2]
Calcium Levulinate [591-64-0]
Calcium Mesoxalate [21085-60-9]
Calcium Methionate [819-17-0]
Calcium Molybdate(VI) [7789-82-4]
Calcium Nitrate [10124-37-5]
Calcium Nitrite [13780-06-8]
Calcium Oleate [142-17-6]
Calcium Oxalate [563-72-4]
Calcium Oxide [1305-78-8]
Calcium Palmitate [542-42-7]
Calcium Pantothenate [137-08-6]
Calcium Permanganate [10118-76-0]
Calcium Peroxide [1305-79-9]
Calcium Phenolsulfonate [127-83-3]
Calcium Phenoxide [5793-84-0]
Calcium Phosphate, Dibasic
  [7757-93-9]
Calcium Phosphate, Monobasic
  [7758-23-8]

Calcium Phosphate, Tribasic
  [7758-87-4]
Calcium Phosphide [1305-99-3]
Calcium Phosphite [21056-98-4]
Calcium Polycarbophil [9003-97-8]
Calcium Propionate [4075-81-4]
Calcium Pyrophosphate [35405-51-7]
Calcium D-Saccharate [5793-88-4]
Calcium Selenide [1305-84-6]
Calcium Silicate [1344-95-2]
Calcium Stearate [1592-23-0]
Calcium Stearyl-2 Lactylate
  [5793-94-2]
Calcium Succinate [140-99-8]
Calcium Sulfate [7778-18-9]
Calcium Sulfide [20548-54-3]
Calcium Sulfite [10257-55-3]
Calcium Tartrate [3164-34-9]
Calcium Thiocyanate [2092-16-2]
Calcium Thioglycollate [814-71-1]
Calcium Thiosulfate [10124-41-1]
Calcium Tungstate(VI) [7790-75-2]
Caldariomycin [465-61-2]
C-Calebassine [7257-29-6]
Californium [7440-71-3]
Calmagite [3147-14-6]
Calomelol [8011-82-3]
Calotropin [1986-70-5]
Calusterone [17021-26-0]
Calycanthine [595-05-1]
Camazepam [36104-80-0]
Cambendazole [26097-80-3]
Camostat [59721-28-7]
Campesterol [474-62-4]
Camphene [79-92-5]
d-Camphocarboxylic Acid
  [18530-30-8]
Camphor [76-22-2]
Camphoric Acid [5394-83-2]
d-Camphorsulfonic Acid [3144-16-9]
Camphotamide [4876-45-3]
Camptothecin [7689-03-4]
Camylofine [54-30-8]
Canadine [522-97-4]
Canavanine [543-38-4]
Candelilla Wax [8006-44-8]
Candicidin [1403-17-4]
Candidin [1405-90-9]
Cannabidiol [13956-29-1]
Cannabinol [521-35-7]
Canrenone [976-71-6]
Canrenoate Potassium* [2181-04-6]
Cantharidin [56-25-7]
Canthaxanthin [514-78-3]
Capobenic Acid [21434-91-3]
Capobenate Sodium* [27276-25-1]
Capreomycin [11003-38-6]
Capreomycin Sulfate* [1405-37-4]
n-Capric Acid [334-48-5]
n-Caproic Acid [142-62-1]
Caproic Aldehyde [66-25-1]
Caprolactam [105-60-2]
Caproyl Chloride [142-61-0]
Caprylene [111-66-0]
Caprylic Acid [124-07-2]
Caprylic Aldehyde [124-13-0]
Capsaicin [404-86-4]
Capsanthin [465-42-9]
Captafol [2425-06-1]
Captan [133-06-2]
Captodiame [486-17-9]
Captodiamine Hydrochloride*
  [904-04-1]
Captopril [62571-86-2]
Capuride [5579-13-5]
Caramiphen Ethanedisulfonate
  [125-86-0]
Caramiphen Hydrochloride
  [125-85-9]
Carazolol [57775-29-8]
Carbachol [51-83-2]

Carbadox [6804-07-5]
Carbamazepine [298-46-4]
Carbamic Acid [463-77-4]
Carbamyl Chloride [463-72-9]
Carbanilic Acid [501-82-6]
Carbanilide [102-07-8]
Carbarsone [121-59-5]
Carbaryl [63-25-2]
Carbazochrome Salicylate
  [13051-01-9]
Carbazole [86-74-8]
9-Carbazoleacetic Acid [524-80-1]
Carbendazim [10605-21-7]
Carbenicillin [4697-36-3]
Carbenicillin Disodium* [4800-94-6]
Carbenoxolone [5697-56-3]
Carbenoxolone Sodium* [7421-40-1]
Carbetapentane [77-23-6]
Carbetapentane Citrate* [23142-01-0]
Carbetidine [469-82-9]
Carbetocin [37025-55-1]
Carbic Anhydride [129-64-6]
Carbidopa [28860-95-9]
Carbimazole [22232-54-8]
Carbinoxamine [486-16-8]
Carbinoxamine Maleate* [3505-38-2]
  Rotoxamine Tartrate* [49746-00-1]
Carbiphene [15687-16-8]
Carbiphene Hydrochloride*
  [467-22-1]
Carbitol® [111-90-0]
Carbobenzoxy Chloride [501-53-1]
Carbocloral [541-79-7]
Carbocysteine [638-23-3]
Carbofuran [1563-66-2]
Carbohydrazide [497-18-7]
γ-Carboline [244-69-9]
Carbon [7440-44-0]
Carbon Dioxide [124-38-9]
Carbon Diselenide [506-80-9]
Carbon Disulfide [75-15-0]
Carbonic Anhydrase [9001-03-0]
Carbon Monoxide [630-08-0]
Carbon Suboxide [504-64-3]
Carbon Tetrachloride [56-23-5]
Carbon Tetrafluoride [75-73-0]
Carbon Tetraiodide [507-25-5]
N,N'-Carbonyldiimidazole [530-62-1]
Carbonyl Fluoride [353-50-4]
Carbophenothion [786-19-6]
Carboplatin [41575-94-4]
Carboprost [35700-23-3]
Carboprost Methyl* [35700-21-1]
Carboprost Tromethamine*
  [58551-69-2]
Carboquone [24279-91-2]
Carbostyril [59-31-4]
Carboxin [5234-68-4]
β-Carboxyaspartic Acid [75898-26-9]
γ-Carboxyglutamic Acid
  [53861-57-7]
Carboxymethylcellulose Sodium
  [9004-32-4]
Carboxypolymethylene [9007-20-9]
Carbromal [77-65-6]
Carbubarb [960-05-4]
Carbutamide [339-43-5]
Carbuterol [34866-47-2]
Carbuterol Hydrochloride*
  [34866-46-1]
Cardiotoxin [11061-96-4]
3-Carene [13466-78-9]
Carfecillin Sodium [21649-57-0]
Carfinate [3567-38-2]
Cargutocin [33605-67-3]
Carindacillin [35531-88-5]
Carisoprodol [78-44-4]
Carlsbad Salt Artificial [8007-49-6]
Carminic Acid [1260-17-9]
Carmofur [61422-45-5]
Carmustine [154-93-8]

Carnauba Wax [8015-86-9]
Carnegine [490-53-9]
Carnidazole [42116-76-7]
Carnitine [461-06-3]
Carnitine (L-Form)* [541-15-1]
Carnosine [305-84-0]
Caro's Acid [7722-86-3]
α-Carotene [7488-99-5]
β-Carotene [7235-40-7]
γ-Carotene [472-93-5]
δ-Carotene [472-92-4]
Carotol [465-28-1]
Caroverine [23465-76-1]
Caroxazone [18464-39-6]
Carpaine [3463-92-1]
Carpetimycin A* [76025-73-5]
Carpetimycin B* [76094-36-5]
Carpetimycin C* [87695-64-5]
Carpetimycin D* [87139-37-5]
Carphenazine [2622-30-2]
Carphenazine Dimaleate* [2975-34-0]
Carpipramine [5942-95-0]
Carpipramine Dihydrochloride Mono-
hydrate* [7075-03-8]
Carprofen [53716-49-7]
Carpronium Chloride [13254-33-6]
Carrageenan [9000-07-1]
Carsalam [2037-95-8]
Cartap [15263-53-3]
Carteolol [51781-06-7]
Carteolol Hydrochloride*
[51781-21-6]
Carthamin [36338-96-2]
Carticaine [23964-58-1]
Carubicin [50935-04-1]
Carubicin Hydrochloride*
[52794-97-5]
Carumonam [87638-04-8]
Carumonam Sodium* [86832-68-0]
Carvacrol [499-75-2]
Carvedilol [72956-09-3]
Carvone [99-49-0]
Caryophyllene [87-44-5]
Carzenide [138-41-0]
Carzinophilin [1403-28-7]
Cascarillin [10118-56-6]
Casein [9000-71-9]
Casimiroedine [5853-02-1]
Casimiroin [477-89-4]
Cassaidine [26296-41-3]
Cassaine [468-76-8]
Cassamine [471-71-6]
Cassella's Acid F [494-44-0]
Castanospermine [79831-76-8]
Catalase [9001-05-2]
Catalposide [6736-85-2]
Catechin [154-23-4]
Catharanthine [2468-21-5]
Cathinone [71031-15-7]
Caulophylline [486-86-2]
Ceanothic Acid [21302-79-4]
Cedrin [6040-62-6]
Cedrol [77-53-2]
Cefaclor [70356-03-5]
Cefadroxil [66592-87-8]
Cefamandole [34444-01-4]
Cefamandole Nafate* [42540-40-9]
Cefatrizine [51627-14-6]
Cefazedone [56187-47-4]
Cefazolin [25953-19-9]
Cefazolin Sodium* [27164-46-1]
Cefbuperazone [76610-84-9]
Cefixime [79350-37-1]
Cefmenoxime [65085-01-0]
Cefmenoxime Hydrochloride*
[75738-58-8]
Cefmetazole [56796-20-4]
Cefmetazole Sodium* [56796-39-5]
Cefminox [84305-41-9]
Cefodizime [69739-16-8]

Cefodizime Disodium Salt*
[86329-79-5]
Cefonicid [61270-58-4]
Cefonicid Sodium* [61270-78-8]
Cefoperazone [62893-19-0]
Cefoperazone Sodium* [62893-20-3]
Ceforanide [60925-61-3]
Cefotaxime [63527-52-6]
Cefotaxime Sodium* [64485-93-4]
Cefotetan [69712-56-7]
Cefotetan Disodium* [74356-00-6]
Cefotiam [61622-34-2]
Cefotiam Hydrochloride*
[66309-69-1]
Cefoxitin [35607-66-0]
Cefoxitin Sodium* [33564-30-6]
Cefpimizole [84880-03-5]
Cefpimizole Sodium* [85287-61-2]
Cefpiramide [70797-11-4]
Cefpiramide Sodium* [74849-93-7]
Cefpodoxime Proxetil [87239-81-4]
Cefroxadine [51762-05-1]
Cefsulodin [62587-73-9]
Cefsulodin Sodium* [52152-93-9]
Ceftazidime [72558-82-8]
Cefteram [82547-58-8]
Ceftezole [26973-24-0]
Ceftibuten [97519-39-6]
Ceftiofur [80370-57-6]
Ceftiofur Sodium* [104010-37-9]
Ceftizoxime [68401-81-0]
Ceftizoxime Sodium* [68401-82-1]
Ceftriaxone [73384-59-5]
Ceftriaxone Sodium* [104376-79-6]
Cefuroxime [55268-75-2]
Cefuroxime Sodium* [56238-63-2]
Cefuzonam [82219-78-1]
Celesticetin [2520-21-0]
Celestin Blue [1562-90-9]
Celiprolol [56980-93-9]
Celiprolol Hydrochloride*
[57470-78-7]
Cellobiose [528-50-7]
Cellocidin [543-21-5]
Celluloid® [8050-88-2]
Cellulose [9004-34-6]
Cellulose Ethyl Hydroxyethyl Ether
[9004-58-4]
Centaurein [35595-03-0]
Cephacetrile Sodium [23239-41-0]
Cephaeline [483-17-0]
Cephalexin [15686-71-2]
Cephalexin Hydrochloride*
[105879-42-3]
Cephaloglycin [3577-01-3]
Cephalonic Acid [18456-04-7]
Cephaloridine [50-59-9]
Cephalosporin C [61-24-5]
Cephalosporin P₁ [28393-42-2]
Cephalothin [153-61-7]
Cephapirin Sodium [24356-60-3]
Cepharanthine [481-49-2]
Cephradine [38821-53-3]
Cerberoside [11005-70-2]
Ceresin [8001-75-0]
Ceric Fluoride [10060-10-3]
Ceric Oxide [1306-38-3]
Ceric Sulfate [13590-82-4]
Cerium [7440-45-1]
Cerous Bromide [14457-87-5]
Cerous Carbonate [537-01-9]
Cerous Chloride [7790-86-5]
Cerous Fluoride [7758-88-5]
Cerous Iodide [7790-87-6]
Cerous Nitrate [10108-73-3]
Cerous Oxalate [139-42-4]
Cerous Sulfate [13454-94-9]
Cerulenin [17397-89-6]
Ceruletide [17650-98-5]
Ceruletide Diethylamine*
[71247-25-1]

Ceruloplasmin [9031-37-2]
Cervicarcin [18700-78-2]
Cesium [7440-46-2]
Cesium Bromide [7787-69-1]
Cesium Carbonate [534-17-8]
Cesium Chloride [7647-17-8]
Cesium Hydroxide [21351-79-1]
Cesium Iodide [7789-17-5]
Cesium Nitrate [7789-18-6]
Cesium Sulfate [10294-54-9]
Cetalkonium Chloride [122-18-9]
Cetamolol [34919-98-7]
Cetamolol Hydrochloride*
[77590-95-5]
Cethexonium Bromide [1794-74-7]
Cetiedil [14176-10-4]
Citiedil Citrate* [16286-69-4]
Cetirizine [83881-51-0]
Cetirizine Hydrochloride*
[83881-52-1]
Cetotiamine [137-76-8]
Cetoxime [25394-78-9]
Cetoxime Hydrochloride*
[22204-29-1]
Cetraric Acid [489-49-6]
Cetraxate [34675-84-8]
Cetraxate Hydrochloride*
[27724-96-5]
Cetrimonium Bromide [57-09-0]
Cetrimonium Stearate [124-23-2]
Cetyl Alcohol [36653-82-4]
Cetyldimethylethylammonium Bro-
mide [124-03-8]
Cetyl Lactate [35274-05-6]
Cetyl Palmitate [540-10-3]
Cetylpyridinium Chloride [123-03-5]
Cevadine [62-59-9]
Cevine [124-98-1]
Chalcomycin [20283-48-1]
Chalcone [94-41-7]
Chalcopyrite [1308-56-1]
D-Chalcose [3150-28-5]
Chamazulene [529-05-5]
Chanoclavine [2390-99-0]
CHAPS [75621-03-3]
CHAPS Hydroxy Analog*
[82473-24-3]
Chartreusin [6377-18-0]
Chaulmoogric Acid [29106-32-9]
Chavicine [495-91-0]
Chavicol [501-92-8]
Cheirolin [505-34-0]
Chelerythrine [34316-15-9]
Chelidonic Acid [99-32-1]
Chelidonine [476-32-4]
Chelidonine (±-Form)* [20267-87-2]
Chelidonine (−)-Form* [88200-01-5]
Chenodiol [474-25-9]
Chimaphilin [482-70-2]
Chimonanthine [5545-89-1]
Chimyl Alcohol [506-03-6]
Chitin [1398-61-4]
Chlophedianol [791-35-5]
Chlophedianol Hydrochloride*
[511-13-7]
Chloracetyl Chloride [79-04-9]
Chloracizine [800-22-6]
Chloral Alcoholate [515-83-3]
Chloral Ammonia [507-47-1]
Chloralantipyrine [603-63-4]
Chloral Betaine [2218-68-0]
Chloral Formamide [515-82-2]
Chloral Hydrate [302-17-0]
α-Chloralose [15879-93-1]
Chloramben [133-90-4]
Chlorambucil [305-03-3]
Chloramine-B [127-52-6]
Chloramine-T [127-65-1]
Chloraminophenamide [121-30-2]
Chloramphenicol [56-75-7]

Denotes derivative of title compound

Chloramphenicol Palmitate [530-43-8]
Chloramphenicol Pantothenate [52993-97-2]
Chloranil [118-75-2]
Chloranilic Acid [87-88-7]
Chlorazanil [500-42-5]
Chlorbenside [103-17-3]
Chlorbenzoxamine [522-18-9]
Chlorbenzoxamine Dihydrochloride* [5576-62-5]
Chlorbetamide [97-27-8]
Chlorbicyclen [2550-75-6]
Chlorcyclizine [82-93-9]
Chlorcyclizine Hydrochloride* [1620-21-9]
Chlordan(e) [57-74-9]
Chlordantoin [5588-20-5]
Chlordecone [143-50-0]
Chlordiazepoxide [58-25-3]
Chlordiazepoxide Hydrochloride* [438-41-5]
Chlordimeform [6164-98-3]
Chlorendic Anhydride [115-27-5]
Chlorfenac [85-34-7]
Chlorfenethol [80-06-8]
Chlorfenvinphos [470-90-6]
Chlorguanide [500-92-5]
Chlorhexadol [3563-58-4]
Chlorhexidine [55-56-1]
Chlorhexidine Dihydrochloride* [3697-42-5]
Chlorhexidine Gluconate* [18472-51-0]
Chloric Acid [7790-93-4]
Chlorimuron Ethyl [90982-32-4]
Chlorinated Lime [7778-54-3]
Chlorindanol [145-94-8]
Chlorine [7782-50-5]
Chlorine Dioxide [10049-04-4]
Chlorine Heptoxide [10294-48-1]
Chlorine Monofluoride [7790-89-8]
Chlorine Monoxide [7791-21-1]
Chlorine Trifluoride [7790-91-2]
Chlorisondamine Chloride [69-27-2]
Chlormadinone Acetate [302-22-7]
Chlormequat Chloride [999-81-5]
Chlormerodrin [62-37-3]
Chlormezanone [80-77-3]
Chlormidazole [3689-76-7]
Chlornaphazine [494-03-1]
Chloroacetaldehyde [107-20-0]
Chloroacetamide [79-07-2]
Chloroacetic Acid [79-11-8]
Chloroacetic Anhydride [541-88-8]
Chloroacetone [78-95-5]
p-Chloroacetophenone [99-91-2]
ω-Chloroacetophenone [532-27-4]
Chloroacetyl Isocyanate [4461-30-7]
N-(3-Chloroallyl)hexaminium Chloride [4080-31-3]
m-Chloroaniline* [108-42-9]
o-Chloroaniline* [95-51-2]
p-Chloroaniline* [106-47-8]
Chloroarsenol [151-07-5]
Chloroazodin [502-98-7]
Chlorobenzene [108-90-7]
p-Chlorobenzenesulfonic Acid [98-66-8]
Chlorobenzilate [510-15-6]
m-Chlorobenzoic Acid [535-80-8]
o-Chlorobenzoic Acid [118-91-2]
p-Chlorobenzoic Acid [74-11-3]
o-Chlorobenzylidenemalononitrile [2698-41-1]
p-Chlorobenzylpseudothiuronium Chloride [544-47-8]
Chlorobutanol [57-15-8]
1-Chloro-2-butene [591-97-9]
3-Chloro-1-butene [563-52-0]
3-Chloro-d-camphor [508-29-2]

4-Chloro-m-cresol [59-50-7]
Chlorocyanohydrin [513-96-2]
1-Chloro-2,4-dinitrobenzene [97-00-7]
2-Chloro-1,3-dinitrobenzene [606-21-3]
β-Chloroethyl Acetate [542-58-5]
N-(2-Chloroethyl)dibenzylamine Hydrochloride [55-43-6]
2-Chloroethyl Vinyl Ether [110-75-8]
Chloroform [67-66-3]
Chlorogenic Acid [327-97-9]
Chlorogenin [562-34-5]
1-Chlorohexane [544-10-5]
α-Chlorohydrin [96-24-2]
Chloromethyl Methyl Ether [107-30-2]
1-Chloro-2-methylpropene [513-37-1]
3-Chloro-2-methylpropene [563-47-3]
1-Chloronaphthalene [90-13-1]
2-Chloronaphthalene [91-58-7]
Chlorophacinone [3691-35-8]
p-Chlorophenacyl Bromide [536-38-9]
Chlorophyll [1406-65-1]
Chloropicrin [76-06-2]
Chloroprednisone [52080-57-6]
Chloroprednisone Acetate* [14066-79-6]
2-Chloroprocaine Hydrochloride [3858-89-7]
β-Chloropropionic Acid [107-94-8]
β-Chloropropionitrile [542-76-7]
6-Chloropurine [87-42-3]
Chloropyramine [59-32-5]
Chloroquine [54-05-7]
Chloroquine Phosphate* [50-63-5]
Chloroselenic Acid [7790-95-6]
N-Chlorosuccinimide [128-09-6]
Chlorosulfonic Acid [7790-94-5]
Chlorothalonil [1897-45-6]
Chlorothen [148-65-2]
Chlorothen Citrate* [148-64-1]
Chlorothiazide [58-94-6]
Chlorothiazide Sodium* [7085-44-1]
Chlorothricin [34707-92-1]
Chlorothymol [89-68-9]
m-Chlorotoluene* [108-41-8]
o-Chlorotoluene* [95-49-8]
p-Chlorotoluene* [106-43-4]
Chlorotrianisene [569-57-3]
Chlorotris(triphenylphosphine)rhodium [14694- 95-2]
Chloroxine [773-76-2]
Chloroxylenol [88-04-0]
Chlorozotocin [54749-90-5]
Chlorphenesin [104-29-0]
Chlorphenesin Carbamate [886-74-8]
Chlorpheniramine [132-22-9]
Chlorpheniramine Maleate* [2438-32-6]
Chlorpheniramine (d-Form)* [25523-97-1]
Chlorpheniramine (d-Form) Maleate* [2438-32-6]
Chlorphenoxamide [3576-64-5]
Chlorphenoxamine [77-38-3]
Chlorphenoxamine Hydrochloride* [562-09-4]
Chlorphentermine [461-78-9]
Chlorphentermine Hydrochloride* [151-06-4]
Chlorproethazine [84-01-5]
Chlorproethazine Hydrochloride* [4611-02-3]
Chlorproguanil [537-21-3]
Chlorproguanil Hydrochloride* [15537-76-5]
Chlorpromazine [50-53-3]

Chlorpromazine Hydrochloride* [69-09-0]
Chlorpropamide [94-20-2]
Chlorpropham [101-21-3]
Chlorprothixene [113-59-7]
Chlorpyrifos [2921-88-2]
Chlorquinaldol [72-80-0]
Chlorsulfuron [64902-72-3]
Chlortetracycline [57-62-5]
Chlortetracycline Hydrochloride* [64-72-2]
Chlorthalidone [77-36-1]
Chlorthenoxazin(e) [132-89-8]
Chlorthion® [500-28-7]
Chlorzoxazone [95-25-0]
Cholane [548-98-1]
Cholanic Acid [25312-65-6]
Cholanthrene [479-23-2]
Cholecystokinin [9011-97-6]
Cholestane [481-21-0]
Cholestanol [80-97-7]
Cholesterol [57-88-5]
Cholestyramine Resin [11041-12-6]
Cholic Acid [81-25-4]
Choline [62-49-7]
Choline Chloride [67-48-1]
Choline Dehydrocholate [4201-78-9]
Choline Dihydrogen Citrate [77-91-8]
Choline Esterase [9001-08-5]
Choline Salicylate [2016-36-6]
Choline Theophyllinate [4499-40-5]
Chondrillasterol [481-17-4]
Chondrocurine [477-58-7]
Chondrofoline [31944-97-5]
Chondroitin Sulfate [9007-28-7]
Chondrosine [499-14-9]
Chonemorphine [4282-07-9]
Chorismic Acid [617-12-9]
Chromic Acetate [1066-30-4]
Chromic Bromide [10031-25-1]
Chromic Carbonate [29689-14-3]
Chromic Chloride [10025-73-7]
Chromic Fluoride [7788-97-8]
Chromic Formate [27115-36-2]
Chromic Hydroxide [1308-14-1]
Chromic Nitrate [13548-38-4]
Chromic Oxide [1308-38-9]
Chromic Phosphate [7789-04-0]
Chromic Potassium Oxalate [14217-01-7]
Chromic Potassium Sulfate [10141-00-1]
Chromic Sulfate [10101-53-8]
Chromium [7440-47-3]
Chromium Carbonyl [13007-92-6]
Chromium Dioxide [12018-01-8]
Chromium Tetrafluoride [10049-11-3]
Chromium Trioxide [1333-82-0]
Chromocarb [4940-39-0]
Chromonar [804-10-4]
Chromonar Hydrochloride* [655-35-6]
Chromotrope 2B [548-80-1]
Chromotropic Acid [148-25-4]
Chromous Acetate [628-52-4]
Chromous Bromide [10049-25-9]
Chromous Chloride [10049-05-5]
Chromous Fluoride [10049-10-2]
Chromous Formate [4493-37-2]
Chromous Oxalate [814-90-4]
Chromous Sulfate [13825-86-0]
Chromyl Chloride [14977-61-8]
Chromyl Fluoride [7788-96-7]
Chrysamminic Acid [517-92-0]
Chrysanthemaxanthin [27780-11-6]
Chrysanthemic Acid [10453-89-1]
Chrysanthenone [473-06-3]
Chrysarobin [491-59-8]
6-Chrysenamine [2642-98-0]
Chrysene [218-01-9]
Chrysergonic Acid [53092-91-4]

Chrysin [480-40-0]
Chrysoidine [532-82-1]
Chrysoidine Citrate* [5909-04-6]
Chrysoidine Free Base* [495-54-5]
Chrysophanic Acid [481-74-3]
Chymopapain [9001-09-6]
Ciafos [2636-26-2]
Cichoriin [531-58-8]
Cicletanine [89943-82-8]
Ciclonicate [53449-58-4]
Ciclopirox [29342-05-0]
Ciclopirox Olamine* [41621-49-2]
Ciclosidomine [66564-16-7]
Cicrotoic Acid [25229-42-9]
Cicutoxin [505-75-9]
Cifenline [53267-01-9]
Cifenline Succinate* [100678-32-8]
Cilastatin [82009-34-5]
Cilastatin Sodium* [81129-83-1]
Cilazapril [92077-78-6]
Cilazaprilat* [90139-06-3]
Cilostazol [73963-72-1]
Cimaterol [54239-37-1]
Cimetidine [51481-61-9]
Cimetidine Hydrochloride*
    [70059-30-2]
Cimetropium Bromide [51598-60-8]
Cimigenol [3779-59-7]
Cinametic Acid [35703-32-3]
Cinamiodyl [1215-70-9]
Cinchomeronic Acid [490-11-9]
Cinchonamine [482-28-0]
Cinchonidine [485-71-2]
Cinchonine [118-10-5]
Cinchophen [132-60-5]
Cinchotoxine [69-24-9]
Cinepazet Maleate [50679-07-7]
Cinepazide [23887-46-9]
Cinepazic Acid* [54063-23-9]
Cinmetacin [20168-99-4]
Cinmethylin [87818-31-3]
Cinnabarine [146-90-7]
Cinnamaldehyde [104-55-2]
Cinnamedrine [90-86-8]
Cinnamic Acid [621-82-9]
Cinnamoyl Chloride [102-92-1]
Cinnamoylcocaine [521-67-5]
Cinnamyl Alcohol [104-54-1]
Cinnamyl Anthranilate [87-29-6]
Cinnamyl Cinnamate [122-69-0]
Cinnarizine [298-57-7]
Cinnoline [253-66-7]
Cinobufotalin [1108-68-5]
Cinoxacin [28657-80-9]
Cinoxate [104-28-9]
Cinromide [58473-74-8]
Ciprofibrate [52214-84-3]
Ciprofloxacin [85721-33-1]
Ciprofloxacin Hydrochloride
    Monohydrate*
    [86393-32-0]
Ciramadol [63269-31-8]
Ciramadol Hydrochloride*
    [63323-46-6]
Cisapride [81098-60-4]
Cisplatin [15663-27-1]
Citalopram [59729-33-8]
Citalopram Hydrobromide*
    [59729-32-7]
Citicoline [987-78-0]
Citiolone [1195-16-0]
Citraconic Acid [498-23-7]
Citral [5392-40-5]
Citramalic Acid [2306-22-1]
β-Citraurin [650-69-1]
Citrazinic Acid [99-11-6]
Citric Acid [77-92-9]
Citrinin [518-75-2]
Citromycetin [478-60-4]
Citronellal [106-23-0]
β-Citronellol [106-22-9]

Citrulline [372-75-8]
Citrullol [1390-93-8]
Citrus Red 2 [6358-53-8]
Civetone [542-46-1]
Clanobutin [30544-61-7]
Clarase® [9001-11-0]
Clarithromycin [81103-11-9]
Clavulanic Acid [58001-44-8]
Clavulanate Potassium* [61177-45-5]
Clazuril [101831-36-1]
Clebopride [55905-53-8]
Clemastine [15686-51-8]
Clemastine Fumarate* [14976-57-9]
Clemizole [442-52-4]
Clenbuterol [37148-27-9]
1,6-Cleve's Acid [119-79-9]
1,7-Cleve's Acid [119-28-8]
Clidanac [28968-07-2]
Clidinium Bromide [3485-62-9]
Clindamycin [18323-44-9]
Clindamycin Hydrochloride Monohy-
    drate* [58207-19-5]
Clinofibrate [30299-08-2]
Clioxanide [14437-41-3]
Clobazam [22316-47-8]
Clobenfurol [3611-72-1]
Clobenoside [29899-95-4]
Clobenzepam [1159-93-9]
Clobenzorex [13364-32-4]
Clobenztropine [5627-46-3]
Clobetasol [25122-41-2]
Clobetasol Propionate* [25122-46-7]
Clobetasone [54063-32-0]
Clobetasone Butyrate* [25122-57-0]
Clobutinol [14860-49-2]
Clobuzarit [22494-47-9]
Clocapramine [47739-98-0]
Clocinizine [298-55-5]
Cloconazole [77175-51-0]
Clocortolone [4828-27-7]
Clocortolone Acetate* [4258-85-9]
Clocortolone Pivalate* [34097-16-0]
Clodronic Acid [10596-23-3]
Clofazimine [2030-63-9]
Clofenamide [671-95-4]
Clofenciclan [5632-52-0]
Clofentezine [74115-24-5]
Clofibrate [637-07-0]
Clofibric Acid [882-09-7]
Cloflucarban [369-77-7]
Clofoctol [37693-01-9]
Cloforex [14261-75-7]
Clomacran [5310-55-4]
Clomestrone [4091-75-2]
Clometacin [25803-14-9]
Clomethiazole [533-45-9]
Clometocillin [1926-49-4]
Clomiphene [911-45-5]
Clomiphene Citrate* [50-41-9]
Clomipramine [303-49-1]
Clomipramine Hydrochloride*
    [17321-77-6]
Clomocycline [1181-54-0]
Clonazepam [1622-61-3]
Clonidine [4205-90-7]
Clonidine Hydrochloride*
    [4205-91-8]
Clonitazene [3861-76-5]
Clonitrate [2612-33-1]
Clopamide [636-54-4]
Clopenthixol [982-24-1]
Cloperastine [3703-76-2]
Clopidol [2971-90-6]
Clopirac [42779-82-8]
Cloprednol [5251-34-3]
Cloprostenol [40665-92-7]
Cloprostenol Sodium* [55028-72-3]
Clopyralid [1702-17-6]
Cloranolol [39563-28-5]
Clorazepate [20432-69-3]

Clorazepate Dipotassium*
    [57109-90-7]
Clorazepate Monopotassium*
    [5991-71-9]
Clorexolone [2127-01-7]
Clorindione [1146-99-2]
Clorophene [120-32-1]
Clorprenaline [3811-25-4]
Clorprenaline Hydrochloride
    Monohydrate*
    [5588-22-7]
Clorsulon [60200-06-8]
Clortermine [10389-73-8]
Clortermine Hydrochloride*
    [10389-72-7]
Closantel [57808-65-8]
Clospirazine [24527-27-3]
Clostebol [1093-58-9]
Clothiapine [2058-52-8]
Clotiazepam [33671-46-4]
Clotrimazole [23593-75-1]
Cloxacillin [61-72-3]
Cloxacillin Benzathine* [23736-58-5]
Cloxacillin Sodium Monohydrate*
    [7081-44-9]
Cloxazolam [24166-13-0]
Cloxyquin [130-16-5]
Clozapine [5786-21-0]
Clupeine [9007-31-2]
Cnicin [24394-09-0]
Cobalt [7440-48-4]
Cobaltic Acetate [917-69-1]
Cobaltic-Cobaltous Oxide
    [1308-06-1]
Cobaltic Fluoride [10026-18-3]
Cobaltic Oxide Monohydrate
    [12016-80-7]
Cobaltic Potassium Nitrite
    [13782-01-9]
Cobaltous Acetate [71-48-7]
Cobaltous Arsenate [7785-24-2]
Cobaltous Bromide [7789-43-7]
Cobaltous Carbonate [513-79-1]
Cobaltous Chloride [7646-79-9]
Cobaltous Chromate(III)
    [13455-25-9]
Cobaltous Cyanide [542-84-7]
Cobaltous Fluoride [10026-17-2]
Cobaltous Formate [544-18-3]
Cobaltous Hydroxide [21041-93-0]
Cobaltous Iodide [15238-00-3]
Cobaltous Nitrate [10141-05-6]
Cobaltous Oxalate [814-89-1]
Cobaltous Oxide [1307-96-6]
Cobaltous Phosphate [13455-36-2]
Cobaltous Potassium Sulfate
    [13596-22-0]
Cobaltous Sulfate [10124-43-3]
Cobaltous Sulfide [1317-42-6]
Cobaltous Thiocyanate [3017-60-5]
Cobamamide [13870-90-1]
Cobrotoxin [12584-83-7]
Cocaethylene [529-38-4]
Cocaine [50-36-2]
Cocarboxylase [154-87-0]
Coclaurine [486-39-5]
Codamine [21040-59-5]
Codeine [76-57-3]
Codeine Acetate* [5913-71-3]
Codeine Hydrobromide* [125-25-7]
Codeine Hydrochloride* [1422-07-7]
Codeine Salicylate* [6020-73-1]
Codeine Methyl Bromide [125-27-9]
Codeine N-Oxide [3688-65-1]
Codeine Phosphate [52-28-8]
Codeine Sulfate [1420-53-7]
Coenzyme A [85-61-0]
Coherin [9044-70-6]
Colchiceine [477-27-0]
Colchicine [64-86-8]
Colestipol [26658-42-4]

Denotes derivative of title compound

Colestipol Hydrochloride*
  [37296-80-3]
Colforsin [66575-29-9]
Colistin [1066-17-7]
Colistimethate Sodium* [8068-28-8]
Colistin Sulfate* [1264-72-8]
Collagenase [9001-12-1]
Collinomycin [27267-69-2]
Colocynthin [1398-78-3]
Colostrokinin [37203-40-0]
Colpormon [1247-71-8]
α-Colubrine* [509-44-4]
β-Colubrine* [509-36-4]
Columbamine [3621-36-1]
Columbin [546-97-4]
Concanavalin A [11028-71-0]
Condurangin [1401-98-5]
Conessine [546-06-5]
Conessine Dihydrobromide*
  [5913-82-6]
Congo Red [573-58-0]
Congressane [2292-79-7]
Conhydrine [495-20-5]
β-Coniceine [538-90-9]
γ-Coniceine [1604-01-9]
Coniferin [531-29-3]
Coniferyl Alcohol [458-35-5]
Coniine [458-88-8]
Coniine Hydrobromide [637-49-0]
Coniine Hydrochloride [555-92-0]
Conquinamine [464-86-8]
Convallamarogenin [16683-27-5]
Convallatoxin [508-75-8]
Convicine [19286-37-4]
Copaene [3856-25-5]
Copaiba [8001-61-4]
Copal [9000-14-0]
Coparaffinate [8001-60-3]
Copper [7440-50-8]
Copper Phthalocyanine [147-14-8]
Coproergostane [511-21-7]
Coprogen [31418-71-0]
Coprostane [481-20-9]
Coprosterol [360-68-9]
Coptine [1391-14-6]
Coptisine [3486-66-6]
Cord Factor [61512-20-7]
Cordycepin [73-03-0]
Coriamyrtin [2571-86-0]
Coroxon [321-54-0]
Corticosterone [50-22-6]
Cortisone [53-06-5]
Cortisone Acetate* [50-04-4]
Cortisone, 21β-Cyclopentanepropio-
  nate [509-00-2]
Cortisone Phosphate [508-95-2]
Cortivazol [1110-40-3]
Cortol [516-38-1]
Cortolone [516-42-7]
Corybulbine [518-77-4]
Corycavamine [521-85-7]
Corycavidine [521-93-7]
Corydaldine [493-49-2]
Corydaline [518-69-4]
Corydine [476-69-7]
Corynanthine [18904-54-6]
Corynanthine [483-10-3]
Corypalmine [6018-40-2]
Corytuberine [517-56-6]
Cosyntropin [16960-16-0]
Cotarnine [82-54-2]
Cotarnine Chloride* [10018-19-6]
Cotarnine Hydrochloride*
  [36647-02-6]
Cotarnine Phthalate* [6190-36-9]
Cotinine [486-56-6]
Cotinine Fumarate* [5695-98-7]
Cotoin [479-21-0]
Coumachlor [81-82-3]
Coumafuryl [117-52-2]
Coumalic Acid [500-05-0]

Coumaphos [56-72-4]
Coumaran [496-16-2]
p-Coumaric Acid [7400-08-0]
Coumarilic Acid [496-41-3]
Coumarin [91-64-5]
Coumarin-3-carboxylic Acid
  [531-81-7]
Coumestrol [479-13-0]
Coumetarol [4366-18-1]
Coumingine [26241-81-6]
Coumithoate [572-48-5]
Creatine [57-00-1]
Creatinine [60-27-5]
Creolin® [12751-04-1]
Creosol [93-51-6]
Creosote, Coal Tar [8001-58-9]
Creosote, Wood [8021-39-4]
m-Cresol [108-39-4]
o-Cresol [95-48-7]
p-Cresol [106-44-5]
o-Cresolphthalein [596-27-0]
Cresol Red [1733-12-6]
m-Cresotic Acid [50-85-1]
o-Cresotic Acid [83-40-9]
p-Cresotic Acid [89-56-5]
m-Cresyl Acetate [122-46-3]
o-Cresyl Acetate [533-18-6]
CRF [9015-71-8]
Crimidine [535-89-7]
Croceic Acid [132-57-0]
Crocetin [27876-94-4]
Crocin [42553-65-1]
Cromolyn [16110-51-3]
Cromolyn Sodium* [15826-37-6]
Cropropamide [633-47-6]
Crotamine [37196-57-9]
Crotamiton [483-63-6]
Crotethamide [6168-76-9]
Crotonaldehyde [4170-30-3]
Crotonic Acid [3724-65-0]
Crotoxin [9007-40-3]
Crotoxyphos [7700-17-6]
Crotyl Alcohol [6117-91-5]
CRP [9007-41-4]
Crufomate [299-86-5]
Cryofluorane [76-14-2]
Cryolite [15096-52-3]
Cryptopine [482-74-6]
Cryptoxanthin [472-70-8]
Cubane [277-10-1]
Cubebin [18423-69-3]
Cumene [98-82-8]
Cumic Acid [536-66-3]
Cumic Alcohol [536-60-7]
Cumidine [99-88-7]
Cuminaldehyde [122-03-2]
Cuminaldehyde Thiosemicarbazone
  [3811-20-9]
Cupferron [135-20-6]
Cupreine [524-63-0]
Cupric Acetate [142-71-2]
Cupric Acetate, Basic [52503-64-7]
Cupric Acetoarsenite [12002-03-8]
Cupric Arsenite [10290-12-7]
Cupric Borate [39290-85-2]
Cupric Bromide [7789-45-9]
Cupric Butyrate [540-16-9]
Cupric Carbonate, Basic [12069-69-1]
Cupric Chlorate [14721-21-2]
Cupric Chloride [7447-39-4]
Cupric Chromate(VI) [13548-42-0]
Cupric Chromite [12018-10-9]
Cupric Citrate [866-82-0]
Cupric Ferrocyanide [13601-13-3]
Cupric Fluoride [7789-19-7]
Cupric Formate [544-19-4]
Cupric Gluconate [527-09-3]
Cupric Glycinate [13479-54-4]
Cupric Hexafluorosilicate
  [12062-24-7]
Cupric Hydroxide [20427-59-2]

Cupric Nitrate [3251-23-8]
Cupric Oleate [1120-44-1]
Cupric Oxalate [814-91-5]
Cupric Oxide [1317-38-0]
Cupric Perchlorate [13770-18-8]
Cupric p-Phenolsulfonate [547-56-8]
Cupric Phosphate [7798-23-4]
Cupric Salicylate [16048-96-7]
Cupric Selenate [15123-69-0]
Cupric Selenide [1317-41-5]
Cupric Selenite [10214-40-1]
Cupric Stearate [660-60-6]
Cupric Sulfate [7758-98-7]
Cupric Sulfate, Basic [1332-14-5]
Cupric Sulfide [1317-40-4]
Cupric Tartrate [815-82-7]
Cupric Tungstate(VI) [13587-35-4]
Cuprobam [7076-63-3]
Cuprous Acetate [598-54-9]
Cuprous Bromide [7787-70-4]
Cuprous Chloride [7758-89-6]
Cuprous Cyanide [544-92-3]
Cuprous Iodide [7681-65-4]
Cuprous Mercuric Iodide
  [13876-85-2]
Cuprous Oxide [1317-39-1]
Cuprous Potassium Cyanide
  [13682-73-0]
Cuprous Selenide [20405-64-5]
Cuprous Sulfide [22205-45-4]
Cuprous Sulfite [13982-53-1]
Cuprous Thiocyanate [1111-67-7]
Cuproxoline [13007-93-7]
Curare [8063-06-7]
C-Curarine I [7168-64-1]
C-Curarine III [6866-93-9]
Curcumin [458-37-7]
Curine [436-05-5]
Curium [7440-51-9]
Curvularin [10140-70-2]
Cuscohygrine [454-14-8]
Cuspareine [442-33-1]
Cusparine [529-92-0]
Cyacetacide [140-87-4]
Cyamelide [462-02-2]
Cyamemazine [3546-03-0]
Cyanamide [420-04-2]
Cyanazine [21725-46-2]
Cyanic Acid [420-05-3]
Cyanidin Chloride [528-58-5]
Cyanoacetamide [107-91-5]
Cyanoacetic Acid [372-09-8]
Cyanofenphos [13067-93-1]
Cyanogen [460-19-5]
Cyanogen Azide [764-05-6]
Cyanogen Bromide [506-68-3]
Cyanogen Chloride [506-77-4]
Cyanogen Iodide [506-78-5]
Cyanuric Acid [108-80-5]
Cycasin [14901-08-7]
Cyclacillin [3485-14-1]
Cyclamic Acid [100-88-9]
Cyclandelate [456-59-7]
Cyclarbamate [5779-54-4]
Cyclazocine [3572-80-3]
Cyclethrin [97-11-0]
Cyclexanone [15301-52-7]
Cyclexedrine [532-52-5]
Cyclic AMP [60-92-4]
Cyclic GMP [7665-99-8]
Cyclizine [82-92-8]
Cyclizine Hydrochloride* [303-25-3]
Cyclobarbital [52-31-3]
Cyclobendazole [31431-43-3]
Cyclobenzaprine [303-53-7]
Cyclobenzaprine Hydrochloride*
  [6202-23-9]
Cyclobutane [287-23-0]
Cyclobutyrol [512-16-3]
Cyclobuxine [2241-90-9]
Cyclocumarol [518-20-7]

Cyclodrine [52109-93-0]
Cyclofenil [2624-43-3]
Cycloguanil [516-21-2]
Cycloguanil Pamoate* [609-78-9]
Cycloheptanone [502-42-1]
Cyclohexane [110-82-7]
Cyclohexanecarboxylic Acid
  [98-89-5]
Cyclohexanol [108-93-0]
Cyclohexanone [108-94-1]
Cyclohexene [110-83-8]
Cycloheximide [66-81-9]
Cyclohexylamine [108-91-8]
Cyclohexyl Bromide [108-85-0]
Cyclohexylcarbinol [100-49-2]
Cyclohexyl Chloride [542-18-7]
2-Cyclohexyl-4,6-dinitrophenol
  [131-89-5]
Cycloleucine [52-52-8]
Cyclomethycaine [139-62-8]
Cyclomethycaine Hydrochloride*
  [537-61-1]
Cyclonite [121-82-4]
Cyclonium Iodide [6577-41-9]
Cyclopentadiene [542-92-7]
Cyclopentamine [102-45-4]
Cyclopentamine Hydrochloride*
  [3459-06-1]
Cyclopentane [287-92-3]
Cyclopentanol [96-41-3]
Cyclopentanone [120-92-3]
1,2-Cyclopentenophenanthrene
  [482-66-6]
Cyclopenthiazide [742-20-1]
Cyclopentobarbital [76-68-6]
Cyclopentolate [512-15-2]
Cyclopentolate Hydrochloride*
  [5870-29-1]
Cyclophosphamide [50-18-0]
Cyclopregnol [465-53-2]
Cyclopropane [75-19-4]
Cyclopropyl Methyl Ether [540-47-6]
Cyclorphan [4163-15-9]
Cycloserine [68-41-7]
Cyclosporin A* [59865-13-3]
Cyclosporin G* [74436-00-3]
Cyclothiazide [2259-96-3]
Cyclovalone [579-23-7]
Cycothiamin(e) [6092-18-8]
Cycrimine Hydrochloride [126-02-3]
Cyfluthrin [68359-37-5]
Cyhalothrin [68085-85-8]
Cyheptamide [7199-29-3]
Cyhexatin [13121-70-5]
Cymarin [508-77-0]
Cymarose [579-04-4]
Cymene [25155-15-1]
Cymiazole [61676-87-7]
Cynanchogenin [6870-10-6]
Cynarin(e) [30964-13-7]
Cyoctol [89672-11-7]
Cypermethrin [52315-07-8]
Cyphenothrin [39515-40-7]
Cyprenorphine [4406-22-8]
Cyprenorphine Hydrochloride*
  [16550-22-4]
Cyproheptadine [129-03-3]
Cyproheptadine Hydrochloride*
  [41354-29-4]
Cyproquinate [19485-08-6]
Cyproterone [2098-66-0]
Cyproterone Acetate* [427-51-0]
Cyromazine [66215-27-8]
Cystamine [51-85-4]
L-Cystathionine [56-88-2]
Cysteamine [60-23-1]
Cysteic Acid [13100-82-8]
Cysteine [52-90-4]
Cysteine Hydrochloride* [52-89-1]
Cystine [56-89-3]
Cytarabine [147-94-4]

Cythioate [115-93-5]
Cytidine [65-46-3]
2'-Cytidylic Acid [85-94-9]
3'-Cytidylic Acid [84-52-6]
Cytisine [485-35-8]
Cytochrome c [9007-43-6]
Cytohemin [19554-22-4]
Cytolipin H [4682-48-8]
Cytosine [71-30-7]
2,4-D [94-75-7]
Dacarbazine [4342-03-4]
Dactinomycin [50-76-0]
Daidzein [486-66-8]
Dalapon [75-99-0]
Daltroban [79094-20-5]
Damar [9000-16-2]
Damascenine [483-64-7]
Daminozide [1596-84-5]
Danazol [17230-88-5]
Dansyl Chloride [605-65-2]
Danthron [117-10-2]
Dantrolene [7261-97-4]
Dantrolene Sodium Hemiheptahy-
  drate* [24868-20-0]
Daphnandrine [1183-76-2]
Daphnetin [486-35-1]
Daphnin [486-55-5]
Daphnoline [479-36-7]
Dapiprazole [72822-12-9]
Dapsone [80-08-0]
Darvan® [9003-24-1]
Datiscetin [480-15-9]
Daucol [887-08-1]
Daunorubicin [20830-81-3]
Daunorubicin Hydrochloride*
  [23541-50-6]
Dauricine [524-17-4]
Dazomet [533-74-4]
2,4-DB [94-82-6]
DBMC [497-39-2]
DCPA [1861-32-1]
DDD (Analytical) [6088-51-3]
DDT [50-29-3]
Deaminooxytocin [113-78-0]
Deanol [108-01-0]
Deanol Aceglumate [3342-61-8]
Deanol Acetamidobenzoate
  [3635-74-3]
Debrisoquin [1131-64-2]
Debrisoquin Sulfate* [581-88-4]
Decaborane(14) [17702-41-9]
Decalin® [91-17-8]
Decamethonium Bromide [541-22-0]
Decamethylcyclopentasiloxane
  [541-02-6]
Decamethylene Glycol [112-47-0]
Decamethyltetrasiloxane [141-62-8]
Dechlorane® Plus [13560-89-9]
Decimemide [14817-09-5]
Decoquinate [18507-89-6]
n-Decyl Alcohol [112-30-1]
Deet [134-62-3]
Defensins [103220-14-0]
Deferoxamine [70-51-9]
Deferoxamine Hydrochloride*
  [1950-39-6]
Deferoxamine Mesylate* [138-14-7]
Defibrotide [83712-60-1]
Deflazacort [14484-47-0]
Defosfamide [3733-81-1]
Deguelin [522-17-8]
Dehydroacetic Acid [520-45-6]
Dehydroascorbic Acid [490-83-5]
7-Dehydrocholesterol [434-16-2]
Dehydrocholic Acid [81-23-2]
Dehydrocholic Acid Sodium*
  [145-41-5]
11-Dehydrocorticosterone [72-23-1]
Dehydroemetine [4914-30-1]
Dehydroergosterol [516-85-8]
3-Dehydroretinal [472-87-7]

7-Dehydrositosterol [521-04-0]
Delapril [83435-66-9]
Delapril Hydrochloride*
  [83435-67-0]
Delmadinone Acetate [13698-49-2]
Delphinidin [528-53-0]
Delphinine [561-07-9]
Delsoline [509-18-2]
Deltamethrin [52918-63-5]
Demanyl Phosphate [6909-62-2]
Demecarium Bromide [56-94-0]
Demeclocycline [127-33-3]
Demeclocycline Hydrochloride*
  [64-73-3]
Demecolcine [477-30-5]
Demegestone [10116-22-0]
Demeton [8065-48-3]
Demexiptiline [24701-51-7]
Denatonium Benzoate [3734-33-6]
Denopamine [71771-90-9]
Denopterin [22006-84-4]
6-Deoxy-L-ascorbic Acid [528-81-4]
Deoxycholic Acid [83-44-3]
Deoxycorticosterone [64-85-7]
Deoxycorticosterone Acetate
  [56-47-3]
Deoxydihydrostreptomycin
  [26086-49-7]
Deoxyepinephrine [501-15-5]
2-Deoxy-D-glucose [154-17-6]
1-Deoxynojirimycin [19103-96-2]
Deoxyribonuclease (Pancreatic)
  [9003-98-9]
D-2-Deoxyribose [533-67-5]
2-Deoxystreptamine [2037-48-1]
Deoxyuridine [951-78-0]
Deprenyl [2323-36-6]
Deptropine Citrate* [2169-75-7]
Dequalinium Acetate [4028-98-2]
Dequalinium Chloride [522-51-0]
Dermostatin [11120-15-3]
Desaspidin [114-43-2]
Desatrine [6746-01-6]
Deserpidine [131-01-1]
Desipramine [50-47-5]
Desipramine Hydrochloride*
  [58-28-6]
Deslanoside [17598-65-1]
Desmopressin [16679-58-6]
Desmosterol [313-04-2]
Desogestrel [54024-22-5]
Desomorphine [427-00-9]
Desonide [638-94-8]
Desosamine [5779-39-5]
Desoximetasone [382-67-2]
6-Desoxy-D-glucosamine [6018-53-7]
11-Desoxy-17-hydroxycorticosterone
  [152-58-9]
4-Desoxypyridoxine Hydrochloride
  [148-51-6]
Desthiobiotin [533-48-2]
Destomycin A [14918-35-5]
Detaxtran [9015-73-0]
Detomidine [76631-46-4]
Deuterium [7782-39-0]
Deuterium Oxide [7789-20-0]
Devarda's Metal [8049-11-4]
Dexamethasone [50-02-2]
Dexamethasone Acetate* [1177-87-3]
Dexamethasone Sodium Phosphate*
  [2392-39-4]
Dexetimide [21888-98-2]
Dexpanthenol [81-13-0]
Dextran [9004-54-0]
Dextranase [9025-70-1]
Dextranomer [56087-11-7]
Dextran Sulfate Sodium [9011-18-1]
Dextri-Maltose® [8006-91-5]
Dextrin [9004-53-9]
Dextroamphetamine Sulfate [51-63-8]
Dextromoramide [357-56-2]

Denotes derivative of title compound

Dextromoramide Tartrate* [2922-44-3]
Dezocine [53648-55-8]
DFDD [603-55-4]
DFDT [475-26-3]
Dhurrin [499-20-7]
Diaboline [509-40-0]
Diacerein [13739-02-1]
Diacetazotol [83-63-6]
Diacetin [25395-31-7]
Diacetonamine [625-04-7]
Diacetone Acrylamide [2873-97-4]
Diacetone Alcohol [123-42-2]
Diacetoneglucose [582-52-5]
Diacetyl [431-03-8]
Diacetyldihydromorphine [509-71-7]
Diacetylmorphine [561-27-3]
Dialifor [10311-84-9]
Diallate [2303-16-4]
Diallylamine [124-02-7]
Diallylcyanamide [538-08-9]
Diamfenetide [36141-82-9]
p-Diaminoazobenzene [538-41-0]
3,5-Diaminobenzoic Acid [535-87-5]
2,6-Diamino-2'-butyloxy-3,5'-azo-pyridine [617-19-6]
4,4'-Diaminodiphenylamine [537-65-5]
p,p'-Diaminodiphenylmethane [101-77-9]
2,4-Diamino-6-hydroxypyrimidine [56-06-4]
2,3-Diaminophenazine [655-86-7]
2,4-Diaminophenol [95-86-3]
2,3-Diaminopropionic Acid [515-94-6]
2,6-Diaminopurine [1904-98-9]
Diamond [7782-40-3]
Diamond Ink [8006-35-7]
Diampromide [552-25-0]
Diamthazole Dihydrochloride [136-96-9]
Diamyl Sodium Sulfosuccinate [922-80-5]
1,2-Dianilinoethane [150-61-8]
Dianisidine [119-90-4]
Diastase of Malt [9000-92-4]
Diathymosulfone [5964-62-5]
Diatretyne I [53318-35-7]
Diatretyne II [3625-68-1]
Diatrizoate Sodium [737-31-5]
Diaveridine [5355-16-8]
Diazepam [439-14-5]
Diazinon® [333-41-5]
Diaziquone [57998-68-2]
Diazoacetic Ester [623-73-4]
Diazoaminobenzene [136-35-6]
p-Diazobenzenesulfonic Acid [305-80-6]
Diazomethane [334-88-3]
6-Diazo-5-oxo-L-norleucine [157-03-9]
5-Diazouracil [2435-76-9]
Diazoxide [364-98-7]
Dibekacin [34493-98-6]
Dibenzalacetone [538-58-9]
1,2:5,6-Dibenzanthracene [53-70-3]
Dibenzepin [4498-32-2]
Dibenzepin Hydrochloride* [315-80-0]
Dibenzoylmethane [120-46-7]
2,3:6,7-Dibenzphenanthrene [222-93-5]
Dibenzylamine [103-49-1]
Dibenzyl Chlorophosphonate [538-37-4]
Dibenzyl Disulfide [150-60-7]
Dibenzyl Phosphite [17176-77-1]
Diborane(6) [19287-45-7]
Diboron Tetrachloride [13701-67-2]
Diboron Tetrahydroxide [13675-18-8]

9,10-Dibromoanthracene [523-27-3]
p-Dibromobenzene [106-37-6]
α,α'-Dibromo-d-camphor [514-12-5]
Dibromochloropropane [96-12-8]
1,2-Dibromo-2,4-dicyanobutane [35691-65-7]
4',5'-Dibromofluorescein [596-03-2]
Dibromogallic Acid [602-92-6]
3,5-Dibromo-4-hydroxybenzenesul-fonic Acid [4232-99-9]
Dibromopropamidine [496-00-4]
Dibromopropamidine Isethionate* [614-87-9]
2,3-Dibromopropene [513-31-5]
2,6-Dibromoquinone-4-chlorimide [537-45-1]
3,5-Dibromosalicylaldehyde [90-59-5]
3,5-Dibromosalicylic Acid [3147-55-5]
2,3-Dibromosuccinic Acid [526-78-3]
3,5-Dibromo-L-tyrosine [300-38-9]
Dibromsalicil [523-88-6]
Dibucaine Hydrochloride [61-12-1]
Dibunate Sodium [39315-52-1]
Dibutoline Sulfate [532-49-0]
n-Dibutylamine [111-92-2]
Di-tert-butyl Ether [6163-66-2]
Di-tert-butyl Malonate [541-16-2]
2,6-Di-tert-butylpyridine [585-48-8]
Dibutyl Succinate [141-03-7]
Di-tert-butyl Succinate [926-26-1]
Dibutyltin Dilaurate [77-58-7]
Dicamba [1918-00-9]
Dicapthon [2463-84-5]
Dicentrine [517-66-8]
Dichlobenil [1194-65-6]
Dichlofenthion [97-17-6]
Dichlofluanid [1085-98-9]
Dichlone [117-80-6]
Dichloralphenazone [480-30-8]
Dichloramine T [473-34-7]
Dichlorisone [7008-26-6]
Dichlorisone Acetate* [79-61-8]
Dichlorisoproterenol [59-61-0]
Dichloroacetic Acid [79-43-6]
1,1-Dichloroacetone [513-88-2]
1,3-Dichloroacetone [534-07-6]
2,2-Dichloroacetyl Chloride [79-36-7]
3,4-Dichloroaniline [95-76-1]
Dichlorobenzalkonium Chloride [102-30-7]
m-Dichlorobenzene [541-73-1]
o-Dichlorobenzene [95-50-1]
p-Dichlorobenzene [106-46-7]
2,2'-Dichlorobenzidine [84-68-4]
3,3'-Dichlorobenzidine [91-94-1]
Dichlorobenzyl Alcohol [1777-82-8]
1,1-Dichloro-2,2-bis(p-chlorophenyl)-ethane [72-54-8]
1,1-Dichloro-2,2-bis(p-ethylphenyl)-ethane [72-56-0]
Dichloro(2-chlorovinyl)arsine [541-25-3]
2,3-Dichloro-5,6-dicyanobenzoqui-none [84-58-2]
Dichlorodifluoromethane [75-71-8]
1,3-Dichloro-5,5-dimethylhydantoin [118-52-5]
sym-Dichloroethyl Ether [111-44-4]
4',5'-Dichlorofluorescein [2320-96-9]
2,6-Dichloroindophenol Sodium [620-45-1]
sym-Dichloromethyl Ether [542-88-1]
Dichlorophen(e) [97-23-4]
Dichlorophenarsine Hydrochloride [536-29-8]
2,4-Dichlorophenol [120-83-2]
2,6-Dichlorophenol [87-65-0]
1,3-Dichloro-2-propanol [96-23-1]
1,3-Dichloropropene [542-75-6]

Dichlororiboflavin [521-96-0]
Dichloroxylenol [133-53-9]
Dichlorphenamide [120-97-8]
Dichlorprop [120-36-5]
Dichlorvos [62-73-7]
Diclobutrazol [75736-33-3]
Diclofenac Sodium [15307-79-6]
Diclofop-Methyl [51338-27-3]
Dicloxacillin [3116-76-5]
Dicloxacillin Sodium Monohydrate* [13412-64-1]
Dicobalt Octacarbonyl [10210-68-1]
Dicofol [115-32-2]
Dicolinium Iodide [382-82-1]
Dicrotophos [141-66-2]
Dicryl [2164-09-2]
Dictamnine [484-29-7]
Dicumarol [66-76-2]
Dicyanine [52260-69-2]
Dicyanodiamide [461-58-5]
Dicyanodiamidine Sulfate [591-01-5]
9-Dicyanomethylene-2,4,7-trinitroflu-orene [1172-02-7]
Dicyclohexylamine [101-83-7]
Dicyclohexylcarbodiimide [538-75-0]
Dicyclomine Hydrochloride [67-92-5]
Didecyldimethylammonium Chloride [7173-51-5]
Dideoxyadenosine [4097-22-7]
Dideoxycytidine [7481-89-2]
Dideoxyinosine [69655-05-6]
Dieldrin [60-57-1]
Dienestrol [84-17-3]
Dienochlor [2227-17-0]
Diethadione [702-54-5]
Diethanolamine [111-42-2]
Diethazine [60-91-3]
Diethazine Hydrochloride* [341-70-8]
Diethylacetic Acid [88-09-5]
Diethylamine [109-89-7]
2-Diethylaminoethanol [100-37-8]
Diethylaniline [91-66-7]
N,N-Diethylbenzhydrylamine [519-72-2]
Diethylberyllium [542-63-2]
Diethylbromoacetamide [511-70-6]
Diethylcarbamazine [90-89-1]
Diethylcarbamazine Citrate* [1642-54-2]
N,N'-Diethylcarbanilide [85-98-3]
Diethyl Carbitol® [112-36-7]
Diethylene Glycol [111-46-6]
Diethylene Glycol Monolaurate [141-20-8]
Diethyl Ketone [96-22-0]
Diethylmagnesium [557-18-6]
Diethyl Maleate [141-05-9]
Diethylmalonic Acid [510-20-3]
Diethyl Oxalate [95-92-1]
Diethylpropion [90-84-6]
Diethylpropion Hydrochloride* [134-80-5]
Diethylsilane [542-91-6]
Diethylstilbestrol [56-53-1]
Diethylstilbestrol Dipropionate [130-80-3]
Diethyl Sulfate [64-67-5]
Diethylzinc [557-20-0]
Difemerine [80387-96-8]
Difenamizole [20170-20-1]
Difenoxin [28782-42-5]
Difenpiramide [51484-40-3]
Diflorasone [2557-49-5]
Diflorasone Diacetate* [33564-31-7]
Difloxacin [98106-17-3]
Difloxacin Hydrochloride* [91296-86-5]
Diflubenzuron [35367-38-5]
Diflucortolone [2607-06-9]
Diflunisal [22494-42-4]

2,4-Difluoroaniline [367-25-9]
p-Difluorobenzene [540-36-3]
4,4'-Difluorodiphenyl [398-23-2]
Difluprednate [23674-86-4]
Digallic Acid [536-08-3]
Digalogenin [6877-35-6]
Diginatigenin [559-57-9]
Diginatin [52589-12-5]
Diginin [467-53-8]
Digitalin [752-61-4]
Digitalose [4481-08-7]
Digitogenin [511-34-2]
Digitonin [11024-24-1]
Digitoxigenin [143-62-4]
Digitoxin [71-63-6]
Digitoxose [527-52-6]
Diglyme [111-96-6]
Digoxigenin [1672-46-4]
Digoxin [20830-75-5]
Dihexyverine [561-77-3]
Dihexyverine Hydrochloride*
 [5588-25-0]
Dihydralazine [484-23-1]
Dihydrazine Sulfate [13464-80-7]
Dihydrocodeine [125-28-0]
Dihydrocodeinone Enol Acetate
 [466-90-0]
Dihydroequilin [3563-27-7]
Dihydroergotamine [511-12-6]
Dihydroergotamine Mesylate*
 [6190-39-2]
Dihydro-β-erythroidine [23255-54-1]
Dihydroisocodeine [795-38-0]
Dihydromorphine [509-60-4]
Dihydroresorcinol [504-02-9]
Dihydrostreptomycin [128-46-1]
Dihydrotachysterol [67-96-9]
Dihydrothebaine [561-25-1]
Dihydrovitamin K₁ [572-96-3]
Dihydroxyacetone [96-26-4]
Dihydroxyaluminum Acetylsalicylate
 [53230-06-1]
Dihydroxyaluminum Aminoacetate
 [13682-92-3]
Dihydroxyaluminum Sodium Carbon-
 ate [539-68-4]
Dihydroxymaleic Acid [526-84-1]
9,10-Dihydroxystearic Acid
 [120-87-6]
Dihydroxytartaric Acid [76-30-2]
2,4-Diiodoaniline [533-70-0]
4',5'-Diiodofluorescein [38577-97-8]
3,5-Diiodosalicylic Acid [133-91-5]
3,5-Diiodothyronine [534-51-0]
3,5-Diiodotyrosine [66-02-4]
Diisoamylamine [544-00-3]
Diisobutyl Sodium Sulfosuccinate
 [127-39-9]
Diisopromine [5966-41-6]
Diisopromine Hydrochloride*
 [24358-65-4]
Diisopropylamine [108-18-9]
Diisopropylamine Dichloroacetate
 [660-27-5]
Diisopropyl Paraoxon [3254-66-8]
Dikegulac [18467-77-1]
Dilazep [35898-87-4]
Dilevalol [75659-07-3]
Dilevalol Hydrochloride*
 [75659-08-4]
Diloxanide [579-38-4]
Diltiazem [42399-41-7]
Diltiazem Hydrochloride*
 [33286-22-5]
Dimecrotic Acid [7706-67-4]
Dimefline [1165-48-6]
Dimefline Hydrochloride*
 [2740-04-7]
Dimefox [115-26-4]
Dimemorfan [36309-01-0]
Dimenhydrinate [523-87-5]

Dimenoxadol [509-78-4]
Dimepheptanol [545-90-4]
Dimercaprol [59-52-9]
2,3-Dimercapto-1-propanesulfonic
 Acid [74-61-3]
Dimestrol [130-79-0]
Dimetacrine [4757-55-5]
Dimetan® [122-15-6]
Dimethadione [695-53-4]
Dimethazan [519-30-2]
Dimethazone [81777-89-1]
Dimethindene [5636-83-9]
Dimethindene Maleate* [3614-69-5]
Dimethirimol [5221-53-4]
Dimethisoquin [86-80-6]
Dimethisterone [79-64-1]
Dimethoate [60-51-5]
Dimethocaine [94-15-5]
Dimethoxanate [477-93-0]
Dimethoxanate Hydrochloride*
 [518-63-8]
Dimethoxane [828-00-2]
1,2-Dimethoxyethane [110-71-4]
2,6-Dimethoxyquinone [530-55-2]
Dimethylacetal [534-15-6]
N,N-Dimethylacetamide [127-19-5]
Dimethylamine [124-40-3]
p-Dimethylaminoazobenzene
 [60-11-7]
p-Dimethylaminobenzaldehyde
 [100-10-7]
p-Dimethylaminobenzalrhodanine
 [536-17-4]
4-(Dimethylamino)benzoic Acid
 [619-84-1]
p-Dimethylaminobenzophenone
 [530-44-9]
N,N-Dimethylaniline [121-69-7]
9,10-Dimethyl-1,2-benzanthracene
 [57-97-6]
5,6-Dimethylbenzimidazole
 [582-60-5]
p,α-Dimethylbenzyl Alcohol
 [536-50-5]
Dimethylberyllium [506-63-8]
2,3-Dimethyl-1,3-butadiene
 [513-81-5]
Dimethylcadmium [506-82-1]
Dimethyl Carbate [39589-98-5]
5,5-Dimethyl-1,3-cyclohexanedione
 [126-81-8]
N,N-Dimethylformamide [68-12-2]
N,N-Dimethylglycine [1118-68-9]
N,N-Dimethylglycine Hydrazide
 Hydrochloride [539-64-0]
Dimethylglyoxime [95-45-4]
1,1-Dimethylhydrazine [57-14-7]
1,2-Dimethylhydrazine [540-73-8]
Dimethylmercury [593-74-8]
Dimethyl(methylene)ammonium
 Iodide
 [33797-51-2]
N,N-Dimethyl-1-naphthylamine
 [86-56-6]
Dimethylolpropionic Acid
 [4767-03-7]
Dimethyl-p-phenylenediamine
 [99-98-9]
Dimethyl Phthalate [131-11-3]
Dimethyl Sulfate [77-78-1]
2,4-Dimethylsulfolane [1003-78-7]
Dimethyl Sulfone [67-71-0]
Dimethyl Sulfoxide [67-68-5]
Dimethylthiambutene [524-84-5]
2,4-Dimethylthiazole [541-58-2]
N,N'-Dimethylthiourea [534-13-4]
N,N-Dimethyltryptamine [61-50-7]
Dimethylzinc [544-97-8]
Dimetilan [644-64-4]
Dimetofrine [22950-29-4]
Dimetridazole [551-92-8]

Diminazene Aceturate [908-54-3]
Dimorpholamine [119-48-2]
Dimoxyline [147-27-3]
Dimsyl [13810-16-7]
β,β'-Dinaphthylamine [532-18-3]
Diniconazole [83657-24-3]
Dinitolmide [148-01-6]
2,4-Dinitroaniline [97-02-9]
2,6-Dinitroaniline [606-22-4]
2,4-Dinitrobenzaldehyde [528-75-6]
2,4-Dinitrobenzenesulfenyl Chloride
 [528-76-7]
3,4-Dinitrobenzoic Acid [528-45-0]
3,5-Dinitrobenzoic Acid [99-34-3]
3,5-Dinitrobenzoyl Chloride
 [99-33-2]
4,4'-Dinitrocarbanilide [587-90-6]
Dinitrocresol [534-52-1]
3,7-Dinitro-5-oxophenothiazine
 [574-81-2]
2,4-Dinitrophenol [51-28-5]
2,5-Dinitrophenol [329-71-5]
2,6-Dinitrophenol [573-56-8]
2,4-Dinitrophenylhydrazine
 [119-26-6]
2,4-Dinitroresorcinol [519-44-8]
5,5'-Dinitrosalicil [528-10-9]
Dinobuton [973-21-7]
Dinocap [39300-45-3]
Dinoseb [88-85-7]
Dinosterol [58670-63-6]
Dinsed [96-62-8]
Diopterin [6807-82-5]
Dioscin [19057-60-4]
Dioscorine [3329-91-7]
Diosgenin [512-04-9]
Diosmetin [520-34-3]
Diosmin [520-27-4]
Diosphenol [490-03-9]
Dioxadrol [6495-46-1]
Dioxadrol Hydrochloride*
 [3666-69-1]
Dioxane [123-91-1]
Dioxaphetyl Butyrate [467-86-7]
Dioxathion [78-34-2]
Dioxethedrine [497-75-6]
Dioxybenzone [131-53-3]
Dioxypyramidon [519-65-3]
2,5-Di-tert-pentylhydroquinone
 [79-74-3]
Diperodon Hydrochloride [537-12-2]
Diphemanil Methylsulfate [62-97-5]
Diphemethoxidine [13862-07-2]
Diphenadione [82-66-6]
Diphenamid [957-51-7]
Diphenane [101-71-3]
Diphenazoline [6703-39-5]
Diphenhydramine [58-73-1]
Diphenhydramine Hydrochloride*
 [147-24-0]
Diphenic Acid [482-05-3]
Diphenicillin Sodium [2289-50-1]
Diphenidol [972-02-1]
Diphenidol Hydrochloride*
 [3254-89-5]
Diphenolic Acid [126-00-1]
Diphenoxylate [915-30-0]
Diphenoxylate Hydrochloride*
 [3810-80-8]
Diphenyl [92-52-4]
Diphenylacetamide [519-87-9]
Diphenylacetic Acid [117-34-0]
Diphenylamine [122-39-4]
Diphenylamine-2,2'-dicarboxylic
 Acid [579-92-0]
N,N'-Diphenylbenzidine [531-91-9]
1,4-Diphenyl-1,3-butadiene
 [886-65-7]
sym-Diphenylcarbazide [140-22-7]
Diphenylcarbazone [538-62-5]
1,1-Diphenylethene [530-48-3]

Denotes derivative of title compound

N-(1,2-Diphenylethyl)nicotinamide [553-06-0]
1,3-Diphenylguanidine [102-06-7]
1,1-Diphenylhydrazine [530-50-7]
Diphenylketene [525-06-4]
Diphenylmagnesium [555-54-4]
Diphenylmethane [101-81-5]
Diphenylmethane-4,4'-disulfonamide [535-66-0]
N,N'-Diphenyl-p-phenylenediamine [74-31-7]
1,1-Diphenyl-2-picrylhydrazyl (Free Radical) [1898-66-4]
α,α-Diphenyl-2-piperidinepropanol [510-07-6]
Diphenylpyraline [147-20-6]
Diphenylpyraline Hydrochloride* [132-18-3]
sym-Diphenylpyrophosphorodiamidic Acid [541-04-8]
Diphenyl Sulfone [127-63-9]
sym-Diphenylthiourea [102-08-9]
Diphetarsone [515-76-4]
Diphosgene [503-38-8]
Dipicrylamine [131-73-7]
Dipin [738-99-8]
Dipipanone [467-83-4]
Dipipanone Hydrochloride* [856-87-1]
Dipiproverine [117-30-6]
Dipivefrin [52365-63-6]
Dipivefrin Hydrochloride* [64019-93-8]
Diploicin [527-93-5]
Diponium Bromide [2001-81-2]
Diprenorphine [14357-78-9]
Dipropalin® [1918-08-7]
Dipropetryn [4147-51-7]
n-Dipropylamine [142-84-7]
Dipropyl Ketone [123-19-3]
5,5-Dipropyl-2,4-oxazolidinedione [512-12-9]
Diprotrizoate Sodium [129-57-7]
Dipyridamole [58-32-2]
α,α'-Dipyridyl [366-18-7]
γ,γ'-Dipyridyl [553-26-4]
Dipyrocetyl [486-79-3]
Dipyrone [5907-38-0]
Diquat Dibromide [85-00-7]
1,3-Di-6-quinolylurea [532-05-8]
Diresorcinol [531-02-2]
Disilane [1590-87-0]
Disodium Dihydrogen Hypophosphate [7782-95-8]
Disodium Phenyl Phosphate [3279-54-7]
Disophenol [305-85-1]
Disopyramide [3737-09-5]
Disopyramide Phosphate* [22059-60-5]
Disparlure [54910-51-9]
Distigmine Bromide [15876-67-2]
Disulfamide [671-88-5]
Disulfiram [97-77-8]
Disulfoton [298-04-4]
Disul-sodium [136-78-7]
Ditazol [18471-20-0]
Dithianone [3347-22-6]
Dithiazanine Iodide [514-73-8]
2,2'-Dithiobis[benzothiazole] [120-78-5]
2,4-Dithiobiuret [541-53-7]
4,4'-Dithiodimorpholine [103-34-4]
3,3-Dithiodipyridine Dihydrochloride [538-45-4]
Dithiosalicylic Acid [527-89-9]
1,4-Dithiothreitol [3483-12-3]
Dithizone [60-10-6]
Ditiocarb Sodium [148-18-5]
1,1-Di-p-tolylethane [530-45-0]
1,2-Di-p-tolylethane [538-39-6]

p-Ditolylmercury [537-64-4]
Diuron [330-54-1]
Divicine [32267-39-3]
Dixanthogen [502-55-6]
Dixyrazine [2470-73-7]
Dizocilpine [77086-21-6]
Dizocilpine Maleate* [77088-22-7]
Dizocilpine (−)-Form* [77086-19-2]
Djenkolic Acid [498-59-9]
DMPA [299-85-4]
Dobesilate Calcium [20123-80-2]
Dobutamine [34368-04-2]
Dobutamine Hydrochloride* [49745-95-1]
Docusate Calcium [128-49-4]
Docusate Sodium [577-11-7]
Docusate Potassium* [7491-09-0]
Dodecahedrane [4493-23-6]
Dodecamethylcyclohexasiloxane [540-97-6]
Dodecamethylpentasiloxane [141-63-9]
1-Dodecanol [112-53-8]
3-(1,3,5,7,9-Dodecapentaenyloxy)-1,2-propanediol [83248-46-8]
Dodecarbonium Chloride [100-95-8]
Dodemorph [1593-77-7]
Dodine [2439-10-3]
Doisynoestrol [15372-34-6]
Doisynolic Acid [482-49-5]
Domesticine [476-71-1]
Domiodol [61869-07-6]
Domiphen Bromide [538-71-6]
Domperidone [57808-66-9]
Donovan's Solution [8012-54-2]
Dopa [63-84-5]
Dopamine [51-61-6]
Dopamine Hydrochloride* [62-31-7]
Dopan [520-09-2]
Dopastin [37134-80-8]
Dopexamine [86197-47-9]
Dopexamine Hydrochloride* [86484-91-5]
Dothiepin [113-53-1]
Dothiepin Hydrochloride* [897-15-4]
Dowicide 9 [13347-42-7]
Doxapram [309-29-5]
Doxapram Monohydrochloride Monohydrate* [7081-53-0]
Doxazosin [74191-85-8]
Doxazosin Mesylate* [77883-43-3]
Doxefazepam [40762-15-0]
Doxenitoin [3254-93-1]
Doxepin [1668-19-5]
Doxepin Hydrochloride* [1229-29-4]
Doxepin (trans-Form Hydrochloride)* [3607-18-9]
Doxifluridine [3094-09-5]
Doxofylline [69975-86-6]
Doxorubicin [23214-92-8]
Doxorubicin Hydrochloride* [25316-40-9]
Doxycycline [17086-28-1]
Doxycycline Monohydrochloride Monohydrate* [24390-14-5]
Doxylamine [469-21-6]
Doxylamine Succinate* [562-10-7]
Drazoxolon [5707-69-7]
Drimenin [2326-89-8]
Drocarbil [900-77-6]
Drofenine [1679-76-1]
Dromostanolone Propionate [521-12-0]
Droperidol [548-73-2]
Droprenilamine [57653-27-7]
Dropropizine [17692-31-8]
Dropropizine L-Form* [99291-24-4]
Drosophilin A [484-67-3]
Drotebanol [3176-03-2]
Droxicam [90101-16-9]
Droxidopa [23651-95-8]

DSIP [62568-57-4]
DTBP [110-05-4]
Dulcin [150-69-6]
Durapatite [1306-06-5]
Durene [95-93-2]
Durohydroquinone [527-18-4]
Duroquinone [527-17-3]
Dyclonine [586-60-7]
Dyclonine Hydrochloride* [536-43-6]
Dydrogesterone [152-62-5]
Dymanthine [124-28-7]
Dymanthine Hydrochloride* [1613-17-8]
Dymixal® [8013-96-5]
Dynorphin [74913-18-1]
Dyphylline [479-18-5]
Dypnone [495-45-4]
Dysprosium [7429-91-6]
Ebimar® [9013-42-7]
Estradiol Enanthate* [4956-37-0]
Estradiol Undecylate* [3571-53-7]
Eburnamonine [474-00-0]
Ecgonidine [484-93-5]
Ecgonine [481-37-8]
Echinenone [432-68-8]
Echinochrome A [517-82-8]
Echinomycin [512-64-1]
Echinopsine [83-54-5]
Echinuline [1859-87-6]
Echitamine [6871-44-9]
Echothiophate Iodide [513-10-0]
Econazole [27220-47-9]
Econazole Nitrate* [68797-31-9]
ECTEOLA-Cellulose [9015-13-8]
Ectylurea [95-04-5]
Edestin [9007-57-2]
Edetate Calcium Disodium [62-33-9]
Edetate Disodium [139-33-3]
Edetate Sodium [64-02-8]
Edetate Trisodium [150-38-9]
Edetic Acid [60-00-4]
Edetic Acid Dipotassium Dihydrate* [25102-12-9]
Edifenphos [17109-49-8]
Edoxudine [15176-29-1]
Edrophonium Chloride [116-38-1]
EEDQ [16357-59-8]
Eflornithine [67037-37-0]
Eflornithine Hydrochloride* [96020-91-6]
Efloxate [119-41-5]
Efrotomycin [56592-32-6]
Eicosamethylnonasiloxane [2652-13-3]
5,8,11,14,17-Eicosapentaenoic Acid [10417-94-4]
Einsteinium [7429-92-7]
Elaidic Acid [112-79-8]
Elaiomycin [23315-05-1]
Elastase [9004-06-2]
Elcatonin [60731-46-6]
Eledoisin [69-25-0]
Element 104 [53850-36-5]
Element 105 [53850-35-4]
Element 106 [54038-81-2]
Element 107 [54037-14-8]
Elenolide [24582-91-0]
Ellagic Acid [476-66-4]
Ellipticine [519-23-3]
Elliptinium Acetate [58337-35-2]
Elliptone [478-10-4]
Elymoclavine [548-43-6]
Embelin [550-24-3]
Embramine [3565-72-8]
Emepronium Bromide [3614-30-0]
Emetamine [483-19-2]
Emetine [483-18-1]
Emodin [518-82-1]
Emorfazone [38957-41-4]
Emylcamate [78-28-4]
Enalapril [75847-73-3]

Enalapril Maleate* [76095-16-4]
Enalaprilat [76420-72-9]
Enallylpropymal [1861-21-8]
Enanthotoxin [20311-78-8]
Encainide [66778-36-7]
Encainide Hydrochloride*
  [66794-74-9]
Endobenzyline Bromide [52080-56-5]
Endosulfan [115-29-7]
Endothall [145-73-3]
Endothelin (human) [117399-94-7]
Endralazine [39715-02-1]
Endralazine Mesylate* [65322-72-7]
Endrin [72-20-8]
Enduracidin [11115-82-5]
Enfenamic Acid [23049-93-6]
Enflurane [13838-16-9]
Enilconazole [35554-44-0]
Enocitabine [55726-47-1]
Enoxacin [74011-58-8]
Enoximone [77671-31-9]
Enoxolone [471-53-4]
Enprofylline [41078-02-8]
Enprostil [73121-56-9]
Enrofloxacin [93106-60-6]
Enterobactin [28384-96-5]
Enterogastrone [9007-67-4]
Enterolactone [78473-71-9]
Enteromycin [3552-16-7]
Entprol [102-60-3]
Enviomycin [33103-22-9]
Enviroxime [72301-79-2]
Eosine I Bluish [548-24-3]
Eosine Yellowish—(YS) [17372-87-1]
Epalrestat [82159-09-9]
Epanolol [86880-51-5]
Eperisone [64840-90-0]
Ephedrine DL-Form* [90-81-3]
Ephedrine DL-Form Hydrochloride*
  [134-71-4]
Ephedrine L-Form Hydrochloride*
  [50-98-6]
Ephedrine L-Form Sulfate*
  [134-72-5]
Pseudoephedrine* [90-82-4]
Pseudoephedrine Hydrochloride*
  [345-78-8]
Pseudoephedrine Sulfate*
  [7460-12-0]
Epichlorohydrin [106-89-8]
Epicholestanol [516-95-0]
Epicholesterol [474-77-1]
Epicillin [26774-90-3]
16-Epiestriol [547-81-9]
Epimestrol [7004-98-0]
Epinephrine [51-43-4]
Epinephrine DL-Form* [329-65-7]
Epinephrine L-Form Bitartrate*
  [51-42-3]
Epinephrine d-Borate* [5579-16-8]
Epiquinidine [572-59-8]
Epiquinine [572-60-1]
Epirizole [18694-40-1]
Epirubicin [56420-45-2]
Epirubicin Hydrochloride*
  [56390-09-1]
Epithiazide [1764-85-8]
Epitiostanol [2363-58-8]
EPN [2104-64-5]
Eprazinone [10402-90-1]
Eprozinol [32665-36-4]
Eptazocine [72522-13-5]
EPTC [759-94-4]
Equilenin [517-09-9]
Equilin [474-86-2]
Equol [531-95-3]
Erbium [7440-52-0]
Erbium Sulfate [13478-49-4]
Erbon [136-25-4]
Erdin [3405-51-4]
Erdmann's Salt [13600-89-0]

Ergocornine [564-36-3]
Ergocorninine [564-37-4]
Ergocristine [511-08-0]
Ergocristinine [511-07-9]
Ergocryptine [511-09-1]
Ergocryptinine [511-10-4]
Ergoflavin [3101-51-7]
Ergometrinine [479-00-5]
Ergonovine [60-79-7]
Ergonovine Maleate* [129-51-1]
Ergosine [561-94-4]
Ergostane [511-20-6]
Ergostanol [6538-02-9]
α-Ergostenol [632-32-6]
β-Ergostenol [632-31-5]
γ-Ergostenol [516-78-9]
Ergosterol [57-87-4]
Ergotamine [113-15-5]
Ergotamine Tartrate* [379-79-3]
Ergotaminine [639-81-6]
Ergothioneine [497-30-3]
Ergotinine [8006-08-4]
Ergotoxine [8006-25-5]
Eriochrome® Black T [1787-61-7]
Eriodictyol [552-58-9]
Eritadenine [23918-98-1]
Erucic Acid [112-86-7]
Erythritol [149-32-6]
Erythritol Anhydride [564-00-1]
Erythrityl Tetranitrate [7297-25-8]
Erythrocentaurin [50276-98-7]
α-Erythroidine [466-80-8]
β-Erythroidine [466-81-9]
Erythromycin [114-07-8]
Erythromycin Ethylsuccinate*
  [41342-53-4]
Erythromycin Acistrate [96128-89-1]
Erythromycin Estolate [3521-62-8]
Erythromycin Glucoheptonate
  [23067-13-2]
Erythromycin Lactobionate
  [3847-29-8]
Erythromycin Propionate [134-36-1]
Erythromycin Stearate [643-22-1]
Erythrophlamine [511-00-2]
Erythrophleine [36150-73-9]
Erythropoietin [11096-26-7]
Erythropterin [7449-03-8]
D-Erythrose [583-50-6]
L-Erythrose [533-49-3]
D-Erythrose 4-Phosphate [585-18-2]
Erythrosine [16423-68-0]
L-Erythrulose [533-50-6]
Esaprazole [64204-55-3]
Escigenin [17806-68-7]
Escin [6805-41-0]
Esculetin [305-01-1]
Esculin [531-75-9]
Eseridine [25573-43-7]
Esmolol [81147-92-4]
Esmolol Hydrochloride* [81161-17-3]
Essential Fatty Acids [11006-87-4]
Estazolam [29975-16-4]
Estradiol [50-28-2]
Estradiol Dipropionate* [113-38-2]
Estradiol Valerate* [979-32-8]
α-Estradiol [57-91-0]
Estradiol Benzoate [50-50-0]
Estradiol 17β-Cypionate [313-06-4]
Estragole [140-67-0]
Estramustine [2998-57-4]
Estramustine Phosphate Sodium*
  [52205-73-9]
Estriol [50-27-1]
Estrone [53-16-7]
Etafedrine [48141-64-6]
Etafedrine Hydrochloride*
  [5591-29-7]
Etafenone [90-54-0]
Etamiphyllin [314-35-2]
Etamycin [299-20-7]

Etaqualone [7432-25-9]
Eterobarb [27511-99-5]
Etersalate [62992-61-4]
Ethacridine [442-16-0]
Ethacridine Lactate* [1837-57-6]
Ethacrynic Acid [58-54-8]
Ethacrynic Acid Sodium* [6500-81-8]
Ethadione [520-77-4]
Ethalfluralin [55283-68-6]
Ethambutol [74-55-5]
Ethambutol Dihydrochloride*
  [1070-11-7]
Ethamivan [304-84-7]
Ethamoxytriphetol [67-98-1]
Ethamsylate [2624-44-4]
Ethane [74-84-0]
Ethanearsonic Acid [507-32-4]
1,2-Ethanedisulfonic Acid [110-04-3]
1,2-Ethanedithiol [540-63-6]
Ethanethiol [75-08-1]
Ethanolamine [141-43-5]
Ethanolamine Oleate* [2272-11-9]
Ethaverine [486-47-5]
Ethaverine Hydrochloride*
  [985-13-7]
Ethchlorvynol [113-18-8]
Ethebenecid [1213-06-5]
Ethenzamide [938-73-8]
Ethephon [16672-87-0]
Ethiazide [1824-58-4]
Ethinamate [126-52-3]
Ethinyl Estradiol [57-63-6]
Ethiodized Oil [8008-53-5]
Ethion [563-12-2]
Ethionamide [536-33-4]
Ethionine [13073-35-3]
Ethiozin [64529-56-2]
Ethirimol [23947-60-6]
Ethisterone [434-03-7]
Ethofumesate [26225-79-6]
Ethoheptazine [77-15-6]
Ethoheptazine Citrate* [2085-42-9]
Ethohexadiol [94-96-2]
Ethopabate [59-06-3]
Ethoprop [13194-48-4]
Ethopropazine [522-00-9]
Ethopropazine Hydrochloride*
  [1094-08-2]
Ethosuximide [77-67-8]
Ethotoin [86-35-1]
Ethoxazene [94-10-0]
Ethoxazene Hydrochloride*
  [2313-87-3]
2-Ethoxyethanol [110-80-5]
2-Ethoxyethyl Acetate [111-15-9]
2-Ethoxynaphthalene [93-18-5]
Ethoxyquin [91-53-2]
Ethoxzolamide [452-35-7]
Ethybenztropine [524-83-4]
Ethyl Acetate [141-78-6]
Ethyl Acetoacetate [141-97-9]
Ethyl Acrylate [140-88-5]
Ethyl Alcohol [64-17-5]
Ethylamine [75-04-7]
Ethyl Aminobenzoate [94-09-7]
N-Ethylamphetamine [457-87-4]
Ethyl Amyl Ketone [106-68-3]
Ethylaniline [103-69-5]
Ethylbenzene [100-41-4]
Ethylbenzhydramine [642-58-0]
Ethyl Benzoate [93-89-0]
Ethyl Benzoylacetate [94-02-0]
α-Ethylbenzyl Alcohol [93-54-9]
Ethylbenzylaniline [92-59-1]
Ethyl Biscoumacetate [548-00-5]
Ethyl Bromide [74-96-4]
Ethyl α-Bromopropionate [535-11-5]
Ethyl tert-Butyl Ether [637-92-3]
Ethyl Butyrate [105-54-4]
Ethyl Caprate [110-38-3]
Ethyl Caproate [123-66-0]

Denotes derivative of title compound

Ethyl Caprylate [106-32-1]
Ethyl β-Carboline-3-carboxylate [74214-62-3]
Ethyl Carbonate [105-58-8]
Ethyl Cellulose [9004-57-3]
Ethyl Chloride [75-00-3]
Ethyl Chloroacetate [105-39-5]
Ethyl Chloroformate [541-41-3]
Ethyl α-Chloropropionate [535-13-7]
Ethyl Cyanoacetate [105-56-6]
Ethyl Dibunate [5560-69-0]
Ethyl Diethylmalonate [77-25-8]
Ethyldimethyl-9-octadecenylammonium Bromide [6458-13-5]
Ethylene [74-85-1]
Ethylene Bromohydrin [540-51-2]
Ethylene Chlorohydrin [107-07-3]
Ethylene Cyanohydrin [109-78-4]
Ethylenediamine [107-15-3]
Ethylene Dibromide [106-93-4]
Ethylene Dichloride [107-06-2]
Ethylene Glycol [107-21-1]
Ethylene Glycol Diacetate [111-55-7]
Ethylene Glycol Monoacetate [542-59-6]
Ethylene Oxide [75-21-8]
Ethylene Thiourea [96-45-7]
Ethylenimine [151-56-4]
Ethylestrenol [965-90-2]
Ethyl Ether [60-29-7]
Ethyl Formate [109-94-4]
2-Ethyl-1-hexanol [104-76-7]
Ethylhydrocupreine [522-60-1]
Ethylidene Chloride [75-34-3]
Ethylidene Diacetate [542-10-9]
Ethylidene Dicoumarol [1821-16-5]
Ethyl Iodide [75-03-6]
Ethyl Isobutyrate [97-62-1]
Ethyl Isothiocyanate [542-85-8]
Ethyl Isovalerate [108-64-5]
Ethyl Lactate [97-64-3]
Ethyl Laurate [106-33-2]
Ethyl Levulinate [539-88-8]
Ethyl Linoleate [544-35-4]
Ethyl Loflazepate [29177-84-2]
N-Ethylmaleimide [128-53-0]
Ethyl Malonate [105-53-3]
Ethylmercuric Chloride [107-27-7]
N-(Ethylmercuri)-p-toluenesulfonanilide [517-16-8]
Ethyl Methanesulfonate [62-50-0]
Ethyl Methyl Ether [540-67-0]
Ethylmethylthiambutene [441-61-2]
Ethylmorphine [76-58-4]
Ethyl Nitrite [109-95-5]
N-Ethyl-N-nitrosourea [759-73-9]
Ethylnorepinephrine [536-24-3]
Ethylnorepinephrine Hydrochloride* [3198-07-0]
Ethyl Oenanthate [106-30-9]
Ethyl Oxalacetate [108-56-5]
Ethylparaben [120-47-8]
Ethyl Pelargonate [123-29-5]
Ethyl Phenylacetate [101-97-3]
Ethyl Phosphorochloridite [589-57-1]
3-Ethyl-4-picoline [529-21-5]
4-Ethyl-2-picoline [536-88-9]
5-Ethyl-2-picoline [104-90-5]
1-Ethyl-3-piperidinol [13444-24-1]
5-Ethyl-5-(1-piperidyl)barbituric Acid [509-87-5]
Ethyl Propionate [105-37-3]
3-Ethylpyridine [536-78-7]
4-Ethylpyridine [536-75-4]
Ethyl Salicylate [118-61-6]
Ethyl Silicate [78-10-4]
Ethylstibamine [1338-98-3]
Ethyl Sulfate [540-82-9]
Ethyl Sulfide [352-93-2]
2-(Ethylsulfonyl)ethanol [513-12-2]
Ethyl Tartrate [87-91-2]

Ethyl Tartrate, Acid [608-89-9]
2-(Ethylthio)ethanol [110-77-0]
Ethyl p-Toluenesulfonate [80-40-0]
Ethyl Vanillin [121-32-4]
Ethynodiol [1231-93-2]
Ethynodiol Diacetate* [297-76-7]
Ethynylbenzene [536-74-3]
Etidocaine [36637-18-0]
Etidronic Acid [2809-21-4]
Etidronic Acid Disodium* [7414-83-7]
Etifelmin [341-00-4]
Etifoxine [21715-46-8]
Etilefrin [709-55-7]
Etiocholane [438-23-3]
Etiocholanic Acid [438-08-4]
Etiocobalamin [27792-36-5]
Etioporphyrin [26608-34-4]
Etiproston [59619-81-7]
Etiroxate [17365-01-4]
Etisazol [7716-60-1]
Etizolam [40054-69-1]
Etodolac [41340-25-4]
Etodroxizine [17692-34-1]
Etofenamate [30544-47-9]
Etofibrate [31637-97-5]
Etofylline [519-37-9]
Etofylline Nicotinate [13425-39-3]
Etoglucid [1954-28-5]
Etomidate [33125-97-2]
Etomidoline [21590-92-1]
Etonitazene [911-65-9]
Etoperidone [52942-31-1]
Etoposide [33419-42-0]
Etorphine [14521-96-1]
Etoxadrol [28189-85-7]
Etoxadrol Hydrochloride* [23239-37-4]
Etozolin [73-09-6]
Etretinate [54350-48-0]
Etrimfos [38260-54-7]
Etryptamine [2235-90-7]
Etryptamine Acetate* [118-68-3]
Etymemazine [523-54-6]
Etymemazine Hydrochloride* [3737-33-5]
β-Eucaine [500-34-5]
Eucaine Hydrochloride* [555-28-2]
Eucalyptol [470-82-6]
Eucalyptus Gum [9000-20-8]
Eucatropine [100-91-4]
Eucatropine Hydrochloride* [536-93-6]
Eugenol [97-53-0]
Euparin [532-48-9]
Eupatorin [855-96-9]
Euprocin [1301-42-4]
Europium [7440-53-1]
Evan's Blue [314-13-6]
Evodiamine [518-17-2]
Exalamide [53370-90-4]
Exaltolide® [106-02-5]
Exifone [52479-85-3]
Exiproben [26281-69-6]
Factor V [9001-24-5]
Factor VII [9001-25-6]
Factor VIII [9001-27-8]
Factor IX [9001-28-9]
Factor X [9001-29-0]
Factor XI [9013-55-2]
Factor XII [9001-30-3]
Factor XIII [9013-56-3]
Fagarine [524-15-2]
Falone® [94-84-8]
Famotidine [76824-35-6]
Famphur [52-85-7]
α-Farnesene [502-61-4]
β-Farnesene [18794-84-8]
Farnesol [4602-84-0]
Fast Green FCF [2353-45-9]
Fazadinium Bromide [49564-56-9]

Febantel [58306-30-2]
Febarbamate [13246-02-1]
Febrifugine [24159-07-7]
Febuprol [3102-00-9]
Feclemine [3590-16-7]
Felbinac [5728-52-9]
Felinine [471-09-0]
Felodipine [72509-76-3]
Felypressin [56-59-7]
Femoxetine [59859-58-4]
Fenadiazole [1008-65-7]
Fenalamide [4551-59-1]
Fenalcomine [34616-39-2]
Fenamiphos [22224-92-6]
Fenapanil [61019-78-1]
Fenarimol [60168-88-9]
Fenbendazole [43210-67-9]
Fenbenicillin [1926-48-3]
Fenbufen [36330-85-5]
Fenbutatin Oxide [13356-08-6]
Fenbutrazate [4378-36-3]
Fencamfamine [1209-98-9]
Fencamfamine Hydrochloride* [2240-14-4]
Fencamine [28947-50-4]
d-Fenchone [4695-62-9]
Fencibutirol [5977-10-6]
Fenclofenac [34645-84-6]
Fenclorac [36616-52-1]
Fenclozic Acid [17969-20-9]
Fendiline [13042-18-7]
Fendosal [53597-27-6]
Fenethazine [522-24-7]
Fenethylline [3736-08-1]
Fenethylline Hydrochloride* [1892-80-4]
Fenfluramine [458-24-2]
Fenfluramine Hydrochloride* [404-82-0]
Fenipentol [583-03-9]
Fenitrothion [122-14-5]
Fenofibrate [49562-28-9]
Fenoldopam [67227-56-9]
Fenoldopam Mesylate* [67227-57-0]
Fenoprofen [31879-05-7]
Fenoterol [13392-18-2]
Fenoverine [37561-27-6]
Fenoxaprop-ethyl [82110-72-3]
Fenoxazoline [4846-91-7]
Fenoxazoline Hydrochloride* [21370-21-8]
Fenoxedil [54063-40-0]
Fenozolone [15302-16-6]
Fenpentadiol [15687-18-0]
Fenpiprane [3540-95-2]
Fenpiprane Hydrochloride* [3329-14-4]
Fenpiverinium Bromide [125-60-0]
Fenpropathrin [39515-41-8]
Fenpropidin [67306-00-7]
Fenpropimorph [67306-03-0]
Fenproporex [15686-61-0]
Fenprostalene [69381-94-8]
Fenquizone [20287-37-0]
Fenspiride [5053-06-5]
Fenspiride Hydrochloride* [5053-08-7]
Fensulfothion [115-90-2]
Fentanyl [437-38-7]
Fentanyl Citrate* [990-73-8]
Fenthion [55-38-9]
Fentiazac [18046-21-4]
Fenticlor [97-24-5]
Fenticonazole [72479-26-6]
Fenticonazole Nitrate* [73151-29-8]
Fentonium Bromide [5868-06-4]
Fenuron [101-42-8]
Fenvalerate [51630-58-1]
Fenvalerate (S,S-Isomer)* [66230-04-4]
Feprazone [30748-29-9]

Ferbam [14484-64-1]
Fermium [7440-72-4]
Ferric Acetate, Basic [10450-55-2]
Ferric and Ammonium Acetate Solution [8006-27-7]
Ferric Bromide [10031-26-2]
Ferric Chloride [7705-08-0]
Ferric Chromate(VI) [10294-52-7]
Ferric Citrate [2338-05-8]
Ferric Ferrocyanide [14038-43-8]
Ferric Fluoride [7783-50-8]
Ferric Formate [555-76-0]
Ferric Fructose [12286-76-9]
Ferric Hydroxide [20344-49-4]
Ferric Hypophosphite [7783-84-8]
Ferriclate Calcium Sodium [34150-62-4]
Ferric Nitrate [10421-48-4]
Ferric Oxide [1309-37-1]
Ferric Oxide, Saccharated [8047-67-4]
Ferric Phosphate [10045-86-0]
Ferric Pyrophosphate [10058-44-3]
Ferric Sodium Edetate [15708-41-5]
Ferric Sodium Pyrophosphate [10045-87-1]
Ferric Subsulfate Solution [1310-45-8]
Ferric Sulfate [10028-22-5]
Ferric Thiocyanate [4119-52-2]
Ferrite [1317-54-0]
Ferritin [9007-73-2]
Ferrocene [102-54-5]
Ferrocholinate [1336-80-7]
Ferroglycine Sulfate [17169-60-7]
Ferrosoferric Oxide [1317-61-9]
Ferrous Bromide [7789-46-0]
Ferrous Carbonate Mass [8030-35-1]
Ferrous Carbonate Saccharated [8001-10-3]
Ferrous Chloride [7758-94-3]
Ferrous Citrate [23383-11-1]
Ferrous Fluoride [7789-28-8]
Ferrous Fumarate [141-01-5]
Ferrous Gluconate [299-29-6]
Ferrous Hydroxide [18624-44-7]
Ferrous Iodide [7783-86-0]
Ferrous Lactate [5905-52-2]
Ferrous Oxalate [516-03-0]
Ferrous Oxide [1345-25-1]
Ferrous Phosphate [14940-41-1]
Ferrous Phosphide [1310-43-6]
Ferrous Selenide [1310-32-3]
Ferrous Succinate [10030-90-7]
Ferrous Sulfate [7720-78-7]
Ferrous Sulfide [1317-37-9]
Ferrous Thiocyanate [6010-09-9]
Fertilysin [1477-57-2]
Fertirelin [38234-21-8]
Fertirelin Acetate* [66002-66-2]
Ferulic Acid [1135-24-6]
Fervenulin [483-57-8]
Fibrin [9001-31-4]
Fibrinogen [9001-32-5]
Fibroblast Growth Factor [62031-54-3]
Fibroblast Growth Factor (aFGF)* [106096-92-8]
Fibroblast Growth Factor (bFGF)* [106096-93-9]
Fibroin [9007-76-5]
Fichtelite [2221-95-6]
Ficin [9001-33-6]
Filicinic Acid [2065-00-1]
Filipin [11078-21-0]
Fipexide [34161-24-5]
Firefly Luciferin [2591-17-5]
Fisetin [528-48-3]
FK-506 [104987-11-3]
Flavaspidic Acid [114-42-1]

Flavine-Adenine Dinucleotide [146-14-5]
Flavipucine [38473-18-6]
Flavone [525-82-6]
Flavopereirine [486-18-0]
Flavoxanthin [512-29-8]
Flavoxate [15301-69-6]
Flavoxate Hydrochloride* [3717-88-2]
Flecainide [54143-55-4]
Flecainide Acetate* [54143-56-6]
Fleroxacin [79660-72-3]
Flindersine [523-64-8]
Floctafenine [23779-99-9]
Flomoxef [99665-00-6]
Flopropione [2295-58-1]
Florantyrone [519-95-9]
Floredil [53731-36-5]
Florfenicol [73231-34-2]
Flosequinan [76568-02-0]
Floxacillin [5250-39-5]
Floxuridine [50-91-9]
Fluacizine [30223-48-4]
Fluanisone [1480-19-9]
Fluazacort [19888-56-3]
Fluazifop-butyl [69806-50-4]
Flubendazole [31430-15-6]
Flubenzimine [37893-02-0]
Fluchloralin [33245-39-5]
Flucloronide [3693-39-8]
Fluconazole [86386-73-4]
Flucythrinate [70124-77-5]
Flucytosine [2022-85-7]
Fludarabine [21679-14-1]
Fludiazepam [3900-31-0]
Fludrocortisone [127-31-1]
Fludrocortisone Acetate* [514-36-3]
Flufenamic Acid [530-78-9]
Fluindione [957-56-2]
Flumazenil [78755-81-4]
Flumecinol [56430-99-0]
Flumedroxone Acetate [987-18-8]
Flumequine [42835-25-6]
Flumethasone [2135-17-3]
Flumethasone Pivalate* [2002-29-1]
Flumethiazide [148-56-1]
Flumethrin [69770-45-2]
Flumetramide [7125-73-7]
Flunarizine [52468-60-7]
Flunarizine Hydrochloride* [30484-77-6]
Flunisolide [3385-03-3]
Flunisolide Acetate* [4533-89-5]
Flunitrazepam [1622-62-4]
Flunixin [38677-85-9]
Flunixin Meglumine* [42461-84-7]
Flunoxaprofen [66934-18-7]
Fluoboric Acid [16872-11-0]
Fluocinolone Acetonide [67-73-2]
Fluocinonide [356-12-7]
Fluocortin Butyl [41767-29-7]
Fluocortolone [152-97-6]
Fluocortolone Caproate* [303-40-2]
Fluometuron [2164-17-2]
9H-Fluorene [86-73-7]
9H-Fluorene-2,7-diamine [525-64-4]
N-2-Fluorenylacetamide [53-96-3]
Fluorescamine [38183-12-9]
Fluorescein [2321-07-5]
Fluorescein Sodium* [518-47-8]
Fluorescin [518-44-5]
Fluoresone [2924-67-6]
Fluoridamid [47000-92-0]
Fluorine [7782-41-4]
Fluorine Dioxide [7783-44-0]
Fluorine Monoxide [7783-41-7]
Fluorine Nitrate [7789-26-6]
Fluorine Perchlorate [10049-03-3]
Fluoroacetamide [640-19-7]
Fluoroacetic Acid [144-49-0]
3-Fluoro-D-alanine [35455-20-0]

p-Fluoroaniline [371-40-4]
Fluorobenzene [462-06-6]
p-Fluorobenzoic Acid [456-22-4]
1-Fluoro-2,4-dinitrobenzene [70-34-8]
Fluoroform [75-46-7]
Fluoromethane [593-53-3]
Fluorometholone [426-13-1]
Fluorometholone Acetate* [3801-06-7]
p-Fluorophenylacetic Acid [405-50-5]
Fluorosalan [4776-06-1]
Fluorosulfonic Acid [7789-21-1]
Fluorouracil [51-21-8]
Fluosilicic Acid [16961-83-4]
Fluosol DA [75216-20-5]
Fluoxetine [54910-89-3]
Fluoxymesterone [76-43-7]
Flupentixol [2709-56-0]
Fluperolone Acetate [2119-75-7]
Fluphenazine [69-23-8]
Fluphenazine Dihydrochloride* [146-56-5]
Fluphenazine Enanthate* [2746-81-8]
Flupirtine [56995-20-1]
Fluprednidene Acetate [1255-35-2]
Fluprednisolone [53-34-9]
Fluproquazone [40507-23-1]
Fluprostenol [40666-16-8]
Fluprostenol Sodium* [55028-71-2]
Flurandrenolide [1524-88-5]
Flurazepam [17617-23-1]
Flurazepam Dihydrochloride* [1172-18-5]
Flurbiprofen [5104-49-4]
Flurogestone Acetate [2529-45-5]
Flurothyl [333-36-8]
Fluroxene [406-90-6]
Fluroxypyr [69377-81-7]
Fluroxypyr 1-Methylheptyl Ester* [81406-37-3]
Flurprimidol [56425-91-3]
Flusilazole [85509-19-9]
Fluspirilene [1841-19-6]
Flutamide [13311-84-7]
Flutazolam [27060-91-9]
Flutoprazepam [25967-29-7]
Flutriafol [76674-21-0]
Flutropium Bromide [63516-07-4]
Fluvalinate [69409-94-5]
Fluvoxamine [54739-18-3]
Folescutol [15687-22-6]
Folic Acid [59-30-3]
Folinic Acid [58-05-9]
Folpet [133-07-3]
Fominoben [18053-31-1]
Fomocaine [17692-39-6]
Fonazine [7456-24-8]
Fonazine Mesylate* [7455-39-2]
Fonofos [944-22-9]
Formaldehyde, Gas [50-00-0]
Formaldehyde Sodium Bisulfite [870-72-4]
Formaldehyde Solution [50-00-0]
Formamide [75-12-7]
Formanilide [103-70-8]
Formic Acid [64-18-6]
Formicin [625-51-4]
Forminitrazole [500-08-3]
Formocortal [2825-60-7]
Formononetin [485-72-3]
Formosulfathiazole [13968-86-0]
Formoterol [73573-87-2]
Formothion [2540-82-1]
Formyldienolone [2454-11-7]
Formyl Fluoride [1493-02-3]
4'-Formylsuccinanilic Acid Thiosemicarbazone [2635-29-2]
N^2-Formylsulfisomidine [795-13-1]
Foscarnet Sodium [63585-09-1]
Fosetyl Al [39148-24-8]

Denotes derivative of title compound

Fosfestrol [522-40-7]
Fosfomycin [23155-02-4]
Fosfosal [6064-83-1]
Fosinopril [98048-97-6]
Fosinopril Diacid* [95399-71-6]
Fosinopril Sodium* [88889-14-9]
Fospirate [5598-52-7]
Fosthietan [21548-32-3]
Fotemustine [92118-27-9]
Francium [7440-73-5]
Frangulin [60529-33-1]
Fraxetin [574-84-5]
Fraxin [524-30-1]
Fraxiparine ® [104521-37-1]
Fredericamycin A [80455-68-1]
Frenolicin [10023-07-1]
Frequentin [29119-03-7]
Friedelin [559-74-0]
Fructose [57-48-7]
DL-Fructose [6035-50-3]
Fructose-1,6-diphosphate [488-69-7]
Fructose-6-phosphate [643-13-0]
FSH [9002-68-0]
Ftaxilide [19368-18-4]
Fucosamine [7577-62-0]
D-Fucose [3615-37-0]
L-Fucose [2438-80-4]
Fucosterol [17605-67-3]
Fucoxanthin [3351-86-8]
Fulvoplumierin [20867-01-0]
Fumagillin [23110-15-8]
Fumaric Acid [110-17-8]
Fumigatin [484-89-9]
Fungichromin [6834-98-6]
Fungisterol [53260-54-1]
Funtumine [474-45-3]
Furaltadone [139-91-3]
Furan [110-00-9]
2-Furanacrylic Acid [539-47-9]
2-Furanacrylonitrile [7187-01-1]
Furazabol [1239-29-8]
Furazolidone [67-45-8]
Furazolium Chloride [5118-17-2]
Furcellaran [9000-21-9]
Furethidine [2385-81-1]
Furfural [98-01-1]
Furfuryl Alcohol [98-00-0]
5-Furfuryl-5-isopropylbarbituric Acid
  [1146-21-0]
Furfurylmethylamphetamine
  [13445-60-8]
α-Furildioxime [522-27-0]
2-Furoic Acid [88-14-2]
Furonazide [3460-67-1]
Furosemide [54-31-9]
Furoyl Chloride [1300-32-9]
Fursultiamine [804-30-8]
Furterene [7761-75-3]
Furtrethonium [7618-86-2]
Furtrethonium Iodide* [541-64-0]
Fusafungine [1393-87-9]
Fusaric Acid [536-69-6]
Fusarubin [1702-77-8]
Fuscin [83-85-2]
Fusidic Acid [6990-06-3]
Fusidic Acid Sodium* [751-94-0]
Fustin [20725-03-5]
Fyrol FR-2® [13674-87-8]
Gabexate [39492-01-8]
Gadolinium [7440-54-2]
Gadopentetic Acid [80529-93-7]
Gadopentetic Acid Dimeglumine Salt*
  [86050-77-3]
Galactaric Acid [526-99-8]
Galactitol [608-66-2]
Galactoflavin [5735-19-3]
D-Galactosamine [7535-00-4]
Galactose [59-23-4]
D-Galacturonic Acid [685-73-4]
Galangin [548-83-4]
Galanthamine [357-70-0]

Galantamine Hydrobromide*
  [1953-04-4]
Galegine [543-83-9]
Galipine [525-68-8]
Gallacetophenone [528-21-2]
Gallamine Triethiodide [65-29-2]
Gallein [2103-64-2]
Gallic Acid [149-91-7]
Methyl Gallate* [99-24-1]
Gallium [7440-55-3]
Gallium Citrate* [41183-64-6]
Gallium Nitrate Anhydrous*
  [13494-90-1]
Gallium Arsenide [1303-00-0]
Gallium Phosphide [12063-98-8]
Gallium Trifluoride [7783-51-9]
Gallocyanine [1562-85-2]
Gallopamil [16662-47-8]
Gamabufotalin [465-11-2]
Gamboge [9000-25-3]
Gambogic Acid [2752-65-0]
Ganciclovir [82410-32-0]
Ganciclovir Sodium* [107910-75-8]
Ganglefene [299-61-6]
Gardol® [137-16-6]
Garryine [561-51-3]
Gaultherin [490-67-5]
Gefarnate [51-77-4]
Geissoschizoline [18397-07-4]
Geissospermine [427-01-0]
Gelatin [9000-70-8]
Gelsemine [509-15-9]
Gemeprost [64318-79-2]
Gemfibrozil [25812-30-0]
Genistein [446-72-0]
Genisteine-Alkaloid [446-95-7]
Genite® [97-16-5]
Gentamicin [1403-66-3]
Gentamicin Sulfate* [1405-41-0]
Gentianine [439-89-4]
Gentian Violet [548-62-9]
Gentiobiose [554-91-6]
Gentiopicrin [20831-76-9]
Gentisic Acid [490-79-9]
Gentisin [437-50-3]
Gentisyl Alcohol [495-08-9]
Gentrogenin [427-28-1]
Geosmin [19700-21-1]
Gepefrine [18840-47-6]
Gephyrotoxin [55893-12-4]
Gepirone [83928-76-1]
Gepirone Hydrochloride*
  [83928-66-9]
Geraniol [106-24-1]
Geranylhydroquinone [10457-66-6]
Germane [7782-65-2]
Germanium [7440-56-4]
Germanium Dichloride [10060-11-4]
Germanium Dioxide [1310-53-8]
Germanium Tetrachloride
  [10038-98-9]
Germanium Tetrafluoride [7783-58-6]
Germine [508-65-6]
Gestodene [60282-87-3]
Gestonorone Caproate [1253-28-7]
Gestrinone [16320-04-0]
Ghatti Gum [9000-28-6]
Gibberellic Acid [77-06-5]
Gibbs Reagent [101-38-2]
Gigantine [32829-58-6]
[6]-Gingerol [23513-14-6]
Giractide [24870-04-0]
Gitalin [1405-76-1]
Gitogenin [511-96-6]
F-Gitonin [28591-01-7]
Gitoxigenin [545-26-6]
Gitoxin [4562-36-1]
Gitoxin Pentaacetate* [7242-04-8]
Gladiolic Acid [478-05-7]
Glafenine [3820-67-5]
Glaucarubin [1448-23-3]

Glaucine [475-81-0]
Gliadin [9007-90-3]
Glibornuride [26944-48-9]
Gliclazide [21187-98-4]
Gliotoxin [67-99-2]
Glipizide [29094-61-9]
Gliquidone [33342-05-1]
Glisoxepid [25046-79-1]
Glucagon [9007-92-5]
Glucametacin [52443-21-7]
Glucamine [488-43-7]
D-Glucaric Acid [87-73-0]
D-Glucoascorbic Acid [528-88-1]
Glucofrangulin [52731-38-1]
α-Glucogallin [53318-36-8]
β-Glucogallin [13405-60-2]
Glucoheptonic Acid [87-74-1]
Gluconic Acid [526-95-4]
Gluconolactone [90-80-2]
Glucosamine [3416-24-8]
Glucose [50-99-7]
Glucose Oxidase [9001-37-0]
Glucose-6-phosphate [56-73-5]
α-Glucose-1-phosphate [59-56-3]
Glucosulfone Sodium [554-18-7]
N⁴-β-D-Glucosylsulfanilamide
  [53274-53-6]
Glucovanillin [494-08-6]
D-Glucuronic Acid [6556-12-3]
β-Glucuronidase [9001-45-0]
D-Glucuronolactone [32449-92-6]
Glutamic Acid [56-86-0]
L-Glutamic Acid 5-Ethyl Ester
  [1119-33-1]
Glutamine [56-85-9]
Glutaraldehyde [111-30-8]
Glutaric Acid [110-94-1]
Glutaronitrile [544-13-8]
Glutathione [70-18-8]
Glutethimide [77-21-4]
Glyburide [10238-21-8]
Glybuthiazol(e) [535-65-9]
Glybuzole [1492-02-0]
Glycarsamide [144-87-6]
Glyceraldehyde [367-47-5]
Glyceraldehyde 3-Phosphate
  [142-10-9]
Glyceric Acid [473-81-4]
Glycerol [56-81-5]
Glycerophosphoric Acid [57-03-4]
Glyceryl p-Aminobenzoate [136-44-7]
Glyceryl Iodide [554-10-9]
Glyceryl Monostearate [31566-31-1]
Glycidol [556-52-5]
Glycine [56-40-6]
Glycine Sulfate [513-29-1]
Glycinin [9007-93-6]
Glycobiarsol [116-49-4]
Glycocholic Acid [475-31-0]
Glycocyamine [352-97-6]
Glycogen [9005-79-2]
Glycol Dilaurate [624-04-4]
Glycolic Acid [79-14-1]
Glycol Salicylate [87-28-5]
Glyconiazide [3691-74-5]
Glycopyrrolate [596-51-0]
Glycosine [6873-15-0]
N-Glycylglycine [556-50-3]
Glycyrrhizic Acid [1405-86-3]
Glyhexamide [451-71-8]
Glymidine [339-44-6]
Glymidine Sodium* [3459-20-9]
Glyodin [556-22-9]
Glyoxal [107-22-2]
Glyoxal-Sodium Bisulfite [517-21-5]
Glyoxylic Acid [298-12-4]
Glyphosate [1071-83-6]
Glyphosine [2439-99-8]
Glypinamide [1228-19-9]
Goitrin [500-12-9]
Gold [7440-57-5]

Gold, Explosive [1332-12-3]
Gold Monochloride [10294-29-8]
Gold Monocyanide [506-65-0]
Gold Monoiodide [10294-31-2]
Gold Monosulfide [1303-60-2]
Gold Monoxide [1303-57-7]
Gold, Radioactive, Colloidal [10043-49-9]
Gold Selenate [10294-32-3]
Gold Selenide [1303-62-4]
Gold Sodium Thiomalate [12244-57-4]
Gold Sodium Thiosulfate [10233-88-2]
Gold Stannate [1345-24-0]
Gold Tribromide [10294-28-7]
Gold Tribromide, Acid [17083-68-0]
Gold Trichloride [13453-07-1]
Gold Trichloride, Acid [16903-35-8]
Gold Tricyanide [535-37-5]
Gold Trihydroxide [1303-52-2]
Gold Trioxide [1303-58-8]
Gold Trisulfide [1303-61-3]
Goserelin [65807-02-5]
Gossyplure [50933-33-0]
Gossypol [303-45-7]
Gougerotin [2096-42-6]
Gramicidin S [113-73-5]
Gramine [87-52-5]
Granaticin [19879-06-2]
Granaticin B* [19879-03-9]
Grandisol [26532-22-9]
Granisetron [109889-09-0]
Granisetron Hydrochloride* [107007-99-8]
Graphite [7782-42-5]
Graphitic Acid [1399-57-1]
Gratiogenin [7067-16-5]
Gratioside [53762-90-6]
Gravitol(e) [6006-09-3]
Grindelic Acid [1438-57-9]
Grisein [1391-82-8]
Griseofulvin [126-07-8]
Griseoviridin [53216-90-3]
Guaiac [9000-29-7]
Guaiacol [90-05-1]
Guaiacol Benzoate [531-37-3]
Guaiacol Carbonate [553-17-3]
Guaiacol Phosphate [563-03-1]
Guaiacol Valerate [531-39-5]
Guaiactamine [15687-23-7]
Guaiapate [852-42-6]
Guaiazulene [489-84-9]
Guaifenesin [93-14-1]
Guaiol [489-86-1]
Guaithylline [5634-38-8]
Guamecycline [16545-11-2]
Guanabenz [5051-62-7]
Guanabenz Acetate* [23256-50-0]
Guanacline [1463-28-1]
Guanacline Sulfate* [1562-71-6]
Guanadrel [40580-59-4]
Guanadrel Sulfate* [22195-34-2]
Guanazodine [32059-15-7]
Guanethidine [55-65-2]
Guanethidine Monosulfate* [645-43-2]
Guanethidine Sulfate* [60-02-6]
Guanfacine [29110-47-2]
Guanfacine Hydrochloride* [29110-48-3]
Guanidine [113-00-8]
Guanidinium Aluminum Sulfate Hexahydrate [10199-21-0]
Guanine [73-40-5]
Guanochlor [5001-32-1]
Guanoclor Sulfate* [551-48-4]
Guanosine [118-00-3]
Guanoxabenz [24047-25-4]
Guanoxan [2165-19-7]

Guanoxan Sulfate* [5714-04-5]
3'-Guanylic Acid [117-68-0]
5'-Guanylic Acid [85-32-5]
Guaran [9000-30-0]
Guar Gum [9000-30-0]
Guinea Green B [4680-78-8]
D-Gulonic Acid [20246-33-7]
L-Gulonic Acid [526-97-6]
D-Gulose [4205-23-6]
L-Gulose [6027-89-0]
Gum Benzoin [9000-05-9]
Gum Tragacanth [9000-65-1]
Gutta-Percha [9000-32-2]
Guvacine [498-96-4]
Gymnemic Acid [1399-64-0]
Gypsogenin [639-14-5]
Hachimycin [1394-02-1]
Hadacidin [689-13-4]
Hafnium [7440-58-6]
Halazepam [23092-17-3]
Halazone [80-13-7]
Halcinonide [3093-35-4]
Halethazole [15599-36-7]
Halimide® [19014-05-2]
Halofantrine [69756-53-2]
Halofantrine Hydrochloride* [36167-63-2]
Halofuginone [55837-20-2]
Halofuginone Hydrobromide* [64924-67-0]
Halometasone [50629-82-8]
Haloperidol [52-86-8]
Halopredone Acetate [57781-14-3]
Haloprogesterone [3538-57-6]
Haloprogin [777-11-7]
Halopropane [679-84-5]
Halostachine [495-42-1]
Halothane [151-67-7]
Haloxazolam [59128-97-1]
Haloxon [321-55-1]
Halquinol [8067-69-4]
Hamamelitannin [469-32-9]
Hamamelose [4573-78-8]
Hamycin [1403-71-0]
Haplophytine [16625-20-0]
Harmaline [304-21-2]
Harmalol [525-57-5]
Harman [486-84-0]
Harmine [442-51-3]
Hasubanonine [1805-85-2]
HCG [9002-61-3]
HCS [11085-36-2]
Hecogenin [467-55-0]
Hectorite [12173-47-6]
Hedaquinium Chloride [4310-89-8]
Hederagenin [465-99-6]
α-Hederin [27013-91-8]
Helenalin [6754-13-8]
Helenynolic Acid [7309-58-2]
Helicin [618-65-5]
Heliosupine [32728-78-2]
Helium [7440-59-7]
Hellebrin [2389-18-4]
Helminthosporal [723-61-5]
Helminthosporol [1619-29-0]
Helveticoside [630-64-8]
Helvolic Acid [29400-42-8]
Hematein [475-25-2]
Hematin [15489-90-4]
Hematoporphyrin [14459-29-1]
Hematoxylin [517-28-2]
Heme [14875-96-8]
Hementin [87041-58-5]
Hemi-Dewar Biphenyl [4655-30-5]
Hemin [16009-13-5]
Hemipyocyanine [528-71-2]
Hemisulfur Mustard [693-30-1]
Hempa [680-31-9]
Heparamine [53260-52-9]
Heparin [9005-49-6]
Hepaxanthin [512-39-0]

HEPES [7365-45-9]
Hepronicate [7237-81-2]
Heptabarbital [509-86-4]
Heptachlor [76-44-8]
Heptaminol [372-66-7]
Heptanal [111-71-7]
Heptanal Sodium Bisulfite [13495-04-0]
n-Heptane [142-82-5]
Heptanoic Acid [111-14-8]
1-Heptanol [111-70-6]
2-Heptanol [543-49-7]
2-Heptanone [110-43-0]
Heptenophos [23560-59-0]
Heptoxime [530-97-2]
D-manno-Heptulose [3615-44-9]
Hercynine [534-30-5]
Herqueinone [26871-30-7]
Hesperetin [520-33-2]
Hesperidin [520-26-3]
Hetacillin [3511-16-8]
Hetacillin Potassium* [5321-32-4]
Hetastarch [9004-62-0]
5-HETE [70608-72-9]
Heteronium Bromide [7247-57-6]
Hetolin® [2390-22-9]
Hexaaminecobalt Trichloride [10534-89-1]
Hexaborane(10) [23777-80-2]
Hexacarbacholine Bromide [306-41-2]
Hexachlorobenzene [118-74-1]
Hexachloroethane [67-72-1]
Hexachlorophene [70-30-4]
Hexacyclonate Sodium [7009-49-6]
Hexadecyl 3-Hydroxy-2-naphthoate [531-84-0]
Hexadimethrine Bromide [9011-04-5]
Hexafluorenium Bromide [317-52-2]
Hexahydroequilenin [517-07-7]
Hexalure [23192-42-9]
Hexamethonium [60-26-4]
Hexamethonium Bromide* [55-97-0]
Hexamethylene Glycol [629-11-8]
Hexamethylolmelamine [531-18-0]
Hexamidine [3811-75-4]
Hexamidine Isethionate* [659-40-5]
n-Hexane [110-54-3]
1,6-Hexanediamine [124-09-4]
1-Hexanol [111-27-3]
Hexapropymate [358-52-1]
Hexazinone [51235-04-2]
Hexazole [4671-03-8]
Hexedine [5980-31-4]
3-Hexen-1-ol [544-12-7]
Hexestrol [84-16-2]
Hexestrol Diphosphate* [14188-82-0]
Hexestrol Bis(β-diethylaminoethyl ether) [2691-45-4]
Hexethal Sodium [144-00-3]
Hexetidine [141-94-6]
Hexobarbital [56-29-1]
Hexobendine [54-03-5]
Hexocyclium Methyl Sulfate [115-63-9]
Hexoprenaline [3215-70-1]
Hexylcaine Hydrochloride [532-76-3]
2-Hexyldecanoic Acid [25354-97-6]
Hexylene Glycol [107-41-5]
Hexyl Methyl Ketone [111-13-7]
4-Hexylresorcinol [136-77-6]
Hexythiazox [78578-05-0]
Hinderin [551-90-6]
Hippuric Acid [495-69-2]
Hirsutic Acid C [3650-17-7]
Hirudin [8001-27-2]
Histaminase [9001-53-0]
Histamine [51-45-6]
Histamine Phosphate* [51-74-1]
Histapyrrodine [493-80-1]
Histidine [7006-35-1]

Denotes derivative of title compound

HN1 [538-07-8]
Holarrhenine [561-22-8]
Holmium [7440-60-0]
Holomycin [488-04-0]
Homarine [445-30-7]
Homatropine [87-00-3]
Homatropine Hydrobromide*
   [51-56-9]
Homatropine Methylbromide*
   [80-49-9]
Homidium [3546-21-2]
Homidium Bromide* [1239-45-8]
Homocamfin [535-86-4]
Homochelidonine [476-33-5]
Homochlorcyclizine [848-53-3]
Homocysteine [6027-13-0]
Homocystine [462-10-2]
Homoeriodictyol [446-71-9]
Homofenazine [3833-99-6]
Homogentisic Acid [451-13-8]
Homonicotinic Acid [501-81-5]
Homosalate [118-56-9]
Homoserine [672-15-1]
Homovanillic Acid [306-08-1]
HON [4439-84-3]
Hopantenic Acid [18679-90-8]
Hordenine [539-15-1]
HPA-23 [59372-48-4]
HQNO [341-88-8]
Humulene [6753-98-6]
Humulon [26472-41-3]
Hyalobiuronic Acid [499-15-0]
Hyaluronic Acid [9004-61-9]
Hyaluronic Acid Sodium*
   [9067-32-7]
Hyaluronidases [9001-54-1]
Hycanthone [3105-97-3]
Hycanthone Mesylate* [23255-93-8]
Hydantoin [461-72-3]
Hydnocarpic Acid [459-67-6]
Hydracarbazine [3614-47-9]
Hydracrylic Acid [503-66-2]
Hydralazine [86-54-4]
Hydralazine Hydrochloride*
   [304-20-1]
Hydrallostane [516-41-6]
Hydramethylnon [67485-29-4]
Hydramitrazine [13957-36-3]
Hydrargaphen [14235-86-0]
Hydrastine [118-08-1]
Hydrastine Hydrochloride*
   [5936-28-7]
Hydrastinine [6592-85-4]
Hydrastinine Hydrochloride*
   [4884-68-8]
Hydrazine [302-01-2]
Hydrazine Dihydrochloride*
   [5341-61-7]
Hydrazine Hydrate [7803-57-8]
Hydrazine Sulfate [10034-93-2]
Hydrazine Tartrate [634-62-8]
4-Hydrazinobenzenesulfonic Acid
   [98-71-5]
2-Hydrazinoethanol [109-84-2]
Hydrazoic Acid [7782-79-8]
Hydrindantin [5103-42-4]
Hydriodic Acid [10034-85-2]
Iodamide Meglumine* [18656-21-8]
Hydrobenzoin [492-70-6]
Hydrobromic Acid [10035-10-6]
Hydrocarbostyril [553-03-7]
Hydrochloric Acid [7647-01-0]
Hydrochlorothiazide [58-93-5]
Hydrocinchonidine [485-64-3]
Hydrocinchonine [485-65-4]
Hydrocinnamic Acid [501-52-0]
Hydrocodone [125-29-1]
Hydrocodone Bitartrate Hydrate*
   [34195-34-1]
Hydrocortamate [76-47-1]

Hydrocortamate Hydrochloride*
   [125-03-1]
Hydrocortisone [50-23-7]
Hydrocortisone Butyrate*
   [13609-67-1]
Hydrocortisone Valerate*
   [57524-89-7]
Hydrocortisone Acetate [50-03-3]
Hydrocortisone Phosphate
   [3863-59-0]
Hydrocortisone Sodium Phosphate*
   [6000-74-4]
Hydrocortisone 21-Sodium Succinate
   [125-04-2]
Hydrocortisone Tebutate [508-96-3]
Hydrocotarnine [550-10-7]
Hydroflumethiazide [135-09-1]
Hydrofluoric Acid [7664-39-3]
Hydrofuramide [494-47-3]
Hydrogen [1333-74-0]
Hydrogen Bromide [10035-10-6]
Hydrogen Chloride [7647-01-0]
Hydrogen Cyanide [74-90-8]
Hydrogen Fluoride [7664-39-3]
Hydrogen Iodide [10034-85-2]
Hydrogen Peroxide [7722-84-1]
Hydrogen Selenide [7783-07-5]
Hydrogen Sulfide [7783-06-4]
Hydrogen Telluride [7783-09-7]
Hydrogen Tetracarbonylferrate(II)
   [17440-90-3]
Hydrohydrastinine [494-55-3]
Hydromorphone [466-99-9]
Hydromorphone Hydrochloride*
   [71-68-1]
Hydrone® [8049-21-6]
Hydroorotic Acid [155-54-4]
Hydroquinidine [1435-55-8]
Hydroquinine [522-66-7]
Hydroquinone [123-31-9]
Hydroxocobalamin [13422-51-0]
Hydroxyamphetamine [1518-86-1]
Hydroxyamphetamine Hydrobromide*
   [306-21-8]
p-Hydroxybenzaldehyde [123-08-0]
p-Hydroxybenzoic Acid [99-96-7]
p-Hydroxybenzylpenicillin Sodium
   [5985-13-7]
α-Hydroxybenzylphosphinic Acid
   [52705-43-8]
4'-Hydroxybutyranilide [101-91-7]
β-Hydroxybutyric Acid [300-85-6]
3-Hydroxycamphor [10373-81-6]
Hydroxychloroquine [118-42-3]
Hydroxychloroquine Sulfate*
   [747-36-4]
1α-Hydroxycholecalciferol
   [41294-56-8]
24-Hydroxycholesterol [474-73-7]
25-Hydroxycholesterol [2140-46-7]
Hydroxycodeinone [508-54-3]
Hydroxydione Sodium [53-10-1]
p-Hydroxyephedrine [365-26-4]
N-Hydroxyethylpromethazine Chloride [2090-54-2]
Hydroxyglutamic Acid [533-62-0]
8-Hydroxy-7-iodo-5-quinolinesulfonic Acid [547-91-1]
8-Hydroxy-7-iodo-5-quinolinesulfonic Acid Compd. with Chloroquine* [7270-12-4]
4-Hydroxyisophthalic Acid
   [636-46-4]
Hydroxylamine [7803-49-8]
Hydroxylamine Hydrochloride*
   [5470-11-1]
Hydroxylamine Sulfate*
   [10039-54-0]
Hydroxylupanine [15358-48-2]

1-(Hydroxymethyl)-5,5-dimethyl-hydantoin [116-25-6]
17-Hydroxy-16-methylene-Δ^6-progesterone [10087-54-4]
5-(Hydroxymethyl)-2-furaldehyde
   [67-47-0]
N-(Hydroxymethyl)nicotinamide
   [3569-99-1]
1-Hydroxy-2-naphthoic Acid
   [86-48-6]
3-Hydroxy-2-naphthoic Acid
   [92-70-6]
4-Hydroxy-19-nortestosterone
   [4721-69-1]
Hydroxypethidine [468-56-4]
Hydroxyphenamate [50-19-1]
N-(4-Hydroxyphenyl)glycine
   [122-87-2]
Hydroxyprocaine [487-53-6]
17α-Hydroxyprogesterone [68-96-2]
17α-Hydroxyprogesterone Caproate
   [630-56-8]
4-Hydroxy-L-proline [51-35-4]
Hydroxypropyl Cellulose [9004-64-2]
Hydroxypropyl Methylcellulose
   [9004-65-3]
8-Hydroxyquinoline [148-24-3]
8-Hydroxyquinoline Sulfate
   [134-31-6]
8-Hydroxy-5-quinolinesulfonic Acid
   [84-88-8]
Hydroxystilbamidine [495-99-8]
Hydroxystilbamidine Isethionate*
   [533-22-2]
Hydroxystreptomycin [6835-00-3]
Hydroxytetracaine [490-98-2]
5-Hydroxytryptophan [56-69-9]
Hydroxyurea [127-07-1]
Hydroxyzine [68-88-2]
Hydroxyzine Dihydrochloride*
   [2192-20-3]
Hydroxyzine Pamoate* [10246-75-0]
Hyenanchin [3484-46-6]
Hygrine [496-49-1]
Hygromycin [6379-56-2]
Hygromycin B [31282-04-9]
Hygrophylline [3573-82-8]
Hymecromone [90-33-5]
Hymecromone O,O-Diethyl Phosphorothioate [299-45-6]
Hyodeoxycholic Acid [83-49-8]
Hyoscyamine [101-31-5]
Hyoscyamine Hydrobromide*
   [306-03-6]
Hyoscyamine Sulfate Dihydrate*
   [6835-16-1]
Hypalon® [9008-08-6]
Hypaphorine [487-58-1]
Hypericin [548-04-9]
Hypochlorous Acid [7790-92-3]
Hypoglycine A [156-56-9]
Hypoglycine B [502-37-4]
Hypophosphoric Acid [7803-60-3]
Hypophosphorous Acid [6303-21-5]
Hypoxanthine [68-94-0]
Ibogaine [83-74-9]
Ibopamine [66195-31-1]
Ibotenic Acid [2552-55-8]
Ibrotamide [466-14-8]
Ibudilast [50847-11-5]
Ibufenac [1553-60-2]
Ibuprofen [15687-27-1]
Ibuproxam [53648-05-8]
Ichthammol [8029-68-3]
Ichthyopterin [490-58-4]
Idebenone [58186-27-9]
Idose [2152-76-3]
Idoxuridine [54-42-2]
Idrocilamide [6961-46-2]
Ifenprodil [23210-56-2]
Ifosfamide [3778-73-2]

Imazamethabenz [81405-85-8]
Imazaquin [81335-37-7]
Imiclopazine [7224-08-0]
Imidazole [288-32-4]
Imidazole Salicylate [36364-49-5]
2-Imidazolidinone [120-93-4]
Imidocarb [27885-92-3]
Imidocarb Dihydrochloride* [5318-76-3]
Iminodiacetic Acid [142-73-4]
4,4'-Iminodicyclohexanecarboxylic Acid [53555-68-3]
Imipenem [74431-23-5]
Imipramine [50-49-7]
Imipramine Hydrochloride* [113-52-0]
Imipramine N-Oxide [6829-98-7]
Imolamine [318-23-0]
Imperatorin [482-44-0]
Imperialine [18059-10-4]
Improsulfan [13425-98-4]
Indaconitine [4491-19-4]
Indalpine [63758-79-2]
Indan [496-11-7]
Indanazoline [40507-78-6]
Indanthrene® [81-77-6]
Indapamide [26807-65-8]
1H-Indazole [271-44-3]
Indecainide [74517-78-5]
Indeloxazine Hydrochloride [65043-22-3]
Indene [95-13-6]
Indenolol [60607-68-3]
Indican (Metabolic Indican) [487-94-5]
Indican (Plant Indican) [487-60-5]
Indigo [482-89-3]
Indigo Carmine [860-22-0]
Indium [7440-74-6]
Indium Antimonide [1312-41-0]
Indium Arsenide [1303-11-3]
Indium Oxide [1312-43-2]
Indium Phosphide [22398-80-7]
Indium Selenide [1312-42-1]
Indium Sulfate [13464-82-9]
Indium Telluride [1312-45-4]
Indium Trichloride [10025-82-8]
Indium Trifluoride [7783-52-0]
Indobufen [63610-08-2]
Indocyanine Green [3599-32-4]
Indole [120-72-9]
Indoleacetic Acid [87-51-4]
Indolebutyric Acid [133-32-4]
Indolmycin [21200-24-8]
3-Indolylacetone [1201-26-9]
Indomethacin [53-86-1]
Indomethacin Sodium Trihydrate* [74252-25-8]
Indoprofen [31842-01-0]
Indoramin [26844-12-2]
Indoramin Hydrochloride* [38821-52-2]
Indospicine [16377-00-7]
Inosine [58-63-9]
Inosine Pranobex [36703-88-5]
Inosinic Acid [131-99-7]
Inositol [87-89-8]
Inositol Monophosphate [573-35-3]
Inositol Niacinate [6556-11-2]
Insularine [549-07-5]
Insulin [9004-10-8]
Insulin Beef* [11070-73-8]
Insulin Human* [11061-68-0]
Insulin Pork* [12584-58-6]
Insulinase [9013-83-6]
Insulin Injection [9004-10-8]
Insulin Zinc Suspension [8049-62-5]
Interferon Alfa-2a* [76543-88-9]
Interferon Alfa-2b* [99210-65-8]
Interferon-γ [82115-62-6]
Interferon gamma-1b* [98059-61-1]

Cys-Tyr-Cys-IFN-γ* [98059-18-8]
Interleukin-2 [85898-30-2]
Inulin [9005-80-5]
Invertase [9001-57-4]
Invert Sugar [8013-17-0]
Iobenzamic Acid [3115-05-7]
Iocarmic Acid [10397-75-8]
Iocarmic Acid Dimeglumine* [54605-45-7]
Iocetamic Acid [16034-77-8]
Iodamide [440-58-4]
Iodic Acid [7782-68-5]
Iodinated Glycerol [5634-39-9]
Iodine [7553-56-2]
Iodine Colloidal [7553-56-2]
Iodine Heptafluoride [16921-96-3]
Iodine Monobromide [7789-33-5]
Iodine Monochloride [7790-99-0]
Iodine Pentafluoride [7783-66-6]
Iodine Pentoxide [12029-98-0]
Iodine Trichloride [865-44-1]
Iodinin [68-81-5]
Iodipamide [606-17-7]
Iodipamide Meglumine* [3521-84-4]
Iodipamide Sodium* [2618-26-0]
Iodized Oil [8001-40-9]
Iodoacetic Acid [64-69-7]
Iodoalphionic Acid [577-91-3]
p-Iodoaniline [540-37-4]
o-Iodoanisole [529-28-2]
Iodobenzene [591-50-4]
o-Iodobenzoic Acid [88-67-5]
Iodochlorhydroxyquin [130-26-7]
Iodofenphos [18181-70-9]
Iodoform [75-47-8]
o-Iodohippurate Sodium [133-17-5]
o-Iodophenol [533-58-4]
p-Iodophenol [540-38-5]
Iodophthalein Sodium [2217-44-9]
Iodophthalein* [386-17-4]
Iodopsin [1415-94-7]
Iodopyracet [300-37-8]
Iodopyrrole [87-58-1]
Iodoquinol [83-73-8]
N-Iodosuccinimide [516-12-1]
3-Iodotyrosine [70-78-0]
Iofetamine I 123 [75917-92-9]
Iofetamine I 123 Hydrochloride* [85068-76-4]
Ioglycamic Acid [2618-25-9]
Iohexol [66108-95-0]
Iomeglamic Acid [25827-76-3]
Iopamidol [60166-93-0]
Iopanoic Acid [96-83-3]
Iopentol [89797-00-2]
Iophendylate [99-79-6]
Iophenoxic Acid [96-84-4]
Iopromide [73334-07-3]
Iopronic Acid [41473-08-9]
Iopydol [5579-92-0]
Iopydone [5579-93-1]
Iothalamic Acid [2276-90-6]
Iothalamic Acid Meglumine* [13087-53-1]
Iothalamic Acid Sodium* [1225-20-3]
Iothion [534-08-7]
Iothiouracil [5984-97-4]
Iotrolan [79770-24-4]
Ioxaglic Acid [59017-64-0]
Ioxaglic Acid Meglumine* [59018-13-3]
Ioxaglic Acid Sodium* [67992-58-9]
Ipodate [5587-89-3]
Ipodate Calcium* [1151-11-7]
Ipodate Sodium* [1221-56-3]
Ipratropium Bromide [22254-24-6]
Ipriflavone [35212-22-7]
Iprindole [5560-72-5]
Iproclozide [3544-35-2]
Iprodione [36734-19-7]
Iproniazid [54-92-2]

Ipronidazole [14885-29-1]
Ipsapirone [95847-70-4]
Ipsapirone Hydrochloride* [92589-98-5]
Iridium [7439-88-5]
Iridium Hexafluoride [7783-75-7]
Iridium Sesquioxide [1312-46-5]
Iridium Trichloride [10025-83-9]
Iridomyrmecin [485-43-8]
Irigenin [548-76-5]
Irisolone [3301-68-6]
Iron [7439-89-6]
α-Irone [79-69-6]
β-Irone [79-70-9]
γ-Irone [79-68-5]
Iron Pentacarbonyl [13463-40-6]
Iron Sorbitex [1338-16-5]
Irsogladine [57381-26-7]
Irsogladine Maleate* [84504-69-8]
Isanic Acid [506-25-2]
Isatide [464-73-3]
Isatin [91-56-5]
Isatropic Acid [596-56-5]
Isaxonine [4214-72-6]
Isazofos [42509-80-8]
Isepamicin [58152-03-7]
Isepamicin Sulfate* [67814-76-0]
Isethionic Acid [107-36-8]
Isoaminile [77-51-0]
Isoamyl Acetate [123-92-2]
Isoamylamine [107-85-7]
Isoamyl Benzoate [94-46-2]
Isoamyl Bromide [107-82-4]
Isoamyl Butyrate [106-27-4]
Isoamyl Chloride [107-84-6]
Isoamyl Cyanide [542-54-1]
Isoamyl Ether [544-01-4]
Isoamyl Formate [110-45-2]
Isoamyl Iodide [541-28-6]
Isoamyl Isovalerate [659-70-1]
Isoamyl Nitrate [543-87-3]
Isoamyl Nitrite [110-46-3]
Isoamyl Phthalate [605-50-5]
Isoamyl Salicylate [87-20-7]
Isoapo-β-erythroidine [2581-07-9]
Isoascorbic Acid [89-65-6]
Isobenzan [297-78-9]
Isoborneol [124-76-5]
Isobornyl Thiocyanoacetate, Technical [115-31-1]
Isobutol [41663-50-7]
Isobutyl Acetate [110-19-0]
Isobutyl Alcohol [78-83-1]
Isobutylamine [78-81-9]
Isobutyl p-Aminobenzoate [94-14-4]
Isobutylbenzene [538-93-2]
Isobutyl Bromide [78-77-3]
Isobutyl n-Butyrate [539-90-2]
Isobutyl Carbamate [543-28-2]
Isobutyl Chloride [513-36-0]
Isobutyl Chlorocarbonate [543-27-1]
Isobutylene [115-11-7]
Isobutyl Ether [628-55-7]
Isobutyl Formate [542-55-2]
Isobutyl Iodide [513-38-2]
Isobutyl Isobutyrate [97-85-8]
Isobutyl Isovalerate [589-59-3]
Isobutyl Mercaptan [513-44-0]
Isobutyl Nitrate [543-29-3]
Isobutyl Nitrite [542-56-3]
Isobutyl Propionate [540-42-1]
Isobutyl Stearate [646-13-9]
Isobutyl Sulfide [592-65-4]
Isobutyl Thiocyanate [591-84-4]
Isobutyl Urethane [539-89-9]
Isobutyraldehyde [78-84-2]
Isobutyric Acid [79-31-2]
Isocarboxazid [59-63-2]
Isochondrodendrine [477-62-3]
Isocil [314-42-1]

Denotes derivative of title compound

Isocinchomeronic Acid [100-26-5]
Isoconazole [27523-40-6]
Isocorybulbine [22672-74-8]
Isocorydine [475-67-2]
Isocorypalmine [53447-14-6]
Isocrotonic Acid [503-64-0]
Isocyanic Acid [75-13-8]
Isodurene [527-53-7]
Isoestradiol [517-04-4]
8-Isoestrone [517-06-6]
Isoetharine [530-08-5]
Isoetharine Hydrochloride*
 [2576-92-3]
Isoeugenol [97-54-1]
Isofenphos [25311-71-1]
Isofezolac [50270-33-2]
Isoflavone [574-12-9]
Isoflupredone [338-95-4]
Isoflupredone Acetate* [338-98-7]
Isoflurane [26675-46-7]
Isoflurophate [55-91-4]
L-Isoglutamine [636-65-7]
Isoladol [530-34-7]
Isolan® [119-38-0]
Isoleucine [73-32-5]
Isolysergic Acid [478-95-5]
Isomaltol [3420-59-5]
Isometamidium Chloride
 [34301-55-8]
Isomethadol [25117-79-7]
Isomethadone [466-40-0]
Isometheptene [503-01-5]
Isometheptene Hydrochloride*
 [6168-86-1]
Isoniazid [54-85-3]
Isoniazid Methanesulfonate
 [13447-95-5]
Isonicotinic Acid [55-22-1]
Isonicotinic Acid Diethylamide
 [530-40-5]
Isonipecotic Acid [498-94-2]
Isonitrosoacetone [306-44-5]
Isonitrosoacetophenone [532-54-7]
Isonixin [57021-61-1]
Isooctane [540-84-1]
Isooctyl Alcohol [26952-21-6]
Isopentyl Alcohol [123-51-3]
Isophane Insulin Suspension
 [9004-17-5]
Isophthalic Acid [121-91-5]
Isophytol [505-32-8]
Isopilosine [491-88-3]
Isopimaric Acid [5835-26-7]
Isoprene [78-79-5]
Isopromethazine [303-14-0]
Isopropalin [33820-53-0]
Isopropamide [71-81-8]
Isopropenyl Acetate [108-22-5]
4-(5-Isopropenyl-2-methyl-1-cyclo-
 penten- 1-yl)-2-butanone [87-45-6]
Isopropyl Acetate [108-21-4]
Isopropyl Acetoacetate [542-08-5]
Isopropylacetone [108-10-1]
Isopropyl Alcohol [67-63-0]
Isopropylamine [75-31-0]
Isopropyl Bromide [75-26-3]
Isopropyl Chloride [75-29-6]
Isopropyl Ether [108-20-3]
Isopropylidene Glycerol [100-79-8]
Isopropyl Iodide [75-30-9]
Isopropyl Myristate [110-27-0]
Isopropyl Nitrite [541-42-4]
Isoproterenol [7683-59-2]
Isoproterenol Sulfate Dihydrate*
 [6700-39-6]
Isoproturon [34123-59-6]
Isopyrocalciferol [474-70-4]
Isoquassin [21293-20-9]
Isoquercitrin [21637-25-2]
Isoquinoline [119-65-3]
Isorubijervine [468-45-1]

Isosafrole [120-58-1]
Isosorbide [652-67-5]
Isosorbide Mononitrate*
 [16051-77-7]
Isosorbide Dinitrate [87-33-2]
β-Isosparteine [24915-04-6]
Isothebaine [568-21-8]
Isothipendyl [482-15-5]
Isothipendyl Hydrochloride*
 [1225-60-1]
Isovaleraldehyde [590-86-3]
Isovaleramide [541-46-8]
Isovaleric Acid [503-74-2]
Isovaleryl Chloride [108-12-3]
Isovaleryl Diethylamide [533-32-4]
2-Isovalerylindane-1,3-dione
 [83-28-3]
Isovaline [595-40-4]
Isoxaben [8255-50-7]
Isoxepac [55453-87-7]
Isoxicam [34552-84-6]
Isoxsuprine [395-28-8]
Isoxsuprine Hydrochloride*
 [579-56-6]
Isradipine [75695-93-1]
Itaconic Acid [97-65-4]
Itraconazole [84625-61-6]
Itramin Tosylate [13445-63-1]
Ivermectin [70288-86-7]
Janus Green B [2869-83-2]
Japan Wax [8001-39-6]
Jasmone [488-10-8]
Jatrorrhizine [3621-38-3]
Javanicin [476-45-9]
Jervine [469-59-0]
Jesaconitine [16298-90-1]
Josamycin [16846-24-5]
Juglone [481-39-0]
Julocrotine [492-87-5]
Kaempferol [520-18-3]
Kainic Acid [487-79-6]
Kallidin [342-10-9]
Kallikrein [9001-01-8]
Kanamycin [8063-07-8]
Kanamycin Sulfate* [25389-94-0]
Kanamycin A* [59-01-8]
Kanamycin A Sulfate* [25389-94-0]
Kanamycin B* [4696-78-8]
Karanjin [521-88-0]
Karaya Gum [9000-36-6]
Karsil [2533-89-3]
Kawain [500-64-1]
Kebuzone [853-34-9]
Keratinase [9025-41-6]
Kermesic Acid [18499-92-8]
Kerosene [8008-20-6]
Ketamine [6740-88-1]
Ketamine Hydrochloride*
 [1867-66-9]
Ketanserin [74050-98-9]
Ketazolam [27223-35-4]
Ketene [463-51-4]
Kethoxal [27762-78-3]
Ketipic Acid [533-76-6]
Ketobemidone [469-79-4]
Ketoconazole [65277-42-1]
α-Ketoglutaric Acid [328-50-7]
2-Keto-L-gulonic Acid [526-98-7]
Ketoprofen [22071-15-4]
11-Ketoprogesterone [516-15-4]
Ketorolac [74103-06-3]
Ketorolac Tromethamine*
 [74103-07-4]
Ketotifen [34580-13-7]
Ketotifen Fumarate* [34580-14-8]
Khellin [82-02-0]
Khellol Glucoside [17226-75-4]
Kinetin [525-79-1]
Kino [8052-20-7]
Kinoprene [42588-37-4]
Kitol [4626-00-0]

Kojic Acid [501-30-4]
Kopsine [559-48-8]
Krebiozen® [9008-19-9]
Krypton [7439-90-9]
Kurchessine [6869-45-0]
Kurcholessine [6869-47-2]
Kyanmethin [461-98-3]
Kynurenic Acid [492-27-3]
Kynurenine [343-65-7]
Labetalol [36894-69-6]
Labetalol Hydrochloride*
 [32780-64-6]
Lachesine [1164-38-1]
Lactaroviolin [85-33-6]
Lactate Dehydrogenase [9001-60-9]
D-Lactic Acid [10326-41-7]
DL-Lactic Acid [598-82-3]
L-Lactic Acid [79-33-4]
Lactic Acid Lactate [617-57-2]
Lactobacillic Acid [19625-10-6]
Lactobionic Acid [96-82-2]
p-Lactophenetide [539-08-2]
Lactose [63-42-3]
Lactucin [1891-29-8]
Lactulose [4618-18-2]
Laminaran [9008-22-4]
Lamotrigine [84507-84-1]
Lankamycin [30042-37-6]
Lanosterol [79-63-0]
Lanthanum [7439-91-0]
Lanthionine [922-55-4]
Lapachol [84-79-7]
Lappaconitine [32854-75-4]
Lapyrium Chloride [6272-74-8]
Lasalocid A [25999-31-9]
Laserpitin [7067-12-1]
Lasiocarpine [303-34-4]
Laudanidine [301-21-3]
Laudanine [85-64-3]
Laudanosine [2688-77-9]
Laudexium Methyl Sulfate
 [3253-60-9]
Laureline [81-38-9]
Laurepukine [34029-94-2]
Lauric Acid [143-07-7]
Laurocapram [59227-89-3]
Lauroguadine [135-43-3]
Laurolinium Acetate [146-37-2]
Laurotetanine [128-76-7]
3-O-Lauroylpyridoxol Diacetate
 [1562-13-6]
Lauryl Bromide [143-15-7]
Lawrencium [22537-19-5]
Lawsone [83-72-7]
Lazurite [1302-83-6]
Lead [7439-92-1]
Lead Acetate [301-04-2]
Lead Antimonate(V) [13510-89-9]
Lead Arsenate [10102-48-4]
Lead Arsenite [10031-13-7]
Lead Azide [13424-46-9]
Lead Borate [10214-39-8]
Lead Bromate [34018-28-5]
Lead Bromide [10031-22-8]
Lead Butyrate [819-73-8]
Lead Chlorate [10294-47-0]
Lead Chloride 7758-95-4]
Lead Chromate(VI) [7758-97-6]
Lead Chromate(VI) Oxide
 [18454-12-1]
Lead Dioxide [1309-60-0]
Lead Fluoride [7783-46-2]
Lead Formate [811-54-1]
Lead Hexafluorosilicate [25808-74-6]
Leadhillite [1319-48-8]
Lead Hydroxide [1311-11-1]
Lead Hypophosphite [10294-58-3]
Lead Iodide [10101-63-0]
Lead Lactate [18917-82-3]
Lead Molybdate(VI)
 [10190-55-3]

Denotes derivative of title compound

Lead Monoxide [1317-36-8]
Lead Nitrate [10099-74-8]
Lead Oleate [1120-46-3]
Lead Oxalate [814-93-7]
Lead Phosphate [7446-27-7]
Lead Selenate [7446-15-3]
Lead Selenite [7488-51-9]
Lead Sesquioxide [1314-27-8]
Lead Sodium Thiosulfate [10101-94-7]
Lead Stearate [1072-35-1]
Lead Subacetate [1335-32-6]
Lead Sulfate [7446-14-2]
Lead Sulfide [1314-87-0]
Lead Telluride [1314-91-6]
Lead Tetraacetate [546-67-8]
Lead Tetrafluoride [7783-59-7]
Lead Tetroxide [1314-41-6]
Lead Thiocyanate [592-87-0]
Lead Tungstate(VI) [7759-01-5]
Lead Vanadate(V) [10099-79-3]
Ledol [577-27-5]
Lefetamine [7262-75-1]
Leiopyrrole [5633-16-9]
Lenacil [2164-08-1]
Lenampicillin [86273-18-9]
Lenperone [24678-13-5]
Lenthionine [292-46-6]
Lentinan [80738-42-7]
Leonurine [24697-74-3]
Lepidine [491-35-0]
Leptodactyline [13957-33-0]
Leptophos [21609-90-5]
Lethane® 60 [301-11-1]
Lethane® 384 [112-56-1]
Letosteine [53943-88-7]
Leucine [61-90-5]
Leucinocaine Mesylate [135-44-4]
Leucite [1302-34-7]
Leucocyanidin [480-17-1]
Leucodrin [14225-07-1]
Leucoglycodrin [13190-88-0]
Leucomycins [1392-21-8]
Leucopterin [492-11-5]
Leuprolide [53714-56-0]
Leuprolide Acetate* [74381-53-6]
Levallorphan [152-02-3]
Levobunolol [47141-42-4]
Levobunolol Hydrochloride* [27912-14-7]
Levodopa [59-92-7]
Levomepate [428-07-9]
Levomethadyl Acetate [1477-40-3]
Levophacetoperane [24558-01-8]
Levopimaric Acid [79-54-9]
Levopropoxyphene [2338-37-6]
Levorphanol [77-07-6]
Levothyroxine Sodium [55-03-8]
Levulinic Acid [123-76-2]
LH [9002-67-9]
LH-RH [9034-40-6]
Lichenin [1402-10-4]
Lidamidine [66871-56-5]
Lidamidine Hydrochloride* [65009-35-0]
Lidocaine [137-58-6]
Lidocaine Hydrochloride Monohydrate* [6108-05-0]
Lidoflazine [3416-26-0]
Light Green SF Yellowish [5141-20-8]
Lignin [9005-53-2]
Lignoceric Acid [557-59-5]
Ligroin [8032-32-4]
Limaprost [74397-12-9]
Limettin [487-06-9]
Limonene [138-86-3]
Limonin [1180-71-8]
Linalool [78-70-6]
Linalyl Acetate [115-95-7]
Linamarin [554-35-8]

Linarin [480-36-4]
Linatine [10139-06-7]
Lincomycin [154-21-2]
Lincomycin Hydrochloride Monohydrate* [7179-49-9]
Lindane [58-89-9]
Lindlar Catalyst [53092-86-7]
Lineatin [65035-34-9]
Linoleic Acid [60-33-3]
Linolenic Acid [463-40-1]
γ-Linolenic Acid [506-26-3]
Linuron [330-55-2]
Liothyronine [6893-02-3]
Liothyronine Sodium* [55-06-1]
Lipase [9001-62-1]
Lipoprotein Lipase [9004-02-8]
Lipotropic Hormone [9035-55-6]
Lisinopril [83915-83-7]
Lisuride [18016-80-3]
Lithium [7439-93-2]
Lithium Acetate [546-89-4]
Lithium Acetylsalicylate [552-98-7]
Lithium Amide [7782-89-0]
Lithium Benzoate [553-54-8]
Lithium Bitartrate [868-16-6]
Lithium Borate [1303-94-2]
Lithium Borohydride [16949-15-8]
Lithium Bromide [7550-35-8]
Lithium Carbonate [554-13-2]
Lithium Chloride [7447-41-8]
Lithium Chromate(VI) [14307-35-8]
Lithium Citrate [919-16-4]
Lithium Dichromate(VI) [13843-81-7]
Lithium Fluoride [7789-24-4]
Lithium Formate [556-63-8]
Lithium Hydride [7580-67-8]
Lithium Hydroxide [1310-65-2]
Lithium Iodide [10377-51-2]
Lithium Nitrate [7790-69-4]
Lithium Oxalate [553-91-3]
Lithium Oxide [12057-24-8]
Lithium Perchlorate [7791-03-9]
Lithium Selenate [15593-52-9]
Lithium Selenite [15593-51-8]
Lithium Silicate [10102-24-6]
Lithium Sulfate [10377-48-7]
Lithium Tetracyanoplatinate(II) [14402-73-4]
Lithocholic Acid [434-13-9]
Lithopone [1345-05-7]
Litmocidin [1392-27-4]
Lobelanidine [552-72-7]
Lobelanine [21511-25-1]
Lobeline [90-69-7]
Lobenzarit [63329-53-3]
Lobenzarit Sodium* [64808-48-6]
Lochnericine [2447-58-7]
Lochneridine [5980-01-8]
Lofentanil [61380-40-3]
Lofepramine [23047-25-8]
Lofepramine Hydrochloride* [26786-32-3]
Lofexidine [31036-80-3]
Lofexidine Hydrochloride* [21498-08-8]
Loflucarban [790-69-2]
Loganin [18524-94-2]
Loline [25161-91-5]
Lomefloxacin [98079-51-7]
Lomefloxacin Hydrochloride* [98079-52-8]
Lomustine [13010-47-4]
Lonazolac [53008-88-1]
Longifolene [475-20-7]
Lonidamine [50264-69-2]
Looplure [14959-86-5]
Loperamide [53179-11-6]
Loperamide Hydrochloride* [34552-83-5]
Lophophorine [17627-78-0]
Lophotoxin [78697-56-0]

Loprazolam [61197-93-1]
Lorajmine [47562-08-3]
Lorajmine Hydrochloride* [40819-93-0]
Loratadine [79794-75-5]
Lorazepam [846-49-1]
Lorcainide [59729-31-6]
Lorcainide Hydrochloride* [58934-46-6]
Lormetazepam [848-75-9]
Lotrifen [66535-86-2]
Lovastatin [75330-75-5]
Loxapine [1977-10-2]
Loxapine Succinate* [27833-64-3]
Loxoprofen [68767-14-6]
Lucanthone Hydrochloride [548-57-2]
Lucensomycin [13058-67-8]
Lucifer Yellow CH [67769-47-5]
Lumazine [487-21-8]
Lumichrome [1086-80-2]
Lumiflavine [1088-56-8]
Luminol [521-31-3]
Lumisterol [474-69-1]
Lunacridine [83-58-9]
Lunacrine [82-40-6]
Lunasine [52306-36-2]
Lunine [518-60-5]
Lunularic Acid [23255-59-6]
Lupeol [545-47-1]
Lupinine [486-70-4]
Luprostiol [73523-00-9]
Lupulon [468-28-0]
Luteolin [491-70-3]
Lutetium [7439-94-3]
2,6-Lutidine [108-48-5]
Lututrin [1407-04-1]
Lyapolate Sodium [25053-27-4]
Lycoctonine [26000-17-9]
Lycoctonine N-Succinylanthranilic Acid Ester* [25867-19-0]
Lycidine [20316-18-1]
Lycofawcine [3175-90-4]
Lycomarasmine [7611-43-0]
Lycopene [502-65-8]
Lycophyll [19891-75-9]
Lycopodine [466-61-5]
Lycoramine [21133-52-8]
Lycorine [476-28-8]
Lycoxanthin [19891-74-8]
Lymecycline [992-21-2]
Lynestrenol [52-76-6]
Lypressin [50-57-7]
Lysalbinic Acid [9006-58-0]
Lysergamide [478-94-4]
Lysergic Acid [82-58-6]
Lysergide [50-37-3]
Lysidine [534-26-9]
Lysine [56-87-1]
Lysine Hydrochloride* [657-27-2]
Lysine Acetylsalicylate [62952-06-1]
L-Lysine L-Glutamate [5408-52-6]
Lysostaphin [9011-93-2]
Lysozyme [9001-63-2]
Lyxoflavine [13123-37-0]
D-Lyxose [1114-34-7]
Mabuterol [56341-08-3]
Maclurin [519-34-6]
Macromerine [2970-95-8]
Maddrell's Salt [10361-03-2]
MADU [840-50-6]
Maduramicin [84878-61-5]
Mafenide [138-39-6]
Mafenide Acetate* [13009-99-9]
Mafenide Hydrochloride* [138-37-4]
Magaldrate [1317-26-6]
Magenta I [632-99-5]
Magnesium [7439-95-4]
Magnesium Acetate [142-72-3]
Magnesium Acetylsalicylate [132-49-0]

Denotes derivative of title compound

Magnesium Amide [7803-54-5]
Magnesium Benzoate [553-70-8]
Magnesium Bisulfate [10028-26-9]
Magnesium Borate [13703-82-7]
Magnesium Bromate [7789-36-8]
Magnesium Bromide [7789-48-2]
Magnesium Carbonate Hydroxide [39409-82-0]
Magnesium Chlorate [10326-21-3]
Magnesium Chloride [7786-30-3]
Magnesium Citrate [3344-18-1]
Magnesium Citrate, Dibasic [144-23-0]
Magnesium Fluoride [7783-40-6]
Magnesium Formate [557-39-1]
Magnesium Germanide [1310-52-7]
Magnesium Hexafluorosilicate [16949-65-8]
Magnesium Hydride [60616-74-2]
Magnesium Hydroxide [1309-42-8]
Magnesium Iodide [10377-58-9]
Magnesium Lactate [18917-93-6]
Magnesium Mandelate [18937-33-2]
Magnesium Nitrate [10377-60-3]
Magnesium Oleate [1555-53-9]
Magnesium Oxalate [547-66-0]
Magnesium Oxide [1309-48-4]
Magnesium Perborate [14635-87-1]
Magnesium Perchlorate [10034-81-8]
Magnesium Permanganate [10377-62-5]
Magnesium Peroxide [1335-26-8]
Magnesium Phosphate, Dibasic [7757-86-0]
Magnesium Phosphate, Monobasic [13092-66-5]
Magnesium Phosphate, Tribasic [7757-87-1]
Magnesium Potassium Selenate [28041-84-1]
Magnesium Pyrophosphate [13446-24-7]
Magnesium Salicylate [18917-89-0]
Magnesium Selenate [14986-91-5]
Magnesium Selenide [1313-04-8]
Magnesium Selenite [15593-61-0]
Magnesium Silicide [22831-39-6]
Magnesium Stannide [1313-08-2]
Magnesium Stearate [557-04-0]
Magnesium Sulfate [7487-88-9]
Magnesium Sulfite [7757-88-2]
Magnesium Thiocyanate [306-61-6]
Magnesium Thiosulfate [10124-53-5]
Magnesium Tungstate(VI) [13573-11-0]
Magneson [74-39-5]
Magnoflorine [2141-09-5]
Magnoline [6859-66-1]
Malachite Green [569-64-2]
Malathion [121-75-5]
Maleamic Acid [557-24-4]
Maleanilic Acid [555-59-9]
Maleic Acid [110-16-7]
Maleic Anhydride [108-31-6]
Maleic Hydrazide [123-33-1]
Maleuric Acid [105-61-3]
Malic Acid [6915-15-7]
Malonic Acid [141-82-2]
Malononitrile [109-77-3]
Malotilate [59937-28-9]
Maltol [118-71-8]
Maltose [69-79-4]
Malvidin Chloride [643-84-5]
Mancozeb [8018-01-7]
Mandelic Acid [90-64-2]
Mandelic Acid Isoamyl Ester [5421-04-5]
Mandelonitrile [532-28-5]
Mandelonitrile Glucoside [138-53-4]
Maneb [12427-38-2]
Manganese [7439-96-5]

Manganese Acetate [638-38-0]
Manganese Borate [12228-91-0]
Manganese Bromide [13446-03-2]
Manganese Carbonate [598-62-9]
Manganese Carbonyl [10170-69-1]
Manganese Chloride [7773-01-5]
Manganese Difluoride [7782-64-1]
Manganese Dioxide [1313-13-9]
Manganese Hypophosphite [10043-84-2]
Manganese Iodide [7790-33-2]
Manganese Nitrate [10377-66-9]
Manganese Oleate [23250-73-9]
Manganese Oxalate [640-67-5]
Manganese Oxide [1317-35-7]
Manganese Phosphate, Dibasic [7782-76-5]
Manganese Pyrophosphate [53731-35-4]
Manganese Selenide [1313-22-0]
Manganese Sesquioxide [1317-34-6]
Manganese Silicate [7759-00-4]
Manganese Sulfate [7785-87-7]
Manganese Sulfide [18820-29-6]
Manganese Trifluoride [7783-53-1]
Mangostin [6147-11-1]
Mannitol [69-65-8]
Mannitol Hexanitrate [15825-70-4]
Mannomustine [551-74-6]
D-Mannose [3458-28-4]
Maprotiline [10262-69-8]
Maprotiline Hydrochloride* [10347-81-6]
Margaric Acid [506-12-7]
Marrubiin [465-92-9]
Matricarin [5989-43-5]
Matrine [519-02-8]
Maytansine [35846-53-8]
Mazindol [22232-71-9]
Mazipredone [60-39-9]
MCPA [94-74-6]
MDMA [42542-10-9]
Mebendazole [31431-39-7]
Mebeverine [3625-06-7]
Mebeverine Hydrochloride* [2753-45-9]
Mebhydroline [524-81-2]
Mebiquine [23910-07-8]
MeBmt [59865-23-5]
Mebrofenin [78266-06-5]
Mebutamate [64-55-1]
Mecamylamine [60-40-2]
Mecamylamine Hydrochloride* [826-39-1]
Mechlorethamine [51-75-2]
Mechlorethamine Oxide Hydrochloride [302-70-5]
Mechlorethamine Hydrochloride* [55-86-7]
Meclizine [569-65-3]
Meclizine Dihydrochloride* [31884-77-2]
Meclocycline [2013-58-3]
Meclocycline Sulfosalicylate* [73816-42-9]
Meclofenamic Acid [644-62-2]
Meclofenamic Acid Sodium* [6385-02-0]
Meclofenoxate [51-68-3]
Mecloqualone [340-57-8]
Mecloralurea [1954-79-6]
Mecloxamine [5668-06-4]
Meconic Acid [497-59-6]
Meconin [569-31-3]
Mecoprop [7085-19-0]
Mecrylate [137-05-3]
Mecysteine Hydrochloride [18598-63-5]
Medazepam [2898-12-6]
Medazepam Hydrochloride* [2898-11-5]

Medetomidine [86347-14-0]
Medetomidine Hydrochloride* [86347-15-1]
Medibazine [53-31-6]
Medicagol [1983-72-8]
Medicarpin [32383-76-9]
Medifoxamine [32359-34-5]
Medifoxamine Fumarate* [16604-45-8]
Medmain [576-11-4]
Medrogestone [977-79-7]
Medroxyprogesterone [520-85-4]
Medroxyprogesterone Acetate* [71-58-9]
Medrylamine [524-99-2]
Medrysone [2668-66-8]
Mefenamic Acid [61-68-7]
Mefenorex [17243-57-1]
Mefenorex Hydrochloride* [5586-87-8]
Mefexamide [1227-61-8]
Mefloquine Hydrochloride [51773-92-3]
Mefluidide [53780-34-0]
Mefruside [7195-27-9]
Megestrol Acetate [595-33-5]
Meglumine Acetrizoate [22154-43-4]
Meglumine Diatrizoate [131-49-7]
Meglutol [503-49-1]
Melamine [108-78-1]
Melanostatin [9083-38-9]
Melarsoprol [494-79-1]
Melatonin [73-31-4]
Meldrum's Acid [2033-24-1]
Melengestrol [5633-18-1]
Melengestrol Acetate* [2919-66-6]
Melezitose [597-12-6]
Melibiose [585-99-9]
Melilotoside [618-67-7]
Melinamide [14417-88-0]
Melinonine A [6801-41-8]
Melitracen [5118-29-6]
Melitracen Hydrochloride* [10563-70-9]
Melittin [20449-79-0]
Mellitic Acid [517-60-2]
Melperone [3575-80-2]
Melphalan [148-82-3]
Memantine [19982-08-2]
Menadiol Diacetate [573-20-6]
Menadiol Dibutyrate [53370-44-8]
Menadiol Diphosphate (Tetrasodium Salt) [131-13-5]
Menadiol Disulfate [29520-22-7]
Menadione [58-27-5]
Menadione Dimethylpyrimidinol Bisulfite [14451-99-1]
Menadione Sodium Bisulfite [130-37-0]
Menadoxime [573-01-3]
Menbutone [3562-99-0]
Mendelevium [7440-11-1]
Menichlopholan [10331-57-4]
Menthol [89-78-1]
l-Menthone [14073-97-3]
Menthyl Acetate [89-48-5]
Menthyl Borate [53370-45-9]
Menthyl Salicylate [89-46-3]
Menthyl Valerate [89-47-4]
Meobentine [46464-11-3]
Meobentine Sulfate* [58503-79-0]
Meparfynol [77-75-8]
Meparfynol Carbamate [302-66-9]
Mepartricin [11121-32-7]
Mepazine [60-89-9]
Mepenzolate Bromide [76-90-4]
Meperidine [57-42-1]
Meperidine Hydrochloride* [50-13-5]
Mephenesin [59-47-2]
Mephenesin Carbamate* [533-06-2]
Mephenhydramine [3572-74-5]

Mephenoxalone [70-07-5]
Mephentermine [100-92-5]
Mephentermine Sulfate* [6190-60-9]
Mephenytoin [50-12-4]
Mephobarbital [115-38-8]
Mephosfolan [950-10-7]
Mepindolol [23694-81-7]
Mepiprazole [20326-12-9]
Mepiquat Chloride [24307-26-4]
Mepitiostane [21362-69-6]
Mepivacaine [96-88-8]
Mepivacaine Hydrochloride*
  [1722-62-9]
Mepixanox [17854-59-0]
Meprednisone [1247-42-3]
Meprobamate [57-53-4]
Meprylcaine Hydrochloride
  [956-03-6]
Meptazinol [54340-58-8]
Mequitazine [29216-28-2]
Meralein Sodium [4386-35-0]
Merbromin [129-16-8]
Mercamphamide [127-50-4]
Mercamphamide Sodium* [124-34-8]
Mercurophylline* [8012-34-8]
2-Mercaptobenzothiazole [149-30-4]
2-Mercaptoethanol [60-24-2]
Mercaptomerin Sodium [21259-76-7]
6-Mercaptopurine [50-44-2]
Mercufenol Chloride [90-03-9]
Mercumallylic Acid [86-36-2]
Mercumatilin Sodium [8018-15-3]
Mercuric Acetate [1600-27-7]
Mercuric Arsenate [7784-37-4]
Mercuric Benzoate [583-15-3]
Mercuric Bromide [7789-47-1]
Mercuric Chloride [7487-94-7]
Mercuric Chloride, Ammoniated
  [10124-48-8]
Mercuric Cyanide [592-04-1]
Mercuric Dichromate(VI)
  [7789-10-8]
Mercuric Fluoride [7783-39-3]
Mercuric Iodate [7783-32-6]
Mercuric Iodide, Red [7774-29-0]
Mercuric Nitrate [10045-94-0]
Mercuric Oleate [1191-80-6]
Mercuric Oxide, Red [21908-53-2]
Mercuric Oxide, Yellow [21908-53-2]
Mercuric Oxycyanide [1335-31-5]
Mercuric Salicylate [5970-32-1]
Mercuric Sodium p-Phenolsulfonate
  [535-55-7]
Mercuric Stearate [645-99-8]
Mercuric Subsulfate [1312-03-4]
Mercuric Succinimide [584-43-0]
Mercuric Sulfate [7783-35-9]
Mercuric Sulfide, Black [1344-48-5]
Mercuric Sulfide, Red [1344-48-5]
Mercuric Thiocyanate [592-85-8]
Mercurophen [52486-78-9]
Mercurous Acetate [631-60-7]
Mercurous Bromide [10031-18-2]
Mercurous Chlorate [10294-44-7]
Mercurous Chloride [10112-91-1]
Mercurous Fluoride [13967-25-4]
Mercurous Iodide [15385-57-6]
Mercurous Nitrate [10415-75-5]
Mercurous Oxide [15829-53-5]
Mercurous Sulfate [7783-36-0]
Mercury [7439-97-6]
Merisoprol Hg 197 [5579-94-2]
Merphyrin [15375-94-7]
Mersalyl [492-18-2]
Mesaconic Acid [498-24-8]
Mesalamine [89-57-6]
Mescaline [54-04-6]
Mesembrine [24880-43-1]
Mesitylene [108-67-8]
Mesityl Oxide [141-79-7]
Mesna [19767-45-4]

Mesoridazine [5588-33-0]
Mesoridazine Besylate* [32672-69-8]
Mesoxalic Acid [473-90-5]
Mesquite Gum [9000-47-9]
Mestanolone [521-11-9]
Mesterolone [1424-00-6]
Mestilbol [7773-60-6]
Mestranol [72-33-3]
Mesulergine [64795-35-3]
Mesulphen [135-58-0]
Metabutoxycaine Hydrochloride
  [550-01-6]
Metachrome Yellow [584-42-9]
Metaclazepam [84031-17-4]
[2.2]Metacyclophane [2319-97-3]
Metalaxyl [57837-19-1]
Metaldehyde [9002-91-9]
Metamfepramone [15351-09-4]
Metamivam [13898-68-5]
Metampicillin [6489-97-0]
Metanephrine [5001-33-2]
Metanilic Acid [121-47-1]
Metanil Yellow [587-98-4]
Metaphanine [1805-86-3]
Metapramine [21730-16-5]
Metaproterenol [586-06-1]
Metaproterenol Sulfate* [5874-97-5]
Metaraminol [54-49-9]
Metaraminol Bitartrate*
  [33402-03-8]
Metaxalone [1665-48-1]
Metazocine [3734-52-9]
Metcaraphen [561-79-5]
Metcaraphen Hydrochloride*
  [1950-31-8]
Meteloidine [526-13-6]
Metepa [57-39-6]
Metergoline [17692-51-2]
Metescufylline [15518-82-8]
Metformin [657-24-9]
Methabenzthiazuron [18691-97-9]
Methacholine Chloride [62-51-1]
Methacrifos [62610-77-9]
Methacrylic Acid [79-41-4]
Methacrylonitrile [126-98-7]
Methacycline [914-00-1]
Methacycline Hydrochloride*
  [3963-95-9]
Methadone Hydrochloride
  [1095-90-5]
Methadyl Acetate [509-74-0]
Methafurylene [531-06-6]
Methallatal [115-56-0]
Methallenestril [517-18-0]
Methallibure [926-93-2]
Methamidophos [10265-92-6]
Methamphetamine [537-46-2]
Metham Sodium [137-42-8]
Methandriol [521-10-8]
Methandrostenolone [72-63-9]
Methane [74-82-8]
Methanearsonic Acid [124-58-3]
Methanesulfonic Acid [124-58-3]
Methanesulfonyl Chloride [124-63-0]
Methanethiol [74-93-1]
Methanol [67-56-1]
Methantheline Bromide [53-46-3]
Methaphenilene [493-78-7]
Methaphenilene Hydrochloride*
  [7084-07-3]
Methapyrilene [91-80-5]
Methapyrilene Fumarate*
  [33032-12-1]
Methapyrilene Hydrochloride*
  [135-23-9]
Methaqualone [72-44-6]
Methaqualone Hydrochloride*
  [340-56-7]
Metharbital [50-11-3]
Methargen [53370-43-7]
Methazolamide [554-57-4]

Methazole [20354-26-1]
Methdilazine [1982-37-2]
Methdilazine Hydrochloride*
  [1229-35-2]
Methenamine [100-97-0]
Methenamine Allyl Iodide
  [36895-62-2]
Methenamine Anhydromethylenecitrate [6190-43-8]
Methenamine Hippurate [5714-73-8]
Methenamine Mandelate [587-23-5]
Methenamine Salicylate [620-34-8]
Methenamine Sulfosalicylate
  [20480-93-7]
Methenamine Tetraiodine
  [12001-65-9]
Methenolone [153-00-4]
Methenolone Acetate* [434-05-9]
Methenolone Enanthate* [303-42-4]
Methestrol [130-73-4]
Methetoin [5696-06-0]
Methicillin Sodium [132-92-3]
Methidathion [950-37-8]
Methimazole [60-56-0]
Methiocarb [2032-65-7]
Methiodal Sodium [126-31-8]
Methionic Acid [503-40-2]
Methionine [63-68-3]
Methionine DL-Form* [59-51-8]
Methionine Hydroxy Analog
  [583-91-5]
Methioprim [588-36-3]
Methiotriazamine [6043-86-3]
Methisazone [1910-68-5]
Methitural [730-68-7]
Methixene [4969-02-2]
Methixene Hydrochloride Monohydrate* [7081-40-5]
Methocarbamol [532-03-6]
Methohexital Sodium [22151-68-4]
Methomyl [16752-77-5]
Methoprene [40596-69-8]
Methopterin [2410-93-7]
Methotrexate [59-05-2]
Methotrimeprazine [60-99-1]
Methoxamine Hydrochloride
  [61-16-5]
Methoxsalen [298-81-7]
Methoxyamine [67-62-9]
Methoxychlor [72-43-5]
Methoxyflurane [76-38-0]
10-Methoxyharmalan [3589-73-9]
3-Methoxy-4-hydroxyphenylglycol
  [534-82-7]
2-(Methoxymethyl)-5-nitrofuran
  [586-84-5]
2-Methoxynaphthalene [93-04-9]
Methoxyphenamine [93-30-1]
Methoxyphenamine Hydrochloride*
  [5588-10-3]
4-Methoxy-m-phenylenediamine
  [615-05-4]
3-(o-Methoxyphenyl)-2-phenylacrylic
  Acid [25333-25-9]
N-(p-Methoxyphenyl)-p-phenylenediamine [101-64-8]
Methoxypromazine [61-01-8]
Methoxypromazine Maleate*
  [3403-42-7]
6-Methoxy-α-tetralone [1078-19-9]
1-Methoxy-3-(trimethylsilyloxy)-1,3-
  butadiene [59414-23-2]
5-Methoxytryptamine [608-07-1]
Methscopolamine Bromide [155-41-9]
Methsuximide [77-41-8]
Methyclothiazide [135-07-9]
Methyl Abietate [127-25-3]
N-Methylacetanilide [579-10-2]
Methyl Acetate [79-20-9]
Methyl Acetoacetate [105-45-3]
Methyl Acetylsalicylate [580-02-9]

Denotes derivative of title compound

Methyl Acrylate [96-33-3]
Methylal [109-87-5]
Methyl Allyl Trisulfide [34135-85-8]
Methylamine [74-89-5]
2-Methylaminoethanol [109-83-1]
p-Methylaminophenol Sulfate
  [55-55-0]
Methylaniline [100-61-8]
Methyl Anthranilate [134-20-3]
2-Methylanthraquinone [84-54-8]
Methylarbutin [6032-32-2]
3-Methylarsacetin [25384-21-8]
Methylbenzethonium Chloride
  [25155-18-4]
Methyl Benzoate [93-58-3]
Methyl Benzoylsalicylate [610-60-6]
α-Methylbenzylamine [98-84-0]
Methyl Blue [28983-56-4]
Methyl Bromide [74-83-9]
2-Methyl-1-butanol [137-32-6]
3-Methyl-2-butanol [598-75-4]
Methyl tert-Butyl Ether [1634-04-4]
Methyl Butyl Ketone [591-78-6]
2-Methyl-3-butyn-2-ol [115-19-5]
Methyl Butyrate [623-42-7]
Methyl Carbamate [598-55-0]
Methyl Carbitol® [111-77-3]
Methyl Carbonate [616-38-6]
Methyl Cellosolve® [109-86-4]
Methyl Cellosolve® Acetate
  [110-49-6]
Methylcellulose [9004-67-5]
Methyl Chloride [74-87-3]
Methyl Chloroacetate [96-34-4]
Methyl Chlorocarbonate [79-22-1]
3-Methylcholanthrene [56-49-5]
Methylconiine [553-75-3]
3'-Methyl-1,2-cyclopentenophenan-
  threne [549-38-2]
5-Methylcytosine [554-01-8]
Methyl Demeton [8022-00-2]
p-Methyldiphenhydramine
  [19804-27-4]
Methyldiphenylamine [552-82-9]
Methyldopa [555-30-6]
Methyldopa Ethyl Ester Hydrochlo-
  ride* [2508-79-4]
3-O-Methyldopa [300-48-1]
Methyleneaminoacetonitrile
  [109-82-0]
4,4'-Methylenebis[2-chloroaniline]
  [101-14-4]
Methylene Blue [61-73-4]
Methylene Bromide [74-95-3]
α-Methylene Butyrolactone
  [547-65-9]
Methylene Chloride [75-09-2]
Methylenedigallic Acid [552-21-6]
5,5'-Methylenedisalicylic Acid
  [122-25-8]
Methylene Iodide [75-11-6]
N-Methylephedrine [552-79-4]
N-Methylepinephrine [554-99-4]
Methylergonovine [113-42-8]
Methylergonovine Maleate*
  [57432-61-8]
Methyl Ether [115-10-6]
Methyl Ethyl Ketone [78-93-3]
α-Methylfentanyl [79704-88-4]
α-Methylfentanyl* [1443-44-3]
Methyl Fluorosulfonate [421-20-5]
Methyl Formate [107-31-3]
N-Methylglucamine [6284-40-8]
N-Methyl-α-L-glucosamine
  [42852-95-9]
α-Methylglucoside [97-30-3]
Methyl Green [14855-76-6]
Methylguanidine [471-29-4]
Methylhexaneamine [105-41-9]
Methylhydrazine [60-34-4]
Methyl Iodide [74-88-4]

Methyl Isobutyrate [547-63-7]
Methyl Isocyanate [624-83-9]
Methyl Isothiocyanate [556-61-6]
Methyl Isovalerate [556-24-1]
Methyl Lactate [547-64-8]
Methyl Linoleate [112-63-0]
Methyl Malonate [108-59-8]
Methyl Methanesulfonate [66-27-3]
3-Methyl-6,7-methylenedioxy-1-
  piperonylisoquinoline [550-73-2]
Methyl N-Methylnipecotate
  [1690-72-8]
N-Methylmyosmine [525-74-6]
Methyl Nicotinate [93-60-7]
Methyl Nitrate [598-58-3]
Methyl p-Nitrobenzenesulfonate
  [6214-20-6]
N-Methyl-N'-nitro-N-nitrosoguani-
  dine [70-25-7]
Methyl Orange [547-58-0]
Methyl Oxalate [553-90-2]
Methylparaben [99-76-3]
Methyl Parathion [298-00-0]
5-Methyl-5-(3-phenanthryl)hydantoin
  [3784- 92-7]
N-Methylphenazonium Methosulfate
  [299-11-6]
Methylphenidate [113-45-1]
Methylphenidate Hydrochloride*
  [298-59-9]
3-Methyl-5-phenylhydantoin
  [6846-11-3]
3'-Methylphthalanilic Acid [85-72-3]
Methylprednisolone [83-43-2]
Methylprednisolone Acetate*
  [53-36-1]
Methylprednisolone Sodium Phos-
  phate* [5015-36-1]
Methylprednisolone Sodium Suc-
  cinate* [2375-03-3]
p-(2-Methylpropenyl)phenol Acetate
  [53370-48-2]
Methyl Propionate [554-12-1]
Methyl Propyl Ether [557-17-5]
Methyl Propyl Ketone [107-87-9]
5-Methylpyrazole-3-carboxylic Acid
  [402-61-9]
Methyl Pyridyl Ketone [350-03-8]
Methyl 4-Pyridyl Ketone Thiosemi-
  carbazone [3115-21-7]
N-Methylpyrroline [554-15-4]
Methyl Red [493-52-7]
Methyl Salicylate [119-36-8]
4'-(Methylsulfamoyl)sulfanilanilide
  [547-53-5]
Methyl Sulfate [75-93-4]
Methyl Sulfide [75-18-3]
N-Methyltaurine [107-68-6]
17-Methyltestosterone [58-18-4]
17α-Methyltestosterone 3-Cyclopentyl
  Enol Ether [67-81-2]
4-Methyl-5-thiazoleethanol
  [137-00-8]
Methyl Thiocyanate [556-64-9]
Methylthiouracil [56-04-2]
Methyltrienolone [965-93-5]
α-Methyl-m-tyrosine [305-96-4]
6-Methyluracil [626-48-2]
Methyl Vinyl Ketone [78-94-4]
Methymycin [497-72-3]
Methysergid(e) [361-37-5]
Methysergide Maleate* [129-49-7]
Methysticin [495-85-2]
Metiazinic Acid [13993-65-2]
Meticrane [1084-65-7]
Metipranolol [22664-75-7]
Metizoline [17692-22-7]
Metizoline Hydrochloride*
  [5090-37-9]
Metobromuron [3060-89-7]
Metochalcone [18493-30-6]

Metoclopramide [364-62-5]
Metoclopramide Monohydrochloride
  Monohydrate* [54143-57-6]
Metocurine Iodide [7601-55-0]
Metofenazate [388-51-2]
Metofoline [2154-02-1]
Metolachlor [51218-45-2]
Metolazone [17560-51-9]
Metomidate [5377-20-8]
Metopimazine [14008-44-7]
Metopon [143-52-2]
Metopon Hydrochloride* [124-92-5]
Metoprolol [37350-58-6]
Metoprolol Tartrate* [56392-17-7]
Metoquinone [622-91-3]
Metoserpate [1178-28-5]
Metoserpate Hydrochloride*
  [1178-29-6]
Metralindole [54188-38-4]
Metribuzin [21087-64-9]
Metrizamide [31112-62-6]
Metrizoic Acid [1949-45-7]
Metrizoic Acid Sodium* [7225-61-8]
Metronidazole [443-48-1]
Metronidazole Hydrochloride*
  [69198-10-3]
Metron S [13946-02-6]
Metsulfuron Methyl [74223-64-6]
Meturedepa [1661-29-6]
Metyrapone [54-36-4]
Metyrapone Ditartrate* [908-35-0]
Metyridine [114-91-0]
Metyrosine [672-87-7]
Metyrosine DL-Form* [620-30-4]
Mevaldic Acid [541-07-1]
Mevalonic Acid [150-97-0]
Mevastatin [73573-88-3]
Mevinphos [7786-34-7]
Mexacarbate [315-18-4]
Mexazolam [31868-18-5]
Mexenone [1641-17-4]
Mexican [52500-59-1]
Mexiletine [31828-71-4]
Mexiletine Hydrochloride*
  [5370-01-4]
Mezlocillin [51481-65-3]
Mezlocillin Sodium* [59798-30-0]
Mianserin [24219-97-4]
Mianserin Hydrochloride*
  [21535-47-7]
Mibolerone [3704-09-4]
Michler's Base [101-61-1]
Michler's Ketone [90-94-8]
Miconazole [22916-47-8]
Miconazole Nitrate* [22832-87-7]
Micranthine [36104-64-0]
Micrococcin P [1392-46-7]
Micronomicin [52093-21-7]
Midazolam [59467-70-8]
Midazolam Hydrochloride*
  [59467-96-8]
Midazolam Maleate* [59467-94-6]
Midodrine [42794-76-3]
Midodrine Hydrochloride*
  [30902-17-9]
Mifentidine [83184-43-4]
Mifepristone [84371-65-3]
Miglitol [72432-03-2]
Mildiomycin [67527-71-3]
Milorganite® [8049-99-8]
Miloxacin [37065-29-5]
Milrinone [78415-72-2]
Mimosine [500-44-7]
Minaprine [25905-77-5]
Minaprine Hydrochloride*
  [25953-17-7]
Mineral Spirits [64475-85-0]
Minocycline [10118-90-8]
Minocycline Hydrochloride*
  [13614-98-7]
Minoxidil [38304-91-5]

Miokamycin [55881-07-7]
Mipafox [371-86-8]
Mirex [2385-85-5]
Miroprofen [55843-86-2]
Misch Metal [8049-20-5]
Misoprostol [59122-46-2]
Mitobronitol [488-41-5]
Mitoguazone [459-86-9]
Mitolactol [10318-26-0]
Mitotane [53-19-0]
Mitoxantrone [65271-80-9]
Mitoxantrone Dihydrochloride*
    [70476-82-3]
Mitragynine [4098-40-2]
Mizoribine [50924-49-7]
Mocimycin [50935-71-2]
Moclobemide [71320-77-9]
Mofebutazone [2210-63-1]
Molindone [7416-34-4]
Molindone Hydrochloride*
    [15622-65-8]
Molsidomine [25717-80-0]
Molybdenum [7439-98-7]
Molybdenum Disulfide [1317-33-5]
Molybdenum Hexafluoride
    [7783-77-9]
Molybdenum Sesquioxide [1313-29-7]
Molybdenum Trioxide [1313-27-5]
Molybdic(VI) Acid [7782-91-4]
Mometasone Furoate [83919-23-7]
Monacetin [26446-35-5]
Monardein Chloride [1329-77-7]
Monensin [17090-79-8]
Monoamine Oxidase [9001-66-5]
Monobenzone [103-16-2]
Monocrotaline [315-22-0]
Monocrotophos [6923-22-4]
Monooctanoin [26402-26-6]
Monophen® [530-77-8]
Monorden [12772-57-5]
Monosodium Glutamate [142-47-2]
Monotropein [5945-50-6]
Montan Wax [8002-53-7]
Montmorillonite [1318-93-0]
Monuron [150-68-5]
Moperone [1050-79-9]
Mopidamol [13665-88-8]
Moprolol [5741-22-0]
MOPS [1132-61-2]
Moquizone [19395-58-5]
Morantel [20574-50-9]
Morantel Tartrate* [26155-31-7]
Morazone [6536-18-1]
Morclofone [31848-01-8]
Moricizine [31883-05-3]
Morin [480-16-0]
Morindin [60450-21-7]
Moroxydine [3731-59-7]
Morphazinamide [952-54-5]
Morphenol [519-56-2]
Morpheridine [469-81-8]
Morphinan [468-10-0]
Morphine [57-27-2]
Morphine Hydrobromide [630-81-9]
Morphine Hydrochloride [52-26-6]
Morphine Methylbromide [125-23-5]
Morphine Mucate [596-19-0]
Morphine Oleate, 20% [6033-05-2]
Morphine N-Oxide [639-46-3]
Morphine Sulfate [64-31-3]
Morpholine [110-91-8]
Morpholine Salicylate [147-90-0]
7-Morpholinomethyltheophylline
    [5089-89-4]
Morphothebaine [478-53-5]
Motilin [52906-92-0]
Motretinide [56281-36-8]
Moveltipril [85856-54-8]
Moveltipril Calcium* [85921-53-5]
Moxalactam [64952-97-2]
Moxalactam Disodium* [64953-12-4]

Moxaverine [10539-19-2]
Moxestrol [34816-55-2]
Moxisylyte [54-32-0]
MPTP [28289-54-5]
Mucochloric Acid [87-56-9]
Mucochloric Anhydride [4419-02-3]
Muconic Acid [505-70-4]
Mupirocin [12650-69-0]
Muramic Acid [1114-41-6]
Muramyl Dipeptide [53678-77-6]
Murexide [3051-09-0]
Murexine [20284-40-6]
Muroctasin [78113-36-7]
Muscalure [27519-02-4]
Muscarine [300-54-9]
Muscazone [2255-39-2]
Muscimol [2763-96-4]
Muscone [541-91-3]
Mustard Gas [505-60-2]
Muzolimine [55294-15-0]
Mycaminose [519-21-1]
Mycarose [6032-92-4]
Mycelianamide [22775-52-6]
Myclobutanil [88671-89-0]
Mycobacillin [18524-67-9]
Mycomycin [544-51-4]
Mycophenolic Acid [24280-93-1]
Mycosamine [527-38-8]
Myoral [645-74-9]
β-Myrcene [123-35-3]
Myricetin [529-44-2]
Myristic Acid [544-63-8]
Myristicin [607-91-0]
Myristyl Alcohol [112-72-1]
Myristyltrimethylammonium Bromide
    [1119-97-7]
Myrophine [467-18-5]
Myrrh [9000-45-7]
Myrtecaine [7712-50-7]
Myrtol [8002-55-9]
Mytatrienediol [5108-94-1]
Myxin [13925-12-7]
Nabam [142-59-6]
Nabilone [51022-71-0]
Nabumetone [42924-53-8]
Nadide [53-84-9]
Nadolol [42200-33-9]
Nadoxolol [54063-51-3]
NADP [53-59-8]
Naepaine [2188-67-2]
Nafamostat [81525-10-2]
Nafarelin [76932-56-4]
Nafarelin Acetate* [86220-42-0]
Nafcillin Sodium [985-16-0]
Nafiverine [5061-22-3]
Nafronyl [31329-57-4]
Nafronyl Oxalate* [3200-06-4]
Naftalofos [1491-41-4]
Naftifine [65472-88-0]
Naftifine Hydrochloride*
    [65473-14-5]
Nalbuphine [20594-83-6]
Nalbuphine Hydrochloride*
    [23277-43-2]
Naled [300-76-5]
Nalidixic Acid [389-08-2]
Nalmefene [55096-26-9]
Nalorphine [62-67-9]
Nalorphine Hydrochloride* [57-29-4]
Nalorphine Dinicotinate [3194-25-0]
Naloxone [465-65-6]
Naloxone Hydrochloride* [357-08-4]
Naloxone Hydrochloride Dihydrate*
    [51481-60-8]
Naltrexone [16590-41-3]
Naltrexone Hydrochloride*
    [16676-29-2]
Nandinine [572-76-9]
Nandrolone [434-22-0]
Nandrolone Decanoate [360-70-3]

Nandrolone p-Hexyloxyphenylpro-
    pionate [52279-57-9]
Nandrolone Phenpropionate [62-90-8]
Nandrolone Propionate [7207-92-3]
Napalm [8031-21-8]
Napelline [5008-52-6]
Naphazoline [835-31-4]
Naphazoline Hydrochloride*
    [550-29-2]
Naphthacene [92-24-0]
Naphthalene [91-20-3]
1-Naphthaleneacetic Acid [86-87-3]
1,8-Naphthalenediamine [479-27-6]
1,6-Naphthalenedisulfonic Acid
    [525-37-1]
2,6-Naphthalenedisulfonic Acid
    [581-75-9]
2,7-Naphthalenedisulfonic Acid
    [92-41-1]
1-Naphthalenesulfonic Acid [85-47-2]
2-Naphthalenesulfonic Acid
    [120-18-3]
1-Naphthalenethiol [529-36-2]
2-Naphthalenethiol [91-60-1]
Naphthalic Acid [518-05-8]
1-Naphthoic Acid [86-55-5]
2-Naphthoic Acid [93-09-4]
1-Naphthol [90-15-3]
2-Naphthol [135-19-3]
1-Naphthol-8-amino-3,6-disulfonic
    Acid [90-20-0]
1-Naphthol-4,8-disulfonic Acid
    [117-56-6]
2-Naphthol-3,6-disulfonic Acid
    [148-75-4]
2-Naphthol-6,8-disulfonic Acid
    [118-32-1]
α-Naphtholphthalein [596-01-0]
1-Naphthol-2-sulfonic Acid
    [567-18-0]
1-Naphthol-4-sulfonic Acid [84-87-7]
2-Naphthol-6-sulfonic Acid [93-01-6]
Naphthol Yellow S [846-70-8]
1,2-Naphthoquinone [524-42-5]
1,4-Naphthoquinone [130-15-4]
Naphthoresorcinol [132-86-5]
2-Naphthoxyacetic Acid [120-23-0]
1-Naphthylamine [134-32-7]
2-Naphthylamine [91-59-8]
1-Naphthylamine-2,7-disulfonic Acid
    [486-54-4]
1-Naphthylamine-4,6-disulfonic Acid
    [85-74-5]
1-Naphthylamine-4,7-disulfonic Acid
    [85-75-6]
1-Naphthylamine-4-sulfonic Acid
    [84-86-6]
1-Naphthylamine-5-sulfonic Acid
    [84-89-9]
1-Naphthylamine-8-sulfonic Acid
    [82-75-7]
2-Naphthylamine-1-sulfonic Acid
    [81-16-3]
2-Naphthylamine-5-sulfonic Acid
    [81-05-0]
2-Naphthyl Benzoate [93-44-7]
N-(1-Naphthyl)ethylenediamine
    [551-09-7]
1-Naphthylisocyanate [86-84-0]
1-Naphthylisothiocyanate [551-06-4]
2-Naphthyl Lactate [93-43-6]
2-(2-Naphthyloxy)ethanol [93-20-9]
1-Naphthyl Salicylate [550-97-0]
2-Naphthyl Salicylate [613-78-5]
Napropamide [15299-99-7]
Naproxen [22204-53-1]
Naproxen Sodium* [26159-34-2]
Naptalam [132-66-1]
Narasin [55134-13-9]
Narbomycin [6036-25-5]
Narceine [131-28-2]

Denotes derivative of title compound

Narcobarbital [125-55-3]
Narcotoline [521-40-4]
Naringenin [480-41-1]
Naringin [10236-47-2]
Natamycin [7681-93-8]
Nauheim Salts (Artificial) [8007-50-9]
Naxagolide [88058-88-2]
Naxagolide Disodium Salt* [100935-99-7]
Nealbarbital [561-83-1]
Neamine [3947-65-7]
Nebularine [550-33-4]
Neburon [555-37-3]
Nedocromil [69049-73-6]
Nedocromil Sodium* [69049-74-7]
Nefopam [13669-70-0]
Nefopam Hydrochloride* [23327-57-3]
Negamycin [33404-78-3]
Negatol® [9011-02-3]
Neoarsphenamine [457-60-3]
Neocembrene [57918-06-6]
Neo-cupferron [1013-20-3]
Neocuproine [484-11-7]
Neodymium [7440-00-8]
Neoergosterol [516-98-3]
Neohesperidin Dihydrochalcone [20702-77-6]
Neomethymycin [497-73-4]
Neomycin [1404-04-2]
Neomycin Sulfate* [1405-10-3]
Neomycin Undecylenate [1406-04-8]
Neon [7440-01-9]
Neopentane [463-82-1]
Neopentyl Alcohol [75-84-3]
Neopentyl Glycol [126-30-7]
Neophyl Chloride [515-40-2]
Neopine [467-14-1]
Neopterin [670-65-5]
Neoquassin [76-77-7]
Neostigmine [59-99-4]
Neostigmine Bromide* [114-80-7]
Neostigmine Methylsulfate* [51-60-5]
Neotetrazolium Chloride [298-95-3]
Neovitamin A [36525-21-0]
Nepetalactone [490-10-8]
Neptunium [7439-99-8]
Nequinate [13997-19-8]
Neriifolin [466-07-9]
Nerol [106-25-2]
Nerolidol [7212-44-4]
Netilmicin [56391-56-1]
Netilmicin Sulfate* [56391-57-2]
Netobimin [88255-01-0]
Netropsin [1438-30-8]
Neuraminic Acid [114-04-5]
Neurine [463-88-7]
Neurotensin [39379-15-2]
Neutral Red [553-24-2]
Niacinamide [98-92-0]
Nialamide [51-12-7]
Niaprazine [27367-90-4]
Nicametate [3099-52-3]
Nicarbazin [330-95-0]
Nicardipine [55985-32-5]
Nicardipine Hydrochloride* [54527-84-3]
Nicergoline [27848-84-6]
Niceritrol [5868-05-3]
Nickel [7440-02-0]
Nickel Acetate [373-02-4]
Nickel Acetylacetonate [3264-82-2]
Nickel Bromide [13462-88-9]
Nickel Carbonate Hydroxide [12607-70-4]
Nickel Carbonyl [13463-39-3]
Nickel Chloride [7718-54-9]
Nickel Cyanide [557-19-7]

Nickel Dimethylglyoxime [13478-93-8]
Nickel Fluoride [10028-18-9]
Nickel Formate [3349-06-2]
Nickel Hydroxide [12054-48-7]
Nickel Iodide [13462-90-3]
Nickel Monoxide [1313-99-1]
Nickel Nitrate [13138-45-9]
Nickel Oxalate [547-67-1]
Nickel Phosphate [10381-36-9]
Nickel Sesquioxide [1314-06-3]
Nickel Sulfate [7786-81-4]
Niclosamide [50-65-7]
Nicoclonate [10571-59-2]
Nicofibrate [31980-29-7]
Nicofuranose [15351-13-0]
Nicomol [27959-26-8]
Nicomorphine [639-48-5]
Nicorandil [65141-46-0]
Nicotelline [494-04-2]
Nicotinamide Ascorbate [1987-71-9]
Nicotine [54-11-5]
Nicotinic Acid [59-67-6]
Nicotinic Acid Benzyl Ester [94-44-0]
Nicotinic Acid Monoethanolamine Salt [3570-15-8]
Nicotinyl Alcohol [100-55-0]
Nidroxyzone [405-22-1]
Nidulin [10089-10-8]
Nifedipine [21829-25-4]
Nifenalol [7413-36-7]
Nifenazone [2139-47-1]
Niflumic Acid [4394-00-7]
Nifuradene [555-84-0]
Nifuraldezone [3270-71-1]
Nifuratel [4936-47-4]
Nifurfoline [3363-58-4]
Nifuroquine [57474-29-0]
Nifuroxazide [965-52-6]
Nifuroxime [6236-05-1]
Nifurpirinol [13411-16-0]
Nifurprazine [1614-20-6]
Nifurtimox [23256-30-6]
Nifurtoinol [1088-92-2]
Nifurzide [39978-42-2]
Nigericin [28380-24-7]
Nihydrazone [67-28-7]
Nikethamide [59-26-7]
Nilutamide [63612-50-0]
Nilvadipine [75530-68-6]
Nimbin [5945-86-8]
Nimbiol [561-95-5]
Nimesulide [51803-78-2]
Nimetazepam [2011-67-8]
Nimidane [50435-25-1]
Nimodipine [66085-59-4]
Nimorazole [6506-37-2]
Nimustine [42471-28-3]
Ninhydrin [485-47-2]
Ninopterin [2179-16-0]
Niobium [7440-03-1]
Niobium Pentachloride [10026-12-7]
Niobium Pentafluoride [7783-68-8]
Niobium Pentoxide [1313-96-8]
Niobium Potassium Oxypentafluoride [17523-77-2]
Nioxime® [492-99-9]
Nipecotic Acid [498-95-3]
Nipradilol [81486-22-8]
Niridazole [61-57-4]
Nisin [1414-45-5]
Nisoldipine [63675-72-9]
Nitarsone [98-72-6]
Nithiazide [139-94-6]
Nitracrine [4533-39-5]
Nitralin [4726-14-1]
Nitramide [7782-94-7]
Nitramine [479-45-8]
Nitranilic Acid [479-22-1]
Nitrapyrin [1929-82-4]
Nitrazepam [146-22-5]

Nitrefazole [21721-92-6]
Nitrendipine [39562-70-4]
Nitric Acid [7697-37-2]
Nitric Acid, Anhydrous [7697-37-2]
Nitric Oxide [10102-43-9]
Nitrilotriacetic Acid [139-13-9]
Nitrin [553-74-2]
Nitroakridin 3582 [6035-39-8]
m-Nitroaniline [99-09-2]
o-Nitroaniline [88-74-4]
p-Nitroaniline [100-01-6]
5-Nitrobarbituric Acid [480-68-2]
Nitrobenzene [98-95-3]
4-Nitrobenzoyl Chloride [122-04-3]
p-Nitrobenzyl Cyanide [555-21-5]
o-Nitrobiphenyl [86-00-0]
p-Nitrobiphenyl [92-93-3]
2-Nitro-1,1-bis(p-chlorophenyl)-propane [117-27-1]
m-Nitrocinnamic Acid [555-68-0]
Nitrodan [962-02-7]
Nitroethane [79-24-3]
Nitrofen [1836-75-5]
Nitrofurantoin [67-20-9]
Nitrofurantoin Sodium* [54-87-5]
Nitrofurazone [59-87-0]
Nitrogen [7727-37-9]
Nitrogen Chloride [10025-85-1]
Nitrogen Dioxide [10102-44-0]
Nitrogen Fluoride [7783-54-2]
Nitrogen Pentoxide [10102-03-1]
Nitrogen Selenide [12033-88-4]
Nitroglycerin [55-63-0]
Nitroguanidine [556-88-7]
Nitrohydrochloric Acid [8007-56-5]
Nitromersol [133-58-4]
Nitromethane [75-52-5]
Nitromide [121-81-3]
Nitron [487-88-7]
1-Nitronaphthalene [86-57-7]
1-Nitro-2-naphthol [550-60-7]
3-Nitropentane [551-88-2]
5-Nitro-o-phenetidine [136-79-8]
Nitrophenide [537-91-7]
m-Nitrophenol [554-84-7]
o-Nitrophenol [88-75-5]
p-Nitrophenol [100-02-7]
p-Nitrophenylacetic Acid [104-03-0]
4-Nitro-o-phenylenediamine [99-56-9]
(4-Nitrophenyl)hydrazine [100-16-3]
o-Nitrophenylpropiolic Acid [530-85-8]
(4-Nitrophenyl)urea [556-10-5]
1-Nitropropane [108-03-2]
2-Nitropropane [79-46-9]
5'-Nitro-2'-propoxyacetanilide [553-20-8]
5-Nitro-2-propoxyaniline [553-79-7]
5-Nitroquinaldic Acid [525-47-3]
3-Nitrosalicylic Acid [85-38-1]
5-Nitrosalicylic Acid [96-97-9]
Nitroscanate [19881-18-6]
N-Nitrosodiethanolamine [1116-54-7]
N-Nitrosodiethylamine [55-18-5]
N-Nitrosodimethylamine [62-75-9]
p-Nitroso-N,N-dimethylaniline [138-89-6]
p-Nitrosodiphenylamine [156-10-5]
N-Nitrosomorpholine [59-89-2]
1-Nitroso-2-naphthol [131-91-9]
4-Nitrosophenol [104-91-6]
N-Nitrosopyrrolidine [930-55-2]
Nitroso-R Salt [525-05-3]
p-Nitrosulfathiazole [473-42-7]
2-Nitro-4-sulfobenzoic Acid [552-23-8]
Nitrosyl Chloride [2696-92-6]
Nitrosyl Fluoride [7789-25-5]
Nitrosylsulfuric Acid [7782-78-7]

Nitrosyl Tetrafluoroborate [14635-75-7]
Nitrourea [556-89-8]
Nitrous Acid [7782-77-6]
Nitrous Oxide [10024-97-2]
Nitrovin [804-36-4]
Nitroxoline [4008-48-4]
Nitroxynil [1689-89-0]
Nitryl Chloride [13444-90-1]
Nitryl Fluoride [10022-50-1]
Nivalenol [23282-20-4]
Nizatidine [76963-41-2]
Nizofenone [54533-85-6]
NMDA [6384-92-5]
NMN [1094-61-7]
Nobelium [10028-14-5]
Nocardamin [26605-16-3]
Nodakenin [495-31-8]
Noformicin [155-38-4]
Nogalamycin [1404-15-5]
Nomifensine [24526-64-5]
Nomifensine Maleate* [32795-47-4]
Nomilin [1063-77-0]
Nonactin [6833-84-7]
2-Nonenal [18829-56-6]
Nonoxynol [26027-38-3]
n-Nonyl Acetate [143-13-5]
n-Nonyl Alcohol [143-08-8]
Nonyl Phenol [25154-52-3]
(p-Nonylphenoxy)acetic Acid [3115-49-9]
Noprylsulfamide [576-97-6]
Noracymethadol [1477-39-0]
Noracymethadol Hydrochloride* [5633-25-0]
Norbolethone [1235-15-0]
Norbormide [991-42-4]
Norcarane [286-08-8]
Norcholanic Acid [511-18-2]
Norcodeine [467-15-2]
Nordazepam [1088-11-5]
Nordefrin Hydrochloride [138-61-4]
Nordefrin L-Form* [829-74-3]
Nordihydroguaiaretic Acid [500-38-9]
Norea [18530-56-8]
Norepinephrine [51-41-2]
Norepinephrine Bitartrate* [69815-49-2]
Norethandrolone [52-78-8]
Norethindrone [68-22-4]
Norethindrone Acetate* [51-98-9]
Norethynodrel [68-23-5]
Norfenefrine [536-21-0]
Norfloxacin [70458-96-7]
Norflurazon [27314-13-2]
Norgesterone [13563-60-5]
Norgestimate [35189-28-7]
Norgestrel [797-63-7]
Norgestrel (−)-Form* [797-64-8]
Norgestrienone [848-21-5]
Norhyoscyamine [537-29-1]
Norleucine [327-57-1]
Norlevorphanol [1531-12-0]
Norlobelanine [6035-31-0]
Normetanephrine [97-31-4]
Normethadone [467-85-6]
Normethandrone [514-61-4]
Normorphine [466-97-7]
Nornicotine [494-97-3]
Nornidulin [33403-37-1]
Norpipanone [561-48-8]
Norpseudoephedrine [36393-56-3]
Nortriptyline [72-69-5]
Nortriptyline Hydrochloride* [894-71-3]
Norvaline [6600-40-4]
Norvinisterone [6795-60-4]
Noscapine [128-62-1]
Noscapine Hydrochloride* [912-60-7]
Noscapine (dl-Form)* [6035-40-1]
Nosiheptide [56377-79-8]

Novembichin [1936-40-9]
Novobiocin [303-81-1]
Novobiocin Sodium* [1476-53-5]
Novoldiamine [140-80-7]
Novonal [512-48-1]
Noxiptilin [3362-45-6]
Noxytiolin [15599-39-0]
NPA Acid [539-13-9]
Nucleocidin [24751-69-7]
Nupharidine [468-89-3]
Nybomycin [30408-30-1]
Nylidrin [447-41-6]
Nylidrin Hydrochloride* [849-55-8]
Nylon 6 [25038-54-4]
Nylon 46 [50327-22-5]
Nystatin [1400-61-9]
Obidoxime Chloride [114-90-9]
Ocimene [29714-87-2]
Octabenzone [1843-05-6]
Octacaine [13912-77-1]
Octacosanol [557-61-9]
Octadecyltrimethylammonium Penta-chlorophenate [488-40-4]
Octafluorocyclobutane [115-25-3]
2,2,3,3,4,4,5,5-Octafluoro-1-pentanol [355-80-6]
Octamethylcyclotetrasiloxane [556-67-2]
Octamethyltrisiloxane [107-51-7]
Octamoxin [4684-87-1]
Octamylamine [502-59-0]
Octane [111-65-9]
Octanohydroxamic Acid [7377-03-9]
1-Octanol [111-87-5]
2-Octanol [123-96-6]
Octaverine [549-68-8]
Octhilinone [26530-20-1]
Octodrine [543-82-8]
Octopamine [104-14-3]
Octotiamine [137-86-0]
Octoxynol [9002-93-1]
Octreotide [83150-76-9]
Octyl Acetate [103-09-3]
n-Octyl Bromide [111-83-1]
sec-Octyl Bromide [557-35-7]
sec-Octyl Iodide [557-36-8]
Octyl Methoxycinnamate [5466-77-3]
Ofloxacin [82419-36-1]
Olaquindox [23696-28-8]
Old Yellow Enzyme [9001-68-7]
Oleandomycin [3922-90-5]
Oleandrin [465-16-7]
Oleanolic Acid [508-02-1]
Oleic Acid [112-80-1]
Sodium Oleate* [143-19-1]
Oleuropein [32619-42-4]
Oleyl Alcohol [143-28-2]
Olibanum [8050-07-5]
Olivacine [484-49-1]
Olivil [2955-23-9]
Olsalazine [15722-48-2]
Olsalazine Sodium* [6054-98-4]
Omeprazole [73590-58-6]
Omoconazole [74512-12-2]
Ondansetron [99614-02-5]
Ondansetron Hydrochloride Dihy-drate* [99614-01-4]
Ontianil [35727-72-1]
Oosporein [475-54-7]
Opianic Acid [519-05-1]
Opiniazide [2779-55-7]
Opipramol [315-72-0]
Opipramol Dihydrochloride* [909-39-7]
Opromazine [969-99-3]
Orange I [523-44-4]
Orange II [633-96-5]
Orange B [15139-76-1]
Orazamide [2574-78-9]
Orcein [1400-62-0]
Orcinol [504-15-4]

Orgotein [9016-01-7]
Oripavine [467-04-9]
Ormosinine [14350-67-5]
Ornidazole [16773-42-5]
Ornipressin [3397-23-7]
Ornithine [70-26-8]
Ornoprostil [70667-26-4]
Orotic Acid [65-86-1]
Orotidine [314-50-1]
Oroxylin A [480-11-5]
Orphenadrine [83-98-7]
Orphenadrine Citrate* [4682-36-4]
o-Orsellinic Acid [480-64-8]
Orthanilic Acid [88-21-1]
Orthocaine [536-25-4]
Orthoformic Acid [463-78-5]
Oryzacidin [1400-67-5]
Oryzalin [19044-88-3]
γ-Oryzanol [11042-64-1]
    Oryzanol (A)* [21238-33-5]
    Oryzanol (B)* [469-36-3]
Osajin [482-53-1]
Osalmid [526-18-1]
Osmaron B [8031-66-1]
Osmium [7440-04-2]
Osmium Hexafluoride [13768-38-2]
Osmium Tetrachloride [10026-01-4]
Osmium Tetroxide [20816-12-0]
Osthole [484-12-8]
Ostruthin [148-83-4]
Ostruthol [642-08-0]
Otobain [3738-01-0]
Ouabagenin [508-52-1]
Ouabain [630-60-4]
Ovex [80-33-1]
Oxaceprol [33996-33-7]
Oxacillin [66-79-5]
Oxacillin Sodium Monohydrate* [7240-38-2]
1,3,4-Oxadiazole [288-99-3]
Oxadiazon [19666-30-9]
Oxaflozane [26629-87-8]
Oxaflumazine [16498-21-8]
Oxalacetic Acid [328-42-7]
Oxalenediuramidoxime [580-52-9]
Oxalic Acid [144-62-7]
Oxalomolybdic Acid [53450-33-2]
Oxalyl Chloride [79-37-8]
Oxamarin [15301-80-1]
Oxamarin Hydrochloride* [6830-17-7]
Oxametacine [27035-30-9]
Oxamic Acid [471-47-6]
Oxamide [471-46-5]
Oxamniquine [21738-42-1]
Oxamyl [23135-22-0]
Oxanamide [126-93-2]
Oxandrolone [53-39-4]
Oxantel [36531-26-7]
Oxantel Pamoate* [68813-55-8]
Oxapropanium Iodide [541-66-2]
Oxaprotiline [56433-44-4]
Oxaprotiline Hydrochloride* [39022-39-4]
Oxaprotiline R(−)-Form* [76496-68-9]
Oxaprotiline R(−)-Form Hydrochlo-ride* [76496-69-0]
Oxaprotiline S(+)-Form* [76496-70-3]
Oxaprotiline S(+)-Form Hydrochlo-ride* [76679-95-3]
Oxaprozin [21256-18-8]
Oxatomide [60607-34-3]
Oxazepam [604-75-1]
Oxazidione [27591-42-0]
Oxazolam [27167-30-2]
Oxeladin [468-61-1]
Oxendolone [33765-68-3]
Oxenin [3230-75-9]
Oxethazaine [126-27-2]

* Denotes derivative of title compound

Oxetorone [26020-55-3]
Oxetorone Fumarate* [34522-46-8]
Oxfendazole [53716-50-0]
Oxibendazole [20559-55-1]
Oxiconazole Nitrate [64211-46-7]
Oxidimethiin [55290-64-7]
Oxiniacic Acid [2398-81-4]
Oxiracetam [62613-82-5]
Oxitropium Bromide [30286-75-0]
Oxolamine [959-14-8]
Oxolinic Acid [14698-29-4]
Oxomemazine [3689-50-7]
Oxonic Acid [937-13-3]
Oxophenarsine Hydrochloride
 [538-03-4]
Oxophenylarsine [637-03-6]
Oxotremorine [70-22-4]
Oxprenolol [6452-71-7]
Oxprenolol Hydrochloride*
 [6452-73-9]
Oxyacanthine [548-40-3]
Oxybenzone [131-57-7]
Oxybutynin Chloride [1508-65-2]
Oxychlorosene [8031-14-9]
Oxychlorosene Sodium* [52906-84-0]
Oxycinchophen [485-89-2]
Oxyclozanide [2277-92-1]
Oxycodone [76-42-6]
Oxycodone Hydrochloride*
 [124-90-3]
4,4'-Oxydi-2-butanol [821-33-0]
10,10'-Oxydiphenoxarsine [58-36-6]
Oxyfedrine [15687-41-9]
Oxyfluorfen [42874-03-3]
Oxygen [7782-44-7]
Oxymesterone [145-12-0]
Oxymetazoline [1491-59-4]
Oxymetazoline Hydrochloride*
 [2315-02-8]
Oxymetholone [434-07-1]
Oxymethurea [140-95-4]
Oxymorphone [76-41-5]
Oxymorphone Hydrochloride*
 [357-07-3]
Oxypendyl [5585-93-3]
Oxypertine [153-87-7]
Oxyphenbutazone [129-20-4]
Oxyphencyclimine [125-53-1]
Oxyphencyclimine Hydrochloride*
 [125-52-0]
Oxyphenisatin Acetate [115-33-3]
Oxyphenonium Bromide [50-10-2]
Oxypinocamphone [10136-65-9]
Oxytetracycline [79-57-2]
Oxytetracycline Hydrochloride*
 [2058-46-0]
Oxythiamine [582-36-5]
Oxythioquinox [2439-01-2]
Oxytocin [50-56-6]
Ozagrel [82571-53-7]
Ozone [10028-15-6]
Paclobutrazol [76738-62-0]
Pactamycin [23668-11-3]
Palitantin [15265-28-8]
Palladium [7440-05-3]
Palladium Chloride [7647-10-1]
Palladium Diacetate [19807-27-3]
Palladium Nitrate [10102-05-3]
Palladium Oxide [1314-08-5]
Palmatine [3486-67-7]
Palmidrol [544-31-0]
Palmitic Acid [57-10-3]
Palustric Acid [1945-53-5]
Palytoxin [77734-91-9]
Pamabrom [606-04-2]
Pamaquine [491-92-9]
Pamidronic Acid [40391-99-9]
Pamidronic Acid Disodium Salt*
 [57248-88-1]
Pamoic Acid [130-85-8]
Pancreatin [8049-47-6]

Pancrelipase [53608-75-6]
Pancuronium Bromide [15500-66-0]
Pankrin [9001-72-3]
Pantetheine [496-65-1]
Pantethine [16816-67-4]
Pantolactone [599-04-2]
Pantothenic Acid [79-83-4]
Papain [9001-73-4]
Papaveraldine [522-57-6]
Papaveretum [8002-76-4]
Papaverine [58-74-2]
Parabanic Acid [120-89-8]
Paraflutizide [1580-83-2]
Paraformaldehyde [30525-89-4]
Paraldehyde [123-63-7]
Paramethadione [115-67-3]
Paramethasone [53-33-8]
Paramethasone Acetate* [1597-82-6]
Paranyline [1729-61-9]
Paranyline Hydrochloride*
 [5585-60-4]
Paraoxon [311-45-5]
Paraquat [4685-14-7]
Parasorbic Acid [10048-32-5]
Parathiazine [290-57-3]
Parathion [56-38-2]
Parathyroid Hormone [9002-64-6]
Parbendazole [14255-87-9]
Parethoxycaine [94-23-5]
Parethoxycaine Hydrochloride*
 [136-46-9]
Pargyline [555-57-7]
Pargyline Hydrochloride* [306-07-0]
Paromomycin [7542-37-2]
Paromomycin Sulfate* [1263-89-4]
Parotin [1392-81-0]
Paroxetine [61869-08-7]
Paroxypropione [70-70-2]
Parsalmide [30653-83-9]
Parthenin [508-59-8]
Parthenolide [20554-84-1]
Partricin [11096-49-4]
Parvaquone [4042-30-2]
Pasiniazide [2066-89-9]
Patchouli Alcohol [5986-55-0]
Patulin [149-29-1]
Pebulate [1114-71-2]
Pecilocin [19504-77-9]
Pectin [9000-69-5]
Pectolinarigenin [520-12-7]
Pederin [27973-72-4]
Pefloxacin [70458-92-3]
Pefloxacin Mesylate* [70458-95-6]
Pelargonic Acid [112-05-0]
Pelargonidin [134-04-3]
Pelletierine [4396-01-4]
Pellitorine [18836-52-7]
Pellotine [83-14-7]
α-Peltatin [568-53-6]
β-Peltatin [518-29-6]
Pemoline [2152-34-3]
Pempidine [79-55-0]
Penamecillin [983-85-7]
Penbutolol [38363-40-5]
Penbutolol Sulfate* [38363-32-5]
Pendimethalin [40487-42-1]
Penethamate Hydriodide [808-71-9]
Penfluridol [26864-56-2]
Penicillamine [52-67-5]
Penicillamine Cysteine Disulfide
 [18840-45-4]
Penicillamine Disulfide [312-10-7]
Penicillanic Acid [87-53-6]
Penicillic Acid [90-65-3]
Penicillinase [9001-74-5]
Penicillin BT [6192-29-6]
Penicillin G Benethamine [751-84-8]
Penicillin G Benzathine [1538-09-6]
Penicillin G Benzhydrylamine
 [1538-11-0]
Penicillin G Calcium [973-53-5]

Penicillin G Hydrabamine
 [3344-16-9]
Penicillin G Potassium [113-98-4]
Penicillin G Procaine [6130-64-9]
Penicillin N [525-94-0]
Penicillin O [87-09-2]
Penicillin O Potassium* [897-61-0]
Penicillin O Sodium* [7177-54-0]
Penicillin S Potassium [525-87-1]
Penicillin V [87-08-1]
Penicillin V Potassium* [132-98-9]
Penicillin V Benzathine [5928-84-7]
Penicillin V Hydrabamine
 [6591-72-6]
Penicilloyl Polylysine [53608-77-8]
Penimepicycline [4599-60-4]
Pentaborane(9) [19624-22-7]
Pentaborane(11) [18433-84-6]
Pentabromoacetone [79-49-2]
Pentacene [135-48-8]
Pentachloroethane [76-01-7]
Pentachlorophenol [87-86-5]
Pentacynium Bis(methyl sulfate)
 [3810-83-1]
3-Pentadecylcatechol [492-89-7]
Pentaerythritol [115-77-5]
Pentaerythritol Chloral [78-12-6]
Pentaerythritol Dichlorohydrin
 [2209-86-1]
Pentaerythritol Tetraacetate
 [597-71-7]
Pentaerythritol Tetranitrate [78-11-5]
Pentagastrin [5534-95-2]
Pentagestrone [7001-56-1]
Pentahomoserine [533-88-0]
Pentamethonium Bromide [541-20-8]
Pentamidine [100-33-4]
Pentamidine Dimethanesulfonate
 [6823-79-6]
Pentamidine Isethionate [140-64-7]
Pentane [109-66-0]
1,5-Pentanediol [111-29-5]
1-Pentanol [71-41-0]
2-Pentanol [6032-29-7]
3-Pentanol [584-02-1]
Pentapiperide [7009-54-3]
Pentapiperium Methylsulfate*
 [7681-80-3]
Pentazocine [359-83-1]
Pentazocine Hydrochloride*
 [64024-15-3]
1-Pentene [109-67-1]
2-Pentene [109-68-2]
2-Pentenylpenicillin Sodium
 [525-86-0]
Pentetate Calcium Trisodium
 [12111-24-9]
Pentetic Acid [67-43-6]
Penthienate Bromide [60-44-6]
Pentifylline [1028-33-7]
Pentigetide [62087-72-3]
Pentobarbital Sodium [57-33-0]
Pentobarbital* [76-74-4]
1-Pentol [105-29-3]
Pentolinium Tartrate [52-62-0]
Pentosan Polysulfate [37300-21-3]
Pentostatin [53910-25-1]
Pentoxifylline [6493-05-6]
Pentoxyl [147-61-5]
Pentrinitrol [1607-17-6]
Pentryl® [4481-55-4]
tert-Pentyl Alcohol [75-85-4]
Pentylenetetrazole [54-95-5]
p-tert-Pentylphenol [80-46-6]
Peonidin [134-01-0]
Peplomycin [68247-85-8]
Peplomycin Sulfate* [70384-29-1]
Pepsin [9001-75-6]
Pepstatin [17891-89-9]
Peptide T [106362-32-7]
Peracetic Acid [79-21-0]

Denotes derivative of title compound

Perazine [84-97-9]
Perbenzoic Acid [93-59-4]
Perchloric Acid [7601-90-3]
Perchloryl Fluoride [7616-94-6]
Perezone [3600-95-1]
Perfluidone [37924-13-3]
Performic Acid [107-32-4]
Pergolide [66104-22-1]
Pergolide Mesylate* [66104-23-2]
Perhexiline [6621-47-2]
Perhexiline Maleate* [6724-53-4]
Pericyazine [2622-26-6]
Perilla Ketone [553-84-4]
Perillaldehyde [2111-75-3]
Perimethazine [13093-88-4]
Perindopril [82834-16-0]
Perindoprilat* [95153-31-4]
Periodic Acid [10450-60-9]
Periodyl [53586-99-5]
Periplocin [13137-64-9]
Periplocymarin [32476-67-8]
Periplogenin [514-39-6]
Perisoxal [2055-44-9]
Perivine [2673-40-7]
Perlapine [1977-11-3]
Permethrin [52645-53-1]
Perphenazine [58-39-9]
Peruvoside [1182-87-2]
Perylene [198-55-0]
Petroleum Benzin [8030-30-6]
Petroselinic Acid [593-39-5]
Petunidin [1429-30-7]
Peucedanin [133-26-6]
Peyonine [19717-25-0]
Pfeiffer's Substance [8015-18-7]
Phalloidin [17466-45-4]
Phanquinone [84-12-8]
Phaseolin [13401-40-6]
Phasin [1392-87-6]
α-Phellandrene [99-83-2]
β-Phellandrene [555-10-2]
Phenacaine Hydrochloride [620-99-5]
Phenacaine Hydrochloride Monohydrate* [6153-19-1]
Phenacemide [63-98-9]
Phenacetin [62-44-2]
Phenacetolin [1340-26-7]
Phenacridane Chloride [3131-08-6]
Phenactropinium Chloride [3784-89-2]
Phenacylamine [613-89-8]
Phenadoxone [467-84-5]
Phenaglycodol [79-93-6]
Phenallymal [115-43-5]
Phenamacide Hydrochloride [31031-74-0]
Phenamet [3819-34-9]
Phenamidine [101-62-2]
Phenamidine Isethionate* [620-90-6]
Phenampromid(e) [129-83-9]
Phenanthrene [85-01-8]
Phenanthrenequinone [84-11-7]
o-Phenanthroline [66-71-7]
Phenarsazine Chloride [578-94-9]
Phenarsone Sulfoxylate [535-51-3]
Phenatine [139-68-4]
Phenazine [92-82-0]
Phenazocine [127-35-5]
Phenazocine Hydrobromide* [1239-04-9]
Phenazopyridine Hydrochloride [136-40-3]
Phenbenzamine [961-71-7]
Phenbutamide [3149-00-6]
Phencarbamide [3735-90-8]
Phencyclidine [77-10-1]
Phencyclidine Hydrochloride* [956-90-1]
Phendimetrazine [634-03-7]
Phendimetrazine Tartrate* [50-58-8]
Phenelzine [51-71-8]

Phenelzine Sulfate* [156-51-4]
Phenesterine [3546-10-9]
Phenetharbital [357-67-5]
Phenethicillin Potassium [132-93-4]
Phenethyl Alcohol [60-12-8]
Phenethylamine [64-04-0]
o-Phenetidine [94-70-2]
p-Phenetidine [156-43-4]
Phenetole [103-73-1]
Pheneturide [90-49-3]
Phenformin [114-86-3]
Phenformin Hydrochloride* [834-28-6]
Phenglutarimide [1156-05-4]
Phenicarbazide [103-03-7]
Phenicin [128-68-7]
Phenindamine [82-88-2]
Phenindamine Tartrate* [569-59-5]
Phenindione [83-12-5]
Pheniprazine [55-52-7]
Pheniramine [86-21-5]
Pheniramine Maleate* [132-20-7]
Phenmedipham [13684-63-4]
Phenmetrazine [134-49-6]
Phenmetrazine Hydrochloride* [1707-14-8]
Phenobarbital [50-06-6]
Phenobarbital Sodium [57-30-7]
Phenobutiodil [554-24-5]
Phenocoll [103-97-9]
Phenoctide [78-05-7]
Phenol [108-95-2]
Phenoldisulfonic Acid [96-77-5]
Phenolphthalein [77-09-8]
Phenolphthalein Sodium [518-51-4]
Phenolphthalin [81-90-3]
Phenolphthalol [81-92-5]
p-Phenolsulfonic Acid [98-67-9]
Phenolsulfonphthalein [143-74-8]
Phenoltetrachlorophthalein [639-44-1]
Phenomorphan [468-07-5]
Phenoperidine [562-26-5]
Phenopyrazone [3426-01-5]
Phenosafranin [81-93-6]
Phenosulfazole [515-54-8]
Phenothiazine [92-84-2]
Phenothrin [26002-80-2]
Phenoxazine [135-67-1]
Phenoxyacetic Acid [122-59-8]
Phenoxyacetyl Cellulose [68332-77-4]
Phenoxybenzamine [59-96-1]
Phenoxybenzamine Hydrochloride* [63-92-3]
2-Phenoxyethanol [122-99-6]
Phenoxypropazine [3818-37-9]
Phenpentermine [434-43-5]
Phenprobamate [673-31-4]
Phenprocoumon [435-97-2]
Phensuximide [86-34-0]
Phentermine [122-09-8]
Phentermine Hydrochloride* [1197-21-3]
Phentetiothalein Sodium [18265-54-8]
Phentolamine [50-60-2]
Phentolamine Hydrochloride* [73-05-2]
Phentolamine Methanesulfonate* [65-28-1]
Phentydrone [634-19-5]
Phenylacetaldehyde [122-78-1]
α-Phenylacetamide [103-81-1]
Phenyl Acetate [122-79-2]
Phenylacetic Acid [103-82-2]
Phenylacetone [103-79-7]
Phenyl Acetylsalicylate [134-55-4]
Phenylalanine [63-91-2]
Phenyl Aminosalicylate [133-11-9]
N-Phenylanthranilic Acid [91-40-7]

2-Phenyl-1H-benzimidazole [716-79-0]
Phenyl Benzoate [93-99-2]
Phenyl Biguanide [102-02-3]
Phenylbutazone [50-33-9]
α-Phenylbutyramide [90-26-6]
Phenyl Carbonate [102-09-0]
2-Phenyl-6-chlorophenol [85-97-2]
4-Phenyl-2-chlorophenol [92-04-6]
α-Phenylcinnamic Acid [3368-16-9]
m-Phenylenediamine [108-45-2]
o-Phenylenediamine [95-54-5]
p-Phenylenediamine [106-50-3]
Phenylephrine Hydrochloride [61-76-7]
Phenylethanolamine [7568-93-6]
Phenyl Ether [101-84-8]
5-(α-Phenylethyl)semioxamazide [93-95-8]
Phenylglyceryl Ether [538-43-2]
N-Phenylglycine [103-01-5]
α-Phenylglycine [69-91-0]
Phenylhydrazine [100-63-0]
Phenylhydrazine Hydrochloride [59-88-1]
Phenylhydroxylamine [100-65-2]
Phenyl Isocyanate [103-71-9]
Phenyl Isothiocyanate [103-72-0]
Phenylmagnesium Chloride [100-59-4]
N-Phenylmaleimide [941-69-5]
Phenylmercuric Acetate [62-38-4]
Phenylmercuric Chloride [100-56-1]
Phenylmercuric Nitrate, Basic [8003-05-2]
Phenylmercury Borate [102-98-7]
Phenylmethylbarbituric Acid [76-94-8]
o-Phenylphenol [90-43-7]
p-Phenylphenol [92-69-3]
Phenyl Phthalate [84-62-8]
Phenylpropanolamine Hydrochloride [154-41-6]
Phenylpropanolamine* [14838-15-4]
Phenylpropylmethylamine [93-88-9]
1-Phenyl-3-pyrazolidinone [92-43-3]
Phenyl Salicylate [118-55-8]
4-Phenylsemicarbazide [537-47-3]
N-Phenylsulfanilic Acid [101-57-5]
Phenyl Sulfide [139-66-2]
Phenylthiourea [103-85-5]
Phenyltoloxamine [92-12-6]
Phenyltrimethylammonium Iodide [98-04-4]
Phenylurea [64-10-8]
Phenylurethan(e) [101-99-5]
Phenyramidol [553-69-5]
Phenyramidol Hydrochloride* [326-43-2]
Phenytoin [57-41-0]
Phenytoin Sodium* [630-93-3]
Phethenylate Sodium [510-34-9]
Phillyrin [487-41-2]
Phloionic Acid [23843-52-9]
Phloionolic Acid [583-86-8]
Phloretin [60-82-2]
Phloridzin [60-81-1]
Phloroglucinol [108-73-6]
Phlorol [90-00-6]
Pholcodine [509-67-1]
Pholedrine [370-14-9]
Phorate [298-02-2]
Phorbol [17673-25-5]
Phorone [504-20-1]
Phosalone [2310-17-0]
Phosfolan [947-02-4]
Phosgene [75-44-5]
Phosmet [732-11-6]
Phosphamidon [13171-21-6]
Phosphine [7803-51-2]
Phosphinothricin [51276-47-2]

Denotes derivative of title compound

Phosphocreatine [67-07-2]
Phosphomolybdic Acid [11104-88-4]
Phosphonium Iodide [12125-09-6]
Phosphoric Acid [7664-38-2]
Phosphoric Acid, Meta [37267-86-0]
Phosphorous Acid [13598-36-2]
Phosphorus [7723-14-0]
Phosphorus Hemitriselenide
  [1314-81-4]
Phosphorus Oxybromide [7789-59-5]
Phosphorus Oxychloride
  [10025-87-3]
Phosphorus Pentabromide
  [7789-69-7]
Phosphorus Pentachloride
  [10026-13-8]
Phosphorus Pentafluoride
  [7647-19-0]
Phosphorus Pentaselenide
  [1314-82-5]
Phosphorus Pentasulfide [1314-80-3]
Phosphorus Pentoxide [1314-56-3]
Phosphorus Sulfochloride [3982-91-0]
Phosphorus Tribromide [7789-60-8]
Phosphorus Trichloride [7719-12-2]
Phosphorus Trifluoride [7783-55-3]
Phosphorus Trioxide [1314-24-5]
Phosphorus Triselenide [1314-86-9]
Phosphorylcholine [107-73-3]
Phosphoserine [407-41-0]
Phosphotungstic Acid [12067-99-1]
Phosvitin [9008-96-2]
Phoxim [14816-18-3]
Phrenosin [586-02-7]
Phthalamide [88-96-0]
Phthalazine [253-52-1]
Phthalic Acid [88-99-3]
Phthalic Anhydride [85-44-9]
Phthalimide [85-41-6]
Phthalofyne [131-67-9]
Phthaloyl Chloride [88-95-9]
Phthalylsulfacetamide [131-69-1]
Phthalylsulfathiazole [85-73-4]
Phthiocol [483-55-6]
Physalaemin [2507-24-6]
Physodic Acid [84-24-2]
Physostigmine [57-47-6]
Physostigmine Salicylate* [57-64-7]
Physostigmine Sulfate* [64-47-1]
Physovenine [6091-05-0]
Phytic Acid [83-86-3]
Phytic Acid Calcium Magnesium
  Salt*
  [3615-82-5]
Phytate Sodium* [7205-52-9]
Phytochlorin [19660-77-6]
Phytofluene [540-05-6]
Phytol [150-86-7]
Piberaline [39640-15-8]
Picadex [99-00-3]
Picein [530-14-3]
Picene [213-46-7]
Picilorex [62510-56-9]
Picloram [1918-02-1]
Picloxydine [5636-92-0]
α-Picoline [109-06-8]
β-Picoline [108-99-6]
γ-Picoline [108-89-4]
Picolinic Acid [98-98-6]
Picoperine [21755-66-8]
Picosulfate Sodium [10040-45-6]
Picotamide [32828-81-2]
Picramic Acid [96-91-3]
Picric Acid [88-89-1]
Picrocrocin [138-55-6]
Picrolichenic Acid [466-34-2]
Picrolonic Acid [550-74-3]
Picromycin [19721-56-3]
Picropodophyllin [477-47-4]
Picrotin [21416-53-5]
Picrotoxin [124-87-8]

Picrotoxinin [17617-45-7]
Picryl Chloride [88-88-0]
Pifarnine [56208-01-6]
Pifoxime [31224-92-7]
Piketoprofen [60576-13-8]
Pildralazine [64000-73-3]
Pilocarpine [92-13-7]
Pilocarpine Hydrochloride* [54-71-7]
Pilocarpine Nitrate* [148-72-1]
Pilocereine [2552-47-8]
Pimaric Acid [127-27-5]
Pimeclone [534-84-9]
Pimefylline [10001-43-1]
Pimelic Acid [111-16-0]
Piminodine [13495-09-5]
Piminodine Esylate* [7081-52-9]
Pimozide [2062-78-4]
Pimpinellin [131-12-4]
Pinacidil [85371-64-8]
Pinacol [76-09-5]
Pinacolone [75-97-8]
Pinaverium Bromide [53251-94-8]
Pinazepam [52463-83-9]
Pindolol [13523-86-9]
Pindone [83-26-1]
α-Pinene [80-56-8]
β-Pinene [127-91-3]
Pinguinain [37288-97-4]
Pinosylvin [22139-77-1]
Pipacycline [1110-80-1]
Pipamazine [84-04-8]
Pipamperone [1893-33-0]
Pipazethate [2167-85-3]
Pipebuzone [27315-91-9]
Pipecolic Acid [535-75-1]
Pipecurium Bromide [52212-02-9]
Pipemidic Acid [51940-44-4]
Pipenzolate Bromide [125-51-9]
Piperacetazine [3819-00-9]
Piperacillin [61477-96-1]
Piperacillin Sodium* [59703-84-3]
Piperazine [110-85-0]
Piperazine Phosphate Monohydrate*
  [18534-18-4]
Piperazine Adipate [142-88-1]
Piperazine Citrate [144-29-6]
2,5-Piperazinedione [106-57-0]
Piperazine Edetate Calcium
  [12002-30-1]
Piperazine Tartrate [133-36-8]
Piperic Acid [5285-18-7]
Piperidine [110-89-4]
Piperidine Phosphate* [767-21-5]
Piperidione [77-03-2]
Piperidolate [82-98-4]
Piperidolate Hydrochloride*
  [129-77-1]
Piperilate [4546-39-8]
Piperilate Hydrochloride*
  [4544-15-4]
Piperine [94-62-2]
Piperitone [89-81-6]
Piperocaine [136-82-3]
Piperocaine Hydrochloride*
  [533-28-8]
Piperonal [120-57-0]
Piperonyl Butoxide [51-03-6]
Piperonylic Acid [94-53-1]
Piperoxan [59-39-2]
Piperylone [2531-04-6]
PIPES [5625-37-6]
PIPES Disodium* [76836-02-7]
PIPES Sodium* [10010-67-0]
Pipobroman [54-91-1]
Piposulfan [2608-24-4]
Pipotiazine [39860-99-6]
Pipotiazine Palmitate* [37517-26-3]
Pipoxolan Hydrochloride
  [18174-58-8]
Pipradrol [467-60-7]
Pipradrol Hydrochloride* [71-78-3]

Piprinhydrinate [606-90-6]
Piprozolin [17243-64-0]
Pipsyl Chloride [98-61-3]
Piracetam [7491-74-9]
Pirarubicin [72496-41-4]
Pirbuterol [38677-81-5]
Pirbuterol Dihydrochloride*
  [38029-10-6]
Pirenoxine [1043-21-6]
Pirenzepine [28797-61-7]
Pirenzepine Dihydrochloride*
  [29868-97-1]
Piretanide [55837-27-9]
Piribedil [3605-01-4]
Piridocaine [87-21-8]
Pirifibrate [55285-45-5]
Pirimicarb [23103-98-2]
Pirimiphos-ethyl [23505-41-1]
Piritramide [302-41-0]
Pirmenol [68252-19-7]
Piroctone [50650-76-5]
Piroctone Olamine* [68890-66-4]
Piroheptine [16378-21-5]
Piromen [9008-99-5]
Piromidic Acid [19562-30-2]
Piroxicam [36322-90-4]
Pirozadil [54110-25-7]
Pirprofen [31793-07-4]
Piscidic Acid [35388-57-9]
Pithecolobine [22368-82-7]
Pivalic Acid [75-98-9]
Pivalylbenzhydrazine [306-19-4]
Pivampicillin [33817-20-8]
Pivampicillin Hydrochloride*
  [26309-95-5]
Pivcefalexin [27726-31-4]
Pizotyline [15574-96-6]
Plafibride [63394-05-8]
Plantisul [1407-93-8]
Plasmin [9001-90-5]
Plasminogen [9001-91-6]
Plasmocid [551-01-9]
Platelet Activating Factor
  [65154-06-5]
Platinic Chloride [16941-12-1]
Platinic Iodide [7790-46-7]
Platinic Oxide [1314-15-4]
Platinous Chloride [10025-65-7]
Platinous Iodide [7790-39-8]
Platinum [7440-06-4]
Platonin [3571-88-8]
Platyphylline [480-78-4]
Plaunotol [64218-02-6]
Plegatil [53597-29-8]
Pleuromutilin [125-65-5]
Pleurotin(e) [1404-23-5]
Plicamycin [18378-89-7]
Plumbagin [481-42-5]
Plumericin [77-16-7]
Plumieride [511-89-7]
Plutonium [7440-07-5]
Podocarpic Acid [5947-49-9]
Pododacric Acid [32630-75-4]
Podophyllinic Acid 2-Ethylhydrazide
  [1508-45-8]
Podophyllotoxin [518-28-5]
Polar® Yellow [6372-96-9]
Poldine Methylsulfate [545-80-2]
Polidexide [9064-92-0]
Polidocanol [9002-92-0]
Polonium [7440-08-6]
Polonium Dioxide [7446-06-2]
Polonium Tetrachloride [10026-02-5]
Poloxalene [9003-11-6]
Poloxamer Forms* [9003-11-6]
Polybenzarsol [54531-52-1]
Polydatin [27208-80-6]
Polyestradiol Phosphate [28014-46-2]
Polyethylene [9002-88-4]
Polyethylene Glycol [25322-68-3]
Polyferose [9009-29-4]

Polygodial [6754-20-7]
Polylysine [25104-18-1]
Polymyxin [1406-11-7]
Polymyxin B Sulfate* [1405-20-5]
Polymyxin B-Methanesulfonic Acid [1407-09-6]
Polypropylene [9003-07-0]
Polysorbate 80 [9005-65-6]
Polytetrafluoroethylene [9002-84-0]
Polythiazide [346-18-9]
Polyvinyl Alcohol [9002-89-5]
Polyvinyl Chloride [9002-86-2]
Ponalrestat [72702-95-5]
Ponasterone A [13408-56-5]
Ponceau 3R [3564-09-8]
Ponceau SX [4548-53-2]
Populin [99-17-2]
Porfiromycin [801-52-5]
Porphyrillic Acid [129-65-7]
Porofor® BSH [80-17-1]
Porphine [101-60-0]
Porphobilinogen [487-90-1]
Porphyrillic Acid [129-65-7]
Porphyropsin [9009-58-9]
Potassium [7440-09-7]
Potassium Acetate [127-08-2]
Potassium Aluminate [1302-63-2]
Potassium p-Aminobenzoate [138-84-1]
Potassium Arsenate [7784-41-0]
Potassium Arsenite [10124-50-2]
Potassium Bicarbonate [298-14-6]
Potassium Bifluoride [7789-29-9]
Potassium Binoxalate [127-95-7]
Potassium Biphthalate [877-24-7]
Potassium Biselenite [7782-70-9]
Potassium Bisulfate [7646-93-7]
Potassium Bisulfide [1310-61-8]
Potassium Bitartrate [868-14-4]
Potassium Borohydride [13762-51-1]
Potassium Borotartrate [12001-68-2]
Potassium Bromate [7758-01-2]
Potassium Bromide [7758-02-3]
Potassium Carbonate [584-08-7]
Potassium Chlorate [3811-04-9]
Potassium Chloride [7447-40-7]
Potassium Chromate(VI) [7789-00-6]
Potassium Citrate [866-84-2]
Potassium Citrate, Monobasic [866-83-1]
Potassium Cobaltous Selenate [28041-86-3]
Potassium Cyanate [590-28-3]
Potassium Cyanide [151-50-8]
Potassium Dichromate(VI) [7778-50-9]
Potassium Dicyanoaurate(I) [13967-50-5]
Potassium Ethyl Sulfate [563-17-7]
Potassium Ferricyanide [13746-66-2]
Potassium Ferrocyanide [13943-58-3]
Potassium Fluoride [7789-23-3]
Potassium Formate [590-29-4]
Potassium Gluconate [299-27-4]
Potassium Glycerophosphate [1319-69-3]
Potassium Guaiacolsulfonate [1321-14-8]
Potassium Hexachloroosmate(IV) [16871-60-6]
Potassium Hexachloroplatinate(IV) [16921-30-5]
Potassium Hexacyanocobaltate(III) [13963-58-1]
Potassium Hexafluoromanganate(IV) [16962-31-5]
Potassium Hexafluorosilicate [16871-90-2]
Potassium Hexafluorozirconate(IV) [16923-95-8]
Potassium Hexathiocyanatoplatinate(IV) [17069-38-4]

Potassium Hydroxide [1310-58-3]
Potassium Hypophosphite [7782-87-8]
Potassium Iodate [7758-05-6]
Potassium Iodide [7681-11-0]
Potassium Manganate(VI) [10294-64-1]
Potassium Metabisulfite [16731-55-8]
Potassium Methyl Sulfate [562-54-9]
Potassium Molybdate(VI) [13446-49-6]
Potassium Nitrate [7757-79-1]
Potassium Nitrite [7758-09-0]
Potassium Nitroprusside [14709-57-0]
Potassium Oleate [143-18-0]
Potassium Osmate(VI) [19718-36-6]
Potassium Oxalate [583-52-8]
Potassium Percarbonate [589-97-9]
Potassium Perchlorate [7778-74-7]
Potassium Periodate [7790-21-8]
Potassium Permanganate [7722-64-7]
Potassium Persulfate [7727-21-1]
Potassium Phenolsulfonate [12167-15-6]
Potassium Phenoxide [100-67-4]
Potassium Phosphate, Dibasic [7758-11-4]
Potassium Phosphate, Monobasic [7778-77-0]
Potassium Phosphate, Tribasic [7778-53-2]
Potassium Phosphite [13492-26-7]
Potassium Picrate [573-83-1]
Potassium Pyroantimonate, Acid [10090-54-7]
Potassium Pyrophosphate [7320-34-5]
Potassium Pyrosulfate [7790-62-7]
Potassium Salicylate [578-36-9]
Potassium Selenate [7790-59-2]
Potassium Selenide [1312-74-9]
Potassium Silicate [1312-76-1]
Potassium Silver Cyanide [506-61-6]
Potassium Sodium Tartrate [304-59-6]
Potassium Sorbate [24634-61-5]
Potassium Stannate(IV) [12142-33-5]
Potassium Stannosulfate [27790-37-0]
Potassium Stearate [593-29-3]
Potassium Sulfate [7778-80-5]
Potassium Sulfide [1312-73-8]
Potassium Sulfite [10117-38-1]
Potassium Sulfobenzoate [53608-78-9]
Potassium Tartrate [921-53-9]
Potassium Tellurate(IV) [7790-58-1]
Potassium Tellurate(VI) [15571-91-2]
Potassium Tetraborate [1332-77-0]
Potassium Tetrabromoaurate(III) [13966-47-7]
Potassium Tetrachloroaurate(III) [13682-61-6]
Potassium Tetrachloroplatinate(II) [10025-99-7]
Potassium Tetracyanomercurate(II) [591-89-9]
Potassium Tetracyanonickelate(II) [14220-17-8]
Potassium Tetracyanoplatinate(II) [562-76-5]
Potassium Tetracyanozincate [14244-62-3]
Potassium Tetrafluoroborate [14075-53-7]
Potassium Tetraiodoaurate(III) [7791-09-8]
Potassium Tetraiodocadmate [584-10-1]
Potassium Tetraiodomercurate(II) [7783-33-7]

Potassium Tetroxalate [127-96-8]
Potassium Thioantimonate(V) [14693-02-8]
Potassium Thiocarbonate [26750-66-3]
Potassium Thiocyanate [333-20-0]
Potassium Thiosulfate [10294-66-3]
Potassium Titanyl Oxalate [14402-67-6]
Potassium Triiodide [7790-42-3]
Potassium Triiodomercurate(II) Solution [22330-18-3]
Potassium Triiodozincate [7790-43-4]
Potassium Tungstate(VI) [7790-60-5]
Potassium Uranate(VI) [7790-63-8]
Potassium Uranyl Nitrate [18078-40-5]
Potassium Uranyl Sulfate [27709-53-1]
Potassium Xanthogenate [140-89-6]
Potassium Zinc Sulfate [13932-17-7]
Potassium Zirconium Sulfate [53608-79-0]
Povidone [9003-39-8]
Povidone-Iodine [25655-41-8]
PPACK [71142-71-7]
Practolol [6673-35-4]
Prajmaline [35080-11-6]
Prajmalium Bitartrate* [2589-47-1]
Pralidoxime Chloride [51-15-0]
Pralidoxime Iodide* [94-63-3]
Pralidoxime Mesylate* [154-97-2]
Pramiverin [14334-40-8]
Pramoxine [140-65-8]
Pramoxine Hydrochloride* [637-58-1]
Pranoprofen [52549-17-4]
Praseodymium [7440-10-0]
Prasterone [53-43-0]
Pratensein [2284-31-3]
Pravastatin Sodium [81131-70-6]
Prazepam [2955-38-6]
Praziquantel [55268-74-1]
Prazosin [19216-56-9]
Prazosin Hydrochloride* [19237-84-4]
Prednicarbate [73771-04-7]
Prednimustine [29069-24-7]
Prednisolone [50-24-8]
Prednisolone Acetate* [52-21-1]
Prednisolone Hemisuccinate* [2920-86-7]
Prednisolone 21-Diethylaminoacetate [5626-34-6]
Prednisolone Sodium Phosphate [125-02-0]
Prednisolone Sodium Succinate [1715-33-9]
Prednisolone Sodium 21-m-Sulfobenzoate [630-67-1]
Prednisolone 21-Stearoylglycolate [5060-55-9]
Prednisolone Tebutate [7681-14-3]
Prednisolone 21-Trimethylacetate [1107-99-9]
Prednisone [53-03-2]
Prednival [15180-00-4]
Prednylidene [599-33-7]
Prednylidene 21-Diethylaminoacetate [6890-42-2]
Pregnane [481-26-5]
Pregnanediol [80-92-2]
3,20-Pregnanedione [128-23-4]
Pregnan-3α-ol-20-one [128-20-1]
4-Pregnene-20,21-diol-3,11-dione [116-56-3]
4-Pregnene-11β,17α,20β,21-tetrol-3-one [116-58-5]
4-Pregnene-17α,20β,21-triol-3,11-dione [116-59-6]

Denotes derivative of title compound

4-Pregnene-17α,20β,21-triol-3-one [128-19-8]
Pregnenolone [145-13-1]
Pregnenolone Methyl Ether [511-26-2]
Prelog-Djerassi Lactone [69056-12-8]
Prenalterol [57526-81-5]
Prenalterol Hydrochloride* [61260-05-7]
Prenoxdiazine Hydrochloride [982-43-4]
Prenylamine [390-64-7]
Prephenic Acid [126-49-8]
Pretilachlor [51218-49-6]
Pridinol [511-45-5]
Prifinium Bromide [4630-95-9]
Prilocaine [721-50-6]
Prilocaine Hydrochloride* [1786-81-8]
Primaperone [1219-35-8]
Primaquine [90-34-6]
Primaquine Phosphate* [63-45-6]
Primeverose [26531-85-1]
Primidone [125-33-7]
Primocarcin [3750-26-3]
Primulaverin [154-61-0]
Primycin [30003-49-7]
Pristane [1921-70-6]
Probarbital [76-76-6]
Probarbital Sodium* [143-82-8]
Probenecid [57-66-9]
Probucol [23288-49-5]
Procainamide Hydrochloride [614-39-1]
Procaine [59-46-1]
Procarbazine [671-16-9]
Procarbazine Hydrochloride* [366-70-1]
Procaterol [72332-33-3]
Procaterol Hydrochloride* [59828-07-8]
Procerin [552-96-5]
Prochloraz [67747-09-5]
Prochlorperazine [58-38-8]
Prochlorperazine Maleate* [84-02-6]
Procodazole [23249-97-0]
Procyclidine [77-37-2]
Procyclidine Hydrochloride* [1508-76-5]
Procymate [13931-64-1]
Procymidone [32809-16-8]
Prodiamine [29092-21-2]
Prodigiosin [82-89-3]
Prodigiosin Hydrochloride* [112373-40-7]
Prodilidine [3734-17-6]
Prodilidine Hydrochloride* [3734-16-5]
Prodipine [31314-38-2]
Prodlure [50767-79-8]
Pro-Drone® [53905-38-7]
Proflavine [92-62-6]
Proflavine Dihydrochloride* [531-73-7]
Proflavine Sulfate* [553-30-0]
Profluralin [26399-36-0]
Progabide [62666-20-0]
Progesterone [57-83-0]
Proglumide [6620-60-6]
Proheptazine [77-14-5]
Proinsulin [9035-68-1]
Prolactin [9002-62-4]
Proline [147-85-3]
Prolintane [493-92-5]
Prolintane Hydrochloride* [1211-28-5]
Prolonium Iodide [123-47-7]
Promazine [58-40-2]
Promazine Hydrochloride* [53-60-1]
Promecarb [2631-37-0]

Promedol [64-39-1]
Promegestone [34184-77-5]
Promethazine [60-87-7]
Promethazine Hydrochloride* [58-33-3]
Promethazine Teoclate* [17693-51-5]
Promethium [7440-12-2]
Prometon [1610-18-0]
Prometryn [7287-19-6]
Promoxolane [470-43-9]
Pronethalol [54-80-8]
Pro-opiomelanocortin [66796-54-1]
Propacetamol [66532-85-2]
Propachlor [1918-16-7]
Propafenone [54063-53-5]
Propafenone Hydrochloride* [34183-22-7]
Propallylonal [545-93-7]
Propamidine [104-32-5]
Propamidine Isethionate* [140-63-6]
Propane [74-98-6]
1-Propanearsonic Acid [107-34-6]
1,3-Propanedithiol [109-80-8]
Propanethial S-Oxide [32157-29-2]
Propanidid [1421-14-3]
Propanil [709-98-8]
Propanocaine [493-76-5]
Propantheline Bromide [50-34-0]
Proparacaine [499-67-2]
Proparacaine Hydrochloride* [5875-06-9]
Propargite [2312-35-8]
Propargyl Alcohol [107-19-7]
Propargyl Chloride [624-65-7]
Propatyl Nitrate [2921-92-8]
Propazine [139-40-2]
Propentofylline [55242-55-2]
Propenzolate [4354-45-4]
Propenzolate Hydrochloride* [1420-03-7]
Properdin [11016-39-0]
Properidine [561-76-2]
Propetamphos [31218-83-4]
Propham [122-42-9]
Propicillin [551-27-9]
Propicillin Potassium* [1245-44-9]
Propiconazole [60207-90-1]
Propineb [12071-83-9]
β-Propiolactone [57-57-8]
Propiolic Acid [471-25-0]
Propiomazine [362-29-8]
Propiomazine Hydrochloride* [1240-15-9]
Propionaldehyde [123-38-6]
Propionamide [79-05-0]
Propionic Acid [79-09-4]
Propionic Anhydride [123-62-6]
Propionitrile [107-12-0]
Propionyl Chloride [79-03-8]
Propionylpromazine [3568-24-9]
Propiophenone [93-55-0]
Propipocaine [3670-68-6]
Propiram [15686-91-6]
Propiram Fumarate* [13717-04-9]
Propivane [132-45-6]
Propizepine [10321-12-7]
Propofol [2078-54-8]
Propoxur [114-26-1]
Propoxycaine Hydrochloride [550-83-4]
Propoxyphene [469-62-5]
Propoxyphene Hydrochloride* [1639-60-7]
Propoxyphene Napsylate* [26570-10-5]
Propoxyphene L-Form Napsylate Monohydrate* [55557-30-7]
Propranolol [525-66-6]
Propranolol Hydrochloride* [318-98-9]
Propyl Acetate [109-60-4]

n-Propyl Alcohol [71-23-8]
Propylamine [107-10-8]
n-Propylbenzene [103-65-1]
Propyl Bromide [106-94-5]
Propyl Butyrate [105-66-8]
Propyl Chloride [540-54-5]
Propyl Chlorocarbonate [109-61-5]
Propyl Docetrizoate [5579-08-8]
Propylene [115-07-1]
Propylene Chlorohydrin [78-89-7]
sec-Propylene Chlorohydrin [127-00-4]
Propylenediamine [78-90-0]
Propylene Dibromide [78-75-1]
Propylene Dichloride [78-87-5]
Propylene Glycol [57-55-6]
Propylene Oxide [75-56-9]
Propyl Ether [111-43-3]
Propyl Formate [110-74-7]
Propyl Gallate [121-79-9]
Propylhexedrine [3595-11-7]
Propylidene Chloride [78-99-9]
Propyl Iodide [107-08-4]
Propyliodone [587-61-1]
n-Propyl Nitrate [627-13-4]
n-Propyl Nitrite [543-67-9]
Propylparaben [94-13-3]
Propyl Propionate [106-36-5]
Propyl Sulfide [111-47-7]
Propylthiouracil [51-52-5]
Propylure [10297-61-7]
Propyphenazone [479-92-5]
Propyromazine [145-54-0]
Propyzamide [23950-58-5]
Proquazone [22760-18-5]
Proscar® [98319-26-7]
Proscillaridin [466-06-8]
Prostacyclin [35121-78-9]
Prostaglandin E₁ [745-65-3]
Prostaglandin E₂ [363-24-6]
Prostaglandin F₂α [551-11-1]
Prostalene [54120-61-5]
Prosultiamine [59-58-5]
Protactinium [7440-13-3]
Protamine Sulfate* [9009-65-8]
Protamine Zinc Insulin Suspension [9004-17-5]
Protheobromine [50-39-5]
Prothipendyl [303-69-5]
Prothipendyl Hydrochloride* [1225-65-6]
Prothrombin [9001-26-7]
Protiofate [58416-00-5]
Protionamide [14222-60-7]
Protizinic Acid [54323-85-2]
Protoanemonin [108-28-1]
Protocatechualdehyde [139-85-5]
Protocatechuic Acid [99-50-3]
Protokosin [10091-70-0]
Protokylol [136-70-9]
Protokylol Hydrochloride* [136-69-6]
Protopine [130-86-9]
Protoporphyrin IX [553-12-8]
Protostephanine [549-28-0]
Protoverine [76-45-9]
Protriptyline [438-60-8]
Protriptyline Hydrochloride* [1225-55-4]
Pro-Urokinase [82657-92-9]
Proxazole [5696-09-3]
Proxazole Citrate* [132-35-4]
Proxibarbal [2537-29-3]
Proxyphylline [603-00-9]
Prozapine [3426-08-2]
Prunetin [552-59-0]
Pseudoaconitine [127-29-7]
Pseudobaptigenin [90-29-9]
Pseudococaine [478-73-9]
Pseudocodeine [466-96-6]
Pseudoconhydrine [140-55-6]

Pseudocumene [95-63-6]
Pseudohecogenin [11005-20-2]
Pseudoionone [141-10-6]
Pseudojervine [36069-05-3]
Pseudomorphine [125-24-6]
Pseudopederin [10352-73-5]
Pseudopelletierine [552-70-5]
Pseudotropine [135-97-7]
Pseudoyohimbine [84-37-7]
Psicofuranine [1874-54-0]
D-Psicose [551-68-8]
Psilocin [520-53-6]
Psilocybin [520-52-5]
PSK® [66455-27-4]
Psoralen [66-97-7]
Psoralidin [18642-23-4]
Psychotrine [7633-29-6]
Pteridine [91-18-9]
Pterocarpin [524-97-0]
Pteroic Acid [119-24-4]
Pteropterin [89-38-3]
Pteroylhexaglutamylglutamic Acid
 [6484-74-8]
Puberulic Acid [99-23-0]
Puberulonic Acid [82-83-7]
Pukateine [81-67-4]
Pulegone [89-82-7]
1H-Purine [120-73-0]
Puromycin [53-79-2]
Purothionin [9009-72-7]
Purpurin [81-54-9]
Purpurogallin [569-77-7]
Putrescine [110-60-1]
Pyocyanine [85-66-5]
Pyracarbolid [24691-76-7]
Pyrantel [15686-83-6]
Pyrantel Pamoate* [22204-24-6]
Pyrantel Tartrate* [33401-94-4]
Pyrathiazine [84-08-2]
Pyrazinamide [98-96-4]
Pyrazine [290-37-9]
2,3-Pyrazinedicarboxylic Acid
 [89-01-0]
Pyrazinoic Acid [98-97-5]
Pyrazole [288-13-1]
2-Pyrazoline [109-98-8]
Pyrazophos [13457-18-6]
Pyrene [129-00-0]
Pyrethrosin [28272-18-6]
Pyridate [55512-33-9]
Pyridazine [289-80-5]
Pyridine [110-86-1]
Pyridine 1-Oxide [694-59-7]
Pyridinium Bromide Perbromide
 [39416-48-3]
Pyridinium Chlorochromate
 [26299-14-9]
Pyridinol Carbamate [1882-26-4]
Pyridofylline [53403-97-7]
Pyridomycin [18791-21-4]
Pyridostigmine Bromide [101-26-8]
Pyridoxal [66-72-8]
Pyridoxal 5-Phosphate [54-47-7]
Pyridoxamine Dihydrochloride
 [524-36-7]
4-Pyridoxic Acid [82-82-6]
Pyridoxine Hydrochloride [58-56-0]
Pyrilamine [91-84-9]
Pyrilamine Maleate* [59-33-6]
Pyrimethamine [58-14-0]
Pyrimidine [289-95-2]
Pyriminil [53558-25-1]
Pyrimithate [5221-49-8]
Pyrinoline [1740-22-3]
Pyrisuccideanol [33605-94-6]
Pyrithiamine [534-64-5]
Pyrithione [1121-30-8]
Pyrithione Zinc* [13463-41-7]
Pyrithyldione [77-04-3]
Pyritinol [10049-83-9]
Pyrocalciferol [128-27-8]

Pyrocarbonic Acid Diethyl Ester
 [1609-47-8]
Pyrocatechol [120-80-9]
Pyrogallol [87-66-1]
Pyrogallol Monoacetate* [1330-51-4]
Pyrogallol Triacetate* [525-52-0]
L-Pyroglutamic Acid [98-79-3]
Pyrolan® [87-47-8]
Pyroligneous Acid [8030-97-5]
Pyromellitic Acid [89-05-4]
Pyronine B [2150-48-3]
Pyronine Y [92-32-0]
Pyrophosphoric Acid [2466-09-3]
Pyrosulfuric Acid [7783-05-3]
Pyrosulfuryl Chloride [7791-27-7]
Pyrovalerone [3563-49-3]
Pyrovalerone Hydrochloride*
 [1147-62-2]
Pyrrobutamine [91-82-7]
Pyrrobutamine Phosphate*
 [135-31-9]
Pyrrocaine [2210-77-7]
Pyrrocaine Hydrochloride*
 [2210-64-2]
1H-Pyrrole [109-97-7]
Pyrrolidine [123-75-1]
2-Pyrrolidone [616-45-5]
3-Pyrroline [109-96-6]
Pyrrolnitrin [1018-71-9]
Pyruvaldehyde [78-98-8]
Pyruvate Decarboxylase [9001-04-1]
Pyruvic Acid [127-17-3]
Pyrvinium Chloride [548-84-5]
Pyrvinium Pamoate [3546-41-6]
Q-Enzyme® [9001-97-2]
Quassin [76-78-8]
Quatrimycin [79-85-6]
Quazepam [36735-22-5]
Quebrachamine [4850-21-9]
Queen Substance [334-20-3]
Quercetagetin [90-18-6]
Quercetin [117-39-5]
Quercimeritrin [491-50-9]
d-Quercitol [488-73-3]
Quercitrin [522-12-3]
Quillaic Acid [631-01-6]
Quinacillin [985-32-0]
Quinacrine [83-89-6]
Quinaldic Acid [93-10-7]
Quinaldine [91-63-4]
Quinaldine Blue [2768-90-3]
Quinaldine Red [117-92-0]
Quinaldine Sulfate [655-76-5]
Quinalizarin [81-61-8]
Quinamine [464-85-7]
Quinapril [85441-61-8]
Quinapril Monohydrochloride Mono-
 hydrate* [90243-99-5]
Quinaprilat* [82768-85-2]
Quinapyramine [20493-41-8]
Quinazoline [253-82-7]
Quinbolone [2487-63-0]
Quinestradiol [1169-79-5]
Quinestrol [152-43-2]
Quinethazone [73-49-4]
Quinfamide [62265-68-3]
Quingestrone [69-25-8]
Quinhydrone [106-34-3]
Quinic Acid [77-95-2]
Quinidine [56-54-2]
Quinidine Gluconate* [7054-25-3]
Quinidine Polygalacturonate
 [7681-28-9]
Quinidine Sulfate [50-54-4]
Quinine [130-95-0]
Quinine Bisulfate [549-56-4]
Quinine Carbonate [146-06-5]
Quinine Dihydrobromide [549-47-3]
Quinine Dihydrochloride [60-93-5]
Quinine Ethylcarbonate [83-75-0]
Quinine Formate [130-90-5]

Quinine Gluconate [4325-25-1]
Quinine Hydriodide [549-50-8]
Quinine Hydrobromide [549-49-5]
Quinine Hydrochloride [130-89-2]
Quinine Iodosulfate [7631-46-1]
Quinine Oleate [10486-11-0]
Quinine Salicylate [750-90-3]
Quinine Sulfate [804-63-7]
Quinine Urea Hydrochloride
 [549-52-0]
Quininic Acid [86-68-0]
Quininone [84-31-1]
Quinizarin [81-64-1]
Quinizarin Green SS [128-80-3]
Quinocide [525-61-1]
Quinoline [91-22-5]
8-Quinolineboronic Acid [86-58-8]
8-Quinolinecarboxylic Acid [86-59-9]
Quinoline Yellow [8004-92-0]
Quinoline Yellow Spirit Soluble
 [8003-22-3]
Quinolinic Acid [89-00-9]
Quinone [106-51-4]
Quinovic Acid [465-74-7]
Quinovin [53516-73-7]
Quinovose [7658-08-4]
Quinoxaline [91-19-0]
Quintozene [82-68-8]
Quinuclidine [100-76-5]
3-Quinuclidinol [1619-34-7]
Quinupramine [31721-17-2]
Quisqualic Acid [52809-07-1]
Quizalofop-Ethyl [76578-14-8]
R-11 [126-15-8]
Racefemine [22232-57-1]
Racemethorphan [510-53-2]
Radicinin [10088-95-6]
Radium [7440-14-4]
Radium Bromide* [10031-23-9]
Radium Chloride* [10025-66-8]
Radon [10043-92-2]
Raffinose [512-69-6]
Rafoxanide [22662-39-1]
Ramifenazone [3615-24-5]
Ramipril [87333-19-5]
Ramiprilat* [87269-97-4]
Raney Nickel® [7440-02-0]
Ranimustine [58994-96-0]
Ranitidine [66357-35-5]
Raubasine [483-04-5]
Raunescine [117-73-7]
Razoxane [21416-87-5]
Reductic Acid [80-72-8]
Reinecke Salt [13573-16-5]
Relaxin [9002-69-1]
Renin [9015-94-5]
Rennin [9001-98-3]
Repirinast [73080-51-0]
Reposal [3625-25-0]
Reproterol [54063-54-6]
Reproterol Hydrochloride*
 [13055-82-8]
Resacetophenone [89-84-9]
Resazurin [550-82-3]
Rescimetol [73573-42-9]
Rescinnamine [24815-24-5]
Reserpic Acid [83-60-3]
Reserpiline [131-02-2]
Reserpine [50-55-5]
Resibufogenin [465-39-4]
Resin Ipomea [9000-34-4]
Resin Jalap [9000-35-5]
Resin Scammony [9000-58-2]
Resistomycin [20004-62-0]
Resodec [9012-13-9]
Resorantel [20788-07-2]
Resorcinol [108-46-3]
β-Resorcylaldehyde [95-01-2]
β-Resorcylic Acid [89-86-1]
Retamine [2122-29-4]
Retene [483-65-8]

Denotes derivative of title compound

Reticuline [485-19-8]
Retinal [116-31-4]
Retine [51848-43-2]
Retinoic Acid [302-79-4]
Retronecine [480-85-3]
Retrorsine [480-54-6]
Rhamnetin [90-19-7]
Rhamnose [3615-41-6]
Rhapontin [155-58-8]
Rheadine [2718-25-4]
Rhein [478-43-3]
Rhenium [7440-15-5]
Rhenium Heptoxide [1314-68-7]
Rhenium Hexafluoride [10049-17-9]
Rhenium Oxychloride [7791-09-5]
Rhenium Trioxide [1314-28-9]
Rhizopterin [119-20-0]
Rhodamine B [81-88-9]
Rhodanilic Acid [1332-53-2]
Rhodanine [141-84-4]
Rhodinal [141-26-4]
Rhodinol [6812-78-8]
Rhodium [7440-16-6]
Rhodium Carbonyl Chloride
[14523-22-9]
Rhodium Chloride [10049-07-7]
Rhododendrin [497-78-9]
Rhodopin [105-92-0]
Rhodopsin [9009-81-8]
Rhodoviolascin [34255-08-8]
Rhodoxanthin [116-30-3]
Rhubarb [8016-55-5]
Rhynchophylline [76-66-4]
Ribavirin [36791-04-5]
α-Ribazole [132-13-8]
Riboflavine [83-88-5]
Riboflavine Phosphate (Sodium)
[130-40-5]
Ribonuclease [9001-99-4]
D-Ribose [50-69-1]
D-Ribose-5-phosphoric Acid
[4300-28-1]
D-Ribulose [488-84-6]
Ricin [9009-86-3]
Ricinine [524-40-3]
Ricinoleic Acid [141-22-0]
Sodium Ricinoleate* [5323-95-5]
Rifamide [2750-76-7]
Rifampin [13292-46-1]
Rifamycin SV [6998-60-3]
Rifaximin [80621-81-4]
Rilmazafone [99593-25-6]
Rilmenidene [54187-04-1]
Rimantadine [13392-28-4]
Rimantadine Hydrochloride*
[1501-84-4]
Rimazolium Metilsulfate
[28610-84-6]
Rimiterol [32953-89-2]
Rimiterol Hydrobromide*
[31842-61-2]
Rimocidin [1393-12-0]
Rioprostil [77287-05-9]
Risocaine [94-12-2]
Ristocetin [1404-55-3]
Ritodrine [26652-09-5]
Ritodrine Hydrochloride*
[23239-51-2]
Robenidine [25875-51-8]
Robenidine Hydrochloride*
[25875-50-7]
Robinin [301-19-9]
Roccellic Acid [29838-46-8]
Rociverine [53716-44-2]
Rokitamycin [74014-51-0]
Rolicyprine [2829-19-8]
Rolipram [61413-54-5]
Rolitetracycline [751-97-3]
Rolitetracycline Nitrate Sesquihy-
drate* [26657-13-6]
Ronidazole [7681-76-7]

Ronifibrate [42597-57-9]
Ronnel [299-84-3]
Rosaprostol [56695-65-9]
Rosaramicin [35834-26-5]
Rose Bengal [11121-48-5]
Rosoxacin [40034-42-2]
Rotenone [83-79-4]
Rotraxate [92071-51-7]
Rottlerin [82-08-6]
Roxarsone [121-19-7]
Roxatidine Acetate [78628-28-1]
Roxithromycin [80214-83-1]
Rubeanic Acid [79-40-3]
Ruberythric Acid [152-84-1]
Rubiadin [117-02-2]
Rubidium [7440-17-7]
Rubidium Bromide [7789-39-1]
Rubidium Chloride [7791-11-9]
Rubidium Hydroxide [1310-82-3]
Rubidium Iodide [7790-29-6]
Rubijervine [79-58-3]
Rubixanthin [3763-55-1]
Rufigallol [82-12-2]
Ruscogenin [472-11-7]
Rutecarpine [84-26-4]
Ruthenium Red [1307-52-4]
Ruthenium Tetroxide [20427-56-9]
Ruthenium Trichloride [10049-08-8]
Rutin [153-18-4]
Rutinose [90-74-4]
Ryania [8047-13-0]
Ryanodine [15662-33-6]
Sabadilla [8051-02-3]
Sabadine [124-80-1]
Saccharin [81-07-2]
L-Saccharopine [997-68-2]
Safranal [116-26-7]
Safrole [94-59-7]
SAICAR [3031-95-6]
Sakuranetin [2957-21-3]
Salacetamide [487-48-9]
Salazosulfadimidine [2315-08-4]
Salicin [138-52-3]
Salicyl Alcohol [90-01-7]
Salicylaldehyde [90-02-8]
Salicylaldoxime [94-67-7]
Salicylamide [65-45-2]
Salicylamide O-Acetic Acid
[25395-22-6]
Salicylanilide [87-17-2]
Salicylhydroxamic Acid [89-73-6]
Salicylic Acid [69-72-7]
4-Salicyloylmorpholine [3202-84-4]
Salicylsulfuric Acid [89-45-2]
Salinazid [495-84-1]
Salinomycin [53003-10-4]
Salmine Sulfate [53597-25-4]
Salsalate [552-94-3]
Salsoline [89-31-6]
Salutaridine [1936-18-1]
Salverine [6376-26-7]
Samandarine [467-51-6]
Samarium [7440-19-9]
Sancycline [808-26-4]
Sanguinarine [2447-54-3]
α-Santalol [115-71-9]
β-Santalol [77-42-9]
Santonic Acid [510-35-0]
α-Santonin [481-06-1]
Saponarin [20310-89-8]
Sapphire [1317-82-4]
Saralasin [34273-10-4]
Saralasin Acetate Hydrate*
[39698-78-7]
Sarcosine [107-97-1]
Sarin [107-44-8]
Sarkomycin [11031-48-4]
Sarmentogenin [76-28-8]
Sarmentose [13484-14-5]
Sarpagine [482-68-8]
Sarsasapogenin [126-19-2]

Sarverogenin [22146-03-8]
Saxitoxin [35554-08-6]
Scabiolide [20055-57-6]
Scandium [7440-20-2]
Scarlet Red [85-83-6]
Schradan [152-16-9]
Schweizer's Reagent [17500-49-1]
Scillabiose [40525-07-3]
Scillaren [11003-70-6]
Scillarenin [465-22-5]
Scilliroside [507-60-8]
Scoparin [301-16-6]
Scoparone [120-08-1]
Scopolamine [51-34-3]
Scopolamine N-Oxide [97-75-6]
Scopoletin [92-61-5]
Scopolin [531-44-2]
Scopoline [487-27-4]
Scotophobin [33579-45-2]
Scutellarein [529-53-3]
Sebacic Acid [111-20-6]
Sebacil [96-01-5]
Sebacoin [96-00-4]
Secnidazole [3366-95-8]
Secobarbital Sodium [309-43-3]
Secretin [1393-25-5]
Securinine [5610-40-2]
Sedecamycin [23477-98-7]
Seidlitz Mixture [8014-63-9]
Selagine [116-28-9]
Selenic Acid [7783-08-6]
Selenious Acid [7783-00-8]
Selenium [7782-49-2]
Selenium Bromide [7789-52-8]
Selenium Chloride [10025-68-0]
Selenium Hexafluoride [7783-79-1]
Selenium Oxide [7446-08-4]
Selenium Oxybromide [7789-51-7]
Selenium Oxychloride [7791-23-3]
Selenium Oxyfluoride [7783-43-9]
Selenium Tetrabromide [7789-65-3]
Selenium Tetrachloride [10026-03-6]
Selenium Tetrafluoride [13465-66-2]
Selenocysteine [10236-58-5]
Selenomethionine [1464-42-2]
Selenomethionine Se75* [1187-56-0]
Semicarbazide Hydrochloride
[563-41-7]
Semioxamazide [515-96-8]
Sempervirine [6882-99-1]
Senecialdehyde [107-86-8]
Senecic Acid [13588-16-4]
Senecionine [130-01-8]
Seneciphylline [480-81-9]
Senegenin [2469-34-3]
Senociclin [4154-10-3]
Serine [56-45-1]
Sermorelin [86168-78-7]
Serotonin [50-67-9]
Serpentine (Alkaloid) [18786-24-8]
Sertraline [79617-96-2]
Sertraline Hydrochloride*
[79559-97-0]
Serum Albumin [9048-49-1]
Sesamex [51-14-9]
Sesamin [607-80-7]
Sesamolin [526-07-8]
Sesin [94-83-7]
Setastine [64294-95-7]
Setastine Hydrochloride*
[59767-13-4]
Sethoxydim [74051-80-2]
Shark Liver Oil [68990-63-6]
Shellac [9000-59-3]
Shellolic Acid [4448-95-7]
Shikimic Acid [138-59-0]
Shionone [10376-48-4]
Showdomycin [16755-07-0]
Siccanin [22733-60-4]
Siduron [1982-49-6]
Sikkimotoxin [18651-67-7]

Silane [7803-62-5]
Silicic Acid [1343-98-2]
Silicon [7440-21-3]
Silicon Carbide [409-21-2]
Silicon Dioxide [7631-86-9]
Silicon Disulfide [13759-10-9]
Silicon Monoxide [10097-28-6]
Silicon Nitride [12033-89-5]
Silicon Tetraacetate [562-90-3]
Silicon Tetrabromide [7789-66-4]
Silicon Tetrachloride [10026-04-7]
Silicon Tetrafluoride [7783-61-1]
Silicotungstic Acid [12520-88-6]
Silver [7440-22-4]
Silver Acetate [563-63-3]
Silver Bromide [7785-23-1]
Silver Carbonate [534-16-7]
Silver Chlorate [7783-92-8]
Silver Chloride [7783-90-6]
Silver Chromate(VI) [7784-01-2]
Silver Citrate [126-45-4]
Silver Cyanide [506-64-9]
Silver Difluoride [7783-95-1]
Silver Fluoride [7775-41-9]
Silver Iodate [7783-97-3]
Silver Iodide [7783-96-2]
Silver Lactate [128-00-7]
Silver Nitrate [7761-88-8]
Silver Nitrate, Toughened
  [8007-31-6]
Silver Nitrite [7783-99-5]
Silver Oxalate [533-51-7]
Silver Oxide [20667-12-3]
Silver(II) Oxide [1301-96-8]
Silver Perchlorate [7783-93-9]
Silver Permanganate [7783-98-4]
Silver Phosphate [7784-09-0]
Silver Picrate [146-84-9]
Silver Protein, Mild [9015-51-4]
Silver Protein, Strong [9008-42-8]
Silver Selenate [7784-07-8]
Silver Selenide [1302-09-6]
Silver Selenite [7784-05-6]
Silver Subfluoride [1302-01-8]
Silver Sulfate [10294-26-5]
Silver Sulfide [21548-73-2]
Silver Tetraiodomercurate(II)
  [7784-03-4]
Silvex [93-72-1]
Simazine [122-34-9]
Simethicone [8050-81-5]
Simetride [154-82-5]
Simetryne® [1014-70-6]
Simfibrate [14929-11-4]
Simvastatin [79902-63-9]
Sinalbin [20196-67-2]
Sinapine [18696-26-9]
Sincalide [25126-32-3]
Sinigrin [3952-98-5]
Sinomenine [115-53-7]
Sisomicin [32385-11-8]
Sisomicin Sulfate* [53179-09-2]
α_1-Sitosterol [474-40-8]
β-Sitosterol [83-46-5]
γ-Sitosterol [83-47-6]
Sizofiran [9050-67-3]
Skatole [83-34-1]
Skimmianine [83-95-4]
Skimmin [93-39-0]
Smilagenin [126-18-1]
Sobrerol [498-71-5]
Soda Lime [8006-28-8]
Sodium [7440-23-5]
Sodium Acetate [127-09-3]
Sodiumacetic Acid Sodium Salt
  [534-12-3]
Sodium Acid Pyrophosphate
  [7758-16-9]
Sodium Alizarinesulfonate [130-22-3]
Sodium Aluminate [1302-42-7]
Sodium Amalgam [11110-52-4]

Sodium Amide [7782-92-5]
Sodium Amylosulfate [9010-01-9]
Sodium Arsanilate [127-85-5]
Sodium Arsenate, Dibasic
  [7778-43-0]
Sodium Arsenite [7784-46-5]
Sodium Arsphenamine [1936-28-3]
Sodium Ascorbate [134-03-2]
Sodium Azide [26628-22-8]
Sodium Benzoate [532-32-1]
Sodium Bicarbonate [144-55-8]
Sodium Bifluoride [1333-83-1]
Sodium Bismuthate(V) [12232-99-4]
Sodium Bisulfate [7681-38-1]
Sodium Bisulfide [16721-80-5]
Sodium Bisulfite [7631-90-5]
Sodium Bitartrate [526-94-3]
Sodium Borate [1330-43-4]
Sodium Borohydride [16940-66-2]
Sodium Bromate [7789-38-0]
Sodium Bromide [7647-15-6]
Sodium Cacodylate [124-65-2]
Sodium Carbonate [497-19-8]
Sodium Cellulose Phosphate
  [9038-41-9]
Sodium Chlorate [7775-09-9]
Sodium Chloride [7647-14-5]
Sodium Chlorite [7758-19-2]
Sodium 6-Chloro-5-nitrotoluene-3-
  sulfonate [535-48-8]
Sodium Chromate(VI) [7775-11-3]
Sodium Chromate(VI), Radioactive
  [10039-53-9]
Sodium Citrate [68-04-2]
Sodium Citrate, Acid [144-33-2]
Sodium Cobaltinitrite [13600-98-1]
Sodium Cyanate [917-61-3]
Sodium Cyanide [143-33-9]
Sodium Cyanoborohydride
  [25895-60-7]
Sodium Diacetate [126-96-5]
Sodium Dichromate(VI) [10588-01-9]
Sodium Dicyanoaurate(I)
  [15280-09-8]
Sodium Dithionate [7631-94-9]
Sodium Ethoxide [141-52-6]
Sodium Ethyl Sulfate [546-74-7]
Sodium Ferricyanide [14217-21-1]
Sodium Ferrocyanide [14101-19-9]
Sodium Fluoborate [13755-29-8]
Sodium Fluoride [7681-49-4]
Sodium Folate [6484-89-5]
Sodium Formaldehydesulfoxylate
  [149-44-0]
Sodium Formate [141-53-7]
Sodium Gluconate [527-07-1]
Sodium Glycerophosphate
  [1334-74-3]
Sodium Hexachloroplatinate(IV)
  [16923-58-3]
Sodium Hexafluorosilicate
  [16893-85-9]
Sodium Hydride [7646-69-7]
Sodium Hydrosulfite [7775-14-6]
Sodium Hydroxide [1310-73-2]
Sodium Hypochlorite [7681-52-9]
Sodium Hypophosphate [13721-43-2]
Sodium Hypophosphite [7681-53-0]
Sodium Iodate [7681-55-2]
Sodium Iodide [7681-82-5]
Sodium Iodide, Radioactive
  [7790-26-3]
Sodium Iodomethamate [519-26-6]
Sodium Isopropyl Xanthate
  [140-93-2]
Sodium Lactate [72-17-3]
Sodium Lauryl Sulfate [151-21-3]
Sodium-Lead Alloy [12740-44-2]
Sodium Metabisulfite [7681-57-4]
Sodium Metaborate [7775-19-1]
Sodium Metaperiodate [7790-28-5]

Sodium Metasilicate [6834-92-0]
Sodium Methoxide [124-41-4]
Sodium Methyl Sulfate [512-42-5]
Sodium Molybdate(VI) [7631-95-0]
Sodium β-Naphthoquinone-4-sulfo-
  nate [521-24-4]
Sodium Nitrate [7631-99-4]
Sodium Nitrite [7632-00-0]
Sodium Nitroprusside [14402-89-2]
Sodium Oxalate [62-76-0]
Sodium Oxide [1313-59-3]
Sodium Oxybate [502-85-2]
Sodium Perborate [7632-04-4]
Sodium Perchlorate [7601-89-0]
Sodium Permanganate [10101-50-5]
Sodium Peroxide [1313-60-6]
Sodium Persulfate [7775-27-1]
Sodium Pertechnetate ^{99m}Tc
  [23288-60-0]
Sodium Phenolsulfonate [1300-51-2]
Sodium Phenoxide [139-02-6]
Sodium Phosphate, Dibasic
  [7558-79-4]
Sodium Phosphate, Monobasic
  [7558-80-7]
Sodium Phosphate, Radioactive
  [8027-28-9]
Sodium Phosphate, Tribasic
  [7601-54-9]
Sodium Phosphite [13708-85-5]
Sodium Phosphomolybdate
  [1313-30-0]
Sodium Phosphotungstate
  [51312-42-6]
Sodium Plumbate(IV) [12201-47-7]
Sodium Polyanetholesulfonate
  [52993-95-0]
Sodium Polymetaphosphate
  [50813-16-6]
Sodium Polystyrene Sulfonate
  [9003-59-2]
Sodium Propionate [137-40-6]
Sodium Pyroantimonate [10049-22-6]
Sodium Rhodizonate [523-21-7]
Sodium Salicylate [54-21-7]
Sodium Selenate [13410-01-0]
Sodium Selenide [1313-85-5]
Sodium Selenite [10102-18-8]
Sodium Sesquicarbonate [533-96-0]
Sodium Silicate [1344-09-8]
Sodium Silicate Solution [1344-09-8]
Sodium Stannate(IV) [12058-66-1]
Sodium Stearate [822-16-2]
Sodium Succinate [150-90-3]
Sodium Sulfate [7757-82-6]
Sodium Sulfide [1313-82-2]
Sodium Sulfite [7757-83-7]
Sodium β-Sulfopropionitrile
  [513-15-5]
Sodium Tartrate [868-18-8]
Sodium Tellurate(IV) [10102-20-2]
Sodium Tellurate(VI) [10101-83-4]
Sodium Tetrachloroaluminate
  [7784-16-9]
Sodium Tetrachloroaurate(III)
  [15189-51-2]
Sodium Tetradecyl Sulfate [139-88-8]
Sodium Tetraphenylborate [143-66-8]
Sodium Tetrathionate [10101-86-7]
Sodium Thioantimonate(V)
  [13776-84-6]
Sodium Thiocarbonate [534-18-9]
Sodium Thioglycolate [367-51-1]
Sodium Thiophosphate [10101-88-9]
Sodium Thiosulfate [7772-98-7]
Sodium Trimetaphosphate
  [7785-84-4]
Sodium Tripolyphosphate [7758-29-4]
Sodium Tungstate(VI) [13472-45-2]
Sodium Uranate(VI) [13721-34-1]
Sodium Vanadate(V) [13718-26-8]

Denotes derivative of title compound

Sofalcone [64506-49-6]
Solan [2307-68-8]
Solanidine [80-78-4]
Solanine [20562-02-1]
Solanocapsine [639-86-1]
Solanone [1937-54-8]
Solasodine [126-17-0]
Solasonine [19121-58-5]
Solasulfone [133-65-3]
Soman [96-64-0]
Somatoliberin [83930-13-6]
Somatostatin [38916-34-6]
Somatotropin [9002-72-6]
Somatotropin (Human)* [12629-01-5]
Songorine [509-24-0]
Sophorabioside [2945-88-2]
Sophoricoside [152-95-4]
Sophorose [534-46-3]
Sorbic Acid [110-44-1]
Sorbic Alcohol [111-28-4]
Sorbinil [68367-52-2]
Sorbitol [50-70-4]
Sorbose [87-79-6]
Sotalol [3930-20-9]
Sotalol Hydrochloride* [959-24-0]
Soterenol [13642-52-9]
Soterenol Monohydrochloride*
    [14816-67-2]
Sozoiodolic Acid [554-71-2]
Sparassol [520-43-4]
Sparsiflorine [2128-61-2]
Sparteine [90-39-1]
Sparteine Sulfate Pentahydrate*
    [6160-12-9]
Spasmolytol [25333-96-4]
Spectinomycin [1695-77-8]
Spectinomycin Dihydrochloride Pen-
    tahydrate* [22189-32-8]
Spermidine [124-20-9]
Spermine [71-44-3]
Spherophysine [25978-54-5]
Sphingosine [123-78-4]
α-Spinasterol [481-18-5]
Spinulosin [85-23-4]
Spiperone [749-02-0]
Spiramycin [8025-81-8]
Spirilene [357-66-4]
Spirit of Ammonia, Aromatic
    [8013-59-0]
Spirit of Camphor [76-22-2]
Spirit of Ether [8013-43-2]
Spirit of Ether Compound
    [8013-44-3]
Spirit of Ethyl Nitrite [109-95-5]
Spirit of Formic Acid [64-18-6]
Spirit of Glyceryl Trinitrate [55-63-0]
Spirogermanium [41992-23-8]
Spirogermanium Dihydrochloride*
    [41992-22-7]
Spironolactone [52-01-7]
Spizofurone [72492-12-7]
Splenin [1416-60-0]
Squalane [111-01-3]
Squalene [111-02-4]
Stachydrine [471-87-4]
Stallimycin [636-47-5]
Stallimycin Hydrochloride*
    [6576-51-8]
Stannic Bromide [7789-67-5]
Stannic Chloride [7646-78-8]
Stannic Chromate(VI) [38455-77-5]
Stannic Fluoride [7783-62-2]
Stannic Iodide [7790-47-8]
Stannic Oxide [18282-10-5]
Stannic Selenide [20770-09-6]
Stannic Selenite [7446-25-5]
Stannic Sulfide [1315-01-1]
Stannous Acetate [638-39-1]
Stannous Bromide [10031-24-0]
Stannous Chloride [7772-99-8]
Stannous Fluoride [7783-47-3]

Stannous Hexafluorozirconate(IV)
    [12419-43-1]
Stannous Iodide [10294-70-9]
Stannous Oxalate [814-94-8]
Stannous Oxide [21651-19-4]
Stannous Pyrophosphate
    [15578-26-4]
Stannous Selenide [1315-06-6]
Stannous Sulfate [7488-55-3]
Stannous Sulfide [1314-95-0]
Stannous Tartrate [815-85-0]
Stanolone [521-18-6]
Stanozolol [10418-03-8]
Starch, Soluble [9005-84-9]
Statine [49642-07-1]
Statolon [11006-77-2]
Stearic Acid [57-11-4]
Stenbolone [5197-58-0]
Stenbolone Acetate* [1242-56-4]
Stepronin [72324-18-6]
Steviol [471-80-7]
Stevioside [57817-89-7]
Stibine [7803-52-3]
Stibocaptate [27279-76-1]
Stibophen [15489-16-4]
Stigmastanol [83-45-4]
Stigmasterol [83-48-7]
Stilbamidine [122-06-5]
Stilbamidine Isethionate* [140-59-0]
Stilbazium Iodide [3784-99-4]
Stilbene [588-59-0]
Stilonium Iodide [2551-76-0]
Stirofos [22248-79-9]
Storax [8023-62-9]
Strepogenin [11000-03-6]
Streptidine [85-17-6]
Streptobiosamine [126-05-6]
Streptokinase [9002-01-1]
Streptolydigin [7229-50-7]
Streptomycin [57-92-1]
Streptomycin Sulfate* [3810-74-0]
Streptomycin B [128-45-0]
Streptonicozid [5667-71-0]
Streptonigrin [3930-19-6]
L-Streptose [13008-73-6]
Streptovaricin [1404-74-6]
Streptovirudin [56833-74-0]
Streptozocin [18883-66-4]
Strigol [11017-56-4]
Strobane® [8001-50-1]
Strontium [7440-24-6]
Strontium Acetate [543-94-2]
Strontium Bromate [14519-18-7]
Strontium Bromide [10476-81-0]
Strontium Carbonate [1633-05-2]
Strontium Chlorate [7791-10-8]
Strontium Chloride [10476-85-4]
Strontium Chromate(VI) [7789-06-2]
Strontium Fluoride [7783-48-4]
Strontium Formate [592-89-2]
Strontium Hydroxide [18480-07-4]
Strontium Iodide [10476-86-5]
Strontium Lactate [29870-99-3]
Strontium Nitrate [10042-76-9]
Strontium Oxalate [814-95-9]
Strontium Oxide [1314-11-0]
Strontium Peroxide [1314-18-7]
Strontium Phosphate, Tribasic
    [7446-28-8]
Strontium Selenate [7446-21-1]
Strontium Sulfate [7759-02-6]
Strontium Sulfide [1314-96-1]
Strophanthidin [66-28-4]
Strophanthin [11005-63-3]
Strophanthobiose [7724-04-1]
Strychnine [57-24-9]
Strychnine Nitrate [66-32-0]
Strychnine N⁶-Oxide [7248-28-4]
Strychnine Phosphate [509-42-2]
Strychnine Sulfate [60-41-3]
Stylopine [7461-02-1]

Styphnic Acid [82-71-3]
Styramate [94-35-9]
Styrene [100-42-5]
Styrene Glycol [93-56-1]
Subathizone [121-55-1]
Suberic Acid [505-48-6]
Substance P [11035-08-8]
Subtilin [1393-38-0]
Succimer [304-55-2]
Succinamide [110-14-5]
Succinanil [83-25-0]
Succinanilic Acid [102-14-7]
Succinic Acid [110-15-6]
Succinic Anhydride [108-30-5]
Succinimide [123-56-8]
Succinonitrile [110-61-2]
Succinyl Chloride [543-20-4]
Succinylcholine Bromide [55-94-7]
Succinylcholine Chloride [71-27-2]
Succinylcholine Iodide [541-19-5]
Succinyl Peroxide [123-23-9]
Succinylsalicylic Acid [578-19-8]
Succinylsulfathiazole [116-43-8]
Succisulfone [547-36-4]
Suclofenide [30279-49-3]
Sucralfate [54182-58-0]
Sucralose [56038-13-2]
Sucrose [57-50-1]
Sucrose Octaacetate [126-14-7]
Sudan III [85-86-9]
Sufentanil [56030-54-7]
Sufentanil Citrate* [60561-17-3]
Sulbactam [68373-14-8]
Sulbenicillin [41744-40-5]
Sulbenox [58095-31-1]
Sulbentine [350-12-9]
Sulconazole [61318-90-9]
Sulconazole Nitrate* [61318-91-0]
Sulfabenz [127-77-5]
Sulfabenzamide [127-71-9]
Sulfabromomethazine [116-45-0]
Sulfacetamide [144-80-9]
Sulfachlorpyridazine [80-32-0]
Sulfachrysoidine [485-41-6]
Sulfacytine [17784-12-2]
Sulfadiazine [68-35-9]
Sulfadiazine Silver Salt*
    [22199-08-2]
Sulfadicramide [115-68-4]
Sulfadimethoxine [122-11-2]
Sulfadoxine [2447-57-6]
Sulfaethidole [94-19-9]
Sulfaguanidine [57-67-0]
Sulfaguanole [27031-08-9]
Sulfalene [152-47-6]
Sulfallate [95-06-7]
Sulfaloxic Acid [14376-16-0]
Sulfamerazine [127-79-7]
Sulfameter [651-06-9]
Sulfamethazine [57-68-1]
Sulfamethizole [144-82-1]
Sulfamethomidine [3772-76-7]
Sulfamethoxazole [723-46-6]
Sulfamethoxypyridazine [80-35-3]
Sulfamethoxypyridazine Acetyl*
    [3568-43-2]
Sulfamethylthiazole [515-59-3]
Sulfametrole [32909-92-5]
Sulfamic Acid [5329-14-6]
Sulfamide [7803-58-9]
Sulfamidochrysoidine [103-12-8]
Sulfamipyrine [129-89-5]
Sulfamoxole [729-99-7]
Sulfanilamide [63-74-1]
Sulfanilamidomethanesulfonic Acid
    Triethanolamine Salt [127-81-1]
4-Sulfanilamidosalicylic Acid
    [6202-21-7]
Sulfanilic Acid [121-57-3]
2-p-Sulfanilylanilinoethanol
    [80-02-4]

p-Sulfanilylbenzylamine [4393-19-5]
Sulfanilyl Fluoride [98-62-4]
N^4-Sulfanilylsulfanilamide [547-52-4]
Sulfanilylurea [547-44-4]
N-Sulfanilyl-3,4-xylamide [120-34-3]
Sulfanitran [122-16-7]
Sulfaperine [599-88-2]
Sulfaphenazole [526-08-9]
Sulfaproxyline [116-42-7]
Sulfapyrazine [116-44-9]
Sulfapyridine [144-83-2]
Sulfapyridine Sodium* [127-57-1]
Sulfaquinoxaline [59-40-5]
Sulfarside [1134-98-1]
Sulfarsphenamine [618-82-6]
Sulfasalazine [599-79-1]
Sulfasomizole [632-00-8]
Sulfasymazine [1984-94-7]
Sulfathiazole [72-14-0]
Sulfathiourea [515-49-1]
Sulfatolamide [1161-88-2]
Sulfazamet [852-19-7]
Sulfazecin [77912-79-9]
Sulfinalol [66264-77-5]
Sulfinalol Hydrochloride*
  [63251-39-8]
Sulfinpyrazone [57-96-5]
4,4'-Sulfinyldianiline [119-59-5]
Sulfiram [95-05-6]
Sulfisomidine [515-64-0]
Sulfisoxazole [127-69-5]
Sulfisoxazole Diolamine* [4299-60-9]
Sulfitocobalamin [15671-27-9]
Sulfoacetic Acid [123-43-3]
Sulfobromophthalein Sodium
  [71-67-0]
Sulfolane [126-33-0]
3-Sulfolene [77-79-2]
Sulfometuron Methyl [74222-97-2]
Sulfonethylmethane [76-20-0]
Sulfoniazide [3691-81-4]
Sulfonmethane [115-24-2]
Sulfonyldiacetic Acid [123-45-5]
p,p'-Sulfonyldianiline-N,N'-digalac-
  toside [1329-37-9]
Sulforaphen [592-95-0]
Sulforidazine [14759-06-9]
Sulfosalicylic Acid [97-05-2]
Sulfotep [3689-24-5]
Sulfoxide [120-62-7]
Sulfoxone Sodium [144-75-2]
Sulfur [7704-34-9]
Sulfur Chloride [10025-67-9]
Sulfur Dioxide [7446-09-5]
Sulfuretin [120-05-8]
Sulfur Hexafluoride [2551-624]
Sulfuric Acid [7664-93-9]
Slfuric Acid Fuming* [8014-95-7]
Sulfur Iodide [1312-15-8]
Sulfurous Acid [7782-99-2]
Sulfur, Pharmaceutical [7704-34-9]
Sulfur Tetrafluoride [7783-60-0]
Sulfur Trioxide [7446-11-9]
Sulfuryl Chloride [7791-25-5]
Sulfuryl Fluoride [2699-79-8]
Sulindac [38194-50-2]
Sulisatin [54935-03-4]
Sulisobenzone [4065-45-6]
Sulmarin [29334-07-4]
Sulmazole [73384-60-8]
Sulmepride [57479-88-6]
Suloctidil [54063-56-8]
Sulphan Blue [129-17-9]
Sulphenone® [80-00-2]
Sulphurenic Acid [1260-08-8]
Sulpiride [15676-16-1]
Sulpiride (l-Form)* [23672-07-3]
Sulprofos [35400-43-2]
Sulprostone [60325-46-4]
Sulthiame [61-56-3]
Sultopride [53583-79-2]

Sultosilic Acid, Piperazine Salt
  [57775-27-6]
Sultroponium [15130-91-3]
Sumatriptan [103628-46-2]
Sumatriptan Succinate*
  [103628-48-4]
Sumatrol [82-10-0]
Sunset Yellow FCF [2783-94-0]
Suprasterol II [562-71-0]
Suprofen [40828-46-4]
Suramin Sodium [129-46-4]
Suriclone [53813-83-5]
Surinamine [537-49-5]
Suxethonium Bromide [111-00-2]
Suxibuzone [27470-51-5]
Swertiamarin [17388-39-5]
Sydnones [50927-09-8]
Symclosene [87-90-1]
Synephrine [94-07-5]
Synhexyl [117-51-1]
Syringaldehyde [134-96-3]
Syringin [118-34-3]
Syrosingopine [84-36-6]
2,4,5-T [93-76-5]
Tabernanthine [83-94-3]
Tabun [77-81-6]
Tachysterol [115-61-7]
Tacrine [321-64-2]
D-Tagatose [87-81-0]
Taglutimide [14166-26-8]
Taka-Diastase [9001-19-8]
Talampicillin [47747-56-8]
Talampicillin Hydrochloride*
  [39878-70-1]
Talastine [16188-61-7]
Talbutal [115-44-6]
Talc [14807-96-6]
Talinolol [57460-41-0]
Tallysomycin [67995-68-0]
Tallysomycin A (Talisomycin)*
  [65057-90-1]
Talniflumate [66898-62-2]
Tamoxifen [10540-29-1]
Tamoxifen Citrate* [54965-24-1]
Tanacetin [1401-54-3]
Tanghinigenin [6875-16-7]
Tanghinin [25390-16-3]
Tannoform [9010-29-1]
Tantalum [7440-25-7]
Tantalum Pentachloride [7721-01-9]
Tantalum Pentafluoride [7783-71-3]
Tantalum Pentoxide [1314-61-0]
Taraxasterol [1059-14-9]
Taraxein [9010-30-4]
Taraxerol [127-22-0]
D-Tartaric Acid [147-71-7]
DL-Tartaric Acid [133-37-9]
L-Tartaric Acid [87-69-4]
meso-Tartaric Acid [147-73-9]
Tartrazine [1934-21-0]
Tartronic Acid [80-69-3]
Taurine [107-35-7]
Taurocholic Acid [81-24-3]
Taxicatin [90-71-1]
Taxine(s) [12607-93-1]
Taxodione [19026-31-4]
Taxol [33069-62-4]
Tazettine [507-79-9]
Taziprinone Hydrochloride
  [68491-57-6]
TCDD [1746-01-6]
Tebuthiuron [34014-18-1]
Technetium [7440-26-8]
Teclozan [5560-78-1]
Tecomanine [6878-83-7]
Tectorigenin [548-77-6]
Teflurane [124-72-1]
Tegafur [17902-23-7]
Teicoplanin [61036-64-4]
Telluric(VI) Acid [7803-68-1]
Tellurium [13494-80-9]

Tellurium Dibromide [7789-54-0]
Tellurium Dichloride [10025-71-5]
Tellurium Dioxide [7446-07-3]
Tellurium Hexafluoride [7783-80-4]
Tellurium Tetrabromide [10031-27-3]
Tellurium Tetrachloride [10026-07-0]
Tellurium Tetraiodide [7790-48-9]
Tellurous Acid [10049-23-7]
Telomycin [19246-24-3]
Temafloxacin [108319-06-8]
Temafloxacin Hydrochloride*
  [105784-61-0]
Temazepam [846-50-4]
Temephos [3383-96-8]
Temocillin [66148-78-5]
Tenidap [100599-27-7]
Teniloxazine [62473-79-4]
Teniposide [29767-20-2]
Tenonitrozole [3810-35-3]
Tenoxicam [59804-37-4]
Tenuazonic Acid [610-88-8]
Tephrosin [76-80-2]
Teprenone [6809-52-5]
Terazosin [63590-64-7]
Terazosin Monohydrochloride Dihy-
  drate* [70024-40-7]
Terbacil [5902-51-2]
Terbinafine [91161-71-6]
Terbium [7440-27-9]
Terbufos [13071-79-9]
Terbutaline [23031-25-6]
Terbutaline Sulfate* [23031-32-5]
Terconazole [67915-31-5]
Terebene [8014-10-6]
Terebic Acid [79-91-4]
Terephthalic Acid [100-21-0]
Terfenadine [50679-08-8]
Terguride [37686-84-3]
Terguride Hydrogen Maleate*
  [37686-85-4]
Teriparatide Acetate [99294-94-7]
Terlipressin [14636-12-5]
Terodiline [15793-40-5]
Terofenamate [29098-15-5]
Terpenylic Acid [26754-48-3]
Terpin [80-53-5]
Terpin Hydrate* [2541-01-6]
α-Terpineol [98-55-5]
Terreic Acid [121-40-4]
Tertatolol [34784-64-0]
α-Terthienyl [1081-34-1]
Testolactone [968-93-4]
Testosterone [58-22-0]
Testosterone 17-Chloral Hemiacetal
  [53608-96-1]
Testosterone 17β-Cypionate [58-20-8]
Testosterone Enanthate [315-37-7]
Testosterone Nicotinate [668-56-4]
Testosterone Phenylacetate
  [5704-03-0]
Testosterone Propionate [57-85-2]
Tetraamminecopper Sulfate
  [14283-05-7]
Tetrabarbital [76-23-3]
Tetrabenazine [58-46-8]
Tetraborane(10) [18283-93-7]
3,4,5,6-Tetrabromo-o-cresol
  [576-55-6]
sym-Tetrabromoethane [79-27-6]
3',3'',5',5''-Tetrabromophenol-
  phthalein [76-62-0]
Tetracaine Hydrochloride [136-47-0]
Tetrachlormethiazide [4267-05-4]
Tetrachloroethane [79-34-5]
Tetrachloroethylene [127-18-4]
3,3',4',5-Tetrachlorosalicylanilide
  [1154-59-2]
Tetracosamethylhendecasiloxane
  [107-53-9]
Tetracyanoethylene [670-54-2]
Tetracycline [60-54-8]

* Denotes derivative of title compound

Tetracycline Hydrochloride*
[64-75-5]
Tetracycline Phosphate Complex*
[1336-20-5]
Tetradecamethylhexasiloxane
[107-52-8]
Tetradifon [116-29-0]
Tetraethylammonium Bromide
[71-91-0]
Tetraethylammonium Chloride
[56-34-8]
Tetraethylammonium Hydroxide
[77-98-5]
Tetraethyllead [78-00-2]
N,N,N',N'-Tetraethylphthalamide
[83-81-8]
Tetraethyl Pyrophosphate [107-49-3]
2,2,3,3-Tetrafluoro-1-propanol
[76-37-9]
Tetraglycine Hydroperiodide
[7097-60-1]
Tetraglyme [143-24-8]
Tetrahydrocortisone [53-05-4]
Tetrahydrofuran [109-99-9]
2,5-Tetrahydrofurandimethanol
[104-80-3]
Tetrahydrofurfuryl Alcohol [97-99-4]
Tetrahydropalmatine [10097-84-4]
Tetrahydropapaveroline [4747-99-3]
Tetrahydropyran [142-68-7]
Tetrahydrozoline [84-22-0]
Tetrahydrozoline Hydrochloride*
[522-48-5]
Tetraiodoethylene [513-92-8]
Tetralin® [119-64-2]
Tetralol [530-91-6]
Tetramethrin [7696-12-0]
Tetramethylammonium Hydroxide
[75-59-2]
Tetramethylammonium Iodide
[75-58-1]
Tetramethyldiaminobutane [111-51-3]
Tetramethylenedisulfotetramine
[80-12-6]
Tetramethyl-p-phenylenediamine
[100-22-1]
Tetramethylurea [632-22-4]
Tetramisole [5036-02-2]
Tetramisole Hydrochloride*
[5086-74-8]
Tetramisole L-Form Hydrochloride*
[16595-80-5]
Tetrandrine [518-34-3]
Tetranectin [109489-77-2]
Tetranitromethane [509-14-8]
Tetrantoin [52094-70-9]
Tetraphenylarsonium Bromide
[507-27-7]
Tetraphenylarsonium Chloride
[507-28-8]
Tetraphosphorus Trisulfide
[1314-85-8]
Tetrasilane [7783-29-1]
Tetrasodium Pyrophosphate
[7722-88-5]
Tetrasulfur Tetranitride [28950-34-7]
Tetrazepam [10379-14-3]
Tetrazolium Blue [1871-22-3]
Tetrodotoxin [4368-28-9]
Tetronasin [75139-06-9]
Tetroquinone [319-89-1]
Tetroxoprim [53808-87-0]
Tevenel® [4302-95-8]
TFM [88-30-2]
Thalicarpine [5373-42-2]
Thalidomide [50-35-1]
Thallium [7440-28-0]
Thallium Acetate [563-68-8]
Thallium Bromide [7789-40-4]
Thallium Carbonate [6533-73-9]
Thallium Chloride [7791-12-0]

Thallium Cyanide [13453-34-4]
Thallium Fluoride [7789-27-7]
Thallium Hydroxide [12026-06-1]
Thallium Iodide [7790-30-9]
Thallium Nitrate [10102-45-1]
Thallium Oxide [1314-12-1]
Thallium Selenate [7446-22-2]
Thallium Selenide [15572-25-5]
Thallium Sesquioxide [1314-32-5]
Thallium Sulfate [7446-18-6]
Thallium Sulfide [1314-97-2]
Thallium Trifluoride [7783-57-5]
Thaumatin [53850-34-3]
Theaflavine [4670-05-7]
Thebaine [115-37-7]
Thebainone [467-98-1]
Thenaldine [86-12-4]
Thenium Closylate [4304-40-9]
3-Thenoic Acid [88-13-1]
Thenyldiamine [91-79-2]
Theobromine [83-67-0]
Theobromine Calcium Salicylate*
[8065-51-8]
Theobromine Sodium Acetate*
[8002-88-8]
Theobromine Sodium Salicylate*
[8048-31-5]
1-Theobromineacetic Acid
[5614-56-2]
Theofibrate [54504-70-0]
Theophylline [58-55-9]
Theophylline Diethanolamine*
[32156-80-2]
Theophylline Ethanolamine*
[573-41-1]
Theophylline Isopropanolamine*
[5600-19-1]
Theophylline Sodium Acetate*
[8002-89-9]
Theophylline Sodium Glycinate*
[8000-10-1]
Thermolysin [9073-78-3]
Thermorubin [11006-83-0]
Thevetin A [37933-66-7]
Thevetose [5675-98-9]
Thiabendazole [148-79-8]
Thiacetazone [104-06-3]
Thialbarbital [467-36-7]
Thiambutene [86-14-6]
Thiamine Disulfide [67-16-3]
Thiamine Hydrochloride [67-03-8]
Thiamine Mononitrate [532-43-4]
Thiamine Phosphoric Acid Ester
Chloride
[532-40-1]
Thiamine Phosphoric Acid Ester
Phosphate Salt [635-97-2]
Thiamine 1,5-Salt [573-09-1]
Thiamine Triphosphoric Acid Ester
[3475-65-8]
Thiamiprine [5581-52-2]
Thiamorpholine [123-90-0]
Thiamphenicol [15318-45-3]
Thiamphenicol DL-Form* [847-25-6]
Thiamylal [77-27-0]
Thiamylal Sodium* [337-47-3]
Thianaphthene [95-15-8]
Thiazesim [5845-26-1]
Thiazesim Hydrochloride*
[3122-01-8]
Thiazinamium Methylsulfate
[58-34-4]
Thiazole [288-47-1]
Thiazolinobutazone [54749-86-9]
Thiazolsulfone [473-30-3]
Thiazol Yellow G [1829-00-1]
Thibenzazoline [6028-35-9]
Thienamycin [59995-64-1]
Thiethylperazine [1420-55-9]
Thiethylperazine Maleate*
[52239-63-1]

Thihexinol [53626-54-3]
Thihexinol Methylbromide*
[7219-91-2]
Thimerfonate Sodium [5964-24-9]
Thimerosal [54-64-8]
Thioacetaldehyde [2765-04-0]
Thioacetamide [62-55-5]
Thioacetic Acid [507-09-5]
Thiobarbital [77-32-7]
Thiobenzyl Alcohol [100-53-8]
Thiobutabarbital [2095-57-0]
Thiocarbamizine [91-71-4]
Thiocarbarsone [120-02-5]
Thiocolchicine [2730-71-4]
Thiocolchicoside* [602-41-5]
Thiocolchicoside* [602-41-5]
Thiocresol [26445-03-4]
Thioctic Acid [62-46-4]
Thiocyanate Sodium [540-72-7]
Thiocyanic Acid [463-56-9]
Thiodicarb [59669-26-0]
2,2'-Thiodiethanol [111-48-8]
Thiodiglycolic Acid [123-93-3]
3,3'-Thiodipropionic Acid [111-17-1]
Thioformamide [115-08-2]
5-Thio-D-glucose [20408-97-3]
Thioglycerol [96-27-5]
Thioglycolic Acid [68-11-1]
Thioguanine [154-42-7]
Thioguanosine [85-31-4]
Thiolactic Acid [79-42-5]
2-Thiolhistidine [13552-61-9]
Thiolutin [87-11-6]
Thiomalic Acid [70-49-5]
Thionalide [93-42-5]
Thionaphthene-2-carboxylic Acid
[6314-28-9]
Thionazin [297-97-2]
Thionine [135-59-1]
Thionyl Bromide [507-16-4]
Thionyl Chloride [7719-09-7]
Thionyl Fluoride [7783-42-8]
Thiopental Sodium [71-73-8]
Thiopeptin [12609-84-6]
Thiophanate [23564-06-9]
Thiophene [110-02-1]
2-Thiophenecarboxylic Acid
[527-72-0]
Thiophenol [108-98-5]
Thiopropazate [84-06-0]
Thioproperazine [316-81-4]
Thioquinox [93-75-4]
Thioredoxin [52500-60-4]
Thioridazine [50-52-2]
Thioridazine Hydrochloride*
[130-61-0]
Thiosalicylic Acid [147-93-3]
Thiosemicarbazide [79-19-6]
Thiosinamine [109-57-9]
1-Thiosorbitol [24531-57-5]
Thiostrepton [1393-48-2]
Thiothiamine [299-35-4]
Thiothixene [5591-45-7]
2-Thiouracil [141-90-2]
Thiourea [62-56-6]
Thioxanthene [261-31-4]
Thioxanthone [492-22-8]
THIP [64603-91-4]
Thiphenamil [82-99-5]
Thiphenamil Hydrochloride*
[548-68-5]
Thiram [137-26-8]
Thomas Phosphate [1306-01-0]
Thonzonium Bromide [553-08-2]
Thonzylamine Hydrochloride
[63-56-9]
Thorium [7440-29-1]
Thorium Chloride [10026-08-1]
Thorium Iodide [7790-49-0]
Thorium Nitrate [13823-29-5]
Thorium Oxide [1314-20-1]

Thorium Sulfate [10381-37-0]
Thorium Tetracyanoplatinate(II) [14481-33-5]
Thozalinone [655-05-0]
Threonine [72-19-5]
D-Threose [95-43-2]
L-Threose [95-44-3]
Thrombin [9002-04-4]
Thrombin, Topical [9002-04-4]
Thromboplastin [9035-58-9]
Thujic Acid [499-89-8]
Thujopsene [470-40-6]
Thulium [7440-30-4]
Thurfyl Nicotinate [70-19-9]
Thymidine [50-89-5]
Thymine [65-71-4]
Thymol [89-83-8]
Thymol Acetate* [528-79-0]
Thymol Carbonate* [552-93-2]
Thymol Blue [76-61-9]
Thymol Iodide [552-22-7]
Thymolphthalein [125-20-2]
Thymomodulin [90803-92-2]
Thymopentin [69558-55-0]
Thymopoietin [60529-76-2]
o-Thymotic Acid [548-51-6]
Thymyl N-Isoamylcarbamate [578-20-1]
Thyropropic Acid [51-26-3]
Thyroprotein [9005-97-4]
Thyroxine [51-48-9]
Tiadenol [6964-20-1]
Tiamenidine [31428-61-2]
Tiamenidine Hydrochloride* [51274-83-0]
Tiamulin [55297-95-5]
Tiamulin Fumarate* [55297-96-6]
Tianeptine [66981-73-5]
Tiapride [51012-32-9]
Tiaprofenic Acid [33005-95-7]
Tiaprost [71116-82-0]
Tiaramide [32527-55-2]
Tiaramide Hydrochloride* [35941-71-0]
Tibezonium Iodide [54663-47-7]
Tibolone [5630-53-5]
Ticarbodine [31932-09-9]
Ticarcillin [34787-01-4]
Ticarcillin Disodium* [4697-14-7]
Ticlopidine [55142-85-3]
Ticlopidine Hydrochloride* [53885-35-1]
Ticrynafen [40180-04-9]
Tiemonium Iodide [144-12-7]
Tigemonam [102507-71-1]
Tigemonam Dicholine* [102916-21-2]
Tiglic Acid [80-59-1]
Tigloidine [495-83-0]
Tigonin [1329-83-5]
Tiliacorine [27073-72-9]
Tilidine [20380-58-9]
Tilorone [27591-97-5]
Tilorone Hydrochloride* [27591-69-1]
Timepidium Bromide [35035-05-3]
Timiperone [57648-21-2]
Timolol [26839-75-8]
Timolol Maleate* [26921-17-5]
Timonacic [444-27-9]
Tin [7440-31-5]
Tinidazole [19387-91-8]
Tinofedrine [66788-41-8]
Tinoridine [24237-54-5]
Tinuvin® P [2440-22-4]
Tiocarlide [910-86-1]
Tioclomarol [22619-35-8]
Tioconazole [65899-73-2]
Tiomesterone [2205-73-4]
Tiopronin [1953-02-2]
Tioxidazole [61570-90-9]

Tioxolone [4991-65-5]
Tipepidine [5169-78-8]
Tiquizium Bromide [71731-58-3]
Tiratricol [51-24-1]
Tiron® [149-45-1]
Tiropramide [55837-29-1]
Titanic(IV) Acid [20338-08-3]
Titanium [7440-32-6]
Titanium Dichloride [10049-06-6]
Titanium Dioxide [13463-67-7]
Titanium Hydride [7704-98-5]
Titanium Isopropylate [546-68-9]
Titanium Sesquisulfate [10343-61-0]
Titanium Sulfate [13825-74-6]
Titanium Tetrabromide [7789-68-6]
Titanium Tetrachloride [7550-45-0]
Titanium Tetrafluoride [7783-63-3]
Titanium Trichloride [7705-07-9]
Titanocene Dichloride [1271-19-8]
Tixocortol [61951-99-3]
Tixocortol Pivalate* [55560-96-8]
Tizanidine [51322-75-9]
TMD [60761-10-6]
Tobramycin [32986-56-4]
Tobramycin Sulfate* [79645-27-5]
Tocainide [41708-72-9]
Tocamphyl [5634-42-4]
Tocol [119-98-2]
β-Tocopherol [148-03-8]
γ-Tocopherol [7616-22-0]
δ-Tocopherol [119-13-1]
ε-Tocopherol [490-23-3]
ζ₁-Tocopherol [1721-51-3]
ζ₂-Tocopherol [493-35-6]
η-Tocopherol [91-86-1]
α-Tocopherol Acid Succinate [4345-03-3]
Todralazine [14679-73-3]
Tofenacin [15301-93-6]
Tofenacin Hydrochloride* [10488-36-5]
Tofisopam [22345-47-7]
Tolan [501-65-5]
Tolazamide [1156-19-0]
Tolazoline [59-98-3]
Tolazoline Hydrochloride* [59-97-2]
Tolboxane [2430-46-8]
Tolbutamide [64-77-7]
Tolbutamide Sodium* [473-41-6]
Tolciclate [50838-36-3]
Tolcyclamide [664-95-9]
Toldimfos Sodium [575-75-7]
Tolfenamic Acid [13710-19-5]
o-Tolidine [119-93-7]
Tolindate [27877-51-6]
Toliprolol [2933-94-0]
Tolmetin [26171-23-3]
Tolmetin Sodium* [64490-92-2]
Tolnaftate [2398-96-1]
Tolonidine [4201-22-3]
Tolonium Chloride [92-31-9]
Toloxatone [29218-27-7]
Toloxychlorinol [6055-48-7]
Tolperisone [728-88-1]
Tolpronine [97-57-4]
Tolpropamine [5632-44-0]
Tolrestat [82964-04-3]
Toltrazuril [69004-03-1]
o-Tolualdehyde [529-20-4]
o-Toluamide [527-85-5]
Toluene [108-88-3]
Toluene 2,4-Diisocyanate [584-84-9]
Toluene-3,4-dithiol [496-74-2]
p-Toluenesulfinic Acid [536-57-2]
p-Toluenesulfonic Acid [104-15-4]
p-Toluenesulfonyl Chloride [98-59-9]
m-Toluidine* [108-44-1]
o-Toluidine* [95-53-4]
p-Toluidine* [106-49-0]
o-Tolunitrile [529-19-1]
p-Tolunitrile [104-85-8]

2-(p-Toluyl)benzoic Acid [85-55-2]
Toluylene Blue [97-26-7]
Tolycaine [3686-58-6]
p-Tolylsulfonylmethylnitrosamide [80-11-5]
Tomatidine [77-59-8]
Tomatine [17406-45-0]
Tonin [53414-68-9]
Torasemide [56211-40-6]
Toremifene [89778-26-7]
Toremifene Citrate* [89778-27-8]
Torularhodin [514-92-1]
Tosufloxacin [108138-46-1]
Tosufloxacin Monohydrate* [107097-79-0]
Tosufloxacin Toluenesulfonate* [115964-29-9]
Toxaphene [8001-35-2]
Toxiferine I [6888-23-9]
Toxoflavin [84-82-2]
Toxohormone [9014-44-2]
Toxopyrimidine [73-67-6]
Toyocamycin [606-58-6]
Tralkoxydim [87820-88-0]
Tramadol [27203-92-5]
Tramadol Hydrochloride* [22204-88-2]
Tramazoline [1082-57-1]
Tramazoline Hydrochloride* [3715-90-0]
Tranexamic Acid [1197-18-8]
Tranid® [15271-41-7]
Tranilast [53902-12-8]
Tranylcypromine [155-09-9]
Tranylcypromine Sulfate* [13492-01-8]
Trapidil [15421-84-8]
Traumatic Acid [6402-36-4]
Traxanox [58712-69-9]
Traxanox Sodium* [70502-82-8]
Trazodone [19794-93-5]
Trazodone Hydrochloride* [25332-39-2]
Trehalose [99-20-7]
Tremetone [4976-25-4]
Tremorine [51-73-0]
Trenbolone [10161-33-8]
Trenbolone Acetate* [10161-34-9]
Trengestone [5192-84-7]
Trepibutone [41826-92-0]
Tretoquinol [18559-59-6]
TRH [24305-27-9]
Triacetin [102-76-1]
Triacetyldiphenolisatin [18869-73-3]
1-Triacontanol [593-50-0]
Triadimefon [43121-43-3]
Triadimenol [55219-65-3]
Triafur [712-68-5]
Triallate [2303-17-5]
Triamcinolone [124-94-7]
Triamcinolone Acetonide [76-25-5]
Triamcinolone Acetonide Disodium Phosphate* [1997-15-5]
Triamcinolone Benetonide [31002-79-6]
Triamcinolone Hexacetonide [5611-51-8]
Triamterene [396-01-0]
s-Triazaborane [6569-51-3]
Triaziquone [68-76-8]
Triazolam [28911-01-5]
1H-1,2,4-Triazole [288-88-0]
Tribenoside [10310-32-4]
Tribromoacetic Acid [75-96-7]
2,4,6-Tribromoaniline [147-82-0]
Tribromo-tert-butyl Alcohol [76-08-4]
2,4,6-Tribromo-m-cresol [4619-74-3]
Tribromoethanol [75-80-9]
2,4,6-Tribromophenol [118-79-6]
1,2,3-Tribromopropane [96-11-7]

* *Denotes derivative of title compound*

Tribromosilane [7789-57-3]
Tribromsalan [87-10-5]
Tributylamine [102-82-9]
Tributyl Phosphate [126-73-8]
Tributyrin [60-01-5]
Tricaine [886-86-2]
Tricarballylic Acid [99-14-9]
Tricetamide [363-20-2]
Trichlorfon [52-68-6]
Trichlormethiazide [133-67-5]
Trichloroacetaldehyde [75-87-6]
Trichloroacetic Acid [76-03-9]
Trichloroacetonitrile [545-06-2]
2,4,6-Trichloroanisole [87-40-1]
1,2,3-Trichlorobenzene [87-61-6]
1,2,4-Trichlorobenzene [120-82-1]
1,3,5-Trichlorobenzene [108-70-3]
α,α,β-Trichloro-n-butyraldehyde
    [76-36-8]
2,3,6-Trichloro-p-cresol [551-77-9]
2,4,6-Trichloro-m-cresol [551-76-8]
4,5,6-Trichloro-o-cresol [551-78-0]
1,1,1-Trichloroethane [71-55-6]
1,1,2-Trichloroethane [79-00-5]
2,2,2-Trichloroethanol [115-20-8]
Trichloroethylene [79-01-6]
Trichlorofluoromethane [75-69-4]
3,4,6-Trichloro-2-nitrophenol
    [82-62-2]
2,4,5-Trichlorophenol [95-95-4]
2,4,6-Trichlorophenol [88-06-2]
1,1,1-Trichloro-2-propanol [76-00-6]
3′,4′,5-Trichlorosalicylanilide
    [642-84-2]
Trichlorosilane [10025-78-2]
2,2′,2″-Trichlorotriethylamine
    [555-77-1]
Trichlorourethan [107-69-7]
Trichodermin [4682-50-2]
Trichosanthin [60318-52-7]
Trichothecin [6379-69-7]
Tricine [5704-04-1]
Triclabendazole [68786-66-3]
Triclobisonium Chloride [79-90-3]
Triclocarban [101-20-2]
Triclodazol [56-28-0]
Triclofenol Piperazine [5714-82-9]
Triclofos [306-52-5]
Triclofos Sodium* [7246-20-0]
Triclopyr [55335-06-3]
Triclosan [3380-34-5]
Tricromyl [85-90-5]
Tridecylbenzene [123-02-4]
Tridemorph [24602-86-6]
Tridihexethyl Iodide [125-99-5]
Tridiphane [58138-08-2]
Trientine [112-24-3]
Trientine Hydrochloride*
    [38260-01-4]
Trietazine [1912-26-1]
Triethanolamine [102-71-6]
Triethylamine [121-44-8]
Triethylammonium Formate
    [585-29-5]
Triethylenediamine [280-57-9]
Triethylene Glycol [112-27-6]
Triethylenemelamine [51-18-3]
Triethylenephosphoramide [545-55-1]
Triethylenethiophosphoramide
    [52-24-4]
Triethyl Phosphate [78-40-0]
Triethyl Phosphine [554-70-1]
Trifenmorph [1420-06-0]
Triflumuron [64628-44-0]
Trifluomeprazine [2622-37-9]
Trifluoperazine [117-89-5]
Trifluoperazine Dihydrochloride*
    [440-17-5]
Trifluoracetic Acid [76-05-1]
Trifluperidol [749-13-3]
Triflupromazine [146-54-3]

Triflupromazine Hydrochloride*
    [1098-60-8]
Trifluralin [1582-09-8]
Trifluridine [70-00-8]
Triflusal [322-79-2]
Triforine [26644-46-2]
Trigentisic Acid [52486-80-3]
Triglyme [112-49-2]
Trigonellamide Chloride [1005-24-9]
Trigonelline [535-83-1]
Trihexyphenidyl Hydrochloride
    [52-49-3]
Trihydrazine Dihydriodide
    [10034-92-1]
Trilobine [6138-73-4]
Trilostane [13647-35-3]
Trimazosin [35795-16-5]
Trimazosin Monohydrochloride
    Monohydrate* [53746-46-6]
Trimebutine [39133-31-8]
Trimecaine [616-68-2]
Trimedlure [12002-53-8]
Trimellitic Acid [528-44-9]
Trimellitic Anhydride [552-30-7]
Trimeprazine [84-96-8]
Trimeprazine Tartrate* [4330-99-8]
Trimetazidine [5011-34-7]
Trimethadione [127-48-0]
Trimethaphan Camsylate [68-91-7]
Trimethidinium Methosulfate
    [7009-82-7]
Trimethobenzamide [138-56-7]
Trimethobenzamide Hydrochloride*
    [554-92-7]
Trimethoprim [738-70-5]
Trimethylamine [75-50-3]
Trimethyl Borate [121-43-7]
N,N-1-Trimethyl-3,3-diphenylpropyl-
    amine [13957-55-6]
Trimethylene Bromide [109-64-8]
Trimethylene Glycol [504-63-2]
Trimethylene Oxide [503-30-0]
Trimethylolmelamine [1017-56-7]
2,4,6-Trimethylpyridine [108-75-8]
Trimethylsilyl Triflate [27607-77-8]
Trimetozine [635-41-6]
Trimetrexate [52128-35-5]
Trimipramine [739-71-9]
Trimipramine Maleate* [521-78-8]
Trimoprostil [69900-72-7]
Trimyristin [555-45-3]
sym-Trinitrobenzene [99-35-4]
2,4,6-Trinitrobenzoic Acid [129-66-8]
2,4,7-Trinitrofluorenone [129-79-3]
Trinitromethane [517-25-9]
2,4,6-Trinitrotoluene [118-96-7]
Triolein [122-32-7]
s-Trioxane [110-88-3]
Trioxsalen [3902-71-4]
Tripalmitin [555-44-2]
Tripamide [73803-48-2]
Triparanol [78-41-1]
Tripelennamine [91-81-6]
Tripelennamine Citrate* [6138-56-3]
Tripelennamine Hydrochloride*
    [154-69-8]
Triphal® [6138-58-5]
Triphenylcarbinol [76-84-6]
Triphenylene [217-59-4]
Triphenylmethane [519-73-3]
Triphenyl Phosphate [115-86-6]
Triphenylphosphine [603-35-0]
Triphenyltetrazolium Chloride
    [298-96-4]
Triphenyltin Hydroxide [76-87-9]
Tripoli [1317-95-9]
Triprolidine [486-12-4]
Triprolidine Monohydrochloride
    Monohydrate* [6138-79-0]
Triptorelin [57773-63-4]
Triptycene [477-75-8]

2,4,6-Tripyridyl-s-triazine
    [3682-35-7]
Tris-BP [126-72-7]
Tris(ethylenediamine)cadmium Dihy-
    droxide [14874-24-9]
Tris(hydroxymethyl)nitromethane
    [126-11-4]
Trisilane [7783-26-8]
Tristearin [555-43-1]
Tristriphenylphosphine Rhodium
    Carbonyl Hydride [17185-29-4]
Trithiocarbonic Acid [594-08-1]
Trithiozine [35619-65-9]
Tritium [10028-17-8]
Tritolyl Phosphate [1330-78-5]
Tri-o-tolyl Phosphate [78-30-8]
Tritoqualine [14504-73-5]
Triuret [556-99-0]
Troclosene Potassium [2244-21-5]
Trofosfamide [22089-22-1]
Troleandomycin [2751-09-9]
Trolnitrate Phosphate [588-42-1]
Tromantadine [53783-83-8]
Tromethamine [77-86-1]
Tropacine [6878-98-4]
Tropacocaine [537-26-8]
Tropaeolin O [547-57-9]
Tropaeolin OO [554-73-4]
Tropane [529-17-9]
Tropenzile [53834-53-0]
Tropic Acid [529-64-6]
Tropicamide [1508-75-4]
Tropine [120-29-6]
Tropine Benzylate [3736-36-5]
Tropylium Bromide [5376-03-4]
Trospium Chloride [10405-02-4]
Troxerutin [7085-55-4]
Troxipide [99777-81-9]
Truxillic Acid [4462-95-7]
Trypan Blue [72-57-1]
Trypan Red [574-64-1]
Tryparsamide [554-72-3]
Trypsin [9002-07-7]
Tryptamine [61-54-1]
Tryptazan [526-30-7]
L-Tryptophan [73-22-3]
Tryptophol [526-55-6]
TSH [9002-71-5]
Tsuduranine [517-97-5]
T-2 Toxin [21259-20-1]
Tuaminoheptane [123-82-0]
Tuaminoheptane Sulfate* [6411-75-2]
Tuberactinomycin [11075-36-8]
Tubercidin [69-33-0]
Tuberin [53643-53-1]
Tubocurarine Chloride [57-94-3]
Tuftsin [9063-57-4]
Tulobuterol [41570-61-0]
Tungsten [7440-33-7]
Tungsten Hexafluoride [7783-82-6]
Tungsten Trioxide [1314-35-8]
Tungstic(VI) Acid [7783-03-1]
Tunichrome B-1 [97689-87-7]
Turanose [547-25-1]
Turicine [515-24-2]
Turks Island Salt [8006-37-9]
ar-Turmerone [38142-58-4]
Tutin [2571-22-4]
Twistane [253-14-5]
Tybamate [4268-36-4]
Tylocrebrine [6879-02-3]
Tylophorine [482-20-2]
Tylosin [1401-69-0]
Tyloxapol [25301-02-4]
Tymazoline [24243-97-8]
Tyramine [51-67-2]
Tyrocidine [8011-61-8]
Tyropanoate Sodium [7246-21-1]
Tyrosinase [9002-10-2]
Tyrosine [60-18-4]
m-Tyrosine [587-33-7]

Tyrothricin [1404-88-2]
Ubenimex [58970-76-6]
Uglow Black Silver [8047-26-5]
Uintahite [12002-43-6]
Ujothion [1219-77-8]
Ultramarine [1317-97-1]
Umbelliferone [93-35-6]
Undecylenic Acid [112-38-9]
Uracil [66-22-8]
Uracil Mustard [66-75-1]
Uramil [118-78-5]
Uranediol [516-51-8]
Uranium [7440-61-1]
Uranium Dioxide [1344-57-6]
Uranium Hexafluoride [7783-81-5]
Uranium Peroxide [19525-15-6]
Uranium Tetrachloride [10026-10-5]
Uranium Tetrafluoride [10049-14-6]
Uranium Trichloride [10025-93-1]
Uranium Trioxide [1344-58-7]
Uranyl Acetate [541-09-3]
Uranyl Chloride [7791-26-6]
Uranyl Nitrate [36478-76-9]
Uranyl Phosphate [18433-48-2]
Uranyl Sulfate [1314-64-3]
Urapidil [34661-75-1]
Urazole [3232-84-6]
Urea [57-13-6]
Urea Hydrochloride [506-89-8]
Urea Hydrogen Peroxide [124-43-6]
Urea Nitrate [513-80-4]
Urease [9002-13-5]
Urea Stibamine [1340-35-8]
Uredepa [302-49-8]
Urethan [51-79-6]
Uric Acid [69-93-2]
Uricase [9002-12-4]
Uridine [58-96-8]
Uridine 5'-Diphosphate [58-98-0]
Uridine Diphosphate Glucose
    [133-89-1]
Uridine 5'-Triphosphate [63-39-8]
5'-Uridylic Acid [58-97-9]
Urochloralic Acid [97-25-6]
Urokinase [9039-53-6]
Urothion [19295-31-9]
Ursodiol [128-13-2]
Ursolic Acid [77-52-1]
Uscharidin [20304-48-7]
Usnic Acid [125-46-2]
Ustilagic Acid [8002-36-6]
Uzarin [20231-81-6]
Vaccenic Acid [693-72-1]
Vacciniin [90-75-5]
Valacidin [53762-92-8]
n-Valeraldehyde [110-62-3]
Valeric Acid, Normal [109-52-4]
Valethamate Bromide [90-22-2]
Valine [72-18-4]
Valinomycin [2001-95-8]
Valnoctamide [4171-13-5]
Valproic Acid [99-66-1]
Valproic Acid Semisodium Salt*
    [76584-70-8]
Valproic Acid Sodium Salt (1:1)*
    [1069-66-5]
Valpromide [2430-27-5]
Vanadium [7440-62-2]
Vanadium Carbonyl [20644-87-5]
Vanadium Pentafluoride [7783-72-4]
Vanadium Pentoxide [1314-62-1]
Vanadium Tetrafluoride [10049-16-8]
Vanadium Trifluoride [10049-12-4]
Vanadium Trioxide [1314-34-7]
Vanadium Trisulfate [13701-70-7]
Vanadium Trisulfide [1315-03-3]
Vanadyl Dichloride [10213-09-9]
Vanadyl Sulfate [27774-13-6]
Vanadyl Trichloride [7727-18-6]
Vancomycin [1404-90-6]

Vancomycin Hydrochloride*
    [1404-93-9]
Vanillic Acid [121-34-6]
Vanillin [121-33-5]
Vanilmandelic Acid [55-10-7]
Vanitiolide [17692-71-6]
Vasicine [6159-55-3]
Vasopressin [9034-50-8]
Veatchine [76-53-9]
Vecuronium Bromide [50700-72-6]
Vellosimine [6874-98-2]
Veralipride [66644-81-3]
Veralkamine [17155-31-6]
Verapamil [52-53-9]
Verapamil Hydrochloride* [152-11-4]
Veratraldehyde [120-14-9]
Veratramine [60-70-8]
Veratric Acid [93-07-2]
Veratridine [71-62-5]
Veratrole [91-16-7]
Verazide [93-47-0]
Verbascose [546-62-3]
Verbenalin [548-37-8]
d-Verbenone [18309-32-5]
Vermiculite [1318-00-9]
Vernadigin [30285-47-3]
Vernamycin B [9040-14-6]
Vernolate [1929-77-7]
Vernolepin [18542-37-5]
Vernolic Acid [503-07-1]
Versalide® [88-29-9]
Versen-Ol® [139-89-9]
Verticine [23496-41-5]
Vetrabutine [3735-45-3]
Vicianin [155-57-7]
Vicianose [14116-69-9]
Vicine [152-93-2]
Vidarabine [5536-17-4]
Vigabatrin [60643-86-9]
Villikinin [9056-49-9]
Viloxazine [46817-91-8]
Viloxazine Hydrochloride*
    [35604-67-2]
Viminol [21363-18-8]
Vinbarbital Sodium [125-44-0]
Vinblastine [865-21-4]
Vinblastine Sulfate* [143-67-9]
Vincamine [1617-90-9]
Vincetoxin [1401-78-1]
Vinclozolin [50471-44-8]
Vincristine [57-22-7]
Vincristine Sulfate* [2068-78-2]
Vindesine [53643-48-4]
Vindesine Sulfate* [59917-39-4]
Vindoline [2182-14-1]
Vinpocetine [42971-09-5]
Vintiamol [26242-33-1]
Vinyl Acetate [108-05-4]
Vinylbital [2430-49-1]
Vinyl Chloride [75-01-4]
Vinyl Ether [109-93-3]
Vinylidene Chloride [75-35-4]
Violacein [548-54-9]
Violaxanthin [126-29-4]
Violuric Acid [87-39-8]
Viomycin [32988-50-4]
Viomycin Sulfate* [37883-00-4]
Viomycin Pantothenate [1401-79-2]
VIP [37221-79-7]
Viquidil [84-55-9]
Viractin [53762-93-9]
Virginiamycin M1* [21411-53-0]
Virginiamycin S1* [23152-29-6]
Viridicatin [129-24-8]
Viridin [3306-52-3]
Viscosin [27127-62-4]
Visnadine [477-32-7]
Visnagin [82-57-5]
Vital Red [574-65-2]
Vitamin A [68-26-8]
Vitamin A₂ [79-80-1]

Vitamin B₁₂ [68-19-9]
Vitamin B₁₂, Radioactive
    [13422-53-2]
Vitamin B₁₂c [20623-13-6]
Vitamin B₁₂r [53762-94-0]
Vitamin B₁₂s [18534-66-2]
Vitamin D₁ [520-91-2]
Vitamin D₂ [50-14-6]
Vitamin D₃ [67-97-0]
Vitamin D₄ [511-28-4]
Vitamin E [59-02-9]
Vitamin E Acetate [58-95-7]
Vitamin K₁ [84-80-0]
Vitamin K₁ Oxide [25486-55-9]
Vitamin K₅ [130-24-5]
Vitamin K₅ N-Acetyl Analog*
    [523-68-2]
Vitamin K₆ [83-68-1]
Vitamin K₇ [83-69-2]
Vitamin K-S(II) [2487-39-0]
Vitamin T [1407-73-4]
Vitamin U [1115-84-0]
Vitride® [22722-98-1]
Voacamine [3371-85-5]
Vomicine [125-15-5]
Vomitoxin [51481-10-8]
VX [50782-69-9]
Warburganal [62994-47-2]
Warfarin [81-81-2]
Warfarin Sodium* [129-06-6]
Water [7732-18-5]
Wieland-Gumlich Aldehyde
    [466-85-3]
Withaferin A [5119-48-2]
Woodward's Reagent K [4156-16-5]
Worenine [38763-29-0]
Wortmannin [19545-26-7]
Xamoterol [81801-12-9]
Xanthan Gum [11138-66-2]
Xanthatin [26791-73-1]
Xanthine [69-89-6]
Xanthinol Niacinate [437-74-1]
Xanthocillin [11042-38-9]
Xanthone [90-47-1]
Xanthophyll [127-40-2]
Xanthopterin [119-44-8]
Xanthosine [146-80-5]
Xanthoxyletin [84-99-1]
Xanthoxylin [90-24-4]
Xanthurenic Acid [59-00-7]
Xanthyletin [553-19-5]
Xenazoic Acid [1174-11-4]
Xenbucin [959-10-4]
Xenon [7440-63-3]
p-Xenylcarbimide [92-95-5]
Xenytropium Bromide [511-55-7]
Xibenolol [81584-06-7]
Xibornol [13741-18-9]
Xipamide [14293-44-8]
Xylazine [7361-61-7]
Xylazine Hydrochloride*
    [23076-35-9]
Xylene [1330-20-7]
Xylenol [1300-71-6]
Xylenol Blue [125-31-5]
Xylidine [1300-73-8]
Xylitol [87-99-0]
Xylometazoline [526-36-3]
Xylometazoline Hydrochloride*
    [1218-35-5]
Xylopropamine [102-31-8]
Xylose [58-86-6]
Xylulose [5962-29-8]
1-Xylylazo-2-naphthol [85-82-5]
Yangonin [500-62-9]
Yellow AB [85-84-7]
Yellow OB [131-79-3]
Yellow Phenolphthalein [8053-05-2]
Yig [12063-56-8]
Ylangene [14912-44-8]
Yohimbine [146-48-5]

Denotes derivative of title compound

allo-Yohimbine [522-94-1]
α-Yohimbine [131-03-3]
Ytterbium [7440-64-4]
Yttrium [7440-65-5]
Zearalenone [17924-92-4]
Zeatin [1637-39-4]
Zeaxanthin [144-68-3]
Zeranol [55331-29-8]
Zidovudine [30516-87-1]
Zimeldine [56775-88-3]
Zimeldine Dihydrochloride Monohydrate* [61129-30-4]
Zinc [7440-66-6]
Zinc Acetate [557-34-6]
Zinc Bromide [7699-45-8]
Zinc Caprylate [557-09-5]
Zinc Carbonate [3486-35-9]
Zinc Chloride [7646-85-7]
Zinc Chromate(VI) Hydroxide [13530-65-9]
Zinc Citrate [546-46-3]
Zinc Cyanide [557-21-1]
Zinc Fluoride [7783-49-5]
Zinc Formate [557-41-5]
Zinc Hexafluorosilicate [16871-71-9]
Zinc Insulin Crystals [8049-62-5]
Zinc Iodate [7790-37-6]
Zinc Iodide [10139-47-6]
Zinc Iodide-Starch [9010-71-3]
Zinc Lactate [16039-53-5]
Zinc Meta-arsenite [10326-24-6]
Zinc Nitrate [7779-88-6]

Zinc Nitride [1313-49-1]
Zinc Nitrite [10102-02-0]
Zinc Oleate [557-07-3]
Zinc Ortho-arsenate [13464-44-3]
Zinc Oxalate [547-68-2]
Zinc Oxide [1314-13-2]
Zinc Perchlorate [13637-61-1]
Zinc Permanganate [23414-72-4]
Zinc Peroxide [1314-22-3]
Zinc p-Phenolsulfonate [127-82-2]
Zinc Phosphate [7779-90-0]
Zinc Phosphide [1314-84-7]
Zinc Propionate [557-28-8]
Zinc Pyrophosphate [7446-26-6]
Zinc Salicylate [16283-36-6]
Zinc Selenate [13597-54-1]
Zinc Selenide [1315-09-9]
Zinc Silicate [13597-65-4]
Zinc Stearate [557-05-1]
Zinc Sulfate [7733-02-0]
Zinc Sulfide [1314-98-3]
Zinc Tannate [8011-65-2]
Zinc Tartrate [551-64-4]
Zinc Telluride [1315-11-3]
Zinc Thiocyanate [557-42-6]
Zinc Valerate [556-38-7]
Zineb [12122-67-7]
Zingerone [122-48-5]
Zinostatin [9014-02-2]
Zipeprol [34758-83-3]
Ziram [137-30-4]
Zirconium [7440-67-7]

Zirconium Chloride [10026-11-6]
Zirconium Fluoride [7783-64-4]
Zirconium Hydride [7704-99-6]
Zirconium Hydroxide [14475-63-9]
Zirconium Iodide [13986-26-0]
Zirconium Nitrate [13746-89-9]
Zirconium Oxide [1314-23-4]
Zirconium Silicate [10101-52-7]
Zirconium Sulfate [14644-61-2]
Zirconyl Acetate [5153-24-2]
Zirconyl Chloride [7699-43-6]
Zoapatanol [71117-51-6]
Zolamine [553-13-9]
Zolamine Hydrochloride* [1155-03-9]
Zolimidine [1222-57-7]
Zolpidem [82626-48-0]
Zolpidem (+)-Tartrate (2:1)* [99294-93-6]
Zomepirac [33369-31-2]
Zomepirac Sodium Dihydrate* [64092-49-5]
Zometapine [51022-73-2]
Zonisamide [68291-97-4]
Zopiclone [43200-80-2]
Zorubicin [54083-22-6]
Zorubicin Hydrochloride* [36508-71-1]
Zotepine [26615-21-4]
Zoxazolamine [61-80-3]
Zygadenine [545-45-9]
Zymosan [9010-72-4]

[50-00-0] Formaldehyde, *4148, 4150*
[50-02-2] Dexamethasone, *2922*
[50-03-3] Hydrocortisone Acetate, *4711*
[50-04-4] Cortisone Acetate*, *2533*
[50-06-6] Phenobarbital, *7201*
[50-10-2] Oxyphenonium Bromide, *6928*
[50-11-3] Metharbital, *5873*
[50-12-4] Mephenytoin, *5741*
[50-13-5] Meperidine Hydrochloride*, *5736*
[50-14-6] Vitamin D_2, *9928*
[50-18-0] Cyclophosphamide, *2753*
[50-19-1] Hydroxyphenamate, *4770*
[50-22-6] Corticosterone, *2532*
[50-23-7] Hydrocortisone, *4710*
[50-24-8] Prednisolone, *7719*
[50-27-1] Estriol, *3659*
[50-28-2] Estradiol, *3653*
[50-29-3] DDT, *2832*
[50-32-8] Benzo[*a*]pyrene, *1113*
[50-33-9] Phenylbutazone, *7248*
[50-34-0] Propantheline Bromide, *7816*
[50-35-1] Thalidomide, *9182*
[50-36-2] Cocaine, *2450*
[50-37-3] Lysergide, *5507*
[50-39-5] Protheobromine, *7901*
[50-41-9] Clomiphene Citrate*, *2384*
[50-42-0] Adiphenine Hydrochloride, *151*
[50-44-2] 6-Mercaptopurine, *5762*
[50-47-5] Desipramine, *2902*
[50-48-6] Amitriptyline, *504*
[50-49-7] Imipramine, *4835*
[50-50-0] Estradiol Benzoate, *3655*
[50-52-2] Thioridazine, *9290*
[50-53-3] Chlorpromazine, *2186*
[50-54-4] Quinidine Sulfate, *8074*
[50-55-5] Reserpine, *8149*
[50-56-6] Oxytocin, *6934*
[50-57-7] Lypressin, *5503*
[50-58-8] Phendimetrazine Tartrate*, *7180*
[50-59-9] Cephaloridine, *1975*
[50-60-2] Phentolamine, *7234*
[50-63-5] Chloroquine Phosphate*, *2163*
[50-65-7] Niclosamide, *6425*
[50-67-9] Serotonin, *8413*
[50-69-1] D-Ribose, *8205*
[50-70-4] Sorbitol, *8680*
[50-71-5] Alloxan, *281*
[50-76-0] Dactinomycin, *2804*
[50-78-2] Aspirin, *873*
[50-81-7] Ascorbic Acid, *855*
[50-85-1] *m*-Cresotic Acid, *2584*
[50-89-5] Thymidine, *9331*
[50-91-9] Floxuridine, *4045*
[50-98-6] Ephedrine L-Form Hydrochloride*, *3561*
[50-99-7] Glucose, *4353*
[51-03-6] Piperonyl Butoxide, *7446*
[51-12-7] Nialamide, *6399*

[51-14-9] Sesamex, *8419*
[51-15-0] Pralidoxime Chloride, *7705*
[51-17-2] Benzimidazole, *1091*
[51-18-3] Triethylenemelamine, *9586*
[51-20-7] 5-Bromouracil, *1430*
[51-21-8] Fluorouracil, *4109*
[51-24-1] Tiratricol, *9391*
[51-26-3] Thyropropic Acid, *9346*
[51-28-5] 2,4-Dinitrophenol, *3274*
[51-34-3] Scopolamine, *8361*
[51-35-4] 4-Hydroxy-L-proline, *4775*
[51-41-2] Norepinephrine, *6612*
[51-42-3] Epinephrine L-Form Bitartrate*, *3569*
[51-43-4] Epinephrine, *3569*
[51-45-6] Histamine, *4640*
[51-48-9] Thyroxine, *9348*
[51-52-5] Propylthiouracil, *7882*
[51-55-8] Atropine, *891*
[51-56-9] Homatropine Hydrobromide*, *4649*
[51-60-5] Neostigmine Methylsulfate*, *6380*
[51-61-6] Dopamine, *3415*
[51-63-8] Dextroamphetamine Sulfate, *2932*
[51-66-1] *p*-Acetanisidine, *43*
[51-67-2] Tyramine, *9743*
[51-68-3] Meclofenoxate, *5660*
[51-71-8] Phenelzine, *7181*
[51-73-0] Tremorine, *9498*
[51-74-1] Histamine Phosphate*, *4640*
[51-75-2] Mechlorethamine, *5655*
[51-77-4] Gefarnate, *4273*
[51-79-6] Urethan, *9789*
[51-83-2] Carbachol, *1780*
[51-85-4] Cystamine, *2783*
[51-98-9] Norethindrone Acetate*, *6614*
[52-01-7] Spironolactone, *8721*
[52-21-1] Prednisolone Acetate*, *7719*
[52-24-4] Triethylenethiophosphoramide, *9588*
[52-26-6] Morphine Hydrochloride, *6188*
[52-28-8] Codeine Phosphate, *2462*
[52-31-3] Cyclobarbital, *2717*
[52-39-1] Aldosterone, *218*
[52-43-7] Allobarbital, *252*
[52-46-0] Apholate, *760*
[52-49-3] Trihexyphenidyl Hydrochloride, *9607*
[52-51-7] Bronopol, *1437*
[52-52-8] Cycloleucine, *2740*
[52-53-9] Verapamil, *9851*
[52-62-0] Pentolinium Tartrate, *7089*
[52-67-5] Penicillamine, *7029*
[52-68-6] Trichlorfon, *9536*
[52-76-6] Lynestrenol, *5501*
[52-78-8] Norethandrolone, *6613*
[52-85-7] Famphur, *3882*
[52-86-8] Haloperidol, *4511*

[52-88-0] Atropine Methylnitrate*, *891*
[52-89-1] Cysteine Hydrochloride*, *2787*
[52-90-4] Cysteine, *2787*
[53-03-2] Prednisone, *7727*
[53-05-4] Tetrahydrocortisone, *9143*
[53-06-5] Cortisone, *2533*
[53-10-1] Hydroxydione Sodium, *4753*
[53-16-7] Estrone, *3660*
[53-18-9] Bietaserpine, *1226*
[53-19-0] Mitotane, *6134*
[53-31-6] Medibazine, *5671*
[53-33-8] Paramethasone, *6977*
[53-34-9] Fluprednisolone, *4119*
[53-36-1] Methylprednisolone Acetate*, *6028*
[53-39-4] Oxandrolone, *6875*
[53-41-8] Androsterone, *673*
[53-43-0] Prasterone, *7710*
[53-46-3] Methantheline Bromide, *5869*
[53-59-8] NADP, *6262*
[53-60-1] Promazine Hydrochloride*, *7793*
[53-70-3] 1,2:5,6-Dibenzanthracene, *2989*
[53-73-6] Angiotensinamide*, *680*
[53-79-2] Puromycin, *7960*
[53-84-9] Nadide, *6259*
[53-86-1] Indomethacin, *4874*
[53-89-4] Benzpiperylon, *1131*
[53-96-3] *N*-2-Fluorenylacetamide, *4083*
[54-03-5] Hexobendine, *4626*
[54-04-6] Mescaline, *5808*
[54-05-7] Chloroquine, *2163*
[54-06-8] Adrenochrome, *162*
[54-11-5] Nicotine, *6434*
[54-21-7] Sodium Salicylate, *8626*
[54-25-1] 6-Azauridine, *920*
[54-30-8] Camylofine, *1743*
[54-31-9] Furosemide, *4221*
[54-32-0] Moxisylyte, *6204*
[54-36-4] Metyrapone, *6083*
[54-42-2] Idoxuridine, *4819*
[54-47-7] Pyridoxal 5-Phosphate, *7992*
[54-49-9] Metaraminol, *5837*
[54-62-6] Aminopterin, *485*
[54-64-8] Thimerosal, *9244*
[54-71-7] Pilocarpine Hydrochloride*, *7395*
[54-80-8] Pronethalol, *7802*
[54-85-3] Isoniazid, *5071*
[54-87-5] Nitrofurantoin Sodium*, *6520*
[54-91-1] Pipobroman, *7451*
[54-92-2] Iproniazid, *4965*
[54-95-5] Pentylenetetrazole, *7097*
[55-03-8] Levothyroxine Sodium, *5351*
[55-06-1] Liothyronine Sodium*, *5388*
[55-10-7] Vanilmandelic Acid, *9840*

Denotes derivative of title compound

[55-18-5] *N*-Nitrosodiethylamine, 6557
[55-21-0] Benzamide, 1067
[55-22-1] Isonicotinic Acid, 5073
[55-38-9] Fenthion, 3945
[55-43-6] *N*-(2-Chloroethyl)dibenzylamine Hydrochloride, 2138
[55-52-7] Pheniprazine, 7197
[55-55-0] *p*-Methylaminophenol Sulfate, 5940
[55-56-1] Chlorhexidine, 2090
[55-63-0] Nitroglycerin, 6528
[55-63-0] Spirit of Glyceryl Trinitrate, 8717
[55-65-2] Guanethidine, 4473
[55-73-2] Bethanidine, 1208
[55-86-7] Mechlorethamine Hydrochloride*, 5656
[55-91-4] Isoflurophate, 5060
[55-94-7] Succinylcholine Bromide, 8845
[55-97-0] Hexamethonium Bromide*, 4609
[55-98-1] Busulfan, 1494
[56-03-1] Biguanide, 1233
[56-04-2] Methylthiouracil, 6048
[56-06-4] 2,4-Diamino-6-hydroxypyrimidine, 2959
[56-10-0] AET, 166
[56-12-2] γ-Aminobutyric Acid, 441
[56-23-5] Carbon Tetrachloride, 1822
[56-25-7] Cantharidin, 1755
[56-28-0] Triclodazol, 9569
[56-29-1] Hexobarbital, 4625
[56-34-8] Tetraethylammonium Chloride, 9134
[56-38-2] Parathion, 6983
[56-40-6] Glycine, 4386
[56-41-7] L-Alanine, 195
[56-45-1] Serine, 8411
[56-47-3] Deoxycorticosterone Acetate, 2883
[56-49-5] 3-Methylcholanthrene, 5967
[56-53-1] Diethylstilbestrol, 3118
[56-54-2] Quinidine, 8072
[56-55-3] 1,2-Benzanthracene, 1069
[56-59-7] Felypressin, 3896
[56-65-5] Adenosine Triphosphate, 145
[56-69-9] 5-Hydroxytryptophan, 4784
[56-72-4] Coumaphos, 2559
[56-73-5] Glucose-6-phosphate, 4355
[56-75-7] Chloramphenicol, 2068
[56-81-5] Glycerol, 4379
[56-84-8] Aspartic Acid, 862
[56-85-9] Glutamine, 4365
[56-86-0] Glutamic Acid, 4363
[56-87-1] Lysine, 5509
[56-88-2] L-Cystathionine, 2784
[56-89-3] Cystine, 2788
[56-94-0] Demecarium Bromide, 2871
[57-00-1] Creatine, 2570
[57-03-4] Glycerophosphoric Acid, 4381
[57-06-7] Allyl Isothiocyanate, 293
[57-08-9] ε-Acetamidocaproic Acid, 38
[57-09-0] Cetrimonium Bromide, 2018
[57-10-3] Palmitic Acid, 6947
[57-11-4] Stearic Acid, 8761
[57-13-6] Urea, 9781
[57-14-7] 1,1-Dimethylhydrazine, 3236

[57-15-8] Chlorobutanol, 2129
[57-22-7] Vincristine, 9891
[57-24-9] Strychnine, 8822
[57-27-2] Morphine, 6186
[57-29-4] Nalorphine Hydrochloride*, 6275
[57-30-7] Phenobarbital Sodium, 7202
[57-33-0] Pentobarbital Sodium, 7087
[57-39-6] Metepa, 5842
[57-41-0] Phenytoin, 7293
[57-42-1] Meperidine, 5736
[57-43-2] Amobarbital, 601
[57-44-3] Barbital, 972
[57-47-6] Physostigmine, 7357
[57-48-7] Fructose, 4185
[57-50-1] Sucrose, 8855
[57-53-4] Meprobamate, 5751
[57-55-6] Propylene Glycol, 7868
[57-57-8] β-Propiolactone, 7832
[57-62-5] Chlortetracycline, 2193
[57-63-6] Ethinyl Estradiol, 3689
[57-64-7] Physostigmine Salicylate*, 7357
[57-66-9] Probenecid, 7760
[57-67-0] Sulfaguanidine, 8879
[57-68-1] Sulfamethazine, 8886
[57-74-9] Chlordan(e), 2079
[57-83-0] Progesterone, 7783
[57-85-2] Testosterone Propionate, 9115
[57-87-4] Ergosterol, 3607
[57-88-5] Cholesterol, 2204
[57-91-0] α-Estradiol, 3654
[57-92-1] Streptomycin, 8786
[57-94-3] Tubocurarine Chloride, 9717
[57-96-5] Sulfinpyrazone, 8926
[57-97-6] 9,10-Dimethyl-1,2-benzanthracene, 3224
[58-00-4] Apomorphine, 776
[58-05-9] Folinic Acid, 4141
[58-08-2] Caffeine, 1635
[58-14-0] Pyrimethamine, 7997
[58-15-1] Aminopyrine, 488
[58-18-4] 17-Methyltestosterone, 6044
[58-20-8] Testosterone 17β-Cypionate, 9111
[58-22-0] Testosterone, 9109
[58-25-3] Chlordiazepoxide, 2082
[58-27-5] Menadione, 5714
[58-28-6] Desipramine Hydrochloride*, 2902
[58-32-2] Dipyridamole, 3354
[58-33-3] Promethazine Hydrochloride*, 7797
[58-34-4] Thiazinamium Methylsulfate, 9234
[58-36-6] 10,10'-Oxydiphenoxarsine, 6914
[58-37-7] Aminopromazine, 479
[58-38-8] Prochlorperazine, 7768
[58-39-9] Perphenazine, 7135
[58-40-2] Promazine, 7793
[58-46-8] Tetrabenazine, 9118
[58-54-8] Ethacrynic Acid, 3669
[58-55-9] Theophylline, 9392
[58-56-0] Pyridoxine Hydrochloride, 7995
[58-61-7] Adenosine, 143
[58-63-9] Inosine, 4880
[58-64-0] Adenosine Diphosphate, 144
[58-73-1] Diphenhydramine, 3308
[58-74-2] Papaverine, 6968
[58-82-2] Bradykinin, 1356
[58-85-5] Biotin, 1244
[58-86-6] Xylose, 9995
[58-89-9] Lindane, 5379

[58-93-5] Hydrochlorothiazide, 4704
[58-94-6] Chlorothiazide, 2169
[58-95-7] Vitamin E Acetate, 9932
[58-96-8] Uridine, 9792
[58-97-9] 5'-Uridylic Acid, 9796
[58-98-0] Uridine 5'-Diphosphate, 9793
[59-00-7] Xanthurenic Acid, 9977
[59-01-8] Kanamycin A*, 5161
[59-02-9] Vitamin E, 9931
[59-05-2] Methotrexate, 5908
[59-06-3] Ethopabate, 3701
[59-23-4] Galactose, 4241
[59-26-7] Nikethamide, 6459
[59-30-3] Folic Acid, 4140
[59-31-4] Carbostyril, 1831
[59-32-5] Chloropyramine, 2162
[59-33-6] Pyrilamine Maleate*, 7996
[59-39-2] Piperoxan, 7448
[59-40-5] Sulfaquinoxaline, 8914
[59-46-1] Procaine, 7763
[59-47-2] Mephenesin, 5737
[59-50-7] 4-Chloro-*m*-cresol, 2133
[59-51-8] Methionine DL-Form*, 5896
[59-52-9] Dimercaprol, 3196
[59-56-3] α-Glucose-1-phosphate, 4356
[59-58-5] Prosultiamine, 7896
[59-61-0] Dichlorisoproterenol, 3036
[59-63-2] Isocarboxazid, 5040
[59-66-5] Acetazolamide, 45
[59-67-6] Nicotinic Acid, 6435
[59-87-0] Nitrofurazone, 6521
[59-88-1] Phenylhydrazine Hydrochloride, 7265
[59-89-2] *N*-Nitrosomorpholine, 6561
[59-92-7] Levodopa, 5344
[59-96-1] Phenoxybenzamine, 7225
[59-97-2] Tolazoline Hydrochloride*, 9430
[59-98-3] Tolazoline, 9430
[59-99-4] Neostigmine, 6380
[60-00-4] Edetic Acid, 3484
[60-01-5] Tributyrin, 9532
[60-02-6] Guanethidine Sulfate*, 4473
[60-09-3] *p*-Aminoazobenzene, 430
[60-10-6] Dithizone, 3383
[60-11-7] *p*-Dimethylaminoazobenzene, 3218
[60-12-8] Phenethyl Alcohol, 7185
[60-13-9] Amphetamine Sulfate*, 616
[60-18-4] Tyrosine, 9747
[60-23-1] Cysteamine, 2785
[60-24-2] 2-Mercaptoethanol, 5760
[60-26-4] Hexamethonium, 4609
[60-27-5] Creatinine, 2571
[60-29-7] Ethyl Ether, 3762
[60-31-1] Acetylcholine Chloride, 81
[60-32-2] ε-Aminocaproic Acid, 442
[60-33-3] Linoleic Acid, 5382
[60-34-4] Methylhydrazine, 6001
[60-35-5] Acetamide, 36
[60-39-9] Mazipredone, 5644
[60-40-2] Mecamylamine, 5654
[60-41-3] Strychnine Sulfate, 8826
[60-44-6] Penthienate Bromide, 7084
[60-46-8] Aminopentamide, 471
[60-51-5] Dimethoate, 3209
[60-54-8] Tetracycline, 9130

Denotes derivative of title compound

[60-56-0] Methimazole, 5892
[60-57-1] Dieldrin, 3093
[60-70-8] Veratramine, 9853
[60-79-7] Ergonovine, 3600
[60-80-0] Antipyrine, 748
[60-81-1] Phloridzin, 7300
[60-82-2] Phloretin, 7299
[60-87-7] Promethazine, 7797
[60-89-9] Mepazine, 5734
[60-91-3] Diethazine, 3098
[60-92-4] Cyclic AMP, 2714
[60-93-5] Quinine Dihydrochloride, 8079
[60-99-1] Methotrimeprazine, 5909
[61-00-7] Acepromazine, 27
[61-01-8] Methoxypromazine, 5923
[61-12-1] Dibucaine Hydrochloride, 3016
[61-16-5] Methoxamine Hydrochloride, 5910
[61-19-8] 5'-Adenylic Acid, 148
[61-24-5] Cephalosporin C, 1976
[61-33-6] Benzylpenicillinic Acid, 1156
[61-50-7] N,N-Dimethyltryptamine, 3251
[61-54-1] Tryptamine, 9705
[61-56-3] Sulthiame, 8974
[61-57-4] Niridazole, 6480
[61-68-7] Mefenamic Acid, 5680
[61-72-3] Cloxacillin, 2414
[61-73-4] Methylene Blue, 5979
[61-75-6] Bretylium Tosylate, 1364
[61-76-7] Phenylephrine Hydrochloride, 7257
[61-78-9] p-Aminohippuric Acid, 454
[61-80-3] Zoxazolamine, 10098
[61-82-5] Amitrole, 506
[61-90-5] Leucine, 5331
[62-31-7] Dopamine Hydrochloride*, 3415
[62-33-9] Edetate Calcium Disodium, 3480
[62-37-3] Chlormerodrin, 2104
[62-38-4] Phenylmercuric Acetate, 7271
[62-44-2] Phenacetin, 7155
[62-46-4] Thioctic Acid, 9255
[62-49-7] Choline, 2207
[62-50-0] Ethyl Methanesulfonate, 3782
[62-51-1] Methacholine Chloride, 5847
[62-53-3] Aniline, 687
[62-54-4] Calcium Acetate, 1643
[62-55-5] Thioacetamide, 9246
[62-56-6] Thiourea, 9299
[62-57-7] α-Aminoisobutyric Acid, 457
[62-59-9] Cevadine, 2025
[62-67-9] Nalorphine, 6275
[62-73-7] Dichlorvos, 3069
[62-75-9] N-Nitrosodimethylamine, 6558
[62-76-0] Sodium Oxalate, 8601
[62-90-8] Nandrolone Phenpropionate, 6283
[62-97-5] Diphemanil Methylsulfate, 3302
[63-05-8] 4-Androstene-3,17-dione, 671
[63-12-7] Benzquinamide, 1133
[63-25-2] Carbaryl, 1789
[63-39-8] Uridine 5'-Triphosphate, 9795
[63-42-3] Lactose, 5221
[63-45-6] Primaquine Phosphate*, 7751

[63-56-9] Thonzylamine Hydrochloride, 9307
[63-68-3] Methionine, 5896
[63-74-1] Sulfanilamide, 8898
[63-75-2] Arecoline, 802
[63-84-3] Dopa, 3414
[63-91-2] Phenylalanine, 7242
[63-92-3] Phenoxybenzamine Hydrochloride*, 7225
[63-98-9] Phenacemide, 7154
[64-02-8] Edetate Sodium, 3482
[64-04-0] Phenethylamine, 7186
[64-10-8] Phenylurea, 7290
[64-17-5] Ethyl Alcohol, 3716
[64-18-6] Formic Acid, 4153
[64-18-6] Spirit of Formic Acid, 8716
[64-19-7] Acetic Acid Glacial, 47
[64-31-3] Morphine Sulfate, 6193
[64-39-1] Promedol, 7795
[64-43-7] Amobarbital Sodium*, 601
[64-47-1] Physostigmine Sulfate*, 7357
[64-55-1] Mebutamate, 5653
[64-65-3] Bemegride, 1036
[64-67-5] Diethyl Sulfate, 3120
[64-69-7] Iodoacetic Acid, 4918
[64-72-2] Chlortetracycline Hydrochloride*, 2193
[64-73-3] Demeclocycline Hydrochloride*, 2872
[64-75-5] Tetracycline Hydrochloride*, 9130
[64-77-7] Tolbutamide, 9432
[64-85-7] Deoxycorticosterone, 2882
[64-86-8] Colchicine, 2470
[65-28-1] Phentolamine Methanesulfonate*, 7234
[65-29-2] Gallamine Triethiodide, 4249
[65-45-2] Salicylamide, 8297
[65-46-3] Cytidine, 2792
[65-49-6] p-Aminosalicylic Acid, 491
[65-71-4] Thymine, 9332
[65-82-7] N-Acetylmethionine, 90
[65-85-0] Benzoic Acid, 1101
[65-86-1] Orotic Acid, 6828
[66-02-4] 3,5-Diiodotyrosine, 3177
[66-22-8] Uracil, 9761
[66-23-9] Acetylcholine Bromide, 80
[66-25-1] Caproic Aldehyde, 1761
[66-27-3] Methyl Methanesulfonate, 6010
[66-28-4] Strophanthidin, 8818
[66-32-0] Strychnine Nitrate, 8823
[66-71-7] o-Phenanthroline, 7169
[66-72-8] Pyridoxal, 7991
[66-75-1] Uracil Mustard, 9762
[66-76-2] Dicumarol, 3080
[66-79-5] Oxacillin, 6858
[66-81-9] Cycloheximide, 2734
[66-97-7] Psoralen, 7944
[67-03-8] Thiamine Hydrochloride, 9222
[67-07-2] Phosphocreatine, 7315
[67-16-3] Thiamine Disulfide, 9221
[67-20-9] Nitrofurantoin, 6520
[67-28-7] Nihydrazone, 6458
[67-43-6] Pentetic Acid, 7083
[67-45-8] Furazolidone, 4210
[67-47-0] 5-(Hydroxymethyl)-2-furaldehyde, 4764
[67-48-1] Choline Chloride, 2208
[67-52-7] Barbituric Acid, 973
[67-56-1] Methanol, 5868
[67-62-9] Methoxyamine, 5912
[67-63-0] Isopropyl Alcohol, 5096

[67-64-1] Acetone, 58
[67-66-3] Chloroform, 2141
[67-68-5] Dimethyl Sulfoxide, 3247
[67-71-0] Dimethyl Sulfone, 3246
[67-72-1] Hexachloroethane, 4601
[67-73-2] Fluocinolone Acetonide, 4076
[67-81-2] 17α-Methyltestosterone 3-Cyclopentyl Enol Ether, 6045
[67-92-5] Dicyclomine Hydrochloride, 3087
[67-95-8] Quingestrone, 8069
[67-96-9] Dihydrotachysterol, 3163
[67-97-0] Vitamin D₃, 9929
[67-98-1] Ethamoxytriphetol, 3674
[67-99-2] Gliotoxin, 4336
[68-04-2] Sodium Citrate, 8549
[68-11-1] Thioglycolic Acid, 9265
[68-12-2] N,N-Dimethylformamide, 3232
[68-19-9] Vitamin B₁₂, 9921
[68-22-4] Norethindrone, 6614
[68-23-5] Norethynodrel, 6615
[68-26-8] Vitamin A, 9918
[68-35-9] Sulfadiazine, 8874
[68-36-0] 1,4-Bis(trichloromethyl)-benzene, 1313
[68-41-7] Cycloserine, 2758
[68-76-8] Triaziquone, 9517
[68-81-5] Iodinin, 4915
[68-88-2] Hydroxyzine, 4786
[68-90-6] Benziodarone, 1093
[68-91-7] Trimethaphan Camsylate, 9621
[68-94-0] Hypoxanthine, 4805
[68-96-2] 17α-Hydroxyprogesterone, 4773
[69-09-0] Chlorpromazine Hydrochloride*, 2186
[69-23-8] Fluphenazine, 4116
[69-24-9] Cinchotoxine, 2291
[69-25-0] Eledoisin, 3502
[69-27-2] Chlorisondamine Chloride, 2101
[69-33-0] Tubercidin, 9714
[69-46-5] Calcium Acetylsalicylate, 1644
[69-52-3] Ampicillin Sodium*, 621
[69-53-4] Ampicillin, 621
[69-57-8] Benzylpenicillin Sodium, 1157
[69-65-8] Mannitol, 5629
[69-72-7] Salicylic Acid, 8301
[69-79-4] Maltose, 5595
[69-89-6] Xanthine, 9968
[69-91-0] α-Phenylglycine, 7263
[69-93-2] Uric Acid, 9790
[70-00-8] Trifluridine, 9599
[70-07-5] Mephenoxalone, 5739
[70-11-1] ω-Bromoacetophenone, 1391
[70-18-8] Glutathione, 4369
[70-19-9] Thurfyl Nicotinate, 9329
[70-22-4] Oxotremorine, 6904
[70-25-7] N-Methyl-N'-nitro-N-nitrosoguanidine, 6017
[70-26-8] Ornithine, 6826
[70-30-4] Hexachlorophene, 4602
[70-34-8] 1-Fluoro-2,4-dinitrobenzene, 4101
[70-43-9] Barthrin, 1014
[70-47-3] Asparagine, 859
[70-49-5] Thiomalic Acid, 9272
[70-51-9] Deferoxamine, 2850
[70-69-9] p-Aminopropiophenone, 482
[70-70-2] Paroxypropione, 6992
[70-78-0] 3-Iodotyrosine, 4937
[71-23-8] n-Propyl Alcohol, 7854

* Denotes derivative of title compound

[71-27-2] Succinylcholine Chloride, 8846
[71-30-7] Cytosine, 2801
[71-36-3] n-Butyl Alcohol, 1540
[71-41-0] 1-Pentanol, 7074
[71-43-2] Benzene, 1074
[71-44-3] Spermine, 8699
[71-48-7] Cobaltous Acetate, 2427
[71-55-6] 1,1,1-Trichloroethane, 9549
[71-58-9] Medroxyprogesterone Acetate*, 5677
[71-62-5] Veratridine, 9855
[71-63-6] Digitoxin, 3146
[71-67-0] Sulfobromophthalein Sodium, 8933
[71-68-1] Hydromorphone Hydrochloride*, 4733
[71-73-8] Thiopental Sodium, 9280
[71-78-3] Pipradrol Hydrochloride*, 7455
[71-81-8] Isopropamide, 5090
[71-91-0] Tetraethylammonium Bromide, 9133
[72-14-0] Sulfathiazole, 8920
[72-17-3] Sodium Lactate, 8586
[72-18-4] Valine, 9818
[72-19-5] Threonine, 9316
[72-20-8] Endrin, 3533
[72-23-1] 11-Dehydrocorticosterone, 2859
[72-33-3] Mestranol, 5819
[72-43-5] Methoxychlor, 5913
[72-44-6] Methaqualone, 5872
[72-48-0] Alizarin, 237
[72-54-8] 1,1-Dichloro-2,2-bis(p-chlorophenyl)ethane, 3049
[72-56-0] 1,1-Dichloro-2,2-bis(p-ethylphenyl)ethane, 3050
[72-57-1] Trypan Blue, 9701
[72-63-9] Methandrostenolone, 5862
[72-69-5] Nortriptyline, 6635
[72-80-0] Chlorquinaldol, 2191
[73-03-0] Cordycepin, 2524
[73-05-2] Phentolamine Hydrochloride*, 7234
[73-09-6] Etozolin, 3845
[73-22-3] L-Tryptophan, 9707
[73-24-5] Adenine, 141
[73-31-4] Melatonin, 5695
[73-32-5] Isoleucine, 5064
[73-40-5] Guanine, 4477
[73-48-3] Bendroflumethiazide, 1045
[73-49-4] Quinethazone, 8067
[73-67-6] Toxopyrimidine, 9482
[74-11-3] p-Chlorobenzoic Acid, 2126
[74-31-7] N,N'-Diphenyl-p-phenylenediamine, 3331
[74-39-5] Magneson, 5578
[74-55-5] Ethambutol, 3672
[74-61-3] 2,3-Dimercapto-1-propanesulfonic Acid, 3197
[74-79-3] Arginine, 805
[74-82-8] Methane, 5863
[74-83-9] Methyl Bromide, 5951
[74-84-0] Ethane, 3676
[74-85-1] Ethylene, 3748
[74-86-2] Acetylene, 84
[74-87-3] Methyl Chloride, 5964
[74-88-4] Methyl Iodide, 6002
[74-89-5] Methylamine, 5938
[74-90-8] Hydrogen Cyanide, 4722
[74-93-1] Methanethiol, 5867
[74-95-3] Methylene Bromide, 5980
[74-96-4] Ethyl Bromide, 3730
[74-98-6] Propane, 7809
[75-00-3] Ethyl Chloride, 3740

[75-01-4] Vinyl Chloride, 9898
[75-03-6] Ethyl Iodide, 3769
[75-04-7] Ethylamine, 3718
[75-05-8] Acetonitrile, 62
[75-07-0] Acetaldehyde, 32
[75-08-1] Ethanethiol, 3680
[75-09-2] Methylene Chloride, 5982
[75-11-6] Methylene Iodide, 5985
[75-12-7] Formamide, 4151
[75-13-8] Isocyanic Acid, 5049
[75-15-0] Carbon Disulfide, 1818
[75-18-3] Methyl Sulfide, 6042
[75-19-4] Cyclopropane, 2755
[75-20-7] Calcium Carbide, 1656
[75-21-8] Ethylene Oxide, 3758
[75-25-2] Bromoform, 1407
[75-26-3] Isopropyl Bromide, 5098
[75-29-6] Isopropyl Chloride, 5099
[75-30-9] Isopropyl Iodide, 5102
[75-31-0] Isopropylamine, 5097
[75-34-3] Ethylidene Chloride, 3766
[75-35-4] Vinylidene Chloride, 9900
[75-36-5] Acetyl Chloride, 79
[75-39-8] Acetaldehyde Ammonia, 33
[75-44-5] Phosgene, 7310
[75-46-7] Fluoroform, 4102
[75-47-8] Iodoform, 4926
[75-50-3] Trimethylamine, 9625
[75-52-5] Nitromethane, 6532
[75-56-9] Propylene Oxide, 7869
[75-58-1] Tetramethylammonium Iodide, 9156
[75-59-2] Tetramethylammonium Hydroxide, 9155
[75-60-5] Cacodylic Acid, 1603
[75-64-9] tert-Butylamine, 1545
[75-65-0] tert-Butyl Alcohol, 1542
[75-66-1] tert-Butyl Mercaptan, 1577
[75-69-4] Trichlorofluoromethane, 9553
[75-71-8] Dichlorodifluoromethane, 3053
[75-73-0] Carbon Tetrafluoride, 1823
[75-80-9] Tribromoethanol, 9525
[75-84-3] Neopentyl Alcohol, 6373
[75-85-4] tert-Pentyl Alcohol, 7096
[75-86-5] Acetone Cyanohydrin, 59
[75-87-6] Trichloroacetaldehyde, 9538
[75-91-2] tert-Butyl Hydroperoxide, 1569
[75-93-4] Methyl Sulfate, 6041
[75-96-7] Tribromoacetic Acid, 9521
[75-97-8] Pinacolone, 7409
[75-98-9] Pivalic Acid, 7482
[75-99-0] Dalapon, 2806
[76-00-6] 1,1,1-Trichloro-2-propanol, 9557
[76-01-7] Pentachloroethane, 7058
[76-03-9] Trichloroacetic Acid, 9539
[76-05-1] Trifluoroacetic Acid, 9595
[76-06-2] Chloropicrin, 2156
[76-08-4] Tribromo-tert-butyl Alcohol, 9523
[76-09-5] Pinacol, 7408
[76-14-2] Cryofluorane, 2608
[76-20-0] Sulfonethylmethane, 8937
[76-22-2] Camphor, 1738
[76-22-2] Spirit of Camphor, 8711
[76-23-3] Tetrabarbital, 9117

[76-24-4] Alloxantin, 282
[76-25-5] Triamcinolone Acetonide, 9512
[76-28-8] Sarmentogenin, 8334
[76-29-9] 3-Bromo-d-camphor, 1402
[76-30-2] Dihydroxytartaric Acid, 3172
[76-36-8] α,α,β-Trichloro-n-butyraldehyde, 9545
[76-37-9] 2,2,3,3-Tetrafluoro-1-propanol, 9139
[76-38-0] Methoxyflurane, 5914
[76-41-5] Oxymorphone, 6922
[76-42-6] Oxycodone, 6912
[76-43-7] Fluoxymesterone, 4113
[76-44-8] Heptachlor, 4576
[76-45-9] Protoverine, 7916
[76-47-1] Hydrocortamate, 4709
[76-49-3] Bornyl Acetate, 1339
[76-53-9] Veatchine, 9844
[76-57-3] Codeine, 2459
[76-58-4] Ethylmorphine, 3785
[76-59-5] Bromthymol Blue, 1435
[76-60-8] Bromcresol Green, 1375
[76-61-9] Thymol Blue, 9334
[76-62-0] 3',3'',5',5''-Tetrabromophenolphthalein, 9122
[76-65-3] Amolanone, 603
[76-64-6] Rhynchophylline, 8198
[76-68-6] Cyclopentobarbital, 2751
[76-74-4] Pentobarbital*, 7087
[76-76-6] Probarbital, 7759
[76-77-7] Neoquassin, 6379
[76-78-8] Quassin, 8037
[76-80-2] Tephrosin, 9082
[76-84-6] Triphenylcarbinol, 9653
[76-87-9] Triphenyltin Hydroxide, 9659
[76-90-4] Mepenzolate Bromide, 5735
[76-93-7] Benzilic Acid, 1089
[76-94-8] Phenylmethylbarbituric Acid, 7275
[77-02-1] Aprobarbital, 782
[77-03-2] Piperidone, 7439
[77-04-3] Pyrithyldione, 8005
[77-06-5] Gibberellic Acid, 4313
[77-07-6] Levorphanol, 5350
[77-09-8] Phenolphthalein, 7208
[77-10-1] Phencyclidine, 7179
[77-14-5] Proheptazine, 7786
[77-15-6] Ethoheptazine, 3698
[77-16-7] Plumericin, 7512
[77-20-3] Alphaprodine, 307
[77-21-4] Glutethimide, 4371
[77-23-6] Carbetapentane, 1798
[77-25-8] Ethyl Diethylmalonate, 3746
[77-26-9] Butalbital, 1502
[77-27-0] Thiamylal, 9231
[77-28-1] Butethal, 1515
[77-32-7] Thiobarbital, 9248
[77-36-1] Chlorthalidone, 2194
[77-37-2] Procyclidine, 7770
[77-38-3] Chlorphenoxamine, 2182
[77-40-7] Bisphenol B, 1312
[77-41-8] Methsuximide, 5928
[77-42-9] β-Santalol, 8322
[77-46-3] Acedapsone, 16
[77-51-0] Isoamine, 4992
[77-52-1] Ursolic Acid, 9802
[77-53-2] Cedrol, 1919
[77-58-7] Dibutyltin Dilaurate, 3025
[77-59-8] Tomatidine, 9470
[77-60-1] Tigogenin, 9367
[77-65-6] Carbromal, 1837
[77-66-7] Acecarbromal, 15
[77-67-8] Ethosuximide, 3704

[77-75-8] Meparfynol, 5731
[77-78-1] Dimethyl Sulfate, 3244
[77-79-2] 3-Sulfolene, 8935
[77-81-6] Tabun, 9001
[77-86-1] Tromethamine, 9684
[77-91-8] Choline Dihydrogen Citrate, 2210
[77-92-9] Citric Acid, 2328
[77-94-1] Butyl Citrate, 1564
[77-95-2] Quinic Acid, 8071
[77-98-5] Tetraethylammonium Hydroxide, 9135
[78-00-2] Tetraethyllead, 9136
[78-05-7] Phenoctide, 7205
[78-10-4] Ethyl Silicate, 3805
[78-11-5] Pentaerythritol Tetranitrate, 7066
[78-12-6] Pentaerythritol Chloral, 7063
[78-28-4] Emylcamate, 3520
[78-30-8] Tri-o-tolyl Phosphate, 9676
[78-34-2] Dioxathion, 3296
[78-40-0] Triethyl Phosphate, 9589
[78-41-1] Triparanol, 9650
[78-44-4] Carisoprodol, 1848
[78-53-5] Amiton, 502
[78-67-1] 2,2'-Azobisisobutyronitrile, 931
[78-70-6] Linalool, 5373
[78-75-1] Propylene Dibromide, 7866
[78-76-2] sec-Butyl Bromide, 1554
[78-77-3] Isobutyl Bromide, 5019
[78-79-5] Isoprene, 5087
[78-81-9] Isobutylamine, 5016
[78-83-1] Isobutyl Alcohol, 5015
[78-84-2] Isobutyraldehyde, 5038
[78-86-4] sec-Butyl Chloride, 1561
[78-87-5] Propylene Dichloride, 7867
[78-89-7] Propylene Chlorohydrin, 7863
[78-90-0] Propylenediamine, 7865
[78-91-1] 2-Aminopropanol, 480
[78-92-2] sec-Butyl Alcohol, 1541
[78-93-3] Methyl Ethyl Ketone, 5991
[78-94-4] Methyl Vinyl Ketone, 6052
[78-95-5] Chloroacetone, 2113
[78-98-8] Pyruvaldehyde, 8030
[78-99-9] Propylidene Chloride, 7874
[79-00-5] 1,1,2-Trichloroethane, 9550
[79-01-6] Trichloroethylene, 9552
[79-03-8] Propionyl Chloride, 7840
[79-04-9] Chloracetyl Chloride, 2054
[79-05-0] Propionamide, 7836
[79-06-1] Acrylamide, 123
[79-07-2] Chloroacetamide, 2109
[79-08-3] Bromoacetic Acid, 1388
[79-09-4] Propionic Acid, 7837
[79-10-7] Acrylic Acid, 124
[79-11-8] Chloroacetic Acid, 2111
[79-14-1] Glycolic Acid, 4394
[79-15-2] N-Bromoacetamide, 1386
[79-17-4] Aminoguanidine, 453
[79-19-6] Thiosemicarbazide, 9292
[79-20-9] Methyl Acetate, 5932
[79-21-0] Peracetic Acid, 7107
[79-22-1] Methyl Chlorocarbonate, 5966
[79-24-3] Nitroethane, 6518
[79-27-6] sym-Tetrabromoethane, 9121
[79-31-2] Isobutyric Acid, 5039

[79-33-4] L-Lactic Acid, 5216
[79-34-5] Tetrachloroethane, 9125
[79-36-7] 2,2-Dichloroacetyl Chloride, 3040
[79-37-8] Oxalyl Chloride, 6867
[79-40-3] Rubeanic Acid, 8256
[79-41-4] Methacrylic Acid, 5849
[79-42-5] Thiolactic Acid, 9269
[79-43-6] Dichloroacetic Acid, 3037
[79-46-9] 2-Nitropropane, 6549
[79-49-2] Pentabromoacetone, 7056
[79-54-9] Levopimaric Acid, 5348
[79-55-0] Pempidine, 7022
[79-57-2] Oxytetracycline, 6931
[79-58-3] Rubijervine, 8264
[79-61-8] Dichlorisone Acetate*, 3035
[79-63-0] Lanosterol, 5232
[79-64-1] Dimethisterone, 3208
[79-68-5] γ-Irone, 4979
[79-69-6] α-Irone, 4977
[79-70-9] β-Irone, 4978
[79-74-3] 2,5-Di-tert-pentylhydroquinone, 3300
[79-80-1] Vitamin A₂, 9919
[79-83-4] Pantothenic Acid, 6964
[79-85-6] Quatrimycin, 8038
[79-90-3] Triclobisonium Chloride, 9567
[79-91-4] Terebic Acid, 9092
[79-92-5] Camphene, 1736
[79-93-6] Phenaglycodol, 7161
[80-00-2] Sulphenone®, 8969
[80-02-4] 2-p-Sulfanilylanilinoethanol, 8902
[80-03-5] Acediasulfone, 17
[80-05-7] Bisphenol A, 1311
[80-06-8] Chlorfenethol, 2086
[80-08-0] Dapsone, 2820
[80-11-5] p-Tolylsulfonylmethylnitrosamide, 9469
[80-12-6] Tetramethylenedisulfotetramine, 9158
[80-13-7] Halazone, 4503
[80-17-1] Porofor® BSH, 7573
[80-32-0] Sulfachlorpyridazine, 8871
[80-33-1] Ovex, 6856
[80-35-3] Sulfamethoxypyridazine, 8890
[80-40-0] Ethyl p-Toluenesulfonate, 3814
[80-46-6] p-tert-Pentylphenol, 7098
[80-49-9] Homatropine Methylbromide*, 4649
[80-50-2] Anisotropine Methylbromide, 701
[80-53-5] Terpin, 9101
[80-56-8] α-Pinene, 7414
[80-58-0] α-Bromobutyric Acid, 1401
[80-59-1] Tiglic Acid, 9365
[80-60-4] α-Aminobutyric Acid, 439
[80-69-3] Tartronic Acid, 9042
[80-72-8] Reductic Acid, 8134
[80-74-0] Acetyl Sulfisoxazole, 100
[80-77-3] Chlormezanone, 2105
[80-78-4] Solanidine, 8660
[80-80-8] Acetosulfone*, 66
[80-92-2] Pregnanediol, 7732
[80-97-7] Cholestanol, 2203
[81-04-9] Armstrong's Acid, 812
[81-05-0] 2-Naphthylamine-5-sulfonic Acid, 6327
[81-07-2] Saccharin, 8282
[81-11-8] Amsonic Acid, 628
[81-13-0] Dexpanthenol, 2924

[81-16-3] 2-Naphthylamine-1-sulfonic Acid, 6326
[81-23-2] Dehydrocholic Acid, 2858
[81-24-3] Taurocholic Acid, 9044
[81-25-4] Cholic Acid, 2206
[81-38-9] Laureline, 5251
[81-54-9] Purpurin, 7962
[81-61-8] Quinalizarin, 8059
[81-64-1] Quinizarin, 8094
[81-67-4] Pukateine, 7954
[81-77-6] Indanthrene®, 4846
[81-81-2] Warfarin, 9950
[81-82-3] Coumachlor, 2556
[81-88-9] Rhodamine B, 8182
[81-90-3] Phenolphthalin, 7210
[81-92-5] Phenolphthalol, 7211
[81-93-6] Phenosafranin, 7218
[82-02-0] Khellin, 5189
[82-05-3] Benzanthrone, 1070
[82-08-6] Rottlerin, 8250
[82-10-0] Sumatrol, 8980
[82-12-2] Rufigallol, 8267
[82-22-4] Anthrimide, 720
[82-24-6] 1-Aminoanthraquinone-2-carboxylic Acid, 429
[82-40-6] Lunacrine, 5473
[82-45-1] 1-Aminoanthraquinone, 428
[82-54-2] Cotarnine, 2551
[82-57-5] Visnagin, 9916
[82-58-6] Lysergic Acid, 5506
[82-62-2] 3,4,6-Trichloro-2-nitrophenol, 9554
[82-66-6] Diphenadione, 3304
[82-68-8] Quintozene, 8108
[82-71-3] Styphnic Acid, 8828
[82-75-7] 1-Naphthylamine-8-sulfonic Acid, 6325
[82-76-8] 1-Anilino-8-naphthalenesulfonate, 690
[82-82-6] 4-Pyridoxic Acid, 7994
[82-83-7] Puberulonic Acid, 7953
[82-88-2] Phenindamine, 7195
[82-89-3] Prodigiosin, 7774
[82-92-8] Cyclizine, 2716
[82-93-9] Chlorcyclizine, 2078
[82-95-1] Buclizine, 1449
[82-98-4] Piperidolate, 7440
[82-99-5] Thiphenamil, 9303
[83-07-8] Ampyrone, 625
[83-12-5] Phenindione, 7196
[83-14-7] Pellotine, 7018
[83-25-0] Succinanil, 8838
[83-26-1] Pindone, 7413
[83-28-3] 2-Isovalerylindane-1,3-dione, 5123
[83-32-9] Acenaphthene, 23
[83-34-1] Skatole, 8503
[83-40-9] o-Cresotic Acid, 2585
[83-43-2] Methylprednisolone, 6028
[83-44-3] Deoxycholic Acid, 2881
[83-45-4] Stigmastanol, 8770
[83-46-5] β-Sitosterol, 8500
[83-47-6] γ-Sitosterol, 8501
[83-48-7] Stigmasterol, 8771
[83-49-8] Hyodeoxycholic Acid, 4794
[83-54-5] Echinopsine, 3472
[83-58-9] Lunacridine, 5472
[83-60-3] Reserpic Acid, 8147
[83-63-6] Diacetazotol, 2940
[83-67-0] Theobromine, 9209
[83-68-1] Vitamin K₆, 9937
[83-69-2] Vitamin K₇, 9938
[83-72-7] Lawsone, 5263
[83-73-8] Iodoquinol, 4935
[83-74-9] Ibogaine, 4806
[83-75-0] Quinine Ethylcarbonate, 8080

*Denotes derivative of title compound

[83-79-4] Rotenone, 8248
[83-81-8] N,N,N',N'-Tetraethyl-phthalamide, 9137
[83-85-2] Fuscin, 4229
[83-86-3] Phytic Acid, 7359
[83-88-5] Riboflavine, 8201
[83-89-6] Quinacrine, 8053
[83-94-3] Tabernanthine, 9000
[83-95-4] Skimmianine, 8505
[83-98-7] Orphenadrine, 6831
[84-01-5] Chlorproethazine, 2184
[84-02-6] Prochlorperazine Male-ate*, 7768
[84-04-8] Pipamazine, 7421
[84-06-0] Thiopropazate, 9286
[84-08-2] Pyrathiazine, 7969
[84-11-7] Phenanthrenequinone, 7168
[84-12-8] Phanquinone, 7148
[84-16-2] Hexestrol, 4621
[84-17-3] Dienestrol, 3094
[84-21-9] 3'-Adenylic Acid, 147
[84-22-0] Tetrahydrozoline, 9150
[84-24-2] Physodic Acid, 7355
[84-26-4] Rutecarpine, 8271
[84-31-1] Quininone, 8093
[84-36-6] Syrosingopine, 8998
[84-37-7] Pseudoyohimbine, 7938
[84-52-6] 3'-Cytidylic Acid, 2794
[84-54-8] 2-Methylanthraquinone, 5943
[84-55-9] Viquidil, 9908
[84-58-2] 2,3-Dichloro-5,6-di-cyanobenzoquinone, 3052
[84-62-8] Phenyl Phthalate, 7278
[84-65-1] Anthraquinone, 717
[84-68-4] 2,2'-Dichlorobenzidine, 3046
[84-74-2] n-Butyl Phthalate, 1586
[84-79-7] Lapachol, 5235
[84-80-0] Vitamin K₁, 9933
[84-82-2] Toxoflavin, 9480
[84-86-6] 1-Naphthylamine-4-sulfonic Acid, 6323
[84-87-7] 1-Naphthol-4-sulfonic Acid, 6311
[84-88-8] 8-Hydroxy-5-quinoline-sulfonic Acid, 4780
[84-89-9] 1-Naphthylamine-5-sulfonic Acid, 6324
[84-96-8] Trimeprazine, 9618
[84-97-9] Perazine, 7108
[84-99-1] Xanthoxyletin, 9975
[85-00-7] Diquat Dibromide, 3359
[85-01-8] Phenanthrene, 7167
[85-02-9] Benzo[f]quinoline, 1115
[85-17-6] Streptidine, 8781
[85-23-4] Spinulosin, 8706
[85-31-4] Thioguanosine, 9267
[85-32-5] 5'-Guanylic Acid, 4484
[85-33-6] Lactaroviolin, 5212
[85-34-7] Chlorfenac, 2085
[85-38-1] 3-Nitrosalicylic Acid, 6553
[85-41-6] Phthalimide, 7347
[85-44-9] Phthalic Anhydride, 7346
[85-47-2] 1-Naphthalenesulfonic Acid, 6295
[85-55-2] 2-(p-Toluyl)benzoic Acid, 9465
[85-61-0] Coenzyme A, 2465
[85-64-3] Laudanine, 5248
[85-66-5] Pyocyanine, 7965
[85-72-3] 3'-Methylphthalanilic Acid, 6027
[85-73-4] Phthalylsulfathiazole, 7351
[85-74-5] 1-Naphthylamine-4,6-disulfonic Acid, 6321

[85-75-6] 1-Naphthylamine-4,7-disulfonic Acid, 6322
[85-82-5] 1-Xylylazo-2-naphthol, 9997
[85-83-6] Scarlet Red, 8349
[85-84-7] Yellow AB, 10003
[85-86-9] Sudan III, 8858
[85-90-5] Tricromyl, 9574
[85-94-9] 2'-Cytidylic Acid, 2793
[85-95-0] Benzestrol, 1082
[85-97-2] 2-Phenyl-6-chlorophe-nol, 7251
[85-98-3] N,N'-Diethylcarbanilide, 3107
[86-00-0] o-Nitrobiphenyl, 6513
[86-12-4] Thenaldine, 9204
[86-14-6] Thiambutene, 9220
[86-21-5] Pheniramine, 7198
[86-22-6] Brompheniramine, 1433
[86-34-0] Phensuximide, 7231
[86-35-1] Ethotoin, 3705
[86-36-2] Mercumallylic Acid, 5764
[86-42-0] Amodiaquin, 602
[86-48-6] 1-Hydroxy-2-naphthoic Acid, 4766
[86-50-0] Azinphos-methyl, 926
[86-54-4] Hydralazine, 4682
[86-55-5] 1-Naphthoic Acid, 6301
[86-56-6] N,N-Dimethyl-1-naph-thylamine, 3240
[86-57-7] 1-Nitronaphthalene, 6535
[86-58-8] 8-Quinolineboronic Acid, 8098
[86-59-9] 8-Quinolinecarboxylic Acid, 8099
[86-60-2] Badische Acid, 952
[86-65-7] Amido-G-Acid, 412
[86-68-0] Quininic Acid, 8092
[86-73-7] 9H-Fluorene, 4081
[86-74-8] Carbazole, 1792
[86-75-9] Benzoxiquine, 1122
[86-80-6] Dimethisoquin, 3207
[86-84-0] 1-Naphthylisocyanate, 6330
[86-87-3] 1-Naphthaleneacetic Acid, 6290
[86-88-4] ANTU, 755
[87-00-3] Homatropine, 4649
[87-08-1] Penicillin V, 7046
[87-09-2] Penicillin O, 7044
[87-10-5] Tribromsalan, 9529
[87-11-6] Thiolutin, 9271
[87-17-2] Salicylanilide, 8299
[87-18-3] 4-tert-Butylphenyl Salic-ylate, 1585
[87-20-7] Isoamyl Salicylate, 5007
[87-21-8] Piridocaine, 7466
[87-28-5] Glycol Salicylate, 4395
[87-29-6] Cinnamyl Anthranilate, 2306
[87-33-2] Isosorbide Dinitrate, 5114
[87-39-8] Violuric Acid, 9904
[87-40-1] 2,4,6-Trichloroanisole, 9541
[87-42-3] 6-Chloropurine, 2161
[87-44-5] Caryophyllene, 1884
[87-45-6] 4-(5-Isopropenyl-2-methyl-1-cyclopenten- 1-yl)-2-butanone, 5092
[87-47-8] Pyrolan®, 8013
[87-51-4] Indoleacetic Acid, 4870
[87-52-5] Gramine, 4440
[87-53-6] Penicillanic Acid, 7032
[87-56-9] Mucochloric Acid, 6208
[87-58-1] Iodopyrrole, 4934
[87-61-6] 1,2,3-Trichlorobenzene, 9542

[87-65-0] 2,6-Dichlorophenol, 3062
[87-66-1] Pyrogallol, 8010
[87-69-4] L-Tartaric Acid, 9039
[87-73-0] D-Glucaric Acid, 4344
[87-74-1] Glucoheptonic Acid, 4349
[87-79-6] Sorbose, 8681
[87-81-0] D-Tagatose, 9005
[87-86-5] Pentachlorophenol, 7059
[87-88-7] Chloranilic Acid, 2072
[87-89-8] Inositol, 4883
[87-90-1] Symclosene, 8993
[87-91-2] Ethyl Tartrate, 3811
[87-99-0] Xylitol, 9992
[88-04-0] Chloroxylenol, 2176
[88-06-2] 2,4,6-Trichlorophenol, 9556
[88-09-5] Diethylacetic Acid, 3099
[88-13-1] 3-Thenoic Acid, 9206
[88-14-2] 2-Furoic Acid, 4219
[88-21-1] Orthanilic Acid, 6835
[88-29-9] Versalide®, 9870
[88-30-2] TFM, 9180
[88-67-5] o-Iodobenzoic Acid, 4923
[88-74-4] o-Nitroaniline, 6504
[88-75-5] o-Nitrophenol, 6541
[88-85-7] Dinoseb, 3282
[88-88-0] Picryl Chloride, 7390
[88-89-1] Picric Acid, 7380
[88-95-9] Phthaloyl Chloride, 7349
[88-96-0] Phthalamide, 7343
[88-99-3] Phthalic Acid, 7345
[89-00-9] Quinolinic Acid, 8102
[89-01-0] 2,3-Pyrazinedicarboxylic Acid, 7972
[89-05-4] Pyromellitic Acid, 8015
[89-31-6] Salsoline, 8308
[89-38-3] Pteropterin, 7950
[89-45-2] Salicylsulfuric Acid, 8303
[89-46-3] Menthyl Salicylate, 5727
[89-47-4] Menthyl Valerate, 5728
[89-48-5] Menthyl Acetate, 5725
[89-56-5] p-Cresotic Acid, 2586
[89-57-6] Mesalamine, 5807
[89-65-6] Isoascorbic Acid, 5009
[89-68-9] Chlorothymol, 2171
[89-73-6] Salicylhydroxamic Acid, 8300
[89-78-1] Menthol, 5723
[89-81-6] Piperitone, 7443
[89-82-7] Pulegone, 7955
[89-83-8] Thymol, 9333
[89-84-9] Resacetophenone, 8143
[89-86-1] β-Resorcylic Acid, 8160
[90-00-6] Phlorol, 7302
[90-01-7] Salicyl Alcohol, 8294
[90-02-8] Salicylaldehyde, 8295
[90-03-9] Mercufenol Chloride, 5763
[90-05-1] Guaiacol, 4457
[90-11-9] 1-Bromonaphthalene, 1413
[90-13-1] 1-Chloronaphthalene, 2149
[90-15-3] 1-Naphthol, 6303
[90-18-6] Quercetagetin, 8043
[90-19-7] Rhamnetin, 8170
[90-20-0] 1-Naphthol-8-amino-3,6-disulfonic Acid, 6305
[90-22-2] Valethamate Bromide, 9816
[90-24-4] Xanthoxylin, 9976
[90-26-6] α-Phenylbutyramide, 7249
[90-29-9] Pseudobaptigenin, 7925
[90-33-5] Hymecromone, 4792
[90-34-6] Primaquine, 7751
[90-39-1] Sparteine, 8692

[90-43-7] o-Phenylphenol, 7276
[90-44-8] Anthrone, 721
[90-45-9] Aminacrine, 418
[90-47-1] Xanthone, 9971
[90-49-3] Pheneturide, 7190
[90-54-0] Etafenone, 3662
[90-59-5] 3,5-Dibromosalicylal-
dehyde, 3011
[90-64-2] Mandelic Acid, 5599
[90-65-3] Penicillic Acid, 7033
[90-69-7] Lobeline, 5432
[90-71-1] Taxicatin, 9045
[90-74-4] Rutinose, 8277
[90-75-5] Vacciniin, 9811
[90-80-2] Gluconolactone, 4351
[90-81-3] Ephedrine DL-Form*,
3561
[90-82-4] Pseudoephedrine*, 3561
[90-84-6] Diethylpropion, 3116
[90-86-8] Cinnamedrine, 2299
[90-89-1] Diethylcarbamazine,
3106
[90-94-8] Michler's Ketone, 6100
[91-00-9] Benzhydrylamine, 1085
[91-01-0] Benzohydrol, 1100
[91-16-7] Veratrole, 9857
[91-17-8] Decalin®, 2839
[91-18-9] Pteridine, 7947
[91-19-0] Quinoxaline, 8107
[91-20-3] Naphthalene, 6289
[91-22-5] Quinoline, 8097
[91-33-8] Benzthiazide, 1134
[91-40-7] N-Phenylanthranilic
Acid, 7244
[91-53-2] Ethoxyquin, 3710
[91-56-5] Isatin, 4985
[91-58-7] 2-Chloronaphthalene,
2150
[91-59-8] 2-Naphthylamine, 6319
[91-60-1] 2-Naphthalenethiol,
6298
[91-63-4] Quinaldine, 8055
[91-64-5] Coumarin, 2563
[91-66-7] Diethylaniline, 3102
[91-71-4] Thiocarbamizine, 9251
[91-75-8] Antazoline, 709
[91-76-9] Benzoguanamine, 1099
[91-79-2] Thenyldiamine, 9207
[91-80-5] Methapyrilene, 5871
[91-81-6] Tripelennamine, 9651
[91-82-7] Pyrrobutamine, 8023
[91-84-9] Pyrilamine, 7996
[91-86-1] η-Tocopherol, 9423
[91-94-1] 3,3'-Dichlorobenzidine,
3047
[92-04-6] 4-Phenyl-2-chlorophe-
nol, 7252
[92-12-6] Phenyltoloxamine, 7287
[92-13-7] Pilocarpine, 7395
[92-24-0] Naphthacene, 6288
[92-28-4] Amido-R-Acid, 414
[92-31-9] Tolonium Chloride, 9444
[92-32-0] Pyronine Y, 8017
[92-41-1] 2,7-Naphthalenedisul-
fonic Acid, 6294
[92-43-3] 1-Phenyl-3-pyrazolidi-
none, 7281
[92-52-4] Diphenyl, 3314
[92-59-1] Ethylbenzylaniline, 3728
[92-61-5] Scopoletin, 8363
[92-62-6] Proflavine, 7780
[92-67-1] p-Biphenylamine, 1248
[92-69-3] p-Phenyl Phenol, 7277
[92-70-6] 3-Hydroxy-2-naphthoic
Acid, 4767
[92-82-0] Phenazine, 7173
[92-84-2] Phenothiazine, 7220
[92-87-5] Benzidine, 1086
[92-93-3] p-Nitrobiphenyl, 6514
[92-95-5] p-Xenylcarbimide, 9982

[93-01-6] 2-Naphthol-6-sulfonic
Acid, 6312
[93-04-9] 2-Methoxynaphthalene,
5918
[93-07-2] Veratric Acid, 9854
[93-09-4] 2-Naphthoic Acid, 6302
[93-10-7] Quinaldic Acid, 8054
[93-14-1] Guaifenesin, 4465
[93-18-5] 2-Ethoxynaphthalene,
3709
[93-20-9] 2-(2-Naphthyloxy)-
ethanol, 6333
[93-30-1] Methoxyphenamine,
5919
[93-35-6] Umbelliferone, 9758
[93-39-0] Skimmin, 8506
[93-42-5] Thionalide, 9273
[93-43-6] 2-Naphthyl Lactate,
6332
[93-44-7] 2-Naphthyl Benzoate,
6328
[93-47-0] Verazide, 9859
[93-51-6] Creosol, 2573
[93-54-9] α-Ethylbenzyl Alcohol,
3727
[93-55-0] Propiophenone, 7842
[93-56-1] Styrene Glycol, 8831
[93-58-3] Methyl Benzoate, 5947
[93-59-4] Perbenzoic Acid, 7109
[93-60-7] Methyl Nicotinate, 6014
[93-71-0] Allidochlor, 250
[93-72-1] Silvex, 8483
[93-75-4] Thioquinox, 9288
[93-76-5] 2,4,5-T, 8999
[93-88-9] Phenylpropylmethylamine, 7280
[93-89-0] Ethyl Benzoate, 3725
[93-95-8] 5-(α-Phenylethyl)-
semioxamazide, 7260
[93-97-0] Benzoic Anhydride, 1102
[93-98-1] Benzanilide, 1068
[93-99-2] Phenyl Benzoate, 7246
[94-02-0] Ethyl Benzoylacetate,
3726
[94-07-5] Synephrine, 8994
[94-09-7] Ethyl Aminobenzoate,
3719
[94-10-0] Ethoxazene, 3706
[94-12-2] Risocaine, 8226
[94-13-3] Propylparaben, 7879
[94-14-4] Isobutyl p-Aminobenzoate, 5017
[94-15-5] Dimethocaine, 3210
[94-16-6] Aminohippurate Sodium*, 454
[94-19-9] Sulfaethidole, 8878
[94-20-2] Chlorpropamide, 2187
[94-23-5] Parethoxycaine, 6987
[94-25-7] Butamben, 1504
[94-26-8] Butylparaben, 1583
[94-35-9] Styramate, 8829
[94-36-0] Benzoyl Peroxide, 1128
[94-41-7] Chalcone, 2028
[94-44-0] Nicotinic Acid Benzyl
Ester, 6436
[94-46-2] Isoamyl Benzoate, 4995
[94-53-1] Piperonylic Acid, 7447
[94-59-7] Safrole, 8287
[94-62-2] Piperine, 7442
[94-63-3] Pralidoxime Iodide*,
7705
[94-67-7] Salicylaldoxime, 8296
[94-70-2] o-Phenetidine, 7187
[94-74-6] MCPA, 5645
[94-75-7] 2,4-D, 2802
[94-82-6] 2,4-DB, 2828
[94-83-7] Sesin, 8422
[94-84-8] Falone®, 3880
[94-96-2] Ethohexadiol, 3699
[95-01-2] β-Resorcylaldehyde,
8159

[95-04-5] Ectylurea, 3478
[95-05-6] Sulfiram, 8928
[95-06-7] Sulfallate, 8882
[95-13-6] Indene, 4851
[95-14-7] 1H-Benzotriazole, 1119
[95-15-8] Thianaphthene, 9232
[95-16-9] Benzothiazole, 1118
[95-25-0] Chlorzoxazone, 2197
[95-38-5] Amine 220®, 419
[95-43-2] D-Threose, 9317
[95-44-3] L-Threose, 9318
[95-45-4] Dimethylglyoxime, 3235
[95-48-7] o-Cresol, 2580
[95-49-8] o-Chlorotoluene*, 2172
[95-50-1] o-Dichlorobenzene, 3044
[95-51-2] o-Chloroaniline*, 2118
[95-53-4] o-Toluidine*, 9462
[95-54-5] o-Phenylenediamine,
7255
[95-55-6] o-Aminophenol, 473
[95-56-7] o-Bromophenol*, 1416
[95-63-6] Pseudocumene, 7929
[95-76-1] 3,4-Dichloroaniline,
3041
[95-86-3] 2,4-Diaminophenol,
2961
[95-92-1] Diethyl Oxalate, 3115
[95-93-2] Durene, 3450
[95-95-4] 2,4,5-Trichlorophenol,
9555
[96-00-4] Sebacoin, 8371
[96-01-5] Sebacil, 8370
[96-11-7] 1,2,3-Tribromopropane,
9527
[96-12-8] Dibromochloropropane,
3003
[96-20-8] 2-Amino-1-butanol, 438
[96-22-0] Diethyl Ketone, 3111
[96-23-1] 1,3-Dichloro-2-propa-
nol, 3063
[96-24-2] α-Chlorohydrin, 2145
[96-26-4] Dihydroxyacetone, 3166
[96-27-5] Thioglycerol, 9264
[96-33-3] Methyl Acrylate, 5935
[96-34-4] Methyl Chloroacetate,
5965
[96-41-3] Cyclopentanol, 2747
[96-45-7] Ethylene Thiourea, 3759
[96-48-0] Butyrolactone, 1596
[96-50-4] 2-Aminothiazole, 494
[96-62-8] Dinsed, 3284
[96-64-0] Soman, 8668
[96-77-5] Phenoldisulfonic Acid,
7207
[96-82-2] Lactobionic Acid, 5219
[96-83-3] Iopanoic Acid, 4944
[96-84-4] Iophenoxic Acid, 4947
[96-88-8] Mepivacaine, 5748
[96-91-3] Picramic Acid, 7379
[96-97-9] 5-Nitrosalicylic Acid,
6554
[97-00-7] 1-Chloro-2,4-dinitro-
benzene, 2135
[97-02-9] 2,4-Dinitroaniline, 3263
[97-05-2] Sulfosalicylic Acid, 8944
[97-11-0] Cyclethrin, 2711
[97-16-5] Genite®, 4283
[97-17-6] Dichlofenthion, 3030
[97-18-7] Bithionol, 1316
[97-23-4] Dichlorophen(e), 3059
[97-24-5] Fenticlor, 3947
[97-25-6] Urochloralic Acid, 9798
[97-26-7] Toluylene Blue, 9466
[97-27-8] Chlorbetamide, 2076
[97-30-3] α-Methylglucoside, 5997
[97-31-4] Normetanephrine, 6627
[97-44-9] Acetarsone, 44
[97-53-0] Eugenol, 3855
[97-54-1] Isoeugenol, 5054
[97-57-4] Tolpronine, 9449
[97-59-6] Allantoin, 246

* Denotes derivative of title compound

[97-62-1] Ethyl Isobutyrate, 3770
[97-64-3] Ethyl Lactate, 3773
[97-65-4] Itaconic Acid, 5130
[97-75-6] Scopolamine N-Oxide, 8362
[97-77-8] Disulfiram, 3370
[97-85-8] Isobutyl Isobutyrate, 5028
[97-99-4] Tetrahydrofurfuryl Alcohol, 9146
[98-00-0] Furfuryl Alcohol, 4215
[98-01-1] Furfural, 4214
[98-04-4] Phenyltrimethylammonium Iodide, 7289
[98-05-5] Benzenearsonic Acid, 1075
[98-06-6] tert-Butylbenzene, 1551
[98-07-7] Benzotrichloride, 1120
[98-08-8] Benzotrifluoride, 1121
[98-09-9] Benzenesulfonyl Chloride, 1080
[98-11-3] Benzenesulfonic Acid, 1078
[98-50-0] Arsanilic Acid, 818
[98-54-4] p-tert-Butylphenol, 1584
[98-55-5] α-Terpineol, 9103
[98-58-8] p-Bromobenzenesulfonyl Chloride, 1395
[98-59-9] p-Toluenesulfonyl Chloride, 9460
[98-61-3] Pipsyl Chloride, 7458
[98-62-4] Sulfanilyl Fluoride, 8904
[98-66-8] p-Chlorobenzenesulfonic Acid, 2122
[98-67-9] p-Phenolsulfonic Acid, 7212
[98-71-5] 4-Hydrazinobenzenesulfonic Acid, 4695
[98-72-6] Nitarsone, 6483
[98-79-3] L-Pyroglutamic Acid, 8012
[98-80-6] Benzeneboronic Acid, 1076
[98-82-8] Cumene, 2619
[98-84-0] α-Methylbenzylamine, 5949
[98-86-2] Acetophenone, 65
[98-87-3] Benzal Chloride, 1064
[98-88-4] Benzoyl Chloride, 1124
[98-89-5] Cyclohexanecarboxylic Acid, 2730
[98-92-0] Niacinamide, 6398
[98-95-3] Nitrobenzene, 6509
[98-96-4] Pyrazinamide, 7970
[98-97-5] Pyrazinoic Acid, 7973
[98-98-6] Picolinic Acid, 7375
[99-00-3] Picadex, 7366
[99-05-8] m-Aminobenzoic Acid, 432
[99-09-2] m-Nitroaniline, 6503
[99-11-6] Citrazinic Acid, 2327
[99-14-9] Tricarballylic Acid, 9534
[99-17-2] Populin, 7571
[99-20-7] Trehalose, 9496
[99-23-0] Puberulic Acid, 7952
[99-24-1] Methyl Gallate*, 4251
[99-32-1] Chelidonic Acid, 2042
[99-33-2] 3,5-Dinitrobenzoyl Chloride, 3270
[99-34-3] 3,5-Dinitrobenzoic Acid, 3269
[99-35-4] sym-Trinitrobenzene, 9639
[99-43-4] Benoxinate, 1056
[99-45-6] Adrenalone, 161
[99-49-0] Carvone, 1883
[99-50-3] Protocatechuic Acid, 7909
[99-56-9] 4-Nitro-o-phenylenediamine, 6544
[99-66-1] Valproic Acid, 9821

[99-73-0] p-Bromophenacyl Bromide, 1415
[99-76-3] Methylparaben, 6021
[99-79-6] Iophendylate, 4946
[99-83-2] α-Phellandrene, 7151
[99-88-7] Cumidine, 2622
[99-90-1] p-Bromoacetophenone, 1390
[99-91-2] p-Chloroacetophenone, 2114
[99-96-7] p-Hydroxybenzoic Acid, 4742
[99-98-9] Dimethyl-p-phenylenediamine, 3242
[100-01-6] p-Nitroaniline, 6505
[100-02-7] p-Nitrophenol, 6542
[100-07-2] p-Anisoyl Chloride, 703
[100-09-4] p-Anisic Acid, 696
[100-10-7] p-Dimethylaminobenzaldehyde, 3219
[100-16-3] (4-Nitrophenyl)hydrazine, 6545
[100-21-0] Terephthalic Acid, 9093
[100-22-1] Tetramethyl-p-phenylenediamine, 9159
[100-26-5] Isocinchomeronic Acid, 5043
[100-33-4] Pentamidine, 7071
[100-37-8] 2-Diethylaminoethanol, 3101
[100-39-0] Benzyl Bromide, 1142
[100-41-4] Ethylbenzene, 3723
[100-42-5] Styrene, 8830
[100-44-7] Benzyl Chloride, 1143
[100-46-9] Benzylamine, 1139
[100-47-0] Benzonitrile, 1107
[100-49-2] Cyclohexylcarbinol, 2737
[100-51-6] Benzyl Alcohol, 1138
[100-52-7] Benzaldehyde, 1065
[100-53-8] Thiobenzyl Alcohol, 9249
[100-55-0] Nicotinyl Alcohol, 6438
[100-56-1] Phenylmercuric Chloride, 7272
[100-59-4] Phenylmagnesium Chloride, 7269
[100-61-8] Methylaniline, 5941
[100-63-0] Phenylhydrazine, 7264
[100-65-2] Phenylhydroxylamine, 7266
[100-66-3] Anisole, 699
[100-67-4] Potassium Phenoxide, 7646
[100-76-5] Quinuclidine, 8109
[100-79-8] Isopropylidene Glycerol, 5101
[100-88-9] Cyclamic Acid, 2707
[100-91-4] Eucatropine, 3854
[100-92-5] Mephentermine, 5740
[100-95-8] Dodecarbonium Chloride, 3404
[100-97-0] Methenamine, 5879
[101-05-3] Anilazine, 685
[101-14-4] 4,4'-Methylenebis[2-chloroaniline], 5978
[101-20-2] Triclocarban, 9568
[101-21-3] Chlorpropham, 2188
[101-26-8] Pyridostigmine Bromide, 7990
[101-27-9] Barban, 969
[101-31-5] Hyoscyamine, 4795
[101-38-2] Gibbs Reagent, 4315
[101-42-8] Fenuron, 3951
[101-53-1] p-Benzylphenol, 1159
[101-57-5] N-Phenylsulfanilic Acid, 7284
[101-60-0] Porphine, 7574
[101-61-1] Michler's Base, 6099
[101-62-2] Phenamidine, 7165

[101-64-4] N-(p-Methoxyphenyl)-p-phenylenediamine, 5922
[101-71-3] Diphenane, 3306
[101-77-9] p,p'-Diaminodiphenylmethane, 2958
[101-81-5] Diphenylmethane, 3329
[101-83-7] Dicyclohexylamine, 3085
[101-84-8] Phenyl Ether, 7259
[101-91-7] 4'-Hydroxybutyranilide, 4745
[101-97-3] Ethyl Phenylacetate, 3794
[101-99-5] Phenylurethan(e), 7291
[102-01-2] Acetoacetanilide, 50
[102-02-3] Phenyl Biguanide, 7247
[102-06-7] 1,3-Diphenylguanidine, 3325
[102-07-8] Carbanilide, 1787
[102-08-9] sym-Diphenylthiourea, 3337
[102-09-0] Phenyl Carbonate, 7250
[102-14-7] Succinanilic Acid, 8839
[102-30-7] Dichlorobenzalkonium Chloride, 3042
[102-31-8] Xylopropamine, 9994
[102-45-4] Cyclopentamine, 2745
[102-54-5] Ferrocene, 3985
[102-60-3] Entprol, 3551
[102-71-6] Triethanolamine, 9581
[102-76-1] Triacetin, 9504
[102-82-9] Tributylamine, 9530
[102-92-1] Cinnamoyl Chloride, 2303
[102-98-7] Phenylmercury Borate, 7274
[103-01-5] N-Phenylglycine, 7262
[103-03-7] Phenicarbazide, 7193
[103-09-3] Octyl Acetate, 6683
[103-12-8] Sulfamidochrysoidine, 8895
[103-16-2] Monobenzone, 6159
[103-17-3] Chlorbenside, 2074
[103-29-7] Bibenzyl, 1219
[103-32-2] Benzylaniline, 1140
[103-33-3] Azobenzene, 930
[103-34-4] 4,4'-Dithiodimorpholine, 3379
[103-41-3] Benzyl Cinnamate, 1144
[103-49-1] Dibenzylamine, 2993
[103-50-4] Benzyl Ether, 1146
[103-65-1] n-Propylbenzene, 7856
[103-69-5] Ethylaniline, 3722
[103-70-8] Formanilide, 4152
[103-71-9] Phenyl Isocyanate, 7267
[103-72-0] Phenyl Isothiocyanate, 7268
[103-73-1] Phenetole, 7189
[103-79-7] Phenylacetone, 7240
[103-81-1] α-Phenylacetamide, 7237
[103-82-2] Phenylacetic Acid, 7239
[103-84-4] Acetanilide, 42
[103-85-5] Phenylthiourea, 7286
[103-88-5] p-Bromoacetanilide, 1387
[103-90-2] Acetaminophen, 40
[103-97-9] Phenocoll, 7204
[104-03-0] p-Nitrophenylacetic Acid, 6543
[104-06-3] Thiacetazone, 9218
[104-14-3] Octopamine, 6679
[104-15-4] p-Toluenesulfonic Acid, 9459
[104-22-3] Benzylsulfamide, 1161
[104-28-9] Cinoxate, 2312
[104-29-0] Chlorphenesin, 2178
[104-31-4] Benzonatate, 1106
[104-32-5] Propamidine, 7808
[104-46-1] Anethole, 675
[104-51-8] n-Butylbenzene, 1549

[104-54-1] Cinnamyl Alcohol, 2305
[104-55-2] Cinnamaldehyde, 2298
[104-57-4] Benzyl Formate, 1148
[104-76-7] 2-Ethyl-1-hexanol, 3764
[104-80-3] 2,5-Tetrahydrofurandimethanol, 9145
[104-85-8] p-Tolunitrile, 9464
[104-90-5] 5-Ethyl-2-picoline, 3798
[104-91-6] 4-Nitrosophenol, 6563
[105-13-5] Anise Alcohol, 695
[105-20-4] Betazole, 1205
[105-29-3] 1-Pentol, 7088
[105-37-3] Ethyl Propionate, 3801
[105-39-5] Ethyl Chloroacetate, 3741
[105-41-9] Methylhexaneamine, 6000
[105-45-3] Methyl Acetoacetate, 5933
[105-46-4] sec-Butyl Acetate, 1536
[105-53-3] Ethyl Malonate, 3779
[105-54-4] Ethyl Butyrate, 3733
[105-56-6] Ethyl Cyanoacetate, 3744
[105-57-7] Acetal, 31
[105-58-8] Ethyl Carbonate, 3738
[105-60-2] Caprolactam, 1762
[105-61-3] Maleuric Acid, 5588
[105-66-8] Propyl Butyrate, 7858
[105-92-0] Rhodopin, 8192
[106-02-5] Exaltolide®, 3867
[106-22-9] β-Citronellol, 2332
[106-23-0] Citronellal, 2331
[106-24-1] Geraniol, 4298
[106-25-2] Nerol, 6387
[106-27-4] Isoamyl Butyrate, 4997
[106-30-9] Ethyl Oenanthate, 3790
[106-31-0] Butyric Anhydride, 1594
[106-32-1] Ethyl Caprylate, 3736
[106-33-2] Ethyl Laurate, 3774
[106-34-3] Quinhydrone, 8070
[106-36-5] Propyl Propionate, 7880
[106-37-6] p-Dibromobenzene, 3001
[106-40-1] p-Bromoaniline, 1392
[106-41-2] p-Bromophenol*, 1416
[106-43-4] p-Chlorotoluene*, 2172
[106-44-5] p-Cresol, 2581
[106-46-7] p-Dichlorobenzene, 3045
[106-47-8] p-Chloroaniline*, 2118
[106-49-0] p-Toluidine*, 9462
[106-50-3] p-Phenylenediamine, 7256
[106-51-4] Quinone, 8103
[106-57-0] 2,5-Piperazinedione, 7434
[106-60-5] δ-Aminolevulinic Acid, 458
[106-68-3] Ethyl Amyl Ketone, 3721
[106-89-8] Epichlorohydrin, 3563
[106-93-4] Ethylene Dibromide, 3753
[106-94-5] Propyl Bromide, 7857
[106-95-6] Allyl Bromide, 286
[106-97-8] Butane, 1507
[106-98-9] 1-Butene, 1513
[106-99-0] 1,3-Butadiene, 1500
[107-01-7] 2-Butene, 1514
[107-02-8] Acrolein, 122
[107-05-1] Allyl Chloride, 287
[107-06-2] Ethylene Dichloride, 3754
[107-07-3] Ethylene Chlorohydrin, 3750
[107-08-4] Propyl Iodide, 7875
[107-10-8] Propylamine, 7855

[107-11-9] Allylamine, 285
[107-12-0] Propionitrile, 7839
[107-13-1] Acrylonitrile, 125
[107-15-3] Ethylenediamine, 3752
[107-18-6] Allyl Alcohol, 284
[107-19-7] Propargyl Alcohol, 7819
[107-20-0] Chloroacetaldehyde, 2108
[107-21-1] Ethylene Glycol, 3755
[107-22-2] Glyoxal, 4405
[107-27-7] Ethylmercuric Chloride, 3780
[107-29-9] Acetaldoxime, 35
[107-30-2] Chloromethyl Methyl Ether, 2146
[107-31-3] Methyl Formate, 5994
[107-32-4] Performic Acid, 7114
[107-34-6] 1-Propanearsonic Acid, 7810
[107-35-7] Taurine, 9043
[107-36-8] Isethionic Acid, 4990
[107-38-0] Arsonoacetic Acid, 838
[107-41-5] Hexylene Glycol, 4631
[107-43-7] Betaine, 1201
[107-44-8] Sarin, 8332
[107-49-3] Tetraethyl Pyrophosphate, 9138
[107-51-7] Octamethyltrisiloxane, 6669
[107-52-8] Tetradecamethylhexasiloxane, 9131
[107-53-9] Tetracosamethylhendecasiloxane, 9128
[107-59-5] tert-Butyl Chloroacetate, 1563
[107-68-6] N-Methyltaurine, 6043
[107-69-7] Trichlorourethan, 9561
[107-73-3] Phosphorylcholine, 7337
[107-82-4] Isoamyl Bromide, 4996
[107-84-6] Isoamyl Chloride, 4998
[107-85-7] Isoamylamine, 4994
[107-86-8] Senecialdehyde, 8399
[107-87-9] Methyl Propyl Ketone, 6032
[107-88-0] 1,3-Butylene Glycol, 1566
[107-89-1] Aldol, 217
[107-91-5] Cyanoacetamide, 2695
[107-92-6] Butyric Acid, 1593
[107-94-8] β-Chloropropionic Acid, 2159
[107-95-9] β-Alanine, 196
[107-97-1] Sarcosine, 8331
[108-01-0] Deanol, 2834
[108-03-2] 1-Nitropropane, 6548
[108-05-4] Vinyl Acetate, 9896
[108-10-1] Isopropylacetone, 5095
[108-12-3] Isovaleryl Chloride, 5121
[108-18-9] Diisopropylamine, 3181
[108-19-0] Biuret, 1319
[108-20-3] Isopropyl Ether, 5100
[108-21-4] Isopropyl Acetate, 5093
[108-22-5] Isopropenyl Acetate, 5091
[108-24-7] Acetic Anhydride, 48
[108-28-1] Protoanemonin, 7907
[108-30-5] Succinic Anhydride, 8841
[108-31-6] Maleic Anhydride, 5586
[108-39-4] m-Cresol, 2579
[108-41-8] m-Chlorotoluene*, 2172
[108-42-9] m-Chloroaniline*, 2118
[108-44-1] m-Toluidine*, 9462
[108-45-2] m-Phenylenediamine, 7254
[108-46-3] Resorcinol, 8158
[108-48-5] 2,6-Lutidine, 5485
[108-56-5] Ethyl Oxalacetate, 3791
[108-59-8] Methyl Malonate, 6009
[108-64-5] Ethyl Isovalerate, 3772

[108-67-8] Mesitylene, 5810
[108-70-3] 1,3,5-Trichlorobenzene, 9544
[108-73-6] Phloroglucinol, 7301
[108-75-8] 2,4,6-Trimethylpyridine, 9632
[108-78-1] Melamine, 5691
[108-80-5] Cyanuric Acid, 2704
[108-85-0] Cyclohexyl Bromide, 2736
[108-86-1] Bromobenzene, 1394
[108-88-3] Toluene, 9455
[108-89-4] γ-Picoline, 7374
[108-90-7] Chlorobenzene, 2121
[108-91-8] Cyclohexylamine, 2735
[108-93-0] Cyclohexanol, 2731
[108-94-1] Cyclohexanone, 2732
[108-95-2] Phenol, 7206
[108-98-5] Thiophenol, 9285
[108-99-6] β-Picoline, 7373
[109-06-8] α-Picoline, 7372
[109-21-7] n-Butyl n-Butyrate, 1556
[109-52-4] Valeric Acid, Normal, 9815
[109-57-9] Thiosinamine, 9293
[109-60-4] Propyl Acetate, 7853
[109-61-5] Propyl Chlorocarbonate, 7860
[109-63-7] Boron Trifluoride Etherate, 1352
[109-64-8] Trimethylene Bromide, 9628
[109-65-9] n-Butyl Bromide, 1553
[109-66-0] Pentane, 7072
[109-67-1] 1-Pentene, 7079
[109-68-2] 2-Pentene, 7080
[109-69-3] n-Butyl Chloride, 1560
[109-73-9] n-Butylamine, 1543
[109-74-0] Butyronitrile, 1597
[109-75-1] Allyl Cyanide, 288
[109-77-3] Malononitrile, 5592
[109-78-4] Ethylene Cyanohydrin, 3751
[109-79-5] n-Butyl Mercaptan, 1575
[109-80-8] 1,3-Propanedithiol, 7811
[109-82-0] Methyleneaminoacetonitrile, 5976
[109-83-1] 2-Methylaminoethanol, 5939
[109-84-2] 2-Hydrazinoethanol, 4696
[109-86-4] Methyl Cellosolve®, 5961
[109-87-5] Methylal, 5936
[109-89-7] Diethylamine, 3100
[109-93-3] Vinyl Ether, 9899
[109-94-4] Ethyl Formate, 3763
[109-95-5] Ethyl Nitrite, 3786
[109-95-5] Spirit of Ethyl Nitrite, 8715
[109-96-6] 3-Pyrroline, 8028
[109-97-7] 1H-Pyrrole, 8025
[109-98-8] 2-Pyrazoline, 7975
[109-99-9] Tetrahydrofuran, 9144
[110-00-9] Furan, 4206
[110-02-1] Thiophene, 9283
[110-04-3] 1,2-Ethanedisulfonic Acid, 3678
[110-05-4] DTBP, 3446
[110-13-4] Acetonylacetone, 63
[110-14-5] Succinamide, 8837
[110-15-6] Succinic Acid, 8840
[110-16-7] Maleic Acid, 5585
[110-17-8] Fumaric Acid, 4200
[110-19-0] Isobutyl Acetate, 5014
[110-27-0] Isopropyl Myristate, 5103
[110-38-3] Ethyl Caprate, 3734

Denotes derivative of title compound

[110-43-0] 2-Heptanone, 4584
[110-44-1] Sorbic Acid, 8677
[110-45-2] Isoamyl Formate, 5001
[110-46-3] Isoamyl Nitrite, 5005
[110-49-6] Methyl Cellosolve®
Acetate, 5962
[110-53-2] n-Amyl Bromide, 637
[110-54-3] n-Hexane, 4613
[110-58-7] n-Amylamine, 630
[110-60-1] Putrescine, 7964
[110-61-2] Succinonitrile, 8843
[110-62-3] n-Valeraldehyde, 9813
[110-66-7] n-Amyl Mercaptan, 648
[110-71-4] 1,2-Dimethoxyethane,
3213
[110-74-7] Propyl Formate, 7871
[110-75-8] 2-Chloroethyl Vinyl
Ether, 2139
[110-77-0] 2-(Ethylthio)ethanol,
3813
[110-80-5] 2-Ethoxyethanol, 3707
[110-82-7] Cyclohexane, 2729
[110-83-8] Cyclohexene, 2733
[110-85-0] Piperazine, 7431
[110-86-1] Pyridine, 7983
[110-88-3] s-Trioxane, 9646
[110-89-4] Piperidine, 7438
[110-91-8] Morpholine, 6194
[110-94-1] Glutaric Acid, 4367
[111-00-2] Suxethonium Bromide, 8989
[111-01-3] Squalane, 8726
[111-02-4] Squalene, 8727
[111-13-7] Hexyl Methyl Ketone,
4632
[111-14-8] Heptanoic Acid, 4581
[111-15-9] 2-Ethoxyethyl Acetate,
3708
[111-16-0] Pimelic Acid, 7401
[111-17-1] 3,3'-Thiodipropionic
Acid, 9261
[111-20-6] Sebacic Acid, 8369
[111-27-3] 1-Hexanol, 4615
[111-28-4] Sorbic Alcohol, 8678
[111-29-5] 1,5-Pentanediol, 7073
[111-30-8] Glutaraldehyde, 4366
[111-42-2] Diethanolamine, 3097
[111-43-3] Propyl Ether, 7870
[111-44-4] sym-Dichloroethyl
Ether, 3055
[111-46-6] Diethylene Glycol, 3109
[111-47-7] Propyl Sulfide, 7881
[111-48-8] 2,2'-Thiodiethanol,
9259
[111-51-3] Tetramethyldiaminobutane, 9157
[111-55-7] Ethylene Glycol Diacetate, 3756
[111-65-9] Octane, 6672
[111-66-0] Caprylene, 1764
[111-70-6] 1-Heptanol, 4582
[111-71-7] Heptanal, 4578
[111-76-2] Butyl Cellosolve®,
1559
[111-77-3] Methyl Carbitol®, 5959
[111-83-1] n-Octyl Bromide, 6684
[111-87-5] 1-Octanol, 6674
[111-90-0] Carbitol®, 1806
[111-92-2] n-Dibutylamine, 3019
[111-96-6] Diglyme, 3148
[112-05-0] Pelargonic Acid, 7013
[112-24-3] Trientine, 9579
[112-27-6] Triethylene Glycol,
9585
[112-30-1] n-Decyl Alcohol, 2847
[112-34-5] Butyl Carbitol®, 1557
[112-36-7] Diethyl Carbitol®,
3108
[112-38-9] Undecylenic Acid, 9760
[112-47-0] Decamethylene Glycol,
2842
[112-49-2] Triglyme, 9604

[112-53-8] 1-Dodecanol, 3402
[112-56-1] Lethane® 384, 5329
[112-63-0] Methyl Linoleate, 6008
[112-72-1] Myristyl Alcohol, 6248
[112-79-8] Elaidic Acid, 3497
[112-80-1] Oleic Acid, 6788
[112-85-6] Behenic Acid, 1033
[112-86-7] Erucic Acid, 3619
[113-00-8] Guanidine, 4475
[113-15-5] Ergotamine, 3609
[113-18-8] Ethchlorvynol, 3683
[113-38-2] Estradiol Dipropionate*,
3653
[113-42-8] Methylergonovine, 5989
[113-45-1] Methylphenidate, 6025
[113-52-0] Imipramine Hydrochloride*, 4835
[113-53-1] Dothiepin, 3419
[113-59-7] Chlorprothixene, 2189
[113-73-5] Gramicidin S, 4439
[113-78-0] Deaminooxytocin, 2833
[113-92-8] Chlorpheniramine Maleate*, 2180
[113-98-4] Penicillin G Potassium,
7041
[114-04-5] Neuraminic Acid, 6392
[114-07-8] Erythromycin, 3626
[114-25-0] Biliverdine, 1236
[114-26-1] Propoxur, 7849
[114-42-1] Flavaspidic Acid, 4027
[114-43-2] Desaspidin, 2899
[114-80-7] Neostigmine Bromide*,
6380
[114-85-2] Bethanidine Sulfate*,
1208
[114-86-3] Phenformin, 7191
[114-90-9] Obidoxime Chloride,
6659
[114-91-0] Metyridine, 6084
[115-02-6] Azaserine, 916
[115-07-1] Propylene, 7862
[115-08-2] Thioformamide, 9262
[115-10-6] Methyl Ether, 5990
[115-11-7] Isobutylene, 5024
[115-17-3] Bromal, 1372
[115-19-5] 2-Methyl-3-butyn-2-ol,
5956
[115-20-8] 2,2,2-Trichloroethanol,
9551
[115-24-2] Sulfonmethane, 8939
[115-25-3] Octafluorocyclobutane,
6666
[115-26-4] Dimefox, 3191
[115-27-5] Chlorendic Anhydride,
2084
[115-29-7] Endosulfan, 3529
[115-31-1] Isobornyl Thiocyanoacetate, Technical, 5012
[115-32-2] Dicofol, 3075
[115-33-3] Oxyphenisatin Acetate,
6927
[115-37-7] Thebaine, 9202
[115-38-8] Mephobarbital, 5742
[115-39-9] Bromphenol Blue, 1434
[115-40-2] Bromcresol Purple, 1376
[115-43-5] Phenallymal, 7162
[115-44-6] Talbutal, 9010
[115-46-8] Azacyclonol, 909
[115-51-5] Ambutonium Bromide,
397
[115-53-7] Sinomenine, 8496
[115-56-0] Methallatal, 5855
[115-61-7] Tachysterol, 9002
[115-63-9] Hexocyclium Methyl
Sulfate, 4627
[115-67-3] Paramethadione, 6976
[115-68-4] Sulfadicramide, 8875
[115-69-5] 2-Amino-2-methyl-1,3-
propanediol, 460
[115-70-8] 2-Amino-2-ethyl-1,3-
propanediol, 451

[115-71-9] α-Santalol, 8321
[115-77-5] Pentaerythritol, 7062
[115-79-7] Ambenonium Chloride,
387
[115-86-6] Triphenyl Phosphate,
9656
[115-90-2] Fensulfothion, 3943
[115-93-5] Cythioate, 2791
[115-95-5] Linalyl Acetate, 5374
[116-06-3] Aldicarb, 216
[116-25-6] 1-(Hydroxymethyl)-5,5-
dimethylhydantoin, 4762
[116-26-7] Safranal, 8286
[116-28-9] Selagine, 8379
[116-29-0] Tetradifon, 9132
[116-30-3] Rhodoxanthin, 8196
[116-31-4] Retinal, 8165
[116-38-1] Edrophonium Chloride,
3487
[116-42-7] Sulfaproxyline, 8911
[116-43-8] Succinylsulfathiazole,
8850
[116-44-9] Sulfapyrazine, 8912
[116-45-0] Sulfabromomethazine,
8869
[116-49-4] Glycobiarsol, 4389
[116-56-3] 4-Pregnene-20,21-diol-
3,11-dione, 7735
[116-58-5] 4-Pregnene-11β,17α,-
20β,21-tetrol-3-one, 7736
[116-59-6] 4-Pregnene-17α,20β,21-
triol-3,11-dione, 7737
[116-63-2] 1-Amino-2-naphthol-4-
sulfonic Acid, 466
[117-02-2] Rubiadin, 8258
[117-10-2] Danthron, 2813
[117-12-4] Anthrarufin, 719
[117-26-0] Bulan®, 1470
[117-27-1] 2-Nitro-1,1-bis(p-
chlorophenyl)propane, 6515
[117-30-6] Dipiproverine, 3343
[117-34-0] Diphenylacetic Acid,
3316
[117-37-3] Anisindione, 698
[117-39-5] Quercetin, 8044
[117-51-1] Synhexyl, 8995
[117-52-2] Coumafuryl, 2557
[117-56-6] 1-Naphthol-4,8-disul-
fonic Acid, 6306
[117-68-0] 3'-Guanylic Acid, 4483
[117-73-7] Raunescine, 8130
[117-80-6] Dichlone, 3032
[117-81-7] Bis(2-ethylhexyl)
Phthalate, 1262
[117-89-5] Trifluoperazine, 9594
[117-92-0] Quinaldine Red, 8057
[118-00-3] Guanosine, 4480
[118-08-1] Hydrastine, 4688
[118-10-5] Cinchonine, 2289
[118-23-0] Bromodiphenhydramine,
1405
[118-32-1] 2-Naphthol-6,8-disul-
fonic Acid, 6308
[118-34-3] Syringin, 8997
[118-42-3] Hydroxychloroquine,
4748
[118-52-5] 1,3-Dichloro-5,5-di-
methylhydantoin, 3054
[118-55-8] Phenyl Salicylate, 7282
[118-56-9] Homosalate, 4660
[118-57-0] Acetaminosalol, 41
[118-58-1] Benzyl Salicylate, 1160
[118-61-6] Ethyl Salicylate, 3804
[118-68-3] Etryptamine Acetate*,
3848
[118-71-8] Maltol, 5594
[118-74-1] Hexachlorobenzene,
4600
[118-75-2] Chloranil, 2071
[118-78-5] Uramil, 9763

Denotes derivative of title compound

[118-79-6] 2,4,6-Tribromophenol, 9526
[118-91-2] o-Chlorobenzoic Acid, 2125
[118-92-3] o-Aminobenzoic Acid, 433
[118-96-7] 2,4,6-Trinitrotoluene, 9643
[119-13-1] δ-Tocopherol, 9419
[119-20-0] Rhizopterin, 8181
[119-24-4] Pteroic Acid, 7949
[119-26-6] 2,4-Dinitrophenylhydrazine, 3277
[119-28-8] 1,7-Cleve's Acid, 2349
[119-29-9] Ambucaine, 393
[119-36-8] Methyl Salicylate, 6038
[119-38-0] Isolan®, 5063
[119-41-5] Efloxate, 3490
[119-44-8] Xanthopterin, 9973
[119-48-2] Dimorpholamine, 3257
[119-53-9] Benzoin, 1103
[119-59-5] 4,4'-Sulfinyldianiline, 8927
[119-61-9] Benzophenone, 1108
[119-64-2] Tetralin®, 9152
[119-65-3] Isoquinoline, 5110
[119-79-9] 1,6-Cleve's Acid, 2348
[119-90-4] Dianisidine, 2970
[119-93-7] o-Tolidine, 9437
[119-96-0] Arsthinol, 840
[119-98-2] Tocol, 9416
[120-02-5] Thiocarbarsone, 9252
[120-05-8] Sulfuretin, 8951
[120-08-1] Scoparone, 8359
[120-12-7] Anthracene, 712
[120-14-9] Veratraldehyde, 9852
[120-18-3] 2-Naphthalenesulfonic Acid, 6296
[120-23-0] 2-Naphthoxyacetic Acid, 6317
[120-29-6] Tropine, 9693
[120-32-1] Clorophene, 2403
[120-34-3] N-Sulfanilyl-3,4-xylamide, 8907
[120-36-5] Dichlorprop, 3068
[120-46-7] Dibenzoylmethane, 2991
[120-47-8] Ethylparaben, 3792
[120-51-4] Benzyl Benzoate, 1141
[120-57-0] Piperonal, 7445
[120-58-1] Isosafrole, 5112
[120-62-7] Sulfoxide, 8946
[120-72-9] Indole, 4869
[120-73-0] 1H-Purine, 7959
[120-78-5] 2,2'-Dithiobis[benzothiazole], 3377
[120-80-9] Pyrocatechol, 8009
[120-82-1] 1,2,4-Trichlorobenzene, 9543
[120-83-2] 2,4-Dichlorophenol, 3061
[120-87-6] 9,10-Dihydroxystearic Acid, 3171
[120-89-8] Parabanic Acid, 6970
[120-92-3] Cyclopentanone, 2748
[120-93-4] 2-Imidazolidinone, 4830
[120-97-8] Dichlorphenamide, 3067
[121-19-7] Roxarsone, 8251
[121-25-5] Amprolium, 623
[121-30-2] Chloraminophenamide, 2067
[121-32-4] Ethyl Vanillin, 3815
[121-33-5] Vanillin, 9839
[121-34-6] Vanillic Acid, 9838
[121-40-4] Terreic Acid, 9104
[121-43-7] Trimethyl Borate, 9626
[121-44-8] Triethylamine, 9582
[121-47-1] Metanilic Acid, 5832
[121-54-0] Benzethonium Chloride, 1083
[121-55-1] Subathizone, 8832

[121-57-3] Sulfanilic Acid, 8901
[121-59-5] Carbarsone, 1788
[121-60-8] N-Acetylsulfanilyl Chloride, 99
[121-61-9] N⁴-Acetylsulfanilamide, 97
[121-62-0] N-Acetylsulfanilic Acid, 98
[121-66-4] 2-Amino-5-nitrothiazole, 469
[121-69-7] N,N-Dimethylaniline, 3223
[121-75-5] Malathion, 5582
[121-79-9] Propyl Gallate, 7872
[121-81-3] Nitromide, 6533
[121-82-4] Cyclonite, 2742
[121-91-5] Isophthalic Acid, 5083
[122-03-2] Cuminaldehyde, 2623
[122-04-3] 4-Nitrobenzoyl Chloride, 6511
[122-06-5] Stilbamidine, 8772
[122-09-8] Phentermine, 7232
[122-10-1] Bomyl®, 1333
[122-11-2] Sulfadimethoxine, 8876
[122-14-5] Fenitrothion, 3922
[122-15-6] Dimetan®, 3200
[122-16-7] Sulfanitran, 8908
[122-18-9] Cetalkonium Chloride, 2009
[122-25-8] 5,5'-Methylenedisalicylic Acid, 5984
[122-32-7] Triolein, 9644
[122-34-9] Simazine, 8485
[122-39-4] Diphenylamine, 3317
[122-42-9] Propham, 7828
[122-46-3] m-Cresyl Acetate, 2587
[122-48-5] Zingerone, 10072
[122-57-6] Benzylideneacetone, 1151
[122-59-8] Phenoxyacetic Acid, 7223
[122-62-3] Bis(2-ethylhexyl) Sebacate, 1263
[122-69-0] Cinnamyl Cinnamate, 2307
[122-78-1] Phenylacetaldehyde, 7236
[122-79-2] Phenyl Acetate, 7238
[122-80-5] p-Aminoacetanilide, 422
[122-87-2] N-(4-Hydroxyphenyl)-glycine, 4771
[122-99-6] 2-Phenoxyethanol, 7226
[123-02-4] Tridecylbenzene, 9575
[123-03-5] Cetylpyridinium Chloride, 2024
[123-08-0] p-Hydroxybenzaldehyde, 4741
[123-11-5] p-Anisaldehyde, 693
[123-19-3] Dipropyl Ketone, 3351
[123-23-9] Succinyl Peroxide, 8848
[123-29-5] Ethyl Pelargonate, 3793
[123-30-8] p-Aminophenol, 474
[123-31-9] Hydroquinone, 4738
[123-33-1] Maleic Hydrazide, 5587
[123-35-3] β-Myrcene, 6243
[123-38-6] Propionaldehyde, 7835
[123-42-2] Diacetone Alcohol, 2944
[123-43-3] Sulfoacetic Acid, 8932
[123-45-5] Sulfonyldiacetic Acid, 8940
[123-47-7] Prolonium Iodide, 7792
[123-51-3] Isopentyl Alcohol, 5081
[123-54-6] Acetylacetone, 75
[123-56-8] Succinimide, 8842
[123-62-6] Propionic Anhydride, 7838
[123-63-7] Paraldehyde, 6975
[123-66-0] Ethyl Caproate, 3735
[123-72-8] Butyraldehyde, 1591
[123-75-1] Pyrrolidine, 8026

[123-76-2] Levulinic Acid, 5352
[123-77-3] Azodicarbonamide, 932
[123-78-4] Sphingosine, 8703
[123-82-0] Tuaminoheptane, 9712
[123-84-2] 1-[(2-Aminoethyl)-amino]-2-propanol, 450
[123-86-4] n-Butyl Acetate, 1535
[123-90-0] Thiamorpholine, 9229
[123-91-1] Dioxane, 3294
[123-92-2] Isoamyl Acetate, 4993
[123-93-3] Thiodiglycolic Acid, 9260
[123-95-5] Butyl Stearate, 1589
[123-96-6] 2-Octanol, 6675
[123-99-9] Azelaic Acid, 921
[124-02-7] Diallylamine, 2951
[124-03-8] Cetyldimethylethylammonium Bromide, 2021
[124-04-9] Adipic Acid, 152
[124-07-2] Caprylic Acid, 1765
[124-09-4] 1,6-Hexanediamine, 4614
[124-13-0] Caprylic Aldehyde, 1766
[124-20-9] Spermidine, 8698
[124-23-2] Cetrimonium Stearate, 2019
[124-28-7] Dymanthine, 3455
[124-34-8] Mercamphamide Sodium*, 5758
[124-38-9] Carbon Dioxide, 1816
[124-40-3] Dimethylamine, 3217
[124-41-4] Sodium Methoxide, 8594
[124-42-5] Acetamidine Hydrochloride, 37
[124-43-6] Urea Hydrogen Peroxide, 9783
[124-58-3] Methanearsonic Acid, 5864
[124-58-3] Methanesulfonic Acid, 5865
[124-63-0] Methanesulfonyl Chloride, 5866
[124-65-2] Sodium Cacodylate, 8540
[124-68-5] 2-Amino-2-methyl-1-propanol, 461
[124-72-1] Teflurane, 9059
[124-76-5] Isoborneol, 5011
[124-80-1] Sabadine, 8281
[124-87-8] Picrotoxin, 7388
[124-90-3] Oxycodone Hydrochloride*, 6912
[124-92-5] Metopon Hydrochloride*, 6071
[124-94-7] Triamcinolone, 9511
[124-98-1] Cevine, 2026
[125-02-0] Prednisolone Sodium Phosphate, 7721
[125-03-1] Hydrocortamate Hydrochloride*, 4709
[125-04-2] Hydrocortisone 21-Sodium Succinate, 4713
[125-15-5] Vomicine, 9946
[125-20-2] Thymolphthalein, 9336
[125-23-5] Morphine Methylbromide, 6189
[125-24-6] Pseudomorphine, 7934
[125-25-7] Codeine Hydrobromide*, 2459
[125-27-9] Codeine Methyl Bromide, 2460
[125-28-0] Dihydrocodeine, 3154
[125-29-1] Hydrocodone, 4708
[125-31-5] Xylenol Blue, 9990
[125-33-7] Primidone, 7753
[125-40-6] Butabarbital*, 1495
[125-44-0] Vinbarbital Sodium, 9886
[125-46-2] Usnic Acid, 9806

* *Denotes derivative of title compound*

[125-51-9] Pipenzolate Bromide, 7428
[125-52-0] Oxyphencyclimine Hydrochloride*, 6926
[125-53-1] Oxyphencyclimine, 6926
[125-55-3] Narcobarbital, 6342
[125-60-0] Fenpiverinium Bromide, 3935
[125-65-5] Pleuromutilin, 7508
[125-84-8] Aminoglutethimide, 452
[125-85-9] Caramiphen Hydrochloride, 1777
[125-86-0] Caramiphen Ethanedisulfonate, 1776
[125-99-5] Tridihexethyl Iodide, 9577
[126-00-1] Diphenolic Acid, 3312
[126-02-3] Cycrimine Hydrochloride, 2763
[126-05-6] Streptobiosamine, 8782
[126-07-8] Griseofulvin, 4453
[126-11-4] Tris(hydroxymethyl)-nitromethane, 9667
[126-12-5] Anileridine Dihydrochloride*, 686
[126-14-7] Sucrose Octaacetate, 8856
[126-15-8] R-11, 8114
[126-17-0] Solasodine, 8665
[126-18-1] Smilagenin, 8508
[126-19-2] Sarsasapogenin, 8338
[126-22-7] Butonate, 1528
[126-27-2] Oxethazaine, 6888
[126-29-4] Violaxanthin, 9902
[126-30-7] Neopentyl Glycol, 6374
[126-31-8] Methiodal Sodium, 5894
[126-33-0] Sulfolane, 8934
[126-45-4] Silver Citrate, 8457
[126-49-8] Prephenic Acid, 7745
[126-52-3] Ethinamate, 3688
[126-72-7] Tris-BP, 9665
[126-73-8] Tributyl Phosphate, 9531
[126-81-8] 5,5-Dimethyl-1,3-cyclohexanedione, 3231
[126-93-2] Oxanamide, 6874
[126-96-5] Sodium Diacetate, 8555
[126-98-7] Methacrylonitrile, 5850
[127-00-4] sec-Propylene Chlorohydrin, 7864
[127-06-0] Acetoxime, 68
[127-07-1] Hydroxyurea, 4785
[127-08-2] Potassium Acetate, 7580
[127-09-3] Sodium Acetate, 8513
[127-17-3] Pyruvic Acid, 8032
[127-18-4] Tetrachloroethylene, 9126
[127-19-5] N,N-Dimethylacetamide, 3216
[127-22-0] Taraxerol, 9033
[127-25-3] Methyl Abietate, 5930
[127-27-5] Pimaric Acid, 7398
[127-29-7] Pseudoaconitine, 7924
[127-31-1] Fludrocortisone, 4059
[127-33-3] Demeclocycline, 2872
[127-35-5] Phenazocine, 7174
[127-39-9] Diisobutyl Sodium Sulfosuccinate, 3179
[127-40-2] Xanthophyll, 9972
[127-48-0] Trimethadione, 9620
[127-50-4] Mercamphamide, 5758
[127-52-6] Chloramine-B, 2065
[127-57-1] Sulfapyridine Sodium*, 8913
[127-60-6] Acediasulfone Sodium*, 17
[127-63-9] Diphenyl Sulfone, 3336
[127-65-1] Chloramine-T, 2066
[127-69-5] Sulfisoxazole, 8930
[127-71-9] Sulfabenzamide, 8868
[127-77-5] Sulfabenz, 8867

[127-79-7] Sulfamerazine, 8884
[127-81-1] Sulfanilamidomethanesulfonic Acid Triethanolamine Salt, 8899
[127-82-2] Zinc p-Phenolsulfonate, 10054
[127-83-3] Calcium Phenolsulfonate, 1697
[127-85-5] Sodium Arsanilate, 8521
[127-91-3] β-Pinene, 7415
[127-95-7] Potassium Binoxalate, 7588
[127-96-8] Potassium Tetroxalate, 7684
[128-00-7] Silver Lactate, 8463
[128-08-5] N-Bromosuccinimide, 1428
[128-09-6] N-Chlorosuccinimide, 2165
[128-12-1] Acetosulfone Sodium, 66
[128-13-2] Ursodiol, 9801
[128-19-8] 4-Pregnene-17α,20β,21-triol-3-one, 7738
[128-20-1] Pregnan-3α-ol-20-one, 7734
[128-23-4] 3,20-Pregnanedione, 7733
[128-27-8] Pyrocalciferol, 8007
[128-37-0] Butylated Hydroxytoluene, 1548
[128-45-0] Streptomycin B, 8787
[128-46-1] Dihydrostreptomycin, 3162
[128-49-4] Docusate Calcium, 3397
[128-53-0] N-Ethylmaleimide, 3778
[128-62-1] Noscapine, 6638
[128-68-7] Phenicin, 7194
[128-76-7] Laurotetanine, 5258
[128-80-3] Quinizarin Green SS, 8095
[128-87-0] Bis(4-amino-1-anthraquinonyl)amine, 1254
[129-00-0] Pyrene, 7977
[129-03-3] Cyproheptadine, 2779
[129-06-6] Warfarin Sodium*, 9950
[129-16-8] Merbromin, 5757
[129-17-9] Sulphan Blue, 8968
[129-20-4] Oxyphenbutazone, 6925
[129-24-8] Viridicatin, 9911
[129-46-4] Suramin Sodium, 8986
[129-49-7] Methysergide Maleate*, 6055
[129-51-1] Ergonovine Maleate*, 3600
[129-57-7] Diprotrizoate Sodium, 3353
[129-63-5] Acetrizoate Sodium, 72
[129-64-6] Carbic Anhydride, 1801
[129-65-7] Porphyrillic Acid, 7576
[129-66-8] 2,4,6-Trinitrobenzoic Acid, 9640
[129-74-8] Buclizine Dihydrochloride*, 1449
[129-77-1] Piperidolate Hydrochloride*, 7440
[129-79-3] 2,4,7-Trinitrofluorenone, 9641
[129-83-9] Phenampromid(e), 7166
[129-89-5] Sulfamipyrine, 8896
[130-01-8] Senecionine, 8402
[130-15-4] 1,4-Naphthoquinone, 6315
[130-16-5] Cloxyquin, 2416
[130-22-3] Sodium Alizarinesulfonate, 8508
[130-24-5] Vitamin K₅, 9936
[130-26-7] Iodochlorhydroxyquin, 4924

[130-37-0] Menadione Sodium Bisulfite, 5716
[130-40-5] Riboflavine Phosphate (Sodium), 8202
[130-61-0] Thioridazine Hydrochloride*, 9290
[130-73-4] Methestrol, 5888
[130-79-0] Dimestrol, 3198
[130-80-3] Diethylstilbestrol Dipropionate, 3119
[130-85-8] Pamoic Acid, 6955
[130-86-9] Protopine, 7912
[130-89-2] Quinine Hydrochloride, 8085
[130-90-5] Quinine Formate, 8081
[130-95-0] Quinine, 8075
[131-01-1] Deserpidine, 2901
[131-02-2] Reserpiline, 8148
[131-03-3] α-Yohimbine, 10013
[131-11-3] Dimethyl Phthalate, 3243
[131-12-4] Pimpinellin, 7406
[131-13-5] Menadiol Diphosphate (Tetrasodium Salt), 5712
[131-28-2] Narceine, 6341
[131-49-7] Meglumine Diatrizoate, 5689
[131-53-3] Dioxybenzone, 3298
[131-54-4] Benzophenone-6, 1109
[131-56-6] Benzoresorcinol, 1117
[131-57-7] Oxybenzone, 6907
[131-67-9] Phthalofyne, 7348
[131-69-1] Phthalylsulfacetamide, 7350
[131-73-7] Dipicrylamine, 3340
[131-74-8] Ammonium Picrate, 575
[131-79-3] Yellow OB, 10004
[131-89-5] 2-Cyclohexyl-4,6-dinitrophenol, 2739
[131-91-9] 1-Nitroso-2-naphthol, 6562
[131-99-7] Inosinic Acid, 4882
[132-13-8] α-Ribazole, 8200
[132-17-2] Benztropine Mesylate, 1135
[132-18-3] Diphenylpyraline Hydrochloride*, 3334
[132-20-7] Pheniramine Maleate*, 7198
[132-21-8] Brompheniramine (d-Form)*, 1433
[132-22-9] Chlorpheniramine, 2180
[132-35-4] Proxazole Citrate*, 7919
[132-45-6] Propivane, 7845
[132-49-0] Magnesium Acetylsalicylate, 5531
[132-57-0] Croceic Acid, 2591
[132-60-5] Cinchophen, 2290
[132-66-1] Naptalam, 6338
[132-69-4] Benzydamine Hydrochloride*, 1136
[132-86-5] Naphthoresorcinol, 6316
[132-89-8] Chlorthenoxazin(e), 2195
[132-92-3] Methicillin Sodium, 5890
[132-93-4] Phenethicillin Potassium, 7184
[132-98-9] Penicillin V Potassium*, 7046
[133-04-0] Asarinin, 848
[133-06-2] Captan, 1771
[133-07-3] Folpet, 4142
[133-09-5] Aminosalicylate Potassium*, 491
[133-11-9] Phenyl Aminosalicylate, 7243
[133-15-3] Aminosalicylate Calcium*, 491

[133-17-5] o-Iodohippurate Sodium, 4927
[133-26-6] Peucedanin, 7144
[133-32-4] Indolebutyric Acid, 4871
[133-36-8] Piperazine Tartrate, 7436
[133-37-9] DL-Tartaric Acid, 9038
[133-53-9] Dichloroxylenol, 3066
[133-58-4] Nitromersol, 6531
[133-60-8] Azosulfamide, 936
[133-65-3] Solasulfone, 8667
[133-67-5] Trichlormethiazide, 9537
[133-89-1] Uridine Diphosphate Glucose, 9794
[133-90-4] Chloramben, 2063
[133-91-5] 3,5-Diiodosalicylic Acid, 3175
[134-01-0] Peonidin, 7099
[134-03-2] Sodium Ascorbate, 8525
[134-04-3] Pelargonidin, 7014
[134-20-3] Methyl Anthranilate, 5942
[134-31-6] 8-Hydroxyquinoline Sulfate, 4779
[134-32-7] 1-Naphthylamine, 6318
[134-36-1] Erythromycin Propionate, 3631
[134-37-2] Amphenidone, 614
[134-49-6] Phenmetrazine, 7200
[134-50-9] Aminacrine Hydrochloride*, 418
[134-53-2] Amprotropine Phosphate, 624
[134-55-4] Phenyl Acetylsalicylate, 7241
[134-58-7] 8-Azaguanine, 912
[134-62-3] Deet, 2848
[134-71-4] Ephedrine DL-Form Hydrochloride*, 3561
[134-72-5] Ephedrine L-Form Sulfate*, 3561
[134-80-5] Diethylpropion Hydrochloride*, 3116
[134-81-6] Benzil, 1087
[134-96-3] Syringaldehyde, 8996
[135-07-9] Methyclothiazide, 5929
[135-09-1] Hydroflumethiazide, 4716
[135-19-3] 2-Naphthol, 6304
[135-20-6] Cupferron, 2625
[135-23-9] Methapyrilene Hydrochloride*, 5871
[135-31-9] Pyrrobutamine Phosphate*, 8023
[135-43-3] Lauroguadine, 5256
[135-44-4] Leucinocaine Mesylate, 5332
[135-48-8] Pentacene, 7057
[135-58-0] Mesulphen, 5821
[135-59-1] Thionine, 9276
[135-67-1] Phenoxazine, 7222
[135-68-2] 4-Amino-4'-chlorodiphenyl, 445
[135-97-7] Pseudotropine, 7937
[135-98-8] sec-Butylbenzene, 1550
[136-25-4] Erbon, 3587
[136-35-6] Diazoaminobenzene, 2981
[136-40-3] Phenazopyridine Hydrochloride, 7175
[136-44-7] Glyceryl p-Aminobenzoate, 4382
[136-46-9] Parethoxycaine Hydrochloride*, 6987
[136-47-0] Tetracaine Hydrochloride, 9123
[136-60-7] n-Butyl Benzoate, 1552
[136-69-6] Protokylol Hydrochloride*, 7911

[136-70-9] Protokylol, 7911
[136-77-6] 4-Hexylresorcinol, 4633
[136-78-7] Disul-sodium, 3372
[136-79-8] 5-Nitro-o-phenetidine, 6538
[136-82-3] Piperocaine, 7444
[136-91-4] Calcium 2-Ethylbutanoate, 1667
[136-95-8] 2-Aminobenzothiazole, 435
[136-96-9] Diamthazole Dihydrochloride, 2967
[137-00-8] 4-Methyl-5-thiazoleethanol, 6046
[137-05-3] Mecrylate, 5667
[137-08-6] Calcium Pantothenate, 1694
[137-16-6] Gardol®, 4267
[137-26-8] Thiram, 9304
[137-30-4] Ziram, 10075
[137-32-5] 2-Methyl-1-butanol, 5952
[137-40-6] Sodium Propionate, 8623
[137-42-8] Metham Sodium, 5860
[137-58-6] Lidocaine, 5359
[137-76-8] Cetotiamine, 2014
[137-86-0] Octotiamine, 6680
[138-14-7] Deferoxamine Mesylate*, 2850
[138-37-4] Mafenide Hydrochloride*, 5525
[138-39-6] Mafenide, 5525
[138-41-0] Carzenide, 1885
[138-52-3] Salicin, 8293
[138-53-4] Mandelonitrile Glucoside, 5602
[138-55-6] Picrocrocin, 7381
[138-56-7] Trimethobenzamide, 9623
[138-58-9] Bismuth Iodosubgallate, 1281
[138-59-0] Shikimic Acid, 8428
[138-61-4] Nordefrin Hydrochloride, 6609
[138-84-1] Potassium p-Aminobenzoate, 7582
[138-86-3] Limonene, 5371
[138-89-6] p-Nitroso-N,N-dimethylaniline, 6559
[138-92-1] Betazole Dihydrochloride*, 1205
[139-02-6] Sodium Phenoxide, 8611
[139-06-0] Calcium Cyclamate, 1665
[139-10-6] Amphetamine Phosphate*, 616
[139-13-9] Nitrilotriacetic Acid, 6499
[139-33-3] Edetate Disodium, 3481
[139-40-2] Propazine, 7822
[139-42-4] Cerous Oxalate, 1995
[139-62-8] Cyclomethycaine, 2741
[139-66-2] Phenyl Sulfide, 7285
[139-68-4] Phenatine, 7172
[139-85-5] Protocatechualdehyde, 7908
[139-88-8] Sodium Tetradecyl Sulfate, 8645
[139-89-9] Versen-Ol®, 9871
[139-91-3] Furaltadone, 4205
[139-93-5] Arsphenamine, 839
[139-94-6] Nithiazide, 6484
[140-11-4] Benzyl Acetate, 1137
[140-22-7] sym-Diphenylcarbazide, 3321
[140-28-3] Benzathine, 1072
[140-29-4] Benzyl Cyanide, 1145
[140-40-9] Aminitrozole, 421
[140-55-6] Pseudoconhydrine, 7928
[140-57-8] Aramite®, 794

[140-59-0] Stilbamidine Isethionate*, 8772
[140-63-6] Propamidine Isethionate*, 7808
[140-64-7] Pentamidine Isethionate*, 7071
[140-65-8] Pramoxine, 7707
[140-67-0] Estragole, 3657
[140-80-7] Novoldiamine, 6642
[140-87-4] Cyacetacide, 2688
[140-88-5] Ethyl Acrylate, 3715
[140-89-6] Potassium Xanthogenate, 7697
[140-93-2] Sodium Isopropyl Xanthate, 8585
[140-95-4] Oxymethurea, 6921
[140-99-8] Calcium Succinate, 1712
[141-00-4] Cadmium Succinate, 1626
[141-01-5] Ferrous Fumarate, 3995
[141-03-7] Dibutyl Succinate, 3023
[141-05-9] Diethyl Maleate, 3113
[141-10-6] Pseudoionone, 7931
[141-20-8] Diethylene Glycol Monolaurate, 3110
[141-22-0] Ricinoleic Acid, 8213
[141-26-4] Rhodinal, 8185
[141-32-2] n-Butyl Acrylate, 1539
[141-43-5] Ethanolamine, 3681
[141-52-6] Sodium Ethoxide, 8560
[141-53-7] Sodium Formate, 8568
[141-62-8] Decamethyltetrasiloxane, 2843
[141-63-9] Dodecamethylpentasiloxane, 3401
[141-66-2] Dicrotophos, 3077
[141-75-3] n-Butyryl Chloride, 1598
[141-78-6] Ethyl Acetate, 3713
[141-79-7] Mesityl Oxide, 5811
[141-82-2] Malonic Acid, 5591
[141-84-4] Rhodanine, 8184
[141-90-2] 2-Thiouracil, 9298
[141-94-6] Hexetidine, 4624
[141-97-9] Ethyl Acetoacetate, 3714
[142-03-0] Aluminum Diacetate, 339
[142-10-9] Glyceraldehyde 3-Phosphate, 4377
[142-17-6] Calcium Oleate, 1690
[142-47-2] Monosodium Glutamate, 6165
[142-59-6] Nabam, 6256
[142-61-0] Caproyl Chloride, 1763
[142-62-1] n-Caproic Acid, 1760
[142-68-7] Tetrahydropyran, 9149
[142-71-2] Cupric Acetate, 2627
[142-72-3] Magnesium Acetate, 5530
[142-73-4] Iminodiacetic Acid, 4832
[142-82-5] n-Heptane, 4580
[142-84-7] n-Dipropylamine, 3350
[142-88-1] Piperazine Adipate, 7432
[142-96-1] n-Butyl Ether, 1568
[143-07-7] Lauric Acid, 5254
[143-08-8] n-Nonyl Alcohol, 6598
[143-13-5] n-Nonyl Acetate, 6597
[143-15-7] Lauryl Bromide, 5260
[143-18-0] Potassium Oleate, 7637
[143-19-1] Sodium Oleate*, 6788
[143-24-8] Tetraglyme, 9141
[143-28-2] Oleyl Alcohol, 6791
[143-33-9] Sodium Cyanide, 8553
[143-50-0] Chlordecone, 2081
[143-52-2] Metopon, 6071
[143-62-4] Digitoxigenin, 3145
[143-66-8] Sodium Tetraphenylborate, 8646

Denotes derivative of title compound

[143-67-9] Vinblastine Sulfate*, 9887
[143-74-8] Phenolsulfonphthalein, 7213
[143-81-7] Butabarbital Sodium, 1495
[143-82-8] Probarbital Sodium*, 7759
[144-00-3] Hexethal Sodium, 4623
[144-02-5] Barbital Sodium*, 972
[144-12-7] Tiemonium Iodide, 9363
[144-14-9] Anileridine, 686
[144-23-0] Magnesium Citrate, Dibasic, 5542
[144-29-6] Piperazine Citrate, 7433
[144-33-2] Sodium Citrate, Acid, 8550
[144-49-0] Fluoroacetic Acid, 4096
[144-55-8] Sodium Bicarbonate, 8528
[144-62-7] Oxalic Acid, 6865
[144-68-3] Zeaxanthin, 10019
[144-75-2] Sulfoxone Sodium, 8947
[144-80-9] Sulfacetamide, 8870
[144-82-1] Sulfamethizole, 8887
[144-83-2] Sulfapyridine, 8913
[144-87-6] Glycarsamide, 4375
[145-12-0] Oxymesterone, 6918
[145-13-1] Pregnenolone, 7739
[145-41-5] Dehydrocholic Acid Sodium*, 2858
[145-54-0] Propyromazine, 7885
[145-73-3] Endothall, 3530
[145-94-8] Chlorindanol, 2094
[146-06-5] Quinine Carbonate, 8077
[146-14-5] Flavine-Adenine Dinucleotide, 4028
[146-22-5] Nitrazepam, 6491
[146-37-2] Laurolinium Acetate, 5257
[146-48-5] Yohimbine, 10011
[146-54-3] Triflupromazine, 9597
[146-56-5] Fluphenazine Dihydrochloride*, 4116
[146-80-5] Xanthosine, 9974
[146-84-9] Silver Picrate, 8473
[146-90-7] Cinnabarine, 2297
[147-14-8] Copper Phthalocyanine, 2515
[147-20-6] Diphenylpyraline, 3334
[147-24-0] Diphenhydramine Hydrochloride*, 3308
[147-27-3] Dimoxyline, 3258
[147-61-5] Pentoxyl, 7093
[147-71-7] D-Tartaric Acid, 9037
[147-73-9] meso-Tartaric Acid, 9040
[147-81-9] Arabinose, 788
[147-82-0] 2,4,6-Tribromoaniline, 9522
[147-85-3] Proline, 7790
[147-90-0] Morpholine Salicylate, 6195
[147-93-3] Thiosalicylic Acid, 9291
[147-94-4] Cytarabine, 2790
[148-01-6] Dinitolmide, 3262
[148-03-8] β-Tocopherol, 9417
[148-18-5] Ditiocarb Sodium, 3384
[148-24-3] 8-Hydroxyquinoline, 4778
[148-25-4] Chromotropic Acid, 2243
[148-51-6] 4-Desoxypyridoxine Hydrochloride, 2913
[148-56-1] Flumethiazide, 4067
[148-64-1] Chlorothen Citrate*, 2168
[148-65-2] Chlorothen, 2168
[148-72-1] Pilocarpine Nitrate*, 7395

[148-75-4] 2-Naphthol-3,6-disulfonic Acid, 6307
[148-79-8] Thiabendazole, 9217
[148-82-3] Melphalan, 5708
[148-83-4] Ostruthin, 6850
[149-15-5] Butacaine Sulfate*, 1496
[149-16-6] Butacaine, 1496
[149-29-1] Patulin, 7002
[149-30-4] 2-Mercaptobenzothiazole, 5759
[149-32-6] Erythritol, 3620
[149-44-0] Sodium Formaldehydesulfoxylate, 8567
[149-45-1] Tiron®, 9392
[149-64-4] N-Butylscopolammonium Bromide, 1588
[149-90-6] Acetylleucine Monoethanolamine, 89
[149-91-7] Gallic Acid, 4251
[150-13-0] p-Aminobenzoic Acid, 434
[150-25-4] Bicine, 1221
[150-38-9] Edetate Trisodium, 3483
[150-59-4] Alverine, 377
[150-60-7] Dibenzyl Disulfide, 2995
[150-61-8] 1,2-Dianilinoethane, 2969
[150-68-5] Monuron, 6169
[150-69-6] Dulcin, 3448
[150-86-7] Phytol, 7362
[150-90-3] Sodium Succinate, 8635
[150-97-0] Mevalonic Acid, 6087
[151-06-4] Chlorphentermine Hydrochloride*, 2183
[151-07-5] Chloroarsenol, 2119
[151-18-8] 3-Aminopropionitrile, 481
[151-21-3] Sodium Lauryl Sulfate, 8587
[151-50-8] Potassium Cyanide, 7607
[151-56-4] Ethylenimine, 3760
[151-67-7] Halothane, 4517
[151-73-5] Betamethasone Sodium Phosphate*, 1202
[152-02-3] Levallorphan, 5342
[152-11-4] Verapamil Hydrochloride*, 9851
[152-16-9] Schradan, 8351
[152-43-2] Quinestrol, 8066
[152-47-6] Sulfalene, 8881
[152-58-9] 11-Desoxy-17-hydroxycorticosterone, 2912
[152-62-5] Dydrogesterone, 3454
[152-72-7] Acenocoumarol, 24
[152-84-1] Ruberythric Acid, 8257
[152-93-2] Vicine, 9880
[152-95-4] Sophoricoside, 8675
[152-97-6] Fluocortolone, 4079
[153-00-4] Methenolone, 5887
[153-18-4] Rutin, 8276
[153-61-7] Cephalothin, 1978
[153-87-7] Oxypertine, 6924
[154-17-6] 2-Deoxy-D-glucose, 2886
[154-21-2] Lincomycin, 5378
[154-23-4] Catechin, 1908
[154-41-6] Phenylpropanolamine Hydrochloride, 7279
[154-42-7] Thioguanine, 9266
[154-61-0] Primulaverin, 7755
[154-68-7] Antazoline Phosphate*, 709
[154-69-8] Tripelennamine Hydrochloride*, 9651
[154-82-5] Simetride, 8487
[154-87-0] Cocarboxylase, 2451
[154-93-8] Carmustine, 1852
[154-97-2] Pralidoxime Mesylate*, 7705

[155-09-9] Tranylcypromine, 9491
[155-38-4] Noformicin, 6590
[155-41-9] Methscopolamine Bromide, 5927
[155-54-4] Hydroorotic Acid, 4735
[155-57-7] Vicianin, 9878
[155-58-8] Rhapontin, 8173
[156-08-1] Benzphetamine, 1130
[156-10-5] p-Nitrosodiphenylamine, 6560
[156-43-4] p-Phenetidine, 7188
[156-51-4] Phenelzine Sulfate*, 7181
[156-56-9] Hypoglycine A, 4801
[156-62-7] Calcium Cyanamide, 1662
[157-03-9] 6-Diazo-5-oxo-L-norleucine, 2984
[192-97-2] Benzo[e]pyrene, 1114
[195-19-7] 3,4-Benzphenanthrene, 1129
[198-55-0] Perylene, 7137
[213-46-7] Picene, 7368
[217-59-4] Triphenylene, 9654
[218-01-9] Chrysene, 2259
[222-93-5] 2,3:6,7-Dibenzphenanthrene, 2992
[244-69-9] γ-Carboline, 1812
[253-14-5] Twistane, 9736
[253-52-1] Phthalazine, 7344
[253-66-7] Cinnoline, 2309
[253-82-7] Quinazoline, 8062
[254-04-6] 1,2-Benzopyran, 1112
[260-94-6] Acridine, 117
[261-31-4] Thioxanthene, 9300
[271-44-3] 1H-Indazole, 4848
[271-89-6] Benzofuran, 1098
[275-51-4] Azulene, 939
[277-10-1] Cubane, 2614
[280-57-9] Triethylenediamine, 9584
[281-23-2] Adamantane, 140
[286-08-8] Norcarane, 6605
[287-23-0] Cyclobutane, 2720
[287-92-3] Cyclopentane, 2746
[288-13-1] Pyrazole, 7974
[288-32-4] Imidazole, 4828
[288-47-1] Thiazole, 9235
[288-88-0] 1H-1,2,4-Triazole, 9519
[288-99-3] 1,3,4-Oxadiazole, 6859
[289-80-5] Pyridazine, 7982
[289-95-2] Pyrimidine, 7998
[290-37-9] Pyrazine, 7971
[290-57-3] Parathiazine, 6982
[292-46-6] Lenthionine, 5321
[297-76-7] Ethynodiol Diacetate*, 3816
[297-78-9] Isobenzan, 5010
[297-97-2] Thionazin, 9275
[298-00-0] Methyl Parathion, 6022
[298-02-2] Phorate, 7305
[298-04-4] Disulfoton, 3371
[298-12-4] Glyoxylic Acid, 4407
[298-14-6] Potassium Bicarbonate, 7586
[298-45-3] Bulbocapnine, 1471
[298-46-4] Carbamazepine, 1783
[298-55-5] Clocinizine, 2366
[298-57-7] Cinnarizine, 2308
[298-59-9] Methylphenidate Hydrochloride*, 6025
[298-81-7] Methoxsalen, 5911
[298-95-3] Neotetrazolium Chloride, 6381
[298-96-4] Triphenyltetrazolium Chloride, 9658
[299-11-6] N-Methylphenazonium Methosulfate, 6024
[299-20-7] Etamycin, 3664
[299-27-4] Potassium Gluconate, 7615

[299-28-5] Calcium Gluconate, 1672
[299-29-6] Ferrous Gluconate, 3996
[299-35-4] Thiothiamine, 9296
[299-45-6] Hymecromone O,O-Diethyl Phosphorothioate, 4793
[299-61-6] Ganglefene, 4263
[299-84-3] Ronnel, 8239
[299-85-4] DMPA, 3394
[299-86-5] Crufomate, 2607
[299-89-8] Acetiamine, 46
[300-08-3] Arecoline Hydrobromide*, 802
[300-37-8] Iodopyracet, 4933
[300-38-9] 3,5-Dibromo-L-tyrosine, 3014
[300-48-1] 3-O-Methyldopa, 5975
[300-54-9] Muscarine, 6219
[300-62-9] Amphetamine, 616
[300-76-5] Naled, 6272
[300-85-6] β-Hydroxybutyric Acid, 4746
[301-04-2] Lead Acetate, 5268
[301-11-1] Lethane® 60, 5328
[301-16-6] Scoparin, 8357
[301-19-9] Robinin, 8230
[301-21-3] Laudanidine, 5247
[302-01-2] Hydrazine, 4691
[302-17-0] Chloral Hydrate, 2061
[302-22-7] Chlormadinone Acetate, 2102
[302-27-2] Aconitine, 113
[302-40-9] Benactyzine, 1037
[302-41-0] Piritramide, 7470
[302-49-8] Uredepa, 9787
[302-66-9] Meparfynol Carbamate, 5732
[302-70-5] Mechlorethamine Oxide Hydrochloride, 5656
[302-72-7] Alanine (DL-Form)*, 195
[302-79-4] Retinoic Acid, 8167
[302-91-0] Allopregnane-3α,11β,-17α,21-tetrol-20- one, 267
[303-14-0] Isopromethazine, 5088
[303-25-3] Cyclizine Hydrochloride*, 2716
[303-34-4] Lasiocarpine, 5245
[303-40-2] Fluocortolone Caproate*, 4079
[303-42-4] Methenolone Enanthate*, 5887
[303-45-7] Gossypol, 4435
[303-49-1] Clomipramine, 2385
[303-53-7] Cyclobenzaprine, 2719
[303-69-5] Prothipendyl, 7902
[303-81-1] Novobiocin, 6641
[304-20-1] Hydralazine Hydrochloride*, 4682
[304-21-2] Harmaline, 4528
[304-55-2] Succimer, 8836
[304-59-6] Potassium Sodium Tartrate, 7660
[304-84-7] Ethamivan, 3673
[305-01-1] Esculetin, 3645
[305-03-3] Chlorambucil, 2064
[305-13-5] Acetamidoeugenol, 39
[305-80-6] p-Diazobenzenesulfonic Acid, 2982
[305-84-0] Carnosine, 1857
[305-85-1] Disophenol, 3365
[305-96-4] α-Methyl-m-tyrosine, 6050
[305-97-5] Anthiolimine, 711
[306-03-6] Hyoscyamine Hydrobromide*, 4795
[306-07-0] Pargyline Hydrochloride*, 6988
[306-08-1] Homovanillic Acid, 4662

[306-19-4] Pivalylbenzhydrazine, 7483
[306-21-8] Hydroxyamphetamine Hydrobromide*, 4740
[306-41-2] Hexacarbacholine Bromide, 4599
[306-44-5] Isonitrosoacetone, 5076
[306-52-5] Triclofos, 9571
[306-53-6] Azamethonium Bromide, 913
[306-60-5] Agmatine, 176
[306-61-6] Magnesium Thiocyanate, 5575
[309-00-2] Aldrin, 219
[309-29-5] Doxapram, 3421
[309-43-3] Secobarbital Sodium, 8374
[311-09-1] Benzoquinonium Chloride, 1116
[311-45-5] Paraoxon, 6979
[312-10-7] Penicillamine Disulfide, 7031
[313-04-2] Desmosterol, 2905
[313-05-3] Azacosterol, 908
[313-06-4] Estradiol 17β-Cypionate, 3656
[313-67-7] Aristolochic Acid, 810
[314-13-6] Evan's Blue, 3863
[314-35-2] Etamiphyllin, 3663
[314-40-9] Bromacil, 1370
[314-42-1] Isocil, 5042
[314-50-1] Orotidine, 6829
[315-18-4] Mexacarbate, 6090
[315-22-0] Monocrotaline, 6160
[315-30-0] Allopurinol, 278
[315-37-7] Testosterone Enanthate, 9112
[315-72-0] Opipramol, 6808
[315-80-0] Dibenzepin Hydrochloride*, 2990
[316-81-4] Thioproperazine, 9287
[317-34-0] Aminophylline, 477
[317-52-2] Hexafluorenium Bromide, 4606
[318-23-0] Imolamine, 4838
[318-98-9] Propranolol Hydrochloride*, 7852
[319-89-1] Tetroquinone, 9177
[320-67-2] Azacitidine, 907
[321-54-0] Coroxon, 2531
[321-55-1] Haloxon, 4519
[321-64-2] Tacrine, 9003
[322-35-0] Benserazide, 1059
[322-79-2] Triflusal, 9600
[326-43-2] Phenyramidol Hydrochloride*, 7292
[327-57-1] Norleucine, 6624
[327-97-9] Chlorogenic Acid, 2142
[328-42-7] Oxalacetic Acid, 6863
[328-50-7] α-Ketoglutaric Acid, 5182
[329-65-7] Epinephrine DL-Form*, 3569
[329-71-5] 2,5-Dinitrophenol, 3275
[330-54-1] Diuron, 3388
[330-55-2] Linuron, 5387
[330-95-0] Nicarbazin, 6402
[331-39-5] Caffeic Acid, 1634
[333-20-0] Potassium Thiocyanate, 7687
[333-36-8] Flurothyl, 4126
[333-41-5] Diazinon®, 2978
[334-20-3] Queen Substance, 8042
[334-48-5] n-Capric Acid, 1759
[334-88-3] Diazomethane, 2983
[337-47-3] Thiamylal Sodium*, 9231
[338-95-4] Isoflupredone, 5058
[338-98-7] Isoflupredone Acetate*, 5058
[339-43-5] Carbutamide, 1839

[339-44-6] Glymidine, 4403
[340-56-7] Methaqualone Hydrochloride*, 5872
[340-57-8] Mecloqualone, 5661
[341-00-4] Etifelmin, 3820
[341-70-8] Diethazine Hydrochloride*, 3098
[341-88-8] HQNO, 4670
[342-10-9] Kallidin, 5158
[343-65-7] Kynurenine, 5207
[345-78-8] Pseudoephedrine Hydrochloride*, 3561
[346-18-9] Polythiazide, 7561
[350-03-8] Methyl Pyridyl Ketone, 6034
[350-12-9] Sulbentine, 8865
[352-21-6] 4-Amino-3-hydroxybutyric Acid, 456
[352-93-2] Ethyl Sulfide, 3809
[352-97-6] Glycocyamine, 4391
[353-50-4] Carbonyl Fluoride, 1826
[355-80-6] 2,2,3,3,4,4,5,5-Octafluoro-1-pentanol, 6667
[356-12-7] Fluocinonide, 4077
[357-07-3] Oxymorphone Hydrochloride*, 6922
[357-08-4] Naloxone Hydrochloride*, 6277
[357-56-2] Dextromoramide, 2933
[357-57-3] Brucine, 1443
[357-66-4] Spirilene, 8709
[357-67-5] Phenetharbital, 7183
[357-70-0] Galanthamine, 4245
[358-52-1] Hexapropymate, 4616
[359-83-1] Pentazocine, 7078
[360-66-7] Androisoxazole, 667
[360-68-9] Coprosterol, 2519
[360-70-3] Nandrolone Decanoate, 6281
[361-37-5] Methysergid(e), 6055
[362-29-8] Propiomazine, 7834
[362-74-3] Bucladesine, 1448
[363-20-2] Tricetamide, 9535
[363-24-6] Prostaglandin E₂, 7893
[364-62-5] Metoclopramide, 6063
[364-98-7] Diazoxide, 2986
[365-26-4] p-Hydroxyephedrine, 4754
[366-18-7] α,α'-Dipyridyl, 3355
[366-70-1] Procarbazine Hydrochloride*, 7764
[367-25-9] 2,4-Difluoroaniline, 3131
[367-47-5] Glyceraldehyde, 4376
[367-51-1] Sodium Thioglycolate, 8650
[369-77-7] Cloflucarban, 2376
[370-14-9] Pholedrine, 7304
[371-40-4] p-Fluoroaniline, 4098
[371-86-8] Mipafox, 6124
[372-09-8] Cyanoacetic Acid, 2696
[372-66-7] Heptaminol, 4577
[372-75-8] Citrulline, 2333
[373-02-4] Nickel Acetate, 6407
[378-44-9] Betamethasone, 1202
[379-79-3] Ergotamine Tartrate*, 3609
[382-45-6] Adrenosterone, 165
[382-67-2] Desoximetasone, 2910
[382-82-1] Dicolinium Iodide, 3076
[386-17-4] Iodophthalein*, 4931
[388-51-2] Metofenazate, 6065
[389-08-2] Nalidixic Acid, 6273
[390-64-7] Prenylamine, 7744
[395-28-8] Isoxsuprine, 5128
[396-01-0] Triamterene, 9515
[398-23-2] 4,4'-Difluorodiphenyl, 3133
[402-61-9] 5-Methylpyrazole-3-carboxylic Acid, 6033

Denotes derivative of title compound

[404-82-0] Fenfluramine Hydrochloride*, 3920
[404-86-4] Capsaicin, 1767
[405-22-1] Nidroxyzone, 6439
[405-50-5] p-Fluorophenylacetic
Acid, 4105
[406-90-6] Fluroxene, 4127
[407-41-0] Phosphoserine, 7338
[409-21-2] Silicon Carbide, 8439
[420-04-2] Cyanamide, 2691
[420-05-3] Cyanic Acid, 2693
[421-20-5] Methyl Fluorosulfonate,
5993
[426-13-1] Fluorometholone, 4104
[427-00-9] Desomorphine, 2907
[427-01-0] Geissospermine, 4275
[427-28-1] Gentrogenin, 4293
[427-51-0] Cyproterone Acetate*,
2781
[428-07-9] Levomepate, 5345
[431-03-8] Diacetyl, 2946
[432-60-0] Allylestrenol, 289
[432-68-8] Echinenone, 3469
[434-03-7] Ethisterone, 3696
[434-05-9] Methenolone Acetate*,
5887
[434-07-1] Oxymetholone, 6920
[434-13-9] Lithocholic Acid, 5423
[434-16-2] 7-Dehydrocholesterol,
2857
[434-22-0] Nandrolone, 6280
[434-43-5] Phenpentermine, 7228
[435-97-2] Phenprocoumon, 7230
[436-05-5] Curine, 2682
[437-38-7] Fentanyl, 3944
[437-50-3] Gentisin, 4291
[437-74-1] Xanthinol Niacinate,
9969
[438-08-4] Etiocholanic Acid, 3824
[438-22-2] Androstane, 668
[438-23-3] Etiocholane, 3823
[438-41-5] Chlordiazepoxide Hydrochloride*, 2082
[438-60-8] Protriptyline, 7917
[439-14-5] Diazepam, 2977
[439-89-4] Gentianine, 4286
[440-17-5] Trifluoperazine Dihydrochloride*, 9594
[440-58-4] Iodamide, 4904
[441-38-3] Benzoin Oxime, 1104
[441-61-2] Ethylmethylthiambutene, 3784
[442-16-0] Ethacridine, 3668
[442-33-1] Cuspareine, 2686
[442-51-3] Harmine, 4531
[442-52-4] Clemizole, 2346
[443-48-1] Metronidazole, 6079
[444-27-9] Timonacic, 9375
[445-30-7] Homarine, 4648
[446-71-9] Homoeriodictyol, 4656
[446-72-0] Genistein, 4281
[446-86-6] Azathioprine, 918
[446-95-7] Genisteine-Alkaloid,
4282
[447-41-6] Nylidrin, 6654
[451-13-8] Homogentisic Acid,
4658
[451-71-8] Glyhexamide, 4402
[452-35-7] Ethoxzolamide, 3711
[454-14-8] Cuscohygrine, 2685
[456-22-4] p-Fluorobenzoic Acid,
4100
[456-59-7] Cyclandelate, 2708
[457-60-3] Neoarsphenamine, 6361
[457-87-4] N-Ethylamphetamine,
3720
[458-24-2] Fenfluramine, 3920
[458-35-5] Coniferyl Alcohol, 2499
[458-37-7] Curcumin, 2681
[458-88-8] Coniine, 2500
[459-67-6] Hydnocarpic Acid, 4679

[459-86-9] Mitoguazone, 6131
[460-19-5] Cyanogen, 2698
[461-06-3] Carnitine, 1856
[461-58-5] Dicyanodiamide, 3082
[461-72-3] Hydantoin, 4678
[461-78-9] Chlorphentermine, 2183
[461-98-3] Kyanmethin, 5205
[462-02-2] Cyamelide, 2689
[462-06-6] Fluorobenzene, 4099
[462-08-8] β-Aminopyridine, 487
[462-10-2] Homocystine, 4655
[462-94-2] Cadaverine, 1608
[463-40-1] Linolenic Acid, 5383
[463-51-4] Ketene, 5177
[463-56-9] Thiocyanic Acid, 9257
[463-72-9] Carbamyl Chloride,
1785
[463-77-4] Carbamic Acid, 1784
[463-78-5] Orthoformic Acid, 6837
[463-82-1] Neopentane, 6372
[463-88-7] Neurine, 6393
[464-41-5] Bornyl Chloride, 1341
[464-72-2] Benzopinacol, 1110
[464-73-3] Isatide, 4984
[464-81-3] Bufotoxin, 1468
[464-85-7] Quinamine, 8060
[464-86-8] Conquinamine, 2505
[465-11-2] Gamabufotalin, 4258
[465-16-7] Oleandrin, 6786
[465-19-0] Bufogenin B, 1464
[465-21-4] Bufalin, 1458
[465-22-5] Scillarenin, 8355
[465-28-1] Carotol, 1863
[465-39-4] Resibufogenin, 8150
[465-42-9] Capsanthin, 1768
[465-53-2] Cyclopregnol, 2754
[465-61-2] Caldariomycin, 1721
[465-65-6] Naloxone, 6277
[465-74-7] Quinovic Acid, 8104
[465-92-9] Marrubiin, 5635
[465-99-6] Hederagenin, 4540
[466-06-8] Proscillaridin, 7889
[466-07-9] Neriifolin, 6386
[466-14-8] Ibrotamide, 4809
[466-34-2] Picrolichenic Acid, 7382
[466-40-0] Isomethadone, 5069
[466-43-3] Atisine, 881
[466-49-9] Aspidospermine, 872
[466-61-5] Lycopodine, 5494
[466-80-8] α-Erythroidine, 3624
[466-81-9] β-Erythroidine, 3625
[466-85-3] Wieland-Gumlich Aldehyde, 9957
[466-90-0] Dihydrocodeinone Enol
Acetate, 3155
[466-96-6] Pseudocodeine, 7927
[466-97-7] Normorphine, 6630
[466-99-9] Hydromorphone, 4733
[467-04-9] Oripavine, 6821
[467-14-1] Neopine, 6376
[467-15-2] Norcodeine, 6607
[467-18-5] Myrophine, 6250
[467-22-1] Carbiphene Hydrochloride*, 1805
[467-36-7] Thialbarbital, 9219
[467-51-6] Samandarine, 8313
[467-53-8] Diginin, 3139
[467-55-0] Hecogenin, 4537
[467-60-7] Pipradrol, 7455
[467-83-4] Dipipanone, 3342
[467-84-5] Phenadoxone, 7160
[467-85-6] Normethadone, 6628
[467-86-7] Dioxaphetyl Butyrate,
3295
[467-98-1] Thebainone, 9203
[468-07-5] Phenomorphan, 7215
[468-10-0] Morphinan, 6185
[468-28-0] Lupulon, 5482
[468-45-1] Isorubijervine, 5111
[468-56-4] Hydroxypethidine, 4769
[468-61-1] Oxeladin, 6885

[468-76-8] Cassaine, 1896
[468-89-3] Nupharidine, 6650
[469-21-6] Doxylamine, 3430
[469-32-9] Hamamelitannin, 4522
[469-36-3] γ-Oryzanol (B)*, 6841
[469-59-0] Jervine, 5145
[469-62-5] Propoxyphene, 7851
[469-79-4] Ketobemidone, 5180
[469-81-8] Morpheridine, 6184
[469-82-9] Carbetidine, 1799
[469-83-0] Cafestol, 1633
[470-40-6] Thujopsene, 9327
[470-43-9] Promoxolane, 7801
[470-82-6] Eucalyptol, 3851
[470-90-6] Chlorfenvinphos, 2087
[471-09-0] Felinine, 3894
[471-25-0] Propiolic Acid, 7833
[471-29-4] Methylguanidine, 5999
[471-34-1] Calcium Carbonate,
1657
[471-35-2] Cacodyl, 1602
[471-46-5] Oxamide, 6871
[471-47-6] Oxamic Acid, 6870
[471-53-4] Enoxolone, 3543
[471-71-6] Cassamine, 1897
[471-80-7] Steviol, 8765
[471-87-4] Stachydrine, 8729
[471-95-4] Bufotalin, 1466
[472-11-7] Ruscogenin, 8270
[472-61-7] Astaxanthin, 877
[472-70-8] Cryptoxanthin, 2612
[472-87-7] 3-Dehydroretinal, 2862
[472-92-4] δ-Carotene, 1862
[472-93-5] γ-Carotene, 1861
[473-06-3] Chrysanthenone, 2256
[473-30-3] Thiazolsulfone, 9237
[473-34-7] Dichloramine T, 3034
[473-41-6] Tolbutamide Sodium*,
9432
[473-42-7] p-Nitrosulfathiazole,
6566
[473-81-4] Glyceric Acid, 4378
[473-90-5] Mesoxalic Acid, 5814
[473-98-3] Betulin, 1212
[474-00-0] Eburnamonine, 3464
[474-07-7] Brazilin, 1362
[474-25-9] Chenodiol, 2044
[474-40-8] α₁-Sitosterol, 8499
[474-43-1] Funtumine, 4204
[474-62-4] Campesterol, 1735
[474-69-1] Lumisterol, 5471
[474-70-4] Isopyrocalciferol, 5107
[474-73-7] 24-Hydroxycholesterol,
4750
[474-77-1] Epicholesterol, 3565
[474-86-2] Equilin, 3582
[475-20-7] Longifolene, 5446
[475-25-2] Hematein, 4553
[475-26-3] DFDT, 2936
[475-31-0] Glycocholic Acid, 4390
[475-54-7] Oosporein, 6805
[475-67-2] Isocorydine, 5046
[475-81-0] Glaucine, 4332
[476-28-8] Lycorine, 5498
[476-32-4] Chelidonine, 2043
[476-33-5] Homochelidonine, 4652
[476-45-9] Javanicin, 5143
[476-66-4] Ellagic Acid, 3508
[476-69-7] Corydine, 2545
[476-70-0] Boldine, 1328
[476-71-1] Domesticine, 3409
[477-27-0] Colchiceine, 2469
[477-30-5] Demecolcine, 2873
[477-32-7] Visnadine, 9915
[477-47-4] Picropodophyllin, 7385
[477-58-7] Chondrocurine, 2215
[477-60-1] Bebeerine, 1024
[477-62-3] Isochondrodendrine,
5041
[477-75-8] Triptycene, 9663
[477-89-4] Casimiroin, 1894

[477-90-7] Bergenin, *1174*
[477-93-0] Dimethoxanate, *3211*
[478-05-7] Gladiolic Acid, *4329*
[478-10-4] Elliptone, *3511*
[478-43-3] Rhein, *8175*
[478-53-5] Morphothebaine, *6197*
[478-60-4] Citromycetin, *2330*
[478-61-5] Berbamine, *1168*
[478-73-9] Pseudococaine, *7926*
[478-84-2] Bromolysergide, *1411*
[478-85-3] Apo-β-erythroidine, *775*
[478-94-4] Lysergamide, *5505*
[478-95-5] Isolysergic Acid, *5065*
[479-00-5] Ergometrinine, *3599*
[479-13-0] Coumestrol, *2565*
[479-18-5] Dyphylline, *3459*
[479-20-9] Atranorin, *885*
[479-21-0] Cotoin, *2553*
[479-22-1] Nitranilic Acid, *6489*
[479-23-2] Cholanthrene, *2200*
[479-27-6] 1,8-Naphthalenediam-
ine, *6291*
[479-36-7] Daphnoline, *2818*
[479-45-8] Nitramine, *6488*
[479-68-5] Broparoestrol, *1438*
[479-81-2] Bietaverine, *1224*
[479-92-5] Propyphenazone, *7884*
[479-98-1] Aucubin, *894*
[480-11-5] Oroxylin A, *6830*
[480-15-9] Datiscetin, *2822*
[480-16-0] Morin, *6179*
[480-17-1] Leucocyanidin, *5334*
[480-30-8] Dichloralphenazone,
3033
[480-36-4] Linarin, *5376*
[480-40-0] Chrysin, *2261*
[480-41-1] Naringenin, *6344*
[480-44-4] Acacetin, *9*
[480-54-6] Retrorsine, *8169*
[480-64-8] *o*-Orsellinic Acid, *6834*
[480-68-2] 5-Nitrobarbituric Acid,
6507
[480-78-4] Platyphylline, *7505*
[480-81-9] Seneciphylline, *8403*
[480-85-3] Retronecine, *8168*
[481-05-0] Artemisin, *843*
[481-06-1] α-Santonin, *8325*
[481-17-4] Chondrillasterol, *2214*
[481-18-5] α-Spinasterol, *8705*
[481-20-9] Coprostane, *2518*
[481-21-0] Cholestane, *2202*
[481-26-5] Pregnane, *7731*
[481-29-8] Epiandrosterone, *3562*
[481-37-8] Ecgonine, *3467*
[481-39-0] Juglone, *5150*
[481-42-5] Plumbagin, *7511*
[481-49-2] Cepharanthine, *1981*
[481-72-1] Aloe-Emodin, *303*
[481-74-3] Chrysophanic Acid,
2263
[482-05-3] Diphenic Acid, *3309*
[482-15-5] Isothipendyl, *5117*
[482-20-2] Tylophorine, *9739*
[482-28-0] Cinchonamine, *2287*
[482-44-0] Imperatorin, *4839*
[482-49-5] Doisynolic Acid, *3408*
[482-53-1] Osajin, *6842*
[482-66-6] 1,2-Cyclopentenophen-
anthrene, *2749*
[482-68-8] Sarpagine, *8336*
[482-70-2] Chimaphilin, *2047*
[482-74-6] Cryptopine, *2611*
[482-89-3] Indigo, *4855*
[482-91-7] Aricine, *809*
[483-04-5] Raubasine, *8129*
[483-10-3] Corynanthine, *2547*
[483-17-0] Cephaeline, *1970*
[483-18-1] Emetine, *3517*
[483-19-2] Emetamine, *3516*
[483-49-8] Berbine, *1172*

[483-54-5] Aurantiogliocladin, *896*
[483-55-6] Phthiocol, *7352*
[483-57-8] Fervenulin, *4012*
[483-63-6] Crotamiton, *2597*
[483-64-7] Damascenine, *2809*
[483-65-8] Retene, *8162*
[483-78-3] Cadalene, *1607*
[484-11-7] Neocuproine, *6364*
[484-12-8] Osthole, *6849*
[484-20-8] Bergapten(e), *1173*
[484-23-1] Dihydralazine, *3152*
[484-29-7] Dictamnine, *3079*
[484-31-1] Apiole (Dill), *766*
[484-49-1] Olivacine, *6794*
[484-67-3] Drosophilin A, *3441*
[484-89-9] Fumigatin, *4201*
[484-93-5] Ecgonidine, *3466*
[485-19-8] Reticuline, *8164*
[485-31-4] Binapacryl, *1237*
[485-35-8] Cytisine, *2795*
[485-41-6] Sulfachrysoidine, *8872*
[485-43-8] Iridomyrmecin, *4972*
[485-47-2] Ninhydrin, *6470*
[485-49-4] Bicuculline, *1223*
[485-64-3] Hydrocinchonidine,
4705
[485-65-4] Hydrocinchonine, *4706*
[485-71-2] Cinchonidine, *2288*
[485-72-3] Formononetin, *4157*
[485-89-2] Oxycinchophen, *6910*
[485-91-6] Allocryptopine, *255*
[486-12-4] Triprolidine, *9661*
[486-16-8] Carbinoxamine, *1804*
[486-17-9] Captodiame, *1772*
[486-18-0] Flavopereirine, *4031*
[486-35-1] Daphnetin, *2816*
[486-39-5] Coclaurine, *2455*
[486-47-5] Ethaverine, *3682*
[486-54-4] 1-Naphthylamine-2,7-
disulfonic Acid, *6320*
[486-55-5] Daphnin, *2817*
[486-56-6] Cotinine, *2552*
[486-66-8] Daidzein, *2805*
[486-70-4] Lupinine, *5479*
[486-79-3] Dipyrocetyl, *3357*
[486-84-0] Harman, *4530*
[486-86-2] Caulophylline, *1914*
[486-89-5] Anagyrine, *660*
[487-06-9] Limettin, *5370*
[487-21-8] Lumazine, *5467*
[487-27-4] Scopoline, *8365*
[487-41-2] Phillyrin, *7296*
[487-48-9] Salacetamide, *8290*
[487-53-6] Hydroxyprocaine, *4772*
[487-58-1] Hypaphorine, *4798*
[487-60-5] Indican (Plant Indican),
4854
[487-79-6] Kainic Acid, *5157*
[487-88-7] Nitron, *6534*
[487-90-1] Porphobilinogen, *7575*
[487-93-4] Bufotenine, *1467*
[487-94-5] Indican (Metabolic Indican), *4853*
[488-04-0] Holomycin, *4647*
[488-10-8] Jasmone, *5141*
[488-40-4] Octadecyltrimethylammonium Pentachlorophenate, *6665*
[488-41-5] Mitobronitol, *6130*
[488-43-7] Glucamine, *4343*
[488-69-7] Fructose-1,6-diphos-
phate, *4187*
[488-73-3] *d*-Quercitol, *8046*
[488-81-3] Adonitol, *157*
[488-84-6] D-Ribulose, *8209*
[489-49-6] Cetraric Acid, *2016*
[489-84-9] Guaiazulene, *4464*
[489-86-1] Guaiol, *4466*
[490-02-8] Aspergillic Acid, *863*
[490-03-9] Diosphenol, *3292*
[490-10-8] Nepetalactone, *6383*

[490-11-9] Cinchomeronic Acid,
2285
[490-23-3] ε-Tocopherol, *9420*
[490-53-9] Carnegine, *1854*
[490-55-1] Amiphenazole, *498*
[490-58-4] Ichthyopterin, *4816*
[490-67-5] Gaultherin, *4272*
[490-79-9] Gentisic Acid, *4290*
[490-83-5] Dehydroascorbic Acid,
2856
[490-98-2] Hydroxytetracaine,
4783
[491-35-0] Lepidine, *5324*
[491-50-9] Quercimeritrin, *8045*
[491-59-8] Chrysarobin, *2257*
[491-67-8] Baicalein, *954*
[491-70-3] Luteolin, *5483*
[491-88-3] Isopilosine, *5085*
[491-92-9] Pamaquine, *6953*
[492-11-5] Leucopterin, *5338*
[492-17-1] 2,4′-Biphenyldiamine,
1249
[492-18-2] Mersalyl, *5805*
[492-22-8] Thioxanthone, *9301*
[492-27-3] Kynurenic Acid, *5206*
[492-38-6] Atropic Acid, *890*
[492-70-6] Hydrobenzoin, *4700*
[492-87-5] Julocrotine, *5151*
[492-89-7] 3-Pentadecylcatechol,
7061
[492-99-9] Nioxime®, *6477*
[493-35-6] ζ₂-Tocopherol, *9422*
[493-49-2] Corydaldine, *2542*
[493-52-7] Methyl Red, *6037*
[493-75-4] Bialamicol, *1217*
[493-76-5] Propanocaine, *7815*
[493-78-7] Methaphenilene, *5870*
[493-80-1] Histapyrrodine, *4641*
[493-92-5] Prolintane, *7791*
[494-03-1] Chlornaphazine, *2107*
[494-04-2] Nicotelline, *6432*
[494-08-6] Glucovanillin, *4359*
[494-44-0] Cassella's Acid F, *1899*
[494-47-3] Hydrofuramide, *4718*
[494-52-0] Anabasine, *655*
[494-55-3] Hydrohydrastinine,
4732
[494-79-1] Melarsoprol, *5694*
[494-97-3] Nornicotine, *6631*
[495-08-9] Gentisyl Alcohol, *4292*
[495-20-5] Conhydrine, *2495*
[495-31-8] Nodakenin, *6589*
[495-42-1] Halostachine, *4516*
[495-45-4] Dypnone, *3460*
[495-48-7] Azoxybenzene, *937*
[495-54-5] Chrysoidine Free Base*,
2262
[495-69-2] Hippuric Acid, *4636*
[495-83-0] Tigloidine, *9366*
[495-84-1] Salinazid, *8304*
[495-85-2] Methysticin, *6056*
[495-91-0] Chavicine, *2038*
[495-99-8] Hydroxystilbamidine,
4781
[496-00-4] Dibromopropamidine,
3008
[496-11-7] Indan, *4844*
[496-16-2] Coumaran, *2560*
[496-41-3] Coumarilic Acid, *2562*
[496-46-8] Acetyleneurea, *87*
[496-49-1] Hygrine, *4788*
[496-65-1] Pantetheine, *6961*
[496-67-3] Bromisovalum, *1385*
[496-74-2] Toluene-3,4-dithiol,
9457
[496-77-5] Butyroin, *1595*
[497-18-1] Carbohydrazide, *1811*
[497-19-8] Sodium Carbonate,
8541
[497-30-3] Ergothioneine, *3611*
[497-39-2] DBMC, *2829*

Denotes derivative of title compound

[497-59-6] Meconic Acid, 5664
[497-72-3] Methymycin, 6053
[497-73-4] Neomethymycin, 6368
[497-75-6] Dioxethedrine, 3297
[497-76-7] Arbutin, 799
[497-78-9] Rhododendrin, 8190
[498-02-2] Apocynin, 772
[498-23-7] Citraconic Acid, 2323
[498-24-8] Mesaconic Acid, 5806
[498-59-9] Djenkolic Acid, 3393
[498-71-5] Sobrerol, 8510
[498-94-2] Isonipecotic Acid, 5075
[498-95-3] Nipecotic Acid, 6478
[498-96-4] Guvacine, 4495
[499-04-7] Arecaidine, 801
[499-12-7] Aconitic Acid, 112
[499-14-9] Chondrosine, 2218
[499-15-0] Hyalobiuronic Acid, 4674
[499-20-7] Dhurrin, 2937
[499-67-2] Proparacaine, 7817
[499-75-2] Carvacrol, 1881
[499-89-8] Thujic Acid, 9325
[500-05-0] Coumalic Acid, 2558
[500-08-3] Forminitrazole, 4155
[500-12-9] Goitrin, 4411
[500-28-7] Chlorthion®, 2196
[500-34-5] β-Eucaine, 3850
[500-38-9] Nordihydroguaiaretic Acid, 6610
[500-42-5] Chlorazanil, 2073
[500-44-7] Mimosine, 6118
[500-55-0] Apoatropine, 770
[500-62-9] Yangonin, 10001
[500-64-1] Kawain, 5167
[500-92-5] Chlorguanide, 2088
[501-15-5] Deoxyepinephrine, 2885
[501-30-4] Kojic Acid, 5197
[501-52-0] Hydrocinnamic Acid, 4707
[501-53-1] Carbobenzoxy Chloride, 1807
[501-65-5] Tolan, 9428
[501-68-8] Beclamide, 1027
[501-81-5] Homonicotinic Acid, 4659
[501-82-6] Carbanilic Acid, 1786
[501-92-8] Chavicol, 2039
[502-37-4] Hypoglycine B, 4802
[502-42-1] Cycloheptanone, 2728
[502-55-6] Dixanthogen, 3390
[502-59-0] Octamylamine, 6671
[502-61-4] α-Farnesene, 3883
[502-65-8] Lycopene, 5492
[502-85-2] Sodium Oxybate, 8603
[502-98-7] Chloroazodin, 2120
[503-01-5] Isometheptene, 5070
[503-07-1] Vernolic Acid, 9868
[503-30-0] Trimethylene Oxide, 9630
[503-38-8] Diphosgene, 3339
[503-40-2] Methionic Acid, 5895
[503-49-1] Meglutol, 5690
[503-64-0] Isocrotonic Acid, 5048
[503-66-2] Hydracrylic Acid, 4681
[503-74-2] Isovaleric Acid, 5120
[504-02-9] Dihydroresorcinol, 3161
[504-15-4] Orcinol, 6819
[504-20-1] Phorone, 7307
[504-29-0] α-Aminopyridine, 486
[504-63-2] Trimethylene Glycol, 9629
[504-64-3] Carbon Suboxide, 1821
[505-32-8] Isophytol, 5084
[505-34-0] Cheirolin, 2040
[505-48-6] Suberic Acid, 8833
[505-60-2] Mustard Gas, 6225
[505-70-4] Muconic Acid, 6210
[505-75-9] Cicutoxin, 2273
[506-03-6] Chimyl Alcohol, 2049
[506-12-7] Margaric Acid, 5634

[506-25-2] Isanic Acid, 4983
[506-26-3] γ-Linolenic Acid, 5384
[506-30-9] Arachidic Acid, 791
[506-32-1] Arachidonic Acid, 792
[506-33-2] Brassidic Acid, 1359
[506-61-6] Potassium Silver Cyanide, 7659
[506-63-8] Dimethylberyllium, 3227
[506-64-9] Silver Cyanide, 8458
[506-65-0] Gold Monocyanide, 4415
[506-66-1] Beryllium Carbide, 1183
[506-68-3] Cyanogen Bromide, 2700
[506-77-4] Cyanogen Chloride, 2701
[506-78-5] Cyanogen Iodide, 2702
[506-80-9] Carbon Diselenide, 1817
[506-82-1] Dimethylcadmium, 3229
[506-89-8] Urea Hydrochloride, 9782
[506-96-7] Acetyl Bromide, 76
[507-02-8] Acetyl Iodide, 88
[507-09-5] Thioacetic Acid, 9247
[507-16-4] Thionyl Bromide, 9277
[507-19-7] tert-Butyl Bromide, 1555
[507-20-0] tert-Butyl Chloride, 1562
[507-25-5] Carbon Tetraiodide, 1824
[507-27-7] Tetraphenylarsonium Bromide, 9166
[507-28-8] Tetraphenylarsonium Chloride, 9167
[507-32-4] Ethanearsonic Acid, 3677
[507-36-8] tert-Amyl Bromide, 638
[507-40-4] tert-Butyl Hypochlorite, 1570
[507-42-6] Bromal Hydrate, 1373
[507-45-9] Amylene Dichloride, 645
[507-47-1] Chloral Ammonia, 2057
[507-60-8] Scilliroside, 8356
[507-61-9] Azafrin, 911
[507-70-0] Borneol, 1338
[507-79-9] Tazettine, 9050
[508-02-1] Oleanolic Acid, 6787
[508-29-2] 3-Chloro-d-camphor, 2132
[508-44-1] Anemonin, 674
[508-52-1] Ouabagenin, 6853
[508-54-3] Hydroxycodeinone, 4752
[508-59-8] Parthenin, 6995
[508-65-6] Germine, 4307
[508-75-8] Convallatoxin, 2508
[508-77-0] Cymarin, 2768
[508-95-2] Cortisone Phosphate, 2535
[508-96-3] Hydrocortisone Tebutate, 4714
[509-00-2] Cortisone, 21β-Cyclopentanepropionate, 2534
[509-14-8] Tetranitromethane, 9164
[509-15-9] Gelsemine, 4277
[509-17-1] Ajacine, 182
[509-18-2] Delsoline, 2868
[509-20-6] Aconine, 110
[509-24-0] Songorine, 8673
[509-36-4] β-Colubrine*, 2487
[509-40-0] Diaboline, 2938
[509-42-2] Strychnine Phosphate, 8825
[509-44-4] α-Colubrine*, 2487
[509-60-4] Dihydromorphine, 3160
[509-67-1] Pholcodine, 7303

[509-71-7] Diacetyldihydromorphine, 2947
[509-74-0] Methadyl Acetate, 5853
[509-77-3] Anilinephthalein, 689
[509-78-4] Dimenoxadol, 3194
[509-86-4] Heptabarbital, 4575
[509-87-5] 5-Ethyl-5-(1-piperidyl)-barbituric Acid, 3800
[509-93-3] Ambrosin, 391
[510-07-6] α,α-Diphenyl-2-piperidinepropanol, 3333
[510-15-6] Chlorobenzilate, 2123
[510-20-3] Diethylmalonic Acid, 3114
[510-25-8] Belladonnine, 1035
[510-34-9] Phethenylate Sodium, 7295
[510-35-0] Santonic Acid, 8324
[510-53-2] Racemethorphan, 8116
[510-90-7] Buthalital Sodium, 1518
[511-00-2] Erythrophlamine, 3633
[511-07-9] Ergocristinine, 3594
[511-08-0] Ergocristine, 3593
[511-09-1] Ergocryptine, 3595
[511-10-4] Ergocryptinine, 3596
[511-12-6] Dihydroergotamine, 3157
[511-13-7] Chlophedianol Hydrochloride*, 2053
[511-18-2] Norcholanic Acid, 6606
[511-20-6] Ergostane, 3602
[511-21-7] Coproergostane, 2516
[511-26-2] Pregnenolone Methyl Ether, 7740
[511-28-4] Vitamin D4, 9930
[511-34-2] Digitogenin, 3143
[511-45-5] Pridinol, 7747
[511-55-7] Xenytropium Bromide, 9983
[511-70-6] Diethylbromoacetamide, 3105
[511-89-7] Plumieride, 7513
[511-96-6] Gitogenin, 4325
[512-04-9] Diosgenin, 3289
[512-12-9] 5,5-Dipropyl-2,4-oxazolidinedione, 3352
[512-15-2] Cyclopentolate, 2752
[512-16-3] Cyclobutyrol, 2721
[512-29-8] Flavoxanthin, 4032
[512-35-6] Benzenesulfonic Anhydride, 1079
[512-39-0] Hepaxanthin, 4572
[512-42-5] Sodium Methyl Sulfate, 8595
[512-48-1] Novonal, 6643
[512-64-1] Echinomycin, 3471
[512-69-6] Raffinose, 8120
[512-85-6] Ascaridole, 852
[513-10-0] Echothiophate Iodide, 3475
[513-12-2] 2-(Ethylsulfonyl)ethanol, 3810
[513-15-5] Sodium β-Sulfopropionitrile, 8639
[513-29-1] Glycine Sulfate, 4387
[513-31-5] 2,3-Dibromopropene, 3009
[513-35-9] Amylene, 644
[513-36-0] Isobutyl Chloride, 5022
[513-37-1] 1-Chloro-2-methylpropene, 2147
[513-38-2] Isobutyl Iodide, 5027
[513-44-0] Isobutyl Mercaptan, 5030
[513-48-4] sec-Butyl Iodide, 1573
[513-53-1] sec-Butyl Mercaptan, 1576
[513-74-6] Ammonium Dithiocarbamate, 539
[513-77-9] Barium Carbonate, 979

[513-78-0] Cadmium Carbonate, 1614

[513-79-1] Cobaltous Carbonate, 2430

[513-80-4] Urea Nitrate, 9784

[513-81-5] 2,3-Dimethyl-1,3-butadiene, 3228

[513-85-9] 2,3-Butylene Glycol, 1567

[513-86-0] Acetoin, 55

[513-88-2] 1,1-Dichloroacetone, 3038

[513-92-8] Tetraiodoethylene, 9151

[513-96-2] Chlorocyanohydrin, 2134

[514-10-3] Abietic Acid, 2

[514-12-5] α,α'-Dibromo-d-camphor, 3002

[514-17-0] Androstane-3β,11β-diol-17-one, 669

[514-36-3] Fludrocortisone Acetate*, 4059

[514-39-6] Periplogenin, 7128

[514-61-4] Normethandrone, 6629

[514-65-8] Biperiden, 1246

[514-73-8] Dithiazanine Iodide, 3376

[514-76-1] Astacin, 875

[514-78-3] Canthaxantin, 1756

[514-92-1] Torularhodin, 9476

[515-24-2] Turicine, 9729

[515-25-3] Betonicine, 1209

[515-30-0] Atrolactic Acid, 889

[515-40-2] Neophyl Chloride, 6375

[515-49-1] Sulfathiourea, 8921

[515-54-8] Phenosulfazole, 7219

[515-59-3] Sulfamethylthiazole, 8891

[515-64-0] Sulfisomidine, 8929

[515-72-0] Barium Benzenesulfonate, 976

[515-76-4] Diphetarsone, 3338

[515-82-2] Chloral Formamide, 2060

[515-83-3] Chloral Alcoholate, 2056

[515-94-6] 2,3-Diaminopropionic Acid, 2962

[515-96-8] Semioxamazide, 8397

[516-02-9] Barium Oxalate, 997

[516-03-0] Ferrous Oxalate, 4000

[516-12-1] N-Iodosuccinimide, 4936

[516-15-4] 11-Ketoprogesterone, 5185

[516-16-5] Allopregnane-3β,11β,-21-triol-20-one, 272

[516-18-7] Asaprol, 847

[516-21-2] Cycloguanil, 2727

[516-38-1] Cortol, 2537

[516-39-2] Allopregnane-3β,11β,-17α,20β,21-pentol, 265

[516-41-6] Hydrallostane, 4683

[516-42-7] Cortolone, 2538

[516-45-0] Allopregnane-3β,17α,-21-triol-11,20-dione, 271

[516-47-2] Allopregnane-3β,17α,-21-triol-20-one, 273

[516-51-8] Uranediol, 9765

[516-53-0] Allopregnane-3β,20β-diol, 261

[516-54-1] Allopregnan-3α-ol-20-one, 274

[516-55-2] Allopregnan-3β-ol-20-one, 275

[516-58-5] Allopregnan-20β-ol-3-one, 277

[516-59-6] Allopregnan-20α-ol-3-one, 276

[516-78-9] γ-Ergostenol, 3606

[516-85-8] Dehydroergosterol, 2861

[516-95-0] Epicholestanol, 3564

[516-98-3] Neoergosterol, 6366

[517-04-4] Isoestradiol, 5051

[517-06-6] 8-Isoestrone, 5052

[517-07-7] Hexahydroequilenin, 4607

[517-09-9] Equilenin, 3581

[517-10-2] Allocholesterol, 253

[517-16-8] N-(Ethylmercuri)-p-toluenesulfonanilide, 3781

[517-18-0] Methallenestril, 5856

[517-21-5] Glyoxal-Sodium Bisulfite, 4406

[517-23-7] α-Acetylbutyrolactone, 77

[517-25-9] Trinitromethane, 9642

[517-28-2] Hematoxylin, 4556

[517-56-6] Corytuberine, 2549

[517-60-2] Mellitic Acid, 5706

[517-66-8] Dicentrine, 3028

[517-69-1] Actinodaphnine, 131

[517-82-8] Echinochrome A, 3470

[517-92-0] Chrysamminic Acid, 2253

[517-97-5] Tsuduranine, 9710

[518-05-8] Naphthalic Acid, 6299

[518-17-2] Evodiamine, 3865

[518-20-7] Cyclocumarol, 2723

[518-28-5] Podophyllotoxin, 7520

[518-29-6] β-Peltatin, 7020

[518-34-3] Tetrandrine, 9162

[518-44-5] Fluorescin, 4087

[518-47-8] Fluorescein Sodium*, 4085

[518-51-4] Phenolphthalein Sodium, 7209

[518-60-5] Lunine, 5475

[518-61-6] Ahistan, 179

[518-63-8] Dimethoxanate Hydrochloride*, 3211

[518-69-4] Corydaline, 2543

[518-75-2] Citrinin, 2329

[518-77-4] Corybulbine, 2539

[518-82-1] Emodin, 3518

[519-02-8] Matrine, 5641

[519-04-0] Anhalonine, 684

[519-05-1] Opianic Acid, 6806

[519-09-5] Benzoylecgonine, 1125

[519-21-1] Mycaminose, 6228

[519-23-3] Ellipticine, 3509

[519-26-6] Sodium Iodomethamate, 8584

[519-30-2] Dimethazan, 3202

[519-34-6] Maclurin, 5519

[519-37-9] Etofylline, 3835

[519-40-4] Aspidinol, 869

[519-44-8] 2,4-Dinitroresorcinol, 3278

[519-56-2] Morphenol, 6183

[519-65-3] Dioxypyramidon, 3299

[519-72-2] N,N-Diethylbenzhydrylamine, 3103

[519-73-3] Triphenylmethane, 9655

[519-87-9] Diphenylacetamide, 3315

[519-88-0] Ambucetamide, 394

[519-95-9] Florantyrone, 4040

[520-07-0] Antipyrine Salicylate, 749

[520-09-2] Dopan, 3416

[520-12-7] Pectolinarigenin, 7010

[520-18-3] Kaempferol, 5156

[520-26-3] Hesperidin, 4591

[520-27-4] Diosmin, 3291

[520-33-2] Hesperetin, 4590

[520-34-3] Diosmetin, 3290

[520-36-5] Apigenin, 763

[520-43-4] Sparassol, 8690

[520-45-6] Dehydroacetic Acid, 2855

[520-52-5] Psilocybin, 7942

[520-53-6] Psilocin, 7941

[520-77-4] Ethadione, 3670

[520-85-4] Medroxyprogesterone, 5677

[520-86-5] Allopregnane-3β,17α,-20β-triol, 270

[520-91-2] Vitamin D$_1$, 9927

[521-04-0] 7-Dehydrositosterol, 2863

[521-10-8] Methandriol, 5861

[521-11-9] Mestanolone, 5816

[521-12-0] Dromostanolone Propionate, 3436

[521-17-5] Androstenediol, 670

[521-18-6] Stanolone, 8753

[521-24-4] Sodium β-Naphthoquinone-4-sulfonate, 8597

[521-31-3] Luminol, 5470

[521-35-7] Cannabinol, 1751

[521-40-4] Narcotoline, 6343

[521-67-5] Cinnamoylcocaine, 2304

[521-74-4] Broxyquinoline, 1441

[521-78-8] Trimipramine Maleate*, 9636

[521-85-7] Corycavamine, 2540

[521-88-0] Karanjin, 5163

[521-93-7] Corycavidine, 2541

[521-96-0] Dichlororiboflavin, 3065

[522-00-9] Ethopropazine, 3703

[522-12-3] Quercitrin, 8047

[522-17-8] Deguelin, 2854

[522-18-9] Chlorbenzoxamine, 2075

[522-24-7] Fenethazine, 3918

[522-27-0] α-Furildioxime, 4218

[522-40-7] Fosfestrol, 4168

[522-48-5] Tetrahydrozoline Hydrochloride*, 9150

[522-51-0] Dequalinium Chloride, 2896

[522-57-6] Papaveraldine, 6966

[522-60-1] Ethylhydrocupreine, 3765

[522-66-7] Hydroquinine, 4737

[522-70-3] Antimycin A$_3$, 747

[522-94-1] allo-Yohimbine, 10012

[522-97-4] Canadine, 1744

[523-21-7] Sodium Rhodizonate, 8625

[523-27-3] 9,10-Dibromoanthracene, 3000

[523-44-4] Orange I, 6812

[523-54-6] Etymemazine, 3849

[523-64-8] Flindersine, 4036

[523-68-2] Vitamin K$_5$ N-Acetyl Analog*, 9936

[523-80-8] Apiole (Parsley), 767

[523-87-5] Dimenhydrinate, 3193

[523-88-6] Dibromsalicil, 3015

[524-15-2] Fagarine, 3879

[524-17-4] Dauricine, 2826

[524-20-9] Armepavine, 811

[524-30-1] Fraxin, 4179

[524-36-7] Pyridoxamine Dihydrochloride, 7993

[524-40-3] Ricinine, 8212

[524-42-5] 1,2-Naphthoquinone, 6314

[524-46-9] Adlumine, 155

[524-63-0] Cupreine, 2626

[524-80-1] 9-Carbazoleacetic Acid, 1793

[524-81-2] Mebhydroline, 5649

[524-83-4] Ethybenztropine, 3712

[524-84-5] Dimethylthiambutene, 3248

[524-96-9] Benzylimidobis(p-methoxyphenyl)methane, 1153

Denotes derivative of title compound

[524-97-0] Pterocarpin, 7948
[524-99-2] Medrylamine, 5678
[525-05-3] Nitroso-R Salt, 6565
[525-06-4] Diphenylketene, 3327
[525-37-1] 1,6-Naphthalenedisulfonic Acid, 6292
[525-47-3] 5-Nitroquinaldic Acid, 6552
[525-52-0] Pyrogallol Triacetate*, 8010
[525-57-5] Harmalol, 4529
[525-61-1] Quinocide, 8096
[525-64-4] 9H-Fluorene-2,7-diamine, 4082
[525-66-6] Propranolol, 7852
[525-68-8] Galipine, 4247
[525-74-6] N-Methylmyosmine, 6013
[525-79-1] Kinetin, 5192
[525-82-6] Flavone, 4030
[525-86-0] 2-Pentenylpenicillin Sodium, 7081
[525-87-1] Penicillin S Potassium, 7045
[525-94-0] Penicillin N, 7043
[526-02-3] Actiphenol, 136
[526-07-8] Sesamolin, 8421
[526-08-9] Sulfaphenazole, 8910
[526-13-6] Meteloidine, 5841
[526-18-1] Osalmid, 6843
[526-30-7] Tryptazan, 9706
[526-31-8] Abrine, 5
[526-35-2] Aloxidone, 305
[526-36-3] Xylometazoline, 9993
[526-55-6] Tryptophol, 9708
[526-78-3] 2,3-Dibromosuccinic Acid, 3013
[526-84-1] Dihydroxymaleic Acid, 3170
[526-94-3] Sodium Bitartrate, 8534
[526-95-4] Gluconic Acid, 4350
[526-97-6] L-Gulonic Acid, 4489
[526-98-7] 2-Keto-L-gulonic Acid, 5183
[526-99-8] Galactaric Acid, 4237
[527-07-1] Sodium Gluconate, 8569
[527-09-3] Cupric Gluconate, 2643
[527-17-3] Duroquinone, 3452
[527-18-4] Durohydroquinone, 3451
[527-38-8] Mycosamine, 6239
[527-52-6] Digitoxose, 3147
[527-53-7] Isodurene, 5050
[527-62-8] 2-Amino-4,6-dichlorophenol, 448
[527-72-0] 2-Thiophenecarboxylic Acid, 9284
[527-73-1] Azomycin, 934
[527-76-4] 4-Amino-2-sulfobenzoic Acid, 493
[527-85-5] o-Toluamide, 9454
[527-89-9] Dithiosalicylic Acid, 3381
[527-93-5] Diploicin, 3345
[528-10-9] 5,5'-Dinitrosalicil, 3279
[528-21-2] Gallacetophenone, 4248
[528-44-9] Trimellitic Acid, 9616
[528-45-0] 3,4-Dinitrobenzoic Acid, 3268
[528-48-3] Fisetin, 4026
[528-50-7] Cellobiose, 1957
[528-53-0] Delphinidin, 2866
[528-58-5] Cyanidin Chloride, 2694
[528-71-2] Hemipyocyanine, 4564
[528-75-6] 2,4-Dinitrobenzaldehyde, 3265
[528-76-7] 2,4-Dinitrobenzenesulfenyl Chloride, 3266
[528-79-0] Thymol Acetate*, 9333
[528-81-4] 6-Deoxy-L-ascorbic Acid, 2880

[528-88-1] D-Glucoascorbic Acid, 4345
[528-92-7] Apronalide, 783
[528-94-9] Ammonium Salicylate, 578
[529-05-5] Chamazulene, 2031
[529-17-9] Tropane, 9689
[529-19-1] o-Tolunitrile, 9463
[529-20-4] o-Tolualdehyde, 9453
[529-21-5] 3-Ethyl-4-picoline, 3796
[529-28-2] o-Iodoanisole, 4921
[529-36-2] 1-Naphthalenethiol, 6297
[529-38-4] Cocaethylene, 2449
[529-44-2] Myricetin, 6244
[529-53-3] Scutellarein, 8367
[529-64-6] Tropic Acid, 9691
[529-86-2] Anthranol, 716
[529-92-0] Cusparine, 2687
[530-08-5] Isoetharine, 5053
[530-14-3] Picein, 7367
[530-31-4] Ammonium Mandelate, 557
[530-34-7] Isoladol, 5062
[530-36-9] 2-Amino-1,2-diphenylethanol, 449
[530-40-5] Isonicotinic Acid Diethylamide, 5074
[530-43-8] Chloramphenicol Palmitate, 2069
[530-44-9] p-Dimethylaminobenzophenone, 3222
[530-45-0] 1,1-Di-p-tolylethane, 3385
[530-48-3] 1,1-Diphenylethene, 3323
[530-50-7] 1,1-Diphenylhydrazine, 3326
[530-55-2] 2,6-Dimethoxyquinone, 3214
[530-62-1] N,N'-Carbonyldiimidazole, 1825
[530-75-6] Acetylsalicylsalicylic Acid, 95
[530-77-8] Monophen®, 6163
[530-78-9] Flufenamic Acid, 4060
[530-85-8] o-Nitrophenylpropiolic Acid, 6546
[530-91-6] Tetralol, 9153
[530-97-2] Heptoxime, 4586
[531-02-2] Diresorcinol, 3361
[531-06-6] Methafurylene, 5854
[531-09-9] p-Bromtripelennamine, 1436
[531-18-0] Hexamethylolmelamine, 4611
[531-29-3] Coniferin, 2498
[531-37-3] Guaiacol Benzoate, 4458
[531-39-5] Guaiacol Valerate, 4461
[531-44-2] Scopolin, 8364
[531-53-3] Azure A, 940
[531-55-5] Azure B, 941
[531-57-7] Azure C, 942
[531-58-8] Cichoriin, 2267
[531-72-6] Arsenamide, 819
[531-73-7] Proflavine Dihydrochloride*, 7780
[531-75-9] Esculin, 3646
[531-81-7] Coumarin-3-carboxylic Acid, 2564
[531-84-0] Hexadecyl 3-Hydroxy-2-naphthoate, 4604
[531-91-9] N,N'-Diphenylbenzidine, 3319
[531-95-3] Equol, 3583
[531-99-7] 5-Acetyl-2-methoxybenzaldehyde, 91
[532-03-6] Methocarbamol, 5903

[532-05-8] 1,3-Di-6-quinolylurea, 3360
[532-11-6] Anethole Trithione, 676
[532-18-3] β,β'-Dinaphthylamine, 3260
[532-27-4] ω-Chloroacetophenone, 2115
[532-28-5] Mandelonitrile, 5601
[532-32-1] Sodium Benzoate, 8527
[532-34-3] Butopyronoxyl, 1529
[532-40-1] Thiamine Phosphoric Acid Ester Chloride, 9224
[532-43-4] Thiamine Mononitrate, 9223
[532-48-9] Euparin, 3857
[532-49-0] Dibutoline Sulfate, 3018
[532-52-5] Cyclexedrine, 2713
[532-54-7] Isonitrosoacetophenone, 5077
[532-55-8] Benzoyl Isothiocyanate, 1126
[532-59-2] Amylocaine Hydrochloride, 650
[532-76-3] Hexylcaine Hydrochloride, 4629
[532-82-1] Chrysoidine, 2262
[533-06-2] Mephenesin Carbamate*, 5737
[533-18-6] o-Cresyl Acetate, 2588
[533-22-2] Hydroxystilbamidine Isethionate*, 4781
[533-28-8] Piperocaine Hydrochloride*, 7444
[533-30-2] 6-Aminobenzothiazole, 436
[533-32-4] Isovaleryl Diethylamide, 5122
[533-45-9] Clomethiazole, 2382
[533-48-2] Desthiobiotin, 2914
[533-49-3] L-Erythrose, 3638
[533-50-6] L-Erythrulose, 3641
[533-51-7] Silver Oxalate, 8467
[533-58-4] o-Iodophenol, 4928
[533-62-0] Hydroxyglutamic Acid, 4756
[533-67-5] D-2-Deoxyribose, 2890
[533-70-0] 2,4-Diiodoaniline, 3173
[533-73-3] 1,2,4-Benzenetriol, 1081
[533-74-4] Dazomet, 2827
[533-76-6] Ketipic Acid, 5179
[533-87-9] Aleuritic Acid, 221
[533-88-0] Pentahomoserine, 7069
[533-96-0] Sodium Sesquicarbonate, 8630
[533-98-2] α-Butylene Dibromide, 1565
[534-00-9] d-Amyl Bromide, 636
[534-07-6] 1,3-Dichloroacetone, 3039
[534-08-7] Iothion, 4953
[534-12-3] Sodiumacetic Acid Sodium Salt, 8514
[534-13-4] N,N'-Dimethylthiourea, 3250
[534-15-6] Dimethylacetal, 3215
[534-16-7] Silver Carbonate, 8453
[534-17-8] Cesium Carbonate, 2003
[534-18-9] Sodium Thiocarbonate, 8649
[534-26-9] Lysidine, 5508
[534-30-5] Hercynine, 4588
[534-33-8] Acetarsone Diethylamine Salt*, 44
[534-46-3] Sophorose, 8676
[534-51-0] 3,5-Diiodothyronine, 3176
[534-52-1] Dinitrocresol, 3272
[534-59-8] n-Butylmalonic Acid, 1574
[534-64-5] Pyrithiamine, 8003

Denotes derivative of title compound

[534-82-7] 3-Methoxy-4-hydroxy-phenylglycol, 5916
[534-84-9] Pimeclone, 7399
[535-11-5] Ethyl α-Bromopropionate, 3731
[535-13-7] Ethyl α-Chloropropionate, 3743
[535-37-5] Gold Tricyanide, 4429
[535-46-6] Benzenestibonic Acid, 1077
[535-48-8] Sodium 6-Chloro-5-nitrotoluene-3-sulfonate, 8546
[535-51-3] Phenarsone Sulfoxylate, 7171
[535-55-7] Mercuric Sodium p-Phenolsulfonate, 5783
[535-65-9] Glybuthiazol(e), 4373
[535-66-0] Diphenylmethane-4,4'-disulfonamide, 3330
[535-75-1] Pipecolic Acid, 7425
[535-80-8] m-Chlorobenzoic Acid, 2124
[535-83-1] Trigonelline, 9606
[535-86-4] Homocamfin, 4651
[535-87-5] 3,5-Diaminobenzoic Acid, 2955
[535-89-7] Crimidine, 2590
[536-08-3] Digallic Acid, 3135
[536-17-4] p-Dimethylaminobenz-alrhodanine, 3220
[536-21-0] Norfenefrine, 6616
[536-24-3] Ethylnorepinephrine, 3789
[536-25-4] Orthocaine, 6836
[536-29-8] Dichlorophenarsine Hydrochloride, 3060
[536-33-4] Ethionamide, 3692
[536-38-9] p-Chlorophenacyl Bromide, 2153
[536-43-6] Dyclonine Hydrochloride*, 3453
[536-50-5] p,α-Dimethylbenzyl Alcohol, 3226
[536-57-2] p-Toluenesulfinic Acid, 9458
[536-60-7] Cumic Alcohol, 2621
[536-66-3] Cumic Acid, 2620
[536-69-6] Fusaric Acid, 4227
[536-74-3] Ethynylbenzene, 3817
[536-75-4] 4-Ethylpyridine, 3803
[536-78-7] 3-Ethylpyridine, 3802
[536-88-9] 4-Ethyl-2-picoline, 3797
[536-93-6] Eucatropine Hydrochloride*, 3854
[536-95-8] p-(Benzylsulfonamido)-benzoic Acid, 1163
[537-01-9] Cerous Carbonate, 1990
[537-12-2] Diperodon Hydrochloride, 3301
[537-17-7] Amanozine, 379
[537-21-3] Chlorproguanil, 2185
[537-26-8] Tropacocaine, 9686
[537-29-1] Norhyoscyamine, 6623
[537-45-1] 2,6-Dibromoquinone-4-chlorimide, 3010
[537-46-2] Methamphetamine, 5859
[537-47-3] 4-Phenylsemicarbazide, 7283
[537-49-5] Surinamine, 8988
[537-61-1] Cyclomethycaine Hydrochloride*, 2741
[537-64-4] p-Ditolylmercury, 3387
[537-65-5] 4,4'-Diaminodiphenyl-amine, 2957
[537-91-7] Nitrophenide, 6539
[538-03-4] Oxophenarsine Hydrochloride, 6902
[538-07-8] HN1, 4644
[538-08-9] Diallylcyanamide, 2952

[538-17-0] Aluminum Thiocyanate, 375
[538-28-3] S-Benzylthiuronium Chloride, 1164
[538-32-9] Benzylurea, 1165
[538-37-4] Dibenzyl Chlorophosphonate, 2994
[538-39-6] 1,2-Di-p-tolylethane, 3386
[538-41-0] p-Diaminoazobenzene, 2954
[538-43-2] Phenylglyceryl Ether, 7261
[538-45-4] 3,3-Dithiodipyridine Dihydrochloride, 3380
[538-51-2] Benzylideneaniline, 1152
[538-58-9] Dibenzalacetone, 2988
[538-62-5] Diphenylcarbazone, 3322
[538-64-7] Benzyl Fumarate, 1149
[538-68-1] Amylbenzene, 635
[538-71-6] Domiphen Bromide, 3411
[538-74-9] Benzyl Sulfide, 1162
[538-75-0] Dicyclohexylcarbodi-imide, 3086
[538-86-3] Benzyl Methyl Ether, 1154
[538-90-9] β-Coniceine, 2496
[538-93-2] Isobutylbenzene, 5018
[539-08-2] p-Lactophenetide, 5220
[539-12-8] p-Anol, 706
[539-13-9] NPA Acid, 6646
[539-15-1] Hordenine, 4666
[539-21-9] Ambazone, 386
[539-30-0] Benzyl Ethyl Ether, 1147
[539-47-9] 2-Furanacrylic Acid, 4207
[539-54-8] Antimony Sodium Thioglycollate, 735
[539-64-0] N,N-Dimethylglycine Hydrazide Hydrochloride, 3234
[539-68-4] Dihydroxyaluminum Sodium Carbonate, 3169
[539-86-6] Allicin, 249
[539-88-8] Ethyl Levulinate, 3775
[539-89-9] Isobutyl Urethane, 5037
[539-90-2] Isobutyl n-Butyrate, 5020
[540-05-6] Phytofluene, 7361
[540-07-8] n-Amyl Caproate, 640
[540-10-3] Cetyl Palmitate, 2023
[540-16-9] Cupric Butyrate, 2633
[540-18-1] n-Amyl Butyrate, 639
[540-36-3] p-Difluorobenzene, 3132
[540-37-4] p-Iodoaniline, 4920
[540-38-5] p-Iodophenol, 4929
[540-42-1] Isobutyl Propionate, 5033
[540-47-6] Cyclopropyl Methyl Ether, 2756
[540-49-8] Acetylene Dibromide, 85
[540-51-2] Ethylene Bromohydrin, 3749
[540-54-5] Propyl Chloride, 7859
[540-59-0] Acetylene Dichloride, 86
[540-61-4] Aminoacetonitrile, 423
[540-63-6] 1,2-Ethanedithiol, 3679
[540-67-0] Ethyl Methyl Ether, 3783
[540-69-2] Ammonium Formate, 548
[540-72-7] Thiocyanate Sodium, 9256
[540-73-8] 1,2-Dimethylhydrazine, 3237
[540-80-7] tert-Butyl Nitrite, 1582

[540-82-9] Ethyl Sulfate, 3808
[540-84-1] Isooctane, 5079
[540-88-5] tert-Butyl Acetate, 1537
[540-92-1] Acetone Sodium Bisulfite, 61
[540-97-6] Dodecamethylcyclo-hexasiloxane, 3400
[541-02-6] Decamethylcyclopenta-siloxane, 2841
[541-04-8] sym-Diphenylpyrophos-phorodiamidic Acid, 3335
[541-07-1] Mevaldic Acid, 6086
[541-09-3] Uranyl Acetate, 9774
[541-15-1] Carnitine L-Form*, 1856
[541-16-2] Di-tert-butyl Malonate, 3021
[541-19-5] Succinylcholine Iodide, 8847
[541-20-8] Pentamethonium Bromide, 7070
[541-22-0] Decamethonium Bromide, 2840
[541-25-3] Dichloro(2-chlorovinyl)-arsine, 3051
[541-28-6] Isoamyl Iodide, 5002
[541-33-3] Butylidene Chloride, 1571
[541-35-5] n-Butyramide, 1592
[541-41-3] Ethyl Chloroformate, 3742
[541-42-4] Isopropyl Nitrite, 5104
[541-43-5] Barium Formate, 987
[541-46-8] Isovaleramide, 5119
[541-48-0] β-Aminobutyric Acid, 440
[541-50-4] Acetoacetic Acid, 51
[541-53-7] 2,4-Dithiobiuret, 3378
[541-58-2] 2,4-Dimethylthiazole, 3249
[541-64-0] Furtrethonium Iodide*, 4225
[541-66-2] Oxapropanium Iodide, 6877
[541-73-1] m-Dichlorobenzene, 3043
[541-79-7] Carbocloral, 1808
[541-88-8] Chloroacetic Anhydride, 2112
[541-91-3] Muscone, 6222
[542-05-2] Acetonedicarboxylic Acid, 60
[542-08-5] Isopropyl Acetoacetate, 5094
[542-10-9] Ethylidene Diacetate, 3767
[542-18-7] Cyclohexyl Chloride, 2738
[542-32-5] α-Aminoadipic Acid, 427
[542-37-0] tert-Amyl Isovalerate, 647
[542-42-7] Calcium Palmitate, 1693
[542-46-1] Civetone, 2337
[542-52-9] n-Butyl Carbonate, 1558
[542-54-1] Isoamyl Cyanide, 4999
[542-55-2] Isobutyl Formate, 5026
[542-56-3] Isobutyl Nitrite, 5032
[542-58-5] β-Chloroethyl Acetate, 2137
[542-59-6] Ethylene Glycol Mono-acetate, 3757
[542-62-1] Barium Cyanide, 983
[542-63-2] Diethylberyllium, 3104
[542-69-8] n-Butyl Iodide, 1572
[542-75-6] 1,3-Dichloropropene, 3064
[542-76-7] β-Chloropropionitrile, 2160

Denotes derivative of title compound

[542-83-6] Cadmium Cyanide, 1616
[542-84-7] Cobaltous Cyanide, 2433
[542-85-8] Ethyl Isothiocyanate, 3771
[542-88-1] sym-Dichloromethyl Ether, 3058
[542-91-6] Diethylsilane, 3117
[542-92-7] Cyclopentadiene, 2744
[543-20-4] Succinyl Chloride, 8844
[543-21-5] Cellocidin, 1958
[543-24-8] Aceturic Acid, 74
[543-27-1] Isobutyl Chlorocarbonate, 5023
[543-28-2] Isobutyl Carbamate, 5021
[543-29-3] Isobutyl Nitrate, 5031
[543-38-4] Canavanine, 1745
[543-49-7] 2-Heptanol, 4583
[543-59-9] Amyl Chloride, 642
[543-63-5] n-Butylmercuric Chloride, 1578
[543-67-9] n-Propyl Nitrite, 7878
[543-80-6] Barium Acetate, 975
[543-81-7] Beryllium Acetate, 1178
[543-82-8] Octodrine, 6678
[543-83-9] Galegine, 4246
[543-87-3] Isoamyl Nitrate, 5004
[543-90-8] Cadmium Acetate, 1612
[543-94-2] Strontium Acetate, 8798
[544-00-3] Diisoamylamine, 3178
[544-01-4] Isoamyl Ether, 5000
[544-10-5] 1-Chlorohexane, 2144
[544-12-7] 3-Hexen-1-ol, 4620
[544-13-8] Glutaronitrile, 4368
[544-16-1] n-Butyl Nitrite, 1581
[544-17-2] Calcium Formate, 1671
[544-18-3] Cobaltous Formate, 2435
[544-19-4] Cupric Formate, 2642
[544-27-4] Arsenoacetic Acid, 836
[544-31-0] Palmidrol, 6946
[544-35-4] Ethyl Linoleate, 3776
[544-40-1] n-Butyl Sulfide, 1590
[544-44-5] Agrocybin, 178
[544-47-8] p-Chlorobenzylpseudothiuronium Chloride, 2128
[544-51-4] Mycomycin, 6237
[544-60-5] Ammonium Oleate, 563
[544-62-7] Batyl Alcohol, 1021
[544-63-8] Myristic Acid, 6246
[544-92-3] Cuprous Cyanide, 2668
[544-97-8] Dimethylzinc, 3252
[545-06-2] Trichloroacetonitrile, 9540
[545-26-6] Gitoxigenin, 4327
[545-45-9] Zygadenine, 10099
[545-47-1] Lupeol, 5478
[545-55-1] Triethylenephosphoramide, 9587
[545-61-9] Ajaconine, 183
[545-80-2] Poldine Methylsulfate, 7529
[545-90-4] Dimepheptanol, 3195
[545-93-7] Propallylonal, 7807
[546-06-5] Conessine, 2492
[546-43-0] Alantolactone, 198
[546-46-3] Zinc Citrate, 10033
[546-62-3] Verbascose, 9860
[546-67-8] Lead Tetraacetate, 5305
[546-68-9] Titanium Isopropylate, 9400
[546-74-7] Sodium Ethyl Sulfate, 8561
[546-88-3] Acetohydroxamic Acid, 54
[546-89-4] Lithium Acetate, 5396
[546-97-4] Columbin, 2489
[547-25-1] Turanose, 9728
[547-36-4] Succisulfone, 8851

[547-44-4] Sulfanilylurea, 8906
[547-52-4] N^4-Sulfanilylsulfanilamide, 8905
[547-53-5] 4'-(Methylsulfamoyl)-sulfanilanilide, 6040
[547-56-8] Cupric p-Phenolsulfonate, 2652
[547-57-9] Tropaeolin O, 9687
[547-58-0] Methyl Orange, 6019
[547-63-7] Methyl Isobutyrate, 6003
[547-64-8] Methyl Lactate, 6007
[547-65-9] α-Methylene Butyrolactone, 5981
[547-66-0] Magnesium Oxalate, 5554
[547-67-1] Nickel Oxalate, 6421
[547-68-2] Zinc Oxalate, 10049
[547-77-3] Allotetrahydrocortisone, 280
[547-81-9] 16-Epiestriol, 3567
[547-91-1] 8-Hydroxy-7-iodo-5-quinolinesulfonic Acid, 4757
[548-00-5] Ethyl Biscoumacetate, 3729
[548-04-9] Hypericin, 4799
[548-24-3] Eosine I Bluish, 3554
[548-37-8] Verbenalin, 9861
[548-40-3] Oxyacanthine, 6906
[548-42-5] Agroclavine, 177
[548-43-6] Elymoclavine, 3512
[548-51-6] o-Thymotic Acid, 9342
[548-54-9] Violacein, 9901
[548-57-2] Lucanthone Hydrochloride, 5463
[548-62-9] Gentian Violet, 4287
[548-68-5] Thiphenamil Hydrochloride*, 9303
[548-73-2] Droperidol, 3437
[548-76-5] Irigenin, 4973
[548-77-6] Tectorigenin, 9057
[548-80-1] Chromotrope 2B, 2242
[548-83-4] Galangin, 4244
[548-84-5] Pyrvinium Chloride, 8033
[548-98-1] Cholane, 2198
[549-07-5] Insularine, 4886
[549-18-8] Amitriptyline Hydrochloride*, 504
[549-28-0] Protostephanine, 7914
[549-38-2] 3'-Methyl-1,2-cyclopentenophenanthrene, 5969
[549-47-3] Quinine Dihydrobromide, 8078
[549-49-5] Quinine Hydrobromide, 8084
[549-50-8] Quinine Hydriodide, 8083
[549-52-0] Quinine Urea Hydrochloride, 8091
[549-56-4] Quinine Bisulfate, 8076
[549-68-8] Octaverine, 6676
[550-01-6] Metabutoxycaine Hydrochloride, 5822
[550-10-7] Hydrocotarnine, 4715
[550-24-3] Embelin, 3513
[550-28-7] Amisometradine, 500
[550-29-2] Naphazoline Hydrochloride*, 6287
[550-33-4] Nebularine, 6353
[550-49-2] Adlumidine, 154
[550-60-7] 1-Nitro-2-naphthol, 6536
[550-73-2] 3-Methyl-6,7-methylenedioxy-1-piperonylisoquinoline, 6011
[550-74-3] Picrolonic Acid, 7383
[550-82-3] Resazurin, 8144
[550-83-4] Propoxycaine Hydrochloride, 7850

[550-97-0] 1-Naphthyl Salicylate, 6334
[551-01-9] Plasmocid, 7495
[551-06-4] 1-Naphthylisothiocyanate, 6331
[551-09-7] N-(1-Naphthyl)ethylenediamine, 6329
[551-11-1] Prostaglandin $F_{2\alpha}$, 7894
[551-16-6] 6-Aminopenicillanic Acid, 470
[551-27-9] Propicillin, 7829
[551-48-4] Guanoclor Sulfate*, 4479
[551-64-4] Zinc Tartrate, 10067
[551-68-8] D-Psicose, 7940
[551-74-6] Mannomustine, 5631
[551-76-8] 2,4,6-Trichloro-m-cresol, 9547
[551-77-9] 2,3,6-Trichloro-p-cresol, 9546
[551-78-0] 4,5,6-Trichloro-o-cresol, 9548
[551-88-2] 3-Nitropentane, 6537
[551-90-6] Hinderin, 4635
[551-92-8] Dimetridazole, 3255
[552-21-6] Methylenedigallic Acid, 5983
[552-22-7] Thymol Iodide, 9335
[552-23-8] 2-Nitro-4-sulfobenzoic Acid, 6567
[552-25-0] Diampromide, 2966
[552-30-7] Trimellitic Anhydride, 9617
[552-33-0] Amidomycin, 413
[552-58-9] Eriodictyol, 3616
[552-59-0] Prunetin, 7923
[552-70-5] Pseudopelletierine, 7936
[552-72-7] Lobelanidine, 5429
[552-79-4] N-Methylephedrine, 5987
[552-82-9] Methyldiphenylamine, 5973
[552-93-2] Thymol Carbonate*, 9333
[552-94-3] Salsalate, 8307
[552-96-5] Procerin, 7766
[552-98-7] Lithium Acetylsalicylate, 5397
[553-03-7] Hydrocarbostyril, 4702
[553-06-0] N-(1,2-Diphenylethyl)-nicotinamide, 3324
[553-08-2] Thonzonium Bromide, 9306
[553-12-8] Protoporphyrin IX, 7913
[553-13-9] Zolamine, 10089
[553-17-3] Guaiacol Carbonate, 4459
[553-19-5] Xanthyletin, 9978
[553-20-8] 5'-Nitro-2'-propoxy-acetanilide, 6550
[553-24-2] Neutral Red, 6395
[553-26-4] γ,γ'-Dipyridyl, 3356
[553-27-5] Aniline Mustard, 688
[553-30-0] Proflavine Sulfate*, 7780
[553-39-9] Allenolic Acid, 247
[553-54-8] Lithium Benzoate, 5399
[553-58-2] Metabutethamine Hydrochloride*, 1517
[553-68-4] Butethamine Hydrochloride*, 1517
[553-69-5] Phenyramidol, 7292
[553-70-8] Magnesium Benzoate, 5533
[553-74-2] Nitrin, 6500
[553-75-3] Methylconiine, 5968
[553-79-7] 5-Nitro-2-propoxyaniline, 6551
[553-84-4] Perilla Ketone, 7118
[553-90-2] Methyl Oxalate, 6020

Denotes derivative of title compound

[553-91-3] Lithium Oxalate, 5415
[554-01-8] 5-Methylcytosine, 5970
[554-10-9] Glyceryl Iodide, 4383
[554-12-1] Methyl Propionate, 6030
[554-13-2] Lithium Carbonate, 5404
[554-15-4] N-Methylpyrroline, 6036
[554-18-7] Glucosulfone Sodium, 4357
[554-24-5] Phenobutiodil, 7203
[554-35-8] Linamarin, 5375
[554-57-4] Methazolamide, 5875
[554-70-1] Triethyl Phosphine, 9590
[554-71-2] Sozoiodolic Acid, 8687
[554-72-3] Tryparsamide, 9703
[554-73-4] Tropaeolin OO, 9688
[554-84-7] m-Nitrophenol, 6540
[554-91-6] Gentiobiose, 4288
[554-92-7] Trimethobenzamide Hydrochloride*, 9623
[554-99-4] N-Methylepinephrine, 5988
[555-10-2] β-Phellandrene, 7152
[555-21-5] p-Nitrobenzyl Cyanide, 6512
[555-28-2] Eucaine Hydrochloride*, 3850
[555-30-6] Methyldopa, 5974
[555-31-7] Aluminum Isopropoxide, 349
[555-32-8] Aluminum Benzoate, 327
[555-35-1] Aluminum Palmitate, 361
[555-37-3] Neburon, 6354
[555-43-1] Tristearin, 9669
[555-44-2] Tripalmitin, 9648
[555-45-3] Trimyristin, 9638
[555-54-4] Diphenylmagnesium, 3328
[555-57-7] Pargyline, 6988
[555-59-9] Maleanilic Acid, 5584
[555-68-0] m-Nitrocinnamic Acid, 6516
[555-75-9] Aluminum Ethoxide, 340
[555-76-0] Ferric Formate, 3966
[555-77-1] 2,2′,2″-Trichlorotriethylamine, 9560
[555-84-0] Nifuradene, 6445
[555-89-5] Bis(p-chlorophenoxy)methane, 1258
[555-92-0] Coniine Hydrochloride, 2502
[556-10-5] (4-Nitrophenyl)urea, 6547
[556-22-9] Glyodin, 4404
[556-24-1] Methyl Isovalerate, 6006
[556-27-4] Alliin, 251
[556-38-7] Zinc Valerate, 10070
[556-50-3] N-Glycylglycine, 4399
[556-52-5] Glycidol, 4385
[556-56-9] Allyl Iodide, 292
[556-61-6] Methyl Isothiocyanate, 6005
[556-63-8] Lithium Formate, 5410
[556-64-9] Methyl Thiocyanate, 6047
[556-67-2] Octamethylcyclotetrasiloxane, 6668
[556-88-7] Nitroguanidine, 6529
[556-89-8] Nitrourea, 6573
[556-91-2] Aluminum tert-Butoxide, 333
[556-99-0] Triuret, 9678
[557-04-0] Magnesium Stearate, 5572

[557-05-1] Zinc Stearate, 10063
[557-07-3] Zinc Oleate, 10047
[557-09-5] Zinc Caprylate, 10029
[557-11-9] Allylurea, 296
[557-17-5] Methyl Propyl Ether, 6031
[557-18-6] Diethylmagnesium, 3112
[557-19-7] Nickel Cyanide, 6413
[557-20-0] Diethylzinc, 3121
[557-21-1] Zinc Cyanide, 10034
[557-24-4] Maleamic Acid, 5583
[557-28-8] Zinc Propionate, 10057
[557-31-3] Allyl Ethyl Ether, 291
[557-34-6] Zinc Acetate, 10026
[557-35-7] sec-Octyl Bromide, 6685
[557-36-8] sec-Octyl Iodide, 6686
[557-39-1] Magnesium Formate, 5544
[557-40-4] Allyl Ether, 290
[557-41-5] Zinc Formate, 10036
[557-42-6] Zinc Thiocyanate, 10069
[557-59-5] Lignoceric Acid, 5364
[557-61-9] Octacosanol, 6664
[559-48-8] Kopsine, 5199
[559-49-9] Annotinine, 705
[559-57-9] Diginatigenin, 3137
[559-70-6] β-Amyrin, 654
[559-74-0] Friedelin, 4184
[560-88-3] Bornyl Salicylate, 1343
[561-07-9] Delphinine, 2867
[561-20-6] Cacotheline, 1604
[561-22-8] Holarrhenine, 4645
[561-25-1] Dihydrothebaine, 3164
[561-27-3] Diacetylmorphine, 2948
[561-48-8] Norpipanone, 6633
[561-51-3] Garryine, 4269
[561-76-2] Properidine, 7826
[561-77-3] Dihexyverine, 3151
[561-78-4] Alphaprodine Hydrochloride*, 307
[561-79-5] Metcaraphen, 5840
[561-83-1] Nealbarbital, 6350
[561-86-4] Brallobarbital, 1357
[561-94-4] Ergosine, 3601
[561-95-5] Nimbiol, 6463
[562-09-4] Chlorphenoxamine Hydrochloride*, 2182
[562-10-7] Doxylamine Succinate*, 3430
[562-26-5] Phenoperidine, 7216
[562-34-5] Chlorogenin, 2143
[562-54-9] Potassium Methyl Sulfate, 7632
[562-71-0] Suprasterol II, 8984
[562-76-5] Potassium Tetracyanoplatinate(II), 7678
[562-81-2] Barium Platinous Cyanide, 1003
[562-90-3] Silicon Tetraacetate, 8445
[563-03-1] Guaiacol Phosphate, 4460
[563-12-2] Ethion, 3691
[563-17-7] Potassium Ethyl Sulfate, 7610
[563-41-7] Semicarbazide Hydrochloride, 8396
[563-47-3] 3-Chloro-2-methylpropene, 2148
[563-52-0] 3-Chloro-1-butene, 2131
[563-63-3] Silver Acetate, 8451
[563-68-8] Thallium Acetate, 9184
[563-72-4] Calcium Oxalate, 1691
[564-00-1] Erythritol Anhydride, 3621
[564-36-3] Ergocornine, 3591
[564-37-4] Ergocorninine, 3592
[565-63-9] Angelic Acid, 678

[565-74-2] α-Bromoisovaleric Acid, 1409
[566-02-9] Allopregnane-3β,21-diol-11,20-dione, 262
[566-41-6] Allopregnane-3β,17α,-20β,21-tetrol, 266
[566-56-3] Allopregnane-3β,20α-diol, 260
[566-57-4] Allopregnane-3α,20β-diol, 259
[566-58-5] Allopregnane-3α,20α-diol, 258
[566-65-4] 3,20-Allopregnanedione, 264
[566-78-9] 21-Acetoxypregnenolone, 70
[567-18-0] 1-Naphthol-2-sulfonic Acid, 6310
[568-02-5] Alizarine Blue, 239
[568-21-8] Isothebaine, 5116
[568-53-6] α-Peltatin, 7019
[568-93-4] Alizarine Orange, 240
[569-31-3] Meconin, 5665
[569-57-3] Chlorotrianisene, 2173
[569-58-4] Aluminon, 320
[569-59-5] Phenindamine Tartrate*, 7195
[569-64-2] Malachite Green, 5581
[569-65-3] Meclizine, 5657
[569-77-7] Purpurogallin, 7963
[570-50-3] Allopregnane-3β,17α,-20α-triol, 269
[570-54-7] Allopregnane-3β,17α-diol-20-one, 263
[572-09-8] Acetobromglucose, 52
[572-48-5] Coumithoate, 2568
[572-59-8] Epiquinidine, 3570
[572-60-1] Epiquinine, 3571
[572-76-9] Nandinine, 6279
[572-96-3] Dihydrovitamin K₁, 3165
[573-01-3] Menadoxime, 5717
[573-09-1] Thiamine 1,5-Salt, 9226
[573-20-6] Menadiol Diacetate, 5710
[573-35-3] Inositol Monophosphate, 4884
[573-41-1] Theophylline Ethanolamine*, 9212
[573-56-8] 2,6-Dinitrophenol, 3276
[573-58-0] Congo Red, 2493
[573-83-1] Potassium Picrate, 7651
[574-12-9] Isoflavone, 5057
[574-64-1] Trypan Red, 9702
[574-65-2] Vital Red, 9917
[574-81-2] 3,7-Dinitro-5-oxophenothiazine, 3273
[574-84-5] Fraxetin, 4178
[574-95-8] Aureothricin, 898
[575-19-9] 6,7-Benzomorphan, 1105
[575-47-3] Amylpenicillin Sodium, 652
[575-74-6] Buclosamide, 1450
[575-75-7] Toldimfos Sodium, 9435
[576-11-4] Medmain, 5675
[576-19-2] Biocytin, 1240
[576-55-6] 3,4,5,6-Tetrabromo-o-cresol, 9120
[576-97-6] Noprylsulfamide, 6601
[577-11-7] Docusate Sodium, 3398
[577-27-5] Ledol, 5313
[577-33-3] Anthrarobin, 718
[577-37-7] Aphylline, 761
[577-48-0] Butamben Picrate*, 1504
[577-91-3] Iodoalphionic Acid, 4919
[578-19-8] Succinylsalicylic Acid, 8849

* Denotes derivative of title compound

[578-20-1] Thymyl *N*-Isoamyl-carbamate, 9343
[578-36-9] Potassium Salicylate, 7655
[578-74-5] Apigetrin, 764
[578-94-9] Phenarsazine Chloride, 7170
[579-04-4] Cymarose, 2769
[579-10-2] *N*-Methylacetanilide, 5931
[579-21-5] Lobelanine, 5430
[579-23-7] Cyclovalone, 2761
[579-38-4] Diloxanide, 3187
[579-56-6] Isoxsuprine Hydrochloride*, 5128
[579-92-0] Diphenylamine-2,2'-dicarboxylic Acid, 3318
[580-02-9] Methyl Acetylsalicylate, 5934
[580-13-2] 2-Bromonaphthalene, 1414
[580-52-9] Oxalenediuramidoxime, 6864
[581-49-7] Anatabine, 661
[581-75-9] 2,6-Naphthalenedisulfonic Acid, 6293
[581-88-4] Debrisoquin Sulfate*, 2837
[582-36-5] Oxythiamine, 6932
[582-52-5] Diacetoneglucose, 2945
[582-60-5] 5,6-Dimethylbenzimidazole, 3225
[583-03-9] Fenipentol, 3921
[583-15-3] Mercuric Benzoate, 5768
[583-39-1] 2-Benzimidazolethiol, 1092
[583-50-6] D-Erythrose, 3637
[583-52-8] Potassium Oxalate, 7639
[583-86-8] Phloionolic Acid, 7298
[583-91-5] Methionine Hydroxy Analog, 5897
[584-02-1] 3-Pentanol, 7076
[584-08-7] Potassium Carbonate, 7599
[584-10-1] Potassium Tetraiodocadmate, 7682
[584-18-9] Acetomeroctol, 57
[584-28-1] Aspidin, 868
[584-42-9] Metachrome Yellow, 5823
[584-43-0] Mercuric Succinimide, 5786
[584-84-9] Toluene 2,4-Diisocyanate, 9456
[584-85-0] Anserine, 708
[585-18-2] D-Erythrose 4-Phosphate, 3639
[585-29-5] Triethylammonium Formate, 9583
[585-48-8] 2,6-Di-*tert*-butylpyridine, 3022
[585-99-9] Melibiose, 5699
[586-02-7] Phrenosin, 7342
[586-06-1] Metaproterenol, 5836
[586-60-7] Dyclonine, 3453
[586-76-5] *p*-Bromobenzoic Acid, 1396
[586-84-5] 2-(Methoxymethyl)-5-nitrofuran, 5917
[587-23-5] Methenamine Mandelate, 5883
[587-33-7] *m*-Tyrosine, 9748
[587-46-2] Benzpyrinium Bromide, 1132
[587-61-1] Propyliodone, 7876
[587-90-6] 4,4'-Dinitrocarbanilide, 3271
[587-98-4] Metanil Yellow, 5833
[588-36-3] Methioprim, 5898

[588-42-1] Trolnitrate Phosphate, 9682
[588-59-0] Stilbene, 8774
[589-15-1] *p*-Bromobenzyl Bromide, 1397
[589-17-3] *p*-Bromobenzyl Chloride, 1398
[589-21-9] *p*-Bromophenylhydrazine, 1417
[589-44-6] 3-Amino-4-hydroxybutyric Acid, 455
[589-57-1] Ethyl Phosphorochloridite, 3795
[589-59-3] Isobutyl Isovalerate, 5029
[589-97-9] Potassium Percarbonate, 7640
[590-01-2] *n*-Butyl Propionate, 1587
[590-28-3] Potassium Cyanate, 7606
[590-29-4] Potassium Formate, 7614
[590-46-5] Betaine Hydrochloride*, 1201
[590-60-3] Amyl Carbamate, Tertiary, 641
[590-63-6] Bethanechol Chloride, 1207
[590-86-3] Isovaleraldehyde, 5118
[590-92-1] β-Bromopropionic Acid, 1421
[591-01-5] Dicyanodiamidine Sulfate, 3083
[591-09-3] Acetyl Nitrate, 92
[591-20-8] *m*-Bromophenol*, 1416
[591-27-5] *m*-Aminophenol, 472
[591-50-4] Iodobenzene, 4922
[591-64-0] Calcium Levulinate, 1684
[591-78-6] Methyl Butyl Ketone, 5955
[591-84-4] Isobutyl Thiocyanate, 5036
[591-89-9] Potassium Tetracyanomercurate(II), 7676
[591-97-9] 1-Chloro-2-butene, 2130
[592-01-8] Calcium Cyanide, 1664
[592-04-1] Mercuric Cyanide, 5772
[592-65-4] Isobutyl Sulfide, 5035
[592-85-8] Mercuric Thiocyanate, 5790
[592-87-0] Lead Thiocyanate, 5308
[592-88-1] Allyl Sulfide, 295
[592-89-2] Strontium Formate, 8806
[592-95-0] Sulforaphen, 8942
[593-26-0] Ammonium Palmitate, 566
[593-29-3] Potassium Stearate, 7664
[593-39-5] Petroselinic Acid, 7142
[593-50-0] 1-Triacontanol, 9506
[593-53-3] Fluoromethane, 4103
[593-74-8] Dimethylmercury, 3238
[594-08-1] Trithiocarbonic Acid, 9671
[595-05-1] Calycanthine, 1731
[595-33-5] Megestrol Acetate, 5687
[595-40-4] Isovaline, 5124
[595-77-7] Algestone, 229
[596-01-0] α-Naphtholphthalein, 6309
[596-03-2] 4',5'-Dibromofluorescein, 3005
[596-19-0] Morphine Mucate, 6190
[596-27-0] *o*-Cresolphthalein, 2582
[596-50-9] Poldine Methylsulfate*, 7529
[596-51-0] Glycopyrrolate, 4397

[596-56-5] Isatropic Acid, 4986
[597-12-6] Melezitose, 5698
[597-71-7] Pentaerythritol Tetraacetate, 7065
[598-31-2] Bromoacetone, 1389
[598-54-9] Cuprous Acetate, 2665
[598-55-0] Methyl Carbamate, 5958
[598-58-3] Methyl Nitrate, 6015
[598-62-9] Manganese Carbonate, 5608
[598-75-4] 3-Methyl-2-butanol, 5953
[598-82-3] DL-Lactic Acid, 5215
[599-04-2] Pantolactone, 6963
[599-33-7] Prednylidene, 7729
[599-79-1] Sulfasalazine, 8917
[599-88-2] Sulfaperine, 8909
[602-41-5] Thiocolchicoside*, 9253
[602-64-2] Anthragallol, 713
[602-81-3] Apomorphine Methylbromide*, 776
[602-92-6] Dibromogallic Acid, 3006
[603-00-9] Proxyphylline, 7921
[603-35-0] Triphenylphosphine, 9657
[603-45-2] Aurin, 899
[603-50-9] Bisacodyl, 1253
[603-55-4] DFDD, 2935
[603-63-4] Chloralantipyrine, 2058
[604-75-1] Oxazepam, 6881
[605-50-5] Isoamyl Phthalate, 5006
[605-65-2] Dansyl Chloride, 2812
[606-04-2] Pamabrom, 6952
[606-17-7] Iodipamide, 4916
[606-21-3] 2-Chloro-1,3-dinitrobenzene, 2136
[606-22-4] 2,6-Dinitroaniline, 3264
[606-58-6] Toyocamycin, 9483
[606-90-6] Piprinhydrinate, 7456
[607-80-7] Sesamin, 8420
[607-91-0] Myristicin, 6247
[608-07-1] 5-Methoxytryptamine, 5926
[608-66-2] Galactitol, 4238
[608-89-9] Ethyl Tartrate, Acid, 3812
[609-78-9] Cycloguanil Pamoate*, 2727
[610-60-6] Methyl Benzoylsalicylate, 5948
[610-88-8] Tenuazonic Acid, 9081
[611-75-6] Bromhexine Hydrochloride*, 1379
[613-78-5] 2-Naphthyl Salicylate, 6335
[613-89-8] Phenacylamine, 7159
[614-39-1] Procainamide Hydrochloride, 7762
[614-87-9] Dibromopropamidine Isethionate*, 3008
[615-05-4] 4-Methoxy-m-phenylenediamine, 5920
[616-05-7] α-Bromo-*n*-caproic Acid, 1403
[616-38-6] Methyl Carbonate, 5960
[616-45-5] 2-Pyrrolidone, 8027
[616-68-2] Trimecaine, 9614
[616-91-1] Acetylcysteine, 82
[617-12-9] Chorismic Acid, 2220
[617-19-6] 2,6-Diamino-2'-butyloxy-3,5'-azopyridine, 2956
[617-57-2] Lactic Acid Lactate, 5217
[618-22-4] Arsacetin, 817
[618-65-5] Helicin, 4544
[618-67-7] Melilotoside, 5701
[618-82-6] Sulfarsphenamine, 8916
[619-84-1] 4-(Dimethylamino)-benzoic Acid, 3221

[620-30-4] Metyrosine DL-Form*, 6085
[620-34-8] Methenamine Salicylate, 5884
[620-45-1] 2,6-Dichloroindophenol Sodium, 3057
[620-90-6] Phenamidine Isethionate*, 7165
[620-99-5] Phenacaine Hydrochloride, 7153
[621-72-7] Bendazol, 1043
[621-82-9] Cinnamic Acid, 2300
[622-91-3] Metoquinone, 6073
[623-42-7] Methyl Butyrate, 5957
[623-73-4] Diazoacetic Ester, 2980
[624-04-4] Glycol Dilaurate, 4393
[624-65-7] Propargyl Chloride, 7820
[624-83-9] Methyl Isocyanate, 6004
[625-04-7] Diacetonamine, 2942
[625-51-4] Formicin, 4154
[626-48-2] 6-Methyluracil, 6051
[627-13-4] n-Propyl Nitrate, 7877
[628-52-4] Chromous Acetate, 2244
[628-55-7] Isobutyl Ether, 5025
[629-11-8] Hexamethylene Glycol, 4610
[630-08-0] Carbon Monoxide, 1820
[630-56-8] 17α-Hydroxyprogesterone Caproate, 4774
[630-60-4] Ouabain, 6854
[630-64-8] Helveticoside, 4551
[630-67-1] Prednisolone Sodium 21-m-Sulfobenzoate, 7723
[630-81-9] Morphine Hydrobromide, 6187
[630-93-3] Phenytoin Sodium*, 7293
[631-01-6] Quillaic Acid, 8049
[631-60-7] Mercurous Acetate, 5792
[631-61-8] Ammonium Acetate, 514
[631-69-6] β-Boswellic Acid, 1354
[632-00-8] Sulfasomizole, 8918
[632-22-4] Tetramethylurea, 9160
[632-31-5] β-Ergostenol, 3605
[632-32-6] α-Ergostenol, 3604
[632-99-5] Magenta I, 5528
[633-03-4] Brilliant Green, 1367
[633-47-6] Cropropamide, 2595
[633-96-5] Orange II, 6813
[634-03-7] Phendimetrazine, 7180
[634-19-5] Phentydrone, 7235
[634-62-8] Hydrazine Tartrate, 4694
[635-41-6] Trimetozine, 9634
[635-65-4] Bilirubin, 1235
[635-97-2] Thiamine Phosphoric Acid Ester Phosphate Salt, 9225
[636-46-4] 4-Hydroxyisophthalic Acid, 4758
[636-47-5] Stallimycin, 8730
[636-54-4] Clopamide, 2391
[636-65-7] L-Isoglutamine, 5061
[637-03-6] Oxophenylarsine, 6903
[637-07-0] Clofibrate, 2374
[637-12-7] Aluminum Stearate, 370
[637-49-0] Coniine Hydrobromide, 2501
[637-58-1] Pramoxine Hydrochloride*, 7707
[637-92-3] Ethyl tert-Butyl Ether, 3732
[638-23-3] Carbocysteine, 1809
[638-38-0] Manganese Acetate, 5605
[638-39-1] Stannous Acetate, 8740
[638-94-8] Desonide, 2908
[638-95-9] α-Amyrin, 653

[639-14-5] Gypsogenin, 4497
[639-43-0] Akuammicine, 189
[639-44-1] Phenoltetrachlorophthalein, 7214
[639-46-3] Morphine N-Oxide, 6192
[639-48-5] Nicomorphine, 6430
[639-81-6] Ergotaminine, 3610
[639-86-1] Solanocapsine, 8662
[639-97-4] Apiose, 768
[640-19-7] Fluoroacetamide, 4095
[640-67-5] Manganese Oxalate, 5617
[641-36-1] Apocodeine, 771
[641-85-0] Allopregnane, 257
[642-08-0] Ostruthol, 6851
[642-15-9] Antimycin A₁, 746
[642-18-2] Alstonine, 314
[642-44-4] Aminometradine, 463
[642-58-0] Ethylbenzhydramine, 3724
[642-72-8] Benzydamine, 1136
[642-75-1] Adrenolutin, 164
[642-83-1] Aceglatone, 20
[642-84-2] 3',4',5-Trichlorosalicylanilide, 9558
[643-13-0] Fructose-6-phosphate, 4188
[643-22-1] Erythromycin Stearate, 3632
[643-60-7] Anhalamine, 682
[643-84-5] Malvidin Chloride, 5596
[644-31-5] Acetozone, 71
[644-62-2] Meclofenamic Acid, 5659
[644-64-4] Dimetilan, 3253
[645-05-6] Altretamine, 318
[645-43-2] Guanethidine Monosulfate*, 4473
[645-74-9] Myoral, 6241
[645-99-8] Mercuric Stearate, 5784
[646-13-9] Isobutyl Stearate, 5034
[650-69-1] β-Citraurin, 2326
[651-06-9] Sulfameter, 8885
[651-43-4] Allopregnane-3β,11β,-17α,21-tetrol-20- one, 268
[652-37-9] Acefylline, 18
[652-67-5] Isosorbide, 5113
[653-03-2] Butaperazine, 1509
[655-00-5] Thozalinone, 9315
[655-35-6] Chromonar Hydrochloride*, 2241
[655-76-5] Quinaldine Sulfate, 8058
[655-86-7] 2,3-Diaminophenazine, 2960
[657-24-9] Metformin, 5845
[657-27-2] Lysine Hydrochloride*, 5509
[659-40-5] Hexamidine Isethionate*, 4612
[659-70-1] Isoamyl Isovalerate, 5003
[660-27-5] Diisopropylamine Dichloroacetate, 3182
[660-60-6] Cupric Stearate, 2658
[664-95-9] Tolcyclamide, 9434
[665-66-7] Amantadine Hydrochloride*, 380
[666-99-9] Agaricic Acid, 174
[668-56-4] Testosterone Nicotinate, 9113
[670-54-2] Tetracyanoethylene, 9129
[670-65-5] Neopterin, 6378
[671-16-9] Procarbazine, 7764
[671-88-5] Disulfamide, 3369
[671-95-4] Clofenamide, 2371
[672-15-1] Homoserine, 4661
[672-87-7] Metyrosine, 6085
[673-31-4] Phenprobamate, 7229
[679-84-5] Halopropane, 4515

[680-31-9] Hempa, 4568
[685-73-4] D-Galacturonic Acid, 4242
[688-37-9] Aluminum Oleate, 357
[689-13-4] Hadacidin, 4500
[692-13-7] Buformin, 1465
[693-30-1] Hemisulfur Mustard, 4565
[693-65-2] n-Amyl Ether, 646
[693-72-1] Vaccenic Acid, 9810
[694-59-7] Pyridine 1-Oxide, 7984
[695-34-1] 2-Amino-4-picoline, 478
[695-53-4] Dimethadione, 3201
[702-54-5] Diethadione, 3096
[709-55-7] Etilefrin, 3822
[709-98-8] Propanil, 7814
[712-68-5] Triafur, 9509
[716-79-0] 2-Phenyl-1H-benzimidazole, 7245
[721-50-6] Prilocaine, 7749
[723-46-6] Sulfamethoxazole, 8889
[723-61-5] Helminthosporal, 4548
[728-88-1] Tolperisone, 9447
[729-99-7] Sulfamoxole, 8897
[730-68-7] Methitural, 5901
[732-11-6] Phosmet, 7311
[737-31-5] Diatrizoate Sodium, 2975
[738-70-5] Trimethoprim, 9624
[738-99-8] Dipin, 3341
[739-71-9] Trimipramine, 9636
[742-20-1] Cyclopenthiazide, 2750
[745-65-3] Prostaglandin E₁, 7892
[747-36-4] Hydroxychloroquine Sulfate*, 4748
[749-02-0] Spiperone, 8707
[749-13-3] Trifluperidol, 9596
[750-90-3] Quinine Salicylate, 8088
[751-19-9] A-Ninopterin, 691
[751-84-8] Penicillin G Benethamine, 7036
[751-94-0] Fusidic Acid Sodium*, 4231
[751-97-3] Rolitetracycline, 8236
[752-61-4] Digitalin, 3140
[759-73-9] N-Ethyl-N-nitrosourea, 3788
[759-94-4] EPTC, 3580
[764-05-6] Cyanogen Azide, 2699
[767-21-5] Piperidine Phosphate*, 7438
[768-94-5] Amantadine, 380
[773-76-2] Chloroxine, 2175
[777-11-7] Haloprogin, 4514
[786-19-6] Carbophenothion, 1827
[790-69-2] Loflucarban, 5440
[791-35-5] Chlophedianol, 2053
[795-13-1] N²-Formylsulfisomidine, 4164
[795-38-0] Dihydroisocodeine, 3159
[797-63-7] Norgestrel, 6621
[797-64-8] Norgestrel (−)-Form*, 6621
[800-22-6] Chloracizine, 2055
[800-24-8] 2,5-Bis(1-aziridinyl)-3,6-bis(2-methoxyethoxy)- 1,4-benzoquinone, 1256
[801-52-5] Porfiromycin, 7572
[804-10-4] Chromonar, 2241
[804-30-8] Fursultiamine, 4223
[804-36-4] Nitrovin, 6576
[804-63-7] Quinine Sulfate, 8089
[807-31-8] Aceperone, 25
[808-26-4] Sancycline, 8316
[808-71-9] Penethamate Hydriodide, 7027
[811-54-1] Lead Formate, 5283
[813-94-5] Calcium Citrate, 1661

* Denotes derivative of title compound

[814-71-1] Calcium Thioglycollate, 1718
[814-80-2] Calcium Lactate, 1683
[814-87-9] Aluminum Oxalate, 358
[814-89-1] Cobaltous Oxalate, 2439
[814-90-4] Chromous Oxalate, 2249
[814-91-5] Cupric Oxalate, 2649
[814-93-7] Lead Oxalate, 5294
[814-94-8] Stannous Oxalate, 8746
[814-95-9] Strontium Oxalate, 8811
[815-78-1] Aluminum Tartrate, 374
[815-82-7] Cupric Tartrate, 2662
[815-85-0] Stannous Tartrate, 8752
[819-17-0] Calcium Methionate, 1686
[819-73-8] Lead Butyrate, 5276
[821-33-0] 4,4'-Oxydi-2-butanol, 6913
[822-16-2] Sodium Stearate, 8634
[826-39-1] Mecamylamine Hydrochloride*, 5654
[828-00-2] Dimethoxane, 3212
[829-74-3] Nordefrin L-Form*, 6609
[830-89-7] Albutoin, 210
[834-12-8] Ametryn, 402
[834-28-6] Phenformin Hydrochloride*, 7191
[835-31-4] Naphazoline, 6287
[840-50-6] MADU, 5523
[841-73-6] Bucolome, 1452
[846-48-0] Boldenone, 1327
[846-49-1] Lorazepam, 5456
[846-50-4] Temazepam, 9074
[846-70-8] Naphthol Yellow S, 6313
[847-25-6] Thiamphenicol DL-Form*, 9230
[848-21-5] Norgestrienone, 6622
[848-53-3] Homochlorcyclizine, 4653
[848-75-9] Lormetazepam, 5458
[849-55-8] Nylidrin Hydrochloride*, 6654
[850-52-2] Altrenogest, 317
[852-19-7] Sulfazamet, 8923
[852-42-6] Guaiapate, 4463
[853-34-9] Kebuzone, 5168
[855-96-9] Eupatorin, 3858
[856-87-1] Dipipanone Hydrochloride*, 3342
[860-22-0] Indigo Carmine, 4856
[865-21-4] Vinblastine, 9887
[865-44-1] Iodine Trichloride, 4914
[866-82-0] Cupric Citrate, 2639
[866-83-1] Potassium Citrate, Monobasic, 7604
[866-84-2] Potassium Citrate, 7603
[868-14-4] Potassium Bitartrate, 7593
[868-16-6] Lithium Bitartrate, 5400
[868-18-8] Sodium Tartrate, 8640
[868-54-2] 2-Amino-1,1,3-tricyanopropene, 495
[870-72-4] Formaldehyde Sodium Bisulfite, 4149
[877-24-7] Potassium Biphthalate, 7589
[882-09-7] Clofibric Acid, 2375
[886-65-7] 1,4-Diphenyl-1,3-butadiene, 3320
[886-74-8] Chlorphenesin Carbamate, 2179
[886-86-2] Tricaine, 9533
[887-08-1] Daucol, 2824
[894-71-3] Nortriptyline Hydrochloride*, 6635
[897-15-4] Dothiepin Hydrochloride*, 3419

[897-61-0] Penicillin O Potassium*, 7044
[900-77-6] Drocarbil, 3434
[904-04-1] Captodiamine Hydrochloride*, 1772
[908-35-0] Metyrapone Ditartrate*, 6083
[908-54-3] Diminazene Aceturate, 3256
[909-39-7] Opipramol Dihydrochloride*, 6808
[910-86-1] Tiocarlide, 9382
[911-45-5] Clomiphene, 2384
[911-65-9] Etonitazene, 3840
[912-60-7] Noscapine Hydrochloride*, 6638
[914-00-1] Methacycline, 5851
[915-30-0] Diphenoxylate, 3313
[915-67-3] Amaranth (Dye), 382
[917-61-3] Sodium Cyanate, 8552
[917-69-1] Cobaltic Acetate, 2422
[918-04-7] Acetaldehyde Sodium Bisulfite, 34
[919-16-4] Lithium Citrate, 5407
[921-53-9] Potassium Tartrate, 7669
[922-55-4] Lanthionine, 5234
[922-80-5] Diamyl Sodium Sulfosuccinate, 2968
[923-06-8] Bromosuccinic Acid, 1427
[926-26-1] Di-tert-butyl Succinate, 3024
[926-93-2] Methallibure, 5857
[930-55-2] N-Nitrosopyrrolidine, 6564
[932-53-6] 6-Azathymine, 919
[937-13-3] Oxonic Acid, 6901
[938-73-8] Ethenzamide, 3685
[941-69-5] N-Phenylmaleimide, 7270
[944-22-9] Fonofos, 4147
[947-02-4] Phosfolan, 7309
[950-10-7] Mephosfolan, 5743
[950-37-8] Methidathion, 5891
[951-78-0] Deoxyuridine, 2892
[952-54-5] Morphazinamide, 6182
[956-03-6] Meprylcaine Hydrochloride, 5752
[956-90-1] Phencyclidine Hydrochloride*, 7179
[957-51-7] Diphenamid, 3305
[957-56-2] Fluindione, 4061
[957-68-6] 7-Aminocephalosporanic Acid, 53
[959-10-4] Xenbucin, 9980
[959-14-8] Oxolamine, 6898
[959-24-0] Sotalol Hydrochloride*, 8682
[960-05-4] Carbubarb, 1838
[961-71-7] Phenbenzamine, 7176
[962-02-7] Nitrodan, 6517
[965-52-6] Nifuroxazide, 6450
[965-90-2] Ethylestrenol, 3761
[965-93-5] Methyltrienolone, 6049
[968-81-0] Acetohexamide, 53
[968-93-4] Testolactone, 9108
[969-99-3] Opromazine, 6810
[972-02-1] Diphenidol, 3311
[973-21-7] Dinobuton, 3280
[973-53-5] Penicillin G Calcium, 7039
[976-71-6] Canrenone, 1753
[977-79-7] Medrogestone, 5676
[979-32-8] Estradiol Valerate*, 3653
[980-71-2] Brompheniramine Maleate*, 1433
[982-24-1] Clopenthixol, 2392
[982-43-4] Prenoxdiazine Hydrochloride, 7743

[983-85-7] Penamecillin, 7024
[985-13-7] Ethaverine Hydrochloride*, 3682
[985-16-0] Nafcillin Sodium, 6266
[985-32-0] Quinacillin, 8052
[987-18-8] Flumedroxone Acetate, 4064
[987-24-6] Betamethasone Acetate*, 1202
[987-78-0] Citicoline, 2321
[990-73-8] Fentanyl Citrate*, 3944
[991-42-4] Norbormide, 6604
[992-21-2] Lymecycline, 5500
[992-59-6] Benzopurpurine 4B, 1111
[997-68-2] L-Saccharopine, 8283
[999-81-5] Chlormequat Chloride, 2103
[1002-89-7] Ammonium Stearate, 582
[1003-78-7] 2,4-Dimethylsulfolane, 3245
[1005-24-9] Trigonellamide Chloride, 9605
[1008-65-7] Fenadiazole, 3898
[1013-20-3] Neo-cupferron, 6363
[1014-70-6] Simetryne®, 8488
[1017-56-7] Trimethylolmelamine, 9631
[1018-71-9] Pyrrolnitrin, 8029
[1028-33-7] Pentifylline, 7085
[1043-21-6] Pirenoxine, 7462
[1050-48-2] Benzilonium Bromide, 1090
[1050-79-9] Moperone, 6170
[1055-55-6] Bunamidine Hydrochloride*, 1475
[1059-14-9] Taraxasterol, 9031
[1063-77-0] Nomilin, 6593
[1066-17-7] Colistin, 2475
[1066-30-4] Chromic Acetate, 2221
[1066-33-7] Ammonium Bicarbonate, 516
[1069-54-1] Bis(1,2-dimethylpropyl)borane, 1261
[1069-55-2] Bucrylate, 1453
[1069-66-5] Valproic Acid Sodium Salt (1:1)*, 9821
[1070-11-7] Ethambutol Dihydrochloride*, 3672
[1070-83-3] tert-Butylacetic Acid, 1538
[1071-83-6] Glyphosate, 4408
[1072-35-1] Lead Stearate, 5300
[1078-19-9] 6-Methoxy-α-tetralone, 5924
[1078-21-3] 4-Amino-3-phenylbutyric Acid, 476
[1081-34-1] α-Terthienyl, 9106
[1082-57-1] Tramazoline, 9486
[1083-57-4] Bucetin, 1445
[1084-65-7] Meticrane, 6058
[1085-98-9] Dichlofluanid, 3031
[1086-80-2] Lumichrome, 5468
[1088-11-5] Nordazepam, 6608
[1088-56-8] Lumiflavine, 5469
[1088-92-2] Nifurtoinol, 6455
[1093-58-9] Clostebol, 2409
[1094-08-2] Ethopropazine Hydrochloride*, 3703
[1094-61-7] NMN, 6585
[1095-90-5] Methadone Hydrochloride, 5852
[1098-60-8] Triflupromazine Hydrochloride*, 9597
[1107-99-9] Prednisolone 21-Trimethylacetate, 7726
[1108-68-5] Cinobufotalin, 2310
[1110-40-3] Cortivazol, 2536
[1110-80-1] Pipacycline, 7420
[1111-39-3] α-Acetyldigitoxin*, 83

[1111-67-7] Cuprous Thiocyanate, 2676
[1111-71-3] Beryllium Formate, 1186
[1111-78-0] Ammonium Carbamate, 527
[1113-38-8] Ammonium Oxalate, 565
[1114-34-7] D-Lyxose, 5516
[1114-41-6] Muramic Acid, 6213
[1114-71-2] Pebulate, 7007
[1115-84-0] Vitamin U, 9942
[1116-54-7] N-Nitrosodiethanol-amine, 6556
[1118-68-9] N,N-Dimethylglycine, 3233
[1119-33-1] L-Glutamic Acid 5-Ethyl Ester, 4364
[1119-34-2] Arginine Hydrochloride*, 805
[1119-97-7] Myristyltrimethylammonium Bromide, 6249
[1120-44-1] Cupric Oleate, 2648
[1120-46-3] Lead Oleate, 5293
[1121-30-8] Pyrithione, 8004
[1131-64-2] Debrisoquin, 2837
[1132-61-2] MOPS, 6173
[1134-47-0] Baclofen, 950
[1134-98-1] Sulfarside, 8915
[1135-24-6] Ferulic Acid, 4011
[1142-70-7] Butallylonal, 1503
[1143-38-0] Anthralin, 714
[1146-21-0] 5-Furfuryl-5-isopropyl-barbituric Acid, 4216
[1146-98-1] Bromindione, 1381
[1146-99-2] Clorindione, 2402
[1147-62-2] Pyrovalerone Hydrochloride*, 8021
[1151-11-7] Ipodate Calcium*, 4958
[1151-15-1] o-(p-Anisoyl)benzoic Acid, 702
[1153-51-1] Androst-16-en-3-ol, 672
[1154-59-2] 3,3',4',5-Tetrachloro-salicylanilide, 9127
[1155-03-9] Zolamine Hydrochloride*, 10089
[1156-05-4] Phenglutarimide, 7192
[1156-19-0] Tolazamide, 9429
[1159-93-9] Clobenzepam, 2358
[1161-88-2] Sulfatolamide, 8922
[1164-38-1] Lachesine, 5210
[1165-48-6] Dimefline, 3190
[1169-79-5] Quinestradiol, 8065
[1172-02-7] 9-Dicyanomethylene-2,4,7-trinitrofluorene, 3084
[1172-18-5] Flurazepam Dihydrochloride*, 4123
[1174-11-4] Xenazoic Acid, 9979
[1177-87-3] Dexamethasone Acetate*, 2922
[1178-28-5] Metoserpate, 6074
[1178-29-6] Metoserpate Hydrochloride*, 6074
[1180-71-8] Limonin, 5372
[1181-54-0] Clomocycline, 2386
[1182-87-2] Peruvoside, 7136
[1183-76-2] Daphnandrine, 2815
[1185-57-5] Ammonium Ferric Citrate, 541
[1187-56-0] Selenomethionine Se75*, 8395
[1191-80-6] Mercuric Oleate, 5778
[1194-65-6] Dichlobenil, 3029
[1195-16-0] Citiolone, 2322
[1197-18-8] Tranexamic Acid, 9487
[1197-21-3] Phentermine Hydrochloride*, 7232
[1197-55-3] p-Aminophenylacetic Acid, 475
[1201-26-9] 3-Indolylacetone, 4873

[1209-98-9] Fencamfamine, 3909
[1210-56-6] Adrenoglomerulotropin, 163
[1211-28-5] Prolintane Hydrochloride*, 7791
[1213-06-5] Ethebenecid, 3684
[1215-70-9] Cinamiodyl, 2284
[1218-35-5] Xylometazoline Hydrochloride*, 9993
[1219-35-8] Primaperone, 7750
[1219-77-8] Ujothion, 9755
[1221-56-3] Ipodate Sodium*, 4958
[1222-57-7] Zolimidine, 10090
[1225-20-3] Iothalamic Acid Sodium*, 4952
[1225-55-4] Protriptyline Hydrochloride*, 7917
[1225-60-1] Isothipendyl Hydrochloride*, 5117
[1225-65-6] Prothipendyl Hydrochloride*, 7902
[1227-61-8] Mefexamide, 5682
[1228-19-9] Glypinamide, 4410
[1229-29-4] Doxepin Hydrochloride*, 3425
[1229-35-2] Methdilazine Hydrochloride*, 5877
[1231-93-2] Ethynodiol, 3816
[1235-15-0] Norbolethone, 6603
[1235-82-1] Biperiden Hydrochloride*, 1246
[1239-04-9] Phenazocine Hydrobromide*, 7174
[1239-29-8] Furazabol, 4209
[1239-45-8] Homidium Bromide*, 4650
[1240-15-9] Propiomazine Hydrochloride*, 7834
[1242-56-4] Stenbolone Acetate*, 8763
[1245-44-9] Propicillin Potassium*, 7829
[1247-42-3] Meprednisone, 5750
[1247-71-8] Colpormon, 2485
[1249-84-9] Azacosterol Dihydrochloride*, 908
[1253-28-7] Gestonorone Caproate, 4309
[1255-35-2] Fluprednidene Acetate, 4118
[1260-08-8] Sulphurenic Acid, 8970
[1260-17-9] Carminic Acid, 1850
[1263-89-4] Paromomycin Sulfate*, 6989
[1264-72-8] Colistin Sulfate*, 2475
[1271-19-8] Titanocene Dichloride, 9407
[1299-86-1] Aluminum Carbide, 335
[1300-32-9] Furoyl Chloride, 4222
[1300-51-2] Sodium Phenolsulfonate, 8610
[1300-71-6] Xylenol, 9989
[1300-73-8] Xylidine, 9991
[1300-81-8] Aluminum β-Naphthol-disulfonate, 353
[1301-16-2] Calcium Iodostearate, 1682
[1301-42-4] Euprocin, 3861
[1301-96-8] Silver(II) Oxide, 8469
[1302-01-8] Silver Subfluoride, 8479
[1302-09-6] Silver Selenide, 8477
[1302-34-7] Leucite, 5333
[1302-42-7] Sodium Aluminate, 8517
[1302-63-2] Potassium Aluminate, 7581
[1302-78-9] Bentonite, 1062
[1302-81-4] Aluminum Sulfide, 373
[1302-82-5] Aluminum Selenide, 367

[1302-83-6] Lazurite, 5265
[1303-00-0] Gallium Arsenide, 4253
[1303-11-3] Indium Arsenide, 4859
[1303-28-2] Arsenic Pentoxide, 827
[1303-32-8] Arsenic Disulfide, 822
[1303-33-9] Arsenic Trisulfide, 834
[1303-34-0] Arsenic Pentasulfide, 826
[1303-35-1] Arsenic Hemiselenide, 823
[1303-36-2] Arsenic Triselenide, 833
[1303-37-3] Arsenic Pentaselenide, 825
[1303-52-2] Gold Trihydroxide, 4430
[1303-57-7] Gold Monoxide, 4418
[1303-58-8] Gold Trioxide, 4431
[1303-60-2] Gold Monosulfide, 4417
[1303-61-3] Gold Trisulfide, 4432
[1303-62-4] Gold Selenide, 4421
[1303-86-2] Boric Anhydride, 1337
[1303-94-2] Lithium Borate, 5401
[1304-28-5] Barium Oxide, 998
[1304-29-6] Barium Peroxide, 1001
[1304-39-8] Barium Selenide, 1004
[1304-40-1] Barium Silicide, 1005
[1304-54-7] Beryllium Nitride, 1191
[1304-56-9] Beryllium Oxide, 1192
[1304-76-3] Bismuth Oxide, 1285
[1304-82-1] Bismuth Telluride, 1303
[1304-85-4] Bismuth Subnitrate, 1298
[1305-62-0] Calcium Hydroxide, 1676
[1305-78-8] Calcium Oxide, 1692
[1305-79-9] Calcium Peroxide, 1696
[1305-84-6] Calcium Selenide, 1708
[1305-99-3] Calcium Phosphide, 1702
[1306-01-0] Thomas Phosphate, 9305
[1306-06-5] Durapatite, 3449
[1306-19-0] Cadmium Oxide, 1621
[1306-23-6] Cadmium Sulfide, 1628
[1306-24-7] Cadmium Selenide, 1625
[1306-25-8] Cadmium Telluride, 1629
[1306-38-3] Ceric Oxide, 1986
[1307-52-4] Ruthenium Red, 8273
[1307-96-6] Cobaltous Oxide, 2440
[1308-06-1] Cobaltic-Cobaltous Oxide, 2423
[1308-14-1] Chromic Hydroxide, 2227
[1308-38-9] Chromic Oxide, 2229
[1308-56-1] Chalcopyrite, 2029
[1309-37-1] Ferric Oxide, 3973
[1309-42-8] Magnesium Hydroxide, 5548
[1309-48-4] Magnesium Oxide, 5555
[1309-60-0] Lead Dioxide, 5281
[1309-64-4] Antimony Trioxide, 743
[1310-32-3] Ferrous Selenide, 4004
[1310-43-6] Ferrous Phosphide, 4003
[1310-45-8] Ferric Subsulfate Solution, 3979
[1310-52-7] Magnesium Germanide, 5545
[1310-53-8] Germanium Dioxide, 4304
[1310-58-3] Potassium Hydroxide, 7625
[1310-61-8] Potassium Bisulfide, 7592
[1310-65-2] Lithium Hydroxide, 5412

Denotes derivative of title compound

[1310-73-2] Sodium Hydroxide, 8575
[1310-82-3] Rubidium Hydroxide, 8262
[1311-11-1] Lead Hydroxide, 5286
[1311-90-6] Ammonium Phosphotungstate, 574
[1312-03-4] Mercuric Subsulfate, 5785
[1312-15-8] Sulfur Iodide, 8954
[1312-41-0] Indium Antimonide, 4858
[1312-42-1] Indium Selenide, 4862
[1312-43-2] Indium Oxide, 4860
[1312-45-4] Indium Telluride, 4864
[1312-46-5] Iridium Sesquioxide, 4970
[1312-73-8] Potassium Sulfide, 7666
[1312-74-9] Potassium Selenide, 7657
[1312-76-1] Potassium Silicate, 7658
[1313-04-8] Magnesium Selenide, 5567
[1313-08-2] Magnesium Stannide, 5571
[1313-13-9] Manganese Dioxide, 5612
[1313-22-0] Manganese Selenide, 5621
[1313-27-5] Molybdenum Trioxide, 6148
[1313-29-7] Molybdenum Sesquioxide, 6147
[1313-30-0] Sodium Phosphomolybdate, 8617
[1313-49-1] Zinc Nitride, 10045
[1313-59-3] Sodium Oxide, 8602
[1313-60-6] Sodium Peroxide, 8607
[1313-82-2] Sodium Sulfide, 8637
[1313-85-5] Sodium Selenide, 8628
[1313-96-8] Niobium Pentoxide, 6475
[1313-99-1] Nickel Monoxide, 6419
[1314-06-3] Nickel Sesquioxide, 6423
[1314-08-5] Palladium Oxide, 6944
[1314-11-0] Strontium Oxide, 8812
[1314-12-1] Thallium Oxide, 9193
[1314-13-2] Zinc Oxide, 10050
[1314-15-4] Platinic Oxide, 7500
[1314-18-7] Strontium Peroxide, 8813
[1314-20-1] Thorium Oxide, 9312
[1314-22-3] Zinc Peroxide, 10053
[1314-23-4] Zirconium Oxide, 10083
[1314-24-5] Phosphorus Trioxide, 7335
[1314-27-8] Lead Sesquioxide, 5298
[1314-28-9] Rhenium Trioxide, 8180
[1314-32-5] Thallium Sesquioxide, 9196
[1314-34-7] Vanadium Trioxide, 9829
[1314-35-8] Tungsten Trioxide, 9724
[1314-41-6] Lead Tetroxide, 5307
[1314-56-3] Phosphorus Pentoxide, 7330
[1314-60-9] Antimony Pentoxide, 730
[1314-61-0] Tantalum Pentoxide, 9028
[1314-62-1] Vanadium Pentoxide, 9826
[1314-64-3] Uranyl Sulfate, 9778
[1314-68-7] Rhenium Heptoxide, 8177
[1314-80-3] Phosphorus Pentasulfide, 7329

[1314-81-4] Phosphorus Hemitriselenide, 7322
[1314-82-5] Phosphorus Pentaselenide, 7328
[1314-84-7] Zinc Phosphide, 10056
[1314-85-8] Tetraphosphorus Trisulfide, 9168
[1314-86-9] Phosphorus Triselenide, 7336
[1314-87-0] Lead Sulfide, 5303
[1314-91-6] Lead Telluride, 5304
[1314-95-0] Stannous Sulfide, 8751
[1314-96-1] Strontium Sulfide, 8817
[1314-97-2] Thallium Sulfide, 9198
[1314-98-3] Zinc Sulfide, 10065
[1315-01-1] Stannic Sulfide, 8739
[1315-03-3] Vanadium Trisulfide, 9831
[1315-04-4] Antimony Pentasulfide, 729
[1315-05-5] Antimony Triselenide, 744
[1315-06-6] Stannous Selenide, 8749
[1315-09-9] Zinc Selenide, 10061
[1315-11-3] Zinc Telluride, 10068
[1317-26-6] Magaldrate, 5527
[1317-33-5] Molybdenum Disulfide, 6145
[1317-34-6] Manganese Sesquioxide, 5622
[1317-35-7] Manganese Oxide, 5618
[1317-36-8] Lead Monoxide, 5291
[1317-37-9] Ferrous Sulfide, 4007
[1317-38-0] Cupric Oxide, 2650
[1317-39-1] Cuprous Oxide, 2671
[1317-40-4] Cupric Sulfide, 2661
[1317-41-5] Cupric Selenide, 2656
[1317-42-6] Cobaltous Sulfide, 2444
[1317-54-0] Ferrite, 3983
[1317-61-9] Ferrosoferric Oxide, 3988
[1317-82-4] Sapphire, 8329
[1317-95-9] Tripoli, 9660
[1317-97-1] Ultramarine, 9757
[1318-00-9] Vermiculite, 9863
[1318-93-0] Montmorillonite, 6168
[1319-48-8] Leadhillite, 5285
[1319-69-3] Potassium Glycerophosphate, 7616
[1319-91-1] Calcium Iodobehenate, 1681
[1321-14-8] Potassium Guaiacolsulfonate, 7617
[1327-39-5] Calcium Aluminosilicate, 1645
[1327-41-9] Aluminum Hydroxychloride, 346
[1327-53-3] Arsenic Trioxide, 832
[1329-37-9] p,p'-Sulfonyldianiline-N,N'-digalactoside, 8941
[1329-77-7] Monardein Chloride, 6154
[1329-83-5] Tigonin, 9368
[1330-20-7] Xylene, 9988
[1330-43-4] Sodium Borate, 8535
[1330-51-4] Pyrogallol Monoacetate*, 8010
[1330-78-5] Tritolyl Phosphate, 9675
[1332-12-3] Gold, Explosive, 4413
[1332-14-5] Cupric Sulfate, Basic, 2660
[1332-52-1] Beryllium Acetate, Basic, 1179
[1332-53-2] Rhodanilic Acid, 8183
[1332-77-0] Potassium Tetraborate, 7672
[1333-38-6] Angelica Lactone, 679
[1333-74-0] Hydrogen, 4719
[1333-82-0] Chromium Trioxide, 2238

[1333-83-1] Sodium Bifluoride, 8529
[1334-74-3] Sodium Glycerophosphate, 8570
[1335-26-8] Magnesium Peroxide, 5559
[1335-31-5] Mercuric Oxycyanide, 5781
[1335-32-6] Lead Subacetate, 5301
[1336-20-5] Tetracycline Phosphate Complex*, 9130
[1336-29-4] Bisacodyl Tannex*, 1253
[1336-80-7] Ferrocholinate, 3986
[1338-16-5] Iron Sorbitex, 4981
[1338-98-3] Ethylstibamine, 3807
[1339-92-0] Basic Aluminum Carbonate Gel, 1015
[1340-26-7] Phenacetolin, 7156
[1340-35-8] Urea Stibamine, 9786
[1341-49-7] Ammonium Bifluoride, 517
[1343-98-2] Silicic Acid, 8437
[1344-09-8] Sodium Silicate, 8631
[1344-09-8] Sodium Silicate Solution, 8632
[1344-28-1] Aluminum Oxide, 359
[1344-28-1] Aluminum Oxide (Brockmann), 360
[1344-36-1] Basic Lead Carbonate, 1016
[1344-48-5] Mercuric Sulfide, Black, 5788
[1344-48-5] Mercuric Sulfide, Red, 5789
[1344-57-6] Uranium Dioxide, 9767
[1344-58-7] Uranium Trioxide, 9773
[1344-95-2] Calcium Silicate, 1709
[1345-04-6] Antimony Trisulfide, 745
[1345-05-7] Lithopone, 5424
[1345-07-9] Bismuth Sulfide, 1301
[1345-24-0] Gold Stannate, 4424
[1345-25-1] Ferrous Oxide, 4001
[1362-42-1] Absinthin, 7
[1390-93-8] Citrullol, 2334
[1391-14-6] Coptine, 2520
[1391-82-8] Grisein, 4452
[1392-21-8] Leucomycins, 5337
[1392-27-4] Litmocidin, 5425
[1392-46-7] Micrococcin P, 6103
[1392-81-0] Parotin, 6990
[1392-87-6] Phasin, 7150
[1393-12-0] Rimocidin, 8224
[1393-25-5] Secretin, 8375
[1393-38-0] Subtilin, 8835
[1393-48-2] Thiostrepton, 9295
[1393-62-0] Abrin, 4
[1393-63-1] Annatto, 704
[1393-64-2] Araroba, 795
[1393-68-6] Bottromycin, 1355
[1393-87-9] Fusafungine, 4426
[1394-02-1] Hachimycin, 4499
[1395-18-2] Azolitmin, 933
[1397-74-6] Acetyltannic Acid, 101
[1397-77-9] Actinorhodine, 135
[1397-84-8] Alazopeptin, 199
[1397-89-3] Amphotericin B, 620
[1398-17-0] Bakankosin, 955
[1398-61-4] Chitin, 2052
[1398-78-3] Colocynthin, 2483
[1399-57-1] Graphitic Acid, 4445
[1399-64-0] Gymnemic Acid, 4496
[1400-61-9] Nystatin, 6658
[1400-62-0] Orcein, 6818
[1400-67-5] Oryzacidin, 6839
[1401-54-3] Tanacetin, 9020
[1401-69-0] Tylosin, 9740
[1401-78-1] Vincetoxin, 9889
[1401-79-2] Viomycin Pantothenate, 9906

Denotes derivative of title compound

[1401-98-5] Condurangin, 2491
[1402-10-4] Lichenin, 5357
[1402-37-5] Actinomycetin, 132
[1402-44-4] Actinomycin F₁, 133
[1402-82-0] Amphomycin, 618
[1403-17-4] Candicidin, 1747
[1403-28-7] Carzinophilin, 1886
[1403-66-3] Gentamicin, 4284
[1403-71-0] Hamycin, 4524
[1404-04-2] Neomycin, 6369
[1404-15-5] Nogalamycin, 6591
[1404-23-5] Pleurotin(e), 7509
[1404-55-3] Ristocetin, 8227
[1404-74-6] Streptovaricin, 8792
[1404-88-2] Tyrothricin, 9749
[1404-90-6] Vancomycin, 9836
[1404-93-9] Vancomycin Hydrochloride*, 9836
[1405-10-3] Neomycin Sulfate*,
6369
[1405-20-5] Polymyxin B Sulfate*,
7550
[1405-37-4] Capreomycin Sulfate*,
1758
[1405-41-0] Gentamicin Sulfate*,
4284
[1405-76-1] Gitalin, 4324
[1405-86-3] Glycyrrhizic Acid, 4401
[1405-87-4] Bacitracin, 948
[1405-88-5] Bacitracin Methylenedisalicylic Acid, 949
[1405-90-9] Candidin, 1748
[1406-04-8] Neomycin Undecylenate, 6370
[1406-11-7] Polymyxin, 7550
[1406-65-1] Chlorophyll, 2155
[1407-04-1] Lututrin, 5486
[1407-09-6] Polymyxin B-Methanesulfonic Acid, 7551
[1407-47-2] Angiotensin, 680
[1407-73-4] Vitamin T, 9941
[1407-85-8] Amphetamine d-Form
Tannate*, 616
[1407-93-8] Plantisul, 7490
[1414-39-7] Albomycin, 204
[1414-45-5] Nisin, 6481
[1415-94-7] Iodopsin, 4932
[1416-60-0] Splenin, 8723
[1420-03-7] Propenzolate Hydrochloride*, 7824
[1420-06-0] Trifenmorph, 9591
[1420-53-7] Codeine Sulfate, 2463
[1420-55-9] Thiethylperazine, 9241
[1421-14-3] Propanidid, 7813
[1421-68-7] Amidephrine Mesylate*,
409
[1422-07-7] Codeine Hydrochloride*, 2459
[1424-00-6] Mesterolone, 5817
[1424-27-7] Acetazolamide Sodium*, 45
[1429-30-7] Petunidin, 7143
[1435-55-8] Hydroquinidine, 4736
[1438-30-8] Netropsin, 6391
[1438-57-9] Grindelic Acid, 4451
[1443-44-3] α-Methylfentanyl*,
5992
[1448-23-3] Glaucarubin, 4331
[1463-28-1] Guanacline, 4470
[1464-42-2] Selenomethionine, 8395
[1476-53-5] Novobiocin Sodium*,
6641
[1477-10-7] Bietamiverine Hydrochloride, 1224
[1477-19-6] Benzarone, 1071
[1477-39-0] Noracymethadol, 6602
[1477-40-3] Levomethadyl Acetate,
5346
[1477-57-2] Fertilysin, 4009
[1480-19-9] Fluanisone, 4047
[1483-07-4] Albizziin, 202

[1491-41-4] Naftalofos, 6269
[1491-59-4] Oxymetazoline, 6919
[1492-02-0] Glybuzole, 4374
[1493-02-3] Formyl Fluoride, 4162
[1501-84-4] Rimantadine Hydrochloride*, 8221
[1503-53-3] 5-Bromosalicylic Acid
Acetate, 1425
[1506-12-3] Butidrine Hydrochloride, 1522
[1508-45-8] Podophyllinic Acid
2-Ethylhydrazide, 7519
[1508-65-2] Oxybutynin Chloride,
6908
[1508-75-4] Tropicamide, 9692
[1508-76-5] Procyclidine Hydrochloride*, 7770
[1518-86-1] Hydroxyamphetamine,
4740
[1524-88-5] Flurandrenolide, 4122
[1531-12-0] Norlevorphanol, 6625
[1538-09-6] Penicillin G Benzathine,
7037
[1538-11-0] Penicillin G Benzhydrylamine, 7038
[1553-60-2] Ibufenac, 4811
[1555-53-9] Magnesium Oleate,
5553
[1562-13-6] 3-O-Lauroylpyridoxol
Diacetate, 5259
[1562-71-6] Guanaclirne Sulfate*,
4470
[1562-85-2] Gallocyanine, 4256
[1562-90-9] Celestin Blue, 1955
[1563-66-2] Carbofuran, 1810
[1580-83-2] Paraflutizide, 6973
[1582-09-8] Trifluralin, 9598
[1590-87-0] Disilane, 3362
[1592-23-0] Calcium Stearate, 1710
[1593-77-7] Dodemorph, 3405
[1596-84-5] Daminozide, 2810
[1597-82-6] Paramethasone Acetate*, 6977
[1600-27-7] Mercuric Acetate, 5766
[1603-91-4] 2-Amino-4-methylthia-
zole, 462
[1604-01-9] γ-Coniceine, 2497
[1605-89-6] Bolasterone, 1326
[1607-17-6] Pentrinitrol, 7094
[1609-47-8] Pyrocarbonic Acid
Diethyl Ester, 8008
[1610-18-0] Prometon, 7799
[1613-17-8] Dymanthine Hydrochloride*, 3455
[1614-20-6] Nifurprazine, 6453
[1617-90-9] Vincamine, 9888
[1618-26-4] Bis[methylthio]methane,
1267
[1619-29-0] Helminthosporol, 4549
[1619-34-7] 3-Quinuclidinol, 8110
[1620-21-9] Chlorcyclizine Hydrochloride*, 2078
[1622-61-3] Clonazepam, 2387
[1622-62-4] Flunitrazepam, 4072
[1633-05-2] Strontium Carbonate,
8801
[1634-04-4] Methyl tert-Butyl Ether,
5954
[1637-39-4] Zeatin, 10018
[1639-60-7] Propoxyphene Hydrochloride*, 7851
[1641-17-4] Mexenone, 6092
[1642-54-2] Diethylcarbamazine
Citrate*, 3106
[1649-18-9] Azaperone, 915
[1661-29-6] Meturedepa, 6082
[1665-48-1] Metaxalone, 5838
[1668-19-5] Doxepin, 3425
[1672-46-4] Digoxigenin, 3149
[1673-06-9] Amphotalide, 619
[1679-76-1] Drofenine, 3435

[1684-42-0] Acranil, 116
[1689-84-5] Bromoxynil, 1431
[1689-89-0] Nitroxynil, 6578
[1690-72-8] Methyl N-Methylnipecotate, 6012
[1695-77-8] Spectinomycin, 8695
[1702-17-6] Clopyralid, 2398
[1702-77-8] Fusarubin, 4228
[1707-14-8] Phenmetrazine Hydrochloride*, 7200
[1715-33-9] Prednisolone Sodium
Succinate, 7722
[1721-51-3] ζ₁-Tocopherol, 9421
[1722-62-9] Mepivacaine Hydrochloride*, 5748
[1729-61-9] Paranyline, 6978
[1733-12-6] Cresol Red, 2583
[1740-22-3] Pyrinoline, 8001
[1746-01-6] TCDD, 9052
[1762-95-4] Ammonium Thiocyanate, 591
[1764-85-8] Epithiazide, 3574
[1777-82-8] Dichlorobenzyl Alcohol,
3048
[1786-81-8] Prilocaine Hydrochloride*, 7749
[1787-61-7] Eriochrome® Black T,
3615
[1794-74-7] Cethexonium Bromide,
2011
[1798-50-1] Azacyclonol Hydrochloride*, 909
[1805-85-2] Hasubanonine, 4533
[1805-86-3] Metaphanine, 5834
[1808-12-4] Bromodiphenhydramine
Hydrochloride*, 1405
[1812-30-2] Bromazepam, 1374
[1821-16-5] Ethylidene Dicoumarol,
3768
[1824-50-6] Benzylhydrochlorothiazide, 1150
[1824-58-4] Ethiazide, 3687
[1829-00-1] Thiazol Yellow G, 9238
[1830-32-6] Azintamide, 927
[1836-75-5] Nitrofen, 6519
[1837-57-6] Ethacridine Lactate*,
3668
[1841-19-6] Fluspirilene, 4131
[1843-05-6] Octabenzone, 6662
[1859-87-6] Echinuline, 3473
[1861-21-8] Enallylpropymal, 3523
[1861-32-1] DCPA, 2830
[1861-40-1] Benfluralin, 1048
[1863-63-4] Ammonium Benzoate,
515
[1867-66-9] Ketamine Hydrochloride*, 5174
[1871-22-3] Tetrazolium Blue, 9173
[1874-54-0] Psicofuranine, 7939
[1882-26-4] Pyridinol Carbamate,
7987
[1891-29-8] Lactucin, 5224
[1892-56-4] Athamantin, 880
[1892-80-4] Fenethylline Hydrochloride*, 3919
[1893-33-0] Pipamperone, 7422
[1897-45-6] Chlorothalonil, 2167
[1898-66-4] 1,1-Diphenyl-2-picryl-
hydrazyl (Free Radical), 3332
[1904-98-9] 2,6-Diaminopurine,
2963
[1910-68-5] Methisazone, 5900
[1912-24-9] Atrazine, 886
[1912-26-1] Trietazine, 9580
[1918-00-9] Dicamba, 3026
[1918-02-1] Picloram, 7370
[1918-08-7] Dipropalin®, 3348
[1918-16-7] Propachlor, 7805
[1921-70-6] Pristane, 7757
[1923-76-8] Bunamiodyl Sodium,
1476

* Denotes derivative of title compound

[1926-48-3] Fenbenicillin, 3905
[1926-49-4] Clometocillin, 2383
[1929-77-7] Vernolate, 9866
[1929-82-4] Nitrapyrin, 6490
[1934-21-0] Tartrazine, 9041
[1936-18-1] Salutaridine, 8309
[1936-28-3] Sodium Arsphenamine, 8524
[1936-40-9] Novembichin, 6640
[1937-54-8] Solanone, 8663
[1945-53-5] Palustric Acid, 6950
[1949-45-7] Metrizoic Acid, 6078
[1950-31-8] Metcaraphen Hydrochloride*, 5840
[1950-39-6] Deferoxamine Hydrochloride*, 2850
[1951-25-3] Amiodarone, 497
[1953-02-2] Tiopronin, 9386
[1953-04-4] Galantamine Hydrobromide*, 4245
[1954-28-5] Etoglucid, 3837
[1954-79-6] Mecloralurea, 5662
[1962-14-7] Acacic Acid, 11
[1976-28-9] Aluminum Nicotinate, 354
[1977-10-2] Loxapine, 5461
[1977-11-3] Perlapine, 7131
[1980-45-6] Benzodepa, 1097
[1982-37-2] Methdilazine, 5877
[1982-49-6] Siduron, 8433
[1983-72-8] Medicagol, 5672
[1984-94-7] Sulfasymazine, 8919
[1986-53-4] Bolandiol Dipropionate*, 1325
[1986-70-5] Calotropin, 1728
[1987-71-9] Nicotinamide Ascorbate, 6433
[1990-29-0] D-Altrose, 319
[1997-15-5] Triamcinolone Acetonide Disodium Phosphate*, 9512
[2001-81-2] Diponium Bromide, 3346
[2001-95-8] Valinomycin, 9819
[2002-29-1] Flumethasone Pivalate*, 4066
[2008-41-5] Butylate, 1546
[2011-67-8] Nimetazepam, 6465
[2013-58-3] Meclocycline, 5658
[2016-36-6] Choline Salicylate, 2212
[2016-63-9] Bamifylline, 965
[2019-68-3] Atrolactamide, 888
[2022-85-7] Flucytosine, 4056
[2030-53-7] Aporeine, 778
[2030-63-9] Clofazimine, 2370
[2032-59-9] Aminocarb, 443
[2032-65-7] Methiocarb, 5893
[2033-24-1] Meldrum's Acid, 5696
[2037-48-1] 2-Deoxystreptamine, 2891
[2037-95-8] Carsalam, 1873
[2043-38-1] Buthiazide, 1519
[2052-01-9] α-Bromoisobutyric Acid, 1408
[2055-44-9] Perisoxal, 7129
[2058-46-0] Oxytetracycline Hydrochloride*, 6931
[2058-52-8] Clothiapine, 2410
[2062-78-4] Pimozide, 7404
[2062-84-2] Benperidol, 1057
[2065-00-1] Filicinic Acid, 4021
[2066-89-9] Pasiniazide, 6999
[2068-78-2] Vincristine Sulfate*, 9891
[2078-54-8] Propofol, 7847
[2079-00-7] Blasticidin S, 1323
[2085-42-9] Ethoheptazine Citrate*, 3698
[2086-83-1] Berberine, 1169
[2090-54-2] N-Hydroxyethylpromethazine Chloride, 4755
[2090-89-3] Butethamine, 1517

[2092-16-2] Calcium Thiocyanate, 1717
[2092-17-3] Barium Thiocyanate, 1010
[2095-57-0] Thiobutabarbital, 9250
[2096-42-6] Gougerotin, 4436
[2098-66-0] Cyproterone, 2781
[2103-64-2] Gallein, 4250
[2104-64-5] EPN, 3576
[2104-96-3] Bromophos, 1419
[2109-73-1] Butacetin, 1497
[2111-75-3] Perillaldehyde, 7119
[2119-75-7] Fluperolone Acetate, 4115
[2122-29-4] Retamine, 8161
[2127-01-7] Clorexolone, 2401
[2128-61-2] Sparsiflorine, 8691
[2135-17-3] Flumethasone, 4066
[2139-47-1] Nifenazone, 6443
[2140-46-7] 25-Hydroxycholesterol, 4751
[2141-09-5] Magnoflorine, 5579
[2150-48-3] Pyronine B, 8016
[2152-34-3] Pemoline, 7021
[2152-44-5] Betamethasone 17-Valerate*, 1202
[2152-56-9] Arabitol, 789
[2152-76-3] Idose, 4818
[2154-02-1] Metofoline, 6066
[2156-27-6] Benproperine, 1058
[2164-08-1] Lenacil, 5318
[2164-09-2] Dicryl, 3078
[2164-17-2] Fluometuron, 4080
[2165-19-7] Guanoxan, 4482
[2167-85-3] Pipazethate, 7423
[2169-75-7] Deptropine Citrate*, 2894
[2179-16-0] Ninopterin, 6471
[2179-37-5] Bencyclane, 1041
[2180-92-9] Bupivacaine, 1485
[2181-04-6] Canrenoate Potassium*, 1753
[2182-14-1] Vindoline, 9893
[2188-67-2] Naepaine, 6263
[2192-20-3] Hydroxyzine Dihydrochloride*, 4786
[2205-73-4] Tiomesterone, 9385
[2207-50-3] Aminorex, 490
[2209-86-1] Pentaerythritol Dichlorohydrin, 7064
[2210-63-1] Mofebutazone, 6141
[2210-64-2] Pyrrocaine Hydrochloride*, 8024
[2210-77-7] Pyrrocaine, 8024
[2217-44-9] Iodophthalein Sodium, 4931
[2218-68-0] Chloral Betaine, 2059
[2221-95-6] Fichtelite, 4019
[2227-17-0] Dienochlor, 3095
[2235-90-7] Etryptamine, 3848
[2240-14-4] Fencamfamine Hydrochloride*, 3909
[2241-90-9] Cyclobuxine, 2722
[2243-76-7] Alizarine Yellow R, 241
[2244-21-5] Troclosene Potassium, 9679
[2255-39-2] Muscazone, 6220
[2259-96-3] Cyclothiazide, 2760
[2272-11-9] Ethanolamine Oleate*, 3681
[2276-90-6] Iothalamic Acid, 4952
[2277-92-1] Oxyclozanide, 6911
[2284-31-3] Pratensein, 7711
[2289-50-1] Diphenicillin Sodium, 3310
[2292-79-7] Congressane, 2494
[2295-58-1] Flopropione, 4039
[2303-16-4] Diallate, 2950
[2303-17-5] Triallate, 9510
[2306-22-1] Citramalic Acid, 2325

[2307-68-8] Solan, 8659
[2310-17-0] Phosalone, 7308
[2312-35-8] Propargite, 7818
[2313-87-3] Ethoxazene Hydrochloride*, 3706
[2315-02-8] Oxymetazoline Hydrochloride*, 6919
[2315-08-4] Salazosulfadimidine, 8291
[2316-64-5] Bromosaligenin, 1426
[2319-97-3] [2.2]Metacyclophane, 5825
[2320-96-9] 4',5'-Dichlorofluorescein, 3056
[2321-07-5] Fluorescein, 4085
[2323-36-6] Deprenyl, 2893
[2326-89-8] Drimenin, 3433
[2338-05-8] Ferric Citrate, 3963
[2338-37-6] Levopropoxyphene, 5349
[2350-32-5] Butoxycaine, 1531
[2353-45-9] Fast Green FCF, 3886
[2363-58-8] Epitiostanol, 3575
[2375-03-3] Methylprednisolone Sodium Succinate*, 6028
[2385-81-1] Furethidine, 4213
[2385-85-5] Mirex, 6126
[2390-22-9] Hetolin®, 4596
[2390-99-0] Chanoclavine, 2033
[2391-03-9] Brompheniramine d-Form Maleate*, 1433
[2392-39-4] Dexamethasone Sodium Phosphate*, 2922
[2398-81-4] Oxiniacic Acid, 6895
[2398-96-1] Tolnaftate, 9442
[2410-93-7] Methopterin, 5907
[2425-06-1] Captafol, 1770
[2429-71-2] Benzo Azurine G, 1095
[2430-27-5] Valpromide, 9822
[2430-46-8] Tolboxane, 9431
[2430-49-1] Vinylbital, 9897
[2435-76-9] 5-Diazouracil, 2985
[2438-32-6] Chlorpheniramine d-Form Maleate*, 2180
[2438-72-4] Bufexamac, 1462
[2438-80-4] L-Fucose, 4193
[2439-01-2] Oxythioquinox, 6933
[2439-10-3] Dodine, 3406
[2439-99-8] Glyphosine, 4409
[2440-22-4] Tinuvin® P, 9381
[2447-54-3] Sanguinarine, 8320
[2447-57-6] Sulfadoxine, 8877
[2447-58-7] Lochnericine, 5434
[2454-11-7] Formyldienolone, 4161
[2463-84-5] Dicapthon, 3027
[2466-09-3] Pyrophosphoric Acid, 8018
[2468-21-5] Catharanthine, 1910
[2469-34-3] Senegenin, 8405
[2470-73-7] Dixyrazine, 3391
[2487-39-0] Vitamin K-S(II), 9939
[2487-63-0] Quinbolone, 8063
[2490-97-3] Aceglutamide, 21
[2493-02-9] p-Bromophenyl Isocyanate, 1418
[2507-24-6] Physalaemin, 7354
[2508-72-7] Antazoline Hydrochloride*, 709
[2508-79-4] Methyldopa Ethyl Ester Hydrochloride*, 5974
[2517-04-6] 2-Azetidinecarboxylic Acid, 923
[2520-21-0] Celesticetin, 1954
[2529-45-5] Flurogestone Acetate, 4125
[2531-04-6] Piperylone, 7449
[2533-89-3] Karsil, 5165
[2537-29-3] Proxibarbal, 7920
[2540-82-1] Formothion, 4160
[2541-01-6] Terpin Hydrate*, 9101
[2550-75-6] Chlorbicyclen, 2077

[2551-62-4] Sulfur Hexafluoride, 8952
[2551-76-0] Stilonium Iodide, 8776
[2552-47-8] Pilocereine, 7397
[2552-55-8] Ibotenic Acid, 4808
[2557-49-5] Diflorasone, 3126
[2571-22-4] Tutin, 9735
[2571-86-0] Coriamyrtin, 2525
[2574-78-9] Orazamide, 6817
[2576-92-3] Isoetharine Hydrochloride*, 5053
[2581-07-9] Isoapo-β-erythroidine, 5008
[2589-47-1] Prajmalium Bitartrate*, 7704
[2591-17-5] Firefly Luciferin, 4025
[2595-97-3] D-Allose, 279
[2606-93-1] Blancophor® R, 1322
[2607-06-9] Diflucortolone, 3129
[2608-24-4] Piposulfan, 7452
[2609-46-3] Amiloride, 417
[2612-33-1] Clonitrate, 2390
[2618-25-9] Ioglycamic Acid, 4939
[2618-26-0] Iodipamide Sodium*, 4916
[2622-26-6] Pericyazine, 7117
[2622-30-2] Carphenazine, 1868
[2622-37-9] Trifluomeprazine, 9593
[2624-43-3] Cyclofenil, 2726
[2624-44-4] Ethamsylate, 3675
[2631-37-0] Promecarb, 7794
[2635-29-2] 4'-Formylsuccinanilic Acid Thiosemicarbazone, 4163
[2636-26-2] Ciafos, 2266
[2642-98-0] 6-Chrysenamine, 2258
[2652-13-3] Eicosamethylnonasiloxane, 3494
[2667-89-2] Bisbentiamine, 1257
[2668-66-8] Medrysone, 5679
[2673-40-7] Perivine, 7130
[2686-47-7] Amphenone B, 615
[2688-77-9] Laudanosine, 5249
[2691-45-4] Hexestrol Bis(β-diethylaminoethyl ether), 4622
[2696-92-6] Nitrosyl Chloride, 6568
[2698-41-1] o-Chlorobenzylidenemalononitrile, 2127
[2699-79-8] Sulfuryl Fluoride, 8960
[2709-56-0] Flupentixol, 4114
[2718-25-4] Rheadine, 8174
[2730-71-4] Thiocolchicine, 9253
[2740-04-7] Dimefline Hydrochloride*, 3190
[2740-52-5] Anagestone, 658
[2746-81-8] Fluphenazine Enanthate*, 4116
[2750-76-7] Rifamide, 8214
[2751-09-9] Troleandomycin, 9681
[2751-68-0] Acetophenazine, 64
[2752-65-0] Gambogic Acid, 4261
[2753-45-9] Mebeverine Hydrochloride*, 5648
[2757-90-6] Agaritine, 175
[2763-96-4] Muscimol, 6221
[2765-04-0] Thioacetaldehyde, 9245
[2768-90-3] Quinaldine Blue, 8056
[2779-55-7] Opiniazide, 6807
[2783-94-0] Sunset Yellow FCF, 8983
[2809-21-4] Etidronic Acid, 3819
[2825-00-5] Aureothin, 897
[2825-60-7] Formocortal, 4156
[2829-19-8] Rolicyprine, 8234
[2834-90-4] 4-Amino-1-naphthol, 465
[2869-83-2] Janus Green B, 5138
[2873-97-4] Diacetone Acrylamide, 2943
[2898-11-5] Medazepam Hydrochloride*, 5669
[2898-12-6] Medazepam, 5669

[2919-66-6] Melengestrol Acetate*, 5697
[2920-86-7] Prednisolone Hemisuccinate*, 7719
[2921-88-2] Chlorpyrifos, 2190
[2921-92-8] Propatyl Nitrate, 7821
[2922-44-3] Dextromoramide Tartrate*, 2933
[2924-67-6] Fluoresone, 4088
[2933-94-0] Toliprolol, 9439
[2945-88-2] Sophorabioside, 8674
[2955-23-9] Olivil, 6797
[2955-38-6] Prazepam, 7713
[2957-21-3] Sakuranetin, 8289
[2970-95-8] Macromerine, 5520
[2971-90-6] Clopidol, 2394
[2975-34-0] Carphenazine Dimaleate*, 1868
[2998-57-4] Estramustine, 3658
[3011-89-0] Aklomide, 188
[3012-65-5] Ammonium Citrate, Dibasic, 534
[3017-60-5] Cobaltous Thiocyanate, 2445
[3031-94-5] AICAR, 180
[3031-95-6] SAICAR, 8288
[3051-09-0] Murexide, 6215
[3060-89-7] Metobromuron, 6061
[3093-35-4] Halcinonide, 4504
[3094-09-5] Doxifluridine, 3426
[3095-65-6] Ammonium Bitartrate, 523
[3099-52-3] Nicametate, 6401
[3101-51-7] Ergoflavin, 3597
[3102-00-9] Febuprol, 3891
[3105-97-3] Hycanthone, 4677
[3115-05-7] Iobenzamic Acid, 4901
[3115-21-7] Methyl 4-Pyridyl Ketone Thiosemicarbazone, 6035
[3115-49-9] (p-Nonylphenoxy)acetic Acid, 6600
[3116-76-5] Dicloxacillin, 3073
[3122-01-8] Thiazesim Hydrochloride*, 9233
[3131-08-6] Phenacridane Chloride, 7157
[3137-73-3] Anagestone Acetate*, 658
[3144-16-9] d-Camphorsulfonic Acid, 1740
[3147-14-6] Calmagite, 1725
[3147-55-5] 3,5-Dibromosalicylic Acid, 3012
[3149-00-6] Phenbutamide, 7177
[3150-28-5] D-Chalcose, 2030
[3164-34-9] Calcium Tartrate, 1716
[3167-49-5] 6-Aminonicotinic Acid, 468
[3175-90-4] Lycofawcine, 5490
[3176-03-2] Drotebanol, 3442
[3194-25-0] Nalorphine Dinicotinate, 6276
[3198-07-0] Ethylnorepinephrine Hydrochloride*, 3789
[3200-06-4] Nafronyl Oxalate*, 6268
[3202-55-9] Benapryzine Hydrochloride*, 1039
[3202-84-2] 4-Salicyloylmorpholine, 8302
[3215-70-1] Hexoprenaline, 4628
[3230-75-9] Oxenin, 6887
[3232-84-6] Urazole, 9780
[3244-88-0] Acid Fuchsin, 103
[3251-23-8] Cupric Nitrate, 2647
[3253-60-9] Laudexium Methyl Sulfate, 5250
[3254-66-8] Diisopropyl Paraoxon, 3183
[3254-89-5] Diphenidol Hydrochloride*, 3311

[3254-93-1] Doxenitoin, 3424
[3264-82-2] Nickel Acetylacetonate, 6408
[3270-71-1] Nifuraldezone, 6446
[3279-54-7] Disodium Phenyl Phosphate, 3364
[3301-68-6] Irisolone, 4974
[3306-52-3] Viridin, 9912
[3329-14-4] Fenpiprane Hydrochloride*, 3934
[3329-91-7] Dioscorine, 3288
[3342-61-8] Deanol Aceglumate, 2835
[3344-16-9] Penicillin G Hydrabamine, 7040
[3344-18-1] Magnesium Citrate, 5541
[3347-22-6] Dithianone, 3375
[3349-06-2] Nickel Formate, 6416
[3351-86-8] Fucoxanthin, 4195
[3354-67-4] Amidephrine, 409
[3362-45-6] Noxiptilin, 6644
[3363-58-4] Nifurfoline, 6448
[3366-95-8] Secnidazole, 8373
[3368-16-9] α-Phenylcinnamic Acid, 7253
[3371-85-5] Voacamine, 9944
[3376-83-8] Biotin l-Sulfoxide, 1245
[3380-34-5] Triclosan, 9573
[3383-96-8] Temephos, 9075
[3385-03-3] Flunisolide, 4071
[3397-23-7] Ornipressin, 6825
[3403-42-7] Methoxypromazine Maleate*, 5923
[3405-51-4] Erdin, 3588
[3416-24-8] Glucosamine, 4352
[3416-26-0] Lidoflazine, 5360
[3420-59-5] Isomaltol, 5066
[3426-01-5] Phenopyrazone, 7217
[3426-08-2] Prozapine, 7922
[3447-95-8] Benfurodil Hemisuccinate, 1051
[3458-28-4] D-Mannose, 5632
[3459-06-1] Cyclopentamine Hydrochloride*, 2745
[3459-20-9] Glymidine Sodium*, 4403
[3459-96-9] Amicarbalide, 405
[3460-67-1] Furonazide, 4220
[3463-92-1] Carpaine, 1866
[3475-65-8] Thiamine Triphosphoric Acid Ester, 9227
[3483-12-3] 1,4-Dithiothreitol, 3382
[3484-46-6] Hyenanchin, 4787
[3485-14-1] Cyclacillin, 2706
[3485-62-9] Clidinium Bromide, 2351
[3486-35-9] Zinc Carbonate, 10030
[3486-66-6] Coptisine, 2522
[3486-67-7] Palmatine, 6945
[3505-38-2] Carbinoxamine Maleate*, 1804
[3511-16-8] Hetacillin, 4592
[3512-87-6] Akuammine, 190
[3521-62-8] Erythromycin Estolate, 3628
[3521-84-4] Iodipamide Meglumine*, 4916
[3538-57-6] Haloprogesterone, 4513
[3540-95-2] Fenpiprane, 3934
[3544-35-2] Iproclozide, 4963
[3546-03-0] Cyamemazine, 2690
[3546-10-9] Phenesterine, 7182
[3546-21-2] Homidium, 4650
[3546-41-6] Pyrvinium Pamoate, 8034
[3552-16-7] Enteromycin, 3550
[3562-84-3] Benzbromarone, 1073
[3562-99-0] Menbutone, 5719
[3563-27-7] Dihydroequilin, 3156
[3563-49-3] Pyrovalerone, 8021

Denotes derivative of title compound

[3563-58-4] Chlorhexadol, 2089
[3564-09-8] Ponceau 3R, 7567
[3565-72-8] Embramine, 3514
[3567-38-2] Carfinate, 1845
[3567-76-8] Aminochlorthenoxazin, 446
[3568-24-9] Propionylpromazine, 7841
[3568-43-2] Sulfamethoxypyridazine Acetyl*, 8890
[3569-99-1] N-(Hydroxymethyl)-nicotinamide, 4765
[3570-15-8] Nicotinic Acid Mono-ethanolamine Salt, 6437
[3571-53-7] Estradiol Undecylate*, 3463
[3571-88-8] Platonin, 7504
[3572-43-8] Bromhexine, 1379
[3572-44-9] Anot, 707
[3572-52-9] Biphenamine, 1247
[3572-60-9] Amidinomycin, 410
[3572-74-5] Mephenhydramine, 5738
[3572-80-3] Cyclazocine, 2710
[3573-82-8] Hygrophylline, 4791
[3575-80-2] Melperone, 5707
[3576-64-5] Chlorphenoxamide, 2181
[3577-01-3] Cephaloglycin, 1973
[3583-64-0] Bumadizon, 1472
[3589-73-9] 10-Methoxyharmalan, 5915
[3590-05-4] Acetyl Sulfamethoxy-pyrazine, 96
[3590-16-7] Feclemine, 3892
[3595-11-7] Propylhexedrine, 7873
[3598-37-6] Acepromazine Male-ate*, 27
[3599-32-4] Indocyanine Green, 4868
[3600-95-1] Perezone, 7112
[3605-01-4] Piribedil, 7465
[3607-18-9] Doxepin (trans-Form Hydrochloride)*, 3425
[3611-72-1] Clobenfurol, 2356
[3614-30-0] Emepronium Bromide, 3515
[3614-47-9] Hydracarbazine, 4680
[3614-69-5] Dimethindene Maleate*, 3204
[3615-24-5] Ramifenazone, 8122
[3615-37-0] D-Fucose, 4192
[3615-41-6] Rhamnose, 8171
[3615-44-9] D-manno-Heptulose, 4587
[3615-82-5] Phytic Acid Calcium Magnesium Salt*, 7359
[3621-36-1] Columbamine, 2488
[3621-38-3] Jatrorrhizine, 5142
[3624-96-2] Bialamicol Hydrochlo-ride*, 1217
[3625-06-7] Mebeverine, 5648
[3625-25-0] Reposal, 8141
[3625-68-1] Diatretyne II, 2974
[3635-74-3] Deanol Acetamidoben-zoate, 2836
[3650-17-7] Hirsutic Acid C, 4637
[3666-69-1] Dioxadrol Hydrochlo-ride*, 3293
[3670-68-6] Propipocaine, 7843
[3679-64-9] Bromosalicylchlorani-lide, 1423
[3682-35-7] 2,4,6-Tripyridyl-s-tri-azine, 9664
[3686-58-6] Tolycaine, 9467
[3688-65-1] Codeine N-Oxide, 2461
[3689-24-5] Sulfotep, 8945
[3689-50-7] Oxomemazine, 6900
[3689-76-7] Chlormidazole, 2106
[3690-04-8] Aminopropylon, 484
[3690-12-8] Bacimethrin, 947

[3691-21-2] Buzepide, 1599
[3691-35-8] Chlorophacinone, 2152
[3691-74-5] Glyconiazide, 4396
[3691-81-4] Sulfoniazide, 8938
[3693-39-8] Flucloronide, 4053
[3697-42-5] Chlorhexidine Dihydro-chloride*, 2090
[3703-76-2] Cloperastine, 2393
[3703-79-5] Bamethan, 964
[3704-09-4] Mibolerone, 6098
[3715-90-0] Tramazoline Hydro-chloride*, 9486
[3717-88-2] Flavoxate Hydrochlo-ride*, 4033
[3724-65-0] Crotonic Acid, 2600
[3731-59-7] Moroxydine, 6181
[3733-81-1] Defosfamide, 2853
[3734-16-5] Prodilidine Hydrochlo-ride*, 7775
[3734-17-6] Prodilidine, 7775
[3734-24-5] Betasine, 1203
[3734-33-6] Denatonium Benzoate, 2877
[3734-52-9] Metazocine, 5839
[3735-45-3] Vetrabutine, 9875
[3735-90-8] Phencarbamide, 7178
[3736-08-1] Fenethylline, 3919
[3736-36-5] Tropine Benzylate, 9694
[3737-09-5] Disopyramide, 3366
[3737-33-5] Etymemazine Hydro-chloride*, 3849
[3738-01-0] Otobain, 6852
[3748-77-4] Bunamidine, 1475
[3750-26-3] Primocarcin, 7754
[3754-19-6] Ambuside, 396
[3763-55-1] Rubixanthin, 8265
[3772-76-7] Sulfamethomidine, 8888
[3778-73-2] Ifosfamide, 4822
[3779-59-7] Cimigenol, 2282
[3784-89-2] Phenactropinium Chlo-ride, 7158
[3784-92-7] 5-Methyl-5-(3-phenan-thryl)hydantoin, 6023
[3784-99-4] Stilbazium Iodide, 8773
[3785-21-5] Butanilicaine, 1508
[3801-06-7] Fluorometholone Ace-tate*, 4104
[3810-35-3] Tenonitrozole, 9079
[3810-74-0] Streptomycin Sulfate*, 8786
[3810-80-8] Diphenoxylate Hydro-chloride*, 3313
[3810-83-1] Pentacynium Bis(methyl sulfate), 7060
[3811-04-9] Potassium Chlorate, 7600
[3811-20-9] Cuminaldehyde Thio-semicarbazone, 2624
[3811-25-4] Clorprenaline, 2404
[3811-56-1] Aminoquinuride, 489
[3811-75-4] Hexamidine, 4612
[3818-37-9] Phenoxypropazine, 7227
[3818-50-6] Bephenium Hydroxy-naphthoate*, 1166
[3818-62-0] Betoxycaine, 1210
[3819-00-9] Piperacetazine, 7429
[3819-34-9] Phenamet, 7164
[3820-67-5] Glafenine, 4330
[3833-99-6] Homofenazine, 4657
[3844-45-9] Brilliant Blue FCF, 1366
[3847-29-8] Erythromycin Lacto-bionate, 3630
[3856-25-5] Copaene, 2510
[3858-89-7] 2-Chloroprocaine Hydrochloride, 2158
[3861-73-2] Anazolene Sodium, 662
[3861-76-5] Clonitazene, 2389

[3863-59-0] Hydrocortisone Phos-phate, 4712
[3900-31-0] Fludiazepam, 4058
[3902-71-4] Trioxsalen, 9647
[3922-90-5] Oleandomycin, 6785
[3930-19-6] Streptonigrin, 8789
[3930-20-9] Sotalol, 8682
[3947-65-7] Neamine, 6351
[3952-98-5] Sinigrin, 8495
[3963-95-9] Methacycline Hydro-chloride*, 5851
[3964-81-6] Azatadine, 917
[3978-86-7] Azatadine Maleate*, 917
[3982-91-0] Phosphorus Sulfochlo-ride, 7331
[4008-48-4] Nitroxoline, 6577
[4028-98-2] Dequalinium Acetate, 2895
[4042-30-2] Parvaquone, 6998
[4044-65-9] Bitoscanate, 1318
[4065-45-6] Sulisobenzone, 8963
[4075-81-4] Calcium Propionate, 1705
[4080-31-3] N-(3-Chloroallyl)hex-aminium Chloride, 2117
[4091-75-2] Clomestrone, 2380
[4093-35-0] Bromopride, 1420
[4097-22-7] Dideoxyadenosine, 3090
[4098-40-2] Mitragynine, 6136
[4119-52-2] Ferric Thiocyanate, 3982
[4147-51-7] Dipropetryn, 3349
[4154-10-3] Senociclin, 8408
[4156-16-5] Woodward's Reagent K, 9962
[4163-15-9] Cyclorphan, 2757
[4170-30-3] Crotonaldehyde, 2599
[4171-13-5] Valnoctamide, 9820
[4201-22-3] Tolonidine, 9443
[4201-78-9] Choline Dehydrocho-late, 2209
[4205-23-6] D-Gulose, 4490
[4205-90-7] Clonidine, 2388
[4205-91-8] Clonidine Hydrochlo-ride*, 2388
[4214-72-6] Isaxonine, 4987
[4232-99-9] 3,5-Dibromo-4-hydr-oxybenzenesulfonic Acid, 3007
[4258-85-9] Clocortolone Acetate*, 2368
[4267-05-4] Tetrachlormethiazide, 9124
[4268-36-4] Tybamate, 9737
[4271-30-1] N-(p-Aminobenzoyl)-glutamic Acid, 437
[4282-07-9] Chonemorphine, 2219
[4299-60-9] Sulfisoxazole Diol-amine*, 8930
[4300-28-1] D-Ribose-5-phosphoric Acid, 8206
[4302-95-8] Tevenel®, 9179
[4304-40-9] Thenium Closylate, 9205
[4310-89-8] Hedaquinium Chloride, 4539
[4317-14-0] Amitriptylinoxide, 505
[4320-30-3] Arginine Glutamate, 806
[4325-25-1] Quinine Gluconate, 8082
[4330-99-8] Trimeprazine Tartrate*, 9618
[4342-03-4] Dacarbazine, 2803
[4345-03-3] α-Tocopherol Acid Succinate, 9424
[4354-45-4] Propenzolate, 7824
[4360-12-7] Ajmaline, 184
[4366-18-1] Coumetarol, 2566
[4368-28-9] Tetrodotoxin, 9175

Denotes derivative of title compound

[4378-36-3] Fenbutrazate, 3908
[4386-35-0] Meralein Sodium, 5755
[4393-19-5] p-Sulfanilylbenzylamine, 8903
[4394-00-7] Niflumic Acid, 6444
[4396-01-4] Pelletierine, 7015
[4403-90-1] Alizarin Cyanine Green F, 238
[4406-22-8] Cyprenorphine, 2777
[4412-09-3] Mucochloric Anhydride, 6209
[4419-39-0] Beclomethasone, 1029
[4438-22-6] Atropine N-Oxide, 892
[4439-84-3] HON, 4663
[4448-95-7] Shellolic Acid, 8427
[4461-30-7] Chloroacetyl Isocyanate, 2116
[4462-95-1] Truxillic Acid, 9700
[4481-08-7] Digitalose, 3142
[4481-55-4] Pentryl®, 7095
[4491-19-4] Indaconitine, 4842
[4493-23-6] Dodecahedrane, 3399
[4493-37-2] Chromous Formate, 2248
[4498-32-2] Dibenzepin, 2990
[4499-40-5] Choline Theophyllinate, 2213
[4533-39-5] Nitracrine, 6485
[4533-89-5] Flunisolide Acetate*, 4071
[4544-15-4] Piperilate Hydrochloride*, 7441
[4546-39-8] Piperilate, 7441
[4548-53-2] Ponceau SX, 7568
[4551-59-1] Fenalamide, 3899
[4562-36-1] Gitoxin, 4328
[4573-78-8] Hamamelose, 4523
[4574-60-1] Atropine Oxide Hydrochloride*, 892
[4589-33-7] Bostrycoidin, 1353
[4599-60-4] Penimepicycline, 7052
[4602-84-0] Farnesol, 3885
[4611-02-3] Chlorproethazine Hydrochloride*, 2184
[4618-18-2] Lactulose, 5225
[4618-41-1] 1-Butyl-3-metanilyl-urea, 1579
[4619-74-3] 2,4,6-Tribromo-m-cresol, 9524
[4626-00-0] Kitol, 5195
[4630-95-9] Prifinium Bromide, 7748
[4655-30-5] Hemi-Dewar Biphenyl, 4562
[4663-83-6] Buramate, 1490
[4670-05-7] Theaflavine, 9201
[4671-03-8] Hexazole, 4618
[4673-18-1] Buphanitine, 1483
[4680-78-8] Guinea Green B, 4487
[4682-36-4] Orphenadrine Citrate*, 6831
[4682-48-8] Cytolipin H, 2800
[4682-50-2] Trichodermin, 9562
[4684-87-1] Octamoxin, 6670
[4685-14-7] Paraquat, 6980
[4695-62-9] d-Fenchone, 3911
[4696-78-8] Kanamycin B*, 5161
[4697-14-7] Ticarcillin Disodium*, 9360
[4697-36-3] Carbenicillin, 1796
[4721-69-1] 4-Hydroxy-19-nortestosterone, 4768
[4726-14-1] Nitralin, 6486
[4747-99-3] Tetrahydropapaveroline, 9148
[4757-55-5] Dimetacrine, 3199
[4767-03-7] Dimethylolpropionic Acid, 3241
[4776-06-1] Fluorosalan, 4106
[4800-94-6] Carbenicillin Disodium*, 1796

[4803-27-4] Anthramycin, 715
[4828-27-7] Clocortolone, 2368
[4846-91-1] Fenoxazoline, 3930
[4850-21-9] Quebrachamine, 8040
[4876-45-3] Camphotamide, 1741
[4884-68-8] Hydrastinine Hydrochloride*, 4689
[4914-30-1] Dehydroemetine, 2860
[4936-47-4] Nifuratel, 6447
[4940-39-0] Chromocarb, 2239
[4945-47-5] Bamipine, 966
[4956-37-0] Estradiol Enanthate*, 3463
[4968-09-6] Algestone Acetonide*, 229
[4969-02-2] Methixene, 5902
[4976-25-4] Tremetone, 9497
[4991-65-5] Tioxolone, 9388
[5001-32-1] Guanochlor, 4479
[5001-33-2] Metanephrine, 5831
[5003-47-4] Betoxycaine Hydrochloride*, 1210
[5003-48-5] Benorylate, 1054
[5008-52-6] Napelline, 6286
[5011-34-7] Trimetazidine, 9619
[5015-36-1] Methylprednisolone Sodium Phosphate*, 6028
[5036-02-2] Tetramisole, 9161
[5051-62-7] Guanabenz, 4469
[5053-06-5] Fenspiride, 3942
[5053-08-7] Fenspiride Hydrochloride*, 3942
[5060-55-9] Prednisolone 21-Stearoylglycolate, 7724
[5061-22-3] Nafiverine, 6267
[5080-50-2] Acetylcarnitine (L-Form Hydrochloride*, 78
[5086-74-8] Tetramisole Hydrochloride*, 9161
[5089-89-4] 7-Morpholinomethyltheophylline, 6196
[5090-37-9] Metizoline Hydrochloride*, 6060
[5103-42-4] Hydrindantin, 4698
[5104-49-4] Flurbiprofen, 4124
[5108-94-1] Mytatrienediol, 6254
[5118-17-2] Furazolium Chloride, 4211
[5118-29-6] Melitracen, 5704
[5119-48-2] Withaferin A, 9960
[5133-19-7] Aloin, 304
[5141-20-8] Light Green SF Yellowish, 5361
[5142-76-7] Bismuth Subacetate, 1294
[5153-24-2] Zirconyl Acetate, 10086
[5169-78-8] Tipepidine, 9389
[5175-83-7] Bismuth Tribromophenate, 1305
[5192-84-7] Trengestone, 9500
[5197-58-0] Stenbolone, 8763
[5205-82-3] Bevonium Methyl Sulfate, 1214
[5221-49-8] Pyrimithate, 8000
[5221-53-4] Dimethirimol, 3206
[5234-68-4] Carboxin, 1832
[5250-39-5] Floxacillin, 4044
[5251-34-3] Cloprednol, 2396
[5285-18-7] Piperic Acid, 7437
[5310-55-4] Clomacran, 2379
[5318-76-3] Imidocarb Dihydrochloride*, 4831
[5321-32-4] Hetacillin Potassium*, 4592
[5323-95-5] Sodium Ricinoleate*, 8213
[5329-14-6] Sulfamic Acid, 8893
[5341-61-7] Hydrazine Dihydrochloride*, 4691
[5355-16-8] Diaveridine, 2976

[5370-01-4] Mexiletine Hydrochloride*, 6094
[5373-42-2] Thalicarpine, 9181
[5376-03-4] Tropylium Bromide, 9696
[5377-20-8] Metomidate, 6069
[5392-40-5] Citral, 2324
[5394-83-2] Camphoric Acid, 1739
[5408-52-6] L-Lysine L-Glutamate, 5511
[5411-22-3] Benzphetamine Hydrochloride*, 1130
[5421-04-5] Mandelic Acid Isoamyl Ester, 5600
[5466-77-3] Octyl Methoxycinnamate, 6687
[5470-11-1] Hydroxylamine Hydrochloride*, 4759
[5486-03-3] Buquinolate, 1489
[5486-77-1] Alloclamide, 254
[5508-58-7] Andrographolide, 666
[5534-09-8] Beclomethasone Dipropionate*, 1029
[5534-95-2] Pentagastrin, 7067
[5536-17-4] Vidarabine, 9881
[5545-89-1] Chimonanthine, 2048
[5560-59-8] Alverine Citrate*, 377
[5560-62-3] Biphenamine Hydrochloride*, 1247
[5560-69-0] Ethyl Dibunate, 3745
[5560-72-5] Iprindole, 4962
[5560-78-1] Teclozan, 9055
[5576-62-5] Chlorbenzoxamine Dihydrochloride*, 2075
[5579-08-8] Propyl Docetrizoate, 7861
[5579-13-5] Capuride, 1774
[5579-16-8] Epinephrine d-Borate*, 3569
[5579-84-0] Betahistine Dihydrochloride*, 1200
[5579-92-0] Iopydol, 4950
[5579-93-1] Iopydone, 4951
[5579-94-2] Merisoprol Hg 197, 5803
[5581-35-1] Amphecloral, 613
[5581-52-2] Thiamiprine, 9228
[5585-60-4] Paranyline Hydrochloride*, 6978
[5585-64-8] Amotriphene, 608
[5585-73-9] Butriptyline Hydrochloride*, 1533
[5585-93-3] Oxypendyl, 6923
[5586-87-8] Mefenorex Hydrochloride*, 5681
[5587-89-3] Ipodate, 4958
[5588-10-3] Methoxyphenamine Hydrochloride*, 5919
[5588-16-9] Althiazide, 316
[5588-20-5] Chlordantoin, 2080
[5588-22-7] Clorprenaline Hydrochloride Monohydrate*, 2404
[5588-25-0] Dihexyverine Hydrochloride*, 3151
[5588-33-0] Mesoridazine, 5813
[5591-22-0] Becanthone Hydrochloride*, 1026
[5591-29-7] Etafedrine Hydrochloride*, 3661
[5591-45-7] Thiothixene, 9297
[5593-20-4] Betamethasone Dipropionate*, 1202
[5598-52-7] Fospirate, 4172
[5600-19-1] Theophylline Isopropanolamine*, 9212
[5610-40-2] Securinine, 8376
[5611-51-8] Triamcinolone Hexacetonide, 9514
[5614-56-2] 1-Theobromineacetic Acid, 9210
[5625-37-6] PIPES, 7450

* Denotes derivative of title compound

[5626-34-6] Prednisolone 21-Diethylaminoacetate, 7720
[5627-46-3] Clobenztropine, 2360
[5630-53-5] Tibolone, 9358
[5632-44-0] Tolpropamine, 9450
[5632-52-0] Clofenciclan, 2372
[5633-14-7] Benzetimide Hydrochloride*, 1084
[5633-16-9] Leiopyrrole, 5316
[5633-18-1] Melengestrol, 5697
[5633-25-0] Noracymethadol Hydrochloride*, 6602
[5634-34-4] Ambuphylline, 395
[5634-38-8] Guaithylline, 4467
[5634-39-9] Iodinated Glycerol, 4906
[5634-42-4] Tocamphyl, 9415
[5636-83-9] Dimethindene, 3204
[5636-92-0] Picloxydine, 7371
[5638-76-6] Betahistine, 1200
[5639-34-9] 1-Amino-2-naphthol-6-sulfonic Acid, 467
[5667-71-0] Streptonicozid, 8788
[5668-06-4] Mecloxamine, 5663
[5675-98-9] Thevetose, 9216
[5695-98-7] Cotinine Fumarate*, 2552
[5696-06-0] Methetoin, 5889
[5696-09-3] Proxazole, 7919
[5697-56-3] Carbenoxolone, 1797
[5704-03-0] Testosterone Phenylacetate, 9114
[5704-04-1] Tricine, 9565
[5707-69-7] Drazoxolon, 3432
[5714-00-1] Acetophenazine Dimaleate*, 64
[5714-04-5] Guanoxan Sulfate*, 4482
[5714-73-8] Methenamine Hippurate, 5882
[5714-82-9] Triclofenol Piperazine, 9570
[5716-20-1] Bamethan Sulfate*, 964
[5728-52-9] Felbinac, 3893
[5735-19-3] Galactoflavin, 4239
[5741-22-0] Moprolol, 6172
[5743-27-1] Calcium Ascorbate, 1648
[5743-29-3] Calcium 3-Aurothio-2-propanol-1-sulfonate, 1649
[5743-34-0] Calcium Borogluconate, 1652
[5779-39-5] Desosamine, 2909
[5779-54-4] Cyclarbamate, 2709
[5786-21-0] Clozapine, 2417
[5793-84-0] Calcium Phenoxide, 1698
[5793-88-4] Calcium D-Saccharate, 1707
[5793-94-2] Calcium Stearyl-2 Lactylate, 1711
[5794-88-7] 5-Bromoanthranilic Acid, 1393
[5798-41-4] Bismuth Potassium Tartrate, 1289
[5798-43-6] Bismuth Sodium Triglycollamate, 1293
[5798-47-0] Bismuth Valerate, Basic, 1306
[5798-49-2] Bis(1-naphthylmethyl)-amine, 1307
[5798-78-7] p-Bromobenzyl Chloroformate, 1399
[5798-79-8] α-Bromobenzyl Cyanide, 1400
[5798-88-9] β-Bromoisovaleric Acid, 1410
[5798-94-7] 5-Bromosalicylhydroxamic Acid, 1424
[5835-26-7] Isopimaric Acid, 5086
[5845-26-1] Thiazesim, 9233

[5853-02-1] Casimiroedine, 1893
[5854-93-3] L-Alanosine, 197
[5868-05-3] Niceritrol, 6405
[5868-06-4] Fentonium Bromide, 3949
[5870-29-1] Cyclopentolate Hydrochloride*, 2752
[5874-97-5] Metaproterenol Sulfate*, 5836
[5875-06-9] Proparacaine Hydrochloride*, 7817
[5892-10-4] Bismuth Subcarbonate, 1295
[5897-76-7] α-(α-Aminopropyl)-benzyl Alcohol, 483
[5902-51-2] Terbacil, 9085
[5905-34-0] Acid Violet 7B, 104
[5905-52-2] Ferrous Lactate, 3999
[5907-38-0] Dipyrone, 3358
[5908-63-4] Baptigenin, 967
[5909-04-6] Chrysoidine Citrate*, 2262
[5913-71-3] Codeine Acetate*, 2459
[5913-82-6] Conessine Dihydrobromide*, 2492
[5928-84-7] Penicillin V Benzathine, 7047
[5936-28-7] Hydrastine Hydrochloride*, 4688
[5942-95-0] Carpipramine, 1869
[5945-50-6] Monotropein, 6166
[5945-86-8] Nimbin, 6462
[5947-49-9] Podocarpic Acid, 7516
[5959-52-4] 3-Amino-2-naphthoic Acid, 464
[5962-29-8] Xylulose, 9996
[5964-24-9] Thimerfonate Sodium, 9243
[5964-62-5] Diathymosulfone, 2972
[5965-33-3] Antimony Potassium Oxalate, 731
[5965-40-2] Allocupreide Sodium, 256
[5966-41-6] Diisopromine, 3180
[5970-32-1] Mercuric Salicylate, 5782
[5972-71-4] Ammonium Bimalate, 518
[5972-72-5] Ammonium Binoxalate, 519
[5972-76-9] Ammonium Caprylate, 526
[5977-10-6] Fencibutirol, 3912
[5978-87-0] D-Araboflavin, 790
[5980-01-8] Lochneridine, 5435
[5980-31-4] Hexedine, 4619
[5984-97-4] Iothiouracil, 4954
[5985-13-7] p-Hydroxybenzylpenicillin Sodium, 4743
[5985-94-2] Apoquinine, 777
[5986-55-0] Patchouli Alcohol, 7001
[5987-82-6] Benoxinate Hydrochloride*, 1056
[5989-43-5] Matricarin, 5640
[5991-71-9] Clorazepate Monopotassium*, 2400
[6000-74-4] Hydrocortisone Sodium Phosphate*, 4712
[6001-51-0] Bis(p-dimethylaminobenzylidene)benzidine, 1260
[6001-97-4] Bis(1-methylamyl) Sodium Sulfosuccinate, 1266
[6006-09-3] Gravitol(e), 4448
[6009-67-2] Amolanone Hydrochloride*, 603
[6010-09-9] Ferrous Thiocyanate, 4008
[6018-19-5] Aminosalicylate Sodium Dihydrate*, 492
[6018-40-2] Corypalmine, 2548

[6018-53-7] 6-Desoxy-D-glucosamine, 2911
[6020-73-1] Codeine Salicylate*, 2459
[6027-13-0] Homocysteine, 4654
[6027-89-0] L-Gulose, 4491
[6028-35-9] Thibenzazoline, 9239
[6032-29-7] 2-Pentanol, 7075
[6032-32-2] Methylarbutin, 5944
[6032-92-4] Mycarose, 6229
[6033-05-2] Morphine Oleate, 20%, 6191
[6035-31-0] Norlobelanine, 6626
[6035-39-8] Nitroakridin 3582, 6502
[6035-40-1] Noscapine (dl-Form)*, 6638
[6035-50-3] DL-Fructose, 4186
[6036-25-5] Narbomycin, 6340
[6040-62-6] Cedrin, 1918
[6043-86-3] Methiotriazamine, 5899
[6054-98-4] Olsalazine Sodium*, 6799
[6055-48-7] Toloxychlorinol, 9446
[6064-83-1] Fosfosal, 4170
[6078-26-8] Bikhaconitine, 1234
[6088-51-3] DDD (Analytical), 2831
[6091-05-0] Physovenine, 7358
[6092-18-8] Cycothiamin(e), 2762
[6108-05-0] Lidocaine Hydrochloride Monohydrate*, 5359
[6117-91-5] Crotyl Alcohol, 2604
[6130-64-9] Penicillin G Procaine, 7042
[6138-56-3] Tripelennamine Citrate*, 9651
[6138-58-5] Triphal®, 9652
[6138-73-4] Trilobine, 9610
[6138-79-0] Triprolidine Monohydrochloride Monohydrate*, 9661
[6147-11-1] Mangostin, 5627
[6153-19-1] Phenacaine Hydrochloride Monohydrate*, 7153
[6159-55-3] Vasicine, 9842
[6160-12-9] Sparteine Sulfate Pentahydrate*, 8692
[6163-66-2] Di-tert-butyl Ether, 3020
[6164-98-3] Chlordimeform, 2083
[6168-76-9] Crotethamide, 2598
[6168-86-1] Isometheptene Hydrochloride*, 5070
[6190-36-9] Cotarnine Phthalate*, 2551
[6190-39-2] Dihydroergotamine Mesylate*, 3157
[6190-43-8] Methenamine Anhydromethylenecitrate, 5881
[6190-60-9] Mephentermine Sulfate*, 5740
[6192-29-6] Penicillin BT, 7035
[6202-21-7] 4-Sulfanilamidosalicylic Acid, 8900
[6202-23-9] Cyclobenzaprine Hydrochloride*, 2719
[6209-17-2] N'-Acetylsulfanilamide Sodium Salt Monohydrate*, 97
[6214-20-6] Methyl p-Nitrobenzenesulfonate, 6016
[6236-05-1] Nifuroxime, 6451
[6272-74-8] Lapyrium Chloride, 5238
[6277-14-1] Acetoxolone, 69
[6284-40-8] N-Methylglucamine, 5995
[6303-21-5] Hypophosphorous Acid, 4804
[6314-28-9] Thionaphthene-2-carboxylic Acid, 9274
[6358-53-8] Citrus Red 2, 2335
[6372-96-9] Polar® Yellow, 7528

Denotes derivative of title compound

[6376-26-7] Salverine, 8310
[6377-18-0] Chartreusin, 2035
[6379-56-2] Hygromycin, 4789
[6379-69-7] Trichothecin, 9564
[6384-92-5] NMDA, 6584
[6385-02-0] Meclofenamic Acid Sodium*, 5659
[6385-58-6] Bithionolate Sodium*, 1316
[6398-98-7] Amodiaquine Dihydrochloride Dihydrate*, 602
[6402-36-4] Traumatic Acid, 9493
[6411-75-2] Tuaminoheptane Sulfate*, 9712
[6452-71-7] Oxprenolol, 6905
[6452-73-9] Oxprenolol Hydrochloride*, 6905
[6458-13-5] Ethyldimethyl-9-octadecenylammonium Bromide, 3747
[6484-52-2] Ammonium Nitrate, 561
[6484-74-8] Pteroylhexaglutamylglutamic Acid, 7951
[6484-89-5] Sodium Folate, 8566
[6489-97-0] Metampicillin, 5830
[6493-05-6] Pentoxifylline, 7092
[6495-46-1] Dioxadrol, 3293
[6500-81-8] Ethacrynic Acid Sodium*, 3669
[6506-37-2] Nimorazole, 6468
[6533-73-9] Thallium Carbonate, 9186
[6533-78-4] Antimony Thioglycollamide, 737
[6536-18-1] Morazone, 6176
[6538-02-9] Ergostanol, 3603
[6556-11-2] Inositol Niacinate, 4885
[6556-12-3] D-Glucuronic Acid, 4360
[6569-51-3] s-Triazaborane, 9516
[6576-51-8] Stallimycin Hydrochloride*, 8730
[6577-41-9] Cyclonium Iodide, 2743
[6591-55-5] Bismuth Oxalate, 1284
[6591-72-6] Penicillin V Hydrabamine, 7048
[6592-85-4] Hydrastinine, 4689
[6600-40-4] Norvaline, 6636
[6620-60-6] Proglumide, 7785
[6621-47-2] Perhexiline, 7116
[6673-35-4] Practolol, 7703
[6700-39-6] Isoproterenol Sulfate Dihydrate*, 5105
[6703-39-5] Diphenazoline, 3307
[6724-53-4] Perhexiline Maleate*, 7116
[6736-85-2] Catalposide, 1907
[6740-88-1] Ketamine, 5174
[6746-01-6] Desatrine, 2900
[6753-98-6] Humulene, 4672
[6754-13-8] Helenalin, 4542
[6754-20-7] Polygodial, 7548
[6793-24-4] Buphanamine, 1482
[6795-60-4] Norvinisterone, 6637
[6801-41-8] Melinonine A, 5703
[6802-93-3] Bipiperidyl Mustard, 1250
[6804-07-5] Carbadox, 1782
[6805-41-0] Escin, 3644
[6807-82-5] Diopterin, 3285
[6809-52-5] Teprenone, 9083
[6812-78-8] Rhodinol, 8186
[6823-79-6] Pentamidine Dimethanesulfonate*, 7071
[6829-98-7] Imipramine N-Oxide, 4836
[6830-17-7] Oxamarin Hydrochloride*, 6868
[6831-14-7] Arborescin, 798
[6833-84-7] Nonactin, 6594

[6834-92-0] Sodium Metasilicate, 8593
[6834-98-6] Fungichromin, 4202
[6835-00-3] Hydroxystreptomycin, 4782
[6835-16-1] Hyoscyamine Sulfate Dihydrate*, 4795
[6846-11-3] 3-Methyl-5-phenylhydantoin, 6026
[6859-66-1] Magnoline, 5580
[6866-93-9] C-Curarine III, 2680
[6869-45-0] Kurchessine, 5203
[6869-47-2] Kurcholessine, 5204
[6870-10-6] Cynanchogenin, 2772
[6871-44-9] Echitamine, 3474
[6873-15-0] Glycosine, 4398
[6874-98-2] Vellosimine, 9846
[6875-16-7] Tanghinigenin, 9021
[6877-35-6] Digalogenin, 3136
[6878-83-7] Tecomanine, 9056
[6878-98-4] Tropacine, 9685
[6879-02-3] Tylocrebrine, 9738
[6882-99-1] Sempervirine, 8398
[6888-11-5] Bietanautine, 1225
[6888-23-9] Toxiferine I, 9479
[6890-42-2] Prednylidene 21-Diethylaminoacetate, 7730
[6893-02-3] Liothyronine, 5388
[6903-12-4] Anabsinthin, 656
[6909-62-2] Demanyl Phosphate, 2870
[6915-15-7] Malic Acid, 5589
[6915-57-1] Bibrocathol, 1220
[6923-22-4] Monocrotophos, 6161
[6940-50-7] p-Bromomandelic Acid, 1412
[6946-29-8] p-Aminosalicylic Acid Hydrazide, 492
[6961-46-2] Idrocilamide, 4820
[6964-20-1] Tiadenol, 9349
[6983-79-5] Bixin, 1320
[6990-06-3] Fusidic Acid, 4231
[6998-60-3] Rifamycin SV, 8217
[7001-56-1] Pentagestrone, 7068
[7004-98-0] Epimestrol, 3568
[7006-35-1] Histidine, 4642
[7008-26-6] Dichlorisone, 3035
[7009-49-6] Hexacyclonate Sodium, 4603
[7009-54-3] Pentapiperide, 7077
[7009-82-7] Trimethidinium Methosulfate, 9622
[7054-25-3] Quinidine Gluconate*, 8072
[7067-12-1] Laserpitin, 5244
[7067-16-5] Gratiogenin, 4446
[7075-03-8] Carpipramine Dihydrochloride Monohydrate*, 1869
[7076-63-3] Cuprobam, 2664
[7081-40-5] Methixene Hydrochloride Monohydrate*, 5902
[7081-44-9] Cloxacillin Sodium Monohydrate*, 2414
[7081-52-9] Piminodine Esylate*, 7403
[7081-53-0] Doxapram Monohydrochloride Monohydrate*, 3421
[7084-07-3] Methaphenilene Hydrochloride*, 5870
[7085-19-0] Mecoprop, 5666
[7085-44-1] Chlorothiazide Sodium*, 2169
[7085-55-4] Troxerutin, 9698
[7097-60-1] Tetraglycine Hydroperiodide, 9140
[7125-73-7] Flumetramide, 4069
[7168-64-1] C-Curarine I, 2679
[7173-51-5] Didecyldimethylammonium Chloride, 3088
[7177-48-2] Ampicillin Trihydrate*, 621

[7177-54-0] Penicillin O Sodium*, 7044
[7179-49-9] Lincomycin Hydrochloride Monohydrate*, 5378
[7181-73-9] Bephenium, 1166
[7187-01-1] 2-Furanacrylonitrile, 4208
[7195-27-9] Mefruside, 5685
[7199-29-3] Cyheptamide, 2766
[7205-52-9] Phytate Sodium*, 7359
[7207-92-3] Nandrolone Propionate, 6284
[7212-44-4] Nerolidol, 6388
[7219-91-2] Thihexinol Methylbromide*, 9242
[7224-08-0] Imiclopazine, 4827
[7225-61-8] Metrizoic Acid Sodium*, 6078
[7229-50-7] Streptolydigin, 8785
[7235-40-7] β-Carotene, 1860
[7236-47-7] β-Benzalbutyramide, 1063
[7237-81-2] Hepronicate, 4574
[7240-38-2] Oxacillin Sodium Monohydrate*, 6858
[7242-04-8] Gitoxin Pentaacetate*, 4328
[7246-07-3] Actinoquinol Sodium*, 134
[7246-20-0] Triclofos Sodium*, 9571
[7246-21-1] Tyropanoate Sodium, 9745
[7247-57-6] Heteronium Bromide, 4595
[7248-28-4] Strychnine N^6-Oxide, 8824
[7257-29-6] C-Calebassine, 1722
[7261-97-4] Dantrolene, 2814
[7262-75-1] Lefetamine, 5314
[7270-12-4] 8-Hydroxy-7-iodo-5-quinolinesulfonic Acid Compd. with Chloroquine*, 4757
[7287-19-6] Prometryn, 7800
[7297-25-8] Erythrityl Tetranitrate, 3622
[7309-58-2] Helenynolic Acid, 4543
[7320-34-5] Potassium Pyrophosphate, 7653
[7361-61-7] Xylazine, 9987
[7365-45-9] HEPES, 4573
[7377-03-9] Octanohydroxamic Acid, 6673
[7400-08-0] p-Coumaric Acid, 2561
[7413-36-7] Nifenalol, 6442
[7414-83-7] Etidronic Acid Disodium*, 3819
[7416-34-4] Molindone, 6142
[7421-40-1] Carbenoxolone Sodium*, 1797
[7429-90-5] Aluminum, 321
[7429-91-6] Dysprosium, 3461
[7429-92-7] Einsteinium, 3496
[7432-25-9] Etaqualone, 3665
[7439-88-5] Iridium, 4968
[7439-89-6] Iron, 4975
[7439-90-9] Krypton, 5202
[7439-91-0] Lanthanum, 5233
[7439-92-1] Lead, 5267
[7439-93-2] Lithium, 5395
[7439-94-3] Lutetium, 5484
[7439-95-4] Magnesium, 5529
[7439-96-5] Manganese, 5604
[7439-97-6] Mercury, 5801
[7439-98-7] Molybdenum, 6144
[7439-99-8] Neptunium, 6384
[7440-00-8] Neodymium, 6365
[7440-01-9] Neon, 6371
[7440-02-0] Nickel, 6406
[7440-02-0] Raney Nickel®, 8124

Denotes derivative of title compound

[7440-03-1] Niobium, 6472
[7440-04-2] Osmium, 6845
[7440-05-3] Palladium, 6940
[7440-06-4] Platinum, 7503
[7440-07-5] Plutonium, 7514
[7440-08-6] Polonium, 7533
[7440-09-7] Potassium, 7579
[7440-10-0] Praseodymium, 7709
[7440-11-1] Mendelevium, 5720
[7440-12-2] Promethium, 7798
[7440-13-3] Protactinium, 7897
[7440-14-4] Radium, 8118
[7440-15-5] Rhenium, 8176
[7440-16-6] Rhodium, 8187
[7440-17-7] Rubidium, 8259
[7440-19-9] Samarium, 8314
[7440-20-2] Scandium, 8348
[7440-21-3] Silicon, 8438
[7440-22-4] Silver, 8450
[7440-23-5] Sodium, 8512
[7440-24-6] Strontium, 8797
[7440-25-7] Tantalum, 9025
[7440-26-8] Technetium, 9054
[7440-27-9] Terbium, 9087
[7440-28-0] Thallium, 9183
[7440-29-1] Thorium, 9308
[7440-30-4] Thulium, 9328
[7440-31-5] Tin, 9376
[7440-32-6] Titanium, 9396
[7440-33-7] Tungsten, 9722
[7440-34-8] Actinium, 129
[7440-35-9] Americium, 401
[7440-36-0] Antimony, 724
[7440-37-1] Argon, 808
[7440-38-2] Arsenic, 820
[7440-39-3] Barium, 974
[7440-40-6] Berkelium, 1175
[7440-41-7] Beryllium, 1177
[7440-42-8] Boron, 1345
[7440-43-9] Cadmium, 1611
[7440-44-0] Carbon, 1814
[7440-45-1] Cerium, 1988
[7440-46-2] Cesium, 2001
[7440-47-3] Chromium, 2234
[7440-48-4] Cobalt, 2421
[7440-50-8] Copper, 2514
[7440-51-9] Curium, 2683
[7440-52-0] Erbium, 3585
[7440-53-1] Europium, 3862
[7440-54-2] Gadolinium, 4235
[7440-55-3] Gallium, 4252
[7440-56-4] Germanium, 4302
[7440-57-5] Gold, 4412
[7440-58-6] Hafnium, 4501
[7440-59-7] Helium, 4546
[7440-60-0] Holmium, 4646
[7440-61-1] Uranium, 9766
[7440-62-2] Vanadium, 9823
[7440-63-3] Xenon, 9981
[7440-64-4] Ytterbium, 10014
[7440-65-5] Yttrium, 10015
[7440-66-6] Zinc, 10025
[7440-67-7] Zirconium, 10076
[7440-68-8] Astatine, 876
[7440-69-9] Bismuth, 1268
[7440-70-2] Calcium, 1642
[7440-71-3] Californium, 1724
[7440-72-4] Fermium, 3955
[7440-73-5] Francium, 4175
[7440-74-6] Indium, 4857
[7446-06-2] Polonium Dioxide, 7534
[7446-07-3] Tellurium Dioxide, 9067
[7446-08-4] Selenium Oxide, 8386
[7446-09-5] Sulfur Dioxide, 8950
[7446-11-9] Sulfur Trioxide, 8958
[7446-14-2] Lead Sulfate, 5302
[7446-15-3] Lead Selenate, 5296
[7446-18-6] Thallium Sulfate, 9197

[7446-21-1] Strontium Selenate, 8815
[7446-22-2] Thallium Selenate, 9194
[7446-25-5] Stannic Selenite, 8738
[7446-26-6] Zinc Pyrophosphate, 10058
[7446-27-7] Lead Phosphate, 5295
[7446-28-8] Strontium Phosphate, Tribasic, 8814
[7446-32-4] Antimony Sulfate, 736
[7446-70-0] Aluminum Chloride, 338
[7447-39-4] Cupric Chloride, 2636
[7447-40-7] Potassium Chloride, 7601
[7447-41-8] Lithium Chloride, 5405
[7449-03-8] Erythropterin, 3636
[7455-39-2] Fonazine Mesylate*, 4146
[7456-24-8] Fonazine, 4146
[7460-12-0] Pseudoephedrine Sulfate*, 3561
[7461-02-1] Stylopine, 8827
[7481-89-2] Dideoxycytidine, 3091
[7487-88-9] Magnesium Sulfate, 5573
[7487-94-7] Mercuric Chloride, 5770
[7488-51-9] Lead Selenite, 5297
[7488-55-3] Stannous Sulfate, 8750
[7488-99-5] α-Carotene, 1859
[7491-09-0] Docusate Potassium*, 3398
[7491-74-9] Piracetam, 7459
[7527-91-5] Acrisorcin, 120
[7527-94-8] Alkofanone, 244
[7535-00-4] D-Galactosamine, 4240
[7542-37-2] Paromomycin, 6989
[7550-35-8] Lithium Bromide, 5403
[7550-45-0] Titanium Tetrachloride, 9404
[7553-56-2] Iodine, 4907
[7553-56-2] Iodine Colloidal, 4908
[7558-79-4] Sodium Phosphate, Dibasic, 8612
[7558-80-7] Sodium Phosphate, Monobasic, 8613
[7568-93-6] Phenylethanolamine, 7258
[7577-62-0] Fucosamine, 4191
[7580-67-8] Lithium Hydride, 5411
[7601-54-9] Sodium Phosphate, Tribasic, 8615
[7601-55-0] Metocurine Iodide, 6064
[7601-89-0] Sodium Perchlorate, 8605
[7601-90-3] Perchloric Acid, 7110
[7611-43-0] Lycomarasmine, 5491
[7616-22-0] γ-Tocopherol, 9418
[7616-94-6] Perchloryl Fluoride, 7111
[7618-86-2] Furtrethonium, 4225
[7631-46-1] Quinine Iodosulfate, 8086
[7631-86-9] Silicon Dioxide, 8440
[7631-90-5] Sodium Bisulfite, 8533
[7631-94-9] Sodium Dithionate, 8558
[7631-95-0] Sodium Molybdate(VI), 8596
[7631-99-4] Sodium Nitrate, 8598
[7632-00-0] Sodium Nitrite, 8599
[7632-04-4] Sodium Perborate, 8604
[7633-29-6] Psychotrine, 7946
[7637-07-2] Boron Trifluoride, 1351
[7646-69-7] Sodium Hydride, 8573
[7646-78-8] Stannic Chloride, 8732
[7646-79-9] Cobaltous Chloride, 2431
[7646-85-7] Zinc Chloride, 10031

[7646-93-7] Potassium Bisulfate, 7591
[7647-01-0] Hydrochloric Acid, 4703
[7647-01-0] Hydrogen Chloride, 4721
[7647-10-1] Palladium Chloride, 6941
[7647-14-5] Sodium Chloride, 8544
[7647-15-6] Sodium Bromide, 8539
[7647-17-8] Cesium Chloride, 2004
[7647-18-9] Antimony Pentachloride, 727
[7647-19-0] Phosphorus Pentafluoride, 7327
[7654-03-7] Benmoxine, 1052
[7658-08-4] Quinovose, 8106
[7664-38-2] Phosphoric Acid, 7318
[7664-39-3] Hydrofluoric Acid, 4717
[7664-39-3] Hydrogen Fluoride, 4723
[7664-41-7] Ammonia, 510
[7664-93-9] Sulfuric Acid, 8953
[7665-99-8] Cyclic GMP, 2715
[7681-11-0] Potassium Iodide, 7628
[7681-14-3] Prednisolone Tebutate, 7725
[7681-28-9] Quinidine Polygalacturonate, 8073
[7681-38-1] Sodium Bisulfate, 8531
[7681-49-4] Sodium Fluoride, 8565
[7681-52-9] Sodium Hypochlorite, 8576
[7681-53-0] Sodium Hypophosphite, 8580
[7681-55-2] Sodium Iodate, 8581
[7681-57-4] Sodium Metabisulfite, 8589
[7681-65-4] Cuprous Iodide, 2669
[7681-76-7] Ronidazole, 8237
[7681-80-3] Pentapiperium Methylsulfate*, 7077
[7681-82-5] Sodium Iodide, 8582
[7681-93-8] Natamycin, 6346
[7683-59-2] Isoproterenol, 5105
[7689-03-4] Camptothecin, 1742
[7696-12-0] Tetramethrin, 9154
[7697-37-2] Nitric Acid, 6495
[7697-37-2] Nitric Acid, Anhydrous, 6496
[7699-43-6] Zirconyl Chloride, 10087
[7699-45-8] Zinc Bromide, 10028
[7700-17-6] Crotoxyphos, 2603
[7704-34-9] Sulfur, 8948
[7704-98-5] Titanium Hydride, 9399
[7704-99-6] Zirconium Hydride, 10079
[7705-07-9] Titanium Trichloride, 9406
[7705-08-0] Ferric Chloride, 3961
[7706-67-4] Dimecrotic Acid, 3189
[7712-50-7] Myrtecaine, 6252
[7716-60-1] Etisazol, 3829
[7718-54-9] Nickel Chloride, 6412
[7719-09-7] Thionyl Chloride, 9278
[7719-12-2] Phosphorus Trichloride, 7333
[7720-78-7] Ferrous Sulfate, 4006
[7721-01-9] Tantalum Pentachloride, 9026
[7722-64-7] Potassium Permanganate, 7643
[7722-76-1] Ammonium Phosphate, Monobasic, 571
[7722-84-1] Hydrogen Peroxide, 4725
[7722-86-3] Caro's Acid, 1858

[7722-88-5] Tetrasodium Pyrophosphate, 9170
[7723-14-0] Phosphorus, 7321
[7724-04-1] Strophanthobiose, 8820
[7726-95-6] Bromine, 1382
[7727-15-3] Aluminum Bromide, 332
[7727-18-6] Vanadyl Trichloride, 9834
[7727-21-1] Potassium Persulfate, 7644
[7727-37-9] Nitrogen, 6522
[7727-43-7] Barium Sulfate, 1006
[7727-54-0] Ammonium Peroxydisulfate, 569
[7732-18-5] Water, 9951
[7733-02-0] Zinc Sulfate, 10064
[7757-79-1] Potassium Nitrate, 7634
[7757-82-6] Sodium Sulfate, 8636
[7757-83-7] Sodium Sulfite, 8638
[7757-86-0] Magnesium Phosphate, Dibasic, 5560
[7757-87-1] Magnesium Phosphate, Tribasic, 5562
[7757-88-2] Magnesium Sulfite, 5574
[7757-93-9] Calcium Phosphate, Dibasic, 1699
[7758-01-2] Potassium Bromate, 7596
[7758-02-3] Potassium Bromide, 7597
[7758-05-6] Potassium Iodate, 7627
[7758-09-0] Potassium Nitrite, 7635
[7758-11-4] Potassium Phosphate, Dibasic, 7647
[7758-16-9] Sodium Acid Pyrophosphate, 8515
[7758-19-2] Sodium Chlorite, 8545
[7758-23-8] Calcium Phosphate, Monobasic, 1700
[7758-29-4] Sodium Tripolyphosphate, 8654
[7758-87-4] Calcium Phosphate, Tribasic, 1701
[7758-88-5] Cerous Fluoride, 1992
[7758-89-6] Cuprous Chloride, 2667
[7758-94-3] Ferrous Chloride, 3992
[7758-95-4] Lead Chloride, 5278
[7758-97-6] Lead Chromate(VI), 5279
[7758-98-7] Cupric Sulfate, 2659
[7759-00-4] Manganese Silicate, 5623
[7759-01-5] Lead Tungstate(VI), 5309
[7759-02-6] Strontium Sulfate, 8816
[7761-75-3] Furterene, 4224
[7761-88-8] Silver Nitrate, 8464
[7772-98-7] Sodium Thiosulfate, 8652
[7772-99-8] Stannous Chloride, 8742
[7773-01-5] Manganese Chloride, 5610
[7773-06-0] Ammonium Sulfamate, 583
[7773-60-6] Mestilbol, 5818
[7774-29-0] Mercuric Iodide, Red, 5776
[7775-09-9] Sodium Chlorate, 8543
[7775-11-3] Sodium Chromate(VI), 8547
[7775-14-6] Sodium Hydrosulfite, 8574
[7775-19-1] Sodium Metaborate, 8590
[7775-27-1] Sodium Persulfate, 8608
[7775-41-9] Silver Fluoride, 8460
[7778-18-9] Calcium Sulfate, 1713

[7778-39-4] Arsenic Acid, 821
[7778-43-0] Sodium Arsenate, Dibasic, 8522
[7778-44-1] Calcium Arsenate, 1646
[7778-50-9] Potassium Dichromate-(VI), 7608
[7778-53-2] Potassium Phosphate, Tribasic, 7649
[7778-54-3] Calcium Hypochlorite, 1677
[7778-54-3] Chlorinated Lime, 2093
[7778-74-7] Potassium Perchlorate, 7641
[7778-77-0] Potassium Phosphate, Monobasic, 7648
[7778-80-5] Potassium Sulfate, 7665
[7779-88-6] Zinc Nitrate, 10044
[7779-90-0] Zinc Phosphate, 10055
[7782-39-0] Deuterium, 2919
[7782-40-3] Diamond, 2964
[7782-41-4] Fluorine, 4090
[7782-42-5] Graphite, 4444
[7782-44-7] Oxygen, 6917
[7782-49-2] Selenium, 8382
[7782-50-5] Chlorine, 2095
[7782-64-1] Manganese Difluoride, 5611
[7782-65-2] Germane, 4301
[7782-68-5] Iodic Acid, 4905
[7782-70-9] Potassium Biselenite, 7590
[7782-76-5] Manganese Phosphate, Dibasic, 5619
[7782-77-6] Nitrous Acid, 6574
[7782-78-7] Nitrosylsulfuric Acid, 6570
[7782-79-8] Hydrazoic Acid, 4697
[7782-87-6] Potassium Hypophosphite, 7626
[7782-89-0] Lithium Amide, 5398
[7782-91-4] Molybdic(VI) Acid, 6149
[7782-92-5] Sodium Amide, 8519
[7782-94-7] Nitramide, 6487
[7782-95-8] Disodium Dihydrogen Hypophosphate, 3363
[7782-99-2] Sulfurous Acid, 8955
[7783-00-8] Selenious Acid, 8381
[7783-03-1] Tungstic(VI) Acid, 9725
[7783-05-3] Pyrosulfuric Acid, 8019
[7783-06-4] Hydrogen Sulfide, 4729
[7783-07-5] Hydrogen Selenide, 4728
[7783-08-6] Selenic Acid, 8380
[7783-09-7] Hydrogen Telluride, 4730
[7783-18-8] Ammonium Thiosulfate, 592
[7783-19-9] Ammonium Selenite, 580
[7783-20-2] Ammonium Sulfate, 584
[7783-21-3] Ammonium Selenate, 579
[7783-22-4] Ammonium Uranate-(VI), 595
[7783-26-8] Trisilane, 9668
[7783-28-0] Ammonium Phosphate, Dibasic, 570
[7783-29-1] Tetrasilane, 9169
[7783-32-6] Mercuric Iodate, 5775
[7783-33-7] Potassium Tetraiodomercurate(II), 7683
[7783-35-9] Mercuric Sulfate, 5787
[7783-36-0] Mercurous Sulfate, 5800
[7783-39-3] Mercuric Fluoride, 5774
[7783-40-6] Magnesium Fluoride, 5543

[7783-41-7] Fluorine Monoxide, 4092
[7783-42-8] Thionyl Fluoride, 9279
[7783-43-9] Selenium Oxyfluoride, 8389
[7783-44-0] Fluorine Dioxide, 4091
[7783-46-2] Lead Fluoride, 5282
[7783-47-3] Stannous Fluoride, 8743
[7783-48-4] Strontium Fluoride, 8805
[7783-49-5] Zinc Fluoride, 10035
[7783-50-8] Ferric Fluoride, 3965
[7783-51-9] Gallium Trifluoride, 4255
[7783-52-0] Indium Trifluoride, 4866
[7783-53-1] Manganese Trifluoride, 5626
[7783-54-2] Nitrogen Fluoride, 6525
[7783-55-3] Phosphorus Trifluoride, 7334
[7783-56-4] Antimony Trifluoride, 741
[7783-57-5] Thallium Trifluoride, 9199
[7783-58-6] Germanium Tetrafluoride, 4306
[7783-59-7] Lead Tetrafluoride, 5306
[7783-60-0] Sulfur Tetrafluoride, 8957
[7783-61-1] Silicon Tetrafluoride, 8448
[7783-62-2] Stannic Fluoride, 8734
[7783-63-3] Titanium Tetrafluoride, 9405
[7783-64-4] Zirconium Fluoride, 10078
[7783-66-6] Iodine Pentafluoride, 4912
[7783-68-8] Niobium Pentafluoride, 6474
[7783-70-2] Antimony Pentafluoride, 728
[7783-71-3] Tantalum Pentachloride, 9027
[7783-72-4] Vanadium Pentafluoride, 9825
[7783-75-7] Iridium Hexafluoride, 4969
[7783-77-9] Molybdenum Hexafluoride, 6146
[7783-79-1] Selenium Hexafluoride, 8385
[7783-80-4] Tellurium Hexafluoride, 9068
[7783-81-5] Uranium Hexafluoride, 9768
[7783-82-6] Tungsten Hexafluoride, 9723
[7783-84-8] Ferric Hypophosphite, 3970
[7783-86-0] Ferrous Iodide, 3998
[7783-90-6] Silver Chloride, 8455
[7783-92-8] Silver Chlorate, 8454
[7783-93-9] Silver Perchlorate, 8470
[7783-95-1] Silver Difluoride, 8459
[7783-96-2] Silver Iodide, 8462
[7783-97-3] Silver Iodate, 8461
[7783-98-4] Silver Permanganate, 8471
[7783-99-5] Silver Nitrite, 8466
[7784-01-2] Silver Chromate(VI), 8456
[7784-03-4] Silver Tetraiodomercurate(II), 8482
[7784-05-6] Silver Selenite, 8478
[7784-07-8] Silver Selenate, 8476
[7784-09-0] Silver Phosphate, 8472

Denotes derivative of title compound

[7784-14-7] Ammonium Tetrachloroaluminate, 589
[7784-16-9] Sodium Tetrachloroaluminate, 8643
[7784-18-1] Aluminum Fluoride, 341
[7784-19-2] Ammonium Hexafluoroaluminate, 549
[7784-21-6] Aluminum Hydride, 344
[7784-22-7] Aluminum Hypophosphite, 347
[7784-23-8] Aluminum Iodide, 348
[7784-25-0] Aluminum Ammonium Sulfate, 325
[7784-30-7] Aluminum Phosphate, 362
[7784-33-0] Arsenic Tribromide, 828
[7784-34-1] Arsenic Trichloride, 829
[7784-35-2] Arsenic Trifluoride, 830
[7784-36-3] Arsenic Pentafluoride, 824
[7784-37-4] Mercuric Arsenate, 5767
[7784-41-0] Potassium Arsenate, 7583
[7784-42-1] Arsine, 837
[7784-45-4] Arsenic Triiodide, 831
[7784-46-5] Sodium Arsenite, 8523
[7785-23-1] Silver Bromide, 8452
[7785-24-2] Cobaltous Arsenate, 2428
[7785-84-4] Sodium Trimetaphosphate, 8653
[7785-87-7] Manganese Sulfate, 5624
[7786-30-3] Magnesium Chloride, 5540
[7786-34-7] Mevinphos, 6089
[7786-81-4] Nickel Sulfate, 6424
[7787-32-8] Barium Fluoride, 986
[7787-35-1] Barium Manganate(VI), 993
[7787-36-2] Barium Permanganate, 1000
[7787-39-5] Barium Sulfite, 1009
[7787-46-4] Beryllium Bromide, 1182
[7787-47-5] Beryllium Chloride, 1184
[7787-49-7] Beryllium Fluoride, 1185
[7787-50-0] Beryllium Potassium Fluoride, 1194
[7787-52-2] Beryllium Hydride, 1187
[7787-53-3] Beryllium Iodide, 1189
[7787-57-7] Bismuth Bromide Oxide, 1271
[7787-58-8] Bismuth Bromide, 1270
[7787-59-9] Bismuth Chloride Oxide, 1274
[7787-60-2] Bismuth Chloride, 1273
[7787-61-3] Bismuth Fluoride, 1276
[7787-62-4] Bismuth Pentafluoride, 1286
[7787-63-5] Bismuth Iodide Oxide, 1280
[7787-64-6] Bismuth Iodide, 1279
[7787-68-0] Bismuth Sulfate, 1300
[7787-69-1] Cesium Bromide, 2002
[7787-70-4] Cuprous Bromide, 2666
[7787-71-5] Bromine Trifluoride, 1384
[7788-96-7] Chromyl Fluoride, 2252
[7788-97-8] Chromic Fluoride, 2225
[7788-98-9] Ammonium Chromate-(VI), 532

[7789-00-6] Potassium Chromate-(VI), 7602
[7789-04-0] Chromic Phosphate, 2230
[7789-06-2] Strontium Chromate-(VI), 8804
[7789-08-4] Ammonium Ferric Chromate, 540
[7789-09-5] Ammonium Dichromate(VI), 538
[7789-10-8] Mercuric Dichromate-(VI), 5773
[7789-17-5] Cesium Iodide, 2006
[7789-18-6] Cesium Nitrate, 2007
[7789-19-7] Cupric Fluoride, 2641
[7789-20-0] Deuterium Oxide, 2920
[7789-21-1] Fluorosulfonic Acid, 4107
[7789-23-3] Potassium Fluoride, 7613
[7789-24-4] Lithium Fluoride, 5409
[7789-25-5] Nitrosyl Fluoride, 6569
[7789-26-6] Fluorine Nitrate, 4093
[7789-27-7] Thallium Fluoride, 9189
[7789-28-8] Ferrous Fluoride, 3994
[7789-29-9] Potassium Bifluoride, 7587
[7789-30-2] Bromine Pentafluoride, 1383
[7789-31-3] Bromic Acid, 1380
[7789-33-5] Iodine Monobromide, 4910
[7789-36-8] Magnesium Bromate, 5536
[7789-38-0] Sodium Bromate, 8538
[7789-39-1] Rubidium Bromide, 8260
[7789-40-4] Thallium Bromide, 9185
[7789-41-5] Calcium Bromide, 1653
[7789-42-6] Cadmium Bromide, 1613
[7789-43-7] Cobaltous Bromide, 2429
[7789-45-9] Cupric Bromide, 2632
[7789-46-0] Ferrous Bromide, 3989
[7789-47-1] Mercuric Bromide, 5769
[7789-48-2] Magnesium Bromide, 5537
[7789-51-7] Selenium Oxybromide, 8387
[7789-52-8] Selenium Bromide, 8383
[7789-54-0] Tellurium Dibromide, 9065
[7789-57-3] Tribromosilane, 9528
[7789-59-5] Phosphorus Oxybromide, 7323
[7789-60-8] Phosphorus Tribromide, 7332
[7789-61-9] Antimony Tribromide, 738
[7789-65-3] Selenium Tetrabromide, 8391
[7789-66-4] Silicon Tetrabromide, 8446
[7789-67-5] Stannic Bromide, 8731
[7789-68-6] Titanium Tetrabromide, 9403
[7789-69-7] Phosphorus Pentabromide, 7325
[7789-74-4] Calcium Fluorophosphate, 1670
[7789-75-5] Calcium Fluoride, 1669
[7789-78-8] Calcium Hydride, 1675
[7789-79-9] Calcium Hypophosphite, 1678
[7789-80-2] Calcium Iodate, 1679

[7789-82-4] Calcium Molybdate-(VI), 1687
[7790-21-8] Potassium Periodate, 7642
[7790-26-3] Sodium Iodide, Radioactive, 8583
[7790-28-5] Sodium Metaperiodate, 8591
[7790-29-6] Rubidium Iodide, 8263
[7790-30-9] Thallium Iodide, 9191
[7790-33-2] Manganese Iodide, 5614
[7790-37-6] Zinc Iodate, 10039
[7790-39-8] Platinous Iodide, 7502
[7790-42-3] Potassium Triiodide, 7690
[7790-43-4] Potassium Triiodozincate, 7692
[7790-44-5] Antimony Triiodide, 742
[7790-46-7] Platinic Iodide, 7499
[7790-47-8] Stannic Iodide, 8735
[7790-48-9] Tellurium Tetraiodide, 9071
[7790-49-0] Thorium Iodide, 9310
[7790-58-1] Potassium Tellurate-(IV), 7670
[7790-59-2] Potassium Selenate, 7656
[7790-60-5] Potassium Tungstate-(VI), 7693
[7790-62-7] Potassium Pyrosulfate, 7654
[7790-63-8] Potassium Uranate(VI), 7694
[7790-69-4] Lithium Nitrate, 5414
[7790-75-2] Calcium Tungstate(VI), 1720
[7790-79-6] Cadmium Fluoride, 1617
[7790-80-9] Cadmium Iodide, 1619
[7790-85-4] Cadmium Tungstate-(VI), 1630
[7790-86-5] Cerous Chloride, 1991
[7790-87-6] Cerous Iodide, 1993
[7790-89-8] Chlorine Monofluoride, 2098
[7790-91-2] Chlorine Trifluoride, 2100
[7790-92-3] Hypochlorous Acid, 4800
[7790-93-4] Chloric Acid, 2091
[7790-94-5] Chlorosulfonic Acid, 2166
[7790-95-6] Chloroselenic Acid, 2164
[7790-98-9] Ammonium Perchlorate, 568
[7790-99-0] Iodine Monochloride, 4911
[7791-03-9] Lithium Perchlorate, 5417
[7791-08-4] Antimony Chloride Oxide, 725
[7791-09-5] Rhenium Oxychloride, 8179
[7791-10-8] Strontium Chlorate, 8802
[7791-11-9] Rubidium Chloride, 8261
[7791-12-0] Thallium Chloride, 9187
[7791-16-4] Antimony Dichlorotrifluoride, 726
[7791-21-1] Chlorine Monoxide, 2099
[7791-23-3] Selenium Oxychloride, 8388
[7791-25-5] Sulfuryl Chloride, 8959
[7791-26-6] Uranyl Chloride, 9775

Denotes derivative of title compound

[7791-27-7] Pyrosulfuryl Chloride, 8020
[7791-29-9] Potassium Tetraiodoaurate(III), 7681
[7798-23-4] Cupric Phosphate, 2653
[7803-49-8] Hydroxylamine, 4759
[7803-51-2] Phosphine, 7313
[7803-52-3] Stibine, 8767
[7803-54-5] Magnesium Amide, 5532
[7803-55-6] Ammonium Vanadate-(V), 599
[7803-57-8] Hydrazine Hydrate, 4692
[7803-58-9] Sulfamide, 8894
[7803-60-3] Hypophosphoric Acid, 4803
[7803-62-5] Silane, 8435
[7803-63-6] Ammonium Bisulfate, 520
[7803-65-8] Ammonium Hypophosphite, 553
[7803-68-1] Telluric(VI) Acid, 9063
[8000-10-1] Theophylline Sodium Glycinate*, 9212
[8000-61-1] Aluminum Subacetate Solution, 371
[8001-08-9] Acerin, 28
[8001-10-3] Ferrous Carbonate Saccharated, 3991
[8001-27-2] Hirudin, 4638
[8001-35-2] Toxaphene, 9478
[8001-39-6] Japan Wax, 5139
[8001-40-9] Iodized Oil, 4917
[8001-50-1] Strobane®, 8796
[8001-58-9] Creosote, Coal Tar, 2574
[8001-60-3] Coparaffinate, 2513
[8001-61-4] Copaiba, 2511
[8001-75-0] Ceresin, 1984
[8002-36-6] Ustilagic Acid, 9807
[8002-53-7] Montan Wax, 6167
[8002-55-9] Myrtol, 6253
[8002-76-4] Papaveretum, 6967
[8002-88-8] Theobromine Sodium Acetate*, 9209
[8002-89-9] Theophylline Sodium Acetate*, 9212
[8003-05-2] Phenylmercuric Nitrate, Basic, 7273
[8003-22-3] Quinoline Yellow Spirit Soluble, 8101
[8004-92-0] Quinoline Yellow, 8100
[8005-78-5] Bismark Brown R, 1264
[8006-08-4] Ergotinine, 3612
[8006-13-1] Aluminum Acetate Solution, 322
[8006-25-5] Ergotoxine, 3613
[8006-27-7] Ferric and Ammonium Acetate Solution, 3959
[8006-28-8] Soda Lime, 8511
[8006-35-7] Diamond Ink, 2965
[8006-37-9] Turks Island Salt, 9731
[8006-44-8] Candelilla Wax, 1746
[8006-91-5] Dextri-Maltose®, 2930
[8007-14-5] Argol, 807
[8007-31-6] Silver Nitrate, Toughened, 8465
[8007-49-6] Carlsbad Salt Artificial, 1849
[8007-50-9] Nauheim Salts (Artificial), 6348
[8007-56-5] Nitrohydrochloric Acid, 6530
[8008-20-6] Kerosene, 5173
[8008-53-5] Ethiodized Oil, 3690
[8011-44-7] Aluminum Boroformate, 330
[8011-61-8] Tyrocidine, 9744

[8011-62-9] Barium Sulfide, Black, 1008
[8011-65-2] Zinc Tannate, 10066
[8011-82-3] Calomelol, 1727
[8012-34-8] Mercurophylline*, 5758
[8012-54-2] Donovan's Solution, 3413
[8012-89-3] Beeswax, 1031
[8013-17-0] Invert Sugar, 4900
[8013-43-2] Spirit of Ether, 8713
[8013-44-3] Spirit of Ether Compound, 8714
[8013-59-0] Spirit of Ammonia, Aromatic, 8710
[8013-61-4] Ammonium Acetate Solution*, 514
[8013-88-5] Calcium Cyanamide Citrated, 1663
[8013-96-5] Dymixal®, 3456
[8014-10-6] Terebene, 9091
[8014-63-9] Seidlitz Mixture, 8378
[8014-95-7] Sulfuric Acid Fuming*, 8953
[8015-18-7] Pfeiffer's Substance, 7146
[8015-86-9] Carnauba Wax, 1853
[8018-01-7] Mancozeb, 5598
[8018-15-3] Mercumatilin Sodium, 5765
[8021-39-4] Creosote, Wood, 2575
[8022-00-2] Methyl Demeton, 5971
[8023-62-9] Storax, 8778
[8025-81-8] Spiramycin, 8708
[8027-28-9] Sodium Phosphate, Radioactive, 8614
[8028-98-6] Acetulan®, 73
[8029-68-3] Ichthammol, 4815
[8030-30-6] Petroleum Benzin, 7141
[8030-35-1] Ferrous Carbonate Mass, 3990
[8030-97-5] Pyroligneous Acid, 8014
[8031-14-9] Oxychlorosene, 6909
[8031-21-8] Napalm, 6285
[8031-66-1] Osmaron B, 6844
[8032-32-4] Ligroin, 5366
[8038-65-1] Ambergris, 389
[8047-13-0] Ryania, 8278
[8047-26-5] Uglow Black Silver, 9753
[8047-67-4] Ferric Oxide, Saccharated, 3974
[8048-31-5] Theobromine Sodium Salicylate*, 9209
[8048-52-0] Acriflavine, 118
[8049-11-4] Devarda's Metal, 2921
[8049-20-5] Misch Metal, 6128
[8049-21-6] Hydrone®, 4734
[8049-47-6] Pancreatin, 6956
[8049-62-5] Insulin Zinc Suspension, 4890
[8049-62-5] Zinc Insulin Crystals, 10038
[8049-99-8] Milorganite®, 6115
[8050-07-5] Olibanum, 6792
[8050-81-5] Simethicone, 8486
[8050-88-2] Celluloid®, 1960
[8051-02-3] Sabadilla, 8280
[8052-16-2] Cactinomycin, 1605
[8052-27-5] Kino, 5193
[8053-05-2] Yellow Phenolphthalein, 10005
[8063-06-7] Curare, 2678
[8063-07-8] Kanamycin, 5161
[8065-36-9] Bufencarb, 1459
[8065-48-3] Demeton, 2875
[8065-51-8] Theobromine Calcium Salicylate*, 9209
[8067-69-4] Halquinol, 4520
[8068-28-8] Colistimethate Sodium*, 2475

[8075-98-7] Ascorbigen, 856
[8255-50-7] Isoxaben, 5125
[9000-01-5] Acacia, 10
[9000-02-6] Amber, 388
[9000-03-7] Ammoniacum, 511
[9000-04-8] Asafetida, 846
[9000-05-9] Gum Benzoin, 4492
[9000-07-1] Carrageenan, 1872
[9000-14-0] Copal, 2512
[9000-16-2] Damar, 2808
[9000-20-8] Eucalyptus Gum, 3853
[9000-21-9] Furcellaran, 4212
[9000-25-3] Gamboge, 4260
[9000-28-6] Ghatti Gum, 4311
[9000-29-7] Guaiac, 4455
[9000-30-0] Guaran, 4485
[9000-30-0] Guar Gum, 4486
[9000-32-2] Gutta-Percha, 4494
[9000-34-4] Resin Ipomea, 8152
[9000-35-5] Resin Jalap, 8153
[9000-36-6] Karaya Gum, 5164
[9000-45-7] Myrrh, 6251
[9000-47-9] Mesquite Gum, 5815
[9000-58-2] Resin Scammony, 8154
[9000-59-3] Shellac, 8426
[9000-65-1] Gum Tragacanth, 4493
[9000-69-5] Pectin, 7009
[9000-70-8] Gelatin, 4276
[9000-71-9] Casein, 1892
[9000-85-5] α-Amylase (Bacterial), 632
[9000-88-8] D-Amino Acid Oxidase, 425
[9000-89-9] L-Amino Acid Oxidase, 426
[9000-90-2] α-Amylase (Swine Pancreas), 633
[9000-91-3] β-Amylase (Sweet Potato), 634
[9000-92-4] Amylase, 631
[9000-92-4] Diastase of Malt, 2971
[9000-95-7] Apyrase, 786
[9001-00-7] Bromelain, 1377
[9001-01-8] Kallikrein, 5159
[9001-03-0] Carbonic Anhydrase, 1819
[9001-04-1] Pyruvate Decarboxylase, 8031
[9001-05-2] Catalase, 1906
[9001-08-5] Choline Esterase, 2211
[9001-09-6] Chymopapain, 2264
[9001-11-0] Clarase®, 2339
[9001-12-1] Collagenase, 2477
[9001-19-8] Taka-Diastase, 9007
[9001-24-5] Factor V, 3871
[9001-25-6] Factor VII, 3872
[9001-26-7] Prothrombin, 7903
[9001-27-8] Factor VIII, 3873
[9001-28-9] Factor IX, 3874
[9001-29-0] Factor X, 3875
[9001-30-3] Factor XII, 3877
[9001-31-4] Fibrin, 4014
[9001-32-5] Fibrinogen, 4015
[9001-33-6] Ficin, 4020
[9001-37-0] Glucose Oxidase, 4354
[9001-45-0] β-Glucuronidase, 4361
[9001-53-0] Histaminase, 4639
[9001-54-1] Hyaluronidases, 4676
[9001-57-4] Invertase, 4899
[9001-60-9] Lactate Dehydrogenase, 5213
[9001-62-1] Lipase, 5389
[9001-63-2] Lysozyme, 5514
[9001-66-5] Monoamine Oxidase, 6158
[9001-68-7] Old Yellow Enzyme, 6784
[9001-72-3] Pankrin, 6960
[9001-73-4] Papain, 6965
[9001-74-5] Penicillinase, 7034
[9001-75-6] Pepsin, 7103

Denotes derivative of title compound

[9001-90-5] Plasmin, 7493
[9001-91-6] Plasminogen, 7494
[9001-97-2] Q-Enzyme®, 8035
[9001-98-3] Rennin, 8139
[9001-99-4] Ribonuclease, 8203
[9002-01-1] Streptokinase, 8784
[9002-04-4] Thrombin, 9319
[9002-07-7] Trypsin, 9704
[9002-10-2] Tyrosinase, 9746
[9002-12-4] Uricase, 9791
[9002-13-5] Urease, 9785
[9002-18-0] Agar, 172
[9002-60-2] ACTH, 127
[9002-61-3] HCG, 4534
[9002-62-4] Prolactin, 7788
[9002-64-6] Parathyroid Hormone, 6984
[9002-67-9] LH, 5353
[9002-68-0] FSH, 4189
[9002-69-1] Relaxin, 8137
[9002-71-5] TSH, 9709
[9002-72-6] Somatotropin, 8672
[9002-84-0] Polytetrafluoroethylene, 7560
[9002-86-2] Polyvinyl Chloride, 7563
[9002-88-4] Polyethylene, 7544
[9002-89-5] Polyvinyl Alcohol, 7562
[9002-91-9] Metaldehyde, 5827
[9002-92-0] Polidocanol, 7532
[9002-93-1] Octoxynol, 6681
[9003-07-0] Polypropylene, 7558
[9003-11-6] Poloxalene, 7536
[9003-11-6] Poloxamer Forms, 7537
[9003-24-1] Darvan®, 2821
[9003-39-8] Povidone, 7700
[9003-59-2] Sodium Polystyrene Sulfonate, 8622
[9003-97-8] Calcium Polycarbophil, 1704
[9003-98-9] Deoxyribonuclease (Pancreatic), 2888
[9004-02-8] Lipoprotein Lipase, 5390
[9004-06-2] Elastase, 3499
[9004-10-8] Insulin, 4887
[9004-10-8] Insulin Injection, 4889
[9004-17-5] Isophane Insulin Suspension, 5082
[9004-17-5] Protamine Zinc Insulin Suspension, 7899
[9004-32-4] Carboxymethylcellulose Sodium, 1835
[9004-34-6] Cellulose, 1961
[9004-53-9] Dextrin, 2931
[9004-54-0] Dextran, 2925
[9004-57-3] Ethyl Cellulose, 3739
[9004-58-4] Cellulose Ethyl Hydroxyethyl Ether, 1963
[9004-61-9] Hyaluronic Acid, 4675
[9004-62-0] Hetastarch, 4593
[9004-64-2] Hydroxypropyl Cellulose, 4776
[9004-65-3] Hydroxypropyl Methylcellulose, 4777
[9004-67-5] Methylcellulose, 5963
[9005-32-7] Alginic Acid, 232
[9005-38-3] Algin, 231
[9005-49-6] Heparin, 4571
[9005-53-2] Lignin, 5363
[9005-65-6] Polysorbate 80, 7559
[9005-79-2] Glycogen, 4392
[9005-80-5] Inulin, 4898
[9005-84-9] Starch, Soluble, 8758
[9005-97-4] Thyroprotein, 9347
[9006-58-0] Lysalbinic Acid, 5504
[9007-03-8] Bifidus Factor, 1230
[9007-12-9] Calcitonin, 1640
[9007-20-9] Carboxypolymethylene, 1836

[9007-28-7] Chondroitin Sulfate, 2217
[9007-31-2] Clupeine, 2418
[9007-40-3] Crotoxin, 2602
[9007-41-4] CRP, 2606
[9007-43-6] Cytochrome c, 2797
[9007-57-2] Edestin, 3479
[9007-64-4] Enterogastrone, 3548
[9007-73-2] Ferritin, 3984
[9007-76-5] Fibroin, 4017
[9007-90-3] Gliadin, 4333
[9007-92-5] Glucagon, 4341
[9007-93-6] Glycinin, 4388
[9008-08-6] Hypalon®, 4797
[9008-19-9] Krebiozen®, 5201
[9008-22-4] Laminaran, 5226
[9008-42-8] Silver Protein, Strong, 8475
[9008-96-2] Phosvitin, 7340
[9008-99-5] Piromen, 7474
[9009-29-4] Polyferose, 7547
[9009-58-9] Porphyropsin, 7577
[9009-65-8] Protamine Sulfate*, 7898
[9009-72-7] Purothionin, 7961
[9009-81-8] Rhodopsin, 8193
[9009-86-3] Ricin, 8211
[9010-01-9] Sodium Amylosulfate, 8520
[9010-29-1] Tannoform, 9024
[9010-30-4] Taraxein, 9032
[9010-71-3] Zinc Iodide-Starch, 10041
[9010-72-4] Zymosan, 10100
[9011-02-3] Negatol®, 6360
[9011-04-5] Hexadimethrine Bromide, 4605
[9011-18-1] Dextran Sulfate Sodium, 2929
[9011-93-2] Lysostaphin, 5513
[9011-97-6] Cholecystokinin, 2201
[9012-13-9] Resodec, 8156
[9013-42-7] Ebimar®, 3463
[9013-55-2] Factor XI, 3876
[9013-56-3] Factor XIII, 3878
[9013-83-6] Insulinase, 4888
[9014-02-2] Zinostatin, 10073
[9014-44-2] Toxohormone, 9481
[9014-67-9] Aloxiprin, 306
[9015-13-8] ECTEOLA-Cellulose, 3477
[9015-51-4] Silver Protein, Mild, 8474
[9015-68-3] L-Asparaginase, 858
[9015-71-8] CRF, 2589
[9015-73-0] Detaxtran, 2916
[9015-94-5] Renin, 8138
[9016-01-7] Orgotein, 6820
[9025-41-6] Keratinase, 5171
[9025-70-1] Dextranase, 2926
[9031-37-2] Ceruloplasmin, 1999
[9034-40-6] LH-RH, 5354
[9034-50-8] Vasopressin, 9843
[9035-55-6] Lipotropic Hormone, 5392
[9035-58-9] Thromboplastin, 9321
[9035-68-1] Proinsulin, 7787
[9037-22-3] Amioca, 496
[9038-41-9] Sodium Cellulose Phosphate, 8542
[9039-53-6] Urokinase, 9799
[9039-61-6] Batroxobin, 1020
[9040-14-6] Vernamycin B, 9865
[9041-92-3] α_1-Antitrypsin, 751
[9041-93-4] Bleomycin Sulfate*, 1324
[9044-70-6] Coherin, 2468
[9046-56-4] Ancrod, 664
[9048-49-1] Serum Albumin, 8417
[9050-67-3] Sizofiran, 8502
[9056-49-9] Villikinin, 9883

[9063-57-4] Tuftsin, 9719
[9064-92-0] Polidexide, 7531
[9067-32-7] Hyaluronic Acid Sodium*, 4675
[9073-78-3] Thermolysin, 9213
[9079-25-8] Amberlite®, 390
[9083-38-9] Melanostatin, 5693
[9087-70-1] Aprotinin, 784
[10001-43-1] Pimefylline, 7400
[10010-67-0] PIPES Sodium*, 7450
[10018-19-6] Cotarnine Chloride*, 2551
[10022-31-8] Barium Nitrate, 995
[10022-50-1] Nitryl Fluoride, 6580
[10023-07-1] Frenolicin, 4182
[10024-97-2] Nitrous Oxide, 6575
[10025-65-7] Platinous Chloride, 7501
[10025-66-8] Radium Chloride*, 8118
[10025-67-9] Sulfur Chloride, 8949
[10025-68-0] Selenium Chloride, 8384
[10025-71-5] Tellurium Dichloride, 9066
[10025-73-7] Chromic Chloride, 2224
[10025-78-2] Trichlorosilane, 9559
[10025-82-8] Indium Trichloride, 4865
[10025-83-9] Iridium Trichloride, 4971
[10025-85-1] Nitrogen Chloride, 6523
[10025-87-3] Phosphorus Oxychloride, 7324
[10025-91-9] Antimony Trichloride, 739
[10025-93-1] Uranium Trichloride, 9772
[10025-99-7] Potassium Tetrachloroplatinate(II), 7675
[10026-01-4] Osmium Tetrachloride, 6847
[10026-02-5] Polonium Tetrachloride, 7535
[10026-03-6] Selenium Tetrachloride, 8392
[10026-04-7] Silicon Tetrachloride, 8447
[10026-07-0] Tellurium Tetrachloride, 9070
[10026-08-1] Thorium Chloride, 9309
[10026-10-5] Uranium Tetrachloride, 9770
[10026-11-6] Zirconium Chloride, 10077
[10026-12-7] Niobium Pentachloride, 6473
[10026-13-8] Phosphorus Pentachloride, 7326
[10026-17-2] Cobaltous Fluoride, 2434
[10026-18-3] Cobaltic Fluoride, 2424
[10028-14-5] Nobelium, 6586
[10028-15-6] Ozone, 6936
[10028-17-8] Tritium, 9674
[10028-18-9] Nickel Fluoride, 6415
[10028-22-5] Ferric Sulfate, 3980
[10028-26-9] Magnesium Bisulfate, 5534
[10030-90-7] Ferrous Succinate, 4005
[10031-13-7] Lead Arsenite, 5271
[10031-18-2] Mercurous Bromide, 5793
[10031-22-8] Lead Bromide, 5275
[10031-23-9] Radium Bromide*, 8118

[10031-24-0] Stannous Bromide, 8741

[10031-25-1] Chromic Bromide, 2222

[10031-26-2] Ferric Bromide, 3960

[10031-27-3] Tellurium Tetrabromide, 9069

[10034-81-8] Magnesium Perchlorate, 5557

[10034-85-2] Hydriodic Acid, 4699

[10034-85-2] Hydrogen Iodide, 4724

[10034-92-1] Trihydrazine Dihydriodide, 9608

[10034-93-2] Hydrazine Sulfate, 4693

[10035-10-6] Hydrobromic Acid, 4701

[10035-10-6] Hydrogen Bromide, 4720

[10038-98-9] Germanium Tetrachloride, 4305

[10039-31-3] Beryllium Selenate, 1196

[10039-53-9] Sodium Chromate(VI), Radioactive, 8548

[10039-54-0] Hydroxylamine Sulfate*, 4759

[10040-45-6] Picosulfate Sodium, 7377

[10042-76-9] Strontium Nitrate, 8810

[10043-01-3] Aluminum Sulfate, 372

[10043-11-5] Boron Nitride, 1348

[10043-35-3] Boric Acid, 1336

[10043-49-9] Gold, Radioactive, Colloidal, 4419

[10043-52-4] Calcium Chloride, 1659

[10043-67-1] Aluminum Potassium Sulfate, 364

[10043-84-2] Manganese Hypophosphite, 5613

[10043-92-2] Radon, 8119

[10045-86-0] Ferric Phosphate, 3975

[10045-87-1] Ferric Sodium Pyrophosphate, 3978

[10045-89-3] Ammonium Ferrous Sulfate, 546

[10045-94-0] Mercuric Nitrate, 5777

[10048-32-5] Parasorbic Acid, 6981

[10048-98-3] Barium Phosphate, Dibasic, 1002

[10048-99-4] Barium Mercuric Iodide, 994

[10049-01-1] Bismuth Phosphate, 1287

[10049-03-3] Fluorine Perchlorate, 4094

[10049-04-4] Chlorine Dioxide, 2096

[10049-05-5] Chromous Chloride, 2246

[10049-06-6] Titanium Dichloride, 9397

[10049-07-7] Rhodium Chloride, 8189

[10049-08-8] Ruthenium Trichloride, 8275

[10049-10-2] Chromous Fluoride, 2247

[10049-11-3] Chromium Tetrafluoride, 2237

[10049-12-4] Vanadium Trifluoride, 9828

[10049-14-6] Uranium Tetrafluoride, 9771

[10049-16-8] Vanadium Tetrafluoride, 9827

[10049-17-9] Rhenium Hexafluoride, 8178

[10049-22-6] Sodium Pyroantimonate, 8624

[10049-23-7] Tellurous Acid, 9072

[10049-25-9] Chromous Bromide, 2245

[10049-83-9] Pyritinol, 8006

[10058-44-3] Ferric Pyrophosphate, 3976

[10060-10-3] Ceric Fluoride, 1985

[10060-11-4] Germanium Dichloride, 4303

[10085-81-1] Benzoctamine Hydrochloride*, 1096

[10087-54-4] 17-Hydroxy-16-methylene-Δ^6-progesterone, 4763

[10088-95-6] Radicinin, 8117

[10089-10-8] Nidulin, 6440

[10090-54-7] Potassium Pyroantimonate, Acid, 7652

[10091-70-0] Protokosin, 7910

[10097-28-6] Silicon Monoxide, 8443

[10097-84-4] Tetrahydropalmatine, 9147

[10099-74-8] Lead Nitrate, 5292

[10099-79-3] Lead Vanadate(V), 5310

[10101-50-5] Sodium Permanganate, 8606

[10101-52-7] Zirconium Silicate, 10084

[10101-53-8] Chromic Sulfate, 2233

[10101-63-0] Lead Iodide, 5288

[10101-83-4] Sodium Tellurate(VI), 8642

[10101-86-7] Sodium Tetrathionate, 8647

[10101-88-9] Sodium Thiophosphate, 8651

[10101-94-7] Lead Sodium Thiosulfate, 5299

[10102-02-0] Zinc Nitrite, 10046

[10102-03-1] Nitrogen Pentoxide, 6526

[10102-05-3] Palladium Nitrate, 6943

[10102-18-8] Sodium Selenite, 8629

[10102-20-2] Sodium Tellurate(IV), 8641

[10102-24-6] Lithium Silicate, 5420

[10102-43-9] Nitric Oxide, 6498

[10102-44-0] Nitrogen Dioxide, 6524

[10102-45-1] Thallium Nitrate, 9192

[10102-48-4] Lead Arsenate, 5270

[10102-68-8] Calcium Iodide, 1680

[10102-71-3] Aluminum Sodium Sulfate, 369

[10108-64-2] Cadmium Chloride, 1615

[10108-73-3] Cerous Nitrate, 1994

[10112-91-1] Mercurous Chloride, 5795

[10114-58-6] Bismark Brown Y, 1265

[10116-22-0] Demegestone, 2874

[10117-38-1] Potassium Sulfite, 7667

[10118-56-6] Cascarillin, 1891

[10118-76-0] Calcium Permanganate, 1695

[10118-90-8] Minocycline, 6121

[10124-36-4] Cadmium Sulfate, 1627

[10124-37-5] Calcium Nitrate, 1688

[10124-41-1] Calcium Thiosulfate, 1719

[10124-43-3] Cobaltous Sulfate, 2443

[10124-48-8] Mercuric Chloride, Ammoniated, 5771

[10124-50-2] Potassium Arsenite, 7584

[10124-53-5] Magnesium Thiosulfate, 5576

[10136-65-9] Oxypinocamphone, 6929

[10137-74-3] Calcium Chlorate, 1658

[10138-04-2] Ammonium Ferric Sulfate, 543

[10139-06-7] Linatine, 5377

[10139-47-6] Zinc Iodide, 10040

[10140-70-2] Curvularin, 2684

[10141-00-1] Chromic Potassium Sulfate, 2232

[10141-05-6] Cobaltous Nitrate, 2438

[10161-33-8] Trenbolone, 9499

[10161-34-9] Trenbolone Acetate*, 9499

[10170-69-1] Manganese Carbonyl, 5609

[10190-55-3] Lead Molybdate(VI), 5290

[10192-30-0] Ammonium Bisulfite, 522

[10196-04-0] Ammonium Sulfite, 588

[10199-21-0] Guanidinium Aluminum Sulfate Hexahydrate, 4476

[10210-64-7] Beryllium Acetylacetonate, 1180

[10210-68-1] Dicobalt Octacarbonyl, 3074

[10213-09-9] Vanadyl Dichloride, 9832

[10214-39-8] Lead Borate, 5273

[10214-40-1] Cupric Selenite, 2657

[10233-88-2] Gold Sodium Thiosulfate, 4423

[10236-47-2] Naringin, 6345

[10236-58-5] Selenocysteine, 8394

[10238-21-8] Glyburide, 4372

[10246-75-0] Hydroxyzine Pamoate*, 4786

[10257-55-3] Calcium Sulfite, 1715

[10262-69-8] Maprotiline, 5633

[10265-92-6] Methamidophos, 5858

[10290-12-7] Cupric Arsenite, 2630

[10294-26-5] Silver Sulfate, 8480

[10294-28-7] Gold Tribromide, 4425

[10294-29-8] Gold Monochloride, 4414

[10294-31-2] Gold Monoiodide, 4416

[10294-32-3] Gold Selenate, 4420

[10294-33-4] Boron Tribromide, 1349

[10294-34-5] Boron Trichloride, 1350

[10294-40-3] Barium Chromate(VI), 982

[10294-44-7] Mercurous Chlorate, 5794

[10294-47-0] Lead Chlorate, 5277

[10294-48-1] Chlorine Heptoxide, 2097

[10294-52-7] Ferric Chromate(VI), 3962

[10294-54-9] Cesium Sulfate, 2008

[10294-58-3] Lead Hypophosphite, 5287

[10294-64-1] Potassium Manganate(VI), 7629

[10294-66-3] Potassium Thiosulfate, 7688

[10294-70-9] Stannous Iodide, 8745

[10297-61-7] Propylure, 7883

[10310-32-4] Tribenoside, 9520

[10311-84-9] Dialifor, 2949

[10318-26-0] Mitolactol, 6132

[10321-12-7] Propizepine, 7846

[10325-94-7] Cadmium Nitrate, 1620

[10326-21-3] Magnesium Chloride, 5539

Denotes derivative of title compound

[10326-24-6] Zinc Meta-arsenite, 10043
[10326-41-7] D-Lactic Acid, 5214
[10331-57-4] Menichlopholan, 5722
[10343-61-0] Titanium Sesquisulfate, 9401
[10347-81-6] Maprotiline Hydrochloride*, 5633
[10352-73-5] Pseudopederin, 7935
[10361-03-2] Maddrell's Salt, 5522
[10361-29-2] Ammonium Carbonate, 528
[10361-37-2] Barium Chloride, 981
[10361-43-0] Bismuth Hydroxide, 1277
[10361-44-1] Bismuth Nitrate, 1282
[10373-81-6] 3-Hydroxycamphor, 4747
[10376-48-4] Shionone, 8429
[10377-48-7] Lithium Sulfate, 5421
[10377-51-2] Lithium Iodide, 5413
[10377-58-9] Magnesium Iodide, 5549
[10377-60-3] Magnesium Nitrate, 5552
[10377-62-5] Magnesium Permanganate, 5558
[10377-66-9] Manganese Nitrate, 5615
[10379-14-3] Tetrazepam, 9172
[10380-31-1] Barium Uranium Oxide, 1013
[10381-36-9] Nickel Phosphate, 6422
[10381-37-0] Thorium Sulfate, 9313
[10389-72-7] Clortermine Hydrochloride*, 2406
[10389-73-8] Clortermine, 2406
[10397-75-8] Iocarmic Acid, 4902
[10402-90-1] Eprazinone, 3577
[10405-02-4] Trospium Chloride, 9697
[10415-75-5] Mercurous Nitrate, 5798
[10417-94-4] 5,8,11,14,17-Eicosapentaenoic Acid, 3495
[10418-03-8] Stanozolol, 8754
[10421-48-4] Ferric Nitrate, 3972
[10450-55-2] Ferric Acetate, Basic, 3957
[10450-60-9] Periodic Acid, 7123
[10453-89-1] Chrysanthemic Acid, 2255
[10457-66-6] Geranylhydroquinone, 4300
[10457-90-6] Bromperidol, 1432
[10476-81-0] Strontium Bromide, 8800
[10476-85-4] Strontium Chloride, 8803
[10476-86-5] Strontium Iodide, 8808
[10486-11-0] Quinine Oleate, 8087
[10488-36-5] Tofenacin Hydrochloride*, 9426
[10534-89-1] Hexaaminecobalt Trichloride, 4597
[10539-19-2] Moxaverine, 6202
[10540-29-1] Tamoxifen, 9019
[10553-31-8] Barium Bromide, 978
[10563-70-9] Melitracen Hydrochloride*, 5704
[10567-69-8] Barium Iodate, 991
[10571-59-2] Nicoclonate, 6426
[10580-02-6] Ammonium Titanium Oxalate, 593
[10588-01-9] Sodium Dichromate-(VI), 8556
[10596-23-3] Clodronic Acid, 2369
[10605-21-7] Carbendazim, 1794
[11000-03-6] Strepogenin, 8780
[11003-38-6] Capreomycin, 1758

[11003-70-6] Scillaren, 8354
[11005-20-2] Pseudohecogenin, 7930
[11005-63-3] Strophanthin, 8819
[11005-70-2] Cerberoside, 1983
[11006-77-2] Statolon, 8760
[11006-83-0] Thermorubin, 9214
[11006-87-4] Essential Fatty Acids, 3650
[11015-37-5] Bambermycins, 962
[11016-39-0] Properdin, 7825
[11017-56-4] Strigol, 8795
[11024-24-1] Digitonin, 3144
[11028-71-0] Concanavalin A, 2490
[11030-71-0] Amanitin, 378
[11031-48-4] Sarkomycin, 8333
[11035-08-8] Substance P, 8834
[11041-12-6] Cholestyramine Resin, 2205
[11042-38-9] Xanthocillin, 9970
[11042-64-1] γ-Oryzanol, 6841
[11051-71-1] Avilamycin, 905
[11061-68-0] Insulin Human*, 4887
[11061-96-4] Cardiotoxin, 1842
[11070-73-8] Insulin Beef*, 4887
[11075-36-8] Tuberactinomycin, 9713
[11076-19-0] Bongkrekic Acid, 1334
[11078-21-0] Filipin, 4022
[11085-36-2] HCS, 4535
[11096-26-7] Erythropoietin, 3635
[11096-49-4] Partricin, 6997
[11104-88-4] Phosphomolybdic Acid, 7316
[11110-52-4] Sodium Amalgam, 8518
[11115-82-5] Enduracidin, 3534
[11120-15-3] Dermostatin, 2897
[11120-25-5] Ammonium Tungstate-(VI), 594
[11121-16-7] Aluminum Borate, 329
[11121-32-7] Mepartricin, 5733
[11121-48-5] Rose Bengal, 8242
[11125-96-5] Burgundy Mixture, 1491
[11138-66-2] Xanthan Gum, 9966
[11141-17-6] Azadirachtin, 910
[12001-65-9] Methenamine Tetraiodine, 5886
[12001-68-2] Potassium Borotartrate, 7595
[12002-03-8] Cupric Acetoarsenite, 2629
[12002-30-1] Piperazine Edetate Calcium, 7435
[12002-43-6] Uintahite, 9754
[12002-53-8] Trimedlure, 9615
[12007-56-6] Calcium Borate, 1651
[12007-58-8] Ammonium Borate, 524
[12016-80-7] Cobaltic Oxide Monohydrate, 2425
[12018-01-8] Chromium Dioxide, 2236
[12018-10-9] Cupric Chromite, 2638
[12026-06-1] Thallium Hydroxide, 9190
[12026-66-3] Ammonium Phosphomolybdate, 573
[12027-06-4] Ammonium Iodide, 554
[12027-67-7] Ammonium Molybdate(VI), 559
[12029-98-0] Iodine Pentoxide, 4913
[12033-38-4] Nitrogen Selenide, 6527
[12033-89-5] Silicon Nitride, 8444
[12040-44-7] Alcian Blue, 211
[12047-27-7] Barium Titanate(IV), 1012
[12048-50-9] Bismuth Tetroxide, 1304

[12054-48-7] Nickel Hydroxide, 6417
[12057-24-8] Lithium Oxide, 5416
[12058-66-1] Sodium Stannate(IV), 8633
[12062-24-7] Cupric Hexafluorosilicate, 2645
[12063-56-8] Yig, 10006
[12063-98-8] Gallium Phosphide, 4254
[12067-99-1] Phosphotungstic Acid, 7339
[12068-69-8] Bismuth Selenide, 1290
[12069-32-8] Boron Carbide, 1346
[12069-69-1] Cupric Carbonate, Basic, 2634
[12071-83-9] Propineb, 7831
[12111-24-9] Pentetate Calcium Trisodium, 7082
[12122-67-7] Zineb, 10071
[12124-97-9] Ammonium Bromide, 525
[12124-99-1] Ammonium Bisulfide, 521
[12125-01-8] Ammonium Fluoride, 547
[12125-02-9] Ammonium Chloride, 531
[12125-08-5] Ammonium Osmium Chloride, 564
[12125-09-6] Phosphonium Iodide, 7317
[12135-76-1] Ammonium Sulfide, 585
[12141-46-7] Aluminum Silicate, 368
[12142-33-5] Potassium Stannate-(IV), 7662
[12167-15-6] Potassium Phenolsulfonate, 7645
[12173-47-6] Hectorite, 4538
[12192-57-3] Aurothioglucose, 901
[12201-47-7] Sodium Plumbate(IV), 8619
[12228-91-0] Manganese Borate, 5606
[12232-99-4] Sodium Bismuthate(V), 8530
[12244-57-4] Gold Sodium Thiomalate, 4422
[12284-76-3] Bismuth Aluminate, 1269
[12286-76-9] Ferric Fructose, 3967
[12419-43-1] Stannous Hexafluorozirconate(IV), 8744
[12427-38-2] Maneb, 5603
[12505-77-0] Boron Monoxide, 1347
[12511-31-8] Aluminum Magnesium Silicate, 352
[12520-88-6] Silicotungstic Acid, 8449
[12584-58-6] Insulin Pork*, 4887
[12584-83-7] Cobrotoxin, 2447
[12607-70-4] Nickel Carbonate Hydroxide, 6410
[12607-92-0] Aceglutamide Aluminum*, 21
[12607-93-1] Taxine(s), 9047
[12609-84-6] Thiopeptin, 9281
[12616-24-9] Ammonium Zirconyl Carbonate, 600
[12629-01-5] Somatotropin (Human)*, 8672
[12650-69-0] Mupirocin, 6212
[12704-90-4] Aurodox, 900
[12740-44-2] Sodium-Lead Alloy, 8588
[12751-04-1] Creolin®, 2572
[12771-68-5] Ancymidol, 665
[12772-35-9] Butirosin, 1523
[12772-57-5] Monorden, 6164

[13007-90-4] Bis(triphenylphosphine)dicarbonylnickel, 1314
[13007-92-6] Chromium Carbonyl, 2235
[13007-93-7] Cuproxoline, 2677
[13008-73-6] L-Streptose, 8790
[13009-99-9] Mafenide Acetate*, 5525
[13010-47-4] Lomustine, 5444
[13011-54-6] Ammonium Sodium Phosphate, 581
[13042-18-7] Fendiline, 3916
[13051-01-9] Carbazochrome Salicylate, 1790
[13055-82-8] Reproterol Hydrochloride*, 8142
[13058-67-8] Lucensomycin, 5464
[13067-93-1] Cyanofenphos, 2697
[13071-79-9] Terbufos, 9088
[13073-35-3] Ethionine, 3693
[13087-53-1] Iothalamic Acid Meglumine*, 4952
[13092-66-5] Magnesium Phosphate, Monobasic, 5561
[13093-88-4] Perimethazine, 7120
[13100-82-8] Cysteic Acid, 2786
[13103-34-9] Boldenone Undecylenate*, 1327
[13121-70-5] Cyhexatin, 2767
[13123-37-0] Lyxoflavine, 5515
[13137-64-9] Periplocin, 7126
[13138-45-9] Nickel Nitrate, 6420
[13171-21-6] Phosphamidon, 7312
[13190-88-0] Leucoglycodrin, 5336
[13194-48-4] Ethoprop, 3702
[13246-02-1] Febarbamate, 3889
[13254-33-6] Carpronium Chloride, 1871
[13289-18-4] Hellebrin, 4547
[13292-46-1] Rifampin, 8215
[13311-84-7] Flutamide, 4132
[13327-32-7] Beryllium Hydroxide, 1188
[13347-42-7] Dowicide 9, 3420
[13356-08-6] Fenbutatin Oxide, 3907
[13364-32-4] Clobenzorex, 2359
[13392-18-2] Fenoterol, 3927
[13392-28-4] Rimantadine, 8221
[13401-40-6] Phaseolin, 7149
[13402-08-9] Acetylpheneturide, 94
[13405-60-2] β-Glucogallin, 4348
[13408-56-5] Ponasterone A, 7566
[13410-01-0] Sodium Selenate, 8627
[13411-16-0] Nifurpirinol, 6452
[13412-64-1] Dicloxacillin Sodium Monohydrate*, 3073
[13422-51-0] Hydroxocobalamin, 4739
[13422-52-1] Aquocobalamin, 787
[13422-53-2] Vitamin B$_{12}$, Radioactive, 9922
[13424-46-9] Lead Azide, 5272
[13425-39-3] Etofylline Nicotinate, 3836
[13425-98-4] Improsulfan, 4841
[13444-24-1] 1-Ethyl-3-piperidinol, 3799
[13444-90-1] Nitryl Chloride, 6579
[13445-60-8] Furfurylmethylamphetamine, 4217
[13445-63-1] Itramin Tosylate, 5132
[13446-03-2] Manganese Bromide, 5607
[13446-24-7] Magnesium Pyrophosphate, 5564
[13446-49-6] Potassium Molybdate(VI), 7633
[13447-95-5] Isoniazid Methanesulfonate, 5072
[13453-07-1] Gold Trichloride, 4427

[13453-34-4] Thallium Cyanide, 9188
[13454-94-9] Cerous Sulfate, 1996
[13455-25-9] Cobaltous Chromate(III), 2432
[13455-36-2] Cobaltous Phosphate, 2441
[13457-18-6] Pyrazophos, 7976
[13462-88-9] Nickel Bromide, 6409
[13462-90-3] Nickel Iodide, 6418
[13463-39-3] Nickel Carbonyl, 6411
[13463-40-6] Iron Pentacarbonyl, 4980
[13463-41-7] Pyrithione Zinc*, 8004
[13463-67-7] Titanium Dioxide, 9398
[13464-44-3] Zinc Ortho-arsenate, 10048
[13464-80-7] Dihydrazine Sulfate, 3153
[13464-82-9] Indium Sulfate, 4863
[13465-66-2] Selenium Tetrafluoride, 8393
[13465-94-6] Barium Nitrite, 996
[13465-95-7] Barium Perchlorate, 999
[13466-78-9] 3-Carene, 1843
[13471-78-8] Beclotiamine, 1030
[13472-45-2] Sodium Tungstate(VI), 8655
[13473-90-0] Aluminum Nitrate, 355
[13477-00-4] Barium Chlorate, 980
[13478-49-4] Erbium Sulfate, 3586
[13478-93-8] Nickel Dimethylglyoxime, 6414
[13479-54-4] Cupric Glycinate, 2644
[13484-14-5] Sarmentose, 8335
[13492-01-8] Tranylcypromine Sulfate*, 9491
[13492-26-7] Potassium Phosphite, 7650
[13494-80-9] Tellurium, 9064
[13494-90-1] Gallium Nitrate Anhydrous*, 4252
[13495-04-0] Heptanal Sodium Bisulfite, 4579
[13495-09-5] Piminodine, 7403
[13510-49-1] Beryllium Sulfate, 1198
[13510-89-9] Lead Antimonate(V), 5269
[13523-86-9] Pindolol, 7412
[13530-57-9] Aluminum Rubidium Sulfate, 365
[13530-65-9] Zinc Chromate(VI) Hydroxide, 10032
[13539-59-8] Apazone, 758
[13548-38-4] Chromic Nitrate, 2228
[13548-42-0] Cupric Chromate(VI), 2637
[13548-43-1] Ammonium Chromic Sulfate, 533
[13552-61-9] 2-Thiolhistidine, 9270
[13560-89-9] Dechlorane® Plus, 2844
[13563-60-5] Norgesterone, 6619
[13573-11-0] Magnesium Tungstate(VI), 5577
[13573-16-5] Reinecke Salt, 8135
[13587-35-4] Cupric Tungstate(VI), 2663
[13588-16-4] Senecic Acid, 8400
[13590-82-4] Ceric Sulfate, 1987
[13596-22-0] Cobaltous Potassium Sulfate, 2442
[13596-46-8] Ammonium Cobaltous Sulfate, 536
[13597-54-1] Zinc Selenate, 10060
[13597-65-4] Zinc Silicate, 10062
[13597-95-0] Beryllium Perchlorate, 1193

[13597-99-4] Beryllium Nitrate, 1190
[13598-36-2] Phosphorous Acid, 7320
[13600-89-0] Erdmann's Salt, 3589
[13600-98-1] Sodium Cobaltinitrite, 8551
[13601-13-3] Cupric Ferrocyanide, 2640
[13601-19-9] Sodium Ferrocyanide, 8563
[13609-67-1] Hydrocortisone Butyrate*, 4710
[13614-98-7] Minocycline Hydrochloride*, 6121
[13637-61-1] Zinc Perchlorate, 10051
[13642-52-9] Soterenol, 8683
[13647-35-3] Trilostane, 9611
[13655-52-2] Alprenolol, 311
[13665-88-8] Mopidamol, 6171
[13669-70-0] Nefopam, 6358
[13674-87-8] Fyrol FR-2®, 4233
[13675-18-8] Diboron Tetrahydroxide, 2999
[13682-61-6] Potassium Tetrachloroaurate(III), 7674
[13682-73-0] Cuprous Potassium Cyanide, 2672
[13682-92-3] Dihydroxyaluminum Aminoacetate, 3168
[13684-63-4] Phenmedipham, 7199
[13698-49-2] Delmadinone Acetate, 2865
[13701-67-2] Diboron Tetrachloride, 2998
[13701-70-7] Vanadium Trisulfate, 9830
[13703-82-7] Magnesium Borate, 5535
[13707-88-5] Alprenolol Hydrochloride*, 311
[13708-85-5] Sodium Phosphite, 8616
[13710-19-5] Tolfenamic Acid, 9436
[13717-04-9] Propiram Fumarate*, 7844
[13718-26-8] Sodium Vanadate(V), 8657
[13718-50-8] Barium Iodide, 992
[13721-34-1] Sodium Uranate(VI), 8656
[13721-43-2] Sodium Hypophosphate, 8579
[13739-02-1] Diacerein, 2939
[13741-18-9] Xibornol, 9985
[13746-66-2] Potassium Ferricyanide, 7611
[13746-89-9] Zirconium Nitrate, 10082
[13755-29-8] Sodium Fluoborate, 8564
[13759-10-9] Silicon Disulfide, 8441
[13762-51-1] Potassium Borohydride, 7594
[13765-19-0] Calcium Chromate(VI), 1660
[13768-38-2] Osmium Hexafluoride, 6846
[13770-18-8] Cupric Perchlorate, 2651
[13776-84-6] Sodium Thioantimonate(V), 8648
[13780-03-5] Calcium Bisulfite, Solution, 1650
[13780-06-8] Calcium Nitrite, 1689
[13782-01-9] Cobaltic Potassium Nitrite, 2426
[13810-16-7] Dimsyl, 3259
[13814-62-5] Cadmium Selenate, 1624

Denotes derivative of title compound

[13820-41-2] Ammonium Platinous Chloride, 577
[13821-06-2] Barium Ferrocyanide, 985
[13823-29-5] Thorium Nitrate, 9311
[13825-74-6] Titanium Sulfate, 9402
[13825-86-0] Chromous Sulfate, 2250
[13838-08-9] Azidamfenicol, 924
[13838-16-9] Enflurane, 3536
[13843-81-7] Lithium Dichromate-(VI), 5408
[13845-17-5] Barium Dithionate, 984
[13862-07-2] Diphemethoxidine, 3303
[13870-90-1] Cobamamide, 2446
[13871-27-7] Beryllium Sodium Fluoride, 1197
[13876-85-2] Cuprous Mercuric Iodide, 2670
[13898-58-3] Benzoylpas, 1127
[13898-68-5] Metamivam, 5829
[13912-77-1] Octacaine, 6663
[13925-12-7] Myxin, 6255
[13931-64-1] Procymate, 7771
[13932-17-7] Potassium Zinc Sulfate, 7698
[13943-58-3] Potassium Ferrocyanide, 7612
[13946-02-6] Metron S, 6080
[13952-84-6] sec-Butylamine, 1544
[13956-29-1] Cannabidiol, 1750
[13957-33-0] Leptodactyline, 5326
[13957-36-3] Hydramitrazine, 4685
[13957-55-6] N,N-1-Trimethyl-3,3-diphenylpropylamine, 9627
[13963-58-1] Potassium Hexacyanocobaltate(III), 7620
[13966-47-7] Potassium Tetrabromoaurate(III), 7673
[13967-25-4] Mercurous Fluoride, 5796
[13967-50-5] Potassium Dicyanoaurate(I), 7609
[13967-90-3] Barium Bromate, 977
[13968-86-0] Formosulfathiazole, 4158
[13982-53-1] Cuprous Sulfite, 2675
[13986-26-0] Zirconium Iodide, 10081
[13993-65-2] Metiazinic Acid, 6057
[13997-19-8] Nequinate, 6385
[14007-64-8] Butethamate, 1516
[14008-44-7] Metopimazine, 6070
[14008-48-1] Bisoxatin Acetate, 1310
[14028-44-5] Amoxapine, 609
[14038-43-8] Ferric Ferrocyanide, 3964
[14051-33-3] Benzetimide, 1084
[14066-79-6] Chloroprednisone Acetate*, 2157
[14073-97-3] l-Menthone, 5724
[14075-53-7] Potassium Tetrafluoroborate, 7680
[14116-69-9] Vicianose, 9879
[14166-26-8] Taglutimide, 9006
[14176-10-4] Cetiedil, 2012
[14188-82-0] Hexestrol Diphosphate*, 4621
[14217-01-7] Chromic Potassium Oxalate, 2231
[14217-21-1] Sodium Ferricyanide, 8562
[14220-17-8] Potassium Tetracyanonickelate(II), 7677
[14221-47-7] Ammonium Ferric Oxalate, 542
[14221-48-8] Ammonium Ferricyanide, 544

[14222-60-7] Protionamide, 7905
[14225-07-1] Leucodrin, 5335
[14235-86-0] Hydrargaphen, 4687
[14244-62-3] Potassium Tetracyanozincate, 7679
[14252-80-3] Bupivacaine Hydrochloride*, 1485
[14255-87-9] Parbendazole, 6985
[14259-45-1] Asperuloside, 866
[14261-75-7] Cloforex, 2378
[14283-05-7] Tetraamminecopper Sulfate, 9116
[14284-36-7] Aluminum Cesium Sulfate, 336
[14286-84-1] Bencyclane Fumarate*, 1041
[14293-44-8] Xipamide, 9986
[14297-87-1] Benzylmorphine, 1155
[14307-33-6] Calcium Dichromate-(VI), 1666
[14307-35-8] Lithium Chromate(VI), 5406
[14334-40-8] Pramiverin, 7706
[14350-67-5] Ormosinine, 6823
[14357-78-9] Diprenorphine, 3347
[14376-16-0] Sulfaloxic Acid, 8883
[14402-67-6] Potassium Titanyl Oxalate, 7689
[14402-70-1] Ammonium Nitroferricyanide, 562
[14402-73-4] Lithium Tetracyanoplatinate(II), 5422
[14402-75-6] Cadmium Potassium Cyanide, 1622
[14402-89-2] Sodium Nitroprusside, 8600
[14417-88-0] Melinamide, 5702
[14437-41-3] Clioxanide, 2354
[14451-99-1] Menadione Dimethylpyrimidinol Bisulfite, 5715
[14457-87-5] Cerous Bromide, 1989
[14459-29-1] Hematoporphyrin, 4555
[14475-63-9] Zirconium Hydroxide, 10080
[14481-29-9] Ammonium Ferrocyanide, 545
[14481-33-5] Thorium Tetracyanoplatinate(II), 9314
[14484-47-0] Deflazacort, 2852
[14484-64-1] Ferbam, 3954
[14504-73-5] Tritoqualine, 9677
[14519-18-7] Strontium Bromate, 8799
[14521-96-1] Etorphine, 3843
[14523-22-9] Rhodium Carbonyl Chloride, 8188
[14556-46-8] Bupranolol, 1486
[14590-13-7] Ammonium Cobaltous Phosphate, 535
[14635-75-7] Nitrosyl Tetrafluoroborate, 6571
[14635-87-1] Magnesium Perborate, 5556
[14636-12-5] Terlipressin, 9097
[14639-94-2] Ammonium Hexafluorogallate, 550
[14639-97-5] Ammonium Tetrachlorozincate, 565
[14639-98-6] Ammonium Pentachlorozincate, 567
[14644-61-2] Zirconium Sulfate, 10085
[14679-73-3] Todralazine, 9425
[14693-02-8] Potassium Thioantimonate(V), 7685
[14694-95-2] Chlorotris(triphenylphosphine)rhodium, 2174
[14698-29-4] Oxolinic Acid, 6899
[14709-57-0] Potassium Nitroprusside, 7636

[14721-21-2] Cupric Chlorate, 2635
[14759-06-9] Sulforidazine, 8943
[14807-96-6] Talc, 9011
[14816-18-3] Phoxim, 7341
[14816-67-2] Soterenol Monohydrochloride*, 8683
[14817-09-5] Decimemide, 2845
[14838-15-4] Phenylpropanolamine*, 7279
[14855-76-6] Methyl Green, 5998
[14860-49-2] Clobutinol, 2363
[14871-79-5] Barium Hypophosphite, 990
[14874-24-9] Tris(ethylenediamine)-cadmium Dihydroxide, 9666
[14875-96-8] Heme, 4558
[14882-18-9] Bismuth Subsalicylate, 1299
[14885-29-1] Ipronidazole, 4966
[14901-08-7] Cycasin, 2705
[14912-44-8] Ylangene, 10007
[14918-35-5] Destomycin A, 2915
[14929-11-4] Simfibrate, 8489
[14940-41-1] Ferrous Phosphate, 4002
[14959-86-5] Looplure, 5449
[14976-57-9] Clemastine Fumarate*, 2345
[14977-61-8] Chromyl Chloride, 2251
[14986-91-5] Magnesium Selenate, 5566
[14992-62-2] Acetylcarnitine, 78
[15096-52-3] Cryolite, 2609
[15123-69-0] Cupric Selenate, 2655
[15130-91-3] Sultroponium, 8977
[15139-76-1] Orange B, 6814
[15176-29-1] Edoxudine, 3486
[15180-00-4] Prednival, 7728
[15180-03-7] Alcuronium Chloride*, 214
[15189-51-2] Sodium Tetrachloroaurate(III), 8644
[15238-00-3] Cobaltous Iodide, 2437
[15263-53-3] Cartap, 1874
[15265-28-8] Palitantin, 6939
[15271-41-7] Tranid®, 9488
[15280-09-8] Sodium Dicyanoaurate-(I), 8557
[15299-99-7] Napropamide, 6336
[15301-40-3] Actinoquinol, 134
[15301-48-1] Bezitramide, 1216
[15301-52-7] Cyclexanone, 2712
[15301-69-6] Flavoxate, 4033
[15301-80-1] Oxamarin, 6868
[15301-93-6] Tofenacin, 9426
[15302-16-6] Fenozolone, 3932
[15304-14-0] Aluminum Hexaurea Sulfate Triiodide, 343
[15307-79-6] Diclofenac Sodium, 3071
[15318-45-3] Thiamphenicol, 9230
[15351-04-9] Becanthone, 1026
[15351-09-4] Metamfepramone, 5828
[15351-13-0] Nicofuranose, 6428
[15358-48-2] Hydroxylupanine, 4760
[15372-34-6] Doisynoestrol, 3407
[15375-94-7] Merphyrin, 5859
[15385-57-6] Mercurous Iodide, 5797
[15421-84-8] Trapidil, 9492
[15477-33-5] Aluminum Chlorate, 337
[15489-16-4] Stibophen, 8769
[15489-90-4] Hematin, 4554
[15500-66-0] Pancuronium Bromide, 6958
[15518-82-8] Metescufylline, 5844
[15537-71-0] N-Acetylpenicillamine, 93

[15537-76-5] Chlorproguanil Hydro-chloride*, 2185
[15571-91-2] Potassium Tellurate-(VI), 7671
[15572-25-5] Thallium Selenide, 9195
[15574-96-6] Pizotyline, 7486
[15578-26-4] Stannous Pyrophos-phate, 8748
[15585-70-3] Bibenzonium Bromide, 1218
[15593-51-8] Lithium Selenite, 5419
[15593-52-9] Lithium Selenate, 5418
[15593-61-0] Magnesium Selenite, 5568
[15599-36-7] Halethazole, 4505
[15599-39-0] Noxytiolin, 6645
[15599-51-6] Apicycline, 762
[15610-76-1] Ammonium Cupric Chloride, 537
[15622-65-8] Molindone Hydrochlo-ride*, 6142
[15662-33-6] Ryanodine, 8279
[15663-27-1] Cisplatin, 2319
[15671-27-9] Sulfitocobalamin, 8931
[15676-16-1] Sulpiride, 8971
[15686-51-8] Clemastine, 2345
[15686-61-0] Fenproporex, 3939
[15686-71-2] Cephalexin, 1971
[15686-83-6] Pyrantel, 7968
[15686-91-6] Propiram, 7844
[15687-16-8] Carbiphene, 1805
[15687-18-0] Fenpentadiol, 3933
[15687-22-6] Folescutol, 4139
[15687-23-7] Guaiactamine, 4462
[15687-27-1] Ibuprofen, 4812
[15687-41-9] Oxyfedrine, 6915
[15699-18-0] Ammonium Nickel Sulfate, 560
[15708-41-5] Ferric Sodium Edetate, 3977
[15722-48-2] Olsalazine, 6799
[15793-40-5] Terodiline, 9098
[15825-70-4] Mannitol Hexanitrate, 5630
[15826-37-6] Cromolyn Sodium*, 2594
[15829-53-5] Mercurous Oxide, 5799
[15876-67-2] Distigmine Bromide, 3368
[15879-93-3] α-Chloralose, 2062
[15972-60-8] Alachlor, 193
[16009-13-5] Hemin, 4563
[16034-77-8] Iocetamic Acid, 4903
[16037-91-5] Antimony Sodium Gluconate, 733
[16039-53-5] Zinc Lactate, 10042
[16048-96-7] Cupric Salicylate, 2654
[16051-77-7] Isosorbide Mononi-trate*, 5113
[16110-51-3] Cromolyn, 2594
[16188-61-7] Talastine, 9009
[16283-36-6] Zinc Salicylate, 10059
[16286-69-4] Citiedil Citrate*, 2012
[16298-90-1] Jesaconitine, 5146
[16320-04-0] Gestrinone, 4310
[16357-59-8] EEDQ, 3488
[16377-00-7] Indospicine, 4877
[16378-21-5] Piroheptine, 7473
[16423-68-0] Erythrosine, 3640
[16498-21-8] Oxaflumazine, 6862
[16545-11-2] Guamecycline, 4468
[16550-22-4] Cyprenorphine Hydro-chloride*, 2777
[16590-41-3] Naltrexone, 6278
[16595-80-5] Tetramisole L-Form Hydrochloride*, 9161
[16604-45-8] Medifoxamine Fuma-rate*, 5674
[16625-20-0] Haplophytine, 4525

[16649-79-9] Calcium N-Carbamo-ylaspartate, 1655
[16662-47-8] Gallopamil, 4257
[16672-87-0] Ethephon, 3686
[16676-29-2] Naltrexone Hydrochlo-ride*, 6278
[16679-58-6] Desmopressin, 2904
[16683-27-5] Convallamarogenin, 2506
[16721-80-5] Sodium Bisulfide, 8532
[16731-55-8] Potassium Metabisul-fite, 7630
[16752-77-5] Methomyl, 5905
[16755-07-0] Showdomycin, 8430
[16773-42-5] Ornidazole, 6824
[16774-21-3] Ammonium Ceric Ni-trate, 529
[16790-49-1] Butazolamide, 1511
[16816-67-4] Pantethine, 6962
[16830-15-2] Asiaticoside, 857
[16846-24-5] Josamycin, 5148
[16853-85-3] Aluminum Lithium Hydride, 351
[16871-60-6] Potassium Hexachloro-osmate(IV), 7618
[16871-71-9] Zinc Hexafluorosilicate, 10037
[16871-90-2] Potassium Hexafluoro-silicate, 7622
[16872-11-0] Fluoboric Acid, 4075
[16893-85-9] Sodium Hexafluorosili-cate, 8572
[16903-35-8] Gold Trichloride, Acid, 4428
[16919-19-0] Ammonium Hexafluo-rosilicate, 552
[16919-58-7] Ammonium Platinic Chloride, 576
[16921-30-5] Potassium Hexachloro-platinate(IV), 7619
[16921-96-3] Iodine Heptafluoride, 4909
[16923-58-3] Sodium Hexachloropla-tinate(IV), 8571
[16923-95-8] Potassium Hexafluoro-zirconate(IV), 7623
[16925-39-6] Calcium Hexafluorosili-cate, 1674
[16925-51-2] Aurothioglycanide, 902
[16940-66-2] Sodium Borohydride, 8537
[16941-10-9] Aluminum Calcium Hydride, 334
[16941-11-0] Ammonium Hexafluo-rophosphate, 551
[16941-12-1] Platinic Chloride, 7498
[16949-15-8] Lithium Borohydride, 5402
[16949-65-8] Magnesium Hexafluo-rosilicate, 5546
[16960-16-0] Cosyntropin, 2550
[16961-83-4] Fluosilicic Acid, 4110
[16962-07-5] Aluminum Borohy-dride, 331
[16962-31-5] Potassium Hexafluoro-manganate(IV), 7621
[17021-26-0] Calusterone, 1730
[17069-38-4] Potassium Hexathiocy-anatoplatinate(IV), 7624
[17083-68-0] Gold Tribromide, Acid, 4426
[17086-28-1] Doxycycline, 3429
[17090-79-8] Monensin, 6157
[17099-70-6] Aluminum Hexafluoro-silicate, 342
[17109-49-8] Edifenphos, 3485
[17125-80-3] Barium Hexafluorosili-cate, 968
[17155-31-6] Veralkamine, 9850
[17162-36-6] Arecoline p-Stibono-benzoic Acid, 803

[17169-60-7] Ferroglycine Sulfate, 3987
[17176-77-1] Dibenzyl Phosphite, 2996
[17185-29-4] Tristriphenylphosphine Rhodium Carbonyl Hydride, 9670
[17194-00-2] Barium Hydroxide, 989
[17226-75-4] Khellol Glucoside, 5190
[17230-88-5] Danazol, 2811
[17243-38-8] Azidocillin, 925
[17243-39-9] Benzoctamine, 1096
[17243-57-1] Mefenorex, 5681
[17243-64-0] Piprozolin, 7457
[17321-77-6] Clomipramine Hydro-chloride*, 2385
[17365-01-4] Etiroxate, 3828
[17372-87-1] Eosine Yellow-ish—(YS), 3555
[17388-39-5] Swertiamarin, 8991
[17397-89-6] Cerulenin, 1997
[17406-45-0] Tomatine, 9471
[17440-83-4] Amiloride Hydrochlo-ride Dihydrate*, 417
[17440-85-6] Beryllium Borohydride, 1181
[17440-90-3] Hydrogen Tetracarbon-ylferrate(II), 4731
[17466-45-4] Phalloidin, 7147
[17500-49-1] Schweizer's Reagent, 8352
[17523-77-2] Niobium Potassium Oxypentafluoride, 6476
[17560-51-9] Metolazone, 6068
[17590-01-1] Amphetaminil, 617
[17598-65-1] Deslanoside, 2903
[17605-67-3] Fucosterol, 4194
[17617-23-1] Flurazepam, 4123
[17617-45-7] Picrotoxinin, 7389
[17627-77-9] Anhalonidine, 683
[17627-78-0] Lophophorine, 5451
[17650-86-1] Amicetin, 406
[17650-98-5] Ceruletide, 1998
[17651-61-5] Adonitoxin, 158
[17673-25-5] Phorbol, 7306
[17692-22-7] Metizoline, 6060
[17692-31-8] Dropropizine, 3439
[17692-34-1] Etodroxizine, 3832
[17692-39-6] Fomocaine, 4145
[17692-51-2] Metergoline, 5843
[17692-71-6] Vanitiolide, 9841
[17693-51-5] Promethazine Teo-clate*, 7797
[17702-41-9] Decaborane(14), 2838
[17754-44-8] Atractyloside, 883
[17784-12-2] Sulfacytine, 8873
[17804-35-2] Benomyl, 1053
[17806-68-7] Escigenin, 3643
[17854-59-0] Mepixanox, 5749
[17891-89-9] Pepstatin, 7104
[17902-23-7] Tegafur, 9060
[17924-92-4] Zearalenone, 10017
[17969-20-9] Fenclozic Acid, 3915
[18016-80-3] Lisuride, 5394
[18046-21-4] Fentiazac, 3946
[18053-31-1] Fominoben, 4144
[18059-10-4] Imperialine, 4840
[18077-77-5] Ammonium Uranium Carbonate, 596
[18078-40-5] Potassium Uranyl Ni-trate, 7695
[18109-80-3] Butamirate, 1505
[18109-81-4] Butamirate Citrate*, 1505
[18174-58-8] Pipoxolan Hydrochlo-ride, 7454
[18181-70-9] Iodofenphos, 4925
[18181-80-1] Bromopropylate, 1422
[18265-54-8] Phentetiothalein Sodi-um, 7233

Denotes derivative of title compound

[18282-10-5] Stannic Oxide, 8736
[18283-93-7] Tetraborane(10), 9119
[18288-22-7] Bismuthine, 1278
[18309-32-5] d-Verbenone, 9862
[18323-44-9] Clindamycin, 2352
[18378-89-7] Plicamycin, 7510
[18397-07-4] Geissoschizoline, 4274
[18423-69-3] Cubebin, 2616
[18433-40-4] Ammonium Uranium Fluoride, 597
[18433-48-2] Uranyl Phosphate, 9777
[18433-84-6] Pentaborane(11), 7055
[18454-12-1] Lead Chromate(VI) Oxide, 5280
[18456-04-7] Cephalonic Acid, 1974
[18464-39-6] Caroxazone, 1865
[18467-77-1] Dikegulac, 3184
[18471-20-0] Ditazol, 3374
[18472-51-0] Chlorhexidine Gluconate*, 2090
[18480-07-4] Strontium Hydroxide, 8807
[18493-30-6] Metochalcone, 6062
[18499-92-8] Kermesic Acid, 5172
[18507-89-6] Decoquinate, 2846
[18524-67-9] Mycobacillin, 6234
[18524-94-2] Loganin, 5441
[18530-30-8] d-Camphocarboxylic Acid, 1737
[18530-56-8] Norea, 6611
[18534-18-4] Piperazine Phosphate Monohydrate*, 7431
[18534-66-2] Vitamin B_{12s}, 9926
[18542-37-5] Vernolepin, 9867
[18559-59-6] Tretoquinol, 9502
[18559-94-9] Albuterol, 209
[18598-63-5] Mecysteine Hydrochloride, 5668
[18624-44-7] Ferrous Hydroxide, 3997
[18642-23-4] Psoralidin, 7945
[18651-67-7] Sikkimotoxin, 8434
[18656-21-8] Iodamide Meglumine*, 4699
[18679-90-8] Hopantenic Acid, 4664
[18683-91-5] Ambroxol, 392
[18691-97-9] Methabenzthiazuron, 5846
[18694-40-1] Epirizole, 3572
[18696-26-9] Sinapine, 8493
[18700-78-2] Cervicarcin, 2000
[18786-24-8] Serpentine (Alkaloid), 8415
[18791-21-4] Pyridomycin, 7989
[18794-84-8] β-Farnesene, 3884
[18820-29-6] Manganese Sulfide, 5625
[18829-56-6] 2-Nonenal, 6595
[18836-52-7] Pellitorine, 7017
[18840-45-4] Penicillamine Cysteine Disulfide, 7030
[18840-47-6] Gepefrine, 4295
[18869-73-3] Triacetyldiphenolisatin, 9505
[18883-66-4] Streptozocin, 8794
[18904-54-6] Corynantheine, 2546
[18917-82-3] Lead Lactate, 5289
[18917-89-0] Magnesium Salicylate, 5565
[18917-91-4] Aluminum Lactate, 350
[18917-93-6] Magnesium Lactate, 5550
[18937-33-2] Magnesium Mandelate, 5551
[19010-79-8] Cadmium Salicylate, 1623
[19014-05-2] Halimide®, 4507
[19026-31-4] Taxodione, 9048
[19044-88-3] Oryzalin, 6840

[19057-60-4] Dioscin, 3286
[19103-96-2] 1-Deoxynojirimycin, 2887
[19121-58-5] Solasonine, 8666
[19216-56-9] Prazosin, 7715
[19237-84-4] Prazosin Hydrochloride*, 7715
[19246-24-3] Telomycin, 9073
[19286-37-4] Convicine, 2509
[19287-45-7] Diborane(6), 2997
[19295-31-9] Urothion, 9800
[19356-17-3] Calcifediol, 1638
[19368-18-4] Ftaxilide, 4190
[19379-90-9] Benzoxonium Chloride, 1123
[19387-91-8] Tinidazole, 9377
[19395-58-5] Moquizone, 6174
[19457-37-5] Batrachotoxinin A, 1019
[19485-08-6] Cyproquinate, 2780
[19504-77-9] Pecilocin, 7008
[19525-15-6] Uranium Peroxide, 9769
[19545-26-7] Wortmannin, 9964
[19554-22-4] Cytohemin, 2799
[19562-30-2] Piromidic Acid, 7475
[19624-22-7] Pentaborane(9), 7054
[19625-10-6] Lactobacillic Acid, 5218
[19660-77-6] Phytochlorin, 7360
[19666-30-9] Oxadiazon, 6860
[19700-21-1] Geosmin, 4294
[19717-25-0] Peyonine, 7145
[19718-36-6] Potassium Osmate(VI), 7638
[19721-56-3] Picromycin, 7384
[19767-45-4] Mesna, 5812
[19793-20-5] Bolandiol, 1325
[19794-93-5] Trazodone, 9495
[19804-27-4] p-Methyldiphenhydramine, 5972
[19807-27-3] Palladium Diacetate, 6942
[19879-03-9] Granaticin B*, 4441
[19879-06-2] Granaticin, 4441
[19881-18-6] Nitroscanate, 6555
[19888-56-3] Fluazacort, 4048
[19891-74-8] Lycoxanthin, 5499
[19891-75-9] Lycophyll, 5493
[19982-08-2] Memantine, 5709
[20004-62-0] Resistomycin, 8155
[20055-57-6] Scabiolide, 8346
[20123-80-2] Dobesilate Calcium, 3395
[20168-99-4] Cinmetacin, 2295
[20170-20-1] Difenamizole, 3123
[20187-55-7] Bendazac, 1042
[20196-67-2] Sinalbin, 8492
[20231-81-6] Uzarin, 9809
[20246-33-7] D-Gulonic Acid, 4488
[20267-87-2] Chelidonine (±-Form)*, 2043
[20283-48-1] Chalcomycin, 2027
[20284-40-6] Murexine, 6216
[20287-37-0] Fenquizone, 3941
[20304-48-7] Uscharidin, 9805
[20310-89-8] Saponarin, 8327
[20311-78-8] Enanthotoxin, 3524
[20316-18-1] Lycodine, 5489
[20326-12-9] Mepiprazole, 5745
[20338-08-3] Titanic(IV) Acid, 9395
[20344-49-4] Ferric Hydroxide, 3969
[20350-15-6] Brefeldin A, 1363
[20354-26-1] Methazole, 5876
[20380-58-9] Tilidine, 9370
[20405-64-5] Cuprous Selenide, 2673
[20408-97-3] 5-Thio-D-glucose, 9263
[20427-56-9] Ruthenium Tetroxide, 8274

[20427-59-2] Cupric Hydroxide, 2646
[20432-69-3] Clorazepate, 2400
[20449-79-0] Melittin, 5705
[20480-93-7] Methenamine Sulfosalicylate, 5885
[20493-41-8] Quinapyramine, 8061
[20548-54-3] Calcium Sulfide, 1714
[20554-84-1] Parthenolide, 6996
[20559-55-1] Oxibendazole, 6891
[20562-02-1] Solanine, 8661
[20574-50-9] Morantel, 6175
[20594-83-6] Nalbuphine, 6271
[20623-13-6] Vitamin B_{12c}, 9924
[20644-87-5] Vanadium Carbonyl, 9824
[20667-12-3] Silver Oxide, 8468
[20684-06-4] Bamifylline Hydrochloride*, 965
[20702-77-6] Neohesperidin Dihydrochalcone, 6367
[20725-03-5] Fustin, 4232
[20770-09-6] Stannic Selenide, 8737
[20788-07-2] Resorantel, 8157
[20816-12-0] Osmium Tetroxide, 6848
[20830-75-5] Digoxin, 3150
[20830-81-3] Daunorubicin, 2825
[20831-76-9] Gentiopicrin, 4289
[20859-73-8] Aluminum Phosphide, 363
[20867-01-0] Fulvoplumierin, 4198
[21018-84-8] Amarogentin, 384
[21040-59-5] Codamine, 2458
[21041-93-0] Cobaltous Hydroxide, 2436
[21041-95-2] Cadmium Hydroxide, 1618
[21056-98-4] Calcium Phosphite, 1703
[21085-60-9] Calcium Mesoxalate, 1685
[21087-64-9] Metribuzin, 6076
[21109-95-5] Barium Sulfide, 1007
[21133-52-8] Lycoramine, 5497
[21187-98-4] Gliclazide, 4335
[21200-24-8] Indolmycin, 4872
[21238-33-5] γ-Oryzanol (A)*, 6841
[21256-18-8] Oxaprozin, 6879
[21259-20-1] T-2 Toxin, 9711
[21259-76-7] Mercaptomerin Sodium, 5761
[21293-20-9] Isoquassin, 5108
[21293-29-8] Abscisic Acid, 6
[21302-79-4] Ceanothic Acid, 1916
[21351-79-1] Cesium Hydroxide, 2005
[21362-69-6] Mepitiostane, 5747
[21363-18-8] Viminol, 9885
[21370-21-8] Fenoxazoline Hydrochloride*, 3930
[21411-53-0] Virginiamycin M1*, 9910
[21416-53-5] Picrotin, 7387
[21416-87-5] Razoxane, 8133
[21434-91-3] Capobenic Acid, 1757
[21466-07-9] Bromofenofos, 1406
[21498-08-8] Lofexidine Hydrochloride*, 5439
[21535-47-7] Mianserin Hydrochloride*, 6097
[21548-32-3] Fosthietan, 4173
[21548-73-2] Silver Sulfide, 8481
[21590-92-1] Etomidoline, 3839
[21609-90-5] Leptophos, 5327
[21637-25-2] Isoquercitrin, 5109
[21645-51-2] Aluminum Hydroxide, 345
[21649-57-0] Carfecillin Sodium, 1844
[21651-19-4] Stannous Oxide, 8747

[21679-14-1] Fludarabine, 4057
[21715-46-8] Etifoxine, 3821
[21721-92-6] Nitrefazole, 6492
[21725-46-2] Cyanazine, 2692
[21730-16-5] Metapramine, 5835
[21738-42-1] Oxamniquine, 6872
[21755-66-8] Picoperine, 7376
[21829-25-4] Nifedipine, 6441
[21888-98-2] Dexetimide, 2923
[21908-53-2] Mercuric Oxide, Red, 5779
[21908-53-2] Mercuric Oxide, Yellow, 5780
[21995-38-0] Ammonium Cerous Sulfate, 530
[22006-84-4] Denopterin, 2879
[22059-60-5] Disopyramide Phosphate*, 3366
[22071-15-4] Ketoprofen, 5184
[22089-22-1] Trofosfamide, 9680
[22103-14-6] Bufeniode, 1460
[22131-35-7] Butalamine, 1501
[22131-79-9] Alclofenac, 212
[22139-77-1] Pinosylvin, 7419
[22146-03-8] Sarverogenin, 8339
[22150-76-1] Biopterin, 1242
[22151-68-4] Methohexital Sodium, 5904
[22154-43-4] Meglumine Acetrizoate, 5688
[22195-34-2] Guanadrel Sulfate*, 4471
[22199-08-2] Sulfadiazine Silver Salt*, 8874
[22204-24-6] Pyrantel Pamoate*, 7968
[22204-29-1] Cetoxime Hydrochloride*, 2015
[22204-53-1] Naproxen, 6337
[22204-88-2] Tramadol Hydrochloride*, 9485
[22205-45-4] Cuprous Sulfide, 2674
[22224-92-6] Fenamiphos, 3901
[22232-54-8] Carbimazole, 1803
[22232-57-1] Racefemine, 8115
[22232-71-9] Mazindol, 5643
[22248-79-9] Stirofos, 8777
[22254-24-6] Ipratropium Bromide, 4960
[22260-51-1] Bromocriptine Mesylate*, 1404
[22263-79-2] Antheridiol, 710
[22298-29-9] Betamethasone Benzoate*, 1202
[22316-47-8] Clobazam, 2355
[22330-18-3] Potassium Triiodomercurate(II) Solution, 7691
[22345-47-7] Tofisopam, 9427
[22368-82-7] Pithecolobine, 7480
[22389-47-0] Aspartame, 861
[22398-80-7] Indium Phosphide, 4861
[22407-74-5] Bisobrin, 1308
[22457-89-2] Benfotiamine, 1049
[22487-42-9] Benapryzine, 1039
[22494-42-4] Diflunisal, 3130
[22494-47-9] Clobuzarit, 2364
[22537-19-5] Lawrencium, 5262
[22573-93-9] Alexidine, 222
[22619-35-8] Tioclomarol, 9383
[22650-86-8] Bismuth Subgallate, 1297
[22661-76-3] Amoproxan, 604
[22662-39-1] Rafoxanide, 8121
[22664-55-7] Metipranolol, 6059
[22672-74-8] Isocorybulbine, 5045
[22722-98-1] Vitride®, 9943
[22733-60-4] Siccanin, 8432
[22760-18-5] Proquazone, 7887
[22775-52-6] Mycelianamide, 6230
[22781-23-3] Bendiocarb, 1044

[22831-39-6] Magnesium Silicide, 5570
[22832-87-7] Miconazole Nitrate*, 6101
[22862-76-6] Anisomycin, 700
[22916-47-8] Miconazole, 6101
[22950-29-4] Dimetofrine, 3254
[22992-10-5] Aluminum Zinc Sulfate, 376
[22994-85-0] Benznidazole, 1094
[23031-25-6] Terbutaline, 9089
[23031-32-5] Terbutaline Sulfate*, 9089
[23047-25-8] Lofepramine, 5438
[23049-93-6] Enfenamic Acid, 3535
[23067-13-2] Erythromycin Glucoheptonate, 3629
[23076-35-9] Xylazine Hydrochloride*, 9987
[23092-17-3] Halazepam, 4502
[23103-98-2] Pirimicarb, 7468
[23110-15-8] Fumagillin, 4199
[23135-22-0] Oxamyl, 6873
[23142-01-0] Carbetapentane Citrate*, 1798
[23152-29-6] Virginiamycin S1*, 9910
[23155-02-4] Fosfomycin, 4169
[23184-66-9] Butachlor, 1498
[23192-42-9] Hexalure, 4608
[23210-56-2] Ifenprodil, 4821
[23214-92-8] Doxorubicin, 3428
[23214-96-2] Alcuronium, 214
[23239-37-4] Etoxadrol Hydrochloride*, 3844
[23239-41-0] Cephacetrile Sodium, 1969
[23239-51-2] Ritodrine Hydrochloride*, 8228
[23249-97-0] Procodazole, 7769
[23250-73-9] Manganese Oleate, 5616
[23255-54-1] Dihydro-β-erythroidine, 3158
[23255-59-6] Lunularic Acid, 5476
[23255-93-8] Hycanthone Mesylate*, 4677
[23256-30-6] Nifurtimox, 6454
[23256-50-0] Guanabenz Acetate*, 4469
[23271-63-8] Amicibone, 407
[23277-43-2] Nalbuphine Hydrochloride*, 6271
[23282-20-4] Nivalenol, 6581
[23288-49-5] Probucol, 7761
[23288-60-0] Sodium Pertechnetate ^{99m}Tc, 8609
[23315-05-1] Elaiomycin, 3498
[23327-57-3] Nefopam Hydrochloride*, 6358
[23383-11-1] Ferrous Citrate, 3993
[23413-80-1] Aluminum Bis(acetylsalicylate), 328
[23414-72-4] Zinc Permanganate, 10052
[23444-65-7] Alkannin, 243
[23465-76-1] Caroverine, 1864
[23477-98-7] Sedecamycin, 8377
[23496-41-5] Verticine, 9873
[23505-41-1] Pirimiphos-ethyl, 7469
[23509-16-2] Batrachotoxin, 1018
[23513-14-6] [6]-Gingerol, 4318
[23541-50-6] Daunorubicin Hydrochloride*, 2825
[23560-59-0] Heptenophos, 4585
[23564-06-9] Thiophanate, 9282
[23593-75-1] Clotrimazole, 2412
[23602-78-0] Benfluorex, 1047
[23605-05-2] α-Antiarin, 722
[23651-95-8] Droxidopa, 3444
[23668-11-3] Pactamycin, 6938

[23672-07-3] Sulpiride (l-Form)*, 8971
[23674-86-4] Difluprednate, 3134
[23694-81-7] Mepindolol, 5744
[23696-28-8] Olaquindox, 6783
[23736-58-5] Cloxacillin Benzathine*, 2414
[23777-80-2] Hexaborane(10), 4598
[23779-99-9] Floctafenine, 4037
[23843-52-9] Phloionic Acid, 7297
[23873-81-6] Benzil Dioxime, 1088
[23887-46-9] Cinepazide, 2293
[23910-07-8] Mebiquine, 5650
[23918-98-1] Eritadenine, 3618
[23930-19-0] Alfaxalone, 226
[23930-37-2] Alfadolone Acetate, 224
[23947-60-6] Ethirimol, 3695
[23950-58-5] Propyzamide, 7886
[23964-58-1] Carticaine, 1878
[24047-25-4] Guanoxabenz, 4481
[24159-07-7] Febrifugine, 3890
[24166-13-0] Cloxazolam, 2415
[24219-97-4] Mianserin, 6097
[24237-54-5] Tinoridine, 9379
[24243-97-8] Tymazoline, 9742
[24279-91-2] Carboquone, 1830
[24280-93-1] Mycophenolic Acid, 6238
[24304-00-5] Aluminum Nitride, 356
[24305-27-9] TRH, 9503
[24307-26-4] Mepiquat Chloride, 5746
[24345-16-2] Apamin, 757
[24356-60-3] Cephapirin Sodium, 1980
[24356-94-3] Algestone Acetophenide, 230
[24358-65-4] Diisopromine Hydrochloride*, 3180
[24390-14-5] Doxycycline Monohydrochloride Monohydrate*, 3429
[24394-09-0] Cnicin, 2419
[24397-89-5] Actinobolin, 130
[24526-64-5] Nomifensine, 6592
[24527-27-3] Clospirazine, 2408
[24531-57-5] 1-Thiosorbitol, 9294
[24558-01-8] Levophacetoperane, 5347
[24582-91-0] Elenolide, 3507
[24602-86-6] Tridemorph, 9576
[24622-52-4] Amixetrine Hydrochloride*, 507
[24622-72-8] Amixetrine, 507
[24634-61-5] Potassium Sorbate, 7661
[24678-13-5] Lenperone, 5320
[24691-76-7] Pyracarbolid, 7966
[24697-74-3] Leonurine, 5323
[24701-51-7] Demexiptiline, 2876
[24751-69-7] Nucleocidin, 6648
[24815-24-5] Rescinnamine, 8146
[24868-20-0] Dantrolene Sodium Hemiheptahydrate*, 2814
[24870-04-0] Giractide, 4322
[24880-43-1] Mesembrine, 5809
[24915-04-6] β-Isosparteine, 5115
[25013-16-5] Butylated Hydroxyanisole, 1547
[25038-54-4] Nylon 6, 6656
[25046-79-1] Glisoxepid, 4339
[25053-27-4] Lyapolate Sodium, 5487
[25057-89-0] Bentazon, 1060
[25102-12-9] Edetic Acid Dipotassium Dihydrate*, 3484
[25104-18-1] Polylysine, 7549
[25117-79-7] Isomethadol, 5068
[25122-41-2] Clobetasol, 2361
[25122-46-7] Clobetasol Propionate*, 2361

Denotes derivative of title compound

[25122-57-0] Clobetasone Butyrate*, 2362
[25126-32-3] Sincalide, 8494
[25152-52-7] Aluminum Antimonide, 326
[25154-52-3] Nonyl Phenol, 6599
[25155-15-1] Cymene, 2770
[25155-18-4] Methylbenzethonium Chloride, 5946
[25161-91-5] Loline, 5442
[25229-42-9] Cicrotoic Acid, 2272
[25301-02-4] Tyloxapol, 9741
[25311-71-1] Isofenphos, 5055
[25312-65-6] Cholanic Acid, 2199
[25316-40-9] Doxorubicin Hydrochloride*, 3428
[25322-68-3] Polyethylene Glycol, 7545
[25332-39-2] Trazodone Hydrochloride*, 9495
[25333-25-9] 3-(o-Methoxyphenyl)-2-phenylacrylic Acid, 5921
[25333-96-4] Spasmolytol, 8693
[25354-97-6] 2-Hexyldecanoic Acid, 4630
[25384-17-2] Allylprodine, 294
[25384-21-8] 3-Methylarsacetin, 5945
[25389-94-0] Kanamycin A Sulfate*, 5161
[25389-94-0] Kanamycin Sulfate*, 5161
[25390-16-3] Tanghinin, 9022
[25394-57-4] Affinin, 167
[25394-75-6] Alstonidine, 313
[25394-78-9] Cetoxime, 2015
[25395-22-6] Salicylamide O-Acetic Acid, 8298
[25395-31-7] Diacetin, 2941
[25486-55-9] Vitamin K₁ Oxide, 9934
[25523-97-1] Chlorpheniramine (d-Form)*, 2180
[25573-43-7] Eseridine, 3647
[25614-03-3] Bromocriptine, 1404
[25635-88-5] AIR, 181
[25655-41-8] Povidone-Iodine, 7701
[25663-23-4] A-Denopterin, 142
[25717-80-0] Molsidomine, 6143
[25803-14-9] Clometacin, 2381
[25808-74-6] Lead Hexafluorosilicate, 5284
[25812-30-0] Gemfibrozil, 4280
[25827-76-3] Iomeglamic Acid, 4941
[25867-19-0] Lycoctonine N-Succinylanthranilic Acid Ester*, 5488
[25875-50-7] Robenidine Hydrochloride*, 8229
[25875-51-8] Robenidine, 8229
[25895-60-7] Sodium Cyanoborohydride, 8554
[25905-77-5] Minaprine, 6119
[25953-17-7] Minaprine Hydrochloride*, 6119
[25953-19-9] Cefazolin, 1925
[25956-17-6] Allura® Red AC, 283
[25967-29-7] Flutoprazepam, 4134
[25978-54-5] Spherophysine, 8701
[25999-31-9] Lasalocid A, 5243
[26000-17-9] Lycoctonine, 5488
[26002-80-2] Phenothrin, 7221
[26020-55-3] Oxetorone, 6889
[26027-38-3] Nonoxynol, 6596
[26086-49-7] Deoxydihydrostreptomycin, 2884
[26097-80-3] Cambendazole, 1733
[26155-31-7] Morantel Tartrate*, 6175
[26159-34-2] Naproxen Sodium*, 6337
[26171-23-3] Tolmetin, 9441

[26225-79-6] Ethofumesate, 3697
[26241-81-6] Coumingine, 2567
[26242-33-1] Vintiamol, 9895
[26281-69-6] Exiproben, 3869
[26296-41-3] Cassaidine, 1895
[26299-14-9] Pyridinium Chlorochromate, 7986
[26309-95-5] Pivampicillin Hydrochloride*, 7484
[26328-53-0] Amoscanate, 606
[26399-36-0] Profluralin, 7781
[26402-26-6] Monooctanoin, 6162
[26445-03-4] Thiocresol, 9254
[26446-35-5] Monacetin, 6152
[26472-41-3] Humulon, 4673
[26530-20-1] Octhilinone, 6677
[26531-85-1] Primeverose, 7752
[26532-22-9] Grandisol, 4442
[26544-34-3] Apiin, 765
[26570-10-5] Propoxyphene Napsylate*, 7851
[26605-16-3] Nocardamin, 6587
[26608-34-4] Etioporphyrin, 3826
[26615-21-4] Zotepine, 10097
[26628-22-8] Sodium Azide, 8526
[26629-87-8] Oxaflozane, 6861
[26644-46-2] Triforine, 9602
[26652-09-5] Ritodrine, 8228
[26657-13-6] Rolitetracycline Nitrate Sesquihydrate*, 8236
[26658-42-4] Colestipol, 2472
[26675-46-7] Isoflurane, 5059
[26750-66-3] Potassium Thiocarbonate, 7686
[26750-81-2] Alibendol, 233
[26754-48-3] Terpenylic Acid, 9100
[26774-90-3] Epicillin, 3566
[26786-32-3] Lofepramine Hydrochloride*, 5438
[26787-78-0] Amoxicillin, 610
[26791-73-1] Xanthatin, 9967
[26807-65-8] Indapamide, 4847
[26839-75-8] Timolol, 9374
[26844-12-2] Indoramin, 4876
[26864-56-2] Penfluridol, 7028
[26871-30-7] Herquinone, 4589
[26921-17-5] Timolol Maleate*, 9374
[26944-48-9] Glibornuride, 4334
[26952-21-6] Isooctyl Alcohol, 5080
[26973-24-0] Ceftezole, 1946
[27013-91-8] α-Hederin, 4541
[27031-08-9] Sulfaguanole, 8880
[27035-30-9] Oxametacine, 6869
[27060-91-9] Flutazolam, 4133
[27073-72-9] Tiliacorine, 9369
[27115-36-2] Chromic Formate, 2226
[27127-62-4] Viscosin, 9914
[27164-46-1] Cefazolin Sodium*, 1925
[27167-30-2] Oxazolam, 6883
[27200-12-0] Ampelopsin, 611
[27203-92-5] Tramadol, 9485
[27208-80-6] Polydatin, 7542
[27214-00-2] Calcium Glycerophosphate, 1673
[27220-47-9] Econazole, 3476
[27223-35-4] Ketazolam, 5176
[27267-69-2] Collinomycin, 2478
[27276-25-1] Capobenate Sodium*, 1757
[27279-76-1] Stibocaptate, 8768
[27314-13-2] Norflurazon, 6618
[27315-91-9] Pipebuzone, 7424
[27367-90-4] Niaprazine, 6400
[27469-53-0] Almitrine, 299
[27470-51-5] Suxibuzone, 8990
[27511-99-5] Eterobarb, 3666
[27519-02-4] Muscalure, 6218
[27523-40-6] Isoconazole, 5044

[27589-33-9] Azosemide, 935
[27591-42-0] Oxazidione, 6882
[27591-69-1] Tilorone Hydrochloride*, 9371
[27591-97-5] Tilorone, 9371
[27607-77-8] Trimethylsilyl Triflate, 9633
[27709-53-1] Potassium Uranyl Sulfate, 7696
[27724-96-5] Cetraxate Hydrochloride*, 2017
[27726-31-4] Pivcefalexin, 7485
[27762-78-3] Kethoxal, 5178
[27774-13-6] Vanadyl Sulfate, 9833
[27780-11-6] Chrysanthemaxanthin, 2254
[27790-37-0] Potassium Stannosulfate, 7663
[27792-36-5] Etiocobalamin, 3825
[27833-64-3] Loxapine Succinate*, 5461
[27848-84-6] Nicergoline, 6404
[27876-94-4] Crocetin, 2592
[27877-51-6] Tolindate, 9438
[27885-92-3] Imidocarb, 4831
[27912-14-7] Levobunolol Hydrochloride*, 5343
[27959-26-8] Nicomol, 6429
[27973-72-4] Pederin, 7011
[28014-46-2] Polyestradiol Phosphate, 7543
[28041-84-1] Magnesium Potassium Selenate, 5563
[28041-86-3] Potassium Cobaltous Selenate, 7605
[28189-85-7] Etoxadrol, 3844
[28272-18-6] Pyrethrosin, 7979
[28289-54-5] MPTP, 6205
[28300-74-5] Antimony Potassium Tartrate, 732
[28380-24-7] Nigericin, 6457
[28384-96-5] Enterobactin, 3547
[28393-42-2] Cephalosporin P₁, 1977
[28395-03-1] Bumetanide, 1473
[28434-01-7] Bioresmethrin, 1243
[28591-01-7] F-Gitonin, 4326
[28610-84-6] Rimazolium Metilsulfate, 8222
[28657-80-9] Cinoxacin, 2311
[28772-56-7] Bromadiolone, 1371
[28782-42-5] Difenoxin, 3124
[28797-61-7] Pirenzepine, 7463
[28860-95-9] Carbidopa, 1802
[28911-01-5] Triazolam, 9518
[28947-50-4] Fencamine, 3910
[28950-34-7] Tetrasulfur Tetranitride, 9171
[28968-07-2] Clidanac, 2350
[28981-97-7] Alprazolam, 310
[28983-56-4] Methyl Blue, 5950
[28994-41-4] o-Benzylphenol, 1158
[29025-14-7] Butropium Bromide, 1534
[29069-24-7] Prednimustine, 7718
[29092-21-2] Prodiamine, 7773
[29094-61-9] Glipizide, 4337
[29098-15-5] Terofenamate, 9099
[29106-32-9] Chaulmoogric Acid, 2037
[29110-47-2] Guanfacine, 4474
[29110-48-3] Guanfacine Hydrochloride*, 4474
[29119-03-7] Frequentin, 4183
[29122-68-7] Atenolol, 879
[29177-84-2] Ethyl Loflazepate, 3777
[29216-28-2] Mequitazine, 5754
[29218-27-7] Toloxatone, 9445
[29334-07-4] Sulmarin, 8964
[29342-05-0] Ciclopirox, 2270
[29393-20-2] Bacilysin, 946

[29400-42-8] Helvolic Acid, 4552
[29520-22-7] Menadiol Disulfate, 5713
[29689-14-3] Chromic Carbonate, 2223
[29714-87-2] Ocimene, 6661
[29767-20-2] Teniposide, 9078
[29838-46-8] Roccellic Acid, 8231
[29868-97-1] Pirenzepine Dihydrochloride*, 7463
[29870-99-3] Strontium Lactate, 8809
[29883-15-6] Amygdalin, 629
[29899-95-4] Clobenoside, 2357
[29908-03-0] S-Adenosylmethionine, 146
[29913-86-8] Amarolide, 385
[29975-16-4] Estazolam, 3651
[30003-49-7] Primycin, 7756
[30042-37-6] Lankamycin, 5230
[30223-48-4] Fluacizine, 4046
[30279-49-3] Suclofenide, 8852
[30285-47-3] Vernadigin, 9864
[30286-75-0] Oxitropium Bromide, 6897
[30299-08-2] Clinofibrate, 2353
[30392-40-6] Bitolterol, 1317
[30392-41-7] Bitolterol Mesylate*, 1317
[30408-30-1] Nybomycin, 6653
[30484-77-6] Flunarizine Hydrochloride*, 4070
[30516-87-1] Zidovudine, 10023
[30525-89-4] Paraformaldehyde, 6974
[30544-47-9] Etofenamate, 3833
[30544-61-7] Clanobutin, 2338
[30560-19-1] Acephate, 26
[30578-37-1] Amezinium Methyl Sulfate, 403
[30653-83-9] Parsalmide, 6993
[30748-29-9] Feprazone, 3953
[30902-17-9] Midodrine Hydrochloride*, 6107
[30924-31-3] Cafaminol, 1632
[30964-13-7] Cynarin(e), 2773
[31002-79-6] Triamcinolone Benetonide, 9513
[31031-74-0] Phenamacide Hydrochloride, 7163
[31036-80-3] Lofexidine, 5439
[31112-62-6] Metrizamide, 6077
[31218-83-4] Propetamphos, 7827
[31224-92-7] Pifoxime, 7392
[31282-04-9] Hygromycin B, 4790
[31314-38-2] Prodipine, 7776
[31329-57-4] Nafronyl, 6268
[31362-50-2] Bombesin, 1332
[31418-71-0] Coprogen, 2517
[31428-61-2] Tiamenidine, 9350
[31430-15-6] Flubendazole, 4050
[31431-39-7] Mebendazole, 5647
[31431-43-3] Cyclobendazole, 2718
[31566-31-1] Glyceryl Monostearate, 4384
[31586-77-3] Bismuth Sodium Tartrate, 1292
[31637-97-5] Etofibrate, 3834
[31677-93-7] Bupropion Hydrochloride*, 1488
[31698-14-3] Ancitabine, 663
[31721-17-2] Quinupramine, 8111
[31774-33-1] Abikoviromycin, 3
[31793-07-4] Pirprofen, 7478
[31828-71-4] Mexiletine, 6094
[31842-01-0] Indoprofen, 4875
[31842-61-2] Rimiterol Hydrobromide*, 8223
[31848-01-8] Morclofone, 6177
[31868-18-5] Mexazolam, 6091
[31879-05-7] Fenoprofen, 3926

[31883-05-3] Moricizine, 6178
[31884-77-2] Meclizine Dihydrochloride*, 5657
[31932-09-9] Ticarbodine, 9359
[31944-97-5] Chondrofoline, 2216
[31980-29-7] Nicofibrate, 6427
[32059-15-7] Guanazodine, 4472
[32156-80-2] Theophylline Diethanolamine*, 9212
[32157-29-2] Propanethial S-Oxide, 7812
[32222-06-3] Calcitriol, 1641
[32267-39-3] Divicine, 3389
[32359-34-5] Medifoxamine, 5674
[32383-76-9] Medicarpin, 5673
[32385-11-8] Sisomicin, 8498
[32421-46-8] Bunaftine, 1474
[32449-92-6] D-Glucuronolactone, 4362
[32467-88-2] ACV, 138
[32476-67-8] Periplocymarin, 7127
[32527-55-2] Tiaramide, 9356
[32619-42-4] Oleuropein, 6790
[32630-75-4] Pododacric Acid, 7517
[32665-36-4] Eprozinol, 3578
[32672-69-8] Mesoridazine Besylate*, 5813
[32728-78-2] Heliosupine, 4545
[32780-64-6] Labetalol Hydrochloride*, 5208
[32791-31-4] C$_{14}$-Aldehyde, 215
[32795-44-1] Acecainide, 14
[32795-47-4] Nomifensine Maleate*, 6592
[32808-51-8] Bucloxic Acid, 1451
[32809-16-8] Procymidone, 7772
[32828-81-2] Picotamide, 7378
[32829-58-6] Gigantine, 4316
[32838-26-9] Butoctamide, 1526
[32854-75-4] Lappaconitine, 5237
[32886-97-8] Amdinocillin Pivoxil, 400
[32887-01-7] Amdinocillin, 399
[32909-92-5] Sulfametrole, 8892
[32953-89-2] Rimiterol, 8223
[32986-56-4] Tobramycin, 9413
[32988-50-4] Viomycin, 9905
[33005-95-7] Tiaprofenic Acid, 9354
[33032-12-1] Methapyrilene Fumarate*, 5871
[33069-62-4] Taxol, 9049
[33089-61-1] Amitraz, 503
[33103-22-9] Enviomycin, 3552
[33125-97-2] Etomidate, 3838
[33178-86-8] Alinidine, 234
[33237-74-0] Aprindine Hydrochloride*, 781
[33245-39-5] Fluchloralin, 4052
[33286-22-5] Diltiazem Hydrochloride*, 3188
[33342-05-1] Gliquidone, 4338
[33369-31-2] Zomepirac, 10092
[33386-08-2] Buspirone Hydrochloride*, 1493
[33401-94-4] Pyrantel Tartrate*, 7968
[33402-03-8] Metaraminol Bitartrate*, 5837
[33403-37-1] Nornidulin, 6632
[33404-78-3] Negamycin, 6359
[33419-42-0] Etoposide, 3842
[33445-15-7] Ammonium Mercuric Chloride, 558
[33564-30-6] Cefoxitin Sodium*, 1938
[33564-31-7] Diflorasone Diacetate*, 3126
[33579-45-2] Scotophobin, 8366
[33605-67-3] Cargutocin, 1846
[33605-94-6] Pyrisuccideanol, 8002
[33629-47-9] Butralin, 1532

[33659-28-8] Calcium Bromolactobionate, 1654
[33665-90-6] Acesulfame, 30
[33671-46-4] Clotiazepam, 2411
[33765-68-3] Oxendolone, 6886
[33797-51-2] Dimethyl(methylene)-ammonium Iodide, 3239
[33817-20-8] Pivampicillin, 7484
[33820-53-0] Isopropalin, 5089
[33996-33-7] Oxaceprol, 6857
[34014-18-1] Tebuthiuron, 9053
[34018-28-5] Lead Bromate, 5274
[34029-94-2] Laurepukine, 5253
[34031-32-8] Auranofin, 895
[34097-16-0] Clocortolone Pivalate*, 2368
[34118-92-8] Acecainide Hydrochloride*, 14
[34123-59-6] Isoproturon, 5106
[34135-85-8] Methyl Allyl Trisulfide, 5937
[34150-62-4] Ferriclate Calcium Sodium, 3971
[34161-24-5] Fipexide, 4024
[34183-22-7] Propafenone Hydrochloride*, 7806
[34184-77-5] Promegestone, 7796
[34195-34-1] Hydrocodone Bitartrate Hydrate*, 4708
[34255-08-8] Rhodoviolascin, 8195
[34273-10-4] Saralasin, 8330
[34301-55-8] Isometamidium Chloride, 5067
[34316-15-9] Chelerythrine, 2041
[34368-04-2] Dobutamine, 3396
[34444-01-4] Cefamandole, 1922
[34493-98-6] Dibekacin, 2987
[34521-09-0] Antimony Sodium Tartrate, 734
[34522-46-8] Oxetorone Fumarate*, 6889
[34524-20-4] Boromycin, 1344
[34552-83-5] Loperamide Hydrochloride*, 5450
[34552-84-6] Isoxicam, 5127
[34580-13-7] Ketotifen, 5187
[34580-14-8] Ketotifen Fumarate*, 5187
[34616-39-2] Fenalcomine, 3900
[34633-34-6] Bifluranol, 1231
[34645-84-6] Fenclofenac, 3913
[34661-75-1] Urapidil, 9779
[34675-84-8] Cetraxate, 2017
[34707-92-1] Chlorothricin, 2170
[34758-83-3] Zipeprol, 10074
[34765-96-3] Alsactide, 312
[34784-64-0] Tertatolol, 9105
[34787-01-4] Ticarcillin, 9360
[34816-55-2] Moxestrol, 6203
[34866-46-1] Carbuterol Hydrochloride*, 1840
[34866-47-2] Carbuterol, 1840
[34911-55-2] Bupropion, 1488
[34915-68-9] Bunitrolol, 1479
[34919-98-7] Cetamolol, 2010
[35035-05-3] Timepidium Bromide, 9372
[35080-11-6] Prajmaline, 7704
[35112-53-9] Barium Thiosulfate, 1011
[35121-78-9] Prostacyclin, 7890
[35189-28-7] Norgestimate, 6620
[35212-22-7] Iprindole, 4961
[35274-05-6] Cetyl Lactate, 2022
[35367-38-5] Diflubenzuron, 3128
[35388-57-9] Piscidic Acid, 7479
[35400-43-2] Sulprofos, 8972
[35405-51-7] Calcium Pyrophosphate, 1706
[35455-20-0] 3-Fluoro-D-alanine, 4097

Denotes derivative of title compound

[35531-88-5] Carindacillin, *1847*
[35554-08-6] Saxitoxin, *8344*
[35554-44-0] Enilconazole, *3537*
[35595-03-0] Centaurein, *1964*
[35604-67-2] Viloxazine Hydrochloride*, *9884*
[35607-66-0] Cefoxitin, *1938*
[35619-65-9] Trithiozine, *9672*
[35691-65-7] 1,2-Dibromo-2,4-dicyanobutane, *3004*
[35700-21-1] Carboprost Methyl*, *1829*
[35700-23-3] Carboprost, *1829*
[35703-32-3] Cinametic Acid, *2283*
[35727-72-1] Ontianil, *6804*
[35795-16-5] Trimazosin, *9612*
[35834-26-5] Rosaramicin, *8241*
[35846-53-8] Maytansine, *5642*
[35898-87-4] Dilazep, *3185*
[35941-65-2] Butriptyline, *1533*
[35941-71-0] Tiaramide Hydrochloride*, *9356*
[36069-05-3] Pseudojervine, *7932*
[36104-64-0] Micranthine, *6102*
[36104-80-0] Camazepam, *1732*
[36141-82-9] Diamfenetide, *2953*
[36150-73-9] Erythrophleine, *3634*
[36167-63-2] Halofantrine Hydrochloride*, *4508*
[36309-01-0] Dimemorfan, *3192*
[36322-90-4] Piroxicam, *7476*
[36330-85-5] Fenbufen, *3906*
[36338-96-2] Carthamin, *1876*
[36364-49-5] Imidazole Salicylate, *4829*
[36393-56-3] Norpseudoephedrine, *6634*
[36478-76-9] Uranyl Nitrate, *9776*
[36504-93-5] Butaclamol, *1499*
[36505-84-7] Buspirone, *1493*
[36508-71-1] Zorubicin Hydrochloride*, *10096*
[36525-21-0] Neovitamin A, *6382*
[36531-26-7] Oxantel, *6876*
[36616-52-1] Fenclorac, *3914*
[36637-18-0] Etidocaine, *3818*
[36647-02-6] Cotarnine Hydrochloride*, *2551*
[36653-82-4] Cetyl Alcohol, *2020*
[36703-88-5] Inosine Pranobex, *4881*
[36734-19-7] Iprodione, *4964*
[36735-22-5] Quazepam, *8039*
[36791-04-5] Ribavirin, *8199*
[36798-79-5] Budralazine, *1457*
[36894-69-6] Labetalol, *5208*
[36895-62-2] Methenamine Allyl Iodide, *5880*
[37025-55-1] Carbetocin, *1800*
[37065-29-5] Miloxacin, *6116*
[37091-65-9] Azlocillin Sodium*, *929*
[37091-66-0] Azlocillin, *929*
[37106-97-1] Bentiromide, *1061*
[37115-32-5] Adinazolam, *150*
[37134-80-8] Dopastin, *3417*
[37148-27-9] Clenbuterol, *2347*
[37196-57-9] Crotamine, *2596*
[37203-40-0] Colostrokinin, *2484*
[37221-79-7] VIP, *9907*
[37267-86-0] Phosphoric Acid, Meta, *7319*
[37288-97-4] Pinguainin, *7418*
[37296-80-3] Colestipol Hydrochloride*, *2472*
[37300-21-3] Pentosan Polysulfate, *7090*
[37321-09-8] Apramycin, *780*
[37332-99-3] Avoparcin, *906*
[37350-58-6] Metoprolol, *6072*
[37517-26-3] Pipotiazine Palmitate*, *7453*

[37517-28-5] Amikacin, *416*
[37517-30-9] Acebutolol, *13*
[37561-27-6] Fenoverine, *3928*
[37640-71-4] Aprindine, *781*
[37661-08-8] Bacampicillin Hydrochloride*, *944*
[37686-84-3] Terguride, *9095*
[37686-85-4] Terguride Hydrogen Maleate*, *9095*
[37693-01-9] Clofoctol, *2377*
[37883-00-4] Viomycin Sulfate*, *9905*
[37893-02-0] Flubenzimine, *4051*
[37895-35-5] Albofungin, *203*
[37924-13-3] Perfluidone, *7113*
[37933-66-7] Thevetin A, *9215*
[38029-10-6] Pirbuterol Dihydrochloride*, *7461*
[38129-37-2] Bicozamycin, *1222*
[38142-58-4] *ar*-Turmerone, *9733*
[38183-12-9] Fluorescamine, *4084*
[38194-50-2] Sulindac, *8961*
[38234-21-8] Fertirelin, *4010*
[38260-01-4] Trientine Hydrochloride*, *9579*
[38260-54-7] Etrimfos, *3847*
[38304-91-5] Minoxidil, *6122*
[38363-32-5] Penbutolol Sulfate*, *7025*
[38363-40-5] Penbutolol, *7025*
[38455-77-5] Stannic Chromate(VI), *8733*
[38473-18-6] Flavipucine, *4029*
[38577-97-8] 4′,5′-Diiodofluorescein, *3174*
[38677-81-5] Pirbuterol, *7461*
[38677-85-9] Flunixin, *4073*
[38763-29-0] Worenine, *9963*
[38821-52-2] Indoramin Hydrochloride*, *4876*
[38821-53-3] Cephradine, *1982*
[38916-34-6] Somatostatin, *8671*
[38957-41-4] Emorfazone, *3519*
[38966-21-1] Aphidicolin, *759*
[39022-39-4] Oxaprotiline Hydrochloride*, *6878*
[39133-31-8] Trimebutine, *9613*
[39148-24-8] Fosetyl Al, *4167*
[39290-85-2] Cupric Borate, *2631*
[39300-45-3] Dinocap, *3281*
[39315-52-1] Dibunate Sodium, *3017*
[39379-15-2] Neurotensin, *6394*
[39409-82-0] Magnesium Carbonate Hydroxide, *5538*
[39416-48-3] Pyridinium Bromide Perbromide, *7985*
[39492-01-8] Gabexate, *4234*
[39515-40-7] Cyphenothrin, *2776*
[39515-41-8] Fenpropathrin, *3936*
[39552-01-7] Befunolol, *1032*
[39562-70-4] Nitrendipine, *6493*
[39563-28-5] Cloranolol, *2399*
[39589-98-5] Dimethyl Carbate, *3230*
[39640-15-8] Piperaline, *7365*
[39698-78-7] Saralasin Acetate Hydrate*, *8330*
[39715-02-1] Endralazine, *3532*
[39718-89-3] Alminoprofen, *298*
[39733-35-2] Ammonium Magnesium Chloride, *556*
[39831-55-5] Amikacin Sulfate*, *416*
[39860-99-6] Pipotiazine, *7453*
[39878-70-1] Talampicillin Hydrochloride*, *9008*
[39978-42-2] Nifurzide, *6456*
[40034-42-2] Rosoxacin, *8247*
[40054-69-1] Etizolam, *3830*
[40164-67-8] Amidochlor, *411*
[40180-04-9] Ticrynafen, *9362*

[40391-99-9] Pamidronic Acid, *6954*
[40487-42-1] Pendimethalin, *7026*
[40507-23-1] Fluproquazone, *4120*
[40507-78-6] Indanazoline, *4845*
[40525-07-3] Scillabiose, *8353*
[40580-59-4] Guanadrel, *4471*
[40596-69-8] Methoprene, *5906*
[40665-92-7] Cloprostenol, *2397*
[40666-16-8] Fluprostenol, *4121*
[40762-15-0] Doxefazepam, *3423*
[40819-93-0] Lorajmine Hydrochloride*, *5454*
[40828-46-4] Suprofen, *8985*
[41078-02-8] Enprofylline, *3544*
[41183-64-6] Gallium Citrate*, *4252*
[41294-56-8] 1α-Hydroxycholecalciferol, *4749*
[41340-25-4] Etodolac, *3831*
[41342-53-4] Erythromycin Ethylsuccinate*, *3626*
[41354-29-4] Cyproheptadine Hydrochloride*, *2779*
[41372-20-7] Apomorphine Hydrochloride*, *776*
[41451-75-6] Bruceantin, *1442*
[41473-09-9] Iopronic Acid, *4949*
[41483-43-6] Bupirimate, *1484*
[41570-61-0] Tulobuterol, *9720*
[41575-94-4] Carboplatin, *1828*
[41621-49-2] Ciclopirox Olamine*, *2270*
[41663-50-7] Isobutol, *5013*
[41708-72-9] Tocainide, *9414*
[41744-40-5] Sulbenicillin, *8863*
[41767-29-7] Fluocortin Butyl, *4078*
[41826-92-0] Trepibutone, *9501*
[41859-67-0] Bezafibrate, *1215*
[41944-01-8] Bismuth Potassium Iodide, *1288*
[41992-22-7] Spirogermanium Dihydrochloride*, *8720*
[41992-23-8] Spirogermanium, *8720*
[42116-76-7] Carnidazole, *1855*
[42200-33-9] Nadolol, *6260*
[42399-41-7] Diltiazem, *3188*
[42408-82-2] Butorphanol, *1530*
[42461-84-7] Flunixin Meglumine*, *4073*
[42471-28-3] Nimustine, *6469*
[42509-80-8] Isazofos, *4988*
[42540-40-9] Cefamandole Nafate*, *1922*
[42542-10-9] MDMA, *5646*
[42553-65-1] Crocin, *2593*
[42576-02-3] Bifenox, *1228*
[42588-37-4] Kinoprene, *5194*
[42597-57-9] Ronifibrate, *8238*
[42739-38-8] Ammonium Valerate, *598*
[42779-82-8] Clopirac, *2395*
[42794-76-3] Midodrine, *6107*
[42835-25-6] Flumequine, *4065*
[42852-95-9] *N*-Methyl-α-L-glucosamine, *5996*
[42864-78-8] Bevantolol Hydrochloride*, *1213*
[42874-03-3] Oxyfluorfen, *6916*
[42924-53-8] Nabumetone, *6258*
[42971-09-5] Vinpocetine, *9894*
[43121-43-3] Triadimefon, *9507*
[43200-80-2] Zopiclone, *10095*
[43210-67-9] Fenbendazole, *3904*
[46464-11-3] Meobentine, *5730*
[46817-91-8] Viloxazine, *9884*
[47000-92-0] Fluoridamid, *4089*
[47141-42-4] Levobunolol, *5343*
[47562-08-3] Lorajmine, *5454*
[47739-98-0] Clocapramine, *2365*
[47747-56-8] Talampicillin, *9008*
[48141-64-6] Etafedrine, *3661*
[49562-28-9] Fenofibrate, *3924*

[49564-56-9] Fazadinium Bromide, 3887
[49642-07-1] Statine, 8759
[49745-95-1] Dobutamine Hydrochloride*, 3396
[49746-00-1] Rotoxamine Tartrate*, 1804
[50264-69-2] Lonidamine, 5447
[50270-33-2] Isofezolac, 5056
[50276-98-7] Erythrocentaurin, 3623
[50327-22-5] Nylon 46, 6657
[50435-25-1] Nimidane, 6466
[50471-44-8] Vinclozolin, 9890
[50594-66-6] Acifluorfen, 105
[50629-82-8] Halometasone, 4510
[50650-76-5] Piroctone, 7472
[50679-07-1] Cinepazet Maleate, 2292
[50679-08-2] Terfenadine, 9094
[50700-49-7] Acetimidoquinone, 49
[50700-72-6] Vecuronium Bromide, 9845
[50767-79-8] Prodlure, 7777
[50782-69-9] VX, 9948
[50813-16-6] Sodium Polymetaphosphate, 8621
[50838-36-3] Tolciclate, 9433
[50847-11-5] Ibudilast, 4810
[50924-49-7] Mizoribine, 6137
[50927-09-8] Sydnones, 8992
[50933-33-0] Gossyplure, 4434
[50935-04-1] Carubicin, 1879
[50935-71-2] Mocimycin, 6138
[50972-17-3] Bacampicillin, 944
[51012-32-9] Tiapride, 9353
[51022-69-6] Amcinonide, 398
[51022-70-9] Albuterol Sulfate*, 209
[51022-71-0] Nabilone, 6257
[51022-73-2] Zometapine, 10093
[51022-98-1] Butirosin Sulfate*, 1523
[51025-84-5] Arbekacin, 797
[51037-30-0] Acipimox, 106
[51218-45-2] Metolachlor, 6067
[51218-49-6] Pretilachlor, 7746
[51235-04-2] Hexazinone, 4617
[51264-14-3] Amsacrine, 627
[51274-83-0] Tiamenidine Hydrochloride*, 9350
[51276-47-2] Phosphinothricin, 7314
[51308-54-4] Buthiobate, 1520
[51312-42-6] Sodium Phosphotungstate, 8618
[51322-75-9] Tizanidine, 9409
[51333-22-3] Budesonide, 1455
[51338-27-3] Diclofop-Methyl, 3072
[51395-42-7] Butedronic Acid, 1512
[51481-10-8] Vomitoxin, 9947
[51481-60-8] Naloxone Hydrochloride Dihydrate*, 6277
[51481-61-9] Cimetidine, 2279
[51481-65-3] Mezlocillin, 6096
[51484-40-3] Difenpiramide, 3125
[51503-61-8] Ammonium Phosphite, 572
[51579-82-9] Amfenac, 404
[51598-60-8] Cimetropium Bromide, 2280
[51627-14-6] Cefatrizine, 1923
[51630-58-1] Fenvalerate, 3952
[51762-05-1] Cefroxadine, 1942
[51773-92-3] Mefloquine Hydrochloride, 5683
[51781-06-7] Carteolol, 1875
[51781-21-6] Carteolol Hydrochloride*, 1875
[51803-78-2] Nimesulide, 6464
[51848-43-2] Retine, 8166
[51940-44-4] Pipemidic Acid, 7427
[52003-58-4] Ammonium Lactate, 555

[52080-56-5] Endobenzyline Bromide, 3527
[52080-57-6] Chloroprednisone, 2157
[52093-21-7] Micronomicin, 6104
[52094-70-9] Tetrantoin, 9165
[52109-93-0] Cyclodrine, 2725
[52128-35-5] Trimetrexate, 9635
[52152-93-9] Cefsulodin Sodium*, 1943
[52205-73-9] Estramustine Phosphate Sodium*, 3658
[52212-02-9] Pipecurium Bromide, 7426
[52214-84-3] Ciprofibrate, 2314
[52239-63-1] Thiethylperazine Maleate*, 9241
[52260-69-2] Dicyanine, 3081
[52279-57-9] Nandrolone p-Hexyloxyphenylpropionate, 6282
[52306-36-2] Lunasine, 5474
[52315-07-8] Cypermethrin, 2775
[52365-63-6] Dipivefrin, 3344
[52443-21-7] Glucametacin, 4342
[52463-83-9] Pinazepam, 7411
[52468-60-7] Flunarizine, 4070
[52479-85-3] Exifone, 3868
[52485-79-7] Buprenorphine, 1487
[52486-78-9] Mercurophen, 5791
[52486-80-3] Trigentisic Acid, 9603
[52500-59-1] Mexicain, 6093
[52500-60-4] Thioredoxin, 9289
[52500-61-5] β-Amino-α-methylphenethyl Alcohol, 459
[52503-64-7] Cupric Acetate, Basic, 2628
[52549-17-4] Pranoprofen, 7708
[52589-12-5] Diginatin, 3138
[52645-53-1] Permethrin, 7132
[52665-69-7] Calcimycin, 1639
[52705-43-8] α-Hydroxybenzylphosphinic Acid, 4744
[52731-38-1] Glucofrangulin, 4346
[52740-16-6] Calcium Arsenite, 1647
[52794-97-5] Carubicin Hydrochloride*, 1879
[52809-07-1] Quisqualic Acid, 8112
[52906-84-0] Oxychlorosene Sodium*, 6909
[52906-92-0] Motilin, 6198
[52918-63-5] Deltamethrin, 2869
[52921-08-1] Betaine Aspartate*, 1201
[52942-31-1] Etoperidone, 3841
[52951-36-7] Bisdequalinium Chloride, 1259
[52951-37-8] Bismuth Ethyl Camphorate, 1275
[52951-38-9] Bismuth Oleate, 1283
[52964-40-6] d-Bornyl α-Bromoisovalerate, 1340
[52993-95-0] Sodium Polyanetholesulfonate, 8620
[52993-97-2] Chloramphenicol Pantothenate, 2070
[53003-10-4] Salinomycin, 8305
[53008-88-1] Lonazolac, 5445
[53022-14-3] d-Bornyl Isovalerate, 1342
[53043-14-4] 6-n-Amyl-m-cresol, 643
[53078-86-7] Arogenic Acid, 814
[53092-86-7] Lindlar Catalyst, 5380
[53092-91-4] Chrysergonic Acid, 2260
[53152-21-9] Buprenorphine Hydrochloride*, 1487
[53164-05-9] Acemetacin, 22
[53179-09-2] Sisomicin Sulfate*, 8498
[53179-11-6] Loperamide, 5450

[53216-90-3] Griseoviridin, 4454
[53230-06-1] Dihydroxyaluminum Acetylsalicylate, 3167
[53251-94-8] Pinaverium Bromide, 7410
[53260-52-9] Heparamine, 4570
[53260-54-1] Fungisterol, 4203
[53267-01-9] Cifenline, 2274
[53274-53-6] N^4-β-D-Glucosylsulfanilamide, 4358
[53318-35-7] Diatretyne I, 2973
[53318-36-8] α-Glucogallin, 4347
[53370-43-7] Methargen, 5874
[53370-44-8] Menadiol Dibutyrate, 5711
[53370-45-9] Menthyl Borate, 5726
[53370-48-2] p-(2-Methylpropenyl)phenol Acetate, 6029
[53370-90-4] Exalamide, 3866
[53403-97-7] Pyridofylline, 7988
[53414-68-9] Tonin, 9472
[53447-14-6] Isocorypalmine, 5047
[53449-58-4] Ciclonicate, 2269
[53450-33-2] Oxalomolybdic Acid, 6866
[53516-73-7] Quinovin, 8105
[53555-68-3] 4,4'-Iminodicyclohexanecarboxylic Acid, 4833
[53558-25-1] Pyriminil, 7999
[53583-79-2] Sultopride, 8975
[53586-99-5] Periodyl, 7124
[53597-25-4] Salmine Sulfate, 8306
[53597-27-6] Fendosal, 3917
[53597-29-8] Plegatil, 7507
[53608-75-6] Pancrelipase, 6957
[53608-77-8] Penicilloyl Polylysine, 7050
[53608-78-9] Potassium Sulfobenzoate, 7668
[53608-79-0] Potassium Zirconium Sulfate, 7699
[53608-96-1] Testosterone 17-Chloral Hemiacetal, 9110
[53626-54-3] Thihexinol, 9242
[53643-48-4] Vindesine, 9892
[53643-53-1] Tuberin, 9716
[53648-05-8] Ibuproxam, 4813
[53648-55-8] Dezocine, 2934
[53678-77-6] Muramyl Dipeptide, 6214
[53684-48-3] Beryllium Potassium Sulfate, 1195
[53684-49-4] Bufetolol, 1461
[53684-61-0] Calcium Ferrous Citrate, 1668
[53714-56-0] Leuprolide, 5341
[53716-44-2] Rociverine, 8232
[53716-49-7] Carprofen, 1870
[53716-50-0] Oxfendazole, 6890
[53731-35-4] Manganese Pyrophosphate, 5620
[53731-36-5] Floredil, 4041
[53746-46-6] Trimazosin Monohydrochloride Monohydrate*, 9612
[53762-90-6] Gratioside, 4447
[53762-92-8] Valacidin, 9812
[53762-93-9] Viractin, 9909
[53762-94-0] Vitamin B_{12r}, 9925
[53778-50-0] Bismuth Sodium Iodide, 1291
[53780-34-0] Mefluidide, 5684
[53783-83-8] Tromantadine, 9683
[53808-87-0] Tetroxoprim, 9178
[53813-83-5] Suriclone, 8987
[53834-53-0] Tropenzile, 9690
[53850-34-3] Thaumatin, 9200
[53850-35-4] Element 105, 3504
[53850-36-5] Element 104, 3503
[53861-57-7] γ-Carboxyglutamic Acid, 1834

Denotes derivative of title compound

[53885-35-1] Ticlopidine Hydrochloride*, 9361
[53897-25-9] Bismuth Butylthiolaurate, 1272
[53902-12-8] Tranilast, 9489
[53905-38-7] Pro-Drone®, 7778
[53910-25-1] Pentostatin, 7091
[53943-88-7] Letosteine, 5330
[54024-22-5] Desogestrel, 2906
[54037-14-8] Element 107, 3506
[54038-81-2] Element 106, 3505
[54063-23-9] Cinepazic Acid*, 2293
[54063-32-0] Clobetasone, 2362
[54063-40-0] Fenoxedil, 3931
[54063-51-3] Nadoxolol, 6261
[54063-53-5] Propafenone, 7806
[54063-54-6] Reproterol, 8142
[54063-56-8] Suloctidil, 8967
[54083-22-6] Zorubicin, 10096
[54110-25-7] Pirozadil, 7477
[54120-61-5] Prostalene, 7895
[54143-55-4] Flecainide, 4034
[54143-56-5] Flecainide Acetate*, 4034
[54143-57-6] Metoclopramide Monohydrochloride Monohydrate*, 6063
[54182-58-0] Sucralfate, 8853
[54187-04-1] Rilmenidine, 8220
[54188-38-4] Metralindole, 6075
[54239-37-1] Cimaterol, 2278
[54323-85-2] Protizinic Acid, 7906
[54340-58-8] Meptazinol, 5753
[54340-62-4] Bufuralol, 1469
[54350-48-0] Etretinate, 3846
[54400-59-8] Butamisole, 1506
[54400-62-3] Butamisole Hydrochloride*, 1506
[54504-70-0] Theofibrate, 9211
[54527-84-3] Nicardipine Hydrochloride*, 6403
[54531-52-1] Polybenzarsol, 7539
[54533-85-6] Nizofenone, 6583
[54605-45-7] Iocarmic Acid Dimeglumine*, 4902
[54663-47-7] Tibezonium Iodide, 9357
[54739-18-3] Fluvoxamine, 4138
[54749-86-9] Thiazolinobutazone, 9236
[54749-90-5] Chlorozotocin, 2177
[54910-51-9] Disparlure, 3367
[54910-89-3] Fluoxetine, 4112
[54935-03-4] Sulisatin, 8962
[54965-21-8] Albendazole, 201
[54965-24-1] Tamoxifen Citrate*, 9019
[55028-70-1] Arbaprostil, 796
[55028-71-2] Fluprostenol Sodium*, 4121
[55028-72-3] Cloprostenol Sodium*, 2397
[55077-30-0] Aclatonium Napadisilate, 109
[55079-83-9] Acitretin, 107
[55096-26-9] Nalmefene, 6274
[55134-13-9] Narasin, 6339
[55142-85-3] Ticlopidine, 9361
[55179-31-2] Bitertanol, 1315
[55219-65-3] Triadimenol, 9508
[55242-55-2] Propentofylline, 7823
[55268-74-1] Praziquantel, 7714
[55268-75-2] Cefuroxime, 1951
[55283-68-6] Ethalfluralin, 3671
[55285-45-5] Pirifibrate, 7467
[55290-64-7] Oxidimethiin, 6893
[55294-15-0] Muzolimine, 6227
[55297-95-5] Tiamulin, 9351
[55297-96-6] Tiamulin Fumarate*, 9351

[55300-29-3] Antrafenine, 754
[55331-29-8] Zeranol, 10022
[55335-06-3] Triclopyr, 9572
[55453-87-7] Isoxepac, 5126
[55512-33-9] Pyridate, 7981
[55557-30-1] Propoxyphene L-Form Napsylate Monohydrate*, 7851
[55560-96-8] Tixocortol Pivalate*, 9408
[55726-47-1] Enocitabine, 3539
[55769-64-7] Butobendine Hydrochloride*, 1524
[55769-65-8] Butobendine, 1524
[55779-18-5] Arprinocid, 816
[55837-14-4] Butaverine, 1510
[55837-18-8] Butibufen, 1521
[55837-20-2] Halofuginone, 4509
[55837-25-7] Buflomedil, 1463
[55837-27-9] Piretanide, 7464
[55837-29-1] Tiropramide, 9393
[55843-86-2] Miroprofen, 6127
[55881-07-7] Miokamycin, 6123
[55893-12-4] Gephyrotoxin, 4296
[55905-53-8] Clebopride, 2344
[55937-99-0] Beclobrate, 1028
[55985-32-5] Nicardipine, 6403
[56030-54-7] Sufentanil, 8860
[56038-13-2] Sucralose, 8854
[56073-10-0] Brodifacoum, 1368
[56087-11-7] Dextranomer, 2928
[56180-94-0] Acarbose, 12
[56187-47-4] Cefazedone, 1924
[56208-01-6] Pifarnine, 7391
[56211-40-6] Torasemide, 9473
[56238-63-2] Cefuroxime Sodium*, 1951
[56281-36-8] Motretinide, 6199
[56287-74-2] Afloqualone, 171
[56341-08-3] Mabuterol, 5517
[56377-79-8] Nosiheptide, 6639
[56390-09-1] Epirubicin Hydrochloride*, 3573
[56391-56-1] Netilmicin, 6389
[56391-57-2] Netilmicin Sulfate*, 6389
[56392-17-7] Metoprolol Tartrate*, 6072
[56420-45-2] Epirubicin, 3573
[56425-91-3] Flurprimidol, 4129
[56430-99-0] Flumecinol, 4063
[56433-44-4] Oxaprotiline, 6878
[56518-41-3] Brodimoprim, 1369
[56592-32-6] Efrotomycin, 3491
[56695-65-9] Rosaprostol, 8240
[56775-88-3] Zimeldine, 10024
[56796-20-4] Cefmetazole, 1929
[56796-39-5] Cefmetazole Sodium*, 1929
[56824-20-5] Amiprilose, 499
[56833-74-0] Streptovirudin, 8793
[56980-93-9] Celiprolol, 1956
[56995-20-1] Flupirtine, 4117
[57021-61-1] Isonixin, 5078
[57109-90-7] Clorazepate Dipotassium*, 2400
[57248-88-1] Pamidronic Acid Disodium Salt*, 6954
[57381-26-7] Irsogladine, 4982
[57432-61-8] Methylergonovine Maleate*, 5989
[57460-41-0] Talinolol, 9012
[57470-78-7] Celiprolol Hydrochloride*, 1956
[57474-29-0] Nifuroquine, 6449
[57475-17-9] Brovincamine, 1440
[57479-88-6] Sulmepride, 8966
[57524-89-7] Hydrocortisone Valerate*, 4710
[57526-81-5] Prenalterol, 7742
[57574-09-1] Amineptine, 420

[57644-54-9] Bismuth Subcitrate Sol (Dried), 1296
[57648-21-2] Timiperone, 9373
[57653-27-7] Droprenilamine, 3438
[57760-36-8] Alborixin, 205
[57773-63-4] Triptorelin, 9662
[57775-27-6] Sultosilic Acid, Piperazine Salt, 8976
[57775-29-8] Carazolol, 1779
[57781-14-3] Halopredone Acetate, 4512
[57801-81-7] Brotizolam, 1439
[57808-65-8] Closantel, 2407
[57808-66-9] Domperidone, 3412
[57817-89-7] Stevioside, 8766
[57837-19-1] Metalaxyl, 5826
[57918-06-6] Neocembrene, 6362
[57938-82-6] Adinazolam Mesylate*, 150
[57982-77-1] Buserelin, 1492
[57982-78-2] Budipine, 1456
[57998-68-2] Diaziquone, 2979
[58001-44-8] Clavulanic Acid, 2342
[58095-31-1] Sulbenox, 8864
[58138-08-2] Tridiphane, 9578
[58152-03-7] Isepamicin, 4989
[58158-77-3] Amantanium Bromide, 381
[58186-27-9] Idebenone, 4817
[58207-19-5] Clindamycin Hydrochloride Monohydrate*, 2352
[58306-30-2] Febantel, 3888
[58337-35-2] Elliptinium Acetate, 3510
[58409-52-2] Albaspidin, 200
[58409-59-9] Bucumolol, 1454
[58416-00-5] Protiofate, 7904
[58473-74-8] Cinromide, 2313
[58503-79-0] Meobentine Sulfate*, 5730
[58551-69-2] Carboprost Tromethamine*, 1829
[58579-51-4] Anagrelide Hydrochloride*, 659
[58581-89-8] Azelastine, 922
[58670-63-6] Dinosterol, 3283
[58712-69-9] Traxanox, 9494
[58786-99-5] Butorphanol Tartrate*, 1530
[58795-03-2] Apalcillin Sodium*, 756
[58798-97-3] Berninamycin, 1176
[58934-46-6] Lorcainide Hydrochloride*, 5457
[58970-76-6] Ubenimex, 9750
[58994-96-0] Ranimustine, 8125
[59017-64-0] Ioxaglic Acid, 4956
[59018-13-3] Ioxaglic Acid Meglumine*, 4956
[59122-46-2] Misoprostol, 6129
[59128-97-1] Haloxazolam, 4518
[59170-23-9] Bevantolol, 1213
[59227-89-3] Laurocapram, 5255
[59277-89-3] Acyclovir, 139
[59338-93-1] Alizapride, 236
[59372-48-4] HPA-23, 4669
[59414-23-2] 1-Methoxy-3-(trimethylsilyloxy)-1,3-butadiene, 5925
[59467-70-8] Midazolam, 6105
[59467-94-6] Midazolam Maleate*, 6105
[59467-96-8] Midazolam Hydrochloride*, 6105
[59619-81-7] Etiproston, 3827
[59669-26-0] Thiodicarb, 9258
[59703-84-3] Piperacillin Sodium*, 7430
[59721-28-7] Camostat, 1734
[59729-31-6] Lorcainide, 5457
[59729-32-7] Citalopram Hydrobromide*, 2320

[59729-33-8] Citalopram, 2320
[59767-13-4] Setastine Hydrochloride*, 8423
[59798-30-0] Mezlocillin Sodium*, 6096
[59804-37-4] Tenoxicam, 9080
[59828-07-8] Procaterol Hydrochloride*, 7765
[59859-58-4] Femoxetine, 3897
[59865-13-3] Cyclosporin A*, 2759
[59865-23-5] MeBmt, 5651
[59917-39-4] Vindesine Sulfate*, 9892
[59937-28-9] Malotilate, 5593
[59995-64-1] Thienamycin, 9240
[60166-93-0] Iopamidol, 4943
[60168-88-9] Fenarimol, 3903
[60200-06-8] Clorsulon, 2405
[60207-90-1] Propiconazole, 7830
[60282-87-3] Gestodene, 4308
[60318-52-7] Trichosanthin, A10
[60325-46-4] Sulprostone, 8973
[60414-06-4] Amiprilose Hydrochloride*, 499
[60450-21-7] Morindin, 6180
[60529-33-1] Frangulin, 4177
[60529-76-2] Thymopoietin, 9339
[60561-17-3] Sufentanil Citrate*, 8860
[60576-13-8] Piketoprofen, 7393
[60607-34-3] Oxatomide, 6880
[60607-68-3] Indenolol, 4852
[60616-74-2] Magnesium Hydride, 5547
[60628-96-8] Bifonazole, 1232
[60643-86-9] Vigabatrin, 9882
[60662-16-0] Binedaline, 1238
[60668-24-8] Alafosfalin, 194
[60719-84-8] Amrinone, 626
[60731-46-6] Elcatonin, 3501
[60761-10-6] TMD, 9410
[60925-61-3] Ceforanide, 1934
[61019-78-1] Fenapanil, 3902
[61036-64-4] Teicoplanin, 9062
[61129-30-4] Zimeldine Dihydrochloride Monohydrate*, 10024
[61177-45-5] Clavulanate Potassium*, 2342
[61197-93-1] Loprazolam, 5453
[61230-25-9] Aplasmomycin, 769
[61260-05-7] Prenalterol Hydrochloride*, 7742
[61270-58-4] Cefonicid, 1932
[61270-78-8] Cefonicid Sodium*, 1932
[61318-90-9] Sulconazole, 8866
[61318-91-0] Sulconazole Nitrate*, 8866
[61336-70-7] Amoxicillin Trihydrate*, 610
[61380-40-3] Lofentanil, 5437
[61413-54-5] Rolipram, 8235
[61422-45-5] Carmofur, 1851
[61477-96-1] Piperacillin, 7430
[61512-20-7] Cord Factor, 2523
[61570-90-9] Tioxidazole, 9387
[61618-27-7] Amfenac Sodium Monohydrate*, 404
[61622-34-2] Cefotiam, 1937
[61676-87-7] Cymiazole, 2771
[61869-07-6] Domiodol, 3410
[61869-08-7] Paroxetine, 6991
[61951-99-3] Tixocortol, 9408
[62031-54-3] Fibroblast Growth Factor, 4016
[62087-72-3] Pentigetide, 7086
[62265-68-3] Quinfamide, 8068
[62473-79-4] Teniloxazine, 9077
[62510-56-9] Picilorex, 7369
[62568-57-4] DSIP, 3445
[62571-86-2] Captopril, 1773

[62587-73-9] Cefsulodin, 1943
[62610-77-9] Methacrifos, 5848
[62613-82-5] Oxiracetam, 6896
[62640-05-5] Ajugarin I*, 187
[62640-06-6] Ajugarin II*, 187
[62640-07-7] Ajugarin III*, 187
[62658-63-3] Bopindolol, 1335
[62666-20-0] Progabide, 7782
[62893-19-0] Cefoperazone, 1933
[62893-20-3] Cefoperazone Sodium*, 1933
[62952-06-1] Lysine Acetylsalicylate, 5510
[62973-76-6] Azanidazole, 914
[62992-61-4] Etersalate, 3667
[62994-47-2] Warburganal, 9949
[63251-39-8] Sulfinalol Hydrochloride*, 8925
[63269-31-8] Ciramadol, 2316
[63323-46-6] Ciramadol Hydrochloride*, 2316
[63329-53-3] Lobenzarit, 5433
[63333-35-7] Bromethalin, 1378
[63358-49-6] Aspoxicillin, 874
[63394-05-8] Plafibride, 7488
[63469-19-2] Apalcillin, 756
[63516-07-4] Flutropium Bromide, 4136
[63527-52-6] Cefotaxime, 1935
[63547-13-7] Adrafinil, 159
[63585-09-1] Foscarnet Sodium, 4166
[63590-64-7] Terazosin, 9084
[63610-08-2] Indobufen, 4867
[63612-50-0] Nilutamide, 6460
[63659-18-7] Betaxolol, 1204
[63659-19-8] Betaxolol Hydrochloride*, 1204
[63675-72-9] Nisoldipine, 6482
[63758-79-2] Indalpine, 4843
[63968-64-9] Artemisinin, 844
[64000-73-3] Pildralazine, 7394
[64019-93-8] Dipivefrin Hydrochloride*, 3344
[64024-15-3] Pentazocine Hydrochloride*, 7078
[64092-49-5] Zomepirac Sodium Dihydrate*, 10092
[64204-55-3] Esaprazole, 3642
[64211-46-7] Oxiconazole Nitrate, 6892
[64218-02-6] Plaunotol, 7506
[64228-81-5] Atracurium Besylate, 884
[64241-34-5] Cadralazine, 1631
[64294-95-7] Setastine, 8423
[64318-79-2] Gemeprost, 4279
[64475-85-0] Mineral Spirits, 6120
[64485-93-4] Cefotaxime Sodium*, 1935
[64490-92-2] Tolmetin Sodium*, 9441
[64506-49-6] Sofalcone, 8658
[64529-56-2] Ethiozin, 3694
[64552-17-6] Butofilolol, 1527
[64603-91-4] THIP, 9302
[64628-44-0] Triflumuron, 9592
[64706-54-3] Bepridil, 1167
[64795-35-3] Mesulergine, 5820
[64808-48-6] Lobenzarit Sodium*, 5433
[64840-90-0] Eperisone, 3558
[64872-76-0] Butoconazole, 1525
[64872-77-1] Butoconazole Nitrate*, 1525
[64902-72-3] Chlorsulfuron, 2192
[64924-67-0] Halofuginone Hydrobromide*, 4509
[64952-97-2] Moxalactam, 6201
[64953-12-4] Moxalactam Disodium*, 6201

[65002-17-7] Bucillamine, 1447
[65009-35-0] Lidamidine Hydrochloride*, 5358
[65035-34-9] Lineatin, 5381
[65043-22-3] Indeloxazine Hydrochloride, 4850
[65057-90-1] Tallysomycin A (Talisomycin)*, 9016
[65085-01-0] Cefmenoxime, 1928
[65141-46-0] Nicorandil, 6431
[65154-06-5] Platelet Activating Factor, 7496
[65271-80-9] Mitoxantrone, 6135
[65277-42-1] Ketoconazole, 5181
[65322-72-7] Endralazine Mesylate*, 3532
[65472-88-0] Naftifine, 6270
[65473-14-5] Naftifine Hydrochloride*, 6270
[65807-02-5] Goserelin, 4433
[65899-73-2] Tioconazole, 9384
[66002-66-2] Fertirelin Acetate*, 4010
[66085-59-4] Nimodipine, 6467
[66104-22-1] Pergolide, 7115
[66104-23-2] Pergolide Mesylate*, 7115
[66108-95-0] Iohexol, 4940
[66148-78-5] Temocillin, 9076
[66195-31-1] Ibopamine, 4807
[66215-27-8] Cyromazine, 2782
[66230-04-4] Fenvalerate (S,S-Isomer)*, 3952
[66264-77-5] Sulfinalol, 8925
[66309-69-1] Cefotiam Hydrochloride*, 1937
[66357-35-5] Ranitidine, 8126
[66455-27-4] PSK®, 7943
[66532-85-2] Propacetamol, 7804
[66535-86-2] Lotrifen, 5459
[66564-16-7] Ciclosidomine, 2271
[66575-29-9] Colforsin, 2473
[66592-87-8] Cefadroxil, 1921
[66644-81-3] Veralipride, 9849
[66711-21-5] Apraclonidine, 779
[66722-44-9] Bisoprolol, 1309
[66734-13-2] Alclometasone Dipropionate*, 213
[66778-36-7] Encainide, 3525
[66788-41-8] Tinofedrine, 9378
[66794-74-9] Encainide Hydrochloride*, 3525
[66796-54-1] Pro-opiomelanocortin, 7803
[66813-51-2] Alexitol Sodium, 223
[66827-12-1] Almagate, 297
[66871-56-5] Lidamidine, 5358
[66898-62-2] Talniflumate, 9017
[66934-18-7] Flunoxaprofen, 4074
[66981-73-5] Tianeptine, 9352
[67037-37-0] Eflornithine, 3489
[67227-56-9] Fenoldopam, 3925
[67227-57-0] Fenoldopam Mesylate*, 3925
[67306-00-7] Fenpropidin, 3937
[67306-03-0] Fenpropimorph, 3938
[67434-14-4] Benoxaprofen, 1055
[67452-97-5] Alclometasone, 213
[67485-29-4] Hydramethylnon, 4684
[67527-71-3] Mildiomycin, 6113
[67747-09-5] Prochloraz, 7767
[67769-47-5] Lucifer Yellow CH, 5466
[67814-76-0] Isepamicin Sulfate*, 4989
[67915-31-5] Terconazole, 9090
[67992-58-9] Ioxaglic Acid Sodium*, 4956
[67995-68-0] Tallysomycin, 9016
[68085-85-8] Cyhalothrin, 2765
[68247-85-8] Peplomycin, 7100

Denotes derivative of title compound

[68252-19-7] Pirmenol, 7471
[68291-97-4] Zonisamide, 10094
[68302-57-8] Amlexanox, 508
[68332-77-4] Phenoxyacetyl Cellulose, 7224
[68359-37-5] Cyfluthrin, 2764
[68367-52-2] Sorbinil, 8679
[68373-14-8] Sulbactam, 8862
[68377-92-4] Arotinolol, 815
[68401-81-0] Ceftizoxime, 1949
[68401-82-1] Ceftizoxime Sodium*, 1949
[68475-42-3] Anagrelide, 659
[68491-57-6] Taziprinone Hydrochloride, 9051
[68630-75-1] Buserelin Acetate*, 1492
[68767-14-6] Loxoprofen, 5462
[68786-66-3] Triclabendazole, 9566
[68797-31-9] Econazole Nitrate*, 3476
[68813-55-8] Oxantel Pamoate*, 6876
[68844-77-9] Astemizole, 878
[68890-66-4] Piroctone Olamine*, 7472
[68990-63-6] Shark Liver Oil, 8425
[69004-03-1] Toltrazuril, 9452
[69047-39-8] Binifibrate, 1239
[69049-73-6] Nedocromil, 6355
[69049-74-7] Nedocromil Sodium*, 6355
[69056-12-8] Prelog-Djerassi Lactone, 7741
[69198-10-3] Metronidazole Hydrochloride*, 6079
[69377-81-7] Fluroxypyr, 4128
[69381-94-8] Fenprostalene, 3940
[69409-94-5] Fluvalinate, 4137
[69558-55-0] Thymopentin, 9338
[69655-05-6] Dideoxyinosine, 3092
[69712-56-7] Cefotetan, 1936
[69739-16-8] Cefodizime, 1931
[69756-53-2] Halofantrine, 4508
[69770-45-2] Flumethrin, 4068
[69806-50-4] Fluazifop-butyl, 4049
[69815-49-2] Norepinephrine Bitartrate*, 6612
[69900-72-7] Trimoprostil, 9637
[69975-86-6] Doxofylline, 3427
[70024-40-7] Terazosin Monohydrochloride Dihydrate*, 9084
[70059-30-2] Cimetidine Hydrochloride*, 2279
[70124-77-5] Flucythrinate, 4055
[70288-86-7] Ivermectin, 5133
[70356-03-5] Cefaclor, 1920
[70356-09-1] Butyl Methoxydibenzoylmethane, 1580
[70384-29-1] Peplomycin Sulfate*, 7100
[70458-92-3] Pefloxacin, 7012
[70458-95-6] Pefloxacin Mesylate*, 7012
[70458-96-7] Norfloxacin, 6617
[70476-82-3] Mitoxantrone Dihydrochloride*, 6135
[70502-82-8] Traxanox Sodium*, 9494
[70608-72-9] 5-HETE, 4594
[70667-26-4] Ornoprostil, 6827
[70797-11-4] Cefpiramide, 1940
[70879-28-6] Alfentanil Hydrochloride*, 227
[71031-15-7] Cathinone, 1912
[71116-82-0] Tiaprost, 9355
[71117-51-6] Zoapatanol, 10088
[71142-71-7] PPACK, 7702
[71195-58-9] Alfentanil, 227
[71247-25-1] Ceruletide Diethylamine*, 1998

[71320-77-9] Moclobemide, 6139
[71439-68-4] Bisantrene Dihydrochloride*, 1255
[71626-11-4] Benalaxyl, 1038
[71675-85-9] Amisulpride, 501
[71731-58-3] Tiquizium Bromide, 9390
[71751-41-2] Abamectin, 1
[71765-30-5] Buthionine Sulfoximine, A3
[71771-90-9] Denopamine, 2878
[71963-77-4] Artemether, 842
[72301-79-2] Enviroxime, 3553
[72324-18-6] Stepronin, 8764
[72332-33-3] Procaterol, 7765
[72420-38-3] Acifran, A1
[72432-03-2] Miglitol, 6110
[72432-10-1] Aniracetam, 692
[72479-26-6] Fenticonazole, 3948
[72492-12-7] Spizofurone, 8722
[72496-41-4] Pirarubicin, 7460
[72509-76-3] Felodipine, 3895
[72522-13-5] Eptazocine, 3579
[72558-82-8] Ceftazidime, 1944
[72702-95-5] Ponalrestat, 7565
[72822-12-9] Dapiprazole, 2819
[72956-09-3] Carvedilol, 1882
[72962-43-7] Brassinolide, 1360
[73080-51-0] Repirinast, 8140
[73121-56-9] Enprostil, 3545
[73151-29-8] Fenticonazole Nitrate*, 3948
[73231-34-2] Florfenicol, 4042
[73334-07-3] Iopromide, 4948
[73384-59-5] Ceftriaxone, 1950
[73384-60-8] Sulmazole, 8965
[73523-00-9] Luprostiol, 5480
[73573-42-9] Rescimetol, 8145
[73573-87-2] Formoterol, 4159
[73573-88-3] Mevastatin, 6088
[73590-58-6] Omeprazole, 6800
[73771-04-7] Prednicarbate, 7717
[73803-48-2] Tripamide, 9649
[73816-42-9] Meclocycline Sulfosalicylate*, 5658
[73963-72-1] Cilostazol, 2277
[74011-58-8] Enoxacin, 3540
[74014-51-0] Rokitamycin, 8233
[74050-98-9] Ketanserin, 5175
[74051-80-2] Sethoxydim, 8424
[74103-06-3] Ketorolac, 5186
[74103-07-4] Ketorolac Tromethamine*, 5186
[74115-24-5] Clofentezine, 2373
[74176-31-1] Alfaprostol, 225
[74191-85-8] Doxazosin, 3422
[74214-62-3] Ethyl β-Carboline-3-carboxylate, 3737
[74222-97-2] Sulfometuron Methyl, 8936
[74223-64-6] Metsulfuron Methyl, 6081
[74252-25-8] Indomethacin Sodium Trihydrate*, 4874
[74258-86-9] Alacepril, 192
[74356-00-6] Cefotetan Disodium*, 1936
[74381-53-6] Leuprolide Acetate*, 5341
[74397-12-9] Limaprost, 5367
[74431-23-5] Imipenem, 4834
[74436-00-3] Cyclosporin G*, 2759
[74512-12-2] Omoconazole, 6801
[74517-78-5] Indecainide, 4849
[74764-40-2] Bepridil Hydrochloride Monohydrate*, 1167
[74849-93-7] Cefpiramide Sodium*, 1940
[74863-84-6] Argatroban, 804
[74913-18-1] Dynorphin, 3458
[75139-06-9] Tetronasin, 9176

[75216-20-5] Fluosol DA, 4111
[75219-46-4] Bestrabucil, 1199
[75330-68-6] Nilvadipine, 6461
[75330-75-5] Lovastatin, 5460
[75558-90-6] Amperozide, 612
[75621-03-3] CHAPS, 2034
[75659-07-3] Dilevalol, 3186
[75659-08-4] Dilevalol Hydrochloride*, 3186
[75695-93-1] Isradipine, 5129
[75736-33-3] Diclobutrazol, 3070
[75738-58-8] Cefmenoxime Hydrochloride*, 1928
[75847-73-3] Enalapril, 3521
[75887-54-6] Arteether, 841
[75898-26-9] β-Carboxyaspartic Acid, 1833
[75917-92-9] Iofetamine I[123], 4938
[76025-73-5] Carpetimycin A*, 1867
[76094-36-5] Carpetimycin B*, 1867
[76095-16-4] Enalapril Maleate*, 3521
[76420-72-9] Enalaprilat, 3522
[76496-68-9] Oxaprotiline R(−)-Form*, 6878
[76496-69-0] Oxaprotiline R(−)-Form Hydrochloride*, 6878
[76496-70-3] Oxaprotiline S(+)-Form*, 6878
[76543-88-9] Interferon Alfa-2a*, 4892
[76568-02-0] Flosequinan, 4043
[76578-14-8] Quizalofop-Ethyl, 8113
[76584-70-8] Valproic Acid Semisodium Salt*, 9821
[76610-84-9] Cefbuperazone, 1926
[76631-46-4] Detomidine, 2917
[76674-21-0] Flutriafol, 4135
[76679-95-3] Oxaprotiline S(+)-Form Hydrochloride*, 6878
[76738-62-0] Paclobutrazol, 6937
[76824-35-6] Famotidine, 3881
[76836-02-7] PIPES Disodium*, 7450
[76932-56-4] Nafarelin, 6265
[76963-41-2] Nizatidine, 6582
[77086-19-2] Dizocilpine (−)-Form*, 3392
[77086-21-6] Dizocilpine, 3392
[77088-22-7] Dizocilpine Maleate*, 3392
[77103-91-4] Acifran (+)-Form*, A1
[77103-92-5] Acifran (−)-Form*, A1
[77175-51-0] Cloconazole, 2367
[77287-05-9] Rioprostil, 8225
[77590-95-5] Cetamolol Hydrochloride*, 2010
[77671-31-9] Enoximone, 3542
[77734-91-9] Palytoxin, 6951
[77883-43-3] Doxazosin Mesylate*, 3422
[77912-79-9] Sulfazecin, 8924
[78110-38-0] Aztreonam, 938
[78113-36-7] Muroctasin, 6217
[78186-34-2] Bisantrene, 1255
[78266-06-5] Mebrofenin, 5652
[78415-72-2] Milrinone, 6117
[78473-71-9] Enterolactone, 3549
[78578-05-0] Hexythiazox, 4634
[78613-35-1] Amorolfine, 605
[78613-38-4] Amorolfine Hydrochloride*, 605
[78628-28-1] Roxatidine Acetate, 8252
[78654-44-1] Amicoumacin A, 408
[78697-56-0] Lophotoxin, 5452
[78718-25-9] Benexate Hydrochloride, 1046
[78755-81-4] Flumazenil, 4062

Denotes derivative of title compound

[79094-20-5] Daltroban, 2807
[79307-93-0] Azelastine Hydrochloride*, 922
[79350-37-1] Cefixime, 1927
[79559-97-0] Sertraline Hydrochloride*, 8416
[79617-96-2] Sertraline, 8416
[79645-27-5] Tobramycin Sulfate*, 9413
[79660-72-3] Fleroxacin, 4035
[79704-88-4] α-Methylfentanyl, 5992
[79770-24-4] Iotrolan, 4955
[79794-75-5] Loratadine, 5455
[79831-76-8] Castanospermine, 1902
[79902-63-9] Simvastatin, 8491
[80214-83-1] Roxithromycin, 8253
[80370-57-6] Ceftiofur, 1948
[80387-96-8] Difemerine, 3122
[80455-68-1] Fredericamycin A, 4181
[80529-93-7] Gadopentetic Acid, 4236
[80621-81-4] Rifaximin, 8218
[80738-42-7] Lentinan, 5322
[80755-51-7] Bunazosin, 1477
[80863-62-3] Alitame, 235
[81098-60-4] Cisapride, 2318
[81103-11-9] Clarithromycin, 2340
[81129-83-1] Cilastatin Sodium*, 2275
[81131-70-6] Pravastatin Sodium, 7712
[81147-92-4] Esmolol, 3648
[81161-17-3] Esmolol Hydrochloride*, 3648
[81335-37-7] Imazaquin, 4826
[81403-68-1] Alfuzosin Hydrochloride*, 228
[81403-80-7] Alfuzosin, 228
[81405-85-8] Imazamethabenz, 4825
[81406-37-3] Fluroxypyr 1-Methylheptyl Ester*, 4128
[81486-22-8] Nipradilol, 6479
[81525-10-2] Nafamostat, 6264
[81584-06-7] Xibenolol, 9984
[81660-57-0] APSAC, 785
[81732-46-9] Bambuterol Hydrochloride*, 963
[81732-65-2] Bambuterol, 963
[81777-89-1] Dimethazone, 3203
[81801-12-9] Xamoterol, 9965
[81982-32-3] Alpiropride, 309
[82009-34-5] Cilastatin, 2275
[82110-72-3] Fenoxaprop-ethyl, 3929
[82115-62-6] Interferon-γ, 4894
[82159-09-9] Epalrestat, 3556
[82219-78-1] Cefuzonam, 1952
[82225-47-6] Ajugarin IV*, 187
[82231-14-9] Ajugarin V*, 187
[82410-32-0] Ganciclovir, 4262
[82419-36-1] Ofloxacin, 6688
[82473-24-3] CHAPS Hydroxy Analog*, 2034
[82547-58-8] Cefteram, 1945
[82560-54-1] Benfuracarb, 1050
[82571-53-7] Ozagrel, 6935
[82626-01-5] Alpidem, 308
[82626-48-0] Zolpidem, 10091
[82657-04-3] Bifenthrin, 1229
[82657-92-9] Pro-Urokinase, 7918
[82768-85-2] Quinaprilat*, A7
[82834-16-0] Perindopril, 7122
[82964-04-3] Tolrestat, 9451
[83150-76-9] Octreotide, 6682
[83184-43-4] Mifentidine, 6108
[83248-46-8] 3-(1,3,5,7,9-Dodecapentaenyloxy)-1,2- propanediol, 3403
[83435-66-9] Delapril, 2864

[83435-67-0] Delapril Hydrochloride*, 2864
[83657-24-3] Diniconazole, 3261
[83712-60-1] Defibrotide, 2851
[83881-51-0] Cetirizine, 2013
[83881-52-1] Cetirizine Hydrochloride*, 2013
[83905-01-5] Azithromycin, 928
[83915-83-7] Lisinopril, 5393
[83919-23-7] Mometasone Furoate, 6151
[83928-66-9] Gepirone Hydrochloride*, 4297
[83928-76-1] Gepirone, 4297
[83930-13-6] Somatoliberin, 8669
[84031-17-4] Metaclazepam, 5824
[84305-41-9] Cefminox, 1930
[84371-65-3] Mifepristone, 6109
[84504-69-8] Irsogladine Maleate*, 4982
[84507-84-1] Lamotrigine, 5228
[84625-61-6] Itraconazole, 5131
[84878-61-5] Maduramicin, 5524
[84880-03-5] Cefpimizole, 1939
[85068-76-4] Iofetamine I¹²³ Hydrochloride*, 4938
[85287-61-2] Cefpimizole Sodium*, 1939
[85320-68-9] Amosulalol, 607
[85371-64-8] Pinacidil, 7407
[85441-61-8] Quinapril, A7
[85509-19-9] Flusilazole, 4130
[85637-73-6] Atrial Natriuretic Factor, 887
[85721-33-1] Ciprofloxacin, 2315
[85856-54-8] Moveltipril, 6200
[85898-30-2] Interleukin-2, 4896
[85921-53-5] Moveltipril Calcium*, 6200
[86050-77-3] Gadopentetic Acid Dimeglumine Salt*, 4236
[86168-78-7] Sermorelin, 8412
[86197-47-9] Dopexamine, 3418
[86220-42-0] Nafarelin Acetate*, 6265
[86273-18-9] Lenampicillin, 5319
[86329-79-5] Cefodizime Disodium Salt*, 1931
[86347-14-0] Medetomidine, 5670
[86347-15-1] Medetomidine Hydrochloride*, 5670
[86386-73-4] Fluconazole, 4054
[86393-32-0] Ciprofloxacin Hydrochloride Monohydrate*, 2315
[86393-37-5] Amifloxacin, 415
[86484-91-5] Dopexamine Hydrochloride*, 3418
[86832-68-0] Carumonam Sodium*, 1880
[86880-51-5] Epanolol, 3557
[87041-58-5] Hementin, 4559
[87139-37-5] Carpetimycin D*, 1867
[87239-81-4] Cefpodoxime Proxetil, 1941
[87269-97-4] Ramiprilat*, 8123
[87333-19-5] Ramipril, 8123
[87638-04-8] Carumonam, 1880
[87695-64-5] Carpetimycin C*, 1867
[87818-31-3] Cinmethylin, 2296
[87820-88-0] Tralkoxydim, 9484
[87848-99-5] Acrivastine, 121
[88036-80-0] Amifloxacin Mesylate*, 415
[88058-88-2] Naxagolide, 6349
[88150-42-9] Amlodipine, 509
[88150-47-4] Amlodipine Maleate*, 509
[88200-01-5] Chelidonine (−)-Form*, 2043
[88255-01-0] Netobimin, 6390
[88426-33-9] Buparvaquone, 1481

[88495-63-0] Artesunate, 845
[88671-89-0] Myclobutanil, 6232
[88889-14-9] Fosinopril Sodium*, 4171
[89672-11-7] Cyoctol, 2774
[89778-26-7] Toremifene, 9474
[89778-27-8] Toremifene Citrate*, 9474
[89797-00-2] Iopentol, 4945
[89943-82-8] Cicletanine, 2268
[90101-16-9] Droxicam, 3443
[90139-06-3] Cilazaprilat*, 2276
[90243-99-5] Quinapril Monohydrochloride Monohydrate*, A7
[90293-01-9] Bifemelane, 1227
[90803-92-2] Thymomodulin, 9337
[90982-32-4] Chlorimuron Ethyl, 2092
[91161-71-6] Terbinafine, 9086
[91296-86-5] Difloxacin Hydrochloride*, 3127
[92071-51-7] Rotraxate, 8249
[92077-78-6] Cilazapril, 2276
[92118-27-9] Fotemustine, 4174
[92285-01-3] Ajoene, 185
[92589-98-5] Ipsapirone Hydrochloride*, 4967
[93106-60-6] Enrofloxacin, 3546
[93413-04-8] Asperlicin, 865
[93413-05-9] Asperlicin E*, 865
[93413-06-0] Asperlicin C*, 865
[93413-07-1] Asperlicin D*, 865
[93413-08-2] Asperlicin B*, 865
[95153-31-4] Perindoprilat*, 7122
[95399-71-6] Fosinopril Diacid*, 4171
[95847-70-4] Ipsapirone, 4967
[96020-91-6] Eflornithine Hydrochloride*, 3489
[96128-89-1] Erythromycin Acistrate, 3627
[97519-39-6] Ceftibuten, 1947
[97689-87-7] Tunichrome B-1, 9727
[97772-42-7] Butedronic Acid Tetrasodium Salt*, 1512
[98048-97-6] Fosinopril, 4171
[98059-18-8] Cys-Tyr-Cys-IFN-γ*, 4894
[98059-61-1] Interferon gamma-1b*, 4894
[98079-51-7] Lomefloxacin, 5443
[98079-52-8] Lomefloxacin Hydrochloride*, 5443
[98106-17-3] Difloxacin, 3127
[98319-26-7] Proscar®, 7888
[99210-65-8] Interferon Alfa-2b*, 4892
[99291-24-4] Dropropizine L-Form*, 3439
[99294-93-6] Zolpidem (+)-Tartrate (2:1)*, 10091
[99294-94-7] Teriparatide Acetate, 9096
[99593-25-6] Rilmazafone, 8219
[99614-01-4] Ondansetron Hydrochloride Dihydrate*, 6802
[99614-02-5] Ondansetron, 6802
[99665-00-6] Flomoxef, 4038
[99751-63-0] AL 721, 191
[99777-81-8] Troxipide, 9699
[100599-27-7] Tenidap, A9
[100678-32-8] Cifenline Succinate*, 2274
[100935-99-7] Naxagolide Disodium Salt*, 6349
[101831-36-1] Clazuril, 2343
[102507-71-1] Tigemonam, 9364
[102916-21-2] Tigemonam Dicholine*, 9364
[103220-14-0] Defensins, 2849
[103628-46-2] Sumatriptan, 8979

* Denotes derivative of title compound

[103628-48-4] Sumatriptan Succinate*, 8979
[104010-37-9] Ceftiofur Sodium*, 1948
[104118-48-1] Adipsin, 153
[104344-23-2] Bisoprolol Fumarate*, 1309
[104376-79-6] Ceftriaxone Sodium*, 1950
[104521-37-1] Fraxiparine ®, 4180
[104987-11-3] FK-506, A5
[105784-61-0] Temafloxacin Hydrochloride*, A8

[105879-42-3] Cephalexin Hydrochloride*, 1971
[106096-92-8] Fibroblast Growth Factor (aFGF)*, 4016
[106096-93-9] Fibroblast Growth Factor (bFGF)*, 4016
[106362-32-7] Peptide T, 7105
[107007-99-8] Granisetron Hydrochloride*, 4443
[107097-79-0] Tosufloxacin Monohydrate*, 9477
[107910-75-8] Ganciclovir Sodium*, 4262

[108138-46-1] Tosufloxacin, 9477
[108319-06-8] Temefloxacin, A8
[108544-30-5] Butedronic acid Technetium Salt*, 1512
[109489-77-2] Tetranectin, 9163
[109889-09-0] Granisetron, 4443
[110101-67-2] U74006F*, 5264
[112373-40-7] Prodigiosin Hydrochloride*, 7774
[115964-29-9] Tosufloxacin Toluenesulfonate*, 9477
[117399-94-7] Endothelin (human), A4

THERAPEUTIC CATEGORY AND
BIOLOGICAL ACTIVITY
INDEX

THERAPEUTIC CATEGORY AND BIOLOGICAL ACTIVITY INDEX

This index, which appears for the first time in The Merck Index, is designed to serve as an additional entry point to the monographs of this edition for which a therapeutic category (THERAP CAT) has been given. It contains listings for discrete chemical entities; plants and plant portions e.g. seeds or leaves, although used medicinally, have generally been excluded. An attempt has been made to standardize the nomenclature used for index terms. Therefore, the heading under which a compound is listed may differ slightly from the therapeutic category found within the monograph.

Biological activities indicating mechanism of action have also been included as index headings. Whenever possible, cross references to appropriate therapeutic categories have been given for these listings. Cross references have also been provided for synonyms of preferred terms or, in some cases, for closely related entries. In addition, many therapeutic category headings have been subclassified according to structural features. A master list of all categories and cross references can be found on pages THER-1 to THER-4.

Monographs are listed alphabetically by title under each heading. Selected derivatives and isomers of title compounds have been listed by generic or trivial name and are referenced to the appropriate monograph by number.

Therapeutic categories have been assigned to reflect the major indications listed in manufacturer's product information, reported in the clinical literature, or published by the USAN Council. Inclusion of a drug in this index does not imply efficacy or endorsement, nor does it imply that the compound is *presently* being marketed. Omission of a compound from a particular category does not imply that the substance is not in use or under investigation for that indication. Additional information on each entry can be found in the cited monograph. Reference to primary sources is encouraged.

MASTER LIST OF HEADINGS

ABORTIFACIENT

ACE-INHIBITOR *see also
Antihypertensive*

α-ADRENERGIC AGONIST *see also
Antihypotensive; Antihypertensive;
Decongestant; Mydriatic*

β-ADRENERGIC AGONIST *see also
Bronchodilator; Cardiotonic;
Tocolytic*

α-ADRENERGIC BLOCKER *see also
Antihypertensive*

β-ADRENERGIC BLOCKER *see also
Antianginal; Antiarrhythmic;
Antiglaucoma; Antihypertensive*

ADRENOCORTICAL STEROID *see
Glucocorticoid; Mineralocorticoid*

ADRENOCORTICAL SUPRESSANT

ADRENOCORTICOTROPIC HORMONE

ALCOHOL DETERRENT

ALDOSE REDUCTASE INHIBITOR

ALDOSTERONE ANTAGONIST *see also
Diuretic*

5-ALPHA REDUCTASE INHIBITOR *see
5α-Reductase Inhibitor*

ANABOLIC

ANALGESIC (DENTAL)

ANALGESIC (NARCOTIC)

ANALGESIC (NON-NARCOTIC)

ANDROGEN

ANESTHETIC (INHALATION)

ANESTHETIC (INTRAVENOUS)

ANESTHETIC (LOCAL)

ANGIOTENSIN CONVERTING EN-
ZYME INHIBITOR *see ACE-Inhibitor*

ANOREXIC

ANTACID

ANTHELMINTIC (CESTODES)

ANTHELMINTIC (NEMATODES)

ANTHELMINTIC (ONCHOCERCA)

ANTHELMINTIC (SCHISTOSOMA)

ANTHELMINTIC (TREMATODES)

ANTIACNE *see also Keratolytic*

ANTIALLERGIC *see also Antihistaminic;
Decongestant; Glucocorticoid*

ANTIALLERGIC (HYPOSENSITIZA-
TION THERAPY)

ANTIALLERGIC (STEROIDAL, NASAL)
see also Glucocorticoid

ANTIALOPECIA AGENT

ANTIAMEBIC

ANTIANDROGEN *see also Antiacne;
Antialopecia; Antineoplastic
(Hormonal)*

ANTIANGINAL *see also Vasodilator
(Coronary)*

ANTIARRHYTHMIC

ANTIARTERIOSCLEROTIC *see also
Antihyperlipoproteinemic*

ANTIARTHRITIC/ANTIRHEUMATIC
*see also Anti-inflammatory (Nonste-
roidal); Glucocorticoid*

ANTIASTHMATIC (NONBRONCHODI-
LATOR) *see also Bronchodilator; Glucocorticoid*

ANTIASTHMATIC (STEROIDAL, IN-
HALANT) *see also Glucocorticoid*

ANTIBACTERIAL (ANTIBIOTIC)
  Aminoglycosides
  Amphenicols
  Ansamycins
  β-Lactams
    Carbapenems
    Cephalosporins
    Cephamycins
    Monobactams
    Oxacephems
    Penicillins
  Lincosamides
  Macrolides
  Polypeptides
  Tetracyclines
  Others

ANTIBACTERIAL (SYNTHETIC)
  2,4-Diaminopyrimidines
  Nitrofurans
  Quinolones and Analogs
  Sulfonamides
  Sulfones
  Others

ANTIBACTERIAL (LEPROSTATIC)

ANTIBACTERIAL (RICKETTSIA) *see
Antirickettsial*

ANTIBACTERIAL (TUBERCULO-
STATIC)

ANTIBACTERIAL ADJUNCT (ENZYME
INHIBITOR)
  β-Lactamase Inhibitors
  Renal Dipeptidase Inhibitors

ANTIBIOTIC *see Antibacterial (Antibiotic); Antifungal (Antibiotic);
Antineoplastic (Antibiotic)*

ANTICANCER *see Antineoplastic*

ANTICHOLELITHOGENIC *see
Cholelitholytic*

ANTICHOLESTEREMIC *see Antihyperlipoproteinemic*

ANTICHOLINERGIC *see also Antispasmodic; Bronchodilator*

ANTICOAGULANT *see also Antithrombotic; Thrombolytic*

ANTICONVULSANT

ANTIDEPRESSANT *see also Antimanic*
  Bicyclics
  Hydrazides/Hydrazines
  Pyrrolidones
  Tetracyclics
  Tricyclics
  Others

ANTIDIABETIC
  Biguanides
  Hormones
  Sulfonylurea Derivatives
  Others

ANTIDIARRHEAL

ANTIDIURETIC

ANTIDOTE (ACETAMINOPHEN
POISONING)

ANTIDOTE (CURARE)

ANTIDOTE (CYANIDE)

ANTIDOTE (FOLIC ACID ANTAGO-
NISTS)

ANTIDOTE (HEAVY METAL POISON-
ING)

ANTIDYSKINETIC *see also
Antiparkinsonian*

ANTIECZEMATIC

ANTIEMETIC

ANTIEPILEPTIC *see Anticonvulsant*

ANTIESTROGEN

ANTIFIBROTIC

ANTIFLATULENT

ANTIFUNGAL (ANTIBIOTIC)
  Polyenes
  Others

ANTIFUNGAL (SYNTHETIC)
  Allylamines
  Imidazoles
  Triazoles
  Others

ANTIGLAUCOMA

ANTIGONADOTROPIN

ANTIGOUT

ANTIHEMORRHAGIC *see Hemostatic*

ANTIHISTAMINIC *see also Antiallergic*
  Alkylamine Derivatives
  Aminoalkyl Ethers
  Ethylenediamine Derivatives
  Piperazines
  Tricyclics
    Phenothiazines
    Others
  Others

ANTIHYPERCHOLESTEROLEMIC *see Antihyperlipoproteinemic*

ANTIHYPERLIPIDEMIC *see Antihyperlipoproteinemic*

ANTIHYPERLIPOPROTEINEMIC
  Aryloxyalkanoic Acid Derivatives
  Bile Acid Sequestrants
  HMG CoA Reductase Inhibitors
  Nicotinic Acid Derivatives
  Thyroid Hormones/Analogs
  Others

ANTIHYPERPHOSPHATEMIC

ANTIHYPERTENSIVE *see also Diuretic*
  Arylethanolamine Derivatives
  Aryloxypropanolamine Derivatives
  Benzothiadiazine Derivatives
  N-Carboxyalkyl(peptide/lactam) Derivatives
  Dihydropyridine Derivatives
  Guanidine Derivatives
  Hydrazines/Phthalazines
  Imidazole Derivatives
  Quaternary Ammonium Compounds
  Quinazolinyl Piperazine Derivatives
  Reserpine Derivatives
  Sulfonamide Derivatives
  Others

ANTIHYPERTHYROID

ANTIHYPOTENSIVE

ANTIHYPOTHYROID

ANTI-INFECTIVE *see Antiseptic/Disinfectant*

ANTI-INFLAMMATORY (GASTRO-INTESTINAL)

ANTI-INFLAMMATORY (NONSTE-ROIDAL) *see also Antiarthritic/Antirheumatic*
  Aminoarylcarboxylic Acid Derivatives
  Arylacetic Acid Derivatives
  Arylbutyric Acid Derivatives
  Arylcarboxylic Acids

  Arylpropionic Acid Derivatives
  Pyrazoles
  Pyrazolones
  Salicylic Acid Derivatives
  Thiazinecarboxamides
  Others

ANTI-INFLAMMATORY (STEROIDAL) *see Glucocorticoid*

ANTILEPROTIC *see Antibacterial (Leprostatic)*

ANTILEUKEMIC *see Antineoplastic*

ANTILIPEMIC *see Antihyperlipoproteinemic*

ANTILIPIDEMIC *see Antihyperlipoproteinemic*

ANTIMALARIAL

ANTIMANIC

ANTIMETHEMOGLOBINEMIC

ANTIMIGRAINE

ANTIMYCOTIC *see Antifungal*

ANTINAUSEANT *see Antiemetic*

ANTINEOPLASTIC
  Alkylating agents
    Alkyl Sulfonates
    Aziridines
    Epoxides
    Ethylenimines and Methylmelamines
    Nitrogen Mustards
    Nitrosoureas
    Others
  Antibiotics
  Antimetabolites
    Folic Acid Analogs
    Purine Analogs
    Pyrimidine Analogs
  Enzymes
  Others

ANTINEOPLASTIC (HORMONAL)
  Androgens
  Antiadrenals
  Antiandrogens
  Antiestrogens
  Estrogens
  LH-RH Analogs
  Progestogens

ANTINEOPLASTIC (RADIATION SOURCE)

ANTINEOPLASTIC ADJUNCT
  Folic Acid Replenisher
  Uroprotective

ANTIOSTEOPOROTIC *see Calcium Regulator*

ANTIPAGETIC *see Calcium Regulator*

ANTIPARKINSONIAN

ANTIPERISTALTIC *see Antidiarrheal*

ANTIPHEOCHROMOCYTOMA

ANTIPNEUMOCYSTIS

ANTIPROSTATIC HYPERTROPHY

ANTIPROTOZOAL (AMEBA) *see Antiamebic*

ANTIPROTOZOAL (GIARDIA)

ANTIPROTOZOAL (HISTOMONAS)

ANTIPROTOZOAL (LEISHMANIA)

ANTIPROTOZOAL (PLASMODIA) *see Antimalarial*

ANTIPROTOZOAL (TRICHOMONAS)

ANTIPROTOZOAL (TRYPANOSOMA)

ANTIPRURITIC

ANTIPSORIATIC

ANTIPSYCHOTIC
  Butyrophenones
  Phenothiazines
  Thioxanthenes
  Other Tricyclics
  Others

ANTIPYRETIC

ANTIRHEUMATIC *see Antiarthritic/Antirheumatic*

ANTIRICKETTSIAL

ANTISEBORRHEIC

ANTISEPTIC/DISINFECTANT
  Alcohols
  Aldehydes
  Dyes
  Guanidines
  Halogens/Halogen Compounds
  Mercurial Compounds
  Nitrofurans
  Peroxides/Permanganates
  Phenols
  Quaternary Ammonium Compounds
  Quinolines
  Silver Compounds
  Others

ANTISPASMODIC *see also Anticholinergic*

ANTISYPHILITIC *see also Antibacterial*

ANTITHROMBOTIC *see also Anticoagulant; Thrombolytic*

ANTITUBERCULAR *see Antibacterial (Tuberculostatic)*

ANTITUMOR see Antineoplastic

ANTITUSSIVE

ANTIULCERATIVE see also Antacid

ANTIUROLITHIC

ANTIVENIN

ANTIVERTIGO see Antiemetic

ANTIVIRAL
  Purines/Pyrimidinones
  Others

ANXIOLYTIC
  Arylpiperazines
  Benzodiazepine Derivatives
  Carbamates
  Others

ASTRINGENT

BENZODIAZEPINE ANTAGONIST

BETA-BLOCKER see β-Adrenergic
  Blocker

BRONCHODILATOR see also Antiasthmatic; Glucocorticoid
  Ephedrine Derivatives
  Quaternary Ammonium Compounds
  Xanthine Derivatives
  Others

CALCIUM CHANNEL BLOCKER see
  also Antianginal; Antihypertensive;
  Vasodilator (Coronary)
  Arylalkylamines
  Dihydropyridine Derivatives
  Piperazine Derivatives
  Others

CALCIUM REGULATOR

CALCIUM SUPPLEMENT see
  Replenishers/Supplements

CANCER CHEMOTHERAPY see Antineoplastic

CAPILLARY PROTECTANT see Vasoprotectant

CARBONIC ANHYDRASE INHIBITOR
  see also Antiglaucoma; Diuretic

CARDIAC DEPRESSANT (ANTIARRHYTHMIC) see Antiarrhythmic

CARDIOTONIC

CATHARTIC see Laxative/Cathartic

CATION-EXCHANGE RESIN see Ion-
  exchange Resin

CCK ANTAGONIST

CENTRAL STIMULANT see CNS Stimulant

CEREBRAL VASODILATOR see Vasodilator (Cerebral)

CHELATING AGENT

CHOLECYSTOKININ ANTAGONIST see
  CCK Antagonist

CHOLELITHOLYTIC AGENT

CHOLERETIC

CHOLINERGIC

CHOLINESTERASE INHIBITOR

CHOLINESTERASE REACTIVATOR

CNS STIMULANT

COGNITION ACTIVATOR see Nootropic

CONTRACEPTIVE (INJECTABLE)

CONTRACEPTIVE (ORAL)

CONTROL OF INTRAOCULAR PRESSURE see also Antiglaucoma

CONVERTING ENZYME INHIBITOR
  see ACE-Inhibitor

CORONARY VASODILATOR see Vasodilator (Coronary)

CYTOPROTECTANT (GASTRIC) see
  also Antiulcerative

DEBRIDING AGENT

DECONGESTANT

DEPIGMENTOR

DERMATITIS HERPETIFORMIS SUPPRESANT

DIAGNOSTIC AID

DIAGNOSTIC AID (NMR CONTRAST
  MEDIUM)

DIAGNOSTIC AID (RADIOACTIVE
  IMAGING AGENT)

DIAGNOSTIC AID (RADIOPAQUE
  MEDIUM)

DIGESTIVE AID

DISINFECTANT see
  Antiseptic/Disinfectant

DIURETIC
  Benzothiadiazine Derivatives
  Organomercurials
  Pteridines
  Purines
  Steroids
  Sulfonamide Derivatives
  Uracils
  Others

DOPAMINE RECEPTOR AGONIST see
  also Antihypertensive; Antimigraine;
  Antiparkinsonian

DOPAMINE RECEPTOR ANTAGONIST
  see also Antiemetic; Antipsychotic

ECTOPARASITICIDE

ELECTROLYTE REPLENISHER see
  Replenishers/Supplements

EMETIC

ENZYME
  Digestive
  Mucolytic
  Penicillin Inactivating
  Proteolytic

ENZYME INDUCER (HEPATIC)

ESTROGEN
  Nonsteroidal
  Steroidal

ESTROGEN ANTAGONIST see Antiestrogen

EXPECTORANT

GASTRIC AND PANCREATIC SECRETION STIMULANT

GASTRIC PROTON PUMP INHIBITOR
  see also Antiulcerative

GASTRIC SECRETION INHIBITOR

GLUCOCORTICOID

α-GLUCOSIDASE INHIBITOR see also
  Antidiabetic

GONAD-STIMULATING PRINCIPLE

GONADOT␣␣PIC HORMONE

GOUT ␣␣␣␣SANT see Antigout

GRC␣␣␣␣ORMONE INHIBITOR

GROWTH HORMONE RELEASING
  FACTOR

GROWTH STIMULANT

HEMATINIC

HEMOLYTIC

HEMOSTATIC

HEPARIN ANTAGONIST

HEPATOPROTECTANT

HISTAMINE H_1-RECEPTOR ANTAGONIST see Antihistaminic

HISTAMINE H_2-RECEPTOR ANTAGONIST see also Antiulcerative

HMG CoA REDUCTASE INHIBITOR see
  also Antihyperlipoproteinemic

HYPNOTIC see Sedative/Hypnotic

HYPOCHOLESTEREMIC see Antihyperlipoproteinemic

HYPOLIPIDEMIC see Antihyperlipoproteinemic

HYPOTENSIVE *see Antihypertensive*

IMMUNOMODULATOR

IMMUNOSUPPRESSANT

INOTROPIC AGENT *see Cardiotonic*

ION EXCHANGE RESIN

KERATOLYTIC *see also Antiacne*

LACTATION STIMULATING HORMONE

LAXATIVE/CATHARTIC

LH-RH AGONIST *see also Antineoplastic; Gonad-Stimulating Principle*

LIPOTROPIC

LOCAL ANESTHETIC *see Anesthetic (Local)*

LUPUS ERYTHEMATOSUS SUPPRESSANT

MAJOR TRANQUILIZER *see Antipsychotic*

MINERALOCORTICOID

MINOR TRANQUILIZER *see Anxiolytic*

MIOTIC

MONOAMINE OXIDASE INHIBITOR *see also Antidepressant; Antihypertensive*

MUCOLYTIC

MUSCLE RELAXANT (SKELETAL)

MUSCLE RELAXANT (SMOOTH) *see Anticholinergic; Antispasmodic; Bronchodilator; Vasodilator*

MYDRIATIC

NARCOTIC ANALGESIC *see Analgesic (Narcotic)*

NARCOTIC ANTAGONIST

NASAL DECONGESTANT *see Decongestant*

NEUROLEPTIC *see Antipsychotic*

NEUROMUSCULAR BLOCKING AGENT *see Muscle Relaxant (Skeletal)*

NEUROPROTECTIVE

NMDA ANTAGONIST *see also Neuroprotective*

NOOTROPIC

NSAID *see Anti-inflammatory (Nonsteroidal)*

OPIOID ANALGESIC *see Analgesic (Narcotic)*

ORAL CONTRACEPTIVE *see Contraceptive (Oral)*

OVARIAN HORMONE

OXYTOCIC

PARASYMPATHOMIMETIC *see Cholinergic*

PEDICULICIDE *see Ectoparasiticide*

PEPSIN INHIBITOR

PERIPHERAL VASODILATOR *see Vasodilator (Peripheral)*

PERISTALTIC STIMULANT

PIGMENTATION AGENT

PLASMA VOLUME EXPANDER

POTASSIUM CHANNEL ACTIVATOR/OPENER *see also Antihypertensive*

PRESSOR AGENT *see Antihypotensive*

PROGESTOGEN

PROLACTIN INHIBITOR

PROSTAGLANDIN/PROSTAGLANDIN ANALOG *see also Abortifacient; Antiulcerative; Oxytocic*

PROTEASE INHIBITOR

PROTON PUMP INHIBITOR *see Gastric Proton Pump Inhibitor*

5α-REDUCTASE INHIBITOR *see also Antiprostatic Hypertrophy*

REPLENISHERS/SUPPLEMENTS
  Calcium
  Electrolyte
  Fluid and Nutrient
  Iodine
  Magnesium
  Phosphorus
  Potassium
  Zinc

RESPIRATORY STIMULANT

REVERSE TRANSCRIPTASE INHIBITOR *see also Antiviral*

SCABICIDE *see Ectoparasiticide*

SCLEROSING AGENT

SEDATIVE/HYPNOTIC *see also Anxiolytic*
  Acyclic Ureides
  Alcohols
  Amides
  Barbituric Acid Derivatives
  Benzodiazepine Derivatives
  Bromides
  Carbamates
  Chloral Derivatives
  Quinazolone Derivatives
  Piperidinediones
  Others

SEROTONIN RECEPTOR AGONIST *see also Anxiolytic*

SEROTONIN RECEPTOR ANTAGONIST *see also Antiemetic; Antihypertensive; Antimigraine*

SEROTONIN UPTAKE INHIBITOR *see also Antidepressant*

SKELETAL MUSCLE RELAXANT *see Muscle Relaxant (Skeletal)*

SOMATOSTATIN ANALOG

SPASMOLYTIC *see Antispasmodic*

STOOL SOFTENER *see Laxative/Cathartic*

SUCCINYLCHOLINE SYNERGIST

SYMPATHOMIMETIC *see α-Adrenergic Agonist; β-Adrenergic Agonist*

THROMBOLYTIC *see also Anticoagulant; Antithrombotic*

THYROID HORMONE *see also Antihypothyroid*

THYROID INHIBITOR *see Antihyperthyroid*

THYROTROPIC HORMONE

TOCOLYTIC

TOPICAL PROTECTANT

TRANQUILIZER *see Antipsychotic; Anxiolytic*

ULTRAVIOLET SCREEN

URICOSURIC *see also Antigout*

VASODILATOR (CEREBRAL)

VASODILATOR (CORONARY) *see also Antianginal*

VASODILATOR (PERIPHERAL)

VASOPRESSOR *see Antihypotensive*

VASOPROTECTANT

VITAMIN/VITAMIN SOURCE
  Antirachitic
  Antiscorbutic
  Antixerophthalmic
  Enzyme Co-factor
  Hematopoietic
  Prothrombogenic
  Others

VULNERARY

WILSON'S DISEASE TREATMENT

XANTHINE OXIDASE INHIBITOR *see also Antigout*

ANABOLIC (continued)

Bolandiol, 1325
Bolasterone, 1326
Clostebol, 2409
Ethylestrenol, 3761
Formyldienolone, 4161
4-Hydroxy-19-nortestosterone, 4768
Methandriol, 5861
Methenolone, 5887
Methyltrienolone, 6049
Nandrolone, 6280
Nandrolone Decanoate, 6281
Nandrolone p-Hexyloxyphenylpropionate, 6282
Nandrolone Phenpropionate, 6283
Nandrolone Propionate, 6284
Norbolethone, 6603
Oxymesterone, 6918
Pizotyline, 7486
Quinbolone, 8063
Stenbolone, 8763
Trenbolone, 9499

ANALGESIC (DENTAL)

Chlorobutanol, 2129
Clove see 2413 and 6718
Eugenol, 3855

ANALGESIC (NARCOTIC)

Alfentanil, 227
Allylprodine, 294
Alphaprodine, 307
Anileridine, 686
Benzylmorphine, 1155
Bezitramide, 1216
Buprenorphine, 1487
Butorphanol, 1530
Clonitazene, 2389
Codeine, 2459
Codeine Methyl Bromide, 2460
Codeine Phosphate, 2462
Codeine Sulfate, 2463
Desomorphine, 2907
Dextromoramide, 2933
Dezocine, 2934
Diampromide, 2966
Dihydrocodeine, 3154
Dihydrocodeinone Enol Acetate, 3155
Dihydromorphine, 3160
Dimenoxadol, 3194
Dimepheptanol, 3195
Dimethylthiambutene, 3248
Dioxaphetyl Butyrate, 3295
Dipipanone, 3342
Eptazocine, 3579
Ethoheptazine, 3698
Ethylmethylthiambutene, 3784
Ethylmorphine, 3785
Etonitazene, 3840
Fentanyl, 3944
Hydrocodone, 4708
Hydromorphone, 4733
Hydroxypethidine, 4769
Isomethadone, 5069
Ketobemidone, 5180
Levorphanol, 5350
Lofentanil, 5437
Meperidine, 5736
Meptazinol, 5753
Metazocine, 5839
Methadone Hydrochloride, 5852
Metopon, 6071
Morphine, 6186
Morphine Derivatives see 6187–6193
Myrophine, 6250
Nalbuphine, 6271

Narceine, 6341
Nicomorphine, 6430
Norlevorphanol, 6625
Normethadone, 6628
Normorphine, 6630
Norpipanone, 6633
Opium, 6809
Oxycodone, 6912
Oxymorphone, 6922
Papaveretum, 6967
Pentazocine, 7078
Phenadoxone, 7160
Phenazocine, 7174
Phenoperidine, 7216
Piminodine, 7403
Piritramide, 7470
Proheptazine, 7786
Promedol, 7795
Properidine, 7826
Propiram, 7844
Propoxyphene, 7851
Sufentanil, 8860
Tilidine, 9370

ANALGESIC (NON-NARCOTIC)

Acetaminophen, 40
Acetaminosalol, 41
Acetanilide, 42
Acetylsalicylsalicylic Acid, 95
Alclofenac, 212
Alminoprofen, 298
Aloxiprin, 306
Aluminum Bis(acetylsalicylate), 328
Aminochlorthenoxazin, 446
2-Amino-4-picoline, 478
Aminopropylon, 484
Aminopyrine, 488
Ammonium Salicylate, 578
Antipyrine, 748
Antipyrine Salicylate, 749
Antrafenine, 754
Apazone, 758
Aspirin, 873
Benorylate, 1054
Benoxaprofen, 1055
Benzpiperylon, 1131
Benzydamine, 1136
p-Bromoacetanilide, 1387
5-Bromosalicylic Acid Acetate, 1425
Bucetin, 1445
Bufexamac, 1462
Bumadizon, 1472
Butacetin, 1497
Calcium Acetylsalicylate, 1644
Carbamazepine, 1783
Carbetidine, 1799
Carbiphene, 1805
Carsalam, 1873
Chloralantipyrine, 2058
Chlorthenoxazin(e), 2195
Choline Salicylate, 2212
Cinchophen, 2290
Ciramadol, 2316
Clometacin, 2381
Cropropamide, 2595
Crotethamide, 2598
Dexoxadrol see 3293
Difenamizole, 3123
Diflunisal, 3130
Dihydroxyaluminum Acetylsalicylate, 3167
Dipyrocetyl, 3357
Dipyrone, 3358
Emorfazone, 3519
Enfenamic Acid, 3535
Epirizole, 3572

Etersalate, 3667
Ethenzamide, 3685
Ethoxazene, 3706
Etodolac, 3831
Felbinac, 3893
Fenoprofen, 3926
Floctafenine, 4037
Flufenamic Acid, 4060
Fluoresone, 4088
Flupirtine, 4117
Fluproquazone, 4120
Flurbiprofen, 4124
Fosfosal, 4170
Gentisic Acid, 4290
Glafenine, 4330
Ibufenac, 4811
Imidazole Salicylate, 4829
Indomethacin, 4874
Indoprofen, 4875
Isofezolac, 5056
Isoladol, 5062
Isonixin, 5078
Ketoprofen, 5184
Ketorolac, 5186
p-Lactophenetide, 5220
Lefetamine, 5314
Loxoprofen, 5462
Lysine Acetylsalicylate, 5510
Magnesium Acetylsalicylate, 5531
Methotrimeprazine, 5909
Metofoline, 6066
Miroprofen, 6127
Morazone, 6176
Morpholine Salicylate, 6195
Naproxen, 6337
Nefopam, 6358
Nifenazone, 6443
5′-Nitro-2′propoxyacetanilide, 6550
Parsalmide, 6993
Perisoxal, 7129
Phenacetin, 7155
Phenazopyridine Hydrochloride, 7175
Phenocoll, 7204
Phenopyrazone, 7217
Phenyl Acetylsalicylate, 7241
Phenyl Salicylate, 7282
Phenyramidol, 7292
Pipebuzone, 7424
Piperylone, 7449
Prodilidine, 7775
Propacetamol, 7804
Propyphenazone, 7884
Proxazole, 7919
Quinine Salicylate, 8088
Ramifenazone, 8122
Rimazolium Metilsulfate, 8222
Salacetamide, 8290
Salicin, 8293
Salicylamide, 8297
Salicylamide O-Acetic Acid, 8298
Salicylsulfuric Acid, 8303
Salsalate, 8307
Salverine, 8310
Simetride, 8487
Sodium Salicylate, 8626
Sulfamipyrine, 8896
Suprofen, 8985
Talniflumate, 9017
Tenoxicam, 9080
Terofenamate, 9099
Tetrandrine, 9162
Tinoridine, 9379
Tolfenamic Acid, 9436
Tolpronine, 9449
Tramadol, 9485
Viminol, 9885
Xenbucin, 9980
Zomepirac, 10092

ANDROGEN

Boldenone, *1327*
Fluoxymesterone, *4113*
Mestanolone, *5816*
Mesterolone, *5817*
Methandrostenolone, *5862*
17-Methyltestosterone, *6044*
17α-Methyltestosterone 3-Cyclo-
  pentyl Enol Ether, *6045*
Norethandrolone, *6613*
Normethandrone, *6629*
Oxandrolone, *6875*
Oxymesterone, *6918*
Oxymetholone, *6920*
Prasterone, *7710*
Stanolone, *8753*
Stanozolol, *8754*
Testosterone, *9109*
Testosterone 17-Chloral
  Hemiacetal, *9110*
Testosterone 17β-Cypionate,
  9111
Testosterone Enanthate, *9112*
Testosterone Nicotinate, *9113*
Testosterone Phenylacetate,
  9114
Testosterone Propionate, *9115*
Tiomesterone, *9385*

ANESTHETIC (INHALATION)

Cyclopropane, *2755*
Enflurane, *3536*
Ethylene, *3748*
Ethyl Ether, *3762*
Fluroxene, *4127*
Halopropane, *4515*
Halothane, *4517*
Isoflurane, *5059*
Methoxyflurane, *5914*
Methyl Propyl Ether, *6031*
Nitrous Oxide, *6575*
Teflurane, *9059*
Tribromoethanol, *9525*
Trichloroethylene, *9552*
Vinyl Ether, *9899*
Xenon, *9981*

ANESTHETIC (INTRAVENOUS)

Acetamidoeugenol, *39*
Alfadolone Acetate, *224*
Alfaxalone, *226*
Buthalital Sodium, *1518*
Etoxadrol, *3844*
Hexobarbital, *4625*
Hydroxydione Sodium, *4753*
Ketamine, *5174*
Methohexital Sodium, *5904*
Midazolam, *6105*
Phencyclidine, *7179*
Propanidid, *7813*
Propofol, *7847*
Thialbarbital, *9219*
Thiamylal, *9231*
Thiobutabarbital, *9250*
Thiopental Sodium, *9280*

ANESTHETIC (LOCAL)

Ambucaine, *393*
Amolanone, *603*
Amylocaine Hydrochloride, *650*
Benoxinate, *1056*
Betoxycaine, *1210*
Biphenamine, *1247*
Bupivacaine, *1485*
Butacaine, *1496*
Butamben, *1504*
Butanilicaine, *1508*

Butethamine, *1517*
Butoxycaine, *1531*
Carticaine, *1878*
2-Chloroprocaine Hydrochloride,
  2158
Cocaethylene, *2449*
Cocaine, *2450*
Cyclomethycaine, *2741*
Dibucaine Hydrochloride, *3016*
Dimethisoquin, *3207*
Dimethocaine, *3210*
Diperodon Hydrochloride, *3301*
Dyclonine, *3453*
Ecgonidine, *3466*
Ecgonine, *3467*
Ethyl Aminobenzoate, *3719*
Ethyl Chloride, *3740*
Etidocaine, *3818*
β-Eucaine, *3850*
Euprocin, *3861*
Fenalcomine, *3900*
Fomocaine, *4145*
Hexylcaine Hydrochloride, *4629*
Hydroxyprocaine, *4772*
Hydroxytetracaine, *4783*
Isobutyl *p*-Aminobenzoate, *5017*
Leucinocaine Mesylate, *5332*
Levoxadrol *see 3293*
Lidocaine, *5359*
Mepivacaine, *5748*
Meprylcaine Hydrochloride, *5752*
Metabutoxycaine Hydrochloride,
  5822
Methyl Chloride, *5964*
Myrtecaine, *6252*
Naepaine, *6263*
Octacaine, *6663*
Orthocaine, *6836*
Oxethazaine, *6888*
Parethoxycaine, *6987*
Phenacaine Hydrochloride, *7153*
Phenol, *7206*
Piperocaine, *7444*
Piridocaine, *7466*
Polidocanol, *7532*
Pramoxine, *7707*
Prilocaine, *7749*
Procaine, *7763*
Propanocaine, *7815*
Proparacaine, *7817*
Propipocaine, *7843*
Propoxycaine Hydrochloride,
  7850
Pseudococaine, *7926*
Pyrrocaine, *8024*
Quinine Urea Hydrochloride,
  8091
Risocaine, *8226*
Salicyl Alcohol, *8294*
Tetracaine Hydrochloride, *9123*
Tolycaine, *9467*
Trimecaine, *9614*
Zolamine, *10089*

ANGIOTENSIN CONVERTING EN-
ZYME INHIBITOR *see ACE-Inhibitor*

ANOREXIC

Aminorex, *490*
Amphecloral, *613*
Amphetamine, *616*
Benzphetamine, *1130*
Chlorphentermine, *2183*
Clobenzorex, *2359*
Cloforex, *2378*
Clortermine, *2406*
Cyclexedrine, *2713*
Dextroamphetamine Sulfate, *2932*
Diethylpropion, *3116*

Diphemethoxidine, *3303*
N-Ethylamphetamine, *3720*
Fenbutrazate, *3908*
Fenfluramine, *3920*
Fenproporex, *3939*
Furfurylmethylamphetamine, *4217*
Levophacetoperane, *5347*
Mazindol, *5643*
Mefenorex, *5681*
Metamfepramone, *5828*
Methamphetamine, *5859*
Norpseudoephedrine, *6634*
Phendimetrazine, *7180*
Phenmetrazine, *7200*
Phenpentermine, *7228*
Phentermine, *7232*
Phenylpropanolamine Hydrochlo-
  ride, *7279*
Picilorex, *7369*

ANTACID

Alexitol Sodium, *223*
Almagate, *297*
Aluminum Hydroxide, *345*
Aluminum Magnesium Silicate,
  352
Aluminum Phosphate, *362*
Azulene, *939*
Basic Aluminum Carbonate Gel,
  1015
Bismuth Aluminate, *1269*
Bismuth Phosphate, *1287*
Bismuth Subgallate, *1297*
Bismuth Subnitrate, *1298*
Calcium Carbonate, *1657*
Dihydroxyaluminum Aminoace-
  tate,
  3168
Dihydroxyaluminum Sodium
  Carbonate, *3169*
Ebimar®, *3463*
Magaldrate, *5527*
Magnesium Carbonate Hydroxide,
  5538
Magnesium Hydroxide, *5548*
Magnesium Oxide, *5555*
Magnesium Peroxide, *5559*
Magnesium Phosphate, Tribasic,
  5562
Magnesium Silicates, *5569*
Polyamine-Methylene Resin, *7538*
Potassium Citrate, *7603*
Sodium Bicarbonate, *8528*

ANTHELMINTIC (CESTODES)

Arecoline, *802*
Aspidin, *868*
Aspidinol, *869*
Dichlorophen(e), *3059*
Embelin, *3513*
Kosin, *5200*
Naphthalene, *6289*
Niclosamide, *6425*
Pelletierine, *7015*
Pelletierine Tannate, *7016*
Quinacrine, *8053*

ANTHELMINTIC (NEMATODES)

Alantolactone, *198*
Amoscanate, *606*
Ascaridole, *852*
Bephenium, *1166*
Bitoscanate, *1318*
Carbon Tetrachloride, *1822*
Carvacrol, *1881*
Cyclobendazole, *2718*
Diethylcarbamazine, *3106*

ANTHELMINTIC (NEMATODES)
(continued)

Diphenane, 3306
Dithiazanine Iodide, 3376
Dymanthine, 3455
Gentian Violet, 4287
4-Hexylresorcinol, 4633
Kainic Acid, 5157
Mebendazole, 5647
2-Naphthol, 6304
Oxantel, 6876
Papain, 6965
Piperazine, 7431
Piperazine Adipate, 7432
Piperazine Citrate, 7433
Piperazine Edetate Calcium, 7435
Piperazine Tartrate, 7436
Pyrantel, 7968
Pyrvinium Pamoate, 8034
α-Santonin, 8325
Stilbazium Iodide, 8773
Tetrachloroethylene, 9126
Tetramisole, 9161
Thiabendazole, 9217
Thymol, 9333
Thymyl N-Isoamylcarbamate, 9343
Triclofenol Piperazine, 9570
Urea Stibamine, 9786

ANTHELMINTIC (ONCHOCERCA)

Ivermectin, 5133
Suramin Sodium, 8986

ANTHELMINTIC (SCHISTOSOMA)

Amoscanate, 606
Amphotalide, 619
Antimony Potassium Tartrate, 732
Antimony Sodium Gluconate, 733
Antimony Sodium Tartrate, 734
Antimony Sodium Thioglycollate, 735
Antimony Thioglycollamide, 737
Becanthone, 1026
Hycanthone, 4677
Lucanthone Hydrochloride, 5463
Niridazole, 6480
Oxamniquine, 6872
Praziquantel, 7714
Stibocaptate, 8768
Stibophen, 8769
Urea Stibamine, 9786

ANTHELMINTIC (TREMATODES)

Anthiolimine, 711
Tetrachloroethylene, 9126

ANTIACNE see also Keratolytic

Algestone Acetophenide, 230
Azelaic Acid, 921
Benzoyl Peroxide, 1128
Cyoctol, 2774
Cyproterone, 2781
Motretinide, 6199
Resorcinol, 8158
Retinoic Acid, 8167
Tetroquinone, 9177

ANTIALLERGIC see also Antihistaminic; Decongestant; Glucocorticoid

Amlexanox, 508
Astemizole, 878
Azelastine, 922
Cromolyn, 2594
Fenpiprane, 3934

Ibudilast, 4810
Nedocromil, 6355
Oxatomide, 6880
Pentigetide, 7086
Repirinast, 8140
Tranilast, 9489
Traxanox, 9494

ANTIALLERGIC (HYPOSENSITIZA-TION THERAPY)

Histamine, 4640
Poison Ivy Extract see 7420
Poison Oak Extract see 7421
Poison Sumac Extract see 7422
Urushiol, 9803

ANTIALLERGIC (STEROIDAL, NASAL)
see also Glucocorticoid

Beclomethasone, 1029
Dexamethasone, 2922
Flunisolide, 4071

ANTIALOPECIA AGENT

Cyoctol, 2774
Minoxidil, 6122

ANTIAMEBIC

Arsthinol, 840
Bialamicol, 1217
Carbarsone, 1788
Cephaeline, 1970
Chlorbetamide, 2076
Chloroquine, 2163
Chlorphenoxamide, 2181
Chlortetracycline, 2193
Dehydroemetine, 2860
Dibromopropamidine, 3008
Diloxanide, 3187
Diphetarsone, 3338
Emetine, 3517
Fumagillin, 4199
Glaucarubin, 4331
Glycobiarsol, 4389
8-Hydroxy-7-iodo-5-quinolinesul-fonic Acid, 4757
Iodochlorhydroxyquin, 4924
Iodoquinol, 4935
Paromomycin, 6989
Phanquinone, 7148
Phenarsone Sulfoxylate, 7171
Polybenzarsol, 7539
Propamidine, 7808
Quinfamide, 8068
Secnidazole, 8373
Sulfarside, 8915
Teclozan, 9055
Tetracycline, 9130
Thiocarbamizine, 9251
Thiocarbarsone, 9252
Tinidazole, 9377

ANTIANDROGEN see also Antiacne; Antialopecia; Antineoplastic (Hormonal)

Bifluranol, 1231
Cyoctol, 2774
Cyproterone, 2781
Delmadinone Acetate, 2865
Flutamide, 4132
Nilutamide, 6460
Oxendolone, 6886

ANTIANGINAL see also Vasodilator (Coronary)

Acebutolol, 13
Alprenolol, 311

Amiodarone, 497
Amlodipine, 509
Arotinolol, 815
Atenolol, 879
Bepridil, 1167
Bevantolol, 1213
Bucumolol, 1454
Bufetolol, 1461
Bufuralol, 1469
Bunitrolol, 1479
Bupranolol, 1486
Carazolol, 1779
Carteolol, 1875
Carvedilol, 1882
Celiprolol, 1956
Cinepazet Maleate, 2292
Diltiazem, 3188
Epanolol, 3557
Felodipine, 3895
Gallopamil, 4257
Imolamine, 4838
Indenolol, 4852
Isosorbide Dinitrate, 5114
Isradipine, 5129
Limaprost, 5367
Mepindolol, 5744
Metoprolol, 6072
Molsidomine, 6143
Nadolol, 6260
Nicardipine, 6403
Nifedipine, 6441
Nifenalol, 6442
Nilvadipine, 6461
Nipradilol, 6479
Nisoldipine, 6482
Nitroglycerin, 6528
Oxprenolol, 6905
Oxyfedrine, 6915
Ozagrel, 6935
Penbutolol, 7025
Pentaerythritol Tetranitrate, 7066
Pindolol, 7412
Pronethalol, 7802
Propranolol, 7852
Sotalol, 8682
Terodiline, 9098
Timolol, 9374
Toliprolol, 9439
Verapamil, 9851

ANTIARRHYTHMIC

Acebutolol, 13
Acecainide, 14
Adenosine, 143
Ajmaline, 184
Alprenolol, 311
Amiodarone, 497
Amoproxan, 604
Aprindine, 781
Arotinolol, 815
Atenolol, 879
Bevantolol, 1213
Bretylium Tosylate, 1364
Bucumolol, 1454
Bufetolol, 1461
Bunaftine, 1474
Bunitrolol, 1479
Bupranolol, 1486
Butidrine Hydrochloride, 1522
Butobendine, 1524
Capobenic Acid, 1757
Carazolol, 1779
Carteolol, 1875
Cifenline, 2274
Cloranolol, 2399
Disopyramide, 3366
Encainide, 3525
Esmolol, 3648

Flecainide, *4034*
Gallopamil, *4257*
Hydroquinidine, *4736*
Indecainide, *4849*
Indenolol, *4852*
Ipratropium Bromide, *4960*
Lidocaine, *5359*
Lorajmine, *5454*
Lorcainide, *5457*
Meobentine, *5730*
Metipranolol, *6059*
Mexiletine, *6094*
Moricizine, *6178*
Nadoxolol, *6261*
Nifenalol, *6442*
Oxprenolol, *6905*
Penbutolol, *7025*
Pindolol, *7412*
Pirmenol, *7471*
Practolol, *7703*
Prajmaline, *7704*
Procainamide Hydrochloride, *7762*
Pronethalol, *7802*
Propafenone, *7806*
Propranolol, *7852*
Pyrinoline, *8001*
Quinidine Polygalacturonate, *8073*
Quinidine Sulfate, *8074*
Quinidine, *8072*
Sotalol, *8682*
Talinolol, *9012*
Timolol, *9374*
Tocainide, *9414*
Verapamil, *9851*
Viquidil, *9908*
Xibenolol, *9984*

ANTIARTERIOSCLEROTIC see also *Antihyperlipoproteinemic*

Pyridinol Carbamate, *7987*

ANTIARTHRITIC/ANTIRHEUMATIC
see also **Anti-inflammatory (Nonste-roidal); Glucocorticoid**

Allocupreide Sodium, *256*
Auranofin, *895*
Aurothioglucose, *901*
Aurothioglycanide, *902*
Azathioprine, *918*
Calcium 3-Aurothio-2-propanol-1-sulfonate, *1649*
Chloroquine, *2163*
Clobuzarit, *2364*
Cuproxoline, *2677*
Diacerein, *2939*
Glucosamine, *4352*
Gold Sodium Thiomalate, *4422*
Gold Sodium Thiosulfate, *4423*
Hydroxychloroquine, *4748*
Kebuzone, *5168*
Lobenzarit, *5433*
Melittin, *5705*
Methotrexate, *5908*
Myoral, *6241*
Penicillamine, *7029*

ANTIASTHMATIC (NONBRONCHODI-LATOR) see also **Bronchodilator; Glucocorticoid**

Amlexanox, *508*
Azelastine, *922*
Cromolyn, *2594*
Ibudilast, *4810*
Ketotifen, *5187*
Nedocromil, *6355*
Oxatomide, *6880*

Tiaramide, *9356*
Traxanox, *9494*

ANTIASTHMATIC (STEROIDAL, IN-HALANT) see also *Glucocorticoid*

Beclomethasone, *1029*
Budesonide, *1455*
Dexamethasone, *2922*
Flunisolide, *4071*
Triamcinolone Acetonide, *9512*

ANTIBACTERIAL (ANTIBIOTIC)

Aminoglycosides
Amikacin, *416*
Apramycin, *780*
Arbekacin, *797*
Bambermycins, *962*
Butirosin, *1523*
Dibekacin, *2987*
Dihydrostreptomycin, *3162*
Fortimicin(s), *4165*
Gentamicin, *4284*
Isepamicin, *4989*
Kanamycin, *5161*
Micronomicin, *6104*
Neomycin, *6369*
Neomycin Undecylenate, *6370*
Netilmicin, *6389*
Paromomycin, *6989*
Ribostamycin, *8208*
Sisomicin, *8498*
Spectinomycin, *8695*
Streptomycin, *8786*
Streptonicozid, *8788*
Tobramycin, *9413*

Amphenicols
Azidamfenicol, *924*
Chloramphenicol, *2068*
Chloramphenicol Palmitate, *2069*
Chloramphenicol Pantothenate, *2070*
Florfenicol, *4042*
Thiamphenicol, *9230*

Ansamycins
Rifamide, *8214*
Rifampin, *8215*
Rifamycin SV, *8217*
Rifaximin, *8218*

β-Lactams

Carbapenems
Imipenem, *4834*

Cephalosporins
Cefaclor, *1920*
Cefadroxil, *1921*
Cefamandole, *1922*
Cefatrizine, *1923*
Cefazedone, *1924*
Cefazolin, *1925*
Cefixime, *1927*
Cefmenoxime, *1928*
Cefodizime, *1931*
Cefonicid, *1932*
Cefoperazone, *1933*
Ceforanide, *1934*
Cefotaxime, *1935*
Cefotiam, *1937*
Cefpimizole, *1939*
Cefpiramide, *1940*
Cefpodoxime Proxetil, *1941*
Cefroxadine, *1942*
Cefsulodin, *1943*
Ceftazidime, *1944*
Cefteram, *1945*
Ceftezole, *1946*
Ceftibuten, *1947*
Ceftizoxime, *1949*
Ceftriaxone, *1950*
Cefuroxime, *1951*

Cefuzonam, *1952*
Cephacetrile Sodium, *1969*
Cephalexin, *1971*
Cephaloglycin, *1973*
Cephaloridine, *1975*
Cephalosporin C, *1976*
Cephalothin, *1978*
Cephapirin Sodium, *1980*
Cephradine, *1982*
Pivcefalexin, *7485*

Cephamycins
Cefbuperazone, *1926*
Cefmetazole, *1929*
Cefminox, *1930*
Cefotetan, *1936*
Cefoxitin, *1938*

Monobactams
Aztreonam, *938*
Carumonam, *1880*
Tigemonam, *9364*

Oxacephems
Flomoxef, *4038*
Moxolactam, *6201*

Penicillins
Amidinocillin, *399*
Amdinocillin Pivoxil, *400*
Amoxicillin, *610*
Ampicillin, *621*
Apalcillin, *756*
Aspoxicillin, *874*
Azidocillin, *925*
Azlocillin, *929*
Bacampicillin, *944*
Benzylpenicillinic Acid, *1156*
Benzylpenicillin Sodium, *1157*
Carbenicillin, *1796*
Carfecillin Sodium, *1844*
Carindacillin, *1847*
Clometocillin, *2383*
Cloxacillin, *2414*
Cyclacillin, *2706*
Dicloxacillin, *3073*
Diphenicillin Sodium, *3310*
Epicillin, *3566*
Fenbenicillin, *3905*
Floxacillin, *4044*
Hetacillin, *4592*
Lenampicillin, *5319*
Metampicillin, *5830*
Methicillin Sodium, *5890*
Mezlocillin, *6096*
Nafcillin Sodium, *6266*
Oxacillin, *6858*
Penamecillin, *7024*
Penethamate Hydriodide, *7027*
Penicillin G Benethamine, *7036*
Penicillin G Benzathine, *7037*
Penicillin G Benzhydrylamine, *7038*
Penicillin G Calcium, *7039*
Penicillin G Hydrabamine, *7040*
Penicillin G Potassium, *7041*
Penicillin G Procaine, *7042*
Penicillin N, *7043*
Penicillin O, *7044*
Penicillin V, *7046*
Penicillin V Benzathine, *7047*
Penicillin V Hydrabamine, *7048*
Penimepicycline, *7052*
Phenethicillin Potassium, *7184*
Piperacillin, *7430*
Pivampicillin, *7484*
Propicillin, *7829*
Quinacillin, *8052*
Sulbenicillin, *8863*
Talampicillin, *9008*
Temocillin, *9076*
Ticarcillin, *9360*

ANTIBACTERIAL (ANTIBIOTIC) (continued)

Lincosamides
Clindamycin, 2352
Lincomycin, 5378

Macrolides
Azithromycin, 928
Carbomycin, 1813
Clarithromycin, 2340
Erythromycin, 3626
Erythromycin Acistrate, 3627
Erythromycin Estolate, 3628
Erythromycin Glucoheptonate,
3629
Erythromycin Lactobionate, 3630
Erythromycin Propionate, 3631
Erythromycin Stearate, 3632
Josamycin, 5148
Leucomycins, 5337
Midecamycins, 6106
Miokamycin, 6123
Oleandomycin, 6785
Primycin, 7756
Rokitamycin, 8233
Rosaramicin, 8241
Roxithromycin, 8253
Spiramycin, 8708
Troleandomycin, 9681

Polypeptides
Amphomycin, 618
Bacitracin, 948
Capreomycin, 1758
Colistin, 2475
Enduracidin, 3534
Enviomycin, 3552
Fusafungine, 4226
Gramicidin(s), 4438
Gramicidin S, 4439
Mikamycin, 6111
Polymyxin, 7550
Polymyxin B-Methanesulfonic
Acid, 7551
Pristinamycin, 7758
Ristocetin, 8227
Teicoplanin, 9062
Thiostrepton, 9295
Tuberactinomycin, 9713
Tyrocidine, 9744
Tyrothricin, 9749
Vancomycin, 9836
Viomycin, 9905
Viomycin Pantothenate, 9906
Virginiamycin, 9910
Zinc Bacitracin, 10027

Tetracyclines
Apicycline, 762
Chlortetracycline, 2193
Clomocycline, 2386
Demeclocycline, 2872
Doxycycline, 3429
Guamecycline, 4468
Lymecycline, 5500
Meclocycline, 5658
Methacycline, 5851
Minocycline, 6121
Oxytetracycline, 6931
Penimepicycline, 7052
Pipacycline, 7420
Rolitetracycline, 8236
Sancycline, 8316
Senociclin, 8408
Tetracycline, 9130

Others
Cycloserine, 2758
Mupirocin, 6212
Tuberin, 9716

ANTIBACTERIAL (SYNTHETIC)

2,4-Diaminopyrimidines
Brodimoprim, 1369
Tetroxoprim, 9178
Trimethoprim, 9624

Nitrofurans
Furaltadone, 4205
Furazolium Chloride, 4211
Nifuradene, 6445
Nifuratel, 6447
Nifurfoline, 6448
Nifurpirinol, 6452
Nifurprazine, 6453
Nifurtoinol, 6455
Nitrofurantoin, 6520

Quinolones and Analogs
Amifloxacin, 415
Cinoxacin, 2311
Ciprofloxacin, 2315
Difloxacin, 3127
Enoxacin, 3540
Fleroxacin, 4035
Flumequine, 4065
Lomefloxacin, 5443
Miloxacin, 6116
Nalidixic Acid, 6273
Norfloxacin, 6617
Ofloxacin, 6688
Oxolinic Acid, 6899
Pefloxacin, 7012
Pipemidic Acid, 7427
Piromidic Acid, 7475
Rosoxacin, 8247
Temafloxacin, A8
Tosufloxacin, 9477

Sulfonamides
Acetyl Sulfamethoxypyrazine, 96
Acetyl Sulfisoxazole, 100
Azosulfamide, 936
Benzylsulfamide, 1161
Chloramine-B, 2065
Chloramine-T, 2066
Dichloramine T, 3034
Formosulfathiazole, 4158
N^2-Formylsulfisomidine, 4164
N^4-β-D-Glucosylsulfanilamide,
4358
Mafenide, 5525
4′-(Methylsulfamoyl)sulfanilanilide,
6040
p-Nitrosulfathiazole, 6566
Noprylsulfamide, 6601
Phthalylsulfacetamide, 7350
Phthalylsulfathiazole, 7351
Salazosulfadimidine, 8291
Succinylsulfathiazole, 8850
Sulfabenzamide, 8868
Sulfacetamide, 8870
Sulfachlorpyridazine, 8871
Sulfachrysoidine, 8872
Sulfacytine, 8873
Sulfadiazine, 8874
Sulfadicramide, 8875
Sulfadimethoxine, 8876
Sulfadoxine, 8877
Sulfaethidole, 8878
Sulfaguanidine, 8879
Sulfaguanol, 8880
Sulfalene, 8881
Sulfaloxic Acid, 8883
Sulfamerazine, 8884
Sulfameter, 8885
Sulfamethazine, 8886
Sulfamethizole, 8887
Sulfamethomidine, 8888
Sulfamethoxazole, 8889
Sulfamethoxypyridazine, 8890
Sulfametrole, 8892
Sulfamidochrysoidine, 8895

Sulfamoxole, 8897
Sulfanilamide, 8898
Sulfanilamidomethanesulfonic
Acid Triethanolamine Salt,
8899
4-Sulfanilamidosalicylic Acid,
8900
N^4-Sulfanilylsulfanilamide,
8905
Sulfanilylurea, 8906
N-Sulfanilyl-3,4-xylamide, 8907
Sulfanitran, 8908
Sulfaperine, 8909
Sulfaphenazole, 8910
Sulfaproxyline, 8911
Sulfapyrazine, 8912
Sulfapyridine, 8913
Sulfasomizole, 8918
Sulfasymazine, 8919
Sulfathiazole, 8920
Sulfathiourea, 8921
Sulfatolamide, 8922
Sulfisomidine, 8929
Sulfisoxazole, 8930

Sulfones
Acedapsone, 16
Acediasulfone, 17
Acetosulfone Sodium, 66
Dapsone, 2820
Diathymosulfone, 2972
Glucosulfone Sodium, 4357
Solasulfone, 8667
Succisulfone, 8851
Sulfanilic Acid, 8901
p-Sulfanilylbenzylamine, 8903
p,p′-Sulfonyldianiline-N,N′digalac-
toside, 8941
Sulfoxone Sodium, 8947
Thiazolsulfone, 9237

Others
Clofoctol, 2377
Hexedine, 4619
Methenamine, 5879
Methenamine Anhydromethylene-
citrate, 5881
Methenamine Hippurate, 5882
Methenamine Mandelate, 5883
Methenamine Sulfosalicylate,
5885
Nitroxoline, 6577
Xibornol, 9985

ANTIBACTERIAL (LEPROSTATIC)

Acedapsone, 16
Acetosulfone Sodium, 66
Clofazimine, 2370
Dapsone, 2820
Diathymosulfone, 2972
Glucosulfone Sodium, 4357
Hydnocarpic Acid, 4679
Solasulfone, 8667
Succisulfone, 8851
Sulfoxone Sodium, 8947

ANTIBACTERIAL (RICKETTSIA) see
Antirickettsial

**ANTIBACTERIAL (TUBERCULO-
STATIC)**

p-Aminosalicylic Acid, 491
p-Aminosalicylic Acid Hydrazide,
492
Benzoylpas, 1127
5-Bromosalicylhydroxamic Acid,
1424
Capreomycin, 1758
Clofazimine, 2370

Cyacetacide, *2688*
Cycloserine, *2758*
Dihydrostreptomycin, *3162*
Enviomycin, *3552*
Ethambutol, *3672*
Ethionamide, *3692*
4'-Formylsuccinanilic Acid
  Thiosemicarbazone, *4163*
Furonazide, *4220*
Glyconiazide, *4396*
Isobutol, *5013*
Isoniazid, *5071*
Isoniazid Methanesulfonate,
  5072
Morphazinamide, *6182*
Opiniazide, *6807*
Parsiniazide, *6999*
Phenyl Aminosalicylate, *7243*
Protionamide, *7905*
Pyrazinamide, *7970*
Rifampin, *8215*
Salinazid, *8304*
Streptomycin, *8786*
Subathizone, *8832*
Sulfoniazide, *8938*
Thiacetazone, *9218*
Tiocarlide, *9382*
Tuberactinomycin, *9713*
Tubercidin, *9714*
Tuberin, *9716*
Verazide, *9859*
Viomycin, *9905*
Viomycin Pantothenate, *9906*

ANTIBACTERIAL ADJUNCT (ENZYME INHIBITOR)

β-Lactamase Inhibitors
  Clavulanic Acid, *2342*
  Sulbactam, *8862*

Renal Dipeptidase Inhibitors
  Cilastatin, *2275*

ANTIBIOTIC *see Antibacterial (Antibiotic); Antifungal (Antibiotic); Antineoplastic (Antibiotic)*

ANTICANCER *see Antineoplastic*

ANTICHOLELITHOGENIC *see Cholelitholytic*

ANTICHOLESTEREMIC *see Antihyperlipoproteinemic*

ANTICHOLINERGIC *see also Antispasmodic; Bronchodilator*

  Adiphenine Hydrochloride, *151*
  Alverine, *377*
  Ambutonium Bromide, *397*
  Aminopentamide, *471*
  Amixetrine, *507*
  Amprotropine Phosphate, *624*
  Anisotropine Methylbromide, *701*
  Apoatropine, *770*
  Atropine, *891*
  Atropine *N*-Oxide, *892*
  Benactyzine, *1037*
  Benapryzine, *1039*
  Benzetimide, *1084*
  Benzilonium Bromide, *1090*
  Benztropine Mesylate, *1135*
  Bevonium Methyl Sulfate, *1214*
  Biperiden, *1246*
  Butropium Bromide, *1534*

N-Butylscopolammonium Bromide, *1588*
Buzepide, *1599*
Camylofine, *1743*
Caramiphen Hydrochloride, *1777*
Chlorbenzoxamine, *2075*
Chlorphenoxamine, *2182*
Cimetropium Bromide, *2280*
Clidinium Bromide, *2351*
Cyclodrine, *2725*
Cyclonium Iodide, *2743*
Cycrimine Hydrochloride, *2763*
Deptropine, *2894*
Dexetimide, *2923*
Dibutoline Sulfate, *3018*
Dicyclomine Hydrochloride, *3087*
Diethazine, *3098*
Difemerine, *3122*
Dihexyverine, *3151*
Diphemanil Methylsulfate, *3302*
N-(1,2-Diphenylethyl)nicotinamide, *3324*
Dipiproverine, *3343*
Diponium Bromide, *3346*
Emepronium Bromide, *3515*
Endobenzyline Bromide, *3527*
Ethopropazine, *3703*
Ethybenztropine, *3712*
Ethylbenzhydramine, *3724*
Etomidoline, *3839*
Eucatropine, *3854*
Fenpiverinium Bromide, *3935*
Fentonium Bromide, *3949*
Flutropium Bromide, *4136*
Glycopyrrolate, *4397*
Heteronium Bromide, *4595*
Hexocyclium Methyl Sulfate, *4627*
Homatropine, *4649*
Hyoscyamine, *4795*
Ipratropium Bromide, *4960*
Isopropamide, *5090*
Levomepate, *5345*
Mecloxamine, *5663*
Mepenzolate Bromide, *5735*
Metcaraphen, *5840*
Methantheline Bromide, *5869*
Methixene, *5902*
Methscopolamine Bromide, *5927*
Octamylamine, *6671*
Oxybutynin Chloride, *6908*
Oxyphencyclimine, *6926*
Oxyphenonium Bromide, *6928*
Pentapiperide, *7077*
Penthienate Bromide, *7084*
Phencarbamide, *7178*
Phenglutarimide, *7192*
Pipenzolate Bromide, *7428*
Piperidolate, *7440*
Piperilate, *7441*
Poldine Methylsulfate, *7529*
Pridinol, *7747*
Prifinium Bromide, *7748*
Procyclidine, *7770*
Propantheline Bromide, *7816*
Propenzolate, *7824*
Propyromazine, *7885*
Scopolamine, *8361*
Scopolamine *N*-Oxide, *8362*
Stilonium Iodide, *8776*
Stramonium, *8779*
Sultroponium, *8977*
Thihexinol, *9242*
Thiphenamil, *9303*
Tiemonium Iodide, *9363*
Timepidium Bromide, *9372*
Tiquizium Bromide, *9390*
Tridihexethyl Iodide, *9577*
Trihexyphenidyl Hydrochloride, *9607*
Tropacine, *9685*
Tropenzile, *9690*
Tropicamide, *9692*

Trospium Chloride, *9697*
Valethamate Bromide, *9816*
Xenytropium Bromide, *9983*

ANTICOAGULANT *see also Antithrombotic; Thrombolytic*

  Acenocoumarol, *24*
  Ancrod, *664*
  Anisindione, *698*
  Bromindione, *1381*
  Clorindione, *2402*
  Coumetarol, *2566*
  Cyclocumarol, *2723*
  Dextran Sulfate Sodium, *2929*
  Dicumarol, *3080*
  Diphenadione, *3304*
  Ethyl Biscoumacetate, *3729*
  Ethylidene Dicoumarol, *3768*
  Fluindione, *4061*
  Heparin, *4571*
  Hirudin, *4638*
  Lyapolate Sodium, *5487*
  Oxazidione, *6882*
  Pentosan Polysulfate, *7090*
  Phenindione, *7196*
  Phenprocoumon, *7230*
  Phosvitin, *7340*
  Picotamide, *7378*
  Tioclomarol, *9383*
  Warfarin, *9950*

ANTICONVULSANT

  Acetylpheneturide, *94*
  Albutoin, *210*
  Aloxidone, *305*
  Aminoglutethimide, *452*
  4-Amino-3-hydroxybutyric Acid, *456*
  Atrolactamide, *888*
  Beclamide, *1027*
  Buramate, *1490*
  Calcium Bromide, *1653*
  Carbamazepine, *1783*
  Cinromide, *2313*
  Clomethiazole, *2382*
  Clonazepam, *2387*
  Decimemide, *2845*
  Diethadione, *3096*
  Dimethadione, *3201*
  Doxenitoin, *3424*
  Eterobarb, *3666*
  Ethadione, *3670*
  Ethosuximide, *3704*
  Ethotoin, *3705*
  Fluoresone, *4088*
  5-Hydroxytryptophan, *4784*
  Lamotrigine, *5228*
  Magnesium Bromide, *5537*
  Magnesium Sulfate, *5573*
  Mephenytoin, *5741*
  Mephobarbital, *5742*
  Metharbital, *5873*
  Methetoin, *5889*
  Methsuximide, *5928*
  5-Methyl-5-(3-phenanthryl)hydantoin, *6023*
  3-Methyl-5-phenylhydantoin, *6026*
  Narcobarbital, *6342*
  Nimetazepam, *6465*
  Nitrazepam, *6491*
  Paramethadione, *6976*
  Phenacemide, *7154*
  Phenetharbital, *7183*
  Pheneturide, *7190*
  Phenobarbital, *7201*
  Phenobarbital Sodium, *7202*
  Phensuximide, *7231*

ANTICONVULSANT (continued)

Phenylmethylbarbituric Acid, 7275
Phenytoin, 7293
Phethenylate Sodium, 7295
Potassium Bromide, 7597
Primidone, 7753
Progabide, 7782
Sodium Bromide, 8539
Solanum, 8664
Strontium Bromide, 8800
Suclofenide, 8852
Sulthiame, 8974
Tetrantoin, 9165
Trimethadione, 9620
Valproic Acid, 9821
Valpromide, 9822
Vigabatrin, 9882
Zonisamide, 10094

ANTIDEPRESSANT see also Antimanic

Bicyclics
Binedaline, 1238
Caroxazone, 1865
Citalopram, 2320
Dimethazan, 3202
Indalpine, 4843
Fencamine, 3910
Indeloxazine Hydrochloride, 4850
Nefopam, 6358
Nomifensine, 6592
Oxitriptan see 4784
Oxypertine, 6924
Paroxetine, 6991
Sertraline, 8416
Thiazesim, 9233
Trazodone, 9495
Zometapine, 10093

Hydrazides/Hydrazines
Benmoxine, 1052
Iproclozide, 4963
Iproniazid, 4965
Isocarboxazid, 5040
Nialamide, 6399
Octamoxin, 6670
Phenelzine, 7181

Pyrrolidones
Cotinine, 2552
Rolicyprine, 8234
Rolipram, 8235

Tetracyclics
Maprotiline, 5633
Metralindole, 6075
Mianserin, 6097
Oxaprotiline, 6878

Tricyclics
Adinazolam, 150
Amitriptyline, 504
Amitriptylinoxide, 505
Amoxapine, 609
Butriptyline, 1533
Clomipramine, 2385
Demexiptiline, 2876
Desipramine, 2902
Dibenzepin, 2990
Dimetacrine, 3199
Dothiepin, 3419
Doxepin, 3425
Fluacizine, 4046
Imipramine, 4835
Imipramine N-Oxide, 4836
Iprindole, 4962
Lofepramine, 5438
Melitracen, 5704
Metapramine, 5835
Nortriptyline, 6635
Noxiptilin, 6644

Opipramol, 6808
Pizotyline, 7486
Propizepine, 7846
Protriptyline, 7917
Quinupramine, 8111
Tianeptine, 9352
Trimipramine, 9636

Others
Adrafinil, 159
Benactyzine, 1037
Bupropion, 1488
Butacetin, 1497
Deanol, 2834
Deanol Aceglumate, 2835
Deanol Acetamidobenzoate, 2836
Dioxadrol, 3293
Etoperidone, 3841
Febarbamate, 3889
Femoxetine, 3897
Fenpentadiol, 3933
Fluoxetine, 4112
Fluvoxamine, 4138
Hematoporphyrin, 4555
Hypericin, 4799
Levophacetoperane, 5347
Medifoxamine, 5674
Minaprine, 6119
Moclobemide, 6139
Oxaflozane, 6861
Piberaline, 7365
Prolintane, 7791
Pyrisuccideanol, 8002
Rubidium Chloride, 8261
Sulpiride, 8971
Sultopride, 8975
Teniloxazine, 9077
Thozalinone, 9315
Tofenacin, 9426
Toloxatone, 9445
Tranylcypromine, 9491
L-Tryptophan, 9707
Viloxazine, 9884
Zimeldine, 10024

ANTIDIABETIC

Biguanides
Buformin, 1465
Metformin, 5845
Phenformin, 7191

Hormones
Glucagon, 4341
Insulin, 4887
Insulin Injection, 4889
Insulin Zinc Suspension, 4890
Isophane Insulin Suspension, 5082
Protamine Zinc Insulin Suspension, 7899
Zinc Insulin Crystals, 10038

Sulfonylurea Derivatives
Acetohexamide, 53
1-Butyl-3-metanilylurea, 1579
Carbutamide, 1839
Chlorpropamide, 2187
Glibornuride, 4334
Gliclazide, 4335
Glipizide, 4337
Gliquidone, 4338
Glisoxepid, 4339
Glyburide, 4372
Glybuthiazol(e), 4373
Glybuzole, 4374
Glyhexamide, 4402
Glymidine, 4403
Glypinamide, 4410
Phenbutamide, 7177
Tolazamide, 9429

Tolbutamide, 9432
Tolcyclamide, 9434

Others
Acarbose, 12
Calcium Mesoxalate, 1685
Miglitol, 6110

ANTIDIARRHEAL

Acetyltannic Acid, 101
Albumin Tannate, 208
Alkofanone, 244
Aluminum Salicylates, Basic, 366
Catechin, 1908
Difenoxin, 3124
Diphenoxylate, 3313
Lidamidine, 5358
Loperamide, 5450
Mebiquine, 5650
Trillium, 9609
Uzarin, 9809

ANTIDIURETIC

Desmopressin, 2904
Felypressin, 3896
Lypressin, 5503
Ornipressin, 6825
Oxycinchophen, 6910
Pituitary, Posterior, 7481
Terlipressin, 9097
Vasopressin, 9843

ANTIDOTE (ACETAMINOPHEN POISONING)

Acetylcysteine, 82
Cysteamine, 2785

ANTIDOTE (CURARE)

Edrophonium Chloride, 3487
Tacrine, 9003

ANTIDOTE (CYANIDE)

p-Aminopropiophenone, 482
Methylene Blue, 5979
Potassium Nitrite, 7635
Sodium Nitrite, 8599
Sodium Thiosulfate, 8652

ANTIDOTE (FOLIC ACID ANTAGONISTS)

Folinic Acid, 4141

ANTIDOTE (HEAVY METAL POISONING)

Albumen, 206
Deferoxamine, 2850
Dimercaprol, 3196
2,3-Dimercapto-1-propanesulfonic Acid, 3197
Ditiocarb Sodium, 3384
Succimer, 8836
Tiopronin, 9386

ANTIDYSKINETIC see also Antiparkinsonian

Amantadine, 380
Clonidine, 2388
Haloperidol, 4511
Pimozide, 7404
Tetrabenazine, 9118
Tiapride, 9353

ANTIECZEMATIC

Evening Primrose Oil, *3864*
γ-Linolenic Acid, *5384*

ANTIEMETIC

Acetylleucine Monoethanolamine, 89
Alizapride, *236*
Benzquinamide, *1133*
Bietanautine, *1225*
Bromopride, *1420*
Buclizine, *1449*
Chlorpromazine, *2186*
Clebopride, *2344*
Cyclizine, *2716*
Dimenhydrinate, *3193*
Diphenidol, *3311*
Domperidone, *3412*
Granisetron, *4443*
Meclizine, *5657*
Methallatal, *5855*
Metoclopramide, *6063*
Metopimazine, *6070*
Nabilone, *6257*
Ondansetron, *6802*
Oxypendyl, *6923*
Pipamazine, *7421*
Piprinhydrinate, *7456*
Prochlorperazine, *7768*
Scopolamine, *8361*
Tetrahydrocannabinols, *9142*
Thiethylperazine, *9241*
Thioproperazine, *9287*
Trimethobenzamide, *9623*

ANTIEPILEPTIC *see Anticonvulsant*

ANTIESTROGEN

Delmadinone Acetate, *2865*
Ethamoxytriphetol, *3674*
Tamoxifen, *9019*
Toremifene, *9474*

ANTIFIBROTIC

Potassium *p*-Aminobenzoate, *7582*

ANTIFLATULENT

Silicones, *8442*
Simethicone, *8486*

ANTIFUNGAL (ANTIBIOTICS)

Polyenes
Amphotericin B, *620*
Candicidin, *1747*
Dermostatin, *2897*
Filipin, *4022*
Fungichromin, *4202*
Hachimycin, *4499*
Hamycin, *4524*
Lucensomycin, *5464*
Mepartricin, *5733*
Natamycin, *6346*
Nystatin, *6658*
Pecilocin, *7008*
Perimycin, *7121*

Others
Azaserine, *916*
Griseofulvin, *4453*
Oligomycins, *6793*
Neomycin Undecylenate, *6370*
Pyrrolnitrin, *8029*
Siccanin, *8432*

Tubercidin, *9714*
Viridin, *9912*

ANTIFUNGAL (SYNTHETIC)

Allylamines
Naftifine, *6270*
Terbinafine, *9086*

Imidazoles
Bifonazole, *1232*
Butoconazole, *1525*
Chlordantoin, *2080*
Chlormidazole, *2106*
Cloconazole, *2367*
Clotrimazole, *2412*
Econazole, *3476*
Enilconazole, *3537*
Fenticonazole, *3948*
Isoconazole, *5044*
Ketoconazole, *5181*
Miconazole, *6101*
Omoconazole, *6801*
Oxiconazole Nitrate, *6892*
Sulconazole, *8866*
Tioconazole, *9384*

Triazoles
Fluconazole, *4054*
Itraconazole, *5131*
Terconazole, *9090*

Others
Acrisorcin, *120*
Amorolfine, *605*
Biphenamine, *1247*
Bromosalicylchloranilide, *1423*
Buclosamide, *1450*
Calcium Propionate, *1705*
Chlorphenesin, *2178*
Ciclopirox, *2270*
Cloxyquin, *2416*
Coparaffinate, *2513*
Diamthazole Dihydrochloride, *2967*
Exalamide, *3866*
Flucytosine, *4056*
Halethazole, *4505*
Hexetidine, *4624*
Loflucarban, *5440*
Nifuratel, *6447*
Potassium Iodide, *7628*
Propionic Acid, *7837*
Pyrithione, *8004*
Salicylanilide, *8299*
Sodium Propionate, *8623*
Sulbentine, *8865*
Tenonitrozole, *9079*
Tolciclate, *9433*
Tolindate, *9438*
Tolnaftate, *9442*
Triacetin, *9504*
Ujothion, *9755*
Undecylenic Acid, *9760*
Zinc Propionate, *10057*

ANTIGLAUCOMA

Acetazolamide, *45*
Befunolol, *1032*
Betaxolol, *1204*
Bupranolol, *1486*
Carteolol, *1875*
Dapiprazole, *2819*
Dichlorphenamide, *3067*
Dipivefrin, *3344*
Epinephrine, *3569*
Levobunolol, *5343*
Methazolamide, *5875*
Metipranolol, *6059*
Pilocarpine, *7395*
Pindolol, *7412*
Timolol, *9374*

ANTIGONADOTROPIN

Danazol, *2811*
Gestrinone, *4310*
Paroxypropione, *6992*

ANTIGOUT

Allopurinol, *278*
Carprofen, *1870*
Colchicine, *2470*
Probenecid, *7760*
Sulfinpyrazone, *8926*

ANTIHEMORRHAGIC *see Hemostatic*

ANTIHISTAMINIC *see also Antiallergic*

Alkylamine Derivatives
Acrivastine, *121*
Bamipine, *966*
Brompheniramine, *1433*
Chlorpheniramine, *2180*
Dimethindene, *3204*
Metron S, *6080*
Pheniramine, *7198*
Pyrrobutamine, *8023*
Thenaldine, *9204*
Tolpropamine, *9450*
Triprolidine, *9661*

Aminoalkyl Ethers
Bietanautine, *1225*
Bromodiphenhydramine, *1405*
Carbinoxamine, *1804*
Clemastine, *2345*
Diphenylpyraline, *3334*
Doxylamine, *3430*
Embramine, *3514*
Medrylamine, *5678*
Mephenhydramine, *5738*
p-Methyldiphenhydramine, *5972*
Orphenadrine, *6831*
Phenyltoloxamine, *7287*
Piprinhydrinate, *7456*
Setastine, *8423*

Ethylenediamine Derivatives
Alloclamide, *254*
p-Bromtripelennamine, *1436*
Chloropyramine, *2162*
Chlorothen, *2168*
Histapyrrodine, *4641*
Methafurylene, *5854*
Methaphenilene, *5870*
Methapyrilene, *5871*
Phenbenzamine, *7176*
Pyrilamine, *7996*
Talastine, *9009*
Thenyldiamine, *9207*
Thonzylamine Hydrochloride, *9307*
Tripelennamine, *9651*
Zolamine, *10089*

Piperazines
Cetirizine, *2013*
Chlorcyclizine, *2078*
Cinnarizine, *2308*
Clocinizine, *2366*
Hydroxyzine, *4786*

Tricyclics

Phenothiazines
Ahistan, *179*
Etymemazine, *3849*
Fenethazine, *3918*
N-Hydroxyethylpromethazine Chloride, *4755*
Isopromethazine, *5088*
Mequitazine, *5754*
Promethazine, *7797*
Pyrathiazine, *7969*
Thiazinamium Methyl Sulfate, *9234*

ANTIHISTAMINIC

Tricyclics (*continued*)

Others

Azatadine, *917*
Clobenzepam, *2358*
Cyproheptadine, *2779*
Deptropine, *2894*
Isothipendyl, *5117*
Loratadine, *5455*
Prothipendyl, *7902*

Others

Antazoline, *709*
Astemizole, *878*
Azelastine, *922*
Cetoxime, *2015*
Clemizole, *2346*
Clobenztropine, *2360*
Diphenazoline, *3307*
Diphenhydramine, *3308*
Mebhydroline, *5649*
Phenindamine, *7195*
Terfenadine, *9094*
Tritoqualine, *9677*

ANTIHYPERCHOLESTEROLEMIC see Antihyperlipoproteinemic

ANTIHYPERLIPIDEMIC see Antihyperlipoproteinemic

ANTIHYPERLIPOPROTEINEMIC

Aryloxyalkanoic Acid Derivatives

Beclobrate, *1028*
Bezafibrate, *1215*
Binifibrate, *1239*
Ciprofibrate, *2314*
Clinofibrate, *2353*
Clofibrate, *2374*
Clofibric Acid, *2375*
Etofibrate, *3834*
Fenofibrate, *3924*
Gemfibrozil, *4280*
Nicofibrate, *6427*
Pirifibrate, *7467*
Ronifibrate, *8238*
Simfibrate, *8489*
Theofibrate, *9211*

Bile Acid Sequesterants

Cholestyramine Resin, *2205*
Colestipol, *2472*
Polidexide, *7531*

HMG CoA Reductase Inhibitors

Lovastatin, *5460*
Pravastatin Sodium, *7712*
Simvastatin, *8491*

Nicotinic Acid Derivatives

Aluminum Nicotinate, *354*
Acipimox, *106*
Niceritrol, *6405*
Nicoclonate, *6426*
Nicomol, *6429*
Oxiniacic Acid, *6895*

Thyroid Hormones/Analogs

Etiroxate, *3828*
Thyropropic Acid, *9346*
Thyroxine, *9348*

Others

Acifran, *A1*
Azacosterol, *908*
Benfluorex, *1047*
β-Benzalbutyramide, *1063*
Carnitine, *1856*
Chondroitin Sulfate, *2217*
Clomestrone, *2380*
Detaxtran, *2916*
Dextran Sulfate Sodium, *2929*

5,8,11,14,17-Eicosapentaenoic
Acid, *3495*
Eritadenine, *3618*
Furazabol, *4209*
Meglutol, *5690*
Melinamide, *5702*
Mytatrienediol, *6254*
Ornithine, *6826*
γ-Oryzanol, *6841*
Pantethine, *6962*
Pentaerythritol Tetraacetate,
7065
α-Phenylbutyramide, *7249*
Pirozadil, *7477*
Probucol, *7761*
β-Sitosterol, *8500*
Sultosilic Acid, Piperazine
Salt, *8976*
Tiadenol, *9349*
Triparanol, *9650*
Xenbucin, *9980*

ANTIHYPERPHOSPHATEMIC

Aluminum Hydroxide, *345*
Aluminum Hydroxychloride, *346*

ANTIHYPERTENSIVE see also Diuretic

Arylethanolamine Derivatives

Amosulalol, *607*
Bufuralol, *1469*
Dilevalol, *3186*
Labetalol, *5208*
Pronethalol, *7802*
Sotalol, *8682*
Sulfinalol, *8925*

Aryloxypropanolamine Derivatives

Acebutolol, *13*
Alprenolol, *311*
Arotinolol, *815*
Atenolol, *879*
Betaxolol, *1204*
Bevantolol, *1213*
Bisoprolol, *1309*
Bopindolol, *1335*
Bunitrolol, *1479*
Bupranolol, *1486*
Butofilolol, *1527*
Carazolol, *1779*
Carteolol, *1875*
Carvedilol, *1882*
Celiprolol, *1956*
Cetamolol, *2010*
Epanolol, *3557*
Indenolol, *4852*
Mepindolol, *5744*
Metipranolol, *6059*
Metoprolol, *6072*
Moprolol, *6172*
Nadolol, *6260*
Nipradilol, *6479*
Oxprenolol, *6905*
Penbutolol, *7025*
Pindolol, *7412*
Propranolol, *7852*
Talinolol, *9012*
Tertatolol, *9105*
Timolol, *9374*
Toliprolol, *9439*

Benzothiadiazine Derivatives

Althiazide, *316*
Bendroflumethiazide, *1045*
Benzthiazide, *1134*
Benzylhydrochlorothiazide, *1150*
Buthiazide, *1519*
Chlorothiazide, *2169*
Chlorthalidone, *2194*
Cyclopenthiazide, *2750*
Cyclothiazide, *2760*

Diazoxide, *2986*
Epithiazide, *3574*
Ethiazide, *3687*
Fenquizone, *3941*
Hydrochlorothiazide, *4704*
Hydroflumethiazide, *4716*
Methyclothiazide, *5929*
Meticrane, *6058*
Metolazone, *6068*
Paraflutizide, *6973*
Polythiazide, *7561*
Tetrachlormethiazide, *9124*
Trichlormethiazide, *9537*

N-Carboxyalkyl(peptide/lactam) Derivatives

Alacepril, *192*
Captopril, *1773*
Cilazapril, *2276*
Delapril, *2864*
Enalapril, *3521*
Enalaprilat, *3522*
Fosinopril, *4171*
Lisinopril, *5393*
Moveltipril, *6200*
Perindopril, *7122*
Quinapril, *A7*
Ramipril, *8123*

Dihydropyridine Derivatives

Amlodipine, *509*
Felodipine, *3895*
Isradipine, *5129*
Nicardipine, *6403*
Nifedipine, *6441*
Nilvadipine, *6461*
Nisoldipine, *6482*
Nitrendipine, *6493*

Guanidine Derivatives

Bethanidine, *1208*
Debrisoquin, *2837*
Guanabenz, *4469*
Guanacline, *4470*
Guanadrel, *4471*
Guanazodine, *4472*
Guanethidine, *4473*
Guanfacine, *4474*
Guanochlor, *4479*
Guanoxabenz, *4481*
Guanoxan, *4482*

Hydrazines/Phthalazines

Budralazine, *1457*
Cadralazine, *1631*
Dihydralazine, *3152*
Endralazine, *3532*
Hydracarbazine, *4680*
Hydralazine, *4682*
Pheniprazine, *7197*
Pildralazine, *7394*
Todralazine, *9425*

Imidazole Derivatives

Clonidine, *2388*
Lofexidine, *5439*
Phentolamine, *7234*
Tiamenidine, *9350*
Tolonidine, *9443*

Quaternary Ammonium Compounds

Azamethonium Bromide, *913*
Chlorisondamine Chloride, *2101*
Hexamethonium, *4609*
Pentacynium Bis(methyl sulfate),
7060
Pentamethonium Bromide, *7070*
Pentolinium Tartrate, *7089*
Phenactropinium Chloride, *7158*
Trimethidinium Methosulfate,
9622

Quinazoline Derivatives

Alfuzosin, *228*
Bunazosin, *1477*

Doxazosin, *3422*
Prazosin, *7715*
Terazosin, *9084*
Trimazosin, *9612*

Reserpine Derivatives
Bietaserpine, *1226*
Deserpidine, *2901*
Rescinnamine, *8146*
Reserpine, *8149*
Syrosingopine, *8998*

Sulfonamide Derivatives
Ambuside, *396*
Clopamide, *2391*
Furosemide, *4221*
Indapamide, *4847*
Quinethazone, *8067*
Tripamide, *9649*
Xipamide, *9986*

Others
Ajmaline, *184*
γ-Aminobutyric Acid, *441*
Bufeniode, *1460*
Chlorthalidone, *2194*
Cicletanine, *2268*
Ciclosidomine, *2271*
Cryptenamine Tannates, *2610*
Fenoldopam, *3925*
Flosequinan, *4043*
Indoramin, *4876*
Ketanserin, *5175*
Mebutamate, *5653*
Mecamylamine, *5654*
Methyldopa, *5974*
Methyl 4-Pyridyl Ketone
    Thiosemicarbazone, *6035*
Metolazone, *6068*
Minoxidil, *6122*
Muzolimine, *6227*
Pargyline, *6988*
Pempidine, *7022*
Pinacidil, *7407*
Piperoxan, *7448*
Primaperone, *7750*
Protoveratrines, *7915*
Raubasine, *8129*
Rescimetol, *8145*
Rilmenidene, *8220*
Saralasin, *8330*
Sodium Nitroprusside, *8600*
Ticrynafen, *9362*
Trimethaphan Camsylate, *9621*
Tyrosinase, *9746*
Urapidil, *9779*

ANTIHYPERTHYROID

2-Amino-4-methylthiazole, *462*
2-Aminothiazole, *494*
Carbimazole, *1803*
3,5-Dibromo-L-tyrosine, *3014*
3,5-Diiodotyrosine, *3177*
Hinderin, *4635*
Iodine, *4907*
Iothiouracil, *4954*
Methimazole, *5892*
Methylthiouracil, *6048*
Propylthiouracil, *7882*
Sodium Perchlorate, *8605*
Thibenzazoline, *9239*
Thiobarbital, *9248*
2-Thiouracil, *9298*

ANTIHYPOTENSIVE

Amezinium Methyl Sulfate, *403*
Angiotensin Amide *see 680*
Dimetofrine, *3254*
Dopamine, *3415*
Etifelmin, *3820*

Etilefrin, *3822*
Gepefrine, *4295*
Metaraminol, *5837*
Midodrine, *6107*
Norepinephrine, *6612*
Pholedrine, *7304*
Synephrine, *8994*

ANTIHYPOTHYROID

Levothyroxine Sodium, *5351*
Liothyronine, *5388*
Thyroid, *9344*
Thyroidin, *9345*
Thyroxine, *9348*
Tiratricol, *9391*
TSH, *9709*

ANTI-INFECTIVE *see Antiseptic/Disinfectant*

ANTI-INFLAMMATORY (GASTRO-INTESTINAL)

Mesalamine, *5807*
Olsalazine, *6799*
Sulfasalazine, *8917*

ANTI-INFLAMMATORY (NONSTE-ROIDAL) *see also Antiarthritic/Antirheumatic*

Aminoarylcarboxylic Acid Derivatives
Enfenamic Acid, *3535*
Etofenamate, *3833*
Flufenamic Acid, *4060*
Isonixin, *5078*
Meclofenamic Acid, *5659*
Mefenamic Acid, *5680*
Niflumic Acid, *6444*
Talniflumate, *9017*
Terofenamate, *9099*
Tolfenamic Acid, *9436*

Arylacetic Acid Derivatives
Acemetacin, *22*
Alclofenac, *212*
Amfenac, *404*
Bufexamac, *1462*
Cinmetacin, *2295*
Clopirac, *2395*
Diclofenac Sodium, *3071*
Etodolac, *3831*
Felbinac, *3893*
Fenclofenac, *3913*
Fenclorac, *3914*
Fenclozic Acid, *3915*
Fentiazac, *3946*
Glucametacin, *4342*
Ibufenac, *4811*
Indomethacin, *4874*
Isofezolac, *5056*
Isoxepac, *5126*
Lonazolac, *5445*
Metiazinic Acid, *6057*
Oxametacin, *6869*
Proglumetacin, *7784*
Sulindac, *8961*
Tiaramide, *9356*
Tolmetin, *9441*
Zomepirac, *10092*

Arylbutyric Acid Derivatives
Bumadizon, *1472*
Butibufen, *1521*
Fenbufen, *3906*
Xenbucin, *9980*

Arylcarboxylic Acids
Clidanac, *2350*
Ketorolac, *5186*
Tinoridine, *9379*

Arylpropionic Acid Derivatives
Alminoprofen, *298*
Benoxaprofen, *1055*
Bucloxic Acid, *1451*
Carprofen, *1870*
Fenoprofen, *3926*
Flunoxaprofen, *4074*
Flurbiprofen, *4124*
Ibuprofen, *4812*
Ibuproxam, *4813*
Indoprofen, *4875*
Ketoprofen, *5184*
Loxoprofen, *5462*
Miroprofen, *6127*
Naproxen, *6337*
Oxaprozin, *6879*
Piketoprofen, *7393*
Pirprofen, *7478*
Pranoprofen, *7708*
Protizinic Acid, *7906*
Suprofen, *8985*
Tiaprofenic Acid, *9354*

Pyrazoles
Difenamizole, *3123*
Epirizole, *3572*

Pyrazolones
Apazone, *758*
Benzpiperylon, *1131*
Feprazone, *3953*
Mofebutazone, *6141*
Morazone, *6176*
Oxyphenbutazone, *6925*
Phenylbutazone, *7248*
Pipebuzone, *7424*
Propyphenazone, *7884*
Ramifenazone, *8122*
Suxibuzone, *8990*
Thiazolinobutazone, *9236*

Salicylic Acid Derivatives
Acetaminosalol, *41*
Aspirin, *873*
Benorylate, *1054*
Bromosaligenin, *1426*
Calcium Acetylsalicylate, *1644*
Diflunisal, *3130*
Etersalate, *3667*
Fendosal, *3917*
Gentisic Acid, *4290*
Glycol Salicylate, *4395*
Imidazole Salicylate, *4829*
Lysine Acetylsalicylate, *5510*
Mesalamine, *5807*
Morpholine Salicylate, *6195*
1-Naphthyl Salicylate, *6334*
Olsalazine, *6799*
Parsalmide, *6993*
Phenyl Acetylsalicylate, *7241*
Phenyl Salicylate, *7282*
Salacetamide, *8290*
Salicylamide *O*-Acetic Acid,
    8298
Salicylsulfuric Acid, *8303*
Salsalate, *8307*
Sulfasalazine, *8917*

Thiazinecarboxamides
Droxicam, *3443*
Isoxicam, *5127*
Piroxicam, *7476*
Tenoxicam, *9080*

Others
ε-Acetamidocaproic Acid, *38*
S-Adenosylmethionine, *146*
3-Amino-4-hydroxybutyric Acid,
    455
Amixetrine, *507*
Bendazac, *1042*
Benzydamine, *1136*
Bucolome, *1452*
Difenpiramide, *3125*
Ditazol, *3374*

ANTI-INFLAMMATORY (NONSTE-ROIDAL)

Others (*continued*)
Emorfazone, *3519*
Guaiazulene, *4464*
Nabumetone, *6258*
Nimesulide, *6464*
Orgotein, *6820*
Oxaceprol, *6857*
Paranyline, *6978*
Perisoxal, *7129*
Pifoxime, *7392*
Proquazone, *7887*
Proxazole, *7919*
Tenidap, *A9*

ANTI-INFLAMMATORY (STEROIDAL)
see Glucocorticoid

ANTILEPROTIC *see Antibacterial (Leprostatic)*

ANTILEUKEMIC *see Antineoplastic*

ANTILIPEMIC *see Antihyperlipo-proteinemic*

ANTILIPIDEMIC *see Antihyperlipo-proteinemic*

ANTIMALARIAL

Acedapsone, *16*
Amodiaquin, *602*
Arteether, *841*
Artemether, *842*
Artemisinin, *844*
Artesunate, *845*
Bebeerine, *1024*
Berberine, *1169*
Chirata, *2051*
Chlorguanide, *2088*
Chloroquine, *2163*
Chlorproguanil, *2185*
Cinchona, *2286*
Cinchonidine, *2288*
Cinchonine, *2289*
Cycloguanil, *2727*
Gentiopicrin, *4289*
Halofantrine, *4508*
Hydroxychloroquine, *4748*
Mefloquine Hydrochloride, *5683*
3-Methylarsacetin, *5945*
Pamaquine, *6953*
Plasmocid, *7495*
Primaquine, *7751*
Pyrimethamine, *7997*
Quinacrine, *8053*
Quinine, *8075*
Quinine Bisulfate, *8076*
Quinine Carbonate, *8077*
Quinine Dihydrobromide, *8078*
Quinine Dihydrochloride, *8079*
Quinine Ethylcarbonate, *8080*
Quinine Formate, *8081*
Quinine Gluconate, *8082*
Quinine Hydriodide, *8083*
Quinine Hydrochloride, *8085*
Quinine Salicylate, *8088*
Quinine Sulfate, *8089*
Quinine Tannate, *8090*
Quinine Urea Hydrochloride, *8091*
Quinocide, *8096*
Quinoline, *8097*
Sodium Arsenate, Dibasic, *8522*

ANTIMANIC

Lithium Acetate, *5396*
Lithium Carbonate, *5404*
Lithium Chloride, *5405*
Lithium Citrate, *5407*
Lithium Sulfate, *5421*

ANTIMETHEMOGLOBINEMIC

Methylene Blue, *5979*

ANTIMIGRAINE

Alpiropride, *309*
Dihydroergotamine, *3157*
Ergocornine, *3591*
Ergocorninine, *3592*
Ergocryptine, *3595*
Ergot, *3608*
Ergotamine, *3609*
Flumedroxone acetate, *4064*
Fonazine, *4146*
Lisuride, *5394*
Methysergid(e), *6055*
Oxetorone, *6889*
Pizotyline, *7486*
Sumatriptan, *8979*

ANTIMYCOTIC *see Antifungal*

ANTINAUSEANT *see Antiemetic*

ANTINEOPLASTIC

Alkylating Agents

Alkyl Sulfonates
Busulfan, *1494*
Improsulfan, *4841*
Piposulfan, *7452*

Aziridines
Benzodepa, *1097*
Carboquone, *1830*
Meturedepa, *6082*
Uredepa, *9787*

Ethylenimines and Methylmelamines
Altretamine, *318*
Triethylenemelamine, *9586*
Triethylenephosphoramide, *9587*
Triethylenethiophosphoramide, *9588*
Trimethylolmelamine, *9631*

Nitrogen Mustards
Chlorambucil, *2064*
Chlornaphazine, *2107*
Cyclophosphamide, *2753*
Estramustine, *3658*
Ifosfamide, *4822*
Mechlorethamine, *5655*
Mechlorethamine Oxide Hydro-chloride, *5656*
Melphalan, *5708*
Novembichin, *6640*
Phenesterine, *7182*
Prednimustine, *7718*
Trofosfamide, *9680*
Uracil Mustard, *9762*

Nitrosoureas
Carmustine, *1852*
Chlorozotocin, *2177*
Fotemustine, *4174*
Lomustine, *5444*
Nimustine, *6469*
Ranimustine, *8125*

Others
Dacarbazine, *2803*
Mannomustine, *5631*
Mitobronitol, *6130*

Mitolactol, *6132*
Pipobroman, *7451*

Antibiotics
Aclacinomycins, *108*
Actinomycin F₁, *133*
Anthramycin, *715*
Azaserine, *916*
Bleomycins, *1324*
Cactinomycin, *1605*
Carubicin, *1879*
Carzinophilin, *1886*
Chromomycins, *2240*
Dactinomycin, *2804*
Daunorubicin, *2825*
6-Diazo-5-oxo-L-norleucine, *2984*
Doxorubicin, *3428*
Epirubicin, *3573*
Mitomycins, *6133*
Mycophenolic Acid, *6238*
Nogalamycin, *6591*
Olivomycins, *6798*
Peplomycin, *7100*
Plicamycin, *7510*
Porfiromycin, *7572*
Puromycin, *7960*
Streptonigrin, *8789*
Streptozocin, *8794*
Tubercidin, *9714*
Ubenimex, *9750*
Zinostatin, *10073*
Zorubicin, *10096*

Antimetabolites

Folic Acid Analogs
Denopterin, *2879*
Methotrexate, *5908*
Pteropterin, *7950*
Trimetrexate, *9635*

Purine Analogs
Fludarabine, *4057*
6-Mercaptopurine, *5762*
Thiamiprine, *9228*
Thioguanine, *9266*

Pyrimidine Analogs
Ancitabine, *663*
Azacitidine, *907*
6-Azauridine, *920*
Carmofur, *1851*
Cytarabine, *2790*
Doxifluridine, *3426*
Enocitabine, *3539*
Floxuridine, *4045*
Fluorouracil, *4109*
Tegafur, *9060*

Enzymes
L-Asparaginase, *858*

Others
Aceglatone, *20*
Amsacrine, *627*
Bestrabucil, *1199*
Bisantrene, *1255*
Carboplatin, *1828*
Cisplatin, *2319*
Defosfamide, *2853*
Demecolcine, *2873*
Diaziquone, *2979*
Eflornithine, *3489*
Elliptinium Acetate, *3510*
Etoglucid, *3837*
Etoposide, *3842*
Gallium Nitrate see *4252*
Hydroxyurea, *4785*
Interferon-α, *4892*
Interferon-β, *4893*
Interferon-γ, *4894*
Interleukin-2, *4896*
Lentinan, *5322*
Lonidamine, *5447*
Mitoguazone, *6131*
Mitoxantrone, *6135*
Mopidamol, *6171*

Nitracrine, *6485*
Pentostatin, *7091*
Phenamet, *7164*
Pirarubicin, *7460*
Podophyllinic Acid
2-Ethylhydrazide, *7519*
Procarbazine, *7764*
PSK®, *7943*
Razoxane, *8133*
Sizofiran, *8502*
Spirogermanium, *8720*
Taxol, *9049*
Teniposide, *9078*
Tenuazonic Acid, *9081*
Triaziquone, *9517*
2,2′,2″-Trichlorotriethylamine,
9560
Urethan, *9789*
Vinblastine, *9887*
Vincristine, *9891*
Vindesine, *9892*

ANTINEOPLASTIC (HORMONAL)

Androgens
Calusterone, *1730*
Dromostanolone Propionate, *3436*
Epitiostanol, *3575*
Mepitiostane, *5747*
Testolactone, *9108*

Antiadrenals
Aminoglutethimide, *452*
Mitotane, *6134*
Trilostane, *9611*

Antiandrogens
Flutamide, *4132*
Nilutamide, *6460*

Antiestrogens
Tamoxifen, *9019*
Toremifene, *9474*

Estrogens
Fosfestrol, *4168*
Hexestrol, *4621*
Polyestradiol Phosphate, *7543*

LH-RH Analogs
Buserelin, *1492*
Goserelin, *4433*
Leuprolide, *5341*
Triptorelin, *9662*

Progestogens
Chlormadinone Acetate, *2102*
Medroxyprogesterone, *5677*
Megestrol Acetate, *5687*
Melengestrol, *5697*

ANTINEOPLASTIC (RADIATION SOURCE)

Americium, *401*
Cobalt, *2421*
[131]I-Ethiodized Oil *see 3690*
Gold, Radioactive, Colloidal,
4419
Radium, *8118*
Radon, *8119*
Sodium Iodide, Radioactive,
8583
Sodium Phosphate, Radioactive,
8614

ANTINEOPLASTIC ADJUNCT

Folic Acid Replenisher
Folinic Acid, *4141*

Uroprotective
Mesna, *5812*

ANTIOSTEOPOROTIC *see Calcium Regulator*

ANTIPAGETIC *see Calcium Regulator*

ANTIPARKINSONIAN

Amantadine, *380*
Benserazide, *1059*
Bietanautine, *1225*
Biperiden, *1246*
Bromocriptine, *1404*
Budipine, *1456*
Carbidopa, *1802*
Deprenyl, *2893*
Dexetimide, *2923*
Diethazine, *3098*
Droxidopa, *3444*
Ethopropazine, *3703*
Ethylbenzhydramine, *3724*
Levodopa, *5344*
Naxagolide, *6349*
Pergolide, *7115*
Piroheptine, *7473*
Pridinol, *7747*
Prodipine, *7776*
Terguride, *9095*
Tigloidine, *9366*
Trihexyphenidyl Hydrochloride,
9607

ANTIPERISTALTIC *see Antidiarrheal*

ANTIPHEOCHROMOCYTOMA

Metyrosine, *6085*
Phenoxybenzamine, *7225*
Phentolamine, *7234*

ANTIPNEUMOCYSTIS

Eflornithine, *3489*
Pentamidine, *7071*
Sulfamethoxazole, *8889*

ANTIPROSTATIC HYPERTROPHY

Gestonorone Caproate, *4309*
Mepartricin, *5733*
Oxendolone, *6886*
Proscar®, *7888*

ANTIPROTOZOAL (AMEBA) *see Antiamebic*

ANTIPROTOZOAL (GIARDIA)

Acranil, *116*
Tinidazole, *9377*

ANTIPROTOZOAL (HISTOMONAS)

Ipronidazole, *4966*

ANTIPROTOZOAL (LEISHMANIA)

Antimony Sodium Gluconate, *733*
Ethylstibamine, *3807*
Hydroxystilbamidine, *4781*
N-Methylglucamine, *5995*
Pentamidine, *7071*
Stilbamidine, *8772*
Urea Stibamine, *9786*

ANTIPROTOZOAL (PLASMODIA) *see Antimalarial*

ANTIPROTOZOAL (TRICHOMONAS)

Acetarsone, *44*
Aminitrozole, *421*
Anisomycin, *700*
Azanidazole, *914*
Forminitrazole, *4155*
Furazolidone, *4210*
Hachimycin, *4499*
Lauroguadine, *5256*
Mepartricin, *5733*
Metronidazole, *6079*
Nifuratel, *6447*
Nifuroxime, *6451*
Nimorazole, *6468*
Secnidazole, *8373*
Silver Picrate, *8473*
Tenonitrozole, *9079*
Tinidazole, *9377*

ANTIPROTOZOAL (TRYPANOSOMA)

Benznidazole, *1094*
Eflornithine, *3489*
Melarsoprol, *5694*
Nifurtimox, *6454*
Oxophenarsine Hydrochloride,
6902
Pentamidine, *7071*
Propamidine, *7808*
Puromycin, *7960*
Quinapyramine, *8061*
Stilbamidine, *8772*
Suramin Sodium, *8986*
Trypan Red, *9702*
Tryparsamide, *9703*

ANTIPRURITIC

Camphor, *1738*
Cyproheptadine, *2779*
Dichlorisone, *3035*
Glycine, *4386*
Halometasone, *4510*
3-Hydroxycamphor, *4747*
Menthol, *5723*
Mesulphen, *5821*
Methdilazine, *5877*
Phenol, *7206*
Polidocanol, *7532*
Risocaine, *8226*
Spirit of Camphor, *8711*
Thenaldine, *9204*
Tolpropamine, *9450*
Trimeprazine, *9618*

ANTIPSORIATIC

Acitretin, *107*
Ammonium Salicylate, *578*
Anthralin, *714*
6-Azauridine, *920*
Bergapten(e), *1173*
Chrysarobin, *2257*
Etretinate, *3846*
Pyrogallol, *8010*

ANTIPSYCHOTIC

Butyrophenones
Benperidol, *1057*
Bromperidol, *1432*
Droperidol, *3437*
Fluanisone, *4047*
Haloperidol, *4511*
Melperone, *5707*
Moperone, *6170*
Pipamperone, *7422*
Spiperone, *8707*
Timiperone, *9373*
Trifluperidol, *9596*

Carvacrol, *1881*
Chloroxylenol, *2176*
Clorophene, *2403*
Creosote, *2575*
Cresol(s), *2578*
p-Cresol, *2581*
Fenticlor, *3947*
Hexachlorophene, *4602*
1-Naphthyl Salicylate, *6334*
2-Naphthyl Salicylate, *6335*
2,4,6-Tribromo-*m*-cresol, *9524*
3',4',5-Trichlorosalicylanilide, *9558*

Quaternary Ammonium Compounds
Amantanium Bromide, *381*
Benzethonium Chloride, *1083*
Benzoxonium Chloride, *1123*
Bisdequalinium Chloride, *1259*
Cetalkonium Chloride, *2009*
Cethexonium Bromide, *2011*
Cetylpyridinium Chloride, *2024*
Dequalinium Acetate, *2895*
Dequalinium Chloride, *2896*
Dodecarbonium Chloride, *3404*
Domiphen Bromide, *3411*
Halimide®, *4507*
Laurolinium Acetate, *5257*
Methylbenzethonium Chloride, *5946*
Phenoctide, *7205*
Tibezonium Iodide, *9357*
Triclobisonium Chloride, *9567*

Quinolines
Aminoquinuride, *489*
Benzoxiquine, *1122*
Broxyquinoline, *1441*
Chloroxine, *2175*
Chlorquinaldol, *2191*
Cloxyquin, *2416*
Ethylhydrocupreine, *3765*
Euprocin, *3861*
Halquinol, *4520*
Hydrastine, *4688*
8-Hydroxyquinoline, *4778*
8-Hydroxyquinoline Sulfate, *4779*
Iodochlorhydroxyquin, *4924*

Silver Compounds
Silver Bromide, *8452*
Silver Fluoride, *8460*
Silver Lactate, *8463*
Silver Nitrate, *8464*
Silver Protein *see 8474 and 8475*

Others
Aluminum Acetate Solution, *322*
Aluminum Subacetate Solution, *371*
Aluminum Sulfate, *372*
3-Amino-4-hydroxybutyric Acid, *455*
Boric Acid, *1336*
Chlorhexidine, *2090*
Chloroazodin, *2120*
m-Cresyl Acetate, *2587*
Cupric Sulfate, *2659*
Dibromopropamidine, *3008*
Ichthammol, *4815*
Negatol®, *6360*
Noxytiolin, *6645*
Ornidazole, *6698*
β-Propiolactone, *7832*
α-Terpineol, *9103*

ANTISPASMODIC *see also Anticholinergic*

Alibendol, *233*
Ambucetamide, *394*
Aminopromazine, *479*

Apoatropine, *770*
Bevonium Methyl Sulfate, *1214*
Bietamiverine, *1224*
Butaverine, *1510*
Butropium Bromide, *1534*
N-Butylscopolammonium Bromide, *1588*
Caroverine, *1864*
Cimetropium Bromide, *2280*
Cinnamedrine, *2299*
Clebopride, *2344*
Coniine Hydrobromide, *2501*
Coniine Hydrochloride, *2502*
Cyclonium Iodide, *2743*
Difemerine, *3122*
Diisopromine, *3180*
Dioxaphetyl Butyrate, *3295*
Diponium Bromide, *3346*
Drofenine, *3435*
Emepronium Bromide, *3515*
Ethaverine, *3682*
Feclemine, *3892*
Fenalamide, *3899*
Fenoverine, *3928*
Fenpiprane, *3934*
Fenpiverinium Bromide, *3935*
Fentonium Bromide, *3949*
Flavoxate, *4033*
Flopropione, *4039*
Gluconic Acid, *4350*
Guaiactamine, *4462*
Hydramitrazine, *4685*
Hymecromone, *4792*
Leiopyrrole, *5316*
Mebeverine, *5648*
Moxaverine, *6202*
Nafiverine, *6267*
Octamylamine, *6671*
Octaverine, *6676*
Pentapiperide, *7077*
Phenamacide Hydrochloride, *7163*
Phloroglucinol, *7301*
Pinaverium Bromide, *7410*
Piperilate, *7441*
Pipoxolan Hydrochloride, *7454*
Pramiverin, *7706*
Prifinium Bromide, *7748*
Properidine, *7826*
Propivane, *7845*
Propyromazine, *7885*
Prozapine, *7922*
Racefemine, *8115*
Rociverine, *8232*
Spasmolytol, *8693*
Stilonium Iodide, *8776*
Sultroponium, *8977*
Tiemonium Iodide, *9363*
Tiquizium Bromide, *9390*
Tiropramide, *9393*
Trepibutone, *9501*
Tricromyl, *9574*
Trifolium, *9601*
Trimebutine, *9613*
N,N-1-Trimethyl-3,3-diphenyl-propylamine, *9627*
Tropenzile, *9690*
Trospium Chloride, *9697*
Xenytropium Bromide, *9983*

ANTISYPHILITIC *see also Antibacterial*

Acetarsone, *44*
Arsacetin, *817*
Arsphenamine, *839*
Bismuth Butylthiolaurate, *1272*
Bismuth Chloride Oxide, *1274*
Bismuth Ethyl Camphorate, *1275*
Bismuth Potassium Tartrate, *1289*
Bismuth Sodium Iodide, *1291*

Bismuth Sodium Tartrate, *1292*
d-Camphocarboxylic Acid, *1737*
Dichlorophenarsine Hydrochloride, *3060*
Ethanearsonic Acid, *3677*
Mercuric Benzoate, *5768*
Mercurous Chloride, *5795*
Sodium Arsanilate, *8521*
Sodium Arsphenamine, *8524*
Sozoiodole-mercury *see 8687*
Sulfarsphenamine, *8916*

ANTITHROMBOTIC *see also Anticoagulant; Thrombolytic*

Anagrelide, *659*
Argatroban, *804*
Cilostazol, *2277*
Daltroban, *2807*
Defibrotide, *2851*
Enoxaparin, *3541*
Fraxiparine®, *4180*
Indobufen, *4867*
Lamoparan, *5227*
Ozagrel, *6935*
Picotamide, *7378*
Plafibride, *7488*
Tedelparin, *9058*
Ticlopidine, *9361*
Triflusal, *9600*

ANTITUBERCULAR *see Antibacterial (Tuberculostatic)*

ANTITUMOR *see Antineoplastic*

ANTITUSSIVE

Alloclamide, *254*
Amicibone, *407*
Benproperine, *1058*
Benzonatate, *1106*
Bibenzonium Bromide, *1218*
Bromoform, *1407*
Butamirate, *1505*
Butethamate, *1516*
Caramiphen Ethanedisulfonate, *1776*
Carbetapentane, *1798*
Chlophedianol, *2053*
Clobutinol, *2363*
Cloperastine, *2393*
Codeine, *2459*
Codeine Methyl Bromide, *2460*
Codeine N-Oxide, *2461*
Codeine Phosphate, *2462*
Codeine Sulfate, *2463*
Cyclexanone, *2712*
Dextromethorphan *see 8116*
Dibunate Sodium, *3017*
Dihydrocodeine, *3154*
Dihydrocodeinone Enol Acetate, *3155*
Dimemorfan, *3192*
Dimethoxanate, *3211*
α,α-Diphenyl-2-piperidinepropanol, *3333*
Dropropizine, *3439*
Drotebanol, *3442*
Eprazinone, *3577*
Ethyl Dibunate, *3745*
Ethylmorphine, *3785*
Fominoben, *4144*
Guaiapate, *4463*
Hydrocodone, *4708*
Isoaminile, *4992*
Levopropoxyphene, *5349*
Morclofone, *6177*
Narceine, *6341*

ANTITUSSIVE (*continued*)

Normethadone, *6628*
Noscapine, *6638*
Oxeladin, *6885*
Oxolamine, *6898*
Pholcodine, *7303*
Picoperine, *7376*
Pipazethate, *7423*
Piperidione, *7439*
Prenoxdiazine Hydrochloride, *7743*
Racemethorphan, *8116*
Taziprinone Hydrochloride, *9051*
Tipepidine, *9389*
Zipeprol, *10074*

ANTIULCERATIVE *see also* **Antacid**

Aceglutamide Aluminum Complex *see 21*
ε-Acetamidocaproic Acid Zinc Salt *see 38*
Acetoxolone, *69*
Arbaprostil, *796*
Benexate Hydrochloride, *1046*
Bismuth Subcitrate Sol (Dried), *1296*
Carbenoxolone, *1797*
Cetraxate, *2017*
Cimetidine, *2279*
Enprostil, *3545*
Esaprazole, *3642*
Famotidine, *3881*
Ftaxilide, *4190*
Gefarnate, *4273*
Guaiazulene, *4464*
Irsogladine, *4982*
Misoprostol, *6129*
Nizatidine, *6582*
Omeprazole, *6800*
Ornoprostil, *6827*
γ-Oryzanol, *6841*
Pifarnine, *7391*
Pirenzepine, *7463*
Plaunotol, *7506*
Ranitidine, *8126*
Rioprostil, *8225*
Rosaprostol, *8240*
Rotraxate, *8249*
Roxatidine Acetate, *8252*
Sofalcone, *8658*
Spizofurone, *8722*
Sucralfate, *8853*
Teprenone, *9083*
Trimoprostil, *9637*
Trithiozine, *9672*
Troxipide, *9699*
Zolimidine, *10090*

ANTIUROLITHIC

Acetohydroxamic Acid, *54*
Allopurinol, *278*
Potassium Citrate, *7603*
Succinimide, *8842*

ANTIVENIN

Lyovac® Antivenin, *5502*

ANTIVERTIGO *see* **Antiemetic**

ANTIVIRAL

Purines/Pyrimidinones
Acyclovir, *139*
Cytarabine, *2790*
Dideoxyadenosine, *3090*
Dideoxycytidine, *3091*

Dideoxyinosine, *3092*
Edoxudine, *3486*
Floxuridine, *4045*
Ganciclovir, *4262*
Idoxuridine, *4819*
Inosine Pranobex, *4881*
MADU, *5523*
Trifluridine, *9599*
Vidarabine, *9881*
Zidovudine, *10023*

Others
Acetylleucine Monoethanolamine, *89*
Amantadine, *380*
Amidinomycin, *410*
Cuminaldehyde Thiosemicarbazone, *2624*
Foscarnet Sodium, *4166*
Interferon-α, *4892*
Interferon-β, *4893*
Interferon-γ, *4894*
Kethoxal, *5178*
Lysozyme, *5514*
Methisazone, *5900*
Moroxydine, *6181*
Podophyllotoxin, *7520*
Ribavirin, *8199*
Rimantadine, *8221*
Stallimycin, *8730*
Statolon, *8760*
Tromantadine, *9683*
Xenazoic Acid, *9979*

ANXIOLYTIC

Arylpiperazines
Buspirone, *1493*
Gepirone, *4297*
Ipsapirone, *4967*

Benzodiazepine Derivatives
Alprazolam, *310*
Bromazepam, *1374*
Camazepam, *1732*
Chlordiazepoxide, *2082*
Clobazam, *2355*
Clorazepate, *2400*
Clotiazepam, *2411*
Cloxazolam, *2415*
Diazepam, *2977*
Ethyl Loflazepate, *3777*
Etizolam, *3830*
Fludiazepam, *4058*
Flutazolam, *4133*
Flutoprazepam, *4134*
Halazepam, *4502*
Ketazolam, *5176*
Lorazepam, *5456*
Loxapine, *5461*
Medazepam, *5669*
Metaclazepam, *5824*
Mexazolam, *6091*
Nordazepam, *6608*
Oxazepam, *6881*
Oxazolam, *6883*
Pinazepam, *7411*
Prazepam, *7713*
Tofisopam, *9427*

Carbamates
Cyclarbamate, *2709*
Emylcamate, *3520*
Hydroxyphenamate, *4770*
Meprobamate, *5751*
Phenprobamate, *7229*
Tybamate, *9737*

Others
Alpidem, *308*
Benzoctamine, *1096*
Captodiamine, *1772*

Chlormezanone, *2105*
Etifoxine, *3821*
Fluoresone, *4088*
Glutamic Acid, *4363*
Hydroxyzine, *4786*
Mecloralurea, *5662*
Mephenoxalone, *5739*
Oxanamide, *6874*
Phenaglycodol, *7161*
Suriclone, *8987*

ASTRINGENT

Albumin Tannate, *208*
Alkannin, *243*
Aluminum Acetate Solution, *322*
Aluminum Acetotartrate, *323*
Aluminum Ammonium Sulfate, *325*
Aluminum Chlorate, *337*
Aluminum Chloride, *338*
Aluminum Hydroxychloride, *346*
Aluminum β-Naphtholdisulfonate, *353*
Aluminum Potassium Sulfate, *364*
Aluminum Sodium Sulfate, *369*
Aluminum Subacetate Solution, *371*
Ammonium Ferric Sulfate, *543*
Baicalein, *954*
Bismuth Oxide, *1285*
Bismuth Subgallate, *1297*
Bismuth Tannate, *1302*
Boric Acid, *1336*
Calcium Hydroxide, *1676*
Cupric Citrate, *2639*
Dichloroacetic Acid, *3037*
Ferric Chloride, *3961*
Formic Acid, *4153*
Gallic Acid, *4251*
Iodic Acid, *4905*
Lead Acetate, *5268*
Methionic Acid, *5895*
Silver Bromide, *8452*
Silver Lactate, *8463*
Sodium Formate, *8568*
Tannic Acid, *9023*
Tannoform, *9024*
Zinc Acetate, *10026*
Zinc Carbonate, *10030*
Zinc Chloride, *10031*
Zinc Iodide, *10040*
Zinc Oxide, *10050*
Zinc Permanganate, *10052*
Zinc Peroxide, *10053*
Zinc p-Phenolsulfonate, *10054*
Zinc Salicylate, *10059*
Zinc Sulfate, *10064*
Zinc Tannate, *10066*

BENZODIAZEPINE ANTAGONIST

Flumazenil, *4062*

BETA-BLOCKER *see* β-*Adrenergic Blocker*

BRONCHODILATOR *see also* **Antiasthmatic; Glucocorticoid**

Ephedrine Derivatives
Albuterol, *209*
Bambuterol, *963*
Bitolterol, *1317*
Carbuterol, *1840*
Clenbuterol, *2347*
Clorprenaline, *2404*
Dioxethedrine, *3297*
Ephedrine, *3561*
Epinephrine, *3569*
Eprozinol, *3578*

Etafedrine, *3661*
Ethylnorepinephrine, *3789*
Fenoterol, *3927*
Formoterol, *4159*
Hexoprenaline, *4628*
Isoetharine, *5053*
Isoproterenol, *5105*
Mabuterol, *5517*
Metaproterenol, *5836*
N-Methylephedrine, *5987*
Pirbuterol, *7461*
Procaterol, *7765*
Protokylol, *7911*
Reproterol, *8142*
Rimiterol, *8223*
Soterenol, *8683*
Terbutaline, *9089*
Tulobuterol, *9720*

Quaternary Ammonium Compounds
Bevonium Methyl Sulfate, *1214*
Flutropium Bromide, *4136*
Ipratropium Bromide, *4960*
Oxitropium Bromide, *6897*

Xanthine Derivatives
Acefylline, *18*
Acefylline Piperazine, *19*
Ambuphylline, *395*
Aminophylline, *477*
Bamifylline, *965*
Choline Theophyllinate, *2213*
Doxofylline, *3427*
Dyphylline, *3459*
Enprofylline, *3544*
Etamiphyllin, *3663*
Etofylline, *3835*
Guaithylline, *4467*
Proxyphylline, *7921*
Theobromine, *9209*
1-Theobromineacetic Acid, *9210*
Theophylline, *9212*

Others
Fenspiride, *3942*
Medibazine, *5671*
Methoxyphenamine, *5919*
Tretoquinol, *9502*

CALCIUM CHANNEL BLOCKER see
also *Antianginal; Antihypertensive; Vaso-*
dilator (Coronary)

Arylalkylamines
Bepridil, *1167*
Diltiazem, *3188*
Fendiline, *3916*
Gallopamil, *4257*
Prenylamine, *7744*
Terodiline, *9098*
Verapamil, *9851*

Dihydropyridine Derivatives
Felodipine, *3895*
Isradipine, *5129*
Nicardipine, *6403*
Nifedipine, *6441*
Nilvadipine, *6461*
Nimodipine, *6467*
Nisoldipine, *6482*
Nitrendipine, *6493*

Piperazine Derivatives
Cinnarizine, *2308*
Flunarizine, *4070*
Lidoflazine, *5360*

Others
Bencyclane, *1041*
Etafenone, *3662*
Perhexiline, *7116*

CALCIUM REGULATOR
Calcifediol, *1638*
Calcitonin, *1640*

Calcitriol, *1641*
Clodronic Acid, *2369*
Dihydrotachysterol, *3163*
Elcatonin, *3501*
Etidronic Acid, *3819*
Ipriflavone, *4961*
Pamidronic Acid, *6954*
Parathyroid Hormone, *6984*
Teriparatide Acetate, *9096*

CALCIUM SUPPLEMENT see *Replenish-*
ers/Supplements

CANCER CHEMOTHERAPY see *Anti-*
neoplastic

CAPILLARY PROTECTANT see *Vaso-*
protectant

CARBONIC ANHYDRASE INHIBITOR
see also *Antiglaucoma; Diuretic*

Acetazolamide, *45*
Butazolamide, *1511*
Dichlorphenamide, *3067*
Diphenylmethane-4,4′-disulfon-
amide, *3330*
Ethoxzolamide, *3711*
Flumethiazide, *4067*
Methazolamide, *5875*

CARDIAC DEPRESSANT (ANTIAR-
RHYTHMIC) see *Antiarrhythmic*

CARDIOTONIC
Acefylline, *18*
Acetyldigitoxins, *83*
2-Amino-4-picoline, *478*
Amrinone, *626*
Benfurodil Hemisuccinate, *1051*
Bucladesine, *1448*
Cerberoside, *1983*
Camphotamide, *1741*
Convallatoxin, *2508*
Cymarin, *2768*
Denopamine, *2878*
Deslanoside, *2903*
Digitalin, *3140*
Digitalis, *3141*
Digitoxin, *3146*
Digoxin, *3150*
Dobutamine, *3396*
Dopamine, *3415*
Dopexamine, *3418*
Enoximone, *3542*
Erythrophleine, *3634*
Fenalcomine, *3900*
Gitalin, *4324*
Gitoxin, *4328*
Glycocyamine, *4391*
Heptaminol, *4577*
Hydrastinine, *4689*
Ibopamine, *4807*
Lanatosides, *5229*
Metamivam, *5829*
Milrinone, *6117*
Neriifolin, *6386*
Oleandrin, *6786*
Ouabain, *6854*
Oxyfedrine, *6915*
Prenalterol, *7742*
Proscillaridin, *7889*
Resibufogenin, *8150*
Scillaren, *8354*
Scillarenin, *8355*
Strophanthin, *8819*

Sulmazole, *8965*
Theobromine, *9209*
Xamoterol, *9965*

CATHARTIC see *Laxative/Cathartic*

CATION-EXCHANGE RESIN see *Ion-*
exchange Resin

CCK ANTAGONIST
Proglumide, *7785*

CENTRAL STIMULANT see *CNS Stimu-*
lant

CEREBRAL VASODILATOR see *Vasodi-*
lator (Cerebral)

CHELATING AGENT

Deferoxamine, *2850*
Ditiocarb Sodium, *3384*
Edetate Calcium Disodium, *3480*
Edetate Disodium, *3481*
Edetate Sodium, *3482*
Edetate Trisodium, *3483*
Penicillamine, *7029*
Pentetate Calcium Trisodium,
7082
Pentetic Acid, *7083*
Succimer, *8836*
Trientine, *9579*

CHOLECYSTOKININ ANTAGONIST see
CCK Antagonist

CHOLELITHOLYTIC AGENT

Chenodiol, *2044*
Methyl *tert*-Butyl Ether, *5954*
Monooctanoin, *6162*
Ursodiol, *9801*

CHOLERETIC

Alibendol, *233*
Anethole Trithion, *676*
Azintamide, *927*
Cholic Acid, *2206*
Cicrotoic Acid, *2272*
Clanobutin, *2338*
Cyclobutyrol, *2721*
Cyclovalone, *2761*
Cynarin(e), *2773*
Dehydrocholic Acid, *2858*
Deoxycholic Acid, *2831*
Dimecrotic Acid, *3189*
α-Ethylbenzyl Alcohol, *3727*
Exiproben, *3869*
Feguprol, *3891*
Fencibutirol, *3912*
Fenipentol, *3921*
Florantyrone, *4040*
Hymecromone, *4792*
Menbutone, *5719*
3-(*o*-Methoxyphenyl)-2-phenyl-
acrylic Acid, *5921*
Metochalcone, *6062*
Moquizone, *6174*
Osalmid, *6843*
Ox Bile Extract, *6884*
4,4′-Oxydi-2-butanol, *6913*
Piprozolin, *7457*
Prozapine, *7922*

CHOLERETIC (continued)

4-Salicyloylmorpholine, 8302
Sincalide, 8494
Taurocholic Acid, 9044
Timonacic, 9375
Tocamphyl, 9415
Trepibutone, 9501
Vanitiolide, 9841

CHOLINERGIC

Aceclidine see 8110
Acetylcholine Bromide, 80
Acetylcholine Chloride, 81
Aclatonium Napadisilate, 109
Benzpyrinium Bromide, 1132
Bethanechol Chloride, 1207
Carbachol, 1780
Carpronium Chloride, 1871
Demecarium Bromide, 2871
Dexpanthenol, 2924
Diisopropyl Paraoxon, 3183
Echothiophate Iodide, 3475
Edrophonium Chloride, 3487
Eseridine, 3647
Furtrethonium, 4225
Isoflurophate, 5060
Methacholine Chloride, 5847
Muscarine, 6219
Neostigmine, 6380
Oxapropanium Iodide, 6877
Physostigmine, 7357
Pyridostigmine Bromide, 7990

CHOLINESTERASE INHIBITOR

Ambenonium Chloride, 387
Distigmine Bromide, 3368
Galanthamine, 4245

CHOLINESTERASE REACTIVATOR

Obidoxime Chloride, 6659
Pralidoxime Chloride, 7705

CNS STIMULANT

Amineptine, 420
Amphetamine, 616
Amphetaminil, 617
Bemegride, 1036
Benzphetamine, 1130
Brucine, 1443
Caffeine, 1635
Chlorphentermine, 2183
Clofenciclan, 2372
Clortermine, 2406
Coca, 2448
Demanyl Phosphate, 2870
Dexoxadrol see 3293
Dextroamphetamine Sulfate, 2932
Diethylpropion, 3116
N-Ethylamphetamine, 3720
Ethamivan, 3673
Etifelmin, 3820
Etryptamine, 3848
Fencamfamine, 3909
Fenethylline, 3919
Fenozolone, 3932
Flurothyl, 4126
Hexacyclonate Sodium, 4603
Homocamfin, 4651
Mazindol, 5643
Mefexamide, 5682
Methamphetamine, 5859
Methylphenidate, 6025
Nikethamide, 6459
Pemoline, 7021
Pentylenetetrazole, 7097
Phendimetrazine, 7180

Phenmetrazine, 7200
Phentermine, 7232
Picrotoxin, 7388
Pipradrol, 7455
Prolintane, 7791
Pyrovalerone, 8021

COGNITION ACTIVATOR see Nootropic

CONTRACEPTIVE (INJECTABLE)

Medroxyprogesterone, 5677
Norethindrone, 6614

CONTRACEPTIVE (ORAL)

Desogestrel, 2906
Ethinyl Estradiol, 3689
Ethynodiol, 3816
Gestodene, 4308
Lynestrenol, 5501
Mestranol, 5819
Norethindrone, 6614
Norethynodrel, 6615
Norgestimate, 6620
Norgestrel, 6621

CONTROL OF INTRAOCULAR PRESSURE see also Antiglaucoma

Apraclonidine, 779

CONVERTING ENZYME INHIBITOR see ACE-Inhibitor

CORONARY VASODILATOR see Vasodilator (Coronary)

CYTOPROTECTANT (GASTRIC) see also Antiulcerative

Aceglutamide Aluminum Complex see 21
Acetoxolone, 69
Benexate Hydrochloride, 1046
Carbenoxolone, 1797
Cetraxate, 2017
Ftaxilide, 4190
Guaiazulene, 4464
Irsogladine, 4982
Plaunotol, 7506
Sofalcone, 8658
Spizofurone, 8722
Sucralfate, 8853
Teprenone, 9083
Troxipide, 9699
Zolimidine, 10090

DEBRIDING AGENT

Collagenase, 2477
Deoxyribonuclease (Pancreatic), 2888
Papain, 6965

DECONGESTANT

Amidephrine, 409
Cafaminol, 1632
Cyclopentamine, 2745
Ephedrine, 3561
Epinephrine, 3569
Fenoxazoline, 3930
Indanazoline, 4845
Metizoline, 6060
Naphazoline, 6287
Nordefrin Hydrochloride, 6609

Octodrine, 6678
Oxymetazoline, 6919
Phenylephrine Hydrochloride, 7257
Phenylpropanolamine Hydrochloride, 7279
Phenylpropylmethylamine, 7280
Propylhexedrine, 7873
Pseudoephedrine see 3561
Tetrahydrozoline, 9150
Tymazoline, 9742
Xylometazoline, 9993

DEPIGMENTOR

Hydroquinine, 4737
Hydroquinone, 4738
Monobenzone, 6159

DERMATITIS HERPETIFORMIS SUPPRESSANT

Dapsone, 2820
Sulfapyridine, 8913

DIAGNOSTIC AID

Alsactide, 312
Americium, 401
p-Aminohippuric Acid, 454
Anazolene Sodium, 662
Arginine, 805
Bentiromide, 1061
Betazole, 1205
Ceruletide, 1998
Congo Red, 2493
Dexamethasone, 2922
Edrophonium Chloride, 3487
Evan's Blue, 3863
Fluorescein, 4085
Galactose, 4241
Glycerol, 4379
Histamine, 4640
Indocyanine Green, 4868
Inulin, 4898
Iodinated Serum Albumin see 8417
Isosulphan Blue see 8968
Mannitol, 5629
Merisoprol Hg 197, 5803
Methacholine Chloride, 5847
Metyrapone, 6083
Oleic Acid, 6788
Penicilloyl Polylysine, 7050
3-Pentadecylcatechol, 7061
Pentagastrin, 7067
Phenolsulfonphthalein, 7213
Phenoltetrachlorophthalein, 7214
Phentolamine, 7234
Piperoxan, 7448
Rose Bengal, 8242
Saralasin, 8330
Sodium Benzoate, 8527
Sodium Chromate(VI), Radioactive, 8548
Sodium Iodide, Radioactive, 8583
Sulfobromophthalein Sodium, 8933
Teriparatide Acetate, 9096
Tolonium Chloride, 9444
TSH, 9709
Tuberculin, 9715
Tubocurarine Chloride, 9717
Vitamin B_{12}, Radioactive, 9922
Xylose, 9995

DIAGNOSTIC AID (NMR CONTRAST MEDIUM)

Gadopentetic Acid, 4236

DIAGNOSTIC AID (RADIOACTIVE IMAGING AGENT)

Butedronic Acid Complex with
  ^{99m}Tc see 1512
Iofetamine ^{123}I, 4938
Pamidronic Acid Complex with
  ^{99m}Tc see 6954
Sodium Pertechnetate ^{99m}Tc,
  8609
Sodium Phosphate, Radioactive,
  8614
Stannous Pyrophosphate Complex
  with ^{99m}Tc see 8748
Succimer Complex with ^{99m}Tc see
  8836
Technetium, 9054
^{133}Xenon see 9981

DIAGNOSTIC AID (RADIOPAQUE MEDIUM)

Acetrizoate Sodium, 72
Barium Sulfate, 1006
Bunamiodyl Sodium, 1476
Cinamiodyl, 2284
Diatrizoate Sodium, 2975
Dimethiodal Sodium, 3205
Diprotrizoate Sodium, 3353
Ethiodized Oil, 3690
Iobenzamic Acid, 4901
Iocarmic Acid, 4902
Iocetamic Acid, 4903
Iodipamide, 4916
Iodized Oil, 4917
Iodoalphionic Acid, 4919
Iodophthalein Sodium, 4931
Iodopyracet, 4933
Ioglycamic Acid, 4939
Iohexol, 4940
Iomeglamic Acid, 4941
Iopamidol, 4943
Iopanoic Acid, 4944
Iopentol, 4945
Iophendylate, 4946
Iophenoxic Acid, 4947
Iopromide, 4948
Iopronic Acid, 4949
Iopydol, 4950
Iopydone, 4951
Iothalamic Acid, 4952
Iotrolan, 4955
Ioxaglic Acid, 4956
Ipodate, 4958
Meglumine Acetrizoate, 5688
Meglumine Diatrizoate, 5689
Methiodal Sodium, 5894
Metrizamide, 6077
Metrizoic Acid, 6078
Monophen®, 6163
Phenobutiodil, 7203
Phentetiothalein Sodium, 7233
Propyl Docetrizoate, 7861
Propyliodone, 7876
Sodium Iodomethamate, 8584
Sozoiodolic Acid, 8687
3′,3″,5′,5″-Tetrabromophenol-
  phthalein, 9122
Thorium Oxide, 9312
Tyropanoate Sodium, 9745

DIGESTIVE AID

α-Amylase (Swine Pancreas), 633
Lipase, 5389
Pancreatin, 6956
Pancrelipase, 6957
Papain, 6965
Pepsin, 7103
Rennin, 8139
Sulpiride, 8971

DISINFECTANT see Antiseptic/Disinfectant

DIURETIC

Benzothiadiazine Derivatives
Althiazide, 316
Bendroflumethiazide, 1045
Benzthiazide, 1134
Benzylhydrochlorothiazide, 1150
Buthiazide, 1519
Chlorothiazide, 2169
Chlorthalidone, 2194
Cyclopenthiazide, 2750
Cyclothiazide, 2760
Epithiazide, 3574
Ethiazide, 3687
Fenquizone, 3941
Hydrochlorothiazide, 4704
Hydroflumethiazide, 4716
Methyclothiazide, 5929
Meticrane, 6058
Metolazone, 6068
Paraflutizide, 6973
Polythiazide, 7561
Tetrachlormethiazide, 9124
Trichlormethiazide, 9537

Organomercurials
Chlormerodrin, 2104
Meralluride, 5756
Mercamphamide, 5758
Mercaptomerin Sodium, 5761
Mercumallylic Acid, 5764
Mercumatilin Sodium, 5765
Mercurous Chloride, 5795
Mersalyl, 5805

Pteridines
Furterene, 4224
Triamterene, 9515

Purines
Acefylline, 18
7-Morpholinomethyltheophylline,
  6196
Pamabrom, 6952
Protheobromine, 7901
Theobromine, 9209

Steroids
Canrenone, 1753
Oleandrin, 6786
Spironolactone, 8721

Sulfonamide Derivatives
Acetazolamide, 45
Ambuside, 396
Azosemide, 935
Bumetanide, 1473
Butazolamide, 1511
Chloraminophenamide, 2067
Clofenamide, 2371
Clopamide, 2391
Clorexolone, 2401
Diphenylmethane-4,4′-disulfonam-
  ide, 3330
Disulfamide, 3369
Ethoxzolamide, 3711
Furosemide, 4221
Indapamide, 4847
Mefruside, 5685
Methazolamide, 5875
Piretanide, 7464
Quinethazone, 8067
Torasemide, 9473
Tripamide, 9649
Xipamide, 9986

Uracils
Aminometradine, 463
Amisometradine, 500

Others
Amanozine, 379
Amiloride, 417

Arbutin, 799
Chlorazanil, 2073
Ethacrynic Acid, 3669
Etozolin, 3845
Hydracarbazine, 4680
Isosorbide, 5113
Mannitol, 5629
Metochalcone, 6062
Muzolimine, 6227
Perhexiline, 7116
Ticrynafen, 9362
Urea, 9781

DOPAMINE RECEPTOR AGONIST see also Antihypertensive; Antimigraine; Antiparkinsonian

Bromocriptine, 1404
Dopexamine, 3418
Fenoldopam, 3925
Ibopamine, 4807
Lisuride, 5394
Naxagolide, 6349
Pergolide, 7115

DOPAMINE RECEPTOR ANTAGONIST see also Antiemetic; Antipsychotic

Amisulpride, 501
Clebopride, 2344
Domperidone, 3412
Metoclopramide, 6063
Sulpiride, 8971

ECTOPARASITICIDE

Amitraz, 503
Benzyl Benzoate, 1141
Carbaryl, 1789
Crotamiton, 2597
DDT, 2832
Dixanthogen, 3390
Isobornyl Thiocyanoacetate,
  Technical, 5012
Lime Sulfurated Solution, 5369
Lindane, 5379
Malathion, 5582
Mercuric Oleate, 5778
Mesulphen, 5821
Sulfur, Pharmaceutical, 8956

ELECTROLYTE REPLENISHER see Replenishers/Supplements

EMETIC

Apocodeine, 771
Apomorphine, 776
Cephaeline, 1970
Ipecac, 4957
Sodium Chloride, 8544
Zinc Acetate, 10026

ENZYME

Digestive
α-Amylase (Swine Pancreas), 633
Lipase, 5389
Pancrelipase, 6957
Pepsin, 7103
Rennin, 8139

Mucolytic
Lysozyme, 5514

Penicillin Inactivating
Penicillinase, 7034

Proteolytic
Collagenase, 2477
Chymopapain, 2264

ENZYME

Proteolytic (*continued*)
Chymotrypsins, 2265
Papain, 6965
Trypsin, 9704

ENZYME INDUCER (HEPATIC)
Flumecinol, 4063

ESTROGEN

Nonsteroidal
Benzestrol, 1082
Broparoestrol, 1438
Chlorotrianisene, 2173
Dienestrol, 3094
Diethylstilbestrol, 3118
Diethylstilbestrol Dipropionate, 3119
Dimestrol, 3198
Fosfestrol, 4168
Hexestrol, 4621
Methallenestril, 5856
Methestrol, 5888

Steroidal
Colpormon, 2485
Conjugated Estrogenic Hormones, 2504
Equilenin, 3581
Equilin, 3582
Estradiol, 3653
Estradiol Benzoate, 3655
Estradiol 17β-Cypionate, 3656
Estriol, 3659
Estrone, 3660
Ethinyl Estradiol, 3689
Mestranol, 5819
Moxestrol, 6203
Mytatrienediol, 6254
Quinestradiol, 8065
Quinestrol, 8066

ESTROGEN ANTAGONIST *see Antiestrogen*

EXPECTORANT
Ambroxol, 392
Ammonium Bicarbonate, 516
Ammonium Carbonate, 528
Bromhexine, 1379
Calcium Iodide, 1680
Carbocysteine, 1809
Guaiacol, 4457
Guaiacol Benzoate, 4458
Guaiacol Carbonate, 4459
Guaiacol Phosphate, 4460
Guaifenesin, 4465
Guaithylline, 4467
Hydriodic Acid, 4699
Iodinated Glycerol, 4906
Potassium Guaiacolsulfonate, 7617
Potassium Iodide, 7628
Sodium Citrate, 8549
Sodium Iodide, 8582
Storax, 8778
Terebene, 9091
Terpin, 9101
Trifolium, 9601

GASTRIC AND PANCREATIC SECRETION STIMULANT
Carnitine, 1856
Ceruletide, 1998
Secretin, 8375
Sincalide, 8494

GASTRIC PROTON PUMP INHIBITOR
*see also **Antiulcerative***
Omeprazole, 6800

GASTRIC SECRETION INHIBITOR
Enterogastrone, 3548
Octreotide, 6682

GLUCOCORTICOID
21-Acetoxypregnenolone, 70
Alclometasone, 213
Algestone, 229
Amcinonide, 398
Beclomethasone, 1029
Betamethasone, 1202
Budesonide, 1455
Chloroprednisone, 2157
Clobetasol, 2361
Clobetasone, 2362
Clocortolone, 2368
Cloprednol, 2396
Corticosterone, 2532
Cortisone, 2533
Cortivazol, 2536
Deflazacort, 2852
Desonide, 2908
Desoximetasone, 2910
Dexamethasone, 2922
Diflorasone, 3126
Diflucortolone, 3129
Difluprednate, 3134
Enoxolone, 3543
Fluazacort, 4048
Flucloronide, 4053
Flumethasone, 4066
Flunisolide, 4071
Fluocinolone Acetonide, 4076
Fluocinonide, 4077
Fluocortin Butyl, 4078
Fluocortolone, 4079
Fluorometholone, 4104
Fluperolone Acetate, 4115
Fluprednidene Acetate, 4118
Fluprednisolone, 4119
Flurandrenolide, 4122
Formocortal, 4156
Halcinonide, 4504
Halometasone, 4510
Halopredone Acetate, 4512
Hydrocortamate, 4709
Hydrocortisone, 4710
Hydrocortisone Acetate, 4711
Hydrocortisone Phosphate, 4712
Hydrocortisone 21-Sodium Succinate, 4713
Hydrocortisone Tebutate, 4714
Mazipredone, 5644
Medrysone, 5679
Meprednisone, 5750
Methylprednisolone, 6028
Mometasone Furoate, 6151
Paramethasone, 6977
Prednicarbate, 7717
Prednisolone, 7719
Prednisolone 21-Diethylaminoacetate, 7720
Prednisolone Sodium Phosphate, 7721
Prednisolone Sodium Succinate, 7722
Prednisolone Sodium 21-*m*-Sulfobenzoate, 7723
Prednisolone 21-Stearoylglycolate, 7724
Prednisolone Tebutate, 7725
Prednisolone 21-Trimethylacetate, 7726
Prednisone, 7727
Prednival, 7728

Prednylidene, 7729
Prednylidene 21-Diethylaminoacetate, 7730
Tixocortol, 9408
Triamcinolone, 9511
Triamcinolone Acetonide, 9512
Triamcinolone Benetonide, 9513
Triamcinolone Hexacetonide, 9514

α-GLUCOSIDASE INHIBITOR *see also Antidiabetic*
Acarbose, 12
Miglitol, 6110

GONAD-STIMULATING PRINCIPLE
Buserelin, 1492
Clomiphene, 2384
Cyclofenil, 2726
Epimestrol, 3568
FSH, 4189
HCG, 4534
LH-RH, 5354

GONADOTROPIC HORMONE
LH, 5353
PMSG, 7515

GOUT SUPPRESSANT *see Antigout*

GROWTH HORMONE INHIBITOR
Octreotide, 6682
Somatostatin, 8671

GROWTH HORMONE RELEASING FACTOR
Sermorelin, 8412

GROWTH STIMULANT
Somatotropin, 8672

HEMATINIC
Ammonium Ferric Citrate, 541
Calcium Ferrous Citrate, 1668
Cobaltous Chloride, 2431
Dextran Iron Complex, 2927
Erythropoietin, 3635
Ferric Albuminate, 3958
Ferric and Ammonium Acetate Solution, 3959
Ferric Citrate, 3963
Ferric Fructose, 3967
Ferriclate Calcium Sodium, 3971
Ferric Oxide, Saccharated, 3974
Ferric Pyrophosphate, 3976
Ferric Sodium Edetate, 3977
Ferritin, 3984
Ferrocholinate, 3986
Ferroglycine Sulfate, 3987
Ferrous Carbonate Mass, 3990
Ferrous Carbonate Saccharated, 3991
Ferrous Citrate, 3993
Ferrous Fumarate, 3995
Ferrous Gluconate, 3996
Ferrous Lactate, 3999
Ferrous Succinate, 4005
Ferrous Sulfate, 4006
Iron Sorbitex, 4981
Liver Extract, 5427
Peptonized Iron, 7106
Polyferose, 7547

HEMOLYTIC

Phenylhydrazine, 7264
Phenylhydrazine Hydrochloride, 7265

HEMOSTATIC

Adrenalone, 161
Adrenochrome, 162
Algin, 231
Alginic Acid, 232
ε-Aminocaproic Acid, 442
Aminochromes, 447
Batroxobin, 1020
Carbazochrome Salicylate, 1790
Carbazochrome Sodium Sulfonate, 1791
Cephalins, 1972
Cotarnine, 2551
Ellagic Acid, 3508
Ethamsylate, 3675
Factor VIII, 3873
Factor IX, 3874
Factor XIII, 3878
Fibrinogen, 4015
1,2-Naphthoquinone, 6314
1-Naphthylamine-4-sulfonic Acid, 6323
Oxamarin, 6868
Oxidized Cellulose, 6894
Styptic Collodion see 2480
Sulmarin, 8964
Thrombin, 9319
Thromboplastin, 9321
Tolonium Chloride, 9444
Tranexamic Acid, 9487
Vasopressin, 9843
Vitamin(s) K$_2$, 9935
Vitamin K$_5$, 9936
Vitamin K-S(II), 9939

HEPARIN ANTAGONIST

Hexadimethrine Bromide, 4605
Protamines, 7898

HEPATOPROTECTANT

S-Adenosylmethionine, 146
Betaine, 1201
Catechin, 1908
Citiolone, 2322
Malotilate, 5593
Orazamide, 6817
Phosphorylcholine, 7337
Protoporphyrin IX, 7913
Silymarin-Group, 8484
Thioctic Acid, 9255
Tiopronin, 9386

HISTAMINE H₁-RECEPTOR ANTAGONIST see Antihistaminic

HISTAMINE H₂-RECEPTOR ANTAGONIST see also Antiulcerative

Cimetidine, 2279
Famotidine, 3881
Nizatidine, 6582
Ranitidine, 8126
Roxatidine Acetate, 8252

HMG CoA REDUCTASE INHIBITOR see also Antihyperlipoproteinemic

Lovastatin, 5460
Pravastatin Sodium, 7712
Simvastatin, 8491

HYPNOTIC see Sedative/Hypnotic

HYPOCHOLESTEREMIC see Antihyperlipoproteinemic

HYPOLIPIDEMIC see Antihyperlipoproteinemic

HYPOTENSIVE see Antihypertensive

IMMUNOMODULATOR

Amiprilose, 499
Bucillamine, 1447
Ditiocarb Sodium, 3384
Inosine Pranobex, 4881
Interferon-γ, 4894
Interleukin-2, 4896
Lentinan, 5322
Muroctasin, 6217
Platonin, 7504
Procodazole, 7769
Tetramisole, 9161
Thymomodulin, 9337
Thymopentin, 9338
Ubenimex, 9750

IMMUNOSUPPRESSANT

Azathioprine, 918
Cyclosporins, 2759
Mizoribine, 6137

INOTROPIC AGENT see Cardiotonic

ION EXCHANGE RESIN

Carbacrylic Resins, 1781
Cholestyramine Resin, 2205
Colestipol, 2472
Polidexide, 7531
Resodec, 8156
Sodium Polystyrene Sulfonate, 8622

KERATOLYTIC see also Antiacne

Benzoyl Peroxide, 1128
Dichloroacetic Acid, 3037
Resorcinol, 8158
Retinoic Acid, 8167
Salicylic Acid, 8301
Tetroquinone, 9177

LACTATION STIMULATING HORMONE

Prolactin, 7788

LAXATIVE/CATHARTIC

Agar, 172
Aloe-emodin, 303
Aloin, 304
Bisacodyl, 1253
Bisoxatin Acetate, 1310
Calcium Polycarbophil, 1704
Calomelol, 1727
Carlsbad Salt Artificial, 1849
Casanthranol, 1887
Castor Oil, 1904
Cellulose Ethyl Hydroxyethyl Ether, 1963
Colocynthin, 2483
Danthron, 2813

Docusate Calcium, 3397
Docusate Sodium, 3398
Emodin, 3518
Frangulin, 4177
Glucofrangulin, 4346
Lactulose, 5225
Magnesium Carbonate Hydroxide, 5538
Magnesium Chloride, 5540
Magnesium Citrate, 5541
Magnesium Hydroxide, 5548
Magnesium Lactate, 5550
Magnesium Phosphate, Dibasic, 5560
Magnesium Sulfate, 5573
Mercurous Chloride, 5795
Mercury Mass, 5802
Oxyphenisatin Acetate, 6927
Petrolatum, Liquid, 7139
Phenolphthalein, 7208
Phenolphthalol, 7211
Phenoltetrachlorophthalein, 7214
Picosulfate Sodium, 7377
Poloxamers, 7537
Potassium Bisulfate, 7591
Potassium Bitartrate, 7593
Potassium Phosphate, Dibasic, 7647
Potassium Sodium Tartrate, 7660
Potassium Sulfate, 7665
Potassium Sulfite, 7667
Potassium Tartrate, 7669
Seidlitz Mixture, 8378
Senna, 8406
Sennoside A & B, 8407
Sodium Phosphate, Dibasic, 8612
Sodium Succinate, 8635
Sodium Sulfate, 8636
Sodium Tartrate, 8640
Sulisatin, 8962
Triacetyldiphenolisatin, 9505
Yellow Phenolphthalein, 10005

LH-RH AGONIST see also Antineoplastic; Gonad-Stimulating Principle

Buserelin, 1492
Goserelin, 4433
Leuprolide, 5341
Nafarelin, 6265
Triptorelin, 9662

LIPOTROPIC

N-Acetylmethionine, 90
Choline Chloride, 2208
Choline Dehydrocholate, 2209
Choline Dihydrogen Citrate, 2210
Inositol, 4883
Lecithin, 5311
Methionine, 5896

LOCAL ANESTHETIC see Anesthetic (Local)

LUPUS ERYTHEMATOSUS SUPPRESSANT

Bismuth Sodium Triglycollamate, 1293
Bismuth Subsalicylate, 1299
Chloroquine, 2163
Hydroxychloroquine, 4748

MAJOR TRANQUILIZER see Antipsychotic

MINERALOCORTICOID

Aldosterone, *218*
Deoxycorticosterone, *2882*
Deoxycorticosterone Acetate, *2883*
Fludrocortisone, *4059*

MINOR TRANQUILIZER *see Anxiolytic*

MIOTIC

Carbachol, *1780*
Physostigmine, *7357*
Pilocarpine, *7395*
Pilocarpus, *7396*

MONOAMINE OXIDASE INHIBITOR
see also Antidepressant; Antihypertensive

Deprenyl, *2893*
Iproclozide, *4963*
Iproniazid, *4965*
Isocarboxazid, *5040*
Moclobemide, *6139*
Octamoxin, *6670*
Pargyline, *6988*
Phenelzine, *7181*
Phenoxypropazine, *7227*
Pivalylbenzhydrazine, *7483*
Prodipine, *7776*
Toloxatone, *9445*
Tranylcypromine, *9491*

MUCOLYTIC

Acetylcysteine, *82*
Bromhexine, *1379*
Carbocysteine, *1809*
Domiodol, *3410*
Letosteine, *5330*
Lysozyme, *5514*
Mecysteine Hydrochloride, *5668*
Mesna, *5812*
Sobrerol, *8510*
Stepronin, *8764*
Tiopronin, *9386*
Tyloxapol, *9741*

MUSCLE RELAXANT (SKELETAL)

Afloqualone, *171*
Alcuronium, *214*
Atracurium Besylate, *884*
Baclofen, *950*
Benzoctamine, *1096*
Benzoquinonium Chloride, *1116*
C-Calebassine, *1722*
Carisoprodol, *1848*
Chlormezanone, *2105*
Chlorphenesin Carbamate, *2179*
Chlorproethazine, *2184*
Chlorzoxazone, *2197*
Curare, *2678*
Cyclarbamate, *2709*
Cyclobenzaprine, *2719*
Dantrolene, *2814*
Decamethonium Bromide, *2840*
Diazepam, *2977*
Eperisone, *3558*
Fazadinium Bromide, *3887*
Flumetramide, *4069*
Gallamine Triethiodide, *4249*
Hexacarbacholine Bromide, *4599*
Hexafluorenium Bromide, *4606*
Idrocilamide, *4820*
Laudexium Methyl Sulfate, *5250*
Leptodactyline, *5326*
Memantine, *5709*
Mephenesin, *5737*

Mephenoxalone, *5739*
Metaxalone, *5838*
Methocarbamol, *5903*
Metocurine Iodide, *6064*
Nimetazepam, *6465*
Orphenadrine, *6831*
Pancuronium Bromide, *6958*
Phenprobamate, *7229*
Phenyramidol, *7292*
Pipecurium Bromide, *7426*
Promoxolane, *7801*
Quinine Sulfate, *8089*
Styramate, *8829*
Succinylcholine Bromide, *8845*
Succinylcholine Chloride, *8846*
Succinylcholine Iodide, *8847*
Suxethonium Bromide, *8989*
Tetrazepam, *9172*
Thiocolchicoside *see 9253*
Tizanidine, *9409*
Tolperisone, *9447*
Tubocurarine Chloride, *9717*
Vecuronium Bromide, *9845*
Zoxazolamine, *10098*

MUSCLE RELAXANT (SMOOTH) *see Anticholinergic; Antispasmodic; Bronchodilator; Vasodilator*

MYDRIATIC

Atropine, *891*
Cyclopentolate, *2752*
Epinephrine, *3569*
Hydroxyamphetamine, *4740*
Phenylephrine Hydrochloride, *7257*
Yohimbine, *10011*

NARCOTIC ANALGESIC *see Analgesic (Narcotic)*

NARCOTIC ANTAGONIST

Amiphenazole, *498*
Cyclazocine, *2710*
Levallorphan, *5342*
Nadide, *6259*
Nalmefene, *6274*
Nalorphine, *6275*
Nalorphine Dinicotinate, *6276*
Naloxone, *6277*
Naltrexone, *6278*

NASAL DECONGESTANT *see Decongestant*

NEUROLEPTIC *see Antipsychotic*

NEUROMUSCULAR BLOCKING AGENT *see Muscle Relaxant (Skeletal)*

NEUROPROTECTIVE

Dizocilpine, *3392*

NMDA RECEPTOR ANTAGONIST *see also Neuroprotective*

Dizocilpine, *3392*

NOOTROPIC

Aceglutamide, *21*
Acetylcarnitine, *78*

Aniracetam, *692*
Bifemelane, *1227*
Exifone, *3868*
Fipexide, *4024*
Idebenone, *4817*
Indeloxazine Hydrochloride, *4850*
Nizofenone, *6583*
Oxiracetam, *6896*
Piracetam, *7459*
Propentofylline, *7823*
Pyritinol, *8006*
Tacrine, *9003*

NSAID *see Anti-inflammatory (Nonsteroidal)*

OPIOID ANALGESIC *see Analgesic (Narcotic)*

ORAL CONTRACEPTIVE *see Contraceptive (Oral)*

OVARIAN HORMONE

Relaxin, *8137*

OXYTOCIC

Carboprost, *1829*
Cargutocin, *1846*
Deaminooxytocin, *2833*
Ergonovine, *3600*
Gemeprost, *4279*
Methylergonovine, *5989*
Oxytocin, *6934*
Pituitary, Posterior, *7481*
Prostaglandin E$_2$, *7893*
Prostaglandin F$_{2\alpha}$, *7894*
Sparteine, *8692*

PARASYMPATHOMIMETIC *see Cholinergic*

PEDICULICIDE *see Ectoparasiticide*

PEPSIN INHIBITOR

Sodium Amylosulfate, *8520*

PERIPHERAL VASODILATOR *see Vasodilator (Peripheral)*

PERISTALTIC STIMULANT

Cisapride, *2318*

PIGMENTATION AGENT

Methoxsalen, *5911*
Trioxsalen, *9647*

PLASMA VOLUME EXPANDER

Serum Albumin, *8417*
Dextran, *2925*
Hetastarch, *4593*
Oxypolygelatin, *6930*

POTASSIUM CHANNEL ACTIVATOR/ OPENER *see also Antihypertensive*

Nicorandil, *6431*
Pinacidil, *7407*

PRESSOR AGENT see *Antihypotensive*

PROGESTOGEN

Allylestrenol, *289*
Anagestone, *658*
Chlormadinone Acetate, *2102*
Delmadinone Acetate, *2865*
Demegestone, *2874*
Desogestrel, *2906*
Dimethisterone, *3208*
Dydrogesterone, *3454*
Ethisterone, *3696*
Ethynodiol, *3816*
Flurogestone Acetate, *4125*
Gestodene, *4308*
Gestonorone Caproate, *4309*
Haloprogesterone, *4513*
17-Hydroxy-16-methylene-
Δ⁶-progesterone, *4763*
17α-Hydroxyprogesterone, *4773*
17α-Hydroxyprogesterone
Caproate, *4774*
Lynestrenol, *5501*
Medrogestone, *5676*
Medroxyprogesterone, *5677*
Megestrol Acetate, *5687*
Melengestrol, *5697*
Norethindrone, *6614*
Norethynodrel, *6615*
Norgesterone, *6619*
Norgestimate, *6620*
Norgestrel, *6621*
Norgestrienone, *6622*
Norvinisterone, *6637*
Pentagestrone, *7068*
Progesterone, *7783*
Promegestone, *7796*
Quingestrone, *8069*
Trengestone, *9500*

PROLACTIN INHIBITOR

Bromocriptine, *1404*
Lisuride, *5394*
Metergoline, *5843*
Terguride, *9095*

**PROSTAGLANDIN/PROSTAGLANDIN
ANALOG** see also *Abortifacient; Anti-
ulcerative; Oxytocic*

Arbaprostil, *796*
Carboprost, *1829*
Enprostil, *3545*
Gemeprost, *4279*
Limaprost, *5367*
Misoprostol, *6129*
Ornoprostil, *6827*
Prostacyclin, *7890*
Prostaglandin E₁, *7892*
Prostaglandin E₂, *7893*
Prostaglandin F₂α, *7894*
Rioprostil, *8225*
Rosaprostol, *8240*
Sulprostone, *8973*
Trimoprostil, *9637*

PROTEASE INHIBITOR

Aprotinin, *784*
Camostat, *1734*
Gabexate, *4234*
Nafamostat, *6264*

PROTON PUMP INHIBITOR see *Gastric
Proton Pump Inhibitor*

5α-REDUCTASE INHIBITOR see also
Antiprostatic Hypertrophy

Proscar®, *7888*

REPLENISHERS/SUPPLEMENTS

Calcium
Calcium Carbonate, *1657*
Calcium Gluconate, *1672*
Calcium Hypophosphite, *1678*
Calcium Lactate, *1683*
Calcium Levulinate, *1684*
Calcium Phosphate, Dibasic,
1699
Calcium Phosphate, Tribasic,
1701

Electrolyte
Calcium Chloride, *1659*
Potassium Chloride, *7601*
Potassium Gluconate, *7615*
Sodium Chloride, *8544*
Sodium Lactate, *8586*

Fluid and Nutrient
Protein Hydrolysates, *7900*

Iodine
Betasine, *1203*
Calcium Iodostearate, *1682*
Methenamine Tetraiodine, *5886*
Periodyl, *7124*
Potassium Iodide, *7628*
Prolonium Iodide, *7792*
Rubidium Iodide, *8263*
Sodium Iodide, *8582*
Strontium Iodide, *8808*

Magnesium
Magnesium Gluconate see *4350*

Phosphorus
Calcium Glycerophosphate, *1673*
Durapatite, *3449*

Potassium
Potassium Bicarbonate, *7586*

Zinc
Zinc Sulfate, *10064*

RESPIRATORY STIMULANT

Almitrine, *299*
Bemegride, *1036*
Carbon Dioxide, *1816*
Cropropamide, *2595*
Crotethamide, *2598*
Dimefline, *3190*
Dimorpholamine, *3257*
Doxapram, *3421*
Ethamivan, *3673*
Fominoben, *4144*
Lobeline, *5432*
Mepixanox, *5749*
Metamivam, *5829*
Nikethamide, *6459*
Picrotoxin, *7388*
Pimeclone, *7399*
Pyridofylline, *7988*
Sodium Succinate, *8635*
Tacrine, *9003*

**REVERSE TRANSCRIPTASE INHIBI-
TOR** see also *Antiviral*

Dideoxyadenosine, *3090*
Dideoxycytidine, *3091*
Dideoxyinosine, *3092*
Foscarnet Sodium, *4166*
Suramin Sodium, *8986*
Zidovudine, *10023*

SCABICIDE see *Ectoparasiticide*

SCLEROSING AGENT

Ethanolamine, *3681*
Ethylamine, *3718*

2-Hexyldecanoic Acid, *4630*
Polidocanol, *7532*
Quinine Bisulfate, *8076*
Quinine Urea Hydrochloride,
8091
Sodium Ricinoleate see *8213*
Sodium Tetradecyl Sulfate, *8645*
Tribenoside, *9520*

SEDATIVE/HYPNOTIC see also *Anxiolytic*

Acylic Ureides
Acecarbromal, *15*
Apronalide, *783*
Bromisovalum, *1385*
Capuride, *1774*
Carbromal, *1837*
Ectylurea, *3478*

Alcohols
Chlorhexadol, *2089*
Ethchlorvynol, *3683*
Meparfynol, *5731*
4-Methyl-5-thiazoleethanol,
6046
tert-Pentyl Alcohol, *7096*
2,2,2-Trichloroethanol, *9551*

Amides
Butoctamide, *1526*
Diethylbromoacetamide, *3105*
Ibrotamide, *4809*
Isovaleryl Diethylamide, *5122*
Niaprazine, *6400*
Tricetamide, *9535*
Trimetozine, *9634*
Zolpidem, *10091*
Zopiclone, *10095*

Barbituric Acid Derivatives
Allobarbital, *252*
Amobarbital, *601*
Aprobarbital, *782*
Barbital, *972*
Brallobarbital, *1357*
Butabarbital Sodium, *1495*
Butalbital, *1502*
Butallylonal, *1503*
Butethal, *1515*
Carbubarb, *1838*
Cyclobarbital, *2717*
Cyclopentobarbital, *2751*
Enallylpropymal, *3523*
5-Ethyl-5-(1-piperidyl)barbituric
Acid, *3800*
5-Furfuryl-5-isopropylbarbituric
Acid, *4216*
Heptabarbital, *4575*
Hexethal Sodium, *4623*
Hexobarbital, *4625*
Mephobarbital, *5742*
Methitural, *5901*
Narcobarbital, *6342*
Nealbarbital, *6350*
Pentobarbital Sodium, *7087*
Phenallymal, *7162*
Phenobarbital, *7201*
Phenobarbital Sodium, *7202*
Phenylmethylbarbituric Acid,
7275
Probarbital, *7759*
Propallylonal, *7807*
Proxibarbal, *7920*
Reposal, *8141*
Secobarbital Sodium, *8374*
Talbutal, *9010*
Tetrabarbital, *9117*
Vinbarbital Sodium, *9886*
Vinylbital, *9897*

Benzodiazepine Derivatives
Brotizolam, *1439*
Doxefazepam, *3423*

SEDATIVE/HYPNOTIC

Benzodiazepine Derivatives (*continued*)
Estazolam, *3651*
Flunitrazepam, *4072*
Flurazepam, *4123*
Haloxazolam, *4518*
Loprazolam, *5453*
Lormetazepam, *5458*
Nitrazepam, *6491*
Quazepam, *8039*
Temazepam, *9074*
Triazolam, *9518*

Bromides
Ammonium Bromide, *525*
Calcium Bromide, *1653*
Calcium Bromolactobionate, *1654*
Lithium Bromide, *5403*
Magnesium Bromide, *5537*
Potassium Bromide, *7597*
Sodium Bromide, *8539*

Carbamates
Amyl Carbamate, Tertiary, *641*
Ethinamate, *3688*
Hexapropymate, *4616*
Meparfynol Carbamate, *5732*
Novonal, *6643*
Trichlorourethan, *9561*

Chloral Derivatives
Carbocloral, *1808*
Chloral Betaine, *2059*
Chloral Formamide, *2060*
Chloral Hydrate, *2061*
Chloralantipyrine, *2058*
Dichloralphenazone, *3033*
Pentaerythritol Chloral, *7063*
Triclofos, *9571*

Piperidinediones
Glutethimide, *4371*
Methyprylon, *6054*
Piperidione, *7439*
Pyrithyldione, *8005*
Taglutimide, *9006*
Thalidomide, *9182*

Quinazolone Derivatives
Etaqualone, *3665*
Mecloqualone, *5661*
Methaqualone, *5872*

Others
Acetal, *31*
Acetophenone, *65*
Aldol, *217*
Ammonium Valerate, *598*
Amphenidone, *614*
d-Bornyl α-Bromoisovalerate, *1340*
d-Bornyl Isovalerate, *1342*
Bromoform, *1407*
Calcium 2-Ethylbutanoate, *1667*
Carfinate, *1845*
α-Chloralose, *2062*
Clomethiazole, *2382*
Cypripedium, *2778*
Doxylamine, *3430*
Etodroxizine, *3832*
Etomidate, *3838*
Fenadiazole, *3898*
Homofenazine, *4657*
Hydrobromic Acid, *4701*
Mecloxamine, *5663*
Menthyl Valerate, *5728*
Opium, *6809*
Paraldehyde, *6975*
Perlapine, *7131*
Propiomazine, *7834*
Rilmazafone, *8219*
Sodium Oxybate, *8603*
Sulfonethylmethane, *8937*
Sulfonmethane, *8939*

SEROTONIN RECEPTOR AGONIST *see also Anxiolytic*

Buspirone, *1493*
Gepirone, *4297*
Ipsapirone, *4967*

SEROTONIN RECEPTOR ANTAGONIST *see also Antiemetic; Antihypertensive; Antimigraine*

Granisetron, *4443*
Ketanserin, *5175*
Methysergid(e), *6055*
Ondansetron, *6802*
Oxetorone, *6889*
Sumatriptan, *8979*

SEROTONIN UPTAKE INHIBITOR *see also Antidepressant*

Femoxetine, *3897*
Fluoxetine, *4112*
Fluvoxamine, *4138*
Indalpine, *4843*
Indeloxazine Hydrochloride, *4850*
Paroxetine, *6991*
Sertraline, *8416*

SKELETAL MUSCLE RELAXANT *see Muscle Relaxant (Skeletal)*

SOMATOSTATIN ANALOG

Octreotide, *6682*

SPASMOLYTIC *see Antispasmodic*

STOOL SOFTENER *see Laxative/Cathartic*

SUCCINYLCHOLINE SYNERGIST

Hexafluorenium Bromide, *4606*

SYMPATHOMIMETIC *see α-Adrenergic Agonist; β-Adrenergic Agonist*

THROMBOLYTIC *see also Anticoagulant; Antithrombotic*

APSAC, *785*
Plasmin, *7493*
Pro-Urokinase, *7918*
Streptokinase, *8784*
Tissue Plasminogen Activator, *9394*
Urokinase, *9799*

THYROID HORMONE *see also Antihypothyroid*

Liothyronine, *5388*
Thyroid, *9344*
Thyroidin, *9345*
Thyroxine, *9348*

THYROID INHIBITOR *see Antihyperthyroid*

THYROTROPIC HORMONE

TRH, *9503*
TSH, *9709*

TOCOLYTIC

Albuterol, *209*
Fenoterol, *3927*
Hexoprenaline, *4628*
Ritodrine, *8228*
Terbutaline, *9089*

TOPICAL PROTECTANT

Allantoin, *246*
Balsam Peru, *959*
Balsam Traumatic, *961*
Bismuth Phosphate, *1287*
Bismuth Subcarbonate, *1295*
Bismuth Subgallate, *1297*
Bismuth Tannate, *1302*
Calamine, *1636*
Collodion, *2480*
Esculin, *3646*
Gum Benzoin, *4492*
Hydroxypropyl Cellulose, *4776*
Pyroxylin, *8022*
Shark Liver Oil, *8425*
Storax, *8778*
Titanium Dioxide, *9398*
Zinc Oxide, *10050*

TRANQUILIZER *see Antipsychotic; Anxiolytic*

ULTRAVIOLET SCREEN

Actinoquinol, *134*
p-Aminobenzoic Acid, *434*
Butyl Methoxydibenzoylmethane, *1580*
β-Carotene, *1860*
Cinoxate, *2312*
4-(Dimethylamino)benzoic Acid, *3221*
Dioxybenzone, *3298*
Lawsone, *5263*
Mexenone, *6092*
Octabenzone, *6662*
Octyl Methoxycinnamate, *6687*
Oxybenzone, *6907*
Sulisobenzone, *8963*

URICOSURIC *see also Antigout*

Benzbromarone, *1073*
Ethebenecid, *3684*
Orotic Acid, *6828*
Oxycinchophen, *6910*
Probenecid, *7760*
Sulfinpyrazone, *8926*
Ticrynafen, *9362*
Zoxazolamine, *10098*

VASODILATOR (CEREBRAL)

Bencyclane, *1041*
Cinnarizine, *2308*
Citicoline, *2321*
Cyclandelate, *2708*
Ciclonicate, *2269*
Diisopropylamine Dichloroacetate, *3182*
Eburnamonine, *3464*
Fenoxedil, *3931*
Flunarizine, *4070*
Ibudilast, *4810*
Ifenprodil, *4821*
Nafronyl, *6268*
Nicametate, *6401*
Nicergoline, *6404*
Nimodipine, *6467*
Papaverine, *6968*
Pentifylline, *7085*

Tinofedrine, *9378*
Vincamine, *9888*
Vinpocetine, *9894*
Viquidil, *9908*

VASODILATOR (CORONARY)

Amotriphene, *608*
Bendazol, *1043*
Benfurodil Hemisuccinate, *1051*
Benziodarone, *1093*
Chloracizine, *2055*
Chromonar, *2241*
Clobenfurol, *2356*
Clonitrate, *2390*
Dilazep, *3185*
Dipyridamole, *3354*
Droprenilamine, *3438*
Efloxate, *3490*
Erythritol, *3620*
Erythrityl Tetranitrate, *3622*
Etafenone, *3662*
Fendiline, *3916*
Floredil, *4041*
Ganglefene, *4263*
Hexestrol Bis(β-diethylaminoethyl ether), *4622*
Hexobendine, *4626*
Itramin Tosylate, *5132*
Khellin, *5189*
Lidoflazine, *5360*
Mannitol Hexanitrate, *5630*
Medibazine, *5671*
Nicorandil, *6431*
Nitroglycerin, *6528*
Pentaerythritol Tetranitrate, *7066*
Pentrinitrol, *7094*
Perhexiline, *7116*
Pimefylline, *7400*
Prenylamine, *7744*
Propatyl Nitrate, *7821*
Pyridofylline, *7988*
Trapidil, *9492*
Tricromyl, *9574*
Trimetazidine, *9619*
Trolnitrate Phosphate, *9682*
Visnadine, *9915*

VASODILATOR (PERIPHERAL)

Aluminum Nicotinate, *354*
Bamethan, *964*
Bencyclane, *1041*
Betahistine, *1200*
Bradykinin, *1356*
Brovincamine, *1440*
Bufeniode, *1460*
Buflomedil, *1463*
Butalamine, *1501*
Cetiedil, *2012*
Ciclonicate, *2269*
Cinepazide, *2293*
Cinnarizine, *2308*
Cyclandelate, *2708*
Diisopropylamine Dichloroacetate, *3182*
Eledoisin, *3502*

Fenoxedil, *3931*
Flunarizine, *4070*
Hepronicate, *4574*
Ifenprodil, *4821*
Inositol Niacinate, *4885*
Isoxsuprine, *5128*
Kallidin, *5158*
Kallikrein, *5159*
Moxisylyte, *6204*
Nafronyl, *6268*
Nicametate, *6401*
Nicergoline, *6404*
Nicofuranose, *6428*
Nicotinyl Alcohol, *6438*
Nylidrin, *6654*
Pentifylline, *7085*
Pentoxifylline, *7092*
Piribedil, *7465*
Prostaglandin E₁, *7892*
Suloctidil, *8967*
Xanthinol Niacinate, *9969*

VASOPRESSOR *see Antihypotensive*

VASOPROTECTANT

Benzarone, *1071*
Bioflavonoids, *1241*
Chromocarb, *2239*
Clobenoside, *2357*
Diosmin, *3291*
Dobesilate Calcium, *3395*
Escin, *3644*
Folescutol, *4139*
Leucocyanidin, *5334*
Metescufylline, *5844*
Quercetin, *8044*
Rutin, *8276*
Troxerutin, *9698*

VITAMIN/VITAMIN SOURCE

Antirachitic
Ergosterol, *3607*
1α-Hydroxycholecalciferol, *4749*
Vitamin D₂, *9928*
Vitamin D₃, *9929*

Antiscorbutic
Ascorbic Acid, *855*
Calcium Ascorbate, *1648*
Nicotinamide Ascorbate, *6433*
Sodium Ascorbate, *8525*

Antixerophthalmic
α-Carotene, *1859*
β-Carotene, *1860*
γ-Carotene, *1861*
Vitamin A, *9918*

Enzyme Co-factor
Acetiamine, *46*
Benfotiamine, *1049*
Bisbentiamine, *1257*
Calcium Pantothenate, *1694*
Cetotiamine, *2014*
Cycothiamin(e), *2762*
Dexpanthenol, *2924*
Fursultiamine, *4223*
Methylol Riboflavine, *6018*
Niacinamide, *6398*

Nicotinamide Ascorbate, *6433*
Nicotinic Acid, *6435*
Nicotinic Acid Monoethanolamine Salt, *6437*
Octotiamine, *6680*
Pantothenic Acid, *6964*
Prosultiamine, *7896*
Pyridoxal 5-Phosphate, *7992*
Pyridoxine Hydrochloride, *7995*
Riboflavine, *8201*
Riboflavine Phosphate (Sodium), *8202*
Thiamine Disulfide, *9221*
Thiamine Hydrochloride, *9222*
Thiamine Mononitrate, *9223*
Vintiamol, *9895*

Hematopoietic
Cobamamide, *2446*
Folic Acid, *4140*
Hydroxocobalamin, *4739*
Sodium Folate, *8566*
Vitamin B₁₂, *9921*
Vitamin B₁₂-Zinc Tannate Complex, *9923*

Prothrombogenic
Dihydrovitamin K₁, *3165*
Menadiol Diacetate, *5710*
Menadiol Dibutyrate, *5711*
Menadiol Diphosphate (Tetrasodium Salt), *5712*
Menadiol Disulfate, *5713*
Menadione, *5714*
Menadione Sodium Bisulfite, *5716*
Menadoxime, *5717*
Vitamin K₁, *9933*
Vitamin K₁ Oxide, *9934*
Vitamin(s) K₂, *9935*
Vitamin K₅, *9936*
Vitamin K-S(II), *9939*

Others
Inositol, *4883*
β-Tocopherol, *9417*
γ-Tocopherol, *9418*
δ-Tocopherol, *9419*
Vitamin E, *9931*
Vitamin E Acetate, *9932*
Vitamin U, *9942*

VULNERARY

Acetylcysteine, *82*
Allantoin, *246*
Asiaticoside, *857*
Cadexomer Iodine, *1609*
Chitin, *2052*
Dextranomer, *2928*
Oxaceprol, *6857*

WILSON'S DISEASE TREATMENT

Penicillamine, *7029*
Trientine, *9579*

XANTHINE OXIDASE INHIBITOR *see also Antigout*

Allopurinol, *278*

FORMULA INDEX

FORMULA INDEX

A

AgBr
Silver Bromide, 8452

AgCl
Silver Chloride, 8455

AgClO₃
Silver Chlorate, 8454

AgClO₄
Silver Perchlorate, 8470

AgF
Silver Fluoride, 8460

AgF₂
Silver Difluoride, 8459

AgI
Silver Iodide, 8462

AgIO₃
Silver Iodate, 8461

AgMnO₄
Silver Permanganate, 8471

AgNO₂
Silver Nitrite, 8466

AgNO₃
Silver Nitrate, 8464

AgO
Silver(II) Oxide, 8469

Ag₂CrO₄
Silver Chromate(VI), 8456

Ag₂F
Silver Subfluoride, 8479

Ag₂HgI₄
Silver Tetraiodomercurate(II), 8482

Ag₂O
Silver Oxide, 8468

Ag₂O₃Se
Silver Selenite, 8478

Ag₂O₄S
Silver Sulfate, 8480

Ag₂O₄Se
Silver Selenate, 8476

Ag₂S
Silver Sulfide, 8481

Ag₂Se
Silver Selenide, 8477

Ag₃O₄P
Silver Phosphate, 8472

AlB₃H₁₂
Aluminum Borohydride, 331

AlBr₃
Aluminum Bromide, 332

AlCl₃
Aluminum Chloride, 338

AlCl₃O₉
Aluminum Chlorate, 337

AlCl₄H₄N
Ammonium Tetrachloroaluminate, 589

AlCl₄Na
Sodium Tetrachloroaluminate, 8643

AlCsO₈S₂
Aluminum Cesium Sulfate, 336

AlF₃
Aluminum Fluoride, 341

AlF₆H₁₂N₃
Ammonium Hexafluoroaluminate, 549

AlF₆Na₃
Cryolite, 2609

AlH₃
Aluminum Hydride, 344

AlH₃O₃
Aluminum Hydroxide, 345

AlH₄Li
Aluminum Lithium Hydride, 351

AlH₄NO₈S₂
Aluminum Ammonium Sulfate, 325

AlH₆O₆P₃
Aluminum Hypophosphite, 347

AlI₃
Aluminum Iodide, 348

AlKO₈S₂
Aluminum Potassium Sulfate, 364

AlN
Aluminum Nitride, 356

AlN₃O₉
Aluminum Nitrate, 355

AlNaO₂
Sodium Aluminate, 8517

AlNaO₈S₂
Aluminum Sodium Sulfate, 369

AlO₄P
Aluminum Phosphate, 362

AlO₈RbS₂
Aluminum Rubidium Sulfate, 365

AlP
Aluminum Phosphide, 363

AlSb
Aluminum Antimonide, 326

Al₂CaH₈
Aluminum Calcium Hydride, 334

Al₂F₁₈Si₃
Aluminum Hexafluorosilicate, 342

Al₂K₂O₄
Potassium Aluminate, 7581

Al₂MgO₈Si₂
Aluminum Magnesium Silicate, 352

Al₂O₃
Aluminum Oxide, 359

Al₂O₅Si
Aluminum Silicate, 368

Al₂O₁₂S₃
Aluminum Sulfate, 372

Al₂O₁₆S₄Zn
Aluminum Zinc Sulfate, 376

Al₂S₃
Aluminum Sulfide, 373

Al₂Se₃
Aluminum Selenide, 367

Al₆Bi₂O₁₂
Bismuth Aluminate, 1269

AsBr₃
Arsenic Tribromide, 828

AsCl₃
Arsenic Trichloride, 829

AsF₃
Arsenic Trifluoride, 830

AsF₅
Arsenic Pentafluoride, 824

AsGa
Gallium Arsenide, 4253

AsHHgO₄
Mercuric Arsenate, 5767

AsHNa₂O₄
Sodium Arsenate, Dibasic, 8522

AsH₂KO₄
Potassium Arsenate, 7583

AsH₃
Arsine, 837

AsH₃O₄
Arsenic Acid, 821

AsI₃
Arsenic Triiodide, 831

AsIn
Indium Arsenide, 4859

As₂Ca₃O₈
Calcium Arsenate, 1646

As₂Co₃O₈
Cobaltous Arsenate, 2428

As₂O₃
Arsenic Trioxide, 832

As₂O₄Zn
Zinc Meta-arsenite, 10043

As₂O₅
Arsenic Pentoxide, 827

As₂O₈Zn₃
Zinc Ortho-arsenate, 10048

As₂S₃
Arsenic Trisulfide, 834

As₂S₅
Arsenic Pentasulfide, 826

As₂Se
Arsenic Hemiselenide, 823

As₂Se₃
Arsenic Triselenide, 833

As₂Se₅
Arsenic Pentaselenide, 825

As₄S₄
Arsenic Disulfide, 822

AuBr₃
Gold Tribromide, 4425

AuBr₄H
Gold Tribromide, Acid, 4426

AuBr₄K
Potassium Tetrabromoaurate(III), 7673

AuCl
Gold Monochloride, 4414

AuCl₃
Gold Trichloride, 4427

AuCl₄H
Gold Trichloride, Acid, 4428

AuCl₄K
Potassium Tetrachloroaurate(III), 7674

AuCl₄Na
Sodium Tetrachloroaurate(III), 8644

AuH₃O₃
Gold Trihydroxide, 4430

AuI
Gold Monoiodide, 4416

AuI₄K
Potassium Tetraiodoaurate(III), 7681

AuNa₃O₆S₄
Gold Sodium Thiosulfate, 4423

Au₂O₃
Gold Trioxide, 4431

Au₂O₁₂Se₃
Gold Selenate, 4420

Au₂S
Gold Monosulfide, 4417

Denotes derivative of title compound

Denotes derivative of title compound

Br_2Cr
Chromous Bromide, 2245
Br_2Cu
Cupric Bromide, 2632
Br_2Fe
Ferrous Bromide, 3989
Br_2Hg
Mercuric Bromide, 5769
Br_2Hg_2
Mercurous Bromide, 5793
Br_2Mg
Magnesium Bromide, 5537
Br_2MgO_6
Magnesium Bromate, 5536
Br_2Mn
Manganese Bromide, 5607
Br_2Ni
Nickel Bromide, 6409
Br_2OS
Thionyl Bromide, 9277
Br_2OSe
Selenium Oxybromide, 8387
Br_2O_6Pb
Lead Bromate, 5274
Br_2O_6Sr
Strontium Bromate, 8799
Br_2Pb
Lead Bromide, 5275
Br_2Se_2
Selenium Bromide, 8383
Br_2Sn
Stannous Bromide, 8741
Br_2Sr
Strontium Bromide, 8800
Br_2Te
Tellurium Dibromide, 9065
Br_2Zn
Zinc Bromide, 10028
Br_3Ce
Cerous Bromide, 1989
Br_3Cr
Chromic Bromide, 2222
Br_3Fe
Ferric Bromide, 3960
Br_3HSi
Tribromosilane, 9528
Br_3OP
Phosphorus Oxybromide, 7323
Br_3P
Phosphorus Tribromide, 7332
Br_3Sb
Antimony Tribromide, 738
Br_4Se
Selenium Tetrabromide, 8391
Br_4Si
Silicon Tetrabromide, 8446
Br_4Sn
Stannic Bromide, 8731
Br_4Te
Tellurium Tetrabromide, 9069
Br_4Ti
Titanium Tetrabromide, 9403
Br_5P
Phosphorus Pentabromide, 7325

C

CAgN
Silver Cyanide, 8458
CAg_2O_3
Silver Carbonate, 8453
CAuN
Gold Monocyanide, 4415
CB_4
Boron Carbide, 1346
$CBaO_3$
Barium Carbonate, 979
CBe_2
Beryllium Carbide, 1183
CBi_2O_5
Bismuth Subcarbonate, 1295
CBrN
Cyanogen Bromide, 2700
$CCaN_2$
Calcium Cyanamide, 1662
$CCaO_3$
Calcium Carbonate, 1657
$CCdO_3$
Cadmium Carbonate, 1614
CClN
Cyanogen Chloride, 2701
CCl_2F_2
Dichlorodifluoromethane, 3053
CCl_2O
Phosgene, 7310
CCl_3F
Trichlorofluoromethane, 9553
CCl_3NO_2
Chloropicrin, 2156
CCl_4
Carbon Tetrachloride, 1822
$CCoO_3$
Cobaltous Carbonate, 2430
CCs_2O_3
Cesium Carbonate, 2003
CCuN
Cuprous Cyanide, 2668
CCuNS
Cuprous Thiocyanate, 2676
CF_2O
Carbonyl Fluoride, 1826
CF_4
Carbon Tetrafluoride, 1823
$CHBr_3$
Bromoform, 1407
$CHCl_3$
Chloroform, 2141
CHFO
Formyl Fluoride, 4162
CHF_3
Fluoroform, 4102
CHI_2NaO_3S
Dimethiodal Sodium, 3205
CHI_3
Iodoform, 4926
$CHKO_2$
Potassium Formate, 7614
$CHKO_3$
Potassium Bicarbonate, 7586
$CHLiO_2$
Lithium Formate, 5410
CHN
Hydrogen Cyanide, 4722
CHNO
Cyanic Acid, 2693
Isocyanic Acid, 5049
CHNS
Thiocyanic Acid, 9257
CHN_3O_6
Trinitromethane, 9642
$CHNaO_2$
Sodium Formate, 8568
$CHNaO_3$
Sodium Bicarbonate, 8528
CH_2AlNaO_5
Dihydroxyaluminum Sodium Carbonate, 3169
CH_2Br_2
Methylene Bromide, 5980
$CH_2CaO_6S_2$
Calcium Methionate, 1686
CH_2ClNO
Carbamyl Chloride, 1785
CH_2Cl_2
Methylene Chloride, 5982
$CH_2Cl_2Na_2O_6P_2$
Clodronic Acid*, 2369
$CH_2Cu_2O_5$
Cupric Carbonate, Basic, 2634
CH_2INaO_3S
Methiodal Sodium, 5894
CH_2I_2
Methylene Iodide, 5985
CH_2N_2
Cyanamide, 2691
Diazomethane, 2983
CH_2O
Formaldehyde, Gas, 4148
CH_2O_2
Formic Acid, 4153
CH_2O_3
Performic Acid, 7114
CH_2S_3
Trithiocarbonic Acid, 9671
$CH_3AsNa_2O_3$
Methanearsonic Acid*, 5864
CH_3BNNa
Sodium Cyanoborohydride, 8554
CH_3Br
Methyl Bromide, 5951
CH_3Cl
Methyl Chloride, 5964
CH_3ClO_2S
Methanesulfonyl Chloride, 5866
CH_3F
Fluoromethane, 4103
CH_3FO_3S
Methyl Fluorosulfonate, 5993
CH_3I
Methyl Iodide, 6002
CH_3KO_4S
Potassium Methyl Sulfate, 7632
CH_3NO
Formamide, 4151
CH_3NO_2
Nitromethane, 6532
CH_3NO_3
Methyl Nitrate, 6015
CH_3NS
Thioformamide, 9262
$CH_3N_3O_3$
Nitrourea, 6573
CH_3NaO
Sodium Methoxide, 8594
CH_3NaO_3S
Sodium Formaldehydesulfoxylate, 8567
CH_3NaO_4S
Formaldehyde Sodium Bisulfite, 4149
Sodium Methyl Sulfate, 8595
CH_4
Methane, 5863
CH_4AsNaO_3
Methanearsonic Acid*, 5864
$CH_4Cl_2O_6P_2$
Clodronic Acid, 2369
CH_4N_2O
Urea, 9781
$CH_4N_2O_2$
Hydroxyurea, 4785
CH_4N_2S
Ammonium Thiocyanate, 591
Thiourea, 9299
$CH_4N_4O_2$
Nitroguanidine, 6529
$CH_4Ni_3O_7$
Nickel Carbonate Hydroxide, 6410
CH_4O
Methanol, 5868
CH_4O_3S
Methanesulfonic Acid, 5865
CH_4O_4S
Methyl Sulfate, 6041
$CH_4O_6S_2$
Methionic Acid, 5895
CH_4S
Methanethiol, 5867

** Denotes derivative of title compound*

CH₅AsO₃
Methanearsonic Acid, 5864
CH₅ClN₂O
Urea Hydrochloride, 9782
CH₅N
Methylamine, 5938
CH₅NO
Methoxyamine, 5912
CH₅NO₂
Ammonium Formate, 548
CH₅NO₃
Ammonium Bicarbonate, 516
CH₅N₃
Guanidine, 4475
CH₅N₃O₄
Urea Nitrate, 9784
CH₅N₃S
Thiosemicarbazide, 9292
CH₆ClN₃O
Semicarbazide Hydrochloride, 8396
CH₆N₂
Methylhydrazine, 6001
CH₆N₂O₂
Ammonium Carbamate, 527
CH₆N₂O₃
Urea Hydrogen Peroxide, 9783
CH₆N₂S₂
Ammonium Dithiocarbamate, 539
CH₆N₄
Aminoguanidine, 453
CH₆N₄O
Carbohydrazide, 1811
CH₁₈AlN₃O₁₄S₂
Guanidinium Aluminum Sulfate
  Hexahydrate, 4476
CIN
Cyanogen Iodide, 2702
CI₄
Carbon Tetraiodide, 1824
CKN
Potassium Cyanide, 7607
CKNO
Potassium Cyanate, 7606
CKNS
Potassium Thiocyanate, 7687
CK₂O₃
Potassium Carbonate, 7599
CK₂S₃
Potassium Thiocarbonate, 7686
CLi₂O₃
Lithium Carbonate, 5404
CMnO₃
Manganese Carbonate, 5608
CNNa
Sodium Cyanide, 8553
CNNaO
Sodium Cyanate, 8552
CNNaS
Thiocyanate Sodium, 9256
CNTl
Thallium Cyanide, 9188
CN₄
Cyanogen Azide, 2699
CN₄O₈
Tetranitromethane, 9164
CNa₂O₃
Sodium Carbonate, 8541
CNa₂S₃
Sodium Thiocarbonate, 8649
CNa₃O₅P
Foscarnet Sodium, 4166
CO
Carbon Monoxide, 1820
CO₂
Carbon Dioxide, 1816
CO₃Sr
Strontium Carbonate, 8801
CO₃Tl₂
Thallium Carbonate, 9186
CO₃Zn
Zinc Carbonate, 10030

CS₂
Carbon Disulfide, 1818
CSe₂
Carbon Diselenide, 1817
CSi
Silicon Carbide, 8439
C₂AgKN₂
Potassium Silver Cyanide, 7659
C₂Ag₂O₄
Silver Oxalate, 8467
C₂AuKN₂
Potassium Dicyanoaurate(I), 7609
C₂AuN₂Na
Sodium Dicyanoaurate(I), 8557
C₂BaN₂
Barium Cyanide, 983
C₂BaN₂S₂
Barium Thiocyanate, 1010
C₂BaO₄
Barium Oxalate, 997
C₂Ca
Calcium Carbide, 1656
C₂CaN₂
Calcium Cyanide, 1664
C₂CaN₂S₂
Calcium Thiocyanate, 1717
C₂CaO₄
Calcium Oxalate, 1691
C₂CdN₂
Cadmium Cyanide, 1616
C₂Cl₂F₄
Cryofluorane, 2608
C₂Cl₂O₂
Oxalyl Chloride, 6867
C₂Cl₃N
Trichloroacetonitrile, 9540
C₂Cl₃NaO₂
Trichloroacetic Acid*, 9539
C₂Cl₄
Tetrachloroethylene, 9126
C₂Cl₄O₂
Diphosgene, 3339
C₂Cl₆
Hexachloroethane, 4601
C₂CoN₂
Cobaltous Cyanide, 2433
C₂CoN₂S₂
Cobaltous Thiocyanate, 2445
C₂CoO₄
Cobaltous Oxalate, 2439
C₂CrO₄
Chromous Oxalate, 2249
C₂CuKN₂
Cuprous Potassium Cyanide, 2672
C₂CuO₄
Cupric Oxalate, 2649
C₂FeN₂S₂
Ferrous Thiocyanate, 4008
C₂FeO₄
Ferrous Oxalate, 4000
C₂HBrClF₃
Halothane, 4517
C₂HBrF₄
Teflurane, 9059
C₂HBr₃O
Bromal, 1372
C₂HBr₃O₂
Tribromoacetic Acid, 9521
C₂HCl₃
Trichloroethylene, 9552
C₂HCl₃O
2,2-Dichloroacetyl Chloride, 3040
Trichloroacetaldehyde, 9538
C₂HCl₃O₂
Trichloroacetic Acid, 9539
C₂HCl₅
Pentachloroethane, 7058
C₂HF₃O₂
Trifluoroacetic Acid, 9595
C₂HKO₄
Potassium Binoxalate, 7588

C₂HNa₃O₆
Sodium Sesquicarbonate, 8630
C₂H₂
Acetylene, 84
C₂H₂AsCl₃
Dichloro(2-chlorovinyl)arsine, 3051
C₂H₂BaO₄
Barium Formate, 987
C₂H₂BeO₄
Beryllium Formate, 1186
C₂H₂Br₂
Acetylene Dibromide, 85
C₂H₂Br₄
sym-Tetrabromoethane, 9121
C₂H₂CaO₂S
Calcium Thioglycollate, 1718
C₂H₂CaO₄
Calcium Formate, 1671
C₂H₂ClNaO₂
Chloroacetic Acid*, 2111
C₂H₂Cl₂
Acetylene Dichloride, 86
Vinylidene Chloride, 9900
C₂H₂Cl₂O
Chloracetyl Chloride, 2054
C₂H₂Cl₂O₂
Dichloroacetic Acid, 3037
C₂H₂Cl₄
Tetrachloroethane, 9125
C₂H₂CoO₄
Cobaltous Formate, 2435
C₂H₂CrO₄
Chromous Formate, 2248
C₂H₂CuO₄
Cupric Formate, 2642
C₂H₂FNaO₂
Fluoroacetic Acid*, 4096
C₂H₂MgO₄
Magnesium Formate, 5544
C₂H₂MoO₇
Oxalomolybdic Acid, 6866
C₂H₂N₂O
1,3,4-Oxadiazole, 6859
C₂H₂Na₂O₂
Sodiumacetic Acid Sodium Salt,
  8514
C₂H₂NiO₄
Nickel Formate, 6416
C₂H₂O
Ketene, 5177
C₂H₂O₂
Glyoxal, 4405
C₂H₂O₃
Glyoxylic Acid, 4407
C₂H₂O₄
Oxalic Acid, 6865
C₂H₂O₄Pb
Lead Formate, 5283
C₂H₂O₄Sr
Strontium Formate, 8806
C₂H₂O₄Zn
Zinc Formate, 10036
C₂H₃AgO₂
Silver Acetate, 8451
C₂H₃AsNa₂O₅.H₂O
Arsonoacetic Acid*, 838
C₂H₃BiO₃
Bismuth Subacetate, 1294
C₂H₃BrO
Acetyl Bromide, 76
C₂H₃BrO₂
Bromoacetic Acid, 1388
C₂H₃Br₃O
Tribromoethanol, 9525
C₂H₃Br₃O₂
Bromal Hydrate, 1373
C₂H₃Cl
Vinyl Chloride, 9898
C₂H₃ClO
Acetyl Chloride, 79
Chloroacetaldehyde, 2108

Denotes derivative of title compound

C₂H₃ClO₂
Chloroacetic Acid, 2111
Methyl Chlorocarbonate, 5966

C₂H₃Cl₃
1,1,1-Trichloroethane, 9549
1,1,2-Trichloroethane, 9550

C₂H₃Cl₃NaO₄P
Triclofos*, 9571

C₂H₃Cl₃O
2,2,2-Trichloroethanol, 9551

C₂H₃Cl₃O₂
Chloral Hydrate, 2061

C₂H₃CuO₂
Cuprous Acetate, 2665

C₂H₃FO₂
Fluoroacetic Acid, 4096

C₂H₃IO
Acetyl Iodide, 88

C₂H₃IO₂
Iodoacetic Acid, 4918

C₂H₃KO₂
Potassium Acetate, 7580

C₂H₃LiO₂
Lithium Acetate, 5396

C₂H₃N
Acetonitrile, 62

C₂H₃NO
Methyl Isocyanate, 6004

C₂H₃NO₃
Oxamic Acid, 6870

C₂H₃NO₄
Acetyl Nitrate, 92

C₂H₃NS
Methyl Isothiocyanate, 6005
Methyl Thiocyanate, 6047

C₂H₃N₃
1H-1,2,4-Triazole, 9519

C₂H₃N₃O₂
Urazole, 9780

C₂H₃NaO₂
Sodium Acetate, 8513

C₂H₃NaO₂S
Sodium Thioglycolate, 8650

C₂H₃O₂Tl
Thallium Acetate, 9184

C₂H₄
Ethylene, 3748

C₂H₄BrNO
N-Bromoacetamide, 1386

C₂H₄Br₂
Ethylene Dibromide, 3753

C₂H₄ClNO
Chloroacetamide, 2109

C₂H₄Cl₂
Ethylene Dichloride, 3754
Ethylidene Chloride, 3766

C₂H₄Cl₂N₆
Chloroazodin, 2120

C₂H₄Cl₂O
sym-Dichloromethyl Ether, 3058

C₂H₄Cl₃NO
Chloral Ammonia, 2057

C₂H₄Cl₃O₄P
Triclofos, 9571

C₂H₄FNO
Fluoroacetamide, 4095

C₂H₄NNaS₂
Metham Sodium, 5860

C₂H₄N₂
Aminoacetonitrile, 423

C₂H₄N₂O₂
Oxamide, 6871

C₂H₄N₂S₂
Rubeanic Acid, 8256

C₂H₄N₄
Amitrole, 506
Dicyanodiamide, 3082

C₂H₄N₄O₂
Azodicarbonamide, 932

C₂H₄Na₂O₈S₂
Glyoxal-Sodium Bisulfite, 4406

C₂H₄O
Acetaldehyde, 32
Ethylene Oxide, 3758

C₂H₄OS
Thioacetic Acid, 9247

C₂H₄O₂
Acetic Acid Glacial, 47
Methyl Formate, 5994

C₂H₄O₂S
Thioglycolic Acid, 9265

C₂H₄O₃
Glycolic Acid, 4394
Peracetic Acid, 7107

C₂H₄O₅S
Sulfoacetic Acid, 8932

C₂H₄S₅
Lenthionine, 5321

C₂H₅AsNa₂O₃
Ethanearsonic Acid*, 3677

C₂H₅AsO₅
Arsonoacetic Acid, 838

C₂H₅Br
Ethyl Bromide, 3730

C₂H₅BrO
Ethylene Bromohydrin, 3749

C₂H₅Cl
Ethyl Chloride, 3740

C₂H₅ClHg
Ethylmercuric Chloride, 3780

C₂H₅ClO
Chloromethyl Methyl Ether, 2146
Ethylene Chlorohydrin, 3750

C₂H₅I
Ethyl Iodide, 3769

C₂H₅KO₄S
Potassium Ethyl Sulfate, 7610

C₂H₅N
Ethylenimine, 3760

C₂H₅NO
Acetaldoxime, 35
Acetamide, 36

C₂H₅NO₂
Acetohydroxamic Acid, 54
Ethyl Nitrite, 3786
Glycine, 4386
Methyl Carbamate, 5958
Nitroethane, 6518

C₂H₅NO₄
Ammonium Binoxalate, 519

C₂H₅NS
Thioacetamide, 9246

C₂H₅N₃O
Formaldehyde Solution*, 4150

C₂H₅N₃O₂
Biuret, 1319
Semioxamazide, 8397

C₂H₅N₃S₂
2,4-Dithiobiuret, 3378

C₂H₅N₅O₃
N-Methyl-N'-nitro-N-nitrosoguan-
idine, 6017

C₂H₅NaO
Sodium Ethoxide, 8560

C₂H₅NaO₃S₂
Mesna, 5812

C₂H₅NaO₄S
Acetaldehyde Sodium Bisulfite, 34
Sodium Ethyl Sulfate, 8561

C₂H₆
Ethane, 3676

C₂H₆AlNO₄
Dihydroxyaluminum Aminoacetate, 3168

C₂H₆AsNaO₂
Sodium Cacodylate, 8540

C₂H₆BaO₈S₂
Methyl Sulfate*, 6041

C₂H₆Be
Dimethylberyllium, 3227

C₂H₆CaO₈S₂
Methyl Sulfate*, 6041

C₂H₆Cd
Dimethylcadmium, 3229

C₂H₆ClO₃P
Ethephon, 3686

C₂H₆Co₅O₁₂
Cobaltous Carbonate*, 2430

C₂H₆Hg
Dimethylmercury, 3238

C₂H₆N₂O
N-Nitrosodimethylamine, 6558

C₂H₆Na₂O₇P₂
Etidronic Acid*, 3819

C₂H₆O
Ethyl Alcohol, 3716
Methyl Ether, 5990

C₂H₆OS
Dimethyl Sulfoxide, 3247
2-Mercaptoethanol, 5760

C₂H₆O₂
Ethylene Glycol, 3755

C₂H₆O₂S
Dimethyl Sulfone, 3246

C₂H₆O₃S
Methyl Methanesulfonate, 6010

C₂H₆O₄S
Dimethyl Sulfate, 3244
Ethyl Sulfate, 3808
Isethionic Acid, 4990

C₂H₆O₆S₂
1,2-Ethanedisulfonic Acid, 3678

C₂H₆S
Ethanethiol, 3680
Methyl Sulfide, 6042

C₂H₆S₂
1,2-Ethanedithiol, 3679

C₂H₆Zn
Dimethylzinc, 3252

C₂H₇AsO₂
Cacodylic Acid, 1603

C₂H₇AsO₃
Ethanearsonic Acid, 3677

C₂H₇ClN₂
Acetamidine Hydrochloride, 37

C₂H₇N
Dimethylamine, 3217
Ethylamine, 3718

C₂H₇NO
Acetaldehyde Ammonia, 33
Ethanolamine, 3681

C₂H₇NO₂
Ammonium Acetate, 514

C₂H₇NO₃S
Taurine, 9043

C₂H₇NS
Cysteamine, 2785

C₂H₇N₃
Methylguanidine, 5999

C₂H₇N₅
Biguanide, 1233

C₂H₈NO₂PS
Methamidophos, 5858

C₂H₈N₂
1,1-Dimethylhydrazine, 3236
1,2-Dimethylhydrazine, 3237
Ethylenediamine, 3752

C₂H₈N₂O
2-Hydrazinoethanol, 4696

C₂H₈N₂O₄
Ammonium Oxalate, 565

C₂H₈O₇P₂
Etidronic Acid, 3819

C₂H₁₄Al₂Mg₆O₂₀·4H₂O
Almagate, 297

C₂HgN₂
Mercuric Cyanide, 5772

C₂HgN₂S₂
Mercuric Thiocyanate, 5790

C₂Hg₂N₂O
Mercuric Oxycyanide, 5781

C₂I₄
Tetraiodoethylene, 9151

C₂K₂O₄
Potassium Oxalate, 7639
C₂K₂O₆
Potassium Percarbonate, 7640
C₂Li₂O₄
Lithium Oxalate, 5415
C₂MgN₂S₂
Magnesium Thiocyanate, 5575
C₂MgO₄
Magnesium Oxalate, 5554
C₂MnO₄
Manganese Oxalate, 5617
C₂N₂
Cyanogen, 2698
C₂N₂Ni
Nickel Cyanide, 6413
C₂N₂PbS₂
Lead Thiocyanate, 5308
C₂N₂S₂Zn
Zinc Thiocyanate, 10069
C₂N₂Zn
Zinc Cyanide, 10034
C₂Na₂O₄
Sodium Oxalate, 8601
C₂NiO₄
Nickel Oxalate, 6421
C₂O₄Pb
Lead Oxalate, 5294
C₂O₄Sn
Stannous Oxalate, 8746
C₂O₄Sr
Strontium Oxalate, 8811
C₂O₄Zn
Zinc Oxalate, 10049
C₃AlN₃S₃
Aluminum Thiocyanate, 375
C₃Al₄
Aluminum Carbide, 335
C₃AuN₃
Gold Tricyanide, 4429
C₃CaO₅
Calcium Mesoxalate, 1685
C₃Ce₂O₉
Cerous Carbonate, 1990
C₃Cl₂KN₃O₃
Troclosene Potassium, 9679
C₃Cl₃N₃O₃
Symclosene, 8993
C₃FeN₃S₃
Ferric Thiocyanate, 3982
C₃HBr₅O
Pentabromoacetone, 7056
C₃H₂ClF₅O
Enflurane, 3536
Isoflurane, 5059
C₃H₂ClNO₂
Chloroacetyl Isocyanate, 2116
C₃H₂Cl₃NO
Chlorocyanohydrin, 2134
C₃H₂N₂
Malononitrile, 5592
C₃H₂N₂O₃
Parabanic Acid, 6970
C₃H₂O₂
Propiolic Acid, 7833
C₃H₂O₅
Mesoxalic Acid, 5814
C₃H₂O₅.H₂O
Mesoxalic Acid*, 5814
C₃H₃BrF₄
Halopropane, 4515
C₃H₃Cl
Propargyl Chloride, 7820
C₃H₃Cl₂NaO₂
Dalapon*, 2806
C₃H₃CrO₆
Chromic Formate, 2226
C₃H₃FeO₆
Ferric Formate, 3966
C₃H₃N
Acrylonitrile, 125

C₃H₃NOS₂
Rhodanine, 8184
C₃H₃NO₂
Cyanoacetic Acid, 2696
C₃H₃NS
Thiazole, 9235
C₃H₃N₃O₂
Azomycin, 934
C₃H₃N₃O₂S
2-Amino-5-nitrothiazole, 469
C₃H₃N₃O₃
Cyamelide, 2689
Cyanuric Acid, 2704
C₃H₄Br₂
2,3-Dibromopropene, 3009
C₃H₄ClN
β-Chloropropionitrile, 2160
C₃H₄Cl₂
1,3-Dichloropropene, 3064
C₃H₄Cl₂F₂O
Methoxyflurane, 5914
C₃H₄Cl₂O
1,1-Dichloroacetone, 3038
1,3-Dichloroacetone, 3039
C₃H₄Cl₂O₂
Dalapon, 2806
C₃H₄Cl₃NO₂
Chloral Formamide, 2060
Trichlorourethan, 9561
C₃H₄F₄O
2,2,3,3-Tetrafluoro-1-propanol, 9139
C₃H₄NNaO₃S
Sodium β-Sulfopropionitrile, 8639
C₃H₄N₂
Imidazole, 4828
Methyleneaminoacetonitrile, 5976
Pyrazole, 7974
C₃H₄N₂O
Cyanoacetamide, 2695
C₃H₄N₂O₂
Hydantoin, 4678
C₃H₄N₂S
2-Aminothiazole, 494
C₃H₄O
Acrolein, 122
Propargyl Alcohol, 7819
C₃H₄O₂
Acrylic Acid, 124
β-Propiolactone, 7832
Pyruvaldehyde, 8030
C₃H₄O₃
Pyruvic Acid, 8032
C₃H₄O₄
Malonic Acid, 5591
C₃H₄O₅
Tartronic Acid, 9042
C₃H₅AgO₃
Silver Lactate, 8463
C₃H₅Br
Allyl Bromide, 286
C₃H₅BrO
Bromoacetone, 1389
C₃H₅BrO₂
β-Bromopropionic Acid, 1421
C₃H₅Br₂Cl
Dibromochloropropane, 3003
C₃H₅Br₃
1,2,3-Tribromopropane, 9527
C₃H₅Cl
Allyl Chloride, 287
C₃H₅ClN₂O₆
Clonitrate, 2390
C₃H₅ClO
Chloroacetone, 2113
Epichlorohydrin, 3563
Propionyl Chloride, 7840
C₃H₅ClO₂
β-Chloropropionic Acid, 2159
Ethyl Chloroformate, 3742
Methyl Chloroacetate, 5965

C₃H₅Cl₃O
1,1,1-Trichloro-2-propanol, 9557
C₃H₅I
Allyl Iodide, 292
C₃H₅KOS₂
Potassium Xanthogenate, 7697
C₃H₅N
Propionitrile, 7839
C₃H₅NO
Acrylamide, 123
Ethylene Cyanohydrin, 3751
C₃H₅NO₂
Isonitrosoacetone, 5076
C₃H₅NO₄
Hadacidin, 4500
C₃H₅NS
Ethyl Isothiocyanate, 3771
Thiocyanic Acid*, 9257
C₃H₅N₃O
Cyacetacide, 2688
C₃H₅N₃O₉
Nitroglycerin, 6528
C₃H₅NaO₂
Sodium Propionate, 8623
C₃H₅NaO₃
Sodium Lactate, 8586
C₃H₆
Cyclopropane, 2755
Propylene, 7862
C₃H₆BrNO₄
Bronopol, 1437
C₃H₆Br₂
Propylene Dibromide, 7866
Trimethylene Bromide, 9628
C₃H₆Cl₂
Propylene Dichloride, 7867
Propylidene Chloride, 7874
C₃H₆Cl₂O
1,3-Dichloro-2-propanol, 3063
C₃H₆FNO₂
3-Fluoro-D-alanine, 4097
C₃H₆I₂O
Iothion, 4953
C₃H₆N₂
3-Aminopropionitrile, 481
2-Pyrazoline, 7975
C₃H₆N₂O
2-Imidazolidinone, 4830
C₃H₆N₂O₂
Cycloserine, 2758
C₃H₆N₂S
Ethylene Thiourea, 3759
C₃H₆N₄O₃
Triuret, 9678
C₃H₆N₆
Melamine, 5691
C₃H₆N₆O₆
Cyclonite, 2742
C₃H₆O
Acetone, 58
Allyl Alcohol, 284
Propionaldehyde, 7835
Propylene Oxide, 7869
Trimethylene Oxide, 9630
C₃H₆OS
Propanethial S-Oxide, 7812
C₃H₆O₂
Ethyl Formate, 3763
Glycidol, 4385
Methyl Acetate, 5932
Propionic Acid, 7837
C₃H₆O₂S
Thiolactic Acid, 9269

Denotes derivative of title compound

C₃H₆O₃
Dihydroxyacetone, 3166
Glyceraldehyde, 4376
Hydracrylic Acid, 4681
D-Lactic Acid, 5214
DL-Lactic Acid, 5215
L-Lactic Acid, 5216
Methyl Carbonate, 5960
s-Trioxane, 9646

C₃H₆O₄
Glyceric Acid, 4378

C₃H₇Br
Isopropyl Bromide, 5098
Propyl Bromide, 7857

C₃H₇CaO₆P
Calcium Glycerophosphate, 1673

C₃H₇Cl
Isopropyl Chloride, 5099
Propyl Chloride, 7859

C₃H₇ClO
Propylene Chlorohydrin, 7863
sec-Propylene Chlorohydrin, 7864

C₃H₇ClO₂
α-Chlorohydrin, 2145

C₃H₇I
Isopropyl Iodide, 5102
Propyl Iodide, 7875

C₃H₇IO₂
Glyceryl Iodide, 4383

C₃H₇K₂O₆P
Potassium Glycerophosphate, 7616

C₃H₇N
Allylamine, 285

C₃H₇NO
Acetoxime, 68
N,N-Dimethylformamide, 3232
Propionamide, 7836

C₃H₇NO₂
L-Alanine, 195
β-Alanine, 196
Formicin, 4154
Isopropyl Nitrite, 5104
1-Nitropropane, 6548
2-Nitropropane, 6549
n-Propyl Nitrite, 7878
Sarcosine, 8331
Urethan, 9789

C₃H₇NO₂S
Cysteine, 2787

C₃H₇NO₂Se
Selenocysteine, 8394

C₃H₇NO₃
n-Propyl Nitrate, 7877
Serine, 8411

C₃H₇NO₅S
Cysteic Acid, 2786

C₃H₇N₃O
N-Ethyl-N-nitrosourea, 3788
Glycocyamine, 4391

C₃H₇N₃O₄
L-Alanosine, 197

C₃H₇NaO₃S₃
2,3-Dimercapto-1-propanesulfonic
Acid*, 3197

C₃H₇NaO₄S
Acetone Sodium Bisulfite, 61

C₃H₇Na₂O₆P
Sodium Glycerophosphate, 8570

C₃H₇O₄P
Fosfomycin, 4169

C₃H₇O₆P
Glyceraldehyde 3-Phosphate, 4377

C₃H₈
Propane, 7809

C₃H₈HgO₂
Merisoprol Hg 197, 5803

C₃H₈IN
Dimethyl(methylene)ammonium
Iodide, 3239

C₃H₈NO₅P
Glyphosate, 4408

C₃H₈NO₆P
Phosphoserine, 7338

C₃H₈N₂OS
Noxytiolin, 6645

C₃H₈N₂O₂
2,3-Diaminopropionic Acid, 2962

C₃H₈N₂O₃
Oxymethurea, 6921

C₃H₈N₂S
N,N'-Dimethylthiourea, 3250

C₃H₈O
Ethyl Methyl Ether, 3783
Isopropyl Alcohol, 5096
n-Propyl Alcohol, 7854

C₃H₈OS₂
Dimercaprol, 3196

C₃H₈O₂
Methylal, 5936
Methyl Cellosolve®, 5961
Propylene Glycol, 7868
Trimethylene Glycol, 9629

C₃H₈O₂S
Thioglycerol, 9264

C₃H₈O₃
Glycerol, 4379

C₃H₈O₃S
Ethyl Methanesulfonate, 3782

C₃H₈O₃S₃
2,3-Dimercapto-1-propanesulfonic
Acid, 3197

C₃H₈S₂
Bis[methylthio]methane, 1267
1,3-Propanedithiol, 7811

C₃H₉AsO₃
1-Propanearsonic Acid, 7810

C₃H₉BO₃
Trimethyl Borate, 9626

C₃H₉N
Isopropylamine, 5097
Propylamine, 7855
Trimethylamine, 9625

C₃H₉NNa₂O₇P₂
Pamidronic Acid*, 6954

C₃H₉NO
2-Aminopropanol, 480
2-Methylaminoethanol, 5939
Trimethylamine*, 9625

C₃H₉NO₃
Ammonium Lactate, 555

C₃H₉NO₃S
N-Methyltaurine, 6043

C₃H₉O₆P
Glycerophosphoric Acid, 4381

C₃H₁₀ClN
Trimethylamine*, 9625

C₃H₁₀N₂
Propylenediamine, 7865

C₃H₁₁Br₂N₃S
AET, 166

C₃H₁₁NO₇P₂
Pamidronic Acid, 6954

C₃H₁₃N₃O₁₀Zr
Ammonium Zirconyl Carbonate,
600

C₃H₁₆N₄O₁₁U
Ammonium Uranium Carbonate,
596

C₃La₂O₉
Lanthanum*, 5233

C₃O₂
Carbon Suboxide, 1821

C₄BaN₄Pt
Barium Platinous Cyanide, 1003

C₄CdK₂N₄
Cadmium Potassium Cyanide, 1622

C₄Cl₂O₄Rh₂
Rhodium Carbonyl Chloride, 8188

C₄F₈
Octafluorocyclobutane, 6666

C₄HCoO₄
Dicobalt Octacarbonyl*, 3074

C₄HI₄N
Iodopyrrole, 4934

C₄H₂Cl₂O₃
Mucochloric Acid, 6208

C₄H₂FeO₄
Ferrous Fumarate, 3995
Hydrogen Tetracarbonylferrate(II),
4731

C₄H₂N₂O₄
Alloxan, 281

C₄H₂N₄O₂
5-Diazouracil, 2985

C₄H₂O₃
Maleic Anhydride, 5586

C₄H₃BrN₂O₂
5-Bromouracil, 1430

C₄H₃FN₂O₂
Fluorouracil, 4109

C₄H₃IN₂OS
Iothiouracil, 4954

C₄H₃KO₈
Potassium Tetroxalate, 7684

C₄H₃N₃O₃S
Forminitrazole, 4155

C₄H₃N₃O₄
Oxonic Acid, 6901
Violuric Acid, 9904

C₄H₃N₃O₅
5-Nitrobarbituric Acid, 6507

C₄H₄As₂Na₂O₄
Arsenoacetic Acid*, 836

C₄H₄Au₂CaO₄S₂
Myoral, 6241

C₄H₄BrNO₂
N-Bromosuccinimide, 1428

C₄H₄Br₂O₄
2,3-Dibromosuccinic Acid, 3013

C₄H₄CaO₄
Calcium Succinate, 1712

C₄H₄CaO₆
Calcium Tartrate, 1716

C₄H₄CdO₄
Cadmium Succinate, 1626

C₄H₄ClNO₂
N-Chlorosuccinimide, 2165

C₄H₄Cl₂O₂
Succinyl Chloride, 8844

C₄H₄Cl₂O₃
Chloroacetic Anhydride, 2112

C₄H₄CuO₆
Cupric Tartrate, 2662

C₄H₄FN₃O
Flucytosine, 4056

C₄H₄F₆O
Flurothyl, 4126

C₄H₄FeO₄
Ferrous Succinate, 4005

C₄H₄INO₂
N-Iodosuccinimide, 4936

C₄H₄KNO₄S
Acesulfame*, 30

C₄H₄KNaO₆
Potassium Sodium Tartrate, 7660

C₄H₄K₂O₄·3H₂O
Succinic Acid*, 8840

C₄H₄K₂O₆
Potassium Tartrate, 7669

C₄H₄N₂
Pyrazine, 7971
Pyridazine, 7982
Pyrimidine, 7998
Succinonitrile, 8843

C₄H₄N₂OS
2-Thiouracil, 9298

C₄H₄N₂O₂
Cellocidin, 1958
Maleic Hydrazide, 5587
Uracil, 9761

C₄H₄N₂O₃
Barbituric Acid, 973

Denotes derivative of title compound

C₄H₄N₆O
8-Azaguanine, 912
C₄H₄NaO₄S₂Sb
Antimony Sodium Thioglycollate, 735
C₄H₄NaO₇Sb
Antimony Sodium Tartrate, 734
C₄H₄Na₂O₄
Sodium Succinate, 8635
C₄H₄Na₂O₆
Sodium Tartrate, 8640
C₄H₄O
Furan, 4206
C₄H₄O₃
Succinic Anhydride, 8841
C₄H₄O₄
Fumaric Acid, 4200
Maleic Acid, 5585
C₄H₄O₅
Oxalacetic Acid, 6863
C₄H₄O₆
Dihydroxymaleic Acid, 3170
C₄H₄O₆Sn
Stannous Tartrate, 8752
C₄H₄O₆Zn
Zinc Tartrate, 10067
C₄H₄S
Thiophene, 9283
C₄H₅BrO₄
Bromosuccinic Acid, 1427
C₄H₅Cl₃O
α,α,β-Trichloro-n-butyraldehyde, 9545
C₄H₅Cl₃O₂
Trichloroacetic Acid*, 9539
C₄H₅F₃O
Fluroxene, 4127
C₄H₅KO₆
Potassium Bitartrate, 7593
C₄H₅LiO₆
Lithium Bitartrate, 5400
C₄H₅N
Allyl Cyanide, 288
Methacrylonitrile, 5850
1H-Pyrrole, 8025
C₄H₅NO₂
Succinimide, 8842
C₄H₅NO₃
Maleamic Acid, 5583
C₄H₅NO₄S
Acesulfame, 30
C₄H₅NS
Allyl Isothiocyanate, 293
Parathiazine, 6982
C₄H₅N₃O
Cytosine, 2801
C₄H₅N₃O₂
6-Azathymine, 919
Mizoribine*, 6137
C₄H₅N₃O₃
Uramil, 9763
C₄H₅NaO₆
Sodium Bitartrate, 8534
C₄H₆
1,3-Butadiene, 1500
C₄H₆As₂O₄
Arsenoacetic Acid, 836
C₄H₆As₆Cu₄O₁₆
Cupric Acetoarsenite, 2629
C₄H₆BaO₄
Barium Acetate, 975
C₄H₆BeO₄
Beryllium Acetate, 1178
C₄H₆CaO₄
Calcium Acetate, 1643
C₄H₆CdO₄
Cadmium Acetate, 1612
C₄H₆CoO₄
Cobaltous Acetate, 2427
C₄H₆CrO₄
Chromous Acetate, 2244

C₄H₆CuO₄
Cupric Acetate, 2627
C₄H₆HgO₄
Mercuric Acetate, 5766
C₄H₆Hg₂O₄
Mercurous Acetate, 5792
C₄H₆MgO₄
Magnesium Acetate, 5530
C₄H₆MnN₂S₄
Maneb, 5603
C₄H₆MnO₄
Manganese Acetate, 5605
C₄H₆N₂Na₂S₄
Nabam, 6256
C₄H₆N₂O₂
Diazoacetic Ester, 2980
Muscimol, 6221
2,5-Piperazinedione, 7434
C₄H₆N₂S
2-Amino-4-methylthiazole, 462
Methimazole, 5892
C₄H₆N₂S₄Zn
Zineb, 10071
C₄H₆N₄O
2,4-Diamino-6-hydroxypyrimidine, 2959
C₄H₆N₄O₂
Acetyleneurea, 87
Divicine, 3389
C₄H₆N₄O₃
Allantoin, 246
C₄H₆N₄O₃S₂
Acetazolamide, 45
C₄H₆N₄O₁₂
Erythrityl Tetranitrate, 3622
C₄H₆NiO₄
Nickel Acetate, 6407
C₄H₆O
Crotonaldehyde, 2599
Methyl Vinyl Ketone, 6052
Vinyl Ether, 9899
C₄H₆O₂
Butyrolactone, 1596
Crotonic Acid, 2600
Diacetyl, 2946
Erythritol Anhydride, 3621
Isocrotonic Acid, 5048
Methacrylic Acid, 5849
Methyl Acrylate, 5935
Vinyl Acetate, 9896
C₄H₆O₂S
3-Sulfolene, 8935
C₄H₆O₃
Acetic Anhydride, 48
Acetoacetic Acid, 51
C₄H₆O₄
Methyl Oxalate, 6020
Succinic Acid, 8840
C₄H₆O₄Pb
Lead Acetate, 5268
C₄H₆O₄Pd
Palladium Diacetate, 6942
C₄H₆O₄S
Thiodiglycolic Acid, 9260
Thiomalic Acid, 9272
C₄H₆O₄S₂
Succimer, 8836
C₄H₆O₄Sn
Stannous Acetate, 8740
C₄H₆O₄Sr
Strontium Acetate, 8798
C₄H₆O₄Zn
Zinc Acetate, 10026
C₄H₆O₅
Malic Acid, 5589
C₄H₆O₆
D-Tartaric Acid, 9037
DL-Tartaric Acid, 9038
L-Tartaric Acid, 9039
meso-Tartaric Acid, 9040

C₄H₆O₆S
Sulfonyldiacetic Acid, 8940
C₄H₆O₆U
Uranyl Acetate, 9774
C₄H₆O₈
Dihydroxytartaric Acid, 3172
C₄H₇AlO₅
Aluminum Diacetate, 339
C₄H₇BrO₂
α-Bromobutyric Acid, 1401
α-Bromoisobutyric Acid, 1408
C₄H₇Br₂Cl₂O₄P
Naled, 6272
C₄H₇Br₃O
Tribromo-tert-butyl Alcohol, 9523
C₄H₇Cl
1-Chloro-2-butene, 2130
1-Chloro-2-methylpropene, 2147
C₄H₇ClO
n-Butyryl Chloride, 1598
2-Chloroethyl Vinyl Ether, 2139
C₄H₇ClO₂
β-Chloroethyl Acetate, 2137
Ethyl Chloroacetate, 3741
Propyl Chlorocarbonate, 7860
C₄H₇Cl₂O₄P
Dichlorvos, 3069
C₄H₇Cl₃N₂O₂
Mecloralurea, 5662
C₄H₇Cl₃O
Chlorobutanol, 2129
C₄H₇Cl₃O₂
Chloral Alcoholate, 2056
C₄H₇FeO₅
Ferric Acetate, Basic, 3957
C₄H₇N
Butyronitrile, 1597
3-Pyrroline, 8028
C₄H₇NO
Acetone Cyanohydrin, 59
2-Pyrrolidone, 8027
C₄H₇NO₂
2-Azetidinecarboxylic Acid, 923
C₄H₇NO₂S
Timonacic, 9375
C₄H₇NO₃
Aceturic Acid, 74
C₄H₇NO₄
Aspartic Acid, 862
Iminodiacetic Acid, 4832
C₄H₇N₃O
Creatinine, 2571
C₄H₇NaOS₂
Sodium Isopropyl Xanthate, 8585
C₄H₇NaO₃
Sodium Oxybate, 8603
C₄H₈
1-Butene, 1513
2-Butene, 1514
Cyclobutane, 2720
Isobutylene, 5024
C₄H₈Br₂
α-Butylene Dibromide, 1565
C₄H₈Cl₂
Butylidene Chloride, 1571
C₄H₈Cl₂O
sym-Dichloroethyl Ether, 3055
C₄H₈Cl₂S
Mustard Gas, 6225
C₄H₈Cl₃O₄P
Trichlorfon, 9536
C₄H₈CuN₂O₄
Cupric Glycinate, 2644
C₄H₈N₂
Lysidine, 5508
C₄H₈N₂O
Allylurea, 296
N-Nitrosopyrrolidine, 6564

Denotes derivative of title compound

*Denotes derivative of title compound

C₄K₂N₄Ni
Potassium Tetracyanonickelate(II), 7677

C₄K₂N₄Pt
Potassium Tetracyanoplatinate(II), 7678

C₄K₂N₄Zn
Potassium Tetracyanozincate, 7679

C₄K₂O₅Ti
Potassium Titanyl Oxalate, 7689

C₄Li₂N₄Pt
Lithium Tetracyanoplatinate(II), 5422

C₄NiO₄
Nickel Carbonyl, 6411

C₅FeK₂N₆O
Potassium Nitroprusside, 7636

C₅FeN₆Na₂O
Sodium Nitroprusside, 8600

C₅FeO₅
Iron Pentacarbonyl, 4980

C₅H₃ClN₄
6-Chloropurine, 2161

C₅H₃ClO₂
Furoyl Chloride, 4222

C₅H₃I₂NO
Iopydone, 4951

C₅H₃NaO₂S
2-Thiophenecarboxylic Acid*, 9284

C₅H₄F₈O
2,2,3,3,4,4,5,5-Octafluoro-1-pentanol, 6667

C₅H₄NNaOS
Pyrithione*, 8004

C₅H₄N₂O₂
Pyrazinoic Acid, 7973

C₅H₄N₂O₄
Nifuroxime, 6451
Orotic Acid, 6828

C₅H₄N₄
1H-Purine, 7959

C₅H₄N₄O
Allopurinol, 278
Hypoxanthine, 4805

C₅H₄N₄O₂
Xanthine, 9968

C₅H₄N₄O₃
Uric Acid, 9790

C₅H₄N₄S
6-Mercaptopurine, 5762

C₅H₄O₂
Furfural, 4214
Protoanemonin, 7907

C₅H₄O₂S
3-Thenoic Acid, 9206
2-Thiophenecarboxylic Acid, 9284

C₅H₄O₃
2-Furoic Acid, 4219

C₅H₅N
Pyridine, 7983

C₅H₅NO
Pyridine 1-Oxide, 7984

C₅H₅NOS
Pyrithione, 8004

C₅H₅NO₂
Mecrylate, 5667

C₅H₅N₃O
Pyrazinamide, 7970

C₅H₅N₃O₃S
Aminitrozole, 421

C₅H₅N₅
Adenine, 141

C₅H₅N₅O
Guanine, 4477

C₅H₅N₅S
Thioguanine, 9266

C₅H₆
Cyclopentadiene, 2744

C₅H₆Br₃N
Pyridinium Bromide Perbromide, 7985

C₅H₆CaN₂O₅
Calcium N-Carbamoylaspartate, 1655

C₅H₆ClCrNO₃
Pyridinium Chlorochromate, 7986

C₅H₆Cl₂N₂O₂
1,3-Dichloro-5,5-dimethylhydantoin, 3054

C₅H₆N₂
α-Aminopyridine, 486
β-Aminopyridine, 487
Glutaronitrile, 4368

C₅H₆N₂OS
Methylthiouracil, 6048

C₅H₆N₂O₂
5-Methylpyrazole-3-carboxylic Acid, 6033
6-Methyluracil, 6051
Thymine, 9332

C₅H₆N₂O₄
Hydroorotic Acid, 4735
Ibotenic Acid, 4808
Maleuric Acid, 5588
Muscazone, 6220

C₅H₆N₆
2,6-Diaminopurine, 2963

C₅H₆Na₄O₁₀P₂
Butedronic Acid*, 1512

C₅H₆O₂
Angelica Lactone, 679
Furfuryl Alcohol, 4215
α-Methylene Butyrolactone, 5981

C₅H₆O₃
Reductic Acid, 8134

C₅H₆O₄
Citraconic Acid, 2323
Itaconic Acid, 5130
Mesaconic Acid, 5806

C₅H₆O₅
Acetonedicarboxylic Acid, 60
α-Ketoglutaric Acid, 5182

C₅H₇NOS
Goitrin, 4411

C₅H₇NO₂
Ethyl Cyanoacetate, 3744

C₅H₇NO₃
Dimethadione, 3201
L-Pyroglutamic Acid, 8012

C₅H₇NO₆
β-Carboxyaspartic Acid, 1833

C₅H₇NS
2,4-Dimethylthiazole, 3249

C₅H₇N₃O
5-Methylcytosine, 5970

C₅H₇N₃O₂
Dimetridazole, 3255

C₅H₇N₃O₄
Azaserine, 916

C₅H₇N₃O₅
Quisqualic Acid, 8112

C₅H₇N₅O
Hydracarbazine, 4680

C₅H₈
Isoprene, 5087

C₅H₈Cl₂O₂
Caldariomycin, 1721

C₅H₈Cl₃NO₃
Carbocloral, 1808

C₅H₈FeN₆O
Ammonium Nitroferricyanide, 562

C₅H₈NNaO₄.H₂O
Monosodium Glutamate, 6165

C₅H₈N₂S₄Zn
Propineb, 7831

C₅H₈N₄O₃S₂
Methazolamide, 5875

C₅H₈N₄O₁₂
Pentaerythritol Tetranitrate, 7066

C₅H₈O
Cyclopentanone, 2748
2-Methyl-3-butyn-2-ol, 5956
Senecialdehyde, 8399

C₅H₈O₂
Acetylacetone, 75
Angelic Acid, 678
Ethyl Acrylate, 3715
Glutaraldehyde, 4366
Isopropenyl Acetate, 5091
Tiglic Acid, 9365

C₅H₈O₃
Levulinic Acid, 5352
Methyl Acetoacetate, 5933

C₅H₈O₄
Glutaric Acid, 4367
Methyl Malonate, 6009

C₅H₈O₅
Citramalic Acid, 2325

C₅H₉BiO₃
Bismuth Valerate, Basic, 1306

C₅H₉BrO₂
α-Bromoisovaleric Acid, 1409
β-Bromoisovaleric Acid, 1410
Ethyl α-Bromopropionate, 3731

C₅H₉ClO
Isovaleryl Chloride, 5121

C₅H₉ClO₂
Ethyl α-Chloropropionate, 3743
Isobutyl Chlorocarbonate, 5023

C₅H₉Cl₂N₃O₂
Carmustine, 1852

C₅H₉IO₃
Domiodol, 3410

C₅H₉N
N-Methylpyrroline, 6036

C₅H₉NO₂
Proline, 7790

C₅H₉NO₂S₂
Cheirolin, 2040

C₅H₉NO₃
δ-Aminolevulinic Acid, 458
4-Hydroxy-L-proline, 4775

C₅H₉NO₃S
Acetylcysteine, 82
Tiopronin, 9386

C₅H₉NO₄
Glutamic Acid, 4363
HON, 4663
NMDA, 6584

C₅H₉NO₄S
Carbocysteine, 1809

C₅H₉NO₅
Hydroxyglutamic Acid, 4756

C₅H₉NS
Isobutyl Thiocyanate, 5036

C₅H₉N₃
Betazole, 1205
Histamine, 4640

C₅H₉N₃O₁₀
Pentrinitrol, 7094

C₅H₁₀
Amylene, 644
Cyclopentane, 2746
1-Pentene, 7079
2-Pentene, 7080

C₅H₁₀ClNO₄
Glutamic Acid*, 4363

C₅H₁₀Cl₂
Amylene Dichloride, 645

C₅H₁₀Cl₂O₂
Pentaerythritol Dichlorohydrin, 7064

C₅H₁₀NNaS₂
Ditiocarb Sodium, 3384

C₅H₁₀N₂O₂S
Methomyl, 5905

C₅H₁₀N₂O₃
Glutamine, 4365
L-Isoglutamine, 5061

C₆H₃Cl₃O

$C_6H_3Cl_3O$
2,4,5-Trichlorophenol, 9555
2,4,6-Trichlorophenol, 9556

$C_6H_3Cl_4N$
Nitrapyrin, 6490

$C_6H_3FN_2O_4$
1-Fluoro-2,4-dinitrobenzene, 4101

$C_6H_3I_2NO_3$
Disophenol, 3365

$C_6H_3I_2NaO_4S.2H_2O$
Sozoiodolic Acid*, 8687

$C_6H_3N_3O_6$
sym-Trinitrobenzene, 9639

$C_6H_3N_3O_7$
Picric Acid, 7380

$C_6H_3N_3O_8$
Styphnic Acid, 8828

$C_6H_4BrClO_2S$
p-Bromobenzenesulfonyl Chloride, 1395

$C_6H_4Br_2$
p-Dibromobenzene, 3001

$C_6H_4Br_2O_4S$
3,5-Dibromo-4-hydroxybenzenesulfonic Acid, 3007

$C_6H_4Br_3N$
2,4,6-Tribromoaniline, 9522

$C_6H_4ClNO_2$
Chloronitrobenzene, 2151

$C_6H_4Cl_2$
m-Dichlorobenzene, 3043

$C_6H_4Cl_2O$
2,4-Dichlorophenol, 3061
2,6-Dichlorophenol, 3062

$C_6H_4Cu_2O_7$
Cupric Citrate, 2639

$C_6H_4F_2$
p-Difluorobenzene, 3132

$C_6H_4HgNNaO_4$
Mercurophen, 5791

$C_6H_4I_2O_4S$
Sozoiodolic Acid, 8687

$C_6H_4NNaO_2$
Nicotinic Acid*, 6435

$C_6H_4N_2O_3S$
p-Diazobenzenesulfonic Acid, 2982

$C_6H_4N_2O_4$
Dinitrobenzene, 3266
2,3-Pyrazinedicarboxylic Acid, 7972

$C_6H_4N_2O_5$
2,4-Dinitrophenol, 3274

$C_6H_4N_2O_6$
2,4-Dinitroresorcinol, 3278

$C_6H_4N_4$
2-Amino-1,1,3-tricyanopropene, 495
Pteridine, 7947

$C_6H_4N_4O_2$
Lumazine, 5467

$C_6H_4N_4O_3S$
Triafur, 9509

$C_6H_4Na_2O_8S_2$
Tiron®, 9392

$C_6H_4O_2$
Quinone, 8103

$C_6H_4O_4$
Coumalic Acid, 2558

$C_6H_4O_6$
Tetroquinone, 9177

$C_6H_5Ag_3O_7$
Silver Citrate, 8457

C_6H_5AsO
Oxophenylarsine, 6903

C_6H_5Br
Bromobenzene, 1394

C_6H_5BrO
Bromophenol, 1416

C_6H_5Cl
Chlorobenzene, 2121

C_6H_5ClHg
Phenylmercuric Chloride, 7272

C_6H_5ClHgO
Mercufenol Chloride, 5763

$C_6H_5ClHgO_2$
Hydroxymercurichlorophenols*, 4761

C_6H_5ClMg
Phenylmagnesium Chloride, 7269

$C_6H_5ClNNaO_2S$
Chloramine-B, 2065

C_6H_5ClO
Chlorophenol, 2154

$C_6H_5ClO_2S$
Benzenesulfonyl Chloride, 1080

$C_6H_5ClO_3S$
p-Chlorobenzenesulfonic Acid, 2122

$C_6H_5Cl_2N$
3,4-Dichloroaniline, 3041

$C_6H_5Cl_2NO$
2-Amino-4,6-dichlorophenol, 448

C_6H_5F
Fluorobenzene, 4099

$C_6H_5F_2N$
2,4-Difluoroaniline, 3131

$C_6H_5GaO_7$
Gallium*, 4252

C_6H_5I
Iodobenzene, 4922

C_6H_5IO
o-Iodophenol, 4928
p-Iodophenol, 4929

$C_6H_5I_2N$
2,4-Diiodoaniline, 3173

C_6H_5KO
Potassium Phenoxide, 7646

$C_6H_5KO_4S$
Potassium Phenolsulfonate, 7645

$C_6H_5K_3O_7$
Potassium Citrate, 7603

$C_6H_5Li_3O_7$
Lithium Citrate, 5407

$C_6H_5NO_2$
Isonicotinic Acid, 5073
Nicotinic Acid, 6435
Nitrobenzene, 6509
4-Nitrosophenol, 6563
Picolinic Acid, 7375

$C_6H_5NO_3$
o-Nitrophenol, 6541
Oxiniacic Acid, 6895

$C_6H_5NO_4$
Citrazinic Acid, 2327

$C_6H_5N_3$
1H-Benzotriazole, 1119

$C_6H_5N_3O_4$
2,4-Dinitroaniline, 3263
2,6-Dinitroaniline, 3264

$C_6H_5N_3O_5$
Picramic Acid, 7379

$C_6H_5N_5O_2$
Xanthopterin, 9973

$C_6H_5N_5O_3$
Leucopterin, 5338

C_6H_5Na
Benzene*, 1074

C_6H_5NaO
Phenol*, 7206
Sodium Phenoxide, 8611

$C_6H_5NaO_3S$
Benzenesulfonic Acid*, 1078

$C_6H_5NaO_4S$
Sodium Phenolsulfonate, 8610

$C_6H_5Na_2O_4P$
Disodium Phenyl Phosphate, 3364

$C_6H_5Na_3O_7$
Sodium Citrate, 8549

C_6H_6
Benzene, 1074

$C_6H_6AsNO_5$
Nitarsone, 6483

$C_6H_6AsNO_6$
Roxarsone, 8251

C_6H_6BrN
p-Bromoaniline, 1392

$C_6H_6Br_2N_2$
1,2-Dibromo-2,4-dicyanobutane, 3004

C_6H_6ClN
Chloroaniline, 2118

$C_6H_6Cl_2N_2O_4S_2$
Dichlorphenamide, 3067

$C_6H_6Cl_6$
Lindane, 5379

C_6H_6FN
p-Fluoroaniline, 4098

$C_6H_6FNO_2S$
Sulfanilyl Fluoride, 8904

$C_6H_6MgO_7$
Magnesium Citrate, Dibasic, 5542

C_6H_6NI
p-Iodoaniline, 4920

$C_6H_6NNa_3O_6$
Nitrilotriacetic Acid*, 6499

$C_6H_6N_2O$
Niacinamide, 6398

$C_6H_6N_2O_2$
6-Aminonicotinic Acid, 468
m-Nitroaniline, 6503

$C_6H_6N_2O_3$
Acipimox, 106

$C_6H_6N_4O_3S$
Niridazole, 6480

$C_6H_6N_4O_4$
2,4-Dinitrophenylhydrazine, 3277
Nitrofurazone, 6521

$C_6H_6N_4O_7$
Ammonium Picrate, 575

$C_6H_6Na_2O_7$
Sodium Citrate, Acid, 8550

C_6H_6O
Phenol, 7206

$C_6H_6O_2$
Hydroquinone, 4738
Pyrocatechol, 8009
Resorcinol, 8158

$C_6H_6O_3$
1,2,4-Benzenetriol, 1081
2-Furoic Acid*, 4219
5-(Hydroxymethyl)-2-furaldehyde, 4764
Isomaltol, 5066
Maltol, 5594
Phloroglucinol, 7301
Pyrogallol, 8010

$C_6H_6O_3S$
Benzenesulfonic Acid, 1078

$C_6H_6O_4$
Kojic Acid, 5197
Muconic Acid, 6210

$C_6H_6O_4S$
p-Phenolsulfonic Acid, 7212

$C_6H_6O_6$
Aconitic Acid, 112
Dehydroascorbic Acid, 2856
Ketipic Acid, 5179

$C_6H_6O_7S_2$
Phenoldisulfonic Acid, 7207

C_6H_6S
Thiophenol, 9285

$C_6H_7AsClNO_2$
Oxophenarsine Hydrochloride, 6902

$C_6H_7AsCl_3NO$
Dichlorophenarsine Hydrochloride, 3060

$C_6H_7AsNNaO_3$
Sodium Arsanilate, 8521

$C_6H_7AsO_3$
Benzenearsonic Acid, 1075

$C_6H_7BHgO_3$
Phenylmercury Borate, 7274

$C_6H_7BO_2$
Benzeneboronic Acid, 1076

Denotes derivative of title compound

Denotes derivative of title compound

Denotes derivative of title compound

Denotes derivative of title compound

C₇H₁₂N₂O₂
Ectylurea, 3478
Heptoxime, 4586

C₇H₁₂N₂O₄
Aceglutamide, 21

C₇H₁₂O
Cycloheptanone, 2728

C₇H₁₂O₂
n-Butyl Acrylate, 1539
Cyclohexanecarboxylic Acid, 2730

C₇H₁₂O₃
Ethyl Levulinate, 3775
Isopropyl Acetoacetate, 5094

C₇H₁₂O₄
n-Butylmalonic Acid, 1574
Diethylmalonic Acid, 3114
Ethyl Malonate, 3779
Pimelic Acid, 7401

C₇H₁₂O₅
Diacetin, 2941

C₇H₁₂O₆
Quinic Acid, 8071

C₇H₁₃BrN₂O₂
Carbromal, 1837

C₇H₁₃N
Quinuclidine, 8109

C₇H₁₃NO
Cycloheptanone*, 2728
3-Quinuclidinol, 8110

C₇H₁₃NO₂
Stachydrine, 8729

C₇H₁₃NO₃
Betonicine, 1209
Turicine, 9729

C₇H₁₃NO₃S
N-Acetylmethionine, 90
N-Acetylpenicillamine, 93

C₇H₁₃NO₃S₂
Bucillamine, 1447

C₇H₁₃NO₄
L-Glutamic Acid 5-Ethyl Ester, 4364

C₇H₁₃N₃O₃S
Oxamyl, 6873

C₇H₁₃NaO₈
Glucoheptonic Acid*, 4349

C₇H₁₃O₅PS
Methacrifos, 5848

C₇H₁₃O₆P
Mevinphos, 6089

C₇H₁₄AsClO₃
Chloroarsenol, 2119

C₇H₁₄BrNO
Ibrotamide, 4809

C₇H₁₄Cl₃NO₄
Chloral Betaine, 2059

C₇H₁₄NO₃PS₂
Phosfolan, 7309

C₇H₁₄NO₅P
Monocrotophos, 6161

C₇H₁₄N₂O₂S
Aldicarb, 216

C₇H₁₄N₂O₄S
L-Cystathionine, 2784

C₇H₁₄N₂O₄S₂
Djenkolic Acid, 3393

C₇H₁₄N₃O₃P
Uredepa, 9787

C₇H₁₄N₃O₄P
Isaxonine*, 4987

C₇H₁₄N₄S₂
Methallibure, 5857

C₇H₁₄O
Cyclohexylcarbinol, 2737
Dipropyl Ketone, 3351
Heptanal, 4578
2-Heptanone, 4584

C₇H₁₄O₂
n-Butyl Propionate, 1587
Ethyl Isovalerate, 3772
Heptanoic Acid, 4581
Isoamyl Acetate, 4993
Isobutyl Propionate, 5033
Propyl Butyrate, 7858
Valeric Acid, Normal*, 9815

C₇H₁₄O₄
D-Chalcose, 2030
Cymarose, 2769
Mycarose, 6229
Sarmentose, 8335

C₇H₁₄O₅
Digitalose, 3142
Thevetose, 9216

C₇H₁₄O₆
α-Methylglucoside, 5997

C₇H₁₄O₇
D-manno-Heptulose, 4587

C₇H₁₄O₈
Glucoheptonic Acid, 4349

C₇H₁₅Cl₂N₂O₂P
Cyclophosphamide, 2753
Ifosfamide, 4822

C₇H₁₅Cl₄N
Novembichin, 6640

C₇H₁₅NO
1-Ethyl-3-piperidinol, 3799

C₇H₁₅NO₂
Emylcamate, 3520
Isobutyl Urethane, 5037

C₇H₁₅NO₃
Carnitine, 1856

C₇H₁₅NO₄S
MOPS, 6173

C₇H₁₅NO₅
N-Methyl-α-L-glucosamine, 5996

C₇H₁₅N₃O₂
Indospicine, 4877

C₇H₁₅N₃O₂S
Cartap, 1874

C₇H₁₅NaO₄S
Heptanal Sodium Bisulfite, 4579

C₇H₁₆
n-Heptane, 4580

C₇H₁₆BrNO₂
Acetylcholine Bromide, 80

C₇H₁₆ClN
Mepiquat Chloride, 5746

C₇H₁₆ClNO₂
Acetylcholine Chloride, 81

C₇H₁₆ClNO₃
Carnitine*, 1856

C₇H₁₆ClN₃O₂S₂
Cartap*, 1874

C₇H₁₆FO₂P
Soman, 8668

C₇H₁₆INO₂
Oxapropanium Iodide, 6877

C₇H₁₆O
1-Heptanol, 4582
2-Heptanol, 4583

C₇H₁₆O₃
Orthoformic Acid*, 6837

C₇H₁₆O₄S₂
Sulfonmethane, 8939

C₇H₁₇AsClNO₃
Chloroarsenol*, 2119

C₇H₁₇ClN₂O₂
Bethanechol Chloride, 1207

C₇H₁₇N
Methylhexaneamine, 6000
Tuaminoheptane, 9712

C₇H₁₇NO₂
Triethylammonium Formate, 9583

C₇H₁₇NO₅
N-Methylglucamine, 5995

C₇H₁₇O₂PS₃
Phorate, 7305

C₇H₁₈NO₇P
Fosfomycin*, 4169

C₇H₁₈NO₈Sb
N-Methylglucamine*, 5995

C₇H₁₉N₃
Spermidine, 8698

C₈Cl₂N₂O₂
2,3-Dichloro-5,6-dicyanobenzoquinone, 3052

C₈Cl₄N₂
Chlorothalonil, 2167

C₈Co₂O₈
Dicobalt Octacarbonyl, 3074

C₈H₂Cl₄O₅
Mucochloric Anhydride, 6209

C₈H₃I₂lNNa₂O₅
Sodium Iodomethamate, 8584

C₈H₃NO₂
Diatretyne II, 2974

C₈H₄AuN₂NaO₂S
Triphal®, 9652

C₈H₄Cl₂O₂
Phthaloyl Chloride, 7349

C₈H₄Cl₆
1,4-Bis(trichloromethyl)benzene, 1313

C₈H₄K₂O₁₂Sb₂.3H₂O
Antimony Potassium Tartrate, 732

C₈H₄N₂S₂
Bitoscanate, 1318

C₈H₄O₃
Phthalic Anhydride, 7346

C₈H₅Cl₃O₂
Chlorfenac, 2085

C₈H₅Cl₃O₃
2,4,5-T, 8999

C₈H₅KO₄
Potassium Biphthalate, 7589

C₈H₅NOS
Benzoyl Isothiocyanate, 1126

C₈H₅NO₂
Agrocybin, 178
Isatin, 4985
Phthalimide, 7347

C₈H₅NO₃
Carsalam, 1873
Diatretyne I, 2973

C₈H₅N₃O₃S₂
Tenonitrozole, 9079

C₈H₆
Ethynylbenzene, 3817

C₈H₆BrClO
p-Chlorophenacyl Bromide, 2153

C₈H₆BrClO₂
p-Bromobenzyl Chloroformate, 1399

C₈H₆BrN
α-Bromobenzyl Cyanide, 1400

C₈H₆Br₂O
p-Bromophenacyl Bromide, 1415

C₈H₆Cl₂O₃
2,4-D, 2802
Dicamba, 3026

C₈H₆F₃N₃O₄S₂
Flumethiazide, 4067

C₈H₆KNO₄S
Indican (Metabolic Indican)*, 4853

C₈H₆N₂
Cinnoline, 2309
Phthalazine, 7344
Quinazoline, 8062
Quinoxaline, 8107

C₈H₆N₂O₂
Fenadiazole, 3898
p-Nitrobenzyl Cyanide, 6512

C₈H₆N₄O₅
Nitrofurantoin, 6520

C₈H₆N₄O₈
Alloxantin, 282

C₈H₆N₆O₁₁
Pentryl®, 7095

C$_8$H$_6$O
 Benzofuran, 1098
C$_8$H$_6$O$_3$
 Piperonal, 7445
C$_8$H$_6$O$_4$
 Isophthalic Acid, 5083
 Phthalic Acid, 7345
 Piperonylic Acid, 7447
 Terephthalic Acid, 9093
C$_8$H$_6$O$_5$
 Fomecins*, 4143
 4-Hydroxyisophthalic Acid, 4758
C$_8$H$_6$O$_6$
 Puberulic Acid, 7952
C$_8$H$_6$S
 Thianaphthene, 9232
C$_8$H$_7$BrO
 p-Bromoacetophenone, 1390
 ω-Bromoacetophenone, 1391
C$_8$H$_7$BrO$_3$
 p-Bromomandelic Acid, 1412
C$_8$H$_7$ClN$_2$O$_2$S
 Diazoxide, 2986
C$_8$H$_7$ClO
 p-Chloroacetophenone, 2114
 ω-Chloroacetophenone, 2115
C$_8$H$_7$ClO$_2$
 p-Anisoyl Chloride, 703
 Carbobenzoxy Chloride, 1807
 m-Chlorobenzoic Acid*, 2124
 p-Chlorobenzoic Acid*, 2126
C$_8$H$_7$Cl$_2$NO$_2$
 Chloramben*, 2063
C$_8$H$_7$Cl$_2$NaO$_5$S
 Disul-sodium, 3372
C$_8$H$_7$Cl$_4$N$_3$O$_4$S$_2$
 Tetrachlormethiazide, 9124
C$_8$H$_7$FO$_2$
 p-Fluorophenylacetic Acid, 4105
C$_8$H$_7$N
 Benzyl Cyanide, 1145
 Indole, 4869
 o-Tolunitrile, 9463
C$_8$H$_7$NO
 Mandelonitrile, 5601
C$_8$H$_7$NO$_2$
 Acetimidoquinone, 49
 Isonitrosoacetophenone, 5077
C$_8$H$_7$NO$_3$
 4-Pyridoxic Acid*, 7994
C$_8$H$_7$NO$_4$
 p-Nitrophenylacetic Acid, 6543
C$_8$H$_7$NO$_4$S
 Indican (Metabolic Indican), 4853
C$_8$H$_7$N$_3$O$_2$
 Luminol, 5470
C$_8$H$_7$N$_3$O$_5$
 Dinitolmide, 3262
 Furazolidone, 4210
C$_8$H$_7$NaO$_3$
 Mandelic Acid*, 5599
C$_8$H$_7$NaO$_4$.H$_2$O
 Dehydroacetic Acid*, 2855
C$_8$H$_8$
 Cubane, 2614
 Styrene, 8830
C$_8$H$_8$AuNOS
 Aurothioglycanide, 902
C$_8$H$_8$BrCl$_2$O$_3$PS
 Bromophos, 1419
C$_8$H$_8$BrNO
 p-Bromoacetanilide, 1387
C$_8$H$_8$ClNO
 Chloroacetanilide, 2110
C$_8$H$_8$ClNO$_3$S
 N-Acetylsulfanilyl Chloride, 99
C$_8$H$_8$Cl$_2$IO$_3$PS
 Iodofenphos, 4925
C$_8$H$_8$Cl$_2$N$_4$
 Guanabenz, 4469

C$_8$H$_8$Cl$_2$N$_4$O
 Guanoxabenz, 4481
C$_8$H$_8$Cl$_2$O
 Dichloroxylenol, 3066
C$_8$H$_8$Cl$_3$N$_3$O$_4$S$_2$
 Clorsulon, 2405
 Trichlormethiazide, 9537
C$_8$H$_8$Cl$_3$O$_3$PS
 Ronnel, 8239
C$_8$H$_8$F$_3$N$_3$O$_4$S$_2$
 Hydroflumethiazide, 4716
C$_8$H$_8$HgN$_2$O$_4$
 Mercuric Succinimide, 5786
C$_8$H$_8$HgO$_2$
 Phenylmercuric Acetate, 7271
C$_8$H$_8$K$_2$O$_5$
 Endothall*, 3530
C$_8$H$_8$NNaO$_4$S
 N-Acetylsulfanilic Acid*, 98
C$_8$H$_8$N$_2$O$_2$
 Phthalamide, 7343
 Ricinine, 8212
C$_8$H$_8$N$_2$O$_2$S$_2$
 Thiolutin, 9271
C$_8$H$_8$N$_2$O$_3$
 Nitroacetanilide, 6501
C$_8$H$_8$N$_2$O$_3$S
 Zonisamide, 10094
C$_8$H$_8$N$_4$
 Hydralazine, 4682
C$_8$H$_8$N$_4$O$_4$
 Nifuradene, 6445
C$_8$H$_8$N$_6$O$_6$
 Murexide, 6215
C$_8$H$_8$Na$_2$O$_5$
 Endothall*, 3530
C$_8$H$_8$O
 Acetophenone, 65
 Coumaran, 2560
 Phenylacetaldehyde, 7236
 o-Tolualdehyde, 9453
C$_8$H$_8$O$_2$
 p-Anisaldehyde, 693
 Benzyl Formate, 1148
 Methyl Benzoate, 5947
 Phenyl Acetate, 7238
 Phenylacetic Acid, 7239
 Toluic Acid, 9461
C$_8$H$_8$O$_3$
 p-Anisic Acid, 696
 m-Cresotic Acid, 2584
 Mandelic Acid, 5599
 Methylparaben, 6021
 Methyl Salicylate, 6038
 Phenoxyacetic Acid, 7223
 Resacetophenone, 8143
 Resorcinol*, 8158
 Vanillin, 9839
C$_8$H$_8$O$_4$
 Dehydroacetic Acid, 2855
 2,6-Dimethoxyquinone, 3214
 Fumigatin, 4201
 Gallacetophenone, 4248
 Homogentisic Acid, 4658
 o-Orsellinic Acid, 6834
 Pyrogallol*, 8010
 Vanillic Acid, 9838
C$_8$H$_8$O$_5$
 Fomecins*, 4143
 Gallic Acid*, 4251
 Spinulosin, 8706
C$_8$H$_9$AsBiNO$_6$
 Glycobiarsol, 4389
C$_8$H$_9$AsNNaO$_5$
 Glycarsamide*, 4375
C$_8$H$_9$Br
 Xylyl Bromide, 9998
C$_8$H$_9$BrO
 Bromophenol*, 1416
C$_8$H$_9$Cl
 Xylyl Chloride, 9999

C$_8$H$_9$ClNO$_5$PS
 Chlorthion®, 2196
 Dicapthon, 3027
C$_8$H$_9$ClN$_4$
 Hydralazine*, 4682
C$_8$H$_9$ClO
 Chloroxylenol, 2176
C$_8$H$_9$Cl$_3$N$_4$O
 Guanoxabenz*, 4481
C$_8$H$_9$FN$_2$O$_3$
 Tegafur, 9060
C$_8$H$_9$FO$_2$S
 Fluoresone, 4088
C$_8$H$_9$HgNaO$_3$S$_2$
 Thimerfonate Sodium, 9243
C$_8$H$_9$I$_2$NO$_3$
 Iopydol, 4950
C$_8$H$_9$NO
 Acetanilide, 42
 Aminoacetophenone, 424
 Phenacylamine, 7159
 α-Phenylacetamide, 7237
 o-Toluamide, 9454
C$_8$H$_9$NO$_2$
 Acetaminophen, 40
 p-Aminophenylacetic Acid, 475
 Methyl Anthranilate, 5942
 N-Phenylglycine, 7262
 α-Phenylglycine, 7263
C$_8$H$_9$NO$_3$
 N-(4-Hydroxyphenyl)glycine, 4771
 Orthocaine, 6836
 Pyridoxal, 7991
C$_8$H$_9$NO$_4$
 4-Pyridoxic Acid, 7994
C$_8$H$_9$NO$_4$S
 N-Acetylsulfanilic Acid, 98
C$_8$H$_9$NO$_5$
 Clavulanic Acid, 2342
C$_8$H$_9$N$_2$NaO$_3$S.H$_2$O
 Sulfacetamide*, 8870
C$_8$H$_9$N$_3$O$_3$
 Anot, 707
C$_8$H$_9$N$_3$O$_4$
 Nicorandil, 6431
C$_8$H$_{10}$
 Ethylbenzene, 3723
 Xylene, 9988
C$_8$H$_{10}$AsNO$_4$
 Arsacetin, 817
C$_8$H$_{10}$AsNO$_5$
 Acetarsone, 44
 Glycarsamide, 4375
C$_8$H$_{10}$AsN$_2$NaO$_4$
 Tryparsamide, 9703
C$_8$H$_{10}$ClN$_3$S
 Tiamenidine, 9350
C$_8$H$_{10}$Cl$_2$N$_2$S
 p-Chlorobenzylpseudothiuronium
  Chloride, 2128
C$_8$H$_{10}$NNaO$_3$S
 Sulbactam*, 8862
C$_8$H$_{10}$NO$_5$PS
 Methyl Parathion, 6022
C$_8$H$_{10}$NO$_6$P
 Pyridoxal 5-Phosphate, 7992
C$_8$H$_{10}$N$_2$O
 p-Aminoacetanilide, 422
 Benzylurea, 1165
 p-Nitroso-N,N-dimethylaniline,
  6559
C$_8$H$_{10}$N$_2$O$_3$
 5-Nitro-o-phenetidine, 6538
 Oxiniacic Acid*, 6895
C$_8$H$_{10}$N$_2$O$_3$S
 N^4-Acetylsulfanilamide, 97
 Sulfacetamide, 8870
 p-Tolylsulfonylmethylnitrosamide,
  9469
C$_8$H$_{10}$N$_2$O$_4$
 Mimosine, 6118

* Denotes derivative of title compound

C$_8$H$_{10}$N$_2$S
Ethionamide, 3692
C$_8$H$_{10}$N$_4$O$_2$
Caffeine, 1635
Enprofylline, 3544
C$_8$H$_{10}$N$_4$O$_5$
Nidroxyzone, 6439
C$_8$H$_{10}$N$_4$S
Methyl 4-Pyridyl Ketone Thiosemi-
carbazone, 6035
C$_8$H$_{10}$N$_6$
Dihydralazine, 3152
C$_8$H$_{10}$O
Benzyl Methyl Ether, 1154
Methylenomycins*, 5986
Phenethyl Alcohol, 7185
Phenetole, 7189
Phlorol, 7302
Xylenol, 9989
C$_8$H$_{10}$O$_2$
Anise Alcohol, 695
Creosol, 2573
2-Phenoxyethanol, 7226
Styrene Glycol, 8831
Veratrole, 9857
C$_8$H$_{10}$O$_3$
Filicinic Acid, 4021
C$_8$H$_{10}$O$_3$S
Benzenesulfonic Acid*, 1078
C$_8$H$_{10}$O$_4$
Penicillic Acid, 7033
C$_8$H$_{10}$O$_5$
Endothall, 3530
C$_8$H$_{10}$O$_8$
Succinyl Peroxide, 8848
C$_8$H$_{11}$BrN$_2$O$_2$
Isocil, 5042
C$_8$H$_{11}$ClN$_2$S
S-Benzylthiuronium Chloride, 1164
C$_8$H$_{11}$ClN$_4$S
Methyl 4-Pyridyl Ketone Thiosemi-
carbazone*, 6035
C$_8$H$_{11}$Cl$_2$N$_3$O$_2$
Uracil Mustard, 9762
C$_8$H$_{11}$Cl$_2$N$_5$S
Tiamenidine*, 9350
C$_8$H$_{11}$Cl$_3$O$_6$
α-Chloralose, 2062
C$_8$H$_{11}$Cl$_3$O$_7$
Urochloralic Acid, 9798
C$_8$H$_{11}$N
N,N-Dimethylaniline, 3223
Ethylaniline, 3722
3-Ethyl-4-picoline, 3796
4-Ethyl-2-picoline, 3797
5-Ethyl-2-picoline, 3798
α-Methylbenzylamine, 5949
Phenethylamine, 7186
2,4,6-Trimethylpyridine, 9632
Xylidine, 9991
C$_8$H$_{11}$NO
Metyridine, 6084
o-Phenetidine, 7187
p-Phenetidine, 7188
Phenylethanolamine, 7258
Tyramine, 9743
C$_8$H$_{11}$NO$_2$
Bucrylate, 1453
Dopamine, 3415
Norfenefrine, 6616
Octopamine, 6679
C$_8$H$_{11}$NO$_3$
Ammonium Mandelate, 557
Norepinephrine, 6612
C$_8$H$_{11}$NO$_5$S
Sulbactam, 8862
C$_8$H$_{11}$N$_2$NaO$_3$
Barbital*, 972
C$_8$H$_{11}$N$_3$O$_6$
6-Azauridine, 920

C$_8$H$_{11}$N$_5$
Phenyl Biguanide, 7247
C$_8$H$_{11}$N$_5$O$_3$
Acyclovir, 139
C$_8$H$_{11}$N$_7$S
Ambazone, 386
C$_8$H$_{12}$ClNO
Allidochlor, 250
Tyramine*, 9743
C$_8$H$_{12}$ClNO$_2$
4-Desoxypyridoxine Hydrochloride,
2913
Dopamine*, 3415
Norfenefrine*, 6616
Octopamine*, 6679
C$_8$H$_{12}$ClNO$_3$
Norepinephrine*, 6612
Pyridoxine Hydrochloride, 7995
C$_8$H$_{12}$FeN$_2$O$_8$.4H$_2$O
Aspartic Acid*, 862
C$_8$H$_{12}$NO$_5$PS$_2$
Cythioate, 2791
C$_8$H$_{12}$N$_2$
Betahistine, 1200
Dimethyl-p-phenylenediamine,
3242
Phenelzine, 7181
C$_8$H$_{12}$N$_2$O$_2$S
Thiobarbital, 9248
C$_8$H$_{12}$N$_2$O$_3$
Barbital, 972
Nicotinic Acid Monoethanolamine
Salt, 6437
Primocarcin, 7754
C$_8$H$_{12}$N$_2$O$_3$S
6-Aminopenicillanic Acid, 470
C$_8$H$_{12}$N$_2$O$_4$S
Pralidoxime Chloride*, 7705
C$_8$H$_{12}$N$_4$
2,2′-Azobisisobutyronitrile, 931
C$_8$H$_{12}$N$_4$O$_3$S
Carnidazole, 1855
C$_8$H$_{12}$N$_4$O$_5$
Azacitidine, 907
Ribavirin, 8199
C$_8$H$_{12}$N$_6$SO$_4$
Dihydralazine*, 3152
C$_8$H$_{12}$N$_6$SO$_4$.2½H$_2$O
Dihydralazine*, 3152
C$_8$H$_{12}$O$_2$
5,5-Dimethyl-1,3-cyclohexanedione,
3231
C$_8$H$_{12}$O$_4$
Diethyl Maleate, 3113
Terpenylic Acid, 9100
C$_8$H$_{12}$O$_5$
Ethyl Oxalacetate, 3791
C$_8$H$_{12}$O$_8$Pb
Lead Tetraacetate, 5305
C$_8$H$_{12}$O$_8$Si
Silicon Tetraacetate, 8445
C$_8$H$_{13}$NO$_2$
Arecoline, 802
Bemegride, 1036
Retronecine, 8168
Scopoline, 8365
C$_8$H$_{13}$NO$_3$
Diethadione, 3096
C$_8$H$_{13}$NO$_5$
Oryzacidin, 6839
C$_8$H$_{13}$N$_2$O$_3$PS
Thionazin, 9275
C$_8$H$_{13}$N$_3$O$_4$S
Tinidazole, 9377
C$_8$H$_{14}$ClNS$_2$
Sulfallate, 8882
C$_8$H$_{14}$ClN$_5$
Atrazine, 886
C$_8$H$_{14}$Cl$_2$N$_2$
Betahistine*, 1200

C$_8$H$_{14}$Cl$_2$N$_2$O$_2$
Pyridoxamine Dihydrochloride,
7993
C$_8$H$_{14}$Cl$_3$O$_5$P
Butonate, 1528
C$_8$H$_{14}$CuO$_4$
Cupric Butyrate, 2633
C$_8$H$_{14}$INO
Furtrethonium*, 4225
C$_8$H$_{14}$MgO$_4$
Butyric Acid*, 1593
C$_8$H$_{14}$NNaO$_3$
ε-Acetamidocaproic Acid*, 38
C$_8$H$_{14}$NO $^+$
Furtrethonium, 4225
C$_8$H$_{14}$N$_2$O
Loline, 5442
C$_8$H$_{14}$N$_2$O$_4$S
Phenelzine*, 7181
C$_8$H$_{14}$N$_3$O$_7$P
AIR, 181
C$_8$H$_{14}$N$_4$NiO$_4$
Nickel Dimethylglyoxime, 6414
C$_8$H$_{14}$N$_4$OS
Metribuzin, 6076
C$_8$H$_{14}$O$_2$S$_2$
Thioctic Acid, 9255
C$_8$H$_{14}$O$_3$
Butyric Anhydride, 1594
C$_8$H$_{14}$O$_4$
Dimethoxane, 3212
Suberic Acid, 8833
Succinic Acid*, 8840
C$_8$H$_{14}$O$_4$Pb
Lead Butyrate, 5276
C$_8$H$_{14}$O$_6$
Ethyl Tartrate, 3811
C$_8$H$_{15}$Cl$_3$O$_3$
Chlorhexadol, 2089
C$_8$H$_{15}$N
β-Coniceine, 2496
γ-Coniceine, 2497
Tropane, 9689
C$_8$H$_{15}$NO
Hygrine, 4788
Pelletierine, 7015
Pseudotropine, 7937
Tropine, 9693
C$_8$H$_{15}$NO$_2$
Methyl N-Methylnipecotate, 6012
Oxanamide, 6874
Tranexamic Acid, 9487
C$_8$H$_{15}$NO$_3$
ε-Acetamidocaproic Acid, 38
C$_8$H$_{15}$NO$_4$
Castanospermine, 1902
C$_8$H$_{15}$N$_3$O$_7$
Streptozocin, 8794
C$_8$H$_{15}$N$_5$O
Noformicin, 6590
Pildralazine, 7394
C$_8$H$_{15}$N$_5$S
Simetryne®, 8488
C$_8$H$_{15}$N$_7$O$_2$S$_3$
Famotidine, 3881
C$_8$H$_{15}$NaO$_2$
Valproic Acid*, 9821
C$_8$H$_{16}$
Caprylene, 1764
C$_8$H$_{16}$NO$_3$PS$_2$
Mephosfolan, 5743
C$_8$H$_{16}$NO$_5$P
Dicrotophos, 3077
C$_8$H$_{16}$N$_2$Na$_2$O$_6$S$_2$
PIPES*, 7450
C$_8$H$_{16}$N$_2$O$_2$S$_2$
4,4′-Dithiodimorpholine, 3379
C$_8$H$_{16}$N$_2$O$_4$S$_2$
Homocystine, 4655
Penicillamine Cysteine Disulfide,
7030

Denotes derivative of title compound

Denotes derivative of title compound

C₉H₈ClN₅S
Tizanidine, 9409

C₉H₈Cl₂O₃
Dichlorprop, 3068

C₉H₈Cl₃NO₂S
Captan, 1771

C₉H₈I₂O₃
3,5-Diiodosalicylic Acid*, 3175

C₉H₈N₂O₂
Pemoline, 7021

C₉H₈N₂O₃S₂
Phenosulfazole, 7219

C₉H₈N₃NaO₂S₂.1½H₂O
Sulfathiazole*, 8920

C₉H₈N₄O₆
Nifurtoinol, 6455

C₉H₈N₈O₂S
Thiamiprine, 9228

C₉H₈O
1,2-Benzopyran, 1112
Cinnamaldehyde, 2298

C₉H₈O₂
Atropic Acid, 890
Cinnamic Acid, 2300

C₉H₈O₃
Carbic Anhydride, 1801
p-Coumaric Acid, 2561

C₉H₈O₄
Acetozone, 71
Aspirin, 873
Caffeic Acid, 1634

C₉H₉AlO₆
Dihydroxyaluminum Acetylsalicylate, 3167

C₉H₉Br₂NO₃
3,5-Dibromo-L-tyrosine, 3014

C₉H₉ClO
Chlorindanol, 2094

C₉H₉ClO₃
MCPA, 5645

C₉H₉Cl₂NO
Propanil, 7814

C₉H₉Cl₂NO₂
Diloxanide, 3187

C₉H₉Cl₂N₃
Clonidine, 2388

C₉H₉Cl₂N₃O
Guanfacine, 4474

C₉H₉Cl₂N₅
Chlorazanil*, 2073

C₉H₉HgNaO₂S
Thimerosal, 9244

C₉H₉I₂NO₃
Betasine, 1203
3,5-Diiodotyrosine, 3177

C₉H₉N
Skatole, 8503

C₉H₉NO
Hydrocarbostyril, 4702

C₉H₉NO₃
Adrenochrome, 162
Adrenolutin, 164
Hippuric Acid, 4636
Salacetamide, 8290

C₉H₉NO₄
Ethyl Nitrobenzoate, 3787
Salicylamide O-Acetic Acid, 8298

C₉H₉N₂NaO₃
p-Aminohippuric Acid*, 454

C₉H₉N₃O₂
Carbendazim, 1794

C₉H₉N₃O₂S₂
Sulfathiazole, 8920
Thiazolsulfone, 9237

C₉H₉N₃S
Amiphenazole, 498

C₉H₉N₄NaO₄
Acefylline*, 18
1-Theobromineacetic Acid*, 9210

C₉H₉N₅
Amanozine, 379
Benzoguanamine, 1099

C₉H₁₀
Indan, 4844

C₉H₁₀Cl₂N₂O
Diuron, 3388

C₉H₁₀Cl₂N₂O₂
Linuron, 5387

C₉H₁₀Cl₂N₄
Apraclonidine, 779

C₉H₁₀Cl₃N₃
Clonidine*, 2388

C₉H₁₀Cl₃N₃O
Guanfacine*, 4474

C₉H₁₀INO₃
3-Iodotyrosine, 4937

C₉H₁₀NO₃PS
Ciafos, 2266

C₉H₁₀N₂
5,6-Dimethylbenzimidazole, 3225

C₉H₁₀N₂O
Aminorex, 490
1-Phenyl-3-pyrazolidinone, 7281

C₉H₁₀N₂O₂
Phenacemide, 7154

C₉H₁₀N₂O₂S
Sulbenox, 8864
Thibenzazoline, 9239

C₉H₁₀N₂O₂S₂
Aureothricin, 898

C₉H₁₀N₂O₃
p-Aminohippuric Acid, 454

C₉H₁₀N₂O₃S₂
Ethoxzolamide, 3711

C₉H₁₀N₂S
Etisazol, 3829

C₉H₁₀N₄O₂S₂
Sulfamethizole, 8887

C₉H₁₀N₄O₃S₂
Sulfametrole, 8892

C₉H₁₀N₄O₄
Acefylline, 18
1-Theobromineacetic Acid, 9210

C₉H₁₀N₆O₅
Orazamide, 6817

C₉H₁₀O
p-Anol, 706
Chavicol, 2039
Cinnamyl Alcohol, 2305
Phenylacetone, 7240
Propiophenone, 7842

C₉H₁₀O₂
Benzyl Acetate, 1137
m-Cresyl Acetate, 2587
o-Cresyl Acetate, 2588
Ethyl Benzoate, 3725
Hydrocinnamic Acid, 4707
Paroxypropione, 6992

C₉H₁₀O₃
Apocynin, 772
Atrolactic Acid, 889
Ethylparaben, 3792
Ethyl Salicylate, 3804
Ethyl Vanillin, 3815
Tropic Acid, 9691
Veratraldehyde, 9852

C₉H₁₀O₄
Flopropione, 4039
Glycol Salicylate, 4395
Homovanillic Acid, 4662
Methylenomycins*, 5986
Syringaldehyde, 8996
Veratic Acid, 9854

C₉H₁₀O₅
Vanilmandelic Acid, 9840

C₉H₁₁BrN₂O₂
Metobromuron, 6061

C₉H₁₁ClN₂O
Monuron, 6169

C₉H₁₁ClN₂S
Etisazol*, 3829

C₉H₁₁ClO₃
Chlorphenesin, 2178

C₉H₁₁Cl₂FN₂O₂S₂
Dichlofluanid, 3031

C₉H₁₁Cl₂N₃O₄S₂
Methyclothiazide, 5929

C₉H₁₁Cl₃NO₃PS
Chlorpyrifos, 2190

C₉H₁₁Cl₃N₄
Apraclonidine*, 779

C₉H₁₁FN₂O₅
Doxifluridine, 3426
Floxuridine, 4045

C₉H₁₁IN₂O₅
Idoxuridine, 4819

C₉H₁₁N
Tranylcypromine, 9491

C₉H₁₁NO
Acetotoluide, 67
p-Aminopropiophenone, 482
Cathinone, 1912
p-Dimethylaminobenzaldehyde, 3219
N-Methylacetanilide, 5931

C₉H₁₁NO₂
p-Acetanisidine, 43
Atrolactamide, 888
4-(Dimethylamino)benzoic Acid, 3221
Ethenzamide, 3685
Ethyl Aminobenzoate, 3719
Phenylalanine, 7242
Phenylurethan(e), 7291

C₉H₁₁NO₃
Adrenalone, 161
Styramate, 8829
Tyrosine, 9747
m-Tyrosine, 9748

C₉H₁₁NO₄
Dopa, 3414
Levodopa, 5344

C₉H₁₁NO₅
Droxidopa, 3444

C₉H₁₁NO₆
Showdomycin, 8430

C₉H₁₁N₃O₄
Ancitabine, 663

C₉H₁₁N₅O₃
Biopterin, 1242

C₉H₁₁N₅O₄
Eritadenine, 3618
Ichthyopterin, 4816
Neopterin, 6378

C₉H₁₂
Cumene, 2619
Mesitylene, 5810
n-Propylbenzene, 7856
Pseudocumene, 7929

C₉H₁₂AsNO₄
3-Methylarsacetin, 5945

C₉H₁₂BNO₄
Epinephrine*, 3569

C₉H₁₂ClNO₃
Adrenalone*, 161

C₉H₁₂ClN₃O₄S₂
Ethiazide, 3687

C₉H₁₂ClO₄P
Heptenophos, 4585

C₉H₁₂Cl₂N₄O
Guanochlor, 4479

C₉H₁₂NO₅PS
Fenitrothion, 3922

C₉H₁₂N₂
Nornicotine, 6631

C₉H₁₂N₂Na₃O₁₅P₃.2H₂O
Uridine 5'-Triphosphate*, 9795

C₉H₁₂N₂O
Fenuron, 3951

Denotes derivative of title compound

FI-21

C₉H₁₂N₂O₂
Dulcin, 3448
C₉H₁₂N₂O₃
5-Nitro-2-propoxyaniline, 6551
C₉H₁₂N₂O₅
Deoxyuridine, 2892
C₉H₁₂N₂O₆
Uridine, 9792
C₉H₁₂N₂S
Protionamide, 7905
C₉H₁₂N₄O₃
Etofylline, 3835
C₉H₁₂N₆
Triethylenemelamine, 9586
C₉H₁₂O
Benzyl Ethyl Ether, 1147
p,α-Dimethylbenzyl Alcohol, 3226
α-Ethylbenzyl Alcohol, 3727
C₉H₁₂O₃
Phenylglyceryl Ether, 7261
C₉H₁₂O₃S
Ethyl p-Toluenesulfonate, 3814
C₉H₁₂O₄
3-Methoxy-4-hydroxyphenylglycol, 5916
C₉H₁₃BrN₂O₂
Bromacil, 1370
Pyridostigmine Bromide, 7990
C₉H₁₃ClN₂O₂
Terbacil, 9085
C₉H₁₃ClN₆
Cyanazine, 2692
C₉H₁₃ClN₆O₂
Nimustine, 6469
C₉H₁₃Cl₂N₃O₂
Dopan, 3416
C₉H₁₃N
Amphetamine, 616
Cumidine, 2622
C₉H₁₃NNaO₂P
Toldimfos Sodium, 9435
C₉H₁₃NO
β-Amino-α-methylphenethyl Alcohol, 459
Gepefrine, 4295
Halostachine, 4516
Hydroxyamphetamine, 4740
Norpseudoephedrine, 6634
C₉H₁₃NO₂
Deoxyepinephrine, 2885
Ecgonidine, 3466
Ethinamate, 3688
Metaraminol, 5837
Pyrithyldione, 8005
Synephrine, 8994
C₉H₁₃NO₃
Epinephrine, 3569
Normetanephrine, 6627
C₉H₁₃N₂O₉P
5'-Uridylic Acid, 9796
C₉H₁₃N₃O
Iproniazid, 4965
C₉H₁₃N₃O₂
Aminometradine, 463
Amisometradine, 500
C₉H₁₃N₃O₃
Dideoxycytidine, 3091
C₉H₁₃N₃O₅
Cytarabine, 2790
Cytidine, 2792
C₉H₁₃N₃O₆
Mizoribine, 6137
C₉H₁₃N₅O₄
Ganciclovir, 4262
C₉H₁₄ClNO
Norpseudoephedrine*, 6634
Phenylpropanolamine Hydrochloride, 7279
C₉H₁₄ClNO₂
Phenylephrine Hydrochloride, 7257

C₉H₁₄ClNO₃
Epinephrine*, 3569
Nordefrin Hydrochloride, 6609
C₉H₁₄Cl₂N₆O₂
Nimustine*, 6469
C₉H₁₄IN
Phenyltrimethylammonium Iodide, 7289
C₉H₁₄N₂
Pheniprazine, 7197
C₉H₁₄N₂O
Phenoxypropazine, 7227
C₉H₁₄N₂O₃
Metharbital, 5873
Probarbital, 7759
C₉H₁₄N₂O₄S
Mafenide*, 5525
C₉H₁₄N₂O₆S
Itramin Tosylate, 5132
C₉H₁₄N₂O₁₂P₂
Uridine 5'-Diphosphate, 9793
C₉H₁₄N₃O₈P
2'-Cytidylic Acid, 2793
3'-Cytidylic Acid, 2794
C₉H₁₄N₄O₃
Carnosine, 1857
Nimorazole, 6468
C₉H₁₄N₄O₄
Molsidomine, 6143
C₉H₁₄N₆SO₄
Dihydralazine*, 3152
C₉H₁₄O
Phorone, 7307
C₉H₁₄OS₃
Ajoene, 185
C₉H₁₄O₆
Triacetin, 9504
C₉H₁₅AlO₉
Aluminum Lactate, 350
C₉H₁₅BrN₂O₃
Acecarbromal, 15
C₉H₁₅Br₆O₄P
Tris-BP, 9665
C₉H₁₅ClN₂
Pheniprazine*, 7197
C₉H₁₅Cl₆O₄P
Fyrol FR-2®, 4233
C₉H₁₅NO
Pseudopelletierine, 7936
C₉H₁₅NO₂
Diacetone Acrylamide, 2943
Piperidione, 7439
3-Quinuclidinol*, 8110
C₉H₁₅NO₃
5,5-Dipropyl-2,4-oxazolidinedione, 3352
Ecgonine, 3467
C₉H₁₅NO₃S
Captopril, 1773
Mycobacidin, 6233
C₉H₁₅N₂O₁₅P₃
Uridine 5'-Triphosphate, 9795
C₉H₁₅N₃O₂
Hercynine, 4588
C₉H₁₅N₃O₂S
Ergothioneine, 3611
C₉H₁₅N₃O₇
Lycomarasmine, 5491
C₉H₁₅N₄O₈P
AICAR, 180
C₉H₁₅N₅O
Minoxidil, 6122
C₉H₁₅N₅O₃
Theophylline*, 9212
C₉H₁₅NaO₃
Hexacyclonate Sodium, 4603
C₉H₁₅O₈P
Bomyl®, 1333
C₉H₁₆ClNO₂
3-Quinuclidinol*, 8110

C₉H₁₆ClN₃O₂
Lomustine, 5444
C₉H₁₆ClN₃O₇
Chlorozotocin, 2177
C₉H₁₆ClN₅
Propazine, 7822
Trietazine, 9580
C₉H₁₆Cl₂N₄
N-(3-Chloroallyl)hexaminium Chloride, 2117
C₉H₁₆NNaO₅
Pantothenic Acid*, 6964
C₉H₁₆NO₄P
Amphetamine*, 616
C₉H₁₆N₂O₂
Apronalide, 783
C₉H₁₆N₂O₃S
Betahistine*, 1200
C₉H₁₆N₄OS
Ethiozin, 3694
Tebuthiuron, 9053
C₉H₁₆O
2-Nonenal, 6595
C₉H₁₆O₄
Azelaic Acid, 921
C₉H₁₇ClN₃O₃PS
Isazofos, 4988
C₉H₁₇IN₄
Methenamine Allyl Iodide, 5880
C₉H₁₇NO
Novonal, 6643
Pelletierine*, 7015
C₉H₁₇NO₂S
Lethane® 384, 5329
C₉H₁₇NO₄
Acetylcarnitine, 78
C₉H₁₇NO₅
Pantothenic Acid, 6964
C₉H₁₇NO₇
Muramic Acid, 6213
C₉H₁₇NO₈
Neuraminic Acid, 6392
C₉H₁₇N₂NaO₆
Betaine*, 1201
C₉H₁₇N₃O₃
Dopastin, 3417
C₉H₁₇N₅S
Ametryn, 402
C₉H₁₈BrNO₂
Methyl N-Methylnipecotate*, 6012
C₉H₁₈ClNO₄
Acetylcarnitine*, 78
C₉H₁₈Cl₃N₂O₂P
Trofosfamide, 9680
C₉H₁₈FeN₃S₆
Ferbam, 3954
C₉H₁₈N₂O₂
Capuride, 1774
C₉H₁₈N₂O₄
Meprobamate, 5751
C₉H₁₈N₃OP
Metepa, 5842
C₉H₁₈N₄
Guanacline, 4470
C₉H₁₈N₄O
Amidinomycin, 410
C₉H₁₈N₆
Altretamine, 318
C₉H₁₈N₆O₆
Hexamethylolmelamine, 4611
C₉H₁₈O₂
n-Amyl Butyrate, 639
Ethyl Oenanthate, 3790
Isoamyl Butyrate, 4997
Isobutyl Isovalerate, 5029
Pelargonic Acid, 7013
C₉H₁₈O₃
n-Butyl Carbonate, 1558
C₉H₁₉ClN₃O₅P
Fotemustine, 4174

Denotes derivative of title compound

C₉H₁₉N
Cyclopentamine, 2745
Isometheptene, 5070
Methylconiine, 5968

C₉H₁₉NO
Isovaleryl Diethylamide, 5122

C₉H₁₉NOS
EPTC, 3580

C₉H₁₉NO₄
Dexpanthenol, 2924

C₉H₂₀Cl₃N₂O₃P
Defosfamide, 2853

C₉H₂₀NO₂⁺
Muscarine, 6219

C₉H₂₀N₄
Guanazodine, 4472

C₉H₂₀N₄O₄
Negamycin, 6359

C₉H₂₀N₄O₄S
Guanacline*, 4470

C₉H₂₀O
n-Nonyl Alcohol, 6598

C₉H₂₁AlO₃
Aluminum Isopropoxide, 349

C₉H₂₁O₂PS₃
Terbufos, 9088

C₉H₂₂N₂
Novoldiamine, 6642

C₉H₂₂N₄O₄S.H₂O
Guanazodine*, 4472

C₉H₂₂O₄P₂S₄
Ethion, 3691

C₉H₂₃INO₃PS
Echothiophate Iodide, 3475

C₉H₂₄I₂N₂O
Prolonium Iodide, 7792

C₁₀Cl₁₀
Dienochlor, 3095

C₁₀Cl₁₀O
Chlordecone, 2081

C₁₀Cl₁₂
Mirex, 6126

C₁₀F₁₈
Fluosol DA*, 4111

C₁₀H₄Cl₂O₂
Dichlone, 3032

C₁₀H₄N₂Na₂O₈S
Naphthol Yellow S, 6313

C₁₀H₅ClN₂
o-Chlorobenzylidenemalononitrile, 2127

C₁₀H₅Cl₇
Heptachlor, 4576

C₁₀H₅NNa₂O₈S₂
Nitroso-R Salt, 6565

C₁₀H₅NaO₅S
Sodium β-Naphthoquinone-4-sulfonate, 8597

C₁₀H₆Cl₂N₂O₂
Pyrrolnitrin, 8029

C₁₀H₆Cl₄O₄
DCPA, 2830

C₁₀H₆Cl₈
Chlordan(e), 2079

C₁₀H₆N₂OS₂
Oxythioquinox, 6933

C₁₀H₆N₂O₄
5-Nitroquinaldic Acid, 6552

C₁₀H₆O₂
1,2-Naphthoquinone, 6314
1,4-Naphthoquinone, 6315

C₁₀H₆O₃
Juglone, 5150
Lawsone, 5263

C₁₀H₆O₄
Chromocarb, 2239
Coumarin-3-carboxylic Acid, 2564

C₁₀H₆O₈
Pyromellitic Acid, 8015

C₁₀H₇Br
1-Bromonaphthalene, 1413
2-Bromonaphthalene, 1414

C₁₀H₇Cl
1-Chloronaphthalene, 2149
2-Chloronaphthalene, 2150

C₁₀H₇Cl₂NO
Chlorquinaldol, 2191

C₁₀H₇Cl₂N₃O
Anagrelide, 659

C₁₀H₇Cl₅O
Tridiphane, 9578

C₁₀H₇F₃O₄
Triflusal, 9600

C₁₀H₇NO₂
1-Nitronaphthalene, 6535
1-Nitroso-2-naphthol, 6562
N-Phenylmaleimide, 7270
Quinaldic Acid, 8054
8-Quinolinecarboxylic Acid, 8099

C₁₀H₇NO₃
Kynurenic Acid, 5206
1-Nitro-2-naphthol, 6536

C₁₀H₇NO₄
Xanthurenic Acid, 9977

C₁₀H₇N₃S
Thiabendazole, 9217

C₁₀H₇NaO
2-Naphthol*, 6304

C₁₀H₇NaO₄S
2-Naphthol-6-sulfonic Acid*, 6312

C₁₀H₈
Azulene, 939
Naphthalene, 6289

C₁₀H₈ClN₃O₂
Drazoxolon, 3432

C₁₀H₈Cl₃N₃O.½H₂O
Anagrelide*, 659

C₁₀H₈NNaO₃S.4H₂O
1-Naphthylamine-4-sulfonic Acid*, 6323

C₁₀H₈NNaO₄S.2½H₂O
1-Amino-2-naphthol-6-sulfonic Acid*, 467

C₁₀H₈N₂
α,α′-Dipyridyl, 3355
γ,γ′-Dipyridyl, 3356

C₁₀H₈N₂O₂S₂
Pyrithione*, 8004

C₁₀H₈N₂O₂S₂Zn
Pyrithione*, 8004

C₁₀H₈N₂O₄
α-Furildioxime, 4218

C₁₀H₈N₄O₃
Nifurprazine, 6453

C₁₀H₈N₄O₃S₂
Nitrodan, 6517

C₁₀H₈N₄O₄
Nitrefazole, 6492

C₁₀H₈N₄O₅
Picrolonic Acid, 7383

C₁₀H₈O
1-Naphthol, 6303
2-Naphthol, 6304

C₁₀H₈OS₃
Anethole Trithione, 676

C₁₀H₈O₂
Naphthoresorcinol, 6316
Tricromyl, 9574

C₁₀H₈O₃
Erythrocentaurin, 3623
Hymecromone, 4792

C₁₀H₈O₃S
1-Naphthalenesulfonic Acid, 6295
2-Naphthalenesulfonic Acid, 6296

C₁₀H₈O₄
Anemonin, 674
Scopoletin, 8363

C₁₀H₈O₄S
Cassella's Acid, 1898
Croceic Acid, 2591
1-Naphthol-2-sulfonic Acid, 6310
1-Naphthol-4-sulfonic Acid, 6311
2-Naphthol-6-sulfonic Acid, 6312

C₁₀H₈O₅
Fraxetin, 4178

C₁₀H₈O₆S₂
Armstrong's Acid, 812
1,6-Naphthalenedisulfonic Acid, 6292
2,6-Naphthalenedisulfonic Acid, 6293
2,7-Naphthalenedisulfonic Acid, 6294

C₁₀H₈O₇S₂
1-Naphthol-4,8-disulfonic Acid, 6306
2-Naphthol-3,6-disulfonic Acid, 6307
2-Naphthol-6,8-disulfonic Acid, 6308

C₁₀H₈O₈S₂
Chromotropic Acid, 2243

C₁₀H₈O₁₀S₂
Sulmarin, 8964

C₁₀H₈S
1-Naphthalenethiol, 6297
2-Naphthalenethiol, 6298

C₁₀H₉ClN₄O₂S
Sulfachlorpyridazine, 8871

C₁₀H₉ClN₄O₃
Nifurprazine*, 6453

C₁₀H₉Cl₂NO
Dicryl, 3078

C₁₀H₉Cl₂NaO₃
2,4-DB*, 2828

C₁₀H₉Cl₄NO₂S
Captafol, 1770

C₁₀H₉Cl₄O₄P
Stirofos, 8777

C₁₀H₉I₃O₃
Phenobutiodil, 7203

C₁₀H₉N
Lepidine, 5324
1-Naphthylamine, 6318
2-Naphthylamine, 6319
Quinaldine, 8055

C₁₀H₉NH₂SO₄
Quinaldine Sulfate, 8058

C₁₀H₉NO
4-Amino-1-naphthol, 465
Echinopsine, 3472

C₁₀H₉NO₂
Carfinate, 1845
Gentianine, 4286
Indoleacetic Acid, 4870
Succinanil, 8838

C₁₀H₉NO₃
Maleanilic Acid, 5584

C₁₀H₉NO₃S
Badische Acid, 952
Cassella's Acid F, 1899
1,6-Cleve's Acid, 2348
1-Naphthylamine-4-sulfonic Acid, 6323
1-Naphthylamine-5-sulfonic Acid, 6324
1-Naphthylamine-8-sulfonic Acid, 6325
2-Naphthylamine-1-sulfonic Acid, 6326
2-Naphthylamine-5-sulfonic Acid, 6327

C₁₀H₉NO₄S
1-Amino-2-naphthol-4-sulfonic Acid, 466
1-Amino-2-naphthol-6-sulfonic Acid, 467

Denotes derivative of title compound

Denotes derivative of title compound

Denotes derivative of title compound

FI-26

* Denotes derivative of title compound

C$_{11}$H$_8$O$_3$
1-Hydroxy-2-naphthoic Acid, 4766
3-Hydroxy-2-naphthoic Acid, 4767
Phthiocol, 7352
Plumbagin, 7511

C$_{11}$H$_8$O$_5$
Purpurogallin, 7963

C$_{11}$H$_9$Cl$_2$NO$_2$
Barban, 969

C$_{11}$H$_9$Cl$_5$O$_3$
Erbon, 3587

C$_{11}$H$_9$FN$_2$O$_3$
Sorbinil, 8679

C$_{11}$H$_9$I$_3$NNaO$_2$
Cinamiodyl*, 2284

C$_{11}$H$_9$I$_3$O$_4$
Iothalamic Acid, 4952

C$_{11}$H$_9$NO$_2$
3-Amino-2-naphthoic Acid, 464

C$_{11}$H$_9$NO$_3$
Quininic Acid, 8092

C$_{11}$H$_9$N$_3$O$_2$
1,2-Naphthoquinone*, 6314

C$_{11}$H$_9$NaO$_5$S
Menadione Sodium Bisulfite, 5716

C$_{11}$H$_{10}$CuN$_2$NaO$_2$S
Allocupreide Sodium, 256

C$_{11}$H$_{10}$FNO$_2$S
Flosequinan, 4043

C$_{11}$H$_{10}$F$_3$NO$_2$
Flumetramide, 4069

C$_{11}$H$_{10}$I$_3$NO$_2$
Cinamiodyl, 2284

C$_{11}$H$_{10}$NNaO$_4$S
Actinoquinol*, 134

C$_{11}$H$_{10}$N$_2$O
Amphenidone, 614

C$_{11}$H$_{10}$N$_2$O$_3$
Phenylmethylbarbituric Acid, 7275

C$_{11}$H$_{10}$N$_2$S
ANTU, 755

C$_{11}$H$_{10}$N$_3$NaO$_2$S.H$_2$O
Sulfapyridine*, 8913

C$_{11}$H$_{10}$N$_4$O$_4$
Carbadox, 1782

C$_{11}$H$_{10}$O
2-Methoxynaphthalene, 5918

C$_{11}$H$_{10}$O$_4$
Limettin, 5370
Scoparone, 8359

C$_{11}$H$_{10}$O$_5$
Gladiolic Acid, 4329

C$_{11}$H$_{10}$O$_6$
Dipyrocetyl, 3357

C$_{11}$H$_{10}$O$_8$S$_2$
Menadiol Disulfate, 5713

C$_{11}$H$_{11}$ClO$_3$
Alclofenac, 212

C$_{11}$H$_{11}$Cl$_2$N$_3$O
Muzolimine, 6227

C$_{11}$H$_{11}$Cl$_4$NO$_2$
Chlorbetamide, 2076

C$_{11}$H$_{11}$F$_3$N$_2$O$_3$
Flutamide, 4132

C$_{11}$H$_{11}$I$_3$O$_3$
Iophenoxic Acid, 4947

C$_{11}$H$_{11}$NO
3-Indolylacetone, 4873
Vitamin K$_5$, 9936
Vitamin K$_7$, 9938

C$_{11}$H$_{11}$NO$_2$
Phensuximide, 7231

C$_{11}$H$_{11}$NO$_4$S
Actinoquinol, 134
Woodward's Reagent K, 9962

C$_{11}$H$_{11}$N$_3$O$_2$S
Sulfapyridine, 8913

C$_{11}$H$_{11}$N$_4$NaO$_2$S
Sulfamerazine*, 8884

C$_{11}$H$_{11}$N$_5$O$_3$S$_2$
Urothion, 9800

C$_{11}$H$_{12}$AsNO$_5$S$_2$
Arsenamide, 819

C$_{11}$H$_{12}$BrNO
Cinromide, 2313

C$_{11}$H$_{12}$ClNO
Vitamin K$_5$*, 9936

C$_{11}$H$_{12}$ClNO$_3$S
Chlormezanone, 2105

C$_{11}$H$_{12}$ClN$_5$
Phenazopyridine Hydrochloride,
7175

C$_{11}$H$_{12}$Cl$_2$N$_2$O
Lofexidine, 5439

C$_{11}$H$_{12}$Cl$_2$N$_2$O$_5$
Chloramphenicol, 2068

C$_{11}$H$_{12}$Cl$_2$O$_3$
2,4-D*, 2802

C$_{11}$H$_{12}$Cl$_3$N
Amphecloral, 613

C$_{11}$H$_{12}$Cl$_4$N$_2$O$_3$
Monuron*, 6169

C$_{11}$H$_{12}$I$_3$NO$_2$
Iopanoic Acid, 4944

C$_{11}$H$_{12}$NO$_4$PS$_2$
Phosmet, 7311

C$_{11}$H$_{12}$N$_2$
Vitamin K$_6$, 9937

C$_{11}$H$_{12}$N$_2$O
Antipyrine, 748
Vasicine, 9842

C$_{11}$H$_{12}$N$_2$O$_2$
Ethotoin, 3705
Fenozolone, 3932
Thozalinone, 9315
L-Tryptophan, 9707

C$_{11}$H$_{12}$N$_2$O$_3$
5-Hydroxytryptophan, 4784

C$_{11}$H$_{12}$N$_2$S
Tetramisole, 9161

C$_{11}$H$_{12}$N$_4$O$_2$
Todralazine, 9425

C$_{11}$H$_{12}$N$_4$O$_2$S
Sulfamerazine, 8884
Sulfaperine, 8909

C$_{11}$H$_{12}$N$_4$O$_3$S
Sulfalene, 8881
Sulfameter, 8885
Sulfamethoxypyridazine, 8890

C$_{11}$H$_{12}$O$_2$
Cinnamic Acid*, 2300
6-Methoxy-α-tetralone, 5924

C$_{11}$H$_{12}$O$_3$
Ethyl Benzoylacetate, 3726
Myristicin, 6247

C$_{11}$H$_{12}$O$_5$
Elenolide, 3507

C$_{11}$H$_{12}$O$_7$
Piscidic Acid, 7479

C$_{11}$H$_{13}$AsN$_2$O$_5$S$_2$
Thiocarbarsone, 9252

C$_{11}$H$_{13}$ClF$_3$N$_3$O$_4$S$_3$
Polythiazide, 7561

C$_{11}$H$_{13}$ClN$_2$S
Tetramisole*, 9161

C$_{11}$H$_{13}$ClN$_4$O$_2$
Todralazine*, 9425

C$_{11}$H$_{13}$ClO
Dowicide 9, 3420

C$_{11}$H$_{13}$Cl$_3$N$_2$O
Lofexidine*, 5439

C$_{11}$H$_{13}$F$_3$N$_2$O$_3$S
Mefluidide, 5684

C$_{11}$H$_{13}$N
Pargyline, 6988

C$_{11}$H$_{13}$NO
β-Benzalbutyramide, 1063

C$_{11}$H$_{13}$NO$_2$
Hydrohydrastinine, 4732
Idrocilamide, 4820

C$_{11}$H$_{13}$NO$_3$
Corydaldine, 2542
Hydrastinine, 4689
4-Salicyloylmorpholine, 8302
Thurfyl Nicotinate, 9329
Toloxatone, 9445

C$_{11}$H$_{13}$NO$_4$
Bendiocarb, 1044
Mephenoxalone, 5739

C$_{11}$H$_{13}$N$_3$O
Ampyrone, 625

C$_{11}$H$_{13}$N$_3$O$_3$S
Sulfamoxole, 8897
Sulfisoxazole, 8930

C$_{11}$H$_{13}$N$_5$O$_5$
Azidamfenicol, 924

C$_{11}$H$_{14}$AsNO$_3$S$_2$
Arsthinol, 840

C$_{11}$H$_{14}$BrN$_2$NaO$_3$
Butallylonal*, 1503

C$_{11}$H$_{14}$ClN
Pargyline*, 6988

C$_{11}$H$_{14}$ClNO
Propachlor, 7805

C$_{11}$H$_{14}$ClNO$_2$
Buclosamide, 1450

C$_{11}$H$_{14}$ClN$_3$O$_4$S$_3$
Althiazide, 316

C$_{11}$H$_{14}$ClN$_5$
Cycloguanil, 2727

C$_{11}$H$_{14}$Cl$_2$N$_2$O$_5$S
Tevenel®, 9179

C$_{11}$H$_{14}$N$_2$
Gramine, 4440

C$_{11}$H$_{14}$N$_2$O
Cytisine, 2795
5-Methoxytryptamine, 5926

C$_{11}$H$_{14}$N$_2$O$_2$
Levulinic Acid*, 5352
Pheneturide, 7190

C$_{11}$H$_{14}$N$_2$O$_3$S
Sulfadicramide, 8875

C$_{11}$H$_{14}$N$_2$O$_4$
5'-Nitro-2'-propoxyacetanilide,
6550

C$_{11}$H$_{14}$N$_2$S
Pyrantel, 7968

C$_{11}$H$_{14}$N$_4$O$_2$
Epirizole, 3572

C$_{11}$H$_{14}$N$_4$O$_4$
Doxofylline, 3427
Tubercidin, 9714

C$_{11}$H$_{14}$O$_2$
n-Butyl Benzoate, 1552

C$_{11}$H$_{14}$O$_3$
Butylparaben, 1583
o-Thymotic Acid, 9342
Zingerone, 10072

C$_{11}$H$_{14}$O$_4$
Dimethyl Carbate, 3230

C$_{11}$H$_{15}$BrN$_2$O$_3$
Butallylonal, 1503
Narcobarbital, 6342

C$_{11}$H$_{15}$ClN$_2$O$_2$
Iproclozide, 4963

C$_{11}$H$_{15}$ClO$_2$
Phenaglycodol, 7161

C$_{11}$H$_{15}$Cl$_2$NO
Dichlorisoproterenol, 3036

C$_{11}$H$_{15}$Cl$_2$N$_5$
Chlorproguanil, 2185

C$_{11}$H$_{15}$NO
Metamfepramone, 5828
Phenmetrazine, 7200

C$_{11}$H$_{15}$NO$_2$
Butamben, 1504
Isobutyl p-Aminobenzoate, 5017
MDMA, 5646
Salsoline, 8308

C$_{11}$H$_{15}$NO$_3$S
Methiocarb, 5893

$C_{11}H_{15}NO_3$
Anhalamine, 682
Hydroxyphenamate, 4770
p-Lactophenetide, 5220
Propoxur, 7849

$C_{11}H_{15}NO_4$
Mephenesin*, 5737
Morpholine Salicylate, 6195

$C_{11}H_{15}NO_4S$
Ethebenecid, 3684

$C_{11}H_{15}NO_5$
Methocarbamol, 5903

$C_{11}H_{15}N_2NaO_2S$
Buthalital Sodium, 1518

$C_{11}H_{15}N_2NaO_3$
Enallylpropymal*, 3523
Vinbarbital Sodium, 9886

$C_{11}H_{15}N_2O_8P$
NMN, 6585

$C_{11}H_{15}N_3O_4$
Pyridinol Carbamate, 7987

$C_{11}H_{15}N_3S$
Cuminaldehyde Thiosemicarbazone, 2624

$C_{11}H_{15}N_5O_5$
Psicofuranine, 7939

$C_{11}H_{16}$
Amylbenzene, 635

$C_{11}H_{16}ClNO$
Clorprenaline, 2404
Metamfepramone*, 5828
Phenmetrazine*, 7200

$C_{11}H_{16}ClNO_2$
MDMA*, 5646

$C_{11}H_{16}ClN_3O_4S_2$
Buthiazide, 1519

$C_{11}H_{16}ClN_5$
Chlorguanide, 2088

$C_{11}H_{16}ClO_2PS_3$
Carbophenothion, 1827

$C_{11}H_{16}Cl_3N_5$
Chlorproguanil*, 2185

$C_{11}H_{16}FN_3O_3$
Carmofur, 1851

$C_{11}H_{16}I_2N_2O_5$
Iodopyracet, 4933

$C_{11}H_{16}N_2O$
Tocainide, 9414

$C_{11}H_{16}N_2O_2$
Aminocarb, 443
Pilocarpine, 7395

$C_{11}H_{16}N_2O_3$
Butalbital, 1502
Enallylpropymal, 3523
Nifenalol, 6442
Talbutal, 9010
Vinylbital, 9897

$C_{11}H_{16}N_2O_3S$
Etozolin*, 3845
Phenbutamide, 7177

$C_{11}H_{16}N_2O_4S$
Thienamycin, 9240

$C_{11}H_{16}N_2O_5$
Edoxudine, 3486

$C_{11}H_{16}N_4O$
Lidamidine, 5358

$C_{11}H_{16}N_4O_4$
Pentostatin, 7091
Razoxane, 8133

$C_{11}H_{16}O$
Fenipentol, 3921
Jasmone, 5141
p-tert-Pentylphenol, 7098

$C_{11}H_{16}O_2$
Butylated Hydroxyanisole, 1547

$C_{11}H_{16}O_3$
d-Camphocarboxylic Acid, 1737

$C_{11}H_{17}ClN_2O$
Tocainide*, 9414

$C_{11}H_{17}ClN_2O_2$
Pilocarpine*, 7395

$C_{11}H_{17}ClN_2O_3$
Nifenalol*, 6442

$C_{11}H_{17}ClN_4O$
Lidamidine*, 5358

$C_{11}H_{17}Cl_2NO.H_2O$
Clorprenaline*, 2404

$C_{11}H_{17}Cl_2N_5$
Chlorguanide*, 2088

$C_{11}H_{17}Cl_3N_2O_2S$
Chlordantoin, 2080

$C_{11}H_{17}N$
N-Ethylamphetamine, 3720
Mephentermine, 5740
Phenpentermine, 7228
Xylopropamine, 9994

$C_{11}H_{17}NO$
Methoxyphenamine, 5919
N-Methylephedrine, 5987
Mexiletine, 6094
Tecomanine, 9056

$C_{11}H_{17}NO_3$
Dimetan®, 3200
Dioxethedrine, 3297
Isoproterenol, 5105
Mescaline, 5808
Metaproterenol, 5836

$C_{11}H_{17}NO_4$
Dimetofrine, 3254

$C_{11}H_{17}N_2NaO_2S$
Thiopental Sodium, 9280

$C_{11}H_{17}N_2NaO_3$
Amobarbital*, 601
Pentobarbital Sodium, 7087

$C_{11}H_{17}N_3O$
Meobentine, 5730

$C_{11}H_{17}N_3O_3$
Emorfazone, 3519
5-Ethyl-5-(1-piperidyl)barbituric
   Acid, 3800

$C_{11}H_{17}N_3O_3S$
1-Butyl-3-metanilylurea, 1579
Carbutamide, 1839

$C_{11}H_{17}N_3O_5$
Carbubarb, 1838
Pilocarpine*, 7395

$C_{11}H_{17}N_3O_8$
Tetrodotoxin, 9175

$C_{11}H_{17}N_5O_2$
Dimethazan, 3202

$C_{11}H_{17}N_5O_3$
Cafaminol, 1632

$C_{11}H_{17}O_4PS_2$
Fensulfothion, 3943

$C_{11}H_{18}BrN_5O_3$
Pamabrom, 6952

$C_{11}H_{18}ClNO$
Mexiletine*, 6094

$C_{11}H_{18}ClNO_3$
Isoproterenol*, 5105
Methoxamine Hydrochloride, 5910

$C_{11}H_{18}ClNO_4$
Dimetofrine*, 3254

$C_{11}H_{18}NO^+$
Leptodactyline, 5326

$C_{11}H_{18}N_2O_2$
p-Aminobenzoic Acid*, 434

$C_{11}H_{18}N_2O_3$
Amobarbital, 601
Pentobarbital Sodium*, 7087

$C_{11}H_{18}N_3O_2^+$
Murexine, 6216

$C_{11}H_{18}N_4O_2$
Pirimicarb, 7468

$C_{11}H_{18}N_4O_6$
Lysine*, 5509

$C_{11}H_{19}NOS$
Octhilinone, 6677

$C_{11}H_{19}NO_3$
d-Camphocarboxylic Acid*, 1737

$C_{11}H_{19}N_3O$
Dimethirimol, 3206
Ethirimol, 3695

$C_{11}H_{19}N_3O_6$
Wildfire Toxin, 9959

$C_{11}H_{19}N_5O_3$
Ambuphylline, 395

$C_{11}H_{20}N_2O_6$
L-Saccharopine, 8283

$C_{11}H_{20}N_2O_6S_2$
Amidephrine*, 409

$C_{11}H_{20}N_3O_3PS$
Pirimiphos-ethyl*, 7469
Pyrimithate, 8000

$C_{11}H_{20}O_2$
Undecylenic Acid, 9760

$C_{11}H_{20}O_4$
Di-tert-butyl Malonate, 3021
Ethyl Diethylmalonate, 3746

$C_{11}H_{20}O_{10}$
Primeverose, 7752
Vicianose, 9879

$C_{11}H_{21}N$
Mecamylamine, 5654

$C_{11}H_{21}NO_8$
Choline Dihydrogen Citrate, 2210

$C_{11}H_{21}N_5S$
Dipropetryn, 3349

$C_{11}H_{22}ClN$
Mecamylamine*, 5654

$C_{11}H_{22}N_2O_4S$
Pantetheine, 6961

$C_{11}H_{22}N_2O_6$
Deanol Aceglumate, 2835

$C_{11}H_{22}N_3O_3P$
Meturedepa, 6082

$C_{11}H_{22}O_2$
n-Amyl Caproate, 640
Ethyl Pelargonate, 3793
n-Nonyl Acetate, 6597

$C_{11}H_{22}O_4$
Monooctanoin, 6162

$C_{11}H_{23}NOS$
Butylate, 1546

$C_{11}H_{23}N_3O_6$
L-Lysine L-Glutamate, 5511

$C_{11}H_{23}N_5O_6$
Arginine Glutamate, 806

$C_{11}H_{23}N_7$
Hydramitrazine, 4685

$C_{11}H_{24}FeNO_{11}$
Ferrocholinate, 3986

$C_{11}H_{25}N$
Metron S, 6080

$C_{11}H_{26}NO_2PS$
VX, 9948

$C_{11}H_{28}Br_2N_2$
Pentamethonium Bromide, 7070

$C_{12}H_4Cl_4Na_2O_2S$
Bithionol*, 1316

$C_{12}H_4Cl_4O_2$
TCDD, 9052

$C_{12}H_5O_{12}N_7$
Dipicrylamine, 3340

$C_{12}H_6Cl_2NNaO_2$
2,6-Dichloroindophenol Sodium, 3057

$C_{12}H_6Cl_2N_2O_6$
Menichlopholan, 5722

$C_{12}H_6Cl_4O_2S$
Bithionol, 1316
Tetradifon, 9132

$C_{12}H_6Cl_4O_3S$
Bithionol*, 1316

$C_{12}H_6I_4O_8S_2Zn.6H_2O$
Sozoiodolic Acid*, 8687

$C_{12}H_6N_2O_2$
Phanquinone, 7148

$C_{12}H_6Na_6O_{12}S_6Sb_2$
Stibocaptate, 8768

Denotes derivative of title compound

$C_{12}H_6O_{12}$
Mellitic Acid, 5706
$C_{12}H_7Br_4O_5P$
Bromofenofos, 1406
$C_{12}H_7Cl_2NO_3$
Nitrofen, 6519
$C_{12}H_7Cl_3O_2$
Triclosan, 9573
$C_{12}H_7NO_4$
Resazurin, 8144
$C_{12}H_7N_3O_5S$
3,7-Dinitro-5-oxophenothiazine,
3273
$C_{12}H_8Cl_2O_2S$
Fenticlor, 3947
$C_{12}H_8Cl_2O_3S$
Genite®, 4283
Ovex, 6856
$C_{12}H_8Cl_6$
Aldrin, 219
$C_{12}H_8Cl_6O$
Dieldrin, 3093
Endrin, 3533
$C_{12}H_8F_2$
4,4'-Difluorodiphenyl, 3133
$C_{12}H_8HgNa_2O_8S_2$
Mercuric Sodium p-Phenolsulfonate,
5783
$C_{12}H_8N_2$
o-Phenanthroline, 7169
Phenazine, 7173
$C_{12}H_8N_2O$
Hemipyocyanine, 4564
$C_{12}H_8N_2O_4$
Iodinin, 4915
$C_{12}H_8N_2O_4S_2$
Nitrophenide, 6539
$C_{12}H_8N_4O_6S$
Nifurzide, 6456
$C_{12}H_8O_4$
Bergapten(e), 1173
Methoxsalen, 5911
Naphthalic Acid, 6299
$C_{12}H_8S_3$
α-Terthienyl, 9106
$C_{12}H_9AsClN$
Phenarsazine Chloride, 7170
$C_{12}H_9ClFN_5$
Arprinocid, 816
$C_{12}H_9ClF_3N_3O$
Norflurazon, 6618
$C_{12}H_9ClO$
2-Phenyl-6-chlorophenol, 7251
$C_{12}H_9ClO_2S$
Sulphenone®, 8969
$C_{12}H_9Cl_2NO_3$
Vinclozolin, 9890
$C_{12}H_9Li_6O_{12}S_3Sb$
Anthiolimine, 711
$C_{12}H_9N$
Carbazole, 1792
$C_{12}H_9NO$
Phenoxazine, 7222
$C_{12}H_9NO_2$
Dictamnine, 3079
o-Nitrobiphenyl, 6513
p-Nitrobiphenyl, 6514
$C_{12}H_9NO_6$
Miloxacin, 6116
$C_{12}H_9NS$
Phenothiazine, 7220
$C_{12}H_9N_2NaO_5S$
Tropaeolin O, 9687
$C_{12}H_9N_3O$
Milrinone, 6117
$C_{12}H_9N_3O_4$
Magneson, 5578
$C_{12}H_9N_3O_5$
Nifuroxazide, 6450
$C_{12}H_9NaO.4H_2O$
o-Phenylphenol*, 7276

$C_{12}H_{10}$
Acenaphthene, 23
Diphenyl, 3314
Hemi-Dewar Biphenyl, 4562
$C_{12}H_{10}AsN_4NaO_7$
NPA Acid*, 6646
$C_{12}H_{10}As_2N_2Na_2O_2$
Sodium Arsphenamine, 8524
$C_{12}H_{10}BaO_6S_2$
Barium Benzenesulfonate, 976
$C_{12}H_{10}BaO_8S_2$
p-Phenolsulfonic Acid*, 7212
$C_{12}H_{10}Ba_3O_{14}.7H_2O$
Citric Acid*, 2328
$C_{12}H_{10}CaO_2$
Calcium Phenoxide, 1698
$C_{12}H_{10}CaO_8S_2$
Calcium Phenolsulfonate, 1697
$C_{12}H_{10}CaO_{10}S_2$
Dobesilate Calcium, 3395
$C_{12}H_{10}Ca_2FeO_{14}$
Calcium Ferrous Citrate, 1668
$C_{12}H_{10}Ca_3O_{14}$
Calcium Citrate, 1661
$C_{12}H_{10}ClN$
4-Amino-4'-chlorodiphenyl, 445
$C_{12}H_{10}ClN_3S$
Thionine, 9276
$C_{12}H_{10}Cl_2N_2$
2,2'-Dichlorobenzidine, 3046
3,3'-Dichlorobenzidine, 3047
$C_{12}H_{10}CuO_8S_2$
Cupric p-Phenolsulfonate, 2652
$C_{12}H_{10}F_3N_3O_2$
Nilutamide, 6460
$C_{12}H_{10}Mg$
Diphenylmagnesium, 3328
$C_{12}H_{10}Mg_3O_{14}$
Magnesium Citrate, 5541
$C_{12}H_{10}N_2$
Azobenzene, 930
Harman, 4530
$C_{12}H_{10}N_2O$
Azoxybenzene, 937
Ethyl β-Carboline-3-carboxylate*,
3737
p-Nitrosodiphenylamine, 6560
$C_{12}H_{10}N_2O_4$
Nifurpirinol, 6452
$C_{12}H_{10}N_2O_5$
Cinoxacin, 2311
$C_{12}H_{10}N_4$
2,3-Diaminophenazine, 2960
$C_{12}H_{10}N_4O_2$
Lumichrome, 5468
$C_{12}H_{10}O$
Phenyl Ether, 7259
o-Phenylphenol, 7276
$C_{12}H_{10}O_2$
Chimaphilin, 2047
1-Naphthaleneacetic Acid, 6290
$C_{12}H_{10}O_3$
2-Naphthoxyacetic Acid, 6317
Spizofurone, 8722
$C_{12}H_{10}O_4$
Acifran, A1
Diresorcinol, 3361
Piperic Acid, 7437
Quinhydrone, 8070
$C_{12}H_{10}O_5S_2$
Benzenesulfonic Anhydride, 1079
$C_{12}H_{10}O_7$
Echinochrome A, 3470
$C_{12}H_{10}O_8S_2Zn$
Zinc p-Phenolsulfonate, 10054
$C_{12}H_{10}O_{14}Zn_3$
Zinc Citrate, 10033
$C_{12}H_{10}S$
Phenyl Sulfide, 7285
$C_{12}H_{10}SO_2$
Diphenyl Sulfone, 3336

$C_{12}H_{11}AsN_4O_7$
NPA Acid, 6646
$C_{12}H_{11}ClN_2O_5S$
Furosemide, 4221
$C_{12}H_{11}ClN_6O_2S_2$
Azosemide, 935
$C_{12}H_{11}Cl_2NO$
Propyzamide, 7886
$C_{12}H_{11}Hg_2NO_4$
Phenylmercuric Nitrate, Basic, 7273
$C_{12}H_{11}I_3N_2O_4$
Iodamide, 4904
Metrizoic Acid, 6078
$C_{12}H_{11}N$
p-Biphenylamine, 1248
Diphenylamine, 3317
$C_{12}H_{11}NOS$
Thionalide, 9273
$C_{12}H_{11}NO_2$
4-Amino-1-naphthol*, 465
Carbaryl, 1789
$C_{12}H_{11}NO_3S$
N-Phenylsulfanilic Acid, 7284
$C_{12}H_{11}NO_4$
Casimiroin, 1894
$C_{12}H_{11}N_2NaO_3$
Phenobarbital Sodium, 7202
$C_{12}H_{11}N_3$
p-Aminoazobenzene, 430
Diazoaminobenzene, 2981
$C_{12}H_{11}N_3O_2$
Furonazide, 4220
$C_{12}H_{11}N_7$
Triamterene, 9515
$C_{12}H_{12}Al_2O_{18}$
Aluminum Tartrate, 374
$C_{12}H_{12}BrN_4NaO_2S.H_2O$
Sulfabromomethazine*, 8869
$C_{12}H_{12}Br_2N_2$
Diquat Dibromide, 3359
$C_{12}H_{12}ClNO_2S$
Dansyl Chloride, 2812
$C_{12}H_{12}ClN_5O_4S$
Chlorsulfuron, 2192
$C_{12}H_{12}I_3N_2NaO_2$
Ipodate*, 4958
$C_{12}H_{12}N_2$
Benzidine, 1086
2,4'-Biphenyldiamine, 1249
1,1-Diphenylhydrazine, 3326
$C_{12}H_{12}N_2O$
Harmalol, 4529
$C_{12}H_{12}N_2OS$
4,4'-Sulfinyldianiline, 8927
$C_{12}H_{12}N_2OS_2$
p-Dimethylaminobenzalrhodanine,
3220
$C_{12}H_{12}N_2O_2$
Tetrantoin, 9165
$C_{12}H_{12}N_2O_2S$
Dapsone, 2820
Enoximone, 3542
Sulfabenz, 8867
$C_{12}H_{12}N_2O_3$
Nalidixic Acid, 6273
Phenobarbital, 7201
$C_{12}H_{12}N_2O_6S_2Zn.4H_2O$
Sulfanilic Acid*, 8901
$C_{12}H_{12}N_4$
Chrysoidine*, 2262
p-Diaminoazobenzene, 2954
$C_{12}H_{12}N_4O_3$
Benznidazole, 1094
$C_{12}H_{12}N_6Na_2O_{10}S_2$
Carumonam*, 1880
$C_{12}H_{12}O$
2-Ethoxynaphthalene, 3709
$C_{12}H_{12}O_2$
2-(2-Naphthyloxy)ethanol, 6333

$C_{12}H_{12}O_5$
Fraxetin*, 4178
Radicinin, 8117

$C_{12}H_{12}O_6$
Pyrogallol*, 8010

$C_{12}H_{13}BrN_4O_2S$
Sulfabromomethazine, 8869

$C_{12}H_{13}ClF_3N_3O_4$
Fluchloralin, 4052

$C_{12}H_{13}ClN_4$
Chrysoidine, 2262
Pyrimethamine, 7997

$C_{12}H_{13}Cl_2N_3$
Alinidine, 234

$C_{12}H_{13}I_3N_2O_2$
Ipodate, 4958

$C_{12}H_{13}I_3N_2O_3$
Iocetamic Acid, 4903
Iomeglamic Acid, 4941

$C_{12}H_{13}N$
N,N-Dimethyl-1-naphthylamine, 3240

$C_{12}H_{13}NO_2$
Indolebutyric Acid, 4871
Methsuximide, 5928

$C_{12}H_{13}NO_2S$
Carboxin, 1832

$C_{12}H_{13}NO_3$
Aniracetam, 692

$C_{12}H_{13}N_2NaO_3$
Cyclopentobarbital*, 2751

$C_{12}H_{13}N_2NaO_5$
2-Cyclohexyl-4,6-dinitrophenol*, 2739

$C_{12}H_{13}N_3$
4,4'-Diaminodiphenylamine, 2957

$C_{12}H_{13}N_3O_2$
Isocarboxazid, 5040
Triaziquone, 9517

$C_{12}H_{13}N_3O_4$
Olaquindox, 6783

$C_{12}H_{13}N_3O_4S_2$
N^4-Sulfanilylsulfanilamide, 8905

$C_{12}H_{13}N_3O_6$
Glyconiazide, 4396

$C_{12}H_{13}N_5Na_2O_9S_2$
Tigemonam*, 9364

$C_{12}H_{13}N_5O_2S$
Sulfamidochrysoidine, 8895

$C_{12}H_{13}N_5O_4$
Toyocamycin, 9483

$C_{12}H_{14}As_2Cl_2N_2O_2$
Arsphenamine, 839

$C_{12}H_{14}CaO_{12}$
Calcium Ascorbate, 1648

$C_{12}H_{14}ClNO_2$
Dimethazone, 3203

$C_{12}H_{14}ClNO_3$
Cotarnine*, 2551

$C_{12}H_{14}Cl_2FNO_4S$
Florfenicol, 4042

$C_{12}H_{14}Cl_2N_2$
Paraquat*, 6980

$C_{12}H_{14}Cl_2O_3$
2,4-D*, 2802

$C_{12}H_{14}Cl_3O_4P$
Chlorfenvinphos, 2087

$C_{12}H_{14}N_2$
Detomidine, 2917
N-(1-Naphthyl)ethylenediamine, 6329
Paraquat, 6980

$C_{12}H_{14}N_2O_2$
Abrine, 5
Mephenytoin, 5741
Methetoin, 5889
Primidone, 7753

$C_{12}H_{14}N_2O_2S_2$
Ujothion, 9755

$C_{12}H_{14}N_2O_3$
Cyclopentobarbital, 2751

$C_{12}H_{14}N_2O_3S$
Tioxidazole, 9387

$C_{12}H_{14}N_2O_4$
5-Furfuryl-5-isopropylbarbituric Acid, 4216

$C_{12}H_{14}N_2O_5$
N-(p-Aminobenzoyl)glutamic Acid, 437
2-Cyclohexyl-4,6-dinitrophenol, 2739

$C_{12}H_{14}N_2O_5P_2$
sym-Diphenylpyrophosphorodiamidic Acid, 3335

$C_{12}H_{14}N_2O_6$
Dinoseb*, 3282

$C_{12}H_{14}N_2O_7$
Nicotinamide Ascorbate, 6433

$C_{12}H_{14}N_2S$
Cymiazole, 2771

$C_{12}H_{14}N_3NaO_4S$
Sulfamipyrine, 8896

$C_{12}H_{14}N_4O_2S$
Sulfamethazine, 8886
Sulfisomidine, 8929

$C_{12}H_{14}N_4O_3S$
4'-Formylsuccinanilic Acid Thiosemicarbazone, 4163
Sulfacytine, 8873
Sulfamethomidine, 8888

$C_{12}H_{14}N_4O_4S$
Sulfadimethoxine, 8876
Sulfadoxine, 8877

$C_{12}H_{14}N_4O_4S_2$
Thiophanate*, 9282

$C_{12}H_{14}N_6O_{10}S_2$
Carumonam, 1880

$C_{12}H_{14}O_2$
p-(2-Methylpropenyl)phenol Acetate, 6029
Precocenes*, 7716

$C_{12}H_{14}O_4$
Apiole (Dill), 766
Apiole (Parsley), 767
Dimecrotic Acid, 3189
Phthalic Acid*, 7345

$C_{12}H_{14}O_5$
Cinametic Acid, 2283

$C_{12}H_{14}O_7$
Piscidic Acid*, 7479

$C_{12}H_{15}AsN_6OS_2$
Melarsoprol, 5694

$C_{12}H_{15}ClNO_4PS_2$
Phosalone, 7308

$C_{12}H_{15}ClN_2$
Detomidine*, 2917

$C_{12}H_{15}ClO_3$
Clofibrate, 2374

$C_{12}H_{15}Cl_2NO$
Karsil, 5165

$C_{12}H_{15}Cl_2NO_5S$
Thiamphenicol, 9230

$C_{12}H_{15}N$
6,7-Benzomorphan, 1105
MPTP, 6205

$C_{12}H_{15}NO_3$
Anhalonine, 684
Carbofuran, 1810
Hydrocotarnine, 4715
Metaxalone, 5838

$C_{12}H_{15}NO_3S$
Vanitiolide, 9841

$C_{12}H_{15}NO_4$
Cotarnine, 2551
Ethopabate, 3701
Flavipucine, 4029

$C_{12}H_{15}N_2NaO_3$
Hexobarbital*, 4625

$C_{12}H_{15}N_2O_3PS$
Phoxim, 7341

$C_{12}H_{15}N_3$
Indanazoline, 4845

$C_{12}H_{15}N_3O_2S$
Albendazole, 201

$C_{12}H_{15}N_3O_2S_2$
Glybuzole, 4374

$C_{12}H_{15}N_3O_3$
Oxibendazole, 6891

$C_{12}H_{15}N_3O_5S$
Amezinium Methyl Sulfate, 403

$C_{12}H_{15}N_5O_3S$
Sulfaguanole, 8880

$C_{12}H_{15}N_5O_9S_2$
Tigemonam, 9364

$C_{12}H_{16}ClNO_3$
Meclofenoxate, 5660

$C_{12}H_{16}ClNO_4$
Cotarnine*, 2551

$C_{12}H_{16}ClN_3$
Indanazoline*, 4845

$C_{12}H_{16}ClN_3O_2S$
Oxythiamine, 6932

$C_{12}H_{16}Cl_2N_2O$
Neburon, 6354

$C_{12}H_{16}Cl_2N_4S$
Beclotiamine, 1030

$C_{12}H_{16}F_3N$
Fenfluramine, 3920

$C_{12}H_{16}N_2$
N,N-Dimethyltryptamine, 3251
Etryptamine, 3848
Fenproporex, 3939

$C_{12}H_{16}N_2O$
Bufotenine, 1467
Caulophylline, 1914
Psilocin, 7941

$C_{12}H_{16}N_2O_3$
Cyclobarbital, 2717
Hexobarbital, 4625

$C_{12}H_{16}N_2O_4$
Betahistine*, 1200

$C_{12}H_{16}N_2S$
Morantel, 6175
Xylazine, 9987

$C_{12}H_{16}N_3O_3P$
Benzodepa, 1097

$C_{12}H_{16}N_3O_3PS_2$
Azinphos-methyl*, 926

$C_{12}H_{16}N_4OS_2$
Thiothiamine, 9296

$C_{12}H_{16}N_4O_2S_2$
Glybuthiazol(e), 4373

$C_{12}H_{16}O_2$
Ibufenac, 4811
Isoamyl Benzoate, 4995
Thymol*, 9333

$C_{12}H_{16}O_3$
Asarones, 849
Guaiacol Valerate, 4461
Isoamyl Salicylate, 5007

$C_{12}H_{16}O_4$
Aspidinol, 869

$C_{12}H_{16}O_4S_2$
Malotilate, 5593

$C_{12}H_{16}O_6S$
Protiofate, 7904

$C_{12}H_{16}O_7$
Arbutin, 799

$C_{12}H_{17}ClF_3N$
Fenfluramine*, 3920

$C_{12}H_{17}ClN_2$
Fenproporex*, 3939

$C_{12}H_{17}ClO_2$
Fenpentadiol, 3933

$C_{12}H_{17}Cl_2NO_3$
2,4-DB*, 2828
Meclofenoxate*, 5660

$C_{12}H_{17}NO$
Deet, 2848
Phendimetrazine, 7180

Denotes derivative of title compound

$C_{12}H_{17}NO_2$
Butacetin, 1497
Ciclopirox, 2270
Promecarb, 7794

$C_{12}H_{17}NO_3$
Anhalonidine, 683
Anisomycin*, 700
Bucetin, 1445
Bufexamac, 1462
Cerulenin, 1997
Ethamivan, 3673
Rimiterol, 8223

$C_{12}H_{17}NO_9$
Norepinephrine*, 6612

$C_{12}H_{17}N_2NaO_2S$
Thiamylal*, 9231

$C_{12}H_{17}N_2NaO_3$
Secobarbital Sodium, 8374

$C_{12}H_{17}N_2O_4P$
Psilocybin, 7942

$C_{12}H_{17}N_3O$
Cimaterol, 2278

$C_{12}H_{17}N_3O_4$
Agaritine, 175

$C_{12}H_{17}N_3O_4 \cdot H_2O$
Imipenem, 4834

$C_{12}H_{17}N_5O_3$
7-Morpholinomethyltheophylline, 6196

$C_{12}H_{17}N_5O_4S$
Thiamine Mononitrate, 9223

$C_{12}H_{17}N_5S$
Methiotriazamine, 5899

$C_{12}H_{17}NaO_7$
Dikegulac*, 3184

$C_{12}H_{18}Be_4O_{13}$
Beryllium Acetate, Basic, 1179

$C_{12}H_{18}BrNO_3$
Rimiterol*, 8223

$C_{12}H_{18}ClN$
Mefenorex, 5681

$C_{12}H_{18}ClNO$
Tulobuterol, 9720

$C_{12}H_{18}ClNO_4$
Methyldopa*, 5974

$C_{12}H_{18}ClN_4O_4PS$
Thiamine Phosphoric Acid Ester Chloride, 9224

$C_{12}H_{18}ClN_5O_3$
Metformin*, 5845

$C_{12}H_{18}Cl_2N_2O$
Clenbuterol, 2347

$C_{12}H_{18}Cl_2N_4OS$
Thiamine Hydrochloride, 9222

$C_{12}H_{18}IN$
Iofetamine I ^{123}, 4938

$C_{12}H_{18}NO_6P$
Diisopropyl Paraoxon, 3183

$C_{12}H_{18}N_2O$
Oxotremorine, 6904
Pivalylbenzhydrazine, 7483

$C_{12}H_{18}N_2O_2$
Mexacarbate, 6090
Nicametate, 6401

$C_{12}H_{18}N_2O_2S$
Thiamylal, 9231

$C_{12}H_{18}N_2O_3$
Nealbarbital, 6350

$C_{12}H_{18}N_2O_3S$
Tolbutamide, 9432

$C_{12}H_{18}N_2O_4$
Midodrine, 6107

$C_{12}H_{18}N_2O_5$
Bacilysin, 946
Hypoglycine B, 4802

$C_{12}H_{18}N_2O_7$
Bicozamycin, 1222

$C_{12}H_{18}N_2O_7S$
N^4-β-D-Glucosylsulfanilamide, 4358

$C_{12}H_{18}N_4O_6S$
Oryzalin, 6840

$C_{12}H_{18}N_{20}$
Isoproturon, 5106

$C_{12}H_{18}Na_5O_{23}S_4Sb$
Stibophen, 8769

$C_{12}H_{18}O$
6-n-Amyl-m-cresol, 643
Propofol, 7847

$C_{12}H_{18}O_2$
4-Hexylresorcinol, 4633

$C_{12}H_{18}O_4$
Butopyronoxyl, 1529

$C_{12}H_{18}O_7$
Dikegulac, 3184

$C_{12}H_{19}BrN_2O_2$
Neostigmine*, 6380

$C_{12}H_{19}ClIN$
Iofetamine I ^{123}*, 4938

$C_{12}H_{19}ClNO_3P$
Crufomate, 2607

$C_{12}H_{19}ClN_2O_4$
Midodrine*, 6107

$C_{12}H_{19}ClN_4O_7P_2S$
Cocarboxylase, 2451

$C_{12}H_{19}Cl_2N$
Mefenorex*, 5681

$C_{12}H_{19}Cl_2NO$
Tulobuterol*, 9720

$C_{12}H_{19}Cl_3N_2O$
Clenbuterol*, 2347

$C_{12}H_{19}Cl_3O_8$
Sucralose, 8854

$C_{12}H_{19}NO$
Etafedrine, 3661

$C_{12}H_{19}NO_2$
Bamethan, 964

$C_{12}H_{19}NO_3$
Macromerine, 5520
Prenalterol, 7742
Terbutaline, 9089

$C_{12}H_{19}NO_4$
Choline Salicylate, 2212

$C_{12}H_{19}N_2NaO_2S_2$
Methitural, 5901

$C_{12}H_{19}N_2NaO_3$
Hexethal Sodium, 4623

$C_{12}H_{19}N_2O_2{}^+$
Neostigmine, 6380

$C_{12}H_{19}N_3O$
Procarbazine, 7764

$C_{12}H_{19}N_4O_{10}P_3S$
Thiamine Triphosphoric Acid Ester, 9227

$C_{12}H_{19}O_2PS_3$
Sulprofos, 8972

$C_{12}H_{20}B_2CaO_{16}$
Calcium Borogluconate, 1652

$C_{12}H_{20}ClNO$
Etafedrine*, 3661

$C_{12}H_{20}ClNO_3$
Prenalterol*, 7742

$C_{12}H_{20}ClN_3O$
Procarbazine*, 7764

$C_{12}H_{20}Cl_4N_2O_2$
Fertilysin, 4009

$C_{12}H_{20}N_2$
Tremorine, 9498

$C_{12}H_{20}N_2O_2$
Aspergillic Acid, 863

$C_{12}H_{20}N_2O_3$
Pirbuterol, 7461
Tetrabarbital, 9117

$C_{12}H_{20}N_2O_3S$
Sotalol, 8682

$C_{12}H_{20}N_2O_4S$
Soterenol, 8683

$C_{12}H_{20}N_4O_2$
Hexazinone, 4617

$C_{12}H_{20}N_4O_8P_2S$
Thiamine Phosphoric Acid Ester Phosphate Salt, 9225

$C_{12}H_{20}N_4O_9S$
Sulfazecin, 8924

$C_{12}H_{20}O_2$
Bornyl Acetate, 1339
Linalyl Acetate, 5374

$C_{12}H_{20}O_4$
Traumatic Acid, 9493

$C_{12}H_{20}O_6$
Diacetoneglucose, 2945

$C_{12}H_{20}O_7$
Citric Acid*, 2328

$C_{12}H_{21}AsN_2O_5$
Acetarsone*, 44

$C_{12}H_{21}ClN_2O_3S$
Sotalol*, 8682

$C_{12}H_{21}ClO_2$
Trimedlure, 9615

$C_{12}H_{21}N$
Memantine, 5709
Rimantadine, 8221

$C_{12}H_{21}NO$
Pimeclone, 7399

$C_{12}H_{21}NO_{11}$
Chondrosine, 2218
Hyalobiuronic Acid, 4674

$C_{12}H_{21}N_2O_3PS$
Diazinon®, 2978

$C_{12}H_{21}N_5O_2S_2$
Nizatidine, 6582

$C_{12}H_{21}N_5O_3$
Cadralazine, 1631

$C_{12}H_{21}NaO_7S$
Diisobutyl Sodium Sulfosuccinate, 3179

$C_{12}H_{22}CaO_4$
Calcium 2-Ethylbutanoate, 1667

$C_{12}H_{22}CaO_{14}$
Calcium Gluconate, 1672

$C_{12}H_{22}ClN$
Memantine*, 5709
Rimantadine*, 8221

$C_{12}H_{22}Cl_2N_2O_3$
Pirbuterol*, 7461

$C_{12}H_{22}CuO_{14}$
Cupric Gluconate, 2643

$C_{12}H_{22}FeO_{14}$
Ferrous Gluconate, 3996

$C_{12}H_{22}MgO_{14} \cdot 2H_2O$
Gluconic Acid*, 4350

$C_{12}H_{22}N_2O_2$
Crotethamide, 2598

$C_{12}H_{22}N_2O_8S_2$
Piposulfan, 7452

$C_{12}H_{22}N_4O_{14}P_4S \cdot H_2O$
Thiamine Triphosphoric Acid Ester*, 9227

$C_{12}H_{22}O$
Geosmin, 4294

$C_{12}H_{22}O_2$
Menthyl Acetate, 5725

$C_{12}H_{22}O_4$
Dibutyl Succinate, 3023
Di-$tert$-butyl Succinate, 3024

$C_{12}H_{22}O_6$
Etoglucid, 3837

$C_{12}H_{22}O_{10}$
Rutinose, 8277
Scillabiose, 8353

* Denotes derivative of title compound

$C_{13}H_{12}N_3NaO_6S$
Cephacetrile Sodium, 1969
$C_{13}H_{12}N_4O$
Diphenylcarbazone, 3322
$C_{13}H_{12}N_4O_2$
Lumiflavine, 5469
$C_{13}H_{12}N_4O_3$
Pyriminil, 7999
$C_{13}H_{12}N_4S$
Dithizone, 3383
$C_{13}H_{12}N_5NaO_5S_2$
Ceftizoxime*, 1949
$C_{13}H_{12}N_8O_4S_3$
Ceftezole, 1946
$C_{13}H_{12}O$
Benzohydrol, 1100
o-Benzylphenol, 1158
Phentydrone, 7235
$C_{13}H_{12}O_2$
Monobenzone, 6159
$C_{13}H_{12}O_3$
Allenolic Acid, 247
Euparin, 3857
2-Naphthyl Lactate, 6332
$C_{13}H_{13}As_2N_2NaO_4S$
Neoarsphenamine, 6361
$C_{13}H_{13}Cl_2N_3O_3$
Iprodione, 4964
$C_{13}H_{13}N$
Benzhydrylamine, 1085
Benzylaniline, 1140
Methyldiphenylamine, 5973
$C_{13}H_{13}NO_2$
Vitamin K_5*, 9936
$C_{13}H_{13}N_3$
1,3-Diphenylguanidine, 3325
Nitrin, 6500
$C_{13}H_{13}N_3O_3$
Cyclobendazole, 2718
$C_{13}H_{13}N_3O_4S$
Proflavine*, 7780
$C_{13}H_{13}N_3O_5S_2$
Succinylsulfathiazole, 8850
$C_{13}H_{13}N_5O_4S$
Sulfachrysoidine, 8872
$C_{13}H_{13}N_5O_5S_2$
Ceftizoxime, 1949
$C_{13}H_{14}ClNO_2$
Pirprofen, 7478
$C_{13}H_{14}Cl_2O_3$
Ciprofibrate, 2314
$C_{13}H_{14}F_3N_3O_4$
Ethalfluralin, 3671
$C_{13}H_{14}N_2$
p,p'-Diaminodiphenylmethane, 2958
Tacrine, 9003
$C_{13}H_{14}N_2O$
Harmaline, 4528
10-Methoxyharmalan, 5915
N-(p-Methoxyphenyl)-p-phenylene-diamine, 5922
Phenyramidol, 7292
$C_{13}H_{14}N_2O_2$
Metomidate, 6069
$C_{13}H_{14}N_2O_2S$
Benzylsulfamide, 1161
p-Sulfanilylbenzylamine, 8903
$C_{13}H_{14}N_2O_3$
Mephobarbital, 5742
$C_{13}H_{14}N_2O_4$
Menadoxime, 5717
$C_{13}H_{14}N_2O_4S_2$
Diphenylmethane-4,4'-disulfon-amide, 3330
Gliotoxin, 4336
$C_{13}H_{14}N_2S$
Metizoline, 6060
$C_{13}H_{14}N_3NaO_4S$
Glymidine*, 4403

$C_{13}H_{14}N_4O$
sym-Diphenylcarbazide, 3321
$C_{13}H_{14}N_4O_3S$
N^2-Formylsulfisomidine, 4164
$C_{13}H_{14}N_4O_4$
Pasiniazide, 6999
$C_{13}H_{14}N_4O_4S$
Acetyl Sulfamethoxypyrazine, 96
Sulfamethoxypyridazine*, 8890
$C_{13}H_{14}O_2$
Tremetone, 9497
$C_{13}H_{14}O_5$
Citrinin, 2329
$C_{13}H_{15}BrN_4O_2$
Brodimoprim, 1369
$C_{13}H_{15}ClN_2$
Tacrine*, 9003
$C_{13}H_{15}ClN_2O$
Phenyramidol*, 7292
$C_{13}H_{15}ClN_2O_2$
Metomidate*, 6069
$C_{13}H_{15}ClN_2S$
Metizoline*, 6060
$C_{13}H_{15}Cl_3N_2O_3$
Chloralantipyrine, 2058
$C_{13}H_{15}NO_2$
Glutethimide, 4371
Pyracarbolid, 7966
Securinine, 8376
$C_{13}H_{15}N_3O_2$
Pyrolan®, 8013
$C_{13}H_{15}N_3O_4S$
Acetyl Sulfisoxazole, 100
Glymidine, 4403
$C_{13}H_{15}N_3O_4S_2$
4'-(Methylsulfamoyl)sulfanilanilide, 6040
$C_{13}H_{15}N_5O_6$
Nifurfoline, 6448
$C_{13}H_{16}ClNO$
Ketamine, 5174
$C_{13}H_{16}ClN_3O_5S_2$
Ambuside, 396
$C_{13}H_{16}Cl_{12}O_8$
Pentaerythritol Chloral, 7063
$C_{13}H_{16}F_3N_3O_4$
Benfluralin, 1048
Trifluralin, 9598
$C_{13}H_{16}HgNNaO_6$
Mersalyl, 5805
$C_{13}H_{16}N_2$
Medetomidine, 5670
Tetrahydrozoline, 9150
$C_{13}H_{16}N_2O$
Adrenoglomerulotropin, 163
Oxantel, 6876
$C_{13}H_{16}N_2O_2$
Aminoglutethimide, 452
Melatonin, 5695
Mofebutazone, 6141
$C_{13}H_{16}N_2O_2S$
Thialbarbital, 9219
$C_{13}H_{16}N_2O_3$
Acetylpheneturide, 94
$C_{13}H_{16}N_3NaO_4S.H_2O$
Dipyrone, 3358
$C_{13}H_{16}N_4$
Mifentidine, 6108
$C_{13}H_{16}N_4O_2$
Diaveridine, 2976
$C_{13}H_{16}N_4O_3S$
Cycothiamin(e), 2762
$C_{13}H_{16}N_4O_6$
Furaltadone, 4205
$C_{13}H_{16}O_2$
Cinnamic Acid*, 2300
R-11, 8114
Tremetone*, 9497
$C_{13}H_{16}O_3$
Precocenes*, 7716

$C_{13}H_{16}O_7$
Helicin, 4544
Vacciniin, 9811
$C_{13}H_{16}O_{10}$
α-Glucogallin, 4347
β-Glucogallin, 4348
$C_{13}H_{17}ClN_2$
Medetomidine*, 5670
Tetrahydrozoline*, 9150
$C_{13}H_{17}ClN_2O_2$
Moclobemide, 6139
$C_{13}H_{17}Cl_2NO$
Ketamine*, 5174
$C_{13}H_{17}F_3N_4O_4$
Prodiamine, 7773
$C_{13}H_{17}KN_2O_4S_2$
Penicillin O*, 7044
$C_{13}H_{17}N$
Deprenyl, 2893
$C_{13}H_{17}NO$
Crotamiton, 2597
$C_{13}H_{17}NO_2$
Alminoprofen, 298
$C_{13}H_{17}NO_3$
Lophophorine, 5451
$C_{13}H_{17}NO_4$
Alibendol, 233
$C_{13}H_{17}NO_7$
Synephrine*, 8994
$C_{13}H_{17}N_2NaO_4S_2$
Penicillin O*, 7044
$C_{13}H_{17}N_3$
Tramazoline, 9486
$C_{13}H_{17}N_3O$
Aminopyrine, 488
$C_{13}H_{17}N_3O_2$
Parbendazole, 6985
$C_{13}H_{17}N_3O_3$
Dioxypyramidon, 3299
$C_{13}H_{17}N_5O_2S$
Sulfasymazine, 8919
$C_{13}H_{17}N_5O_8S_2$
Aztreonam, 938
$C_{13}H_{18}Br_2N_2O$
Ambroxol, 392
$C_{13}H_{18}ClF_3N_2O$
Mabuterol, 5517
$C_{13}H_{18}ClN$
Deprenyl*, 2893
$C_{13}H_{18}ClNO$
Bupropion, 1488
Solan, 8659
$C_{13}H_{18}ClNO_2$
Cloforex, 2378
$C_{13}H_{18}ClNO_4$
Alclofenac*, 212
$C_{13}H_{18}ClN_3.H_2O$
Tramazoline*, 9486
$C_{13}H_{18}ClN_3O_3S$
Glypinamide, 4410
$C_{13}H_{18}ClN_3O_4S_2$
Cyclopenthiazide, 2750
$C_{13}H_{18}Cl_2N_2O_2$
Melphalan, 5708
$C_{13}H_{18}Cl_2N_4$
Mifentidine*, 6108
$C_{13}H_{18}N_2$
Medmain, 5675
$C_{13}H_{18}N_2O$
Fenoxazoline, 3930
$C_{13}H_{18}N_2O_2$
Lenacil, 5318
$C_{13}H_{18}N_2O_3$
Heptabarbital, 4575
$C_{13}H_{18}N_2O_4S_2$
Penicillin O, 7044
$C_{13}H_{18}N_2O_5$
Phenoxypropazine*, 7227
$C_{13}H_{18}N_4O_3$
Methenamine Salicylate, 5884
Pentoxifylline, 7092

Denotes derivative of title compound

$C_{13}H_{18}N_4O_6S$
Methenamine Sulfosalicylate, 5885
$C_{13}H_{18}O_2$
Ibuprofen, 4812
$C_{13}H_{18}O_3$
Mandelic Acid Isoamyl Ester, 5600
$C_{13}H_{18}O_5S$
Ethofumesate, 3697
$C_{13}H_{18}O_7$
Methylarbutin, 5944
Salicin, 8293
$C_{13}H_{19}Br_2ClN_2O$
Ambroxol*, 392
$C_{13}H_{19}ClN_2O$
Butanilicaine, 1508
Fenoxazoline*, 3930
$C_{13}H_{19}ClN_2O_5S_2$
Mefruside, 5685
$C_{13}H_{19}Cl_2F_3N_2O$
Mabuterol*, 5517
$C_{13}H_{19}Cl_2NO$
Bupropion*, 1488
$C_{13}H_{19}Cl_2NO_2$
Cloranolol, 2399
$C_{13}H_{19}NO$
Diethylpropion, 3116
$C_{13}H_{19}NO_2$
Bufencarb, 1459
Carnegine, 1854
Dioscorine, 3288
Exalamide, 3866
Ibuproxam, 4813
$C_{13}H_{19}NO_2S$
Isobornyl Thiocyanoacetate, Technical, 5012
$C_{13}H_{19}NO_3$
Gigantine, 4316
Metamivam, 5829
Pellotine, 7018
Viloxazine, 9884
$C_{13}H_{19}NO_4S$
Probenecid, 7760
$C_{13}H_{19}NO_7$
Gepefrine*, 4295
$C_{13}H_{19}NO_8$
Metaraminol*, 5837
$C_{13}H_{19}N_3O_4$
Dipropalin®, 3348
Pendimethalin, 7026
$C_{13}H_{19}N_3O_6S$
Nitralin, 6486
$C_{13}H_{19}N_4O_{12}P$
SAICAR, 8288
$C_{13}H_{19}N_5 \cdot H_2O$
Pinacidil, 7407
$C_{13}H_{20}ClNO$
Diethylpropion*, 3116
$C_{13}H_{20}ClNO_2$
Phenamacide Hydrochloride, 7163
$C_{13}H_{20}ClNO_3$
Viloxazine*, 9884
$C_{13}H_{20}Cl_2N_2O_2$
2-Chloroprocaine Hydrochloride, 2158
$C_{13}H_{20}Cl_3NO_2$
Cloranolol*, 2399
$C_{13}H_{20}N_2O$
Prilocaine, 7749
$C_{13}H_{20}N_2O_2$
Butethamine, 1517
Dropropizine, 3439
Procaine, 7763
$C_{13}H_{20}N_2O_3$
Hydroxyprocaine, 4772
$C_{13}H_{20}N_2O_3S$
Carticaine, 1878
Etozolin, 3845
$C_{13}H_{20}N_2O_4$
Deanol Acetamidobenzoate, 2836
$C_{13}H_{20}N_2O_6$
Actinobolin, 130

$C_{13}H_{20}N_4O_2$
Pentifylline, 7085
$C_{13}H_{20}N_4O_3$
Ciclosidomine, 2271
$C_{13}H_{20}N_4O_7$
Methenamine Anhydromethylenecitrate, 5881
$C_{13}H_{20}N_6O_4$
Acefylline Piperazine, 19
$C_{13}H_{20}O$
Ionone, 4942
4-(5-Isopropenyl-2-methyl-1-cyclopenten-1-yl)-2-butanone, 5092
Pseudoionone, 7931
$C_{13}H_{20}O_3$
Febuprol, 3891
$C_{13}H_{20}O_8$
Pentaerythritol Tetraacetate, 7065
$C_{13}H_{21}ClN_2O$
Prilocaine*, 7749
$C_{13}H_{21}ClN_2O_2$
Butethamine*, 1517
Procaine*, 7763
$C_{13}H_{21}ClN_2O_3S$
Carticaine*, 1878
$C_{13}H_{21}ClN_4O_3$
Ciclosidomine*, 2271
$C_{13}H_{21}N$
2,6-Di-tert-butylpyridine, 3022
$C_{13}H_{21}NO_2$
Guaiactamine, 4462
Tigloidine, 9366
Toliprolol, 9439
$C_{13}H_{21}NO_3$
Albuterol, 209
Isoetharine, 5053
Moprolol, 6172
$C_{13}H_{21}NO_4$
Meteloidine, 5841
$C_{13}H_{21}N_3O_3$
Carbuterol, 1840
$C_{13}H_{22}ClNO_2$
Toliprolol*, 9439
$C_{13}H_{22}ClNO_3$
Isoetharine*, 5053
Moprolol*, 6172
$C_{13}H_{22}ClN_3O$
Procainamide Hydrochloride, 7762
$C_{13}H_{22}ClN_3O_3$
Carbuterol*, 1840
$C_{13}H_{22}ClN_5O_2$
Etamiphyllin, 3663
$C_{13}H_{22}NO_3PS$
Fenamiphos, 3901
$C_{13}H_{22}N_2$
Dicyclohexylcarbodiimide, 3086
$C_{13}H_{22}N_2O$
Norea, 6611
$C_{13}H_{22}N_2O_4$
Salicylamide O-Acetic Acid*, 8298
$C_{13}H_{22}N_2O_6S$
Neostigmine*, 6380
$C_{13}H_{22}N_4O_3S$
Ranitidine, 8126
$C_{13}H_{22}O$
Solanone, 8663
$C_{13}H_{23}ClN_4O_3S$
Ranitidine*, 8126
$C_{13}H_{23}NO_9$
Streptobiosamine, 8782
$C_{13}H_{24}N_2O$
Cuscohygrine, 2685
$C_{13}H_{24}N_2O_2$
Cropropamide, 2595
$C_{13}H_{24}N_3O_3PS$
Pirimiphos-ethyl, 7469
$C_{13}H_{24}N_4O_3$
Melanostatin*, 5693

$C_{13}H_{24}N_4O_3S$
Bupirimate, 1484
Timolol, 9374
$C_{13}H_{24}O$
TMD, 9410
$C_{13}H_{24}O_9$
Strophanthobiose, 8820
$C_{13}H_{24}O_{10}$
Gaultherin*, 4272
$C_{13}H_{25}B_5N_2O_{12}$
Procaine*, 7763
$C_{13}H_{25}N_3O_8S_2$
Sulfanilamidomethanesulfonic Acid Triethanolamine Salt, 8899
$C_{13}H_{26}N_2O_3$
Elaiomycin, 3498
$C_{13}H_{26}N_2O_4$
Tybamate, 9737
$C_{13}H_{29}N$
Octamylamine, 6671
$C_{13}H_{32}I_2N_2O$
Plegatil, 7507
$C_{13}H_{33}Br_2N_3$
Azamethonium Bromide, 913
$C_{14}H_4N_2O_2S_2$
Dithianone, 3375
$C_{14}H_4N_4O_{12}$
Chrysamminic Acid, 2253
$C_{14}H_6ClF_3NNaO_5$
Acifluorfen*, 105
$C_{14}H_6O_8$
Ellagic Acid, 3508
$C_{14}H_7Br_3F_3N_3O_4$
Bromethalin, 1378
$C_{14}H_7ClF_3NO_5$
Acifluorfen, 105
$C_{14}H_7NO_6$
Alizarine Orange, 240
$C_{14}H_7NaO_7S$
Sodium Alizarinesulfonate, 8516
$C_{14}H_8Br_2$
9,10-Dibromoanthracene, 3000
$C_{14}H_8Br_2F_3NO_2$
Fluorosalan, 4106
$C_{14}H_8Br_2O_4$
Dibromsalicil, 3015
$C_{14}H_8ClNNa_2O_4$
Lobenzarit*, 5433
$C_{14}H_8Cl_2N_4$
Clofentezine, 2373
$C_{14}H_8N_2Na_2O_6$
Olsalazine*, 6799
$C_{14}H_8N_2O_6$
Nifuroquine, 6449
$C_{14}H_8N_2O_8$
5,5'-Dinitrosalicil, 3279
$C_{14}H_8N_2S_4$
2,2'-Dithiobis[benzothiazole], 3377
$C_{14}H_8O_2$
Anthraquinone, 717
Morphenol, 6183
Phenanthrenequinone, 7168
$C_{14}H_8O_4$
Alizarin, 237
Anthrarufin, 719
Danthron, 2813
Quinizarin, 8094
$C_{14}H_8O_5$
Anthragallol, 713
Purpurin, 7962
$C_{14}H_8O_6$
Quinalizarin, 8059
$C_{14}H_8O_8$
Rufigallol, 8267
$C_{14}H_9ClF_2N_2O_2$
Diflubenzuron, 3128
$C_{14}H_9ClN_2O_3S$
Tenidap, A9
$C_{14}H_9Cl_2F_3N_2O$
Cloflucarban, 2376

Denotes derivative of title compound

Denotes derivative of title compound

$C_{14}H_{14}N_2O_6S_2$
Amsonic Acid, 628
$C_{14}H_{14}N_3NaO_3S$
Methyl Orange, 6019
$C_{14}H_{14}N_3NaO_5S_2$
Acetosulfone Sodium, 66
$C_{14}H_{14}N_4O$
Phenamidine, 7165
$C_{14}H_{14}N_4O_2S$
Cambendazole, 1733
$C_{14}H_{14}N_4O_8S_2$
Dinsed, 3284
$C_{14}H_{14}N_8O_4S_3$
Cefazolin, 1925
$C_{14}H_{14}O$
Benzyl Ether, 1146
$C_{14}H_{14}O_2$
Hydrobenzoin, 4700
$C_{14}H_{14}O_3$
Euparin*, 3857
2-Isovalerylindane-1,3-dione, 5123
Kawain, 5167
Naproxen, 6337
Pindone, 7413
$C_{14}H_{14}O_4$
Phthalofyne, 7348
$C_{14}H_{14}S$
Benzyl Sulfide, 1162
$C_{14}H_{14}S_2$
Dibenzyl Disulfide, 2995
$C_{14}H_{15}ClN_2$
Naphazoline*, 6287
$C_{14}H_{15}ClN_4$
Zometapine, 10093
$C_{14}H_{15}Cl_2N$
Chlornaphazine, 2107
$C_{14}H_{15}Cl_2NS$
Ticlopidine*, 9361
$C_{14}H_{15}N$
Dibenzylamine, 2993
$C_{14}H_{15}NO$
2-Amino-1,2-diphenylethanol, 449
$C_{14}H_{15}NO_5$
Folescutol, 4139
$C_{14}H_{15}N_3$
o-Aminoazotoluene, 431
p-Dimethylaminoazobenzene, 3218
$C_{14}H_{15}N_3O_2$
Indolmycin, 4872
$C_{14}H_{15}N_5O$
Endralazine, 3532
$C_{14}H_{15}N_5O_6S$
Metsulfuron Methyl, 6081
$C_{14}H_{15}O_2PS_2$
Edifenphos, 3485
$C_{14}H_{15}O_3P$
Dibenzyl Phosphite, 2996
$C_{14}H_{16}$
Chamazulene, 2031
$C_{14}H_{16}As_2N_2O_6$
Diphetarsone, 3338
$C_{14}H_{16}CaN_6O_8S_2$
Isoniazid Methanesulfonate*, 5072
$C_{14}H_{16}ClNO_5$
Folescutol*, 4139
$C_{14}H_{16}ClN_3O_2$
Triadimefon, 9507
$C_{14}H_{16}ClN_3O_4S_2$
Cyclothiazide, 2760
$C_{14}H_{16}ClO_5PS$
Coumaphos, 2559
$C_{14}H_{16}ClO_6P$
Coroxon, 2531
$C_{14}H_{16}Cl_2N_4O_3$
Obidoxime Chloride, 6659
$C_{14}H_{16}Cl_2O_2$
Fenclorac, 3914
$C_{14}H_{16}Cl_6O_5$
Toloxychlorinol, 9446
$C_{14}H_{16}F_3N_3O_4$
Profluralin, 7781

$C_{14}H_{16}I_2O_3$
Monophen®, 6163
$C_{14}H_{16}N_2$
1,2-Dianilinoethane, 2969
o-Tolidine, 9437
$C_{14}H_{16}N_2O_2$
Dianisidine, 2970
Etomidate, 3838
Rolicyprine, 8234
$C_{14}H_{16}N_2O_3$
Nadoxolol, 6261
Phenetharbital, 7183
$C_{14}H_{16}N_2O_3S$
2-p-Sulfanilylanilinoethanol, 8902
$C_{14}H_{16}N_2O_4$
Taglutimide, 9006
$C_{14}H_{16}N_4$
Budralazine, 1457
$C_{14}H_{16}N_4O$
Ethoxazene, 3706
$C_{14}H_{16}N_4O_3$
Piromidic Acid, 7475
$C_{14}H_{16}O_3$
Kawain*, 5167
$C_{14}H_{16}O_9$
Bergenin, 1174
$C_{14}H_{17}ClNO_4PS_2$
Dialifor, 2949
$C_{14}H_{17}ClN_2O_3$
Nadoxolol*, 6261
$C_{14}H_{17}ClN_2O_3S$
Clorexolone, 2401
$C_{14}H_{17}ClN_4O$
Ethoxazene*, 3706
$C_{14}H_{17}I_3N_2O_2$
Ipodate*, 4958
$C_{14}H_{17}NO_2$
3-Quinuclidinol*, 8110
$C_{14}H_{17}NO_3$
EEDQ, 3488
$C_{14}H_{17}NO_4$
Chromocarb*, 2239
$C_{14}H_{17}NO_6$
Indican (Plant Indican), 4854
Mandelonitrile Glucoside, 5602
$C_{14}H_{17}NO_7$
Dhurrin, 2937
$C_{14}H_{17}NS_2$
Dimethylthiambutene, 3248
$C_{14}H_{17}N_2NaO_3$
Methohexital Sodium, 5904
$C_{14}H_{17}N_3O_9$
6-Azauridine*, 920
$C_{14}H_{17}N_5O_3$
Pipemidic Acid, 7427
$C_{14}H_{17}O_5PS$
Hymecromone O,O-Diethyl Phosphorothioate, 4793
$C_{14}H_{18}CaN_3Na_3O_{10}$
Pentetate Calcium Trisodium, 7082
$C_{14}H_{18}ClKN_2O_4S_2$
Penicillin S Potassium, 7045
$C_{14}H_{18}ClN$
Picilorex, 7369
$C_{14}H_{18}ClNO_2$
Indeloxazine Hydrochloride, 4850
3-Quinuclidinol*, 8110
$C_{14}H_{18}ClN_3O_2$
Triadimenol, 9508
$C_{14}H_{18}ClN_3S$
Chlorothen, 2168
$C_{14}H_{18}F_3NO$
Oxaflozane, 6861
$C_{14}H_{18}N_2O$
Ibudilast, 4810
Propyphenazone, 7884
$C_{14}H_{18}N_2O_2$
Hypaphorine, 4798
Parsalmide, 6993

$C_{14}H_{18}N_2O_3$
Physovenine, 7358
Reposal, 8141
$C_{14}H_{18}N_2O_4$
α-Ribazole, 8200
$C_{14}H_{18}N_2O_5$
Aspartame, 861
$C_{14}H_{18}N_2O_6S$
Carpetimycins*, 1867
$C_{14}H_{18}N_2O_7$
Dinobuton, 3280
$C_{14}H_{18}N_2O_9S_2$
Carpetimycins*, 1867
$C_{14}H_{18}N_4O_3$
Benomyl, 1053
Trimethoprim, 9624
$C_{14}H_{18}N_4O_4S_2$
Thiophanate, 9282
$C_{14}H_{18}N_6O$
2,6-Diamino-2'-butyloxy-3,5'-azopyridine, 2956
$C_{14}H_{18}O_4$
Cinoxate, 2312
$C_{14}H_{18}O_7$
Picein, 7367
$C_{14}H_{18}O_8$
Glucovanillin, 4359
$C_{14}H_{19}BrO_9$
Acetobromglucose, 52
$C_{14}H_{19}ClF_3NO$
Oxaflozane*, 6861
$C_{14}H_{19}ClN_4$
Amprolium, 623
$C_{14}H_{19}Cl_2N$
Picilorex*, 7369
$C_{14}H_{19}Cl_2NO_2$
Chlorambucil, 2064
$C_{14}H_{19}Cl_2N_3S$
Chlorothen*, 2168
$C_{14}H_{19}NO$
Ethoxyquin, 3710
$C_{14}H_{19}NO_2$
Levophacetoperane, 5347
Methylphenidate, 6025
Piperoxan, 7448
$C_{14}H_{19}NO_4$
Anisomycin, 700
$C_{14}H_{19}NO_4S$
Furtrethonium*, 4225
Trithiozine, 9672
$C_{14}H_{19}NO_5$
Trimetozine, 9634
$C_{14}H_{19}NO_8$
p-Phenetidine*, 7188
$C_{14}H_{19}N_2NaO_4S$
2-Pentenylpenicillin Sodium, 7081
$C_{14}H_{19}N_3O$
Methafurylene, 5854
Oxolamine, 6898
Ramifenazone, 8122
$C_{14}H_{19}N_3S$
Methapyrilene, 5871
Thenyldiamine, 9207
$C_{14}H_{19}N_5O_4S_3$
Sulfatolamide, 8922
$C_{14}H_{19}O_6P$
Crotoxyphos, 2603
$C_{14}H_{20}$
Congressane, 2494
$C_{14}H_{20}Br_2N_2$
Bromhexine, 1379
$C_{14}H_{20}Br_2N_4O$
Pyrithiamine, 8003
$C_{14}H_{20}ClNO_2$
Alachlor, 193
Levophacetoperane*, 5347
$C_{14}H_{20}ClN_3O_3S$
Clopamide, 2391
$C_{14}H_{20}Cl_2N_4$
Amprolium*, 623

Denotes derivative of title compound

$C_{14}H_{20}Cl_6N_2$
Chlorisondamine Chloride, 2101

$C_{14}H_{20}GdN_3O_{10}$
Gadopentetic Acid, 4236

$C_{14}H_{20}IN_3O_8$
Nitroxynil*, 6578

$C_{14}H_{20}N_2O$
Pyrrocaine, 8024
Siduron, 8433
Tymazoline, 9742

$C_{14}H_{20}N_2O_2$
Bunitrolol, 1479
Etryptamine*, 3848
Pindolol, 7412
Piridocaine, 7466

$C_{14}H_{20}N_2O_3$
Propacetamol, 7804

$C_{14}H_{20}N_2O_3S$
Tolcyclamide, 9434

$C_{14}H_{20}N_2O_6S$
p-Methylaminophenol Sulfate, 5940

$C_{14}H_{20}N_2O_8S_2$
Paraquat*, 6980

$C_{14}H_{20}N_3O_5PS$
Pyrazophos, 7976

$C_{14}H_{20}N_4O$
Imolamine, 4838

$C_{14}H_{20}N_4O_3$
Methenamine Mandelate, 5883

$C_{14}H_{20}N_4O_7S_2$
Netobimin, 6390

$C_{14}H_{20}O_2$
Butibufen, 1521

$C_{14}H_{20}O_3$
Nudic Acids*, 6649

$C_{14}H_{20}O_4$
Frequentin, 4183

$C_{14}H_{20}O_8$
Taxicatin, 9045

$C_{14}H_{21}BO_2$
Tolboxane, 9431

$C_{14}H_{21}Br_2ClN_2$
Bromhexine*, 1379

$C_{14}H_{21}Br_2NO_2$
Spasmolytol, 8693

$C_{14}H_{21}ClN_2O_2$
Bunitrolol*, 1479

$C_{14}H_{21}ClN_2O_3$
Propacetamol*, 7804

$C_{14}H_{21}ClN_4O$
Imolamine*, 4838

$C_{14}H_{21}NO_2$
4-(Dimethylamino)benzoic Acid*, 3221

$C_{14}H_{21}NO_8$
6-Desoxy-D-glucosamine*, 2911

$C_{14}H_{21}N_2NaO_4S$
Amylpenicillin Sodium, 652

$C_{14}H_{21}N_3O_2S$
Sumatriptan, 8979

$C_{14}H_{21}N_3O_3$
Oxamniquine, 6872

$C_{14}H_{21}N_3O_3S$
Tolazamide, 9429

$C_{14}H_{21}N_3O_4$
Butralin, 1532

$C_{14}H_{21}N_3O_4S$
Sulmepride, 8966

$C_{14}H_{21}N_3O_5$
Leonurine, 5323

$C_{14}H_{21}N_3O_6S$
Penicillin N, 7043

$C_{14}H_{21}N_5O_6S.H_2O$
Serotonin*, 8413

$C_{14}H_{22}BrN_3O_2$
Bromopride, 1420

$C_{14}H_{22}ClNO$
Clobutinol, 2363

$C_{14}H_{22}ClNO_2$
Amylocaine Hydrochloride, 650
Bupranolol, 1486
Meprylcaine Hydrochloride, 5752

$C_{14}H_{22}ClNO_3$
Mecoprop*, 5666

$C_{14}H_{22}ClN_3O_2$
Metoclopramide, 6063

$C_{14}H_{22}Cl_2N_2O_6S_4$
Clomethiazole*, 2382

$C_{14}H_{22}N_2O$
Lidocaine, 5359
Octacaine, 6663

$C_{14}H_{22}N_2O_2$
Naepaine, 6263

$C_{14}H_{22}N_2O_3$
Atenolol, 879
Bucolome, 1452
Practolol, 7703
Trimetazidine, 9619

$C_{14}H_{22}N_2O_3S$
Piprozolin, 7457

$C_{14}H_{22}N_2O_4S_2$
Penicillin BT, 7035

$C_{14}H_{22}N_2O_7S$
Rimazolium Metilsulfate, 8222

$C_{14}H_{22}O$
C_{14}-Aldehyde, 215
Irone, 4976

$C_{14}H_{22}O_4$
Palitantin, 6939

$C_{14}H_{23}BrClN_3O_2$
Bromopride*, 1420

$C_{14}H_{23}ClN_2O.H_2O$
Lidocaine*, 5359

$C_{14}H_{23}Cl_2NO$
Clobutinol*, 2363

$C_{14}H_{23}Cl_2NO_2$
Bupranolol*, 1486

$C_{14}H_{23}Cl_2N_3O_2.H_2O$
Metoclopramide*, 6063

$C_{14}H_{23}NO$
Affinin, 167

$C_{14}H_{23}NO_2$
Piroctone, 7472

$C_{14}H_{23}NO_4$
4,4'-Iminodicyclohexanecarboxylic Acid, 4833

$C_{14}H_{23}N_3O_{10}$
Pentetic Acid, 7083

$C_{14}H_{24}BrClN_3O_2.H_2O$
Bromopride*, 1420

$C_{14}H_{24}CaN_4O_8$
Piperazine Edetate Calcium, 7435

$C_{14}H_{24}Cl_2N_2O_3$
Trimetazidine*, 9619

$C_{14}H_{24}Cl_3N_3O_2.H_2O$
Metoclopramide*, 6063

$C_{14}H_{24}N_2O_3$
Ciclopirox*, 2270

$C_{14}H_{24}N_2O_5$
Pirbuterol*, 7461

$C_{14}H_{24}N_2O_7$
Spectinomycin, 8695

$C_{14}H_{25}HgNO_5$
Mercamphamide, 5758

$C_{14}H_{25}NO$
Pellitorine, 7017

$C_{14}H_{25}N_3O_4S$
Alitame, 235

$C_{14}H_{25}N_3O_6S$
ACV, 138

$C_{14}H_{25}N_4NaO_{11}P_2$
Citicoline*, 2321

$C_{14}H_{25}NaO_7S$
Diamyl Sodium Sulfosuccinate, 2968

$C_{14}H_{26}CaO_{16}$
Glucoheptonic Acid*, 4349

$C_{14}H_{26}Cl_2N_2$
Bipiperidyl Mustard, 1250

$C_{14}H_{26}Cl_2N_2O_7.5H_2O$
Spectinomycin*, 8695

$C_{14}H_{26}MgO_{16}$
Glucoheptonic Acid*, 4349

$C_{14}H_{26}N_4O_{11}P_2$
Citicoline, 2321

$C_{14}H_{26}O_2$
Looplure, 5449

$C_{14}H_{26}O_4$
Sebacic Acid*, 8369

$C_{14}H_{27}NO_6$
Amiprilose, 499
Pempidine*, 7022

$C_{14}H_{28}ClNO_6$
Amiprilose*, 499

$C_{14}H_{28}O_2$
Ethyl Laurate, 3774
Myristic Acid, 6246

$C_{14}H_{29}NaSO_4$
Sodium Tetradecyl Sulfate, 8645

$C_{14}H_{30}Br_2N_2O_4$
Succinylcholine Bromide, 8845

$C_{14}H_{30}Cl_2N_2O_4$
Succinylcholine Chloride, 8846

$C_{14}H_{30}I_2N_2O_4$
Succinylcholine Iodide, 8847

$C_{14}H_{30}O$
Myristyl Alcohol, 6248

$C_{14}H_{30}O_2$
Decamethylene Glycol*, 2842

$C_{14}H_{30}O_2S_2$
Tiadenol, 9349

$C_{14}H_{32}N_2O_4$
Entprol, 3551

$C_{14}H_{42}O_5Si_6$
Tetradecamethylhexasiloxane, 9131

$C_{15}H_8O_5$
Coumestrol, 2565

$C_{15}H_8O_6$
Rhein, 8175

$C_{15}H_9BrO_2$
Bromindione, 1381

$C_{15}H_9ClO_2$
Clorindione, 2402

$C_{15}H_9FO_2$
Fluindione, 4061

$C_{15}H_9NO_4$
1-Aminoanthraquinone-2-carboxylic Acid, 429

$C_{15}H_{10}BrClN_4S$
Brotizolam, 1439

$C_{15}H_{10}ClF_3N_2O_3$
Triflumuron, 9592

$C_{15}H_{10}ClI_2NO_3$
Clioxanide, 2354

$C_{15}H_{10}ClN_3O_3$
Clonazepam, 2387

$C_{15}H_{10}Cl_2N_2O_2$
Lonidamine, 5447
Lorazepam, 5456

$C_{15}H_{10}I_4NNaO_4$
Levothyroxine Sodium, 5351
Thyroxine*, 9348

$C_{15}H_{10}O_2$
Flavone, 4030
Isoflavone, 5057
2-Methylanthraquinone, 5943
Phenindione, 7196

$C_{15}H_{10}O_4$
Chrysin, 2261
Chrysophanic Acid, 2263
Daidzein, 2805
Rubiadin, 8258

$C_{15}H_{10}O_5$
Aloe-Emodin, 303
Apigenin, 763
Baicalein, 954
Emodin, 3518
Galangin, 4244
Genistein, 4281
Sulfuretin, 8951

C₁₅H₁₀O₆ → $C_{15}H_{10}O_6$

Baptigenin, 967
Datiscetin, 2822
Fisetin, 4026
Kaempferol, 5156
Luteolin, 5483
Scutellarein, 8367

$C_{15}H_{10}O_7$
Morin, 6179
Quercetin, 8044

$C_{15}H_{10}O_8$
Myricetin, 6244
Quercetagetin, 8043

$C_{15}H_{11}ClF_3NO_4$
Oxyfluorfen, 6916

$C_{15}H_{11}ClN_2O$
Mecloqualone, 5661
Nordazepam, 6608

$C_{15}H_{11}ClN_2O_2$
Oxazepam, 6881

$C_{15}H_{11}ClO_2$
Clobenfurol, 2356

$C_{15}H_{11}ClO_5$
Pelargonidin, 7014

$C_{15}H_{11}ClO_6$
Cyanidin Chloride, 2694

$C_{15}H_{11}ClO_7$
Delphinidin, 2866

$C_{15}H_{11}I_3NNaO_4$
Liothyronine*, 5388

$C_{15}H_{11}I_3O_4$
Thyropropic Acid, 9346

$C_{15}H_{11}I_4NO_4$
Thyroxine, 9348

$C_{15}H_{11}NO_2$
Viridicatin, 9911

$C_{15}H_{11}NO_5$
Bostrycoidin, 1353

$C_{15}H_{11}N_2NaO_2$
Phenytoin*, 7293

$C_{15}H_{11}N_3$
Nicotelline, 6432

$C_{15}H_{11}N_3O_3$
Nitrazepam, 6491

$C_{15}H_{12}ClNO_2$
Carprofen, 1870

$C_{15}H_{12}Cl_2O_3$
Sesin, 8422

$C_{15}H_{12}I_2O_3$
Iodoalphionic Acid, 4919

$C_{15}H_{12}I_3NO_4$
Liothyronine, 5388

$C_{15}H_{12}NNaO_3.H_2O$
Amfenac*, 404

$C_{15}H_{12}N_2O$
Carbamazepine, 1783

$C_{15}H_{12}N_2O_2$
Phenopyrazone, 7217
Phenytoin, 7293

$C_{15}H_{12}N_2O_3$
Hydrofuramide, 4718

$C_{15}H_{12}N_6O_4$
Rhizopterin, 8181

$C_{15}H_{12}O$
Chalcone, 2028

$C_{15}H_{12}O_2$
Dibenzoylmethane, 2991
α-Phenylcinnamic Acid, 7253

$C_{15}H_{12}O_3$
2-(p-Toluyl)benzoic Acid, 9465

$C_{15}H_{12}O_4$
o-(p-Anisoyl)benzoic Acid, 702
Methyl Benzoylsalicylate, 5948
Phenyl Acetylsalicylate, 7241

$C_{15}H_{12}O_5$
Naringenin, 6344

$C_{15}H_{12}O_6$
Eriodictyol, 3616
Fustin, 4232
5,5′-Methylenedisalicylic Acid, 5984

$C_{15}H_{12}O_8$
Ampelopsin, 611

$C_{15}H_{12}O_{10}$
Methylenedigallic Acid, 5983

$C_{15}H_{13}ClI_3NO_4$
Liothyronine*, 5388

$C_{15}H_{13}ClNNaO_3.2H_2O$
Zomepirac*, 10092

$C_{15}H_{13}ClN_2$
Chlormidazole, 2106

$C_{15}H_{13}ClN_2O_5$
Gallocyanine, 4256

$C_{15}H_{13}Cl_2NO_2$
2-Nitro-1,1-bis(p-chlorophenyl)-propane, 6515

$C_{15}H_{13}Cl_2N_5$
Robenidine, 8229

$C_{15}H_{13}FO_2$
Flurbiprofen, 4124

$C_{15}H_{13}I_2NO_4$
3,5-Diiodothyronine, 3176

$C_{15}H_{13}NO$
N-2-Fluorenylacetamide, 4083

$C_{15}H_{13}NO_2S$
Metiazinic Acid, 6057

$C_{15}H_{13}NO_3$
Amfenac, 404
Ketorolac, 5186
3′-Methylphthalanilic Acid, 6027
Pranoprofen, 7708

$C_{15}H_{13}NO_3S_2$
Epalrestat, 3556

$C_{15}H_{13}NO_4$
Acetaminosalol, 41

$C_{15}H_{13}N_3O_2S$
Fenbendazole, 3904

$C_{15}H_{13}N_3O_3S$
Oxfendazole, 6890

$C_{15}H_{13}N_3O_4S$
Piroxicam, 7476

$C_{15}H_{14}ClNO_3$
Zomepirac, 10092

$C_{15}H_{14}ClN_3O_4S.H_2O$
Cefaclor, 1920

$C_{15}H_{14}ClN_3O_4S_3$
Benzthiazide, 1134

$C_{15}H_{14}Cl_2N_2$
Chlormidazole*, 2106

$C_{15}H_{14}Cl_2N_4O_6$
Dichlororiboflavin, 3065

$C_{15}H_{14}Cl_3N_5$
Robenidine*, 8229

$C_{15}H_{14}FN_3O_3$
Flumazenil, 4062

$C_{15}H_{14}F_3N_3O_4S_2$
Bendroflumethiazide, 1045

$C_{15}H_{14}NNaO_3.2H_2O$
Tolmetin*, 9441

$C_{15}H_{14}NO_2PS$
Cyanofenphos, 2697

$C_{15}H_{14}N_2Na_2O_6S_2$
Ticarcillin*, 9360

$C_{15}H_{14}N_2O$
Doxenitoin, 3424

$C_{15}H_{14}N_4O_2S$
Sulfaphenazole, 8910

$C_{15}H_{14}N_4O_6S_2$
Ceftibuten, 1947

$C_{15}H_{14}O$
Lactaroviolin, 5212

$C_{15}H_{14}O_3$
Equol, 3583
Fenoprofen, 3926
Guaiacol*, 4457
Lapachol, 5235
Mexenone, 6092

$C_{15}H_{14}O_4$
Lunularic Acid, 5476
Menadiol Diacetate, 5710
Menbutone, 5719
Peucedanin, 7144
Xanthoxyletin, 9975
Yangonin, 10001

$C_{15}H_{14}O_5$
Benzophenone-6, 1109
Guaiacol Carbonate, 4459
Methysticin, 6056
Phloretin, 7299

$C_{15}H_{14}O_6$
Catechin, 1908
Javanicin, 5143
Plumericin, 7512

$C_{15}H_{14}O_7$
Fusarubin, 4228
Leucocyanidin, 5334

$C_{15}H_{15}ClN_2O_4S$
Xipamide, 9986

$C_{15}H_{15}ClN_4O_6S$
Chlorimuron Ethyl, 2092

$C_{15}H_{15}F_3N_2O_2$
Flurprimidol, 4129

$C_{15}H_{15}I_3NNaO_3$
Bunamiodyl Sodium, 1476

$C_{15}H_{15}NO$
p-Dimethylaminobenzophenone, 3222

$C_{15}H_{15}NO_2$
Apo-β-erythroidine, 775
Enfenamic Acid, 3535
Isoapo-β-erythroidine, 5008
Mefenamic Acid, 5680

$C_{15}H_{15}NO_3$
Tolmetin, 9441

$C_{15}H_{15}NO_3S$
Adrafinil, 159

$C_{15}H_{15}NO_6$
Ascorbigen, 856

$C_{15}H_{15}N_3O$
Ethacridine, 3668

$C_{15}H_{15}N_3O_2$
Methyl Red, 6037

$C_{15}H_{15}N_3O_3$
Verazide, 9859

$C_{15}H_{15}N_5O_4$
Etofylline Nicotinate, 3836

$C_{15}H_{16}ClN_3S$
Azure B, 941
Tolonium Chloride, 9444

$C_{15}H_{16}Cl_3N_3O_2$
Prochloraz, 7767

$C_{15}H_{16}N_2Na_2O_8S_3$
Noprylsulfamide, 6601

$C_{15}H_{16}N_2O$
Benmoxine, 1052
Phenatine, 7172

$C_{15}H_{16}N_2O_2$
Ancymidol, 665

$C_{15}H_{16}N_2O_3S$
N-Sulfanilyl-3,4-xylamide, 8907

$C_{15}H_{16}N_2O_6S_2$
Ticarcillin, 9360

$C_{15}H_{16}N_4O_5S$
Sulfometuron Methyl, 8936

$C_{15}H_{16}N_5NaO_6S_2$
Cefpodoxime Proxetil*, 1941

$C_{15}H_{16}N_6O$
Amicarbalide, 405

$C_{15}H_{16}N_7NaO_5S_3$
Cefmetazole*, 1929

$C_{15}H_{16}O_2$
Bisphenol A, 1311
Nabumetone, 6258

$C_{15}H_{16}O_3$
Osthole, 6849

Denotes derivative of title compound

C₁₅H₁₆O₅
$C_{15}H_{16}O_5$
Fuscin, 4229
Lactucin, 5224
Vernolepin, 9867

$C_{15}H_{16}O_6$
Picrotoxinin, 7389

$C_{15}H_{16}O_8$
Leucodrin, 5335
Skimmin, 8506

$C_{15}H_{16}O_9$
Cichoriin, 2267
Daphnin, 2817
Esculin, 3646

$C_{15}H_{17}BrN_2O_2$
Benzpyrinium Bromide, 1132

$C_{15}H_{17}Br_2NO_2$
Bromoxynil*, 1431

$C_{15}H_{17}ClN_4$
Myclobutanil, 6232
Neutral Red, 6395

$C_{15}H_{17}Cl_2N_3O$
Diniconazole, 3261

$C_{15}H_{17}Cl_2N_3O_2$
Propiconazole, 7830

$C_{15}H_{17}FN_4O_2$
Flupirtine, 4117

$C_{15}H_{17}FN_4O_3$
Enoxacin, 3540

$C_{15}H_{17}FN_4O_3 \cdot 1\frac{1}{2}H_2O$
Enoxacin*, 3540

$C_{15}H_{17}F_2N_6NaO_7S_2$
Flomoxef*, 4038

$C_{15}H_{17}HgNO_2S$
N-(Ethylmercuri)-p-toluenesulfon-
anilide, 3781

$C_{15}H_{17}I_3NNaO_3$
Tyropanoate Sodium, 9745

$C_{15}H_{17}N$
Ethylbenzylaniline, 3728

$C_{15}H_{17}NO_4$
Actiphenol, 136

$C_{15}H_{17}NS_2$
Tipepidine, 9389

$C_{15}H_{17}N_3O$
Cetoxime, 2015
Metralindole, 6075

$C_{15}H_{17}N_5O_6S_2$
Cefpodoxime Proxetil*, 1941

$C_{15}H_{17}N_7O_5S_3$
Cefmetazole, 1929

$C_{15}H_{17}NaO_3$
Loxoprofen*, 5462

$C_{15}H_{17}NaO_3S$
Guaiazulene*, 4464

$C_{15}H_{18}$
Cadalene, 1607
Guaiazulene, 4464

$C_{15}H_{18}ClN_3O$
Cetoxime*, 2015
Metralindole*, 6075

$C_{15}H_{18}ClN_3O_3S$
Tiaramide, 9356

$C_{15}H_{18}Cl_2N_2O_3$
Oxadiazon, 6860

$C_{15}H_{18}Cl_6N_2O_5$
Dichloralphenazone, 3033

$C_{15}H_{18}F_2N_6O_7S_2$
Flomoxef, 4038

$C_{15}H_{18}I_3NO_5$
Iopronic Acid, 4949

$C_{15}H_{18}N_2O$
Selagine, 8379

$C_{15}H_{18}N_2O_6$
Binapacryl, 1237

$C_{15}H_{18}N_4O_5$
Mitomycins*, 6133

$C_{15}H_{18}N_6O_2$
Pimefylline, 7400

$C_{15}H_{18}O_2$
Procerin, 7766

$C_{15}H_{18}O_3$
Ambrosin, 391
Loxoprofen, 5462
α-Santonin, 8325
Xanthatin, 9967

$C_{15}H_{18}O_4$
Artemisin, 843
Helenalin, 4542
Parthenin, 6995

$C_{15}H_{18}O_5$
Coriamyrtin, 2525

$C_{15}H_{18}O_6$
Cedrin, 1918
Tutin, 9735

$C_{15}H_{18}O_7$
Hyenanchin, 4787
Picrotin, 7387

$C_{15}H_{18}O_8$
Melilotoside, 5701

$C_{15}H_{19}BrN_2O_5$
Mebrofenin, 5652

$C_{15}H_{19}ClN_4$
Toluylene Blue, 9466

$C_{15}H_{19}Cl_2N_3O$
Diclobutrazol, 3070

$C_{15}H_{19}Cl_2N_3O_3S$
Tiaramide*, 9356

$C_{15}H_{19}F_3N_2S$
Ticarbodine, 9359

$C_{15}H_{19}NO$
Furfurylmethylamphetamine, 4217
Pronethalol, 7802

$C_{15}H_{19}NO_2$
Tropacocaine, 9686

$C_{15}H_{19}NS_2$
Ethylmethylthiambutene, 3784

$C_{15}H_{19}N_3OS$
Butamisole, 1506

$C_{15}H_{19}N_3O_5$
Carboquone, 1830

$C_{15}H_{19}N_5O_4S$
Endralazine*, 3532

$C_{15}H_{20}ClN_3O$
Paclobutrazol, 6937

$C_{15}H_{20}ClN_3OS$
Butamisole*, 1506

$C_{15}H_{20}FNO$
Primaperone, 7750

$C_{15}H_{20}NO_7Sb$
Arecoline p-Stibonobenzoic Acid,
803

$C_{15}H_{20}N_2$
Indalpine, 4843

$C_{15}H_{20}N_2O$
Anagyrine, 660

$C_{15}H_{20}N_2O_2$
Fenspiride, 3942

$C_{15}H_{20}N_2O_3$
Pifoxime, 7392

$C_{15}H_{20}N_2O_4S$
Acetohexamide, 53

$C_{15}H_{20}N_2O_6S$
Pyrantel*, 7968

$C_{15}H_{20}N_2S$
Methaphenilene, 5870

$C_{15}H_{20}N_6O_5$
Alazopeptin, 199

$C_{15}H_{20}O$
ar-Turmerone, 9733

$C_{15}H_{20}O_2$
Alantolactone, 198
Periplanones*, 7125

$C_{15}H_{20}O_3$
Arborescin, 798
Illudins*, 4824
Parthenolide, 6996
Perezone, 7112
Periplanones*, 7125

$C_{15}H_{20}O_4$
Abscisic Acid, 6
Hirsutic Acid C, 4637
Illudins*, 4824
Santonic Acid, 8324
Tanacetin, 9020

$C_{15}H_{20}O_6$
Shellolic Acid, 8427
Vomitoxin, 9947

$C_{15}H_{20}O_7$
Nivalenol, 6581

$C_{15}H_{21}ClN_2O_2$
Amidochlor, 411
Fenspiride*, 3942

$C_{15}H_{21}ClN_2S$
Methaphenilene*, 5870

$C_{15}H_{21}Cl_2FN_2O_3$
Fluroxypyr*, 4128

$C_{15}H_{21}F_3N_2O_2$
Fluvoxamine, 4138

$C_{15}H_{21}N$
Fencamfamine, 3909

$C_{15}H_{21}NO$
Eptazocine, 3579
Metazocine, 5839

$C_{15}H_{21}NO_2$
Ciclonicate, 2269
β-Eucaine, 3850
Indenolol, 4852
Ketobemidone, 5180
Meperidine, 5736
Naxagolide, 6349
Prodilidine, 7775
Tolpronine, 9449

$C_{15}H_{21}NO_3$
Hydroxypethidine, 4769

$C_{15}H_{21}NO_4$
Metalaxyl, 5826

$C_{15}H_{21}NO_4S$
Furtrethonium*, 4225

$C_{15}H_{21}N_3O$
Primaquine, 7751
Quinocide, 8096

$C_{15}H_{21}N_3OS$
Zolamine, 10089

$C_{15}H_{21}N_3O_2$
Physostigmine, 7357

$C_{15}H_{21}N_3O_2S_3$
Arotinolol, 815

$C_{15}H_{21}N_3O_3$
Eseridine, 3647

$C_{15}H_{21}N_3O_3S$
Gliclazide, 4335

$C_{15}H_{21}N_5O_3$
Methenamine Hippurate, 5882

$C_{15}H_{22}ClN$
Fencamfamine*, 3909

$C_{15}H_{22}ClNO_2$
Indenolol*, 4852
Ketobemidone*, 5180
Meperidine*, 5736
Metolachlor, 6067
Prodilidine*, 7775

$C_{15}H_{22}ClN_3OS$
Zolamine*, 10089

$C_{15}H_{22}ClN_3O_2S_3$
Arotinolol*, 815

$C_{15}H_{22}IN_3O_8$
Nitroxynil*, 6578

$C_{15}H_{22}N_2O$
Mepivacaine, 5748

$C_{15}H_{22}N_2O_2$
Mepindolol, 5744

$C_{15}H_{22}N_2O_3$
Tolycaine, 9467

$C_{15}H_{22}N_2O_4$
Troxipide, 9699

$C_{15}H_{22}N_2O_6$
Lysine Acetylsalicylate, 5510
Nipradilol, 6479

Denotes derivative of title compound

C₁₅H₂₂N₄O₃
Propentofylline, 7823
C₁₅H₂₂N₆O₅S
S-Adenosylmethionine, 146
C₁₅H₂₂O
Vetivones, 9874
C₁₅H₂₂O₂
Drimenin, 3433
Helminthosporal, 4548
Polygodial, 7548
C₁₅H₂₂O₃
3-(1,3,5,7,9-Dodecapentaenyloxy)-
1,2- propanediol, 3403
Gemfibrozil, 4280
Warburganal, 9949
C₁₅H₂₂O₅
Artemisinin, 844
C₁₅H₂₂O₆
Sesamex, 8419
C₁₅H₂₂O₉
Aucubin, 894
C₁₅H₂₃ClN₂O
Mepivacaine*, 5748
C₁₅H₂₃ClN₂O₃
Tolycaine*, 9467
C₁₅H₂₃ClO₄S
Aramite®, 794
C₁₅H₂₃N
Prolintane, 7791
C₁₅H₂₃NO
Meptazinol, 5753
C₁₅H₂₃NO₂
Alprenolol, 311
Ciramadol, 2316
Nupharidine, 6650
C₁₅H₂₃NO₃
Oxprenolol, 6905
Parethoxycaine, 6987
C₁₅H₂₃NO₄
Cycloheximide, 2734
C₁₅H₂₃NO₆
Phenpentermine*, 7228
C₁₅H₂₃N₃O₂
Acecainide, 14
C₁₅H₂₃N₃O₃S
Amdinocillin, 399
C₁₅H₂₃N₃O₄
Isopropalin, 5089
C₁₅H₂₃N₃O₄S
Cyclacillin, 2706
Sulpiride, 8971
C₁₅H₂₄
Cadinenes, 1610
Caryophyllene, 1884
Copaene, 2510
α-Farnesene, 3883
β-Farnesene, 3884
Humulene, 4672
Longifolene, 5446
Thujopsene, 9327
Ylangene, 10007
C₁₅H₂₄ClN
Prolintane*, 7791
C₁₅H₂₄ClNO
Meptazinol*, 5753
C₁₅H₂₄ClNO₂
Alprenolol*, 311
C₁₅H₂₄ClNO₃
Oxprenolol*, 6905
C₁₅H₂₄ClN₃O₂
Acecainide*, 14
C₁₅H₂₄NO₄PS
Isofenphos, 5055
C₁₅H₂₄N₂O
Aphylline, 761
Lupanine, 5477
Matrine, 5641
Trimecaine, 9614
C₁₅H₂₄N₂O₂
Hydroxylupanine, 4760

C₁₅H₂₄N₂O₃
Hydroxytetracaine, 4783
Mefexamide, 5682
C₁₅H₂₄N₂O₄S
Tiapride, 9353
C₁₅H₂₄N₂O₁₇P₂
Uridine Diphosphate Glucose, 9794
C₁₅H₂₄N₄O₂S₂
Prosultiamine, 7896
C₁₅H₂₄N₄O₅S
Sulfisoxazole*, 8930
C₁₅H₂₄O
Butylated Hydroxytoluene, 1548
DBMC, 2829
α-Santalol, 8321
β-Santalol, 8322
C₁₅H₂₄O₂
Helminthosporol, 4549
C₁₅H₂₄O₅
Artemisinin*, 844
C₁₅H₂₅BrO₂
d-Bornyl α-Bromoisovalerate, 1340
C₁₅H₂₅ClNO₂
Xibenolol*, 9984
C₁₅H₂₅ClN₂O₂
Tetracaine Hydrochloride, 9123
C₁₅H₂₅ClN₂O₄S
Tiapride*, 9353
C₁₅H₂₅ClN₄O₂S₂
Prosultiamine*, 7896
C₁₅H₂₅Cl₂N₃OS
Diamthazole Dihydrochloride, 2967
C₁₅H₂₅NO₂
Xibenolol, 9984
C₁₅H₂₅NO₃
Metoprolol, 6072
C₁₅H₂₆N₂
Genisteine-Alkaloid, 4282
β-Isosparteine, 5115
Sparteine, 8692
C₁₅H₂₆N₂O
Retamine, 8161
C₁₅H₂₆O
Carotol, 1863
Cedrol, 1919
Farnesol, 3885
Guaiol, 4466
Ledol, 5313
Nerolidol, 6388
Patchouli Alcohol, 7001
C₁₅H₂₆O₂
d-Bornyl Isovalerate, 1342
Daucol, 2824
C₁₅H₂₆O₆
Tributyrin, 9532
C₁₅H₂₇NO₂S
Lethane® 60, 5328
C₁₅H₂₇NO₉S₃
Improsulfan*, 4841
C₁₅H₂₈NNaO₃
Gardol®, 4267
C₁₅H₂₈N₂O₄S.5H₂O
Sparteine*, 8692
C₁₅H₂₈O₂
Exaltolide®, 3867
Menthyl Valerate, 5728
C₁₅H₂₉N₇O₆
Hydramitrazine*, 4685
C₁₅H₃₂N₄O₅
Fortimicin(s)*, 4165
C₁₅H₃₃NO₂
Dodine, 3406
C₁₆H₅N₅O₆
9-Dicyanomethylene-2,4,7-trinitro-
fluorene, 3084
C₁₆H₇N₂NaO₅
Pirenoxine*, 7462
C₁₆H₈N₂Na₂O₈S₂
Indigo Carmine, 4856
C₁₆H₈N₂O₅
Pirenoxine, 7462

C₁₆H₈O₆
Medicagol, 5672
C₁₆H₉N₃Na₂O₁₀S₂
Chromotrope 2B, 2242
C₁₆H₉N₄Na₃O₉S₂
Tartrazine, 9041
C₁₆H₁₀
Pyrene, 7977
C₁₆H₁₀ClN₃
Lotrifen, 5459
C₁₆H₁₀Cl₂O₇
Erdin, 3588
C₁₆H₁₀Cl₄O₅
Diploicin, 3345
C₁₆H₁₀LiNO₂.8H₂O
Cinchophen*, 2290
C₁₆H₁₀N₂Na₂O₇S₂
Sunset Yellow FCF, 8983
C₁₆H₁₀N₂O₂
Indigo, 4855
C₁₆H₁₀O₅
Pseudobaptigenin, 7925
C₁₆H₁₀O₇
Laccaic Acid*, 5209
Porphyrillic Acid, 7576
C₁₆H₁₀O₈
Kermesic Acid, 5172
C₁₆H₁₁ClK₂N₂O₄
Clorazepate*, 2400
C₁₆H₁₁ClN₄
Estazolam, 3651
C₁₆H₁₁NO₂
Benzoxiquine, 1122
Cinchophen, 2290
C₁₆H₁₁NO₃
Oxycinchophen, 6910
C₁₆H₁₁N₂NaO₄S
Orange I, 6812
Orange II, 6813
C₁₆H₁₁N₃O₅S
Droxicam, 3443
C₁₆H₁₂ClFN₂O
Fludiazepam, 4058
C₁₆H₁₂ClKN₂O₄
Clorazepate*, 2400
C₁₆H₁₂ClNO₂
Cinchophen*, 2290
C₁₆H₁₂ClNO₃
Benoxaprofen, 1055
C₁₆H₁₂Cl₂N₂O₂
Lormetazepam, 5458
C₁₆H₁₂FNO₃
Flunoxaprofen, 4074
C₁₆H₁₂FN₃O₃
Flubendazole, 4050
Flunitrazepam, 4072
C₁₆H₁₂N₂O₄
Isatide, 4984
C₁₆H₁₂O₃
Anisindione, 698
C₁₆H₁₂O₄
Chrysin*, 2261
Formononetin, 4157
Isoxepac, 5126
C₁₆H₁₂O₅
Acacetin, 9
Emodin*, 3518
Genistein*, 4281
Oroxylin A, 6830
Prunetin, 7923
C₁₆H₁₂O₆
Acetylsalicylsalicylic Acid, 95
Diosmetin, 3290
Hematein, 4553
Pratensein, 7711
Tectorigenin, 9057
C₁₆H₁₂O₇
Rhamnetin, 8170
C₁₆H₁₃ClN₂O
Diazepam, 2977
Mazindol, 5643

Denotes derivative of title compound

$C_{16}H_{13}ClN_2O_2$
Clobazam, 2355
Temazepam, 9074
$C_{16}H_{13}ClN_2O_4$
Clorazepate, 2400
$C_{16}H_{13}ClN_2O_4S$
Suclofenide, 8852
$C_{16}H_{13}ClO_6$
Peonidin, 7099
$C_{16}H_{13}ClO_7$
Petunidin, 7143
$C_{16}H_{13}Cl_2NO_4$
Quinfamide, 8068
$C_{16}H_{13}Cl_3N_2OS$
Tioconazole, 9384
$C_{16}H_{13}F_2N_3O$
Flutriafol, 4135
$C_{16}H_{13}I_3N_2O_3$
Iobenzamic Acid, 4901
$C_{16}H_{13}NO_3S$
1-Anilino-8-naphthalenesulfonate, 690
$C_{16}H_{13}N_3$
Yellow AB, 10003
$C_{16}H_{13}N_3O_3$
Mebendazole, 5647
Nimetazepam, 6465
$C_{16}H_{14}$
1,4-Diphenyl-1,3-butadiene, 3320
$C_{16}H_{14}CaO_6$
Mandelic Acid*, 5599
$C_{16}H_{14}ClN_3O$
Chlordiazepoxide, 2082
$C_{16}H_{14}Cl_2O_3$
Chlorobenzilate, 2123
$C_{16}H_{14}Cl_2O_4$
Diclofop-Methyl, 3072
$C_{16}H_{14}FN_3O$
Afloqualone, 171
$C_{16}H_{14}F_3NO_3S$
Tolrestat, 9451
$C_{16}H_{14}MgO_6$
Magnesium Mandelate, 5551
$C_{16}H_{14}N_2O$
Glycosine, 4398
Methaqualone, 5872
$C_{16}H_{14}N_2O_2$
Miroprofen, 6127
$C_{16}H_{14}N_2O_3$
Bendazac, 1042
$C_{16}H_{14}N_2O_4$
Amlexanox, 508
Nybomycin, 6653
$C_{16}H_{14}N_2O_6S$
Phthalylsulfacetamide, 7350
$C_{16}H_{14}N_7NaO_5S_4.2H_2O$
Cefuzonam*, 1952
$C_{16}H_{14}O$
Dypnone, 3460
$C_{16}H_{14}O_2$
Benzyl Cinnamate, 1144
$C_{16}H_{14}O_3$
Fenbufen, 3906
Ketoprofen, 5184
3-(o-Methoxyphenyl)-2-phenyl-acrylic Acid, 5921
$C_{16}H_{14}O_4$
Imperatorin, 4839
Medicarpin, 5673
$C_{16}H_{14}O_5$
Aspirin*, 873
Brazilin, 1362
Sakuranetin, 8289
$C_{16}H_{14}O_6$
Hematoxylin, 4556
Hesperetin, 4590
Homoeriodictyol, 4656
$C_{16}H_{15}ClN_2$
Medazepam, 5669
$C_{16}H_{15}ClN_2O$
Methaqualone*, 5872

$C_{16}H_{15}ClN_2OS$
Clotiazepam, 2411
$C_{16}H_{15}Cl_2NO_2$
Bulan®, 1470
$C_{16}H_{15}ClN_3O$
Chlordiazepoxide*, 2082
$C_{16}H_{15}Cl_3O_2$
Methoxychlor, 5913
$C_{16}H_{15}CrN_6S_4$
Rhodanilic Acid, 8183
$C_{16}H_{15}F_2N_3Si$
Flusilazole, 4130
$C_{16}H_{15}F_3O$
Flumecinol, 4063
$C_{16}H_{15}N$
Dizocilpine, 3392
$C_{16}H_{15}NO$
Cyheptamide, 2766
$C_{16}H_{15}NO_2$
Cinnamyl Anthranilate, 2306
$C_{16}H_{15}NO_3$
Ftaxilide, 4190
$C_{16}H_{15}N_2NaO_6S_2$
Cephalothin*, 1978
$C_{16}H_{15}N_3O_5$
Opiniazide, 6807
$C_{16}H_{15}N_3O_7S$
Sulfaloxic Acid, 8883
$C_{16}H_{15}N_4NaO_8S$
Cefuroxime*, 1951
$C_{16}H_{15}N_5O_7S_2$
Cefixime, 1927
$C_{16}H_{15}N_7O_5S_4$
Cefuzonam, 1952
$C_{16}H_{16}$
[2.2]Metacyclophane, 5825
$C_{16}H_{16}ClNO_2$
Nicoclonate, 6426
$C_{16}H_{16}ClNO_3$
Fenoldopam, 3925
Nicofibrate, 6427
$C_{16}H_{16}ClNO_4S$
Daltroban, 2807
$C_{16}H_{16}ClN_3O_3S$
Indapamide, 4847
Metolazone, 6068
$C_{16}H_{16}ClN_3O_3S.\frac{1}{2}H_2O$
Indapamide*, 4847
$C_{16}H_{16}Cl_6N_2O_2$
Triclofenol Piperazine, 9570
$C_{16}H_{16}KN_5O_4S$
Azidocillin*, 925
$C_{16}H_{16}NO_6P$
Naftalofos, 6269
$C_{16}H_{16}N_2Na_2O_7S_2$
Sulbenicillin*, 8863
Temocillin*, 9076
$C_{16}H_{16}N_2OS$
Ahistan, 179
$C_{16}H_{16}N_2O_2$
Isolysergic Acid, 5065
Lysergic Acid, 5506
Rugulovasines, 8268
$C_{16}H_{16}N_2O_4$
Phenmedipham, 7199
$C_{16}H_{16}N_2O_4S$
Acedapsone, 16
$C_{16}H_{16}N_2O_5S$
Succisulfone, 8851
$C_{16}H_{16}N_2O_6S_2$
Cephalothin, 1978
$C_{16}H_{16}N_3NaO_4S$
Cephalexin, 1971
$C_{16}H_{16}N_3NaO_7S_2$
Cefoxitin*, 1938
$C_{16}H_{16}N_4$
Stilbamidine, 8772
$C_{16}H_{16}N_4O$
Hydroxystilbamidine, 4781
$C_{16}H_{16}N_4O_2S$
Sulfazamet, 8923

$C_{16}H_{16}N_4O_8S$
Cefuroxime, 1951
$C_{16}H_{16}N_5NaO_4S$
Azidocillin*, 925
$C_{16}H_{16}N_5NaO_7S_2$
Cefotaxime*, 1935
$C_{16}H_{16}O_2$
Xenbucin, 9980
$C_{16}H_{16}O_3$
Parvaquone, 6998
$C_{16}H_{16}O_5$
Alkannin, 243
$C_{16}H_{17}BrClN_3O_3$
Halofuginone, 4509
$C_{16}H_{17}BrN_2$
Zimeldine, 10024
$C_{16}H_{17}ClN_2O$
Tetrazepam, 9172
$C_{16}H_{17}Cl_2NO_3$
Nicofibrate*, 6427
$C_{16}H_{17}KN_2O_4S$
Penicillin G Potassium, 7041
$C_{16}H_{17}KN_2O_5S$
Penicillin V*, 7046
$C_{16}H_{17}NO$
Diphenamid, 3305
$C_{16}H_{17}NO_3$
Normorphine, 6630
$C_{16}H_{17}NO_4$
Lunine, 5475
Lycorine, 5498
Tetrahydropapaveroline, 9148
$C_{16}H_{17}N_2NaO_4S$
Benzylpenicillin Sodium, 1157
$C_{16}H_{17}N_2NaO_5S$
p-Hydroxybenzylpenicillin Sodium, 4743
$C_{16}H_{17}N_3O$
Lysergamide, 5505
$C_{16}H_{17}N_3O_4$
Anthramycin, 715
$C_{16}H_{17}N_3O_4S$
Cephalexin, 1971
$C_{16}H_{17}N_3O_5S.H_2O$
Cefadroxil, 1921
$C_{16}H_{17}N_3O_7S_2$
Cefoxitin, 1938
$C_{16}H_{17}N_5O_4S$
Azidocillin, 925
$C_{16}H_{17}N_5O_7S_2$
Cefotaxime, 1935
$C_{16}H_{17}N_9O_5S_2$
Cefteram, 1945
$C_{16}H_{17}N_9O_5S_3$
Cefmenoxime, 1928
$C_{16}H_{18}$
1,1-Di-p-tolylethane, 3385
1,2-Di-p-tolylethane, 3386
$C_{16}H_{18}Br_2ClN_3O_3$
Halofuginone*, 4509
$C_{16}H_{18}ClN$
Clobenzorex, 2359
$C_{16}H_{18}ClN_3O_4.H_2O$
Cephalexin*, 1971
$C_{16}H_{18}ClN_3S$
Methylene Blue, 5979
$C_{16}H_{18}CrN_7S_4$
Rhodanilic Acid*, 8183
$C_{16}H_{18}FN_3O_3$
Norfloxacin, 6617
$C_{16}H_{18}KN_3O_4S$
Ampicillin*, 621
$C_{16}H_{18}N_2$
Agroclavine, 177
Metapramine, 5835
Nomifensine, 6592
$C_{16}H_{18}N_2O$
Amphenone B, 615
Elymoclavine, 3512
$C_{16}H_{18}N_2O_3$
Isopilosine, 5085

C₁₆H₁₈N₂O₄S

$C_{16}H_{18}N_2O_4S$
Benzylpenicillinic Acid, 1156
Sulfaproxyline, 8911

$C_{16}H_{18}N_2O_5S$
Penicillin V, 7046

$C_{16}H_{18}N_2O_7S_2$
Sulbenicillin, 8863
Temocillin, 9076

$C_{16}H_{18}N_2S$
Fenethazine, 3918

$C_{16}H_{18}N_4O_2$
Nialamide, 6399
Piribedil, 7465

$C_{16}H_{18}NaN_3O_4S$
Ampicillin*, 621

$C_{16}H_{18}O_2$
Bisphenol B, 1312

$C_{16}H_{18}O_9$
Chlorogenic Acid, 2142
Scopolin, 8364

$C_{16}H_{18}O_{10}$
Fraxin, 4179

$C_{16}H_{19}BrCl_2N_2.H_2O$
Zimeldine*, 10024

$C_{16}H_{19}BrN_2$
Brompheniramine, 1433

$C_{16}H_{19}ClN_2$
Chlorpheniramine, 2180

$C_{16}H_{19}ClN_2O$
Carbinoxamine, 1804

$C_{16}H_{19}ClO_2$
Clidanac, 2350

$C_{16}H_{19}ClO_3$
Bucloxic Acid, 1451

$C_{16}H_{19}Cl_2N$
N-(2-Chloroethyl)dibenzylamine
Hydrochloride, 2138
Clobenzorex*, 2359

$C_{16}H_{19}FN_4O_3$
Amifloxacin, 415

$C_{16}H_{19}N$
Lefetamine, 5314

$C_{16}H_{19}NO_2$
Medifoxamine, 5674

$C_{16}H_{19}NO_2S$
Teniloxazine, 9077

$C_{16}H_{19}NO_3$
α-Erythroidine, 3624
β-Erythroidine, 3625
Isoladol, 5062
Lunacrine, 5473

$C_{16}H_{19}NO_4$
Benzoylecgonine, 1125

$C_{16}H_{19}NO_5$
Peyonine, 7145

$C_{16}H_{19}N_3$
Fenapanil, 3902

$C_{16}H_{19}N_3O_3$
Febrifugine, 3890

$C_{16}H_{19}N_3O_4S$
Ampicillin, 621
Cephradine, 1982

$C_{16}H_{19}N_3O_5S$
Amoxicillin, 610
Cefroxadine, 1942

$C_{16}H_{19}N_3O_6$
Mitomycins*, 6133

$C_{16}H_{19}N_3S$
Isothipendyl, 5117
Prothipendyl, 7902

$C_{16}H_{20}BrN_3$
p-Bromtripelennamine, 1436

$C_{16}H_{20}ClN$
Lefetamine*, 5314

$C_{16}H_{20}ClN_3$
Chloropyramine, 2162

$C_{16}H_{20}ClN_3O_3S$
Tripamide, 9649

$C_{16}H_{20}ClN_3S$
Isothipendyl*, 5117

$C_{16}H_{20}NNaO_8S$
1-Naphthylamine-4-sulfonic Acid*, 6323

$C_{16}H_{20}N_2$
Benzathine, 1072
Pheniramine, 7198

$C_{16}H_{20}N_2O$
Chanoclavine, 2033

$C_{16}H_{20}N_2O_2$
Perisoxal, 7129

$C_{16}H_{20}N_2O_3$
Imazamethabenz, 4825

$C_{16}H_{20}N_2O_4S_2$
Pyritinol, 8006

$C_{16}H_{20}N_2O_5$
Eterobarb, 3666

$C_{16}H_{20}N_4O_2$
Apazone, 758

$C_{16}H_{20}N_4O_3S$
Torasemide, 9473

$C_{16}H_{20}N_4O_5$
Porfiromycin, 7572

$C_{16}H_{20}N_4O_6$
Diaziquone, 2979

$C_{16}H_{20}N_7NaO_7S_3.7H_2O$
Cefminox*, 1930

$C_{16}H_{20}O_5$
Curvularin, 2684

$C_{16}H_{20}O_6P_2S_3$
Temephos, 9075

$C_{16}H_{20}O_9$
Gentiopicrin, 4289

$C_{16}H_{21}BrClN_3$
p-Bromtripelennamine*, 1436

$C_{16}H_{21}ClN_4$
Mepiprazole, 5745

$C_{16}H_{21}Cl_2N_3$
Chloropyramine*, 2162

$C_{16}H_{21}HgN_6NaO_7$
Meralluride*, 5756

$C_{16}H_{21}N$
Morphinan, 6185

$C_{16}H_{21}NO$
Norlevorphanol, 6625

$C_{16}H_{21}NO_2$
HQNO, 4670
Propranolol, 7852

$C_{16}H_{21}NO_3$
Annotinine, 705
Dihydro-β-erythroidine, 3158
Homatropine, 4649
Norhyoscyamine, 6623
Rolipram, 8235

$C_{16}H_{21}NO_4$
Befunolol, 1032

$C_{16}H_{21}NS_2$
Thiambutene, 9220

$C_{16}H_{21}N_3$
Tripelennamine, 9651

$C_{16}H_{21}N_3O_4S$
Epicillin, 3566

$C_{16}H_{21}N_3O_8S$
Cephalosporin C, 1976

$C_{16}H_{21}N_5O_2$
Alizapride, 236

$C_{16}H_{21}N_7O_7S_3$
Cefminox, 1930

$C_{16}H_{22}BrNO_3$
Homatropine*, 4649

$C_{16}H_{22}ClNO_2$
Propranolol*, 7852

$C_{16}H_{22}ClNO_4$
Befunolol*, 1032

$C_{16}H_{22}ClN_3$
Tripelennamine*, 9651

$C_{16}H_{22}ClN_3O_4$
Plafibride, 7488

$C_{16}H_{22}ClN_5O_2$
Alizapride*, 236

$C_{16}H_{22}Cl_2N_2O_4S_2.H_2O$
Pyritinol*, 8006

$C_{16}H_{22}FNO$
Melperone, 5707

$C_{16}H_{22}HgN_6O_7$
Meralluride, 5756

$C_{16}H_{22}NNaO_6$
Capobenic Acid*, 1757

$C_{16}H_{22}N_2$
Lycodine, 5489

$C_{16}H_{22}N_2O_3$
Procaterol, 7765

$C_{16}H_{22}N_2O_3S$
Glyhexamide, 4402

$C_{16}H_{22}N_2O_4$
Antipyrine*, 748

$C_{16}H_{22}N_2O_6$
2,5-Bis(1-aziridinyl)-3,6-bis(2-
methoxyethoxy)- 1,4-benzoqui-
none, 1256

$C_{16}H_{22}N_2O_6S$
Morantel*, 6175

$C_{16}H_{22}N_4O_2$
Aminopropylon, 484

$C_{16}H_{22}N_4O_4$
Tetroxoprim, 9178

$C_{16}H_{22}N_4O_4S$
Acetiamine, 46

$C_{16}H_{22}N_6O_4$
TRH, 9503

$C_{16}H_{22}O_2$
Geranylhydroquinone, 4300

$C_{16}H_{22}O_3$
Fencibutirol, 3912
Homosalate, 4660

$C_{16}H_{22}O_4$
n-Butyl Phthalate, 1586

$C_{16}H_{22}O_6$
Trepibutone, 9501

$C_{16}H_{22}O_8$
Coniferin, 2498

$C_{16}H_{22}O_{10}$
Swertiamarin, 8991

$C_{16}H_{22}O_{11}$
Monotropein, 6166

$C_{16}H_{23}AsN_2O_7$
Drocarbil, 3434

$C_{16}H_{23}ClFNO$
Melperone*, 5707

$C_{16}H_{23}ClN_2O_2$
Alloclamide, 254

$C_{16}H_{23}ClN_2O_3.½H_2O$
Procaterol*, 7765

$C_{16}H_{23}ClN_4O$
Thonzylamine Hydrochloride, 9307

$C_{16}H_{23}ClN_4O_4S$
Acetiamine*, 46

$C_{16}H_{23}Cl_3N_4$
Mepiprazole*, 5745

$C_{16}H_{23}I_3N_2O_8$
Meglumine Acetrizoate, 5688

$C_{16}H_{23}NO$
Dezocine, 2934
Pyrovalerone, 8021
Tolperisone, 9447

$C_{16}H_{23}NO_2$
Alphaprodine, 307
Bufuralol, 1469
Ethoheptazine, 3698
Etoxadrol, 3844
Piperocaine, 7444
Properidine, 7826

$C_{16}H_{23}NO_3$
Acetamidoeugenol, 39

$C_{16}H_{23}NO_6$
Capobenic Acid, 1757
Monocrotaline, 6160

$C_{16}H_{23}NO_8$
Bakankosin, 955

$C_{16}H_{23}N_3O_4$
Gabexate, 4234

$C_{16}H_{23}NaO_5$
Exiproben*, 3869

Denotes derivative of title compound

C₁₆H₂₄ClNO
Pyrovalerone*, 8021
Tolperisone*, 9447
C₁₆H₂₄ClNO₂
Alphaprodine*, 307
Bufuralol*, 1469
Etoxadrol*, 3844
Hexylcaine Hydrochloride, 4629
C₁₆H₂₄HgO₃
Acetomeroctol, 57
C₁₆H₂₄NO₅⁺
Sinapine, 8493
C₁₆H₂₄N₂
Isoaminile, 4992
Xylometazoline, 9993
C₁₆H₂₄N₂O
Oxymetazoline, 6919
C₁₆H₂₄N₂O₂
Molindone, 6142
N,N,N',N'-Tetraethylphthalamide, 9137
C₁₆H₂₄N₂O₃
Carteolol, 1875
C₁₆H₂₄N₂O₄
Ubenimex, 9750
C₁₆H₂₄N₂O₄S
2-Amino-4-picoline*, 478
C₁₆H₂₄N₂O₅
Tricetamide, 9535
C₁₆H₂₄N₂O₆
Pyrisuccideanol, 8002
C₁₆H₂₄N₂O₆S
Phenylethanolamine*, 7258
C₁₆H₂₄N₁₀O₄
Aminophylline, 477
C₁₆H₂₄O₄
Brefeldin A, 1363
C₁₆H₂₄O₅
Exiproben, 3869
C₁₆H₂₄O₇
Rhododendrin, 8190
C₁₆H₂₅ClN₂
Xylometazoline*, 9993
C₁₆H₂₅ClN₂O
Oxymetazoline*, 6919
C₁₆H₂₅ClN₂O₃
Carteolol*, 1875
C₁₆H₂₅HgNNa₂O₆S
Mercaptomerin Sodium, 5761
C₁₆H₂₅NO
Lycopodine, 5494
C₁₆H₂₅NO₂
Butethamate, 1516
Cyclexanone, 2712
Gravitol(e), 4448
Thymyl N-Isoamylcarbamate, 9343
Tramadol, 9485
C₁₆H₂₅NO₂S
Tertatolol, 9105
C₁₆H₂₅NO₃
Moxisylyte, 6204
C₁₆H₂₅NO₄
Esmolol, 3648
Floredil, 4041
C₁₆H₂₅N₂NaO₅S
Cilastatin*, 2275
C₁₆H₂₅N₃O
Propiram, 7844
C₁₆H₂₅N₃O₅
Xamoterol, 9965
C₁₆H₂₅N₇O₈
Gougerotin, 4436
C₁₆H₂₆ClNO
Butidrine Hydrochloride, 1522
C₁₆H₂₆ClNO₂
Cyclexanone*, 2712
Tramadol*, 9485
C₁₆H₂₆ClNO₂S
Tertatolol*, 9105
C₁₆H₂₆ClNO₃
Moxisylyte*, 6204

C₁₆H₂₆ClNO₄
Esmolol*, 3648
Floredil*, 4041
C₁₆H₂₆N₂O₂
Dimethocaine, 3210
C₁₆H₂₆N₂O₃
Proparacaine, 7817
C₁₆H₂₆N₂O₄
Cetamolol, 2010
C₁₆H₂₆N₂O₅S
Cilastatin, 2275
C₁₆H₂₆O₂
2,5-Di-tert-pentylhydroquinone, 3300
C₁₆H₂₆O₅
Artemether, 842
C₁₆H₂₆O₇
Picrocrocin, 7381
C₁₆H₂₇ClN₂O₃
Proparacaine*, 7817
Propoxycaine Hydrochloride, 7850
C₁₆H₂₇ClN₂O₄
Cetamolol*, 2010
C₁₆H₂₇N₃O₈
Dinoseb*, 3282
C₁₆H₂₈NO₃Zn
ε-Acetamidocaproic Acid*, 38
C₁₆H₂₈N₂O₂
Tromantadine, 9683
C₁₆H₂₈N₄O₄S
Biocytin, 1240
C₁₆H₂₈O₂
Cyoctol, 2774
Hydnocarpic Acid, 4679
Prodlure, 7777
C₁₆H₂₉ClN₂O₂
Tromantadine*, 9683
C₁₆H₂₉NO₅
Butoctamide*, 1526
C₁₆H₂₉N₃O₈
Diethylcarbamazine*, 3106
C₁₆H₂₉NaO₇S
Bis(1-methylamyl) Sodium Sulfosuccinate, 1266
C₁₆H₃₀MgO₄
Valproic Acid*, 9821
C₁₆H₃₀N₂O₃
Piroctone*, 7472
C₁₆H₃₀O
Muscone, 6222
C₁₆H₃₀O₄Zn
Zinc Caprylate, 10029
C₁₆H₃₁NaO₄
Valproic Acid*, 9821
C₁₆H₃₂O₂
2-Hexyldecanoic Acid, 4630
Myristic Acid*, 6246
Palmitic Acid, 6947
C₁₆H₃₂O₄
Diethylene Glycol Monolaurate, 3110
C₁₆H₃₂O₅
Aleuritic Acid, 221
C₁₆H₃₃BiO₄S
Bismuth Butylthiolaurate, 1272
C₁₆H₃₄Br₂N₂O₄
Suxethonium Bromide, 8989
C₁₆H₃₄I₂N₂O₂
Dicolinium Iodide, 3076
C₁₆H₃₄O
Cetyl Alcohol, 2020
C₁₆H₃₅NO₂
Ammonium Palmitate, 566
C₁₆H₃₈Br₂N₂
Decamethonium Bromide, 2840
C₁₆H₄₂I₇N₈O₁₆
Tetraglycine Hydroperiodide, 9140
C₁₇H₉NO₄
Alizarine Blue, 239
C₁₇H₁₀Cl₂N₄O₂
Clazuril, 2343

C₁₇H₁₀F₆N₄S
Flubenzimine, 4051
C₁₇H₁₀N₂O₅
Phenocoll*, 7204
C₁₇H₁₀O
Benzanthrone, 1070
C₁₇H₁₀O₄
Fluorescamine, 4084
C₁₇H₁₁ClF₄N₂S
Quazepam, 8039
C₁₇H₁₁NO₇
Aristolochic Acid, 810
C₁₇H₁₂BrFN₂O₃
Ponalrestat, 7565
C₁₇H₁₂Br₂O₃
Benzbromarone, 1073
C₁₇H₁₂ClF₃N₂O
Halazepam, 4502
C₁₇H₁₂ClNO₂S
Fentiazac, 3946
C₁₇H₁₂Cl₂N₂O
Fenarimol, 3903
C₁₇H₁₂Cl₂N₄
Triazolam, 9518
C₁₇H₁₂Cl₂O₇
Erdin*, 3588
C₁₇H₁₂I₂O₃
Benziodarone, 1093
C₁₇H₁₂O₂
2-Naphthyl Benzoate, 6328
C₁₇H₁₂O₃
1-Naphthyl Salicylate, 6334
2-Naphthyl Salicylate, 6335
C₁₇H₁₂O₆
Aflatoxins B*, 168
Irisolone, 4974
C₁₇H₁₂O₇
Aflatoxins G*, 169
Aflatoxins M*, 170
C₁₇H₁₃ClN₂O₃
Lonazolac, 5445
C₁₇H₁₃ClN₄
Alprazolam, 310
C₁₇H₁₃N₃O₅S₂
Phthalylsulfathiazole, 7351
C₁₇H₁₄
1,2-Cyclopentenophenanthrene, 2749
C₁₇H₁₄BrFN₂O₂
Haloxazolam, 4518
C₁₇H₁₄ClFN₂O₃
Doxefazepam, 3423
C₁₇H₁₄Cl₂N₂O₂
Cloxazolam, 2415
C₁₇H₁₄N₂
Ellipticine, 3509
Flavopereirine, 4031
Olivacine, 6794
C₁₇H₁₄N₂O₃
Rosoxacin, 8247
C₁₇H₁₄N₂O₅S
Calmagite, 1725
C₁₇H₁₄O
Dibenzalacetone, 2988
C₁₇H₁₄O₃
Benzarone, 1071
C₁₇H₁₄O₅
Coumafuryl, 2557
Pterocarpin, 7948
C₁₇H₁₄O₆
Aflatoxins B*, 168
Pectolinarigenin, 7010
C₁₇H₁₄O₇
Aflatoxins G*, 169
Aflatoxins M*, 170
C₁₇H₁₅ClN₄S
Etizolam, 3830
C₁₇H₁₅ClO₇
Malvidin Chloride, 5596
C₁₇H₁₅Cl₃N₂O₂
Triclodazol, 9569

Denotes derivative of title compound

FI-43

Denotes derivative of title compound

C$_{18}$H$_{10}$I$_6$N$_2$O$_7$
Ioglycamic Acid, 4939

C$_{18}$H$_{10}$O$_6$
Hydrindantin, 4698

C$_{18}$H$_{12}$
1,2-Benzanthracene, 1069
3,4-Benzphenanthrene, 1129
Chrysene, 2259
Naphthacene, 6288
Triphenylene, 9654

C$_{18}$H$_{12}$AlN$_3$O$_6$
Aluminum Nicotinate, 354

C$_{18}$H$_{12}$Cl$_{12}$
Dechlorane® Plus, 2844

C$_{18}$H$_{12}$NNaO$_3$
Naptalam*, 6338

C$_{18}$H$_{12}$N$_2$O$_2$
Xanthocillin*, 9970

C$_{18}$H$_{12}$N$_2$O$_3$
Xanthocillin*, 9970

C$_{18}$H$_{12}$N$_2$O$_4$
Xanthocillin*, 9970

C$_{18}$H$_{12}$N$_6$
2,4,6-Tripyridyl-s-triazine, 9664

C$_{18}$H$_{12}$O$_4$
Karanjin, 5163

C$_{18}$H$_{13}$ClFN$_3$
Midazolam, 6105

C$_{18}$H$_{13}$ClN$_2$O
Pinazepam, 7411

C$_{18}$H$_{13}$N
6-Chrysenamine, 2258

C$_{18}$H$_{13}$NNa$_2$O$_8$S$_2$
Picosulfate Sodium, 7377

C$_{18}$H$_{13}$NO$_3$
Naptalam, 6338

C$_{18}$H$_{13}$N$_3$O
Rutecarpine, 8271

C$_{18}$H$_{14}$CaO$_8$
Calcium Acetylsalicylate, 1644

C$_{18}$H$_{14}$ClFN$_2$O$_3$
Ethyl Loflazepate, 3777

C$_{18}$H$_{14}$Cl$_2$FN$_3$
Midazolam*, 6105

C$_{18}$H$_{14}$Cl$_2$N$_5$NaO$_6$S$_3$
Cefazedone*, 1924

C$_{18}$H$_{14}$Cl$_4$N$_2$O
Isoconazole, 5044
Miconazole, 6101

C$_{18}$H$_{14}$Cl$_4$N$_4$O$_4$
Oxiconazole Nitrate, 6892

C$_{18}$H$_{14}$F$_3$N$_5$O$_4$S
Toltrazuril, 9452

C$_{18}$H$_{14}$MgO$_8$
Magnesium Acetylsalicylate, 5531

C$_{18}$H$_{14}$N$_2$Na$_2$O$_7$S$_2$
Ponceau SX, 7568

C$_{18}$H$_{14}$N$_2$Na$_2$O$_8$S$_2$
Allura® Red AC, 283

C$_{18}$H$_{14}$N$_2$O$_2$
5-Methyl-5-(3-phenanthryl)hydantoin, 6023

C$_{18}$H$_{14}$N$_3$NaO$_3$S
Metanil Yellow, 5833
Tropaeolin OO, 9688

C$_{18}$H$_{14}$N$_4$Na$_2$O$_{10}$S$_3$
Azosulfamide, 936

C$_{18}$H$_{14}$N$_4$O$_5$S
Sulfasalazine, 8917

C$_{18}$H$_{14}$O$_3$
Cinnamic Acid*, 2300

C$_{18}$H$_{14}$O$_8$
Succinylsalicylic Acid, 8849

C$_{18}$H$_{15}$AlO$_9$
Aluminum Bis(acetylsalicylate), 328

C$_{18}$H$_{15}$ClN$_2$O
Cloconazole, 2367

C$_{18}$H$_{15}$ClN$_4$
Phenosafranin, 7218

C$_{18}$H$_{15}$Cl$_2$N$_5$O$_5$S$_3$
Cefazedone, 1924

C$_{18}$H$_{15}$Cl$_3$N$_2$O
Econazole, 3476

C$_{18}$H$_{15}$Cl$_3$N$_2$S
Sulconazole, 8866

C$_{18}$H$_{15}$Cl$_4$N$_3$O$_4$
Isoconazole*, 5044
Miconazole*, 6101

C$_{18}$H$_{15}$NO$_3$
Oxaprozin, 6879

C$_{18}$H$_{15}$O$_4$P
Triphenyl Phosphate, 9656

C$_{18}$H$_{15}$P
Triphenylphosphine, 9657

C$_{18}$H$_{16}$
3'-Methyl-1,2-cyclopentenophenanthrene, 5969

C$_{18}$H$_{16}$ClNO$_5$
Fenoxaprop-ethyl, 3929

C$_{18}$H$_{16}$ClN$_3$O$_4$
Econazole*, 3476

C$_{18}$H$_{16}$Cl$_2$N$_2$O
Cloconazole*, 2367

C$_{18}$H$_{16}$Cl$_2$N$_2$O$_2$
Mexazolam, 6091

C$_{18}$H$_{16}$Cl$_3$N$_3$O$_3$S
Sulconazole*, 8866

C$_{18}$H$_{16}$N$_2$
N,N'-Diphenyl-p-phenylenediamine, 3331

C$_{18}$H$_{16}$N$_2$O
1-Xylylazo-2-naphthol, 9997

C$_{18}$H$_{16}$N$_2$O$_3$
Citrus Red 2, 2335

C$_{18}$H$_{16}$N$_2$O$_6$S
8-Hydroxyquinoline Sulfate, 4779

C$_{18}$H$_{16}$N$_4$O$_6$S
Quinacillin, 8052

C$_{18}$H$_{16}$N$_6$Na$_2$O$_8$S$_3$
Cefonicid*, 1932

C$_{18}$H$_{16}$N$_8$Na$_2$O$_7$S$_3$·3½H$_2$O
Ceftriaxone*, 1950

C$_{18}$H$_{16}$OSn
Triphenyltin Hydroxide, 9659

C$_{18}$H$_{16}$O$_2$
Cinnamyl Cinnamate, 2307

C$_{18}$H$_{16}$O$_3$
Ipriflavone, 4961
Phenprocoumon, 7230

C$_{18}$H$_{16}$O$_4$
Benzyl Fumarate, 1149
Isatropic Acid, 4986
Truxillic Acid, 9700

C$_{18}$H$_{16}$O$_7$
Eupatorin, 3858
Usnic Acid, 9806

C$_{18}$H$_{16}$O$_8$
Irigenin, 4973

C$_{18}$H$_{17}$ClNNaO$_4$
Clanobutin*, 2338

C$_{18}$H$_{17}$ClN$_2$O
Oxazolam, 6883

C$_{18}$H$_{17}$ClO$_6$
Monorden, 6164

C$_{18}$H$_{17}$FN$_2$O
Fluproquazone, 4120

C$_{18}$H$_{17}$I$_4$NO$_4$
Etiroxate, 3828

C$_{18}$H$_{17}$NO$_2$
Aporeine, 778

C$_{18}$H$_{17}$NO$_3$
Indobufen, 4867
Pukateine, 7954

C$_{18}$H$_{17}$NO$_4$
Actinodaphnine, 131
Laurepukine, 5253

C$_{18}$H$_{17}$NO$_5$
Tranilast, 9489

C$_{18}$H$_{18}$
Retene, 8162

C$_{18}$H$_{18}$BrClN$_2$O
Metaclazepam, 5824

C$_{18}$H$_{18}$ClI$_4$NO$_4$
Etiroxate*, 3828

C$_{18}$H$_{18}$ClNOS
Zotepine, 10097

C$_{18}$H$_{18}$ClNO$_4$
Clanobutin, 2338

C$_{18}$H$_{18}$ClNO$_5$
Etofibrate, 3834

C$_{18}$H$_{18}$ClNS
Chlorprothixene, 2189

C$_{18}$H$_{18}$ClN$_3$O
Loxapine, 5461

C$_{18}$H$_{18}$ClN$_3$S
Clothiapine, 2410

C$_{18}$H$_{18}$F$_3$NO$_2$
Flufenamic Acid*, 4060

C$_{18}$H$_{18}$F$_3$NO$_4$
Etofenamate, 3833

C$_{18}$H$_{18}$N$_2$
Cifenline, 2274

C$_{18}$H$_{18}$N$_2$O
Demexiptiline, 2876
Proquazone, 7887

C$_{18}$H$_{18}$N$_2$O$_4$
Antipyrine Salicylate, 749

C$_{18}$H$_{18}$N$_6$O$_5$S$_2$
Cefamandole, 1922
Cefatrizine, 1923

C$_{18}$H$_{18}$N$_6$O$_8$S$_3$
Cefonicid, 1932

C$_{18}$H$_{18}$N$_8$O$_7$S$_3$
Ceftriaxone, 1950

C$_{18}$H$_{18}$O$_2$
Dienestrol, 3094
Equilenin, 3581

C$_{18}$H$_{18}$O$_4$
Enterolactone, 3549
Metochalcone, 6062

C$_{18}$H$_{18}$O$_6$
Samaderins*, 8312

C$_{18}$H$_{18}$O$_7$
Frenolicin, 4182

C$_{18}$H$_{19}$BrCl$_2$N$_2$O
Metaclazepam*, 5824

C$_{18}$H$_{19}$ClN$_2$O
Demexiptiline*, 2876

C$_{18}$H$_{19}$ClN$_4$
Clozapine, 2417

C$_{18}$H$_{19}$Cl$_2$NO$_4$
Felodipine, 3895

C$_{18}$H$_{19}$Cl$_2$N$_3$O
Loxapine*, 5461

C$_{18}$H$_{19}$F$_3$N$_2$S
Triflupromazine, 9597

C$_{18}$H$_{19}$N
Benzoctamine, 1096

C$_{18}$H$_{19}$NOS
Tolindate, 9438

C$_{18}$H$_{19}$NO$_2$
Apocodeine, 771

C$_{18}$H$_{19}$NO$_3$
Morphothebaine, 6197
Oripavine, 6821
Tsuduranine, 9710

C$_{18}$H$_{19}$NO$_4$
Hydroxycodeinone, 4752

C$_{18}$H$_{19}$N$_3$O
Ondansetron, 6802

C$_{18}$H$_{19}$N$_3$O$_2$
Diacetazotol, 2940

C$_{18}$H$_{19}$N$_3$O$_6$S
Cephaloglycin, 1973

C$_{18}$H$_{20}$BrNO$_2$
Apomorphine*, 776

C$_{18}$H$_{20}$ClF$_3$N$_2$S
Triflupromazine*, 9597

C$_{18}$H$_{20}$ClN
Benzoctamine*, 1096

C$_{18}$H$_{20}$ClN$_3$O·2H$_2$O
Ondansetron*, 6802

Denotes derivative of title compound

$C_{18}H_{20}ClN_3O_6S_2$
Sporidesmins*, 8724

$C_{18}H_{20}Cl_2$
1,1-Dichloro-2,2-bis(p-ethylphenyl)ethane, 3050

$C_{18}H_{20}Cl_2N_8$
Bismark Brown Y, 1265

$C_{18}H_{20}FN_3O_4$
Ofloxacin, 6688

$C_{18}H_{20}N_2$
Mianserin, 6097

$C_{18}H_{20}N_2O_6$
Nitrendipine, 6493

$C_{18}H_{20}N_2S$
Methdilazine, 5877
Pyrathiazine, 7969

$C_{18}H_{20}N_4O_2$
Nitracrine, 6485

$C_{18}H_{20}Na_4O_8P_2$
Fosfestrol*, 4168

$C_{18}H_{20}O_2$
Diethylstilbestrol, 3118
Equilin, 3582

$C_{18}H_{20}O_8S_2$
Diethylstilbestrol*, 3118

$C_{18}H_{21}BrN_2OS$
Propiomazine*, 7834

$C_{18}H_{21}ClN_2$
Chlorcyclizine, 2078
Clomacran, 2379
Mianserin*, 6097

$C_{18}H_{21}ClN_2S$
Methdilazine*, 5877
Pyrathiazine*, 7969

$C_{18}H_{21}ClN_4O_7$
Chrysoidine*, 2262

$C_{18}H_{21}ClO_4$
Barthrin, 1014

$C_{18}H_{21}KN_2O_5S$
Propicillin*, 7829

$C_{18}H_{21}NO$
Azacyclonol, 909
Pipradrol, 7455

$C_{18}H_{21}NO_3$
Codeine, 2459
Hydrocodone, 4708
Metopon, 6071
Morphine*, 6186
Neopine, 6376
Pseudocodeine, 7927
Thebainone, 9203

$C_{18}H_{21}NO_4$
Codeine N-Oxide, 2461
Oxycodone, 6912

$C_{18}H_{21}NO_5$
Protokylol, 7911
Tazettine, 9050

$C_{18}H_{21}N_3O$
Dibenzepin, 2990

$C_{18}H_{21}N_3O_4.H_2O$
Ethacridine, 3668

$C_{18}H_{22}BrNO$
Embramine, 3514

$C_{18}H_{22}BrNO_3$
Morphine Methylbromide, 6189

$C_{18}H_{22}BrNO_3S$
Heteronium Bromide, 4595

$C_{18}H_{22}ClNO$
Chlorphenoxamine, 2182
Phenoxybenzamine, 7225

$C_{18}H_{22}ClNO.H_2O$
Azacyclonol*, 909
Pipradrol*, 7455

$C_{18}H_{22}ClNO_4$
Oxycodone*, 6912

$C_{18}H_{22}ClNO_5$
Protokylol*, 7911

$C_{18}H_{22}ClNO_6$
Clofibric Acid*, 2375

$C_{18}H_{22}ClN_3O$
Dibenzepin*, 2990

$C_{18}H_{22}Cl_2N_2$
Chlorcyclizine*, 2078

$C_{18}H_{22}Cl_2N_4O_2.H_2O$
Nitracrine*, 6485

$C_{18}H_{22}I_3N_3O_8$
Metrizamide, 6077

$C_{18}H_{22}N_2$
Cyclizine, 2716
Desipramine, 2902

$C_{18}H_{22}N_2OS$
Methoxypromazine, 5923

$C_{18}H_{22}N_2O_2$
Carazolol, 1779

$C_{18}H_{22}N_2O_2S$
Oxomemazine, 6900

$C_{18}H_{22}N_2O_5S$
Propicillin, 7829

$C_{18}H_{22}N_2S$
Diethazine, 3098
Trimeprazine, 9618

$C_{18}H_{22}N_4O_7$
Galactoflavin, 4239

$C_{18}H_{22}O_2$
Dihydroequilin, 3156
Estrone, 3660
Hexestrol, 4621
8-Isoestrone, 5052
Trenbolone, 9499

$C_{18}H_{22}O_3$
Methallenestril, 5856

$C_{18}H_{22}O_4$
Nordihydroguaiaretic Acid, 6610

$C_{18}H_{22}O_5$
Zearalenone, 10017

$C_{18}H_{22}O_8P_2$
Fosfestrol, 4168

$C_{18}H_{22}O_{11}$
Asperuloside, 866

$C_{18}H_{23}BrClNO$
Embramine*, 3514

$C_{18}H_{23}ClN_2$
Cyclizine*, 2716
Desipramine*, 2902

$C_{18}H_{23}ClN_2O_2$
Phenacaine Hydrochloride, 7153

$C_{18}H_{23}ClN_2O_2S$
Oxomemazine*, 6900

$C_{18}H_{23}ClN_2S$
Diethazine*, 3098

$C_{18}H_{23}Cl_2NO$
Chlorphenoxamine*, 2182
Phenoxybenzamine*, 7225

$C_{18}H_{23}N$
Tolpropamine, 9450
N,N-1-Trimethyl-3,3-diphenylpropylamine, 9627

$C_{18}H_{23}NO$
Bifemelane, 1227
Mephenhydramine, 5738
p-Methyldiphenhydramine, 5972
Orphenadrine, 6831
Racefemine, 8115

$C_{18}H_{23}NO_2$
Medrylamine, 5678

$C_{18}H_{23}NO_3$
Dihydrocodeine, 3154
Dihydroisocodeine, 3159
Dobutamine, 3396
Isoxsuprine, 5128

$C_{18}H_{23}NO_4$
Cocaethylene, 2449
Denopamine, 2878

$C_{18}H_{23}NO_5$
Seneciphylline, 8403

$C_{18}H_{23}N_3O_4S$
Phentolamine*, 7234

$C_{18}H_{23}N_3O_5$
Methafurylene*, 5854

$C_{18}H_{23}N_3O_5S$
Acediasulfone*, 17

$C_{18}H_{23}N_5NaO_8P$
Bucladesine*, 1448

$C_{18}H_{23}N_5O_2$
Fenethylline, 3919

$C_{18}H_{23}N_5O_5$
Reproterol, 8142

$C_{18}H_{23}N_9O_4S_3$
Cefotiam, 1937

$C_{18}H_{24}BrNO_3S$
Bretylium Tosylate, 1364

$C_{18}H_{24}BrNO_4$
Methscopolamine Bromide, 5927

$C_{18}H_{24}ClNO$
Bifemelane*, 1227
Orphenadrine*, 6831

$C_{18}H_{24}ClNO_3$
Dobutamine*, 3396
Isoxsuprine*, 5128

$C_{18}H_{24}ClN_2OP_4$
Clomacran*, 2379

$C_{18}H_{24}ClN_5O_2$
Fenethylline*, 3919

$C_{18}H_{24}ClN_5O_5$
Reproterol*, 8142

$C_{18}H_{24}FN_3O_6S.2H_2O$
Pefloxacin*, 7012

$C_{18}H_{24}INO_2S$
Tiemonium Iodide, 9363

$C_{18}H_{24}I_3N_3O_8$
Iopromide, 4948

$C_{18}H_{24}NO_7P$
Codeine Phosphate, 2462

$C_{18}H_{24}N_2O_3$
Julocrotine, 5151

$C_{18}H_{24}N_2O_4$
Isoxaben, 5125

$C_{18}H_{24}N_2O_4S$
Tranylcypromine*, 9491

$C_{18}H_{24}N_2O_5.2H_2O$
Enalaprilat, 3522

$C_{18}H_{24}N_2O_5S$
Amosulalol, 607

$C_{18}H_{24}N_2O_6$
Dinocap, 3281

$C_{18}H_{24}N_2O_7$
Scopolamine*, 8361

$C_{18}H_{24}N_4O$
Granisetron, 4443

$C_{18}H_{24}N_5O_8P$
Bucladesine, 1448

$C_{18}H_{24}O_2$
Estradiol, 3653
α-Estradiol, 3654
Hexahydroequilenin, 4607
Isoestradiol, 5051
Nimbiol, 6463

$C_{18}H_{24}O_3$
Doisynolic Acid, 3408
16-Epiestriol, 3567
Estriol, 3659

$C_{18}H_{24}O_8P_2$
Hexestrol*, 4621

$C_{18}H_{25}ClN_2O_5S$
Amosulalol*, 607

$C_{18}H_{25}ClN_4$
Granisetron*, 4443

$C_{18}H_{25}Cl_2N_9O_4S_3$
Cefotiam*, 1937

$C_{18}H_{25}N$
Dimemorfan, 3192

$C_{18}H_{25}NO$
Cyclazocine, 2710
Racemethorphan, 8116

$C_{18}H_{25}NO_2$
Allylprodine, 294

$C_{18}H_{25}NO_3$
Levomepate, 5345

$C_{18}H_{25}NO_5$
Senecionine, 8402

$C_{18}H_{25}NO_6$
Retrorsine, 8169

C₁₉H₁₅ClN₄
Triphenyltetrazolium Chloride, 9658

$C_{19}H_{15}ClN_4$
Triphenyltetrazolium Chloride, 9658
$C_{19}H_{15}ClO_4$
Coumachlor, 2556
$C_{19}H_{15}Cl_3O_5$
Nornidulin, 6632
$C_{19}H_{15}F_3N_4O_3$
Tosufloxacin, 9477
$C_{19}H_{15}KO_4$
Warfarin*, 9950
$C_{19}H_{15}NO_2$
Cinchophen*, 2290
$C_{19}H_{15}NO_4$
3-Methyl-6,7-methylenedioxy-1-piperonylisoquinoline, 6011
$C_{19}H_{15}NO_6$
Acenocoumarol, 24
$C_{19}H_{15}NaO_4$
Warfarin*, 9950
$C_{19}H_{16}$
Triphenylmethane, 9655
$C_{19}H_{16}ClFN_2O$
Flutoprazepam, 4134
$C_{19}H_{16}ClFN_3NaO_5S.H_2O$
Floxacillin*, 4044
$C_{19}H_{16}ClF_3N_4O_3$
Tosufloxacin*, 9477
$C_{19}H_{16}ClNO_4$
Clometacin, 2381
Indomethacin, 4874
$C_{19}H_{16}ClN_5$
Phenformin*, 7191
$C_{19}H_{16}Cl_2N_3NaO_5S.H_2O$
Dicloxacillin*, 3073
$C_{19}H_{16}N_2$
Sempervirine, 8398
$C_{19}H_{16}N_2Na_2O_7S_2$
Ponceau 3R, 7567
$C_{19}H_{16}N_2O$
Difenpiramide, 3125
$C_{19}H_{16}N_5NaO_7S_3$
Ceftiofur*, 1948
$C_{19}H_{16}O$
Triphenylcarbinol, 9653
$C_{19}H_{16}O_4$
Warfarin, 9950
$C_{19}H_{16}O_5$
Efloxate, 3490
$C_{19}H_{17}ClFN_3O_5S$
Floxacillin, 4044
$C_{19}H_{17}ClN_2O$
Prazepam, 7713
$C_{19}H_{17}ClN_2O_4$
Glafenine, 4330
Oxametacine, 6869
Quizalofop-Ethyl, 8113
$C_{19}H_{17}ClN_3NaO_5S.H_2O$
Cloxacillin*, 2414
$C_{19}H_{17}Cl_2N_3O_5S$
Dicloxacillin, 3073
$C_{19}H_{17}Cl_3N_2S$
Butoconazole, 1525
$C_{19}H_{17}NNa_2O_7$
Nedocromil*, 6355
$C_{19}H_{17}NOS$
Tolnaftate, 9442
$C_{19}H_{17}NO_3$
Cusparine, 2687
$C_{19}H_{17}NO_4$
Stylopine, 8827
$C_{19}H_{17}NO_7$
Nedocromil, 6355
$C_{19}H_{17}N_3O$
Evodiamine, 3865
$C_{19}H_{17}N_3O_4S_2$
Cephaloridine, 1975
$C_{19}H_{17}N_5O_2$
Nafamostat, 6264
$C_{19}H_{17}N_5O_5S$
Salazosulfadimidine, 8291

$C_{19}H_{17}N_5O_7S_3$
Ceftiofur, 1948
$C_{19}H_{17}N_6NaO_6S_2$
Cefamandole*, 1922
$C_{19}H_{18}CaN_2O_9$
Calcium Acetylsalicylate*, 1644
$C_{19}H_{18}ClFN_2O_3$
Flutazolam, 4133
$C_{19}H_{18}ClN_3O_2$
Camazepam, 1732
$C_{19}H_{18}ClN_3O_5S$
Cloxacillin, 2414
$C_{19}H_{18}ClN_5$
Adinazolam, 150
$C_{19}H_{18}ClN_5O_7S_3$
Ceftiofur*, 1948
$C_{19}H_{18}Cl_3N_3O_3S$
Butoconazole*, 1525
$C_{19}H_{18}N_2O_3$
Kebuzone, 5168
$C_{19}H_{18}N_3NaO_5S.H_2O$
Oxacillin*, 6858
$C_{19}H_{18}N_6O_6$
Nicarbazin, 6402
$C_{19}H_{18}N_7NaO_6$
Sodium Folate, 8566
$C_{19}H_{18}O_7$
Benfurodil Hemisuccinate, 1051
Gardenins*, 4265
$C_{19}H_{18}O_8$
Atranorin, 885
Gardenins*, 4265
$C_{19}H_{18}O_9$
Gardenins*, 4265
$C_{19}H_{19}N$
Phenindamine, 7195
$C_{19}H_{19}NOS$
Ketotifen, 5187
$C_{19}H_{19}NO_3$
Laureline, 5251
$C_{19}H_{19}NO_4$
Bulbocapnine, 1471
Domesticine, 3409
Nandinine, 6279
$C_{19}H_{19}NO_6$
Etersalate, 3667
$C_{19}H_{19}N_3O_5S$
Oxacillin, 6858
$C_{19}H_{19}N_3O_6$
Nilvadipine, 6461
$C_{19}H_{19}N_7O_6$
Folic Acid, 4140
$C_{19}H_{20}ClNO_4$
Bezafibrate, 1215
$C_{19}H_{20}ClNO_5$
Ronifibrate, 8238
$C_{19}H_{20}ClN_3$
Clemizole, 2346
$C_{19}H_{20}FNO_3$
Paroxetine, 6991
$C_{19}H_{20}F_3NO_2$
Benfluorex, 1047
$C_{19}H_{20}F_3NO_4$
Fluazifop-butyl, 4049
$C_{19}H_{20}F_3N_3O_3$
Niflumic Acid*, 6444
$C_{19}H_{20}N_2$
Mebhydroline, 5649
$C_{19}H_{20}N_2O$
Vellosimine, 9846
$C_{19}H_{20}N_2O_2$
Phenylbutazone, 7248
$C_{19}H_{20}N_2O_3$
Amphotalide, 619
Ditazol, 3374
Oxyphenbutazone, 6925
$C_{19}H_{20}N_2O_4$
Antipyrine*, 748
$C_{19}H_{20}N_6O$
Imidocarb, 4831

$C_{19}H_{20}N_8O_5$
Aminopterin, 485
$C_{19}H_{20}O_2$
Cyclofenil*, 2726
$C_{19}H_{20}O_9$
Cervicarcin, 2000
$C_{19}H_{20}O_{10}$
Khellol Glucoside, 5190
$C_{19}H_{21}ClF_3NO_2$
Benfluorex*, 1047
$C_{19}H_{21}ClN_2OS$
Chloracizine, 2055
Halethazole, 4505
$C_{19}H_{21}ClN_4O_5$
Theofibrate, 9211
$C_{19}H_{21}FN_2O_6$
Flupirtine*, 4117
$C_{19}H_{21}F_3N_2S$
Trifluomeprazine, 9593
$C_{19}H_{21}KO_6$
Gibberellic Acid*, 4313
$C_{19}H_{21}N$
Nortriptyline, 6635
Protriptyline, 7917
$C_{19}H_{21}NO$
Doxepin, 3425
$C_{19}H_{21}NO_3$
Isothebaine, 5116
Nalorphine, 6275
Thebaine, 9202
$C_{19}H_{21}NO_4$
Boldine, 1328
Corytuberine, 2549
Laurotetanine, 5258
Naloxone, 6277
Salutaridine, 8309
$C_{19}H_{21}NS$
Dothiepin, 3419
Pizotyline, 7486
$C_{19}H_{21}N_3$
Perlapine, 7131
$C_{19}H_{21}N_3O$
Talastine, 9009
Zolpidem, 10091
$C_{19}H_{21}N_3O_5$
Isradipine, 5129
$C_{19}H_{21}N_3S$
Cyamemazine, 2690
$C_{19}H_{21}N_5O_2$
Pirenzepine, 7463
$C_{19}H_{21}N_5O_4$
Prazosin, 7715
$C_{19}H_{22}BrNO_3$
Nalorphine*, 6275
$C_{19}H_{22}ClN$
Nortriptyline*, 6635
Protriptyline*, 7917
$C_{19}H_{22}ClNO$
Doxepin*, 3425
$C_{19}H_{22}ClNO_3$
Nalorphine*, 6275
$C_{19}H_{22}ClNO_4$
Naloxone*, 6277
$C_{19}H_{22}ClNS$
Dothiepin*, 3419
$C_{19}H_{22}ClN_3O$
Talastine*, 9009
$C_{19}H_{22}ClN_5O$
Trazodone, 9495
$C_{19}H_{22}ClN_5O_4$
Prazosin*, 7715
$C_{19}H_{22}Cl_2N_6O$
Imidocarb*, 4831
$C_{19}H_{22}FN_3O$
Azaperone, 915
$C_{19}H_{22}FN_3O_3$
Enrofloxacin, 3546
$C_{19}H_{22}KN_3O_4S$
Hetacillin*, 4592
$C_{19}H_{22}N_2$
Triprolidine, 9661

C₁₉H₂₂N₂O
$C_{19}H_{22}N_2O$
Cinchonidine, 2288
Cinchonine, 2289
Cinchotoxine, 2291
Eburnamonine, 3464
Noxiptilin, 6644

$C_{19}H_{22}N_2OS$
Acepromazine, 27
Thiazesim, 9233

$C_{19}H_{22}N_2O_2$
Apoquinine, 777
Cupreine, 2626
Sarpagine, 8336
Wieland-Gumlich Aldehyde, 9957

$C_{19}H_{22}N_2O_3$
Bumadizon, 1472

$C_{19}H_{22}N_2O_3S$
Dimethoxanate, 3211

$C_{19}H_{22}N_2O_6S$
Penamecillin, 7024

$C_{19}H_{22}N_2S$
Mepazine, 5734

$C_{19}H_{22}O_2$
Mestilbol, 5818

$C_{19}H_{22}O_3$
Doisynoestrol, 3407
Ostruthin, 6850

$C_{19}H_{22}O_4$
Menadiol Dibutyrate, 5711

$C_{19}H_{22}O_6$
Gibberellic Acid, 4313
Strigol, 8795

$C_{19}H_{22}O_7$
Samaderins*, 8312

$C_{19}H_{23}ClN_2$
Clomipramine, 2385
Homochlorcyclizine, 4653

$C_{19}H_{23}ClN_2 \cdot H_2O$
Triprolidine*, 9661

$C_{19}H_{23}ClN_2O$
Noxiptilin*, 6644

$C_{19}H_{23}ClN_2OS$
Thiazesim*, 9233

$C_{19}H_{23}ClN_2O_2S$
Pyridate, 7981

$C_{19}H_{23}ClN_2S$
Chlorproethazine, 2184

$C_{19}H_{23}ClO_2$
Clomestrone, 2380

$C_{19}H_{23}Cl_2N_3O$
Trazodone*, 9495

$C_{19}H_{23}Cl_3N_5O_2$
Pirenzepine*, 7463

$C_{19}H_{23}I_2NO_2$
Bufeniode, 1460

$C_{19}H_{23}NO$
Cinnamedrine, 2299
Diphenylpyraline, 3334

$C_{19}H_{23}NO_3$
Armepavine, 811
Biphenamine, 1247
Dihydrothebaine, 3164
Ethylmorphine, 3785
Oxyfedrine, 6915

$C_{19}H_{23}NO_4$
Cinnamoylcocaine, 2304
Reticuline, 8164
Sinomenine, 8496

$C_{19}H_{23}NO_5$
Metaphanine, 5834
Tretoquinol, 9502

$C_{19}H_{23}N_3$
Amitraz, 503
Binedaline, 1238

$C_{19}H_{23}N_3O$
Benzydamine, 1136

$C_{19}H_{23}N_3O_2$
Ergometrinine, 3599
Ergonovine, 3600

$C_{19}H_{23}N_3O_4S$
Hetacillin, 4592

$C_{19}H_{23}N_4O_6PS$
Benfotiamine, 1049

$C_{19}H_{23}N_5O_3$
Trimetrexate, 9635

$C_{19}H_{23}N_5O_3S$
Ipsapirone, 4967

$C_{19}H_{24}BrNO_3$
Codeine Methyl Bromide, 2460

$C_{19}H_{24}BrNS_2$
Tiquizium Bromide, 9390

$C_{19}H_{24}ClNO$
Diphenylpyraline*, 3334
Mecloxamine, 5663

$C_{19}H_{24}ClNO_3$
Biphenamine*, 1247

$C_{19}H_{24}ClNO_3 \cdot 2H_2O$
Ethylmorphine*, 3785

$C_{19}H_{24}ClNO_5$
Tretoquinol*, 9502

$C_{19}H_{24}ClN_3$
Binedaline*, 1238

$C_{19}H_{24}ClN_3O$
Benzydamine*, 1136

$C_{19}H_{24}ClN_5O_3S$
Ipsapirone*, 4967

$C_{19}H_{24}Cl_2N_2$
Clomipramine*, 2385
Homochlorcyclizine*, 4653

$C_{19}H_{24}Cl_2N_2S$
Chlorproethazine*, 2184

$C_{19}H_{24}F_6N_2O_5$
Flecainide*, 4034

$C_{19}H_{24}N_2$
Bamipine, 966
Histapyrrodine, 4641
Imipramine, 4835

$C_{19}H_{24}N_2O$
Aminopentamide, 471
Cinchonamine, 2287
Hydrocinchonidine, 4705
Hydrocinchonine, 4706
Imipramine N-Oxide, 4836

$C_{19}H_{24}N_2OS$
Methotrimeprazine, 5909
Phencarbamide, 7178

$C_{19}H_{24}N_2O_2$
Conquinamine, 2505
Praziquantel, 7714
Quinamine, 8060
Salverine, 8310

$C_{19}H_{24}N_2O_3$
Dilevalol, 3186
Labetalol, 5208

$C_{19}H_{24}N_2O_4$
Formoterol, 4159

$C_{19}H_{24}N_2O_6$
Ketorolac*, 5186

$C_{19}H_{24}N_2S$
Ethopropazine, 3703

$C_{19}H_{24}N_4O_2$
Pentamidine, 7071

$C_{19}H_{24}O_2$
Methyltrienolone, 6049

$C_{19}H_{24}O_3$
Adrenosterone, 165
Testolactone, 9108

$C_{19}H_{24}O_5$
Trichothecin, 9564

$C_{19}H_{24}O_7$
Samaderins*, 8312

$C_{19}H_{25}ClN_2$
Histapyrrodine*, 4641
Imipramine*, 4835

$C_{19}H_{25}ClN_2O$
Imipramine N-Oxide*, 4836

$C_{19}H_{25}ClN_2OS$
N-Hydroxyethylpromethazine Chloride, 4755
Phencarbamide*, 7178

$C_{19}H_{25}ClN_2O_3$
Labetalol*, 5208

$C_{19}H_{25}ClN_2S$
Ethopropazine*, 3703

$C_{19}H_{25}Cl_3N_2$
Homochlorcyclizine*, 4653

$C_{19}H_{25}F_3N_2O_6$
Fluvoxamine*, 4138

$C_{19}H_{25}NO$
Ethylbenzhydramine, 3724
Levallorphan, 5342

$C_{19}H_{25}NO_2$
Nylidrin, 6654

$C_{19}H_{25}NO_4$
Tetramethrin, 9154

$C_{19}H_{25}NO_{10}$
Vicianin, 9878

$C_{19}H_{25}N_3$
Picoperine, 7376

$C_{19}H_{25}N_3O_2S_2$
Fonazine, 4146

$C_{19}H_{25}N_3S$
Aminopromazine, 479

$C_{19}H_{25}N_5O_4$
Terazosin, 9084

$C_{19}H_{26}BrNO$
Bibenzonium Bromide, 1218

$C_{19}H_{26}BrNO_4$
Oxitropium Bromide, 6897

$C_{19}H_{26}ClNO$
Ethylbenzhydramine*, 3724

$C_{19}H_{26}ClNO_2$
Nylidrin*, 6654

$C_{19}H_{26}ClN_3$
Picoperine*, 7376

$C_{19}H_{26}ClN_5O_4 \cdot 2H_2O$
Terazosin*, 9084

$C_{19}H_{26}Cl_2N_2$
Bamipine*, 966

$C_{19}H_{26}I_3N_3O_9$
Iohexol, 4940

$C_{19}H_{26}N_2$
Quebrachamine, 8040

$C_{19}H_{26}N_2O$
Geissoschizoline, 4274

$C_{19}H_{26}N_2O_4S_2$
Thiazinamium Methylsulfate, 9234

$C_{19}H_{26}N_2S$
Pergolide, 7115

$C_{19}H_{26}N_6O_6$
Xanthinol Niacinate, 9969

$C_{19}H_{26}O_2$
4-Androstene-3,17-dione, 671
Boldenone, 1327

$C_{19}H_{26}O_3$
Allethrins*, 248
Epimestrol, 3568

$C_{19}H_{26}O_4S$
Propargite, 7818

$C_{19}H_{26}O_{12}$
Gaultherin, 4272

$C_{19}H_{27}ClO_2$
Clostebol, 2409

$C_{19}H_{27}NO$
Pentazocine, 7078

$C_{19}H_{27}NO_3$
Tetrabenazine, 9118

$C_{19}H_{27}NO_4$
Drotebanol, 3442

$C_{19}H_{27}N_5$
Dapiprazole, 2819

$C_{19}H_{27}N_5O_3$
Bunazosin, 1477

$C_{19}H_{27}N_5O_4$
Alfuzosin, 228

$C_{19}H_{27}NaO_5S$
Prasterone*, 7710

$C_{19}H_{27}NaO_8$
Artesunate*, 845

$C_{19}H_{28}BrNO_3$
Glycopyrrolate, 4397

$C_{19}H_{28}ClNO$
Pentazocine*, 7078

Denotes derivative of title compound

Denotes derivative of title compound

$C_{20}H_{16}O_4$
Phenolphthalin, 7210

$C_{20}H_{16}O_5$
Psoralidin, 7945

$C_{20}H_{16}O_6$
Anthralin*, 714
Elliptone, 3511
Viridin, 9912

$C_{20}H_{17}ClN_2O_3$
Ketazolam, 5176

$C_{20}H_{17}Cl_3N_2O_2$
Omoconazole, 6801

$C_{20}H_{17}Cl_3O_5$
Nidulin, 6440

$C_{20}H_{17}FO_3S$
Sulindac, 8961

$C_{20}H_{17}F_3N_2O_4$
Floctafenine, 4037

$C_{20}H_{17}NO_6$
Adlumidine, 154
Bicuculline, 1223

$C_{20}H_{17}N_3Na_2O_9S_3$
Acid Fuchsin, 103

$C_{20}H_{18}ClNO_6$
Ochratoxins*, 6660

$C_{20}H_{18}Cl_3N_3O_5$
Omoconazole*, 6801

$C_{20}H_{18}NO_4{}^+$
Berberine, 1169

$C_{20}H_{18}N_2O$
N-(1,2-Diphenylethyl)nicotinamide, 3324

$C_{20}H_{18}N_6Na_2O_9S$
Moxalactam*, 6201

$C_{20}H_{18}O_2Sn$
Triphenyltin Hydroxide*, 9659

$C_{20}H_{18}O_3$
Phenolphthalol, 7211

$C_{20}H_{18}O_4$
Cyclocumarol, 2723
Phaseolin, 7149

$C_{20}H_{18}O_6$
Asarinin, 848
Sesamin, 8420

$C_{20}H_{18}O_7$
Sesamolin, 8421

$C_{20}H_{18}O_9$
Cetraric Acid, 2016
Frangulin*, 4177

$C_{20}H_{19}NO_3$
Oxazidione, 6882

$C_{20}H_{19}NO_4$
Dizocilpine*, 3392

$C_{20}H_{19}NO_5$
Chelidonine, 2043
Papaveraldine, 6966
Protopine, 7912

$C_{20}H_{19}NO_6$
Ochratoxins*, 6660

$C_{20}H_{19}NO_8S$
Berberine*, 1169

$C_{20}H_{20}$
Dodecahedrane, 3399

$C_{20}H_{20}ClN_3$
Magenta I, 5528

$C_{20}H_{20}ClN_5O_3S_2$
Suriclone, 8987

$C_{20}H_{20}Cl_2MgO_6$
Clofibric Acid*, 2375

$C_{20}H_{20}NO_4{}^+$
Columbamine, 2488
Jatrorrhizine, 5142

$C_{20}H_{20}N_2O_2$
Feprazone, 3953

$C_{20}H_{20}N_2O_3$
Elliptinium Acetate, 3510

$C_{20}H_{20}N_2O_5$
Antipyrine*, 748

$C_{20}H_{20}N_2O_6$
Antipyrine*, 748

$C_{20}H_{20}N_6Na_2O_7S_4$
Cefodizime*, 1931

$C_{20}H_{20}N_6O_7S_4$
Cefodizime, 1931

$C_{20}H_{20}N_6O_9S$
Moxalactam, 6201

$C_{20}H_{20}N_8Na_2O_5$
Methotrexate*, 5908

$C_{20}H_{20}O_4$
Otobain, 6852

$C_{20}H_{20}O_6$
Cubebin, 2616

$C_{20}H_{20}O_7$
Herqueinone, 4589

$C_{20}H_{20}O_9$
Gardenins*, 4265

$C_{20}H_{20}O_{14}$
Hamamelitannin, 4522

$C_{20}H_{21}AlCl_2O_7$
Clofibric Acid*, 2375

$C_{20}H_{21}CaN_7O_7.5H_2O$
Folinic Acid*, 4141

$C_{20}H_{21}ClN_2O_4$
Fipexide, 4024

$C_{20}H_{21}ClO_4$
Fenofibrate, 3924

$C_{20}H_{21}FN_2O$
Citalopram, 2320

$C_{20}H_{21}F_3N_2OS$
Fluacizine, 4046

$C_{20}H_{21}N$
Cyclobenzaprine, 2719

$C_{20}H_{21}NOS$
Tolciclate, 9433

$C_{20}H_{21}NOS_2$
Tinofedrine, 9378

$C_{20}H_{21}NO_2$
Moxaverine, 6202

$C_{20}H_{21}NO_3$
Dimefline, 3190
Galipine, 4247
Mepixanox, 5749

$C_{20}H_{21}NO_4$
Canadine, 1744
Dicentrine, 3028
Papaverine, 6968

$C_{20}H_{21}NO_5$
Repirinast, 8140

$C_{20}H_{21}N_3O_3$
Moquizone, 6174

$C_{20}H_{21}N_7NaO_6S_2$
Ceforanide*, 1934

$C_{20}H_{21}N_7O_6$
Methopterin, 5907
Ninopterin, 6471

$C_{20}H_{21}N_7O_6S_2$
Ceforanide, 1934

$C_{20}H_{22}ClF_3N_2OS$
Fluacizine*, 4046

$C_{20}H_{22}ClN$
Cyclobenzaprine*, 2719
Pyrrobutamine, 8023

$C_{20}H_{22}ClNOS_2$
Tinofedrine*, 9378

$C_{20}H_{22}ClNO_2$
Moxaverine*, 6202

$C_{20}H_{22}ClNO_3$
Dimefline*, 3190

$C_{20}H_{22}ClNO_4$
Papaverine*, 6968

$C_{20}H_{22}ClN_3O$
Amodiaquin, 602

$C_{20}H_{22}ClN_3O_3$
Moquizone*, 6174

$C_{20}H_{22}ClN_5O_3S$
Adinazolam*, 150

$C_{20}H_{22}Cl_2N_2O_4$
Fipexide*, 4024

$C_{20}H_{22}N_2$
Azatadine, 917

$C_{20}H_{22}N_2O_2$
Akuammicine, 189
Gelsemine, 4277
Quininone, 8093

$C_{20}H_{22}N_2O_3$
Perivine, 7130

$C_{20}H_{22}N_2O_4$
Nomifensine*, 6592

$C_{20}H_{22}N_2S$
Mequitazine, 5754

$C_{20}H_{22}N_4O$
Difenamizole, 3123

$C_{20}H_{22}N_4O_5$
Camostat, 1734

$C_{20}H_{22}N_4O_6S$
Febantel, 3888

$C_{20}H_{22}N_4O_{10}S$
Cefuroxime*, 1951

$C_{20}H_{22}N_5NaO_6S$
Azlocillin*, 929

$C_{20}H_{22}N_8O_5$
A-Ninopterin, 691
Methotrexate, 5908

$C_{20}H_{22}O_2$
Norgestrienone, 6622

$C_{20}H_{22}O_3$
Butyl Methoxydibenzoylmethane, 1580

$C_{20}H_{22}O_6$
Columbin, 2489

$C_{20}H_{22}O_8$
Polydatin, 7542
Populin, 7571

$C_{20}H_{23}BrN_2OS$
Propyromazine, 7885

$C_{20}H_{23}BrN_2O_4$
Brompheniramine*, 1433

$C_{20}H_{23}ClN_2O_4$
Chlorpheniramine*, 2180

$C_{20}H_{23}ClN_2O_5$
Carbinoxamine*, 1804

$C_{20}H_{23}ClN_4O_2$
Clonitazene, 2389

$C_{20}H_{23}ClO_3$
Beclobrate, 1028

$C_{20}H_{23}N$
Amitriptyline, 504
Maprotiline, 5633

$C_{20}H_{23}NO$
Amitriptylinoxide, 505
Oxaprotiline, 6878

$C_{20}H_{23}NO_2$
Amolanone, 603
Dioxadrol, 3293

$C_{20}H_{23}NO_3$
Benalaxyl, 1038

$C_{20}H_{23}NO_4$
Corydine, 2545
Corypalmine, 2548
Dihydrocodeinone Enol Acetate, 3155
Isocorydine, 5046
Isocorypalmine, 5047
Naltrexone, 6278

$C_{20}H_{23}NO_5$
Cyproquinate, 2780

$C_{20}H_{23}NO_6$
Medifoxamine*, 5674

$C_{20}H_{23}NO_6S$
Teniloxazine*, 9077

$C_{20}H_{23}NS$
Methixene, 5902

$C_{20}H_{23}N_2O{}^+$
C-Curarine III, 2680

$C_{20}H_{23}N_3O_2$
Bitertanol, 1315

$C_{20}H_{23}N_3O_4$
Aminopyrine*, 488
Epanolol, 3557

$C_{20}H_{23}N_5O_6S$
Azlocillin, 929

Denotes derivative of title compound

$C_{20}H_{23}N_7O_7$
Folinic Acid, 4141
$C_{20}H_{24}BrN_3O$
Bromolysergide, 1411
$C_{20}H_{24}ClN$
Amitriptyline*, 504
Maprotiline*, 5633
$C_{20}H_{24}ClNO$
Cloperastine, 2393
Oxaprotiline*, 6878
$C_{20}H_{24}ClNO_2$
Dioxadrol*, 3293
Metofoline, 6066
$C_{20}H_{24}ClNO_4$
Dihydrocodeinone Enol Acetate*, 3155
Naltrexone*, 6278
$C_{20}H_{24}ClNS$
Methixene*, 5902
$C_{20}H_{24}ClN_3O_2$
Clebopride, 2344
$C_{20}H_{24}ClN_3S$
Prochlorperazine, 7768
$C_{20}H_{24}Cl_2N_{10}$
Picloxydine, 7371
$C_{20}H_{24}Cl_3N_3O.2H_2O$
Amodiaquin*, 602
$C_{20}H_{24}I_2O_2$
Thymol Iodide, 9335
$C_{20}H_{24}NO_4{}^+$
Magnoflorine, 5579
$C_{20}H_{24}N_2$
Dimethindene, 3204
$C_{20}H_{24}N_2O$
Indecainide, 4849
$C_{20}H_{24}N_2OS$
Propiomazine, 7834
Propionylpromazine, 7841
$C_{20}H_{24}N_2O_2$
Epiquinidine, 3570
Quinidine, 8072
Quinine, 8075
Viquidil, 9908
$C_{20}H_{24}N_2O_2S$
Hycanthone, 4677
$C_{20}H_{24}N_2O_3$
Lochneridine, 5435
$C_{20}H_{24}N_2O_4$
Metoquinone, 6073
Pheniramine*, 7198
$C_{20}H_{24}N_2O_5$
Diamfenetide, 2953
$C_{20}H_{24}N_2O_6$
Nisoldipine, 6482
$C_{20}H_{24}O_2$
Dimestrol, 3198
Ethinyl Estradiol, 3689
$C_{20}H_{24}O_3$
Estrone*, 3660
Trenbolone*, 9499
$C_{20}H_{24}O_4$
Crocetin, 2592
$C_{20}H_{24}O_6$
Crown Ethers*, 2605
$C_{20}H_{24}O_7$
Olivil, 6797
$C_{20}H_{24}O_9$
Nodakenin, 6589
$C_{20}H_{25}BrN_2O_2$
Quinine Hydrobromide, 8084
$C_{20}H_{25}ClN_2O$
Indecainide*, 4849
$C_{20}H_{25}ClN_2OS$
Lucanthone Hydrochloride, 5463
Propiomazine*, 7834
Propionylpromazine*, 7841
$C_{20}H_{25}ClN_2O_2$
Quinine Hydrochloride, 8085
$C_{20}H_{25}ClN_2O_5$
Amlodipine, 509

$C_{20}H_{25}ClN_2O_7$
Carbinoxamine*, 1804
$C_{20}H_{25}Cl_2NO$
Cloperastine*, 2393
$C_{20}H_{25}FN_4O$
Niaprazine, 6400
$C_{20}H_{25}IN_2O_2$
Quinine Hydriodide, 8083
$C_{20}H_{25}N$
Fenpiprane, 3934
Prodipine, 7776
$C_{20}H_{25}NO$
Diphemethoxidine, 3303
α,α-Diphenyl-2-piperidinepropanol, 3333
Normethadone, 6628
Pridinol, 7747
$C_{20}H_{25}NOS$
Thiphenamil, 9303
$C_{20}H_{25}NO_2$
Cuspareine, 2686
Femoxetine, 3897
Fomocaine, 4145
Propanocaine, 7815
$C_{20}H_{25}NO_3$
Benactyzine, 1037
Difemerine, 3122
Dimenoxadol, 3194
$C_{20}H_{25}NO_4$
Codamine, 2458
Laudanidine, 5247
Laudanine, 5248
$C_{20}H_{25}N_2O^+$
Macusines*, 5521
$C_{20}H_{25}N_3O$
Lysergide, 5507
Prodigiosin, 7774
$C_{20}H_{25}N_3O_2$
Methylergonovine, 5989
$C_{20}H_{25}N_3S$
Perazine, 7108
$C_{20}H_{26}Br_2N_2O_2$
Quinine Dihydrobromide, 8078
$C_{20}H_{26}ClN$
Prodipine*, 7776
$C_{20}H_{26}ClNO$
Normethadone*, 6628
Pridinol*, 7747
$C_{20}H_{26}ClNOS$
Thiphenamil*, 9303
$C_{20}H_{26}ClNO_2$
Adiphenine Hydrochloride, 151
Femoxetine*, 3897
$C_{20}H_{26}ClNO_3$
Benactyzine*, 1037
Difemerine*, 3122
Lachesine, 5210
$C_{20}H_{26}ClN_3O_7S$
Chlorothen*, 2168
$C_{20}H_{26}Cl_2N_2O_2$
Quinine Dihydrochloride, 8079
$C_{20}H_{26}Cl_4N_{10}$
Picloxydine*, 7371
$C_{20}H_{26}INO_3$
Ethylmorphine*, 3785
$C_{20}H_{26}N_2$
Dimetacrine, 3199
Trimipramine, 9636
$C_{20}H_{26}N_2O$
Ibogaine, 4806
Tabernanthine, 9000
$C_{20}H_{26}N_2O_2$
Ajmaline, 184
Hydroquinidine, 4736
Hydroquinine, 4737
$C_{20}H_{26}N_2O_5S$
Alacepril, 192
$C_{20}H_{26}N_2O_6S.4H_2O$
Quinidine*, 8072
$C_{20}H_{26}N_2O_6S.7H_2O$
Quinine Bisulfate*, 8076

$C_{20}H_{26}N_2O_6S$
Quinine Bisulfate, 8076
$C_{20}H_{26}N_2S$
Etymemazine, 3849
$C_{20}H_{26}N_4O$
Lisuride, 5394
$C_{20}H_{26}N_4OS$
Oxypendyl, 6923
$C_{20}H_{26}N_4O_2$
Hexamidine, 4612
$C_{20}H_{26}O$
3-Dehydroretinal, 2862
$C_{20}H_{26}O_2$
Benzestrol, 1082
Methestrol, 5888
Norethindrone, 6614
Norethynodrel, 6615
$C_{20}H_{26}O_3$
Taxodione, 9048
$C_{20}H_{26}O_5$
Allethrins*, 248
$C_{20}H_{26}O_7$
Cnicin, 2419
$C_{20}H_{27}BrN_2O$
Ambutonium Bromide, 397
$C_{20}H_{27}ClN_2O_2$
Hydroquinidine*, 4736
$C_{20}H_{27}ClN_2S$
Etymemazine*, 3849
$C_{20}H_{27}Cl_2N_3O_9$
Chloramphenicol Pantothenate, 2070
$C_{20}H_{27}N$
Alverine, 377
Terodiline, 9098
$C_{20}H_{27}NO$
Cyclorphan, 2757
$C_{20}H_{27}NO_2$
Fenalcomine, 3900
Vetrabutine, 9875
$C_{20}H_{27}NO_3$
Tralkoxydim, 9484
Trilostane, 9611
$C_{20}H_{27}NO_4$
Bevantolol, 1213
$C_{20}H_{27}NO_4S$
Sulfinalol, 8925
$C_{20}H_{27}NO_5$
Buquinolate, 1489
Chromonar, 2241
$C_{20}H_{27}NO_{11}$
Amygdalin, 629
$C_{20}H_{27}N_3O_5$
Cilazapril*, 2276
$C_{20}H_{27}N_3O_6$
Febarbamate, 3889
$C_{20}H_{27}N_3O_7S$
Succisulfone*, 8851
$C_{20}H_{27}N_3O_8$
Oxolamine*, 6898
$C_{20}H_{27}N_5O_2$
Cilostazol, 2277
$C_{20}H_{27}N_5O_3$
Bamifylline, 965
$C_{20}H_{27}N_5O_5S$
Glisoxepid, 4339
$C_{20}H_{27}N_7O_6$
Diminazene Aceturate*, 3256
$C_{20}H_{28}BrN$
Emepronium Bromide, 3515
$C_{20}H_{28}BrNO_3$
Endobenzyline Bromide, 3527
$C_{20}H_{28}ClN$
Terodiline*, 9098
$C_{20}H_{28}ClNO_2$
Fenalcomine*, 3900
Vetrabutine*, 9875
$C_{20}H_{28}ClNO_4$
Bevantolol*, 1213
$C_{20}H_{28}ClNO_4S$
Sulfinalol*, 8925

Denotes derivative of title compound

$C_{20}H_{28}ClNO_5$
Chromonar*, 2241
$C_{20}H_{28}ClN_5O_3$
Bamifylline*, 965
$C_{20}H_{28}Cl_2N_4OS$
Oxypendyl*, 6923
$C_{20}H_{28}Cl_4N_2O_4$
Teclozan, 9055
$C_{20}H_{28}INO$
Ethylbenzhydramine*, 3724
$C_{20}H_{28}I_3N_3O_9$
Iopentol, 4945
$C_{20}H_{28}N_2O_3$
Oxyphencyclimine, 6926
$C_{20}H_{28}N_2O_4$
Naproxen*, 6337
$C_{20}H_{28}N_2O_5$
Enalapril, 3521
$C_{20}H_{28}N_4O$
Terguride, 9095
$C_{20}H_{28}N_4O_8S_2$
Stilbamidine*, 8772
$C_{20}H_{28}N_6O_2$
Fencamine, 3910
$C_{20}H_{28}N_6O_4S$
Debrisoquin*, 2837
$C_{20}H_{28}N_6O_8S$
Guanoxan*, 4482
$C_{20}H_{28}O$
Lynestrenol, 5501
Retinal, 8165
Vitamin A_2, 9919
$C_{20}H_{28}O_2$
Ethynodiol, 3816
Methandrostenolone, 5862
Norgesterone, 6619
Norvinisterone, 6637
Retinoic Acid, 8167
$C_{20}H_{28}O_3$
Cafestol, 1633
Cinerins*, 2294
Mytatrienediol, 6254
$C_{20}H_{28}O_3S$
Ethyl Dibunate, 3745
$C_{20}H_{28}O_4$
Marrubiin, 5635
$C_{20}H_{28}O_5$
Pododacric Acid, 7517
$C_{20}H_{28}O_6$
Amarolide, 385
Phorbol, 7306
$C_{20}H_{28}O_6S$
Tiaprost, 9355
$C_{20}H_{28}O_{13}$
Primulaverin, 7755
$C_{20}H_{29}ClN_6O_2$
Fencamine*, 3910
$C_{20}H_{29}FO_3$
Fluoxymesterone, 4113
$C_{20}H_{29}NO_3$
Propenzolate, 7824
$C_{20}H_{29}NO_5S$
Latrunculins*, 5246
$C_{20}H_{29}NO_6S$
Sultroponium, 8977
$C_{20}H_{29}N_3O_5$
Propiram*, 7844
$C_{20}H_{29}N_3O_5S_3$
Fonazine*, 4146
$C_{20}H_{29}N_3O_7$
Amicoumacin A, 408
$C_{20}H_{29}N_5O_3$
Urapidil, 9779
$C_{20}H_{29}N_5O_6$
Trimazosin, 9612
$C_{20}H_{30}BrNO_3$
Ipratropium Bromide, 4960
$C_{20}H_{30}ClNO_3$
Propenzolate*, 7824
$C_{20}H_{30}ClN_3O_2$
Dibucaine Hydrochloride, 3016

$C_{20}H_{30}ClN_5O_6 \cdot H_2O$
Trimazosin*, 9612
$C_{20}H_{30}N_2O_2$
Dipiproverine, 3343
Furazabol, 4209
$C_{20}H_{30}N_2O_3$
Morpheridine, 6184
$C_{20}H_{30}N_2O_3S_2$
Pergolide*, 7115
$C_{20}H_{30}N_2O_5S$
Benfuracarb, 1050
$C_{20}H_{30}O$
Neovitamin A, 6382
Vitamin A, 9918
$C_{20}H_{30}O_2$
Abietic Acid, 2
5,8,11,14,17-Eicosapentaenoic Acid, 3495
Hepaxanthin, 4572
Isopimaric Acid, 5086
Levopimaric Acid, 5348
Methenolone, 5887
17-Methyltestosterone, 6044
Mibolerone, 6098
Norethandrolone, 6613
Oxendolone, 6886
Oxenin, 6887
Palustric Acid, 6950
Pimaric Acid, 7398
Stenbolone, 8763
$C_{20}H_{30}O_3$
Leukotrienes*, 5339
Oxymesterone, 6918
Steviol, 8765
$C_{20}H_{30}O_5$
Andrographolide, 666
Taxicins*, 9046
$C_{20}H_{30}O_6$
Taxicins*, 9046
$C_{20}H_{31}NO_2$
Drofenine, 3435
Metcaraphen, 5840
$C_{20}H_{31}NO_2S$
Cetiedil, 2012
$C_{20}H_{31}NO_3$
Carbetapentane, 1798
$C_{20}H_{31}NO_6S$
Tetrabenazine*, 9118
$C_{20}H_{31}NO_7$
Heliosupine, 4545
$C_{20}H_{31}NaO_5$
Prostacyclin*, 7890
$C_{20}H_{32}$
Neocembrene, 6362
$C_{20}H_{32}ClNO$
Procyclidine*, 7770
Trihexyphenidyl Hydrochloride, 9607
$C_{20}H_{32}ClNO_2$
Drofenine*, 3435
$C_{20}H_{32}Cl_2N_2O_2$
Dipiproverine*, 3343
$C_{20}H_{32}Cl_2N_2O_3$
Morpheridine*, 6184
$C_{20}H_{32}N_2O_6S$
Ephedrine*, 3561
$C_{20}H_{32}N_6O_4S$
Bethanidine*, 1208
$C_{20}H_{32}N_6O_{12}S_2$
Glutathione*, 4369
$C_{20}H_{32}O$
Ethylestrenol, 3761
$C_{20}H_{32}O_2$
Arachidonic Acid, 792
Etiocholanic Acid, 3824
Mestanolone, 5816
Mesterolone, 5817
Methandriol, 5861
$C_{20}H_{32}O_3$
Grindelic Acid, 4451
5-HETE, 4594

$C_{20}H_{32}O_4$
Leukotrienes*, 5339
$C_{20}H_{32}O_5$
Grayanotoxins*, 4449
Prostacyclin, 7890
Prostaglandin E_2, 7893
Thromboxanes*, 9323
$C_{20}H_{33}NO$
Fenpropimorph, 3938
$C_{20}H_{33}NO_3$
Ganglefene, 4263
Oxeladin, 6885
$C_{20}H_{33}NO_6S$
Pentapiperide*, 7077
$C_{20}H_{33}N_3O_3$
Talinolol, 9012
$C_{20}H_{33}N_3O_4$
Celiprolol, 1956
$C_{20}H_{34}AuO_9PS$
Auranofin, 895
$C_{20}H_{34}ClNO_3$
Ganglefene*, 4263
$C_{20}H_{34}ClN_3O_4$
Celiprolol*, 1956
$C_{20}H_{34}O_2$
Plaunotol, 7506
$C_{20}H_{34}O_4$
Aphidicolin, 759
Zoapatanol, 10088
$C_{20}H_{34}O_5$
Prostaglandin E_1, 7892
Prostaglandin $F_{2\alpha}$, 7894
$C_{20}H_{34}O_6$
Grayanotoxins*, 4449
Thromboxanes*, 9323
$C_{20}H_{35}ClO_4S$
Oxychlorosene, 6909
$C_{20}H_{35}NOS$
Suloctidil, 8967
$C_{20}H_{35}NO_2$
Dihexyverine, 3151
$C_{20}H_{35}NO_7$
Ibuprofen*, 4812
$C_{20}H_{35}NO_{13}$
Validamycins*, 9817
$C_{20}H_{36}CaNO_5 \cdot \frac{1}{2}H_2O$
Hopantenic Acid*, 4664
$C_{20}H_{36}ClNO_2$
Dihexyverine*, 3151
$C_{20}H_{36}N_6O$
Lauroguadine, 5256
$C_{20}H_{36}O_2$
Ethyl Linoleate, 3776
$C_{20}H_{36}O_6$
Crown Ethers*, 2605
$C_{20}H_{37}KO_7S$
Docusate Sodium*, 3398
$C_{20}H_{37}NO_3$
Rociverine, 8232
$C_{20}H_{37}N_3O_{13}$
Destomycin A, 2915
Hygromycin B, 4790
$C_{20}H_{37}NaO_7S$
Docusate Sodium, 3398
$C_{20}H_{38}BrNO_2$
Diponium Bromide, 3346
$C_{20}H_{38}Cl_2N_6O \cdot H_2O$
Lauroguadine*, 5256
$C_{20}H_{38}N_4O_4$
Dimorpholamine, 3257
$C_{20}H_{38}O_2$
Oleic Acid*, 6788
$C_{20}H_{39}NO_3$
Dodemorph*, 3405
$C_{20}H_{40}N_2O_{12}$
Hexamethonium*, 4609
$C_{20}H_{40}N_6O_8S$
Guanadrel*, 4471
$C_{20}H_{40}O$
Isophytol, 5084
Phytol, 7362

Denotes derivative of title compound

C₂₁H₂₄O₇
Visnadine, 9915
C₂₁H₂₄O₉
Rhapontin, 8173
C₂₁H₂₄O₁₀
Phloridzin, 7300
C₂₁H₂₅BrN₂O₃
Brovincamine, 1440
C₂₁H₂₅ClN₂O₃
Cetirizine, 2013
C₂₁H₂₅ClN₂O₄S
Tianeptine, 9352
C₂₁H₂₅ClO₂
Trengestone, 9500
C₂₁H₂₅ClO₅
Chloroprednisone, 2157
Cloprednol, 2396
C₂₁H₂₅Cl₂N₃O₃
Fominoben*, 4144
C₂₁H₂₅Cl₃FN₃O
Flurazepam*, 4123
C₂₁H₂₅FN₂O₂
Fluanisone, 4047
C₂₁H₂₅N
Melitracen, 5704
Terbinafine, 9086
C₂₁H₂₅NO₂
Piperidolate, 7440
C₂₁H₂₅NO₃
Nalmefene, 6274
Piperilate, 7441
C₂₁H₂₅NO₄
Corybulbine, 2539
Glaucine, 4332
Isocorybulbine, 5045
Tetrahydropalmatine, 9147
C₂₁H₂₅NO₅
Demecolcine, 2873
Diacetyldihydromorphine, 2947
C₂₁H₂₅NO₇S₂.H₂O
Tipepidine*, 9389
C₂₁H₂₅N₂O₆P
Strychnine Phosphate, 8825
C₂₁H₂₅N₃O₃S
Pipazethate, 7423
C₂₁H₂₅N₅O₈S₂
Mezlocillin, 6096
Nafamostat*, 6264
C₂₁H₂₅O₄N
Protostephanine, 7914
C₂₁H₂₆BrNO₃
Mepenzolate Bromide, 5735
Methantheline Bromide, 5869
C₂₁H₂₆ClN
Melitracen*, 5704
C₂₁H₂₆ClNO
Clemastine, 2345
C₂₁H₂₆ClNO₂
Piperidolate*, 7440
C₂₁H₂₆ClNO₃
Piperilate*, 7441
C₂₁H₂₆ClN₃OS
Perphenazine, 7135
C₂₁H₂₆ClN₃O₃S
Pipazethate*, 7423
C₂₁H₂₆Cl₂F₃N₃S
Trifluoperazine*, 9594
C₂₁H₂₆Cl₂N₈
Bismark Brown R, 1264
C₂₁H₂₆Cl₂O
Clofoctol, 2377
C₂₁H₂₆Cl₂O₄
Dichlorisone, 3035
C₂₁H₂₆N₂OS₂
Mesoridazine, 5813
C₂₁H₂₆N₂O₂S₂
Sulforidazine, 8943

C₂₁H₂₆N₂O₃
Corynanthine, 2547
Pseudoyohimbine, 7938
Vincamine, 9888
Yohimbine, 10011
allo-Yohimbine, 10012
α-Yohimbine, 10013
C₂₁H₂₆N₂O₄
Quinine Formate, 8081
C₂₁H₂₆N₂O₇
Nimodipine, 6467
C₂₁H₂₆N₂S₂
Thioridazine, 9290
C₂₁H₂₆N₄O₇S₂
Cefatrizine*, 1923
C₂₁H₂₆N₄O₈S
Camostat*, 1734
C₂₁H₂₆O₂
Altrenogest, 317
Cannabinol, 1751
Gestodene, 4308
Mestranol, 5819
C₂₁H₂₆O₃
Acitretin, 107
Buparvaquone, 1481
Moxestrol, 6203
Octabenzone, 6662
C₂₁H₂₆O₅
Prednisone, 7727
C₂₁H₂₆O₁₂
Plumieride, 7513
C₂₁H₂₆O₁₃
Leucoglycodrin, 5336
C₂₁H₂₇ClN₂O
Pyronine B, 8016
C₂₁H₂₇ClN₂O₂
Hydroxyzine, 4786
C₂₁H₂₇ClN₂O₃
Vincamine*, 9888
Yohimbine*, 10011
C₂₁H₂₇ClN₂S₂
Thioridazine*, 9290
C₂₁H₂₇Cl₃N₂O₃
Cetirizine*, 2013
C₂₁H₂₇FO₅
Fluprednisolone, 4119
Isoflupredone, 5058
C₂₁H₂₇FO₆
Triamcinolone, 9511
C₂₁H₂₇N
Budipine, 1456
Butriptyline, 1533
Pramiverin, 7706
Prozapine, 7922
C₂₁H₂₇NO
Benproperine, 1058
Diphenidol, 3311
Isomethadone, 5069
C₂₁H₂₇NO₂
Etafenone, 3662
Ifenprodil, 4821
C₂₁H₂₇NO₃
Benapryzine, 1039
Propafenone, 7806
C₂₁H₂₇NO₄
Laudanosine, 5249
Nalbuphine, 6271
C₂₁H₂₇NO₄S
Diphemanil Methylsulfate, 3302
C₂₁H₂₇NO₅
Hasubanonine, 4533
C₂₁H₂₇NO₁₀
Pseudococaine*, 7926
C₂₁H₂₇N₃O₂
Methysergid(e), 6055
C₂₁H₂₇N₃O₅
Pyrilamine*, 7996
C₂₁H₂₇N₃O₆
Casimiroedine, 1893
C₂₁H₂₇N₃O₇S
Bacampicillin, 944

C₂₁H₂₇N₅O₄S
Glipizide, 4337
C₂₁H₂₇N₅O₇S
Aspoxicillin, 874
C₂₁H₂₇N₅O₉S₂
Cefpodoxime Proxetil, 1941
C₂₁H₂₇N₇O₁₄P₂
Nadide, 6259
C₂₁H₂₇Na₂O₈P
Prednisolone Sodium Phosphate, 7721
C₂₁H₂₈BrFO₂
Haloprogesterone, 4513
C₂₁H₂₈BrNO₂
Adiphenine Hydrochloride*, 151
C₂₁H₂₈BrNO₃
Benactyzine*, 1037
C₂₁H₂₈BrNO₄
Cimetropium Bromide, 2280
C₂₁H₂₈ClN
Budipine*, 1456
Butriptyline*, 1533
Pramiverin*, 7706
Prozapine*, 7922
C₂₁H₂₈ClNO
Diphenidol*, 3311
Methadone Hydrochloride, 5852
C₂₁H₂₈ClNO₂
Etafenone*, 3662
C₂₁H₂₈ClNO₃
Benapryzine*, 1039
Propafenone*, 7806
C₂₁H₂₈ClNO₄
Nalbuphine*, 6271
C₂₁H₂₈ClN₃O₇S
Bacampicillin*, 944
C₂₁H₂₈Cl₃N₃O₂
Acranil, 116
C₂₁H₂₈F₃N₃O₇
Flunixin*, 4073
C₂₁H₂₈N₂O
Diampromide, 2966
C₂₁H₂₈N₂O₂
Aspidospermine*, 872
Ethylhydrocupreine, 3765
C₂₁H₂₈N₂O₅
Doxylamine*, 3430
Trimethobenzamide, 9623
C₂₁H₂₈N₂S₂
Buthiobate, 1520
C₂₁H₂₈N₇O₁₇P₃
NADP, 6262
C₂₁H₂₈O₂
Demegestone, 2874
Dydrogesterone, 3454
Ethisterone, 3696
Norgestrel, 6621
Tibolone, 9358
C₂₁H₂₈O₃
Cyclethrin, 2711
Estradiol*, 3653
11-Ketoprogesterone, 5185
Pyrethrins*, 7978
C₂₁H₂₈O₄
11-Dehydrocorticosterone, 2859
Formyldienolone, 4161
C₂₁H₂₈O₅
Aldosterone, 218
Cinerins*, 2294
Cortisone, 2533
Prednisolone, 7719
C₂₁H₂₈O₈
Scabiolide, 8346
C₂₁H₂₉ClN₂O₂
Ethylhydrocupreine*, 3765
C₂₁H₂₉ClN₂O₅
Trimethobenzamide*, 9623
C₂₁H₂₉ClO₃
Clostebol*, 2409
C₂₁H₂₉ClO₆S
Luprostiol, 5480

* *Denotes derivative of title compound*

$C_{22}H_{14}$
1,2:5,6-Dibenzanthracene, 2989
2,3:6,7-Dibenzphenanthrene, 2992
Pentacene, 7057
Picene, 7368
$C_{22}H_{14}Cl_2I_2N_2O_2$
Closantel, 2407
$C_{22}H_{16}Cl_2O_4S$
Tioclomarol, 9383
$C_{22}H_{16}N_4Na_2O_9S_2$
Orange B, 6814
$C_{22}H_{16}N_4O$
Sudan III, 8858
$C_{22}H_{16}O_6$
Resistomycin, 8155
$C_{22}H_{16}O_8$
Ethyl Biscoumacetate, 3729
$C_{22}H_{17}ClFN_3O_4$
Midazolam*, 6105
$C_{22}H_{17}ClN_2$
Clotrimazole, 2412
$C_{22}H_{18}Cl_2FNO_3$
Cyfluthrin, 2764
$C_{22}H_{18}N_2$
Bifonazole, 1232
$C_{22}H_{18}O_4$
o-Cresolphthalein, 2582
$C_{22}H_{18}O_7$
Justicidins*, 5154
$C_{22}H_{19}Br$
Broparoestrol, 1438
$C_{22}H_{19}Br_2NO_3$
Deltamethrin, 2869
$C_{22}H_{19}Cl_2NO_3$
Cypermethrin, 2775
$C_{22}H_{19}N$
Bis(1-naphthylmethyl)amine, 1307
$C_{22}H_{19}NO_4$
Bisacodyl, 1253
$C_{22}H_{19}N_4NaO_8S_2$
Cefsulodin*, 1943
$C_{22}H_{20}N_2O_2$
Piketoprofen, 7393
$C_{22}H_{20}N_4O_8S_2$
Cefsulodin, 1943
$C_{22}H_{20}O_{10}$
Granaticin, 4441
$C_{22}H_{20}O_{13}$
Carminic Acid, 1850
$C_{22}H_{21}ClN_2O_2$
Piketoprofen*, 7393
$C_{22}H_{21}ClN_2O_8$
Meclocycline, 5658
$C_{22}H_{21}KN_2O_5S$
Fenbenicillin*, 3905
$C_{22}H_{21}NO_2$
Benzylimidobis(p-methoxyphenyl)-
methane, 1153
$C_{22}H_{22}ClNO_6$
Ochratoxins*, 6660
$C_{22}H_{22}FN_3O_2$
Droperidol, 3437
$C_{22}H_{22}FN_3O_3$
Ketanserin, 5175
$C_{22}H_{22}N_2O_5S$
Fenbenicillin, 3905
$C_{22}H_{22}N_2O_8$
Methacycline, 5851
$C_{22}H_{22}N_6O_7S_2$
Ceftazidime, 1944
$C_{22}H_{22}N_6O_7S_2.5H_2O$
Ceftazidime*, 1944
$C_{22}H_{22}N_8$
Bisantrene, 1255
$C_{22}H_{22}O_4$
Dienestrol*, 3094
$C_{22}H_{22}O_5$
Cyclovalone, 2761

$C_{22}H_{22}O_8$
β-Peltatin, 7020
Picropodophyllin, 7385
Podophyllotoxin, 7520
$C_{22}H_{22}O_9$
Formononetin*, 4157
$C_{22}H_{22}O_{10}$
Prunetin*, 7923
$C_{22}H_{22}O_{11}$
Scoparin, 8357
Tectorigenin*, 9057
$C_{22}H_{23}ClN_2O_2$
Loratadine, 5455
$C_{22}H_{23}ClN_2O_8$
Chlortetracycline, 2193
Methacycline*, 5851
$C_{22}H_{23}F_2NO_2$
Lenperone, 5320
$C_{22}H_{23}F_4NO_2$
Trifluperidol, 9596
$C_{22}H_{23}NO_3$
Fenpropathrin, 3936
$C_{22}H_{23}NO_4$
Nequinate, 6385
$C_{22}H_{23}NO_6$
Aureothin, 897
$C_{22}H_{23}NO_7$
Noscapine, 6638
$C_{22}H_{23}N_3O_6S_2$
Amsacrine*, 627
$C_{22}H_{23}N_3O_9$
Aluminon, 320
$C_{22}H_{24}ClF_2NO_2$
Lenperone*, 5320
$C_{22}H_{24}ClF_4NO_2$
Trifluperidol*, 9596
$C_{22}H_{24}ClN_3O$
Azelastine, 922
$C_{22}H_{24}ClN_3OS_2$
Clospirazine, 2408
$C_{22}H_{24}ClN_3O_5$
Loxapine*, 5461
$C_{22}H_{24}ClN_5O_2$
Domperidone, 3412
$C_{22}H_{24}Cl_2N_2O_8$
Chlortetracycline*, 2193
$C_{22}H_{24}Cl_2N_8$
Bisantrene*, 1255
$C_{22}H_{24}FN_3OS$
Timiperone, 9373
$C_{22}H_{24}FN_3O_2$
Benperidol, 1057
$C_{22}H_{24}N_2O_2$
Acrivastine, 121
$C_{22}H_{24}N_2O_3$
Colubrines, 2487
$C_{22}H_{24}N_2O_4$
Alstonidine, 313
Cifenline*, 2274
Kopsine, 5199
Vomicine, 9946
$C_{22}H_{24}N_2O_8$
Quatrimycin, 8038
Tetracycline, 9130
$C_{22}H_{24}N_2O_8.H_2O$
Doxycycline, 3429
$C_{22}H_{24}N_2O_9$
Oxytetracycline, 6931
$C_{22}H_{24}N_4O_7S_3$
Thiamine 1,5-Salt, 9226
$C_{22}H_{24}O_8$
Lophotoxin, 5452
$C_{22}H_{24}O_9$
Podophyllic Acids, 7518
$C_{22}H_{24}O_{10}$
Sakuranetin*, 8289
$C_{22}H_{25}ClN_2OS$
Clopenthixol, 2392
$C_{22}H_{25}ClN_2O_2$
Tetracycline*, 9130

$C_{22}H_{25}ClN_2O_8$
Doxycycline*, 3429
$C_{22}H_{25}Cl_2N_3O$
Azelastine*, 922
$C_{22}H_{25}N$
Piroheptine, 7473
$C_{22}H_{25}NO_2$
Lobelanine, 5430
Tropacine, 9685
$C_{22}H_{25}NO_3$
Tropine Benzylate, 9694
$C_{22}H_{25}NO_4$
Dimoxyline, 3258
$C_{22}H_{25}NO_5$
Corycavidine, 2541
$C_{22}H_{25}NO_5S$
Thiocolchicine, 9253
$C_{22}H_{25}NO_6$
Colchicine, 2470
$C_{22}H_{25}N_3O$
Benzpiperylon, 1131
Indoramin, 4876
$C_{22}H_{25}N_3O_4S$
Moricizine, 6178
$C_{22}H_{26}BrNO_3$
Clidinium Bromide, 2351
$C_{22}H_{26}ClFO_4$
Clobetasone, 2362
$C_{22}H_{26}ClN$
Piroheptine*, 7473
$C_{22}H_{26}ClNO_3$
Pipoxolan Hydrochloride, 7454
$C_{22}H_{26}ClN_3O$
Indoramin*, 4876
$C_{22}H_{26}ClN_3O_4S$
Moricizine*, 6178
$C_{22}H_{26}FNO_2$
Moperone, 6170
$C_{22}H_{26}F_3N_3OS$
Fluphenazine, 4116
$C_{22}H_{26}N_2O_2$
Vinpocetine, 9894
$C_{22}H_{26}N_2O_3$
Corynantheine, 2546
$C_{22}H_{26}N_2O_4$
Akuammine, 190
Aricine, 809
Lochnericine*, 5434
Tofisopam, 9427
$C_{22}H_{26}N_2O_4S$
Diltiazem, 3188
$C_{22}H_{26}N_2O_5S$
Methoxypromazine*, 5923
$C_{22}H_{26}N_4$
Calycanthine, 1731
Chimonanthine, 2048
$C_{22}H_{26}N_4O_2S$
Thiazolinobutazone, 9236
$C_{22}H_{26}O_3$
Bioresmethrin, 1243
$C_{22}H_{26}O_5$
Colpormon, 2485
$C_{22}H_{26}O_{12}$
Catalposide, 1907
$C_{22}H_{27}ClFNO_2$
Moperone*, 6170
$C_{22}H_{27}ClF_2O_5$
Halometasone, 4510
$C_{22}H_{27}ClF_2O_5.H_2O$
Halometasone*, 4510
$C_{22}H_{27}ClN_2O$
Lorcainide, 5457
$C_{22}H_{27}ClN_2O_3$
Lorajmine, 5454
$C_{22}H_{27}ClN_2O_4S$
Diltiazem*, 3188
$C_{22}H_{27}ClO_3$
Cyproterone, 2781
$C_{22}H_{27}Cl_3N_2OS$
Clopenthixol*, 2392

Denotes derivative of title compound

C$_{22}$H$_{27}$NO
Ethybenztropine, 3712
Phenazocine, 7174

C$_{22}$H$_{27}$NO$_2$
Amineptine, 420
Danazol, 2811
Lobeline, 5432

C$_{22}$H$_{27}$NO$_3$
Dioxaphetyl Butyrate, 3295

C$_{22}$H$_{27}$NO$_4$
Corydaline, 2543

C$_{22}$H$_{27}$NO$_5$
Racefemine*, 8115

C$_{22}$H$_{27}$NO$_9$·2½H$_2$O
Hydrocodone*, 4708

C$_{22}$H$_{27}$N$_2$O$_3$$^+$
Macusines*, 5521
Melinonine A, 5703

C$_{22}$H$_{27}$N$_3$O$_2$
Caroverine, 1864

C$_{22}$H$_{27}$N$_3$O$_3$S$_2$
Metopimazine, 6070

C$_{22}$H$_{27}$N$_3$O$_5$
Physostigmine*, 7357

C$_{22}$H$_{27}$N$_3$O$_6$
Eseridine*, 3647

C$_{22}$H$_{27}$N$_3$O$_6$S
Pivcefalexin, 7485

C$_{22}$H$_{27}$N$_3$O$_7$S
Griseoviridin, 4454

C$_{22}$H$_{27}$N$_9$O$_4$
Stallimycin, 8730

C$_{22}$H$_{27}$N$_9$O$_7$S$_2$
Cefteram*, 1945

C$_{22}$H$_{28}$BrN
Prifinium Bromide, 7748

C$_{22}$H$_{28}$BrNO
Ethybenztropine*, 3712
Phenazocine*, 7174

C$_{22}$H$_{28}$BrNO$_3$
Benzilonium Bromide, 1090
Pipenzolate Bromide, 7428

C$_{22}$H$_{28}$ClFO$_4$
Clobetasol, 2361
Clocortolone, 2368

C$_{22}$H$_{28}$ClNO
Ethybenztropine*, 3712
Setastine, 8423

C$_{22}$H$_{28}$ClNO$_2$
Amineptine*, 420
Lobeline*, 5432

C$_{22}$H$_{28}$ClN$_3$O$_4$
Diperodon Hydrochloride, 3301

C$_{22}$H$_{28}$ClN$_3$O$_6$S
Pivcefalexin*, 7485

C$_{22}$H$_{28}$ClNaO$_6$
Cloprostenol*, 2397

C$_{22}$H$_{28}$Cl$_2$F$_3$N$_3$OS
Fluphenazine*, 4116

C$_{22}$H$_{28}$Cl$_2$N$_2$O
Lorcainide*, 5457

C$_{22}$H$_{28}$Cl$_2$N$_2$O$_3$
Lorajmine*, 5454

C$_{22}$H$_{28}$FNa$_2$O$_8$P
Betamethasone*, 1202
Dexamethasone*, 2922

C$_{22}$H$_{28}$F$_2$O$_4$
Diflucortolone, 3129

C$_{22}$H$_{28}$F$_2$O$_5$
Diflorasone, 3126
Flumethasone, 4066

C$_{22}$H$_{28}$N$_2$O
Buzepide, 1599
Fentanyl, 3944

C$_{22}$H$_{28}$N$_2$O$_2$
Anileridine, 686
Encainide, 3525

C$_{22}$H$_{28}$N$_2$O$_2$S
Becanthone, 1026
Perimethazine, 7120

C$_{22}$H$_{28}$N$_2$O$_3$
Voacamine*, 9944

C$_{22}$H$_{28}$N$_2$O$_4$
Rhynchophylline, 8198

C$_{22}$H$_{28}$N$_2$O$_5$
Ketoprofen*, 5184
Mycelianamide, 6230
Reserpic Acid, 8147

C$_{22}$H$_{28}$N$_4$O$_3$
Etonitazene, 3840

C$_{22}$H$_{28}$N$_4$O$_5$
Nitroakridin 3582, 6502

C$_{22}$H$_{28}$N$_4$O$_6$
Mitoxantrone, 6135

C$_{22}$H$_{28}$N$_9$NaO$_9$S$_2$
Cefbuperazone*, 1926

C$_{22}$H$_{28}$O$_3$
Canrenone, 1753
17-Hydroxy-16-methylene-Δ^6-
  progesterone, 4763
Norethindrone*, 6614

C$_{22}$H$_{28}$O$_5$
Estradiol*, 3653
Meprednisone, 5750
Prednylidene, 7729
Pyrethrins*, 7978

C$_{22}$H$_{28}$O$_6$
Isoquassin, 5108
Quassin, 8037

C$_{22}$H$_{29}$BrN$_2$O
Fenpiverinium Bromide, 3935

C$_{22}$H$_{29}$ClN$_2$O$_2$
Encainide*, 3525

C$_{22}$H$_{29}$ClN$_2$O$_2$S
Becanthone*, 1026

C$_{22}$H$_{29}$ClO$_5$
Alclometasone, 213
Beclomethasone, 1029

C$_{22}$H$_{29}$ClO$_6$
Cloprostenol, 2397

C$_{22}$H$_{29}$Cl$_2$NO
Setastine*, 8423

C$_{22}$H$_{29}$FO$_4$
Desoximetasone, 2910
Fluocortolone, 4079
Fluorometholone, 4104

C$_{22}$H$_{29}$FO$_5$
Betamethasone, 1202
Dexamethasone, 2922
Paramethasone, 6977

C$_{22}$H$_{29}$NO$_2$
Levopropoxyphene, 5349
Lobelanidine, 5429
Noracymethadol, 6602
Propoxyphene, 7851

C$_{22}$H$_{29}$NO$_4$S
Benztropine Mesylate, 1135

C$_{22}$H$_{29}$NO$_5$
Trimebutine, 9613

C$_{22}$H$_{29}$NO$_7$S
Poldine Methylsulfate, 7529

C$_{22}$H$_{29}$NO$_9$
Dihydrocodeine*, 3154

C$_{22}$H$_{29}$N$_2$O$_4$$^+$
Echitamine, 3474

C$_{22}$H$_{29}$N$_3$O$_6$S
Pivampicillin, 7484

C$_{22}$H$_{29}$N$_3$S$_2$
Thiethylperazine, 9241

C$_{22}$H$_{29}$N$_7$O$_5$
Puromycin, 7960

C$_{22}$H$_{29}$N$_9$O$_6$
Diminazene Aceturate, 3256

C$_{22}$H$_{29}$N$_9$O$_9$S$_2$
Cefbuperazone, 1926

C$_{22}$H$_{29}$Na$_2$O$_8$P
Methylprednisolone*, 6028

C$_{22}$H$_{30}$ClNO$_2$
Noracymethadol*, 6602
Propoxyphene*, 7851

C$_{22}$H$_{30}$ClN$_3$O$_6$S
Pivampicillin*, 7484

C$_{22}$H$_{30}$Cl$_2$N$_4$O$_6$
Mitoxantrone*, 6135

C$_{22}$H$_{30}$Cl$_2$N$_{10}$
Chlorhexidine, 2090

C$_{22}$H$_{30}$FO$_8$P
Dexamethasone*, 2922

C$_{22}$H$_{30}$INO
Stilonium Iodide, 8776

C$_{22}$H$_{30}$KO$_4$
Canrenone*, 1753

C$_{22}$H$_{30}$N$_2$
Aprindine, 781

C$_{22}$H$_{30}$N$_2$O
Pirmenol, 7471

C$_{22}$H$_{30}$N$_2$O$_2$
Aspidospermine, 872
Eprozinol, 3578

C$_{22}$H$_{30}$N$_2$O$_2$S
Sufentanil, 8860

C$_{22}$H$_{30}$N$_4$O$_2$S$_2$
Thioproperazine, 9287

C$_{22}$H$_{30}$O
Desogestrel, 2906

C$_{22}$H$_{30}$O$_2$
Promegestone, 7796

C$_{22}$H$_{30}$O$_3$
Siccanin, 8432

C$_{22}$H$_{30}$O$_5$
Jasmolins*, 5140
Methylprednisolone, 6028

C$_{22}$H$_{30}$O$_6$
Neoquassin, 6379

C$_{22}$H$_{31}$ClN$_2$
Aprindine*, 781

C$_{22}$H$_{31}$ClN$_2$O
Pirmenol*, 7471

C$_{22}$H$_{31}$NO$_3$
Amicibone, 407
Songorine, 8673

C$_{22}$H$_{31}$NO$_5$S
Latrunculins*, 5246

C$_{22}$H$_{31}$NO$_6$
Pentapiperide*, 7077

C$_{22}$H$_{31}$N$_3$O$_5$
Cinepazide, 2293

C$_{22}$H$_{31}$N$_3$O$_5$·H$_2$O
Cilazapril, 2276

C$_{22}$H$_{32}$Br$_2$N$_4$O$_4$
Distigmine Bromide, 3368

C$_{22}$H$_{32}$ClNO$_3$
Oxybutynin Chloride, 6908

C$_{22}$H$_{32}$Cl$_2$N$_2$O$_2$
Eprozinol*, 3578

C$_{22}$H$_{32}$Cl$_4$N$_{10}$
Chlorhexidine*, 2090

C$_{22}$H$_{32}$IN$_3$O$_4$S
Penethamate Hydriodide, 7027

C$_{22}$H$_{32}$N$_2$O$_2$
Dopexamine, 3418

C$_{22}$H$_{32}$N$_2$O$_5$
Benzquinamide, 1133

C$_{22}$H$_{32}$N$_2$O$_5$S
Estrone*, 3660

C$_{22}$H$_{32}$N$_2$O$_6$
Hexoprenaline, 4628

C$_{22}$H$_{32}$N$_2$O$_7$
Isoaminile*, 4992

C$_{22}$H$_{32}$N$_2$O$_{10}$
Synephrine*, 8994

C$_{22}$H$_{32}$O$_2$
Synhexyl, 8995

C$_{22}$H$_{32}$O$_3$
Medroxyprogesterone, 5677
Medrysone, 5679
Stenbolone*, 8763
Testosterone Propionate, 9115

C$_{22}$H$_{32}$O$_4$
Algestone*, 229

Denotes derivative of title compound

$C_{23}H_{28}ClN_3O_5S$
Glyburide, 4372
$C_{23}H_{28}ClN_3O_7$
Minocycline*, 6121
$C_{23}H_{28}Cl_2O_5$
Dichlorisone*, 3035
$C_{23}H_{28}F_3N_3OS$
Homofenazine, 4657
$C_{23}H_{28}F_3NaO_6$
Fluprostenol*, 4121
$C_{23}H_{28}I_2O_4$
Hinderin, 4635
$C_{23}H_{28}N_2O$
Leiopyrrole, 5316
$C_{23}H_{28}N_2O_3$
Bopindolol, 1335
$C_{23}H_{28}N_2O_4$
Quinine Ethylcarbonate, 8080
$C_{23}H_{28}N_2O_5$
Reserpiline, 8148
$C_{23}H_{28}N_2O_5S$
Methotrimeprazine*, 5909
$C_{23}H_{28}O_6$
Enprostil, 3545
Prednisone*, 7727
$C_{23}H_{28}O_{15}$
Purpurogallin*, 7963
$C_{23}H_{29}ClFN_3O_4$
Cisapride, 2318
$C_{23}H_{29}ClFN_3O_4.H_2O$
Cisapride*, 2318
$C_{23}H_{29}ClN_2O$
Leiopyrrole*, 5316
$C_{23}H_{29}ClO_4$
Chlormadinone Acetate, 2102
$C_{23}H_{29}FO_6$
Isoflupredone*, 5058
$C_{23}H_{29}F_2N_3O$
Amperozide, 612
$C_{23}H_{29}F_3O_6$
Fluprostenol, 4121
$C_{23}H_{29}NO$
Norpipanone, 6633
$C_{23}H_{29}NO_2$
Phenadoxone, 7160
$C_{23}H_{29}NO_3$
Fenbutrazate, 3908
Phenoperidine, 7216
$C_{23}H_{29}NO_{11}$
Morphine Mucate, 6190
$C_{23}H_{29}NO_{12}$
Hygromycin, 4789
$C_{23}H_{29}N_3O$
Opipramol, 6808
$C_{23}H_{29}N_3O_2$
Etomidoline, 3839
Oxypertine, 6924
$C_{23}H_{29}N_3O_2S$
Acetophenazine, 64
$C_{23}H_{29}N_3O_2S_2$
Thiothixene, 9297
$C_{23}H_{30}BrNO_3$
Piperilate*, 7441
Propantheline Bromide, 7816
$C_{23}H_{30}ClNO$
Norpipanone*, 6633
$C_{23}H_{30}ClNO_2$
Phenadoxone*, 7160
$C_{23}H_{30}ClNO_3$
Fenbutrazate*, 3908
Phenoperidine*, 7216
$C_{23}H_{30}ClN_3O$
Quinacrine, 8053
$C_{23}H_{30}Cl_2F_3N_3OS$
Homofenazine*, 4657
$C_{23}H_{30}Cl_2NNa_2O_6P$
Estramustine*, 3658
$C_{23}H_{30}Cl_3N_3O_2S$
Thiopropazate*, 9286
$C_{23}H_{30}N_2O$
α-Methylfentanyl, 5992

$C_{23}H_{30}N_2O_2$
Piminodine, 7403
$C_{23}H_{30}N_2O_4$
Mitragynine, 6136
Pholcodine, 7303
$C_{23}H_{30}N_4O_2$
Phenylbutazone*, 7248
$C_{23}H_{30}N_4O_3.H_2O$
Oxyphenbutazone*, 6925
$C_{23}H_{30}O_3$
Etretinate, 3846
Melengestrol, 5697
$C_{23}H_{30}O_6$
Cortisone*, 2533
Fenprostalene, 3940
Prednisolone*, 7719
$C_{23}H_{30}O_7$
Sarverogenin, 8339
$C_{23}H_{31}ClN_2O_3$
Etodroxizine, 3832
$C_{23}H_{31}Cl_2NO_3$
Estramustine, 3658
$C_{23}H_{31}Cl_2N_3O_7$
Opipramol*, 6808
$C_{23}H_{31}FO_5$
Flurogestone Acetate, 4125
$C_{23}H_{31}FO_6$
Fludrocortisone*, 4059
$C_{23}H_{31}IN_2O$
Buzepide*, 1599
$C_{23}H_{31}NO_2$
Levomethadyl Acetate, 5346
Methadyl Acetate, 5853
Motretinide, 6199
$C_{23}H_{31}NO_3$
Norgestimate, 6620
$C_{23}H_{31}NO_7$
Levallorphan*, 5342
$C_{23}H_{31}NO_7S$
Bevonium Methyl Sulfate, 1214
Sulprostone, 8973
$C_{23}H_{32}Cl_2NO_6P$
Estramustine*, 3658
$C_{23}H_{32}Cl_3N_3O.2H_2O$
Quinacrine*, 8053
$C_{23}H_{32}N_2O_2S$
Tiocarlide, 9382
$C_{23}H_{32}N_2O_3$
Zipeprol, 10074
$C_{23}H_{32}N_2O_5$
Ramipril, 8123
$C_{23}H_{32}N_2O_8$
Trichostatin(s)*, 9563
$C_{23}H_{32}N_6O_{14}$
Polyoxins*, 7553
$C_{23}H_{32}O_2$
Dimethisterone, 3208
Medrogestone, 5676
$C_{23}H_{32}O_3$
Estradiol*, 3653
Quinestradiol, 8065
$C_{23}H_{32}O_4$
Deoxycorticosterone Acetate, 2883
17α-Hydroxyprogesterone*, 4773
$C_{23}H_{32}O_5$
Tanghinigenin, 9021
$C_{23}H_{32}O_6$
Hydrocortisone Acetate, 4711
Strophanthidin, 8818
$C_{23}H_{32}O_7$
Strophanthidin*, 8818
$C_{23}H_{33}IN_2O$
Isopropamide, 5090
$C_{23}H_{33}N_2O_2$ +
Prajmaline, 7704
$C_{23}H_{33}N_3O_5$
Aminopyrine*, 488
$C_{23}H_{33}N_3O_8$
Proxazole*, 7919
$C_{23}H_{34}Cl_2N_2O_3$
Zipeprol*, 10074

$C_{23}H_{34}NO_5P$
Fosinopril*, 4171
$C_{23}H_{34}O_3$
Testosterone*, 9109
$C_{23}H_{34}O_4$
21-Acetoxypregnenolone, 70
Digitoxigenin, 3145
Uzarin*, 9809
$C_{23}H_{34}O_5$
Alfadolone Acetate, 224
Digoxigenin, 3149
Gitoxigenin, 4327
Mevastatin, 6088
Periplogenin, 7128
Sarmentogenin, 8334
$C_{23}H_{34}O_6$
Ajugarins*, 187
Diginatigenin, 3137
$C_{23}H_{34}O_8$
Ouabagenin, 6853
$C_{23}H_{35}NO_5$
Bencyclane*, 1041
$C_{23}H_{35}NaO_7$
Pravastatin Sodium, 7712
$C_{23}H_{36}N_2O_2$
Proscar®, 7888
$C_{23}H_{36}N_4O_5S_3$
Octotiamine, 6680
$C_{23}H_{36}N_4O_{10}S_2$
Pentamidine*, 7071
$C_{23}H_{36}N_6O_5S$
Argatroban, 804
$C_{23}H_{36}O_3$
Dromostanolone Propionate, 3436
$C_{23}H_{37}NO_5S$
Leukotrienes*, 5339
$C_{23}H_{37}NO_6$
Tocamphyl, 9415
$C_{23}H_{37}N_5O_6S$
Etamiphyllin*, 3663
$C_{23}H_{38}O$
Teprenone, 9083
$C_{23}H_{38}O_2$
Norcholanic Acid, 6606
$C_{23}H_{38}O_3$
Pregnanediol*, 7732
$C_{23}H_{38}O_4$
Trimoprostil, 9637
$C_{23}H_{38}O_5$
Gemeprost, 4279
$C_{23}H_{38}O_6$
Ornoprostil, 6827
$C_{23}H_{39}NO_4$
Perhexiline*, 7116
$C_{23}H_{41}ClN_2O$
Dodecarbonium Chloride, 3404
$C_{23}H_{42}ClNO_2$
Benzoxonium Chloride, 1123
$C_{23}H_{42}N_2$
Chonemorphine, 2219
$C_{23}H_{42}N_2O_{12}$
Pentolinium Tartrate, 7089
$C_{23}H_{45}N_5O_{14}$
Paromomycin, 6989
$C_{23}H_{46}$
Muscalure, 6218
$C_{23}H_{46}N_6O_{13}$
Neomycin*, 6369
$C_{23}H_{47}N_5O_{18}S$
Paromomycin*, 6989
$C_{24}H_{16}As_2O_3$
10,10'-Oxydiphenoxarsine, 6914
$C_{24}H_{16}O_{12}$
Laccaic Acid*, 5209
$C_{24}H_{19}NO_5$
Oxyphenisatin Acetate, 6927
$C_{24}H_{19}NO_6$
Bisoxatin Acetate, 1310
$C_{24}H_{19}N_3O_5S$
Piroxicam*, 7476

Denotes derivative of title compound

C$_{24}$H$_{20}$AsBr
Tetraphenylarsonium Bromide, 9166

C$_{24}$H$_{20}$AsCl
Tetraphenylarsonium Chloride, 9167

C$_{24}$H$_{20}$BNa
Sodium Tetraphenylborate, 8646

C$_{24}$H$_{20}$Cl$_2$N$_2$OS
Fenticonazole, 3948

C$_{24}$H$_{20}$I$_6$N$_4$O$_8$
Iocarmic Acid, 4902

C$_{24}$H$_{20}$N$_2$
N,N'-Diphenylbenzidine, 3319

C$_{24}$H$_{20}$N$_4$O
Scarlet Red, 8349

C$_{24}$H$_{21}$Cl$_2$N$_3$O$_4$S
Fenticonazole*, 3948

C$_{24}$H$_{21}$Cl$_6$O$_6$P
Falone®, 3880

C$_{24}$H$_{21}$I$_6$N$_5$O$_8$
Ioxaglic Acid, 4956

C$_{24}$H$_{23}$N$_3$O$_6$S
Talampicillin, 9008

C$_{24}$H$_{24}$CaI$_6$N$_4$O$_4$
Ipodate*, 4958

C$_{24}$H$_{24}$ClN$_3$O$_6$S
Talampicillin*, 9008

C$_{24}$H$_{25}$ClN$_6$O$_6$S
Loprazolam*, 5453

C$_{24}$H$_{25}$NO$_3$
Benzylmorphine, 1155
Cyphenothrin, 2776

C$_{24}$H$_{25}$NO$_4$
Flavoxate, 4033

C$_{24}$H$_{26}$BrN$_3$O$_3$
Nicergoline, 6404

C$_{24}$H$_{26}$ClNO$_3$
Benzylmorphine*, 1155

C$_{24}$H$_{26}$ClNO$_4$
Flavoxate*, 4033

C$_{24}$H$_{26}$MgO$_8$
Dimecrotic Acid*, 3189

C$_{24}$H$_{26}$N$_2$O$_4$
Carvedilol, 1882

C$_{24}$H$_{26}$N$_2$O$_6$
Suxibuzone, 8990

C$_{24}$H$_{26}$N$_8$O$_9$
Diopterin, 3285

C$_{24}$H$_{26}$O$_6$
Mangostin, 5627

C$_{24}$H$_{26}$O$_{13}$
Centaurein, 1964
Irigenin*, 4973

C$_{24}$H$_{27}$N
Prenylamine, 7744

C$_{24}$H$_{27}$NO$_4$
Tylocrebrine, 9738
Tylophorine, 9739

C$_{24}$H$_{28}$BiN$_4$Na$_7$O$_{25}$
Bismuth Sodium Triglycollamate, 1293

C$_{24}$H$_{28}$ClNO$_4$
Phenactropinium Chloride, 7158

C$_{24}$H$_{28}$ClN$_5$O$_3$
Dimenhydrinate, 3193

C$_{24}$H$_{28}$FN$_3$O
Spirilene, 8709

C$_{24}$H$_{28}$N$_2$O$_4$
Dimethindene*, 3204

C$_{24}$H$_{28}$N$_2$O$_5$S
Propiomazine*, 7834

C$_{24}$H$_{28}$N$_4$O$_6$
Cotinine*, 2552

C$_{24}$H$_{28}$O$_4$
Diethylstilbestrol Dipropionate, 3119

C$_{24}$H$_{29}$BrFNO$_3$
Flutropium Bromide, 4136

C$_{24}$H$_{29}$ClN$_2$O$_9$
Amlodipine*, 509

C$_{24}$H$_{29}$ClO$_4$
Cyproterone*, 2781

C$_{24}$H$_{29}$Cl$_2$FO$_5$
Flucloronide, 4053

C$_{24}$H$_{29}$FN$_2$O$_2$
Aceperone, 25

C$_{24}$H$_{29}$FO$_6$
Fluprednidene Acetate, 4118

C$_{24}$H$_{29}$NO
Phenomorphan, 7215

C$_{24}$H$_{29}$NO$_4$
Ethaverine, 3682

C$_{24}$H$_{29}$N$_3$O$_6$
Methylergonovine*, 5989

C$_{24}$H$_{29}$N$_5$O$_8$S
Doxazosin*, 3422

C$_{24}$H$_{30}$BrNO$_4$
Tropenzile*, 9690

C$_{24}$H$_{30}$BrN$_7$O$_3$
Pyrilamine*, 7996

C$_{24}$H$_{30}$CaN$_4$O$_6$
Cyclobarbital*, 2717

C$_{24}$H$_{30}$ClFO$_5$
Clocortolone*, 2368

C$_{24}$H$_{30}$ClNO$_4$
Ethaverine*, 3682

C$_{24}$H$_{30}$ClN$_2$O$_7$
Clebopride*, 2344

C$_{24}$H$_{30}$FNa$_2$O$_9$P
Triamcinolone Acetonide*, 9512

C$_{24}$H$_{30}$F$_2$O$_6$
Fluocinolone Acetonide, 4076

C$_{24}$H$_{30}$N$_2$O$_2$
Doxapram, 3421

C$_{24}$H$_{30}$N$_2$O$_2$S
Piperacetazine, 7429

C$_{24}$H$_{30}$N$_2$O$_4$
Trimipramine*, 9636

C$_{24}$H$_{30}$N$_2$O$_5$
Vindoline*, 9893

C$_{24}$H$_{30}$N$_2$O$_8$
Podophyllinic Acid 2-Ethylhydrazide, 7519

C$_{24}$H$_{30}$N$_4$O$_5$
Lisuride*, 5394

C$_{24}$H$_{30}$O$_4$
17-Hydroxy-16-methylene-Δ^6-progesterone*, 4763

C$_{24}$H$_{30}$O$_7$
Athamantin, 880

C$_{24}$H$_{30}$O$_8$
Desaspidin, 2899
Flavaspidic Acid, 4027

C$_{24}$H$_{30}$O$_9$
T-2 Toxin, 9711

C$_{24}$H$_{31}$ClN$_2$O$_2$·H$_2$O
Doxapram*, 3421

C$_{24}$H$_{31}$FO$_6$
Betamethasone*, 1202
Dexamethasone*, 2922
Flunisolide, 4071
Fluperolone Acetate, 4115
Paramethasone*, 6977
Triamcinolone Acetonide, 9512

C$_{24}$H$_{31}$F$_3$O$_4$
Flumedroxone Acetate, 4064

C$_{24}$H$_{31}$NO
Dipipanone, 3342

C$_{24}$H$_{31}$NO$_8$
Orphenadrine*, 6831

C$_{24}$H$_{31}$N$_3$OS
Butaperazine, 1509

C$_{24}$H$_{31}$N$_3$O$_2$S
Carphenazine, 1868

C$_{24}$H$_{32}$ClFO$_5$
Halcinonide, 4504

C$_{24}$H$_{32}$N$_2$O$_2$
Eprazinone, 3577

C$_{24}$H$_{32}$N$_2$O$_5$
Metoserpate, 6074

C$_{24}$H$_{32}$N$_2$O$_9$
Enalapril*, 3521

C$_{24}$H$_{32}$N$_2$O$_{10}$
Cinepazet Maleate, 2292

C$_{24}$H$_{32}$N$_2$O$_{12}$S
p,p'-Sulfonyldianiline-N,N'-digalactoside, 8941

C$_{24}$H$_{32}$N$_2$O$_{14}$
Pyrisuccideanol*, 8002

C$_{24}$H$_{32}$N$_4$O$_5$
Terguride*, 9095

C$_{24}$H$_{32}$O$_2$
Quinbolone, 8063

C$_{24}$H$_{32}$O$_4$
Estradiol*, 3653
Ethynodiol*, 3816
Megestrol Acetate, 5687
Resibufogenin, 8150
Scillarenin, 8355

C$_{24}$H$_{32}$O$_4$S
Spironolactone, 8721

C$_{24}$H$_{32}$O$_6$
Desonide, 2908

C$_{24}$H$_{32}$O$_7$
Etiproston, 3827

C$_{24}$H$_{33}$ClN$_2$O$_5$
Metoserpate*, 6074

C$_{24}$H$_{33}$FO$_6$
Flurandrenolide, 4122

C$_{24}$H$_{33}$N
Droprenilamine, 3438

C$_{24}$H$_{33}$NO$_3$
Nafronyl, 6268

C$_{24}$H$_{33}$N$_3$O$_2$S
Dixyrazine, 3391

C$_{24}$H$_{33}$N$_3$O$_3$S$_2$
Pipotiazine, 7453

C$_{24}$H$_{33}$NaO$_5$
Dehydrocholic Acid*, 2858

C$_{24}$H$_{33}$NaO$_7$S
Bufalin*, 1458

C$_{24}$H$_{34}$ClN
Droprenilamine*, 3438

C$_{24}$H$_{34}$Cl$_2$N$_2$O$_2$
Eprazinone*, 3577

C$_{24}$H$_{34}$N$_2$Na$_2$O$_{18}$S$_3$
Glucosulfone Sodium, 4357

C$_{24}$H$_{34}$N$_2$O
Bepridil, 1167

C$_{24}$H$_{34}$N$_2$O$_2$
Euprocin, 3861

C$_{24}$H$_{34}$N$_8$O$_4$S$_2$
Thiamine Disulfide, 9221

C$_{24}$H$_{34}$O$_4$
Algestone*, 229
Bufalin, 1458
Medroxyprogesterone*, 5677

C$_{24}$H$_{34}$O$_4$S$_2$
Tiomesterone, 9385

C$_{24}$H$_{34}$O$_5$
Bufogenin B, 1464
Dehydrocholic Acid, 2858
Gamabufotalin, 4258

C$_{24}$H$_{34}$O$_7$
Ajugarins*, 187

C$_{24}$H$_{35}$ClN$_2$O·H$_2$O
Bepridil*, 1167

C$_{24}$H$_{35}$NO$_5$
Batrachotoxinin A, 1019
Decoquinate, 2846

C$_{24}$H$_{36}$ClN$_3$O$_{10}$
Zorubicin*, 10096

C$_{24}$H$_{36}$N$_2$O$_9$S
Celesticetin, 1954

C$_{24}$H$_{36}$O$_3$
Anagestone*, 658
Nabilone, 6257

C$_{24}$H$_{36}$O$_4$
Bolandiol*, 1325

C$_{24}$H$_{36}$O$_5$
Lovastatin, 5460

FI-62

Denotes derivative of title compound

$C_{25}H_{36}O_3$
Estradiol*, 3653
Nandrolone*, 6280
$C_{25}H_{36}O_4$
Cephalonic Acid, 1974
$C_{25}H_{36}O_6$
Hydrocortisone*, 4710
$C_{25}H_{36}O_{10}$
Glaucarubin, 4331
$C_{25}H_{38}N_2O$
Bunamidine, 1475
$C_{25}H_{38}O_2$
17α-Methyltestosterone 3-Cyclo-
pentyl Enol Ether, 6045
$C_{25}H_{38}O_5$
Simvastatin, 8491
$C_{25}H_{38}O_7$
Laserpitin, 5244
$C_{25}H_{39}ClN_2O$
Bunamidine*, 1475
$C_{25}H_{39}NO_5$
Cassamine, 1897
$C_{25}H_{39}NO_6$
Erythrophlamine, 3633
$C_{25}H_{40}N_2O_6S$
Leukotrienes*, 5339
$C_{25}H_{40}O_2S$
Mepitiostane, 5747
$C_{25}H_{40}O_4$
Pregnanediol*, 7732
$C_{25}H_{41}NO_7$
Delsoline, 2868
Lycoctonine, 5488
$C_{25}H_{41}NO_9$
Aconine, 110
$C_{25}H_{42}N_2O$
Cyclobuxine, 2722
$C_{25}H_{43}NO_7$
Methymycin, 6053
Neomethymycin, 6368
$C_{25}H_{43}NO_{18}$
Acarbose, 12
$C_{25}H_{43}N_{13}O_9$
Tuberactinomycin*, 9713
$C_{25}H_{43}N_{13}O_{10}$
Enviomycin, 3552
Viomycin, 9905
$C_{25}H_{43}N_{13}O_{11}$
Tuberactinomycin*, 9713
$C_{25}H_{44}N_2$
Kurchessine, 5203
$C_{25}H_{44}N_2O$
Azacosterol, 908
$C_{25}H_{44}N_2O_2$
Kurcholessine, 5204
$C_{25}H_{44}N_{14}O_7$
Capreomycin*, 1758
$C_{25}H_{44}N_{14}O_8$
Capreomycin*, 1758
$C_{25}H_{45}FeN_6O_8$
Deferoxamine*, 2850
$C_{25}H_{45}NO_9$
Pederin, 7011
$C_{25}H_{46}BrNO_2$
Amantanium Bromide, 381
$C_{25}H_{46}ClN$
Cetalkonium Chloride, 2009
$C_{25}H_{46}Cl_2N_2O$
Azacosterol*, 908
$C_{25}H_{47}NO_8$
Carboprost*, 1829
$C_{25}H_{48}N_6O_8$
Deferoxamine, 2850
$C_{25}H_{49}ClN_6O_8$
Deferoxamine*, 2850
$C_{26}H_{16}N_3Na_3O_{10}S_3$
Anazolene Sodium, 662
$C_{26}H_{18}CuN_4O_2$
Myxin*, 6255
$C_{26}H_{19}NO_{12}$
Laccaic Acid*, 5209

$C_{26}H_{21}NO_6$
Triacetyldiphenolisatin, 9505
$C_{26}H_{22}ClF_3N_2O_3$
Fluvalinate, 4137
$C_{26}H_{22}O_2$
Benzopinacol, 1110
$C_{26}H_{23}F_2NO_4$
Flucythrinate, 4055
$C_{26}H_{23}F_3N_4O_6S.H_2O$
Tosufloxacin*, 9477
$C_{26}H_{24}N_4O_8$
Valacidin, 9812
$C_{26}H_{25}Cl_3N_3O$
Hetolin®, 4596
$C_{26}H_{25}N_2NaO_6S$
Carindacillin*, 1847
$C_{26}H_{25}N_3O_3S$
Fenoverine, 3928
$C_{26}H_{25}N_3O_{11}$
Tunichrome B-1, 9727
$C_{26}H_{26}Cl_4N_2O$
Hetolin®*, 4596
$C_{26}H_{26}F_2N_2$
Flunarizine, 4070
$C_{26}H_{26}N_2O_6S$
Carindacillin, 1847
$C_{26}H_{27}ClN_2$
Clocinizine, 2366
$C_{26}H_{27}ClN_2O$
Lofepramine, 5438
$C_{26}H_{27}NO_{10}$
Carubicin, 1879
$C_{26}H_{28}ClNO$
Clomiphene, 2384
Phenacridane Chloride, 7157
Toremifene, 9474
$C_{26}H_{28}ClNO_{10}$
Carubicin*, 1879
$C_{26}H_{28}ClN_3$
Pyrvinium Chloride, 8033
$C_{26}H_{28}Cl_2F_2N_2$
Flunarizine*, 4070
$C_{26}H_{28}Cl_2N_2O$
Lofepramine*, 5438
$C_{26}H_{28}Cl_2N_4O_4$
Ketoconazole, 5181
$C_{26}H_{28}N_2$
Cinnarizine, 2308
$C_{26}H_{28}N_2O_2$
Fenproporex*, 3939
$C_{26}H_{28}O_{14}$
Apiin, 765
$C_{26}H_{29}F_2N_7$
Almitrine, 299
$C_{26}H_{29}NO$
Tamoxifen, 9019
$C_{26}H_{29}NO_3$
Amotriphene, 608
$C_{26}H_{29}N_3O_6$
Nicardipine, 6403
$C_{26}H_{30}ClN_3O_6$
Nicardipine*, 6403
$C_{26}H_{30}ClN_5O_3$
Piprinhydrinate, 7456
$C_{26}H_{30}Cl_2F_3NO$
Halofantrine, 4508
$C_{26}H_{30}Na_2O_9$
Estriol*, 3659
$C_{26}H_{30}O_8$
Limonin, 5372
Physodic Acid, 7355
$C_{26}H_{31}Cl_2N_5O_3$
Terconazole, 9090
$C_{26}H_{31}Cl_3F_3NO$
Halofantrine*, 4508
$C_{26}H_{32}ClFO_5$
Clobetasone*, 2362
$C_{26}H_{32}F_2O_7$
Diflorasone*, 3126
Fluocinonide, 4077

$C_{26}H_{32}F_3N_3O_2S$
Oxaflumazine, 6862
$C_{26}H_{32}MgN_6O_8S_2$
Dipyrone*, 3358
$C_{26}H_{32}N_2O_5$
Delapril, 2864
$C_{26}H_{32}N_2O_8$
Tritoqualine, 9677
$C_{26}H_{32}O_4$
Bixin*, 1320
$C_{26}H_{33}ClN_2O_5$
Delapril*, 2864
$C_{26}H_{33}FO_7$
Flunisolide*, 4071
$C_{26}H_{33}NO_4$
Cyprenorphine, 2777
$C_{26}H_{33}NO_9$
Trimebutine*, 9613
$C_{26}H_{33}NaO_8$
Methylprednisolone*, 6028
$C_{26}H_{34}ClNO_4$
Cyprenorphine*, 2777
$C_{26}H_{34}O_4$
Methestrol*, 5888
$C_{26}H_{34}O_7$
Cinobufotalin, 2310
Fumagillin, 4199
$C_{26}H_{35}FO_5$
Fluocortin Butyl, 4078
$C_{26}H_{35}NO_4$
Diprenorphine, 3347
$C_{26}H_{35}NO_7$
Alverine*, 377
Nafronyl*, 6268
$C_{26}H_{35}N_3O_9$
Cinepazide*, 2293
$C_{26}H_{35}O_4$
Trenbolone*, 9499
$C_{26}H_{36}ClNO_4$
Diprenorphine*, 3347
$C_{26}H_{36}N_2O_4$
Bisobrin, 1308
$C_{26}H_{36}N_2O_9$
Antimycin A_3, 747
Quinidine*, 8072
Quinine Gluconate, 8082
$C_{26}H_{36}O_3$
Estradiol 17β-Cypionate, 3656
$C_{26}H_{36}O_6$
Bufotalin, 1466
Prednisolone 21-Trimethylacetate,
7726
Prednival, 7728
$C_{26}H_{37}ClN_4O_6S_2.H_2O$
Penicillin O*, 7044
$C_{26}H_{38}Cl_2N_{10}O_4$
Chlorhexidine*, 2090
$C_{26}H_{38}O_2$
Quingestrone, 8069
$C_{26}H_{38}O_3$
Pentagestrone, 7068
$C_{26}H_{38}O_4$
Gestonorone Caproate, 4309
4-Hydroxy-19-nortestosterone*,
4768
Lupulon, 5482
$C_{26}H_{38}O_5S$
Tixocortol*, 9408
$C_{26}H_{38}O_6$
Hydrocortisone*, 4710
$C_{26}H_{39}ClN_2O_4$
Mazipredone, 5644
$C_{26}H_{39}NO_9S$
Cetiedil*, 2012
$C_{26}H_{39}NO_{10}$
Carbetapentane*, 1798
$C_{26}H_{40}O_3$
Testosterone Enanthate, 9112
$C_{26}H_{40}O_4$
Methandriol*, 5861

Denotes derivative of title compound

C26H41Br2NO4
  Pinaverium Bromide, 7410
C26H41NO
  Melinamide, 5702
C26H41NO10
  Oxeladin*, 6885
C26H42O9
  Pseudomonic Acids*, 7933
C26H43NO6
  Glycocholic Acid, 4390
C26H44NNaO7S
  Taurocholic Acid*, 9044
C26H44N2O10S
  Albuterol*, 209
C26H44O2
  Tocol, 9416
C26H44O8
  Pseudomonic Acids*, 7933
C26H44O9
  Mupirocin, 6212
C26H45NO7S
  Taurocholic Acid, 9044
C26H49NO3S
  Cetrimonium Bromide*, 2018
C26H50O4
  Bis(2-ethylhexyl) Sebacate, 1263
  Glycol Dilaurate, 4393
C26H52N6O11S
  Deferoxamine*, 2850
C26H54NO7P
  Platelet Activating Factor*, 7496
C26H56N10
  Alexidine, 222
C27H20N4O
  Pyrinoline, 8001
C27H20O12
  Collinomycin, 2478
C27H22Cl2N4
  Clofazimine, 2370
C27H24AlN3O15S3
  8-Hydroxyquinoline Sulfate*, 4779
C27H24N2O9
  Albofungin, 203
C27H28Br2O5S
  Bromthymol Blue, 1435
C27H29IN2
  Dicyanine, 3081
C27H29NO10
  Daunorubicin, 2825
  Pirozadil, 7477
C27H29NO11
  Doxorubicin, 3428
  Epirubicin, 3573
C27H30ClNO10
  Daunorubicin*, 2825
C27H30ClNO11
  Doxorubicin*, 3428
  Epirubicin*, 3573
C27H30Cl2O6
  Mometasone Furoate, 6151
C27H30N2O5
  Quinine Salicylate, 8088
C27H30N4O
  Oxatomide, 6880
C27H30O5S
  Thymol Blue, 9334
C27H30O6
  Sofalcone, 8658
C27H30O14
  Glucofrangulin, 4346
  Kaempferol*, 5156
  Morindin, 6180
  Sophorabioside, 8674
C27H30O16
  Rutin, 8276
  Saponarin, 8327
C27H31ClN2O
  Chlorbenzoxamine, 2075
C27H31ClO15
  Cyanidin Chloride*, 2694
  Pelargonidin*, 7014

C27H31ClO16
  Cyanidin Chloride*, 2694
C27H31ClO17
  Delphinidin*, 2866
C27H31N2NaO6S2
  Sulphan Blue, 8968
C27H32ClNO
  Triparanol, 9650
C27H32N2O4S3
  Mesoridazine*, 5813
C27H32N2O7
  Bopindolol*, 1335
C27H32N4O8
  Pyridomycin, 7989
C27H32O8
  Verrucarins*, 9869
C27H32O9
  Verrucarins*, 9869
C27H32O14
  Naringin, 6345
C27H33Cl3N2O
  Chlorbenzoxamine*, 2075
C27H33NO3
  Ethamoxytriphetol, 3674
  Prenylamine*, 7744
C27H33NO10S
  Thiocolchicine*, 9253
C27H33N3O6S
  Gliquidone, 4338
C27H33N3O8
  Rolitetracycline, 8236
C27H33N9O15P2
  Flavine-Adenine Dinucleotide, 4028
C27H34F2O7
  Difluprednate, 3134
C27H34N2O4S
  Brilliant Green, 1367
C27H34N2O7
  Lofentanil*, 5437
C27H34N4O
  Piritramide, 7470
C27H34N4O11·1½H2O
  Rolitetracycline*, 8236
C27H34O3
  Nandrolone Phenpropionate, 6283
  Testosterone Phenylacetate, 9114
C27H34O4
  Bixin*, 1320
C27H34O8
  Verrucarins*, 9869
C27H34O9
  Verrucarins*, 9869
C27H34O11
  Phillyrin, 7296
C27H35BrClN3
  Methyl Green, 5998
C27H35NO5
  Etorphine*, 3843
C27H35NO8
  Sedecamycin, 8377
C27H35N5O7S
  Endorphins*, 3528
C27H36ClFO5
  Clocortolone*, 2368
C27H36ClNO5
  Etorphine*, 3843
C27H36F2O5
  Diflucortolone*, 3129
C27H36F2O6
  Flumethasone*, 4066
C27H36O8
  Prednicarbate, 7717
C27H37FO6
  Betamethasone*, 1202
C27H38N2O4
  Verapamil, 9851
C27H38N2O8
  Bunaftine*, 1474
  Prajmaline*, 7704
C27H38O3
  Norethindrone*, 6614

C27H38O4
  Azafrin, 911
C27H38O6
  Prednisolone Tebutate, 7725
C27H39ClN2O4
  Verapamil*, 9851
C27H39NO2
  Veratramine, 9853
C27H39NO3
  Jervine, 5145
C27H39NO6
  Prednisolone 21-Diethylamino-
    acetate, 7720
C27H40ClNO6
  Prednisolone 21-Diethylamino-
    acetate*, 7720
C27H40N2O2
  Pifarnine, 7391
C27H40O
  Neoergosterol, 6366
C27H40O3
  Hexadecyl 3-Hydroxy-2-naphtho-
    ate, 4604
  Nandrolone*, 6280
  Testosterone 17β-Cypionate, 9111
C27H40O4
  Gentrogenin, 4293
  17α-Hydroxyprogesterone Caproate,
    4774
C27H40O6
  Hydrocortisone Tebutate, 4714
C27H41NO6
  Hydrocortamate, 4709
C27H42ClNO
  Phenoctide, 7205
C27H42ClNO
  Benzethonium Chloride, 1083
C27H42ClNO6
  Hydrocortamate*, 4709
C27H42Cl2N2O6
  Chloramphenicol Palmitate, 2069
C27H42FeN9O12
  Ferrichromes*, 3968
C27H42O3
  Diosgenin, 3289
C27H42O4
  Convallamarogenin, 2506
  Hecogenin, 4537
  Pseudohecogenin, 7930
  Ruscogenin, 8270
C27H43NO
  Solanidine, 8660
C27H43NO2
  Isorubijervine, 5111
  Rubijervine, 8264
  Solasodine, 8665
  Veralkamine, 9850
C27H43NO3
  Imperialine, 4840
C27H43NO7
  Zygadenine, 10099
C27H43NO8
  Cevine, 2026
  Germine, 4307
C27H43NO9
  Protoverine, 7916
C27H44O
  7-Dehydrocholesterol, 2857
  Desmosterol, 2905
  Vitamin D3, 9929
C27H44O2
  Calcifediol, 1638
  Gefarnate, 4273
  1α-Hydroxycholecalciferol, 4749
C27H44O3
  Calcitriol, 1641
  Sarsasapogenin, 8338
  Smilagenin, 8508
  Tigogenin, 9367

C$_{27}$H$_{44}$O$_4$
Chlorogenin, 2143
Digalogenin, 3136
Gitogenin, 4325
C$_{27}$H$_{44}$O$_5$
Digitogenin, 3143
C$_{27}$H$_{44}$O$_6$
Ecdysones*, 3465
Ponasterone A, 7566
C$_{27}$H$_{44}$O$_7$
Ecdysones*, 3465
C$_{27}$H$_{45}$NO$_2$
Tomatidine, 9470
C$_{27}$H$_{45}$NO$_3$
Verticine, 9873
C$_{27}$H$_{46}$Cl$_5$NO
Octadecyltrimethylammonium
Pentachlorophenate, 6665
C$_{27}$H$_{46}$N$_2$O$_2$
Solanocapsine, 8662
C$_{27}$H$_{46}$O
Allocholesterol, 253
Cholesterol, 2204
Epicholesterol, 3565
C$_{27}$H$_{46}$O$_2$
24-Hydroxycholesterol, 4750
25-Hydroxycholesterol, 4751
δ-Tocopherol, 9419
η-Tocopherol, 9423
C$_{27}$H$_{47}$ClN$_2$O$_3$
Lapyrium Chloride*, 5238
C$_{27}$H$_{48}$
Cholestane, 2202
Coprostane, 2518
C$_{27}$H$_{48}$N$_6$O$_9$
Nocardamin, 6587
C$_{27}$H$_{48}$O
Cholestanol, 2203
Coprosterol, 2519
Epicholestanol, 3564
C$_{27}$H$_{49}$N$_7$O$_{17}$
Streptomycin B, 8787
C$_{27}$H$_{50}$N$_6$O$_8$
Deferoxamine*, 2850
C$_{28}$H$_{14}$N$_2$O$_4$
Indanthrene®, 4846
C$_{28}$H$_{15}$NO$_4$
Anthrimide, 720
C$_{28}$H$_{17}$N$_3$O$_4$
Bis(4-amino-1-anthraquinonyl)-
amine, 1254
C$_{28}$H$_{18}$O$_4$
α-Naphtholphthalein, 6309
C$_{28}$H$_{19}$N$_5$Na$_2$O$_6$S$_4$
Thiazol Yellow G, 9238
C$_{28}$H$_{20}$CaN$_2$O$_8$.5H$_2$O
Benzoylpas*, 1127
C$_{28}$H$_{20}$N$_2$Na$_2$O$_8$S$_2$
Alizarin Cyanine Green F, 238
C$_{28}$H$_{22}$Cl$_2$FNO$_3$
Flumethrin, 4068
C$_{28}$H$_{22}$N$_2$O$_2$
Quinizarin Green SS, 8095
C$_{28}$H$_{22}$N$_4$Na$_2$O$_8$S$_2$
Blancophor® R, 1322
C$_{28}$H$_{24}$Br$_2$N$_6$
Fazadinium Bromide, 3887
C$_{28}$H$_{25}$N$_6$NaO$_{10}$S$_2$
Cefpimizole*, 1939
C$_{28}$H$_{26}$ClN$_7$
Isometamidium Chloride, 5067
C$_{28}$H$_{26}$N$_6$O$_{10}$S$_2$
Cefpimizole, 1939
C$_{28}$H$_{27}$ClF$_5$NO
Penfluridol, 7028
C$_{28}$H$_{28}$N$_2$O$_2$
Difenoxin, 3124
C$_{28}$H$_{29}$ClN$_2$O$_2$
Difenoxin*, 3124
C$_{28}$H$_{29}$F$_2$N$_3$O
Pimozide, 7404

C$_{28}$H$_{29}$NO$_4$
Bephenium*, 1166
C$_{28}$H$_{30}$N$_2$O$_8$
Azatadine*, 917
C$_{28}$H$_{30}$O$_4$
Thymolphthalein, 9336
C$_{28}$H$_{30}$O$_{14}$.3H$_2$O
Pseudobaptigenin*, 7925
C$_{28}$H$_{31}$ClN$_2$O$_3$
Rhodamine B, 8182
C$_{28}$H$_{31}$FN$_4$O
Astemizole, 878
C$_{28}$H$_{31}$NO$_5$
Bitolterol, 1317
C$_{28}$H$_{31}$NO$_9$
Rhodomycins*, 8191
C$_{28}$H$_{31}$N$_3$O$_6$
Hepronicate, 4574
C$_{28}$H$_{31}$NaO$_9$S
Prednisolone Sodium 21-m-Sulfo-
benzoate, 7723
C$_{28}$H$_{32}$ClN$_3$O$_8$S
Prochlorperazine*, 7768
C$_{28}$H$_{32}$FNO$_6$
Dexamethasone*, 2922
C$_{28}$H$_{32}$IN$_3$S$_2$
Tibezonium Iodide, 9357
C$_{28}$H$_{32}$NaO$_9$S
Prednisolone*, 7719
C$_{28}$H$_{32}$O$_{14}$
Acacetin*, 9
Linarin, 5376
C$_{28}$H$_{32}$O$_{15}$
Diosmin, 3291
C$_{28}$H$_{33}$ClN$_2$
Buclizine, 1449
C$_{28}$H$_{33}$ClO$_{16}$
Peonidin*, 7099
C$_{28}$H$_{33}$ClO$_{17}$
Petunidin*, 7143
C$_{28}$H$_{33}$NO$_2$
Xenbucin*, 9980
C$_{28}$H$_{33}$N$_5$O$_{11}$
Butamben*, 1504
C$_{28}$H$_{34}$N$_2$O$_2$
Carbiphene, 1805
C$_{28}$H$_{34}$N$_2$O$_3$
Denatonium Benzoate, 2877
C$_{28}$H$_{34}$O$_8$
Uliginosins*, 9756
C$_{28}$H$_{34}$O$_9$
Nomilin, 6593
C$_{28}$H$_{34}$O$_{15}$
Hesperidin, 4591
C$_{28}$H$_{35}$ClN$_2$O$_2$
Carbiphene*, 1805
C$_{28}$H$_{35}$Cl$_3$N$_2$
Buclizine*, 1449
C$_{28}$H$_{35}$FO$_7$
Amcinonide, 398
C$_{28}$H$_{35}$FO$_9$
Triamcinolone Acetonide*, 9512
C$_{28}$H$_{35}$N$_3$O$_7$
Virginiamycin*, 9910
C$_{28}$H$_{36}$N$_2$O$_4$
Psychotrine, 7946
C$_{28}$H$_{36}$N$_2$O$_8$
Fentanyl*, 3944
C$_{28}$H$_{36}$O$_6$
Clinofibrate, 2353
C$_{28}$H$_{36}$O$_8$
Uliginosins*, 9756
C$_{28}$H$_{36}$O$_{11}$
Bruceantin, 1442
C$_{28}$H$_{36}$O$_{15}$
Neohesperidin Dihydrochalcone,
6367
C$_{28}$H$_{37}$ClN$_2$O$_4$
Viminol*, 9885
C$_{28}$H$_{37}$ClN$_4$O
Clocapramine, 2365

C$_{28}$H$_{37}$ClO$_7$
Alclometasone*, 213
Beclomethasone*, 1029
C$_{28}$H$_{37}$FO$_7$
Betamethasone*, 1202
Dexamethasone*, 2922
C$_{28}$H$_{37}$F$_2$N$_7$O$_6$S$_2$
Almitrine*, 299
C$_{28}$H$_{37}$N$_5$O$_7$
Endorphins*, 3528
C$_{28}$H$_{38}$BrNO$_4$
Butropium Bromide, 1534
C$_{28}$H$_{38}$N$_2$O$_4$
Cephaeline, 1970
C$_{28}$H$_{38}$N$_2$O$_6$
Simetride, 8487
C$_{28}$H$_{38}$N$_2$O$_9$S
Sufentanil*, 8860
C$_{28}$H$_{38}$N$_4$
Carpipramine, 1869
C$_{28}$H$_{38}$N$_4$O$_8$
Pactamycin, 6938
C$_{28}$H$_{38}$O$_6$
Withaferin A, 9960
C$_{28}$H$_{38}$O$_7$
Bongkrekic Acid, 1334
Prednival*, 7728
C$_{28}$H$_{38}$O$_{19}$
Cellobiose*, 1957
Sucrose Octaacetate, 8856
C$_{28}$H$_{39}$Cl$_3$N$_4$O.H$_2$O
Clocapramine*, 2365
C$_{28}$H$_{39}$FO$_5$
Fluocortolone*, 4079
C$_{28}$H$_{39}$FO$_6$
Dexamethasone*, 2922
C$_{28}$H$_{39}$NO$_6$
Prednylidene 21-Diethylamino-
acetate, 7730
C$_{28}$H$_{40}$ClNO$_6$
Prednylidene 21-Diethylamino-
acetate*, 7730
C$_{28}$H$_{40}$Cl$_2$N$_4$O.H$_2$O
Carpipramine*, 1869
C$_{28}$H$_{40}$N$_2$O$_2$
Bialamicol, 1217
C$_{28}$H$_{40}$N$_2$O$_5$
Gallopamil, 4257
C$_{28}$H$_{40}$N$_2$O$_9$
Antimycin A$_1$, 746
C$_{28}$H$_{40}$O$_4$
Pentagestrone*, 7068
C$_{28}$H$_{40}$O$_7$
Diginin, 3139
C$_{28}$H$_{41}$ClN$_2$O$_5$
Gallopamil*, 4257
C$_{28}$H$_{41}$FNO$_6$
Dexamethasone*, 2922
C$_{28}$H$_{41}$N$_3$O$_3$
Oxethazaine, 6888
Tiropramide, 9393
C$_{28}$H$_{41}$N$_3$O$_8$
Detoxin Complex*, 2918
C$_{28}$H$_{42}$ClN$_3$O$_3$
Tiropramide*, 9393
C$_{28}$H$_{42}$Cl$_4$N$_4$O$_2$
Ambenonium Chloride, 387
C$_{28}$H$_{42}$N$_2$O$_5$
Fenoxedil, 3931
C$_{28}$H$_{42}$O
Dehydroergosterol, 2861
C$_{28}$H$_{42}$O$_2$
ε-Tocopherol, 9420
C$_{28}$H$_{42}$O$_6$
Cynanchogenin, 2772
C$_{28}$H$_{43}$ClN$_2$O$_5$
Fenoxedil*, 3931
C$_{28}$H$_{44}$ClNO$_2$
Methylbenzethonium Chloride,
5946

Denotes derivative of title compound

$C_{28}H_{44}O$
Ergosterol, 3607
Fungisterol, 4203
Isopyrocalciferol, 5107
Lumisterol, 5471
Pyrocalciferol, 8007
Suprasterol II, 8984
Tachysterol, 9002
Vitamin D_2, 9928
$C_{28}H_{44}O_3$
Nandrolone Decanoate, 6281
$C_{28}H_{46}O$
Dihydrotachysterol, 3163
Ergosterol*, 3607
Vitamin D_4, 9930
$C_{28}H_{47}NO_4S$
Tiamulin, 9351
$C_{28}H_{47}NO_7$
Narbomycin, 6340
$C_{28}H_{47}NO_8$
Picromycin, 7384
$C_{28}H_{48}O$
Campesterol, 1735
α-Ergostenol, 3604
β-Ergostenol, 3605
γ-Ergostenol, 3606
$C_{28}H_{48}O_2$
β-Tocopherol, 9417
γ-Tocopherol, 9418
ζ_2-Tocopherol, 9422
$C_{28}H_{48}O_6$
Brassinolide, 1360
$C_{28}H_{50}$
Coproergostane, 2516
Ergostane, 3602
$C_{28}H_{50}N_2O_4$
Carpaine, 1866
$C_{28}H_{50}O$
Ergostanol, 3603
$C_{28}H_{58}NO_7P$
Platelet Activating Factor*, 7496
$C_{28}H_{58}O$
Octacosanol, 6664
$C_{29}H_{24}N_4O_8$
Niceritrol, 6405
$C_{29}H_{24}O_{12}$
Theaflavine, 9201
$C_{29}H_{25}N_3O_5$
Nicomorphine, 6430
$C_{29}H_{27}ClN_2O_{14}S$
Meclocycline*, 5658
$C_{29}H_{27}NO_4S_2$
Tipepidine*, 9389
$C_{29}H_{30}O_{13}$
Amarogentin, 384
$C_{29}H_{31}F_2N_3O$
Fluspirilene, 4131
$C_{29}H_{31}N_3O_4S$
Penicillin G Benzhydrylamine, 7038
$C_{29}H_{32}N_2O_7S_3$
Phencarbamide*, 7178
$C_{29}H_{32}O_{13}$
Etoposide, 3842
$C_{29}H_{33}ClN_2O_2$
Loperamide, 5450
$C_{29}H_{33}FO_6$
Betamethasone*, 1202
$C_{29}H_{33}N_9O_{12}$
Pteropterin, 7950
$C_{29}H_{34}Cl_2N_2O_2$
Loperamide*, 5450
$C_{29}H_{34}O_6$
Tribenoside, 9520
$C_{29}H_{34}O_{15}$
Pectolinarigenin*, 7010
$C_{29}H_{35}ClO_{17}$
Malvidin Chloride*, 5596
$C_{29}H_{35}NO_2$
Mifepristone, 6109
$C_{29}H_{35}NO_8$
Deptropine*, 2894

$C_{29}H_{35}NO_8S$
Bitolterol*, 1317
$C_{29}H_{36}N_2O_4$
Emetamine, 3516
$C_{29}H_{36}O_4$
Algestone Acetophenide, 230
$C_{29}H_{37}NO_5$
Cytochalasins*, 2796
$C_{29}H_{37}N_3O_6$
Calcimycin, 1639
$C_{29}H_{38}ClFO_8$
Formocortal, 4156
$C_{29}H_{38}F_3N_3O_2S$
Fluphenazine*, 4116
$C_{29}H_{38}N_2O_4$
Dehydroemetine, 2860
$C_{29}H_{38}N_4O_6S.H_2O$
Penicillin G Procaine, 7042
$C_{29}H_{38}N_4O_9$
Pipacycline, 7420
$C_{29}H_{38}N_4O_{10}$
Lymecycline, 5500
$C_{29}H_{38}N_8O_8$
Guamecycline, 4468
$C_{29}H_{38}O_9$
Uscharidin, 9805
$C_{29}H_{39}N_3O_2$
Echinuline, 3473
$C_{29}H_{40}Cl_2N_2O_4$
Dehydroemetine*, 2860
$C_{29}H_{40}Cl_2N_8O_8$
Guamecycline*, 4468
$C_{29}H_{40}N_2O_4$
Emetine, 3517
$C_{29}H_{40}O_6$
Cortisone, 21β-Cyclopentanepro-
pionate, 2534
$C_{29}H_{40}O_9$
Calotropin, 1728
$C_{29}H_{41}NO_4$
Buprenorphine, 1487
$C_{29}H_{42}ClNO_4$
Buprenorphine*, 1487
$C_{29}H_{42}Cl_2N_2O_4$
Emetine*, 3517
$C_{29}H_{42}N_6O_9$
Amicetin, 406
$C_{29}H_{42}N_6O_{19}S_5$
S-Adenosylmethionine*, 146
$C_{29}H_{42}O_5$
Antheridiol, 710
$C_{29}H_{42}O_9$
Helveticoside, 4551
$C_{29}H_{42}O_{10}$
Adonitoxin, 158
Convallatoxin, 2508
$C_{29}H_{42}O_{11}$
α-Antiarin, 722
$C_{29}H_{44}O_2$
ζ_1-Tocopherol, 9421
$C_{29}H_{44}O_3$
Estradiol*, 3653
$C_{29}H_{44}O_{12}$
Ouabain, 6854
$C_{29}H_{45}N_3O_9S_2$
Pentacynium Bis(methyl sulfate),
7060
$C_{29}H_{47}NO_6$
Coumingine, 2567
$C_{29}H_{47}NO_8$
Sabadine, 8281
$C_{29}H_{48}NO_6$
Choline Dehydrocholate, 2209
$C_{29}H_{48}O$
Chondrillasterol, 2214
7-Dehydrositosterol, 2863
Fucosterol, 4194
α-Spinasterol, 8705
Stigmasterol, 8771
$C_{29}H_{48}O_3$
Testosterone*, 9109

$C_{29}H_{50}O$
β-Sitosterol, 8500
γ-Sitosterol, 8501
$C_{29}H_{50}O_2$
Vitamin E, 9931
$C_{29}H_{52}O$
Stigmastanol, 8770
$C_{30}H_{16}O_8$
Hypericin, 4799
$C_{30}H_{18}Al_2O_{21}S_6$
Aluminum β-Naphtholdisulfonate,
353
$C_{30}H_{21}FeN_3O_{15}{}^{3-}$
Enterobactin*, 3547
$C_{30}H_{21}NO_9$
Fredericamycin A, 4181
$C_{30}H_{23}BrO_4$
Bromadiolone, 1371
$C_{30}H_{24}N_4O_{10}$
Nicofuranose, 6428
$C_{30}H_{26}CaO_6.2H_2O$
Fenoprofen*, 3926
$C_{30}H_{26}F_6N_4O_2$
Antrafenine, 754
$C_{30}H_{26}MgO_8$
Menbutone*, 5719
$C_{30}H_{26}O_{14}$
Ergoflavin, 3597
$C_{30}H_{27}N_3O_{15}$
Enterobactin, 3547
$C_{30}H_{28}N_2Na_4O_{14}S_5$
Solasulfone, 8667
$C_{30}H_{28}N_6O_6S_4$
Verticillins*, 9872
$C_{30}H_{28}N_6O_7S_4$
Verticillins*, 9872
$C_{30}H_{28}N_6O_7S_5$
Verticillins*, 9872
$C_{30}H_{28}O_8$
Rottlerin, 8250
$C_{30}H_{30}N_4$
Bis(p-dimethylaminobenzylidene)-
benzidine, 1260
$C_{30}H_{30}O_8$
Gossypol, 4435
$C_{30}H_{31}ClN_6$
Janus Green B, 5138
$C_{30}H_{32}N_2O_2$
Diphenoxylate, 3313
$C_{30}H_{33}O_8FNa$
Dexamethasone*, 2922
$C_{30}H_{34}BrNO_3$
Xenytropium Bromide, 9983
$C_{30}H_{34}O_{13}$
Picrotoxin, 7388
$C_{30}H_{35}F_2N_3O$
Lidoflazine, 5360
$C_{30}H_{36}O_9$
Nimbin, 6462
$C_{30}H_{37}N_3O_8S_2$
Thiethylperazine*, 9241
$C_{30}H_{37}N_5O_5$
Ergosine, 3601
$C_{30}H_{38}N_4O_{11}$
Apicycline, 762
$C_{30}H_{40}Cl_2N_4$
Dequalinium Chloride, 2896
$C_{30}H_{40}O_2$
β-Citraurin, 2326
$C_{30}H_{40}O_6$
Absinthin, 7
Anabsinthin, 656
$C_{30}H_{41}FO_7$
Triamcinolone Hexacetonide, 9514
$C_{30}H_{42}N_2O_{15}S_2$
Sinalbin, 8492
$C_{30}H_{42}O_8$
Proscillaridin, 7889
$C_{30}H_{44}K_2O_{16}S_2$
Atractyloside, 883

C₃₀H₄₄N₂O₁₀
$C_{30}H_{44}N_2O_{10}$
Hexobendine, 4626

$C_{30}H_{44}O_3$
Boldenone*, 1327

$C_{30}H_{44}O_9$
Cymarin, 2768
Peruvoside, 7136

$C_{30}H_{44}O_{10}$
Vernadigin, 9864

$C_{30}H_{45}ClO_6$
Senegenin, 8405

$C_{30}H_{45}NNaO_7P$
Fosinopril*, 4171

$C_{30}H_{46}Cl_2N_2O_{10}$
Hexobendine*, 4626

$C_{30}H_{46}NO_7P$
Fosinopril, 4171

$C_{30}H_{46}N_2O_{14}S_2$
Aclatonium Napadisilate, 109

$C_{30}H_{46}N_4O_8S$
Mepindolol*, 5744

$C_{30}H_{46}O_4$
Enoxolone, 3543
Gypsogenin, 4497

$C_{30}H_{46}O_5$
Ceanothic Acid, 1916
Quillaic Acid, 8049
Quinovic Acid, 8104

$C_{30}H_{46}O_8$
Neriifolin, 6386
Periplocymarin, 7127

$C_{30}H_{47}N_3O_9S$
Leukotrienes*, 5339

$C_{30}H_{48}N_2O_2$
Hexestrol Bis(β-diethylaminoethyl ether), 4622

$C_{30}H_{48}O_3$
β-Boswellic Acid, 1354
Nandrolone*, 6280
Oleanolic Acid, 6787
Ursolic Acid, 9802

$C_{30}H_{48}O_4$
Gratiogenin, 4446
Hederagenin, 4540

$C_{30}H_{48}O_5$
Acacic Acid, 11
Cimigenol, 2282
Escigenin, 3643

$C_{30}H_{49}N_9O_9$
Thymopentin, 9338

$C_{30}H_{50}$
Squalene, 8727

$C_{30}H_{50}Cl_2N_2O_2$
Hexestrol Bis(β-diethylaminoethyl ether)*, 4622

$C_{30}H_{50}O$
α-Amyrin, 653
β-Amyrin, 654
Friedelin, 4184
Lanosterol, 5232
Lupeol, 5478
Shionone, 8429
α₁-Sitosterol, 8499
Taraxasterol, 9031
Taraxerol, 9033

$C_{30}H_{50}O_2$
Betulin, 1212

$C_{30}H_{51}B$
Longifolene*, 5446

$C_{30}H_{52}O$
Dinosterol, 3283

$C_{30}H_{52}O_{26}$
Verbascose, 9860

$C_{30}H_{53}NO_{11}$
Benzonatate, 1106

$C_{30}H_{57}BO_3$
Menthyl Borate, 5726

$C_{30}H_{58}N_8O_{17}$
Dihydrostreptomycin*, 3162

$C_{30}H_{60}I_3N_3O_3$
Gallamine Triethiodide, 4249

$C_{30}H_{62}$
Squalane, 8726

$C_{30}H_{62}O$
1-Triacontanol, 9506

$C_{30}H_{66}N_4O_8S$
Dibutoline Sulfate, 3018

$C_{31}H_{23}BrO_3$
Brodifacoum, 1368

$C_{31}H_{27}N_3O_5$
Nalorphine Dinicotinate, 6276

$C_{31}H_{28}O_{14}$
Ergochrysins, 3590

$C_{31}H_{29}N_5O_4$
Asperlicin, 865

$C_{31}H_{32}N_4O_2$
Bezitramide, 1216

$C_{31}H_{34}BrNO_4$
Fentonium Bromide, 3949

$C_{31}H_{35}N_2NaO_{11}$
Novobiocin*, 6641

$C_{31}H_{35}N_3O_4S$
Penicillin G Benethamine, 7036

$C_{31}H_{36}ClN_3O_5S$
Metofenazate, 6065

$C_{31}H_{36}IN_3$
Stilbazium Iodide, 8773

$C_{31}H_{36}N_2O_8$
Raunescine, 8130

$C_{31}H_{36}N_2O_{11}$
Novobiocin, 6641

$C_{31}H_{37}N_3O_{10}S$
Acetophenazine*, 64

$C_{31}H_{39}ClN_2O_{11}$
Etodroxizine*, 3832

$C_{31}H_{39}N_5O_5$
Ergocornine, 3591
Ergocorninine, 3592

$C_{31}H_{41}ClFNO_3$
Haloperidol*, 4511

$C_{31}H_{42}N_2O_6$
Batrachotoxin, 1018

$C_{31}H_{44}N_2O_{10}$
Dilazep, 3185

$C_{31}H_{44}O_7$
Milbemycins*, 6112

$C_{31}H_{44}O_8$
Proscillaridin*, 7889

$C_{31}H_{46}Cl_2N_2O_{10}$
Dilazep*, 3185

$C_{31}H_{46}O_2$
Vitamin K₁, 9933

$C_{31}H_{46}O_3$
Vitamin K₁ Oxide, 9934

$C_{31}H_{47}NaO_6$
Fusidic Acid*, 4231

$C_{31}H_{48}N_{10}O_6$
Metformin*, 5845

$C_{31}H_{48}Na_2O_8P_2$
Dihydrovitamin K₁*, 3165

$C_{31}H_{48}O_2$
Dihydrovitamin K₁, 3165

$C_{31}H_{48}O_2S_2$
Probucol, 7761

$C_{31}H_{48}O_4$
Testosterone*, 9109

$C_{31}H_{48}O_6$
Fusidic Acid, 4231

$C_{31}H_{50}O_4$
Sulphurenic Acid, 8970

$C_{31}H_{51}NO_9$
Rosaramicin, 8241

$C_{31}H_{52}O_3$
Vitamin E Acetate, 9932

$C_{31}H_{55}N_3O_6$
Enocitabine, 3539

$C_{32}H_{16}CuN_8$
Copper Phthalocyanine, 2515

$C_{32}H_{19}N_6Na_5O_{15}S_5$
Trypan Red, 9702

$C_{32}H_{22}N_6Na_2O_6S_2$
Congo Red, 2493

$C_{32}H_{24}O_{12}$
Thermorubin*, 9214

$C_{32}H_{26}MgO_6$
3-(o-Methoxyphenyl)-2-phenyl-acrylic Acid*, 5921

$C_{32}H_{26}O_{14}$
Actinorhodine, 135

$C_{32}H_{30}O_{14}$
Chrysergonic Acid, 2260
Secalonic Acids, 8372

$C_{32}H_{32}Ag_2N_4O_4S$
Diathymosulfone*, 2972

$C_{32}H_{32}N_2O_{10}$
Cotarnine*, 2551

$C_{32}H_{32}O_{13}S$
Teniposide, 9078

$C_{32}H_{32}O_{14}$
Chartreusin, 2035

$C_{32}H_{34}CaN_4O_8S_2$
Penicillin G Calcium, 7039

$C_{32}H_{34}CaN_4O_{10}S_2$
Penicillin V*, 7046

$C_{32}H_{34}N_4O_4S$
Diathymosulfone, 2972

$C_{32}H_{35}ClN_{18}O_{10}S_6$
Cefmenoxime*, 1928

$C_{32}H_{36}CaCl_2O_6$
Bucloxic Acid*, 1451

$C_{32}H_{36}ClNO_8$
Clomiphene*, 2384
Toremifene*, 9474

$C_{32}H_{37}NO_5S$
Levopropoxyphene*, 5349

$C_{32}H_{37}NO_5S.H_2O$
Propoxyphene*, 7851

$C_{32}H_{37}NO_8$
Tamoxifen*, 9019

$C_{32}H_{37}NO_{12}$
Pirarubicin, 7460

$C_{32}H_{38}N_2O_5$
Cortivazol, 2536

$C_{32}H_{38}N_2O_8$
Deserpidine, 2901

$C_{32}H_{38}N_4$
Etioporphyrin, 3826

$C_{32}H_{39}NO_{11}S$
Noscapine*, 6638

$C_{32}H_{39}N_3O_9S$
Butaperazine*, 1509

$C_{32}H_{40}BrN_5O_5$
Bromocriptine, 1404

$C_{32}H_{40}N_2O_5S_2$
Trimethaphan Camsylate, 9621

$C_{32}H_{41}NO_2$
Terfenadine, 9094

$C_{32}H_{41}N_5O_5$
Ergocriptine, 3595
Ergocriptinine, 3596

$C_{32}H_{44}F_3N_3O_2S$
Fluphenazine*, 4116

$C_{32}H_{44}N_2O_8$
Lappaconitine, 5237

$C_{32}H_{44}N_2O_9$
Streptolydigin, 8785

$C_{32}H_{44}O_8$
Cucurbitacins*, 2617

$C_{32}H_{44}O_{12}$
Scilliroside, 8356

$C_{32}H_{45}FO_7$
Betamethasone*, 1202

$C_{32}H_{46}N_8O_6S_2$
Thiamine Disulfide*, 9221

$C_{32}H_{46}O_8$
Cucurbitacins*, 2617

$C_{32}H_{46}O_{10}$
Tanghinin, 9022

$C_{32}H_{48}N_2O_{10}$
Bisobrin*, 1308
Butobendine, 1524

$C_{32}H_{48}O_5$
Acetoxolone, 69

Denotes derivative of title compound

$C_{32}H_{48}O_7$
Milbemycins*, 6112
$C_{32}H_{48}O_9$
Neriifolin*, 6386
Oleandrin, 6786
$C_{32}H_{49}NO_9$
Cevadine, 2025
$C_{32}H_{50}Cl_2N_2O_{10}$
Butobendine*, 1524
$C_{32}H_{51}NO_8S$
Tiamulin*, 9351
$C_{32}H_{52}Br_2N_4O_4$
Demecarium Bromide, 2871
$C_{32}H_{52}O_2$
Taraxasterol*, 9031
$C_{32}H_{55}BrN_4O$
Thonzonium Bromide, 9306
$C_{32}H_{56}N_4O_8$
Sporidesmolides*, 8725
$C_{32}H_{58}N_2O_7S$
CHAPS, 2034
$C_{32}H_{58}N_2O_8S$
CHAPS*, 2034
$C_{32}H_{62}CaO_4$
Calcium Palmitate, 1693
$C_{32}H_{64}O_2$
Cetyl Palmitate, 2023
$C_{32}H_{64}O_4Sn$
Dibutyltin Dilaurate, 3025
$C_{33}H_{24}Hg_2O_6S_2$
Hydrargaphen, 4687
$C_{33}H_{25}N_3O_3$
Norbormide, 6604
$C_{33}H_{34}N_4O_6$
Biliverdine, 1236
$C_{33}H_{35}N_5O_5$
Ergotamine, 3609
Ergotaminine, 3610
$C_{33}H_{36}N_4O_6$
Bilirubin, 1235
$C_{33}H_{37}N_5O_5$
Dihydroergotamine, 3157
$C_{33}H_{38}N_2O_8$
Rescimetol, 8145
$C_{33}H_{40}N_2O_9$
Deserpidine*, 2901
Reserpine, 8149
$C_{33}H_{40}O_{19}$
Robinin, 8230
$C_{33}H_{42}O_{19}$
Troxerutin, 9698
$C_{33}H_{43}FO_6$
Betamethasone*, 1202
$C_{33}H_{44}BrN_5O_8S$
Bromocriptine*, 1404
$C_{33}H_{44}N_4O_{10}$
Riboflavine*, 8201
$C_{33}H_{44}O_8$
Helvolic Acid, 4552
$C_{33}H_{45}NO_9$
Delphinine, 2867
$C_{33}H_{46}Bi_2O_{11}$
d-Camphocarboxylic Acid*, 1737
$C_{33}H_{46}O_4$
Nandrolone p-Hexyloxyphenylpropionate, 6282
$C_{33}H_{47}NO_{13}$
Natamycin, 6346
$C_{33}H_{48}O_7$
Milbemycins*, 6112
$C_{33}H_{49}NO_8$
Pseudojervine, 7932
$C_{33}H_{50}O_8$
Cephalosporin P_1, 1977
$C_{33}H_{53}NO_8$
Imperialine*, 4840
$C_{33}H_{54}O_5$
α-Tocopherol Acid Succinate, 9424
$C_{33}H_{55}NO_8$
Verticine*, 9873

$C_{33}H_{57}N_3O_9$
Enniatins*, 3538
$C_{33}H_{58}N_4O_8$
Sporidesmolides*, 8725
$C_{34}H_{24}CaCl_2N_4O_4$
Lonazolac*, 5445
$C_{34}H_{24}N_4Na_2O_{10}S_2$
Benzo Azurine G, 1095
$C_{34}H_{24}N_6Na_4O_{14}S_4$
Evan's Blue, 3863
Trypan Blue, 9701
$C_{34}H_{25}N_6Na_3O_9S_3$
Vital Red, 9917
$C_{34}H_{26}N_6Na_2O_6S_2$
Benzopurpurine 4B, 1111
$C_{34}H_{30}N_2O_6S$
Pyrantel*, 7968
$C_{34}H_{32}ClFeN_4O_4$
Hemin, 4563
$C_{34}H_{32}FeN_4O_4$
Heme, 4558
$C_{34}H_{32}N_2O_5$
Micranthine, 6102
$C_{34}H_{32}N_4Na_2O_4$
Protoporphyrin IX*, 7913
$C_{34}H_{32}N_4O_9$
Nicomol, 6429
$C_{34}H_{32}N_6O_4$
1,3-Diphenylguanidine*, 3325
$C_{34}H_{33}FeN_4O_5$
Hematin, 4554
$C_{34}H_{34}HgN_4Na_2O_6$
Merphyrin, 5804
$C_{34}H_{34}N_4O_4$
Protoporphyrin IX, 7913
$C_{34}H_{35}N_3O_{10}$
Zorubicin, 10096
$C_{34}H_{36}N_2O_6$
Pseudomorphine, 7934
$C_{34}H_{36}N_4O_6$
Phytochlorin, 7360
$C_{34}H_{38}N_2O_4$
Nafiverine, 6267
$C_{34}H_{38}N_4O_6$
Hematoporphyrin, 4555
$C_{34}H_{39}ClN_4O_6.H_2O$
Hematoporphyrin*, 4555
$C_{34}H_{40}N_2O_{10}S$
Morphine Sulfate, 6193
$C_{34}H_{40}O_{12}$
Filixic Acids*, 4023
$C_{34}H_{41}N_5O_8S$
Dihydroergotamine*, 3157
$C_{34}H_{42}N_2O_4$
Belladonnine, 1035
$C_{34}H_{44}F_3N_3O_{10}S$
Oxaflumazine*, 6862
$C_{34}H_{44}N_2O_8S.H_2O$
Belladonnine*, 1035
$C_{34}H_{46}ClN_3O_{10}$
Maytansine, 5642
$C_{34}H_{46}Cl_2N_2$
Hedaquinium Chloride, 4539
$C_{34}H_{46}N_4O_4$
Dequalinium Acetate, 2895
$C_{34}H_{46}O_8$
Helvolic Acid*, 4552
$C_{34}H_{47}NO_{10}$
Indaconitine, 4842
$C_{34}H_{47}NO_{11}$
Aconitine, 113
$C_{34}H_{48}I_6N_4O_9$
Iodipamide*, 4916
$C_{34}H_{48}N_2O_9$
Ajacine, 182
$C_{34}H_{48}N_2O_{10}S.H_2O$
Atropine*, 891
$C_{34}H_{48}N_2O_{10}S.2H_2O$
Hyoscyamine*, 4795
$C_{34}H_{48}Na_2O_7$
Carbenoxolone*, 1797

$C_{34}H_{50}Cl_2N_4O_2$
Benzoquinonium Chloride, 1116
$C_{34}H_{50}O_7$
Carbenoxolone, 1797
$C_{34}H_{51}NO_{13}$
Tetrin*, 9174
$C_{34}H_{51}NO_{14}$
Tetrin*, 9174
$C_{34}H_{54}Cl_2N_{10}O_{14}$
Chlorhexidine*, 2090
$C_{34}H_{54}O_8$
Lasalocid A, 5243
$C_{34}H_{56}CuN_6O_{14}S_4$
Cuproxoline, 2677
$C_{34}H_{56}N_2O_{12}$
Metoprolol*, 6072
$C_{34}H_{57}BrN_2O_4$
Vecuronium Bromide, 9845
$C_{34}H_{59}N_3O_9$
Enniatins*, 3538
$C_{34}H_{60}N_4O_8$
Sporidesmolides*, 8725
$C_{34}H_{63}N_5O_9$
Pepstatin, 7104
$C_{35}H_{34}N_2O_5$
Trilobine, 9610
$C_{35}H_{36}N_2O_6$
Daphnoline, 2818
$C_{35}H_{38}Cl_2N_8O_4$
Itraconazole, 5131
$C_{35}H_{39}N_5O_5$
Ergocristine, 3593
Ergocristinine, 3594
$C_{35}H_{41}N_9O_9$
Bietanautine, 1225
$C_{35}H_{42}FNO_8$
Triamcinolone Benetonide, 9513
$C_{35}H_{42}N_2O_9$
Rescinnamine, 8146
$C_{35}H_{42}N_2O_{11}$
Syrosingopine, 8998
$C_{35}H_{42}O_9$
Taxicins*, 9046
$C_{35}H_{42}O_{12}$
Filixic Acids*, 4023
$C_{35}H_{44}O_{16}$
Azadirachtin, 910
$C_{35}H_{45}Cl_2NO_6$
Prednimustine, 7718
$C_{35}H_{48}N_8O_{11}S$
Phalloidin, 7147
$C_{35}H_{48}N_{10}O_{15}$
DSIP, 3445
$C_{35}H_{49}NO_{12}$
Jesaconitine, 5146
$C_{35}H_{53}FeN_6O_{13}$
Coprogen, 2517
$C_{35}H_{53}NO_3$
Vitamin E*, 9931
$C_{35}H_{54}O_8$
Tetronasin, 9176
$C_{35}H_{54}O_{14}$
Uzarin, 9809
$C_{35}H_{55}N_9O_{16}$
Peptide T, 7105
$C_{35}H_{56}O_{14}$
Chalcomycin, 2027
$C_{35}H_{58}O_{11}$
Filipin*, 4022
$C_{35}H_{58}O_{12}$
Fungichromin, 4202
$C_{35}H_{59}Al_3N_{10}O_{14}$
Aceglutamide*, 21
$C_{35}H_{60}Br_2N_2O_4$
Pancuronium Bromide, 6958
$C_{35}H_{61}NO_{12}$
Oleandomycin, 6785
$C_{35}H_{62}Br_2N_4O_4.2H_2O$
Pipecurium Bromide, 7426
$C_{35}H_{64}NO_{16}P$
Oleandomycin*, 6785

$C_{36}H_{36}N_2O_5$
Tiliacorine, 9369
Trilobine*, 9610

$C_{36}H_{37}ClO_{17}$
Monardein Chloride, 6154

$C_{36}H_{37}ClO_{18}$
Delphinidin*, 2866

$C_{36}H_{38}N_2O_6$
Bebeerine, 1024
Chondrocurine, 2215
Curine, 2682
Daphnandrine, 2815
Isochondodendrine, 5041

$C_{36}H_{40}N_2O_6$
Magnoline, 5580

$C_{36}H_{42}Br_2N_2$
Hexafluorenium Bromide, 4606

$C_{36}H_{44}N_2O_{10}S$
Codeine Sulfate, 2463

$C_{36}H_{44}O_{12}$
Filixic Acids*, 4023

$C_{36}H_{45}N_5O_8S$
Ergoloid Mesylates*, 3598

$C_{36}H_{48}N_2O_{10}$
Lycoctonine*, 5488

$C_{36}H_{48}N_2O_{12}$
Rhodomycins*, 8191

$C_{36}H_{51}NO_{11}$
Bikhaconitine, 1234
Veratridine, 9855

$C_{36}H_{51}NO_{12}$
Pseudoaconitine, 7924

$C_{36}H_{52}O_{13}$
Scillaren*, 8354

$C_{36}H_{52}O_{15}$
Hellebrin, 4547

$C_{36}H_{53}NO_{13}$
Lucensomycin, 5464

$C_{36}H_{54}N_6O_{14}$
Xamoterol*, 9965

$C_{36}H_{55}NO_{12}$
Desatrine, 2900

$C_{36}H_{56}O_8$
Phorbol*, 7306

$C_{36}H_{56}O_9$
Quinovin*, 8105

$C_{36}H_{56}O_{10}$
Quinovin*, 8105

$C_{36}H_{56}O_{13}$
Periplocin, 7126

$C_{36}H_{56}O_{14}$
Digitalin, 3140

$C_{36}H_{57}BiO_{12}$
Bismuth Ethyl Camphorate, 1275

$C_{36}H_{60}N_2O_8S$
Penbutolol*, 7025

$C_{36}H_{60}O_2$
Vitamin A*, 9918

$C_{36}H_{60}O_{30}$
Cyclodextrins*, 2724

$C_{36}H_{61}NaO_{11}$
Monensin*, 6157

$C_{36}H_{62}N_4O_8S$
Butacaine*, 1496

$C_{36}H_{62}O_{11}$
Monensin, 6157

$C_{36}H_{63}N_3O_9$
Enniatins*, 3538

$C_{36}H_{64}N_2O_{12}$
Bisoprolol*, 1309

$C_{36}H_{66}BaO_4$
Oleic Acid*, 6788

$C_{36}H_{66}CaO_4$
Calcium Oleate, 1690

$C_{36}H_{66}CuO_4$
Cupric Oleate, 2648

$C_{36}H_{66}HgO_4$
Mercuric Oleate, 5778

$C_{36}H_{66}MgO_4$
Magnesium Oleate, 5553

$C_{36}H_{66}O_4Zn$
Zinc Oleate, 10047

$C_{36}H_{68}CaI_2O_4$
Calcium Iodostearate, 1682

$C_{36}H_{70}CaO_4$
Calcium Stearate, 1710

$C_{36}H_{70}CuO_4$
Cupric Stearate, 2658

$C_{36}H_{70}HgO_4$
Mercuric Stearate, 5784

$C_{36}H_{70}MgO_4$
Magnesium Stearate, 5572

$C_{36}H_{70}O_4Zn$
Zinc Stearate, 10063

$C_{36}H_{74}Cl_2N_2$
Triclobisonium Chloride, 9567

$C_{37}H_{27}N_3Na_2O_9S_3$
Methyl Blue, 5950

$C_{37}H_{34}N_2Na_2O_9S_3$
Brilliant Blue FCF, 1366
Light Green SF Yellowish, 5361

$C_{37}H_{34}N_2Na_2O_{10}S_3$
Fast Green FCF, 3886

$C_{37}H_{35}N_2NaO_6S_2$
Guinea Green B, 4487

$C_{37}H_{36}N_3NaO_6S_2$
Acid Violet 7B, 104

$C_{37}H_{38}N_2O_6$
Cepharanthine, 1981

$C_{37}H_{40}N_2O_6$
Berbamine, 1168
Chondrofoline, 2216
Oxyacanthine, 6906

$C_{37}H_{40}N_4O_7$
Haplophytine, 4525

$C_{37}H_{42}Cl_2N_2O_6$
Tubocurarine Chloride, 9717

$C_{37}H_{42}N_2O_7$
Hydrocortisone*, 4710

$C_{37}H_{45}NO_{12}$
Rifamycins*, 8216

$C_{37}H_{47}NO_{12}$
Rifamycin SV, 8217

$C_{37}H_{48}I_6N_6O_{18}$
Iotrolan, 4955

$C_{37}H_{49}N_7O_9S$
Pentagastrin, 7067

$C_{37}H_{61}NO_9$
Pavoninin-5, 7003

$C_{37}H_{67}NO_{13}$
Erythromycin, 3626

$C_{37}H_{77}NO_2$
Cetrimonium Stearate, 2019

$C_{38}H_{28}Cl_2N_8$
Neotetrazolium Chloride, 6381

$C_{38}H_{30}NiO_2P_2$
Bis(triphenylphosphine)dicarbonyl-nickel, 1314

$C_{38}H_{36}CaN_4O_4$
Phenylbutazone*, 7248

$C_{38}H_{40}N_2O_6$
Insularine, 4886

$C_{38}H_{42}CaN_4O_6.\frac{1}{2}H_2O$
Bumadizon*, 1472

$C_{38}H_{42}N_2O_6$
Tetrandrine, 9162

$C_{38}H_{42}N_8O_6S_2$
Bisbentiamine, 1257

$C_{38}H_{44}N_2O_6$
Dauricine, 2826

$C_{38}H_{44}O_8$
Gambogic Acid, 4261

$C_{38}H_{48}N_4O_{11}$
Perisoxal*, 7129

$C_{38}H_{51}NO_4$
Myrophine, 6250

$C_{38}H_{52}Br_2N_2O_4.4H_2O$
Belladonnine*, 1035

$C_{38}H_{54}I_6N_6O_{18}$
Iocarmic Acid*, 4902

$C_{38}H_{54}O_{13}$
Colocynthin, 2483

$C_{38}H_{58}CaN_4O_{10}S_2$
Moveltipril*, 6200

$C_{38}H_{59}FO_6$
Dexamethasone*, 2922

$C_{38}H_{60}N_2O_{10}S_2$
Caramiphen Ethanedisulfonate, 1776

$C_{38}H_{60}O_{18}$
Stevioside, 8766

$C_{38}H_{61}I_2N_3S_3$
Platonin, 7504

$C_{38}H_{69}NO_{13}$
Clarithromycin, 2340

$C_{38}H_{72}N_2O_{12}$
Azithromycin, 928

$C_{38}H_{84}N_{10}O_{34}S_5$
Sisomicin*, 8498

$C_{39}H_{44}ClN_3O_{13}S$
Metofenazate*, 6065

$C_{39}H_{46}Cl_2N_2O_6$
Tubocurarine Chloride*, 9717

$C_{39}H_{47}NO_{14}$
Rifamycins*, 8216
Streptovaricin*, 8792

$C_{39}H_{49}NO_{14}$
Rifamycins*, 8216

$C_{39}H_{49}NO_{16}$
Nogalamycin, 6591

$C_{39}H_{53}N_3O_9$
Bietaserpine, 1226

$C_{39}H_{56}N_6O_5S$
Lazaroids*, 5264

$C_{39}H_{59}Cl_2NO_2$
Phenesterine, 7182

$C_{39}H_{61}NO_{14}$
Rimocidin, 8224

$C_{40}H_{32}Cl_2N_8O_2$
Tetrazolium Blue, 9173

$C_{40}H_{44}N_4O^{2+}$
C-Curarine I, 2679

$C_{40}H_{46}N_4O_2^{2+}$
Toxiferine I, 9479

$C_{40}H_{48}I_2N_2O_6$
Metocurine Iodide, 6064

$C_{40}H_{48}N_4O_2^{2+}$
C-Calebassine, 1722

$C_{40}H_{48}N_4O_3$
Geissospermine, 4275

$C_{40}H_{48}O_4$
Astacin, 875

$C_{40}H_{49}NO_{14}$
Streptovaricin*, 8792

$C_{40}H_{50}N_4O_6S_2$
Trimeprazine*, 9618

$C_{40}H_{50}N_4O_8S.2H_2O$
Quinine Sulfate*, 8089

$C_{40}H_{50}N_4O_8S$
Quinidine Sulfate, 8074
Quinine Sulfate, 8089

$C_{40}H_{50}O_2$
Rhodoxanthin, 8196

$C_{40}H_{51}NO_{13}$
Streptovaricin*, 8792

$C_{40}H_{51}NO_{14}$
Streptovaricin*, 8792

$C_{40}H_{51}NO_{15}$
Streptovaricin*, 8792

$C_{40}H_{52}O_2$
Canthaxanthin, 1756
Torularhodin, 9476

$C_{40}H_{52}O_4$
Astaxanthin, 877

$C_{40}H_{54}O$
Echinenone, 3469

Denotes derivative of title compound

$C_{40}H_{56}$
 α-Carotene, 1859
 β-Carotene, 1860
 γ-Carotene, 1861
 δ-Carotene, 1862
 Lycopene, 5492
$C_{40}H_{56}O$
 Cryptoxanthin, 2612
 Lycoxanthin, 5499
 Rubixanthin, 8265
$C_{40}H_{56}O_2$
 Lycophyll, 5493
 Xanthophyll, 9972
 Zeaxanthin, 10019
$C_{40}H_{56}O_3$
 Capsanthin, 1768
 Chrysanthemaxanthin, 2254
 Flavoxanthin, 4032
$C_{40}H_{56}O_4$
 Violaxanthin, 9902
$C_{40}H_{58}Cl_2N_4$
 Bisdequalinium Chloride, 1259
$C_{40}H_{58}O$
 Rhodopin, 8192
$C_{40}H_{58}O_4$
 γ-Oryzanol*, 6841
$C_{40}H_{60}BNaO_{14}$
 Aplasmomycin, 769
$C_{40}H_{60}N_4O_{10}$
 Bufotoxin, 1468
$C_{40}H_{60}O_2$
 Kitol, 5195
$C_{40}H_{63}N_3O_4S_2$
 Pipotiazine*, 7453
$C_{40}H_{64}O_{12}$
 Nonactin, 6594
$C_{40}H_{66}N_6$
 Ormosinine, 6823
$C_{40}H_{66}O_{10}$
 Venturicidins*, 9848
$C_{40}H_{67}NO_{14}$
 Leucomycins*, 5337
$C_{40}H_{68}$
 Phytofluene, 7361
$C_{40}H_{68}N_4O_{12}$
 Amidomycin, 413
$C_{40}H_{68}O_{11}$
 Nigericin, 6457
$C_{40}H_{71}NO_{14}$
 Erythromycin Propionate, 3631
$C_{40}H_{74}CaO_{14}S_2$
 Docusate Calcium, 3397
$C_{40}H_{92}N_{10}O_{34}S_5$
 Micronomicin*, 6104
$C_{41}H_{39}ClO_{21}$
 Delphinidin*, 2866
$C_{41}H_{46}N_4O_5$
 Quinine Carbonate, 8077
$C_{41}H_{47}Cl_2NO_6$
 Bestrabucil, 1199
$C_{41}H_{48}N_2O_8$
 Thalicarpine, 9181
$C_{41}H_{52}N_4O_{12}\cdot2H_2O$
 Formoterol*, 4159
$C_{41}H_{52}N_6O_7$
 Ergonovine*, 3600
$C_{41}H_{56}O_2$
 Vitamin(s) K$_2$*, 9935
$C_{41}H_{58}FeN_9O_{20}$
 Ferrichromes*, 3968
$C_{41}H_{60}O_4$
 γ-Oryzanol*, 6841
$C_{41}H_{63}NO_{14}$
 Protoveratrines*, 7915
$C_{41}H_{63}NO_{15}$
 Protoveratrines*, 7915
$C_{41}H_{64}O_8$
 Prednisolone 21-Stearoylglycolate, 7724
$C_{41}H_{64}O_{13}$
 Digitoxin, 3146

$C_{41}H_{64}O_{14}$
 Digoxin, 3150
 Gitoxin, 4328
$C_{41}H_{64}O_{15}$
 Diginatin, 3138
$C_{41}H_{65}NO_{15}$
 Midecamycins*, 6106
$C_{41}H_{66}O_{12}$
 α-Hederin, 4541
 Nonactin*, 6594
$C_{41}H_{67}NO_{11}$
 Venturicidins*, 9848
$C_{41}H_{67}NO_{15}$
 Midecamycins*, 6106
 Troleandomycin, 9681
$C_{41}H_{76}N_2O_{15}$
 Roxithromycin, 8253
$C_{42}H_{27}AlF_9N_3O_6$
 Flufenamic Acid*, 4060
$C_{42}H_{30}N_6O_{12}$
 Inositol Niacinate, 4885
$C_{42}H_{38}O_{20}$
 Sennoside A & B, 8407
$C_{42}H_{45}N_3O_7$
 Pamaquine*, 6953
$C_{42}H_{46}N_4O_8S$
 Strychnine Sulfate, 8826
$C_{42}H_{48}N_6O_8$
 Zolpidem*, 10091
$C_{42}H_{49}Cl_2N_5O_{16}$
 Senociclin, 8408
$C_{42}H_{51}NO_{15}$
 Aclacinomycins*, 108
$C_{42}H_{52}N_6O_{10}$
 Ergonovine*, 3600
$C_{42}H_{53}NO_{15}$
 Aclacinomycins*, 108
 Streptovaricin*, 8792
$C_{42}H_{53}NO_{16}$
 Streptovaricin*, 8792
$C_{42}H_{54}ClNO_{15}$
 Aclacinomycins*, 108
$C_{42}H_{54}N_6O_4S_2$
 Aminopromazine*, 479
$C_{42}H_{58}O_6$
 Fucoxanthin, 4195
$C_{42}H_{59}N_2O_{20}P_3$
 Glaucine*, 4332
$C_{42}H_{60}K_2O_{16}$
 Glycyrrhizic Acid*, 4401
$C_{42}H_{60}O_2$
 Rhodoviolascin, 8195
$C_{42}H_{62}N_8O_7S$
 Bottromycin*, 1355
$C_{42}H_{62}O_{16}$
 Glycyrrhizic Acid, 4401
$C_{42}H_{64}O_{19}$
 Thevetin A, 9215
$C_{42}H_{65}N_{11}O_{12}$
 Cargutocin, 1846
$C_{42}H_{65}N_{13}O_{10}$
 Saralasin, 8330
$C_{42}H_{66}O_{14}$
 Digoxin*, 3150
$C_{42}H_{66}O_{18}$
 Cerberoside, 1983
$C_{42}H_{67}NO_{15}$
 Carbomycin*, 1813
$C_{42}H_{67}NO_{16}$
 Carbomycin*, 1813
$C_{42}H_{68}O_{12}$
 Nonactin*, 6594
$C_{42}H_{68}O_{14}$
 Gratioside, 4447
$C_{42}H_{69}NO_{15}$
 Josamycin, 5148
 Rokitamycin, 8233
$C_{42}H_{69}NaO_{11}$
 Salinomycin*, 8305
$C_{42}H_{70}O_{11}$
 Salinomycin, 8305

$C_{42}H_{70}O_{35}$
 Cyclodextrins*, 2724
$C_{42}H_{72}O_{16}$
 Lankamycin, 5230
$C_{42}H_{84}CaCl_8N_{14}O_{24}$
 Streptomycin*, 8786
$C_{42}H_{84}N_{14}O_{36}S_3$
 Streptomycin*, 8786
$C_{42}H_{88}N_{14}O_{36}S_3$
 Dihydrostreptomycin*, 3162
$C_{42}H_{92}N_{10}O_{34}S_5$
 Netilmicin*, 6389
$C_{43}H_{42}O_{22}$
 Carthamin, 1876
$C_{43}H_{47}N_2NaO_6S_2$
 Indocyanine Green, 4868
$C_{43}H_{49}N_7O_{10}$
 Virginiamycin*, 9910
$C_{43}H_{51}N_3O_{11}$
 Rifaximin, 8218
$C_{43}H_{52}N_4O_5$
 Voacamine, 9944
$C_{43}H_{55}N_5O_7$
 Vindesine, 9892
$C_{43}H_{57}N_5O_{11}S$
 Vindesine*, 9892
$C_{43}H_{58}N_2O_{13}$
 Rifamide, 8214
$C_{43}H_{58}N_4O_{12}$
 Rifampin, 8215
$C_{43}H_{60}N_2O_{12}$
 Mocimycin, 6138
$C_{43}H_{62}N_2O_{12}$
 Mocimycin*, 6138
$C_{43}H_{65}N_{11}O_{12}S_2$
 Deaminooxytocin, 2833
$C_{43}H_{66}N_{12}O_{12}S_2$
 Oxytocin, 6934
$C_{43}H_{66}O_{14}$
 Acetyldigitoxins, 83
$C_{43}H_{66}O_{15}$
 Digoxin*, 3150
$C_{43}H_{70}O_{12}$
 Nonactin*, 6594
$C_{43}H_{72}O_{11}$
 Narasin, 6339
$C_{43}H_{74}N_2O_{14}$
 Spiramycin*, 8708
$C_{43}H_{75}NO_{16}$
 Erythromycin*, 3626
$C_{43}H_{78}N_6O_{13}$
 Muroctasin, 6217
$C_{44}H_{43}ClN_2O_8$
 Hydroxyzine*, 4786
$C_{44}H_{50}Cl_2N_4O_2$
 Alcuronium*, 214
$C_{44}H_{50}N_4O_2{}^{2+}$
 Alcuronium, 214
$C_{44}H_{56}N_2O_8S$
 Lobeline*, 5432
$C_{44}H_{62}N_2O_{12}$
 Aurodox, 900
$C_{44}H_{62}N_8O_{11}$
 Etamycin, 3664
$C_{44}H_{64}O_{24}$
 Crocin, 2593
$C_{44}H_{69}NO_{12}$
 FK-506, A5
$C_{44}H_{72}O_{11}$
 Oligomycins*, 6793
$C_{44}H_{81}NO_{21}$
 Erythromycin Glucoheptonate, 3629
$C_{44}H_{84}CaI_2O_4$
 Calcium Iodobehenate, 1681
$C_{45}H_{42}NO_{15}$
 Erythromycin Stearate, 3632
$C_{45}H_{44}Cl_2N_{10}O_6$
 Cycloguanil*, 2727
$C_{45}H_{54}N_8O_{10}$
 Mikamycin*, 6111

Denotes derivative of title compound

Denotes derivative of title compound

*Denotes derivative of title compound

CaH₂
Calcium Hydride, 1675

CaH₂O₂
Calcium Hydroxide, 1676

CaH₄O₄P₂
Calcium Hypophosphite, 1678

CaH₄O₈P₂
Calcium Phosphate, Monobasic, 1700

CaI₂
Calcium Iodide, 1680

CaI₂O₆
Calcium Iodate, 1679

CaMn₂O₈
Calcium Permanganate, 1695

CaMoO₄
Calcium Molybdate(VI), 1687

CaN₂O₄
Calcium Nitrite, 1689

CaN₂O₆
Calcium Nitrate, 1688

CaO
Calcium Oxide, 1692

CaO₂
Calcium Peroxide, 1696

CaO₃S
Calcium Sulfite, 1715

CaO₄S
Calcium Sulfate, 1713

CaO₄W
Calcium Tungstate(VI), 1720

CaS
Calcium Sulfide, 1714

CaS₂O₃
Calcium Thiosulfate, 1719

CaSe
Calcium Selenide, 1708

Ca₂O₇P₂
Calcium Pyrophosphate, 1706

Ca₃O₈P₂
Calcium Phosphate, Tribasic, 1701

Ca₃P₂
Calcium Phosphide, 1702

CdCl₂
Cadmium Chloride, 1615

CdF₂
Cadmium Fluoride, 1617

CdH₂O₂
Cadmium Hydroxide, 1618

CdI₂
Cadmium Iodide, 1619

CdI₄K₂
Potassium Tetraiodocadmate, 7682

CdN₂O₆
Cadmium Nitrate, 1620

CdO
Cadmium Oxide, 1621

CdO₄S
Cadmium Sulfate, 1627

CdO₄Se
Cadmium Selenate, 1624

CdO₄W
Cadmium Tungstate(VI), 1630

CdS
Cadmium Sulfide, 1628

CdSe
Cadmium Selenide, 1625

CdTe
Cadmium Telluride, 1629

CeCl₃
Cerous Chloride, 1991

CeF₃
Cerous Fluoride, 1992

CeF₄
Ceric Fluoride, 1985

CeH₄NO₈S₂
Ammonium Cerous Sulfate, 530

CeH₈N₈O₁₈
Ammonium Ceric Nitrate, 529

CeI₃
Cerous Iodide, 1993

CeN₃O₉
Cerous Nitrate, 1994

CeO₂
Ceric Oxide, 1986

CeO₈S₂
Ceric Sulfate, 1987

Ce₂O₁₂S₃
Cerous Sulfate, 1996

ClCs
Cesium Chloride, 2004

ClCu
Cuprous Chloride, 2667

ClF
Chlorine Monofluoride, 2098

ClFO₃
Perchloryl Fluoride, 7111

ClFO₄
Fluorine Perchlorate, 4094

ClF₃
Chlorine Trifluoride, 2100

ClH
Hydrogen Chloride, 4721

ClHO
Hypochlorous Acid, 4800

ClHO₃
Chloric Acid, 2091

ClHO₃S
Chlorosulfonic Acid, 2166

ClHO₃Se
Chloroselenic Acid, 2164

ClHO₄
Perchloric Acid, 7110

ClH₂HgN
Mercuric Chloride, Ammoniated, 5771

ClH₄N
Ammonium Chloride, 531

ClH₄NO
Hydroxylamine*, 4759

ClH₄NO₄
Ammonium Perchlorate, 568

ClI
Iodine Monochloride, 4911

ClK
Potassium Chloride, 7601

ClKO₃
Potassium Chlorate, 7600

ClKO₄
Potassium Perchlorate, 7641

ClLi
Lithium Chloride, 5405

ClLiO₄
Lithium Perchlorate, 5417

ClNO
Nitrosyl Chloride, 6568

ClNO₂
Nitryl Chloride, 6579

ClNa
Sodium Chloride, 8544

ClNaO
Sodium Hypochlorite, 8576

ClNaO₂
Sodium Chlorite, 8545

ClNaO₃
Sodium Chlorate, 8543

ClNaO₄
Sodium Perchlorate, 8605

ClOSb
Antimony Chloride Oxide, 725

ClO₂
Chlorine Dioxide, 2096

ClO₃Re
Rhenium Oxychloride, 8179

ClRb
Rubidium Chloride, 8261

ClTl
Thallium Chloride, 9187

Cl₂Co
Cobaltous Chloride, 2431

Cl₂Cr
Chromous Chloride, 2246

Cl₂CrO₂
Chromyl Chloride, 2251

Cl₂Cu
Cupric Chloride, 2636

Cl₂CuO₆
Cupric Chlorate, 2635

Cl₂CuO₈
Cupric Perchlorate, 2651

Cl₂F₃Sb
Antimony Dichlorotrifluoride, 726

Cl₂Fe
Ferrous Chloride, 3992

Cl₂Ge
Germanium Dichloride, 4303

Cl₂H₆N₂Pt
Cisplatin, 2319

Cl₂Hg
Mercuric Chloride, 5770

Cl₂Hg₂
Mercurous Chloride, 5795

Cl₂Hg₂O₆
Mercurous Chlorate, 5794

Cl₂Mg
Magnesium Chloride, 5540

Cl₂MgO₆
Magnesium Chlorate, 5539

Cl₂MgO₈
Magnesium Perchlorate, 5557

Cl₂Mn
Manganese Chloride, 5610

Cl₂Ni
Nickel Chloride, 6412

Cl₂O
Chlorine Monoxide, 2099

Cl₂OS
Thionyl Chloride, 9278

Cl₂OSe
Selenium Oxychloride, 8388

Cl₂OV
Vanadyl Dichloride, 9832

Cl₂OZr
Zirconyl Chloride, 10087

Cl₂O₂S
Sulfuryl Chloride, 8959

Cl₂O₂U
Uranyl Chloride, 9775

Cl₂O₅S₂
Pyrosulfuryl Chloride, 8020

Cl₂O₆Pb
Lead Chlorate, 5277

Cl₂O₆Sr
Strontium Chlorate, 8802

Cl₂O₇
Chlorine Heptoxide, 2097

Cl₂O₈Zn
Zinc Perchlorate, 10051

Cl₂Pb
Lead Chloride, 5278

Cl₂Pd
Palladium Chloride, 6941

Cl₂Pt
Platinous Chloride, 7501

Cl₂S₂
Sulfur Chloride, 8949

Cl₂Se₂
Selenium Chloride, 8384

Cl₂Sn
Stannous Chloride, 8742

Cl₂Sr
Strontium Chloride, 8803

Cl₂Te
Tellurium Dichloride, 9066

Cl₂Ti
Titanium Dichloride, 9397

Cl₂Zn
Zinc Chloride, 10031

Cl₃CoH₁₈N₆
Hexaaminecobalt Trichloride, 4597

Cl₃Cr
Chromic Chloride, 2224

Denotes derivative of title compound

Cl₃Fe
Ferric Chloride, 3961
Cl₃HSi
Trichlorosilane, 9559
Cl₃H₄MgN
Ammonium Magnesium Chloride, 556
Cl₃I
Iodine Trichloride, 4914
Cl₃In
Indium Trichloride, 4865
Cl₃Ir
Iridium Trichloride, 4971
Cl₃N
Nitrogen Chloride, 6523
Cl₃OP
Phosphorus Oxychloride, 7324
Cl₃OV
Vanadyl Trichloride, 9834
Cl₃P
Phosphorus Trichloride, 7333
Cl₃PS
Phosphorus Sulfochloride, 7331
Cl₃Rh
Rhodium Chloride, 8189
Cl₃Ru
Ruthenium Trichloride, 8275
Cl₃Sb
Antimony Trichloride, 739
Cl₃Ti
Titanium Trichloride, 9406
Cl₃U
Uranium Trichloride, 9772
Cl₄CuH₈N₂
Ammonium Cupric Chloride, 537
Cl₄Ge
Germanium Tetrachloride, 4305
Cl₄H₈HgN₂
Ammonium Mercuric Chloride, 558
Cl₄H₈N₂Pt
Ammonium Platinous Chloride, 577
Cl₄H₈N₂Zn
Ammonium Tetrachlorozincate, 590
Cl₄K₂Pt
Potassium Tetrachloroplatinate(II), 7675
Cl₄Os
Osmium Tetrachloride, 6847
Cl₄Po
Polonium Tetrachloride, 7535
Cl₄Se
Selenium Tetrachloride, 8392
Cl₄Si
Silicon Tetrachloride, 8447
Cl₄Sn
Stannic Chloride, 8732
Cl₄Te
Tellurium Tetrachloride, 9070
Cl₄Th
Thorium Chloride, 9309
Cl₄Ti
Titanium Tetrachloride, 9404
Cl₄U
Uranium Tetrachloride, 9770
Cl₄Zr
Zirconium Chloride, 10077
Cl₅H₁₂N₃Zn
Ammonium Pentachlorozincate, 567
Cl₅Nb
Niobium Pentachloride, 6473
Cl₅P
Phosphorus Pentachloride, 7326
Cl₅Sb
Antimony Pentachloride, 727
Cl₅Ta
Tantalum Pentachloride, 9026
Cl₆H₂Pt
Platinic Chloride, 7498
Cl₆H₈N₂Os
Ammonium Osmium Chloride, 564

Cl₆H₈N₂Pt
Ammonium Platinic Chloride, 576
Cl₆H₄₂N₁₄O₂Ru₃
Ruthenium Red, 8273
Cl₆K₂Os
Potassium Hexachloroosmate(IV), 7618
Cl₆K₂Pt
Potassium Hexachloroplatinate(IV), 7619
Cl₆Na₂Pt
Sodium Hexachloroplatinate(IV), 8571
CoCr₂O₄
Cobaltous Chromate(III), 2432
CoF₂
Cobaltous Fluoride, 2434
CoF₃
Cobaltic Fluoride, 2424
CoHO₂
Cobaltic Oxide Monohydrate, 2425
CoH₂O₂
Cobaltous Hydroxide, 2436
CoH₄NO₄P
Ammonium Cobaltous Phosphate, 535
CoH₈N₂O₈S₂
Ammonium Cobaltous Sulfate, 536
CoH₁₀N₇O₈
Erdmann's Salt, 3589
CoI₂
Cobaltous Iodide, 2437
CoK₂O₈S₂
Cobaltous Potassium Sulfate, 2442
CoK₂O₈Se₂
Potassium Cobaltous Selenate, 7605
CoK₃N₆O₁₂
Cobaltic Potassium Nitrite, 2426
CoN₂O₆
Cobaltous Nitrate, 2438
CoN₆Na₃O₁₂
Sodium Cobaltinitrite, 8551
CoO
Cobaltous Oxide, 2440
CoO₄S
Cobaltous Sulfate, 2443
CoS
Cobaltous Sulfide, 2444
Co₃O₄
Cobaltic-Cobaltous Oxide, 2423
Co₃O₈P₂
Cobaltous Phosphate, 2441
CrCuO₄
Cupric Chromate(VI), 2637
CrF₂
Chromous Fluoride, 2247
CrF₂O₂
Chromyl Fluoride, 2252
CrF₃
Chromic Fluoride, 2225
CrF₄
Chromium Tetrafluoride, 2237
CrH₃O₃
Chromic Hydroxide, 2227
CrH₄NO₈S₂
Ammonium Chromic Sulfate, 533
CrH₈N₂O₄
Ammonium Chromate(VI), 532
CrKO₈S₂
Chromic Potassium Sulfate, 2232
CrK₂O₄
Potassium Chromate(VI), 7602
CrLi₂O₄
Lithium Chromate(VI), 5406
CrN₃O₉
Chromic Nitrate, 2228
CrNa₂O₄
Sodium Chromate(VI), 8547
CrO₂
Chromium Dioxide, 2236

CrO₃
Chromium Trioxide, 2238
CrO₄P
Chromic Phosphate, 2230
CrO₄Pb
Lead Chromate(VI), 5279
CrO₄S
Chromous Sulfate, 2250
CrO₄Sr
Strontium Chromate(VI), 8804
CrPb₂O₅
Lead Chromate(VI) Oxide, 5280
Cr₂CuO₄
Cupric Chromite, 2638
Cr₂FeH₄NO₈
Ammonium Ferric Chromate, 540
Cr₂H₈N₂O₇
Ammonium Dichromate(VI), 538
Cr₂HgO₇
Mercuric Dichromate(VI), 5773
Cr₂K₂O₇
Potassium Dichromate(VI), 7608
Cr₂Li₂O₇
Lithium Dichromate(VI), 5408
Cr₂Na₂O₇
Sodium Dichromate(VI), 8556
Cr₂O₃
Chromic Oxide, 2229
Cr₂O₈Sn
Stannic Chromate(VI), 8733
Cr₂O₁₂S₃
Chromic Sulfate, 2233
Cr₃Fe₂O₁₂
Ferric Chromate(VI), 3962
CsHO
Cesium Hydroxide, 2005
CsI
Cesium Iodide, 2006
CsNO₃
Cesium Nitrate, 2007
Cs₂O₄S
Cesium Sulfate, 2008
CuF₂
Cupric Fluoride, 2641
CuF₆Si
Cupric Hexafluorosilicate, 2645
CuFeS₂
Chalcopyrite, 2029
CuH₂O₂
Cupric Hydroxide, 2646
CuH₁₂N₄O₄S
Tetraamminecopper Sulfate, 9116
CuI
Cuprous Iodide, 2669
CuN₂O₆
Cupric Nitrate, 2647
CuO
Cupric Oxide, 2650
CuO₃Se
Cupric Selenite, 2657
CuO₄S
Cupric Sulfate, 2659
CuO₄Se
Cupric Selenate, 2655
CuO₄W
Cupric Tungstate(VI), 2663
CuS
Cupric Sulfide, 2661
CuSe
Cupric Selenide, 2656
Cu₂HgI₄
Cuprous Mercuric Iodide, 2670
Cu₂O
Cuprous Oxide, 2671
Cu₂O₃S
Cuprous Sulfite, 2675
Cu₂S
Cuprous Sulfide, 2674
Cu₂Se
Cuprous Selenide, 2673

* Denotes derivative of title compound

Fe$_5$O$_{12}$Y$_3$
  Yig, 10006

G

GaN$_3$O$_9$
  Gallium*, 4252
GaP
  Gallium Phosphide, 4254
Gd$_2$O$_3$
  Gadolinium*, 4235
GeH$_4$
  Germane, 4301
GeMg$_2$
  Magnesium Germanide, 5545
GeO$_2$
  Germanium Dioxide, 4304

H

HI
  Hydrogen Iodide, 4724
HIO$_3$
  Iodic Acid, 4905
HKO
  Potassium Hydroxide, 7625
HKO$_3$Se
  Potassium Biselenite, 7590
HKO$_4$S
  Potassium Bisulfate, 7591
HKS
  Potassium Bisulfide, 7592
HK$_2$O$_3$P
  Potassium Phosphite, 7650
HK$_2$O$_4$P
  Potassium Phosphate, Dibasic, 7647
HLi
  Lithium Hydride, 5411
HLiO
  Lithium Hydroxide, 5412
HMgO$_4$P
  Magnesium Phosphate, Dibasic, 5560
HMnO$_4$P
  Manganese Phosphate, Dibasic, 5619
HNO$_2$
  Nitrous Acid, 6574
HNO$_3$
  Nitric Acid, 6495
HNO$_5$S
  Nitrosylsulfuric Acid, 6570
HN$_3$
  Hydrazoic Acid, 4697
HNa
  Sodium Hydride, 8573
HNaO
  Sodium Hydroxide, 8575
HNaO$_3$S
  Sodium Bisulfite, 8533
HNaO$_4$S
  Sodium Bisulfate, 8531
HNaS
  Sodium Bisulfide, 8532
HNa$_2$O$_3$P
  Sodium Phosphite, 8616
HNa$_2$O$_4$P
  Sodium Phosphate, Dibasic, 8612
HORb
  Rubidium Hydroxide, 8262
HOTl
  Thallium Hydroxide, 9190
HO$_6$PU
  Uranyl Phosphate, 9777

H$_2$KO$_2$P
  Potassium Hypophosphite, 7626
H$_2$KO$_4$P
  Potassium Phosphate, Monobasic, 7648
H$_2$LiN
  Lithium Amide, 5398
H$_2$Mg
  Magnesium Hydride, 5547
H$_2$MgO$_2$
  Magnesium Hydroxide, 5548
H$_2$MgO$_8$S$_2$
  Magnesium Bisulfate, 5534
H$_2$MoO$_4$
  Molybdic(VI) Acid, 6149
H$_2$NNa
  Sodium Amide, 8519
H$_2$N$_2$O$_2$
  Nitramide, 6487
H$_2$NaO$_2$P
  Sodium Hypophosphite, 8580
H$_2$NaO$_4$P
  Sodium Phosphate, Monobasic, 8613
H$_2$Na$_2$O$_6$P$_2$
  Disodium Dihydrogen Hypophosphate, 3363
H$_2$Na$_2$O$_7$P$_2$
  Sodium Acid Pyrophosphate, 8515
H$_2$NiO$_2$
  Nickel Hydroxide, 6417
H$_2$O
  Water, 9951
H$_2$O$_2$
  Hydrogen Peroxide, 4725
H$_2$O$_2$Sr
  Strontium Hydroxide, 8807
H$_2$O$_3$Se
  Selenious Acid, 8381
H$_2$O$_3$Te
  Tellurous Acid, 9072
H$_2$O$_4$Pb$_3$
  Lead Hydroxide, 5286
H$_2$O$_4$S
  Sulfuric Acid, 8953
H$_2$O$_4$Se
  Selenic Acid, 8380
H$_2$O$_4$W
  Tungstic(VI) Acid, 9725
H$_2$O$_5$S
  Caro's Acid, 1858
H$_2$O$_7$S$_2$
  Pyrosulfuric Acid, 8019
H$_2$S
  Hydrogen Sulfide, 4729
H$_2$Se
  Hydrogen Selenide, 4728
H$_2$Te
  Hydrogen Telluride, 4730
H$_2$Ti
  Titanium Hydride, 9399
H$_3$N
  Ammonia, 510
H$_3$NO
  Hydroxylamine, 4759
H$_3$NO$_3$S
  Sulfamic Acid, 8893
H$_3$O$_2$P
  Hypophosphorous Acid, 4804
H$_3$O$_3$P
  Phosphorous Acid, 7320
H$_3$O$_4$P
  Phosphoric Acid, 7318
H$_3$P
  Phosphine, 7313
H$_3$Sb
  Stibine, 8767
H$_4$IN
  Ammonium Iodide, 554
H$_4$IP
  Phosphonium Iodide, 7317

H$_4$MgN$_2$
  Magnesium Amide, 5532
H$_4$MgO$_8$P$_2$
  Magnesium Phosphate, Monobasic, 5561
H$_4$MnO$_4$P$_2$
  Manganese Hypophosphite, 5613
H$_4$NO$_3$V
  Ammonium Vanadate(V), 599
H$_4$N$_2$
  Hydrazine, 4691
H$_4$N$_2$O$_2$S
  Sulfamide, 8894
H$_4$N$_2$O$_3$
  Ammonium Nitrate, 561
H$_4$O$_4$P$_2$Pb
  Lead Hypophosphite, 5287
H$_4$O$_4$Zr
  Zirconium Hydroxide, 10080
H$_4$O$_6$P$_2$
  Hypophosphoric Acid, 4803
H$_4$O$_7$P$_2$
  Pyrophosphoric Acid, 8018
H$_4$O$_{40}$SiW$_{12}$
  Silicotungstic Acid, 8449
H$_4$Si
  Silane, 8435
H$_5$F$_2$N
  Ammonium Bifluoride, 517
H$_5$IO$_6$
  Periodic Acid, 7123
H$_5$NNaO$_4$P
  Ammonium Sodium Phosphate, 581
H$_5$NO$_3$S
  Ammonium Bisulfite, 522
H$_5$NO$_4$S
  Ammonium Bisulfate, 520
H$_5$NS
  Ammonium Bisulfide, 521
H$_6$NO$_2$P
  Ammonium Hypophosphite, 553
H$_6$NO$_4$P
  Ammonium Phosphate, Monobasic, 571
H$_6$N$_2$O
  Hydrazine Hydrate, 4692
H$_6$N$_2$O$_3$S
  Ammonium Sulfamate, 583
H$_6$N$_2$O$_4$S
  Hydrazine Sulfate, 4693
H$_6$O$_6$Te
  Telluric(VI) Acid, 9063
H$_6$Si$_2$
  Disilane, 3362
H$_8$N$_2$NiO$_8$S$_2$
  Ammonium Nickel Sulfate, 560
H$_8$N$_2$O$_3$S
  Ammonium Sulfite, 588
H$_8$N$_2$O$_3$S$_2$
  Ammonium Thiosulfate, 592
H$_8$N$_2$O$_3$Se
  Ammonium Selenite, 580
H$_8$N$_2$O$_4$S
  Ammonium Sulfate, 584
H$_8$N$_2$O$_4$Se
  Ammonium Selenate, 579
H$_8$N$_2$O$_6$S
  Hydroxylamine*, 4759
H$_8$N$_2$O$_7$U$_2$
  Ammonium Uranate(VI), 595
H$_8$N$_2$O$_8$S$_2$
  Ammonium Peroxydisulfate, 569
H$_8$N$_2$S
  Ammonium Sulfide, 585
H$_8$Si$_3$
  Trisilane, 9668
H$_9$N$_2$O$_3$P
  Ammonium Phosphite, 572
H$_9$N$_2$O$_4$P
  Ammonium Phosphate, Dibasic, 570

Denotes derivative of title compound

Denotes derivative of title compound

Denotes derivative of title compound

O₄PbS
Lead Sulfate, 5302
O₄PbSe
Lead Selenate, 5296
O₄PbW
Lead Tungstate(VI), 5309
O₄Pb₃
Lead Tetroxide, 5307
O₄Ru
Ruthenium Tetroxide, 8274
O₄SSn
Stannous Sulfate, 8750
O₄SSr
Strontium Sulfate, 8816
O₄STl₂
Thallium Sulfate, 9197
O₄SZn
Zinc Sulfate, 10064
O₄SeSr
Strontium Selenate, 8815
O₄SeTl₂
Thallium Selenate, 9194
O₄SeZn
Zinc Selenate, 10060
O₄SiZn₂
Zinc Silicate, 10062
O₄SiZr
Zirconium Silicate, 10084
O₄U
Uranium Peroxide, 9769
O₅P₂
Phosphorus Pentoxide, 7330
O₅STi
Titanium Sulfate, 9402
O₅SV
Vanadyl Sulfate, 9833
O₅Sb₂
Antimony Pentoxide, 730
O₅Ta₂
Tantalum Pentoxide, 9028
O₅V₂
Vanadium Pentoxide, 9826
O₆PbV₂
Lead Vanadate(V), 5310

O₆SU
Uranyl Sulfate, 9778
O₆Se₂Sn
Stannic Selenite, 8738
O₇P₂Sn₂
Stannous Pyrophosphate, 8748
O₇P₂Zn₂
Zinc Pyrophosphate, 10058
O₇Re₂
Rhenium Heptoxide, 8177
O₈P₂Pb₃
Lead Phosphate, 5295
O₈P₂Sr₃
Strontium Phosphate, Tribasic, 8814
O₈P₂Zn₃
Zinc Phosphate, 10055
O₈S₂Th
Thorium Sulfate, 9313
O₈S₂Zr
Zirconium Sulfate, 10085
O₁₂S₃Sb₂
Antimony Sulfate, 736
O₁₂S₃Ti₂
Titanium Sesquisulfate, 9401
O₁₂S₃V₂
Vanadium Trisulfate, 9830

P

P₂S₅
Phosphorus Pentasulfide, 7329
P₂Se₃
Phosphorus Hemitriselenide, 7322
P₂Se₅
Phosphorus Pentaselenide, 7328
P₂Zn₃
Zinc Phosphide, 10056
P₄S₃
Tetraphosphorus Trisulfide, 9168
P₄Se₃
Phosphorus Triselenide, 7336

PbS
Lead Sulfide, 5303
PbTe
Lead Telluride, 5304

S

SSn
Stannous Sulfide, 8751
SSr
Strontium Sulfide, 8817
STl₂
Thallium Sulfide, 9198
SZn
Zinc Sulfide, 10065
S₂Si
Silicon Disulfide, 8441
S₂Sn
Stannic Sulfide, 8739
S₃Sb₂
Antimony Trisulfide, 745
S₃V₂
Vanadium Trisulfide, 9831
S₅Sb₂
Antimony Pentasulfide, 729
Sb₂Se₃
Antimony Triselenide, 744
SeSn
Stannous Selenide, 8749
SeTl₂
Thallium Selenide, 9195
SeZn
Zinc Selenide, 10061
Se₂Sn
Stannic Selenide, 8737

T

TeZn
Zinc Telluride, 10068

Denotes derivative of title compound

CROSS INDEX OF NAMES

CROSS INDEX OF NAMES

The Cross Index of Names refers the reader to monograph numbers with three types of citations:

Ethacrynic Acid, 3643 (format for titles)
Ethacrynate Sodium *see* 3643 (format for all synonyms)
CVP *see* MISC-2 (format for entries in Miscellaneous Tables Section)

Company name included in citation denotes trademark ownership:

Edecrin [Merck & Co.] *see* 3643

An asterisk (*) preceding an entry signifies that the name does not appear in the monograph.

(−)H 80/62 *see* 7742
1C50 *see* 3784
2-DG *see* 2886
2-M-4-A *see* 39
3C Antibiotic *see* 6255
3-MC *see* 5967
3Y9 *see* 3880
4A65 *see* 4831
4-C-32 *see* 9361
5-FC *see* 4056
8 AL *see* 6405
21P *see* 9603
25-HCC *see* 1638
33T57 *see* 5900
84L *see* 3106
#96H60 *see* 4519
101-E *see* 6323
101-G *see* 6323
190 F *see* 44
191C49 *see* 9220
194-B *see* 10089
217 MI *see* 3475
291.C.51 *see* 4155
295.C.51 *see* 9661
309F *see* 8986
338C48 *see* 3248
374 JL *see* 6638
378C48 *see* 3342
501 P *see* 4357
516 MD *see* 2308
554L *see* 9677
574-Chromogen *see* 4572
606 *see* 839
611C55 *see* 9205
6-12 [Union Carbide] *see* 3699
683 M *see* 6898
688-A *see* 7225
693 B *see* 3807
710 F *see* 7495
722 D *see* 7477
730-CERM *see* 604
746 CE *see* 3577
00836 *see* 5870
844 *see* 7087
864T *see* 4841
921 C *see* 9577
933F *see* 7448
106-7 *see* 2758
1080 [Monsanto] *see* 4096
1081 *see* 4095
1162 F *see* 8898
1314 Th *see* 3692
1358F *see* 2820
1399F *see* 16
1489 RB *see* 7427
1497 CB *see* 7841
1522 CB *see* 27
1589 MRB *see* 7012
1589 RB *see* 7012
1600 Antibiotic *see* 6989
1609 RB *see* 7460
1633 Labaz *see* 3017
1665 RB *see* 797
1678 CB *see* 7834
1717 *see* 7410

1766 *see* 2182
02-115 *see* 9012
2249 F *see* 6877
III-2318 *see* 5734
2329 Labaz *see* 1093
2601 A *see* 2186
3024 CERM *see* 10074
3123 L *see* 7960
4114 Th *see* 7203
4311/b Ciba *see* 6025
53-32 C *see* 9361
5512-M *see* 709
6029-M *see* 1487
#6063 *see* 45
6315-S *see* 4038
7432-S *see* 1947
8053 CB *see* 9230
8102 CB *see* 965
10040 *see* 2090
10275-S *see* 3575
10364-S *see* 5747
10,580 *see* 2727
12494 Hoechst *see* 3935
27-400 *see* 2759
29060-LE *see* 9887
31252-S *see* 7129
36801 *see* 3821
47657 *see* 780
69276 MD *see* 9445
81723 Hfu *see* 9351
1703-18B *see* 4789
450191-S *see* 8219
710674-S *see* 2367
A₃ *see* 9099
A 16 *see* 394
A 21 *see* 5105
A 65 *see* 7470
A 66 *see* 7200
A-101 *see* 6608
A 118 *see* 8977
A 124 *see* 8693
A 145 *see* 7793
A 272 *see* 6793
A 300-I *see* 9563
A 350 *see* 446
A 363 *see* 443
A 585 *see* 8976
A 820 *see* 1532
A 3823A *see* 6157
A 4696 *see* 126
A 4942 *see* 4822
A 5610 *see* 922
A 8103 *see* 7451
A 19120 *see* 6988
A 20968 *see* 7452
A 23187 *see* 1639
A 29622 *see* 4166
A 35957 *see* 317
A 38414 [Enzyme] *see* 664
A 41-304 *see* 2910
A 43818 *see* 5341
A 46745 *see* 4310
A 56268 *see* 2340
A 56619 *see* 3127
A 60969 *see* 9477

A 61827 *see* 9477
A 62254 *see* A8
A 63004 *see* A8
A 64730 *see* 9477
α-0817185 *see* 8218
Aα *see* 3952
AA 149 *see* 9501
AA 673 *see* 508
Aa 5648 *see* 8936
Aacifemine [Aaciphar] *see* 3659
AAF *see* 4083
AA-protein *see* 651
Aarane [Syntex] *see* 2594
Aararre [Syntex] *see* 2594
AAT *see* 751
A1AT *see* 751
AAtrex [Geigy] *see* 886
AB-100 *see* 9787
AB-103 *see* 1097
AB-132 *see* 6082
AB-206 *see* 6116
ABA *see* 6
ABA 663 *see* 2278
Abacin [Isola-Ibi] *see* 8889
Abadol [Specia] *see* 494
Abaktal [Lek] *see* 7012
Abalyn *see* 5930
Abamectin, 1
Abapresin [Polfa] *see* 4473
Abar [Velsicol] *see* 5327
Abasin [Bayer] *see* 15
Abate [Am. Cyanamid] *see* 9075
Abbocillin-DC [Abbott] *see* 7042
Abbocillin V [Abbott] *see* 7048
Abbocin [Abbott] *see* 6931
Abboflox [Abbott] *see* 4044
Abbokinase [Abbott] *see* 9799
Abbolexin *see* 7292
Abboticine [Abbott] *see* 3632
Abbott 16900 *see* 9059
Abbott 30400 *see* 7021
Abbott 34842 *see* 1504
Abbott 36581 *see* 1505
Abbott 38414 *see* 664
Abbott 43326 *see* 1875
Abbott 43818 *see* 5341
Abbott 44747 *see* 4165
Abbott 45975 *see* 9084
Abbott 46811 *see* 1943
Abbott 48999 *see* 1937
Abbott 50192 *see* 1928
Abbott 50711 *see* 9821
Abbott 56619 *see* 3127
Abboxapam [Abbott] *see* 6881
Abbsa [Sanko] *see* 9447
ABC 12/3 *see* 3427
Abensanil *see* 40
Abequito [Am. Cyanamid] *see* 6466
Aberel [McNeil] *see* 8167
Abicoviromycin *see* 3
Δ 6,8(14)-Abietadienoic Acid *see* 5348
Abietic Acid, 2
Abietic Anhydride *see* 8245
Abietin *see* 2498
Abiguanil *see* 8879

Acetylsalicylic Acid Calcium Salt *see* 1644
Acetylsalicylic Acid Calcium Salt Complex with Urea *see* 1644
Acetylsalicylic Acid Methyl Ester *see* 5934
Acetylsalicylsalicylic Acid, 95
Acetylsalol *see* 7241
5-Acetylspiro[benzofuran-2(3*H*),1'-cyclopropan]-3-one *see* 8722
Acetyl Sulfamethoxypyrazine, 96
4'-(Acetylsulfamyl)phthalanilic Acid *see* 7350
N¹-Acetylsulfanilamide *see* 8870
N⁴-Acetylsulfanilamide, 97
2-(N¹-Acetylsulfanilamido)-3-methoxypyrazine *see* 96
N-Acetylsulfanilic Acid, 98
N-Acetylsulfanilyl Chloride, 99
N¹-Acetyl-6-(sulfanilylmetanilamido)-sodium *see* 66
Acetyl Sulfisoxazole, 100
Acetyltanghinigenin *see* 9021
Acetyltannic Acid, 101
6-(Acetylthio)-8-[[2-[[(4-amino-2-methyl-5-pyrimidinyl)methyl]formylamino]-1-(2-hydroxyethyl)-1-propenyl]dithio]octanoic Acid Methyl Ester *see* 6680
3-Acetylthio-4-[(4-amino-2-methyl-5-pyrimidinyl)methyl-N-formylamino]-3-pentenyl Acetate *see* 46
S-(3-Acetylthio-7-carbomethoxyheptylthio)thiamine *see* 6680
7-(Acetylthio)-17-hydroxy-3-oxopregn-4-ene-21-carboxylic Acid γ-Lactone *see* 8721
(S)-N-[1-[3-(Acetylthio)-2-methyl-1-oxopropyl]-L-prolyl]-L-phenylalanine *see* 192
1-(D-3-Acetylthio-2-methylpropanoyl)-L-prolyl-L-phenylalanine *see* 192
Acetylthymol *see* 9333
N-Acetyl-*m*-toluidine *see* 67
Acetyl-4-trifluoromethylsalicylic Acid *see* 9600
1-[6-[(3-Acetyl-2,4,6-trihydroxy-5-methylphenyl)methyl]-5,7-dihydroxy-2,2-dimethyl-2*H*-1-benzopyran-8-yl]-3-phenyl-2-propen-1-one *see* 8250
N-Acetyl-N-(2,4,6-triiodo-3-aminophenyl)-β-aminoisobutyric Acid *see* 4903
1-Acetyl-1,5,5-trimethyl-2-phenylsemioxamazide *see* 3299
Acetylveratroylbikhaconine *see* 1234
N-Acetyl-Wieland-Gumlich Aldehyde *see* 2938
Acexamic Acid *see* 38
Acezide [Duncan, Flockhart] *see* 1773
Ac-globulin *see* 3871
Achillea, 102
Achilleic Acid *see* 112
Achless [Tatsumi] *see* 4060
Achletin [Toyama] *see* 9537
Achro [Lederle] *see* 9130
Achromycin [Lederle] *see* 9130
Achromycin V [Lederle] *see* 9130
Acibilin [Exa] *see* 2279
1,2,4-Acid *see* 466
Acid Acriflavine *see* 118
Acid Ammonium Carbonate *see* 516
Acid Ammonium Fluoride *see* 517
Acid Ammonium Purpurate *see* 6215
Acid Ammonium Sulfate *see* 520
Acid Blue 74 *see* 4856
Acid Calcium Phosphate *see* 1700
Acid Fuchsin, 103
Acidic Fibroblast Growth Factor *see* 4016

Acid Magenta *see* 103
Acid Magnesium Citrate *see* 5542
Acid Magnesium Phosphate *see* 5561
Acid Magnesium Sulfate *see* 5534
Acid Methyl Sulfate *see* 6041
Acidogen *see* 4363
Acidogen Nitrate *see* 9784
Acidol [Winthrop] *see* 1201
Acidomycin *see* 6233
Acidoride [Abbott] *see* 4363
Acid Platinic Chloride *see* 7498
Acid Potassium Phthalate *see* 7589
Acid Potassium Tartrate *see* 7593
Acid Quinine Hydrochloride *see* 8079
Acid Quinine Sulfate *see* 8076
Acid Roseine *see* 103
Acid Rubin *see* 103
Acid Sodium Phosphate *see* 8613
Acid Trypaflavine *see* 118
Acidulin [Lilly] *see* 4363
Acidum Acetylsalicylicum *see* 873
Acidum Fenclozicum *see* 3915
Acid Violet 7B, 104
Acid Yellow 3 *see* 8100
Acid Yellow D *see* 9688
Acid Yellow S *see* 6313
Acifluorfen, 105
Acifran, A1
Aciglumin *see* 4363
Aci-Jel [Ortho-Cilag] *see* 47
Acillin [Bentex] *see* 621
Aciloc [Orion] *see* 2279
Acimetion *see* 5896
Acimetten [Kwizda] *see* 873
Acinil [GEA] *see* 2279
Acinitrazole *see* 421
Acino [IMA] *see* 3920
Acintol C *see* 9013
Acipen-V *see* 7046
Acipimox, 106
Acistrate *see* MISC-3
Acitretin, 107
Acket *see* 8297
ACL-59 [Monsanto] *see* 9679
ACL-85 [Monsanto] *see* 8993
Aclacinomycin A *see* 108
Aclacinomycin B *see* 108
Aclacinomycins, 108
Aclacinon [Sanraku] *see* 108
Aclaplastin [Behringwerke] *see* 108
Aclarubicin *see* 108
Aclarubicin Hydrochloride *see* 108
Aclatonium Napadisilate, 109
Aclor *see* 4363
Aclosone [Unicet] *see* 213
Aclovate [Glaxo] *see* 213
Aclovir [Lek] *see* 139
Acnegel [Warrick] *see* 1128
Acnestrol [Dermik] *see* 1438
Acnosan *see* 9650
ACNU [Asta] *see* 6469
Acocantherin *see* 6854
Acodeen [Hommel] *see* 1505
Acolen *see* 2858
Acon [Endo] *see* 9918
Aconine, 110
Aconite, 111
Aconitic Acid, 112
Aconitine, 113
Aconitine, Amorphous, 114
Aconitum Ferox, 115
Aconitysat [Arzneimittelwerk VEB] *see* 110
Acopyrine *see* 748
"Acorn Sugar" *see* 8046
Acortan [Ferring] *see* 127
Acorto *see* 127
Acosterina *see* 9650
Acozid *see* 8304
ACP 332 *see* 6338
ACPC *see* 2740

ACPM-629 *see* 2063
Acquat CDAC *see* 2009
Acquinite *see* 2156
Acraconitine *see* 7924
Acraldehyde *see* 122
Acramine Yellow *see* 418
Acranil® [Bayer], 116
Acrasin *see* 2714
Acrex [Murphy] *see* 3280
Acrichine *see* 8053
Acricid [Hoechst] *see* 1237
9-Acridinamine *see* 418
Acridine, 117
3,6-Acridinediamine *see* 7780
4'-(9-Acridinylamino)methanesulfon-*m*-anisidide *see* 627
N-[4-(9-Acridinylamino)-3-methoxyphenyl]methanesulfonamide *see* 627
Acriflavine, 118
Acriflavine Hydrochloride *see* 118
Acrilan®, 119
Acrinol *see* 3668
Acriquine *see* 8053
Acrisorcin, 120
Acritet [Stauffer] *see* 125
Acrivastine, 121
Acrizane Chloride [Abbott] *see* 7157
Acrofollin *see* 3653
Acrol *see* 2961
Acrolactine *see* 3668
Acrolein, 122
Acronize [Am. Cyanamid] *see* 2193
α-Acrose *see* 4186
Acrosoxacin *see* 8247
Acrylaldehyde *see* 122
Acrylamide, 123
Acrylic Acid, 124
Acrylic Acid *n*-Butyl Ester *see* 1539
Acrylic Acid Ethyl Ester *see* 3715
Acrylic Acid Methyl Ester *see* 5935
Acrylic Aldehyde *see* 122
Acrylonitrile, 125
ACS *see* 750
Actaea *see* 2281
Actal [Winthrop] *see* 223
Actamer [Monsanto] *see* 1316
Actaplanins, 126
Actasal [Purdue Frederick] *see* 2212
Actase [Ortho] *see* 7493
Actedron *see* 616
Actellic [ICI] *see* 7469
Actellifog [ICI] *see* 7469
Actemin *see* 616
Acterol [Lappe] *see* 6468
ACTH, 127
Acthar [Armour Pharm.] *see* 127
Acthiazidum *see* 3687
ACTH-β-lipotropin Common Precursor *see* 7803
Actholain *see* 2550
Acthormon [Shionogi] *see* 4322
Actidil [Burroughs-Wellcome] *see* 9661
Actidilon *see* 9661
Actidione [Upjohn] *see* 2734
Actifed-C [Burroughs Wellcome] *see* 4465
Actigall [Ciba-Geigy] *see* 9801
Actigam [Daltan] *see* 4356
Actilin *see* 6369
Actilyse [Boehringer, Ing.] *see* 9394
Actimide [Tobishi] *see* 2446
Actin, 128
F-Actin *see* 128
G-Actin *see* 128
α-Actin *see* 128
β-Actin *see* 128
γ-Actin *see* 128
Actinamin [Daiichi] *see* 1871
Actinium, 129
Actinium K *see* 4175

Actinium X see 8118
Actinobolin, 130
Actinodaphnine, 131
Actinolite see 851
Actinomycetin, 132
Actinomycin IV see 2804
Actinomycin A$_{IV}$ see 2804
Actinomycin C see 1605
Actinomycin C$_1$ see 2804
Actinomycin C$_2$ see 1605
Actinomycin C$_3$ see 1605
Actinomycin D see 2804
Actinomycin F$_1$, 133
Actinomycin I$_1$ see 2804
Actinomycin KS4 see 133
Actinomycin-[thr-val-pro-sar-meval]
    see 2804
Actinomycin X$_1$ see 2804
Actinon see 8119
Actinoquinol, 134
Actinorhodine, 135
Actinospectacin see 8695
Actiphenol, 136
Actithiazic Acid see 6233
Activated Aluminum Oxide see 360
Activated Charcoal see 1815
Activated 7-Dehydrocholesterol see
    9929
Activated Ergosterol see 9928
Active Amyl Alcohol see 5952
Active Lipid see 191
"Active Methionine" see 146
Activin see 6283
Activin A see 137
Activin AB see 137
Activins, 137
Activol see 474
Actocortin see 4712
Actol [Beecham] see 6444
Actomyosin see 128
Acton [Ferring] see 127
Actonar see 127
Actosin [Daiichi Seiyaku] see 1448
Actosolv [Behringwerke] see 9799
Actospar [Sandoz] see 8692
Actozine 1037
Actrapid [Novo] see 4889
Actriol [Organon] see 3567
Actybaryte [Labaz] see 1006
Actylamide see 8290
Acupan [Riker] see 6358
Acutil-S [ISF] see 21
Acutran [Mallinckrodt] see 613
ACV, 138
Acycloguanosine see 139
Acyclovir, 139
Acylanid [Sandoz] see 83
Acylcolaminoformylmethylpyridinium
    Chloride see 9759
N-(Acylcolaminoformylmethyl)pyri-
    dinium Chloride see 5238
Acylpyrin [Spofa] see 873
AD-106 see 2272
AD-810 see 10094
Ada see 3168
ADAH see 1255
Adalat(e) [Bayer] see 6441
Adalin [Bayer; Winthrop] see 1837
1-Adamantanamine see 380
Adamantane, 140
2-(1'-Adamantanecarbonyloxy)ethyl-
    dimethyldecylammonium Bromide
    see 381
N-1-Adamantyl-N-[2-(dimethylami-
    no)ethoxy]acetamide see 9683
Adamite see 10048
Adams' Catalyst see 7500
Adamsite see 7170
Adanon Hydrochloride [Winthrop] see
    5852
Adapin [Pennwalt] see 3425

Adapress [Lagap] see 6441
Adaptinol [Bayer] see 9972
Adazine [Upjohn] see 9597
ADCA see 1255
Adcortyl [Squibb] see 9511
Adcortyl-A see 9512
Addisomnol [Synochem] see 1837
Adecut [Takeda] see 2864
Adelir [Nagase] see 9751
Ademetionine see 146
Ademide [Toyo Jozo] see 2446
Ademin(e) see 9515
Ademol [Squibb] see 4067
Adenaron [Kowa] see 1791
Adenex [ICN] see 855
Adenine, 141
Adenine Arabinoside see 9881
Adenine-D-ribose-phosphate-phos-
    phate-D-ribose-nicotinamide see
    6259
Adenine Riboside see 143
Adenocard [Medco] see 143
Adenock [Shiraimatsu] see 278
Adenogen see 1790
Adenohypophyseal Growth Hormone
    see 8672
Adenohypophysial Luteotropin see
    7788
A-Denopterin, 142
Adenosine, 143
Adenosine Cyclic 3',5'-(Hydrogen
    Phosphate) see 2714
Adenosine 3',5'-Cyclic Monophos-
    phate see 2714
Adenosine 3',5'-Cyclic Phosphate see
    2714
Adenosine Diphosphate, 144
Adenosinediphosphoric Acid see 144
Adenosine 3'-Monophosphate see 147
Adenosine 3',5'-Monophosphate see
    2714
Adenosine 5'-Monophosphate see 148
Adenosine-3'-monophosphoric Acid
    see 147
Adenosine-5'-monophosphoric Acid
    see 148
Adenosine Phosphate see 148
Adenosine 3',5'-Phosphate see 2714
Adenosine-3'-phosphoric Acid see
    147
Adenosine-5'-phosphoric Acid see
    148
Adenosine 5'-Pyrophosphoric Acid
    see 144
Adenosine 5'-(Tetrahydrogen Tri-
    phosphate) see 145
Adenosine 5'-(Trihydrogen Diphos-
    phate) see 144
Adenosine 5'-(Trihydrogen Diphos-
    phate) 2'-(Dihydrogen Phosphate)
    5'→5'-Ester with 3-(Aminocarbon-
    yl)-1-β-D-ribofuranosylpyridinium
    Hydroxide Inner Salt see 6262
Adenosine 5'-(Trihydrogen Diphos-
    phate) 5'→5'-Ester with 3-(Amino-
    carbonyl)-1-β-D-ribofuranosylpyri-
    dinium, Hydroxide, Inner Salt see
    6259
Adenosine 5'-(Trihydrogen Diphos-
    phate) 5'→5'-Ester With Ribofla-
    vine see 4028
Adenosine Triphosphate, 145
Adenosine 5'-Triphosphoric Acid see
    145
Adenosyl-B$_{12}$ see 2446
S-Adenosylmethionine, 146
Adenyl [Auclair] see 148
Adenylate Cyclase see 2714
Adenylic Acid b see 147
h-Adenylic Acid see 147
t-Adenylic Acid see 148

3'-Adenylic Acid, 147
5'-Adenylic Acid, 148
5'-Adenylphosphoric Acid see 144
Adephos see 145
Adepril see 504
Adeps see 5239
Adepsine Oil see 7139
Adergon see 4287
Adermine Hydrochloride see 7995
Adermykon see 2178
Aderoxal see 7992
Adesitrin [Erba] see 6528
Adetol see 145
Adetphos [Kowa] see 145
Adhatoda, 149
Adhatodai see 149
AD/here see 5667
Adiab see 4372
Adiaben see 2187
Adiazine see 8874
ADIC see MISC-2
Adicillin see 7043
Adifax [Servier] see 3920
Adigal [Beiersdorf] see 5229
Adinazolam, 150
Adinazolam Mesylate see 150
Adipan see 616
Adiparthrol [Riker] see 3720
Adipex-P [Lemmon] see 7232
Adiphenine Hydrochloride, 151
Adipic Acid, 152
Adipic Acid Di(3-carboxy-2,4,6-tri-
    iodoanilide) see 4916
Adipic Ketone see 2748
Adipiodone see 4916
Adipokinetic Hormone see 5392
Adipomin [Streuli] see 3920
Adiposettin [Reiss] see 6634
3,3'-(Adipoyldiimino)bis[2,4,6-tri-
    iodobenzoic Acid] see 4916
5,5'-(Adipoyldiimino)bis[2,4,6-tri-
    iodo-N-methylisophthalamic Acid]
    see 4902
Adiprazine see 7432
Adipsin [Heilmittelwerk] see 6012
Adipsin, 153
N,N'-Adipylbis(3-amino-2,4,6-tri-
    iodobenzoic Acid) see 4916
Adisné see 9212
Adiuretin SD [Spofa] see 2904
Adlone [Pharmascience] see 3868
Adlumidine, 154
Adlumine, 155
Adnephrine [Winthrop] see 3569
Adnexol [Protina] see 4815
Adobacillin [Tobishi] see 621
Adobiol [Yoshitomi] see 1461
Adofeed [Lead Chem.] see 4124
Adoisine see 9950
Adomal [Malesci] see 3130
AdoMet see 146
Adona [Tanabe Seiyaku] see 1791
Adonigen see 156
Adonilen see 156
Adonisid see 156
Adonis vernalis, 156
Adonite see 157
Adonitol, 157
Adonitoxin, 158
Adonival see 156
D-Adonose see 8209
Adopon see 1743
Adovern [Roche] see 156
ADP see 144 and 6954
Adphen [Ferndale] see 7180
ADR-033 see 9649
Adrafinil, 159
Adral [Lifepharma] see 6450
Adran [Takata] see 4812
Adrechros [Toho Iyaku] see 1791
Adrenal see 3569

Adrenal Cortical Extract, 160
Adrenalin [Parke, Davis] see 3569
Adrenaline Oxidase see 6158
Adrenalone, 161
Adrenamine see 3569
Adrenine see 3569
Adrenochrome, 162
Adrenochrome Monosemicarbazone
   Sodium Salicylate Complex see 1790
Adrenochrome Semicarbazone Compd
   with Sodium Salicylate see 1790
Adrenocorticotrop(h)ic Hormone of
   the Pituitary Gland see 127
Adrenocorticotrop(h)in see 127
Adrenoglomerulotropin, 163
Adrenolutin, 164
Adrenone see 161
Adrenor see 6612
Adrenosem [Beecham] see 1790
Adrenosem Salicylate see 1790
Adrenosterone, 165
Adrenoxyl [Nordmark] see 162
Adreson see 2533
Adrestat-F [Organon] see 1790
Adrevil [Zyma] see 1501
Adriablastina [Farmitalia] see 3428
Adriacin [Kyowa] see 3428
Adriamicina [Farmochim. Ital.] see
   5851
Adriamycin [Farmitalia] see 3428
Adriamycin (former generic name) see
   3428
Adriamycinone see 3428
Adrianol [Anasco] see 7257
Adriblastina [Farmitalia] see 3428
Adrin [Merck & Co.] see 3569
Adroyd [Parke, Davis] see 6920
Adrucil [Adria] see 4109
Adulsa see 149
Adumbran [Thomae] see 6881
Adurix [Benzon] see 2391
Advil [Whitehall] see 4812
AE-17 see 8990
Aedurid [Robugen] see 3486
Aequamen [Promonta] see 1200
Aerbron [Angelini Francesco] see 7919
Aerobin [Farmasa] see 9212
Aerolin [Glaxo] see 209
Aerolone [Lilly] see 5105
Aeromatt see 1657
Aeropax see 8486
Aeropent [Fisons] see 7071
Aeroseb-D [Herbert] see 2922
Aerosol AY see 2968
Aerosol IB see 3179
Aerosol MA see 1266
Aerosol OT see 3398
Aerosporin [Burroughs Wellcome] see
   7550
Aerotrol [Abbott] see 5105
Aerrane [Anaquest] see 5059
Aerugipen [Wlfing] see 9360
Aescigenin see 3643
Aescin see 3644
Aescusan see 3644
AET, 166
Aether Oenanthicus see 3790
Aethoform see 3719
Aethon see 6837
Aethylis Chloridum see 3740
Aetina see 3692
Aetiocholanic Acid see 3824
Aetoxisclerol [Dexo] see 7532
AF 438 see 6898
AF 634 see 7919
AF 983 see 1042
AF 1161 see 9495
AF 1191 see 3841
AF 1890 see 5447
AF 2071 see 4710

AF 2139 see 2819
Afalon [Hoechst] see 5387
Afatin see 2932
Afaxin [Winthrop] see 9918
Affinin, 167
Affinitins see 5312
Affirm [Merck & Co.] see 1
aFGF see 4016
aFGF-1 see 4016
aFGF-2 see 4016
Afibrin [Lederle] see 442
AFI-Ftalyl see 7351
AFI-phyllin see 3459
AFI-Tiazin see 7220
Afko-Hist see 7996
Aflatoxicol see 168
Aflatoxin B$_1$ see 168
Aflatoxin B$_2$ see 168
Aflatoxin G$_1$ see 169
Aflatoxin G$_2$ see 169
Aflatoxin M$_1$ see 170
Aflatoxin M$_2$ see 170
Aflatoxins B, 168
Aflatoxins G, 169
Aflatoxins M, 170
Aflix [Sandoz] see 4160
Afloben [Esseti] see 1136
Aflodac [Janus] see 8961
Afloqualone, 171
Aflorix [Gramon] see 6101
Afloxan [Rotta] see 7784
Afluon [Asta] see 5543
Afonilum [Minden] see 9212
Afos [Tiber] see 4169
Afrazine [Kirby-Warrick] see 6919
African Saffron see 1877
Afrin [Schering] see 6919
Afrinol [Schering] see 3561
Afsillin [Squibb] see 7042
Aftate [Plough] see 9442
Afugan [Hoechst] see 7976
Afungil [EGYT] see 2191
Afungin [Decenta] see 8865
Afwillite see 1709
Afzelechin see 1908
AG 629 see 8722
AG 34-55 see 2284
Agar, 172
Agar-agar see 172
Agaric, 173
Agaric Acid see 174
Agaricic Acid, 174
Agaricin [Houd] see 174
Agaricinic Acid see 174
Agarin see 6221
Agaritine, 175
Agaropectin see 172
Agarose see 172
Agate see 8440
Agbayun see 6125
Agedal [Bayer] see 6644
Agedoite see 859
Agene see 6523
Agent Orange see 9052
AGEPC see 7496
Ageratochromene see 7716
Agerite see 6159
Ageroplas [Serono] see 3374
Agestal [Femada] see 5677
Agglutinin see 4
Agglutinins see 5312
Agifutol S [Kyorin] see 4369
Agilene see 7544
Agiolan see 9918
Agit [Midy] see 3157
Agkistrodon Rhodostoma Venom
   Protease see 664
Agkistrodon Serine Proteinase see 664
Aglumin see 3675
Agmatine, 176
Agnin see 5231

Agnolin see 5231
Agofell [Janssen] see 3180
Agofollin see 3653
Agolanid [Sandoz] see 3150
Agon [Hoechst] see 3895
Agoniadin see 7513
Agontan [Knoll] see 3177
Agotan see 2290
Agozol see 7744
Agr 1240 see 6119
Agradil [Vita] see 9849
Agram [Inava] see 610
Agréal [Delagrange] see 9849
Agribon [Roche] see 8876
Agricultural Limestone see 5368
Agrimek [Merck & Co.] see 1
Agri-Strep [Merck & Co.] see 8786
Agritan see 2832
Agritol see 945
Agritox [M & B] see 5645
Agroclavine, 177
Agrocybin, 178
Agromicina see 9130
Agropyrum see 9673
Agrosan GN see 7271
Agrothrin [Sumitomo] see 2775
Agroxone [ICI] see 5645
Agrozyme see 632
Agrypnal see 7201
Agstone see 5368
Ague Tree see 8340
Agurin see 9209
AH-42 see 5871
AH-289 see 2078
AH-2250 see 1485
AH-3365 see 209
AH-5158A see 5208
AH-8165 see 3887
AH-19065 see 8126
AHA see 54
Ahanon see 9009
AHB-DKB see 797
AHCTL see 2322
AHF see 3873
AHG see 3873
Ahistan, 179
AHMHA see 8759
AHP 200 see 6857
AHPrBP see 6954
AHR-85 see 5903
AHR-233 see 5739
AHR-438 see 5838
AHR-504 see 4397
AHR-619 see 3421
AHR-857 see 8885
AHR-2277 see 5320
AHR-3018 see 758
AHR-3053 see 1809
AHR-3070-C see 6063
AHR-5850D see 404
A-hydroCort [Abbott] see 4713
AI 3-36206 see 7778
AI 27303 see 2010
AIBN see 931
Aicamin [Crinos] see 6817
AICA Orotate see 6817
AICAR, 180
Aicorat [Mack, Illert.] see 6817
Aida [Paraphar] see 4738
Aiglonyl [Fumouze] see 8971
Aimax [Ayerst] see 5857
AIR, 181
Airbron [BDH] see 82
Airoform see 1281
Airogen see 1281
Airol [Roche] see 8167
Airum [Promeco] see 3927
Aisemide see 4221
Aizumycin see 1222
Ajacine, 182
Ajaconine, 183

Alfadolone Acetate, 224
Alfadryl "Spofa" see 5738
Alfa-interferon see 4892
Alfamox [Alfa] see 610
Alfaprostol, 225
Alfapsin [Choay] see 2265
Alfarol [Chugai] see 4749
Alfason [Thomae] see 4710
Alfaspoven [Alfa] see 1971
Alfathesin [Glaxo] see 226
Alfatil [Lilly] see 1920
Alfatrofin see 127
Alfavet [Hoffmann-La Roche] see 225
Alfaxalone, 226
Alfenamin see 4060
Alfenta [Janssen] see 227
Alfentanil, 227
Alfentanil Hydrochloride see 227
Alferon [Interferon Sciences] see 4892
Alficetin [Argentia] see 2475
Alficetyn [Allen & Hanburys] see 2068
Alfide [OM] see 8990
Alflorone [Merck & Co.] see 4059
Alfone see 244
Alfonic Ethoxylates [Conoco] see 7554
Alfospas [Rotta/Rorer] see 9393
Alfoten [Synthelabo] see 228
Alfuzosin, 228
Algafan see 7851
Algamon see 8297
Algamon Soluble see 8298
Algedrate see 345
Algeril [Bayer] see 7844
Algestone, 229
Algestone Acetonide see 229
Algestone Acetophenide, 230
Algiamida see 8297
Algidon see 5852
Algil see 5736
Algin, 231
Alginic Acid, 232
Alginic Acid Sodium Salt see 231
Alginodia see 3358
Alginor [De Angeli] see 2280
Alglyn [Brayten] see 3168
Algobaz [Labaz] see 9921
Algocalmin see 3358
Algocetil [Francia] see 8961
Algocor [Ravizza] see 4257
Algocor see 1093
Algolysin see 5852
Algopent [Alfa] see 7078
Algylen see 9552
Alibendol, 233
Alidase [Searle] see 4676
Alidine see 686
Aliette [Rhône-Poulenc] see 4167
Alimemazine see 9618
Alimet [Monsanto] see 5897
Alinam [Lucien] see 2105
Alinamin see 7896
Alinamin F [Takeda] see 4223
Alindapril see 2864
Alindor see 7248
Alinidine, 234
Aliporina [Asla] see 1975
Alisactide see 312
Alisobumal see 1502
Alitame, 235
Alius [Scharper] see 4146
Alival [Hoechst] see 6592
Alivin see 7027
Alizapride, 236
Alizarin, 237
Alizarin Blue R see 239
Alizarinbordeaux see 8059
Alizarin Cyanine Green F, 238
Alizarine Blue, 239
Alizarine Carmine see 8516

Alizarine Orange, 240
Alizarine S see 8516
Alizarine Yellow C see 4248
Alizarine Yellow GG see 5823
Alizarine Yellow R, 241
Alizarine Yellow RW see 241
β-2-Alizarin Primeveroside see 8257
Alizarin Violet see 4250
Alkacitron see 8550
Alkagel see 345
Alkalovert [Klein] see 7359
Alkanet, 242
Alkanet Extract see 243
Alkanna see 242
Alkanna Red see 243
Alkannin, 243
Alkannin Paper see 242
Alkathene [ICI] see 7544
Alkavervir see 9858
Alkeran [Burroughs Wellcome] see 5708
Alkiron see 6048
Alkofanone, 244
Alkron see 6983
Alkyd Resins, 245
1-O-Alkyl-2-acetyl-sn-glycero-3-phosphorylcholine see 7496
Alkylaluminum Halide see 324
Alkylaluminum Sesquihalide see 324
Allantoin, 246
Allantoxanic Acid see 6901
Alledryl see 3308
Allegan see 4375
Allegron [Dista] see 6635
Allenolic Acid, 247
Alleoside A see 4551
Alleract [Burroughs Wellcome] see 9661
Allerclor [Fellows-Testagar] see 2180
Allercorb [Plessner] see 855
Allercur [Roerig] see 2346
Allergam [C.F.T.S.] see 4837
Allergefon [Lafon] see 1804
Allergen see 3334
Allergin see 3308
Allergisan [Pharmacia] see 2180
Allergocrom [Ursapharm] see 2594
Allerkif [Edmond] see 5187
Alleron see 6983
Allerplus [Simes] see 9094
Allethrin I see 248
Allethrin II see 248
Allethrins, 248
Allethrolone Ester of Chrysanthemumdicarboxylic Acid Monomethyl Ester see 248
Allethrolone Ester of Chrysanthemummonocarboxylic Acid see 248
Allicin, 249
Allidochlor, 250
Allie [Du Pont] see 6081
Alliin, 251
Allium see 4268
Allobarbital, 252
Allobarbitone see 252
Alloca [Upjohn] see 6827
Allocaine see 7763
Allocholesterol, 253
Alloclamide, 254
Allocor [Kwizda] see 5229
Allocryptopine, 255
Allocupreide Sodium, 256
Allodene see 616
Allodihydro F see 4683
Allodihydrohydrocortisone see 4683
Allofenyl see 7162
Alloferin [Roche] see 214
Allohydroxyproline see 4775
Allohydroxy-D-proline Betaine see 9729

Allohydroxy-L-proline Betaine see 9729
Alloid [Kyosei] see 231
L(+)-Alloisoleucine see 5064
Allomaleic Acid see 4200
Allomelanins see 5692
Allomethadione see 305
Allomones see 7294
Allophycocyanins see 7353
Allopregnane, 257
Allopregnane-3α,20α-diol, 258
Allopregnane-3α,20β-diol, 259
Allopregnane-3β,20α-diol, 260
Allopregnane-3β,20β-diol, 261
Allopregnane-3β,21-diol-11,20-dione, 262
Allopregnane-3β,17α-diol-20-one, 263
3,20-Allopregnanedione, 264
Allopregnane-3β,11β,17α,20β,21-pentol, 265
Allopregnane-3β,17α,20β,21-tetrol, 266
Allopregnane-3α,11β,17α,21-tetrol-20-one, 267
Allopregnane-3β,11β,17α,21-tetrol-20-one, 268
Allopregnane-3β,17α,20α-triol, 269
Allopregnane-3β,17α,20β-triol, 270
Allopregnane-3α,17α,21-triol-11,20-dione see 280
Allopregnane-3β,17α,21-triol-11,20-dione, 271
Allopregnane-11β,17α,21-triol-3,20-dione see 4683
Allopregnane-3β,11β,21-triol-20-one, 272
Allopregnane-3β,17α,21-triol-20-one, 273
Allopregnan-3α-ol-20-one, 274
Allopregnan-3β-ol-20-one, 275
Allopregnan-20α-ol-3-one, 276
Allopregnan-20β-ol-3-one, 277
Allo-Puren [Klinge-Nattermann] see 278
Allopurinol, 278
Allopydin [Chugai] see 212
β-D-Allopyranose see 279
Allorphine see 6275
Allose [Kyosei] see 231
D-Allose, 279
Allotelluric Acid see 9063
3α-Allotetrahydrocortisol see 267
Allotetrahydrocortisone, 280
Allotropal [Heyl] see 5731
Alloxan, 281
Alloxan 5-Oxime see 9904
Alloxantin, 282
Alloxazine Mononucleotide see 8202
Alloyohimbine see 10012
Allozym [Sawai] see 278
Allspice see 7402
Alltox see 9478
D-Allulose see 7940
Allural [Nativelle] see 278
Allura® Red AC, 283
Alluval [Berlin-Chemie] see 1385
Allvoran [TAD] see 3071
Ally [Du Pont] see 6081
Allyl Alcohol, 284
Allylamine, 285
1-Allyl-6-amino-3-ethyl-2,4(1H,3H)-pyrimidinedione see 463
1-Allyl-6-amino-3-ethyluracil see 463
p-Allylanisole see 3657
Allylbarbital see 1502
Allyl Bromide, 286
5-Allyl-5-(2-bromoallyl)barbituric Acid see 1357
5-Allyl-5-sec-butylbarbituric Acid see 9010

Alphacillin [Merck & Co.] *see* 7484
Alpha Cobione [Merck & Co.] *see* 4739
Alpha D3 [Teva] *see* 4749
Alphadione *see* 226
Alphadolone Acetate *see* 224
Alphadril "Spofa" *see* 5738
Alphadrol [Upjohn] *see* 4119
Alphadryl "Spofa" *see* 5738
Alpha-hypophamine *see* 6934
Alpha-interferon *see* 4892
Alphakil [Rentokil] *see* 2062
Alphalin [Lilly] *see* 9918
Alpha-methyldopa *see* 5974
Alphamide *see* 8903
Alphamin [SS Pharm.] *see* 2345
Alphamine [Centerchem] *see* 6107
Alphanaphthol *see* 6303
Alpha-pipradrol *see* 7455
Alphaprodine, 307
AlphaRedisol [Merck & Co.] *see* 4739
Alpha-Ruvite [Savage] *see* 4739
Alphasol AY *see* 2968
Alphasol IB *see* 3179
Alphasol MA *see* 1266
Alphasol OT *see* 3398
Alphasone *see* 229
Alphasone Acetonide *see* 229
Alphasone Acetophenide *see* 230
Alphaxalone *see* 226
Alphenal *see* 7162
Alphenate *see* 7162
Alphol *see* 6334
Alphozone *see* 8848
Alpidem, 308
Alpiny [SS Pharm.] *see* 40
Alpiropride, 309
Alplax [Gador] *see* 310
Alprazolam, 310
Alprenolol, 311
Alpress LP [Pfizer] *see* 7715
Alprostadil *see* 7892
AL-protein *see* 651
Alredase [Am. Home] *see* 9451
Alrheumat [Bayer] *see* 5184
Alrheumun [Bayropharm] *see* 5184
Alsactide, 312
Alsadorm [Woelm] *see* 3430
Alsanate [Dainippon] *see* 4273
Alserin [Merck & Co.] *see* 8149
Alsol [Athenstaedt] *see* 323
Alstonia Cortex *see* 3373
Alstonidine, 313
Alstonine, 314
Alsystin [Bayer] *see* 9592
Altabactina *see* 4205
Altacite [Roussel-UCLAF] *see* 5538
Altafur [Norwich] *see* 4205
Altaite *see* 5304
Altan *see* 2813
Altapin [Marion] *see* 3419
Altat [Teikoku Hormone] *see* 8252
Alternagel [Stuart] *see* 345
Altex [Cenci] *see* 8721
Althea, 315
Altheine *see* 859
Althesin [Glaxo] *see* 226
Althiazide, 316
Altiazem [Lusofarmaco] *see* 3188
Alticina [Squibb] *see* 7184
Altilev [Squibb] *see* 6635
Altim [Roussel-UCLAF] *see* 2536
Altimina [Miquel] *see* 3910
Altimol [E. Merck] *see* 6492
Altinil [Squibb] *see* 9233
Altiopril *see* 6200
Altizide *see* 316
Alto [Kyosei] *see* 231
Altodor [Delalande] *see* 3675
Altolat *see* 8034
Altosid SR-10 [Zoecon] *see* 5906

Altracin [A. L. Labs] *see* 948
Altrenogest, 317
Altretamine, 318
Altrigen [Ethigen] *see* 191
D-Altropyranose *see* 319
D-Altrose, 319
Alubasine *see* 3168
Alucol [Dorsey] *see* 345
Aluctyl *see* 350
Aludrin [Boehringer, Ing.] *see* 5105
Aludrine *see* 5105
Aludrox [Wyeth] *see* 345
Aludyal [Sandoz] *see* 345
Alufibrate *see* 2375
Aluline [Steinhard] *see* 278
Alum *see* 364
Alum Flour *see* 364
Alumina *see* 359
Aluminium *see* 321
Aluminon, 320
Aluminum, 321
Aluminum Acetate Solution, 322
Aluminum Acetotartrate, 323
Aluminum Acetylsalicylate N.F *see* 328
Aluminum Alkyls, 324
Aluminum Aminoacetate (Basic) *see* 3168
Aluminum Ammonium Chloride *see* 589
Aluminum Ammonium Sulfate, 325
Aluminum Antimonide, 326
Aluminum Aspirin *see* 328
Aluminum Benzoate, 327
Aluminum Bis(acetylsalicylate), 328
Aluminum Bismuth Oxide *see* 1269
Aluminum Borate, 329
Aluminum Boroformate, 330
Aluminum Boro-formicicum *see* 330
Aluminum Borohydride, 331
Aluminum Bromide, 332
Aluminum *tert*-Butoxide, 333
Aluminum Calcium Hydride, 334
Aluminum Calcium Silicate *see* 1645
Aluminum Carbide, 335
Aluminum Cesium Sulfate, 336
Aluminum Chlorate, 337
Aluminum Chloride, 338
Aluminum Chlorohydrate *see* 346
Aluminum Chlorohydroxide *see* 346
Aluminum Diacetate, 339
Aluminum Diacetylsalicylate *see* 328
Aluminum Diaspirin *see* 328
Aluminum Dihydroxyaminoacetate *see* 3168
Aluminum Ethoxide, 340
Aluminum Ethylate *see* 340
Aluminum Flufenamate *see* 4060
Aluminum Fluoride, 341
Aluminum Fluosilicate *see* 342
Aluminum Glycinate *see* 3168
Aluminum Hexacarbamide Sulfate Triiodide *see* 343
Aluminum Hexafluorosilicate, 342
Aluminum Hexaurea Sulfate Triiodide, 343
Aluminum Hydrate *see* 345
Aluminum Hydride, 344
Aluminum Hydroxide, 345
Aluminum Hydroxyacetate *see* 339
Aluminum Hydroxychloride, 346
Aluminum Hypophosphite, 347
Aluminum Iodide, 348
Aluminum Isopropoxide, 349
Aluminum Isopropylate *see* 349
Aluminum Lactate, 350
Aluminum Lithium Hydride, 351
Aluminum Magnesium Hydroxide (AlMg(OH)$_7$) Monohydrate *see* 5527
Aluminum Magnesium Silicate, 352

Aluminum β-Naphtholdisulfonate, 353
Aluminum Nicotinate, 354
Aluminum Nitrate, 355
Aluminum Nitride, 356
Aluminum Oleate, 357
Aluminum Orthophosphate *see* 362
Aluminum Oxalate, 358
Aluminum Oxide, 359
Aluminum Oxide (Brockmann), 360
Aluminum Palmitate, 361
Aluminum Phosphate, 362
Aluminum Phosphide, 363
Aluminum Potassium Oxide *see* 7581
Aluminum Potassium Sulfate, 364
Aluminum Rubidium Sulfate, 365
Aluminum Salicylates, Basic, 366
Aluminum Selenide, 367
Aluminum Silicate, 368
Aluminum Silicofluoride *see* 342
Aluminum Sodium Carbonate Hexitol Complex *see* 223
Aluminum Sodium Carbonate Hydroxide *see* 3169
Aluminum Sodium Chloride *see* 8643
Aluminum Sodium Oxide *see* 8517
Aluminum Sodium Sulfate, 369
Aluminum Stearate, 370
Aluminum Subacetate *see* 339
Aluminum Subacetate Solution, 371
Aluminum Sulfate, 372
Aluminum Sulfide, 373
Aluminum Sulfocyanate *see* 375
Aluminum Tartrate, 374
Aluminum Tetrahydroborate *see* 331
Aluminum Thiocyanate, 375
Aluminum Trifluoride *see* 341
Aluminum Trihydrate *see* 345
Aluminum Tris(ethyl Phosphite) *see* 4167
Aluminum Tris(*O*-ethylphosphonate) *see* 4167
Aluminum Tristearate *see* 370
Aluminum Zinc Sulfate, 376
Alum Meal *see* 364
Alumnol [Winthrop] *see* 353
Alum Root *see* 4299
Alundum *see* 359
Alunex [Steinhard] *see* 2180
Alunite *see* 354
Alunogenite *see* 372
Alunozal *see* 366
Alupent [Geigy; Boehringer, Ing.] *see* 5836
Aluphos *see* 362
Alupram [Steinhard] *see* 2977
Alurate [Roche] *see* 782
Alurate Sodium [Roche] *see* 782
Alure *see* 343
Alurene *see* 2169
"Alutiae" *see* 7503
Alutyl *see* 2290
Aluwets [Stiefel] *see* 338
Aluzine [Steinhard] *see* 4221
Alven [FIRMA] *see* 9520
Alveograf [Sterling] *see* 3449
Alverine, 377
Alvinine [Wampole] *see* 1247
Alvo [Wyeth; Taisho] *see* 6879
Alvodine [Winthrop] *see* 7403
Alvonal MR [Gödecke] *see* 2768
Alvyl [Nobel] *see* 7562
Alyrane [Anaquest] *see* 3536
Alysine [Merrell] *see* 8626
Alzinox [SM & P] *see* 3168
Am 109 *see* 5901
AM-715 *see* 6617
AM-725 *see* 7012
AM-833 *see* 4035
AMA-1080 *see* 1880
Amabevan *see* 1788

Amadil see 40
Amadou see 173
Amal see 601
Amanin see 378
Amanitin, 378
Amanozine, 379
Amantadine, 380
Amantanium Bromide, 381
Amantol [Rotta] see 381
Amaranth (Dye), 382
Amaranth (Plant), 383
Amarogentin, 384
Amarolide, 385
Amaromycin see 7384
Amarsan see 44
Amarylline see 5498
Amasulin [Takeda] see 1880
Amasust see 601
Amatine [Roberts] see 6107
Amaze [Mobay] see 5055
Amazolon [Sawai] see 380
Ambacamp [Upjohn] see 944
Ambamide see 5525
Ambathizon see 9218
Ambaxin [Upjohn] see 944
Ambazone, 386
Amben see 434
Ambenonium Chloride, 387
Amber, 388
Amber Acid see 8840
Ambergris, 389
Amberlite®, 390
Ambilhar [Ciba] see 6480
Ambinon [Organon] see 4534
Ambivalon [Nattermann] see 505
Amblosin [Hoechst] see 621
Amblygonite see 5395
Amboclorin see 2064
Ambodryl [Parke, Davis] see 1405
Ambox [Hoechst] see 1237
Ambracyn see 9130
Ambramicina see 9130
Ambramycin see 9130
Ambraveine [Lepetit] see 7420
Ambra-Vena [Lepetit] see 7420
Ambrocef [Lusofarmaco] see 1980
Ambrosia see 6713
Ambrosin, 391
Ambroxol, 392
Ambrunate [Rhodia] see 6057
Ambucaine, 393
Ambucetamide, 394
Ambuphylline, 395
Ambush [ICI] see 7132
Ambuside, 396
Ambuterol see 5517
Ambutonium Bromide, 397
Ambutoxate see 393
Ambutyrosin see 1523
Ambylan [Lilly] see 780
Amcap [Circle] see 621
AMCHA see 9487
Amchem-65-81B see 2063
Amchem -70-25 see 1532
Amchlor [Cooper] see 531
Amcide see 583
Amciderm [E. Merck] see 398
Amcill [Parke, Davis] see 621
Amcill-S [Parke, Davis] see 621
Amcinonide, 398
Amco see 7558
AMD see 5974
Amdinocillin, 399
Amdinocillin Pivoxil, 400
Amdro [Am. Cyanamid] see 4684
Amebacilin see 4199
Ameban see 1788
Amebarsin see 3338
Amebarsone [Lilly] see 1788
Amebil see 4924
Ame-Boots see 3187

Amechol [Moorsfields] see 5847
Amedel [Dainippon] see 7451
Ameisensäure (German) see 4153
Amekrin [Parke-Davis] see 627
Ameliaroside see 7367
Amen [Carnrick] see 5677
Amenide [Sterling] see 8068
Amenox [Sterling] see 8068
Amenyl see 2102
Amerfil see 7558
Americaine [Arnar-Stone] see 3719
American Cyanamid 3911 see 7305
American Cyanamid 4124 see 3027
American Cyanamid 12880 see 3209
American Cyanamid 18133 see 9275
American Cyanamid 38023 see 3882
American Elder see 8315
American Hellebore see 9858
American Horsemint see 6153
American Indian Hemp see 774
American Nightshade Root see 7363
American Penicillin see 1157
American Saffron see 1877
American Spikenard see 793
American Valerian see 2778
American Veratrum see 9858
American Wormseed see 6713
Americium, 401
Ametazole see 1205
Amethocaine Hydrochloride see 9123
Amethone [Abbott] see 603
A-Methopterin see 5908
Amethopterin see 5908
Amethyst see 8440
Ametox see 8652
Ametrex [Makhteshim-Agan] see 402
Ametriodinic Acid see 4904
Ametryn, 402
Ametryne see 402
Ametycine [Choay] see 6133
Amex [Amchem] see 1532
Amezinium Methyl Sulfate, 403
Amfebutamon(e) see 1488
Amfecloral see 613
Amfenac, 404
Amfepramone see 3116
d-Amfetasul [Pitman-Moore] see 2932
Amfipen [Brocades] see 621
Amfomycin see 618
Amianthus see 851
Amiben [Amchem] see 2063
Amibiarson see 1788
Amibufen [Harris] see 4812
Amicar [Lederle] see 442
Amicarbalide, 405
Amicardine see 5189
Amicetin, 406
Amicibone, 407
Amicos [Banyu] see 2344
Amicoumacin A, 408
Amidalgon see 3295
Amidate [Abbott] see 3838
Amidazine see 3692
Amidazophen see 488
Amidephrine, 409
Amidephrine Mesylate see 409
Amide PP see 6398
N-Amidino-3,5-diamino-6-chloro-
  pyrazinamide see 417
N-Amidino-3,5-diamino-6-chloro-
  pyrazinecarboxamide see 417
N-Amidino-2-(2,6-dichlorophenyl)-
  acetamide see 4474
N-(2'-Amidinoethyl)-3-aminocyclo-
  pentanecarboxamide see 410
2-[N-(2-Amidinoethyl)carbamoyl]-5-
  iminopyrrolidine see 6590
N'-(2-Amidinoethyl)-4-(2-guanidino-
  acetamido)-1,1'-dimethyl-N,4'-bi-
  [pyrrole-2-carboxamide] see 6391

N-(2-Amidinoethyl)-5-imino-2-pyr-
  rolidinecarboxamide see 6590
N-Amidinoglycine see 4391
Amidinoguanidine see 1233
1-Amidinohydrazono-4-thiosemicarb-
  azono-2,5-cyclohexadiene see 386
Amidinomycin, 410
6-Amidino-2-naphthyl-4-guanidino-
  benzoate see 6264
L-6-Amidinonorleucine see 4877
7-m-Amidinophenyldiazoamino-2-
  amino-10-ethyl-9-phenylphenan-
  thridinium Chloride see 5067
8-[3-(m-Amidinophenyl)-2-triazeno]-
  3-amino-5-ethyl-6-phenylphenan-
  thridinium Chloride see 5067
N-Amidinosarcosine see 2570
N¹-Amidinosulfanilamide see 8879
2-Amidino-1,2,3,4-tetrahydroisoqui-
  noline see 2837
Amidochlor, 411
Amidocyanogen see 2691
Amidofebrin see 488
Amido-G-Acid, 412
Amidol see 2961
Amidolacetate see 5853
Amidomycin, 413
Amidonal [Madaus] see 781
Amidon Hydrochloride see 5852
Amidoprocain see 7762
Amidopyrazoline see 488
Amidopyrine see 488
Amido-R-Acid, 414
Amidosulfonic Acid see 8893
1-(p-Amidosulfonylphenyl)-2-thiapi-
  peridine 2,2-Dioxide see 8974
Amidozol see 8918
Amidryl see 3308
Amid-Sal [Glenwood] see 8297
AMIF-72 see 1547
Amifloxacin, 415
Amifloxacin Mesylate see 415
Amifur [Norwich] see 6521
Amigen see 7900
Amikacin, 416
Amikal [Gea] see 417
Amikapron [Kabi] see 9487
Amikin [Bristol] see 416
Amiklin [Bristol] see 416
Amilan [Toyo] see 6656
Amilar [Du Pont] see 7546
Amilco [Norton] see 417
Amiloride, 417
Amilyt see 1116
Amimetilina see 7917
Amimycin see 6785
Aminacrine, 418
Aminacyl see 491
Aminarsone see 1788
Aminazine see 2186
Amine 220 [Carbide & Carbon Chem.],
  419
Amine Oxidase see 6158
Amineptine, 420
Amin-Glaukosan [Woelm] see 4640
Aminicotin see 6398
Aminitrozole, 421
p-Aminoacetanilide, 422
4'-Aminoacetanilide see 422
Aminoacetic Acid see 4386
Aminoacetonitrile, 423
α-Amino-p-acetophenetide see 7204
2-Amino-p-acetophenetidide see 7204
Aminoacetophenetidine see 7204
Aminoacetophenone, 424
m-Aminoacetophenone see 424
o-Aminoacetophenone see 424
p-Aminoacetophenone see 424
2-Aminoacetophenone see 7159
3'-Aminoacetophenone see 424
α-Aminoacetophenone see 7159

2,2'-[[[4-(Aminocarbonyl)phenyl]arsinidene]bis(thio)]bisacetic Acid *see* 819

3-(Aminocarbonyl)-1-(5-O-phosphono-β-D-ribofuranosyl)pyridinium Hydroxide Inner Salt *see* 6585

(D-4-Amino-4-carboxybutyl)penicillinic Acid *see* 7043

2-Amino-2-carboxyethyl 2-Amino-2-carboxyethanethiosulfonate *see* 2789

S-(2-Amino-2-carboxyethyl)cysteine *see* 5234

2-Amino-3-O-(D-1'-carboxyethyl)-2-deoxy-D-glucose *see* 6213

2-Amino-3-O-(1-carboxyethyl)-2-deoxy-D-glucose *see* 6213

3-[(2-Amino-2-carboxyethyl)dithio]-D-valine *see* 7030

(R)-S-(2-Amino-2-carboxyethyl)-L-homocysteine *see* 2784

7β-(2-D-Amino-2-carboxyethylthioacetamido)-7α-methoxy-3-[[(1-methyl-1H-tetrazol-5-yl)thio]methyl]-3-cephem-4-carboxylic Acid *see* 1930

[6R-[6α,7α,7(S*)]]-7-[[[(2-Amino-2-carboxyethyl)thio]acetyl]amino]-7-methoxy-3-[[(1-methyl-1H-tetrazol-5-yl)thio]methyl]-8-oxo-5-thia-1-azabicyclo[4.2.0]oct-2-ene-2-carboxylic Acid *see* 1930

[R-[R*,S*(Z)]]-7-[(2-Amino-2-carboxyethyl)thio]-2-[[(2,2-dimethylcyclopropyl)carbonyl]amino]-2-heptenoic Acid *see* 2275

6-[(2-Amino-2-carboxyethyl)thio]-5-hydroxy-7,9,11,14-eicosatetraenoic Acid *see* 5339

α-Amino-1-carboxy-4-hydroxy-2,5-cyclohexadiene-1-propanoic Acid *see* 814

p-Amino-p'-(carboxymethylamino)-diphenyl Sulfone *see* 17

1-[(4-Amino-4-carboxy-1-oxobutyl)-amino]-D-proline *see* 5377

[6R-(6α,7α)]-7-[[[4-(2-Amino-1-carboxy-2-oxoethylidene)-1,3-dithietan-2-yl]carbonyl]amino]-7-methoxy-3-[[(1-methyl-1H-tetrazol-5-yl)thio]methyl]-8-oxo-5-thia-1-azabicyclo[4.2.0]oct-2-ene-2-carboxylic Acid *see* 1936

[2S-[2α,5α,6β(S*)]]-6-[(5-Amino-5-carboxy-1-oxopentyl)amino]-3,3-dimethyl-7-oxo-4-thia-1-azabicyclo[3.2.0]heptane-2-carboxylic Acid *see* 7043

(S)-N-[N-(5-Amino-5-carboxy-1-oxopentyl)-L-cysteinyl]-D-valine *see* 138

N-(5-Amino-5-carboxypentyl)-L-glutamic Acid *see* 8283

3-Amino-N-(α-carboxyphenethyl)-succinamic Acid N-Methyl Ester *see* 861

4-Amino-4'-β-carboxypropionylaminodiphenylsulfone *see* 8851

4-Amino-4'-(β-carboxypropionylamino)phenylsulfonylbenzene *see* 8851

Z-3-[[[4-(3-Amino-3-carboxypropoxy)phenyl](hydroxyimino)acetyl]amino]-4-(4-hydroxyphenyl)-2-oxo-1-azetidineacetic Acid *see* 6588

S-(3-Amino-3-carboxypropyl)-S-butylsulfoximine *see* A3

(3-Amino-3-carboxypropyl)dimethyl Sulfonium Chloride *see* 9942

(3-Amino-3-carboxypropyl)methylphosphinic Acid *see* 7314

5'-[(3-Amino-3-carboxypropyl)methylsulfonio]-5'-deoxyadenosine Hydroxide, Inner Salt *see* 146

6-Amino-3-carboxypyridine *see* 468

6-(D-5-Amino-5-carboxyvaleramido)-3,3-dimethyl-7-oxo-4-thia-1-azabicyclo[3.2.0]heptane-2-carboxylic Acid *see* 7043

7-(D-5-Amino-5-carboxyvaleramido)-3-(hydroxymethyl)-8-oxo-5-thia-1-azabicyclo[4.2.0]oct-2-ene-2-carboxylic Acid Acetate *see* 1976

N-[N-(L-5-Amino-5-carboxyvaleryl)-L-cysteinyl]-D-valine *see* 138

Aminocardol *see* 477

7-Aminocephalosporanic Acid, 444

Aminochinuride *see* 489

2-Amino-4-(p-chloroanilino)-s-triazine *see* 2073

4-Amino-6-chloro-m-benzenedisulfonamide *see* 2067

4-Amino-6-chloro-1,3-benzenedisulfonamide *see* 2067

6-Amino-4-chlorobenzene-1,3-disulfonamide *see* 2067

4-Amino-2-chlorobenzoic Acid 2-(Diethylamino)ethyl Ester Monohydrochloride *see* 2158

2-Amino-5-chlorobenzoxazole *see* 10098

p-Amino-p'-chlorobiphenyl *see* 445

4-Amino-5-chloro-N-[2-(diethylamino)ethyl]-o-anisamide *see* 6063

4-Amino-5-chloro-N-[2-(diethylamino)ethyl]-2-methoxybenzamide *see* 6063

4-Amino-3-chloro-α-[[(1,1-dimethylethyl)amino]methyl]-5-(trifluoromethyl)benzenemethanol *see* 5517

4-Amino-4'-chlorodiphenyl, 445

6-Amino-2-(2-chloroethyl)-2,3-dihydro-4H-1,3-benzoxazin-4-one *see* 446

cis-4-Amino-5-chloro-N-[1-[3-(p-fluorophenoxy)propyl]-3-methoxy-4-piperidinyl]-o-anisamide *see* 2318

cis-4-Amino-5-chloro-N-[1-[3-(4-fluorophenoxy)propyl]-3-methoxy-4-piperidinyl]-2-methoxybenzamide *see* 2318

4-Amino-5-chloro-2-methoxy-N-(β-diethylaminoethyl)benzamide *see* 6063

4-Amino-5-chloro-2-methoxy-N-[1-(phenylmethyl)-4-piperidinyl]benzamide *see* 2344

γ-Amino-β-(p-chlorophenyl)butyric Acid *see* 950

4-Amino-N-(6-chloro-3-pyridazinyl)benzenesulfonamide *see* 8871

1-(4'-Amino-3'-chloro-5'-trifluoromethylphenyl)-2-tert-butylaminoethanol *see* 5517

Aminochlorthenoxazin, 446

Aminochromes, 447

6-Aminochrysene *see* 2258

o-Aminocinnamic Acid Lactam *see* 1831

p-Aminoclonidine *see* 779

4-Aminocumene *see* 2622

1-(4-Amino-o-cyanophenyl)-2-isopropylaminoethanol *see* 2278

4-Amino-5-cyano-7-(D-ribofuranosyl)-7H-pyrrolo[2,3-d]pyrimidine *see* 9483

7-[D-2-Amino-2-(1,4-cyclohexadienyl)acetamide]-3-methoxy-3-cephem-4-carboxylic Acid *see* 1942

7-[D-2-Amino-2-(1,4-cyclohexadienyl)acetamido]desacetoxycephalosporanic Acid *see* 1982

6-[D-α-Amino-2-(1,4-cyclohexadien-1-yl)acetamido]penicillanic Acid *see* 3566

6-[(Amino-1,4-cyclohexadien-1-ylacetyl)amino]-3,3-dimethyl-7-oxo-4-thia-1-azabicyclo[3.2.0]heptane-2-carboxylic Acid *see* 3566

7-[(Amino-1,4-cyclohexadien-1-ylacetyl)amino]-3-methoxy-8-oxo-5-thia-1-azabicyclo[4.2.0]oct-2-ene-2-carboxylic Acid *see* 1942

[6R-[6α,7β(R*)]]-7-[(Amino-1,4-cyclohexadien-1-ylacetyl)amino]-3-methyl-8-oxo-5-thia-1-azabicyclo-[4.2.0]oct-2-ene-2-carboxylic Acid *see* 1982

D-α-Amino-(1,4-cyclohexadien-1-yl)methylpenicillin *see* 3566

Aminocyclohexane *see* 2735

6-(1-Aminocyclohexanecarboxamido)-penicillanic Acid *see* 2706

6-[[(1-Aminocyclohexyl)carbonyl]amino]-3,3-dimethyl-7-oxo-4-thia-1-azabicyclo[3.2.0]heptane-2-carboxylic Acid *see* 2706

(1-Aminocyclohexyl)penicillin *see* 2706

1-Aminocyclopentanecarboxylic Acid *see* 2740

Aminodal *see* 18

2-Amino-5-(3,7,11,15,19,23,27,31,35,-39-decamethyl-2,6,10,14,18,22,26,-30,34,38-tetracontadecaenyl)-3-methoxy-6-methyl-2,5-cyclohexadiene-1,4-dione *see* 8194

2-Amino-2-deoxy-D-galactose *see* 4240

1-Amino-1-deoxy-D-glucitol *see* 4343

O-3-Amino-3-deoxy-α-D-glucopyranosyl-(1→6)-O-[6-amino-6-deoxy-α-D-glucopyranosyl-(1→4)]-N¹-(4-amino-2-hydroxy-1-oxobutyl)-2-deoxy-D-streptamine *see* 416

O-3-Amino-3-deoxy-α-D-glucopyranosyl-(1→6)-O-[6-amino-6-deoxy-α-D-glucopyranosyl-(1→4)]-2-deoxy-D-streptamine *see* 5161

O-4-Amino-4-deoxy-α-D-glucopyranosyl-(1→8)-O-(8R)-2-amino-2,3,7-trideoxy-7-(methylamino)-D-glycero-α-D-allo-octodialdo-1,5:8,4-di-pyranosyl-(1→4)-2-deoxy-D-streptamine *see* 780

O-2-Amino-2-deoxy-α-D-glucopyranosyl-(1→4)-O-[3-deoxy-3-(methylamino)-α-D-xylopyranosyl-(1→6)]-2-deoxy-D-streptamine *see* 4284

(S)-O-6-Amino-6-deoxy-α-D-glucopyranosyl-(1→4)-O-[3-deoxy-4-C-methyl-3-(methylamino)-β-L-arabinopyranosyl-(1→6)]-N¹-(3-amino-2-hydroxy-1-oxopropyl)-2-deoxy-D-streptamine *see* 4989

O-2-Amino-2-deoxy-α-D-glucopyranosyl-(1→4)-O-[O-2,6-diamino-2,6-dideoxy-β-L-idopyranosyl-(1→3)-β-D-ribofuranosyl-(1→5)]-2-deoxy-D-streptamine *see* 6989

(S)-O-3-Amino-3-deoxy-α-D-glucopyranosyl-(1→6)-O-[2,6-diamino-2,3,4,6-tetradeoxy-α-D-erythro-hexopyranosyl-(1→4)]-N¹-(4-amino-2-hydroxy-1-oxobutyl)-2-deoxy-D-streptamine *see* 797

O-3-Amino-3-deoxy-α-D-glucopyranosyl-(1→6)-O-[2,6-diamino-2,3,4,6-tetradeoxy-α-D-erythro-hexopyranosyl-(1→4)]-2-deoxy-D-streptamine *see* 2987

2-Amino-6-(1'-methyl-4'-nitro-5'-imidazolyl)mercaptopurine *see* 9228

2-Amino-6-[(1-methyl-4-nitroimidazol-5-yl)thio]purine *see* 9228

(*E*)-2-Amino-4-[2-(1-methyl-5-nitroimidazol-2-yl)vinyl]pyrimidine *see* 914

(±)-*erythro*-2-Amino-3-methylpentanoic Acid *see* 5064

(+)-*threo*-2-Amino-3-methylpentanoic Acid *see* 5064

2-Amino-3-methylpentanoic Acid *see* 5064

2-Amino-4-methylpentanoic Acid *see* 5331

4-Amino-4-methyl-2-pentanone *see* 2942

β-Amino-α-methylphenethyl Alcohol, 459

7-[*o*-(Aminomethyl)phenylacetamido]-3-[[[1-(carboxymethyl)-1*H*-tetrazol-5-yl]thio]methyl]-3-cephem-4-carboxylic Acid *see* 1934

7-[[[2-(Aminomethyl)phenyl]acetyl]-amino]-3-[[[1-(carboxymethyl)-1*H*-tetrazol-5-yl]thio]methyl]-8-oxo-5-thia-1-azabicyclo[4.2.0]oct-2-ene-2-carboxylic Acid *see* 1934

4-Amino-*N*-(3-methyl-1-phenyl-1*H*-pyrazol-5-yl)benzenesulfonamide *see* 8923

2-Amino-4-methylphosphinobutyric Acid *see* 7314

1-Amino-2-methylpropane *see* 5016

2-Amino-2-methylpropane *see* 1545

2-Amino-2-methyl-1,3-propanediol, 460

2-Amino-2-methylpropanoic Acid *see* 457

2-Amino-2-methyl-1-propanol, 461

2-Amino-2-methyl-2-propanol 8-Bromotheophyllinate *see* 6952

4-Amino-*N*[10]-methylpteroylglutamic Acid *see* 5908

4-Amino-9-methylpteroylglutamic Acid *see* 691

2-Amino-4-methylpyridine *see* 478

4-Amino-2-methyl-5-pyrimidinemethanol *see* 9482

4-Amino-5-methyl-2(1*H*)-pyrimidinone *see* 5970

4-Amino-*N*-(4-methyl-2-pyrimidinyl)benzenesulfonamide *see* 8884

4-Amino-*N*-(5-methyl-2-pyrimidinyl)benzenesulfonamide *see* 8909

N-[(4-Amino-2-methyl-5-pyrimidinyl)methyl]-*N*-[2-[(2-benzoylvinyl)-thio]-4-hydroxy-1-methyl-1-butenyl]formamide *see* 9895

3-[(4-Amino-2-methyl-5-pyrimidinyl)methyl]-5-(2-chloroethyl)-4-methylthiazolium Chloride *see* 1030

N'-[(4-Amino-2-methyl-5-pyrimidinyl)methyl]-*N*-(2-chloroethyl)-*N*-nitrosourea *see* 6469

8-[[2-[*N*-[(4-Amino-2-methyl-5-pyrimidinyl)methyl]formamido]-1-(2-hydroxyethyl)propenyl]dithio]-6-mercaptooctanoic Acid Methyl Ester S(or 6)-Acetate *see* 6680

1-[(4-Amino-2-methyl-5-pyrimidinyl)methyl]-3-(2-hydroxyethyl)-2-methylpyridinium Bromide Monohydrobromide *see* 8003

3-[(4-Amino-2-methyl-5-pyrimidinyl)methyl]-5-(2-hydroxyethyl)-4-methyl-2(3*H*)-thiazolethione *see* 9296

3-[(4-Amino-2-methyl-5-pyrimidinyl)methyl]-5-(2-hydroxyethyl)-4-methylthiazolium Chloride Monohydrochloride *see* 9222

N-[(4-Amino-2-methyl-5-pyrimidinyl)methyl]-*N*-(4-hydroxy-2-mercapto-1-methyl-1-butenyl)formamide *O*,*S*-Diacetate *see* 46

N-[(4-Amino-2-methyl-5-pyrimidinyl)methyl]-*N*-[4-hydroxy-1-methyl-2-[(3-oxo-3-phenyl-1-propenyl)-thio]-1-butenyl]formamide *see* 9895

N-[(4-Amino-2-methyl-5-pyrimidinyl)methyl]-*N*-[4-hydroxy-1-methyl-2-(propyldithio)-1-butenyl]form-amide *see* 7896

N-[(4-Amino-2-methyl-5-pyrimidinyl)methyl]-*N*-[4-hydroxy-1-methyl-2-[[(tetrahydro-2-furanyl)methyl]-dithio]-1-butenyl]formamide *see* 4223

3-[(4-Amino-2-methyl-5-pyrimidinyl)methyl]-5-[2-[[hydroxy(phosphonooxy)phosphinyl]oxy]ethyl]-4-methylthiazolium Chloride *see* 2451

3-[(4-Amino-2-methyl-5-pyrimidinyl)methyl]-5-[2-(phosphonooxy)ethyl]thiazolium Chloride *see* 9224

3-[(4-Amino-2-methyl-5-pyrimidinyl)methyl] 4-Methyl-5-(4,6,6-trihydroxy-3,5-dioxa-4,6-diphosphahex-1-yl)thiazolium Chloride, *P*,*P*'-Dioxide *see* 2451

N-[(4-Amino-2-methyl-5-pyrimidinyl)methyl]-*N*-[1-(2-oxo-1,3-oxathian-4-ylidene)ethyl]formamide *see* 2762

3-(4-Amino-2-methylpyrimidyl-5-methyl)-4-methyl-5-(β-hydroxyethyl)thiazolium Nitrate *see* 9223

2-Aminomethylpyrrol-3-acetic Acid 4-Propionic Acid *see* 7575

2-Amino-4-(methylseleno)butanoic Acid *see* 8395

α-Amino-γ-(methylseleno)butyric Acid *see* 8395

4-Amino-*N*-(5-methyl-1,3,4-thiadiazol-2-yl)benzenesulfonamide *see* 8887

2-Amino-4-methylthiazole, 462

4-Amino-*N*-(4-methyl-2-thiazole)-benzenesulfonamide *see* 8891

2-Amino-4-methylthiobutanoic Acid *see* 5896

2-Amino-4-(methylthio)butyric Acid *see* 5896

4-Amino-2-methylthio-5-pyrimidinemethanol *see* 5898

2-Amino-3-methylvaleric Acid *see* 5064

2-Amino-4-methylvaleric Acid *see* 5331

α-Amino-β-methylvaleric Acid *see* 5064

α-(Aminomethyl)vanillyl Alcohol *see* 6627

Aminometradine, 463

Aminometramide *see* 463

Aminomux [Gador] *see* 6954

Aminomycin *see* 7121

1-Aminonaphthalene *see* 6318

2-Aminonaphthalene *see* 6319

3-Amino-2-naphthalenecarboxylic Acid *see* 464

1-Amino-2,7-naphthalenedisulfonic Acid *see* 6320

3-Amino-2,7-naphthalenedisulfonic Acid *see* 414

4-Amino-1,6-naphthalenedisulfonic Acid *see* 6322

4-Amino-1,7-naphthalenedisulfonic Acid *see* 6321

7-Amino-1,3-naphthalenedisulfonic Acid *see* 412

2-Amino-1-naphthalenesulfonic Acid *see* 6326

4-Amino-1-naphthalenesulfonic Acid *see* 6323

5-Amino-1-naphthalenesulfonic Acid *see* 6324

5-Amino-2-naphthalenesulfonic Acid *see* 2348

6-Amino-1-naphthalenesulfonic Acid *see* 6327

6-Amino-2-naphthalenesulfonic Acid *see* 1899

7-Amino-1-naphthalenesulfonic Acid *see* 952

7-Amino-2-naphthalenesulfonic Acid *see* 1899

8-Amino-1-naphthalenesulfonic Acid *see* 6325

8-Amino-2-naphthalenesulfonic Acid *see* 2349

4-Amino-1-naphthalenol *see* 465

3-Amino-2-naphthoic Acid, 464

4-Amino-1-naphthol, 465

8-Amino-1-naphthol-3,6-disulfonic Acid *see* 6305

1-Amino-2-naphthol-4-sulfonic Acid, 466

1-Amino-2-naphthol-6-sulfonic Acid, 467

1-Amino-2-(α-naphthylamino)ethane *see* 6329

Aminonat *see* 7900

6-Aminonicotinic Acid, 468

2-Amino-5-(5-nitro-2-furyl)-1,3,4-thiadiazole *see* 9509

3-Amino-6-[2-(5-nitro-2-furyl)vinyl]-pyridazine *see* 6453

2-Amino-4-nitro-1-propoxybenzene *see* 6551

L-2-Amino-3-[(*N*-nitroso)hydroxylamino]propionic Acid *see* 197

2-Amino-5-nitrothiazole, 469

3-Amino-5-nitro-*o*-toluamide *see* 707

2-Amino-4-octadecene-1,3-diol *see* 8703

[5*R*-(5α,11α,13*S**)]-13-Amino-5,6,7,-8,9,10,11,12-octahydro-5-methyl-5,11-methanobenzocyclodecen-3-ol *see* 2934

Aminooxamide *see* 8397

Aminooxoacetic Acid *see* 6870

Aminooxoacetic Acid Hydrazide *see* 8397

Aminooxoacetic Acid [(5-Nitro-2-furanyl)methylene]hydrazide *see* 6446

(*Z*)-4-Amino-4-oxo-2-butenoic Acid *see* 5583

4-Amino-2-oxo-1,2-dihydropyrimidine *see* 2801

1-(4-Amino-4-oxo-3,3-diphenylbutyl)-1-methylpiperidinium Bromide *see* 3935

N-[2-[(2-Amino-2-oxoethyl)amino]-2-carboxyethyl]-L-aspartic Acid *see* 5491

[4-[(2-Amino-2-oxoethyl)amino]phenyl]arsonic Acid Monosodium Salt *see* 9703

(*E*)-8-Amino-8-oxo-2-octene-4,6-diynoic Acid *see* 2973

α-Amino-2-oxo-4-oxazoline-5-acetic Acid *see* 6220

5-Amino-4-oxopentanoic Acid *see* 458

3α-Amino-20-oxo-5α-pregnane *see* 4204

Ammonium Metavanadate *see* 599
Ammonium Molybdate(VI), 559
Ammonium Molybdophosphate *see* 573
Ammonium Muriate *see* 531
Ammonium α-Naphthylnitrosohydroxylamine *see* 6363
Ammonium Nickel Sulfate, 560
Ammonium Nitrate, 561
Ammonium Nitratocerate(IV) *see* 529
Ammonium Nitroferricyanide, 562
Ammonium Nitroprusside *see* 562
Ammonium Oleate, 563
Ammonium Osmium Chloride, 564
Ammonium Oxalate, 565
Ammonium Oxodioxalatotitanate(IV) *see* 593
Ammonium Palmitate, 566
Ammonium Paramolybdate *see* 559
Ammonium Paratungstate *see* 594
Ammonium Pentachlorozincate, 567
Ammonium Pentacyanonitrosylferrate(III) *see* 562
Ammonium Perchlorate, 568
Ammonium Peroxydisulfate, 569
Ammonium Persulfate *see* 569
Ammonium Phenate *see* 7206
Ammonium *p*-Phenolsulfonate *see* 7212
Ammonium Phosphate, Dibasic, 570
Ammonium Phosphate, Monobasic, 571
Ammonium Phosphite, 572
Ammonium Phosphomolybdate, 573
Ammonium Phosphorus Hexafluoride *see* 551
Ammonium Phosphotungstate, 574
Ammonium Phosphowolframate *see* 574
Ammonium Picrate, 575
Ammonium Picronitrate *see* 575
Ammonium Platinic Chloride, 576
Ammonium Platinochloride *see* 577
Ammonium Platinous Chloride, 577
Ammonium Polysulfide Soln *see* 586
Ammonium Purpurate *see* 6215
Ammonium Reineckate *see* 8135
Ammonium Rhodanide *see* 591
Ammonium Rhodanilate *see* 8183
Ammonium Salicylate, 578
Ammonium Salt of Aurintricarboxylic Acid *see* 320
Ammonium Selenate, 579
Ammonium Selenite, 580
Ammonium Silicofluoride *see* 552
Ammonium Sodium Phosphate, 581
Ammonium Stearate, 582
Ammonium Sulfamate, 583
Ammonium Sulfate, 584
Ammonium Sulfhydrate *see* 521
Ammonium Sulfhydrate Soln *see* 587
Ammonium Sulfide, 585
Ammonium Sulfide Solution, Red, 586
Ammonium Sulfide Solution, Yellow, 587
Ammonium Sulfite, 588
Ammonium Sulfobituminate *see* 4815
Ammonium Sulfocarbamate *see* 539
Ammonium Sulfocyanate *see* 591
Ammonium Sulfocyanide *see* 591
Ammonium Sulfoichthyolate *see* 4815
Ammonium Tetraborate *see* 524
Ammonium Tetrachloroaluminate, 589
Ammonium Tetrachlorocuprate(II) *see* 537
Ammonium Tetrachlorodiaquocuprate-(II) *see* 537
Ammonium Tetrachloromercurate(II) *see* 558

Ammonium Tetrachloroplatinate(II) *see* 577
Ammonium Tetrachlorozincate, 590
Ammonium Tetrathiocyanodiammonochromate *see* 8135
Ammonium Thiocyanate, 591
Ammonium Thiosulfate, 592
Ammonium Titanium Oxalate, 593
Ammonium Tricarbonatozirconate *see* 600
Ammonium Trioxalatoferrate(III) *see* 542
Ammonium Tungstate(VI), 594
Ammonium Tungstophosphate *see* 574
Ammonium Uranate(VI), 595
Ammonium Uranium Carbonate, 596
Ammonium Uranium Fluoride, 597
Ammonium Valerate, 598
Ammonium Valerianate *see* 598
Ammonium Vanadate(V), 599
Ammonium Zirconyl Carbonate, 600
Ammonyx DME [Onyx] *see* 2021
Ammonyx G *see* 2009
Ammonyx T *see* 2009
Ammophyllin *see* 477
A.M.N. *see* 8202
Amnestrogen [Squibb] *see* 2504
Amoban [Rhone-Poulenc] *see* 10095
Amobarbital, 601
Amocaine *see* 603
Amocilline [Inpharzam] *see* 610
Amodex [Robert et Carrière] *see* 610
Amodiaquin, 602
Amoenol *see* 4924
Amoksiklav [Lek] *see* 2342
Amolanone, 603
Amolin [Takeda] *see* 610
Amopenixin [Hishiyama] *see* 610
Amoproxan, 604
Amorolfine, 605
Amorphan [Heumann] *see* 6634
Amosamine *see* 406
Amoscanate, 606
Amosene [Ferndale] *see* 5751
Amosite *see* 851
Amosulalol, 607
Amosyt [AB Leo] *see* 3193
Amotril *see* 2374
Amotriphene, 608
Amovane [Rhône-Poulenc] *see* 10095
Amoxanox *see* 508
Amoxapine, 609
Amoxi [Beecham] *see* 610
Amoxibiotic [Aristochimica] *see* 610
Amoxicillin, 610
Amoxidal [De Angeli] *see* 610
Amoxidin [Lagap] *see* 610
Amoxil [Beecham] *see* 610
Amoxillat [Azuchemie] *see* 610
Amoxipen [Lorenzini] *see* 610
Amoxi-Wolff [Wolff] *see* 610
Amoxycillin *see* 610
Amoxypen [Grünenthal] *see* 610
AMP [Barnes-Hind] *see* 148
AMP-3 *see* 9179
A5MP *see* 148
3′,5′-AMP *see* 2714
Ampazine *see* 7793
AMPC *see* 610
Ampecyclal [Sarget] *see* 4577
Ampelopsin, 611
Ampeloptin *see* 611
Amperil [Geneva] *see* 621
Amperozide, 612
Amphecloral, 613
Amphedroxyn [Lilly] *see* 5859
Amphenidone, 614
Amphenone *see* 615
Amphenone B, 615
Amphetamine, 616
d-Amphetamine Sulfate *see* 2932

Amphetaminil, 617
Amphibole *see* 851
Amphicol *see* 2068
Amphigene *see* 5333
Amphihydroxynaphthyl-β-propionic Acid *see* 247
Amphojel [Wyeth] *see* 345
Ampho-Moronal [Heyden] *see* 620
Amphomycin, 618
Amphotalide, 619
Amphotericin B, 620
Amphothalide *see* 619
Amphozone [Squibb] *see* 620
Ampi-Bol [Beecham] *see* 621
Ampichel [Rachelle] *see* 621
Ampicillin, 621
Ampicillin A *see* 621
Ampicillin B *see* 621
Ampicillin (5-Methyl-2-oxo-1,3-dioxolen-4-yl)methyl Ester *see* 5319
Ampicillin 1-Oxo-1,3-dihydroisobenzofuran-3-yl Ester *see* 9008
Ampicillin Pivaloyloxymethyl Ester *see* 7484
Ampicin [Bristol] *see* 621
Ampicina [Sigma-Tau] *see* 621
Ampikel [Dreikehl] *see* 621
Ampilag [Dumex] *see* 621
Ampilar [Lagap] *see* 621
Ampimed [Medix] *see* 621
Ampinova *see* 621
Ampin-penicillin *see* 7042
Ampipenin *see* 621
Ampi-Tablinen [Sanorania] *see* 621
Ampliactil *see* 2186
Amplicain *see* 6663
Amplictil *see* 2186
Amplidione [Dausse] *see* 6882
Ampligen, 622
Ampligram [Hermes] *see* 1975
Amplin [Winston] *see* 621
Amplisom [I.S.O.M.] *see* 621
Amplit [Daiichi] *see* 5438
Amplital [Farmitalia] *see* 621
Amplivix [Sigma-Tau] *see* 1093
Amprol [Merck & Co.] *see* 623
Amprolium, 623
Amprol Plus [Merck & Co.] *see* 3701
Amprotropine Phosphate, 624
Ampy-Penyl [Proto] *see* 621
Ampyrone, 625
Amrinone, 626
AMS *see* 583
m-AMSA *see* 627
Amsacrine, 627
Amsebarb *see* 601
Amsidine [Parke-Davis/Warner-Lambert] *see* 627
Amsidyl [Parke-Davis] *see* 627
Amsonate *see* MISC-4
Amsonic Acid, 628
Amstat [Lederle] *see* 9487
Amsubit *see* 4815
Amsustain *see* 2932
Amudane *see* 4453
Amuno [Merck & Co.] *see* 4874
Amurex *see* 5896
Amycor [Medicia] *see* 1232
Amygdalic Acid *see* 5599
Amygdalin, 629
Amygdalinic Acid *see* 5599
Amygdalose *see* 4288
Amygdaloside *see* 629
Amylacetic Ester *see* 4993
dl-sec-Amyl Alcohol *see* 7075
n-Amyl Alcohol *see* 7074
tert-Amyl Alcohol *see* 7096
Amyl Alcohol (Commercial) *see* 4230
n-Amylamine, 630
2-*n*-Amyl-aminoethyl *p*-Aminobenzoate *see* 6263

Amylase, 631
α-Amylase (Aspergillus Oryzae) see 9007
α-Amylase (Bacterial), 632
α-Amylase (Swine Pancreas), 633
β-Amylase (Sweet Potato), 634
γ-Amylases see 631
Amylbenzene, 635
n-Amylbenzene see 635
d-Amyl Bromide, 636
d-pri-act-Amyl Bromide see 636
n-Amyl Bromide, 637
tert-Amyl Bromide, 638
n-Amyl Butyrate, 639
n-Amyl Caproate, 640
Amyl Carbamate, Tertiary, 641
Amylcarbinol see 4615
Amyl Chloride, 642
n-Amyl Chloride see 642
6-n-Amyl-m-cresol, 643
Amyleine Hydrochloride see 650
Amylene, 644
α-n-Amylene see 7079
β-n-Amylene see 7080
Amylene Dichloride, 645
tert-Amylene Dichloride see 645
Amylene Hydrate see 7096
n-Amyl Ether, 646
Amyl Ethyl Ketone see 3721
3-Amyl-1-hydroxy-6,6,9-trimethyl-
  6H-dibenzo[b,d]pyran see 1751
Amyl Isovalerate see 5003
tert-Amyl Isovalerate, 647
n-Amyl Mercaptan, 648
Amylmethylcarbinol see 4583
Amyl Nitrite, 649
Amylobarbitone see 601
Amylocaine Hydrochloride, 650
Amylodextrin see 8758
Amylogen see 8758
Amyloid, 651
Amyloid Substance see 651
Amylo-Liquifase see 632
Amylopectin see 496 and 8757
Amylopsin see 633
Amylose see 8757
Amyl Oxide see 646
Amylpenicillin Sodium, 652
p-tert-Amylphenol see 7098
4-[3-(4-tert-Amylphenyl)-2-methyl-
  propyl]-2,6-dimethylmorpholine see
  605
Amyl Phthalate see 5006
Amylsine see 6263
Amyl Thioalcohol see 648
Amylum see 8757
Amyl Valerate see 5003
α-Amyrenol see 653
β-Amyrenol see 654
α-Amyrin, 653
β-Amyrin, 654
Amyron see 8407
Amytal [Lilly] see 601
Amytal Sodium [Lilly] see 601
AN 1 [Voight] see 617
AN 148 see 5852
AN 1022 see 5921
AN 1317 see 7120
AN 1324 see 4374
Anabactyl [Beecham] see 1796
Anabasi [Pierrel] see 2446
Anabasine, 655
Anabet [Squibb] see 6260
Anabiol [Searle] see 1325
Anabloc [Irbi] see 7292
Anaboleen [Badarznei] see 8753
Anabolex see 8753
Anabolicum Vister see 8063
Anabsinthin, 656
Anabsynthin see 656
Anacardic Acid, 657

Anacardiol [IBI] see 5829
Anacardone see 6459
Anacetin [Philips Roxane] see 2068
Anacobin [BDH] see 9921
Anacyclin [Ciba] see 5501
Anadonis Green see 2229
Anador [Logeais] see 6282
Anadrol [Syntex] see 6920
Anadur [AB Leo] see 6282
Anafebrina see 488
Anaflex [Geistlich] see 7552
Anaflogistico see 2922
Anaflon see 40
Anafranil [Geigy] see 2385
Anafung see 9568
Anagestone, 658
Anagrelide, 659
Anagyrine, 660
Anahaemin [Bardin] see 5427
Anahist [Warner-Lambert] see 9307
Analate [Winston] see 5565
Analeptin see 8994
Analetil see 9137
Analexin [Mallinckrodt] see 7292
Analgesine [Elder] see 748
Analgin see 3358
Analud [Unifa] see 3953
Analux [Polfa] see 5660
Anamidol [Iwaki] see 6918
Anamycin [Chephasaar] see 3626
Ananase [Rorer] see 1377
Anandron [Roussel-UCLAF] see 6460
Ananxyl [Synthelabo] see 308
Anapolon [Syntex] see 6920
Anaprel see 8146
Anaprotin see 8753
Anaprox [Syntex] see 6337
Anaptivan [Help] see 1951
Anarcon see 6275
Anarel [Cutter] see 4471
Anarexol [Merck & Co.] see 2779
Anasclerol [Fardeco] see 9888
Anasterone see 6920
Anastil [Eberth] see 4457
Anastress see 5751
Anatabine, 661
Anatase see 9398
Anatensol [Squibb] see 4116
Anatola [Parke, Davis] see 9918
Anatran [Tobishi] see 9537
Anatrofin see 8763
Anatropin [Ortho] see 658
Anaus see 9623
Anautine see 3193
Anauxite see 368
Anavar [Searle] see 6875
Anavenol [ICI] see 6333
Anayodin see 4757 and 8582
Anazid see 6999
Anazolene Sodium, 662
Ancef [SK & F] see 1925
Anceron [Essex] see 1029
Anchoic Acid see 921
Anchovyxanthin see 10019
Anchusa see 242
Anchusa Acid see 243
Anchusin see 243
Anchusin Paper see 242
Ancillin [SK & F] see 3310
Ancitabine, 663
Anco [Kanoldt] see 4812
Ancobon [Roche] see 4056
Ancolan see 5657
Ancoron [Libbs] see 497
Ancortone [Merck & Co.] see 7727
Ancotil [Roche] see 4056
Ancrod, 664
Ancylol [Am. Cyanamid] see 3365
Ancymidol, 665
Ancytabine see 663
Ancyte [Abbott] see 7452

Andalusite see 368
Andantol [Homburg] see 5117
Andanton [Lacer] see 5117
Andaxin see 5751
Andere [Toyama] see 1465
Andergin [ISOM] see 6101
Andiamine [Polfa] see 4626
Andion [Gea] see 1029
Andirine see 8988
Andolex [Riker] see 1136
Andractim [Besins-Iscovesco] see 8753
Andrade Indicator see 103
Andramine see 3193
Andriol [Organon] see 9109
Androcur [Schering AG] see 2781
Androdiol see 5861
Androdurin [Ayerst] see 9109
Androfluorene see 4113
Androfluorone see 4113
Androfurazanol see 4209
Andrographolide, 666
Androisoxazole, 667
Androlone see 8753
Andromedotoxin see 4449
Androsan see 6044
Androsan (Amps.) see 9115
1,4-Androstadien-17β-ol-3-one see
  1327
Androstalone [Roussel] see 5816
Androstanazole see 8754
Androstane, 668
5α-Androstane see 668
5β-Androstane see 3823
5β-Androstane-17β-carboxylic Acid
  see 3824
Androstane-3β,11β-diol-17-one, 669
Androstanolone see 8753
Androstan-3(α)-ol-17-one see 673
Androstan-17β-ol-3-one see 8753
3β-Androstanol-17-one see 3562
Androstenediol, 670
Androst-5-ene-3β,17β-diol see 670
Δ5-Androstene-3β,17β-diol see 670
4-Androstene-3,17-dione, 671
Δ4-Androstene-17β-propionate-3-one
  see 9115
Androst-4-ene-3,11,17-trione see 165
Androst-16-en-3-ol, 672
Δ16-Androsten-3-ol see 672
Δ4-Androsten-17β-ol-3-one see 9109
Δ5-Androsten-3β-ol-17-one see 7710
Androsterolo [Pierrel] see 4113
Androsterone, 673
cis-Androsterone see 673
Androtardyl see 9112
Androteston-M see 5861
Androtest P see 9115
Androtex see 671
Androviron [Schering AG] see 5817
Anectine Chloride [Burroughs Well-
  come] see 8846
Anekain [Pliva] see 1485
Anelmid [Lilly] see 3376
Anemolin [Toyama] see 610
Anemone Camphor see 674
Anemonin see 674 and 7907
Anergan see 3918
Anergen see 3918
Anergex [Mulford] see 7526
Anertan see 9115
Anertan (Tabl) see 6044
Anestacon [Webcon] see 5359
Anesthesin [Abbott] see 3719
Anesthesol see 7763
Anesthetic Ether see 3762
Anesthone [Parke, Davis] see 3719
Anestil see 7763
Anetamin [Sankyo] see 8603
Anethaine see 9123
Anethole, 675

Anetholesulfonic Acid Sodium Salt
  Polymer see 8620
Anethole Trithione, 676
Aneural [Wyeth] see 5751
Aneurimec [Pharmec] see 7896
Aneurin-AS [A. S. Biol. Prod.] see
  9222
Aneurine Disulfide see 9221
Aneurine Hydrochloride see 9222
Aneurine Mononitrate see 9223
Aneurin-1,5-salt see 9226
Anexate [Roche] see 4062
ANF see 887
AN Factor see 1245
Anfamon [Fatol] see 3116
Anflagen [Ohta] see 4812
Anflam [Cox] see 4710
Angel Dust see 7179
Angelica, 677
Angelic Acid, 678
Angelica Lactone, 679
Δ¹-Angelica Lactone see 679
Δ²-Angelica Lactone see 679
Angeline see 8988
Angeli's Sulfone see 4357
Angelite see 362
Angex [Janssen] see 5360
Angibid [Meyer] see 6528
Angilol [DDSA] see 7852
Angimuth see 1737
Anginal [Yamanouchi] see 3354
Anginin [Banyu] see 7987
Anginine [Burroughs Wellcome] see
  6528
Anginosan [Promed] see 3334
Anginyl [Marion] see 3188
Angiociclan [IBI] see 1041
Angio-Conray [Mallinckrodt] see 4952
Angio-Contrix "48" see 4952
Angiodarona [Riker] see 497
Angiogenin, A2
Angiografin [Schering AG] see 5689
Angiolingual [Atmos] see 6528
Angiomin [Manetti] see 9969
Angionorm [Farmasan] see 3157
Angiopac [UCB] see 9888
Angiophtal [Merck & Co.] see 2239
Angiotensin, 680
Angiotensin I see 680
Angiotensin II see 680
Angiotensinase see 680
Angiotensin II Aspartic-β-amide
  5-Valine see 680
β-Angiotensin-I Converting Enzyme
  (Formerly) see 9472
Angiotensinogen see 680 and 8138
Angiotonin see 680
Angioxine [Roussel-UCLAF] see
  7987
Angitet [Purdue Frederick] see 7066
Angitrit [Leo Pharm.] see 9682
Angium [Roche] see 1469
Angizem [Inverni] see 3188
Anglesite see 5302
Angolon see 4838
Angopril [CERM] see 1167
Angorin see 6528
Angormin see 7744
Angostura Bark, 681
Anguifugan see 3376
Angustmycin C see 7939
Anhalamine, 682
Anhaline see 4666
Anhalonidine, 683
Anhalonine, 684
Anhiba [Hokuriku] see 40
Anhistan [Nippon Zoki] see 2345
Anhydr Butylchloral see 9545
Anhydr Chloral see 9538
Anhydr Hydrofluoric Acid see 4723

"Anhydride" of Ammonium Carbonate
  see 527
Anhydrite see 1713
2,2'-Anhydro-(1β-D-arabinofuranos-
  yl)cytosine see 663
Anhydroara C see 663
Anhydro-4,4'-bis(diethylamino)tri-
  phenylmethanol-2'',4''-disulfonic
  Acid Monosodium Salt see 8968
N¹,N¹-Anhydrobis(β-hydroxyethyl)-
  biguanide see 6181
4-[[[2''',3''-Anhydro]-O-3,6-di-
  deoxy-α-L-erythro-hexopyranos-4-
  ulos-1-yl-(1 → 4)-O-2,6-dideoxy-α-
  L-lyxo-hexopyranosyl-(1 → 4)-2,3,6-
  trideoxy-3-(dimethylamino)-α-L-
  lyxo-hexopyranosyl]oxy]-2-ethyl-
  1,2,3,4,6,11-hexahydro-2,5,7-tri-
  hydroxy-6,11-dioxo-1-naphthacene-
  carboxylic Acid Methyl Ester see
  108
Anhydroecgonin see 3466
Anhydrogitalin see 4328
Anhydroglucochloral see 2062
10-(1',5'-Anhydroglucosyl)-aloe-
  emodin-9-anthrone see 304
Anhydrohydroxynorprogesterone see
  6614
Anhydrohydroxyprogesterone see
  3696
Anhydrol [Dermal] see 338
Anhydron [Lilly] see 2760
Anhydrone see 5557
Anhydro-3,3,3',3'-tetramethyl-1,1'-
  bis(4-sulfobutyl)-4,5,4',5'-dibenzo-
  indotricarbocyanine Hydroxide In-
  ner Salt Sodium Salt see 4868
Anhydrotrimellitic Acid see 9617
Anhydrous Alcohol see 3716
Anhydrous Boric Acid see 1337
Anhydrous Gypsum see 1713
Anhydrous Hydriodic Acid see 4724
Anhydrous Hydrobromic Acid see
  4720
Anhydrous Hydrochloric Acid see
  4721
"Anhydrous" Potassium Acid Sulfate
  see 7654
Anhydrous Sulfate of Lime see 1713
Anifed [Zoja] see 6441
Anilazine, 685
Anileridine, 686
Aniline, 687
Aniline Green see 5581
Aniline Mustard, 688
Aniline Oil see 687
Anilinephthalein, 689
p-Anilinesulfonamide see 8898
Aniline-m-sulfonic Acid see 5832
o-Anilinesulfonic Acid see 6835
p-Anilinesulfonic Acid see 8901
Aniline Tribromide see 9522
Aniline Violet see 4287
Aniline Yellow see 430
Anilinoacetic Acid see 7262
Anilinoazobenzene see 2981
Anilinobenzene see 687
p-Anilinobenzenesulfonic Acid see
  7284
2-Anilinobenzoic Acid see 7244
Anilinoformylhydrazine see 7283
4'-Anilino-8-hydroxy-1,1'-azonaph-
  thalene-3,5',6-trisulfonic Acid Trisodium Salt see 662
1-Anilino-8-naphthalenesulfonate,
  690
8-Anilino-1-naphthalenesulfonic Acid
  see 690
m-[(p-Anilinophenyl)azo]benzenesul-
  fonic Acid Sodium Salt see 5833

p-[(p-Anilinophenyl)azo]benzenesul-
  fonic Acid Sodium Salt see 9688
1-(3-Anilinopropyl)-4-phenylisonipe-
  cotic Acid Ethyl Ester see 7403
Anilotic Acid see 6554
Animal Charcoal see 1815
Animal Coniine see 1608
Animal Galactose Factor see 6828
Animal Starch see 4392
Anime (Soft Copal) see 2512
A-Ninopterin, 691
Aniobi see 1291
Aniprime [Syntex Diamond] see 4066
Aniracetam, 692
p-Anisaldehyde, 693
Anise, 694
Anise Alcohol, 695
Anise Camphor see 675
Anise Seed see 694
p-Anisic Acid, 696
Anisic Aldehyde see 693
Anisidine, 697
m-Anisidine see 697
o-Anisidine see 697
p-Anisidine see 697
Anisindione, 698
Anisole, 699
Anisomycin, 700
Anisotropine Methylbromide, 701
Anisoylated Plasminogen Strepto-
  kinase Activator Complex see 785
o-(p-Anisoyl)benzoic Acid, 702
p-Anisoyl Chloride, 703
1-p-Anisoyl-2-pyrrolidinone see 692
Anistadin [Maruko] see 9537
Anistreplase see 785
Anisyl Alcohol see 695
3-(p-Anisyl)-4,5-dithiacyclopent-2-
  ene-1-thione see 676
2-(p-Anisyl)-1,3-indandione see 698
3-(p-Anisyl)trithione see 676
ANIT see 6331
Ankebin [Volpino] see 3924
Ankilostin see 9126
Ankyrin see 8696
Annalin see 1713
Annatto, 704
Annidalin see 9335
Annotinine, 705
Annotta see 704
Anobesina see 2739
Anobial see 9558
Anodynine see 748
Anodynon see 3740
Anohist [Warner-Lambert] see 9307
p-Anol, 706
Anone see 2732
Anopridine see 7403
Anoprolin [Nippon Shoji] see 278
Anorex see 3116
Anorthite see 1645
Anot, 707
Anovlar 21 [Schering AG] see 6614
Anoxin see 4138
ANP see 887
ANP 235 see 5660
ANP 3624 see 9362
Anparton [Sanwa] see 2374
Anprolene see 3758
Anquil [Janssen] see 1057
ANS see 690
Ansaid [Upjohn] see 4124
Ansar 170 [Ansul] see 5864
Ansar 184 [Ansul] see 5864
Ansar 529 [Ansul] see 5864
Ansar 8100 [Ansul] see 5864
Ansatin [Ono] see 4060
Ansepron [Fuso] see 7229
Anseren [Geigy] see 5176
Anserine, 708
Ansiacal see 2082

Ansieten [Exa] see 5176
Ansilan [Lek] see 5669
Ansimar [ABC] see 3427
Ansiolin [Scharper] see 2977
Ansmin [SS Pharm.] see 3311
Ansolysen Bitartrate [Wyeth] see 7089
Ansolysen Tartrate [Wyeth] see 7089
Anspor [SK & F] see 1982
Antabuse [Ayerst; Tosse] see 3370
Antadix see 3370
Antadol see 7248
Antadril see 3307
Antagonate [Dome] see 2180
Antagosan [Behringwerke] see 784
Antalka see 4363
Antallergan see 7996
Antallin see 3480
Antalvic see 7851
Antalzyme see 5514
Antamin see 3308
Antamine see 7996
Antapentan [Ayerst] see 7180
Antastan see 709
Antasten see 709
Antaxone [Simes] see 6278
Antazoline, 709
Antebor see 8870
Antees [Oriental] see 1798
Antegan [Merck & Co.] see 2779
Antelobine see 4534
Antemin [Streuli] see 3193
Antemovis see 8413
Antepan [Henning] see 9503
Antepar [Burroughs Wellcome] see 7433
Antèparsine see 4534
Antepsin [Ayerst] see 8853
Antergan see 7176
Antergan Hydrochloride see 7176
Anterior Pituitary Growth Hormone see 8672
Anterior Pituitary Luteotropin see 7788
Anteron [Schering AG] see 7515
Antex-490 [Leo Pharm.] see 7515
Anthalazine [Bowman] see 7431
Anthelcide EQ [Norden] see 6891
Anthelin see 803
Anthelone see 3492
Anthelone E see 3548
Anthelone U see 3492
Anthelvet [McNeil] see 9161
Anthen [Byk-Gulden] see 7776
Antheridiol, 710
Anthesterin see 9031
Anthio see 4160
Anthiolimine, 711
Anthiomaline [Specia] see 711
Anthion see 7644
Anthiphen see 3059
Anthisan [M&B] see 7996
Anthol see 803
Anthon see 9536
Anthophyllite see 851
Anthorine see 881
Anthracene, 712
Anthracene Brown see 713
9,10-Anthracenedicarboxaldehyde Bis[(4,5-dihydro-1H-imidazol-2-yl)hydrazone] see 1255
9,10-Anthracenedicarboxaldehyde Bis(2-imidazolin-2-ylhydrazone) see 1255
9,10-Anthracenedione see 717
1,2,10-Anthracenetriol see 718
1,8,9-Anthracenetriol Triacetate see 714
9-Anthracenol see 716
9(10H)-Anthracenone see 721
Anthra-Derm [Dermik] see 714
Anthragallic Acid see 713

Anthragallol, 713
Anthralin, 714
Anthramycin, 715
Anthranilic Acid see 433
Anthranilic Acid Cinnamyl Ester see 2306
Anthranilic Acid 2-(2-Piperidyl)ethyl Ester see 7466
3-Anthraniloylalanine see 5207
Anthranol, 716
Anthraquinone, 717
9,10-Anthraquinone see 717
6,6'-(1,4-Anthraquinonylenediimino)-di-m-toluenesulfonic Acid Disodium Salt see 238
Anthrarobin, 718
Anthrarufin, 719
1,2,10-Anthratriol see 718
Anthrimide, 720
9-Anthrol see 716
Anthrone, 721
Anthropodesoxycholic Acid see 2044
Antiangor [I.S.O.M.] see 2241
α-Antiarin, 722
Antib see 9218
Antibason see 6048
Antibiocin [Dorsch] see 7046
Antibiotic 1037 see 9483
Antibiotic 5879 see 1222
Antibiotic 6640 see 8498
Antibiotic 67-694 see 8241
Antibiotic A 246 see 4202
Antibiotic A 300 see 9563
Antibiotic A 5283 see 6346
Antibiotic A 23187 see 1639
Antibiotic A-28086 Factor A see 6339
Antibiotic AB 206 see 6116
Antibiotic Ab 651 see 1867
Antibiotic B-41 see 6112
Antibiotic B-41A1 see 6112
Antibiotic B-41A3 see 6112
Antibiotic B41D see 6112
Antibiotic C-19393 H₂ see 1867
Antibiotic C-19393 S₂ see 1867
Antibiotic DE-3936 see 5448
Antibiotic FI 1163 see 5464
Antibiotic G-6302 see 8924
Antibiotic K 178 see 6457
Antibiotic KA- 6643-A see 1867
Antibiotic KA-6643-B see 1867
Antibiotic KW-1062 see 6104
Antibiotic LA-7017 see 7510
Antibiotic M139603 see 9176
Antibiotic MA 144A1 see 108
Antibiotic MA 144B1 see 108
Antibiotic MYC 8003 see 6138
Antibiotic No. 899 see 9910
Antibiotic S 14750A see 205
Antibiotic SF 837 see 6106
Antibiotic SF 837A3 see 6106
Antibiotic T-2636A see 8377
Antibiotic WR 141 see 4441
Antibiotic X-464 see 6457
Antibiotic X-5108 see 900
Antibiotic X-465A see 2035
Antibiotic X-537A see 5243
Antibiotic X-14868A Ammonium Salt see 5524
Antibiotic YL 704B1 see 6106
Antibiotique EF 185 see 6369
Anticanitic Vitamin see 434
Anticarie see 4600
Antichlor see 8652
Antichoc Hipmag see 5576
Anti-chromotrichia Factor see 434
Antideprin see 4835
Antidiar [Armour] see 345
Antidipsin see 3684
Antidiuretic Hormone see 9843
Antidrasi see 3067
Antierythrite see 3620

Antietanol see 3370
Antifebrin see 42
Antifoam A [Dow] see 8486
Antiformin see 8577
"Antifungin" see 5535
Antigens, 723
Antigestil see 3118
Antigorite see 5569
Anti-Gray-hair Factor see 434
Antihemophilic Factor A see 3873
Antihemophilic Factor B see 3874
Antihemophilic Globulin see 3873
Antihemorrhagic Vitamin see 9933
Antihistal see 709
Antihypertensive Polar Renomedullary Lipid see 7496
Anti-hypo see 7640
Anti-infective Vitamin see 9918
Anti-inflammatory Hormone see 4710
Anti-JH see 7716
Antikrein [Teikoku Zoki] see 784
Antilirium [Forest] see 7357
Antilon [Lakeside] see 3666
Antimalarine see 7495
Antimicina see 5071
Antiminth [Roerig] see 7968
Antimonial Saffron see 729
Antimonic "Acid" see 730
Antimonic Oxide see 730
Antimonic Sulfide see 729
Antimonous Fluoride see 741
Antimonous Sulfate see 736
Antimonous Sulfide see 745
Antimony, 724
Antimony Chloride Fluoride see 726
Antimony Chloride Oxide, 725
Antimony Chloride Solution see 740
Antimony Dichlorofluoride see 726
Antimony Dichlorotrifluoride, 726
"Antimony Dimercaptosuccinate" see 8768
Antimony Fluoride see 741
Antimony Glance see 745
Antimony Gluconate Complex Sodium Salt see 733
Antimony Gluconate Sodium see 733
Antimony Hydride see 8767
Antimony Oxychloride see 725
Antimony Pentachloride, 727
Antimony Pentafluoride, 728
Antimony Pentasulfide, 729
Antimony Pentoxide, 730
Antimony Potassium Oxalate, 731
Antimony Potassium Tartrate, 732
Antimony Pyrocatechol Sodium Disulfonate see 8769
Antimony Red see 729
"Antimony Salt" see 731
Antimony Sodium Gluconate, 733
Antimony Sodium Oxide L(+)-Tartrate see 734
Antimony Sodium Tartrate, 734
Antimony Sodium Thioacetate see 735
Antimony Sodium Thioglycollate, 735
Antimony Sulfate, 736
Antimony Sulfide see 745
Antimony Thioglycollamide, 737
Antimony Thioglycollic Acid Triamide see 737
Antimony Tribromide, 738
Antimony Trichloride, 739
Antimony Trichloride Solution, 740
Antimony Trifluoride, 741
Antimony Trifluorodichloride see 726
Antimony Triiodide, 742
Antimony Trioxide, 743
Antimony Triselenide, 744
Antimony Trisulfate see 736
Antimony Trisulfide, 745
Antimosan see 8769
Antimycin see 2329

Antimycin A$_1$, 746
Antimycin A$_3$, 747
Antin see 7287
Antinonnin see 3272
Antinosin see 4931
Antipar [Farmitalia] see 3098
Antiparkin [Arzneimittelwerk VEB]
see 3724
Antipellagra Vitamin see 6435
Antipernicin see 9921
Antipernicious Anemia Principle see
9921
Antiplasmin see 7493
Antipyrine, 748
Antipyrine Amygdalate see 748
Antipyrine Chloral Hydrate see 2058
Antipyrine 2-Hydroxy-2-methylbut-
yrate see 748
Antipyrine Salicylate, 749
(Antipyrinylamino)methanesulfonic
Acid Sodium Salt see 8896
N-(Antipyrinyl)-2-(dimethylamino)-
propionamide see 484
(Antipyrinylmethylamino)methanesul-
fonic Acid Sodium Salt see 3358
N-Antipyrinylnicotinamide see 6443
Antirad see 166
Antiradon see 166
Antireticular Cytotoxic Serum, 750
Antirex see 3487
Antirobe [Upjohn] see 2352
Antirrhinin see 2694
Antisacer [Wander] see 7293
Antiscorbutic Vitamin see 855
Antisep see 8116
Antisepsin see 1387
Antiserotonin see 5675
Antispasmin see 377
Antisterility Vitamin see 9931
Antistin [Ciba] see 709
Antistine [Ciba] see 709
Anti-Stress [Sintyal] see 5731
Antitanil see 3163
Anti-tetany Substance 10 see 3163
Antithermin see 5352
Antithrombin III see 4571
α$_1$-Antitrypsin, 751
Antituxil-Z [Ghimas] see 10074
Antiulcera Master [Coli] see 2075
Antivariz see 3681
Antivenin (Crotalidae) Polyvalent, 752
Antiverm see 7220
Antivert [Roerig] see 5657
Antivirin, 753
Antivom [Ritsert] see 3334
Antiweinsäure (German) see 9040
Antixerophthalmic Vitamin see 9918
Antlerite see 2660
Antodyne see 7261
Antopen [Meiji] see 6127
Antophysin see 4534
Antoral [Recordati] see 9357
Antorphine see 6275
Antostab see 7515
Antra [Globopharm] see 6800
Antracol [Bayer] see 7831
Antraderm [Brocades] see 714
Antrafenine, 754
Antrancine 8 [Jandekker] see 1548
Antrancine 12 [Jandekker] see 1547
Antrapurol see 2813
Antrenyl [Ciba] see 6928
Antrycide see 8061
Antrypol see 8986
ANTU, 755
Antuitrin-Growth see 8672
Antuitrin S [Parke, Davis] see 4534
Antulcus see 6926
Anturan [Ciba-Geigy] see 8926
Anturane [Geigy] see 8926
Anturano [Ciba-Geigy] see 8926

Anturat see 755
Anuspiramin see 7248
Anvene [Searle] see 6254
Anvitoff [Knoll] see 9487
Anxon [Beecham] see 5176
Anzief [Nippon Chemiphar] see 278
AO-12 see 6029
Aolan [Beiersdorf] see 3665
Aolept [Bayer] see 7117
A.O.P. see 70
Aoral see 9918
AP2 see 8487
AP-14 see 3123
AP-43 see 603
Ap-67 see 2195
AP-1288 see 1039
6-APA see 470
Apacil see 491
Apacizin see 492
Apacizina [Farmitalia] see 492
Apalcillin, 756
Apamide [Ames] see 40
Apamin, 757
APAP see 40
Aparasin see 5379
Aparkan see 9607
Aparkazin see 3098
Apas see 491
Apascil see 5751
Apatef [ICI] see 1936
Apaurin see 2977
Apazone, 758
APC see 7155
Apegmone [Oberval] see 9383
Aperdor [Tokyo Tanabe] see 9425
Apertase see 4676
Apesan see 1848
Apetinil [Medial] see 3720
Apeton see 8753
Apex 462-5 see 9665
Apexol [Pfizer] see 9918
APF-1 see 9752
Aphamite see 6983
Aphanin see 3469
Aphidicolin, 759
Aphilan R [UCEPHA] see 1449
Apholate, 760
Aphox [ICI] see 7468
Aphoxide see 9587
Aphrodine see 10011
Aphrodyne [Star] see 10011
Aphtiria [Debat] see 5379
Aphylline, 761
A1PI see 751
Apicycline, 762
Apigenin, 763
Apigenin-7-apiosylglucoside see 765
Apigenin-7-D-glucoside see 764
Apigenin-4'-methyl Ether see 9
Apigetrin, 764
Apihepar [Apia] see 8484
Apiin, 765
Apilak see 8254
Apinol see 5724
7-[(2-O-D-Apio-β-D-furanosyl-β-D-
glucopyranosyl)oxy]-5-hydroxy-2-
(4-hydroxyphenyl)-4H-1-benzo-
pyran-4-one see 765
6-O-(D-Apiofuranosyl)-1,6,8-trihydr-
oxy-3-methylanthraquinone see
4177
Apiol see 767
Apiole (Dill), 766
Apiole (Parsley), 767
Apioline see 767
Apiose, 768
D-Apiose see 768
Apioside see 765
Apiquel [McNeil] see 490
Apiracohl [Kyowa] see 9425
Apirazin see 2195

A.P.L. [Ayerst] see 4534
Aplace [Kyorin] see 9699
Aplactan [Eisai] see 2308
Aplakil see 6881
Aplasmomycin, 769
Aplexal [Taiyo] see 2308
Aplisol [Parke, Davis] see 9715
Apllobal [Fujisawa] see 311
Aplonidine see 779
APM see 861
APNPS see 8908
APO see 9587
APO see MISC-2
Apoatropine, 770
Apocard [Esteve] see 4034
Apocodeine, 771
Apocretin [Kyowa] see 3822
Apocupreine see 777
Apocynamarin see 8818
Apocynin, 772
Apocynum androsaemifolium, 773
Apocynum cannabinum, 774
Apodol [Squibb] see 686
Apo-β-erythroidine, 775
Apoferritin see 3984
Apohemocyanin see 4566
Apoidina [Parke, Davis] see 4534
Apolan [Stockli] see 2374
Apollo [FBC] see 2373
Apomorphine, 776
Aponal [Boehringer, Mann.] see 3425
Aponal see 641
Aponeuron [Apogepha] see 617
Aponorin [Kodama] see 9537
Apopen [Apothekernes] see 7046
Apophedrin see 7258
Apoquinine, 777
Aporeine, 778
Aporheine see 778
6aβ-Aporphine-10,11-diol see 776
Apostavit see 9918
Apo-Sulfatrim [Apotex] see 8889
Apoterin [Seiko] see 8146
Apotomin [Kowa] see 2308
3α,16α-Apovincaminic Acid Ethyl
Ester see 9894
Apozepam [Apothekernes] see 2977
Appertex [Janssen] see 2343
Apple Acid see 5589
Apple Oil see 5003
Apple Of Peru see 8779
Apraclonidine, 779
Apralan [Lilly] see 780
Apramycin, 780
Apranax [Laroche Navarron] see 6337
Aprecon see 4992
Apresoline [Ciba] see 4682
Aprical [Rentschler] see 6441
Apride [Kodama] see 9425
Apridol see 5731
Aprindine, 781
Aprinox [Borden] see 1045
APRL see 7496
Aprobal [Fujisawa] see 311
Aprobarbital, 782
Aprobarbital Sodium see 782
Aprobit see 4755
Aprocarb see 7849
Apronai see 783
Apronalide, 783
Aprotinin, 784
Aprozal see 782
APSAC, 785
Apsifen [APS] see 4812
Apsin VK see 7046
Apsolol [APS] see 7852
Aptine [Pharma-Stern] see 311
Aptol Duriles [Astra] see 311
Apulonga see 278
Apurin [De Haas] see 278
Apurol [Siegfried] see 278

Apurone [Riker] see 4065
APY-606 [Yoshitomi] see 2408
Apyrase, 786
Apyron [Wülfing] see 5531
AQ 110 see 9502
Aqua Ammonia see 513
Aquacare [Herbert] see 9781
Aquacillin [Riker] see 7042
Aquacycline [Syntex] see 6931
Aquadrate [Eaton] see 9781
Aquafortis see 6495
Aquakay see 5714
Aqualin [Shell] see 122
Aquamarine see 1177
AquaMEPHYTON [Merck & Co.] see 9933
Aquamollin see 3482
Aquamox [Lederle] see 8067
Aquamycetin [Winzer] see 2068
Aquamycin see 1958
Aquaphor [Beiersdorf] see 9986
Aquareduct [Azuchemie] see 8721
Aqua Regia see 6530
Aquaretic [Azuchemie] see 417
Aquarius [Merck & Co.] see 4704
Aquasuspen see 7042
Aquatag [Tutag] see 1134
Aquatensen [Mallinckrodt] see 5929
Aquathol [Pennwalt] see 3530
Aquathol K [Pennwalt] see 3530
Aquedux [Delagrange] see 2371
Aquex [Sandoz] see 2391
Aquinone see 5714
Aquirel see 2760
Aquocobalamin, 787
Aquocobamide see 787
Aquo-Trinitrosan [E. Merck] see 6528
AR 12008 see 9492
Ara-A see 9881
Arabinitol see 789
9-β-D-Arabinofuranosyladenine Monohydrate see 9881
1-β-D-Arabinofuranosylcytosine see 2790
N-(1-β-D-Arabinofuranosyl-1,2-di-hydro-2-oxo-4-pyrimidinyl)docos-anamide see 3539
9-β-D-Arabinofuranosyl-2-fluoro-adenine see 4057
9-β-D-Arabinofuranosyl-2-fluoro-9H-purin-6-amine see 4057
9-β-D-Arabinofuranosyl-9H-purine-6-amine Monohydrate see 9881
6-O-α-L-Arabinopyranosyl-D-glucose see 9879
Arabinose, 788
L-Arabinose see 788
6-(α-L-Arabinosido)-D-glucose see 9879
Arabinosyladenine see 9881
Arabite see 789
Arabitin [Sankyo] see 2790
Arabitol, 789
D-Araboascorbic Acid see 5009
D-Araboflavin, 790
Ara-C see 2790
Arachic Acid see 791
Arachidic Acid, 791
Arachidonic Acid, 792
Arachis Oil see 7006
Aracytidine see 2790
Aracytine [Upjohn] see 2790
Aragonite see 1657
Aralen [Winthrop] see 2163
Aralia, 793
Aramine [Merck & Co.] see 5837
Aramite®, 794
Arantoick see 9447
"Arariba" see 795
Araroba, 795

Arasan [Du Pont] see 9304
Arasena-A [Mochida] see 9881
Arathane see 3281
Aratron see 794
Arbacet [Upjohn] see 796
Arbaprostil, 796
Arbekacin, 797
Arborescin, 798
Arborine see 4398
Arbor Vitae see 9324
Arbotect [Merck & Co.] see 9217
Arbuse see 9953
Arbutin, 799
Arbutoside see 799
Arbuz see 6965
Arcacil see 7046
Arcadine see 1037
Arcalion [Servier] see 9221
Arcanum Duplicatum see 7665
Arcasin [Engelhard KG) see 7046
Archil see 6818
Archin see 3518
Arcilla see 807
Arcoban [Arcum] see 5751
Arcomonol Tablets see 6141
Arcospectron [Arco, Switzerland] see 6931
Arcton [ICI] see 2140
Arcton 11 [ICI] see 9553
Arcton 12 [ICI] see 3053
Arcton 114 [ICI] see 2608
Arcylate [Hauck] see 8307
Ardall (from plant sources) see 8626
Ardésyl [Beytout] see 9942
Ardex see 2932
Ardeydorm [Ardeypharm] see 9707
AR-DF 26 see 4338
Ardine [Antibioticos] see 610
Ardisic Acid B see 1174
Arduan [Gedeon Richter] see 7426
Areca, 800
Arecaidine, 801
Arecaine see 801
Arecaline see 802
Arechin [Polfa] see 2163
Arecholine see 802
Arecoline, 802
Arecoline-acetarsol see 3434
Arecoline-acetarsone see 3434
Arecoline p-Stibonobenzoic Acid, 803
Arelix [Hoechst] see 7464
Arelon [Hoechst] see 5106
A-Rest [Elanco] see 665
Aretit [Hoechst] see 3282
Arfonad [Roche] see 9621
Arg (IUPAC Abbrev.) see 805
Argatroban, 804
Argentic Fluoride see 8459
Argentic Oxide see 8469
Argentite see 8481
Argentous Chlorate see 8454
Argentous Fluoride see 8460
Argentous Oxide see 8468
Argentous Sulfide see 8481
Argentum Vitellinatum see 8474
Argentum Vitellinum see 8474
Argil see 807
Argilla see 5162
Argilla Vini see 807
Arginine, 805
Arginine L-Aspartate see 862
Arginine Glutamate, 806
5-L-Arginine-27-glycine-33-L-isoleu-cine-34-L-serine-42-glycine-puro-thionin A I (Reduced) see 7961
8-D-Arginine-1-(3-mercaptopropanoic Acid) Vasopressin see 2904
5-L-Arginine-6-L-threonine-18-L-serine-26-L-serine-27-L-threonine-42-glycine-purothionin A I (Re-duced) see 7961

Arginine Vasopressin see 9843
N-(N-(N-(N²-L-Arginyl-L-lysyl)-L-α-aspartyl)-L-valyl)-L-tyrosine see 9338
Argipidine see 804
Argipressin see 9843
Argisal see 8474
Argivene [Gray] see 805
Argobyl [Specia] see 4039
Argol, 807
Argon, 808
Argun [Merckle] see 4874
Argun see 212
Argyn see 8474
Argyrol [SM & P] see 8474
Arheol see 6765
Aribine see 4530
Aricine, 809
Aricyl [Winthrop] see 838
Arilin [Wolff] see 6079
Aristamid [Nordmark] see 8929
Aristochin see 8077
Aristocort [Lederle] see 9511
Aristoderm [Lederle] see 9512
Aristol see 9335
Aristolochic Acid, 810
Aristolochic Acid-I see 810
Aristolochine see 810
Aristololactams see 810
Aristoquin [Lederle] see 8077
Aristoquinine see 8077
Aristosol [Lederle] see 9512
Aristospan [Lederle] see 9514
Arkitropin see 4649
ARL see 815
AR-L 115BS see 8965
Arlef [Parke, Davis] see 4060
Arlidin [USV] see 6654
Arliflav see 1241
Arlitene [Chinoin] see 6204
Arlytene see 6204
Armazal see 2688
Armepavine, 811
Armoise see 8
Armol see 5940
Armophylline [Armour-Montagu] see 9212
Armstrong's Acid, 812
Armyl [Armour Pharm.] see 5500
Arnaudon's Green see 2230
Arnica, 813
Arnica Flowers see 813
Arnosulfan see 8876
Arnotta see 704
Arobon see 5436
Aroclor [Monsanto] see 7541
Aroclor 1242 see 7541
Aroclor 1254 see 7541
Aroclor 1260 see 7541
Aroft [Tanabe] see 171
Arofuto [Tanabe] see 171
Arogenate see 814
Arogenic Acid, 814
Aropax [Beecham] see 6991
Arotinolol, 815
Arovit [Roche] see 9918
Aroxine see 4155
Arpezine see 7431 and 7433
Arpicolin [R. P. Drugs] see 7770
Arpimycin [R. P. Drugs] see 3626
Arpocox [Merck & Co.] see 816
ARPPRN see 6259
Arprinocid, 816
Arquad 10 [Armak] see 3088
Arquad 16 Stearate see 2019
Arquel [Parke, Davis] see 5659
Arresten [Nippon Shinyaku] see 6058
Arret [Janssen] see 5450
Arret see 2212
Arrhenal see 5864
Arrivo [FMC] see 2775

Arrow Wood *see* 4176
Arsacetin, 817
Arsacol [Arsac] *see* 9801
Arsambide *see* 1788
Arsamin *see* 8521
Arsaminol *see* 839
Arsanilic Acid, 818
Arsanilic Acid Sodium Salt *see* 8521
Arsanyl [Taiyo] *see* 4273
Arsaphen *see* 44
Arsaphenan *see* 44
Arseclor *see* 3060
Arsecodile *see* 8540
Arsemétine *see* 3517
Arsen (German) *see* 820
Arsenamide, 819
Arsenic, 820
Arsenic Acid, 821
Arsenic Acid Anhydride *see* 827
Arsenical Soln *see* 7585
Arsenic Chloride Solution *see* 835
Arsenic Disulfide, 822
Arsenic Hemiselenide, 823
Arsenic Pentafluoride, 824
Arsenic Pentaselenide, 825
Arsenic Pentasulfide, 826
Arsenic Pentoxide, 827
Arsenic Sesquioxide *see* 832
Arsenic Sulfide *see* 822
Arsenic Tribromide, 828
Arsenic Trichloride, 829
Arsenic Trifluoride, 830
Arsenic Trihydride *see* 837
Arsenic Triiodide, 831
Arsenic Trioxide, 832
Arsenic Triselenide, 833
Arsenic Trisulfide, 834
Arsenic Yellow *see* 834
Arsenik *see* 820
Arsenious Acid Copper(2+) Salt (1:1) *see* 2630
Arsenious Acid Solution, 835
Arsenious Selenide *see* 833
Arseno 39 *see* 6902
Arsenoacetic Acid, 836
4,4'-Arsenobis(2-aminophenol) Dihydrochloride *see* 839
4,4'-Arsenobis[2-aminophenol] Disodium Salt *see* 8524
Arsenodiacetic Acid *see* 836
Arsenolite *see* 832
Arsenosan *see* 6902
Arsenosobenzene *see* 6903
Arsenous Acid *see* 832
Arsenous Acid Anhydride *see* 832
Arsenous Oxide *see* 832
Arsenous Selenide *see* 833
Arsenoxide *see* 6902
Arsenphenolamine Hydrochloride *see* 839
Arsevan *see* 6361
Arsicodile *see* 8540
Arsine, 837
Arsinyl *see* 5864
Arsion *see* 2119
Arsobal [Specia] *see* 5694
Arsonic Acid Copper(2+) Salt (1:1) *see* 2630
Arsonoacetic Acid, 838
[(5-Arsono-2-hydroxyphenyl)amino]-methanesulfinic Acid Disodium Salt *see* 7171
p-Arsonophenylurea *see* 1788
Arsphenamine, 839
Arsphenamine Methylenesulfoxylic Acid Sodium Salt *see* 6361
Arsphenamine Sodium *see* 8524
Arsthinol, 840
Arsycodile *see* 8540
Arsynal *see* 5864
Artabotrine *see* 5046

Artam *see* 2290
Artane [Lederle] *see* 9607
Artate [Nippon Chemiphar] *see* 2308
Arteannuin *see* 844
Arteether, 841
Artegodan [Artesan] *see* 6968
Artemether, 842
Artemisin, 843
Artemisine *see* 844
Artemisinin, 844
Arteoptic [Novopharma] *see* 1875
Arteparon [Luitpold] *see* 2217
Arterenol [Hoechst] *see* 6612
Arteriohom [Weisskopf] *see* 2375
Arteriovinca [Farma-Leporil] *see* 9888
Arterium V [ICI] *see* 6427
Arterocoline *see* 81
Arterolo [Nuovo Cons.] *see* 2380
Artes [Sumitomo] *see* 5702
Artesunate, 845
Artesunic Acid *see* 845
Artevil *see* 2374
Artex [Servier] *see* 9105
Arthaxan [Atmos] *see* 6258
Arthrex [Sandoz] *see* 7887
Arthrisin [Sandoz] *see* 4120
Arthriticine *see* 7431
Arthrobid [Merck & Co.] *see* 8961
Arthrocine [Merck & Co.] *see* 8961
Arthrodont [Veyron-Froment] *see* 3543
Arthropan [Purdue Frederick] *see* 2212
Arthrosan *see* 2290
Artificial Cognac Essence *see* 3790
Artificial Essential Oil of Almond *see* 1065
Artificial Ethyl Oenanthate *see* 3790
Artificial Lanthanite *see* 5233
Artificial Oil of Ants *see* 4214
Artil [Hoechst] *see* 5126
Artilan [Steinhard] *see* 9607
Artisone Acetate [Wyeth] *see* 70
Artisone-Wyeth *see* 6028
Artolon *see* 5751
Artomycin *see* 9130
Artonil *see* 9430
Artosin [Boehringer, Mann.] *see* 9432
Artracin [DDSA] *see* 4874
Artrene [Irbi] *see* 4812
Artribid [Merck & Co.] *see* 8961
Artrichin *see* 2163
Artril 300 [Farmasa] *see* 4812
Artrinovo [Llorns, Spain] *see* 4874
Artrivia [Lifasa] *see* 4874
Artrizin *see* 7248
Artrobione *see* 2212
Artrodar [Proter] *see* 2939
Artrolasi [Lenza] *see* 6820
Artromialgina *see* 3357
Artrosilene [Dompè] *see* 5184
Artroxicam [Coli] *see* 7476
Artume *see* 1712
ARTZ [Seikagaku] *see* 4675
Arumel [SS Pharm.] *see* 4109
Arumil [Sharp & Dohme] *see* 417
Arusa *see* 149
Arusal *see* 1848
Arvigol [Zyma] *see* 2357
Arvin [Toyford] *see* 664
Arvynol [Pfizer] *see* 3683
Arwin [Knoll] *see* 664
Arylam *see* 1789
*Arythmol *see* 7806
Arzene [Am. Cyanamid] *see* 6903
AS *see* 651
AS 101 *see* 818
Asa *see* 1833
A.S.A. [Lilly] *see* 873
ASA-226 *see* 2073
5-ASA *see* 5807

Asabaine *see* 5869
Asacol [Tillotts] *see* 5807
Asacolitin [Roehm] *see* 5807
Asafetida, 846
Asafoetida *see* 846
Asamedol [Maruko] *see* 3662
Asant *see* 846
Asaprol, 847
Asarabacca Camphor *see* 849
Asarin *see* 849
Asarinin, 848
l-Asarinin *see* 848
α-Asarone *see* 849
β-Asarone *see* 849
Asarones, 849
Asarum, 850
Asarum Camphor *see* 849
Asatard [De Angeli] *see* 873
Asbestos, 851
ASC *see* 99
Ascabin *see* 1141
Ascabiol [May & Baker] *see* 1141
Ascal *see* 1644
Ascaridole, 852
Ascaril *see* 7431
Ascarisin *see* 852
Ascarite [Baker] *see* 851
Ascaryl [Wander] *see* 4633
Ascensil *see* 478
Asceptichrome *see* 5757
Ascharite *see* 5535
Asclepias, 853
Asclepias syriaca, 854
Ascorbic Acid, 855
L-Ascorbic Acid *see* 855
Ascorbic Acid Calcium Salt *see* 1648
L-Ascorbic Acid Mixt. with 3-Pyridinedicarboxamide *see* 6433
Ascorbic Acid Nicotinamide Complex *see* 6433
Ascorbic Acid Sodium Derivative *see* 8525
Ascorbicin [Squibb] *see* 8525
Ascorbigen, 856
Ascorbigen A *see* 856
Ascorbigen B *see* 856
Ascorbin [Lakeside] *see* 8525
Ascorin *see* 855
Ascorteal *see* 855
Ascorvit *see* 855
Ascotoxin *see* 1363
Ascuron *see* 8847
Asebotoxin *see* 4449
Asellacrin [Calbiochem] *see* 8672
Asendin [Lederle] *see* 609
Asepsin *see* 1387
Aseptamide [Merminod] *see* 4503
Aseptorid *see* 8922
ASH *see* 163
Asiatic Acid *see* 857
Asiaticoside, 857
Asil *see* 4815
Asiprenol *see* 5105
Aska-Rid *see* 1621
Askensil *see* 478
ASL-279 *see* 3415
ASL-601 *see* 14
ASL-8052 *see* 3648
Aslos [Kotani] *see* 1798
Asmalar *see* 5105
Asmaten [Riker] *see* 8223
Asmaterol [Lusofarmaco] *see* 8142
Asmaven [APS] *see* 209
Asn (IUPAC Abbrev.) *see* 859
L-Asnase *see* 858
ASP-47 *see* 8945
Asp (IUPAC Abbrev.) *see* 862
Aspara *see* 862
Aspara K *see* 862
Asparagic Acid *see* 862
L-Asparaginase, 858

Asparaginate Calcium see 862
Asparagine, 859
D-β-Asparagine see 859
L-β-Asparagine see 859
L-Asparagine Amidohydrolase see 858
1-Asparagine-5-valine-angiotensin II
 see 680
Asparaginic Acid see 862
Asparaginsäure (German) see 862
Asparagus, 860
Asparamide see 859
Aspartame, 861
Aspartat [Debat] see 862
Aspartic Acid, 862
Aspartic Acid β-Amide see 859
3-(L-Aspartyl-D-alaninamido)-2,2,4,4-
 tetramethylthietane see 235
L-Aspartyl-D-alanine-N-(2,2,4,4-tet-
 ramethylthietan-3-yl)amide see 235
N-L-α-Aspartyl-L-phenylalanine
 1-Methyl Ester see 861
N²-[1-[N-(N-L-α-Aspartyl-L-seryl)-L-
 α-aspartyl]-L-prolyl]-L-arginine see
 7086
L-α-Aspartyl-N-(2,2,4,4-tetramethyl-
 3-thietanyl)-D-alaninamide see 235
Aspasan [Hoechst] see 3934
ASPC see 874
Aspegic [Egic] see 5510
Aspenil [Chemil] see 610
Aspenon [Mitsui] see 781
Aspergillic Acid, 863
Aspergillin, 864
Aspergillus Diastase see 9007
Asperlicin, 865
Asperule Absolute see 9961
Asperuloside, 866
Asphalt, 867
Asphaltum see 867
Asphocalcium see 855
Aspiculamycin see 4436
Aspidin, 868
Aspidinol, 869
Aspidium, 870
Aspidol [Piam] see 5510
Aspidosperma, 871
Aspidospermine, 872
Aspirin, 873
Aspirin Aluminum see 328
Aspirin Lysine Salt see 5510
Aspisol [Bayer] see 5510
Aspogen [Eaton] see 3168
Aspoxicillin, 874
Aspro [Nicholas] see 873
Assaren [Permamed] see 3071
Assert [Am. Cyanamid] see 4825
Assiprenol see 5105
Assugrin see 2707
Assur see 6181
Assure [Nissan] see 8113
Asta C 4898 see 3185
Astacene see 875
Astacin, 875
Astaril see 3807
Astatine, 876
Astaxanthin, 877
Asta Z 4942 see 4822
Astemisan [Zdravlje] see 878
Astemizole, 878
Astenile [Recordati] see 7710
Asteric [Cooper] see 873
Asterin see 2694
Asterol Dihydrochloride [Roche] see
 2967
Asteromycin see 4436
Asthenthilo [Thilo] see 3146
Asthmalitan [Kettelhack] see 5053
Asthma Weed see 5431
Asthmolysin [Kade] see 3459
Astiban see 8768
Astix CMPP [Agrotec] see 5666

Astmamasit [Showa Yakuhin] see
 3459
Astomin [Yamanouchi] see 3192
Astonin-H [E. Merck] see 4059
Astragalin see 5156
Astridine [Astra] see 5114
Astringen see 346
Astrobain see 6854
Astrobot [Arnolds] see 3069
Astrocar see 6459
Astroderm [Medix] see 3035
Astrolin see 748
Astromicin see 4165
Astrophyllin see 3459
Astryl see 4375
Asturidon [Cilag] see 1495
Astyn see 3478
Asuccin see 8840
Asucrol see 2187
[ASU¹,⁷]-E-CT see 3501
Asuntol [Bayer] see 2559
Asuro [Nippon Kayaku] see 2929
Asverin [Tanabe Seiyaku] see 9389
Asymmetrical Trimethylbenzene see
 7929
AT-7 see 4602
AT-10 see 3163
AT-17 see 3192
AT-101 see 5113
AT-327 see 9389
AT-2266 see 3540
ATA see 506
Atabrine Dihydrochloride [Winthrop]
 see 8053
Atapren [Sumitomo] see 9425
Ataractan see 909
Atarax [Pfizer] see 4786
Atav see 9918
ATBAC see 1757
ATC [Medial] see 9375
Atcotibine see 5071
Ateben [Chem-Sintyal] see 6635
Atebrin Hydrochloride see 8053
Ateculon [Nippon Chemiphar] see
 2374
AteHexal [Hexal] see 879
Atelor [Roche] see 2967
Atem [Chiesi] see 4960
Atemorin see 5731
Atempol [Norgine] see 5731
Atenen [Tsuruhara] see 145
Atenezol [Tsuruhara] see 45
Atenol [CT] see 879
Atenolol, 879
Atenos [UCB] see 9720
Atenos see 1799
Atensil see 7394
Atensin see 5737
Atensine [Berk] see 2977
Aterax see 4786
Aterian see 8879
Ateriosan see 2374
Ateroid [Mack, Illert.] see 2217
Aterosan [Lancet] see 7987
Atgard [Shell] see 3069
Athamantin see 880
Atheran see 2380
Atheroitin see 2217
Atherolip see 2375
Atherolipin [Pharma Schwarz] see
 2375
Atherophylline see 7988
Atheropront [Mack, Illert.] see 2374
Athrombin-K [Perdue Frederick] see
 9950
Athrombon see 7196
Athymil [Organon] see 6097
Athyromazole see 1803
Atilen [Spofa] see 2977
Atipi see 145
Atirin [Intersint] see 1925

Atisine, 881
Ativan [Wyeth] see 5456
Atlacide [Chipman] see 8543
Atlansil [Roemmers] see 497
Atlas G-2133 [Atlas Powder] see 7532
Atlas G-3705 [Atlas Powder] see 7532
Atmosgen [Maruko] see 9447
Atmosphere, 882
Atocin see 2290
Atock [Yamanouchi] see 4159
Atophan [Warner-Chilcott] see 2290
Atophanyl [Warner-Chilcott] see 2290
Atoquinol see 2290
Atosil [Bayer] see 7797
Atover [Oti] see 7987
Atoxicocaine see 7763
Atoxyl see 8521
Atoxylic Acid see 818
ATP see 145
ATP-dependent Proteolytic Factor
 see 9752
Atractyl see 5600
Atractylin (C₃₀ Glucoside) see 883
Atractyloside, 883
Atracurium Besylate, 884
Atral see 3360
Atranex [Makhteshim-Agan] see 886
Atranoric Acid see 885
Atranorin, 885
Atravet [Ayerst] see 27
Atraxin see 5751
Atrazine, 886
Atrial Natriuretic Factor, 887
Atrican see 9079
Atrilon 5 see 7821
Atrimycon [Bioindustria] see 7904
Atrinal [Roche] see 3184
Atriopeptigen see 887
Atriopeptin see 887
Atriphos see 145
Atro-Dote [Hart-Delta] see 891
Atrol see 2834
Atrolactamide, 888
Atrolactic Acid, 889
Atromid [Ayerst] see 2374
Atromidin see 2374
Atromid-S [Ayerst] see 2374
Atropamine see 770
Atrophate [Burns-Biotec] see 891
Atropic Acid, 890
Atropine, 891
Atropine Aminoxide see 892
Atropine Hyperduric see 891
Atropine N-Oxide, 892
Atropisol [CooperVision] see 891
Atropyltropeine see 770
Atroscine see 8361
Atrosed [Burns-Biotec] see 891
Atrovent [Boehringer, Ing.] see 4960
Attacins, 893
Attar of Rose see 6762
Attentil [Ravizza] see 4024
Atumin [Merrell] see 3087
Aturbal see 7192
Aturban(e) [Ciba] see 7192
Aturgyl [Dausse] see 3930
Atussil [Squibb] see 1798
Atysmal [Schaefer] see 3704
Aubépine see 2569
Aucubin, 894
Aucuboside see 894
Audax [Napp] see 2212
Audes [Nippon Kayaku] see 2321
Augmentin [Beecham] see 2342
Auligen see 3390
Aulin [Robin] see 6464
Aulinogen [Boehringer, Ing.] see 3390
Auranofin, 895
Aurantex [Glaxo] see 226
Aurantiin see 6345
Aurantiogliocladin, 896

Aurcoloid [Abbott] see 4419
Aurcoscan-198 [Abbott] see 4419
Aureine see 8402
Aurelic Acid see 7510
Aureociclina see 2193
Aureocina see 2193
Aureolic Acid see 7510
Aureomycin [Lederle] see 2193
Aureotan [Byk-Gulden] see 901
Aureothin, 897
Aureothricin, 898
Aureotope [Squibb] see 4419
Auric Bromide see 4425
Auric Chloride see 4427
Auric Cyanide see 4429
Auric Hydroxide see 4430
Auricidine see 4423
Auric Oxide see 4431
Auric Selenate see 4420
Auric Selenide see 4421
Auric Sulfide see 4432
Auriculin see 887
Aurin, 899
Aurinol see 4266
Auripigment see 834
Aurochlorohydric Acid see 4428
Aurocidin see 4423
Aurodox, 900
Aurolin see 4423
α-Auromercaptoacetanilide see 902
Auropex see 4423
Auropin see 4423
Aurorix [Roche] see 6139
Aurosan see 4423
Aurothioglucose, 901
Aurothioglycanide, 902
Aurothioglycolanilide see 902
Aurothioglycolic Acid Anilide see 902
Aurothiol see 9652
Aurothion see 4423
Aurothiosulfate Natrium see 4423
Aurothiosulfate Sodium see 4423
Aurous Chloride see 4414
Aurous Cyanide see 4415
Aurous Iodide see 4416
Aurous Oxide see 4418
Aurous Stannate see 4424
Aurous Sulfide see 4417
Aurumine see 901
Aurum Paradoxum see 9064
Ausocef [Ausonia] see 1971
Ausovit B$_1$ [Oscar] see 7896
Austracol see 2068
Australian Fever Bark see 3373
Australian Fever Tree see 3852
Austrapen see 621
Austrapine see 8149
Austrawolf see 8131
Austrian Cinnabar see 5280
Autan see 2848
Autoprothrombin II see 3874
Autumn Crocus see 2471
Auxiloson [Thomae] see 2922
Auxisone [Boehringer, Ing.] see 2922
Auxit [Heyden] see 1379
AV see 753
AV see MISC-2
AV-290 see 906
Ava-ava see 5166
Avacan [Asta] see 1743
Avadex see 2950
Avadex BW [Monsanto] see 9510
Avadyl see 1743
Avagai see 5869
Avan [Takeda] see 4817
Avantyl see 6635
Avatec [Roche] see 5243
Avazyme [Wampole] see 2265
Avenein see 4359
Avenin see 1892
Aventyl [Lilly] see 6635

Avermectin A$_{1a/b}$ see 903
Avermectin A$_{2a/b}$ see 903
Avermectin B$_1$ see 1
Avermectin B$_{1a/b}$ see 903
Avermectin B$_{2a/b}$ see 903
Avermectins, 903
Averon-1 [Alfar] see 1978
Avertin [Winthrop] see 9525
Avibon see 9918
Avicalm [Squibb] see 6074
Avicel see 1961
Avicol see 8108
Avicol [Troponwerke] see 2183
Avicol SL [Troponwerke] see 2378
Avidin, 904
Avigilen [Efeka] see 7459
Avil [Albert-Roussel] see 7198
Avilamycin, 905
Avilamycin A see 905
Avilamycin C see 905
Avinar [Armour Pharm.] see 9787
Aviochina see 8914
Aviral [Medici Domus] see 4881
Avisco Rayon [FMC] see 8132
Avita see 9918
Avitol see 9918
Avlane [Roussel-UCLAF] see 5453
Avlocardyl [ICI] see 7852
Avloclor [ICI] see 2163
Avloprocil see 7042
Avlosulfon [Ayerst] see 2820
Avlosulphone see 2820
AVM see 903
Avogadrite see 7680
Avolin see 3243
Avomec [Merck & Co.] see 1
Avomine see 7797
Avoparcin, 906
Avornin see 4177
Avosyl see 5737
Avotan [Am. Cyanamid] see 906
Avoxin [Krka] see 4138
Avoxyl see 5737
Avrazor [Spofa] see 6824
AW 105 843 see 6270
AW 14'2333 see 7131
Awelysin [Arzneimittelwerk VEB] see
8784
AX 250 [Durachemie] see 610
Axeen [Hommel] see 7920
Axer [Alfa] see 6337
Axerol see 9918
Axerophthal see 8165
Axerophthol see 9918
Axetil see MISC-3 and 4
Axid [Lilly] see 6582
Axiomin [Promeco] see 3841
Axion see 4739
Axiquel [McNeil] see 9820
Axiten see 5653
Axlon [Albert-Roussel] see 4739
Axoril [Glaxo] see 1951
Axungia Porci see 5239
Axuris see 4287
AY 5406-1 see 1037
AY 5710 see 5527
AY 6108 see 621
AY 6608 see 7067
AY 8682 see 2766
AY 21011 see 7703
AY 23028 see 1499
AY 24031 see 5354
AY 24236 see 3831
AY 25650 see 9662
AY 25712 see A1
AY 27773 see 9451
AY 62014 see 1533
AY 62021 see 2392
AY 62022 see 5676
AY 64043 see 7852
Ayeramate see 5751

Ayerlucil see 8887
Ayermicina [Ayerst] see 5337
Ayfactin see 6997
Ayfivin see 948
Aygestin [Ayerst] see 6614
AZ 8 see 4464
10-Azaanthracene see 117
5-Aza-10-arsenaanthracene Chloride
see 7170
1-Azabicyclo[2.2.2]octane see 8109
1-Azabicyclo[2.2.2]octan-3-ol see
8110
5-(1-Azabicyclo[2.2.2]oct-3-yl)-
10,11-dihydro-5H-dibenz[b,f]-
azepine see 8111
10-(1-Azabicyclo[2.2.2]oct-3-ylmeth-
yl)-10H-phenothiazine see 5754
1-(3-Azabicyclo[3.3.0]oct-3-yl)-3-(p-
tolylsulfonyl)urea see 4335
Azacitidine, 907
Azacort see 2852
Azacortid [Lepetit] see 4048
Azacosterol, 908
Azactam [Squibb] see 938
Azacyclonol, 909
2-(1'-Azacyclooctyl)ethylguanidine
see 4473
1-Azacyclooct-2-ylmethylguanidine
see 4472
Azacyclopropane see 3760
5-Azacytidine see 907
Azadirachtin, 910
9-Azafluorene see 1792
Azafrin, 911
8-Azaguanine, 912
Azalomycin M see 6457
Azalone [SM & P] see 709
Azamethone see 913
Azamethonium Bromide, 913
4-Aza-5-(N-methyl-4-piperidinylid-
ene)-10,11-dihydro-5H-dibenzo-
[a,d]cycloheptene see 917
Azameton see 913
Azamin 4B see 1111
Azamune [Penn] see 918
Azanidazole, 914
Azanin [Tanabe] see 918
Azantac [Glaxo] see 8126
5-Azaorotic Acid see 6901
Azapen [Pfizer] see 5890
Azaperone, 915
1-Azaphenothiazine-10-carboxylic
Acid 2-(2-Piperidinoethoxy)ethyl
Ester see 7423
Azapren [Robins] see 758
Azapropazone see 758
Azaribine see 920
Azaron see 9651
Azaserine, 916
5α,20β$_F$,22α$_F$, 25β$_F$,27-azaspirostan-
3β-ol see 9470
Δ^5-20β$_F$,22α$_F$,25α$_F$,27-azaspirosten-
3β-ol see 8665
Azasterol see 908
Azatadine, 917
6-Azathymine, 919
6-Azauracil Riboside see 920
6-Azauridine, 920
AZ 8 Beris [Weimer] see 4464
Azelaic Acid, 921
Azelastine, 922
Azene [Endo] see 2400
Azepinamide see 4410
Azeptin [Eisai] see 922
2-Azetidinecarboxylic Acid, 923
Azidamfenicol, 924
Azidin [USSR] see 3256
D-(−)-threo-2-Azidoacetamido-1-p-
nitrophenyl-1,3-propanediol see 924
Azidoamphenicol see 924

α-Azidobenzylpenicillin *see* 925
Azidocillin, 925
3'-Azido-3'-deoxythymidine *see* 10023
2-Azido-*N*-[2-hydroxy-1-(hydroxy-methyl)-2-(4-nitrophenyl)ethyl]-acetamide *see* 924
6-[D-α-Azidophenylacetamido]peni-cillanic Acid *see* 925
6-[(Azidophenylacetyl)amino]-3,3-dimethyl-7-oxo-4-thia-1-azabicy-clo[3.2.0]heptane-2-carboxylic Acid *see* 925
Azidothymidine *see* 10023
Azimethylene *see* 2983
Azimidobenzene *see* 1119
Azindole *see* 1091
Azinepurine *see* 7947
Azinphos-ethyl *see* 926
Azinphos-methyl, 926
Azintamide, 927
Azipranone *see* 9051
Aziridine *see* 3760
Aziridinylbenzoquinone *see* 2979
1-Aziridinylphosphonitrile Trimer *see* 760
Azithromycin, 928
Azlin [Miles] *see* 929
Azlocillin, 929
Azmacort [Rorer] *see* 9514
Azoangin [Permicutan] *see* 2262
p-Azoaniline *see* 2954
Azobenzene, 930
Azobenzide *see* 930
Azobenzol *see* 930
4,4'-Azobisbenzenamine *see* 2954
1,1'-Azobiscarbamide *see* 932
Azobiscarbonamide *see* 932
Azobiscarboxamide *see* 932
α,α'-Azobis[chloroformamidine] *see* 2120
1,1'-Azobisformamide *see* 932
3,3'-Azobis(6-hydroxybenzoic Acid) *see* 6799
2,2'-Azobisisobutyronitrile, 931
1,1'-Azobis[3-methyl-2-phenylimid-azo[1,2-*a*]pyridinium] Dibromide *see* 3887
2,2'-Azobis[2-methylpropanenitrile] *see* 931
5,5'-Azobis(salicylic Acid) *see* 6799
Azochloramide [Pennwalt] *see* 2120
Azo Compd No. 4 *see* 8872
5-Azocytosine *see* 907
4,4'-Azodianiline *see* 2954
Azodicarbonamide, 932
Azodicarboxamide *see* 932
α,α'-Azodiisobutyronitrile *see* 931
Azodisal *see* 6799
Azodisal Sodium *see* 6799
Azodolen *see* 2940
Azodrin [Shell] *see* 6161
Azohel [Permicutan] *see* 2262
Azoksodon *see* 7021
Azol *see* 474
Azole *see* 8025
Azolid [USV] *see* 7248
Azolitmin, 933
Azolmen [Menarini] *see* 1232
Azolmetazin *see* 8886
Azoman *see* 4618
Azomycin, 934
Azonam [Krka] *see* 938
Azone [Nelson] *see* 5255
Azoniaspiro(3α-benziloyloxynortro-pane-8,1'-pyrrolidine) Chloride *see* 9697
Azoniaspiro(3α-diphenylglycoloyloxy-nortropan-8,1'-pyrrolidine) Chlo-ride *see* 9697
Azophenylene *see* 7173

Azopyrin *see* 8917
Azoran [Searle] *see* 918
Azosemide, 935
Azosulfamide, 936
Azothioprine *see* 918
Azovan Blue *see* 3863
Azoxodone *see* 7021
Azoxybenzene, 937
Azoxybenzide *see* 937
AZQ *see* 2979
AZT *see* 10023
Azthreonam *see* 938
Aztreon [Squibb] *see* 938
Aztreonam, 938
Azubromaron [Azupharma] *see* 1073
Azucaps *see* 7191
Azudimidine [Pharmacia] *see* 8291
Azudoxat [Azuchemie] *see* 3429
Azuglucon *see* 4372
Azulene, 939
Azulfidine [Pharmacia] *see* 8917
Azulon [Homburg] *see* 4464
Azupentat [Azuchemie] *see* 7092
AzUR *see* 920
Azure I *see* 5977
Azure II *see* 5977
Azure II Eosin *see* 5977
Azure A, 940
Azure B, 941
Azure Blue II *see* 5977
Azure C, 942
Azurene [Cilag-Chemie] *see* 1432
Azurite *see* 2634
Azusalen [Ohta] *see* 939
Azutranquil [Azuchemie] *see* 6881
B 9 *see* 2810
B 360 *see* 6992
B 436 *see* 7744
B 518 *see* 2753
B 577 *see* 3833
B 612 *see* 2853
B 663 *see* 2370
B 995 *see* 2810
B 1500 *see* 5685
B 2310 *see* 6182
B 2311 *see* 6182
B 4576 *see* 7781
B 11420 *see* 4949
B 15000 *see* 4943
B 66256 *see* 9779
B 68138 *see* 3901
BA *see* 1334
BA 168 *see* 5439
Ba 253 *see* 6897
BA 253-BR-L *see* 6897
BA 598BR *see* 4136
Ba 1355 *see* 7178
Ba 2758 *see* 2886
Ba 5473 *see* 6928
BA 7205 *see* 2359
BA 7602-06 *see* 9017
Ba 10370 *see* 8871
Ba 18189 *see* 3303
Ba 21401 *see* 9520
Ba 29038 *see* 1327
Ba 29837 *see* 2850
BA 30803 *see* 1096
Ba 32644 *see* 6480
Ba 34276 *see* 5633
Ba 34647 *see* 950
Ba 39089 *see* 6905
BA 180265 *see* 203
BAAM [Upjohn] *see* 503
Bab-O *see* 8993
Babbitt Metal, 943
Babesan *see* 3360
Babesin [Hoechst] *see* 3256
Babidium *see* 4650
Baburan *see* 3360
Bacacil [Pfizer] *see* 944
Bacampicillin, 944

Bacampicine [Upjohn] *see* 944
Bacancosin *see* 955
Bacarate [Tutag] *see* 7180
Baccidal [Kyorin] *see* 6617
Bacfeed [Fujisawa] *see* 1222
Bachelor's Buttons *see* 6652
Baciferm *see* 10027
Bacifurane [Meram] *see* 6450
Bacillin *see* 946
Bacillus Calmette-Guérin *see* 1023
Bacillus Thermoproteolyticus Neutral Proteinase *see* 9213
Bacillus thuringiensis, 945
Bacilysin, 946
Bacimethrin, 947
Bacitracin, 948
Bacitracin A *see* 948
Bacitracin Methylenebis[2-hydroxy-benzoate] *see* 949
Bacitracin Methylenedisalicylate *see* 949
Bacitracin Methylenedisalicylic Acid, 949
Bacitracin Zinc Complex *see* 10027
Bacitracin Zinc Salt *see* 10027
Baclofen, 950
Baclon [Medica] *see* 950
BACOP *see* MISC-2
Bacterial Vitamin H¹ *see* 434
Bacterio-opsin *see* 951
Bacteriopsins *see* 6811
Bacteriorhodopsin, 951
Bacteron [Ciba-Geigy] *see* 1222
Bacticlens *see* 2090
Bactidan [Recordati] *see* 3540
Bactocill [Beecham] *see* 6858
Bactoderm [Beecham] *see* 6212
Bactol *see* 4924
Bactopen [Beecham] *see* 2414
Bactospeine *see* 945
Bactramin [Nippon Roche] *see* 8889
Bactramyl [Boots-Dacour] *see* 7475
Bactrim [Roche] *see* 8889
Bactroban [Beecham] *see* 6212
Bactromin *see* 8889
Bactrovet [Pitman-Moore] *see* 8876
Bactylan *see* 491
Baddeleyite *see* 10083
Badil [Bayer] *see* 4287
Badional *see* 8921
Badische Acid, 952
Bad Nauheim Salts (Artificial) *see* 6348
BAF *see* 4895
Bafhameritin-M [Hishiyama] *see* 5680
Bagasse, 953
Bagodryl *see* 3308
Bagolax *see* 5963
Bagrosin-Natrium [Cassella-Riedel] *see* 6023
Bahama White Wood *see* 1749
BAHS *see* 1526
Baicalein, 954
Bajaten [Volpino] *see* 4847
Bakankoside *see* 955
Bakankosin, 955
Bakash *see* 149
Baking Soda *see* 8528
Bakontal *see* 1006
Baktar [Shionogi] *see* 8889
Bakthane *see* 945
BAL [HW & D] *see* 3196
Balan [Lilly] *see* 1048
Balanophorin *see* 654
Balarsen [Endo] *see* 840
Baldex [Bausch & Lomb] *see* 2922
Balfin [Lilly] *see* 1048
Balm of Gilead *see* 958
"Balm of Gilead" *see* 956
Balsam Canada, 956
Balsam Capivi *see* 2511

Balsam Copaiba see 2511
Balsam of Fir see 956
Balsam of Gilead see 958
Balsam Gurjun, 957
Balsam Mecca, 958
Balsam Peru, 959
Balsam Tolu, 960
Balsam Traumatic, 961
Balsam Tree see 5636
Balsán-Katél see 958
Baltic Amber see 388
Baluvet [Veterinaria] see 366
Bambec [Astra] see 963
Bambermycins, 962
Bamboo Curare see 2678
Bambuterol, 963
Bamethan, 964
Bamifylline, 965
Bamiphylline see 965
Bamipine, 966
Bamo 400 [Misemer] see 5751
Banabins see 7490
Banabin-Sintyal see 2105
Banamine [Schering] see 4073
Banan [Sankyo] see 1941
Banana Oil see 4993
Bancaris [Burroughs Wellcome] see 9205
Bandol [Squibb] see 1805
Banewort see 1034
Banflex [OJ & F] see 6831
Banicol see 2009
Banisterine see 4531
Banistyl [M & B] see 4146
Banlene [Fisons] see 3026
Banminth [Pfizer] see 7968
Bannal see 8358
Banner [Ciba-Geigy] see 7830
Banocide [Wellcome] see 3106
Bantenol [Janssen] see 5647
Bantex see 5759
Banthine Bromide [Searle] see 5869
Banthionine see 5896
Bantogen see 7046
Bantron see 5432
Bantu see 755
Banvel D [Velsicol] see 3026
BAPP see MISC-2
Baptigenin, 967
Baptisia, 968
Baptitoxine see 2795
BAQD 10 see 2896
Barakshin see 4815
Baratol [Wyeth] see 4876
Barazan [Merck & Co.] see 6617
Barbaloin see 304
Barbamate see 969
Barbamil see 601
Barbamyl see 601
Barban, 969
Barbane see 969
Barbasco, 970
Barbenyl see 7201
Barberry Bark, 971
Barbexaclone see 7873
Barbiphenyl see 7201
Barbipil [Conal] see 7201
Barbital, 972
Barbital Sodium see 972
Barbitone see 972
Barbitone Sodium see 972
Barbituric Acid, 973
Barbonin [Knoll] see 3682
Barbosec [Rowell] see 8374
Bardac 2250/2280 [Lonza] see 3088
Bardana see 5236
Bareon [Hokuriku] see 5443
Baridol see 1006
Barite see 1006
Baritop [Kaigen] see 1006
Barium, 974

Barium Acetate, 975
Barium Benzenesulfonate, 976
Barium Benzoate see 1101
Barium Bromate, 977
Barium Bromide, 978
Barium Carbonate, 979
Barium Chlorate, 980
Barium Chloride, 981
Barium Chromate(VI), 982
Barium Citrate see 2328
Barium Cyanide, 983
Barium Cyanoplatinate(II) see 1003
Barium Dioxide see 1001
Barium Dithionate, 984
Barium Diuranate see 1013
Barium Ethyl Sulfate see 3808
Barium Ferrocyanide, 985
Barium Fluoride, 986
Barium Fluosilicate see 988
Barium Formate, 987
Barium Hexacyanoferrate(II) see 985
Barium Hexafluorosilicate, 988
Barium Hydrate see 989
Barium Hydroxide, 989
Barium Hypophosphite, 990
Barium "Hyposulfate" see 984
Barium Hyposulfite see 1011
Barium Iodate, 991
Barium Iodide, 992
Barium Lactate see 5215
Barium Manganate(VI), 993
Barium Manganate(VII) see 1000
Barium Mercuric Iodide, 994
Barium Metatitanate see 1012
Barium Methyl Sulfate see 6041
Barium Monoxide see 998
Barium Nitrate, 995
Barium Nitrite, 996
Barium Oleate see 6788
Barium Oxalate, 997
Barium Oxide, 998
Barium Perchlorate, 999
Barium Permanganate, 1000
Barium Peroxide, 1001
Barium p-Phenolsulfonate see 7212
Barium Phosphate, Dibasic, 1002
Barium Platinocyanide see 1003
Barium Platinous Cyanide, 1003
Barium Propionate see 7837
Barium Protoxide see 998
Barium Rhodanide see 1010
Barium Selenide, 1004
Barium Silicide, 1005
Barium Silicofluoride see 988
Barium Sulfate, 1006
Barium Sulfide, 1007
Barium Sulfide, Black, 1008
Barium Sulfite, 1009
Barium Sulfocyanate see 1010
Barium Sulfocyanide see 1010
Barium Superoxide see 1001
Barium Taurocholate see 9044
Barium Tetracyanoplatinate(II) see 1003
Barium Tetraiodomercurate(II) see 994
Barium Thiocyanate, 1010
Barium Thiosulfate, 1011
Barium Titanate(IV), 1012
Barium Uranate(VI) see 1013
Barium Uranium Oxide, 1013
Barizin [Lek] see 6403
Barnetil [Delagrange] see 8975
Barnotil [Vita] see 8975
Baron [Dow] see 3587
Baros [Horii] see 8486
Baros Camphor see 1338
Barosma Camphor see 3292
Barosmin see 3291
Barosperse [Mallinckrodt] see 1006
Barpental see 7087

Barquinol see 4924
Barricade [Shell] see 2775
Barrier [Falk] see 1296
Barthrin, 1014
Baryta Yellow see 982
Barytes see 1006
BAS 083 see 5746
BAS 238F see 3405
BAS 352F see 9890
BAS 392H see 4052
BAS 3460 see 1794
BAS 9052 see 8424
BAS 42100F see 3938
BAS 67054 see 1794
BAS 85559X see 5746
Basagran [BASF] see 1060
Basalin [BASF] see 4052
Basaljel [Wyeth] see 1015
Basanite [BASF] see 3282
Basecil see 6048
Basedol see 494
Base L see 5490
Basergin see 3600
Basethyrin see 6048
Basfapon B [BASF] see 2806
Basforin [Compo] see 9602
Basham's Mixture see 3959
Basic Aluminum Acetate see 339
Basic Aluminum Carbonate Gel, 1015
Basic Aluminum Chloride see 346
Basic Aluminum Salicylates see 366
Basic Antimony Chloride see 725
Basic Bismuth Bromide see 1271
Basic Bismuth Chloride see 1274
Basic Bismuth Iodide see 1280
Basic Bismuth Salicylate see 1299
Basic Bismuth Valerate see 1306
Basic Chromic Sulfates see 2233
Basic Cupric Chromates see 2637
Basic Dextran see 2916
Basic Ferric Sulfate Soln see 3979
Basic Fibroblast Growth Factor see 4016
Basic Lead Carbonate, 1016
Basic Lead Chromate see 5280
Basic Lead Hydroxide see 5286
Basic Mercuric Sulfate see 5785
Basi-Cop see 2660
Basic Quinine Sulfate see 8089
Basic Zirconium Chloride see 10087
Basodexan [Rohm] see 9781
Basofortina [Sandoz] see 5989
Basolan see 5892
Basswood, 1017
Basta [Hoechst] see 7314
Bastard Saffron see 1877
Bastiverit see 4372
Basudin [Geigy] see 2978
BAT see 9694
Batidrol [Métadier] see 714
Batrachotoxin, 1018
Batrachotoxinin A, 1019
Batrachotoxinin A 20-(2,4-Dimethyl-1H-pyrrole-3-carboxylate) see 1018
Batrafen [Hoechst] see 2270
Batroxobin, 1020
Batyl Alcohol, 1021
Baukal [Bruschetti] see 7884
Baum's Acid see 6310
Bauxite see 359
Bavenite see 1645
Bavistin [BASF] see 1794
BAX [McKesson] see 3308
BAX 1400Z see 3201
BAX 1526 see 2264
BAX 2739Z see 965
Baxacor [Mack, Illert.] see 3662
Baxan [Bristol-Myers] see 1921
Baxo [Toyama] see 7476
BAY 1470 see 9987
Bay 1521 see 6644

BAY 2502 see 6454
Bay 4503 see 7844
Bay 5621 see 7341
Bay 6681 F see 9507
BAY 9002 see 6269
BAY 9010 see 7849
BAY 19639 see 3371
BAY 25/141 see 3943
BAY 34727 see 2266
BAY 39007 see 7849
Bay 44646 see 443
Bay 68138 see 3901
BAY 70143 see 1810
Bay 77488 see 7341
Bay 92114 see 5055
Bay 94337 see 6076
BAY a 1040 see 6441
BAY b 4343 b see 1869
BAY b 5097 see 2412
BAY e 5009 see 6493
Bay e 6905 see 929
Bay e 9736 see 6467
Bay f 1353 see 6096
Bay g 5421 see 12
Bay g 2821 see 6227
Bay h 4502 see 1232
Bay h 5757 see 3888
Bay k 5552 see 6482
BAY m 1099 see 6110
BAY n 5595 see 2887
Bay o 6893 see 8225
Bay o 9867 see 2315
Bay q 7821 see 4967
Bayberry Bark, 1022
Baycain [Bayer] see 9467
Baycaron [Bayer] see 5685
Baycid [Bayer] see 3945
Baycillin [Bayer] see 7829
Baycipen [Bayer] see 6096
Baycor [Bayer] see 1315
Baycovin see 8008
Baycox [Bayer] see 9452
Bayer 205 see 8986
Bayer 1219 see 9618
Bayer 1362 see 1509
Bayer 1420 see 7813
Bayer 2353 see 6425
Bayer 3231 see 9517
Bayer 5312 see 3692
Bayer 5630 see 6079
Bayer 5633 see 5846
Bayer 8169 see 2875
Bayer 9015 see 5722
Bayer 9051 see 9161
Bayer 16259 see 926
Bayer 17147 see 926
Bayer 21/116 see 5971
Bayer 21/199 see 2559
Bayer 25820 see 6269
Bayer 29493 see 3945
Bayer 36205 see 6933
Bayer 37344 see 5893
Bayer 41831 see 3922
Bayer 46131 see 7831
Bayer 47531 see 3031
Bayer 68138 see 3901
Bayer 71628 see 5858
Bayer 74283 see 5846
Bayer 78418 see 3485
Bayer A 128 see 784
Bayer A 173 see 7768
Bayer E 39 Soluble see 1256
Bayer E 393 see 8945
Bayerite see 359
Bayer L 13/59 see 9536
Bayer S 5660 see 3922
Bayer's Acid see 2591
BAY FCR 1272 see 2764
Baygnostil see 7203
Baygon [Bayer] see 7849
BAY KWG 0519 see 9508

BAY KWG 0599 see 1315
Bayleton [Bayer] see 9507
Bayluscid see 6425
Baymicin [Bayer] see 8498
Baymix [Chemagro] see 2559
Baymycard [Bayer] see 6482
BAY NTN 9306 see 8972
Bayo-n-ox [Bayer] see 6783
Bayotensin [Bayer] see 6493
Baypen [Bayer] see 6096
Baypress [Bayer] see 6493
Bayrena see 8885
Bayrogel [Bayropharm] see 3833
BAY SIR 8514 see 9592
BAY SMY 1500 see 3694
Baytan [Bayer] see 9508
Baytex [Parke, Davis] see 3945
Baythion [Bayer] see 7341
Baythroid [Bayer] see 2764
Bayticol [Bayer] see 4068
Baytinal [Bayer] see 1518
Baytril [Mobay] see 3546
BAY VA 1470 see 9987
BAY VA 5387 see 3829
BAY Va 9391 see 6783
Bay Vh 5757 see 3888
Bay Vi 9142 see 9452
Bay Vp 2674 see 3546
BBB see 4023
B.B.C. see 1400
BB-K8 [Bristol] see 416
BB Powder see 7661
BC 16 see 4599
BC 48 see 2871
BC 51 see 3368
BC 105 see 7486
BC 681 see 3488
BC 757 see 7508
Bc Conjugate see 7951
B-cell Activating Factor see 4895
β-Cell Differentiation Factor see 4893
BCG, 1023
BCM see 5631
BCM see 1794
BCME see 3058
BCNU see 1852
BCP see 1452
BCVPP see MISC-2
BD 40 A see 4159
BDH 312 see 5737
B.D.P.E. see 1438
Beacillin see 7037
Beamette see 7558
Beaprine [Aspro-Nicholas] see 1873
Bearberry see 9808
Bearberry Bark see 1889
Bear's Weed see 3617
Bearwood see 1889
Bebate see 1202
Bebeerine, 1024
d-Bebeerine see 1024
l-Bebeerine see 2682
Bebeeru Bark, 1025
Beben [Warner-Lambert] see 1202
Becantal see 3017
Becantex see 3017
Becanthone, 1026
Becantone see 1026
Becaptan [Labaz] see 2785
Becenun [Bristol-Myers] see 1852
Bécilan see 7995
Beclacin [Kaigai] see 1029
Beclamide, 1027
Beclipur [Siegfried] see 1028
Beclobrate, 1028
Becloforte [Allen & Hanburys] see 1029
Beclomet [Orion] see 1029
Beclomethasone, 1029
Beclorhinol [Declimed] see 1029
Beclosclerin [Siegfried] see 1028

Beclotiamine, 1030
Becloval [Valeas] see 1029
Beclovent [Glaxo] see 1029
Becodisks [Allen & Hanburys] see 1029
Beconase [Allen & Hanburys] see 1029
Beconasol [Glaxo] see 1029
Becort [Rachelle] see 1202
Becotide [Allen & Hanburys] see 1029
Bedermin [Damor] see 1202
Bedoce [Lincoln] see 9921
Bedodeka see 9921
Bedome [Merck & Co.] see 9222
Bedoz see 9921
Bedranol [Lagap] see 7852
Bedriol [Andromaco] see 1232
Bee Bread see 7848
Beechwood Creosote see 2575
Beeswax, 1031
Beet Sugar see 8855
Befeniol see 1166
Befizal [Oberval] see 1215
Beflavine see 8201
Befunolol, 1032
Begiolan see 9222
Behenic Acid, 1033
N(4)-Behenoyl-1-β-D-arabinofuranos-ylcytosine see 3539
Behenoylcytosine Arabinoside see 3539
Behepan see 9921
Behyd [Kyorin] see 1150
Bei 1293 see 9986
Bekadid see 4708
Bekanamycin see 5161
Belamarine see 5498
Belfacillin see 5890
Belfene [Lafarge] see 3334
Bellacristin see 1035
Belladonna, 1034
Belladonnine, 1035
Belladonnine Bis[bromoethylate] Tetrahydrate see 1035
Belladine see 2685
Bellasthman see 5105
Belmark [Shell] see 3952
Beloc [Astra] see 6072
Belosin see 1743
Belseren [Mead Johnson] see 2400
Belt see 2079
Belustine [Roger Bellon] see 5444
Bemaphate see 2163
Bémarsal [Specia] see 3338
Bemarside see 8915
Bemegride, 1036
Bemidone see 4769
Bemperil [Sidus] see 8967
Benacol [Cenci] see 3087
Benactyzine, 1037
Benadon [Roche] see 7995
Benadryl [Parke, Davis] see 3308
Ben-a-hist see 709
Benalaxyl, 1038
Benalgin [Polfa] see 1136
Benambax [Rhône-Poulenc] see 7071
Benapen see 7036
Benapryzine, 1039
Benazoline (obsolete) see 6060
Bencef [Fargal-Pharmasint] see 7485
Bence-Jones Proteins, 1040
Bencyclane, 1041
Bendacort see 4710
Bendazac, 1042
Bendazol, 1043
Bendazole see 1043
Bendazolic Acid see 1042
Bendectin [Merrell Dow] see 3430
Bendiocarb, 1044
Bendioxide see 1060
Bendogen [Lagap] see 1208

Bendopa [ICN] see 5344
Bendralan [Bristol] see 7184
Bendrofluazide see 1045
Bendroflumethiazide, 1045
Benecardin see 5189
Benedorm see 8005
Benefex [Makhteshim-Agan] see 1048
Benefin see 1048
Benemid [Merck & Co.] see 7760
Benerva [Roche] see 9222
Benesal see 8297
Benethamine Penicillin G see 7036
Benetolin see 7036
Benexate-CD see 1046
Benexate Hydrochloride, 1046
Benfluorex, 1047
Benfluralin, 1048
Benfluramate see 1047
Benfofen [Plantorgan] see 3071
Benfotiamine, 1049
Benfuracarb, 1050
Benfuran [Kaken] see 1032
Benfurodil Hemisuccinate, 1051
Bengal Isinglass see 172
Benicot see 6398
Benirol see 1066
Benisone [Warner-Chilcott] see 1202
Benitoite see 9396
Benlate [Du Pont] see 1053
Benmoxine, 1052
Benne Oil see 8418
Benocten [Medinova] see 3308
Benodaine [Merck & Co.] see 7448
Benodin see 3308
Benomyl, 1053
Benoquin see 6159
Benoral [Winthrop] see 1054
Benortan [Winthrop] see 1054
Benorylate, 1054
Benovocylin see 3655
Benoxaprofen, 1055
Benoxil [Santen] see 1056
Benoxinate, 1056
Benoxyl [Stiefel] see 1128
Benozil [Kyowa] see 4123
Benperidol, 1057
Benproperine, 1058
Benserazide, 1059
Bensylyt see 7225
Bent [USV] see 2082
Bentazon, 1060
Bentazone see 1060
Bentelan [Glaxo] see 1202
Bentex [Steinhard] see 9607
Benthiozone see 9218
Bentiromide, 1061
Bentomine [Darby] see 3087
Bentonite, 1062
Bentonyl see 9682
Bentos [Kaken] see 1032
Bentox [Kaken] see 1032
Bentrofene [Cilag-Chemie] see 17
Bentyl Hydrochloride [Merrell] see 3087
Bentylol Hydrochloride see 3087
Benuride [Bengue] see 7190
Benuron [Bristol] see 1045
Ben-u-ron [Bene-Arzneimittel] see 40
Benylan see 3308
Benylate [Breon] see 1141
Benylin DM [Parke, Davis] see 8116
Benzacyl [Wander] see 1127
Benzagel 10 [Rorer] see 1128
Benzaidin see 1208
Benzaknen [Alcon] see 1128
Benzalacetone see 1151
Benzalacetophenone see 2028
Benzalaniline see 1152
β-Benzalbutyramide see 1063
Benzal Chloride, 1064
Benzaldehyde, 1065

Benzaldehyde Cyanohydrin see 5601
Benzaldehyde FFC see 1065
Benzaldehyde Green see 5581
Benzal Green see 5581
Benzalin [Shionogi] see 6491
Benzalkonium Chloride, 1066
Benzamide, 1067
Benzamidoacetic Acid see 4636
DL-α-Benzamid-p-[2-(diethylamino)-ethyl]-N,N-dipropylhydrocin-namamide see 9393
DL-4-Benzamido-N,N-dipropylglu-taramic Acid see 7785
(S)-p-(α-Benzamido-p-hydroxyhydro-cinnamamido)benzoic Acid see 1061
3-[2-(4-Benzamidopiperidino)ethyl]-indole see 4876
4-Benzamidosalicylic Acid see 1127
Benzamine see 3850
Benzamine Blue see 9701
Benzamizole see 5125
Benzamon see 4225
Benzanilide, 1068
Benz[a]anthracene see 1069
1,2-Benzanthracene, 1069
2,3-Benzanthracene see 6288
7H-Benz[de]anthracen-7-one see 1070
Benzanthrene see 1069
Benzanthrone, 1070
Benzantin see 3308
Benzapas [Dorsey] see 1127
Benzarone, 1071
Benzathine, 1072
Benzathine Benzylpenicillin see 7047
Benzathine Penicillin G see 7037
Benzathine Penicillin V see 7047
1-Benzazine see 8097
2-Benzazine see 5110
Benzazoline see 9430
Benzbromarone, 1073
Benzcarbimine see 1097
Benzchlorpropamide see 1027
3,4-Benzchrysene see 7368
Benzcurine Iodide see 4249
Benzcyclan see 1041
Benzedrex [SK & F] see 7873
Benzedrine [SK & F] see 616
Benzehist [Pharmex] see 3308
Benzenamine see 687
Benzene, 1074
Benzeneacetaldehyde see 7236
Benzeneacetamide see 7237
Benzeneacetic Acid see 7239
Benzeneacetic Acid α-Ethyl-2-(di-ethylamino)ethyl Ester see 1516
Benzeneacetic Acid Ethyl Ester see 3794
Benzeneacetonitrile see 1145
Benzenearsonic Acid, 1075
Benzeneazoaniline see 2981
Benzeneazobenzene see 930
Benzeneboronic Acid, 1076
Benzene Boronic Anhydride see 1076
Benzenecarbonyl Chloride see 1124
Benzenecarboperoxoic Acid see 7109
Benzenecarbothioic Acid S-[2-[[(4-Amino-2-methyl-5-pyrimidinyl)-methyl]formylamino]-1-[2-(phos-phonooxy)ethyl]-1-propenyl] Ester see 1049
Benzenecarboxylic Acid see 1101
Benzene Chloride see 2121
1,2-Benzenediamine see 7255
1,3-Benzenediamine see 7254
1,4-Benzenediamine see 7256
1,2-Benzenedicarbonyl Dichloride see 7349
1,2-Benzenedicarboxamide see 7343
1,2-Benzenedicarboxylic Acid see 7345

1,3-Benzenedicarboxylic Acid see 5083
1,4-Benzenedicarboxylic Acid see 9093
1,2-Benzenedicarboxylic Acid Bis(2-ethylhexyl) Ester see 1262
1,2-Benzenedicarboxylic Acid Bis(3-methylbutyl) Ester see 5006
1,2-Benzenedicarboxylic Acid Dibutyl Ester see 1586
1,2-Benzenedicarboxylic Acid Di-methyl Ester see 3243
1,2-Benzenedicarboxylic Acid Diphen-yl Ester see 7278
1,2-Benzenedicarboxylic Acid Mono-(1-ethyl-1-methyl-2-propynyl) Ester see 7348
1,2-Benzenediol see 8009
1,3-Benzenediol see 8158
1,4-Benzenediol see 4738
Benzeneethanamine see 7186
Benzeneethanol see 7185
Benzenehexacarboxylic Acid see 5706
γ-Benzene Hexachloride see 5379
Benzenemethanamine see 1139
Benzenemethanethiol see 9249
Benzenemethanol see 1138
Benzenepropanoic Acid see 4707
Benzenepropanol Carbamate see 7229
Benzenestibonic Acid, 1077
4-(2-Benzenesulfinylethyl)-1,2-di-phenylpyrazolidine-3,5-dione see 8926
Benzenesulfohydrazide see 7573
2-Benzenesulfonamido-5-tert-butyl-1,3,4-thiadiazole see 4374
2-Benzenesulfonamido-5-(β-methoxy-ethoxy)pyrimidine see 4403
Benzene Sulfonechloride see 1080
Benzenesulfonic Acid, 1078
Benzenesulfonic Acid Barium Salt see 976
Benzenesulfonic (Acid) Chloride see 1080
Benzenesulfonic Acid 2,4-Dichloro-phenyl Ester see 4283
Benzenesulfonic Acid Hydrazide see 7573
Benzenesulfonic Anhydride, 1079
N-Benzenesulfonyl-N'-butylurea see 7177
Benzenesulfonyl Chloride, 1080
1,2,4,5-Benzenetetracarboxylic Acid see 8015
Benzenethiol see 9285
1,2,4-Benzenetricarboxylic Acid see 9616
1,2,3-Benzenetriol see 8010
1,2,4-Benzenetriol, 1081
1,3,5-Benzenetriol see 7301
2,2',2''-[1,2,3-Benzenetriyltris-(oxy)]tris[N,N,N-triethylethanamini-um] Triiodide see 4249
N,N'-Benzenyl-o-phenylenediamine see 7245
Benzenyl Trichloride see 1120
Benzerial [Houdé] see 4481
Benzestrol, 1082
Benzetamophylline see 965
Benzethacil see 7037
Benzethonium Chloride, 1083
Benzetimide, 1084
Benzhexol Chloride see 9607
Benzhormovarine see 3655
Benzhydramine see 3308
Benzhydrol see 1100
Benzhydrylamine, 1085
Benzhydrylamine Penicillin see 7038
N-Benzhydryl-N'-trans-cinnamyl-piperazine see 2308

Benzhydryl 2-Diethylaminoethyl Ether *see* 3724

O-Benzhydryldimethylaminoethanol *see* 3308

O-Benzhydryldimethylaminoethanol Bis(theophylline 7-acetate) *see* 1225

O-Benzhydryldimethylaminoethanol 8-Chlorotheophyllinate *see* 3193

2-Benzhydryl β-Dimethylaminoethyl Ether *see* 7287

(*N*-Benzhydryl)(*N*′-methyl)diethylenediamine *see* 2716

N-Benzhydryl-*N*′-methylpiperazine *see* 2716

2-(Benzhydryloxy)-*N*,*N*-diethylethylamine *see* 3724

2-(Benzhydryloxy)-*N*,*N*-dimethylethylamine *see* 3308

2-(Benzhydryloxy)-*N*,*N*-dimethylethylamine Bis(theophylline 7-Acetate) *see* 1225

2-(Benzhydryloxy)-*N*,*N*-dimethylethylamine 8-Chlorotheophyllinate *see* 3193

2-Benzhydryloxymethyl-2-imidazoline *see* 3307

4-Benzhydryloxy-*N*-methylpiperidine *see* 3334

4-(Benzhydryloxy)-1-methylpiperidine *see* 3334

2-(Benzhydryloxy)triethylamine *see* 3724

2-Benzhydryl-1-piperidineethanol *see* 3303

1-Benzhydryl-4-piperonylpiperazine *see* 5671

2-(Benzhydrylsulfinyl)acetohydroxamic Acid *see* 159

Benzidazol *see* 9430

Benzidine, 1086

Benzil, 1087

Benzil Dioxime, 1088

Benzilic Acid, 1089

Benzilic Acid β-Diethylaminoethyl Ester *see* 1037

Benzilic Acid Ester with 2-(Hydroxymethyl)-1,1-dimethylpiperidinium Methyl Sulfate *see* 1214

Benzilic Acid, 1-Ethyl-3-piperidyl Ester Methyl Bromide *see* 7428

Benzilic Acid 2-(Ethylpropylamino)ethyl Ester *see* 1039

Benzilic Acid 1-Ethyl-3-pyrrolidinyl Ester Ethyl Bromide *see* 1090

Benzilic Acid *N*-β-Fluoroethylnortropine Ester Methobromide *see* 4136

Benzilic Acid 6-Methoxytropine Ester *see* 9690

Benzilic Acid 1-Piperidineethanol Ester *see* 7441

Benzilic Acid 3α-Tropanyl Ester *see* 9694

Benzilonium Bromide, 1090

3-Benziloyloxy-1-azabicyclo[2.2.2]octane Methobromide *see* 2351

3-Benziloyloxy-1,1-diethylpyrrolidinium Bromide *see* 1090

(8*r*)-3α-Benziloyloxy-8-(2-fluoroethyl)-8-methyl-8-azoniabicyclo-[3.2.1]octane Bromide *see* 4136

2-Benziloyloxymethyl-1,1-dimethylpyrrolidinium Methyl Sulfate *see* 7529

3α-Benziloyloxyspiro(nortropane-8,1′-pyrrolidinium) Chloride *see* 9697

Benzimidazole, 1091

2-Benzimidazolecarbamic Acid Methyl Ester *see* 1794

1*H*-Benzimidazole-2-propanoic Acid *see* 7769

2-Benzimidazolethiol, 1092

1*H*-Benzimidazol-2-ylcarbamic Acid Methyl Ester *see* 1794

β-(2-Benzimidazolyl)propionic Acid *see* 7769

4-(2-Benzimidazolyl)thiazole *see* 9217

Benziminazole *see* 1091

Benzin *see* 7141

Benzin (German) *see* 4270

Benzindamine *see* 1136

Benzinoform [Mallinckrodt] *see* 1822

Benziodarone, 1093

1,2-Benzisothiazol-3(2*H*)-one 1,1-Dioxide *see* 8282

Benzisotriazole *see* 1119

1,2-Benzisoxazole-3-methanesulfonamide *see* 10094

Benzite *see* 9639

Benzitramide *see* 1216

6,7-Benzmorphan *see* 1105

Benznidazole, 1094

Benzoaric Acid *see* 3508

Benzo Azurine G, 1095

Benzo Blue *see* 9701

Benzocaine *see* 3719

Benzoctamine, 1096

Benzodepa, 1097

Benzodiapin [Lisapharma] *see* 2082

7,8-Benzo-1,3-diazaspiro[4.5]decane-2,4-dione *see* 9165

1,2-Benzodiazine *see* 2309

1,3-Benzodiazine *see* 8062

1,4-Benzodiazine *see* 8107

2,3-Benzodiazine *see* 7344

1,3-Benzodiazole *see* 1091

Benzodioxane *see* 7448

(1,4-Benzodioxan-2-ylmethyl)guanidine *see* 4482

1,3-Benzodioxole-5-carboxaldehyde *see* 7445

1,3-Benzodioxole-5-carboxylic Acid *see* 7447

5-[4-(1,3-Benzodioxolol-5-yloxy)-tetrahydro-1*H*,3*H*-furo[3,4-*c*]furan-1-yl]-1,3-benzodioxole *see* 8421

6-[2-(1,3-Benzodioxol-5-yl)ethenyl]-5,6-dihydro-4-methoxy-2*H*-pyran-2-one *see* 6056

2-[2-(1,3-Benzodioxol-5-yl)ethyl]-4-methoxyquinoline *see* 2687

3-(1,3-Benzodioxol-5-yl)-7-hydroxy-4*H*-1-benzopyran-4-one *see* 7925

1-(1,3-Benzodioxol-5-ylmethyl)-4-[[(4-chlorophenoxy)acetyl]piperazine *see* 4024

1-(1,3-Benzodioxol-5-ylmethyl)-4-(diphenylmethyl)piperazine *see* 5671

4-[2-[[2-(1,3-Benzodioxol-5-yl)-1-methylethyl]amino]-1-hydroxyethyl]-1,2-benzenediol *see* 7911

5-(1,3-Benzodioxol-5-ylmethyl)-7-methyl-1,3-dioxolo[4,5-*g*]isoquinoline *see* 6011

10-[[4-(1,3-Benzodioxol-5-ylmethyl)-1-piperazinyl]acetyl]-10*H*-phenothiazine *see* 3928

2-[4-(1,3-Benzodioxol-5-ylmethyl)-1-piperazinyl]pyrimidine *see* 7465

1-(1,3-Benzodioxol-5-ylmethyl)-4-(3,7,11-trimethyl-2,6,10-dodecatrienyl)piperazine *see* 7391

1-[5-(1,3-Benzodioxol-5-yl)-1-oxo-2,4-pentadienyl]piperidine *see* 7442

1-[5-(1,3-Benzodioxol-5-yl)-1-oxo-2,4-pentadienyl]piperidine *see* 2038

trans-(−)-3-[(1,3-Benzodioxol-5-yloxy)methyl]-4-(4-fluorophenyl)-piperidine *see* 6991

5-(1,3-Benzodioxol-5-yl)-2,4-pentadienoic Acid *see* 7437

9α-(1,3-Benzodioxol-5-yl)-6,7,8,9-tetrahydro-7α,8β-dimethylnaphtho-[1,2-*d*]-1,3-dioxole *see* 6852

9-(1,3-Benzodioxol-5-yl)-4,6,7-trimethoxynaphtho[2,3-*c*]furan-1(3*H*)-one *see* 5154

Benzodol *see* 7539

Benzoestrofol *see* 3655

Benzofoline *see* 3655

Benzofuran, 1098

2-Benzofurancarboxylic Acid *see* 2562

4-(4-Benzofurazanyl)-1,4-dihydro-2,6-dimethyl-3,5-pyridinedicarboxylic Acid Methyl 1-Methylethyl Ester *see* 5129

3-Benzofuro[3,2-*c*][1]benzoxepin-6(12*H*)-ylidene-*N*,*N*-dimethyl-1-propanamine *see* 6889

Benzofurodil *see* 1051

2-Benzofuryl-*p*-chlorophenylcarbinol *see* 2356

Benzoglyoxaline *see* 1091

Benzoguanamine, 1099

Benzo-Gynoestryl [Roussel-UCLAF] *see* 3655

Benzohydrol, 1100

Benzoic Acid, 1101

Benzoic Acid [1-[4-[(3-Amino-2,3,6-trideoxy-α-L-*lyxo*-hexopyranosyl)-oxy]-1,2,3,4,6,11-hexahydro-2,5,12-trihydroxy-7-methoxy-6,11-dioxo-2-napthacenyl]ethylidene]hydrazide *see* 10096

Benzoic Acid Anhydride *see* 1102

Benzoic Acid Benzyl Ester *see* 1141

Benzoic Acid Butyl Ester *see* 1552

Benzoic Acid 2,4-Dichlorophenoxyethyl Ester *see* 8422

Benzoic Acid Ethyl Ester *see* 3725

Benzoic Acid Hydrazide 3-Hydrazone with Daunorubicin *see* 10096

Benzoic Acid Methyl Ester *see* 5947

Benzoic Acid Phenyl Ester *see* 7246

Benzoic Acid 2-(1-Phenylethyl)hydrazide *see* 1052

Benzoic Acid Phenylmethyl Ester *see* 1141

Benzoic Acid *p*-Sulfamide *see* 1885

Benzoic Aldehyde *see* 1065

Benzoic Anhydride, 1102

Benzoic Sulfimide *see* 8282

Benzoin, 1103

Benzoin Oxime, 1104

Benzol *see* 1074

Benzolin *see* 3016

Benzoline *see* 5366

Benzometan [Castejon] *see* 1131

6,7-Benzomorphan, 1105

Benzo[*b*]naphthacene *see* 7057

Benzonaphthol *see* 6328

Benzonatate, 1106

Benzone [Hyde Chemical] *see* 7248

Benzonitrile, 1107

Benzononatine *see* 1106

Benzoparadiazine *see* 8107

Benzo[*c*]phenanthrene *see* 1129

Benzo[*def*]phenanthrene *see* 7977

Benzophenone, 1108

Benzophenone-1 *see* 1117

Benzophenone-3 *see* 6907

Benzophenone-4 *see* 8963

Benzophenone-6, 1109

Benzophenone-8 *see* 3298

Benzophenone-10 *see* 6092

Benzophenone-12 *see* 6662

Benzopinacol, 1110

Benzopurpurine 4B, 1111

1,2-Benzopyran, 1112

2*H*-1-Benzopyran, 1112

α-5:6-Benzopyran *see* 1112

2*H*-1-Benzopyran-2-one *see* 2563

Benzylpenicillin Potassium see 7041
Benzylpenicillin Procaine see 7042
Benzylpenicillin Sodium, 1157
Benzylpenicilloyl Polylysine see 7050
o-Benzylphenol, 1158
p-Benzylphenol, 1159
4-(o-Benzylphenoxy)-N-methylbutyl-
  amine see 1227
1-[2-(2-Benzylphenoxy)-1-methyleth-
  yl]piperidine see 1058
Benzylphenylamine see 1140
α-(N-Benzyl-N-phenylamino)acet-
  amidoxime see 2015
p-Benzylphenyl Carbamate see 3306
2-Benzylphenyl β-Dimethylaminoethyl Ether see 7287
N-Benzyl-2-phenylethylamine Salt of
  Benzylpenicillin see 7036
N-Benzyl-N-phenylpyrrolidinoethylamine see 4641
p-Benzylphenylurethan see 3306
Benzyl Phosphite see 2996
Benzyl Phosphorochloridate see 2994
1-Benzyl-4-picolinoylpiperazine see
  7365
2-(4-Benzylpiperidino)-1-(4-hydroxy-
  phenyl)-1-propanol see 4821
N-(1'-Benzyl-4'-piperidyl)-2-meth-
  oxy-4-amino-5-chlorobenzamide
  see 2344
(S)-(+)-2-(1-Benzyl-4-piperidyl)-2-
  phenylglutarimide see 2923
2-(1-Benzyl-4-piperidyl)-2-phenyl-
  glutarimide see 1084
(+)-3-(1-Benzyl-4-piperidyl)-3-
  phenylpiperidine-2,6-dione see 2923
1-Benzyl-2-pivalylhydrazine see 7483
N-Benzyl-N-pyrrolidinoethylaniline
  see 4641
Benzyl-Rodiuran [Boehringer, Ing.]
  see 1045
Benzyl Salicylate, 1160
Benzyl Salicylate trans-4-(Guanidinomethyl)cyclohexanecarboxylate
  Hydrochloride see 1046
Benzylsulfamide, 1161
N⁴-Benzylsulfanilamide see 1161
Benzyl Sulfide, 1162
p-(Benzylsulfonamido)benzoic Acid,
  1163
5-Benzyl-1,2,3,4-tetrahydro-2-meth-
  yl-γ-carboline see 5649
5-Benzyl-2,3,4,5-tetrahydro-2-meth-
  yl-1H-pyrido[4,3-b]indole see 5649
3-[(Benzylthio)methyl]-6-chloro-2H-
  1,2,4-benzothiadiazine-7-sulfon-
  amide 1,1-Dioxide see 1134
3-[(Benzylthio)methyl]-6-chloro-7-
  sulfamoyl-2H-benzo-1,2,4-thiadi-
  azine 1,1-Dioxide see 1134
S-Benzylthiuronium Chloride, 1164
3-Benzyl-6-trifluoromethyl-3,4-di-
  hydro-7-sulfamoyl-2H-1,2,4-benzo-
  thiadiazine 1,1-Dioxide see 1045
1-Benzyl-2-trimethylacetylhydrazine
  see 7483
Benzylurea, 1165
Benzyl Viologen see 9903
Benzyrin [Yoshitomi] see 1136
Benzytol see 2176
Beocid-Isoptal see 8870
BEP see MISC-2
Bepadin [Carter Wallace] see 1167
Bepanthen [Roche] see 2924
Bephenium, 1166
Bephenium Embonate see 1166
Beprane [Riker/3M] see 7852
Bepridil, 1167
Béprochine see 6953
Béprocin see 1257
Bequin see 9222

Berachin [Tanabe] see 9720
Béradia see 4300
Beraunite see 3975
Berbamine, 1168
Berberine, 1169
Berberine Bisulfate see 1169
Berberis, 1170
Berberis Aristata, 1171
Berberis Bark see 971
Berbine, 1172
Berculon A [ICI] see 9218
Berenil see 3256
Bergacef [Berganon] see 1922
Bergamol see 5374
Bergaptan see 1173
Bergapten(e), 1173
Bergenin, 1174
Bergenit see 1174
Berin see 9222
Berkatens [Berk] see 9851
Berkazon see 9218
Berkfuran see 6520
Berkfurin see 6520
Berkmycen see 6931
Berkolol [Berk] see 7852
Berkomine see 4835
Berkozide [Berk] see 1045
Berlicid see 8885
Berlin Blue see 3964
Bernacaine see 7763
Berninamycin, 1176
Berninamycin A see 1176
Berninamycin B see 1176
Bernstein see 388
Bernsteinsäure (German) see 8840
Bernsteinsäureanhydrid (German) see
  8841
Berocillin [Boehringer, Ing.] see 7484
Berofor Alpha 2 [Boehringer, Ing.] see
  4892
Berolase [Roche] see 2451
Beromycin [Boehringer, Ing.] see 7046
Beromycin 400 [Boehringer, Ing.] see
  7046
Beronald [Kowa] see 4221
Berotec [Boehringer, Ing.] see 3927
Berry Alder see 4176
Berry Wax see 1022
Bersen see 394
Bertholite see 2095
Berubi see 9921
Berubigen [Upjohn] see 9921
Beryl see 1177
Beryllia see 1192
Beryllium, 1177
Beryllium Acetate, 1178
Beryllium Acetate, Basic, 1179
Beryllium Acetylacetonate, 1180
Beryllium Borohydride, 1181
Beryllium Bromide, 1182
Beryllium Carbide, 1183
Beryllium Chloride, 1184
Beryllium Diethyl see 3104
Beryllium Dimethyl see 3227
Beryllium Fluoride, 1185
Beryllium Formate, 1186
Beryllium Hydride, 1187
Beryllium Hydroxide, 1188
Beryllium Iodide, 1189
Beryllium Nitrate, 1190
Beryllium Nitride, 1191
Beryllium Oxide, 1192
Beryllium Oxide Acetate see 1179
Beryllium Perchlorate, 1193
Beryllium Potassium Fluoride, 1194
Beryllium Potassium Sulfate, 1195
Beryllium Selenate, 1196
Beryllium Sodium Fluoride, 1197
Beryllium Sulfate, 1198

Beryllium Tetrahydroborate(1−) see
  1181
Berzelianite see 2673
Berzeline see 2673
Besilate see MISC-4
Besnoline [Kotobuki] see 9447
Bespar [Bristol] see 1493
Bessisterol see 8705
Bestatin [Nippon Kayaku] see 9750
Bestcall [Takeda] see 1928
Beston [Tanabe Seiyaku] see 1257
Bestrabucil, 1199
Besuric [Labaz] see 1073
Beta-alanine see 196
Betabactyl [Beecham-Wulfing] see
  2342
Betabion Hydrochloride [E. Merck]
  see 9222
Betabion Mononitrate [E. Merck] see
  9223
Betabloc [Siphar] see 1522
Betacaine see 3850
Beta-Cardone [Duncan-Flockhart] see
  8682
Betacef [Firma] see 1938
Betacetylmethadol see 5853
Beta-Chlor [Mead-Johnson] see 2059
Betacor [Ayerst] see 2010
Beta-Corlan see 1202
Betacyamine see 1201
Betadexamethasone see 1202
Betadid [Hoechst] see 1445
Betadine [Purdue Frederick] see 7701
Betadival [Fardeco] see 1202
Betadran [Logeais] see 1486
Betadrenol [Pharma Schwarz] see
  1486
Betafedrina see 2932
Betafluorene [Lepetit] see 1202
Betagan [Allergan] see 5343
Betagon [Schering] see 5744
Betahistine, 1200
Beta-hypophamine see 9843
Betaine, 1201
Betaine Hydrazide Hydrochloride see
  4323
Betaine Viologen see 9903
Beta-Intensain [Cassella-Riedel] see
  6442
Beta-interferon see 4893
Betaisodona [Mundipharma] see 7701
Betalin-12 [Lilly] see 9921
Betaling [Tanabe] see 1208
Betalin S [Lilly] see 9222
Betaloc [Astra] see 6072
Betalone [Lepetit] see 5750
Betamethasone, 1202
Betamethasone 21-(Dihydrogen Phos-
  phate) Disodium Salt see 1202
Betamox [Norbrook] see 610
Betanal [Schering AG] see 7199
Betanaphthol see 6304
Betanaphthol Benzoate see 6328
Betanaphthol Orange see 6813
Betanaphthoxyethanol see 6333
Beta-Neg [Ellem] see 7852
Betanidol [Tanabe Seiyaku] see 1208
Betanol [Dulcis] see 6059
Betapace [Bristol-Myers] see 8682
Betapar [Parke, Davis] see 5750
Betapen [Bristol] see 7036
Betapen VK [Bristol] see 7046
d-Betaphedrine see 2932
Betapindol [Helvepharm] see 7412
Betapred [Schering] see 5750
Betapressin [Hoechst] see 7025
Betaprodine see 307
Betaprone [OJ & F] see 7832
Betarin [Beta] see 2446
Betaserc [Duphar] see 1200
Betaseron [Triton Biosci.] see 4893

Betasine, 1203
Betasinum see 1203
Betasolon see 1202
Betasyamine see 1201
Beta-Tablinen [Sanorania] see 7852
Beta-Timelets [Temmler] see 7852
Betatron [Francia] see 7896
Betaxin [Winthrop] see 9222
Betaxina [Beta] see 6273
Betaxolol, 1204
Betazed see 7248
Betazine see 1203
Betazole, 1205
BETE see 9694
Betel, 1206
Betel Nuts see 800
Bethanechol Chloride, 1207
Bethanidine, 1208
Bethiazine see 9222
Bethrodine see 1048
Beth Root see 9609
Betim [Leo Pharm.] see 9374
Betnelan [Glaxo] see 1202
Betnesol Injectable [Glaxo] see 1202
Betnesol Tablets [Glaxo] see 1202
Betnesol-V [Glaxo] see 1202
Betneval [Glaxo] see 1202
Betnovate [Glaxo] see 1202
Betol see 6335
Betolvex [Dumex] see 9921
Betonicine, 1209
Betoptic [Alcon] see 1204
Betoptima [Alcon] see 1204
Betoxycaine, 1210
Betriol [Boehringer, Ing.] see 1479
Betsovet [Bristol] see 1202
Betula, 1211
Betula Oil see 6038
Betulin, 1212
Betulinol see 1212
Betulol see 1212
Betuloside see 8190
Bevantolol, 1213
Bevatine-12 [Dorsey] see 9921
Beveno [Fischer] see 2761
Bevidox [Abbott] see 9921
Bevitam [Merrell] see 9923
Bevitex see 9222
Bevonium Methyl Sulfate, 1214
Bewon see 9222
Bexedan [Smit] see 2078
Bexide see 3390
Bexii [Conal] see 9921
Bexil see 9921
Bexol see 3297
Bextasol [Glaxo] see 1202
Bexton [Dow] see 7805
Bezafibrate, 1215
Bezalip [MCP] see 1215
Bezatol [Kissei] see 1215
Bezitramide, 1216
BFE 60 see 1032
BFGF see 4016
des 1-15 BFGF see 4016
α-Bgt see 1478
BH 6 see 6659
BHA see 1547
BH-AC see 3539
Bhang see 1752
BHB see 1534
γ-BHC see 5379
Bhimsaim Camphor see 1338
B Hormone see 6206
BHT see 1548
Biacetyl see 2946
Bialamicol, 1217
Bialaphos see 7314
Bialatan [Geigy] see 5851
Bialcol [Ciba-Geigy] see 1123
Biallylamicol see 1217
Bialzepam [Bial] see 2977

Biarison [Sandoz] see 7887
Biarsan [Sandoz] see 7887
Biazolina [Panthox & Burck] see 1925
Bibenzal see 8774
Bibenzene 3314
o,o'-Bibenzoic Acid see 3309
Bibenzonium Bromide, 1218
Bibenzoyl see 1087
Bibenzyl, 1219
Bibenzylidene see 8774
Bibiru Bark see 1025
Bibrocathin see 1220
Bibrocathol, 1220
Bicarbamimide see 9780
Bicarnesine [Labaz] see 1856
Bichloracetic Acid see 3037
Bichol [Sana] see 6843
Bichromate of Soda see 8556
Bicillin [Wyeth] see 7037
Bicillin V see 7047
Bicine, 1221
Biciron [Basotherm] see 9486
Bickie-mol see 40
BiCNU [Bristol] see 1852
Bicol see 6884
Bicol [Sigma Tau] see 2721
Bicol [Wampole] see 1253
Bicolon [Wm. R. Warner] see 8486
Bicolorin see 3646
Bicor [KabiVitrum] see 9098
Bicortone [Merck & Co.] see 7727
Bicozamycin, 1222
Bicuculline, 1223
Bicyclo[4.4.0]decane see 2839
Bicyclo-[0.3.5]-deca-1,3,5,7,9-penta-
    ene see 939
Bicyclo-[5.3.0]-deca-2,4,6,8,10-penta-
    ene see 939
Bicyclo[4.1.0]heptane see 6605
(endo,endo)-Bicyclo[2.2.1]hept-5-ene-
    2,3-dicarboxylic Acid Dimethyl
    Ester see 3230
endo-cis-Bicyclo[2.2.1]hept-5-ene-
    2,3-dicarboxylic Anhydride see 1801
3-Bicyclo[2.2.1]hept-5-en-2-yl-6-
    chloro-3,4-dihydro-2H-1,2,4-benzo-
    thiadiazine-7-sulfonamide 1,1-Di-
    oxide see 2760
2-[(Bicyclo[2.2.1]hept-5-en-2-ylhydr-
    oxyphenylacetyl)oxy]-N,N,N-tri-
    methylethanaminium Bromide see
    3527
α-Bicyclo[2.2.1]hept-5-en-2-yl-α-
    phenyl-1-piperidinepropanol see
    1246
1-Bicycloheptenyl-1-phenyl-3-piperi-
    dinopropanol see 1246
[1,1'-Bicyclohexyl]-1-carboxylic Acid
    2-(Diethylamino)ethyl Ester Hydro-
    chloride see 3087
[1,1'-Bicyclohexyl]-1-carboxylic Acid
    2-(1-Piperidinyl)ethyl Ester see
    3151
Bicyclomycin see 1222
5-Bicyclo[3.2.1]oct-2-en-3-yl-5-eth-
    ylbarbituric Acid see 8141
5-Bicyclo[3.2.1]oct-2-en-3-yl-5-eth-
    yl-2,4,6(1H,3H,5H)-pyrimidine-
    trione see 8141
Bidizole see 8918
Bidocef [Ciba-Geigy] see 1921
Bidrin [Shell] see 3077
Bieberite see 2443
Biebrich Scarlet Red see 8349
Bietamiverine, 1224
Bietanautine, 1225
Bietaserpine, 1226
Biethylene see 1500
Bifemelane, 1227
Bifenox, 1228
Bifenthrin, 1229

Bifex [Provet] see 7849
Bifidus Factor, 1230
Bifiteral [Philips Roxane] see 5225
Bifluranol, 1231
Bifonazole, 1232
Biformyl see 4405
Biforon [Meiji] see 1465
Bigitalin see 4328
Biglumide see 9006
Biguanide, 1233
Bigunal [Nikken] see 1465
Bihypnal see 3033
[Δ²,²'-Biindoline]-3,3'-dione see 4855
Bijosal see 3175
Bi-Ketolan see 6884
Bikhaconitine, 1234
Bikhroot see 115
Biklin [Bristol; Grünenthal] see 416
Bilagen see 9415
Bilagol [Janssen] see 3180
Bilamid(e) [Cilag-Chemie] see 4765
Bilarcil [Bayer] see 9536
Bilcolic [Unifa] see 4792
Bilcrine [Bellon] see 5921
Bildux [Perrier] see 9841
Bilein see 6884
Biletan [Gador] see 9255
Bilevon see 4602
Bilevon-M see 5722
Bilibyk [Byk-Gulden] see 4901
Bilicante [Oryx] see 4792
Bilicholan see 6884
Bilidren see 2858
Biligen see 3912
Biligrafin [Schering AG] see 4916
Biligram [Schering AG] see 4939
Bilijodon-Natrium see 4944
Bilimiro [Byk-Gulden; Bracco] see
    4949
Bilimiron [Bracco] see 4949
Bilimix [Gero] see 2721
Bilineurine see 2207
Biliognost see 4919
Biliphorine see 9415
Bilirubin, 1235
Bilirubin IXα see 1235
Biliselectan see 4919
Biliton [Berlin-Chemie] see 2858
Bilitrast see 4931
Biliverdine, 1236
Bilopac see 9745
Bilopaque [Winthrop] see 9745
Biloptin [Schering AG] see 4958
Bilordyl [Fisopharma] see 9212
Bilostat [Pennwalt] see 2858
Biloxazol see 1315
Biltricide [Bayer] see 7714
Bimaran [Roux-Ocefa] see 9495
Bimethadol see 3195
Bimethyl see 3676
2,2'-Bimorphine see 7934
Binapacryl, 1237
β,β-Binaphthyleneethene see 7368
Binazine [Polfa] see 9425
Bindan see 7196
Bindazac see 1042
Binedaline, 1238
Binifibrate, 1239
Biniwas [Wassermann] see 1239
Binodaline see 1238
Binodrenal see 6612
Binotal [Bayer] see 621
Binova [Gentili] see 7896
Binovum [Ortho-Cilag] see 6614
Biobamate see 5751
Biobenzyfuroline [Sumitomo] see 1243
Biocefalin [Benvegna] see 8006
Biochanin A see 4281
Biochanin B see 4157
Biociclin [Del Saz & Filippini] see
    1951

6,7-Bis(cyclopropylmethoxy)-4-hydr-oxy-3-quinolinecarboxylic Acid Ethyl Ester see 2780

Bis-DEAE-fluorenone see 9371

N,N'-Bis(dehydroabietyl)ethylenedi-amine Bis(phenoxymethylpenicillin) see 7048

N,N'-Bis(dehydroabietyl)ethylenedi-amine Dipenicillin G see 7040

1,6-Bisdehydro-6-chloro-17α-acet-oxyprogesterone see 2865

α-Bisdehydrodoisynolic Acid Methyl Ether see 3407

Bisdequalinium Chloride, 1259

α,α-Bis(3,5-dibromo-4-hydroxyphen-yl)-α-hydroxy-o-toluenesulfonic Acid, γ-Sultone see 1434

3,3-Bis(3,5-dibromo-4-hydroxyphen-yl)-1(3H)-isobenzofuranone see 9122

α,α-Bis(3,5-dibromo-4-hydroxy-o-tolyl)-α-hydroxytoluenesulfonic Acid, γ-Sultone see 1375

N,N'-Bis(dichloroacetyl)-N,N'-bis(2-ethoxyethyl)-1,4-bis(aminomethyl)-benzene see 9055

N,N'-Bis(dichloroacetyl)-1,8-octa-methylenediamine see 4009

Bis[3,4-dichloro-2(5)-furanonyl] Ether see 6209

2,6-Bis(diethanolamino)-4,8-dipiperi-dinopyrimido-[5,4-d]pyrimidine see 3354

2,6-Bis(diethanolamino)-4-piperidino-pyrimido[5,4-d]pyrimidine see 6171

Bis[S-(diethoxyphosphinothioyl)mer-capto]methane see 3691

Bis[1-(2-diethylaminoethoxycarbon-yl)-1-phenylcyclopentane] Ethane-disulfonate see 1776

4,4'-Bis(β-diethylaminoethoxy)-α,β-diethyldiphenylethane see 4622

2,7-Bis[2-(diethylamino)ethoxy]-9H-fluoren-9-one see 9371

6,7-Bis[2-(diethylamino)ethoxy]-4-methyl-2H-1-benzopyran-2-one see 6868

6,7-Bis[2-(diethylamino)ethoxy]-4-methylcoumarin see 6868

3,4-Bis[p-(β-diethylaminoethoxy)-phenyl]hexane see 4622

N,N'-Bis[2-diethylaminoethyl]ox-amide Bis[2-chlorobenzyl Chloride] see 387

2,4-Bis(diethylamino)-6-hydrazino-s-triazine see 4685

3,3'-Bis[(diethylamino)methyl]-5,5'-di-2-propenyl-[1,1'-biphenyl]-4,4'-diol see 1217

1,3-Bis(diethylamino)-2-(α-phenyl-α-cyclohexylmethyl)propane see 3892

2,5-Bis(3-diethylaminopropylamino)-benzoquinone Bis(benzyl Chloride) see 1116

4,6-Bis(diethylamino)-1,3,5-triazin-2(1H)-one Hydrazone see 4685

3,6-Bis(diethylamino)xanthylium Chloride see 8016

Bis-O,O-diethylphosphoric Anhydride see 9138

Bis(diethylthiocarbamoyl) Disulfide see 3370

Bis(diethylthiocarbamoyl)sulfide see 8928

Bis(diethylthiocarbamyl) Disulfide see 3370

N,N'-Bis[3-(4,5-dihydro-1H-imida-zol-2-yl)phenyl]urea see 4831

1,8-Bis(3,4-dihydro-3,5,7-trihydroxy-2H-1-benzopyran-2-yl)-3,4,6-tri-hydroxy-5H-benzocyclohepten-5-one see 9201

Bis[4,5-dihydroxy-1,3-benzenedisul-fonato(4—)-O⁴,O⁵]-antimonate(5—) Pentasodium Heptahydrate see 8769

2,3-Bis(3,4-dihydroxybenzyl)butane see 6610

Bis[μ-[2,3-dihydroxybutanedioato-(4—)-01,02:03,04]]-diantimonate dipotassium trihydrate see 732

2,2-Bisdihydroxymethyl-1,3-propane-diol Tetranitrate see 7066

N,N'-Bis[2-(3,4-dihydroxyphenyl)-2-hydroxyethyl]hexamethylenediamine see 4628

(1α,3α,4α,5β)-1,3-Bis[[3-(3,4-dihydr-oxyphenyl)-1-oxo-2-propenyl]oxy]-4,5-dihydroxycyclohexanecarboxylic Acid see 2773

N,N'-Bis(2,3-dihydroxypropyl)-5-[N-(2,3-dihydroxypropyl)acetamido]-2,4,6-triiodoisophthalamide see 4940

N,N'-Bis(2,3-dihydroxypropyl)-5-[N-(2-hydroxy-3-methoxypropyl)acet-amido]-2,4,6-triiodoisophthalamide see 4945

N,N'-Bis(2,3-dihydroxypropyl)-2,4,6-triiodo-5-(2-methoxyacetamido)-N-methylisophthalamide see 4948

N,N'-Bis(2,3-dihydroxypropyl)-2,4,6-triiodo-5-[(methoxyacetyl)amino]-N-methyl-1,3-benzenedicarbox-amide see 4948

Bis(dimethylamido)fluorophosphate see 3191

Bis(dimethylamido)phosphoryl Fluo-ride see 3191

N,N'-Bis(p-dimethylaminobenzal)-benzidine see 1260

4,4'-Bis(dimethylamino)benzophenone see 6100

Bis(p-dimethylaminobenzylidene)-benzidine, 1260

Bis(4-dimethylaminobenzylidene)-p,p'-diaminodiphenyl see 1260

Bis[6-(dimethylamino)-2-[2-(2,5-di-methyl-1-phenylpyrrol-3-yl)vinyl]-1-methylquinolinium] 4,4'-Methyl-enebis(3-hydroxy-2-naphthoate) see 8034

Bis[2-dimethylaminoethyl]succinate Bis[ethobromide] see 8989

Bis[2-dimethylaminoethyl]succinate Bis[methobromide] see 8845

Bis[2-dimethylaminoethyl]succinate Bis[methochloride] see 8846

Bis(β-dimethylaminoethyl)succinate Bis(methyl Iodide) see 8847

Bisdimethylaminofluorophosphine Oxide see 3191

4,7-Bis(dimethylamino)-1,4,4a,5,5a,6,-11,12a-octahydro-3,10,12,12a-tetra-hydroxy-1,11-dioxo-2-naphthacene-carboxamide see 6121

3,7-Bis(dimethylamino)phenazathi-onium Chloride see 5979

3,7-Bis(dimethylamino)phenothiazin-5-ium Chloride see 5979

Bis(p-dimethylaminophenyl)methane see 6099

Bis[4-(dimethylamino)phenyl]methan-one see 6100

N,N'-Bis[[4-(dimethylamino)phenyl]-methylene][1,1'-biphenyl]-4,4'-di-amine see 1260

N-[4-[Bis[4-(dimethylamino)phenyl]-methylene]-2,5-cyclohexadien-1-ylidene]-N-methylmethanaminium Chloride see 4287

Bis[p-(dimethylamino)phenyl]phenyl-methylium Chloride see 5581

3β,20α-Bis(dimethylamino)-5-preg-nene see 5203

10-[2,3-Bis(dimethylamino)propyl]-phenothiazine see 479

3,6-Bis(dimethylamino)xanthylium Chloride see 8017

[Bis(2,2-dimethyl-1-aziridinyl)phos-phinyl]carbamic Acid Ethyl Ester see 6082

Bis(dimethylcarbamodithioato-S,S')-zinc see 10075

1-[Bis(3',5'-N,N-dimethylcarbamoyl-oxy)phenyl]-2-N-tert-butylamino-ethanol see 963

Bis(dimethyldithiocarbamato)zinc see 10075

4-[3,5-Bis(1,1-dimethylethyl)-4-hydr-oxyphenoxy]-3,5-diiodobenzenepro-panoic Acid see 4635

2,4-Bis(1,1-dimethylethyl)-5-methyl-phenol see 2829

2,6-Bis(1,1-dimethylethyl)-4-methyl-phenol see 1548

3,6-Bis(1,1-dimethylethyl)-1-naph-thalenesulfonic Acid Ethyl Ester see 3745

3,6-Bis(1,1-dimethylethyl)-1-naph-thalenesulfonic Acid Sodium Salt Mixt. with 3,7-Bis(1,1-dimethyl-ethyl)-1-naphthalenesulfonic Acid Sodium Salt see 3017

Bis(1,1-Dimethylethyl) Peroxide see 3446

2,6-Bis(1,1-dimethylethyl)pyridine see 3022

Bis(dimethylglyoximato)nickel see 6414

2,5-Bis(1,1-dimethylpropyl)-1,4-benz-enediol see 3300

Bis(1,2-dimethylpropyl)borane, 1261

2,5-Bis(1,1-dimethylpropyl)hydro-quinone see 3300

Bis(dimethylthiocarbamoyl) Disulfide see 9304

Bis(dimethylthiocarbamyl) Disulfide see 9304

(±)-1,2-Bis(3,5-dioxopiperazinyl)-propane see 8133

1,2-Bis[2-(2,3-epoxypropoxy)ethoxy]-ethane see 3837

O,S-Bis(ethoxycarbonyl)thiamine see 2014

1,2-Bis(3-ethoxycarbonyl-2-thioure-ido)benzene see 9282

N,N'-Bis(ethoxyethyl)-N,N'-bis(di-chloroacetyl)-1,4-xylylenediamine see 9055

N¹,N²-Bis(p-ethoxyphenyl)acetamidine Hydrochloride see 7153

N,N'-Bis(4-ethoxyphenyl)ethanimid-amide Monohydrochloride see 7153

2,4-Bis(ethylamino)-6-chloro-s-tri-azine see 8485

2,4-Bis(ethylamino)-6-(methylthio)-s-triazine see 8488

2,4-Bis(ethylamino)-6-(methylthio)-1,3,5-triazine see 8488

Bis(ethylenimido)phosphorylurethane see 9787

Bis[2-ethylhexyl]calcium Sulfosuc-cinate see 3397

N,N'''-Bis(2-ethylhexyl)-3,12-di-imino-2,4,11,13-tetraazatetradecan-ediimidamide see 222

2,6-Bis(2-ethylhexyl)hexahydro-7a-methyl-1H-imidazo[1,5-c]imidazole see 4619

1,3-Bis(2-ethylhexyl)hexahydro-5-methyl-5-pyrimidinamine see 4624

1,3-Bis(β-ethylhexyl)-5-methyl-5-
aminohexahydropyrimidine see 4624
3,7-Bis(2-ethylhexyl)-5-methyl-1,3,7-
triazabicyclo[3.3.0]octane see 4619
Bis(2-ethylhexyl) Phthalate, 1262
Bis(2-ethylhexyl) Sebacate, 1263
Bis(2-ethylhexyl)sodium Sulfosuccinate see 3398
Bis[ethyl(2-hydroxyethyl)dimethyl-
ammonium] Sulfate Bis(dibutylcarbamate) see 3018
2,2-Bis(p-ethylphenyl)-1,1-dichloro-
ethane see 3050
Bis[1-ethylquinoline-(2)]trimethinecy-
anine Chloride see 8056
2,2-Bis(ethylsulfonyl)butane see 8937
2,2-Bis(ethylsulfonyl)propane see
8939
N-[4-[Bis[4-[ethyl(3-sulfophenyl)-
amino]phenyl]methylene]-2,5-cyclo-
hexadien-1-ylidene]-N-methylmeth-
anaminium Hydroxide Inner Salt
Monosodium Salt see 104
Bisethylxanthogen see 3390
Bisexovis see 670
1-[4,4-Bis(4-fluorophenyl)butyl]-4-[4-
chloro-3-(trifluoromethyl)phenyl]-
4-piperidinol see 7028
1-[4,4-Bis(p-fluorophenyl)butyl]-4-(4-
chloro-α,α,α-trifluoro-m-tolyl)-4-
piperidinol see 7028
4-[4,4-Bis(4-fluorophenyl)butyl]-N-
(2,6-dimethylphenyl)-1-piperazine-
acetamide see 5360
4-[4,4-Bis(4-fluorophenyl)butyl]-N-
ethyl-1-piperazinecarboxamide see
612
1-(4,4-Bis(4-fluorophenyl)butyl]-4-
hydroxy-4-(3-trifluoromethyl-4-
chlorophenyl)piperidine see 7028
8-[4,4-Bis(4-fluorophenyl)butyl]-1-
phenyl-1,3,8-triazaspiro[4.5]decan-
4-one see 4131
4-[4,4-Bis(p-fluorophenyl)butyl]-1-
piperazineaceto-2′,6′-xylidide see
5360
1-[1-[4,4-Bis(4-fluorophenyl)butyl]-4-
piperidinyl]-1,3-dihydro-2H-benz-
imidazol-2-one see 7404
1-[1-[4,4-Bis(p-fluorophenyl)butyl]-4-
piperidyl]-2-benzimidazolinone see
7404
(E)-1-[Bis(p-fluorophenyl)-methyl]-4-
cinnamylpiperazine see 4070
(E)-1-[Bis(4-fluorophenyl)methyl]-4-
(3-phenyl-2-propenyl)piperazine see
4070
6-[4-[Bis(4-fluorophenyl)methyl]-1-
piperazinyl]-N,N′-di-2-propenyl-
1,3,5-triazine-2,4-diamine see 299
1-[[Bis(4-fluorophenyl)methylsilyl]-
methyl]-1H-1,2,4-triazole see 4130
Bis(4-fluorophenyl)methyl(1H-1,2,4-
triazol-1-ylmethyl)silane see 4130
1,4-Bis(1-formamido-2,2,2-trichloro-
ethyl)piperazine see 9602
2,2′-Bis[8-formyl-1,6,7-trihydroxy-
5-isopropyl-3-methylnaphthalene]
see 4435
3,5-Bis(β-D-glucopyranosyloxy)-7-
hydroxy-2-(4-hydroxy-3-methoxy-
phenyl)-1-benzopyrylium Chloride
see 7099
3,5-Bis(β-D-glucopyranosyloxy)-7-
hydroxy-2-(4-hydroxyphenyl)-1-
benzopyrylium Chloride see 7014
5,5′-Bis(β-D-glucopyranosyloxy)-
9,9′,10,10′-tetrahydro-4,4′-dihydr-
oxy-10,10′-dioxo[9,9′-bianthra-
cene]-2,2′-dicarboxylic Acid see
8407

3,5-Bis(glucosyloxy)-4′,7-dihydroxy-
3′,5′-dimethoxyflavylium Chloride
see 5596
3,5-Bis(glucosyloxy)-3′,4′,5′,7-tetra-
hydroxyflavylium Chloride see 2866
Bis(glycinato)copper see 2644
Bisguadine [Sterwin] see 222
1,3-Bis[4-guanylphenyl]triazene Diaceturate see 3256
Bish see 115
Bishma see 115
trans-(±)-2,3-Bis(3′-hydroxybenzyl)-
γ-butyrolactone see 3549
Bis(p-hydroxybenzylidene)ethylene
Isocyanide see 9970
Bis[3-hydroxybutyl] Ether see 6913
Bis[2-hydroxy-5-chlorophenyl]sulfide
see 3947
Bishydroxycoumarin (rescinded) see
3080
2,2-Bis[4′-hydroxy-3′-coumarinyl]-
ethyl Methyl Ether see 2566
Bis-3,3′-(4-hydroxycumarinyl)acetic
Acid Ethyl Ester see 3729
Bis(2-hydroxy-3,5-dichlorophenyl)-
sulfide see 1316
2-[3,4-Bis(2-hydroxyethoxy)phenyl]-
3-[[6-O-(6-deoxy-α-L-mannopyran-
osyl)-β-D-glucopyranosyl]oxy]-5-
hydroxy-7-(2-hydroxyethoxy)-4H-
1-benzopyran-4-one see 9698
Bis(hydroxyethyl)amine see 3097
N,N-Bis(hydroxyethyl)aminoacetic
Acid see 1221
2-[Bis(β-hydroxyethyl)amino]-4,5-
diphenyloxazole see 3374
Bis(hydroxyethyl)ammonium 3,5-
Diiodo-4-pyridone-N-acetate see
4933
N,N-Bis(2-hydroxyethyl)glycine see
1221
Bis[hydroxyethylpoly(ethyleneoxy)-
ethyl]polypropyleneglycol see 7536
Bis(hydroxyethyl)sulfide see 9259
[7,12-Bis(1-hydroxyethyl)-3,8,13,17-
tetramethyl-21H,23H-porphine-
2,18-dipropanoato(4−)-N²¹,N²²,-
N²³,N²⁴]mercurate(2−) Disodium see
5804
7,12-Bis(1-hydroxyethyl)-3,8,13,17-
tetramethyl-21H,23H-porphine-
2,18-dipropanoic Acid see 4555
7,12-Bis(1-hydroxyethyl)-3,8,13,17-
tetramethyl-2,18-porphinedipropi-
onic Acid see 4555
1,10-Bis(2-hydroxyethylthio)decane
see 9349
(S)-N,N′-Bis[2-hydroxy-1-(hydroxymethyl)ethyl]-5-[(2-hydroxy-1-oxo-
propyl)amino]-2,4,6-triiodo-1,3-
benzenedicarboxamide see 4943
(S)-N,N′-Bis[2-hydroxy-1-(hydroxymethyl)ethyl]-2,4,6-triiodo-5-lact-
amidoisophthalamide see 4943
Bis[(3-hydroxy-4-hydroxymethyl-2-
methyl-5-pyridyl)methyl] Disulfide
see 8006
Bis(4-hydroxyiminomethylpyridinium-
1-methyl) Ether Dichloride see 6659
2,6-Bis(4-hydroxy-3-methoxybenzyl-
idene)cyclohexanone see 2761
1,7-Bis(4-hydroxy-3-methoxyphenyl)-
1,6-heptadiene-3,5-dione see 2681
Bis(2-hydroxy-4-methoxyphenyl)-
methanone see 1109
1,3-Bis(hydroxymethyl)-2-benzimida-
zolinethione see 9239
2,2-Bis(hydroxymethyl)-1-butanol
Trinitrate see 7821

Bis(4-hydroxymethyl-5-hydroxy-6-
methyl-3-pyridylmethyl) Disulfide
see 8006
Bis[p-(4-hydroxy-2-methyl-5-iso-
propylphenylazo)phenyl] Sulfone see
2972
3,3-Bis[4-hydroxy-2-methyl-5-(1-
methylethyl)phenyl]-1(3H)-isoben-
zofuranone see 9336
3,3-Bis(4-hydroxy-3-methylphenyl)-
1(3H)-isobenzofuranone see 2582
2,2-Bis(hydroxymethyl)-1,3-propane-
diol see 7062
2,2-Bis(hydroxymethyl)propionic Acid
see 3241
d-N,N′-Bis(1-hydroxymethylpropyl)-
ethylenediamine see 3672
2,5-Bis(hydroxymethyl)tetrahydro-
furan see 9145
N,N′-Bis(hydroxymethyl)urea see
6921
3,3-Bis(4-hydroxynaphthalenyl)-
1(3H)-isobenzofuranone see 6309
Bis(6-hydroxy-2-naphthyl) Disulfide
see 2831
3,3-Bis(4-hydroxy-1-naphthyl)phthal-
ide see 6309
Bis(2-hydroxy-5-nitrophenyl)ethane-
dione see 3279
Bis(4-hydroxy-2-oxo-2H-1-benzo-
pyran-3-yl)acetic Acid Ethyl Ester
see 3729
3,3-Bis(p-hydroxyphenyl)-3H-2,1-
benzoxathiole 1,1-Dioxide see 7213
2,2-Bis(4-hydroxyphenyl)butane see
1312
1,4-Bis(p-hydroxyphenyl)-2,3-diisoni-
trilo-1,3-butadiene see 9970
3,4-Bis(p-hydroxyphenyl)-2,4-hexa-
diene see 3094
meso-3,4-Bis(p-hydroxyphenyl)-n-
hexane see 4621
3,4-Bis(p-hydroxyphenyl)-3-hexene
see 3118
Bis(4-hydroxyphenyl)-(2-hydroxy-
methylphenyl)methane see 7211
3,3-Bis(4-hydroxyphenyl)-1(3H)-iso-
benzofuranone see 7208
3,3-Bis(4-hydroxyphenyl)-1(3H)-iso-
benzofuranone Disodium Salt see
7209
2-[Bis(4-hydroxyphenyl)methyl]benz-
enemethanol see 7211
2-[Bis(4-hydroxyphenyl)methyl]benz-
oic Acid see 7210
o-[Bis(p-hydroxyphenyl)methyl]benzyl
Alcohol see 7211
4-[Bis(4-hydroxyphenyl)methylene]-
2,5-cyclohexadien-1-one see 899
3,3-Bis(p-hydroxyphenyl)-7-methyl-
2-indolinone Bis(hydrogen sulfate)
(Ester) see 8962
4,4-Bis[4′-hydroxyphenyl]pentanoic
Acid see 3312
3,3-Bis(p-hydroxyphenyl)phthalide
see 7208
2,2-Bis(4-hydroxyphenyl)propane see
1311
3,3-Bis(4-hydroxyphenyl)-4,5,6,7-
tetraiodo-1(3H)-isobenzofuranone
Disodium Salt see 7233
γ,γ-Bis-(p-hydroxyphenyl)valeric
Acid see 3312
1,4-Bis(3-hydroxypropionyl)pipera-
zine Dimethanesulfonate see 7452
N,N′-(Bis-ω-hydroxypropyl)homo-
piperazine 3,4,5-Trimethoxybenzo-
ate (Diester) see 3185
Bis-(1-hydroxy-2(1H)-pyridinethio-
nato-O,S)zinc see 8004
Bisibutiamine see 9221

D-Bis(*N*-pantothenyl-β-aminoethyl) Disulfide *see* 6962
Bispecia *see* 1272
Bis(pentachloro-2,4-cylopentadien-1-yl) *see* 3095
Bis(2,4-pentanedionato)beryllium *see* 1180
Bis(2,4-pentanedionato-*O,O* ′)nickel *see* 6408
Bis(2,4-pentanediono)nickel(II) *see* 6408
p-Bis(perchloromethyl)benzene *see* 1313
Bisphenabid [Dow] *see* 7761
Bisphenol A, 1311
Bisphenol B, 1312
Bis(phenoxarsin-10-yl)ether *see* 6914
Bis(10-phenoxarsinyl)oxide *see* 6914
Bis(10-phenoxarsyl)oxide *see* 6914
Bis(phenylmercuri)methylenedinaphthalenesulfonate *see* 4687
1,1′-Bis(phenylmethyl)-4,4′-bipyridinium *see* 9903
Bis(phenylmethyl) Disulfide *see* 2995
N,N ′-Bis(phenylmethyl)-1,2-ethanediamine *see* 1072
Bis(γ-phenylpropyl)ethylamine *see* 377
4,4′-Bis(3-phenylureido)-2,2′-stilbenedisulfonic Acid Disodium Salt *see* 1322
N,*N*-Bis(phosphonomethyl)glycine *see* 4409
N,N ′-Bis(3-picolyl)-4-methoxyisophthalamide *see* 7378
Bis(2-pyridylthio)zinc 1,1′-Dioxide *see* 8004
2,6-Bis(*p*-1-pyrrolidinylstyryl)-1-ethylpyridinium Iodide *see* 8773
Bis(6-quinolyl)urea *see* 3360
[[4-[Bis[4-[(sulfophenyl)amino]phenyl]methylene]-2,5-cyclohexadien-1-ylidene]amino]benzenesulfonic Acid Disodium Salt *see* 5950
Bissy Nuts *see* 5198
Bisteril *see* 7175
Bis-*N,N,N* ′,*N* ′-tetramethylphosphorodiamidic Anhydride *see* 8351
Biston [Spofa] *see* 1783
Bistovol *see* 44
1,1′-[2,2-Bis[(2,2,2-trichloro-1-hydroxyethoxy)methyl]-1,3-propanediyl-bis(oxy)]bis[2,2,2-trichloroethanol] *see* 7063
Bis(3,5,6-trichloro-2-hydroxyphenyl)-methane *see* 4602
1,4-Bis(trichloromethyl)benzene, 1313
Bis(2,4,5-trichlorophenol) Piperazine *see* 9570
Bis(trifluoroethyl) Ether *see* 4126
Bis(2,2,2-trifluoroethyl) Ether *see* 4126
1,5-Bis(α,α,α-trifluoro-*p*-tolyl)-1,4-pentadiene-3-one (1,4,5,6-Tetrahydro-5,5-dimethyl-2-pyrimidinyl)-hydrazone *see* 4684
2,2′-Bis[1,6,7-trihydroxy-3-methyl-5-isopropyl-8-aldehydonaphthalene] *see* 4435
Bistrimate [SM & P] *see* 1293
N,N ′-Bis[3-(3,4,5-trimethoxybenzoyl-oxy)propyl]homopiperazine *see* 3185
1,4-Bis[3-(3,4,5-trimethoxybenzoyl-oxy)propyl]perhydro-1,4-diazepine *see* 3185
1,3-Bis(trimethylamino)-2-propanol Diiodide *see* 7792
α,ω-Bis(trimethylammonium)hexane *see* 4609
α,ω-Bis(trimethylammonium)pentane Dibromide *see* 7070

N,N ′-Bis[3-trimethylammoniumphenoxycarbonyl]-*N,N* ′-dimethyldecamethylenediamine Dibromide *see* 2871
Bistrimin *see* 7287
Bis(triphenylphosphine)dicarbonylnickel, 1314
Bistrium Bromide [Squibb] *see* 4609
Bistrium Chloride [Squibb] *see* 4609
Bistyryl *see* 3320
Bisvanil [Boehringer, Ing.] *see* 7463
Bitertanol, 1315
Bitevan *see* 9921
Bithin *see* 1316
Bithiodine *see* 9389
Bithionol, 1316
Bithionolate Sodium *see* 1316
Bithymol Diiodide *see* 9335
Bitin-S [Tanabe Seiyaku] *see* 1316
Bitolterol, 1317
Bitolterol Mesylate *see* 1317
Bitoscanate, 1318
Bitrex [Macfarlan Smith] *see* 2877
Bitriben *see* 1313
Bitrop [Boehringer, Ing.] *see* 4960
α-Bitter Acid *see* 4673
β-Bitter Acid *see* 5482
Bitter-almond-oil Camphor *see* 1103
Bitter Apple *see* 2482
Bitter Ash *see* 3856 and 8036
Bitter-bloom *see* 1965
Bitter Cucumber *see* 2482
Bitter Gourd *see* 2482
Bitter Herb *see* 1967
Bitter Root *see* 773 and 4285
Bitter Salts *see* 5573
Bitter Stick *see* 2051
Bittersweet *see* 3447
Bitter Wintergreen *see* 2046
Bitter Wood *see* 8036
Bitulan [Adroka] *see* 4815
Bitumen *see* 867
Bituminol *see* 4815
Bitumol *see* 4815
Bituvitan [Hishiyama] *see* 8201
Biuno *see* 9222
Biuret, 1319
Biuretamidine Sulfate *see* 3083
Bivatin *see* 9222
Biverm *see* 7220
Bivinyl *see* 1500
Bivita *see* 9222
Bivitasi *see* 2451
Bixin, 1320
Bizolin [Philips Roxane] *see* 7248
Bizolin 200 [Philips Roxane] *see* 7248
BL-5 *see* 2723
BL-139 *see* 471
BL-191 *see* 7092
BL-4162A *see* 659
BL-5583 *see* 9274
Black Ash *see* 1008
Black Balsam *see* 959
Blackberry Bark *see* 8266
Black Cohosh *see* 2281
Black Copper Oxide *see* 2650
Black Dogwood *see* 4176
Black Hawk *see* 9877
Black Henbane *see* 4796
Black Indian Hemp *see* 774
Black Iron Oxide *see* 3988
Black Lead *see* 4444
Black Leaf 40 *see* 6434
Black Manganese Oxide *see* 5612
Black Nickel Oxide *see* 6423
Black Pepper *see* 7101
Black Phosphorus *see* 7321
Black Root *see* 5325
Black Sea Sterol *see* 3283
Black Snake Root *see* 2281
Blackstrap Molasses, 1321

Black-tang *see* 4196
Black Uranium Oxide *see* 9767
Black and White Bleaching Cream [Plough] *see* 4738
Bladafum [Bayer] *see* 8945
Bladan [Bayer] *see* 9138
Bladder Fucus *see* 4196
Bladderon [Nippon Shinyaku] *see* 4033
Bladder Pod *see* 5431
Bladder-wrack *see* 4196
Bladex [Shell] *see* 2692
Blanc De Perle *see* 1274
Blanc d'Espagne *see* 1274
Blanc Fixe *see* 1006
Blancol C *see* 1322
Blancophor® R, 1322
Blankophor R *see* 1322
Blascorid [Guidotti] *see* 1058
Blasticidin S, 1323
Blasting Gelatin *see* 6528
Blasting Oil *see* 6528
Blastmycin *see* 747
Blattanex *see* 7849
Blätteralkohol *see* 4620
Blaud's Mass *see* 3990
Blausäure (German) *see* 4722
Blazer [Rohm & Haas] *see* 105
Blazing Star *see* 220 and 4550
Bleached Beeswax *see* 1031
Bleached Yellow Wax *see* 1031
Bleaching Powder *see* 2093
Bled [Poli] *see* 2269
Bleiweiss (German) *see* 1016
Bleminol [Desitin] *see* 278
Blenoxane [Bristol] *see* 1324
Bleo [Nippon Kayaku] *see* 1324
Bleomycin A₂ *see* 1324
Bleomycins, 1324
Bleph-10 [Allergan] *see* 8870
Blex [ICI] *see* 7469
Blistering Beetle *see* 1754
Blistering Collodion *see* 2481
Blistering Fly *see* 1754
Bloat Guard [SK & F] *see* 7536
Bloc [Elanco] *see* 3903
Blocadren [Merck & Co.] *see* 9374
Blockade [Sandoz] *see* 7773
Blockaine Hydrochloride *see* 7850
Blocklin L [Shionogi] *see* 7412
Blood-coagulation Factor II *see* 7903
Blood-coagulation Factor III *see* 9321
Blood-coagulation Factor V *see* 3871
Blood-coagulation Factor VII *see* 3872
Blood-coagulation Factor VIII *see* 3873
Blood-coagulation Factor IX *see* 3874
Blood-coagulation Factor X *see* 3875
Blood-coagulation Factor XI *see* 3876
Blood-coagulation Factor XII *see* 3877
Blood-coagulation Factor XIII *see* 3878
Bloodroot *see* 8319
Blood Sugar *see* 4353
Bloodwort *see* 1967
Blotic [Sandoz] *see* 7827
Blow Gas *see* 7779
Blox [Biomed Foscama] *see* 5450
Bloxanth [Calmic] *see* 278
BL-P 1322 *see* 1980
BL-S 578 *see* 1921
BL-S640 *see* 1923
BL-S786 *see* 1934
Blue Asbestos *see* 851
Blue Chamomile Oil *see* 6710
Blue Cohosh *see* 1915
Blue Gas *see* 9952
Blue-gum Tree *see* 3852
Blue Mass *see* 5802

Blue Pill see 5802
Bluestone see 2659
Blue Tetrazolium see 9173
Blue Verdigris see 2628
Blue Vitriol see 2659
Blusalt see 8544
Blutene Chloride [Abbott] see 9444
Bluton [Morishita] see 4812
BM 02015 see 9473
BM 13505 see 2807
BM 14190 see 1882
BM 15075 see 1215
BM 51052 see 1779
BMC see 1794
BMP see MISC-2
BMY 13805 see 4297
BMY 13805-1 see 4297
BMY 26538-01 see 659
B-mycin see 1355
BN 1270 see 2268
B-Nine [Uniroyal] see 2810
BO 714 see 2322
Bobierrite see 5562
Boc-β-Ala-Try-Met-Asp-Phe(NH₂)
see 7067
B.O.E.A. see 3729
Boehmite see 359
Boettger's Paper see 242
Boforsin (obsolete) see 2473
Bog Bean see 5729
Bogomolets' Serum see 750
Bogoserum see 750
Bohrium see 3504
BOL-148 see 1411
Bolandiol, 1325
Bolasterone, 1326
BOLD see MISC-2
Boldea see 1329
Boldenone, 1327
Boldenone Undecylenate see 1327
Boldine, 1328
Boldine Dimethyl Ether see 4332
Boldo, 1329
Boldoa see 1329
Boldu see 1329
Boldus see 1329
Bole, Armenian, 1330
Bolecic Acid see 4983
Boleko Oil, 1331
Boletic Acid see 4200
Bolfortan see 9113
Bolstar [Bayer] see 8972
Bolus Alba see 5162
Bolus Armena see 1330
Bolus Rubra see 1330
Bolvidon [Organon] see 6097
Bombesin, 1332
N-(4-Bomophenyl)-2,6-dihydroxy-
benzamide see 8157
Bomyl®, 1333
Bonabol [Sawai] see 5680
Bonacid [Lilly] see 6441
Bonadorm see 3033
Bonaid [Norwich] see 1489
Bonamid [Heilmittelwerke] see 917
Bonamine [Roerig] see 5657
Bonapar see 7292
Bonapicillin [Taiyo] see 621
Bonare see 6881
Bonasanit [Weimer] see 7995
Bonatranquan [Recip] see 5456
Bondou see 8341
"Bone Ash" see 1701
Bonefos [Oy Star] see 2369
Boneset see 3859
Bongast see 1297
Bongkrekic Acid, 1334
Bonicor see 7901
Bonifen see 8006
Bonine [Pfizer] see 5657
Bonlam [Merck & Co.] see 1733

Bonoform see 9125
Bonomycin [Pfizer] see 8316
Bonopen [Belupo] see 5830
Bonpac [Hoei] see 4146
Bonpyrin see 3358
Bontril [Carnrick] see 2932
Bonyl [Erco] see 6337
Bonzi [ICI] see 6937
Bonzol [Tokyo Tanabe] see 2811
Bopindolol, 1335
Boracic Acid see 1336
Borated Cream of Tartar see 7595
Borax see 8535
Borax Glass see 8535
Borazane see 9516
Borazine see 9516
Borazole see 9516
Borazon see 1348
Bor-Cefazol [Proter] see 1925
Bordeaux Mixture see 2660
Boric Acid, 1336
Boric Acid Trimethyl Ester see 9626
Boric Anhydride, 1337
Boric Oxide see 1337
Bor-Ind [Borromeo] see 4875
endo-2-Bornanol see 1338
exo-2-Bornanol see 5011
2-Bornanone see 1738
Borneo Camphor see 1338
Borneol, 1338
Borneol Acetate see 1339
Borneol α-Bromoisovalerate see 1340
Borneol Isovalerate see 1342
Borneol Salicylate see 1343
Bornite see 2029
Bornyl Acetate, 1339
Bornyl Alcohol see 1338
d-Bornyl α-Bromoisovalerate, 1340
d-Bornyl 2-Bromo-3-methylbutyrate
see 1340
Bornyl Chloride, 1341
d-Bornyl Isovalerate, 1342
Bornyl Salicylate, 1343
Bornyval see 1342
Borobutane see 9119
Borocaine see 7763
Boroethane see 2997
Borofax [Burroughs Wellcome] see
1336
Borofluoric Acid see 4075
Borohexane see 4598
Boromycin, 1344
Boron, 1345
Boron Carbide, 1346
Boron Chloride see 2998
Boron Fluoride Etherate see 1352
Boron Fluoride Ethyl Ether see 1352
Boron Monoxide, 1347
Boron Nitride, 1348
Boron Oxide see 1337
Boron Sesquioxide see 1337
Boron Tribromide, 1349
Boron Trichloride, 1350
Boron Trifluoride, 1351
Boron Trifluoride Etherate, 1352
Boron Trioxide see 1337
Bostrycoidin, 1353
β-Boswellic Acid, 1354
Bothrops Atrox Serine Proteinase see
1020
Bothrops Venom Proteinase see 1020
Botogenin see 4293
Botrilex see 8108
Botropase [Ravizza] see 1020
Bottromycin, 1355
Bottromycin A₂ Acid Methyl Ester see
1355
Bouncing Bet see 8326
Bourbonal see 3815
Bovanide [Merck & Co.] see 8121
Bovatec see 5243

Bovicam [Merck & Co.] see 1733
Boviclox [Squibb] see 2414
Bovilene [Syntex] see 3940
Bovizole [Merck & Co.] see 9217
Bo-Xan see 9972
BP 662 see 4024
BPAA see 3893
BPAS see 1127
B.P.C. 1954 see 3850
BPM see 1250
BPPS see 7818
BQ 22-708 see 3532
BQC Reagent see 3010
BR see 951
BR 700 see 3946
BRA see 8160
Brace [Ciba-Geigy] see 4988
Bradilan [Mundipharma] see 6428
Bradophen [Ciba-Geigy] see 1123
Bradosol Bromide [Ciba-Geigy] see
3411
Bradykinin, 1356
Bradyl [Lafon] see 6261
Brainine see 2834
Brain Sugar see 4241
Brallobarbital, 1357
Branching Factor see 8035
Brandy, 1358
Brandy Mint see 7102
Branigen [Glaxo] see 78
Brantur [Dong-A] see 6119
Brasilin see 1362
Brassel [Schiapparelli] see 2321
Brassicol see 8108
Brassidic Acid, 1359
Brassinolide, 1360
Bratenol [Elmu] see 7467
Braunite see 5604
Braunol [Braun Melsungen] see 7701
Braunosan H [Braun Melsungen] see
7701
Bravo [Diamond] see 2167
Brayera, 1361
Brazilin, 1362
Brazil Powder see 795
Brazil Wax see 1853
Brazil Wood see 7133
Bredinin [Toyo Jozo] see 6137
Bredon [Organon] see 6898
Brefeldin A, 1363
Brek [Irbi] see 5450
Brelomax [Abbott; Hollister] see 9720
Bremax [Abbott] see 9720
Bremen Blue see 2634
Bremen Green see 2634
Bremfol see 6471
Bremil see 4704
Brenal [Tanabe Seiyaku] see 5660
Brendil [Daiichi] see 2293
Brentan [Janssen] see 6101
Brenzschleimsäure (German) see 4219
Brenztraubensäure (German) see 8032
Breokinase [Breon] see 9799
Breon see 7563
Brestan see 9659
Brethine [Ciba-Geigy] see 9089
Bretol [Fine Organics] see 2021
Bretylan [Burroughs Wellcome] see
1364
Bretylate [Burroughs Wellcome] see
1364
Bretylium Tosylate, 1364
Bretylol [Arnar-Stone] see 1364
Brevetoxin A see 1365
Brevetoxin B see 1365
Brevetoxin C see 1365
Brevetoxins, 1365
Brevibloc [Am. Crit. Care] see 3648
Brevicid [Arzneimittelwerk VEB] see
9970
Brevicidin [Penick] see 9744

Brevicon [Syntex] see 6614
Brevidil E see 8989
Brevidil M see 8845
Brevifolin see 9976
Brevimytal Sodium [Lilly] see 5904
Brevinarcon see 9250
Brevinor [Syntex] see 6614
Brevital [Lilly] see 5904
Brevital Sodium [Lilly] see 5904
Brevium see 7897
Brexin [Chiesi] see 7476
Bricanyl [Draco] see 9089
Bricef [Bristol-Banyu] see 1923
Bridal see 7176
Brietal Sodium [Lilly] see 5904
Brigade [FMC] see 1229
Brij [ICI] see 7554
Brij 58 [ICI] see 7554
Brilliant Blue FCF, 1366
Brilliant Congo R see 9917
Brilliant Cotton Blue see 5950
Brilliant Green, 1367
Brilliant Vital Red see 9917
Brimstone see 8948
Brinaldix [Sandoz] see 2391
Briofil [Alfa] see 965
Bripadon [Anphar] see 4088
Briplatin [Bristol-Myers] see 2319
Brisfirina [Bristol] see 1980
Brispen see 3073
Bristab [Bristol] see 4716
Bristaciclina [Bristol] see 9130
Bristacin [Bristol] see 8236
Bristacin-A [Bristol] see 8236
Bristagen [Bristol] see 4284
Bristamin [Bristol] see 7287
Bristamox [Bristol] see 610
Bristamycin [Bristol] see 3632
Bristocef [Bristol] see 1980
Bristopen [Bristol] see 6858
Bristuric see 1045
Bristurin [Bristol] see 4716
Bristuron [Bristol] see 1045
Britacil see 621
Britai [Bristol-Banyu] see 2350
Britiazim [Thames] see 3188
British Anti-Lewisite see 3196
British Gum see 2931
BRL 284 see 7829
BRL 804 see 4592
BRL 1241 see 5890
BRL 1288 see 1039
BRL 1341 see 621
BRL 1400 see 6858
BRL 1621 see 2414
BRL 1702 see 3073
BRL 2039 see 4044
BRL 2064 see 1796
BRL 2288 see 9360
BRL 2333 see 610
BRL 2351 see 925
BRL 3475 see 1844
BRL 8988 see 9008
BRL 13856 see 2395
BRL 14777 see 6258
BRL 17421 see 9076
BRL 26921 see 785
BRL 29060 see 6991
BRL 43694A see 4443
BRL 4910A see 6212
Brocadopa [Brocades-Stheeman] see 5344
Brocalcin see 1654
Brocasipal see 6831
Brochantite see 2660
Brocide see 3754
Brocillin see 7829
Brocsil see 7184
Brodiar [Intervet] see 1441
Brodifacoum, 1368
Brodimoprim, 1369

Broflex [Bio-Medical] see 9607
Brolene see 3008
Broline Drops [Rhône-Poulenc] see 7808
Broline Ointment [Rhône-Poulenc] see 3008
Brolitène [Lab. Medicia] see 4820
Bromacil, 1370
Bromadal see 1837
Bromadiolone, 1371
Bromadryl [Spofa] see 3514
Bromal, 1372
Bromal Hydrate, 1373
Bromallylene see 286
Bromanautine see 1405
Bromat see 2018
Bromated Camphor see 1402
Bromazepam, 1374
Bromazine see 1405
Bromchlophos see 6272
Bromcholitin see 4332
Bromcresol Green, 1375
Bromcresol Purple, 1376
Bromdiphenhydramine see 1405
Bromelain, 1377
Bromelia see 3709
Bromelin see 1377
Brometazepam see 5824
Bromethalin, 1378
Bromethol see 9525
Brometone see 9523
Bromex see 6272
Bromex [Makhteshim-Agan] see 3030
Bromhexine, 1379
Bromic Acid, 1380
Bromic Ether see 3730
Brominated Biphenyls see 7540
Bromindione, 1381
Bromine, 1382
Bromine Cyanide see 2700
Bromine Pentafluoride, 1383
Bromine Trifluoride, 1384
Brominil [Amchem] see 1431
Bromisoval see 1385
Bromisovalum, 1385
N-Bromoacetamide, 1386
p-Bromoacetanilide, 1387
4'-Bromoacetanilide see 1387
Bromoacetic Acid, 1388
Bromoacetone, 1389
p-Bromoacetophenone, 1390
ω-Bromoacetophenone, 1391
2-Bromoallyl Bromide see 3009
α-Bromoallyl Bromide see 3009
5-(2-Bromoallyl)-5-sec-butylbarbituric Acid see 1503
5-(2-Bromoallyl)-5-isopropylbarbituric Acid see 7807
5-(2-Bromoallyl)-5-isopropyl-1-methylbarbituric Acid see 6342
5-Bromo-2-aminobenzoic Acid see 1393
Bromoanilide see 1387
p-Bromoaniline, 1392
4-Bromoaniline see 1392
p-Bromoanisole see 1416
5-Bromoanthranilic Acid, 1393
Bromoantifebrin see 1387
Bromoaspirin see 1425
Bromoauric Acid see 4426
Bromo-Benadryl see 1405
Bromobenzene, 1394
α-Bromobenzeneacetonitrile see 1400
4-Bromobenzeneamine see 1392
p-Bromobenzenesulfonyl Chloride, 1395
β-(p-Bromobenzhydryloxy)ethyldimethylamine see 1405
p-Bromobenzoic Acid, 1396
p-Bromobenzyl Bromide, 1397
p-Bromobenzyl Chloride, 1398

p-Bromobenzyl Chloroformate, 1399
α-Bromobenzyl Cyanide, 1400
2-[(p-Bromobenzyl)(2-dimethylaminoethyl)amino]pyridine see 1436
(o-Bromobenzyl)ethyldimethylammonium p-Toluenesulfonate see 1364
3-[3-(4'-Bromo[1,1'-biphenyl]-4-yl)-3-hydroxy-1-phenylpropyl]-4-hydroxy-2H-1-benzopyran-2-one see 1371
3-[3-(4'-Bromo[1,1'-biphenyl]-4-yl)-1,2,3,4-tetrahydro-1-naphthalenyl]-4-hydroxy-2H-1-benzopyran-2-one see 1368
3-Bromo-d-2-bornanone see 1402
1-Bromo-4-(bromomethyl)benzene see 1397
2-Bromo-2-(bromomethyl)glutaronitrile see 3004
2-Bromo-2-(bromomethyl)pentanedinitrile see 3004
2-Bromo-1-(4-bromophenyl)ethanone see 1415
4-Bromo-α-(4-bromophenyl)-α-hydroxybenzeneacetic Acid 1-Methyl Ethyl Ester see 1422
1-Bromobutane see 1553
2-Bromobutane see 1554
Bromobutanedioic Acid see 1427
2-Bromobutanoic Acid see 1401
5-Bromo-3-sec-butyl-6-methyluracil see 1370
dl-2-Bromobutyric Acid see 1401
α-Bromobutyric Acid, 1401
3-Bromo-d-camphor, 1402
3α-Bromo-d-camphor see 1402
3β-Bromo-d-camphor see 1402
α'-Bromo-d-camphor see 1402
α-Bromo-d-camphor see 1402
α-Bromo-n-caproic Acid, 1403
p-Bromocarbanil see 1418
2-Bromo-4'-chloroacetophenone see 2153
α-Bromo-p-chloroacetophenone see 2153
7-Bromo-6-chlorofebrifugine see 4509
7-Bromo-6-chloro-3-[3-(3-hydroxy-2-piperidinyl)-2-oxopropyl]-4(3H)-quinazolinone see 4509
(±)-trans-7-Bromo-6-chloro-3-[3-(3-hydroxy-2-piperidyl)acetonyl]-4(3H)-quinazolinone see 4509
1-Bromo-4-(chloromethyl)benzene see 1398
7-Bromo-5-(2-chlorophenyl)-2,3-dihydro-2-(methoxymethyl)-1-methyl-1H-1,4-benzodiazepine see 5824
2-Bromo-1-(4-chlorophenyl)ethanone see 2153
5-Bromo-N-(4-chlorophenyl)-2-hydroxybenzamide see 1423
2-Bromo-4-(o-chlorophenyl)-9-methyl-6H-thieno[3,2-f]-s-triazolo-[4,3-a][1,4]diazepine see 1439
2-Bromo-4-(2-chlorophenyl)-9-methyl-6H-thieno[3,2-f][1,2,4]triazolo-[4,3-a][1,4]diazepine see 1439
8-Bromo-6-(o-chlorophenyl)-1-methyl-4H-s-triazolo[3,4-c]thieno[2,3-e]-1,4-diazepine see 1439
5-Bromo-4'-chlorosalicylanilide see 1423
p-Bromo-α-chlorotoluene see 1398
Bromochlorotrifluoroethane see 4517
2-Bromo-2-chloro-1,1,1-trifluoroethane see 4517
Bromocriptine, 1404
Bromocyclohexane see 2736
Bromocyl see 1424

Cross Index of Names

O-(4-Bromo-2,5-dichlorophenyl)-
O,O-dimethylphosphorothioate see
1419
O-(4-Bromo-2,5-dichlorophenyl)
O-Methyl Phenylphosphonothioate
see 5327
(8β)-2-Bromo-9,10-didehydro-N,N-
diethyl-6-methylergoline-8-carbox-
amide see 1411
2-Bromo-2,2-diethylacetamide see
3105
Bromodiethylacetylcarbamide see
1837
Bromodiethylacetylurea see 1837
2-Bromo-N,N-diethyl-D-lysergamide
see 1411
2-Bromo-6β,9-difluoro-11β,17,21-tri-
hydroxypregna-1,4-diene-3,20-
dione 17,21-Diacetate see 4512
8-Bromo-3,7-dihydro-1,3-dimethyl-
1H-purine-2,6-dione Compd with
2-Amino-2-methyl-1-propanol (1:1)
see 6952
(3α,14β,16α)-11-Bromo-14,15-di-
hydro-14-hydroxyeburnamenine-
14-carboxylic Acid Methyl Ester
see 1440
7-Bromo-1,3-dihydro-5-(2-pyridinyl)-
2H-1,4-benzodiazepin-2-one see
1374
5-Bromo-N,2-dihydroxybenzamide
see 1424
4'-Bromo-2,6-dihydroxybenzanilide
see 8157
4-[(2-Bromo-4,5-dimethoxyphenyl)-
methyl]-4-[2-[2-(6,6-dimethylbicy-
clo[3.1.1]hept-2-yl)ethoxy]ethyl]-
morpholinium Bromide see 7410
5-[(4-Bromo-3,5-dimethoxyphenyl)-
methyl]-2,4-pyrimidinediamine see
1369
2-[p-Bromo-α-(2-dimethylaminoeth-
yl)benzyl]pyridine see 1433
α-Bromo-β-dimethylpropanoylurea
see 1385
N¹-(5-Bromo-4,6-dimethyl-2-pyrimi-
dinyl)sulfanilamide see 8869
5-Bromo-4,6-dimethyl-2-sulfanilami-
dopyrimidine see 8869
Bromodiphenhydramine, 1405
1-(2-Bromo-1,2-diphenylethenyl)-4-
ethylbenzene see 1438
2-[1-(4-Bromodiphenyl)ethoxy]-N,N-
dimethylethylamine see 3514
α-Bromo-α,β-diphenyl-β-(p-ethylphenyl)ethylene see 1438
1-Bromododecane see 5260
Bromoeosine see 3555
2-Bromoergocryptine see 1404
2-Bromo-α-ergokryptin see 1404
Bromoethane see 3730
2-Bromoethanol see 3749
β-Bromoethyl Alcohol see 3749
2-Bromo-2-ethylbutanamide see 3105
(α-Bromo-α-ethylbutyryl)carbamide
see 1837
(α-Bromo-α-ethylbutyryl)urea see
1837
trans-3-Bromo-N-ethylcinnamamide
see 2313
2-Bromo-N-ethyl-N,N-dimethylbenz-
enemethanaminium 4-Methylbenz-
enesulfonate see 1364
2-Bromo-2-ethylisovaleramide see
4809
2-Bromo-2-ethyl-3-methylbutanamide
see 4809
1-Bromo-2-(4-ethylphenyl)-1,2-di-
phenylethene see 1438
1-Bromo-2-(p-ethylphenyl)-1,2-di-
phenylethylene see 1438

Bromofenofos, 1406
Bromofluoresceic Acid see 3555
3-(4-Bromo-2-fluorobenzyl)-4-oxo-
3H-phthalazin-1-ylacetic Acid see
7565
3-[(4-Bromo-2-fluorophenyl)methyl]-
3,4-dihydro-4-oxo-1-phthalazine-
acetic Acid see 7565
10-Bromo-11b-(2-fluorophenyl)-2,3,-
7,11b-tetrahydrooxazolo[3,2-d]
[1,4]benzodiazepin-6(5H)-one see
4518
17-Bromo-6-fluoropregn-4-ene-3,20-
dione see 4513
17α-Bromo-6α-fluoroprogesterone see
4513
Bromoform, 1407
2-Bromohexanoic Acid see 1403
4-Bromo-α-hydroxybenzeneacetic
Acid see 1412
5-Bromo-2-hydroxybenzenemethanol
see 1426
5-Bromo-2-hydroxybenzyl Alcohol
see 1426
(5'α)-2-Bromo-12'-hydroxy-2'-(1-
methylethyl)-5'-(2-methylpropyl)-
ergotaman-3',6',18-trione see 1404
α-(5-Bromo-4-hydroxy-m-tolyl)-α-
(3-bromo-5-methyl-4-oxo-2,5-
cyclohexadien-1-ylidene)-o-toluene-
sulfonic Acid see 1376
2-Bromoisobutane see 1555
α-Bromoisobutyric Acid, 1408
1-Bromo-4-isocyanatobenzene see
1418
2-Bromoisopentane see 638
α-Bromo-α-isopropylbutyramide see
4809
5-Bromo-3-isopropyl-6-methyluracil
see 5042
α-Bromoisovaleric Acid, 1409
β-Bromoisovaleric Acid, 1410
(α-Bromoisovaleryl)urea see 1385
Bromol see 9526
Bromolate [Roner] see 4363
Bromo-LSD see 1411
D-2-Bromolysergic Acid Diethylamide
see 1411
Bromolysergide, 1411
p-Bromomandelic Acid, 1412
[[[(3-Bromomesityl)carbamoyl]meth-
yl]imino]diacetic Acid see 5652
Bromomethane see 5951
(Bromomethyl)benzene see 1142
1-Bromo-3-methylbenzene see 1429
p-Bromo-α-methylbenzhydryl 2-Di-
methylaminoethyl Ether see 3514
L-1-Bromo-2-methylbutane see 636
1-Bromo-2-methylbutane see 636
1-Bromo-3-methylbutane see 4996
2-Bromo-2-methylbutane see 638
2-Bromo-3-methylbutanoic Acid see
1409
3-Bromo-3-methylbutanoic Acid see
1410
2-Bromo-3-methylbutanoic Acid 1,7,-
7-Trimethylbicyclo[2.2.1]hept-2-yl
Ester see 1340
2-Bromo-3-methylbutyric Acid Ester
with d-Borneol see 1340
1-(Bromomethyl)-3-methylbenzene
see 9998
5-Bromo-6-methyl-3-(1-methyleth-
yl)-2,4(1H,3H)-pyrimidinedione see
5042
5-Bromo-6-methyl-3-(1-methylprop-
yl)-2,4(1H,3H)-pyrimidinedione see
1370
5-Bromo-6-methyl-3-(1-methylprop-
yl)uracil see 1370

2-[(p-Bromo-α-methyl-α-phenylbenzyl)oxy]-N,N-dimethylethylamine see
3514
1-Bromo-2-methylpropane see 5019
2-Bromo-2-methylpropane see 1555
2-Bromo-2-methylpropanoic Acid see
1408
1-Bromonaphthalene, 1413
2-Bromonaphthalene, 1414
α-Bromonaphthalene see 1413
β-Bromonaphthalene see 1414
Bromone [Lipha] see 1371
8β-[(5-Bromonicotinoyloxy)methyl]-
1,6-dimethyl-10α-methoxyergoline
see 6404
2-Bromo-2-nitro-1,3-propanediol see
1437
β-Bromo-β-nitrotrimethyleneglycol
see 1437
1-Bromooctane see 6684
2-Bromooctane see 6685
1-Bromopentane see 637
p-Bromophenacyl Bromide, 1415
p-Bromophenetole see 1416
Bromophenol, 1416
2-Bromophenol see 1416
3-Bromophenol see 1416
4-Bromophenol see 1416
Bromophenophos see 1406
N-(4-Bromophenyl)acetamide see
1387
α-Bromophenylacetonitrile see 1400
2-(p-Bromo-α-phenylbenzyloxy)-N,N-
dimethylethylamine see 1405
4-Bromophenylcarbimide see 1418
p-Bromophenylcarbonimide see 1418
γ-(4-Bromophenyl)-N,N-dimethyl-2-
pyridinepropanamine see 1433
(Z)-3-(4-Bromophenyl)-N,N-dimeth-
yl-3-(3-pyridinyl)-2-propen-1-
amine see 10024
2-(4-Bromophenyl)-1,3-dioxohydrin-
dene see 1381
1-(4-Bromophenyl)ethanone see 1390
2-Bromo-1-phenylethanone see 1391
(E)-3-(3-Bromophenyl)-N-ethyl-2-
propenamide see 2313
p-Bromophenylglycolic Acid see 1412
p-Bromophenylhydrazine, 1417
3-[α-[p-(p-Bromophenyl)-β-hydroxy-
phenethyl]benzyl]-4-hydroxycou-
marin see 1371
4-[4-(p-Bromophenyl)-4-hydroxypip-
eridino]-4'-fluorobutyrophenone
see 1432
4-[4-(4-Bromophenyl)-4-hydroxy-1-
piperidinyl]-1-(4-fluorophenyl)-1-
butanone see 1432
2-(4-Bromophenyl)-1H-indene-
1,3(2H)-dione see 1381
p-Bromophenyl Isocyanate, 1418
3-(p-Bromophenyl)-1-methoxy-1-
methylurea see 6061
N-[(4-Bromophenyl)methyl]-N',N'-
dimethyl-N-2-pyridinyl-1,2-ethane-
diamine see 1436
1-(p-Bromophenyl)-1-phenyl-1-(2-
dimethylaminoethoxy)ethane see
3514
2-[1-(4-Bromophenyl)-1-phenylethox-
y]-N,N-dimethylethanamine see
3514
[2-(1-p-Bromophenyl-1-phenylethox-
y)ethyl]dimethylamine see 3514
2-[(4-Bromophenyl)phenylmethoxy]-
N,N-dimethylethanamine see 1405
(Z)-3-(4'-Bromophenyl)-3-(3''-pyr-
idyl)dimethylallylamine see 10024
1-(p-Bromophenyl)-1-(2-pyridyl)-3-
dimethylaminopropane see 1433

3-(p-Bromophenyl)-3-(2-pyridyl)-
N,N-dimethylpropylamine see 1433
Bromophin see 776
Bromophos, 1419
Bromophos-ethyl see 1419
Bromophos-methyl see 1419
Bromopride, 1420
1-Bromopropane see 7857
2-Bromopropane see 5098
3-Bromopropanoic Acid see 1421
2-Bromopropanoic Acid Ethyl Ester
see 3731
1-Bromo-2-propanone see 1389
3-Bromo-1-propene see 286
5-(2-Bromo-2-propenyl)-5-(1-methyl-
ethyl)-2,4,6(1H,3H,5H)pyrimidine-
trione see 7807
5-(2-Bromo-2-propenyl)-1-methyl-5-
(1-methylethyl)-2,4,6(1H,3H,5H)-
pyrimidinetrione see 6342
5-(2-Bromo-2-propenyl)-5-(1-methyl-
propyl)-2,4,6(1H,3H,5H)pyrimidine-
trione see 1503
5-(2-Bromo-2-propenyl)-5-(2-propen-
yl)-2,4,6(1H,3H,5H)-pyrimidine-
trione see 1357
β-Bromopropionic Acid, 1421
Bromopropylate, 1422
3-Bromopropylene see 286
7-Bromo-5-(2-pyridyl)-3H-1,4-
benzodiazepin-2(1H)-one see 1374
5-Bromo-2,4(1H,3H)-pyrimidinedione
see 1430
1-Bromo-2,5-pyrrolidinedione see
1428
4'-Bromo-γ-resorcylanilide see 8157
Bromosalicylchloranilide, 1423
5-Bromosalicylhydroxamic Acid,
1424
5-Bromosalicylic Acid Acetate, 1425
N-5-Bromosalicyloyl-p-chloroaniline
see 1423
Bromosaligenin, 1426
5-Bromosaligenin see 1426
Bromosuccinic Acid, 1427
N-Bromosuccinimide, 1428
5-Bromosulfamethazine see 8869
Bromosulfophthalein see 8933
2-Bromo-1,1,1,2-tetrafluoroethane
see 9059
1-Bromo-2,2,3,3-tetrafluoropropane
see 4515
3-Bromo-1,1,2,2-tetrafluoropropane
see 4515
8-Bromotheophylline Compd with
2-Amino-2-methyl-1-propanol (1:1)
see 6952
Bromotiren see 3014
Bromotoluene, 1429
m-Bromotoluene see 1429
o-Bromotoluene see 1429
p-Bromotoluene see 1429
3-Bromotoluene see 1429
α-Bromotoluene see 1142
ω-Bromotoluene see 1142
α-Bromo-α-tolunitrile see 1400
N-(3-Bromo-2,4,6-trimethylacetanil-
ide)iminodiacetic Acid see 5652
3-Bromo-1,7,7-trimethylbicyclo-
[2.2.1]heptan-2-one see 1402
N-[2-[(3-Bromo-2,4,6-trimethylphen-
yl)amino]-2-oxoethyl]-N-(carboxymethyl)glycine see 5652
5-Bromouracil, 1430
4-(6-Bromoveratryl)-4-[2-[2-(6,6-
dimethyl-2-norpinyl)ethoxy]ethyl]-
morpholinium Bromide see 7410
cis-11-Bromovincamine see 1440
α-Bromo-m-xylene see 9998
ω-Bromo-m-xylene see 9998
Bromoxynil, 1431

Bromperidol, 1432
Brompheniramine, 1433
Bromphenol Blue, 1434
Bromphenphos see 1406
Bromsalizol [HW & D] see 1426
Bromsulfophthalein [HW & D] see
8933
Bromsulphalein Sodium see 8933
Brom-Tetragnost see 8933 and 9122
Bromth [Brayten] see 7996
Bromthalein [E. Merck] see 8933
Bromthymol Blue, 1435
p-Bromtripelennamine, 1436
Bromural [Knoll] see 1385
Bromuvan see 1385
Bromvaletone see 1385
Bronalin [Byk-Liprandi] see 4628
Broncaspin [Bayer] see 873
Broncatar [Pulitzer] see 6898
Bronchocillin see 7027
Bronchodil [Keymer] see 8142
Broncholin [Kaken] see 5517
Broncholysin see 82 and 4628
Bronchopront [Mack] see 392
Bronchoretard [Klinge] see 9212
Bronchoselectan [Schering AG] see 72
Bronchospasmin [Homburg] see 8142
Broncocor [Irbi] see 7461
Broncocur [Irbi] see 7461
Broncon [Wakamoto] see 2404
Broncopen see 7027
Broncoplus [Sigma-Tau] see 8764
Broncovaleas [Valeas] see 209
Bronkephrine [Breon] see 3789
Bronkodyl [Breon] see 9212
Bronner's Acid see 1899
Bronopol, 1437
Bronosol [Green Cross] see 1437
Bronsecur [SK & F] see 1840
Brontine [Brocades-Stheeman] see
2894
Brontyl see 7921
Brookite see 9398
Broom see 8358
Broparoestrol, 1438
Brosalamid see 1424
Brotizolam, 1439
Brotopon [Pfizer Taito] see 4511
Brovalol see 1340
Brovalurea see 1385
Brovel [Lepetit] see 3578
Brovincamine, 1440
"Brown Acetate of Lime" see 1643
Brown Mustard see 6224
Broxalax see 1253
Broxil [Beecham] see 7184
Broxolin [Breon] see 4389
Broxykinolin see 1441
Broxynil see 1431
Broxyquinoline, 1441
BRS 640 see 1356
Bruceantin see 1442
Brucine, 1443
Brucite see 5548
Brufaneuxol see 488
Brufanic [Taiyo Yakuko] see 4812
Brufen [Boots] see 4812
Brufort [Lampugnani] see 4812
Bruisewort see 8326
Brulidine [M & B] see 3008
Brumetidina [Bruschettini] see 2279
Brumixol [Bruschettini] see 2270
Brunac [Bruschettini] see 82
Bruneomycin see 8789
Brunet Saxifrage see 7405
Brushite see 1699
Bruxicam [Bruschettini] see 7476
Bruzem [Bruschettini] see 3188
Bryamycin [Bristol] see 9295
Bryonia, 1444
Bryony see 1444

BS 572 see 2708
BS 4231 see 4339
BS 5930 see 6831
BS 6987 see 2894
BS 7331 see 9426
BS 100-141 see 4474
BSF-2 4893
B.S.G. [Lemmon] see 1297
BSM 906M see 2709
BSP see 8933
BT see 945 and 9173
BT 621 see 9425
BTC see 1066
BTC 1010 [Onyx] see 3088
BTE see 9694
B-Telve see 9921
BTPABA see 1061
BTS see 1316
BTS 18322 see 4124
BTS 27419 see 503
BTS 40542 see 7767
BTS 49465 see 4043
B-Twelv [Sherman] see 9921
BTX see 1365
BTX-B see 1365
BTX-C see 1365
BU-2231 see 9016
BU-2231A see 9016
BU-2231B see 9016
Buban [Burroughs Wellcome] see 1475
Bubarbital Sodium [Philips Roxane]
see 1495
Buburone [Towa Yakuhin] see 4812
Bucarban see 1839
Buccalsone see 4713
Buccastem [Reckitt & Colman] see
7768
Bucco see 1446
Bucetalon [Isei] see 1445
Bucetin, 1445
Buchu, 1446
Buchu Camphor see 3292
Buchu Resin see 3291
Bucillamine, 1447
Buck Bean see 5729
Buckminsterfullerene see 1814
Buckthorn Bark see 4176 and 8172
Bucu see 1446
Bucladesine, 1448
Buclamase [Rystan] see 633
Buclifen [Pfizer] see 1449
Buclina [Usafarma] see 1449
Buclizine, 1449
Buclosamide, 1450
Bucloxic Acid, 1451
Bucloxonic Acid see 1451
Bucolome, 1452
Bucrilate, 1453
Bucrol see 1839
Bucrylate, 1453
Buctril [M & B] see 1431
Bucumarol [Sankyo] see 1454
Bucumolol, 1454
Buddleoflavonoloside see 5376
Budeson [Fujisawa] see 1455
Budesonide, 1455
Budipine, 1456
Budoform [Dolder] see 4924
Budralazine, 1457
Bueno [Diamond] see 5864
Bufalin, 1458
Bufedil [Abbott; Pharma-Linie] see
1463
Bufedon [Cosmopharma] see 6654
Bufemid [Lederle] see 3906
Bufencarb, 1459
Bufeniode, 1460
Bufetolol, 1461
Bufexamac, 1462
Buflan [Pierrel] see 1463
Buflomedil, 1463

Bufogenin B, 1464
Bufon see 3118
Bufonamin [Kaken] see 1465
Bufopto Homatrocel [Softcon] see 4649
Bufor [Bofors] see 6405
Buformin, 1465
Bufotalin, 1466
Bufotalin 3-Suberoylarginine Ester see 1468
Bufotenine, 1467
Bufotoxin, 1468
Bufuralol, 1469
Bufylline (rescinded USAN) see 395
Bugbane see 2281
Bugleweed see 5496
Bugwort see 2281
Bukarban see 1839
Buksamin see 456
Buku see 1446
Bulan®, 1470
Bulbocapnine, 1471
Bulbonin [Sankyo] see 1465
Bulbus Scillae see 8728
Bull Nettle see 8664
Bumadizon, 1472
Bumaflex [Byk-Liprandi] see 1472
Bumetanide, 1473
Bumex [Roche] see 1473
Buminate see 8417
Buna see 1500
Buna S [I. G. Farben] see 8345
Bunaftine, 1474
Bunaiod see 1476
Bunamidine, 1475
Bunamiodyl Sodium, 1476
Bunaphtide see 1474
Bunaphtine see 1474
Bunapsilate see MISC-4
Bunazosin, 1477
Bundlin B see 8377
Bunetzone see 7248
α-Bungarotoxin see 1478
β-Bungarotoxin see 1478
Bungarotoxins, 1478
Buniodyl see 1476
Bunitrolol, 1479
l-Bunolol see 5343
Bunsenite see 6419
Bunt-cure see 4600
Bunte Salts, 1480
Bunt-no-more see 4600
Buparvaquone, 1481
Bupatol [Gedeon Richter] see 964
Buphanamine, 1482
Buphanitine, 1483
Buphedrin [Tatsumi] see 6654
Buphenine see 6654
Bupirimate, 1484
Bupivacaine, 1485
Bupranol see 1486
Bupranolol, 1486
Buprenex [Norwich] see 1487
Buprenorphine, 1487
Bupropion, 1488
Buquinolate, 1489
Buramate, 1490
Burdock see 5236
Burgodin [Janssen] see 1216
Burgundy Mixture, 1491
Burinex [Leo Pharm.] see 1473
Burning Bush see 3856
Burnt Alum see 364
Burnt Ammonium Alum see 325
Burnt Lime see 1692
Burnt Sugar see 1775
Burnt Sugar Coloring see 1775
Buronil [Ferrosan] see 5707
Burow's Solution see 322
Bursine see 2207
Burtonite 44 see 4212

Burtonite V-7-E see 4486
Buscapina see 1588
Buscol see 1588
Buscolamin [Tokyo Hosei] see 1588
Buscolysin [Pharmachim] see 1588
Buscopan [Boehringer, Ing.] see 1588
Buserelin, 1492
Busodium [Truxton] see 1495
Busotran [Galen] see 1495
Buspar [Mead Johnson] see 1493
Buspinol [Zdravlje] see 1493
Buspirone, 1493
Busulfan, 1494
Busulphan see 1494
Butabar [Flar] see 1495
Butabarbital Sodium, 1495
Butabarbitone Sodium see 1495
Butabarpal Sodium [Philadelphia Labs.] see 1495
Butabon see 1495
Butacaine, 1496
Butacetin, 1497
Butachlor, 1498
Butacide [Fairfield] see 7446
Butaclamol, 1499
Butacote [Ciba-Geigy] see 7248
1,3-Butadiene, 1500
α,γ-Butadiene see 1500
1,3-Butadiene-1,4-dicarboxylic Acid see 6210
Butadiene Diepoxide see 3621
Butadiene Dioxide see 3621
1,1'-(1,3-Butadiene-1,4-diyl)bisbenzene see 3320
Butadiene Sulfone see 8935
Butadion see 7248
Butak [Lemmon] see 1495
Buta-Kay [King] see 1495
Butalamine, 1501
Butalan [Lannett] see 1495
Butalbital, 1502
Butalex [Coopers] see 1481
Butalgin see 5852
Butalix [Vale] see 1495
Butallylonal, 1503
Butamben, 1504
Butamide see 1511
Butamirate, 1505
Butamisole, 1506
Butamiverine see 1510
Butamyrate see 1505
Butanal see 1591
Butanamide see 1592
1-Butanamine see 1543
2-Butanamine see 1544
Butane, 1507
n-Butane see 1507
Butanediamide see 8837
1,4-Butanediamine see 7964
1,4-Butanedicarboxylic Acid see 152
Butanedinitrile see 8843
Butanedioic Acid see 8840
Butanedioic Acid Bis(1,1-dimethylethyl) Ester see 3024
Butanedioic Acid Calcium Salt see 1712
Butanedioic Acid Dibutyl Ester see 3023
Butanedioic Acid 2-(Dimethylamino)-ethyl [5-Hydroxy-4-(hydroxymethyl)-6-methyl-3-pyridinyl]methyl Ester see 8002
Butanedioic Acid Mono[(4-butyl-3,5-dioxo-1,2-diphenyl-4-pyrazolidinyl]methyl Ester see 8990
[3R-(3α,5aβ,6β,8aβ,9α,10β,12β,-12aR*)]-Butanedioic Acid Mono-(decahydro-3,6,9-trimethyl-3,12-epoxy-12H-pyrano[4,3-j]-1,2-benzodioxepin-10-yl) Ester see 845

Butanedioic Acid Mono[1-[5-(2,5-dihydro-5-oxo-3-furanyl)-3-methyl-2-benzofuranyl]ethyl] Ester see 1051
Butanedioic Acid Mono(2,2-dimethylhydrazide) see 2810
Butanedioic Acid Mono[3-[(2-ethylhexyl)amino]-1-methyl-3-oxopropyl] Ester see 1526
Butanedioic Anhydride see 8841
Butane-1,3-diol see 1566
1,3-Butanediol see 1566
2,3-Butanediol see 1567
1,4-Butanediol Dimethanesulfonate Esters see 1494
Butanedione see 3621
2,3-Butanedione see 2946
2,3-Butanedionedioxime see 3235
Butanedioyl Chloride see 8844
1,1'-(1,4-Butanediyl)bis[1,2,3,4-tetrahydro-6,7-dimethoxyisoquinoline] see 1308
Butanenitrile see 1597
1,2,3,4-Butanetetrol see 3620
(R*,S*)-1,2,3,4-Butanetetroltetranitrate see 3622
1-Butanethiol see 1575
2-Butanethiol see 1576
Butanex [Makhteshim-Agan] see 1498
Butanilicaine, 1508
Butanimide see 8842
Butanoic Acid see 1593
Butanoic Acid Anhydride see 1594
1-Butanoic Acid-26-L-aspartic Acid-27-L-valine-29-L-alanine-1,7-dicarbacalcitonin (Salmon) see 3501
Butanoic Acid Butyl Ester see 1556
Butanoic Acid Ethyl Ester see 3733
1-Butanoic Acid-7-glycine-1,6-dicarbaoxytocin see 1846
Butanoic Acid Methyl Ester see 5957
1-Butanoic Acid-2-(O-methyl-L-tyrosine)-1-carbaoxytocin see 1800
Butanoic Acid Pentyl Ester see 639
Butanoic Acid 1,2,3-Propanetriyl Ester see 9532
Butanoic Acid Propyl Ester see 7858
Butanoic Acid 2,2,2-Trichloro-1-(dimethoxyphosphinyl)ethyl Ester see 1528
1-Butanol see 1540
2-Butanol see 1541
1,2-Butanolide see 1596
1,4-Butanolide see 1596
2,3-Butanolone see 55
2-Butanone see 5991
Butanotic [Medical Arts] see 1495
Butanoyl Chloride see 1598
Butaperazine, 1509
Butaphyllamine [Merrell] see 395
Butapirazol see 7248
Butased [Schlicksup] see 1495
Butatensin [Benvegna] see 5653
Butatran [Palmedico] see 1495
Butaverine, 1510
Butazem [Zemmer] see 1495
Butazolamide, 1511
Butazolidin [Geigy] see 7248
Butazonic [Taiyo Yakuko] see 6925
Butedrin see 964
Butedronic Acid, 1512
Butelline see 1496
2-Butenal see 2599
i-Butene, 1513
2-Butene, 1514
Butenedioic Acid see 5585
(E)-2-Butenedioic Acid see 4200
(E)-2-Butenedioic Acid Bis(phenylmethyl) Ester see 1149
(Z)-2-Butenedioic Acid, Diethyl Ester see 3113

cis-Butenedioic Anhydride see 5586
3-Butenenitrile see 288
cis-2-Butenoic Acid see 5048
trans-2-Butenoic Acid see 2600
2-Buten-1-ol see 2604
3-Buten-2-one see 6052
Δ^3-2-Butenone see 6052
β-Butenonitrile see 288
(4R)-4-[(E)-2-Butenyl]-4,N-dimethyl-L-threonine see 5651
3-(2-Butenylidene)-2-carboxy-α-(hydroxymethylene)-1,4-cyclopenta-diene-1-acetic Acid δ-Lactone Methyl Ester see 4198
(E,E)-7-(2-Butenylidene)-1,7-di-hydro-1-oxocyclopenta[c]pyran-4-carboxylic Acid Methyl Ester see 4198
Buteprate see MISC-3
Buterazine [Daiichi] see 1457
Butesin [Abbott] see 1504
Butesin Picrate [Abbott] see 1504
Butethal, 1515
Butethamate, 1516
Butethamine, 1517
Butethanol see 9123
Butex [Phillips] see 1495
Butformin see 1465
Buthalital Sodium, 1518
Buthalitone Sodium see 1518
Buthiazide, 1519
Buthiobate, 1520
Buthionine Sulfoximine, A3
Buthoid [Merrell] see 395
Butibufen, 1521
Buticaps [McNeil] see 1495
Butidiona see 7248
Butidrine Hydrochloride, 1522
Butilate see 1546
Butilopan [Juste] see 1521
Butionine Sulfoximine see A3
Butirosin, 1523
Butirosin A see 1523
Butirosin B see 1523
Butisol Sodium [McNeil] see 1495
Butixirate see 9980
Butizide see 1519
Butobarbital see 1515
Butobarbitone see 1515
Butoben [Merck & Co.] see 1583
Butobendine, 1524
Butoconazole, 1525
Butoctamide, 1526
Butofilolol, 1527
Butoform see 1504
Butolan [Bayer] see 3306
Butolen see 3306
Butonate, 1528
Butopyronoxyl, 1529
Butorphanol, 1530
Butox [Distrivet] see 2869
Butoxone see 2828
4'-tert-Butoxyacetanilide see 1497
2-Butoxy-4-aminobenzoic Acid β-Di-ethylaminoethyl Ester see 393
3-Butoxy-4-aminobenzoic Acid 2-(Di-ethylamino)ethyl Ester see 1056
2-Butoxy-3-aminobenzoic Acid β-Di-ethylaminoethyl Ester Hydrochlo-ride see 5822
4-Butoxybenzoic Acid 2-(Diethyl-amino)ethyl Ester see 1531
8-(p-Butoxybenzyl)-3α-hydroxy-1αH,5αH-tropanium Bromide (−)-Tropate see 1534
l-[1-(p-n-Butoxybenzyl)hyoscyamini-um] Bromide see 1534
Butoxycaine, 1531
1-(3-Butoxy-2-carbamoyloxypropyl)-5-phenyl-5-ethylbarbituric Acid see 3889

N-[N-[N-[N-(N-tert-Butoxycarbonyl-β-alanyl)-L-tryptophanyl]-L-methi-onyl]-L-aspartyl]-L-phenylalan-inamide see 7067
2-Butoxy-N-(2-diethylaminoethyl)-cinchoninamide Hydrochloride see 3016
2-Butoxy-N-[2-(diethylamino)ethyl]-4-quinolinecarboxamide Monohy-drochloride see 3016
2-Butoxyethanol see 1559
2-(2-Butoxyethoxy)ethanol see 1557
α-[2-(2-Butoxyethoxy)ethoxy]-4,5-methylenedioxy-2-propyltoluene see 7446
5-[[2-(2-Butoxyethoxy)ethoxy]meth-yl]-6-propyl-1,3-benzodioxole see 7446
2-[2-(Butoxy)ethoxy]ethyl Ester of Thiocyanic Acid see 5329
4-Butoxy-N-hydroxybenzeneacet-amide see 1462
1-(3-Butoxy-2-hydroxypropyl)-5-ethyl-5-phenylbarbituric Acid Car-bamate see 3889
N-(Butoxymethyl)-2-chloro-2',6'-diethylacetanilide see 1498
N-(Butoxymethyl)-2-chloro-N-(2,6-diethylphenyl)acetamide see 1498
2-(4-Butoxyphenoxy)-N-(2,5-dieth-oxyphenyl)-N-[2-(diethylamino)-ethyl]acetamide see 3931
1-Butoxy-3-phenoxy-2-propanol see 3891
4-[3-(4-Butoxyphenoxy)propyl]-morpholine see 7707
2-(p-Butoxyphenyl)acetohydroxamic Acid see 1462
2-(p-Butoxyphenyl)-N-[2-(diethylami-no)ethyl]-2',5'-diethoxyacetanilide see 3931
[3(S)-endo]-8-[(4-Butoxyphenyl)meth-yl]-3-(3-hydroxy-1-oxo-2-phenyl-propoxy)-8-methyl-8-azoniabicy-clo[3.2.1]octane Bromide see 1534
p-Butoxyphenyl γ-Morpholinopropyl Ether see 7707
4-Butoxyphenyl Piperidineethyl Ke-tone see 3453
1-(4-Butoxyphenyl)-3-(1-piperidinyl)-1-propanone see 3453
4-Butoxy-β-piperidinopropiophenone see 3453
4-n-Butoxy-β-(1-piperidyl)propio-phenone see 3453
3-[(6-Butoxy-3-pyridinyl)azo]-2,6-pyridinediamine see 2956
2-Butoxy-2'-thiocyanodiethyl Ether see 5329
1-Butoxy-α-(2-thiocyanoethoxy)-ethane see 5329
Butoz [Hamilton] see 7248
Butralin, 1532
Butrate [Kay] see 1495
Butriptylene see 1533
Butriptyline, 1533
Butropium Bromide, 1534
Butte [Scrip] see 1495
Butter of Antimony see 739
Butter of Arsenic see 829
Buttercup Yellow see 10032
Butterfly Weed see 853
Butternut see 5149
Butter Yellow see 3218
Butter of Zinc see 10031
β-BuTX see 1478
Butydrine Hydrochloride see 1522
n-Butyl Acetate, 1535
sec-Butyl Acetate, 1536
tert-Butyl Acetate, 1537

21-tert-Butylacetate-9α-fluoro-11β-hydroxy-16α,17α-(isopropylidene-dioxy)pregna-1,4-diene-3,20-dione see 9514
tert-Butylacetic Acid, 1538
n-Butyl Acrylate, 1539
Butyl Alcohol see 1540
n-Butyl Alcohol, 1540
sec-Butyl Alcohol, 1541
tert-Butyl Alcohol, 1542
n-Butylamine, 1543
sec-Butylamine, 1544
tert-Butylamine, 1545
ω-n-Butylaminoacetic Acid 2-Methyl-6-chloroanilide see 1508
1-(Butylaminoacetylamino)-2-chloro-6-methylbenzene see 1508
N-(Butylaminoacetyl)-6-chloro-o-toluidine see 1508
Butyl Aminobenzoate see 1504
n-Butyl p-Aminobenzoate see 1504
4-(Butylamino)benzoic Acid 2-(Di-methylamino)ethyl Ester Hydro-chloride see 9123
p-Butylaminobenzoic Acid ω-O-Meth-ylnonaethyleneglycol Ester see 1106
4-(Butylamino)benzoic Acid 3,6,9,12,-15,18,21,24,27-Nonaoxaoctacos-1-yl Ester see 1106
p-Butylaminobenzoyl-2-dimethylami-noethanol Hydrochloride see 9123
N-[(Butylamino)carbonyl]benzenesul-fonamide see 7177
[1-[(Butylamino)carbonyl]-1H-benz-imidazol-2-yl]carbamic Acid Methyl Ester see 1053
N-[(Butylamino)carbonyl]-4-methyl-benzenesulfonamide see 9432
2-(Butylamino)-6'-chloro-o-aceto-toluidide see 1508
1-tert-Butylamino-3-(2-chloro-5-methylphenoxy)-2-propanol see 1486
2-(Butylamino)-N-(2-chloro-6-meth-ylphenyl)acetamide see 1508
(±)-2-(tert-Butylamino)-3'-chloro-propiophenone see 1488
1-(tert-Butylamino)-3-[(6-chloro-m-tolyl)oxy]-2-propanol see 1486
(S)-1-(tert-Butylamino)-3-(o-cyclo-pentylphenoxy)-2-propanol see 7025
1-(tert-Butylamino)-3-(2,5-dichloro-phenoxy)-2-propanol see 2399
3-(tert-Butylamino)-3-(2',3'-dimeth-ylphenoxy)-2-propanol see 9984
4-Butylamino-2-hydroxybenzoic Acid 2-Dimethylaminoethyl Ester see 4783
(−)-1-tert-Butylamino-2-hydroxy-3-(2'-cyclopentylphenoxy)propane see 7025
2-(2-tert-Butylamino-1-hydroxyeth-yl)-7-ethylbenzofuran see 1469
(±)-5-[2-(tert-Butylamino)-1-hydr-oxyethyl]-m-phenylene Bis(dimeth-ylcarbamate) see 963
2-(tert-Butylamino)-1-(4-hydroxy-3-hydroxymethylphenyl)ethanol see 209
2-Butylamino-1-p-hydroxyphenyl-ethanol see 964
o-[3-(tert-Butylamino)-2-hydroxypro-poxy]benzonitrile see 1479
5-[3-(tert-Butylamino)-2-hydroxypro-poxy]-3,4-dihydrocarbostyril see 1875
(−)-5-[3-(tert-Butylamino)-2-hydr-oxypropoxy]-3,4-dihydro-1(2H)-naphthalenone see 5343

Cross Index of Names

Butylethylcarbamothioic Acid *S*-Propyl Ester see 7007
N-Butyl-*N*-ethyl-2,6-dinitro-4-trifluoromethylaniline see 1048
N-Butyl-*N*-ethyl-2,6-dinitro-4-(trifluoromethyl)benzenamine see 1048
tert-Butyl Ethyl Ether see 3732
5-Butyl-5-ethyl-2,4,6(1*H*,3*H*,5*H*)-pyrimidinetrione see 1515
5-*sec*-Butyl-5-ethyl-2-thiobarbituric Acid see 9250
N-Butyl-*N*-ethyl-α,α,α-trifluoro-2,6-dinitro-*p*-toluidine see 1048
Butyl 6α-Fluoro-11β-hydroxy-16α-methyl-3,20-dioxopregna-1,4-dien-21-oate see 4078
S-(*n*-Butyl)homocysteine Sulfoximine see A3
tert-Butyl Hydroperoxide, 1569
n-Butyl *p*-Hydroxybenzoate see 1583
5-Butyl-5-(2-hydroxyethyl)barbituric Acid Carbamate see 1838
6-*sec*-Butyl-1-hydroxy-3-isobutyl-2(1*H*)-pyrazinone see 863
4-Butyl-4-(hydroxymethyl)-1,2-diphenyl-3,5-pyrazolidinedione Hydrogen Succinate (Ester) see 8990
[7(S)-(1α,2β,4β,5α,7β)]-9-Butyl-7-(3-hydroxy-1-oxo-2-phenylpropoxy)-9-methyl-3-oxa-9-azoniatricyclo-[3.3.1.0²,⁴]nonane Bromide see 1588
4-Butyl-1-(4-hydroxyphenyl)-2-phenyl-3,5-pyrazolidinedione see 6925
4-Butyl-2-(*p*-hydroxyphenyl)-1-phenyl-3,5-pyrazolidinedione see 6925
t-Butyl Hypochlorite see 1570
tert-Butyl Hypochlorite, 1570
16,l7-Butylidenebis(oxy)-11,21-dihydroxypregna-1,4-diene-3,20-dione see 1455
Butylidene Chloride, 1571
p,*p*'-*sec*-Butylidenediphenol see 1312
6-*tert*-Butyl-3-(2-imidazolin-2-yl-methyl)-2,4-dimethylphenol see 6919
N-Butylimidodicarbonimidic Diamide see 1465
n-Butyl Iodide, 1572
sec-Butyl Iodide, 1573
6-*sec*-Butyl-3-isobutylpyrazinol 1-Oxide see 1455
2-[(3-Butyl-1-isoquinolinyl)oxy]-*N*,*N*-dimethylethanamine see 3207
n-Butylmalonic Acid, 1574
Butylmalonic Acid Mono(1,2-diphenylhydrazide) see 1472
n-Butyl Mercaptan, 1575
sec-Butyl Mercaptan, 1576
tert-Butyl Mercaptan, 1577
p-Butylmercaptobenzhydryl β-Dimethylaminoethyl Sulfide see 1772
Butylmercaptomethylpenicillin see 7035
n-Butylmercuric Chloride, 1578
Butyl Mesityl Oxide Oxalate see 1529
*N*¹-Butyl-*N*²-metanilylcarbamide see 1579
1-Butyl-3-metanilylurea, 1579
Butyl Methoxydibenzoylmethane, 1580
4-*tert*-Butyl-4'-methoxydibenzoyl-methane see 1580
2-*tert*-Butyl-4-methoxyphenol see 1547
3-*tert*-Butyl-4-methoxyphenol see 1547
N-*tert*-Butyl-1-methyl-3,3-diphenyl-propylamine see 9098
tert-Butyl Methyl Ether see 5954
tert-Butyl Methyl Ketone see 7409

4-Butyl-4-[(4-methyl-1-piperazinyl)-methyl]-1,2-diphenyl-3,5-pyrazolidinedione see 7424
2-*sec*-Butyl-2-methyl-1,3-propanediol Dicarbamate see 5653
N-Butyl-2-methyl-2-propyl-1,3-propanediol Dicarbamate see 9737
2-*sec*-Butyl-2-methyltrimethylene-carbamate see 5653
Butylmin [Nippon Kayaku] see 1588
n-Butyl Nitrite, 1581
tert-Butyl Nitrite, 1582
Butyl-Nor-Sympatol see 964
(±)-3α-*tert*-Butyl-2,3,4,4aβ-8,9,-13bα,14-octahydro-1*H*-benzo[6,7]-cyclohepta[1,2,3-*de*]pyrido[2,1-*a*]-isoquinolin-3-ol see 1499
Butylon [Grelan] see 1445
2-[*p*-(Butyloxy)phenyl]acetohydroxamic Acid see 1462
Butylparaben, 1583
Butyl Parasept see 1583
Butylphen see 1584
p-*tert*-Butylphenol, 1584
2-(*p*-*tert*-Butylphenoxy)cyclohexyl Propargyl Sulfite see 7818
2-(*p*-*tert*-Butylphenoxy)isopropyl 2-Chloroethyl Sulfite see 794
4-Butyl-1-phenyl-3,5-dioxopyrazolidine see 6141
1-(*p*-*tert*-Butylphenyl)-4-[4'-(α-hydroxydiphenylmethyl)-1'-piperidyl]-butanol see 9094
α-(*p*-*tert*-Butylphenyl)-4-(α-hydroxy-α-phenylbenzyl)-1-piperidinebutanol see 9094
α-Butyl-α-phenyl-1*H*-imidazole-1-propanenitrile see 3902
cis-4-[3-(4-*tert*-Butylphenyl)-2-methylpropyl]-2,6-dimethylmorpholine see 3938
1-[3-(*p*-*tert*-Butylphenyl)-2-methyl-propyl]piperidine see 3937
Butyl β-Phenyl-1-piperidinepropionate see 1510
4-Butyl-1-phenyl-3,5-pyrazolidinedione see 6141
4-*tert*-Butylphenyl Salicylate, 1585
1-Butyl-3-(phenylsulfonyl)urea see 7177
n-Butyl Phthalate, 1586
5-Butylpicolinic Acid see 4227
dl-*N*-*n*-Butylpipecolic Acid 2,6-Xylidide see 1485
dl-1-Butyl-2',6'-pipecoloxylidide see 1485
dl-1-*n*-Butylpiperidine-2-carboxylic Acid 2,6-Dimethylanilide see 1485
Butyl β-Piperidinohydrocinnamate see 1510
n-Butyl β-(*N*-Piperidyl)-β-phenylpropionate see 1510
Butylpropanedioic Acid see 1574
Butylpropanedioic Acid Mono(1,2-diphenylhydrazide) see 1472
n-Butyl Propionate, 1587
5-Butyl-2-pyridinecarboxylic Acid see 4227
Butylscopolamine [Kyoritsu] see 1588
Butylscopolamine Bromide see 1588
N-Butylscopolammonium Bromide, 1588
Butyl Stearate, 1589
1-Butyl-3-sulfanilylurea see 1839
n-Butyl Sulfide, 1590
N-(5-*tert*-Butyl-1,3,4-thiadiazol-2-yl)benzenesulfonamide see 4374
1-(5-*tert*-Butyl-1,3,4-thiadiazol-2-yl)-1,3-dimethylurea see 9053
*N*¹-(5-*tert*-Butyl-1,3,4-thiadiazol-2-yl)sulfanilamide see 4373

6-[[(Butylthio)acetyl]amino]-3,3-dimethyl-7-oxo-4-thia-1-azabicyclo[3.2.0]heptane-2-carboxylic Acid see 7035
sec-Butyl Thioalcohol see 1576
Butylthiobutane see 1590
p-Butylthiodiphenylmethyl 2-Dimethylaminoethyl Sulfide see 1772
2-(Butylthio)dodecanoic Acid Bismuth Basic Salt see 1272
Butylthiomethylpenicillin see 7035
2-[*p*-(Butylthio)-α-phenylbenzylthio]-*N*,*N*-dimethylethylamine see 1772
2-[[[4-(Butylthio)phenyl]phenylmethyl]thio]-*N*,*N*-dimethylethanamine see 1772
1-Butyl-3-(*p*-tolylsulfonyl)urea see 9432
N-*n*-Butyl-*N*'-tosylurea see 9432
1-*t*-Butyl-2-(1,2,4-triazol-1-yl)-2-(2',4'-dichlorobenzyl)ethanol see 3070
Butyl 2-[4-(5-Trifluoromethyl-2-pyridyloxy)phenoxy]propionate see 4049
1-Butyl-2-(2,6-xylylcarbamoyl)piperidine see 1485
2-Butynediamide see 1958
1,1'-(2-Butyne-1,4-diyl)bispyrrolidine see 9498
Butynorate see 3025
Butyn Sulfate [Abbott] see 1496
1,1'-(2-Butynylene)dipyrrolidine see 9498
Butyrac [Union Carbide] see 2828
Butyraldehyde, 1591
n-Butyramide, 1592
3-Butyramido-α-ethyl-2,4,6-triiodocinnamic Acid Sodium Salt see 1476
3-Butyramido-α-ethyl-2,4,6-triiodohydrocinnamic Acid Sodium Salt see 9745
5'-Butyramido-2'-(2-hydroxy-3-isopropylaminopropoxy)acetophenone see 13
5-Butyramido-1,3,4-thiadiazole-2-sulfonamide see 1511
Butyrchloral see 9545
Butyric Acid, 1593
n-Butyric Acid see 1593
1-Butyric Acid-6-(L-2-aminobutyric Acid)-7-glycineoxytocin see 1846
Butyric Acid Butyl Ester see 1556
Butyric Acid Ester with Dimethyl (2,2,2-Trichloro-1-hydroxyethyl)-phosphonate see 1528
Butyric Acid Ethyl Ester see 3733
Butyric Acid, Lead Salt see 5276
1-Butyric Acid-2-[3-(*p*-methoxyphenyl)-L-alanine]oxytocin see 1800
Butyric Acid Nitrile see 1597
Butyric Anhydride, 1594
γ-Butyrobetaine Chloride Methyl Ester see 1871
Butyroin, 1595
Butyrolactone, 1596
γ-Butyrolactone see 1596
Butyrone see 3351
Butyronitrile, 1597
N-Butyroyl-*p*-aminophenol see 4745
3-(3-Butyrylamino-2,4,6-triiodophenyl)-2-ethylacrylic Acid Sodium Salt see 1476
Butyryl Chloride, 1598
n-Butyryl Chloride, 1598
Butyrylcholinesterase see 2211
3'-[(5-Butyryl-2,4-dihydroxy-3,3-dimethyl-6-oxo-1,4-cyclohexadien-1-yl)methyl]-2',6'-dihydroxy-4'-methoxybutyrophenone see 2899

Calcium Benzylpenicillinate see 7039
Calcium Bichromate see 1666
Calcium Biphosphate see 1700
Calcium Bisulfite, Solution, 1650
Calcium Borate, 1651
Calcium Borogluconate, 1652
Calcium Bromide, 1653
Calcium Bromolactobionate, 1654
Calcium N-Carbamoylaspartate, 1655
Calcium Carbide, 1656
Calcium Carbimide see 1662
Calcium Carbolate see 1698
Calcium Carbonate, 1657
Calcium Chel 330 see 7082
Calcium Chlorate, 1658
Calcium Chloride, 1659
Calcium Chromate(VI), 1660
Calcium Chrome Yellow see 1660
Calcium Citrate, 1661
Calcium Creosotate see 2575
Calcium Cyanamide, 1662
Calcium Cyanamide Citrated, 1663
Calcium Cyanide, 1664
Calcium Cyclamate, 1665
Calcium Cyclohexanesulfamate see 1665
Calcium Cyclohexylsulfamate see 1665
Calcium-dependent Regulator Protein see 1726
Calcium Diborogluconate see 1652
Calcium Dichromate(VI), 1666
Calcium D(+)-N-(2,4-Dihydroxy-3,3-dimethylbutyryl)-β-alaninate see 1694
Calcium Dioctyl Sulfosuccinate see 3397
Calcium Dioxide see 1696
Calcium Diphosphate see 1706
Calcium Disodium Edetate see 3480
Calcium Disodium Ethylenediamine-tetraacetate see 3480
Calcium Disodium (Ethylenedinitrilo)-tetraacetate see 3480
Calcium Disodium Versenate [Riker] see 3480
Calcium Diuretin see 9209
Calcium Dobesilate see 3395
Calcium 2-Ethylbutanoate, 1667
Calcium Ethyl Sulfate see 3808
Calcium Ferrous Citrate, 1668
Calcium Fluoride, 1669
Calcium Fluorophosphate, 1670
Calcium Fluosilicate see 1674
Calcium Folinate see 4141
Calcium Formate, 1671
Calcium Galactogluconate Bromide see 1654
Calcium Glucoheptonate see 4349
Calcium Glucomonocarbonate see 4349
Calcium Gluconate, 1672
Calcium Glucosemonocarbonate see 4349
Calcium Glycerinophosphate see 1673
Calcium Glycerophosphate, 1673
Calcium Hexafluorosilicate, 1674
Calcium Homopantothenate see 4664
Calcium Hydrate see 1676
Calcium Hydride, 1675
Calcium Hydroxide, 1676
Calcium 2-Hydroxy-1-naphthalene-sulfonate see 847
Calcium Hypochlorite, 1677
Calcium Hypophosphite, 1678
Calcium Hyposulfite see 1719
Calcium Iodate, 1679
Calcium Iodide, 1680
Calcium Iodobehenate, 1681
Calcium Iodostearate, 1682
Calcium Ipodate see 4958

Calcium Ketomalonate see 1685
Calcium Lactate, 1683
Calcium Lactobionate see 5219
Calcium Levulinate, 1684
Calcium Lysinate see 5509
Calcium Mandelate see 5599
Calcium [Mercaptoacetato(2−)-O,S]-aurate(1−) (1:2) see 6241
Calcium Mesoxalate, 1685
Calcium Methionate, 1686
Calcium Methyl Sulfate see 6041
Calcium Molybdate(VI), 1687
Calcium Monofluorophosphate see 1670
Calcium Monohydrogen Phosphate see 1699
Calcium β-Naphthol-α-monosulfonate see 847
Calcium 2-Naphthol-1-sulfonate see 847
Calcium Nitrate, 1688
Calcium Nitrite, 1689
Calcium Oleate, 1690
Calcium Orthophosphate, Basic see 3449
Calcium Oxalate, 1691
Calcium Oxide, 1692
Calcium Oxomalone see 1685
"Calcium Oxychloride" see 2093
Calcium Oxysulfide Solution see 5369
Calcium Palmitate, 1693
Calcium Pantothenate, 1694
Calcium Pectolith see 1709
Calcium Penicillin G see 7039
Calcium Permanganate, 1695
Calcium Peroxide, 1696
Calcium Phenate see 1698
Calcium Phenolate see 1698
Calcium Phenolsulfonate, 1697
Calcium Phenoxide, 1698
Calcium Phenylate see 1698
Calcium Phosphate, Dibasic, 1699
Calcium Phosphate Hydroxide see 3449
Calcium Phosphate, Monobasic, 1700
Calcium Phosphate, Tetrabasic see 9305
Calcium Phosphate, Tribasic, 1701
Calcium Phosphide, 1702
Calcium Phosphite, 1703
Calcium Phosphoglycerate see 1673
Calcium Polycarbophil, 1704
Calcium Propionate, 1705
Calcium Pyroborate see 1651
Calcium Pyrophosphate, 1706
Calcium Rhodanate see 1717
Calcium D-Saccharate, 1707
Calcium Selenide, 1708
Calcium Silicate, 1709
Calcium Silicofluoride see 1674
Calcium Stearate, 1710
Calcium Stearyl-2 Lactylate, 1711
Calcium Stelate see 1711
Calcium Succinate, 1712
Calcium Sulfate, 1713
Calcium Sulfide, 1714
Calcium Sulfite, 1715
Calcium Sulfocarbolate see 1697
Calcium Sulfocyanate see 1717
Calcium Sulfomethylate see 6041
Calcium Sulfophenolate see 1697
Calcium Sulfovinate see 3808
"Calcium Superphosphate" see 1700
Calcium Tartrate, 1716
Calcium Tetraborate see 1651
Calcium Tetrahydroaluminate see 334
Calcium Thiocyanate, 1717
Calcium Thioglycollate, 1718
Calcium Thiosulfate, 1719
Calcium Trisodium Pentetate see 7082
Calcium Tungstate(VI), 1720

Calcium Ureidosuccinate see 1655
Calcort [Lepetit] see 2852
Calcotheobromine see 9209
Calcreose [Pennwalt] see 2575
Caldariomycin, 1721
Calderol [Upjohn] see 1638
Caldon [Hoechst] see 3282
C-Calebassine, 1722
Calendula, 1723
Calepsin [Orion] see 1783
Calglucon see 1672
Calgon see 8621
Calheptose see 4349
Calibène [Carrion] see 8990
Calico Yellow see 6179
Californit [Merckle] see 6925
Californium, 1724
Caliment [Apotex] see 7476
Calioben see 1681
Calisaya Bark see 2286
Calixin [BASF] see 9576
Callicrein see 5159
Callidin I see 1356
Callistephin see 7014
Callusolve [Dermal] see 1066
Calmabel see 377
Calmagite, 1725
Calmalone [Cassenne] see 2709
Calmatel (Aerosol) [Almirall] see 7393
Calmatel (Cream) [Almirall] see 7393
Calmax see 5751
Calmaxid [Lilly] see 6582
Calmday [Will-Pharma] see 6608
Calmeran see 909
Calmipan see 4465
Calmiren see 5751
Calmoden [Berk] see 2082
Calmodid see 4708
Calmodulin, 1726
Calmonal [Heyden] see 5657
Calmpose [Ranbaxy] see 2977
Calmus see 1637
Calnathal see 1798
Calnegyt [Toyo Jozo] see 4472
Calogreen [Mallinckrodt] see 5795
Calomel see 5795
Calomelol, 1727
Calomide [Yamanouchi] see 2446
Calonat see 2922
Calorose see 4900
Calosen [Apotex] see 6337
Calotropin, 1728
Calpanate [Consumers Vitamin] see 1694
Calphosan [Carlton] see 1673
Calpol [Calmic] see 40
Calscorbat [Biosedra] see 855
Calscorbate [Cole] see 855
Calsekin [Mect] see 4146
Calsmin [Upjohn] see 6491
Calsol see 3482
Calsyn [Armour-Montagu] see 1640
Calsynar [Armour] see 1640
Calthor [Ayerst] see 2706
Caltidren [Liphar] see 1875
Calumba, 1729
Calurin [Dorsey] see 1644
Calusterone, 1730
Calvisken [Sandoz-Sankyo] see 7412
Calx see 1692
Calycanthine, 1731
Calystigine see 6945
CaM see 1726
CAM-AQ1 see 602
Camazepam, 1732
Cambendazole, 1733
Cambenzole [Merck & Co.] see 1733
Cambet [Merck & Co.] see 1733
Cambilene [Fisons] see 3026
Cambogia see 4260
Camcolit [Norgine] see 5404

Camcopot [Camden] *see* 7601
Camdelate *see* 5599
Camganiba *see* 5597
Camite *see* 1400
Camoform [Parke, Davis] *see* 1217
Camolar [Parke, Davis] *see* 2727
Camomile *see* 2032
Camont [Klinge] *see* 6441
Camoquin [Parke, Davis] *see* 602
Camostat, 1734
Camostat Mesylate *see* 1734
cAMP *see* 2714
Campel [Farmitalia Carlo Erba] *see* 2239
Campesterol, 1735
endo-2-Camphanol *see* 1338
exo-2-Camphanol *see* 5011
2-Camphanone *see* 1738
Camphechlor *see* 9478
Camphene, 1736
Camphetamide *see* 1741
Camphidonium *see* 9622
Camphobismol *see* 1737
d-Camphocarboxylic Acid, 1737
Camphol *see* 1338
Camphophyline [Millot] *see* 3663
Camphor, 1738
d-3-Camphorcarboxylic Acid *see* 1737
Camphoric Acid, 1739
Camphoric Acid 1-(*p*,α-Dimethylbenzyl) Ester, Compd with 2,2'-Imidodiethanol (1:1) *see* 9415
d-Camphoric Acid Ethyl Ester Bismuth Salt *see* 1275
Camphor Monobromated *see* 1402
Camphor Monochlorated *see* 2132
Camphor Solubilized *see* 1737
d-Camphorsulfonic Acid, 1740
10-Camphorsulfonic Acid *see* 1740
β-Camphorsulfonic Acid *see* 1740
Camphostene *see* 6929
Camphostyl *see* 1740
Camphosulfonyl-*N*-methylpyridine-β-diethylcarboxamide *see* 1741
Camphotamide, 1741
Camphramine *see* 1741
Camphydryl *see* 1737
Campolon [Winthrop; Bayer] *see* 5427
Camporit *see* 1677
Camposan *see* 3686
Campovit *see* 5427
Campoviton 6 *see* 7995
Camptothecin, 1742
Camsellite *see* 5535
Camsilate *see* MISC-4
Camsylate *see* 1740
Camylofine, 1743
Camyna [Boehringer, Ing.] *see* 9388
Canada Snakeroot *see* 850
Canada Turpentine *see* 956
Canadian Hemp *see* 774
Canadine, 1744
Canadol *see* 5366
Cananga Oil *see* 10008
Canary Dextrin *see* 2931
Canavanine, 1745
Canchalagua *see* 1966
Candamide *see* 5404
Candelilla Wax, 1746
Candeptin [Julius Schmid] *see* 1747
Canderel [Searle] *see* 861
Candex [Dome] *see* 6658
Candicidin, 1747
Candicidin D *see* 1747
Candidin, 1748
Candidinin *see* 1748
Candidoin *see* 1748
Candimon [Ayerst] *see* 1747
Candio-Hermal [Hermal] *see* 6658
Candiolin *see* 4187
Candleberry Bark *see* 1022

Canella, 1749
Canescine *see* 2901
Canesten [Bayer] *see* 2412
Cane Sugar *see* 8855
Canex [Pitman-Moore] *see* 8248
Canferon [Takeda] *see* 4892
Canfodion [Gentili] *see* 8116
Canfoxil *see* 1737
Canifug [Wolff] *see* 2412
Cannabidiol, 1750
Cannabinol, 1751
Cannabis, 1752
Cannabiscetin *see* 6244
Cannogenin α-L-Thevetoside *see* 7136
Canocenta [Byk Gulden] *see* 5459
Canogard [Shell] *see* 3069
Canopar [Cooper] *see* 9205
Canrenone, 1753
Cantabilin [Formenti] *see* 4792
Cantabiline *see* 4792
Cantan [Hoechst] *see* 855
Cantaxin [Winthrop] *see* 855
Cantharides, 1754
Cantharides Camphor *see* 1755
Cantharidin, 1755
Cantharis Vesicatoria see 1754
Canthaxanthin, 1756
Cantil [Lakeside] *see* 5735
Canton's Phosphorus *see* 1714
Cantor [Clin Midy] *see* 6119
Cantralax [Ferrosan] *see* 1887
Cantrex *see* 5161
Cantricin [Corvi] *see* 4202
Cantril *see* 5735
Cantrodifene (obsolete) *see* 6555
Canutillo *see* 3560
Caocobre [Sandoz] *see* 2671
Caomet [Simes] *see* 9751
Caoutchouc *see* 8255
CAP *see* MISC-2
Caparlem *see* 5153
Caparol [Geigy] *see* 7800
Caparside *see* 819
Caparsolate [Abbott] *see* 819
Capastat [Lilly] *see* 1758
Capazine *see* 7768
Capben [Comm. Solvents] *see* 1757
Capen [Phoenix] *see* 9386
Caperase *see* 1906
Capisten [Kissei] *see* 5184
Capitol [Dermal] *see* 1066
Capitrol [Westwood] *see* 2175
Capitus [Berk] *see* 3704
Capla [Wallace Labs.] *see* 5653
Caplenal [Berk] *see* 278
Capmul 8210 [Stokely-Van Camp] *see* 6162
Capnoidine *see* 154
Capobenic Acid, 1757
Capostatin *see* 1758
Capoten [Squibb] *see* 1773
Cap-O-Tran [Croyden-Browne] *see* 5751
Capozide [Squibb] *see* 1773
Capquin [Ciba-Geigy] *see* 2163
Capralense [Choay] *see* 442
Capramol [Choay] *see* 442
Capreomycin, 1758
Capreomycin IA *see* 1758
Capreomycin IB *see* 1758
Capreomycin IIB *see* 1758
Capreomycins IIA *see* 1758
n-Capric Acid, 1759
Caprin [Sinclair] *see* 873
Caprine *see* 6624
Caproaldehyde *see* 1761
Caproate *see* MISC-3
Caprocid *see* 442
Caprocin [Lilly] *see* 1758
Caprodat *see* 1848
n-Caproic Acid, 1760

n-Caproic Acid *n*-Amyl Ester *see* 640
Caproic Aldehyde, 1761
Caprokol [Merck & Co.] *see* 4633
Caprolactam, 1762
ε-Caprolactam *see* 1762
Caprolan [Allied] *see* 6656
Caprolin *see* 1758
Capromycin *see* 1758
Caprosem *see* 9110
Caproyl Chloride, 1763
Caprylaldehyde *see* 1766
Caprylene, 1764
Caprylic Acid, 1765
Caprylic Acid Ammonium Salt *see* 526
Caprylic Acid Monoglyceride *see* 6162
Caprylic Alcohol *see* 6674
Caprylic Aldehyde, 1766
Caprylohydroxamic Acid *see* 6673
Capsaicin, 1767
Capsanthin, 1768
Capsebon [Pitman-Moore] *see* 1628
Capsicum, 1769
Captafol, 1770
Captagon [Homburg] *see* 3919
Captan, 1771
Captax *see* 5759
Captea [Theraplix] *see* 1773
Captin [Krewel] *see* 40
Captodiam *see* 1772
Captodiamine, 1772
Captodramin *see* 1772
Captolane [Theraplix] *see* 1773
Captopril, 1773
Captoril [Sankyo] *see* 1773
Capture [FMC] *see* 1229
Capuride, 1774
Capval [Dreluso] *see* 6638
Carace [Morson] *see* 5393
Carachol *see* 2858
Caracurine VII *see* 9957
Caradrin [Asta] *see* 7889
Carafate [Marion] *see* 8853
Caramel, 1775
Caramiphen Ethanedisulfonate, 1776
Caramiphen Hydrochloride, 1777
Caraway, 1778
Carazolol, 1779
Carbachol, 1780
Carbacrylic Resins, 1781
Carbacylamine Resins *see* 1781
Carbadipimidine *see* 1869
Carbadox, 1782
Carbamaldehyde *see* 4151
Carbamamidine *see* 4475
Carbamate *see* 3954
Carbamate of 2-(Hydroxymethyl)-2-methylpentyl Ester of Butylcarbamic Acid *see* 9737
Carbamate of (2-Hydroxypropyl)trimethylammonium Chloride *see* 1207
Carbamazepine, 1783
Carbamazine *see* 3106
Carbamic Acid, 1784
Carbamic Acid *p*-Benzylphenyl Ester *see* 3306
Carbamic Acid 2-*sec*-Butyl-2-methyltrimethylene Ester *see* 5653
Carbamic Acid 3-(*p*-Chlorophenoxy)-2-hydroxypropyl Ester *see* 2179
Carbamic Acid 1-Cyclohexylpropyl Ester *see* 7771
Carbamic Acid 1,1-Dimethylpropyl Ester *see* 1641
Carbamic Acid Ester with 5-Butyl-5-(2-hydroxyethyl)barbituric Acid *see* 1838
Carbamic Acid Ester with 5-(β-Hydroxyethyl)-5-butylmalonylurea *see* 1838
Carbamic Acid Ethyl Ester *see* 9789

Carbocaina [Pierrel] see 5748
Carbocaine Hydrochloride [Winthrop] see 5748
Carbocalcitonin see 3501
Carbocholine see 1780
Carbochromen see 2241
Carbocit [CT] see 1809
Carbocloral, 1808
Carbocromen see 2241
Carbocysteine, 1809
Carbodicyclohexylimide see 3086
Carbodiimide see 2691
N-(1S-Carbethoxy-3-phenylpropyl)-S-alanyl-cis,endo-2-azabicyclo-[3.3.0]octane-3S-carboxylic Acid see 8123
Carboethoxyphthalazinohydrazine see 9425
Carbofil see 1704
Carbofos see 5582
Carbofuran, 1810
Carbohydrazide, 1811
Carbolic Acid see 7206
5-Carboline see 1812
β-Carboline see 3737
γ-Carboline, 1812
Carbolith [Winley Morris] see 5404
Carbolithium [IFI] see 5404
Carbomer see 1836
Carbomethene see 5177
2α-Carbomethoxy-3β-benzoxytropane see 7926
2β-Carbomethoxy-3β-benzoxytropane see 2450
Carbomethoxyibogaine see 9944
2-Carbomethoxy-1-methylvinyl Dimethyl Phosphate see 6089
Carbomix [Penn] see 1815
Carbomycin, 1813
Carbomycin A see 1813
Carbomycin B see 1813
Carbon, 1814
Carbon, Activated see 1815
Carbon, Amorphous, 1815
Carbonate Dihydratase see 1819
Carbonate Hydro-lyase see 1819
[Carbonato(1−)-O]dihydroxyaluminum Monosodium Salt see 3169
[Carbonato(2)]heptahydroxy(aluminum)trimagnesium Dihydrate see 297
Carbon Bisulfide see 1818
Carbon Black see 1815
Carbon, Decolorizing see 1815
Carbon Dioxide, 1816
Carbon Diselenide, 1817
Carbon Disulfide, 1818
Carbon Hexachloride see 4601
Carbonic Acid, Aluminum-magnesium Complex see 297
Carbonic Acid Bis(2-methoxyphenyl) Ester see 4459
Carbonic Acid 2-sec-Butyl-4,6-dinitrophenyl Isopropyl Ester see 3280
Carbonic Acid Calcium Salt (1:1) see 1657
Carbonic Acid Dibutyl Ester see 1558
Carbonic Acid Diethyl Ester see 3738
Carbonic Acid Dimethyl Ester see 5960
Carbonic Acid Diphenyl Ester see 7250
Carbonic Acid Gas see 1816
Carbonic Acid Guaiacol Ether see 4459
Carbonic Acid 1-Methylethyl 2-(1-Methylpropyl)-4,6-dinitrophenyl Ester see 3280
Carbonic Anhydrase, 1819
Carbonic Anhydrase Inhibitor No. 6063 see 45

Carbonic Anhydride see 1816
Carbonic Dichloride see 7310
Carbonic Difluoride see 1826
Carbonic Dihydrazide see 1811
Carbon Monoxide, 1820
Carbonochloridic Acid (4-Bromophenyl)methyl Ester see 1399
Carbonochloridic Acid Ethyl Ester see 3742
Carbonochloridic Acid Methyl Ester see 5966
Carbonochloridic Acid 2-Methylpropyl Ester see 5023
Carbonochloridic Acid Phenylmethyl Ester see 1807
Carbonochloridic Acid Propyl Ester see 7860
Carbonochloridic Acid Trichloromethyl Ester see 3339
Carbonodithioic Acid O-Ethyl Ester Potassium Salt see 7697
Carbonotrithioic Acid see 9671
Carbon Pernitride see 2699
Carbon Selenide see 1817
Carbon Suboxide, 1821
Carbon Tetrachloride, 1822
Carbon Tetrafluoride, 1823
Carbon Tetraiodide, 1824
1,1'-Carbonylbis-1H-imidazole see 1825
8,8'-[Carbonylbis[imino-3,1-phenylenecarbonylimino(4-methyl-3,1-phenylene)carbonylimino]]bis-1,3,5-naphthalenetrisulfonic Acid Hexasodium Salt see 8986
Carbonyl Chloride see 7310
Carbonyldiamide see 9781
N,N'-Carbonyldiimidazole, 1825
3,3'-(Carbonyldiimino)bisbenzenecarboximidamide see 405
Carbonyldiurea see 9678
Carbonyl Fluoride, 1826
Carbonylhydrotris(triphenylphosphine)rhodium see 9670
Carbonylsalicylamide see 1873
Carbophenothion, 1827
Carboplatin, 1828
Carbopol see 1836
Carboprost, 1829
Carboprost Methyl see 1829
Carboprost Trometamol see 1829
Carboquone, 1830
Carboraffin see 1815
Carbo-Resin [Lilly] see 1781
Carbose D see 1835
Carb-O-Sep [Rohm & Haas] see 1788
Carbostesin [Woelm] see 1485
Carbostibamide see 9786
Carbostyril, 1831
Carbothiamine see 2762
Carbothrone see 721
Carbowax [Union Carbide] see 7545
Carboxin, 1832
N-Carboxy-β-alanyl-L-tryptophyl-L-methionyl-L-aspartylphenyl-L-alaninamide N-tert-Butyl Ester see 7067
1-Carboxy-2-amino-9-hydroxymethylphenoxazin-3-one see 2297
D-8β-[(Carboxyamino)methyl]-1,6-dimethylergoline I Benzyl Ester see 5843
N-Carboxyaniline see 1786
α-Carboxy-o-anisic Acid Compd with Antipyrine see 748
β-Carboxyaspartic Acid, 1833
N-[p-(o-Carboxybenzamido)benzenesulfonyl]acetamide see 7350
p-Carboxybenzenesulfo-di-N-ethylamide see 3684

p-Carboxybenzenesulfonamide see 1885
p-Carboxybenzenesulfondichloroamide see 4503
[(4-Carboxy-2-benzimidazolyl)thio]gold see 9652
N-(o-Carboxybenzoyl)sulfacetamide see 7350
α-Carboxybenzylpenicillin see 1796
α-Carboxybenzylpenicillin Phenyl Ester Sodium Salt see 1844
d-3-Carboxy-2-bornanone see 1737
(2S,3aS,7aS)-1-[(S)-N-[(S)-1-Carboxybutyl]alanyl]hexahydro-2-indolinecarboxylic Acid 1-Ethyl Ester see 7122
d-3-Carboxy-2-camphanone see 1737
α-Carboxycaproyl-N,N'-diphenylhydrazine see 1472
N-(2-Carboxycaproyl)hydrazobenzene see 1472
[6R-[6α,7β(7R*)]]-1-[[2-Carboxy-7-[[[[(5-carboxy-1H-imidazol-4-yl)-carbonyl]amino]phenylacetyl]amino]-8-oxo-5-thia-1-azabicyclo-[4.2.0]oct-2-en-3-yl]methyl]-4-(2-sulfoethyl)pyridinium Hydroxide, Inner Salt see 1939
1-[(6R,7R)-2-Carboxy-7-[(R)-2-(5-carboxy-4-imidazolylcarboxamido)-2-phenylacetamido]-8-oxo-5-thia-1-azabicyclo[4.2.0]oct-2-en-3-yl-methyl]pyridino-4-ethylsulfonate see 1939
18-Carboxy-20-(carboxymethyl)-8-ethenyl-13-ethyl-2,3-dihydro-3,7,-12,17-tetramethyl-21H,23H-porphine-2-propanoic Acid see 7360
2-Carboxy-3-carboxymethyl-4-isopropenylpyrrolidine see 5157
p-Carboxy-N,N-diethylbenzenesulfonamide see 3684
α-(5-Carboxy-1,2-dihydro-2-hydroxyphenoxy)acrylic Acid see 2220
D-N-Carboxydihydro-1-methyllysergamine I Benzyl Ester see 5843
(S)-α-Carboxy-2,3-dihydro-N,N,N-trimethyl-2-thioxo-1H-imidazole-4-ethanaminium Hydroxide Inner Salt see 3611
1-Carboxy-4,5-dihydroxy-1,3-cyclohexylenebis-(3,4-dihydroxycinnamate) see 2773
2-Carboxy-3,4-dimethoxybenzal Isonicotinylhydrazone see 6807
1-(2-Carboxy-3,4-dimethoxybenzylidene)-2-isonicotinoylhydrazine see 6807
1-Carboxy-7-(dimethylamino)-3,4-dihydroxyphenoxazin-5-ium Chloride see 4256
N-(2-Carboxy-3,3-dimethyl-7-oxo-4-thia-1-azabicyclo[3.2.0]hept-6-yl)-2-phenylmalonamic Acid see 1796
N-(2-Carboxy-3,3-dimethyl-7-oxo-4-thia-1-azabicyclo[3.2.0]hept-6-yl)-2-phenylmalonamic Acid 1-(5-Indanyl) Ester see 1847
N-(2-Carboxy-3,3-dimethyl-7-oxo-4-thia-1-azabicyclo[3.2.0]hept-6-yl)-2-phenylmalonamic Acid 1-Phenyl Ester Sodium Salt see 1844
N-(2-Carboxy-3,3-dimethyl-7-oxo-4-thia-1-azabicyclo[3.2.0]hept-6-yl)-3-thiophenemalonamic Acid see 9360
2-Carboxy-1,1-dimethylpyrrolidinium Hydroxide Inner Salt see 8729
(3R-trans)-3-[(1-Carboxyethenyl)-oxy]-4-hydroxy-1,5-cyclohexadiene-1-carboxylic Acid see 2220

2-(2-Carboxyethyl)benzimidazole *see* 7769

3-*O*-α-Carboxyethyl-D-glucosamine *see* 6213

17α-(2-Carboxyethyl)-17β-hydroxyandrosta-4,6-dien-3-one Lactone *see* 1753

17α-(2-Carboxyethyl)-17β-hydroxy-3-oxoandrosta-4,6-diene Lactone *see* 1753

3-Carboxy-1-ethyl-7-methyl-1,8-naphthyridin-4-one *see* 6273

2-[4-(1-Carboxyethyl)phenyl]-1-isoindolinone *see* 4875

2-(2-Carboxyethyl)-1,2,3,4-tetrahydro-6,7-dimethoxy-2-methyl-1-veratrylisoquinolinium Benzenesulfonate Pentamethylene Ester *see* 884

6-(1-Carboxyethyl)-3,4,5,6-tetrahydro-3,5-dimethyl-2-pyranone *see* 7741

L-γ-Carboxyglutamic Acid *see* 1834

γ-Carboxyglutamic Acid, 1834

γ-Carboxy-L-glutamic Acid *see* 1834

β-Carboxyglutaric Acid *see* 9534

Carboxyhemoglobin *see* 4567

3-Carboxy-4-hydroxybenzenesulfonic Acid *see* 8944

N-[*S*-[1-(4-Carboxy-1-hydroxybutyl)-2,4,6,9-pentadecatetraenyl]-L-cysteinyl]glycine *see* 5339

N-[*S*-[1-(4-Carboxy-1-hydroxybutyl)-2,4,6,9-pentadecatetraenyl]-*N*-L-γ-glutamyl-L-cysteinyl]glycine *see* 5339

1-Carboxy-4-hydroxy-2,5-cyclohexadiene-1-pyruvic Acid *see* 7745

L-(8*S*)-β-(1-Carboxy-4-hydroxy-2,5-cyclohexadien-1-yl)alanine *see* 814

cis-2-Carboxy-4-hydroxy-1,1-dimethylpyrrolidinium Hydroxide Inner Salt *see* 9729

trans-2-Carboxy-4-hydroxy-1,1-dimethylpyrrolidinium Hydroxide, Inner Salt *see* 1209

2α-Carboxy-3β-hydroxy-*A*(1)-norlup-20(29)-en-28-oic Acid *see* 1916

1-Carboxy-4-hydroxy-α-oxo-2,5-cyclohexadiene-1-propanoic Acid *see* 7745

7β-[2-Carboxy-2-(4-hydroxyphenyl)-acetamido]-7α-methoxy-3-[[(1-methyl-1*H*-tetrazol-5-yl)thio]methyl]-1-oxa-1-dethia-3-cephem-4-carboxylic Acid *see* 6201

7-[[Carboxy(4-hydroxyphenyl)acetyl]-amino]-7-methoxy-3-[[(1-methyl-1*H*-tetrazol-5-yl)thio]methyl]-8-oxo-5-oxa-1-azabicyclo[4.2.0]oct-2-ene-2-carboxylic Acid *see* 6201

5-[(3-Carboxy-4-hydroxyphenyl)(3-carboxy-4-oxo-2,5-cylohexadien-1-ylidene)methyl]-2-hydroxybenzoic Acid Triammonium Salt *see* 320

3-(3-Carboxy-4-hydroxyphenyl)-2-phenyl-4,5-dihydro-3*H*-benz[e]-indole *see* 3917

(3-Carboxy-2-hydroxypropyl)trimethylammonium Hydroxide, Inner Salt *see* 1856

(3-Carboxy-2-hydroxypropyl)trimethylammonium Hydroxide Inner Salt Acetate *see* 78

3-Carboxy-5-hydroxy-1-*p*-sulfophenyl-4-*p*-sulfophenylazopyrazole Trisodium Salt *see* 9041

3-Carboxy-2-hydroxy-*N*,*N*,*N*-trimethyl-1-propanaminium Hydroxide, Inner Salt *see* 1856

7-β-[D-(−)-α-(4-Carboxyimidazole-5-carboxamido)phenylacetamido]-3-(4-β-sulfoethylpyridinium)methyl-3-cephem-4-carboxylic Acid *see* 1939

(1-Carboxy-2-imidazol-4-ylethyl)trimethyl Ammonium Hydroxide Inner Salt *see* 4588

α-Carboxylase *see* 8031

3-[[2-(Carboxylatomethoxy)benzoyl]-amino]-2-methoxypropyl]hydroxymercurate(1−) Sodium *see* 5805

[3-[[(3-Carboxylato-2,2,3-trimethylcyclopentyl)carbonyl]amino]-2-methoxypropyl][mercaptoacetato-(2−)-*O*,*S*]mercurate(2−) Disodium-(*T*-4) *see* 5761

[1-Carboxy-2-[2-mercaptoimidazol-4(or 5)-yl]ethyl]trimethylammonium Hydroxide, Inner Salt *see* 3611

2′-Carboxymethoxy-4,4′-bis(3-methyl-2-butenyloxy)chalcone *see* 8658

N-[(6*R*,7*R*)-2-Carboxy-7-methoxy-3-[[(1-methyl-1*H*-tetrazol-5-yl)thio]-methyl]-8-oxo-5-oxa-1-azabicyclo-[4.2.0]oct-2-en-7-yl]-2-(*p*-hydroxyphenyl)malonamic Acid *see* 6201

4-Carboxymethylamino-4′-aminodiphenylsulfone *see* 17

4-Carboxymethylbiphenyl *see* 3893

N-Carboxymethyl-*N*,*N*-bis(methylenephosphonic Acid)amine *see* 4409

Carboxymethylcellulose Sodium, 1835

S-Carboxymethylcysteine *see* 1809

S-(Carboxymethyl)-L-cysteine *see* 1809

4-*O*-(Carboxymethyl)-1-deoxy-1,4-dihydro-4-hydroxy-1-oxorifamycin γ-Lactone *see* 8216

3,3′-Carboxymethylene Bis(4-hydroxycoumarin)ethyl Ester *see* 3729

(*E*)-13-(Carboxymethylene)-14α-methyl-7-oxopodocarpan-16-oic Acid 13-[2-(Dimethylamino)ethyl] methyl Ester *see* 1897

2-Carboxy-4-(1-methylethenyl)-3-pyrrolidineacetic Acid *see* 5157

Carboxymethyl Ether Cellulose Sodium Salt *see* 1835

N-(Carboxymethyl)glycine *see* 4832

N-(Carboxymethyl)-*N*′-(2-hydroxyethyl)-*N*,*N*′-ethylenediglycine Trisodium Salt *see* 9871

[[(Carboxymethyl)imino]bis(ethylenenitrilo)]tetraacetic Acid *see* 7083

[[(Carboxymethyl)imino]bis(ethylenenitrilo)]tetraacetic Acid Calcium Complex Trisodium Salt *see* 7082

Carboxymethylmenadione Monoxime Ammonium Salt *see* 5717

N-(γ-Carboxymethylmercaptomercuri-β-methoxy)propylcamphoramic Acid Disodium Salt *see* 5761

[*R*-[*R**,*S**-(*E*,*Z*,*Z*,*E*,*E*,*Z*,*E*)]]-20-(Carboxymethyl)-6-methoxy-2,5,17-trimethyl-2,4,8,10,14,18,20-docosaheptaenedioic Acid *see* 1334

3-Carboxymethyl-17-methoxy-6,18,-21-trimethyldocosa-2,4,8,12,14,18,-20-heptaenedioic Acid *see* 1334

3-Carboxymethyl-5-(2-methylcinnamylidene)rhodanine *see* 3556

2-Carboxy-5-methylpyrazine 4-Oxide *see* 106

2-Carboxy-1-methylpyridinium Hydroxide, Inner Salt *see* 4648

3-Carboxy-1-methylpyridinium Hydroxide Inner Salt *see* 9606

4-*O*-(Carboxymethyl)rifamycin *see* 8216

(Carboxymethyl)sodium Sodium Salt *see* 8514

Carboxymethyltheophylline *see* 18

3-[(Carboxymethyl)thio]alanine *see* 1809

2-[2-[(Carboxymethyl)thio]ethyl]-4-thiazolidinecarboxylic Acid 2-Ethyl Ester *see* 5330

[(Carboxymethyl)thio]gold Calcium Salt *see* 6241

(Carboxymethyl)trimethylammonium Chloride *see* 1201

(Carboxymethyl)trimethylammonium Chloride Hydrazide *see* 4323

(Carboxymethyl)trimethylammonium Hydroxide *see* 1201

(Carboxymethyl)trimethylammonium Hydroxide Inner Salt *see* 1201

N-Carboxy-3-morpholinosydnonimine Ethyl Ester *see* 6143

2-Carboxy-4-[2′-(5′-nitrofuryl)]-quinoline 1-Oxide *see* 6449

[3-(3-Carboxy-2-oxo-2*H*-1-benzopyran-8-yl)-2-methoxypropyl]hydroxymercurate(1−) Hydrogen *see* 5764

[3-(3-Carboxy-2-oxo-2*H*-1-benzopyran-8-yl)-2-methoxypropyl]hydroxymercury Sodium Salt Compd with Theophylline *see* 5765

20β-Carboxy-11-oxo-30-norolean-12-en-3β-yl-2-*O*-β-D-glucopyranuronosyl-α-D-glucopyranosiduronic Acid *see* 4401

21-(3-Carboxy-1-oxopropoxy)-11β,17-dihydroxypregna-1,4-diene-3,20-dione Monosodium Salt *see* 7722

3-(3-Carboxy-1-oxopropoxy)-11-oxoolean-12-en-29-oic Acid *see* 1797

21-(3-Carboxy-1-oxopropoxy)-5β-pregnane-3,20-dione Sodium Salt *see* 4753

[3-[[[(3-Carboxy-1-oxopropyl)amino]carbonyl]amino]-2-methoxypropyl]hydroxymercury, Mixture with 3,7-Dihydro-1,3-dimethyl-1*H*-purine-2,6-dione *see* 5756

1-[[2-Carboxy-8-oxo-7-[(2-thienylacetyl)amino]-5-thia-1-azabicyclo-[4.2.0]oct-2-en-3-yl]methyl]pyridinium Hydroxide Inner Salt *see* 1975

2-(5-Carboxypentyl)-4-thiazolidone *see* 6233

Carboxyphen *see* 2932

6-(2-Carboxy-2-phenylacetamido)-3,3-dimethyl-7-oxo-4-thia-1-azabicyclo[3.2.0]heptane-2-carboxylic Acid 6-(5-Indanyl Ester) *see* 1847

6-(α-Carboxyphenylacetamido)penicillanic Acid *see* 1796

6-[(Carboxyphenylacetyl)amino]-3,3-dimethyl-7-oxo-4-thia-1-azabicyclo[3.2.0]heptane-2-carboxylic Acid *see* 1796

2-[(2-Carboxyphenyl)amino]-4-chlorobenzoic Acid *see* 5433

4-[(2-Carboxyphenyl)amino]-7-chloroquinoline α-Monoglyceride *see* 4330

N-(2-Carboxyphenyl)-4-chloroanthranilic Acid *see* 5433

N-[9-(2-Carboxyphenyl)-6-(diethylamino)-3*H*-xanthen-3-ylidene]-*N*-ethylethanaminium Chloride *see* 8182

9-(*o*-Carboxyphenyl)-6-hydroxy-3-isoxanthenone *see* 4085

9-(*o*-Carboxyphenyl)-6-hydroxy-3*H*-xanthen-3-one *see* 4085

4′-Carboxyphenylmethanesulfonanilide *see* 1163

o-Carboxyphenyl Phosphate *see* 4170

Cross Index of Names

(2S,3aS,6aS)-1-[(S)-N-[(S)-1-Carboxy-3-phenylpropyl]alanyl]octahydrocyclopenta[b]pyrrole-2-carboxylic Acid 1-Ethyl Ester see 8123
1-(N-1-Carboxy-3-phenylpropyl)-L-alanyl-L-proline Dihydrate see 3522
1-[N-[(S)-1-Carboxy-3-phenylpropyl]-L-alanyl]-L-proline 1'-Ethyl Ester see 3521
(S)-2-[(S)-N-[(S)-1-Carboxy-3-phenylpropyl]alanyl]-1,2,3,4-tetrahydro-3-isoquinolinecarboxylic Acid 1-Ethyl Ester see A7
(1S,9S)-9-[[(S)-1-Carboxy-3-phenylpropyl]amino]octahydro-10-oxo-6H-pyridazino[1,2-a][1,2]diazepine-1-carboxylic Acid 9-Ethyl Ester Monohydrate see 2276
(S)-1-[N²-(1-Carboxy-3-phenylpropyl)-L-lysyl]-L-proline Dihydrate see 5393
4-Carboxyphenylstibonic Acid N-Methyltetrahydropyridinecarboxylic Acid Methyl Ester see 803
[(o-Carboxyphenyl)thio]ethylmercury Sodium Salt see 9244
Carboxypolymethylene, 1836
4-(β-Carboxypropionylamino)-4'-aminodiphenyl Sulfone see 8851
p-[(3-Carboxypropionyl)amino]benzaldehyde Thiosemicarbazone see 4163
3-O-(β-Carboxypropionyl)-11-oxo-18β-olean-12-en-30-oic Acid see 1797
[3-[3-(3-Carboxypropionyl)ureido]-2-methoxypropyl](theophyllinato)mercury see 5756
2-[4-(1-Carboxypropyl)phenyl]-1-isoindolinone see 4867
(3-Carboxypropyl)trimethylammonium Chloride Methyl Ester see 1871
3-Carboxypyridine N-Oxide see 6895
N-(D-2-Carboxy-1-pyrrolidinyl)-L-glutamine see 5377
6-[[(3-Carboxy-2-quinoxalinyl)carbonyl]amino]-3,3-dimethyl-7-oxo-4-thia-1-azabicyclo[3.2.0]heptane-2-carboxylic Acid see 8052
3-Carboxy-2-quinoxalinylpenicillin see 8052
Carboxysulfamidochrysoidine see 8872
3-Carboxy-6,7,8,9-tetrahydro-1,6-dimethyl-4-oxo-4H-pyrido[1,2-a]pyrimidinium Methyl Sulfate, Ethyl Ester see 8222
(6S)-6-[2-Carboxy-2-(3-thienyl)acetamido]-6-methoxypenicillanic Acid see 9076
6-[D(−)-α-Carboxy-3-thienylacetamido]penicillanic Acid see 9360
6-[((Carboxy-3-thienylacetyl)amino]-3,3-dimethyl-7-oxo-4-thia-1-azabicyclo[3.2.0]heptane-2-carboxylic Acid see 9360
6-[((Carboxy-3-thienylacetyl)amino]-6-methoxy-3,3-dimethyl-7-oxo-4-thia-1-azabicyclo[3.2.0]heptane-2-carboxylic Acid see 9076
α-Carboxy-3-thienylmethylpenicillin see 9360
α-Carboxy-N,N,N-trimethyl-1H-imidazole-4-ethanaminium Hydroxide Inner Salt see 4588
α-Carboxy-N,N,N-trimethyl-1H-indole-3-ethanaminium Hydroxide Inner Salt see 4798
1-Carboxy-N,N,N-trimethylmethanaminium Chloride see 1201

1-Carboxy-N,N,N-trimethylmethanaminium Hydroxide Inner Salt see 1201
1-Carboxy-N,N,N-trimethylmethanaminium Hydroxide Inner Salt Compd with 2,2,2-Trichloro-1,1-ethanediol (1:1) see 2059
2-Carboxy-1,1,6-trimethylpiperidinium Iodide Diethyl(2-hydroxyethyl)methylammonium Iodide Ester see 3076
3-Carboxy-N,N,N-trimethyl-1-propanaminium Chloride Methyl Ester see 1871
6-Carboxyuridine see 6829
Carboxyvinyl Polymer see 1836
Carbrital [Parke, Davis] see 7087
Carbromal, 1837
Carbromide see 3105
Carbubarb, 1838
Carbutamide, 1839
Carbuten [Kalopharma] see 5653
Carbuterol, 1840
Carbyne [Spencer] see 969
Carcholin [Merck & Co.] see 1780
Carcinil [Abbott] see 5341
Cardace [Hoechst-Roussel] see 8123
Cardamine see 6459
Cardamist see 6528
Cardamom Seed, 1841
Cardelmycin see 6641
Cardene [Syntex] see 6403
Cardiacap [Consolidated] see 7066
Cardiagutt [Engelhard KG] see 9851
Cardiamid see 6459
Cardiazol [Knoll] see 7097
Cardibeltin [Pharma Schwarz] see 9851
Cardiem [Marion] see 3188
Cardigin [Merrell] see 3146
Cardilate [Burroughs Wellcome] see 3622
Cardiloid see 3622
Cardimon see 6459
Cardine [Marshall] see 9915
Cardinol [CP] see 7852
Cardinophyllin see 1886
Cardio 10 [Nicholas] see 5114
Cardiofilina see 477
Cardiogen [Mediolanum] see 1856
Cardiografin [Squibb] see 5689
Cardio-Green [HW & D] see 4868
Cardio-Khellin see 5189
Cardiolan [Tosi] see 3150
Cardiolipol [Gremy-Longuet] see 6405
Cardiomin see 887
Cardiomone (Na Salt) [Endo] see 148
Cardion [Nippon Chemiphar] see 7889
Cardionatrin see 887
Cardioquin [Purdue Frederick] see 8073
Cardiorythmine see 184
Cardiosteril [Fresenius] see 3415
Cardiotoxin, 1842
Cardiovanil see 3673
Cardis [Iwaki] see 5114
Cardisan see 5132
Carditin see 7744
Carditoxin [Hungary] see 3146
Cardivix see 1093
Cardizem [Marion] see 3188
Cardomec [Merck & Co.] see 5133
Cardophylin see 477
Cardophyllin see 477
Cardovar [Pfizer] see 9612
Cardoxin [RAFA] see 3354
Cardrase [Upjohn] see 3711
Carduben [Madaus] see 9915
Cardura [Pfizer] see 3422
Carduran [Pfizer] see 3422
Carecin [Zensei Yakuhin] see 2308

Carena [Delagrange] see 477
3-Carene, 1843
Δ³-Carene see 1843
Carfecillin Sodium, 1844
Carfenil [Chugai] see 5433
Carfinate, 1845
Carfonal [Lafon] see 4041
Cargentos see 8474
Carguto [Tanabe Seiyaku] see 2878
Cargutocin, 1846
Carica see 6969
Caricaxanthin see 2612
Caricide [Am. Cyanamid] see 3106
Caridan see 6926
Caridian [Schering AG] see 5744
Caridorol [Sankyo Zoki] see 7852
Carinamide see 1163
Carindacillin, 1847
Carindapen [Pfizer] see 1847
Carisano [Roedler] see 4268
Carisoma [Wallace Labs.] see 1848
Carisoprodate see 1848
Carisoprodol, 1848
Carlsbad Salt Artificial, 1849
Carlytene [Dedieu] see 6204
Carmazon [Nikken] see 7889
Carmethose [Ciba-Geigy] see 1835
Carmine see 1850
Carminic Acid, 1850
Carminomycin see 1879
Carminomycin I see 1879
Carmofur, 1851
Carmol HC [Ingram] see 4711
Carmubris [Bristol] see 1852
Carmurit see 3706
Carmustine, 1852
Carnacid-Cor [Tad] see 4328
Carnallite see 7579
Carnauba Wax, 1853
Carnegine, 1854
Carnicor [Sigma-Tau] see 1856
Carnidazole, 1855
Carnigen [Albert Roussel] see 4754
Carnitene [Sigma-Tau] see 1856
Carnitine, 1856
Carnitine Acetyl Ester see 78
Carnitor [Sigma-Tau] see 1856
Carnosine, 1857
Carnotite see 9823
Caro's Acid, 1858
Carob Flour see 5436
Carofur [Boehringer, Mann.] see 6453
Caroid [Breon] see 6965
Carolina Pink see 8704
Caronamide see 1163
Carony Bark see 681
Carophyll Red [Roche] see 1756
Carotaben see 1860
Carotaben Plus see 1756
α-Carotene, 1859
β-Carotene, 1860
β,β-Carotene, 1860
γ-Carotene, 1861
δ-Carotene, 1862
ε,ψ-Carotene, 1862
ψ,ψ-Carotene see 5492
all trans-β-Carotene-3,3'-diol see 10019
β,β-Carotene-3,3'-diol see 10019
β,ε-Carotene-3,3'-diol see 9972
ψ,ψ-Carotene-16,16'-diol see 5493
β,β-Carotene-4,4'-dione see 1756
ψ,ψ-Carotene-16-ol see 5499
β,β-Carotene-3,3',4,4'-tetrone see 875
(3R)-β,β-Caroten-3-ol see 2612
(3R)-β,ψ-Caroten-3-ol see 8265
β-Caroten-3-ol see 2612
β,β-Caroten-4-one see 3469
Carotol, 1863
Caroverine, 1864

Caroxazone, 1865
Carpaine, 1866
Carpene [Agrimont] see 3406
Carpetimycins, 1867
Carphenazine, 1868
Carphenol [Bayer] see 3306
Carpidine see 5085
Carpiline see 5085
Carpipramine, 1869
Carprofen, 1870
Carpronium Chloride, 1871
Carrageen see 1872
Carrageenan, 1872
Carrageenin see 1872
Carrbutabarb [Century] see 1495
Carrel-Dakin Soln see 8578
Carrier [Chiesi] see 1856
Carsalam, 1873
Cartap, 1874
Carteol [Chauvin-Blache] see 1875
Carteolol, 1875
Carthamic Acid see 1876
Carthamin, 1876
Carthamus, 1877
Carticaine, 1878
Cartric [Sanwa] see 8146
Cartrol [Otsuka] see 1875
Carubicin, 1879
Carubinose see 5632
Carudol [Franc. Therap.] see 7248
Carumonam, 1880
Carvacrol, 1881
Carvacron [Taiyo] see 9537
Carvanil see 5114
Carvasin [Wyeth] see 5114
Carvedilol, 1882
Carvol see 1883
6,8-Carvomenthenediol see 8510
Carvone, 1883
Carwin [ICI] see 9965
Carylderm [Napp] see 1789
Caryolysine [Delagrange] see 5655
Caryophyllene, 1884
α-Caryophyllene see 4672
β-Caryophyllene see 1884
γ-Caryophyllene see 1884
Caryophyllic Acid see 3855
Caryophyllin see 6787
Caryophyllus see 2413
Carzenide, 1885
Carzinophilin, 1886
Carzinophilin A see 1886
Carzonal [Tobishi] see 9060
Casakol [Upjohn] see 1887
Casamino Acids see 1892
Casanthranol, 1887
Casanthranol A see 1887
Casanthranol B see 1887
Casantin [Cassella-Riedel] see 3098
Casca Bark see 8341
Cascapride [Cascan] see 1420
Cascara Amara, 1888
Cascara Sagrada, 1889
Cascarilla, 1890
Cascarillin, 1891
Cascarin see 4177
Cascarosides see 1889
Caseanine see 9147
Casein, 1892
Cas-Evac see 1889
Cashoo see 1909
Casil see 339
Casimiroedine, 1893
Casimiroin, 1894
Casmalon see 2709
Casoron [Thompson Hayward] see 3029
Casoron-133 [Thompson Hayward] see 3029
Cassaidine, 1895
Cassaine, 1896

Cassamine, 1897
Cassella 4489 see 2241
Cassella's Acid, 1898
Cassella's Acid F, 1899
Cassel's Green see 993
Cassia Fistula, 1900
Cassia Pods see 1900
Cassia Pulp see 1900
Cassic Acid see 8175
Cassiopeium see 5484
Cassiterite see 8736
Castanea, 1901
Castanospermine, 1902
Castle's Intrinsic Factor, 1903
Castor Oil, 1904
Castor Oil, Hydrogenated, 1905
Castorwax see 1905
Castrix see 2590
Cataclot [Ono] see 6935
Catalase, 1906
Catalin [Senju] see 7462
Catalpin see 1907
Catalposide, 1907
Catanil [DeAngeli] see 2187
Catapres [Boehringer, Ing.] see 2388
Catapresan [Boehringer, Ing.] see 2388
Catapyrin see 463
Catarase [Cooper] see 2265
Cataria see 1913
Catatrol [ICI] see 9884
Catavin C see 855
Catechin, 1908
Catechinic Acid see 1908
Catechol see 1908 and 8009
Catecholase see 9746
Catechu see 4259
Catechu Black, 1909
Catechuic Acid see 1908
Catenulin see 6989
Catergen [Zyma] see 1908
Catharanthine, 1910
Cathepsin C see 1911
Cathepsin D see 1911
Cathepsin G see 1911
Cathepsins, 1911
Cathine see 6634
Cathinone, 1912
Cathocin [Merck & Co.] see 6641
Cathomycin [Merck & Co.] see 6641
Catmint see 1913
Catnep, 1913
Catnip see 1913
Catolep [Sumitomo] see 4874
Catovit see 7791
Catral see 7197
Catron [Lakeside] see 7197
Catroniazid see 7197
Cat's Hair see 3860
Caudaline [Exa] see 9361
Caulophylline, 1914
Caulophyllum, 1915
Caulosapogenin see 4540
Caustic Alcohol see 8560
Caustic Barley see 8280
Caustic Baryta see 989
Caustic Potash see 7625
Caustic Soda see 8575
Causyth [Causyth] see 7884
CAV see MISC-2
Cav-Ecol see 9954
Cavinton [Gedeon Richter] see 9894
Cavitands see 2341
Cavodil see 7197
Cavonyl see 2717
Cayenne Pepper see 1769
Cay Note see 3560
Caytine [Lakeside] see 7911
CB 11 see 7160
CB 154 see 1404
CB 154 Mesylate see 1404

CB 302 see 3967
CB 304 see 920
CB 311 see 8672
CB 313 see 6134
CB 337 see 5690
CB 804 see 1451
CB 1048 see 2107
CB 1314 see 6062
CB 1348 see 2064
CB 2041 see 1494
CB 3025 see 5708
CB 3026 see 5708
CB 3697 see 8115
CB 4261 see 9172
CB 4306 see 2400
CB 8061 see 7090
CB 30038 see 6119
CBDCA see 1828
CBS see 1296
CBV see MISC-2
CC see 9929
CC-2481 see 2718
CCA see 5433
CCC [BASF] see 2103
CCC see 1663
β-CCE see 3737
CCI 15641 see 1951
CCK 179 see 3598
CCK-PZ see 2201
CCK C-Terminal Octapeptide see 8494
CCNU [Lundbeck] see 5444
CCT see 2762
CD-68 see 2079
CD-3400 see 8145
CDA see 2021
CDAA see 250
CDC [Weddel] see 2044
CDEC see 8882
CDM see 2083
CDP-choline see 2321
CDR see 1726
CE 3624 see 9362
Cealysin see 9631
Ceanothic Acid, 1916
Cebera [Bouchara] see 233
Cebesine [Chauvin-Blache] see 1056
Cebicure [Merck & Co.] see 855
Cebid [Winston] see 855
Cebion [E. Merck] see 855
Cebitate [Merck & Co.] see 8525
Cebrogen [Walker] see 4365
Cebroton [Sancarlo] see 2321
Cebrum [SIFA] see 2082
Cebutid [Boots-Dacour] see 4124
Cecenu [Medac] see 5444
Ceclor [Lilly] see 1920
Cecon [Abbott] see 855
Cecropins, 1917
Cedad [Recordati] see 1037
Cedar Camphor see 1919
Cedar Leaf Oil see 6707
Cedilanid [Sandoz] see 5229
Cedilanid-D [Sandoz] see 2903
Cedin (Aerosol) [Lyssia] see 5071
Cedocard [Tillotts] see 5114
Cedol [Tiber] see 1922
8βH-Cedran-8-ol see 1919
Cedrin, 1918
Cedrol, 1919
Cedro Oil see 6740
Cedulamin see 5883
Cedur [Boehringer, Mann.] see 1215
CeeNU [Bristol] see 5444
Ceepryn [Merrell] see 2024
Cefacidal [Allard] see 1925
Cefaclor, 1920
Cefadol [Nippon Shinyaku] see 3311
Cefa-Drops [Fort Dodge] see 1921
Cefadros [Proter] see 1971
Cefadroxil, 1921

Cefadyl [Bristol] see 1980
Cefa-Iskia [Iskia] see 1971
Cefa-Lak [Bristol] see 1980
Cefalex [Von Boch] see 7485
Cefaloridin see 1975
Cefalotin see 1978
Cefaloto [Lifepharma] see 1971
Cefam [Magis] see 1922
Cefamandole, 1922
Cefamar [Firma] see 1951
Cefamedin [Fujisawa] see 1925
Cefamezin [Fujisawa] see 1925
Cefamox [Bristol] see 1921
Cefaperos [Allard] see 1923
Cefatrexyl [Bristol] see 1980
Cefatriaxone see 1950
Cefatrizine, 1923
Cefazedone, 1924
Cefazil [Giustini] see 1925
Cefazina [Chemil] see 1925
Cefazolin, 1925
Cefbuperazone, 1926
Cefibacter [Rubio] see 1971
Cefiran [Pierrel] see 1922
Cefixime, 1927
Cefizox [SmithKline] see 1949
Ceflorin [Glaxo] see 1975
Cefmax [TAP] see 1928
Cefmenoxime, 1928
Cefmetazole, 1929
Cefmetazon [Sankyo] see 1929
Cefminox, 1930
Cefobid [Pfizer] see 1933
Cefobine [Pfizer] see 1933
Cefobis [Pfizer] see 1933
Cefodie [ISF-Italseber; ICAR Leo] see 1932
Cefodizime, 1931
Cefomonil [TAP] see 1943
Cefonicid, 1932
Cefoperazone, 1933
Cefoprim [Esseti] see 1951
Ceforal [Farmoffer] see 1921
Ceforanide, 1934
Cefosint [Proter] see 1933
Cefossim [Coli] see 1951
Cefotan [Yamanouchi] see 1936
Cefotax [Chugai] see 1935
Cefotaxime, 1935
Cefotetan, 1936
Cefotiam, 1937
Cefotrizin [Firma] see 1923
Cefoxitin, 1938
Cefpimizole, 1939
Cefpiramide, 1940
Cefpiran [Sumitomo] see 1940
Cefpodoxime see 1941
Cefpodoxime Proxetil, 1941
Cefracycline Suspension [Frosst] see 9130
Cefracycline Tablets [Frosst] see 9130
Cefradex [Ausonia] see 1982
Cefradin see 1982
Cefrag [Magis] see 1982
Cefro [Sankyo] see 1982
Cefroxadine, 1942
Cefspan [Fujisawa] see 1927
Cefsulodin, 1943
Ceftazidime, 1944
Ceftenon [Biochemie] see 1936
Cefteram, 1945
Cefteram Pivoxil see 1945
Ceftetrame see 1945
Ceftezole, 1946
Ceftibuten, 1947
Ceftim [Bonomelli] see 1944
Ceftin [Glaxo] see 1951
Ceftiofur, 1948
Ceftix [Boehringer, Mann.] see 1949
Ceftizoxime, 1949
Ceftriaxone, 1950

Cefumax [Locatelli] see 1951
Cefurex [Sarm] see 1951
Cefurin [Magis] see 1951
Cefuroxime, 1951
Cefuroxime Axetil see 1951
Cefuzonam, 1952
Cegiolan see 855
Ceglunat [DDR] see 5229
Ceglution [Ariston] see 5404
Celadigal [Beiersdorf] see 5229
Celaskon [Spofa] see 855
Cela W-524 see 9602
Celbenin [Beecham] see 5890
Celectol [Rorer] see 1956
Celeport [Mitsubishi; Eisai] see 1227
Celery Seed, 1953
Celestan [Byk-Essex] see 1202
Celestan-V [Byk-Essex] see 1202
Celeste [Andromaco] see 9511
Celestene [Cetrane] see 1202
Celesticetin, 1954
Celestin Blue, 1955
Celestine see 8816
Celestite see 8816
Celestoderm-V [Schering] see 1202
Celestone [Schering] see 1202
Celevac [WB Pharm] see 5963
Celex [Aristochimica] see 1982
Celexane [Rhone-Poulenc] see 8987
Celin see 855
Celiomycin see 9905
Celiprolol, 1956
Celiptium [Sanofi] see 3510
Celite see 4878
Cellative [Tobishi] see 5660
Cellidrin [Hennig] see 278
Cellobiose, 1957
β-Cellobiose see 1957
Cellocidin, 1958
Celloidin see 8022
Cellolax see 1835
Cellon see 9125
Cellophane, 1959
Cellose see 1957
Cellosolve see 3707
Cellosolve Acetate see 3708
Cellothyl [Warner-Chilcott] see 5963
Cellucon [Medo] see 5963
Celluflex see 9675
Celluloid®, 1960
Cellulose, 1961
Cellulose Acetates, 1962
Cellulose-ECTEOLA see 3477
Cellulose Ethyl Ether see 3739
Cellulose Ethyl Hydroxyethyl Ether, 1963
Cellulose 2-Hydroxypropyl Ether see 4776
Cellulose 2-Hydroxypropyl Methyl Ether see 4777
Cellulose Methyl Ether see 5963
Cellulose Nitrate see 8022
Cellulose Phenoxyacetate see 7224
Cellulosic Acid see 6894
Cellumeth [Conal] see 5963
Celmidol [Tobishi] see 3311
Cel-O-Brandt see 1835
Celocurine see 8847
Celontin [Parke, Davis] see 5928
Celoslin [Hoechst] see 1946
Celospor [Ciba-Geigy] see 1969
Celphos [Excel] see 363
Celpillina see 5890
Celtect [Kyowa] see 6880
Celupan [Lacer] see 6278
Cemado [Farmochim. Ital.] see 1922
Cemandil [SIT] see 1922
Cembrene A see 6362
Cemix [Takeda] see 1928
Cenalene-M [Central Pharm.] see 7097

Cenazol see 7097
Cenetone see 855
Cenocort [Central Pharm.] see 9511
Cenolate [Abbott] see 8525
Cenomycin [Daiichi] see 1938
Censedal [M & B] see 6350
Censpar [Bristol-Myers] see 1493
Centaurein, 1964
Centaurin see 2419
Centaury, American, 1965
Centaury, Chilean, 1966
Centaury, Minor, 1967
Centaury, Spiked, 1968
Centedrin see 6025
Centelase Dermatologico [Laroche Navarron] see 857
Centractil see 7793
Centractyl see 7793
Centralgin [Amino] see 5736
Centralgol [Valpan] see 7920
Centralgyl see 7920
Centrallasite see 1709
Centrax [Parke, Davis] see 7713
Centrine [Bristol] see 471
Centrophenoxine see 5660
Centroton [Wander] see 8021
Centurina [Merck & Co.] see 7484
Centyl [Leo Pharm.] see 1045
Ceolat [Kali-Chemie] see 8486
Ceosunin [Kyowa] see 1998
CEPA see 3686
Cepacilina see 7037
Cepacol [Merrell] see 2024
Cepaloridin [Glaxo] see 1975
Cepalorin see 1975
Cepan [IBI] see 1936
Cepaverin [Globopharm] see 6968
Cepazine [Glaxo] see 1951
CEPH see 9425
CEPHA [GAF] see 3686
Cephacetrile Sodium, 1969
Cephaeline, 1970
Cephaeline Methyl Ether see 3517
Cephalexin, 1971
Cephalins, 1972
Cephaloglycin, 1973
Cephalonic Acid, 1974
Cephaloridine, 1975
Cephalosporin C, 1976
Cephalosporin C_A see 1976
Cephalosporin C_C see 1976
Cephalosporin N see 7043
Cephalosporin P_1, 1977
Cephalothin, 1978
Cephamycins, 1979
Cephapirin Sodium, 1980
Cepharanthine, 1981
Cephation [Meiji] see 1978
Cephos [CT] see 1921
Cephradine, 1982
Cephrol see 2332
Cephulac [Merrell] see 5225
Ceporacin [Torii] see 1978
Ceporan [Glaxo] see 1975
Ceporex [Glaxo] see 1971
Ceporexine [Glaxo] see 1971
Ceporin [Glaxo] see 1975
Cepovenin [Hoechst; Glaxo] see 1978
Cepticol [Banyu] see 1923
Cequartyl see 1066
Cer [Nippon Glaxo] see 1975
Ceractin [Wyeth-Ayerst] see 9894
Ceradon [Takeda] see 1937
Ceramide-β-lactoside see 2800
Ceraphyl 28 see 2022
Cerargyrite see 8450
Cerasynt 660 [Van Dyk] see 7555
Cerberigenin see 3145
Cerberin see 6386
Cerberoside, 1983
Cercobin [Pennwalt] see 9282

Cercobin-M *see* 9282
Cereb [Ohta] *see* 2321
Cerebid [Saron] *see* 6968
Cerebolan [Tobishi] *see* 2308
Cerebon [Isis Chemie] *see* 5660
Cerebro [Sidus] *see* 8967
Cerebroforte [Azuchemie] *see* 7459
Cerebron *see* 7342
Cerebronylsphingosylglucosidogalactoside *see* 2800
Cerebrose *see* 4241
Cerebrostenediol *see* 4750
Cerebrosterol *see* 4750
Cerebroxine [CCP] *see* 9888
Cereclor *see* 6972
Ceregulart [Kaken] *see* 2977
Cerekinon [Tanabe] *see* 9613
Cereon *see* 855
Cerepar [Merckle] *see* 5344
Cerepar [Mepha] *see* 2308
Cerepax [Promeco] *see* 9074
Ceresan Dry *see* 7271
Ceresan M *see* 3781
Ceresan Slaked Lime *see* 7271
Ceresan Wet *see* 3780
Ceresin, 1984
Cerespan [USV] *see* 6968
Cerevon *see* 4005
Cerfenil [Chugai] *see* 5433
Cergem [Searle] *see* 4279
Cergona *see* 855
Ceria *see* 1986
Ceric Ammonium Nitrate *see* 529
Ceric Fluoride, 1985
Ceric Oxide, 1986
Ceric Sulfate, 1987
Cerin *see* 1984
Cerium, 1988
CERM 1766 *see* 6861
CERM 1978 *see* 1167
CERM 10137 *see* 9443
Cero [Dexter] *see* 2102
Cer-O-Cillin Sodium *see* 7044
Cerocral [Funai] *see* 4821
Cerone [Union Carbide] *see* 3686
Cerosin *see* 1984
Cerous Ammonium Sulfate *see* 530
Cerous Benzoate *see* 1101
Cerous Bromide, 1989
Cerous Carbonate, 1990
Cerous Chloride, 1991
Cerous Fluoride, 1992
Cerous Iodide, 1993
Cerous Nitrate, 1994
Cerous Oxalate, 1995
Cerous Sulfate, 1996
Certinal *see* 474
Certomycin [Byk-Essex] *see* 6389
Cerubidin [M & B] *see* 2825
Cérubidine [Rhône-Poulenc] *see* 2825
Cerucal [Arzneimittelwerk VEB] *see* 6063
Cerulein *see* 1998
Cerulen [Adria] *see* 1998
Cerulenin, 1997
Ceruletide, 1998
Ceruloplasmin, 1999
Ceruse *see* 1016
Cerussa *see* 1016
Cerussite *see* 5267
Cervagem(e) [M&B] *see* 4279
Cervicarcin, 2000
Cervilaxin *see* 8137
Cerviprost [Organon] *see* 7893
Cervoxan [Sorbio] *see* 3464
Cesamet [Lilly] *see* 6257
C'esar [Roussel-UCLAF] *see* 4634
Cescan-131 [Abbott] *see* 2004
Cescorbat *see* 855
Cesium, 2001
Cesium Alum *see* 336

Cesium Bromide, 2002
Cesium Carbonate, 2003
Cesium Chloride, 2004
Cesium Hydrate *see* 2005
Cesium Hydroxide, 2005
Cesium Iodide, 2006
Cesium Nitrate, 2007
Cesium Sulfate, 2008
Cesol [E. Merck] *see* 7714
Cesplon [Esteve] *see* 1773
Cesporan [Errekappa] *see* 1982
Cestarsol *see* 3434
Cestocid *see* 6425
Cetab *see* 2018
Cetaceum *see* 8697
Cetacillin *see* 7829
Cetadol [Rybar] *see* 40
Cetain *see* 7763
Cetal [Parke, Davis] *see* 9888
Cetalkonium Chloride, 2009
Cetamid *see* 855
Cetamide [Alcon] *see* 8870
Cetamin *see* 3308
Cetamium [Merrell] *see* 2024
Cetamolol, 2010
Cetampin [Scarium] *see* 621
Cetapril [Dainippon] *see* 192
Cetats [Fine Organics] *see* 2018
Cetavlon [Ayerst] *see* 2018
Ceteareth *see* 7554
Cetebe [Stroschein] *see* 855
Cetemican *see* 855
Ceteth *see* 7554
Ceteth-20 *see* 7554
Cethexonium Bromide, 2011
Cethylose *see* 1835 and 5963
Cethytin [Ascher] *see* 5963
Cetiedil, 2012
Cetina [Robeco] *see* 8697
Cetiprin [Kabi] *see* 3515
Cetirizine, 2013
Cetol *see* 2009
Cetomacrogol 1000 *see* 7554
Cetosanol [Sanol] *see* 5229
Cetotiamine, 2014
Cetoxime, 2015
Cetraphylline [Cetrane] *see* 9212
Cetraric Acid, 2016
Cetrarin *see* 2016
Cetraxate, 2017
Cetrexin [Leciva] *see* 5660
Cetrimide *see* 2018
Cetrimonium Bromide, 2018
Cetrimonium Stearate, 2019
Cetsim [Saint-Germain] *see* 209
Cetyl Alcohol, 2020
Cetylamine *see* 2018
α-Cetylcitric Acid *see* 174
Cetyldimethylbenzylammonium Chloride *see* 2009
N-Cetyl-N,N-dimethyl-2-cyclohexanolammonium Bromide *see* 2011
Cetyldimethylethylammonium Bromide, 2021
Cetyldimethyl(2-hydroxycyclohexyl)ammonium Bromide *see* 2011
Cetylic Acid *see* 6947
Cetyl Lactate, 2022
Cetyl[2-[(p-methoxybenzyl)-2-pyrimidinylamino]ethyl]dimethylammonium Bromide *see* 9306
Cetyl Palmitate, 2023
Cetylpyridinium Chloride, 2024
Cetyltrimethylammonium Bromide *see* 2018
Cetyltrimethylammonium Stearate *see* 2019
Cevadilla *see* 8280
Cevadine, 2025
Cevalin [Lilly] *see* 855

(3β,4α,16β)-Cevane-3,4,12,14,16,17,-20-heptol 3-Acetate *see* 8281
Cevane-3,6,20-triol *see* 9873
Cevanol [ICI] *see* 1037
Cevatine *see* 855
Cevex [Merrell] *see* 855
Cevimin *see* 855
Cevine, 2026
Ce-Vi-Sol [Mead Johnson] *see* 855
Cevitamic Acid *see* 855
Cevitamin *see* 855
Cevitan *see* 855
Cevitex *see* 855
Cewin [Winthrop] *see* 855
Cex [Glaxo Fuji; Nippon Glaxo] *see* 1971
Ceylon Isinglass *see* 172
CEZ *see* 1925
CF *see* 4141
CFCs *see* 2140
C-Film [Arun] *see* 6596
CFP *see* MISC-2
CFPMV *see* MISC-2
CFPQ *see* 3546
CFPT *see* MISC-2
CG 113 *see* 7746
CG 201 *see* 1214
CG 315E *see* 9485
CG 635 *see* 3828
CG 1283 *see* 6126
CGA 10832 *see* 7781
CGA 12223 *see* 4988
CGA 18731 *see* 5106
CGA 20168 *see* 5848
CGA 23654 *see* 6555
CGA 24705 *see* 6067
CGA 26 423 *see* 7746
CGA 45156 *see* 9258
CGA 48988 *see* 5826
CGA 50439 *see* 2771
CGA 64250 *see* 7830
CGA 72662 *see* 2782
CGA 89317 *see* 9566
CG B3Q *see* 1946
CGMP *see* 2715
CGP *see* 4535
CGP-2175 *see* 6072
CGP-4540 *see* 606
CGP-7174/E *see* 1943
CGP-7760B *see* 7742
CGP-8426 *see* 8852
CGP-9000 *see* 1942
CGP-12103A *see* 6878
CGP-12104A *see* 6878
CGP-14221/E *see* 1937
CH 800 *see* 3946
CH 846 *see* 3464
CH 3635 (formerly) *see* 9573
CHAD *see* MISC-2
Chaetomidin *see* 6805
Chalcanthite *see* 2659
Chalcedony *see* 8440
Chalcocite *see* 2674
Chalcomenite *see* 2657
Chalcomycin, 2027
Chalcone, 2028
Chalcopyrite, 2029
D-Chalcose, 2030
Chalkone *see* 2028
Chalkopyrite *see* 2029
Chamazulene, 2031
Chamber Crystals *see* 6570
Chameleon Mineral *see* 7643
CHAMOCA *see* MISC-2
Chamomile, 2032
Champaca Camphor *see* 4466
Champacol *see* 4466
Chanchalagua *see* 1966
Channel Black *see* 1815
Channing's Soln *see* 7691
Chanoclavin-I *see* 2033

Chanoclavine, 2033
Chanoclavine II see 2033
CHAP-5 see MISC-2
CHAPS, 2034
CHAPSO see 2034
Charas see 1752
Charcot-Neumann Crystals see 8699
Chartarin see 2035
Chartreusin, 2035
Chat see 5188
Chaulmestrol see 2037
Chaulmoogra Oil, 2036
Chaulmoogric Acid, 2037
Chavicine, 2038
Chavicinic Acid see 7437
Chavicol, 2039
Chavicol Methyl Ether see 3657
Chebutan see 5168
Cheeseflower see 5590
Chefarox see 345
Cheirolin, 2040
Cheladrate [Pharmex] see 3481
Chelafer see 3986
Chelafrin see 3569
Chelaplex III see 3481
Chelen see 3740
Chelerythrine, 2041
ψ-Chelerythrine see 8320
Chelidonic Acid, 2042
Chelidonine, 2043
Chel-Iron [Kinney] see 3986
Chemcef [Chemil] see 1935
Chemestrogen see 1082
Chem-Fish [Tifa] see 7446
Chemical 109 see 755
Chemical Mace see 2115
Chemicetina [Erba] see 2068
Chemifluor [Chemipharm] see 8565
Chemiofuran see 6520
Chemipen [Squibb] see 7184
Chemocide PK see 7879
Chemodyn see 1161
Chemofuran see 6521
Chemotrim [R. P. Drugs] see 8889
Chemox DN [Blue Spruce] see 3282
Chemox PE [Blue Spruce] see 3282
Chemox Selective [Blue Spruce] see 3282
Chem Rice see 7814
Chendol [Weddel] see 2044
Chenic Acid see 2044
Chenix [Rowell] see 2044
Chenocedon [Tillotts] see 2044
Chenocol [Caber] see 2044
Chenodeoxycholic Acid see 2044
Chenodex [Roussel-UCLAF] see 2044
Chenodiol, 2044
Chenofalk [Falk] see 2044
Chenopodiol [Pohl] see 6713
Chenoposan see 6713
Chenoposetten see 6713
Chenosäure [Falk] see 2044
Chenossil [Guiliani] see 2044
Chephalotin [Lilly] see 1978
Chepirol see 5168
Cheque [Upjohn] see 6098
Chessylite see 2634
Chestnut see 1901
Chetazolidin [Zeria] see 5168
Chetil see 5168
Chevreul's Salt see 2675
CHF see MISC-2
Chibro-Amuno [Chibret] see 4874
Chibro Pilocarpine [Chibret] see 7395
Chibro-Rifamycin [Chibret] see 8217
Chick Antidermatitis Factor see 6964
Chicle, 2045
Chile Saltpeter see 8598
Chimaphila, 2046
Chimaphilin, 2047
Chimonanthine, 2048

Chimyl Alcohol, 2049
China Bark see 8050
China Clay see 5162
Chinacrin Hydrochloride see 8053
China Green see 5581
China Oil see 959
"China White" see 2948 and 5992
China Wood Oil see 9721
Chinese Anise see 8756
Chinese Blistering Flies see 6240
Chinese Blue see 3964
Chinese Cantharides see 6240
Chinese Ginger see 4243
Chinese Isinglass see 172
Chinese Pea see 8684
Chinese Red see 5789
Chinese Seasoning see 6165
Chinese Wax, 2050
Chinic Acid see 8071
Chinicine see 9908
Chinidin-Duriles [Astra] see 8072
Chiniofon see 4757
Chinocide see 8096
Chinoform see 4924
Chinofungin [Chinoin] see 9442
Chinoleine see 8097
Chinomethionat(e) see 6933
Chinosol see 4779
Chinothionat see 9288
Chinova Acid see 8104
Chinovic Acid see 8104
Chinovin see 8105
Chinovose see 8106
CHIPCO-26019 [Rhône Poulenc] see 4964
Chirata, 2051
Chirayita see 2051
Chiretta see 2051
Chitin, 2052
Chitosamine see 4352
Chitosan see 2052
Chittem Bark see 1889
Chittim Bark see 1889
Chlo-Amine [Hollister-Stier] see 2180
Chlochinate see 4757
Chlomycol see 2068
Chloor-hexaviet see 4609
Chlophedianol, 2053
Chlophenadione see 2402
Chloquinate see 4757
Chloracetone see 2113
Chloracetyl Chloride, 2054
Chloracizine, 2055
Chloracon see 1027
Chloractil [DDSA] see 2186
Chloracysin see 2055
Chloral see 9538
Chloral Alcoholate, 2056
Chloralamide see 2060
Chloral Ammonia, 2057
Chloralantipyrine, 2058
Chloral Betaine, 2059
Chloraldurat [Pohl] see 2061
Chloral Ethylalcoholate see 2056
Chloral Formamide, 2060
Chloral Hydrate, 2061
Chloral Hydrocyanide see 2134
Chlorallylene see 287
Chloralodol see 2089
Chloralosane see 2062
α-Chloralose, 2062
β-Chloralose see 2062
Chloral-urethane see 1808
Chloramben, 2063
Chlorambucil, 2064
Chloramex see 2068
Chloramfilin see 2068
Chloramide see 2060
Chloramine see 2066
Chloramine-B, 2065
Chloramine-T, 2066

Chloraminophenamide, 2067
Chloraminophene see 2064
Chloramiphene see 2384
Chloramphenicol, 2068
Chloramphenicol Arginine Succinate see 2068
Chloramphenicol Calcium Pantothenate see 2070
Chloramphenicol Palmitate, 2069
Chloramphenicol Pantothenate, 2070
Chloramphenicol Pantothenate Complex see 2070
Chloramphenicol Sodium Pantothenate see 2070
Chloramphenicol Succinate Compd with Rolitetracycline see 8408
Chloramsaar [Chephasaar] see 2068
Chloranautine see 3193
Chloranil, 2071
Chloranilic Acid, 2072
Chlorapatite see 7321
Chlorarsen see 3060
Chlorarsol see 3060
Chlorasept 2000 see 2090
Chloraseptic [Norwich] see 7206
Chloraseptine [Walker] see 2066
Chlorasol [EVSCO] see 2068
Chlorazanil, 2073
Chlorazene [Wisconsin Pharmacal] see 2066
Chlorazin [Streuli] see 2186
Chlorazine (Russian) see 2727
Chlorazinil see 2073
Chlorazodin see 2120
Chlorazol Yellow 2G [ICI] see 9238
Chlorazone see 2066
Chlorbenside, 2074
Chlorbenzilat see 2123
Chlorbenzoxamine, 2075
5-Chlorbenzoxazolin-2-one see 2197
Chlorbenzoxyethamine see 2075
Chlorbetamide, 2076
Chlorbicyclen, 2077
Chlorbismol see 1274
Chlorbutol see 2129
Chlorcosane see 6972
Chlorcyclizine, 2078
Chlordan(e), 2079
Chlordantoin, 2080
Chlordecone, 2081
Chlorderazin see 2186
Chlordiazachel [Rachelle] see 2082
Chlordiazepoxide, 2082
Chlordimeform, 2083
Chloreal see 8993
Chlorendic Anhydride, 2084
Chloresium [Rystan] see 2155
Chlorethiazol see 2382
Chlorethyl see 3740
Chloretone [Parke, Davis] see 2129
Chlorex see 3055
Chlorfenac, 2085
Chlorfenethol, 2086
Chlorfenson see 6856
Chlorfenvinphos, 2087
Chlorguanide, 2088
Chlorguanide Triazine see 2727
Chlorhexadol, 2089
Chlorhexidine, 2090
Chlorhydrol [Armour Pharm.] see 346
Chloric Acid, 2091
Chloricol [EVSCO] see 2068
"Chloride of Lime" see 2093
Chloridin see 7997
Chlor-IFC see 2188
Chlorimipramine see 2385
Chlorimpiphenine see 4827
Chlorimuron Ethyl, 2092
Chlorinat see 969
Chlorinated Biphenyls see 7541
Chlorinated Camphene see 9478

2-Chloroethyl Alcohol see 3750

2-Chloro-4-ethylamino-6-isopropyl-amine-s-triazine see 886

6-Chloro-2-(ethylamino)-4-methyl-4-phenyl-4H-3,1-benzoxazine see 3821

2-[[4-Chloro-6-(ethylamino)-1,3,5-triazin-2-yl]amino]-2-methylpro-panenitrile see 2692

2-[[4-Chloro-6-(ethylamino)-s-tri-azin-2-yl]amino]-2-methylpropioni-trile see 2692

2-Chloroethyl N,N-Bis(2-chloroethyl)-N′-(3-hydroxypropyl)phosphorodi-amidate see 2853

β-Chloroethyl β-(p-tert-Butylphen-oxy)-α-methylethyl Sulfite see 794

3-(2-Chloroethyl)-2-[(2-chloroethyl)-amino]tetrahydro-2H-1,3,2-oxaza-phosphorin-2-oxide see 4822

N-(2-Chloroethyl)-N′-cyclohexyl-N-nitrosourea see 5444

1-(2-Chloroethyl)-3-cyclohexyl-1-nitrosourea see 5444

N-(2-Chloroethyl)dibenzylamine Hydrochloride, 2138

6-Chloro-3-ethyl-3,4-dihydro-2H-1,2,4-benzothiadiazine-7-sulfon-amide 1,1-Dioxide see 3687

2-(2-Chloroethyl)-2,3-dihydro-4H-1,3-benzoxazin-4-one see 2195

2-(β-Chloroethyl)-2,3-dihydro-4-keto-(benzo-1,3-oxazine) see 2195

2-(β-Chloroethyl)-2,3-dihydro-4-oxo-6-amino-1,3-benzoxazine see 446

2-(β-Chloroethyl)-2,3-dihydro-4-oxo-(benzo-1,3-oxazine) see 2195

6-Chloro-3-ethyl-3,4-dihydro-7-sulfamoylbenzo-1,2,4-thiadiazine 1,1-Dioxide see 3687

N-(2-Chloroethyl)-2,6-dinitro-N-propyl-4-(trifluoromethyl)benzen-amine see 4052

Chloroethylene see 9898

Chloroethylene Polymer see 7563

7-Chloro-4-[4-(N-ethyl-N-β-hydr-oxyethylamino)-1-methylbutyl-amino]quinoline see 4748

7-Chloro-4-[4-[ethyl(2-hydroxyeth-yl)amino]-1-methylbutylamino]-quinoline see 4748

7-Chloro-4-[5-(N-ethyl-N-2-hydr-oxyethylamino)-2-pentyl]aminoqui-noline see 4748

2-Chloroethyl 2-Hydroxyethyl Sulfide see 4565

β-Chloroethyl β-Hydroxyethyl Thio-ether see 4565

Chloroethylmercury see 3780

2-Chloro-6′-ethyl-N-(2-methoxy-1-methylethyl)acet-o-toluidide see 6067

2-Chloroethyl 1-Methyl-4-(p-tert-butylphenoxy)ethyl Sulfite see 794

6-Chloro-N-ethyl-N′-(1-methyleth-yl)-1,3,5-triazine-2,4-diamine see 886

α-Chloro-2′-ethyl-6′-methyl-N-(1-methyl-2-methoxyethyl)acetanilide see 6067

N-(2-Chloroethyl)-N-(1-methyl-2-phenoxyethyl)benzenemethanamine see 7225

N-(2-Chloroethyl)-N-(1-methyl-2-phenoxyethyl)benzylamine see 7225

6-Chloro-N-ethyl-4-methyl-4-phenyl-4H-3,1-benzoxazin-2-amine see 3821

2-Chloro-N-(2-ethyl-6-methylphen-yl)-N-(2-methoxy-1-methylethyl)-acetamide see 6067

5-(2-Chloroethyl)-4-methylthiazole see 2382

2-[[[(2-Chloroethyl)nitrosoamino]-carbonyl]amino]-2-deoxy-D-glucose see 2177

[1-[[[(2-Chloroethyl)nitrosoamino]-carbonyl]amino]ethyl]phosphonic Acid Diethyl Ester see 4174

1-(2-Chloroethyl)-1-nitroso-3-(D-glucos-2-yl)urea see 2177

2-[3-(2-Chloroethyl)-3-nitrosour-eido]-2-deoxy-D-glucopyranose see 2177

6-[3-(2-Chloroethyl)-3-nitrosour-eido]-6-deoxy-α-D-glucopyranoside see 8125

1-[N-(2-Chloroethyl)-N-nitrosour-eido]ethylphosphonic Acid Diethyl Ester see 4174

1-Chloro-3-ethyl-1-penten-4-yl-3-ol see 3683

5-Chloro-3-ethylpent-1-yn-4-en-3-ol see 3683

Chloroethylphenamide see 1027

N-(2-Chloroethyl)-N-(phenylmethyl)-benzenemethanamine Hydrochloride see 2138

(2-Chloroethyl)phosphonic Acid see 3686

7-Chloro-2-ethyl-6-sulfamoyl-1,2,-3,4-tetrahydro-4-quinazolinone see 8067

7-Chloro-2-ethyl-1,2,3,4-tetrahydro-4-oxo-6-quinazolinesulfonamide see 8067

7-Chloro-2-ethyl-1,2,3,4-tetrahydro-4-oxo-6-sulfamoylquinazoline see 8067

5-Chloroethylthiamine see 1030

2-(2-Chloroethylthio)ethanol see 4565

N-(2-Chloroethyl)-α,α,α-trifluoro-2,6-dinitro-N-propyl-p-toluidine see 4052

(2-Chloroethyl)trimethylammonium Chloride see 2103

2-Chloroethyl Vinyl Ether, 2139

2-Chloroflumethasone see 4510

9-(2-Chloro-6-fluorobenzyl)adenine see 816

Chlorofluorocarbons, 2140

9-Chloro-6α-fluoro-11β,21-dihydr-oxy-16α-methylpregna-1,4-diene-3,20-dione see 2368

(11β,16β)-21-Chloro-9-fluoro-11,17-dihydroxy-16-methylpregna-1,4-diene-3,20-dione see 2361

21-Chloro-9α-fluoro-11β-hydroxy-16α,17α-isopropylidenedioxy-4-pregnene-3,20-dione see 4504

21-Chloro-9-fluoro-11-hydroxy-16,17-[(1-methylethylidene)bis-(oxy)]pregn-4-ene-3,20-dione see 4504

(16β)-21-Chloro-9-fluoro-17-hydr-oxy-16-methylpregna-1,4-diene-3,11,20-trione see 2362

9-Chloro-6α-fluoro-16α-methyl-1,4-pregnadiene-11β,21-diol-3,20-dione see 2368

7-Chloro-5-(2-fluorophenyl)-1,3-dihydro-3-hydroxy-1-(2-hydroxy-ethyl)-2H-1,4-benzodiazepin-2-one see 3423

7-Chloro-5-(2-fluorophenyl)-1,3-di-hydro-1-methyl-2H-1,4-benzodi-azepin-2-one see 4058

7-Chloro-5-(2-fluorophenyl)-2,3-dihydro-2-oxo-1H-1,4-benzodiaze-pine-3-carboxylic Acid Ethyl Ester see 3777

7-Chloro-5-(2-fluorophenyl)-1,3-dihydro-1-(2,2,2-trifluoroethyl)-2H-1,4-benzodiazepine-2-thione see 8039

6-Chloro-3-[(4-Fluorophenyl)methyl]-3,4-dihydro-2H-1,2,4-benzothiadi-azine-7-sulfonamide 1,1-Dioxide see 6973

8-Chloro-6-(2-fluorophenyl)-1-meth-yl-4H-imidazo[1,5-a][1,4]benzodi-azepine see 6105

6-[3-(2-Chloro-6-fluorophenyl)-5-methyl-4-isoxazolecarboxamido]-penicillanic Acid see 4044

6-[[[3-(2-Chloro-6-fluorophenyl)-5-methyl-4-isoxazolyl]carbonyl]-amino]-3,3-dimethyl-7-oxo-4-thia-1-azabicyclo[3.2.0]heptane-2-carb-oxylic Acid see 4044

3-(2-Chloro-6-fluorophenyl)-5-meth-yl-4-isoxazolylpenicillin see 4044

9-[(2-Chloro-6-fluorophenyl)methyl]-9H-purin-6-amine see 816

10-Chloro-11b-(2-fluorophenyl)-2,3,7,11b-tetrahydro-7-(2-hydroxy-ethyl)oxazolo[3,2-d][1,4]benzodi-azepin-6(5H)-one see 4133

21-Chloro-9-fluoro-11β,16α,17-tri-hydroxypregn-4-ene-3,20-dione Cyclic 16,17-Acetal with Acetone see 4504

Chlorofolin see 2155

Chloroform, 2141

Chloroformamide see 1785

Chloroformic Acid Benzyl Ester see 1807

Chloroformic Acid p-Bromobenzyl Ester see 1399

Chloroformic Acid Ethyl Ester see 3742

Chloroformic Acid Trichloromethyl Ester see 3339

Chloroformyl Chloride see 7310

Chlorofos see 9536

4-Chloro-N-furfuryl-5-sulfamoylan-thranilic Acid see 4221

4-Chloro-N-(2-furylmethyl)-5-sulfa-moylanthranilic Acid see 4221

Chlorogenic Acid, 2142

Chlorogenin, 2143

Chloroguanide see 2088

Chlorohemin see 4563

2-Chloro-1-heptene-1-arsonic Acid see 2119

(2-Chloro-1-heptenyl)arsonic Acid see 2119

4-Chloro-N-[[(hexahydro-1H-azepin-1-yl)amino]carbonyl]benzenesulfon-amide see 4410

9-Chloro-2,3,5a,6,10b,11-hexahydro-10b,11-dihydroxy-7,8-dimethoxy-2,3,6-trimethyl-3,11a-epidithio-11aH-pyrazino[1′,2′:1,5]pyrrolo[2,-3-b]indole-1,4-dione see 8724

4-Chloro-N-(endo-hexahydro-4,7-methanoisoindolin-2-yl)-3-sulfa-moylbenzamide see 9649

10-Chloro-2,3,5,6,7,11b-hexahydro-2-methyl-11b-phenylbenzo[6,7]-1,4-diazepino[5,4-b]oxazol-6-one see 6883

1-Chlorohexane, 2144

9-Chloro-9-[p-(hexyloxy)phenyl]-10-methylacridan see 7157

α-Chlorohydrin, 2145

4-Chloro-17β-hydroxyandrost-4-en-3-one see 2409

4-Chloro-2-hydroxybenzoic Acid n-Butylamide see 1450

5-Chloro-2-hydroxybenzoxazole see 2197

2-[[p-(p-Chlorophenyl)benzyl]oxy]-2-methylpropionic Acid see 2364

3-[(p-Chloro-α-phenylbenzyl)oxy]-tropane see 2360

[2-[4-(p-Chloro-α-phenylbenzyl)-1-piperazinyl]ethoxy]acetic Acid see 2013

(3-Chlorophenyl)carbamic Acid 4-Chloro-2-butynyl Ester see 969

(3-Chlorophenyl)carbamic Acid 1-Methylethyl Ester see 2188

p-Chlorophenyl p-Chlorobenzenesulfonate see 6856

2-(2-Chlorophenyl)-2-(4-chlorophenyl)-1,1-dichloroethane see 6134

α-(2-Chlorophenyl)-α-(4-chlorophenyl)-5-pyrimidinemethanol see 3903

3-[3-(4-Chlorophenyl)-1-(5-chloro-2-thienyl)-3-hydroxypropyl]-4-hydroxy-2H-1-benzopyran-2-one see 9383

N-(4-Chlorophenyl)-N′-[4-chloro-3-(trifluoromethyl)phenyl]urea see 2376

4-Chloro-α-phenyl-o-cresol see 2403

1-(p-Chlorophenyl)cyclohexyl β-Diethylaminoethyl Ether see 2372

trans-5-(4-Chlorophenyl)-N-cyclohexyl-4-methyl-2-oxo-3-thiazolidinecarboxamide see 4634

2-[[1-(4-Chlorophenyl)cyclohexyl]oxy]-N,N-diethylethanamine see 2372

2-[[1-(p-Chlorophenyl)cyclohexyl]oxy]triethylamine see 2372

3-(4-Chlorophenyl)-5-cyclopropyl-2-methylpyrrolidine see 7369

1-p-Chlorophenyl-2,4-diamino-6,6-dimethyl-1,6-dihydro-1,3,5-triazine see 2727

N-(p-Chlorophenyl)-2,4-diamino-s-triazine see 2073

(±)-1-[4-(4-Chlorophenyl)-2-[(2,6-dichlorophenyl)thio]butyl]-1H-imidazole see 1525

N-(4-Chlorophenyl)-N′-(3,4-dichlorophenyl)urea see 9568

2-(p-Chlorophenyl)-1-[p-(2-diethylaminoethoxy)phenyl]-1-(p-tolyl)-ethanol see 9650

(2-Chlorophenyl)[2-[2-[(diethylamino)methyl]-1H-imidazol-1-yl]-5-nitrophenyl]methanone see 6583

1-(4-Chlorophenyl)-3-(2,6-difluorobenzoyl)urea see 3128

1-(4-Chlorophenyl)-2-[[3-(10,11-dihydro-5H-dibenz[b,f]azepin-5-yl)-propyl]methylamino]ethanone see 5438

1-(4-Chlorophenyl)-1,6-dihydro-6,6-dimethyl-1,3,5-triazine-2,4-diamine see 2727

5-(4-Chlorophenyl)-2,3-dihydro-5-hydroxy-5H-imidazo[2,1-a]isoindole see 5643

5-(4-Chlorophenyl)-2,5-dihydro-3H-imidazo[2,1-a]isoindol-5-ol see 5643

(±)-3-(4-Chlorophenyl)-1,3-dihydro-6-methylfuro[3,4-c]pyridin-7-ol see 2268

6-(2-Chlorophenyl)-2,4-dihydro-2-[(4-methyl-1-piperazinyl)methylene]-8-nitro-1H-imidazo[1,2-a]-[1,4]benzodiazepin-1-one see 5453

5-(2-Chlorophenyl)-1,3-dihydro-7-nitro-2H-1,4-benzodiazepin-2-one see 2387

(4-Chlorophenyl)[3,5-dimethoxy-4-[2-(4-morpholinyl)ethoxy]phenyl]methanone see 6177

1-p-Chlorophenyl-2,3-dimethyl-4-dimethylamino-2-butanol see 2363

1-(3-Chlorophenyl)-2-[(1,1-dimethylethyl)amino]-1-propanone see 1488

[2-(4-Chlorophenyl)-1,1-dimethylethyl]carbamic Acid Ethyl Ester see 2378

γ-(4-Chlorophenyl)-N,N-dimethyl-2-pyridinepropanamine see 2180

1-(4-Chlorophenyl)-2,5-dimethyl-1H-pyrrole-3-acetic Acid see 2395

(2RS,3RS)-1-(4-chlorophenyl)-4,4-dimethyl-2-(1H-1,2,4-triazol-1-yl)-pentan-3-ol see 6937

N′-(4-Chlorophenyl)-N,N-dimethylurea see 6169

N-(4-Chlorophenyl)-2,6-dioxocyclohexanecarbothioamide see 6804

1-[(o-Chlorophenyl)diphenylmethyl]-imidazole see 2412

1-[(2-Chlorophenyl)diphenylmethyl]-1H-imidazole see 2412

1-(4-Chlorophenyl)ethanone see 2114

2-Chloro-1-phenylethanone see 2115

5-(2-Chlorophenyl)-7-ethyl-1,3-dihydro-1-methyl-2H-thieno[2,3-e]-1,4-diazepin-2-one see 2411

4-(2-Chlorophenyl)-2-ethyl-9-methyl-6H-thieno[3,2-f][1,2,4]triazolo[4,3-a][1,4]diazepine see 3830

5-(4-Chlorophenyl)-6-ethyl-2,4-pyrimidinediamine see 7997

4-[[α-(p-Chlorophenyl)-5-fluoro-2-hydroxybenzylidene]amino]butyramide see 7782

4-[[(4-Chlorophenyl)(5-fluoro-2-hydroxyphenyl)methylene]amino]-butanamide see 7782

4-[[α-(p-Chlorophenyl)-5-fluorosalicylidene]amino]butyramide see 7782

β-(4-Chlorophenyl)GABA see 950

p-Chlorophenyl α-Glyceryl Ether see 2178

3-Chloro-7-D-(2-phenylglycinamido)-3-cephem-4-carboxylic Acid Monohydrate see 1920

4-(2-Chlorophenylhydrazono)-3-methyl-5(4H)-isoxazolone see 3432

4-(4-Chlorophenyl)-4-hydroxy-N,N-dimethyl-α,α-diphenyl-1-piperidinebutanamide see 5450

4-(p-Chlorophenyl)-4-hydroxy-N,N-dimethyl-α,α-diphenyl-1-piperidinebutyramide see 5450

N-(β-o-Chlorophenyl-β-hydroxyethyl)isopropylamine see 2404

4-[4-(p-Chlorophenyl)-4-hydroxypiperidino]-4′-fluorobutyrophenone see 4511

4-[4-(4-Chlorophenyl)-4-hydroxy-1-piperidinyl]-1-(4-fluorophenyl)-1-butanone see 4511

2-(p-Chlorophenyl)-1,3-indandione see 2402

2-(4-Chlorophenyl)indan-1,3-dione see 2402

2-(4-Chlorophenyl)-1H-indene-1,3(2H)-dione see 2402

1-(p-Chlorophenyl)isobutyl Nicotinate see 6426

1-(o-Chlorophenyl)-2-isopropylaminoethanol see 2404

1-(p-Chlorophenyl)-5-isopropylbiguanide see 2088

p-Chlorophenylisopropylcarbinol Nicotinate see 6426

N¹-p-Chlorophenyl-N⁵-isopropyldiguanide see 2088

Chlorophenylmagnesium see 7269

Chlorophenylmercury see 7272

1-[2-[(4-Chlorophenyl)methoxy]-2-(2,4-dichlorophenyl)ethyl]-1H-imidazole see 3476

1-[1-[2-[(3-Chlorophenyl)methoxy]-phenyl]ethenyl]-1H-imidazole see 2367

(±)-2-(2-Chlorophenyl)-2-(methylamino)cyclohexanone see 5174

1-(o-Chlorophenyl)-2-methyl-2-aminopropane see 2406

1-(p-Chlorophenyl)-2-methyl-2-aminopropane see 2183

2-(4-Chlorophenyl)-α-methyl-5-benzoxazoleacetic Acid see 1055

2-(4-Chlorophenyl)-3-methyl-2,3-butanediol see 7161

2-[(4-Chlorophenyl)methyl]-N,N-diethyl-5-nitro-1H-benzimidazole-1-ethanamine see 2389

(R*,R*)-(±)-β-[(4-Chlorophenyl)-methyl]-α-(1,1-dimethylethyl)-1H-1,2,4-triazole-1-ethanol see 6937

2-[(2-Chlorophenyl)methyl]-4,4-dimethyl-3-isoxazolidinone see 3203

N-[(4-Chlorophenyl)methyl]-N′,N′-dimethyl-N-2-pyridinyl-1,2-ethanediamine see 2162

[(2-Chlorophenyl)methylene]propanedinitrile see 2127

N-(4-Chlorophenyl)-N′-(1-methylethyl)imidodicarbonimidic Diamide see 2088

N-(4-Chlorophenyl)-N-[1-(1-methylethyl)-4-piperidinyl]benzeneacetamide see 5457

4-[(4-Chlorophenyl)methyl]-2-(hexahydro-1-methyl-1H-azepin-4-yl)-1(2H)-phthalazinone see 922

6-[3-(o-Chlorophenyl)-5-methyl-4-isoxazolecarboxamido]penicillanic Acid see 2414

6-[[[3-(2-Chlorophenyl)-5-methyl-4-isoxazolyl]carbonyl]amino]-3,3-dimethyl-7-oxo-4-thia-1-azabicyclo-[3.2.0]heptane-2-carboxylic Acid see 2414

[3-(o-Chlorophenyl)-5-methyl-4-isoxazolyl]penicillin see 2414

2-(4-Chlorophenyl)-3-methyl-4-metathiazanone 1,1-Dioxide see 2105

N-[(2-Chlorophenyl)methyl]-α-methylbenzeneethanamine see 2359

1-[(4-Chlorophenyl)methyl]-2-methyl-1H-benzimidazole see 2106

2-(4-Chlorophenyl)-4-methyl-2,4-pentanediol see 3933

4-Chloro-2-(phenylmethyl)phenol see 2403

(±)-2-[4-[(4-Chlorophenyl)methyl]-phenoxy]-2-methylbutanoic Acid Ethyl Ester see 1028

3-Chloro-N-(phenylmethyl)propanamide see 1027

1-(o-Chlorophenyl)-2-methyl-2-propylamine see 2406

1-(p-Chlorophenyl)-2-methyl-2-propylamine see 2183

1-(3-Chlorophenyl)-4-[2-(5-methyl-1H-pyrazol-3-yl)ethyl]piperazine see 5745

1-[(4-Chlorophenyl)methyl]-2-(1-pyrrolidinylmethyl)-1H-benzimidazole see 2346

3-(2-Chlorophenyl)-2-methyl-4(3H)-quinazolinone see 5661

5-[(2-Chlorophenyl)methyl]-4,5,6,7-tetrahydrothieno[3,2-c]pyridine see 9361

1-[2-[[(4-Chlorophenyl)methyl]thio]-2-(2,4-dichlorophenyl)ethyl]-1H-imidazole see 8866

4-[(7-Chloro-4-quinolinyl)amino]-2-[(diethylamino)methyl]phenol see 602

2-[[4-[(7-Chloro-4-quinolinyl)amino]pentyl]ethylamino]ethanol see 4748

N^4-(7-Chloro-4-quinolinyl)-N^1,N^1-diethyl-1,4-pentanediamine see 2163

2-[(7-Chloro-4-quinolyl)amino]benzoic Acid α-Glyceride see 4330

4-[(7-Chloro-4-quinolyl)amino]-α-(diethylamino)-o-cresol see 602

N-(7-Chloro-4-quinolyl)anthranilic Acid 2,3-Dihydroxypropyl Ester see 4330

2-[4-[(6-Chloro-2-quinoxalinyl)oxy]phenoxy]propanoic Acid Ethyl Ester see 8113

6-Chlororthoxenol see 7251

Chlorosal see 2169

5-Chlorosalicylic Acid 3′,4′-Dichloroanilide see 9558

5-Chlorosalicyloyl-(o-chloro-p-nitranilide) see 6425

Chloroselenic Acid, 2164

N-Chlorosuccinimide, 2165

1-(4-Chloro-3-sulfamoylbenzamido)-2,6-dimethylpiperidine see 2391

N-(4-Chloro-3-sulfamoylbenzenesulfonyl)-N-methyl-2-furfurylamine see 5685

6-Chloro-7-sulfamoyl-2H-1,2,4-benzothiadiazine 1,1-Dioxide see 2169

6-Chloro-7-sulfamoyl-2H-1,2,4-benzothiadiazin-2-ylsodium 1,1-Dioxide see 2169

6-Chloro-7-sulfamoyl-3-benzyl-3,4-dihydro-1,2,4-benzothiadiazine 1,1-Dioxide see 1150

6-Chloro-7-sulfamoyl-3-benzylthiomethyl-2H-1,2,4-benzothiadiazine 1,1-Dioxide see 1134

3-(4′-Chloro-3′-sulfamoylphenyl)-3-hydroxyphthalimidine see 2194

1-(2-Chloro-4-sulfamoylphenyl)-3-phenylsuccinimide see 8852

4-Chloro-5-sulfamoyl-2′,6′-salicyloxylidide see 9986

6-Chloro-7-sulfamyl-1,2,4-benzothiadiazine 1,1-Dioxide see 2169

6-Chloro-7-sulfamyl-3,4-dihydro-1,2,4-benzothiadiazine 1,1-Dioxide see 4704

4-Chloro-5-sulfamylsalicyloyl-2′,6′-dimethylanilide see 9986

3-Chloro-6-sulfanilamidopyridazine see 8871

Chlorosulfonic Acid, 2166

Chlorosulfonic Anhydride see 8020

Chlorosulfuric Acid see 2166

Chlorosulthiadil see 4704

4-Chlorotestosterone see 2409

7-Chlorotetracycline see 2193

8-Chloro-1a,14,15,15a-tetrahydro-9,11-dihydroxy-14-methyl-6H-oxireno[e][2]benzoxacyclotetradecin-6,12(7H)-dione see 6164

6-Chloro-2,3,4,5-tetrahydro-1-(4-hydroxyphenyl)-1H-3-benzazepine-7,8-diol see 3925

7-Chloro-1,2,3,4-tetrahydro-2-methyl-3-(2-methylphenyl)-4-oxo-6-quinazolinesulfonamide see 6068

7-Chloro-1,2,3,4-tetrahydro-2-methyl-4-oxo-3-o-tolyl-6-quinazolinesulfonamide see 6068

10-Chloro-2,3,7,11b-tetrahydro-2-methyl-11b-phenyloxazolo[3,2-d]-[1,4]benzodiazepin-6(5H)-one see 6883

7-Chloro-1,2,3,4-tetrahydro-4-oxo-2-phenyl-6-quinazolinesulfonamide see 3941

9-Chloro-7-(1H-tetrazol-5-yl)-5H-[1]benzopyrano[2,3-b]pyridin-5-one see 9494

9-Chloro-7-(5-1H-tetrazolyl)-5-oxo-5H-[1]benzopyrano[2,3-b]pyridine see 9494

2-Chloro-5-(2H-tetrazol-5-yl)-N^4-2-thenylsulfanilamide see 935

2-Chloro-5-(1H-tetrazol-5-yl)-4-[(2-thienylmethyl)amino]benzenesulfonamide see 935

Chlorothalonil, 2167

Chlorothen, 2168

Chlorothene see 9549

5-Chloro-3-(2-thenoyl)-2-oxindole-1-carboxamide see A9

5-(4′-Chloro-2′-thenylamino-5′-sulfamoylphenyl)tetrazole see 935

2-[(5-Chloro-2-thenyl)(2-dimethylaminoethyl)amino]pyridine see 2168

Chlorothenylpyramine see 2168

Chlorothiamine see 1030

Chlorothiazide, 2169

1-[2-[(2-Chloro-3-thienyl)methoxy]-2-(2,4-dichlorophenyl)ethyl]-1H-imidazole see 9384

N-[(5-Chloro-2-thienyl)methyl]-N′,N′-dimethyl-N-2-pyridinyl-1,2-ethanediamine see 2168

N-5-Chloro-2-thienylmethyl-N′,N′-dimethyl-N-2-pyridylethylenediamine see 2168

Chlorothion see 2196

3-(2-Chloro-9H-thioxanthen-9-ylidene)-N,N-dimethyl-1-propanamine see 2189

4-[3-(2-Chloro-9H-thioxanthen-9-ylidene)propyl]-1-piperazineethanol see 2392

Chlorothricin, 2170

Chlorothricolide see 2170

Chlorothymol, 2171

4-Chlorothymol see 2171

6-Chlorothymol see 2171

(4-Chloro-o-toloxy)acetic Acid see 5645

Chlorotoluene, 2172

m-Chlorotoluene see 2172

o-Chlorotoluene see 2172

p-Chlorotoluene see 2172

α-Chlorotoluene see 1143

5-Chlorotoluene-2,4-disulfonamide see 3369

(N-Chloro-p-toluenesulfonamido)sodium see 2066

2-(2-Chloro-p-toluidino)-2-imidazoline see 9443

N-(3-Chloro-o-tolyl)anthranilic Acid see 9436

N′-(4-Chloro-o-tolyl)-N,N-dimethylformamidine see 2083

(±)-2-[(4-Chloro-o-tolyl)oxy]propionic Acid see 5666

Chlorotrianisene, 2173

2-Chloro-6-(trichloromethyl)pyridine see 6490

2-Chloro-1-(2,4,5-trichlorophenyl)vinyl Dimethyl Phosphate see 8777

6-Chloro-N,N,N′-triethyl-1,3,5-triazine-2,4-diamine see 9580

1-Chloro-2,2,2-trifluoroethyl Difluoromethyl Ether see 5059

2-Chloro-1,1,2-trifluoroethyl Difluoromethyl Ether see 3536

2-Chloro-N-[[[4-(trifluoromethoxy)phenyl]amino]carbonyl]benzamide see 9592

5-[2-Chloro-4-(trifluoromethyl)phenoxy]-2-nitrobenzoic Acid see 105

N-[2-Chloro-4-(trifluoromethyl)phenyl]-DL-valine Cyano(3-phenoxyphenyl)methyl Ester see 4137

3-(2-Chloro-3,3,3-trifluoro-1-propenyl)-2,2-dimethylcyclopropanecarboxylic Acid Cyano(3-phenoxyphenyl)methyl Ester see 2765

[1α,3α(Z)]-(±)-3-(2-Chloro-3,3,3-trifluoro-1-propenyl)-2,2-dimethylcyclopropanecarboxylic Acid (2-Methyl[1,1′-biphenyl]-3-yl)methyl Ester see 1229

2-Chloro-α,α,α-trifluoro-p-tolyl-3-ethoxy-4-nitrophenyl Ether see 6916

7-Chloro-11,17,21-trihydroxy-16-methylpregna-1,4-diene-3,20-dione see 213

9-Chloro-11,17,21-trihydroxy-16-methylpregna-1,4-diene-3,20-dione see 1029

(11β)-6-Chloro-11,17,21-trihydroxypregna-1,4,6-triene-3,20-dione see 2396

7-Chloro-2′,4,6-trimethoxy-6′-methylspiro[benzofuran-2(3H),1′-[2]-cyclohexene]-3,4′-dione see 4453

endo-2-Chloro-1,7,7-trimethylbicyclo-[2.2.1]heptane see 1341

3-Chloro-1,7,7-trimethylbicyclo-[2.2.1]heptan-2-one see 2132

2-Chloro-N,N,N-trimethylethanaminium Chloride see 2103

2-Chloro-N,N,6-trimethyl-4-pyrimidinamine see 2590

2-Chloro-1,3,5-trinitrobenzene see 7390

Chlorotrioxochromate(1−), Hydrogen, Compd with Pyridine (1:1) see 7986

Chlorotris(p-methoxyphenyl)ethylene see 2173

Chlorotris(triphenylphosphine)rhodium, 2174

1-(o-Chlorotrityl)imidazole see 2412

Chlorovinylarsine Dichloride see 3051

2-Chlorovinyldichloroarsine see 3051

4′-Chloroxenylamine see 445

Chioroxine, 2175

Chloroxyl see 2290

α-Chloro-m-xylene see 9999

ω-Chloro-m-xylene see 9999

Chloroxylenol, 2176

p-Chloro-m-xylenol see 2176

2-Chloro-m-xylenol see 2176

4-Chloro-3,5-xylenol see 2176

Chlorozotocin, 2177

Chlorparacide see 2074

Chlorphenamide see 2371

Chlorphenamidine see 2083

Chlorphenamine see 2180

Chlorphencyclan see 2372

Chlorphenesin, 2178

Chlorphenesin Carbamate, 2179

Chlorpheniramine, 2180

d-Chlorpheniramine see 2180

Chlorphenoxamide see 2181

Chlorphenoxamine, 2182

Chlorphentermine, 2183

Chlorphthalidolone see 2194

Chlorpiprazine see 7135

Chlorpiprozine see 7135

Chlorproethazine, 2184

Chlorproguanil, 2185

Chlorpromados [Pharma Holz] see 2186

Chlor-Promanyl [Maney] see 2186

Chlorpromazine, 2186

Chlorpromazine Sulfoxide see 6810

Chlorpropamide, 2187
Chlorpropham, 2188
Chlorprophenpyridamine see 2180
Chlorprothixene, 2189
Chlorpyrifos, 2190
Chlorpyrifos-ethyl see 2190
Chlorpyrifos-methyl see 2190
Chlorquinaldol, 2191
Chloroquinol see 4520
Chlorsulfonamidodihydrobenzothiadi-
   azine Dioxide see 4704
Chlorsulfuron, 2192
Chlorsulphacide see 2074
Chlortetracycline, 2193
Chlor-Tetragnost see 7214
Chlorthalidone, 2194
Chlorthal-methyl see 2830
Chlorthalonil see 2167
Chlorthenoxazin(e), 2195
Chlorthiepin see 3529
Chlorthion®, 2196
Chlor-Trimeton [Schering] see 2180
Chlor-Tripolon [Schering] see 2180
Chlorurit see 2169
Chlorvescent see 7601
Chloryl Anesthetic see 3740
Chlorylen [Schering] see 9552
Chlorzide [Foy] see 4704
Chlorzoxazone, 2197
Chlosudimeprimyl see 2391
Chlotride [Merck & Co.] see 2169
ChlVPP see MISC-2
CHO see MISC-2
Chocolax [Elder] see 7208
Choisine [Rhône Poulenc] see 7366
Cholagon see 2858
Cholaic Acid see 9044
Cholalic Acid see 2206
3-[(3-Cholamidopropyl)dimethylam-
   monio]-1-propanesulfonate see 2034
Cholan-DH [Pennwalt] see 2858
Cholane, 2198
5β-Cholane see 2198
Cholanic Acid, 2199
5β-Cholan-24-oic Acid see 2199
Cholanorm [Grünenthal] see 2044
Cholanthrene, 2200
Cholatol [Upjohn] see 6884
Cholaxine see 8680
Cholebrine [Nicholas] see 4903
Cholecalciferol see 9929
Cholecystokinin, 2201
Cholecystokinin-pancreozymin see
   2201
Cholecystokinin C-Terminal Octapep-
   tide see 8494
Choledyl [Warner-Chilcott] see 2213
Cholegnostyl see 9122
Cholegrafin [Squibb] see 4916
Choleic Acids see 2881
Cholepatin [AB Leo] see 2858
Cholepulvis see 4931
Cholergol [Giulini] see 6828
Cholesolvin [Yoshitomi] see 8489
Cholestabyl [Holphar] see 2472
Cholesta-5,7-dien-3-ol see 2857
3β-Cholesta-5,24-dien-3-ol see 2905
Cholestane, 2202
5β-Cholestane see 2518
Cholestanol, 2203
5α-Cholestan-3α-ol see 3564
5β-Cholestan-3β-ol see 2519
β-Cholestanol see 2203
ε-Cholestanol see 3564
Cholest-5-ene-3,25-diol see 4751
Cholest-5-ene-3β,24-diol see 4750
Cholest-5-ene-3β,24α-diol see 4750
Cholest-5-ene-3β,24β-diol see 4750
Δ5-Cholestene-3β,25-diol see 4751
Cholest-4-en-3β-ol see 253
Cholest-5-en-3α-ol see 3565

Cholest-5-en-3β-ol see 2204
Cholest-5-en-3β-ol 4-[Bis(2-chloro-
   ethyl)amino]benzeneacetate see
   7182
Cholesterin see 2204
Cholesterol, 2204
Cholesteryl p-Bis(2-chloroethyl)-
   aminophenylacetate see 7182
Cholestyramine Resin, 2205
Choletec [Squibb] see 5652
Cholexamin [Kyorin] see 6429
Cholibil [Takeda] see 9501
Cholic Acid, 2206
Choligen see 4765
Cholimil [Takeda] see 4903
Choline, 2207
Choline Bromide Hexamethylenedi-
   carbamate see 4599
Choline Bromide α-Phenyl-5-nor-
   bornene-2-glycolate see 3527
Choline Chloride, 2208
Choline Chloride Carbamate see 1780
Choline Chloride Dihydrogen Phos-
   phate see 7337
Choline Chloride Phosphate see 7337
Choline Chloride Succinate (2:1) see
   8846
Choline Cytidine 5'-Pyrophosphate
   (Ester) see 2321
Choline Dehydrocholate, 2209
Choline Dichloride see 2103
Choline Dihydrogen Citrate, 2210
Choline Esterase, 2211
Choline 1,5-Naphthalenedisulfonate
   (2:1), Dilactate, Diacetate see 109
Choline Orotate see 6828
Choline Phosphate Chloride see 7337
Choline Phosphoric Acid Ester (Chlo-
   ride) see 7337
"Choline Plasmalogen" see 7492
Choline Salicylate, 2212
Choline Salicylic Acid Salt see 2212
Choline Succinate Dichloride see 8846
Choline Theophyllinate, 2213
Cholinfall [Tanabe Seiyaku] see 5902
Cholinophylline see 2213
Cholit-Ursan [Fresenius] see 9801
Chologon see 2858
Cholografin [Squibb] see 4916
Cholonerton [Dolorgiet] see 4792
Cholospect see 4916
Choloxin [Baxter] see 9348
N-Choloyltaurine see 9044
Cholspasmin [Lipha] see 4792
Cholumbrin see 4931
Cholybar [Warner-Lambert] see 2205
N-Cholylglycine see 4390
Cholyltaurine see 9044
d-Chondocurine see 2215
Chondodendrine see 1024
Chondrillasterol, 2214
Chondrocurine, 2215
Chondrofoline, 2216
Chondroitin Sulfate, 2217
Chondroitin Sulfate A see 2217
Chondroitin Sulfate B see 2217
Chondroitin Sulfate C see 2217
Chondroitin 4-Sulfate see 2217
Chondroitin 6-Sulfate see 2217
Chondroitin 4-Sulfate Disodium Salt
   see 2217
Chondroitinsulfuric Acid see 2217
Chondrosamine see 4240
Chondrosine, 2218
Chonemorphine, 2219
Chonsurid see 2217
CHOP see MISC-2
CHOP-B see MISC-2
Chop Nut see 7356
Choragon [Ferring] see 4534
Choriogonin see 4534

Chorionic Gonadotropin see 4534
Chorionic Growth Hormone-prolactin
   see 4535
Chorismic Acid, 2220
Chothyn [Tilden-Yates] see 2210
CHP-Depot see 7873
CHQ see 4520
Chrisanol see 1649
Christmas Factor see 3874
Chromargyre see 5757
Chromatophorotropic Hormone see
   6206
Chrome Alum see 2232
Chrome Alum Ammonium see 533
Chrome Green see 2229
1,2-Chromene see 1112
3-Chromene see 1112
Chrome Ocher see 2229
Chrome Oxide Green see 2229
Chrome Red see 5280
Chrome Yellow see 5279
Chromia see 2229
Chromic Acetate, 2221
Chromic Acid see 2238
Chromic Acid Iron(3+) Salt (3:2) see
   3962
Chromic Acid Lead(2+) Salt (1:2) see
   5280
Chromic Ammonium Sulfate see 533
Chromic Anhydride see 2238
Chromic Bromide, 2222
Chromic Carbonate, 2223
Chromic Chloride, 2224
Chromic Fluoride, 2225
Chromic Formate, 2226
Chromic Hydroxide, 2227
Chromic Nitrate, 2228
Chromic Oxide, 2229
Chromic Oxide Gel see 2227
Chromic Oxide, Hydrous see 2227
Chromic Phosphate, 2230
Chromic Potassium Oxalate, 2231
Chromic Potassium Sulfate, 2232
Chromic Sulfate, 2233
Chromite see 2234
Chromitope Sodium [Squibb] see 8548
Chromium, 2234
Chromium Carbonyl, 2235
Chromium Dioxide, 2236
Chromium Dioxychloride see 2251
Chromium Hexacarbonyl see 2235
Chromium Hydroxide see 2227
Chromium Lead Oxide see 5280
Chromium Oxyfluoride see 2252
Chromium Sesquioxide see 2229
Chromium Tetrafluoride, 2237
Chromium Trioxide, 2238
Chromocarb, 2239
Chromomycin A₃ see 2240
Chromomycinone see 7510
Chromomycins, 2240
Chromonar, 2241
2-Chromonecarboxylic Acid see 2239
Chromotrichia Factor see 434
Chromotrope 2B, 2242
Chromotropic Acid, 2243
Chromous Acetate, 2244
Chromous Bromide, 2245
Chromous Chloride, 2246
Chromous Fluoride, 2247
Chromous Formate, 2248
Chromous Oxalate, 2249
Chromous Sulfate, 2250
Chromyl Chloride, 2251
Chromyl Fluoride, 2252
Chronogest [Intervet] see 4125
Chronogyn [Winthrop] see 2811
Chrono-Indocid [Merck & Co.] see
   4874
Chrysammic Acid see 2253
Chrysamminic Acid, 2253

Chrysanol see 1649
Chrysanthemaxanthin, 2254
Chrysanthemic Acid, 2255
Chrysanthemin see 2694
N-(Chrysanthemoxymethyl)-1-cyclo-
  hexene-1,2-dicarboximide see 9154
Chrysanthemumdicarboxylic Acid
  Monomethyl Ester Pyrethrolone
  Ester see 7978
Chrysanthemumic Acid see 2255
Chrysanthemummonocarboxylic Acid
  see 2255
Chrysanthemummonocarboxylic Acid
  6-Chloropiperonyl Ester see 1014
Chrysanthemummonocarboxylic Acid
  Ester With 3-(2-Cyclopenten-1-yl)-
  2-methyl-4-oxo-2-cyclopenten-1-ol
  see 2711
Chrysanthemummonocarboxylic Acid
  Pyrethrolone Ester see 7978
Chrysanthenone, 2256
Chrysarobin, 2257
Chrysatropic Acid see 8363
Chrysazin see 2813
Chrysazin-3-carboxylic Acid see 8175
6-Chrysenamine, 2258
Chrysene, 2259
Chrysenex see 2258
6-Chrysenylamine see 2258
Chrysergonic Acid, 2260
Chrysidenon 1438 see 2261
Chrysin, 2261
Chrysoberyl see 1177
Chrysogen see 6288
Chrysoidine, 2262
Chrysoidine Orange see 2262
Chrysoidine Y see 2262
Chrysoine see 9687
Chrysomykine see 2193
Chrysophanein see 2263
Chrysophanic Acid, 2263
Chrysophaniin see 2263
Chrysophanol see 2263
Chrysotile see 5569
Chrysotile see 851
6-Chrysylamine see 2258
Chrytemin [Fujinaga] see 4835
CHX-3101 see 8695
CHX-3311 see 2790
CHX-3673 see 508
Chymar [Armour Pharm.] see 2265
Chymetin [BDH] see 2265
Chymex [Adria] see 1061
Chymodiactin [Smith] see 2264
Chymolase see 2265
Chymopapain, 2264
Chymosin see 8139
Chymotase (Tabl.) see 2265
Chymotrypsins, 2265
Chymozym [Ika] see 2265
CI-336 see 1808
CI-379 see 1090
CI-395 see 7179
CI-416 see 9570
CI-440 see 4060
CI-473 see 5680
CI-501 see 2727
CI-556 see 16
CI-581 see 5174
CI-583 see 5659
CI-588 see 3355
CI-633 see 2354
CI-673 see 9881
CI-719 see 4280
CI-775 see 1213
CI-781 see 10093
CI-825 see 7091
CI-845 see 7471
CI-880 see 627
CI-898 see 9635
CI-904 see 2979

CI-906 see A7
CI-912 see 10094
CI-919 see 3540
CI-928 see A7
C.I. 10316 see 6313
C.I. 11000 see 430
C.I. 11020 see 3218
C.I. 11050 see 5138
C.I. 11160 see 431
C.I. 11270 see 2262
C.I. 11380 see 10003
C.I. 11390 see 10004
C.I. 12140 see 9997
C.I. 12156 see 2335
C.I. 13020 see 6037
C.I. 13025 see 6019
C.I. 13065 see 5833
C.I. 13080 see 9688
C.I. 13390 see 662
C.I. 14025 see 5823
C.I. 14030 see 241
C.I. 14130 see 6799
C.I. 14270 see 9687
C.I. 14600 see 6812
C.I. 14645 see 3615
C.I. 14700 see 7568
C.I. 15510 see 6813
C.I. 15985 see 8983
C.I. 16035 see 283
C.I. 16155 see 7567
C.I. 16185 see 382
C.I. 16575 see 2242
C.I. 18950 see 7528
C.I. 19140 see 9041
C.I. 19235 see 6814
C.I. 19540 see 9238
C.I. 21000 see 1265
C.I. 21010 see 1264
C.I. 21010:1 see 1264
C.I. 22120 see 2493
C.I. 22850 see 9702
C.I. 23500 see 1111
C.I. 23570 see 9917
C.I. 23850 see 9701
C.I. 23860 see 3863
C.I. 24140 see 1095
C.I. 26100 see 8858
C.I. 26105 see 8349
C.I. 37500 see 6304
C.I. 40600 see 1322
C.I. 40850 see 1756
C.I. 42000 see 5581
C.I. 42040 see 1367
C.I. 42045 see 8968
C.I. 42053 see 3886
C.I. 42085 see 4487
C.I. 42090 see 1366
C.I. 42095 see 5361
C.I. 42510 see 5528
C.I. 42555 see 4287
C.I. 42590 see 5998
C.I. 42685 see 103
C.I. 42745 see 104
C.I. 42780 see 5950
C.I. 43800 see 899
C.I. 45005 see 8017
C.I. 45010 see 8016
C.I. 45170 see 8182
C.I. 45350 see 4085
C.I. 45350:1 see 4085
C.I. 45365 see 3056
C.I. 45370:1 see 3005
C.I. 45380 see 3555
C.I. 45400 see 3554
C.I. 45425 see 3174
C.I. 45425:1 see 3174
C.I. 45430 see 3640
C.I. 45440 see 3242
C.I. 45445 see 4250
C.I. 47000 see 8101
C.I. 47005 see 8100

C.I. 49410 see 9466
C.I. 50040 see 6395
C.I. 50200 see 7218
C.I. 51030 see 4256
C.I. 51050 see 1955
C.I. 51400 see 5211
C.I. 52000 see 9276
C.I. 52002 see 942
C.I. 52005 see 940
C.I. 52010 see 941
C.I. 52015 see 5979
C.I. 52040 see 9444
C.I. 57000 see 4248
C.I. 58000 see 237
C.I. 58005 see 8516
C.I. 58015 see 240
C.I. 58050 see 8094
C.I. 58205 see 7962
C.I. 58500 see 8059
C.I. 58600 see 8267
C.I. 61565 see 8095
C.I. 61570 see 238
C.I. 67410 see 239
C.I. 73000 see 4855
C.I. 73015 see 4856
C.I. 74160 see 2515
C.I. 74240 see 211
C.I. 75140 see 1876
C.I. 75240 see 5519
C.I. 75280 see 1362
C.I. 75410 see 7962
C.I. 75450 see 5209
C.I. 75460 see 5172
C.I. 75470 see 1850
C.I. 75500 see 5150
C.I. 75530 see 243
C.I. 75620 see 4026
C.I. 75660 see 6179
C.I. 76050 see 5920
C.I. 76076 see 7256
C.I. 77007 see 9757
C.I. 77085 see 822
C.I. 77103 see 982
C.I. 77199 see 1628
C.I. 77223 see 1660
C.I. 77266 see 1815
C.I. 77288 see 2229
C.I. 77350 see 2428
C.I. 77357 see 2426
C.I. 77360 see 2441
C.I. 77402 see 2671
C.I. 77410 see 2629
C.I. 77482 see 4424
C.I. 77505 see 3962
C.I. 77510 see 3964
C.I. 77578 see 5307
C.I. 77597 see 1016
C.I. 77600 see 5279
C.I. 77766 see 5789
C.I. 77800 see 8273
C.I. 77947 see 10050
C.I. Acid Blue 1 see 8968
C.I. Acid Blue 9 see 1366
C.I. Acid Blue 74 see 4856
C.I. Acid Blue 92 see 662
C.I. Acid Blue 93 see 5950
C.I. Acid Green 3 see 4487
C.I. Acid Green 5 see 5361
C.I. Acid Green 25 see 238
C.I. Acid Orange 5 see 9688
C.I. Acid Orange 6 see 9687
C.I. Acid Orange 7 see 6813
C.I. Acid Orange 11 see 3005
C.I. Acid Orange 20 see 6812
C.I. Acid Orange 52 see 6019
C.I. Acid Orange 137 see 6814
C.I. Acid Red 2 see 6037
C.I. Acid Red 27 see 382
C.I. Acid Red 51 see 3640
C.I. Acid Red 87 see 3555
C.I. Acid Red 91 see 3554

C.I. Acid Red 94 see 8242
C.I. Acid Red 95 see 3174
C.I. Acid Red 176 see 2242
C.I. Acid Violet 19 see 103
C.I. Acid Violet 25 see 104
C.I. Acid Yellow 1 see 6313
C.I. Acid Yellow 3 see 8100
C.I. Acid Yellow 23 see 9041
C.I. Acid Yellow 36 see 5833
C.I. Acid Yellow 40 see 7528
C.I. Acid Yellow 73 see 4085
C.I. Azoic Coupling Component 1 see 6304
C.I. Basic Blue 9 see 5979
C.I. Basic Blue 17 see 9444
C.I. Basic Brown 1 see 1265
C.I. Basic Brown 4 see 1264
C.I. Basic Green 1 see 1367
C.I. Basic Green 4 see 5581
C.I. Basic Orange 2 see 2262
C.I. Basic Red 5 see 6395
C.I. Basic Violet 3 see 4287
C.I. Basic Violet 10 see 8182
C.I. Basic Violet 14 see 5528
C.I. Developer 5 see 6304
C.I. Direct Blue 8 see 1095
C.I. Direct Blue 14 see 9701
C.I. Direct Blue 53 see 3863
C.I. Direct Red 2 see 1111
C.I. Direct Red 28 see 2493
C.I. Direct Red 34 see 9917
C.I. Direct Yellow 9 see 9238
C.I. Fluorescent Brightener 30 see 1322
C.I. Food Blue 1 see 4856
C.I. Food Blue 2 see 1366
C.I. Food Blue 3 see 8968
C.I. Food Green 1 see 4487
C.I. Food Green 3 see 3886
C.I. Food Red 1 see 7568
C.I. Food Red 6 see 7567
C.I. Food Red 14 see 3640
C.I. Food Red 17 see 283
C.I. Food Yellow 3 see 8983
C.I. Food Yellow 4 see 9041
C.I. Food Yellow 8 see 9687
C.I. Ingrain Blue 1 see 211
C.I. Mordant Black 11 see 3615
C.I. Mordant Blue 10 see 4256
C.I. Mordant Blue 14 see 1955
C.I. Mordant Orange I see 241
C.I. Mordant Orange 14 see 240
C.I. Mordant Red 3 see 8516
C.I. Mordant Red 11 see 237
C.I. Mordant Violet 26 see 8059
C.I. Mordant Yellow 1 see 5823
C.I. Mordant Yellow 5 see 6799
C.I. Natural Brown 1 see 4026
C.I. Natural Brown 7 see 5150
C.I. Natural Red 3 see 5172
C.I. Natural Red 4 see 1850
C.I. Natural Red 8 see 7962
C.I. Natural Red 16 see 7962
C.I. Natural Red 20 see 243
C.I. Natural Red 24 see 1362
C.I. Natural Red 25 see 5209
C.I. Natural Red 26 see 1876
C.I. Natural Yellow 8 see 6179
C.I. Natural Yellow 11 see 5519 and 6179
C.I. Oxidation Base 12 see 5920
C.I. Pigment Blue 15 see 2515
C.I. Pigment Blue 27 see 3964
C.I. Pigment Blue 29 see 9757
C.I. Pigment Blue 66 see 4855
C.I. Pigment Brown 9 see 2640
C.I. Pigment Green 17 see 2229
C.I. Pigment Green 21 see 2629
C.I. Pigment Red 83 see 237
C.I. Pigment Red 105 see 5307
C.I. Pigment Red 106 see 5789

C.I. Pigment Red 109 see 4424
C.I. Pigment Violet 14 see 2441
C.I. Pigment White 1 see 1016
C.I. Pigment White 4 see 10050
C.I. Pigment White 5 see 5424
C.I. Pigment Yellow 31 see 982
C.I. Pigment Yellow 33 see 1660
C.I. Pigment Yellow 34 see 5279
C.I. Pigment Yellow 36 see 10032
C.I. Pigment Yellow 39 see 822
C.I. Pigment Yellow 40 see 2426
C.I. Pigment Yellow 45 see 3962
C.I. Solvent Brown 12 see 1264
C.I. Solvent Green 3 see 8095
C.I. Solvent Orange 3 see 2262
C.I. Solvent Orange 7 see 9997
C.I. Solvent Orange 32 see 3056
C.I. Solvent Red 23 see 8858
C.I. Solvent Red 24 see 8349
C.I. Solvent Red 72 see 3005
C.I. Solvent Red 73 see 3174
C.I. Solvent Red 80 see 2335
C.I. Solvent Yellow 1 see 430
C.I. Solvent Yellow 2 see 3218
C.I. Solvent Yellow 3 see 431
C.I. Solvent Yellow 5 see 10003
C.I. Solvent Yellow 6 see 10004
C.I. Solvent Yellow 33 see 8101
C.I. Solvent Yellow 94 see 4085
C.I. Vat Blue 1 see 4855
Ciafos, 2266
Ciamin see 855
Cianatil see 2690
Cianurina see 5772
Ciatyl [Troponwerke] see 2392
Ciba 570 see 7312
Ciba 2059 see 4080
Ciba 3126 see 6061
Ciba 5968 see 4682
Ciba-8514 see 2083
Ciba 9295 see 913
Ciba 10370 see 8871
Ciba 11925 see 7148
CIBA 32644-Ba see 6480
CIBA 36278-Ba see 1969
Cibacalcin (Human Synthetic) [Ciba] see 1640
Cibacthen [Ciba] see 127
Cibazol [Ciba] see 8920
Cibenzoline see 2274
Ciberon [Taisho] see 1804
Cibian [Yamanouchi] see 4892
Cichorigenin see 3645
Cichoriin, 2267
Ciclacillin see 2706
Cicletanide see 2268
Cicletanine, 2268
Ciclincaf [Medici] see 8408
Ciclobendazole see 2718
Ciclobiotic [Beta] see 5851
Cicloche [Novag] see 2270
Ciclolysal see 5500
Ciclonicate, 2269
Ciclonium Iodide see 2743
Ciclopirox, 2270
Ciclopirox Olamine see 2270
Cicloral see 1839
Ciclosidomine, 2271
Ciclosporin see 2759
Ciclotate see MISC-4
Cicloven [Agips] see 7987
Cicrotoic Acid, 2272
Cicutine see 2500
Cicutoxin, 2273
Cidal see 8297
Cidamex see 45
Cidandopa [Cidán] see 5344
Cidex [Arbrook] see 4366
Cidifos [Neopharmed] see 2321
Cidocetine see 2068

Cidomycin [Roussel-UCLAF] see 4284
Cidoxepin see 3425
Cidrex see 4704
Cifenline, 2274
Ciflox [Bayer] see 2315
CIG see 4018
Cignolin [Bayer] see 714
Cigthranol [Bayer] see 714
Cilag 61 see 9631
Cilastatin, 2275
Cilastatin Sodium see 2275
Cilazapril, 2276
Cilazaprilat see 2276
Cilest [Cilag] see 6620
Cilicaine see 7042
Cilifor [C.E.P.A.] see 1975
Cillenta [Ayerst] see 7037
Cilleral [Bristol] see 621
Cilligen [Sigma] see 7050
Cillimycin [Wyeth] see 5378
Ciloprin [Cilag] see 17
Cilostazol, 2277
Cimadon see 7403
Cimagel [Usines Fournier-Cimag] see 7532
Cimal [AL] see 2279
Cimaterol, 2278
Cimedone see 8667
Cimet [Lancet] see 2279
Cimetag [Cehasol] see 2279
Cimetidine, 2279
Cimetropium Bromide, 2280
Cimetum [Sintyal] see 2279
Cimicifuga, 2281
Cimicifugol see 2282
Cimigenol, 2282
Cinametic Acid, 2283
Cinamiodil see 2284
Cinamiodyl, 2284
Cinaperazine [Yakult] see 2308
Cinarine see 2773
Cinazyn [Italchimici] see 2308
Cincaine see 3016
Cinch [Du Pont] see 2296
Cinchamidine see 4705
Cinchocaine see 3016
Cinchol see 8500
Cincholepidine see 5324
Cincholic Acid β-D-Quinovoside see 8105
Cinchomeronic Acid, 2285
Cinchona, 2286
Cinchona Bark see 2286
Cinchonamine, 2287
Cinchonan-6′,9-diol see 2626
Cinchonan-9-ol see 2288 and 2289
Cinchonicine see 2291
Cinchonidine, 2288
Cinchonine, 2289
Cinchophen, 2290
Cinchotine see 4706
Cinchotoxine, 2291
Cinchovatine see 809 and 2288
Cinco F U [Montedison] see 4109
Cinconal see 2290
Cinconifine see 4706
Cincosal see 2290
Cincuental [Nemi] see 9888
Cindomet [Chiesi] see 2295
Cinene see 5371
Cineole see 3851
Cinepazet Maleate, 2292
Cinepazic Acid Ethyl Ester Maleate see 2292
Cinepazide, 2293
Cinerin I see 2294
Cinerin II see 2294
Cinerins, 2294
Cinmetacin, 2295
Cinmethylin, 2296

Cinnabar see 5789
Cinnabarine, 2297
Cinnacet [Schwarzhaupt] see 2308
Cinnageron [Streuli] see 2308
Cinnaloid [Pfizer] see 8146
Cinnamal see 2298
Cinnamaldehyde, 2298
Cinnamedrine, 2299
Cinnamein see 1144
Cinnamene see 8830
Cinnamic Acid, 2300
trans-Cinnamic Acid Benzyl Ester see 1144
Cinnamic Acid Cinnamyl Ester see 2307
Cinnamic Alcohol see 2305
Cinnamic Aldehyde see 2298
Cinnamic Anhydride see 2300
Cinnamin [Nippon Chemiphar] see 758
Cinnamol see 8830
Cinnamon, Ceylon, 2301
Cinnamonin see 6233
Cinnamon, Saigon, 2302
Cinnamon Wood see 8340
Cinnamoyl Chloride, 2303
Cinnamoylcocaine, 2304
Cinnamoylecgonine Methyl Ester see 2304
1-Cinnamoyl-5-methoxy-2-methylin-dole-3-acetic Acid see 2295
Cinnamoylmethylecgonine see 2304
1-Cinnamoyl-2-methyl-5-methoxy-3-indolylacetic Acid see 2295
Cinnamyl Alcohol, 2305
Cinnamyl o-Aminobenzoate see 2306
Cinnamyl Anthranilate, 2306
1-Cinnamyl-4-benzhydrylpiperazine see 2308
1-Cinnamyl-4-(4-chlorobenzhydryl)-piperazine see 2366
1-Cinnamyl-4-(p-chloro-α-phenyl-benzyl)piperazine see 2366
Cinnamyl Cinnamate, 2307
Cinnamylcocaine see 2304
1-Cinnamyl-4-(di-p-fluorobenzhy-dryl)piperazine see 4070
1-Cinnamyl-4-diphenylmethylpiper-azine see 2308
1-trans-Cinnamyl-4-diphenylmethyl-piperazine see 2308
N-Cinnamylephedrine see 2299
α-[1-(N-Cinnamyl-N-methylamino)-ethyl]benzyl Alcohol see 2299
Cinnamyl Methyl Ketone see 1151
(E)-N-Cinnamyl-N-methyl-1-naph-thalenemethylamine see 6270
Cinnarizine, 2308
Cinnipirine see 2308
Cinnoline, 2309
Cinnoxicam see 7476
Cinnyl Cinnamate see 2307
CINO-40 [Tutag] see 9511
Cinobac [Lilly] see 2311
Cinobufotalin, 2310
Cinolone [Pierrel] see 9511
Cinopal [Lederle] see 3906
Cinopenil [Hoechst] see 5890
Cinopol [Lederle] see 3906
Cinoxacin, 2311
Cinoxate, 2312
Cin-Quin [Rowell] see 8074
Cinromide, 2313
CiNU [Bristol] see 5444
Ciodrin [Shell] see 2603
CIPC see 2188
Cipca see 855
Cipionate see MISC-4
Ciplamin H see 4739
Ciplamycetin see 2068
Cipractin [Andromaco] see 2779

Cipralan [Roche; UPSA] see 2274
Cipro [Bayer] see 2315
Ciprobay [Bayer] see 2315
Ciprofibrate, 2314
Ciprofloxacin, 2315
Ciprol [Sterling] see 2314
Ciproxan [Bayer] see 2315
Ciproxin [Bayer] see 2315
Ciradol [Wyeth] see 2316
Ciramadol, 2316
Cirantin [Indian] see 4591
Circair [Circle] see 3459
Circanol [Kettelhack] see 3598
Circleton [IBI] see 8967
Circolene [Inverni Della Beffa] see 8129
Circubid [W. F. Merchant] see 3682
Circulen "Kyorin" see 145
Circuletin see 5159
Circulin A see 2317
Circulin B see 2317
Circulins, 2317
Circupon [Troponwerke] see 3822
Cirotyl [Parke, Davis] see 6927
Cirpon see 5751
Cirrocolina see 2210
Cisapride, 2318
Cismaplat [Mack, Illert.] see 2319
Cisordinol [Lundbeck] see 2392
Cisplatin, 2319
Cisplatyl [Roger Bellon] see 2319
Cistobil [Bracco] see 4944
Citalopram, 2320
Citanest [Astra Pharm.] see 7749
Citarin [Bayer] see 9161
Citexal see 5872
Citicoline, 2321
Citifar [Lafare] see 2321
Citilat [CT] see 6441
Citiolase [Roussel-Maestretti] see 2322
Citiolone, 2322
Citireuma [CT] see 8961
Citizeta [CT] see 10074
Citnatin see 8549
Citobaryum see 1006
Citodon see 4625
Citofur [Lusofarmaco] see 9060
Citol see 474
Citopan see 4625
Citoplatino [Rhône-Poulenc] see 2319
Citosarin [Toyo Jozo] see 2706
Citoxid [Disprovent] see 6268
Citracholine see 2210
Citraconic Acid, 2323
Citral, 2324
Citramalic Acid, 2325
Citramin see 5881
Citrated Caffeine see 1635
Citrated Calcium Carbimide see 1663
Citraurin see 2326
β-Citraurin, 2326
Citrazinic Acid, 2327
Citresia see 5542
Citric Acid, 2328
Citric Acid Diammonium Salt see 534
Citric Acid Tributyl Ester see 1564
Citric Acid Trilithium Salt see 5407
Citrical [Shire] see 1657
Citridic Acid see 112
Citrinin, 2329
Citromycetin, 2330
Citronellal, 2331
α-Citronellal see 8185
Citronellol see 2332
l-Citronellol see 8186
α-Citronellol see 8186
β-Citronellol, 2332
Citronin A see 6313
Citrophen see 7188
Citropten see 5370

Citrosodine see 8549
Citrostadienol see 8499
Citrovorum Factor see 4141
Citrullamon [Südmedica] see 7293
Citrulline, 2333
Citrullol, 2334
Citrus Flavonoid Compounds see 1241
Citrus Red 2, 2335
Citrylideneacetone see 7931
Civet, 2336
Civetone, 2337
CL 68 see 2368
Cl 337 see 916
Cl 636 see 8873
CL 1388R see 4471
CL 1848C see 3844
CL 11344 see 7981
CL 12625 see 6346
CL 13494 see 8890
Cl 13900 see 7960
Cl 14377 see 5908
CL 19823 see 9511
CL 26691 see 2791
CL 34433 see 9514
CL 34699 see 398
CL 36010 see 8067
CL 47031 see 7309
CL 62362 see 5461
CL 64475 see 4173
CL 67772 see 609
CL 71563 see 5461
CL 81588 see 906
CL 82204 see 3906
CL 112302 see 1487
CL 118523 see 1952
CL 206214 see 1506
CL 206576 see 8864
CL 216942 see 1255
CL 222293 see 4825
Cl 227193 see 7430
Cl 232315 see 6135
CL 251931 see 1952
CL 263780 see 2278
CL 273703 see 5524
CL 284635 see 1927
Cl 67310465 see 7091
Clabber see 10010
Cladinose see 6229
Claforan [Hoechst] see 1935
Clamoxyl [Beecham] see 610
Clam Poison see 8344
Clanobutin, 2338
Clantin [Schering] see 5455
Clanzol [Orfi Farma] see 2344
Claradin [Nicholas] see 873
Clarase®, 2339
Claresan see 2375
Claretin-12 see 9921
Clarified Butter see 4312
Clarin [Pfizer] see 4571
Claripex [Usafarma] see 2374
Clarithromycin, 2340
Claritin [Schering-Plough] see 5455
Clarityne [Schering] see 5455
Clarmil see 9431
Clarphoril see 9431
Clarvisan see 7462
Classic [DuPont] see 2092
Clast [Meiji Seika] see 2344
Clasteon [Gentili] see 2369
Clathrates, 2341
Clathromycin [Taisho] see 2340
Clauden [Luitpold] see 9321
Claudetite, 832
Clausthalite see 8382
Clavacin see 7002
Clavatin see 7002
Claversal [SmithKline] see 5807
Claviformin see 7002
Clavitol see 4448
Claviton see 9577

Clavulanic Acid, 2342
Clayton Yellow *see* 9238
Clazuril, 2343
Clearing Factor *see* 5390
Clearnal [Yoshitomi] *see* 9494
Clearteck [Penreco] *see* 7139
Clebopride, 2344
Cleboril [Almirall-Omega] *see* 2344
Clédial [Anphar-Rolland] *see* 5674
Clefamide *see* 2181
Cleiton [Sankyo Zoki] *see* 4710
Cleland's Reagent [Calbiochem] *see* 3382
Clemanil [Kyoritsu Yamagata] *see* 2345
Clemastine, 2345
Clemizole, 2346
Clenbuterol, 2347
Clenil-A [Chiesi] *see* 1029
Cleocin [Upjohn] *see* 2352
Cleofil [Ciba] *see* 3303
Cleprid [Recordati] *see* 2344
Clera [Person & Covey] *see* 6287
Cleregil *see* 2835
Cleridium 150 [Millot-Solac] *see* 3354
1,6-Cleve's Acid, 2348
1,7-Cleve's Acid, 2349
Clexane [Rhône-Poulenc] *see* 8987
Clexon [Wellcome] *see* 6998
Cliacil [Hoechst] *see* 7046
Clidanac, 2350
Clidinium Bromide, 2351
Clift [Knoll] *see* 7889
Clin *see* 8626
Clindamycin, 2352
Clinestrol *see* 3119
Clinicide [DeWitt] *see* 1789
Clinimycin *see* 6931
Clinimycin (rescinded) *see* 2352
Clinitar [Smith & Nephew] *see* 2420
Clinium [McNeil] *see* 5360
Clinoenstatite *see* 5569
Clinofibrate, 2353
Clinolamide *see* 5382
Clinoril [Merck & Co.] *see* 8961
Clinovir [Upjohn] *see* 5677
Clinoxan [Midy] *see* 9172
Clinozoisite *see* 1645
Cliocinizine *see* 2366
Clionasterol *see* 8501
Clioquinol *see* 4924
Clioxanide, 2354
CLIP *see* 7803
Clipper [ICI] *see* 6937
Cliquinol *see* 4924
Cliradon [Ciba] *see* 5180
Clistin [McNeil] *see* 1804
Clisundac [Lagap] *see* 8961
Clitizina *see* 4220
Cl₂MDP *see* 2369
Clobazam, 2355
Clobenfurol, 2356
Clobenoside, 2357
Clobenzepam, 2358
Clobenzorex, 2359
Clobenztropine, 2360
Cloberat [Negroni] *see* 2374
Clobesol [Glaxo] *see* 2361
Clobetasol, 2361
Clobetasone, 2362
Clobren-SF [Morishita] *see* 2374
Clobutinol, 2363
Clobuzarit, 2364
Clocapramine, 2365
Clocarpramine *see* 2365
Clocete [Toyama] *see* 5660
Clocinizine, 2366
Cloconazole, 2367
Clocortolone, 2368
Cloderm [Ortho] *see* 2368
Clodronate Disodium *see* 2369

Clodronic Acid, 2369
Clofazimine, 2370
Clofedanol *see* 2053
Clofekton [Yoshitomi] *see* 2365
Clofenamide, 2371
Clofenciclan, 2372
Clofenotane *see* 2832
Clofenpyride *see* 6427
Clofentezine, 2373
Clofibrate, 2374
Clofibric Acid, 2375
Clofinit [Gentili] *see* 2374
Cloflucarban, 2376
Clofoctol, 2377
Cloforex, 2378
Clomacran, 2379
Clomag *see* 2375
Clomestrone, 2380
Clometacillin *see* 2383
Clometacin, 2381
Clometazin *see* 2381
Clomethacillin *see* 2383
Clomethiazole, 2382
Clometocillin, 2383
Clomid [Merrell] *see* 2384
Clomidazole *see* 2106
Clomifene *see* 2384
Clomiphene, 2384
Clomipramine, 2385
Clomivid [Draco] *see* 2384
Clomocycline, 2386
Clomphid *see* 2384
Clonazepam, 2387
Clonidine, 2388
Clonistada [Stadapharm] *see* 2388
Clonitazene, 2389
Clonitrate, 2390
Clonopin [Roche] *see* 2387
Clont [Bayer] *see* 6079
Clopamide, 2391
Clopane [Lilly] *see* 2745
Clopenthixol, 2392
α-Clopenthixol *see* 2392
Cloperastine, 2393
Clophen [Bayer] *see* 7541
Clopidol, 2394
Clopindol *see* 2394
Clopinerin [Nippon Shoji] *see* 2404
Clopirac, 2395
Clopiran [Continental Pharma] *see* 2395
Clopixol [Lundbeck] *see* 2392
Clopoxide *see* 2082
Cloprane [Ibis] *see* 8238
Cloprednol, 2396
Clopromate [Purdue Frederick] *see* 6063
Cloprostenol, 2397
Clopyralid, 2398
Cloquinate *see* 4757
Cloradryn [Recordati] *see* 2396
Cloramfen [Sclavo] *see* 2068
Cloramficin *see* 2068
Cloramicol *see* 2068
Cloramin [Simes] *see* 5655
Cloranolol, 2399
Clorarsen *see* 3060
Clorazepate, 2400
Clorazepate Dipotassium *see* 2400
Clorazil [Sandoz] *see* 2417
Clorazolam *see* 9518
Clordelazin *see* 2186
Clorevan [Evans] *see* 2182
Clorexolone, 2401
Clorfentermina *see* 2183
Clorina [Heyden] *see* 2066
Clorindione, 2402
Clorociclin [Panthox & Burck] *see* 8408
Clorocyn *see* 2068
Clorolifarina *see* 2069

Cloromilen *see* 2068
Cloromisan *see* 2068
Cloronaftina *see* 2107
Clorophene, 2403
Cloropiril *see* 2180
Clorox *see* 8576
Clorpactin [Guardian] *see* 6909
Clorpactin WCS [Guardian] *see* 6909
Clorpactin XCB [Guardian] *see* 6909
Clorprenaline, 2404
Clorsulon, 2405
Clortermine, 2406
Clortetrin [Fargal] *see* 2872
Closantel, 2407
Closilate *see* MISC-4
Closina *see* 2758
Clospirazine, 2408
Clostebol, 2409
Clostilbegyt [EGYT] *see* 2384
Clostridiopeptidase A *see* 2477
Closylate *see* 2122
Clotam [Medica] *see* 9436
Clothiapine, 2410
Clotiamina [Squibb] *see* 9222
Clotiamine *see* 1030
Clotiazepam, 2411
Clotride [Merck & Co.] *see* 2169
Clotrimazole, 2412
Clotrox *see* 9650
Clout [Scott] *see* 5864
Clove, 2413
Clove Oil *see* 6718
Cloxacillin, 2414
Cloxapen [Beecham] *see* 2414
Cloxazolam, 2415
Cloxiquine *see* 2416
Cloxypen [Allard] *see* 2414
Cloxyquin, 2416
Clozan [Roerig] *see* 2411
Clozapine, 2417
Clozaril [Sandoz] *see* 2417
Clozic [ICI] *see* 2364
Club-moss Spores *see* 5495
Clupeine, 2418
CLY-503 *see* 8489
Clysodrast [Barnes-Hind] *see* 1253
CM 6805 *see* 1527
CM 6912 *see* 3777
CM 8282 *see* 6801
CM 9155 *see* 3134
CM 31-916 *see* 1948
CMC [Stuart] *see* 1835
CME 74770 *see* 9602
C-Meton [SS Pharm] *see* 2180
CMF *see* MISC-2
CMFP *see* MISC-2
CMFVP *see* MISC-2
CMME *see* 2146
C-MOPP *see* MISC-2
2'-CMP *see* 2793
3'-CMP *see* 2794
CMPP *see* 5666
CMT *see* 1922
CMU *see* 6169
CN *see* 2115
CN 3123 *see* 8897
CN 15757 *see* 916
CN 11-2936 *see* 7773
Cnicin, 2419
Co II *see* 6262
CO 61 *see* 6857
CO/1063 *see* 3642
CoA *see* 2465
Coal Oil *see* 7140
Coal Tar, 2420
Coal Tar Creosote *see* 2574
Co-amilozide *see* 417
COAP *see* MISC-2
Coapt [Ethicon] *see* 5667
Coaxin [Tobishi] *see* 1978

Cobactin see 6235
Cobadex see 4710
Cobalamin see 9921
Cobalex see 4739
Cobalin see 9921
Cobalin-H [Paines & Byrne] see 4739
Cobalion [Houdé] see 2446
Cobalt, 2421
Cobaltamin S [Wakamoto] see 2446
Cobalt Bloom see 2428
Cobalt Carbonate Hydroxide see 2430
Cobalt Carbonyl Hydride see 3074
Cobalt Chromite see 2432
Cobalt Cyanide see 2433
Cobalt Dibromide see 2429
Cobalt Dichloride see 2431
Cobalt Difluoride see 2434
Cobalt Diiodide see 2437
Cobalt Hydrocarbonyl see 3074
Cobalt Hydroxide Oxide see 2425
Cobaltic Acetate, 2422
Cobaltic-Cobaltous Oxide, 2423
Cobaltic Fluoride, 2424
Cobaltic Hydroxide see 2425
Cobaltic Oxide Monohydrate, 2425
Cobaltic Potassium Nitrite, 2426
Cobaltite see 2421
Cobaltmethyl-5,6-dimethylbenzimida-
  zolecobalamin see 9921
Cobalto-cobaltic Oxide see 2423
Cobalt Octacarbonyl see 3074
Cobaltosic Oxide see 2423
Cobaltous Acetate, 2427
Cobaltous Ammonium Phosphate see
  535
Cobaltous Ammonium Sulfate see 536
Cobaltous Arsenate, 2428
Cobaltous Bromide, 2429
Cobaltous Carbonate, 2430
Cobaltous Carbonate Basic see 2430
Cobaltous Chloride, 2431
Cobaltous Chromate(III), 2432
Cobaltous Cyanide, 2433
Cobaltous Fluoride, 2434
Cobaltous Formate, 2435
Cobaltous Hydroxide, 2436
Cobaltous Iodide, 2437
Cobaltous Nitrate, 2438
Cobaltous Oxalate, 2439
Cobaltous Oxide, 2440
Cobaltous Phosphate, 2441
Cobaltous Potassium Sulfate, 2442
Cobaltous Rhodanide see 2445
Cobaltous Sulfate, 2443
Cobaltous Sulfide, 2444
Cobaltous Sulfocyanate see 2445
Cobaltous Thiocyanate, 2445
Cobalt Potassium Cyanide see 7620
Cobalt Spar see 2430
Cobalt Tetracarbonyl see 3074
Cobalt Trifluoride see 2424
Cobalt Yellow see 2426
Cobamamide, 2446
Cobamamidum see 2446
Cobamic Acid see 9921
Cobamide see 9921
Cobamin see 9921
Cobamine see 9921
Coban [Lilly] see 6157
Cobantril [Pfizer] see 7968
Cobanzyme [Bouchara] see 2446
Cobazymase [Bouchara] see 2446
Cobefrin Hydrochloride [Winthrop]
  see 6609
Coben [Takeda] see 7376
Coben P [Takeda] see 7376
Cobinamide see 9921
Cobinamide Acetate Phosphate 3'-
  Ester with 5,6-Dimethyl-1-α-D-
  ribofuranosylbenzimidazole Inner
  Salt see 4739

Cobinamide Co-methyl Deriv. Hydr-
  oxide Dihydrogen Phosphate (Ester)
  Inner Salt 3'-Ester with 5,6-Di-
  methyl-1-α-D-ribofuranosyl-1H-
  benzimidazole see 9921
Cobinamide Cyanide Phosphate 3'-
  Ester with 5,6-Dimethyl-1-α-D-
  ribofuranosylbenzimidazole Inner
  Salt see 9921
Cobinamide Dicyanide see 3825
Cobinamide Dihydroxide Dihydrogen
  Phosphate (Ester), Mono(inner Salt),
  3'-Ester with 5,6-Dimethyl-1-α-D-
  ribofuranosyl-1H-benzimidazole see
  4739
Cobinamide Dihydroxide, Monohy-
  drate, Dihydrogen Phosphate Ester,
  Mono(inner Salt), 3'-Ester with 5,6-
  Dimethyl-1-α-D-ribofuranosyl-1H-
  benzimidazole see 787
Cobinamide, 3'-Ester with 5,6-Di-
  methyl-1-α-D-ribofuranosylbenz-
  imidazole, Co-(5'-deoxyadenosine-
  5') Deriv., Dihydrogen
  Phosphate (Ester), Inner Salt see
  2446
Cobinamide Hydride Hydroxide,
  Dihydrogen Phosphate Ester, Inner
  Salt 3'-Ester with 5,6-Dimethyl-1-
  α-D-ribofuranosyl-1H-benzimid-
  azole see 9926
Cobinamide Hydroxide Nitrite Salt,
  Dihydrogen Phosphate Ester, Inner
  Salt, 3'-Ester with 5,6-Dimethyl-1-
  α-D-ribofuranosyl-1H-benzimid-
  azole see 9924
Cobinamide Sulfite Phosphate 3'-
  Ester with 5,6-Dimethyl-1-α-D-
  ribofuranosylbenzimidazole see 8931
Cobinic Acid see 9921
Cobiona [Esteve] see 6880
Cobione [Merck & Co.] see 9921
Cobrentin Methanesulfonate [Merck &
  Co.] see 1135
Cobrotoxin, 2447
Cobutolin [Cox] see 209
Cobyric Acid see 9921
Cobyrinic Acid see 9921
Coca, 2448
Cocaethylene, 2449
Cocaic Acid see 9700
β-Cocaic Acid see 9700
Cocaine, 2450
l-Cocaine see 2450
β-Cocaine see 2450
Cocaine Muriate see 2450
Cocalose [Maruko] see 2451
Cocarbina see 2451
Cocarboxylase, 2451
Cocciden [Sankyo] see 1030
Coccoclase see 8913
Cocculin see 7388
Cocculus, 2452
Cocculus Indicus see 2452
Cochineal, 2453
Cocillana, 2454
Coclanoline see 8164
Coclaurine, 2455
Cocoa, 2456
Cocoa Butter see 9208
Cocoa Shells see 1601
Coconut Oil, 2457
Coco-Quinine [Lilly] see 8089
Coculine see 8496
Codamine, 2458
Codecarboxylase see 7992
Codehydrase II see 6262
Codehydrogenase see 6259
Codehydrogenase II see 6262
Codeigene see 2461
Codeine, 2459

β-Codeine see 6376
ψ-Codeine see 7927
Codeine Methyl Bromide, 2460
Codeine N-Oxide, 2461
Codeine Phosphate, 2462
Codeine Sulfate, 2463
Codelcortone [Merck & Co.] see 7719
Codelcortone-T.B.A. [Merck & Co.]
  see 7725
Codelsol [Merck & Co.] see 7721
Co-dergocrine Mesylate see 3598
Codethyline see 3785
Codhydrine see 3154
Codicept [Sanol] see 2459
Cod Liver Oil, 2464
Codons see 8204
Codroxomin [OJ & F] see 4739
Codylin see 7303
Coenzyme I see 6259
Coenzyme II see 6262
Coenzyme A, 2465
Coenzyme B₁₂ see 2446
Coenzyme Q₁₀ see 9751
Coenzyme Q-199 see 9751
Coenzyme R see 1244
Coenzymes Q see 9751
Coeruleolactite see 362
Co-Ervonum [Glaxo] see 5687
Coffearine see 9606
Coffee Bean Oil see 2467
Coffee Beans see 2466
Coffee, Green, 2466
Coffee Oil, 2467
Coffeine see 1635
Coflavinase see 8202
Cofosfolactamines see 4169
Co-Fram [Abbott] see 8897
Co-galactoisomerase see 9794
Cogentin [Merck & Co.] see 1135
Cogentinol [Pharma-Stern] see 1135
Cogesic [Mead Johnson] see 7775
Cognac Oil, Synthetic see 3790
Cogomycin see 4202
Coherin, 2468
Co I see 6259
Cola see 5198
Colace [Mead Johnson] see 3398
Colalin see 2206 and 6884
Colamine see 3681
Colascor [DHA] see 855
Colaspase see 858
Colcamyl see 9253
Colcemid [Ciba] see 2873
Colchamine see 2873
Colchiceine, 2469
Colchicine, 2470
Colchicine-binding Protein see 9718
Colchicum Corm, 2471
Coldan see 6287
Coldrin [Nippon Shinyaku] see 2053
Colecalciferol see 9929
Colectril [Merck & Co.] see 417
Coleflux [Finadiet] see 7457
Colemanite see 1651
Colenormol see 6913
Colepan [Panther-Osfa] see 2721
Colepax see 4944
Colepur [Draco] see 1441
Colerainite see 352
Colesterinex [Prodes] see 7987
Colestid [Upjohn] see 2472
Colestipol, 2472
Colestyramin see 2205
Coletrast see 4919
Coletyl see 1780
Colfarit [Bayer] see 873
Colforsin, 2473
Colicine see 2474
Colicins, 2474
Colic Root see 220, 3287 and 4243

Colidosan see 8213
Colifoam [Trommsdorf] see 4711
Colifos see 7337
Colimune [Fisons] see 2594
Colimycin see 2475
Coliopan [Eisai] see 1534
Colipar [UCEPHA] see 1441
Coliquifilm [Pharm-Allergan] see 2129
Colisone [Frosst] see 7727
Colisticina see 2475
Colistimethate Sodium see 2475
Colistin, 2475
Colite [Nippon Chemiphar] see 2321
Colitiazolo see 7351
Collagen, 2476
Collagenase, 2477
Collastin see 6818
sym-Collidine see 9632
2,4,6-Collidine see 9632
α-Collidine see 3797
α,γ,α'-Collidine see 9632
β-Collidine see 3796
γ-Collidine see 9632
Collinomycin, 2478
Collinsonia, 2479
Colliron I.V. see 3974
Collodion, 2480
Collodion, Cantharidal, 2481
Collodion Cotton see 8022
Collodion Wool see 8022
Colloidal Bismuth Subcitrate see 1296
Colloidal Calomel see 1727
Colloidal Gold—^{198}Au see 4419
Colloxylin see 8022
Collubiazol see 8872
Collumol see 345
Collunosol see 9555
Collunovar see 6361
Colme [Lasa] see 1663
Colocynth, 2482
Colocynthin, 2483
Colofac [Duphar] see 5648
Colofoam [Stafford-Miller] see 4711
Cologel [Lilly] see 5963
Cologne Yellow see 5279
Colombo see 1729
Colonorm [Mundipharma] see 8407
Colophony see 8245
Colo-Pleon [Henning] see 8917
Colostrokinin, 2484
Colpermin [Tillotts] see 6758
Colpogyn [Angelini] see 3659
Colpogynon see 2485
Colpormon, 2485
Colpovis [SIT] see 8065
Colpro [Dagra] see 5676
Colprone [Ayerst] see 5676
Colrex Expectorant [Rowell] see 4465
Coltericin [Quimica Argentia] see 5161
Coltirot see 9749
Coltramyl see 9253
Coltrax see 9253
Coltromyl see 9253
Coltsfoot, 2486
α-Colubrine see 2487
β-Colubrine see 2487
Colubrines, 2487
Columbamine, 2488
Columbin, 2489
Columbite see 6472
Columbium see 6472
Columbium Pentachloride see 6473
Columbium Pentafluoride see 6474
Columbium Pentoxide see 6475
Columbium Potassium Oxypentafluoride see 6476
Colvasone [Norbrook] see 2922
Colybar [Warner-Lambert] see 2205

Coly-Mycin [Warner-Chilcott] see 2475
Colyonal [Mochida] see 2929
Colza Oil see 8127
Combantrin [Pfizer] see 7968
Combat [Cyanamid] see 4684
Combec [Tokyo Tanabe] see 4060
Combelen [Bayer] see 7841
Combetin see 8819
Combot Equine [Bayvet] see 9536
Comelian [Kowa] see 3185
Cometamine [Yamanouchi] see 2762
Co-methylcobalamin see 9921
Comfolax [Searle] see 3398
Comite [Uniroyal] see 7818
COMLA see MISC-2
Command [FMC] see 3203
Commetamin see 2762
Common Mallow see 5590
Common Salt see 8544
Common Sundew see 3440
Common Verdigris see 2628
Comox [Norton] see 8889
COMP see MISC-2
Compactin see 6088
Compazine [SK & F] see 7768
Compd 118 see 219
Compd 338 see 2123
Compd 711 see 219
Compd 3-120 see 8914
Compd 47-83 see 2716
Compd 08958 see 2763
Compd 22/190 see 2196
Compd 47-282 see 2078
Compd 48/268 see 8845
Compendium [Polifarma] see 1374
Compitox [M & B] see 5666
Compitox Plus [M & B] see 5666
Complamex [Wülfing] see 9969
Complamin [Italchimici] see 9969
Complemix see 3398
Complexone see 3482
Compocillin [Abbott] see 7040
Compocillin-V [Abbott] see 7048
Compocillin VK [Ross] see 7046
Component B$_{1a}$ see 5133
Component B$_{1b}$ see 5133
Compound 20 see 5250
Compound 42 see 9950
Compound 88R see 794
Compound 269 see 3533
Compound 347 see 3536
Compound 469 see 5059
Compound 497 see 3093
Compound 545 see 4940
Compound 1080 see 4096
Compound 4072 see 2087
Compound 43-663 see 4481
Compound 64716 see 2311
Compound 72500 see 4129
Compound 79891 see 6339
Compound 81929 see 3396
Compound 83405 see 1922
Compound 83846 see 781
Compound 90459 see 1055
Compound 99170 see 781
Compound 99638 see 1920
Compound 109514 see 6257
Compound 112531 see 9892
Compound 180/442 see 3549
Compound B see 2532
Compound F-2 see 10017
Compral see 9561
Compralgyl see 9561
Comprecin [Warner-Lambert] see 3540
Compudose 365 [Elanco] see 3653
ConA see 2490
Conadil [Riker] see 8974
Conalbumin see 9490
Concanavalin A, 2490

Concemin see 855
"Concentrated Opium" see 6967
Conceplan [Grünenthal] see 6614
Conceptrol [Ortho] see 6596
Conceral [Takeda] see 4010
Conco NI [Continental Chem.] see 6596
Conco NI-90 [Continental Chem.] see 6596
Conco NIX-100 [Continental Chem.] see 6681
Concor [E. Merck] see 1309
Concordin [Merck & Co.] see 7917
Condol see 9928
Condor Vine see 2491
Condrosulf [IBSA] see 2217
Conducton [Klinge] see 1779
Condurangin, 2491
Condurangogenin A see 2491
Condyline [Brocades] see 7520
Condylox [pHarma Medica] see 7520
Cone Flower see 3468
Conessine, 2492
Conestron [Wyeth] see 2504
Conflictan [Riom] see 6861
Confortid [Dumex] see 4874
Congo Blue see 9701
Congocidine see 6391
Congo Red, 2493
Congressane, 2494
Conhydrine, 2495
β-Coniceine, 2496
γ-Coniceine, 2497
Conicine see 2500
Coniferin, 2498
Coniferyl Alcohol, 2499
Coniine, 2500
Coniine Hydrobromide, 2501
Coniine Hydrochloride, 2502
Conium Fruit, 2503
Conjugated Estrogenic Hormones, 2504
Conjugol see 3660
Conjuncain [Mann] see 1056
Conludag [Astra-Syntex] see 6614
Conmel see 3358
Connettivina [Fidia] see 4675
Conoco C-50 see 8559
Conoco C-60 see 8559
Conoco SD 40 see 8559
Conoderm [C-Vet] see 6101
Conofite [Pitman-Moore] see 6101
Conotrane see 4687
Conovid E [Searle] see 6615
Conquinamine, 2505
Conquinine see 8072
Conray [Mallinckrodt] see 4952
Conray-400 [Mallinckrodt] see 4952
Conselt [Sana] see 2404
Consolan [Beecham] see 6258
Constaphyl [Bristol] see 3073
Consumptive's Weed see 3617
Cont see 6079
Contalax [Riker] see 1253
Contamex [Beecham-Wülfing] see 5176
Contaverm see 7220
Contax see 6927
Conteben [Riker] see 9218
Contenton [Dauelsberg] see 380
Contergan see 9182
Continal see 7087
Contomin see 2186
Contopheron see 9932
Contralin see 3370
Contrapar [Burroughs Wellcome] see 5178
Contrathion see 7705
Contratuss see 5349
Contravul [Riker] see 8974
Contrheuma Retard [Spitzner] see 873

Contristamine see 2182
Contrix "28" see 4952
Control-Om see 5739
Convacard [Madaus] see 2507
Convallamarin see 2506
Convallamarogenin, 2506
Convallan [Gödecke] see 2507
Convallaria, 2507
Convallaria Glycosides see 2507
Convallaton see 2508
Convallatoxigenin see 8818
Convallatoxin, 2508
Convallen see 2507
Convalyt see 2507
Convasid see 2507
Convenil [Hommel] see 1516
Convicine, 2509
Convulex [Byk-Gulden] see 9821
Coolspan [Hishiyama] see 8971
Coomassie Blue [Ayerst] see 662
Coomassie Blue Medicinal [Ayerst]
    see 662
Coomassie Blue RL [Ayerst] see 662
Coopex see 2531
CO-ORD [Baxter] see 210
COP see MISC-2
Copaene, 2510
α-Copaene see 2510
Copaiba, 2511
Copal, 2512
Coparaffinate, 2513
Coparogin [Nippon Chemiphar] see
    9060
COP-BLAM see MISC-2
Copharcilin [Cophar] see 621
Copinal [Vinas] see 38
Copirene see 5168
COPP see MISC-2
Copper, 2514
Copper Acetate Arsenite see 2629
Copperas see 4006
Copper Carbonate Hydroxide see 2634
Copper-chromium Oxide see 2638
Copper DOS see 2677
Copper Glance see 2674
Copper Hydrate see 2646
Copper Hydroxide Sulfate see 2660
Copper Phthalocyanine, 2515
Copper Pyrites see 2029
Copper-Sandoz [Sandoz] see 2671
Copper Sulfate Dibasic see 2660
Copper Sulfate Tribasic see 2660
Coppertone [Schering] see 4660
Copra Oil see 2457
Copren [Fuso] see 4369
Coproergostane, 2516
Coprogen, 2517
Coprol see 3398
Coprostane, 2518
3β-Coprostanol see 2519
Coprostenol see 253
4:5-Coprosten-3-ol see 253
Coprosterol, 2519
Copsamine [Durst] see 7996
Coptin [Pfizer] see 8874
Coptine, 2520
Coptis, 2521
Coptisine, 2522
Coracon see 6459
Coracten [SmithKline] see 6441
Coractiv N [Phyteia] see 6459
Coradon see 7996
Corafurone see 5189
Co-ral [Chemagro] see 2559
Coralgil see 4622
Coralgina see 4622
Corallin see 899
Coralox see 2531
Coramedan [Medice] see 3146
Coramine [Ciba] see 6459
Coranormol see 7097

Corathiem [Ohta] see 2308
Co-Rax [Prentiss] see 9950
Corax see 2082
Corazole see 7097
Corbadrine see 6609
Corbasil see 6609
Corbel [BASF] see 3938
Corchorgenin see 8818
Corchorin see 8818
Corchsularin see 8818
Cordabromin [Homburg] see 7901
Cordalin [Homburg] see 3835
Cordan [IBI] see 3916
Cordanum [Arzneimittelwerk VEB]
    see 9012
Cordarex [Labaz] see 497
Cordarone [Labaz; Sigma-Tau] see
    497
Cordarone X [Labaz] see 497
Cordes Vas [Ichthyol] see 8167
Cord Factor, 2523
Cordiamin see 6459
Cordianine see 246
Cordicant [Mundipharma] see 6441
Cordilox [Pfizer] see 9851
Cordiomon see 4536
Cordioxil see 3150
Corditrine [Roger Bellon] see 6528
Cordium [Riom] see 1167
Cordoval [Tempelhof] see 4328
Cordoxene [Laroche-Navarron] see
    3900
Cordran see 4122
Cordycepic Acid see 5629
Cordycepin, 2524
Cordycepin-5'-triphosphate see 2524
9-Cordyceposidoadenine see 2524
Coredamin see 7744
Corediol [Meiji] see 6459
Corein 2R see 1955
Coreminal [Mitsui] see 4133
Corenalin [Kaken] see 2321
Co-Renitec [Merck & Co.] see 3521
Coretal [Polfa] see 6905
Corflazine [Daltan] see 5360
Corgal [Pliva] see 4257
Corgard [Squibb] see 6260
Corglykon see 2508
Corhormon see 4536
Coriamyrtin, 2525
Coriander, 2526
Coriandrol see 5373
Coriantin see 4534
Coriban [Burroughs Wellcome] see
    2953
Corid [Merck & Co.] see 623
Cori Ester see 4356
Corilagin see 9023
Corindolan [Schering AG] see 5744
Coriphate [Tokyo Tanabe] see 4076
Coristin [San Carlo] see 3598
Coritat [Green Cross] see 6616
Cork, 2527
Corlan see 4713
Corlan [Glaxo] see 4710
Corlin see 2533
Corlopam [SKB] see 3925
Corlutin see 7783
Corlutina see 7783
Corluvite see 7783
Cormed [Reiss] see 6459
Cormelian [Asta] see 3185
Cormid see 6459
Cornetite see 2653
Cornin see 9861
Cornocentin see 3600
Corn Oil, 2528
Cornox [Boots] see 5645
Cornox RK [Boots] see 3068
Corn Silk see 10016
Corn Steep Liquor, 2529

Corn Steep Water see 2529
Corn Sugar see 4353
Cornus, 2530
Corodane see 2079
Corodenin see 134
Corodilan [Meiji] see 3662
Coronarin see 3459
Coronarine [NEGMA] see 3354
Coronin see 5189
Coro-Nitro [Boehringer, Mann.] see
    6528
Corontin see 7744
Corotrend [Siegfried] see 6441
Corotrope [Winthrop] see 6117
Corovliss [Boehringer, Mann.] see
    5114
Coroxon, 2531
Corozate see 10075
Corpax see 7744
Corps De Pfeiffer see 7146
Cor-Puren [Klinge-Nattermann] see
    3150
Corpus Luteum Hormone see 7783
Corrigast [Searle] see 7816
Corrigen see 6786
Corrin see 9921
Corrole see 9921
Corrosive Mercury Chloride see 5770
Corrosive Sublimate see 5770
Corsair [Rhône-Poulenc] see 7132
Corsodyl [ICI] see 2090
Corson [Takeda] see 2922
Corstiline see 127
Cortadren [Schering] see 2533
Cortaid [Upjohn] see 4711
Cortalone [Halsey] see 7719
Cortancyl [Roussel] see 7727
Cortate [Schering] see 2883
Cortazac see 4710
Cort-Dome see 4710
Cortef [Upjohn] see 4710
Cortelan [Glaxo] see 2533
Cortenil [Hoechst] see 2883
Cortensor see 4577
Cortes [Taisho] see 4711
Cortesan see 2883
Cortex Angosturae see 681
Cortex Cuspariae see 681
Cortexilar see 4066
Cortexolone see 2912
Cortexone see 2882
Cortexone Acetate see 2883
Cor-Theophyllin [Paramedical] see
    3459
Cortical [ION] see 160
Corti-Clyss [Vifor] see 7719
Corticoliberin(e) see 2589
Corticopan [Ist. Chim. Inter.] see 160
Cortico-Sol see 7723
Corticosterone, 2532
Corticotrophin see 127
Corticotrophin Zinc Hydroxide Sus-
    pension see 127
Corticotropin see 127
α1-24-Corticotropin see 2550
β1-24-Corticotropin see 2550
Corticotropin-like Intermediate Lobe
    Peptide see 7803
Corticotropin-releasing Factor see
    2589
Corticotropin-releasing Hormone see
    2589
Cortiden see 6977
Cortidin [Crinos] see 160
Cortidyn see 160
Cortifar see 2883
Cortifoam [Reed & Carnrick] see 4710
Cortigen see 2883
Cortilet see 4104
Cortine Naturelle [Laroche Navarron]
    see 160

Cortiphate Injectable [Baxter] see 4712
Cortiphyson [Promonta] see 127
Cortiplastol [Medici] see 4076
Cortiron [Schering AG] see 2883
Cortisol see 4710
Cortisol Acetate see 4711
Cortisol 21-(N,N-Diethyl)glycinate see 4709
Cortisol 21-Ester with N,N-Diethyl-glycine see 4709
Cortisone, 2533
Δ¹-Cortisone see 7727
Cortisone Acetate see 2533
Cortisone, 21β-Cyclopentanepropionate, 2534
Cortisone, 21β-Cyclopentylpropionate see 2534
Cortisone 21-Dihydrogen Phosphate see 2535
Cortisone Phosphate, 2535
21-Cortisonephosphoric Acid see 2535
Cortisone 17-Valerate see 4710
Cortistab [Boots] see 2533
Cortisumman see 2922
Cortisural see 160
Cortisyl Artriona [Roussel-UCLAF] see 2533
Cortivazol, 2536
Cortivis [Vister] see 2883
Cortixyl see 2883
Cortocin-F [Eisai] see 4156
Cortogen [Schering] see 2533
Cortol, 2537
α-Cortol see 2537
β-Cortol see 2537
Cortolone, 2538
Cortone [Merck & Co.] see 2533
Cortril [Pfizer] see 4710
Cortrophin [Organon] see 127
Cortrophin-Z see 127
Cortrosinta [Organon] see 2550
Cortrosyn [Organon] see 2550
Corundum see 359
Corvasal [Hoechst] see 6143
Corvasol see 7097
Corvasymton see 8994
Corvaton [Cassella-Riedel] see 6143
Corvitol see 6459
Corvotone see 6459
Corwin [ICI] see 9965
Corybulbine, 2539
Corycavamine, 2540
Corycavidine, 2541
Corycavine see 2540
Corydaldine, 2542
Corydaline, 2543
Corydalis, 2544
Corydalis-G see 2539
Corydine, 2545
Corylophyline see 4354
Corynantheine, 2546
Corynanthidine see 10013
Corynanthine, 2547
Corynine see 10011
Corynomycolic Acids see 6236
Corypalmine, 2548
Corystibin see 8769
Corytuberine, 2549
Cosaldon [Albert-Roussel] see 7085
Cosaprin see 98
Cosbiol [Laserson] see 8726
Coscopin [BDH] see 6638
Coscotabs see 6638
Coslan see 5680
Cosmegen [Merck & Co.] see 2804
Cosmoline [E. F. Houghton] see 7138
Cosmopen see 7041
Cosmosin [Nippon Lederle] see 1952
Cospanon [Eisai] see 4039
Cosprin [Glenbrook] see 873

Cossmetin see 764
Cossym [Egnaro] see 1798
Costimulator see 4896
Cosulfa see 8871
Cosulid [Ciba] see 8871
Cosuric [DDSA] see 278
Cosylan [Warner-Lambert] see 8116
Cosyntropin, 2550
Cotarnine, 2551
Cotarnine Chloride see 2551
Cotarninium Chloride see 2551
Cotazym [Organon] see 6957
Co-tetroxazine see 9178
Cothera Syrup [Ayerst] see 3211
Co-thromboplastin see 3872
Cotinazin [Pfizer] see 5071
Cotinine, 2552
Cotnion-methyl [Makhteshim-Agan] see 926
Cotofor [Ciba-Geigy] see 3349
Cotoin, 2553
Cotoran [Ciba] see 4080
Cotrane [Midy] see 3211
Co-trimazine see 8874
Co-trimoxazole see 8889
Cotrim-Puren [Klinge] see 8889
Cottonex [Makhteshim-Agan] see 4080
Cotton Red 4B see 1111
Cotton-root Bark, 2554
Cottonseed Oil, 2555
Cotunnite see 5278
Couch Grass see 9673
Coughwort see 2486
Coumachlor, 2556
Coumadin [Endo] see 9950
Coumafos see 2559
Coumafuryl, 2557
Coumalic Acid, 2558
Coumaphos, 2559
Coumaphos Oxygen Analog see 2531
Coumaran, 2560
p-Coumaric Acid, 2561
Coumarilic Acid, 2562
Coumarin, 2563
Coumarin-3-carboxylic Acid, 2564
cis-o-Coumarinic Acid Lactone see 2563
Coumarinic Anhydride see 2563
Coumarone see 1098
Coumarone-2-carboxylic Acid see 2562
Coumestrol, 2565
Coumetarol, 2566
Coumingine, 2567
Coumithoate, 2568
Counter [Am. Cyanamid] see 9088
Courlene [Courtaulds] see 7544
Courlene PY [Courtaulds] see 7558
Cousso see 1361
Covalan [Dausse] see 4139
Covatine [Bailly] see 1772
Covatix [Gödecke] see 1772
Covellite see 2661
Coverine [Astier] see 2268
Coversyl [Servier] see 7122
Covit see 9921
Covitol see 9931
Co-waldenase see 9794
Cow Clover see 9601
Cow's Milk see 6114
Cowrie see 2512
Coxigon [Lilly] see 1055
Coxistac [Pfizer] see 8305
Coxistat see 6521
Coxytrol [Squibb] see 2780
Coyden [Dow] see 2394
Cozymase see 6259
Cozyme [Travenol] see 2924
CP 172AP see 2395
CP 556S see 8967

CP 1001 see 2258
CP 1044 J3 see 1462
CP 4742 see 8882
CP 6343 see 250
CP 10423-16 see 7968
CP 10423-18 see 7968
CP 12009-18 see 6175
CP 12299-1 see 7715
CP 12574 see 9377
CP 14445 see 6876
CP 14445-16 see 6876
CP 15336 see 2950
CP 15464 see 1847
CP 15464-2 see 1847
CP 15639-2 see 1796
CP 16171 see 7476
CP 19106-1 see 9612
CP 23426 see 9510
CP 24314-1 see 7461
CP 31393 see 7805
CP 34089 see 8973
CP 41845 see 4409
CP 45634 see 8679
CP 45899 see 8862
CP 45899-2 see 8862
CP 50144 see 193
CP 51974-1 see 8416
CP 52640-2 see 1933
CP 53619 see 1498
CP 54802 see 235
CP 62993 see 928
CP 66248 see A9
CPA see 2781
CPC 10997 see 2774
CPDC see 2319
C.P.H. [Cutter] see 7900
CPIB see 2374
Cpiron see 3995
C-Quens see 2102
CR 242 see 7785
CR 604 see 7784
CR 605 see 9393
CR/662 see 9389
CR 1639 see 3281
Crab-E-Rad see 5864
Crag 974 see 2827
Crag Fruit Fungicide 341 [Union Carbide] see 4404
Crag Herbicide-1 [Union Carbide] see 3372
Crag Sesone [Union Carbide] see 3372
Cramp Bark see 9876
Crampol see 94
Cranberry Tree see 9876
Cranesbill see 4299
Crapinon [Sanzen] see 7440
Crasnitin [Bayer] see 858
Crataegus, 2569
Crataegus-Kreussler see 2569
Cratecil see 7201
Craviten [Polfa] see 1524
CRD-401 see 3188
C-Reactive Protein see 2606
Creamalin [Winthrop] see 345
Cream of Tartar see 7593
Creasote see 2575
Creatergyl [Midy] see 7315
Creatine, 2570
Creatine Phosphate see 7315
Creatinephosphoric Acid see 7315
Creatinine, 2571
Cremesone [Dalin] see 4710
Cremorin see 345
Cremor Tartari see 7593
Creolin®, 2572
Creolin-Pearson see 2572
Creon [Duphar] see 6956
Creosedin [Osiris] see 1374
Creosol, 2573
Creosote, Coal Tar, 2574
Creosote, Wood, 2575

Creosotic Acid, 2576
Crepasin [Hoei] see 7744
Cresatin [Merck & Co.] see 2587
Cresatin Metacresylacetate [Merck & Co.] see 2587
Cresatin-Sulzberger [Merck & Co.] see 2587
Crescefel [Ayerst] see 6884
Crescormon [Kabi] see 8672
Creslan®, 2577
Cresol(s), 2578
m-Cresol, 2579
o-Cresol, 2580
p-Cresol, 2581
o-Cresol Acetate see 2588
m-Cresol Acetic Acid Ester see 2587
Cresolase see 9746
o-Cresolphthalein, 2582
Cresol Red, 2583
o-Cresolsulfonphthalein see 2583
m-Cresotic Acid, 2584
o-Cresotic Acid, 2585
p-Cresotic Acid, 2586
2,3-Cresotic Acid see 2585
2,4-Cresotic Acid see 2584
2,5-Cresotic Acid see 2586
γ-Cresotic Acid see 2584
m-Cresotinic Acid see 2584
o-Cresotinic Acid see 2585
p-Cresotinic Acid see 2586
Cresoxydiol see 5737
Cresoxypropanediol see 5737
Crestabolic see 5861
Crestanil [Crest] see 5751
Crestmoreite see 1709
Crestomycin see 6989
m-Cresyl Acetate, 2587
o-Cresyl Acetate, 2588
o-Cresyl Glycerol Ether see 5737
o-Cresylic Acetate see 2588
Cresylic Acid see 2578
o-Cresylic Acid see 2580
Cresylol [Norden] see 2578
CRF, 2589
CRH see 2589
Crimidine, 2590
Crinovaryl see 3660
Crinovyl [Rhovyl] see 7563
Crinuryl [Assia] see 3669
Crisalbine see 4423
Criseociclina see 9130
Criseocil see 7052
Crisinar [Rubio] see 895
Crisofin [Allergan] see 895
Crispin [Kowa] see 9485
Cristallovar see 3660
Cristalomicina see 5161
Cristaloxine see 4324
Cristapen see 7041
Cristapurat see 3146
Cristerona MB see 8753
Cristobalite see 8440
Crithidia Factor see 6378
CRL 40028 see 159
Croceic Acid, 2591
Crocetin, 2592
Crocidolite see 851
Crocin, 2593
α-Crocin see 2593
Crocoite see 5279
Croconazole see 2367
Crocus see 8285
Crodimyl see 9574
Cromadrenal see 162
Cromaton see 5427
Cromoci see 2797
Cromoglycic Acid see 2594
Cromolyn, 2594
Cromolyn Sodium see 2594
Cromonalgina see 9574
Cromosan see 5530

Cromosil see 162
Cronassial [Fidia] see 4264
Cronetal see 3370
Cronil [Farmigea] see 3478
Cronizat [Farmitalia] see 6582
Cronodione see 7196
Cronolone [Searle] see 4125
Cropotex [Bayer] see 4051
Cropropamide, 2595
Crotaline see 6160
Crotamine, 2596
Crotamitex [Troponwerke] see 2597
Crotamiton, 2597
Crotethamide, 2598
Crotonaldehyde, 2599
cis-Crotonaldehyde see 2599
Crotonchloral see 9545
Crotonic Acid, 2600
α-Crotonic Acid see 2600
Crotonic Aldehyde see 2599
Croton Oil, 2601
Croton Oil Factor A₁ see 7306
Crotonyl Alcohol see 2604
α-(N'-Crotonyl-N'-ethyl)amino-N,N-dimethylbutyramide see 2598
Crotonyl-N-ethyl-o-toluidine see 2597
α-(N'-Crotonyl-N'-propyl)amino-N,N-dimethylbutyramide see 2595
Crotothane [M & B] see 3281
Crotoxin, 2602
Crotoxyphos, 2603
Crottle see 2618
Crotyl Alcohol, 2604
Crotyl Chloride see 2130
Crovaril see 6925
Crovicina [Viti] see 8408
18-Crown-6 see 2605
Crown Compounds see 2605
Crown Ethers, 2605
Croysulfone see 2820
Croysulphone see 2820
CRP, 2606
Crude Chrysarobin see 795
Crude Cream of Tartar see 807
Crude Oil see 7140
Crude Opium see 6809
Crude Potassium Bitartrate see 807
Crufomate, 2607
Crustecdysone see 3465
Crylène [Auclair] see 7077
Cryofluorane, 2608
Cryogenine see 7193
Cryolite, 2609
Cryptates see 2605
Cryptenamine Tannates, 2610
Cryptocavine see 2611
Cryptocillin [Hoechst] see 6858
Cryptohalite see 552
Cryptopine, 2611
Cryptoxanthin, 2612
Cryptoxanthol see 2612
Crystallinic Acid see 6641
Crystallins, 2613
Crystallized Verdigris see 2627
Crystallose [Tenneco] see 8282
Crystal Violet see 4287
Crystamin [Armour Pharm.] see 9921
Crystapen see 7041
Crystex see 8948
Crysticillin [Squibb] see 7042
Crystodigin [Lilly] see 3146
Crystoids [Merck & Co.] see 4633
Crystoserpine [Dorsey] see 8149
Crytion see 4423
CS see 2127
CS-359 see 1454
CS-370 see 2415
CS-386 see 6091
CS-430 see 4518
CS-500 see 6088
CS-514 see 7712

CS-600 see 5462
CS-684 see 7506
CS-807 see 1941
CS-847 see 969
CS-1170 see 1929
CSA see 2217
CSA see 1873
CSP see 4018
C-Stuff see 4692
CT see 1030
CT-848 see 6896
CT-1341 see 226
C.T.A.B. see 2018
CTR 6110 see 6517
CTR 6669 see 1794
C-tre [Chemioterapico] see 1757
CTZ see 1946
CU 32-085 see 5820
Cubane, 2614
Cube Alum see 364
Cubeb, 2615
Cubebin, 2616
Cubic Niter see 8598
Cuca see 2448
Cucoline see 8496
Cucurbitacin B see 2617
Cucurbitacin E see 2617
Cucurbitacins, 2617
Cudbear, 2618
Cuemid [Merck & Co.] see 2205
Cuivasal [IDC] see 5359
Cujec [TVL] see 2677
Culpen [Beecham] see 4044
Cultar [ICI] see 6937
Culver's Root see 5325
Cu-lyt see 2667
Cumaldehyde see 2623
Cumaldehyde Thiosemicarbazone see 2624
Cumaran see 2560
Cumarin see 2563
Cumarone see 1098
Cumarote C [Towa Yakuhin] see 4792
Cumelon [Teijin] see 8249
Cumene, 2619
3-p-Cumenyl-1,1-dimethylurea see 5106
2-[(o-Cumenyloxy)methyl]-2-imidazo-line see 3930
Cumertilin Sodium [Endo] see 5765
Cumetharol see 2566
Cumethoxaethane see 2566
Cumic Acid, 2620
Cumic Alcohol, 2621
Cumidine, 2622
Cuminal see 2623
Cuminaldehyde, 2623
Cuminaldehyde Thiosemicarbazone, 2624
Cuminal Thiosemicarbazone see 2624
Cuminic Acid see 2620
Cuminol see 2621
Cuminyl Alcohol see 2621
Cumol [Cutter] see 2619
Cumopyran [Abbott] see 2723
Cumopyrin see 2723
Cumotocopherol see 9417
Cupferron, 2625
Cupralene [Specia] see 256
Cupralylnatrium see 256
Cupralylsodium see 256
Cuprammonium Sulfate see 9116
Cupreine, 2626
Cuprelon see 256
Cuprenil [Polfa] see 7029
Cupreol see 8500
Cuprex [Beecham] see 2648
Cupriaseptol see 2652
Cupric Acetate, 2627
Cupric Acetate, Basic, 2628
Cupric Acetoarsenite, 2629

Cupric Aminoacetate see 2644
Cupric Ammonium Chloride see 537
Cupric Arsenite, 2630
Cupric Benzoate see 1101
Cupric Bis[8-hydroxyquinoline Di-(diethylammonium Sulfonate)] see 2677
Cupric Borate, 2631
Cupric Bromide, 2632
Cupric Butyrate, 2633
Cupric Carbonate, Basic, 2634
Cupric Chlorate, 2635
Cupric Chloride, 2636
Cupric Chromate(III) see 2638
Cupric Chromate(VI), 2637
Cupric Chromite, 2638
Cupric Citrate, 2639
Cupric Ferrocyanide, 2640
Cupric Ferrous Sulfide see 2029
Cupric Fluoride, 2641
Cupric Fluosilicate see 2645
Cupric Formate, 2642
Cupric Gluconate, 2643
Cupric Glycinate, 2644
Cupric Hexacyanoferrate(II) see 2640
Cupric Hexafluorosilicate, 2645
Cupric Hydroxide, 2646
Cupricin see 2668
Cupric Lactate see 5215
Cupric Nitrate, 2647
Cupric Oleate, 2648
Cupric Oxalate, 2649
Cupric Oxide, 2650
Cupric Perchlorate, 2651
Cupric p-Phenolsulfonate, 2652
Cupric Phosphate, 2653
Cupric Salicylate, 2654
Cupric Selenate, 2655
Cupric Selenide, 2656
Cupric Selenite, 2657
Cupric Silicofluoride see 2645
Cupric Stearate, 2658
Cupric Subacetate see 2628
Cupric Subcarbonate see 2634
Cupric Subsulfate see 2660
Cupric Sulfate, 2659
Cupric Sulfate, Ammoniated see 9116
Cupric Sulfate, Basic, 2660
Cupric Sulfide, 2661
Cupric Sulfocarbolate see 2652
Cupric Tartrate, 2662
Cupric Tungstate(VI), 2663
Cupric Wolframate see 2663
Cuprid [Merck & Co.] see 9579
Cuprimine [Merck & Co.] see 7029
Cuprimyl [M&B] see 2677
Cuprimyxin see 6255
Cuprion see 256
Cuprite see 2671
Cuprobam, 2664
Cuprobamé see 2664
Cuprocitrol see 2639
Cupro-cupric Sulfate see 2675
Cupron [Ayerst] see 1104
Cuprothiosinamine m-Benzoate Sodium see 256
Cuprous Acetate, 2665
Cuprous Bromide, 2666
Cuprous Chloride, 2667
Cuprous Cyanide, 2668
Cuprous Dimethyldithiocarbamate, Cuprous Chloride Complex see 2664
Cuprous Iodide, 2669
Cuprous Mercuric Iodide, 2670
Cuprous Oxide, 2671
Cuprous Potassium Cyanide, 2672
Cuprous Selenide, 2673
Cuprous Sulfide, 2674
Cuprous Sulfite, 2675
Cuprous Sulfocyanate see 2676
Cuprous Sulfocyanide see 2676

Cuprous Tetraiodomercurate(II) see 2670
Cuprous Thiocyanate, 2676
Cuproxat [Lentia] see 2660
Cuproxoline, 2677
Curacit see 8847
Curamil [Hoechst] see 7976
Curan-17-ol see 4274
Curantyl [Arzneimittelwerk VEB] see 3354
Curare, 2678
C-Curarine I, 2679
C-Curarine III, 2680
Curarin-HAF [Ethicon] see 9717
Curatin [Pfizer] see 3425
Curatrem [Merck & Co.] see 2405
Curcuma Longa see 9732
Curcumin, 2681
Curethyl see 5427
Curine, 2682
Curium, 2683
Curled Dock see 8269
Curling Factor see 4453
Curocef [Glaxo] see 1951
Curosajin [Mohan] see 4653
Curoxim [Glaxo-Hoechst] see 1951
Curretab [Reid-Provident] see 5677
Curtacain see 9123
Curtacrat see 2569
Curvularin, 2684
Curythan see 5737
Cuscohygrine, 2685
Cus Cus Oil see 6778
Cuscutin see 1174
Cuskhygrine see 2685
Cuspareine, 2686
Cusparia Bark see 681
Cusparine, 2687
Cusso see 1361
Cutch see 1909
Cutheparine [Biosedra] see 4571
Cuthizone see 2624
Cutinolone Simple [Labaz] see 9512
Cutisan [Chantereau] see 9568
Cutisterol [Farmitalia] see 4156
Cutizon see 2624
Cutless [Elanco] see 4129
Cuttle-fish Bone see 8409
Cut-weed see 4196
Cuvalit [Schering AG] see 5394
Cuxacillin [TAD] see 610
CV-2619 see 4817
CV-3317 see 2864
CV-58903 see 9979
CVF see MISC-2
C-Vimin [Astra] see 855
CVK see 7184
CVP see 2087
CVP see MISC-2
C.V.P. [USV] see 1241
CY-116 see 442
CY-216 see 4180
Cyacetacide, 2688
Cyamelide, 2689
Cyamemazine, 2690
Cyamepromazine see 2690
Cyamopsis Gum see 4486
Cyanacethydrazide see 2688
Cyanacethydrazine see 2688
Cyanacetylhydrazide see 2688
Cyanamide, 2691
"Cyanamide" see 1662
Cyanazide see 2688
Cyanazine, 2692
Cyanein see 1363
Cyanic Acid, 2693
Cyanic Acid Sodium Salt see 8552
(+)-Cyanidanol-3 see 1908
Cyanidanon-3-methyl Ether 1625 see 4656

Cyanidanon 4'-Methyl Ether 1626 see 4590
Cyanidenolon 1522 see 8044
Cyanidenolon-7-methyl Ether 1537 see 8170
Cyanidenon 1470 see 5483
Cyanidenon-4'-methyl Ether 1479 see 3290
Cyanidin Chloride, 2694
Cyanidol see 1908
Cyanin see 2694
Cyanite see 368
Cyanizide see 2688
Cyanoacetamide, 2695
Cyanoacetic Acid, 2696
Cyanoacetic Acid Ethyl Ester see 3744
Cyanoacetic Acid Hydrazide see 2688
Cyanoacetic Ester see 3744
Cyanoacetohydrazide see 2688
Cyanoacetonitrile see 5592
2-Cyanoacrylic Acid Isobutyl Ester see 1453
2-Cyanoacrylic Acid Methyl Ester see 5667
Cyanobenzene see 1107
Cyanocobalamin see 9921
Cyanocobalaminzinc Tannate Complex see 9923
N-Cyanodiallylamine see 2952
2-Cyano-1,4-dihydro-6-methyl-4-(3-nitrophenyl)-3,5-pyridinedicarboxylic Acid 3-Methyl 5-(1-Methylethyl) Ester see 6461
2-Cyano-10-(3-dimethylamino-2-methylpropyl)phenothiazine see 2690
3-Cyano-5-dimethylamino-3-phenyl-2-methylhexane see 4992
α-Cyanodiphenylmethane see 3316
4-[2-[(5-Cyano-5,5-diphenylpentyl)dimethylammonio]ethyl]-4-methylmorpholinium Bis(methyl Sulfate) see 7060
N^1-(5-Cyano-5,5-diphenylpentyl)-N^1,N^1,N^2-trimethylethylene-1-ammonium-2-morpholinium Bis(methyl Sulfate) see 7060
1'-(3-Cyano-3,3-diphenylpropyl)-[1,4'-bipiperidine]-4'-carboxamide see 7470
1-(3-Cyano-3,3-diphenylpropyl)-4-(2-oxo-3-propionyl-1-benzimidazolinyl)piperidine see 1216
1-(3-Cyano-3,3-diphenylpropyl)-4-phenylisonipecotic Acid see 3124
1-(3-Cyano-3,3-diphenylpropyl)-4-phenylisonipecotic Acid Ethyl Ester see 3313
1-(3-Cyano-3,3-diphenylpropyl)-4-phenyl-4-piperidinecarboxylic Acid see 3124
1-(3-Cyano-3,3-diphenylpropyl)-4-phenyl-4-piperidinecarboxylic Acid Ethyl Ester see 3313
1-[1-(3-Cyano-3,3-diphenylpropyl)-4-piperidinyl]-1,3-dihydro-3-(1-oxopropyl)-2H-benzimidazol-2-one see 1216
1-[1-(3-Cyano-3,3-diphenylpropyl)-4-piperidyl]-3-propionyl-2-benzimidazolinone see 1216
2α-Cyano-4α,5α-epoxyandrostan-17β-ol-3-one see 9611
2-Cyanoethanesulfonic Acid Sodium Salt see 8639
Cyanoethydrazide see 2688
(±)-N-2-Cyanoethylamphetamine see 3939
Cyanoethylene see 125
Cyanofenphos, 2697

(*R,S*)-α-Cyano-4-fluoro-3-phenoxy-
benzyl-(1*R,S*)-*cis,trans*-3-(2,2-di-
chlorovinyl)-2,2-dimethylcyclopro-
panecarboxylate *see* 2764
Cyanogas [Am. Cyanamid] *see* 1664
Cyanogen, 2698
Cyanogenamide *see* 2691
Cyanogen Azide, 2699
Cyanogen Bromide, 2700
Cyanogen Chloride, 2701
Cyanogen Iodide, 2702
Cyanogran *see* 8553
Cyanoguanidine *see* 3082
7-Cyano-2-heptene-4,6-diynoic Acid
see 2974
2-Cyano-10-[3-(4-hydroxypiperidi-
no)propyl]phenothiazine *see* 7117
Cyanomethane *see* 62
Cyanomethylamine *see* 423
N-Cyano-*N*′-methyl-*N*′′-[2-[[(5-
methyl-1*H*-imidazol-4-yl)methyl]-
thio]ethyl]guanidine *see* 2279
7-[[[(Cyanomethyl)thio]acetyl]amino]-
7-methoxy-3-[[(1-methyl-1*H*-tetra-
zol-5-yl)thio]methyl]-8-oxo-5-thia-
1-azabicyclo[4.2.0]oct-2-ene-2-
carboxylic Acid *see* 1929
α-Cyano-3-phenoxybenzyl α-(4-Chlo-
rophenyl)isovalerate *see* 3952
α-Cyano-3-phenoxybenzyl-2-(4-chlo-
rophenyl)-3-methylbutyrate *see*
3952
(*RS*)-α-cyano-3-phenoxybenzyl (1*R*)-
cis,trans-Chrysanthemate *see* 2776
(*S*)-α-Cyano-3-phenoxybenzyl (1*R*)-
cis-3-(2,2-dibromovinyl)-2,2-di-
methylcyclopropane Carboxylate *see*
2869
(±)-α-Cyano-3-phenoxybenzyl-(±)-
cis,trans-3-(2,2-dichlorovinyl)-2,2-
dimethylcyclopropane Carboxylate
see 2775
α-Cyano-*m*-phenoxybenzyl 2,2-Di-
methyl-3-(2-methylpropenyl)cyclo-
propanecarboxylate *see* 2776
α-Cyano-3-phenoxybenzyl 2,2,3,3-
Tetramethylcyclopropanecarboxyl-
ate *see* 3936
1-(2-Cyanophenoxy)-2-hydroxy-3-
tert-butylaminopropane *see* 1479
1-(2-Cyanophenoxy)-3-β-(4-hydroxy-
phenylacetamido)ethylamino-2-pro-
panol *see* 3557
N-[2-[[3-(2-Cyanophenoxy)-2-hydr-
oxypropyl]amino]ethyl]-4-hydroxy-
benzeneacetamide *see* 3557
Cyano(3-phenoxyphenyl)methyl 4-
Chloro-α-(1-methylethyl)benzene-
acetate *see* 3952
(±)-Cyano-(3-phenoxyphenyl)methyl
(+)-4-(Difluoromethoxy)-α-(1-
methylethyl)benzeneacetate *see*
4055
Cyanophos *see* 2266
2-Cyano-2-propenoic Acid Methyl
Ester *see* 5667
2-Cyano-2-propenoic Acid 2-Methyl-
propyl Ester *see* 1453
Cyanopsin, 2703
(±)-*N*-Cyano-*N*′-4-pyridinyl-*N*′′-
(1,2,2-trimethylpropyl)guanidine
Monohydrate *see* 7407
o-Cyanotoluene *see* 9463
ω-Cyanotoluene *see* 1145
Cyanox [Sumitomo] *see* 2266
Cyantin [Lederle] *see* 6520
Cyanuric Acid, 2704
Cyanurotriamide *see* 5691
Cyasorb UV 9 (Obsolete) [Am. Cyan-
amid] *see* 6907

Cyasorb UV 24 (Obsolete) [Am.
Cyanamid] *see* 3298
Cyasorb UV 284 (Obsolete) [Am.
Cyanamid] *see* 8963
Cybis [Breon] *see* 6273
Cybolt [Am. Cyanamid] *see* 4055
Cycasin, 2705
Cyclacillin, 2706
Cycladiene [Bruneau] *see* 3094
Cyclaine [Merck & Co.] *see* 4629
Cyclamate Calcium *see* 1665
Cyclamate Sodium *see* 2707
Cyclamic Acid, 2707
Cyclamide *see* 53 and 9434
Cyclamin [Wyeth] *see* 9681
Cyclamin *see* 5596
Cyclamycin [Wyeth] *see* 9681
Cyclan *see* 1665
Cyclandelate, 2708
Cyclanon *see* 4266
Cyclapen [Wyeth] *see* 2706
Cyclarbamate, 2709
Cyclazenin *see* 4470
Cyclazocine, 2710
Cyclergine [Deglaude] *see* 2708
Cycletanide *see* 2268
Cyclethrin, 2711
Cyclexanone, 2712
Cyclexedrine, 2713
Cyclic Adenosine 3′,5′-Monophos-
phate *see* 2714
Cyclic AMP, 2714
Cyclic GMP, 2715
Cyclic Guanosine 3′,5′-Monophos-
phate *see* 2715
Cyclic Methylene (4-Chloro-*o*-tolyl)-
dithioimidocarbonate *see* 6466
Cyclic Nucleotide Phosphodiesterases
see 2714
Cyclic Propylene (Diethoxyphosphin-
yl)dithioimidocarbonate *see* 5743
Cyclizine, 2716
Cycloamyloses *see* 2724
Cycloartenyl Ferulate *see* 6841
Cyclobarbital, 2717
Cyclobarbitone *see* 2717
Cyclobendazole, 2718
Cyclobenzaprine, 2719
Cycloblastin *see* 2753
Cyclobutane, 2720
1,1-Cyclobutanedicarboxylic Acid
Platinum Complex *see* 1828
(−)-*N*-Cyclobutylmethyl-3,14-di-
hydroxymorphinan *see* 1530
17-(Cyclobutylmethyl)-4,5-epoxymor-
phinan-3,6,14-triol *see* 6271
N-Cyclobutylmethyl-14-hydroxydihy-
dronormorphine *see* 6271
17-(Cyclobutylmethyl)morphinan-
3,14-diol *see* 1530
Cyclobutyrol, 2721
Cyclobuxine, 2722
Cyclo-C [Cohjin] *see* 663
Cyclocarbothiamine *see* 2762
Cyclocort [Lederle] *see* 398
Cyclocumarol, 2723
*O*²,²′-Cyclocytidine *see* 663
2,2′-*O*-Cyclocytidine *see* 663
1,2-Cyclodecanedione *see* 8370
1-Cyclodecanol-2-one *see* 8371
α-Cyclodextrin *see* 2724
β-Cyclodextrin *see* 2724
γ-Cyclodextrin *see* 2724
Cyclodextrins, 2724
4-Cyclododecyl-2,6-dimethylmorpho-
line *see* 3405
Cyclodol *see* 9607
Cyclodorm *see* 2717
Cyclodrine *see* 2725
Cycloestrol *see* 4621

Cyclofenil, 2726
Cycloform *see* 5017
Cyclogest [Collins] *see* 7783
Cycloglucans *see* 2724
Cycloglycylglycine *see* 7434
Cycloguanil, 2727
Cycloguanil Embonate *see* 2727
Cyclogyl [Alcon] *see* 2752
Cycloheptaamylose *see* 2724
9-Cycloheptadecen-1-one *see* 2337
1,2-Cycloheptanedione Dioxime *see*
4586
Cycloheptanone, 2728
Cycloheptatrienocarbonium Bromide
see 9696
Cycloheptatrienylium Bromide *see*
9696
5-(1-Cyclohepten-1-yl)-5-ethylbarbi-
turic Acid *see* 4575
5-(1-Cyclohepten-1-yl)-5-ethyl-2,4,6-
(1*H*,3*H*,5*H*)-pyrimidinetrione *see*
4575
Cyclohexaamylose *see* 2724
1,4-Cyclohexadienedione *see* 8103
2,5-Cyclohexadiene-1,4-dione *see*
8103
2,5-Cyclohexadiene-1,4-dione Compound with 1,4-Benzenediol (1:1) *see*
8070
Cyclohexanamine *see* 2735
Cyclohexane, 2729
Cyclohexanecarbinol *see* 2737
N-[3-(*N*-Cyclohexanecarbonyl-D-
alanylthio)-2-methylpropanoyl]-L-
proline *see* 6200
Cyclohexanecarboxylic Acid, 2730
1,3-Cyclohexanedione *see* 3161
1,2-Cyclohexanedione Dioxime *see*
6477
Cyclohexanehexol *see* 4883
1,2,3,4,5,6-Cyclohexanehexolphos-
phoric Acid *see* 7359
Cyclohexanehexyl Hexaphosphate *see*
7359
Cyclohexanemethanol *see* 2737
1,2,3,4,5-Cyclohexanepentol *see* 8046
Cyclohexanesulfamic Acid *see* 2707
Cyclohexanesulfamic Acid Calcium
Salt *see* 1665
Cyclohexanol, 2731
1-Cyclohexanol-α-butyric Acid *see*
2721
Cyclohexanone, 2732
Cyclohexatriene *see* 1074
Cyclohexene, 2733
5-(2-Cyclohexen-1-yl)-5-allyl-2-thio-
barbituric Acid *see* 9219
5-(2-Cyclohexen-1-yl)dihydro-5-(2-
propenyl)-2-thioxo-4,6(1*H*,5*H*)-
pyrimidinedione *see* 9219
5-Cyclohexenyl-3,5-dimethylbarbitur-
ic Acid *see* 4625
5-(1-Cyclohexen-1-yl)-1,5-dimethyl-
barbituric Acid *see* 4625
5-(1-Cyclohexen-1-yl)-1,5-dimethyl-
2,4,6(1*H*,3*H*,5*H*)-pyrimidinetrione
see 4625
5-(1-Cyclohexen-1-yl)-5-ethylbarbi-
turic Acid *see* 2717
5-(1-Cyclohexen-1-yl)-5-ethyl-2,4,6-
(1*H*,3*H*,5*H*)-pyrimidinetrione *see*
2717
Cycloheximide, 2734
Cyclohexitol *see* 4883
3-Cyclohexyl-1-(*p*-acetylphenylsulfo-
nyl)urea *see* 53
Cyclohexylamine, 2735
N-[2-[4-[[[(Cyclohexylamino)carbo-
nyl]amino]sulfonyl]phenyl]ethyl]-5-
methylpyrazinecarboxamide *see*
4337

2-Cyclopenten-2,3-diol-1-one *see* 8134

2-Cyclopentene-1-tridecanoic Acid *see* 2037

2-Cyclopentene-1-undecanoic Acid *see* 4679

1,2-Cyclopentenophenanthrene, 2749

3-(2-Cyclopentenyl)-2-methyl-4-oxo-2-cyclopentenyl Ester of Chrysan-themummonocarboxylic Acid *see* 2711

2-(1-Cyclopenten-1-yl)-2-[2-(4-mor-pholinyl)ethyl]cyclopentanone *see* 2712

17β-(1-Cyclopenten-1-yloxy)andro-sta-1,4-dien-3-one *see* 8063

5-(2-Cyclopenten-1-yl)-5-(2-propen-yl)-2,4,6(1*H*,3*H*,5*H*)-pyrimidinetri-one *see* 2751

2-(Δ-1′-Cyclopentenyl)-2-(β-tetrahy-droparoxazinoethyl)cyclopentanone *see* 2712

D-13-(2-Cyclopenten-1-yl)tridecanoic Acid *see* 2037

11-(2-Cyclopenten-1-yl)undecanoic Acid *see* 4679

Cyclopenthiazide, 2750

Cyclopentobarbital, 2751

Cyclopentolate, 2752

Cyclopentyl Alcohol *see* 2747

3-[(Cyclopentylhydroxyphenylacetyl)-oxy]-1,1-dimethylpyrrolidinium Bromide *see* 4397

2-[(Cyclopentylhydroxy-2-thienyl-acetyl)oxy]-*N*,*N*-diethyl-*N*-methyl-ethanaminium Bromide *see* 7084

α-Cyclopentylmandelic Acid Ester with 3-Hydroxy-1,1-dimethylpyrro-lidinium Bromide *see* 4397

1-Cyclopentyl-2-methylaminopropane *see* 2745

3-Cyclopentylmethyl-6-chloro-7-sul-famyl-3,4-dihydro-1,2,4-benzothia-diazine 1,1-Dioxide *see* 2750

17β-(3-Cyclopentyl-1-oxopropoxy)-androst-4-en-3-one *see* 9111

(16α,17β)-3-(Cyclopentyloxy)estra-1,3,5(10)-trien-16,17-diol *see* 8065

3-(Cyclopentyloxy)-17-hydroxypreg-na-3,5-dien-20-one *see* 7068

4-[3-(Cyclopentyloxy)-4-methoxy-phenyl]-2-pyrrolidinone *see* 8235

(17β)-3-(Cyclopentyloxy)-17-methyl-androsta-3,5-dien-17-ol *see* 6045

3-(Cyclopentyloxy)-19-nor-17α-preg-na-1,3,5(10)-trien-20-yn-17-ol *see* 8066

3-(Cyclopentyloxy)pregna-3,5-dien-20-one *see* 8069

(*S*)-1-(2-Cyclopentylphenoxy)-3-[(1,1-dimethylethyl)amino]-2-propanol *see* 7025

α-Cyclopentyl-α-phenyl-1-piperidine-propanol Hydrochloride *see* 2763

17β-(2-Cyclopentylpropionyloxy)-4-hydroxyestr-4-en-3-one *see* 4768

Cyclophosphamide, 2753

Cyclophosphane *see* 2753

Cyclopon *see* 4266

Cyclopregnol, 2754

Cyclopropane, 2755

2-Cyclopropylamino-4,6-diamino-*s*-triazine *see* 2782

[5-(Cyclopropylcarbonyl)-1*H*-benz-imidazol-2-yl]carbamic Acid Methyl Ester *see* 2718

1-Cyclopropyl-7-(4-ethyl-1-pipera-zinyl)-6-fluoro-1,4-dihydro-4-oxo-3-quinolinecarboxylic Acid *see* 3546

1-Cyclopropyl-6-fluoro-1,4-dihydro-4-oxo-7-(1-piperazinyl)-3-quino-linecarboxylic Acid *see* 2315

21-Cyclopropyl-7α-(2-hydroxy-3,3-dimethyl-2-butyl)-6,14-*endo*-ethano-6,7,8,14-tetrahydrooripavine *see* 1487

21-Cyclopropyl-7α-[(*S*)-1-hydroxy-1,2,2-trimethylpropyl]-6,14-*endo*-ethano-6,7,8,14-tetrahydrooripavine *see* 1487

1-[4-[2-(Cyclopropylmethoxy)ethyl]-phenoxy]-3-[(1-methylethyl)amino]-2-propanol *see* 1204

α-Cyclopropyl-α-(4-methoxyphenyl)-5-pyrimidinemethanol *see* 665

α-Cyclopropyl-4-methoxy-α-(pyrimi-din-5-yl)benzyl Alcohol *see* 665

N-(Cyclopropylmethyl)-7,8-dihydro-7α-(1-hydroxy-1-methylethyl)-*O*⁶-methyl-6,14-*endo*-ethenonormor-phine *see* 2777

[5α,7α(*S*)]-17-(Cyclopropylmethyl)-α-(1,1-dimethylethyl)-4,5-epoxy-18,19-dihydro-3-hydroxy-6-meth-oxy-α-methyl-6,14-ethenomorphi-nan-7-methanol *see* 1487

N-(Cyclopropylmethyl)-2,6-dinitro-*N*-propyl-4-(trifluoromethyl)benzen-amine *see* 7781

(5α,7α)-17-(Cyclopropylmethyl)-4,5-epoxy-18,19-dihydro-3-hydroxy-6-methoxy-α,α-dimethyl-6,14-etheno-morphinan-7-methanol *see* 3347

17-(Cyclopropylmethyl)-4,5-epoxy-3,14-dihydroxymorphinan-6-one *see* 6278

17-(Cyclopropylmethyl)-4,5-epoxy-3-hydroxy-6-methoxy-α,α-dimethyl-6,14-ethenomorphinan-7-methanol *see* 2777

8-(Cyclopropylmethyl)-6β,7β-epoxy-3α-hydroxy-1α*H*,5α*H*-tropanium Bromide (−)-(*S*)-Tropate *see* 2280

(5α)-17-(Cyclopropylmethyl)-4,5-epoxy-6-methylenemorphinan-3,14-diol *see* 6274

N-(Cyclopropylmethyl)-6,14-*endo*-etheno-7α-(2-hydroxy-2-propyl)-tetrahydronororipavine *see* 2777

N-(Cyclopropylmethyl)-6,14-*endo*-ethenotetrahydronororipavine *see* 2777

Cyclopropyl Methyl Ether, 2756

3-(Cyclopropylmethyl)-1,2,3,4,5,6-hexahydro-6,11-dimethyl-2,6-meth-ano-3-benzazocin-8-ol *see* 2710

N-Cyclopropylmethyl-14-hydroxydi-hydromorphinone *see* 6278

[7(*S*)-(1α,2β,4β,5α,7β)]-9-(Cycloprop-ylmethyl)-7-(3-hydroxy-1-oxo-2-phenylpropoxy)-9-methyl-3-oxa-9-azoniatricyclo[3.3.1.0²,⁴]nonane Bromide *see* 2280

N-(Cyclopropylmethyl)-19-methylnor-orvinol *see* 3347

17-(Cyclopropylmethyl)morphinan-3-ol *see* 2757

1-(Cyclopropylmethyl)-5-phenyl-7-chloro-1*H*-1,4-benzodiazepin-2(3*H*)-one *see* 7713

N-Cyclopropylmethyl-*N*-*n*-propyl-4-trifluoromethyl-2,6-dinitroaniline *see* 7781

N-Cyclopropylmethylscopolamine Bromide *see* 2280

N-(Cyclopropylmethyl)-α,α,α-triflu-oro-2,6-dinitro-*N*-propyl-*p*-tolui-dine *see* 7781

21-Cyclopropyl-6,7,8,14-tetrahydro-7α-(1-hydroxy-1-methylethyl)-6,14-*endo*-ethanooripavine *see* 3347

N-Cyclopropyl-1,3,5-triazine-2,4,6-triamine *see* 2782

Cyclo-Prostin [Upjohn] *see* 7890

Cyclorphan, 2757

Cyclosa [Nourypharma] *see* 2906

Cyclosal *see* 4651

Cyclosan *see* 5795

Cyclosan *see* 4799

Cycloserine, 2758

Cyclospasmol [Brocades; Ives-Cameron] *see* 2708

Cyclosporin A *see* 2759

Cyclosporin B *see* 2759

Cyclosporin C *see* 2759

Cyclosporin D *see* 2759

Cyclosporin G *see* 2759

Cyclosporine *see* 2759

Cyclosporins, 2759

Cyclostin [Farmitalia] *see* 2753

Cyclosulfyne *see* 7818

Cyclothiazide, 2760

Cyclotrimethylenetrinitramine *see* 2742

Cyclotriphosphazenes *see* 2341

Cyclovalone, 2761

Cyclo-Werrol [Werrol] *see* 4799

Cycobemin *see* 9921

Cycocel [Am. Cyanamid] *see* 2103

Cycogan [Makhteshim-Agan] *see* 2103

Cycolamin *see* 9921

Cycostat [Am. Cyanamid] *see* 8229

Cycothiamin(e), 2762

Cycotiamine *see* 2762

Cycrimine Hydrochloride, 2763

Cycrin [Ayerst] *see* 5677

Cydonia Seed *see* 8064

Cydril [Tutag] *see* 616

Cyfen [Am. Cyanamid] *see* 3922

Cyfluthrin, 2764

Cyfos [Mead Johnson] *see* 4822

Cyfoxylate *see* 2764

Cygon [Am. Cyanamid] *see* 3209

Cygro [Am. Cyanamid] *see* 5524

Cyhalothrin, 2765

Cyheptamide, 2766

Cyhexatin, 2767

Cyklodorm *see* 2717

Cyklokapron [Kabi] *see* 9487

Cyklosal *see* 2745

Cykobeminet *see* 9921

Cylan [Am. Cyanamid] *see* 7309

Cylert [Abbott] *see* 7021

Cylphenicol [Trent] *see* 2068

Cymarigenin *see* 8818

Cymarin, 2768

Cymarose, 2769

Cymbi [Dolorgiet] *see* 621

Cymbush [ICI] *see* 2775

Cymene, 2770

m-Cymene *see* 2770

o-Cymene *see* 2770

p-Cymene *see* 2770

2-*p*-Cymenol *see* 1881

3-*p*-Cymenol *see* 9333

Cymerine [Tokyo Tanabe] *see* 8125

Cymevan [Syntex] *see* 4262

Cymevene [Syntex] *see* 4262

Cymiazole, 2771

Cymidon *see* 5180

m-Cym-5-yl Methylcarbamate *see* 7794

Cynanchin *see* 9889

Cynanchogenin, 2772

Cynarin(e), 2773

Cynaron *see* 5896

Cynaroside *see* 5483

Cynem [Am. Cyanamid] *see* 9275

Cynisin see 2419
Cynoglossophine see 4545
Cynomel [SK & F] see 5388
Cynotoxin see 8818
Cyoctol, 2774
Cyolane [Am. Cyanamid] see 7309
CYP see 2697
Cypercare [Virbac] see 2775
Cyperkill [Mitchell Cotts] see 2775
Cypermethrin, 2775
Cypersect [Mitchell Cotts] see 2775
Cyphenothrin, 2776
Cypip [Am. Cyanamid] see 3106
Cyprenorphine, 2777
Cypress Camphor see 1919
Cyprex see 3406
Cypripedium, 2778
Cyproheptadine, 2779
Cyprome Ether see 2756
Cypromin [Lakeside] see 8234
Cyproquinate, 2780
Cyproquinidate see 2780
Cyprostat [Keymer] see 2781
Cyproterone, 2781
Cyproxyquine see 2780
Cyren A [Bayer] see 3118
Cyren B [Bayer] see 3119
Cyromazine, 2782
Cyrpon [Troponwerke] see 5751
Cys (IUPAC Abbrev.) see 2787
Cyscholin [Kanto] see 2321
Cystamin see 5879
Cystamine, 2783
L-Cystathionine, 2784
Cysteamine, 2785
Cysteic Acid, 2786
Cysteine, 2787
L-Cysteine see 2787
L-Cysteine Methyl Ester Hydrochloride see 5668
L-Cysteine Thioacetal of Formaldehyde see 3393
[N-(N-L-Cysteinyl-L-tyrosyl)-L-cysteinyl]-interferon γ (Human Lymphocyte Protein Moiety Reduced) see 4894
Cystine, 2788
Cystine S,S-Dioxide see 2789
Cystine S,S'-Dioxide see 2789
Cystine S-Dioxides, 2789
Cystine Disulfoxide see 2789
sym-L-Cystine Disulfoxide see 2789
L-Cystine Thiosulfonate see 2789
Cystit [Heyden] see 6520
Cysto-Conray [Mallinckrodt] see 4952
Cystogen see 5879
Cystografin [Squibb] see 5689
Cystokon [Mallinckrodt] see 72
Cystorelin [Abbott] see 5354
Cystospaz [Alcon] see 4795
Cystural see 3706
Cytacon [Glaxo] see 9921
Cytadren [Ciba-Geigy] see 452
Cytamen [Glaxo] see 9921
Cytarabine, 2790
Cyten [Am. Cyanamid] see 3922
Cythioate, 2791
Cythion [Am. Cyanamid] see 5582
Cytidine, 2792
Cytidine Diphosphate Choline Ester see 2321
Cytidine-2'-monophosphate see 2793
Cytidine-3'-monophosphate see 2794
Cytidine-2'-phosphate see 2793
Cytidine-3'-phosphate see 2794
2'-Cytidinephosphoric Acid see 2793
3'-Cytidinephosphoric Acid see 2794
Cytidine 5'-(Trihydrogen Diphosphate) Mono[2-(trimethylammonio)-ethyl] Ester Hydroxide Inner Salt see 2321

Cytidylic Acid a see 2793
Cytidylic Acid b see 2794
2'-Cytidylic Acid, 2793
3'-Cytidylic Acid, 2794
Cytisine, 2795
Cytiton [USSR] see 2795
Cytobin [Norden] see 5388
Cytobion [E. Merck, W. Ger.] see 9921
Cytochalasin B see 2796
Cytochalasins, 2796
Cytochrome c, 2797
Cytochrome c₁ see 2797
Cytochrome c₂ see 2797
Cytochrome c₃ see 2797
Cytochrome e: see 2797
Cytochromes P₄₅₀, 2798
Cytocym see 9321
Cytoflav see 8202
Cytofol [Lappe] see 4140
Cytohemin, 2799
Cytolipin H, 2800
Cytomel [SK & F] see 5388
Cytomine [Darby] see 5388
Cytonal [Berlin-Chemie] see 4168
Cytophosphane see 2753
Cytorest [Mochida] see 2797
Cytosar [Upjohn] see 2790
Cytosine, 2801
β-Cytosine Arabinoside see 2790
Cytosine Riboside see 2792
1-(1'-Cytosinyl)-4-[L-3'-amino-5'-(1''-N-methylguanidino)-valerylamino]-1,2,3,4-tetradeoxy-β-D-erythro-hex-2-enuronic Acid see 1323
Cytostatin see 4621
2'-Cytosylic Acid see 2793
3'-Cytosylic Acid see 2794
Cytotec [Searle] see 6129
Cytotect [Biotest] see 4837
Cytovene [Syntex] see 4262
Cytoxan [Mead Johnson] see 2753
Cytoxin-anticoytotoxin Serum see 750
Cytozyme see 9321
Cytrol [Am. Cyanamid] see 506
Cytrolane [Am. Cyanamid] see 5743
CYVADIC see MISC-2
CZ see 1886
D 20 see 2727
D 25-Antimykotikum see 3947
D 32 see 9984
D 40 [Armour Pharm.] see 8584
D 40TA see 3651
D 41 see 9683
D 50 see 7970
D 65MT see 310
D 109 see 4629
D 138 see 6620
D 145 see 5709
D 201 see 5117
D 206 see 7902
D 254 see 7423
D 301 see 1123
D 365 see 9851
D 563 see 6915
D 600 see 4257
D 706 see 6923
D 860 see 9432
D 1959 see 8142
D 2083 see 2908
D 4018 see 3544
D 8955 see 9378
D 9998 see 4117
2,4-D, 2802
DA 398 see 3572
DA 688 see 4273
DA 708 see 9059
DA 759 see 5914
DA 2370 see 3953
DA 3177 see 2280

DA 4577 see 6108
2,4-DAA see 5920
DAAO see 425
Dabco see 9584
Dabroson [Hoyer] see 278
Dabylen [SCHI-WA] see 3308
DAC-2787 [Diamond] see 2167
Dacarbazine, 2803
Dacarel [Roemmers] see 6403
Dacatic [Orion] see 2803
Daconate [Diamond] see 5864
Daconil 2787 [Diamond] see 2167
Dacoren [Rhodia] see 3598
Dacortilen see 7729
Dacortin see 7727
DACPM see 5978
Dacron [Du Pont] see 7546
Dacthal see 2830
Dactil [Lakeside] see 7440
Dactin see 3054
Dactinomycin, 2804
DADA see 3182
DADDS see 16
Dadex see 2932
Dadibutol see 3672
DADPS see 2820
Dafalgan [UPSA] see 40
Daflon [Servier] see 3291
Daflon-500 Mg see 3291
Dafnegin [Poli] see 2270
Dagenan [Merck & Co.] see 8913
Dagenan Chloride [Merck & Co.] see 99
Dagger [Am. Cyanamid] see 4825
Dagutan see 10010
Dahi see 10010
Dahl's Acid see 6327
Dahl's Acid II see 6321
Dahl's Acid III see 6322
Dahlin see 4898
Daidzein, 2805
Daidzin see 2805
Daipisate [Taisho] see 7441
Dairena see 8885
Daital [Alter] see 3667
Daktar [Janssen] see 6101
Daktarin [Janssen] see 6101
Dalacin T see 2352
Dalacin C [Upjohn] see 2352
Dalactine [Upjohn] see 2352
Dalapon, 2806
Dalaron [OJ & F] see 2922
Dal-E-Rad [Vineland] see 5864
Dalf (Obsolete) see 6022
Dalgan [Wyeth] see 2934
Dalgol see 5731
Dalmadorm [Roche] see 4123
Dalmane [Roche] see 4123
Dalmate [Nippon Roche] see 4123
Dalmatian Insect Powder see 7980
Dalnate [USV] see 9438
Dalpac see 1548
Daltroban, 2807
Dalyde [HW & D] see 3011
Dalysep [Syntex] see 8881
Dalzic [Rhein-Pharma] see 7703
Damar, 2808
Damascenine, 2809
Dambose see 4883
Dametin [E. Merck] see 2860
Damfin [Ciba-Geigy] see 5848
Damide [SIT] see 4847
Daminozide, 2810
Dammar see 2808
DAMP see 1253
D-AMP-3 see 9179
DAN-603 see 8962
DAN 2163 see 501
Danabol see 5862
Danantizol see 5892
Danazol, 2811

Dandelion *see* 9030
Daneral [Hoechst] *see* 7198
Danex [Makhteshim-Agan] *see* 9536
Danilon [Esteve] *see* 8990
Danilone [Frosst; Schieffelin] *see* 7196
Danish Agar *see* 4212
Danishefsky's Diene *see* 5925
Danitol [Sumitomo] *see* 3936
Danizol *see* 6079
Danka [Mediolanum] *see* 3439
Danocrine [Winthrop] *see* 2811
Danol [Winthrop] *see* 2811
Danoval [Krka] *see* 2811
Dansyl Chloride, 2812
Dantamacrin [Rhm] *see* 2814
Danten [McKesson] *see* 7293
Danthron, 2813
Dantrium [Eaton] *see* 2814
Dantrium [Bago] *see* 1041
Dantrix [SIT] *see* 2814
Dantrolene, 2814
Dantromin *see* 7021
Dantron *see* 2813
Danylen [Temmler] *see* 3116
Daonil [Hoechst] *see* 4372
Daphnandrine, 2815
Daphnetin, 2816
Daphnetin 7-β-D-Glucoside *see* 2817
Daphnin, 2817
Daphnoline, 2818
Dapiprazole, 2819
Dapocel *see* 3182
Dapotum [Heyden] *see* 4116
Dapotum D [Heyden] *see* 4116
Dapsone, 2820
DAPT *see* 498
Daptazile *see* 498
Daptazole [Nicholas] *see* 498
Daquin [Riker] *see* 2073
DAR *see* 2939
Daramin *see* 8282
Darammon *see* 531
Daranide [Merck & Co.] *see* 3067
Darapram *see* 7997
Daraprim [Burroughs Wellcome] *see* 7997
Darbid [SK & F] *see* 5090
D'Arcet Metal—Fusible *see* 1268
Darcil [Wyeth] *see* 7184
Darenthin [Burroughs Wellcome] *see* 1364
Daricol *see* 6926
Daricon [Pfizer] *see* 6926
Darlan *see* 2821
Darmol [Omegin] *see* 7208
Darotol *see* 2155
Dartal [Searle] *see* 9286
Dartalan [Searle] *see* 9286
Dartranol [Ester] *see* 7248
Darvan®, 2821
Darvilen [Schering] *see* 1936
Darvisul [Burroughs Wellcome] *see* 2976
Darvon [Lilly] *see* 7851
Darvon-N [Lilly] *see* 7851
Darvyl *see* 5509
Dasanit [Chemagro] *see* 3943
Daserol *see* 5737
Daskil [Wander] *see* 6435
Dasten [Duncan] *see* 6880
DAT [Wander] *see* 46
Datanil *see* 9382
DATC *see* 2950 and 9382
Datiscetin, 2822
Datril [Bristol-Myers] *see* 40
Datura, 2823
Daturine *see* 4795
Daucol, 2824
Daunoblastina [Farmitalia] *see* 2825
Daunomycin *see* 2825
Daunomycinone *see* 2825

Daunorubicin, 2825
Daunosamine *see* 2825 and 3428
Dauricine, 2826
Davainex *see* 3025
Davistar [Uriach] *see* 2375
Davitamon C *see* 855
Davitamon-K [Organon] *see* 5710
Davitin [Ives] *see* 9928
Davosin [Parke, Davis] *see* 8890
Davosin Suspension *see* 8890
Davoxin [Lakeside] *see* 3150
DAV Ritter [Ritter] *see* 2904
Daxauten [Kettelhack] *see* 7744
Daxolin [Dome] *see* 5461
Dayfen *see* 3334
Dazomet, 2827
Dazzle *see* 8576
DB 32 *see* 8905
DB 87 *see* 6040
2,4-DB, 2828
DBC *see* 2446
DBcAMP *see* 1448
DBCP *see* 3003
DBD *see* 6132
DBED *see* 1072
DBED Penicillin *see* 7037
DBI [USV] *see* 7191
DBI-TD *see* 7191
DBM *see* 6130
DBMC, 2829
DBP *see* 1586
DBPCl *see* 2994
D Bretard *see* 7191
DBS *see* 3015
D & C Blue No. 6 *see* 4855
D & C Green No. 5 *see* 238
D & C Green No. 6 *see* 8095
D & C Orange No. 4 *see* 6813
D & C Orange No. 5 *see* 3005
D & C Orange No. 8 *see* 3056
D & C Orange No. 10 *see* 3174
D & C Orange No. 11 *see* 3174
D & C Red No. 17 *see* 8858
D & C Red No. 19 *see* 8182
D & C Red No. 22 *see* 3555
D & C Yellow No. 7 *see* 4085
D & C Yellow No. 8 *see* 4085
D & C Yellow No. 10 *see* 8100
D & C Yellow No. 11 *see* 8101
DC 826 *see* 1631
DC 2797 *see* 1448
DCA *see* 2883 and 3037
DCB *see* 3047
DCC *see* 3086
DCCI *see* 3086
DCCK [Rentschler] *see* 3598
DCEE *see* 3055
DCET *see* 2014
DCF *see* 7091
2'-DCF *see* 7091
DCH-21 *see* 3869
DCI *see* 3036
DClMDP *see* 2369
DCMO *see* 1832
DCMX *see* 3066
DCNU *see* 2177
d(COMOT) *see* 1800
3,6-DCP *see* 2398
DCPA, 2830
DCPC *see* 2086
DCPM *see* 1258
D-D [Shell] *see* 3064
DD 3480 *see* 9373
DDA *see* 3090
DdAdo *see* 3090
DDAVP [Ferring] *see* 2904
DDC *see* 3384
DdCyd *see* 3091
DDD *see* 3090
o,p'-DDD *see* 6134
p,p'-DDD *see* 3049

DDD (Analytical), 2831
DDI *see* 3092
DdIno *see* 3092
DDP *see* 2319
cis-DDP *see* 2319
DDQ *see* 1477 and 3052
DDS *see* 2820
DDT, 2832
p,p'-DDT *see* 2832
DDTC *see* 3384
DDVP *see* 3069
DE-019 *see* 1447
Deacetylandromedotoxin *see* 4449
Deacetylanhydroandromedotoxin *see* 4449
Deacetylanisomycin *see* 700
1-Deacetyldiaboline *see* 9957
Deacetyllanatoside C *see* 2903
N-Deacetyl-*N*-methylcolchicine *see* 2873
Dead-burned Gypsum *see* 1713
Deadly Nightshade *see* 1034
Deadopa [DeAngeli] *see* 5344
DEAE-cellulose *see* 1961
DEAE-dextran *see* 2916
DEAE-Sephadex [Pharmacia] *see* 7531
Deal Pine *see* 9956
Deamino-2-*O*-methyltyrosine-1-carbaoxytocin *see* 1800
Deaminooxytocin, 2833
Deanase [Consolidated Chem.] *see* 2888
Deandros [Farmochimica] *see* 7710
Deaner [Riker] *see* 2836
Deanol, 2834
Deanol Aceglumate, 2835
Deanol Acetamidobenzoate, 2836
Deapasil *see* 491
Deapril-ST [Mead Johnson] *see* 3598
Death's Herb *see* 1034
7-Deazaadenosine *see* 9714
Deba *see* 972
Debecacin *see* 2987
Débékacyl [Roger Bellon] *see* 2987
Debenal *see* 8874
Debenal M *see* 8884
Debeone *see* 7191
Debetrol *see* 9348
Debinyl *see* 7191
Deblaston [Madaus] *see* 7427
Debricin [J & J] *see* 4020
Debridat [Jouveinal; Sigma-Tau] *see* 9613
Debrisan [Pharmacia] *see* 2928
Debrisoquin, 2837
Debrisorb [Deutsche Pharmacia] *see* 2928
Debroxide [Alcon] *see* 1128
Dec *see* 2839
Decabid [Lilly] *see* 4849
Decaborane(14), 2838
Decaboron Tetradecahydride *see* 2838
Decacarbonyldimanganese *see* 5609
Decachlor *see* 3095
1,1',2,2',3,3',4,4',5,5'-Decachlorobi-2,4-cyclopentadien-1-yl *see* 3095
1,1a,3,3a,4,5,5,5a,5b,6-Decachloro-octahydro-1,3,4-metheno-2*H*-cyclo-buta[*cd*]pentalen-2-one *see* 2081
Decacil *see* 2082
Decacortin *see* 2922
Decaderm [Merck & Co.] *see* 2922
Decadron [Merck & Co.] *see* 2922
Decadron-LA [Merck & Co.] *see* 2922
Decadron TBA [Merck & Co.] *see* 2922
Deca-Durabol *see* 6281
Deca-Durabolin [Organon] *see* 6281
Deca-Hybolin [Hyrex] *see* 6281

Decahydro-3,5,1,7-[1,2,3,4]butane-
tetraylnaphthalene see 2494
Decahydro-6β,9aα-dihydroxy-5aβ-
methyl-3,9-bis(methylene)naphtho-
[1,2-b]furan-2(3H)-one see 9020
1,6,7,8,9,11a,12,13,14,14a-Decahydro-
1,13-dihydroxy-6-methyl-4H-cyclo-
pent[f]oxacyclotridecin-4-one see
1363
(1R)-1,2,3,4,4a,4bα,5,9,10,10aα-Deca-
hydro-1,4aβ-dimethyl-7-(1-methyl-
ethyl)-1α-phenanthrenecarboxylic
Acid see 5348
[1R-(1α,4aβ,4bα,10aα)]-1,2,3,4,4a,4b,-
5,6,10,10a-Decahydro-1,4a-dimeth-
yl-7-(1-methylethyl)-1-phenan-
threnecarboxylic Acid see 2
[1R-(1α,4aβ,10aα)]-1,2,3,4,4a,5,6,9,-
10,10a-Decahydro-1,4a-dimethyl-7-
(1-methylethyl)-1-phenanthrenecar-
boxylic Acid see 6950
3a,4,5,6a,7,7a,8,11a,11b,11c-Decahy-
dro-5-hydroxy-2,10-dimethoxy-
3,8,11a,11c-tetramethylphenanthro-
[10,1-bc]pyran-1,11-dione see 6379
Decahydro-2-hydroxy-3a,5-dimethyl-
3-methylenecyclopenta[4,5]penta-
leno[1,6a-b]oxirene-5-carboxylic
Acid see 4637
1,2,3,4,4a,4b,7,9,10,10a-Decahydro-2-
hydroxy-2,4b-dimethyl-7-oxo-1-
phenanthrenepropionic Acid δ-Lactone see 9108
1,2,3,4,7a,10,11,11a,12,13-Decahy-
dro-14-(1-hydroxyethyl)-2,11a-di-
methyl-7H-9,11b-epoxy-13a,5a-
propenophenanthro[2,1-f][1,4]-
oxazepine-9,12(8H)-diol see 1019
3-[2-[Decahydro-6-hydroxy-5-(hydroxymethyl)-5,8a-dimethyl-2-meth-
ylene-1-naphthalenyl]ethylidene]-
dihydro-4-hydroxy-2(3H)-furanone
see 666
[3R-(3α,5aβ,6β,8aβ,9α,10α,12β,-
12aR*)]-Decahydro-10-methoxy-
3,6,9-trimethyl-3,12-epoxy-12H-
pyrano[4,3-j]-1,2-benzodioxepin see
842
(2aα,4aβ,5β,6β,8aα,12bβ,12cβ,12dβ)-
(−)-2a,3,4,4a,5,6,7,8a,12b,12c-
Decahydro-6-methyl-2H-5,12d-
ethanofuro[4',3',2':4,10]anthra[9,1-
bc]oxepin-2,9,12-trione see 7509
Decahydronaphthalene see 2839
Decahydro-2-oxo-1,3-bis(phenylmethyl)thieno[1',2':1,2]thieno[3,4]-
imidazol-5-ium, Salt with (+)-7,7-
Dimethyl-2-oxobicyclo[2.2.1]hep-
tane-1-methanesulfonic Acid (1:1)
see 9621
1,1a,1b,4,4a,7a,7b,8,9,9a-Decahydro-
4a,7b,9,9a-tetrahydroxy-3-(hydroxymethyl)-1,1,6,8-tetramethyl-5H-
cyclopropa[3,4]benz[1,2-e]azulen-5-
one see 7306
[4aS-(4aα,6aα,11bα,13aR*,13bα)]-
1,2,3,4,4a,5,6,6a,11b,13b-Decahy-
dro-4,4,6a,9-tetramethyl-13H-ben-
zo[a]furo[2,3,4-mn]xanthen-11-ol
see 8432
Decahydro-1,1,4,7-tetramethyl-1H-
cycloprop[e]azulen-4-ol see 5313
3,4,5,6,7,8,9,10,11,12-Decahydro-
7,14,16-trihydroxy-3-methyl-1H-2-
benzoxacyclotetradecin-1-one see
10022
Decahydro-4a,7,9-trihydroxy-2-meth-
yl-6,8-bis(methylamino)-4H-py-
rano[2,3-b][1,4]benzodioxin-4-one
see 8695

Decahydro-4,8,8-trimethyl-9-methyl-
ene-1,4-methanoazulene see 5446
Decahydro-1,1,4a-trimethyl-2-naph-
thalenol see 9410
Decalin®, 2839
Decalix [Parmed] see 2922
Decamethonium Bromide, 2840
Decamethrin see 2869
Decamethylcyclopentasiloxane, 2841
1,1'-Decamethylenebis[4-aminoquin-
aldinium Acetate] see 2895
1,1'-Decamethylenebis[4-amino-
quinaldinium Chloride] see 2896
Decamethylene-α-ω-bis[1-(3',4'-di-
methoxybenzyl)-1,2,3,4-tetrahydro-
6,7-dimethoxy-2-methylisoquinoli-
nium Methosulfate] see 5250
Decamethylenebis[m-dimethylamino-
phenyl N-Methylcarbamate] Dimethobromide see 2871
Decamethylenebis[N-methylcarbamic
Acid m-Dimethylaminophenyl Es-
ter] Bromomethylate see 2871
Decamethylenebis[1,2,3,4-tetrahydro-
6,7-dimethoxy-1-(3,4-dimethoxy-
benzyl)-2-methylisoquinolinium
Methyl Sulfate] see 5250
2,2'-Decamethylenebis[1,2,3,4-tetra-
hydro-6,7-dimethoxy-2-methyl-1-
veratrylisoquinolinium Methyl Sul-
fate] see 5250
Decamethylenebis[trimethylammo-
nium Bromide] see 2840
N¹,N¹'-Decamethylene-N⁴,N⁴'-deca-
methylenebis[4-aminoquinaldinium
Chloride] see 1259
1,1'-Decamethylene-4,4'-(1,10-deca-
methylenediimino)bis[quinaldinium
Chloride] see 1259
2,2'-(Decamethylenedithio)diethanol
see 9349
Decamethylene Glycol, 2842
Decamethyltetrasiloxane, 2843
Decamine see 2896
Decanedioic Acid see 8369
Decanedioic Acid Bis(2-ethylhexyl)
Ester see 1263
1,10-Decanediol see 2842
1,1'-(1,10-Decanediyl)bis-[4-amino-
2-methylquinolinium Chloride] see
2896
1,1'-(1,10-Decanediyl)bis[4-amino-2-
methylquinolinium Diacetate] see
2895
2,2'-(1,10-Decanediyl)bis[1-[(3,4-
dimethoxyphenyl)methyl]-1,2,3,4-
tetrahydro-6,7-dimethoxy-2-meth-
ylisoquinolium] Bis(methyl Sulfate)
see 5250
3,3'-[1,10-Decanediylbis[(methylimino)carbonyloxy]]bis[N,N,N-trimeth-
ylbenzenaminium]dibromide see
2871
2,2'-[1,10-Decanediylbis(thio)]bis-
ethanol see 9349
Decanoic Acid see 1759
Decanoic Acid Ethyl Ester see 3734
1-Decanol see 2847
Decapeptyl [Ferring] see 9662
Decaprednil [Dorsch] see 7719
Decapryn Succinate [Merrell] see 3430
Decaps see 9928
(3-Decarbamoyloxy)-3-hydroxyven-
turicidin A see 9848
Decarboxycysteine see 2785
Decarboxycystine see 2783
Decardil [Trenker] see 3150
Decasone [Merck & Co.] see 2922
Decaspir [Pulitzer] see 3942
Decaspiride see 3942

2,4,6,8-Decatetraenedioic Acid Mono-
[4-(1,2-epoxy-1,5-dimethyl-4-hex-
enyl)-5-methoxy-1-oxaspiro[2.5]-
oct-6-yl] Ester see 4199
2,4,6,8-Decatetraenedioic Acid Mono-
[5-methoxy-4-[2-methyl-3-(3-
methyl-2-butenyl)oxiranyl]-1-oxa-
spiro[2.5]oct-6-yl] Ester see 4199
Deccotane see 1544
Deccox [Merrell] see 2846
1-Decene-1,10-dicarboxylic Acid see
9493
Decentan [E. Merck] see 7135
Dechlorane see 6126
Dechlorane® Plus [Hooker], 2844
Decholin [Dome] see 2858
Decholin Sodium [Dome] see 2858
Decicain see 9123
Decimemide, 2845
Decis [Roussel-UCLAF] see 2869
Deckor see 7563
Declid see 9760
Declinax [Roche] see 2837
Declomycin [Lederle] see 2872
Decme [Spitzner] see 3598
Decoderm Sine Gentamycin [E.
Merck] see 4118
Decolorized Phenolphthalein see 7210
Decomoton [Aesculaap] see 1800
Decontractyl see 5737
Decoquinate, 2846
Decorenone [Lifepharma] see 9751
Decorpa [Norgine] see 4486
Decortancyl see 7727
Decortilen [E. Merck] see 7729
Decortilen Soluble [E. Merck] see
7730
Decortin [E. Merck] see 7727
Decortin [Schieffelin] see 2883
Decortin H [E. Merck] see 7719
Decortisyl [Roussel] see 7727
Decosterone see 2883
Decrelip [Ferrer] see 4280
Decreten [Dumex] see 7412
Decril [Mediolanum] see 3598
Dectancyl see 2922
Decumbin see 1363
n-Decyl Alcohol, 2847
N-Decyl-N,N-dimethyl-1-decanamini-
um Chloride see 3088
Decyl(2-hydroxyethyl)dimethylammo-
nium Bromide 1-Adamantanecarb-
oxylate see 381
2-Decyl-3-(5-methylhexyl)oxirane see
3367
4-(Decyloxy)-3,5-dimethoxybenzam-
ide see 2845
6-Decyloxy-7-ethoxy-4-hydroxy-3-
quinolinecarboxylic Acid Ethyl
Ester see 2846
DEDC see 3384
Dedevap [Bayer] see 3069
Dedrogyl [Roussel-UCLAF] see 1638
DeDTC see 3384
Dedyl see 3182
DeeO see 4354
Dee-Ron see 9928
Deer's Tongue see 5355
Deet, 2848
Defekton [Yoshitomi] see 1869
Defencin [Bristol] see 5128
Defenidol see 3311
Defensins, 2849
Deferoxamine, 2850
Defibrase [Pharma-Zentrale] see 1020
Defibrinotide see 2851
Defibrotide, 2851
Deficol [Vangard] see 1253
Défiltran [Jouveinal] see 45
Deflamene [Farmitalia] see 4156
Deflamon [SPA] see 6079

Deflan [Guidotti] see 2852
Deflazacort, 2852
Deflexol see 10098
Deflorin [Glaxo] see 1975
Defluina [Simes] see 3598
De-Fol-Ate [Pennwalt] see 8543
Defosfamide, 2853
Degalol [Ames] see 2881
Deganol see 2872
Degener's Indicator see 7156
Deglycobleomycin see 1324
De Graafina see 3655
Degranol see 5631
Dégryp [Houd] see 254
Deguelia Root see 2898
Deguelin, 2854
Dehacodin see 3154
Dehistin see 9651
Dehychol [Rexall] see 2858
Dehydol [Henkel] see 7554
Dehydrated Alcohol see 3716
Dehydrite see 5557
Dehydroacetic Acid, 2855
trans-Dehydroandrosterone see 7710
Dehydroascorbic Acid, 2856
Dehydrobenzperidol [J & J] see 3437
Dehydrobilirubin see 1236
11-Dehydro C see 280
7-Dehydrocholesterol, 2857
24-Dehydrocholesterol see 2905
Dehydrocholic Acid, 2858
Dehydrocholic Acid Salt of Choline
  see 2209
Dehydrocollinusin see 5154
11-Dehydrocorticosterone, 2859
Δ^1-Dehydrocortisol see 7719
Δ^1-Dehydrocortisone see 7727
Dehydroemetine, 2860
2-Dehydroemetine see 2860
2,3-Dehydroemetine see 2860
Dehydroepiandrosterone see 7710
Dehydroergosterol, 2861
1-Dehydro-9α-fluorohydrocortisone
  see 5058
Dehydrofolliculinic Acid see 3407 Δ^1-
  Dehydrohydrocortisone see 7719
11-Dehydro-17-hydroxycorticoster-
  one see 2533
6-Dehydro-17α-hydroxy-16-methyl-
  eneprogesterone see 4763
Dehydroiopanoic Acid see 2284
Dehydroisoandrosterone see 7710
2,3-Dehydroisoemetine see 2860
6-Dehydro-6-methyl-17α-acetoxypro-
  gesterone see 5687
1-Dehydro-16α-methyl-9α-fluorohy-
  drocortisone see 2922
1-Dehydro-6α-methylhydrocortisone
  see 6028
1-Dehydro-17α-methyltestosterone
  see 5862
Dehydronivalenol see 9947
Dehydroperillic Acid see 9325
6-Dehydro-retro-progesterone see
  3454
3-Dehydroretinal, 2862
Dehydroretinol see 9919
7-Dehydrositosterol, 2863
1-Dehydrotestololactone see 9108
Dehydrotestosterone see 1327
1-Dehydrotestosterone 17-Cyclopent-
  1'-enyl Ether see 8063
6-Dehydrotestosterone-17α-propionic
  Acid γ-Lactone see 1753
Deidrocolico Vita see 2858
Dejo see 3376
Dekacort see 2922
Dekadin see 2896
DeKalin see 2839
Dekamin see 2896
Dekelmin see 6084

Dekortin see 7727
Dekrysil see 3272
DEL 1267 see 6063
Delacillin [Sankyo] see 610
Delacurarine [Squibb] see 9717
Deladroxone (obsolete) see 230
Delalutin [Squibb] see 4774
Delan see 3375
Delapril, 2864
Delatestryl [Squibb] see 9112
Delaxin [Ferndale] see 5903
Delcaine see 7926
Delcortin see 7727
Delcosine see 2868
Delesan see 8486
Delestrec [Squibb] see 3653
Delestrogen [Squibb] see 3653
Delfen [Ortho] see 6596
Delgesic [KarlsPharma] see 5510
Delgesic [Rhône Poulenc] see 873
Delimmun [Delalande/Newport] see
  4881
Delinal [Merrell] see 7824
Delipid [Coop. Farm.] see 9349
Delmadinone Acetate, 2865
Delmeson [Hoechst] see 4104
Delnav [Hercules] see 3296
Delonal [Byk-Essex] see 213
m-Delphene see 2848
Delphicort [Lederle] see 9511
Delphidenolon 1575 see 6244
Delphimix [Cyanamid] see 3071
Delphin see 2866
Delphinic Acid see 5120
Delphinidin, 2866
Delphinidol see 2866
Delphinin see 2866
Delphinine, 2867
Delphinium see 5242
Delphoside see 2866
Delpregnin [Novo] see 5687
Delsoline, 2868
Delsterol see 9929
Delsym [McNeil] see 8116
Delta-Corlin see 7727
Delta-Cortef [Upjohn] see 7719
Delta-Cortelan see 7727
Deltacortene Beta see 5750
Deltacortenolo see 7719
Deltacortisone see 7727
Deltacortone [Merck & Co.] see 7727
Deltacortril [Pfizer] see 7719
Deltacortril DA see 7720
Delta E see 7727
Delta F see 7719
Deltafluorene [Lepetit] see 2922
Delta Gluconolactone see 4351
Deltalin [Lilly] see 9928
Deltamethrin, 2869
Deltamine see 7021
Deltamycin A$_4$ see 1813
Deltan [Berna] see 3247
Delta Prenovis see 7727
Delta Sleep Factor see 3445
Delta Sleep-inducing Peptide (Rabbit)
  see 3445
Delta Sleep Peptide see 3445
Deltasone [Upjohn] see 7727
Deltastab [Boots] see 7719
Delta-1-testololactone see 9108
Deltathione [Tobishi] see 4369
Deltisolone see 7719
Deltisona B see 5750
Deltisone see 7727
Deltoin [Sandoz] see 5889
Deltra [Merck & Co.] see 7727
Delursan [Houd] see 9801
Delvex [Lilly] see 3376
Delvinal Sodium [Merck & Co.] see
  9886
Delvomycin see 6138

Delvoprim [Gist-Brocades] see 8874
Demanol Aceglumate see 2835
Demanyl Phosphate, 2870
Demasorb [Squibb] see 3247
Demavet [Squibb] see 3247
Demecarium Bromide, 2871
Demeclocycline, 2872
Demecolcine, 2873
Demegestone, 2874
Demelon see 6280
Demerol Hydrochloride [Winthrop]
  see 5736
Demeso [Merck & Co.] see 3247
Demethocaine [Elder] see 8116
6-Demethoxyageratochromene see
  7716
2-Demethoxy-2-glucosidoxythiocol-
  chicine see 9253
5-O-Demethylavermectin A$_{1a}$ and
  5-O-Demethyl-25-de(1-methylprop-
  yl)-25-(1-methylethyl)avermectin
  A$_{1a}$ (4:1) see 1
1-Demethylcalycanthidine see 2048
Demethylchlortetracycline (obsolete)
  see 2872
4-O-Demethyldaunorubicin see 1879
(6R,25R)-5-O-Demethyl-28-deoxy-
  6,28-epoxy-25-(1-methylethyl)mil-
  bemycin B see 6112
(6R,25R)-5-O-Demethyl-28-deoxy-
  6,28-epoxy-25-methylmilbemycin B
  see 6112
6-Demethyl-6-deoxytetracycline see
  8316
S-Demethyl-3'-depropyl-S-[2-[(2-
  hydroxybenzoyl)oxy]ethyl]-7-O-
  methyllincomycin see 1954
40-Demethyl-3,7-dideoxo-3,7-dihydr-
  oxy-N see 5733
5-O-Demethyl-22,23-dihydroaver-
  mectin A$_{1a}$ see 5133
Demethyldihydrothebaine Acetate
  (Ester) see 3155
Demethyldopan see 9762
(6S,7S,8R,12S,15R,16R,19S,22S,-
  23R,24S,26R,27S)-4-Demethylene-
  22,24-dimethyltetronomycin see
  9176
4'-Demethylepipodophyllotoxin 9-
  [4,6-O-Ethylidene-β-D-glucopyran-
  oside] see 3842
4'-Demethylepipodophyllotoxin 9-
  (4,6-O-2-Thenylidene-β-D-gluco-
  pyranoside) see 9078
4'-Demethylepipodophyllotoxin-β-D-
  thenylidine Glucoside see 9078
Demethylhomopterocarpin see 5673
5-O-Demethyl-25-de(1-methylprop-
  yl)-22,23-dihydro-25-(1-methyl-
  ethyl)avermectin A$_{1a}$ see 5133
6-Demethylmevinolin see 6088
26-Demethyloligomycin A see 6793
N-Demethylorphenadrine see 9426
4-Demethyl-3-oxovobasan-17-oic
  Acid Methyl Ester see 7130
O^3-Demethylthebaine see 6821
N-Demethylthiolutin see 4647
Demeton, 2875
Demeton-methyl see 5971
Demetraciclina [Bios] see 2872
Demetrin [Gödecke] see 7713
Demexiptiline, 2876
Demidone see 4769
Demigran [Leo Pharm] see 4064
Democracin [Inter-Alia] see 9130
Demolox [Lederle] see 609
Demon [ICI] see 2775
Demorphan Hydrobromide see 8116
Demosulfan see 8909
Demotil see 3302
Demoxytocin see 2833

O-3-Deoxy-4-C-methyl-3-(methyl-amino)-β-L-arabinopyranosyl-(1→6)-O-[2,6-diamino-2,3,4,6-tetradeoxy-α-D-*glycero*-hex-4-enopyranosyl-(1→4)]-2-deoxy-D-streptamine *see* 8498

O-3-Deoxy-4-C-methyl-3-(methyl-amino)-β-L-arabinopyranosyl-(1→6)-O-[2,6-diamino-2,3,4,6-tetradeoxy-α-D-*erythro*-hexopyra-nosyl-(1→4)]-2-deoxy-D-strept-amine *see* 4284

2-Deoxy-2-[[(methylnitrosoamino)-carbonyl]amino]-D-glucopyranose *see* 8794

2-Deoxy-2-(3-methyl-3-nitrosourei-do)-D-glucopyranose *see* 8794

4-Deoxy-4'-methylpyrido[1',2'-1,2]-imidazo[5,4-c]rifamycin SV *see* 8218

[2R-(2α,3β,4α,5β,6α,11β,13α,14α)]-11-[(6-Deoxy-3-C-methyl-2,3,4-tri-O-methyl-α-L-mannopyranosyl)-oxy]-4-(dimethylamino)-3,4,5,6,9,-11,12,13,14,16-decahydro-3,5,8,10,-13-pentahydroxy-6,13-dimethyl-9,16-dioxo-2,6-epoxy-2H-naphtha-ceno[1,2-b]oxocin-14-carboxylic Acid Methyl Ester *see* 6591

Deoxynivalenol *see* 9947

1-Deoxynojirimycin, 2887

Deoxynybomycin *see* 6653

12-Deoxyoligomycin A *see* 6793

9-Deoxy-9-oxoleucomycin V 3-Acetate 4^B-(3-Methylbutanoate) *see* 1813

9-Deoxy-9-oxoleucomycin V 3,4^B-Dipropanoate *see* 6106

9-Deoxy-9-oxo-1,8-secocinchonine *see* 2291

α-6-Deoxyoxytetracycline Monohy-drate *see* 3429

1-(2-Deoxy-α-D-*erythro*-pentofurano-syl)-5-fluorouracil *see* 4045

1-(2-Deoxy-α-D-*erythro*-pentofurano-syl)-5-iodouracil *see* 4819

(R)-3-(2-Deoxy-β-D-*erythro*-pento-furanosyl)-3,6,7,8-tetrahydroimida-zo[4,5-d][1,3]diazepin-8-ol *see* 7091

1-(2-Deoxy-α-D-*erythro*-pentofuran-syl)uracil *see* 2892

1-(2-Deoxy-β-D-*erythro*-pentofurano-syl)uracil *see* 2892

D-*erythro*-2-Deoxypentose *see* 2890

2-Deoxy-D-*erythro*-pentose *see* 2890

12-Deoxypicromycin *see* 6340

1-(2-Deoxy-β-D-ribofuranosyl)-5-fluorouracil *see* 4045

1-(β-D-5'-Deoxyribofuranosyl)-5-fluorouracil *see* 3426

1-(2-Deoxy-α-D-ribofuranosyl)-5-iodouracil *see* 4819

1-(2-Deoxy-β-D-ribofuranosyl)-5-iodouracil *see* 4819

1-(2-Deoxy-β-D-ribofuranosyl)-5-methyluracil *see* 9331

1-(2-Deoxy-α-D-ribofuranosyl)uracil *see* 2892

1-(2-Deoxy-β-D-ribofuranosyl)uracil *see* 2892

Deoxyribonuclease (Pancreatic), 2888

Deoxyribonucleic Acid, 2889

D-2-Deoxyribose, 2890

(±)-15-Deoxy-(16RS)-16-hydroxy-16-methyl-PGE1 Methyl Ester *see* 6129

1-[4-Deoxy-4-(sarcosyl,D-seryl)-amino-β-D-glucopyranuronamide]-cytosine *see* 4436

2-Deoxystreptamine, 2891

2'-Deoxy-5-(trifluoromethyl)uridine *see* 9599

11-Deoxy-11α,16,16-trimethyl-PGE$_2$ *see* 9637

Deoxyuridine, 2892

2'-Deoxyuridine *see* 2892

Depakene [Abbott] *see* 9821

Depakin [Sigma-Tau] *see* 9821

Dépakine [Labaz] *see* 9821

Depakote [Abbott] *see* 9821

Dépamide [Labaz; Sigma-Tau] *see* 9822

Depamine [Berk] *see* 7029

Deparal *see* 9929

Deparkin *see* 3098

Deparon [Aron] *see* 2876

Depas [Yoshitomi] *see* 3830

Depasan [Giulini] *see* 8692

DEPC *see* 8008

Depen [Wallace Labs.] *see* 7029

Depepsen [Searle] *see* 8520

D-Epifrin [Allergan] *see* 3344

Depersolone [Gedeon Richter] *see* 5644

Depigman [Hermal] *see* 6159

Depil [Calmic] *see* 1718

Depinar [Armour Pharm.] *see* 9923

Depixol [Lundbeck] *see* 4114

Depleil *see* 9124

Depo-Cer-O-Cillin Chloroprocaine [Upjohn] *see* 7044

Depocid *see* 8910

Depocillin [Brocades] *see* 7042

Depo-Clinovir [Upjohn] *see* 5677

Depocolin-S [Ohta] *see* 7913

Depocural *see* 2346

Depoestradiol [Upjohn] *see* 3656

Depofemin [Hoechst] *see* 3656

Depogamma [Ankermann] *see* 4739

Depo-Insulin *see* 7899

Depo-Medrate [Upjohn] *see* 6028

Depo-Medrol [Upjohn] *see* 6028

Depo-Medrone [Upjohn] *see* 6028

Depomide *see* 8897

Deponit [Pharma Schwarz] *see* 6528

Depo-Provera [Upjohn] *see* 5677

Deporone [Hyrex-Key] *see* 5677

Deposal [Benvegna] *see* 8302

Depostat [Schering AG] *see* 4309

Deposulin [Upjohn] *see* 7899

Depo-testosterone *see* 9111

Depot-Glumorin *see* 5159

Depot-Novadral [Diwag] *see* 6616

Depotocin [Spofa] *see* 1800

Depot-Oestromenine [E. Merck] *see* 3198

Depot-Oestromon *see* 3198

Depot-Salicyl [Fischer] *see* 6195

Depovernil *see* 8890

Depovirin [Hoechst] *see* 9111

Depracer [Lepori] *see* 3841

Deprancol [Substancia] *see* 7851

Deprelin *see* 8843

Deprenalin *see* 2893

Deprenil *see* 2893

Deprenyl, 2893

Depressan [Apogepha] *see* 4682

Depressan (Tabl) [Apogepha] *see* 3152

Depressan (Amp) [Apogepha] *see* 3152

Depressin [UPSA] *see* 7846

Depressin [Lentia] *see* 4609

Depresym *see* 3419

Depreton *see* 6035

Deprex [Mowatt & Moore] *see* 504

Depridol *see* 5852

Deprinol *see* 4835

Depromic *see* 7851

Depsococaine *see* 7926

Deptropine, 2894

Dequadin Acetate [Allen & Hanburys] *see* 2895

Dequadin Chloride [Allen & Han-burys] *see* 2896

Dequafungan [Kreussler] *see* 2896

Dequalinium Acetate, 2895

Dequalinium Chloride, 2896

Dequavagyn [Kreussler] *see* 2896

Dequavet *see* 2896

Deracil *see* 9298

Deracyn [Upjohn] *see* 150

Deralbine [Andromaco] *see* 6101

Deralin [ABIC] *see* 7852

Derantel [Nippon Chemiphar] *see* 1971

De-Rat Concentrate [Rentokil] *see* 9928

Deratol [Brewer] *see* 9928

Derbac [International] *see* 1789

Derbac-M [International] *see* 5582

Dereuma *see* 446

Derfon *see* 3116

Dergotamine [Abbott] *see* 3157

Derizene [Hollister-Stier] *see* 7293

D-Ergotox Forte L.U.T. [Pharmafrid] *see* 3598

Dergramin *see* 2922

Dermabet [Taro] *see* 1202

Dermacid *see* 5759

Dermacort [Rowell] *see* 4710

Dermadex [Alconox] *see* 4602

Dermaflor [Brocchieri] *see* 3126

Dermafur [Eaton] *see* 4211

Dermagan *see* 2940

Dermairol [Roche] *see* 8167

Dermalar *see* 4076

Dermaplus [Ripari-Gero] *see* 4077

Dermasorb *see* 3247

Dermastatin [Cooper] *see* 2897

Dermatan Sulfate *see* 2217

Dermatol [Hoechst] *see* 1297

Dermaton [Burroughs Wellcome] *see* 2087

Dermatop [Cassella] *see* 7717

Dermevan *see* 4930

Dermolate [Schering] *see* 4710

Dermolen [Syntex] *see* 4710

Dermonistat [Ortho] *see* 6101

Dermosol [Iwaki] *see* 1202

Dermostatin, 2897

Dermostatin A *see* 2897

Dermostatin B *see* 2897

Dermotricine [Roger Bellon] *see* 9749

Dermoval *see* 2361

Dermovaleas [Valeas] *see* 1202

Dermovate [Glaxo] *see* 2361

Dermoxin [Glaxo] *see* 2361

Dermoxinale [Glaxo] *see* 2361

Deronil [Schering] *see* 2922

Derosal [Hoechst] *see* 1794

Derride *see* 3511

Derringer [Fairfield] *see* 7446

Derris Root, 2898

DES *see* 3118

Desacchromin Dispersion *see* 7474

Desace [Dexo] *see* 2903

Desacetylbufotalin *see* 1464

Desacetyldigilanide C *see* 2903

Desacetylvinblastine Amide *see* 9892

Desaci [Simes] *see* 2903

Desagiybuzole *see* 4374

1-Desamino-8-D-arginine Vasopressin *see* 2904

1-Desamino-1-monocarba-[2-tyr-(OMe)]-OT *see* 1800

Desaminooxytocin *see* 2833

Desanden [Max Ritter] *see* 1128

Desaspidin, 2899

Desatrine, 2900

Desatrine-3-(N,N-diethylaminoace-tate) *see* 2900

Deschlorobiomycin *see* 9130

Desclidium [S.P.R.E.T.] *see* 9908

Descocin [Kanto] see 9230
Descorterone see 2883
Descotone see 2883
Desdemin [Vitacain] see 4221
Déselmine see 3376
Deseptyl see 8898
Deseril [Sandoz] see 6055
Désernil-Sandoz [Sandoz] see 6055
Deserol see 1405
Deseronil see 2922
Deserpidine, 2901
Desfedrin see 5859
Desferal [Ciba] see 2850
Desferricoprogen see 2517
Desferrioxamine B see 2850
Desferrioxamine Mesylate see 2850
Desglucotransvaaline see 7889
Des-Gly ^{10}-NH$_2$-LH-RH-ethylamide see 4010
Desicol [Parke, Davis] see 6884
Desipramine, 2902
Desitriptilina see 6635
Desiver [USV] see 5427
Deslanoside, 2903
Desmel [Ciba-Geigy] see 7830
Desmesterol see 2905
11-Desmethoxyreserpine see 2901
Desmethylamitriptyline see 6635
N-Desmethylcodeine see 6607
Desmethyldiazepam see 6608
Desmethyldopan see 9762
Desmethylemetine see 1970
Desmethylimipramine see 2902
Desmethylmethadone see 6628
Desmethylmorphine see 6630
Desmethylnarcotine see 6343
5-O-Desmethyltangeretin see 4265
Desmofosfamide see 2853
Desmopressin, 2904
Desmospray [Ferring] see 2904
Desmosterol, 2905
Desmycosin see 9740
2,4-DES-Na see 3372
Desogestrel, 2906
Desomedine see 4612
Desomorphine, 2907
Desonide, 2908
Desopam [Mochida] see 9611
Desosamine, 2909
Desoximetasone, 2910
Desoxiribon see 2889
Desoxostrophanthidin see 7128
S-(5'-Desoxyadenosin-5'-yl)-L-methionine see 146
4-Desoxyadermin Hydrochloride see 2913
Desoxyalizarin see 718
2-Desoxy-D-altromethylose see 3147
Desoxycholic Acid see 2881
Desoxycorticosterone see 2882
Desoxycorticosterone Acetate see 2883
11-Desoxycortisone see 2912
Desoxycortone see 2882
Desoxycortone Acetate see 2883
14-Desoxy-14-[(2-diethylaminoethyl)-mercaptoacetoxy]mutilin see 9351
d-Desoxyephedrine see 5859
Desoxyepinephrine see 2885
Desoxyfed see 5859
21-Desoxy-9α-fluoro-6α-methylpred-nisolone see 4104
6-Desoxy-D-glucosamine, 2911
D-arabino-2-Desoxyhexose see 2886
11-Desoxy-17-hydroxycorticosterone, 2912
Desoxymethasone see 2910
6-Desoxy-6-methylenenaltrexone see 6274
21-Desoxy-6α-methyl-9α-fluoropred-nisolone see 4104

Desoxymycin [Kaken] see 2884
Desoxyn [Abbott] see 5859
3-Desoxynorlutin see 5501
2-Desoxyphenobarbital see 7753
4-Desoxypyridoxine Hydrochloride, 2913
5-Desoxyquercetin see 4026
Desoxyribonuclease (Pancreatic) see 2888
Desoxyribonucleic Acid see 2889
Desoxyribose see 2890
DESP see 3942
Despacilina [Squibb] see 7042
Desposal see 6195
De-Squaman [Hermal] see 8004
Dessin [Union Carbide] see 3280
Desthiobiotin, 2914
Destim [Central Pharm.] see 5859
Destolit [Sandoz S.A.R.L.] see 9801
Destomycin A, 2915
Destonate 20 [Meiji] see 2915
Destriol see 3659
Destrone see 3660
Destun [3M] see 7113
N-Desulfoheparin see 4570
Desuric [Labaz; Sigma-Tau] see 1073
Desyrel [Mead Johnson] see 9495
m-DETA see 2848
Detal see 3272
Detalup [Bayer] see 9928
Detamide see 2848
Detantol [Eisai] see 1477
Detaril see 7455
Detaxtran, 2916
Detensiel [Merck-Clevenot] see 1309
Detergent Alkylate #5 see 9575
Dethylandiamine see 9207
Dethyrona [Baxter] see 9348
Detia [Freyberg] see 363
Deticene [Roger Bellon] see 2803
Detigon [F.B.A. Pharmaceuticals] see 2053
DET MS [Rentschler] see 3157
Detomidine, 2917
Detoxepa [Ayerst] see 9375
Detoxin C$_1$ see 2918
Detoxin C$_{a1}$ see 2918
Detoxin D$_1$ see 2918
Detoxin Complex, 2918
Detoxinine (β,3-Dihydroxy-2-pyrroli-dinepropanoic Acid) see 2918
Detraine see 7815
Detravis [Vis] see 2872
Detrovel [Geigy] see 39
Détryptoréline see 9662
Dettol [Reckitt & Colman] see 2176
Detulin [Woelm] see 9932
Detyroxin see 9348
Deumacard see 7097
Deursil [Giuliani] see 9801
Deuterium, 2919
Deuterium Oxide, 2920
Devarda's Alloy see 2921
Devarda's Metal, 2921
Devaricin see 4630
Devegan [Hoechst] see 44
Develin [Gödecke] see 7851
Deverol [Waldheim] see 8721
Devil's Apple see 8779
Devil's Dung see 846
Devincan [Gedeon Richter] see 9888
Devonium [Perga] see 7484
Devrinol [Stauffer] see 6336
Devryl [SK & F] see 2379
Dexabene [Merckle] see 2922
Dexacilina [Squibb] see 3566
Dexacillin [Squibb] see 3566
Dexacortal [Pharmacia] see 2922
Dexa-Cortidelt [Roussel-UCLAF] see 2922
Dexacortin [Streuli] see 2922

Dexa-Cortisyl see 2922
Dexafarma [Llano] see 2922
Dexal [Pulitzer] see 5184
Dexalone see 2932
Dexa-Mamallet [Showa] see 2922
Dexambutol [Sobio] see 3672
Dexameth [USV] see 2922
Dexamethasone, 2922
Dexamethasone tert-Butylacetate see 2922
Dexamethasone 21-(Dihydrogen Phosphate) Disodium Salt see 2922
Dexamethasone 21-(4-Pyridinecarboxylate) see 2922
Dexamisole see 9161
Dexampex [Lemmon] see 2932
Dexamphetamine see 2932
Dexapos [Ursapharm] see 2922
Dexa-Scheroson [Schering AG] see 2922
Dexa-sine [Thilo] see 2922
Dexasone see 2922
Dexawin [Winthrop] see 9230
Dexbenzetimide see 2923
Dexbrompheniramine see 1433
Dexchlorpheniramine see 2180
Dexcyanidanol see 1908
Dexedrine Sulfate [SK & F] see 2932
Dexetimide, 2923
Dexfenfluramine see 3920
Dexide [Fargal] see 2916
Dexindoprofen see 4875
Dexinolon [Desitin] see 2922
Dexinoral [Desitin] see 2922
Dexium [Delalande] see 3395
Dexivacaine see 5748
Dexnorgestrel (obsolete) see 6621
Dexnorgestrel Acetime see 6620
Dexol see 8604
Dexoval [Vale] see 5859
Dexoxadrol Hydrochloride see 3293
Dexpanthenol, 2924
Dextelan [Glaxo] see 2922
Dexten see 2932
Dextran, 2925
Dextran 40 see 2925
Dextran 70 see 2925
Dextran 75 see 2925
Dextranase, 2926
Dextran 2-(Diethylamino)ethyl 2-[[2-(Diethylamino)ethyl]diethylammonio]ethyl Ether Chloride Hydrochloride Epichlorohydrin Crosslinked see 7531
Dextran 2-(Diethylamino)ethyl Ether see 2916
Dextran 2,3-Dihydroxypropyl 2-Hydroxy-1,3-propanediyl Ethers see 2928
Dextran Iron Complex, 2927
Dextranomer, 2928
Dextran Sulfate Sodium, 2929
Dextran Sulfuric Acid Ester Sodium Salt see 2929
Dextrarine see 2929
Dextraven [Fisons] see 2925
Dextri-Maltose® see 2930
Dextrin, 2931
Dextroamphetamine Sulfate, 2932
Dextrobenzetimide see 2923
Dextrocaine see 7926
Dextrocamphoric Acid see 1739
Dextrofenfluramine see 3920
Dextroid see 9348
Dextromethorphan Hydrobromide see 8116
Dextromoramide, 2933
Dextronic Acid see 4350
Dextropimaric Acid see 7398
Dextropropoxyphene see 7851
Dextropur see 4353

Dextrorotatory Lactic Acid *see* 5216
Dextrorphan *see* 5350
Dextrose *see* 4353
Dextrosol *see* 4353
Dextrosulphenidol *see* 9230
Dextrotartaric Acid *see* 9039
Dextrothyroxine Sodium *see* 9348
Dexulate *see* 2929
Dezocine, 2934
Dezone [Tutag] *see* 2922
DF 118 *see* 3154
DFDD, 2935
DFDT, 2936
DFMO *see* 3489
DFOM *see* 2850
DFP *see* 5060
5'-DFUR *see* 3426
5'-DFUrd *see* 3426
DFV *see* 3129
DH 245 *see* 4209
DH 581 *see* 7761
DHA *see* 2855
DHA-S *see* 7710
DHA-S [Dow] *see* 2855
DHAD *see* 6135
DHAQ *see* 6135
DHBE *see* 6913
DHC [Napp] *see* 3154
1,25-DHCC *see* 1641
DH-codeine *see* 3154
DHE-45 [Sandoz] *see* 3157
DHIC *see* 3159
DHPG *see* 4262
DHSM *see* 3162
Dhurrin, 2937
Dia-basan *see* 4372
Diabechlor *see* 2187
Diabefagos *see* 5845
Diaben *see* 9432
Diabenal *see* 2187
Diabesan *see* 9432
Diabeta [Hoechst] *see* 4372
Diabetamid *see* 9432
Diabetoral [Boehringer, Mann.] *see* 2187
Diabetosan *see* 5845
Diabewas *see* 9429
Diabex *see* 5845
Diabinese [Pfizer] *see* 2187
Diabis *see* 7191
Diaboline, 2938
Diaboral [Erba] *see* 9434
Diabrin [USV] *see* 1465
Diabuton *see* 9432
Diacarb *see* 45
Diacepin *see* 2977
Diacerein, 2939
Diacerhein *see* 2939
3,5-Diacetamido-2,4,6-triiodobenzoic Acid Methylglucamine Salt *see* 5689
3,5-Diacetamido-2,4,6-triiodobenzoic Acid Sodium Salt *see* 2975
α,5-Diacetamido-2,4,6-triiodo-*m*-toluic Acid *see* 4904
Diacetatopalladium(II) *see* 6942
Diacetatozirconic Acid *see* 10086
Diacetazotol, 2940
Diacethiamine *see* 46
Diacetic Acid *see* 51
Diacetin, 2941
1,2-Diacetin *see* 2941
1,3-Diacetin *see* 2941
Diacetonamine, 2942
Diacetone Acrylamide, 2943
Diacetone Alcohol, 2944
Diacetoneglucose, 2945
Diacetone-2-ketogulonic Acid *see* 3184
Diacetone-2-oxo-L-gulonic Acid *see* 3184
Diacetotoluide *see* 2940

4,4'-Diacetoxybenzhydrylidenecyclohexane *see* 2726
2,3-Diacetoxybenzoic Acid *see* 3357
1,8-Diacetoxy-3-carboxyanthraquinone *see* 2939
Diacetoxydiphenylisatin *see* 6927
2-(4,4'-Diacetoxydiphenylmethyl)-pyridine *see* 1253
(4,4'-Diacetoxydiphenyl)(2-pyridyl)-methane *see* 1253
3α,17β-Diacetoxy-2β,16β-dipiperidino-5α-androstane Dimethobromide *see* 6958
3,16α-Diacetoxy-Δ^{1,3,5}-estratrien-17-one *see* 2485
1,1-Diacetoxyethane *see* 3767
3β,17β-Diacetoxy-17α-ethynyl-4-estrene *see* 3816
16α,21-Diacetoxy-9α-fluoro-11β,17α-dihydroxy-1,4-pregnadiene-3,20-dione *see* 9511
1,4-Diacetoxy-2-methylnaphthalene *see* 5710
Di(acetoxyphenyl)oxindole *see* 6927
3α,20α-Diacetoxypregnane *see* 7732
Diacetyl, 2946
Diacetylaminoazotoluene *see* 2940
4,4'-Diacetylaminodiphenyl Sulfone *see* 16
3-(Diacetylamino)-2,4,6-triiodobenzoic Acid Propyl Ester *see* 7861
Diacetylcholine Dichloride *see* 8846
Diacetylcholine Diiodide *see* 8847
Diacetylcholine Iodide *see* 8847
Diacetyldapsone *see* 16
N,N'-Diacetyl-4,4'-diaminodiphenyl Sulfone *see* 16
4,4'-Diacetyldiaminodiphenyl Sulfone *see* 16
Diacetyldihydromorphine, 2947
2,6-Diacetyl-7,9-dihydroxy-8,9b-dimethyl-1,3(2*H*,9b*H*)-dibenzofurandione *see* 9806
Diacetyldihydroxydiphenylisatin *see* 6927
Diacetyldioxime *see* 3235
Diacetyldioxyphenylisatin *see* 6927
Diacetyldiphenolisatin *see* 6927
α,β-Diacetylethane *see* 63
2,5-Di-*O*-acetyl-D-glucaro-1,4:6,3-dilactone *see* 20
2,5-Di-*O*-acetyl-D-glucosaccharo-1,4:6,3-dilactone *see* 20
Diacetylhydroxyphenylisatin *see* 6927
Diacetylmethane *see* 75
9,3''-Diacetylmidecamycin *see* 6123
Diacetylmorphine, 2948
Diacetylpyrocatechol-3-carboxylic Acid *see* 3357
Diacetylrhein *see* 2939
Diacetyltannic Acid *see* 101
O,S-Diacetylthiamine *see* 46
N,N-Diacetyl-*o*-tolyazo-*o*-toluidine *see* 2940
Diacid *see* 1837
α,γ-Diacipiperazine *see* 7434
Diacta *see* 8879
Di-Actane [Millot-Solac] *see* 6268
Diactol *see* 9928
Diacycline [Bristol] *see* 9130
Diacylglycerol Lipase *see* 5390
Di-Ademil [Squibb] *see* 4716
Diadilan *see* 7196
Diadol [Tilden-Yates] *see* 252
Di-Adreson *see* 7727
Di-Adreson-F *see* 7719
Di-Adreson-F-aquosum *see* 7722
Diadril [Pliva] *see* 5657
Diafen [Riker] *see* 3334
Diafusor [Fabre] *see* 6528
Diaginol [M & B] *see* 72

Diagnorenol *see* 5894
Diakarmon *see* 8680
Dial [Ciba] *see* 252
Dialar [Lagap] *see* 2977
Dial-a-gesic [Borden] *see* 40
Dialicor [Kissei] *see* 3662
Dialifor, 2949
Dialifos *see* 2949
Diallate, 2950
Diallylamine, 2951
5,5-Diallylbarbituric Acid *see* 252
6,6'-Diallyl-α,α'-bis(diethylamino)-4,4'-bi-*o*-cresol *see* 1217
5,5'-Diallyl-α,α'-bis(diethylamino)-*m,m*'-bitolyl-4,4'-diol *see* 1217
Diallylbis(nortoxiferine) *see* 214
N,N-Diallyl-2-chloroacetamide *see* 250
Diallylcyanamide, 2952
Diallyl Ether *see* 290
Diallylnortoxiferine *see* 214
N,N'-Diallylnortoxiferinium *see* 214
N,N'-Diallylnortoxiferinium Dichloride *see* 214
Diallyl Sulfide *see* 295
Diallyltoxiferine *see* 214
Dialuramide *see* 9763
Diamantane *see* 2494
Diamantane (obsolete) *see* 140
Diamarin *see* 3193
Diamfenetide, 2953
Diamicron [Servier] *see* 4335
3,3'-Diamidinocarbanilide *see* 405
4,4'-Diamidinodiazoaminobenzene Diaceturate *see* 3256
4,4'-Diamidino-α,ω-diphenoxyhexane *see* 4612
4,4'-Diamidino-α,ω-diphenoxypentane *see* 7071
4,4'-Diamidino-α,ω-diphenoxypropane *see* 7808
4,4'-Diamidinodiphenyl Ether *see* 7165
N,N'-Di(*m*-amidinophenyl)urea *see* 405
4,4'-Diamidinostilbene *see* 8772
N,N'-Diamidinostreptamine *see* 8781
Diamine Blue *see* 9701
2,6-Diamine-5-hydroxy-4(3*H*)-pyrimidinone *see* 3389
Diamine Oxidase *see* 4639
Diamine Penicillin *see* 7037
2,8-Diaminoacridine *see* 7780
3,6-Diaminoacridine *see* 7780
3,5-Diamino-*N*-(aminoiminomethyl)-6-chloropyrazinecarboxamide *see* 417
3,5-Diamino-2-[[4-(aminosulfonyl)-phenyl]azo]benzoic Acid *see* 8872
2,4-Diaminoanisole *see* 5920
4,4'-Diamino-1,1'-anthrimide *see* 1254
p-Diaminoazobenzene, 2954
4,4'-Diaminoazobenzene *see* 2954
2,4-Diaminoazobenzene Hydrochloride *see* 2262
2,4-Diaminoazobenzene Hydrochloride Citrate *see* 2262
2,4-Diaminoazobenzene-4'-sulfonamide *see* 8895
m-Diaminobenzene *see* 7254
o-Diaminobenzene *see* 7255
p-Diaminobenzene *see* 7256
3,5-Diaminobenzoic Acid, 2955
2,4-Diamino-5-(4-bromo-3,5-dimethoxybenzyl)pyrimidine *see* 1369
2,6-Diamino-2'-butoxy-3,5'-azopyridine *see* 2956
2',6'-Diamino-2-butoxy-5,5'-azopyridine *see* 2956

Diamond Green Bx *see* 5581
Diamond Green G *see* 1367
Diamond Green P Extra *see* 5581
Diamond Ink, 2965
Diamorphine *see* 2948
Diamox [Lederle] *see* 45
Diamphenethide *see* 2953
Diampromide, 2966
Diampron [M & B] *see* 405
Diamthazole Dihydrochloride, 2967
Diamyceline [Diamant] *see* 2106
Diamyl Ether *see* 646
2,5-Di-*tert*-amylhydroquinone *see* 3300
Diamyl Sodium Sulfosuccinate, 2968
Dianabol [Ciba] *see* 5862
Dianat *see* 3026
Diancina [Septa] *see* 7484
Diandron *see* 7710
Diane 35 [Schering] *see* 2781
Dianette [Schering] *see* 2781
1,4:3,6-Dianhydro-D-glucitol *see* 5113
1,4:3,6-Dianhydro-D-glucitol Dinitrate *see* 5114
1,4:3,6-Dianhydrosorbitol *see* 5113
1,4:3,6-Dianhydrosorbitol 2,5-Dinitrate *see* 5114
Dianil Blue *see* 9701
o,p '-Dianiline *see* 1249
1,4-Dianilinobenzene *see* 3331
1,2-Dianilinoethane, 2969
Dianin's Compound *see* 2341
Dianisidine, 2970
3,3'-Dianisolebis[4,4'-(3,5-diphenyl)-tetrazolium Chloride] *see* 9173
3,4-Dianisyl-3-hexene *see* 3198
2,2-Di-*p*-anisyl-1,1,1-trichloroethane *see* 5913
Dianthraquinonylamine *see* 720
Dianthrimide *see* 720
Diantimony Trioxide *see* 743
Diaparene Chloride [Homemakers' Products] *see* 5946
Diaperos [Merieux] *see* 7177
Diaphene [Stecker] *see* 9529
Diaphenylsulfone *see* 2820
Diapid [Sandoz] *see* 5503
8,8'-Diapocarotenedioic Acid *see* 2592
8,8'-Diapo-ψ,ψ-carotenedioic Acid Bis(6-O-β-D-glucopyranosyl-β-D-glucopyranosyl) Ester *see* 2593
6,6'-Diapo-ψ,ψ-carotenedioic Acid Monomethyl Ester *see* 1320
Diaquone *see* 2813
Diarconal [Recordati] *see* 9490
Diarétyl [SEDAPH] *see* 5650
Diarlidan [Vinas] *see* 6450
Diarsed *see* 3313
4,4'-(1,2-Diarsenediyl)bis[2-aminophenol] Dihydrochloride *see* 839
[1,2-Diarsenediylbis[(6-hydroxy-3,1-phenylene)imino]]bismethanesulfonic Acid Disodium Salt *see* 8916
Diarsenic Pentasulfide *see* 826
Diarsenoacetic Acid *see* 836
Diart [Boehringer, Mann.] *see* 935
Diasan *see* 9218
Diasmen *see* 632
Diasone [Abbott] *see* 8947
Diaspasmyl *see* 7885
Diaspirin *see* 8849
Diaspore *see* 359
Diastal [Bayropharm] *see* 1460
Diastase of Malt, 2971
Diastase Vera *see* 6956
Diastatin *see* 6658
Diaster [Diamant] *see* 2536
Diasulfa *see* 8876
Diasulfon [Streuli] *see* 9432
Diasulfyl *see* 8876

Diatensec [Searle] *see* 8721
Diathymosulfone, 2972
Diatomaceous Earth *see* 4878
Diatox *see* 2972
Diatrast *see* 4933
Diatretyne I, 2973
Diatretyne II, 2974
Diatretyne Amide *see* 2973
Diatretyne Nitrile *see* 2974
Diatrin *see* 5870
Diatrin Base [Warner-Chilcott] *see* 5870
Diatrizoate Meglumine *see* 5689
Diatrizoate Methylglucamine *see* 5689
Diatrizoate Sodium, 2975
Diatropic Acid *see* 4986
Diaveridine, 2976
1,4-Diazabicyclo[2.2.2]octane *see* 9584
Diazachel (obsolete) [Rachelle] *see* 2082
20,25-Diazacholesterol *see* 908
1,3-Diaza-2,4-cyclopentadiene *see* 4828
8,10-Diaza-6-hydroxy-5-methylene-1-(2-methyl-1,2,3-trihydroxypropyl)-2-oxabicyclo[4.2.2]decan-7,9-dione *see* 1222
Diazald [Aldrich] *see* 9469
Diazamine Golden Yellow T *see* 9238
1,2-Diazanaphthalene *see* 2309
3,6-Diazaoctane-1,8-diamine *see* 9579
Diazasterol *see* 908
Diazemuls [Kabivitrum] *see* 2977
Diazenedicarboxamide *see* 932
Diazepam, 2977
Diazil *see* 8886
1,2-Diazine *see* 7982
1,3-Diazine *see* 7998
1,4-Diazine *see* 7971
Diazinon, 2978
Diaziquone, 2979
2,5-Diaziridinyl-3,6-bis(ethoxycarbonylamino)-1,4-benzoquinone *see* 2979
Diazoacetic Ester, 2980
O-Diazoacetyl-L-serine *see* 916
Diazoaminobenzene, 2981
4,4'-(Diazoamino)dibenzamidine Diaceturate *see* 3256
2,2'-[(Diazoamino)di-*p*-phenylene]-bis[6-methyl-7-benzothiazolesulfonic Acid] Disodium Salt *see* 9238
p-Diazobenzenesulfonic Acid, 2982
5-Diazo-2,4-dioxopyrimidine *see* 2985
Diazol [Makhteshim-Agan] *see* 2978
1,2-Diazole *see* 7974
1,3-Diazole *see* 4828
Diazoline *see* 5649
Diazomethane, 2983
Diazon *see* 8947
6-Diazo-5-oxo-L-norleucine, 2984
5-Diazo-2,4(1H,3H)-pyrimidinedione *see* 2985
"Diazoresorcinol" *see* 8144
5-Diazouracil, 2985
Diazoxide, 2986
Diazyl *see* 8874
Dibactil [Roemmers] *see* 8897
Dibasic Sodium Phosphate *see* 8612
Dibasol *see* 1043
Dibasole *see* 1043
Dibein *see* 7191
Dibekacin, 2987
Dibenamine Hydrochloride [SK & F] *see* 2138
Dibencil *see* 7037
Dibencillin *see* 7037
Dibencozide *see* 2446
Dibendrin *see* 3308

Dibenyline [SK&F] *see* 7225
Dibenzalacetone, 2988
Dibenz[*a,h*]anthracene *see* 2989
Dibenz[*de,kl*]anthracene *see* 7137
1,2:5,6-Dibenzanthracene, 2989
5H-Dibenz[*b,f*]azepine-5-carboxamide *see* 1783
4-[3-(5H-Dibenz[*b,f*]azepin-5-yl)-propyl]-1-piperazineethanol *see* 6808
Dibenzcozamide *see* 2446
Dibenzepin, 2990
Dibenzheptropine *see* 2894
1,2,3,4-Dibenznaphthalene *see* 9654
2,3,6,7-Dibenzoanthracene *see* 7057
4-[3-(5H-Dibenzo[*b,f*]azepin-5-yl)-propyl]-1-(2-hydroxyethyl)piperazine *see* 6808
Dibenzo-18-crown-6 *see* 2605
Dibenzo[*a,d*][1,4]cycloheptadiene-5-carboxamide *see* 2766
5H-Dibenzo[*a,d*]cyclohepten-5-one O-[2-(Methylamino)ethyl]oxime *see* 2876
3-(5H-Dibenzo[*a,d*]cyclohepten-5-ylidene)-N,N-dimethyl-1-propanamine *see* 2719
4-(5H-Dibenzo[*a,d*]cyclohepten-5-ylidene)-1-methylpiperidine *see* 2779
2,3,11,12-Dibenzo-1,4,7,10,13,16-hexaoxacyclooctadeca-2,11-diene *see* 2605
Dibenzoparadiazine *see* 7173
Dibenzopenthiophene *see* 9300
Dibenzo[*a,i*]phenanthrene *see* 7368
Dibenzo[*b,h*]phenanthrene *see* 2992
Dibenzopyrazine *see* 7173
Dibenzo[*b,e*]pyridine *see* 117
Dibenzo-γ-pyrone *see* 9971
Dibenzopyrrole *see* 1792
Dibenzothiazine *see* 7220
2,2'-Dibenzothiazyl Disulfide *see* 3377
3-Dibenzo[*b,e*]thiepin-11(6H)-ylidene-N,N-dimethyl-1-propanamine *see* 3419
Dibenzothiopyran *see* 9300
3-Dibenz[*b,e*]oxepin-11(6H)-ylidene-N,N-dimethyl-1-propanamine *see* 3425
Dibenzoxin *see* 6644
Dibenzoyl *see* 1087
Dibenzoylmethane, 2991
Dibenzoyl Peroxide *see* 1128
1,2,7,8-Dibenzphenanthrene *see* 7368
2,3:6,7-Dibenzphenanthrene, 2992
β,β'-Dibenzphenanthrene *see* 2992
Dibenzthiazyl Disulfide *see* 3377
Dibenzthione *see* 8865
Dibenzyl *see* 1219
Dibenzylamine, 2993
N,N-Dibenzylaminoethyl Chloride Hydrochloride *see* 2138
N,N-Dibenzyl-β-chloroethylamine Hydrochloride *see* 2138
Dibenzylchlorophosphate *see* 2994
Dibenzyl Chlorophosphonate, 2994
N,N'-Dibenzyl-γ,γ'-dipyridylium *see* 9903
Dibenzyl Disulfide, 2995
Dibenzyl Ether *see* 1146
N,N'-Dibenzylethylenediamine *see* 1072
N,N'-Dibenzylethylenediamine Bis-[benzylpenicillin] *see* 7037
Dibenzylethylenediamine Dipenicillin G *see* 7037
Dibenzyl Fumarate *see* 1149
Dibenzyl Hydrogen Phosphite *see* 2996
Dibenzylidene Acetone *see* 2988

Dibenzyline see 7225
d-3,4-(1′,3′-Dibenzyl-2′-ketoimid-azolido)-1,2-trimethylenethiophani-um d-Camphorsulfonate see 9621
4,6-Dibenzyl-5-oxo-1-thia-4,6-diaza-tricyclo[6.3.0.0³·⁷]undecanium (+)-β-Camphorsulfonate see 9621
Dibenzyl Phosphite, 2996
Dibenzylphosphoryl Chloride see 2994
Dibenzylsulfide see 1162
3,5-Dibenzyltetrahydro-2H-1,3,5-thiadiazine-2-thione see 8865
Dibenzyran [Röhm] see 7225
Dibestil (formerly) [Breon] see 3119
N,N′-Dibetaine-γ,γ′-dipyridylium see 9903
Dibetos [Kodama] see 1465
Dibiraf see 7191
Dibondrin see 3308
Diborane(6), 2997
Diboron Hexahydride see 2997
Diboron Tetrachloride, 2998
Diboron Tetrahydroxide, 2999
Dibotin see 7191
Dibrom [Chevron] see 6272
Dibromated Camphor see 3002
p,α-Dibromoacetophenone see 1415
2,4′-Dibromoacetophenone see 1415
9,10-Dibromoanthracene see 3000
p-Dibromobenzene, 3001
4,4′-Dibromobenzilic Acid Isopropyl Ester see 1422
d-3,3-Dibromo-2-bornanone see 3002
3,5-Dibromo-N-(4-bromophenyl)-2-hydroxybenzamide see 9529
1,2-Dibromobutane see 1565
2,3-Dibromobutanedioic Acid see 3013
α,α′-Dibromo-d-camphor, 3002
[2,7-Dibromo-9-(o-carboxyphenyl)-6-hydroxy-3-oxo-3H-xanthen-4-yl]-hydroxymercury Disodium Salt see 5757
2,6-Dibromo-N-chloro-p-benzoquin-oneimine see 3010
2,6-Dibromo-4-(chloroimino)-2,5-cyclohexadien-1-one see 3010
Dibromochloropropane, 3003
1,2-Dibromo-3-chloropropane see 3003
2,6-Dibromo-N-chloroquinonimine see 3010
5,5′-Dibromo-o-cresolsulfonphthalein see 1376
2,6-Dibromo-4-cyanophenol see 1431
3,5-Dibromo-Nα-cyclohexyl-Nα-meth-yltoluene-α,2-diamine see 1379
2′,2′′-Dibromo-4′,4′′-diamidino-1,3-diphenoxypropane see 3008
1,2-Dibromo-2,4-dicyanobutane, 3004
1,6-Dibromo-1,6-dideoxydulcitol see 6132
1,6-Dibromo-1,6-dideoxygalactitol see 6132
1,6-Dibromo-1,6-dideoxy-D-mannitol see 6130
5,5′-Dibromo-2,2′-dihydroxybenzil see 3015
5,5′-Dibromo-2,2′-dihydroxybenz-oyl see 3015
4′,5′-Dibromo-3′,6′-dihydroxy-2′,7′-dinitrospiro[isobenzofuran-1(3H),9′-[9H]xanthen]-3-one Diso-dium Salt see 3554
(2′7′-Dibromo-3′,6′-dihydroxy-3-oxospiro[isobenzofuran-1(3H),9′-[9H]xanthen]-4′-yl)hydroxymercu-ry Disodium Salt see 5757

4′,5′-Dibromo-3′,6′-dihydroxyspiro-[isobenzofuran-1(3H),9′-[9H]xanth-en]-3-one see 3005
4′,5′-Dibromo-2′,7′-dinitrofluores-cein Disodium Salt see 3554
Dibromodulcit see 6132
Dibromodulcitol see 6132
sym-Dibromoethane see 3753
1,2-Dibromoethane see 3753
1,2-Dibromoethene see 85
3-(2,2-Dibromoethenyl)-2,2-dimethyl-cyclopropanecarboxylic Acid Cy-ano(3-phenoxyphenyl)methyl Ester see 2869
sym-Dibromoethylene see 85
1,2-Dibromoethylene see 85
4,5-Dibromo-3,6-fluorandiol see 3005
4′,5′-Dibromofluorescein, 3005
Dibromogallic Acid, 3006
2,6-Dibromogallic Acid see 3006
1,5-Dibromohexamethylpentaneam-monium see 7070
3,5-Dibromo-2-hydroxybenzaldehyde see 3011
3,5-Dibromo-4-hydroxybenzenesulf-onic Acid, 3007
3,5-Dibromo-2-hydroxybenzoic Acid see 3012
3,5-Dibromo-4-hydroxybenzonitrile see 1431
3-(3,5-Dibromo-4-hydroxybenzoyl)-2-ethylbenzofuran see 1073
Dibromohydroxymercurifluorescein Disodium Salt see 5757
β-(3,5-Dibromo-4-hydroxyphenyl)-alanine see 3014
3,5-Dibromo-4-hydroxyphenyl Cyan-ide see 1431
3,5-Dibromo-4-hydroxyphenyl 2-Eth-yl-3-benzofuranyl Ketone see 1073
(3,5-Dibromo-4-hydroxyphenyl)(2-ethyl-3-benzofuranyl)methanone see 1073
5,7-Dibromo-8-hydroxyquinoline see 1441
3,5-Dibromo-2-hydroxy-N-[3-(triflu-oromethyl)phenyl]benzamide see 4106
Dibromol [Trommsdorff] see 3007
Dibromomannitol see 6130
Dibromomethane see 5980
3,5-Dibromo-2-methoxybenzyl Alco-hol β-Diethylaminoethyl Ether see 8693
2-(3,5-Dibromo-2-methoxybenzyl-oxy)triethylamine see 8693
2-[(3,5-Dibromo-2-methoxyphenyl)-methoxy]-N,N-diethylethanamine see 8693
2,6-Dibromophenol-4-sulfonic Acid see 3007
Dibromopropamidine, 3008
1,2-Dibromopropane see 7866
1,3-Dibromopropane see 9628
α,γ-Dibromopropane see 9628
ω,ω′-Dibromopropane see 9628
2,3-Dibromo-1-propanol Phosphate-(3:1) see 9665
2,3-Dibromopropene, 3009
2,3-Dibromopropylene, 3009
5,7-Dibromo-8-quinolinol see 1441
2,6-Dibromoquinone-4-chlorimide, 3010
5,5′-Dibromosalicil see 3015
3,5-Dibromosalicylaldehyde, 3011
3,5-Dibromosalicylic Acid, 3012
erythro-2,3-Dibromosuccinic Acid see 3013
sym-Dibromosuccinic Acid see 3013
threo-2,3-Dibromosuccinic Acid see 3013

2,3-Dibromosuccinic Acid, 3013
α,α′-Dibromosuccinic Acid see 3013
Dibromotetraaquochromium Bromide Dihydrate see 2222
3,3′-Dibromothymolsulfonphthalein see 1435
p,α-Dibromotoluene see 1397
3,5-Dibromo-3′-trifluoromethylsali-cylanilide see 4106
3,5-Dibromo-α,α,α-trifluoro-m-salic-ylotoluidide see 4106
2,6-Dibromo-3,4,5-trihydroxybenzoic Acid see 3006
d-3,3-Dibromo-1,7,7-trimethylbicy-clo[2.2.1]heptan-2-one see 3002
3,5-Dibromo-L-tyrosine, 3014
Dibromsalan see 9529
Dibromsalicil, 3015
Dibrosal, 3015
Dibucaine Hydrochloride, 3016
Dibuline Sulfate [Merck & Co.] see 3018
Dibunafon see 3017
Dibunate Ethyl see 3745
Dibunate Sodium, 3017
Dibutalin see 1532
Dibutamide [Upjohn] see 394
Dibutil [Bayer] see 3703
Dibutoline Sulfate, 3018
n-Dibutylamine, 3019
2-[[(Dibutylamino)carbonyl]oxy]-N-ethyl-N,N-dimethylethanaminium Sulfate (2:1) see 3018
γ-(Dibutylamino)-1,3-dichloro-6-(tri-fluoromethyl)-9-phenanthrenepro-panol see 4508
5-[[2-(Dibutylamino)ethyl]amino]-3-phenyl-1,2,4-oxadiazole see 1501
α-(Dibutylamino)-4-methoxybenzene-acetamide see 394
α-Dibutylamino-α-(p-methoxyphen-yl)acetamide see 394
3-(Dibutylamino)-1-propanol 4-Ami-nobenzoate see 1496
Dibutylaminopropyl-p-aminobenzoate see 1496
Dibutylbis(lauroyloxy)tin see 3025
Dibutylbis[(1-oxododecyl)oxy]stan-nane see 3025
N,N-Dibutyl-1-butanamine see 9530
(2-Dibutyl-carbamyloxyethyl)dimeth-ylethylammonium Sulfate see 3018
Dibutyl Carbonate see 1558
2,6-Di-tert-butyl-p-cresol see 1548
4,6-Di-tert-butyl-m-cresol see 2829
N,N′-Dibutyl-N,N′-dicarboxyethyl-ene Diaminemorpholide see 3257
N,N′-Dibutyl-N,N′-dicarboxymor-pholideethylenediamine see 3257
Di-tert-butyl Ether, 3020
n-Dibutyl Ether see 1568
N,N-Dibutyl-4-(hexyloxy)-1-naphtha-lenecarboximidamide see 1475
N,N-Dibutyl-4-hexyloxy-1-naphth-amidine see 1475
Di-tert-butyl Malonate, 3021
2,6-Di-tert-butyl-4-methylphenol see 1548
2,7-Di-tert-butylnaphthalene-4-sul-fonic Acid Ethyl Ester see 3745
3,6-Di-tert-butyl-1-naphthalenesul-fonic Acid Ethyl Ester see 3745
2,7-(And 2,6-)Di-tert-butylnaphtha-lene-4-sulfonic Acid Sodium Salt see 3017
3,6-(And 3,7-)Di-tert-butyl-1-naph-thalenesulfonic Acid Sodium Salt see 3017
Di-tert-butyl Peroxide see 3446

N,N-Dibutyl-N'-(3-phenyl-1,2,4-oxa-
diazol-5-yl)-1,2-ethanediamine *see*
1501
Dibutyl Phthalate *see* 1586
2,6-Di-*tert*-butylpyridine, 3022
Dibutyl Succinate, 3023
Di-*n*-butyl Succinate *see* 3023
Di-*tert*-butyl Succinate, 3024
Dibutyl Sulfide *see* 1590
Dibutyltin Dilaurate, 3025
*N*⁶,2'-*O*-Dibutyryladenosine 3',5'-
Cyclic Monophosphate *see* 1448
*N*⁶,2'-*O*-Dibutyryl CAMP *see* 1448
DIC *see* 2803
DICA *see* 5447
Dicacodyl *see* 1602
1,3-Dicaffeoylquinic Acid *see* 2773
1,5-Dicaffeylquinic Acid *see* 2773
Dicain *see* 9123
Dicalcium Orthophosphate *see* 1699
Dicamba, 3026
Dicamoylmethane *see* 5653
Dicaptan *see* 3027
Dicapthon, 3027
Dicaptol *see* 3196
Dicarbam [BASF] *see* 1789
**2,2-Dicarbamoyloxymethyl-3-methyl-
pentane** *see* 5653
2,2-Di(carbamoyloxymethyl)pentane
see 5751
1,3-Dicarbamylurea *see* 9678
S-(1,2-Dicarbethoxyethyl) *O*,*O*-Di-
methyldithiophosphate *see* 5582
O,*S*-Dicarbethoxythiamine *see* 2014
Dicarbonic Acid Diethyl Ester *see*
8008
**Dicarbonylbis(triphenylphosphine)-
nickel** *see* 1314
**Dicarbonyldi(triphenylphosphino)-
nickel** *see* 1314
Di-μ-carbonylhexacarbonyldicobalt
see 3074
Dicarbosulf *see* 9258
4,4'-Dicarboxydicyclohexylamine *see*
4833
**3,3'-Dicarboxy-4,4'-dihydroxyazo-
benzene** *see* 6799
**2,2'-[(1,2-Dicarboxy-1,2-ethanediyl)-
bis(thio)]bis-1,3,2-dithiastibolane-
4,5-dicarboxylic Acid Hexasodium
Salt** *see* 8768
**4,5-Di(2-carboxyethyl)-1,3,6,7-tetra-
methyl-2,8-divinylbilatriene** *see*
1236
**1,3-Di(2-carboxy-4-oxochromen-5-
yloxy)propan-2-ol** *see* 2594
**2,3-Dicarboxypropane-1,1-diphos-
phonic Acid** *see* 1512
Dicentrine, 3028
Diceplon [Yoshitomi] *see* 2408
Dicestal *see* 3059
Dicetamin [Shionogi] *see* 2014
Dicetel [Latema] *see* 7410
Dicethiamin *see* 2014
Dichlobenil, 3029
Dichlobutrazol *see* 3070
Dichlofenthion, 3030
Dichlofluanid, 3031
Dichlone, 3032
Dichloralantipyrine *see* 3033
Dichloralphenazone, 3033
Dichloramine T, 3034
Dichloran *see* 3042
Dichloren [Ciba] *see* 5655
Dichlorethanoic Acid *see* 3037
Di-chloricide *see* 3045
Dichloricide Aerosol [Merck & Co.]
see 8796
**Dichloricide Mothproofer [Merck &
Co.]** *see* 8796
Dichlorine Heptoxide *see* 2097

Dichlorine Monoxide *see* 2099
Dichlorisone, 3035
Dichlorisoproterenol, 3036
Dichlorman [Shell] *see* 3069
D-*d*-*threo*-2-Dichloroacetamido-1-(4-
methylsulfonyl)-1,3-propanediol *see*
9230
D(−)-*threo*-2-(Dichloroacetamido)-1-
(*p*-nitrophenyl)-3-pantothenyloxy-
1-propanol *see* 2070
D(−)-*threo*-2-Dichloroacetamido-1-*p*-
nitrophenyl-1,3-propanediol *see*
2068
**Dichloroacet-4-hydroxy-N-methylan-
ilide** *see* 3187
**N-Dichloroacet-4-hydroxy-N-methyl-
anilide** *see* 3187
Dichloroacetic Acid, 3037
**Dichloroacetic Acid Compd with N-(1-
Methylethyl)-2-propanamine (1:1)**
see 3182
**Dichloroacetic Acid Diisopropylam-
monium Salt** *see* 3182
sym-Dichloroacetic Anhydride *see*
2112
sym-Dichloroacetone *see* 3039
uns-Dichloroacetone *see* 3038
1,1-Dichloroacetone, 3038
1,3-Dichloroacetone, 3039
α,α-Dichloroacetone *see* 3038
α,γ-Dichloroacetone *see* 3039
2,2-Dichloroacetyl Chloride, 3040
**1-(Dichloroacetyl)-6-(2-furoyloxy)-
1,2,3,4-tetrahydroquinoline** *see* 8068
D-*threo*-N-Dichloroacetyl-1-*p*-nitro-
phenyl-2-amino-1,3-propanediol *see*
2068
S-2,3-Dichloroallyl Diisopropylthiocarbamate *see* 2950
2,4-Dichloro-6-aminophenol *see* 448
4,6-Dichloro-o-aminophenol *see* 448
**4-[(Dichloroamino)sulfonyl]benzoic
Acid** *see* 4503
**9α,11β-Dichloro Analog of Predniso-
lone** *see* 3035
**3,4-Dichloroanilide of 5-Chlorosali-
cylic Acid** *see* 9558
3,4-Dichloroaniline, 3041
**2-(2,6-Dichloroanilino)-1,3-diazacy-
clopentene-(2)** *see* 2388
2-(2,6-Dichloroanilino)-2-imidazoline
see 2388
[*o*-(2,6-Dichloroanilino)phenyl]acetic
Acid Sodium Salt *see* 3071
3,6-Dichloro-o-anisic Acid *see* 3026
Dichloroazodicarbonamidine *see* 2120
Dichlorobenzalkonium Chloride, 3042
m-Dichlorobenzene,** 3043
o-Dichlorobenzene,** 3044
p-Dichlorobenzene,** 3045
1,2-Dichlorobenzene *see* 3044
1,3-Dichlorobenzene *see* 3043
3,4-Dichlorobenzeneamine *see* 3041
**4,5-Dichloro-1,3-benzenedisulfonam-
ide** *see* 3067
2,4-Dichlorobenzenemethanol *see*
3048
2,2'-Dichlorobenzidine, 3046
3,3'-Dichlorobenzidine, 3047
**4,4'-Dichlorobenzilic Acid Ethyl
Ester** *see* 2123
2,6-Dichlorobenzonitrile *see* 3029
**2,6-Dichloro-*p*-benzoquinone-4-chlo-
roimine** *see* 4315
**2,6-Dichloro-1,4-benzoquinone-4-(4-
hydroxyanil) Sodium** *see* 3057
Dichlorobenzyl Alcohol, 3048
**N-(2,4-Dichlorobenzyl)-N-(2-hydr-
oxyethyl)dichloroacetamide** *see*
2076

**N-(2,6-Dichlorobenzylidene)-N'-
amidinohydrazine** *see* 4469
**[(2,6-Dichlorobenzylidene)amino]-
guanidine** *see* 4469
**1-[(2,6-Dichlorobenzylidene)amino]-
3-hydroxyguanidine** *see* 4481
**1-(2,4-Dichlorobenzyl)-1H-indazole-
3-carboxylic Acid** *see* 5447
**2,2'-Dichloro[1,1'-biphenyl]-4,4'-
diamine** *see* 3046
**3,3-Dichloro-(1,1'-biphenyl)-4,4'-
diamine** *see* 3047
3,3'-Dichloro-4,4'-biphenyldiamine
see 3047
**1,1-Dichloro-2,2-bis(*p*-chlorophenyl)-
ethane,** 3049
**Dichlorobis(η⁵-2,4-cyclopentadien-1-
yl)titanium** *see* 9407
**1,1-Dichloro-2,2-bis(*p*-ethylphenyl)-
ethane,** 3050
**1,1-Dichloro-2,2-bis(*p*-fluorophenyl)-
ethane** *see* 2935
1,1-Dichlorobutane *see* 1571
**2,4-Dichloro-6-(o-chloroanilino)-s-
triazine** *see* 685
**1-[2,4-Dichloro-β-[(*p*-chlorobenzyl)-
oxy]phenethyl]imidazole** *see* 3476
**(±)-1-[2,4-Dichloro-β-[(*p*-chloroben-
zyl)thio]phenethyl]imidazole** *see*
8866
**2,6-Dichloro-4-(chloroimino)-2,5-
cyclohexadien-1-one** *see* 4315
**2,4-Dichloro-α-(chloromethylene)-
benzyl Alcohol Diethyl Phosphate**
see 2087
**(Z)-1-[2,4-Dichloro-β-[2-(*p*-chloro-
phenoxy)ethoxy]-α-methylstyryl]-
imidazole** *see* 6801
**1,1-Dichloro-2-(o-chlorophenyl)-2-(*p*-
chlorophenyl)ethane** *see* 6134
**4,6-Dichloro-N-(2-chlorophenyl)-
1,3,5-triazin-2-amine** *see* 685
**1-[2,4-Dichloro-β-[(2-chloro-3-then-
yl)oxy]phenethyl]imidazole** *see* 9384
Dichloro(2-chlorovinyl)arsine *see* 3051
α,3-Dichloro-4-cyclohexylbenzene-
acetic Acid *see* 3914
α,*m*-Dichloro-*p*-cyclohexylphenylace-
tic Acid *see* 3914
**(1S-*trans*)-2,2-Dichloro-1,3-cyclopen-
tanediol** *see* 1721
**2-[4-(2,2-Dichlorocyclopropyl)phen-
oxy]isobutyric Acid** *see* 2314
**2-[4-(2,2-Dichlorocyclopropyl)phen-
oxy]-2-methylpropanoic Acid** *see*
2314
**N,N''-Dichlorodiazenedicarboximid-
amide** *see* 2120
**1,3-Dichloro-α-[2-(dibutylamino)eth-
yl]-6-(trifluoromethyl)-9-phenan-
threnemethanol** *see* 4508
**2,2-Dichloro-N-(2,4-dichlorobenzyl)-
N-(2-hydroxyethyl)acetamide** *see*
2076
**1-[2,4-Dichloro-β-[(2,4-dichlorobenz-
yl)oxy]phenethyl]imidazole** *see* 6101
**1-[2,4-Dichloro-β-[(2,6-dichlorobenz-
yl)oxy]phenethyl]imidazole** *see* 5044
**3,5-Dichloro-6-[(3,5-dichloro-6-hydr-
oxy-4-methoxy-o-tolyl)oxy]-4,2-
cresotic Acid ε-Lactone** *see* 3345
**3,5-Dichloro-N-(3,4-dichlorophenyl)-
2-hydroxybenzamide** *see* 9127
**2,2-Dichloro-N-[(2,4-dichlorophenyl)-
methyl]-N-(2-hydroxyethyl)acet-
amide** *see* 2076
**2,3-Dichloro-5,6-dicyanobenzoqui-
none,** 3052
**Dichlorodi-π-cyclopentadienyltitani-
um** *see* 9407

Dichloromethyl Methyl Ketone *see* 3038

2-[(2,6-Dichloro-3-methylphenyl)-amino]benzoic Acid *see* 5659

2-[(2,6-Dichloro-3-methylphenyl)-amino]benzoic Acid Ethoxymethyl Ester *see* 9099

5,7-Dichloro-2-methyl-8-quinolinol *see* 2191

3′,4′-Dichloro-2-methylvaleranilide *see* 5165

Dichloromonoxide *see* 2099

2,3-Dichloro-1,4-naphthalenedione *see* 3032

2,3-Dichloro-1,4-naphthoquinone *see* 3032

2,4-Dichloro-1-(4-nitrophenoxy)-benzene *see* 6519

2′,5-Dichloro-4′-nitrosalicylanilide *see* 6425

2,3-Dichloro-4-oxo-2-butenoic Acid *see* 6208

7-[[(3,5-Dichloro-4-oxo-1(4H)-pyri-dinyl)acetyl]amino]-3-[[(5-methyl-1,3,4-thiadiazol-2-yl)thio]methyl]-8-oxo-5-thia-1-azabicyclo[4.2.0]-oct-2-ene-2-carboxylic Acid *see* 1924

Dichlorooxozirconium *see* 10087

Dichlorophen(e), 3059

Dichlorophenarsine Hydrochloride, 3060

2,4-Dichlorophenol, 3061

2,6-Dichlorophenol, 3062

2,4-Dichlorophenol Benzenesulfonate *see* 4283

2,6-Dichlorophenol-indophenol Sodi-um *see* 3057

(2,4-Dichlorophenoxy)acetic Acid *see* 2802

2-(2,4-Dichlorophenoxy)benzeneacetic Acid *see* 3913

4-(2,4-Dichlorophenoxy)butanoic Acid *see* 2828

4-(2,4-Dichlorophenoxy)butyric Acid *see* 2828

1-(2,5-Dichlorophenoxy)-3-[(1,1-dimethylethyl)amino]-2-propanol *see* 2399

2-(2,4-Dichlorophenoxy)ethanol Benzoate *see* 8422

2-(2,4-Dichlorophenoxy)ethanol Hydrogen Sulfate Sodium Salt *see* 3372

2-(2,4-Dichlorophenoxy)ethanol Phosphite (3:1) *see* 3880

[[2-(2,6-Dichlorophenoxy)ethyl]-amino]guanidine *see* 4479

2,4-Dichlorophenoxyethyl Benzoate *see* 8422

2-[1-(2,6-Dichlorophenoxy)ethyl]-4,5-dihydro-1H-imidazole *see* 5439

2-[2-(2,4-Dichlorophenoxy)ethyl]hy-drazinecarboximidamide *see* 4479

2,4-Dichlorophenoxyethyl Hydrogen Sulfate Sodium Salt *see* 3372

2-[1-(2,6-Dichlorophenoxy)ethyl]-2-imidazoline *see* 5439

Di-(p-chlorophenoxy)methane *see* 1258

5-(2,4-Dichlorophenoxy)-2-nitrobenz-oic Acid Methyl Ester *see* 1228

2-[4-(2,4-Dichlorophenoxy)phenoxy]-propanoic Acid Methyl Ester *see* 3072

[o-(2,4-Dichlorophenoxy)phenyl]acetic Acid *see* 3913

2-(2,4-Dichlorophenoxy)propanoic Acid *see* 3068

2-(2,4-Dichlorophenoxy)propionic Acid *see* 3068

[(2,6-Dichlorophenyl)acetyl]guanidine *see* 4474

2-[(2,6-Dichlorophenyl)amino]benz-eneacetic Acid Monosodium Salt *see* 3071

2-[(2,6-Dichlorophenyl)amino]-2-imidazoline *see* 2388

2,4-Dichlorophenyl Benzenesulfonate *see* 4283

1-(3′,4′-Dichlorophenyl)-3-(4′-chlo-rophenyl)urea *see* 9568

1-[2-(2,4-Dichlorophenyl)-2-[(2,4-dichlorophenyl)methoxy]ethyl]-1H-imidazole *see* 6101

1-[2-(2,4-Dichlorophenyl)-2-[(2,6-dichlorophenyl)methoxy]ethyl]-1H-imidazole *see* 5044

O-2,4-Dichlorophenyl O,O-Diethyl Phosphorothioate *see* 3030

1,6-Di(4′-chlorophenyldiguanido)-hexane *see* 2090

4-(2,3-Dichlorophenyl)-1,4-dihydro-2,6-dimethyl-3,5-pyridinedicarbox-ylic Acid Ethyl Methyl Ester *see* 3895

N-(2,6-Dichlorophenyl)-4,5-dihydro-N-2-propenyl-1H-imidazol-2-amine *see* 234

3-(3,5-Dichlorophenyl)-1,5-dimethyl-3-azabicyclo[3.1.0]hexane-2,4-dione *see* 7772

N-(3,5-Dichlorophenyl)-1,2-dimethyl-1,2-cyclopropanedicarboximide *see* 7772

1-(2,4-Dichlorophenyl)-4,4-dimethyl-2-(1,2,4-triazol-1-yl)pentan-3-ol *see* 3070

(E)-1-(2,4-Dichlorophenyl)-4,4-di-methyl-2-(1,2,4-triazol-1-yl)-1-penten-3-ol *see* 3261

N′-(3,4-Dichlorophenyl)-N,N-dimeth-ylurea *see* 3388

3-(3,5-Dichlorophenyl)-5-ethenyl-5-methyl-2,4-oxazolidinedione *see* 9890

N-(3,5-Dichlorophenyl)-N′-(4-fluoro-phenyl)thiourea *see* 5440

N-[β-(3,4-Dichlorophenyl)-β-hydr-oxyethyl]isopropylamine *see* 3036

(Z)-1-(2,4-Dichlorophenyl)-2-(1H-imidazol-1-yl)ethanone O-[(2,4-Dichlorophenyl)methyl]oxime Mononitrate *see* 6892

1-(2,4-Dichlorophenyl)-2-(N-imid-azolyl)ethyl-4-phenylthiobenzyl Ether *see* 3948

1-(3,4-Dichlorophenyl)-2-isopropyl-aminoethanol *see* 3036

N¹-3,4-Dichlorophenyl-N⁵-isopropyl-biguanide *see* 2185

1-(3,4-Dichlorophenyl)-5-isopropylbi-guanide *see* 2185

N¹-3,4-Dichlorophenyl-N⁵-isopropyl-diguanide *see* 2185

N-(3,4-Dichlorophenyl)methacrylam-ide *see* 3078

6-[[(3,4-Dichlorophenyl)methoxyacet-yl]amino]-3,3-dimethyl-7-oxo-4-thia-1-azabicyclo[3.2.0]heptane-2-carboxylic Acid *see* 2383

2,4-Dichlorophenyl 3-(Methoxycarbo-nyl)-4-nitrophenyl Ether *see* 1228

N′-(3,4-Dichlorophenyl)-N-methoxy-N-methylurea *see* 5387

3-(3,4-Dichlorophenyl)-1-methyl-1-n-butylurea *see* 6354

(R*,R*)-(±)-β-[(2,4-Dichlorophenyl)-methyl]-α-(1,1-dimethylethyl)-1H-1,2,4-triazole-1-ethanol *see* 3070

(E)-(±)-β-[(2,4-Dichlorophenyl)meth-ylene]-α-(1,1-dimethylethyl)-1H-1,2,4-triazole-1-ethanol *see* 3261

2-[(2,6-Dichlorophenyl)methylene]hy-drazinecarboximidamide *see* 4469

2-[(2,6-Dichlorophenyl)methylene]-N-hydroxyhydrazinecarboximidamide *see* 4481

3-(3,5-Dichlorophenyl)-N-(1-methyl-ethyl)-2,4-dioxo-1-imidazolidine-carboxamide *see* 4964

N-(3,4-Dichlorophenyl)-N′-(1-meth-ylethyl)imidodicarbonimidic Diam-ide *see* 2185

1-[(2,4-Dichlorophenyl)methyl]-1H-indazole-3-carboxylic Acid *see* 5447

O-(2,4-Dichlorophenyl) O-Methyl Isopropylphosphoramidothioate *see* 3394

6-[3-(2,6-Dichlorophenyl)-5-methyl-4-isoxazolecarboxamido]penicillanic Acid *see* 3073

6-[[[3-(2,6-Dichlorophenyl)-5-methyl-4-isoxazolyl]carbonyl]amino]-3,3-dimethyl-7-oxo-4-thia-1-azabicy-clo[3.2.0]heptane-2-carboxylic Acid *see* 3073

3-(2,6-Dichlorophenyl)-5-methyl-4-isoxazolylpenicillin *see* 3073

2-(3,4-Dichlorophenyl)-4-methyl-1,2,4-oxadiazolidine-3,5-dione *see* 5876

N-(3,4-Dichlorophenyl)-2-methyl-pentanamide *see* 5165

N-(3,4-Dichlorophenyl)-2-methyl-2-propenamide *see* 3078

2-[(2,4-Dichlorophenyl)methyl]-4-(1,1,3,3-tetramethylbutyl)phenol *see* 2377

3-(3,5-Dichlorophenyl)-5-methyl-5-vinyloxazolidine-2,4-dione *see* 9890

2,4-Dichlorophenyl p-Nitrophenyl Ether *see* 6519

1-[2-(2,4-Dichlorophenyl)-2-[[4-(phenylthio)phenyl]methoxy]ethyl]-1H-imidazole *see* 3948

N-(3,4-Dichlorophenyl)propanamide *see* 7814

1-[2-(2,4-Dichlorophenyl)-2-(2-prop-enyloxy)ethyl]-1H-imidazole *see* 3537

N-(3,4-Dichlorophenyl)propionamide *see* 7814

1-[[2-(2,4-Dichlorophenyl)-4-propyl-1,3-dioxolan-2-yl]methyl]-1H-1,2,4-triazole *see* 7830

(1S-cis)-4-(3,4-Dichlorophenyl)-1,2,-3,4-tetrahydro-N-methyl-1-naph-thalenamine *see* 8416

(1S,4S)-4-(3,4-Dichlorophenyl)-1,2,-3,4-tetrahydro-N-methyl-1-naph-thylamine *see* 8416

α-(2,4-Dichlorophenyl)-4-(1,1,3,3-tetramethylbutyl)-o-cresol *see* 2377

1-[2,4-Dichloro-β-[[p-(phenylthio)-benzyl]oxy]phenethyl]imidazole *see* 3948

2,4-Dichloro-4′-phenylthio-(N-imid-azolylmethyl)dibenzyl Ether *see* 3948

6-(2,3-Dichlorophenyl)-1,2,4-triazine-3,5-diamine *see* 5228

6-(2,5-Dichlorophenyl)-1,3,5-triazine-2,4-diamine *see* 4982

cis-1-[4-[[2-(2,4-Dichlorophenyl)-2-(1H-1,2,4-triazol-1-ylmethyl)-1,3-dioxolan-4-yl]methoxy]phenyl]-4-(1-methylethyl)piperazine *see* 9090

4-[4-[4-[4-[[2-(2,4-Dichlorophenyl)-2-(1*H*-1,2,4-triazol-1-ylmethyl)-1,3-dioxolan-4-yl]methoxy]phenyl]-1-piperazinyl]phenyl]-2,4-dihydro-2-(1-methylpropyl)-3*H*-1,2,4-triazol-3-one 5131

2-(3,5-Dichlorophenyl)-2-(2,2,2-trichloroethyl)oxirane *see* 9578

Di(*p*-chlorophenyl)trichloromethylcarbinol *see* 3075

Dichlorophos *see* 3069

3,6-Dichloropicolinic Acid *see* 2398

9α,11β-Dichloro-1,4-pregnadiene-17α,21-diol-3,20-dione *see* 3035

Dichloroprop *see* 3068

1,1-Dichloropropane *see* 7874

1,2-Dichloropropane *see* 7867

2,2-Dichloropropanoic Acid *see* 2806

2,2-Dichloropropanoic Acid 2-(2,4,5-Trichlorophenoxy)ethyl Ester *see* 3587

1,3-Dichloro-2-propanol, 3063

1,3-Dichloro-2-propanol Phosphate (3:1) *see* 4233

1,1-Dichloro-2-propanone *see* 3038

1,3-Dichloro-2-propanone *see* 3039

1,3-Dichloropropene, 3064

3′,4′-Dichloropropionanilide *see* 7814

α,α-Dichloropropionic Acid *see* 2806

2,2-Dichloropropionic Acid 2-(2,4,5-Trichlorophenoxy)ethyl Ester *see* 3587

1,3-Dichloropropylene *see* 3064

α,γ-Dichloropropylene *see* 3064

3,6-Dichloro-2-pyridinecarboxylic Acid *see* 2398

2,4′-Dichloro-α-(pyrimidin-5-yl)benzhydryl Alcohol *see* 3903

5,7-Dichloro-8-quinaldinol *see* 2191

5,7-Dichloro-8-quinolinol *see* 2175

5,7-Dichloro-8-quinolinol Mixt with 5-Chloro-8-quinolinol and 7-Chloro-8-quinolinol *see* 4520

2,6-Dichloroquinone Chloroimide *see* 4315

6,7-Dichloro-9-ribitylisoalloxazine *see* 3065

7,8-Dichloro-10-ribitylisoalloxazine *see* 3065

Dichlororiboflavin, 3065

Dichlorosal *see* 4704

p-(Dichlorosulfamoyl)benzoic Acid *see* 4503

Dichlorotetraaquochromium Chloride Dihydrate *see* 2224

Dichlorotetracarbonyldirhodium *see* 8188

1,2-Dichloro-1,1,2,2-tetrafluoroethane *see* 2608

6,7-Dichloro-1,2,3,5-tetrahydroimidazo[2,1-*b*]quinazolin-2-one *see* 659

3,5-Dichlorotetrahydro-2,4,6-trioxo-*s*-triazin-1(2*H*)-yl Potassium *see* 9679

7,8-Dichloro-10-(D-*ribo*-2,3,4,5-tetrahydroxypentyl)isoalloxazine *see* 3065

[2,3-Dichloro-4-(2-thenoyl)phenoxy]acetic Acid *see* 9362

[2,3-Dichloro-4-(2-thienylcarbonyl)phenoxy]acetic Acid *see* 9362

[2,3-Dichloro-4-(2-thiophenecarbonyl)phenoxy]acetic Acid *see* 9362

α,α-Dichlorotoluene *see* 1064

N,N-Dichloro-*p*-toluenesulfonamide *see* 3034

N-(2,6-Dichloro-*m*-tolyl)anthranilic Acid *see* 5659

N-(2,6-Dichloro-*m*-tolyl)anthranilic Acid Ethoxymethyl Ester *see* 9099

1,3-Dichloro-1,3,5-triazine-2,4,6(1*H*,-3*H*,5*H*)-trione Potassium Salt *see* 9679

4,4′-Dichloro-α-(trichloromethyl)-benzhydrol *see* 3075

2,2′-Dichlorotriethylamine *see* 4644

4,4′-Dichloro-3-(trifluoromethyl)-carbanilide *see* 2376

1-(1,3-Dichloro-6-trifluoromethyl-9-phenanthryl)-3-di(*n*-butyl)amino-propanol *see* 4508

Dichlorovos *see* 3069

Dichloroxide *see* 2099

Dichloroxylenol, 3066

2,4-Dichloro-3,5-xylenol *see* 3066

Dichlorphenamide, 3067

Dichlorprop, 3068

Dichlor-Stapenor [Bayer] *see* 3073

Dichlorvos, 3069

Dichlotride [Merck & Co.] *see* 4704

β-Dichroine *see* 3890

Dichronic [Sana] *see* 3071

Dichystrolum *see* 3163

Dickite *see* 368

Diclobenin [Chassot] *see* 3071

Diclobutrazol, 3070

Diclocil [Bristol] *see* 3073

Diclofenac Sodium, 3071

Diclofop-Methyl, 3072

Dicloguamine *see* 4982

Diclondazolic Acid *see* 5447

Diclo-Phlogont [Azuchemie] *see* 3071

Diclo-Puren [Klinge-Nattermann] *see* 3071

Diclord [Wakamoto] *see* 3071

Dicloreum [Alfa] *see* 3071

Diclotride [Merck & Co.] *see* 4704

Dicloxacillin, 3073

Dico *see* 3154

Dicobalt Octacarbonyl, 3074

Dicodid [Knoll] *see* 4708

Dicodrine [Westerfield] *see* 4708

Dicoferin *see* 6450

Dicofol, 3075

Dicoline *see* 3076

Dicolinium Iodide, 3076

Dicopac [Amersham] *see* 9922

Dicophane *see* 2832

Dicopper Tetraiodomercurate(2−) *see* 2670

Dicorantil [Silva Araujo-Roussel] *see* 3366

Dicortol *see* 7719

Dicoumarin *see* 3080

Dicoumarol *see* 3080

Dicoumoxyl [Labaz] *see* 2566

Dicroden *see* 4596

Dicromil [Organon] *see* 2906

Dicrotalic Acid *see* 5690

Dicrotophos, 3077

Dicryl, 3078

Dictamine *see* 3079

Dictamnine, 3079

Dictycide [Fort Dodge] *see* 2688

Dictyzide *see* 2688

Dicumacyl *see* 3729

Dicumarol, 3080

Dicumol *see* 3080

Dicumoxane [Boots] *see* 2566

Dicuprene [UCB] *see* 2677

Dicurone *see* 4362

Dicyan *see* 2698

Dicyanine, 3081

2,2′-Dicyano-2,2′-azopropane *see* 931

β,β-Dicyano-*o*-chlorostyrene *see* 2127

Dicyanodiamide, 3082

Dicyanodiamidine Sulfate, 3083

2,3-Dicyano-1,4-dithiaanthraquinone *see* 3375

sym-Dicyanoethane *see* 8843

Dicyanomethane *see* 5592

9-Dicyanomethylene-2,4,7-trinitrofluorene, 3084

1,3-Dicyano-2,4,5,6-tetrachlorobenzene *see* 2167

Dicyclidine *see* 7772

Dicyclohexano-18-crown-6 *see* 2605

Dicyclohexylacetic Acid 2-Piperidinoethyl Ester *see* 3151

Dicyclohexylamine, 3085

Dicyclohexylcarbodiimide, 3086

"Dicyclohexyl-18-crown-6" *see* 2605

2-(2,2-Dicyclohexylethyl)piperidine *see* 7116

1,1-Dicyclohexyl-2-(2-piperidyl)ethane *see* 7116

Dicyclomine Hydrochloride, 3087

Dicyclopentadiene *see* 2744

Dicyclopentadienyliron *see* 3985

Dicyclopentylacetic Acid β-Diethylaminoethyl Ester Ethobromide *see* 3346

2-[(Dicyclopentylacetyl)oxy]-N,N,N-triethylethanaminium Bromide *see* 3346

2-[N-(Dicyclopropylmethyl)amino]-oxazoline *see* 8220

N-(Dicyclopropylmethyl)-4,5-dihydro-2-oxazolamine *see* 8220

(2α,7α,8R,19α,22E)-7,19:8,19-Dicyclo-9,10-secoergosta-5(10),22-dien-2-ol *see* 8984

Dicycloverin Hydrochloride *see* 3087

Dicynene [Travenol] *see* 3675

Dicynone [OM] *see* 3675

Dicysteine *see* 2788

Didakene *see* 9126

Didandin [Boots] *see* 3304

Didecyldimethylammonium Chloride, 3088

4′,5′-Didehydro-4,5-*retro*-β,β-carotene-3,3′-dione *see* 8196

3′,4′-Didehydro-β,ψ-caroten-16′-oic Acid *see* 9476

(16R,19E)-19,20-Didehydro-16-[(10β,13β,21β)-23-deoxy-21,22-dihydro-11-oxa-12,14-secostrychnidin-10-yl]corynan-17-oic Acid Methyl Ester *see* 4275

2′,3′-Didehydro-2′-deoxyverrucarin A *see* 9869

9,10-Didehydro-N,N-diethyl-6-methylergoline-8β-carboxamide *see* 5507

3,10-Didehydro-10,11-dihydrocinchonan-6′,9-diol *see* 777

12,13-Didehydro-13,14-dihydro-α-erythroidine *see* 3625

11,12-Didehydro-7,10-dihydro-10-hydroxyretinol *see* 6887

(3β)-1,6-Didehydro-14,17-dihydro-3-methoxy-16(15*H*)-oxaerythrinan-15-one *see* 3158

(16E,20β)-16,17-Didehydro-9,17-dimethoxycorynan-16-carboxylic Acid Methyl Ester *see* 6136

16,17-Didehydro-10,11-dimethoxy-19-methyloxayohimban-16-carboxylic Acid Methyl Ester *see* 8148

(E)-16,17-Didehydro-9,17-dimethoxy-17,18-seco-20α-yohimban-16-carboxylic Acid Methyl Ester *see* 6136

8,9-Didehydro-6,8-dimethylergoline *see* 177

2,3-Didehydro-6,7-epoxyaspermidine-3-carboxylic Acid Methyl Ester *see* 5434

19,20-Didehydro-17,18-epoxycuran-17-ol *see* 9957

6,7-Didehydro-4,5α-epoxy-3,6-dimethoxy-17-methylmorphinan *see* 3164

Di(2-ethylhexyl) Phthalate see 1262
Di(2-ethylhexyl) Sebacate see 1263
Diethyl-2-[1-hydroxy-1-cyclopentyl-1-(2-thienyl)acetoxy]ethylmethyl-ammonium Bromide see 7084
1,1-Diethyl-3-[(hydroxydiphenylacetyl)oxy]pyrrolidinium Bromide see 1090
Diethyl(2-hydroxyethyl)methylammonium Bromide α-Cyclopentyl-2-thiopheneglycolate see 7084
Diethyl(2-hydroxyethyl)methylammonium Bromide Xanthene-9-carboxylate see 5869
Diethyl(2-hydroxyethyl)methylammonium 3-Methyl-2-phenylvalerate Bromide see 9816
Diethyl(2-hydroxyethyl)methylammonium α-Phenylcyclohexaneglycolate Bromide see 6928
dl-13β,17α-Diethyl-17β-hydroxygon-4-en-3-one see 6603
N,N-Diethyl-4-hydroxy-3-methoxybenzamide see 3673
O,O-Diethyl O-7-Hydroxy-3,4-tetramethylenecoumarinyl Phosphorothioate see 2568
4,4'-(1,2-Diethylidene-1,2-ethanediyl)bisphenol see 3094
4,4'-(Diethylideneethylene)diphenol see 3094
N,N-Diethyl-N'-2-indanyl-N'-phenyl-1,3-propanediamine see 781
N,N-Diethylisonicotinamide see 5074
Diethyl 2-Isopropyl-4-methyl-6-pyrimidyl Thionophosphate see 2978
O,O-Diethyl O-2-Isopropyl-4-methyl-6-pyrimidyl Thiophosphate see 2978
N,N-Diethylisovaleramide see 5122
Diethyl Ketone, 3111
N,N-Diethyl-D-lysergamide see 5507
Diethylmagnesium, 3112
Diethyl Maleate, 3113
Diethyl Malonate see 3779
Diethylmalonic Acid, 3114
Diethylmalonylurea see 972
N,N-Diethyl-2-methoxy-4-allylphenoxyacetamide see 39
N,N-Diethyl-2-(2-methoxyphenoxy)-ethanamine see 4462
N,N-Diethyl-2-[2-methoxy-4-(2-propenyl)phenoxy]acetamide see 39
N¹,N¹-Diethyl-N⁴-(6-methoxy-8-quinolinyl)-1,4-pentanediamine see 6953
N,N-Diethyl-N'-(6-methoxy-8-quinolinyl)-1,3-propanediamine see 7495
α,α'-Diethyl-4'-methoxy-4-stilbenol see 5818
2-[[2-(Diethylmethylammonio)ethoxy]carbonyl]-1,1,6-trimethylpiperidinium Diiodide see 3076
5,5-Diethyl-1-methylbarbituric Acid see 5873
N,N-Diethyl-3-methylbenzamide see 2848
N,N-Diethyl-3-methylbutyramide see 5122
Diethyl Methyl Carbinol Urethan see 3520
O,O-Diethyl O-(4-Methyl-7-coumarinyl) Phosphorothioate see 4793
O,O-Diethyl O-(4-Methyl-7-coumarinyl) Thiophosphate see 4793
N,N-Diethyl-1-methyl-3,3-di-2-thienylallylamine see 9220
N,N-Diethyl-N'-[(8α)-6-methylergolin-8-yl]urea see 9095
1,1-Diethyl-3-(D-6-methylisoergolen-8-yl)urea see 5394

N,N-Diethyl-N-methyl-2-[(3-methyl-1-oxo-2-phenylpentyl)oxy]ethanaminium Bromide see 9816
N,N-Diethyl-α-methyl-10H-phenothiazine-10-ethanamine see 3703
N,N-Diethyl-2-[2-(2-methyl-5-phenyl-1H-pyrrol-1-yl)phenoxy]-ethanamine see 5316
N,N-Diethyl-N-methyl-2-[[4-[4-(phenylthio)phenyl]-3H-1,5-benzodiazepin-2-yl]thio]ethanaminium Iodide see 9357
Diethylmethyl[2-[[4-[p-(phenylthio)-phenyl]-3H-1,5-benzodiazepin-2-yl]thio]ethyl]ammonium Iodide see 9357
N,N-Diethyl-4-methyl-1-piperazine-carboxamide see 3106
3,3-Diethyl-5-methyl-2,4-piperidinedione see 6054
4,4'-(1,2-Diethyl-3-methyl-1,3-propanediyl)bisphenol see 1082
5,5-Diethyl-1-methyl-2,4,6(1H,3H,-5H)-pyrimidinetrione see 5873
O,O-Diethyl O-[p-(Methylsulfinyl)-phenyl] Phosphorothioate see 3943
N,N'-Diethyl-6-(methylthio)-1,3,5-triazine-2,4-diamine see 8488
N,N-Diethyl-5-methyl-[1,2,4]triazolo-[1,5-a]pyrimidin-7-amine see 9492
N,N-Diethyl-N-methyl-2-[2-(trimethylammonio)ethoxy]-1-propanaminium Diiodide see 7507
O,O-Diethyl O-(4-Methylumbelliferone) Phosphorothioate see 4793
N,N-Diethyl-N-methyl-2-[(9H-xanthen-9-ylcarbonyl)oxy]ethanaminium Bromide see 5869
N,N-Diethyl-2-(1-naphthalenyloxy)-propanamide see 6336
O,O-Diethyl O-Naphthaloximide Phosphate see 6269
N,N-Diethyl-2-(1-naphthyloxy)propionamide see 6336
N,N-Diethylnicotinamide see 6459
Diethyl-p-nitrophenyl Monothiophosphate see 6983
Diethyl p-Nitrophenyl Phosphate see 6979
O,O-Diethyl O-p-Nitrophenyl Phosphorothioate see 6983
Diethylnitrosamine see 6557
Diethylolamine see 3097
Diethylolglycine see 1221
Diethyl Oxalacetate see 3791
Diethyl Oxalate, 3115
Diethyl Oxide see 3762
Diethyl Oxydiformate see 8008
N¹,N¹-Diethyl-1,4-pentanediamine see 6642
2,2-Diethyl-4-pentenamide see 6643
N,N-Diethyl-10H-phenothiazine-10-ethanamine see 3098
α,α-Diethylphenylacetic Acid 2-(2-Diethylaminoethoxy)ethyl Ester see 6885
5,5-Diethyl-1-phenylbarbituric Acid see 7183
N,N-Diethyl-α-phenylbenzenemethanamine see 3103
Di(p-ethylphenyl)dichloroethane see 3050
N,N-Diethyl-3-phenyl-1,2,4-oxadiazole-5-ethanamine see 6898
N,N-Diethyl-3-(1-phenylpropyl)-1,2,4-oxadiazole-5-ethanamine see 7919
5,5-Diethyl-1-phenyl-2,4,6(1H,3H,-5H)-pyrimidinetrione see 7183
Diethylphosphorous Acid Chloride see 3795

Diethyl Phthalate see 7345
3,3-Diethyl-2,4-piperidinedione see 7439
Diethylpropanedioic Acid see 3114
Diethylpropanedioic Acid Diethyl Ester see 3746
Diethylpropion, 3116
2,6-Diethyl-N-(2'-n-propoxyethyl)-chloroacetanilide see 7746
O,O-Diethyl-O-(2-pyrazinyl) Phosphorothioate see 9275
N,N-Diethyl-3-pyridinecarboxamide see 6459
3,3-Diethyl-2,4-(1H,3H)pyridinedione see 8005
5,5-Diethyl-2,4,6(1H,3H,5H)-pyrimidinetrione see 972
Diethyl Pyrocarbonate see 8008
N,N-Diethylrifomycin B Amide see 8214
Diethyl Sebacate see 8369
Diethylsilane, 3117
α,α'-Diethylstilbenediol see 3118
α,α'-Diethyl-4,4'-stilbenediol Diphosphoric Acid Ester see 4168
α,α'-Diethyl-4,4'-stilbenediol Dipropionyl Ester see 3119
Diethylstilbestrol, 3118
Diethylstilbestrol Diphosphate see 4168
Diethylstilbestrol 4,4'-Diphosphoric Ester see 4168
Diethylstilbestrol Dipropionate, 3119
Diethylstilbestrol Monomethyl Ether see 5818
Diethylstilbestryl Diphosphate see 4168
p-(Diethylsulfamoyl)benzoic Acid see 3684
4,6-Diethyl-2-sulfanilamido-1,3,5-triazine see 8919
Diethyl Sulfate, 3120
Diethyl Sulfide see 3809
Diethyl Sulfide 2,2'-Dicarboxylic Acid see 9261
Diethylsulfondimethylmethane see 8939
Diethylsulfonmethylethylmethane see 8937
Diethyl Tartrate see 3811
5,5-Diethyl-2,3,5,6-tetrahydro-4H-1,3-oxazine-2,4-dione see 3096
1,8-Diethyl-1,3,4,9-tetrahydropyrano-[3,4-b]indole-1-acetic Acid see 3831
N,N-Diethyl-N-[2-[4-(1,1,3,3-tetramethylbutyl)phenoxy]ethyl]benzenemethanaminium Chloride see 7205
N,N-Diethyl-N,N',N',N'-tetramethyl-N,N'-(2-methyl-3-oxapentamethylene)bis[ammonium Iodide] see 7507
3,3'-Diethylthiadicarbocyanine Iodide see 3376
Diethylthiambutene see 9220
5,5-Diethyl-2-thiobarbituric Acid see 9248
O,O-Diethylthiophosphoric Ester of 3,4-Tetramethyleneumbelliferone see 2568
N,N-Diethyl-m-toluamide see 2848
N¹-(4,6-Diethyl-s-triazin-2-yl)sulfanilamide see 8919
O,O-Diethyl O-3,5,6-Trichloro-2-pyridyl Phosphorothioate see 2190
1,1'-Diethyl-2,2'-trimethinequinocyanine Chloride see 8056
O,O-Diethyl S-(2-Trimethylammoniumethyl)phosphorothioate Iodide see 3475
5,5-Diethyl-2,4,6-trioxo-1-phenylhexahydropyrimidine see 7183

9-(3,3-Diethylureido)-4,6,6a,7,8,9-
hexahydro-7-methylindolo[4,3-*f*,*g*]-
quinoline *see* 5394
N,N-Diethylvanillamide *see* 3673
Diethyl Xanthogenate *see* 3390
Diethylzinc, 3121
Dietreen *see* 8777
Dietrol *see* 7180
Difacil Hydrochloride *see* 151
2-Difarnesyl-3-methyl-1,4-naphtho-
quinone *see* 9935
Difaterol [Andreu] *see* 1215
Difemerine, 3122
Difenamizole, 3123
Difenax [Zambeletti] *see* 3125
Difenidolin [Taiyo] *see* 3311
Difenoxilic Acid *see* 3124
Difenoxin, 3124
Difenoxylic Acid *see* 3124
Difenpiramide, 3125
Diferuloylmethane *see* 2681
Difflam [Carnegie] *see* 1136
Diffolisterol *see* 3655
Diffu-K [Delagrange] *see* 7601
Diffumal [Malesci] *see* 9212
Diffusin [Ortho] *see* 4676
Diffusing Factor *see* 4676
Difhydan [AB Leo] *see* 7293
Diflorasone, 3126
Difloxacin, 3127
Diflubenzuron, 3128
Diflucan [Pfizer] *see* 4054
Diflucortolone, 3129
Diflunisal, 3130
N,N'-Di-9H-fluoren-9-yl-N,N,N',N'-
tetramethyl-1,6-hexanediaminium
Dibromide *see* 4606
2,4-Difluoroaniline, 3131
2,4-Difluorobenzenamine *see* 3131
p-Difluorobenzene, 3132
4,4'-Difluoro-1,1'-biphenyl *see* 3133
2,4-Difluoro-α,α-bis(1H-1,2,4-triazol-
1-ylmethyl)benzyl Alcohol *see* 4054
1,1-Difluoro-2,2-dichloroethyl Methyl
Ether *see* 5914
Difluorodichloromethane *see* 3053
erythro-3,3'-Difluoro-4,4'-dihydr-
oxy-α-ethyl-α'-methyldibenzyl *see*
1231
6,9-Difluoro-11,21-dihydroxy-16,17-
[(1-methylethylidene)bis(oxy)]-preg-
na-1,4-diene-3,20-dione *see* 4076
6,9-Difluoro-11,21-dihydroxy-16-
methylpregna-1,4-diene-3,20-dione
see 3129
Difluorodioxochromium *see* 2252
4,4'-Difluorodiphenyl, 3133
Difluorodiphenyldichloroethane *see*
2935
Difluorodiphenyltrichloroethane *see*
2936
6,8-Difluoro-1-(2-fluoroethyl)-1,4-
dihydro-7-(4-methyl-1-piperazin-
yl)-4-oxo-3-quinolinecarboxylic
Acid *see* 4035
2',4'-Difluoro-4-hydroxy-[1,1'-bi-
phenyl]-3-carboxylic Acid *see* 3130
2',4'-Difluoro-4-hydroxy-3-biphen-
ylcarboxylic Acid *see* 3130
2',4'-Difluoro-4-hydroxy-[1',1-di-
phenyl]-3-carboxylic Acid *see* 3130
6α,9α-Difluoro-16α-hydroxypredniso-
lone 16,17-Acetonide *see* 4076
6α,9α-Difluoro-16α,17α-isopropylid-
enedioxy-1,4-pregnadiene-3,20-
dione *see* 4076
4-(Difluoromethoxy)-α-(1-methyleth-
yl)benzeneacetic Acid Cyano(3-
phenoxyphenyl)methyl Ester *see*
4055

6α,9α-Difluoro-16α-methyl-1-de-
hydrocorticosterone *see* 3129
2-(Difluoromethyl)-DL-ornithine *see*
3489
α-Difluoromethylornithine *see* 3489
6α,9α-Difluoro-16α-methylpredniso-
lone *see* 4066
6α,9α-Difluoro-16β-methylpredniso-
lone *see* 3126
6α,9α-Difluoro-16α-methyl-1,4-preg-
nadiene-11β,21-diol-3,20-dione *see*
3129
6α,9α-Difluoro-16β-methyl-Δ 1,4-preg-
nadiene-11β,17α,21-triol-3,20-dione
see 3126
7-β-Difluoromethylthioacetamido-7α-
methoxy-3-[[1-(2-hydroxyethyl)-
1H-tetrazol-5-yl]thiomethyl]-1-
oxa-3-cephem-4-carboxylic Acid
see 4038
(6R-*cis*)-7-[[[(Difluoromethyl)thio]-
acetyl]amino]-3-[[[1-(2-hydroxyeth-
yl)-1H-tetrazol-5-yl]thio]methyl]-
7-methoxy-8-oxo-5-oxa-1-azabicy-
clo[4.2.0]oct-2-ene-2-carboxylic
Acid *see* 4038
2-(2,4-Difluorophenyl)-1,3-bis(1H-
1,2,4-triazol-1-yl)propan-2-ol *see*
4054
1-[4,4-Di(4-fluorophenyl)butyl]-4-
[(2,6-dimethylanilinocarbonyl)meth-
yl]piperazine *see* 5360
1-[4,4-Di-(4-fluorophenyl)butyl]-4-
(2-oxo-1-benzimidazolinyl)piperi-
dine *see* 7404
1-(2,4-Difluorophenyl)-6-fluoro-1,4-
dihydro-7-(3-methyl-1-piperazin-
yl)-4-oxo-3-quinolinecarboxylic
Acid *see* A8
5-(2,4-Difluorophenyl)salicylic Acid
see 3130
α-(2,4-Difluorophenyl)-α-(1H-1,2,4-
triazol-1-ylmethyl)-1H-1,2,4-tri-
azole-1-ethanol *see* 4054
6α,9α-Difluoroprednisolone-21-ace-
tate-17-butyrate *see* 3134
6α,9α-Difluoro-11β,16α,17,21-tetra-
hydroxypregna-1,4-diene-3,20-di-
one Cyclic 16,17-Acetal with Acetone *see* 4076
6α,9α-Difluoro-11β,16α,17,21-tetrahy-
droxypregna-1,4-diene-3,20-dione,
Cyclic 16,17-Acetal with Acetone,
21-Acetate *see* 4077
(RS)-2,4'-difluoro-α-(1H-1,2,4-tri-
azol-1-ylmethyl)benzhydryl Alcohol
see 4135
6,9-Difluoro-11,17,21-trihydroxy-16-
methylpregna-1,4-diene-3,20-dione
see 4066
(6α,11β,16β)-6,9-Difluoro-11,17,21-
trihydroxy-16-methylpregna-1,4-
diene-3,20-dione *see* 3126
6α,9-Difluoro-11β,17,21-trihydroxy-
pregna-1,4-diene-3,20-dione 21-
Acetate 17-Butyrate *see* 3134
Difluprednate, 3134
Diflupyl *see* 5060
Diflurex [Anphar] *see* 9362
Difluron *see* 3128
Difmecor [UCM-Difme] *see* 3916
Difmedol [UCM-Difme] *see* 6925
Difolatan *see* 1770
Difolliculine *see* 3655
Diforene *see* 2836
Diformyl *see* 4405
2,3-Diformyl-6-methoxy-5-methyl-
benzoic Acid *see* 4329
1,3-Diformylpropane *see* 4366
Difosfocin [Magis] *see* 2321
Difosfonal [Benedetti] *see* 2369

Difral [Yamanouchi; Upjohn; Sumito-
mo Seiyaku] *see* 3126
Di-2-furanylethanedione Dioxime *see*
4218
N,N'-Difurfurylidene-2-furanmeth-
anediamine *see* 4718
Digacin [Beiersdorf] *see* 3150
Digallic Acid, 3135
"Digallic Acid" *see* 9023
m-Digallic Acid *see* 3135
Digalogenin, 3136
Digenic Acid *see* 5157
Digenin *see* 5157
Di-gentiobiose Ester of Crocetin *see*
2593
Digerent Polifarma [Polifarma] *see*
9613
Digicor [Hennig] *see* 3146
Digifortis [Parke, Davis] *see* 3141
Digilanide A *see* 5229
Digilanide B *see* 5229
Digilanide C *see* 5229
Digilanide D *see* 5229
Digilong [Boehringer, Mann.] *see* 3146
Digimed [Trommsdorf] *see* 3146
Digimerck [E. Merck] *see* 3146
Digin *see* 4325
Diginatigenin, 3137
Diginatin, 3138
Diginigenin *see* 3139
Diginin, 3139
Diginorgin *see* 3140
Diginose *see* 3139
3β-(Diginosyloxy)-12α,20α-epoxy-
14β,17α-pregn-5-ene-11,15-dione
see 3139
Digipural [Schaper & Brümmer] *see*
3146
Digisidin [Winthrop] *see* 3146
Digitalin, 3140
Digitalin, Crystalline *see* 3146
Digitaline Nativelle [Nativelle; Savage]
see 3146
Digitalinum True *see* 3140
Digitalinum Verum *see* 3140
Digitalis, 3141
Digitalose, 3142
Digitin *see* 3144
Digitoflavone *see* 5483
Digitogenin, 3143
Digitonin, 3144
Digitophyllin *see* 3146
Digitora [Upjohn] *see* 3141
Digitoxigenin, 3145
Digitoxin, 3146
Digitoxose, 3147
Diglycine *see* 4832
Diglycolic Acid Bis(2,4,6-triiodo-3-
carboxanilide) *see* 4939
Diglycol Laurate *see* 3110
3,3'-(Diglycoloyldiimino)bis(2,4,6-
triiodobenzoic Acid) *see* 4939
Diglycolyldiamide *see* 7434
Diglyme, 3148
Dignonitrat [Dignos] *see* 5114
Dignover [Dignos] *see* 9851
Digoxigenin, 3149
Digoxin, 3150
Diguanide *see* 1233
2,4-Diguanidino-1-dodecyloxybenzene
see 5256
2,4-Diguanidino-1-lauryloxybenzene
see 5256
2,4-Diguanidinophenyl Dodecyl Ether
see 5256
2,4-Diguanidinophenyl Lauryl Ether
see 5256
1,3-Diguanido-2,4,5,6-cyclohexanetet-
rol *see* 8781
Diguanyl *see* 2088

Dihydrohydroxymorphinone *see* 6922
Dihydro-14-hydroxymorphinone *see* 6922
1,2-Dihydro-6-hydroxy-2-oxo-4-pyridinecarboxylic Acid *see* 2327
3,4-Dihydro-3-(4-hydroxyphenyl)-2*H*-1-benzopyran-7-ol *see* 3583
Dihydro-8-(hydroxyphenylmethyl)-4-[(1-methyl-1*H*-imidazol-5-yl)methyl]-2(3*H*)-furanone *see* 5085
3,7-Dihydro-1-(2-hydroxypropyl)-3,7-dimethyl-1*H*-purine-2,6-dione *see* 7901
3,7-Dihydro-7-(2-hydroxypropyl)-1,3-dimethyl-1*H*-purine-2,6-dione *see* 7921
1,2-Dihydro-12-hydroxysenecionan-11,16-dione *see* 7505
Dihydro-14-hydroxy-6β-thebainol 4-Methyl Ether *see* 3442
5,6-Dihydro-2-hydroxy-3,9,10-trimethoxydibenzo[*a,g*]quinolizinium *see* 2488
5,6-Dihydro-3-hydroxy-2,9,10-trimethoxydibenzo[*a,g*]quinolizinium *see* 5142
9,10-Dihydro-5-hydroxy-4,8,8-trimethyl-2*H*,4*H*-benzo[1,2-*b*:4,3-*c* ']dipyran-2,6(8*H*)-dione *see* 4229
(3*R-trans*)-4,6-Dihydro-8-hydroxy-3,4,5-trimethyl-6-oxo-3*H*-2-benzopyran-7-carboxylic Acid *see* 2329
2,3-Dihydro-3α-hydroxytropidine *see* 9693
3-[(4,5-Dihydro-1*H*-imidazol-2-yl)methyl]-6-(1,1-dimethylethyl)-2,4-dimethylphenol *see* 6919
3-[[(4,5-Dihydro-1*H*-imidazol-2-yl)methyl](4-methylphenyl)amino]phenol *see* 7234
1,6-Dihydro-6-iminopurine *see* 141
3,6-Dihydro-6-iminopurine *see* 141
2,3-Dihydro-1*H*-indene *see* 4844
N-(2,3-Dihydro-1*H*-inden-2-yl)-*N* ',*N* '-diethyl-*N*-phenyl-1,3-propanediamine *see* 781
N-(2,3-Dihydro-1*H*-inden-4-yl)-4,5-dihydro-1*H*-imidazol-2-amine *see* 4845
(*S*)-*N*-(2,3-Dihydro-1*H*-inden-2-yl)-*N*-[*N*-[1-(ethoxycarbonyl)-3-phenylpropyl]-L-alanyl]glycine *see* 2864
6-[[3-[(2,3-Dihydro-1*H*-inden-5-yl)oxy]-1,3-dioxo-2-phenylpropyl]amino]-3,3-dimethyl-7-oxo-4-thia-1-azabicyclo[3.2.0]heptane-2-carboxylic Acid *see* 1847
8,13-Dihydroindolo[2',3':3,4]pyrido[2,1-*b*]quinazolin-5(7*H*)-one *see* 8271
2,3-Dihydro-5-iodo-2-thioxo-4(1*H*)-pyrimidinone *see* 4954
Dihydroisocodeine, 3159
2,3-Dihydro-2-isopropenyl-5-benzofuranyl Methyl Ketone *see* 9497
3,4-Dihydro-2(1*H*)-isoquinolinecarboxamidine *see* 2837
3,4-Dihydro-2(1*H*)-isoquinolinecarboximidamide *see* 2837
1,2-Dihydro-2-ketobenzisosulfonazole *see* 8282
9,10α-Dihydrolisuride *see* 9095
Dihydromenperformon *see* 3653
(6a*R-cis*)-6a,11a-Dihydro-9-methoxy-6*H*-benzofuro[3,2-*c*][1]benzopyran-3-ol *see* 5673
(8α,9*R*)-10,11-Dihydro-6'-methoxycinchonan-9-ol *see* 4737
(9*S*)-10,11-Dihydro-6'-methoxycinchonan-9-ol *see* 4736

(6a*R-cis*)-6a,12a-Dihydro-3-methoxy-6*H*-[1,3]dioxolo[5,6]benzofuro-[3,2-*c*][1]benzopyran *see* 7948
3,4-Dihydromethoxyharman *see* 5915
7,8-Dihydro-4-methoxy-6-methyl-1,3-dioxolo[4,5-*g*]isoquinolinium Chloride *see* 2551
3,9-Dihydro-8-methoxy-9-methyl-2-(1-methylethyl)furo[2,3-*b*]quinolin-4(2*H*)-one *see* 5473
1,2-Dihydro-4-methoxy-1-methyl-2-oxonicotinonitrile *see* 8212
1,2-Dihydro-4-methoxy-1-methyl-2-oxo-3-pyridinecarbonitrile *see* 8212
3,4-Dihydro-2-methoxy-2-methyl-4-phenyl-2*H*,5*H*-pyrano[3,2-*c*][1]-benzopyran-5-one *see* 2723
4,9-Dihydro-6-methoxy-1-methyl-3*H*-pyrido[3,4-*b*]indole *see* 5915
4,9-Dihydro-7-methoxy-1-methyl-3*H*-pyrido[3,4-*b*]indole *see* 4528
3,4-Dihydro-6-methoxy-1(2*H*)-naphthalenone *see* 5924
5,8-Dihydro-5-methoxy-8-oxo-1,3-dioxolo[4,5-*g*]quinoline-7-carboxylic Acid *see* 6116
5,6-Dihydro-4-methoxy-6-(2-phenylethenyl)-2*H*-pyran-2-one *see* 5167
10,11-Dihydro-5-[3-(methylamino)-propyl]-5*H*-dibenz[*b,f*]azepine *see* 2902
10,11-Dihydro-5-(3-methylaminopropylidene)-5*H*-dibenzo[*a,d*][1,4]-cycloheptene *see* 6635
1,2-Dihydro-3-methylbenz[*j*]aceanthrylene *see* 5967
4,5-Dihydro-2-[(2-methylbenzo[*b*]-thien-3-yl)methyl]-1*H*-imidazole *see* 6060
3,4-Dihydro-6-methyl-2*H*-1-benzothiopyran-7-sulfonamide 1,1-Dioxide *see* 6058
5,6-Dihydro-14-methylbis[1,3]benzodioxolo[5,6-*a*:5',6'-*g*]quinolizinium *see* 9963
1,3-Dihydro-7-methyl-3,3-bis[4-(sulfooxy)phenyl]-2*H*-indol-1-one *see* 8962
10,11-Dihydro-6'-(3-methylbutoxy)-cinchonan-9-ol *see* 3861
Dihydro-5-(1-methylbutyl)-5-[2-(methylthio)ethyl]-2-thioxo-4,6-(1*H*,5*H*)-pyrimidinedione Monosodium Salt *see* 5901
Dihydro-5-(1-methylbutyl)-5-(2-propenyl)-2-thioxo-4,6(1*H*,5*H*)-pyrimidinedione *see* 9231
16,17-Dihydro-17-methyl-15*H*-cyclopenta[*a*]phenanthrene *see* 5969
10,11-Dihydro-*N*-methyl-5*H*-dibenz-[*b,f*]azepine-5-propanamine *see* 2902
10,11-Dihydro-*N*-methyl-5*H*-dibenzo[*a,d*]cycloheptene-Δ^5,γ-propylamine *see* 6635
10,11-Dihydro-5-methyl-5*H*-dibenzo-[*a,d*]cyclohepten-5,10-imine *see* 3392
5,6-Dihydro-1-methyl-5,6-dioxo-2-indolinesulfonic Acid 5-Semicarbazone Sodium Salt *see* 1791
3-[(1,4-Dihydro-3-methyl-1,4-dioxo-2-naphthalenyl)thio]propanoic Acid *see* 9939
3-(1,4-Dihydro-3-methyl-1,4-dioxo-2-naphthylthio)propionic Acid *see* 9939
Dihydro-5-methylene-2(3*H*)-furanone *see* 679
4,5-Dihydro-3-methylene-2(3*H*)-furanone *see* 5981

1-[2,3-Dihydro-2-(1-methylethenyl)-5-benzofuranyl]ethanone *see* 9497
(*R*)-1-[2,3-Dihydro-2-(1-methylethyl)-5-benzofuranyl]ethanone *see* 9497
4,5-Dihydro-2-[[2-(1-methylethyl)-phenoxy]methyl]-1*H*-imidazole *see* 3930
4,5-Dihydro-2-methyl-1*H*-imidazole *see* 5508
1,3-Dihydro-1-methyl-2*H*-imidazole-2-thione *see* 5892
10,11-Dihydro-5-methyl-10-(methylamino)-5*H*-dibenz[*b,f*]azepine *see* 5835
cis-(±)-Dihydro-4'-methyl-4-(methylamino)spiro[benz[*cd*]indole-5(1*H*),2'(5'*H*)-furan]-5'-one *see* 8268
7,10-Dihydro-10-methyl-8-(1-methylethyl)-1,3-dioxolo[4,5-*h*]furo-[2,3-*b*]quinolin-6(8*H*)-one *see* 5475
2-[4,5-Dihydro-4-methyl-4-(1-methylethyl)-5-oxo-1*H*-imidazol-2-yl)-4(and 5)-Methylbenzoic Acid Methyl Ester *see* 4825
2-[4,5-Dihydro-4-methyl-4-(1-methylethyl)-5-oxo-1*H*-imidazol-2-yl)-3-quinolinecarboxylic Acid *see* 4826
4,5-Dihydro-2-[[5-methyl-2-(1-methylethyl)phenoxy]methyl]-1*H*-imidazole *see* 9742
1,3-Dihydro-4-methyl-5-[4-(methylthio)benzoyl]-2*H*-imidazol-2-one *see* 3542
2,4-Dihydro-5-methyl-4-nitro-2-(4-nitrophenyl)-3*H*-pyrazol-3-one *see* 7383
1,3-Dihydro-1-methyl-7-nitro-5-phenyl-2*H*-1,4-benzodiazepin-2-one *see* 6465
2,3-Dihydro-8-methylnortropidine *see* 9689
1,6-Dihydro-2-methyl-6-oxo-(3,4'-bipyridine)-5-carbonitrile *see* 6117
N-(4,5-Dihydro-4-methyl-5-oxo-1,2-dithiolo[4,3-*b*]pyrrol-6-yl)acetamide *see* 9271
N-(4,5-Dihydro-4-methyl-5-oxo-1,2-dithiolo[4,3-*b*]pyrrol-6-yl)propanamide *see* 898
3-[[(2,5-Dihydro-4-methyl-5-oxo-2-furanyl)oxy]methylene]-3,3a,4,5,6,-7,8,8b-octahydro-5-hydroxy-8,8-dimethyl-2*H*-indeno[1,2-*b*]furan-2-one *see* 8795
3,7-Dihydro-3-methyl-1-(5-oxohexyl)-7-propyl-1*H*-purine-2,6-dione *see* 7823
2-(1,2-Dihydro-1-methyl-2-oxo-3*H*-indol-3-ylidene)hydrazinecarbothioamide *see* 5900
4,5-Dihydro-5-methyl-4-oxo-5-phenyl-2-furancarboxylic Acid *see* A1
1,2-Dihydro-6-methyl-2-oxo-5-(4-pyridinyl)nicotinonitrile *see* 6117
3,4-Dihydro-2-methyl-4-oxo-*N*-2-pyridyl-2*H*-1,2-benzothiazine-3-carboxamide 1,1-Dioxide *see* 7476
3-[(1,4-Dihydro-2-methyl-4-oxo-5-pyrimidinyl)methyl]-5-(2-hydroxyethyl)-4-methylthiazolium Chloride *see* 6932
1,4-Dihydro-1-methyl-4-oxoquinoline *see* 3472
3,4-Dihydro-2-methyl-4-oxo-3-*o*-tolylquinazoline *see* 5872
3,6-Dihydro-α-[(2-methylphenoxy)methyl]-1(2*H*)-pyridineethanol *see* 9449

5,6-Dihydro-2-methyl-3-(phenyl-carbamoyl)-4H-pyran *see* 7966

5,6-Dihydro-2-methyl-N-phenyl-1,4-oxathiin-3-carboxamide *see* 1832

3,4-Dihydro-6-methyl-N-phenyl-2H-pyran-5-carboxamide *see* 7966

5,11-Dihydro-11-[(4-methyl-1-piperazinyl)acetyl]-6H-pyrido[2,3-b]-[1,4]benzodiazepin-6-one *see* 7463

6,11-Dihydro-11-(1-methyl-4-piperidinylidene)-5H-benzo[5,6]cyclohepta[1,2-b]pyridine *see* 917

4,9-Dihydro-4-(1-methyl-4-piperidinylidene)-10H-benzo[4,5]cyclohepta[1,2-b]thiophen-10-one *see* 5187

1,2-Dihydro-2-(1-methyl-4-piperidinyl)-5-phenyl-4-(phenylmethyl)-3H-pyrazol-3-one *see* 1131

Dihydro-5-(2-methylpropyl)-5-(2-propenyl)-2-thioxo-4,6(1H,5H)-pyrimidinedione Monosodium Salt *see* 1518

3,4-Dihydro-6-methyl-2H-pyran-5-carboxanilide *see* 7966

S-5,6-Dihydro-6-methyl-2H-pyran-2-one *see* 6981

[2-(3,6-Dihydro-4-methyl-1(2H)-pyridinyl)ethyl]guanidine *see* 4470

3,4-Dihydro-6-methyl-9H-pyrido-[3,4-b]indol-7-ol *see* 4529

4,9-Dihydro-1-methyl-3H-pyrido-[3,4-b]indol-7-ol *see* 4529

3-(4,5-Dihydro-1-methyl-1H-pyrrol-2-yl)pyridine *see* 6013

1a,7a-Dihydro-7a-methyl-1a-(3,7,11,-15-tetramethyl-2-hexadecenyl)-naphth[2,3-b]oxirene-2,7-dione *see* 9934

2,3-Dihydro-3-methyl-2-thioxo-1H-imidazole-1-carboxylic Acid Ethyl Ester *see* 1803

2,3-Dihydro-6-methyl-2-thioxo-4(1H)-pyrimidinone *see* 6048

3,4-Dihydro-2-methyl-2-(4,8,12-trimethyltridecyl)-2H-1-benzopyran-6-ol *see* 9416

5,6-Dihydromocimycin *see* 6138

Dihydromorphine, 3160

Dihydromorphinone *see* 4733

2,3-Dihydro-1-[(4-morpholinyl)acetyl]-3-phenyl-4(1H)-quinazolinone *see* 6174

Dihydromyricetin *see* 611

4,5-Dihydro-2-(1-naphthalenylmethyl)-1H-imidazole *see* 6287

Dihydrone *see* 6912

Dihydroneopine *see* 3154

Dihydronicotinamide Adenine Dinucleotide Phosphate Diaphorase *see* 6784

4,5-Dihydro-β-nicotyrine *see* 6013

2,3-Dihydro-4-nitro-2,3-dioxo-9,10-secostrychnidin-10-oic Acid *see* 1604

6,7-Dihydro-3-(5-nitro-2-furanyl)-5H-imidazo[2,1-b]thiazol-4-ium Chloride *see* 4211

1,3-Dihydro-7-nitro-5-phenyl-2H-1,4-benzodiazepin-2-one *see* 6491

4,5-Dihydroorotic Acid *see* 4735

9,10-Dihydro-9-oxoanthracene *see* 721

2,3-Dihydro-3-oxobenzisosulfonazole *see* 8282

3,4-Dihydro-1-oxo-1H-2-benzopyran-5-carboxaldehyde *see* 3623

6,11-Dihydro-11-oxodibenz[b,e]oxepin-2-acetic Acid *see* 5126

N-(4,5-Dihydro-5-oxo-1,2-dithiolo-[4,3-b]pyrrol-6-yl)acetamide *see* 4647

N-(4,5-Dihydro-5-oxo-1,2-dithiolo-[4,3-b]pyrrol-6-yl)-N-methylformamide *see* 4647

2-(1,3-Dihydro-3-oxo-2H-indol-2-ylidene)-1,2-dihydro-3H-indol-3-one *see* 4855

(±)-4-(1,3-Dihydro-1-oxo-2H-isoindol-2-yl)-α-ethylbenzeneacetic Acid *see* 4867

4-(1,3-Dihydro-1-oxo-2H-isoindol-2-yl)-α-methylbenzeneacetic Acid *see* 4875

2-(1,3-Dihydro-3-oxo-5-sulfo-2H-indol-2-ylidene)-2,3-dihydro-3-oxo-1H-indole-5-sulfonic Acid Disodium Salt *see* 4856

4,5-Dihydro-5-oxo-4-[(4-sulfo-1-naphthalenyl)azo]-1-(4-sulfophenyl)-1H-pyrazole-3-carboxylic Acid 3-Ethyl Ester Disodium Salt *see* 6814

4,5-Dihydro-5-oxo-1-(4-sulfophenyl)-4-[(4-sulfophenyl)azo]-1H-pyrazole-3-carboxylic Acid Trisodium Salt *see* 9041

Dihydropentaborane(9) *see* 7055

2-[(1,2-Dihydro-5,6,17,19,21-pentahydroxy-23-methoxy-2,4,12,16,18,-20,22-heptamethyl-1,11-dioxo-2,7-(epoxypentadeca[1,11,13]trienimino)naphtho[2,1-b]furan-9-yl)oxy]-N,N-diethylacetamide 21-Acetate *see* 8214

5-(4,5-Dihydro-2-phenyl-3H-benz[e]-indol-3-yl)-2-hydroxybenzoic Acid *see* 3917

5-(4,5-Dihydro-2-phenyl-3H-benz[e]-indol-3-yl)salicylic Acid *see* 3917

4,5-Dihydro-2-(phenylmethyl)-1H-imidazole *see* 9430

3,4-Dihydro-3-(phenylmethyl)-6-(trifluoromethyl)-2H-1,2,4-benzothiadiazine-7-sulfonamide 1,1-Dioxide *see* 1045

4,5-Dihydro-5-phenyl-2-oxazolamine *see* 490

4,5-Dihydro-N-phenyl-N-(phenylmethyl)-1H-imidazole-2-methanamine *see* 709

2,3-Dihydro-1,4-phthalazinedione Dihydrazone *see* 3152

3,7-Dihydro-3-propyl-1H-purine-2,6-dione *see* 3544

2,3-Dihydro-6-propyl-2-thioxo-4(1H)pyrimidinone *see* 7882

Dihydropsychotrine *see* 1970

3,7-Dihydro-1H-purine-2,6-dione *see* 9968

7,9-Dihydro-1H-purine-2,6,8(3H)-trione *see* 9790

1,7-Dihydro-6H-purin-6-one *see* 4805

4,5-Dihydro-1H-pyrazole *see* 7975

1,5-Dihydro-4H-pyrazolo[3,4-d]-pyrimidin-4-one *see* 278

4',5'-Dihydropyrethrin I *see* 5140

4',5'-Dihydropyrethrin II *see* 5140

1,2-Dihydro-3,6-pyridazinedione *see* 5587

2,5-Dihydro-1H-pyrrole *see* 8028

3,4-Dihydropyrrole-2,5-dione *see* 8842

Dihydro-3-pyrroline-2,5-dione *see* 8842

Dihydroqinghaosu *see* 844

Dihydroqinghaosu Ethyl Ether *see* 841

Dihydroqinghaosu Hemisuccinate *see* 845

Dihydroqinghaosu Methyl Ether *see* 842

Dihydroquinidine *see* 4736

Dihydroquinine *see* 4737

3,4-Dihydro-2-quinolinol *see* 4702

3,4-Dihydro-2(1H)-quinolinone *see* 4702

Dihydro-α-quinolone *see* 4702

10,11-Dihydro-5-(3-quinuclidinyl)-5H-dibenz[b,f]azepine *see* 8111

Dihydroresorcinol, 3161

Dihydrorhengol *see* 7061

Dihydro-β-sitosterol *see* 8770

3',4'-Dihydrospiro[imidazolidine-4,2'(1'H)-naphthalene]-2,5-dione *see* 9165

22:23-Dihydrostigmasterol *see* 8500

Dihydrostreptomycin, 3162

3,4-Dihydro-7-sulfamyl-6-trifluoromethyl-1,2,4-benzothiadiazine 1,1-Dioxide *see* 4716

Dihydrotachysterol, 3163

4-Dihydrotestosterone *see* 8753

[2S-[2α,9β,9(R*),9aβ]]-6,7-Dihydro-7-[[2,3,9,9a-tetrahydro-9-hydroxy-2-(2-methylpropyl)-3-oxo-1H-imidazo[1,2-a]indol-9-yl]methyl]-quinazolino[3,2-a][1,4]benzodiazepine-5,13-dione *see* 865

4,5-Dihydro-N-(5,6,7,8-tetrahydro-1-naphthalenyl)-1H-imidazol-2-amine *see* 9486

4,5-Dihydro-2-(1,2,3,4-tetrahydro-1-naphthalenyl)-1H-imidazole *see* 9150

6a,7-Dihydro-3,4,6a,10-tetrahydroxy-benz[b]indeno[1,2-d]pyran-9(6H)-one *see* 4553

9,10-Dihydro-3,5,6,8-tetrahydroxy-1-methyl-9,10-dioxo-2-anthracene-carboxylic Acid *see* 5172

5,6-Dihydro-2,3,9,10-tetramethoxydibenzo[a,g]quinolizinium *see* 6945

6,7-Dihydro-1,2,3,10-tetramethoxy-7-(methylamino)benzo[a]heptalen-9(5H)-one *see* 2873

3,4-Dihydro-2,5,7,8-tetramethyl-2-(4,8,12-trimethyl-3,7,11-tridecatrienyl)-2H-1-benzopyran-6-ol *see* 9421

3,4-Dihydro-2,5,7,8-tetramethyl-2-(4,8,12-trimethyltridecyl)-2H-1-benzopyran-6-ol *see* 9931

3,4-Dihydro-2,5,7,8-tetramethyl-2-(4,8,12-trimethyltridecyl)-2H-1-benzopyran-6-ol Acetate *see* 9932

Dihydrothebaine, 3164

Dihydrotheelin *see* 3653

2,5-Dihydrothiophene 1,1-Dioxide *see* 8935

4-[4-(2,3-Dihydro-2-thioxo-1H-benzimidazol-1-yl)-1-piperidinyl]-1-(4-fluorophenyl)-1-butanone *see* 9373

2,3-Dihydro-2-thioxo-4(1H)-pyrimidinone *see* 9298

3,4-Dihydro-6-(trifluoromethyl)-2H-1,2,4-benzothiadiazine-7-sulfonamide 1,1-Dioxide *see* 4716

3,4-Dihydro-3,6,9-trihydroxy-7-methoxy-3-methyl-1H-naphtho-[2,3-c]pyran-5,10-dione *see* 4228

(E,E)-6',7'-Dihydro-4,9,9'-trihydroxy-6-methoxy-3'-(1,3-pentadienyl)spiro[2H-benz(f)indene-2,8'-(8H)cyclopent(g)isoquinoline]-1,1',3,5,8(2'H)-pentone *see* 4181

8,9-Dihydro-4,6,7a-trihydroxy-5-methoxy-1,8,8,9-tetramethyl-3H-phenaleno[1,2-b]furan-3,7(7aH)-dione *see* 4589

(2R-trans)-2,3-Dihydro-3,5,7-trihydroxy-2-(3,4,5-trihydroxyphenyl)-4H-1-benzopyran-4-one *see* 611

10,11-Dihydro-*N*,*N*,*β*-trimethyl-5*H*-dibenz[*b*,*f*]azepine-5-propanamine *see* 9636

10,11-Dihydro-*N*,*N*,*β*-trimethyl-5*H*-dibenzo[*a*,*d*]cycloheptene-5-propanamine *see* 1533

3,7-Dihydro-1,3,7-trimethyl-8-[[2-[methyl(1-methyl-2-phenylethyl)-amino]ethyl]amino]-1*H*-purine-2,6-dione *see* 3910

3,7-Dihydro-1,3,7-trimethyl-1*H*-purine-2,6-dione *see* 1635

3,4-Dihydro-2,5,8-trimethyl-2-(4,8,-12-trimethyl-3,7,11-tridecatrienyl)-2*H*-1-benzopyran-6-ol *see* 9420

3,4-Dihydro-2,5,7-trimethyl-2-(4,8,-12-trimethyltridecyl)-2*H*-1-benzopyran-6-ol *see* 9422

3,4-Dihydro-2,5,8-trimethyl-2-(4,8,-12-trimethyltridecyl)-2*H*-1-benzopyran-6-ol *see* 9417

3,4-Dihydro-2,7,8-trimethyl-2-(4,8,-12-trimethyltridecyl)-2*H*-1-benzopyran-6-ol *see* 9418

22:23-Dihydrovitamin D₂ see 9930

Dihydrovitamin K₁, 3165

Dihydroxyacetone, 3166

2′,4′-Dihydroxyacetophenone *see* 8143

Dihydroxy(acetylsalicylato)aluminum *see* 3167

2(*R*),3(*R*)-Dihydroxy-4-(9-adenyl)-butyric Acid, (D-*erythro*-Form) *see* 3618

Dihydroxyaluminum Acetylsalicylate, 3167

Dihydroxyaluminum Aminoacetate, 3168

Dihydroxyaluminum Aspirin *see* 3167

Dihydroxyaluminum Sodium Carbonate, 3169

1,3-Dihydroxy-2-amino-4-octadecene *see* 8703

3,11-Dihydroxyandrostan-17-one *see* 669

4,4′-(3α,17β-Dihydroxy-5α-androstan-2β,16β-ylene)bis(1,1-dimethyl-piperazinium)dibromide Diacetate Dihydrate *see* 7426

1,1′-(3α,17β-Dihydroxy-5α-androstan-2β,16β-ylene)bis[1-methylpiperidinium] Dibromide Diacetate *see* 6958

1,2-Dihydroxy-9,10-anthracenedione *see* 237

1,4-Dihydroxy-9,10-anthracenedione *see* 8094

1,5-Dihydroxy-9,10-anthracenedione *see* 719

1,8-Dihydroxy-9,10-anthracenedione *see* 2813

1,8-Dihydroxy-9(10*H*)-anthracenone *see* 714

3,4-Dihydroxyanthranol *see* 718

1,2-Dihydroxyanthraquinone *see* 237

1,4-Dihydroxyanthraquinone *see* 8094

1,5-Dihydroxyanthraquinone *see* 719

1,8-Dihydroxyanthraquinone *see* 2813

1,8-Dihydroxyanthraquinone-3-carboxylic Acid *see* 8175

4,5-Dihydroxyanthraquinone-2-carboxylic Acid *see* 8175

1,8-Dihydroxyanthrone *see* 714

1,2-Dihydroxybenz[*j*]aceanthrylene *see* 2200

2,4-Dihydroxybenzaldehyde *see* 8159

3,4-Dihydroxybenzaldehyde *see* 7908

N,2-Dihydroxybenzamide *see* 8300

m-Dihydroxybenzene *see* 8158

p-Dihydroxybenzene *see* 4738

1,2-Dihydroxybenzene *see* 8009

2,5-Dihydroxybenzeneacetic Acid *see* 4658

2,4-Dihydroxybenzenecarbonal *see* 8159

3,4-Dihydroxybenzenecarbonal *see* 7908

2,4-Dihydroxybenzenecarboxylic Acid *see* 8160

1,2-Dihydroxybenzene-3,5-disulfonic Acid Disodium Salt *see* 9392

4,5-Dihydroxy-1,3-benzenedisulfonic Acid Disodium Salt *see* 9392

2,5-Dihydroxybenzenesulfonic Acid Calcium Salt *see* 3395

2,5-Dihydroxybenzenesulfonic Acid Compd with *N*-Ethylethanamine *see* 3675

2-(4,4′-Dihydroxybenzhydryl)benzyl Alcohol *see* 7211

4-(*p*,*p* ′-Dihydroxybenzhydrylidene)-2,5-cyclohexadien-1-one *see* 899

3,9-Dihydroxy-6*H*-benzofuro[3,2-c]-[1]benzopyran-6-one *see* 2565

2,4-Dihydroxybenzoic Acid *see* 8160

2,5-Dihydroxybenzoic Acid *see* 4290

3,4-Dihydroxybenzoic Acid *see* 7908

2-6-Dihydroxybenzoic Acid 4′-Bromoanilide *see* 8157

4,5-Dihydroxybenzoic Acid Monogallate *see* 3135

2,4-Dihydroxybenzophenone *see* 1117

6,7-Dihydroxy-2*H*-1-benzopyran-2-one *see* 3645

7,8-Dihydroxy-2*H*-1-benzopyran-2-one *see* 2816

2,5-Dihydroxybenzyl Alcohol *see* 4292

L-α-(3,4-Dihydroxybenzyl)-α-hydrazinopropionic Acid Monohydrate *see* 1802

2-(3,4-Dihydroxybenzylidene)-6-hydroxy-3(2*H*)-benzofuranone *see* 8951

1-(3,4-Dihydroxybenzyl)-1,2,3,4-tetrahydro-6,7-isoquinolinediol *see* 9148

5,5′-Dihydroxy-5,5′-bibarbituric Acid *see* 282

2,2′-Dihydroxy-[2,2′-biindan]-1,1′,-3,3′-tetrone *see* 4698

2,2′-Dihydroxy-[2,2′-bi-1*H*-indene]-1,1′,3,3′-(2*H*,2′*H*)-tetrone *see* 4698

3,3′-Dihydroxy-[3,3′-biindoline]-2,2′-dione *see* 4984

5,5′-Dihydroxy-[5,5′-bipyrimidine]-2,2′,4,4′,6,6′(1*H*,1′*H*,3*H*,3′*H*,5*H*,-5′*H*)-hexone *see* 282

2,6-Dihydroxy-5-bis[2-chloroethyl]-aminopyramidine *see* 9762

1,4-Dihydroxy-5,8-bis[[2-[(2-hydroxyethyl)amino]ethyl]amino]-9,10-anthracenedione *see* 6135

1,4-Dihydroxy-5,8-bis[[2-[(2-hydroxyethyl)amino]ethyl]amino]-9,10-anthraquinone *see* 6135

8-[(Dihydroxybismutho)oxy]-6-methylquinoline *see* 5650

3,14-Dihydroxybufa-20,22-dienolide *see* 1458

3β,14-Dihydroxybufa-4,20,22-trienolide *see* 8355

1,3-Dihydroxybutane *see* 1566

2,3-Dihydroxybutane *see* 1567

L-2,3-Dihydroxybutanedioic Acid *see* 9039

[*R*-(*R**,*R**)]-2,3-Dihydroxybutanedioic Acid *see* 9039

[*S*-(*R**,*R**)]-2,3-Dihydroxybutanedioic Acid *see* 9037

2,3-Dihydroxybutanedioic Acid *see* 9038

2,3-Dihydroxybutanedioic Acid Aluminum Salt *see* 374

2,3-Dihydroxybutanedioic Acid Calcium Salt *see* 1716

2,3-Dihydroxybutanedioic Acid Copper Salt *see* 2662

(*R*)-2,3-Dihydroxybutanedioic Acid Diethyl Ester *see* 3811

(*R*)-2,3-Dihydroxybutanedioic Acid Monoammonium Salt *see* 523

(*R*)-2,3-Dihydroxybutanedioic Acid Monoethyl Ester *see* 3812

(*Z*)-2,3-Dihydroxy-2-butenedioic Acid *see* 3170

3,14-Dihydroxycard-20(22)-enolide *see* 3145

3β,14-Dihydroxy-5α-card-20(22)-enolide *see* 9809

(3*R*,3′*R*)-Dihydroxy-β-carotene *see* 10019

3,3′-Dihydroxy-β,β-carotene-4,4′-dione *see* 877

(3*R*,3′*S*,5′*R*)-3,3′-Dihydroxy-β,*k*-caroten-6′-one *see* 1768

10β,13-Dihydroxycedr-8-ene-12,15-dioic Acid *see* 8427

3,20-Dihydroxycevan-6-one *see* 4840

3α,6α-Dihydroxy-5β-cholanic Acid *see* 4794

3α,7α-Dihydroxy-5β-cholanic Acid *see* 2044

(3α,5β,6α)-3,6-Dihydroxycholan-24-oic Acid *see* 4794

(3α,5β,7α)-3,7-Dihydroxycholan-24-oic Acid *see* 2044

(3α,5β,7β)-3,7-Dihydroxycholan-24-oic Acid *see* 9801

(3α,5β,12α)-3,12-Dihydroxy-5-cholan-24-oic Acid *see* 2881

1α,25-Dihydroxycholecalciferol *see* 1641

21,25-Dihydroxycholecalciferol *see* 1641

3,4-Dihydroxycinnamic Acid *see* 1634

3,4-Dihydroxycinnamic Acid 1-Carboxy-4,5-dihydroxy-1,3-cyclohexylene Ester *see* 2773

3-(3,4-Dihydroxycinnamoyl)quinic Acid *see* 2142

6,7-Dihydroxycoumarin *see* 3645

7,8-Dihydroxycoumarin *see* 2816

6,7-Dihydroxycoumarin 6-Glucoside *see* 3646

6,7-Dihydroxycoumarin-7-glucoside *see* 2267

7,8-Dihydroxycoumarin 7-β-D-Glucoside *see* 2817

7′,6-Dihydroxycoumarino(3′,4′,3,2)-coumarone *see* 2565

1,2-Dihydroxy-1,2-cyclobutanediacrylic Acid Di-γ-lactone *see* 674

5,6-Dihydroxy-5-cyclohexene-1,2,-3,4-tetrone Disodium Salt *see* 8625

2,3-Dihydroxy-2-cyclopenten-1-one *see* 8134

3,3′-Dihydroxydibutyl Ether *see* 6913

1,3-Dihydroxy-2,2-dichlorocyclopentane *see* 1721

2,2′-Dihydroxy-5,5′-dichlorodiphenylmethane *see* 3059

2,2′-Dihydroxy-5,5′-dichlorodiphenyl Sulfide *see* 3947

2,2′-Dihydroxydiethylamine *see* 3097

4,4′-Dihydroxy-α,β-diethyldiphenylethane *see* 4621

4,4′-Dihydroxy-α,β-diethylstilbene *see* 3118

3,3′-Dihydroxy-3,3′-dihydroisoindigo *see* 4984

6,7-Dihydroxy-1-(3,4-dihydroxybenz-yl)-1,2,3,4-tetrahydroisoquinoline *see* 9148

(3',6'-Dihydroxy-2',7'-diiodospiro-[3*H*-2,1-benzoxanthiole-3,9'-[9*H*]-xanthen]-4'-yl)hydroxymercury *S*,*S*-Dioxide Monosodium Salt *see* 5755

3',6'-Dihydroxy-4',5'-diiodospiro-[isobenzofuran-1(3*H*),9'-[9*H*]xan-then]-3-one *see* 3174

3,3'-Dihydroxy-4,4'-diketo-β-caro-tene *see* 877

1,11-Dihydroxy-2,10-dimethoxyapor-phine *see* 2549

2,6-Dihydroxy-3,5-dimethoxyapor-phine *see* 1328

2,2'-Dihydroxy-4,4'-dimethoxyben-zophenone *see* 1109

5,7-Dihydroxy-4',6-dimethoxyflavone *see* 7010

6β,14-Dihydroxy-3,4-dimethoxy-N-methylmorphinan *see* 3442

7',12'-Dihydroxy-6,6'-dimethoxy-2,2',2'-trimethyltubocuraranium Chloride Hydrochloride *see* 9717

3,3'-Dihydroxy-5,5'-dimethyl-2,2'-bi-*p*-benzoquinone *see* 7194

2,2'-Dihydroxy-4,4'-dimethyl[bi-1,4-cyclohexadien-1-yl]-3,3',6,6'-tetrone *see* 7194

4-(2,4-Dihydroxy-3,3-dimethylbutyr-amido)butyric Acid *see* 4664

2,4-Dihydroxy-3,3-dimethylbutyric Acid γ-Lactone *see* 6963

D(+)-N-(2,4-Dihydroxy-3,3-dimeth-ylbutyryl)-β-alanine *see* 6964

α,γ-Dihydroxy-β,β-dimethylbutyryl-β-alanyl-β-aminoethanethiol *see* 6961

3,5-Dihydroxy-4,4-dimethyl-2,5-cyclohexadien-1-one *see* 4021

(11α,13E,15S,17S)-11,15-Dihydroxy-17,20-dimethyl-6,9-dioxoprost-13-en-1-oic Acid Methyl Ester *see* 6827

Dihydroxydimethyldiphenylmethane-disulfonic Acid Polymer *see* 6360

1,3-Dihydroxydimethyl Ketone *see* 3166

trans-1,4-Dihydroxy-2-(3,7-dimethyl-2,6-octadienyl)benzene *see* 4300

(R)-N-(2,4-Dihydroxy-3,3-dimethyl-1-oxobutyl)-β-alanine *see* 6964

N-(2,4-Dihydroxy-3,3-dimethyl-1-oxobutyl)-β-alanine Calcium Salt *see* 1694

[2R-[(1R*),2R*,3R*]]-N-(2,4-Dihydr-oxy-3,3-dimethyl-1-oxobutyl)-β-alanine 2-[(Dichloroacetyl)amino]-3-hydroxy-3-(4-nitrophenyl)propyl Ester *see* 2070

4-[(2,4-Dihydroxy-3,3-dimethyl-1-oxobutyl)amino]butanoic Acid *see* 4664

3,5-Dihydroxy-4,4-dimethyl-2-(1-oxobutyl)-6-[[2,4,6-trihydroxy-3-methyl-5-(1-oxobutyl)phenyl]meth-yl]-2,5-cyclohexadien-1-one *see* 4027

(2E,11α,13E,15S,17S)-11,15-Dihydr-oxy-17,20-dimethyl-9-oxoprosta-2,13-dien-1-oic Acid *see* 5367

11,15-Dihydroxy-16-16-dimethyl-9-oxoprosta-2,13-dien-1-oic Acid Methyl Ester *see* 4279

5,7-Dihydroxy-2,2-dimethyl-6-(2,4,6-trihydroxy-3-methyl-5-acetylbenz-yl)-8-cinnamoyl-1,2-chromene *see* 8250

2,2'-Dihydroxy-6,6'-dinaphthyl Di-sulfide *see* 2831

2,2'-Dihydroxy-1,1'-dinaphthylmeth-ane-3,3'-dicarboxylic Acid *see* 6955

2,2'-Dihydroxy-5,5'-dinitrobenzil *see* 3279

2,5-Dihydroxy-3,6-dinitro-*p*-benzo-quinone *see* 6489

2,5-Dihydroxy-3,6-dinitro-2,5-cyclo-hexadiene-1,4-dione *see* 6489

2,5-Dihydroxy-3,6-dinitroquinone *see* 6489

11β,17α-Dihydroxy-3,20-dioxo-21-(4-methyl-1-piperazinyl)pregna-1,4-diene Hydrochloride *see* 5644

(Z)-6β,16β-Dihydroxy-3,7-dioxo-29-nor-8α,9β,13α,14β-dammara-1,17(20),24-trien-21-oic Acid Di-acetate *see* 4552

3β,21-Dihydroxy-11,20-dioxo-5α-pregnane *see* 262

(11β)-11,21-Dihydroxy-3,20-dioxo-pregn-4-en-18-al *see* 218

4,4'-Dihydroxy-γ,δ-diphenyl-β,δ-hexadiene *see* 3094

4,4'-Dihydroxy-γ,δ-diphenylhexane *see* 4621

4,4'-Dihydroxydiphenylmethane-3,3'-dicarboxylic Acid *see* 5984

2,2'-Dihydroxy-N-(4,5-diphenyloxa-zol-2-yl)diethylamine *see* 3374

threo-2,3-Dihydroxy-1,4-dithiol-butane *see* 3382

Dihydroxydurene *see* 3451

5,12-Dihydroxy-6,8,10,14-eicosatetra-enoic Acid *see* 5339

3,17-Dihydroxyestratriene *see* 3654

3,16α-Dihydroxyestra-1,3,5(10)-trien-17-one Diacetate *see* 2485

3β,17β-Dihydroxyestr-4-ene *see* 1325

4,17β-Dihydroxyestr-4-en-3-one *see* 4768

Dihydroxyestrin *see* 3653

1,2-Dihydroxy-1,2-ethanedisulfonic Acid Disodium Salt *see* 4406

N,N-Di(hydroxyethyl)aminoacetic Acid *see* 1221

17-(1,2-Dihydroxyethyl)androstane-3,17-diol *see* 266

17-(1,2-Dihydroxyethyl)androstane-3,11,17-triol *see* 265

17-(1,2-Dihydroxyethyl)androsten-3-one-11,17-diol *see* 7736

17-(1,2-Dihydroxyethyl)-Δ⁴-androst-en-3-on-17α-ol *see* 7738

α,β-Dihydroxyethylbenzene *see* 8831

1,2-Dihydroxyethylenedicarboxylic Acid *see* 3170

(5Z,13E)-(8R,9S,11R,12R)-9,11-Di-hydroxy-15,15-ethylenedioxy-16-phenoxy-17,18,19,20-tetranorpros-tadienoic Acid *see* 3827

8-(1,2-Dihydroxyethyl)-4-[4-(D-glu-copyranosyloxy)phenyl]-9-hydroxy-1,7-dioxaspiro[4.4]nonane-2,6-dione *see* 5336

Di(hydroxyethyl)glycine *see* 1221

8-(1,2-Dihydroxyethyl)-9-hydroxy-4-(4-hydroxyphenyl)-1,7-dioxaspiro-[4.4]nonane-2,6-dione *see* 5335

Di-(2-hydroxyethyl)nitrosamine *see* 6556

5,7-Dihydroxyflavone *see* 2261

3',6'-Dihydroxyfluoran *see* 4085

11β,17β-Dihydroxy-9α-fluoro-17α-methyl-4-androsten-3-one *see* 4113

(3β,5α,25R)-3,26-Dihydroxyfurost-20(22)-en-12-one *see* 7930

4,6-Dihydroxy-2-(β-D-glucosido)-β-(*p*-hydroxyphenyl)propiophenone *see* 7300

Dihydroxy(glycinato)aluminum *see* 3168

2,2'-Dihydroxy-3,3',5,5',6,6'-hexa-chlorodiphenylmethane *see* 4602

1,6-Dihydroxyhexane *see* 4610

5,7-Dihydroxy-3-(3-hydroxy-4,5-dimethoxyphenyl)-6-methoxy-4*H*-1-benzopyran-4-one *see* 4973

γ,4-Dihydroxy-2-(6-hydroxy-1-hept-enyl)-4-cyclopentanecrotonic Acid λ-Lactone *see* 1363

3,5-Dihydroxy-4-(3-hydroxy-4-meth-oxyhydrocinnamoyl)phenyl-2-O-(6-deoxy-α-L-mannopyranosyl)-β-D-glucopyranoside *see* 6367

5,7-Dihydroxy-2-(3-hydroxy-4-meth-oxyphenyl)-4*H*-1-benzopyran-4-one *see* 3290

5,7-Dihydroxy-3-(3-hydroxy-4-meth-oxyphenyl)-4*H*-1-benzopyran-4-one *see* 7711

3,4-Dihydroxy-1-[1-hydroxy-2-(methylamino)ethyl]benzene *see* 3569

1,8-Dihydroxy-3-(hydroxymethyl)-9,10-anthracenedione *see* 303

1,8-Dihydroxy-3-(hydroxymethyl)-anthraquinone *see* 303

1,8-Dihydroxy-3-hydroxymethyl-10-(6-hydroxymethyl-3,4,5-trihydroxy-2-pyranyl)anthrone *see* 304

5,8-Dihydroxy-2-(hydroxymethyl)-6-methoxy-3-(2-oxopropyl)-1,4-naph-thalenedione *see* 4228

1,7-Dihydroxy-9-hydroxymethyl-3-methyl-4,8-dibenzofurandicarbox-ylic Acid (8,9)-γ-Lactone *see* 7576

7-[3,5-Dihydroxy-2-(3-hydroxy-3-methyl-1-octenyl)cyclopentyl]-5-heptenoic Acid *see* 1829

(S)-5,8-Dihydroxy-2-(1-hydroxy-4-methyl-3-pentenyl)-1,4-naphtha-lenedione *see* 243

(—)-5,8-Dihydroxy-2-(1-hydroxy-4-methyl-3-pentenyl)-1,4-naphthoqui-none *see* 243

3,5-Dihydroxy-α-[[(*p*-hydroxy-α-methylphenethyl)amino]methyl]-benzyl Alcohol *see* 3927

7-[3,5-Dihydroxy-2-(3-hydroxy-1-octenyl)cyclopentyl]-5-heptenoic Acid *see* 7894

7-[3,5-Dihydroxy-2-(3-hydroxy-4-phenoxy-1-butenyl)cyclopentyl]-4,5-heptadienoic Acid Methyl Ester *see* 3940

5,7-Dihydroxy-2-(4-hydroxyphenyl)-4*H*-1-benzopyran-4-one *see* 763

5,7-Dihydroxy-3-(4-hydroxyphenyl)-4*H*-1-benzopyran-4-one *see* 4281

5,7-Dihydroxy-3-(4-hydroxyphenyl)-6-methoxy-4*H*-1-benzopyran-4-one *see* 9057

2,3-Dihydroxy-2-[(4-hydroxyphenyl)-methyl]butanedioic Acid *see* 7479

3,4-Dihydroxy-N-[3-(4-hydroxyphen-yl)-1-methylpropyl]-β-phenylethyl-amine *see* 3396

2,4-Dihydroxy-N-(3-hydroxypropyl)-3,3-dimethylbutanamide *see* 2924

D(+)-α,γ-Dihydroxy-N-(3-hydroxy-propyl)-β,β-dimethylbutyramide *see* 2924

7-[3,5-Dihydroxy-2-[3-hydroxy-4-(3-thienyloxy)-1-butenyl]cyclopentyl]-5-heptenoic Acid *see* 9355

7-[3,5-Dihydroxy-2-[3-hydroxy-4-[3-(trifluoromethyl)phenoxy]-1-buten-yl]cyclopentyl]-5-heptenoic Acid *see* 4121

2,2-Dihydroxy-1,3-indanedione *see* 6470

2,2-Dihydroxy-1*H*-indene-1,3(2*H*)-dione *see* 6470

2-[(5,7-Dihydroxy-8-isobutyryl-2,2-dimethyl-2*H*-1-benzopyran-6-yl)methyl]-3,5-dihydroxy-6-isobutyryl-4,4-dimethyl-2,5-cyclohexadien-1-one *see* 9756

3'-[(2,4-Dihydroxy-5-isobutyryl-3,3-dimethyl-6-oxo-1,4-cyclohexadien-1-yl)methyl]-2',4',6'-trihydroxy-2-methyl-5'-(3-methyl-2-butenyl)propiophenone *see* 9756

4',7-Dihydroxyisoflavan *see* 3583

4',7-Dihydroxyisoflavone *see* 2805

2,6-Dihydroxyisonicotinic Acid *see* 2327

3,4-Dihydroxy-α-[(isopropylamino)-methyl]benzyl Alcohol *see* 5105

3,5-Dihydroxy-α-[(isopropylamino)-methyl]benzyl Alcohol *see* 5836

3,4-Dihydroxy-α-[1-(isopropylamino)propyl]benzyl Alcohol *see* 5053

β,β'-Dihydroxyisopropyl Chloride *see* 2145

11β,21-Dihydroxy-16α,17-isopropylidenedioxy-1,4-pregnadiene-3,20-dione *see* 2908

5,6-Dihydro-2-(2,6-xylidino)-4*H*-1,3-thiazine *see* 9987

Dihydroxymaleic Acid, 3170

Dihydroxymalonic Acid *see* 5814

2,4-Dihydroxy-*N*-[3-[(2-mercaptoethyl)amino]-3-oxopropyl]-3,3-dimethylbutanamide *see* 6961

2,4-Dihydroxy-*N*-[2-[(2-mercaptoethyl)carbamoyl]ethyl]-3,3-dimethylbutyramide *see* 6961

(11β)-11,17-Dihydroxy-21-mercaptopregn-4-ene-3,20-dione *see* 9408

2,11-Dihydroxy-10-methoxyaporphine *see* 6197

α,4-Dihydroxy-3-methoxybenzeneacetic Acid *see* 9840

2,2'-Dihydroxy-4-methoxybenzophenone *see* 3298

2,6-Dihydroxy-4-methoxybenzophenone *see* 2553

7,8-Dihydroxy-6-methoxy-2*H*-1-benzopyran-2-one *see* 4178

3,17-Dihydroxy-16-(methoxycarbonyl)-4-methyl-2,4(1*H*)-cyclo-3,4-secoakuammilanium *see* 3474

7,8-Dihydroxy-6-methoxycoumarin *see* 4178

7,8-Dihydroxy-6-methoxycoumarin-8-β-D-glucoside *see* 4179

4,4'-Dihydroxy-3,3'-(2-methoxyethylidene)dicoumarin *see* 2566

4',5-Dihydroxy-7-methoxyflavanone *see* 8289

4',5-Dihydroxy-7-methoxyflavanone 5-Glucoside *see* 8289

5,7-Dihydroxy-4'-methoxyflavone *see* 9

5,7-Dihydroxy-6-methoxyflavone *see* 6830

5,7-Dihydroxy-4'-methoxyflavone-D-glucosido-L-rhamnoside *see* 5376

4',5-Dihydroxy-7-methoxyisoflavone *see* 7923

5,7-Dihydroxy-4'-methoxyisoflavone *see* 4281

5,7-Dihydroxy-6-methoxy-2-(4-methoxyphenyl)-4*H*-1-benzopyran-4-one *see* 7010

1,8-Dihydroxy-3-methoxy-6-methylanthraquinone *see* 3518

5,8-Dihydroxy-6-methoxy-3-methyl-2-aza-9,10-anthraquinone *see* 1353

6,9-Dihydroxy-7-methoxy-3-methylbenz[g]isoquinoline-5,10-dione *see* 1353

2,5-Dihydroxy-3-methoxy-6-methyl-*p*-benzoquinone *see* 8706

2',6'-Dihydroxy-4'-methoxy-3'-methyl-1-butyrophenone *see* 869

2,5-Dihydroxy-3-methoxy-6-methyl-2,5-cyclohexadiene-1,4-dione *see* 8706

2-[[2,6-Dihydroxy-4-methoxy-3-methyl-5-(1-oxobutyl)phenyl]methyl]-3,5-dihydroxy-4,4-dimethyl-6-(1-oxobutyl)-2,5-cyclohexadien-1-one *see* 868

5,8-Dihydroxy-6-methoxy-2-methyl-3-(2-oxopropyl)-1,4-naphthalenedione *see* 5143

1-(2,6-Dihydroxy-4-methoxy-3-methylphenyl)-1-butanone *see* 869

3,17-Dihydroxy-11β-methoxy-19-nor-17α-pregna-1,3,5-trien-20-yne *see* 6203

2-[[2,4-Dihydroxy-6-methoxy-3-(1-oxobutyl)phenyl]methyl]-3,5-dihydroxy-4,4-dimethyl-6-(1-oxobutyl)-2,5-cyclohexadien-1-one *see* 2899

1,2-Dihydroxy-3-(2-methoxyphenoxy)propane *see* 4465

5,7-Dihydroxy-2-(4-methoxyphenyl)-4*H*-1-benzopyran-4-one *see* 9

5,7-Dihydroxy-6-methoxy-2-phenyl-4*H*-1-benzopyran-4-one *see* 6830

2-(3,4-Dihydroxy-5-methoxyphenyl)-3,5-bis(β-D-glucopyranosyloxy)-7-hydroxy-1-benzopyrylium Chloride *see* 7143

(2,6-Dihydroxy-4-methoxyphenyl)phenylmethanone *see* 2553

2-(3,4-Dihydroxy-5-methoxyphenyl)-3,5,7-trihydroxy-1-benzopyrylium Chloride *see* 7143

3,6-Dihydroxy-4-methoxy-2,5-toluquinone *see* 8706

3,6-Dihydroxy-5-methoxy-*p*-toluquinone *see* 8706

1,7-Dihydroxy-3-methoxy-9*H*-xanthen-9-one *see* 4291

4,7-Dihydroxy-2-methoxyxanthone *see* 4291

3',4'-Dihydroxy-2-(methylamino)acetophenone *see* 161

3,4-Dihydroxy-α-methylaminoacetophenone *see* 161

3,4-Dihydroxy-α-[(methylamino)methyl]benzyl Alcohol *see* 3569

(±)-3,4-Dihydroxy-α-[(methylamino)methyl]benzyl Alcohol 3,4-Dipivalate *see* 3344

4,17-Dihydroxy-17-methylandrost-4-en-3-one *see* 6918

1,3-Dihydroxy-2-methyl-9,10-anthracenedione *see* 8258

1,8-Dihydroxy-3-methyl-9,10-anthracenedione *see* 2263

1,3-Dihydroxy-2-methylanthraquinone *see* 8258

1,8-Dihydroxy-3-methylanthraquinone *see* 2263

1,8-Dihydroxy-3-methyl-9-anthrone *see* 2257

2,4-Dihydroxy-6-methylbenzenecarboxylic Acid *see* 6834

2,4-Dihydroxy-6-methylbenzoic Acid *see* 6834

2,6-Dihydroxy-4-methyl-5-bis[2-chloroethyl]aminopyrimidine *see* 3416

3,9-Dihydroxy-2-(3-methyl-2-butenyl)-6*H*-benzofuro[3,2-*c*][1]benzopyran-6-one *see* 7945

6,7-Dihydroxy-4-methylcoumarin Disulfate *see* 8964

1,1-Dihydroxymethylcyclopentane *N,N'*-Diphenylcarbamate *see* 2709

3,4-Dihydroxy-2-methylenebutanoic Acid 2,3,3a,4,5,8,9,11a-Octahydro-10-(hydroxymethyl)-6-methyl-3-methylene-2-oxocyclodeca[*b*]furan-4-yl Ester *see* 2419

3β,15α-Dihydroxy-24-methylene-lanost-8-en-21-oic Acid *see* 8970

11,21-Dihydroxy-16,17-[(1-methylethylidene)bis(oxy)]pregna-1,4-diene-3,20-dione *see* 2908

2,2-Dihydroxymethyl-*n*-octanol Trinicotinate *see* 4574

11α,17β-Dihydroxy-17-methyl-3-oxoandrosta-1,4-diene-2-carboxaldehyde *see* 4161

4,17β-Dihydroxy-17α-methyl-3-oxoandrost-4-ene *see* 6918

(5Z,11α,13E,15R)-11,15-Dihydroxy-15-methyl-9-oxoprosta-5,13-dien-1-oic Acid *see* 796

(11α,13E)-(±)-11,16-Dihydroxy-16-methyl-9-oxoprost-13-en-1-oic Acid Methyl Ester *see* 6129

8,9-Dihydroxy-2-methyl-4-oxo-4*H*,5*H*-pyrano[3,2-*c*][1]benzopyran-10-carboxylic Acid *see* 2330

3,5-Dihydroxy-3-methylpentanoic Acid *see* 6087

1,2-Dihydroxy-3-(2-methylphenoxy)propane *see* 5737

7,20-Dihydroxy-16-methyl-10-phenyl-24-oxo[14]cytochalasa-6(12),13,-21-triene-1,23-dione *see* 2796

11β,17-Dihydroxy-21-(4-methyl-1-piperazinyl)pregna-1,4-diene-3,20-dione Hydrochloride *see* 5644

11β,17-Dihydroxy-21-(4-methyl-1-piperazinyl)-Δ¹-progesterone Hydrochloride *see* 5644

(E)-3β,7β-Dihydroxy-14α-methylpodocarpane-Δ¹³,ᵅ-acetic Acid 2-(Dimethylamino)ethyl Ester *see* 1895

17,21-Dihydroxy-16β-methylpregna-1,4-diene-3,11,20-trione *see* 5750

11β,21-Dihydroxy-2'-methyl-5'βH-pregna-1,4-dieno[17,16-*d*]oxazole-3,20-dione 2l-Acetate *see* 2852

2,3-Dihydroxy-2-methylpropanoic Acid 6-[(Acetyloxy)methyl]-2,3,3a,-4,7,8,11,11a-octahydro-10-methyl-3-methylene-2-oxocyclodeca[*b*]-furan-4-yl Ester *see* 8346

2,4-Dihydroxy-5-methylpyrimidine *see* 9332

Dihydroxy(6-methyl-8-quinolinolato)bismuth *see* 5650

3,5-Dihydroxy-6-methyl-1,2,4-triazine *see* 919

N,N'-Dihydroxymethylurea *see* 6921

3,5-Dihydroxy-3-methylvaleric Acid *see* 6087

β,δ-Dihydroxy-β-methylvaleric Acid *see* 6087

β,δ-Dihydroxy-β-methyl-δ-valerolactone *see* 6086

1,5-Dihydroxy-2-methyl-6-[(6-*O*-β-D-xylopyranosyl-β-D-glucopyranosyl)oxy]-9,10-anthracenedione *see* 6180

6,7-Dihydroxy-4-(morpholinomethyl)coumarin *see* 4139

N-[[6-(2,3-Dimethoxypropyl)tetrahy-
dro-4-hydroxy-5,5-dimethyl-2H-
pyran-2-yl]methoxymethyl]tetrahy-
dro-α-hydroxy-2-methoxy-5,6-
dimethyl-4-methylene-2H-pyran-2-
acetamide see 7011
N-[[6-(2,3-Dimethoxypropyl)tetrahy-
dro-4-hydroxy-5,5-dimethyl-2H-
pyran-2-yl]methoxymethyl]tetrahy-
dro-2-methoxy-5,6-dimethyl-4-
methylene-2H-pyran-2-glycolamide
see 7011
N'-(5,6-Dimethoxy-4-pyrimidinyl)-
sulfanilamide see 8877
N¹-(2,6-Dimethoxy-4-pyrimidinyl)-
sulfanilamide see 8876
2,6-Dimethoxyquinone, 3214
2,3-Dimethoxystrychnidin-10-one see
1443
10,11-Dimethoxystrychnine see 1443
2,4-Dimethoxy-6-sulfanilamido-1,3-
diazine see 8876
2,6-Dimethoxy-4-sulfanilamidopyri-
midine see 8876
Dimethoxytetraethylene Glycol see
9141
9-[4,5-Dimethoxy-2-[(1,2,3,4-tetrahy-
dro-6,7-dimethoxy-2-methyl-1-
isoquinolinyl)methyl]phenoxy]-5,6,-
6a,7-tetrahydro-1,2,10-trimethoxy-
6-methyl-4H-dibenzo[de,g]quinoline
see 9181
6,7-Dimethoxy-3-(5,6,7,8-tetrahydro-
4-hydroxy-6-methyl-1,3-dioxolo[4,-
5-g]isoquinolin-5-yl)-1(3H)-isoben-
zofuranone see 6343
[S-(R*,S*)]-6,7-Dimethoxy-3-(5,6,7,-
8-tetrahydro-4-methoxy-6-methyl-
1,3-dioxolo[4,5-g]isoquinolin-5-yl)-
1(3H)-isobenzofuranone see 6638
6,7-Dimethoxy-3-(5,6,7,8-tetrahydro-
6-methyl-1,3-dioxolo[4,5-g]isoqui-
nolin-5-yl)-1(3H)-isobenzofuranone
see 4688
6,7-Dimethoxy-1-(3,4,5-triethoxy-
phenyl)isoquinoline see 6676
(3β,16β,17α,18β,20α)-11,17-Dimeth-
oxy-18-[(3,4,5-trimethoxybenzoyl)-
oxy]yohimban-16-carboxylic Acid
Methyl Ester see 8149
6,7-Dimethoxy-1-veratroylisoquino-
line see 6966
6,7-Dimethoxy-1-veratrylisoquinoline
see 6968
Dimethpyrindene see 3204
Dimethulene see 2031
Dimethyl see 3676
Dimethylacetal, 3215
N,N-Dimethylacetamide, 3216
Dimethylacetone see 3111
3,3-Dimethylacrolein see 8399
β,β-Dimethylacrolein see 8399
N¹-Dimethylacroylsulfanilamide see
8875
cis-2,3-Dimethylacrylic Acid see 678
trans-2,3-Dimethylacrylic Acid see
9365
3,3-Dimethylacrylic Acid 2-sec-Butyl-
4,6-dinitrophenyl Ester see 1237
3,5-Dimethyl-1-adamantanamine see
5709
6,7-Dimethylalloxazine see 5468
7,8-Dimethylalloxazine see 5468
2-Dimethylallyl-5,9-dimethyl-2'-
hydroxybenzomorphan see 7078
N-3,3-Dimethylallylguanidine see
4246
Dimethylamidoethoxyphosphoryl
Cyanide see 9001
α-(Dimethylamidooxalyl)-β,β-methyl-
acetylphenylhydrazine see 3299

Dimethylamine, 3217
(Dimethylamino)acetaldehyde Diphenyl Acetal see 5674
(Dimethylamino)acetic Acid see 3233
10-(Dimethylamino)acetyl-10H-
phenothiazine see 179
Dimethylamino-analgesine see 488
4-(Dimethylamino)antipyrine see 488
4-(Dimethylamino)antipyrine Compd
with 5,5-Diethylbarbituric Acid (1:1)
see 7146
p-Dimethylaminoazobenzene, 3218
p-Dimethylaminobenzaldehyde, 3219
p-Dimethylaminobenzalrhodanine,
3220
4-Dimethylaminobenzenecarbonal see
3219
4-(Dimethylamino)benzoic Acid, 3221
p-Dimethylaminobenzophenone, 3222
5-[p-(Dimethylamino)benzylidene]-
rhodanine see 3220
3-Dimethylamino-1,1-bis(2-thienyl)-
1-butene see 3248
3-[[(Dimethylamino)carbonyl]oxy]-1-
methylpyridinium Bromide see 7990
3-[[(Dimethylamino)carbonyl]oxy]-1-
(phenylmethyl)pyridinium Bromide
see 1132
3-[[(Dimethylamino)carbonyl]oxy]-
N,N,N-trimethylbenzenaminium see
6380
N-[1-[(Dimethylamino)carbonyl]-
propyl]-N-ethyl-2-butenamide see
2598
N-[1-[(Dimethylamino)carbonyl]-
propyl]-N-propyl-2-butenamide see
2595
3β-(Dimethylamino)con-5-enine see
2492
3β-(Dimethylamino)con-5-enine-12β-
ol see 4645
(4-Dimethylaminocyclohexyl)di-2-
thienylmethanol see 9242
7-Dimethylamino-6-demethyl-6-
deoxytetracycline see 6121
6-Dimethylamino-9-[3-deoxy-3-(p-
methoxy-L-phenylalanylamino)-β,D-
ribofuranosyl]-β-purine see 7960
2-Dimethylamino-6-(β-diethylaminoethoxy)benzothiazole Dihydrochloride see 2967
4-(Dimethylamino)-1,2-dihydro-1,5-
dimethyl-2-phenyl-3H-pyrazol-3-
one see 488
[4-[p-(Dimethylamino)-α-[p-(dimethylamino)phenyl]benzylidene]-2,5-
cyclohexadien-1-ylidene]dimethyl-
ammonium Chloride Ethobromide
see 5998
2-(Dimethylamino)-1,1-dimethylethyl
Benzilate see 3122
4-(Dimethylamino)-3,5-dimethylphe-
nol Methylcarbamate (Ester) see
6090
4-Dimethylamino-2,3-dimethyl-1-
phenyl-3-pyrazolin-5-one see 488
6-(Dimethylamino)-2-[2-(2,5-dimeth-
yl-1-phenyl-1H-pyrrol-3-yl)ethen-
yl]-1-methylquinolinium Chloride
see 8033
6-(Dimethylamino)-2-[2-(2,5-dimeth-
yl-1-phenyl-1H-pyrrol-3-yl)ethen-
yl]-1-methylquinolinium, Salt with
4,4'-Methylenebis[3-hydroxy-2-
naphthalenecarboxylic Acid] (2:1)
see 8034
6-(Dimethylamino)-2-[2-(2,5-dimeth-
yl-1-phenyl-3-pyrryl)vinyl]-1-
methylquinolinium Salt of 2,2'-
Dihydroxy-1,1'-dinaphthylmeth-
ane-3,3'-dicarboxylic Acid see 8034

2-(Dimethylamino)-5,6-dimethyl-4-
pyrimidinyl Dimethylcarbamate see
7468
1-Dimethylamino-2,3-dioxamethyl-
enepropane Methiodide see 6877
3-Dimethylamino-1,1-diphenylbutane
see 9627
4-Dimethylamino-2,2-diphenylbutyr-
amide Ethyl Bromide see 397
2-(Dimethylamino)-1,2-diphenyldieth-
yl Ether Methyl Bromide see 1218
2-Dimethylamino-4,4-diphenyl-5-
heptanol see 3195
6-(Dimethylamino)-4,4-diphenyl-3-
heptanol see 3195
6-(Dimethylamino)-4,4-diphenyl-3-
heptanol Acetate (Ester) see 5853
6-(Dimethylamino)-4,4-diphenyl-3-
heptanol Acetate (Ester) see 5346
6-Dimethylamino-4,4-diphenyl-3-
heptanone Hydrochloride see 5852
1-Dimethylamino-3,3-diphenyl-4-
hexanone see 6628
6-(Dimethylamino)-4,4-diphenyl-3-
hexanone see 6628
α-l-4-Dimethylamino-1,2-diphenyl-3-
methyl-2-butanol Propionate see
5349
6-Dimethylamino-4,4-diphenyl-5-
methyl-3-hexanone see 5069
(+)-4-Dimethylamino-1,2-diphenyl-
3-methyl-2-propionyloxybutane see
7851
2-(Dimethylamino)-N-(1,3-diphenyl-
1H-pyrazol-5-yl)propanamide see
3123
4-Dimethylamino-2,2-diphenylvaler-
amide see 471
3-Dimethylamino-1,1-di(2'-thienyl)-
but-1-ene see 3248
2-(Dimethylamino)ethanol see 2834
2-(Dimethylamino)ethanol p-Acetami-
dobenzoate see 2836
Dimethylaminoethanol Acetyl-L-glutamate see 2835
2-Dimethylaminoethanol Dihydrogen
Phosphate see 2870
β-Dimethylaminoethanol Diphenylmethyl Ether see 3308
1-(Dimethylaminoethoxyacetamido)-
adamantane see 9683
N-[(2-Dimethylaminoethoxy)benzyl]-
3,4,5-trimethoxybenzamide see 9623
1-(β-Dimethylaminoethoxy)-3-n-
butylisoquinoline see 3207
5-(2-Dimethylaminoethoxy)carvacrol
Acetate see 6204
2-(2-Dimethylaminoethoxy)-N,N-di-
ethylpropylamine Dimethiodide see
7507
2-(2-Dimethylaminoethoxy)diphenyl-
methane see 7287
α-(2-Dimethylaminoethoxy)diphenyl-
methane see 3308
β-Dimethylaminoethoxyethyl Pheno-
thiazine-10-carboxylate see 3211
5-[β-(Dimethylamino)ethoxyimino]-
10,11-dihydro-5H-dibenzo[a,d]-
cycloheptene see 6644
4-(2-Dimethylaminoethoxy)-5-iso-
propyl-2-methylphenyl Acetate see
6204
2-[α-(2-Dimethylaminoethoxy)-α-
methylbenzyl]pyridine see 3430
4-(2-Dimethylaminoethoxy)-2-meth-
yl-5-isopropylphenyl Acetate see
6204
4-[2-(Dimethylamino)ethoxy]-2-meth-
yl-5-(1-methylethyl)phenol Acetate
(Ester) see 6204

10-(2-Dimethylamino-2-methylethyl)-10H-pyrido[3,2-b][1,4]benzothiazine *see* 5117

N-[2-[[[5-[(Dimethylamino)methyl]-2-furanyl]methyl]thio]ethyl]-N'-methyl-2-nitro-1,1-ethenediamine *see* 8126

7-(Dimethylamino)-3-(methylimino)-3H-phenothiazine Hydrochloride *see* 941

3-(Dimethylaminomethyl)indole *see* 4440

8-[(Dimethylamino)methyl]-7-methoxy-3-methylflavone *see* 3190

8-[(Dimethylamino)methyl]-7-methoxy-3-methyl-2-phenyl-4H-1-benzopyran-4-one *see* 3190

8-Dimethylaminomethyl-7-methoxy-3-methyl-2-phenylchromone *see* 3190

2-[(Dimethylamino)methyl]-1-(3-methoxyphenyl)cyclohexanol *see* 9485

1-(Dimethylaminomethyl)-1-methyl-propyl Benzoate Hydrochloride *see* 650

4-(Dimethylamino)-3-methylphenol Methylcarbamate (Ester) *see* 443

4-[4-(Dimethylamino)-N-methyl-L-phenylalanine]virginamycin S₁ *see* 6111

10-(3-Dimethylamino-2-methylpropyl)-3-cyanophenothiazine *see* 2690

5-(3-Dimethylamino-2-methylpropyl)-10,11-dihydro-5H-dibenz[b,f]-azepine *see* 9636

5-(3-Dimethylamino-2-methylpropyl)-10,11-dihydro-5H-dibenz[a,d]-cycloheptene *see* 1533

10-[3-(Dimethylamino)-2-methylpropyl]-2-ethylphenothiazine *see* 3849

5-(3-Dimethylamino-2-methylpropyl)-iminodibenzyl *see* 9636

3-Dimethylamino-7-methyl-1,2-(n-propylmalonyl)-1,2-dihydro-1,2,4-benzotriazine *see* 758

10-(3-Dimethylamino-2-methylpropyl)-2-methoxyphenothiazine *see* 5909

10-[3-(Dimethylamino)-2-methylpropyl]phenothiazine *see* 9618

10-[3-(Dimethylamino)-2-methylpropyl]-10H-phenothiazine-2-carbonitrile *see* 2690

10-(3-Dimethylamino-2-methylpropyl)phenothiazine 5,5-Dioxide *see* 6900

5-(Dimethylamino)-9-methyl-2-propyl-1H-pyrazolo[1,2-a][1,2,4]benzotriazine-1,3(2H)-dione *see* 758

10-[3-(Dimethylamino)-2-methylpropyl]-2-(trifluoromethyl)phenothiazine *see* 9593

α-(Dimethylaminomethyl)protocatechuyl Alcohol *see* 5988

N-[2-[[[2-[(Dimethylamino)methyl]-4-thiazolyl]methyl]thio]ethyl]-N'-methyl-2-nitro-1,1-ethenediamine *see* 6582

α-[(Dimethylamino)methyl]veratryl Alcohol *see* 5520

1-Dimethylaminonaphthalene *see* 3240

5-(Dimethylamino)-1-naphthalenesulfonyl Chloride *see* 2812

4-(Dimethylamino)-1,4,4a,5,5a,6,11,-12a-octahydro-3,5,6,10,12,12a-hexahydroxy-6-methyl-1,11-dioxo-2-naphthacenecarboxamide *see* 6931

4-(Dimethylamino)-1,4,4a,5,5a,6,11,-12a-octahydro-3,6,10,12,12a-penta-hydroxy-N-[[4-(2-hydroxyethyl)-1-piperazinyl]methyl]-6-methyl-1,11-dioxo-2-naphthacenecarboxamide *see* 7420

4-(Dimethylamino)-1,4,4a,5,5a,6-11,12a-octahydro-3,6,10,12,12a-pentahydroxy-6-methyl-1,11-dioxo-2-naphthacenecarboxamide *see* 9130

[4R-(4α,4aβ,5aβ,6α,12aβ)]-4-(Dimethylamino)-1,4,4a,5,5a,6,11,12a-octa-hydro-3,6,10,12,12a-pentahydroxy-6-methyl-1,11-dioxo-2-naphtha-cenecarboxamide *see* 8038

4-(Dimethylamino)-1,4,4a,5,5a,6,11,-12a-octahydro-3,5,10,12,12a-penta-hydroxy-6-methyl-1,11-dioxo-2-naphthacenecarboxamide Monohydrate *see* 3429

α-[4-(Dimethylamino)-1,4,4a,5,5a,6,-11,12a-octahydro-3,6,10,12,12a-pentahydroxy-6-methyl-1,11-dioxo-2-naphthacenecarboxamido]-4-(2-hydroxyethyl)-1-piperazineacetic Acid *see* 762

N⁶-[[[[4-(Dimethylamino)-1,4,4a,5,-5a,6,11,12a-octahydro-3,6,10,12,-12a-pentahydroxy-6-methyl-1,11-dioxo-2-naphthacenyl]carbonyl]-amino]methyl]-L-lysine *see* 5500

4-(Dimethylamino)-1,4,4a,5,5a,6,11,-12a-octahydro-3,6,10,12,12a-penta-hydroxy-6-methyl-1,11-dioxo-N-(1-pyrrolidinylmethyl)-2-naphtha-cenecarboxamide *see* 8236

[4S-(4α,4aα,5α,5aα,12aα)]-4-Dimethyl-amino-1,4,4a,5,5a,6,11,12a-octa-hydro-3,5,10,12,12a-pentahydroxy-6-methylene-1,11-dioxo-2-naphtha-cenecarboxamide *see* 5851

4-(Dimethylamino)-1,4,4a,5,5a,6,11,-12a-octahydro-3,10,12,12a-tetrahy-droxy-1,11-dioxo-2-naphthacene-carboxamide *see* 8316

(Dimethylamino)oxoacetic Acid 2-Acetyl-2-methyl-1-phenylhydrazide *see* 3299

Dimethylaminophenazone *see* 488

4-[[(4-Dimethylamino)phenyl]azo]-benzenesulfonic Acid Sodium Salt *see* 6019

2-[[4-(Dimethylamino)phenyl]azo]-benzoic Acid *see* 6037

3-Dimethylamino-4-phenyl-4-carbethoxy-Δ¹-cyclohexene *see* 9370

2-(Dimethylamino)-1-phenyl-3-cyclo-hexene-1-carboxylic Acid Ethyl Ester *see* 9370

4-[[4-(Dimethylamino)phenyl][4-(di-methylimino)-2,5-cyclohexadien-1-ylidene]methyl]-N-ethyl-N,N-di-methylbenzenaminium Bromide Chloride *see* 5998

Dimethylaminophenyldimethylpyrazolone *see* 488

2-[2-[4-(Dimethylamino)phenyl]ethen-yl]-1-ethylquinolinium Iodide *see* 8057

α-(p-Dimethylaminophenylethylene)-quinoline Ethiodide *see* 8057

7-[4-(Dimethylamino)phenyl]-N-(β-D-glucopyranosyloxy)-4,6-dimethyl-7-oxo-2,4-heptadienamide *see* 9563

7-[4-(Dimethylamino)phenyl]-N-hydr-oxy-4,6-dimethyl-7-oxo-2,4-hepta-dienamide *see* 9563

(11β,17β)-11-[4-(Dimethylamino)-phenyl]-17-hydroxy-17-(1-propyn-yl)estra-4,9-dien-3-one *see* 6109

5-[[4-(Dimethylamino)phenyl]methyl-ene]-2-thioxo-4-thiazolidinone *see* 3220

2-(Dimethylamino)-5-phenyl-2-oxa-zolin-4-one *see* 9315

2-(Dimethylamino)-5-phenyl-4(5H)-oxazolone *see* 9315

[4-(Dimethylamino)phenyl]phenyl-methanone *see* 3222

N-[4-[[4-(Dimethylamino)phenyl]-phenylmethylene]-2,5-cyclohexadi-en-1-ylidene]-N-methylmethanami-nium Chloride *see* 5581

2-Dimethylamino-1-phenylpropanol *see* 5987

2-(Dimethylamino)-1-phenylpropa-none *see* 5828

11β-[4-(N,N-Dimethylamino)phenyl]-17α-(prop-1-ynyl)-Δ⁴,⁹-estradiene-17β-ol-3-one *see* 6109

3-Dimethylamino-1-phenyl-1-p-tolyl-propane *see* 9450

4-[2-(Dimethylamino)propionamido]-antipyrine *see* 484

4-[α-(Dimethylamino)propionamido]-2,3-dimethyl-1-phenyl-3-pyrazolin-5-one *see* 484

2-(Dimethylamino)propiophenone *see* 5828

α-(Dimethylamino)propiophenone *see* 5828

9-[[3-(Dimethylamino)propyl]amino]-1-nitroacridine *see* 6485

10-(2-Dimethylaminopropyl)-1-aza-phenothiazine *see* 5117

10-(γ-Dimethylaminopropyl)-1-aza-phenothiazine *see* 7902

5-(γ-Dimethylaminopropyl)-3-chloro-iminodibenzyl *see* 2385

N-(3-Dimethylaminopropyl)-3-chloro-phenothiazine *see* 2186

10-(γ-Dimethylaminopropyl)-3-chlor-phenothiazine 9-Oxide *see* 6810

2-[3-(Dimethylamino)propyl]-8,8-di-ethyl-2-aza-8-germaspiro[4.5]-decane *see* 8720

5-(3-Dimethylaminopropyl)-10,11-di-hydro-5H-dibenz[b,f]azepine *see* 4835

5-[3-(Dimethylamino)propyl]-10,11-dihydro-5H-dibenz[b,f]azepine 5-Oxide *see* 4836

6-[2-(Dimethylamino)propyl]-1,6-di-hydro-5H-pyrido[2,3-b][1,5]benzo-diazepin-5-one *see* 7846

10-[3-(Dimethylamino)propyl]-9,9-dimethylacridan *see* 3199

10-[2-(Dimethylamino)propyl]-N,N-dimethyl-10H-phenothiazine-2-sul-fonamide *see* 4146

10-(2-Dimethylaminopropyl)-3-di-methylsulfamidophenothiazine *see* 4146

β-[2-(Dimethylamino)propyl]-α-ethyl-β-phenylbenzeneethanol *see* 3195

[S-(R*,R*)]-β-[2-(Dimethylamino)-propyl]-α-ethyl-β-phenylbenzene-ethanol Acetate (Ester) *see* 5346

β-[2-(Dimethylamino)propyl]-α-ethyl-β-phenylbenzeneethanol Acetate (Ester) *see* 5853

1-[3-(Dimethylamino)propyl]-1-(4-fluorophenyl)-1,3-dihydro-5-iso-benzofurancarbonitrile *see* 2320

1-[3-(Dimethylamino)propyl]-1-(4-fluorophenyl)-5-phthalancarbonitrile *see* 2320

5-[3-(Dimethylamino)propyl]-6,7,8,9,-10,11-hexahydro-5H-cyclooct[b]in-dole *see* 4962

Cross Index of Names

Cross Index of Names

3,7-Dimethyl-1,3,6-octatriene *see* 6661

3,7-Dimethyl-1,3,7-octatriene *see* 6661

3,7-Dimethyl-6-octenal *see* 2331

3,7-Dimethyl-7-octenal *see* 8185

S-(—)-3,7-Dimethyl-7-octen-1-ol *see* 8186

2,6-Dimethyl-2-octen-8-ol *see* 2332

3,7-Dimethyl-6-octen-1-ol *see* 2332

Dimethylolpropionic Acid, 3241

2,3-Dimethyl-2-oxabicyclo[2.2.1]-heptane-2,3-dicarboxylic Anhydride *see* 1755

Dimethyl Oxalate *see* 6020

5,5-Dimethyl-2,4-oxazolidinedione *see* 3201

N^1-[(4,5-Dimethyl-2-oxazolyl)amidino]sulfanilamide *see* 8880

N^1-(4,5-Dimethyl-2-oxazolyl)sulfanilamide *see* 8897

1-(4,5-Dimethyloxazol-2-yl)-3-sulfanilylguanidine *see* 8880

7,7-Dimethyl-2-oxobicyclo[2.2.1]-heptane-1-methanesulfonic Acid *see* 1740

21-(3,3-Dimethyl-1-oxobutoxy)-11,17-dihydroxypregna-1,4-diene-3,20-dione *see* 7725

21-(3,3-Dimethyl-1-oxobutoxy)-9-fluoro-11-hydroxy-16,17-[(1-methylethylidene)bis(oxy)]pregna-1,4-diene-3,20-dione *see* 9514

N-(1,1-Dimethyl-3-oxobutyl)acrylamide *see* 2943

N-(1,1-Dimethyl-3-oxobutyl)-2-propenamide *see* 2943

3-[2-(3,5-Dimethyl-2-oxocyclohexyl)-2-hydroxyethyl]glutarimide *see* 2734

4-[2-(3,5-Dimethyl-2-oxocyclohexyl)-2-hydroxyethyl]-2,6-piperidinedione *see* 2734

8,8-Dimethyl-2-oxo-9,10-dihydro-2H,8H-benzo[1,2-b:3,4-b´]dipyran-9,10-diyl-10-acetate-9-(α-methylbutyrate) *see* 9915

3,7-Dimethyl-1-(5-oxohexyl)-1H,3H-purin-2,6-dione *see* 7092

3,3-Dimethyl-7-oxo-6-[(1-oxo-3-hexenyl)amino]-4-thia-1-azabicyclo-[3.2.0]heptane-2-carboxylic Acid Monosodium Salt *see* 7081

3,3-Dimethyl-7-oxo-6-[(1-oxohexyl)-amino]-4-thia-1-azabicyclo[3.2.0]-heptane-2-carboxylic Acid Monosodium Salt *see* 652

3,3-Dimethyl-7-oxo-6-[[[[(2-oxo-1-imidazolidinyl)carbonyl]amino]phenylacetyl]amino]-4-thia-1-azabicyclo[3.2.0]heptane-2-carboxylic Acid *see* 929

[2S-(2α,5α,6β)]-3,3-Dimethyl-7-oxo-6-[(1-oxo-2-phenoxybutyl)amino]-4-thia-1-azabicyclo[3.2.0]heptane-2-carboxylic Acid *see* 7829

3,3-Dimethyl-7-oxo-6-[(1-oxo-2-phenoxypropyl)amino]-4-thia-1-azabicyclo[3.2.0]heptane-2-carboxylic Acid Potassium Salt *see* 7184

15-[(3,4-Dimethyl-1-oxo-2-pentenyl)-oxy]-13,20-epoxy-3,11,12-trihydroxy-2,16-dioxopicras-3-en-21-oic Acid Methyl Ester *see* 1442

12-[(3,4-Dimethyl-1-oxo-2-pentenyl)-oxy]-3,8,14-trihydroxypregn-5-en-20-one *see* 2772

3,3-Dimethyl-7-oxo-6-[(phenoxyacetyl)amino]-4-thia-1-azabicyclo-[3.2.0]heptane-2-carboxylic Acid *see* 7046

[2S-(2α,5α,6β)]-3,3-Dimethyl-7-oxo-6-[(phenoxyacetyl)amino]-4-thia-1-azabicyclo[3.2.0]heptane-2-carboxylic Acid Compd with N,N´-Bis-(phenylmethyl)-1,2-ethanediamine (2:1) *see* 7047

[2S-(2α,5α,6β)]-3,3-Dimethyl-7-oxo-6-[(phenoxyacetyl)amino]-4-thia-1-azabicyclo[3.2.0]heptane-2-carboxylic Acid Compd with [4S-(4α,4aα,5aα,6β,12aα)]-4-(Dimethylamino)-1,4,4a,5,5a,6,11,12a-octahydro-3,6,10,12,12a-pentahydroxy-N-[[4-(2-hydroxyethyl)-1-piperazinyl]methyl]-6-methyl-1,11-dioxo-2-naphthacenecarboxamide (1:1) *see* 7052

3,3-Dimethyl-7-oxo-6-(2-phenoxy-2-phenylacetamido)-4-thia-1-azabicyclo[3.2.0]heptane-2-carboxylate *see* 3905

3,3-Dimethyl-7-oxo-6-[(phenoxyphenylacetyl)amino]-4-thia-1-azabicyclo[3.2.0]heptane-2-carboxylic Acid *see* 3905

3,3-Dimethyl-7-oxo-6-[(phenylacetyl)amino]-4-thia-1-azabicyclo-[3.2.0]heptane-2-carboxylic Acid *see* 1156

3,3-Dimethyl-7-oxo-6-[(phenylacetyl)amino]-4-thia-1-azabicyclo-[3.2.0]heptane-2-carboxylic Acid (Acetyloxy)methyl Ester *see* 7024

[2S-(2α,5α,6β)]-3,3-Dimethyl-7-oxo-6-[(phenylacetyl)amino]-4-thia-1-azabicyclo[3.2.0]heptane-2-carboxylic Acid Calcium Salt (2:1) *see* 7039

[2S-(2α,5α,6β)]-3,3-Dimethyl-7-oxo-6-[(phenylacetyl)amino]-4-thia-1-azabicyclo[3.2.0]heptane-2-carboxylic Acid Compd with N,N´-Bis-(phenylmethyl)-1,2-ethanediamine *see* 7037

[2S-(2α,5α,6β)]-3,3-Dimethyl-7-oxo-6-[(phenylacetyl)amino]-4-thia-1-azabicyclo[3.2.0]heptane-2-carboxylic Acid Compd with 2-(Diethylamino)ethyl 4-Aminobenzoate (1:1) Monohydrate *see* 7042

[2S-(2α,5α,6β)]-3,3-Dimethyl-7-oxo-6-[(phenylacetyl)amino]-4-thia-1-azabicyclo[3.2.0]heptane-2-carboxylic Acid Compd with α-Phenylbenzenemethanamine (1:1) *see* 7038

3,3-Dimethyl-7-oxo-6-[(phenylacetyl)amino]-4-thia-1-azabicyclo-[3.2.0]heptane-2-carboxylic Acid Compd with N-(Phenylmethyl)-benzeneethanamine (1:1) *see* 7036

3,3-Dimethyl-7-oxo-6-[(phenylacetyl)amino]-4-thia-1-azabicyclo-[3.2.0]heptane-2-carboxylic Acid 2-(Diethylamino)ethyl Ester Monohydriodide *see* 7027

[2S-(2α,5α,6β)]-3,3-Dimethyl-7-oxo-6-[(phenylacetyl)amino]-4-thia-1-azabicyclo[3.2.0]heptane-2-carboxylic Acid Monopotassium Salt *see* 7041

[2S-(2α,5α,6β)]-3,3-Dimethyl-7-oxo-6-[(phenylacetyl)amino]-4-thia-1-azabicyclo[3.2.0]heptane-2-carboxylic Acid Monosodium Salt *see* 1157

6-(2,2-Dimethyl-5-oxo-4-phenyl-1-imidazolidinyl)-3,3-dimethyl-7-oxo-4-thia-1-azabicyclo[3.2.0]heptane-2-carboxylic Acid *see* 4592

6-(2,2-Dimethyl-5-oxo-4-phenyl-1-imidazolidinyl)penicillanic Acid *see* 4592

[2S-[2α,5α,6β(S*)]]-3,3-Dimethyl-7-oxo-6-[(phenylsulfoacetyl)amino]-4-thia-1-azabicyclo[3.2.0]heptane-2-carboxylic Acid *see* 8863

3,3-Dimethyl-7-oxo-6-[[(2-propenylthio)acetyl]amino]-4-thia-1-azabicyclo[3.2.0]heptane-2-carboxylic Acid *see* 7044

(11B)-21-(2,2-Dimethyl-1-oxopropoxy)-11,17-dihydroxypregna-1,4-diene-3,20-dione *see* 7726

2-(2,2-Dimethyl-1-oxopropyl)-1H-indene-1,3(2H)-dione *see* 7413

endo-8,8-Dimethyl-3-[(1-oxo-2-propylpentyl)oxy]-8-azoniabicyclo-[3.2.1]octane Bromide *see* 701

(11β)-21-[(2,2-Dimethyl-1-oxopropyl)thio]-11,17-dihydroxypregn-4-ene-3,20-dione *see* 9408

17S,20-Dimethyl-6-oxoprostaglandin E₁ Methyl Ester *see* 6827

3,3-Dimethyl-7-oxo-4-thia-1-azabicyclo[3.2.0]heptane-2-carboxylic Acid *see* 7032

(2S-cis)-3,3-Dimethyl-7-oxo-4-thia-1-azabicyclo[3.2.0]heptane-2-carboxylic Acid 4,4-Dioxide *see* 8862

Dimethyloxychinizin *see* 748

Dimethyloxyquinazine *see* 748

Dimethyl Parathion *see* 6022

7,8-Dimethyl-10-(D-galacto-2,3,4,5,6-pentahydroxyhexyl)benzo[g]pteridine-2,4(3H,10H)-dione *see* 4239

7,8-Dimethyl-10-(D-galacto-2,3,4,5,6-pentahydroxyhexyl)isoalloxazine *see* 4239

16,16-Dimethyl-*trans*-Δ²-PGE₁ Methyl Ester *see* 4279

2,9-Dimethyl-o-phenanthroline *see* 6364

2,9-Dimethyl-1,10-phenanthroline *see* 6364

d-N,α-Dimethylphenethylamine *see* 5859

dl-N,β-Dimethylphenethylamine *see* 7280

α,α-Dimethylphenethylamine *see* 7232

Dimethylphenol *see* 9989

3,3´-Dimethylphenolphthalein *see* 2582

N,N-Dimethyl-10H-phenothiazine-10-ethanamine *see* 3918

N,N-Dimethyl-10H-phenothiazine-10-propanamine *see* 7793

1-(2´,6´-Dimethylphenoxy)-2-aminopropane *see* 6094

5-(2,5-Dimethylphenoxy)-2,2-dimethylpentanoic Acid *see* 4280

N,N-Dimethyl-N-(2-phenoxyethyl)-benzenemethanaminium *see* 1166

N,N-Dimethyl-N-(2-phenoxyethyl)-1-dodecanaminium Bromide *see* 3411

Dimethyl(2-phenoxyethyl)-2-thenylammonium p-Chlorobenzenesulfonate *see* 9205

N,N-Dimethyl-N-(2-phenoxyethyl)-2-thiophenemethanaminium Salt with 4-Chlorobenzenesulfonic Acid (1:1) *see* 9205

5-(3,5-Dimethylphenoxymethyl)-2-oxazolidinone *see* 5838

1-(2,6-Dimethylphenoxy)-2-propanamine *see* 6094

Dimethylphenylamine *see* 3223

2-[(2,3-Dimethylphenyl)amino]benzoic Acid *see* 5680

2-[[(2,6-Dimethylphenyl)amino]carbonyl]benzoic Acid *see* 4190

2-(2,6-Dimethylphenylamino)-4H-5,6-dihydro-1,3-thiazine *see* 9987

Cross Index of Names

Di(phenylpropyl)ethylamine *see* 377
1-(3,3-Diphenylpropyl)hexahydro-1*H*-azepine *see* 7922
1-(3,3-Diphenylpropyl)hexamethylen-imine *see* 7922
N-(3,3-Diphenylpropyl)-α-methyl-benzylamine *see* 3916
(±)-*N*-(3,3-Diphenylpropyl)-α-meth-ylcyclohexaneethylamine *see* 3438
N-(3,3-Diphenylpropyl)-α-methyl-phenethylamine *see* 7744
1-(3,3-Diphenylpropyl)piperidine *see* 3934
Diphenylpyraline, 3334
Diphenylpyraline 8-Chlorotheophyl-linate *see* 7456
1,4-Diphenyl-3,5-pyrazolidinedione *see* 7217
Diphenyl (γ-Pyridyl)carbinol *see* 909
Diphenylpyrilene *see* 3334
sym-Diphenylpyrophosphorodiamidic Acid, 3335
Diphenylsulfide *see* 7285
Diphenyl Sulfone, 3336
5,5-Diphenyltetrahydroglyoxalin-4-one *see* 3424
Diphenylthioacetic Acid S-(2-Dieth-ylaminoethyl) Ester *see* 9303
Diphenylthiocarbamic Acid S-[2-(Diethylamino)ethyl] Ester *see* 7178
Diphenylthiocarbazone *see* 3383
Diphenylthiolacetic Acid 2-Diethyl-aminoethyl Ester *see* 9303
N,*N*'-Diphenylthiourea *see* 3337
sym-Diphenylthiourea, 3337
1,3-Diphenyl-1-triazene *see* 2981
5,5-Diphenyl-3-(2,2,2-trichloro-1-hydroxyethyl)-4-imidazolidinone *see* 9569
2,2-Diphenyl-1-(2,4,6-trinitrophenyl)-hydrazyl *see* 3332
N,*N*'-Diphenylurea *see* 1787
sym-Diphenylurea *see* 1787
1,3-Diphenylurea *see* 1787
Di(phenylurethan) of 1-Piperidinep-ropane-2,3-diol Hydrochloride *see* 3301
α-(4-Diphenylyl)butyric Acid *see* 9980
4-Diphenylylethylacetic Acid *see* 9980
Diphepanol *see* 3333
Diphergan [Polfa] *see* 7797
Diphesatin *see* 6927
Diphetarsone, 3338
Diphone *see* 2820
Diphos [Boehringer, Mann.] *see* 3819
Diphosgene, 3339
(Diphosphonomethyl)butanedioic Acid *see* 1512
(Diphosphonomethyl)succinic Acid *see* 1512
Diphosphopyridine Nucleotide *see* 6259
Diphosphoric Acid *see* 8018
Diphosphoric Acid Tetraethyl Ester *see* 9138
Diphosphoric Acid Tetrapotassium Salt *see* 7653
Diphosphoric Acid Tin(2+) Salt (1:2) *see* 8748
Diphosphoric Acid Zinc Salt (1:2) *see* 10058
Diphosphorus Pentoxide *see* 7330
Diphosphorus Trioxide *see* 7335
Diphylline *see* 2043
Diphyllin Methyl Ether *see* 5154
Dipicin [Doctors] *see* 8215
Dipicrylamine, 3340
Dipidolor [Janssen] *see* 7470
Dipin, 3341
Dipipanone, 3342

2β,16β-Dipiperidino-5α-androstane-3α,17β-diol Diacetate Dimetho-bromide *see* 6958
2,2',2'',2'''-[(4,8-Di-1-piperidinyl-pyrimido[5,4-*d*]pyrimidine-2,6-diyl)dinitrilo]tetrakisethanol *see* 3354
Dipiperon [Janssen] *see* 7422
Dipiproverine, 3343
Dipirin *see* 488
1-(3',4'-Dipivaloyloxyphenyl)-2-methylamino-1-ethanol *see* 3344
Dipivalyl Epinephrine *see* 3344
Dipivefrin, 3344
Diploicin, 3345
Diplosal *see* 8307
Diplosal Acetate *see* 95
Dipolyoxyethylated Polypropyleneg-lycol Ether *see* 7536
Diponium Bromide, 3346
Dipotassium Bis[ethanedioato(2−)-*O*,*O*']oxotitanate(2−) *see* 7689
Dipotassium Hexakis(thiocyanato-S)-platinate(2−) *see* 7624
Dipotassium Hydrogen Phosphate *see* 7647
Dipotassium Pentakis(cyano-C)nitro-sylferrate(2−) *see* 7636
Dipotassium Phosphate *see* 7647
Dipotassium Tetraiodocadmate(2−) *see* 7682
Dipotassium Tetrakis(cyano-C)mer-curate(2−) *see* 7676
Dipotassium Tetrakis(cyano-C)nickel-ate(2−) *see* 7677
Dipotassium Tetrakis(cyano-C)platin-ate(2−) *see* 7678
Dipotassium Tetrakis(cyano-C)zinc-ate(2−) *see* 7679
Diprazin *see* 5088
Diprenorphine, 3347
Diprivan [ICI] *see* 7847
Diproderm [Schering] *see* 1202
Diprolene [Schering] *see* 1202
Dipropalin®, 3348
Di-2-propenylamine *see* 2951
Di-2-propenylcyanamide *see* 2952
5,5-Di-2-propenyl-2,4,6(1*H*,3*H*,5*H*)-pyrimidinetrione *see* 252
Dipropetryn, 3349
Diprophos [Schering] *see* 1202
Diprophylline *see* 2043
3β,17β-Dipropionyloxy-4-estrene *see* 1325
Dipropylacetamide *see* 9822
Di-*n*-propylacetic Acid *see* 9821
n-Dipropylamine, 3350
4-(Dipropylamino)-3,5-dinitrobenz-enesulfonamide *see* 7760
4-[(Dipropylamino)sulfonyl]benzoic Acid *see* 7760
5-Dipropylamino-α,α,α-trifluoro-4,6-dinitro-*o*-toluidine *see* 7773
Dipropylcarbamothioic Acid S-Ethyl Ester *see* 3580
Dipropylcarbamothioic Acid S-Propyl Ester *see* 9866
Dipropyl 3,4-Dihydroxy-2,5-thio-phenedicarboxylate *see* 7904
N,*N*-Dipropyl-2,6-dinitro-4-methyl-aniline *see* 3348
N,*N*-Dipropyl-2,6-dinitro-4-trifluoro-methylaniline *see* 9598
*N*³,*N*³-Di-*n*-propyl-2,4-dinitro-6-tri-fluoromethyl-1,3-phenylenediamine *see* 7773
Dipropyl Ether *see* 7870
Dipropyl Ketone, 3351
5,5-Dipropyl-2,4-oxazolidinedione, 3352

p-(Dipropylsulfamoyl)benzoic Acid *see* 7760
p-(Dipropylsulfamyl)benzoic Acid *see* 7760
Dipropyl Sulfide *see* 7881
Dipropylthiocarbamic Acid S-Ethyl Ester *see* 3580
Dipropyl Thiocarbamic Acid S-Propyl Ester *see* 9866
Diprosis [Byk-Essex] *see* 1202
Diprosone [Schering] *see* 1202
Diprotrizoate Sodium, 3353
Dipsan [Lederle] *see* 1663
Dipterex [Bayer] *see* 9536
Dipyridamole, 3354
Dipyridan [Hokuriku] *see* 3354
α-[3-(Di-2-pyridinylmethylene)-1,4-cyclopentadien-1-yl]-α-2-pyridinyl-2-pyridinemethanol *see* 8001
2,2'-Dipyridyl *see* 3355
4,4'-Dipyridyl *see* 3356
α,α'-Dipyridyl, 3355
γ,γ'-Dipyridyl, 3356
Di-2-pyridyl-(6,6-di-2-pyridylfulven-2-yl)methanol *see* 8001
3,3-Dipyridyl Disulfide Dihydrochlo-ride *see* 3380
3-(Di-2-pyridylmethylene)-α,α-di-2-pyridyl-1,4-cyclopentadiene-1-methanol *see* 8001
2,4-Di(3-pyridyl)pyridine *see* 6432
2,4-Di(β-pyridyl)pyridine *see* 6432
Dipyrine *see* 488
Dipyrithione *see* 8004
Dipyrocetyl, 3357
Dipyrone, 3358
1,4-Dipyrrolidino-2-butyne *see* 9498
21-[4-(2,6-Di-1-pyrrolidinyl-4-pyr-imidinyl)-1-piperazinyl]-16α-meth-ylpregna-1,4,9(11)-triene-3,20-dione Monomethanesulfonate *see* 5264
Diquat Dibromide, 3359
Diquel *see* 3849
Diquinine Carbonate *see* 8077
Di-Quinol *see* 4935
Diquinol [Parke, Davis] *see* 3682
N,*N*'-Di-6-quinolinylurea *see* 3360
sym-Di-(6-quinolyl)urea *see* 3360
1,3-Di-6-quinolylurea, 3360
6,6'-Diquinolylurea *see* 3360
Dira [Kakenyaku] *see* 8721
Diralgan [Roussel-UCLAF] *see* 4037
Dirame [Schering] *see* 7844
Dirax [Daiichi] *see* 4902
Direma *see* 4704
Diresorcinol, 3361
Direxiode [Delalande] *see* 4935
Dirgotarl [Horita] *see* 3157
Dirhenium Heptoxide *see* 8177
Dirimal [Lilly] *see* 6840
Dirnate [Merck & Co.] *see* 1885
Dirocide [Squibb] *see* 3106
Dironyl [Schering AG] *see* 9095
Dirox [Winthrop] *see* 40
Dirythmin SA [Astra] *see* 3366
Disadine D.P. [Stuart] *see* 7701
Disalcid [Riker] *see* 8307
Disalgesic [Kettelhack Riker] *see* 8307
Disalicylic Acid *see* 8307
Disalunil *see* 4704
Disamide [BDH] *see* 3369
Disarim *see* 2082
Discase [Travenol] *see* 2264
Dis-Cinil [Lusofarmaco] *see* 6913
Discoid [Sagitta] *see* 4221
Disdolen [Uriach] *see* 4170
Disepron [Yoshitomi] *see* 2408
Diseptal B *see* 6040
Diseptal C *see* 8905
Disgren [Uriach] *see* 9600

Disiamylborane *see* 1261
Disilane, 3362
Disilicane *see* 3362
Disilicoethane *see* 3362
Disilicon Hexahydride *see* 3362
Disilver Fluoride *see* 8479
Disipal *see* 6831
Di-Sipidin *see* 7481
Diskin *see* 6913
Disocarban *see* 9382
Disoderm [Schering] *see* 3035
Disodium Acetarsenate *see* 838
Disodium Arsenate *see* 8522
Disodium Arsenoacetate *see* 836
Disodium Arsonoacetate *see* 838
Disodium Azodisalicylate *see* 6799
Disodium Citrate *see* 8550
Disodium Cromoglycate *see* 2594
Disodium 3,3'-Diamino-4,4'-dihydr-oxyarsenobenzene N-Dimethylene-sulfonate *see* 8916
Disodium *p,p*'-Diaminodiphenylsul-fone-*N,N*'-diglucose Sulfonate *see* 4357
Disodium *o*-Dianisidinediazobis(1-naphthol-4-sulfonate) *see* 1095
Disodium Dihydrogen (1-Hydroxyeth-ylidene)bis[phosphonate] *see* 3819
Disodium Dihydrogen Hypophosphate, 3363
Disodium Dihydrogen Pyrophosphate *see* 8515
Disodium Dihydrogen Subphosphate *see* 3363
Disodium-1,2-dihydroxybenzene-3,5-disulfonate *see* 9392
*N*⁴-(Disodium 1,3-Disulfo-3-phenyl-propyl)sulfanilamide *see* 6601
Disodium 4,4'-Disulfoxydiphenyl-(2-pyridyl)methane *see* 7377
Disodium Edathamil *see* 3481
Disodium Edetate *see* 3481
Disodium Ethylenebis[dithiocarbam-ate] *see* 6256
Disodium Ethylenediaminetetraacetate *see* 3481
Disodium Formaldehydesulfoxylate-diaminodiphenylsulfone *see* 8947
Disodium Hexachloroplatinate(2−) *see* 8571
Disodium Hydrogen Citrate *see* 8550
Disodium Hydrogen Phosphate *see* 8612
Disodium 5,5'-Indigotin Disulfonate *see* 4856
Disodium *N*-Methyl-3,5-diiodo-4-pyridone-2,6-dicarboxylate *see* 8584
Disodium Monomethanearsonate *see* 5864
Disodium Orthophosphate *see* 8612
Disodium Pentaiodobismuthate(2−) *see* 1291
Disodium Phenoltetrabromophthalein Sulfonate *see* 8933
Disodium Phenyl Phosphate, 3364
Disodium *p*-(γ-Phenylpropylamino)-benzenesulfonamide-α,γ-disulfonate *see* 6601
Disodium 1-Phenyl-3-*p*-sulfamoylani-lino-1,3-propanedisulfonate *see* 6601
Disodium Phosphate *see* 8612
Disodium Prednisolone 21-Phosphate *see* 7721
Disodium Pyrocatechol-3,5-disulfon-ate *see* 9392
Disodium Succinate *see* 8635
Disodium 2-(4'-Sulfamylphenylazo)-7-acetamido-1-hydroxynaphtha-lene-3,6-disulfonate *see* 936

Disodium[sulfonylbis(*p*-phenylenimi-no)]dimethanesulfinate *see* 8947
Disodium *o*-Tolidinediazobis(1-naph-thylamine-4-sulfonate) *see* 1111
Diso-Duriles [Astra] *see* 3366
Disomer [Schering] *see* 1433
Disonate *see* 3398
Disophenol, 3365
Disoprivan [ICI] *see* 7847
Disoprofol *see* 7847
Disopromine *see* 3180
Disopyramide, 3366
Disoquin *see* 4935
Disorat [Boehringer, Mann.] *see* 6059
Disorlon [Nativelle] *see* 5114
Disotat [Isis-Chemie] *see* 3182
Dispadol *see* 5736
Dispan *see* 5345
Disparlure, 3367
Dispas [Ankerfarm] *see* 3151
Dispermin *see* 7431
Disphex [ICI] *see* 7701
Dispranol *see* 7064
Dispril *see* 1644
Disprin *see* 1644
Disprol [Reckitt & Colman] *see* 40
Dispronil [Liberman] *see* 3881
Disrupt [Hercon] *see* 3367
Dissenten [SPA] *see* 5450
Distaclor [Lilly] *see* 1920
Distakaps V-K *see* 7046
Distamine [Dista] *see* 7029
Distamycin A *see* 8730
Distaquaine *see* 7042
Distaquaine V *see* 7046
Distaquaine V-K [Dista] *see* 7046
Distaval *see* 9182
Distigmine Bromide, 3368
Distilbene [UCEPHA] *see* 3118
Distivit (B₁₂ Peptide) *see* 9921
Disto-5 [Cogla] *see* 1316
Distobram [Dista] *see* 9413
Distol 8 *see* 3482
Distraneurin [Pharma-Stern] *see* 2382
Distyryl *see* 3320
Distyryl Ketone *see* 2988
Disulfamide, 3369
1,3-Disulfamyl-4,5-dichlorobenzene *see* 3067
Disulfan *see* 8905
Disulfatozirconic Acid *see* 10085
3,3'-Disulfinyldialanine *see* 2789
Disulfiram, 3370
Disulfoton, 3371
Disulfur Dichloride *see* 8949
Disulfuric Acid *see* 8019
Disulfuric Acid Dipotassium Salt *see* 7654
Disulfur Pentoxydichloride *see* 8020
Disulfuryl Chloride *see* 8020
Disulon [Specia] *see* 8905
Disulone *see* 2820
Disulphamide *see* 3369
Disulphine Blue *see* 8968
Disul-sodium, 3372
Disyncran *see* 5877
Disynformon *see* 3660
Di-Syntramine *see* 3087
Di-Syston [Chemagro] *see* 3371
Dita Bark, 3373
Ditaine *see* 3474
Ditan *see* 3329
Ditaven [Cascan] *see* 3146
Ditazol, 3374
Diteftin *see* 4223
Ditetrazolium Chloride *see* 9173
Dithane 945 [Rohm & Haas] *see* 5598
Dithane D-14 *see* 6256
Dithane LF [La Littorale] *see* 5598
Dithane M-22 [Rohm & Haas] *see* 5603

Dithane M-45 [Rohm & Haas] *see* 5598
Dithane Z-78 [Rohm & Haas] *see* 10071
1,4-Dithiaanthraquinone-2,3-dicarbo-nitrile *see* 3375
δ-[3-(1,2-Dithiacyclopentyl)]pentanoic Acid *see* 9255
Dithianone, 3375
1,2-Dithia-5,8,11,14,17-pentaazacy-cloeicosane Cyclic Peptide Deriv *see* 6682
2,6-Dithia-1,3,5,7-tetraazaadaman-tane 2,2,6,6-Tetraoxide *see* 9158
2,6-Dithia-1,3,5,7-tetraazatricyclo-[3.3.1.1³,⁷]decane 2,2,6,6-Tetraoxide *see* 9158
Dithiazanine Iodide, 3376
α-[1-[(3,3-Di-3-thienylallyl)amino]-ethyl]benzyl Alcohol *see* 9378
α,α'-Dithienyl-4-dimethylaminocy-clohexyl Carbinol *see* 9242
3-(Di-2-thienylmethylene)-5-meth-oxy-1,1-dimethylpiperidinium Bromide *see* 9372
3-(Di-2-thienylmethylene)-1-methyl-piperidine *see* 9389
3-(Di-2-thienylmethylene)-5-methyl-*trans*-quinolizidinium Bromide *see* 9390
trans-3-(Di-2-thienylmethylene)octa-hydro-5-methyl-2*H*-quinolizinium Bromide *see* 9390
α-[1-[(3,3-Di-3-thienyl-2-propenyl)-amino]ethyl]benzenemethanol *see* 9378
1,3-Dithietan-2-ylidenephosphorami-dic Acid Diethyl Ester *see* 4173
Dithio *see* 8945
4,4'-Dithiobis[2-aminobutanoic Acid] *see* 4655
4,4'-Dithiobis[2-aminobutyric Acid] *see* 4655
3,3'-Dithiobis(2-aminopropanoic Acid) *see* 2788
2,2'-Dithiobis[benzothiazole], 3377
N,N'-[Dithiobis[2-[2-(benzoyloxy)-ethyl]-1-methyl-2,1-ethenediyl]]bis-[*N*-[(4-amino-2-methyl-5-pyrimi-dinyl)methyl]formamide] *see* 1257
N,N'-[Dithiobis[1-[(carboxymethyl)-carbamoyl]ethylene]]diglutamine *see* 4369
2,2'-Dithiobisethanamine *see* 2783
N,N'-[Dithiobis[2,1-ethanediylimino-(3-oxo-3,1-propanediyl)]]bis[2,4-di-hydroxy-3,3-dimethylbutanamide] *see* 6962
2,2'-Dithiobis[ethylamine] *see* 2783
N,N'-[Dithiobis(ethyleneiminocarbon-ylethylene)]bis(2,4-dihydroxy-3,3-dimethylbutyramide) *see* 6962
N,N'-[Dithiobis[2-(2-hydroxyethyl)-1-methyl-2,1-ethenediyl]]bis[*N*-[(4-amino-2-methyl-5-pyrimidinyl)-methyl]formamide] *see* 9221
N,N'-[Dithiobis[2-(2-hydroxyethyl)-1-methylvinylene]]bis[*N*-[(4-amino-2-methyl-5-pyrimidinyl)methyl]-formamide] Dibenzoate *see* 1257
3,3'-[Dithiobis(methylene)]bis[5-hydroxy-6-methyl-4-pyridinemeth-anol] *see* 8006
4,4'-Dithiobis[morpholine] *see* 3379
6,6'-Dithiobis-2-naphthalenol *see* 2831
6,6'-Dithiobis(2-naphthol) *see* 2831
3,3'-[Dithiobis[pyridine] Dihydrochlo-ride *see* 3380
2,2'-Dithiobispyridine 1,1'-Dioxide *see* 8004

Cross Index of Names

Dithiobis[thioformic Acid] *O,O*-Di-
ethyl Ester *see* 3390
3,3'-Dithiobis[valine] *see* 7031
2,4-Dithiobiuret, 3378
Dithiocarb *see* 3384
Dithiocarbamic Acid Monoammonium
Salt *see* 539
Dithiocarbonic Acid Cyclic *S,S*-(6-
Methyl-2,3-quinoxalinediyl) Ester
see 6933
Dithiocarbonic Anhydride *see* 1818
Dithiodemeton *see* 3371
β,β'-Dithiodialanine *see* 2788
4,4'-Dithiodimorpholine, 3379
6,6'-Dithiodi-2-naphthol *see* 2831
3,3-Dithiodipyridine Dihydrochloride,
3380
3,3'-Dithiodivaline *see* 7031
Dithioethyleneglycol *see* 3679
1,2-Dithioglycerol *see* 3196
Dithioglycolyl *p*-Arsenobenzamide *see*
819
"Dithiol" *see* 9457
1,2-Dithiolane-3-pentanoic Acid *see*
9255
1,2-Dithiolane-3-valeric Acid *see*
9255
1,3-Dithiolan-2-ylidenephosphorami-
dic Acid Diethyl Ester *see* 7309
5-[3-(1,2-Dithiolanyl)]pentanoic Acid
see 9255
5-(1,2-Dithiolan-3-yl)valeric Acid *see*
9255
1,3-Dithiolo[4,5-*b*]quinoxaline-2-
thione *see* 9288
2,3-Dithiolpropanesulfonic Acid *see*
3197
1,3-Dithiol-2-ylidenepropanedioic
Acid Bis(1-methylethyl) Ester *see*
5593
Dithion [Agrimont] *see* 2568
Dithione *see* 8945
Dithionous Acid Disodium Salt *see*
8558
Dithiooxamide *see* 8256
Dithiophos *see* 8945
Dithiophosphoric Acid *O,O* '-Dimeth-
yl-*S*-[(5-methoxy-1,3,4-thiadiazol-
2(3*H*)-one-3-yl)methyl] Ester *see*
5891
Dithiophosphoric Acid *O,O* '-Dimeth-
yl-*S*-[(2-methoxy-1,3,4-thiadiazol-
5(4*H*)-on-4-yl)methyl] Ester *see*
5891
Dithiopropylthiamine *see* 7896
Dithiosalicylic Acid, 3381
Dithiosystox [Bayer] *see* 3371
1,4-Dithiothreitol, 3382
Dithiotrimethyleneglycol *see* 7811
Dithizone, 3383
Dithranol *see* 714
Dithrocream [Dermal] *see* 714
Dithymol Diiodide *see* 9335
Diticyl *see* 9650
Ditin Diphosphate *see* 8748
Ditin Pyrophosphate *see* 8748
Ditiocarb Sodium, 3384
Dition [Agrimont] *see* 2568
Ditiovit [IBI] *see* 7896
1,4-Di-*p*-toluidinoanthraquinone *see*
8095
Ditolyldiazo-3,6-disulfo-β-naphthyl-
amine-β-naphthylamine-6-sulfonic
Acid Sodium Salt *see* 9917
asym-Di-*p*-tolylethane *see* 3385
sym-Di-*p*-tolylethane *see* 3386
1,1-Di-*p*-tolylethane, 3385
1,2-Di-*p*-tolylethane, 3386
α,α-Di-*p*-tolylethane *see* 3385
α,β-Di-*p*-tolylethane *see* 3386
p-Ditolylmercury, 3387

Ditrazin *see* 3106
1,4-Di(2,2,2-trichloro-1-formamido-
ethyl)piperazine *see* 9602
Di[tri-(2-methyl-2-phenylpropyl)tin]-
oxide *see* 3907
Ditripentat [Heyl] *see* 7082
Ditropan [Marion] *see* 6908
Ditropyl Isatropate *see* 1035
Ditrosol *see* 3272
Ditubin [Schering] *see* 5071
Diucardin [Ayerst] *see* 4716
Diucardyn Sodium [Ayerst] *see* 5761
Diucen [Central Pharm.] *see* 1134
Diulo [Searle] *see* 6068
Diumax [Cusi-Norte] *see* 7464
Diu-melusin [Melusin Schwarz] *see*
4704
Diural [Apothokermes] *see* 4221
Diurapid [Boehringer, Mann.] *see* 935
Diurazine *see* 2073
Diuresal *see* 2169
Diurese [Am. Urologicals] *see* 9537
Diureticum-Holzinger [Holzinger] *see*
45
Diuretin [Knoll] *see* 9209
Diurex [Makhteshim-Agan] *see* 3388
Diurexan [E. Merck] *see* 9986
Diuril [Merck & Co.] *see* 2169
Diurilix [Theraplix] *see* 2169
Diuril Lyovac [Merck & Co.] *see* 2169
Diurite *see* 2169
Diuriwas [Wassermann] *see* 45
Diuron, 3388
Diurone *see* 2104
Diursal *see* 5805
Diutazol *see* 45
Divalproex Sodium *see* 9821
Divanil *see* 2761
Divanillalcyclohexanone *see* 2761
2,6-Divanillylidenecyclohexanone *see*
2761
Divanon (obsolete) *see* 2761
Divarine [Tutag] *see* 3358
Divasil *see* 8469
Divegal [Waldheim] *see* 3157
Divercillin [Ascher] *see* 621
Diverine [Courtois] *see* 3151
Diviator *see* 7750
Divicine, 3389
Divicine 5-Glucoside *see* 9880
Divicine-β-glucoside *see* 9880
Dividol [Zambon] *see* 9885
Diviminol *see* 9885
Divinyl *see* 1500
Divinylene Oxide *see* 4206
Divinylene Sulfide *see* 9283
Divinylenimine *see* 8025
Divinyl Ether *see* 9899
Divinyl Oxide *see* 9899
Divipan [Makhteshim-Agan] *see* 3069
Divit Urto *see* 9928
Dixanthogen, 3390
Dixarit [Boehringer, Ing.] *see* 2388
Dixeran [Lundbeck] *see* 5704
Dixiben [Benvegna] *see* 6273
Dixina *see* 3150
Dixnalate [Sana] *see* 4273
Di-*m*-xylylene *see* 5825
m-Dixylylene *see* 5825
N,N-Di-(2,4-xylyliminomethyl)meth-
ylamine *see* 503
Dixyrazine, 3391
Dizan [Elanco] *see* 3376
Dizocilpine, 3392
Dizol *see* 498
DJ-1461 *see* 1457
DJ-7041 *see* 6217
Djenkolic Acid, 3393
DK-7419 *see* 804
DKB *see* 2987
DKP *see* 7647

DL-152 *see* 1226
DL-832 *see* 4332
DL-3117 *see* 4955
DL-8280 *see* 6688
DL-458-IT *see* 2852
DL-717-IT *see* 5459
DLP [Dawbarn] *see* 7558
DLP-787 *see* 7999
DM *see* 7170
DMAA *see* 5709
DMAC *see* 3216
DMAE *p*-Acetamidobenzoate *see*
2836
DMC *see* 2086 and 7029
DMDP *see* 2369
DMDT *see* 5913
DMDZ *see* 6608
DMF *see* 3232
DMFA *see* 3232
DMG *see* 3233
DMGG *see* 5845
DMN *see* 6558
DMNA *see* 6558
DMO *see* 3201
DMP *see* 3243
DMPA, 3394
DMPA [IMC] *see* 3241
DMPS *see* 3197
DMS *see* 3244 and 8836
DMS-70 *see* 3247
DMS-90 *see* 3247
DMSA *see* 8836
DMSO *see* 3247
DMSO₂ *see* 3246
DMT *see* 3251 and 9093
DMTT *see* 2827
DN *see* 3272
DN-289 *see* 3282
DNA *see* 2889
A-DNA *see* 2889
B-DNA *see* 2889
Z-DNA *see* 2889
DNase I *see* 2888
DNBP *see* 3282
DNC *see* 3271 and 3272
DNFB *see* 4101
DNOC *see* 3272
DNOCHP *see* 2739
DNOCP *see* 3281
DNP *see* 3365
DNS-55 *see* 3279
DNTP *see* 6983
DO-14 *see* 7818
Dobell's Soln *see* 8536
Dobendan [Merrell] *see* 2024
Doberol [Boehringer, Ing.] *see* 9439
Dobesilate Calcium, 3395
Dobesin [Pharmacia] *see* 3116
Dobetin *see* 9921
Dobren [Ravizza] *see* 8971
Doburil [Boehringer, Ing.] *see* 2760
Dobutamine, 3396
Dobutrex [Lilly] *see* 3396
Doca [Organon] *see* 2883
Docelan *see* 4739
Docemine [Roussel-UCLAF] *see* 9921
Docetrizoate Propyl *see* 7861
Docevita [Boizot] *see* 4739
Docibin [Merrell] *see* 9921
Docigram [Endopharm] *see* 9921
Dociton [Rhein-Pharma] *see* 7852
Docivit [Robisch] *see* 9921
Doclizid-T *see* 2073
6,7,8,9,10,11,12,13,14,15,16,17,24,25,-
26,27,28,29,30,31,32,33-Docosahy-
dro-35,37-dimethyl-5,34:18,23-di-
ethenodibenzo[*b,r*][1,5,16,20]tetra-
azacyclotriacontine-23,34-diium
Dichloride 1259
Docosanoic Acid *see* 1033
trans-13-Docosenoic Acid *see* 1359

Dopamet [Merck & Co.] see 5974
Dopamine, 3415
Dopan, 3416
Dopar see 5344
Doparkine [Armstrong] see 5344
Doparl [Kyowa] see 5344
threo-Dopaserine see 3444
Dopasol [Daiichi] see 5344
Dopastat [Parke, Davis] see 3415
Dopastin, 3417
Dopaston [Sankyo] see 5344
Dopastral [Astra] see 5344
Dopatec [Labatec] see 5974
Dopazinol see 6349
Dopegyt see 5974
Dopergin [Schering] see 5394
Dopexamine, 3418
Dopexamine Hydrochloride see 3418
Dopom see 4473
Dopram [Robins] see 3421
Doprin [SK & F] see 5344
Dops [Sumitomo] see 3444
L-DOPS see 3444
L-threo-DOPS see 3444
Doracil [Gador] see 5681
Doral see 9928
Doralese [Bridge] see 4876
Dorantamin [Dorsey] see 7996
Doraphen [Cenci] see 7851
Doraxamin [Dorsey] see 3168
Dorbane [Riker] see 2813
Dorcalm [Frere] see 2062
Dorcostrin [Dorsey] see 2883
Dorevane see 7834
Dorico see 4625
Dorico Soluble see 4625
Doridamina [Angelini] see 5447
Doriden [Ciba] see 4371
Doriden-Sed see 4371
Dorinamin see 1136
Dorison see 5731
Dormalest see 5731
Dormalin [Schering-Plough] see 8039
Dormate [Wallace] see 5653
Dorme [A.V.P.] see 7797
Dormicum [Roche] see 6105
Dormidin see 5731
Dormigen see 5731
Dormigene [Pharmacobel] see 1385
Dormigoa [Scheurich] see 5872
Dormin see 6 and 5871
Dorminal see 601
Dormiphen see 5731
Dormiral see 7201
Dormison [Schering] see 5731
Dormodor [Roche] see 4123
Dormogen [Spofa] see 5872
Dormonal see 972
Dormonoct [Roussel] see 5453
Dormosan see 5731
Dormovit see 4216
Dormutil see 5872
Dormwell see 3033
Dormytal see 601
Dornavac [Merck & Co.] see 2888
Dornwal [Maltbie] see 614
Dorsacaine [Dorsey] see 1056
Dorsallin "A.R." [Dorsey] see 7042
Dorsedin see 5872
Dorsiflex see 5739
Dorsilon see 5739
Doryl [E. Merck] see 1780
Doryx [Parke, Davis] see 3429
Dosberotec [Boehringer, Ing.] see 3927
Dosulepin see 3419
Dosulfin [Geigy] see 8911
Dothiepin, 3419
Double-mycin [Heyl] see 3162
Double Thiosulfate of Gold And Sodium see 4423
Dovenix [Specia] see 6578

Dow 1329 see 3394
Dowco 118 see 3394
Dowco 132 see 2607
Dowco 179 see 2190
Dowco 213 [Dow] see 2767
Dowco 214 see 2190
Dowco 217 [Dow] see 4172
Dowco 233 see 9572
Dowco 290 see 2398
Dowco 356 see 9578
Dowco 433 see 4128
Dowex 1-X2-Cl see 2205
Dowfax 9N [Dow] see 6596
Dowfax 9N9 [Dow] see 6596
Dowfume W 85 see 3753
Dow General [Dow] see 3282
Dowicide A [Dow] see 7276
Dowicide B [Dow] see 9555
Dowicide G [Dow] see 7059
Dowicide 1 [Dow] see 7276
Dowicide 2 [Dow] see 9555
Dowicide 9, 3420
Dowicide Q [Dow] see 2117
Dowicide 2S [Dow] see 9556
Dowlap see 9554
Dowmycin E [Dow] see 3632
Dowpen V-K [Dow] see 7046
Dowpon see 2806
Dow Selective [Dow] see 3282
Doxans [Schiapparelli] see 3423
Doxapram, 3421
Doxatet [Cox Continental] see 3429
Doxazosin, 3422
Doxazosin Mesylate see 3422
Doxefazepam, 3423
Doxenitoin, 3424
Doxephrin see 5859
Doxepin, 3425
Doxergan see 6900
Doxidan [Hoechst] see 3397
Doxifluridine, 3426
Doxigalumicina [Galup] see 3429
Doxinate [Hoechst] see 3398
Doxitard [Mack, Illert.] see 3429
Doxium [Carrion] see 3395
Doxofylline, 3427
Doxol see 3398
Doxophylline see 3427
Doxorubicin, 3428
Doxy-II (Caps) [USV] see 3429
Doxychol see 6884
Doxycycline, 3429
Doxycycline Hyclate see 3429
Doxylamine, 3430
Doxylar [Lagap] see 3429
Doxy-Puren [Klinge-Nattermann] see 3429
Doxy-Tablinen [Sanorania] see 3429
Doxytem [Temmler] see 3429
Doyle [Tanabe Seiyaku] see 874
Dozic [R. P. Drugs] see 4511
2,4-DP see 3068
DPA see 3312 and 7814
DPA Sodium see 9821
DPD see 1512
DPE see 3344
DPN see 6259
DPPD see 3331
DPPH see 3332
D-Pron [Minnesota Pharm.] see 3358
DPX 1410 see 6873
DPX 3674 see 4617
DPX 4189 see 2192
DPX 5648 see 8936
DPX F6025 see 2092
DPX H6573 see 4130
DPX T6376 see 6081
DPX Y5893-9 see 4634
DPX Y6202 see 8113
DQ 2466 see 1882
DQV K see 7046

Dracanyl [Astra] see 9089
Dracylic Acid see 1101
Draganon [Nippon Roche] see 692
Dragnet [FMC] see 7132
Dragon's Blood, 3431
Drakeol [Penreco] see 7139
Dramamine [Searle] see 3193
Dramarin see 3193
Dramcillin-S [White] see 7184
Dramocen [Central Pharm.] see 3193
Dramyl see 3193
Drapolene [Calmic] see 1066
Drapolex see 1066
Drastinetten [Herbrand] see 5872
Drat [M & B] see 2152
Draza [Bayer] see 5893
Drazifon [Roussel-UCLAF] see 9051
Drazine [Smith & Nephew] see 7227
Drazoxolon, 3432
DRC 1201 see 4903
Drenamist see 3569
Drenaren see 9650
Drenison [Lilly] see 4122
Drenusil [Pfizer] see 7561
Dribazil see 6843
Driclor [Stiefel] see 338
Dricol [Bristol] see 409
Dridase [Pharm. Arzneimittel] see 6908
Dri-Die [Fairfield] see 8437
Dridol see 3437
Dried Barium Hydroxide see 989
Dried Calcium Sulfate see 1713
Dried Cupric Sulfate see 2659
Dried Egg White see 206
Dried Ferrous Sulfate see 4006
Dried Gypsum see 1713
Dried Magnesium Sulfate see 5573
Dried Zinc Sulfate see 10064
Drierite see 1713
Drimenin, 3433
Drimyl [Cassenne] see 3832
Dri-Na see 8588
Drinalfa [Squibb] see 5859
Drinox see 4576
Drinupal see 2088
Driol see 6843
Driol-Labaz see 6843
Drisdol [Winthrop] see 9928
Drixin see 6919
Drocarbil, 3434
Drocode see 3154
Drocort [Lilly] see 4122
Drofenine, 3435
Drogenil [Schering AG] see 4132
Drolban [Lilly] see 3436
Droleptan [Le Brun] see 3437
Drometil see 936
Dromilac see 4650
Dromisol [Merck & Co.] see 3247
Dromoran [Roche] see 5350
Dromostanolone Propionate, 3436
Dronabinol see 9142
Droncit [Bayvet] see 7714
Drop Chalk see 1657
Droperidol, 3437
Droprenilamine, 3438
Droprenylamine see 3438
Dropropizine, 3439
Drosera, 3440
Drosophilin A, 3441
Drosophilin B see 7508
Drostanolone Propionate see 3436
Drosteakard [Plantorgan] see 9851
Drotebanol, 3442
Droxarol [Pharmos] see 1462
Droxaryl [Continental Pharma] see 1462
Droxicam, 3443
Droxidopa, 3444
Droxol [Bernabo] see 2082

Droxomin [Tutag] see 4739
Droxone (obsolete) [Squibb] see 230
Drumstick see 1900
Dry-Clox [Bristol] see 2414
Dry Ice see 1816
Drylin [Merckle] see 8889
Drynap [Wako] see 8588
Dryobalanops Camphor see 1338
Dryophantin see 7963
Dryptal [Berk] see 4221
DS 103-282 see 9409
DSCG see 2594
DSF see 6195
DSIP, 3445
DSMA see 5864
DSP see 8612
DSS see 3398
DST see 3162
DT-327 see 2391
D-Tamin Retard L.U.T. [Pharmafrid]
  see 3157
DTBP, 3446
DTC see 3384
DTF see 3084
DTIC see 2803
DTIC-Dome [Dome] see 2803
DTMC see 3075
DTPA see 7083
DTPT see 7896
D-Tracetten see 9928
DU see 2985
DU 1219 see 192
DU 21220 [Philips-Duphar] see 8228
DU 23000 see 4138
DU 112307 see 3128
Duact [Burroughs Wellcome] see 121
Dual [Ciba-Geigy] see 6067
Dualar [Armour Pharm.] see 1097
Duatok [Am. Cyanamid] see 8920
Dubimax [Oberval] see 6268
Dubnium see 3503
Duboisine see 4795
"Dubos Crude Crystals" see 9749
Ducobee [Breon] see 9921
Ducobee-Hy [Breon] see 4739
Dufalone [Merck & Co.] see 3080
Dufaston see 3689
Dufaston [ISM] see 3454
Dufrenite see 3975
Dugro [Merck & Co.] see 8237
Duhnul-balsan see 958
Dulasi [Dukron] see 8967
Dulcamara, 3447
Dulcin, 3448
Dulcion [Dulcis] see 3598
Dulcite see 4238
Dulcitol see 4238
Dulcolan see 1253
Dulcolax [Thomae] see 1253
Dulcose see 4238
Duloctil [Searle] see 8967
Dulsivac see 3398
Dumitone see 2820
Dumocyclin see 9130
Dumopranol [Dumex] see 7852
Duncaine see 5359
Duodecibin see 9921
Duodin see 4708
Duofas [Merck & Co.] see 8121
Duogastral [ISM] see 7463
Duogastrone [Merrell] see 1797
Duolax [Vangard] see 2813
Duolip [Merckle] see 9211
Duosan see 3548
Duoscorb [Upsher-Smith] see 855
Duosol [Interdelta] see 1296
Duotal see 4459
Dupéran [Cassenne] see 2381
Duphacid [Duphar] see 3128
Duphacycline [Duphar] see 6931
Duphafral D₃ 1000 see 9929

Duphalac [Duphar] see 5225
Duphar see 9132
Duphaston [Roxane] see 3454
Duphenicol [Duphar] see 2068
Duplamin [Bruschettini] see 7797
Duplosan KV [BASF] see 5666
Duponol see 4266
Du Pont 634 see 5318
Du Pont Herbicide 326 see 5387
Du Pont Herbicide 732 see 9085
Du Pont Herbicide 976 see 1370
Duprene see 6377
Dura AL [Durachemie] see 278
Dura AX [Durachemie] see 610
Durabetason [Durachemie] see 1202
Durabol see 6283
Durabolin [Organon] see 6283
Durabolin-O [Organon] see 3761
Duracebrol [Durachemie] see 6404
Duracef [Bristol] see 1921
Duracide [Endura] see 7446
Duracillin [Lilly] see 7042
Duraclamid see 6063
Duracreme [LRC] see 6596
Duracroman [Durachemie] see 2594
Duradoce see 4739
Duradoxal [Durachemie] see 3429
Duraflex [McNeil] see 4069
Durafurid [Durachemie] see 4221
Duragel [LRC] see 6596
Duragentam [Durachemie] see 4284
Duraglucon [Durachemie] see 4372
Duralta-12 [Merck & Co.] see 4739
Duramax see 873
Durametacin [Durachemie] see 4874
Duramipress [Durachemie] see 7715
Duramucal [Durachemie] see 392
Duranest [Astra] see 3818
Duranifin [Durachemie] see 6441
Duranitrat [Durachemie] see 5114
Duranol [Elan] see 7852
Durapaediat [Durachemie] see 3626
Durapatite, 3449
Durapental [Durachemie] see 7092
Duraphat [Woelm] see 8565
Duraphyl [Teikoku Seiyaku] see 9212
Duraphyllin [Durachemie] see 9212
Durapindol [Durachemie] see 7412
Durapirenz [Durachemie] see 7463
Duraprox [Wyeth-Ayerst] see 6879
Duraprost [Am. Home] see 6879
Duraquin [Parke, Davis] see 8072
Duraset see 6027
Duraset 20W see 6027
Dura-Silymarin see 8484
Duraspiron [Durachemie] see 8721
Dura-Tab see 8072
Duratrimet [Durachemie] see 8889
Duravolten [Durachemie] see 3071
Durazanil [Durachemie] see 1374
Durazepam [Durachemie] see 6881
Durenat [Bayer; Schering A.G.] see
  8885
Durene, 3450
Duretic see 5929
Duricef [Mead Johnson] see 1921
Durohydroquinone, 3451
Durol see 3450
Durolax see 1253
Duromine [Riker] see 7232
Duromorph [L.A.B.] see 6186
Duronitrin see 9682
Duropenin see 7037
Duroprocin see 8888
Duroquinone, 3452
Durox see 8890
Durrax [Dermik] see 4786
Dursban [Dow] see 2190
Dusodril [Roland] see 6268
Duspatal [Duphar] see 5648
Duspatalin [Duphar] see 5648

Dutch Drops see 6776
Dutch Liquid see 3754
Dutch Oil see 6776
Du-ter see 9659
Duvadilan [Duphar] see 5128
Duvaline [Almirall] see 7987
Duvoid [Norwich] see 1207
Duxima [Dukron] see 1951
DV [Merrell-Dow] see 3094
DV 1 see 5650
DV 714 see 5316
DV 1006 see 2017
DVC see 2761
D₃-Vicotrat see 9929
DW 61 see 4033
DW 3418 see 2692
Dwale see 1034
Dwarf Bay see 6095
DX.C see 2918
Dya-Tron [Myers-Carter] see 3358
Dybar [Du Pont] see 3951
Dybenal see 3048
Dycholium see 2858
Dycill [Beecham] see 3073
Dyclone [Dow] see 3453
Dyclonine, 3453
Dydrogesterone, 3454
Dyer's Alkanet see 242
Dyer's Saffron see 1877
Dyflos see 5060
Dyfonate [Stauffer] see 4147
Dygratyl [Ferrosan] see 3163
Dykon see 8555
Dylate [Unimed] see 2390
Dylene see 8830
Dyloform see 3689
Dylox [Chemagro] see 9536
Dymadon see 40
Dymanthine, 3455
Dymelor [Lilly] see 53
Dymid [Lilly] see 3305
Dymion [Pulitzer] see 9751
Dymixal®, 3456
Dynacaine see 8024
DynaCirc [Sandoz] see 5129
Dynacoryl see 6459
Dynacrine [Sandoz] see 5129
Dynafac see 2019
Dynalin Feed Premix [Squibb] see
  9351
Dynalin Injectable [Squibb] see 9351
Dynalin Soluble Powder [Squibb] see
  9351
Dynaltone see 3042
Dynamisan [Sandoz] see 862
Dynamite see 6528
Dynamos [Siphar] see 3150
Dynamutilin [Squibb] see 9351
Dynamyxin [Pfizer] see 7551
Dynapen [Bristol] see 3073
Dynaphylline [Welcker-Lyster] see 19
Dynarsan see 44
Dynatra [Simes] see 3415
Dyna-Zina see 4835
Dynel [Union Carbide] see 6140
Dynel®, 3457
Dyneric [Merrell] see 2384
Dynese [Galen] see 5527
Dynium Chloride see 3042
Dynorphin, 3458
Dynorphin₁.₁₃ see 3458
1-13-Dynorphin (Pig) see 3458
Dynothel [Henning] see 9348
Dynovas [Mayrand] see 6968
Dyodin see 4935
Dyphylline, 3459
Dypnone, 3460
Dyprin see 5896
Dyren [SK & F] see 9515
Dyrene [Chemagro] see 685
Dyrenium [SK & F] see 9515

Dyrex *see* 9536
Dyscural *see* 3682
Dysect [Deosan] *see* 2775
Dysedon *see* 6900
Dysentulin *see* 4389
Dyskinébyl [Saunier] *see* 6913
Dysmalgine *see* 8115
Dysmenalgit N [Krewel] *see* 6337
Dyspamet [Bridge] *see* 2279
Dyspas [Byk-Gulden] *see* 3087
Dysprosia *see* 3461
Dysprosium, 3461
Dystrophin, 3462
Dytac *see* 9515
Dytransin *see* 4811
E 3 *see* 5210
E 36U31 *see* 9083
E 141 *see* 3675
E 212 *see* 9483
E 250 *see* 2893
E 265 *see* 9485
E 600 *see* 6979
E 601 *see* 6022
E 605 *see* 6983
E 607 *see* 5238
E 614 *see* 9649
E 643 *see* 1477
E 646 *see* 3558
E 838 *see* 4793
E 0659 *see* 922
E 0671 *see* 9083
E 0687 *see* 1227
E 1059 *see* 2875
E 2663 *see* 1061
E 3128 *see* 3443
E 3314 *see* 4576
EACA Kabi *see* 442
EACS *see* 442
Eagle Vine *see* 2491
EAK *see* 3721
Eaklite *see* 1709
Early Bird [Mentholatum] *see* 7968
Earthnut *see* 7005
Earthnut Oil *see* 7006
Earth Wax *see* 1984
Easter Flower *see* 7956
Eastern Poison Oak *see* 7526
East Indian Balmony *see* 2051
"East Indian Copaiba" *see* 957
East Indian Sandalwood Oil *see* 6765
East India Root *see* 4243
Eastman No. 1361 *see* 8057
Eatan [Desitin] *see* 6491
Eau Celeste *see* 9116
Eau de Javelle *see* 4800
Eau de Labarraque *see* 8576
Eazaminum *see* 3098
EB 382 *see* 298
EB 1856 *see* 233
Ebalin [Allergopharma] *see* 1433
E-Base [Barr] *see* 3626
Ebert-Merz α-Acid *see* 6294
Ebert-Merz β-Acid *see* 6293
Ebesal [Hoechst] *see* 256
Ebimar®, 3463
Ebivit *see* 9929
Ebrantil [Byk-Gulden] *see* 9779
Ebrimycin [Chinoin] *see* 7756
Ebucin *see* 1672
Ebufac [DDSA] *see* 4812
Eburnal [Chiesi] *see* 3464
Eburnamenine-14-carboxylic Acid
  Ethyl Ester *see* 9894
Eburnamenin-14(15*H*)-one *see* 3464
(±)-Eburnamenin-14(15*H*)-one *see*
  3464
3α,16α-Eburnamenin-14(15*H*)-one *see*
  3464
Eburnamonine, 3464
Ebutol *see* 3672
E.C. 1.4.3.6 *see* 4639

E.C. 3.5.1.1 *see* 858
E.C. 3.2.1.11 *see* 2926
E.C. 3.4.24.4 *see* 9213
Ecarazine *see* 9425
Ecasolv [Lepetit] *see* 4571
Ecazide [Squibb] *see* 1773
Ecboline *see* 3613
Eccothal [Centerchem] *see* 9197
α-Ecdysone *see* 3465
β-Ecdysone *see* 3465
Ecdysones, 3465
Ecdysterone *see* 3465
ECGF *see* 4016
Ecgonidine, 3466
Ecgonine, 3467
Ecgonine Benzoate *see* 1125
Ecgonine Cinnamate Methyl Ester *see*
  2304
Ecgonine Ethyl Ester Benzoate (Ester)
  see 2449
Ecgonine Methyl Ester Benzoate *see*
  2450
Echinacea, 3468
Echinacin *see* 3468
Echinenone, 3469
Echinochrome A, 3470
Echinomycin, 3471
Echinopsine, 3472
Echinuline, 3473
epi-Echinuline *see* 3473
Echitamine, 3474
Echothiophate Iodide, 3475
Echujetin *see* 3145
Eclabron [U.S. Ethicals] *see* 4467
Eclipse Red *see* 1111
ECM [Grove] *see* 873
Ecobutazone *see* 7248
Ecoderm [Corvi] *see* 213
Ecodipin [Ecosol] *see* 6441
Ecodox [Pulitzer] *see* 3429
Ecodurex [Ecosol] *see* 417
Ecofenac [Ecosol] *see* 3071
Ecofrol *see* 9932
Ecolid [Ciba] *see* 2101
Ecolid Chloride [Ciba] *see* 2101
Econ [Cole] *see* 9932
Econazole, 3476
Ecosporina [Ecobi] *see* 1982
Ecostatin [F.A.I.R.] *see* 3476
Ecothiopate Iodide *see* 3475
Ecotrin [SK & F] *see* 873
Ecoval 70 *see* 1202
Ecovent [Ecosol] *see* 209
ECP [Upjohn] *see* 3656
Ecristidine *see* 4642
Ecstasy *see* 5646
ECTEOLA-Cellulose, 3477
Ectiban [ICI] *see* 7132
Ectimar [Bayer] *see* 3829
Ectoral [Pitman-Moore] *see* 8239
Ectylurea, 3478
Ecuanil [Wyeth] *see* 5751
Ecylert *see* 7021
Eczecidin *see* 4924
Eczederm [Quinoderm] *see* 1636
ED *see* 3816
Edalene [Pharmuka] *see* 2279
Edathamil *see* 3484
Edathamil Calcium Disodium *see* 3480
Edathamil Disodium *see* 3481
EDB *see* 3753
EDC *see* 3754
EDDP *see* 3485
Edecril [Merck & Co.] *see* 3669
Edecrin [Merck & Co.] *see* 3669
Edemex [Savage] *see* 1134
Eden-psich *see* 2082
Edestin, 3479
Edetate *see* MISC 3
Edetate Calcium Disodium, 3480

Edetate Dipotassium *see* 3484
Edetate Disodium, 3481
Edetate Sodium, 3482
Edetate Trisodium, 3483
Edetic Acid, 3484
Edetic Acid Calcium Disodium Salt
  see 3480
Edetic Acid Disodium Salt *see* 3481
Edetic Acid Sodium Iron Salt *see* 3977
Edetic Acid Tetrasodium Salt *see* 3482
Edetic Acid Trisodium Salt *see* 3483
EDGF-II *see* 4016
Edifenphos, 3485
Ediphenphos *see* 3485
Edisilate *see* MISC 4
Edisylate *see* MISC 3
Edolan [Lepetit] *see* 3831
Edornat [Albert-Roussel] *see* 4754
Edoxudine, 3486
EDPA *see* 3820
Edpetiline *see* 4840
Edrophone Bromide *see* 3487
Edrophonium Chloride, 3487
Edrul [Bayer] *see* 6227
EDTA *see* 3484
EDTA Calcium *see* 3480
EDTA Dipotassium *see* 3484
EDTA Disodium *see* 3481
EDTA Tetrasodium *see* 3482
EDTA Trisodium *see* 3483
EDU *see* 3486
Edurid [Robugen] *see* 3486
EEDQ, 3488
E.E.S. [Abbott] *see* 3626
EFA *see* 3650
Efalexin [Maipe] *see* 1971
Efamol [Efamol] *see* 3864
Efcorbin *see* 4710
Efcorlin *see* 4710
Efcortelan [Glaxo] *see* 4710
EF-Cortelan *see* 4710
EF-Cortelan Soluble *see* 4713
Efcortelin *see* 4710
Efcortesol *see* 4712
Efektolol [Efeka] *see* 7852
E-Ferol *see* 9932
Eferox [Efeka] *see* 5351
Effectin [Winthrop; Shionogi] *see*
  1317
Effederm [Sauba] *see* 8167
Effekton [Efeka] *see* 3071
Effilone *see* 5828
Effluderm [Pharm-Allergan] *see* 4109
Efflumidex [Allergan] *see* 4104
Effontil *see* 3822
Effortil [Boehringer, Ing.] *see* 3822
Effusan *see* 3272
Efisol [Roland] *see* 2896
Eflornithine, 3489
Efloxate, 3490
Efo-Dine [Fougera] *see* 7701
Efortil *see* 3822
Efosite Al *see* 4167
Efpenix [Toyo Jozo] *see* 610
Efrane [Abbott] *see* 3536
Efrotomycin, 3491
Efroxine [Pennwalt] *see* 5859
Eftapan [Merckle] *see* 3577
Efudex [Roche] *see* 4109
Efudix [Roche] *see* 4109
Efuranol [Pfizer Taito] *see* 4835
Egacene *see* 4795
Egalin *see* 8131
Egazil Duretter *see* 4795
EGF *see* 3492
EGF-URO *see* 3492
EGF-Urogastrone, 3492
Egg Albumin *see* 6855
Egg Oil, 3493
Egg White *see* 206
EGIS 2062 *see* 8423

Eglen [Tatsumi] see 2308
Egmol see 7211
Egressin see 9343
EGYT 201 see 1041
EGYT 341 see 9427
EGYT 475 see 7365
EGYT 739 see 4472
EGYT 1050 see 2845
EGYT 2062 see 8423
EHDP [Monsanto] see 3819
Ehrlich 5 see 6902
Ehrlich 594 see 44
Ehrlich 606 see 839
Ehrlich's Reagent see 3219
EI 3911 see 7305
EI 47031 see 7309
EI 47470 see 5743
Eicosamethylnonasiloxane, 3494
Eicosanoic Acid see 791
all-cis-5,8,11,14,17-Eicosapentaenoic
    Acid see 3495
5,8,11,14,17-Eicosapentaenoic Acid,
    3495
5,8,11,14-Eicosatetraenoic Acid see
    792
Eikonogen see 467
Einalon S [Kodama] see 4511
EinsAlpha [Thomae] see 4749
Einsteinium, 3496
Eismycin [Beecham-Wulfing] see 6212
Ejibil see 3727
Eka-cesium see 4175
Ekamet [Sandoz] see 3847
Eka-tantalum see 3504
Ekilan see 5739
Ekko [Fleming] see 7293
Ekomine [Hoechst] see 891
Ekonal [Yoshitomi] see 6583
Eksmin [Sumitomo] see 7132
Ektafos [Ciba] see 3077
Ektebin [Bayer] see 7905
Ektophanol [Gremy-Longuet] see
    2290
Ektyl see 3478
Ekvacillin [Astra] see 2414
EL 103 see 9053
EL 107 see 5125
EL 110 see 1048
EL 119 see 6840
EL 161 see 3671
EL 179 see 5089
EL 222 see 3903
EL 466 see 7467
EL 500 see 4129
EL 531 see 665
EL 614 see 1378
EL 857 see 780
EL 974 see 9359
EL 857/820 see 780
Elaidic Acid, 3497
Elaiomycin, 3498
Elamine [Interchemical] see 7900
Elamol [Brocades-Stheeman] see 9426
Elan [Abbott] see 7292
Elan [Valeas] see 6615
Elanone [Robins] see 5320
Elanpres see 5974
Elantan [Pharma Schwarz] see 5114
Elarzone [Dausse] see 7424
Elastase, 3499
Elasterin [Phoenix] see 3924
Elastin, 3500
Elastonon see 616
Elaszym [Eisai] see 3499
α-Elaterin see 2617
Elavil [Merck & Co.] see 504
Elayl see 3748
Elbrol [Pfleger] see 7852
Elbrus [Roemmers] see 5669
Elcatonin, 3501
Elcitonin [Toyo Jozo] see 3501

Elcosal see 3168
Elcosine see 8929
Eldéprine [Unicet] see 2893
Eldepryl [Britannia] see 2893
Elder see 8315
Elderfield Pyrimidine Mustard see
    3416
Eldisine [Lilly] see 9892
Eldopal [Brocades-Stheeman] see 5344
Eldopaque [Elder] see 4738
Eldopar [Weifa] see 5344
Eldopatec [Labatec] see 5344
Eldoquin [Elder] see 4738
Eldoral see 3800
Eldrin see 8276
Eleagol see 7515
Elecampane see 4897
Elecampane Camphor see 198
Elecol [Andromaco] see 8803
Elecor see 7744
Electan [Eutherapie] see 7122
Electrocortin [Ciba] see 218
Eledoisin, 3502
Element 104, 3503
Element 105, 3504
Element 106, 3505
Element 107, 3506
Elen [Yamanouchi] see 4850
Elenium [Polfa] see 2082
Elenolide, 3507
Eleparon [Luitpold] see 2217
Eleudron see 8920
Eleuthera Bark see 1890
Elfwort see 4897
Elgetol see 3272
Elics [Senju] see 508
Elidin [Endo] see 3202
Elieten [Nippon Kayaku] see 6063
Elimin [Sumitomo] see 6465
Eliminoxy see 2300
Elinol see 4116
Elipten [Ciba] see 452
Elisal see 8974
Elixicon [Cooper] see 9212
Elkapin [Gödecke] see 3845
Elkosil see 8929
Elkosin [Ciba] see 8929
Ellagic Acid, 3508
Ellatun [Basotherm] see 9486
Ellipticine, 3509
Elliptinium Acetate, 3510
Elliptone, 3511
Elmarin see 2186
Elmedal [Thiemann] see 7248
Elmetacin [Luitpold] see 4874
Elmifarma [Farmitalia] see 7366
Elobromol [Chinoin] see 6132
Elocon [Schering-Plough] see 6151
Elodrine see 4716
Eloisin [Farmitalia] see 3502
Elon see 5940
Elorine Chloride [Lilly] see 7770
Elorine Sulfate [Lilly] see 7770
Elrodorm see 4371
Elspar [Merck & Co.] see 858
Elsyl [Lakeside] see 6060
Eltrianyl [Saarstickstoff-Fatol] see
    8889
Eltroxin [Glaxo] see 5351
Elvanol [Du Pont] see 7562
Elvaron [Bayer] see 3031
Elvetil [Maggioni] see 8776
Elymoclavine, 3512
Elyzol [Dumex] see 6079
Elzogram [Lilly] see 1925
EM-923 see 4283
EMA-CO see MISC-2
Emamin see 9623
Emanation see 8119
Emanil see 4819
EMB see 3672

Embacetin see 2068
Embafume see 5951
Embanox see 1547
Embarin [Merckle] see 278
Embark [3M] see 5684
EMBAY 8440 see 7714
Embazin see 8914
Embden Ester see 4188
"Embelic Acid" see 3513
Embelin, 3513
Embequin [M & B] see 4935
EMB-Fatol [Saarstickstoff-Fatol] see
    3672
Embichen see 5655
Embichin 7 see 6640
Embikhin 7 see 6640
Embikhine see 5655
Embinal see 972
Embiol see 9921
Embonate see MISC-4
Embonic Acid see 6955
Embramine, 3514
Embutal see 7087
Embutox see 2828
Emcol E-607 [Witco] see 5238
Emcol E-607S [Witco] see 5238
Emconcor [E. Merck] see 1309
Emcor [E. Merck] see 1309
Emcyt [Roche] see 3658
EMD 9806 see 7706
EMD 15700 see 6492
EMD 16923 see 5745
EMD 30087 see 1924
EMD 33 512 see 1309
EMD 34946 see 5480
Emdabol [E. Merck] see 9385
Emdabolin [Chugai] see 9385
Emdisterone see 3436
Emedan see 1839
Emedyl see 3193
Emelent see 7768
Emepride [Roche] see 1420
Emepronium Bromide, 3515
Emerald see 1177
Emerald Green see 1367 and 2629
Emerest 2640 [Emery] see 7555
Emerest 2672 [Emery] see 7555
Emerest 2600 Series [Emery] see 7555
Emergil [Labaz] see 4114
Emericid see 5448
Emersol 132 see 8761
Emes see 3193
Emesazine see 2308
Emeside [L.A.B.] see 3704
Emetamine, 3516
Emete-Con [Pfizer] see 1133
Emethibutin see 3784
Emetic Herb see 5431
Emeticon [Pfizer] see 1133
Emetine, 3517
Emetine Hydrochloride see 3517
Emetiral see 7768
Emeto-Na see 734
Emex [Archifar] see 1791
Emicholin see 2321
Emilene see 5525
Eminase [Beecham] see 785
Emivan [USV] see 3673
Emmolic Acid see 1916
Emociclina see 9921
Emodin [Upjohn] see 4812
Emodin, 3518
Emodin-l-rhamnoside see 4177
Emoren [Wassermann] see 6888
Emorfazone, 3519
Emorhalt [Bayropharm] see 9487
Emoril [Roemmers] see 1420
Emotival [Armstrong] see 5456
Emovate [Glaxo] see 2362
Empecid [Bayer/Yoshitomi] see 2412
Emperal [Orion] see 6063

Empirin [Burroughs Wellcome] see 873
EMPP see 3558
Empyreumatic Oil Of Juniper see 5153
EMQ see 3710
EMS see 3782
Emtexate see 7029
Emtexate [Nordic] see 5908
Emtryl see 3255
Emtrylvet [Rhodia] see 3255
EMU see 3626
Emulphogene BC [GAF] see 7554
Emulphor [GAF] see 7555
Emulsept (Obsolete) see 5238
Emulsion 212 [Apex] see 4233
Emulsynt [Van Dyk] see 7555
E-Mycin [Protea] see 3632
E-Mycin [Upjohn] see 3626
E-Mycin E [Upjohn] see 3626
Emylcamate, 3520
Emyrenil [Emyfar] see 6899
EN 141 see 5148
EN 313 see 6178
EN 1010 see 8024
EN 1530 see 6277
EN 1639A see 6278
EN 1661L see 1308
EN 1733A see 6142
EN 2234A see 6271
EN 18133 see 9275
Enadel [Pfizer] see 2415
Enalapril, 3521
Enalaprilat, 3522
Enalaprilic Acid see 3522
Enallachrome see 3646
Enallylpropymal, 3523
Enantate see MISC-4
Enanthal see 4578
Enanthaldehyde see 4578
Enanthate see MISC-3
Enanthic Acid see 4581
Enanthic Alcohol see 4582
Enanthotoxin, 3524
Enap [Krka] see 3521
Enapren [Merck & Co.] see 3521
Enargite see 2514
Enarmon see 9115
Enavid [Searle] see 6615
Enbol Base [Chugai] see 8006
Encainide, 3525
Encaprin [Procter & Gamble] see 873
Encare [Norwich] see 6596
Encephabol [E. Merck] see 8006
Encetrop [Siegfried] see 7459
Enclomiphene see 2384
Encordin [E. Merck] see 7136
Encorton [Polfa] see 7727
Endak [Madaus] see 1875
Endecril see 3669
Endep [Roche] see 504
Endiandric Acid A see 3526
Endiandric Acid B see 3526
Endiandric Acid C see 3526
Endiandric Acids, 3526
Endoamylases see 631
Endobenziline Bromide see 3527
Endobenzyline Bromide, 3527
Endobulin [Immuno] see 4837
Endocaine [Endo] see 8024
Endociclina [Del Saz & Filippini] see 4169
Endocistobil [Bracco] see 4916
Endocorion see 4534
Endodextranases see 2926
Endo E Dompé see 9932
Endofolliculina see 3660
Endogenous Pyrogen see 4895
Endografin [Schering AG] see 4916
Endojodin [Bayer] see 7792
Endokolat [Weisskopf] see 1253

Endolat see 5736
Endolin [Endo] see 7021
3-(1,4-Endomethylenecyclohexane-2,3-endo-cis-dicarboximido)piperidine-2,6-dione see 9006
cis-3,6-Endomethylene-Δ^4-tetrahydrophthalic Acid Dimethyl Ester see 3230
3,6-Endomethylene-1,2,3,6-tetrahydro-cis-phthalic Anhydride see 1801
3,6-Endomethylene-Δ^4-tetrahydrophthalic Anhydride see 1801
Endomixin [Lusofarmaco] see 6369
Endopancrine see 4889
Endo-Paractol [Homburg] see 8486
Endophenolphthalein see 6927
Endophleban [Rentschler] see 3157
Endopituitrina [ISM] see 6934
α-Endorphin see 3528
α-neo-Endorphin see 3528
β-Endorphin see 3528
β-neo-Endorphin see 3528
γ-Endorphin see 3528
Endorphins, 3528
Endosan [Hoechst] see 1237
Endosulfan, 3529
Endothal see 3530
Endothall, 3530
Endothelial Cell Growth Factor see 4016
Endothelin, A4
Endothelin-1 see A4
Endothelin-2 see A4
Endothelin-3 see A4
Endotoxins, 3531
Endoxan [Asta] see 2753
3,6-Endoxohexahydrophthalic Acid see 3530
Endralazine, 3532
Endrate Disodium [Abbott] see 3481
Endrate Tetrasodium [Abbott] see 3482
Endrin, 3533
Enduracidin, 3534
Enduracidin A Hydrochloride see 3534
Enduracidin B Hydrochloride see 3534
Endural [Arcolab] see 4221
Endurance [Sandoz] see 7773
Enduron [Abbott] see 5929
Enduronum [Abbott] see 5929
Endydol [Guidotti] see 873
E.N.E. see 3789
Enelfa [Dolorgiet] see 40
Enerbol [Polfa] see 8006
Energofit [Roleca] see 4321
Energona [Maurer] see 6616
Eneril see 40
Enesol see 5864
Enfenamic Acid, 3535
Enflurane, 3536
Engemycin [Gist-Brocades] see 6931
Englate see 9212
"English" Aconitine see 7924
English Hawthorn see 2569
English White see 1657
Enheptin [Am. Cyanamid] see 469
Enheptin-A [Am. Cyanamid] see 421
Enhexymal see 4625
Enibomal see 6342
Enicol (Capsules) see 2068
Enide [Tuco] see 3305
Enidin see 5596
Enidran see 6843
Enidrel [Syncro] see 6881
Enidrel see 6615
Enilconazole, 3537
Enin see 5596
Enirant [Desitin] see 3598
Enisyl [Person & Covey] see 5509

Enkade [Mead Johnson] see 3525
Enkaid [Mead Johnson] see 3525
Enkalene [Kunstzijde] see 7546
Enkalon [Am. Enka] see 6656
Enkephalins see 3528
Enlon [Anaquest] see 3487
Ennds see 2155
Enniatin A see 3538
Enniatin B see 3538
Enniatin C see 3538
Enniatins, 3538
Enocitabine, 3539
Enol Luteovis see 8069
3-Enolpyruvic Ether of trans-3,4-Dihydroxycyclohexa-1,5-diene Carboxylic Acid see 2220
Enoram [Roger Bellon] see 3540
Enovid see 6615
Enoxacin, 3540
Enoxaparin, 3541
Enoxen [Zambeletti] see 3540
Enoximone, 3542
Enoxolone, 3543
Enprofylline, 3544
Enprostil, 3545
Enradin [Takeda] see 3534
Enramycin see 3534
Enrofloxacin, 3546
Enrumay [Cooper] see 7996
E.N.S. see 3789
Enseal Potassium Chloride [Lilly] see 7601
Ensidon [Geigy] see 6808
Ensign [Yamanouchi] see 2321
Enstamine see 5870
Enstar [Zoecon] see 5194
Enstatite see 5569
ENT 987 see 9304
ENT 1122 see 3282
ENT 7796 see 5379
ENT 14250 see 7446
ENT 14874 see 10071
ENT 15108 see 6983
ENT 16225 see 3093
ENT 16273 see 8945
ENT 16519 see 794
ENT 17034 see 5582
ENT 17251 see 3533
ENT 17292 see 6022
ENT 17588 see 8013
ENT 20218 see 2848
ENT 20852 see 1431
ENT 22014 see 926
ENT 22374 see 6089
ENT 22879 see 3296
ENT 23233 see 926
ENT 23347 see 3371
ENT 23648 see 3075
ENT 23969 see 1789
ENT 24042 see 7305
ENT 24105 see 3691
ENT 24482 see 3077
ENT 24717 see 2603
ENT 24727 see 3281
ENT 24833 see 1333
ENT 24988 see 6272
ENT 25445 see 506
ENT 25515 see 7312
ENT 25540 see 3945
ENT 25567 see 6269
ENT 25580 see 9275
ENT 25640 see 2791
ENT 25644 see 3882
ENT 25647 see 3394
ENT 25705 see 7311
ENT 25715 see 3922
ENT 25719 see 6126
ENT 25760 see 5718
ENT 25784 see 443
ENT 25793 see 1237
ENT 25830 see 7309

4α,9-Epoxycevane-3β,4,6α,7α,14,15α,-16β,20-octol 6,7-Diacetate 15-(2-Methylbutanoate) see 2900

6β,6aβ-Epoxy-2,3,3aα,3b,4,5,6,6a,7,-7aα-decahydro-5β-hydroxy-2β,3bβ-dimethyl-4-methylene-1H-cyclopenta[a]pentalene-2-carboxylic Acid see 4637

5,8-Epoxy-5,8-dihydro-β,ε-carotene-3,3'-diol see 2254 and 4032

5,6-Epoxy-5,6-dihydroretinol see 4572

(4α,5α,17β)-4,5-Epoxy-3,17-dihydroxyandrost-2-ene-2-carbonitrile see 9611

7β,8-Epoxy-3β,14-dihydroxy-5β-card-20(22)-enolide see 9021

1,10-Epoxy-6,8-dihydroxygermacra-4,11(13)-dien-12-oic Acid 12,8-Lactone Acetate see 7979

15,16-Epoxy-6β,9-dihydroxy-8βH-labda-13(16),14-dien-19-oic Acid γ-Lactone see 5635

4,5-Epoxy-3,14-dihydroxy-17-methylmorphinan-6-one see 6922

20,24-Epoxy-3,25-dihydroxy-9-methyl-19-nor-9β-lanost-5-en-11-one see 4446

4,5-Epoxy-3,14-dihydroxy-17-(2-propenyl)morphinan-6-one see 6277

(5Z,9α,11α,13E,15S)-6,9-Epoxy-11,15-dihydroxyprosta-5,13-dien-1-oic Acid see 7890

6',7-Epoxy-6,12'-dimethoxy-2'-methyl-1'α-oxyacanthan see 9610

4a,9a-Epoxy-3-(2,3-epoxybutyryl)-1,2,3,4,4a,9a-hexahydro-1,3,4,5,10-pentahydroxy-2-methylanthrone see 2000

3α,9α-Epoxy-14β,18β-(epoxyethano-N-methylimino)-5β-pregna-7,16-diene-3β,11α,20α-triol see 1019

3α,9α-Epoxy-14β,18β-(epoxyethano-N-methylimino)-5β-pregna-7,16-diene-3β,11α,20α-triol, 20α-Ester with 2,4-Dimethylpyrrole-3-carboxylic Acid see 1018

cis-10,11-Epoxy-7-ethyl-3,11-dimethyl-trans,trans-2,6-tridecadienoic Acid see 5155

2,3-Epoxy-2-ethylhexanamide see 6874

(8r)-6β,7β-Epoxy-8-ethyl-3α-hydroxy-1αH,5αH-tropanium Bromide (−)-Tropate see 6897

β,17-Epoxy-α-(3-ethylidene-1,2,3,4,-6,7,12,12b-octahydroindole[2,3-a]-quinolizin-2-yl)-curan-1-propanoic Acid Methyl Ester see 4275

4,4a-Epoxy-5-ethylidene-2,3,4,4a-tetrahydro-5H-1-pyridine see 3

(3β,23β)-17,23-Epoxy-3-(β-D-glucopyranosyloxy)veratraman-11-one see 7932

5β,20-Epoxy-1,2α,4,7β,10β,13α-hexahydroxytax-11-en-9-one 4,10-Diacetate 2-Benzoate 13-Ester with (2R,3S)-N-Benzoyl-3-phenylisoserine see 9049

(3β,5β,5β,15β)-14,15-Epoxy-3-hydroxy-5-bufa-20,22-dienolide see 8150

8,10-Epoxy-8-hydroxy-3,4-dimethoxy-17-methylhasubanan-7-one see 5834

4,5-Epoxy-3-hydroxy-5,17-dimethylmorphinan-6-one see 6071

4,5α-Epoxy-6β-hydroxy-germacra-1(10),11(13)-dien-12-oic Acid γ-Lactone see 6996

1,10-Epoxy-6β-hydroxy-1β,5β,7α-guaian-3-en-12-oic Acid γ-Lactone see 798

[5α,7α(R)]-4,5-Epoxy-3-hydroxy-6-methoxy-α,17-dimethyl-α-propyl-6,14-ethenomorphinan-7-methanol see 3843

4,5-Epoxy-14-hydroxy-3-methoxy-17-methylmorphinan-6-one see 6912

4,5-Epoxy-3-hydroxy-17-methylmorphinan-6-one see 4733

6β,7β-Epoxy-3α-hydroxy-8-methyl-1αH,5αH-tropanium Bromide Tropate (Ester) see 5927

4,5-Epoxy-17-hydroxy-3-oxoandrostane-2-carbonitrile see 9611

5,6-Epoxy-3-hydroxy-p-toluquinone see 9104

12,13-Epoxy-4-hydroxytrichothec-9-en-8-one Crotonate see 9564

3α,6α-Epoxy-7β-hydroxytropane see 8365

(3β,23β)-17,23-Epoxy-3-hydroxyveratraman-11-one see 5145

9,13-Epoxylabd-7-en-15-oic Acid see 4451

1,8-Epoxy-p-menthane see 3851

Epoxymethamine Bromide see 5927

4,5-Epoxy-3-methoxy-17-methylmorphinan-6-ol see 3154

4,5α-Epoxy-3-methoxy-17-methylmorphinan-6β-ol see 3159

4,5-Epoxy-3-methoxy-17-methylmorphinan-6-one see 4708

6',7-Epoxy-6-methoxy-2-methyloxyacanthan-12'-ol see 6102

(5α,6α)-4,5-Epoxy-17-methylmorphinan-3,6-diol see 3160

4,5-Epoxy-17-methylmorphinan-3,6-diol Diacetate (Ester) see 2947

4,5a-Epoxy-17-methylmorphinan-3-ol see 2907

cis-7,8-Epoxy-2-methyloctadecane see 3367

cis-12,13-Epoxyoctadec-cis-9-enoic Acid see 9868

16α,21α-Epoxyolean-12-ene-3β,22α,-23,28-tetrol see 3643

12,13-Epoxyoleic Acid see 9868

12,13-Epoxy-4-[(1-oxo-2-butenyl)-oxy]trichothec-9-en-8-one see 9564

(2R,3S)-2,3-Epoxy-4-oxo-7E,10E-dodecadienamide see 1997

2,3-Epoxy-4-oxo-7,10-dodecadienoylamide see 1997

1,3-Epoxypropane see 9630

2,3-Epoxy-1-propanol see 4385

(−)-(1R,2S)-(1,2-Epoxypropyl)phosphonic Acid see 4169

[1β,2α,11β,12α,15β(S)]-11,20-Epoxy-1,2,11,12-tetrahydroxy-15-(2-hydroxy-2-methyl-1-oxobutoxy)picras-3-en-16-one see 4331

12,13-Epoxy-3,4,7,15-tetrahydroxytrichothec-9-en-8-one see 6581

12,13-Epoxytrichothec-9-ene-3,4,8,-15-tetrol 4,15-Diacetate 8-(3-Methylbutanoate) see 9711

12,13-Epoxytrichothec-9-en-4-ol Acetate see 9562

14,15β-Epoxy-3β,5,16β-trihydroxy-5β-bufa-20,22-dienolide 16-Acetate see 2310

15,16-Epoxy-1β,4,12-trihydroxy-5,9-dimethyl-17,18-dinor-8βH,9βH,10α-labda-2,13(16),14-triene-19,20-dioic Acid 19,1:20,12-Dilactone see 2489

7β,8-Epoxy-3β,11α,14-trihydroxy-12-oxo-5β-card-20(22)-enolide see 8339

5,6-Epoxy-4,22,27-trihydroxy-1-oxoergosta-2,24-dien-26-oic Acid δ-Lactone see 9960

12,13-Epoxy-3,7,15-trihydroxytrichothec-9-en-8-one see 9947

cis-10,11-Epoxy-3,7,11-trimethyl-trans,trans-2,6-tridecadienoic Acid Methyl Ester see 5155

3β,7β-Epoxy-1βH,5βH-tropan-6α-ol see 8365

6β,7β-Epoxy-1αH,5αH-tropan-3α-ol (−)-Tropate see 8361

6β,7β-Epoxy-1αH,5αH-tropan-3α-ol (−)-Tropate 8-Oxide see 8362

6β,7β-Epoxy-3α-tropanyl S-(−)-Tropate see 8361

6,7-Epoxytropine Tropate see 8361

Eppy [Barnes-Hind] see 3569

Eprazin see 7970

Eprazinone, 3577

Eprex [Cilag] see 3635

Eprofil see 9217

Eprolin-S [Lilly] see 9931

Eprozinol, 3578

Epsamon [Emser] see 442

Epsikapron [Kabi] see 442

Epsilan see 9931

Epsilan-M see 9932

Epsilcapramin see 442

Epsilon-aminocaproic Acid see 442

Epsomite see 5573

Epsom Salts see 5573

Epsyl [Exa] see 7457

Eptam [Stauffer] see 3580

Eptastatin see 7712

Eptazocine, 3579

EPTC, 3580

Eptoin see 7293

Epuric [Stago] see 278

Equal [Searle] see 861

Equanil [Wyeth; Clin-Comar-Byla] see 5751

Equibar [Biogalenique] see 5974

Equiben [Merck & Co.] see 1733

Equibral [Ravizza] see 2082

Equicol [Burns-Bio] see 4465

Equigard [Shell] see 3069

Equigel [Shell] see 3069

Equilase see 1906

Equilenin, 3581

Equilibrin [Nattermann] see 505

Equilin, 3582

Equilium see 1845

Equimate see 4121

Equine Cyonin see 7515

Equine Gonadotrop(h)in see 7515

Equinil see 5751

Equipalazone [Arnolds] see 7248

Equipax see 7771

Equipertine see 6924

Equipose see 4786

Equiproxen [Syntex] see 6337

Equipur [Fresenius] see 9888

Equisetic Acid see 112

Equitac [SK & F] see 6891

Equizole [Merck & Co.] see 9217

Equol, 3583

Equron [Squibb] see 4675

Eqvalan [Merck & Co.] see 5133

ER 115 see 9074

ER 5461 see 7781

Erabutoxin A see 3584

Erabutoxin B see 3584

Erabutoxin C see 3584

Erabutoxins, 3584

Eracine [Winthrop] see 8247

Eradacil [Winthrop] see 8247

Eradacin [Sterling] see 8247

Eradex [Chemagro] see 9288

Eradicane [ICI] see 3580

Eraldin [ICI] see 7703

Erantin *see* 7851
Erasis [Orion] *see* 3627
Erasol *see* 5655
Eratrex [Bristol] *see* 3632
Eraverm (Tabl.) [Asta] *see* 7431
Eraverm (Syrup) [Asta] *see* 7431
Erazon [Krka] *see* 7476
Erbalax-N [Erba] *see* 4346
Erbia *see* 3585
Erbium, 3585
Erbium Sulfate, 3586
Erbocain [Heilit] *see* 4145
Erbon, 3587
Erbumine *see* MISC-3
Ercefurol *see* 6450
Ercefuryl [Robert et Carrière] *see* 6450
Erco-Fer *see* 3995
Ercoquin *see* 4748
Ercotina *see* 7816
Erdin, 3588
Erdmann's Salt, 3589
Erebile *see* 2858
Eremeyevite *see* 329
Eremfat [Saarstickstoff] *see* 8215
4βH,5α-Eremophila-1(10),7(11)-dien-2-one *see* 9874
Eremursine *see* 4666
Ergadenylic Acid *see* 148
Ergamine [Burroughs Wellcome] *see* 4640
Ergamisol [Janssen] *see* 9161
Ergate *see* 3609
Ergenyl [Labaz] *see* 9821
Ergine *see* 5505
Ergobasine *see* 3600
Ergocalciferol *see* 9928
Ergocalm [Efeka] *see* 5458
Ergochrome AA(2,2') *see* 8372
Ergochrome CC(2,2') *see* 3597
Ergochrysin A *see* 3590
Ergochrysin B *see* 3590
Ergochrysins, 3590
Ergoclavine *see* 3601
Ergoclavinine *see* 3601
Ergocornine, 3591
Ergocorninine, 3592
Ergocristine, 3593
Ergocristinine, 3594
Ergocryptine, 3595
Ergocryptinine, 3596
Ergodesit [Graf] *see* 3598
Ergoflavin, 3597
Ergohydrin [Streuli] *see* 3598
Ergoklinine *see* 3600
Ergokryptine *see* 3595
Ergokryptinine *see* 3596
Ergoloid Mesylates, 3598
Ergomar [Fisons] *see* 3609
Ergometrine *see* 3600
Ergometrinine, 3599
Ergomimet [Klinge] *see* 3157
Ergonovine, 3600
Ergonovinine *see* 3599
Ergont [Desitin] *see* 3157
Ergoplus [Klinge] *see* 3598
Ergorone *see* 9928
Ergosine, 3601
Ergosinine *see* 3601
Ergostane, 3602
5β-Ergostane *see* 2516
Ergostanol, 3603
Ergostan-3-ol *see* 3603
Ergostat [Parke, Davis] *see* 3609
Ergosta-5,7,9(11),22-tetraen-3β-ol *see* 2861
Ergosta-5:6,7:8,22:23-trien-3-ol *see* 3607
Ergosta-6,8,22-trien-3β-ol *see* 4203
(3β,22E)-Ergosta-5,7,22-trien-3-ol *see* 3607

9β-Ergosta-5,7,22-trien-3β-ol *see* 5107
9β,10α-Ergosta-5,7,22-trien-3β-ol *see* 5471
10α-Ergosta-5,7,22-trien-3β-ol *see* 8007
Ergost-7-en-3β-ol *see* 3606
Ergost-8(14)-en-3β-ol *see* 3604
Ergost-14-en-3β-ol *see* 3605
(24R)-Ergost-5-en-3β-ol *see* 1735
α-Ergostenol, 3604
β-Ergostenol, 3605
γ-Ergostenol, 3606
Ergosterin *see* 3607
Ergosterol, 3607
9β-Ergosterol *see* 5107
Ergostetrine *see* 3600
Ergot, 3608
Ergotamine, 3609
Ergotaminine, 3610
Ergotartrate *see* 3609
Ergotex [Dumex] *see* 3157
Ergothioneine, 3611
L(+)-Ergothionine *see* 3611
Ergotinine, 3612
Ergotocine *see* 3600
Ergoton-A [Azusa] *see* 3609
Ergotonin [Streuli] *see* 3157
Ergotoxine, 3613
Ergotrate [Lilly] *see* 3600
Ergotrate-H [Lilly] *see* 3600
Ergotrate Maleate [Lilly] *see* 3600
Eridan *see* 2977
Erigeron, 3614
Erinitrit *see* 8599
Eriochrome® Black T, 3615
Eriodictin *see* 3616
Eriodictyol, 3616
Eriodictyon, 3617
Eriodictyonone *see* 4656
Erion *see* 8053
Eriosept [Kreussler] *see* 2896
Eriscel [Rachelle] *see* 3628
Erisimin *see* 4551
Erispan [Sumitomo] *see* 4058
Eritadenine, 3618
Eritrocina [Abbott] *see* 3626
Eritroger [Isnardi] *see* 3628
Eritrone *see* 9921
Ermetrine [Organon] *see* 3600
Ermysin [Britannia] *see* 3626
Eromycin *see* 3628
Erpalfa [Intes] *see* 2790
Errolon *see* 4221
Ertron *see* 9928
Ertuban *see* 5071
Erucic Acid, 3619
Ervasil [Gödecke] *see* 352
ERYC [Parke, Davis] *see* 3626
Erycen [Berk] *see* 3626
Erycin *see* 3626
Erycinum [Schering AG] *see* 3626
Erycorbin *see* 5009
Erycytol *see* 9921
Ery Derm [Abbott] *see* 3626
Eryliquid *see* 3626
Erymax [Parke, Davis] *see* 3626
Erypar [Parke, Davis] *see* 3632
Eryped [Abbott] *see* 3626
Erypo [Cilag] *see* 3635
Eryprim [Scarium] *see* 3632
Erysan *see* 2107
Erysimin *see* 4551
Ery-Tab [Abbott] *see* 3626
Erythorbic Acid *see* 5009
Erythrene *see* 1500
Erythricine *see* 4286
Erythrite *see* 2428 and 3620
Erythritol, 3620
meso-Erythritol *see* 3620
Erythritol Anhydride, 3621

Erythritol Tetranitrate *see* 3622
Erythrityl Tetranitrate, 3622
Erythrocentaurin, 3623
Erythrocin [Abbott] *see* 3632
Erythrocin Lactobionate [Abbott] *see* 3630
Erythro ES [Sanko] *see* 3626
Erythrogenic Acid *see* 4983
Erythroglucin *see* 3620
D-Erythrohexulose *see* 7940
Erythro-Holz [Pharma Holz] *see* 3626
α-Erythroidine, 3624
β-Erythroidine, 3625
Erythrol *see* 3620
Erythrol Tetranitrate *see* 3622
Erythromast 36 *see* 3626
Erythromid [Abbott] *see* 3626
Erythromycin, 3626
Erythromycin A *see* 3626
Erythromycin 2'-Acetate Octadeca-noate (Salt) *see* 3627
Erythromycin Acistrate, 3627
Erythromycin Estolate, 3628
Erythromycin Gluceptate *see* 3629
Erythromycin Glucoheptonate, 3629
Erythromycin Glucoheptonic Acid Salt *see* 3629
Erythromycin Lactobionate, 3630
Erythromycin 9-[O-[(2-Methoxyeth-oxy)methyl]oxime] *see* 8253
Erythromycin Octadecanoate (Salt) *see* 3632
Erythromycin 2'-Propanoate *see* 3631
Erythromycin 2'-Propanoate Dodecyl Sulfate (Salt) *see* 3628
Erythromycin Propionate, 3631
Erythromycin Propionate Lauryl Sul-fate *see* 3628
Erythromycin Stearate, 3632
Erythroped [Abbott] *see* 3626
Erythrophlamine, 3633
Erythrophleine, 3634
Erythropoiesis Stimulating Factor *see* 3635
Erythropoietin, 3635
Erythropterin, 3636
Erythro S [Sanko] *see* 3632
D-Erythrose, 3637
L-Erythrose, 3638
D-Erythrose 4-Phosphate, 3639
4-D-Erythrosephosphoric Acid *see* 3639
Erythrosine, 3640
Erythrosine B *see* 3640
Erythrosine BS *see* 3640
Erythrosine Extra Yellowish *see* 3174
Erythrotin *see* 9921
Erythroxylon *see* 2448
L-Erythrulose, 3641
Erytrarco [Arco] *see* 3628
Erytrarsin *see* 1602
ES 132 *see* 1218
ES 771 *see* 7400
ES 902 *see* 7400
Esafosfina *see* 4187
Esametina *see* 4609
Esanin *see* 9994
Esantene *see* 4885
Esaprazole, 3642
Esbatal [Burroughs Wellcome] *see* 1208
Esbecythrin *see* 2869
Esbericard [Schaper & Brümmer] *see* 2569
Esberidin [Schaper & Brümmer] *see* 9888
Esberiven [Schaper & Brümmer] *see* 5700
Esbuphon [Schaper & Brümmer] *see* 6616
Escalol 106 [Van Dyk] *see* 4382

Escalol 506 [Van Dyk] see 3221
Escalol 507 [Van Dyk] see 3221
Escatin see 160
Eschatin [Parke, Davis] see 160
Eschenmoser's Salt see 3239
Escigenin, 3643
Escin, 3644
α-Escin see 3644
β-Escin see 3644
Esclama [Montedison] see 6468
Esclebin [Boizot] see 6617
Esclerosina see 3681
Escobedin see 911
Escorpal [Bayer] see 7178
EsCort [Hoechst-Roussel] see 7717
Escosyl see 160
Escre [SS Pharm.] see 2061
Esculetin, 3645
Esculetin Dimethyl Ether see 8359
Esculin, 3646
Esculoside see 3646
Esdragol see 3657
Eseridine, 3647
Eserine see 7357
Eserine Aminoxide see 3647
Eserine Oxide see 3647
ESF see 3635
Esfar see 1451
Esfenvalerate see 3952
Esiclene [LPB Braglia] see 4161
Esidrex [Ciba] see 4704
Esidrix [Ciba] see 4704
Esilate see MISC-4
Esilgan [Takeda] see 3651
Esimil [Ciba] see 4473
Esinol [Toyama] see 3626
Eskabarb [SK & F] see 7201
Eskacef [SK & F] see 1982
Eskacillin [SK & F] see 7041
Eskacillin V [SK & F] see 7046
Eskadiazine [SK & F] see 8874
Eskalin V [SK & F] see 9910
Eskalith [SK & F] see 5404
Eskaserp [SK & F] see 8149
Eskazine [SK & F] see 9594
Eskazinyl [SK & F] see 9594
Eskel [SK & F] see 5189
Eskima [Mochida] see 3495
Esmail [Gedeon Richter] see 5669
Esmarin [E. Merck, W. Ger.] see 9537
Esmind [Otsuka] see 2186
Esmolol, 3648
Esocalm see 3391
Esoderm [Napp] see 5379
Esoiodine see 7792
Esomedine see 4612
Esomid Chloride [Ciba-Geigy] see 4609
Esophotrast [Barnes-Hind] see 1006
Esorb see 9931
Esparin see 7793
Esparto Wax, 3649
Espasmo-Gemora see 1510
Espectinomicina see 8695
Esperal [Thersa] see 3370
Esperan [Toyama] see 2743
Esperson [Hoechst] see 2910
Espinomycin A see 6106
Espiran [Fardeco] see 3942
Espril see 6399
Espyre [SS Pharm.] see 3358
Esquinon [Sankyo] see 1830
Essence of Mirbane see 6509
Essence de Niobe see 3725
Essence of Niobe see 5947
Essence of Rose see 6762
Essential Fatty Acids, 3650
"Essential Oil of Birch Wood" see 1252
"Essential Salt of Lemons" see 7684
Essigsäure Tonerde see 371

Essitol [Engelhard] see 323
Estazolam, 3651
Esteed [Warren-Teed] see 3689
Ester 25 see 6979
S-Ester with O,O-Dimethyl Phosphorothioate see 5582
Estergel [Merck & Co.] see 5103
Ester Gums, 3652
Esteron 44 [Dow] see 2802
Esterone 245 [Dow] see 8999
Estigyn see 3689
Estil [Reiss] see 39
Estilben see 3119
Estimulocel [Lafarquim] see 7769
Estinerval see 7181
Estinyl [Schering] see 3689
Estolate see MISC-3 and 4
Estomicina [Bergamon] see 3628
Estomycin see 6989
Eston-B see 3655
Eston-E see 3689
Estonmite see 6856
Estopen see 7027
Estoral see 5726
Estrace [Mead Johnson] see 3653
Estracyt [AB Leo] see 3658
Estradep see 3656
Estraderm [Ciba-Geigy] see 3653
Estradiol, 3653
cis-Estradiol see 3653
8α-Estradiol see 5051
α-Estradiol, 3654
α-Estradiol (obsolete) see 3653
β-Estradiol see 3653
Estradiol Benzoate, 3655
β-Estradiol 3-Benzoate see 3655
Estradiol 3-Bis(2-chloroethyl)carbamate see 3658
Estradiol 17β-Cyclopentanepropionate see 3656
Estradiol 17β-Cyclopentylpropionate see 3656
Estradiol 17β-Cypionate, 3656
Estradiol Enanthate see 3653
Estradiol Phosphate Polymer see 7543
Estradiol Undecylate see 3653
Estradurin [Ayerst] see 7543
Estragole, 3657
Estragon see 9036
Estramustine, 3658
1,3,5:10,6,8-Estrapentaen-3-ol-17-one see 3581
Estra-1,3,5(10),7-tetraene-3,17-diol see 3156
1,3,5,7-Estratetraen-3-ol-17-one see 3582
Estra-1,3,5(10)-triene-3,17α-diol see 3654
Estra-5,7,9-triene-3β,17β-diol see 4607
1,3,5-Estratriene-3,17α-diol see 3654
8α-Estra-1,3,5(10)-triene-3,17β-diol see 5051
(17β)-Estra-1,3,5(10)-triene-3,17-diol see 3653
(17β)-Estra-1,3,5(10)-triene-3,17-diol 3-Benzoate see 3655
(17β)-Estra-1,3,5(10)-triene-3,17-diol 3-Benzoate 17-[[4-[4-[Bis(2-chloroethyl)amino]phenyl]-1-oxobutoxy]-acetate] see 1199
Estra-1,3,5(10)-triene-3,17-diol 3-[Bis(2-chloroethyl)carbamate] see 3658
Estra-1,3,5(10)triene-3,17β-diol 3-[N,N-Bis(2-chloroethyl)carbamate] see 3658
(17β)-Estra-1,3,5(10)-triene-3,17-diol 17-Cyclopentanepropanoate see 3656

Estra-1,3,5(10)-triene-3,16,17-triol see 3659
Estra-1,3,5(10)-triene-3,16β,17β-triol see 3567
1,3,5-Estratriene-3β,16α,17β-triol see 3659
Δ1,3,5-Estratriene-3,16β,17β-triol see 3567
1,3,5-Estratrien-3-ol-17-one see 3660
4,9,11-Estratrien-17β-ol-3-one see 9499
Estr-4-ene-3β,17β-diol see 1325
4-Estren-17β-ol-3-one see 6280
Estrex [Syntex] see 2865
Estrifol [Premo] see 2504
Estriol, 3659
Estriol 3-Cyclopentyl Ether see 8065
Estriol Succinate see 3659
Estrobene [Ayerst] see 3118
Estrobene DP [Ayerst] see 3119
Estrodienol see 3094
Estrol see 3660
Estrone, 3660
8α-Estrone see 5052
Estropipate see 3660
Estroral see 3094
Estrosol [Russian] see 3069
Estrosyn [Cooper] see 3118
Estrovis [Gödecke] see 8066
Estrovite see 3653
Estrugenone [Kremers-Urban] see 3660
Estrumate [ICI] see 2397
Estrusol [SM & P] see 3660
Estulic [Sandoz] see 4474
Esucos [UCB] see 3391
Esylate see MISC-3
Esyntin see 3724
ET-1 see A4
ET-495 see 7465
Etabus see 3370
Etacortin [Hermal] see 4118
Etadrol [Farmitalia] see 4119
Etafedrine, 3661
Etafenone, 3662
Etafillina [Delalande] see 19
Etakridin see 3668
Etalate see 3718
Etalontin [Parke, Davis] see 6614
Etalpha [AB Leo] see 4749
Etambol see 3672
Etambro see 9133
Etamiphyllin, 3663
Etamiphylline Methesculetol see 5844
Etamon Chloride [Parke, Davis] see 9134
Etamsylate see 3675
Etamycin, 3664
Etamycin A see 3664
Etanautine see 1225 and 3724
Etaphydel [Delalande] see 19
Etaphylline [Delalande] see 19
Etapiam [Piam] see 3672
Etaqualone, 3665
Etard's Salt see 2675
Etavit see 9931
ETBE see 3732
Etching Ink see 2965
Eterilate see 3667
Eterobarb, 3666
Eterobarbital see 3666
Etersalate, 3667
Eterylate see 3667
Ethabid [Meyer] see 3682
Ethacridine, 3668
Ethacrynate Sodium see 3669
Ethacrynic Acid, 3669
Ethadione, 3670
Ethal see 2020
Ethalfluralin, 3671
Ethambutol, 3672

Ethambutol Isoniazid Methanesulfonate see 5013
Ethamicort see 4709
Ethamide [Allergan] see 3711
Ethamivan, 3673
Ethamolin [Glaxo] see 3681
Ethamoxytriphetol, 3674
Ethamsylate, 3675
Ethanal see 32
Ethanamidine Hydrochloride see 37
Ethanamine see 3718
Ethane, 3676
Ethanearsonic Acid, 3677
Ethanedial see 4405
Ethanediamide see 6871
1,2-Ethanediamine see 3752
Ethanedinitrile see 2698
[Ethanedioato(1—)-*O*]hydroxydioxomolybdenum see 6866
[Ethanedioato(2—)-*O,O* ']trioxomolybdate(2—) Dihydrogen see 6866
Ethanedioic Acid see 6865
Ethanedioic Acid Anhydride with Antimonic Acid (3:1) Tripotassium Salt see 731
Ethanedioic Acid Barium Salt see 997
Ethanedioic Acid Calcium Salt see 1691
Ethanedioic Acid Copper Salt see 2649
Ethanedioic Acid Diamide see 6871
Ethanedioic Acid Diammonium Salt see 565
Ethanedioic Acid Diethyl Ester see 3115
Ethanedioic Acid Dimethyl Ester see 6020
Ethanedioic Acid Disodium Salt see 8601
Ethanedioic Acid Magnesium Salt see 5554
Ethanedioic Acid Silver Salt see 8467
Ethanedioic Acid Strontium Salt see 8811
1,2-Ethanediol see 3755
1,1-Ethanediol Diacetate see 3767
1,2-Ethanediol Diacetate see 3756
1,2-Ethanediol Monoacetate see 3757
Ethanedioyl Dichloride see 6867
1,2-Ethanedisulfonic Acid, 3678
Ethanedithioamide see 8256
1,2-Ethanedithiol, 3679
1,1'-(1,2-Ethanediyl)bisbenzene see 1219
N,N '-1,2-Ethanediylbis[*N*-butyl-4-morpholinecarboxamide] see 3257
[[1,2-Ethanediylbis[carbamodithioato]](2—)]manganese see 5603
[[1,2-Ethanediylbis(carbamodithioato)](2—)]manganese Mixt. with [[1,2-Ethanediylbis(carbamodithioato)](2—)]zinc see 5598
[[1,2-Ethanediylbis[carbamodithioato]](2—)]zinc see 10071
1,2-Ethanediylbiscarbamodithioic Acid Disodium Salt see 6256
[[*N,N* '-1,2-Ethanediylbis[*N*-(carboxymethyl)glycinato]](4—)-*N,N* ',*O,O* ',-*O*ᴺ,*O*ᴺ']calciate(2—)disodium see 3480
[[*N,N* '-Ethanediylbis[*N*-(carboxymethyl)glycinato]](4—)]-*N,N* ',*O,*-*O* ',*O*ᴺ,*O*ᴺ-ferrate(1—) Sodium see 3977
[[*N,N* '-1,2-Ethanediylbis[*N*-(carboxymethyl)glycinato]](4—)-*N,N* ',*O,O* ',-*O*ᴺ,*O*ᴺ]-(OC-6-21)calciate (2—) Dihydrogen Compd with Piperazine (1:1) see 7435
N,N '-1,2-Ethanediylbis[*N*-(carboxymethyl)glycine] see 3484

N,N '-1,2-Ethanediylbis[*N*-(carboxymethyl)glycine] Disodium Salt see 3481
N,N '-1,2-Ethanediylbis[*N*-(carboxymethyl)glycine] Tetrasodium Salt see 3482
N,N '-1,2-Ethanediylbis[*N*-(carboxymethyl)glycine] Trisodium Salt see 3483
[1,2-Ethanediylbis(imino-4,1-phenylene)]bis[arsonic Acid] Disodium Salt see 3338
1,1'-(1,2-Ethanediyl)bis[4-methylbenzene] see 3386
N,N '-1,2-Ethanediylbis[3-nitrobenzenesulfonamide] see 3284
2,2'-[1,2-Ethanediylbis(oxy)]bisethanol see 9585
2,2'-(1,2-Ethanediyldiimino)bis-1-butanol see 3672
1,1',1'',1'''-(1,2-Ethanediyldinitrilo)tetrakis[2-propanol] see 3551
Ethane-1-hydroxy-1,1-diphosphonic Acid see 3819
Ethanenitrile see 62
Ethaneperoxoic Acid see 7107
Ethanethioamide see 9246
Ethanethioic Acid see 9247
Ethanethioic Acid *S*-[1-[2-(Acetyloxy)ethyl]-2-[[(4-amino-2-methyl-5-pyrimidinyl)methyl]formylamino]-1-propenyl] Ester see 46
Ethanethiol, 3680
Ethanimidamide Hydrochloride see 37
Ethanion see 1667
Ethanite see 9268
Ethanol see 3716
Ethanol Aluminum Salt see 340
Ethanolamine, 3681
Ethanolamine Oxiniacate see 6895
"Ethanolamine Plasmalogen" see 7492
Ethanoylaminoethanoic Acid see 74
Ethaphene see 8994
Ethavan [Monsanto] see 3815
Ethaverine, 3682
Ethbenzamide see 3685
Ethchlorvynol, 3683
Ethchlorvynol see 3683
Ethebenecid, 3684
Ethene see 3748
2,2'-(1,2-Ethenediyl)bis[5-aminobenzenesulfonic Acid] see 628
1,1'-(1,2-Ethenediyl)bis[benzene] see 8774
4,4'-(1,2-Ethenediyl)bisbenzenecarboximidamide see 8772
2,2'-(1,2-Ethenediyl)bis[5-[[(phenylamino)carbonyl]amino]benzenesulfonic Acid] Disodium Salt see 1322
Ethene Homopolymer see 7544
Ethenesulfonic Acid Homopolymer Sodium Salt see 5487
Ethenetetracarbonitrile see 9129
Ethenol Homopolymer see 7562
Ethenone see 5177
Ethenylamidine Hydrochloride see 37
2-(5-Ethenyl-1-azabicyclo[2.2.2]oct-2-yl)-1*H*-indole-3-ethanol see 2287
8a-(5-Ethenyl-1-azabicyclo[2.2.2]oct-2-yl)-2,3,8,8a-tetrahydro-3a*H*-furo-[2,3-*b*]indol-3a-ol see 8060
Ethenylbenzene see 8830
5-Ethenyl-3,4-dihydro-1*H*-pyrano[3,-4-*c*]pyridin-1-one see 4286
7-Ethenyl-1,2,3,4,4a,4b,5,6,7,8,10,-10a-dodecahydro-1,4a,7-trimethyl-1-phenanthrenecarboxylic Acid see 5086

7-Ethenyl-1,2,3,4,4a,4b,5,6,7,9,10,-10a-dodecahydro-1,4a,7-trimethyl-1-phenanthrenecarboxylic Acid see 7398
5-Ethenyl-6-(β-D-glucopyranosyloxy)-5,6-dihydro-1*H*,3*H*-pyrano[3,-4-*c*]pyran-1-one see 4289
4-Ethenyl-3-(β-D-glucopyranosyloxy)-3,4,4a,5,6,7-hexahydro-8*H*-pyrano[3,4-*c*]pyridin-8-one see 955
[4a*R*-(4aα,5β,6α)]-5-Ethenyl-6-(β-D-glucopyranosyloxy)-4,4a,5,6-tetrahydro-4a-hydroxy-1*H*,3*H*-pyrano-[3,4-*c*]pyran-1-one see 8991
1,1'-Ethenylidenebis[benzene] see 3323
5-Ethenyl-5-(1-methylbutyl)-2,4,6-(1*H*,3*H*,5*H*)-pyrimidinetrione see 9897
5a-Ethenyloctahydro-4-hydroxy-3,9-bis(methylene)-2*H*-furo[2,3-*f*][2]-benzopyran-2,8(3*H*)-dione see 9867
(*S*)-5-Ethenyl-2-oxazolidinethione see 4411
Ethenyloxyethene see 9899
3-(3-Ethenyl-4-piperidinyl)-1-(6-methoxy-4-quinolinyl)-1-propanone see 9908
(*cis*)-3-(3-Ethenyl-4-piperidinyl)-1-(4-quinolinyl)-1-propanone see 2291
1-Ethenyl-2-pyrrolidinone Homopolymer Compd with Iodine see 7701
1-Ethenyl-2-pyrrolidinone Polymers see 7700
Ethenzamide, 3685
Ethephon, 3686
Ether see 3762
Ether Chloratus see 3740
Ether Hydrochloric see 3740
Ether Muriatic see 3740
Etherylate see 3667
Ethiazide, 3687
Ethidium see 4650
Ethidol see 3689
Ethimide see 3692
Ethinamate, 3688
Ethinazone see 3665
Ethine see 84
Ethinyl Estradiol, 3689
17-Ethinylestradiol see 3689
17α-Ethinylestradiol 3-Cyclopentyl Ether see 8066
Ethinylestrenol see 5501
17α-Ethinyl-17β-hydroxyestr-4-ene see 5501
Ethinyl-Oestradiol [Roussel-UCLAF] see 3689
17α-Ethinyltestosterone see 3696
Ethinyl Trichloride see 9552
Ethiodan see 4946
Ethiodized Oil, 3690
Ethiodol [Savage] see 3690
Ethion, 3691
Ethionamide, 3692
Ethioniamide see 3692
Ethionine, 3693
Ethiops Iron see 3988
Ethiops Mineral see 5788
Ethiozin, 3694
Ethirimol, 3695
Ethisterone, 3696
Ethizone see 8832
Ethmosine see 6178
Ethmozin see 6178
Ethmozine [USSR] see 6178
Ethnine [Purdue Frederick] see 7303
Ethobrom see 9525
Ethocaine see 7763
Ethocel see 3739
Ethodin [Winthrop] see 3668

Cross Index of Names

Ethodryl [BDH] *see* 3106
Ethofat [Armak] *see* 7555
Ethofibrate *see* 3834
Ethofumesate, 3697
Ethoglucid *see* 3837
Ethoheptazine, 3698
Ethohexadiol, 3699
Ethol *see* 2020
Ethomids®, 3700
Ethopabate, 3701
Ethopropazine, 3703
Ethoprophos *see* 3702
Ethosperse [Glyco] *see* 7554
Ethosuximide, 3704
Ethotoin, 3705
Ethotrimeprazine *see* 3849
Ethovan *see* 3815
Ethoxazene, 3706
2-Ethoxy-4-acetamidobenzoic Acid
  Methyl Ester *see* 3701
p-Ethoxyacetanilide *see* 7155
7-Ethoxy-3,9-acridinediamine *see*
  3668
1-Ethoxy-2-amino-4-nitrobenzene
  see 6538
4-Ethoxyaniline *see* 7188
2-Ethoxybenzamide *see* 3685
2-Ethoxybenzenamine *see* 7187
4-Ethoxybenzenamine *see* 7188
Ethoxybenzene *see* 7189
2-Ethoxybenzenecarbonamide *see*
  3685
4-Ethoxybenzoic Acid 2-(Diethylamino)ethyl Ester *see* 6987
6-Ethoxy-2-benzothiazolesulfonamide
  see 3711
2-*p*-Ethoxybenzyl-1-(2-diethylamino-
  ethyl)-5-nitrobenzimidazole *see*
  3840
[2S-[1-[R*,(R*)],2α,3aβ,7aβ]]-1-[2-
  [[1-(Ethoxycarbonyl)butyl]amino]-
  1-oxopropyl]octahydro-1H-indole-
  2-carboxylic Acid *see* 7122
N-Ethoxycarbonyl-2-ethoxy-1,2-
  dihydroquinoline *see* 3488
1-Ethoxycarbonyl-3-methyl-2-thio-4-
  imidazoline *see* 1803
N-(Ethoxycarbonyl)-3-(4-morpholin-
  yl)sydnone Imine *see* 6143
18-[[4-[(Ethoxycarbonyl)oxy]-3,5-
  dimethoxybenzoyl]oxy]-11,17-
  dimethoxyyohimban-16-carboxylic
  Acid Methyl Ester *see* 8998
1'-Ethoxycarbonyloxyethyl 6-(D-α-
  Aminophenylacetamido)penicillanate
  see 944
(11β)-17-[(Ethoxycarbonyl)oxy]-11-
  hydroxy-21-(1-oxopropoxy)pregna-
  1,4-diene-3,20-dione *see* 7717
N-[N-[(S)-1-(Ethoxycarbonyl)-3-
  phenylpropyl]-L-alanyl]-N-(indan-
  2-yl)glycine *see* 2864
(S)-1-[N-[1-(Ethoxycarbonyl)-3-phen-
  ylpropyl]-L-alanyl]-L-proline *see*
  3521
[1S-[1α,9α(R*)]]-9-[[1-(Ethoxycarbonyl)-3-phenylpropyl]amino]octa-
  hydro-10-oxo-6H-pyridazino[1,2-
  a][1,2]diazepine-1-carboxylic Acid
  Monohydrate *see* 2276
[2S-[1[R*(R*)],2α,3aβ,6aβ]]-1-[2-[[1-
  (Ethoxycarbonyl)-3-phenylpropyl]-
  amino]-1-oxopropyl]octahydrocy-
  clopenta[b]pyrrole-2-carboxylic
  Acid *see* 8123
[3S-[2-[R*(R*)]],3R*]-2-[2-[[1-Eth-
  oxycarbonyl-3-phenylpropyl]ami-
  no]-1-oxopropyl]-1,2,3,4-tetrahy-
  dro-3-isoquinolinecarboxylic Acid
  see A7

11-[N-(Ethoxycarbonyl)-4-piperidyli-
  dene]-8-chloro-6,11-dihydro-5H-
  benzo[5,6]cyclohepta[1,2-b]pyridine
  see 5455
3-(Ethoxycarbonyl)-6,7,8,9-tetrahy-
  dro-1,6-dimethyl-4-oxo-4H-pyrido-
  [1,2-a]pyrimidinium Methyl Sulfate
  see 8222
p-Ethoxychrysoidine *see* 3706
[3R-(3α,5aβ,6β,8aβ,9α,10α,12β,-
  12aR*)]-10-Ethoxydecahydro-3,6,9-
  trimethyl-3,12-epoxy-12H-pyrano-
  [4,3-j]-1,2-benzodioxepin *see* 841
2-Ethoxy-6,9-diaminoacridine *see*
  3668
p-Ethoxy-2,4-diaminoazobenzene *see*
  3706
3-Ethoxy-N,N-diethyl-4-hydroxy-
  benzamide *see* 5829
6'-Ethoxy-10,11-dihydro-8α-cincho-
  nan-9R-ol *see* 3765
2-Ethoxy-2,3-dihydro-3,3-dimethyl-
  5-benzofuranol Methanesulfonate
  see 3697
6-Ethoxy-1,2-dihydro-2,2,4-trimeth-
  ylquinoline *see* 3710
3-Ethoxy-1,1-dihydroxy-2-butanone
  see 5178
Ethoxydiphenylacetic Acid 2-Dimeth-
  ylaminoethyl Ester *see* 3194
Ethoxyethane *see* 3762
2-Ethoxyethanol, 3707
2-Ethoxyethanol Acetate *see* 3708
2-(2-Ethoxyethoxy)ethanol *see* 1806
5-[1-[2-(2-Ethoxyethoxy)ethoxy]eth-
  oxy]-1,3-benzodioxole *see* 8419
2-Ethoxyethyl Acetate, 3708
2,2'-[1-(1-Ethoxyethyl)-1,2-ethanedi-
  ylidene]bishydrazinecarbothioamide
  see 5178
2-Ethoxyethyl *p*-Methoxycinnamate
  see 2312
O-(6-Ethoxy-2-ethyl-4-pyrimidinyl)-
  phosphorothioic Acid O,O-Dimethyl
  Ester *see* 3847
3-Ethoxy-4-hydroxybenzaldehyde *see*
  3815
3-Ethoxy-4-hydroxybenzoic Acid
  Diethylamide *see* 5829
p-Ethoxy-N-(β-hydroxybutyryl)ani-
  line *see* 1445
2-[1-(Ethoxyimino)butyl]-5-[2-(ethylthio)propyl]-3-hydroxy-2-cyclohex-
  en-1-one *see* 8424
2-[1-(Ethoxyimino)propyl]-3-hydr-
  oxy-5-mesitylcyclohex-2-en-1-one
  see 9484
2-[1-(Ethoxyimino)propyl]-3-hydr-
  oxy-5-(2,4,6-trimethylphenyl)-2-
  cyclohexen-1-one *see* 9484
p-[N-(α-Ethoxy-β-keto-β-*para*-bi-
  phenyl)ethylamino]benzoic Acid *see*
  9979
β-Ethoxy-α-ketobutyraldehyde *see*
  5178
Ethoxylated Fatty Acid Esters *see*
  7555
Ethoxylated Fatty Alcohols *see* 7554
1-(4-Ethoxy-3-methoxybenzyl)-6,7-
  dimethoxy-3-methylisoquinoline *see*
  3258
(Ethoxymethyl)benzene *see* 1147
2-[[Ethoxy[(1-methylethyl)amino]-
  phosphinothioyl]oxy]benzoic Acid
  1-Methylethyl Ester *see* 5055
9-(Ethoxymethyl)-4-formyl-3,8-di-
  hydroxy-1,6-dimethyl-11-oxo-11H-
  dibenzo[b,e][1,4]dioxepin-7-carbox-
  ylic Acid *see* 2016

2-Ethoxy-N-methyl-N-[2-(methylphenethylamino)ethyl]-2,2-diphen-
  ylacetamide *see* 1805
α-Ethoxy-N-methyl-N-[2-[methyl(2-
  phenylethyl)amino]ethyl]-α-phenyl-
  benzeneacetamide *see* 1805
4-Ethoxy-2-methyl-5-(4-morpholin-
  yl)-3(2H)-pyridazinone *see* 3519
2-Ethoxy-2-methylpropane *see* 3732
2-Ethoxynaphthalene, 3709
6-[[(2-Ethoxy-1-naphthalenyl)carbon-
  yl]amino]-3,3-dimethyl-7-oxo-4-
  thia-1-azabicyclo[3.2.0]heptane-2-
  carboxylic Acid Monosodium Salt
  see 6266
6-(2-Ethoxy-1-naphthamido)penicillin
  Sodium Salt *see* 6266
2-Ethoxy-5-nitrobenzenamine *see*
  6538
Ethoxyol [Emery] *see* 7554
3-Ethoxy-2-oxobutyraldehyde Hydrate *see* 5178
2-[2-[(2-Ethoxy-2-oxoethyl)thio]eth-
  yl]-4-thiazolidinecarboxylic Acid
  see 5330
3-[4-(β-Ethoxyphenethyl)-1-piper-
  azinyl]-2-methylpropiophenone *see*
  3577
2-[(2-Ethoxyphenoxy)methyl]morpho-
  line *see* 9884
2-(2-Ethoxyphenoxymethyl)tetrahy-
  dro-1,4-oxazine *see* 9884
N-(4-Ethoxyphenyl)acetamide *see*
  7155
4-[(4-Ethoxyphenyl)azo]-1,3-benzene-
  diamine *see* 3706
4-[(*p*-Ethoxyphenyl)azo]-*m*-phenyl-
  enediamine *see* 3706
α-Ethoxy-α-phenylbenzeneacetic Acid
  2-(Dimethylamino)ethyl Ester *see*
  3194
4-Ethoxy-7-phenyl-3,5-dioxa-6-aza-
  4-phosphaoct-6-ene-8-nitrile 4-
  Sulfide *see* 7341
3-[4-(2-Ethoxy-2-phenylethyl)-1-
  piperazinyl]-2-methyl-1-phenyl-1-
  propanone *see* 3577
N-(4-Ethoxyphenyl)-3-hydroxybutan-
  amide *see* 1445
N-(4-Ethoxyphenyl)-2-hydroxypro-
  panamide *see* 5220
2-[(4-Ethoxyphenyl)methyl]-N,N-di-
  ethyl-5-nitro-1H-benzimidazole-1-
  ethanamine *see* 3840
p-(α-Ethoxy-*p*-phenylphenacylami-
  do)benzoic Acid *see* 9979
p-[(α-Ethoxy-*p*-phenylphenacyl)-
  amino]benzoic Acid *see* 9979
(4-Ethoxyphenyl)urea *see* 3448
3-Ethoxy-1-propene *see* 291
Ethoxyquin, 3710
2-Ethoxy-1(2H)-quinolinecarboxylic
  Acid Ethyl Ester *see* 3488
8-Ethoxy-5-quinolinesulfonic Acid
  see 134
Ethoxyzolamide *see* 3711
Ethoxzolamide, 3711
Ethrane [Anaquest] *see* 3536
Ethrel [Amchem] *see* 3686
Ethril [Squibb] *see* 3632
Ethryn [Faulding] *see* 3632
Ethy 11 [Cassenne] *see* 3689
Ethybenztropine, 3712
Ethyl Acetate, 3713
Ethylacetic Acid *see* 1593
Ethyl Acetoacetate, 3714
Ethyl Acetone *see* 6032
Ethyl Acrylate, 3715
Ethyl Adipate *see* 152
Ethyl Adrianol *see* 3822
Ethyl Alcohol, 3716

16-Ethyl-17-hydroxyestr-4-en-3-one *see* 6886

Ethyl 1-[2-(2-Hydroxyethoxy)ethyl]-4-phenylisonipecotate *see* 1799

7-[2-[Ethyl(2-hydroxyethyl)amino]-ethyl]-3,7-dihydro-1,3-dimethyl-8-(phenylmethyl)-1*H*-purine-2,6-dione *see* 965

Ethyl Hydroxyethyl Cellulose *see* 1963

Ethyl(2-hydroxyethyl)dimethylammonium Chloride Benzilate *see* 5210

Ethyl 2-Hydroxyethyl Ether Cellulose *see* 1963

1-Ethyl-7-hydroxy-2-methyl-1,2,3,4,-4a,9,10,10a-octahydrophenanthrene-2-carboxylic Acid *see* 3408

1-Ethyl-3-hydroxy-1-methylpiperidinium Bromide Benzilate *see* 7428

2-Ethyl-2-(hydroxymethyl)-1,3-propanediol Trinitrate *see* 7821

1-[[2-[Ethyl(2-hydroxy-2-methylpropyl)amino]ethyl]amino]-4-methyl-9*H*-thioxanthen-9-one *see* 1026

N-Ethyl-α-(hydroxymethyl)-*N*-(4-pyridinylmethyl)benzeneacetamide *see* 9692

20-Ethyl-4-(hydroxymethyl)-1,6,14,-16-tetramethoxyaconitane-7,8-diol *see* 5488

17α-Ethyl-17-hydroxy-4-norandrosten-3-one *see* 6613

17α-Ethyl-17-hydroxy-19-norandrost-4-en-3-one *see* 6613

9-Ethyl-7-(3-hydroxy-1-oxo-2-phenylpropoxy)-9-methyl-3-oxa-9-azoniatricyclo[3.3.1.0 2,4]nonane Bromide *see* 6897

(*E*)-14-Ethyl-13-hydroxy-3,5,7,9,13-pentamethyl-6-[[3,4,6-trideoxy-3-(dimethylamino)-β-D-*xylo*-hexopyranoside]oxy]oxacyclotetradec-11-ene-2,4,10-trione *see* 7384

β-Ethyl-β-hydroxyphenethyl Carbamate *see* 4770

β-Ethyl-β-hydroxyphenethyl Carbamic Acid Ester *see* 4770

α-Ethyl-1-hydroxy-4-phenylcyclohexaneacetic Acid *see* 3912

Ethyl(*m*-hydroxyphenyl)dimethylammonium Chloride *see* 3487

Ethyl *p*-Hydroxyphenyl Ketone *see* 6992

Ethyl 4-(*m*-Hydroxyphenyl)-1-methylisonipecotate *see* 4769

1-Ethyl-3-hydroxypiperidine *see* 3799

Ethyl α-hydroxypropionate *see* 3773

2-[6-[Ethyl(2-hydroxypropyl)amino]-3-pyridazinyl]hydrazinecarboxylic Acid Ethyl Ester *see* 1631

4-Ethyl-4-hydroxy-1*H*-pyrano[3',4':6,7]indolizino[1,2-*b*]quinoline-3,14(4*H*,12*H*)-dione *see* 1742

3-Ethyl-7-hydroxy-2,8,12,16-tetramethyl-5,13-dioxo-9-[[3,4,6-trideoxy-3-(dimethylamino)-β-D-*xylo*-hexopyranosyl]oxy]-4,17-dioxabicyclo[14.1.0]heptadec-14-ene-10-acetaldehyde *see* 8241

α-Ethyl-3-hydroxy-2,4,6-triiodobenzenepropanoic Acid *see* 4947

α-Ethyl-3-hydroxy-2,4,6-triiodohydrocinnamic Acid *see* 4947

α-Ethyl-β-(3-hydroxy-2,4,6-triiodophenyl)propionic Acid *see* 4947

3,3'-Ethylidenebis[4-hydroxy-2*H*-1-benzopyran-2-one] *see* 3768

3,3'-Ethylidenebis(4-hydroxycoumarin) *see* 3768

1,1'-Ethylidenebis[4-methylbenzene] *see* 3385

3,3-Ethylidenebis(4-oxycoumarin) *see* 3768

Ethylidene Chloride, 3766

24-Ethylidenecholest-5-en-3β-ol *see* 4194

Ethylidene Diacetate, 3767

Ethylidene Dicoumarol, 3768

Ethylidene Diethyl Ether *see* 31

Ethylidene Dimethyl Ether *see* 3215

9-[(4,6-*O*-Ethylidene-β-D-glucopyranosyl)oxy]-5,8,8a,9-tetrahydro-5-(4-hydroxy-3,5-dimethoxyphenyl)-furo[3',4':6,7]naphtho[2,3-*d*]-1,3-dioxol-6(5a*H*)-one *see* 3842

5-Ethylidene-2-hydroxy-2,3-dimethylhexanedioic Acid *see* 8400

Ethylidenehydroxylamine *see* 35

7-Ethylidene-1a,2,3,7-tetrahydrocyclopent[*b*]oxireno[*c*]pyridine *see* 3

[3aS-(3*E*,3aα,4aβ,7aβ,9a*R**,9bβ)]-3-Ethylidene-3,3a,7a,9b-tetrahydro-2-oxo-2*H*,4a*H*-1,4,5-trioxadicyclopent[*a*,*hi*]indene-7-carboxylic Acid Methyl Ester *see* 7512

α-Ethyl-1*H*-indole-3-ethanamine *see* 3848

3-Ethyl-12*H*-indolo[2,3-*a*]quinolizin-5-ium *see* 4031

Ethyl Iodide, 3769

Ethyl 10-(*p*-Iodophenyl)hendecanoate *see* 4946

Ethyl 10-(*p*-Iodophenyl)undecylate *see* 4946

5-Ethyl-5-isoamylbarbituric Acid *see* 601

Ethylisobutrazine *see* 3849

Ethyl Isobutyrate, 3770

α-Ethylisonicotinoylthioamide *see* 3692

5-Ethyl-5-isopentylbarbituric Acid *see* 601

O-Ethyl *O*-2-Isopropoxycarbonylphenyl Isopropylphosphoramidothioate *see* 5055

5-Ethyl-5-isopropylbarbituric Acid *see* 7759

α-Ethyl-α-isopropyl-α-bromoacetamide *see* 4809

Ethyl Isothiocyanate, 3771

2-Ethylisothionicotinamide *see* 3692

3-Ethylisothionicotinamide *see* 3692

Ethyl Isovalerate, 3772

Ethyl Lactate, 3773

Ethyl Laurate, 3774

Ethyl Levulinate, 3775

Ethyl Linoleate, 3776

Ethyl Loflazepate, 3777

Ethyl Maleate *see* 3113

N-Ethylmaleimide, 3778

Ethyl Malonate, 3779

Ethyl Mercaptan *see* 3680

Ethyl(4-mercaptobenzenesulfonato-*S* 4)mercury Sodium Salt *see* 9243

Ethyl[2-mercaptobenzoato(2−)-*O*,*S*]mercurate(1−) Sodium *see* 9244

3-Ethylmercapto-10-(1'-methylpiperazinyl-4'-propyl)phenothiazine *see* 9241

Ethylmercuric Chloride, 3780

N-(Ethylmercuri)-*p*-toluenesulfonanilide, 3781

Ethyl Mesylate *see* 3782

N-Ethyl-*N*-methallyl-4-trifluoromethyl-2,6-dinitroaniline *see* 3671

Ethyl Methanesulfonate, 3782

Ethyl Methanesulfonic Acid *see* 3782

Ethylmethiambutene *see* 3784

β-Ethyl-6-methoxy-α,α-dimethyl-2-naphthalenepropionic Acid *see* 5856

4-[Ethyl(*p*-methoxy-α-methylphenethyl)amino]butyl 3,4-Dimethoxybenzoate *see* 5648

4-[1-Ethyl-2-(4-methoxyphenyl)-1-butenyl]phenol *see* 5818

N-Ethyl-9-(4-methoxy-2,3,6-trimethylphenyl)-3,7-dimethyl-2,4,6,8-nonatetraenamide *see* 6199

Ethyl-β-methylallylthiobarbituric Acid *see* 5855

5-Ethyl-5-(2-methylallyl)-2-thiobarbituric Acid *see* 5855

3-Ethylmethylamino-1,1-di(2'-thienyl)but-1-ene *see* 3784

α-[1-(Ethylmethylamino)ethyl]benzenemethanol *see* 3661

l-α-[1-(Ethylmethylamino)ethyl]benzyl Alcohol *see* 3661

α-Ethyl-β-[2-(methylamino)propyl]-β-phenylbenzeneethanol Acetate *see* 6602

N-Ethyl-α-methylbenzeneethanamine *see* 3720

5-Ethyl-5-(1-methyl-1-butenyl)barbituric Acid Sodium Salt *see* 9886

5-Ethyl-5-(1-methyl-1-butenyl)-2,4,6(1*H*,3*H*,5*H*)-pyrimidinetrione Sodium Salt *see* 9886

5-Ethyl-5-(3-methylbutyl)-2,4,6(1*H*,-3*H*,5*H*)-pyrimidinetrione *see* 601

5-Ethyl-5-(1-methylbutyl)-2,4,6(1*H*,-3*H*,5*H*)-pyrimidinetrione Monosodium Salt *see* 7087

5-Ethyl-5-(1-methylbutyl)-2-thiobarbituric Acid Sodium Salt *see* 9280

3-Ethyl-5-methyl-1,4-dihydro-2,6-dimethyl-4-(3-nitrophenyl)-3,5-pyridinedicarboxylate *see* 6493

3-Ethyl-2-methyl-5-dimethylaminoindole *see* 5675

4-Ethyl-4-methyl-2,6-dioxopiperidine *see* 1036

N-Ethyl-*N*-methyl-4,4-di-2-thienyl-3-buten-2-amine *see* 3784

(17α)-13-Ethyl-11-methylene-18,19-dinorpregn-4-en-20-yn-17-ol *see* 2906

1-Ethyl-6,7-methylenedioxy-4(1*H*)-oxocinnoline-3-carboxylic Acid *see* 2311

1-Ethyl-6,7-methylenedioxy-4-quinolone-3-carboxylic Acid *see* 6899

Ethyl Methylene Phosphorodithioate *see* 3691

4,4'-(1-Ethyl-2-methyl-1,2-ethanediyl)bis[2-fluorophenol] *see* 1231

Ethyl Methyl Ether, 3783

N-Ethyl-*N*'-(1-methylethyl)-6-(methylthio)-1,3,5-triazine-2,4-diamine *see* 402

5-Ethyl-5-(1-methylethyl)-2,4,6(1*H*,-3*H*,5*H*)-pyrimidinetrione *see* 7759

3-Ethyl-3-methylglutarimide *see* 1036

7-Ethyl-2-methyl-4-hendecanol Sulfate Sodium Salt *see* 8645

Ethyl Methyl Ketone *see* 5991

O-Ethyl *O*-[4-(Methylmercapto)phenyl]-*S*-*n*-propylphosphorothionothiolate *see* 8972

1-Ethyl-2-methyl-7-methoxy-1,2,3,4-tetrahydrophenanthryl-1-carboxylic Acid *see* 3407

21-Ethyl-4-methyl-16-methylene-7,20-cycloveatchane-1,12-15-triol *see* 6286

Ethyl 3-Methyl-4-(methylthio)phenyl (1-Methylethyl)phosphoramidate *see* 3901

Ethyl-*N*-methylmorphinium Iodide *see* 3785

Eunerpan [Nordmark] see 5707
Euneryl see 7201
Eunoctal [I.S.H.] see 601
Eunoctin [Gedeon Richter] see 6491
Euonymit see 4238
Euonymus, 3856
Euparen(e) [Bayer] see 3031
Euparin, 3857
Eupatal see 198
Eupatin II see 8941
Eupatorin, 3858
Eupatorium, 3859
Eupaverin [E. Merck] see 6202
Eupaverin see 6011
Eupaverina [Bracco] see 6202
Euphorbia, 3860
Euphozid see 4965
Euphthalmine [Warner-Chilcott] see 3854
Euphyllin [Byk-Gulden] see 477
Euphylline L.A. [Valpan] see 9212
Euphylong [Byk Gulden] see 9212
Euplit [Desitin] see 504
Eupnéron [Lyocentre] see 3578
Euporphin see 776
Eupractone [Baxter] see 3201
Euprax [Baxter] see 210
Eupressyl [Valpan] see 9779
Euprocin, 3861
Euprovasin [Magis] see 7852
Euquinine see 8080
Euradal [Lacer] see 1309
Eurax [Geigy] see 2597
Euraxil [Thomae] see 2597
Eureceptor [Zambon] see 2279
EureCor [Kade] see 5114
Eurekene see 9821
Euresol [Knoll] see 8158
Eurex [Labaz] see 7715
Eurinol see 9537
Eurodin [Takeda] see 3651
Eurodopa [Europharma] see 5344
Eurogale [Europa] see 4792
European White Birch see 1211
Europia see 3862
Europium, 3862
Eurosan [Mepha] see 2977
Eusaprim [Wellcome] see 8889
Eusmanid see 1208
Eusovit [Woelfer] see 9932
Euspiran see 5105
Eustrophinum see 8819
Eutagen see 6912
Euteberol [Merckle] see 8721
Eutensin [Hoechst] see 4221
Eutensol see 4473
Euthatal see 7087
Euthyrox [E. Merck] see 5351
Eutocol see 3653
Eutonyl [Abbott] see 6988
Eutrit [Takeda] see 9992
Euvaderm [Gödecke] see 1202
Euvasal [Selvi] see 8967
Euvernil [Heyden] see 8906
Euvestin see 3119
Euvifor [UCB] see 7459
Euvitol [Allen & Hanburys] see 3909
E-VA-16 see 4845
Evacalm [Unimed] see 2977
Evaclin see 5062
Evacort [Evans] see 4710
Evadene [Ayerst] see 1533
Evadol [Clin-Comar-Byla] see 5062
Evadyne [Ayerst] see 1533
Evan's Blue, 3863
Evanol [Almirall] see 7377
Evansite see 362
Evasprin see 7292
Evazol [Ravensberg] see 2896
Evening Primrose Oil, 3864
Eventin [Minden] see 2713

Everfree [Cooper] see 6517
Everninic Acid Methyl Ester see 8690
Evik [Ciba-Geigy] see 402
E-Vimin see 9931
Evinopon [Bros] see 3071
Evion [E. Merck] see 9931
Evipal [Winthrop] see 4625
Evipal Sodium [Winthrop] see 4625
Evipan [Bayer] see 4625
Evipan Sodium [Bayer] see 4625
Evipherol see 9931 and 9932
Evital [Sandoz] see 6618
Evodiamine, 3865
Evonogenin see 3145
Evoxin [Sterling] see 3412
Evramicina see 9681
Evramycin [Wyeth] see 9681
Ewer-Pick Acid see 6292
EX 4810 see 396
Ex 4883 see 8234
Ex 12-095 see 3666
Exacin [Toyo Jozo] see 4989
Exacyl [Choay] see 9487
Exal see 9887
Exalamide, 3866
Exalgin see 5931
Exaltolide®, 3867
Examen see 5427
Exandron [Chantal] see 2774
Exangit see 1106
Exaprazole see 3642
Exceglan [Dainippon] see 10094
Excegram [Dainippon] see 10094
Excenel [Upjohn] see 1948
EXD see 3390
Exdol [Frosst] see 40
Exelderm [ICI] see 8866
Exelgyn [Cassenne] see 1525
Exelmin [Lederle] see 7433
Exhirud [Plantorgan] see 4638
Exhirudine see 4638
Exhoran see 3370
Exifone, 3868
Exiproben, 3869
Exirel [Pfizer] see 7461
Exitelite see 743
Exlan see 2577
Exlutena [Organon] see 5501
Exluton(a) [Organon] see 5501
Exmigra see 3609
ExNa [Robins] see 1134
Exoamylases see 631
Exocorpol [Green Cross] see 7537
Exoderil [Sandoz] see 6270
Exodextranases see 2926
Exodor-Grun see 2155
Exofene see 4602
Exolan see 714
(±)-2-Exo-(2-methylbenzyloxy)-1-
   methyl-4-isopropyl-7-oxabicyclo-
   [2.2.1]heptane see 2296
Exonal [Toyama] see 9060
Exopan [Ciba] see 2712
Exopin see 7433
Exopon [Ciba] see 2712
Exorbin [Ayerst] see 7538
Exosalt [Bayer] see 1134
Exosulfonyl see 8851
Exotancain [DDR] see 7843
Exotherm Termil [Diamond] see 2167
Exotoxins, 3870
EXP-105-1 see 380
EXP-126 see 8221
EXP-999 see 6070
Expandex [Comm. Solvents] see 2925
Expansine see 7002
Expar [Coopers] see 7132
Experimental Insecticide No. 269 see
   3533
Exponcit [Arzneimittelwerk VEB] see
   6634

Expressed Almond Oil see 6770
Exrheudon N [Neos] see 7248
Exsel [Herbert] see 8390
Exsiccated Alum see 364
Exsiccated Ammonium Alum see 325
Exsiccated Ferrous Sulfate see 4006
Exsiccated Sodium Phosphate see
   8612
Extacol [Nikken] see 7229
Ext. D & C Orange 3 see 6812
Ext. D & C Red No. 14 see 9997
Ext. D & C Red No. 15 see 7567
Ext. D & C Yellow No. 1 see 5833
Ext. D & C Yellow No. 7 see 6313
Ext. D & C Yellow No. 9 see 10003
Ext. D & C Yellow No. 10 see 10004
Extencilline see 7037
Exterol [Dermal] see 9783
Extracort [Basotherm] see 9511
Extramycin [Bayer] see 8498
Extranase [Rorer] see 1377
Extrinsic Factor see 9921
Extrinsic Plasminogen Activator see
   9394
Exuril see 2169
Exypaque [Nyegaard] see 4940
Eye-Cort [Mallard] see 4710
Eye-derived Growth Factor-II see
   4016
Eyekas [Showa Shinyaku] see 8201
Eye-Sul [Mallard] see 8870
E-Z-HD [Heyden] see 1006
E-Z-Paque [E-Z-Em Co.] see 1006
F 4 see 914
F 139 [Thuron] see 1528
F 151 see 5854
F 190 see 44
F 440 see 2814
F 1500 see 8851
F 1983 see 8021
F 1991 see 1053
F 2559 see 4249
F 6060 see 2726
F 6066 see 2726
F 6113 see 4224
FA see 4097
FA 402 see 3949
FA 2071 see 4964
2-FAA see 4083
Fabahistin [Bayer] see 5649
Fabianol [Ciba-Geigy] see 416
Fabrol [Geigy] see 82
FAC see MISC-2
F Acid see 1898
Factitious Air see 6575
Factor I see 4015
Factor II see 7903
Factor III see 9321
Factor V, 3871
Factor VII, 3872
Factor VIII, 3873
Factor IX, 3874
Factor IX Complex (Human) see 3874
Factor X, 3875
Factor XI, 3876
Factor XII, 3877
Factor XIII, 3878
Factorate [Armour Pharm.] see 3873
Factor B see 3825
Factor T see 9941
Factrel [Ayerst] see 5354
FAD see 4028 and 8201
Fademin [Chugai] see 4028
Fado [Errekappa] see 1922
Fadormir see 5872
Faecla see 7593
Faecula see 7593
Fagarasterol see 5478
Fagarine, 3879
α-Fagarine see 255
β-Fagarine see 8505

γ-Fagarine see 3879
Fagarol see 8420
Fagine see 2207
Fairy Gloves see 3141
FAK III see 4164
Falapen [Frosst] see 7041
Falcopen V see 7047
Falicain see 7843
Falignost [Fahlberg-List] see 4941
Falimint see 6550
Falmonox [Winthrop] see 9055
Falomesin [Chugai] see 1946
Falone®, 3880
False Hellebore see 156
False Indigo see 968
False Saffron see 1877
False Unicorn see 4550
False White Cedar see 9324
False Winter's Bark see 1749
Falvin [Farmades] see 3948
FAM see MISC-2
Famet see 8887
Famodil [Sigma-Tau] see 3881
Famophos see 3882
Famosan [Alkaloid] see 3881
Famosept see 7274
Famotidine, 3881
Famoxal [Silanes] see 3881
Famphur, 3882
Fanasil [Roche] see 8877
Fangorex [Siegfried] see 6801
Fanodormo see 2717
Fanosin [Abello] see 3881
Fansidar see 5683
Fansimef see 5683
Fantorin see 8769
Fanzil [Roche] see 8877
FAP see MISC-2
2-F-araA see 4057
2-F-Ara-AMP see 4057
Farecef [Lafare] see 1933
Faredina [Lafare] see 1975
Faremicin [Lafare] see 4169
Fareston [Farmos] see 9474
Faretrizin [Lafare] see 1923
Farexin [Lafare] see 1971
Fargan see 7797
Far-Go [Monsanto] see 9510
Farial [Nordmark] see 4845
Farlutal [Farmitalia] see 5677
Farmacyrol [Farmaryn] see 3094
Farmicetina [Farmitalia] see 2068
Farmidril [Farmitalia] see 5256
Farmiglucin [Farmitalia] see 6989
Farminosidin [Farmitalia] see 6989
Farmiserina [Farmitalia] see 2758
Farmitalia 204/122 see 8881
Farmolisina see 3358
Farmorubicin [Farmitalia Carlo Erba]
    see 3573
Farmoxin [Erba] see 1938
Farnesene see 3883
α-Farnesene, 3883
β-Farnesene, 3884
Farnesol, 3885
Farnoquinone see 9935
Fas-Cile 200 see 5751
Fasigin [Pfizer] see 9377
Fasigyn [Pfizer] see 9377
Fasinex [Ciba-Geigy] see 9566
Fastac [Agrishell] see 2775
Fast Green see 5581
Fast Green FCF, 3886
Fast Green J see 1367
Fastin [Beecham] see 7232
Faston [Duphar] see 3193
Fast Scarlet see 1111
Fastum [Menarini] see 5184
Fast Yellow see 9688
Fasupond see 6634
Fat Ponceau R see 8349

Fatroximin [Fatro] see 8218
all-cis-Fatty Acid 20:5 Omega-3 see
    3495
Faustan [East Germany] see 2977
Faverin [Duphar] see 4138
Favistan [Asta] see 5892
Fazadinium Bromide, 3887
Fazadon [Duncan, Flockhart] see 3887
Fazol [Fournier Frères] see 5044
Fazor [Uniroyal] see 5587
FB/2 see 3359
FB 5097 see 2412
FBA 1420 see 7813
FBA 1464 see 4470
FBA 4503 see 7844
FC 54 see 233
FC 1157a see 9474
FC 3001 see 9354
FCCs see 2140
F-Cortef [Upjohn] see 4059
FCR 1272 see 2764
FCRC-A48 see 4181
FDA 20 see 4111
FDA 1541 see 3580
FDC see 4111
FD & C Blue No. 1 see 1366
FD & C Blue No. 2 see 4856
FD & C Green 1 see 4487
FD & C Green No. 2 see 5361
FD & C Green No. 3 see 3886
FD & C Orange I see 6812
FD & C Red No. 1 see 7567
FD & C Red No. 2 see 382
FD & C Red No. 3 see 3640
FD & C Red No. 4 see 7568
FD & C Red No. 32 see 9997
FD & C Red No. 40 see 283
FD & C Yellow No. 1 see 6313
FD & C Yellow No. 3 see 10003
FD & C Yellow No. 4 see 10004
FD & C Yellow No. 5 see 9041
FD & C Yellow No. 6 see 8983
FDNB see 4101
Featherfew see 4013
Featherfoil see 4013
Febantel, 3888
Febarbamate, 3889
Febramine see 2015
Febrifugine, 3890
Febrilix see 40
Febrinina see 488
Febuprol, 3891
Fecarb [Ayerst] see 3990
Feclemine, 3892
Fectrim [DDSA] see 8889
Fedacilina [Fedal] see 5830
Fedan [ICI] see 6783
Fedibaretta see 7183
Feguanide see 7191
Feinalmin [Sanko] see 4835
Felacrinos see 2788
Felben [Künzler] see 3308
Felbinac, 3893
Feldene [Pfizer] see 7476
Felicur [Asche] see 3727
Felinine, 3894
Felison [Sigurta] see 4123
Felitrope see 3653
Fellozine [Fellows-Testagar] see 7797
Felmane see 4123
Felodipine, 3895
Felosan see 4765
Felviten [Grünenthal] see 676
Felypressin, 3896
Femadol see 7851
Femadon [Dolorgiet] see 4812
Femergin see 3609
Femestral see 3653
Femestrone see 3655
Femestrone Inj. see 3660
Femidyn see 3660

Feminone [Upjohn] see 3689
Femodene [Schering] see 4308
Femovan [Schering AG] see 4308
Femoxetine, 3897
Femstat [Syntex] see 1525
Femulen [Searle] see 3816
Fenac see 2085
Fenacilin see 7046
Fenactil see 2186
Fenadiazole, 3898
Fenadone see 5852
Fenalamide, 3899
Fenalcomine, 3900
Fenallymal see 7162
Fenamate [Meiji] see 404
Fenamiphos, 3901
Fenamisal see 7243
Fenamizol see 498
Fenapanil, 3902
Fenarimol, 3903
Fenarol [Winthrop] see 2105
Fenarsone see 1788
Fenase see 6425
Fenasprate see 1054
Fenate see 2927
Fenatin see 7172
Fenazil see 7797
Fenazol [Hokuriku Seiyaku] see 4060
Fenazolina [Polfa] see 709
Fenazox [Meiji Seika] see 404
Fenazoxine see 6358
Fenbendazole, 3904
Fenbenicillin, 3905
Fenbid [SK&F] see 4812
Fenbrac [Rorer] see 3914
Fen-Bridal see 5088
Fenbufen, 3906
Fenbutamide see 7177
Fenbutatin Oxide, 3907
Fenbutrazate, 3908
Fencamfamine, 3909
Fencamine, 3910
Fencarbamide see 7178
Fenchlorphos see 8239
d-Fenchone, 3911
Fencibutirol, 3912
Fenclofenac, 3913
Fenclor [Caffaro] see 7541
Fenclorac, 3914
Fenclozic Acid, 3915
Fendilar [SPA] see 3916
Fendiline, 3916
Fendizoate see MISC-4
Fendosal, 3917
Fenergan see 7797
Fenesterin see 7182
Fenestrin see 7182
Fenethazine, 3918
Fenethylline, 3919
Feneticilline see 7184
Fenfluramine, 3920
Fenformin see 7191
Fenhydren see 7196
Fenibutazona see 7248
Fenibutol see 7248
Fenicol see 3727
Fenicol [Alcon] see 2068
Fenidrone see 6910
Fenigam see 476
Fenigama see 476
Fenilbutina see 7248
Fenilfar [Farmila] see 7257
Fenilin see 7196
Fenilor [UCB] see 1441
Fenipentol, 3921
Fenistil [Zyma] see 3204
Fenitrothion, 3922
Fennel, 3923
Fenobrate [Gramon] see 3924
Fénocycline see 3407
Fenocylin see 3407

Fenofibrate, 3924
Fenolactine see 5220
Fenoldopam, 3925
Fenolovo see 9659
Fenoprofen, 3926
Fenopron [Dista] see 3926
Fenoprop see 8483
Fenormin see 7191
Fenospen see 7046
Fenostil [Zyma] see 3204
Fenoterol, 3927
Fenotone see 7248
Fenoverine, 3928
Fenoverm see 7220
Fenoxaprop-ethyl, 3929
Fenoxazoline, 3930
Fenoxedil, 3931
Fenoximone see 3542
Fenoxypen [Novo] see 7046
Fenoxypropazine see 7227
Fenozolone, 3932
Fenpentadiol, 3933
Fenpidon see 3342
Fenpipramide Methobromide see 3935
Fenpiprane, 3934
Fenpiverinium Bromide, 3935
Fenpropanate see 3936
Fenpropathrin, 3936
Fenpropidin, 3937
Fenpropimorph, 3938
Fenproporex, 3939
Fenproporex Retard Bottu see 3939
Fenprostalene, 3940
Fenpyrate see 7981
Fenquizone, 3941
Fenspiride, 3942
Fensulfothion, 3943
Fental [Kanebo] see 9060
Fentanest [Sankyo] see 3944
Fentanyl, 3944
Fentazin [Allen & Hanburys] see 7135
Fenthion, 3945
Fentiazac, 3946
Fentiazin see 7220
Fenticlor, 3947
Fenticonazole, 3948
Fentin Acetate see 9659
Fentin Hydroxide see 9659
Fentonium Bromide, 3949
Fentrinol [Mead Johnson] see 409
Fenugreek, 3950
Fenuron, 3951
Fenuron TCA see 3951
Fenvalerate, 3952
Fenylhist [Mallard] see 3308
Fenyramidol see 7292
Feojectin see 3974
Feosol [SK & F] see 4006
Feospan [SK & F] see 4006
Feostat [Westerfield] see 3995
Feprazone, 3953
Fepron [Lilly] see 3926
Feprona [Lilly] see 3926
Feraconitine see 7924
Ferbam, 3954
Ferbeck see 3954
Fergon [Breon] see 3996
Fergon 500 [Alfar] see 1971
Fergusonite see 10015
Fer-in-Sol [Mead Johnson] see 4006
Ferlucon see 3996
Fermate see 3954
Fermcozyme see 4354
Fermentation Amyl Alcohol see 5081
Fermentation Butyl Alcohol see 5015
Fermentation L. Casei Factor see 7950
Fermine see 3243
Fermium, 3955
Fernambuco see 7133
Fernasan see 9304
Fernex [ICI] see 7469

Fernisone [Ferndale] see 7727
Fernos [ICI] see 7468
Ferodin SL see 7777
Fero-Gradumet [Abbott] see 4006
Ferolactan [Bioindustria] see 7788
Feromax see 4006
Feron [Toray] see 4893
Feroritard [Nikken] see 4006
Feroton [Maney] see 3995
Ferradow see 3954
Ferredoxins, 3956
Ferric Acetate, Basic, 3957
Ferric Albuminate, 3958
Ferric Alum see 543
Ferric and Ammonium Acetate Solution, 3959
Ferric Ammonium Chromate see 540
Ferric Ammonium Citrate see 541
Ferric Ammonium Oxalate see 542
Ferric Ammonium Sulfate see 543
Ferric Bromide, 3960
Ferric Chloride, 3961
Ferric Chromate(VI), 3962
Ferric Citrate, 3963
Ferric Dimethyldithiocarbamate see 3954
Ferric Enterobactin see 3547
Ferric Ferrocyanide, 3964
Ferric Ferrous Oxide see 3988
Ferric Fluoride, 3965
Ferric Formate, 3966
Ferric Fructose, 3967
Ferric Gallotannate see 3981
Ferric Hexacyanoferrate(II) see 3964
Ferrichrome see 3968
Ferrichrome A see 3968
Ferrichromes, 3968
Ferrichrysin see 3968
Ferric Hydroxide, 3969
Ferric Hydroxide Oxide see 3969
Ferric Hypophosphite, 3970
Ferriclate Calcium Sodium, 3971
Ferric Monosodium Ethylenediaminetetraacetate see 3977
Ferric Nitrate, 3972
Ferric Oxide, 3973
Ferric Oxide, Saccharated, 3974
Ferric Persulfate see 3980
Ferric Phosphate, 3975
Ferric Pyrophosphate, 3976
Ferricrocin see 3968
Ferric Sesquioxide see 3973
Ferric Sesquisulfate see 3980
Ferric Sodium Edetate, 3977
Ferric Sodium Pyrophosphate, 3978
Ferric Subsulfate Solution, 3979
Ferric Sulfate, 3980
Ferric Sulfocyanate see 3982
Ferric Sulfocyanide see 3982
Ferric Tannate, 3981
Ferric Tersulfate see 3980
Ferric Thiocyanate, 3982
Ferricytochrome c see 2797
Ferriheme Chloride see 4563
Ferriheme Hydroxide see 4554
Ferrihemoglobin see 5878
Ferrimycins see 2850
Ferrioxamine B see 2850
Ferriporphyrin Chloride see 4563
Ferriporphyrin Hydroxide see 4554
Ferriprotoporphyrin Basic see 4554
Ferriprotoporphyrin Chloride see 4563
Ferrirhodin see 3968
Ferrirubin see 3968
Ferrite, 3983
Ferritin, 3984
Ferritose [Calbiochem] see 3967
Ferrivenin see 3974
Ferrocal see 1668
Ferrocene, 3985
Ferrocholinate, 3986

Ferrocytochrome c see 2797
Ferrofolin [Farmades] see 3984
Ferrofume [Nordic] see 3995
Ferroglycine Sulfate, 3987
Ferroglycine Sulfate Complex see 3987
Ferro-Gradumet [Abbott] see 4006
Ferroheme see 4558
"Ferroin" see 7169
Ferrol [Chemil] see 3984
Ferrolip [Flint] see 3986
Ferromyn [Calmic] see 4005
Ferron see 4757
Ferronat see 3995
Ferrone [Wolf] see 3995
Ferronicum [Sandoz] see 3996
Ferronord [Cooper] see 3987
Ferroprotoporphyrin see 4558
Ferrosanol [Sanol] see 3987
Ferrosoferric Oxide, 3988
Ferrospinel see 3983
Ferrosprint [Poli] see 3984
Ferrostrane see 3977
Ferrostrene see 3977
Ferrotemp [Medix] see 3995
Ferrous Aminoacetosulfate see 3987
Ferrous Ammonium Sulfate see 546
Ferrous Aspartate see 862
Ferrous Bromide, 3989
Ferrous Calcium Citrate see 1668
Ferrous Carbonate Mass, 3990
Ferrous Carbonate Saccharated, 3991
Ferrous Chloride, 3992
Ferrous Citrate, 3993
Ferrous Fluoride, 3994
Ferrous Fumarate, 3995
Ferrous Gluconate, 3996
Ferrous Glucuronate see 4362
Ferrous Hydroxide, 3997
Ferrous Iodide, 3998
Ferrous Lactate, 3999
Ferrous Oxalate, 4000
Ferrous Oxide, 4001
Ferrous Phosphate, 4002
Ferrous Phosphide, 4003
Ferrous Selenide, 4004
Ferrous Succinate, 4005
Ferrous Sulfate, 4006
Ferrous Sulfate Glycine Complex see 3987
Ferrous Sulfide, 4007
Ferrous Sulfocyanate see 4008
Ferrous Sulfocyanide see 4008
Ferrous Thiocyanate, 4008
Ferrox see 4000
Ferroxidase see 1999
Ferrum [Green Cross] see 3995
Fersamal [Glaxo] see 3995
Fer-Sul [Pharmex] see 4006
Fertagyl [Intervet] see 5354
Ferti-Cept [Elanco] see 4534
Fertilvit see 9932
Fertilysin, 4009
Fertinorm [Serono] see 4189
Fertiral [Hoechst] see 5354
Fertirelin, 4010
Fertodur [Schering AG] see 2726
Ferulic Acid, 4011
3-Feruloylquinic Acid see 2142
Ferum Hausmann see 3974
Fervenulin, 4012
FES see 10017
Fesofor [SK & F] see 4006
Fesotyme [Elder] see 4006
Fespan [SK & F] see 4006
Festamoxin [Shionogi] see 6201
Festucine see 5442
Fevarin [Duphar] see 4138
Feverall see 3358
Feverfew, 4013
Fever-producing Substances see 8011

Fevonil [Century] see 3358
Feximac [Nicholas] see 1462
FG 4963 see 3897
FG 5111 see 5707
FG 5606 see 612
FG 7051 see 6991
FGF see 4016
FHD-3 see 4515
FI 106 see 3428
FI 1163 see 5464
FI 5853 see 6989
FI 6341 see 4156
FI 6426 see 8730
FI 6654 see 1865
FI 6714 see 6404
FI 6934 see 1998
Fiasone see 7719
Fiber A see 6822
Fiber V [Du Pont] see 7546
Fiber X-51 see 2577
Fiber X-54 see 2577
Fiblaferon [Rentschler] see 4893
Fibocil [Lilly] see 781
Fibonel [Pharma Investi] see 3881
Fiboran [Christiaens] see 781
Fibrase [SK & F] see 7090
Fibravyl [Rhovyl] see 7563
Fibrin, 4014
Fibrin-i see 4014
Fibrin-s see 4014
Fibrinase see 3878
Fibrindex see 9320
Fibrinogen, 4015
Fibrinokinase see 9394
Fibrinolysin see 7493
Fibrin-stabilizing Factor see 3878
Fibroblast Growth Factor, 4016
Fibroblast Interferon see 4893
Fibrogamin [Behringwerke] see 3878
Fibroin, 4017
Fibronectins, 4018
Fibrotan see 4687
Ficam [Fisons] see 1044
Fichtelite, 4019
Ficin, 4020
Ficoid [Fisons] see 4079
Ficortril [Pfizer] see 4710
Ficusin see 7944
Ficus Protease see 4020
Ficus Proteinase see 4020
FIF see 4893
Filair [Riker] see 9089
Filaribits [Norden] see 3106
Filarsen see 3060
Filazine see 3106
Fildesin [Shionogi] see 9892
Filicic Acid see 4023
Filicin see 4023
Filicinic Acid, 4021
Filicinsäure (German) see 4021
Filimarisin see 4022
Filipin, 4022
Filipin III see 4022
Filixic Acids, 4023
Filix Mas (B.P.) see 870
Filixsäure (German) see 4023
Filon [Berk] see 3908
Filoral see 2213
Filtrawhite [Petroleum Spec.] see 7139
Filtrax [Biomed. Foscama] see 7427
Filtrolatum [Petroleum Spec.] see 7138
Filtrosoft [Petroleum Spec.] see 7138
Finadyne [Fisons] see 4073
Finalin [Yamanouchi] see 1037
Finaplix [Hoechst] see 9499
*Finasteride see 7888
Finaten [Finadiet] see 4144
Finimal [Mepros] see 40
Finlepsin see 1783
Finquel [Ayerst] see 9533

Finuret see 4716
Fiobrol see 2195
Fipexide, 4024
Firefly Luciferin, 4025
Firemaster BP-6 [Michigan Chem.] see 7540
Firemaster LV-T 23P [Michigan Chem.] see 9665
Firemaster T 23P [Michigan Chem.] see 9665
Firmacef [Firma] see 1925
Firon [Beard] see 3995
Fir-wood Oil see 6761
Fisalamine see 5807
Fischer's Yellow see 2426
Fisetin, 4026
Fish-berry see 2452
Fish Glue see 4991
Fisidenolon 1521 see 4026
Fisiodar [Gentili] see 2939
Fisostina see 4225
Fivent [Fisopharma] see 2594
FK 027 see 1927
FK 235 see 6461
FK 506, A5
FK 749 see 1949
FK 1160 see 9356
FL 113 see 4961
FL 1039 see 400
FL 1060 see 399
Flabelline see 5890
Flagecidin [Pfizer] see 700
Flagentyl [Rhône-Poulenc] see 8373
Flagyl [Searle] see 6079
Flagyl I.V. [Searle] see 6079
Flake Lead see 1016
Flamanil [Salvoxyl-Wander] see 7392
Flamasone [Norbrook] see 7719
Flamazine [Smith & Nephew] see 8874
Flamitajin [Daisan/Iwaki] see 4028
Flammazine [Duphar] see 8874
Flammex AP see 9665
Flammex T 23P see 9665
Flanax [Syntex] see 6337
Flanin F [Tokyo Tanabe] see 4028
Flatistine [Sauba] see 1856
Flavacidin see 652
Flavan [Millot-Solac] see 5334
3,3',4,4',5,7-Flavanhexol see 5334
3,3',4',5,7-Flavanpentol see 1908
Flavaspidic Acid, 4027
Flavaxin [Winthrop] see 8201
Flavicin see 652
Flavin-adenine Dinucleotide see 8201
Flavine-Adenine Dinucleotide, 4028
Flavine Mononucleotide see 8201 and 8202
Flavine Yellow Shade see 8047
Flavipucine, 4029
Flavitan see 4028
Flavolutan see 7783
Flavomycin [Hoechst] see 962
Flavone, 4030
7-Flavone Ethyl Hydroxyacetate see 3490
7-Flavonoxyacetic Acid Ethyl Ester see 3490
Flavopereirine, 4031
Flavophospholipol see 962
Flavoquine [Roussel-UCLAF] see 602
Flavoxanthin, 4032
Flavoxate, 4033
Flavugal [Tempelhof] see 2761
Flavurol see 5757
Flaxedil [Lederle; Davis & Geck] see 4249
Flaxseed see 5385
FLC 1374 see 9353
Fleabane see 3614
Flea Seed see 7489
Flebopex [Profarma] see 3291

Flebosan [Dukron] see 9520
Flebosmil [Bouchara] see 3291
Flebosten [Bonomelli] see 3291
Flebotropin [Bago] see 3291
Flecainide, 4034
Flectadol [Maggioni] see 5510
Flectar see 9980
Flectron [Shell] see 2775
Fleishmilchsäure see 5216
Flemun [Intermuti] see 8500
Flenac [Norwich] see 3913
Fleroxacin, 4035
Flexal see 1848
Flexartal [Clin-Comar-Byla] see 1848
Flexazone see 7248
Flexeril [Merck & Co.] see 2719
Flexiban [Merck & Co.] see 2719
Flexible Collodion see 2480
Flexidor [Elanco] see 5125
Flexilon [McNeil] see 10098
Flexin [McNeil] see 10098
Flindersine, 4036
Flint see 8440
Flobacin [Sigma Tau/Glaxo] see 6688
Flo-Cillin Aqueous [Bristol] see 7042
Floctafenine, 4037
Floganol [Zyma] see 2357
Flogar [UCB] see 6869
Flogencyl [S.A.R.E.P.] see 3644
Flogene [Polifarma] see 3946
Floginax [Farmochim. Ital.] see 6337
Flogitolo see 6925
Flogobene [Farge] see 7476
Flogobron [Intersint] see 6898
Flogoril see 6925
Flogos [Gentili] see 8990
Flogovital [Bago] see 6464
Floionic Acid see 7297
Flolan [Wellcome] see 7890
Flomoxef, 4038
Flonatril [Specia] see 2401
Flopion [Kyoritsu] see 4039
Flopropione, 4039
Florantyrone, 4040
Floraquin [Searle] see 4935
Floredil, 4041
Florel [Union Carbide] see 3686
Flores Martis see 3961
Florfenicol, 4042
Florid [Mochida] see 6101
Floridin see 4197
Floridin [Coli] see 1975
Florimycin see 9905
Florinef [Squibb] see 4059
Floripavine see 8309
Florisil see 5569
Flormidal [Galenika] see 6105
Florocid see 8565
Florone [Upjohn] see 3126
Floropipamide see 7422
Floropryl [Merck & Co.] see 5060
Flosequinan, 4043
Flosequinon see 4043
Flosin [Erba] see 4875
Flosint [Erba] see 4875
Flou [ELEA] see 7919
Flovacil [Andromaco] see 3130
Flowering Dogwood see 2530
Flowers of Antimony see 743
Flowers of Tin see 8736
Flowers of Zinc see 10050
Flowery-headed Spurge see 3860
Floxacillin, 4044
Floxacin [Merck & Co.] see 6617
Floxapen [Beecham] see 4044
Floxicam [Menarini] see 5127
Floxin [J & J] see 6688
Floxuridine, 4045
Floxyfral [Solvay] see 4138
Fluacizine, 4046
Fluagel see 345

Fluamine see 5845
Fluanisone, 4047
Fluanxol [Lundbeck] see 4114
Fluanxol Dépot [Lundbeck] see 4114
Fluaton [Tubi Lux] see 4104
Fluatox [Inpharzam] see 82
Fluazacort, 4048
Fluazifop-butyl, 4049
Flubendazole, 4050
Flubenisolone see 1202
Flubenol [Janssen] see 4050
Flubenzimine, 4051
Flubron [SS Pharm] see 4136
Flubuperone see 5707
Fluchloralin, 4052
Flucinom [Schering] see 4132
Fluclorolone Acetonide see 4053
Flucloronide, 4053
Flucloxacillin see 4044
Fluconazole, 4054
Flucort [Syntex] see 4066
Flucythrinate, 4055
Flucytosine, 4056
Fludarabine, 4057
Fludarene [Merck & Co.] see 2239
Fludemil see 4067
Fluderma [Farmitalia] see 4156
Fludestrin [Heyden] see 9108
Fludex [Biopharma] see 4847
Fludiazepam, 4058
Fludilat [Thiemann] see 1041
Fludrocortisone, 4059
Fludrocortone [Merck & Co.] see 4059
Fludroxycortide see 4122
Fluellite see 341
Flufenamic Acid, 4060
Flugeral [Italfarmaco] see 4070
Flugeril [Byk-Essex] see 4132
Flugestone Acetate see 4125
Fluibil [Zambon] see 2044
Fluibron [Chiesi] see 392
Fluidane see 1381
Fluiden [Lafare] see 3942
Fluidil [Adria] see 2760
Fluifort [Dompé] see 1809
Fluimucetin [Zambon] see 82
Fluimucil [Zambon] see 82
Fluindione, 4061
Fluitran [Shionogi] see 9537
Fluixol [Ripari-Gero] see 392
Flukanide [Merck & Co.] see 8121
Flukiver [Janssen] see 2407
Flumadine [Roche] see 8221
Flumamine see 5845
Flumarin [Shionogi] see 4038
Flumark [Dainippon] see 3540
Flumazenil, 4062
Flumazepil see 4062
Flumecinol, 4063
Flumedroxone Acetate, 4064
Flumen see 2169
Flumequine, 4065
Flumesil see 1045
Flumethasone, 4066
Flumethiazide, 4067
Flumetholon [Santen] see 4104
Flumethone see 6977
Flumethrin, 4068
Flumetramide, 4069
Flumidin [Kabi] see 6181
Flumoperone see 9596
Flumoxal [Janssen] see 4050
Flumoxane [Lebrun] see 4050
Flunagen [Gentili] see 4070
Flunarizine, 4070
Flunarl [Kyowa Hakko] see 4070
Fluniget [Merck & Co.] see 3130
Flunisolide, 4071
Flunitrazepam, 4072
Flunixin, 4073
Flunoxaprofen, 4074

Fluoboric Acid, 4075
Fluocinil [Pradel] see 4076
Fluocinolide (obsolete) see 4077
Fluocinolide Acetate (obsolete) see 4077
Fluocinolone Acetonide, 4076
Fluocinolone Acetonide Acetate see 4077
Fluocinonide, 4077
Fluocortin Butyl, 4078
Fluocortolone, 4079
Fluocortolone 21-Caproate see 4079
Fluocortolone Trimethylacetate see 4079
Fluodonil [Pirri] see 3130
Fluodrocortisone see 4059
Fluohydric Acid see 4717
Fluohydric Acid Gas see 4723
Fluohydrisone see 4059
Fluohydrocortisone see 4059
Fluomazina see 9597
Fluometuron, 4080
Fluon see 7560
Fluonid [Marion] see 4076
Fluonilid [Continental Pharma] see 5440
Fluophosgene see 1826
Fluophosphoric Acid Di(dimethyl-amide) see 3191
Fluoracisine see 4046
Fluoracizine see 4046
Fluoracyzine see 4046
Fluorakil 100 see 4095
3',6'-Fluorandiol see 4085
Fluorandrenolone see 4122
β-(8-Fluoranthoyl)propionic Acid see 4040
Fluorapatite see 7321
9H-Fluorene, 4081
9H-Fluorene-2,7-diamine, 4082
N-2-Fluorenylacetamide, 4083
N-9H-Fluoren-2-ylacetamide see 4083
4-(9H-Fluoren-9-ylidenemethyl)benz-enecarboximidamide see 6978
α-Fluoren-9-ylidene-p-toluamidine see 6978
Fluorescamine, 4084
Fluorescein, 4085
Fluorescein Paper, 4086
Fluorescent Blue see 5211
Fluorescin, 4087
Fluorescite see 4085
Fluorescyanine see 4816
Fluoresone, 4088
Fluorets [Akorn] see 4085
Fluoridamid, 4089
Fluorine, 4090
Fluorine Dioxide, 4091
Fluorine Monoxide, 4092
Fluorine Nitrate, 4093
Fluorine Oxide see 4092
Fluorine Perchlorate, 4094
Fluoristan see 8743
Fluor-i-strip see 4085
Fluorite see 1669
Fluormetholon see 4104
Fluormone see 2922
Fluoroacetamide, 4095
4'-Fluoro-4-(4-acetamidomethyl-4-phenylpiperido)butyrophenone see 25
Fluoroacetic Acid, 4096
Fluoroacetic Acid Amide see 4095
4'-C-Fluoroadenosine 5'-Sulfamate see 6648
3-Fluoro-D-alanine, 4097
C₃-Fluoroalcohol see 9145
p-Fluoroaniline, 4098
2-Fluoro-9-β-D-arabinofuranosyl-adenine see 4057

4-Fluorobenzenamine see 4098
Fluorobenzene, 4099
4-Fluorobenzeneacetic Acid see 4105
p-Fluorobenzoic Acid, 4100
5-(p-Fluorobenzoyl)-2-benzimidazole-carbamic Acid Methyl Ester see 4050
[5-(4-Fluorobenzoyl)-1H-benzimid-azol-2-yl]carbamic Acid Methyl Ester see 4050
3-[2-[4-(4-Fluorobenzoyl)-1-piperi-dinyl]ethyl]-2,4[1H,3H]-quinazo-linedione see 5175
4-[4-(4-Fluorobenzoyl)-1-piperidin-yl]-1-(4-fluorophenyl)-1-butanone see 5320
1-[γ-(4-Fluorobenzoyl)propyl]-4-acetamidomethyl-4-phenylpiperidine see 25
1'-[3-(p-Fluorobenzoyl)propyl]-[1,4'-bipiperidine]-4'-carboxamide see 7422
1-(3-p-Fluorobenzoylpropyl)-4-p-chlorophenyl-4-hydroxypiperidine see 4511
1-(3'-p-Fluorobenzoylpropyl)-4-hydroxy-4-p-tolylpiperidine see 6170
1-(3'-p-Fluorobenzoylpropyl)-4-hydroxy-4-(3''-trifluoromethyl-phenyl)piperidine see 9596
N-[1-[3-(p-Fluorobenzoyl)propyl]-4-phenylpiperidin-4-ylmethyl]acet-amide see 25
N-[[1-[3-(p-Fluorobenzoyl)propyl]-4-phenyl-4-piperidyl]methyl]acet-amide see 25
8-[3-(p-Fluorobenzoyl)propyl]-1-phenyl-1,3,8-triazaspiro[4.5]decan-4-one see 8707
1-[γ-(4-Fluorobenzoyl)propyl]-4-piperidinopiperidine-4-carboxamide see 7422
1-[1-[3-(p-Fluorobenzoyl)propyl]-4-piperidyl]-2-benzimidazolinone see 1057
1-[1-[3-(4-Fluorobenzoyl)propyl]-4-piperidyl]-2,3-dihydrobenzimid-azole-2-thione see 9373
1-[1-[3-(4-Fluorobenzoyl)propyl]-4-piperidyl]-2-mercaptobenzimidazole see 9373
1-[3-(4-Fluorobenzoyl)propyl]-4-(2-pyridyl)piperazine see 915
1-[1-[3-(p-Fluorobenzoyl)propyl]-1,2,3,6-tetrahydro-4-pyridyl]-2-benzimidazolinone see 3437
1-(p-Fluorobenzyl)-2-[[1-(p-methoxy-phenethyl)-4-piperidyl]amino]benz-imidazole see 878
2-(2-Fluoro-4-biphenylyl)propionic Acid see 4124
2-(2-Fluoro-4-bromobenzyl)-1,2-dihydro-1-oxophthalazin-4-ylacetic Acid see 7565
6α-Fluoro-17α-bromoprogesterone see 4513
9α-Fluoro-21-chloro-11β,16α,17α-trihydroxypregn-4-ene-3,20-dione 16,17-Acetonide see 4504
9α-Fluorocortisol see 4059
C-Fluorocurarine see 2680
5-Fluorocytosine see 4056
6α-Fluoro-1-dehydrohydrocortisone see 4119
5-Fluoro-2'-deoxy-β-uridine see 4045
6α-Fluorodexamethasone see 4066
6α-Fluoro-9α,11β-dichloro-1,4-preg-nadiene-16α,17,21-triol-3,20-dione 16,17-Acetonide see 4053

6-Fluoro-1,4-dihydro-1-(methylamino)-7-(4-methyl-1-piperazinyl)-4-oxo-3-quinolinecarboxylic Acid *see* 415

(±)-9-Fluoro-2,3-dihydro-3-methyl-10-(4-methyl-1-piperazinyl)-7-oxo-7*H*-pyrido[1,2,3-*de*]-1,4-benzoxazine-6-carboxylic Acid *see* 6688

9-Fluoro-6,7-dihydro-5-methyl-1-oxo-1*H*,5*H*-benzo[*ij*]quinolizine-2-carboxylic Acid *see* 4065

8-Fluoro-5,6-dihydro-5-methyl-6-oxo-4*H*-imidazo[1,5-*a*][1,4]benzodiazepine-3-carboxylic Acid Ethyl Ester *see* 4062

(*S*)-6-Fluoro-2,3-dihydrospiro[4*H*-1-benzopyran-4,4′-imidazolidine]-2′,5′-dione *see* 8679

6α-Fluoro-11β,21-dihydroxy-16α,17-isopropylidenedioxy-Δ¹,⁴-pregnadiene-3,20-dione *see* 4071

9α-Fluoro-11β,21-dihydroxy-16α,17α-isopropylidenedioxy-1,4-pregnadiene-3,20-dione *see* 9512

9-Fluoro-11β,17α-dihydroxy-17(*S*)-lactoylandrosta-1,4-dien-3-one 17β-Acetate *see* 4115

9-Fluoro-11,17-dihydroxy-17-methylandrost-4-en-3-one *see* 4113

6-Fluoro-11,21-dihydroxy-16,17-[(1-methylethylidene)bis(oxy)]pregna-1,4-diene-3,20-dione *see* 4071

9-Fluoro-11,21-dihydroxy-16,17-[1-methylethylidenebis(oxy)]pregna-1,4-diene-3,20-dione *see* 9512

6α-Fluoro-11β,21-dihydroxy-16α,17-[(1-methylethylidene)bis(oxy)]-pregn-4-ene-3,20-dione *see* 4122

6-Fluoro-11,21-dihydroxy-16-methylpregna-1,4-diene-3,20-dione *see* 4079

(6α,11β)-9-Fluoro-11,17-dihydroxy-6-methylpregna-1,4-diene-3,20-dione *see* 4104

9-Fluoro-11,21-dihydroxy-16-methylpregna-1,4-diene-3,20-dione *see* 2910

9-Fluoro-11β,21-dihydroxy-2′-methyl-5′β*H*-pregna-1,4-dieno[17,16-*d*]-oxazole-3,20-dione 21-Acetate *see* 4048

9-Fluoro-11β,17-dihydroxypregn-4-ene-3,20-dione-17-acetate *see* 4125

9-Fluoro-11β,17-dihydroxyprogesterone 17-Acetate *see* 4125

1-Fluoro-2,4-dinitrobenzene, 4101

Fluoroethanoic Acid 4096

(*endo,syn*)-8-(2-Fluoroethyl)-3-[(hydroxydiphenylacetyl)oxy]-8-methyl-8-azoniabicyclo[3.2.1]octane Bromide *see* 4136

(8*r*)-8-(2-Fluoroethyl)-3α-hydroxy-1α*H*,5α*H*-tropanium Bromide Benzilate *see* 4136

N-β-Fluoroethylnortropine Benzilate Methobromide *see* 4136

Fluorofen *see* 9597

Fluoroflex *see* 7560

4′-Fluoro-4-[4-(*p*-fluorobenzoyl)-piperidino]butyrophenone *see* 5320

6-Fluoro-1-(4-fluorophenyl)-1,4-dihydro-7-(4-methyl-1-piperazinyl)-4-oxo-3-quinolinecarboxylic Acid *see* 3127

Fluoroform, 4102

Fluoroformylon *see* 4156

Fluorogesarol *see* 2936

5-Fluoro-*N*-hexyl-3,4-dihydro-2,4-dioxo-1(2*H*)-pyrimidinecarboxamide *see* 1851

9α-Fluorohydrocortisone *see* 4059

(±)-5-Fluoro-2-(2-hydroxy-3-*t*-butylaminopropoxy)butyrophenone *see* 1527

4′-Fluoro-4-(4-hydroxy-4-*p*-chlorophenylpiperidino)butyrophenone *see* 4511

9α-Fluoro-17-hydroxycorticosterone *see* 4059

Δ¹-9α-Fluoro-16α-hydroxyhydrocortisone *see* 9511

(6α,11β,16α)-6-Fluoro-11-hydroxy-16-methyl-3,20-dioxopregna-1,4-dien-21-oic Acid Butyl Ester *see* 4078

p-Fluoro-4-(4′-hydroxy-4′-*p*-methylphenylpiperidino)butyrophenone *see* 6170

9α-Fluoro-11β-hydroxy-17α-methyltestosterone *see* 4113

9α-Fluoro-16α-hydroxyprednisolone *see* 9511

9α-Fluoro-16α-hydroxyprednisolone Acetonide *see* 9512

9α-Fluoro-16α-hydroxyprednisolone 16α,17α-Acetonide 21-(β-Benzoylamino)isobutyrate *see* 9513

p-Fluoro-4-(4′-hydroxy-4′-*p*-tolyl-piperidino)butyrophenone *see* 6170

p-Fluoro-4-[4′-hydroxy-4′-(3″-trifluoromethyl)phenyl]piperidinobutyrophenone *see* 9596

4′-Fluoro-4-[4-hydroxy-4-(α,α,α-trifluoro-*m*-tolyl)piperidino]butyrophenone *see* 9596

9α-Fluoro-16α,17-isopropylidenedioxyprednisolone *see* 9512

6α-Fluoro-16α,17-isopropylidenedioxy-4-pregnene-11β,21-diol-3,20-dione *see* 4122

Fluoromar [Ohio Med.] *see* 4127

Fluoromethane, 4103

Fluorometholone, 4104

4′-Fluoro-4-[4-(*o*-methoxyphenyl)-1-piperazinyl]butyrophenone *see* 4047

Fluoromethylbenzene *see* 4108

2-Fluoro-α-methyl[1,1′-biphenyl]-4-acetic Acid *see* 4124

2-Fluoro-α-methyl-4-biphenylacetic Acid *see* 4124

9α-Fluoro-16α-methyl-Δ¹-corticosterone *see* 2910

6α-Fluoro-16α-methyl-1-dehydrocorticosterone *see* 4079

9α-Fluoro-16α-methyl-17-desoxyprednisolone *see* 2910

9α-Fluoro-16-methyleneprednisolone 21-Acetate *see* 4118

9α-Fluoro-16-methylene-Δ¹,⁴-pregnadiene-11β,17,21-triol-3,20-dione 21-Acetate *see* 4118

cis-5-Fluoro-2-methyl-1-[*p*-(methylsulfinyl)benzylidene]indene-3-acetic Acid *see* 8961

(*Z*)-5-Fluoro-2-methyl-1-[[4-(methylsulfinyl)phenyl]methylene]-1*H*-indene-3-acetic Acid *see* 8961

7-Fluoro-1-methyl-3-(methylsulfinyl)-4(1*H*)-quinolinone *see* 4043

4′-Fluoro-4-(4-methylpiperidino)-butyrophenone *see* 5707

6α-Fluoro-16α-methylprednisolone *see* 6977

9α-Fluoro-16α-methylprednisolone *see* 2922

9α-Fluoro-16β-methylprednisolone *see* 1202

6α-Fluoro-16α-methyl-Δ¹,⁴-pregnadiene-11β,21-diol-3,20-dione *see* 4079

Fluorophene [Stecker] *see* 4106

p-Fluorophenylacetic Acid, 4105

5-(2-Fluorophenyl)-1,3-dihydro-1-methyl-7-nitro-2*H*-1,4-benzodiazepin-2-one *see* 4072

p-Fluorophenyl Ethyl Sulfone *see* 4088

α-(2-Fluorophenyl)-α-(4-fluorophenyl)-1*H*-1,2,4-triazole-1-ethanol *see* 4135

3-Fluoro-4-phenylhydratropic Acid *see* 4124

1-(4-Fluorophenyl)-4-[4-hydroxy-4-(4-methylphenyl)-1-piperidinyl]-1-butanone *see* 6170

2-(*p*-Fluorophenyl)-1,3-indandione *see* 4061

1-(4-Fluorophenyl)-4-[4-(2-methoxyphenyl)-1-piperazinyl]-1-butanone *see* 4047

(+)-2-(4-Fluorophenyl)-α-methyl-5-benzoxazoleacetic Acid *see* 4074

(−)-*trans*-4-(*p*-Fluorophenyl)-3-[[3,4-(methylenedioxy)phenoxy]methyl]-piperidine *see* 6991

1-[(4-Fluorophenyl)methyl]-*N*-[1-[2-(4-methoxyphenyl)ethyl]-4-piperidinyl]-1*H*-benzimidazol-2-amine *see* 878

4-(4-Fluorophenyl)-7-methyl-1-(1-methylethyl)-2(1*H*)-quinazolinone *see* 4120

1-(4-Fluorophenyl)-4-(4-methyl-1-piperidinyl)-1-butanone *see* 5707

1′-[4-(4-Fluorophenyl)-4-oxobutyl]-[1,4′-bipiperidine]-4′-carboxamide *see* 7422

N-[[1-[4-(4-Fluorophenyl)-4-oxobutyl]-4-phenyl-4-piperidinyl]methyl]-acetamide *see* 25

8-[4-(4-Fluorophenyl)-4-oxobutyl]-1-phenyl-1,3,8-triazaspiro[4.5]decan-4-one *see* 8707

1-[1-[4-(*p*-Fluorophenyl)-4-oxobutyl]piperidin-4-yl]-2-benzimidazolinone *see* 1057

1-[1-[4-(4-Fluorophenyl)-4-oxobutyl]-4-piperidinyl]-1,3-dihydro-2*H*-benzimidazol-2-one *see* 1057

1-[1-[4-(4-Fluorophenyl)-4-oxobutyl]-1,2,3,6-tetrahydro-4-pyridinyl]-1,3-dihydro-2*H*-benzimidazol-2-one *see* 3437

1-[1-[4-(*p*-Fluorophenyl)-4-oxobutyl]-1,2,3,6-tetrahydro-4-pyridyl]-2-benzimidazolinone *see* 3437

8-[4-(4-Fluorophenyl)-3-pentenyl]-1-phenyl-1,3,8-triazaspiro[4.5]decan-4-one *see* 8709

1-(*p*-Fluorophenyl)-4-(4-phenyl-4-acetamidomethylpiperidino)-1-butanone *see* 25

N-[3-[4-(*p*-Fluorophenyl)-1-piperazinyl]-1-methylpropyl]nicotinamide *see* 6400

N-[3-[4-(4-Fluorophenyl)-1-piperazinyl]-1-methylpropyl]-3-pyridinecarboxamide *see* 6400

1-(*p*-Fluorophenyl)-4-(4-piperidino-4-carbamoylpiperidino)-1-butanone *see* 7422

1-(4-Fluorophenyl)-4-(1-piperidinyl)-1-butanone *see* 7750

1-(4-Fluorophenyl)-4-[4-(2-pyridinyl)-1-piperazinyl]-1-butanone *see* 915

4′-Fluoro-4-piperidinobutyrophenone *see* 7750

4′-Fluoro-4-[*N*-[4-(*N*-piperidino)-4-carbamido]piperidino]butyrophenone *see* 7422

Fluoroplex [Allergan] *see* 4109

6α-Fluoroprednisolone *see* 4119

9-Fluoroprednisolone *see* 5058
6α-Fluoro-1,4-pregnadiene-11β,17α,-21-triol-3,20-dione *see* 4119
4'-Fluoro-4-[4-(2-pyridyl)-1-piperazinyl]butyrophenone *see* 915
5-Fluoro-2,4(1*H*,3*H*)-pyrimidinedione *see* 4109
Fluoros [Dieckmann] *see* 8565
Fluorosalan, 4106
Fluorosilicic Acid *see* 4110
(+)-(4*S*)-6-Fluorospiro[chroman-4,4'-imidazolidine]-2',5'-dione *see* 8679
4'-Fluoro-5'-*O*-sulfamoyladenosine *see* 6648
9-(4-Fluoro-5-*O*-sulfamoylpentofuranosyl)adenine *see* 6648
Fluorosulfonic Acid, 4107
Fluorosulfuric Acid *see* 4107
Fluorosulfuric Acid Methyl Ester *see* 5993
5-Fluoro-1-(tetrahydro-2-furanyl)-2,4(1*H*,3*H*)-pyrimidinedione *see* 9060
5-Fluoro-1-(tetrahydro-2-furyl)uracil *see* 9060
9-Fluoro-11,16,17,21-tetrahydroxy-pregna-1,4-diene-3,20-dione *see* 9511
6α-Fluoro-11β,16α,17,21-tetrahydroxypregna-1,4-diene-3,20-dione Cyclic 16,17-Acetal with Acetone *see* 4071
9α-Fluoro-11β,16α,17,21-tetrahydroxypregna-1,4-diene-3,20-dione Cyclic 16,17-Acetal with Acetone *see* 9512
9-Fluoro-11β,16α,17,21-tetrahydroxypregna-1,4-diene-3,20-dione Cyclic 16,17-Acetal with Acetone, 21-(3,3-Dimethylbutyrate) *see* 9514
9-Fluoro-11β,16α,17,21-tetrahydroxypregna-1,4-diene-3,20-dione Cyclic 16,17-Acetal with Acetone 21-Ester with *N*-Benzoyl-2-methyl-β-alanine *see* 9513
9-Fluoro-11β,16α,17,21-tetrahydroxypregna-1,4-diene-3,20-dione Cyclic 16,17-Acetal with Cyclopentanone, 21-Acetate *see* 398
6α-Fluoro-11β,16α,17,21-tetrahydroxypregn-4-ene-3,20-dione Cyclic 16,17-Acetal with Acetone *see* 4122
6α-Fluoro-11β,16α,17,21-tetrahydroxyprogesterone Cyclic 16,17-Acetal with Acetone *see* 4122
Fluorothiamphenicol *see* 4042
4-Fluoro-4-[4-(2-thioxo-1-benzimidazolinyl)piperidino]butyrophenone *see* 9373
Fluorotoluene, 4108
Fluorotrichloromethane *see* 9553
9α-Fluoro-11β,17,21-trihydroxy-16-methylenepregna-1,4-diene-3,20-dione 21-Acetate *see* 4118
6α-Fluoro-11β,17,21-trihydroxy-16α-methylpregna-1,4-diene-3,20-dione *see* 6977
9-Fluoro-11,17,21-trihydroxy-16-methylpregna-1,4-diene-3,20-dione *see* 1202
(11β,16α)-9-Fluoro-11,17,21-trihydroxy-16-methylpregna-1,4-diene-3,20-dione *see* 2922
9α-Fluoro-11β,17α,21-trihydroxy-21-methylpregna-1,4-diene-3,20-dione 21-Acetate *see* 4115
6-Fluoro-11,17,21-trihydroxypregna-1,4-diene-3,20-dione *see* 4119
9-Fluoro-11,17,21-trihydroxypregna-1,4-diene-3,20-dione *see* 5058

(11β)-9-Fluoro-11,17,21-trihydroxy-pregn-4-ene-3,20-dione *see* 4059
Fluorouracil, 4109
2-Fluorovidarabine *see* 4057
Fluorspar *see* 1669
Fluosilicic Acid, 4110
Fluosol DA, 4111
Fluostigmine *see* 5060
Fluosulfonic Acid *see* 4107
Fluotestin [Roter] *see* 4113
Fluothane [Ayerst] *see* 4517
Fluovitef [Italfarmaco] *see* 4076
Fluoxetine, 4112
Fluoxymesterone, 4113
Flupen [Alfa] *see* 4044
Flupenthixol *see* 4114
Flupentixol, 4114
Fluperolone Acetate, 4115
Fluphenazine, 4116
Flupirtine, 4117
Fluprednidene Acetate, 4118
Fluprednisolone, 4119
Fluprednylidene 21-Acetate *see* 4118
Fluproquazone, 4120
Fluprostenol, 4121
Fluprowit [Thiemann] *see* 82
Fluracil *see* 4109
Flura-Drops *see* 8565
Fluram [Roche] *see* 4084
Flurandrenolide, 4122
Flurandrenolone *see* 4122
Flurandrenolone Acetonide (obsolete) *see* 4122
Flurazepam, 4123
Flurazepam Hydrochloride *see* 4123
Flurbiprofen, 4124
Fluril *see* 4109
Flurobate [Texas Pharmacal] *see* 1202
Fluroblastin [Farmitalia] *see* 4109
Flurofen [Boots] *see* 4124
Flurogestone Acetate, 4125
Fluropryl *see* 5060
Flurothyl, 4126
Fluro Uracil *see* 4109
Fluroxene, 4127
Fluroxypyr, 4128
Flurprimidol, 4129
Flusilazole, 4130
Fluspirilene, 4131
Flustar [Firma] *see* 3130
Flutamide, 4132
Flutazolam, 4133
Flutex [Syosset] *see* 9512
Flutoprazepam, 4134
Flutra *see* 9537
Flutriafen *see* 4135
Flutriafol, 4135
Flutron [Nippon Roche] *see* 3426
Flutropium Bromide, 4136
Fluvalinate, 4137
Fluvean [Kowa] *see* 4076
Fluvermal [Janssen-Le Brun] *see* 4050
Fluversin [Searle] *see* 8967
Fluvin *see* 4704
Fluvisco [Searle] *see* 8967
Fluvoxamine, 4138
Fluxarten [Beecham] *see* 4070
Fluxema *see* 1041
Fluzilazol *see* 4130
Fluzon [Taisho] *see* 4076
Flypel *see* 2848
FMC 1240 *see* 3691
FMC 9260 *see* 9154
FMC 30980 *see* 2775
FMC 33297 *see* 7132
FMC 45498 *see* 2869
FMC 54800 *see* 1229
FMC 57020 *see* 3203
FML [Allergan] *see* 4104
FMN *see* 8201
Fobex *see* 1037

Focusan *see* 9442
Fodrin *see* 8696
Folacin *see* 4140
Folaemin [O.P.G.] *see* 4140
Folbex [Ciba-Geigy] *see* 2123
Folbex VA [Ciba-Geigy] *see* 1422
Folcodal [Syncro] *see* 2308
Foldine [Specia] *see* 4140
Foldox [Sidus] *see* 9613
Folescutol, 4139
Folettes [Fawns & McAllan] *see* 4140
Folex [Adria] *see* 5908
Foliamin *see* 4140
Folic Acid, 4140
Folic Acid Sodium Salt *see* 8566
Folicet [Mission] *see* 4140
Folidol [Bayer] *see* 6983
Folidol-M [Bayer] *see* 6022
Foligan [Henning] *see* 278
Folikrin *see* 3660
Folinerin *see* 6786
Folinic Acid, 4141
Folipac *see* 4140
Folipex *see* 3660
Folisan *see* 3660
Folithion *see* 3922
Follestrine *see* 3660
Follicle-stimulating Hormone *see* 4189
Follicormon (Ampuls) *see* 3655
Follicular Hormone *see* 3660
Follicular Hormone Hydrate *see* 3659
Folliculin *see* 3660
Follicunodis *see* 3660
Follidrin (Ampuls) *see* 3655
Follidrin (Tablets) *see* 3660
Follitropin *see* 4189
Follutein [Squibb] *see* 4534
Fologenon *see* 7783
Folosan *see* 8108
Folpet, 4142
Folsan [Kali-Chemie] *see* 4140
Folsäure *see* 4140
Folvite [Lederle] *see* 4140
Fomecin A *see* 4143
Fomecin B *see* 4143
Fomecins, 4143
Fominoben, 4144
Fomocaine, 4145
Fonatol *see* 3118
Fonazine, 4146
Fonazine Mesylate *see* 4146
Fonderma [Biosedra] *see* 8004
Fongarex [Sanofi] *see* 6801
Fonlipol [Lafon] *see* 9349
Fonofos [Pulitzer] *see* 4169
Fonofos, 4169
Fontamide *see* 8921
Fontarsan *see* 6902
Fontarsol *see* 3060
Fontego [Polifarma] *see* 1473
Fontex [Lilly] *see* 4112
Fontilix [Diamant] *see* 6058
Fonurit *see* 45
Fonzylane [Orsymonde] *see* 1463
Food of The Gods *see* 846
Food Orange 8 *see* 1756
Food Yellow 13 *see* 8100
Foralamin [Eaton] *see* 5854
Forane [Anaquest] *see* 5059
Forapin [Mack, Illert.] *see* 5705
Fordiuran [Thomae] *see* 1473
Fordonal [Almirall] *see* 9236
Forene [Abbott] *see* 5059
Forenol [Roemmers] *see* 6444
Forhistal [Ciba] *see* 3204
Foriod *see* 4931
Forit *see* 6924
Formagene *see* 6974
Formal *see* 5936
Formaldehyde Acetamide *see* 4154

Formaldehyde Dimethyl Acetal *see* 5936
Formaldehyde, Gas, 4148
α,α'-Formaldehyde Glycerol *see* 4380
α,β-Formaldehyde Glycerol *see* 4380
Formaldehyde Semicarbazone *see* 4150
Formaldehyde Sodium Bisulfite, 4149
Formaldehyde Sodium Sulfoxylate *see* 8567
Formaldehyde Solution, 4150
Formaldehyde-sulfathiazole *see* 4158
Formaldehydesulfoxylic Acid Sodium Salt *see* 8567
Formalin *see* 4150
Formamide, 4151
2-Formamido-5-nitrothiazole *see* 4155
Formanilide, 4152
Formanol *see* 5881
Formic Acid, 4153
Formic Acid Ammonium Salt *see* 548
Formic Acid Benzyl Ester *see* 1148
Formic Acid Beryllium Salt *see* 1186
Formic Acid Ethyl Ester *see* 3763
Formic Acid Methyl Ester *see* 5994
Formic Acid Phenylmethyl Ester *see* 1148
Formic Acid Propyl Ester *see* 7871
Formic Aldehyde *see* 4148
Formicin, 4154
N-Formimidoylthienamycin Monohydrate *see* 4834 and 9240
Formin *see* 5879
Forminitrazole, 4155
Formin Salicylate *see* 5884
Formison *see* 5731
Formo-Cibazol [Ciba] *see* 4158
Formocortal, 4156
Formol *see* 4150
Formononetin, 4157
Formononetol *see* 4157
Formophthaloylsulfanilyl Urea *see* 8883
Formosa Camphor *see* 1738
Formosa Oil of Camphor *see* 6702
Formosulfathiazole, 4158
Formoterol, 4159
Formothion, 4160
Formula 1 *see* 3717
Formula 2B *see* 3717
Formula 3A *see* 3717
Formula 6B *see* 3717
Formula 12A *see* 3717
Formula 13A *see* 3717
Formula 19 *see* 3717
Formula 20 *see* 3717
Formula 23A *see* 3717
Formula 28 *see* 3717
Formula 28A *see* 3717
Formula 30 *see* 3717
Formula 32 *see* 3717
Formula 33 *see* 3717
Formula 35A *see* 3717
Formula 39C *see* 3717
Formula 44 *see* 3717
4'-Formylacetanilide Thiosemicarbazone *see* 9218
3-Formylamino-4-hydroxy-α-[*N*-[1-methyl-2-(*p*-methoxyphenyl)ethyl]-aminomethyl]benzyl Alcohol *see* 4159
Formylaniline *see* 4152
3-*C*-Formyl-5-deoxy-L-lyxofuranose *see* 8790
Formyldienolone, 4161
5-Formyl-3,4-dihydro-1*H*-2-benzo-pyran-1-one *see* 3623
5-Formyl-3,4-dihydroisocoumarin *see* 3623

3-Formyl-2,4-dihydroxy-6-methyl-benzoic Acid 3-Hydroxy-4-(methoxycarbonyl)-2,5-dimethylphenyl Ester *see* 885
6-Formyl-2,3-dimethoxybenzoic Acid *see* 6806
Formyl Fluoride, 4162
Formylformic Acid *see* 4407
16-Formylgitoxin *see* 4324
Formyl Hydroperoxide *see* 7114
N-Formyl-*N*-hydroxyaminoacetic Acid *see* 4500
N-Formyl-*N*-hydroxyglycine *see* 4500
2-Formyl-11α-hydroxy-Δ¹-methyl-testosterone *see* 4161
*N*α-Formyl Melittin *see* 5705
N-Formyl *trans-p*-Methoxystyryl-amine *see* 9716
2-Formyl-17α-methylandrosta-1,4-diene-11α,17β-diol-3-one *see* 4161
S-(*N*-Formyl-*N*-methylcarbamoyl-methyl) *O,O*-Dimethyl Phosphorodithioate *see* 4160
1-Formyl-4-methyl-7-isopropenyl-azulene *see* 5212
2-Formyl-1-methylpyridinium Chloride Oxime *see* 7705
4-Formylphenol *see* 4741
4-(1-Formyl-1-propenyl)-3,4-dihydro-2-oxo-2*H*-pyran-5-carboxyl-ic Acid Methyl Ester *see* 3507
Formylpteroic Acid *see* 8181
2-Formylquinoxaline-1,4-dioxide Carbomethoxyhydrazone *see* 1782
4'-Formylsuccinanilic Acid Thiosemicarbazone, 4163
Formylsulfamethine *see* 4164
*N*²-Formylsulfisomidine, 4164
5-Formyl-5,6,7,8-tetrahydrofolic Acid *see* 4141
5-Formyl-5,6,7,8-tetrahydropteroyl-L-glutamic Acid *see* 4141
Foromacidin *see* 8708
Foromacidin A *see* 8708
Foromacidin B *see* 8708
Foromacidin C *see* 8708
Forpen *see* 7041
Forskolin *see* 2473
Forstan [Bayer] *see* 6933
Forsterite *see* 5569
Forsythin *see* 7296
Fortalgesic [Winthrop] *see* 7078
Fortalin [Winthrop] *see* 7078
Fortam [Glaxo] *see* 1944
Fortamine *see* 2180
Fortasec [Esteve] *see* 5450
Fortaz [Glaxo] *see* 1944
Fortecortin [E. Merck] *see* 2922
Forthane [Lilly] *see* 6000
Forticef [Sasse] *see* 1982
Fortigro [Pfizer] *see* 1782
Fortimicin A *see* 4165
Fortimicin B *see* 4165
Fortimicin(s), 4165
Fortizyme [Breon] *see* 633
Fortodyl *see* 9928
Fortombrine M [Dagra] *see* 5688
Fortoshade M *see* 5688
Fortracin [A.L. Labs.] *see* 948
Fortral [Winthrop] *see* 7078
Fortrel [Celanese] *see* 7546
Fortrol [Shell] *see* 2692
Fortum [Glaxo] *see* 1944
Fortunan [Steinhard] *see* 4511
Forturf *see* 2167
Fortuss [GP] *see* 3154
Foscarnet Sodium, 4166
Foscavir [Astra] *see* 4166
Foschlor *see* 9536
Fosenopril *see* 4171
Fosetyl Al, 4167

Fosfakol *see* 6979
Fosfalugel [De Angeli] *see* 362
Fosferno [Plant Protection] *see* 6983
Fosfestrol, 4168
Fosfobiotic [Bergamon] *see* 4169
Fosfocin [Crinos] *see* 4169
Fosfocina [C.E.P.A.] *see* 4169
Fosfocolina [Afarit] *see* 7337
Fosfogram [Firma] *see* 4169
Fosfomycin, 4169
Fosfomycin Trometamol *see* 4169
Fosfonomycin *see* 4169
Fosforal [Sis-Ter] *see* 4169
Fosforina B₁₂ [Francia] *see* 7338
Fosfosal, 4170
Fosfotricina [Italfarmaco] *see* 4169
Foshagite *see* 1709
Foshallasite *see* 1709
Fosinopril, 4171
Fosmicin [Meiji] *see* 4169
Fospirate, 4172
Fossil Flour *see* 4878
Fossyol [Merckle] *see* 6079
Fosten [Serono] *see* 784
Fosthietan, 4173
Fostion MM [Agrimont] *see* 3209
Fotemustine, 4174
Fouadin *see* 8769
Fourneau 190 *see* 44
Fourneau 710 *see* 7495
Fourneau 933 *see* 7448
Fourneau 1500 *see* 8851
Fovane [Pfizer] *see* 1134
Fowler's Soln *see* 7585
Foxglove *see* 3141
Fox Green *see* 4868
Foximin [Caber] *see* 4169
Foy [Ono] *see* 4234
Foy 305 *see* 1734
Foypan [Ono] *see* 1734
FPL 670 *see* 2594
FPL 59002 *see* 6355
FPL 59002KP *see* 6355
FPL 60278 *see* 3418
FPL 60278AR *see* 3418
FPRMeCl *see* 7702
α-FPT-*dec see* 4114
FR 02A *see* 3491
FR 10123 *see* 1946
FR 13479 *see* 1949
FR 17027 *see* 1927
FR 34235 *see* 6461
FR 900506 *see* A5
Frabel *see* 6925
Fraction P *see* 2851
Frademicina *see* 5378
Fradiomycin *see* 6369
Fragivil *see* 1071
Fragivix [Labaz; Sigma-Tau] *see* 1071
Fragmin [KabiVitrum] *see* 9058
Framycetin *see* 6369
Framygen *see* 6369
Francephane *see* 1959
Francital [Francia] *see* 4169
Francium, 4175
Frandol [Yamanouchi] *see* 5114
Frangula, 4176
Frangula Emodin *see* 3518
Frangulic Acid *see* 3518
Frangulin, 4177
Frangulin A *see* 4177
Frangulin B *see* 4177
Franguloside *see* 4177
Frankincense *see* 6792
Franklinite *see* 10025
Franocide [Burroughs Wellcome] *see* 3106
Franroze [Hishiyama] *see* 9060
Frantin *see* 1166
Fraquinol *see* 6369
Fratol *see* 4096

Fraxetin, 4178
Fraxetin-8-glucoside *see* 4179
Fraxin, 4179
Fraxiparine ® [Choay], 4180
Fraxoside *see* 4179
Frazalon [Daiichi] *see* 4209
Fredericamycin A, 4181
Free Benzylpenicillin *see* 1156
Free Penicillin II *see* 1156
Free Penicillin G *see* 1156
Freeuril *see* 1134
Frekentine *see* 3116
Frekven [Ferrosan] *see* 7852
Frénactil [Clin-Comar-Byla] *see* 1057
Frenactyl [Janssen] *see* 1057
Frenantol [Laroche Navarron] *see* 6992
Frenapyl [Troponwerke] *see* 2378
Frenasma [ISF] *see* 2594
French Chalk *see* 9011
French Green *see* 2629
French Purple *see* 2618
French Saffron *see* 8285
French Verdigris *see* 2628
Frenohypon *see* 6992
Frenolicin, 4182
Frenolon [EGYT] *see* 6065
Frenolyse [Specia] *see* 9487
Frenopect [Hefa-Frenon] *see* 392
Frenoton *see* 909
Frenquel [Merrell] *see* 909
Frentirox *see* 5892
Freon [Du Pont] *see* 2140
Freon 11 [Du Pont] *see* 9553
Freon 12 [Du Pont] *see* 3053
Freon 14 [Du Pont] *see* 1823
Freon 114 [Du Pont] *see* 2608
Freon C318 [Du Pont] *see* 6666
Frequentic Acid *see* 2330
Frequentin, 4183
Frescon *see* 9591
Fresmin *see* 9921
Friar's Balsam *see* 961
Friar's Cowl *see* 111
Friedelan-3-one *see* 4184
Friedelin, 4184
D:A-Friedooleanan-3-one *see* 4184
D-Friedoolean-14-en-3β-ol *see* 9033
D:A-Friedo-18,19-secolup-19-en-3-one *see* 8429
Frigen [Hoechst] *see* 2140
Frigen 11 [Hoechst] *see* 9553
Frigen 12 [Hoechst] *see* 3053
Frigen 114 [Hoechst] *see* 2608
Frigol [Sherwood] *see* 7139
Fringanor [Sobio] *see* 7180
Frisium [Hoechst] *see* 2355
Froben [Boots] *see* 4124
Frone [Serono] *see* 4893
Frucote [Elanco] *see* 1544
Fructergyl [Robert et Carrière] *see* 4187
β-D-Fructofuranose 1,3,4,6-Tetra-3-pyridinecarboxylate *see* 6428
β-D-Fructofuranosidases *see* 4899
β-D-Fructofuranosyl O-α-D-Galactopyranosyl-(1→6)-O-α-D-galactopyranosyl-(1→6)-O-α-D-galactopyranosyl-(1→6)-α-D-glucopyranoside *see* 9860
β-D-Fructofuranosyl-O-α-D-galactopyranosyl-(1→6)-α-D-glucopyranoside *see* 8120
β-D-Fructofuranosyl-α-D-glucopyranoside *see* 8855
β-D-Fructofuranosyl-α-D-glucopyranoside Octakis(hydrogen sulfate) Aluminum Complex *see* 8853
Fructose, 4185
D-Fructose *see* 4185
DL-Fructose, 4186

β-D-Fructose *see* 4185
D-Fructose 1,6-Bis(dihydrogen phosphate) *see* 4187
D-Fructose 6-(Dihydrogen phosphate) *see* 4188
Fructose-1,6-diphosphate, 4187
1,6-D-Fructosediphosphoric Acid *see* 4187
D-Fructose Iron(3+)-contg Complex, Potassium Salt (2:1) *see* 3967
Fructose Monophosphate *see* 4188
Fructose-6-phosphate, 4188
D-Fructose-6-phosphoric Acid *see* 4188
Fructose 1,3,4,6-Tetranicotinate *see* 6428
β-h-Fructosidases *see* 4899
Fructosteril *see* 4185
Frugalan [Diamant] *see* 4217
Fruitone-N [Amchem] *see* 6290
Fruit Sugar *see* 4185
Frumin AL *see* 3371
Frumin G *see* 3371
Frusemide *see* 4221
Frusemin [Toho] *see* 4221
Frusetic [Unimed] *see* 4221
Frusid [DDSA] *see* 4221
FSF *see* 3878
FSH, 4189
FSR-3 *see* 5071
FT 207 [Taiho] *see* 9060
Ftalazol *see* 7351
Ftalicetimida *see* 7350
Ftalofyne *see* 7348
Ftaxilide, 4190
F3TDR *see* 9599
Ftoracizine *see* 4046
Ftorafur [Medexport] *see* 9060
Ftorocort [Gedeon Richter] *see* 9512
FTPA *see* 4111
5-FU *see* 4109
Fuadin [Winthrop] *see* 8769
Fua-Med [MED] *see* 6520
Fuchsin(e) Acid *see* 103
Fuchsine *see* 5528
Fucidin [Sigma-Tau] *see* 4231
Fucidina [Leo Pharm.] *see* 4231
Fucidine [Leo Pharm.; Thomae] *see* 4231
Fucidin Intertulle [Leo Pharm.] *see* 4231
Fucithalmic [Leo] *see* 4231
Fuclasin *see* 10075
Fucosamine, 4191
D-Fucose, 4192
L-Fucose, 4193
Fucostanol *see* 8770
Fucosterol, 4194
Fucoxanthin, 4195
Fucus, 4196
FUDR [Roche] *see* 4045
Fugerel [Byk-Essex] *see* 4132
Fugillin [Upjohn] *see* 4199
Fugoa [Scheurich] *see* 6634
Fugu Poison *see* 9175
Fuklasin *see* 10075
Fulaid [Takeda] *see* 9060
Fulcin [ICI] *see* 4453
Fuldazin [Sumitomo] *see* 6116
Fulfeel [Kyorin] *see* 9060
Ful-glo *see* 4085
Fulgram [ABC] *see* 6617
Fuller's Earth, 4197
Fuller's Herb *see* 8326
Fullsafe [Ohta] *see* 4060
"Fulminating Gold" *see* 4413
Fulsix [Tatsumi] *see* 4221
Fuluminol [Tatsumi] *see* 2345
Fuluvamide [Kanto] *see* 4221
Fulvicin [Schering] *see* 4453
Fulvoplumierin, 4198

Fumadil B *see* 4199
Fumafer [Labaz] *see* 3995
Fumagillin, 4199
Fumar F *see* 3995
Fumaric Acid, 4200
Fumaric Acid Dibenzyl Ester *see* 1149
Fumarin *see* 2557
Fumarine *see* 7912
Fumazone *see* 3003
Fumidil [Abbott] *see* 4199
Fumigacin *see* 4552
Fumigatin, 4201
Fumigrain *see* 125
Fuming Liquid Arsenic *see* 829
Fuming Spirit of Libavius *see* 8732
Fumiron [Knoll] *see* 3995
Fundal [Schering] *see* 2083
Funduscein [SM & P] *see* 4085
Fundyl [Roger Bellon] *see* 3545
Fungacetin [Blair] *see* 9504
Fungarest [Janssen] *see* 5181
Fungibacid [Asche] *see* 9384
Fungichromin, 4202
Fungicidin *see* 6658
Fungifos [Basotherm] *see* 9433
Fungilin [Squibb] *see* 620
Fungimycin *see* 7121
Funginex [Celamerck] *see* 9602
Fungiplex [Hermal] *see* 8865
Fungisdin [Isdin] *see* 6101
Fungistat [Janssen] *see* 9090
Fungisterol, 4203
Fungistop [Schering] *see* 9442
Fungizone [Squibb] *see* 620
Fungoral [Janssen] *see* 5181
Funtumine, 4204
Furachel [Rachelle] *see* 6520
Furacin [Eaton] *see* 6521
Furacinetten *see* 6521
Furacoccid *see* 6521
Furacrylic Acid *see* 4207
Furadan [FMC] *see* 1810
Furadantin [Eaton] *see* 6520
Furadantine MC [Eaton; Norwich] *see* 6520
Furadoine *see* 6520
Furadonine *see* 6520
Furadroxyl [Eaton] *see* 6439
Furafluor [Green Cross] *see* 9060
2-Furalacetic Acid *see* 4207
Furalan [Lannett] *see* 6520
2-Furaldehyde *see* 4214
Furaltadone, 4205
Furamazone [Eaton] *see* 6446
Furamide [Clin-Comar-Byla] *see* 3187
Furamon *see* 4225
Furamterene *see* 4224
Furan, 4206
Furanace [Dainippon] *see* 6452
2-Furanacrylic Acid, 4207
2-Furanacrylonitrile, 4208
2-Furancarbinol *see* 4215
Furancarbonyl Chloride *see* 4222
2-Furancarboxaldehyde *see* 4214
2-Furancarboxylic Acid *see* 4219
2-Furancarboxylic Acid 1-(Dichloroacetyl)-1,2,3,4-tetrahydro-6-quinolinyl Ester *see* 8068
2,5-Furandione *see* 5586
N_1-(2'-Furanidyl)-5-fluorouracil *see* 9060
2-Furanmethanol *see* 4215
Furanol *see* 4225
Furantoin *see* 6520
1-(2-Furanyl)-N,N'-bis(2-furanylmethylene)methanediamine *see* 4718
9-(3-Furanyl)decahydro-4-hydroxy-4a,10a-dimethyl-1,4-etheno-3H,7H-benzo[1,2-c:3,4-c']dipyran-3,7-dione *see* 2489

1-*O*-(4-*O*-β-D-Galactopyranosyl-β-D-
glucopyranosyl)ceramide *see* 2800
4-*O*-β-D-Galactopyranosyl-D-glucose
see 5221
6-*O*-α-D-Galactopyranosyl-D-glucose
see 5699
N-[1-[(β-D-Galactopyranosyloxy)-
methyl]-2-hydroxy-3-heptadecen-
yl]-2-hydroxytetracosanamide *see*
7342
Galactoquin *see* 8073
Galactosaccharic Acid *see* 4237
D-Galactosamine, 4240
Galactose, 4241
4-β-D-Galactosido-D-fructose *see*
5225
4-(β-D-Galactosido)-D-gluconic Acid
see 5219
4-(β-D-Galactosido)-D-glucose *see*
5221
6-(α-D-Galactosido)-D-glucose *see*
5699
4-*O*-β-D-Galactosyl-D-fructose *see*
5225
D-Galacturonic Acid, 4242
Galamila [Delta-Chemie] *see* 1694
Galanga, 4243
Galangal *see* 4243
Galangin, 4244
Galantamine *see* 4245
Galanthamine, 4245
Galanthidine *see* 5498
Galatone *see* 4396
Galatur [Wyeth] *see* 4962
Galben [Farmoplant] *see* 1038
Galcodine [Galen] *see* 2462
Galecron [Ciba] *see* 2083
Galegine, 4246
Galena *see* 5303
Galenphol [Galen] *see* 7303
Galfer [Galen] *see* 3995
Galβ1→4Glcβ1→1cer *see* 2800
Galipine, 4247
Galla *see* 6651
Gallacetophenone, 4248
Gallaldehyde 3,5-Dimethyl Ether *see*
8996
Gallamine Triethiodide, 4249
(Gallato)hydroxyiodobismuth *see* 1281
Gallein, 4250
Gallepronin [Taiyo] *see* 4039
Gallic Acid, 4251
Gallic Acid Bismuth Basic Salt *see*
1297
Gallic Acid 5,6-Dihydroxy-3-carb-
oxyphenyl Ester *see* 3135
Gallic Acid 3-Monogallate *see* 3135
Gallic Acid Propyl Ester *see* 7872
Gallicin *see* 4251
Gallimycin [Abbott] *see* 3632
Gallium, 4252
Gallium Arsenide, 4253
Gallium Phosphide, 4254
Gallium Trifluoride, 4255
Gallobromol *see* 3006
Gallocatechin *see* 1908
Gallochrome *see* 5757
Gallocyanine, 4256
Gallodesoxycholic Acid *see* 2044
Gallopamil, 4257
Gallotannic Acid *see* 9023
Gallotannin *see* 9023
Gallotox *see* 7271
Galloxon [Burroughs Wellcome] *see*
4519
m-Galloylgallic Acid *see* 3135
1-Galloyl-α-D-glucose *see* 4347
1-Galloyl-β-D-glucose *see* 4348
(Galloyloxy)hydroxyiodobismuthine
see 1281
4-Galloylpyrogallol *see* 3868

Galls *see* 6651
GalN *see* 4240
Galoxone *see* 4519
Galphol [Galen] *see* 7303
Galpseud [Galen] *see* 3561
Galuteolin *see* 5483
Gamabufagin *see* 4258
Gamabufogenin *see* 4258
Gamabufotalin, 4258
Gamanil [E. Merck] *see* 5438
Gamaquil [Siegfried] *see* 7229
Gamarex [Causyth] *see* 441
Gamasol 90 *see* 3247
Gamastan [Cutter] *see* 4837
Gamatran *see* 377
Gambir, 4259
Gambir Catechu *see* 4259
Gamboge, 4260
Gambogic Acid, 4261
Gamefar *see* 6953
Gamene [Barnes-Hind] *see* 5379
Gamibetal [ISF] *see* 456
Gamimune [Cutter] *see* 4837
Gamiso [Texas Pharmacal] *see* 5379
Gamma Benzene Hexachloride *see*
5379
Gammabulin [Immuno] *see* 4837
Gammacorten [Ciba] *see* 2922
Gammagard [Travenol] *see* 4837
Gammagee [Merck & Co.] *see* 4837
Gamma Globulins *see* 4837
Gammagrippyl *see* 5872
Gamma Hexachlor *see* 5379
Gamma-interferon *see* 4894
Gammajust 50 [Horita] *see* 6841
Gammalin [ICI] *see* 5379
Gammalon [Daiichi] *see* 441
Gamma OH *see* 8603
Gamma-OZ [Kanebo] *see* 6841
Gamma-pipradrol *see* 909
Gammar [Armour] *see* 4837
Gammariza [Toyo] *see* 6841
Gammatsul [Nippon Chemiphar] *see*
6841
Gamma-vinyl GABA *see* 9882
Gammexane [ICI] *see* 5379
Gamolenic Acid *see* 5384
Gamonil [E. Merck] *see* 5438
Gamophen [Ethicon] *see* 4602
Ganal [Byk-Gulden] *see* 3920
Ganasag [Squibb] *see* 3256
Ganciclovir, 4262
Gangesol *see* 9537
Ganglefene, 4263
Gangleron [USSR] *see* 4263
Gangliosides, 4264
Gangliostat *see* 4609
Ganidan *see* 8879
Ganja *see* 1752
Ganlion *see* 913
Ganocide [ICI] *see* 3432
Ganor [Thomae] *see* 3881
Ganphen [Tutag] *see* 7797
Gansil *see* 2066
Gantanol [Roche] *see* 8889
Gantaprim [Ausonia] *see* 8889
Gantrim [Geymonat] *see* 8889
Gantrisin [Roche] *see* 8930
Gantrisin Acetyl [Roche] *see* 100
Gantrosan *see* 8930
GAP A *see* 4018
Garamycin [Schering] *see* 4284
Garantose *see* 8282
Garasin [Wakamoto] *see* 1971
Garasol [Schering] *see* 4284
Gardan [Hoechst] *see* 7884
Gardenal *see* 7201
Gardenal Sodium *see* 7202
Garden Chamomile *see* 2032
Gardenin A *see* 4265
Gardenin B *see* 4265

Gardenin C *see* 4265
Gardenin D *see* 4265
Gardenin E *see* 4265
Gardenins, 4265
Garden Lavender *see* 5261
Garden Rosemary *see* 8244
Garden Tox [Geigy] *see* 2978
Gardepanyl *see* 7201
Gardinol *see* 4266
Gardinol Type Detergents, 4266
Gardol®, 4267
Gardona [Shell] *see* 8777
Gardrin(e) [Syntex] *see* 3545
Garget *see* 7363
Gargilon *see* 2895 and 2896
Gargon [Squibb] *see* 9295
Garlic, 4268
Garlon [Dow] *see* 9572
Garmian [Fuso] *see* 964
Garnierite *see* 6406
Garosamine *see* 4284 and 8498
Garranil [Aristegui] *see* 1773
Garrathion [Stauffer] *see* 1827
Garryine, 4269
GAs *see* 4314
Gas Black *see* 1815
GASH *see* 4476
Gaslon [Nippon Shinyaku] *see* 4982
Gasoline, 4270
Gasstenon [Tatsumi] *see* 4039
Gaster [Yamanouchi] *see* 3881
Gasteril [Ripari-Gero] *see* 7463
Gastomax [Brocchieri] *see* 2075
Gastramine *see* 1205
Gastrax [Asche] *see* 6582
Gastrese [Robins] *see* 6063
Gastrhéma *see* 1903
Gastridan [Merck & Co.] *see* 3881
Gastridene [Bernabo] *see* 7785
Gastridin [Merck & Co.] *see* 3881
Gastrins, 4271
Gastripon *see* 9983
Gastrobid [Napp] *see* 6063
Gastrodiagnost [E. Merck] *see* 7067
Gastrofrenal [ISF-Italseber] *see* 2594
Gastrografin [Squibb; Schering AG]
see 5689
Gastromax [Farmitalia Carlo Erba]
see 6063
Gastromet [Sigurta] *see* 2279
Gastron [Winthrop] *see* 5869
Gastronerton [Dolorgiet] *see* 6063
Gastropen [CEPA] *see* 3881
Gastropidil *see* 5735
Gastropin *see* 9983
Gastrosedan *see* 5869
Gastrosil [Heumann] *see* 6063
Gastro-Tablinen [Sanorania] *see* 6063
Gastrotem [Temmler] *see* 6063
Gastro-Timelets [Temmler] *see* 6063
Gastrozepin [Thomae] *see* 7463
Gastrurol [Gibipharma] *see* 7475
Gastuloric *see* 4363
Gatalone [Barnes-Hind] *see* 4396
Gaultherin, 4272
Gaultherioside *see* 4272
GB *see* 8332
GB 1 *see* 1365
GB 2 *see* 1365
GB 94 *see* 6097
GBE *see* 4320
GC 1189 *see* 2081
GC 2466 *see* 6209
GC 3707 *see* 1333
GD *see* 8668
Gd-DTPA *see* 4236
GEA 6414 *see* 9436
Geangin [Gea] *see* 9851
Geapur [Gea] *see* 278
Geatrim-Boli [Gea] *see* 8874
Gebutox [Hoechst] *see* 3282

Gecolate [Summit] see 4465
Gedamycin Methyl Ester see 5733
Gefanil [Sumitomo] see 4273
Gefarnate, 4273
Gefarnil see 4273
Gefarnyl see 4273
Gefulcer [Ohta] see 4273
Gehlenite see 1645
Geigy 867 see 8907
Geigy Rodenticide Exp. 332 see 2556
Geissoschizoline, 4274
Geissospermine, 4275
Gelatin, 4276
Gelbin see 1660
Gelée Royale (French) see 8254
Gelfoam [Upjohn] see 4276
Gelocatil [Gelos] see 40
Gelomyrtol [Pohl] see 6253
Gelose see 172
Gelosedine [Bayer] see 7178
Gelovermin see 4633
Gelsemine, 4277
Gelseminic Acid see 8363
Gelsemium, 4278
Gelstaph [Beecham] see 2414
Gelumina see 345
Gelusil [Warner-Chilcott] see 352
Gelvatol [Shawinigan] see 7562
Gemalgene see 9552
Gemeprost, 4279
Gemfibrozil, 4280
Gemonil [Abbott] see 5873
Gemora see 1510
Genabol [Wyeth] see 6603
Genacort (Lotion) see 4710
Genatropine see 892
Gendon see 8131
Generlac [My K] see 5225
Génésérine 3 [Amido] see 3647
Genetron [Allied-Signal] see 2140
Genetron 12 [Allied-Signal] see 3053
Genicide see 9971
Geniphene see 9478
Genisis [Organon] see 2504
Genistein, 4281
Genisteine-Alkaloid, 4282
Genistein-4'-glucoside see 8675
Genistein-4'-glucosidorhamnoside see 8674
Genisteol see 4281
Genistin see 4281
Genite®, 4283
Genite 883 see 6856
Genitol 923 see 4283
Genkodein see 2461
Genocodeine see 2461
Géno-cristaux Gremy see 9109
Genogris [Vita] see 7459
Genol see 5940
Genomorphine see 6192
Genophyllin see 477
Genoptic [Allergan] see 4284
Genoscopolamine see 8362
Genostrychnine see 8824
Genotonorm [Kabi] see 8672
Genotropin [Sumitomo] see 8672
Gentacin [Schering/Shionogi] see 4284
Gentak [Akorn] see 4284
Gentalline [Unilabo] see 4284
Gentalyn [Essex] see 4284
Gentamicin, 4284
Gentamicin A see 4284
Gentamicin C_1 see 4284
Gentamicin C_{1a} see 4284
Gentamicin C_2 see 4284
Gentamicin C_{2b} see 4284
Gentamicin D see 4284
Gentamycin see 4284
Gentersal [Ortho] see 6596
Gentian, 4285

Gentianic Acid see 4291
Gentianin see 4291
Gentianine, 4286
Gentian Violet, 4287
Gentiaverm see 4287
Gentibioptal [Farmila] see 4284
Genticin [DDSA] see 4284
Gentiin see 4291
Gentinatre see 4290
Gentiobiose, 4288
Gentiopicrin, 4289
Gentiopicroside see 4289
Gentisic Acid, 4290
Gentisin, 4291
Gentisine U.C.B. see 4290
Gentisod see 4290
Gentisyl Alcohol, 4292
Gentocin [Schering] see 4284
Gentogram [Merck-Clevenot] see 4284
Gent-Ophtal [Winzer] see 4284
Gentran [Travenol] see 2925
Gentran 40 [Travenol] see 2925
Gentran 75 [Travenol] see 2925
Gentrogenin, 4293
Genurin [Recordati] see 4033
Geocillin [Roerig] see 1847
Geodin see 3588
Geoffroyine see 8988
Geomycin [Pliva] see 6931
Geon [Goodrich] see 7563
Geopen [Roerig] see 1796
Geosmin, 4294
Geotricyn see 7052
Gepefrine, 4295
Gephyrotoxin, 4296
Gepirone, 4297
Geranial see 2324
Geraniol, 4298
Geranium, 4299
Geranyl-1,4-benzenediol see 4300
Geranyl Farnesylacetate see 4273
Geranylgeranylacetone see 9083
Geranylhydroquinone, 4300
Geratacaca see 5597
Gerdaxyl [Gerda] see 5674
Geref [Serono] see 8412
Gerfil see 7558
Gerhardite see 2647
Germalgene see 9552
German Chamomile see 5639
Germane, 4301
German Fungus see 173
Germanic Acid see 4304
Germanin [Bayer] see 8986
Germanium, 4302
Germanium Dichloride, 4303
Germanium Dioxide, 4304
Germanium Hydride see 4301
Germanium Tetrachloride, 4305
Germanium Tetrafluoride, 4306
Germapect [Thiemann] see 1798
Germiciclin [Mendelejeff] see 5851
Germine, 4307
Germinol see 1066
Germitol see 1066
Gernebcin [Lilly] see 9413
Gerobit [Gerot] see 5859
Gerodyl see 7455
Gerolin [CT] see 2321
Gerontex H3 see 4498
Gerontine see 8699
Geroquinol see 4300
Gerostop [Merckle] see 6680
Gerot-Epilan [Gerot] see 5741
Gerovital see 4498
Gerovital H3 see 4498
Gesafloc see 9580
Gesafram [Ciba-Geigy] see 7799
Gesagard [Geigy] see 7800
Gesamil [Geigy] see 7822
Gesapax [Ciba-Geigy] see 402

Gesapon [Ciba-Geigy] see 2832
Gesaprim [Geigy] see 886
Gesarex [Ciba-Geigy] see 2832
Gesarol [Geigy] see 2832
Gesatop [Geigy] see 8485
Gestafortin [E. Merck] see 2102
Gestageno see 4773
Gestamestrol [Hermal] see 2102
Gestanin [Organon] see 289
Gestanol see 289
Gestanon [Organon] see 289
Gestanyn see 289
Gestapuran see 5677
Gestapuron see 5677
Gestasol Dry [Merrell-National] see 4534
Gestatron [AB Leo] see 3454
Gestodene, 4308
Gestone [Paines & Byrne] see 7783
Gestonorone Caproate, 4309
Gestoral see 3696
Gestormone see 7783
Gestovis see 7068
Gestrinone, 4310
Gestron see 7783
Gestronol Caproate see 4309
Gestyl see 7515
Gevatran [Anphar-Rolland] see 6268
Gevelina see 7826
Gevex see 9932
Gevilon [Warner-Lambert] see 4280
Gewacalm [Heilmittelwerke] see 2977
Gewalan [Lannacher] see 6430
Gewazol see 7097
GEWO 399 see 6999
Gexane [Pennwalt] see 5379
G-Farlutal [Farmitalia] see 5677
GGA see 9083
GH see 8672
GH 3 see 4498
Ghatti Gum, 4311
Ghee see 4312
Ghi, 4312
Ghimacef [Firma] see 1923
GH-RF see 8669
GH-RH see 8669
GH-RIF see 8671
Giant Yam see 10000
Giardil see 4210
Giarlam see 4210
Gibbane see 4314
Gibberellane see 4314
Gibberellic Acid, 4313
Gibberellin A_3 see 4313
Gibberellins, 4314
Gibberellin X see 4313
Gibbsite see 359
Gibbs Reagent, 4315
Gibicef [Gibipharma] see 1951
Gibrel [Merck & Co.] see 4313
Gichtex [Gerot] see 278
Gifblaar Poison see 4096
Giganten [Troponwerke] see 2308
Gigantine, 4316
Gilemal [Chinoin] see 4372
Gilsonite see 9754
Giltex see 8621
Gilucor "Nitro" see 6528
Gilurytmal [Giulini] see 184
Gilutensin [Giulini] see 3820
Gina see 7821
Ginamate see 806
Ginapect see 7821
Ginarsol see 44
Gindarine see 9147
Gineflavir [Crosara] see 6079
Ginestrene see 3689
Ginger, 4317
[6]-Gingerol, 4318
Gingicain M [Hoechst] see 9123
Gingilli Oil see 8418

Cross Index of Names

Gingko *see* 4319
Ginkgo, 4319
Ginkgo Biloba Extract, 4320
Ginkgoic Acid *see* 657
Ginkgolides *see* 4319
Ginkogink [Biogalenique] *see* 4320
Ginoden [Schering] *see* 4308
Ginorite *see* 1651
Ginseng, 4321
Ginsenosides *see* 4321
Giquel [Danal] *see* 7816
Giractide, 4322
Girard Reagent P *see* 4323
Girard Reagent T *see* 4323
Girard Reagents, 4323
Girard's Reagent D *see* 3234
Girard's Rubiazol *see* 8872
Gismondite *see* 1645
Gitaligin [Schering] *see* 4324
Gitalin, 4324
Gitaloxin *see* 4324
Githagenin *see* 4497
Gitogenin, 4325
Gitogenin β-Lycotetraoside *see* 4326
Gitonin *see* 4326
F-Gitonin, 4326
Gitoxigenin, 4327
Gitoxin, 4328
Gittalun [Thomae] *see* 3430
Give-Tan [Givaudan] *see* 2312
Giv-Gard DXN [Givaudan] *see* 3212
Gix *see* 2936
Gjellebaekite *see* 1709
GL-7 *see* 4395
GLA *see* 5384
Gla *see* 1834
Glacial Phosphoric Acid *see* 7319
Gladiolic Acid, 4329
Glafenine, 4330
Glamidolo [Angelini Francesco] *see* 2819
Glandin [TAD] *see* 7894
Glanduantin-Ch *see* 4534
Glandubolin *see* 3660
Glanil [Mekos] *see* 2308
Glaphenine *see* 4330
Glassy Sodium Metaphosphate *see* 8621
Glauber's Salt *see* 8636
Glaucarubin, 4331
Glaucine, 4332
Glaucon [Alcon] *see* 3569
Glauconex [Thilo] *see* 1032
Glaucotensil [Farmila] *see* 3711
Glauco-Viskin [Wander] *see* 7412
Glaudin [Sifi] *see* 8110
Glauline *see* 6059
Glaumeba [Merck & Co.] *see* 4331
Glaupax *see* 45
Glaurin *see* 3110
Glausyn [Benzon] *see* 6059
Glauvent *see* 4332
Glazidim [Glaxo] *see* 1944
Glean [Du Pont] *see* 2192
Glentonin-retard [Glenwood] *see* 5114
Gliadin, 4333
Glianimon [Troponwerke] *see* 1057
Glibenclamide *see* 4372
Glibenese [Roerig] *see* 4337
Gliben-Puren N *see* 4372
Glibornuride, 4334
Glicerinformal [Calipe] *see* 4380
Gliclazide, 4335
Glidiabet [Ferrer] *see* 4372
Glifan *see* 4330
Glifanan [Albert-Roussel] *see* 4330
Gliguanid *see* 5845
Glimicron [Dainippon] *see* 4335
Glimidstada *see* 4372
Glior *see* 3424

Gliorosein *see* 896
Gliotoxin, 4336
Glipasol *see* 4373
Glipizide, 4337
Gliporal [Grossmann] *see* 1465
Gliquidone, 4338
Glisoxepid, 4339
Gln (IUPAC Abbrev.) *see* 4365
Globacillin [Astra] *see* 925
Globaline *see* 9140
Globenicol *see* 2068
Globin, 4340
Globociclina [Importex] *see* 5851
Globucid *see* 8878
Globularicitrin *see* 8276
Globulin G₁ *see* 5514
γ-Globulins *see* 4837
Glofil-131 [Abbott] *see* 4952
Glomycin *see* 6931
Glonoin *see* 6528
Glosso-Sterandryl *see* 6044
Gloxazone *see* 5178
GLQ 223 *see* A10
Glu (IUPAC Abbrev) *see* 4363
Gluborid [Grünenthal] *see* 4334
Glucagon, 4341
Glucal *see* 1672
Glucametacin, 4342
Glucametacine *see* 4342
Glucamethacin *see* 4342
Glucamine, 4343
D-Glucamine *see* 4343
α-Glucan Branching Glycosyltransferase *see* 8035
α-1,6-Glucan 6-Glucanohydrolase *see* 2926
α-1,4-Glucan 4-Glucanohydrolases *see* 631
α-1,4-Glucan Maltohydrolases *see* 631
Glucantim [Farmitalia] *see* 5995
Glucantime [Specia] *see* 5995
D-Glucaric Acid, 4344
D-Glucaric Acid Calcium Salt *see* 1707
D-Glucaric Acid Di-γ-lactone Diacetate *see* 20
Glucaron [Chugai] *see* 20
Glucazide *see* 4396
Gluceptate *see* MISC-3 and 4
Gluceptate Calcium *see* 4349
Gluceptate Sodium *see* 4349
Glucid *see* 8282
Glucidoral [Servier] *see* 1839
Glucinan [Anphar] *see* 5845
Glucinium *see* 1177
D-Glucitol *see* 8680
Glucitol Iron Complex, Compd with Citric Acid *see* 4981
Glucoamylases *see* 631
D-Glucoascorbic Acid, 4345
Glucobay [Bayer] *see* 12
Glucobiogen *see* 1672
Glucochloral *see* 2062
α-D-Glucochloralose *see* 2062
Gluco-Ferrum [Van Pelt & Brown] *see* 3996
Glucofos *see* 4187
Glucofrangulin, 4346
Glucofrangulin A *see* 4346
Glucofren [Cophar] *see* 1839
D-Glucofuranurono-6,3-lactone *see* 4362
Glucogallic Acid *see* 4348
α-Glucogallin, 4347
β-Glucogallin, 4348
Glucoheptonic Acid, 4349
α-Glucoheptonic Acid *see* 4349
Glucolin *see* 4353
Glucomag [Amfre-Grant] *see* 4350
D-Glucomethylose *see* 8106

Glucomonocarbonic Acid *see* 4349
Gluconiazide *see* 4396
Gluconic Acid, 4350
D-Gluconic Acid *see* 4350
Gluconic Acid Antimony Sodium Derivative *see* 733
D-Gluconic Acid Calcium Salt (2:1) *see* 1672
D-Gluconic Acid Cyclic 4,5-Ester with Boric Acid Calcium Salt (2:1) *see* 1652
D-Gluconic Acid 4-*O*-β-D-Galactopyranosyl Calcium Salt (2:1) Compd with Calcium Bromide (CaBr₂) (1:1) *see* 1654
D-Gluconic Acid δ-Lactone *see* 4351
Gluconic Acid Potassium Salt *see* 7615
Gluconic Acid Quinidine Salt *see* 8072
Gluconic Acid Quinine Salt *see* 8082
Gluconic Acid Sodium Salt *see* 8569
Glucono Delta Lactone *see* 4351
Gluconolactone, 4351
Gluconsan K [Kayaku] *see* 7615
Glucoperiplocymarin *see* 7126
Glucophage [Homburg] *see* 5845
Glucopostin [Boehringer, Mann.] *see* 7191
Glucoproscillaridin A *see* 8354
β-D-Glucopyranose Aerodehydrogenase *see* 4354
β-D-Glucopyranose 6-Benzoate *see* 9811
α-D-Glucopyranose-1-gallate *see* 4347
β-D-Glucopyranose-1-gallate *see* 4348
α-D-Glucopyranose-1-phosphate *see* 4356
α-D-Glucopyranose 1-(3,4,5-Trihydroxybenzoate) *see* 4347
β-D-Glucopyranose 1-(3,4,5-Trihydroxybenzoate) *see* 4348
4-(β-D-Glucopyranosylamino)benzenesulfonamide *see* 4358
α-D-Glucopyranosyl Bromide 2,3,4,6-Tetraacetate *see* 52
2-*O*-β-D-Glucopyranosylcucurbitacin E *see* 2483
(3β,5β,12β)-3-[(*O*-β-D-Glucopyranosyl-(1→4)-*O*-2,6-dideoxy-β-D-*ribo*-hexopyranosyl-(1→4)-*O*-2,6-dideoxy-β-D-*ribo*-hexopyranosyl-(1→4)-2,6-dideoxy-β-D-*ribo*-hexopyranosyl)oxy]-12,14-dihydroxycard-20(22)-enolide *see* 2903
7-α-D-Glucopyranosyl-9,10-dihydro-3,5,6,8-tetrahydroxy-1-methyl-9,10-dioxo-2-anthracenecarboxylic Acid *see* 1850
8-β-D-Glucopyranosyl-5,7-dihydroxy-2-(4-hydroxy-3-methoxyphenyl)-4*H*-1-benzopyran-4-one *see* 8357
10-Glucopyranosyl-1,8-dihydroxy-3-(hydroxymethyl)-9(10*H*)-anthracenone *see* 304
α-D-Glucopyranosyl-β-D-fructofuranoside *see* 8855
O-α-D-Glucopyranosyl-(1→3)-β-D-fructofuranosyl-α-D-glucopyranoside *see* 9496
3-*O*-α-D-Glucopyranosyl-D-fructose *see* 9728
α-D-Glucopyranosyl-α-D-glucopyranoside *see* 9496
3-[(*O*-β-D-Glucopyranosyl-(1→6)-*O*-D-glucopyranosyl-(1→4)-6-deoxy-3-*O*-methyl-α-L-glucopyranosyl)oxy]-14-hydroxycard-20(22)-enolide *see* 1983

3-[(O-β-D-Glucopyranosyl-(1 → 6)-O-
D-glucopyranosyl-(1 → 4)-6-deoxy-
3-O-methyl-α-L-glucopyranosyl)-
oxy]-14-hydroxy-19-oxocard-
20(22)-enolide see 9215
[(6-O-β-D-Glucopyranosyl-β-D-gluco-
pyranosyl)oxy]benzeneacetonitrile
see 629
3β-[6-O-β-D-Glucopyranosyl-β-D-
glucopyranosyl)oxy]-14-hydroxy-
5α-card-20(22)-enolide see 9809
6-β-D-Glucopyranosyl-7-(β-D-glucopyranosyloxy)-5-hydroxy-2-(4-
hydroxyphenyl)-4H-1-benzopyran-
4-one see 8327
13-[(2-O-β-D-Glucopyranosyl-α-D-
glucopyranosyl)oxy]kaur-16-en-18-
oic Acid β-D-Glucopyranosyl Ester
see 8766
6-β-D-Glucopyranosyl-2-[[3-β-D-
glucopyranosyl-2,3,4-trihydroxy-5-
[3-(4-hydroxyphenyl)-1-oxo-2-pro-
penyl]-6-oxo-1,4-cyclohexadien-1-
yl]methylene]-5,6-dihydroxy-4-[3-
(4-hydroxyphenyl)-1-oxo-2-propen-
yl]-4-cyclohexene-1,3-dione see
1876
2-O-β-D-Glucopyranosyl-α-D-glucose
see 8676
4-O-α-D-Glucopyranosyl-D-glucose
see 5595
4-O-β-D-Glucopyranosyl-D-glucose
see 1957
6-O-β-D-Glucopyranosyl-D-glucose
see 4288
4'-(β-D-Glucopyranosyloxy)aceto-
phenone see 7367
2-(β-D-Glucopyranosyloxy)benzalde-
hyde see 4544
α-(β-D-Glucopyranosyloxy)benzene-
acetonitrile see 5602
7-(β-D-Glucopyranosyloxy)-2H-1-
benzopyran-2-one see 8506
1-(β-D-Glucopyranosyloxy)-4a,7a-
dihydro-4'-(1-hydroxyethyl)-5'-
oxospiro[cyclopenta[c]pyran-7(1H),-
2'(5'H)-furan]-4-carboxylic Acid
Methyl Ester see 7513
3-(β-D-Glucopyranosyloxy)-5,7-di-
hydroxy-2-(4-hydroxyphenyl)-1-
benzopyrylium Chloride see 7014
1-[2-(β-D-Glucopyranosyloxy)-4,6-
dihydroxyphenyl]-3-(4-hydroxy-
phenyl)-1-propanone see 7300
1-(β-D-Glucopyranosyloxy)-1,4a,5,6,-
7,7a-hexahydro-6-hydroxy-7-meth-
ylcyclopenta[c]pyran-4-carboxylic
Acid Methyl Ester see 5441
1-(β-D-Glucopyranosyloxy)-1,4a,5,6,-
7,7a-hexahydro-7-methyl-5-oxo-
cyclopenta[c]pyran-4-carboxylic
Acid Methyl Ester see 9861
(S)-α-(β-D-Glucopyranosyloxy)-4-
hydroxybenzeneacetonitrile see
2937
6-(β-D-Glucopyranosyloxy)-7-hydr-
oxy-2H-1-benzopyran-2-one see
3646
7-(β-D-Glucopyranosyloxy)-6-hydr-
oxy-2H-1-benzopyran-2-one see
2267
7-β-D-Glucopyranosyloxy-8-hydroxy-
2H-1-benzopyran-2-one see 2817
7-(β-D-Glucopyranosyloxy)-5-hydr-
oxy-2-(3-hydroxy-4-methoxyphen-
yl)-3,6-dimethoxy-4H-1-benzo-
pyran-4-one see 1964
7-(β-D-Glucopyranosyloxy)-5-hydr-
oxy-2-(4-hydroxyphenyl)-4H-1-
benzopyran-4-one see 764

β-D-Glucopyranosyloxy-L-p-hydroxymandelonitrile see 2937
8-(β-D-Glucopyranosyloxy)-7-hydr-
oxy-6-methoxy-2H-1-benzopyran-
2-one see 4179
4-(β-D-Glucopyranosyloxy)-3-meth-
oxybenzaldehyde see 4359
7-(β-D-Glucopyranosyloxy)-6-meth-
oxy-2H-1-benzopyran-2-one see
8364
2-[1-(β-D-Glucopyranosyloxy)-1-
methylethyl]-2,3-dihydro-7H-furo-
[3,2-g][1]benzopyran-7-one see
6589
7-[(β-D-Glucopyranosyloxy)methyl]-
4-methoxy-5H-furo[3,2-g][1]benzo-
pyran-5-one see 5190
2-(β-D-Glucopyranosyloxy)-2-meth-
ylpropanenitrile see 5375
3-[4-(β-D-Glucopyranosyloxy)phen-
yl]-5,7-dihydroxy-4H-1-benzo-
pyran-4-one see 8675
1-[4-(β-D-Glucopyranosyloxy)phen-
yl]ethanone see 7367
3-[2-(β-D-Glucopyranosyloxy)phen-
yl]-2-propenoic Acid see 5701
1-(β-D-Glucopyranosyloxy)-1,4a,7,7a-
tetrahydro-7-hydroxy-7-(hydroxymethyl)cyclopenta[c]pyran-4-carb-
oxylic Acid see 6166
4-(β-D-Glucopyranosyloxy)-2,6,6-
trimethyl-1-cyclohexene-1-carbox-
aldehyde see 7381
4-O-β-D-Glucopyranosyl-L-rhamnose
see 8353
3-O-(β-D-Glucopyranosyluronic
Acid)-2-amino-2-deoxy-D-glucose
see 4674
β-D-Glucopyranosyluronic Acid 2-
Deoxy-2-amino-D-galactose see
2218
Glucoremed [Econerica] see 4372
D-Glucosaccharic Acid see 4344
Glucosaccharonic Acid see 5009
Glucosamine, 4352
Glucosan see 4349
Glucosan Transglycosylase see 8035
Glucoscan [Du Pont] see 4349
Glucose, 4353
D-Glucose see 4353
D-Glucose 6-(Dihydrogen phosphate)
see 4355
Glucosemonocarboxylic Acid see 4349
Glucose Oxidase, 4354
Glucose-6-phosphate, 4355
α-Glucose-1-phosphate, 4356
Glucose-6-phosphoric Acid see 4355
α-Glucose-1-phosphoric Acid see
4356
α-Glucosidases see 4899
N-D-Glucoside of N-Cinnamoyl-N-
methylhistamine see 1893
3-(α-D-Glucosido)-D-fructose see
9728
4-(α-D-Glucosido)-D-glucose see 5595
4-(β-D-Glucosido)-D-glucose see 1957
6-(β-D-Glucosido)-D-glucose see 4288
(α-D-Glucosido)-α-D-glucoside see
9496
7-Glucosido-8-hydroxycoumarin see
2817
3-(β-Glucosido)indole see 4854
α-n-Glucosidoinvertases see 4899
Glucosidorhamnose see 8353
N⁴-β-d-Glucosidosulfanilamide see
4358
Glucosulfone Sodium, 4357
β-D-Glucosyloxyazoxymethane see
2705
7-(Glucosyloxy)coumarin see 8506

7-(Glucosyloxy)-3',5-dihydroxy-
4',5',6-trimethoxyisoflavone see
4973
8-(Glucosyloxy)-6-methoxyumbelli-
ferone see 4179
3-(Glucosyloxy)-4',5,7-trihydroxy-
3',5'-dimethoxyflavylium Chloride
see 5596
N⁴-β-D-Glucosylsulfanilamide, 4358
(1-D-Glucosylthio)gold see 901
Gluco-Tablinen [Beiersdorf] see 4372
Glucotard [M.C.P.] see 4486
α-D-Glucothiopyranose see 9263
Glucotrol [Pfizer] see 4337
Glucovanillin, 4359
Glucoxy see 4362
Glucurolactone see 4362
Glucurone see 4362
D-Glucuronic Acid, 4360
D-Glucuronic Acid γ-Lactone see
4362
D-Glucuronic Acid γ-Lactone 1-[(4-
Pyridinylcarboxyl)hydrazone] see
4396
β-Glucuronidase, 4361
D-Glucuronolactone, 4362
D-Glucuronolactone Isonicotinoylhy-
drazone see 4396
Gludiase [Roger Bellon] see 4374
Glufosinate see 7314
Glufosinate-ammonium see 7314
Glukagon see 4341
Glukor Injection [Hyrex] see 4534
Glumal [Kyowa] see 21
Glumamycin see 618
Glumin see 4365
Glumorin see 5159
Glupax [Erco] see 45
Gluquinate see 8072
Glurenorm [Thomae] see 4338
Gluronazide see 4396
Gluside see 8282
Glusulase [Endo] see 4361
Glutacid see 4363
Glutacyl see 6165
Glutamic Acid, 4363
Glutamic Acid 1-Amide see 5061
Glutamic Acid 5-Amide see 4365
Glutamic Acid Compd with L-Arginine
see 806
L-Glutamic Acid 5-Ethyl Ester, 4364
L-Glutamic Acid γ-Ethyl Ester see
4364
L-Glutamic Acid 5-[2-[4-(Hydroxy-
methyl)phenyl]hydrazide] see 175
Glutamic Acid Lactam see 8012
L-Glutamic Acid L-Lysine Salt see
5511
Glutamic Acid Monosodium Salt
Monohydrate see 6165
Glutamicine see 4029
Glutamidin see 4363
Glutamine, 4365
2-L-Glutamine-6-L-asparaginealytesin
see 1332
Glutaminic Acid see 4363
Glutaminol see 4363
Glutamycin see 4029
3-γ-Glutamyl-D-alanylamino-3-
methoxyazetidin-2-one-1-sulfonic
Acid see 8924
(3R)-3-(γ-D-Glutamyl-D-alanylamino)-3-methoxy-2-oxoazetidine-1-
sulfonic Acid see 8924
γ-L-Glutamyl-α-amino-β-(2-methyl-
enecyclopropyl)propionic Acid Dipeptide see 4802
1-[N-(γ-L-Glutamyl)amino]-D-proline
see 5377
N-(N-L-γ-Glutamyl-L-cysteinyl)gly-
cine see 4369

β-N-[γ-L(+)-Glutamyl]-4-hydroxy-
  methylphenylhydrazine *see* 175
γ-L-**Glutamylhypoglycine** *see* 4802
D-γ-**Glutamyl-N-(3-methoxy-2-oxo-
  1-sulfo-3-azetidinyl)-D-alaninamide**
  see 8924
N-L-γ-**Glutamyl-3-(2-methylenecy-
  clopropyl)alanine** *see* 4802
Glutan-HCl [Lederle] *see* 4363
Glutaral *see* 4366
Glutaraldehyde, 4366
Glutargin *see* 806
Glutaric Acid, 4367
Glutaric Acid Dinitrile *see* 4368
Glutaric Dialdehyde *see* 4366
Glutarol [Dermal] *see* 4366
Glutaronitrile, 4368
ε-N-(L-**Glutar-2-yl)-L-lysine** *see* 8283
Glutasin [McNeil] *see* 4363
Glutathin [Mochida] *see* 4369
Glutathiol *see* 4369
Glutathion [Nichiiko] *see* 4369
Glutathione, 4369
L-**Glutathione** *see* 4369
Glutathione-SH *see* 4369
Glutaton *see* 4363
Glutavene [Cooper] *see* 6165
Gluten, 4370
Glutestere *see* 4364
Glutethimide, 4371
Glutimic Acid *see* 8012
Glutiminic Acid *see* 8012
Glutinal [Sankyo] *see* 4369
Glutril [Roche] *see* 4334
Gly (IUPAC Abbrev.) *see* 4386
[**Gly**⁷, Asu¹,⁶]oxytocin *see* 1846
Glybenzcyclamide *see* 4372
Glybrom [CooperVision] *see* 7996
Glyburide, 4372
Glybuthiazol(e), 4373
Glybuthizol *see* 4373
Glybuzole, 4374
Glycamine *see* 4343
Glycarsamide, 4375
Glyceraldehyde, 4376
Glyceraldehyde 3-Phosphate, 4377
Glyceric Acid, 4378
Glyceric Aldehyde *see* 4376
Glycerin *see* 4379
Glycerine *see* 4379
D-**Glycero-D-*gulo*-heptonic Acid
  Compd with Erythromycin (1:1)** *see*
  3629
Glycerol, 4379
Glycerol Diacetate *see* 2941
Glycerol α,γ-**Dichlorohydrin** *see* 3063
sym-**Glycerol Dichlorohydrin** *see* 3063
Glycerol Dimethylketal *see* 5101
Glycerol Formal, 4380
Glycerol Guaiacolate *see* 4465
Glycerol α-(2-Methoxyphenyl) Ether
  see 4465
Glycerol α-Monochlorohydrin *see*
  2145
**Glycerol Mono(2-methoxyphenyl)
  Ether** *see* 4465
Glycerol Nitric Acid Triester *see* 6528
Glycerol Tribromohydrin *see* 9527
Glycerophosphoric Acid, 4381
α-**Glycerophosphoric Acid** *see* 4381
β-**Glycerophosphoric Acid** *see* 4381
α-**Glycerophosphoric Acid Calcium
  Salt** *see* 1673
β-**Glycerophosphoric Acid Calcium
  Salt** *see* 1673
Glycerose *see* 4376
Glyceryl *p*-**Aminobenzoate,** 4382
Glycerylaminophenaquine *see* 4330
**Glyceryl 2-(*p*-Chlorophenoxyisobutyr-
  ate-1,3-dinicotinate** *see* 1239
Glyceryl Diacetate *see* 2941

α-**Glyceryl Guaiacol Ether** *see* 4465
Glyceryl Guaiacyl Ether *see* 4465
Glyceryl Iodide, 4383
Glyceryl Monoacetate *see* 6152
Glyceryl Monocaprylate *see* 6162
DL-**Glyceryl-1-mono-octanoate** *see*
  6162
Glyceryl Monostearate, 4384
Glyceryl *o*-**Tolyl Ether** *see* 5737
Glyceryl Triacetate *see* 9504
Glyceryl Tributyrate *see* 9532
Glyceryl Trimyristate *see* 9638
Glyceryl Trinitrate *see* 6528
Glyceryl Trioleate *see* 9644
Glyceryl Tripalmitate *see* 9648
Glyceryl Tripetroselinate *see* 7142
Glyceryl Tristearate *see* 9669
Glycidol, 4385
Glycin *see* 4771
(**Glycinato-N,O)dihydroxyaluminum**
  see 3168
Glycine, 4386
Glycine, Aluminum Salt *see* 3168
**7-Glycine-1,6-aminosuberic Acid-
  oxytocin** *see* 1846
Glycine Anhydride *see* 7434
1-Glycine-18-L-argininamide-α⁽¹⁻¹⁸⁾-
  corticotropin *see* 4322
Glycine Betaine *see* 1201
Glycine Copper Complex *see* 2644
Glycine-ferrous Sulfate Complex *see*
  3987
Glycine Nitrile *see* 423
Glycine *p*-**Phenetidide** *see* 7204
Glycine Sulfate, 4387
Glycinin, 4388
Glycinonitrile *see* 423
Glycobiarsol, 4389
Glycocholic Acid, 4390
Glycocoll *see* 4386
Glycocoll Betaine *see* 1201
Glycocoll-copper *see* 2644
Glycocoll-*p*-phenetidide *see* 7204
Glycocyamine, 4391
Glycodex [Burns-Bio] *see* 4465
Glycodiazine *see* 4403
Glycodine *see* 7303
Glycogen, 4392
Glycogenic Acid *see* 4350
Glycolande [Delalande] *see* 4372
Glycol Benzylcarbamate *see* 1490
Glycol Bromohydrin *see* 3749
Glycol Chlorohydrin *see* 3750
Glycol Cyanohydrin *see* 3751
Glycol Diacetate *see* 3756
Glycol Dilaurate, 4393
Glycoleucine *see* 6624
Glycolic Acid, 4394
**Glycolic Acid 8-Ester with Octahy-
  dro-5,8-dihydroxy-4,6,9,10-tetra-
  methyl-6-vinyl-3a,9-propano-3a*H*-
  cyclopentacycloocten-1(4*H*)-one** *see*
  7508
Glycol-monoacetin *see* 3757
N-**Glycoloylarsanilic Acid** *see* 4375
N-**Glycoloylarsanilic Acid Bismuth
  Deriv** *see* 4389
**3-Glycoloyl-1,2,3,4,6,11-hexahydro-
  3,5,12-trihydroxy-10-methoxy-
  6,11-dioxo-1-naphthacenyl-3-ami-
  no-2,3,6-tridioxy-**α-L-**arabino-
  hexopyranoside** *see* 3573
Glycol Salicylate, 4395
Glycoluril *see* 87
p-**Glycolylaminobenzenearsonic Acid**
  see 4375
p-**Glycolylaminophenylarsinic Acid**
  see 4375
N-**Glycolylarsanilic Acid** *see* 4375
Glycolylurea *see* 4678
Glyconiazide, 4396

Glyconic Acid *see* 4350
Glyconon [DDSA] *see* 9432
Glyconormal *see* 4403
Glycophene *see* 4964
Glycoprotein G *see* 9322
Glycoprotein P *see* A6
Glycopyrrolate, 4397
Glycopyrronium Bromide *see* 4397
Glycosine, 4398
Glycosthène [AB Leo] *see* 4386
7-D-Glycosylapigenin *see* 764
**8-Glycosyl-4',5,7-trihydroxy-3'-
  methoxyflavone** *see* 8357
Glycotauro [HW & D] *see* 6884
Glycyclamide *see* 9434
4-Glycylaminophenetol *see* 7204
4-Glycylaminophenol Ethyl Ether *see*
  7204
N-**Glycylglycine,** 4399
Glycylglycine Lactam *see* 7434
N-[N-(N-**Glycylglycyl)glycyl]-8-**L-
  lysinevasopressin *see* 9097
Nα-**Glycylglycylglycylvasopressin** *see*
  9097
Glycylpressin [Ferring] *see* 9097
Glycyrrhetic Acid *see* 3543
Glycyrrhetic Acid Acetate *see* 69
Glycyrrhetic Acid Hydrogen Succinate
  see 1797
**18β-Glycyrrhetic Acid Hydrogen
  Succinate** *see* 1797
18β-Glycyrrhetinic Acid *see* 3543
Glycyrrhetinic Acid Glycoside *see*
  4401
Glycyrrhiza, 4400
Glycyrrhizic Acid, 4401
Glycyrrhizin *see* 4401
Glycyrrhizinic Acid *see* 4401
Glydiazinamide *see* 4337
Glyferro [Hek] *see* 3987
Glyfyllin *see* 3459
Glyhexamide, 4402
Glykin *see* 9694
Glykocellon *see* 1835
Glykresin *see* 5737
Glyme *see* 3213
Glymidine, 4403
Glymol *see* 7139
Glyodin, 4404
Glyotol *see* 5737
Glyoxal, 4405
**Glyoxal Compound with Sodium Bi-
  sulfite** *see* 4406
Glyoxaldiureine *see* 87
Glyoxalic Acid *see* 4407
Glyoxaline *see* 4828
Glyoxaline-5-alanine *see* 4642
Glyoxal-Sodium Bisulfite, 4406
Glyoxyldiureide *see* 246
Glyoxylic Acid, 4407
Glypesin [Uzara-Werk] *see* 4624
Glyphen *see* 7191
Glyphenarsine *see* 9703
Glyphosate, 4408
Glyphosine, 4409
Glyphylline *see* 3459
Glypinamide, 4410
Glypressin [Ferring] *see* 9097
Glysal *see* 4395
Glysennid [Sandoz] *see* 8407
Glytheonate [SM & P] *see* 9212
Glyvenol [Ciba] *see* 9520
GMP *see* 4484
3',5'-GMP *see* 2715
Gnoscopine *see* 6638
Go-560 *see* 3889
Gö 687 *see* 3845
Gö 919 *see* 7457
Go 1261 C *see* 9370
Goal [Rohm & Haas] *see* 6916
Goa Powder *see* 795

GOBAB see 455
Godalax [Pfleger] see 1253
Gödecke 3282 see 3845
Goethite see 3969
Goetsch's Vitamin see 9941
Goitrin, 4411
Gokilaht [Sumitomo] see 2776
Golarsyl see 44
Gold, 4412
Gold-bloom see 1723
Gold Colloid [198]Au see 4419
Gold Cyanide see 4415
Golden Antimony Sulfide see 729
Golden Apple Seed see 8064
Golden Ragwort see 8401
Golden Seal see 4690
Gold, Explosive, 4413
Goldinodox see 900
Goldinomycin see 900
Gold Monochloride, 4414
Gold Monocyanide, 4415
Gold Monoiodide, 4416
Gold Monosulfide, 4417
Gold Monoxide, 4418
Gold Orange see 6019
Gold Oxide see 4431
Gold Potassium Bromide see 7673
Gold Potassium Chloride see 7674
Gold Potassium Cyanide see 7609
Gold Potassium Iodide see 7681
Gold, Radioactive, Colloidal, 4419
Gold Selenate, 4420
Gold Selenide, 4421
Gold Sesquioxide see 4431
Gold Sodium Chloride see 8644
Gold Sodium Cyanide see 8557
Gold Sodium Thiomalate, 4422
Gold Sodium Thiosulfate, 4423
Gold Stannate, 4424
Gold Thioglucose see 901
Goldthread see 2521
Gold-tin Precipitate see 4424
Gold-tin Purple see 4424
Gold Tribromide, 4425
Gold Tribromide, Acid, 4426
Gold Trichloride, 4427
Gold Trichloride, Acid, 4428
Gold Tricyanide, 4429
Gold Trihydroxide, 4430
Gold Trioxide, 4431
Gold Trisulfide, 4432
Gold Yellow see 9687
Gonacrine see 118
Gonadoliberin see 5354
Gonadorelin see 5354
Gonadorelin Acetate see 5354
Gonadotraphon F.S.H. see 7515
Gonadotraphon L.H. see 4534
Gonadotropin-releasing Factor see 5354
Gonadyl see 7515
Gonadyl-Chorionic see 4534
Gonak [Akorn] see 4777
Gonan see 4534
Gondafon [Schering] see 4403
Goniosol [CooperVision] see 4777
Gonorcin [Farmas] see 6617
Gonosan see 5167
Gonyaulax Toxin see 8344
"Good" Buffers see 4573
Good-Rite Nix see 8585
Gooroo Nuts see 5198
Gormon see 7515
Goserelin, 4433
Gossypine see 2207
Gossypitrin see 8045
Gossyplure, 4434
Gossypol, 4435
Gossypose see 8120
Gotensin [Warner-Lambert] see 5343
Gougerotin, 4436

Gourd Curare see 2678
Government Rubber Styrene see 8345
Goyl see 44
GP 45840 see 3071
G-Proteins, 4437
GR/1214 see 2362
GR 2/234 see 226
GR 2/925 see 2361
GR 20263 see 1944
GR 2/1574 see 224
GR 38032F see 6802
GR 43175 see 8979
GR 43175C see 8979
Graafina see 3655
Gradient [Polifarma] see 4070
Gradocycline see 8408
Grafestrol see 3118
Graham's Salt see 8621
Grains of Paradise see 1841
Gramalil [Fujisawa] see 9353
Gramaxin [Boehringer, Mann.] see 1925
Gramicidin(s), 4438
Gramicidin C (Soviet) see 4439
Gramicidin D (Dubos) see 4438
Gramicidin S, 4439
Gramicidin S (Soviet) see 4439
Gramine, 4440
Graminis see 9673
Graminon [Ciba-Geigy] see 5106
Gramipan [Mayoly-Spindler] see 2896
Gram-Micina [Lagap] see 4169
Grammite see 1709
Gramoderm [Schering] see 4438
Gramoxone [ICI] see 6980
Grampenil [Argentia] see 621
Gramplus [Chiesi] see 2377
Granaticin, 4441
Granaticin B see 4441
Granatum see 7564
Grandaxin [EGYT] see 9427
Grandisol, 4442
Grandlure see 4442
Granisetron, 4443
Granosan see 3780
Granudoxy [Leurquin] see 3429
Granular see 5864
Granulestin see 5311
Grape Sugar see 4353
Graphite, 4444
Graphite Oxide see 4445
Graphitic Acid, 4445
Graphitic Oxide see 4445
Graphol see 5940
Graslan [Elanco] see 9053
Grasp [ICI] see 9484
Gratibain see 6854
Gratiogenin, 4446
Gratioside, 4447
Gratus Strophanthin see 6854
Gravimun [Sachsisches Serumwerk] see 4534
Gravitol(e), 4448
Gravocain [Winthrop] see 5359
Gravol [Horner] see 3193
"Gray Acetate of Lime" see 1643
Grayanotoxane-3,5,6,10,14,16-hexol see 4449
Grayanotoxane-3,5,6,10,14,16-hexol 14-Acetate see 4449
Grayanotox-10(20)-ene-3,5,6,14,16-pentol see 4449
Grayanotoxin I see 4449
Grayanotoxin II see 4449
Grayanotoxin III see 4449
Grayanotoxins, 4449
GR-C507/75 see 6802
Greek Hay see 3950
Green Broom see 8358
Green Cinnabar see 2229
Greenhartin see 5235

Greenheart see 1025
Green Hellebore see 9858
Green Hydroquinone see 8070
"Green Nickel Oxide" see 6417
Greenockite see 1628
Green Oxide of Chromium see 2229
Green Rouge see 2229
Green Verdigris see 2628
Green Vitriol see 4006
Grenade [ICI] see 2765
Grey Arsenic see 820
GRF see 8669
GRF(1-29)NH$_2$ see 8412
Griffith's Zinc White see 5424
Grifomin see 477
Grifulvin [McNeil] see 4453
Grilon [Fibron] see 6656
Grindelia, 4450
Grindelic Acid, 4451
Grinsil [Argentia] see 610
Gripenin-O [Fujisawa] see 1844
Grisactin [Ayerst] see 4453
Griséfuline [Sanofi] see 4453
Grisein, 4452
Griseofulvin, 4453
Griseoviridin, 4454
Grisovin [Glaxo] see 4453
Gris-PEG [Dorsey] see 4453
GR-M see 6377
Grocreme [Grossmann] see 2896
Grodurex [Grossmann] see 417
Grorm [Serono] see 8672
Grossularite see 1645
Ground Apple see 2032
Ground Holly see 2046
Ground Lily see 9609
Groundnut see 7005
Groundnut Oil see 7006
Grovex [Groves] see 7446
Growth Hormone see 8672
Growth Hormone-release Inhibiting Factor see 8671
Growth Hormone-releasing Factor see 8669
Growth Hormone-releasing Hormone see 8669
Grozyme see 632
GR-S see 8345
Grysio [Ayerst] see 4453
GS-385 see 8852
GS-1339 see 3455
GS-2989 see 5658
GS-3065 see 3429
GS-6244 see 1782
GS-13005 see 5891
GS-13332 see 3253
GS-16068 see 3349
GS-19851 see 1422
GS-23654 see 6555
G Salt see 6308
GSH see 4369
GSSG see 4369
G-strophanthidin see 6853
G-strophanthin see 6854
GT-41 see 1494
GT-92 see 2725
GT-1012 see 7704
GTN [Farillon] see 6528
GTP Binding Proteins see 4437
G-Tril see 3889
Guacetisal see 873
Guaiac, 4455
Guaiac Alcohol see 4466
Guaiac-Copper Sulfate Paper, 4456
Guaiacol, 4457
Guaiacol Benzoate, 4458
Guaiacol Carbonate, 4459
Guaiacol Carbonic Acid Neutral Ester see 4459
Guaiacol Glyceryl Ether see 4465

Guaiacol Glyceryl Ether Carbamate see 5903
Guaiacol Phosphate, 4460
Guaiacol Valerate, 4461
Guaiactamine, 4462
Guaiacum see 4455
Guaiacuran see 4465
Guaiacyl Glyceryl Ether see 4465
Guaiamar see 4465
Guaiapate, 4463
Guaiaspir [Lampugnani] see 873
Guaiazulene, 4464
S-Guaiazulene see 4464
Guaiazulene Soluble see 4464
Guaifenesin, 4465
Guaiol, 4466
Guaiphenesin see 4465
Guaithylline, 4467
Guajol see 4466
Guamecycline, 4468
Guamide see 8879
Guanabenz, 4469
Guanacline, 4470
Guanadrel, 4471
Guanamprazine see 417
Guanatol [Lilly] see 2088
Guanazodine, 4472
Guanazolo see 912
Guaneran [Burroughs Wellcome] see 9228
Guanethidine, 4473
Guanfacine, 4474
Guanicil [Cilag] see 8879
Guanidine, 4475
Guanidineacetic Acid see 4391
Guanidine, Compd with Aluminum Sulfate see 4476
Guanidinium Aluminum Sulfate Hexahydrate, 4476
4-Guanidino-1-butanol Syringate see 5323
N-(2-Guanidinoethyl)-4-methyl-Δ³-piperidine see 4470
N-(2-Guanidinoethyl)-4-methyl-1,2,3,6-tetrahydropyridine see 4470
1-(2-Guanidinoethyl)octahydroazocine see 4473
1-(2-Guanidinoethyl)-1,2,3,6-tetrahydro-4-picoline see 4470
2-Guanidinomethyl-1,4-benzodioxan see 4482
α-Guanidinomethylheptamethylenimine see 4472
Guanidoacetic Acid see 4391
Guanine, 4477
Guanine Nucleotide Binding Proteins see 4437
Guanine Riboside see 4480
Guanine Riboside-3-phosphoric Acid see 4483
Guanine Riboside-5-phosphoric Acid see 4484
Guano, 4478
Guanochlor, 4479
Guanosine, 4480
Guanosine Cyclic 3′,5′-(Hydrogen Phosphate) see 2715
Guanosine 3′,5′-Cyclic Monophosphate see 2715
Guanosine 3′,5′-Cyclic Phosphate see 2715
Guanosine 3′-Monophosphate see 4483
Guanosine 3′,5′-Monophosphate see 2715
Guanosine 5′-Monophosphate see 4484
Guanosine 5′-Phosphate see 4484
Guanoxabenz, 4481
Guanoxan, 4482
Guantal see 3325

Guanylate Cyclase see 2715
9-(p-Guanylbenzal)fluorene see 6978
9-(p-Guanylbenzylidene)fluorene see 6978
Guanylguanidine see 1233
Guanylhydrazine see 453
Guanylic Acid b see 4483
3′-Guanylic Acid, 4483
5′-Guanylic Acid, 4484
N¹-Guanylsulfanilamide see 8879
Guanylurea Sulfate see 3083
Guaran, 4485
Guaranine see 1635
Guardian [Boehringer, Ing.] see 4055
Guarem [Rybar] see 4486
Guar Flour see 4486
Guar Gum, 4486
Guarina [Norgine] see 4486
Guastil [Biohorm-Uriach] see 8971
Guatambuinine see 6794
Guayanesin see 4465
Guayule see 8255
Gubernal [Geigy] see 311
Guesapon [Ciba-Geigy] see 2832
Guethine see 4473
Guiatuss [Sheraton] see 4465
Guicitrina [Perga] see 621
Guidazide see 4396
Guinea Green B, 4487
Guinea-pig-anti-stiffness Factor see 8771
Gujaphenyl [DDD] see 4457
L-Gulitol see 8680
Gulliostin [Taiyo] see 3354
D-Gulonic Acid, 4488
L-Gulonic Acid, 4489
D-Gulose, 4490
L-Gulose, 4491
Gum Ammoniac see 511
Gum Arabic see 10
Gumbaral [Homburg Degussa] see 146
Gum Benjamin see 4492
Gum Benzoin, 4492
Gum Camphor see 1738
Gum Copal see 2512
Gum Cyamopsis see 4486
Gum Damar see 2808
Gum Ghatti see 4311
Gum Guaiac see 4455
Gum Karaya see 5164
Gum Kino see 5193
Gum Opium see 6809
Gum Plant see 3617
Gum-plant (of California) see 4450
Gum Quince Seed see 8064
Gum-resin Myrrh see 6251
Gum Thus see 6792 and 9734
Gum Tragacanth, 4493
Gum Wood see 3852
Guntrin [Zensei] see 6841
Gunyl [Hogapharm] see 4457
G.U.-Pen [Pfizer] see 1847
Guronsan see 4362
Guronsan Fe [Chugai] see 4362
Guru Nuts see 5198
Gusathion A [Bayer] see 926
Gusathion M [Bayer] see 926
Guthion [Bayer] see 926
Gutron [OSSW] see 6107
Guttalax [De Angeli] see 7377
Gutta-Percha, 4494
Guvacine, 4495
GV see 4454
GVG see 9882
Gy-bon [Ciba-Geigy] see 8488
GYKI-41099 see 2399
Gymnemic Acid, 4496
Gymnemin see 4496
Gynamide [Merminod] see 4503
Gynamousse see 6931
Gynasan [Bastian] see 3659

Gyneclorina [Heyden] see 2066
Gynécormone [Nigy] see 3655
Gynefollin see 3094
Gyne-Lotrimin [Schering] see 2412
Gyne-Merfen [Zyma] see 7274
Gynera [Schering AG] see 4308
Gynergen [Sandoz] see 3609
Gynergon see 3653
Gynesine see 9606
Gyn-Hydralin [Lefrancq] see 4386
Gynipral [Chemie-Linz] see 4628
Gynocardia Oil see 2036
Gynochrome see 5757
Gyno-Daktarin [Janssen] see 6101
Gynoestryl [Roussel-UCLAF] see 3653
Gynofon [Endopancrine] see 421
Gynokhellan see 5189
Gynol II [Ortho-Cilag] see 6596
Gynolett [Labopharma] see 3689
Gyno-Monistat [Cilag-Chemie] see 6101
Gynomyk [Cassenne] see 1525
Gyno-Pevaryl [Cilag-Chemie] see 3476
Gynoplix [Théraplix] see 44
Gynorest [Mead Johnson] see 3454
Gyno-Sterosan [Geigy] see 2191
Gyno-Terazol [Cilag] see 9090
Gynotherax see 2191
Gyno-Travogen [Schering AG] see 5044
Gyno-Trosyd [Pfizer] see 9384
Gynovlar [Schering] see 6614
Gypsogenin, 4497
Gypsophilasapogenin see 4497
Gypsum see 1713
Gyramid [Parke-Davis] see 3540
Gyrolite see 1709
H3®, 4498
H 33 see 3891
H 115 see 2106
H 133 see 3029
H 321 see 5893
H 365 see 6992
H 610 see 3909
H 990 see 6919
H 1032 see 6060
H 3292 see 3366
H 3452 see 2726
H 3625 see 2536
H 3749 see 7987
H 3774 see 233
H 4007 see 5745
H 4723 see 2355
H 56/28 see 311
(—)H 80/62 see 7742
H 93/26 see 6072
cis-H 102/09 see 10024
H 133/22 see 7742
H 154/82 see 3895
H 168/68 see 6800
HA 106 see 9622
Haarlem Oil see 5153 and 6776
HABA-DKB see 797
Habekacin see 797
Hachimycin, 4499
H Acid see 6305
HAD see MISC-2
Hadacidin, 4500
Haelan [Dista] see 4122
Haemodyn see 7700
Haemofort see 4006
Haemostop Injection see 6314
Haertolan see 6328
Haflutan [Cassella-Riedel] see 2371
Hafnium, 4501
Hagan Phosphate see 8621
Hageman Factor see 3877
Hahnium see 3504
Haiprex [Riker] see 5882

Halamid see 2066
Halane see 3054
Halarsol see 3060
Halazepam, 4502
Halazone, 4503
Halbmond [Much] see 3308
Halciderm [Squibb] see 4504
Halcimat [Heyden] see 4504
Halcinonide, 4504
Halcion [Upjohn] see 9518
Halcort [F.A.I.R.] see 4504
Haldol [McNeil] see 4511
Haldol Decanoate [Janssen] see 4511
Haldrate see 6977
Haldrone [Lilly] see 6977
Haletazole see 4505
Halethazole, 4505
Halfan [SKB] see 4508
Halfa Wax see 3649
Halibut Liver Oil, 4506
Halidor [EGYT] see 1041
Halimide®, 4507
Halinone [USV] see 1381
Halite see 8544
Haloanisone see 4047
Haloart [Taiho] see 4512
Halofantrine, 4508
Halofuginone, 4509
Halog [Squibb; Heyden] see 4504
Halogabide see 7782
Halometasone, 4510
Halomonth [Dainippon] see 4511
Halon see 3053
Haloperidol, 4511
Halopredone Acetate, 4512
Haloprogesterone, 4513
Haloprogin, 4514
Halopropane, 4515
Halopyramine see 2162
Halospor [Ciba-Geigy] see 1937
Halostachine, 4516
Halosten [Shionogi] see 4511
Halotestin [Upjohn] see 4113
Halotex [Mead Johnson] see 4514
Halothane, 4517
Haloxazolam, 4518
Haloxon, 4519
Halquinol, 4520
Halquivet see 4520
Haltran [Upjohn] see 4812
Hamaméliode P [Oberlin] see 5334
Hamamelis, 4521
Hamamelitannin, 4522
Hamamelose, 4523
Hamarin [Nicholas] see 278
Hamburg Blue see 3964
Hämovannid [Bastian] see 4885
Hamycin, 4524
Hamycin A see 4524
Hansolar [Parke, Davis] see 16
Haocolin [Fuso] see 2321
HAPA-B see 4989
Hapadex [Schering] see 6390
Haplophytine, 4525
Haptens, 4526
Haptocil see 8913
Haptocil [Cilag-Chemie] see 2718
Haptoglobins, 4527
Harden-Young Ester see 4187
Hard Paraffin see 6971
Harlem Oil see 5153
Harmaline, 4528
Harmalol, 4529
Harmalol Methyl Ether see 4528
Harman, 4530
Harmar [Zemmer] see 7851
Harmidine see 4528
Harmine, 4531
Harmogen [Abbott] see 3660
Harmonin see 5751
Harmonyl [Abbott] see 2901

Harodase see 4676
Harrical see 5114
Hartol see 5751
Hartshorn see 528
Harvade [Uniroyal] see 6893
Harvamine see 7996
Harvatrate [Tilden-Yates] see 891
Harzol [Hoyer] see 8500
Hasach see 1752 and 4532
Hasethrol see 7066
Hashish, 4532
Hasubanan see 4533
Hasubanonine, 4533
Hatchett's Brown see 2640
Haurymellin [Haury] see 5845
Hauser's Salt see 10085
Hausmannite see 5618
Havapen see 7024
Havidote [Haver-Lockhart] see 3484
Haw Apple see 2569
Haworth Lignan see 5362
Haws see 2569
Hawthorn see 2569
Haynon [R. P. Drugs] see 2180
Hayo see 2448
Hazol see 6919
HB 419 see 4372
HBA see 3866
HBBL see 3549
HBF 386 see 1605
HBK see 797
Hc45 [Crookes] see 4711
HC-58 see 3501
HC-064 see 6458
HC-20-511 see 5187
HCFU see 1851
HCG, 4534
γ-HCH see 5379
HCS, 4535
HE-69 see 6137
He 781 see 9178
Head & Shoulders [Procter & Gamble]
    see 8004
Healon [Pharmacia] see 4675
Healonid [Pharmacia] see 4675
Heartcin [Ohta] see 9751
Heartgard 30 [Merck & Co.] see 5133
Heart Muscle Extract, 4536
Heavy Hydrogen see 2919
Heavy Spar see 1006
Heavy Water see 2920
Hebanil see 2186
Hebaral see 4623
Hebucol see 2721
Hecogenin, 4537
Hectorite, 4538
Hedaquinium Chloride, 4539
Hedoma see 7053
Hederagenin, 4540
α-Hederin see 4541
Hedex see 40
Hedonal [Bayer] see 5666
Hedonal see 2802
Hedonal DP [Bayer] see 3068
Hedulin [Merrell] see 7196
Hefasolon [Hefa-Frenon] see 7719
Heferol [Alkaloid] see 3995
Heitrin [Abbott] see 9084
Hekbilin [Hek] see 2044
Helenalin, 4542
Helenien see 9972
Helenin see 198
Helenynolic Acid, 4543
Helfergin [Promonta] see 5660
Helfo-Dopa [Helfenberg] see 5344
Helgotan see 9024
Helianthine B see 6019
Helicin, 4544
Helicocerin see 1997
Helicon see 873
Heliophan [Greef] see 4660

Heliosupine, 4545
Heliotropin see 7445
Helium, 4546
Helixin C see 6457
Helixin (the Saponin) see 4541
Hellebrigenin Glucorhamnoside see
    4547
Hellebrin, 4547
Hellipidyl [Hellwig] see 491
Helmatac [SK & F] see 6985
Helmet Flower see 8368
Helmetina see 7220
Helmex [Roerig] see 7968
Helmezine see 7433
Helmifren see 7431
Helminal see 5157
Helminthosporal, 4548
Helminthosporol, 4549
Helmirone see 4519
Helmitol [Winthrop] see 5881
Helogaphen [Spitzner] see 2082
Helonias, 4550
Helvecyclin [Helvepharm] see 9130
Helveprim [Helvepharm] see 8889
Helvetia Blue see 5950
Helveticoside, 4551
Helvolic Acid, 4552
Hem see 4558
Hemabate [Upjohn] see 1829
Hemanthine see 1483
Hematein, 4553
Hematin, 4554
Hematin-protein see 2797
Hematite see 3973
Hematoporphyrin, 4555
Hematoporphyrin IX see 4555
Hematoporphyrinmercury Disodium
    Salt see 5804
Hematoxiline see 4556
Hematoxylin, 4556
Hematoxylon, 4557
Heme, 4558
Hemel see 318
Hementin, 4559
Hemerven [Ciba-Geigy] see 3291
Hemerythrin, 4560
Hemicelluloses, 4561
Hemi-Daonil [Hoechst] see 4372
Hemi-Dewar Biphenyl, 4562
Hemiglobin see 5878
Hemin, 4563
Hemineurin [Debat] see 2382
Heminevrin [Astra Pharm.] see 2382
Hemipyocyanine, 4564
Hemisine see 3569
Hemisulfur Mustard, 4565
Hemlock see 2503
Hemoantin [Lepetit] see 8967
Hemo-B-Doze see 9921
Hemocaprol [Delagrange] see 442
Hémoclar [Clin Midy] see 7090
Hemocuron [Takeda] see 9520
Hemocyanins, 4566
Hemodal [Hoechst] see 5716
Hemodex [Central Biokhim.] see 2925
Hemofil [Hyland] see 3873
Hemoglobin, 4567
Hemolidione see 7196
Hemometina [Cusi] see 3517
Hemomin see 9921
Hemo-Pak [J & J] see 6894
Hemoplex (obsolete) [Cutter] see 3874
Hemopoietine see 3635
Hemostasin see 3569
Hemostatin see 3569
Hemotrope [Andromaco] see 1501
Hempa, 4568
Henbane see 4796
10-Hendecenoic Acid see 9760
Heneicosafluorotripropylamine see
    4111

Henequem *see* 8497
Henna, 4569
HEOD *see* 3093
Hepacholine *see* 2208
Hepadial [Biocodex] *see* 3189
Hepadist *see* 4602
Hepagin *see* 7160
Hepagon *see* 9921
Hepalande [Delalande] *see* 5719
Hépalidine [Medial] *see* 9375
Hepalon *see* 5427
Hepa-Merz *see* 6826
Heparamine, 4570
Hepar Calcis *see* 1714
Héparégène *see* 9375
Heparexine *see* 7337
Heparides *see* 4570
Heparin, 4571
β-Heparin *see* 2217
Heparinic Acid *see* 4571
Heparin Sodium *see* 4571
Hepar Sulfuris *see* 7578
Hepartest *see* 8933
Hepasil *see* 3912
Hepasynthyl *see* 9415
Hepation [Nippon Chemiphar] *see* 5593
Hepatocatalase *see* 1906
Hepatopron *see* 5427
Hepatoxane *see* 9415
Hepavis *see* 9921
Hepaxanthin, 4572
Hepbisul *see* 4579
Hepcovite [Endo] *see* 9921
HEPES, 4573
Hepin *see* 442
Hepol *see* 5427
Heporal *see* 676
HEPP *see* 7086
Heprinar [Armour] *see* 4571
Hepronicate, 4574
Hepsal [Weddel] *see* 4571
Heptabarb *see* 4575
Heptabarbital, 4575
Heptachlor, 4576
1H-1,4,5,6,7,8,8-Heptachloro-3a,4,7,-
7a-tetrahydro-4,7-methanoindene *see* 4576
Heptadecanoic Acid *see* 5634
(E,E,E)-(−)-8,10,12-Heptadecatriene-4,6-diyne-1,14-diol *see* 2273
2,8,10-Heptadecatriene-4,6-diyne-1,14-diol *see* 3524
2-(8-Heptadecenyl)-4,5-dihydro-1H-imidazole-1-ethanol *see* 419
2-(8-Heptadecenyl)-2-imidazoline-1-ethanol *see* 419
2-Heptadecyl-4,5-dihydro-1H-imidazole Monoacetate *see* 4404
5-Heptadecylene-1-carboxylic Acid *see* 7142
2-Heptadecylglyoxalidine Acetate *see* 4404
3-(1,3-Heptadienyl)-5,6-dihydroxy-2-(hydroxymethyl)cyclohexanone *see* 6939
6-(1,3-Heptadienyl)-3,4-dihydro-2-oxocyclohexanecarboxaldehyde *see* 4183
Heptadon Hydrochloride *see* 5852
Heptadorm *see* 4575
1,1,2,2,3,3,3-Heptafluoro-N,N-bis-(heptafluoropropyl)-1-propanamine *see* 4111
Heptaldehyde *see* 4578
Heptaldehyde Sodium Bisulfite *see* 4579
Heptalgin *see* 7160
Heptalin *see* 7160
2-(1-N,N-Heptamethylenimino)ethyl-guanidine *see* 4473

(all-E)-2-(3,7,11,15,19,23,27-Hepta-methyl-2,6,10,14,18,22,26-octacosa-heptaenyl)-3-methyl-1,4-naphtha-lenedione *see* 9935
2-(3,7,11,15,19,23,27-Heptamethyl-2,6,10,14,18,22,26-octacosahepta-enyl)-3-methyl-1,4-naphthoquinone *see* 9935
Heptamine *see* 9712
Heptaminol, 4577
Heptamul *see* 4576
Heptanal, 4578
Heptanal Sodium Bisulfite, 4579
2-Heptanamine *see* 9712
n-Heptane, 4580
1,7-Heptanedicarboxylic Acid *see* 921
Heptanedioic Acid *see* 7401
Heptanoic Acid, 4581
Heptanoic Acid Ethyl Ester *see* 3790
1-Heptanol, 4582
2-Heptanol, 4583
Heptanon [Pliva] *see* 5852
2-Heptanone, 4584
4-Heptanone *see* 3351
17β-Heptanoyloxy-1-methyl-5α-androst-1-en-3-one *see* 5887
Heptazone *see* 7160
Heptedrine *see* 9712
D-arabino-Hept-2-enonic Acid γ-Lactone *see* 4345
Heptenophos, 4585
Heptin *see* 9712
Hept-a-myl [Delalande] *see* 4577
n-Heptoic Acid *see* 4581
Heptone *see* 7160
D-glycero-D-gulo-Heptonic Acid *see* 4349
Heptoxime, 4586
D-manno-Heptulose, 4587
n-Heptyl Alcohol *see* 4582
Heptylaldehyde *see* 4578
2-Heptyl-4-hydroxyquinoline N-Oxide *see* 4670
19-Heptyl-10-hydroxy-1,5,10,14-tetraazacyclononadecan-15-one *see* 7480
n-Heptylic Acid *see* 4581
2,2'-[3-[(3-Heptyl-4-methyl-2(3H)-thiazolylidene)ethylidene]-1-prop-ene-1,3-diyl]bis(3-heptyl-4-methyl-thiazolium) Diiodide *see* 7504
Heptylon [Delalande] *see* 4577
2-Heptyl-4-quinolinol 1-Oxide *see* 4670
Hepzide [Merck & Co.] *see* 6484
Héraclène [Mauchant] *see* 2446
Heraclin *see* 1173
Heraldium [Solac] *see* 5662
Herapathite *see* 8086
Herbadox [Am. Cyanamid] *see* 7026
Herban [Hercules] *see* 6611
Herbesser [Tanabe Seiyaku] *see* 3188
Herbisan [Roberts Chemicals] *see* 3390
Hercules 528 *see* 3296
Hercules 3956 *see* 9478
Hercules 7531 *see* 6611
Hercules 14503 *see* 2949
Herculon [Hercules] *see* 7558
Hercynine, 4588
Herkol *see* 3069
Hermesetas *see* 8282
Hermophényl *see* 5783
Heroin *see* 2948
Herperal [Farmitalia] *see* 8730
Herpes-Gel [Master] *see* 4819
Herplex [Allergan] *see* 4819
Herqueinone, 4589
Herzberg's Paper *see* 2493
Herzo [Toho] *see* 7889
Herzolan *see* 4536

HES *see* 4593
6-H.E.S. [Morishita] *see* 4593
Hesofen *see* 5731
Hesotanol *see* 3836
Hesotin *see* 3836
Hespan [McGaw] *see* 4593
Hespander [Kyorin] *see* 4593
Hesperetin, 4590
Hesperetin 7-Rhamnoglucoside *see* 4591
Hesperetin-7-rutinoside *see* 4591
Hesperidin, 4591
Hestar [Otsuka] *see* 4593
Hestat [Otsuka] *see* 4593
Hestrium Chloride *see* 4609
Hestsol [Green Cross] *see* 4593
Hetacillin, 4592
Hetacin K [Bristol] *see* 4592
Hetaphenone *see* 3662
Hetastarch, 4593
5-HETE, 4594
Heteroauxin *see* 4870
Heterocodeine *see* 6186
Heteronium Bromide, 4595
Heterophylline *see* 809
Hetol *see* 1313
Hetolin®, 4596
Hetrazan [Lederle] *see* 3106
Hetrazeen [Heterochemical] *see* 5715
Hetrum Bromide [Lilly] *see* 4595
Heulandite *see* 1645
Hevyteck [Penreco] *see* 7139
Hexaaminecobalt Trichloride, 4597
Hexaaquochromium Triacetate *see* 2221
Hexaaquochromium Tribromide *see* 2222
Hexaaquochromium Trichloride *see* 2224
1,4,7,10,13,16-Hexaazacyclotricosane Cyclic Peptide Deriv *see* 3501
Hexabendin *see* 4626
Hexabetalin *see* 7995
Hexabione Hydrochloride *see* 7995
Hexabolan [Phartec] *see* 9499
Hexaborane(10), 4598
Hexaboron Decahydride *see* 4598
Hexabrix [Guerbet] *see* 4956
2,2',4,4',5,5'-Hexabromobiphenyl *see* 7540
Hexa-CAF *see* MISC-2
Hexacarbacholine Bromide, 4599
Hexachlorobenzene, 4600
1,2,3,4,7,7-Hexachlorobicyclo[2.2.1]-2-heptene-5,6-bisoxymethylene Sulfite *see* 3529
1,4,5,6,7,7-Hexachloro-endo-bicyclo-[2.2.1]hept-5-ene-2,3-dicarboxylic Anhydride *see* 2084
1,4,5,6,7,7-Hexachloro-2,3-bis(chlo-romethyl)bicyclo[2.2.1]hept-5-ene *see* 2077
1,2,3,4,7,7-Hexachloro-5,6-bis(chlo-romethyl)-2-norbornene *see* 2077
1α,2α,3β,4α,5α,6β-Hexachlorocyclo-hexane *see* 5379
Hexachloroendomethylenetetrahydro-phthalic Anhydride *see* 2084
1,2,3,4,10,10-Hexachloro-6,7-epoxy-1,4,4a,5,6,7,8,8a-octahydro-endo,-endo-1,4:5,8-dimethanonaphthalene *see* 3533
1,2,3,4,10,10-Hexachloro-6,7-epoxy-1,4,4a,5,6,7,8,8a-octahydro-endo,-exo-1,4:5,8-dimethanonaphthalene *see* 3093
Hexachloroethane, 4601
1,2,3,4,10,10-Hexachloro-1,4,4a,5,8,-8a-hexahydro-1,4:5,8-dimethano-naphthalene *see* 219

(4α,4aβ,9bβ)-(±)-N-(1,2,3,4,4a,9b-
Hexahydro-8,9b-dimethyl-3-oxo-4-
dibenzofuranyl)-4-methyl-1-pipera-
zinepropanamide Dihydrochloride
see 9051
1,2,3,4,5,6-Hexahydro-6,11-dimeth-
yl-3-(2-phenethyl)-2,6-methano-3-
benzazocin-8-ol see 7174
Hexahydro-1,3-dimethyl-4-phenyl-
1H-azepin-4-ol Propanoate (Ester)
see 7786
(1S,3R,7S,8S,8aR)-1,2,3,7,8,8a-Hexa-
hydro-3,7-dimethyl-8-[2-[(2R,4R)-
tetrahydro-4-hydroxy-6-oxo-2H-
pyran-2-yl]ethyl]-1-naphthalenyl
(S)-2-Methylbutyrate see 5460
2,4,5,7,12b,12c-Hexahydro-1H-[1,3]-
dioxolo[4,5-j]pyrrolo[3,2,1-de]-
phenanthridine-1,2-diol see 5498
Hexahydro-2,6-dioxo-4-pyrimidine-
carboxylic Acid see 4735
4,4',5,5',6,6'-Hexahydrodiphenic
Acid 2,6,2',6'-Dilactone see 3508
Hexahydro-α,α-diphenyl-1H-azepine-
1-butanamide see 1599
Hexahydroequilenin, 4607
[1aR-(1aα,2aβ,3β,6β,6aβ,8aS*,8bβ,-
9S*)]-Hexahydro-2a-hydroxy-9-(1-
hydroxy-1-methylethyl)-8b-methyl-
3,6-methano-8H-1,5,7-trioxacyclo-
penta[ij]cycloprop[a]azulene-4,8-
(3H)-dione see 7387
2,3,4,7,8,8a-Hexahydro-4-hydroxy-8-
(hydroxymethyl)-8-methyl-1H-
3a,7-methanoazulene-3,6-dicarbox-
ylic Acid see 8427
3,3a,4,5,9a,9b-Hexahydro-4-hydroxy-
9-(hydroxymethyl)-6-methyl-3-
methyleneazuleno[4,5-b]furan-2,7-
dione see 5224
1,3,4,9,10,10a-Hexahydro-6-hydroxy-
2H-10,4a-iminoethanophenanthrene
see 6625
2,3,6a,8,9,9a-Hexahydro-9a-hydroxy-
4-methoxycyclopenta[c]furo[3',2':
4,5]furo[2,3-h][1]benzopyran-1,11-
dione see 170
11,12,13,14,15,16-Hexahydro-3-hydr-
oxy-13-methyl-17H-cyclopenta[a]-
phenanthren-17-one see 3581
[1aR-(1aα,2aβ,3β,6β,6aβ,8aS*,8bβ,-
9R*)]-Hexahydro-2a-hydroxy-8b-
methyl-9-(1-methylethenyl)-3,6-
methano-8H-1,5,7-trioxacyclo-
penta[ij]cycloprop[a]azulene-4,8-
(3H)-dione see 7389
[1aS-(1aα,1bβ,2β,5β,6aβ,7β,7aα,8S*)]-
Hexahydro-1b-hydroxy-6a-methyl-
8-(1-methylethenyl)spiro[2,5-meth-
ano-7H-oxireno[3,4]cyclopent[1,2-
d]oxepin-7,2'-oxiran]-3(2H)-one
see 2525
(+)-trans-1a,2,3,4a,5,6-Hexahydro-
9-hydroxy-4-propyl-4H-naphth-
[1,2-b]-1,4-oxazine see 6349
2,3,4,4a,10,10a-Hexahydro-6-hydr-
oxy-1,1,4a,7-tetramethyl-9(1H)-
phenanthrenone see 6463
(4bS-trans)-4b,5,6,7,8,8a-Hexahydro-
4-hydroxy-8,8,8-trimethyl-2-(1-
methylethyl)-3,9-phenanthrenedione
see 9048
1,2,3,4,4a,7-Hexahydro-1-hydroxy-
α,4a,8-trimethyl-7-oxo-2-naphthal-
eneacetic Acid γ-Lactone see 8325
1,2,3,9,10,10a-Hexahydro-10,4a(4H)-
iminoethanophenanthrene see 6185
(4aR)-1,3,4,9,10,10aα-Hexahydro-2H-
10α,4aα-(iminoethano)phenanthrene
see 6185

1,3,4,9,10,10a-Hexahydro-2H-10,4a-
iminoethanophenanthren-6-ol see
6625
Hexahydroisonicotinic Acid see 5075
2,3,4,5,8,8a-Hexahydro-3-isopropyl-
6,8a-dimethyl-3a(1H)-azulenol see
1863
1,2,3,4,5,6-Hexahydro-2,6-methano-
3-benzazocine see 1105
3-(Hexahydro-4,7-methanoindan-5-
yl)-1,1-dimethylurea see 6611
1,2,3,4,5,6-Hexahydro-1,5-methano-
8H-pyrido-[1,2-a][1,5]diazocin-8-
one see 2795
2,3,6aα,8,9,9aα-Hexahydro-4-meth-
oxycyclopenta[c]furo[3',2':4,5]furo-
[2,3-h][1]benzopyran-1,11-dione see
168
3,4,7aα,9,10,10aα-Hexahydro-5-
methoxy-1H,12H-furo[3',2':4,5]-
furo[2,3-h]pyrano[3,4-c][1]benzo-
pyran-1,12-dione see 169
Hexahydro-4-(5-methoxyheptyl)-
2(1H)-pentalenone see 2774
4a,5,9,10,11,12-Hexahydro-3-meth-
oxy-11-methyl-6H-benzofuro[3a,3,-
2-ef][2]benzazepin-6-ol see 4245
dl-cis-1,2,3,9,10,10a-Hexahydro-6-
methoxy-11-methyl-4H-10,4a-
iminoethanophenanthrene see 8116
dl-cis-1,3,4,9,10,10a-Hexahydro-6-
methoxy-11-methyl-2H-10,4a-
iminoethanophenanthrene see 8116
5b,6,7,12b,13,14-Hexahydro-13-
methyl[1,3]benzodioxolo[5,6-c]-1,3-
dioxolo[4,5-i]phenanthridin-6-ol see
2043
1,2,3,4,10,14b-Hexahydro-2-methyl-
dibenzo[c,f]pyrazino[1,2-a]azepine
see 6097
Hexahydro-N-methyl-2,4-methano-
4H-furo[3,2-b]pyrrol-3-amine see
5442
Hexahydro-4-methyl-2,5-methano-
2H-furo[3,2-b]pyrrol-6-ol see 8365
1,2,3,4,5,6-Hexahydro-3-methyl-1,5-
methano-8H-pyrido[1,2-a][1,5]di-
azocin-8-one see 1914
7-[1,2,6,7,8,8a-Hexahydro-2-methyl-
8-(methylbutyryloxy)naphthyl]-3-
hydroxyheptan-5-olide see 6088
3,3a,4,7,8,8a-Hexahydro-7-methyl-3-
methylene-6-(3-oxo-1-butenyl)-2H-
cyclohepta[b]furan-2-one see 9967
Hexahydro-1-methyl-4-phenyl-1H-
azepine-4-carboxylic Acid Ethyl
Ester see 3698
2,3,4,4a,5,6-Hexahydro-12-methyl-
1H-5,10b-propano-1,7-phenanthro-
line see 5489
Hexahydronicotinic Acid see 6478
Hexahydro-2-oxo-1H-thieno[3,4-d]-
imidazole-4-pentanoic Acid see
1244
[3aS-(3aα,4β,5β,6aα)]-Hexahydro-2-
oxo-1H-thieno[3,4-d]imidazole-4-
pentanoic Acid 5-Oxide see 1245
cis-Hexahydro-2-oxo-1H-thieno[3,4]-
imidazole-4-valeric Acid see 1244
N6-[5-(Hexahydro-2-oxo-1H-thieno-
[3,4-d]imidazol-4-yl)-1-oxopentyl]-
L-lysine see 1240
(−)-1,3,4,9,10,10a-Hexahydro-11-
phenethyl-2H-10,4a-iminoethano-
phenanthren-6-ol see 7215
Hexahydrophenol see 2731
Hexahydropicolinic Acid see 7425
(4aR)-trans-3,4,4a,5,6,10b-Hexahy-
dro-4-propyl-2H-naphth[1,2-b]-1,4-
oxazin-9-ol see 6349

4,5,7,8,9,12-Hexahydro-11H-pyrano-
[3,4-d]pyrrolo[3,2,1-jk][1]benzaze-
pin-11-one see 775
Hexahydropyrazine see 7431
Hexahydropyridine see 7438
Hexahydro-1,3,4,5-tetrahydroxybenz-
oic Acid see 8071
[3aS-(3aα,5α,8α,9α,11β,13bβ,15S*)]-
3,3a,5,8,11,13b-Hexahydro-7,8,12,-
15-tetrahydroxy-5,9-dimethyl-8,11-
ethanofuro[2,3-e]naphtho[2,3-c:6,7-
c']dipyran-2,6,13(9H)-trione see
4441
(S)-9,11,12,13,13a,14-Hexahydro-2,-
3,6,7-tetramethoxydibenzo[f,h]-
pyrrolo[1,2-b]isoquinoline see 9739
9,11,12,13,13a,14-Hexahydro-2,3,5,6-
tetramethoxydibenzo[f,h]pyrrolo-
[1,2-b]isoquinoline see 9738
1,10,19,22,23,24-Hexahydro-2,7,13,-
17-tetramethyl-1,19-dioxo-3,18-
divinylbiline-8,12-dipropionic Acid
see 1235
Hexahydrothymol see 5723
Hexahydro-s-triazaborine see 9516
Hexahydro-4-[3-[2-(trifluoromethyl)-
phenothiazin-10-yl]propyl]-1H-1,4-
diazepine-1-ethanol see 4657
Hexahydro-1b,6,8-trihydroxy-6a-
methyl-8-(1-methylethenyl)spiro-
[2,5-methano-7H-oxireno[3,4]cyclo-
pent[1,2-d]oxepin-7,2'-oxiran]-
3(2H)-one see 4787
[1S-[1α(βS*,δS*),2α,6α,8β(R*),8aα]]-
1,2,6,7,8,8a-Hexahydro-β,δ,6-trihy-
droxy-2-methyl-8-(2-methyl-1-
oxobutoxy)-1-naphthaleneheptanoic
Acid Monosodium Salt see 7712
Hexahydro-α,3a,5-trimethyl-6,8-
dioxo-1,4-methanoindan-1-acetic
Acid see 8324
1,2,3,4,5,6-Hexahydro-3,6,11-trimeth-
yl-2,6-methano-3-benzazocin-8-ol
see 5839
2,3,4,4a,9,9a-Hexahydro-2,4aα,9α-
trimethyl-1,2-oxazino[6,5-b]indol-
6-ol Methylcarbamate see 3647
5,6,6a,7,9a,9b-Hexahydro-1,4a,7-
trimethyl-3H-oxireno[8,8a]azuleno-
[4,5-b]furan-8(4aH)-one see 798
[4R-(4α,7β,7aβ)]-1,2,3,4,7,7a-Hexa-
hydro-4,7-trimethyl-6H-2-pyrin-
din-6-one see 9056
(3aS-cis)-1,2,3,3a,8,8a-Hexahydro-
1,3a,8-trimethylpyrrolo[2,3-b]indol-
5-ol Methylcarbamate (Ester) see
7357
Hexahydro-1,3,5-trinitro-1,3,5-tri-
azine see 2742
5-[(Hexahydro-2,4,6-trioxo-5-pyrimi-
dinyl)imino]-2,4,6(1H,3H,5H)-pyri-
midinetrione Monoammonium Salt
see 6215
1,2,3,5,6,7-Hexahydroxy-9,10-anthra-
cenedione see 8267
1,2,3,5,6,7-Hexahydroxyanthraquin-
one see 8267
2,3,3',4,4',5'-Hexahydroxybenzo-
phenone see 3868
1β,3β,5,11α,14,19-Hexahydroxy-5β-
card-20(22)-enolide see 6853
2,3,14,20,22,25-Hexahydroxycholest-
7-en-6-one see 3465
Hexahydroxycyclohexane see 4883
1,1',6,6',7,7'-Hexahydroxy-5,5'-di-
isopropyl-3,3'-dimethyl[2,2'-bi-
naphthalene]-8,8'-dicarboxaldehyde
see 4435

Hexestrol Bis(β-diethylaminoethyl ether), 4622
Hexestrol 4,4'-Diphosphoric Ester see 4621
Hexethal Sodium, 4623
Hexetidine, 4624
Hexetone see 4651
Hexigel [Warner-Lambert] see 4624
Hexobarbital, 4625
Hexobarbital Soluble see 4625
Hexobarbitone see 4625
Hexobendine, 4626
Hexobion [E. Merck] see 7995
Hexobutyramide see 1526
Hexocil [Parke-Davis] see 4624
Hexocyclium Methyl Sulfate, 4627
L-threo-2,3-Hexodiulosonic Acid γ-Lactone see 2856
Hexoestrol see 4621
Hexofen see 5731
Hexogen see 2742
Hexomedine [Rhodia; Théraplix] see 4612
Hexone see 5095
Hexone Chloride see 4609
Hexopal [Winthrop] see 4885
Hexoprenaline, 4628
Hexopropynate see 4616
Hexoral [Gödecke] see 4624
Hexosan see 4602
Hexose Diphosphate see 4187
Hexose Monophosphate see 4188
Hexose Phosphate see 4188
Hextol [Hoechst] see 7823
Hextril see 4624
D-lyxo-Hexulose see 9005
L-xylo-2-Hexulosonic Acid see 5183
Hexydaline see 5883
n-Hexyl Alcohol see 4615
4-Hexyl-1,3-benzenediol see 4633
4-Hexyl-1,3-benzenediol Compd with 9-Acridinamine (1:1) see 120
Hexylcaine Hydrochloride, 4629
1-(n-Hexylcarbamoyl)-5-fluorouracil see 1851
n-Hexyl Chloride see 2144
2-Hexylcyclopropanedecanoic Acid see 5218
2-Hexyldecanoic Acid, 4630
1-Hexyl-3,7-dihydro-3,7-dimethyl-1H-purine-2,6-dione see 7085
4-Hexyl-1,3-dihydroxybenzene see 4633
1-Hexyl-3,7-dimethylxanthine see 7085
Hexylene Glycol, 4631
2-Hexyl-5-hydroxycyclopentane-heptanoic Acid see 8240
2-Hexyl-2-(hydroxymethyl)-1,3-propanediol Trinicotinate see 4574
Hexylmethylcarbinol see 6675
Hexyl Methyl Ketone, 4632
2-(Hexyloxy)benzamide see 3866
2-[3-(Hexyloxy)-2-hydroxypropoxy]-benzoic Acid see 3869
9-(p-Hexyloxyphenyl)-10-methyl-9-acridanyl Chloride see 7157
9-[4-(Hexyloxy)phenyl]-10-methyl-acridinium Chloride see 7157
17β-[3-[4-(Hexyloxy)phenyl]-1-oxo-propoxy]estr-4-en-3-one see 6282
4-Hexylresorcinol, 4633
3-Hexyl-7,8,9,10-tetrahydro-6,6,9-trimethyl-6H-dibenzo[b,d]pyran-1-ol see 8995
1-Hexyltheobromine see 7085
Hexythiazox, 4634
Heyden 611 [Tenneco] see 8769
HF see 3877
HF 1854 see 2417
HF 1927 see 2990

HF 2333 see 7131
HGF see 4341 and 4893
HG-factor see 4341
HGH see 8672
HGRF see 8669
HH 184 see 7920
HH 197 see 1505
HHDN see 219
Hi-Alazin see 9442
Hibanil [Specia] see 2186
Hibenzate see MISC-4
Hiberna see 7797
Hibernal [AB Leo] see 2186
Hibernon [Diwag] see 1436
Hibernyl see 7303
Hi-Bestrol [Boyle] see 3118
Hibiclens [Stuart] see 2090
Hibicon [Lederle] see 1027
Hibidil [ICI] see 2090
Hibiscrub [ICI] see 2090
Hibitane [Ayerst; ICI] see 2090
Hibon see 8201
Hibschite see 1645
Hichillos [Kotani] see 5168
Hiconcil [Allard] see 610
Hicoseen [Klimitschek] see 1516
Hidacian see 2688
Hidantal see 7293
Hi-Deratol see 9928
Hidrasonil see 5071
Hidro-Colisona see 4710
Hidroestron see 3655
Hidroferol [Juventus] see 1638
Hidroronol see 4704
Hi-Enterol see 4924
Hiestrone see 3660
Hi-Fresmin [Takeda] see 2446
High Bush Cranberry see 9876
High Mallow see 5590
Higueroxyl Delabarre see 4020
Hijuven [Eisai] see 9931
Hilactan [Kyoritsu] see 2308
Hillebrandite see 1709
Hilong [Banyu] see 6881
Himecol [Kissei] see 4792
Hinderin, 4635
Hinosan [Bayer] see 3485
Hiochic Acid see 6087
Hiohex Chloride [Keith-Victor] see 4609
Hioxyl [Quinoderm] see 4725
Hipberries see 8243
Hipercilina see 7041
Hipertan [Gramon] see 6107
Hiphyllin see 3459
Hipocolestina see 9650
Hipoftalin see 4682
Hippo see 4957
Hippodin see 4927
Hippramine [Riker] see 5882
Hippuran I 131 [Mallinckrodt] see 4927
Hippuric Acid, 4636
Hipputope [Squibb] see 4927
Hiprex [Merrell] see 5882
Hirathiol see 4815
Hiropon see 5859
Hirsutic Acid C, 4637
Hirudex see 4638
Hirudin, 4638
His (IUPAC Abbrev.) see 4642
Hiserpia [Bowman] see 8149
Hislosine [Toho] see 1804
Hismanal [Janssen] see 878
Hispril [SK & F] see 3334
Histabromamine see 1405
Histabutizine see 1449
Histabutyzine see 1449
Histacap see 7996
Histacuran see 2346
Histadur [Cooper] see 2180

Histalen [Len-Tag] see 2180
Histalog [Lilly] see 1205
Histalon see 7996
Histamen [Polipharma] see 878
Histametizine see 5657
Histaminase, 4639
Histamine, 4640
Histamine Deaminase see 4639
Histamine Oxidase see 4639
Histaminos [Lesvi] see 878
Histan [Cooper] see 7996
Histanorm [Ozothine] see 4190
Histantin [Burroughs Wellcome] see 2078
Histantin (Richter) see 179
Histaphen(e) [UCB] see 5678
Histapon [Endo] see 4640
Histapyran see 7996
Histapyrrodine, 4641
Histasan see 7996
Histaspan [USV] see 2180
Histatex see 7996
Histazine see 709
Histazol [Krka] see 878
Histidine, 4642
L-Histidine see 4642
Histidine-betaine see 4588
Histidine Trimethylbetaine see 4588
Histimin see 1205
Histionex [Pennwalt] see 7287
Histocarb [Rohm & Haas] see 1788
Histomibal see 3187
Histones, 4643
Histosol [Norden] see 7996
Histostab [Boots] see 709
Histrionicotoxin D see 4296
Histryl [SK&F] see 3334
Histyn [Hässle] see 3334
Hi-Ti [Kyowa] see 5538
Hitocobamin-M [Hishiyama] see 9921
Hitodesterol see 8705
Hittorf's Phosphorus see 7321
Hive Dross see 7848
Hiwolfia [Bowman] see 8131
HI-Z [Otsuka] see 6841
HK 137 see 3916
HK 256 see 7743
HL 255 see 1381
HL 362 see 2473
HL 2186 see 9009
HL 5746 see 2186
HL 8731 see 7507
3-HMC see 3737
HMD see 1802
HME see 3510
HMF see 4764
HMG see 5690
HMGA see 5690
HMM see 318
HMPA see 4568
HMPT see 4568
HMS [Allergan] see 5679
HMT see 5879
HMTA see 5879
HN1, 4644
HN2 see 5655
HN3 see 9560
HO-2,474 see 2936
Hoarhound see 4667
Hoban see 5427
Hodag 40-S [Hodag] see 7555
Hoe 045 see 1878
HOE 118 see 7464
HOE 280 see 6688
Hoe 285 see 7823
HOE 296 see 2270
Hoe 296V see 8157
HOE 304 see 2910
HOE 433 see 312
HOE 440 see 9350
Hoe 498 see 8123

Hoe 661 see 7314
Hoe 760 see 8252
HOE 766 see 1492
HOE 777 see 7717
HOE 881v see 3904
HOE 893d see 7025
HOE 984 see 6592
HOE 2784 see 1237
Hoe 2810 see 5387
HOE 2873 see 7976
Hoe 2904 see 3282
Hoe 2982 see 4585
Hoe 13764 see 7966
Hoe 16410 see 5106
Hoe 17411 see 1794
HOE 23408 see 3072
HOE 33171 see 3929
HOE 36801 see 3821
Hoe 39866 see 7314
HOE 39-893d see 7025
Hoe 40045 see 1878
Hoechst 10446 see 4769
Hoechst 10495 see 6633
Hoechst 10582 see 6628
Hoechst 10600 see 7160
Hoechst 10682 see 3333
Hoechst 10720 see 5180
Hoechst 10805 see 3342
Hoechst 10820 see 5852
Hoegrass [Hoechst] see 3072
Hoelon [Hoechst] see 3072
Hoffmann's Anodyne see 8714
Hoffmann's Drops see 8713
HOG see 7179
Hog's Bean see 4796
Hoggar N [Stada] see 3430
Hogival see 3660
Hogpax [Ferrosan] see 612
Hogweed see 8358
Hokunalin [Hokuriku] see 9720
Holarrhenine, 4645
Holbamate see 5751
Holland Balsam see 5153
Holligold see 1723
Holly-leaved Barberry see 1170
Holmia see 4646
Holmium, 4646
Holocaine Hydrochloride [Winthrop] see 7153
Holodorm see 5872
Holomycin, 4647
Holopon [Byk-Gulden] see 5927
Holoxan [Lucien] see 4822
Homandren (Tabl) see 6044
Homandren (Amps.) see 9115
Homapin [Mission Pharmacal] see 4649
Homarine, 4648
Homatrisol [Cooper] see 4649
Homatropine, 4649
Homburg 814 see 3919
Homidium, 4650
Homoalanin-4-yl(methyl)phosphinic Acid see 7314
Homoarterenol Hydrochloride see 6609
Homocaine see 2449
Homocamfin, 4651
Homochelidonine, 4652
α-Homochelidonine see 4652
Homochlorcyclizine, 4653
Homoclomin see 4653
Homocodeine see 7303
Homocysteine, 4654
Homocysteine S-Ethyl Ether see 3693
Homocystine, 4655
Homoeriodictyol, 4656
Homofenazine, 4657
Homogentisic Acid, 4658
Homoginin [Zeria] see 4653
Homomenthyl Salicylate see 4660

Homomycin see 4789
Homomyrtenyl β-(Diethylamino)ethyl Ether see 6252
2-Homomyrtenyloxy-1-(diethylamino)ethane see 6252
Homonal [SS Pharm.] see 5525
Homonicotinic Acid, 4659
Homoolan see 40
D-Homo-17a-oxaandrosta-1,4-diene-3,17-dione see 9108
D-Homopantothenic Acid see 4664
Homoproline see 7425
Homorestar [Ohta] see 4653
Homosalate, 4660
m-Homosalicylic Acid see 2584
o-Homosalicylic Acid see 2585
p-Homosalicylic Acid see 2586
Homoserine, 4661
Homosteron see 9109
Homosulfamine see 5525
4-Homosulfanilamide see 5525
4-Homosulfanilamide Salt with 1-Sulfanilyl-2-thiourea see 8922
Homotrilobine see 9610
Homovanillic Acid, 4662
HON, 4663
Honduras Balsam see 959
Honduras Bark see 1888
Honvan [W.B. Pharm.] see 4168
Honvol [Horner] see 4168
Hopantenate Calcium see 4664
Hopantenic Acid, 4664
Hopate [Tanabe] see 4664
Hopeite see 10055
Hops, 4665
Hordenine, 4666
Horehound, 4667
Hormantoxone see 5427
Hormezon [Tobishi] see 1202
Hormocardiol see 4536
Hormodin [Merck & Co.] see 4871
Hormoestrol see 4621
Hormofemin [Medo] see 3094
Hormoflaveine see 7783
Hormofollin see 3660
17-Hormoforin see 7710
Hormogynon see 3655
Hormoluton see 7783
Hormomed [Merckle] see 3659
Hormone A see 710
Hormonisene see 2173
Hormoteston see 9115
Hormovarine [Clin-Comar-Byla] see 3660
Hornbest [Shizuoka] see 2321
Horn Silver see 8450
Horse Balm see 2479
Horseheal see 4897
Horse Nettle see 8664
Horse-radish, 4668
Horseweed see 3614
Hostacaine [Hoechst] see 1508
Hostacortin see 7727
Hostacortin H see 7719
Hostacyclin [Hoechst] see 9130
Hostaginan see 7744
Hostalival [Hoechst] see 6592
Hostaquick [Hoechst] see 4585
Hourbese [Delta] see 7180
HP 129 see 3917
HP 549 see 5126
HPA-23, 4669
HPC see 6910
HPEK-1 see 9177
5-HPETE see 4594
HpGRF see 8669
HpGRF(1-29)NH₂ see 8412
HPL see 4535
HPMF see 3549
HPP see 278
HPTH 1-34 Acetate see 9096

HQ 495 see 171
HQNO, 4670
HR see 9698
HR-158 see 5453
HR-221 see 1931
HR-376 see 2355
HR-756 see 1935
HRF [Ayerst] see 5354
HRS-16 see 3095
HS-592 see 2345
HS-902 see 9390
131I-HSA see 8417
HSP 2986 see 7706
HSR-902 see 9390
5-HT see 8413
HTH see 1677
L-5HTP see 4784
5-HTP see 4784
HTX D see 4296
HTZ see 4634
Huanghuahaosu see 844
Humafac [Parke, Davis] see 3873
Humagel [Parke, Davis] see 6989
Human Actrapid [Novo] see 4889
Human Chorionic Gonadotrop(h)in see 4534
Human Chorionic Somatomammotropin see 4535
Human EGF-URO see 3492
Human Growth Hormone see 8672
Human Growth Hormone-releasing Factor(1-29)amide see 8412
Human IgE Pentapeptide see 7086
Human Monotard [Novo] see 4890
Human Pancreatic Somatoliberin-(1-29)amide see 8412
Human Placental Lactogen see 4535
Human Serum Growth Factor see 7497
Humatin [Parke, Davis] see 6989
Humatrope [Lilly] see 8672
Humedil [Bio-Mar] see 1131
Humegon [Organon] see 4189
Humic Acids, 4671
Huminsulin [Lilly] see 4887
Humorsol [Merck & Co.] see 2871
Humoryl [Delalande] see 9445
Humulene, 4672
α-Humulene see 4672
β-Humulene see 4672
Humulin [Lilly] see 4887
Humulin I [Lilly] see 5082
Humulin S [Lilly] see 4889
Humulon, 4673
Hungarian Chamomile see 5639
Hungarian Chamomile Oil see 6710
Hustazol [Takeda] see 2393
HVB see 4710
HWA 285 see 7823
2-HxG see 1221
HY-185 see 1808
Hyacin see 2866
Hyacinthin see 7236
Hyadur [Grünenthal] see 3247
Hyaenanchin see 4787
Hyalase [C.P. Pharm.] see 4676
Hyalgan [Fidia] see 4675
Hyalidase see 4676
Hyalobiuronic Acid, 4674
Hyalovet [Fidia] see 4675
Hyalozima see 4676
Hyaluronic Acid, 4675
Hyaluronidases, 4676
Hyamate [Xttrium] see 1490
Hyamine 1622 see 1083
Hyamine 10X [Rohm & Haas] see 5946
Hyanit [Burnus; Angelopharm] see 9781
Hyasmonta see 4676
Hyason see 4676

Hyasorb see 7041
Hyazyme [Abbott] see 4676
Hybenzate see MISC-3
Hybolin Decanoate [Hyrex] see 6281
Hybridoma Growth Factor see 4893
Hybridomas see 4837
Hybrin see 855
Hycanthone, 4677
Hycholin [Ayerst] see 7077
Hyclate see MISC-3
Hyclorate [Funai] see 2374
Hycobal [Eisai] see 2446
Hycolal see 345
Hycorace [OJ & F] see 4713
Hycozid see 5071
Hydac [Hoechst] see 3895
Hydantoin, 4678
Hydantol [Fujinaga-Sankyo] see 7293
Hydeltra [Merck & Co.] see 7719
Hydeltrasol [Merck & Co.] see 7721
Hydeltra-T.B.A. [Merck & Co.] see 7725
Hydeltrone-T.B.A. [Merck & Co.] see 7725
Hydergine [Sandoz] see 3598
Hydnocarpate Sodium see 4679
Hydnocarpic Acid, 4679
Hydnocarpus Oil see 2036
Hydnocarpylacetic Acid see 2037
Hydol see 4716
Hydoxamin see 9921
Hydrabamine Penicillin G see 7040
Hydrabamine Penicillin V see 7048
Hydrabamine Phenoxymethyl Penicillin see 7048
Hydracarbazine, 4680
Hydracillin see 7042
Hydracrylic Acid, 4681
Hydracrylic Acid β-Lactone see 7832
Hydracrylonitrile see 3751
Hydralazine, 4682
Hydrallostane, 4683
Hydramethylnon, 4684
Hydramitrazine, 4685
Hydramycin [Sankyo] see 3429
Hydrangea, 4686
Hydrangin see 9758
Hydraphen see 4687
Hydrargaphen, 4687
Hydrargyrum see 5801
Hydrastine, 4688
l-β-Hydrastine see 4688
Hydrastinine, 4689
Hydrastis, 4690
Hydrated Alumina see 345
Hydrated Cupric Oxide see 2646
Hydrated Ferric Oxide see 3969
Hydrated Magnesium-aluminum-iron Silicate see 9863
Hydrazid see 5071
Hydrazine, 4691
Hydrazine Acid Tartrate see 4694
Hydrazine Anhydrous see 4691
Hydrazine Bitartrate see 4694
Hydrazinecarbothioamide see 9292
Hydrazinecarboxamide Monohydrochloride see 8396
Hydrazinecarboximidamide see 453
Hydrazine Dihydriodide see 9608
Hydrazine Hydrate, 4692
Hydrazine Hydrogen Tartrate see 4694
Hydrazine Sulfate, 4693
Hydrazine Tartrate, 4694
Hydrazine Yellow see 9041
Hydrazinium Sulfate see 4693
Hydrazinobenzene see 7264
4-Hydrazinobenzenesulfonic Acid, 4695
2-Hydrazino-4,6-bis(diethylamino)-1,3,5-triazine see 4685

3-Hydrazino-6-carbamoylpyridazine see 4680
S-α-Hydrazino-3,4-dihydroxy-α-methylbenzenepropanoic Acid Monohydrate see 1802
(−)-L-α-Hydrazino-3,4-dihydroxy-α-methylhydrocinnamic Acid Monohydrate see 1802
2-Hydrazinoethanol, 4696
3-Hydrazino-6-[(2-hydroxypropyl)-methylamino]pyridazine see 7394
α-Hydrazino-α-methyl-β-(3,4-dihydroxyphenyl)propionic Acid Monohydrate see 1802
2-Hydrazinooctane see 6670
1-(2-Hydrazino-2-oxoethyl)pyridinium Chloride see 4323
1-Hydrazinophthalazine see 4682
3-Hydrazinopyridazine-6-carboxamide see 4680
6-Hydrazino-3-pyridazinecarboxamide see 4680
(±)-1-[(6-Hydrazino-3-pyridazinyl)-methylamino]-2-propanol see 7394
2-Hydrazino-N,N,N-trimethyl-2-oxoethanaminium Chloride see 4323
Hydrazodicarbonimide see 9780
Hydrazoic Acid, 4697
Hydrazonium Sulfate see 4693
Hydrea [Squibb] see 4785
Hydrenox [Boots] see 4716
Hydridocarbonyltris(triphenylphosphine)rhodium(I) see 9670
Hydridocobalamin see 9926
Hydril see 4704
Hydrindantin, 4698
Hydrindene see 4844
Hy-Drine [Zemmer] see 1134
Hydriodic Acid, 4699
Hydrion [Robert et Carrière] see 396
Hydrionic [Upjohn] see 4363
Hydro-Adreson see 4710
Hydro-Aquil see 4704
Hydrobenzoin, 4700
Hydrobromic Acid, 4701
Hydrobromic Ether see 3730
Hydrocarbostyril, 4702
Hydrocerussite see 1016
Hydrochlorbenzethylamine see 3832
Hydrochloric Acid, 4703
Hydrochloroauric Acid see 4428
Hydrochlorothiazide, 4704
Hydrocinchonidine, 4705
Hydrocinchonine, 4706
Hydrocinnamic Acid, 4707
Hydrocodin see 3154
Hydrocodone, 4708
Hydrocomp [Beiersdorf] see 417
Hydroconchinine see 4736
Hydrocort [Ferring] see 4710
Hydrocortamate, 4709
Hydrocortisat see 4711
Hydrocortisone, 4710
Δ¹-Hydrocortisone see 7719
Hydrocortisone Acetate, 4711
Hydrocortisone tert-Butylacetate see 4714
Hydrocortisone 21-Diethylaminoacetate see 4709
Hydrocortisone 21-(Dihydrogen Phosphate) see 4712
Hydrocortisone 21-(3,3-Dimethylbutanoic Acid Ester) see 4714
Hydrocortisone 21-(3,3-Dimethylbutyrate) see 4714
Hydrocortisone 21-β,β-Dimethylbutyrate see 4714
Hydrocortisone Free Alcohol see 4710
Hydrocortisone Hemisuccinate Sodium Salt see 4713

Hydrocortisone Phosphate, 4712
21-Hydrocortisonephosphoric Acid see 4712
Hydrocortisone 21-Sodium Succinate, 4713
Hydrocortisone TBA see 4714
Hydrocortisone Tebutate, 4714
Hydrocortisone Valerate see 4710
Hydrocortistab [Boots] see 4711
Hydrocortisyl see 4710
Hydrocortone [Merck & Co.] see 4710
Hydrocortone Acetate [Merck & Co.] see 4711
Hydrocortone Sodium Phosphate [Merck & Co.] see 4712
Hydrocortone-TBA [Merck & Co.] see 4714
Hydrocotarnine, 4715
Hydrocupreine see 2626
Hydrocupreine Ethyl Ether see 3765
Hydrocupreine Isopentyl Ether see 3861
Hydrocyanic Acid see 4722
Hydrocyanite see 2659
Hydrodeltalone see 7719
Hydrodeltisone see 7719
Hydro-Diuril [Merck & Co.] see 4704
Hydroflumethiazide, 4716
Hydrofluoboric Acid see 4075
Hydrofluoric Acid, 4717
Hydrofluoric Acid Gas see 4723
Hydrofluosilicic Acid see 4110
α-Hydroformamine Cyanide see 5976
Hydrofuramide, 4718
Hydrogen, 4719
Hydrogen Arsenide see 837
Hydrogen Azide see 4697
Hydrogen Bis(benzenamine-N)tetrakis(thiocyanato-N)chromate(1−) see 8183
Hydrogen Bromide, 4720
Hydrogen Chloride, 4721
[Hydrogen Citrato(3−)]triaquoiron, Choline Salt see 3986
Hydrogen Cyanamide see 2691
Hydrogen Cyanate see 2693
Hydrogen Cyanide, 4722
Hydrogen Dioxide see 4725
Hydrogen Dioxide Soln see 4726
Hydrogen Fluoride, 4723
Hydrogen Hexachloroplatinate(IV) see 7498
Hydrogen Hexafluorosilicate see 4110
Hydrogen Iodide, 4724
Hydrogen Oxide see 9951
Hydrogen Peroxide, 4725
Hydrogen Peroxide Carbamide see 9783
Hydrogen Peroxide Solution 3%, 4726
Hydrogen Peroxide Solution 30%, 4727
Hydrogen Selenide, 4728
Hydrogen Sulfide, 4729
Hydrogen Telluride, 4730
Hydrogen Tetrabromoaurate(1−) see 4426
Hydrogen Tetracarbonylferrate(II), 4731
Hydrogen Tetrachloroaurate(1−) see 4428
Hydrogen Tetrafluoroborate see 4075
Hydrogen Thiocyanate see 9257
Hydrogen Tribromide Compd with Pyridine (1:1) see 7985
Hydrogen Trioxooxalatomolybdate-(VI) see 6866
Hydro-Giene [Water Science] see 2669
Hydrogrisevit see 4739
Hydrohydrastinine, 4732

α-Hydro-ω-hydroxypoly(oxy-1,2-ethanediyl) see 7545

α-Hydro-ω-hydroxypoly(oxyethylene)poly(oxypropylene)poly(oxyethylene) Block Copolymers see 7537

Hydrokon see 4708

Hydroled [Lederle] see 4221

Hydro-long [Sanorania] see 2194

Hydrolose [Upjohn] see 5963

Hydrolum see 345

Hydro-Magma [Merck & Co.] see 5548

Hydromedin [Merck & Co.] see 3669

Hydromorphone, 4733

Hydromox [Lederle] see 8067

Hydrone®, 4734

Hydronitric Acid see 4697

Hydronol [Stuart] see 5113

Hydronsan see 4396

Hydroorotic Acid, 4735

Hydroperoxide see 4725

5-Hydroperoxy-6,8,11,14-eicosatetraenoic Acid see 4594

Hydropyrin see 5397

Hydroquinidine, 4736

Hydroquinine, 4737

Hydroquinol see 4738

Hydroquinone, 4738

Hydroquinone Benzyl Ether see 6159

Hydroquinone Calcium Sulfonate see 3395

Hydroquinone-β-D-glucopyranoside see 799

Hydroquinone Glucose see 799

Hydroquinone Monobenzyl Ether see 6159

Hydro-rapid [Sanorania] see 4221

Hydroresorcinol see 3161

Hydroretrocortine see 7719

Hydrosaluric [Merck & Co.] see 4704

Hydrosarpan [Eutherapie] see 8129

Hydrosilicofluoric Acid see 4110

"Hydrosulfuric Acid" see 4729

Hydrotalcite see 5538

Hydro-Tamin [Ratiopharm] see 3157

Hydrothide see 4704

Hydrothiobicin see 4163

Hydroton see 2194

Hydrotrichlorothiazide see 9537

Hydrotricine see 9749

Hydrourushiol see 7061

Hydrovit see 4739

Hydroxamethocaine see 4783

Hydroxobase [Hommel] see 9921

Hydroxocobalamin, 4739

Hydroxocobemine see 4739

N-Hydroxyacetamide see 54

p-Hydroxyacetanilide see 40

4′-Hydroxyacetanilide see 40

Hydroxyacetic Acid see 4394

Hydroxyacetic Acid 6-Ethenyldecahydro-5-hydroxy-4,6,9,10-tetramethyl-1-oxo-3a,9-propano-3aH-cyclopentacycloocten-8-yl Ester see 7508

p-Hydroxyacetophenone-D-glucoside see 7367

3-Hydroxy-21-acetoxy-5-pregnen-20-one see 70

[[4-[(Hydroxyacetyl)amino]phenyl]arsonato(1−)]oxobismuth see 4389

[4-[(Hydroxyacetyl)amino]phenyl]arsonic Acid see 4375

4-Hydroxy-N-acetylproline see 6857

4-Hydroxyaflatoxin B₁ see 170

4-Hydroxyaflatoxin B₂ see 170

β-Hydroxyalanine see 8411

2-Hydroxy-6-alkylbenzoic Acids see 657

17-Hydroxy-17α-allyl-4-estrene see 289

l-3-Hydroxy-N-allylmorphinan see 5342

γ-Hydroxy-β-aminobutyric Acid see 455

α-Hydroxy-β-aminodibenzyl see 449

1-Hydroxy-2-aminodiphenylethane see 449

m-Hydroxy-α-(1-aminoethyl)benzyl Alcohol see 5837

5-Hydroxy-3-(β-aminoethyl)indole see 8413

3-Hydroxy-5-aminomethylisoxazole see 6221

6-(p-Hydroxy-α-aminophenylacetamido)penicillanic Acid see 610

α-Hydroxy-β-aminopropylbenzene Hydrochloride see 7279

Hydroxyamphetamine, 4740

p-Hydroxyampicillin see 610

17-Hydroxyandrosta-1,4-dien-3-one see 1327

3α-Hydroxy-5α-androstan-17-one see 673

3α-Hydroxy-5β-androstan-17-one see 8011

3α-Hydroxy-17-androstanone see 673

3β-Hydroxy-5α-androstan-17-one see 3562

3β-Hydroxy-17-androstanone see 3562

17-Hydroxyandrostan-3-one see 8753

17β-Hydroxy-3-androstanone see 8753

3α-Hydroxy-5α-androst-16-ene see 672

3-Hydroxyandrost-5-en-17-one see 7710

17β-Hydroxyandrost-4-en-3-one see 9109

p-Hydroxyaniline see 474

2-Hydroxyaniline see 473

3-Hydroxyaniline see 472

p-Hydroxyanilinoacetic Acid see 4771

o-Hydroxyanisole see 4457

9-Hydroxyanthracene see 716

1-Hydroxy-2-anthraquinonyl 6-O-β-D-Xylopyranosyl-β-D-glucopyranoside see 8257

Hydroxyapatite see 3449

p-Hydroxybenzaldehyde, 4741

2-Hydroxybenzaldehyde see 8295

2-Hydroxybenzaldehyde Oxime see 8296

o-Hydroxybenzal Isonicotinylhydrazone see 8304

2-Hydroxybenzamide see 8297

N-Hydroxybenzenamine see 7266

Hydroxybenzene see 7206

α-Hydroxybenzeneacetic Acid see 5599

endo-(±)-α-Hydroxybenzeneacetic Acid 8-Methyl-8-azabicyclo[3.2.1]-oct-3-yl Ester see 4649

α-Hydroxybenzeneacetic Acid 3-Methylbutyl Ester see 5600

α-Hydroxybenzeneacetic Acid Monoammonium Salt see 557

α-Hydroxybenzeneacetic Acid 1,2,2,6-Tetramethyl-4-piperidinyl Ester see 3854

α-Hydroxybenzeneacetic Acid 3,3,5-Trimethylcyclohexyl Ester see 2708

α-Hydroxybenzeneacetonitrile see 5601

4-p-Hydroxybenzeneazo-1-p-chloro-o-sulfophenyl-3-methyl-5-hydroxypyrazole Toluene-p-sulfonyl Ester Sodium Salt see 7528

2-Hydroxybenzenecarbodithioic Acid see 3381

4-Hydroxy-1,3-benzenedicarboxylic Acid see 4758

4-Hydroxy-m-benzenedisulfonic Acid see 7207

4-Hydroxy-1,3-benzenedisulfonic Acid see 7207

2-Hydroxybenzenemethanol see 8294

4-Hydroxybenzenesulfonic Acid see 7212

p-Hydroxybenzenesulfonic Acid Calcium Salt see 1697

p-Hydroxybenzenesulfonic Acid Copper Salt see 2652

Hydroxybenzenesulfonic Acid Potassium Salt see 7645

Hydroxybenzenesulfonic Acid Sodium Salt see 8610

p-Hydroxybenzenesulfonic Acid Zinc Salt see 10054

2-Hydroxybenzhydroxamic Acid see 8300

(2-Hydroxybenzoato-O¹)oxobismuth see 1299

6-Hydroxy-5-benzofuranacrylic Acid δ-Lactone see 7944

p-Hydroxybenzoic Acid, 4742

2-Hydroxybenzoic Acid see 8301

2-Hydroxybenzoic Acid Acetate 2-Carboxyphenyl Ester see 95

2-Hydroxybenzoic Acid 4-(Acetylamino)phenyl Ester see 41

2-Hydroxybenzoic Acid Bismuth (3+) Salt, Basic see 1299

4-Hydroxybenzoic Acid Butyl Ester see 1583

2-Hydroxybenzoic Acid 2-Carboxyphenyl Ester see 8307

2-Hydroxybenzoic Acid Compd with 1,2-Dihydro-1,5-dimethyl-2-phenyl-3H-pyrazol-3-one (1:1) see 749

2-Hydroxybenzoic Acid Compd with 1H-Imidazole (1:1) see 4829

2-Hydroxybenzoic Acid Compd with Morpholine (1:1) see 6195

2-Hydroxybenzoic Acid Copper Salt see 2654

2-Hydroxybenzoic Acid 4-(1,1-Dimethylethyl)phenyl Ester see 1585

2-Hydroxybenzoic Acid Ethyl Ester see 3804

4-Hydroxybenzoic Acid Ethyl Ester see 3792

p-Hydroxybenzoic Acid Ethyl Ester 6-Guanidinohexanoate see 4234

2-Hydroxybenzoic Acid 2-Hydroxyethyl Ester see 4395

2-Hydroxybenzoic Acid Magnesium Salt see 5565

2-Hydroxybenzoic Acid Methyl Ester see 6038

4-Hydroxybenzoic Acid Methyl Ester see 6021

2-Hydroxybenzoic Acid 5-Methyl-2-(1-methylethyl)cyclohexyl Ester see 5727

2-Hydroxybenzoic Acid Monoammonium Salt see 578

2-Hydroxybenzoic Acid Monosodium Salt see 8626

2-Hydroxybenzoic Acid Monosodium Salt Compd With 2-(1,2,3,6-Tetrahydro-3-hydroxy-1-methyl-6-oxo-5H-indol-5-ylidene)hydrazinecarboxamide (1:1) see 1790

2-Hydroxybenzoic Acid 1-Naphthalenyl Ester see 6334

2-Hydroxybenzoic Acid 2-Naphthalenyl Ester see 6335

2-(Hydroxymethyl)-1,1-dimethylpi-
peridinium Methyl Sulfate Benzilate
see 1214

2-Hydroxymethyl-1,1-dimethylpyrro-
lidinium Methyl Sulfate Benzilate
see 7529

4-(Hydroxymethyl)-1,3-dioxolane *see*
4380

9-Hydroxy-2-methylellipticinium
Acetate *see* 3510

7-Hydroxy-5′,6′-methylenedioxy-
benzofurano(3′,2′:3,4)coumarin *see*
5672

7-Hydroxy-11,12-(methylenedioxy)-
coumestan *see* 5672

7-Hydroxy-3′,4′-(methylenedioxy)-
isoflavone *see* 7925

1-Hydroxy-6,7-methylenedioxy-2-
methyl-1,2,3,4-tetrahydroisoquino-
line *see* 4689

2-Hydroxymethylene-17α-methylan-
drostan-17β-ol-3-one *see* 6920

2-Hydroxymethylene-17α-methyl-
dihydrotestosterone *see* 6920

2-Hydroxymethylene-17α-methyl-
17β-hydroxy-5α-androstan-3-one
see 6920

17-Hydroxy-16-methylenepregna-4,6-
diene-3,20-dione *see* 4763

17-Hydroxy-16-methylene-Δ⁶-pro-
gesterone, 4763

17β-Hydroxy-17-methylestra-4,9,11-
trien-3-one *see* 6049

17β-Hydroxy-17-methylestr-4-en-3-
one *see* 6629

1-[6-Hydroxy-2-(1-methylethenyl)-5-
benzofuranyl]ethanone *see* 3857

2-(2-Hydroxy-1-methylethenyl)-5-
methylcyclopentanecarboxylic Acid
Delta Lactone *see* 6383

4-[1-Hydroxy-2-[(1-methylethyl)-
amino]butyl]-1,2-benzenediol *see*
5053

4-[1-Hydroxy-2-[(1-methylethyl)-
amino]ethyl]-1,2-benzenediol *see*
5105

5-[1-Hydroxy-2-[(1-methylethyl)-
amino]ethyl]-1,3-benzenediol *see*
5836

N-[4-[1-Hydroxy-2-[(1-methylethyl)-
amino]ethyl]phenyl]methanesulfon-
amide *see* 8682

4-[2-Hydroxy-3-[(1-methylethyl)-
amino]propoxy]benzeneacetamide
see 879

4-[2-Hydroxy-3-[(1-methylethyl)-
amino]propoxy]benzenepropanoic
Acid Methyl Ester *see* 3648

1-[7-[2-Hydroxy-3-[(1-methylethyl)-
amino]propoxy]-2-benzofuranyl]-
ethanone *see* 1032

(S)-4-[2-Hydroxy-3-[(1-methylethyl)-
amino]propoxy]phenol *see* 7742

N-[4-[2-Hydroxy-3-[(1-methylethyl)-
amino]propoxy]phenyl]acetamide
see 7703

4-[2-Hydroxy-3-[(1-methylethyl)-
amino]propoxy]-2,3,6-trimethylphe-
nol 1-Acetate *see* 6059

3-(1-Hydroxy-1-methylethyl)glutaric
Acid γ-Lactone *see* 9100

N-[α-(Hydroxymethyl)ethyl]-D-
lysergamide *see* 3600

12′-Hydroxy-2′(1-methylethyl)-5′α-
(2-methylpropyl)ergotaman-3′,6′,-
18-trione *see* 3595

12′-Hydroxy-2′-(1-methylethyl)-
5′α-(2-methylpropyl)-8α-ergot-
aman-3′,6′,18-trione *see* 3596

12′-Hydroxy-2′-(1-methylethyl)-5′-
(phenylmethyl)ergotaman-3′,6′,18-
trione *see* 3593

12′-Hydroxy-2′-(1-methylethyl)-
5′α-(phenylmethyl)-8α-ergotaman-
3′,6′,18-trione *see* 3594

(1-Hydroxy-1-methylethyl)succinic
Acid γ-Lactone *see* 9092

5-Hydroxymethyl-2-formylfuran *see*
4764

5-(Hydroxymethyl)-2-furaldehyde,
4764

2-Hydroxymethylfuran *see* 4215

5-(Hydroxymethyl)-2-furancarbonal
see 4764

5-(Hydroxymethyl)-2-furancarboxal-
dehyde *see* 4764

5-(Hydroxymethyl)-2-furfural *see*
4764

3-Hydroxy-3-methylglutaraldehydic
Acid *see* 6086

3-Hydroxy-3-methylglutaric Acid *see*
5690

3-C-(Hydroxymethyl)-D-glyceroaldo-
tetrose *see* 768

2-Hydroxymethyl-3-hydroxy-6-(1-
hydroxy-2-tert-butylaminoethyl)-
pyridine *see* 7461

2-Hydroxymethyl-5-hydroxy-γ-
pyrone *see* 5197

3-Hydroxy-1-methyl-5,6-indoline-
dione *see* 162

3-Hydroxy-1-methyl-1,5,6-indoline-
dione Semicarbazone Compd with
Sodium Salicylate *see* 1790

4-Hydroxymethyl-2-iodomethyl-1,3-
dioxolane *see* 3410

6-Hydroxymethyl-2-isopropylamino-
methyl-7-nitro-1,2,3,4-tetrahydro-
quinoline *see* 6872

8-(Hydroxymethyl)-4-isopropyl-1,7-
dimethylbicyclo[3.2.1]oct-6-ene-6-
carboxaldehyde *see* 4549

4-Hydroxy-3-(5-methyl-3-isoxazo-
locarbamyl)-2-methyl-2H-1,2-
benzothiazine 1,1-Dioxide *see* 5127

2-Hydroxy-4-(methylmercapto)butyr-
ic Acid *see* 5897

2-Hydroxymethyl-5-methoxyfurano-
chrome Glucoside *see* 162

3-(Hydroxymethyl)-7-methoxy-8-
oxo-7-[2-(2-thienyl)acetamido]-5-
thia-1-azabicyclo[4.2.0]oct-2-ene-
2-carboxylic Acid Carbamate
(Ester) *see* 1938

[2S-(2R*,3S*,4S*,6E)]-3-Hydroxy-4-
methyl-2-(methylamino)-6-octenoic
Acid *see* 5651

α-(Hydroxymethyl)-α-methylbenz-
eneacetic Acid 8-Methyl-8-azabicy-
clo[3.2.1]oct-3-yl Ester *see* 5345

17α-Hydroxy-6-methyl-16-methyl-
enepregna-4,6-diene-3,20-dione *see*
5697

2-Hydroxy-6-methyl-3-(1-methyl-
ethyl)benzoic Acid *see* 9342

2-Hydroxy-3-methyl-6-(1-methyleth-
yl)-2-cyclohexen-1-one *see* 3292

4-Hydroxy-2-methyl-N-(5-methyl-3-
isoxazolyl)-2H-1,2-benzothiazine-
3-carboxamide 1,1-Dioxide *see*
5127

3β-(Hydroxymethyl)-2α-methyl-4β-
[9-methyl-9H-pyrido[3,4-b]indol-
1-yl)methyl]-2H-pyran-5-carboxyl-
ic Acid Methyl Ester *see* 313

5-(Hydroxymethyl)-3-(3-methylphen-
yl)-2-oxazolidinone *see* 9445

N-Hydroxy-α-methyl-4-(2-methyl-
propyl)benzeneacetamide *see* 4813

12′-Hydroxy-2′-methyl-5′α-(2-
methylpropyl)ergotaman-3′,6′,18-
trione *see* 3601

5-Hydroxymethyl-6-methyl-2,4(1H,-
3H)-pyrimidinedione *see* 7093

N-(Hydroxymethyl)-N′-methylthio-
urea *see* 6645

1-(Hydroxymethyl)-3-methyl-2-thio-
urea *see* 6645

5-Hydroxymethyl-4-methyluracil *see*
7093

5-Hydroxymethyl-6-methyluracil *see*
7093

(−)-3-Hydroxy-N-methylmorphinan
see 5350

2-Hydroxy-3-methyl-1,4-naphtha-
lenedione *see* 7352

5-Hydroxy-2-methyl-1,4-naphtha-
lenedione *see* 7511

2-Hydroxy-3-methyl-1,4-naphtho-
quinone *see* 7352

5-Hydroxy-2-methyl-1,4-naphtho-
quinone *see* 7511

N-(Hydroxymethyl)nicotinamide,
4765

D-7-[(4-Hydroxy-6-methylnicotin-
amido)-4-hydroxyphenylacetamido]-
3-(1-methyltetrazol-5-yl)thiometh-
ylcephem-4-carboxylic Acid *see*
1940

7-[(R)-2-(4-Hydroxy-6-methylnico-
tinamido)-2-(p-hydroxyphenyl)acet-
amido]-3-[[(1-methyl-1H-tetrazol-
5-yl)thio]methyl]-8-oxo-5-thia-1-
azabicyclo[4.2.0]oct-2-ene-2-carb-
oxylic Acid *see* 1940

3-(Hydroxymethyl)-1-[[(5-nitro-2-
furanyl)methylene]amino]-2,4-imid-
azolidinedione *see* 6455

3-(Hydroxymethyl)-1-[(5-nitrofurfur-
ylidene)amino]hydantoin *see* 6455

6-(Hydroxymethyl)-2-[2-(5-nitro-2-
furyl)vinyl]pyridine *see* 6452

2-(Hydroxymethyl)-2-nitro-1,3-pro-
panediol *see* 9667

17β-Hydroxy-17-methyl-19-noran-
drosta-4,9,11-trien-3-one *see* 6049

(7α,17α)-17-Hydroxy-7-methyl-19-
norpregn-5(10)-en-20-yn-3-one *see*
9358

17-Hydroxy-17-methyl-2-oxaandro-
stan-3-one *see* 6875

(1R)-3-Hydroxy-4-methyl-7-oxabicy-
clo[4.1.0]hept-3-ene-2,5-dione *see*
9104

(2β,4α,15α)-15-Hydroxy-2-[[2-O-(3-
methyl-1-oxobutyl)-3,4-di-O-sulfo-
β-D-glucopyranosyl]oxy]-19-nor-
kaur-16-en-18-oic Acid Dipotassium Salt *see* 883

7-Hydroxy-4-methyl-2-oxo-3-
chromene *see* 4792

1-(8-Hydroxy-6-methyl-1-oxo-2,4,6-
dodecatrienyl)-2-pyrrolidinone *see*
7008

3-Hydroxy-3-methyl-5-oxopentanoic
Acid *see* 6086

6-Hydroxy-7-methyl-9-oxopodo-
carpane *see* 6463

(E)-3β-Hydroxy-14α-methyl-7-oxo-
podocarpane-Δ¹³,α-acetic Acid 2-
(Dimethylamino)ethyl Ester *see*
1896

3-(Hydroxymethyl)-8-oxo-7-[2-(4-
pyridylthio)acetamido]-5-thia-1-
azabicyclo[4.2.0]oct-2-ene-2-carb-
oxylic Acid Acetate Monosodium
Salt *see* 1980

3-(Hydroxymethyl)-8-oxo-7-[2-(2-thienyl)acetamido]-5-thia-1-azabicyclo[4.2.0]oct-2-ene-2-carboxylic Acid Acetate *see* 1978

3-Hydroxy-3-methylpentanedioic Acid *see* 5690

4-Hydroxy-4-methyl-2-pentanone *see* 2944

2-(1-Hydroxy-4-methyl-3-pentenyl)-5,8-dihydroxy-1,4-naphthoquinone *see* 243

1-Hydroxy-3-methyl-2-penten-4-yne *see* 7088

1-Hydroxy-5-methylphenazinium Hydroxide Inner Salt *see* 7965

dl-*p*-Hydroxy-α-methylphenethylamine *see* 4740

L-3-[(β-Hydroxy-α-methylphenethyl)amino]-3′-methoxypropiophenone *see* 6915

4-Hydroxy-α-[1-[(1-methyl-2-phenoxyethyl)amino]ethyl]benzenemethanol *see* 5128

p-Hydroxy-α-[1-[(1-methyl-2-phenoxyethyl)amino]ethyl]benzyl Alcohol *see* 5128

p-Hydroxy-*N*-(1-methyl-2-phenoxyethyl)norephedrine *see* 5128

3-Hydroxy-α-methylphenylalanine *see* 6050

4-Hydroxy-α-methylphenylalanine *see* 6085

1-(1-Hydroxy-4-methyl-2-phenylazo)-2-naphthol-4-sulfonic Acid *see* 1725

2-(2′-Hydroxy-5′-methylphenyl)-benzotriazole *see* 9381

[*R*-(*R**,*S**)]-3-[(2-Hydroxy-1-methyl-2-phenylethyl)amino]-1-(3-methoxyphenyl)-1-propanone *see* 6915

2-(Hydroxymethyl)phenyl-β-D-glucopyranoside *see* 8293

2-(Hydroxymethyl)phenyl-β-D-glucopyranoside 6-Benzoate *see* 7571

12′-Hydroxy-2′-methyl-5′α-(phenylmethyl)ergotaman-3′,6′,18-trione *see* 3609

12′-Hydroxy-2′-methyl-5′α-(phenylmethyl)-8α-ergotaman-3′,6′,18-trione *see* 3610

4-Hydroxy-α-[1-[(1-methyl-3-phenylpropyl)amino]ethyl]benzenemethanol *see* 6654

p-Hydroxy-α-[1-[(1-methyl-3-phenylpropyl)amino]ethyl]benzyl Alcohol *see* 6654

5-[1-Hydroxy-2-[(1-methyl-3-phenylpropyl)amino]ethyl]salicylamide *see* 5208

p-Hydroxy-*N*-(1-methyl-3-phenylpropyl)norephedrine *see* 6654

2-Hydroxy-5-[[(4-methylphenyl)-sulfonyl]oxy]benzenesulfonic Acid Compd with Piperazine (1:1) *see* 8976

3-Hydroxy-2-methyl-5-[(phosphonooxy)methyl]-4-pyridinecarboxaldehyde *see* 7992

6-Hydroxy-2-methyl-2-phytylchroman *see* 9416

[2*R*-(2α,3β,4α,5β)]-2-(Hydroxymethyl)-3,4,5-piperidinetriol *see* 2887

17-Hydroxy-6-methylpregna-4,6-diene-3,20-dione Acetate *see* 5687

(6α)-17-Hydroxy-6-methylpregn-4-ene-3,20-dione *see* 5677

11-Hydroxy-6-methylpregn-4-ene-3,20-dione *see* 5679

17-Hydroxy-6-methylpregn-4-en-20-one *see* 658

11β-Hydroxy-6α-methylprogesterone *see* 5679

17α-Hydroxy-6α-methylprogesterone *see* 5677

1-Hydroxymethylpropane *see* 5015

2-Hydroxy-2-methylpropanenitrile *see* 59

N-[α-(Hydroxymethyl)propyl]-D-lysergamide *see* 5989

N-[1-(Hydroxymethyl)propyl]-1-methyl-*d*-(+)-lysergamide *see* 6055

N-[α-(Hydroxymethyl)propyl]-1-methyl-D-lysergamide *see* 6055

1-Hydroxy-6-(1-methylpropyl)-3-(2-methylpropyl)-2(1*H*)-pyrazinone *see* 863

17β-Hydroxy-6-methyl-17-(1-propynyl)androst-4-en-3-one *see* 3208

12-Hydroxy-*N*-methylpseudostrychnine *see* 9946

3-Hydroxy-2-methyl-4*H*-pyran-4-one *see* 5594

2-Hydroxy-5-methylpyrazolo[1,5-*a*]pyrimidine-6-carboxylic Acid Ethyl Ester, *O*-Ester with *O,O*-Diethyl Phosphorthioate *see* 7976

3-Hydroxymethylpyridine *see* 6438

N-(Hydroxymethyl)-3-pyridinecarboxamide *see* 4765

3-Hydroxymethylpyridine *p*-Chlorophenoxyisobutyrate *see* 6427

5-Hydroxy-6-methyl-3,4-pyridinedicarbinol Hydrochloride *see* 7995

5-Hydroxy-6-methyl-3,4-pyridinedimethanol Hydrochloride *see* 7995

3-Hydroxy-1-methylpyridinium Bromide Dimethylcarbamate *see* 7990

3-Hydroxy-1-methylpyridinium Bromide Hexamethylenebis[methylcarbamate] *see* 3368

4-Hydroxy-2-methyl-*N*-2-pyridinyl-2*H*-1,2-benzothiazine-3-carboxamide 1,1-Dioxide *see* 7476

[6*R*-[6α,7β(*R**)]]-7-[[[[(4-Hydroxy-6-methyl-3-pyridinyl)carbonyl]amino](4-hydroxyphenyl)acetyl]amino]-3-[[(1-methyl-1*H*-tetrazol-5-yl)thio]methyl]-8-oxo-5-thia-1-azabicyclo[4.2.0]oct-2-ene-2-carboxylic Acid *see* 1940

4-Hydroxy-2-methyl-*N*-2-pyridinyl-2*H*-thieno[2,3-*e*]-1,2-thiazine-3-carboxamide 1,1-Dioxide *see* 9080

3-Hydroxy-2-methyl-4-pyrone *see* 5594

3-Hydroxy-2-methyl-γ-pyrone *see* 5594

3-Hydroxy-1-methylquinuclidinium Bromide Benzilate *see* 2351

2-*C*-(Hydroxymethyl)-D-ribofuranose 2′,5-Digallate *see* 4522

2-*C*-(Hydroxymethyl)-D-ribose *see* 4523

17-Hydroxy-4-methylsarpaganium *see* 5521

4-Hydroxy-19-methyl-16,19-secostrychnidine-10,16-dione *see* 9946

2-Hydroxy-2-methylsuccinic Acid *see* 2325

4-Hydroxy-17α-methyltestosterone *see* 6918

2-Hydroxy-4-(methylthio)-butanoic Acid *see* 5897

2-Hydroxy-4-(methylthio)butyric Acid *see* 5897

5-(Hydroxymethyl)-3-*m*-tolyl-2-oxazolidinone *see* 9445

2-Hydroxy-2-methyl-4-(2,2,2-trichloro-1-hydroxyethoxy)pentane *see* 2089

(2*R*,3*R*,4*R*,5*S*)-2-Hydroxymethyl-3,4,5-trihydroxypiperidine *see* 2887

2-Hydroxy-13-methyl-3,9,10-trimethoxyberbine *see* 5045

3-Hydroxy-13-methyl-2,9,10-trimethoxyberbine *see* 2539

(*E,Z,E*)-7-Hydroxymethyl-3,11,15-trimethyl-2,6,10,14-hexadecatetraen-1-ol *see* 7506

1-Hydroxy-4-methyl-6-(2,4,4-trimethylpentyl)-2(1*H*)-pyridinone *see* 7472

3α-Hydroxy-8-methyl-1α*H*,5α*H*-tropanium Bromide 2-Propylvalerate *see* 701

3-Hydroxy-α-methyl-L-tyrosine *see* 5974

2-[4-(Hydroxymethylureidosulfonyl)-phenylcarbamoyl]benzoic Acid *see* 8883

β-Hydroxy-β-methyl-δ-valerolactone *see* 6087

17-Hydroxy-13-methyl-17α-vinyl-1,2,3,6,7,8,9,10,11,12,13,14,16,17-tetradecahydro-15*H*-cyclopenta[*a*]-phenanthren-3-one *see* 6637

(−)-3-Hydroxymorphinan *see* 6625

2-Hydroxy-5-(morpholinothiocarbonyl)anisole *see* 9841

Hydroxymycin *see* 6989

α-Hydroxynaphthalene *see* 6303

β-Hydroxynaphthalene *see* 6304

3-Hydroxy-2-naphthalenecarboxylic Acid *see* 4767

3-Hydroxy-2-naphthalenecarboxylic Acid Hexadecyl Ester *see* 4604

2-Hydroxy-1,4-naphthalenedione *see* 5263

5-Hydroxy-1,4-naphthalenedione *see* 5150

2-Hydroxynaphthalene-6,8-disulfonic Acid *see* 6308

3-Hydroxy-2,7-naphthalenedisulfonic Acid *see* 6307

4-Hydroxy-1,5-naphthalenedisulfonic Acid *see* 6306

7-Hydroxy-1,3-naphthalenedisulfonic Acid *see* 6308

2-Hydroxynaphthalenedisulfonic Acid Aluminum Salt *see* 353

6-Hydroxy-2-naphthalenepropanoic Acid *see* 247

2-Hydroxy-6-naphthalenepropionic Acid *see* 247

1-Hydroxy-2-naphthalenesulfonic Acid *see* 6310

4-Hydroxy-1-naphthalenesulfonic Acid *see* 6311

6-Hydroxy-2-naphthalenesulfonic Acid *see* 6312

7-Hydroxy-1-naphthalenesulfonic Acid *see* 2591

7-Hydroxy-2-naphthalenesulfonic Acid *see* 1898

2-Hydroxy-1-naphthalenesulfonic Acid Calcium Salt *see* 847

4-[(2-Hydroxy-1-naphthalenyl)azo]benzenesulfonic Acid Monosodium Salt *see* 6813

4-[(4-Hydroxy-1-naphthalenyl)azo]benzenesulfonic Acid Monosodium Salt *see* 6812

N-Hydroxynaphthalimide Diethyl Phosphate *see* 6269

1-Hydroxy-2-naphthoic Acid, 4766

3-Hydroxy-2-naphthoic Acid, 4767

2-Hydroxy-1,4-naphthoquinone *see* 5263

5-Hydroxy-1,4-naphthoquinone *see* 5150

1-(*p*-Hydroxyphenyl)-2-aminoethanol *see* 6679

1-(*m*-Hydroxyphenyl)-2-amino-1-propanol *see* 5837

α-(*m*-Hydroxyphenyl)-β-aminopropanol *see* 5837

4-Hydroxy-5-[[4-(phenylamino)-5-sulfo-1-naphthalenyl]azo]-2,7-naphthalenedisulfonic Acid Trisodium Salt *see* 662

(4-Hydroxyphenyl)arsonic Acid Polymer with Formaldehyde *see* 7539

2-Hydroxy-*N*-phenylbenzamide *see* 8299

α-Hydroxy-α-phenylbenzeneacetic Acid *see* 1089

α-Hydroxy-α-phenylbenzeneacetic Acid 2-(Diethylamino)ethyl Ester *see* 1037

α-Hydroxy-α-phenylbenzeneacetic Acid 2-(Dimethylamino)-2-methylpropyl Ester *see* 3122

α-Hydroxy-α-phenylbenzeneacetic Acid 2-(Ethylpropylamino)ethyl Ester *see* 1039

α-Hydroxy-α-phenylbenzeneacetic Acid 6-Methoxy-8-methyl-8-azabicyclo[3.2.1]oct-3-yl Ester *see* 9690

endo-α-Hydroxy-α-phenylbenzeneacetic Acid 8-Methyl-8-azabicyclo[3.2.1]oct-3-yl Ester *see* 9694

α-Hydroxy-α-phenylbenzeneacetic Acid 2-(1-Piperidinyl)ethyl Ester *see* 7441

p-Hydroxyphenyl Benzyl Ether *see* 6159

3α-Hydroxy-8-(*p*-phenylbenzyl)-1α*H*,5α*H*-tropanium Bromide (±)-Tropate *see* 9983

N-(4-Hydroxyphenyl)butanamide *see* 4745

p-Hydroxyphenylbutazone *see* 6925

1-(*p*-Hydroxyphenyl)-2-butylaminoethanol *see* 964

2-Hydroxy-2-phenylbutyl Carbamate *see* 4770

3-Hydroxy-4-phenylcarbostyril *see* 9911

3-Hydroxy-2-phenylcinchoninic Acid *see* 6910

α-(1-Hydroxy-4-phenylcyclohexyl)-butyric Acid *see* 3912

1-Hydroxy-α-phenylcyclopentaneacetic Acid 2-Diethylaminoethyl Ester *see* 2725

1-Hydroxy-α-phenylcyclopentaneacetic Acid 2-(Dimethylamino)ethyl Ester *see* 2752

2-(*p*-Hydroxyphenyl)-5,7-dihydroxychromone *see* 763

O-(4-Hydroxyphenyl)-3,5-diiodotyrosine *see* 3176

(−)-(*R*)-1-(4-Hydroxyphenyl)-2-(3,4-dimethoxyphenethylamino)ethanol *see* 2878

(3-Hydroxyphenyl)dimethylethylammonium Chloride *see* 3487

m-Hydroxyphenylethanolamine *see* 6616

p-Hydroxyphenylethanolamine *see* 6679

2-*p*-Hydroxyphenylethylamine *see* 9743

α-(*m*-Hydroxyphenyl)-β-(ethylamino)ethanol *see* 3822

2-Hydroxy-2-phenylethyl Carbamate *see* 8829

N-(*p*-Hydroxyphenylethyl)-4-hydroxynorephedrine *see* 8228

N-[2-(*p*-Hydroxyphenyl)ethyl]-*N*-[2-(*p*-hydroxyphenyl)-2-hydroxy-1-methylethyl]amine *see* 8228

2-[6-(2-Hydroxy-2-phenylethyl)-1-methyl-2-piperidinyl]-1-phenylethanone *see* 5432

3-Hydroxy-*N*-(2-phenylethyl)morphinan *see* 7215

4-Hydroxyphenyl-β-D-glucopyranoside *see* 799

N-(4-Hydroxyphenyl)glycine, 4771

1-(4-Hydroxyphenyl)-1-hydroxy-2-butylaminoethane *see* 964

1-(*p*-Hydroxyphenyl)-2-[[β-hydroxy-β-(3′,5′-dihydroxyphenyl)]ethyl]-aminopropane *see* 3927

1-(4-Hydroxyphenyl)-2-[2-(4-hydroxyphenyl)ethylamino]propanol *see* 8228

p-Hydroxyphenylisopropylamine *see* 4740

β-(*p*-Hydroxyphenyl)isopropylmethylamine *see* 7304

o-Hydroxyphenylmercuric Chloride *see* 5763

7-(4-Hydroxyphenyl)-9-methoxy-8*H*-1,3-dioxolo[4,5-g][1]benzopyran-8-one *see* 4974

3-*p*-Hydroxyphenyl-4-*p*-methoxyphenyl-3-hexene *see* 5818

l-1-Hydroxy-1-phenyl-2-methylaminoethane *see* 4516

1-(4-Hydroxyphenyl)-2-methylaminoethanol *see* 8994

l-1-(*m*-Hydroxyphenyl)-2-methylaminoethanol Hydrochloride *see* 7257

α-(*p*-Hydroxyphenyl)-β-methylaminopropane *see* 7304

1-(4-Hydroxyphenyl)-2-methylaminopropanol *see* 4754

[(2-Hydroxyphenyl)methylene]hydrazide 4-Pyridinecarboxylic Acid *see* 8304

4-(*m*-Hydroxyphenyl)-1-methylisonipecotic Acid Ethyl Ester *see* 4769

1-(*p*-Hydroxyphenyl)-2-(1-methyl-2-phenoxyethylamino)-1-propanol *see* 5128

α-(4-Hydroxyphenyl)-β-methyl-4-(phenylmethyl)-1-piperidineethanol *see* 4821

1-(*p*-Hydroxyphenyl)-2-(1′-methyl-3′-phenylpropylamino)-1-propanol *see* 6654

4-(3-Hydroxyphenyl)-1-methyl-4-piperidinecarboxylic Acid Ethyl Ester *see* 4769

4-(*m*-Hydroxyphenyl)-1-methyl-4-piperidyl Ethyl Ketone *see* 5180

1-[4-(3-Hydroxyphenyl)-1-methyl-4-piperidyl]-1-propanone *see* 5180

(±)-4-[2-[[3-(4-Hydroxyphenyl)-1-methylpropyl]amino]ethyl]-1,2-benzenediol *see* 3396

(±)-4-[2-[[3-(*p*-Hydroxyphenyl)-1-methylpropyl]amino]ethyl]pyrocatechol *see* 3396

3-(4-Hydroxyphenyl)-1-methylpropyl β-D-Glucopyranoside *see* 8190

2-(*o*-Hydroxyphenyl)-1,3,4-oxadiazole *see* 3898

α-(*p*-Hydroxyphenyl)-α-(4-oxo-2,5-cyclohexadien-1-ylidene)-*o*-toluic Acid *see* 7208

3α-Hydroxy-8-(*p*-phenylphenacyl)-1α*H*,5α*H*-tropanium Bromide (−)-Tropate *see* 3949

1-(*p*-Hydroxyphenyl)-2-phenyl-4-butylpyrazolidine-3,5-dione *see* 6925

N-Hydroxy-2-[phenyl(phenylmethyl)-amino]ethanimidamide *see* 2015

β-(*p*-Hydroxyphenyl)phloropropiophenone *see* 7299

m-Hydroxyphenylpropanolamine *see* 5837

1-(4-Hydroxyphenyl)-1-propanone *see* 6992

3-(4-Hydroxyphenyl)-2-propenoic Acid *see* 2561

2-Hydroxy-2-phenylpropionamide *see* 888

α-Hydroxy-α-phenylpropionamide *see* 888

2-Hydroxy-2-phenylpropionic Acid *see* 889

α-Hydroxy-α-phenylpropionic Acid *see* 889

dl-1-*p*-Hydroxyphenyl-2-propylamine *see* 4740

L-(1-Hydroxy-1-phenyl-2-propylamino)-1-(*m*-methoxyphenyl)-1-propanone *see* 6915

4-Hydroxy-3-(1-phenylpropyl)-2*H*-1-benzopyran-2-one *see* 7230

γ-(*p*-Hydroxyphenyl)-α-propylene *see* 2039

1-[γ-Hydroxy-γ-phenylpropyl]-4-phenyl-4-carbethoxypiperidine *see* 7216

1-(3-Hydroxy-3-phenylpropyl)-4-phenylisonipecotic Acid Ethyl Ester *see* 7216

1-(3-Hydroxy-3-phenylpropyl)-4-phenyl-4-piperidinecarboxylic Acid Ethyl Ester *see* 7216

3-Hydroxy-2-phenyl-4-quinolinecarboxylic Acid *see* 6910

3-Hydroxy-4-phenyl-2(1*H*)-quinolinone *see* 9911

N-(*p*-Hydroxyphenyl)salicylamide *see* 6843

3-[(Hydroxyphenyl-2-thienylacetyl)-oxy]-1,1-dimethylpyrrolidinium Bromide *see* 4595

4-[3-Hydroxy-3-phenyl-3-(2-thienyl)propyl]-4-methylmorpholinium Iodide *see* 9363

2-[*N*-(*m*-Hydroxyphenyl)-*p*-toluidinomethyl]imidazoline *see* 7234

3-(4-Hydroxyphenyl)-1-(2,4,6-trihydroxyphenyl)-1-propanone *see* 7299

β-(*p*-Hydroxyphenyl)-2,4,6-trihydroxypropiophenone *see* 7299

(*m*-Hydroxyphenyl)trimethylammonium Bromide, Decamethylenebis(methylcarbamate) *see* 2871

2-Hydroxy-3-(phosphonooxy)propanal *see* 4377

2-Hydroxy-3-pinanone *see* 6929

2-Hydroxypinocamphone *see* 6929

4-(Hydroxy-2-piperidinylmethyl)-1,2-benzenediol *see* 8223

3-[3-(3-Hydroxy-2-piperidinyl)-2-oxopropyl]-4(3*H*)-quinazolinone *see* 3890

10-[3-(4-Hydroxy-1-piperidinyl)propyl]-10*H*-phenothiazine-2-carbonitrile *see* 7117

3-[3-(3-Hydroxy-2-piperidyl)acetonyl]-4(3*H*)-quinazolinone *see* 3890

2-Hydroxypiperitone *see* 3292

Hydroxypiracetam *see* 6896

12-Hydroxypodocarpa-8,11,13-trien-16-oic Acid *see* 7516

Hydroxypolyethoxydodecane *see* 7532

16α-Hydroxyprednisolone-16α,17-acetonide *see* 2908

3-Hydroxypregnane-11,20-dione *see* 226

Hypnodil [Janssen] *see* 6069
Hypnodin [Takeda] *see* 7131
Hypnofon *see* 9569
Hypnogène *see* 972
Hypnomidate [Janssen] *see* 3838
Hypnon [Sumitomo] *see* 6465
Hypnone *see* 65
Hypnorex [Delalande] *see* 5404
Hypnovel [Roche] *see* 6105
"Hypo" *see* 8652
Hypoboric Acid *see* 2999
Hypochlorous Acid, 4800
Hypochlorous Acid 1,1-Dimethylethyl
  Ester *see* 1570
Hypochlorous Anhydride *see* 2099
Hypochylin *see* 4363
Hypocrine [Tanabe Seiyaku] *see* 5354
Hypoglycin *see* 4801
Hypoglycin A *see* 4801
Hypoglycine A, 4801
Hypoglycine B, 4802
Hyponitrous Acid Anhydride *see* 6575
β-Hypophamine *see* 9843
Hypophosphoric Acid, 4803
Hypophosphorous Acid, 4804
Hypophthalin *see* 4682
Hypophyseal Growth Hormone *see*
  8672
Hypostamine *see* 9677
Hypostat [Herbrand] *see* 6992
Hyposterol *see* 7249
Hyposulfite of Gold and Sodium *see*
  4423
Hypothiazide *see* 4704
Hypovase [Pfizer] *see* 7715
Hypoxanthine, 4805
Hypoxanthine Riboside *see* 4880
Hypoxanthine Riboside-5-phosphoric
  Acid *see* 4882
Hypoxanthosine *see* 4880
Hyprenan [Astra] *see* 7742
Hypromellose *see* 4777
Hyprotigen [Magaw] *see* 7900
Hyproval P.A. [Tutag] *see* 4774
Hyptor *see* 5872
Hyrazin [Kowa] *see* 9230
Hyryl *see* 8202
Hyskon [Pharmacia] *see* 2925
Hysron H [Farmitalia] *see* 5677
Hystryl *see* 3334
Hytakerol [Winthrop] *see* 3163
Hyton *see* 7021
Hytracin [Dainabot] *see* 9084
Hytrin [Abbott] *see* 9084
Hytrinex [Abbott] *see* 9084
Hytrol O *see* 2732
Hyvar [Du Pont] *see* 1370
Hyvar, General Weed Killer [Du Pont]
  see 5042
Hyxobamine *see* 4739
Hyzyd [Mallinckrodt] *see* 5071
I-612 *see* 8238
I-2586 *see* 8885
I.A. 307 *see* 66
IAA *see* 4870
Ia-But [Inter-Alia] *see* 7248
Ial [Fidia] *see* 4675
Iangene [Farmochim. Ital.] *see* 8967
Iatroneural *see* 9594
IBC *see* 1453
IBCA *see* 1453
IBD [IPG] *see* 5114
Ibenzmethyzin *see* 7764
Ibiamox [IBI] *see* 610
IBI-C83 *see* 8240
Ibidomide *see* 5208
Ibifur *see* 4205
Ibilex [IBI] *see* 1971
Ibinolo [IBI] *see* 879
Ibiotyzil *see* 1037
Ibistacin [IBI] *see* 8208

Ibition *see* 9248
Ibogaine, 4806
Ibopamine, 4807
Ibotenic Acid, 4808
Ibrotamide, 4809
Ibu-Attritin *see* 4812
Ibudilast, 4810
Ibudros [Manetti Roberts] *see* 4813
Ibufenac, 4811
Ibumetin [Benzon] *see* 4812
Ibunac [Kaken] *see* 4811
Ibuprocin [Nisshin] *see* 4812
Ibuprofen, 4812
Ibuproxam, 4813
Ibu-slo [Rona] *see* 4812
Ibustrin [Farmitalia] *see* 4867
Ibutid [Dumex] *see* 4812
Ibutop [Chefaro] *see* 4812
Ibylcaine *see* 1517
Icacine [Bristol] *see* 2987
Iceland Moss, 4814
Ice Spar *see* 2609
Ichden *see* 4815
Ichtammon *see* 4815
Ichthadone *see* 4815
Ichthalum *see* 4815
Ichthammol, 4815
Ichthammonium *see* 4815
Ichthium *see* 4815
Ichthosan *see* 4815
Ichthosauran *see* 4815
Ichthosulfol *see* 4815
Ichthymall [Mallinckrodt] *see* 4815
Ichthynat [Tenneco] *see* 4815
Ichthyocolla *see* 4991
Ichthyol [Stiefel] *see* 4815
Ichthyopon *see* 4815
Ichthyopterin, 4816
Ichthysalle *see* 4815
Ichtopur [Ichthyol] *see* 4815
ICI 122,378 *see* 769
ICI 350 *see* 446
ICI 29661 *see* 8000
ICI 32865 *see* 3837
ICI 33828 *see* 5857
ICI 35868 *see* 7847
ICI 38174 *see* 7802
ICI 45520 *see* 7852
ICI 45763 *see* 9439
ICI 46474 *see* 9019
ICI 47699 *see* 9019
ICI 48213 *see* 2726
ICI 50123 *see* 7067
ICI 50172 *see* 7703
ICI 54450 *see* 3915
ICI 55052 *see* 6385
ICI 55897 *see* 2364
ICI 58834 *see* 9884
ICI 59118 *see* 8133
ICI 66082 *see* 879
ICI 69653 *see* 759
ICI 72222 *see* 2010
ICI 80008 *see* 4121
ICI 80996 *see* 2397
ICI 81008 *see* 4121
ICI 118587 *see* 9965
ICI 118630 *see* 4433
ICI 128436 *see* 7565
ICI 139603 *see* 9176
ICI 141292 *see* 3557
ICI 156834 *see* 1936
Icipen *see* 7046
ICI-PP 333 *see* 6937
ICN-1229 *see* 8199
Iconyl *see* 4771
Icoral B *see* 5837
ICRF 159 *see* 8133
ICSH *see* 5353
Ictéryl *see* 5719
ID 540 *see* 4058
ID 1937 *see* 4134

IDA *see* 4832
Idaein *see* 2694
Idalon [Nippon Roussel] *see* 4037
Idaltim [Lutetia] *see* 2536
Idarac [Diamant] *see* 4037
Idebenone, 4817
Idein *see* 2694
Idexur *see* 4819
Idocyl Novum *see* 8626
Ido-K *see* 5716
Idomethine [Kowa] *see* 4874
Idonor [Roger] *see* 7488
Idonyx *see* 4930
Idorese *see* 2067
Idose, 4818
Idoxene [Spodefell] *see* 4819
Idoxo-B$_{12}$ *see* 4739
Idoxuridine, 4819
Idrocilamide, 4820
Idroestril *see* 3118
Idrogriseovit *see* 4739
Idrolone [Maggioni] *see* 3941
Idro P$_2$ [Maggioni] *see* 8964
Idro P$_3$ [Maggioni] *see* 6868
Idrotiobicina *see* 4163
IDU *see* 4819
Idulea *see* 4819
Idulian [Unilabo] *see* 917
Idu Oculos [Minsa] *see* 4819
IDUR *see* 4819
Iduridin [Ferring] *see* 4819
L-Iduronic Acid *see* 4360
IF *see* 1903
Ifenec [Italfarmaco] *see* 3476
Ifenprodil, 4821
Ifex [Bristol] *see* 4822
IFN *see* 4891
IFN-α *see* 4892
IFN-α$_2$ *see* 4892
IFN-αA *see* 4892
IFN-β *see* 4893
IFN-β$_1$ *see* 4893
IFN-β$_2$ *see* 4893
IFN-γ *see* 4894
Ifosfamide, 4822
IFP *see* 4379
Ifrasarl [Showa Shinyaku] *see* 2779
Ig's *see* 4837
IgA *see* 4837
IgD *see* 4837
IgE *see* 4837
Igepal CA [GAF] *see* 6681
Igepal CA-630 [GAF] *see* 6681
Igepal CO [GAF] *see* 6596
Igepal CO-630 [GAF] *see* 6596
IGF-I *see* 8670
IGF-II *see* 8670
IgG *see* 4837
IgM *see* 4837
Ignatia, 4823
Ignatius Bean *see* 4823
Ignotine *see* 1857
Igralin [Zeria] *see* 9230
IHMS [Daiichi] *see* 5072
Ikaclomine [Ika] *see* 2384
Ikaran [Fabre] *see* 3157
Iktorivil [Roche] *see* 2387
IL-1 *see* 4895
IL-2 *see* 4896
IL-17803A *see* 13
^{123}I-labeled IMP *see* 4938
Ilbion *see* 9218
Ilcocillin P *see* 7042
Ildamen [Homburg] *see* 6915
α-Ile *see* 5064
Ile (IUPAC Abbrev.) *see* 5064
Ilentazol *see* 7351
Ileogastrone *see* 3548
Iletin [Lilly] *see* 4889
Ileu *see* 5064
ILeu *see* 5064

Iliadin [E. Merck] see 6919
Iliren [Hoechst] see 9355
Ilixathin see 8276
Illcut [Nippon Zoki] see 9425
Illoxan [Hoechst] see 3072
Illudin M see 4824
Illudin S see 4824
Illudins, 4824
Ilmenite see 9398
Ilopan [Warren-Teed] see 2924
Ilosone [Lilly] see 3628
Ilotycin [Lilly] see 3626
Ilotycin Gluceptate [Lilly] see 3629
Iloxan [Hoechst] see 3072
Ilozyme [Adria] see 6957
Ilvin [E. Merck] see 1433
^{123}I-M123 see 4938
Imadyl [Roche] see 1870
Imadyl [obsolete] see 4640
Imagon [Astra] see 2163
Imagotan [Sandoz] see 8943
Imakol [Rhodia] see 6900
Imap [McNeil] see 4131
Imavate [Robins] see 4835
Imaverol [Janssen] see 3537
Imazalil see 3537
Imazamethabenz, 4825
Imazamethabenz Methyl see 4825
Imazaquin, 4826
Imazethabenz see 4825
Imbaral [Sharp & Dohme] see 8961
Imbretil [Burroughs Wellcome] see 4599
Imbrilon [Berk] see 4874
Imbun [Merckle] see 6925
Imdur [Astra] see 5114
Imecromone see 4792
Imeson [Desitin] see 6491
Imesonal see 8374
Imex [Merz] see 9130
Imexim [Cimex] see 8889
Imferon [Lakeside] see 2927
IMI 28 see 3573
Imiclopazine, 4827
Imidamine see 709
Imidan [Stauffer] see 7311
Imidazole, 4828
2,4-(3H,5H)-Imidazoledione see 4678
1H-Imidazole-4-ethanamine see 4640
4-Imidazoleethylamine see 4640
5-Imidazoleethylamine see 4640
Imidazole Salicylate, 4829
Imidazoletrione see 6970
2,4-Imidazolidinedione see 4678
2-Imidazolidinethione see 3759
Imidazolidinetrione see 6970
2-Imidazolidinone, 4830
D-α-[(Imidazolidin-2-on-1-yl)carbon-
  ylamino]benzylpenicillin see 929
2-Imidazolidone see 4830
Imidazoline-2-thiol see 3759
1-(2-Δ²-Imidazolinyl)-2,2-diphenylcy-
  clopropane see 2274
N-(2-Imidazolin-2-yl)-N-(4-indanyl)-
  amine see 4845
2-Imidazolinylmethyl Benzhydryl
  Ether see 3307
β-(4-Imidazolyl)acrylcholine see 6216
2-(4-Imidazolyl)ethylamine see 4640
(E)-4-(Imidazol-1-ylmethyl)cinnamic
  Acid see 6935
(E)-3-[4-(1H-Imidazol-1-ylmethyl)-
  phenyl]-2-propenoic Acid see 6935
2-[[3-(1H-Imidazol-4-yl)-1-oxo-2-
  propenyl]oxy]-N,N,N-trimethyleth-
  anaminium see 6216
N-(p-Imidazol-4-ylphenyl)-N'-iso-
  propylformamidine see 6108
N-[4-(1H-Imidazol-4-yl)phenyl]-N'-
  (1-methylethyl)methanimidamide
  see 6108

p-Imidazo[1,2-a]pyridin-2-ylhydra-
  tropic Acid see 6127
4-Imidazo[1,2-a]pyridin-2-yl-α-meth-
  ylbenzeneacetic Acid see 6127
2-[p-(2-Imidazo[1,2-a]pyridyl)phen-
  yl]propionic Acid see 6127
7H-Imidazo[4,5-d]pyrimidine see
  7959
Imido [Roche] see 4640
Imidocarb, 4831
Imidodicarbonic Diamide see 1319
Imidodicarbonimidic Diamide see
  1233
Imidol [Yoshitomi] see 4835
Imidole see 8025
ImIFN see 4894
Imilanyle [Takata] see 4835
Iminazole see 4828
1,1'-Iminobis[4-amino-9,10-anthra-
  cenedione] see 1254
1,1'-Iminobis-9,10-anthracenedione
  see 720
2,2'-Iminobis[benzoic Acid] see 3318
2,2'-Iminobisethanol see 3097
3,3'-Iminobis-1-propanol Dimethanesulfonate (Ester) see 4841
Iminodiacetic Acid, 4832
1,1'-Iminodianthraquinone see 720
4,4'-Iminodicyclohexanecarboxylic
  Acid, 4833
Iminodiethanoic Acid see 4832
2,2'-Iminodiethanol see 3097
3-Imino-7-(methylamino)-3H-pheno-
  thiazine Hydrochloride see 942
N-[Imino(phosphonoamino)methyl]-N-
  methylglycine see 7315
β-(5-Imino-2-pyrrolidinecarboxami-
  do)propamidine see 6590
Iminourea see 4475
Imipemide see 4834
Imipenem, 4834
Imipramine, 4835
Imipramine N-Oxide, 4836
Imiprex [Dumex] see 4836
Imiprin see 4835
Imizad Equine Injection [Burroughs
  Wellcome] see 4831
Imizin see 4835
Imizocarb see 4831
Imizol see 4831
Immenoctal see 8374
Immobilon [Reckitt & Colman] see
  3843
Immuglobin [Savage] see 4837
Immune Globulins see 4837
Immune IFN see 4894
Immune Protein P5 see 893
Immuneron [Biogen] see 4894
Immunoglobulins, 4837
Immunol [Sarm] see 9161
Immunosome see 5391
Immunox [Cilag] see 9338
Imnudorm see 5731
Imodium [Janssen] see 5450
Imolamine, 4838
Imosec [Janssen] see 5450
Imotryl [Cassenne] see 1136
Imovance [Rhône-Poulenc] see 10095
Imovane [Rhône Poulenc] see 10095
IMP see 4938
IMP see 4882
^{123}I-labeled IMP see 4938
Impact [ICI] see 4135
Imperacin [ICI] see 6931
Imperan see 6063
Imperatorin, 4839
Imperial Green see 2629
Imperialine, 4840
Impral [Tanabe Seiyaku] see 2265
Impromen [Janssen] see 1432
Improsulfan, 4841

Impruvol see 1548
Imunovir [Edwin Burgess] see 4881
Imunoviral [Newport] see 4881
Imuran [Burroughs Wellcome] see 918
Imurek [Burroughs Wellcome] see 918
Imurel [Burroughs Wellcome] see 918
Imuthiol see 3384
IN 511 see 7292
Inabrin [Upjohn] see 4812
Inacid [Merck & Co.] see 4874
Inacilin [Inibsa] see 7484
Inactin [Promonta] see 9250
Inactive Limonene see 5371
Inadine [J & J] see 7701
Inaktin see 9250
Inalone O [Lampugnani] see 1029
Inalone R [Lampugnani] see 1029
Inamycin [Hoechst] see 6641
Inapetyl [Upjohn] see 1130
Inapsine [McNeil] see 3437
Inbestan [Maruko] see 2345
Incafolic see 4140
Incazan [USSR] see 6075
Incidal see 5649
Incoran see 7744
Incorporation Factor see 4379
Incortin see 2533
Incortin-H see 4710
Indacin [Merck & Co.] see 4874
Indaconitine, 4842
Indaflex [Lampugnani] see 4847
Indalapril see 2864
Indaliton [Geigy] see 2402
Indalone see 1529
Indalpine, 4843
Indamol [Armour Pharm.] see 4847
Indan, 4844
Indanal [Takeda] see 2350
Indanazoline, 4845
Indanthrene®, 4846
1,2,3-Indantrione Monohydrate see
  6470
1-(5-Indanyl) N-(2-Carboxy-3,3-
  dimethyl-7-oxo-4-thia-1-azabicy-
  clo[3.2.0]hept-6-yl)-2-phenylmalo-
  namate see 1847
O-(5-Indanyl) m,N-Dimethylthiocarb-
  anilate see 9438
α-(5-Indanyloxycarbonyl)benzylpeni-
  cillin see 1847
6-[2-(5-Indanyloxycarbonyl)phenyl-
  acetamido]-3,3-dimethyl-7-oxo-4-
  thia-1-azabicyclo[3.2.0]heptane-2-
  carboxylic Acid see 1847
Indapamide, 4847
1H-Indazole, 4848
3-Indazolealanine see 9706
Indecainide, 4849
Indeloxazine Hydrochloride, 4850
Indema see 7196
Indene, 4851
Indenolol, 4852
(±)-1-[Inden-4(or 7)-Yloxy]-3-(isopropylamino)-2-propanol see 4852
1-[1H-Inden-4(or 7)-Yloxy]-3-[(1-
  methylethyl)amino]-2-propanol see
  4852
(R,S)-2-[(7-Indenyloxy)methyl]mor-
  pholine Hydrochloride see 4850
2-[(1H-Inden-7-yloxy)methyl]morph-
  oline Hydrochloride see 4850
Inderal [Ayerst] see 7852
Inderex [ICI] see 7852
Inderite see 5535
Indian Aconite see 115
Indian Apple see 7521
Indian Arrow Wood see 3856
Indian Balm see 9609
Indian Balsam see 959
Indian Barberry see 1171
Indian Berry see 2452

Indian Cannabis *see* 1752
Indian Dogbane *see* 774
Indian Ginger *see* 850
Indian Grass Oil *see* 6733
Indian Gum *see* 4311
Indian Hemp *see* 1752
Indian Laburnum *see* 1900
Indian Melissa Oil *see* 6741
Indian Oil of Verbena *see* 6741
Indian Physic *see* 774
Indian Pink *see* 8704
Indian Poke *see* 9858
Indian Tobacco *see* 5431
Indian Tragacanth *see* 5164
Indian Turmeric *see* 4690
"Indian Yellow" *see* 2426
India Rubber *see* 8255
Indican (Metabolic Indican), 4853
Indican (Plant Indican), 4854
Indic Fluoride *see* 4866
Indigo, 4855
Indigo Blue *see* 4855
Indigo Carmine, 4856
Indigo Copper *see* 2661
Indigotin *see* 4855
Indigotine *see* 4856
Indigo Weed *see* 968
Indium, 4857
Indium Antimonide, 4858
Indium Arsenide, 4859
Indium Chloride *see* 4865
Indium Fluoride *see* 4866
Indium Oxide, 4860
Indium Phosphide, 4861
Indium Selenide, 4862
Indium Sesquioxide *see* 4860
Indium Sulfate, 4863
Indium Telluride, 4864
Indium Trichloride, 4865
Indium Trifluoride, 4866
Indobloc [Homburg] *see* 7852
Indobufen, 4867
Indocid [Merck & Co.] *see* 4874
Indocin [Merck & Co.] *see* 4874
Indocollyre [Chauvin] *see* 4874
Indocyanine Green, 4868
Indocybin [Sandoz] *see* 7942
Indoklon [Ohio Med.] *see* 4126
Indolacin [Sumitomo] *see* 2295
Indole, 4869
Indoleacetic Acid, 4870
1*H*-Indole-3-acetic Acid *see* 4870
1*H*-Indole-3-butanoic Acid *see* 4871
Indolebutyric Acid, 4871
Indole-3-butyric Acid *see* 4871
Indole-2,3-dione *see* 4985
1*H*-Indole-3-ethanamine *see* 9705
3-Indoleethanol *see* 9708
Indolin [Guidi] *see* 1136
2,3-Indolinedione *see* 4985
Indolmycin, 4872
1*H*-Indol-3-ol Hydrogen Sulfate Ester *see* 4853
3-Indolylacetone, 4873
l-β-3-Indolylalanine *see* 9707
4-(3-Indolyl)butyric Acid *see* 4871
2-Indolyl(3)-ethanol *see* 9708
2-(3-Indolyl)ethyl Alcohol *see* 9708
β-Indolylethyl Alcohol *see* 9708
2-(3-Indolyl)ethylamine *see* 9705
5-(1-Indol-3-ylethyl)-2-(methylami-no)-2-oxazolin-4-one *see* 4872
5-[1-(1*H*-Indol-3-yl)ethyl]-2-(methyl-amino)-4(5*H*)-oxazolone *see* 4872
4-[2-(3-Indolyl)ethyl]piperidine *see* 4843
N-[1-[2-(1*H*-Indol-3-yl)ethyl]-4-pi-peridinyl]benzamide *see* 4876
1*H*-Indol-3-yl-β-D-glucopyranoside *see* 4854

2-*C*-(1*H*-Indol-3-ylmethyl)-β-L-*lyxo*-3-hexulofuranosonic Acid γ-Lactone Mixt with 2-*C*-(1*H*-Indol-3-ylmeth-yl)-β-L-*xylo*-3-hexulofuranosonic Acid γ-Lactone *see* 856
1-(1*H*-Indol-4-yloxy)-3-[(1-methyl-ethyl)amino]-2-propanol *see* 7412
Indol-3-yl Potassium Sulfate *see* 4853
Indol-3-yl-2-propanone *see* 4873
Indol-3-yl Sulfate *see* 4853
Indomed *see* 4874
Indomee [Merck & Co.] *see* 4874
Indomethacin, 4874
Indomethacin Glucosamide *see* 4342
Indomethine [Teika Seiyaku] *see* 4874
Indomod [Benzon] *see* 4874
Indon [Parke, Davis] *see* 7196
Indonaphthene *see* 4851
Indo-Phlogont [Azuchemie] *see* 4874
Indoprofen, 4875
Indoptic [Merck & Co.] *see* 4874
Indoramin, 4876
Indorektal [Sanorania] *see* 4874
Indorm *see* 7834
Indospicine, 4877
Indo-Tablinen [Sanorania] *see* 4874
Indoxamic Acid *see* 6869
Indoxen [Sigma-Tau] *see* 4874
Indoxyl-β-D-glucoside *see* 4854
3-Indoxylsulfuric Acid *see* 4853
Indunox [UCB] *see* 3832
Indusil [Diamant; Recordati] *see* 2446
Inestra [Merck & Co.] *see* 3689
INF *see* 4220
INF 1837 *see* 4060
INF 3355 *see* 5680
INF 4668 *see* 5659
Infacol [Pharmax] *see* 8486
Infectomycin [Heyden] *see* 610
Inferno *see* 502
Infertine *see* 8166
Infiltrase [Armour Pharm.] *see* 4676
Infiltrina [Heyden] *see* 3247
Inflamase [Cooper] *see* 7721
Inflamen [Hokuriku] *see* 1377
Inflanefran [Pharm-Allergan] *see* 7719
Inflatine *see* 5432
Inflazon [Taisho] *see* 4874
Informational RNA *see* 8204
Infrocin [Frosst] *see* 4874
Infron [Whittier] *see* 9928
Infusorial Earth, 4878
INH *see* 5071
INH-Burgthal [Conzen] *see* 5071
INH-G *see* 4396
Inhibace [Roche] *see* 2276
Inhibin, 4879
Inhibostamin [Swiss Pharma] *see* 9677
Inhiston *see* 7198
Inicardio *see* 6459
Inimur [ICN] *see* 6447
Iniprol [Choay] *see* 784
Inmetal *see* 601
Innovace [Merck & Co.] *see* 3521
Innovar *see* 3437
Innoxalon [Sanko] *see* 6273
Inocor [Sterling] *see* 626
Inofal [Sandoz] *see* 8943
Inokiton [Nippon Yakuhin] *see* 9751
Inolin [Tanabe Seiyaku] *see* 9502
Inopamil [Simes] *see* 4807
Inophylline [Millot] *see* 477
Inosie [Morishita] *see* 4880
Inosine, 4880
Inosine:dimethylaminoisopropanol Acetamidobenzoate (1:3) *see* 4881
Inosine, Mono[4-(acetylamino)benzo-ate] (Salt), Compd with 1-(Dimeth-ylamino)-2-propanol (1:3) *see* 4881
Inosine Pranobex, 4881

Inosinic Acid, 4882
t-Inosinic Acid *see* 4882
5-Inosinic Acid *see* 4882
5'-Inosinic Acid *see* 4882
Inosiplex *see* 4881
Inosite *see* 4883
Inositol, 4883
i-Inositol *see* 4883
meso-Inositol *see* 4883
myo-Inositol *see* 4883
myo-Inositol 1-(Dihydrogen Phos-phate) *see* 4884
myo-Inositol Hexakis(dihydrogen Phosphate) *see* 7359
Inositol Hexanicotinate *see* 4885
meso-Inositol Hexanicotinate *see* 4885
Inositolhexaphosphoric Acid *see* 7359
myo-Inositol Hexa-3-pyridinecarb-oxylate *see* 4885
Inositol Monophosphate, 4884
Inositol Niacinate, 4885
Inostral [Syntex] *see* 2594
Inotrex (obsolete) [Lilly] *see* 3396
Inovan [Kyowa] *see* 3415
Inoven [Janssen] *see* 4812
INPC *see* 7828
INPEA *see* 6442
Inpea [Selvi] *see* 6442
Inplacen [Merz] *see* 7487
Insane Root *see* 4796
Insariotoxin *see* 9711
Insecticide 1179 *see* 5905
Insecticide 3960-X14 [B.F. Goodrich] *see* 8796
Insecticide ACC 4124 *see* 3027
Insecticide No. 497 *see* 3093
Insecticide No. 4049 *see* 5582
Insibrin [Byk-Liprandi] *see* 3598
Insidon [Geigy] *see* 6808
Insoluble Anhydrite *see* 1713
Insoluble Cyanuric Acid *see* 2689
Insoluble Salumin *see* 366
Insomin [Orion] *see* 6491
Insoral [Roger Bellon] *see* 7191
Inspir [Vitrum] *see* 82
Insubeta *see* 4900
Insulamin [Iwaki] *see* 1465
Insular [Armour Pharm.] *see* 4889
Insularine, 4886
Insulatard [Nordisk] *see* 5082
Insulin, 4887
Insulin ^{131}I *see* 4887
Insulinase, 4888
Insulin (Emp) *see* 4887
Insulin Injection, 4889
Insulin-like Growth Factor *I* *see* 8670
Insulin-like Growth Factor II *see* 8670
Insulin Novo Lente *see* 4890
Insulin (Prb) *see* 4887
Insulin Protamine Zinc *see* 7899
Insulin Ultracard [Novo] *see* 4890
Insulin Zinc Protamine *see* 7899
Insulin Zinc Suspension, 4890
Insulin Zinc Suspension (Amorphous) *see* 4890
Insulin Zinc Suspension (Crystalline) *see* 4890
Insulin Zinc Suspension Extended *see* 4890
Insulin Zinc Suspension Prompt *see* 4890
Insulton *see* 5741
Insulyl *see* 4889
Insulyl-Retard *see* 7899
Insumin [Kyorin] *see* 4123
Insuven [Berenguer Beneyto] *see* 3291
Intal [Fisons] *see* 2594
Inteban SP [Sumitomo] *see* 4874
Integerrimine *see* 8402
Integrin [Sterling] *see* 6924
Intenkordin [Polfa] *see* 2241

Intensain [Abbott; Cassella-Riedel] see 2241
Intensopan [Sandoz] see 1441
Intercept [Ortho] see 6596
Interceptor [Isnardi] see 6820
Interferon, 4891
Interferon-α, 4892
Interferon-β, 4893
Interferon-γ, 4894
Intergravin-orales [Werfft-Chemie] see 1659
Interleukin-1, 4895
Interleukin-2, 4896
Interleukin-6 see 4893
Intermedin(e) see 6206
Intermigran [Sanol] see 7852
Internal Antiseptic No. 307 see 66
Internally Compensated Tartaric Acid see 9040
Interomycetine see 2068
Interstitial Cell Stimulating Hormone see 5353
Intestiazol see 7351
Intestibar [Spofa] see 1006
Intestin-Euvernil [Heyden] see 8883
Intocostrin [Squibb] see 9717
Intrabilix see 4916
Intrabutazone see 7248
Intracaine see 6987
Intracort see 4713
Intradex [Glaxo] see 2925
Intraformazol see 4158
Intraglobin [Biotest] see 4837
Intraheptol see 5427
Intramycetin [Parke, Davis] see 2068
Intranarcon see 9219
Intrapan [USV] see 2924
Intrasporin [Torlan] see 1975
Intrastigmina [Lusofarmaco] see 6380
Intraval Sodium [M & B] see 9280
Intrazone [Arnolds] see 7248
Intrinsic Factor see 1903
Introcar [Coup] see 6441
Introl [Fisons] see 2594
Intromene [Teikoku Kagaku] see 9537
Intron [Tika] see 3205
Intron A [Schering] see 4892
Introna [Schering] see 4892
Intropin [Arnar-Stone] see 3415
Intybin see 5224
Inula, 4897
Inula Camphor see 198
Inulin, 4898
Inutral [Galenika] see 4887
Invasin see 4676
Invenol [Hoechst] see 1839
Inversine [Merck & Co.] see 5654
Invertase, 4899
Invertin see 4899
Invertogen [Zirkulin] see 4900
Invert Sugar, 4900
Invesol see 4900
Investin [Sagitta] see 3429
Invisi-Gard see 7849
Inyoite see 1651
Iobac [West Chemical] see 4930
Iobenzamic Acid, 4901
Iocarmic Acid, 4902
Iocetamic Acid, 4903
Iodafilina [B.O.I.] see 3663
Iodamide, 4904
Iodamide Meglumine see 4904
Iodaphyline [Millot] see 3663
Iodeikon [Mallinckrodt] see 4931
Iodic Acid, 4905
Iodic Anhydride see 4913
Iodinated Glycerol, 4906
Iodinated (131I) Human Serum Albumin see 8417
Iodinated p-Toluidine Polyvinylpyrrolidone see 9448

Iodine, 4907
Iodine Colloidal, 4908
Iodine Cyanide see 2702
Iodine Heptafluoride, 4909
Iodine Monobromide, 4910
Iodine Monochloride, 4911
Iodine Pentafluoride, 4912
Iodine Pentoxide, 4913
Iodine-polyvinylpyrrolidone Complex see 7701
Iodine Trichloride, 4914
Iodinin, 4915
Iodipamide, 4916
Iodisan see 7792
Iodistol see 9335
Iodized Oil, 4917
Iodoacetic Acid, 4918
Iodoalphionic Acid, 4919
p-Iodoaniline, 4920
o-Iodoanisole, 4921
Iodobenzene, 4922
4-Iodobenzenesulfonyl Chloride see 7458
o-Iodobenzoic Acid, 4923
N-(2-Iodobenzoyl)glycine Monosodium Salt see 4927
Iodobil see 4919
1-Iodobutane see 1572
2-Iodobutane see 1573
Iodochlorhydroxyquin, 4924
Iodochlorohydroxyquinoline see 4924
Iodochloroxyquinoline see 4924
5-Iodo-2'-deoxyuridine see 4819
3-Iodo-1,2-dihydroxypropane see 4383
Iododocosanoic Acid Calcium Salt see 1681
Iodoenterol see 4924
Iodoethane see 3769
2-(1-Iodoethyl)-1,3-dioxolane-4-methanol see 4906
2-(2-Iodoethyl)-1,3-dioxolane-4-methanol see 4906
Iodofenphos, 4925
Iodoform, 4926
Iodoformine see 5886
Iodognost see 4931
Iodogorgoic Acid see 3177
o-Iodohippurate Sodium, 4927
4-(3-Iodo-4-hydroxyphenoxy)-3,5-diiodophenylalanine see 5388
β-[4-(3'-Iodo-4'-hydroxyphenoxy)-3,5-diiodophenyl]propionic Acid see 9346
m-Iodo-o-hydroxyquinolineanasulfonic Acid see 4757
7-Iodo-8-hydroxyquinoline-5-sulfonic Acid see 4757
1-(ω-p-Iodo-131I-benzylpolyethylene)-2-pyrrolidinone see 9448
(±)-p-Iodo-123I-N-isopropyl-α-methylphenethylamine see 4938
4-Iodo-iota-methylbenzenedecanoic Acid Ethyl Ester see 4946
Iodol see 4934
Iodomethane see 6002
Iodomethanesulfonic Acid Sodium Salt see 5894
1-Iodo-3-methylbutane see 5002
2-(Iodomethyl)-1,3-dioxolane-4-methanol see 3410
(±)-4-(Iodo-131I)-α-methyl-N-(1-methylethyl)benzeneethanamine see 4938
1-Iodo-2-methylpropane see 5027
2-Iodooctadecanoic Acid Calcium Salt see 1682
2-Iodooctane see 6686
Iodopanoic Acid see 4944
Iodopaque [Labaz] see 72
Iodophene see 4931

Iodophene Sodium see 4931
o-Iodophenol, 4928
p-Iodophenol, 4929
p-Iodophenyl Sulfonyl Chloride see 7458
Iodophors, 4930
Iodophthalein see 4931
Iodophthalein Sodium, 4931
1-Iodopropane see 7875
2-Iodopropane see 5102
3-Iodo-1,2-propanediol see 4383
3-Iodo-1-propene see 292
3-Iodopropylene see 292
γ-Iodopropyleneglycol see 4383
2,3-(2-Iodopropylidenedioxy)propanol see 4906
2,3-(3-Iodopropylidenedioxy)propanol see 4906
Iodopropylidene Glycerol see 4906
3-Iodo-2-propynyl 2,4,5-Trichlorophenyl Ether see 4514
Iodopsin, 4932
Iodopyracet, 4933
Iodopyracet Compound Solution see 4933
Iodopyracet Concentrated Solution see 4933
Iodopyracet Injection see 4933
Iodopyrrole, 4934
Iodoquinine Sulfate see 8086
Iodoquinol, 4935
Iodorayoral see 4931
Iodosol see 9335
Iodosorb [Perstorp; Stuart] see 1609
N-Iodosuccinimide, 4936
Iodosulfane see 8954
Iodothiouracil see 4954
5-Iodo-2-thiouracil see 4954
Iodothymol see 9335
Iodothyrin see 9345
Iodotope [Squibb] see 8583
3-Iodotyrosine, 4937
Iodoxyl see 8584
Iodtetragnost see 4931
Ioduril see 8582
Iofetamine 123I, 4938
Ioglycamic Acid, 4939
Iohexol, 4940
Iomapidol see 4943
Iomeglamic Acid, 4941
Ionamin [Pennwalt] see 7232
Ionol CP [Shell] see 1548
Ionone, 4942
Ionophore X-4537A see 5243
Iopamidol, 4943
Iopamiro [Bracco] see 4943
Iopamiron [Schering] see 4943
Iopanoic Acid, 4944
Iopentol, 4945
Iophendylate, 4946
Iophenoxic Acid, 4947
Iopidine [Alcon] see 779
Ioprep [J & J] see 4930
Iopromide, 4948
Iopronic Acid, 4949
Iopropane see 4953
Iopydol, 4950
Iopydone, 4951
Iopyracil see 4933
Ioquin [Abbott] see 4935
Iosalide [Schering] see 5148
Iosan [West Chemical] see 4930
Iosol see 9335
Iothalamic Acid, 4952
Iothion, 4953
Iothiouracil, 4954
Iothymol see 9335
Iotrol see 4955
Iotrolan, 4955
Iotrovist [Schering AG] see 4955
Ioxaglate Meglumine see 4956

p-Isobutoxybenzoic Acid 3-(Diethylamino)-1,2-dimethylpropyl Ester see 4263

p-Isobutoxybenzoic Acid α,β-Dimethyl-γ-diethylaminopropyl Ester see 4263

1-Isobutoxy-2-pyrrolidino-3-N-benzylanilinopropane see 1167

3-Isobutoxy-2-pyrrolidino-N-phenyl-N-benzylpropylamine see 1167

Isobutyl Acetate, 5014

Isobutyl Alcohol, 5015

Isobutylaldehyde see 5038

5-Isobutyl-5-allylbarbituric Acid see 1502

Isobutylamine, 5016

Isobutyl p-Aminobenzoate, 5017

2-(Isobutylamino)ethanol p-Aminobenzoate (Ester) see 1517

2-(Isobutylamino)ethyl p-Aminobenzoate see 1517

Isobutylbenzene, 5018

Isobutyl Bromide, 5019

3-Isobutyl-6-sec-butyl-2-hydroxypyrazine 1-Oxide see 863

Isobutyl n-Butyrate, 5020

Isobutyl Carbamate, 5021

Isobutyl Carbinol see 5081

Isobutylcarbylamine see 4994

Isobutyl Chloride, 5022

Isobutyl Chlorocarbonate, 5023

Isobutyl Chloroformate see 5023

Isobutyl Cyanoacrylate see 1453

E,E-N-Isobutyl-2,4-decadienamide see 7017

N-Isobutyldeca-trans-2-cis-6-trans-8-trienamide see 167

N-Isobutyl-2,6,8-decatrienamide see 167

Isobutyl 1,4-Dihydro-5-methoxycarbonyl-2,6-dimethyl-4-(2-nitrophenyl)-3-pyridinecarboxylate see 6482

Isobutylene, 5024

Isobutyl Ether, 5025

Isobutyl Formate, 5026

p-Isobutylhydratropic Acid see 4812

Isobutylhydrochlorothiazide see 1519

Isobutyl Iodide, 5027

Isobutyl Isobutyrate, 5028

Isobutyl Isovalerate, 5029

Isobutyl Kelo-form see 5017

Isobutyl Mercaptan, 5030

Isobutyl Methyl 1,4-Dihydro-2,6-dimethyl-4-(o-nitrophenyl)-3,5-pyridinedicarboxylate see 6482

Isobutyl Nitrate, 5031

Isobutyl Nitrite, 5032

(p-Isobutylphenyl)acetic Acid see 4811

2-(4-Isobutylphenyl)butyric Acid see 1521

2-(4-Isobutylphenyl)propionic Acid see 4812

dl-2-(4-Isobutylphenyl)propionohydroxamic Acid see 4813

Isobutyl Propionate, 5033

Isobutyl Stearate, 5034

Isobutyl Sulfide, 5035

Isobutyl Sulfocyanate see 5036

Isobutyl Thiocyanate, 5036

p-Isobutyl-α-toluic Acid see 4811

Isobutyltrimethylmethane see 5079

Isobutyl Urethane, 5037

Isobutyl Valerate see 5029

Isobutyraldehyde, 5038

Isobutyric Acid, 5039

Isobutyric Aldehyde see 5038

3-Isobutyryl-2-isopropylpyrazolo-[1,5-a]pyridine see 4810

O-Isobutyrylthiamine Disulfide see 9221

Isocaine [Columbus] see 5017

Isocaine-Asid see 7763

Isocaine-Heisler see 7763

Isocalm [Kaken] see 9447

Isocapronitrile see 4999

Isocaramidine see 2837

Isocarboxazid, 5040

Isocaryophyllene see 1884

Isochavicine see 7442

Isochavicinic Acid see 7437

Isochinol [Pharmakos] see 3207

Isochlorothion see 3027

Isochondrodendrine, 5041

Isochrysene see 9654

Isocid see 5071

Isocil, 5042

Isocillin [Hoechst] see 7046

Isocinchomeronic Acid, 5043

Isococaine see 7926

Isocolin [Isola-Ibi] see 7337

Isocolumbin see 2489

Isoconazole, 5044

Iso-Cornox [FBC] see 5666

Isocorybulbine, 5045

Isocorydine, 5046

Isocorypalmine, 5047

Isocothane see 3281

Isocotin see 5071

Isocrin [Warner-Chilcott] see 6927

Isocrotonic Acid, 5048

Isocrotyl Chloride see 2147

Isocyanatobenzene see 7267

4-Isocyanato-1,1'-biphenyl see 9982

Isocyanatomethane see 6004

1-Isocyanatonaphthalene see 6330

Isocyanic Acid, 5049

Isocyanic Acid 4-Biphenylyl Ester see 9982

Isocyanic Acid Methyl Ester see 6004

Isodiamylamine see 3178

Isodianisylethanolamine see 5062

Isodin [Tosi] see 6881

Isodine [Blair] see 7701

Isodiprene see 1843

Isodormid see 783

Isodrin see 219

Isodulcit see 8171

Isodurene, 5050

Isoendoxan see 4822

d-Isoephedrine see 3561

Isoestradiol, 5051

8-Isoestradiol-17β see 5051

8-Isoestrone, 5052

Isoetam [Ferrer] see 5013

Isoetarine see 5053

Isoetham see 5013

Isoetharine, 5053

Isoeugenol, 5054

Isofenphos, 5055

Isofezolac, 5056

4',7-Isoflavandiol see 3583

Isoflavone, 5057

Isofluorphate see 5060

Isoflupredone, 5058

Isoflurane, 5059

Isoflurophate, 5060

Isoglaucon [Boehringer, Ing.] see 2388

L-Isoglutamine, 5061

Isoindazole see 4848

1H-Isoindole-1,3(2H)-dione see 7347

Isoinokosterone see 3465

Iso-Iodeikon see 7233

Iso-K [San Carlo] see 5184

Isoket [Sanol] see 5114

Isoladol, 5062

Isolait [Elder] see 5128

Isolan®, 5063

Isoleucine, 5064

Isolevin see 5105

Isolobelanine see 6626

α-Isolupanine see 5477

Isolyn see 5071

Isolysergic Acid, 5065

IsoMack [Mack, Illert.] see 5114

Isomaltol, 5066

Isomenyl [Kaken] see 5105

Isomeride [Ardix] see 3920

Isomerine see 2180

Isometamidium Chloride, 5067

Isomethadol, 5068

Isomethadone, 5069

Isomethazine see 5088

Isometheptene, 5070

Isomist see 5105

Isomycin [Werfft] see 7907

Isomycomycin see 6237

Isomyn see 616

Isomytal see 601

Isonal [Roussel-UCLAF] see 5742

Isonal (Swedish) see 782

Isonaphthoic Acid see 6302

Isonaphthol see 6304

Isonex see 5071

Isoniazid, 5071

Isoniazid 4-Aminosalicylate see 6999

Isoniazid Mesylate see 5072

Isoniazid Methanesulfonate, 5072

Isonicazide see 5071

Isonicid see 5071

Isonicotan see 5071

Isonicotinic Acid, 5073

Isonicotinic Acid 2-[2-(Benzylcarbamoyl)ethyl]hydrazide see 6399

Isonicotinic Acid Diethylamide, 5074

Isonicotinic Acid Hydrazide see 5071

Isonicotinic Acid Hydrazide p-Aminosalicylate see 6999

Isonicotinic Acid Hydrazide Hydrazone with Glucuronic Acid Lactone see 4396

Isonicotinic Acid Hydrazide, Hydrazone with Streptomycin, Sulfate (Salt) (2:3) see 8788

Isonicotinic Acid Hydrazide Methanesulfonic Acid see 5072

Isonicotinic Acid 2-Isopropylhydrazide see 4965

Isonicotinic Acid α-Methylfurfurylidenehydrazide see 4220

Isonicotinic Acid Salicylidenehydrazide see 8304

Isonicotinic Acid m-Sulfobenzylidene Hydrazide see 8938

Isonicotinic Acid 2-Sulfomethylhydrazide see 5072

Isonicotinic Acid Veratrylidenehydrazide see 9859

N-Isonicotinoyl-N'-[β-(N-benzylcarboxamido)ethyl]hydrazide see 6399

Isonicotinoylhydrazine see 5071

Isonicotinoylhydrazone of D-Glucuronic Acid Lactone see 4396

Isonicotinoyl Hydrazone of m-Sulfobenzaldehyde see 8938

1-Isonicotinoyl-2-isopropylhydrazine see 4965

1-Isonicotinoyl-2-salicylidenehydrazine see 8304

1-Isonicotinoyl-2-veratrylidenehydrazine see 9859

Isonicotinylhydrazine see 5071

Isonicotinyl Hydrazonotoluene-m-sulfonic Acid see 8938

1-Isonicotinyl-2-isopropylhydrazine see 4965

Isonidrin see 5071

Isonilex see 5071

Isonindon see 5071

Isonipecaine see 5736

Isonipecotic Acid, 5075

Isonitrosoacetone, 5076

Isonitrosoacetophenone, 5077
5-Isonitrosobarbituric Acid *see* 9904
N-[*p*-(1-Isonitrosoethyl)phenoxyacet-
  yl]piperidine *see* 7392
β-Isonitrosopropane *see* 68
Isonixin, 5078
Isonizide *see* 5071
Isonootkatone *see* 9874
Isonorin [SM & P] *see* 5105
Isonovobiocin *see* 6641
Isooctane, 5079
Isooctyl Alcohol, 5080
p-Isooctylpolyoxyethylenephenol
  Formaldehyde Polymer *see* 9741
Isoolean-14-en-3β-ol *see* 9033
Iso-oosporein *see* 6805
Isopaque [Winthrop] *see* 6078
Isopar *see* 2513
Isopedine *see* 7826
Isopelletierine *see* 7015
4-(2-Isopentenyl)-1,2-diphenyl-3,5-
  pyrazolidinedione *see* 3953
Isopentyl Alcohol, 5081
sec-Isopentyl Alcohol *see* 5953
Isopentylamine *see* 4994
Isopentyl 5,6-Dihydro-7,8-dimethyl-
  4,5-dioxo-4*H*-pyrano[3,2-*c*]quinol-
  ine-2-carboxylate *see* 8140
N-Isopentyl-1,5-dimethylhexylamine
  see 6671
Isopentylhydrocupreine *see* 3861
Isopentyl Nitrite *see* 5005
α-[(Isopentyloxy)methyl]-4-morpho-
  lineethanol 3,4,5-Trimethoxybenzo-
  ate *see* 604
1-[β-(Isopentyloxy)phenethyl]pyrroli-
  dine *see* 507
Isopentyl 2-Phenylglycinate Hydro-
  chloride *see* 7163
Isopestox [Fisons] *see* 6124
Isophane Insulin *see* 5082
Isophane Insulin Injection *see* 5082
Isophane Insulin Suspension, 5082
Isophen [Knoll] *see* 5859
Isophenergan *see* 5088
Isophenethanol *see* 6442
Isophenphos *see* 5055
Isophosphamide *see* 4822
Isophrin Hydrochloride *see* 7257
Isophthalic Acid, 5083
Isophytol, 5084
Isopilocarpine *see* 7395
Isopilosine, 5085
Isopimaric Acid, 5086
Isopiperine *see* 7442
Isopiperinic Acid *see* 7437
IsoPPC *see* 7828
Isopral *see* 9557
10α-Isopregnenone *see* 3454
Isoprenaline *see* 5105
Isoprene, 5087
Isoprinosin [Newport] *see* 4881
Isoprinosina [Delalande/Newport] *see*
  4881
Isoprinosine [Rohm Pharma/New-
  port] *see* 4881
Isoprofenamine *see* 2404
Isopromedol *see* 7795
Isopromethazine, 5088
Isopropalin, 5089
Isopropanol *see* 5096
Isopropene Cyanide *see* 5850
Isopropenyl Acetate, 5091
4-Isopropenyl-1-cyclohexene-1-carb-
  oxaldehyde *see* 7119
2-Isopropenyl-2,3-dihydro-5-acetyl-
  benzofuran *see* 9497
2-Isopropenyl-6-methoxy-5-benzo-
  furanyl Methyl Ketone *see* 3857

7-Isopropenyl-4-methyl-1-azulene-
  carboxaldehyde *see* 5212
cis-(+)-2-Isopropenyl-1-methylcyclo-
  butaneethanol *see* 4442
4-(5-Isopropenyl-2-methyl-1-cyclo-
  penten-1-yl)-2-butanone, 5092
Isopropenylnitrile *see* 5850
Isoprophenamine *see* 2404
N¹-(4-Isopropoxybenzoyl)-*p*-amino-
  benzenesulfonamide *see* 8911
N¹-(*p*-Isopropoxybenzoyl)sulfanil-
  amide *see* 8911
5-Isopropoxycarbonylaminothiabend-
  azole 1733
5-Isopropoxycarbonylamino-2-(4-
  thiazolyl)benzimidazole *see* 1733
(*E*)-*O*-2-Isopropoxycarbonyl-1-meth-
  ylvinyl *O*-Methyl Ethylphosphoramidothioate *see* 7827
1-(Isopropoxycarbonyloxy)ethyl (6*R*,-
  7*R*)-7-[2-(2-Amino-4-thiazolyl)-
  (*Z*)-2-(methoxyimino)acetamido]-3-
  methoxymethyl-3-cephem-4-carb-
  oxylate *see* 1941
(±)-1-[*p*-(2-Isopropoxyethoxymeth-
  yl)phenoxy]-3-(isopropylamino)-2-
  propanol *see* 1309
(±)-1-[[α-(2-Isopropoxyethoxy)-*p*-
  tolyl]oxy]-3-(isopropylamino)-2-
  propanol *see* 1309
7-Isopropoxyisoflavone *see* 4961
Isopropoxymethylphosphoryl Fluoride
  see 8332
7-Isopropoxy-3-phenyl-4*H*-1-benzo-
  pyran-4-one *see* 4961
7-Isopropoxy-3-phenylchromone *see*
  4961
o-Isopropoxyphenyl *N*-Methylcarb-
  amate *see* 7849
2-Isopropoxypropane *see* 5100
p-Isopropoxy-*N*-sulfanilylbenzamide
  see 8911
Isopropydrin *see* 5105
Isopropyl Acetate, 5093
Isopropylacetic Acid *see* 5120
Isopropyl Acetoacetate, 5094
Isopropylacetone, 5095
Isopropyl Alcohol, 5096
Isopropylamine, 5097
4-Isopropylaminoantipyrine *see* 8122
(±)-1-(Isopropylamino)-3-[*p*-(cyclopropylmethoxyethyl)phenoxy]-2-
  propanol *see* 1204
4-Isopropylamino-2,3-dimethyl-1-
  phenyl-3-pyrazolin-5-one *see* 8122
1-(Isopropylamino)-2-hydroxy-3-[*o*-
  (allyloxy)phenoxy]propane *see* 6905
(±)-1-(Isopropylamino)-3-[*p*-(β-
  methoxyethyl)phenoxy]-2-propanol
  see 6072
1-(Isopropylamino)-3-(*o*-methoxy-
  phenoxy)-2-propanol *see* 6172
Isopropylaminomethyl-(3,4-dihydr-
  oxyphenyl)carbinol *see* 5105
2-Isopropylamino-6-methylheptane
  see 6080
1-(Isopropylamino)-3-[(2-methylin-
  dol-4-yl)oxy]-2-propanol *see* 5744
α-[(Isopropylamino)methyl]-2-naph-
  thalenemethanol *see* 7802
α-[(Isopropylamino)methyl]-*p*-nitro-
  benzyl Alcohol *see* 6442
α-(Isopropylaminomethyl)protocate-
  chuyl Alcohol *see* 5105
2-Isopropylamino-1-(2-naphthyl)eth-
  anol *see* 7802
1-(Isopropylamino)-3-(1-naphthyl-
  oxy)-2-propanol *see* 7852
Isopropylaminophenazone *see* 8122
9-[3-(Isopropylamino)propyl]-9-(aminocarboyl)fluorene *see* 4849

α-(1-Isopropylaminopropyl)protocate-
  chuyl Alcohol *see* 5053
2-(Isopropylamino)pyrimidine *see*
  4987
1-(Isopropylamino)-3-(*m*-tolyloxy)-2-
  propanol *see* 9439
p-Isopropylaniline *see* 2622
4-Isopropylantipyrine *see* 7884
Isopropylarterenol *see* 5105
p-Isopropylbenzaldehyde *see* 2623
p-Isopropylbenzaldehyde Thiosemicarbazone *see* 2624
Isopropylbenzene 2619
p-Isopropylbenzoic Acid *see* 2620
3-Isopropyl-1*H*-2,1,3-benzothiadi-
  azin-4(3*H*)-one 2,2-Dioxide *see*
  1060
Isopropyl 4-(2,1,3-Benzoxadiazol-4-
  yl)-1,4-dihydro-5-methoxycarbon-
  yl-2,6-dimethyl-3-pyridinecarbox-
  ylate *see* 5129
p-Isopropylbenzyl Alcohol *see* 2621
p-Isopropylbenzylidene Thiosemicarbazone *see* 2624
Isopropyl Bromide, 5098
5-Isopropyl-5-(β-bromoallyl)-*N*-
  methylbarbituric Acid *see* 6342
N-4-Isopropylcarbamoylbenzyl-*N* '-
  methylhydrazine *see* 7764
3-Isopropylcarbamylsulfonamido-4-
  (3'-methylphenyl)aminopyridine *see*
  9473
Isopropyl Carbanilate *see* 7828
Isopropylcarbinol *see* 5015
Isopropyl Chloride, 5099
N-Isopropyl-α-chloroacetanilide *see*
  7805
Isopropyl [4'-(*p*-Chlorobenzoyl)-2-
  phenoxy-2-methyl]propionate *see*
  3924
Isopropyl-*m*-chlorocarbanilate *see*
  2188
Isopropyl *N*-(3-Chlorophenyl)carb-
  amate *see* 2188
Isopropyl-*o*-cresol *see* 1881
Isopropyl-*m*-cresyl Ester of Isoamylcarbamic Acid *see* 9343
Isopropyl 6-Cyano-5-methoxycarbon-
  yl-2-methyl-4-(3-nitrophenyl)-1,4-
  dihydropyridine-3-carboxylate *see*
  6461
Isopropyl 4,4'-Dibromobenzilate *see*
  1422
N-Isopropyl-β-dihydroxyphenyl-β-
  hydroxyethylamine *see* 5105
7-Isopropyl-1,4-dimethylazulene *see*
  4464
5-Isopropyl-3,8-dimethyl-1-azulene-
  sulfonic Acid Sodium Salt *see* 4464
N-Isopropyl-1,5-dimethylhexylamine
  see 6080
4-Isopropyl-1,6-dimethylnaphthalene
  see 1607
4-Isopropyl-2,3-dimethyl-1-phenyl-3-
  pyrazolin-5-one *see* 7884
(1*R*,2*S*,6*S*,7*S*,8*S*)-(−)-8-Isopropyl-
  1,3-dimethyltricyclo[4.4.0.0²,⁷]*dec*-3-
  ene *see* 2510
(1*S*,2*R*,6*R*,7*R*,8*S*)-(+)-8-Isopropyl-
  1,3-dimethyltricyclo[4.4.0.0²,⁷]*dec*-3-
  ene *see* 10007
Isopropyl 2,4-Dinitro-6-*sec*-butyl-
  phenyl Carbonate *see* 3280
4-Isopropyl-2,6-dinitro-*N*,*N*-dipropyl-
  aniline *see* 5089
N-Isopropyl-4,4-diphenylcyclohexyl-
  amine *see* 7706
1-Isopropyl-4,4-diphenylpiperidine
  see 7776
Isopropyl Ether, 5100

Isoviral [Lenza] see 4881
Isovist [Nihon Schering] see 4955
Isovitamin C see 5009
Isovon see 5105
Isovue [Squibb] see 4943
Isovyl [Rhovyl] see 7563
Isoxaben, 5125
Isoxal [Shionogi] see 7129
Isoxepac, 5126
Isoxicam, 5127
Isoxsuprine, 5128
Isoxyl see 9382
Isoyohimbine see 10013
Isozid [Saarstickstoff] see 5071
Isozyd see 5071
Ispenoral [Chephasaar] see 7046
Isphamycin see 2193
Isradipine, 5129
Isrodipine see 5129
Issium [Farmochim. Ital.] see 4070
Isteropac E.R. [Bracco] see 4904
Istin see 2813
Istizin [Winthrop] see 2813
Istonil [Nippon Chemiphar/Siegfried] see 3199
Isuprel [Winthrop] see 5105
Isupren see 5105
Iszilin see 4889
ITA-104 see 7488
Itaconic Acid, 5130
Italchin see 8053
Italprid [Prophin] see 9353
Itamidone [Farmitalia] see 488
ITF-182 see 4829
Itobarbital see 1502
Itraconazole, 5131
Itramin Tosylate, 5132
Itridal see 2717
Itrol see 8457
Itrop [Boehringer, Ing.] see 4960
Itrumil [Ciba] see 4954
Ituran [Promonta] see 6520
IUDR see 4819
Iuvacor [Inverni] see 9751
Ivaugan [Voigt] see 4704
Ivermectin, 5133
Iversal [Bayer] see 386
Iversine [Merck & Co.] see 5654
Iviron see 3974
Ivomec [Merck & Co.] see 5133
Ivosit [Hoechst] see 3282
Iwalexin [Iwaki] see 1971
Ixbut, 5134
Ixertol [Rottendorf] see 2308
Ixodin, 5135
Ixoten [Asta] see 9680
Ixprim [Siegfried] see 1238
Izaberizin [Toho Iyaku] see 2308
J-38 see 4160
Jaborandi see 7396
Jaclacin [Lundbeck] see 108
Jacodine see 8403
Jacutin [Hermal] see 5379
Jadit [Hoechst] see 1450
Jaguar see 4486
Jaikin [Basotherm] see 8535
Jalap, 5136
Jalovis see 4676
Jamaica Pepper see 7402
Jamboo see 5137
Jambul, 5137
Jamestown Weed see 8779
Jamylene see 3398
Janimine [Abbott] see 4835
Janus Green B, 5138
Japan Agar see 172
Japan Camphor see 1738
Japanese Oil of Camphor see 6702
Japan Isinglass see 172
Japan Tallow see 5139
Japan Wax, 5139

Jasmolin I see 5140
Jasmolin II see 5140
Jasmolins, 5140
Jasmone, 5141
Jateorrhizine see 5142
Jatroneural [Rohm & Haas] see 9594
Jatropur [Rohm & Haas] see 9515
Jatrorrhizine, 5142
Jaundice Berry see 971
Jaune Brilliant see 1628
Javanicin, 5143
Java Pepper see 2615
Java Plum see 5137
Javelle Water see 4800
JB 11 see 9635
JB 251 see 7911
JB 305 see 7440
JB 323 see 7428
JB 516 see 7197
JB 8181 see 2902
JD 91 see 3520
JDL 464 see 9473
Jectofer [Astra] see 4981
Jecto-Sal [Mallard] see 9291
Jeffox [Jefferson] see 7545
Jefron [Pitman-Moore] see 7547
Jellin [Grünenthal] see 4076
Jenacaine see 7763
Jen-Diril [Jenkins] see 4704
Jequirity see 4
Jeremejevite see 329
Jerusalem Artichoke, 5144
Jerusalem Tea see 6713
Jerva Acid see 2042
Jervasic Acid see 2042
Jervine, 5145
Jesaconitine, 5146
Jesuit's Balsam see 2511
Jesuit's Bark see 2286
Jesuit's Tea see 5637
Jetrium [Hek] see 2933
Jeweler's Rouge see 3973
Jexin [Duncan, Flockhart] see 9717
JH see 5155
C-17 JH see 5155
C-18 JH see 5155
Jilkon see 4245
Jimpson Weed see 8779
Jimson Weed see 8779
JL 991 see 9117
JL 1078 see 3151
J-Liberty [J. Pharmacal] see 2082
JM8 see 1828
JO 1016 see 9408
Jodairol see 4927
Jodid [E. Merck] see 7628
Jodobil see 4919
Jodo-Metil-Fillina [Malesci] see 3663
Jodomiron [Bracco] see 4904
Jodozoat Meglumin [Dagra] see 5688
Joghurt see 10010
Johannisbrotmehl see 5436
Jojoba Oil, 5147
Jomezol see 5892
Jomybel [Sarva] see 5148
Jonctum [Merrell] see 6857
Jonit [Hoechst] see 1318
Jopamiro [Gerot] see 4943
Jordan Almond see 301
Joristen see 7124
Josacine [Spret-Mauchant] see 5148
Josamina [Novag] see 5148
Josamy [Yamanouchi] see 5148
Josamycin, 5148
Josaxin [Yamanouchi] see 5148
JP 992 see 1047
Jubalon see 1805
Judean Pitch see 867
Judolor [Woelm] see 4223
Juglans, 5149
Juglone, 5150

Julin's Carbon Chloride see 4600
Julocrotine, 5151
Julodin [Takeda] see 3651
Jumbul see 5137
Jumex [Chinoin] see 2893
Junipene see 5446
Juniper Berries, 5152
Juniperberry Oil see 6737
Juniper Tar, 5153
Justamil see 8897
Justelmin see 7435
Justicidin A see 5154
Justicidin B see 5154
Justicidins, 5154
Juvamycetin see 2068
Juvason see 7727
Juvastigmin [Dolder] see 6380
Juvela [Eisai] see 9932
Juvela Nicotinate [Eisai] see 9931
Juvenile Hormones, 5155
Juvenimicin A₃ see 8241
Juvocaine see 7763
Juvoxin see 8885
K III see 3272
K IV see 3272
K 17 see 9182
K 31 [Kyorin] see 6429
K 33 see 9124
K 351 see 6479
K 386 see 9434
K 1875 see 1258
K 1900 see 6468
K 2004 see 9006
K 2680 see 3839
K 3712 see 6177
K 3917 see 9074
K 3920 see 4867
K 4024 see 4337
K 4277 see 4875
K 6451 see 6856
K 9321 see 106
K 10033 see 3891
K 11941 see 225
K 22023 see 3394
Ka 2547 see 5824
KABI 925 see 3520
Kabi 2165 see 9058
Kabiglobulin [KabiVitrum] see 4837
Kabikinase [Kabi] see 8784
Kabipenin see 7042
Kadaya see 5164
Kaemferitrin see 5156
Kaempferol, 5156
Kaempferol 3-Robinoside 7-Rhamnoside see 8230
Kaergona see 5714
Kafocin [Lilly] see 1973
Kafylox [Bristol] see 7586
Kainic Acid, 5157
L₅-xylo-Kainic Acid see 5157
α-Kainic Acid see 5157
Kairomones see 7294
Kaleorid [Leo Pharm.] see 7601
Kalex see 3482
Kalgan see 7181
Kalgut [Tanabe Seiyaku] see 2878
Kalignost see 8646
Kalimozan [Nikken] see 7615
Kalinite see 364
Kalitabs see 7601
Kalium see 7579
Kalium-Duriles [Astra] see 7601
Kalle's Acid see 6320
Kallidin, 5158
Kallidin I see 1356
Kallidin II see 5158
Kallidin-9 see 1356
Kallidin-10 see 5158
Kallidinogenase see 5159
Kallikrein, 5159
Kalma [Fresenius] see 9707

Kalmocaps [Italnysco] *see* 2082
Kalmopyrin *see* 1644
Kalodil *see* 3182
Kalsetal *see* 1644
Kalutein [Tatsumi] *see* 2404
Kalymin *see* 7990
Kamala, 5160
Kamassin *see* 8040
Kamaver [Engelhard KG] *see* 2068
Kameela *see* 5160
Kamila *see* 5160
Kaminax [Ital. Res.] *see* 416
Kammerer's Porphyrin *see* 7913
Kamoran [Lilly] *see* 126
Kampfstoff "Lost" *see* 6225
Kamycin *see* 5161
Kamynex *see* 5161
Kanabristol [Bristol] *see* 5161
Kanacedin *see* 5161
Kanamycin, 5161
Kanamycin A *see* 5161
Kanamycin B *see* 5161
Kanamycin C *see* 5161
Kanamytrex [Bristol] *see* 5161
Kanaqua [Andromaco] *see* 5161
Kanasig *see* 5161
Kanatrol [Lusofarmaco] *see* 5161
Kanechlor [Kanegafuchi] *see* 7541
Kanendomycin [Meiji] *see* 5161
Kanendos [Crinos] *see* 5161
Kanescin [Torlan] *see* 5161
Kanicin *see* 5161
Kankohso 101 *see* 7504
Kannasyn *see* 5161
Kannit [Kanto] *see* 9992
Kano [Pierrel] *see* 5161
Kanochol [Nakano] *see* 6843
Kanone [Beecham] *see* 5714
Kanrenol [SPA] *see* 1753
Kantec [Daiichi] *see* 5593
Kantrex [Bristol] *see* 5161
Kantrexil [Bristol] *see* 5161
Kantrox [Bristol] *see* 5161
Kaolin, 5162
Kaolinite *see* 368
Kaon [Warren-Teed] *see* 7615
Kaon-Cl [Adria] *see* 7601
Kaoxidin *see* 8850
Kaparlem *see* 5153
Kapilin *see* 5710
Kapilon [Glaxo] *see* 5710
Kapilon Injectable *see* 5717
Kappadione [Lilly] *see* 5712
Kappati [Erbamont] *see* 2987
Kappaxan *see* 5710
Kappaxin *see* 5714
Kapron [Klin] *see* 6656
Karanjin, 5163
Karanum *see* 5711
Karathane [Rohm & Haas] *see* 3281
Karaya Gum, 5164
Karbam Black *see* 3954
Karbam White *see* 10075
Karbinon [Badrial] *see* 6314
Kardiamed [Medice] *see* 3150
Kardin *see* 9682
Kareon *see* 5714
Karidium *see* 8565
Karil [Sandoz] *see* 1640
Karion *see* 8680
Karion [Gedeon Richter] *see* 7399
Karmex *see* 3388
Karmex Monuron Herbicide *see* 6169
Karminomycin *see* 1879
Karsil, 5165
Karsivan [Hoechst] *see* 7823
Karstenite *see* 1713
Karsulphan *see* 8916
Kasimid *see* 9430
Kaskay [Ayerst] *see* 7601
"Kastle-Meyer Reagent" *see* 7210

KAT 256 *see* 2363
Katadolon [Degussa] *see* 4117
Kata-Lipid [IBI] *see* 1063
Katapyrin [Endopharm] *see* 463
Katchung Oil *see* 7006
Katen [Spofa] *see* 6094
Kathon [Rohm & Haas] *see* 6677
Katilo *see* 5164
Katine *see* 6634
Kativ-G *see* 5714
Katlex [Iwaki] *see* 4221
Katonil [Kali-Chemie] *see* 2104
Katorin *see* 7615
Katoseran [Hishiyama] *see* 2308
Katovit [Thomae] *see* 7791
Kaurie *see* 2512
Kautschin *see* 5371
Kava, 5166
Kavahin *see* 6056
Kavain *see* 5167
Kava-kava *see* 5166
Kavatin *see* 6056
Kavitan [Ayerst] *see* 5716
Kawa *see* 5166
Kawain, 5167
Kayback [Ayerst] *see* 7601
Kay-Cee-L [Geistlich] *see* 7601
Kaydol [Sonneborn] *see* 7139
Kayexalate [Breon/Winthrop] *see* 8622
Kayhydrin *see* 3165
Kayklot *see* 5714
Kayquinone [Abbott] *see* 5714
Kayvisyn *see* 9936
Kayvite *see* 5710
KB-95 *see* 1131
KB-227 *see* 9486
KB-509 *see* 4134
KB-1585 *see* 5319
KBT 1585 *see* 5319
KC-404 *see* 4810
KC-2547 *see* 5824
KC-9147 *see* 9433
K-Contin [Napp] *see* 7601
KD-136 *see* 4511
KE *see* 2533
Kéal [Sinbio] *see* 8853
Kebilis [Roussel] *see* 2044
Kebuzone, 5168
Kedacillin [Takeda] *see* 8863
Kefadol [Lilly] *see* 1922
Kefamin [Lilly] *see* 1944
Kefandol [Lilly] *see* 1922
Kefazim [Lilly] *see* 1944
Kefenid [SIT] *see* 5184
Kefglycin [Lilly] *see* 1973
Kefir Fungi, 5169
Kefir Grains *see* 5169
Kefir Seeds *see* 5169
Keflet [Lilly] *see* 1971
Keflex [Lilly] *see* 1971
Keflin [Lilly] *see* 1978
Keflodin [Lilly] *see* 1975
Keflordin (obsolete) [Lilly] *see* 1975
Keforal [Lilly] *see* 1971
Kefoxina [CT] *see* 1923
Kefroxil [Wharton] *see* 1921
Kefspor [Lilly] *see* 1975
Keftab [Lilly] *see* 1971
Kefurox [Lilly] *see* 1951
Kefzol [Lilly] *see* 1925
Keiperazon [Kaken] *see* 1926
Kelamin *see* 5189
Kelecin *see* 5311
Kelene *see* 3740
Kelfer *see* 3971
Kelferon [MCP] *see* 3987
Kelfiprim [Farmitalia Carlo Erba] *see* 8881
Kelfizina [Abbott] *see* 8881
Kelfizine W [Farmitalia] *see* 8881

Kelgin *see* 231
Kelicor *see* 5189
Kelicorin *see* 5189
Kellin *see* 5189
Kelnac [Sankyo] *see* 7506
Kelocyanor [Laroche Navarron] *see* 3484
Keloid *see* 5189
Kelox [Elder] *see* 9177
Kelp [Omni] *see* 878
Kelpware *see* 4196
Kelthane [Rohm & Haas] *see* 3075
Keltrol F [Kelco] *see* 9966
Kelzan [Kelco] *see* 9966
Kemadrin [Burroughs Wellcome] *see* 7770
Kemi [Otsuka] *see* 7852
Kemicetine [Erba] *see* 2068
Kemithal [ICI] *see* 9219
Kemsol [Horner] *see* 3247
Kenacort [Squibb] *see* 9511
Kenacort-A [Squibb] *see* 9512
Kenalog [Squibb] *see* 9512
Kenaquart *see* 9512
Kendall's Compound A *see* 2859
Kendall's Compound B *see* 2532
Kendall's Compound C *see* 267
Kendall's Compound D *see* 265
Kendall's "Compound E" *see* 2533
Kendall's Compound F *see* 4710
Kendall's Compound G *see* 271
Kendall's Compound H *see* 262
Kendall's Desoxy Compound B *see* 2882
Kephalins *see* 1972
Kephina [Horiuchi] *see* 5184
Kephrine *see* 161
Kepinol [Pfleger] *see* 8889
Kepone *see* 2081
Keracyanin *see* 2694
Keralyt [Bristol-Myers] *see* 8301
Keramik *see* 3116
Keramin *see* 3116
Keraphen *see* 4931
Keratin, 5170
Keratinamin [Kowa] *see* 9781
Keratinase, 5171
Kerb [Rohm & Haas] *see* 7886
Kerecid *see* 4819
Kerlone [Synthelabo] *see* 1204
Kermesic Acid, 5172
Kernechtrot *see* 6395
Kerocaine *see* 7763
Kerosene, 5173
Kerosine *see* 5173
Kertasin *see* 3822
Keselan [Sumitomo] *see* 4511
Kesint [Proter] *see* 1951
Kessar [Farmitalia] *see* 9019
Kessazulen *see* 4464
Kesso-Bamate [McKesson] *see* 5751
Kestrone [Cooper] *see* 3660
Ket [Irbi] *see* 5175
Ketaject [Bristol] *see* 5174
Ketalar [Parke, Davis] *see* 5174
Ketalgin Hydrochloride *see* 5852
Ketaman *see* 7816
Ketamine, 5174
Ketanest *see* 5174
Ketanrift [Ohta] *see* 278
Ketanserin, 5175
Ketas [Kyorin] *see* 4810
Ketaset [Bristol] *see* 5174
Ketason *see* 5168
Ketavet [Bristol] *see* 5174
Ketazolam, 5176
Ketazone [Beytout] *see* 5168
Ketene, 5177
Keteocort [Desitin] *see* 7727
Kethamed *see* 7021
Kethoxal, 5178

Ketipic Acid, 5179
Ketobemidone, 5180
Ketobun-A [Isei] see 278
Ketobutane-Jade [Toho] see 5168
β-Ketobutyranilide see 50
2-Ketobutyric Acid see 51
3-Ketobutyric Acid see 51
4-Keto-β-carotene see 3469
Ketochromin see 3166
Ketoconazole, 5181
Ketocycloheptane see 2728
Ketocyclopentane see 2748
Ketoderm [Janssen] see 5181
Ketodestrin see 3660
(dl)-9-Keto-11α,15α-dihydroxy-16-
   phenoxy-17,18,19,20-tetranorpro-
   sta-4,5,13-trans-trienoic Acid
   Methyl Ester see 3545
17-(1-Ketoethyl)androstane-3,17-diol
   see 263
17β-(1-Ketoethyl)-Δ⁵-androsten-3β-ol
   see 7739
Ketogan [Lundbeck] see 5180
Ketogestin [Upjohn] see 5185
Ketogin [Lundbeck] see 5180
3-Keto-D-glucoheptonofuranolactone
   see 4345
α-Ketoglutaric Acid, 5182
β-Ketoglutaric Acid see 60
2-Keto-L-gulonic Acid, 5183
Ketoheptamethylene see 2728
D-manno-Ketoheptose see 4587
Ketohexamethylene see 2732
2-Ketohexamethylenimine see 1762
D-erythro-3-Ketohexonic Acid Lactone see 5009
D-ribo-2-Ketohexose see 7940
Ketohydroxyestrin see 3660
17-(1-Keto-2-hydroxyethyl)androst-
   ane-3,17-diol-11-one see 271
17-(1-Keto-2-hydroxyethyl)androst-
   ane-3,11,17-triol see 267
17-(1-Keto-2-hydroxyethyl)-Δ⁴-
   androsten-3,11-dione see 2859
17-(1-Keto-2-hydroxyethyl)-4-
   androsten-17α-ol-3-one see 2912
3-[β-Keto-γ-(3-hydroxy-2-piperidyl)-
   propyl]-4-quinazolone see 3890
Ketoisdin [Isdin] see 5181
Ketomalonic Acid see 5814
Ketomalonic Acid Calcium Salt see
   1685
γ-Keto-β-methoxy-δ-methylene-Δᵅ-
   hexenoic Acid see 7033
1-Keto-6-methoxy-1,2,3,4-tetrahy-
   dronaphthalene see 5924
5-Keto-2-methoxy-5,6,7,8-tetrahy-
   dronaphthalene see 5924
Ketopentamethylene see 2748
D-erythro-2-Ketopentose see 8209
Ketophenylbutazone see 5168
Ketoprofen, 5184
11-Ketoprogesterone, 5185
Ketopron [Alcon] see 5184
β-Ketopropane see 58
2-Ketopropionaldehyde see 8030
α-Ketopropionic Acid see 8032
2-Ketopyrrolidine see 8027
2-Ketopyrrolidine-1-ylacetamide see
   7459
Ketorolac, 5186
Ketoscilium [Zambon] see 3949
Ketosuccinic Acid see 6863
1-Keto-3-(3'-sulfamyl-4'-chloro-
   phenyl)-3-hydroxyisoindoline see
   2194
L-3-Ketothreohexuronic Acid Lactone
   see 855
Ketotifen, 5187
17-Ketotrilostane see 9611

2-Keto-1,7,7-trimethylnorcamphane
   see 1738
Keuten see 3017
Kevadon [Merrell] see 9182
Kevadon [Lemonier] see 5184
Kevopril [Pharmuka] see 8111
Kew Tree see 4319
Keyhole-limpet Hemocyanin see 4566
Keypyrone [Key] see 3358
Key-tusscapine [Hyrex-Key] see 6638
KF 868 see 5517
KH3 [Schwarzhaupt] see 4498
Kharophen [Burroughs Wellcome] see
   44
Kharsivan see 839
Khas Khas Oil see 6778
Khat, 5188
Khelfren see 5189
Khellin, 5189
Khellinin see 5190
Khellol Glucoside, 5190
Khinocyde see 8096
Khloratsizin see 2055
Khus Oil see 6778
K-IAO see 7615
Kidira [ICN] see 3157
Kiditard [Delandale] see 8072
Kidon [Ono] see 4422
Kidrolase [Specia] see 858
Kieselguhr see 4878
Kieserite see 5573
Kif see 1752 and 4532
Kiku Oil, 5191
Killax see 9138
Kilmicen [Farmitalia] see 9433
Kilmite 40 see 9138
Kimopsin [Eisai] see 2265
KI-N [Mallinckrodt] see 7628
Kinaden see 4676
Kinalysin [Merck & Co.] see 8784
Kinavosyl [Riker] see 5737
Kinecid see 8885
Kinedak [Ono] see 3556
Kineorl [Showa Shinyaku] see 9447
Kinetin, 5192
Kinetin-Schering see 4676
Kinevac [Squibb] see 8494
King's Gold see 834
King's Yellow see 834 and 5279
Kinic Acid see 8071
Kinichron [Biochimica] see 8072
Kinidin [Astra] see 8072
Kinidrin see 8072
Kino, 5193
Kinoprene, 5194
Kinotomin [Toa Eiyo] see 2345
Kino-yellow see 5519
Kinupril [Pharmuka] see 8111
Kipca, Oil Soluble see 5714
Kipca, Water Soluble see 5712
Kirocid see 8885
Kiron see 8885
α-Kirondrin see 4331
Kir Richter [Lepetit] see 784
Kirromycin see 6138
Kisselo-mleko see 10010
Kitasamycin see 5337
Kitol, 5195
K-Ject see 9933
KL 255 see 1486
KL 373 see 1246
Klamar [Maggioni] see 4463
Klavikordal [Baer] see 6528
Klebcil [Beecham] see 5161
Kleenodyne see 4930
Klemidox see 2346
KLH see 4566
Klimax E [Fink] see 3659
Klimicin [Lek] see 2352
Klimoral [AB Leo] see 3659

Klinit [Eisai] see 9992
Klinium [Janssen] see 5360
Klinomycin [Am. Cyanamid] see 6121
Klion [Gedeon Richter] see 6079
Klismacort [Benechemie] see 7719
Kloben Neburon [Du Pont] see 6354
Klockmannite see 2656
Klonopin [Roche] see 2387
Klor-Con [Upsher-Smith] see 7601
Klorita see 2068
Klorpromex [Dumex] see 2186
Klort [Lemmon] see 5751
Klot [Warren-Teed] see 9444
Klotogen [Abbott] see 5716
Klottone [MET] see 5714
Klucel [Hercules] see 4776
K-Lyte [Mead Johnson] see 7586
KM 2210 see 1199
Knight's Spur see 5242
Knob Root see 2479
Knoll H₇₅ see 7304
Knollide see 7628
K-Norm [Pennwalt] see 7601
Ko 592 see 9439
Kö 1173 see 6094
Kö 1366 see 1479
Koate [Speywood] see 3873
Koaxin see 5714
Kobaton see 3248
Kochite see 368
Kodel®, 5196
Kodocytochalasin-1 see 2796
Koettigite see 10048
Koglucoid see 8131
Koji see 9007
Kojic Acid, 5197
Kola, 5198
Kolikodal see 4708
Kolklot see 5714
Kollateral-forte [Permicutan] see 6202
Kollerdormfix see 2717
Kollidon see 7700
Kolpolyn [Organon] see 3689
Kolpon [Organon] see 3660
Kolton (Tabl.) [Promonta] see 7456
Koltonal see 7456
Kolton (Jelly) [Promonta] see 3334
Kombetin [Boehringer, Mann.] see
   8819
Kombiquens [Novo] see 5687
Kompensan [Roerig] see 3169
Komplexon see 3482
Konakion [Roche] see 9933
Kondurangin see 2491
Konesta [Akzo] see 9539
Koninckite see 3975
Konlax [Nippon Shinyaku] see 7747
Kon Oil see 5518
Kontexin [AB Leo] see 7279
Konyne [Cutter] see 3874
Kopsine, 5199
Korazole see 7097
Korbutone [Nippon Glaxo] see 1029
Kordafen [Polfa] see 6441
Korglykon see 2508
Korlan [Dow] see 8239
Koroseal [Goodrich] see 7563
Korotrin see 4534
Korum [Geneva] see 40
Kosin, 5200
Koso see 1361
Kosso see 1361
K-Othrine [Roussel] see 2869
Kotoite see 5535
Kouso see 1361
Kousso see 1361
KPABA see 7582
KPB see 5168
31K-precursor see 7803
Kraton [Shell] see 8345
Krebiozen®, 5201

Krebon see 1465
Kremol [Sherwood] see 7139
Kremoline [Sherwood] see 7138
Kren see 4668
Kresatin see 2587
KRESTIN [Kureha] see 7943
Kriplex [Alfa] see 3071
Kriptin [Whitehall] see 7996
Kristallose see 8282
Krizanol 1649
Kronitex see 9675
Kronocin see 7768
Kryogenin see 7193
Kryolith see 2609
Kryptocur [Hoechst] see 5354
Krypton, 5202
Krypton Difluoride see 5202
Kryptosterol see 5232
Krysid see 755
KS4 see 133
K-strophanthin see 8819
K-strophanthin-α see 2768
K-strophanthoside see 8819
K-Tab [Abbott] see 7601
K-Thrombyl [Roussel-UCLAF] see 5714
KU 54 see 9699
Kubacron [Kayaku] see 9537
KUE 13032c see 3031
Kujimycin B see 5230
Kukoline see 8496
Kullo see 5164
Kupferkies see 2029
Kuratsit see 8847
Kurchatovium see 3503
Kurchessine, 5203
Kurchiline see 5203
Kurchiphyllamine see 5203
Kurchiphylline see 5203
Kurcholessine, 5204
Kurnakovite see 5535
Kuromanin see 2694
Kuromatsuene see 5446
Kuron see 8483
Kusnarin [Toho Iyaku] see 6273
Kusso see 1361
Kusum Oil see 5518
Kuteera see 5164
Kuthison see 2624
Kutizon see 2624
Ku-Zyme HP [Kremers-Urban] see 6957
K-Vitan see 5714
K-Vitrat see 9936
KW-110 see 21
KW-1062 see 6104
KWD-2183 see 963
Kwell [Reed & Carnrick] see 5379
Kyamepromazine see 2690
Kyanmethin, 5205
Kyanol see 687
Kylar [Uniroyal] see 2810
Kylit [Taiho] see 9992
Kymo-trypure see 2265
Kynex [Lederle] see 8890
Kynex Acetyl [Lederle] see 8890
Kynurenic Acid, 5206
Kynurenine, 5207
Kyocristine [Kyorin] see 9891
Kyorin AP2 see 8487
Kyurinett [Zensei] see 9619
L 67 see 7749
L 105 (Alfa) see 8218
L 105 (Lederle) see 1952
L 105SV see 8218
L141 see 3893
L 1102 see 8302
L 1573 see 2785
L 1633 see 3017
L 2103 see 4616

L 2197 see 1071
L 2214 see 1073
L 2329 see 1093
L 3428 see 497
L 5458 see 2852
L 6257 see 6889
L 6400 see 4048
L 12717 see 5459
L 34314 see 3305
L 35355 see 3348
L 36352 see 9598
L 647339 see 6349
La III see 2977
LA 1 see 6491
LA 956 see 7021
LA 1211 see 4838
LA 1221 see 1501
La 6023 see 5845
LAAM see 5346
LAAO see 426
Lab see 8139
Labazene [Labaz] see 9821
Labazyl (obsolete) see 8290
Labelol [Elea] see 5208
Labetalol, 5208
(R,R)-Labetalol see 3186
Lab Ferment (German) see 8139
Labican [Boniscontro] see 2082
LaBID [Baylor] see 9212
Labile Factor see 3871
Labitan [Labinca] see 3185
Labophylline [L.A.B.] see 9212
Labosept [L.A.B.] see 2896
Labrocol [Lagap] see 5208
Labroda [Specia] see 4039
Labrodax [Specia] see 4039
Labyrin see 2308
Lac see 8426
LAC-43 see 1485
Lacalmin [Tatsumi] see 8721
Lacca see 8426
Lacca Coerulea see 5426
Laccaic Acid, 5209
Laccaic Acid A see 5209
Laccaic Acid A₁ see 5209
Laccaic Acid B see 5209
Laccaic Acid C see 5209
Laccaic Acid D see 5209
Lacca Musica see 5426
Lacdeen [Tsuruhara] see 8721
Lac Dye see 5209
Lacflavin [Towa Yakuhin] see 8201
Lachesine, 5210
Lacmoid, 5211
Lacmus see 5426
Laco [Maney] see 1253
Lacolin see 8586
Lacotein [Christina] see 7900
Lacretin [Tokyo Tanabe] see 2345
Lacril [Allergan] see 4777
Lacrimin [Santen] see 1056
Lacrisert [Merck & Co.] see 4776
Lacrypos P.O.S. [P.O.S.] see 2217
Lactacidogen see 4188
β-Lactamase see 7034
Lactamin (obsolete) see 7900
Lactamine [Daisan] see 7744
Lactaminic Acid see 8431
Lactaroviolin, 5212
Lactate Dehydrogenase, 5213
d-Lactic Acid see 5216
D-Lactic Acid, 5214
D(−)-Lactic Acid see 5214
DL-Lactic Acid, 5215
l-Lactic Acid see 5214
L-Lactic Acid, 5216
L(+)-Lactic Acid see 5216
DL-Lactic Acid Ammonium Salt see 555
Lactic Acid Cetyl Ester see 2022
Lactic Acid Hexadecyl Ester see 2022

Lactic Acid Lactate, 5217
Lactic Acid β-Naphthyl Ester see 6332
Lactic Dehydrogenase see 5213
Lactinium [Roland] see 6828
Lactobacillamide see 5218
Lactobacillic Acid, 5218
Lactobacillus Bifidus Factor see 1230
Lactobacillus Bifidus Growth Factor see 1230
Lactobacillus Bulgaricus Factor see 5266
Lactobacillus Lactis Dorner Factor see 9921
Lactobionic Acid, 5219
Lactobionic Acid Calcium Salt, Compd with Calcium Bromide see 1654
Lactobionic δ-Lactone see 5219
Lactoferrin see 9490
Lactoflavine see 8201
Lactogen see 7788
Lactogen (Human Placental) see 4535
Lactogenic Hormone (Placental Human) see 4535
Lactol see 6332
Lactonaphthol see 6332
p-Lactophenetide, 5220
p-Lactophenetidide see 5220
Lactophenin see 5220
Lactose, 5221
Lactosylceramide see 2800
Lactotransferrin see 9490
Lactovagan [Schwartzhaupt] see 5215
2-(Lactoyloxy)propanoic Acid see 5217
Lactucarium—"German", 5223
Lactucarium—"French", 5222
Lactucerin see 9031
α-Lactucerol see 9031
Lactucin, 5224
Lactucon see 9031
Lactucopicrin see 5224
Lactuflor [Chephesar] see 5225
Lactulose, 5225
Lactyl-p-phenetidin-N-(p-ethoxyphenyl)lactamide see 5220
Lacumin see 5734
Ladakamycin see 907
Ladogal [Winthrop] see 2811
Ladropen [Berk] see 4044
Lady's Slipper see 2778
Laetrile see 629
Laetrile® see 629
Laevilac [Wander] see 5225
Laevoral [Laevosan] see 4185
Laevosan [Laevosan] see 4185
Laevoxin see 5351
Laevulinic Acid see 5352
LAF see 4895
Lagistase [Labaz] see 3508
Lagosin see 4202
Laguncurin see 5519
Laki-Lorand Factor see 3878
Lamar [Tokyo Tanabe] see 9060
Lamasil [Sandoz] see 9086
Lamasine [Bristol-Myers] see 627
Lambdamycin see 2035
Lambeth see 7558
Lamb Mint see 7102
Lambratene (formerly) [Cilag Italiano] see 2785
Lambril see 9430
Lamdiol see 3653
Lamictal [Wellcome] see 5228
Lamidon [Kowa] see 4812
Laminaran see 5226
Laminaran Hydrogen Sulfate see 5226
Laminarin see 5226
Lamisil see 9086
Lamoparan, 5227
Lamoryl [Leo Pharm.] see 4453

Lamotrigine, 5228
Lamoxactam [Lilly] see 6201
Lamp Black see 1815
Lampit [Bayer] see 6454
Lampocef [Lampugnani] see 1925
Lampomandol [Von Boch] see 1922
Lamprecid see 9180
Lamprecid 2770 see 9180
Lampren(e) [Geigy] see 2370
Lampsporin [Von Boch] see 1951
Lampterol see 4824
Lamra [Merckle] see 2977
Lamuran [Boehringer, Mann.] see 8129
Lanacordin see 3150
Lanacort [Combe] see 4711
Lanadigenin see 3149
Lanain see 5231
Lanalin see 5231
Lanarkite see 5302
Lanatilin [Heilmittelwerk] see 3150
Lanatosides, 5229
Lanatoxin [Beiersdorf] see 3146
Lancetina [Aandersen] see 4169
Landamycine [Delalande] see 8208
Landomycin see 6785
Landrax see 2797
Landruma [Landerln] see 6444
Landsen [Sumitomo] see 2387
Lanesin see 5231
Lanesta [Breon] see 2094
Laneth see 7554
Lanettes [Esta] see 7554
Lanex [Nor-Am] see 4080
Lanexat [Roche] see 4062
Langite see 2660
Langoran [Merrell] see 5114
Laniazid [Lannett] see 5071
Lanichol see 5231
Lanicor [Boehringer, Mann.] see 3150
Lanimerck (Ampuls) [E. Merck] see 2903
Lanimerck (Suppositories) [E. Merck] see 5229
Laniol see 5231
Lanirapid [Yamanouchi] see 3150
Lanitop [Boehringer, Mann.] see 3150
Lankacidin A see 8377
Lankacidin C14-Acetate see 8377
Lankamycin, 5230
Lankavose see 2030
Lannate [Du Pont] see 5905
Lanolin, 5231
Lanosta-8,24-dien-3-ol see 5232
Lanosterol, 5232
Lanoxin [Burroughs Wellcome] see 3150
Lansfordite see 5538
Lantadin [Lepetit] see 2852
Lantanon [Ravasini] see 6097
Lanthana see 5233
Lanthanum, 5233
Lanthanum Sesquioxide see 5233
Lanthanum Trioxide see 5233
Lanthionine, 5234
Lanum [Merck & Co.] see 5231
Lanvis [Burroughs Wellcome] see 9266
Lanzyme [Nissui] see 5514
Lapachic Acid see 5235
Lapachol, 5235
Lapav [Federal] see 6968
Lapis Lazuli see 5265
Lappa, 5236
Lappaconitine, 5237
Lapudrine [ICI] see 2185
Lapyrium Chloride, 5238
Laracor [Lagap] see 6905
Laractone [Lagap] see 8721
Laradopa [Roche] see 5344
Laraflex [Lagap] see 6337

Laragon [Roemmers] see 8484
Larapam [Lagap] see 7476
Laratrim [Lagap] see 8889
Larch Agaric see 173
Larch Turpentine see 9847
Lard, 5239
Lard, Benzoinated, 5240
Lard-factor see 9918
Lard Oil, 5241
Larex see 7552
Largactil [SK & F] see 2186
Largaktyl see 2186
Large Fennel see 3923
Large-flowered Cereus see 1606
Largon [Wyeth] see 7834
Lariam [Hoffmann-La Roche] see 5683
Laricic Acid see 174
Laricin see 2498
Laridal [Cusi] see 878
Laristine [Roche] see 4642
Larixin [Toyama] see 1971
Larixinic Acid see 5594
Lark's-claw see 5242
Lark's-heel see 5242
Larkspur, 5242
Larnite see 1709
Larocaine [Roche] see 3210
Larocin (obsolete) [Roche] see 610
Larodopa [Roche] see 5344
Laroscorbine [Roche] see 855
Larostidin [Roche] see 4642
Larotid [Roche] see 610
Laroxyl [Roche] see 504
Larten see 5751
Larumen see 3434
Larvacide 100 see 2156
Larvadex [Ciba-Geigy] see 2782
Larvatrol see 945
Larvin [Rhône Poulenc] see 9258
LAS see 5510
LAS 11871 see 9236
Lasalocid A, 5243
Laser [Tosi] see 6337
Laserdil [Laser] see 5114
Laserpitin, 5244
Lasilix [Hoechst] see 4221
Lasiocarpine, 5245
Lasix [Hoechst] see 4221
Lasma [Pharmax] see 9212
Lasso [Monsanto] see 193
Lastet [Nippon Kayaku] see 3842
Lasurite see 5265
LAT-A see 5246
LAT-B see 5246
Latamoxef see 6201
Latibon [Bayer] see 3098
Latrunculin A see 5246
Latrunculin B see 5246
Latrunculins, 5246
Latumcidin see 3
Laubanite see 1645
Laudanidine, 5247
dl-Laudanidine see 5248
Laudanine, 5248
l-Laudanine see 5247
Laudanosine, 5249
Laudexium Methyl Sulfate, 5250
Laudicon see 4733
Laudissine see 5250
Laudolissin see 5250
Laughing Gas see 6575
Laumontite see 1645
Laurabolin V see 6280
Lauracycline see 9130
Laurel Berry Oil see 5252
Laurel Camphor see 1738
Laureline, 5251
Laurel Oil, 5252
Laurent's Acid see 6324
Laurepukine, 5253

Laureth see 7554
Laureth 9 see 7554
Lauric Acid, 5254
Lauric Acid Ester with 2-Hydroxyethyl Thiocyanate see 5328
Lauric Acid Ester with Pyridoxol Diacetate (Ester) see 5259
Lauric Acid 2-Thiocyanatoethyl Ester see 5328
Lauril see MISC-4
Laurilsulfate see MISC-4
Laurocapram, 5255
Laurodin see 5257
Lauroguadine, 5256
Laurolinium Acetate, 5257
Lauromacrogol 400 see 7554
Lauromicina [Dukron] see 3628
Lauron [Endo] see 902
Laurostearic Acid see 5254
Laurotetanine, 5258
N-(Lauroylcolaminoformylmethyl)-pyridinium Chloride see 5238
5-Lauroyloxy-6-methyl-3,4-pyridine-dimethanol Diacetate see 5259
3-Lauroyloxy-2-picoline-4,5-dimeth-anol Diacetate see 5259
3-O-Lauroylpyridoxol Diacetate, 5259
N-Lauroylsarcosine Sodium Salt see 4267
Lauryl Alcohol see 3402
Lauryl Bromide, 5260
Lauryl Sulfate Salt of the Propionic Acid Ester of Erythromycin see 3628
Lausit [Showa Yakuhin] see 4874
Lautarite see 1679
Lauth's Violet see 9276
Lavender, 5261
Laverin [Lemmon] see 3682
Lawrencite see 3992
Lawrencium, 5262
Lawsone, 5263
Lawsonite see 1645
Laxadin see 1253
Laxagen see 9505
Laxagetten [Tempelhof] see 1253
Laxanin N see 1253
Laxenna see 8407
Laxidogol [Agpharm] see 7377
Laxin see 7208
Laxitex [Andreu] see 8962
Laxoberal [Thomae] see 7377
Laxoberon [Boehringer, Ing.] see 7377
Laxo-Isatin see 6927
Laxonalin see 1310
Laxorex see 1253
Layor Carang see 172
Lazaroids, 5264
Lazurite, 5265
LB 46 see 7412
LB 502 see 4221
LBF, 5266
LC 2 see 5062
LC 33 see 5322
LC 44 see 4114
LCB 29 see 4820
LCR see 9891
LD 935 see 3343
LD 2988 see 4139
LD 3394 see 3932
LD 3612 see 6973
LD 4610 see 6882
LD 4644 see 7424
LDA see 3181
Lead, 5267
Lead Acetate, 5268
Lead Antimonate(V), 5269
Lead Arsenate, 5270
Lead Arsenite, 5271
Lead Azide, 5272

Lead Benzoate see 1101
Lead Borate, 5273
Lead Bromate, 5274
Lead Bromide, 5275
Lead Butyrate, 5276
Lead Chlorate, 5277
Lead Chloride, 5278
Lead Chromate(VI), 5279
Lead Chromate(VI) Oxide, 5280
Lead Difluoride see 5282
Lead Dioxide, 5281
Lead Fluoride, 5282
Lead Fluosilicate see 5284
Lead Formate, 5283
Lead Hexafluorosilicate, 5284
Leadhillite, 5285
Lead Hydroxide, 5286
Lead Hypophosphite, 5287
Lead Iodide, 5288
Lead Lactate, 5289
Lead Metavanadate see 5310
Lead Molybdate(VI), 5290
Lead Monosubacetate see 5301
Lead Monoxide, 5291
Lead Nitrate, 5292
Lead Oleate, 5293
Lead Orthoplumbate see 5307
Lead Oxalate, 5294
Lead Oxide Brown see 5281
Lead Oxide Hydrate see 5286
Lead Oxide Red see 5307
Lead Oxide Yellow see 5291
Lead Peroxide see 5281
Lead Phosphate, 5295
Lead Protoxide see 5291
Lead Selenate, 5296
Lead Selenite, 5297
Lead Sesquioxide, 5298
Lead Silicofluoride see 5284
Lead-sodium Alloy see 8588
Lead Sodium Hyposulfite see 5299
Lead Sodium Thiosulfate, 5299
Lead Stearate, 5300
Lead Subacetate, 5301
Lead Subcarbonate see 1016
Lead Sulfate, 5302
Lead Sulfide, 5303
Lead Sulfocyanate see 5308
Lead Superoxide see 5281
Lead Telluride, 5304
Lead Tetraacetate, 5305
Lead Tetraethyl see 9136
Lead Tetrafluoride, 5306
Lead Tetroxide, 5307
Lead Thiocyanate, 5308
Lead Trioxide see 5298
Lead Tungstate(VI), 5309
Lead Vanadate(V), 5310
Leaf Alcohol see 4620
Leaf Green see 2229
Lealgin see 7216
Leandin see 2688
Leanol [Yoshitomi] see 4628
Lebaycid see 3945
Leben Raib see 10010
Leblon [DeAngeli] see 7463
Lebrufen [Lederle] see 4812
Lecasol [Kaken] see 2345
Lecedil [Zdravlje] see 3881
Lecibis [Andromaco] see 1166
Lecibral [Nezel] see 6403
Lecithin, 5311
Lecithol see 5311
Lectins, 5312
Lectopam [Roche] see 1374
Ledakrin [Polfa] see 6485
Ledclair [Sinclair] see 3480
Ledercillin [Lederle] see 7042
Ledercillin VK [Lederle] see 7046
Ledercort [Lederle] see 9511
Ledercort D [Lederle] see 9512

Lederdopa [Lederle] see 5974
Lederfen [Am. Cyanamid] see 3906
Lederglib [Lederle] see 4372
Lederkyn [Lederle] see 8890
Lederlon [Lederle] see 9514
Ledermycin [Lederle] see 2872
Lederpam [Lederle] see 6881
Lederplatin [Lederle] see 2319
Lederspan [Lederle] see 9514
Ledol, 5313
Ledopa [Lepetit] see 5344
Ledopur [Lederle] see 278
Ledosten see 3096
"Ledum Camphor" see 5313
Lefax [Asche] see 8486
Lefcar [Glaxo] see 1856
Lefetamine, 5314
Lefron [Kyorin] see 9699
Leftose [Nippon Shinyaku] see 5514
Legalon [Madaus] see 8484
Legederm [Essex] see 213
Legential see 4290
Leghemoglobin, 5315
Legoglobin see 5315
Legumex D see 2828
Legumin see 1892
Lehydan [AB Leo] see 7293
LeIF see 4892
Leioplegil [Diathera] see 5316
Leiopyrrole, 5316
Leiormone see 9843
Leipzig Yellow see 5279
Lemascorb [Ascher] see 855
Lemazide [Lemmon] see 1134
Lembrol see 2977
Lemoflur see 8565
Lemonol see 4298
Lemon Peel, 5317
Lemon Walnut see 5149
Lemon Yellow see 982
Lemoran see 5350
Lenacil, 5318
Lenampicillin, 5319
Lenamycin see 1958
Lendorm [Boehringer, Ing.] see 1439
Lendormin [Boehringer, Ing.] see 1439
Lenetran [Lakeside] see 5739
Lenicet see 339
Lenigallol [Knoll] see 8010
Lenirit [Bonomelli] see 4711
Lenisarin see 3390
Lenital [Besins-Iscovesco] see 6528
Lenopect [Draco] see 7423
LenoxiCaps [Wellcome] see 3150
Lenoxin [Deutsche Wellcome] see 3150
Lenperone, 5320
Lentac see 8890
Lentagran [Chemie Linz] see 7981
Lentamid see 8909
Lentard [Novo] see 4890
Lente Iletin see 4890
Lente Insulin see 4890
Lenthionine, 5321
Lenticillin see 7042
Lentin [Dow] see 1780
Lentinacin see 3618
Lentinan, 5322
Lentisk see 5636
Lentizol [Warner] see 504
Lentobetic see 7191
Lentogest [Samil] see 4774
Lento-Kalium [Robin] see 7601
Lentonitrina [Pierrel] see 6528
Lentopenil see 7037
Lentor [Parke, Davis] see 3540
Lentotran [Farm. Patria] see 2082
Lentrat see 7066
Lentysine see 3618
Lenzacef [Lenza] see 1982
Leo 640 see 5438

Leo 1031 see 7718
Leocillin [Leo Pharm.] see 7027
Leodrine [Leo Pharm.] see 4716
Leo K [Leo] see 7601
Leonurine, 5323
Leopard's Bane see 813
Leostesin [Leo Pharm.] see 5359
Lepargylic Acid see 921
Lepasen see 491
Lepetown see 5751
Lepicron [Ciba-Geigy] see 9258
Lepidine, 5324
Lepidocrocite see 3969
Lepidolite see 5395
Lepidopterans: see 1917
Lepitoin Sodium see 7293
Leponex [Wander] see 2417
Lepotex [Dorsey] see 2417
Lepsiral see 7753
Leptanal see 3944
Leptandra, 5325
Leptazol see 7097
Leptilan [Ciba-Geigy] see 9821
Leptodactyline, 5326
Leptophos, 5327
Leptryl [Roger Bellon] see 7120
Lergefin [Larma] see 1804
Lergigan see 7797
Lergine Chloride see 7770
Lergitin see 7176
Lergoban [Riker] see 3334
Lergopenin see 2346
Leritine [Merck & Co.] see 686
Leron [Bayer] see 4470
Lertus [Exa] see 5184
Lescodil [Zdravlje] see 6403
Lesidrin see 6062
Lespedin see 5156
Lespenephryl [Endopharm; Fraysse] see 5156
Lesser Centaury see 1967
Lestid [Upjohn] see 2472
Lethalaire G-58 see 6856
Lethane® 60, 5328
Lethane® 384, 5329
Lethelmin see 7220
Lethidrone [Burroughs Wellcome] see 6275
Letosteine, 5330
LETS see 4018
Letter [Armour Pharm.] see 5351
"Lettuce Opium" see 5223
Letusin see 5349
Leu⁵-E see 3528
Leu (IUPAC Abbrev.) see 5331
Leucaenine see 6118
Leucaenol see 6118
Leucarsone see 1788
Leucenine see 6118
Leucenol see 6118
Leuchtenbergite see 352
Leucine, 5331
Leucine⁵-enkephalin see 3528
6-D-Leucine-9-(N-ethyl-L-prolin-
  amide)-10-deglycinamideluteinizing
  Hormone-releasing Factor (Pig) see 5341
Leucinocaine Mesylate, 5332
Leucite, 5333
Leuco-4 see 141
Leucoalizarin see 718
Leucocyanidin, 5334
Leucocyanidol see 5334
Leucodinine see 6250
Leucodrin, 5335
Leucogen [Bayer] see 858
Leucoglycodrin, 5336
Leucoharmine see 4531
Leucokinin see 9719
Leucoline see 8097
Leucomycin A₁ see 5337

Cross Index of Names

Leucomycin A₃ see 5148 and 5337
Leucomycin V 3-Acetate 4ᴮ-(3-Methylbutanoate) see 5148
Leucomycin V 4B-Butanoate 3B-Propanoate see 8233
Leucomycin V 3ᴮ,9-Diacetate 3,4ᴮ-Dipropanoate see 6123
Leucomycin V 3,4ᴮ-Dipropanoate see 6106
Leucomycins, 5337
Leucophor R see 1322
Leucopterin, 5338
Leucosar [Adria] see 4141
Leucotrofina [Ellem] see 9337
Leucovorin see 4141
Leucovorin [Lederle] see 4141
(D-Leu⁶)-des-Gly¹⁰-LH-RH-ethylamide see 5341
Leu⁵-Enkephalin see 3528
Leukaemomycin C see 2825
Leukeran [Burroughs Wellcome] see 2064
Leukerin see 5762
Leukochthol see 4815
Leukocyte Interferon see 4892
Leukocytic Endogenous Mediator see 4895
Leukomycin [Bayer] see 2068
Leukomycin N [Bayer] see 924
Leukotriene A see 5339
Leukotriene A₄ see 5339
Leukotriene B see 5339
Leukotriene B₄ see 5339
Leukotriene C see 5339
Leukotriene C₁ see 5339
Leukotriene C₄ see 5339
Leukotriene D see 5339
Leukotriene D₄ see 5339
Leukotriene E see 5339
Leukotriene E₄ see 5339
Leukotrienes, 5339
Leunase [Kyowa] see 858
Leupeptin Ac-LL see 5340
Leupeptin Pr-LL see 5340
Leupeptins, 5340
Leuprolide, 5341
Leuprorelin see 5341
Leurocristine see 9891
Levacecarnine Hydrochloride see 78
Levadone [Burroughs Wellcome] see 5852
Levallorphan, 5342
Levamfetamine see 616
Levamisole see 9161
Levamphetamine see 616
Levanil [Upjohn] see 3478
Levant Wormseed see 8323
Levanxene [Erba] see 9074
Levanxol [Erba] see 9074
Levarterenol see 6612
Levarterenol Bitartrate see 6612
Levasole [Pitman-Moore] see 9161
Levatol [Lilly] see 7025
Levaxin see 5351
Levium [Sodelco] see 2977
Levius [Farmitalia] see 873
Levlen [Berlex] see 6621
Levo-alpha-acetylmethadol see 5346
Levo-BC 2627 see 1530
Levobunolol, 5343
Levocarbinoxamine see 1804
Levocarnil [Sigma-Tau] see 1856
Levocarnitine see 1856
Levociclina U.M. [Archifar] see 8408
Levocitrol [Firmenich] see 2332
Levocycline see 8408
Levodopa, 5344
Levo-Dromoran see 5350
Levodropropizine see 3439
Levoglutamina see 4365
Levomepate, 5345

Levomeprazine see 5909
Levomepromazine see 5909
Levomethadyl Acetate, 5346
Levomethorphan Hydrobromide see 8116
Levomicetina see 2068
Levomoprolol see 6172
Levomycetin see 2068
Levonordefrin see 6609
Levonorgestrel see 6621
Levopa [ICN] see 5344
Levophacetoperane, 5347
Levophed [Winthrop] see 6612
Levopimaric Acid, 5348
Levopraid [Ravizza] see 8971
Levoprome [Lederle] see 5909
Levopropoxyphene, 5349
Levopropoxyphene Napsylate see 5349
Levopropylcillin see 7829
Levoprotiline see 6878
Levorenine see 3569
Levorin see 1747
Levorin A₂ see 1747
Levorotatory Lactic Acid see 5214
Levorphan see 5350
Levorphanol, 5350
Lévospasme [Dausse] see 3343
Levospasmol see 3343
Levosulpiride see 8971
Levotartaric Acid see 9037
Levotensin [Simes] see 6172
Levothroid [Armour] see 5351
Levothyl see 5852
Levothym [Karlspharma] see 4784
Levothyrox [Merck-Clevenot] see 5351
Levothyroxine Sodium, 5351
Levotuss [Dompe] see 3439
Levoxadrol Hydrochloride see 3293
Levoxan [Cutter] see 3293
Levsin [Kremers-Urban] see 4795
Levugen see 4185
Levulinic Acid, 5352
Levulinic Acid Calcium Salt see 1684
Levulose see 4185
Levynite see 1645
Lewisite see 3051
Lexatol see 7257
Lexibiotico [Llano] see 1971
Lexinor [Astra] see 6617
Lexomil [Roche] see 1374
Lexone [Du Pont] see 6076
Lexotan [Roche] see 1374
Lexotanil [Roche] see 1374
LF 178 see 3924
LFA 2043 see 4964
LG 152 see 5674
LG 11457 see 3662
LG 30158 see 8232
LH, 5353
LH-RF see 5354
LH-RH, 5354
LH-RH/FSH-RH see 5354
Liatris, 5355
Libanil [APS] see 4372
Libethenite see 2653
Libexin [Chinoin] see 7743
Libigen [Savage] see 4534
Libiolan [L.I.B.S.] see 5751
Libratar [UCEPHA] see 2075
Libritabs [Roche] see 2082
Librium [Roche] see 2082
Licabile see 3892
Licaran see 3892
Licarbin see 9415
Licareol see 5373
Licarpin [Allergan] see 7395
Licheniformins, 5356
Lichenin, 5357
Licorice see 4400

Lidamidine, 5358
Lidanar [Sandoz] see 5813
Lidanil [Sandoz] see 5813
Lidaprim [Chemie Linz] see 8892
Lidarral [Rorer] see 5358
Liden see 5069
Lidepran see 5347
Lidex [Syntex] see 4077
Lidocaine, 5359
Lidoflazine, 5360
Lidol see 5736
Lidone [Abbott] see 6142
Lidothesin see 5359
Life [SIT] see 8006
Lifeampil [Lifepharma] see 621
Lifène see 7231
Life Root see 8401
Lifril [Kissei] see 9060
Light Green N see 5581
Light Green SF Yellowish, 5361
Light Oil of Camphor see 6702
Light Spar see 1713
Lignans, 5362
Lignavet [C-Vet] see 5359
Lignin, 5363
Lignite Wax see 6167
Lignocaine see 5359
Lignocaine Benzyl Benzoate see 2877
Lignoceric Acid, 5364
Lignum Vitae, 5365
Ligroin, 5366
Ligustrin see 8997
Likinozym [Kaigai] see 5514
Likuden [Hoechst] see 4453
Lilacillin [Takeda] see 8863
Lilacin see 8997
Lilly 01526 see 5088
Lilly 35483 see 2760
Lilly 36352 see 9598
Lilly 53838 see 3926
Lilly 69323 see 3926
Lilly 99170 see 781
Lily of the Valley see 2507
Liman [Kali-Chemie] see 9080
Limaprost, 5367
Limarsol see 44
Limas [Taisho] see 5404
Lima Wood see 7133
Limbial see 6881
Limclair [Sinclair] see 3483
Lime see 1692
Limestone, 5368
Lime Sulfurated Solution, 5369
Limethasone [Green Cross] see 2922
Limettin, 5370
Limit [Bock] see 7180
Limit [Monsanto] see 411
Limonene, 5371
Limonin, 5372
Limonite see 3969
Limonoic Acid 3,19:16,17-Dilactone see 5372
Limonoic Acid Di-δ-lactone see 5372
LIN-1418 see 8975
Linalol see 5373
Linalool, 5373
Linalyl Acetate, 5374
Linamarin, 5375
Linarigenin-glucoside see 5376
Linarin, 5376
Linaris [R.A.N.] see 8889
Linatine, 5377
Linaxar see 8829
Lincocin [Upjohn] see 5378
Lincolcina see 5378
Lincoln Bean see 8684
Lincolnensin see 5378
Lincomix see 5378
Lincomycin, 5378
Linctussal see 3017
Lindafor [Rhône Poulenc] see 5379

Lindane, 5379
Lindatox see 5379
Linden Tree see 1017
Lindlar Catalyst, 5380
Lindol see 9675
Lindoxyl [Lindopharm] see 392
Linear Gramicidins see 4438
Lineatin, 5381
4,6,6-Lineatin see 5381
Lingraine [Bayer] see 3609
Lingran [Winthrop] see 3609
Linnaeite see 2421
Linodil [Breon] see 4885
Linoleic Acid, 5382
9,12-Linoleic Acid see 5382
Linolenic Acid, 5383
α-Linolenic Acid see 5383
γ-Linolenic Acid, 5384
Linolexamide see 5382
Linolic Acid see 5382
Linoral [Pharmacia] see 3689
Linostil [Siegfried] see 3199
Linseed, 5385
Linseed Oil, 5386
Lintex see 6425
Linton [Yoshitomi] see 4511
Lintrin see 6328
Linum see 5385
Linurex [Makhteshim-Agan] see 5387
Linuron, 5387
Linyl see 7232
Liometacen [Chiesi] see 4874
Liomycin [Daiichi] see 3429
Lion's Tooth see 9030
Lioplacentyl [Maestretti] see 7487
Lioresal [Ciba] see 950
Liosol see 9980
Liothyronine, 5388
Liothyronine Sodium see 5388
Lioxone see 4659
Lipal see 7554 and 7555
Lipal 39S [PVO Intl.] see 7555
Lipal 400S [PVO Intl.] see 7555
Lipal 9LA [PVO Intl.] see 7554
Lipalt [Von Boch] see 2916
Lipamone see 3094
Lipan see 3272
Lipanor [Winthrop] see 2314
Lipanthyl [Fournier; Holphar] see 3924
Lipantil [Fournier] see 3924
Liparon see 2834
Lipase, 5389
Lipavlon [Avlon] see 2374
Liphadione [Lipha] see 2152
Lipidax [UCB-Smit] see 3924
Lipidil [Ibirn] see 3924
Lipidium [SEDAPH] see 6426
Lipid-mobilizing Hormone see 5392
Lipiodol [Guerbet] see 4917
Liple [Green Cross] see 7892
Lipoclar [Farmacosmici] see 3924
Lipoclin [Sumitomo] see 2353
Lipocol [Lipo] see 7554
Lipocol C-20 [Lipo] see 7554
Lipodel [Delalande] see 6962
Lipofene [Selvi] see 3924
Lipoglutaren [Ausonia] see 5690
Lipo-Hepin [Riker] see 4571
Lipo-Hepinette [Riker] see 4571
α-Lipoic Acid see 9255
Lipoicin see 9255
Lipo-Lutin [Parke, Davis] see 7783
Lipolytic Hormone see 5392
Lipo-Merz [Merz] see 3834
Liponeurina [COSMA] see 7896
Lipopeg [Lipo] see 7555
Lipopeg 4-S [Lipo] see 7555
Lipoprotein Lipase, 5390
Liposana [Farmaroma] see 9980
Liposit [SIT] see 3924

Liposolvin [Tosi] see 8489
Liposomes, 5391
Liposom Forte [Fidia] see 5391
Lipotril see 2208
Lipotrophin see 5392
Lipotropic Hormone, 5392
Lipotropin see 5392
β-Lipotropin (61-91) see 3528
β-Lipotropin C-Fragment see 3528
Lipozid [Pierrel] see 4280
Liprinal [Bristol] see 2374
Lipsin [Fournier] see 3924
Liptan [Kowa] see 4812
Lipur [Substantia] see 4280
Liquaemin Sodium [Organon] see 4571
Liquamar [Organon] see 7230
Liquid Butter of Antimony see 740
Liquid Paraffin see 7139
Liquid Rosin see 9013
Liquid Silver see 5801
Liquifilm [Allergan] see 7562
Liquiphene see 7271
Liquémin [Roche] see 4571
Liquoid [Roche] see 8620
Liquor Aluminii Aceticotartarici see 323
Liquorice see 4400
LIR 1660 see 9849
Liranol [Wyeth] see 7793
Lironox see 2802
Lirotil see 6609
Lisacef [Lisapharm] see 1982
Lisacort [Fellows-Testagar] see 7727
Lisagal see 6927
Lisaglucon [Farmasa] see 4372
Liserdol [Farmitalia] see 5843
Lisergan see 3918
Lisidonil [Ciba] see 4685
Lisil [Lenza] see 1809
Lisino [Essex] see 5455
Lisinopril, 5393
Lisium [Brunton] see 2090
Liskantin [Desitin] see 7753
Liskonum [SK & F] see 5404
Lisomucil [Synthelabo] see 1809
Lisozima [SPA] see 5514
Lispamol [Specia] see 479
Lispine [Nippon Chemiphar] see 3366
Lissamine Green SF [ICI] see 5361
Lissapol see 4266
Lissephen see 5737
Lissolamine V see 2018
Listenon see 8846
Listica [Armour Pharm.] see 4770
Listomin S [Lion] see 1526
Listrocol [Farmitalia] see 2773
Lisuride, 5394
Litalir [Heyden] see 4785
Litarex [Weddel] see 5407
Litec [Sandoz] see 7486
Lithamide see 5398
Lithane [Roerig] see 5404
Litharge see 5291
Lithiodiisopropylamine see 3181
Lithiophor [Vifor] see 5421
Lithium, 5395
Lithium Acetate, 5396
Lithium Acetylsalicylate, 5397
Lithium Alanate see 351
Lithium Aluminohydride see 351
Lithium Aluminum Hydride see 351
Lithium Amide, 5398
Lithium Antimoniothiomalate see 711
Lithium Antimony Thiomalate see 711
Lithium Benzoate, 5399
Lithium Biborate see 5401
Lithium Bichromate see 5408
Lithium Bitartrate, 5400
Lithium Borate, 5401
Lithium Borohydride, 5402

Lithium Bromide, 5403
Lithium Carbonate, 5404
Lithium Chloride, 5405
Lithium Chromate(VI), 5406
Lithium Citrate, 5407
Lithium Dichromate(VI), 5408
Lithium Diisopropylamide see 3181
Lithium-Duriles [Pharma-Stern] see 5421
Lithium Fluoride, 5409
Lithium Formate, 5410
Lithium Hydrate see 5412
Lithium Hydride, 5411
Lithium Hydroxide, 5412
Lithium Iodide, 5413
Lithium Metasilicate see 5420
Lithium Nitrate, 5414
Lithium Oxalate, 5415
Lithium Oxide, 5416
Lithium Perchlorate, 5417
Lithium Platinocyanide see 5422
Lithium Salicylate see 8301
Lithium Selenate, 5418
Lithium Selenite, 5419
Lithium Silicate, 5420
Lithium Sulfate, 5421
Lithium Tetraborate see 5401
Lithium Tetracyanoplatinate(II), 5422
Lithium Tetrahydroaluminate see 351
Lithium Tetrahydroborate see 5402
Lithobid [Rowell] see 5404
Lithocholic Acid, 5423
Lithographic Stone see 5368
Lithol see 4815
Lithonate [Rowell] see 5404
Lithonate S [Rowell] see 5407
Lithopone, 5424
Lithostat [Uro-Research] see 54
Lithotabs [Rowell] see 5404
Liticon [Aristochimica] see 7078
Litlure A see 7777
Litmocidin, 5425
Litmomycin see 4441
Litmopyrine see 5397
Litmus, 5426
Litraderm [Declimed] see 4711
Litsoeine see 5258
Little Water see 9945
Litursol [Alfa] see 9801
Livalfa [Mitsubishi Kasei] see 9677
Livazone [Specia] see 9218
Liver Extract, 5427
Liver *Lactobacillus Casei* Factor see 4140
Liver of Lime see 1714
Liver Starch see 4392
Liver of Sulfur see 7578
Livetins, 5428
Livial [Akzo] see 9358
Liviatin see 3429
Liviclina [Sierochimica] see 1925
Livonal see 3272
Lixacol [Cusi] see 5807
Lixil [AB Leo] see 1473
LJ 48 see 5668
LJ 206 see 1809
LJC 10141 see 3893
LJC 10305 see 1952
LL 1530 see 6261
LL 1558 see 9349
LL 1656 see 1463
LL-AV290 see 906
LLD Factor see 9921
LLF see 3878
Lloncefal [Castillon] see 1975
Llonexina [Castillon] see 1971
LM 91 see 2152
LM 123 see 4061
LM 192 see 9908
LM 208 see 8111
LM 209 see 5754

LM 280 see 4902
LM 637 see 1371
LM 2717 see 2355
LM 2909 see 2876
LM 5008 see 4843
LM 22102 see 5056
LMD see 2925
LMWD see 2925
LN 107 see 1438
Lobak see 2105
Lobamine see 5896
Lobelanidine, 5429
Lobelanine, 5430
Lobelia, 5431
Lobelidine see 5432
Lobeline, 5432
α-Lobeline see 5432
Lobenzarit, 5433
Lobeton see 5432
Lobidan see 5432
Lobron see 5432
Lobufen [Lederle] see 4812
Lobulantina see 7515
Locabiotal [Servier] see 4226
Locacorten (formerly) [Ciba] see 4066
Localyn [Syntex] see 4076
Locapred [Fabre] see 2908
Lochnericine, 5434
Lochneridine, 5435
Lochnerinine see 5434
Locoid [Gist-Brocades] see 4710
Locorten [Ciba] see 4066
Locron [Hoechst] see 346
Loctidon [Lepetit] see 8967
Locton [Lepetit] see 8967
Locula see 8870
Locust Bean Gum, 5436
Loderix [EGIS] see 8423
Lodine [Ayerst] see 3831
Lodopin [Fujisawa] see 10097
Lodosin [Merck & Co.] see 1802
Lodosyn [Merck & Co.] see 1802
Loestrin [Parke, Davis] see 6614
Lofentanil, 5437
Lofepramine, 5438
Lofetensin [Nattermann] see 5439
Lofexidine, 5439
Loflucarban, 5440
Lofoxin [Locatelli] see 4169
Lofton [Abbott] see 1463
Loftran [Beecham] see 5176
Loftyl [Abbott] see 1463
Logan [Ist. Chim. Int.] see 2321
Loganin, 5441
Logical [Armstrong] see 9821
Logwood see 4557
Lokalison F [Dorsch] see 2922
Lokarin see 3194
Loline, 5442
Lomadine [M & B] see 7165
Lomapect [TAD] see 7743
Lombristop [Septa] see 9217
Lomefloxacin, 5443
Lomexin [Recordati] see 3948
Lomidine [Rhône-Poulenc] see 7071
Lomir [Sandoz] see 5129
Lomotil [Searle] see 3313
Lomper [Esteve] see 5647
Lomudal [Fisons] see 2594
Lomudas [Fisons] see 2594
Lomupren [Fisons in England] see 5105
Lomupren [Fisons in Germany] see 2594
Lomuspray [Fisons] see 2594
Lomustine, 5444
LON 798 see 4474
Lonacol [Bayer] see 10071
Lonavar [Searle] see 6875
Lonazolac, 5445
Londomin [Taiyo] see 8998

Londomycin see 5851
Longacid see 3684
Longacilina see 7037
Longasa [Squibb] see 873
Longasulf see 8885
Longatin see 6638
Longatren [Bayer] see 925
Longdigox [Trommsdorf] see 3150
Longestrol [Laroche Navarron] see 1438
Longheparin see 4571
Longicid [Chassot] see 3106
Longicil see 7037
Longifene [Sarva] see 1449
Longifolene, 5446
α-Longilobine see 8403
β-Longilobine see 8169
Longum [Farmitalia] see 8881
Lonidamine, 5447
Loniten [Upjohn] see 6122
Lonmiel [Teikoku] see 1046
Lonolox [Upjohn] see 6122
Lonomycins, 5448
Lontrel [Dow] see 2398
Looplure, 5449
Looser [Kaken] see 1486
Lo/Ovral [Wyeth] see 6621
Lopantrol [Janssen] see 5457
Lopatol [Ciba-Geigy] see 6555
Lopemid [Gentili] see 5450
Lopemin [Dainippon] see 5450
Loperamide, 5450
Loperyl [Zambeletti] see 5450
Lophophorine, 5451
Lophotoxin, 5452
Lopid [Warner-Lambert] see 4280
Lopirin [Squibb] see 1773
LOPP see MISC-2
Lopramine see 5438
Loprazolam, 5453
Lopremone (Rescinded USAN) see 9503
Lopresor [Ciba-Geigy] see 6072
Lopress [Tutag] see 4682
Lopressor [Ciba] see 6072
Lopril [Squibb] see 1773
Loprox [Hoechst] see 2270
Lopurin [Boots] see 278
Lora (formerly) [Wallace Labs.] see 2089
Lorajmine, 5454
Loramet [Wyeth-Pharma] see 5458
Loranil [Winthrop] see 1026
Lorasolid [Dumex] see 5456
Loratadine, 5455
Lorax [Fontoura-Wyeth] see 5456
Lorazepam, 5456
Lorcainide, 5457
Lorelco [Dow] see 7761
Loretin see 4757
Lorexane [ICI] see 5379
Lorfan [Roche] see 5342
Loricin [Sigma Tau] see 8862
Loridine [Lilly] see 1975
Lorinal [Arnar-Stone] see 2061
Lorinden [Polfa] see 4066
Lorisal [Seatrace] see 5565
Lorivox [Srbolek] see 5457
Lormetazepam, 5458
Lormin [Lilly] see 2102
Loromisin see 2068
Lorothidol see 1316
Lorox [Du Pont] see 5387
Lorphen [Geneva] see 2180
Lorsban [Dow] see 2190
Lorsilan [Belupo] see 5456
Lorusil [Bayer] see 479
Losalen [Ciba] see 4066
Losantin see 1677
Losec [Astra] see 6800
Lospoven [Hoechst] see 1978

Loticort see 4104
Lotrial [Roemmers] see 3521
Lotrifen, 5459
Lotrimin [Delbay] see 2412
Loturine see 4530
Lotusate [Winthrop] see 9010
Loubarb see 1495
Louisiana Long Pepper see 1769
Louisiana Sport Pepper see 1769
Lovastatin, 5460
Love-lies-bleeding see 383
Lovenox [Pharmuka] see 3541
Loverine [Yamagata] see 2922
Loviscol [Robins] see 1809
Lowgan [Yamanouchi] see 607
Lowpres [Chugai] see 6200
Lowpstron [Maruko] see 4221
Loxacor [Merrell-Dow] see 5439
Loxapac [Lederle] see 5461
Loxapine, 5461
Loxeen [Maruko/Tobishi] see 7747
Loxen [Sandoz] see 6403
Loxitane [Lederle] see 5461
Loxitane C [Lederle] see 5461
Loxon [Burroughs Wellcome] see 4519
Loxonin [Sankyo] see 5462
Loxoprofen, 5462
Loxuran [EGYT] see 3106
Lozol [USV] see 4847
LPH see 5392
LPH (61-55) see 3528
LPH (61-76) see 3528
LPH (61-77) see 3528
LPH (61-91) see 3528
β-LPH see 5392
γ-LPH see 5392
LRF see 5354
LRH see 5354
LS 74-783 see 4167
LS 121 see 6268
L-S 519 see 7463
L-S 519-C12 see 7463
LSA₂-L₂ see MISC-2
LSD [Sandoz] see 5507
LSD-25 see 5507
LT 1 see 8845
LT 31-200 see 1335
LTA₄ see 5339
LTB₄ see 5339
LTC₄ see 5339
LTD₄ see 5339
LTE₄ see 5339
LTG see 5228
LTH see 7788
LTs see 5339
LTX see 5452
Lu 5-110 see 4114
Lu 10-171 see 2320
LU 1631 see 403
Lubalix [Lubapharm] see 2415
Lubergal see 7201
Lubricort [Texas Pharmacal] see 4710
Lubrokal see 7201
Lubrol PX [ICI] see 7532
Lucaine [Pennwalt] see 7466
Lucamide see 3685
Lucanthone Hydrochloride, 5463
Lucayan [Corvi] see 9370
Lucelan [Krka] see 1493
Lucensomycin, 5464
Luciculine see 6286
Lucidil [Smith & Nephew] see 1037
Lucidol see 1128
Lucidril [Dainippon] see 5660
Luciferin, 5465
D-(−)-Luciferin see 4025
Lucifer Yellow CH, 5466
Lucinite see 362
Lucite see 5849
Lucofen [Wm. R. Warner] see 2183
Lucorteum Oral see 3696

Lucorteum Sol *see* 7783
Lucosil *see* 8887
Lucrin [Abbott] *see* 5341
Ludiomil [Ciba] *see* 5633
Lufyllin [Mallinckrodt] *see* 3459
Lugacin [Lagap] *see* 4284
LüH6 *see* 6659
Lukadin [San Carlo] *see* 416
Luliberin *see* 5354
Lullamin [Reed & Carnrick] *see* 5871
Lumazine, 5467
Lumbrical *see* 7431
Lumichrome, 5468
Lumiflavine, 5469
Luminal [Winthrop] *see* 7201
Luminal Sodium [Winthrop] *see* 7202
Luminol, 5470
Lumirelax *see* 5903
Lumisol RV *see* 1322
9α-Lumista-5,7,22-trien-3β-ol *see* 8007
Lumisterol, 5471
9α-Lumisterol *see* 8007
Lumitens [Sarbach] *see* 9986
Lumopaque [Winthrop] *see* 9745
Lumota [Thomae] *see* 756
Lunacridine, 5472
Lunacrine, 5473
Lunal *see* 151
Lunamin *see* 4616
Lunamycin *see* 4824
Lunar Caustic *see* 8465
Lunargen *see* 8474
Lunasine, 5474
Lunetoron [Sankyo] *see* 1473
Lunine, 5475
Lunipax [Beecham] *see* 4123
Lunis [Valeas] *see* 4071
Lunularic Acid, 5476
Luostyl [UPSA] *see* 3122
Lupanine, 5477
Lup-20(29)-ene-3,28-diol *see* 1212
Lup-20(30)-ene-3β,28-diol *see* 1212
Lup-20(29)-en-3β-ol *see* 5478
Lupeol, 5478
Lupinidine *see* 8692
Lupinine, 5479
(−)-Lupinine *see* 5479
epi-Lupinine *see* 5479
l-Lupinine *see* 5479
l-epi-Lupinine *see* 5479
Lupolen [BASF] *see* 7544
Lupron [TAP] *see* 5341
Luprostiol, 5480
α-Lupulic Acid *see* 4673
β-Lupulic Acid *see* 5482
Lupulin, 5481
Lupulon, 5482
Luride-SF [Hoyt] *see* 8565
Luridine *see* 2207
Luronase *see* 4676
Lurselle [Merrell] *see* 7761
Lutalyse [Upjohn] *see* 7894
Luteanine *see* 5046
Lutecium *see* 5484
Lutein *see* 9972
Luteinizing Hormone *see* 5353
Luteinizing Hormone-releasing Factor *see* 5354
Luteinizing Hormone-releasing Hormone *see* 5354
Luteoantine *see* 4189
Luteocobaltic Chloride *see* 4597
Luteodyn *see* 7783
Luteogan *see* 7783
Luteohormone *see* 7783
Luteol *see* 7783
Luteolas [Serono] *see* 3816
Luteolin, 5483
Luteolin-4′-methyl Ether *see* 3290
Luteonorm [Serono] *see* 3816

Luteosan *see* 7783
Luteotropic Hormone *see* 7788
Luteotropin *see* 7788
Luteovis *see* 7783
Luteran [Cassenne] *see* 2102
Lutestral [Cassenne] *see* 2102
Lutetium, 5484
Lutex *see* 7783
α,α′-Lutidin *see* 5485
2,6-Lutidine, 5485
"β-Lutidine" *see* 3802
Lutidon *see* 7783
Lutidon Oral *see* 3696
Lutionex [Roussel-UCLAF] *see* 2874
Lutocyclin [Ciba] *see* 3696
Lutocyclin M [Ciba] *see* 7783
Lutocylin *see* 7783
Lutocylol [Ciba] *see* 3696
Lutoform *see* 7783
Lutogyl (Tabl) [Roussel-UCLAF] *see* 3696
Lutogyl (Inj) [Roussel-UCLAF] *see* 7783
Luto-Metrodiol *see* 3816
Lutoral [Midy] *see* 5677
Lutoral [Schieffelin] *see* 3696
Lutoral [Syntex] *see* 2102
Lutormone *see* 4534
Lutrapulse [Ortho] *see* 6620
Lutrelef [Ferring] *see* 5354
Lutren *see* 7783
Lutrexin [HW & D] *see* 5486
Lutromone [Endo] *see* 7783
Lututrin, 5486
Luvatren [Cilag-Chemie] *see* 6170
Luvistin [Boehringer, Ing.] *see* 4641
Luxazone *see* 2922
Luxoben [Chinoin] *see* 9353
Luxon *see* 4519
LVD *see* 2925
Ly 141B *see* 7115
Ly 12735 *see* 6201
LY 048740 *see* 905
LY 61017 *see* 3893
LY 061188 *see* 1971
LY 099094 *see* 9892
LY 109514 *see* 6257
LY 110140 *see* 4112
LY 122772 *see* 3553
Ly 127809 *see* 7115
LY 135837 *see* 4849
LY 139037 *see* 6582
Lyamine [Merck & Co.] *see* 5509
Lyapolate Sodium, 5487
Lycaconitine *see* 5488
Lycanol *see* 4403
Lycedan *see* 148
Lycine *see* 1201
Lycoctonine, 5488
Lycodine, 5489
Lycofawcine, 5490
Lycomarasmine, 5491
Lycopene, 5492
(*all-trans*)-Lycopene-16,16′-diol *see* 5493
(*all-trans*)-Lycopen-16-ol *see* 5499
Lycopersicin *see* 9471
Lycophyll, 5493
Lycopodine, 5494
Lycopodium, 5495
Lycopodium Seed (Spores) *see* 5495
Lycopus, 5496
Lycoramine, 5497
Lycoremine *see* 4245
Lycorine, 5498
Lycoxanthin, 5499
Lycra *see* 8688
Lydol *see* 5736
Lyeton [Von Boch] *see* 9801
Lyman [Lubapharm] *see* 306
Lymecycline, 5500

Lymethol *see* 9415
Lymphazurin [Hirsch] *see* 8968
Lymphoblastoid Interferon *see* 4892
Lymphochin *see* 688
Lymphocin *see* 688
Lymphocyte Activating Factor *see* 4895
Lymphocyte Mitogenic Factor *see* 4896
Lymphoquin *see* 688
Lymphotoxin *see* 9411
Lynamine *see* 5189
Lyndiol [Organon] *see* 5501
Lynestrenol, 5501
Lynoral [Organon] *see* 3689
Lyobex [Lappe] *see* 6638
Lyogen [Promonta] *see* 4116
Lyomethyl [Bouchara] *see* 9921
Lyovac® Antivenin, 5502
Lyovac Diuril [Merck & Co.] *see* 2169
Lyovac Sodium Edecrin [Merck & Co.] *see* 3669
Lyovit-H *see* 4739
Lyphocin [LyphoMed] *see* 9836
Lypressin, 5503
Lys (IUPAC Abbrev.) *see* 5509
Lysalbinic Acid, 5504
Lysalgo [Schiapparelli] *see* 5680
Lysanxia [Substantia] *see* 7713
Lysbex *see* 1218
Lyseen *see* 7747
Lysenyl [Spofa] *see* 5394
Lysergamide, 5505
Lysergan *see* 3918
Lysergic Acid, 5506
Lysergic Acid Amide *see* 5505
D-Lysergic Acid (+)-Butanolam-ide-(2) *see* 5989
D-Lysergic Acid Diethylamide *see* 5507
d-Lysergic Acid-*dl*-hydroxybutylam-ide-2 *see* 5989
D-Lysergic Acid D-Propanolamide *see* 3599
D-Lysergic Acid L-2-Propanolamide *see* 3600
Lysergide, 5507
Lysergsäure Diethylamid *see* 5507
Lysibex *see* 1218
Lysidine, 5508
Lysine, 5509
Lysine Acetylsalicylate, 5510
L-Lysine L-Glutamate, 5511
DL-Lysine Mono[2-(acetyloxy)benzo-ate] *see* 5510
Lysine Monosalicylate Acetate *see* 5510
8-L-Lysinevasopressin *see* 5503
N-Lysinomethyltetracycline *see* 5500
Lysivane [M & B] *see* 3703
Lysmucol [Schering] *see* 8510
Lysobex *see* 1218
Lysodren [Calbiochem] *see* 6134
Lysofon [Lafon] *see* 320
Lysomiol *see* 4536
"Lysophosphatidal Choline" *see* 7492
"Lysophosphatidal Ethanolamine" *see* 7492
Lysoplasmalogens *see* 7492
Lysortine [Adrian-Marinier] *see* 5509
Lysosomes, 5512
Lysostaphin, 5513
Lysozyme, 5514
Lyspafen [Cilag-Chemie] *see* 3124
Lyspamin [Cilag-Chemie] *see* 3324
Lyspamin Forte [Cilag-Chemie] *see* 3324
Lyssipoll [Lyssia] *see* 3334
Lysthenon [Lentia] *see* 8846
Lysuride *see* 5394
Lysuron [Laevosan] *see* 278

N^2-L-**Lysylbradykinin** see 5158
Lyteca [Westerfield] see 40
Lytensium see 7070
Lytispasm [Medical] see 701
Lyxoflavine, 5515
L-**Lyxoflavine** see 5515
D-**Lyxose,** 5516
α-D-**Lyxose** see 5516
LZ 544 see 4596
M 2H see 1526
M 14 see 8214
M 19-Q see 8233
M 71 see 1524
M 99 see 3843
M 115 see 8989
M 141 see 8695
M 144 see 888
M 183 see 3843
M 285 see 2777
M 1028 see 4514
M 4209 see 1813
M 4365A2 see 8241
M 4888 see 2088
M 5050 see 3347
M 5943 see 2185
M 7555 see 8061
M 9834 see 1038
M 10580 see 2727
M 73101 see 3519
M 139603 see 9176
Mabertin [Sidus] see 9074
Mabuterol, 5517
MAC see MISC-2
Macasirool [Hisiyama] see 4221
Macassar Oil, 5518
Machete [Monsanto] see 1498
Machiline see 2455
Mackreazid see 2688
Macleyine see 7912
Maclicine [Christiaens] see 3073
Maclurin, 5519
Macmiror [Poli] see 6447
Macocyn [Mack, Illert.] see 6931
Macodyn see 6931
Macol [Mazer] see 7554
MACOP-B see MISC-2
Macphenicol [Nakataki] see 9230
Macquer's Salt see 7583
Macrabin see 9921
Macrobin see 2409
Macrodantin [Eaton] see 6520
Macrodex [Pharmacia] see 2925
Macrodiol see 3653
Macrogol see 7545
Macrogol Fatty Acid Esters see 7555
Macrogol Fatty Alcohol Ethers see 7554
Macrogol Nonylphenyl Ether see 6596
Macromerine, 5520
Macrophage Activating Factor see 4894
Macrophage-derived Growth Factor see 4016
Macroscan-131 [Abbott] see 8417
Macrose see 2925
Maculotoxin see 9175
Macusine A see 5521
Macusine B see 5521
Macusine C see 5521
Macusines, 5521
MAD see 5861
Madar [Ravizza] see 6608
Maddrell's Salt, 5522
Madecassol [Laroche Navarron] see 857
Madelen [Finadiet] see 6824
Maderan [Ciba-Geigy] see 8892
Madlexin [Meiji] see 1971
Madopar [Roche] see 1059
Madribon [Roche] see 8876
Madrine [Langley] see 5859

MADU, 5523
Maduramicin, 5524
Maduramycin see 5524
MAF see 4894
Mafatate [Torii] see 5525
Mafenide, 5525
Magainins, 5526
Magaldrate, 5527
Magan [Warren-Teed] see 5565
Magcal see 5555
Magdelate see 5551
Magell see 6095
Magenta see 5528
Magenta I, 5528
Maggioni 1559 see 9980
Maghemite see 3973
Maghen [Caber] see 7463
Magic Acid [Cationics] see 4107
Magic Methyl [Aldrich] see 5993
Magic Tan see 3166
Magisal see 5531
Magistery of Bismuth see 1298
Maglite [Merck & Co.] see 5555
Magmilor [Poli] see 6447
Magnacort [Pfizer] see 4709
Magnamycin [Pfizer] see 1813
Magnamycin A see 1813
Magnamycin B see 1813
Magnesia see 5555
Magnesia Usta see 5555
Magnesite see 5538
Magnesium, 5529
Magnesium Acetate, 5530
Magnesium Acetylsalicylate, 5531
Magnesium Aluminate Hydrate see 5527
Magnesium Aluminum Silicate see 352
Magnesium Amide, 5532
Magnesium Ammonium Chloride see 556
Magnesium Aspirin see 5531
Magnesium Benzoate, 5533
Magnesium Biphosphate see 5561
Magnesium Bisulfate, 5534
Magnesium Borate, 5535
Magnesium Bromate, 5536
Magnesium Bromide, 5537
Magnesium Bromoglutamate see 4363
Magnesium Butyrate see 1593
Magnesium Carbonate Hydroxide, 5538
Magnesium Chlorate, 5539
Magnesium Chloride, 5540
Magnesium Citrate, 5541
Magnesium Citrate, Dibasic, 5542
Magnesium Citrate Sol see 5542
Magnesium Clofibrate see 2375
Magnesium Diamide see 5532
Magnesium Diethyl see 3112
Magnesium Dioxide see 5559
Magnesium Diphenyl see 3328
Magnesium Fluoride, 5543
Magnesium Fluosilicate see 5546
Magnesium Formate, 5544
Magnesium Germanide, 5545
Magnesium Glucoheptonate see 4349
Magnesium Glucomonocarbonate see 4349
Magnesium Gluconate see 4350
Magnesium Glucosemonocarbonate see 4349
Magnesium Glutamate Hydrobromide see 4363
Magnesium Heparinate see 4571
Magnesium Hexafluorosilicate, 5546
Magnesium Hydrate see 5548
Magnesium Hydride, 5547
Magnesium Hydrogen Phosphate see 5560

Magnesium Hydrogen Sulfate see 5534
Magnesium Hydroxide, 5548
Magnesium Hyposulfite see 5576
Magnesium Iodide, 5549
Magnesium Lactate, 5550
Magnesium Mandelate, 5551
Magnesium Mesotrisilicate see 5569
Magnesium Metasilicate see 5569
Magnesium Nitrate, 5552
Magnesium Oleate, 5553
Magnesium Orthosilicate see 5569
Magnesium Oxalate, 5554
Magnesium Oxide, 5555
Magnesium Pemoline see 7021
Magnesium Perborate, 5556
Magnesium Perchlorate, 5557
Magnesium Perhydrol see 5559
Magnesium Permanganate, 5558
Magnesium Peroxide, 5559
Magnesium Phosphate, Dibasic, 5560
Magnesium Phosphate, Monobasic, 5561
Magnesium Phosphate, Tribasic, 5562
Magnesium Potassium Selenate, 5563
Magnesium Pyrophosphate, 5564
Magnesium Salicylate, 5565
Magnesium Selenate, 5566
Magnesium Selenide, 5567
Magnesium Selenite, 5568
Magnesium Silicates, 5569
Magnesium Silicide, 5570
Magnesium Silicofluoride see 5546
Magnesium Stannide, 5571
Magnesium Stearate, 5572
Magnesium Sulfate, 5573
Magnesium Sulfite, 5574
Magnesium Sulfocyanate see 5575
Magnesium Superoxol see 5559
Magnesium Thiocyanate, 5575
Magnesium Thiosulfate, 5576
Magnesium Trisilicate see 5569
Magnesium Tungstate(VI), 5577
Magneson, 5578
Magnespirin see 5531
Magnetic Iron Oxide see 3988
Magnetite see 3988
Magnetkies see 4007
Magnevist [Schering AG] see 4236
Magnipen [Clin-Comar-Byla] see 5830
Magnofenyl see 6910
Magnoflorine, 5579
Magnogene see 5540
Magnoline, 5580
Magnophenyl see 6910
Magnopyrol see 3358
Magnosil [Polfa] see 5569
Magnosulf [Normark] see 5576
Magrene see 3116
Magrilon [Sintyal] see 5643
Magsalyl see 8626
Ma Huang see 3559
Maidenhair Tree see 4319
Maikohis [Nichiiko] see 2345
Maintasone [Owen] see 4710
Maiorad [Rotta] see 9393
Maipedopa [Maipe] see 5344
Maitansine see 5642
Maizena [CPC] see 8757
Maize Oil see 2528
Majeptil [Rhône-Poulenc] see 9287
Majudin see 1173
Makarol [Mallinckrodt] see 3118
Maki [Lipha] see 1371
Makon see 6596
Makrocef [Krka] see 1935
Malabar Nut see 149
Malachite see 2634
Malachite Green, 5581
Malachite Green G see 1367

Malagride *see* 44
Malamar 50 *see* 5582
Malaquin [Dolder] *see* 2163
Malaspray *see* 5582
Malathion, 5582
Malathon (obsolete) *see* 5582
Malayan Camphor *see* 1338
Malazide *see* 5587
Malazol *see* 305
Malcotran [Pennwalt] *see* 4649
Maldocil [Cilag-Chemie] *see* 8985
Maleamic Acid, 5583
Maleanilic Acid, 5584
Male Fern *see* 870
Maleic Acid, 5585
Maleic Acid, Diethyl Ester *see* 3113
Maleic Acid Hydrazide *see* 5587
Maleic Acid Monoamide *see* 5583
Maleic Anhydride, 5586
Maleic Hydrazide, 5587
Male Shield-fern *see* 870
Malestrone (Tabl) *see* 6044
Malestrone (Amps) *see* 9109
Maleuric Acid, 5588
Malexil [Ferrosan] *see* 3897
Maleylurea *see* 5588
Maliasin [Knoll] *see* 7873
Malic Acid, 5589
l-Malic Acid Monoammonium Salt
 see 518
Malidone [Schering] *see* 305
Malilum *see* 252
Maliner Kren *see* 4668
Malipuran [Scheurich] *see* 1462
Malix [Lagap] *see* 4372
Malix *see* 3529
Mallebrin [Krewel] *see* 337
Mallermin-F [Taiyo] *see* 2345
Mallophene [Mallinckrodt] *see* 7175
Mallorol [Sandoz] *see* 9290
Mallotoxin *see* 8250
Mallow, 5590
Malocide *see* 7997
Malogen L.A. [OJ & F] *see* 9112
Malol *see* 9802
Malonal *see* 972
Malonamide Nitrile *see* 2695
Malonic Acid, 5591
Malonic Acid Cyclic Isopropylidene
 Ester *see* 5696
Malonic Acid Di-*tert*-butyl Ester *see*
 3021
Malonic Acid Ethyl Ester Nitrile *see*
 3744
Malonic Ester *see* 3779
Malonic Mononitrile *see* 2696
Malononitrile, 5592
Malononitrile Dimer *see* 495
Malononitrile Hydrazide *see* 2688
5,5′-[Malonylbis(methylimino)]bis[*N*,-
 N ′-bis[2,3-dihydroxy-1-(hydroxy-
 methyl)propyl]-2,4,6-triiodoiso-
 phthalamide] *see* 4955
Malonylurea *see* 973
Malotilate, 5593
Maltin *see* 2971
Maltobiose *see* 5595
Maltol, 5594
Maltonic Acid *see* 4350
Maltos [Otsuka] *see* 5595
Maltose, 5595
Malt Sugar *see* 5595
Maltyl [Diabetylingesellschaft] *see*
 6063
Malvidin Chloride, 5596
Malvin *see* 5596
Malvoside *see* 5596
Malysol *see* 1036
Mamalexin [Showa Yakuhin] *see* 1971
Mamallet-A [Showa] *see* 488
Mamiesan [Kyowa] *see* 3087

Mammex *see* 6521
Mammotropin *see* 7788
Manaca, 5597
Manacan *see* 5597
Manchurian Bean *see* 8684
Mancona Bark *see* 8341
Mancozeb, 5598
Mandacon [Conal] *see* 5883
Mandamina *see* 5883
Mandarin G *see* 6813
Mandastat [Am. Urologicals] *see* 5883
Mandaverm [Asta] *see* 5600
7-D-Mandelamido-3-(1-methyl-1,2,-
 3,4-tetrazole-5-thiomethyl)-Δ³-
 cephem-4-carboxylic Acid *see* 1922
7-D-Mandelamido-3-[[(1-methyl-1*H*-
 tetrazol-5-yl)thio]methyl]-3-ceph-
 em-4-carboxylic Acid *see* 1922
7-Mandelamido-3-[[(1-methyl-1*H*-
 tetrazol-5-yl)thio]methyl]-8-oxo-5-
 thia-1-azabicyclo[4.2.0]oct-2-ene-
 2-carboxylic Acid *see* 1922
(6*R*,7*R*)-7-[(*R*)-Mandelamido]-8-oxo-
 3-[[[1-(sulfomethyl)-1*H*-tetrazol-5-
 yl]thio]methyl]-5-thia-1-azabicy-
 clo[4.2.0]oct-2-ene-2-carboxylic
 Acid *see* 1932
Mandelamine [Warner-Chilcott] *see*
 5883
Mandelic Acid, 5599
dl-Mandelic Acid *see* 5599
Mandelic Acid Ammonium Salt *see*
 557
Mandelic Acid Isoamyl Ester, 5600
Mandelic Acid Isopentyl Ester *see*
 5600
Mandelic Acid Nitrile *see* 5601
Mandelic Acid β-1,2,2,6-Tetramethyl-
 4-piperidyl Ester *see* 3854
Mandelic Acid 3,3,5-Trimethylcyclo-
 hexyl Ester *see* 2708
Mandelonitrile, 5601
Mandelonitrile-β-gentiobioside *see*
 629
Mandelonitrile Glucoside, 5602
D-Mandelonitrile-β-D-glucosido-6-β-
 D-glucoside *see* 629
Mandelonitrile β-Glucuronide *see* 629
Mandelonitrile Vicianoside *see* 9878
β-4-Mandeloyloxy-1,2,2,6-tetrameth-
 ylpiperidine *see* 3854
Mandelyltropeine *see* 4649
Mandenol *see* 3776
Mandokef [Lilly] *see* 1922
Mandol [Lilly] *see* 1922
Mandolsan [San Carlo] *see* 1922
Mandoz *see* 5883
Mandrake Root *see* 7521
Mandrax [Roussel-UCLAF] *see* 5872
Mandrozep [Henk] *see* 2977
Mandurin *see* 5883
Maneb, 5603
Maneon [Poli] *see* 420
Manexin *see* 5630
Manganblende *see* 5625
Manganese, 5604
Manganese Acetate, 5605
Manganese Benzoate *see* 1101
Manganese Binoxide *see* 5612
Manganese Borate, 5606
Manganese Bromide, 5607
Manganese Carbonate, 5608
Manganese Carbonyl, 5609
Manganese Chloride, 5610
Manganese Dibromide *see* 5607
Manganese Dichloride *see* 5610
Manganese Difluoride, 5611
Manganese Diiodide *see* 5614
Manganese Dioxide, 5612

Manganese Ethylenebis(dithiocarbam-
 ate) (Polymeric) Complex with Zinc
 Salt *see* 5598
Manganese Fluoride *see* 5611
Manganese Green *see* 993
Manganese Hypophosphite, 5613
Manganese Iodide, 5614
Manganese Monosulfide *see* 5625
Manganese Nitrate, 5615
Manganese Oleate, 5616
Manganese Oxalate, 5617
Manganese Oxide, 5618
Manganese Peroxide *see* 5612
Manganese Phosphate, Dibasic, 5619
Manganese Pyrophosphate, 5620
Manganese Selenide, 5621
Manganese Sesquioxide, 5622
Manganese Silicate, 5623
Manganese Sulfate, 5624
Manganese Sulfide, 5625
Manganese Superoxide *see* 5612
Manganese Trifluoride, 5626
Manganic Fluoride *see* 5626
Manganite *see* 5622
Manganjustite *see* 5623
Manganomanganic Oxide *see* 5618
Manganosite *see* 5604
Manganous Chloride *see* 5610
Manganous Ethylenebis[dithiocarbam-
 ate] *see* 5603
Manganous Fluoride *see* 5611
Mangostin, 5627
Manicol *see* 5629
Maninil [Arzneimittelwerk, VEB] *see*
 4372
Maniol [Morishita] *see* 3311
Manna, 5628
Manna Sugar *see* 5629
Mannidex *see* 5629
Mannite *see* 5629
Mannitol, 5629
D-Mannitol *see* 5629
Mannitol Hexanitrate, 5630
Mannitol Nitrate *see* 5630
Mannitol Nitrogen Mustard *see* 5631
Mannitrin *see* 5630
L-Mannomethylose *see* 8171
Mannomustine, 5631
D-Mannose, 5632
Mannosidostreptomycin *see* 8787
Mannosylstreptomycin *see* 8787
Manoplax [Boots] *see* 4043
Mansil [Pfizer] *see* 6872
Mansonil (Vet Feed-mix) *see* 6425
Manta *see* 5906
Mantadan [De Angeli] *see* 380
Mantadine [DuPont] *see* 380
Mantadix [Théraplix] *see* 380
Man-Tan *see* 3166
Mantomide [Winthrop] *see* 2076
Mantropine *see* 5883
Manvene *see* 6254
Manzate [Du Pont] *see* 5603
Manzate 200 [Du Pont] *see* 5598
Manzeb *see* 5598
Manzin 80 [Crystal Chem.] *see* 5598
MAO *see* 6158
MAOA *see* 5872
Maolate [Upjohn] *see* 2179
Maon [Takeda] *see* 8722
MAP *see* 5677
Mapeg [Mazer] *see* 7555
Mapeg S 40 [Mazer] *see* 7555
Mapeg 400 MS [Mazer] *see* 7555
Mapharsal *see* 6902
Mapharsen [Parke, Davis] *see* 6902
Mapharside *see* 6902
Maphenide *see* 5525
MAPO *see* 5842
Maposol [Procida] *see* 5860
Mappine *see* 1467

Maprofix see 4266
Maprotiline, 5633
Maranhist see 7996
Maratan [Ravizza] see 1310
Marathon [Velsicol] see 7773
Marazine [Geneva] see 2186
Marbadal [Bayer] see 8922
Mar-Bate [Mardale] see 5751
Marboran [Burroughs Wellcome] see 5900
Marcain [BDH] see 1485
Marcaina [Pierrel] see 1485
Marcaine [Winthrop] see 1485
Marcoeritrex see 3628
Marcumar [Roche] see 7230
Maretin [Bayer] see 6269
Marevan [Duncan, Flockhart] see 9950
Marezine [Burroughs Wellcome] see 2716
Marfanil see 5525
Marfanil Salt of Badional see 8922
Margaric Acid, 5634
Margarite see 1645
Margosan-O [Vikwood] see 6356
Margosa Oil see 6357
Maricolene see 2721
Marignac's Salt see 7663
Marigold see 1723
Marihuana see 1752
Marijuana see 1752
Marinco C [Merck & Co.] see 5538
Marinco H [Merck & Co.] see 5548
Marindinin see 5167
Marineurina [Farmetrusca] see 7896
Marinol [Unimed] see 9142
Marisilan [Wakamoto] see 621
Markweed see 7525
Marlate [Du Pont] see 5913
Marmelosin see 5635
Marme's Reagent see 7682
Marogen [Chugai-Upjohn] see 3635
Marplan [Roche] see 5040
Marrubiin, 5635
Marsh Gas see 5863
Marshite see 2669
Marshmallow see 315
Marsh Trefoil see 5729
Marsilid [Roche] see 4965
Marsin see 7200
Marsthine [Towa Yakuhin] see 2345
Martec [CPH] see 7520
Martos-10 [Otsuka] see 5595
Martricin see 9749
Marucotol [Maruko] see 5660
Marvelon 150/30 [Akzo] see 2906
Marvinol see 7563
Mary-bud see 1723
Maryland Pink see 8704
Marzine [Burroughs Wellcome] see 2716
Masacin [Boehringer, Mann.] see 7765
Mascagnite see 584
Maschitt [Showa Shinyaku] see 4704
Masdil [Esteve] see 3188
Masdiol see 5861
Masletine [Sioe] see 2345
Masmoran [Pfizer] see 4786
Massicot see 5291
Masterfen [Dompe] see 8985
Masterid [Grünenthal] see 3436
Masteril [Syntex] see 3436
Masterone [Syntex] see 3436
Mastic, 5636
Mastiche see 5636
Mastiphen [Intervet] see 2068
Mastisol see 5636
Mastix see 5636
Matacil [Bayer] see 443

Mataperro see 2491
Maté, 5637
Matenon [Upjohn] see 6098
Matico, 5638
Matricaria, 5639
Matricarin, 5640
Matridine see 5641
Matridin-15-one see 5641
Matrine, 5641
Matrol see 2102
Matromicina see 9681
Matromycin [Pfizer] see 6785
MATS see 5937
Matulane [Roche] see 7764
Mavrik [Zoecon] see 4137
Maxair [Riker] see 7461
Maxeran [Nordic] see 6063
Maxforce [Cyanamid] see 4684
Maxibolin [Organon] see 3761
Maxicaine see 6987
Maxicam [Parke, Davis] see 5127
Maxicortex [Manetti Roberts] see 160
Maxidex [Alcon] see 2922
Maxifen [Sharp & Dohme] see 7484
Maxiflor [Herbert] see 3126
Maxilase see 633
Maximed [Sharp & Dohme] see 7917
Maxipen [Pfizer] see 7184
l-Maxipen Potassium see 7184
Maxitate [Pennwalt] see 5630
Maxiton see 2932
Maxivate [Westwood] see 1202
Maxolon [Beecham] see 6063
Maxomat [Choay] see 8672
Maxulvet see 8876
Max-Uric [Labinca] see 1073
May Apple see 7521
May Blossom see 2507
Maycor [Parke, Davis] see 5114
Maydol see 2528
Mayeptil [Rhône-Poulenc] see 9287
Maygace [Bristol] see 5687
May Lily see 2507
May Pops see 7000
Maytansine, 5642
Mazildene [Farmochim. Ital.] see 5643
Mazindol, 5643
Mazipredone, 5644
Mazola [CPC] see 2528
Mazun see 10010
M & B 125 see 1161
M & B 693 see 8913
M & B 736 see 7165
M & B 744 see 8772
M & B 760 see 8920
M & B 782 see 7808
M & B 800 see 7071
M & B 2050A see 7089
M & B 2207 see 8845
M & B 2210 see 8989
M & B 4180A see 5067
M & B 4486 see 7022
M & B 5062A see 405
M & B 8430 see 2401
M & B 15497 see 2846
M & B 17803A see 13
MB 10064 see 1431
MB 10731 see 1431
MBA see 5655
M-BACOD see MISC-2
MBC see 1794
MBD see MISC-2
MBLA see 5702
MBR 8251 see 7113
MBR 12325 see 5684
MBT see 5759
MBTS see 3377
MBU see 5846
MC 838 see 6200
MC 4379 see 1228
MCA see 2111

MCAA see 5454
MCF see MISC-2
MCI 2016 see 1227
MCI 9038 see 804
M-Cillin [Misemer] see 7041
MCMN see 8237
McN 485 see 10098
McN 742 see 490
McN 1025 see 6604
McN 1210 see 8001
McN 1546 see 4069
McN 2559 see 9441
McN 2559-21-98 see 9441
McN-JR 4584 see 1057
McN-JR 7094 see 5360
McN-JR 8299 see 9161
McN-JR 15403-11 see 3124
McN-R 73-Z see 1804
McN-R 726-47 see 7529
MCNU see 8125
McN-X-181 see 9820
MCP see 5645
MCPA, 5645
MCPP see 5666
MCP-ratiopharm [Ratiopharm] see 6063
MD 76 [Mallinckrodt] see 5689
MD 141 see 3675
MD 805 see 804
MD 6134 see 9950
MD 6753 see 2292
MD 67350 see 2293
MD 690276 see 9445
M-Det see 2848
MDGF see 4016
MDL 507 see 9062
MDL 9918 see 9094
MDL 14042A see 5439
MDL 17043 see 3542
MDL 71754 see 9882
MDMA, 5646
MDMH see 4762
MDP see 6214
MDP-Lys(L18) see 6217
MDS [Kowa] see 2929
Me 3625 see 5722
MEA see 2785
Meadow Anemone see 7956
Meadow Clover see 9601
Meadow Crocus see 2471
Meadow Saffron see 2471
Measurin [Winthrop-Breon] see 873
Meat Sugar see 4883
Meaverin [Woelm] see 5748
MEB 6447 see 9507
Mebadin [Glaxo] see 2860
Meballymal Sodium see 8374
Mebaral [Winthrop] see 5742
Mebendazole, 5647
Mebenvet [Janssen] see 5647
Meberyt see 3632
Mebeverine, 5648
Mebhydroline, 5649
Mebinol [Erba] see 2181
Mebiquine, 5650
MeBmt, 5651
Mebrofenin, 5652
Mebron [Daiichi] see 3572
Mebrophenhydramine see 3514
Mebryl [SK & F] see 3514
Mebumal Sodium see 7087
Mebutamate, 5653
Mebutina [Formenti] see 5653
3-MECA see 5967
Mecadox [Pfizer] see 1782
Mecalmin [Yoshitomi] see 3311
Mecamine see 5654
Mecamylamine, 5654
Mecca-galls see 6651
Mechlorethamine, 5655

Mechlorethamine Oxide Hydrochloride, 5656
Mechlorprop see 5666
Mechothane see 1207
Meciclin [Citobios] see 2872
Mecilex [C.E.P.A.] see 1971
Mecillinam see 399
Meclan [Ortho] see 5658
Meclastine see 2345
Meclizine, 5657
Meclocycline, 5658
Mecloderm see 5658
Meclofenamic Acid, 5659
Meclofenoxane see 5660
Meclofenoxate, 5660
Meclomen [Parke, Davis] see 5659
Meclophenamic Acid see 5659
Meclopran [Lagap] see 6063
Mecloqualone, 5661
Mecloralurea, 5662
Meclosorb [Basotherm] see 5658
Mecloxamine, 5663
Meclozine see 5657
Meclutin [ABC] see 5658
Mecobalamin see 9921
Mecocyanin see 2694
Mecodin see 5852
Mecodrin see 616
Mecomec [PBI/Gordon] see 5666
Meconic Acid, 5664
Meconin, 5665
Meconinic Acid Lactone see 5665
Mecopar see 5666
Mecoprop, 5666
Mecoprop-P see 5666
Mecoral see 2089
Mecostrin Chloride see 9717
Mecrylate, 5667
Mectizan [Merck & Co.] see 5133
Mecysteine Hydrochloride, 5668
Medaject [Neda] see 7763
Medan [Kowa] see 1200
Medapsol see 8927
Medaron see 4210
Medazepam, 5669
Medazepol [Farmasa] see 5669
Medemanol see 5630
Medemycin [Meiji] see 6106
Mederel [Riom] see 604
Medetomidine, 5670
Medialan LL-33 see 4267
Medialan LL-99 see 4267
Mediator [Servier] see 1047
Mediaven [Medial] see 6314
Mediaxal [Servier] see 1047
Medibazine, 5671
Mediben see 3026
Medicagol, 5672
Medicarpin, 5673
Medicef [Profarmi] see 1982
Medichol see 2068
Medicil [Medici] see 6177
"Medicinal Chrysophanic Acid" see 2257
Medicoal [Lundbeck] see 1815
Medicort [Care] see 4710
Medifenac [Medici] see 212
Medifoxamine, 5674
Medifuran see 4205
Medigoxin see 3150
Medilave [Farillon] see 2024
Medilla [Omega] see 4792
Medinal [Warner-Chilcott] see 972
Medio-Contrix "38" see 4952
Medipectol see 1218
Medipren [McNeil] see 4812
Medizinc [Medics] see 10064
Medmain, 5675
Medodorm [Medo] see 2089
Medomet [DDSA] see 5974
Medomin [Geigy] see 4575

Medopa [Kaigi/Nippon Kayaku] see 5974
Medopren [Malesci] see 5974
Medoxim [Medici] see 1951
Medphalan see 5708
Medrate [Upjohn] see 6028
Medrocort [Upjohn] see 5679
Medrogestone, 5676
Medroglutaric Acid see 5690
Medrol [Upjohn] see 6028
Medrol Stabisol [Upjohn] see 6028
Medrone [Upjohn] see 6028
Medroxyprogesterone, 5677
Medrylamine, 5678
Medrysone, 5679
Meerschaum see 5569
Meerzwiebel see 8728
Mefamide [Bayer] see 5525
Mefedina [Farmitalia] see 5736
Mefenal see 8886
Mefenamic Acid, 5680
Mefenorex, 5681
Mefexadyne [Anphar] see 5682
Mefexamide, 5682
Mefloquine Hydrochloride, 5683
Mefluidide, 5684
Mefoxin [Merck & Co.] see 1938
Mefoxitin [Sharp & Dohme] see 1938
Mefruside, 5685
Mega see 5653
Megabion (Indian) see 9921
Megabion (Japanese) see 5861
Megace [Mead Johnson] see 5687
Megacef [Beytout] see 1982
Megacillin Oral [Grünenthal] see 7046
Megacillin Suspension [Frosst] see 7037
Megacillin Tablets [Frosst] see 7041
Megacins, 5686
Megaclor [Pharmax] see 2386
Megallate see MISC-4
Megalovel see 9921
Megamycine [C.R.E.A.T.] see 5851
Megapen see 7042
Megaphen [Bayer] see 2186
Megasedan [Andreu] see 5669
Megasul [Am. Cyanamid] see 6539
Megestat [Lappe] see 5687
Megestrol Acetate, 5687
Megimide [Abbott] see 1036
Meglum [Bago] see 9161
Meglumine see 5995
Meglumine Acetrizoate, 5688
Meglumine Amidotrizoate see 5689
Meglumine Diatrizoate, 5689
Meglumine Iothalamate see 4952
Meglutol, 5690
Megrin [Yoshitomi] see 4574
Meguan see 5845
Meicelin [Meiji Seika] see 1930
Meilax [Meiji Seika] see 3777
Meionite see 1645
MEK see 5991
Mekamine see 5654
Meksamin see 5926
Meladinine [Basotherm] see 5911
Meladrazine see 4685
Melamine, 5691
Melaminsulfone see 8896
2-(4-Melamin-2-ylphenyl)-4-hydroxymethyl-1,3-dithia-2-arsolane see 5694
Melampyrin see 4238
Melampyrite see 4238
Melampyrum see 4238
Melaniline see 3325
Melanins, 5692
Melanocyte-stimulating Hormone see 6206

Melanocyte-stimulating Hormone-release Inhibiting Factor see 5693
Melanophore-affecting Hormone see 6206
Melanophore Dilating Hormone see 6206
Melanophore Expanding Hormone see 6206
Melanophore Hormone see 6206
Melanophore-stimulating Hormone see 6206
Melanosome-dispersing Hormone see 6206
Melanostatin, 5693
Melanostatin I (Ox) see 5693
Melanotropic Hormone see 6206
Melanotropic Inhibiting Factor see 5693
Melanotropin see 6206
α-Melanotropin see 6206
β-Melanotropin see 6206
γ-Melanotropin see 6206
Melanterite see 4006
Melanthigenin see 4540
Melarsoprol, 5694
Melatonin, 5695
Mel B see 5694
Melbex [Lilly] see 6238
Meldane [Chemagro] see 2559
Meldrum's Acid, 5696
Melengestrol, 5697
Meletin see 8044
Melex [Sankyo] see 6091
Melezitose, 5698
Melfalan see 5708
Melfax [Apotex] see 8126
Melibiose, 5699
Melidorm [Asta] see 5663
Melilot, 5700
Melilotoside, 5701
Melin see 8276
Melinamide, 5702
Melinonine A, 5703
Melinonine G see 4031
Melipramin [EGYT] see 4835
Melissyl Alcohol see 9506
Melitase [Berk] see 2187
Melitose see 8120
Melitoxin see 3080
Melitracen, 5704
Melitriose see 8120
Melittin, 5705
Melittin I see 5705
Melixeran [Lusofarma] see 5704
Mellaril [Sandoz] see 9290
Melleretten see 9290
Melleril see 9290
Mellic Acid see 5706
Mellitic Acid, 5706
Mellitin see 5705
Mellitoxin see 4787
Melon Tree see 6969
Melopat [Deiglmayr] see 1200
Meloxine [Upjohn] see 5911
Melperone, 5707
Melphalan, 5708
Melprex [Am. Cyanamid] see 3406
Melsedin [Boots] see 5872
Melsomin see 5872
Meltatox [BASF] see 3405
Meltrol [USV] see 7191
Melubrin see 8896
Melysin [Takeda] see 400
Memantine, 5709
Memcozine see 8876
Memento 400 [Volpino] see 7427
Memine see 7303
Memoq [Gödecke] see 6404
Memphenesin see 5737
Menacor [Menarini] see 2356
Menadiol Diacetate, 5710

Menadiol Dibutyrate, 5711
Menadiol Diphosphate (Tetrasodium Salt), 5712
Menadiol Disulfate, 5713
Menadiol Sodium Diphosphate see 5712
Menadiol Tetrasodium Diphosphate see 5712
Menadione, 5714
Menadione Carboxymethoxime Ammonium Salt see 5717
Menadione Dimethylpyrimidinol Bisulfite, 5715
Menadione Diphosphate Tetrasodium Salt see 5712
Menadione Sodium Bisulfite, 5716
Menadoxime, 5717
Menamin [Chugai] see 5184
Menaphthone see 5714
Menaphthone Carboxymethoxime Ammonium Salt see 5717
Menaquinone 6 see 9935
Menaquinone 7 see 9935
Menaquinones see 9935
Menazon, 5718
Menazone see 4220
Menbutone, 5719
Mencortex [Menarini] see 160
Mendelevium, 5720
Mendiaxon [Roland] see 4792
Mendon [Dainippon] see 2400
Mendrin see 3533
Menetryl see 3661
Menformon [Organon] see 3660
Menhaden Oil, 5721
Menhydrinate see 3193
Menichlopholan, 5722
Menitazine [Towa Yakuhin] see 1200
Menocil [Cilag-Chemie] see 490
Menolyn [Arcum] see 3689
Menonasal see 9123
Menopatol [Nippon Chemiphar] see 9447
Menophase [Syntex] see 5819
Menotropins see 4189
Menova [E. Merck] see 2102
Menrium [Roche] see 2082
Mensiso [Menarini] see 8498
Menstridyl see 2102
p-Mentha-1,5-diene see 7151
p-Mentha-1(7),2-diene see 7152
p-Mentha-1,8-diene see 5371
p-Mentha-6,8-dien-2-one see 1883
(3R,4R)-2-p-Mentha-1,8-dien-3-yl-5-pentylresorcinol see 1750
p-Menthane-1,8-diol see 9101
3-p-Menthanol see 5723
l-p-Menthan-3-one see 5724
2-[(p-Mentha-1,3,5-trien-2-yloxy)methyl]-2-imidazoline see 9742
p-Menth-6-ene-2,8-diol see 8510
1-p-Menthene-6,8-diol see 8510
p-Menth-1-en-8-ol see 9103
1-p-Menthen-2-ol-3-one see 3292
m-Menth-6-en-5-one see 4651
p-Menth-1-en-3-one see 7443
R-(+)-p-Menth-4(8)-en-3-one see 7955
Menthol, 5723
l-Menthol see 5723
l-Menthone, 5724
Menthyl Acetate, 5725
Menthyl Borate, 5726
Menthyl Salicylate, 5727
Menthyl Valerate, 5728
Mentium [Guidotti] see 8002
Menyanthes, 5729
Meobentine, 5730
Meonine [Ives] see 5896
Meothrin [Shell] see 3936
MEP see 3922

Mepacrine see 8053
Mepantin see 5751
Meparfynol, 5731
Meparfynol Carbamate, 5732
Mepartricin, 5733
Mepartricin A see 5733
Mepartricin B see 5733
Mepasin see 5734
Mepavlon [ICI] see 5751
Mepazine, 5734
Mepedyl see 7456
Mepenicycline see 7052
Mepenzolate Bromide, 5735
Meperidine, 5736
Mephabutazon see 7248
Mephacyclin [Mepha] see 9130
Mephedine see 5736
Mephenamin [Boehringer, Mann.] see 6831
Mephenesin, 5737
Mephenhydramine, 5738
Mephenon see 5852
Mephenoxalone, 5739
Mephentermine, 5740
Mephenytoin, 5741
Mepherol see 5737
Mephesin see 5737
Mephine see 5740
Mephobarbital, 5742
Mephosfolan, 5743
Mephson see 5737
Mephyton [Merck & Co.] see 9933
Mephyton DK [Merck & Co.] see 3165
Mepiben see 3334
Mepicaton [Pharmaton] see 5748
Mepiciclina see 7420
Mepicycline see 7420
Mepicycline Phenoxymethylpenicillinate see 7052
Mepidium [Recordati] see 9372
Mepindolol, 5744
Mepiprazole, 5745
Mepiquat Chloride, 5746
Mepirizole see 3572
Mepitiostane, 5747
Mepivacaine, 5748
Mepivastesin [Espe] see 5748
Mepixanox, 5749
Mepixanthone see 5749
Meposed see 5751
Meprane Dipropionate [Reed & Carnrick] see 5888
Mepred [Savage] see 6028
Meprednisone, 5750
Meprin see 5751
Meprindon see 5751
Meprobamate, 5751
Meproban see 5751
Meprocompren [E. Merck; Boehringer, Mann.] see 5751
Meprofen [Agips] see 5184
Meprol see 5751
Meproscillarin see 7889
Meprosin see 5751
Meprospan [Wallace Labs.] see 5751
Meprotabs [Wallace Labs.] see 5751
Meprotan see 5751
Meprotil [Bruner-Tillman] see 5751
Meprylcaine Hydrochloride, 5752
Meptazinol, 5753
Meptid [Wyeth] see 5753
Meptin [Otsuka] see 7765
Meptran [Reid-Provident] see 5751
Mepyramine see 7996
Mepyrapone see 6083
Mepyren see 7996
Mequelon [Frosst] see 5872
Mequin [Lemmon] see 5872
Mequitazine, 5754
Mequiverine see 9908
MER-17 see 909

MER-25 see 3674
MER-27 see 6978
MER-29 see 9650
Meractinomycin see 2804
Meradan(e) [USSR] see 8221
Meraklon [Montecatini] see 7558
Meralein Sodium, 5755
Meralen [Merrell] see 4060
Meralluride, 5756
Meralop [ISF] see 2694
Meralops [Dulcis] see 2694
Meratonic see 7455
Meratran (formerly) [Merrell] see 7455
Merbak [Schieffelin] see 57
Merbentul [Merrell] see 2173
Merbentyl [Merrell] see 3087
Merbromin, 5757
Mercaleukin see 5762
Mercamine see 2785
Mercamphamide, 5758
Mercaptamine see 2785
Mercaptoacetamide Antimony Derivative see 737
2-Mercaptoacetanilide S-Gold Derivative see 902
Mercaptoacetic Acid see 9265
Mercaptoacetic Acid Antimony Derivative Sodium Salt see 735
Mercaptoacetic Acid Calcium Derivative see 1718
Mercaptoacetic Acid Calcium Salt Gold Derivative see 6241
Mercaptoacetic Acid Sodium Salt see 8650
β-Mercaptoalanine see 2787
Mercaptoarsenol see 840
2-Mercaptobenzimidazole see 1092
2-Mercaptobenzimidazole-1,3-dimethylol see 9239
2-Mercaptobenzoic Acid see 9291
2-Mercaptobenzothiazole, 5759
Mercaptobenzthiazyl Ether see 3377
Mercaptobutanedioic Acid see 9272
Mercaptobutanedioic Acid Antimony-(3+) Lithium Salt (3:1:6) see 711
Mercaptobutanedioic Acid Monogold-(1+) Sodium Salt see 4422
Mercaptodiacetic Acid see 9260
Mercaptodimethur see 5893
Mercaptoethane see 3680
2-Mercaptoethanesulfonic Acid Sodium Salt see 5812
2-Mercaptoethanol, 5760
β-Mercaptoethanol see 5760
β-Mercaptoethylamine see 2785
(2-Mercaptoethyl)trimethylammonium Iodide O,O-Diethyl Phosphorothioate see 3475
2-Mercaptohistidine see 9270
2-Mercapto-4-hydroxypyrimidine see 9298
2-Mercaptoimidazoline see 3759
N-(2-Mercaptoisobutyryl)-L-cysteine see 1447
Mercaptomerin Sodium, 5761
Mercaptomethane see 5867
N-(2-Mercapto-2-methyl-1-oxoprop-yl)-L-cysteine see 1447
(S)-1-(3-Mercapto-2-methyl-1-oxo-propyl)-L-proline see 1773
N-(Mercaptomethyl)phthalimide S-(O,O-Dimethyl Phosphorodithioate) see 7311
N-(2-Mercapto-2-methylpropanoyl)-L-cysteine see 1447
(2S)-1-(3-Mercapto-2-methylpropion-yl)-L-proline see 1773
N-[1-[(S)-3-Mercapto-2-methylpro-pionyl]-L-prolyl]-3-phenyl-L-ala-nine Acetate (Ester) see 192

1-Mercaptonaphthalene *see* 6297
2-Mercapto-*N*-2-naphthalenylacet-
  amide *see* 9273
2-Mercaptonapthalene *see* 6298
N-(2-Mercapto-1-oxopropyl)glycine
  see 9386
(2-Mercapto-*N*-phenylacetamidato-
  O,S)gold *see* 902
Mercaptophos [Bayer] *see* 3945
Mercaptophos *see* 2875
3-Mercapto-1,2-propanediol *see* 9264
2-Mercaptopropanoic Acid *see* 9269
α-Mercaptopropanoic Acid *see* 9269
1-(3-Mercaptopropanoic Acid)-8-D-
  arginine Vasopressin *see* 2904
1-(3-Mercaptopropanoic Acid)oxy-
  tocin *see* 2833
2-Mercaptopropionic Acid *see* 9269
N-(2-Mercaptopropionyl)glycine *see*
  9386
α-Mercaptopropionylglycine *see* 9386
N-(2-Mercaptopropionyl)glycine
  2-Thiophenecarboxylate (Ester) *see*
  8764
6-Mercaptopurine, 5762
2-Mercaptopyridine 1-Oxide *see* 8004
2-Mercapto-4(1*H*)-pyrimidinone *see*
  9298
2-Mercapto-4-pyrimidone *see* 9298
Mercaptosuccinic Acid *see* 9272
Mercaptosuccinic Acid Antimonate-
  (III) Hexalithium Salt *see* 711
Mercaptosuccinic Acid *S*-Antimony
  Derivative Lithium Salt *see* 711
Mercaptosuccinic Acid Diethyl Ester
  see 5582
Mercaptothion *see* 5582
3-Mercapto-D-valine *see* 7029
Mercaptyl [Knoll] *see* 7029
Mercardan [Parke, Davis] *see* 5756
Mercate "5" *see* 5009
Mercate "20" *see* 5009
Mercazole *see* 5892
Mercazolyl *see* 5892
Mercilon [Organon] *see* 2906
Mercloran [Parke, Davis] *see* 2104
Mercodinone *see* 4708
Mercoral *see* 2104
Mercufenol Chloride, 5763
Mercuhydrin [Lakeside] *see* 5756
Mercumallylic Acid, 5764
Mercumallylic Acid Theophylline
  Sodium *see* 5765
Mercumatilin Sodium, 5765
Mercuramide *see* 5805
Mercuranine *see* 5757
Mercuretin *see* 5756
Mercuric Acetate, 5766
Mercuric Ammonium Chloride *see* 558
Mercuric Arsenate, 5767
Mercuric Barium Iodide *see* 994
Mercuric Benzoate, 5768
Mercuric Bromide, 5769
Mercuric Cacodylate *see* 1603
Mercuric Chloride, 5770
Mercuric Chloride, Ammoniated,
  5771
Mercuric Cuprous Iodide *see* 2670
Mercuric Cyanide, 5772
Mercuric Dichromate(VI), 5773
Mercuric Fluoride, 5774
Mercuric Imidosuccinate *see* 5786
Mercuric Iodate, 5775
Mercuric Iodide, Red, 5776
Mercuric Nitrate, 5777
Mercuric Oleate, 5778
Mercuric Oxide, Red, 5779
Mercuric Oxide, Yellow, 5780
Mercuric Oxycyanide, 5781
Mercuric Potassium Cyanide *see* 7676
Mercuric Potassium Iodide *see* 7683

Mercuric Potassium Iodide Soln *see*
  7691
Mercuric Salicylate, 5782
Mercuric Silver Iodide *see* 8482
Mercuric Sodium *p*-Phenolsulfonate,
  5783
Mercuric Stearate, 5784
Mercuric Subsulfate, 5785
Mercuric Succinimide, 5786
Mercuric Sulfate, 5787
Mercuric Sulfide, Black, 5788
Mercuric Sulfide, Red, 5789
Mercuric Sulfocyanate *see* 5790
Mercuric Sulfocyanide *see* 5790
Mercuric Thiocyanate, 5790
Mercuri-hematoporphyrin Disodium
  Salt *see* 5804
Mercurin *see* 5758
Mercuriphenoldisulfonate Sodium *see*
  5783
Mercurius Vitae *see* 725
Mercurochrome [HW & D] *see* 5757
Mercurochrome-220 Soluble *see* 5757
Mercurocol *see* 5757
Mercurome *see* 5757
Mercurophage *see* 5757
Mercurophen, 5791
Mercurophyllin *see* 5758
Mercurothiolate *see* 9244
Mercurous Acetate, 5792
Mercurous Bromide, 5793
Mercurous Chlorate, 5794
Mercurous Chloride, 5795
Mercurous Fluoride, 5796
Mercurous Iodide, 5797
Mercurous Nitrate, 5798
Mercurous Oxide, 5799
Mercurous Sulfate, 5800
Mercury, 5801
Mercury Amide Chloride *see* 5771
Mercury Ammonium Chloride *see*
  5771
Mercury Bichloride *see* 5770
Mercury Bichromate *see* 5773
Mercury Biniodide *see* 5776
Mercury Bisulfate *see* 5787
Mercury Cyanide Oxide *see* 5781
Mercury Difluoride *see* 5774
Mercury, Dimethyl *see* 3238
Mercury Mass, 5802
Mercury Monochloride *see* 5795
Mercury Oxide Black *see* 5799
Mercury Oxide Sulfate *see* 5785
Mercury Oxonium Sulfate *see* 5785
Mercury Perchloride *see* 5770
Mercury Pernitrate *see* 5777
Mercury Protochloride *see* 5795
Mercury Protoiodide *see* 5797
Mercury Protonitrate *see* 5798
Mercury And Sodium Phenolsulfonate
  see 5783
Mercury Subchloride *see* 5795
Mercury Subsalicylate *see* 5782
Mercusal *see* 5805
Meregon [Malesci] *see* 1474
Mereprine [Merrell] *see* 3430
Meresa [Dolorgiet] *see* 8971
Merfamin *see* 9244
Merfen [Zyma] *see* 7274
Mergital LM 11 [Siphon] *see* 7532
Merilid *see* 2104
Merinax *see* 4616
Merislon [Eisai] *see* 1200
Merisoprol Hg 197, 5803
Merital [Hoechst] *see* 6592
Meritin *see* 394
Merlum [Merck & Co.] *see* 345
Merocet [Merrell Dow] *see* 2024
Merodicein [HW & D] *see* 5755
Meropenin *see* 7046
Meroxyl [Wander] *see* 4287

Meroxylan [Wander] *see* 4287
Merpan *see* 1771
Merphalan *see* 5708
Merphenyl Nitrate *see* 7273
Merphyrin, 5804
Merprane [Squibb] *see* 5803
Merpress [Merck & Co.] *see* 6433
Merprodem *see* 5872
Mersalin *see* 5805
Mersalyl, 5805
Mersolite *see* 7271
Mertax *see* 5759
Mertect [Merck & Co.] *see* 9217
Mertestate *see* 9109
Merthiolate [Lilly] *see* 9244
Mertionin *see* 5896
Mertorgan *see* 9244
Mervaldin [Lannett] *see* 5737
Mervan [Cont. Pharma; Warner-
  Lambert] *see* 212
Mervastrept *see* 2884
Merxin [Merck & Co.] *see* 1938
Merzonin *see* 9244
Mesaconic Acid, 5806
Mesalamine, 5807
Mesalazine *see* 5807
Mesantoin [Sandoz] *see* 5741
Mesasal [SK & F] *see* 5807
Mescaline, 5808
Mescaline Oxidase *see* 6158
Mescopil *see* 5927
Mesembrine, 5809
Mesentol *see* 3704
Mesidicaine *see* 9614
Mesilate *see* MISC-4
Mesitylene, 5810
Mesityl Oxide, 5811
Mesityl Oxide (1-Phthalazinyl)hydra-
  zone *see* 1457
Mesmar *see* 5751
Mesna, 5812
Mesnex [Asta] *see* 5812
Mesocaine *see* 9614
Mesoetioporphyrin *see* 3826
Mesoinosite *see* 4883
Mesokain *see* 9614
Mesonex *see* 4885
Mesontoin *see* 5741
Mesopin [Endo] *see* 4649
Mesoridazine, 5813
Mesoridazine Besylate *see* 5813
Mesotal [Galma] *see* 4885
Mesotartaric Acid *see* 9040
Mesothorium I *see* 8118
Mesothorium II *see* 129
Mesoxalic Acid, 5814
Mesoxalic Acid Calcium Salt *see* 1685
Mesoxalylcarbamide *see* 281
Mesoxalylurea *see* 281
Mesoxan *see* 1685
Mesoyohimbine *see* 10013
Mespafin [Merckle] *see* 3429
Mesquite Gum, 5815
Messenger RNA *see* 8204
Mestanolone, 5816
Mestenediol *see* 5861
Mesterolone, 5817
Mestilbol, 5818
Mestinon Bromide [Roche] *see* 7990
Mestoranum [Schering AG] *see* 5817
Mestranol, 5819
Mesudin *see* 5525
Mesudrin *see* 5525
Mesulergine, 5820
Mesulfa *see* 8884
Mesulid *see* 6464
Mesural *see* 2082
Mesurol [Chemagro] *see* 5893
Mesuximide *see* 5928
Mesylate *see* MISC-3

Met^5-E *see* 3528
Metab-Auxil *see* 7062
Met (IUPAC Abbrev.) *see* 5896
Metabenzthiazuron *see* 5846
Metabolin *see* 9222
Metabolite 4B *see* 1641
Metabutethamine Hydrochloride *see* 1517
Metabutoxycaine Hydrochloride, 5822
Metacaine *see* 9533
Metacetaldehyde *see* 5827
Metachlor *see* 193
Metachloral *see* 9538
Metachrome Yellow, 5823
Metacide [Chemagro] *see* 6022
Metaclazepam, 5824
Metacortandracin *see* 7727
Metacortandralone *see* 7719
Metacresol Acetate *see* 2587
Metacresylacetate-Sulzberger *see* 2587
Metacycline *see* 5851
[2.2]Metacyclophane, 5825
Metadee *see* 9928
Meta-Delphene [Hercules] *see* 2848
Metadiazine *see* 7998
Metadin *see* 7455
Metadomus [Med. Domus] *see* 5851
Metadorm *see* 5872
Metaformaldehyde *see* 9646
Metahydrin [Lakeside] *see* 9537
Metalaxyl, 5826
Metalcaptase [Heyl] *see* 7029
β_1-Metal-combining Protein *see* 9490
Metaldehyde, 5827
Metalkamate *see* 1459
Metallibure *see* 5857
Metallic Arsenic *see* 820
Metallum Problematum *see* 9064
Metalutin [Parke, Davis] *see* 6629
Metamfepramone, 5828
Metamfepyramone *see* 5828
Metamide [Protea] *see* 6063
Metamin [Takeda] *see* 4114
Metamine *see* 9682
Metaminodiazepoxide *see* 2082
Metamivam, 5829
Metamizol *see* 3358
Metampicillin, 5830
Metam Sodium *see* 5860
Metandiol *see* 5861
Metandren [Ciba] *see* 6044
Metanephrine, 5831
Metanilic Acid, 5832
Metanil Yellow, 5833
N^1-Metanilyl-N^2-*n*-butylurea *see* 1579
Metanite *see* 891
Metanyl *see* 5869
Metaoxedrin *see* 7257
Metaphanine, 5834
Metaphen [Abbott] *see* 6531
Metaphos *see* 6022
Metaphosphoric Acid *see* 7319
Metaphyllin *see* 477
Metaplexan [Badarznei] *see* 5754
Metapramine, 5835
Metaprel [Dorsey] *see* 5836
Metaproterenol, 5836
Metapyrin *see* 8896
Metaradrine *see* 5837
Metaraminol, 5837
Metarsenobillon *see* 8916
Metasclene *see* 9650
Metaspas [Leeming] *see* 3151
Metasqualene *see* 9650
Metastab [Boots] *see* 6028
Metastigmin [Star] *see* 6380
Meta Sympatol *see* 7257
Meta-Synephrine Hydrochloride *see* 7257
Meta-Systox *see* 5971
Metathion *see* 3922
Metatone [Yoshitomi/Green Cross] *see* 9077
Metatyrosine *see* 9748
Metavariscite *see* 362
Metaxalone, 5838
Metaxan *see* 5869
Metaxanin *see* 5826
Metazepium Iodide *see* 1599
Metazocine, 5839
Metazolo *see* 5892
Metcaraphen, 5840
Metebanyl [Sankyo] *see* 3442
Metelilachlor *see* 6067
Meteloidine, 5841
Metembonate *see* MISC-4
Metenarin [Teikoku Zoki] *see* 5989
Metendiol *see* 5861
Metenix [Hoechst] *see* 6068
Met^5-Enkephalin *see* 3528
Méténolone *see* 5887
Meteorex *see* 8486
Metepa, 5842
Meterazine *see* 7768
Meterfer [Sinclair] *see* 3995
Metergoline, 5843
Metescufylline, 5844
Metflorylthiazidine *see* 4716
Metformin, 5845
Metformin Pamoate *see* 5845
"Meth" *see* 5859
Methabenzthiazuron, 5846
Methacetin *see* 43
Methacetone *see* 3111
Methacholine Chloride, 5847
Methacide *see* 9455
Methacolimycin *see* 2475
Methacrifos, 5848
Methacrylic Acid, 5849
Methacrylonitrile, 5850
Methacycline, 5851
Methaderm [Taiho] *see* 2922
Methadol *see* 3195
Methadone Hydrochloride, 5852
Methadren(e) *see* 5988
Methadyl Acetate, 5853
Methafluoridamid *see* 5684
Methaform *see* 2129
Methafrone *see* 5189
Methafurylene, 5854
Methagon [Corvel] *see* 4066
Methallatal, 5855
Methallenestril, 5856
Methallibure, 5857
Methallylchloride *see* 2148
α-Methallyl Chloride *see* 2131
β-Methallyl Chloride *see* 2148
γ-Methallyl Chloride *see* 2130
1-Methallyl-3-methyl-6-aminotetrahydro-2,4-pyrimidinedione *see* 500
Methalutin *see* 6629
Methamidophos, 5858
Methaminodiazepoxide *see* 2082
Methampex [Lemmon] *see* 5859
Methamphetamine, 5859
Methampicillin *see* 5830
Methampyrone *see* 3358
Metham Sodium, 5860
Methanabol *see* 5861
Methanal *see* 4148
Methanamide *see* 4151
Methanamine *see* 5938
Methandienone *see* 5862
Methandiol *see* 5861
Methandriol, 5861
Methandrostenolone, 5862
Methane, 5863
Methanearsonic Acid, 5864
Methanedicarboxylic Acid *see* 5591
Methanedisulfonic Acid *see* 5895
Methanedisulfonic Acid Calcium Salt *see* 1686
Methaneperoxoic Acid *see* 7114
Methanesulfonic Acid, 5865
Methanesulfonic Acid Ethyl Ester *see* 3782
Methanesulfonic Acid Methyl Ester *see* 6010
Methanesulfonic Acid Tetramethylene Ester *see* 1494
Methanesulfonyl Chloride, 5866
***N,N'*-Methanetetraylbiscyclohexanamine** *see* 3086
Methanethiol, 5867
Methanetriol *see* 6837
Methanide *see* 5869
Methanol, 5868
***O*-(1,4-Methano-1,2,3,4-tetrahydro-6-naphthyl)-*N*-methyl-*N*-(*m*-tolyl)-thiocarbamate** *see* 9433
Methantheline Bromide, 5869
Methanthine Bromide *see* 5869
Methaphenilene, 5870
Methapoxide *see* 5842
Methapyrilene, 5871
Methaqualone, 5872
Metharbital, 5873
Methargen, 5874
Methasan *see* 10075
Methased *see* 5872
β-Methasone *see* 1202
Methazol [Park Plaza Pharm.] *see* 8887
Methazolamide, 5875
Methazole, 5876
Met Hb *see* 5878
Methbipyranone *see* 6083
Methdilazine, 5877
Methedrine [Burroughs Wellcome] *see* 5859
Methemoglobin, 5878
Methenamine, 5879
Methenamine Allyl Iodide, 5880
Methenamine Anhydromethylenecitrate, 5881
Methenamine Hippurate, 5882
Methenamine Mandelate, 5883
Methenamine Salicylate, 5884
Methenamine Sulfosalicylate, 5885
Methenamine Tetraiodide *see* 5886
Methenamine Tetraiodine, 5886
Methenolone, 5887
Methenolone Enanthate *see* 5887
***N,N'*-Methenyl-*o*-phenylenediamine** *see* 1091
Metheph [Napp] *see* 5987
Methergin [Sandoz] *see* 5989
Methergine [Sandoz] *see* 5989
Methergoline *see* 5843
Methescufylline *see* 5844
Methestrol, 5888
Methetharimide *see* 1036
Methetoin, 5889
Methexenyl *see* 4625
Methexenyl Sodium *see* 4625
Methforylthiazidine *see* 4716
Met-HGH *see* 8672
Methiacil *see* 6048
Methiazic Acid *see* 6057
Methiazinic Acid *see* 6057
Methicil *see* 6048
Methicillin Sodium, 5890
Methidathion, 5891
Methilanin *see* 5896
Methimazole, 5892
Methiocarb, 5893
Methiocil *see* 6048
Methiodal Sodium, 5894
Methionamine *see* 90

Methionic Acid, 5895
Methionine, 5896
Methionine⁵-enkephalin see 3528
Methionine Hydroxy Analog, 5897
Methionyl Human Growth Hormone see 8672
N²-L-Methionyl-1-139-interferon γ (Human Lymphocyte Protein Moiety Reduced) see 4894
Methioplegium see 9621
Methioprim, 5898
Methiotriazamine, 5899
Methioturiate see 5901
Methisazone, 5900
Methisoprinol see 4881
Methitural, 5901
Methium Chloride see 4609
Methixart [Fuso] see 5902
Methixene, 5902
Methobenzmorphan see 5839
Methocarbamol, 5903
Methocel [Dow] see 5963
Methocel HG [Dow] see 4777
Methocillin-S see 2414
Methofadin see 8888
Methohexital Sodium, 5904
Methohexitone Sodium see 5904
Methoin see 5741
Methomidate see 6069
Methomyl, 5905
Methone see 3231
Methophenazine see 6065
Methopholine see 6066
Methoplain [Kowa Yakuhin] see 5974
Methoprene, 5906
Methopromazine see 5923
Methopterin, 5907
Methopyrapone see 6083
Methorphan see 8116
Methorphinan see 5350
Methosarb [Upjohn] see 1730
Methose see 4186
Methoserpidine see 2901
Methostan [Schering] see 5861
Methotrexate, 5908
Methotrimeprazine, 5909
Methoxa-Dome see 5911
Methoxadone see 5739
Methoxamine Hydrochloride, 5910
Methoxone [ICI] see 5645
Methoxsalen, 5911
6-Methoxyacacetin see 7010
p-Methoxyacetanilide see 43
[[2-[(Methoxyacetyl)amino]-4-(phenylthio)phenyl]carbonimidoyl]biscarbamic Acid Dimethyl Ester see 3888
5-Methoxyacetylamino-2,4,6-triiodoisophthalic Acid [(2,3-Dihydroxy-N-methylpropyl)-(2,3-dihydroxypropyl)]diamide see 4948
2-Methoxy-6-allylphenol Diethylaminoethyl Ether see 4448
2-Methoxy-4-allylphenoxyacetic Acid N,N-Diethylamide see 39
Methoxyamine, 5912
3-Methoxy-4-(4'-aminobenzenesulfonamido)-1,2,5-thiadiazole see 8892
4-Methoxy-4'-aminodiphenylamine see 5922
6-Methoxy-8-(4-aminopentylamino)-quinoline see 8096
2,2'-[3-Methoxy-4'-amyl-5'-methyl-5-(2''-pyrryl)]dipyrrylmethene see 7774
3-Methoxyaniline see 697
4-Methoxy-6-[β-(p-anisyl)vinyl]-α-pyrone see 10001
10-Methoxyaporphine-2,11-diol see 6197

10-Methoxy-6aβ-aporphin-11-ol see 771
4-Methoxy-Benadryl see 5678
4-Methoxybenzaldehyde see 693
2-Methoxybenzenamine see 697
3-Methoxybenzenamine see 697
4-Methoxybenzenamine see 697
Methoxybenzene see 699
4-Methoxy-1,3-benzenediamine see 5920
4-Methoxybenzenemethanol see 695
p-Methoxybenzhydryl β-Dimethylaminoethylether see 5678
β-(p-Methoxybenzhydryloxy)ethyldimethylamine see 5678
4-Methoxybenzoic Acid see 696
2-(4-Methoxybenzoyl)benzoic Acid see 702
4-Methoxybenzoyl Chloride see 703
1-(4-Methoxybenzoyl)-2-pyrrolidinone see 692
p-Methoxybenzyl Alcohol see 695
1-(p-Methoxybenzyl)-2,3-dimethylguanidine see 5730
N-p-Methoxybenzyl-N',N'-dimethyl-N-α-pyridylethylenediamine see 7996
α-(α-Methoxybenzyl)-4-(β-methoxyphenethyl)-1-piperazineethanol see 10074
(p-Methoxybenzyl)tartaric Acid see 7479
1-Methoxy-2,2-bis[4-hydroxy-3-coumarinyl]ethane see 2566
2-Methoxy-4,6-bis(isopropylamino)-s-triazine see 7799
6-Methoxy-N,N'-bis(1-methylethyl)-1,3,5-triazine-2,4-diamine see 7799
4-Methoxy-N,N'-bis(3-pyridinylmethyl)-1,3-benzenedicarboxamide see 7378
4-Methoxy-N,N'-bis(3-pyridylmethyl)isophthalamide see 7378
2-(Methoxycarbonylamino)benzimidazole see 1794
2-[[[(Methoxycarbonyl)amino][[2-nitro-5-(propylthio)phenyl]amino]methylene]amino]ethanesulfonic Acid see 6390
2-[[[(Methoxycarbonyl)amino][[2-nitro-5-(propylthio)phenyl]imino]methyl]amino]ethanesulfonic Acid see 6390
3-Methoxycarbonyl-6-n-butyl-7-benzyloxy-4-oxoquinoline see 6385
N-Methoxycarbonyl-N'-[2-nitro-5-(propylthio)phenyl]-N''-2-(ethylsulfonic Acid)guanidine see 6390
1-Methoxycarbonyl-1-propen-2-yl Dimethyl Phosphate see 6089
Methoxychlor, 5913
N-[4-(β-(2-Methoxy-5-chlorobenzamido)ethyl)benzosulfonyl]-N'-cyclohexylurea see 4372
N¹-[4-[β-(2-Methoxy-5-chlorobenzoylamino)ethyl]benzenesulfonyl]-N²-cyclohexylurea see 4372
6'-Methoxycinchonan-9-ol see 8075
6'-Methoxycinchonan-9-ol see 8072
(9R)-6'-Methoxycinchonan-9-ol see 3570
(9S)-6'-Methoxycinchonan-9-ol see 3571
6'-Methoxycinchonan-9-ol Ethyl Carbonate (Ester) see 8080
6'-Methoxycinchonan-9-ol Monohydrochloride see 8085
6'-Methoxycinchonan-9-ol Sulfate see 8074
(8α)-6'-Methoxycinchonan-9-one see 8093

6-Methoxycinchoninic Acid see 8092
p-Methoxycinnamic Acid 2-Ethoxyethyl Ester see 2312
Methoxyconiferine see 8997
9-Methoxycorynantheidine see 6136
2-Methoxy-p-cresol see 2573
4-Methoxy-2,6-cresotic Acid Methyl Ester see 8690
9-Methoxycrinan-1α,3α-diol see 1483
Methoxycyclopropane see 2756
Methoxy-DDT see 5913
10-Methoxydeserpidine see 2901
Methoxydichlorobenzoate see 3026
2-Methoxy-3,6-dichlorobenzoic Acid see 3026
6-(α-Methoxy-3,4-dichlorophenylacetamido)penicillanic Acid see 2383
8-Methoxydictamnine see 3879
6-Methoxy-8-(3-diethylaminopropylamino)quinoline see 7495
[3-Methoxy-4-[(N,N-diethylcarbamido)methoxy]phenyl]acetic Acid n-Propyl Ester see 7813
6-Methoxy-3,4-dihydro-1(2H)-naphthalenone see 5924
4-Methoxy-2,2'-dihydroxybenzophenone see 3298
3-Methoxy-10-(3-dimethylamino-2-methylpropyl)phenothiazine see 5909
2-Methoxy-10-(3'-dimethylaminopropyl)phenothiazine see 5923
3-Methoxy-10-(3'-dimethylaminopropyl)phenothiazine see 5923
2-Methoxy-N,α-dimethylbenzeneethanamine see 5919
5-Methoxy-8,8-dimethyl-2H,8H-benzo[1,2-b:5,4-b']dipyran-2-one see 9975
7-Methoxy-2,2-dimethyl-2H-1-benzopyran see 7716
7-Methoxy-2,2-dimethylchromene see 7716
(8β)-10-Methoxy-1,6-dimethylergoline-8-methanol 5-Bromo-3-pyridinecarboxylate (Ester) see 6404
1-(8-Methoxy-4,8-dimethylnonyl)-4-(1-methylethyl)benzene see 7778
o-Methoxy-N,α-dimethylphenethylamine see 5919
(±)-7-Methoxy-α,10-dimethyl-10H-phenothiazine-2-acetic Acid see 7906
7-Methoxy-α,10-dimethylphenothiazine-3-acetic Acid see 7906
2-Methoxy-N,N-dimethyl-10H-phenothiazine-10-propanamine see 5923
2-Methoxy-3,5-dimethyl-6-[tetrahydro-4-[2-methyl-3-(4-nitrophenyl)-2-propenylidene]-2-furanyl]-4H-pyran-4-one see 897
Methoxydiuron see 5387
Methoxydon(e) see 5739
3-Methoxy-17-epiestriol see 3568
3-Methoxyestra-1,3,5(10)-triene-16,17-diol see 3568
Methoxyethane see 3783
2-Methoxyethanol see 5961
2-Methoxyethanol Acetate see 5962
2-(2-Methoxyethoxy)ethanol see 5959
N-[5-(2-Methoxyethoxy)-2-pyrimidinyl]benzenesulfonamide see 4403
2-Methoxyethyl 1,4-Dihydro-5-(isopropoxycarbonyl)-2,6-dimethyl-4-(3-nitrophenyl)-3-pyridinecarboxylate see 6467
3,3'-(2-Methoxyethylidene)bis[4-hydroxy-2H-1-benzopyran-2-one] see 2566

6-Methoxytropine Benzilate *see* 9690
5-Methoxytryptamine, 5926
3-Methoxy-L-tyrosine *see* 5975
6-Methoxyumbelliferone *see* 8363
Methoxyverapamil *see* 4257
6-Methoxy-α-(5-vinyl-2-quinuclidin-yl)-4-quinolinemethanol *see* 3570 and 8072
Methral [Pfizer] *see* 4115
Methrazone [W. B. Pharm.] *see* 3953
Methscopolamine Bromide, 5927
Methscopolamine Nitrate *see* 8361
Methsuximide, 5928
Methural *see* 6921
Methyclothiazide, 5929
Methycobal [Eisai] *see* 9921
Methyl Abietate, 5930
Methylacetaldehyde *see* 7835
Methyl 4-Acetamido-2-ethoxybenzo-ate *see* 3701
3-Methyl-4-acetaminophenylarsanilic Acid *see* 5945
m-Methylacetanilide *see* 67
N-Methylacetanilide, 5931
Methyl Acetate, 5932
Methylacetic Acid *see* 7837
Methylacetic Anhydride *see* 7838
Methyl Acetoacetate, 5933
N-Methyl-2-acetonylpyrrolidine *see* 4788
Methylacetopyronone *see* 2855
6-α-Methyl-17α-acetoxyprogesterone *see* 5677
3-Methyl-4-acetylaminophenylarson-ate *see* 5945
Methyl Acetylsalicylate, 5934
β-Methylacrolein *see* 2599
Methyl Acrylate, 5935
α-Methylacrylic Acid *see* 5849
β-Methylacrylic Acid *see* 2600
α-Methylacrylonitrile *see* 5850
α-Methyl-1-adamantanemethylamine *see* 8221
N-Methyladrenaline *see* 5988
3-*O*-Methyladrenaline *see* 5831
Methylal, 5936
Methylal Acetamide *see* 4154
2-Methylalanine *see* 457
3-Methyl-*N*,α-(β-alanyl)-L-histidine *see* 708
Methyl Alcohol *see* 5868
Methyl Aldehyde *see* 4148
2-(*p*-Methylallylaminophenyl)propion-ic Acid *see* 298
α-Methylallyl Chloride *see* 2131
β-Methylallylchloride *see* 2148
γ-Methylallyl Chloride *see* 2130
13β-Methyl-17α-allyl-Δ⁴,⁹,¹¹-gonatri-ene-17β-ol-3-one *see* 317
N-Methyl-5-allyl-5-isopropylbarbi-turic Acid *see* 3523
α-*dl*-1-Methyl-5-allyl-5-(1-methyl-2-pentynyl)barbituric Acid Sodium Salt *see* 5904
Methyl Allyl Trisulfide, 5937
Methylamine, 5938
N-Methylaminoacetic Acid *see* 8331
2-(Methylamino)-*m*-anisic Acid Methyl Ester *see* 2809
Methyl 2-Aminobenzoate *see* 5942
2-(4-Methylaminobutoxy)diphenyl-methane *see* 1227
N-[[(Methylamino)carbonyl]oxy]-ethanimidothioic Acid Methyl Ester *see* 5905
4-Methylaminocetopyrocatechol *see* 161
2-(Methylamino)-2-(2-chlorophenyl)-cyclohexanone *see* 5174
2-(Methylamino)-2-deoxy-α-L-gluco-pyranose *see* 5996

5-(Methylamino)-2′-deoxyuridine *see* 5523
3-Methylamino-7-dimethylamino-phenazathonium Chloride *see* 941
4-Methylamino-1,5-dimethyl-2-phen-yl-3-pyrazolone Sodium Methane-sulfonate *see* 3358
α-*dl*-6-(Methylamino)-4,4-diphenyl-3-heptanol Acetate *see* 6602
N-Methylaminodithioformic Acid Sodium Salt *see* 5860
N-Methylaminoethane Sodium Sulfon-ate *see* 6043
2-Methylaminoethanesulfonic Acid *see* 6043
β-Methylaminoethane-α-sulfonic Acid *see* 6043
Methylaminoethanoic Acid *see* 8331
2-Methylaminoethanol, 5939
Methylaminoethanolcatechol *see* 3569
p-Methylaminoethanolphenol *see* 8994
m-Methylaminoethanolphenol Hydro-chloride *see* 7257
p-Methylaminoethanolphenol Tartrate *see* 8994
4-[2-(Methylamino)ethyl]-1,2-benz-enediol *see* 2885
α-[1-(Methylamino)ethyl]benzene-methanol *see* 3561
α-[1-(Methylamino)ethyl]benzyl Alco-hol *see* 3561
α-(1-Methylaminoethyl)-*p*-hydroxy-benzyl Alcohol *see* 4754
N-Methylaminoethyl 2-Methylbenz-hydryl Ether *see* 9426
4-[2-(Methylamino)ethyl]-*o*-phenyl-ene Diisobutyrate *see* 4807
2-[2-(Methylamino)ethyl]pyridine *see* 1200
4-[2-(Methylamino)ethyl]pyrocatechol *see* 2885
2-Methyl-6-aminoheptane *see* 6678
2-Methyl-6-amino-2-heptanol *see* 4577
6-Methyl-2-amino-6-heptanol *see* 4577
Methyl 3-Amino-4-hydroxybenzoate *see* 6836
2-Methyl-4-amino-1-hydroxynaph-thalene *see* 9936
3-Methyl-4-amino-1-hydroxynaph-thalene *see* 9938
β-Methylamino-α-(4-hydroxyphenyl)-ethyl Alcohol *see* 8994
1-(1-Methylamino-2-hydroxy-3-propyl)dibenzo[*b*,*e*]bicyclo[2.2.2.]-octadiene *see* 6878
α-Methylamino-β-(3-indole)propionic Acid *see* 5
2-Methylamino-5α-(β-indolyl)ethyl-2-oxazolin-4-one *see* 4872
2-Methylaminoisocamphane *see* 5654
3-Methylaminoisocamphane *see* 5654
β-(Methylamino)lpropylcyclopentane *see* 2745
Methyl α-Amino-β-mercaptopropion-ate Hydrochloride *see* 5668
N-Methylaminomethanethionothiolic Acid Sodium Salt *see* 5860
α-[(Methylamino)methyl]benzene-methanol *see* 4516
l-α-(Methylaminomethyl)benzyl Alco-hol *see* 4516
1-(Methylaminomethyl)dibenzo[*b*,*e*]-bicyclo[2.2.2]octadiene *see* 1096
9-(Methylaminomethyl)-9,10-dihydro-9,10-ethanoanthracene *see* 1096
α-[(Methylamino)methyl]-9,10-ethanoanthracene-9(10*H*)-ethanol *see* 6878

6-Methylamino-2-methylheptene *see* 5070
Methylaminomethyl 4-Hydroxyphenyl Carbinol *see* 8994
1-Methylamino-2-methyl-2-phenyl-ethane *see* 7280
2-Methylamino-2-methyl-1-phenyl-propane *see* 5740
α-(Methylaminomethyl)vanillyl Alco-hol *see* 5831
2-Methyl-4-amino-1-naphthol *see* 9936
3-Methyl-4-amino-1-naphthol *see* 9938
N-[4-(6-Methylamino-7-nitro-2-thia-5-aza-6-heptene-1-yl)-2-thiazolyl-methyl]-*N*,*N*-dimethylamine *see* 6582
p-Methylaminophenol Sulfate, 5940
2-Methylamino-1-phenylethanol *see* 4516
1-Methylamino-2-phenylpropane *see* 7280
DL-*threo*-2-(Methylamino)-1-phenyl-propan-1-ol *see* 3561
L-*erythro*-2-(Methylamino)-1-phenyl-propan-1-ol *see* 3561
2-Methylamino-1-phenyl-1-propanol *see* 3561
1-Methyl-4-amino-*N*-phenyl-*N*-(2-thenyl)piperidine *see* 9204
1-(3-Methylaminopropyl)dibenzo[*b*,*e*]-bicyclo[2.2.2]octadiene *see* 5633
7-(3-Methylaminopropyl)-1,2:5,6-di-benzocycloheptatriene *see* 7917
5-(3-Methylaminopropyl)-5*H*-di-benzo[*a*,*d*]cycloheptene *see* 7917
9-(γ-Methylaminopropyl)-9,10-di-hydro-9,10-ethanoanthracene *see* 5633
5-(α-Methylaminopropylidene)di-benzo[*a*,*d*]cyclohepta[1,4]diene *see* 6635
N-(3-Methylaminopropyl)iminobi-benzyl *see* 2902
5-(γ-Methylaminopropyl)iminodi-benzyl *see* 2902
4-[2-(Methylamino)propyl]phenol *see* 7304
Methylaminopterin *see* 5908
4-Methyl-2-aminopyridine *see* 478
6-(2-Methyl-6-amino-5-pyrimidinyl)-5-aza-5-formyl-4-methyl-3-benz-oylthio-3-hexenyl Phosphate *see* 1049
2-(2-Methyl-4-aminopyrimidin-5-yl)-methylformamido-5-hydroxy-2-penten-3-yl Propyl Disulfide *see* 7896
1-(2-Methyl-4-amino-5-pyrimidyl)-methyl-2-methyl-3-hydroxyethyl-pyridinium Bromide Hydrobromide *see* 8003
3-[(2-Methyl-4-amino-5-pyrimidyl)-methyl]-4-methyl-5-(β-hydroxyeth-yl)thiothiazol-2-one *see* 9296
N-Methyl-*N*-(3-amino-2,4,6-triiodo-phenyl)glutaramic Acid *see* 4941
3β-Methylamino-2,2,3-trimethylbicy-clo[2.2.1]heptane *see* 5654
6*S*-[6α(2*S**,3*S**),8β(*R**),9β,11α]-5-(Methylamino)-2-[[3,9,11-trimeth-yl-8-[1-methyl-2-oxo-2-(1*H*-pyrrol-2-yl)ethyl]-1,7-dioxaspiro-[5.5]undec-2-yl]methyl]-4-benzoxa-zolecarboxylic Acid *see* 1639
2-Methylamino-3,3,3-trimethylnor-bornane *see* 5654
d-*N*-Methylamphetamine *see* 5859
Methyl Amyl Ketone *see* 4584

Cross Index of Names

3-Methylbutanoic Acid Methyl Ester
see 6006
(1R-endo)-3-Methylbutanoic Acid
1,7,7-Trimethylbicyclo[2.2.1]hept-2-
yl Ester see 1342
2-Methyl-1-butanol, 5952
2-Methyl-2-butanol see 7096
3-Methyl-1-butanol see 5081
3-Methyl-2-butanol, 5953
2-Methyl-2-butanol Carbamate see
641
3-Methyl-1-butanol Nitrate see 5004
3-Methyl-2-butenal see 8399
2-Methyl-2-butene see 644
(E)-2-Methyl-2-butenedioic Acid see
5806
2-Methyl-2-butenedioic Acid see
2323
(E)-2-Methyl-2-butenoic Acid see
9365
(Z)-2-Methyl-2-butenoic Acid see 678
3-Methyl-2-butenoic Acid 2-sec-But-
yl-4,6-dinitrophenyl Ester see 1237
[1R-[1α,3aα,4β(Z),6β,8β(Z),8aβ]]-2-
Methyl-2-butenoic Acid Decahy-
dro-1,6-dihydroxy-3a,6-dimethyl-1-
(1-methylethyl)-5-oxo-4,8-azulene-
diyl Ester see 5244
2-Methyl-2-butenoic Acid 7-[[2,3-
Dihydroxy-2-(1-hydroxyethyl)-3-
methyl-1-oxobutoxy]methyl]-2,3,-
5,7a-tetrahydro-1H-pyrrolizin-1-yl
Ester see 4545
2-Methyl-2-butenoic Acid 7-[[2,3-
Dihydroxy-2-(1-methoxyethyl)-3-
methyl-1-oxobutoxy]methyl]-2,3,-
5,7a-tetrahydro-1H-pyrrolizin-1-yl
Ester see 5245
2-Methyl-2-butenoic Acid 6α,7α-
Dihydroxy-8-methyl-8-azabicyclo-
[3.2.1]oct-3β(E)-yl Ester see 5841
2-Methyl-2-butenoic Acid 2-Hydr-
oxy-2-methyl-1-[[(7-oxo-7H-furo-
[3,2-g][1]benzopyran-4-yl)oxy]-
methyl]propyl Ester see 6851
2-Methyl-2-butenoic Acid [1α,3α(E),-
5α]- 8-Methyl-8-azabicyclo[3.2.1]-
oct-3-yl Ester see 9366
3-Methyl-2-butenoic Acid 2-(1-
Methylpropyl)-4,6-dinitrophenyl
Ester see 1237
6-(3-Methylbut-2-enyl)coumestrol
see 7945
4-(3-Methyl-2-butenyl)-1,2-diphenyl-
3,5-pyrazolidinedione see 3953
(3-Methyl-2-butenyl)guanidine see
4246
8-(3-Methyl-2-butenyl)herniarin see
6849
3-(3-Methyl-2-butenyl)-1,2,3,4,5,6-
hexahydro-6,11-dimethyl-2,6-meth-
ano-3-benzazocin-8-ol see 7078
9-[(3-Methyl-2-butenyl)oxy]-7H-
furo[3,2-g][1]benzopyran-7-one see
4839
[5-[(3-Methyl-2-butenyl)oxy]-2-[3-
[4-[(3-methyl-2-butenyl)oxy]phen-
yl]-1-oxo-2-propenyl]phenoxy]-
acetic Acid see 8658
1-[2-(3-Methylbutoxy)-2-phenyleth-
yl]pyrrolidine see 507
N,N'-[4-(3-Methylbutoxy)phenyl]-
thiourea see 9382
3-Methylbutylamine see 4994
Methyl 5-Butyl-2-benzimidazolecarb-
amate see 6985
Methyl 1-(Butylcarbamoyl)-2-benz-
imidazolecarbamate see 1053
Methyl tert-Butyl Ether, 5954
Methyl Butyl Ketone, 5955

5-(1-Methylbutyl)-5-[2-(methylthio)-
ethyl]-2-thiobarbituric Acid Sodium
Salt see 5901
2-Methyl-2-sec-butyl-1,3-propanediol
Dicarbamate see 5653
5-(1-Methylbutyl)-5-(2-propenyl)-
2,4,6(1H,3H,5H)-pyrimidinetrione
Monosodium Salt see 8374
5-(1-Methylbutyl)-5-vinylbarbituric
Acid see 9897
2-Methyl-3-butyn-2-ol, 5956
Methyl Butyrate, 5957
2-Methylbutyric Acid 9-Ester with
9,10-Dihydro-9,10-dihydroxy-8,8-
dimethyl-2H,8H-benzo[1,2-b:3,4-
b']Dipyran-2-one Acetate see 9915
3-(α-Methylbutyryloxy)-4-acetoxy-
3,4-dihydroseseline see 9915
8α-(3-Methylbutyryloxy)-4β,15-di-
acetoxyscirp-9-en-3α-ol see 9711
Methyl Carbamate, 5958
Methylcarbamic Acid m-Cym-5-yl
Ester see 7794
Methyl Carbamic Acid 2,3-Dihydro-
2,2-dimethyl-7-benzofuranyl Ester
see 1810
Methylcarbamic Acid 4-(Dimethylamino)-m-tolyl Ester see 443
Methylcarbamic Acid 4-(Dimethylamino)-3,5-xylyl Ester see 6090
Methylcarbamic Acid m-(1-Ethyl-
propyl)phenyl Ester Mixt. with
m-(1-Methylbutyl)phenyl Ester see
1459
Methylcarbamic Acid 2,3-(Isopropyli-
denedioxy)phenyl Ester see 1044
Methylcarbamic Acid 4-(Methylthio)-
3,5-xylyl Ester see 5893
Methyl Carbamic Acid 1-Naphthyl
Ester see 1789
Methylcarbamodithioic Acid Sodium
Salt see 5860
1-Methyl-2-[(carbamoyloxy)methyl]-
5-nitroimidazole see 8237
N-[(Methylcarbamoyl)oxy]thioacet-
imidic Acid Methyl Ester see 8374
Methyl N-Carbamyl-N'-(2-chloroeth-
yl)-N'-nitroso-6-amino-6-deoxy-α-
D-glucopyranoside see 8125
Methyl Carbethoxysyringoyl Reserpate see 8998
Methyl O-(O'-Carbethoxysyringoyl)-
reserpate see 8998
1-Methyl-3-carbethoxy-2-thioglyoxa-
lone see 1803
Methyl Carbitol®, 5959
2-Methyl-β-carboline see 4530
3-Methyl-4-carboline see 4530
2-Methyl-6-carbomethoxy-N-diethyl-
aminoacetanilide see 9467
Methyl Carbonate, 5960
4H-3-Methylcarboxamide-3-benzoxa-
zine-2-one see 1865
4-Methyl-5-(ω-carboxyamyl)imidazo-
lidone-2 see 2914
17β-(1-Methyl-3-carboxypropyl)etio-
cholane see 2199
17β-(1-Methyl-3-carboxypropyl)etio-
cholane-3α,7α-diol see 2044
17β-(1-Methyl-3-carboxypropyl)etio-
cholane-3α,7β-diol see 9801
17β-(1-Methyl-3-carboxypropyl)etio-
cholane-3α,12α-diol see 2881
17β-(1-Methyl-3-carboxypropyl)etio-
cholane-3α,7α,12α-triol see 2206
17β-(1-Methyl-3-carboxypropyl)etio-
cholan-3α-ol see 5423
Methylcatechol see 4457
Methyl Cellosolve®, 5961
Methyl Cellosolve® Acetate, 5962
Methylcellulose, 5963

β-Methylchalcone see 3460
Methyl Chemosept see 6021
Methyl Chloride, 5964
Methyl Chloroacetate, 5965
N-[2-(α-Methyl-p-chlorobenzhydryloxy)ethyl]hexamethyleneimine see
8423
N-Methyl-N'-(4-chlorobenzhydryl)-
piperazine see 2078
Methyl m-Chlorobenzoate see 2124
Methyl p-Chlorobenzoate see 2126
3-Methyl-7-chloro-1,2,4-benzothiadi-
azine 1,1-Dioxide see 2986
[2-Methyl-3-(p-chlorobenzoyl)-6-
methoxyindol-1-yl]acetic Acid see
2381
N-Methyl-N-(4-chlorobenzoylmeth-
yl)-3-(10,11-dihydro-5H-dibenzo-
[b,f]azepin-5-yl)propylamine see
5438
2-Methyl-1-(p-chlorobenzyl)benz-
imidazole see 2106
(±)-2-Methyl-2-[p-(p'-chlorobenz-
yl)phenoxy]butyric Acid Ethyl Ester
see 1028
Methyl Chlorocarbonate, 5966
Methyl 6-[[[(2-Chloroethyl)nitroso-
amino]carbonyl]amino]-6-deoxy-α-
D-glucopyranoside see 8125
4-Methyl-5-(β-chloroethyl)thiazole
see 2382
Methylchloroform see 9549
Methyl Chloroformate see 5966
Methyl Chloromethyl Ether see 2146
3-Methyl-4-chlorophenol see 2133
2-Methyl-4-chlorophenoxyacetic Acid
see 5645
N-(2-Methyl-3-chlorophenyl)anthra-
nilic Acid see 9436
trans-4-Methyl-5-(4-chlorophenyl)-3-
cyclohexylcarbamoyl-2-thiazolidone
see 4634
1-Methyl-6-o-chlorophenyl-8-ethyl-
4H-s-triazolo[3,4-c]thieno[2,3-e]-
1,4-diazepine see 3830
3-Methyl-4-[(2-chlorophenyl)hydra-
zone]-4,5-isoxazoledione see 3432
3-Methyl-4-(o-chlorophenylhydrazono)-5-isoxazolone see 3432
[5-Methyl-3-(o-chlorophenyl)-4-isox-
azolyl]penicillin see 2414
2-Methyl-4-(p-chlorophenyl)-2,4-
pentanediol see 3933
2-Methyl-3-(2-chlorophenyl)-4-quin-
azolone see 5661
(2S-trans)-Methyl 7-Chloro-6,7,8-tri-
deoxy-6-[[(1-methyl-4-propyl-2-
pyrrolidinyl)carbonyl]amino]-1-
thio-L-threo-α-D-galacto-octopyra-
noside see 2352
3-Methylcholanthrene, 5967
20-Methylcholanthrene see 5967
3-Methylchromone see 9574
3-Methylchrysazin see 2263
5-[(E),β-Methylcinnamylidene]-4-
oxo-2-thioxo-3-thiazolidineacetic
Acid see 3556
Methylcobaz [Labaz] see 9921
Methylcoffanolamine see 1632
6α-Methylcompactin see 5460
Methylconiine, 5968
13-Methyl-ψ-coptisine see 9963
3-Methylcrotonaldehyde see 8399
(E)-2-Methylcrotonic Acid see 9365
3-Methylcrotonic Acid 2-sec-Butyl-
4,6-dinitrophenyl Ester see 1237
γ-Methyl-α,β-crotonolactone see 679
γ-Methyl-β,γ-crotonolactone see 679
Methyl 7-Crotonylidenecyclopenta[c]-
pyran-1-(7H)-one-4-carboxylate
see 4198

α-Methylene Butyrolactone, 5981
γ-Methylene-γ-butyrolactone *see* 679
[4-(Methylenebutyryl)-2,3-dichloro-phenoxy]acetic Acid *see* 3669
Methylene Chloride, 5982
Methylene Cyanide *see* 5592
24-Methylenecycloartanyl Ferulate *see* 6841
1,2-Methylenecyclohexane *see* 6605
α-(2,5-*endo*-Methylene-Δ³-cyclo-hexenyl)mandelic Acid β-Dimethyl-aminoethyl Ester Methyl Bromide *see* 3527
(3β)-24-Methylene-9,19-cyclolano-stan-3-ol 3-(4-Hydroxy-3-meth-oxyphenyl)-2-propenoate *see* 6841
2-Methylenecyclopropanealanine *see* 4801
4,4′-Methylenedianiline *see* 2958
Methylene Dichloride *see* 5982
Methylenedigallic Acid, 5983
4,4′-Methylenedi(3-hydroxy-2-naph-thoic Acid) *see* 6955
Methylenedinaphthalenesulfonic Acid Disilver Salt *see* 5874
Methylenedinaphthylenesulfonic Acid Bisphenylmercuri Salt *see* 4687
1,2-Methylenedioxyaporphine *see* 778
1,2-(Methylenedioxy)aporphin-11-ol *see* 7954
3,4-(Methylenedioxy)benzaldehyde *see* 7445
3,4-Methylenedioxybenzoic Acid *see* 7447
2-[4-(3,4-Methylenedioxybenzyl)-piperazino]pyrimidine *see* 7465
1,2-Methylenedioxy-9,10-dimethoxy-aporphine *see* 3028
5,6-Methylenedioxy-2,3-dimethyl-4-(3′,4′-methylenedioxyphenyl)-1,2,-3,4-tetrahydronaphthalene *see* 6852
1,2-Methylenedioxy-11-hydroxyapor-phine *see* 7954
1,2-Methylenedioxy-9-hydroxy-10-methoxynoraporphine *see* 131
3,4-Methylenedioxymethamphetamine *see* 5646
3,4-Methylenedioxy-8-methoxy-10-nitro-1-phenanthrenecarboxylic Acid *see* 810
6,7-Methylenedioxy-1-methoxy-4-oxo-1,4-dihydroquinoline-3-carb-oxylic Acid *see* 6116
1,2-(Methylenedioxy)-4-[2-(octylsul-finyl)propyl]benzene *see* 8946
2-(3,4-Methylenedioxyphenoxy)-3,6,9-trioxaundecane *see* 8419
N-[β-(3,4-Methylenedioxyphenyl)iso-propyl]-β-(3,4-dihydroxyphenyl)-β-hydroxyethylamine *see* 7911
N-[2-(3,4-Methylenedioxyphenyliso-propyl]norepinephrine *see* 7911
5-(3,4-Methylenedioxyphenyl)-2,4-pentadienoic Acid *see* 7437
1,2-(Methylenedioxy)-4-propenyl-benzene *see* 5112
[3,4-(Methylenedioxy)-6-propylbenz-yl] Butyl Diethyleneglycol Ether *see* 7446
6-(3′,4′-Methylenedioxystyryl)-4-methoxy-5,6-dihydro-2H-pyran-2-one *see* 6056
α-Methylene-diphenylmethane *see* 3323
5,5′-Methylenedisalicylic Acid, 5984
Methyleneditannin *see* 9024
3,3′-Methylenedithiobis(2-amino-propanoic Acid) *see* 3393
3,3′-(Methylenedithio)dialanine *see* 3393

16-Methylene-9α-fluoroprednisolone 21-Acetate *see* 4118
5-Methylene-2(5H)-furanone *see* 7907
α,α′-Methylene Glycerin *see* 4380
α,β-Methylene Glycerin *see* 4380
6-Methylene-5-hydroxytetracycline *see* 5851
Methylene Iodide, 5985
Methylenemalononitrile Copolymer with Vinyl Acetate *see* 2821
3-Methylene-6-(1-methylethyl)cyclo-hexene *see* 7152
[1R*,2R*5S*,6E,10R*]-(±)-8-Meth-ylene-5-(1-methylethyl)spiro(11-oxabicyclo[8.1.0]undec-6-ene-2,2′-oxiran)-3-one *see* 7125
11,12-Methyleneoctadecanoic Acid *see* 5218
Methylene Oxide *see* 4148
(R)-2-Methylene-3-oxocyclopentane-carboxylic Acid *see* 8333
5-Methylene-2-oxodihydrofuran *see* 7907
6-Methyleneoxytetracycline *see* 5851
16-Methyleneprednisolone *see* 7729
16-Methyleneprednisolone 21-Dieth-ylaminoacetate *see* 7730
Methylenesuccinic Acid *see* 5130
16-Methylene-11β,17α,21-trihydroxy-pregna-1,4-diene-3,20-dione *see* 7729
Methylenomycin A *see* 5986
Methylenomycin B *see* 5986
Methylenomycins, 5986
N-Methylephedrine, 5987
N-Methylephedrine Camsylate *see* 5987
N-Methylephedrone *see* 5828
Methyl 18-Epi-O-methylreserpate *see* 6074
N-Methylepinephrine, 5988
3-O-Methylepinephrine *see* 5831
O-Methyl-18-epireserpic Acid Methyl Ester *see* 6074
Methylergobasine *see* 5989
Methylergobrevin [Arzneimittelwerk, VEB] *see* 5989
Methylergol Carbamide *see* 5394
Methylergometrine *see* 5989
Methylergonovine, 5989
6-O-Methylerythromycin *see* 2340
4-Methylesctietindisulfonic Acid *see* 8964
β-Methylesculetin *see* 8363
4-Methylesculetin Bis(hydrogen Sulf-ate) *see* 8964
4-Methylesculetin-6,7-disulfuric Ester *see* 8964
17α-Methyl-4,9,11-estratrien-17β-ol-3-one *see* 6049
Methylestrenolone *see* 6629
[[(1-Methyl-1,2-ethanediyl)bis[carb-amodithioato]](2—)]zinc *see* 7831
4,4′-(1-Methyl-1,2-ethanediyl)bis-2,6-piperazinedione *see* 8133
2,2′-(1-Methyl-1,2-ethanediylidene)-bis[hydrazinecarboximidamide] *see* 6131
1,1′-[(Methylethanediylidene)dini-trilo]diguanidine *see* 6131
N-Methyl-9,10-ethanoanthracene-9(10H)-methanamine *see* 1096
N-Methyl-9,10-ethanoanthracene-9(10H)-propanamine *see* 5633
Methylethene *see* 7862
4-(1-Methylethenyl)-1-cyclohexene-1-carboxaldehyde *see* 7119
Methyl Ether, 5990
1-[4-[[2-(1-Methylethoxy)ethoxy]-methyl]phenoxy]-3-[(1-methyleth-yl)amino]-2-propanol *see* 1309

2-(1-Methylethoxy)phenol Methyl-carbamate *see* 7849
7-(1-Methylethoxy)-3-phenyl-4H-1-benzopyran-4-one *see* 4961
N-[[(1-Methylethyl)amino]carbonyl]-4-[(3-methylphenyl)amino]-3-pyri-dinesulfonamide *see* 9473
1-[(1-Methylethyl)amino]-3-[(2-meth-yl-1H-indol-4-yl)oxy]-2-propanol *see* 5744
(±)-α-[[(1-Methylethyl)amino]meth-yl]-4-nitrobenzenemethanol *see* 6442
1-[(1-Methylethyl)amino]-3-(3-meth-ylphenoxy)-2-propanol *see* 9439
1-[(1-Methylethyl)amino]-3-(1-naph-thalenyloxy)-2-propanol *see* 7852
2-Methylethylamino-1-phenyl-1-propanol *see* 3661
1-[(1-Methylethyl)amino]-3-[2-(2-propenyloxy)phenoxy]-2-propanol *see* 6905
1-[(1-Methylethyl)amino]-3-[2-(2-propenyl)phenoxy]-2-propanol *see* 311
9-[3-[(1-Methylethyl)amino]propyl]-9H-fluorene-9-carboxamide *see* 4849
4-(1-Methylethyl)benzaldehyde *see* 2623
4-(1-Methylethyl)benzenamine *see* 2619
(1-Methylethyl)benzene *see* 2619
4-(1-Methylethyl)benzenemethanol *see* 2621
4-(1-Methylethyl)benzoic Acid *see* 2620
3-(1-Methylethyl)-1H-2,1,3-benzo-thiadiazin-4(3H)-one 2,2-Dioxide *see* 1060
Methylethylbromomethane *see* 1554
(1-Methylethyl)carbamic Acid 2-[[(Aminocarbonyl)oxy]methyl]-2-methylpentyl Ester *see* 1848
Methyl Ethyl Carbinol *see* 1541
2-Methyl-3-ethyl-5-dimethylamino-indole *see* 5675
Methylethyldimethylaminomethyl-carbinol Benzoyl Ester Hydrochlo-ride *see* 650
4-(1-Methylethyl)-2,6-dinitro-N,N-dipropylbenzenamine *see* 5089
N-(1-Methylethyl)-4,4-diphenylcyclo-hexanamine *see* 7706
1-(1-Methylethyl)-4,4-diphenylpiperi-dine *see* 7776
Methylethylene *see* 7862
Methyl Ethyl Ether *see* 3783
1-Methylethyl 2-[[Ethoxy[(1-methyl-ethyl)amino]phosphinothioyl]oxy]-benzoate *see* 5055
(E)-1-Methylethyl 3-[[(Ethylamino)-methoxyphosphinothioyl]oxy]-2-butenoate *see* 7827
sym-Methylethylethylene *see* 7080
β,β-Methylethylglutarimide *see* 1036
3-Methyl-6-[7-ethyl-4-hydroxy-3,5-dimethyl-6-oxo-7-[5-ethyl-3-meth-yl-5-(5-ethyl-5-hydroxy-6-methyl-2-tetrahydropyranyl)-2-tetrahydro-furyl]heptyl]salicylic Acid *see* 5243
1-Methyl-3-ethyl-3-(m-hydroxyphen-yl)hexahydro-1H-azepine *see* 5753
16α,17-[(1-Methylethylidene)bis-(oxy)]pregn-4-ene-3,20-dione *see* 229
4,4′-(1-Methylethylidene)bisphenol *see* 1311
4,4′-[(1-Methylethylidene)bis(thio)]-bis[2,6-bis(1,1-dimethylethyl)-phenol] *see* 7761

4-[2-(1-Methyl-5-nitro-1H-imidazol-2-yl)ethenyl]-2-pyrimidinamine *see* 914

[2-(2-Methyl-5-nitro-1H-imidazol-1-yl)ethyl]carbamothioic Acid O-Methyl Ester *see* 1855

6-(1-Methyl-4-nitro-5-imidazolyl)-mercaptopurine *see* 918

1-(2-Methyl-5-nitroimidazol-1-yl)-2-propanol *see* 8373

6-[(1-Methyl-4-nitro-1H-imidazol-5-yl)thio]-1H-purin-2-amine *see* 9228

6-[(1-Methyl-4-nitro-1H-imidazol-5-yl)thio]-1H-purine *see* 918

2-Methyl-4-nitro-1-(4-nitrophenyl)-1H-imidazole *see* 6492

3-Methyl-4-nitro-1-(p-nitrophenyl)-2-pyrazolin-5-one *see* 7383

N-Methyl-N'-nitro-N-nitrosoguanidine, 6017

5-Methyl-2-nitro-7-oxa-8-mercurabicyclo[4.2.0]octa-1,3,5-triene *see* 6531

[2-Methyl-5-nitrophenolato(2−)-C⁶,O¹]mercury *see* 6531

3-Methyl-5-[(p-nitrophenyl)azo]-rhodanine *see* 6517

3-Methyl-5-[(p-nitrophenyl)azo]-2-thio-2,4-thiazolidinedione *see* 6517

3-Methyl-5-[(4-nitrophenyl)azo]-2-thioxo-4-thiazolidinone *see* 6517

Methyl [N'-[2-Nitro-5-(propylthio)-phenyl]-N-(2-sulfoethyl)amidino]-carbamate *see* 6390

N-Methyl-N-nitrosomethanamine *see* 6558

N-Methyl-N-nitroso-N'-nitroguanidine *see* 6017

N-Methyl-N-nitroso-p-toluenesulfonamide *see* 9469

2-Methyl-N-[4-nitro-3-(trifluoromethyl)phenyl]propanamide *see* 4132

3-O-Methylnoradrenaline *see* 6627

α-Methylnoradrenaline Hydrochloride *see* 6609

3-O-Methylnorepinephrine *see* 6627

α-Methylnorepinephrine Hydrochloride *see* 6609

17-Methyl-19-norpregna-4,9-diene-3,20-dione *see* 2874

17α-Methyl-19-nor-Δ⁴,⁹-pregnadiene-3,20-dione *see* 2874

17α-Methyl-Δ⁹-19-norprogesterone *see* 2874

17-Methyl-18-nor-16,28-secosolanida-5,12-diene-3β,16β-diol *see* 9850

Methylnortestosterone *see* 6629

17α-Methyl-19-nortestosterone *see* 6629

Methyloctenylamine *see* 5070

N-Methylol-7-chlortetracycline *see* 2386

Methylol Dimethylhydantoin *see* 4762

Methyl Oleate *see* 6788

Methylol Riboflavine, 6018

Methylolsulfonic Acid Sodium Salt *see* 4149

(Methyl-ONN-azoxy)methyl β-D-Glucopyranoside *see* 2705

Methyl Orange, 6019

Methyl Orthoformate *see* 6837

Methyl Orthovalerate *see* 6838

Methyl Oxalate, 6020

Methyloxan [Nippon Shoji] *see* 5902

6-Methyl-1,2,3-oxathiazin-4(3H)-one 2,2-Dioxide *see* 30

N-Methyloxazepam *see* 9074

Methyl cis-10,11-Oxido-3,11-dimethyl-7-ethyltrideca-trans,trans-2,6-dienoate *see* 5155

Methyloxirane *see* 7869

Methyloxirane Polymer with Oxirane *see* 7536

Methyl Oxirane Polymers, Polymer with Oxirane *see* 7537

(2R-cis)-(3-Methyloxiranyl)phosphonic Acid *see* 4169

(2R-cis)-(3-Methyloxiranyl)-phosphonic Acid Compd with 2-Amino-2-(hydroxymethyl)-1,3-propanediol (1:1) *see* 4169

2-(3-Methyl-1-oxobutyl)-1H-indene-1,3(2H)-dione *see* 5123

N-(2-Methyl-1-oxobutyl)-L-phenylalanine 1-[3-(Acetyloxy)-1-(2-amino-3-methyl-1-oxobutyl)-2-pyrrolidinyl]-2-carboxyethyl Ester *see* 2918

N-[(3-Methyl-1-oxobutyl)-L-valyl-L-valyl-4-amino-3-hydroxy-6-methylheptanoyl-L-alanyl]-4-amino-3-hydroxy-6-methylheptanoic Acid *see* 7104

Methyl-5-oxocandicidin D Methyl Ester Cyclic 15,19-Hemiacetal *see* 5733

α-Methyl-4-[(2-oxocyclopentyl)methyl]benzeneacetic Acid *see* 5462

4-[2-Methyl-4-oxo-3,3-diphenyl-4-(1-pyrrolidinyl)butyl]morpholine *see* 2933

6-Methyl-2-oxo-1,3-dithio[4,5-b]-quinoxaline *see* 6933

N-Methyl-N-(1-oxododecyl)glycine Sodium Salt *see* 4267

3-Methyl-1-(5-oxohexyl)-7-propylxanthine *see* 7823

5-Methyl-2-oxo-4-imidazolidineca-proic Acid *see* 2914

(4R-cis)-5-Methyl-2-oxo-4-imidazolidinehexanoic Acid *see* 2914

[[(3-Methyl-4-oxo-1(4H)-naphthalenylidene)amino]oxy]acetic Acid Ammonium Salt *see* 5717

N-[2-(2-Methyl-4-oxopentyl)]acrylamide *see* 2943

3-Methyl-4-oxo-2-phenyl-4H-1-benzopyran-8-carboxylic Acid 2-(1-Piperidinyl)ethyl Ester *see* 4033

[1R-(exo,exo)]-8-Methyl-3-[(1-oxo-3-phenyl-2-propenyl)oxy]-8-azabicyclo[3.2.1]octane-2-carboxylic Acid Methyl Ester *see* 2304

3-Methyl-4-oxo-5-piperidino-Δ²,ᵅ-thiazolidineacetic Acid Ethyl Ester *see* 3845

[3-Methyl-4-oxo-5-(1-piperidinyl)-2-thiazolidinylidene]acetic Acid Ethyl Ester *see* 3845

2α-Methyl-17β-(1-oxopropoxy)-5α-androstan-3-one *see* 3436

4-Methyl-3-[[1-oxo-2-(propylamino)-propyl]amino]-2-thiophenecarboxylic Acid Methyl Ester *see* 1878

(17β)-17-Methyl-17-(1-oxopropyl)-estra-4,9-dien-3-one *see* 7796

3-Methyl-4-[(1-oxopropyl)phenylamino]-1-(2-phenylethyl)-4-piperidinecarboxylic Acid Methyl Ester *see* 5437

10-(2-Methyl-1-oxo-2-pyrrolidinylethyl)phenothiazine Methobromide *see* 7885

[2S-[2α(E),3β,4β,5α(2E,4S*,5R*)]]-9-[[3-Methyl-1-oxo-4-[tetrahydro-3,4-dihydroxy-5-(5-hydroxy-4-methyl-2-hexenyl)-2H-pyran-2-yl]-2-butenyl]oxy]nonanoic Acid *see* 7933

[2S-[2α(E),3β,4β,5α[2R*,3R*(1R*,-2R*)]]]-9-[[3-Methyl-1-oxo-4-[tetrahydro-3,4-dihydroxy-5-[[3-(2-hydroxy-1-methylpropyl)oxiranyl]-methyl]-2H-pyran-2-yl]-2-butenyl]oxy]nonanoic Acid *see* 6212

[2S-[2α[E(E)],3β,4β,5α[2R*,3R*-(1R*,2R*)]]]-9-[[3-Methyl-1-oxo-4-[tetrahydro-3,4-dihydroxy-5-[[3-(2-hydroxy-1-methylpropyl)oxiranyl]methyl]-2H-pyran-2-yl]-2-butenyl]oxy]-4-nonenoic Acid *see* 7933

Methylparaben, 6021

Methylparafynol *see* 5731

Methyl Parasept [Tenneco] *see* 6021

Methyl Parathion, 6022

Methylpartricin *see* 5733

N-Methylpelletierine *see* 7015

2-Methyl-2,4-pentanediol *see* 4631

4-Methylpentanenitrile *see* 4999

3-Methyl-3-pentanol Carbamate *see* 3520

4-Methyl-2-pentanone *see* 5095

4-Methyl-3-penten-2-one *see* 5811

4-Methyl-3-penten-2-one (1-Phthalazinyl)hydrazone *see* 1457

3-Methyl-2-(2-pentenyl)-2-cyclopenten-1-one *see* 5141

3-Methyl-2-penten-4-yn-1-ol *see* 7088

3-Methyl-3-pentyl Carbamate *see* 3520

5-Methyl-2-propylphenol *see* 643

Methylpentynol *see* 5731

3-Methyl-1-pentyn-3-ol *see* 5731

Methylpentynol Carbamate *see* 5732

3-Methyl-1-pentyn-3-ol Carbamate *see* 5732

3-Methyl-1-pentyn-3-yl Acid Phthalate *see* 7348

Methylperidol *see* 6170

Methylperone *see* 5707

(15R)-15-Methyl-PGE₂ *see* 796

(15S)-15-Methyl PGF₂ₐ *see* 1829

5-Methyl-5-(3-phenanthrenyl)-2,4-imidazolidinedione *see* 6023

5-Methyl-5-(3-phenanthryl)hydantoin, 6023

5-Methylphenazinium Methyl Sulfate *see* 6024

N-Methylphenazonium Methosulfate, 6024

4-Methyl-α-phenethyl Alcohol *see* 3226

dl-α-Methylphenethylamine *see* 616

d-α-Methylphenethylamine Sulfate *see* 2932

7-[2-[(α-Methylphenethyl)amino]-ethyl]theophylline *see* 3919

3-[(α-Methylphenethyl)amino]propionitrile *see* 3939

N-[2-(Methylphenethylamino)propyl]-propionanilide *see* 2966

(α-Methylphenethyl)hydrazine *see* 7197

N-(α-Methylphenethyl)nicotinamide *see* 7172

N-(α-Methylphenethyl)-2-phenylglycinonitrile *see* 617

(−)-cis-3-Methyl-1-phenethyl-4-(N-phenylpropionamido)isonipecotic Acid Methyl Ester *see* 5437

N-[1-(α-Methylphenethyl)-4-piperidyl]propionanilide *see* 5992

Methylphenidan *see* 6025

Methylphenidate, 6025

Methyl Phenidylacetate *see* 6025

Methylphenobarbital *see* 5742

2-Methylphenol *see* 2580

3-Methylphenol *see* 2579

α-Methyl-4-(2-thienylcarbonyl)benz-
eneacetic Acid see 8985
γ-Methylthio-α-aminobutyric Acid
see 5896
2-Methylthio-4,6-bis(isopropylamino)-s-triazine see 7800
2-Methylthio-4,6-bis(monoethylamino)-s-triazine see 8488
N-Methylthiocarbamoyl-N'-[(1-meth-
ylallyl)thiocarbamoyl]hydrazine see
5857
6-Methylthiochroman-7-sulfonamide
1,1-Dioxide see 6058
Methyl Thiocyanate, 6047
4-Methylthio-3,5-dimethylphenyl
N-Methylcarbamate see 5893
2-Methylthio-4-ethylamino-6-iso-
propylamino-s-triazine see 402
O-[[N-[N'-(1-Methylthioethylidene-
iminooxycarbonyl)-N'-methylami-
nosulfenyl]-N-methylcarbamoyl]]-S-
methylacetohydroximate see 9258
5-(2-Methylthioethyl)-5-(1-methylbu-
tyl)-2-thiobarbituric Acid Sodium
Salt see 5901
5-(2-Methylthioethyl)-5-(2-pentyl)-2-
thiobarbituric Acid Sodium Salt see
5901
5-[(Methylthio)methyl]-3-[[(5-nitro-
2-furanyl)methylene]amino]-2-oxa-
zolidinone see 6447
5-[(Methylthio)methyl]-3-[(5-nitro-
furfurylidene)amino]-2-oxazolidi-
none see 6447
8-[(Methylthio)methyl]-6-propylergo-
line see 7115
Methylthioninium Chloride see 5979
1-(Methylthio)propyl 1-Propenyl Disulfide see 846
Methylthiouracil, 6048
4-Methyl-2-thiouracil see 6048
6-Methyl-2-thiouracil see 6048
1-Methyl-3-(9H-thioxanthen-9-yl-
methyl)piperidine see 5902
4-(Methylthio)-3,5-xylyl Methylcarbamate see 5893
D,L-α-Methylthyroxine Ethyl Ester
see 3828
5-Methyltocol see 9420
7-Methyltocol see 9423
8-Methyltocol see 9419
1-Methyl-5-p-toluoylpyrrole-2-acetic
Acid see 9441
N-Methyl-2-[α-(2-tolylbenzyl)oxy]-
ethylamine see 9426
Methyl 3-(m-Tolylcarbamoyloxy)-
phenylcarbamate see 7199
Methyl-p-tolylcarbinol see 3226
Methyl p-Tolylcarbinol Camphorate,
Diethanolamine Salt see 9415
2-Methyl-6-p-tolyl-2-hepten-4-one
see 9733
2-Methyl-3-o-tolyl-4(3H)-quinazoli-
none see 5872
2-Methyl-3-o-tolyl-6-sulfamyl-7-
chloro-1,2,3,4-tetrahydro-4-quin-
azolinone see 6068
6-Methyl-1,2,4-triazine-3,5(2H,4H)-
dione see 919
α-Methyl-N-(2,2,2-trichloroethyli-
dene)benzeneethanamine see 613
α-Methyl-N-(2,2,2-trichloroethyli-
dene)phenethylamine see 613
2-Methyl-4-(2,2,2-trichloro-1-hydr-
oxyethoxy)-2-pentanol see 2089
N-Methyl-N'-(2,2,2-trichloro-1-
hydroxyethyl)urea see 5958
α-Methyltricyclo[3.3.1.1^{3,7}]decane-1-
methanamine see 8221
Methyltrienolone, 6049

2-Methyl-3-(β,β,β-trifluoroethylthio-
methyl)-6-chloro-7-sulfamyl-3,4-
dihydro-1,2,4-benzothiadiazine 1,1-
Dioxide see 7561
2-(2-Methyl-3-trifluoromethylani-
lino)nicotinic Acid see 4073
2-[[α-Methyl-m-(trifluoromethyl)-
phenethyl]amino]ethanol Benzoate
(Ester) see 1047
(±)-N-Methyl-γ-[4-(trifluoromethyl)phenoxy]benzenepropanamine see
4112
N-Methyl-3-(p-trifluoromethylphenoxy)-3-phenylpropylamine see 4112
2-[[2-Methyl-3-(trifluoromethyl)-
phenyl]amino]-3-pyridinecarboxylic
Acid see 4073
2-[[1-Methyl-2-[3-(trifluoromethyl)-
phenyl]ethyl]amino]ethanol Benzoate (Ester) see 1047
N-[4-Methyl-3-[[(trifluoromethyl)-
sulfonyl]amino]phenyl]acetamide
see 4089
1-Methyl-3-[4-[p-[(trifluoromethyl)-
thio]phenoxy]-m-tolyl]-s-triazine-
2,4,6(1H,3H,5H)-trione see 6055
16-Methyl-1,11α,16RS-trihydroxy-
prost-13E-en-9-one see 8225
N-Methyl-N-(2,4,6-triiodo-3-amino-
phenyl)glutaramidic Acid see 4941
N-Methyltrilobine see 9610
Methyl 3,4,5-Trimethoxycinnamoyl
Reserpate see 8146
5-Methyl-6-[[(3,4,5-trimethoxyphen-
yl)amino]methyl]-2,4-quinazolinedi-
amine see 9635
Methyl N-Trimethyl-γ-aminobutyrate
Chloride see 1871
2-Methyl-4-(2,6,6-trimethyl-1-cyclo-
hexen-1-yl)-3-butenal see 215
Methyltrimethylene Glycol see 1566
2-Methyl-2-(4,8,12-trimethyltridec-
yl)-6-chromanol see 9416
1-Methyl-2,4,6-trinitrobenzene see
9643
6α-Methyl-11β,17α,21-triol-1,4-
pregnadiene-3,20-dione see 6028
1-Methyl-4-[3,3,3-tris(4-chlorophen-
yl)-1-oxopropyl]piperazine see 4596
1-Methyl-4-[3,3,3-tris(p-chlorophenyl)propionyl]piperazine see 4596
8-Methyltropinium Bromide 2-Prop-
ylvalerate see 701
N-Methyl-L-tryptophan see 5
α-Methyltyramine see 4740
α-Methyl-m-tyramine see 4295
N-Methyltyrosine see 8988
α-Methyltyrosine see 6085
α-Methyl-L-tyrosine see 6085
α-Methyl-m-tyrosine, 6050
α-Methyl-p-tyrosine see 6085
(2-O-Methyltyrosine)deamino-1-car-
baoxytocin see 1800
4-Methylumbelliferone see 4792
β-Methylumbelliferone see 4792
4-Methylumbelliferone O,O-Diethyl
Phosphorothioate see 4793
4-Methyluracil see 6051
5-Methyluracil see 9332
6-Methyluracil, 6051
Methylurethane see 5958
Methylustin see 6440
trans-8-Methyl-N-vanillyl-6-nonen-
amide see 1767
N-Methyl-β-vinyldiacetonalkamine
Mandelate see 3854
Methyl Vinyl Ketone, 6052
13α-Methyl-13-vinylpodocarp-8(14)-
ene-15-oic Acid see 7398
13β-Methyl-13-vinylpodocarp-7-ene-
15-oic Acid see 5086

Methyl Viologen (2+) see 6980
Methyl Vitamin B$_{12}$ see 9921
1-Methyl-2-(2,6-xylyloxy)ethylamine
see 6094
N-Methyl-N'-2,4-xylyl-N-(N-2,4-
xylylformimidoyl)formamidine see
503
Methyl Yellow see 3218
Methylzineb see 7831
Methymycin, 6053
Methypranol see 6059
Methyprylon, 6054
Methyridine see 6084
Methysergid(e), 6055
Methysergide Maleate see 6055
Methysticin, 6056
Metiazic Acid see 6057
Metiazinic Acid, 6057
Meticlorpindol see 2394
Meticortelone [Schering] see 7719
Meticortelone Soluble [Schering] see
7722
Meticorten [Schering] see 7727
Meticrane, 6058
Metidione see 5861
Metifex [Albert Roussel] see 3668
Metiguanide [Erba] see 5845
Metilar see 6977
Metilcaf [Ital Suisse] see 8408
Metildigoxin see 3150
Metildiolo see 5861
Metilenbiotic [Coli] see 5851
Metilon [Daiichi] see 3358
Metilsulfate see MISC-4
Metina [Pierrel] see 1856
Metione see 5896
Metipranolol, 6059
Metirosine see 6085
Metizoline, 6060
Metligine [Taisho] see 6107
Metmercapturon see 5893
Metobromuron, 6061
Metochalcone, 6062
Metoclol [Toyama] see 6063
Metoclopramide, 6063
Metocobil [Vita] see 6063
Metocryst see 5861
Metocurine Iodide, 6064
Metodik [Labinca] see 878
Metofane [Pitman-Moore] see 5914
Metofenazate, 6065
Metofoline, 6066
Metol see 5940
Metolachlor, 6067
Metolazone, 6068
Metolquizolone see 5872
Metomidate, 6069
Meton see 4609
Metopimazine, 6070
Metopirone [Ciba] see 6083
Metopon, 6071
Metoprolol, 6072
Metopyrone see 6083
Metoquine see 8053
Metoquinone, 6073
Metoros [Geigy] see 6072
Metoryl see 4639
Metoserpate, 6074
Metosomin [Maruishi] see 9447
Metosyn [ICI] see 4077
Metox [Steinhard] see 6063
Metoxadone see 5739
Metoxidon see 8876
Metracin [EF] see 2279
Metralindole, 6075
Metramac see 502
Metramid [Nicholas] see 6063
Metranil see 7066
Metrasil see 8900
Metrazole see 7097
Metreton [Schering] see 7721

Metribuzin, 6076
Metrifonate see 9536
Metrisone see 6028
Metrizamide, 6077
Metrizoate Sodium see 6078
Metrizoic Acid, 6078
Metrodin [Serono] see 4189
Metrodiol [Clin-Comar-Byla] see 3816
MetroGel [Curatek] see 6079
Metrolag [Lagap] see 6079
Metrolyl [Lagap] see 6079
Metron see 6022
Metronidazole, 6079
Metron (Japanese) see 6080
Metron S, 6080
Metropin [Benzon] see 3724
Metropine [Pennwalt] see 891
Metroprione see 6083
Metrulen see 3816
Metsol [Muro] see 5836
Metsulfuron Methyl, 6081
Metubine Iodide [Lilly] see 6064
Metuclazepam see 5824
Meturedepa, 6082
Metycaine [Lilly] see 7444
Metyrapone, 6083
Metyridine, 6084
Metyrosine, 6085
Meusicort see 4710
Mevacor [Merck & Co.] see 5460
Mevaldic Acid, 6086
Mevalolactone see 6087
Mevalon [Guidotti] see 5690
Mevalonic Acid, 6087
Mevalonic Lactone see 6087
Mevalotin [Sankyo] see 7712
Mevasine [Merck & Co.] see 5654
Mevastatin, 6088
Mevinacor [Merck & Co.] see 5460
Mevinolin see 5460
Mevinphos, 6089
Mevlor [Merck & Co.] see 5460
Mexacarbate, 6090
Mexamine see 5926
Mexate [Bristol] see 5908
Mexazolam, 6091
Mexenone, 6092
Mexephenamide see 5682
Mexicain, 6093
Mexican Scammony (Root) see 4959
Mexican Tea see 6713
Mexiletine, 6094
Mexitil [Boehringer, Ing.] see 6094
Mexocine [Specia] see 2872
Meyerhofferite see 1651
Mezaton see 7257
Mezcaline see 5808
Mezepan [Hosbon] see 5669
Mezereon see 6095
Mezereum, 6095
Mezineb see 7831
Mezlin [Miles] see 6096
Mezlocillin, 6096
Mezolin [Merck & Co.] see 4874
MG 143 see 8964
MG 322 see 3307
MG 500 see 4163
MG 624 see 8776
MG 652 see 6868
MG 1559 see 9980
MG 2552 see 2402
MG 2555 see 1381
MG 4833 see 3912
MG 5454 see 4463
MG 5771 see 9980
MG 8926 see 3438
MG 13054 see 3941
MG 13608 see 3410
MGA [Upjohn] see 5697
MGK-11 see 8114

MGK Repellent 11 [McLaughlin] see 8114
Mg 5-Sulfat [Artesan] see 5573
M. H. see 5804
MH see 5587
MH 532 see 7229
MHA [Monsanto] see 5897
MHIP see 9439
MHPG see 5916
Mi 85 see 758
Mi 85Di see 758
Miacalcic see 1640
Miadone see 5852
Mianine see 2066
Mianserin, 6097
Miaquin [Parke, Davis] see 602
Miarsenol see 6361
Miazine see 7998
Miazole see 4828
Mibolerone, 6098
MIC see 6004
Micatex see 9524
Micatin [J & J] see 6101
Michler's Base, 6099
Michler's Ketone, 6100
Michrome No. 24 see 211
Michrome No. 66 see 1955
Michrome No. 226 see 6395
Micinovo [Andreu] see 5830
Micloretin see 2068
Micoclorina see 2068
Micofugal [Biopharma]] see 3476
Micofur [Norwich] see 6451
Micogin [Crosara] see 3476
Micol see 2018
Miconal Ecobi [Ecobi] see 6101
Miconazole, 6101
Micoren see 2595 and 2598
Micoserina see 2758
Micoxolamina [Domp] see 2270
Micranthine, 6102
Micrest [Beecham] see 3118
Micridium Chloride [J & J] see 7157
Micrin see 2895 and 2896
Microbar [Max Ritter] see 1006
Microbin see 2126
Micro-Cell see 1709
Microcetina see 2068
Microcide see 4354
Microcidin see 6304
Microcillin [Bayer] see 1796
Microcline see 7579
Micrococcin see 6103
Micrococcin P, 6103
Microcosmic Salt see 581
Micro-Dee [Diamond] see 9929
Micro K [E. Merck] see 7601
Microlite see 9025
Microlut [Schering AG] see 6621
Micromerol see 9802
Micromet see 8621
Micromite [Uniroyal] see 3128
Micronase [Upjohn] see 4372
Micronett [J & J] see 6614
Micronomicin, 6104
Micronor [Ortho] see 6614
Micronovum [Cilag-Chemie] see 6614
Micropaque [Nicholas] see 1006
Micropenin [Kabi] see 6858
Microtin [AB Leo] see 1978
Microtrast [Nicholas] see 1006
Microtrim [Chephasaar] see 8889
Microtubules see 9718
Microval [Wyeth] see 6621
Microx [Pennwalt] see 6068
Mictine [Searle] see 463
Mictone [Werner] see 1207
Mictrol [KabiVitrum] see 9098
Micturin [KaviVitrum] see 9098
Micturol [KabiVitrum] see 9098
Midafenone see 6583

Midalgan see 6014
Midamine [Roberts] see 6107
Midamor [Merck & Co.] see 417
Midantan see 380
Midarine see 8846
Midazolam, 6105
Midécacine [Clin Midy] see 6106
Midecamycin A$_1$ see 6106
Midecamycin A$_3$ see 6106
Midecamycins, 6106
Midecin [Farmaka] see 6106
Midelid [Beta] see 7785
Midicel [Parke, Davis] see 8890
Midikel see 8890
Midodrine, 6107
Midol [Glenbrook] see 2299
Midoxin [Minerva] see 3429
Midronal [Delalande] see 2308
Midsummer Daisy see 4013
Mielucin [Simes] see 1494
MIF see 5693
MIF-I see 5693
Mifegyne [Roussel-UCLAF] see 6109
Mifentidine, 6108
Mifepristone, 6109
Miforon see 3187
Mifurol [Mitsui] see 1851
Miglitol, 6110
Migristène [Specia] see 4146
MIH see 7764
Mijal [Juste] see 1521
Mikametan [Mikasa] see 4874
Mikamycin, 6111
Mikamycin A see 6111 and 9910
Mikamycin B see 6111
Mikamycin I$_A$ see 6111
Mikavir [Salus] see 416
Mikedimide [Panray] see 1036
Mikelan [Otsuka] see 1875
Miketorin [Mitsui] see 504
Mikrotsid see 4354
Milban [Mallinckrodt] see 3405
Milbedoce see 9921
Milbemycin D see 6112
Milbemycin α_1 see 6112
Milbemycin β_1 see 6112
Milbemycins, 6112
Milchsäure (German) see 5215
D-Milchsäure (German) see 5214
L-Milchsäure see 5216
Mil-Col [ICI] see 3432
Milcurb [ICI] see 3206
Milcurb Super [ICI] see 3695
Mild Aconitine see 114
Mildex see 3281
Mildiomycin, 6113
Mildison [Brocade] see 4710
Mildmen [Icar] see 2082
Mild Mercury Chloride see 5795
Mild Protargin see 8474
Mild Silver Protein see 8474
Milestrol [SM & P] see 3118
Milezin [Spofa] see 5909
Milfoil see 102
Milfoil Oil see 6782
Milgo [ICI] see 3695
Milhéparine [Millot] see 3663
Milibis [Winthrop] see 4389
Milid [Rotta] see 7785
Milide [Beytout] see 7785
Milk, 6114
Milk Ipecac see 773
Milk Sugar see 5221
Milkweed see 854
Millafol [Parke, Davis] see 4140
Millerite see 6406
Miller Nu Set (Hormone Spray) see 8483
Millevit [Nordmark] see 9921
Millicaine [Corbiere] see 1210
Millicorten [Ciba] see 2922

Cross Index of Names

Milligynon [Schering] see 6614
Millinese [Lampugnani] see 2187
Millithrol [Nippon Kayaku] see 6528
Millon's Base see 5801
Millophyline see 3663
Miloderme [Schering-Essex] see 213
Milogard [Geigy] see 7822
Milontin [Parke, Davis] see 7231
Milorganite®, 6115
Miloxacin, 6116
Milrinone, 6117
Milstem [ICI] see 3695
Miltaun [Mack, Illert.] see 5751
Miltax [Watanabe] see 5184
Miltown [Wallace Labs.] see 5751
Mimedran [Esteve] see 8976
Mimetite see 5267
Mimosine, 6118
Minacalm [Tobishi] see 9447
Minacide see 7794
Mina D₂ see 9928
Minalfene [Bouchara] see 298
Minaphil see 477
Minaprine, 6119
Mincard [Searle] see 463
Mindiab [Farmitalia] see 4337
Mindolic Acid see 2381
Minedil [Formenti] see 4659
Minelcin [Parke, Davis] see 1090
Minelsin [Parke, Davis] see 1090
Mineral Blue see 3964
Mineral Carbon see 4444
Mineral Green see 2629
Mineral Oil see 7139 and 7140
Mineral Orange see 5307
Mineral Pitch see 867
Mineral Red see 5307
Mineral Spirits, 6120
Mineral Wax see 1984
Mineral White see 1713
Minetoin see 7293
Minicid see 3169
Minidiab [Erba] see 4337
Minihep [Leo Pharm.] see 4571
Miniluteolas [Serono] see 3816
Minilyn [Organon] see 5501
Minims [Smith & Nephew] see 7721
Mini-Pe [Astra-Syntex] see 6614
"Mini-pill" see 6614
Miniplanor [Kodama] see 278
Minipress [Pfizer] see 7715
Minirin [Ferring] see 2904
Minitran [3M Riker] see 6528
Minium see 5307
Minocin [Lederle] see 6121
Minocycline, 6121
Minocyn [Lederle] see 6121
Minolip [Master] see 1047
Minomycin [Lederle] see 6121
Minovlar [Schering] see 6614
Minoxidil, 6122
Minoximen [Menarini] see 6122
Minozinan see 5909
Minprog [Upjohn] see 7892
Minprostin E2 [Upjohn] see 7893
Mint see 8694
Mintacol see 6979
Mintec [SKF] see 6758
Mintezol [Merck & Co.] see 9217
Mintic [Ayerst] see 6084
Minulet [Wyeth] see 4308
Minuric see 1073
Minus see 231
Minus [Federal] see 7180
Minusin [Doetsch, Grether] see 6634
Minzil see 2169
Minzolum [Sharp & Dohme] see 9217
Mioblock [Sankyo] see 6958
Miocaina see 4465
Miocamycin [Meiji] see 6123
Miocard [Millot-Roux] see 497

Miochol [Cooper] see 81
Miodar see 7292
Miodaron [Alcon] see 497
Miokamycin, 6123
Miokon Sodium [Mallinckrodt] see 3353
Miolaxene [Lepetit] see 5903
Miolene [Lusofarmaco] see 8228
Miolisodal see 1848
Mional [Eisai] see 3558
Miopropan [Bernabo] see 9613
Mioril see 1848
Mio-Sed see 2105
Miostat [Alcon] see 1780
Mioticol see 3183
Miotisal A see 6979
Miotolon [Daiichi] see 4209
Mioton [Pharmakhim] see 1767
Mipafox, 6124
Mipax see 3243
Mirabilite see 8636
Miracil D see 5463
Miracle Fruit see 6125
Miracol see 5463
Miraculin, 6125
Miradol [Mitsui] see 8971
Miradon [Schering] see 698
Miral [Ciba-Geigy] see 4988
Miralin [Miralin] see 6125
Miranax [Astra-Syntex] see 6337
Mirapront [Mack, Illert.] see 7232
Mirbanil [Boehringer, Ing.] see 8971
Mircol [Pharmuka] see 5754
Miretilan [Sandoz] see 3532
Mirex, 6126
Mirfat [Merckle] see 4221
Miridacin [Taiho] see 7784
Mirion see 5886
Mirlon see 6656
Miromorfalil see 6275
Mirontin [Parke, Davis] see 7231
Miropinic Acid see 5086
Miroprofen, 6127
Mirsol [Permamed] see 10074
Mirvan A [Cont. Pharma] see 212
Misch Metal, 6128
Miso see 8684
Misoprostol, 6129
Mistabron [UCB] see 5812
Mistabronco [UCB] see 5812
Mistarel see 5105
Mistral [M & B] see 3938
Misulban [Techni-Pharma] see 1494
Misulvan [Bernabo] see 8971
Mitaban [Upjohn] see 503
Mitac [Boots] see 503
Mitanoline [Toyo Pharmar] see 7747
Mitarson [Asta] see 2853
Mithracin [Pfizer] see 7510
Mithramycin see 7510
Mitigal see 5821
Mitigan [Makhteshim-Agan] see 3075
Mitis Green see 2629
Mitobronitol, 6130
Mitocin-C [Bristol] see 6133
Mitocor [Zambon] see 9751
Mitoguazone, 6131
Mitolac [Bristol-Myers] see 6132
Mitolactol, 6132
Mitomen see 5656
Mitomycin A see 6133
Mitomycin B see 6133
Mitomycin C see 6133
Mitomycins, 6133
Mitopodozide see 7519
Mitoquinones see 9751
Mitosan see 1494
Mitotane, 6134
Mitox see 2074
Mitoxana [W. B. Pharm.] see 4822
Mitoxantrone, 6135

Mitragynine, 6136
Mitramycin see 7510
Mitran see 6856
Mitrinermine see 8198
Mitrolan [Robins] see 1704
Mitronal [Searle] see 2308
Mixed Rare Earth Metal see 6128
Mizoribine, 6137
MJ 505 see 7292
MJ 1992 see 8683
MJ 1996 see 409
MJ 1999 see 8682
MJ 5190 see 409
MJ 9067 see 3525
MJ 10061 see 1073
MJ 13805 see 4297
MJF 9325 see 4822
MJF 11567-3 see 1921
MJF 12264 see 9060
MJF 12637 see 8967
MK-130 see 2719
MK-135 see 2205
MK-188 see 10022
MK-191 see 7484
MK-208 see 3881
MK-231 see 8961
MK-240 see 7917
MK-264 see 4138
MK-302 see 816
MK-306 see 1938
MK-351 see 5974
MK-360 see 9217
MK-366 see 6617
MK-401 see 2405
MK-421 see 3521
MK-422 see 3522
MK-458 see 6349
MK-486 see 1802
MK-521 see 5393
MK-538 see 7565
MK-595 see 3669
MK-621 see 3491
MK-647 see 3130
MK-650 see 2536
MK-733 see 8491
MK-781 see 6085
MK-787 see 4834
MK-791 see 2275
MK-801 see 3392
MK-803 see 5460
MK-870 see 417
MK-905 see 1733
MK-906 see 7888
MK-933 see 5133
MK-936 see 1
MK-950 see 9374
MK-955 see 4169
MK-965 see 858
MK-990 see 8121
ML 236 B see 6088
ML 1024 see 9211
MM 4450 see 6795
MM 13902 see 6795
MM 14151 see 2342
MM 17880 see 6795
MM 22380 see 6795
MM 22381 see 6795
MM 22382 see 6795
MM 22383 see 6795
MMC see 6133
MMH see 6001
4 MMPD see 5920
4-MMPDS see 5920
MMS see 6010
MMSC see 9942
α-MMT see 6050
MN-1695 see 4982
MNE see 6404
MNNG see 6017
MNPA see 6337
MN-10T see 9096

MO-911 see 6988
MOB see 6907
Moban [Endo] see 6142
Mobenol see 9432
Mobidin [Ascher] see 5565
Mobiflex [Roche] see 9080
Mobilan [Galen] see 4874
Mobisyl [Ascher] see 9581
Mobutazon see 6141
Mobuzon see 6141
MOCA see 5978
Mocap [Mobil] see 3702
Mocimycin, 6138
Moclobemide, 6139
Moctanin [Ascot] see 6162
Mod [Irbi] see 3412
Modacin [Tanabe] see 1944
Modacor [I.S.H.] see 6915
Modacrylic Fibers, 6140
Modalina see 9594
Modamide [Merck & Co.] see 417
Modane [Adria] see 2813
Modatrop [Nordmark] see 7228
Modecate [Squibb] see 4116
Modenol see 1519
Moderatan [Thranol] see 3116
Moderil [Pfizer] see 8146
Modicard [Riker] see 4034
Modicare see 3411
Modicon [Ortho] see 6614
Modified Dakin's Soln see 8578
Modimmunal [Ravizza] see 4881
Modinal see 4266
Modirax see 4616
Moditen (Tabl. or Elixir) [Squibb] see 4116
Moditen Enanthate [Squibb] see 4116
Moditen-Retard [Squibb] see 4116
Modown [Mobil/Rhône-Poulenc] see 1228
Modrasone [Kirby-Warrick] see 213
Modrastane [Sterling] see 9611
Modrenal [Winthrop] see 9611
Modulan see 5231
Modulor see 3116
Modumate [Abbott] see 806
Moduretic [Merck & Co.] see 417
Moduretik [Frosst] see 417
Modustatina [Midy] see 8671
Modutrol [Reed & Carnrick] see 7441
Moebiquin [Consolidated Midland] see 4935
Moenocinol see 962
Moenomycin see 962
Moenomycin A see 962
Mofebutazone, 6141
Mofedione see 6882
Mofenar [C.E.P.A.] see 1462
Mofesal [Medice] see 6141
Mogadan [Roche] see 6491
Mogadon [Roche] see 6491
Moheptan see 5852
Mohr's Salt see 546
Mohrus [Hisamitsu] see 5184
Molatoc see 3398
Molcer [Wallace Labs.] see 3398
Moldamin see 7037
Molded Silver Nitrate see 8465
Molecular Sieves see 2341
Molevac [Parke, Davis] see 8034
Molindone, 6142
Molipaxin [Roussel] see 9495
Mol-Iron [Schering] see 4006
Mol-Iron (obsolete) see 6147
Molivate [Glaxo] see 2362
Mollinox see 5872
Molofac see 3398
Moloid [Südmedica] see 5630
Molsidolat [Hoechst] see 6143
Molsidomine, 6143
Molybdenite see 6145

Molybdenum, 6144
Molybdenum Disulfide, 6145
Molybdenum Hexafluoride, 6146
Molybdenum Sesquioxide, 6147
Molybdenum Trioxide, 6148
Molybdic(VI) Acid, 6149
Molybdic Acid, 85%, 6150
Molybdic Anhydride see 6148
Molybdophosphoric Acid see 7316
Molycor-R [Mepha] see 6616
Molyhibit 100 [Amax] see 8596
Molysite see 3961
MOM see 6123
Momentol [Med. Prod. Quim.] see 8889
Momentum [Much] see 40
Mometasone Furoate, 6151
Momicine [Morrith] see 6106
MON-0573 see 4408
MON-2139 see 4408
MON-4621 see 411
Monacetin, 6152
Monacolin K see 5460
Monacrin [Winthrop] see 418
Monactin see 6594
Monalium Hydrate [Ayerst] see 5527
Monapen [Fujisawa] see 9360
Monarch [SS Pharm.] see 278
Monarda, 6153
Monardaein Chloride see 6154
Monardein Chloride, 6154
Monardin see 7014
Monargan [Evans] see 44
Mon-Arsone see 3677
Monase [Upjohn] see 3848
Monasirup [Arznei Mller-Rorer] see 675
Monaspor [Ciba] see 1943
Monazan see 6141
Monazol see 4771
Mondamin [CPC] see 8757
Mondus [Labinca] see 4070
Monel®, 6155
Monellins, 6156
Monensic Acid (obsolete) see 6157
Monensin, 6157
Monetite see 1699
Monicor [Fabre] see 5114
Moniflagon see 9079
Monistat [Cilag-Chemie] see 6101
Monit [Stuart] see 5114
Monitan [Ives] see 7559
Monitor [Chevron] see 5858
Monkshood see 111
Monoacetin see 6152
Monoacetylneriifolin see 6386
N^1-Monoacetyl Sulfisoxazole see 100
Monoamine Oxidase, 6158
Mono-Attritin [Atmos] see 4812
Monobactam see 938 and 8924
Monobasic Lead Acetate see 5301
Mono-Baycuten see 2412
Monobenzone, 6159
Monobenzyl Hydroquinone see 6159
Monobromoacetanilide see 1387
Monobromobenzene see 1394
Monobromoethane see 3730
2-Monobromoisovalerylurea see 1385
Monobromomethane see 5951
Monobromosuccinic Acid see 1427
Monobutyl see 6141
Monocaine [Novocol] see 1517
Monocalcium Orthophosphate see 1700
Monocalcium Phosphate see 1700
Monocalcium Tetrasodium Bis[penta-aquatetra-μ-hydroxy[D-gluconato-(4−)]dioxotriferrate(3−)] see 3971
α-Monocaprylin see 6162
Mono-Cedocard [Tillotts] see 5114
Monochloracetone see 2113

Monochlorethane see 3740
Monochloroacetaldehyde see 2108
Monochloroacetic Acid see 2111
Monochloroacetic Acid Anhydride see 2112
Monochloroacetone see 2113
17-Monochloroacetylajmaline see 5454
Monochlorobenzene see 2121
α-Monochlorohydrin see 2145
Monochloromethyl Ether see 2146
Monochlorphenamide see 2371
Monocid [SK & F] see 1932
Monocidur [SK & F] see 1932
Monoclair [Hennig] see 5114
Monoclate [Rorer] see 3873
Monoclonal Antibodies see 4837
Monocor [Am. Cyanamid] see 1309
Monocortin [Grünenthal] see 6977
Monocortin S [Grünenthal] see 6977
Monocron [Makhteshim-Agan] see 6161
Monocrotaline, 6160
Monocrotophos, 6161
Monodie [Schering AG] see 4308
Mono[2-(dimethylamino)ethyl]phosphoric Acid Ester see 2870
Monodral Bromide [Winthrop] see 7084
5,6-Monoepoxyvitamin A see 4572
Monoethanolamine see 3681
Monoethanolamine DL-Acetylleucinate see 89
Monoethanolamine Nicotinate see 6437
Monoethanolamine Salt of α-Acetamidoisocaproic Acid see 89
Monoethylamine see 3718
Monoethyl Tartrate see 3812
Monoferrous Acid Citrate Monohydrate see 3993
Monofluoroacetamide see 4095
Monofluorophytosterols see 7364
Mono-Gesic [Central] see 8307
Monoglycerol p-Aminobenzoate see 4382
Mono-Glycocard [Rheingold] see 3146
Monoglycol Salicylate see 4395
Monoglyme see 3213
Monogynol B see 5478
Monohydrated Selenium Dioxide see 8381
Monohydroxyaluminum Bis(acetylsalicylate) see 328
Monohydroxyaluminum Diacetylsalicylate see 328
Mono(2-hydroxybenzoate)-1H-imidazole see 4829
Monohydroxymercuridiiodoresorcinsulfonphthalein Sodium Salt see 5755
Mono-[2-hydroxy-5-[[(4-methylphenyl)sulfonyl]oxy]benzenesulfonate]-piperazine see 8976
Monoiodotyrosine see 4937
Mono-Kay see 9933
Monoket [Chiesi] see 5114
Monolene see 450
Monolupine see 660
Mono Mack [Mack, Illert.] see 5114
Monomestrol see 5818
Monomethylamine see 5938
Monomethyl-p-aminophenol Sulfate see 5940
Monomethylaniline see 5941
Monomethylarsinic Acid see 5864
Monomethyldiaminodiphenazothionium Chloride see 942
o-Monomethyldiphenhydramine see 6831

Monomethylhydrazine see 6001
Monomethyl Mercury see 3238
Monomethyloldimethylhydantoin see 4762
Monomethyl Sulfate see 6041
Monomethylthionine Chloride see 942
Monomycin [Grünenthal] see 3626
Monomycin A see 6989
Mononitrogen Monoxide see 6498
Mononuclear Cell Factor see 4895
Monooctadecyl Ether of Glycerol see 1021
Monooctanoin, 6162
Monopar [Burroughs Wellcome] see 8773
Monoparin [Weddel] see 4571
Monopen see 7041
Monophen®, 6163
Monophenolase see 9746
Monophenoloxidase see 9746
Monophenylbutazone see 6141
Monophos see 616
Monopotassium Citrate see 7604
Monopotassium Phosphate see 7648
Monopril [Squibb] see 4171
Monopropionylerythromycin see 3631
Monorden, 6164
Monores [Valeas] see 2347
Monorhein see 8175
Monorheumetten [Nadrol] see 6141
Monosilane see 8435
Monosodium Glutamate, 6165
Monosodium L-Glutamate Monohydrate see 6165
Monosodium Methanearsonate see 5864
Monosodium Orthophosphate see 8613
Monosodium N-Phenylglycinamide-p-arsonate see 9703
Monosorb [Unicet] see 5114
Monospan [Santen] see 1588
Monostearin see 4384
Monosulfiram see 8928
Monotard [Novo] see 4890
Monotheamin [Lilly] see 9212
Monothioethyleneglycol see 5760
α-Monothioglycerol see 9264
Monotrim [Duphar] see 9624
Monotropein, 6166
Monotropitin see 4272
Monotropitoside see 4272
Monovent [Lagap] see 9089
Monoverin [Cascan] see 7706
Monoxone [Plant Protection] see 2111
Monoxychlorosene see 6909
Monozol see 3118
Monsel's Soln see 3979
Montamed see 8310
Montan Wax, 6167
Monteban [Lilly] see 6339
Montmorillonite, 6168
Montrel [Dow] see 2607
Montricin see 5733
Monuril [Zambon] see 4169
Monuron, 6169
Monuron TCA see 6169
Monydrin [Draco] see 7279
Monzal [Thomae] see 9875
Monzaldon see 9875
Moogrol [Burroughs Wellcome] see 2037
5-MOP see 1173
8-MOP see 5911
Mopazine see 5923
Moperone, 6170
Mopidamol, 6171
Moplen [Novamont] see 7558
Mopral [Astra] see 6800
Moprolol, 6172
MOPP see MISC-2

MOPS, 6173
Moquizone, 6174
Moracizine see 6178
D-Moramide see 2933
Moranoline see 2887
Morantel, 6175
Moranyl see 8986
Morazone, 6176
Morbam see 5751
Morbicid see 4150
Morclofone, 6177
Mordant Rouge see 339
Mordant Violet 25 see 4250
Mordant Yellow 3R see 241
Morena [Kettelhack] see 3157
Morepen [Lab. Morejon] see 621
Morestan [Bayer; Chemagro] see 6933
Morfazinamide see 6182
Morial [Takeda] see 6143
Moricizine, 6178
Morin, 6179
Morinamide see 6182
Morindin, 6180
β-Morindin see 6180
Morindone see 6180
Moringine see 1139
Morintannic Acid see 5519
Moriperan [Morishita] see 6063
Morison's Paste see 5573
Moritannic Acid see 5519
Morkit [Bayer] see 717
Mornidine see 7421
Morning-glory see 4959
Morocide [Boots] see 1237
Moronal [Heyden] see 6658
Morosan [Dolder] see 2977
Moroxydine, 6181
Morpan T [A.B.M. Chem.] see 6249
Morphazinamide, 6182
Morphenol, 6183
Morpheridine, 6184
Morphia see 6186
Morphina see 6186
Morphinan, 6185
Morphinan-3-ol see 6625
Morphine, 6186
Morphine Bis(nicotinate) see 6430
Morphine Bis(pyridine-3-carboxylate) see 6430
Morphine Dinicotinate see 6430
Morphine Ester with Nicotinic Acid see 6430
Morphine Hydrobromide, 6187
Morphine Hydrochloride, 6188
Morphine Hyperduric see 6190
Morphine Methylbromide, 6189
Morphine 3-Methyl Ether see 2459
Morphine Monomethyl Ether see 2459
Morphine Mucate, 6190
Morphine Oleate, 20%, 6191
Morphine Oxide see 6192
Morphine N-Oxide, 6192
Morphine Sulfate, 6193
Morphium see 6186
Morphodone see 7160
Morpholine, 6194
4-Morpholinecarboximidoylguanidine see 6181
Morpholine, N,N'-Disulfide see 3379
4-Morpholinepropanesulfonic Acid see 6173
Morpholine Salicylate, 6195
(−)-3-Morpholino-4-(3-tert-butyl-amino-2-hydroxypropoxy)-1,2,5-thiadiazole see 9374
4-Morpholino-2,2-diphenylbutyric Acid Ethyl Ester see 3295
1-(β-N-Morpholinoethyl)-1-cyclopentenylcyclopentane-2-one see 2712

3-(2-Morpholinoethyl)morphine see 7303
N-2-Morpholinoethyl-5-nitroimidazole see 6468
Morpholinoethyl Norpethidine see 6184
1-(2-Morpholinoethyl)-4-phenylisonipecotic Acid Ethyl Ester see 6184
4-Morpholinomethylescutol see 4139
3-(Morpholinomethyl)-1-[(5-nitrofur-furylidene)amino]hydantoin see 6448
5-Morpholinomethyl-3-(5-nitrofur-furylideneamino)-2-oxazolidinone see 4205
N-Morpholinomethylpyrazinamide see 6182
7-Morpholinomethyltheophylline, 6196
3-(N-Morpholino)propanesulfonic Acid see 6173
10-(3-Morpholinopropionyl)phenothiazine-2-carbamic Acid Ethyl Ester see 6178
γ-Morpholinopropyl 4-n-Butoxyphenyl Ether see 7707
4-(Morpholinothiocarbonyl)guaiacol see 9841
6-(4-Morpholinyl)-4,4-diphenyl-3-heptanone see 7160
3-[2-(4-Morpholinyl)ethyl]morphine see 7303
β-Morpholinylethylmorphine see 7303
1-(2-N-Morpholinylethyl)-5-nitro-imidazole see 6468
1-[2-(Morpholinyl)ethyl]-4-phenyl-4-piperidinecarboxylic Acid Ethyl Ester see 6184
3-(4-Morpholinylmethyl)-1-[[(5-nitro-2-furanyl)methylene]amino]-2,4-imidazolidinedione see 6448
5-(4-Morpholinylmethyl)-3-[[(5-nitro-2-furanyl)methylene]amino]-2-oxazolidinone see 4205
2-(4-Morpholinylmethyl)-2-phenyl-1H-indene-1,3(2H)-dione see 6882
N-(4-Morpholinylmethyl)pyrazine-carboxamide see 6182
[10-[3-(4-Morpholinyl)-1-oxopropyl]-10H-phenothiazin-2-yl]carbamic Acid Ethyl Ester see 6178
d-Morpholylmethyldiphenylbutyryl-pyrrolidine see 2933
Morphosan see 6189
Morphothebaine, 6197
Morsydomine see 6143
Mortopal see 9138
Moryl see 1780
Mosaic Gold see 8739
Mosatil see 3480
Moscontin [Sarget] see 6193
Moségor [Wander] see 7486
Mosidal see 5855
Mosquito Plant see 7053
Mossbunker Oil see 5721
Moss Starch see 5357
Motazomin [Takeda] see 6143
Motiax [Neopharmed] see 3881
Motilex [Guidotti] see 2344
Motilin, 6198
Motilium [Janssen] see 3412
Motilyn [Abbott] see 2924
Motolon see 5872
Motovar [OTW] see 3655
Motox [Tenneco] see 9478
Motretinide, 6199
Motrin [Upjohn] see 4812
Mountain Balm see 3617
Mountain Grape see 1170
Mountain Tobacco see 813
Mouse Antialopecia Factor see 4883

Mouse-bane see 111
Mouse EGF-URO see 3492
Movecil [Carlo Erba] see 7987
Movellan (Tabl.) [Asta] see 8824
Movellan (Amps.) [Asta] see 8824
Moveltipril, 6200
Movens [Inverni] see 5659
Movergan [Homburg] see 2893
Moviol [Hoechst] see 7562
Movirene see 3357
Movyl [Montecatini] see 7563
Moxadil [Lederle] see 609
Moxal [Roger Bellon] see 610
Moxalactam, 6201
Moxalactam [Lilly] see 6201
Moxaline [Mead Johnson] see 610
Moxam [Lilly] see 6201
Moxaverine, 6202
Moxestrol, 6203
Moxisylyte, 6204
Moxyl see 6204
MP 11 see 7131
MP 12 see 4190
MP 302 see 4956
MP 620 see 4903
6MP see 5762
8-MP see 5911
MPB see 5715
M.P. Chlorcaps T.D. see 2180
MPMP see 5734
MPP + see 6205
α-MPT see 6085
MPTP, 6205
MPV-785 see 5670
MQPA see 804
MRD-108 see 7455
MRIH see 5693
MRL-41 see 2384
MS-222 [Sandoz] see 9533
Ms 752 see 5463
MS 4101 see 4133
MSA see 8670
MS Antifoam M [Midsil] see 8486
MS Contin [Chinoin] see 6193
MSG see 6165
MSH, 6206
MSH-release Inhibiting Factor see
5693
MSH-release Inhibiting Hormone see
5693
MSMA see 5864
MST-1 Continus [Napp] see 6193
MST 10 Mundipharma [Mundi-
pharma] see 6193
MST 30 Mundipharma [Mundi-
pharma] see 6193
MT 141 see 1930
L-α-MT see 6085
MTB 51 see 5869
MTBE see 5954
L-3-MTO see 5975
MTQ see 5872
MTU see 6048
MTX see 5908
4-MU see 4792
Mucalan see 4992
Mucara [Ives] see 5164
Mucic Acid see 4237
Muciclar [Substantia] see 1809
Mucinol [Kali-Chemie] see 676
Mucins, 6207
Mucitux [Riom] see 3577
Muclox [Sigma Tau] see 3881
Muco-Burg [Burg] see 392
Mucocedyl [Kettelhack Riker] see 82
Mucochloric Acid, 6208
Mucochloric Anhydride, 6209
Mucocis [Coli/Crosara] see 1809
Mucoclear [Mundipharma] see 392
Mucodyne [USV] see 1809
Mucofluid [UCB] see 5812

Mucolase [Lampugnani] see 1809
Mucolator [Allard] see 82
Mucolex [Warner] see 1809
Mucolitico [Maggioni] see 3410
Mucolysin [Proter] see 9386
Mucolyticum [Lappe] see 82
Mucomycin see 5500
Mucomyst [Mead Johnson] see 82
Muconic Acid, 6210
Muconomycin A see 9869
Muconomycin B see 9869
Mucopront [Mack, Illert.] see 1809
Mucorama [Boehringer, Mann.] see
7279
Muco Sanigen [Beecham-Wulfing] see
82
Mucosolvan [Thomae] see 392
Mucosolvin [Berlin-Chemie] see 82
Mucotab [Pharmakon] see 1809
Mucovent [Byk-Gulden] see 392
Mucret [Astra] see 82
Muira Puama, 6211
Mukinyl [Searle] see 1809
Mullite see 368
Mulsiferol see 9928
Mulsopaque [Alcon] see 4946
Multergan see 9234
Multezin see 9234
Multhiomycin see 6639
Multifuge [Blue Line] see 7433
Multifungin [Knoll] see 1423
Multilind [F.A.I.R.] see 6658
Multimycine see 2475
Multiplication Stimulating Activity
see 8670
Multum [Chephasaar] see 2082
Mundisal [Purdue Frederick] see 2212
Munobal [Hoechst] see 3895
Mupirocin, 6212
Muracil [Organon] see 6048
Muramic Acid, 6213
Muramidase see 5514
Muramyl Dipeptide, 6214
Murazyme [Prospa] see 5514
Murel [Ayerst] see 9816
Murexan see 9763
Murexide, 6215
Murexine, 6216
Muriacite see 1713
Muriamic see 4363
Muriatic Acid see 4703
Murillo Bark see 8050
Muroctasin, 6217
Murrayin see 8364
Musaril [Mack-Midy] see 9172
Muscalm [Nippon Kayaku] see 9447
Muscalure, 6218
Muscarine, 6219
Muscatox [Bayer] see 2559
Muscazone, 6220
Muscimol, 6221
Muscle Adenylic Acid see 148
Muscle Inosinic Acid see 4882
Muscone, 6222
Musco-Ril see 9253
Musculamine see 8699
Musculax [Nippon Organon] see 9845
Muskone see 6222
Musk Root see 8981
Musks, 6223
Mussel Poison see 8344
Mustard, Black, 6224
Mustard Chlorohydrin see 4565
Mustard Gas, 6225
Mustard, White, 6226
Mustargen Hydrochloride [Merck &
Co.] see 5655
Mustine Hydrochloride see 5655
Mustron see 5656
Mutabase [Schering] see 2986
Mutamycin [Bristol] see 6133

Mutesa see 6888
Muthesa [Wyeth] see 6888
Muthmann's Liquid see 9121
Mutton Suet see 8859
Muzolimine, 6227
MV-678 see 7778
MVPP see MISC-2
MY 41-6 see 6993
My 301 see 4465
MY 5116 see 8140
Myacine see 6369
Myacyne [Schur] see 6369
Myagen [Upjohn] see 1326
Myalex [ICI] see 3915
Myambutol [Lederle] see 3672
Myanesin see 5737
Myanol see 5737
Myarsenol see 8916
Myasul see 8890
Myavan [Warner-Chilcott] see 9446
Mybasan see 5071
My-B-Den [Ames] see 148
MYC 8003 see 6138
Mycaminose, 6228
Mycanden [Schering] see 4514
Mycardol [Winthrop] see 7066
Mycarose, 6229
Mycelex-G [Miles] see 2412
Mycelianamide, 6230
Mycetins, 6231
Mychel [Rachelle] see 2068
Mycifradin [Upjohn] see 6369
Mycil [BDH] see 2178
Mycilan [Théraplix] see 4514
Mycinol see 2068
Mycivin see 5378
Myclobutanil, 6232
Mycobacidin, 6233
Mycobacillin, 6234
Mycobactic Acid see 6235
Mycobactin P see 6235
Mycobactins, 6235
Mycobactyl see 4396
Mycoban see 1705 and 8623
Mycobutol see 3672
Mycofug [Hermal] see 2412
Mycoin C₃ see 7002
Mycolic Acids, 6236
Mycolutein see 897
(6-O-Mycolyl-α-D-glucopyranosyl)-
6-O-mycolyl-α-D-glucopyranoside
see 2523
Mycomycin, 6237
Mycophenolic Acid, 6238
Mycophyt [Mycofarm] see 6346
Mycosamine, 6239
Mycose see 9496
Mycospor [Bayer] see 1232
Mycosporan [Bayer] see 1232
Mycosporin [Bayer] see 2412
Mycostatin [Squibb] see 6658
Mycoster [Fabre] see 2270
Mycotoxin T-2 see 9711
Mydantane see 380
Mydecamycin see 6106
Mydetone see 9447
Mydfrin [Alcon] see 7257
Mydocalm [TAD] see 9447
Mydplegic [Cooper] see 2752
Mydriacyl [Alcon] see 9692
Mydrial [Winzer] see 9743
Mydriatine see 7279
Mydrilate [W. B. Pharm.] see 2752
Myebrol [Kyorin] see 6130
Myeleukon see 1494
Myelobromol [Chinoin; Lentia] see
6130
Myelographin see 4958
Myeloleukon see 1494
Myelosan see 1494
Myelotrast [Winthrop] see 4902

Cross Index of Names

Myfungar [Siegfried] see 6892
Mygdalon [DDSA] see 6063
Myk [Cassenne] see 8866
Mykostin see 9928
Mykrox [Fisons] see 6068
Mylabris, 6240
Mylar [Du Pont] see 7546
Mylaxen [Mallinckrodt] see 4606
Mylepsin [ICI] see 7753
Mylepsinum [Rhein-Pharma] see 7753
Myleran [Burroughs Wellcome] see 1494
Mylicon [Stuart] see 8486
Mylipen see 7042
Mylis [Kanebo] see 7710
Mylocon see 8486
Mylodorm see 601
Mylofanol see 2290
Mylone see 2827
Mylosar [Upjohn] see 907
Mylosul see 8890
Mylproin [ICI] see 9821
Mynosedin [Toho Kogyo] see 4812
Myocaine see 4465
Myocardone see 4536
Myocholine see 1207
Myochrysine [Merck & Co.] see 4422
Myocord [Szabo] see 879
Myocrisin [M & B] see 4422
Myodetensine see 5737
Myodigin see 3146
Myodil see 4946
Myoflex [Warren-Teed] see 9581
Myoglycerin see 6528
Myohematin see 2797
Myolastan [Clin-Comar-Byla] see 9172
Myolysin see 5737
Myonal [Eisai] see 3558
Myopan see 5737
Myopone see 9954
Myoral, 6241
Myordil [Winthrop] see 608
Myorexon [ICI] see 5114
Myosalvarsan see 8916
Myoscain see 4465
Myoserol see 5737
Myosin, 6242
Myospan [Lagap] see 950
Myospaz [Conal] see 8829
Myostibin see 733
Myoston [Martinet] see 148
Myoten [Central Pharm.] see 5737
Myotolon [Daiichi] see 4209
Myotonine Chloride [Glenwood] see 1207
Myoxam [Menarini] see 6106
Myoxane [Ascher] see 5737
Myoxanthin see 3469
Myprozine [Am. Cyanamid] see 6346
α-Myrcene see 6243
β-Myrcene, 6243
Myreth see 7554
Myrica see 1022
Myricetin, 6244
Myricitrin see 6244
Myricodine see 6250
Myricyl Alcohol see 9506
Myristica, 6245
Myristic Acid, 6246
Myristicin, 6247
Myristin see 9638
Myristyl Alcohol, 6248
Myristyl Benzylmorphine see 6250
Myristyltrimethylammonium Bromide, 6249
Myrj [ICI] see 7555
Myrj 45 [ICI] see 7555
Myrj 52 [ICI] see 7555
Myrj 52S [ICI] see 7555
Myrocodine see 6250

Myronate Potassium see 8495
Myrophine, 6250
Myrophinium see 6250
Myrrh, 6251
Myrtecaine, 6252
Myrticolorin see 8276
Myrtillin-a see 2866
Myrtle Wax see 1022
Myrtol, 6253
Mysalfon [Spofa] see 9095
Myser [Nikken] see 3134
Mysoline [Ayerst] see 7753
Mysuran [Winthrop] see 387
Mytab [Fine Organics] see 6249
Mytatrienediol, 6254
Mytelase [Winthrop] see 387
Mytolon [Winthrop] see 1116
My-trans see 5751
Myvizone see 9218
Myxin, 6255
Myxoviromycin see 410
MZ 144 see 8222
MZ 0780 see 8222
M-Zyme see 5171
N-5′ see 9489
N-399 see 9983
N-553 see 9447
N-714 see 2189
N-746 see 2392
N-869 see 5860
N-0252 see 5255
N-0500 see 6349
N-2790 see 4147
N-7009 see 4114
Na III see 3820
NA 22 [DuPont] see 3759
NA 66 see 7399
NA 73 see 4634
NA 97 see 6958
NA 274 see 1379
NA 872 see 392
NA 8318 see 5125
NAA see 6290
N.A.B. see 6361
NAB 365 see 2347
Nabadial see 5861
Nabam, 6256
NAB 365Cl see 2347
Nabilone, 6257
Nabolin see 5862
Nabumetone, 6258
Nabuser [Bayer] see 6258
NAC [Mead Johnson] see 82
Nacconate 100 see 9456
Nacid [Shionogi] see 5538
NaClex [Glaxo] see 4716
Nacom [Merck & Co.] see 1802
Nactate see 7529
Nacton [McNeil] see 7529
NAD see 6259
Nadeine see 3154
Nadex [Zyma-Galen] see 8002
Nadic Anhydride see 1801
Nadide, 6259
Nadisan [Boehringer, Mann.] see 1839
Nadolol, 6260
Nadone see 2732
Nadoxolol, 6261
NADP, 6262
NADPH₂ Diaphorase see 6784
Naepaine, 6263
Nafamostat, 6264
Nafamostat Mesilate see 6264
Nafamstat see 6264
Nafarelin, 6265
Nafcil [Bristol] see 6266
Nafcillin Sodium, 6266
Naferon [Sclavo] see 4893
Nafiverine, 6267
Nafrine see 6919
Nafronyl, 6268

Naftalofos, 6269
Naftazone see 6314
Naftidan see 6267
Naftidrofuryl see 6268
Naftifine, 6270
Naftifungin see 6270
Naftin [Sandoz] see 6270
Naftopen [KN Gistfabriek] see 6266
NAG see 6265
Naganin see 8986
Naganol see 8986
Nagemid [Dexo] see 1433
Nagravon see 9921
Na¹³¹I see 8583
Naismeritin [Hishiyama] see 9447
Naixan [Tanabe/Syntex] see 6337
Nalador [Schering AG] see 8973
Nalbuphine, 6271
Nalcrom [Fisons] see 2594
Nalcron [Fisons] see 2594
Nalde [Mead Johnson] see 409
Naled, 6272
Nalfon [Lilly] see 3926
Nalgesic [Lilly] see 3926
Nalidicron [Toyo Shinyaku] see 6273
Nalidixic Acid, 6273
Nalitucsan [Hishiyama] see 6273
D-Nal(2)⁶-LHRH see 6265
Nalline [Merck & Co.] see 6275
Nalmefene, 6274
Nalmetrene see 6274
Nalone [Winthrop] see 6277
Nalorex [Robin] see 6278
Nalorphine, 6275
Nalorphine Bis(nicotinate) see 6276
Nalorphine Dinicotinate, 6276
Naloxone, 6277
Nalpen [Beecham] see 925
Nalpen G see 1157
Naltrexone, 6278
Nalutron see 7783
Namphen [Toyo Iyaku] see 5680
Namuron see 2717
NAN see 8431
NANA see 8431
Nanbacine [Fournier Freres] see 9985
Nancimycin see 8216
Nandinine, 6279
Nandrolin [Tutag] see 6283
Nandrolone, 6280
Nandrolone Decanoate, 6281
Nandrolone p-Hexyloxyphenylpropionate, 6282
Nandrolone Laurate see 6280
Nandrolone Phenpropionate, 6283
Nandrolone Propionate, 6284
Nandron [Morishita] see 3519
Nankor [Dow] see 8239
NANM see 6275
Nanormon [Hormon-Chemie] see 8672
Nantenine see 3409
Nantokite see 2667
Naotin see 6435
NAPA [Arnar-Stone] see 14
Napacetin [Toyama] see 4812
Napadisilate see MISC-4
Napageln [Lederle] see 3893
Napalm, 6285
Napelline, 6286
Napellonine see 8673
Naphazoline, 6287
Naphcon [Alcon] see 6287
Naphtha see 7141
Naphthacene, 6288
Naphthacetol see 465
Naphthalane see 2839
1-Naphthalenamine see 6318
2-Naphthalenamine see 6319
Naphthalene, 6289
1-Naphthaleneacetic Acid, 6290

Narcozep [Roche] see 4072
Narcyl see 6341
Nardelzine [Substantia] see 7181
Nardil [Warner-Chilcott] see 7181
Nargoline [Specia] see 6404
Naridan see 6926
Narigix [Taiyo] see 6273
Naringenin, 6344
Naringenin-7-rhamnoglucoside see 6345
Naringen 7-Methyl Ether see 8289
Naringetol see 6344
Naringin, 6345
Narkolan see 9525
Narkothion see 9250
Narone [Ulmer] see 3358
Narphen [Smith & Nephew] see 7174
Narsis [Sumitomo] see 5669
Nartate see 3358
Nasalcrom [Fisons] see 2594
Nasalide [Syntex] see 4071
Nasan [Beecham Wulfing] see 9150
Nasdol see 9115
Nasemo see 6425
Nasivin [E. Merck] see 6919
Nasmil [Lusofarmaco] see 2594
Nastenon see 6920
NAT-333 see 3942
Natacillin [Banyu] see 4592
Natacyn [Alcon] see 6346
Natamycin, 6346
Naticardina see 8073
Natil see 2708
Natirene 25 see 3369
Native Calcium Sulfate see 1713
Natol see 7249
Natrascorb [USV] see 8525
Natrilix [Servier] see 4847
Natrionex [Glutan] see 45
Natriphene see 7276
Natrite see 8541
Natrium see 8512
Natrol see 1292
Natron see 8541
Natto see 8684
Natulan [Roche] see 7764
Natural Calcium Carbonate see 5368
Natural Gas, 6347
Natural Tartaric Acid see 9039
Natural Trehalose see 9496
Naturetin [Squibb] see 1045
Naturine [AB Leo] see 1045
Naturon see 5929
Natyl [Nativelle] see 3354
Naucaine see 7763
Nauheim Salts (Artificial), 6348
Naumannite see 8382
Nauseton [Lepetit] see 9623
Nausidol see 7421
Nausilen [Inverni] see 236
Nautamine [Delagrange] see 1225
Nautazine see 2716
Nauzelin [Janssen] see 3412
Navadel [Hercules] see 3296
Navadyl see 1743
Navane [Roerig] see 9297
Navanide see 434
Navaron (obsolete) see 9297
Navicalm see 5657
Navidrex [Ciba] see 2750
Navidrix [Ciba] see 2750
Navisin see 6919
Navolin see 4349
Naxagolide, 6349
Naxamide [Mead Johnson] see 4822
Naxcel [Upjohn] see 1948
Naxen [Syntex] see 6337
Naxidine [Lilly] see 6582
Naxofem [Farmitalia] see 6468
Naxogin [Erba] see 6468
NBS see 1428

NC 14 see 779
NC 45 see 9845
NC 123 see 5813
NC 302 see 8113
NC 1318 see 9446
NC 1667 see 9580
NC 1968 see 7121
NC 6897 see 1044
NC 8438 see 3697
NC 21314 see 2373
NCI 96683 see 8113
NCI-C56462 see 6160
NCS see 10073
NDEA see 6557
NDELA see 6556
2'NDG see 4262
NDGA see 6610
NDR 263 see 7824
NDR 304 see 3745
NDR 5998A see 3942
Nealbarbital, 6350
Nealbarbitone see 6350
Neallymal see 6350
Neamine, 6351
Neamoxyl [Arcopharma] see 610
Neantine [Givaudan] see 7345
Neargal see 3334
Neatsfoot Oil, 6352
Neazina see 8886
Neazolin see 8930
Nebactam [Specia] see 938
Nebcin [Lilly] see 9413
Neberk [Fuji K.K./Morishita] see 9060
Nebicina [Lilly] see 9413
Nebralin [Gayoso Wellcome] see 9094
Nebramycin see 9413
Nebramycin Factor 2 see 780
Nebramycin Factor 6 see 9413
Nebularine, 6353
Neburon, 6354
Necatorina see 1822
Necic Acids see 8400
"Necine" Bases see 8168
Nectadon [Merck & Co.] see 6638
Nectocyd [Pfizer] see 3376
Nedius [Carlo Erba] see 4875
Nedocromil, 6355
Needle Antimony see 745
Neem, 6356
Neem Oil, 6357
Nefco see 6521
Nefopam, 6358
Nefrix see 4704
Nefrolan [M & B] see 2401
Nefrosul [Riker] see 8871
Neftin see 4210
Nefurofan [Maruko] see 8721
Negamycin, 6359
Negatan [Savage] see 6360
Negatol® [Valpan], 6360
Negaxid [Sigma-Tau] see 400
NegGram [Winthrop] see 6273
Negram see 6273
Neguvon [Bayer] see 9536
Neguvon A [Bayer] see 9536
Nehydrin [TAD] see 3598
Nektrohan [ICN] see 278
Nelbon [Sankyo] see 6491
Nema [Parke, Davis] see 9126
Nemacide VC-13 [Mobil] see 3030
Nemacur [Baychem] see 3901
Nemadital see 7433
Nemafax [M & B] see 9282
Nemafos [Am. Cyanamid] see 9275
Nemafume see 3003
Nemagon see 3003
Nemapan see 9217
Nem-A-Tak [Am. Cyanamid] see 4173
Nematolyt see 6965

Nemazine see 7220
Nembutal [Abbott] see 7087
Nembutal Calcium [Abbott] see 7087
Nemestran [Hoechst Pharma] see 4310
Nemex see 1166
Nemicide [ICI] see 9161
Nemispor [Farmoplant] see 5598
Nemural [Winthrop] see 3434
Nendrin see 3533
Neo-Absentol see 3670
Neoamyl Alcohol see 6373
Neo-Antergan [Merck & Co.] see 7996
Neoantimosan see 8769
Neo-Arsoluin see 6361
Neoarsphenamine, 6361
Neo-Arsycodile see 5864
Neo-Atromid see 2374
Neo-Avagal see 5927
Neobar [E. Merck] see 1006
Neobenadol see 3308
Neo-Benodine see 5972
Neo-Betalin 12 [Lilly] see 4739
Neobiosamine B see 6369
Neobiosamine C see 6369
Neobrettin [Norbrook] see 6369
Neo-Bridal see 7996
Neocaine see 7763
Neocarcinostatin see 10073
Neo-Cardiamine see 9137
Neocardyl see 1272
Neocarzinostatin see 10073
Neocarzinostatin K see 10073
Neo-Cebicure see 5009
Neo-Cebitate see 5009
Neocefal [Gibipharma] see 1922
Neocembrene, 6362
Neocembrene A see 6362
Neochanin see 4157
Neocid [Geigy] see 2832
Neocidol [Ciba-Geigy] see 2978
Neoclym [Poli] see 2726
Neo-Codema see 4704
Neocon 1/35 [Ortho-Cilag] see 6614
Neo-Corovas see 7066
Neo-cupferron, 6363
Neocuproine, 6364
Neo-Cytamen [Duncan-Flockhart] see 4739
Neodaian [Nichiiko] see 9221
Neodalit [Hässle] see 2990
Neodecyllin [Penick] see 6370
Neo-Dema see 2169
Neo-Devomit see 2716
Neodigitalis see 3141
Neo-Dioxanin [Boehringer, Ing.] see 3150
Neo Dohyfral D₃ see 9929
Neodorm [Minden] see 7087
Neo-Douxan see 6538
Neodrenal see 5105
Neodrol [Pfizer] see 8753
Neodymium, 6365
Neodyne [Merrell] see 3745
Neo-Epinine [Burroughs Wellcome] see 5105
Neoergosterol, 6366
Neo-Erycinum [Schering AG] see 3628
Neoesserin (Tabl.) [Isis-Chemie] see 6380
Neoesserin (Amp.) [Arzneimittelwerk VEB] see 6380
Neo-Estrone see 3689
Neofamid see 5525
Neo-Farmadol see 6925
Neofemergen see 3600
Neo-Ferrum see 3974
Neo-Fluimucil [Inpharzam] see 82
Neoflumen see 4704
Neofocin [Medici] see 4169
Neo-Fulcin see 4453

Neogama see 8971
Neogel [Homburg] see 1797
Neogest [Schering] see 6621
Neo-Gilurytmal [Giulini] see 7704
Neo-Hepatex see 5427
Neohesperidin DHC see 6367
Neohesperidin Dihydrochalcone, 6367
Neohetramine Hydrochloride see 9307
Neo-Hibernex see 7793
Neo-Hombreol [Organon] see 9115
Neo-Hombreol-M [Organon] see 6044
Neohydrazid see 2688
Neohydrin [Lakeside] see 2104
Neo-Iopax R [Schering] see 8584
Neo-Iopax Sodium [Schering] see 8584
Neo-Iscotin [Daiichi] see 5072
Neoisocodeine see 7927
Neo-Istafene see 5657
Neoisuprel see 5053
Neolamin [Nippon Kayaku] see 9221
Neolate see 6369
Neolexina [Asla] see 1971
Neolignans see 5362
Neolinarin see 7010
Neoloid [Lederle] see 1904
Neolutin Depositum see 230
Neo-Macrabin see 4739
Neomagnol [Chinoin] see 2065
Neomas see 6369
Neo-mercazole [Aspro-Nicholas] see 1803
Neomestine [Taiyo] see 4146
Neo-Metantyl see 7816
Neo-Methidin see 5896
Neo-Methiodal see 4933
Neomethymycin, 6368
Neomethynolide see 6368
Neomin see 6369
Neomix [Tuco] see 6369
Neomycin, 6369
Neomycin A see 6351
Neomycin B see 6369
Neomycin C see 6369
Neomycin E see 6989
Neomycin Undecylenate, 6370
Neomyson [Eisai] see 9230
Neon, 6371
Neo-Naclex [Glaxo] see 1045
Neonal [Abbott] see 1515
Neonicotine see 655
Neo-Nilorex [A.V.P.] see 7180
Neo-Octon see 6671
Neo-Oestranol I see 3118
Neo-Oestranol II see 3119
Neo-Oxypaat see 8034
Neopax [I.F.C.I.] see 7377
Neopenil see 7027
Neopentane, 6372
Neopentanol see 6373
Neopentyl Alcohol, 6373
Neopentyl Glycol, 6374
Neopenyl see 2346
Neoperidole [Kyowa Hakko] see 4511
Neophryn [Winthrop] see 7257
Neophyl Chloride, 6375
Neopine, 6376
Neoplatin [Mead Johnson] see 2319
Neo-Ponden [Serono] see 667
Neoprene, 6377
Neopres [Boehringer, Ing.] see 2271
Neoproc see 7042
Neoprontosil [Winthrop] see 936
Neo-Protostan [Standex] see 7900
Neoprotoveratrine see 7915
Neopsicaine see 7926
Neopterin, 6378
Neo-Pynamin [Sumitomo] see 9154
Neopyrithiamine see 8003
Neoquassin, 6379
Neo-Rojamin see 4739

Neoron [Ciba-Geigy] see 1422
Neosalvarsan [Hoechst] see 6361
Neosanamid II see 8905
Neoscan [Medi-Physics] see 4252
Neo-Skiodan see 4933
Neospasmina [Biofarmacoterapico] see 3151
Neospiran see 9137
Neostene [SM & P] see 5861
Neostenovasan see 3459
Neosteron see 5861
Neostibosan see 3807
Neostigmin [AB Leo] see 6380
Neostigmine, 6380
Neoston [Beiersdorf] see 212
Neo-Strepsan see 8920
Neostreptal see 8876
Neosulf see 6369
Neosulfine see 5821
Neo-Synephrine Hydrochloride [Winthrop] see 7257
Neo-T see 6381
Neoteben [Bayer] see 5071
Neo-Tenebryl see 4933
Neotetrazolium Blue see 6381
Neotetrazolium Chloride, 6381
Neothesin see 7444
Neothyl see 6031
Neothylline [Lemmon] see 3459
Neo-Thyreostat [Herbrand] see 1803
Neotigason [Roche] see 107
Neotilina see 3459
Neo-Tizide (Tabl.) see 5072
Neo-Tizide (Amp.) see 5072
Neotocopherol see 9417
Neoton [Schiapparelli] see 7315
Neotran see 1258
Neotropin see 2956
Neo-Uliron see 6040
Neo-Urofort see 2073
Neo-Vasophylline see 3459
Neoviridogrisein IV see 3664
Neoviridogriseins see 3664
Neovitamin A, 6382
Neoxin see 5071
Neo-Zine see 7200
Neozine [Rhodia] see 5909
"Nepal" Aconitine see 7924
Nepaline see 7924
Nepenthe [Evans] see 6186
Nepetalactone, 6383
Nepetalic Acid see 6383
Nephramid [Bitterfeld] see 45
Nephridine see 3569
Nephril [Pfizer] see 7561
Nephroflow [Medi-Physics] see 4927
Nephrotest [Cassella-Riedel] see 454
Nepresol (Tabl.) [Ciba-Geigy] see 3152
Nepresol (Amp.) [Ciba-Geigy] see 3152
Népressol [Ciba-Geigy] see 3152
Neprotin see 5142
Neptal [Rhône-Poulenc] see 5805
Neptall [Rhône-Poulenc] see 13
Neptazane [Lederle] see 5875
Neptunium, 6384
Nequinate, 6385
Neral see 2324
Neraval [Schering] see 5901
Neravan [Promonta] see 1495
Nerbowdine see 1483
Nerdipina [Ferrer] see 6403
Nerfactor [Ipsen] see 4987
Nericur [Schering] see 1128
Neriifolin, 6386
Neriine see 2492
Neriodin [Nagase] see 3071
Neriolin see 6786
Nerisona [Schering AG] see 3129
Nerisone [Schering AG] see 3129

Nerobol see 5862
Nerol, 6387
Nerolidol, 6388
Nerolin "New" see 3709
Nerolin "Old" see 5918
Neroli Oil, Artificial see 5942
Nervacton see 1037
Nervanaid B see 3482
Nerve Growth Factor see 6397
Nerve Root see 2778
Nervonus see 5751
Nesacaine [Pennwalt] see 2158
Nesdonal Sodium [Specia] see 9280
Nesontil see 6881
Nespor [Farmoplant] see 5603
Neston see 5896
Nethalide see 7802
Nethamine [Merrell] see 3661
Netillin [Kirby Warrick] see 6389
Netilmicin, 6389
Netilyn [Schering] see 6389
Netobimin, 6390
Netrin [Geigy] see 5840
Netromicine [Unilabo] see 6389
Netromycin [Schering] see 6389
Netropsin, 6391
Netrosylla [Bayer] see 3829
Netsusarin see 488
Nettacin [Essex] see 6389
Neuberg Ester see 4188
Neuchlonic [Taiyo] see 6491
Neucolis [Nippon Shinyaku] see 2321
Neuer [Daiichi] see 2017
Neulactil [M & B] see 7117
Neuleptil [Farmitalia; Specia] see 7117
Neumandin see 5071
Neumolisina see 3765
Neupentedrin see 8994
Neuquinone [Eisai] see 9751
Neuracen [Promonta] see 1027
Neuractil see 5909
Neuractiv [Ciba] see 6896
Neuralex [Millot] see 1052
Neuraminic Acid, 6392
Neuraxin see 5903
Neuridine see 8699
Neurine, 6393
Neuriplege [Genevrier] see 2184
Neurobarb see 7201
Neurocalm [Efeka] see 9707
Neurocil [Bayer] see 5909
Neurofort see 2073
Neurogard [Merck & Co.] see 3392
Neurolene [Magis] see 6592
Neuroleptone see 1037
Neurolytril [Dorsch] see 2977
Neuromet [ISF] see 6896
Neuronal see 3105
Neuronika see 873
Neuroprocin see 3478
Neurosedyn see 9182
Neurosin see 1673
Neurosterone [BDH] see 2754
Neurostop see 1049
Neurotensin, 6394
Neurotensin(ox) Triacetate(salt) see 6394
Neuroton see 4465
Neuroton [Nuovo Cons.] see 2321
Neurotoxin see 2602
Neurotrast see 4946
Neurotropan [Itting] see 2210
Neurvit [Opoatma] see 7896
Neuryl see 4603
Neustab [Boots] see 9218
Neutase [Sawai] see 5514
Neuthion [Senju] see 4369
Neutrafil see 3459
Neutral Acriflavine see 118
Neutral Ammonium Chromate see 532
Neutral Ammonium Fluoride see 547

Nidanthel [Cooper] see 6517
Nidantin [Sasse] see 6899
Nidaton see 5071
Nidaxin [Chassot] see 5677
Nidazol [Steinhard] see 6079
Nidrafur see 6458
Nidran [Sankyo] see 6469
Nidrane [Phoenix] see 1027
Nidrel [Specia] see 6493
Nidroxyzone, 6439
Nidulin, 6440
Nielsbohrium see 3504
Nieraline see 3569
Nifedicor [Schiapparelli] see 6441
Nifedin [Gentili] see 6441
Nifedipine, 6441
Nifelan [Elan] see 6441
Nifelat [Sidus] see 6441
Nifenalol, 6442
Nifenazone, 6443
Niflan [Yoshitomi] see 7708
Niflumic Acid, 6444
Nifluril see 6444
Nifluril Suppositories [UPSA] see 6444
Nifos T [Monsanto] see 9138
Nifran [Yoshitomi] see 7708
Niftolid see 4132
Nifulidone see 4210
Nifuradene, 6445
Nifuraldezone, 6446
Nifuratel, 6447
Nifurfoline, 6448
Nifuroquine, 6449
Nifuroxazide, 6450
Nifuroxime, 6451
Nifurpirinol, 6452
Nifurprazine, 6453
Nifurtimox, 6454
Nifurtoinol, 6455
Nifurzide, 6456
Nifuzon see 6521
Nigakilactone D see 8037
Nigalax see 1253
Nigelline see 2809
Nigericin, 6457
Niggerhead see 10000
Night-blooming Cereus see 1606
Nigrin [Pfizer] see 8789
NIH-4185 see 9220
NIH-4542 see 3248
NIH-5145 see 3784
NIH-7274 see 7215
NIH-7440 see 294
NIH-7519 see 7174
NIH-7539 see 6625
NIH-7590 see 7403
NIH-7667 see 6602
NIH-8805 see 1487
Nihydrazone, 6458
Nikardin see 6459
Nikethamide, 6459
Nikion see 1045
Nikofezon see 6443
Nikoform see 4765
Nikozid see 5071
Nilatil [USV] see 5132
Nilergex [ICI] see 5117
Nilevar [Searle] see 6613
Nilhistin see 5870
Nilodin see 5463
Niltuvin [EGYT] see 6438
Nilutamide, 6460
Nilvadipine, 6461
Nilverm [J & J] see 9161
Nilyph [Astra] see 3544
Nim see 6356
Nimaol see 6670
Nimbecetin see 5156
Nimbin, 6462
Nimbiol, 6463

Nimed see 6464
Nimelan see 6276
Nimergoline see 6404
Nimesulide, 6464
Nimetazepam, 6465
Nimicor [Formenti] see 6403
Nimidane, 6466
Nimodipine, 6467
Nim Oil see 6357
Nimorazole, 6468
Nimotop [Bayer] see 6467
Nimrod [ICI] see 1484
Nimustine, 6469
Ninhydrin, 6470
Ninopterin, 6471
Nio-A-Let [Nion] see 9918
Niobium, 6472
Niobium Pentachloride, 6473
Niobium Pentafluoride, 6474
Niobium Pentoxide, 6475
Niobium Potassium Oxypentafluoride, 6476
Nioform see 4924
Nionate [Nion] see 3996
Niopam [E. Merck] see 4943
Nioxime®, 6477
Niozymin see 6398
Nip [Nichay] see 7563
Nipagin A see 3792
Nipagin M see 6021
Nipantiox 1-F [Nipa] see 1547
Nipasol see 7879
Nipaxon [AB Leo] see 6638
Nipecotan see 686
Nipecotic Acid, 6478
Niperyt see 7066
Nipodal see 7768
Nippas see 491
Nipradilol, 6479
Nipradolol, 6479
Niprazina [Italfarmaco] see 6443
Nipride [Roche] see 8600
Nipruss [Pharma Schwarz] see 8600
Niran [Monsanto] see 6983
Niran [Tamogan] see 2079
Nirexon see 3330
Niridazole, 6480
Nirvan see 5909
Nirvanil see 9820
Nirvotin see 1845
NIS see 4936
Nisentil [Roche] see 307
Nisidana [Geigy] see 6808
Nisin, 6481
Nisintel see 307
Nisoldipine, 6482
Nisolone see 7719
Nisotin see 3692
Nissorun [Nippon Soda] see 4634
Nisulfazole [Breon] see 6566
Nitadon see 5071
Nitalapram see 2320
Nitan see 7021
Nitarsone, 6483
Niter see 7634
Nithiazide, 6484
Nithiocyamine see 606
Niticolin [Morishita] see 2321
Nitobanil [Kyowa Yakuhin] see 9060
Nitogenin see 3289
Nitoman [Roche] see 9118
Niton see 8119
Nitorol [Eisai] see 5114
Nitossil [Zyma] see 2393
Nitracrine, 6485
Nitradisc [Searle] see 6528
Nitrados [Berk] see 6491
Nitraldone see 4205
Nitramide, 6487

Nitramine, 6488
Nitran [Riker] see 6528
Nitranilic Acid, 6489
m-Nitraniline see 6503
o-Nitraniline see 6504
p-Nitraniline see 6505
Nitranitol [Merrell] see 5630
Nitranol see 9682
Nitraphen see 6519
Nitrapyrin, 6490
2-Nitratoethylaminotoluene-p-sulfonate see 5132
(Nitrato-O)phenylmercury see 7273
Nitrazepam, 6491
Nitrefazole, 6492
Nitrendipine, 6493
Nitrenes, 6494
Nitrenpax see 6491
Nitretamin [Squibb] see 9682
Nitric Acid, 6495
Nitric Acid, Anhydrous, 6496
Nitric Acid Concentrated see 6495
Nitric Acid, Fuming, 6497
Nitric Acid Methyl Ester see 6015
Nitric Acid Propyl Ester see 7877
Nitric Anhydride see 6526
Nitric Oxide, 6498
Nitriderm [Geigy] see 6528
5,5'-Nitrilodibarbituric Acid Monoammonium Salt see 6215
Nitrilotriacetic Acid, 6499
Nitrilotriacetic Acid Bismuth Complex Sodium Salt see 1293
2,2',2''-Nitrilotrisethanol see 9581
2,2'2''-Nitrilotrisethanol Trinitrate (Ester) Phosphate (1:2) (Salt) see 9682
Nitrimidazine see 6468
Nitrin, 6500
Nitritocobalamin see 9924
Nitro-tabl see 9682
Nitroacetanilide, 6501
m-Nitroacetanilide see 6501
o-Nitroacetanilide see 6501
p-Nitroacetanilide see 6501
Nitro Acid Sulfite see 6570
Nitroacridine 3582 see 6502
Nitroakridin 3582, 6502
3-Nitroalizarin see 240
4-Nitro-2-aminophenetole see 6538
1-Nitro-3-amino-4-phenyl Ethyl Ether see 6538
m-Nitroaniline, 6503
o-Nitroaniline, 6504
p-Nitroaniline, 6505
Nitroanisole, 6506
m-Nitroanisole see 6506
o-Nitroanisole see 6506
p-Nitroanisole see 6506
5-Nitrobarbituric Acid, 6507
Nitrobenzaldehyde, 6508
m-Nitrobenzaldehyde see 6508
o-Nitrobenzaldehyde see 6508
p-Nitrobenzaldehyde see 6508
3-Nitrobenzenamine see 6503
Nitrobenzene, 6509
4-Nitrobenzeneacetic Acid see 6543
4-Nitrobenzeneacetonitrile see 6512
p-Nitrobenzenearsonic Acid see 6483
p-Nitrobenzeneazochromotropic Acid Sodium Salt see 2242
p-Nitrobenzeneazosalicylic Acid see 241
4-Nitro-1,2-benzenediamine see 6544
p-Nitrobenzenesulfonic Acid Methyl Ester see 6016
Nitrobenzoic Acid, 6510
m-Nitrobenzoic Acid see 6510
o-Nitrobenzoic Acid see 6510
p-Nitrobenzoic Acid see 6510
Nitrobenzol see 6509

2-Nitro-*N*-(phenylmethyl)-1*H*-imid-
azole-1-acetamide see 1094
1-(4-Nitrophenyl)-2-methyl-4-nitro-
imidazole see 6492
3-(3-Nitrophenyl)-2-propenoic Acid
see 6516
o-Nitrophenylpropiolic Acid, 6546
3-(2-Nitrophenyl)-2-propynoic Acid
see 6546
N-(4-Nitrophenyl)-*N*'-(3-pyridinyl-
methyl)urea see 7999
4'-[(*p*-Nitrophenyl)sulfamoyl]acet-
anilide see 8908
2-(*p*-Nitrophenylsulfonamido)thiazole
see 6566
2-(4-Nitrophenylsulfonamido)thiazole
see 6566
(4-Nitrophenyl)urea, 6547
Nitropress [Abbott] see 8600
NitroPRN [Warner-Chilcott] see 6528
1-Nitropropane, 6548
2-Nitropropane, 6549
5'-Nitro-2'-propoxyacetanilide, 6550
5-Nitro-2-propoxyaniline, 6551
5-Nitro-2-propoxybenzenamine see
6551
N-(5-Nitro-2-propoxyphenyl)acet-
amide see 6550
1,1'-(2-Nitropropylidene)bis[4-chlo-
robenzene] see 6515
5-Nitro-2,4,6(1*H*,3*H*,5*H*)-pyrimidine-
trione see 6507
5-Nitroquinaldic Acid, 6552
5-Nitroquinaldinic Acid see 6552
5-Nitro-2-quinolinecarboxylic Acid
see 6552
5-Nitro-8-quinolinol see 6577
Nitrorectal [Pohl] see 6528
Nitroretard [Dumex] see 6528
3-Nitrosalicylic Acid, 6553
5-Nitrosalicylic Acid, 6554
Nitroscanate, 6555
Nitrose [Fort Dodge] see 6570
Nitrosigma [Sigma-Tau] see 6528
2,2'-Nitrosiminodiethanol see 6556
Nitrosocobalamin see 9924
N-Nitrosodiethanolamine, 6556
N-Nitrosodiethylamine, 6557
N-Nitrosodimethylamine, 6558
p-Nitroso-*N*,*N*-dimethylaniline, 6559
p-Nitrosodiphenylamine, 6560
N-(2-Nitrosohydroxylamino-3-meth-
ylbutyl)crotonamide see 3417
2,2'-(Nitrosoimino)bisethanol see
6556
N-Nitrosomorpholine, 6561
4-Nitrosomorpholine see 6561
1-Nitroso-2-naphthalenol see 6562
Nitroso-β-naphthol see 6562
1-Nitroso-2-naphthol, 6562
Nitrosonium Tetrafluoroborate see
6571
4-Nitrosophenol, 6563
4-Nitroso-*N*-phenylbenzenamine see
6560
N-Nitrosophenylhydroxylamine
Ammonium Salt see 2625
N-Nitrosopyrrolidine, 6564
1-Nitrosopyrrolidine see 6564
Nitrosorbon [Pohl-Boskamp] see 5114
Nitroso-R Salt, 6565
Nitrososulfuric Acid see 6570
Nitrostat [Parke, Davis] see 6528
p-Nitrosulfathiazole, 6566
2-Nitro-4-sulfobenzoic Acid, 6567
Nitrosulfonic Acid see 6570
Nitrosyl Borofluoride see 6571
Nitrosyl Chloride, 6568
Nitrosyl Fluoborate see 6571
Nitrosyl Fluoride, 6569
Nitrosyl Hydrogen Sulfate see 6570

Nitrosyl Sulfate see 6570
Nitrosylsulfuric Acid, 6570
Nitrosyl Tetrafluoroborate, 6571
Nitrothiamidazol see 6480
5-Nitro-2-thiazolamine see 469
N-(5-Nitro-2-thiazolyl)acetamide see
421
p-Nitro-*N*-2-thiazolylbenzenesulfon-
amide see 6566
N-(5-Nitro-2-thiazolyl)formamide see
4155
1-(5-Nitro-2-thiazolyl)-2-imidazolidi-
none see 6480
1-(5-Nitro-2-thiazolyl)-2-oxotetrahy-
droimidazole see 6480
N-(5-Nitro-2-thiazolyl)-2-thiophene-
carboxamide see 9079
5-Nitro-2-thiophenecarboxylic Acid
[3-(5-Nitro-2-furanyl)-2-propenyli-
dene]hydrazide see 6456
5-Nitro-2-thiophenecarboxylic Acid
[3-(5-Nitro-2-furyl)allylidene]-
hydrazide see 6456
Nitrotoluene, 6572
m-Nitrotoluene see 6572
o-Nitrotoluene see 6572
p-Nitrotoluene see 6572
p-Nitro-α-toluic Acid see 6543
p-Nitro-α-tolunitrile see 6512
2-Nitro-3,4,6-trichlorophenol see
9554
4'-Nitro-3'-trifluoromethylisobutyr-
anilide see 4132
4-Nitro-3-(trifluoromethyl)phenol see
9180
2-[Nitro(2,4,6-trinitrophenyl)amino]-
ethanol Nitrate (Ester) see 7095
Nitrourea, 6573
Nitrous Acid, 6574
Nitrous Acid Butyl Ester see 1581
Nitrous Acid *tert*-Butyl Ester see 1582
Nitrous Acid 1,1-Dimethyl Ethyl
Ester see 1582
Nitrous Acid Ethyl Ester see 3786
Nitrous Acid Isopropyl Ester see 5104
Nitrous Acid 1-Methylethyl Ester see
5104
Nitrous Acid *n*-Propyl Ester see 7878
Nitrous Acid Sodium Salt see 8599
Nitrous Ether see 3786
Nitrous Oxide, 6575
Nitrovin, 6576
Nitrox 80 see 6022
Nitroxanthic Acid see 7380
Nitroxoline, 6577
Nitroxy Fluoride see 4093
Nitroxyl Chloride see 6579
Nitroxylsulfuric Acid see 6570
Nitroxynil, 6578
Nitroxynil Eglumine see 6578
Nitroxynil Meglumine see 6578
Nitrozell Retard [Byk-Gulden] see
6528
Nitrozone [Century] see 6521
Nitrumon [Simes] see 1852
Nitryl Chloride, 6579
Nitryl Fluoride, 6580
Nitryl Hypofluorite see 4093
Nitux [Inpharzam] see 6177
Nivadil [Fujisawa] see 6461
Nivadipine see 6461
Nivaldipine see 6461
Nivalenol, 6581
Nivalin [Sofia] see 4245
Nivaquine [M & B] see 2163
Nivaquine B see 2163
Nivelona see 3302
Nivemycin [Boots] see 6369
Nivitin see 8680
Nivocilin [Larma] see 3429
Nivoman see 9597

Nix [Burroughs Wellcome] see 7132
Nixyn [Hermes S.A.] see 5078
Nizatidine, 6582
Nizax [Lilly] see 6582
Nizaxid [Lilly] see 6582
Nizin [Broemmel] see 8901
Nizofenone, 6583
Nizoral [Janssen] see 5181
NK-19 see 7504
NK-421 see 9750
NK-631 see 7100
NK-1006 see 5161
NKK 105 see 5593
NMDA, 6584
NMN, 6585
NMOR see 6561
No. 220 Sol see 5757
No. 356 see 2383
Noah's Ark see 2778
Noan [Ravizza] see 2977
Nobacter [Chantereau] see 9568
Nobecutan see 9304
Nobedon see 40
Nobedorm see 5872
Nobelium, 6586
Nobfelon [Toho Iyaku] see 4812
Nobfen [Toho/Funai] see 4812
Nobgen [Kanebo] see 4812
Nobitocin S see 6934
N-Oblivon see 5732
Nobrium [Roche] see 5669
Nocardamin, 6587
Nocardic Acids see 6236
Nocardicin A see 6588
Nocardicin B see 6588
Nocardicin(s), 6588
Nocardomycolic Acids see 6236
Nocertone [Labaz] see 6889
Noctal [Cassella-Riedel] see 7807
Noctamid [Schering AG] see 5458
Noctan see 6054
Noctazepam [Brenner] see 6881
Noctec [Squibb] see 2061
Noctesed [Unimed] see 6491
Noctilene see 5872
Noctivane see 4625
Noctivane Sodium see 4625
Noctone [Gea] see 8126
Noctosom see 4123
Nodakenetin Glucoside see 6589
Nodakenin, 6589
Nodapton see 4397
No-Doz [Bristol] see 1635
Noformicin, 6590
Noformycin see 6590
Nogalamycin, 6591
L-Nogalose see 6591
Nogédal [Théraplix] see 6644
Nogexan [Farmex] see 1838
Nogos [Ciba-Geigy] see 3069
Nogram [Winthrop] see 6273
Noin [Essex Nippon] see 4850
Noleptan [Thomae] see 4144
Nolicin [Krka] see 6617
Nolipax [Foscama] see 3924
Noltam [Lederle] see 9019
Noludar [Roche] see 6054
Nolvadex [ICI] see 9019
Nolvasan [Fort Dodge] see 2090
Nomersan see 9304
Nometan see 7432
Nometine see 7421
Nomifensine, 6592
Nomilin, 6593
Nonactic Acid see 6594
Nonactin, 6594
Nonaethyleneglycol Monomethyl
Ether *p*-*n*-Butylaminobenzoate see
1106
Nonalol see 6598
Nonalupine see 5115

Nonanedioic Acid *see* 921
Nonanoic Acid *see* 7013
Nonanoic Acid Ethyl Ester *see* 3793
1-Nonanol *see* 6598
Nonanol Acetate *see* 6597
2-Nonenal, 6595
trans-2-Nonenaldehyde *see* 6595
Nonflamin [Yoshitomi] *see* 9379
Nonipol NO [Sankyo] *see* 6596
Nonisol [Ciba-Geigy] *see* 7555
Nonoic Acid *see* 7013
Nonoxinol *see* 6596
Nonoxynol, 6596
Nonoxynol-9 *see* 6596
Nonoxynol-11 *see* 6596
Nonplesin [Santen] *see* 7747
Nonulosaminic Acids *see* 8431
n-Nonyl Acetate, 6597
n-Nonyl Alcohol, 6598
Nonylcarbinol *see* 2847
n-Nonyl Ethanoate *see* 6597
Nonylic Acid *see* 7013
Nonyl Phenol, 6599
(*p*-Nonylphenoxy)acetic Acid, 6600
Nonylphenoxypolyethoxyethanol *see* 6596
α-(4-Nonylphenyl)-ω-hydroxypoly-
  (oxy-1,2-ethanediyl) *see* 6596
Nonylphenyl Polyethyleneglycol Ether
  see 6596
Nootron [Biosintetica] *see* 7459
Nootrop [UCB] *see* 7459
Nootropyl [UCEPHA] *see* 7459
Nopalcol [Diamond] *see* 7555
Nopil [Mepha] *see* 8889
Nopinene *see* 7415
Nopoxamine *see* 6252
No-Press *see* 5653
Nopron [Carrion] *see* 6400
Noprylsulfamide, 6601
Noptil *see* 7201
NO-PYR *see* 6564
Noracin [Chew] *see* 6617
Noracyclin [Ciba] *see* 5501
Noracymethadol, 6602
Noradrenaline *see* 6612
Noralutin *see* 6614
Noraminopyrine Methanesulfonate
  Sodium *see* 3358
Noranat [Labinca] *see* 4847
19-Norandrosta-4,9,11-trien-17β-ol-
  3-one *see* 9499
19-Norandrost-4-ene-4,17β-diol-3-
  one *see* 4768
Norandrostenolone Decanoate *see*
  6281
19-Nor-Δ⁴-androsten-17β-ol-3-one
  β-Phenylpropionate *see* 6283
19-Nor-Δ⁴-androsten-17β-ol-3-one
  Propionate *see* 6284
Norantoin *see* 6026
Noravid [Roussel Maestretti] *see* 2851
Noraxin [TP] *see* 6617
Norbide [Norton] *see* 1346
Norbiline [Fournier Frères] *see* 7922
Norbolethone, 6603
Norboral *see* 1839
Norbormide, 6604
cis-5-Norbornene-2,3-dicarboxylic
  Acid Dimethyl Ester *see* 3230
cis-endo-5-Norbornene-2,3-dicarb-
  oxylic Anhydride *see* 1801
α-5-Norbornen-2-yl-α-phenyl-1-
  piperidinepropanol *see* 1246
Norcamphane *see* 3909
Norcarane, 6605
Norcassamidine *see* 3634
Norcholanic Acid, 6606
24-Norcholan-23-oic Acid *see* 6606
Norcodeine, 6607
Norcolut [Köbányai] *see* 6614

Norcuron [Organon] *see* 9845
Norcycline *see* 8316
Nordaz [Bouchara] *see* 6608
Nordazepam, 6608
Nordefrin Hydrochloride, 6609
Norden *see* 6679
2′-Nor-2′-deoxyguanosine *see* 4262
Nordette [Wyeth] *see* 6621
Nordialex [Pharmacodex] *see* 4335
Nordiazepam *see* 6608
Nordicort *see* 4713
Nordihydroguaiaretic Acid, 6610
Norditropin [Nordisk] *see* 8672
Nordox [Norton] *see* 3429
Nor-Durandron *see* 6280
Norea, 6611
Norephedrane *see* 616
Nor-ψ-ephedrine *see* 6634
ψ-Norephedrine *see* 6634
dl-Norephedrine Hydrochloride *see*
  7279
Norepinephrine, 6612
19-Norergosta-5,7,9,22-tetraen-3β-ol
  see 6366
Norethandrolone, 6613
Norethindrone, 6614
19-Norethisterone *see* 6614
Norethynodrel, 6615
19-Nor-17α-ethynylandrosten-17β-
  ol-3-one *see* 6614
19-Nor-17α-ethynyl-17β-hydroxy-4-
  androsten-3-one *see* 6614
19-Nor-17α-ethynyltestosterone *see*
  6614
Norfemac [Nordic] *see* 1462
Norfen [Morishita] *see* 6679
Norfenefrine, 6616
Norfin *see* 6275
Norflex [Riker] *see* 6831
Norfloxacin, 6617
Norflurazon, 6618
Norgamen *see* 9375
Norgan *see* 4708
Norgesic [Kettelhack] *see* 4395
Norgesterone, 6619
Norgestimate, 6620
Norgeston [Schering] *see* 6621
Norgestrel, 6621
D-Norgestrel *see* 6621
Norgestrienone, 6622
Norgine *see* 232
Norglycin [Upjohn] *see* 9429
Norhomoepinephrine Hydrochloride
  see 6609
19-Nor-17β-hydroxy-3-ketoandros-
  tene 17-Phenylpropionate *see* 6283
Norhyoscyamine, 6623
Noriday [Syntex] *see* 6614
Noridyl *see* 9515
Norimin [Syntex] *see* 6614
Norimipramine *see* 2902
Norinyl-1 [Syntex] *see* 6614
Norinyl 1 + 35 [Syntex] *see* 6614
Norisodrine [Abbott] *see* 5105
dl-Norisoephedrine *see* 459
Noristerat [Schering AG] *see* 6614
Norit [Norit N.V.] *see* 1815
Noritren [Lundbeck] *see* 6635
Norizalpinin *see* 4244
Norkel *see* 5189
Norlaudanosoline *see* 9148
Norlestrin [Parke, Davis] *see* 6614
Norleucine, 6624
Norlevorphanol, 6625
Norlobelanine, 6626
Norlongandron *see* 6280
Norlutate [Parke, Davis] *see* 6614
Norluten [Parke, Davis] *see* 6614
Norlutin [Parke, Davis] *see* 6614
Norlutin-A [Parke, Davis] *see* 6614
Norluton [Parke, Davis] *see* 6614

Normabrain [Cassella-Riedel] *see*
  7459
Normadate [Glindia] *see* 5208
Normal Butyl Thioalcohol *see* 1575
Normal Cyanuric Acid *see* 2704
Normal Lead Acetate *see* 5268
Norma-oestren [Norma] *see* 3689
Normase [Molteni] *see* 5225
Normastigmin [Sigmapharm] *see* 6380
Normatensyl [Théraplix] *see* 4680
Normenon [Syntex] *see* 2102
Normet [Gedeon Richter] *see* 2374
Normetandrone *see* 6629
Normetanephrine, 6627
Normethadone, 6628
Normethandrolone *see* 6629
Normethandrone, 6629
Normetic [Abbott] *see* 417
Normicina [E. Merck] *see* 6106
Normi-Nox [Herbrand] *see* 5872
Normison [Wyeth] *see* 9074
Normix [Alfa] *see* 8218
Normoc [Merckle] *see* 1374
Normocytin [Lederle] *see* 9921
Normodyne [Schering] *see* 5208
Normoglucina *see* 7191
Normolipol [Delagrange] *see* 2374
Normonal [Eisai] *see* 9649
Normorix [Benzon] *see* 417
Normorphine, 6630
Normorphine 3-Methyl Ether *see*
  6607
Normoson *see* 3683
Normosterol *see* 7065
Normotiroide *see* 462
Normud [Astra] *see* 10024
Normurat [Grünenthal] *see* 1073
Nornicotine, 6631
Nornidulin, 6632
Noroclox DC [Willows Francis] *see*
  2414
Norodin [Endo] *see* 5859
Norodine [Norbrook] *see* 8874
Noroxin(e) [Merck & Co.] *see* 6617
Noroxylin *see* 954
Norpace [Searle] *see* 3366
Norphen (Ampules) [Byk-Gulden] *see*
  6679
Norphenylephrine *see* 6616
Norphytane *see* 7757
Norpipanone, 6633
Norpramin [Lakeside] *see* 2902
(17α)-19-Norpregna-1,3,5(10)-trien-
  20-yne-3,17-diol *see* 3689
Norpregneninolone *see* 6614
19-Nor-17α-pregn-4-en-17-ol *see*
  3761
(3β,17α)-19-Norpregn-4-en-20-yne-
  3,17-diol *see* 3816
(17α)-19-Norpregn-4-en-20-yn-17-ol
  see 5501
Nor-Progestelea *see* 6637
Norpropandrolate *see* 1325
Norpseudoephedrine, 6634
Nor-QD [Syntex] *see* 6614
Norquen [Syntex] *see* 5819
Norquentiel [SK & F] *see* 6614
Norsulfasol *see* 8920
Norsulfazole *see* 8920
Norsympatol *see* 6679
Norsynephrine *see* 6679
Nortesto *see* 6284
19-Nortestosterone *see* 6280
19-Nortestosterone Cyclohexylpropionate *see* 6280
19-Nortestosterone Decanoate *see*
  6281
19-Nortestosterone Furylpropionate
  see 6280
19-Nortestosterone Hexahydrobenzo-
  ate *see* 6280

Nortestosterone Hexoxyphenylpropionate *see* 6282
19-Nortestosterone-3-(*p*-hexyloxyphenyl)propionate *see* 6282
19-Nortestosterone β-Phenylpropionate *see* 6283
19-Nortestosterone Propionate *see* 6284
North American Sumach *see* 8978
Northern Pine *see* 9956
North And South American Antisnakebite Serum *see* 752
Norticon [Nihon] *see* 7441
Nortimil [Chiesi] *see* 2902
Nortran [Norden] *see* 9593
Nortrilen [Troponwerke] *see* 6635
Nortriptyline, 6635
Nortron [Fisons] *see* 3697
1α*H*,5α*H*-Nortropan-3α-ol (−)-Tropate *see* 6623
Noruron *see* 6611
Norval [Bencard] *see* 6097
Norvaline, 6636
7-L-Norvaline Cyclosporin A *see* 2759
Norvedan [Wyeth] *see* 3946
Norvinisterone, 6637
Norvinodrel *see* 6619
Norxacin [Siam Bheasach] *see* 6617
Norybol-19 *see* 6284
Norzepine *see* 6635
Norzetam [Albert-Farma] *see* 7459
Noscapal [Aspro-Nicholas] *see* 6638
Noscapalin *see* 6638
Noscapine, 6638
Nosiheptide, 6639
Nosim [Richet] *see* 5114
Nosophen *see* 4931
Nosophene Sodium *see* 4931
Nospan *see* 9737
Nostal [Ames] *see* 3478
Nostyn [Ames] *see* 3478
Nosydrast *see* 4933
Nosylan *see* 4933
Notair [DeAngeli] *see* 7021
Notandron *see* 5861
Notaral [Wyeth] *see* 7041
Notatin *see* 4354
Notensil [Fisons] *see* 27
Notézine [Specia] *see* 3106
Nothiazine *see* 5734
Notul [Tosi/Euromedical] *see* 2279
Nova [Rohm & Haas] *see* 6232
Novacetyl *see* 5531
Novacid *see* 3358
Novacrysin *see* 4423
Novacyl [Lematte Et Boinot] *see* 6895
Novadox [Sernagiotto] *see* 3429
Novadral [Diwag] *see* 6616
Novafed [Dow] *see* 3561
Novakol *see* 9218
Noval [Yamanouchi] *see* 5687
Novaldin [Winthrop] *see* 3358
Novalgin [Hoechst] *see* 3358
Novamidon [Specia] *see* 488
Novamin *see* 3193
Novamin [Bristol] *see* 416
Novantrone [Lederle] *see* 6135
Novapirina [Zyma] *see* 3071
Novarsan *see* 6361
Novarsenobenzol *see* 6361
Novarsenobillon *see* 6361
Novartrina [Manzoni] *see* 6176
Novasmasol *see* 5836
Novastan [Mitsubishi] *see* 804
Novastat *see* 188 and 8908
Novatec [Merck & Co.] *see* 5393
Novatrin [Ayerst] *see* 4649
Novatropine *see* 4649
Novazam [Genevrier] *see* 2977
Novazole [Merck & Co.] *see* 1733
Novecyl *see* 8297

Novedrin *see* 3661
Novembichin, 6640
Novemina *see* 3358
Noveril [Morishita] *see* 2990
Noverme [Gist Brocades] *see* 5647
Novesine [Wander] *see* 1056
Novestrol [Rorer] *see* 3689
Novex *see* 3947
Noviben [Merck & Co.] *see* 1733
Novicet [Schwarzhaupt] *see* 9888
Novicodin *see* 3154
Novidium *see* 4650
Novidorm [Sintyal] *see* 9518
Novidroxin [Saarstickstoff-Fatol] *see* 4739
Noviform [Heyden] *see* 1220
Novil *see* 3358
Novismuth *see* 1298
Novobiocin, 6641
Novocain [Winthrop] *see* 7763
Novocainamid *see* 7762
Novocamid [Hoechst] *see* 7762
Novocebrin [Homburg] *see* 9378
Novocillin *see* 1157
Novocol *see* 4460
Novodigal [Beiersdorf] *see* 3150
Novodil [Chauvin-Blache] *see* 2708
Novodolan [Roussel-UCLAF] *see* 4037
Novodrin *see* 5105
Novoembichin *see* 6640
Novofluen [Engelhard KG] *see* 3598
Novoform *see* 1220
Novofosfan *see* 9435
Novogent N [Temmler] *see* 4812
Novohetramin *see* 9307
Novohydrin [Lakeside] *see* 396
Novolaudon *see* 4733
Novoldiamine, 6642
Novol Ketone *see* 6642
Novomazina *see* 2186
Novomycetin *see* 2068
Novon [SK & F] *see* 3587
Novonal, 6643
Novo-Nastizol A [Bago] *see* 878
Novophone *see* 2820
Novoridazine [Novopharm] *see* 9290
Novorin [Polfa] *see* 9993
Novosed [Neo Bologna] *see* 2082
Novoseptale *see* 8891
Novosparol [Hépatrol] *see* 1224
Novospasmin *see* 1743
Novotrone *see* 8947
Novotusil [Inpharzam] *see* 2393
Novoxapin [Ester] *see* 3425
Novrad [Lilly] *see* 5349
Novurit [Chinoin] *see* 5758
Novydrine *see* 616
Noxaben [Unifa] *see* 3073
Noxal *see* 3370
Noxibiol *see* 6577
Noxigram [Firma] *see* 2311
Noxiptilin, 6644
Noxiptyline *see* 6644
Noxiurotan *see* 7436
Noxyflex-S [Geistlich] *see* 6645
Noxythiolin *see* 6645
Noxytiolin, 6645
Nozinan [Rhône-Poulenc] *see* 5909
NP *see* 6539
NP 13 *see* 1036
NP 55 *see* 8424
NP 113 *see* 4881
NP 297 *see* 5682
NPA *see* 7704
NPA-3 *see* 6338
NPA Acid, 6646
NPAB *see* 7704
NPA Sodium *see* 6646
NPH 1320 *see* 1431
NPH Iletin *see* 5082

NPH Insulin *see* 5082
NPT 10381 *see* 4881
NPYR *see* 6564
NRC 910 *see* 4964
NRDC 107 *see* 1243
NRDC 143 *see* 7132
NRDC 149 *see* 2775
NRDC 161 *see* 2869
NSC-185 *see* 2734
NSC-1026 *see* 2740
NSC-1771 *see* 9304
NSC-2101 *see* 8251
NSC-3053 *see* 2804
NSC-3364 *see* 4022
NSC-3590 *see* 4141
NSC-5159 *see* 2035
NSC-5366 *see* 6638
NSC-5547 *see* 3455
NSC-6470 *see* 6445
NSC-8806 *see* 5708
NSC-9706 *see* 9586
NSC-10023 *see* 7727
NSC-11905 *see* 5235
NSC-13875 *see* 318
NSC-14279 *see* 7996
NSC-15200 *see* 4252
NSC-15780 *see* 629
NSC-17777 *see* 7292
NSC-19893 *see* 4109
NSC-20264 *see* 148
NSC-21626 *see* 7832
NSC-23436 *see* 3416
NSC-23759 *see* 9108
NSC-25154 *see* 7451
NSC-25614 *see* 7348
NSC-25855 *see* 9375
NSC-26805 *see* 3782
NSC-27640 *see* 4045
NSC-28693 *see* 6160
NSC-30152 *see* 3201
NSC-30689 *see* 6131
NSC-32519 *see* 3089
NSC-32946 *see* 6131
NSC-34462 *see* 9762
NSC-35051 *see* 5708
NSC-37095 *see* 9787
NSC-37096 *see* 1097
NSC-39084 *see* 918
NSC-39415 *see* 3411
NSC-39470 *see* 1202
NSC-40725 *see* 5113
NSC-43193 *see* 8754
NSC-45383 *see* 8789
NSC-45388 *see* 2803
NSC-47439 *see* 4119
NSC-47774 *see* 7452
NSC-49171 *see* 8307
NSC-52644 *see* 8024
NSC-52947 *see* 6938
NSC-55975 *see* 3568
NSC-60584 *see* 8963
NSC-64375 *see* 1133
NSC-68982 *see* 4469
NSC-69856 *see* 10073
NSC-70731 *see* 5378
NSC-70845 *see* 6591
NSC-71047 *see* 9647
NSC-75520 *see* 9599
NSC-77518 *see* 2977
NSC-77625 *see* 9515
NSC-78502 *see* 5658
NSC-79037 *see* 5444
NSC-82116 *see* 5178
NSC-83653 *see* 380
NSC-84223 *see* 8603
NSC-85998 *see* 8794
NSC-91523 *see* 7852
NSC-102627 *see* 4841
NSC-102816 *see* 907
NSC-104800 *see* 6132
NSC-107429 *see* 2710

[2-[2-(p-Octylcresoxy)ethoxy]ethyl]-
dimethylbenzylammonium Chloride
see 5946
Octylene *see* 1764
Octylene Glycol *see* 3699
sec-Octyl Iodide, 6686
2-Octyl-4-isothiazolin-3-one *see* 6677
2-Octyl-3(2H)-isothiazolone *see* 6677
Octyl Methoxycinnamate, 6687
β-p-tert-Octylphenoxyethyldiethyl-
benzylammonium Chloride *see* 7205
Octylphenoxy Polyethoxyethanol *see*
6681
5-[2-(Octylsulfinyl)propyl]-1,3-
benzodioxole *see* 8946
n-Octylsulfoxide of Isosafrole *see*
8946
Ocufen [Allergan] *see* 4124
Ocusert Pilo [Alza] *see* 7395
Ocu-Vinc [Alcon] *see* 9888
Ocytocin *see* 6934
ODA-914 *see* 2833
Odanon [Towa Yakuhin] *see* 1791
Odemase [Azuchemie] *see* 4221
ODGF *see* 7497
Odontalg [Giovanardi] *see* 5359
Odorigenin *see* 9809
Odylen [Winthrop] *see* 5821
Oedemex [Mepha] *see* 4221
Oekolp [Kade] *see* 3659
Oenanthal *see* 4578
Oenanthaldehyde *see* 4578
Oenanthic Acid *see* 4581
Oenanthic Ether *see* 3790
Oenanthol *see* 4578
Oenanthotoxin *see* 3524
Oenanthylic Acid *see* 4581
Oesipos *see* 5231
Oestergon *see* 3653
Oestradiol Monobenzoate *see* 3655
Oestrasid *see* 3094
Oestrin *see* 3660
Oestriol *see* 3659
Oestrodiene *see* 3094
Oestroform *see* 3660
Oestroform [BDH] *see* 3655
Oestrogel [Besins-Iscovesco] *see* 3653
Oestrogenine *see* 3118
Oestromenin *see* 3118
Oestromensyl *see* 3118
Oestromon [E. Merck] *see* 3118
Oestrone *see* 3660
Oestroperos *see* 3660
Oestroral *see* 3094
Off *see* 2848
Oflocet [Roussel] *see* 6688
Oflocin [Glaxo] *see* 6688
Ofloxacin, 6688
Ofloxacine *see* 6688
Oftalfrine *see* 7257
Oftanol [Bayer] *see* 5055
Ogen [Abbott] *see* 3660
Ogostal [Lilly] *see* 1758
Ogyline [Roussel] *see* 6622
OHB₁₂ *see* 4739
1α-OH-CC *see* 4749
OH-Duphar *see* 4739
Ohlexin [Ohta] *see* 1971
Ohton 3248
Oil of Allspice *see* 6760
Oil of Amber, Rectified, 6689
Oil of American Wormseed *see* 6713
Oil of Angelica, 6690
Oil of Anise, 6691
Oil Anise, Japanese, 6692
Oil of Anthemis *see* 6711
Oil of Arbor Vitae *see* 6779
Oil of Asarum, 6693
Oil of Balm, 6694
Oil of Basil, 6695
Oil of Bay, 6696

Oil Bergamot, 6697
Oil of Bitter Almond, 6698
Oil of Bitter Orange, 6699
Oil of Cade *see* 5153
Oil of Cajeput, 6700
Oil of Calamus, 6701
Oil of Camphor, Rectified, 6702
Oil of Canada Fleabane *see* 6730
Oil of Canada Snakeroot *see* 6693
Oil of Caraway, 6703
Oil of Cardamom, 6704
Oil of Cascarilla, 6705
Oil of Cashew Nut Shell, 6706
Oil of Cassia *see* 6715
Oil of Cedar Leaf, 6707
Oil of Cedar Wood, 6708
Oil of Celery, 6709
Oil of Chamomile—German, 6710
Oil of Chamomile—Roman, 6711
Oil of Champaca, 6712
Oil of Chenopodium, 6713
Oil of Cherry Laurel, 6714
Oil of Chinese Cinnamon *see* 6715
Oil of Cinnamon, 6715
Oil of Cinnamon, Ceylon, 6716
Oil of Citronella, 6717
Oil of Clove, 6718
Oil of Copaiba, 6719
Oil of Coriander, 6720
Oil of Crispmint *see* 6768
Oil of Cubeb, 6721
Oil of Cumin, 6722
Oil of Curled Mint *see* 6768
Oil of Cypress, 6723
Oil of Dill, 6724
Oil of Dwarf Pine Needles, 6725
Oil of Egg Yolk *see* 3493
Oil of Erigeron *see* 6730
Oil of Eucalyptus, 6726
Oil of Fennel, 6727
Oil of Fir, 6728
Oil of Fir—Siberian, 6729
Oil of Fleabane, 6730
Oil of Garlic, 6731
"Oil Garlic" *see* 295
Oil of Geranium, 6732
Oil of Geranium—East Indian, 6733
Oil of Ginger, 6734
Oil of Grapes *see* 3790
Oil Green *see* 2229
Oil of Hedeoma 6755
Oil of Hops, 6735
Oil of Hyssop, 6736
Oil of Jojoba *see* 5147
Oil of Juniper, 6737
Oil of Juniper Tar *see* 5153
Oil of Juniper Wood, 6738
Oil of Lavender, 6739
Oil of Lemon, 6740
Oil of Lemon Balm *see* 6694
Oil of Lemon Grass, 6741
Oil of Levant Wormseed, 6742
Oil of Linaloe, 6743
Oil of Mace *see* 6748
Oil of Marjoram, 6744
Oil of Melissa Balm *see* 6694
Oil of Mirbane *see* 6509
Oil of Monarda *see* 6153
Oil of Mountain Pine *see* 6725
Oil of Mustard, Expressed, 6745
Oil of Myrcia *see* 6696
Oil of Myristica *see* 6749
Oil of Myrtle, 6746
Oil of Neroli *see* 6751
Oil of Niaouli, 6747
Oil of Niobe *see* 5947
Oil Nut *see* 5149
Oil of Nutmeg, Expressed, 6748
Oil of Nutmeg, Volatile, 6749
Oil of Orange, 6750
Oil of Orange Flowers, 6751

Oil of Origanum, 6752
Oil of Palma Christi *see* 1904
Oil of Parsley, 6753
Oil of Patchouli, 6754
Oil of Pelargonium Geranium *see*
6732
Oil of Pennyroyal—American, 6755
Oil of Pennyroyal—European, 6756
Oil of Pepper, 6757
Oil of Peppermint, 6758
Oil of Pettigrain, 6759
Oil of Pimenta, 6760
Oil of Pimento *see* 6760
Oil of Pine *see* 6729
Oil of Pine Needles, 6761
Oil of Pulegium *see* 6756
Oil Red *see* 8858
Oil Red XO *see* 9997
Oil of Rice Bran *see* 8210
Oil of Rose, 6762
Oil of Rose Geranium *see* 6732
Oil of Rosemary, 6763
Oil of Rue, 6764
Oil of Santal, 6765
Oil of Sassafras, 6766
Oil of Savin, 6767
Oil Scarlet *see* 8858
Oil of Scotch Fir *see* 6761
Oil of Silver Fir *see* 6728
Oil of Silver Pine *see* 6728
Oil of Spearmint, 6768
Oil of Spike, 6769
Oil of Sweet Almond, 6770
Oil of Sweet Bay, 6771
Oil of Sweet Flag *see* 6701
Oil Sweet Orange *see* 6750
Oil of Tansy, 6772
Oil of Thuja *see* 6779
Oil of Thyme, 6773
Oil of Turpentine, 6774
Oil of Turpentine, Rectified, 6775
Oil of Turpentine, Sulfurated, 6776
Oil of Valerian, 6777
Oil of Vetiver, 6778
Oil of Vitriol *see* 8953
Oil White Birch *see* 1251
Oil of White Cedar, 6779
Oil of Wild Marjoram *see* 6752
Oil of Wine *see* 3790
Oil of Wine, "Heavy", 6780
Oil of Wormwood, 6781
Oil of Yarrow, 6782
Oizine *see* 7982
OK 174 *see* 1050
Okenite *see* 1709
Oksafenamide *see* 6843
Oksilidin *see* 8110
Oktadin *see* 4473
Oktatensin *see* 4473
Oktatenzin *see* 4473
Oktaverine *see* 6676
OKY-046 *see* 6935
Olamin [Sigamed] *see* 2308
Olamine *see* MISC-3 and 4
Olane [Avi-Sun] *see* 7558
Olaquindox, 6783
Olaxin [SK & F] *see* 2379
Olbemox [Farmitalia] *see* 106
Olbetam [Farmitalia] *see* 106
Olcadil [Sandoz] *see* 2415
Oldren [Roemmers] *see* 6068
Old Tuberculin *see* 9715
Old Yellow Enzyme, 6784
Oleandocetine *see* 9681
Oleandomycin, 6785
Oleandomycin Triacetate Ester *see*
9681
Oleandrin, 6786
Olean-12-en-3β-ol *see* 654
Oleanol *see* 6787
Oleanolic Acid, 6787

Oleate Of Mercury *see* 5778
Olefiant Gas *see* 3748
Oleic Acid, 6788
Oleic Acid Aluminum Salt *see* 357
Oleic Acid Ammonium Salt *see* 563
Oleic Acid Bismuth Salt *see* 1283
Oleic Acid Calcium Salt *see* 1690
Oleic Acid Lead Salt *see* 5293
Oleic Acid Potassium Salt *see* 7637
Olein *see* 9644
Oleochrysine *see* 1649
Oleocreosote *see* 2575
Oleoresin of Aspidium, 6789
Oleo Stock *see* 9014
Oleovitamin A *see* 9918
Oleovitamin D$_2$ *see* 9928
Oleovitamin D$_3$ *see* 9929
Oleptan [Bender] *see* 4144
Olestra *see* 8857
Oleth *see* 7554
Oleum *see* 8953
Oleum Abietis *see* 6729
Oleum Andropogonis Muricati *see* 6778
Oleum Rusci *see* 1251
Oleum Vitis Viniferae *see* 3790
Oleuropein, 6790
Oleyl Alcohol, 6791
Olibanum, 6792
Olicard [Giulini] *see* 5114
Olicin *see* 9681
Oligomycin A *see* 6793
Oligomycin B *see* 6793
Oligomycin C *see* 6793
Oligomycin D *see* 6793
Oligomycins, 6793
Olimpen *see* 7052
Olimplex [Vaillant-Defresne] *see* 3151
Olitensol *see* 9430
Olivacine, 6794
Olivanic Acids, 6795
Olive Oil, 6796
Oliver [Sanzen] *see* 6841
Olive Spurge *see* 6095
Olivil, 6797
Olivin *see* 6798
Olivin [Lek] *see* 3521
Olivomycin A *see* 6798
Olivomycins, 6798
Olmagran [Heyden] *see* 4716
Olmelin *see* 4281
Olmifon [Lafon] *see* 159
Ololiuqui *see* 4959
Olothorb *see* 7559
Olsalazine, 6799
Olymp [Du Pont] *see* 4130
Olynth [Sasse] *see* 9993
OM-518 *see* 5739
OM-805 *see* 804
Omadine *see* 8004
Omadine Disulfide [Olin] *see* 8004
Omaflora [Olin] *see* 4696
Omaine *see* 2873
Omal *see* 9556
Omca [Heyden] *see* 4116
Omcilon *see* 9511
Omcilon-A *see* 9512
OMD *see* 5975
OM-dopa *see* 5975
OMDS *see* 8004
Omegamycin *see* 9130
Omepral [Fujisawa] *see* 6800
Omeprazole, 6800
Omeral [Simes] *see* 6172
Omeril [Bayer] *see* 5649
Omite [Uniroyal Chem.] *see* 7818
Omnes [Fumouze] *see* 6447
Omnibex *see* 7232
Omnipaque [Nyegaard; Schering AG] *see* 4940
Omnipen [Wyeth] *see* 621

Omnipen-N [Wyeth] *see* 621
Omnisan [Squibb] *see* 3566
Omnizole [Merck & Co.] *see* 9217
Omnopon [Roche] *see* 6967
Omnyl *see* 5872
Omoconazole, 6801
Omperan [Taiho] *see* 8971
OMS 29 *see* 1789
OMS 45 *see* 3922
OMS 115 *see* 3394
OMS 658 *see* 1419
OMS 659 *see* 1419
OMS 968 *see* 4160
OMS 1804 *see* 3128
OMS 3023 *see* 3952
Omsat [Boehringer, Mann.] *see* 8889
Omtan *see* 5010
ONB *see* 6513
Oncol [Otsuka] *see* 1050
Oncotrex [Warner-Lambert] *see* 9635
Oncovin [Lilly] *see* 9891
Ondansetron, 6802
Ondena [Bayer] *see* 2825
Ondogyne [Roussel-UCLAF] *see* 2726
Ondonid [Roussel-UCLAF] *see* 2726
One-Alpha [Leo Pharm.] *see* 4749
One-Iron *see* 3995
Onion Oil, 6803
Oniria [Lifepharma] *see* 8039
Onkotin *see* 2925
Ono-802 *see* 4279
ONO-1206 *see* 5367
ONO-1308 *see* 6827
Ono-2235 *see* 3556
Onokrein P [Ono] *see* 5159
Ononin *see* 4157
Onquinin [Ono] *see* 784
Onsukil [Grünenthal] *see* 7765
Ontianil, 6804
Ontosein [Diagnostic Data] *see* 6820
Onychomal [Hermal] *see* 9781
Onyxide *see* 3747
Oöcyan *see* 1236
Ooporphyrin *see* 7913
Oosporein, 6805
OP-1206 *see* 5367
Opacin *see* 4931
Opacist E. R. [Bracco] *see* 4904
Opacoron [Cilag-Chemie] *see* 72
Opal *see* 8437
Opalene [Théraplix] *see* 9634
Opalmon [Ono] *see* 5367
Opalwax *see* 1905
Oparenol *see* 4933
OPC 1005 *see* 1875
OPC-2009 *see* 7765
OPC-13013 *see* 2277
Opclor [Parke, Davis] *see* 2068
Opcon [Bausch & Lomb] *see* 6287
Opegan [Santen] *see* 4675
Operidine [Janssen] *see* 7216
OPG *see* 6930
Ophiobolin D *see* 1974
Ophtagram [Chauvin-Blache] *see* 4284
Ophtalmokalixan [Bristol] *see* 5161
Ophtamedine [De Bournonville] *see* 4612
Ophthaine [Squibb] *see* 7817
Ophthalamin (obsolete) *see* 9918
Ophthalgan [Wyeth-Ayerst] *see* 4379
Ophthalmadine [SAS] *see* 4819
Ophthetic *see* 7817
Ophthochlor [Parke, Davis] *see* 2068
Ophtocortin [Winzer] *see* 5679
Ophtorenin [Winzer] *see* 1486
Ophtosol [Winzer] *see* 1379
Opian *see* 6638
Opianic Acid, 6806
Opianine *see* 6638

Opianyl *see* 5665
Opilon [Parke, Davis] *see* 6204
Opiniazide, 6807
Opino [Bayropharma] *see* 6654
Opinsul *see* 8890
Opipramol, 6808
Opiran [Cassenne] *see* 7404
Opium, 6809
Opobalsam *see* 960
Opocarbyl *see* 1815
Oposim [Richet] *see* 7852
Opren [Lilly] *see* 1055
Opridan [Locatelli] *see* 1420
Opromazine, 6810
Opsins, 6811
α$_2$-Opsonins *see* 4018
Op-Sulfa [Broemmel] *see* 8870
Optal *see* 7854
Optanox *see* 9897
Op-Thal-Zin [Alcon] *see* 10064
Optically Active Amyl Bromide *see* 636
Opticlox [Norbrook] *see* 2414
Opticrom [Fisons] *see* 2594
Opticron [Fisons] *see* 2594
Optidase [Solac] *see* 1906
Optimax [E. Merck] *see* 9707
Optimil [Wallace Labs.] *see* 5872
Optimine [Schering] *see* 917
Optimycin [Biochemie; Sandoz] *see* 5851
Optinoxan [Robisch] *see* 5872
Optipect [Thiemann] *see* 2896
Optipen *see* 7184
Optipress [Wellcome] *see* 1875
Optisulin Long [Hoechst] *see* 4887
Optium [Disprovent] *see* 610
Optojod *see* 8687
Optoquine *see* 3765
Optovit-A *see* 9918
Optovit-E [Hermes] *see* 9932
Optraex *see* 10064
Opturem [Kade] *see* 4812
Opyrin [Taisho] *see* 4060
Orabet [Upjohn] *see* 9432
Orabilex [Fougera] *see* 1476
Orabilix [Guerbet] *see* 1476
Orabiotic *see* 8226
Orabolin [Organon] *see* 3761
Oracaine Hydrochloride *see* 5752
Oracef [Lilly] *see* 1971
Oracéfal [Bristol] *see* 1921
Oracillin [Théraplix] *see* 7046
α-Oracillin *see* 7184
Oracil-VK *see* 7046
Oracocin [Tobishi] *see* 1971
Oracon [Mead Johnson] *see* 3208
Oradexon [Organon] *see* 2922
Oradian [Chinoin] *see* 2187
Oradiol [Mallinckrodt] *see* 3689
Oradol [Ciba-Geigy] *see* 3411
Oraflex [Lilly] *see* 1055
Orafuran *see* 6520
Oragallin *see* 927
Oragest *see* 5677
Orageston [Organon] *see* 289
Oragrafin-Calcium *see* 4958
Oragrafin-Sodium [Squibb] *see* 4958
Oragulant *see* 3304
Orahexal [Siegfried] *see* 2090
Oralcid *see* 44
Oraldene [Wm. R. Warner] *see* 4624
Oralin *see* 9432
Oralopen [Bayer] *see* 7184
Ora-Lutin [Parke, Davis] *see* 3696
Oramid *see* 8297
Oramorph [Boehringer, Ing.] *see* 6193
Oranabol [Farmitalia] *see* 6918
Orange I, 6812
Orange II, 6813

Cross Index of Names

Orange III see 6019
Orange IV see 9688
Orange B, 6814
Orange Crush [Cyanamid] see 1255
Orange GS see 9688
Orange N see 9688
Orange Peel, Bitter, 6815
Orange Peel, Sweet, 6816
Orange Root see 4690
Oranil [Arzneimittelwerk VEB] see 1839
Oranixon [Organon] see 5737
Orap [Janssen] see 7404
Orapen see 7046
Orarsan [Boots] see 44
Orasone [Rowell] see 7727
Oraspor [Ciba-Geigy] see 1942
Orasthin [Hoechst] see 6934
Orasulin see 1839
Oratestin [Hoechst] see 4113
Ora-Testryl [Squibb] see 4113
Oratrast [Barnes-Hind] see 1006
Oratren [Bayer] see 7046
Oratrol [Alcon] see 3067
Oravue [Squibb] see 4949
Orazamide, 6817
Orbenin [Beecham] see 2414
Orbenin Dry Cow [Beecham] see 2414
Orbicilina [Bago] see 621
Orbicin [Pfizer/Mack, Illert.] see 2987
Orbinamon [Pfizer] see 9297
Orbit [Ciba-Geigy] see 7830
Orcanette see 242
Orcein, 6818
Orchil see 2618
Orchisterone-M [Merck & Co.] see 6044
Orchisterone-P [Merck & Co.] see 9115
Orcin see 6819
Orcinol, 6819
Orcinolcarboxylic Acid see 6834
Orciprenaline see 5836
Ordeal Bark see 8341
Ordeal Bean see 7356
Ordiflazine see 5360
Ordimel [Lilly] see 53
Ordinary Lactic Acid see 5215
Ordinary Tartaric Acid see 9039
Ordinator [Dausse] see 3932
Oregon Grape Root see 1170
Orencil see 7038
Oreoselone Methyl Ether see 7144
Oresol see 4465
Oreson see 4465
Orestol see 3119
Orestralyn [McNeil] see 3689
Oretic [Abbott] see 4704
Oreton [Schering] see 9115
Oreton-F [Schering] see 9109
Oreton-M [Schering] see 6044
ORF 10131 see 6620
ORF 11676 see 6274
ORF 15244 see 9338
ORF 15817 see 3486
ORF 15927 see 8225
Orfenso see 6633
Orfiril [Desitin] see 9821
ORG-817 see 3568
Org-2969 see 2906
Org-10172 see 5227
Orgabolin (obsolete) [Organon] see 3761
Orgaboral see 3761
Orgadrone [Sankyo] see 2922
Orgametil see 5501
Orgametril [Organon] see 5501
Organidin [Wampole] see 4906
Organoderm [Mundipharma] see 5582
Orgasteron [Organon] see 6629
Orgastyptin [Organon] see 3659

Orgasuline [Organon] see 4887
Orgatrax [Organon] see 4786
Org GB 94 see 6097
Org NA 97 see 6958
Org NC 45 see 9845
Org OD 14 see 9358
Orgotein, 6820
Oricillin [Grünenthal] see 7829
Oriconazole see 5131
Oricur see 2104
Oriens [Asche] see 69
Oriental Berry see 2452
Orientomycin see 2758
Orifungal M [Janssen] see 5181
Orimeten [Ciba] see 452
Orimon see 7220
Orinase [Upjohn] see 9432
Oriodide [Abbott] see 8583
Orion see 9511
Oripavine, 6821
Orisul [Ciba] see 8910
Orisulf see 8910
Orix [Biomedica] see 6441
Orizaba Jalap Root see 4959
γ-Orizanol see 6841
Orkanet see 242
Orlon®, 6822
Ormeta [Tanabe Seiyaku] see 862
Ormetein (rescinded) see 6820
Ormosanine see 6823
21-Ormosanin-20-yl Panamine see 6823
Ormosinine, 6823
Ornicetil [Logeais; Nordmark] see 5182
Ornid see 1364
Ornidal [Selvi] see 6824
Ornidazole, 6824
Ornidyl [Merrell Dow] see 3489
Ornipressin, 6825
Ornithine, 6826
L-Ornithine L-Aspartate see 862
Ornithine α-Ketoglutarate see 5182
Ornithine Vasopressin see 9843
8-L-Ornithinevasopressin see 6825
Ornitrol [Searle] see 908
Ornoprostil, 6827
Orn(8)-vasopressin see 6825
Orobetina [Bergamon] see 7896
Orobronze [Applipharm] see 1756
Orocillin see 7046
Orofar [Zyma] see 1123
Orofungin [SK & F] see 5733
Oroken [Pharmuka] see 1927
Oronol [Schering] see 901
Oropur see 6828
Orosulfan see 8909
Orotic Acid, 6828
Orotic Acid Compd with 5(or 4)-Aminoimidazole-4(or 5)-Carbox-amide (1:1) see 6817
Orotidine, 6829
Oroturic see 6828
Orotyl see 6828
Orovermol [Brasil] see 9161
Oroxine see 5351
Oroxylin see 6830
Oroxylin A, 6830
Orparan see 862 and 6826
Orphenadrine, 6831
Orphol [Mack-Midy] see 3598
Orpidan [Heumann] see 2073
Orpiment see 834
Orpizin see 2073
Orquisteron [Merck & Co.] see 9109
Orquisteron-E [Merck & Co.] see 9112
Orquisteron-P [Merck & Co.] see 9115
Orris, 6832
Orris Root Oil, 6833

Orsanil [Orion] see 9290
Orseilles see 2618
o-Orsellinic Acid, 6834
Orsellinic Acid Methyl Ester 4-Meth-yl Ether see 8690
Orsile see 1045
Orsin see 7256
Orstanorm [Sandoz] see 3157
Orsudan see 5945
Orsulon see 8913
Ortacrone [Pharmainvesti] see 497
Ortal Sodium see 4623
Ortazol see 4210
Ortédrine see 616
Ortensan [Cimex] see 40
Orthanilic Acid, 6835
Orthene [Chevron] see 26
Orthesin see 3719
Ortho 5353 see 1459
Ortho 9006 see 5858
Ortho 12420 see 26
Orthoarsenic Acid see 821
[Orthoborato(3—)-O]phenylmercur-ate(2—) Dihydrogen see 7274
Orthoboric Acid see 1336
Orthocaine, 6836
Orthocide-406 see 1771
Orthoclase see 7579
Orthocoll see 7617
Ortho-Creme [Ortho] see 6596
Orthodiazine see 7982
Ortho-Dibrom [Chevron] see 6272
Orthodichlorobenzene see 3044
Orthoform [Winthrop] see 6836
Orthoformic Acid, 6837
Orthoform New see 6836
Ortho-Gynest [Cilag] see 3659
Ortho-Klor see 2079
Ortho-Mite see 794
Ortho-Novin 1/50 [Ortho-Cilag] see 6614
Ortho-Novum 1/35 [Ortho] see 6614
Ortho-Novum 1/50 [Ortho] see 6614
Ortho-Novum 7/7/7 [Ortho] see 6614
Orthophosphoric Acid see 7318
Orthophthalic Acid Didiethylamide see 9137
Ortho Spotless [Ortho] see 3261
Orthotelluric Acid see 9063
Orthotitanic Acid see 9395
Orthotran see 6856
Orthovaleric Acid, 6838
Orthoxenol see 7276
Orthoxine [Upjohn] see 5919
Orthoxycol see 4708
Orticalm [Squibb] see 8149
Ortin see 9682
Ortisporina [Turro] see 1971
Ortizon see 9783
Ortodrinex see 5919
Ortonal see 5872
Ortrel [Ortho] see 6620
Ortyn Retard see 1090
Orudis [M & B] see 5184
Orugesic [Rhône-Poulenc] see 5184
Oruvail [M & B] see 5184
Orvagil see 6079
Oryvita [Ijaku] see 6841
Oryzaal [Sankei] see 6841
Oryzacidin, 6839
Oryzacidin A see 6839
Oryzalin, 6840
γ-Oryzanol, 6841
OS-1897 see 3003
OS-2046 see 6089
Osacyl see 491
Osadrin see 7217
Osajin, 6842
Osalmid, 6843
Osarsal see 44
Osarsol [Rosse] see 44

Osbil [Upjohn; M & B] see 4901
Oscine see 8365
Oscorel [SK & F] see 5184
Osiren [Hoechst] see 8721
Osmarins see 6845
Osmaron B, 6844
Osmic Acid see 6848
Osmitrol [Baxter] see 5629
Osmium, 6845
Osmium Ammonium Chloride see 564
Osmium Hexafluoride, 6846
Osmium Octafluoride: see 6846
Osmium Potassium Chloride see 7618
Osmium Tetrachloride, 6847
Osmium Tetroxide, 6848
Osmosal see 5629
Osmosin [Merck & Co.] see 4874
Osnervan [Burroughs Wellcome] see 7770
Ospamox [Biochemie] see 610
Ospen [Sandoz] see 7047
Ospen see 7046
Ospeneff [Sandoz] see 7046
Ospolot [Bayer] see 8974
Ossalin [Chemipharm] see 8565
Ossazone see 2195
Ossein see 2476
Ossian [Bioindustria] see 6899
Ossin [Sulzbach-Neuweiler] see 8565
Ossipirina see 2195
Ossiten [Robin] see 2369
Ossopan [Labaz] see 3449
Ostac [Boehringer, Mann.] see 2369
Ostelin see 9928
Osten [Takeda] see 4961
Ostensin [Wyeth] see 9622
Osteofluor [Merck-Clevenot] see 8565
Osteol [Yamanouchi] see 4273
Osteosarcoma-derived Growth Factor see 7497
Osthole, 6849
Ostreogrycin A see 9910
Ostreogrycin B see 6111
Ostreogrycin B₁ see 9865
Ostreogrycin B₂ see 9865
Ostrocilline see 7047
Ostro-Primolut [Schering AG] see 6614
Ostruthin, 6850
Ostruthol, 6851
Osvan see 1066
Osvarsan see 44
Osyritin see 8276
Osyritrin see 8276
Osyrol [Hoechst] see 8721
Otavite see 1614
Oterben see 9432
Otifuril see 4205
Otobain, 6852
Otobite see 6852
Otodyne see 10089
Otokalixin see 5161
Otriven [Ciba] see 9993
Otrivin [Ciba] see 9993
Otrix [Ciba] see 9993
Otrun see 2835
Otto of Rose see 6762
OU-1308 see 6827
Ouabagenin, 6853
Ouabain, 6854
Ourari see 2678
Oust [Du Pont] see 8936
Ovaban [Schering] see 5687
Ovahormon see 3653
Ovalbumin, 6855
Ovalyse [Upjohn] see 4010
Ovanon [Nourypharma] see 5501
Ovaras [Serono] see 3816
Ovarelin see 5354
Ovasterol [Merck & Co.] see 3653
Ovasterol-B [Merck & Co.] see 3655

Ovastol see 5819
Ovazyme see 4354
Ovcon [Mead Johnson] see 6614
Ovesterin see 3659
Ovestin [Organon] see 3659
Ovex, 6856
Ovex [Mead Johnson] see 5687
Ovex [AFI] see 3660
Ovex B [AB Leo] see 3655
Ovicox [Hoechst] see 8305
Ovifollin see 3660
Ovin [Mead Johnson] see 3208
Oviol [Nourypharma] see 2906
Ovisot see 81
Ovitelmin [Cilag-Chemie] see 5647
Ovochlor see 6856
Ovocyclin [Ciba] see 3653
Ovocyclin-P [Ciba] see 3653
Ovocyclin Benzoate [Ciba] see 3655
Ovocyclin Dipropionate [Ciba] see 3653
Ovocyclin-M [Ciba] see 3655
Ovocyclin-MB [Ciba] see 3655
Ovocylin [Ciba] see 3653
Ovoester see 877
Ovol [Armour Pharm.] see 8486
Ovoresta [Organon] see 5501
Ovotox see 6856
Ovotran see 6856
Ovotransferrin see 9490
Ovo-Vinces [Wolff] see 3659
Ovral [Wyeth] see 6621
Ovrette [Wyeth] see 6621
O-V Statin [Squibb] see 6658
Ovulen see 3816
Ovysmen [Ortho-Cilag] see 6614
Oxabel [Sarva] see 6858
7-Oxabicyclo[2.2.1]heptane-2,3-di-carboxylic Acid see 3530
Oxaceprol, 6857
Oxacillin, 6858
Oxacycline [Crookes Vet.] see 6931
Oxacyclohexadecan-2-one see 3867
Oxacyclotetradecane Erythromycin Deriv see 8253
1,3,4-Oxadiazole, 6859
2-(1,3,4-Oxadiazol-2-yl)phenol see 3898
Oxadiazon, 6860
Oxaflozane, 6861
Oxaflumazine, 6862
Oxaflumine [Diamant] see 6862
Oxafuradene (rescinded) see 6445
Oxaine [Wyeth] see 6888
Oxalacetic Acid, 6863
Oxalaldehyde see 4405
Oxalamide see 6871
Oxaldin [Daiichi] see 6688
Oxalenediuramidoxime, 6864
Oxalic Acid, 6865
Oxalic Acid Bismuth Salt see 1284
Oxalic Acid Diamide see 6871
Oxalic Acid Diethyl Ester see 3115
Oxalic Acid Dinitrile see 2698
Oxalid [USV] see 6925
Oxaloacetic Ester see 3791
Oxalodiacetic Acid see 5179
Oxalomolybdic Acid, 6866
[Oxalylbis(iminoethylene)]bis[(o-chlorobenzyl)diethylammonium Chloride] see 387
Oxalyl Chloride, 6867
Oxalylurea see 6970
Oxamarin, 6868
Oxametacine, 6869
Oxamic Acid, 6870
Oxamic Acid Hydrazide see 8397
Oxamide, 6871
Oxamidic Acid see 6870
Oxaminozoline see 8220
Oxammonium Hydrochloride see 4759

Oxammonium Sulfate see 4759
Oxamniquine, 6872
Oxamycin [Merck & Co.] see 2758
Oxamyl, 6873
Oxanamide, 6874
Oxandrolone, 6875
Oxanid [Steinhard] see 6881
Oxantel, 6876
Oxantel Embonate see 6876
Oxaphenamide see 6843
Oxaphor see 4747
Oxapium Iodide see 2743
Oxapro [Wyeth] see 6879
Oxapropanium Iodide, 6877
Oxaprotiline, 6878
Oxaprozin, 6879
Oxa-Puren [Klinge-Nattermann] see 6881
Oxarmin see 6898
Oxatets see 6931
Oxatomide, 6880
Oxatone [Baxter] see 3166
Oxazacort see 2852
Oxazepam, 6881
Oxazidione, 6882
Oxazocilline see 6858
Oxazolam, 6883
Oxazolazepam see 6883
Ox Bile Extract, 6884
Oxcord [Alcon] see 6441
Oxedix [Labaz] see 6889
Oxedrine see 8994
Oxeladin, 6885
Oxendolone, 6886
Oxenin, 6887
Oxepinac see 5126
Oxetacaine see 6888
Oxetane see 9630
2-Oxetanone see 7832
Oxethazaine, 6888
Oxethazine see 6888
Oxetorone, 6889
Oxeze [Astra] see 3544
Oxfendazole, 6890
Oxiamine [Made] see 4880
Oxiarsolan see 6902
Oxibendazole, 6891
Oxibuprokain see 1056
Oxibutinina Hydrochloride see 6908
Oxichlorochine see 4748
Oxiclipine see 7824
Oxiconazole Nitrate, 6892
Oxidimethiin, 6893
Oxidized Cellulose, 6894
Oxidized Glutathione see 4369
Oxikon see 6912
Oxilapine see 5461
Oxilin [Allergan] see 6919
1-[p-(1-Oximidoethyl)phenoxyacetyl]-piperidine see 7392
Oxine see 4778
Oxine Sulfate see 4779
Oxiniacic Acid, 6895
Oxinofen see 6910
Oxinorm [Zambeletti] see 6820
Oxipendyl see 6923
Oxipethidine see 4769
Oxiracetam, 6896
Oxirane see 3758
Oxiranemethanol see 4385
Oxistat [Glaxo] see 6892
Oxitol see 3707
Oxitriptan see 4784
Oxitropium Bromide, 6897
Oxivent [Boehringer, Ing.] see 6897
Oxlopar [Parke, Davis] see 6931
Oxoacetic Acid see 4407
3-(3-Oxo-7α-acetylthio-17β-hydr-oxy-4-androsten-17α-yl)propionic Acid γ-Lactone see 8721
2-Oxo-acid Carboxy-lyase see 8031

Cross Index of Names

Oxobemin *see* 4739

α-Oxobenzeneacetaldehyde Aldoxime *see* 5077

β-Oxobenzenepropanoic Acid Ethyl Ester *see* 3726

2-Oxo-2*H*-1-benzopyran-3-carboxylic Acid *see* 2564

4-Oxo-4*H*-1-benzopyran-2-carboxylic Acid *see* 2239

2-Oxo-2*H*-1,3-benzoxazine-3(4*H*)-acetamide *see* 1865

γ-Oxo-[1,1′-biphenyl]-4-butanoic Acid *see* 3906

2-Oxo-3,3-bis[*p*-aminophenyl]butane *see* 615

Oxoboi [B.O.I.] *see* 6899

d-2-Oxo-3-bornanecarboxylic Acid *see* 1737

2-Oxo-10-bornanesulfonic Acid *see* 1740

2-Oxobutane *see* 5991

Oxobutanedioic Acid *see* 6863

Oxobutanedioic Acid Diethyl Ester *see* 3791

3-Oxobutanoic Acid *see* 51

3-Oxobutanoic Acid Ethyl Ester *see* 3714

3-Oxobutanoic Acid Methyl Ester *see* 5933

3-Oxobutanoic Acid 1-Methylethyl Ester *see* 5094

N-(1-Oxobutyl)adenosine Cyclic 3′,-5′-(Hydrogen Phosphate) 2′-Butanoate *see* 1448

4-(3-Oxobutyl)-1,2-diphenyl-3,5-pyrazolidinedione *see* 5168

δ-Oxo-α-butylene *see* 6052

d-2-Oxo-3-camphanecarboxylic Acid *see* 1737

4-Oxo-β-carotene *see* 3469

4-Oxo-2-(β-chloroethyl)-2,3-dihydrobenzo-1,3-oxazine *see* 2195

4-Oxo-4*H*-chromene-2-carboxylic Acid *see* 2239

N-(4-Oxo-2,5-cyclohexadien-1-ylidene)acetamide *see* 49

[(4-Oxo-2,5-cyclohexadien-1-ylidene)amino]guanidine Thiosemicarbazone *see* 386

(±)-*p*-[(2-Oxocyclopentyl)methyl]-hydratropic Acid *see* 5462

(*E*)-9-Oxo-2-decenoic Acid *see* 8042

17β-[(1-Oxodecyl)oxy]estr-4-en-3-one *see* 6281

17α-Oxo-D-homo-1,4-androstadiene-3,17-dione *see* 9108

9-Oxo-11α,15α-dihydroxy-17*S*,20-dimethylprosta-*trans*-2,*trans*-13-dienoic Acid *see* 5367

6-Oxo-17*S*,20-dimethyl-PGE₁ Methyl Ester *see* 6827

16-Oxoeburnane *see* 3464

Oxoethanoic Acid *see* 4407

1-Oxo-2-[*p*-[(α-ethyl)carboxymethyl]phenyl]isoindoline *see* 4867

γ-Oxo-8-fluoranthenebutanoic Acid *see* 4040

Oxoglurate *see* MISC-4

2-Oxoglutaric Acid *see* 5182

3-Oxoglutaric Acid *see* 60

3-Oxo-L-gulofuranolactone (Enol Form) *see* 855

2-Oxo-L-gulonic Acid *see* 5183

Oxogulonic Acid Diacetonide *see* 3184

17-[(1-Oxoheptyl)oxy]androst-4-en-3-one *see* 9112

2-Oxohexamethylenimine *see* 1762

D-*erythro*-3-Oxohexonic Acid Lactone *see* 5009

1-(5-Oxohexyl)-3,7-dimethylxanthine *see* 7092

1-(5′-Oxohexyl)-3-methyl-7-propylxanthine *see* 7823

17-[(1-Oxohexyl)oxy]-19-norpregn-4-ene-3,20-dione *see* 4309

17-[(1-Oxohexyl)oxy]pregn-4-ene-3,20-dione *see* 4774

1-(5-Oxohexyl)theobromine *see* 7092

Oxo(hydrogen *N*-Glycoloylarsanilato)bismuth *see* 4389

3-Oxo-17β-hydroxy-1,4-androstadiene *see* 1327

3-(3-Oxo-17β-hydroxy-4,6-androstadien-17α-yl)propionic Acid γ-Lactone *see* 1753

3-Oxo-17β-hydroxyandrostane *see* 8753

2-Oxo-3-isobutyl-9,10-dimethoxy-1,2,3,4,6,7-hexahydro-11b*H*-benzo-[*a*]quinolizine *see* 9118

p-(1-Oxo-2-isoindolinyl)hydratropic Acid *see* 4875

(±)-2-[*p*-(1-Oxo-2-isoindolinyl)phenyl]butyric Acid *see* 4867

2-[4-(1-Oxo-2-isoindolinyl)phenyl]propionic Acid *see* 4875

Oxolamine, 6898

Oxolamine [Arcum] *see* 4739

Oxole *see* 4206

Oxolinic Acid, 6899

Oxomalonic Acid *see* 5814

Oxomalonic Acid Calcium Salt *see* 1685

Oxomemazine, 6900

Oxomethane *see* 4148

γ-Oxo-4-methoxy-1-naphthalenebutyric Acid *see* 5719

1-Oxo-2-[*p*-[(α-methyl)carboxymethyl]phenyl]isoindoline *see* 4875

Oxonic Acid, 6901

3-(1-Oxo-4,7-nonadienyl)oxiranecarboxamide *see* 1997

28-Oxooligomycin *A see* 6793

[(5-Oxo-1,3,2-oxathiostibolan-2-yl)-thio]acetic Acid Sodium Salt *see* 735

1-[2-Oxo-2-[[2-[(1-oxododecyl)oxy]-ethyl]amino]ethyl]pyridinium Chloride *see* 5238

2-Oxopentanedioic Acid *see* 5182

2-Oxo-1,5-pentanedioic Acid *see* 5182

3-Oxopentanedioic Acid *see* 60

4-Oxopentanoic Acid *see* 5352

4-Oxopentanoic Acid Calcium Salt *see* 1684

4-Oxopentanoic Acid Ethyl Ester *see* 3775

Oxophenarsine Hydrochloride, 6902

4-Oxo-4-(phenylamino)butanoic Acid *see* 8839

(*Z*)-4-Oxo-4-(phenylamino)-2-butenoic Acid *see* 5584

Oxophenylarsine, 6903

[(4-Oxo-2-phenyl-4*H*-1-benzopyran-7-yl)oxy]acetic Acid Ethyl Ester *see* 3490

3-Oxo-*N*-phenylbutanamide *see* 50

5-Oxo-*N*-(2-phenylcyclopropyl)-2-pyrrolidinecarboxamide *see* 8234

Oxo-[[(1-phenylethyl)amino]acetic Acid Hydrazide *see* 7260

(17β)-(1-Oxo-3-phenylpropoxy)estr-4-en-3-one *see* 6283

11-Oxoprogesterone *see* 5185

5-Oxo-L-proline *see* 8012

5-Oxo-L-prolyl-L-histidyl-L-prolinamide *see* 9503

5-Oxo-L-prolyl-L-histidyl-L-tryptophyl-L-seryl-L-tyrosyl-3-(2-naphthyl)-D-alanyl-L-leucyl-L-arginyl-L-prolylglycinamide *see* 6265

2-Oxopropanal *see* 8030

2-Oxopropanal 1-Oxime *see* 5076

Oxopropanedioic Acid *see* 5814

2-Oxopropanoic Acid *see* 8032

17β-(1-Oxopropoxy)estr-4-en-3-one *see* 6284

2-Oxo-1,2*H*-pyran-5-carboxylic Acid *see* 2558

4-Oxo-1,4-pyran-2,6-dicarboxylic Acid *see* 2042

4-Oxo-4*H*-pyran-2,6-dicarboxylic Acid *see* 2042

2-Oxopyrrolidine *see* 8027

2-Oxo-1-pyrrolidineacetamide *see* 7459

5-Oxo-2-pyrrolidinecarboxylic Acid *see* 8012

1-[2-Oxo-2-(1-pyrrolidinyl)ethyl]-4-[1-oxo-3-(3,4,5-trimethoxyphenyl)-2-propenyl]piperazine *see* 2293

1-[4-[2-Oxo-2-(1-pyrrolidinyl)ethyl]-1-piperazinyl]-3-(3,4,5-trimethoxyphenyl)-2-propen-1-one *see* 2293

1-(2-Oxo-1-pyrrolidinyl)-4-(1-pyrrolidinyl)-2-butyne *see* 6904

Oxo(salicylato)bismuth *see* 1299

Oxosuccinic Acid *see* 6863

1-Oxo-3-(3-sulfamyl-4-chlorophenyl)-3-hydroxyisoindoline *see* 2194

5-Oxo-4-[(4-sulfo-1-naphthyl)azo]-1-(*p*-sulfophenyl)-2-pyrazoline-3-carboxylic Acid 3-Ethyl Ester Disodium Salt *see* 6814

2-Oxo-1,2,3,4-tetrahydroquinoline *see* 4702

8-Oxo-7-[(1*H*-tetrazol-1-ylacetyl)-amino]-3-[(1,3,4-thiadiazol-2-yl-thio)methyl]-5-thia-1-azabicyclo-[4.2.0]oct-2-ene-2-carboxylic Acid *see* 1946

4-Oxo-2-thiazolidinehexanoic Acid *see* 6233

4-Oxo-4-[4-[(2-thiazolylamino)sulfonyl]phenyl]amino]butanoic Acid *see* 8850

N-[1-Oxo-2-[(2-thienylcarbonyl)-thio]propyl]glycine *see* 8764

4-Oxo-2-thionothiazolidine *see* 8184

9-Oxothioxanthene *see* 9301

Oxotremorine, 6904

4-[1-Oxo-3-(3,4,5-trimethoxyphenyl)-2-propenyl]-1-piperazineacetic Acid Ethyl Ester (*Z*)-2-Butenedioate(1:1) *see* 2292

22-Oxovincaleukoblastine *see* 9891

9-Oxoxanthene *see* 9971

Oxpentifylline *see* 7092

Oxphylline [Amido] *see* 3835

Oxprenolol, 6905

Oxsoralen Ultra [ICN] *see* 5911

Oxtriphylline *see* 2213

Oxucide *see* 7433

Oxurasin *see* 7432

Oxy-5 [USV] *see* 1128

Oxyacanthine, 6906

OxyBan *see* 4354

Oxybenzene *see* 7259

Oxybenzone *see* 7206

Oxybenzone, 6907

Oxybenzopyridine *see* 4778

Oxybiocycline *see* 6931

Oxybiotic [Star Pharm.] *see* 6931

β,β′-Oxybis[*p*-acetophenetidide] *see* 2953

1,1′-Oxybisbenzene *see* 7259

4,4′-Oxybisbenzenecarboximidamide *see* 7165

1,1′-Oxybis[butane] *see* 1568

4,4′-Oxybis[2-butanol] *see* 6913

1,1′-Oxybis[2-chloroethane] *see* 3055

Oxybis[chloromethane] *see* 3058

5,5′-Oxybis[3,4-dichloro-2(5*H*)-furanone] *see* 6209

1,1′-Oxybisethane *see* 3762

N,N'-[Oxybis(2,1-ethanediyloxy-4,1-phenylene)]bisacetamide *see* 2953
2,2'-Oxybisethanol *see* 3109
1,1'-Oxybisethene *see* 9899
1,1'-Oxybis(2-ethoxy)ethane *see* 3108
Oxybismethane *see* 5990
1,1'-Oxybis[2-methoxyethane] *see* 3148
1,1'-Oxybis[3-methylbutane] *see* 5000
1,1'-[Oxybis(methylene)]bis[benzene] *see* 1146
1,1'-[Oxybis(methylene)]bis[4-(hydroxyimino)methyl]pyridinium Dichloride *see* 6659
3,3'-[Oxybis(methylenecarbonylimino)]bis[2,4,6-triiodobenzoic Acid] *see* 4939
1,1'-Oxybis[2-methylpropane] *see* 5025
2,2'-Oxybis[2-methylpropane] *see* 3020
3,3'-[Oxybis[(1-oxo-2,1-ethanediyl)imino]]bis[2,4,6-triiodobenzoic Acid] *see* 4939
1,1'-Oxybispentane *see* 646
10,10'-Oxybis-10H-phenoxarsine *see* 6914
1,1'-Oxybispropane *see* 7870
2,2'-Oxybispropane *see* 5100
3,3'-Oxybis-1-propene *see* 290
1,1'-Oxybis[2,2,2-trifluoroethane] *see* 4126
Oxybuprocaine *see* 1056
Oxybutynin Chloride, 6908
Oxycaine *see* 4772
Oxycamphor *see* 4747
Oxycel [Parke, Davis] *see* 6894
Oxychelidonic Acid *see* 5664
Oxychinolin *see* 4778
Oxychloroquine *see* 4748
Oxychlorosene, 6909
5-Oxychlorpromazine *see* 6810
Oxycholin [Blue Line] *see* 2858
Oxycinchophen, 6910
Oxyclipine *see* 7824
Oxyclozanide, 6911
Oxycodone, 6912
Oxycon *see* 6912
Oxycycline *see* 6931
Oxydapatit *see* 1701
N,N-(Oxydiacetyl)bis[3-amino-2,4,6-triiodobenzoic Acid] *see* 4939
Oxydiazepam *see* 9074
Oxydiazol *see* 5876
4,4'-Oxydibenzamidine *see* 7165
4,4'-Oxydi-2-butanol, 6913
2,2'-Oxydiethanol *see* 3109
Oxydiformic Acid Diethyl Ester *see* 8008
1,1'-(Oxydimethylene)bis[4-formylpyridinium]dichloride Dioxime *see* 6659
Oxydimethylquinizine *see* 748
10,10'-Oxydiphenoxarsine, 6914
N-Oxyd-Lost-HCl *see* 5656
Oxydol *see* 4726
Oxydon *see* 6931
Oxy-Dumocyclin *see* 6931
Oxyephedrin *see* 4754
Oxyethylated Tertiary Octylphenol Formaldehyde Polymer *see* 9741
Oxyethylated Tertiary Octylphenol-polymethylene Polymer *see* 9741
Oxyethylene Oxypropylene Polymer *see* 7536
8-(β-Oxyethyl)methylaminocaffeine *see* 1632
Oxyethyltheophylline *see* 3835
Oxyfedrine, 6915
Oxyflavil *see* 3490

Oxyfluorfen, 6916
Oxygen, 6917
Oxygen Fluoride *see* 4092
Oxygeron [Will-Pharma] *see* 9888
Oxyhemerythrin *see* 4560
Oxyhemocyanin *see* 4566
Oxyhemoglobin *see* 4567
Oxyjavanicin *see* 4228
Oxyject [Diamond] *see* 6931
Oxy-L [Woelm] *see* 1128
Oxylag [Lagap] *see* 6931
Oxylan [Burroughs Wellcome] *see* 3306
Oxylidine *see* 8110
Oxylone [Upjohn] *see* 4104
Oxylupanine *see* 4760
Oxymesterone, 6918
Oxymestrone *see* 6918
Oxymetazoline, 6919
Oxymethebanol *see* 3442
Oxymetholone, 6920
Oxymethurea, 6921
Oxymethylene *see* 4148
Oxymethyleneurea *see* 7552
Oxymorphone, 6922
Oxymycin *see* 6931
Oxyneurine *see* 1201
Oxypaat *see* 7432
Oxypan *see* 6931
Oxypangam [Sanorania] *see* 3182
Oxypendyl, 6923
Oxypertine, 6924
Oxypetidin *see* 4769
Oxyphedrine *see* 6915
Oxyphenbutazone, 6925
Oxyphencyclimine, 6926
Oxyphenisatin Acetate, 6927
Oxyphenonium Bromide, 6928
Oxyphyllin [Astra] *see* 9212
Oxypinocamphone, 6929
Oxypinone, 6929
Oxypolygelatin, 6930
Oxyprocain *see* 4772
Oxypropylated Cellulose *see* 4776
Oxyquinoline *see* 4778
Oxyquinoline Sulfate *see* 4779
Oxystin [Arzneimittelwerk VEB] *see* 6934
Oxytetracid *see* 6931
Oxytetracycline, 6931
Oxytheonyl *see* 3835
Oxythiamine, 6932
Oxythioquinox, 6933
Oxytocin, 6934
Oxytrimethylline *see* 2213
Oxyzin (Tabl.) *see* 7432
Oxyzin (Syrup) *see* 7433
OYE *see* 6784
OZ *see* 6841
γ-OZ *see* 6841
Ozagrel, 6935
Ozolinone *see* 3845
Ozone, 6936
P₄₅₀ *see* 2798
P2S *see* 7705
P7 *see* 5256
p28ˢⁱˢ *see* 7497
P 50 *see* 621
P 071 *see* 2013
P 113 *see* 8330
P 165 *see* 916
P 253 *see* 3334
P 286 *see* 4956
P 350 *see* 2269
P 391 *see* 5734
P 607 *see* 2187
P 638 *see* 7960
P 652 *see* 4145
P 725 *see* 7108
P 1011 *see* 3073
P 1134 *see* 7407

P 1496 *see* 10022
P 1742 *see* 4115
P 1779 *see* 316
P 2105 *see* 3574
P 2647 *see* 1133
P 3693A *see* 3425
P 4000 *see* 6551
P 4241 *see* 4827
P 4599 *see* 3425
P 7138 *see* 6452
P 71-0129 *see* 3917
P 720549 *see* 5126
PA 93 *see* 6641
PA 94 *see* 2758
PA 105 *see* 6785
PA 248 *see* 7829
Pa 155A *see* 4872
PAA-701 *see* 1217
Paarlan [Lilly] *see* 5089
PABA *see* 434
Pabanol [Elder] *see* 434
Pabestrol D *see* 3119
Pabracort *see* 4711
PAC *see* MISC-2
P.A.C. [Upjohn] *see* 491
Pacatal [Warner-Chilcott] *see* 5734
Pacatol *see* 5734
Pacemo *see* 40
Pacetyn *see* 3478
Paceum [Orion] *see* 2977
Pacilan [Daltan] *see* 1655
Pacinol *see* 4116
Pacinox [McNeil] *see* 1774
Pacitane *see* 9607
Pacitran [Squibb] *see* 6074
Pacitran [Lafi] *see* 2977
Pacitron [Berk] *see* 9707
Paclobutrazol, 6937
Pactamycin, 6938
Pacyl [Warner-Lambert] *see* 5127
Padan [Takeda] *see* 1874
Padimate A *see* 3221
Padimate O *see* 3221
Padisal *see* 9234
Padophène *see* 7220
Padreatin *see* 5159
Padrin *see* 7748
Padukrein [Bayropharm] *see* 5159
Padutin [Winthrop] *see* 5159
Paederine *see* 7011
ψ-Paederine *see* 7935
Paediathrocin [Abbott] *see* 3626
PAF *see* 7496
C₁₆-PAF *see* 7496
C₁₈-PAF *see* 7496
PAF-acether *see* 7496
Pagano-Cor [Helopharm] *see* 3662
Paginol [Jugoremedija] *see* 7025
Pagitane Hydrochloride [Lilly] *see* 2763
PAH *see* 454
Paint White *see* 1298
Paka Oil *see* 5518
Palacrin *see* 8053
Palafer [M & M] *see* 3995
Palafuge *see* 3306
Palaprin [Nicholas] *see* 306
Palatinol A [Advanced Solvents] *see* 7345
Palatinol M *see* 3243
Palatone [Dow] *see* 5594
Palavale [Otsuka] *see* 3476
Palcin [Sumitomo] *see* 756
Paldesic [R.P. Drugs] *see* 40
Pale Catechu *see* 4259
Pale Gentian *see* 4285
Palerol *see* 9690
Palerol (Pelerol) [Sandoz] *see* 3358 and 7449
Palestrol *see* 3118
Palfium [Purdue Frederick] *see* 2933

Palinum see 2717
Palitantin, 6939
Paliuroside see 8276
Palladium, 6940
Palladium(II) Acetate see 6942
Palladium Chloride, 6941
Palladium Diacetate, 6942
Palladium Monoxide see 6944
Palladium Nitrate, 6943
Palladium Oxide, 6944
Palladous Chloride see 6941
Palladous Nitrate see 6943
Palladous Oxide see 6944
Pallidin [E. Merck] see 8909
Pallimerck see 8909
Palmarosa Oil see 6733
Palmatine, 6945
Palm Butter see 6948
Palmidrol, 6946
Palmita [Sanwa] see 7229
Palmitic Acid, 6947
Palmitic Acid Aluminum Salt see 361
Palmitic Acid Ammonium Salt see 566
Palmitic Acid Calcium Salt see 1693
Palmitic Acid Hexadecyl Ester see 2023
Palmitin see 9648
Palmityl Alcohol see 2020
Palmofen [Zambon] see 4169
Palm Oil, from Fruit, 6948
Palm Oil, from Seed, 6949
Palohex see 4885
Palonyl [Palmedico] see 3689
Palosein [Diagnostic Data] see 6820
Palphium see 2933
Paludrine [Ayerst; ICI] see 2088
Palusil see 2088
Palustric Acid, 6950
Palux [Taisho] see 7892
Palytoxin, 6951
Palytoxin (C51-55 Hemiacetal) see 6951
L-PAM see 5708
2-PAM see 7705
Pamabrom, 6952
Pamacyl see 491
Pamaquine, 6953
Pamaquine Embonate see 6953
Pamaquine Naphthoate see 6953
2-PAM Chloride see 7705
Pamedon(e) see 3342
Pameion [Simes] see 6968
Pamelor [Sandoz] see 6635
Pamidronic Acid, 6954
Pamine Bromide [Upjohn] see 5927
Pamisyl [Parke, Davis] see 491
Pamisyl Sodium [Parke, Davis] see 491
Pamoate see MISC-3
Pamocil [Lancet] see 610
Pamoic Acid, 6955
Pamovin [Frosst] see 8034
Pamprin see 6952
Panacef [Lilly] see 1920
Panacid [Dainippon] see 7475
Panacur [Hoechst] see 3904
Panadol see 40
Panaldine [Daiichi] see 9361
Panaleve [Leo] see 40
Panama Bark see 8050
Panamicin [Gramon] see 2987
Panamine see 6823
Panaquilins see 4321
Panasone [Norbrook] see 2922
Panasorb [Winthrop] see 40
Panatus [Krka] see 1505
Panax see 4321
Panaxosides see 4321
Panazon [Toyama] see 6576
Panclar [Lematte & Boinot] see 2870
Pancodine see 6912

Pancoral [Eisai] see 3921
Pancorton see 160
Pancrease [Ortho-Cilag] see 6957
Pancrease see 6956
Pancreatic Basic Trypsin Inhibitor see 784
Pancreatic Desoxyribonuclease see 2888
Pancreatic Dornase see 2888
Pancreatic Trypsin Inhibitor [Kunitz] see 784
Pancreatin, 6956
Pancreatopeptidase E see 3499
Pancrelipase, 6957
Pancreozymin see 2201
Pancreozymin C-Terminal Octapeptide see 8494
Pancrex Vet [Paines & Byrne] see 6956
Pancuronium Bromide, 6958
Pandermite see 1651
Pandrocine see 6045
Panectyl see 9618
Panediol see 5751
Panergon [Mack, Illert.] see 6968
Panets see 40
Panex [Mallard] see 40
Panflavin [Hoechst] see 118
Panformin [Shionogi] see 1465
Panfungol [Esteve] see 5181
Pangamic Acid, 6959
Pangerin see 3195
Pangram [Virbac] see 4284
Panheprin [Abbott] see 4571
Panimit [Nattermann] see 1486
Panimycin [Meiji] see 2987
Panithal see 8297
Pankrin, 6960
Pankrotanon [Hausmann] see 6956
Panlomyc [Janssen] see 9090
Panmycin [Upjohn] see 9130
Panmycin Phosphate [Upjohn] see 9130
Panofen [Ormont] see 40
Panolid [Sandoz] see 3712
Panophylline [Panray] see 9212
Panoral [Lilly] see 1920
Panosine see 5714
Panoxolin see 6884
PanOxyl [Stiefel] see 1128
Panparnit [Geigy] see 1777
Panpurol (Tabl.) [Nippon Shinyaku] see 7441
Panpurol (Inj.) [Nippon Shinyaku] see 7441
Panrone [Panray] see 9218
Pansporin [Takeda] see 1937
Pansporine [Cassenne-Takeda] see 1937
Pantalgine [UCB] see 5736
Pantelmin [J & J] see 5647
Pantenyl [Kay] see 2924
Pantestone [Organon] see 9109
Pantetheine, 6961
Pantethine, 6962
Pantetina [Maggioni] see 6962
Panthecin [Sawai] see 6962
Pantheline [Protea] see 7816
Panthenol see 2924
Pantherine see 6221
Panthesin [Sandoz] see 5332
Panthoderm [USV] see 2924
Panthoject [USV] see 6964
Pantholin [Lilly] see 1694
Pantocaine [Hoechst] see 9123
Pantocid see 4503
Pantofenicol [Pluriquimica] see 2070
Pantogam see 4664
Pantoic Acid γ-Lactone see 6963
Pantoic Lactone see 6963
Pantolactone, 6963

Pantomicina [Abbott] see 3632
Pantomin [Daiichi] see 6962
Pantonyl [Invenex] see 2924
Pantopaque [Lafayette] see 4946
Pantopon [Roche] see 6967
Pantosediv see 9182
Pantosin [Daiichi] see 6962
Pantostrep see 3162
Pantothenic Acid, 6964
Pantothenic Acid Calcium Salt see 1694
Pantothenic Acid Ester with Chloramphenicol see 2070
Pantothenic Acid Ester with 2,2-Dichloro-N-[β-hydroxy-α-(hydroxymethyl)-p-nitrophenethyl]acetamide see 2070
Pantothenol see 2924
Pantothenyl Alcohol see 2924
N-(Pantothenyl)-β-aminoethanethiol see 6961
Pantothenylol see 2924
Pantovernil [Heyden] see 2068
Pantoyl Lactone see 6963
N-Pantoyl-3-propanolamine see 2924
Panurin see 4704
Panwarfin [Abbott] see 9950
Panzid [Duncan] see 1944
Papacon [Consolidated Midland] see 6968
Papain, 6965
Papaveraldine, 6966
Papaveretum, 6967
Papaverine, 6968
Papaw see 6969
Papaya, 6969
Papayotin see 6965
Paper Red 4B [Crompton & Knowles] see 1111
Papetherine [Lincoln] see 3682
Papital T.R. [Zemmer] see 6968
Papoose Root see 1915
PAPP see 482
Parabanic Acid, 6970
Parabencil see 3306
Parabis [Dow] see 3059
Parabolan [Negma] see 9499
Parabolin see 3116
Parabromdylamine see 1433
(+)-Parabromdylamine see 1433
Paracain see 7763
Paracarbinoxamine see 1804
Paracetaldehyde see 6975
Paracetamol see 40
Parachloramine see 5657
Parachlorometacresol see 2133
Parachlorometaxylenol see 2176
Parachlorophenol see 2154
Paracide see 3045
Paracipan see 491
B-Paracipan see 1127
Paracodin [Knoll] see 3154
Paracort [Parke, Davis] see 7727
Paracortol [Parke, Davis] see 7719
Paradiazine see 7971
Paradichlorobenzene see 3045
Paradione [Abbott] see 6976
Paradise Plant see 6095
Paradormalene see 5871
Paraffin, 6971
Paraffin Chlorinated, 6972
Paraffin Jelly see 7138
Paraffin Oil see 7139
Paraffin Wax see 6971
Paraflex [McNeil] see 2197
Paraflu [Dainippon] see 4060
Paraflutizide, 6973
Paraform see 6974
Paraformaldehyde, 6974
Paraguay Tea see 5637
Parahexyl see 8995

Paral [OJ & F] see 6975
Paralactic Acid see 5216
Paralaudin see 2947
Paraldehyde, 6975
Paralergin [Vita] see 878
Paralest see 9607
Paralgin see 3358
Paralkan see 1066
Paralytic Shellfish Poison see 8344
Paramandelic Acid see 5599
Paramelaconite see 2650
Paramenyl see 5525
Paramethadione, 6976
Paramethasone, 6977
Paramezone see 6977
Paramibe [UCEPHA] see 1441
Paramicina [Raglionieri] see 6989
Paramidin [Takeda] see 1452
Paraminol see 434
Paraminyl see 7996
Paraminyl Maleate [Cooper] see 7996
Paramisan see 491
Paramorfan see 3160
Paramorphan see 3160
Paramorphine see 9202
Paramoth [Esquire] see 3045
Paramycin see 491
Paranephrin see 3569
Paraniazide see 6999
Paranol see 474
Paranyline, 6978
Paraoxon, 6979
Paraoxypropiophenone see 6992
Para-Pas [Glenwood] see 491
Paraphos see 6983
Paraplatin [Bristol-Myers] see 1828
Paraquat, 6980
Paraquat I see 6980
Parasal [Panray] see 491
Parasalicil see 491
Parasalindon see 491
Parasan see 1037
Parasepiolite see 5569
Parasiticol see 168
Parasorbic Acid, 6981
Paraspen [Fisons] see 40
Parasympatol see 8994
Paratartaric Acid see 9038
Paratect [Pfizer] see 6175
Paratensiol [Latma] see 8148
Parathar [Rorer] see 9096
Parathesin see 3719
Parathiazan see 9229
Parathiazine see 6982 and 7969
Parathion, 6983
Parathion-methyl see 6022
Parathorm [Hormone-Chemie] see 6984
Parathormone see 6984
Para-thor-mone [Lilly] see 6984
Parathyroid Hormone, 6984
Paravermin see 7436
Parawollastonite see 1709
Paraxenol see 7277
Paraxin [Boehringer, Mann.] see 2068
Paraxin Succinate A [Yamanouchi] see 2068
Para-zene [Reliable Chem.] see 3045
Parazine see 7433
Parazol see 8895
Parazone see 9218
Parbendazole, 6985
Parbetan [Parke, Davis] see 1202
Parbocyl-Rev [Elder] see 8626
Parda [Parke, Davis] see 5344
Pardisol see 3703
Pardroyd [Parke, Davis] see 6920
Paredrine [SK & F] see 4740
Paredrinex [SK & F] see 4740
Paredrinol [SK & F] see 7304
Pareira, 6986

Pareira Brava see 6986
Parelan [Elan] see 40
Parenabol [Ciba] see 1327
Parenamine [Winthrop] see 7900
Parenogen [Cutter] see 4015
Parenzyme [Merrell] see 9704
Parenzymol see 9704
Parest [Parke, Davis] see 5872
Parethoxycaine, 6987
Parfenac [Lederle] see 1462
Parfenal [Cyanamid] see 1462
Parfuran [Parke, Davis] see 6520
Pargin [Gibipharma] see 3476
Pargonyl see 6989
Pargyline, 6988
Paricina [Archifar] see 6989
Parid see 7431
Parietic Acid see 8175
Parietin see 3518
Parietrope [Biosedra] see 8044
Parigenin see 8338
Parinase [Upjohn] see 4410
Paris Blue see 3964
Paris Green see 2629
Paris Red see 5307
Paris White see 1657
Paris Yellow see 5279
Paritane [Berk] see 6905
Parkan see 9607
Parkazin see 3098
Parkemed [Parke, Davis] see 5680
Parkin [Yoshitomi] see 3703
Parkinane Retard [Lederle] see 9607
Parkinsan see 9607
Parkinsan [Byk-Gulden] see 1456
Park Lily see 2507
Parkopan see 9607
Par KS-12 [Hommel; Laroze] see 7747
Parlay [ICI] see 6937
Parlef [Parke, Davis] see 4060
Parlodel [Sandoz] see 1404
Parlodion [Mallinckrodt] see 8022
Parmal [Central Pharm.] see 7996
Parmid [Lagap] see 6063
Parminal see 5872
Parmol see 40
Parnate [SK & F] see 9491
Parnitene see 9491
Parodyne see 748
Paroidin [Parke, Davis] see 6984
Paroleine see 7139
Paromomycin, 6989
Paromomycin I see 6989
Parotin, 6990
Paroven [Zyma] see 9698
Paroxetine, 6991
Paroxon see 6992
Paroxyl see 44
Paroxypropione, 6992
Parpanit [Geigy] see 1777
Parphezein see 3703
Parphezin see 3703
Parpon see 1037
Parrot Green see 2629
Parsal [Midy] see 6993
Parsalmide, 6993
Parsidol [Warner-Chilcott; Specia] see 3703
Parsitan see 3703
Parsley Apiole see 767
Parsley Camphor see 767
Parsley Fruit see 6994
Parsley Seed, 6994
Parsley Seed Oil see 6753
Parsol 1789 [Givaudan] see 1580
Parsol A [Givaudan] see 1580
Parsol MCX [Givaudan] see 6687
Parsol MOX [Givaudan] see 6687
Parsotil see 3703

Partel [Lilly] see 3376
Partergin see 5989
Parterol see 3163
Parthenicin see 6995
Parthenin, 6995
Parthenolide, 6996
Partocon [Ferring] see 6934
Partrex [Parmed] see 9130
Partricin, 6997
Partricin A see 6997
Partricin B see 6997
Partricin Methyl Ester see 5733
Partusisten [Boehringer, Ing.] see 3927
Parvaquone, 6998
Parvex [Upjohn] see 7366
Parvolex [Duncan, Flockhart] see 82
Parzate [Du Pont] see 10071
Parzate Liquid see 6256
Parzone [Mallinckrodt] see 3154
PAS see 491
Pasaden [Homburg] see 4657
Pasalin [Takeda] see 3123
Pasalon [Bayer] see 491
Pasalon-Rakeet see 491
PAS-C [Hellwig] see 491
Pascorbic [Hellwig] see 491
Pasetocin [Kyowa] see 610
Pasid see 491
Pasiniazide, 6999
Paskalium [Glenwood] see 491
Paskate [Lilly] see 491
Pasmed Sodium see 491
Pasmicina see 491
Pasmus [Daiichi] see 4039
Pasnodia see 491
Pasolac see 491
Pasolind see 40
Pasolind N [Stada] see 40
Paspaline-P see 2796
Paspertin [Kali-Chemie] see 6063
Pasque Flower see 7956
Passiflora, 7000
Passiflorin see 4530
Passion Flower see 7000
Passion Vine see 7000
Passodico see 491
PAS Sodium Dihydrate see 491
Pastaron [Sato] see 9781
Pasura Calcium [Dorsey] see 491
Patchouli Alcohol, 7001
Patchouli Camphor see 7001
Patchouli Oil see 6754
Patent Alum see 372
Paternoite see 5535
Pathilon [Lederle] see 9577
Pathocidin see 912
Pathocil [Wyeth] see 3073
Pathomycin [Byk-Essex] see 8498
Patoran [Ciba] see 6061
Patrol [ICI] see 3937
Patronite see 9823
Patrovina see 151
Pattonex [Makhteshim-Agan] see 6061
Patulin, 7002
Paucimycin see 6989
Pavabid [Marion] see 6968
Pavacap [Reid-Provident] see 6968
Pavacen [Central Pharm.] see 6968
Pavadel [Canright] see 6968
Pavagen [Darby] see 6968
Pavagrant [Amfre-Grant] see 6968
Pavakey [Key] see 6968
Pavased [Mallard] see 6968
Pavatest [Fellows-Testagar] see 6968
Pavecef [IBP] see 1922
Paveril [Lilly] see 3258
Paviin see 4179
Pavoninin-5, 7003
Pavoninins see 7003

Pavulon [Organon] see 6958
Paxadon [Steinhard] see 7995
Paxanol see 2834
Paxate [Mead Johnson] see 2977
Paxel [Elliott-Marion] see 2977
Paxeladine [Beaufour] see 6885
Paxidorm see 5872
Paxilon [Fisons] see 5876
Paxipam [Schering] see 4502
Paxistil see 4786
Paxisyn [Synthetic] see 6491
Paxital [Warner-Chilcott] see 5734
Paxofen [Steinhard] see 4812
Pay-Off [Am. Cyanamid] see 4055
Payzone [Am. Cyanamid] see 6576
Pazital [Andromaco] see 5669
PB see 8282
PB 89 see 4144
PB 868Cl see 5517
PBB see 4023
PBB's see 7540
PB-NOX [Penick-Bio] see 7446
PBP see 4023
PBU see 7190
PBZ [Ciba] see 9651
PC [Squibb] see 7315
PC 1 see 3921
PC 904 see 756
PC 1238 see 3527
PC 1421 see 7429
(3P-2C) see 7433
PCBs see 7541
PCC see 7986
PCNB see 8108
P-Component see 651
PCP see 7179
PCP see 7059
PD 93 see 7475
PD 107779 see 3540
PDB see 3045
3-PDC see 7061
PDDB see 3411
PDGA see 3285
PDGF see 7497
P-DHP see 230
P-DMEA see 2870
PDP see 6122
PDX Chloride see 7531
PE see MISC-2
PEA see 7186
Peach Oil, Expressed, 7004
Peak IV see 1641
Peanut, 7005
Peanut Oil, 7006
Pearl Ash see 7599
Pearl White see 1274
Pear Oil see 4993
Peau d'Espagne see 3725
PEBC see 7007
β-PEBG see 7191
Pebulate, 7007
Pecazine see 5734
Pecilocin, 7008
Pecitrol Veinogène see 1241
Pecnon [Sanken] see 5168
Pecram [Zyma] see 477
Pect [Rentschler] see 392
Pectamol [BDH] see 6885
Pectamon see 6885
Pectenine see 1854
Pectin, 7009
Pectin Sugar see 788
Pectipront see 407
Pectobloc [Siegfried] see 7412
Pectolin see 7303
Pectolinarigenin, 7010
Pectolinarin see 7010
Pectolitan [Kettelhack] see 2053
Pectox [Nattermann] see 1809
Pedameth [OJ & F] see 5896
Pederin, 7011

Pederine see 7011
ψ-Pederine see 7935
PEDG see 7191
Pediamycin [Ross] see 3626
Pedipen see 7046
Pedisafe [Sagitta] see 2412
Peflacine [Roger Bellon] see 7012
Pefloxacin, 7012
Pefloxacine see 7012
Pefloxacine Mesylate see 7012
PEG see 7545
Peganine see 9842
Peganone [Abbott] see 3705
PEG Fatty Acid Esters see 7555
Pegosperse [Glyco] see 7555
Pegosperse 400-MS [Glyco] see 7555
Pegosperse 1750-MS [Glyco] see 7555
Pegu Catechu see 1909
Peimine see 9873
Peiminoside see 9873
Pelargidanon 1602 see 6344
Pelargidenolon 1497 see 5156
Pelargidenon 1449 see 763
Pelargonic Acid, 7013
Pelargonidin, 7014
Pelargonidin-3-(6-p-coumaroyl)gluco-
  sido-5-glucoside Chloride see 6154
Pelargonin see 7014
Pelazid see 5071
Pelentan see 3729
Pelerol see 9690
Peligal see 5821
Pellagra Preventive Factor see 6435
Pelletierine, 7015
Pelletierine Tannate, 7016
Pellidol see 2940
Pellitorine, 7017
Pellotine, 7018
Pelmine see 6398
Pelonin Amide see 6398
Pelosine see 1024
PELS see 3628
Pelson see 6491
α-Peltatin, 7019
β-Peltatin, 7020
β-Peltatin A see 7020
Peltophorin see 1174
Peluces [Isei] see 4511
Pemal [Benzon] see 3704
Pemix [Prodes] see 7477
Pemoline, 7021
Pempidil [Farmigea] see 7022
Pempidine, 7022
Pempiten [I.G.I.] see 7022
Pen A [Pfizer] see 621
Pen A/N [Pfizer] see 621
Pen-200 see 7184
Pen-A-Brasive see 1157
Penadur [Wyeth] see 7037
Penagen [Genethic] see 7046
Penaldic Acids, 7023
Penamecillin, 7024
Penamox [Beecham] see 610
Penaquacaine G see 7042
Penavlon V see 7046
Penbar [Vangard] see 7087
Pen-Bristol [Grnenthal] see 621
Penbritin see 621
Penbritin-S [Ayerst] see 621
Penbrock [Beecham] see 621
Penbutolol, 7025
Pencal see 1798
Pencal [Tatsumi] see 1646
Penchlorol see 7059
Pencompren see 7046
Pencompren [E. Merck] see 7046
Pen-Di-Ben see 7037
Pendimethalin, 7026
Pendiomid [Ciba] see 913
Penditan see 7037
Pendramine [E. Merck] see 7029

Penemve see 7184
Penetek see 7062
Penethamate Hydriodide, 7027
Penethecillin see 7027
Penetracyne see 7052
Pen-Fifty see 7042
Penfluridol, 7028
Pengitoxin see 4328
Penglobe [Astra] see 944
Penialmen [Prodes] see 621
Penicals see 7046
Penicidin see 7002
Penicillamine, 7029
Penicillamine Cysteine Disulfide,
  7030
Penicillamine Disulfide, 7031
Penicillanic Acid, 7032
Penicillanic Acid 1,1-Dioxide see
  8862
Penicillanic Acid Sulfone see 8862
Penicillic Acid, 7033
Penicillin see 1157
Penicillin-152 see 7184
Penicillin-152 Potassium see 7184
Penicillin 356 see 2383
Penicillin Aminodiphenylmethane Salt
  see 7038
Penicillinase, 7034
Penicillin AT see 7044
Penicillin BT, 7035
Penicillin Dihydro F Sodium see 652
Penicillin G Benethamine, 7036
Penicillin G Benzathine, 7037
Penicillin G Benzhydrylamine, 7038
Penicillin G Calcium, 7039
Penicillin G Compd with 2-(Diethyl-
  amino)ethyl p-Aminobenzoate
  Monohydrate see 7042
Penicillin G Compd with N,N'-Di-
  benzylethylenediamine (2:1) see
  7037
Penicillin G 2-Diethylaminoethyl
  Ester Hydriodide see 7027
Penicillin G Hydrabamine, 7040
Penicillin G Hydroxymethyl Ester
  Acetate see 7024
Penicillin G Potassium, 7041
Penicillin G Procaine, 7042
Penicillin G Salt of N,N'-Dibenzyl-
  ethylenediamine see 7037
Penicillin G Sodium see 1157
Penicillin MV see 7184
Penicillin N, 7043
Penicillin O, 7044
Penicillin O Potassium see 7044
Penicillin P-12 see 6858
Penicillin Phenoxymethyl see 7046
Penicillin S Potassium, 7045
Penicillin V, 7046
Penicillin V Benzathine, 7047
Penicillin V Compd with Mepicycline
  see 7052
Penicillin V DBED see 7047
Penicillin V Hydrabamine, 7048
Penicilloic Acids, 7049
Penicilloyl Polylysine, 7050
Penicin see 470
Péniciline [Delagrange] see 621
Penidryl see 7038
Penidural [Wyeth] see 7037
Penidure see 7037
Penilaryn see 1157
Penillic Acids, 7051
Peniltetra see 7052
Penimepiciclina see 7052
Penimepicycline, 7052
Penin see 470
Peniplus see 7184
Penistaph see 5890
Penitardon [Woelm] see 6654
Penitracin [Penick] see 948

Penmestrol see 6045
Penncap M [Pennwalt] see 6022
Penncozeb [Pennwalt] see 5598
Pennyroyal, 7053
Pen-Oral see 7046
Penorale [Lusofarmaco] see 7184
Penotrane [WB Pharm.] see 4687
Penova see 7184
Penoxalin see 7026
Penphene see 9478
Penplenum [Bristol] see 4592
Penplus [Farmalabor] see 4044
Pensanate see 7441
Pensig see 7184
Pen-Sint [Cophar] see 3073
Penspek (formerly) see 3905
Penstabil [Chemapol] see 621
Penstapho [Bristol] see 6858
Penstaphocid [Bristol] see 6858
Pensyn [Upjohn] see 621
Penta see 7059
Penta-O-acetylgitoxin see 4328
Pentaaqua[D-gluconato(4−)-O²,-O⁴,O⁵]tetra-μ-hydroxydioxotri-ferrate(3−) Calcium Sodium (2:1:4) see 3971
Penta-O-benzylquercetin see 8044
Pentaborane(9), 7054
Pentaborane(11), 7055
Pentaboron Nonahydride see 7054
Pentaboron Undecahydride see 7055
Pentabromoacetone, 7056
1,1,1,3,3-Pentabromo-2-propanone see 7056
Pentac [Zoecon] see 3095
Pentacarbonyliron see 4980
Pentacarboxymethyl Diethylenetri-amine see 7083
Pentacard [Byk] see 5114
Pentacarinat [M & B] see 7071
Pentacene, 7057
Pentachlorin see 2832
3,3′,5,5′,6-Pentachloro-2,2′-dihydr-oxybenzanilide see 6911
Pentachloroethane, 7058
3,3′,5,5′,6-Pentachloro-2′-hydroxy-salicylanilide see 6911
Pentachloronitrobenzene see 8108
Pentachlorophenol, 7059
Pentacin see 7082
Pentacyclo[4.2.0.0²,⁵.0³,⁸.0⁴,⁷]octane see 2614
Pentacyclo[7.3.1.1⁴,¹².0²,⁷.0⁶,¹¹]-tetradecane see 2494
Pentacynium Bis(methyl sulfate), 7060
Pentacyone Mesylate see 7060
6-(8-Pentadecenyl)-2-hydroxybenzoic Acid see 657
3-Pentadecyl-1,2-benzenediol see 7061
3-Pentadecylcatechol, 7061
6-Pentadecyl-2-hydroxybenzoic Acid see 657
3-Pentadecylpyrocatechol see 7061
6-Pentadecylsalicylic Acid see 657
Pentadoll [Showa Shinyaku] see 2404
Pentadorm see 5731
Pentaerythritol, 7062
Pentaerythritol Chloral, 7063
Pentaerythritol Dichlorohydrin, 7064
Pentaerythritol Tetraacetate, 7065
Pentaerythritol Tetranicotinate see 6405
Pentaerythritol Tetranitrate, 7066
Pentaerythritol Trinitrate see 7094
Pentaerythrityl Tetraacetate see 7065
Pentaerythrityl Tetranitrate see 7066
Pentafin see 7066
Pentagastrin, 7067
Pentagestrone, 7068

Pentagin [Sankyo] see 7078
Pentagit [VEB Ysat] see 4328
Pentagonal Dodecahedrane see 3399
Pentahomoserine, 7069
2,3′,4,4′,6-Pentahydroxybenzophen-one see 5519
Pentahydroxycaproic Acid see 4350
2,3,14,22,25-Pentahydroxycholest-7-en-6-one see 3465
(2β,3β,5β,22R)-2,3,14,20,22-Pentahy-droxycholest-7-en-6-one see 7566
2′,3,4′,5,7-Pentahydroxyflavone see 6179
3,3′,4′,5,7-Pentahydroxyflavone see 8044
3,3′,4′,5,7-Pentahydroxyflavone-3-glucoside see 5109
3,3′,4′,5,7-Pentahydroxyflavone-7-D-glucoside see 8045
3,3′,4′,5,7-Pentahydroxyflavone-3-rutinoside see 8276
3,3′,4′,5,7-Pentahydroxyflavylium Chloride see 2694
3,3′,4′,5,7-Pentahydroxy-5′-meth-oxyflavylium Chloride see 7143
5,6,21,23,25-Pentahydroxy-27-meth-oxy-2,4,11,16,20,22,24,26-octameth-yl-2,7-(epoxypentadeca[1,11,13]tri-enimino)benzofuro[4,5-e]pyrido-[1,2-a]benzimidazole-1,15(2H)-dione 25-Acetate see 8218
3,3′,4′,5,7-Pentahydroxy-2-phenyl-benzopyrylium Chloride see 2694
3β,11β,17,20β,21-Pentahydroxy-5α-pregnane see 265
Pentakis(N²-acetyl-L-glutaminato)-tetrahydroxytrialuminum see 21
3,3′,4′,5,7-Pentakis(benzyloxy)fla-vone see 8044
Pentakis(cyano-C)nitrosylferrate(2−) Disodium see 8600
Pentalgina [Pierrel] see 7078
Pentalin see 7058
Pentam 300 [LyphoMed] see 7071
Pentaméthazène see 913
Pentamethazene Dibromide see 913
Pentamethonium Bromide, 7070
N,N,N′,N′-3-Pentamethyl-N,N′-di-ethyl-3-azapentylene-1,5-diammo-nium Dibromide see 913
Pentamethylene see 2746
Pentamethylene-1,5-bis(1-methylpyr-rolidinium) Hydrogen Tartrate see 7089
Pentamethylenebis[trimethylammoni-um Bromide] see 7070
Pentamethylenediamine see 1608
4,4′-(Pentamethylenedioxy)dibenz-amidine see 7071
Pentamethylene Glycol see 7073
β,β-Pentamethylene-γ-hydroxybutyric Acid Sodium Salt see 4603
Pentamethylene Oxide see 9149
1,5-Pentamethylenetetrazole see 7097
Pentamethylpararosaniline Chloride see 4287
1,2,2,6,6-Pentamethylpiperidine see 7022
Pentamidine, 7071
Pentamidine Mesylate see 7071
Pentamin [Boyle] see 913
Pentamycin see 4202
Pentanal see 9813
1-Pentanamine see 630
Pentane, 7072
n-Pentane see 7072
Pentanedial see 4366
1,5-Pentanediamine see 1608
Pentane-1,1-dicarboxylic Acid see 1574

1,5-Pentanedicarboxylic Acid see 7401
3,3-Pentanedicarboxylic Acid see 3114
Pentanedinitrile see 4368
Pentanedioic Acid see 4367
1,5-Pentanediol, 7073
2,4-Pentanedione see 75
2,4-Pentanedione Nickel Complex see 6408
1,1′-(1,5-Pentanediyl)bis[1-methyl-pyrrolidinium] Salt with [R-(R*,-R*)]-2,3-Dihydroxybutanedioic Acid (1:2) see 7089
4,4′-[1,5-Pentanediylbis(oxy)]bis-benzenecarboximidamide see 7071
2,2′-[1,5-Pentanediylbis[oxy(3-oxo-3,1-propanediyl)]]bis[1-[(3,4-di-methoxyphenyl)methyl]-1,2,3,4-tetrahydro-6,7-dimethoxy-2-meth-ylisoquinolinium] Dibenzenesul-fonate see 884
xylo-Pentane-1,2,3,4,5-pentol see 9992
1,2,3,4,5-Pentanepentol see 789
1-Pentanethiol see 648
Pentanitrine see 7066
Pentanoic Acid see 9815
Pentanoic Acid Ammonium Salt see 598
Pentanoic Acid Zinc Salt see 10070
tert-Pentanol see 7096
1-Pentanol, 7074
2-Pentanol, 7075
3-Pentanol, 7076
2-Pentanone see 6032
3-Pentanone see 3111
Pentanyl see 3944
2,5,8,11,14-Pentaoxapentadecane see 9141
Pentaphen see 7098
Pentaphene see 2992
Pentapiperide, 7077
Pentapiperide Methylsulfate see 7077
Pentapiperium Methylsulfate see 7077
Pentapyrrolidinium Bitartrate see 7089
Pentasa [Ferring] see 5807
Pentasodium 4,4′-[(3-Sulfo-4,4′-bi-phenylene)bis(azo)]bis(3-amino-2,7-naphthalenedisulfonate) see 9702
Pentasodium Triphosphate see 8654
1,2,3,5,6-Pentathiepane see 5321
Pentavalent Sodium Antimonyl Gluco-nate see 733
Pentazocine, 7078
Pentcillin [Toyama] see 7430
Pentedrin see 8994
Pentek see 7062
1-Pentene, 7079
2-Pentene, 7080
2-Pentenylpenicillin Sodium, 7081
Δ-β,γ-Pentenylpenicillin Sodium see 7081
Pentetate Calcium Trisodium, 7082
Pentethylcyclanone see 2712
Pentetic Acid, 7083
Pentetrazole see 7097
Penthamil [Geigy] see 7082
Penthienate Bromide, 7084
Penthiobarbital Sodium see 9280
Penthonium see 7070
Penthrane [Abbott] see 5914
Penthrit see 7066
Penticort [Lederle] see 398
Pentid see 7041
Pentifylline, 7085
Pentigetide, 7086
Pentilium see 7089
Pentione see 5092
Pentitrate see 7066

Cross Index of Names

Pentlandite see 6406
Pentobarbital Calcium see 7087
Pentobarbital Sodium, 7087
Pentobarbitone Sodium see 7087
Pentofuryl [Karlspharma] see 6450
1-Pentol, 7088
Pentolinium Tartrate, 7089
Pentolonium Bitartrate see 7089
Pentone see 7087
Pentorex see 7228
Pentosalen see 4839
Pentosan Polysulfate, 7090
Pentostam [Burroughs Wellcome] see 733
Pentostatin, 7091
Pentothal Sodium [Abbott] see 9280
Pentovis [Parke, Davis] see 8065
Pentoxifylline, 7092
Pentoxiverin see 1798
Pentoxyl, 7093
Pentoxyverine see 1798
Pentoyl [Morishita] see 3519
Pentral 80 see 7066
Pentrane see 5914
Pentrex see 621
Pentrexyl [UCB] see 621
Pentrinitrol, 7094
Pentrite see 7066
Pentritol [Armour Pharm.] see 7066
Pentrium [Roche] see 2082
Pentryate see 7066
Pentryl®, 7095
D-erythro-2-Pentulose see 8209
threo-Pentulose see 9996
Pentyde [Immunetech] see 7086
Pentyl see 7087
Pentyl Alcohol see 7074
tert-Pentyl Alcohol, 7096
Pentylamine see 630
2-(Pentylamino)ethanol 4-Aminobenz-oate (Ester) see 6263
2-n-Pentylaminoethyl p-Aminobenzo-ate see 6263
Pentylbenzene see 635
Pentylcarbinol see 4615
6-Pentyl-m-cresol see 643
Pentylenetetrazole, 7097
Pentyl Ether see 646
Pentyl Hexanoate see 640
tert-Pentyl Isovalerate see 647
11-(3-Pentyloxiranyl)-9-undecenoic Acid see 9868
p-tert-Pentylphenol, 7098
Pentymal see 601
Pen-Vee [Wyeth] see 7046
Pen-Vee K [Wyeth] see 7046
Penvikal see 7046
Peonidin, 7099
Peonin see 7099
PEP see 7543
Pepcid [Merck & Co.] see 3881
Pepcidina [Merck & Co.] see 3881
Pepcidine [Merck & Co.] see 3881
Pepdine [Merck & Co.] see 3881
Pepdul [Merck & Co.] see 3881
Pepleo Injection [Nippon Kayaku] see 7100
Pepleomycin see 7100
Peplomycin, 7100
Pepo see 7958
Pepper, 7101
Pepperidge Bush see 971
Peppermint, 7102
Peppermint Camphor see 5723
Pepsamar see 345
Pepsdol see 4363
Pepsin, 7103
Pepstatin, 7104
Pepstatin A see 7104
Pepstatin B see 7104
Pepstatin C see 7104

Peptan [Vianex] see 3881
Peptard [Riker] see 4795
Peptarom [Fresenius] see 9801
Peptavlon [Ayerst] see 7067
Peptide T, 7105
Peptinimid [Iromedica] see 3704
Peptol [Horner] see 2279
Peptonized Iron, 7106
Per-Abrodil see 4933
Peracetic Acid, 7107
Peracon [Kali-Chemie] see 4992
Peraemon see 9921
Peragit see 9607
Perandren M [Ciba] see 9109
Perandren (Inj) [Ciba] see 9115
Perandren (Lozenges) [Ciba] see 6044
Perandren Phenylacetate [Ciba] see 9114
Perandrone A see 9109
Peraprin [Taiyo] see 6063
Perazil [Burroughs Wellcome] see 2078
Perazine, 7108
Perazyl see 2078
Perbenzoic Acid, 7109
Percaine see 3016
Percapyl see 2104
Perchloracap [Mallinckrodt] see 7641
Perchloric Acid, 7110
Perchloric Anhydride see 2097
Perchlorobenzene see 4600
Perchloroethane see 4601
Perchloroethylene see 9126
Perchloromethane see 1822
Perchloron see 1677
Perchloropentacyclo[5.2.1.0 ^{2,6}.0 ^{3,9}.0 ^{5,8}]-decane see 6126
Perchloryl Fluoride, 7111
Perclene see 9126
Percoccide see 8884
Percodan [Endo] see 6912
Percorten [Ciba] see 2883
Percotol see 2883
Percutacrine Androgénique see 9109
Percutacrine Luteinique see 7783
Percutacrine Oestrogénique Iscovesco see 3118
Percutol [Reckitt & Colman] see 6528
Perderm [Schering-Plough] see 213
Perdilatal [Smith & Nephew] see 6654
Perdipina [Sandoz] see 6403
Perdipine [Syntex] see 6403
Perdolat [Inca] see 9370
Perdolat [Knoll AG; Schering-Plough] see 7029
Perduren see 9268
Perebral [Beytout] see 2708
Perebron [Angelini Francesco] see 6898
Peregal ST see 7700
Pereirine see 4274
Peremesin see 5657
Peremine [Chinoin] see 4640
Perenan [Millot-Solac] see 3598
Perenex [ICI] see 2671
Perenum [Delalande] see 9445
Perequil [Lepetit] see 5751
Perezone, 7112
Perfan [Dow] see 3542
Perfekthion [BASF] see 3209
Perflan [Lilly] see 9053
Perfluamine see 4111
Perfluidone, 7113
Perflunafene see 4111
Perfluoroacetic Acid see 9595
Perfluorocyclobutane see 6666
Perfluorodecalin see 4111
Perfluorotripropylamine see 4111
Performic Acid, 7114
Perfusamine [Medi-Physics] see 4938
Pergitral see 7066

Perglottal see 6528
Pergolide, 7115
Pergonal [Serono] see 4189
Pergotime see 2384
Perhexilene see 7116
Perhexiline, 7116
Perhydrit see 9783
N-(2-Perhydroazocin-1-ylethyl)guani-dine see 4473
Perhydrol-Urea see 9783
Perhydronaphthalene see 2839
Perhydrosqualene see 8726
Peri Acid see 6325
Periactin [Merck & Co.] see 2779
Periactinol [Merck & Co.] see 2779
Perichthol see 4815
Periciazine see 7117
Periciclina see 2872
Periclase see 5555
Periclor [Ives] see 7063
Peri-Colace [Mead Johnson] see 3398
Peri-Colace see 1887
Pericyazine, 7117
Peridamol [Fran. Thérap.] see 3354
Peridex [Procter & Gamble] see 2090
Peridon [Italchimici] see 3412
Peridys [Pharmuka] see 3412
Perifadil [Sterling] see 8925
Perifunal [Aristegui] see 7488
Perigen [Wellcome] see 7132
Perilax [Nordex] see 1253
Perilla Ketone, 7118
Perillaldehyde, 7119
l-Perillaldehyde α-anti-Oxime see 7119
l-Perillaldehyde α-syn-Oxime see 7119
Perillartine see 7119
"Perilla Sugar" see 7119
Perimetazine see 7120
Perimethazine see 7120
Perimycin, 7121
Perimycin A see 7121
Perin [Endo] see 7435
Perindopril, 7122
Perindoprilat see 7122
Periodic Acid, 7123
Periodyl, 7124
Periograf [Sterling] see 3449
Periplanone A see 7125
Periplanone B see 7125
Periplanones, 7125
Periplobiose see 8820
Periplocin, 7126
Periplocoside see 7126
Periplocymarin, 7127
Periplogenin, 7128
Periplum [Italfarmaco] see 6467
Peripress [Pfizer] see 7715
Perisalol [Mitsubishi] see 6431
Perisoxal, 7129
Peristil [Recofarma] see 6174
Peristim [Mead Johnson] see 1887
Periston [Bayer] see 7700
Peritol [EGYT] see 2779
Peritrate [Warner-Chilcott] see 7066
Perityl see 7066
Perivine, 7130
Perizin [Provet] see 2559
Perlapine, 7131
Perlatan see 3660
Perlatanol see 3653
Perlinganit [Pharma Schwarz] see 6528
Perlon see 6656
Perlopal see 5731
Perlutex [Leo Pharm.] see 5677
Permanent Yellow see 982
Permanganic Acid Potassium Salt see 7643
Permapen [Pfizer] see 7037

Permastril [Cassenne] see 3436
Permax [Lilly] see 7115
Permease see 4676
Permethanoic Acid see 7114
Permethrin, 7132
Permiran [Fran. Thérap] see 9908
Permitil [Schering] see 4116
Permonid [Roche] see 2907
Pernaemon [Organon] see 5427
Pernaemyl see 5427
Pernaevit see 9921
Pernambuco, 7133
Pernazene see 9742
Pernexin see 5427
Perniciosan see 5427
Pernipur see 9921
Pernocton see 1503
Pernoston see 1503
Pernovin see 7195
Perocan [Toyo Jozo] see 4992
Peroidin see 7641
Perolysen [M & B] see 7022
Peronine see 1155
Peronine Myristate see 6250
Perovex see 3689
Perovskite see 9398
Peroxidases, 7134
Peroxidin [Pliva] see 6820
1,4-Peroxido-*p*-menthene-2 see 852
Peroxinorm [Grünenthal] see 6820
Peroxyacetic Acid see 7107
Peroxybenzoic Acid see 7109
Peroxydex [Virbac] see 1128
Peroxyformic Acid see 7114
Peroxymonosulfuric Acid see 1858
Perparin [Chemopuro] see 3682
Perperine see 3682
Perphenan see 7135
Perphenazine, 7135
Perphenazine 3,4,5-Trimethoxybenzo-
    ate see 6065
Perquietil see 5751
Per-Radiographol see 4933
Perrhenyl Chloride see 8179
Persadox [Texas Pharmacal] see 1128
Persa-gel [Ortho] see 1128
Persantine [Geigy] see 3354
Persedon [Roche] see 8005
Persian Bark see 1889
Persian Berries see 4176
Persian Insect Powder see 7980
Persian Red see 5280
Persic Oil see 7004
Persio see 2618
Persisten see 3096
Persistol Hö 1/193 see 9586
Persolv [Lepetit] see 9799
Perspex see 5849
Persulfuric Acid see 1858
Pertestis see 9111
Perthane see 3050
Pertix see 1505 and 1516
Pertofran [Geigy] see 2902
Pertofrane [USV] see 2902
Pertoxil [Violani] see 2363
Pertranquil see 5751
Pertrombon see 3768
Pertscan see 8609
Peruvian Balsam see 959
Peruvian Bark see 2286
Peruviol see 6388
Peruvoside, 7136
Perval [Lepetit] see 9888
Pervetral [Homburg] see 6923
Pervincamine [Dausse] see 9888
Pervitin [Temmler] see 5859
Perviton see 7172
Pervone [Millot] see 9888
Perycit [Bofors] see 6405
Perylene, 7137
PES see 5487

Peson [Hoechst] see 5487
Pesos [Valeas] see 3920
Pestox III [Fisons] see 8351
Pestox XIV (obsolete) [Fisons] see
    3191
Pestox XV [Fisons] see 6124
PET see 7546
Petalite see 5395
Pet Derm III [Beecham] see 2922
Peteha see 7905
Peter-Kal see 7601
Peterphyllin see 477
Pethidine see 5736
Petidiol see 3670
Petidion [Gerot] see 3670
Petidon see 9620
Petimin see 5569
Petinimid see 3704
Petinutin [Parke, Davis] see 5928
Petisan [Gerot] see 3670
Petit Muguet see 9961
PETN see 7066
Petnidan [Desitin] see 3704
Petrichloral see 7063
Petrin [Parke, Davis] see 7094
Petrisul see 8890
Petrohol see 5096
Petrolatum, 7138
Petrolatum, Liquid, 7139
Petrol (British) see 4270
Petroleum, 7140
Petroleum Benzin, 7141
Petroleum Jelly see 7138
Petroleum Naphtha see 7141
Petroleum Spirits see 6120
Petroselic Acid see 7142
Petroselinic Acid, 7142
Petrosulpho see 4815
Petty-morrel see 793
Petunidin, 7143
Petunidol see 7143
Petunin see 7143
Petzite see 4412
Peucedanin, 7144
Pevaryl [Cilag] see 3476
Pexid [Merrell] see 7116
Peyocactine see 4666
Peyonine, 7145
Peyote see 5808
Peyrone's Chloride see 2319
Peyrone's Salt see 2319
Pezetamid [Hefa-Frenon] see 7970
PF 38 see 4233
PF 1593 see 1473
P-FAD see 4354
Pfeiffer's Substance, 7146
Pfiklor [Pfizer] see 7601
Pfizer-E [Pfizer] see 3632
Pfizerpen VK [Pfizer] see 7046
PFT Roche [Roche] see 1061
PG see 7872
PGA see 4140
PGE₁ see 7892
PGE₂ see 7893
PGF₂ₐ see 7894
PGG₂ see 7890 and 9323
PGH₂ see 7890
PGI₂ see 7890
P-Glycoprotein, A6
P-gp see A6
PGX see 7890
PH 60-40 see 3128
Ph 137 see 2566
PH 1882 see 1406
Phacetoperane see 5347
Phacetur see 7154
Phaenthine see 9162
Phaeomelanins see 5692
Phaeva [Schering AG] see 4308
Phalloidin, 7147
Phalloidine see 7147

Phaltan see 4142
Phanchinone see 7148
Phanodorm [Winthrop] see 2717
Phanodorn [Winthrop] see 2717
Phanquinone, 7148
Phanquone see 7148
Phanurane [Specia] see 1753
Pharlon see 4774
Pharmorubicin [Farmitalia] see 3573
Phasal [Lagap] see 5404
Phaseolin, 7149
Phaseollin see 7149
Phaseolunatin see 5375
Phaseomannite see 4883
Phasil [Reed & Carnrick] see 8486
Phasin, 7150
Phasins see 5312
Phazyme [Stafford-Miller] see 8486
Ph BC see 3921
Ph CJ 91B see 6799
PHD see 9323
Phe (IUPAC Abbrev.) see 7242
Phebuzine see 7248
α-Phellandrene, 7151
β-Phellandrene, 7152
Phe³-Lys⁸-oxytocin see 5503
Phe²-Lys⁸-vasopressin see 3896
Phemeride see 1083
Phemerol Chloride [Parke, Davis] see
    1083
Phemithyn [Flint] see 1083
Phemiton [Pliva] see 5742
Phenacaine Hydrochloride, 7153
Phenacemide, 7154
Phenacetin, 7155
Phenacetolin, 7156
Phenacetylurea see 7154
Phenacide see 9478
Phenacite see 1177
Phenacridane Chloride, 7157
Phenactropinium Chloride, 7158
Phenacylamine, 7159
Phenacyl Bromide see 1391
Phenacyl Chloride see 2115
8-Phenacylhomatropinium Chloride
    see 7158
N-Phenacyl-O-*dl*-mandelyltropinium
    Chloride see 7158
Phenadone Hydrochloride see 5852
Phenadoxone, 7160
Phenaglycodol, 7161
Phenallymal, 7162
Phenallymalum see 7162
Phenalzine see 7181
Phenamacide Hydrochloride, 7163
Phenamet, 7164
Phenamidine, 7165
Phenamin [Nycomed] see 2180
Phenamine see 7204
Phenamiphos see 3901
Phenamizole see 498
Phenampromid(e), 7166
9,10-Phenanthraquinone see 7168
Phenanthrene, 7167
9,10-Phenanthrenedione see 7168
Phenanthrenequinone, 7168
Phenanthro[4,5-*bcd*]furan-3-ol see
    6183
o-Phenanthroline, 7169
1,10-Phenanthroline see 7169
4,5-Phenanthroline see 7169
4,7-Phenanthroline-5,6-dione see
    7148
4,7-Phenanthroline-5,6-quinone see
    7148
5-(3'-Phenanthryl)-5-methylhydan-
    toin see 6023
Phenantoin see 5741
Phenapronil see 3902
Phenarsazine Chloride, 7170
Phenarsone Sulfoxylate, 7171

Phenatine, 7172
Phenatox see 9478
Phenazacillin see 4592
Phenazarsine Chloride see 7170
Phenazine see 7180
Phenazine, 7173
2,3-Phenazinediamine see 2960
1,6-Phenazinediol 5,10-Dioxide see 4915
Phenazine Methosulfate see 6024
1-Phenazinol see 4564
Phenazocine, 7174
Phenazodine see 7175
Phenazoline see 709
Phenazone see 748
Phenazone Salicylate see 749
Phenazopyridine Hydrochloride, 7175
Phenazoxine see 7222
Phenbenicillin see 3905
Phenbenzamine, 7176
Phenbutamide, 7177
Phenbutrazate see 3908
Phencamine see 3910
Phencarbamide, 7178
Phencarbamide Napadisilate see 7178
Phencen [Central Pharm.] see 7797
Phencyclidine, 7179
Phendextro see 2180
Phendimetrazine, 7180
Phenecyclamine see 3892
Phenedrine see 616
Phénégic see 7220
Phenelzine, 7181
[ν-Phenenyltris(oxyethylene)]tris[triethylammonium Triiodide] see 4249
Phenergan [Wyeth] see 7797
Phenesin see 1516
Phenesterine, 7182
Phenetamid see 7249
Phenetamine see 3892
Phenetharbital, 7183
Phenethazinum see 3918
Phenethicillin Potassium, 7184
Phenethyl Alcohol, 7185
Phenethylamine, 7186
4-[2-[[6-(Phenethylamino)hexyl]amino]ethyl]pyrocatechol see 3418
N-Phenethyl Anthranilic Acid see 3535
Phenethylazocine see 7174
4-p-Phenethylazo-m-phenylenediamine see 3706
1-Phenethylbiguanide see 7191
Phenethyldiguanide see 7191
N'-β-Phenethylformamidinyliminourea see 7191
(2-Phenethyl)hydrazine see 7181
N-(1-Phenethyl-4-piperidinyl)-N-phenylpropionamide see 3944
N-(1-Phenethyl-4-piperidyl)propionanilide see 3944
o-Phenetidine, 7187
p-Phenetidine, 7188
Phenetin [Alpinapharm] see 1516
p-Phenetolcarbamide see 3448
Phenetole, 7189
Phenetsal see 41
Pheneturide, 7190
p-Phenetylurea see 3448
Phenformin, 7191
Phenglutarimide, 7192
Phenhydan (Tabl) [Desitin] see 7293
Phenhydan (Inj) [Desitin] see 7293
Phenic Acid see 7206
Phenicarbazide, 7193
Phenicazone see 6443
Phenicin, 7194
Phenicol see 3727
Phenidiemal see 7183
Phenidone see 7281
Phenidylate see 6025

Phenigam see 476
Phenigama see 476
Phenindamine, 7195
Phenindione, 7196
Pheniodol see 4919
Pheniprazine, 7197
Pheniramine, 7198
Phenisatin see 9505
Phenisobromolate see 1422
Phenmedipham, 7199
Phenmerzyl Nitrate [Pitman-Moore] see 7273
Phenmetrazine, 7200
Phenmiazine see 8062
Phenobal see 7201
Phenobarbital, 7201
Phenobarbital Compd with 1-Propylhexedrine see 7873
Phenobarbital Sodium, 7202
Phenobarbitone see 7201
Phenobenzorphan see 7174
Phenobutiodil, 7203
Phenoclor [Prodelec] see 7541
Phenocoll, 7204
Phenoctide, 7205
Phenoctyl see 7834
α-Phenodiazine see 2309
β-Phenodiazine see 7344
Phenodin see 4554
Phenododecinium Bromide see 3411
Phenodyn see 7181
Phenokoll see 7204
Phenol, 7206
Phenolactine see 5220
Phenolate Sodium see 7206
Phenoldisulfonic Acid, 7207
1-Phenol-2,4-disulfonic Acid see 7207
Phenolphthalein, 7208
Phenolphthalein Sodium, 7209
Phenolphthalin, 7210
Phenolphthalol, 7211
Phenol Potassium Salt see 7646
Phenol Red see 7213
Phenol Sodium see 8611
p-Phenolsulfonic Acid, 7212
1-Phenol-4-sulfonic Acid Zinc Salt see 10054
Phenolsulfonphthalein, 7213
Phenoltetrachlorophthalein, 7214
Phenoltetraiodophthalein Sodium see 7233
Phenomorphan, 7215
Phenonyl see 7201
Phenopenicillin see 7046
Phenoperidin see 7216
Phenoperidine, 7216
Phenopropazine see 3703
Phenopryldiasulfone Sodium see 8667
Phenopyrazone, 7217
Phenopyridine see 4778
Phenoquin see 2290
Phenosafranin, 7218
Phenosal see 41
Phenosulfazole, 7219
Phenothiazine, 7220
10H-Phenothiazine see 7220
10H-Phenothiazine-10-carboxylic Acid 2-[2-(Dimethylamino)ethoxy]ethyl Ester see 3211
[1-(10-Phenothiazinylmethyl)ethyl]-2-hydroxyethyldimethylammonium Chloride see 4755
N-[β-(10-Phenothiazinyl)propyl]trimethylammonium Methyl Sulfate see 9234
Phenothrin, 7221
Phenoverm see 7220
Phenovis see 7220
Phenoxadrine see 7287
Phenoxazine, 7222
Phenoxazole see 7021

Phenoxazoline see 3930
Phenoxene [Dow] see 2182
Phenoxethol see 7226
Phenoxetol see 7226
Phenoxine see 2182
Phenoxur see 7220
6-Phenoxyacetamidopenicillanic Acid see 7046
Phenoxyacetic Acid, 7223
Phenoxyacetyl Cellulose, 7224
Phenoxybenzamine, 7225
3-Phenoxybenzyl cis,trans-Chrysanthemate see 7221
m-Phenoxybenzyl (±)-cis,trans-3-(2,2-Dichlorovinyl)-2,2-dimethylcyclopropanecarboxylate see 7132
α-Phenoxybenzylpenicillin see 3905
6-(α-Phenoxybutyramido)penicillanic Acid see 7829
3-Phenoxy-α-cyanobenzylchrysanthemate see 2776
Phenoxyethanoic Acid see 7223
2-Phenoxyethanol, 7226
(β-Phenoxyethyl)dimethyldodecylammonium Bromide see 3411
α-Phenoxyethylpenicillinic Acid Potassium Salt see 7184
(α-Phenoxyethyl)penicillin Potassium see 7184
3'-Phenoxy-4'-fluoro-α-cyanobenzyl 2,2-Dimethyl-3-[2-(4-chlorophenyl)-2-chlorovinyl]cyclopropane Carboxylate see 4068
(±)-m-Phenoxyhydratropic Acid see 3926
N-Phenoxyisopropyl-N-benzyl-β-chloroethylamine see 7225
Phenoxymethylpenicillin see 7046
Phenoxymethylpenicillinic Acid see 7046
4-[3-[4-(Phenoxymethyl)phenyl]propyl]morpholine see 4145
6-(α-Phenoxy-α-phenylacetamido)penicillanate see 3905
6-(α-Phenoxyphenylacetamido)penicillanic Acid see 3905
3-(Phenoxyphenyl)methyl (±)-cis,-trans-3-(2,2-Dichloroethenyl)-2,2-dimethylcyclopropanecarboxylate see 7132
α-dl-2-(3-Phenoxyphenyl)propionic Acid see 3926
3-Phenoxy-1,2-propanediol see 7261
Phenoxypropazine, 7227
6-(α-Phenoxypropionamido)penicillanic Acid Potassium Salt see 7184
2-(3-Phenoxy-2-propylamino)-1-(p-hydroxyphenyl)-1-propanol see 5128
α-Phenoxypropylpenicillin see 7829
4-Phenoxy-3-(1-pyrrolidinyl)-5-sulfamoylbenzoic Acid see 7464
4-[3-(α-Phenoxy-p-tolyl)propyl]morpholine see 4145
Phenozin see 10054
Phenpentermine, 7228
Phenpiazine see 8107
Phenprobamate, 7229
Phenprocoumon, 7230
Phenpromethamine see 7280
Phenpropionate see MISC-3
Phensuximide, 7231
Phentanyl see 3944
Phentermine, 7232
Phentetiothalein Sodium, 7233
Phentolamine, 7234
Phentonium Bromide see 3949
Phentydrone, 7235
Phenurone see 7154
Phenvalerate see 3952
Phenychol see 3727

Phenylmercury Acetate see 7271
Phenylmercury Borate, 7274
Phenylmercury Chloride see 7272
Phenylmercury 2,2'-Dinaphthylmeth-
ane-3,3'-disulphonate see 4687
Phenylmercury Methylenedinaphthalenesulfonate see 4687
Phenylmethane see 9455
Phenylmethanol see 1138
α-Phenyl-o-methoxycinnamic Acid
see 5921
4-(Phenylmethoxy)phenol see 6159
Phenyl-p-methoxyphenylcarbinyl
Dimethylaminoethyl Ether see 5678
Phenylmethylamine see 1139
4-[(Phenylmethyl)amino]benzenesul-
fonamide see 1161
l-Phenyl(methylaminomethyl)carbinol
see 4516
1-Phenyl-2-methylaminopropane see
5859
1-Phenyl-2-methylaminopropanol see
3561
Phenylmethylbarbituric Acid, 7275
N-(Phenylmethyl)benzenemethan-
amine see 2993
2-(Phenylmethyl)-1H-benzimidazole
see 1043
2-Phenyl-3-methyl-3-butylamine see
7228
1-[1-(Phenylmethyl)butyl]pyrrolidine
see 7791
(Phenylmethyl)carbamic Acid 2-Hydr-
oxyethyl Ester see 1490
1-Phenyl-5-methyl-8-chloro-1,2,4,5-
tetrahydro-2,4-dioxo-3H-1,5-
benzodiazepine see 2355
N-(Phenylmethylene)benzenamine see
1152
α-(Phenylmethylene)benzeneacetic
Acid see 7253
Phenylmethylimidazoline see 9430
[[1-(Phenylmethyl)-1H-indazol-3-yl]-
oxy]acetic Acid see 1042
Phenyl Methyl Ketone see 65
2-Phenyl-3-methylpentanoic Acid
1-Methyl-4-piperidyl Ester see
7077
2-(Phenylmethyl)phenol see 1158
4-(Phenylmethyl)phenol see 1159
4-(Phenylmethyl)phenol Carbamate
see 3306
α-Phenyl-α-(1-methyl-2-phenyl)eth-
ylaminoacetonitrile see 617
1-Phenyl-3-methyl-5-pyrazolyl Dimethylcarbamate see 8013
1-(Phenylmethyl)-4-(2-pyridinylcarb-
onyl)piperazine see 7365
4-[[(Phenylmethyl)sulfonyl]amino]-
benzoic Acid see 1163
2-Phenyl-3-methyltetrahydro-1,4-
oxazine see 7200
(Phenylmethyl)urea see 1165
2-Phenyl-3-methylvaleric Acid β-
(Diethylamino)ethyl Ester Bromomethylate see 9816
2-Phenyl-2-morpholinomethylindane-
1,3-dione see 6882
Phenyl Mustard Oil see 7268
α-Phenyl-5-norbornene-2-glycolic
Acid 2-Dimethylaminoethyl Ester
Methyl Bromide see 3527
Phenylone see 748
Phenyl PAS see 7243
4-[(E,E)-5'-Phenylpenta-2',4'-dien-
1'-yl]tetracyclo[5,4,0,0²,⁵,0³,⁹]undec-
10-ene-8-carboxylic Acid see 3526
1-Phenylpentane see 635
Phenylpentanol see 3921
Phenylperi Acid see 690

N-(4-Phenylphenacyl)-1-hyoscyami-
nium Bromide see 3949
Phenyl Phenacyl Ketone see 2991
o-Phenylphenol, 7276
p-Phenylphenol, 7277
4-Phenyl-1-[3-(phenylamino)propyl]-
4-piperidinecarboxylic Acid Ethyl
Ester see 7403
1-Phenyl-3-[(4'-phenyl-4'-carbeth-
oxy)piperidino]-1-propanol see 7216
γ-Phenyl-N-(1-phenylethyl)benzene-
propanamine see 3916
N-Phenyl-N-[1-(2-phenylethyl)-4-
piperidinyl]propanamide see 3944
α-Phenyl-α-N-(1-phenylisopropyl)-
aminoacetonitrile see 617
α-Phenyl-α-(β-phenylisopropylamino)acetonitrile see 617
3-Phenyl-1'-(phenylmethyl)-[3,4'-
bipiperidine]-2,6-dione see 2923
3-Phenyl-3-[1-(phenylmethyl)-4-pi-
peridinyl]-2,6-piperidinedione see
1084
N-Phenyl-N-(phenylmethyl)-1-pyrrol-
idineethanamine see 4641
Phenylphosphonothioic Acid O-(4-
Bromo-2,5-dichlorophenyl) O-
Methyl Ester see 5327
Phenylphosphonothioic Acid O-(4-Cy-
anophenyl) O-Ethyl Ester see 2697
Phenylphosphonothioic Acid O-Ethyl
Ester O-Ester with p-Hydroxyben-
zonitrile see 2697
Phenylphosphonothioic Acid O-Ethyl
O-(4-Nitrophenyl) Ester see 3576
Phenyl Phthalate, 7278
3-(4-Phenyl-1-piperazinyl)-1,2-prop-
anediol see 3439
α-Phenyl-1-piperidineacetic Acid
2-(Diethylamino)ethyl Ester see
1224
α-Phenyl-2-piperidineacetic Acid
Methyl Ester see 6025
α-Phenyl-2-piperidineacetic Acid
2-(1-Piperidinyl)ethyl Ester see
3343
α-Phenyl-2-piperidinemethanol Acetate see 5347
β-Phenyl-1-piperidinepropanoic Acid
Butyl Ester see 1510
β-Phenyl-1-piperidinepropionic Acid
Butyl Ester see 1510
N-Phenyl-N-[2-(1-piperidinyl)ethyl]-
2-pyridinemethanamine see 7376
α-Phenyl-α-(2-piperidyl)acetic Acid
Methyl Ester see 6025
1-Phenyl-1-(2'-piperidyl)-1-acetoxy-
methane see 5347
Phenylpiperone see 3342
Phenylprenazone see 3953
1-Phenylpropane see 7856
1-Phenyl-1-propanol see 3727
Phenylpropanolamine Hydrochloride,
7279
l-Phenyl-2-propanone see 7240
1-Phenyl-1-propanone see 7842
3-Phenyl-2-propenal see 2298
3-Phenyl-2-propenoic Acid see 2300
3-Phenyl-2-propenoic Acid Phenylmethyl Ester see 1144
3-Phenyl-2-propenoic Acid 3-Phenyl-
2-propenyl Ester see 2307
3-Phenyl-2-propen-1-ol see 2305
3-Phenyl-2-propenoyl Chloride see
2303
3-Phenyl-2-propenyl Anthranilate see
2306
5-Phenyl-5-(2-propenyl)-2,4,6(1H,-
3H,5H)-pyrimidinetrione see 7162
3-Phenylpropionic Acid see 4707
β-Phenylpropionic Acid see 4707

4-Phenyl-4-propionoxy-1,3-dimethyl-
azacycloheptane see 7786
1-Phenylpropyl Alcohol see 3727
7-(3-Phenyl-2-propylaminoethyl)-
theophylline see 3919
γ-Phenylpropyl Carbamate see 7229
d-N-(1-Phenyl-2-propyl)-2-chloro-
benzylamine see 2359
N-(3'-Phenyl-2'-propyl)-1,1-diphen-
yl-3-propylamine see 7744
3-(1-Phenylpropyl)-4-hydroxycouma-
rin see 7230
Phenylpropylmethylamine, 7280
N-[2-(1-Phenylpropyl)]-2,2,2-trichlo-
roethylidenimine see 613
Phenylpseudohydantoin see 7021
6-Phenyl-2,4,7-pteridinetriamine see
9515
1-Phenyl-3-pyrazolidinone, 7281
1-Phenyl-3-pyrazolidone see 7281
N'-(1-Phenylpyrazol-5-yl)sulfanilam-
ide see 8910
1-Phenyl-1-(2-pyridyl)-3-dimethyl-
aminopropane see 7198
(±)-1-Phenyl-1-(2-pyridyl)-4-(cis-
2,6-dimethyl-1-piperidyl)butanol
see 7471
3-Phenyl-3-(2-pyridyl)-N,N-dimeth-
ylpropylamine see 7198
1-[2-[Phenyl(2-pyridylmethyl)amino]-
ethyl]piperidine see 7376
Phenyl-2-pyridylmethyl-β-N,N-di-
methylaminoethyl Ether see 3430
N-Phenyl-N-(2-pyridylmethyl)-2-pi-
peridinoethylamine see 7376
1-Phenyl-1H-pyrrole-2,5-dione see
7270
1-Phenyl-2,5-pyrrolidinedione see
8838
Phenylpyrrolidinopentane see 7791
1-Phenyl-2-pyrrolidylpentane see
7791
2-Phenyl-4-quinolinecarboxylic Acid
see 2290
N-Phenylsalicylamide see 8299
Phenyl Salicylate, 7282
3-Phenylsalicylic Acid 2-Diethylami-
noethyl Ester see 1247
1-Phenylsemicarbazide see 7193
4-Phenylsemicarbazide, 7283
Phenyl Sodium see 1074
4-Phenylspiro[furan-2(3H),1'(3'H)-
isobenzofuran]-3,3'-dione see 4084
4-Phenylspiro[furan-2(3H),1'-phtha-
lan]-3,3'-dione see 4084
Phenylstibinic Acid see 1077
Phenylstibonic Acid see 1077
Phenyl Styryl Ketone see 2028
N-Phenylsuccinamic Acid see 8839
N-Phenylsuccinimide see 8838
1-(α-Phenylsuccinimido)-2-chloro-4-
sulfamoylbenzene see 8852
1-Phenyl-3-p-sulfamoylanilino-1,3-
propanedisulfonic Acid Disodium
Salt see 6601
N¹-Phenylsulfanilamide see 8867
1-Phenyl-5-sulfanilamidopyrazole see
8910
N-Phenylsulfanilic Acid, 7284
3-Phenyl-3-sulfanilylpropiophenone
see 244
Phenyl Sulfide, 7285
[5-(Phenylsulfinyl)-1H-benzimidazol-
2-yl]carbamic Acid Methyl Ester
see 6890
5-Phenylsulfinyl-2-carbomethoxyami-
nobenzimidazole see 6890
Phenylsulfohydrazide see 7573
Phenyl Sulfone see 3336
4-(Phenylsulfoxyethyl)-1,2-diphenyl-
3,5-pyrazolidinedione see 8926

Cross Index of Names

Phosphorodithioic Acid S-[(tert-Butyl-thio)methyl] O,O-Diethyl Ester see 9088

Phosphorodithioic Acid S-[2-Chloro-1-(1,3-dihydro-1,3-dioxo-2H-isoindol-2-yl)ethyl] O,O-Diethyl Ester see 2949

Phosphorodithioic Acid S-[(6-Chloro-2-oxo-3(2H)-benzoxazolyl)methyl] O,O-Diethyl Ester see 7308

Phosphorodithioic Acid S-[[(4-Chlorophenyl)thio]methyl] O,O-Diethyl Ester see 1827

Phosphorodithioic Acid, O,O-Diethyl Ester, S-Ester with 6-Chloro-3-(mercaptomethyl)-2-benzoxazolinone see 7308

Phosphorodithioic Acid O,O-Diethyl S-[2-(Ethylthio)ethyl] Ester see 3371

Phosphorodithioic Acid O,O-Diethyl S-[(Ethylthio)methyl] Ester see 7305

Phosphorodithioic Acid S-[(1,3-Dihydro-1,3-dioxo-2H-isoindol-2-yl)methyl] O,O-Dimethyl Ester see 7311

Phosphorodithioic Acid O,O-Dimethyl Ester S-Ester with N-Formyl-2-mercapto-N-methylacetamide see 4160

Phosphorodithioic Acid O,O-Dimethyl Ester, Ester with 2-Mercapto-N-methylacetamide see 3209

Phosphorodithioic Acid O,O-Dimethyl Ester, S-Ester with 3-Mercapto-methyl-1,2,3-benzotriazin-4(3H)-one see 926

Phosphorodithioic Acid O,O-Dimethyl Ester S-Ester with 4-(Mercapto-methyl)-2-methoxy-Δ²-1,3,4-thiadiazolin-5-one see 5891

Phosphorodithioic Acid O,O-Dimethyl Ester S-Ester with N-(Mercapto-methyl)phthalimide see 7311

Phosphorodithioic Acid S-[[(1,1-Dimethylethyl)thio]methyl] O,O-Diethyl Ester see 9088

Phosphorodithioic Acid O,O-Dimethyl S-[2-(Methylamino)-2-oxoethyl] Ester see 3209

Phosphorodithioic Acid O,O-Dimethyl S-[(4-Oxo-1,2,3-benzotriazin-3(4H)-yl)methyl] Ester see 926

Phosphorodithioic Acid S,S'-1,4-Dioxane-2,3-diyl O,O,O',O'-Tetraethyl Ester see 3296

Phosphorodithioic Acid O-Ethyl S,S-Diphenyl Ester see 3485

Phosphorodithioic Acid O-Ethyl S,S-Dipropyl Ester see 3702

Phosphorodithioic Acid S-[2-(Formyl-methylamino)-2-oxoethyl] O,O-Dimethyl Ester see 4160

Phosphorodithioic Acid S-[(5-Meth-oxy-2-oxo-1,3,4-thiadiazol-3(2H)-yl)methyl] O,O-Dimethyl Ester see 5891

Phosphorodithioic Acid S,S'-Methyl-ene O,O,O',O'-Tetraethyl Ester see 3691

Phosphorodithoic Acid O,O-Diethyl Ester S-Ester with N-(2-Chloro-1-mercaptoethyl)phthalimide see 2949

Phosphorofluoridic Acid Bis(1-meth-ylethyl) Ester see 5060

Phosphorofluoridic Acid Calcium Salt see 1670

Phosphorofluoridic Acid Diisopropyl Ester see 5060

Phosphorothioic Acid O-[4-(Amino-sulfonyl)phenyl] O,O-Dimethyl Ester see 2791

Phosphorothioic Acid O-(4-Bromo-2,5-dichlorophenyl) O,O-Dimethyl Ester see 1419

Phosphorothioic Acid O-[5-Chloro-1-(1-methylethyl)-1H-1,2,4-triazol-3-yl] O,O-Diethyl Ester see 4988

Phosphorothioic Acid O-(3-Chloro-4-methyl-2-oxo-2H-1-benzopyran-7-yl) O,O-Diethyl Ester see 2559

Phosphorothioic Acid O-(2-Chloro-4-nitrophenyl) O,O-Dimethyl Ester see 3027

Phosphorothioic Acid O-(3-Chloro-4-nitrophenyl) O,O-Dimethyl Ester see 2196

Phosphorothioic Acid O-(4-Cyano-phenyl) O,O-Dimethyl Ester see 2266

Phosphorothioic Acid O-(2,5-Dichloro-4-iodophenyl) O,O-Dimethyl Ester see 4925

Phosphorothioic Acid O-2,4-Dichloro-phenyl O,O-Diethyl Ester see 3030

Phosphorothioic Acid O,O-Diethyl O-[2-(Ethylthio)ethyl] Ester Mixture with O,O-Diethyl S-[2-(Ethyl-thio)ethyl]phosphorothioate see 2875

Phosphorothioic Acid O,O-Diethyl O-[6-Methyl-2-(1-methylethyl)-4-pyrimidinyl] Ester see 2978

Phosphorothioic Acid O,O-Diethyl O-[4-(Methylsulfinyl)phenyl] Ester see 3943

Phosphorothioic Acid O,O-Diethyl O-(4-Nitrophenyl) Ester see 6983

Phosphorothioic Acid O,O-Diethyl O-Pyrazinyl Ester see 9275

Phosphorothioic Acid O,O-Diethyl O-(7,8,9,10-Tetrahydro-6-oxo-6H-dibenzo[b,d]pyran-3-yl) Ester see 2568

Phosphorothioic Acid O,O-Diethyl O-(3,5,6-Trichloro-2-pyridinyl) Ester see 2190

Phosphorothioic Acid O-[2-(Dimeth-ylamino)-6-methyl-4-pyrimidinyl] O,O-Diethyl Ester see 8000

Phosphorothioic Acid O-[4-[(Dimeth-ylamino)sulfonyl]phenyl] O,O-Dimethyl Ester see 3882

Phosphorothioic Acid O,O-Dimethyl Ester O,O-Diester with 4,4'-Thio-diphenol see 9075

Phosphorothioic Acid O,O-Dimethyl Ester O-Ester with p-Hydroxybenz-enesulfonamide see 2791

Phosphorothioic Acid O,O-Dimethyl ester, O-Ester with p-Hydroxyben-zonitrile see 2266

Phosphorothioic Acid O,O-Dimethyl Ester, O-Ester with p-Hydroxy-N,N-dimethylbenzenesulfonamide see 3882

Phosphorothioic Acid O,O-Dimethyl O-[3-Methyl-4-(methylthio)phenyl] Ester see 3945

Phosphorothioic Acid O,O-Dimethyl O-(3-Methyl-4-nitrophenyl) Ester see 3922

Phosphorothioic Acid O,O-Dimethyl O-(4-Nitrophenyl) Ester see 6022

Phosphorothioic Acid, O,O-Dimethyl O-(2,4,5-Trichlorophenyl)ester see 8239

Phosphorothioic Acid O-[2-(Ethyl-thio)ethyl] O,O-Dimethyl Ester Mixture with S-[2-(Ethylthio)ethyl] O,O-Dimethyl Phosphorothioate see 5971

Phosphorous Acid, 7320
Phosphorous Bromide see 7332
Phosphorous Chloride see 7333
Phosphor-roesslerite see 5560
Phosphorsalz see 581
Phosphorus, 7321
Phosphorus Chloride see 7324
Phosphorus Hemitriselenide, 7322
Phosphorus Oxybromide, 7323
Phosphorus Oxychloride, 7324
Phosphorus Pentabromide, 7325
Phosphorus Pentachloride, 7326
Phosphorus Pentafluoride, 7327
Phosphorus Pentaselenide, 7328
Phosphorus Pentasulfide, 7329
Phosphorus Pentoxide, 7330
Phosphorus Perbromide see 7325
Phosphorus Perchloride see 7326
Phosphorus Persulfide see 7329
Phosphorus Sesquisulfide see 9168
Phosphorus Sulfochloride, 7331
Phosphorus Tribromide, 7332
Phosphorus Trichloride, 7333
Phosphorus Trifluoride, 7334
Phosphorus Trioxide, 7335
Phosphorus Triselenide, 7336
Phosphoryl Bromide see 7323
Phosphoryl Chloride see 7324
Phosphorylcholine, 7337
Phosphorylcholine Chloride see 7337
Phosphoryldimethylaminoethanol see 2870
Phosphoryldimethylcolamine see 2870
Phosphoryldimethylethanolamine see 2870
O-Phosphoryl-4-hydroxy-N,N-di-methyltryptamine see 7942
Phosphoryl Tribromide see 7323
Phosphoserine, 7338
O-Phospho-DL-serine see 7338
Phosphosiderite see 3975
Phosphothion see 5582
Phosphotope [Squibb] see 8614
Phosphotungstic Acid, 7339
Phostoxin [Degesch] see 363
Phosvel [Velsicol] see 5327
Phosvitin, 7340
Photine R see 1322
Photobiline see 4931
Photodyn see 4555
Photoglycine see 4771
Photol [Merck & Co.] see 5940
Photophor see 1702
Photopsins see 6811
Photo-Rex see 5940
Photosensitizer 101 see 7504
Phoxim, 7341
Phrenazol see 7097
Phrenosin, 7342
Phrenotropin see 7902
Phrilon [Phrixwerke] see 6656
Phthalamide, 7343
Phthalamodine see 2194
Phthalamudine see 2194
Phthalazine, 7344
1(2H)-Phthalazinone (1,3-Dimethyl-2-butenylidene)hydrazone see 1457
1(2H)-Phthalazinone Hydrazone see 4682
3-(1-Phthalazinyl)carbazic Acid Ethyl Ester see 9425
2-(1-Phthalazinyl)hydrazinecarboxylic Acid Ethyl Ester see 9425
Phthalic Acid, 7345
m-Phthalic Acid see 5083
p-Phthalic Acid see 9093

o-Phthalic Acid Bis[diethylamide] *see* 9137
Phthalic Acid Diamide *see* 7343
Phthalic Acid Dibutyl Ester *see* 1586
Phthalic Acid Dimethyl Ester *see* 3243
Phthalic Acid 1-Ethyl-1-methyl-2-propynyl Ester *see* 7348
Phthalic Acid Potassium Acid Salt *see* 7589
Phthalic Anhydride, 7346
Phthalic 2,6-Dimethylanilide *see* 4190
Phthalidyl D-α-Aminobenzylpenicillanate *see* 9008
Phthalidyl 2-(3-Trifluoromethylanilino)nicotinate *see* 9017
Phthalidyl 2-(α,α,α-Trifluoro-*m*-toluidino)nicotinate *see* 9017
Phthalimide, 7347
3-Phthalimidoglutarimide *see* 9182
α-Phthalimidoglutarimide *see* 9182
Phthalin *see* 7210
(SP-4-1)-[29*H*,31*H*-Phthalocyaninato(2−)-*N*²⁹,*N*³⁰,*N*³¹,*N*³²]copper *see* 2515
Phthalofyne, 7348
Phthalophos *see* 6269
Phthalophos (USSR) *see* 7311
Phthaloyl Chloride, 7349
N-Phthaloylglutamimide *see* 9182
Phthaloylsulfacetamide *see* 7350
Phthalthrin *see* 9154
2-(*N*⁴-Phthalylaminobenzenesulfonamido)thiazole *see* 7351
N-Phthalylglutamic Acid Imide *see* 9182
Phthalylsulfacetamide, 7350
Phthalylsulfacetimide *see* 7350
2-(*N*⁴-Phthalylsulfanilamido)thiazole *see* 7351
Phthalylsulfathiazole, 7351
Phthalylsulfonazole *see* 7351
Phthiocol, 7352
Phtorazisin [Medexport] *see* 4046
Phycite *see* 3620
Phycobilin *see* 7353
Phycobiliproteins, 7353
Phycobilisomes *see* 7353
Phycocyanins *see* 7353
Phycoerythrins *see* 7353
Phygon *see* 3032
Phygon Paste *see* 3032
Phygon XL *see* 3032
Phylcardin *see* 477
Phylletten [Arznei Müller-Rorer] *see* 2896
Phyllindon *see* 477
Phyllocontin [Napp] *see* 477
Phyllocormin N *see* 3835
α-Phyllohydroquinone *see* 3165
Phylloquinone *see* 9933
Phyllyrin *see* 7296
Phyol *see* 8672
Phyomone [ICI] *see* 6290
Phyone *see* 8672
Physalaemin, 7354
Physalemin *see* 7354
Physalien *see* 10019
Physalin *see* 10019
Physcione *see* 3518
Physeptone [Calmic] *see* 5852
Physex [Leo Pharm.] *see* 4534
Physiomycine *see* 5851
Physodalin *see* 7355
Physodic Acid, 7355
Physostab *see* 4534
Physostigma, 7356
Physostigmine, 7357
Physostigmine Aminoxide *see* 3647
Physostol *see* 7357
Physovenine, 7358

Physpan [Savage] *see* 9212
Phytar [Ansul] *see* 1603
Phytat D. B. [Daniel-Brunet] *see* 7359
Phytic Acid, 7359
Phytin *see* 7359
Phytoagglutinins *see* 5312
Phytoalexins *see* 7944
Phytochlorin, 7360
Phytochlorin(e) e *see* 7360
Phytochrome *see* 7353
Phytoecdysones *see* 3465
Phytofluene, 7361
Phytogermine *see* 9931
Phytohemagglutinin *see* 5312
Phytol, 7362
Phytolacca, 7363
Phytomelin *see* 8276
Phytomenadione *see* 9933
Phytomonic Acid *see* 5218
Phytonadiol *see* 3165
Phytonadione *see* 9933
α-Phytosterol *see* 8500
Phytosterols, 7364
3-Phytylmenadione *see* 9933
Piavetrin *see* 7431
Piazofolina *see* 6182
Piazolin *see* 6182
Piberaline, 7365
Picadex, 7366
Piceid *see* 7542
Picein, 7367
Picene, 7368
Piceol *see* 7367
Piceoside *see* 7367
Picfume [Dow] *see* 2156
Picilorex, 7369
Picloram, 7370
Picloxydine, 7371
4-Picoline *see* 7374
α-Picoline, 7372
β-Picoline, 7373
γ-Picoline, 7374
Picolinic Acid, 7375
γ-Picolinic Acid *see* 5073
Picolinic Acid N-Methylbetaine *see* 4648
7-(β-3′-Picolylaminoethyl)theophylline *see* 7400
2-Picolylidenebis(*p*-phenyl Sodium Sulfate) *see* 7377
4,4′-(2-Picolylidene)bis(phenylsulfuric Acid) Disodium Salt *see* 7377
N-(2-Picolyl)-N-phenyl-N-(2-piperidinoethyl)amine *see* 7376
Picoperidamine *see* 7376
Picoperine, 7376
Pico-Salax [Ferring] *see* 7377
Picosulfate Sodium, 7377
Picosulfol *see* 7377
Picotamide, 7378
Picragol [Wyeth] *see* 8473
Picramic Acid, 7379
Picrasmin *see* 5108
Picric Acid, 7380
Picric Acid Ammonium Salt *see* 575
Picric Acid Silver Derivative *see* 8473
Picrocine *see* 2909
Picrocrocin, 7381
Picrolichenic Acid, 7382
Picrolichenin *see* 7382
Picrolonic Acid, 7383
Picromycin, 7384
Picronitric Acid *see* 7380
Picropodophyllic Acid Hydrazide *see* 7518
Picropodophyllin, 7385
Picropodophyllinic Acid *see* 7518
Picropodophyllinic Acid 2-Ethylhydrazide *see* 7519
Picropodophyllinic Acid Hydrazide *see* 7518

Picropodophyllinic Acid Lactone *see* 7385
Picrorhiza, 7386
Picrotin, 7387
Picrotol *see* 8473
Picrotoxin, 7388
Picrotoxinin, 7389
Picryl Chloride, 7390
Picryimethylnitramine *see* 6488
Picrylnitromethylamine *see* 6488
Pictol [Mallinckrodt] *see* 5940
PID *see* 7196
Pidilat [Giulini] *see* 6441
Pidorubicin *see* 3573
Pierami [Pierrel] *see* 416
Pietil [Quimica] *see* 6899
Pifarnine, 7391
Pifatidine *see* 8252
Pifazin [Selvi] *see* 7391
Pifoxime, 7392
Piglet Pro-Gen V *see* 8521
Pigmentation Hormone *see* 6206
Pigmex *see* 6159
Pigweed *see* 383
PIH *see* 7197
Piketoprofen, 7393
Pikromycin *see* 7384
Pilagan [Allergan] *see* 7395
Pil-Digis [Lakeside] *see* 3141
Pildralazine, 7394
Piliophen *see* 4931
Pill-bearing Spurge *see* 3860
Pilocarpine, 7395
β-Pilocarpine *see* 7395
Pilocarpus, 7396
Pilocereine, 7397
Pilofrin [Allergan] *see* 7395
Pilomiotin [Cooper] *see* 7395
Pilopine HS [Alcon] *see* 7395
Piloral [Nippon Kayaku] *see* 2345
Pilosine, 5085
Pilot [Nissan] *see* 8113
Pilovisc [Cooper] *see* 7395
Pilzcin [Shionogi] *see* 2367
Pimadin *see* 7403
Pimafucin [Basotherm] *see* 6346
Pimaric Acid, 7398
d-Pimaric Acid *see* 7398
l-Pimaric Acid *see* 5348
α-Pimaric Acid *see* 7398
β-Pimaric Acid *see* 5348
Pimaricin *see* 6346
Pimavecort [Mycofarm] *see* 6369
Pimeclone, 7399
Pimeiyiline, 7400
Pimelic Acid, 7401
Pimelic Ketone *see* 2732
Pimenta, 7402
Pimephylline, 7400
Pimexone [Formenti] *see* 5749
Piminodine, 7403
Pimozide, 7404
Pimpernel *see* 7405
Pimpinella, 7405
Pimpinellin, 7406
Pinacidil, 7407
Pinacol, 7408
Pinacolin *see* 7409
Pinacolone, 7409
Pinacoloxymethylphosphoryl Fluoride *see* 8668
Pinacolyl Methylphosphonofluoridate *see* 8668
Pinacone *see* 7408
Pinacyanol Chloride *see* 8056
Pinakon *see* 4631
Pinang *see* 800
Pinaverium Bromide, 7410
Pinazepam, 7411
Pinbetol [Dolorgiet] *see* 7412
Pincets [Merrell] *see* 7431

Pindac [Lilly] see 7407
Pindione see 7196
Pindolol, 7412
Pindone, 7413
Pinene see 7414
2-Pinene see 7414
α-Pinene, 7414
β-Pinene, 7415
Pinene Hydrochloride see 1341
2-Pinen-4-one see 9862
2-Pinen-7-one see 2256
Pine Oil, 7416
Pineroro [Maruko] see 3311
Pine Tar, 7417
Pine Tar Oil see 9035
Pine Tulip see 2046
Pinguinain, 7418
Pinkroot see 8704
Pinnoite see 5535
Pinol Hydrate see 8510
Pinorubicin [Sanraku] see 7460
Pinosylvin, 7419
Pinozan see 7433
Pinrou [Century] see 7433
Pinsirup [Merrell] see 7431
Pinus Montana Oil see 6725
Pinus Pumilio Oil see 6725
PIO see 7021
Pioxol see 7021
Pipacycline, 7420
Pipadone [Burroughs Wellcome] see 3342
Pipadox see 7432
Pipamazine, 7421
Pipamperone, 7422
Pipanol [Winthrop] see 9607
Pipazethate, 7423
Pipeacid [SIT] see 7427
Pipebuzone, 7424
Pipecolic Acid, 7425
Pipecolinic Acid see 7425
Pipecurium Bromide see 7426
Pipecuronium Bromide see 7426
Pipedac [Mediolanum] see 7427
Pipedase [Cristalfarma] see 7427
Pipemid [Gentili] see 7427
Pipemidic Acid, 7427
Pipenale [Fuso] see 7441
Pipenzolate Bromide, 7428
Pipenzolate Methylbromide see 7428
Pipenzolone Bromide see 7428
Piperacetazine, 7429
Piperacillin, 7430
Piperamic Acid see 7427
Piperate see 7436
Piperaverm (Tabl.) see 7431
Piperaverm (Syrup) see 7433
Piperazate [Leeming] see 7431
Piperazidine see 7431
Piperazine, 7431
Piperazine Adipate, 7432
Piperazine-1,4-bis(N,N'-diethylene-phosphonediamide) see 3341
Piperazine-N,N'-bis(2-ethanesulfonic Acid) see 7450
1,4-Piperazinebis(ethanesulfonic Acid) see 7450
Piperazine Calcium Edathamil see 7435
Piperazine Calcium Edetate see 7435
1-Piperazinecarbodithioic Acid see 7366
1-Piperazinecarbodithioic Acid Betaine see 7366
1-Piperazinecarbodithioic Acid Inner Salt see 7366
Piperazine Carbon Disulfide Complex see 7366
Piperazine Citrate, 7433
1,4-Piperazinediethanesulfonic Acid see 7450

1,4-Piperazinediethanol Bis(α-methyl-1-naphthaleneacetate) see 6267
[R-(R*,R*)]-Piperazine-2,3-dihydroxybutanedioate (1:1) see 7436
2,5-Piperazinedione, 7434
Piperazine Dithiocarbamate see 7366
1,4-Piperazinediylbis[bis(1-aziridinyl)phosphine Oxide] see 3341
1,1'-[1,4-Piperazinediylbis(imidocarbonyl)]bis[3-(p-chlorophenyl)guanidine] see 7371
N,N'-[1,4-Piperazinediylbis(2,2,2-trichloroethylidene)]bisformamide see 9602
Piperazine Edetate Calcium, 7435
Piperazine Estrone Sulfate see 3660
Piperazine 2-Hydroxy-1,2,3-propanetricarboxylate (3:2) see 7433
Piperazine Salt of Bis(2,4,5-trichlorophenol) see 9570
Piperazine Tartrate, 7436
Piperazine Theophylline-7-acetate see 19
Piperazine Theophylline Ethanoate see 19
N-(1-Piperazinylacetyl)cyclohexylamine see 3642
1-Piperazinyl-4-methylenecarbonylcyclohexylamine see 3642
Piperic Acid, 7437
Piperidic Acid see 441
Piperidine, 7438
2-Piperidinecarboxylic Acid see 7425
3-Piperidinecarboxylic Acid see 6478
4-Piperidinecarboxylic Acid see 5075
2,2'-(2,6-Piperidinediyl)bis[1-phenylethanone] see 6626
2-Piperidineethanol 2-Aminobenzoate (Ester) see 7466
2-Piperidineethanol Anthranilate see 7466
1-Piperidineethanol Benzilate see 7441
3-Piperidino-4'-butoxypropiophenone see 3453
5-[3-(4-Piperidino-4-carbamoylpiperidino)propyl]-10,11-dihydro-5H-dibenz[b,f]azepine see 1869
6-Piperidino-2,4-diaminopyrimidine 3-Oxide see 6122
2-(1-Piperidino)-4,4-diphenyl-5-heptanone see 3342
6-Piperidino-4,4-diphenylheptan-3-one see 3342
1-Piperidino-3,3-diphenyl-4-hexanone see 6633
3-Piperidino-1,1-diphenyl-1-propanol see 7747
2-(2-Piperidinoethoxy)ethyl 10H-Pyrido[3,2-b][1,4]benzothiadiazine-10-carboxylate see 7423
2-(2-Piperidinoethoxy)ethyl 10-Thia-1,9-diazaanthracene-10-carboxylate see 7423
2-(1-Piperidino)ethyl Benzilate see 7441
β-Piperidinoethyl-4-butoxyphenyl Ketone see 3453
2-Piperidinoethyl 3-Methylflavone-8-carboxylate see 4033
2-Piperidinoethyl 3-Methyl-4-oxo-2-phenyl-4H-1-benzopyran-8-carboxylate see 4033
Piperidinoethyl α-Piperidinophenylacetate see 3343
β-Piperidinoethyl 4-Propoxyphenyl Ketone see 7843
N-(2-Piperidinoethyl)-N-(2-pyridylmethyl)aniline see 7376
4-Piperidino-4'-fluorobutyrophenone see 7750

3-(2-Piperidino-1-hydroxyethyl)-5-phenylisoxazole see 7129
α-Δ³-Piperidino-β-hydroxy-γ-o-toloxypropane see 9449
2-Piperidinomethyl-1,4-benzodioxan see 7448
2-(Piperidinomethyl)cyclohexanone see 7399
1-Piperidino-2-methyl-3-(p-tolyl)-3-propanone see 9447
3-Piperidino-1-phenyl-1-bicycloheptenyl-1-propanol see 1246
3-Piperidino-1-phenyl-(Δ⁵-bicyclo-[2.2.1]hepten-2-yl)-1-propanol see 1246
3-Piperidino-1,2-estranediol Dicarbanilate Hydrochloride see 3301
1-Piperidino-2-(N-propionylanilino)-propane see 7166
3-Piperidino-4'-propoxypropiophenone see 7843
2,2',2'',2'''-[(4-Piperidinopyrimido-[5,4-d]-pyrimidine-2,6-diyl)dinitrilo]tetraethanol see 6171
1-Δ³-Piperidino-3-o-toloxypropan-2-ol see 9449
(R*,S*)-(±)-α-2-Piperidinyl-2,8-bis-(trifluoromethyl)-4-quinolinemethanol Monohydrochloride see 5683
3-[2-(4-Piperidinyl)ethyl]-1H-indole see 4843
N-(2-Piperidinylmethyl)-2,5-bis(2,2,2-trifluoroethoxy)benzamide see 4034
2-(1-Piperidinylmethyl)cyclohexanone see 7399
3-(1-Piperidinyl)-1,2-propanediol Bis(phenylcarbamate) Monohydrochloride see 3301
1-(2-Piperidinyl)-2-propanone see 7015
3-(1-Piperidinyl)-1-(4-propoxyphenyl)-1-propanone see 7843
3-(2-Piperidinyl)pyridine see 655
6-(1-Piperidinyl)-2,4-pyrimidinediamine 3-Oxide see 6122
2,2',2'',2'''-[[4-(1-Piperidinyl)pyrimido[5,4-d]pyrimidine-2,6-diyl]-dinitrilo]tetrakisethanol see 6171
Piperidione, 7439
Piperidolate, 7440
Piperidyl Amidone see 3342
α-(2-Piperidyl)benzhydrol see 7455
α-(4-Piperidyl)benzhydrol see 909
DL-erythro-α-2-Piperidyl-2,8-bis(trifluoromethyl)-4-quinolinemethanol Monohydrochloride see 5683
4'-(1-Piperidylcarbonylmethoxy)-acetophenone Oxime see 7747
3-(1-Piperidyl)-1-cyclohexyl-1-phenyl-1-propanol Hydrochloride see 9607
6-(1-Piperidyl)-4,4-diphenyl-3-hexanone see 6633
3-(N-Piperidyl)-1,1-diphenyl-1-propanol see 7747
β-(2-Piperidyl)ethyl o-Aminobenzoate see 7466
β-Piperidylethyl Benzilate see 7441
2-(1-Piperidyl)ethyl p-Butoxyphenyl Ketone see 3453
2-(β-N-Piperidylethyl)-4,4-diphenyl-1,3-dioxolan-5-one Hydrochloride see 7454
2-(1-Piperidylmethyl)-1,4-benzodioxan see 7448
3-(1-Piperidyl)-1,2-propanediol Dicarbanilate Hydrochloride see 3301
Piperilate, 7441
Piperine, 7442
Piperinic Acid see 7437
Piperinsäure (German) see 7437

Piperitone, 7443
Piperocaine, 7444
Piperonal, 7445
Piperonyl see 7422
Piperonylaldehyde see 7445
Piperonyl Butoxide, 7446
Piperonylic Acid, 7447
10-[(4-Piperonyl-1-piperazinyl)acetyl]phenothiazine see 3928
2-(4-Piperonyl-1-piperazinyl)pyrimidine see 7465
1-Piperonyl-4-(3,7,11-trimethyl-2,6,10-dodecatrienyl)piperazine see 7391
Piperoxan, 7448
(Z,Z)-1-Piperoylpiperidine see 2038
1-Piperoylpiperidine see 7442
Piperylone, 7449
PIPES, 7450
Pipethanate see 7441
Pipitzahoic Acid see 7112
Pipizan Citrate [Merck & Co.] see 7433
Pipobroman, 7451
Piportil [Specia] see 7453
Piportil L4 [Specia] see 7453
Piportil M2 [Specia] see 7453
Piposulfan, 7452
Pipothiazine see 7453
Pipotiazine, 7453
Pipotiazine Palmitate see 7453
Pipotiazine Undecylenate see 7453
Pipoxolan Hydrochloride, 7454
Pipracid (Syrup) [Universal Drug] see 7433
Pipracil [Am. Cyanamid] see 7430
Pipradol see 7455
Pipradrol, 7455
Pipram [Roger Bellon] see 7427
Pipril [Lederle] see 7430
Piprinhydrinate, 7456
Piproxen see 6337
Piprozolin, 7457
Pipsissewa see 2046
Pipsyl Chloride, 7458
Piptal [Lakeside] see 7428
Piptelate [Vale] see 7433
Pipurin [Brocchieri] see 7427
Piracetam, 7459
Piraldina [Bracco] see 7970
Piralgo [Schiapparelli] see 6443
Piramox [Radiumpharma] see 610
Piranver [Usafarma] see 7968
Pirarreumol "B" see 7248
Pirarubicin, 7460
Pirbuterol, 7461
Pirecin [Yoshitomi] see 7919
Pirefar [Farmakos] see 7463
Pirem [Sasse] see 1840
Pirenoxine, 7462
Pirenzepine, 7463
Piretanide, 7464
Pirevan see 3360
Pirexyl [Pharmacia] see 1058
Pirfenoxone see 7462
Piria's Acid see 6323
Piribedil, 7465
Piribenzil Methyl Sulfate see 1214
Piricardio [Nagel] see 478
Piricef [CT] see 1980
Piridazol see 8913
Piridocaine, 7466
Piridol see 488
Piridolan [AB Leo] see 7470
Pirifibrate, 7467
Pirilène [Lepetit] see 7970
Pirimicarb, 7468
Pirimiphos-ethyl, 7469
Pirimiphos-methyl see 7469
Pirimor [ICI] see 7468
Pirinitramide see 7470

Piristerol [Janus] see 4659
Pirisudanol see 8002
Piriton [Allen & Hanburys; West-Ward] see 2180
Piritramide, 7470
Pirmazin [Cilag-Chemie] see 8886
Pirmenol, 7471
Piroan [Towa Yakuhin] see 3354
Pirocrid [Théraplix] see 7906
Piroctone, 7472
Pirodal [ISF] see 7475
Piroflex [Lagap] see 7476
Piroheptine, 7473
Piromen, 7474
Piromidic Acid, 7475
Piroxicam, 7476
Piroxicam Cinnamate see 7476
Piroxina see 2195
Pirozadil, 7477
Pirprofen, 7478
Pirroxil [SIT] see 7459
Piscarol see 4815
Piscidic Acid, 7479
Pisciol see 4815
Pistachia Galls see 5636
Pitayine see 8072
Pitchblende see 9767
Pithecolobine, 7480
Pitocin [Parke, Davis] see 6934
Pitocin-Buccal [Parke, Davis] see 6934
Piton-S see 6934
Pitressin [Parke, Davis] see 9843
Pituamin see 7481
Pituitary Growth Hormone see 8672
Pituitary Lactogenic Hormone see 7788
Pituitary Lipotropic Hormone see 5392
Pituitary, Posterior, 7481
Pituitrin [Parke, Davis] see 7481
Pivacef [Firma] see 7485
Pivadorm [Specia] see 1385
Pival see 7413
Pivalate see MISC-3 and 4
Pivaldione see 7413
Pivalephrine [Santen] see 3344
Pivalic Acid, 7482
Pivalic Acid 2-Benzylhydrazide see 7483
Pivalone [Jouveinal] see 9408
N^1-Pivaloyl-N^2-benzylhydrazine see 7483
2-Pivaloyl-1,3-indandione see 7413
Pivaloyloxymethyl D-α-Aminobenzylpenicillinate see 7484
Pivaloyloxymethyl Ampicillinate see 7484
Pivaloyloxymethyl 6-[(Hexahydro-1H-azepin-1-yl)methyleneamino]penicillanate see 400
Pivalylbenzhydrazine, 7483
Pivalyl Indandione see 7413
2-Pivalyl-1,3-indandione see 7413
Pivalyl Valone see 7413
Pivalyn see 7413
Pivamdinocillin see 400
Pivampicillin, 7484
Pivatil [Merck & Co.] see 7484
Pivcefalexin, 7485
Pivcephalexin see 7485
Pivhydrazine see 7483
Pivmecillinam see 400
Pivoxetil see MISC-4
Pivoxil see MISC-3 and 4
Pix [BASF] see 5746
Pixalbol [Knoll] see 2420
Pixifenide see 7392
Pizotifan see 7486
Pizotifen see 7486
Pizotyline, 7486

PJ 185 see 5450
PK 10169 see 3541
PKhNB see 8108
PK-Merz [Merz] see 380
Placenta, 7487
Placentafil [Fisons] see 7487
Placidex see 5739
Placidon see 5751
Placidyl [Abbott] see 3683
Placitate see 5751
Plac Out [Bernabo] see 2090
Plactamin [Morishita] see 7744
Plactidil [Samil] see 7378
Plafibride, 7488
Planadalin see 1837
Planate [ICI] see 2397
Planavin [Shell] see 6486
Planavit C see 855
Plancol [Chugai] see 4710
Planetol see 5940
Planocaine see 7763
Planochrome see 5757
Planofix see 6290
Planoform see 1504
Planomide see 8920
Planomycin see 4012
Planovin [Novo] see 5687
Plantago Seed, 7489
Plant Amine Oxidase see 6158
Plantisul, 7490
Planum [Farmitalia] see 9074
Plaquenil Sulfate [Winthrop] see 4748
Plasdone see 7700
Plasil [Lepetit] see 6063
Plasma (in Physics), 7491
Plasmalogens, 7492
Plasmasteril [Fresenius] see 4593
Plasma Thromboplastin Antecedent see 3876
Plasma Thromboplastin Component see 3874
Plasma Trypsinogen see 7494
Plasmil [Lepetit] see 6063
Plasmin, 7493
Plasminogen, 7494
Plasminokinase see 8784
Plasmochin see 6953
Plasmocid, 7495
Plasmoquine see 6953
Plasmosan see 7700
Plastenan [Choay; Italfarmaco] see 38
Plaster of Paris see 1713
Platamine [Montedison] see 2319
Platelet Activating Factor, 7496
Platelet-Derived Growth Factor, 7497
Platet [Nicholas] see 873
Plath-Lyse [Genevrier] see 3059
Platiblastin [Farmitalia] see 2319
Platifillin see 7505
Platilon [Toyo] see 7544
Platinex [Bristol] see 2319
Platinic Ammonium Chloride see 576
Platinic Chloride, 7498
Platinic Iodide, 7499
Platinic Oxide, 7500
Platinic Potassium Chloride see 7619
Platinic Potassium Thiocyanate see 7624
Platinol [Bristol] see 2319
Platinous Ammonium Chloride see 577
Platinous Barium Cyanide see 1003
Platinous Chloride, 7501
Platinous Iodide, 7502
Platinous Lithium Cyanide see 5422
Platinous Potassium Chloride see 7675
Platinous Potassium Cyanide see 7678
Platinous Thorium Cyanide see 9314
Platinoxan [Degussa] see 2319
Platinum, 7503

cis-Platinum II see 2319
Platinum Diiodide see 7502
Platinum Dioxide see 7500
Platistin [Farmitalia] see 2319
Platonin, 7504
Platonin J see 7504
Platosin [Nordic] see 2319
Plattnerite see 5281
Platyphylline, 7505
Plaunotol, 7506
Plausitin [Erba] see 6177
Plavolex [Wyeth] see 2925
Plazolite see 1645
Ple-1053 see 935
Plebilin see 6884
Plecyamin see 9921
Plegangin see 5654
Plegatil, 7507
Plegicil [Clin-Comar-Byla] see 27
Plegine [Ayerst] see 7180
Plegomazin see 2186
Plemocil [Farmitalia] see 2773
Plenastril [Protochemie] see 6920
Plendil [Astra] see 3895
Pleniron see 3987
Plenur [Lasa] see 5404
Pleocide [Merck & Co.] see 421
Plesium [Chiesi] see 1420
Plesmet [Coates & Cooper] see 3987
Plessy's Green see 2230
Pletaal [Otsuka] see 2277
Pletil [Andromaco] see 9377
Pleurisy Root see 853
Pleuromutilin, 7508
Pleurotin(e), 7509
Plexamine [Biodica] see 4642
Plexan see 5427
Plexiglas see 5849
Plexol 201 see 1263
Plicamycin, 7510
Plictran [Dow] see 2767
Plitican [Delagrange] see 236
Pluchine see 1201
Plumbagin, 7511
Plumbago see 4444
Plumbic Fluoride see 5306
Plumbous Fluoride see 5282
Plumbous Oxide see 5291
Plumbous Plumbate see 5298
Plumericin, 7512
Plumieride, 7513
Pluracol E [BASF-Wyandotte] see 7545
Plurazol see 8913
Plurexid [Fluxine] see 2090
Pluronic F68 see 7537
Pluronic L101 [BASF-Wyandotte] see 7537
Pluronic L62LF see 7537
Pluronics [Wyandotte] see 7537
Pluropen [Boehringer, Ing.] see 8484
Pluryle [Leo Pharm.] see 1045
Plusuril see 1045
Plutonium, 7514
PLV-2 [Sandoz] see 3896
PM-185184 see 8373
PMA see 7271
PMAC see 7271
PMAS see 7271
P-Mega-Tablinen [Sanorania] see 7046
PMF see MISC-2
P.M.F. see 4687
PMP see 5123
P.M.S. see 7515
PMSG, 7515
PN see 8029
PN 200-110 see 5129
PN 205-033 see 5129
PN 205-034 see 5129
PNB see 6514

Pneumopent [Fisons] see 7071
Pneumorel [Stroder] see 3942
P.O. 12 [Biothérax] see 3543
Poast [BASF] see 8424
Pocan see 7363
Podocarpic Acid, 7516
Pododacric Acid, 7517
Podofilox see 7520
Podophyllic Acids, 7518
Podophyllin see 7523
Podophyllinic Acid see 7518
Podophyllinic Acid 2-Ethylhydrazide, 7519
Podophyllinic Acid Hydrazide see 7518
Podophyllinic Acid Lactone see 7520
Podophyllotoxin, 7520
Podophyllum, 7521
Podophyllum, Indian, 7522
Podophyllum Resin, 7523
POE Fatty Acid Esters see 7555
Pogy Oil see 5721
Poi, 7524
Poison Black Cherry see 1034
Poison Elder see 7527
Poison Hemlock see 2503
Poison Ivy, 7525
Poison Nut see 6652
Poison Oak, 7526
Poison Parsley see 2503
Poison Sumac, 7527
Poison Tobacco see 4796
Poison Vine see 7525
Poke Root see 7363
L-Polamidon [Hoechst] see 5852
Polaramine [Schering] see 2180
Polaris [Monsanto] see 4409
Polaronil see 2180
Polaronil (Germany) [Byk-Essex] see 2180
Polar® Yellow, 7528
Polcortolon [Polfa] see 9511
Poldine Methylsulfate, 7529
Poleon [Sumitomo] see 6273
Pole Reagent Paper, 7530
Polibutin [Juste] see 9613
Policar MZ [Visplant] see 5598
Policar S [Visplant] see 5598
Policort see 2922
Polidexide, 7531
Polidocanol, 7532
Polik see 4514
Polinalin see 488
Poliseptil see 8920
Polishing Powder see 8736
Polistin T-Caps [Trommsdorff] see 1804
Poliuron [Lepetit] see 1045
Polival [Andromaco] see 9217
Polivasal [Coli] see 8967
Polixima [Cristalfarma] see 1951
Pollakisu [Kodama] see 6908
Pollucite see 2001
Polmiror [Poli] see 6447
Polonium, 7533
Polonium Dioxide, 7534
Polonium Tetrachloride, 7535
Poloxalene, 7536
Poloxalkol (obsolete) see 7537
Poloxamer 188 see 7537
Poloxamer 331 see 7537
Poloxamer 182LF see 7537
Poloxamers, 7537
Poly see 8654
Polyacrylonitrile see 6822
Polyamide see 6655
Polyamine-Methylene Resin, 7538
Polyanetholesulfonic Acid Sodium Salt see 7527
Polyanhydroglucuronic Acid see 6894
Polybar [E-Z-EM] see 1006

Polybenzarsol, 7539
Polybrene see 4605
Polybrominated Biphenyls, 7540
Polybromobiphenyls see 7540
Polycell see 1835
Polychlorinated Biphenyls, 7541
Poly(2-chloro-1,3-butadiene) see 6377
Polychlorocamphene see 9478
Polychloroprene see 6377
Polychol [Croda] see 7554
Polychrome see 3646
Polycidal see 8881
Polycidine see 2896
Polycillin [Bristol] see 621
Polycillin-N [Bristol] see 621
Polycondensed Hexahydroxytricarb-oxytriphenylmethane see 9603
Polycycline [Bristol] see 9130
Polydatin, 7542
Poly[2-(diethylamino)ethyl] Poly-glycerylene Dextran Hydrochloride see 7531
Polyestradiol Phosphate, 7543
Polyetherin A see 6457
Polyethylene, 7544
Polyethylene Glycol, 7545
Polyethylene Glycol 200 see 7545
Polyethylene Glycol 400 see 7545
Polyethylene Glycol 600 see 7545
Polyethylene Glycol 1500 see 7545
Polyethylene Glycol 4000 see 7545
Polyethylene Glycol 6000 see 7545
Polyethylene Glycol Esters of Fatty Acids see 7555
Polyethylene Glycol Fatty Alcohol Ethers see 7554
Polyethylene Glycol p-Isooctylphenyl Ether see 6681
Polyethyleneglycol 1000 Monocetyl Ether see 7554
Polyethylene Glycol Monododecyl Ether see 7532
Polyethylene Glycol 400 Monostearate see 7555
Polyethyleneglycols Mono(nonylphen-yl) Ether see 6596
Polyethylene Oxide Sorbitan Mono-oleate see 7559
Polyethylenepolypropylene Glycols, Polymers see 7537
Polyethylene Sodium Sulfonate see 5487
Polyethylene Terephthalates, 7546
Polyferon [Biogen] see 4894
Polyferose, 7547
Poly-G [Olin] see 7545
Polyglucin see 2925
Polyglycol E [Dow] see 7545
Polygodial, 7548
Polygris [Byk-Essex] see 4453
Poly[imino-1,4-butanediylimino(1,6-dioxo-1,6-hexanediyl)] see 6657
Poly(iminocarbonylpentamethylene) see 6656
Poly[imino(1-oxo-1,6-hexanediyl)] see 6656
Poly I . Poly (C$_{12}$,U) see 622
Polylysine, 7549
Polymannuronic Acid see 232
Polymeric Sodium Metaphosphate see 5522
Polymer of N,N,N',N'-Tetramethyl-hexamethylenediamine and Tri-methylene Bromide see 4605
Polymetatelluric Acid see 9063
Poly[methylenedi(hydroxymethyl)-urea] see 7552
Polymox [Bristol] see 610
Polymyxin, 7550
Polymyxin B see 7550
Polymyxin B$_1$ see 7550

Cross Index of Names

Polymyxin B$_2$ *see* 7550
Polymyxin D *see* 7550
Polymyxin D$_1$ *see* 7550
Polymyxin D$_2$ *see* 7550
Polymyxin E *see* 2475 and 7550
Polymyxin E$_1$ *see* 2475
Polymyxin E$_2$ *see* 2475
Polymyxin F *see* 7550
Polymyxin K *see* 7550
Polymyxin M *see* 7550
Polymyxin P *see* 7550
Polymyxin S$_1$ *see* 7550
Polymyxin T$_1$ *see* 7550
Polymyxin B-Methanesulfonic Acid, 7551
Polymyxin B-*N,N,N,N,N*-Pentakis-(methanesulfonic Acid) *see* 7551
Polymyxin B N-Sulfomethyl Deriv Sodium Salt *see* 7551
Polynoxylin, 7552
Polynoxyline *see* 7552
Polyoxin A *see* 7553
Polyoxin B *see* 7553
Polyoxin AL *see* 7553
Polyoxins, 7553
Poly[1-(2-oxo-1-pyrrolidinyl)ethylene] *see* 7700
Polyoxyaluminum Acetylsalicylate *see* 306
Polyoxyethylene Alcohols, 7554
Polyoxyethylene Fatty Acid Esters, 7555
Polyoxyethylene Lauryl Ether *see* 7532
Polyoxyethylene(*n*)nonylphenyl Ether *see* 6596
Poly(oxyethylene)-poly(oxypropylene)-poly(oxyethylene) Polymer *see* 7536
Polyoxyethylene (20) Sorbitan Monooleate *see* 7559
Polyoxyl 8 Stearate *see* 7555
Polyoxyl 40 Stearate *see* 7555
Polyoxymethylene *see* 6974
Polyoxymethyleneurea *see* 7552
Polypeptide β_1 *see* 9340
Polyphenolase *see* 9746
Polyphenoloxidase *see* 9746
Polyphlogin *see* 2290
Polyphosphazenes, 7556
Polyphosphoric Acid, 7557
Polypropylene, 7558
Polyram Z [BASF] *see* 10071
Polyribosome *see* 8207
Poly(rI) . Poly[r(C$_{12}$U)$_n$] *see* 622
Polysaccharide B-1459 *see* 9966
Polysaccharide C 16 *see* 3463
Polysilon *see* 8486
Polysome *see* 8207
Polysorbate 80, 7559
Polystat [Salsbury] *see* 8908
Polystichalbin *see* 200
Polystichin *see* 868
Polystichocitrin *see* 4027
Polystictine *see* 2297
Polystyrene *see* 8830
Polysulfide Rubber *see* 9268
Polytef *see* 7560
Polytergent B [Olin] *see* 6596
Polytergent G [Union Carbide] *see* 6681
Poly-Tergent J [Olin] *see* 7554
Polytetrafluoroethylene, 7560
Polytetrafluoroethylene Resin *see* 7560
Poly(tetramethyleneadipamide) *see* 6657
Poly(*N,N,N ',N '*-tetramethyl-*N*-trimethylenehexamethylenediammonium Dibromide) *see* 4605
Polythene [Du Pont] *see* 7544

Polythiazide, 7561
Polytrim [Wellcome] *see* 7550
Polytrin [Ciba-Geigy] *see* 2775
Polyvidone *see* 7700
Polyvinyl Alcohol, 7562
Polyvinyl Chloride, 7563
Polyvinylpyrrolidone *see* 7700
Polyvinylpyrrolidone-iodine Complex *see* 7701
Polyviol [Wacker] *see* 7562
Pomarsol *see* 9304
POMC *see* 7803
Pomegranate, 7564
Ponalar [Adenylchemie] *see* 5680
Ponalid [Sandoz] *see* 3712
Ponalrestat, 7565
Ponasterone A, 7566
Ponasterones *see* 7566
Ponceau 3R, 7567
Ponceau SX, 7568
Poncuronium Bromide (rescinded USAN) *see* 6958
Poncyl *see* 4453
Ponderal [Servier] *see* 3920
Ponderax [Selpharm] *see* 3920
Ponderex [Robins] *see* 3920
Pondex [Chinoin] *see* 7021
Pondimin [Robins] *see* 3920
Pondinil [Roche] *see* 5681
Pondinol *see* 5681
Pondocil [Leo Pharm.] *see* 7484
Pondocillin [Leo Pharm.] *see* 7484
Pondocillina [Sigma-Tau] *see* 7484
Ponecil [Tutag] *see* 621
Ponoxylan [Berk] *see* 7552
Ponsinomycin *see* 6123
Ponsital *see* 4827
Ponstan [Parke, Davis] *see* 5680
Ponstel [Parke, Davis] *see* 5680
Ponstil [Parke, Davis] *see* 5680
Ponstyl *see* 5680
Pontacyl Fast Blue R *see* 662
Pontal [Parke, Davis] *see* 5680
Pontalin *see* 2076
Pontamine White BR *see* 1322
Ponticin *see* 8173
Pontocaine Hydrochloride [Winthrop] *see* 9123
P.O.P. [Armour Pharm.] *see* 6992
Poppy Capsules, 7569
Poppy Heads *see* 7569
Poppy Oil, 7570
Poppy-seed Oil *see* 7570
Populin, 7571
Populnetin *see* 5156
Populoside *see* 7571
Poquil [Parke, Davis] *see* 8034
Porcelain Clay *see* 5162
Porcine Hormone Secretin *see* 8375
Porcine Motilin *see* 6198
Porfiromycin, 7572
Porofor-57 *see* 931
Porofor® BSH [Bayer], 7573
Porphin *see* 7574
Porphine, 7574
21*H*,23*H*-Porphine *see* 7574
Porphobilinogen, 7575
Porphyrillic Acid, 7576
Porphyrins *see* 7574
Porphyropsin, 7577
Por 8 Sandoz [Sandoz] *see* 6825
Portamycin [Upjohn] *see* 8785
Portyn [Parke, Davis] *see* 1090
Posdel [UCB] *see* 1449
Posedrine [Aron] *see* 1027
Posorutin [Ursapharm] *see* 9698
Possipione [Recordati] *see* 6992
Postacton *see* 5503
Postafen Salve [UCB] *see* 5678
Postafen Supp [UCB] *see* 5657
Postafen Tabl [UCB] *see* 1449

Pot *see* 1752
Potaba [Glenwood] *see* 7582
Potable Aqua [Wisconsin Pharmacal] *see* 9140
Potasan *see* 4793
Potash Sulfurated, 7578
Potasoral [Warren-Teed] *see* 7615
Potassa *see* 7625
Potassio-cuprous Cyanide *see* 2672
Potassium, 7579
Potassium Acetate, 7580
Potassium Acid Arsenate *see* 7583
Potassium Acid Carbonate *see* 7586
Potassium Acid Fluoride *see* 7587
Potassium Acid Oxalate *see* 7588
Potassium Acid Phosphate *see* 7648
Potassium Acid Phthalate *see* 7589
Potassium Acid Sulfate *see* 7591
Potassium Acid Tartrate *see* 7593
Potassium Alum *see* 364
Potassium Aluminate, 7581
Potassium *p*-Aminobenzoate, 7582
Potassium *para*-Aminobenzoate *see* 7582
Potassium Antimonate(V) *see* 7652
Potassium Antimonyltartrate *see* 732
Potassium Arsenate, 7583
Potassium Arsenite, 7584
Potassium Arsenite Solution, 7585
Potassium Atractylate *see* 883
Potassium Auribromide *see* 7673
Potassium Aurichloride *see* 7674
Potassium Auric Iodide *see* 7681
Potassium Aurocyanide *see* 7609
Potassium Benzoate *see* 1101
Potassium Benzylpenicillinate *see* 7041
Potassium Biborate *see* 7672
Potassium Bicarbonate, 7586
Potassium Bichromate *see* 7608
Potassium Bifluoride, 7587
Potassium Binoxalate, 7588
Potassium Biphosphate *see* 7648
Potassium Biphthalate, 7589
Potassium Bis(cyano-C)argentate(1 −) *see* 7659
Potassium Bis(cyano-C)cuprate(1 −) *see* 2672
Potassium Biselenite, 7590
Potassium Bismuthotartrate *see* 1289
Potassium Bismuth Tartrate *see* 1289
Potassium Bismuthyl Tartrate *see* 1289
Potassium Bisulfate, 7591
Potassium Bisulfide, 7592
Potassium Bitartrate, 7593
Potassium Borate *see* 7672
Potassium Borofluoride *see* 7680
Potassium Borohydride, 7594
Potassium Borotartrate, 7595
Potassium Bromate, 7596
Potassium Bromide, 7597
Potassium Cadmium Iodide *see* 7682
Potassium Canrenoate *see* 1753
Potassium Carbacrylic Resin, 7598
Potassium Carbolate *see* 7646
Potassium Carbonate, 7599
Potassium Chlorate, 7600
Potassium Chloride, 7601
Potassium γ-Chlorocrotylmercaptomethylpenicillinate *see* 7045
Potassium Chloroplatinate *see* 7619 and 7675
Potassium Chromate(VI), 7602
Potassium Chromic Oxalate *see* 2231
Potassium Chromic Sulfate *see* 2232
Potassium Citrate, 7603
Potassium Citrate, Monobasic, 7604
Potassium Cobalticyanide *see* 7620
Potassium Cobaltihexacyanide *see* 7620

PP 148 see 6980
PP 149 see 3695
PP 175 see 5718
PP 211 see 7469
PP 296 see 3070
PP 333 see 6937
PP 383 see 2775
PP 450 see 4135
PP 511 see 7469
PP 557 see 7132
PP 581 see 1368
PP 588 see 1484
PP 604 see 9484
PP 675 see 3206
PP 781 see 3432
PP 910 see 6980
PPACK, 7702
PPD see 9715
P. P. Factor see 6435
^{99m}Tc-PPI see 8748
PPL see 7050
^{99m}Tc-PPY see 8748
PR 66 see 7449
Practolol, 7703
Praecicalm [Molimin] see 7087
Praeciglucon see 4372
Praecirheumin [Molimin] see 7248
Praenitron see 9682
Praequine see 6953
Pragman see 9450
Pragmazone [UPSA] see 9495
Pragmoline [Specia] see 80
Praiden [Italchemie] see 1420
Prajmaline, 7704
Prajmaline Bitartrate see 7704
Prajmalium see 7704
Pralidoxime Chloride, 7705
Prälumin see 2717
Pramidex [Berk] see 9432
Pramiel [Teisan/Nagase] see 6063
Pramil [Hässle] see 8374
Pramindole (obsolete) see 4962
Pramitol [Ciba-Geigy] see 7799
Pramiverin, 7706
Pramiverine see 7706
Pramocaine see 7707
Pramoxine, 7707
Prandiol [Bottu] see 3354
Pranone [Schering] see 3696
Pranoprofen, 7708
Prano-Puren [Klinge] see 7852
Pranosina [Newport] see 4881
Pranosine [Newport] see 4881
Prantal [Schering] see 3302
Präparat 5968 see 4682
Präparat 9295 see 913
Praseodymium, 7709
Prasterone, 7710
Pratensein, 7711
Pravachol [Squibb] see 7712
Pravacilin [Recordati] see 5830
Pravastatin Sodium, 7712
Pravidel [Sandoz] see 1404
Pravocaine Hydrochloride see 7850
Praxadium [Bottu] see 6608
Praxilene [Oberval] see 6268
Praxis [Lisapharma] see 4875
Praxiten see 6881
Praxiten SP [Wyeth] see 397
Prazene [Parke, Davis] see 7713
Prazepam, 7713
Prazine [Wyeth] see 7793
Prazinil [Specia] see 1869
Praziquantel, 7714
Prazol [Corvi] see 3642
Prazosin, 7715
Prean [Chemil] see 5653
Prebediolone Acetate see 70
Precef [Bristol] see 1934
Preceptin [Ortho] see 6681
Precipitated Bismuth see 1268

Precipitated Calcium Carbonate see 1657
Precipitated Calcium Sulfate see 1713
Precipitated Chalk see 1657
Precipitated Silica see 8437
Precipitated Sulfur see 8956
Precipité Blanc see 5795
Precocene I see 7716
Precocene II see 7716
Precocenes, 7716
Precortal see 7727
Precortancyl see 7719
Precortilon [Heilit] see 7719
PreCortisyl see 7719
Precursor to ACTH-LPH-β-endor-phin see 7803
Predalon [Organon] see 4534
Predalone T.B.A. [OJ & F] see 7725
Predalon-S see 7515
Predef [Upjohn] see 5058
Predenema [Pharmax] see 7719
Predfoam [Pharmax] see 7723
Pred Forte [Allergan] see 7719
Pred Mild [Allergan] see 7719
Prednacinolone see 2908
Prednelan see 7719
Prednesol [Glaxo] see 7721
Prednicarbate, 7717
Prednicen [Central Pharm.] see 7719
Predniliderm [Ferring] see 7719
Prednilonga [Dorsch] see 7727
Prednimustine, 7718
Predniretard see 7719
Prednisolamate see 7720
Prednisolone, 7719
Prednisolone 21-[4-[p-[Bis(2-chloro-ethyl)amino]phenyl]butyrate] see 7718
Prednisolone 2l-tert-Butylacetate see 7725
Prednisolone 21-Diethylaminoacetate, 7720
Prednisolone 21-N,N-Diethylglycinate see 7720
Prednisolone 17-Ethylcarbonate 21-Propionate see 7717
Prednisolone Phosphate Disodium see 7721
21-Prednisolonephosphoric Acid Disodium Salt see 7721
Prednisolone Sodium Phosphate, 7721
Prednisolone Sodium Succinate, 7722
Prednisolone Sodium 21-m-Sulfoben-zoate, 7723
Prednisolone Steaglate see 7724
Prednisolone 21-Stearoylglycolate, 7724
Prednisolone Tebutate, 7725
Prednisolone 21-Trimethylacetate, 7726
Prednisolone 17-Valerate see 7728
Prednisone, 7727
Prednival, 7728
Prednylidene, 7729
Prednylidene 21-Diethylaminoacetate, 7730
Predonine see 7719
Predsol [Glaxo] see 7721
Predsolan see 7721
Prefamone [Dexo] see 3116
Prefaxil see 9430
Preferid [Astra] see 1455
Prefrin [Allergan] see 7257
Pregard [Ciba-Geigy] see 7781
Preglandin [Ono] see 4279
Pregna-1,4-diene-11β,21-diol-3,20-dione[17α,16α-d]-2'-methyloxazo-line 21-Acetate see 2852
1,4-Pregnadiene-17α,21-diol-3,11,20-trione see 7727

(9β,10α)-Pregna-4,6-diene-3,20-dione see 3454
10α-Pregna-4,6-diene-3,20-dione see 3454
1,4-Pregnadiene-3,20-dione-11β,17α,-21-triol see 7719
Δ^{1,4}-Pregnadiene-16-methylene-11β,17α,21-triol-3,20-dione see 7729
1,4-Pregnadiene-11β,17α,21-triol-3,20-dione see 7719
Pregna-2,4-dien-20-yno[2,3-d]isox-azol-17-ol see 2811
Pregnancy Urine Extract see 4534
Pregnane, 7731
5α-Pregnane see 257
5β-Pregnane see 7731
Pregnanediol, 7732
(3α,5β,20S)-Pregnane-3,20-diol see 7732
5α-Pregnane-3β,20β-diol see 261
5α-Pregnane-3α,20β-diol see 259
5α-Pregnane-3α,20α-diol see 258
5α-Pregnane-3β,20α-diol see 260
3,20-Pregnanedione, 7733
5α-Pregnane-3,20-dione see 264
(5β)-Pregnane-3,20-dione see 7733
11,20-Pregnanedione-3α,17α,21-triol see 9143
(3α,5β,20S)-Pregnane-3,11,17,20,-21-pentol see 2537
5α-Pregnane-3β,11β,17,20β,21-pentol see 265
5β-Pregnane-3α,11β,17,20α,21-pentol see 2537
5β-Pregnane-3α,11β,17,20β,21-pentol see 2537
Pregnane-3,17,20,21-tetrol see 266
3α,17α,20α,21-Pregnanetetrol-11-one see 2538
5α-Pregnane-3β,17,20α-triol see 269
5α-Pregnane-3β,17,20β-triol see 270
3α,17α,21-Pregnanetriol-11,20-dione see 9143
Pregnan-3α-ol-20-one, 7734
Pregnan-3β-ol-20-one see 7734
Pregnant Mare Serum Gonadotrop(h)-in see 7515
4-Pregnene-11β,21-diol-3,20-dione see 2532
4-Pregnene-16α,17α-diol-3,2-dione see 229
4-Pregnene-17α,21-diol-3,20-dione see 2912
4-Pregnene-20,21-diol-3,11-dione, 7735
5-Pregnene-3β,21-diol-20-one 21-Acetate see 70
Δ^{4}-Pregnene-17α,21-diol-3,11,20-trione see 2533
Pregn-4-ene-3,20-dione see 7783
Δ^{4}-Pregnene-3,20-dione see 7783
4-Pregnene-11β,17α,20β,21-tetrol-3-one, 7736
4-Pregnene-11β,17α,21-triol-3,20-dione see 4710
4-Pregnene-17α,20β,21-triol-3,11-dione, 7737
4-Pregnene-17α,20β,21-triol-3-one, 7738
17α-Pregnenetriolone see 7738
Pregn-4-ene-3,11,20-trione see 5185
4-Pregnene-3,11,20-trione-17α,21-diol, 21β-Cyclopentanepropionate see 2534
Pregneninolone see 3696
4-Pregnen-17α-ol-3,20-dione see 4773
4-Pregnen-21-ol-3,20-dione see 2882
Pregnenolone, 7739
Δ^{5}-Pregnen-3β-ol-20-one see 7739

Pregnenolone Methyl Ether, 7740
Δ⁴-Pregnen-21-ol-3,11,20-trione *see* 2859
17α-Pregn-4-en-20-yno[2,3-*d*]isoxazol-17-ol *see* 2811
Pregnesin [E. Merck] *see* 4534
Pregnyl [Organon] *see* 4534
Pregova [Ortho] *see* 4189
Prehemataminic Acid *see* 6392
Prelis [Brunnengräber] *see* 6072
Prelital *see* 1495
Prelog-Djerassi Lactone, 7741
Preludin *see* 7200
Premarin [Ayerst] *see* 2504
Premarrubiin *see* 5635
Premerge [Dow] *see* 3282
Premocillin *see* 7042
Prempar [Duphar] *see* 8228
Prenalterol, 7742
Prenazone *see* 3953
Prenolone *see* 7719
Prenormine [ICI-Farma] *see* 879
Prenoxdiazine Hydrochloride, 7743
Prent [Bayer] *see* 13
Prentox [Prentiss] *see* 7446
Prenylamine, 7744
4-Prenyl-1,2-diphenyl-3,5-dioxopyrazolidine *see* 3953
4-Prenyl-1,2-diphenyl-3,5-pyrazolidinedione *see* 3953
Prepalin *see* 9918
PrePar [Thomae; Duphar] *see* 8228
Preparation K *see* 3390
Prepared Calcium Carbonate *see* 1657
Prepared Chalk *see* 1657
Pre-Pen [Kremers-Urban] *see* 7050
Prephenic Acid, 7745
Prepidil [Upjohn] *see* 7893
PreproANF *see* 887
Prepulsid [J & J] *see* 2318
Pres [Dieckmann] *see* 3521
Presamine [USV] *see* 4835
Pre-Sate [Warner-Lambert] *see* 2183
Prescal [Ciba-Geigy] *see* 5129
Presdate [Alfa] *see* 5208
Presfersul *see* 4006
Presidal *see* 7060
Presidon *see* 8005
Presinol [Bayer] *see* 5974
Pressalolo [Locatelli] *see* 5208
Pressamina [Zambeletti] *see* 3254
Pressionorm [Merrell Pharma] *see* 4295
Pressomin Hydrochloride *see* 5910
Pressonex [Winthrop] *see* 5837
Pressorol [Baxter] *see* 5837
Pressunic [Unipharm] *see* 3152
Pressural [Polifarma] *see* 4847
Prestociclina *see* 7052
PreSun [Westwood] *see* 434
Presuren *see* 4753
Prethcamide *see* 2595 and 2598
Pretilachlor, 7746
Pretiron *see* 9709
Pretonine [Arkodex] *see* 4784
Pretor [Hoechst] *see* 1935
Pretyrosine *see* 814
Prevangor *see* 7066
Preventol G-D *see* 3059
Previscan [Nativelle] *see* 4061
Prexan [LaFare] *see* 6337
Prexidil [Bioindustria] *see* 6122
PR-G 138-Cl *see* 2271
Priadel [Delalande] *see* 5404
Priamide [Janssen] *see* 5090
Priatin *see* 7515
Priaxim [Ravizza] *see* 4074
Priceite *see* 1651
Pricortin [Primedics] *see* 7719
Pridinol, 7747
Prifinium Bromide, 7748

Prilagin [Gambar] *see* 6101
Prilocaine, 7749
Prilon [Cassenne] *see* 6898
Primabalt RP [Primedics] *see* 4739
Primacaine Hydrochloride *see* 5822
Primaclone *see* 7753
Primacor [Sterling] *see* 6117
Primal *see* 386
Primalan [Pharmuka] *see* 5754
Primamycin *see* 4524
Primantron *see* 7515
Primaperone, 7750
Primaquine, 7751
Primary Ammonium Phosphate *see* 571
Primary Calcium Phosphate *see* 1700
Primary Isoamyl Alcohol *see* 5081
Primary Magnesium Phosphate *see* 5561
Primary Sodium Phosphate *see* 8613
Primatol [Ciba-Geigy] *see* 7799
Primatol A [Geigy] *see* 886
Primatol S [Geigy] *see* 8485
Primaverine *see* 7706
Primaxin [Merck & Co.] *see* 2275 and 4834
Primbactam [Menarini] *see* 938
Primeral [Master] *see* 6337
Primeverose, 7752
Primicid [ICI] *see* 7469
Primidone, 7753
Primobolan-Depot [Schering AG] *see* 5887
Primobolan Tablets [Schering AG] *see* 5887
Primocarcin, 7754
Primocort *see* 2883
Primocortan *see* 2883
Primofax [Rhône-Poulenc] *see* 6639
Primofol *see* 3653
Primogonyl *see* 4534
Primogyn *see* 3689
Primogyn B *see* 3655
Primogyn C *see* 3689
Primogyn I *see* 3655
Primogyn M *see* 3689
Primoline [Darby] *see* 7753
Primolut [Schering AG] *see* 7783
Primolut C [Schering AG] *see* 3696
Primolut N [Schering AG] *see* 6614
Primolut-Nor [Schering AG] *see* 6614
Primonabol *see* 5887
Primonabol Depot *see* 5887
Primosiston [Schering] *see* 6614
Primotest *see* 9109
Primoteston [Schering] *see* 9112
Primperan [Delagrange] *see* 6063
Primulaverin, 7755
Primulidin *see* 5596
Primulin *see* 5596
Primun [Inca] *see* 4072
Primycin, 7756
Prinadol [SK & F] *see* 7174
Prinalgin [Berk] *see* 212
Princep [Geigy] *see* 8485
Prince's Feather *see* 383
Prince's Pine *see* 2046
Princillin [Squibb] *see* 621
Principen [Squibb] *see* 621
Principen/N [Squibb] *see* 621
Prinivil [Merck & Co.] *see* 5393
Prinodolol *see* 7412
Prinzone [Squibb] *see* 8871
Priodax [Schering] *see* 4919
Prioderm [Coates & Cooper] *see* 5582
Priomicina [San Carlo] *see* 4169
Pripsen [Purdue Frederick] *see* 7431
Priscol [Ciba] *see* 9430
Priscoline [Ciba] *see* 9430
Prisilidene *see* 307
Pristacin *see* 2024

Pristane, 7757
Pristinamycin, 7758
Pristinamycin I_A *see* 6111
Pristinamycin I_B *see* 9865
Pristinamycin I_C *see* 9865
Pristinamycin II_A *see* 9910
Privadol [Roland Marie] *see* 4330
Privaprol [Lepetit] *see* 5459
Privenal *see* 4625
Privine Hydrochloride [Ciba] *see* 6287
PRN [Bristol] *see* 7287
Pro (IUPAC Abbrev.) *see* 7790
Proaccelerin *see* 3871
Pro-Actidil [Wellcome] *see* 9661
Pro-Air [Warner-Lambert] *see* 7765
Pro-Amid [Amid] *see* 7915
Proaqua [Reid-Provident] *see* 1134
Proasma *see* 5919
Proazamine *see* 7797
Probamyl *see* 5751
Proban [Am. Cyanamid] *see* 2791
Pro-Banthine [Searle] *see* 7816
Probarbital, 7759
Probe [Velsicol] *see* 5876
Probecid *see* 7760
Proben *see* 7760
Probenecid, 7760
Probese-P *see* 7200
Probilin [Gödecke; Warner-Lambert] *see* 7457
Probolin [Tennessee Pharm.] *see* 5861
Probon [Chinoin] *see* 8222
Probonal [Farmos] *see* 8222
Probucol, 7761
Pro-Bumin [Lederle] *see* 8417
Probutylin [Rorer] *see* 7763
Procadil [Recordati] *see* 7765
Procainamide Hydrochloride, 7762
Procaine, 7763
Procaine Amide Hydrochloride *see* 7762
Procaine Benzylpenicillinate *see* 7042
Procaine Penicillin G *see* 7042
Procalmadiol *see* 5751
Procalmidol *see* 5751
Procamide *see* 7762
Procanodia *see* 7042
Procan-SR [Parke, Davis] *see* 7762
Procapan [Panray] *see* 7762
Procaptan [Biopharma] *see* 7122
Procarbazine, 7764
Procardia [Pfizer] *see* 6441
Procardyl *see* 7762
Procasil *see* 7882
Procaterol, 7765
Procerin, 7766
Processine [Sanko] *see* 2308
Procetofen *see* 3924
Procetofene *see* 3924
Procetoken [Bernabo] *see* 3924
Prochloraz, 7767
Prochlorpemazine *see* 7768
Prochlorperazine, 7768
Procholon [Squibb] *see* 2858
Procholon Sodium [Squibb] *see* 2858
Prociclide [Crinos] *see* 2851
Proclival [Houdé] *see* 1460
Proclorperazine *see* 7768
Procoagulo *see* 5712
Procodazole, 7769
Procolon [Sandoz] *see* 9408
Proconazole, 7830
Proconvertin *see* 3872
Procorum [Knoll] *see* 4257
Proctocort [Rowell] *see* 4710
Proctodon [Rowell] *see* 3301
Proctofoam [Reed & Carnrick] *see* 7707
Procyanidins *see* 1908
Procyclidine, 7770

Procyclomin *see* 3087
Procymate, 7771
Procymidone, 7772
Procytox [Horner] *see* 2753
Pro-Dafalgan [UPSA] *see* 7804
Prodasone [Upjohn] *see* 5677
Prodectin [Gedeon Richter] *see* 7987
Prodel *see* 3454
Pro-Diaban [Bayer; Schering AG] *see* 4339
Prodiamine, 7773
Prodiax [Merck & Co.] *see* 7565
Prodigiosin, 7774
Prodigiosine *see* 7774
Prodilidine, 7775
α-Prodine *see* 307
β-Prodine *see* 307
Prodipine, 7776
Prodixamon *see* 7816
Prodlure, 7777
Pro-Dorm [Delagrange] *see* 5872
Pro Dorm *see* 5456
Prodox [Upjohn] *see* 4773
Prodoxol [Wm. R. Warner] *see* 6899
Prodromine *see* 7303
Pro-Drone®, 7778
Producer Gas, 7779
Producil [Merck & Co.] *see* 3491
Pro-Entra *see* 9661
Profamina *see* 616
Profasi [Interlabo] *see* 4534
Profasi HP [Serono] *see* 4534
Profecundin *see* 9931
Profemin [Yamanouchi] *see* 4221
Profenamine *see* 3703
Profenid [Specia] *see* 5184
Profenil [SM & P] *see* 377
Profenine *see* 377
Profenon [Morishita] *see* 4039
Profenone *see* 6992
Proferrin [Merck & Co.] *see* 3974
Profetamine Phosphate [Clark & Clark] *see* 616
Profibrinolysin *see* 7494
Profilate [Alpha Ther.] *see* 3873
Proflavine, 7780
Proflavine Hemisulfate *see* 7780
Proflax [Merck & Co.] *see* 9374
Profluralin, 7781
Profoliol [Schering] *see* 3653
Proformiphen *see* 7229
Profundol *see* 9010
Progabide, 7782
Progallin P *see* 7872
Proganol *see* 8475
Proge [Mochida] *see* 4774
Progekan *see* 7783
Progeril [Midy] *see* 3598
Progesic [Lilly] *see* 3926
Progestasert [Alza/M & B] *see* 7783
Progesterol [Organon] *see* 7783
Progesterone, 7783
Progesterone Cyclopentyl-3-enol Ether *see* 8069
Progestin *see* 7783
Progestogel [Nourypharma] *see* 7783
Progestol [Bruco] *see* 7783
Progestone *see* 7783
Progestoral *see* 3696
Proglicem [Byk-Essex] *see* 2986
Proglumetacin, 7784
Proglumide, 7785
Proglycem [Schering] *see* 2986
Proguanil *see* 2088
Proglyut [Schering AG] *see* 3689
Progynon-B [Schering AG] *see* 3655
Progynon Benzoate [Schering AG] *see* 3655
Progynon C [Schering AG] *see* 3689
Progynon-DH [Schering AG] *see* 3653

Progynon-DP [Schering AG] *see* 3653
Progynon M [Schering AG] *see* 3689
Progynova [Schering AG] *see* 3653
Prohalone [Ayerst] *see* 4513
Proheptatriene *see* 2719
Proheptazine, 7786
Proinsulin, 7787
Prokarbol *see* 3272
Prokayvit [BDH] *see* 5714
Prokayvit Oral *see* 5710
Proketazine [Wyeth] *see* 1868
Prokrein [Tobishi] *see* 5159
Prolactin, 7788
Proladone [Boots] *see* 6912
Prolamins, 7789
Prolan [CSC] *see* 6515
Prolan [Bayer] *see* 4534
Prolastin [Cutter] *see* 751
Prolate [Stauffer] *see* 7311
Prolax *see* 5737
Prolene [Industrial Rayon] *see* 7558
Proleukin [Cetus] *see* 4896
Prolex [Makhteshim-Agan] *see* 7805
Prolex *see* 5427
Prolidon (Tabl) *see* 3696
Prolidon (Inj) *see* 7783
Proline, 7790
Prolintane, 7791
Prolixan [Siegfried; Logeais] *see* 758
Prolixin [Squibb] *see* 4116
Prolixin Decanoate [Squibb] *see* 4116
Prolixin Enanthate [Squibb] *see* 4116
Prolonium Iodide, 7792
Proloprim [Burroughs Wellcome] *see* 9624
Proluton [Schering AG] *see* 7783
Proluton C [Schering AG] *see* 3696
Proluton Depot [Schering AG] *see* 4774
L-Prolyl-L-leucylglycinamide *see* 5693
Proma [Armour Pharm.] *see* 2186
ProMACE *see* MISC-2
ProMACE-CytaBOM *see* MISC-2
ProMACE-MOPP *see* MISC-2
Promacetin [Parke, Davis] *see* 66
Promacid [Knoll] *see* 2186
Promacortine *see* 6028
Promactil *see* 2186
Promanide *see* 4357
Promantine *see* 7797
Promaquid [Rhône-Poulenc] *see* 4146
Promassol [DDR] *see* 386
Promassolax *see* 6927
Promate [Truett] *see* 5751
Promaxol [Esteve] *see* 7765
Promazil *see* 2186
Promazine, 7793
Promecarb, 7794
Promecon [Endopharm] *see* 1133
Promedol, 7795
α-Promedol *see* 7795
γ-Promedol *see* 7795
Promegestone, 7796
Promethazine, 7797
Promethegan [Parmed] *see* 7797
Promethestrol *see* 5888
γ-Promethestrol *see* 5888
Promethium, 7798
Prometon, 7799
Prometone *see* 7799
Prometryn, 7800
Promid [Kaken] *see* 7785
Promide [Kempthorne Prosser] *see* 7762
Promidione *see* 4964
Promin [Parke, Davis] *see* 4357
Prominal *see* 5742
Promine [Lacer] *see* 8166
Promine *see* 7797
Promintic [Ayerst] *see* 6084

Promit [Knoll; Braun Melsungen] *see* 2925
Promizole [Parke, Davis] *see* 9237
Promotil [Badrial] *see* 7791
Promotin [Nippon Chemiphar] *see* 5159
Promoxolane, 7801
Promulsin *see* 8761
Promwill *see* 7793
Pronamid *see* 7886
Prondol [Wyeth] *see* 4962
Pronestyl Hydrochloride [Squibb] *see* 7762
Pronethalol, 7802
Proneurin [Importex] *see* 7896
Pronilin [Akzo] *see* 5480
Pronison *see* 7727
Pronox [Reiss] *see* 2717
Prontalbin *see* 8898
Prontalgin [SPA] *see* 4812
Prontamid [SIT] *see* 8870
Prontobario [Bracco] *see* 1006
Prontoglucal *see* 4358
Prontoglukal *see* 4358
Prontosil *see* 8895
Prontosil III *see* 8872
Prontosil Album *see* 8898
Prontosil Flavum *see* 8895
Prontosil Rubrum *see* 8895
Prontosil S *see* 936
Prontosil Soluble *see* 936
Prontylin *see* 8898
Pronzin Rubrum *see* 8895
Proopiocortin *see* 7803
Pro-opiomelanocortin, 7803
Pro u-PA *see* 7918
Propacetamol, 7804
Propachlor, 7805
Propacil *see* 7882
Propacin *see* 9230
Propaderm [Allen & Hanburys] *see* 1029
1,2-Propadiene-1,3-dione *see* 1821
Propadrine Hydrochloride [Merck & Co.] *see* 7279
Propaesin *see* 8226
Propafenone, 7806
Propahexal [Hexal] *see* 7852
Propal *see* 5105
Propallylonal, 7807
Pro-Pam [Protea] *see* 2977
Propamidine, 7808
Propaminodiphen *see* 7706
Propanal *see* 7835
Propanamide, 7836
1-Propanamine *see* 7855
2-Propanamine *see* 5097
Propane, 7809
1-Propanearsonic Acid, 7810
1,2-Propanediamine *see* 7865
1,3-Propanedicarboxylic Acid *see* 4367
Propane-diethyl Sulfone *see* 8939
Propanedinitrile *see* 5592
Propanedioic Acid *see* 5591
Propanedioic Acid Bis(1,1-dimethylethyl) Ester *see* 3021
Propanedioic Acid Diethyl Ester *see* 3779
Propanedioic Acid Dimethyl Ester *see* 6009
1,2-Propanediol *see* 7868
1,3-Propanediol *see* 9629
1,3-Propanediol Bis[α-(p-chlorophenoxy)isobutyrate] *see* 8489
1,3-Propanediol Bis[2-(4-chlorophenoxy)-2-methylpropionate] *see* 8489
1,2-Propanediol-3-(p-chlorophenoxy)-1-carbamate *see* 2179
1,3-Propanedithiol, 7811

4,4'-[1,3-Propanediylbis(oxy)]bis-benzenecarboximidamide *see* 7808
4,4'-[1,3-Propanediylbis(oxy)]bis(3-bromobenzenecarboximidamide) *see* 3008
Propanenitrile *see* 7839
Propanethial *S*-Oxide, 7812
1,2,3-Propanetricarboxylic Acid *see* 9534
1,2,3-Propanetriol *see* 4379
1,2,3-Propanetriol 1-(4-Aminobenzoate) *see* 4382
1,2,3-Propanetriol Diacetate *see* 2941
1,2,3-Propanetriol Monoacetate *see* 6152
1,2,3-Propanetriol, Mono(dihydrogen Phosphate) Calcium Salt (1:1) *see* 1673
1,2,3-Propanetriol Mono(dihydrogen Phosphate) Disodium Salt *see* 8570
1,2,3-Propanetriol Triacetate *see* 9504
1,2,3-Propanetriol Trinitrate *see* 6528
Propanidid, 7813
Propanil, 7814
Propanocaine, 7815
Propanoic Acid *see* 7837
Propanoic Acid Anhydride *see* 7838
Propanoic Acid Butyl Ester *see* 1587
Propanoic Acid Ethyl Ester *see* 3801
Propanoic Acid Methyl Ester *see* 6030
Propanoic Acid Propyl Ester *see* 7880
Propanoic Anhydride *see* 7838
1-Propanol *see* 7854
2-Propanol *see* 5096
2-Propanol Aluminum Salt *see* 349
β-Propanolamine *see* 480
Propanolide *see* 7832
2-Propanol Nitrite *see* 5104
2-Propanone *see* 58
Propanone 1-Oxime *see* 5076
2-Propanone Oxime *see* 68
Propanoyl Chloride *see* 7840
Propantheline Bromide, 7816
Propaphenin *see* 2186
Proparacaine, 7817
Propargil *see* 7818
Propargite, 7818
Propargyl Alcohol, 7819
Propargyl Chloride, 7820
Propargylic Acid *see* 7833
2-Propargyloxy-5-amino-*N*-butyl-benzamide *see* 6993
Propasa [Merck & Co.] *see* 491
Propat [Hishiyama] *see* 9425
Propatyl Nitrate, 7821
Propavan *see* 7834
Propax *see* 6881
Propaxoline *see* 7919
Propazine, 7822
Propazol *see* 7769
Propazone *see* 3352
Propazyl *see* 8226
Pro-Pen [Merck & Co.] *see* 7042
2-Propenal *see* 122
2-Propenamide *see* 123
2-Propen-1-amine *see* 285
1-Propene *see* 7862
trans-1-Propene-1,2-dicarboxylic Acid *see* 5806
1-Propene Homopolymer *see* 7558
2-Propenenitrile *see* 125
Propene Oxide *see* 7869
2-Propene-1-sulfinothioic Acid *S*-2-Propenyl Ester *see* 249
1-Propene-1,2,3-tricarboxylic Acid *see* 112
2-Propenoic Acid *see* 124
2-Propenoic Acid Butyl Ester *see* 1539
2-Propenoic Acid Ethyl Ester *see* 3715

2-Propenoic Acid Methyl Ester *see* 5935
1-Propenol-3 *see* 284
2-Propen-1-ol *see* 284
1-Propen-2-ol Acetate *see* 5091
Propentofylline, 7823
1-Propen-2-yl Acetate *see* 5091
2-Propenylacrylic Acid *see* 8677
3-[[(2-Propenylamino)thioxomethyl]-amino]benzoic Acid Monocopper-(1+) Monosodium Salt *see* 256
p-Propenylanisole *see* 675
5-(1-Propenyl)-1,3-benzodioxole *see* 5112
5-(2-Propenyl)-1,3-benzodioxole *see* 8287
17-(2-Propenyl)estr-4-en-17-ol *see* 289
4-Propenylguaiacol *see* 5054
17-(2-Propenyl)morphinan-3-ol *see* 5342
p-Propenylphenol *see* 706
4-(1-Propenyl)phenol *see* 706
4-(2-Propenyl)phenol *see* 2039
2-(1-Propenyl)piperidine *see* 2496
N-2-Propenyl-2-propen-1-amine *see* 2951
2-Propenyl 3-(2-Propenylsulfinyl)-1-propenyl Disulfide *see* 185
3-(2-Propenylsulfinyl)-ʟ-alanine *see* 251
(2-Propenyl)thiourea *see* 9293
N-2-Propenylurea *see* 296
Propenzolate, 7824
Properdin, 7825
Propericiazine *see* 7117
Properidine, 7826
Propesin *see* 8226
Propetamphos, 7827
Propethonum Iodide *see* 9577
Propham, 7828
Prophelan *see* 377
Prophenal *see* 7162
Prophenatin [Hishiyama] *see* 3071
Propheniramine *see* 7198
Prophenpyridamine *see* 7198
Prophylux [Henning] *see* 7852
Propicillin, 7829
Propicol *see* 3183
Propiconazole, 7830
Propildazine *see* 7394
Propine [Allergan] *see* 3344
Propineb, 7831
Propiocine [Roussel] *see* 3631
Propiocine Enfant [Roussel] *see* 3628
Propiodal *see* 7792
Propiodone *see* 7876
Propiokan *see* 9115
β-Propiolactone, 7832
Propiolic Acid, 7833
Propiomazine, 7834
Propionaldehyde, 7835
Propionamide, 7836
Propione *see* 3111
Propionic Acid, 7837
Propionic Acid Amide *see* 7836
Propionic Acid Butyl Ester *see* 1587
Propionic Acid Calcium Salt *see* 1705
Propionic Acid 1,2-Dimethyl-3-phenyl-3-pyrrolidinyl Ester *see* 7775
Propionic Acid Sodium Salt *see* 8623
Propionic Anhydride, 7838
Propionitrile, 7839
β-Propionolactone *see* 7832
Propionopyrrothine *see* 898
4-Propionoxy-1,3-dimethyl-4-phenyl-hexamethylenimine *see* 7786
Propionylbenzene *see* 7842
Propionyl Chloride, 7840
3-Propionyl-10-dimethylaminoiso-propylphenothiazine *see* 7834

3-Propionyl-10-(γ-dimethylamino-propyl)phenothiazine *see* 7841
Propionylerythromycin *see* 3631
Propionylerythromycin Lauryl Sulfate *see* 3628
2-Propionyl-10-[γ-(*N*'-β-hydroxy-ethylpiperazino)propyl]phenothi-azine *see* 1868
3''-Propionylleucomycin A5 *see* 8233
p-Propionylphenol *see* 6992
N-Propionyl-2-(1-piperidinoisoprop-yl)aminopyridine *see* 7844
Propionylpromazine, 7841
Propionylpromethazine *see* 7834
N-Propionyl-*N*-(2-pyridyl)-1-piperi-dino-2-aminopropane *see* 7844
Propiophenone, 7842
Propiopromazine *see* 7841
Propipocaine, 7843
Propiram, 7844
Propisamine *see* 616
Propitan *see* 7422
Propitocaine *see* 7749
Propivane, 7845
Propizepine, 7846
Proplex [Hyland] *see* 3874
Propofol, 7847
Propolis, 7848
Propolis Balsam *see* 7848
Propolis Resin *see* 7848
Propolis Wax *see* 7848
Proponesin *see* 9449
Proponex-Plus [Shell] *see* 5666
Propox [Phoenix] *see* 7851
Propoxur, 7849
1-Propoxy-2-acetamino-4-nitrobenz-ene *see* 6550
1-Propoxy-2-amino-4-nitrobenzene *see* 6551
5-Propoxy-2-benzimidazolecarbamic Acid Methyl Ester *see* 6891
(5-Propoxy-1*H*-benzimidazol-2-yl)-carbamic Acid Methyl Ester *see* 6891
(6-Propoxy-2-benzothiazolyl)carb-amic Acid Methyl Ester *see* 9387
Propoxycaine Hydrochloride, 7850
5-Propoxy-2-(carbomethoxyamino)-benzimidazole *see* 6891
Propoxychel [Rachelle] *see* 7851
N-Propoxyethyl-*N*-chloroacetyl-2,6-diethylaniline *see* 7746
Propoxyphene, 7851
d-Propoxyphene *see* 7851
l-Propoxyphene *see* 5349
4-Propoxyphenyl Piperidine Ethyl Ketone *see* 7843
4-*n*-Propoxy-β-(1-piperidyl)propio-phenone *see* 7843
Propranolol, 7852
Propranur [Henning] *see* 7852
Propulm [Chiesi] *see* 7765
Propycil *see* 7882
Propyl Acetate, 7853
Propylacetic Acid *see* 9815
N-Propylajmaline *see* 7704
*N*⁴-Propylajmalinium *see* 7704
n-Propyl Alcohol, 7854
Propylaldehyde *see* 7835
Propylamine, 7855
Propyl *p*-Aminobenzoate *see* 8226
β-(*n*-Propylamino)-β,β-dimethylethyl Benzoate Hydrochloride *see* 5752
α-Propylamino-2-methylpropionani-lide *see* 7749
2-(Propylamino)-*o*-propionotoluidide *see* 7749
3-Propylamino-α-propionylamino-2-carbomethoxy-4-methylthiophene *see* 1878

Protoheme *see* 4558
Protoheme IX *see* 4558
Protokosin, 7910
Protokylol, 7911
Protolipan [Millet] *see* 3924
Protolysate [Mead Johnson] *see* 7900
Protomin *see* 4357
Protona [Grémy-Longuet] *see* 8753
Protopam Chloride [Ayerst] *see* 7705
Protopam Iodide [Ayerst] *see* 7705
Protopam Methanesulfonate [Ayerst] *see* 7705
Protophenicol *see* 2068
Protopine, 7912
Protoporphyrin IX, 7913
Protopyrin *see* 3685
(+)-Protoquercitol *see* 8046
Protosol *see* 3166
Protostephanine, 7914
Protostib *see* 5995
Protoveratrine A *see* 7915
Protoveratrine B *see* 7915
Protoveratrines, 7915
Protoverine, 7916
Protoverine 6,7-Diacetate 15(*l*)-2'-Methylbutyrate *see* 2900
Protoverine 6,7-Diacetate 15-(2-Methylbutyrate) *see* 2900
Protoxyl *see* 8521
Protran [Halsey] *see* 5751
Protriptyline, 7917
Protropin [Genentech] *see* 8672
Pro-UK *see* 7918
Pro-Urokinase, 7918
Prourokinase (Enzyme-activating) *see* 7918
Proustite *see* 8450
Provamycin *see* 8708
Provas [Sanol Schwarz] *see* 1463
Provatene [Microbio] *see* 1860
Provell Maleate [Lilly] *see* 7915
Provent [Wellcome] *see* 9212
Proventil [Schering] *see* 209
Provera [Upjohn] *see* 5677
Proverine *see* 377
Provest *see* 5677
Provigan [Reid-Provident] *see* 7797
Proviodine *see* 7701
Proviron [Schering AG] *see* 5817
Proviscol Wax *see* 8761
Provitamin D$_3$ *see* 2857
Provitar [Clin-Comar-Byla] *see* 6875
Provitina [Promonta] *see* 9929
Provocholine [Roche] *see* 5847
Prower Factor *see* 3875
Prowl [Am. Cyanamid] *see* 7026
Proxagesic [Tutag] *see* 7851
Proxazocain *see* 7707
Proxazole, 7919
Proxen [Grünenthal; Syntex] *see* 6337
Proxetil *see* MISC-4
Proxibarbal, 7920
Proxibarbital *see* 7920
Proxil [Rorer] *see* 7784
Proxol [Tuco] *see* 9536
Proxymetacaine *see* 7817
Proxyphylline, 7921
Proxy-Retardoral [Artesan] *see* 7921
Prozac [Lilly] *see* 4112
Prozapine, 7922
Prozil *see* 2186
Prozinex [Makhteshim-Agan] *see* 7822
Prulaurasin *see* 5602
Prulet [Mission Pharmacal] *see* 6927
Prunasin *see* 5602
Prunetin, 7923
Prunetol *see* 4281
Prunitrin *see* 7923
Prunol [Kremers-Urban] *see* 9802
Prunusetin *see* 7923

Pruralgan [Pharmacia] *see* 3207
Pruralgin [Pharmacia] *see* 3207
Prussian Blue *see* 3964
Prussic Acid *see* 4722
Pryleugan [Arzneimittelwerk VEB] *see* 4835
PS 207 *see* 9124
Pseudoaconitine, 7924
Pseudobaptigenin, 7925
Pseudobaptisin *see* 7925
Pseudobatrachotoxin *see* 1019
Pseudobutylbenzene *see* 1551
Pseudo-butylene *see* 1514
Pseudocef [Grünenthal] *see* 1943
Pseudochelerythrine *see* 8320
Pseudocholestane *see* 2518
Pseudocinchonine *see* 4706
Pseudococaine, 7926
Pseudocodeine, 7927
Pseudoconhydrine, 7928
Pseudocubebin *see* 8420
Pseudocumene, 7929
Pseudocumol *see* 7929
Pseudodigitoxin *see* 4328
d-Pseudoephedrine *see* 3561
Pseudoergostane *see* 2516
Pseudofructose *see* 7940
Pseudohecogenin, 7930
Pseudohyoscyamine *see* 6623
Pseudoionone, 7931
Pseudojervine, 7932
Pseudolibethenite *see* 2653
Pseudomalachite *see* 2653
trans-Pseudomonic Acid *see* 6212
Pseudomonic Acid I *see* 7933
Pseudomonic Acid A *see* 6212
Pseudomonic Acid B *see* 7933
Pseudomonic Acid C *see* 7933
Pseudomonic Acid D *see* 7933
Pseudomonic Acids, 7933
Pseudomonil [Ciba-Geigy] *see* 1943
Pseudomorphine, 7934
Pseudonorephedrine *see* 6634
Pseudopaederine *see* 7935
Pseudopederin, 7935
Pseudopederine *see* 7935
Pseudopelletierine, 7936
Pseudopunicine *see* 7936
3-Pseudotropanol *see* 7937
Pseudotropine, 7937
Pseudotropine Benzoate *see* 9686
Pseudo-wollastonite *see* 1709
Pseudoyohimbine, 7938
Psicaine *see* 7926
Psicaine N *see* 7926
Psichial [RIT] *see* 2082
Psicofuranine, 7939
9β-D-Psicofuranosyladenine *see* 7939
9-β-D-Psicofuranosyl-9*H*-purin-6-amine *see* 7939
Psicopax [Vemaco] *see* 5456
Psicoperidol *see* 9596
Psicosan [Vermont] *see* 2082
D-Psicose, 7940
Psico-Soma [Ferrer] *see* 4363
Psicosoma [Trommsdorff] *see* 4363
Psicosterone *see* 7710
Psicoterina [Farmindustria] *see* 2082
Psicronizer [Albert-Pharma] *see* 6592
Psigodal [E. Merck] *see* 5745
Psilocin, 7941
Psilocybin, 7942
Psilocyn *see* 7941
Psiquium [Sintofarma] *see* 5669
PSK® [Kureha], 7943
Psoil *see* 9388
Psoraderm [Goupil] *see* 1173
Psoradrate [Eaton] *see* 714
Psoralen, 7944
Psoralidin, 7945
Psoralon [Hermal] *see* 246

Psoralon-MOP [Hermal] *see* 5911
Psoriacide [Doak] *see* 714
Psorigel [Alcon] *see* 2420
P.S.P. *see* 7213
Psychamine A 66 *see* 7200
Psychedrine *see* 616
Psychoperidol *see* 9596
Psychosan *see* 909
Psycho-Soma [Boots-Dacour; Ferrer; Roner] *see* 4363
Psychostyl *see* 6635
Psychoton *see* 616
Psychotrine, 7946
Psychoverlan [Verla] *see* 4363
Psyllium Seed *see* 7489
Psymod [Pitman-Moore] *see* 7429
Psyquil [Heyden] *see* 9597
PT-14 *see* 7466
(1P-1T) *see* 7436
PTA *see* 3876
PTC *see* 3874
Pteridine, 7947
2,4(1*H*,3*H*)-Pteridinedione *see* 5467
Pterin HB$_2$ *see* 1242
Pterocarpin, 7948
Pterofen *see* 9515
Pteroic Acid, 7949
Pterolyl-γ-glutamyl-γ-glutamyl-glutamic Acid *see* 7950
Pterophene *see* 9515
Pteropterin, 7950
Pteroyldiglutamic Acid *see* 3285
Pteroyldi-γ-glutamylglutamic Acid *see* 7950
Pteroylglutamic Acid *see* 4140
Pteroyl-α-glutamylglutamic Acid *see* 3285
N-[*N*-(*N*-Pteroyl-γ-glutamyl)-γ-glutamyl]glutamic Acid *see* 7950
Pteroylheptaglutamic Acid *see* 7951
Pteroylhexaglutamylglutamic Acid, 7951
Pteroyltriglutamic Acid *see* 7950
PTFE *see* 7560
PTGA *see* 7950
PTH *see* 6984
Ptimal *see* 9620
PTO *see* 8004
PTX *see* 6951
Ptychotis Oil *see* 186
PU-239 *see* 1090
Puberulic Acid, 7952
Puberulonic Acid, 7953
Puccoon Root *see* 8319
Pudding Pipe *see* 1900
Pudding-stick *see* 1900
Puerzym [Firma] *see* 7103
PUK *see* 7918
Pukateine, 7954
Pularin *see* 4571
Pulegone, 7955
Pulmadil [Riker] *see* 8223
Pulmaxil N *see* 7027
Pulmicort [Astra] *see* 1455
Pulmidol *see* 7861
PulmiDur [Astra] *see* 9212
Pulmo 500 *see* 7027
Pulmoclase [UCB] *see* 1809
Pulmo-Timelets [Temmler] *see* 9212
Pulonon [Yamanouchi] *see* 7806
Pulsamin *see* 3822
Pulsan [Yamanouchi] *see* 4852
Pulsar [Medosan] *see* 2916
Pulsatilla, 7956
Pulsatilla Camphor *see* 674
Pulsoton *see* 4740
Pulsotyl *see* 7304
Puma [Hoechst] *see* 3929
Pumice, 7957
Pumpellyite *see* 1645
Pumpkin Seed, 7958

Cross Index of Names

Punch [Du Pont] see 4130
Punicin see 7014
Punicine see 7015
Punicine Tannate see 7016
Punktyl [Krewel] see 5456
Puragel see 4276
Puralin see 9304
Purantix [Wander] see 2368
Purapuridine see 8665
Purapurine see 8666
Purecal [Wyandotte] see 1657
Pureline [Sherwood] see 7138
Purerin [Kowa] see 5883
Purgaceen see 6927
Purging Agaric see 173
Purging Cassia see 1900
Purified Araroba see 2257
Purified Goa Powder see 2257
Purified Oxgall see 6884
Purified Ozokerite see 1984
Purified Protein Derivative see 9715
Purified Rayon see 8132
Purim [Laphal] see 7475
1H-Purin-6-amine see 141
1H-Purine, 7959
1H-Purine-2,6-diamine see 2963
2,6(1H,3H)-Purinedione see 9968
6-Purinethiol see 5762
Puri-Nethol [Burroughs Wellcome]
  see 5762
Purinethol [Burroughs Wellcome] see
  5762
Purine-2,6,8-triol see 9790
Purine-2,6,8(1H,3H,9H)-trione see
  9790
Purin-6(1H)-one see 4805
Purochin [Sclavo] see 9799
Purocyclina see 9130
Purodigin [Wyeth] see 3146
Puromycin, 7960
Purophyllin see 7921
Purosin-TC [Tatsumi] see 7889
Purostrophan see 6854
Purothionin, 7961
Purothionin A I see 7961
Purothionin A II see 7961
α₁-Purothionin see 7961
α₂-Purothionin see 7961
β-Purothionin see 7961
Puroverin(e) [Sandoz] see 7915
Purple of Cassius see 4424
Purple Clover see 9601
Purple Foxglove see 3141
Purpurea Glycoside C see 2903
Purpurid [Promonta] see 3146
Purpurin, 7962
Purpurogallin, 7963
Purpurosamine see 4284
Pursennid see 8407
Purshiana Bark see 1889
Pusilline see 5115
Putrescine, 7964
Putty Powder see 8736
PVA see 7562
PVB see MISC-2
PVC see 7563
PVK see 7046
P.V.P. see 7700
PVP-1 see 7701
PX-917 see 9675
Pycaril see 6436
Pycazide see 5071
Pydirone [Breon] see 3358
Pydrin [Shell] see 3952
Pyelectan see 8584
Pyelokon-R [Mallinckrodt] see 72
Pyelosil see 4933
Pygnoforton [Plantorgan] see 5334
Pykaryl see 6436
Pyknolepsinum [Rhein-Pharma] see
  3704

Pylapron [Kyorin] see 7852
Pylen [Toyo] see 7544
Pylumbrin see 4933
Pynamin [Sumitomo] see 248
Pynastin [Maruko] see 7412
Pynosect [Mitchell Cotts] see 7132
Pyocefal [Cassenne-Takeda] see 1943
Pyocianil [Farmitalia] see 1796
Pyocyanine, 7965
Pyoktanin see 4287
Pyopen [Beecham] see 1796
Pyostacine see 7758
Pyoxanthose see 4564
Pyra see 7996
Pyracarbolid, 7966
β-Pyracine see 7994
Pyracort D [Lemmon] see 7257
Pyradone see 488
Pyrafat [Saarstickstoff-Fatol] see 7970
Pyrahexyl see 8995
Pyralcid see 8884
Pyralene [Prodelec] see 7541
Pyralgin [Savage] see 3358
Pyralin see 1960
Pyramal [Roxane] see 7996
Pyramidon [Winthrop] see 488
Pyramidon Bicamphorate [Winthrop]
  see 488
Pyramidon Salicylate [Winthrop] see
  488
Pyramin see 9482
Pyramine see 9482
Pyran, 7967
1,2-Pyran see 7967
1,4-Pyran see 7967
2H-Pyran see 7967
4H-Pyran see 7967
α-Pyran see 7967
γ-Pyran see 7967
Pyranisamine see 7996
Pyrantel, 7968
Pyrantel Embonate see 7968
Pyranton see 2944
Pyrargyrite see 8450
Pyrasanone see 7248
Pyrathiazine, 7969
Pyrathyn [Flint] see 5871
Pyrazinamide, 7970
Pyrazine, 7971
Pyrazinecarboxamide see 7970
Pyrazine Carboxylamide see 7970
Pyrazinecarboxylic Acid see 7973
2,3-Pyrazinedicarboxylic Acid, 7972
Pyrazinemonocarboxylic Acid see
  7973
Pyrazinobutazone see 7248
Pyrazinoic Acid, 7973
Pyrazinoic Acid Amide see 7970
Pyrazino[2,3-d]pyrimidine see 7947
N¹-2-Pyrazinylsulfanilamide see 8912
Pyrazogin see 8896
Pyrazole, 7974
1H-Pyrazole-3-ethanamine see 1205
2-Pyrazoline, 7975
1H-Pyrazolo[3,4-d]pyrimidin-4-ol see
  278
Pyrazon see 7248
Pyrazophos, 7976
Pyrcon [Jenapharm] see 8034
Pyrene, 7977
Pyresyn [Penick] see 248
Pyrethrin I see 7978
Pyrethrin II see 7978
Pyrethrins, 7978
Pyrethrosin, 7979
Pyrethrum Flowers, 7980
Pyriamid see 8913
Pyribenzamine [Ciba] see 9651
Pyricarbate see 7987
Pyricardyl see 6459
Pyricidin see 5071

Pyrictal see 7183
Pyridacil see 7175
Pyridamal-100 [Bel-Mar] see 2180
Pyridate, 7981
Pyridazine, 7982
Pyridiate [Darby] see 7175
Pyridin [Du Pont] see 3952
2-Pyridinamine see 486
3-Pyridinamine see 487
2H-Pyrid[4,3-b]indole see 1812
Pyridine, 7983
3-Pyridineacetic Acid see 4659
2-Pyridine Aldoxime Methiodide see
  7705
2-Pyridine Aldoxime Methyl Chloride
  see 7705
3-Pyridinecarboxamide see 6398
4-(3-Pyridinecarboxamido)-2,3-di-
  methyl-1-phenyl-5-pyrazolone see
  6443
Pyridine-β-carboxylic Acid see 6435
o-Pyridinecarboxylic Acid see 7375
2-Pyridinecarboxylic Acid see 7375
3-Pyridinecarboxylic Acid see 6435
4-Pyridinecarboxylic Acid see 5073
γ-Pyridinecarboxylic Acid see 5073
3-Pyridinecarboxylic Acid Aluminum
  Salt see 354
Pyridine-β-carboxylic Acid Benzyl
  Ester see 6436
3-Pyridinecarboxylic Acid 2,2-Bis-
  [[(3-pyridinylcarbonyl)oxy]methyl]-
  1,3-propanediyl Ester see 6405
4-Pyridinecarboxylic Acid [(2-Carb-
  oxy-3,4-dimethoxyphenyl)methyl-
  ene]hydrazide see 6807
3-Pyridinecarboxylic Acid 2-[2-(4-
  Chlorophenoxy)-2-methyl-1-oxo-
  propoxy]ethyl Ester see 3834
3-Pyridinecarboxylic Acid 2-[2-(4-
  Chlorophenoxy)-2-methyl-1-oxo-
  propoxy]-1,3-propanediyl Ester see
  1239
3-Pyridinecarboxylic Acid 3-[2-(4-
  Chlorophenoxy)-2-methyl-1-oxo-
  propoxy]propyl Ester see 8238
3-Pyridinecarboxylic Acid 1-(4-
  Chlorophenyl)-2-methylpropyl Ester
  see 6426
3-Pyridinecarboxylic Acid Compd
  with 2-Aminoethanol (1:1) see 6437
3-Pyridinecarboxylic Acid Compd
  with 3,7-Dihydro-7-[2-hydroxy-3-
  [(2-hydroxyethyl)methylamino]-
  propyl]-1,3-dimethyl-1H-purine-
  2,6-dione (1:1) see 9969
Pyridine-3-carboxylic Acid Diethylamide see 6459
Pyridine-4-carboxylic Acid Diethylamide see 5074
3-Pyridinecarboxylic Acid 2-(Diethylamino)ethyl Ester see 6401
4-Pyridinecarboxylic Acid [(3,4-Di-
  methoxyphenyl)methylene]hydrazide
  see 9859
4-Pyridinecarboxylic Acid [1-(2-
  Furanyl)ethylidene]hydrazide see
  4220
3-Pyridinecarboxylic Acid 2-Hexyl-
  2-[[(3-pyridinylcarbonyl)oxy]meth-
  yl]-1,3-propanediyl Ester see 4574
4-Pyridinecarboxylic Acid Hydrazide
  see 5071
4-Pyridinecarboxylic Acid Hydrazide,
  Hydrazone with O-2-Deoxy-2-
  (methylamino)-α-L-glucopyranosyl-
  (1→2)-O-5-deoxy-3-C-formyl-α-L-
  lyxofuranosyl-(1→4)-N,N'-bis-
  (aminoiminomethyl)-D-streptamine
  Sulfate (2:3) (Salt) see 8788

4-Pyridinecarboxylic Acid Hydrazide Mono(4-amino-2-hydroxybenzoate) see 6999

3-Pyridinecarboxylic Acid (2-Hydroxy-1,3-cyclohexanediylidene)tetrakis(methylene) Ester see 6429

3-Pyridinecarboxylic Acid Hydroxymethylamide see 4765

3-Pyridinecarboxylic Acid Methyl Ester see 6014

4-Pyridinecarboxylic Acid 2-(1-Methylethyl)hydrazide see 4965

Pyridine-3-carboxylic Acid N-Methylolamide see 4765

3-Pyridinecarboxylic Acid 1-Oxide see 6895

4-Pyridinecarboxylic Acid 2-[3-Oxo-3-[(phenylmethyl)amino]propyl]-hydrazide see 6399

3-Pyridinecarboxylic Acid Phenylmethyl Ester see 6436

4-Pyridinecarboxylic Acid 2-(Sulfomethyl)hydrazide see 5072

4-Pyridinecarboxylic Acid 2-(Sulfomethyl)hydrazide Compd with [S-(R*,R*)]-2,2′-(1,2-Ethanediyldiimino)bis(1-butanol) (2:1) see 5013

4-Pyridinecarboxylic Acid [(3-Sulfophenyl)methylene]hydrazide see 8938

3-Pyridinecarboxylic Acid 2-(1,2,3,6-Tetrahydro-1,3-dimethyl-2,6-dioxo-7H-purin-7-yl)ethyl Ester see 3836

3-Pyridinecarboxylic Acid 3,3,5-Trimethylcyclohexyl Ester see 2269

2,3-Pyridinedicarboxylic Acid see 8102

2,5-Pyridinedicarboxylic Acid see 5043

3,4-Pyridinedicarboxylic Acid see 2285

2,6-Pyridinedimethanol Bis(methylcarbamate) (Ester) see 7987

2,6-Pyridinedimethanol Bis(3,4,5-trimethoxybenzoate) see 7477

2,6-Pyridinedimethanol Mono-p-chlorophenoxyisobutyrate see 7467

3-Pyridinemethanol see 6438

Pyridine N-Oxide see 7984

Pyridine 1-Oxide, 7984

2-Pyridinethiol 1-Oxide see 8004

Pyridinium Bromide Perbromide, 7985

Pyridinium Chlorochromate, 7986

2,3-Pyridino-(5′,6′)-5,6-benzo-4-(3″-dimethylaminopropyl)-1,4-thiazine see 7902

Pyridinol Carbamate, 7987

α-[(2-Pyridinylamino)methyl]benzenemethanol see 7292

N-2-Pyridinyl-[1,1′-biphenyl]-4-acetamide see 3125

3-Pyridinylcarbonimidodithioic Acid Butyl [4-(1,1-Dimethylethyl)phenyl]methyl Ester see 1520

17-[(3-Pyridinylcarbonyl)oxy]androst-4-en-3-one see 9113

2,6-Pyridinylenebis[methyl-N-methylcarbamate] see 7987

1-(3-Pyridinyl)ethanone see 6034

2-[1-(4-Pyridinyl)ethylidene]hydrazinecarbothioamide see 6035

4,4′-(2-Pyridinylmethylene)bisphenol Bis(hydrogen Sulfate) (Ester) Disodium Salt see 7377

Pyridipca see 7995

Pyridium [Warner-Lambert] see 7175

10H-Pyrido[3,2-b][1,4]benzothiadiazine-10-carboxylic Acid 2-(2-Piperidinoethoxy)ethyl Ester see 7423

4-[3-(10H-Pyrido[3,2-b][1,4]benzothiazin-10-yl)propyl]-1-piperazineethanol see 6923

Pyridofylline, 7988

5H-Pyrido[4,3-b]indole see 1812

9H-Pyrido[3,4-b]indole-3-carboxylic Acid Ethyl Ester see 3737

9H-Pyrido[3,4-b]indole-3-methanol see 3737

Pyridomycin, 7989

2-Pyridone-3-carboxylic Acid 2,6-Xylidide see 5078

Pyridostigmine Bromide, 7990

Pyridox [Oxford] see 7995

Pyridoxal, 7991

Pyridoxal 5-Monophosphoric Acid Ester see 7992

Pyridoxal 5-Phosphate, 7992

Pyridoxamine Dihydrochloride, 7993

4-Pyridoxic Acid, 7994

Pyridoxine p-Chlorophenoxyisobutyrate see 2375

Pyridoxine-5-disulfide see 8006

Pyridoxine Hydrochloride, 7995

Pyridoxinium Chloride see 7995

Pyridoxol Compd with 7-(2-Hydroxyethyl)theophylline Hydrogen Sulfate see 7988

Pyridoxol 3-[2-(Dimethylamino)ethyl Succinate] see 8002

Pyridoxol Hydrochloride see 7995

4-(Pyridyl-2-amidosulfonyl)-3′-carboxy-4′-hydroxyazobenzene see 8917

N-(2-Pyridyl)-N-(p-bromobenzyl)-N′,N′-dimethylethylenediamine see 1436

β-Pyridylcarbinol see 6438

3-[α-(2′-Pyridyl)ethyl]-2-(β-dimethylaminoethyl)indene see 3204

[2-(2-Pyridyl)ethyl]methylamine see 1200

7-[2-[(3-Pyridylmethyl)amino]ethyl]-theophylline see 7400

1-[2-[N-(2-Pyridylmethyl)anilino]ethyl]piperidine see 7376

4,4′-(2-Pyridylmethylene)bisphenol Diacetate see 1253

4,4′-(2-Pyridylmethylene)diphenolbis(hydrogen Sulfate) (Ester) Disodium Salt see 7377

N-3-Pyridylmethyl-N′-p-nitrophenylurea see 7999

N-(2-Pyridylmethyl)-N-phenyl-N-2-(piperidinoethyl)amine see 7376

β-Pyridyl-α-N-methylpyrrolidine see 6434

2-(3-Pyridyl)piperidine see 655

2-(3-Pyridyl)pyrrolidine see 6631

trans-1-(2-Pyridyl)-3-pyrrolidino-1-p-tolylprop-1-ene see 9661

5-[4-(2-Pyridylsulfamoyl)phenylazo]-2-hydroxybenzoic Acid see 8917

5-[p-(2-Pyridylsulfamoyl)phenylazo]-salicylic Acid see 8917

N1-2-Pyridylsulfanilamide see 8913

2-(3-Pyridyl)-1,2,3,6-tetrahydropyridine see 661

N-(α-Pyridyl)-N-(β-thenyl)-N′,N′-dimethylethylenediamine see 9207

7-[α-(4-Pyridylthio)acetamido]cephalosporanic Acid Sodium Salt see 1980

Pyrikappl [Isei] see 8971

Pyril [Maurry Biol.] see 3358

Pyrilamine, 7996

Pyrilax [Berlin-Chemie] see 1253

Pyrilgin [Savage] see 3358

Pyrimal see 8874

Pyrimal M see 8884

Pyrimethamine, 7997

Pyrimidine, 7998

2,4-Pyrimidinediol see 9761

2,4(1H,3H)-Pyrimidinedione see 9761

Pyrimidine-4′,5′:2,3-pyrazine see 7947

2,4,5,6(1H,3H)-Pyrimidinetetrone see 281

2,4,5,6(1H,3H)-Pyrimidinetetrone 5-Oxime see 9904

2,4,6(1H,3H,5H)-Pyrimidinetrione see 973

8-[4-[4-(2-Pyrimidinyl)-1-piperazinyl]butyl]-8-azaspiro[4.5]decane-7,9-dione see 1493

2-[4-[4-(2-Pyrimidinyl)-1-piperazinyl]butyl]-1,2-benzisothiazolin-3(2H)-one 1,1-Dioxide see 4967

N1-2-Pyrimidinylsulfanilamide see 8874

Pyrimido[4,5-b]pyrazine see 7947

N-(2-Pyrimidyl)-N-(p-methoxybenzyl)-N′,N′-dimethyl-N′-cetyl-N′-bromoethylenediamine see 9306

1-(2″-Pyrimidyl)-4-(methylene-3′,4′-dioxybenzyl)piperazine see 7465

1-(2-Pyrimidyl)-4-piperonylpiperazine see 7465

N1-2-Pyrimidylsulfanilamide see 8874

Pyriminil, 7999

Pyrimithate, 8000

Pyrinamine [Erba] see 9651

Pyrinex [Makhteshim-Agan] see 2190

Pyrinistab see 5871

Pyrinistol see 5871

Pyrinoline, 8001

Pyrinuron see 7999

Pyripyridium see 7175

Pyrisept see 2024

Pyrisuccideanol see 8002

Pyrithiamine, 8003

Pyrithione, 8004

Pyrithioxin see 8006

Pyrithyldione, 8005

Pyritinol, 8006

Pyrizidin [Warner-Chilcott] see 5071

Pyro [Baxter] see 9170

Pyroace [Fujisawa] see 8029

Pyroacetic Ether see 58

Pyrocalciferol, 8007

Pyrocarbonic Acid Diethyl Ester, 8008

Pyrocatechin see 8009

Pyrocatechol, 8009

Pyrocatechol Dimethyl Ether see 9857

o-Pyrocatechuic Acid Diacetate see 3357

Pyrochlore see 6472

Pyrodextrin see 2931

Pyrodifenium Bromide see 7748

Pyrogallic Acid see 8010

Pyrogallol, 8010

Pyrogallol 1,2,3-(Diethylaminoethyl Ether) Tris(ethyl Iodide) see 4249

Pyrogallolphthalein see 4250

Pyrogastrone [Winthrop] see 1797

Pyrogen-free Water see 9951

Pyrogens, 8011

PyroGlu see 8012

L-Pyroglutamic Acid, 8012

Pyroglutamylhistidylprolinamide see 9503

Pyrojec [Maurry Biol.] see 3358

Pyrola see 2046

Pyrolan®, 8013

Pyroligneous Acid, 8014

Pyroligneous Vinegar see 8014

Pyrolusite see 5612

Pyromellitic Acid, 8015

Pyromen [Baxter] see 7474

Pyromijin see 7992

Pyromorphite *see* 5267
Pyromucic Acid *see* 4219
Pyromucic Aldehyde *see* 4214
α-Pyrone-5-carboxylic Acid *see* 2558
Pyronil [Lilly] *see* 8023
Pyronine B, 8016
Pyronine G *see* 8017
Pyronine Y, 8017
Pyrope *see* 352
Pyrophosphoric Acid, 8018
Pyrophosphoric Acid Tetraethyl Ester *see* 9138
Pyrophyllite *see* 368
Pyroplasmin *see* 3360
Pyroracemic Acid *see* 8032
Pyrosal *see* 748
Pyrostib *see* 8769
Pyrosulfuric Acid, 8019
Pyrosulfuryl Chloride, 8020
Pyrovalerone, 8021
Pyroxylin, 8022
Pyrrhotine *see* 4007
Pyrrobutamine, 8023
Pyrrocaine, 8024
Pyrrocycline-N *see* 8236
Pyrrodiazole *see* 9519
Pyrrolamidol *see* 2933
Pyrrolazote [Upjohn] *see* 7969
Pyrrolazote Abergic [Upjohn] *see* 7969
1*H*-Pyrrole, 8025
Pyrrolidine, 8026
1-Pyrrolidineaceto-2',6'-xylidide *see* 8024
2-Pyrrolidinecarboxylic Acid *see* 7790
2,5-Pyrrolidinedione *see* 8842
1-Pyrrolidinoaceto-2,6-dimethylani-lide *see* 8024
N-(β-Pyrrolidinoethyl)phenothiazine *see* 7969
N-(Pyrrolidinomethyl)tetracycline *see* 8236
α-Pyrrolidino-*p*-methylvalerophenone *see* 8021
2-Pyrrolidinone *see* 8027
2-Pyrrolidinoneacetamide *see* 7459
2-(1-Pyrrolidinyl)-2',6'-acetoxyl-idide *see* 8024
1-(1-Pyrrolidinyl)butyl *p*-Tolyl Ketone *see* 8021
1-[4-(1-Pyrrolidinyl)-2-butynyl]-2-pyrrolidinone *see* 6904
1-[(1-Pyrrolidinylcarbonyl)methyl]-4-(3,4,5-trimethoxycinnamoyl)piper-azine *see* 2293
10-[2-(1-Pyrrolidinyl)ethyl]-10*H*-phenothiazine *see* 7969
N-(1-Pyrrolidinylmethyl)tetracycline *see* 8236
10-[2-(1-Pyrrolidinyl)propionyl]phe-nothiazine Methobromide *see* 7885
3-(2-Pyrrolidinyl)pyridine *see* 6631
trans-2-[3-(1-Pyrrolidinyl)-1-*p*-tolyl-propenyl]pyridine *see* 9661
(*E*)-6-[(*E*)-3-(1-Pyrrolidinyl)-1-*p*-tolylpropenyl]-2-pyridineacrylic Acid *see* 121
4-(1-Pyrrolidinyl)-1-(2,4,6-trimeth-oxyphenyl)-1-butanone *see* 1463
2-Pyrrolidone, 8027
α-Pyrrolidone *see* 8027
2-Pyrrolidoneacetamide *see* 7459
2-Pyrrolidone-5-carboxylic Acid *see* 8012
10-[2-(1-Pyrrolidyl)ethyl]phenothi-azine *see* 7969
N-Pyrrolidylethyl-N-phenylbenzyl-amine *see* 4641
3-Pyrroline, 8028
Pyrrolnitrin, 8029
Pyrrolylene *see* 1500

Pyrro[*b*]monazole *see* 4828
Pyrroxate [Upjohn] *see* 7969
Pyruvaldehyde, 8030
Pyruvaldehyde Bis(amidinohydrazone) *see* 6131
Pyruvaldoxime *see* 5076
Pyruvate Decarboxylase, 8031
Pyruvate Oxidation Factor *see* 9255
Pyruvic Acid, 8032
Pyruvic Carboxylase *see* 8031
Pyruvic Decarboxylase *see* 8031
Pyrvinium Chloride, 8033
Pyrvinium Embonate *see* 8034
Pyrvinium Pamoate, 8034
Py-tetrahydroserpentine *see* 8129
PZC *see* 7135
Q-137 *see* 3050
Q-275 *see* 9751
QCD 84924 *see* 9142
QD 10733 *see* 4116
Q-Enzyme®, 8035
Q-Gas *see* 4546
QHS *see* 844
QIDamp [Mallinckrodt] *see* 621
Qinghao *see* 844
Qinghaosu *see* 844
Qing Hau Sau *see* 844
QNB *see* 8110
Q-Pam [Quantum] *see* 2977
Q.T. (Quick Tan) *see* 3166
Quaalude [Rorer] *see* 5872
Quadracycline [Bristol] *see* 9130
Quadrafos *see* 8621
Quadrol [Wyandotte] *see* 3551
Quaker Buttons *see* 6652
Quamaquil *see* 7229
Quamonium *see* 2018
Quaname *see* 5751
Quanil *see* 5751
Quantalan [Lappe] *see* 2205
Quantrel [Pfizer] *see* 7968
Quantril [Roerig] *see* 1133
Quarcyl [Winthrop] *see* 4375
Quartz *see* 8440
Quarzan [Roche] *see* 2351
Quasar [Brocchieri] *see* 9751
Quassia, 8036
Quassin, 8037
Quat *see* 5188
Quaternium-7 *see* 5238
Quatrachlor [Pitman-Moore] *see* 1083
Quatrex [Virax] *see* 9130
Quatrimycin, 8038
Quazepam, 8039
Quazium [Essex] *see* 8039
Quebrachamine, 8040
Quebrachine *see* 10011
Quebracho *see* 871
Quebracho Colorado, 8041
Quebrachol *see* 8500
Queen Bee Jelly *see* 8254
Queen's Root *see* 8775
Queensland Asthma Weed *see* 3860
Queen Substance, 8042
Queletox *see* 3945
Quelicin Chloride [Abbott] *see* 8846
Quellada [Stafford-Miller] *see* 5379
Quemicetina *see* 2068
Quensyl *see* 4748
Quercetagetin, 8043
Quercetagitrin *see* 8043
Quercetin, 8044
Quercetin-3-glucoside *see* 5109
Quercetin-7-D-glucoside *see* 8045
Quercetin-3-L-rhamnoside *see* 8047
Quercetin-3-rutinoside *see* 8276
Quercimelin *see* 8047
Quercimeritrin, 8045
d-Quercitol, 8046
Quercitrin, 8047
Quercitroside *see* 8047

Quercus, 8048
Querton 210CL [Kemanord] *see* 3088
Quesil [EGYT] *see* 2191
Questex *see* 3482
Questran [Mead Johnson] *see* 2205
Quiactin [Merrell] *see* 6874
Quick [Rhône-Poulenc] *see* 2152
Quick Grass *see* 9673
Quicklime *see* 1692
Quicksilver *see* 5801
Quide [Dow] *see* 7429
Quiescin *see* 8149
Quiess [OJ&F] *see* 4786
Quietim [Nativelle] *see* 4784
Quilan [Lilly] *see* 1048
Quilene [Ayerst] *see* 7077
Quilibrex [Isnardi] *see* 6881
Quillaic Acid, 8049
Quillaja, 8050
Quillaja Sapogenin *see* 8049
Quillaja Saponin, 8051
Quillay Bark *see* 8050
Quilonorm [SmithKline] *see* 5396
Quilonorm-retard [Smith-Kline] *see* 5404
Quilonum [Smith Kline Dauelsberg] *see* 5396
Quilonum-retard [Smith Kline Dauelsberg] *see* 5404
Quimar [Armour Pharm.] *see* 2265
Quimoral *see* 2265
Quimotrase [Farmila] *see* 2265
Quinacillin, 8052
Quinacrine, 8053
Quinadome [Dome] *see* 4935
Quinaglute *see* 8072
Quinalbarbitone Sodium *see* 8374
Quinaldic Acid, 8054
Quinaldine, 8055
Quinaldine Blue, 8056
Quinaldine Red, 8057
Quinaldine Sulfate, 8058
Quinaldinic Acid *see* 8054
Quinaldofur *see* 6449
Quinalizarin, 8059
Quinalspan *see* 8374
Quinambicide *see* 4924
Quinamin [Merrell-National] *see* 8089
Quinamine, 8060
Quinamm [Merrell Dow] *see* 8089
Quinapril, A7
Quinaprilat *see* A7
Quinapyramine, 8061
Quinate [Knoll] *see* 8089
Quinazoline, 8062
Quinbisan [Protea] *see* 8076
Quinbolone, 8063
Quince Seed, 8064
Quine [Rowell] *see* 8089
Quinestradiol, 8065
Quinestradol *see* 8065
Quinestrol, 8066
Quinethazone, 8067
Quinexin [Robin] *see* 1054
Quinfamide, 8068
Quingestrone, 8069
Quinhydrone, 8070
Quinic Acid, 8071
Quinic Acid 1,5-Dicaffeic Ester *see* 2773
Quinicardine [Fougera] *see* 8074
Quinicine *see* 9908
Quinidex [Robins] *see* 8074
Quinidine, 8072
α-Quinidine *see* 2288
Quinidine Bisulfate *see* 8072
Quinidine Polygalacturonate, 8073
Quinidine Sulfate, 8074
Quiniduran *see* 8072
Quinine, 8075
β-Quinine *see* 8072

Quinine Acid Hydrobromide see 8078
Quinine Bimuriate see 8079
Quinine Bisulfate, 8076
Quinine "Bromide" see 8084
Quinine Carbonate, 8077
Quinine Chloride see 8085
Quinine Dibromide see 8078
Quinine Dichloride see 8079
Quinine Dihydrobromide, 8078
Quinine Dihydrochloride, 8079
Quinine Ethylcarbonate, 8080
Quinine Formate, 8081
Quinine Gluconate, 8082
Quinine Hydriodide, 8083
Quinine Hydrobromide, 8084
Quinine Hydrochloride, 8085
Quinine "Iodide" see 8083
Quinine Iodosulfate, 8086
Quinine Monohydrochloride see 8085
Quinine Muriate see 8085
Quinine Oleate, 8087
Quinine Salicylate, 8088
Quinine Sulfate, 8089
Quinine Tannate, 8090
Quinine Urea Hydrochloride, 8091
Quininic Acid, 8092
Quininone, 8093
Quiniofon see 4757
Quinisocaine see 3207
Quinitex [Brenner] see 8074
Quinizarin, 8094
Quinizarin Green SS, 8095
Quinnone [Dermohr] see 4738
Quinocide, 8096
Quinofen see 2290
Quinofop-ethyl see 8113
Quinoform see 4924 and 8081
Quinol see 4738
Quinoline, 8097
8-Quinolineboronic Acid, 8098
2-Quinolinecarboxylic Acid see 8054
8-Quinolinecarboxylic Acid, 8099
Quinoline Yellow, 8100
Quinoline Yellow A see 8101
Quinoline Yellow Base see 8101
Quinoline Yellow Spirit Soluble, 8101
Quinolinic Acid, 8102
2-Quinolinol see 1831
8-Quinolinol see 4778
8-Quinolinol Benzoate (Ester) see
  1122
8-Quinolinol Sulfate see 4779
2(1H)-Quinolinone see 1831
2-(2-Quinolinyl)-1H-indene-1,3-(2H)-
  dione see 8101
2(1H)-Quinolone see 1831
Quinolor [Squibb] see 4520
Quinomethionate see 6933
Quinomycin A see 3471
Quinone, 8103
p-Quinone see 8103
Quinone Monoxime see 6563
Quinone Oxime see 6563
Quinophan see 2290
Quinophenol see 4778
Quinophthalone see 8101
Quinora see 8074
Quinosol see 4779
Quinothionate see 9288
Quinotoxine see 9908
Quinotoxol see 9908
Quinova-bitter see 8105
Quinovaic Acid see 8104
Quinovatine see 809
Quinovic Acid, 8104
Quinovic Acid β-D-Glucoside see 8105
Quinovic Acid β-D-Quinovoside see
  8105
Quinovin, 8105
Quinovose, 8106
Quinovose 3-Methyl Ether see 9216

Quinoxaline, 8107
"Quinoxaline Antibiotic" see 3471
"Quinoxaline Antibiotic Complex" see
  9645
2,3-Quinoxalinedithiol Cyclic Trithiocarbonate see 9288
3-(2-Quinoxalinylmethylene)carbazic
  Acid Methyl Ester N,N'-Dioxide
  see 1782
(2-Quinoxalinylmethylene)hydrazine-
  carboxylic Acid Methyl Ester N,N'-
  Dioxide see 1782
N¹-(2-Quinoxalinyl)sulfanilamide see
  8914
N¹-(2-Quinoxalyl)sulfanilamide see
  8914
Quinoxyl [Burroughs Wellcome] see
  4757
Quinsan [Prosana] see 8089
Quintozene, 8108
Quintrate see 7066
Quinuclidine, 8109
3-Quinuclidinol, 8110
3-Quinuclidinyl Acetate see 8110
3-Quinuclidinyl Benzilate see 8110
3-Quinuclidinyl Benzoate see 8110
10-(3-Quinuclidinylmethyl)phenothi-
  azine see 5754
Quinupramine, 8111
Quinuronium Sulfate see 3360
Quipenyl see 6953
Quisqualic Acid, 8112
L-Quisqualic Acid see 8112
Quitaxon [Boehringer, Mann.] see
  3425
Quival [Schering] see 1704
Quixalin [Squibb] see 4520
Quixalud see 4520
Quizalofop-Ethyl, 8113
Quotane [SK & F] see 3207
QZ-2 see 5872
R 11, 8114
R 48 see 2107
R 79 see 5090
R 100 see 397
R 131 see 7177
R 242 see 8969
R 381 see 3908
R 400 see 6989
R 445 see 6176
R 658 see 1599
R 738 see 6358
R 798 see 8223
R 802 see 4065
R 805 see 6464
R 818 see 4034
R 875 see 2933
R 1132 see 3313
R 1303 see 1827
R 1504 see 7311
R 1513 see 926
R 1582 see 926
R 1607 see 9866
R 1608 see 3580
R 1625 see 4511
R 1658 see 6170
R 1707 see 4330
R 1881 see 6049
R 1910 see 1546
R 1929 see 915
R 2028 see 4047
R 2113 see 2910
R 2323 see 4310
R 2453 see 2874
R 2498 see 9596
R 2858 see 6203
R 3248 see 25
R 3345 see 7422
R 3365 see 7470
R 3746 see 1941
R 3763 see 1941

R 3959 see 2381
R 4263 see 3944
R 4318 see 4037
R 4584 see 1057
R 4749 see 3437
R 4845 see 1216
R 4929 see 1084
R 5020 see 7796
R 6128 see 4131
R 6238 see 7404
R 6700 see 5010
R 7315 see 6069
R 7465 see 6336
R 7904 see 5360
R 8299 see 9161
R 11333 see 1432
R 12563 see 9161
R 12564 see 9161
R 14827 see 3476
R 14889 see 6101
R 15403 see 3124
R 15454 see 5044
R 16341 see 7028
R 16470 see 2923
R 16659 see 3838
R 17147 see 2718
R 17635 see 5647
R 17889 see 4050
R 18553 see 5450
R 23979 see 3537
R 25061 see 8985
R 25831 see 1855
R 30730 see 8860
R 31520 see 2407
R 33800 see 8860
R 33812 see 3412
R 34995 see 5437
R 35443 see 6880
R 39209 see 227
R 41400 see 5181
R 41468 see 5175
R 42470 see 9090
R 43512 see 878
R 51211 see 5131
R 51619 see 2318
R 67408 see 3913
R 152450 see 4135
RA₂ see 9560
RA 8 see 3354
RA 233 see 6171
Rabalan see 7350
Rabon [Shell] see 8777
Race-acetylmethadol see 5853
Racefemine, 8115
Racemethionine see 5896
Racemethorphan, 8116
Racemic Acid see 9038
Racemic Desoxy-nor-ephedrine see
  616
Racemic Ephedrine see 3561
Racemic Lactic Acid see 5215
Racemic Mandelic Acid see 5599
Racemic Propoxyphene see 7851
Racemic Tartaric Acid see 9038
Racemomycin A see 8791
Racemomycins see 8791
Racemorphan see 5350
Raceophenidol see 9230
Racephedrine see 3561
Racephedrine Hydrochloride see 3561
Racephedrine Sulfate see 3561
Racephen see 616
Racephenicol see 9230
Rachromate-51 [Abbott] see 8548
R Acid see 6307
Racobalamin⁶⁰ [Abbott] see 9922
Radanil [Roche] see 1094
Radapon [Dow] see 2806
Radar [ICI] see 7830
Rad-e-cate [Vineland] see 8540
Raddeamine see 4840

Radecol see 6438
Radedorm see 6491
Radical Weed see 8664
Radicicol see 6164
Radicinin, 8117
Radioactive Colloidal Gold see 4419
Radioactive Cyanocobalamin see 9922
Radioactive Sodium Phosphate see 8614
Radiocaps-131 [Abbott] see 8583
Radiocyanocobalamin see 9922
Radio-gold (¹⁹⁸Au) Colloid see 4419
Radiographol see 5894
Radio-Iodinated Insulin see 4887
Radio-iodinated (¹³¹I) Tolpovidone see 9448
Radiopaque [Schering] see 1006
Radiorose Bengal Sodium see 8242
Radiostol see 9928
Radiotetrane see 4931
Radium, 8118
Radium F see 7533
Radium Emanation see 8119
Radon, 8119
Radon Fluoride see 8119
Radonin see 8876
Radsterin see 9928
Rafamebin see 4935
Raffinose, 8120
Rafoxanide, 8121
Ragadan [Hoechst] see 4585
Ralabol [IMC] see 10022
Ralfen see 8131
Ralgro [IMC] see 10022
Rally [Rohm & Haas] see 6232
Ralone [ICI] see 10022
RALs see 10017
Rametin [Bayer] see 6269
Ramifenazone, 8122
Ramipril, 8123
Ramiprilat see 8123
Ramodar [Wyeth] see 3831
R/AMP see 8215
Ramrod [Monsanto] see 7805
Ramycin see 4231
Rancinamycin IV see 7908
Randa [Nippon Kayaku] see 2319
Randolectil [Bayer] see 1509
Randonos see 4534
Randox see 250
Ranestol [Parke, Davis] see 1213
Ranestol (obsolete) [Parke, Davis] see 9570
Raney Nickel®, 8124
Raney Nickel Catalyst see 8124
Rangasil 400 [Ciba-Geigy] see 7478
Raniben [Firma] see 8126
Ranide [Merck & Co.] see 8121
Ranidil [Glaxo] see 8126
Ranimustine, 8125
Raniplex [Fournier-Dijon] see 8126
Ranitidine, 8126
Ranomustine see 8125
Ranoroc [Dieckmann] see 7248
Rantudil [Troponwerke] see 22
Ranunculin see 7907
Ranvil [Gentili] see 6403
Raovin [Abbott] see 9448
Rapacodin [Knoll] see 3154
Rapenton [Thomae] see 6171
Rapeseed Oil, 8127
Raphanin see 8942
Raphanus Rusticanus see 4668
Raphetamine Phosphate [Pennwalt] see 616
Rapicidin see 9744
Rapidase see 632
Rapifen [Janssen] see 227
Rapinovet [Coopers] see 7847
Rapostan [Mepha] see 6925
Rapynogen [Maruko] see 7852

Rarical [Ortho] see 1668
Rasapen see 4668
Raslogin [Brunner] see 9178
Raspberry, 8128
Raspite see 5309
Rastinon [Hoechst] see 9432
Ratak+ [ICI] see 1368
Ratanhine see 8988
Rat Antispectacled Eye Factor see 4883
Rathimed N [Pfleger] see 6079
Raticate see 6604
Ratilan [Ciba-Geigy] see 2556
Rattlesnake Root see 8404
Rattrack see 755
Raubasine, 8129
Raucolyt see 8131
Raudixin [Squibb] see 8131
Rauhimbine see 2547
Raulfin see 8131
Raumanon see 5071
Raunescine, 8130
Raunormine [Penick] see 2901
Raunova see 8998
Raupine see 8336
Raurixin see 8131
Rau-sed [Squibb] see 8149
Rauserpol see 8131
Rautensine [Dorsey] see 8131
Rauverid [Westerfield] see 8131
Rauwiloid [Riker] see 8131
Rauwoldin [Westerfield] see 8131
Rauwolfia serpentina, 8131
Rauwolfine see 184
Rauwolscine see 10013
Ra-Valeas see 8131
Ravenil [Caber] see 7987
Raviac [Lipha] see 2152
Ravocaine Hydrochloride [Winthrop] see 7850
Ravyon [Makhteshim-Agan] see 1789
Rawotal see 8131
Raybar [Damancy] see 1006
Raylina [Robert] see 610
Rayon, 8132
Raythesin [Raymer] see 8226
Razoxane, 8133
Razoxin [ICI] see 8133
RB 1509 see 5444
RC 61-91 see 4821
RC-27109 see 6450
RCL I see 8211
RCL II see 8211
RCL III see 8211
RCL IV see 8211
RD 292 see 3933
RD 1572 see 4650
RD 4593 see 5666
RD 13621 see 4812
RDX see 2742
RE 4355 see 6272
Re 1-0185 see 3490
Reacid see 2688
Reacthin see 127
Reactivan see 3909
Reactrol see 2346
Readex see 9288
Reagin [Baliarda] see 2321
Realgar see 822
Realphene see 44
Reasec [J & J] see 3313
Reazide see 2688
Rebriden [Sandoz] see 5129
Rebugen [Dessy] see 4812
Rec 7-0040 see 4033
Rec 7-0267 see 3190
Rec 14-0127 see 6174
Rec-15/0691 see 9357
Rec 15/1476 see 3948
Recacid [Recip] see 4363
Recanescine see 2901

Receptal [Hoechst] see 1492
Recetan [Simes] see 1522
Recheton see 5168
Recidol [Lampugnani] see 4812
Recipavrin see 9627
Reclaim [Dow] see 2398
Recognan [Toyo Jozo] see 2321
Recolip [Benzon] see 2374
Reconin [Toyama] see 2345
Reconox see 7220
Recordil [Recordati] see 3490
Recosen see 4536
Rectadione see 7196
Rectalad Enema [Wallace] see 3398
Recthormone Oestradiol see 3655
Recthormone Testosterone see 9115
Rectodelt [Trommsdorff] see 7727
Rectoid [Pharmacia] see 4710
Rectovalone [Jouveinal] see 9408
Redamina see 9921
Red Arsenic Glass see 822
Red Arsenic Sulfide see 822
Red Bole see 1330
Red Clover see 9601
Red Copper Oxide see 2671
Redeptin [SK & F] see 4131
Redergin [Lek] see 3598
Red Gum see 3853
Redicilin [Rocador] see 621
Redisol [Merck & Co.] see 9921
Redisol H [Merck & Co.] see 4739
Red Lead see 5307
Red Lead Chromate see 5280
Red Liquor see 339
Red Mustard see 6224
Red No. 2 see 382
Red Oil see 9730
Redomex see 504
Red Orpiment see 822
Redoxon [Roche] see 855
Red Phosphorus see 7321
Red Precipitate see 5779
Red Prussiate of Potash see 7611
Red Puccoon see 8319
Red Quebracho see 8041
Red Root see 8319
Redroot see 383
Red Sandalwood see 8342
Redskin see 293
Red Tetrazolium see 9658
Reduced Hematin see 4558
Reduced Nicotinamide Adenine Dinucleotide Phosphate Dehydrogenase see 6784
Reduced Ninhydrin see 4698
Reductic Acid, 8134
Reducto [Arcum] see 7180
Reductol [Ayerst] see A1
Reducymol [Elanco] see 665
Reduform [Kodipharm] see 6634
Redul [Schering] see 4403
Redupresin [Thilo] see 3711
Red Uranium Oxide see 9773
Redusterol see 3737
Red-water Tree Bark see 8341
Redwood see 7133
Reelon [Sanwa] see 7475
Reevon [Reeves] see 7544
Refagan see 5649
Refined Solvent Naphtha see 5366
Refkas [Maruko] see 3626
Reflux see 5883
Refobacin [E. Merck] see 4284
Refosporin [E. Merck] see 1924
Refugal [Bayer] see 2053
Refungine see 8865
Regacholyl see 4767
Regaine [Upjohn] see 6122
Regelan [Rhein-Pharma] see 2374
Regenit [Hoechst] see 7717
Regenon [Temmler] see 3116

Regianin see 5150
Regitine [Ciba] see 7234
Reglan [Robins] see 6063
Regletin [Teikoku Zoki] see 311
Reglone [ICI] see 3359
Regolax [Corvi] see 7211
Regonol [Organon] see 7990
Regular Iletin [Lilly] see 4889
Regulipid [Heilit] see 2375
Regulox see 5587
Regulton [Nordmark / Knoll] see 403
Regulus of Antimony see 724
Regumate [Hoechst] see 317
Regutol [Schering] see 3398
Rehibin see 9603
Rehibin [Thames] see 2726
Reicaf [San Carlo] see 8408
Reichstein's Substance A see 265
Reichstein's Substance C see 267
Reichstein's Substance D see 271
Reichstein's Substance "Dehydro-C" see 280
Reichstein's Substance E see 7736
Reichstein's "Substance Fa" see 2533
Reichstein's Substance G see 165
Reichstein's Substance H see 2532
Reichstein's Substance J see 270
Reichstein's Substance K see 266
Reichstein's Substance L see 263
Reichstein's Substance M see 4710
Reichstein's Substance N see 262
Reichstein's Substance O see 269
Reichstein's Substance P see 273
Reichstein's Substance Q see 2882
Reichstein's Substance R see 272
Reichstein's "Substance S" see 2912
Reichstein's Substance T see 7735
Reichstein's Substance U see 7737
Reichstein's Substance V see 268
Reidamine [Reid-Provident] see 3193
Reinecke Salt, 8135
Reissert Compounds, 8136
Rekawan [Giulini] see 7601
Rela [Schering] see 1848
Relact [Lemonier] see 6491
Relafen [Beecham] see 6258
Relan Beta see 1045
Relane [Cutter] see 3293
Relanium [Polfa] see 2977
Relasom [RAFA] see 1848
Relaspium [Isei] see 9697
Relaxan see 4249
Relaxar see 5737
Relaxil see 5737
Relaxil G see 4465
Relaxin, 8137
Relbapiridina see 8913
Relbazon see 8895
Reldan [Dow] see 2190
Releasin (formerly) [Warner-Chilcott] see 8137
Relefact LH-RH [Hoechst] see 5354
Relefact TRH [Hoechst] see 9503
Relestrid see 5903
Reliberan [Geymonat] see 2082
Relicor see 3662
Relifen [Beecham] see 6258
Relifex [Beecham] see 6258
Reliveran see 6063
Relvene [Pharmascience] see 9698
Remantadin(e) see 8221
Remark [Nippon Zoki] see 1200
Remeflin [Recordati] see 3190
Remestan [Wyeth] see 9074
Remicyclin [Schaper & Brmmer] see 9130
Remid [TAD] see 278
Reminitrol [Schafer] see 6528
Remivox [Janssen] see 5457
Remnos [DDSA] see 6491

Removine [Brocades-Stheeman] see 3193
Remsed [Endo] see 7797
Renacor [Merck & Co.] see 5393
Renafur [Eaton] see 6445
Renaglandin see 3569
Renaleptine see 3569
Renalina see 3569
Renarcol see 5737
Renascin [Mack, Illert.] see 9931
Renasul see 8887
Rencal [Squibb] see 7359
Renelate see 5883
Renese [Pfizer] see 7561
Renex 20 [ICI] see 7555
Renex 600's [Atlas] see 6596
Renex 698 [Atlas] see 6596
Rengasil [Ciba-Geigy] see 7478
Renilan [Mochida] see 1939
Renin, 8138
Renin Substrate see 680
Renitec [Merck & Co.] see 3521
Reniten [Merck & Co.] see 3521
Renivace [Merck & Co.] see 3521
Rennase see 8139
Rennet see 8139
Rennin, 8139
Renoform [Schering] see 3569
Renografin [Squibb] see 5689
Reno M [Squibb] see 5689
Renoquid [Warner-Lambert] see 8873
Ren-O-sal see 8251
Renostypticin see 3569
Renostyptin see 3569
Renosulfan see 8930
Renovist [Squibb] see 5689
Renovue-65 see 4904
Renovue-Dip [Squibb] see 4904
Renselin see 9760
Rentylin [Rentschler] see 7092
Reocorin see 7744
Reomax [Bioindustria] see 3669
Reomucil [Tosi] see 1809
Reorganin see 4465
Reostral see 5751
Reoxyl [Byk-Gulden] see 4626
Reparil see 3644
Repel [Wisconsin Pharmacal] see 2848
Repeltin [Bayer] see 9618
Repidose [Agrovet] see 6890
Repirinast, 8140
Repocal [Desitin] see 7087
Repodral see 8769
Repoise [Robins] see 1509
Repone K see 7601
Reposal, 8141
Reposamal see 8141
Reprodin [E. Merck] see 5480
Reproterol, 8142
Reptilase [Richter; Knoll] see 1020
Reptilase R see 1020
Repulson [Mochida] see 784
Resacetophenone, 8143
Resantin see 3935
Resazoin see 8144
Resazurin, 8144
Resbenzophenone see 1117
Resbuthrin [Wellcome] see 1243
Rescaloid see 8146
Rescimetol, 8145
Rescinnamine, 8146
Rescisan see 8146
Rescufolin [Nordic] see 4141
Rescuvolin [Medac] see 4141
Resectisol [Am. McGaw] see 5629
Reserpex see 8149
Reserpic Acid, 8147
Reserpiline, 8148
Reserpine, 8149
Reserpinine see 8146
Reserpinolic Acid see 8147

Reserpoid see 8149
Resibufogenin, 8150
Resilin, 8151
Resimatil [Labaz] see 7753
Resinat [Merrell] see 7538
Resin Benjamin see 4492
Resin Benzoin see 4492
Resin Copal see 2512
Resin Damar see 2808
Resin Guaiac see 4455
Resin Ipomea, 8152
Resin Jalap, 8153
Resin Kino see 5193
Resin of Mexican Scammony see 8152
Resin Scammony, 8154
Resin Tolu see 960
Resistab [Bristol] see 9307
Resistamine see 9651
Resistoflex [Resistoflex] see 7562
Resistomycin, 8155
Resistomycin [Bayer] see 5161
Resistopen [Squibb] see 6858
Resitan see 9816
Resitox [Bayer] see 2559
Resivit [Oberlin] see 5334
Resloom M 75 [Monsanto] see 4611
(+)-trans-Resmethrin see 1243
Resmit [Shionogi] see 5669
Resochin [Bayer] see 2163
Resodec, 8156
Resolvable Tartaric Acid see 9038
Resonium A [Winthrop] see 8622
Résoquine see 2163
Resorantel, 8157
Resorcin see 8158
Resorcin Blue see 5211
Resorcinol, 8158
Resorcinolphthalein see 4085
Resorcinol Phthalein Sodium see 4085
Resorcinolphthalin see 4087
Resorcinol Yellow see 9687
β-Resorcylaldehyde, 8159
Resorcylam see 8157
β-Resorcylic Acid, 8160
Resorcylic Acid Lactones see 10017
Resotren [Bayer] see 4757
Resovist [Schering AG] see 4236
Respacal [UCB] see 9720
Respaire (obsolete) [Bristol] see 82
Respbid [Boehringer, Ing.] see 9212
Respenyl see 4465
Respicort [Mundipharma] see 9512
Respifral see 5105
Respigon see 8150
Respilene [Lyocentre; Sigma-Tau] see 10074
Respirase [Gibipharma] see 10074
Respiride [Schiapparelli] see 3942
Resporisan [Tsuruhara] see 8148
Restandol [Organon] see 9109
Restanolon [Isei] see 2404
Restar [Sumitomo] see 4134
Restas [Kanebo] see 4134
Restenacht [Kabi] see 3515
Restenil [Kabi] see 5751
Restetal [Kabi] see 3520
Rest-On see 5871
Restoril [Sandoz] see 9074
Restrol [Central Pharm.] see 3094
Restropin [Frosst] see 5927
Restryl see 5871
Resulfon see 8879
Resveratrol-3-β-mono-D-glucoside see 7542
Resyl [Ciba] see 4465
Retabolil see 6281
Re 82-TAD-15 see 858
Retalon see 3094
Retalon-Lingual see 4621
Retalon Oleosum see 4621
Retalon-Oral see 3094

Retamine, 8161
Retarcyl [Gallier] see 8302
Retarcyl [Purdue Frederick] see 6195
Retardo see 7191
Retardon [Chassot] see 8909
Retasulfin see 8890
Retcin [DDSA] see 3626
Retene, 8162
Retens [Wassermann] see 3429
Retensin see 4249
Reticulin (the Antibiotic) see 4782
Reticulin (the Protein), 8163
Reticuline, 8164
Reticulogen see 5427
Retin-A [J & J] see 8167
Retina-derived Endothelial Cell
    Growth Factors see 4016
Retinal, 8165
Retinal 2 see 2862
Retinaldehyde see 8165
Retine, 8166
Retinene₁ see 8165
Retinene 2 see 2862
Retinoic Acid, 8167
Retinol see 8246 and 9918
Retinol₂ see 9919
Retolen [Elmu] see 878
Retozide see 5071
Retractyl [Rhovyl] see 7563
Retrangor see 1093
Retrocortine see 7727
Retroid [Roche] see 9500
Retrone see 3454
Retronecine, 8168
Retrorsine, 8169
Retrorsine N-Oxide see 8169
Retrovir [Burroughs Wellcome] see
    10023
Reublonil see 1131
Reudene [ABC] see 7476
Reudo see 7248
Reudox see 7248
Reufenac [Tosi] see 212
Reuflos [Scharper] see 3130
Reumachlor see 2163
Reumagrip see 2195
Reumalon [Tosi] see 6910
Reumatox see 6141
Reumofene [Von Boch] see 4875
Reumofil [Ital. Res.] see 8961
Reumyl [Lenza] see 8961
Reupiron see 6443
Reutol [Errekappa] see 9441
Reuxen [Syntex] see 6337
REV-6000A see 2864
Reverin see 8236
Reversil [Angelini] see 2819
Reversol [Organon] see 3487
Revertina see 5654
Revistel [Delagrange] see 309
Revivon [Reckitt & Colman] see 3347
Revonal [E. Merck] see 5872
Rexan see 2105
Rexigen [Bago] see 2359
Rexitene [LPB] see 4469
Rexort [Cassenne] see 2321
Rexulfa see 8909
Reychler's Acid see 1740
Rezifilm [Squibb] see 9304
Rezipas [Squibb] see 491
RF 46-790 see 4120
RF 2535 see 4612
RG 270 see 4941
R-Gene [Cutter] see 805
RGH-1106 see 7426
RGH-3332 see 4063
RGH-4405 see 9894
RH-8 see 3535
RH-315 see 7886
RH-787 see 7999
RH-893 see 6677

RH-2161 see 3902
RH-2915 see 6916
RH-3866 see 6232
RH-6201 see 105
Rhabarberone see 303
Rhamnetin, 8170
β-Rhamnocitrin see 8170
Rhamnol see 8500
Rhamnolutein see 5156
Rhamnose, 8171
L-Rhamnose see 8171
3β-Rhamnosido-14β-hydroxy-Δ⁴,²⁰,²²-
    bufatrienolide see 7889
6-O-α-L-Rhamnosyl-D-glucose see
    8277
Rhamnoxanthin see 4177
Rhamnus cathartica, 8172
Rhaponticin see 8173
Rhapontin, 8173
Rheadine, 8174
Rheic Acid see 8175
Rhein, 8175
Rhematan see 2290
Rhenic Anhydride see 8180
Rhenium, 8176
Rhenium Blue see 8180
Rhenium Chloride Oxide see 8179
Rhenium Heptoxide, 8177
Rhenium Hexafluoride, 8178
Rhenium Oxychloride, 8179
Rhenium Trioxide, 8180
Rhenium Trioxide Chloride see 8179
Rhenocain see 4783
Rheochrysidin see 3518
Rheomacrodex [Pharmacia] see 2925
Rheotran [Pharmachem] see 2925
Rheumacin LA [CP Pharm.] see 4874
Rheumapax [Oryx] see 6925
Rheumatism Root see 3287
Rheumatism Weed see 773 and 2046
Rheumatol [Tosse] see 1472
Rheumatrex [Lederle] see 5908
Rheum Emodin see 3518
Rheumibis [IBIS] see 22
Rheumin see 2290
Rheumon Gel [Troponwerke; Bayer]
    see 3833
Rheumox [Robins] see 758
Rhex "Hobein" [Hobein] see 5737
Rhinalair [Fabre] see 3561
Rhinalar [Syntex] see 4071
Rhinanthin see 894
Rhinantin see 6287
Rhinaspray see 9486
Rhinathiol [Joulli] see 1809
Rhinetten [Arzneimittelwerk VEB] see
    1632
Rhinocort [Astra] see 1455
Rhinofrenol see 6919
Rhinogutt [Thomae] see 9486
Rhinolitan [Kettelhack] see 6919
Rhinoperd see 6287
Rhinopront [Mack, Illert.] see 9150
Rhinoptil [Promonta] see 1632
Rhinospray [Thomae] see 9486
Rhizopterin, 8181
Rhocya see 7687
Rhodacryst see 9921
Rhodalline see 9293
Rhodamine B, 8182
Rhodanic Acid see 8184
Rhodanilic Acid, 8183
Rhodanine, 8184
Rhodanwasserstoffsäure (German) see
    9257
Rhodeose see 4192
Rhodialothan [Rhodia] see 4517
Rhodiatox see 6983
Rhodinal, 8185
Rhodine see 873
Rhodinol, 8186

β-Rhodinol see 2332
Rhodium, 8187
Rhodium Carbonyl Chloride, 8188
Rhodium Chloride, 8189
Rhodium Trichloride see 8189
Rhodizite see 2001
Rhodochrosite see 5608
Rhododendrin, 8190
Rhodol see 5940
Rhodomycin A see 8191
Rhodomycin B see 8191
β-Rhodomycin II see 8191
Rhodomycins, 8191
Rhodonite see 5623
Rhodopin, 8192
Rhodopsin, 8193
Rhodoquine see 7495
Rhodoquinone, 8194
Rhodoquinone-10 see 8194
Rhodotoxin see 4449
Rhodoviol [Rhône-Poulenc] see 7562
Rhodoviolascin, 8195
Rhodoxanthin, 8196
Rhoeadine see 8174
Rhombinin see 660
Rhomex see 7433
Rhothane see 3049
Rhovyl [Rhovyl] see 7563
Rhubarb, 8197
Rhubarb Yellow see 8175
Rhudane see 4930
Rhumalgan [Lagap] see 3071
Rhynchophylline, 8198
Rhyncophylline see 8198
Rhythminal [Teikoku; Nikken] see
    9984
Rhythmy [Shionogi] see 8219
Riabal [Fujisawa] see 7748
Riacen [Chiesi] see 7476
Riball [Mitsui] see 278
Ribavirin, 8199
α-Ribazole, 8200
Ribena see 855
Ribex [Formenti] see 3439
Ribipca see 8201
Ribitol see 157
Ribo [Tennessee Pharm.] see 8202
Ribo-Azauracil [Spofa] see 920
D-2-Ribodesose see 2890
Riboflavin 5'-(Dihydrogen Phosphate)
    Monosodium Salt see 8202
Riboflavine, 8201
Riboflavine 5'-Adenosine Diphosphate
    see 4028
Riboflavine-5-phosphate see 8201
Riboflavine 5'-Phosphate Ester
    Monosodium Salt see 8202
Riboflavine Phosphate (Sodium), 8202
Riboflavine 5'-(Trihydrogen Diphos-
    phate) 5'→5'-Ester with Adenosine
    see 4028
D-Ribofuranose-5-phosphoric Acid
    see 8206
9-β-D-Ribofuranosidoadenine see 143
1-β-D-Ribofuranosylcytosine see 2792
9-β-D-Ribofuranosylhypoxanthine see
    4880
2-β-D-Ribofuranosylmaleimide see
    8430
3-β-D-Ribofuranosylorotic Acid see
    6829
9-β-D-Ribofuranosyl-9H-purin-6-
    amine see 143
9-β-D-Ribofuranosyl-9H-purine see
    6353
9-β-D-Ribofuranosyl-9H-purine-2,6-
    diol see 9974
9-β-D-Ribofuranosyl-9H-purine-2,6-
    (1H,3H)-dione see 9974

N-(9-β-D-Ribofuranosyl-9H-purin-6-yl)butyramide Cyclic 3',5'-(Hydrogen Phosphate) 2'-Butyrate *see* 1448
3-β-D-Ribofuranosyl-1H-pyrrole-2,5-dione *see* 8430
7-β-D-Ribofuranosyl-7H-pyrrolo-[2,3-d]pyrimidin-4-amine *see* 9714
2-β-D-Ribofuranosyl-1,2,4-triazine-3,5(2H,4H)-dione *see* 920
1-β-D-Ribofuranosyl-1H-1,2,4-triazole-3-carboxamide *see* 8199
1-β-D-Ribofuranosyluracil *see* 9792
9-β-D-Ribofuranosyl Xanthine *see* 9974
D-**Ribohexulose** *see* 7940
Ribomycine [Delalande] *see* 8208
Ribonosine [Toyo Jozo] *see* 4880
Ribonuclease, 8203
Ribonucleic Acid, 8204
Riboract [Taiho] *see* 8201
D-**Ribose,** 8205
D-**Ribose-5-phosphate** *see* 8206
D-**Ribose-5-phosphoric Acid,** 8206
Ribosomal RNA *see* 8204
Ribosome, 8207
Ribostamin [Delalande] *see* 8208
Ribostamycin, 8208
Ribovis [Tsuruhara] *see* 8201
Ribrain [Endopharm] *see* 1200
D-**Ribulose,** 8209
RIC 272 *see* 5872
Ricainide *see* 4849
Ricamycin [Toyo Jozo] *see* 8233
Rice Bran Oil, 8210
Rice Oil *see* 8210
Richweed *see* 2479
Ricidine *see* 8212
Ricin, 8211
Ricin D *see* 8211
Ricinine, 8212
Ricinoleic Acid, 8213
Ricinolsulfuric Acid *see* 8213
Ricinstearolic Acid Diiodide *see* 7124
Ricinus Oil *see* 1904
Rickamicin *see* 8498
Ricketon *see* 9929
Ricridene [Lipha] *see* 6456
RI$_n$. r(C$_{12}$U)$_n$ *see* 622
Ricycline *see* 9130
Rid *see* 2830
Ridaura [SK & F] *see* 895
Ridauran [SK & F] *see* 895
Ridazin [Taro] *see* 9290
Ridect Pour-on [Beecham] *see* 7132
Ridene [Syntex] *see* 6403
Ridinol *see* 7747
Ridomil [Ciba-Geigy] *see* 5826
Ridzol [Merck & Co.] *see* 8237
Riegel's Paper *see* 2493
Rifa [Grünenthal] *see* 8215
Rifacol [Wassermann] *see* 8218
Rifadin [Dow] *see* 8215
Rifadine [Lepetit] *see* 8215
Rifaldazine *see* 8215
Rifaldin [Lepetit] *see* 8215
Rifamicine SV *see* 8217
Rifamide, 8214
Rifampicin *see* 8215
Rifampin, 8215
Rifamycin AG *see* 8216
Rifamycin AMP *see* 8215
Rifamycin B *see* 8216
Rifamycin B Diethylamide *see* 8214
Rifamycin O *see* 8216
Rifamycin S *see* 8216
Rifamycin SV, 8217
Rifamycin X *see* 8216
Rifamycins, 8216
Rifaprodin [Prodes] *see* 8215
Rifaxidin *see* 8218

Rifaximin, 8218
Rifit [Ciba-Geigy] *see* 7746
Rifloc Retard [Merrell] *see* 5114
Rifobac [Liade] *see* 8215
Rifocin [Lepetit] *see* 8217
Rifocina M [Dow] *see* 8214
Rifocin M [Lepetit] *see* 8214
Rifocyn [Lepetit] *see* 8217
Rifoldin [Merrell Dow] *see* 8215
Rifoldine [Lepetit] *see* 8215
Rifomycins *see* 8216
Rifomycin SV *see* 8217
Riforal [Liade] *see* 8215
Rigedal *see* 5114
Rigelon [Dojin] *see* 9230
Rigidyl *see* 3724
Rikamycin *see* 8233
Rikavarin [Toyo Jozo] *see* 9487
Riker 548 *see* 9535
Riker 52G *see* 784
Rilansyl *see* 2105
Rilaquil *see* 2105
Rilassol *see* 2105
Rilaten [Guidotti] *see* 8232
Rilmazafone, 8219
Rilmenidene, 8220
Rimactan [Ciba-Geigy] *see* 8215
Rimadyl [Roche] *see* 1870
Rimagina [Farmos] *see* 8222
Rimantadine, 8221
Rimaon *see* 3668
Rimatil [Santen] *see* 1447
Rimazole [Cheil Sugar] *see* 2412
Rimazolium Metilsulfate, 8222
Rimidin(e) [Elanco] *see* 3903
Rimifon [Roche] *see* 5071
Rimiterol, 8223
Rimitsid *see* 5071
Rimocidin, 8224
Rimso-50 [Res. Industries] *see* 3247
Rinatec [Boehringer, Ing.] *see* 4960
Rinderon-DP [Shionogi] *see* 1202
Rinder-vasopressin *see* 9843
Rindex [Sidus] *see* 5851
Rinesal [Kissei] *see* 1971
Rineton [Sanwa] *see* 9512
Ringworm Powder *see* 795
Rinlaxer [Taisho] *see* 2179
Rino-Clenil [Chiesi] *see* 1029
Rinogutt [Boehringer, Ing.] *see* 9486
Rintal [Bayvet] *see* 3888
Riogen *see* 7271
Riogon *see* 4534
Riol [Toa Eiyo] *see* 9060
Riomitsin *see* 6931
Riopan [Ayerst] *see* 5527
Rioprostil, 8225
Ripason *see* 5427
Ripcord [Shell] *see* 2775
Ripercol [Am. Cyanamid] *see* 9161
Ripirin [Kabi] *see* 3515
Ripon *see* 5527
Riporest [Formenti] *see* 5872
Riposon *see* 5731
Ripresil [Erba] *see* 2446
Ririlim *see* 1136
Riripen [Daiichi] *see* 1136
Risa-131-H [Abbott] *see* 8417
Risamal [Yoshitomi] *see* 2318
Risatarun [Ravensberg] *see* 2835
Rise [Yoshitomi] *see* 2411
Riself [Gibipharma] *see* 5739
Riseptin *see* 3042
Risocaine, 8226
Risolid [Dumex] *see* 2082
Risordan [Théraplix] *see* 5114
Ristat *see* 8251
Ristocetin, 8227
Ristogen [Kowa Yakuhin] *see* 4060
Ristomycin *see* 8227
Ristomycin A *see* 8227

Ristomycin B *see* 8227
Riston *see* 8227
Ristosamine *see* 126
Risulfasens *see* 8909
Risumic [Dainippon] *see* 403
RIT 1140 *see* 762
Ritacol [Schiapparelli] *see* 8218
Ritalin *see* 6025
Ritalmex [Alkaloida] *see* 6094
Ritmalan [Roche] *see* 2274
Ritmocardyl [Bago] *see* 497
Ritmodan [Maestretti] *see* 3366
Ritmos [Inverni] *see* 184
Ritmos Elle [Inverni] *see* 5454
Ritmusin [Gebro] *see* 781
Ritodrine, 8228
RIV-2093 *see* 309
Rivadescin *see* 8131
Rivanol [Winthrop] *see* 3668
Rivasin *see* 8149
Riversideite *see* 1709
Rivotril [Roche] *see* 2387
Rixaben [RIT] *see* 2383
Rizaben [Kissei] *see* 9489
Rize [Yoshitomi] *see* 2411
Rizen [Puropharma] *see* 2411
Rizinsan K2 A2 *see* 4401
RL-50 *see* 6165
RMI 17043 *see* 3542
RMI-71754 *see* 9882
RMI 71782 *see* 3489
RMI 83027 *see* 8234
RNA *see* 8204
mRNA *see* 8204
rRNA *see* 8204
sRNA *see* 8204
tRNA *see* 8204
RNase *see* 8203
Ro 1-5130 *see* 7990
Ro 1-5431 *see* 5350
Ro 1-5470 *see* 8116
Ro 1-6794 *see* 5350
Ro 1-7788 *see* 8116
Ro 1-7977 *see* 8836
Ro 1-9569 *see* 9118
Ro 1-5431/7 *see* 5350
Ro 1-5470/5 *see* 8116
Ro 1-5470/6 *see* 8116
Ro 2-0404 *see* 244
Ro 2-2453 *see* 2967
Ro 2-2985 *see* 5243
Ro 2-3198 *see* 3487
Ro 2-3308 *see* 8110
Ro 2-3773 *see* 2351
Ro 2-7113 *see* 294
Ro 2-9578 *see* 9623
Ro 2-9757 *see* 4109
Ro 2-9915 *see* 4056
Ro 3-4787 *see* 1469
Ro 4-1575 *see* 504
Ro 4-1577 *see* 2719
Ro 4-1634 *see* 7483
Ro 4-3780 *see* 8167
Ro 4-3816 *see* 214
Ro 4-4393 *see* 8877
Ro 4-4602 *see* 1059
Ro 4-5360 *see* 6491
Ro 4-6467 *see* 7764
Ro 4-8347 *see* 9500
Ro 4-1544/6 *see* 8768
Ro 4-1778/1 *see* 6066
Ro 5-0831 *see* 5040
Ro 5-2180 *see* 6608
Ro 5-2807 *see* 2977
Ro 5-3350 *see* 1374
Ro 5-3438 *see* 4058
Ro 5-4023 *see* 2387
Ro 5-4200 *see* 4072
Ro 5-5345 *see* 9074
Ro 5-6901 *see* 4123
Ro 5-0810/1 *see* 9567

Ro 5-3307/1 see 2837
Ro 6-4563 see 4334
Ro 7-0207 see 6824
Ro 7-1051 see 1094
Ro 7-1554 see 4966
Ro 7-6145 see 3184
RO 7-4488/1 see 6255
Ro 03-7008 see 194
Ro 07-6102 see 4133
Ro 10-1670 see 107
Ro 10-5970 see 1369
Ro 10-6338 see 1473
Ro 10-9070 see 399
Ro 10-9359 see 3846
Ro 11-780 see 8222
Ro 11-1163 see 6139
Ro 11-1430 see 6199
Ro 12-0068 see 9080
Ro 13-1042 see 5457
Ro 13-5057 see 692
Ro 13-8996 see 6892
Ro 13-9904/001 see 1950
Ro 14-3169/000 see 3938
Ro 14-4767/002 see 605
Ro 15-1788 see 4062
Ro 17-2301 see 1880
Ro 19-5247 see 1945
Ro 20-7234 see 4084
Ro 20-5720/000 see 1870
Ro 21-5535 see 1641
Ro 21-5998 see 5683
Ro 21-6937 see 9637
Ro 21-8837 see 3658
Ro 21-9738 see 3426
Ro 21-3981/001 see 6105
Ro 21-3981/003 see 6105
Ro 22-7796 see 2274
Ro 22-8181 see 4892
Ro 22-9000 see 225
Ro 23-6240 see 4035
Ro 31-2848 see 2276
Ro 31-3113 see 2276
Roaccutane [Roche] see 8167
Ro-Ampen [Rowell] see 621
Robac 22 [Robinson Bros.] see 3759
Robalate [Robins] see 3168
Robamate [Robinson] see 5751
Robamol [Cenci] see 5903
Robamox [Robins] see 610
Robane [Robeco] see 8726
Robanul see 4397
Robaxin [Robins] see 5903
Robecote [Robeco] see 8425
Robenidine, 8229
Robenogatope [Squibb] see 8242
Robenz [Am. Cyanamid] see 8229
Robenzidene see 8229
Robigenin see 5156
Robinin, 8230
Robinul [Robins] see 4397
Robiocina [Pradel] see 6641
Robisellin see 5071
Robison Ester see 4355
Robitussin [Robins] see 4465
Robizone-V [Robins] see 7248
Robuoy [Robeco] see 7757
Rocaltrol [Roche] see 1641
Roccal [Winthrop] see 1066
Roccellic Acid, 8231
Rocefin [Roche] see 1950
Rocephin(e) [Roche] see 1950
Rochelle Salt see 7660
Ro-Chlorozide [Robinson] see 2169
Rocillin see 7184
Rociverine, 8232
Rock Oil see 7140
Rock Salt see 8544
Rocornal [Deut. Hydrierwerk] see 9492
Rocosenin see 4536
Ro-cycline [Rowell] see 9130

Rodalon see 1066
Rodameb [Specia] see 3338
Rodex [ABCO] see 9950
Rodilone see 16
Rodinal see 474
Rodipal see 3703
Rodiuran see 4716
Rody [Sumitomo] see 3936
(+)-Roemerine see 778
Roeridorm [Roerig] see 3683
Roferon-A [Roche] see 4892
Roflual [Roche] see 8221
Rogaine [Upjohn] see 6122
Rogitine [Ciba] see 7234
Rogojski's Salt see 2675
Rogor [Agrimont] see 3209
Rogue [Monsanto] see 7814
Rohrbach's Soln see 994
Rohydra [Robinson] see 3308
Ro-Hydrazide [Robinson] see 4704
Rohypnol [Roche] see 4072
Roidenin [Showa Shinyaku] see 4812
Roinin [Mohan] see 7744
Roipnol [Roche] see 4072
Rökan [Intersan] see 4320
Rokitamycin, 8233
Rolazote see 7969
Roldiol [Robinson] see 3689
Rolicton [Searle] see 500
Rolicypram see 8234
Rolicyprine, 8234
Rolipram, 8235
Rolitetracycline, 8236
Rolitetracycline Chloramphenicol
  Succinate see 8408
Roman Chamomile see 2032
Roman Vitriol see 2659
Romensin [Elanco] see 6157
Romet [Mitsubishi] see 8140
Rometin see 4924
Romicil see 6785
Romilar Hydrobromide [Sauter] see 8116
Romotal [Duncan Flockhart] see 9003
Romparkin see 9607
Rompun (also the Hydrochloride)
  [Bayer] see 9987
Rona-Phyllin [Rona] see 9212
Rondar [Wyeth] see 6881
Rondase see 4676
Rondimen [Homburg] see 5681
Rondomycin [Wallace] see 5851
Ronfenil [Zeria] see 2068
Rongalite see 8567
Rongalite C see 8567
Roniacol [Roche] see 6438
Roniacol Tartrate [Roche] see 6438
Ronicol see 6438
Ronidase see 4676
Ronidazole, 8237
Ronifibrate, 8238
Ronilan [BASF] see 9890
Ronnel, 8239
Ronok [Ono] see 6827
Ronoprost see 6827
Ronstar [Rhône-Poulenc] see 6860
Rontyl [Leo Pharm.] see 4716
Ronyl see 7021
Rootone [Amchem] see 6290
ROP 500F see 4964
Ropredlone [Robinson] see 7719
Roptazol see 4210
Roquessine see 2492
Rorer 148 see 5872
Roridin C see 9562
Roridins see 9869
Rosal [IBI] see 8240
Rosamicin see 8241
Rosamit see 5259
Rosampline [Rosa-Phytopharma] see 621

Rosaniline Hydrochloride see 5528
Rosapin see 2899
Rosaprostol, 8240
Rosaramicin, 8241
Roscoelite see 9823
Roscosulf [Syntex] see 8876
Rose Bengal, 8242
Rose Bengale B see 8242
Rose Bengal Sodium I 131 see 8242
Rose Hips, 8243
Rosemary, 8244
Rosemide [Ono] see 4221
Rosenstiehl's Green see 993
Rose-pink see 1965
Rosimon-Neu [Ravensberg] see 6176
Rosin, 8245
Rosin Oil, 8246
Rosinol see 8246
p-Rosolic Acid see 899
Rosoxacin, 8247
Rostil [Ortho] see 8225
Ro-Sulfiram [Robinson] see 3370
Rotacide [Fairfield] see 7446
Rotenone, 8248
Rotersept [Roterpharma] see 2090
Rotesar [Argentina] see 964
Rotondin [Casasco] see 3920
Rotoxamine see 1804
Rotraxate, 8249
Rottlerin, 8250
Rotundine see 9147
Rougoxin see 3150
Roulone [Roussel-UCLAF] see 5872
Round-leafed Sundew see 3440
Round-up [Monsanto] see 4408
Rouqualone see 5872
Rovamicina [Farmalabor] see 8708
Rovamycin [Rhodia] see 8708
Rovral [Rhône-Poulenc] see 4964
Rowamyelin see 8702
Rowapraxin [Rowa] see 7454
Rowasa [Rowell] see 5807
Roxadil [Sterling] see 8247
Roxane [Hoechst] see 8252
Roxanthin Red 10 [Roche] see 1756
Roxarsone, 8251
Roxatidine Acetate, 8252
Roxen [Sandoz] see 6403
Roxenan [UPSA] see 7369
Roxicam [Gramon] see 7476
Roxiden [Pulitzer] see 7476
Roximycin [Kyorin] see 3429
Roxin [Hoechst] see 8252
Roxinil see 8131
Roxinoid see 8149
Roxion [Cela] see 3209
Roxit [Hoechst] see 8252
Roxithromycin, 8253
Roxomicina see 3628
Royal Jelly, 8254
Royal Jelly Acid see 8254
Royline see 5488
Rozol see 2152
RP 40 see 6601
RP 46 see 1161
RP 143 see 7700
RP 177 see 7845
RP 866 see 8053
RP 2090 see 8920
RP 2168 see 5995
RP 2255 see 8921
RP 2259 see 4373
RP 2275 see 8879
RP 2339 see 7176
RP 2512 see 7071
RP 2591 see 3060
RP 2632 see 8884
RP 2740 see 5870
RP 2786 see 7996
RP 2831 see 4577
RP 2987 see 3098

RP 3015 see 3918
RP 3203 see 4933
RP 3276 see 7793
RP 3277 see 7797
RP 3356 see 3703
RP 3359 see 2088
RP 3377 see 2163
RP 3389 see 5088 and 7797
RP 3554 see 9234
RP 3668 see 8667
RP 3697 see 4249
RP 3735 see 5463
RP 3799 see 3106
RP 3828 see 479
RP 4207 see 9218
RP 4460 see 5088
RP 4482 see 8915
RP 4560 see 2186
RP 4632 see 5923
RP 4753 see 7997
RP 4763 see 3338
RP 4909 see 2184
RP 5015 see 5071
RP 5337 see 8708
RP 6140 see 7768
RP 6171 see 619
RP 6484 see 3849
RP 6549 see 9618
RP 6847 see 6900
RP 7044 see 5909
RP 7162 see 9636
RP 7204 see 2690
RP 7293 see 7758
RP 7452 see 89
RP 7522 see 8890
RP 7746 see 9593
RP 7843 see 9287
RP 7891 see 4374
RP 8228 see 5347
RP 8595 see 3255
RP 8599 see 4146
RP 8823 see 6079
RP 8908 see 7117
RP 9159 see 7120
RP 9671 see 6639
RP 9715 see 2719
RP 9921 see 784
RP 9965 see 6070
RP 10192 see 2872
RP 11974 see 7308
RP 12222 see 6045
RP 12833 see 2401
RP 13907 see 4039
RP 14539 see 8373
RP 16091 see 6057
RP 16272 see 1431
RP 17623 see 6860
RP 18429 see 9487
RP 19366 see 7453
RP 19551 see 7453
RP 19552 see 7453
RP 19560 see 5835
RP 19583 see 5184
RP 22050 see 10096
RP 22410 see 4339
RP 26019 see 4964
RP 27267 see 10095
RP 31264 see 8987
RR 32705 see 6793
RS-1301 see 2865
RS-1320 see 4071
RS-2252 see 4053
RS-3540 see 6337
RS-3650 see 6337
RS-3999 see 4071
RS-4691 see 2396
RS-8858 see 6890
RS-9390 see 7895
RS-21592 see 4262
RS-35887 see 1525
RS-37619 see 5186

RS-44872 see 8866
RS-44872-00-10-3 see 8866
RS-69216 see 6403
RS-84043 see 3940
RS-84135 see 3545
RS-94991-298 see 6265
RT see 9658
RTCA see 8199
RU 486 see 6109
RU 965 see 8253
RU 2267 see 317
RU 2323 see 4310
RU 5020 see 7796
RU 15060 see 9354
RU 15750 see 4037
RU 19110 see 4509
RU 20201 see 9051
RU 22974 see 2869
RU 23908 see 6460
Ru 24756 see 1935
RU 28965 see 8253
RU 31158 see 5453
RU 38486 see 6109
RU 39780 see 1238
RU 43-715 see 7887
Rubber, 8255
Rubeanic Acid, 8256
Rubene see 6288
Ruberythric Acid, 8257
Ruberythrinic Acid see 8257
Rubesol [Central Pharm.] see 9921
Rubiadin, 8258
Rubian see 8257
Rubianic Acid see 8257
Rubiazol see 8872
Rubiazol I see 8895
Rubiazol II see 8872
Rubiazol IV see 8872
Rubiazol C see 8872
Rubichloric Acid see 866
Rubidazone [Rhône-Poulenc] see 10096
Rubidium, 8259
Rubidium Alum see 365
Rubidium Bromide, 8260
Rubidium Chloride, 8261
Rubidium Hydroxide, 8262
Rubidium Iodide, 8263
Rubidomycin see 2825
Rubigan [Elanco] see 3903
Rubigervine see 8264
Rubijervine, 8264
Rubimycin [Nikken] see 6106
Rubinorm [Ifi] see 8261
Rubivitan [Bayer] see 9921
Rubixanthin, 8265
Rubozinc [Labcatal] see 4350
Rubramin [Squibb] see 9921
Rubratope-57 see 9922
Rubratope⁶⁰ [Squibb] see 9922
Rubriment [Nordmark] see 6436
Rubripca see 9921
Rubrocitol see 9921
Rubroglioclladin see 896
α-Rubromycin see 2478
Rubus, 8266
Ruby [Spencer] see 6425
Ruby Arsenic see 822
Ruby Wood see 8342
Rucaina see 5359
Rudilin [Darby] see 6654
Rudotel [Arzneimittelwerk VEB] see 5669
RU-EF-Tb see 5071
Ruelene [Dow] see 2607
Rufigallic Acid see 8267
Rufigallol, 8267
Rufol see 8887
Rugulovasine A see 8268
Rugulovasine B see 8268
Rugulovasines, 8268

Rulid [Roussel] see 8253
Rumasal see 328
Rumatral [Wander] see 306
Rumensin [Lilly] see 6157
Rumex, 8269
Ruocid see 8879
Rusa Oil see 6733
Rusat see 1171
Ruscogenin, 8270
Ruscorectal [Jouveinal] see 8270
Ruswut see 1171
Rusyde [CP] see 4221
Rutabion see 8276
Rutaecarpine see 8271
Rutamycin see 6793
Rutecarpine, 8271
Rutenol see 6502
Rutergan see 3918
Rutgers 612 see 3699
Ruthenium, 8272
Ruthenium(VIII) Oxide see 8274
Ruthenium Oxychloride Ammoniated see 8273
Ruthenium Red, 8273
Ruthenium Tetroxide, 8274
Ruthenium Trichloride, 8275
Ruthenium Violet see 8273
Rutherfordium see 3503
Ruticina [Bernabo] see 5830
Rutile see 9398
Rutin, 8276
Rutinose, 8277
Rutonal see 7275
Rutoside see 8276
Rutozyd see 8276
Ruven [Mepha] see 9698
RV-12424 see 4074
Rx 67408 see 3913
RX 5050M see 3347
RX 6029-M see 1487
Ryanex see 8278
Ryania, 8278
Ryanicide [Penick] see 8278
Ryanodine, 8279
Ryanodol 3-(1H-Pyrrole-2-carbox-ylate) see 8279
Rycarden [Syntex] see 6403
Rycopel [Rycovet] see 2775
Rydar [Cutter] see 3293
Rydene [Syntex] see 6403
Rydex [Sandoz] see 7773
Rydrin [Kodama] see 6654
Ryegonovin [Morishita] see 5989
Rynacrom [Fisons] see 2594
Ryomycin see 6931
Rythmarone [Leurquin] see 497
Rythmatine [Burroughs Wellcome] see 5730
Rythminal [Teikoku; Nikken] see 9984
Rythmodan [Roussel-UCLAF] see 3366
Rythmodul [Albert-Roussel] see 3366
Rythmol [Helopharm] see 7806
Rytmonorm [Knoll] see 7806
Ryzelan see 6840
R-3-ZON see 7248
S7 see 3947
S 8 [Medinova] see 3308
S94 see 4116
S 222 see 3374
S 314 see 4226
S 486 see 6426
S 640P see 1923
S 596 see 815
S 768 see 3920
S 780 see 1047
S 805 see 5461
S 940 see 6269
S 992 see 1047
S 1320 see 1455
S 1358 see 1520

S 1520 *see* 4847
S 1530 *see* 6465
S 1574 *see* 9352
S 1694 *see* 420
S 1702 *see* 4335
S 1752 *see* 3945
S 1844 *see* 3952
S 1942 *see* 1419
S 2225 *see* 1419
S 2-676 *see* 9165
S 23/46 *see* 702
S 2395 *see* 9105
S 2539 *see* 7221
S 2620 *see* 299
S 2703 *see* 2776
S 3151 *see* 7132
S 3206 *see* 3936
S 3308L *see* 3261
S 3308-10 *see* 3261
S 3341 *see* 8220
S 3500 *see* 1519
S 4084 *see* 2266
S 4087 *see* 2697
S 5602 *see* 3952
S 5602α *see* 3952
S 5682 *see* 3291
S 6059 *see* 6201
S 6810 *see* 4894
S 7131 *see* 7772
S 8527 *see* 2353
S 9490 *see* 7122
S 9490-3 *see* 7122
S 9780 *see* 7122
S 10036 *see* 4174
S 50022 *see* 4322
S 73 4118 *see* 7464
S 77 0777 *see* 7717
SA *see* 9751
SA 79 *see* 7806
SA96 *see* 1447
SA 97 *see* 4653
Sa 267 *see* 3346
SA 504 *see* 9372
Sabadilla, 8280
Sabadine, 8281
Sabalamin [Santen] *see* 2446
Sabari *see* 5657
Sabatine *see* 8281
Sabidal [Zyma] *see* 2213
Sabiden [Szabo] *see* 7242
Sabril [Merrell-Dow] *see* 9882
Sabromin [Sandoz] *see* 1440
Sab Simplex [Parke, Davis] *see* 8486
Saccharase *see* 4899
Saccharated Iron *see* 3974
Saccharic Acid *see* 4344
Saccharin, 8282
Saccharin Ammonium *see* 8282
Saccharin Insoluble *see* 8282
Saccharinol *see* 8282
Saccharinose *see* 8282
Saccharin Sodium *see* 8282
Saccharol *see* 8282
Saccharolactic Acid *see* 4237
Saccharolactone *see* 4344
L-Saccharopine, 8283
Saccharose *see* 8855
Saccharosonic Acid *see* 5009
Sacerno [EGYT] *see* 5741
Sacophan *see* 8116
Sacred Bark *see* 1889
Sadamin [Polfa] *see* 9969
Sadopan *see* 4266
Saferon *see* 7106
Safersan *see* 7366
Saffan [Glaxo] *see* 226
Safflor Carmine *see* 1876
Safflor Red *see* 1876
Safflower *see* 1877
Safflower Oil, 8284
Saffron, 8285

Saffron-bitter *see* 7381
Saffron Indian *see* 9732
Safranal, 8286
Safranin B Extra *see* 7218
Safrole, 8287
Safrotin [Sandoz] *see* 7827
Sagamicin [Kyowa] *see* 6104
Sagamicin (formerly) *see* 6104
Sagatal *see* 7087
Sage *see* 8311
Sagisal [Sagitta] *see* 8721
Sagittol [Sagitta] *see* 7852
SaH 42548 *see* 5643
SAICAR, 8288
Saiodin *see* 1681
Saisan [ICI] *see* 3432
Saizen [Serono] *see* 8672
Sajodin *see* 1681
Sakuranetin, 8289
Sakuranin *see* 8289
Salacetamide, 8290
Salacetin *see* 873
Sal Acetosella *see* 7588 and 7684
Saladjidi *see* 4815
Salamid [Phillips Roxane] *see* 8297
Salamid [Ferrosan] *see* 2067
Sal Ammoniac *see* 531
Salanil [Sato Yakuhin] *see* 4273
Salazolon *see* 749
Salazopyrin [Pharmacia] *see* 8917
Salazosulfadimidine, 8291
Salazosulfapyridine *see* 8917
Sal Barnit *see* 10066
Salbulin [Riker] *see* 209
Salbumol [Glaxo] *see* 209
Salbutamol *see* 209
Salbutine [Paucort] *see* 209
Salbuvent [Nycomed] *see* 209
Salcatonin *see* 1640
Salcetogen *see* 873
Salco *see* 2371
Salcyl *see* 8303
Salcylix *see* 8303
Saldox *see* 8296
Sal Enixum *see* 7591
Salep, 8292
Sal Ethyl [Parke, Davis] *see* 3804
Saletin *see* 873
Salflex [Carnrick] *see* 8307
Saliamin *see* 8297
Salicain [Hoechst] *see* 4783
Salicin, 8293
Salicin Benzoate *see* 7571
Salicinerein *see* 7367
Salicoside *see* 8293
N-Salicoylaminophenol *see* 6843
Salicylacetylsalicylic Acid *see* 95
Salicyl Alcohol, 8294
Salicyl Alcohol Glucoside *see* 8293
Salicylaldehyde, 8295
Salicylaldehyde β-D-Glucoside *see* 4544
Salicylaldoxime, 8296
Salicylamide, 8297
Salicylamide O-Acetic Acid, 8298
Salicylamide o-Ethyl Ether *see* 3685
Salicylanilide, 8299
Salicylazosulfadimidine *see* 8291
Salicylazosulfamethazine *see* 8291
Salicylazosulfapyridine *see* 8917
Salicylhydroxamic Acid, 8300
Salicylic Acid, 8301
Salicylic Acid Acetate *see* 873
Salicylic Acid Acetate Calcium Salt *see* 1644
Salicylic Acid Acetate, Ester with 4-Hydroxyacetanilide *see* 1054
Salicylic Acid, Acid Sulfate *see* 8303
Salicylic Acid Basic Bismuth Salt *see* 1299
Salicylic Acid Benzyl Ester *see* 1160

Salicylic Acid Bimolecular Ester *see* 8307
Salicylic Acid Bornyl Ester *see* 1343
Salicylic Acid *p-tert*-Butylphenyl Ester *see* 1585
Salicylic Acid Choline Salt *see* 2212
Salicylic Acid Dihydrogen Phosphate *see* 4170
Salicylic Acid Ethyl Ester *see* 3804
Salicylic Acid Isopropyl Ester O-Ester with O-Ethyl Isopropylphosphor-amidothioate *see* 5055
Salicylic Acid Monoammonium Salt *see* 578
Salicylic Acid Sulfuric Acid Ester *see* 8303
Salicylic Acid 3,3,5-Trimethylcyclo-hexyl Ester *see* 4660
Salicylic Aldehyde *see* 8295
Salicylic Ether *see* 3804
Salicyl Morpholide *see* 8302
Salicyloxysalicylic Acid *see* 8307
4-Salicyloylmorpholine, 8302
Salicyl Phosphate *see* 4170
Salicylsalicylic Acid *see* 8307
Salicylsulfonic Acid *see* 8944
Salicylsulfuric Acid, 8303
Salicyl-Vasogen [Pearson] *see* 578
Salicyl Yellow *see* 5823
Saliformin *see* 5884
Saligenin *see* 8294
Saligenin-β-D-glucopyranoside *see* 8293
Saligenol *see* 8294
Salimid *see* 2750
Salimol *see* 8887
Salinaphthol *see* 6335
Salinazid, 8304
Saliniazid *see* 8304
Salinidol *see* 8299
Salinigrin *see* 7367
Salinomycin, 8305
Saliphenazon *see* 749
Salipran [Bottu] *see* 1054
Salipur *see* 8297
Salipurpol *see* 6344
Salipyrazolon *see* 749
Salipyrine *see* 749
Salisan [DePree] *see* 2169
Salit [Tenneco] *see* 1343
Saliuzid *see* 6807
Salizell [Byk-Gulden] *see* 8297
Salizell (Amp.) *see* 8298
Salizid [Warner-Chilcott] *see* 8304
Salizolo *see* 4829
Salmiac *see* 531
Salmine Sulfate, 8306
Salmotonin [Yamanouchi] *see* 1640
Salocin [Hoechst] *see* 8305
Salocoll *see* 7204
Salofalk [Falk] *see* 5807
Salol *see* 7282
Saloop *see* 8340
Salophen [Bayer] *see* 41
Salpetersäure (German) *see* 6495
Salpix [Ortho] *see* 72
Sal Polychrestum *see* 7665
Salrin *see* 8297
Salsalate, 8307
Sal Soda *see* 8541
Salsoline, 8308
Salt *see* 8544
Salt Cake *see* 8636
Salthion *see* 8899
"Salt of Lemon" *see* 7588
Saltpeter *see* 7634
Salt of Phosphorus *see* 581
Saltron *see* 2371
Salt of Saturn *see* 5268
Salt of Sorrel *see* 7588 and 7684
Salt of Tartar *see* 7599

Scammony Root, 8347
Scandia see 8348
Scandicain [Bofors] see 5748
Scandine [Zambon] see 4807
Scandium, 8348
Scarlet Berry see 3447
Scarlet Red, 8349
SCE-129 see 1943
SCE-963 see 1937
SCE-1365 see 1928
Scepter [Am. Cyanamid] see 4826
Sch 1000 see 4960
Sch 3940 see 7135
Sch 4831 see 1202
Sch 7056 see 120
Sch 10144 see 9442
Sch 10159 see 9571
Sch 10649 see 917
Sch 11460 see 1202
Sch 12041 see 4502
Sch 13475 see 8498
SCH 13521 see 4132
Sch 14714 see 4073
Sch 14947 see 8241
SCH 15719W see 5208
Sch-16134 see 8039
Sch 18020W see 1029
SCH 19927 see 3186
Sch 20569 see 6389
Sch-21420 see 4989
Sch 21480 see 9387
Sch 22219 see 213
Sch 25298 see 4042
Sch 28316Z see 4852
Sch-29851 see 5455
SCH-30500 see 4892
Sch-32088 see 6151
Sch 32481 see 6390
SCH 39720 see 1947
Schaeffer's α-Acid see 6310
Schaeffer's β-Acid see 6312
Schaeffer's Salt see 6312
Schardinger Dextrins see 2724
Scheele's Green see 2630
Scheelite see 1720 and 5309
Schemergen [Azusa] see 7248
Schering 34615 see 7794
Schering 36268 see 2083
Schering 38584 see 7199
Scherisolon see 7719
Scherofluron [Schering AG] see 4059
Scheroson see 2533
Scheroson F [Schering AG] see 4710
Schiff Bases, 8350
Schistomide [Specia] see 619
Schizophyllan see 8502
Schleimsäure (German) see 4237
Schlippe's Salt see 8648
Schmiedeberg's Digitalin see 3140
Schöllkopf's Acid see 6306
Schönbein-Pagenstecher's Paper see 4456
Schradan, 8351
Schultenite see 5270
Schwefelstickstoff see 9171
Schweine-Vasopressin see 5503
Schweinfurt Green see 2629
Schweizer's Reagent, 8352
Scifluorfen see 105
Scillabiose, 8353
Scillacrist [Drago] see 7889
Scilla "Didier" see 7889
Scillaren, 8354
Scillaren A see 8354
Scillaren B see 8354
Scillarenin, 8355
Scillarenin 3β-Rhamnoside see 7889
Scilliroside, 8356
Scinnamina see 8146
Sclane see 9650
Sclerosol [Res. Med.] see 3247

Scobro [Ono] see 1588
Scobron [Mohan] see 1588
Scobutil [Rumanian] see 1588
Scoke see 7363
Scolaban [Cooper] see 1475
Scolecite see 1645
Scoline [Duncan Flockhart] see 8846
Scombrine see 7898
Scop [Ciba-Geigy] see 8361
Scoparin, 8357
Scoparius, 8358
Scoparone, 8359
Scoparoside see 8357
Scopine Tropate see 8361
Scopoderm TTS [Geigy] see 8361
Scopola, 8360
Scopolamine, 8361
l-Scopolamine see 8361
Scopolamine Aminoxide see 8362
Scopolamine Bromobutylate see 1588
Scopolamine-N-butyl Bromide see 1588
Scopolamine Methobromide see 5927
Scopolamine Methyl Bromide see 5927
Scopolamine N-Oxide, 8362
Scopolammonium Bromide see 8361
Scopoletin, 8363
Scopolin, 8364
Scopoline, 8365
Scopos see 8361
Scorbacid see 855
Scorbu-C see 855
Scorprin [Willows] see 8874
Scotch Broom see 8358
Scotcil [Scott-Cord] see 7041
Scotine [Unimed] see 2552
Scotophobin, 8366
Scotopsins see 6811
Scourge [Penick-Bio] see 7446
SCP see 8542
S.C.T.Z. see 2382
Scu-PA see 7918
Scurenaline see 3569
Scurocaine see 7763
Scuroforme see 1504
Scutellarein, 8367
Scutellaria, 8368
Scutellarin see 8367
Scutl see 7271
SD 709 see 3199
SD 1601 see 6172
SD 1750 see 3069
SD 3562 see 3077
SD 4294 see 2603
SD 4402 see 5010
SD 7859 see 2087
SD 8447 see 8777
SD 9129 see 6161
SD 104.19 see 7885
SD 11831 see 6486
SD 14114 see 3907
SD 15418 see 2692
SD 270-31 see 6862
SD 271-12 see 2359
SD 41706 see 3936
SD 43775 see 3952
SD 95481 see 2296
S-Dimidine see 8886
SDMH see 3237
SDS see 8587
Sdt 91 see 8769
Sdt 1041 see 9218
Se 780 see 1047
SE 1520 see 4847
SE 2395 see 9105
Sea-Legs see 5657
Sea-oak see 4196
Sea Onion see 8728
Searlequin see 4935
Sea Salt see 8544

Sea-wrack see 4196
Sebacic Acid, 8369
Sebacil, 8370
Sebacil [Bayer] see 7341
Sebaclen [Denver] see 1247
Sebacoin, 8371
Sebacoin [Fumouze] see 1247
Sebar [Vangard] see 8374
Sebatrol [Schering] see 4132
Sebercim [ISF-Italseber] see 6617
Sebizon [Schering] see 8870
Secaclavine see 2033
Secale Cornutum see 3608
Secalip [Fournier] see 3924
Secalonic Acid A see 8372
Secalonic Acid B see 8372
Secalonic Acid C see 8372
Secalonic Acid D see 8372
Secalonic Acid E see 8372
Secalonic Acid F see 8372
Secalonic Acid G see 8372
Secalonic Acids, 8372
Secalysat [Ysatfabrik] see 2551
Secbutobarbitone Sodium see 1495
Seccidin [Nippon Kayaku] see 7744
Secergan [Astra] see 7834
Secholex [Pharmacia] see 7531
Sechvitan see 7992
Seclar [Andromaco] see 1027
Secletan [Ipsen] see 2268
Seclin [Panthox & Burck] see 3151
Seclodin [Much] see 4812
Secnidazole, 8373
Secobarbital Sodium, 8374
(1α,3β,5Z,7E)-9,10-Secocholesta-5,7,10(19)-triene-1,3-diol see 4749
(3β,5Z,7E)-9,10-Secocholesta-5,7,10-(19)-triene-3,25-diol see 1638
(1α,3β,5Z,7E)-9,10-Secocholesta-5,7,10(19)-triene-1,3,25-triol see 1641
9,10-Secocholesta-5,7,10(19)-trien-3-ol see 9929
(6E,22E)-9,10-Secoergosta-5(10),6,8,-22-tetraen-3-ol see 9002
9,10-Secoergosta-5,7,10(19),22-tetra-en-3-ol see 9928
9,10-Secoergosta-5,7,10(19)-trien-3-ol see 9930
9,10-Secoergosta-5,7,22-trien-3β-ol see 3163
Seconal Sodium [Lilly] see 8374
Secondary Ammonium Phosphate see 570
Secondary Barium Phosphate see 1002
Secondary Calcium Phosphate see 1699
Secondary Caprylic Alcohol see 6675
Secondary Magnesium Phosphate see 5560
Secondary Propyl Alcohol see 5096
Secondary Sodium Phosphate see 8612
Secotil see 6810
Secrebil [Isnardi] see 7457
Secretin, 8375
Secretine [Sinbio] see 8375
Secretin-Kabi [Kabi] see 8375
Secretin [Ox] see 8375
Secretin [Pig] see 8375
Secrosteron see 3208
Sectorgel [Hisamitsu] see 5184
Sectral [M & B] see 13
Securinan-11-one see 8376
Securinine, 8376
Securit [Pierrel] see 5456
Securon [Knoll] see 9851
Securopen [Bayer] see 929
Securpres [Poli] see 4852
Sedafamen see 7180
Sedaform see 2129

Sedalande [J & J] see 4047
Sedalis see 9182
Sedamyl [Riker] see 15
Sedanoct [Woelm] see 9707
Sedantoinal [Sandoz] see 5741
Sedapain [Nichiiko] see 3579
Sedapam [Duncan, Flockhart] see 2977
Sedaplus [Chephasaar] see 3430
Sedapran [Kowa] see 7713
Sedaquin [Therapharm] see 5872
Sedaraupin see 8149
Sedatine see 748
Sedatival [Raffo] see 5456
Sedatromin [Takata] see 2308
Sedazil see 5751
Sedecamycin, 8377
Sedemesis see 1995
Sédeval see 972
Sednotic see 601
Sedobex see 1218
Sedolatan see 7744
Sedoneural see 8539
Sedopretten [Schoening] see 3308
Sedor see 3033
Sedormid [Roche] see 783
Sedothyron see 9248
Sedotussin [UCB] see 1798
Sedral [Banyu] see 1921
Sedrena [Daiichi] see 9607
Sedulon [Roche] see 7439
Sedural see 7175
Sedutain see 8374
Seduxen [Gedeon Richter] see 2977
Sefacin [Schering] see 1975
Seffein see 1789
Seflenyl [Ciba-Geigy] see 7478
Sefona see 4757
Sefril [Squibb] see 1982
Seglor [Perrier] see 3157
Segontin [Albert Roussel] see 7744
Segosin see 5894
Segurex [Ricar] see 1473
Seidlitz Mixture, 8378
Seignette Salt see 7660
Sekin [Sintesa] see 2393
Sekisanine see 9050
Sekisanoline see 9050
Sekretolin [Hoechst] see 8375
Sekundal-D [Woelm] see 3308
Selacryn [SK & F] see 9362
Selagine, 8379
Selagnine see 5495
Selapin [Teikoku/Nikken] see 9984
Selbex [Eisai] see 9083
Seldane [Merrell Dow] see 9094
Sel de Fordos et Gelis see 4423
Selectol [Chemie Linz] see 1956
Selectomycin [Grnenthal] see 8708
Selegiline see 2893
Selemicina [Italchemi] see 4169
Selenex see 8390
Selenic Acid, 8380
Seleninyl Bromide see 8387
Seleninyl Chloride see 8388
Seleninyl Fluoride see 8389
Selenious Acid, 8381
Selenious Anhydride see 8386
Selenite see 1713
Selenium, 8382
Selenium Bromide, 8383
Selenium Chloride, 8384
Selenium Cysteine see 8394
Selenium Dioxide see 8386
Selenium Hexafluoride, 8385
Selenium Hydride see 4728
Selenium Monobromide see 8383
Selenium Monochloride see 8384
Selenium Nitride see 6527
Selenium Oxide, 8386
Selenium Oxybromide, 8387

Selenium Oxychloride, 8388
Selenium Oxyfluoride, 8389
Selenium Sulfides, 8390
Selenium Tetrabromide, 8391
Selenium Tetrachloride, 8392
Selenium Tetrafluoride, 8393
Selenkupfer see 2673
Selenocysteine, 8394
Selenocystine see 8394
Selenomethionine, 8395
Selenous Acid see 8381
3-Selenyl-L-alanine see 8394
Selenyl Bromide see 8387
Selenyl Fluoride see 8389
Selepam [Schering-Plough] see 8039
Seles Beta [Farmitalia] see 879
Selexid (Tabl.) [Leo Pharm.] see 400
Selexidin [Leo Pharm.] see 399
Selexid (Susp) [Leo Pharm.] see 400
Selezen [Italfarmaco] see 4829
Seligocidin see 3550
Selinon see 3272
Sellagen see 5569
Sellaite see 5543
Selobloc [Lagap] see 879
Seloken [Hässle] see 6072
Selopral [Hässle] see 6072
Sel-O-Rinse [USV] see 8390
Selsun [Abbott] see 8390
Selvigon [Homburg] see 7423
Semap [Janssen] see 7028
Sembrina see 5974
Semen Amomi see 7402
Semen Cydonia see 8064
Semesan [Du Pont] see 4761
Semicarbazide Hydrochloride, 8396
Semicid [Bottu] see 6596
Semi-Daonil [Hoechst] see 4372
Semi-Euglucon [Boehringer, Mann.] see 4372
Semi-Gliben-Puren N see 4372
Semikon [Beecham] see 5871
Semilente Iletin [Lilly] see 4890
Semi-mustard Gas see 4565
Seminose see 5632
Semioxamazide, 8397
Semisulfur Mustard see 4565
Semi-synthetic Human Insulin see 4887
Semitard [Novo] see 4890
Semopen [Beecham] see 7184
Semoxydrine [Beecham] see 5859
Sempervirene see 8398
Sempervirine, 8398
Semprex [Burroughs Wellcome] see 121
Senarmontite see 743
Sencephalin [Takeda] see 1971
Sencor [Bayer] see 6076
Sencoral see 6076
Sendoxan [Pharmacia] see 2753
Sendran see 7849
Seneca Oil see 7140
Seneca Snakeroot see 8404
Senecialdehyde, 8399
Senecic Acid, 8400
Senecio, 8401
Senecioaldehyde see 8399
Senecioic Acid 2-sec-Butyl-4,6-dini-trophenyl Ester see 1237
Senecionine, 8402
N¹-Senecioylsulfanilamide see 8875
Seneciphylline, 8403
Senega, 8404
Senega Snakeroot see 8404
Senegenin, 8405
Seniramin [Hokuriku] see 8998
Senna, 8406
Sennoside A & B, 8407
Senociclin, 8408
Sensaval see 6635

Sensibion see 4555
Sensidyn [Medica] see 2180
Sensit [Thiemann] see 3916
Sensival see 6635
Sentiloc [Benzon] see 1213
Seotal [Lilly] see 8374
Sepamit [Kanebo] see 6441
Sepan [Janssen] see 2308
Sepatren [Sumitomo] see 1940
Sepazon [Sankyo] see 2415
Sephadex 2-(Diethylamino)ethyl Ether see 7531
Sepia, 8409
Sepiolite see 5569
Sepiomelanin, 8410
Seponver [Ethnor] see 2407
Sepsinol see 4205
Septalan [Polfa] see 246
Septazine [Specia] see 1161
Septicid see 5525
Septicol see 2068
Septiphene see 2403
Septipulmon see 8913
Septivon-Lavril [Porche-Lavril] see 9568
Septocillin see 7046
Septopal [E. Merck] see 4284
Septoplix see 8898
Septosan [Mallard] see 8895
Septotan see 4687
Septra [Burroughs Wellcome] see 8889
Septrim [Burroughs Wellcome] see 8889
Septural [Grünenthal] see 7475
Sequamycin see 8708
Sequens see 2102
Sequestrene see 3482
Sequestrene NA 2 [Ciba-Geigy] see 3481
Sequestrene NA 3 see 3483
Ser (IUPAC Abbrev.) see 8411
Seradix see 4871
Seragon [Ferring] see 7515
Seragonin see 7515
Seral see 5731
Serax [Wyeth] see 6881
D-Ser (Bu¹)⁶Azgly¹⁰-gonadorelin see 4433
D-Ser(Bu¹)⁶Azgly¹⁰-luliberin see 4433
Serc [Unimed] see 1200
Serecor [Houdé-I.S.H.] see 4736
Serefrex [Janssen] see 5175
Serenace [Searle] see 4511
Serenal see 6883
Serenal [Sankyo] see 6881
Serenase [Lusofarmaco] see 4511
Serenesil [Abbott] see 3683
Serenid see 6881
Serenium [Gedeon Richter] see 5669
Serenium [Squibb] see 3706
Serentil [Boehringer, Ing.] see 5813
Seren Vita [Vita] see 2082
Serepax [Wyeth] see 6881
Serepress [Formenti] see 5175
Sereprile [Vita] see 9353
Seresta see 6881
Serfin [Parke, Davis] see 8149
Sergetyl [Vaillant-Defresne] see 3849
Sericin see 4017
Sericosol N see 4380
Seriel [Fabre] see 9427
Seril see 5751
Serine, 8411
L-Serine Diazoacetate(ester) see 916
Serine Dihydrogen Phosphate (Ester) see 7338
Serine Phosphate see 7338
DL-Serine 2-[(2,3,4-Trihydroxyphen-yl)methyl] Hydrazide see 1059
Sermaka [Lilly] see 4122
Sermion [Farmitalia] see 6404

Sermorelin, 8412
Sernevin [Toho] see 8971
Sernyl see 7179
Sernylan see 7179
Seroden [Allen & Hanburys] see 9218
Serogan see 7515
Serolfia [Ascher] see 8149
Seromycin [Lilly] see 2758
Serono-Bagren [Interlabo] see 1404
Serophene [Serono] see 2384
Seroten see 504
Serotinex see 2374
Serotonin, 8413
Serotropin see 7515
Seroxat [Beecham] see 6991
Serozyme see 7903
Serpanray see 8149
Serpasil [Ciba] see 8149
Serpasol see 8149
Serpate [Vale] see 8149
Serpen see 8149
Serpentaria, 8414
Serpentine see 851 and 5569
Serpentine (Alkaloid), 8415
Serpiloid see 8149
Serpina see 8131
Serpine see 8149
Serral see 3118
Sertan see 7753
Sertraline, 8416
Serum Albumin, 8417
Serum Gonadotrop(h)in see 7515
Serum Lactic Dehydrogenase see 5213
Serum Prothrombin Conversion Ac-
  celerator see 3872
Serum Transferrin see 9490
Serum Tryptase see 7493
Servicof [Servipharm] see 8116
Servisone [Lederle] see 7727
L-Seryl-L-aspartyl-L-asparaginyl-L-
  asparaginyl-L-glutaminyl-L-gluta-
  minylglycyl-L-lysyl-L-seryl-L-ala-
  nyl-L-glutaminyl-L-glutaminylgly-
  cylglycyl-L-tyrosinamide see 8366
N-(DL-Seryl)-N'-(2,3,4-trihydroxy-
  benzyl) Hydrazine see 1059
SES see 3372
Sesame Oil, 8418
Sesamex, 8419
Sesamin, 8420
Sesamolin, 8421
Sesden [Tanabe] see 9372
Sesin, 8422
Sesone see 3372
Sesoxane see 8419
Sestron see 377
Sestron Base see 377
Setastine, 8423
Setazine [Merck & Co.] see 1161
Sethadil see 8878
Sethotope [Squibb] see 8395
Sethoxydim, 8424
Sethyl see 4649
Setonil see 2977
Setran see 5751
Setrol [Baxter] see 6926
Settima [IMA] see 7713
Seven Barks see 4686
Sevicaine see 7763
Sevin [Union Carbide] see 1789
Sevinol see 4116
Sevinon see 9760
Sexadieno [AB Leo] see 2726
Sex-Ex see 4498
Sexocretin see 3118
Sexovid [Ferrosan] see 2726
SF-277 see 2939
SF-837 see 6106
SF-1293 see 7314
SF 86-327 see 9086
SF-733 Antibiotic see 8208

S-2703 Forte see 2776
SG-75 see 6431
Sgd 301-76 see 6892
Sgd 12878 see 6801
SGD 24774 see 1028
Sgd-Scha 1059 see 1238
S-GI see 2887
SH 100 see 2743
SH 261 see 3727
SH 406 see 9095
SH 437 see 4955
SH-582 see 4309
SH 617L see 4958
SH 714 see 2781
SH 717 see 4403
SH 770 see 4079
SH 818 see 2368
SH 863 see 2368
SH 881 see 2781
SH 926 see 4904
SH 80881 see 2781
Shadocol see 4931
Shark Liver Oil, 8425
SHB-286 see 8973
SH B 331 see 4308
SHCH-58 see 2624
SH-E-222 see 5744
Sheep Berry see 9877
Shekanin see 9057
Shellac, 8426
Shellolic Acid, 8427
Sheridanite see 352
Sherolatum [Sherwood] see 7138
Sherpa [Rhône-Poulenc] see 2775
Shield [Dow] see 2398
Shigatox [Fawns & McAllan] see 8879
Shikalkin see 243
Shikimic Acid, 8428
Shikonin see 243
Shiomarin [Shionogi] see 6201
Shionone, 8429
Shiosol [Shionogi] see 4422
Shirlan Extra see 8299
Shisonin A see 2694
SH K 203 see 4078
SHL 451 A see 4236
Shock-Ferol see 9928
Showdomycin, 8430
Showersan see 4930
Shoxin see 6604
Siagoside see 4264
O-Sialic Acid see 8431
Sialic Acids, 8431
Sibelium [Janssen/J & J] see 4070
Siberian Pine Needle Oil see 6729
Sibol see 3118
Sibutol see 1315
Sicarol [Hoechst] see 7966
Siccanin, 8432
Sicoclor [Gramon] see 3910
Sicorten [Ciba-Geigy] see 4510
Sideramines see 2850
Siderin Yellow see 3962
Siderite see 4975
Sideromycins see 2850
Siderophilin see 9490
Siderophores see 6235
Sideros [Inverni] see 3984
Siderotil see 4006
Sideryl [Dandoy] see 862
Siduron, 8433
Sieromicin [Sierochimica] see 7420
Sigacalm [Siegfried] see 6881
Sigadoxin [Siegfried] see 3429
Sigamopen [Siegfried] see 610
Sigapedil [Siegfried] see 3626
Sigaprim [Siegfried] see 8889
Sigmacef [Sigma Tau] see 7485
Sigmacort see 4710
Sigmadyn see 7021
Sigmafon [Fontana] see 5653

Sigmaform see 1305
Sigmart [Chugai] see 6431
Signopam see 9074
Sikkimotoxin, 8434
Silain [Robins] see 8486
Silamox [Karlspharma] see 610
Silane, 8435
Silanols, 8436
Silarin [Boehringer, Mann.] see 8484
Silbephylline [Berk] see 3459
Silberol see 8457
Silbesan [Atmos] see 2163
Silene see 1709
Silentium see 8116
Silepar [Ibirn] see 8484
Silibrin see 2082
Silica see 8440
Silica Gel see 8437
Silicane see 8435
Siliceous Earth see 4878
Silicic Acid, 8437
Silicic Acid Tetraethyl Ester see 3805
Silicic Anhydride see 8440
Silicobromoform see 9528
Silicobutane see 9169
Silicochloroform see 9559
Silicoethane see 3362
Silicofluoric Acid see 4110
Silicon, 8438
Silicon Bromide see 8446
Silicon Carbide, 8439
Silicon Chloride see 8447
Silicon Dioxide, 8440
Silicon Disulfide, 8441
Silicones, 8442
Silicon Fluoride see 8448
Silicon Monoxide, 8443
Silicon Nitride, 8444
Silicon Tetraacetate, 8445
Silicon Tetrabromide, 8446
Silicon Tetrachloride, 8447
Silicon Tetrafluoride, 8448
Silicon Tetrahydride see 8435
Silicopropane see 9668
Silicotungstic Acid, 8449
Silicristin see 8484
Silidianin see 8484
Siligaz [Menley & James] see 8486
Silirex [Lampugnani] see 8484
Silital see 2416
Silkweed see 854
Sillimanite see 368
Silliver [Abbott] see 8484
Silomat [Thomae] see 2363
Silopentol [Ciba-Geigy] see 6885
Silo San [ICI] see 7469
Silubin see 1465
Silvadene [Marion] see 8874
Silver, 8450
Silver Acetate, 8451
Silver Balsam see 5153
Silver Benzoate see 1101
Silver Bromide, 8452
Silver Carbonate, 8453
Silver Chlorate, 8454
Silver Chloride, 8455
Silver Chromate(VI), 8456
Silver Citrate, 8457
Silver Cyanide, 8458
Silver Difluoride, 8459
Silver Dinaphthylmethane Disulfonate
  see 5874
Silver Drops see 5153
Silver Fluoride, 8460
Silver Iodate, 8461
Silver Iodide, 8462
Silver Lactate, 8463
Silver Leaf see 8775
Silver Manganate(VII) see 8471
Silver Methylenebis(2-naphthyl-3-
  sulfonate) see 5874

Soln Potassium Iodohydrargyrate *see* 7691
Solocalm [Microsules] *see* 7476
Solodelf *see* 9512
Solon [Taisho] *see* 8658
Solone [Fawns & McAllan] *see* 7719
Solosin [Cassella] *see* 9212
Solozone *see* 8607
Sol Phenobarbital *see* 7202
Sol Phenobarbitone *see* 7202
Solprene [Phillips Petr.] *see* 8345
Solprin *see* 1644
Solprin [Reckitt & Colman] *see* 873
Solprofen [Phoenix] *see* 4812
Solpyron [Beecham-Wülfing] *see* 873
Solubacter [Chantereau] *see* 9568
Solu-Biloptin [Schering] *see* 4958
Soluble Anhydrite *see* 1713
Soluble Aspirin *see* 1644
Soluble Barbital *see* 972
Soluble Cream of Tartar *see* 7595
Soluble Ferric Citrate *see* 541
Soluble Fluorescein *see* 4085
Soluble Glass *see* 8631
Soluble Gun Cotton *see* 8022
Soluble Indigo Blue *see* 4856
Soluble Iodophthalein *see* 4931
Soluble Pentobarbital *see* 7087
Soluble Potash Glass *see* 7658
Soluble Potash Water Glass *see* 7658
Soluble RNA *see* 8204
Soluble Saccharin *see* 8282
Soluble Sulfacetamide *see* 8870
Soluble Sulfadiazine *see* 8874
Soluble Sulfamerazine *see* 8884
Soluble Sulfapyridine *see* 8913
Soluble Sulfathiazole *see* 8920
Soluble Tartar *see* 7669
Solucin *see* 6601
Solucort *see* 7721
Solu-Cortef [Upjohn] *see* 4713
Soludagenan *see* 8913
Solu-Decadron [Merck & Co.] *see* 2922
Solu-Decortin-H *see* 7722
Solu-Dilar [Cassenne] *see* 6977
Solufilin *see* 3459
Solufontamide *see* 8921
Solu-Forte-Cortin [E. Merck] *see* 2922
Solufyllin *see* 3459
Soluglaucit *see* 6979
Solu-Glyc *see* 4713
Solumédine *see* 8884
Solu-Medrol [Upjohn] *see* 6028
Solupemid [Recordati] *see* 7427
Solupred [Houdé] *see* 7723
Solupred *see* 7719
Solu-Predalone [OJ & F] *see* 7721
Solupsan [UPSA] *see* 1644
Soluran *see* 2371
Solurol [Delalande] *see* 8990
Soluseptasine *see* 6601
Soluseptazine *see* 6601
Solusetazine *see* 6601
Solusol *see* 3398
Solustibosan *see* 733
Solutedarol *see* 9512
Solutin [Lindopharm] *see* 9089
Solutrast [Byk-Gulden] *see* 4943
Solutrat [Labinca] *see* 9801
Solutricine [Roger Bellon] *see* 9749
Solvan [Simes] *see* 3304
Solvar [Shawinigan] *see* 7562
Solvay Soda *see* 8541
Solvazinc [Thames] *see* 10064
Solvent Blue 8 *see* 5979
Solvent Naphtha *see* 5366
Solvezink [Draco] *see* 10064
Solvirex *see* 3371
Solvolip [Knoll] *see* 3939

Solyusurmin *see* 733
Soma [Wallace Labs.] *see* 1848
Somacton *see* 8672
Somadril [Dumex] *see* 1848
Somagerol [Efeka] *see* 5456
Somagest [Riom] *see* 507
Somalgen [Bago] *see* 9017
Somalgit *see* 1848
Soman, 8668
Somatocrinin *see* 8669
Somatofalk [Falk] *see* 8671
Somatoliberin, 8669
Somatomedin *see* 8670
Somatomedin A *see* 8670
Somatomedin C *see* 8670
Somatomedins, 8670
Somatonorm [KabiVitrum] *see* 8672
Somatostatin, 8671
Somatotropic Hormone *see* 8672
Somatotropin, 8672
Somatotropin Release Inhibiting Factor *see* 8671
Somatrem *see* 8672
Somatropin *see* 8672
Somatyl [Rolland] *see* 1201
Somazina [Ferrer] *see* 2321
Somberol *see* 5872
Sombrevin [Gedeon Richter] *see* 7813
Sombucaps *see* 4625
Sombulex [Riker] *see* 4625
Somelin [Sankyo] *see* 4518
Somilan [BDH] *see* 2059
Sominat [Merrell] *see* 3033
Somio *see* 2062
Somipront [Mack, Illert.] *see* 3247
Somlan [Sintyal] *see* 4123
Somnafac [Cooper] *see* 5872
Somnal *see* 601
Somnalert [Warren-Teed] *see* 4625
Somnased [Duncan, Flockhart] *see* 6491
Somnatrol [Takeda] *see* 3651
Somnesin [Central Pharm.] *see* 5731
Somnibel [UCB] *see* 6491
Somnipron *see* 782
Somnite [Norgine] *see* 6491
Somnium *see* 5872
Somnomed *see* 5872
Somnopentyl [Pitman-Moore] *see* 7087
Somnos [Merck & Co.] *see* 2061
Somnovit [Hosbon] *see* 5453
Somnurol [Synochem] *see* 1385
Somonal *see* 7201
Somophyllin [Fisons] *see* 477
Somsanit *see* 8603
Sonaform *see* 2717
Sonal *see* 5872
Sonalan [Lilly] *see* 3671
Sonalen [Lilly] *see* 3671
Sonapax [Polfa] *see* 9290
Sonate *see* 8521
Sonazine [Cord] *see* 2186
Sonbutal *see* 1503
Sone [Fawns & McAllan] *see* 7727
Sonebon *see* 6491
Soneryl [M & B] *see* 1515
Songar [Valeas] *see* 9518
Songorine, 8673
Sonifilan [Kaken] *see* 8502
Sonilyn [Mallinckrodt] *see* 8871
Sonimen *see* 2082
Sonin [Lipha] *see* 5453
Soni-Slo [Lipha] *see* 5114
Sonistan *see* 7087
Sonnolin [DIMA] *see* 6491
Sonora *see* 5815
Sopental *see* 7087
Sophiamin [Kyowa Yakuhin] *see* 2082
Sophium [Fabre] *see* 4320
Sophocarpidine *see* 5641

Sophorabioside, 8674
Sophoretin *see* 8044
Sophoricoside, 8675
Sophorin *see* 8276
Sophorine *see* 2795
Sophorose, 8676
Sopor [Arnar-Stone] *see* 5872
Soprol [Lederle] *see* 1309
Soprontin [Knoll] *see* 27
Sorbangil *see* 5114
Sorbic Acid, 8677
Sorbic Acid Potassium Salt *see* 7661
Sorbic Alcohol, 8678
Sorbichew [Stuart] *see* 5114
Sorbic Oil *see* 6981
Sorbicolan *see* 8680
Sorbidilat [Globopharm] *see* 5114
Sorbid SA [Stuart] *see* 5114
Sorbilande *see* 8680
Sorbin *see* 8681
Sorbinil, 8679
Sorbinose *see* 8681
Sorbiperan [Oberval] *see* 6063
Sorbislo [Rona] *see* 5114
Sorbistat [Ono] *see* 3556
Sorbit [Marion] *see* 8680
Sorbitan Mono-9-octadecenoate Poly-(oxy-1,2-ethanediyl) Derivs *see* 7559
Sorbitan Mono-oleate Polyoxyethylene *see* 7559
Sorbitol, 8680
D-Sorbitol *see* 8680
Sorbitrate [Stuart] *see* 5114
Sorbo *see* 8680
Sorbol *see* 8680
Sorboquel [Schering] *see* 1704
Sorbosan *see* 6202
Sorbose, 8681
L-Sorbose *see* 8681
Sorbostyl *see* 8680
Sorbsan [Ross Fraser] *see* 232
Sordenac [Lundbeck] *see* 2392
Sordinol [Ayerst] *see* 2392
Sorensen's Phosphate *see* 8612
Sörensen's Potassium Phosphate *see* 7648
Sorensen's Sodium Phosphate *see* 8612
Sorethytan (20) Mono-oleate *see* 7559
Sorgoa [Scheurich] *see* 9442
Soriatane [Roche] *see* 107
Soricin [Merrell] *see* 8213
Soridermal [Specia; Rhône-Poulenc] *see* 6057
Soriflor [Thomae] *see* 3126
Soripal [Specia] *see* 6057
Sorlate *see* 7559
Sormetal *see* 3480
Sorot [Ravensberg] *see* 2896
Sorquad [Tutag] *see* 5114
Sorquetan [Basotherm] *see* 9377
"Sosan" *see* 4974
Sosigon *see* 7078
Sosol [McKesson] *see* 8930
Sospitan [Kali-Chemie] *see* 7987
Sostenil [Pfizer] *see* 9888
Sostril [Cascan] *see* 8126
Sotacor [Bristol] *see* 8682
Sotal [Gramon] *see* 9389
Sotalex [Mead Johnson; Lappe] *see* 8682
Sotalol, 8682
Soterenol, 8683
Sotyl [Pitman-Moore] *see* 7087
Soudan Coffee *see* 5198
Souframine *see* 7220
Sour-spine *see* 7087
Sovcaine *see* 3016
Soventol [Knoll] *see* 966
Soverin *see* 5872

Cross Index of Names

Sowberry *see* 971
Sowell *see* 5751
Soxisol *see* 8930
Soxo [Sutliff & Case] *see* 8930
Soxomide [Upjohn] *see* 8930
Soxysympamine [Ferndale] *see* 5859
Soya Bean *see* 8684
Soybean, 8684
Soybean Agglutinin *see* 5312
Soybean Oil, 8685
Soy Sauce, 8686
Sozoiodole-mercury *see* 8687
Sozoiodole-sodium *see* 8687
Sozoiodole-zinc *see* 8687
Sozoiodolic Acid, 8687
SP *see* 8834
SP 54 [Bene-Arzneimittel] *see* 7090
Sp 281 *see* 9875
Sp 732 *see* 7791
SP 1103 *see* 9154
SPA [Santen] *see* 5314
SPA-S132 *see* 6997
SPA-S160 *see* 5733
SPA-S-222 *see* 5733
SPA-S510 *see* 7476
Spabucol [Lagap] *see* 9613
Spacolin [Roxane] *see* 377
Spactin [Meiji] *see* 1599
Spaderizine [Kotobuki] *see* 2308
Spametrin-M [Sanzen] *see* 5989
Spamol *see* 479
Spamorin [Tanabe Seiyaku] *see* 4039
Spandex, 8688
Spanestrin [Savage] *see* 3689
Spanish Fly *see* 1754
"Spanish Fly" *see* 1755
Spanish Grass Wax *see* 3649
Spanish Saffron *see* 8285
Spanish Tea *see* 6713
Spanish White *see* 1298
Span-K *see* 7601
Spanon *see* 2083
Spanor [Pharmascience] *see* 3429
Spantin [Deutsche Pharmacia] *see* 7921
Spantol [Nippon Chemiphar] *see* 7229
Span® and Tween®, 8689
Sparassol, 8690
Sparicon [Yamanouchi] *see* 1588
Sparine [Wyeth] *see* 7793
Sparsamycin A *see* 9714
Sparsiflorine, 8691
Spartakon [ICI] *see* 9161
Spartase [Wyeth] *see* 862
Spartein-Asal [Asal] *see* 8692
Sparteine, 8692
l-Sparteine *see* 8692
Spartocin [Ayerst] *see* 8692
Spartocine [UCB] *see* 862
Spartopan [Pailluseau] *see* 8692
Spartrix [Janssen] *see* 1855
Spasfon-Lyoc [Lafon] *see* 7301
Spasmalex *see* 3151
Spasmamide [Schering AG] *see* 3899
Spasmaparid [Nordmar] *see* 1224
Spasmaverine [Ethica] *see* 377
Spasmedal [Medinova] *see* 5736
Spasmentral [Janssen] *see* 1084
Spasmex [Pfleger] *see* 9697
Spasmexan *see* 3892
Spasmipront *see* 5872
Spasmisolvina [Dessy] *see* 1224
Spasmium [Donau] *see* 1864
Spasmocan *see* 1743
Spasmocromona *see* 9574
Spasmocyclon [Kettelhack] *see* 2708
Spasmodex [Crinex] *see* 3151
Spasmodolin [Grossman] *see* 5736
Spasmodolisina *see* 7826
Spasmo-Dolviran [Bayer] *see* 7178
Spasmol *see* 5600

Spasmolyn *see* 5737
Spasmolysin *see* 7921
Spasmolytin *see* 151
Spasmolytol, 8693
Spasmonal [Arzneimittelwerk VEB] - *see* 3343
Spasmo-Nit [Stroschein] *see* 6968
Spasmo-Paparid [Nordmark] *see* 1224
Spasmophen *see* 6928
Spasmopriv [Paillusseau] *see* 3928
Spasmoril *see* 4039
Spasmostenyl *see* 5600
Spasmoxal(e) *see* 3295
Spasuret [Asche] *see* 4033
Spathulatine *see* 5115
Spatomac *see* 1037
Spatonin *see* 3106
SPCA *see* 3872
SPC 97D *see* 925
SPE *see* 8857
SPE 2792 *see* 698
Spearmint, 8694
Specifin [Bergamon] *see* 6273
Spectacillin [Sandoz] *see* 3566
Spectam *see* 8695
Spectamedryn [Allergan] *see* 5679
Spectamine [Medi-Physics] *see* 4938
Spectazole [Ortho] *see* 3476
Spectinomycin, 8695
Spectogard [Diamond] *see* 8695
Spectracide *see* 2978
Spectra-Sorb UV 9 [Am. Cyanamid] *see* 6907
Spectra-Sorb UV 24 [Am. Cyanamid] *see* 3298
Spectra-Sorb UV 284 [Am. Cyanamid] *see* 8963
Spectra-Sorb UV 531 [Am. Cyanamid] *see* 6662
Spectrin, 8696
Spectrobact [UCB] *see* 9681
Spectrobid [Pfizer] *see* 944
Spectrum [Sigma-Tau] *see* 1944
Speda *see* 9897
"Speed" *see* 5859
Spenitol *see* 6181
Spergon *see* 2071
Spermaceti, 8697
Spermidine, 8698
Spermine, 8699
Spermine Oxidase *see* 6158
Spermine Phosphate *see* 8699
Sperm Oil, 8700
Spermwax [Robeco] *see* 8697
SPG *see* 8502
Sphaerocobaltite *see* 2430
Sphaerophysine *see* 8701
Sphalerite *see* 10065
Sphene *see* 9396
Spheroidine *see* 9175
Spheromycin *see* 6641
Spherophysine, 8701
4-Sphingenine *see* 8703
Sphingomyelins, 8702
Sphingosine, 8703
SP-I *see* 7519
Spice Berry *see* 793
Spiclomazine *see* 2408
Spigelia, 8704
Spignet *see* 793
Spike [Lilly] *see* 9053
Spikenard *see* 793
Spilanthol *see* 167
Spinacane *see* 8726
Spinacene *see* 8727
α-Spinasterin *see* 8705
α-Spinasterol, 8705
Spindle Tree *see* 3856
Spinnaker [Shell] *see* 9508
Spinulosin, 8706
Spiperone, 8707

Spira [RMB] *see* 8708
Spiractin *see* 7399
Spiramin [Mitsui] *see* 9487
Spiramycin, 8708
Spiramycin I *see* 8708
Spiramycin II *see* 8708
Spiramycin III *see* 8708
Spiramycin Adipate *see* 8708
Spirein *see* 4544
Spiretic [DDSA] *see* 8721
Spirilene, 8709
Spirilloxanthin *see* 8195
Spirit of Ammonia, Aromatic, 8710
Spirit of Ants *see* 8716
Spirit of Camphor, 8711
Spirit of Chloroform, 8712
Spirit of Ether, 8713
Spirit of Ether Compound, 8714
Spirit of Ethyl Nitrite, 8715
Spirit of Formic Acid, 8716
Spirit of Glonoin *see* 8717
Spirit of Glyceryl Trinitrate, 8717
"Spirit of Hartshorn" *see* 513
Spirit of Hartshorn, Aromatic *see* 8710
Spirit of Mindererus *see* 514
Spirit of Nitroglycerin *see* 8717
Spirit of Nitrous Ether *see* 8715
Spirit of Peppermint, 8718
Spirit of Spearmint, 8719
Spirit of Trinitroglycerin *see* 8717
"Spirit of Turpentine" *see* 6774
Spiro-32 [Unimed] *see* 8720
Spirocid [Hoechst] *see* 44
Spirocort [Astra] *see* 1455
Spiroctan [MCP] *see* 8721
Spirodon *see* 9165
Spiroform *see* 7241
Spirofulvin *see* 4453
Spirogermanium, 8720
Spirolone [Berk] *see* 8721
Spironolactone, 8721
Spiropent [Thomae; Boehringer, Ing.] *see* 2347
Spiropitan [Janssen] *see* 8707
Spirosal *see* 4395
(3β,5α,22β,25S)-Spirosolan-3-ol *see* 9470
Spirosol-5-en-3β-ol *see* 8665
(3β,22α,25R)-Spirosol-5-en-3-yl O-6-Deoxy-α-L-mannopyranosyl-(1→2)-O-[β-D-glucopyranosyl-(1→3)]-β-D-galactopyranoside *see* 8666
Spirostan-2,3-diol *see* 4325
(25R)-5α-Spirostan-3β,6α-diol *see* 2143
(25R)-5α-Spirostan-3β,15β-diol *see* 3136
(25R)-Spirostan-3β-ol *see* 8508
(25R)-5α-Spirostan-3β-ol *see* 9367
(25S)-Spirostan-3β-ol *see* 8338
(25R)-5α-Spirostan-2α,3β,15β-triol *see* 3143
(25R)-Spirost-5-ene-1β,3β-diol *see* 8270
(25R)-Spirost-5-en-3β-ol *see* 3289
20α,22a,25D-Spirost-5-en-3β-ol-12-one *see* 4293
(25R)-Spirost-5-en-3β-yl O-6-Deoxy-α-L-mannopyranosyl-(1→2)-O-[6-deoxy-α-L-mannopyranosyl-(1→4)]-β-D-glucopyranoside *see* 3286
Spiro-Tablinen [Sanorania] *see* 8721
Spizef [Grünenthal] *see* 1937
Spizofurone, 8722
Splendil [Astra] *see* 3895
Splendor [ICI-Zeltia] *see* 9484
Splenin *see* 9339
Splenin, 8723

Split Nut see 7356
Splotin [Taiyo Yakuko] see 8971
Spodumene see 5395
Spofa 325 see 5738
Spofadazine see 8890
Spoloven see 2922
Spondyril [Dorsch] see 7248
Spondyvit 500 [Efeka] see 9932
Spongoadenosine see 9881
Sponsin [Farmasan] see 3598
Spontin see 8227
Spoonwood see 5160
Sporamin [Hishiyama] see 1588
Sporanos [Janssen] see 5131
Sporanox [Janssen] see 5131
Sporeine see 945
Sporidesmin see 8724
Sporidesmin A see 8724
Sporidesmins, 8724
Sporidesmolide I see 8725
Sporidesmolide II see 8725
Sporidesmolide III see 8725
Sporidesmolide IV see 8725
Sporidesmolides, 8725
Sporidyn [Cyanamid] see 1937
Sporiline see 9442
Sporostacin [Ortho] see 2080
Sporostatin see 4453
Sportak [FBC] see 7767
Spotless [Sumitomo] see 3261
Spotted Alder see 4521
Spotted Cowbane see 2503
Spotted Hemlock see 2503
Spotton [Chemagro] see 3945
Spreading Dogbane see 773
Spreading Factor see 4676
Sprecur [Hoechst] see 1492
Sprout-Nip [PPG] see 2188
Spurge Flax see 6095
Spurred Rye see 3608
S.Q. [Merck & Co.] see 8914
SQ 1489 see 9304
SQ 4918 see 4116
SQ 9453 see 3247
SQ 9538 see 9108
SQ 10269 see 1805
SQ 10496 see 9233
SQ 10733 see 4116
SQ 11302 see 3566
SQ 11436 see 1982
SQ 11725 see 6260
SQ 13050 see 3476
SQ 13396 see 4943
SQ 14055 see 9351
SQ 14225 see 1773
SQ 15101 see 230
SQ 15860 see 4402
SQ 16144 see 4116
SQ 16150 see 3653
SQ 16401 see 4520
SQ 18566 see 4504
SQ 19844 see 8494
SQ 21983 see 4949
SQ 22022 see 1982
SQ 22947 see 9351
SQ 26776 see 938
SQ 27519 see 4171
SQ 28555 see 4171
SQ 30213 see 9364
SQ 30836 see 9364
SQ 31000 see 7712
Squalane, 8726
Squalene, 8727
Squalidine see 8402
Square-stem Rose-gentian see 1965
Squaw Bush see 9876
Squaw Mint see 7053
Squaw Root see 1915
Squaw Weed see 8401
Squill, 8728
Squirrel Corn see 2544

SR 406 see 1771
SR 720-22 see 6068
SRA 3886 see 3901
SRA 5172 see 5858
SRA 12869 see 5055
SRG 95213 see 2986
SRIF see 8671
SRIF-A see 8671
SRIF-14 see 8671
SRIF-25 see 8671
SRIF-28 see 8671
Srilane [Medicia] see 4820
SRS see 5339
SRS-A see 5339
SS 578 see 4935
ST-21 see 7485
S.T. 37 [Merck & Co.] see 4633
ST52-Asta see 4168
St-155 see 2388
ST-200 see 78
ST-374 see 3910
ST-567 see 234
ST-1085 see 6107
ST-1191 see 3841
ST-1396 see 1956
ST-1512 see 4628
ST-2121 see 3579
St-4331 see 4789
ST-7090 see 4626
ST-9067 see 927
STA 307 see 9385
Stabicillin see 7046
Stabilene [Auclair] see 3729
Stabillin V-K [Boots] see 7046
Stabinol [Horner] see 2187
Stabisol [Upjohn] see 1299
Stablon [Servier] see 9352
Stachydrine, 8729
Stacillin [Schiapparelli] see 2342
Stadacain [Stada] see 1531
Stadadorm see 601
Stadalax see 1253
Stadol [Bristol-Myers] see 1530
Stafac [SK & F] see 9910
Staff [Sigma-Tau] see 3841
Stafoxil [Brocades] see 4044
Stag Bush see 9877
Staggerweed see 5242
Stagid [Stago] see 5845
Stagural [Stada-Chemie] see 6616
Stakane [Dausse] see 754
Stalleril [Pharmacal] see 9290
Stallimycin, 8730
Stam [Rohm & Haas] see 7814
Stamine [Tutag] see 7996
Stampede [Rohm & Haas] see 7814
Stampen [Beecham] see 3073
Stanaprol [Pfizer] see 8753
Stanazol see 8754
Stangen see 7996
Stangyl (Tabl) [Specia] see 9636
Stangyl (Amp) [Specia] see 9636
Stanilo [Upjohn] see 8695
Stannic Anhydride see 8736
Stannic Bromide, 8731
Stannic Chloride, 8732
Stannic Chromate(VI), 8733
Stannic Fluoride, 8734
Stannic Iodide, 8735
Stannic Oxide, 8736
Stannic Selenide, 8737
Stannic Selenite, 8738
Stannic Sulfide, 8739
Stannochlor see 8742
Stannous Acetate, 8740
Stannous Bromide, 8741
Stannous Chloride, 8742
Stannous Fluoride, 8743
Stannous Fluozirconate(IV) see 8744
Stannous Hexafluorozirconate(IV), 8744

Stannous Iodide, 8745
Stannous Oxalate, 8746
Stannous Oxide, 8747
Stannous Pyrophosphate, 8748
Stannous Selenide, 8749
Stannous Sulfate, 8750
Stannous Sulfide, 8751
Stannous Tartrate, 8752
Stanolene [Standard Oil] see 7138
Stanolind [Standard Oil] see 7138
Stanolone, 8753
Stanomycetin see 2068
Stanozol see 8754
Stanozolol, 8754
Stanquinate see 4935
Stanyl [DSM] see 6657
Stapenor [Bayer] see 6858
Staphcil [Lederle] see 4044
Staphcillin [Bristol] see 5890
Staphisagria, 8755
Staphlipen [Lederle] see 4044
Staphobristol-250 [Bristol] see 2414
Staphybiotic [Delagrange] see 2414
Staphylex [Beecham] see 4044
Staphylomycin see 9910
Staphylomycin M₁ see 9910
Staphylomycin S see 9910
Staphylomycine [SK & F] see 9910
Stapyocine [Specia] see 7758
Starane [Wacker-Dow] see 4128
Star Anise, 8756
Starazine see 7793
Starch, 8757
Starch Gum see 2931
Starch 2-Hydroxyethyl Ether see 4593
Starch, Soluble, 8758
Star Grass see 220
Starwort see 220 and 4550
Stas-Hustenlöser [Stada] see 392
Staticin [Westwood] see 3626
Staticin [Merck & Co.] see 1163
Statil [ICI] see 7565
Statimo [Merck & Co.] see 1790
Statine, 8759
Statobex [Lemmon] see 7180
Statocin [Yoshitomi] see 1846
Statolon, 8760
Statomin [Bowman] see 7996
Statran [Delagrange] see 3520
Statyl [Ayerst] see 6385
Stauffer 2061 see 7007
Stauroderm see 4123
Stavesacre see 8755
Stayban [Yoshitomi] see 4124
Staycept [Syntex] see 6596
Stazepine [Polfa] see 1783
St. Bartholomew's Tea see 5637
Steaglate see MISC-4
Steareth see 7554
Stearic Acid, 8761
Stearic Acid Aluminum Salt see 370
Stearic Acid Ammonium Salt see 582
Stearic Acid Calcium Salt see 1710
Stearic Acid, Ester with Lactate of Lactic Acid, Calcium Salt see 1711
Stearic Acid Lead Salt see 5300
Stearic Acid Potassium Salt see 7664
Stearic Acid Sodium Salt see 8634
Stearin see 9669
Stearodine see 1682
Stearyl Alcohol, 8762
Steatite see 9011
Steclin [Squibb] see 9130
Stecsolin [Squibb] see 6931
Stediril [Wyeth] see 6621
Steinbühl Yellow see 982
Stelabid [SK & F] see 5090
Steladone [Ciba-Geigy] see 2087
Stelazine [SK & F] see 9594
Stellamicina see 3628

Stellarid [Zambeletti] see 7889
Stellerite see 1645
Stemetil [M & B] see 7768
Stemex [Syntex] see 6977
Stemphylone see 8117
Stenandiol see 670
Stenbolone, 8763
Stenediol [Organon] see 5861
Stenobolone (rescinded USAN) see 8763
Stenol see 8762
Stenolon see 5862
Stenorol [Roussel] see 4509
Stenosine see 5864
Stenovasan see 477
Stephanine see 5834
Stepin [Basotherm] see 9388
Stepronin, 8764
Steranabol [Farmitalia] see 2409
Steranabol Long-acting see 4768
Steranabol Ritardo see 4768
Sterandryl (Amps.) see 9115
Sterane [Pfizer] see 7719
Sterathal see 7350
Sterax [Alcon] see 2908
Sterazine see 8874
Stercobilin see 9797
Stercofuge see 232
Stercorin see 2519
Sterculia see 5164
Sterecyt [Pharmaleo] see 7718
Stereocyt [Roger Bellon] see 7718
Stereomycine [Toraude; Ayerst] see 5337
Stereon [Firestone] see 8345
Sterilate see 4624
Sterilon see 2090
Sterinor [Heumann] see 9178
Sterisil [Wm. R. Warner] see 4624
Sterisol see 4624
Sterisol (Domestic) [Warner-Lambert] see 4619
Sterocort see 7729
Steroderm [De Angeli] see 2908
Sterogyl [Roussel-UCLAF] see 9928
Sterolone [Rowell] see 7719
Sterosan [Geigy] see 2191
Sterox [Monsanto] see 6596
Sterox CD [Monsanto] see 7555
Steroxin [Geigy] see 2191
Sterretite see 362
Sterwin 904 see 222
Ster-Zac [Hough; Hoseason] see 9573
Stesolid [Weddel] see 2977
Stesolin see 2977
Stevacin see 6931
Steviol, 8765
Stevioside, 8766
Steviosin see 8766
STH see 8672
Stibanate see 733
Stibanose see 733
Stibatin see 733
"Stibic" Anhydride see 730
2,2',2''-[Stibilidynetris(thio)]tris-
   butanedioic Acid Hexalithium Salt
   see 711
Stibine, 8767
Stibinol see 733
Stibium see 724
Stibnite see 745
Stibocaptate, 8768
p-Stibonobenzoic Acid Methyl 1,2,5,6-
   Tetrahydro-1-methylnicotinate see 803
Stibophen, 8769
Stibosamine see 3807
Stibron [Iwaki] see 1588
Stibunal see 734
Stickstoff Lost see 9560
Stickstofflost [Ebewe] see 5655

Stickstoffwasserstoffsäure (German) see 4697
Stiedex [Stiefel] see 2910
Stiemycin [Stiefel] see 3626
Stigmasta-5,7-dien-3β-ol see 2863
Stigmasta-5,24(28)-dien-3-ol see 4194
(3β,5α,22E)-Stigmasta-7,22-dien-3-ol see 8705
(3β,5α,22E,24R)-Stigmasta-7,22-dien-3-ol see 2214
(3β,22E)-Stigmasta-5,22-dien-3-ol see 8771
Stigmastanol, 8770
(3β,5α)-Stigmastan-3-ol see 8770
(3β)-Stigmast-5-en-3-ol see 8500
(3β,24S)-Stigmast-5-en-3-ol see 8501
Δ⁵-Stigmasten-3β-ol see 8500
Stigmasterol, 8771
Stigmenene Bromide see 1132
Stigmonene Bromide [Warner-Chil-
   cott] see 1132
St. Ignatius' Bean see 4823
Stilalgin see 5737
Stilamin [Serono] see 8671
Stilbamidine, 8772
Stilbazium Iodide, 8773
Stilbene, 8774
[Stilbene-(4,4')]bis[ω-phenylurea]-
   2,2'-disulfonic Acid Disodium Salt
   see 1322
Stilbene-α-carboxylic Acid see 7253
4,4'-Stilbenedicarboxamidine see 8772
E-3,5-Stilbenediol see 7419
Stilbestrol see 3118
Stilbestrol Dimethyl Ether see 3198
Stilbestrol Diphosphate see 4168
Stilbestronate see 3119
Stilbetin [Squibb] see 3118
Stilboefral see 3118
Stilboestroform see 3118
Stilboestrol see 3118
Stilboestrol DP see 3119
Stilbofax see 3119
Stilbostatin [Taro] see 4168
Stilciclina see 9130
Stilkap see 3118
Stillacor [Wolff] see 3150
Stillingia, 8775
Stilnox [Synthelabo] see 10091
Stilny [Will Pharma] see 6608
Stilonium Iodide, 8776
Stilpalmitate see 3118
Stilphostrol [Dome] see 4168
Stilronate see 3119
Stimamizol [J & J] see 9161
Stimol [Biocodex] see 2333
Stimolag Fortis [Lagap] see 7455
Stimovul [Organon] see 3568
Stimsen [Lederle] see 9315
Stimul see 7021
Stimulexin [Robins] see 3421
Stimulin see 6459
Stimulina see 4365
Stimu-T.S.H. [Roussel] see 9503
Stinerval see 7181
Stinkweed see 8779
Stipolac [Burroughs Wellcome] see 4931
Stiptanon [Organon] see 3659
Stirofos, 8777
Stivane [Beaufour] see 8002
StL 1106 see 4118
Stock Guard [Cyanamid UK] see 4055
Stoddard Solvent see 6120
Stolzite see 5309
Stomacain see 6888
Stomp [Am. Cyanamid] see 7026
Stone Oil see 4815
Stone-root see 2479
Stopetyl see 3370

Stopspot [ICI] see 7272
Storax, 8778
Storinal see 1325
Storksbill see 4299
Stovaine see 650
Stovarsol [M & B; Specia] see 44
Stovarsolan see 44
Stoxil [SK & F] see 4819
Stozzon-Chlorophyll see 2155
STPP see 8654
Strabolene see 6283
Straderm [ITA] see 4077
Straminol [Cilag-Chemie] see 3404
Stramonium, 8779
Stratene [Innothera; Sigma-Tau] see 2012
Strawberry Tree see 3856
Strazide see 8788
Strengite see 3975
Strepogenin, 8780
Streptase [Behringwerke] see 8784
Streptidine, 8781
Streptobiosamine, 8782
Streptobrettin [Norbrook] see 8786
Streptocid Album see 8898
Streptocide [Evans] see 8895
Streptocide [Merck & Co.] see 8898
Streptococcal Fibrinolysin see 8784
Streptogramin, 8783
Streptogramin A see 8783 and 9910
Streptogramin B see 6111 and 8783
Streptohydrazid [Pfizer] see 8788
Streptokinase, 8784
Streptolin see 8791
Streptolydigin, 8785
Streptomagma [Wyeth] see 3162
Streptomycin, 8786
Streptomycin A see 8786
Streptomycin B, 8787
Streptomycin Hydrochloride see 8786
Streptomycin Hydrochloride-calcium
   Chloride Complex see 8786
Streptomycin Sulfate see 8786
Streptomyclidene Isonicotinyl Hydra-
   zine Sulfate see 8788
Streptoniazide see 8788
Streptonicozid, 8788
Streptonigrin, 8789
Streptonivicin see 6641
Streptorex see 8786
L-Streptose, 8790
Streptothenat [Grünenthal] see 8786
Streptothricin VI see 8791
Streptothricin F see 8791
Streptothricins, 8791
Streptovaricin, 8792
Streptovaricin A see 8792
Streptovaricin B see 8792
Streptovaricin C see 8792
Streptovaricin D see 8792
Streptovaricin E see 8792
Streptovaricin F see 8792
Streptovaricin G see 8792
Streptovaricin J see 8792
Streptovarycin see 8792
Streptovirudin see 8793
Streptozocin, 8794
Streptozon see 8895
Streptozotocin see 8794
Stresam [Beaufour] see 3821
Strese & Hofmann's Hectorite see 4538
Stresnil [Janssen] see 915
Stresson [Boehringer, Ing.] see 1479
Streunex see 5379
Striadyne see 145
Striatran [Merck & Co.] see 3520
Strictylon see 6287
Strigol, 8795
Striped Alder see 4521
Strobane®, 8796

Strobane-T [Tenneco] see 9478
Strodival [Herbert] see 6854
Stromba [Winthrop] see 8754
Strombaject [Winthrop] see 8754
Strongid [Pfizer] see 7968
Strong Protargin see 8475
Strong Protein Silver see 8475
Strong Silver Protein see 8475
Strontia see 8812
Strontianite see 8801
Strontium, 8797
Strontium Acetate, 8798
Strontium Bromate, 8799
Strontium Bromide, 8800
Strontium Carbonate, 8801
Strontium Chlorate, 8802
Strontium Chloride, 8803
Strontium Chromate(VI), 8804
Strontium Fluoride, 8805
Strontium Formate, 8806
Strontium Hydrate see 8807
Strontium Hydride see 8797
Strontium Hydroxide, 8807
Strontium Iodide, 8808
Strontium Lactate, 8809
Strontium Monoxide see 8812
Strontium Nitrate, 8810
Strontium Nitride see 8797
Strontium Oxalate, 8811
Strontium Oxide, 8812
Strontium Peroxide, 8813
Strontium Phosphate, Tribasic, 8814
Strontium Selenate, 8815
Strontium Sulfate, 8816
Strontium Sulfide, 8817
Strontolac [Wyeth] see 8809
Strophadogenin see 8818
Strophanthidin, 8818
Strophanthidin α-L-Rhamnoside see 2508
Strophanthin, 8819
Strophanthobiose, 8820
Strophanthus, 8821
Strophoperm see 6854
Structum [Grémy-Longuet] see 2217
Strumacil see 6048
Strumazol see 5892
Strychnidin-10-one see 8822
Strychnine, 8822
Strychnine Nitrate, 8823
Strychnine N^6-Oxide, 8824
Strychnine Phosphate, 8825
Strychnine Sulfate, 8826
C-Strychnotoxine see 1722
Stryphnonasal see 161
Stryphnone see 161
STS [Durst] see 8645
Stuart Factor see 3875
Stuart-Prower Factor see 3875
Stugeron [Janssen] see 2308
Sturine see 7898
Stutgeron [Janssen] see 2308
Stutgin see 2308
STX see 8344
Stylomycin see 7960
Stylophorin see 2043
Stylopine, 8827
Styphnic Acid, 8828
Stypnone see 161
Styptic Collodion see 2480
Stypticin [Merck & Co.] see 2551
Styptirenal see 3569
Styptol see 2551
Styquin see 1506
Styracin see 2307
Styramate, 8829
Styrax see 8778
Styrene, 8830
Styrene-butadiene Rubber see 8345
Styrene Glycol, 8831
Styrol see 8830

Styrolene see 8830
Styryl Carbinol see 2305
5-Styrylresorcinol see 7419
SU 88 see 8658
SU 3088 see 2101
Su 3118 see 8998
Su 4885 see 6083
Su 5864 see 4473
Su 6187 see 1519
Su 8341 see 2750
SU 9064 see 6074
Su 10568 see 2406
Su 18137 see 2780
SU 21524 see 7478
Suacron [Praemix] see 1779
Suanovil [Biokema] see 8708
Suanozil [Farmitalia] see 8708
Suavitil [Merck & Co.] see 1037
Subamycin see 9130
Subathizone, 8832
Sub-boric Acid see 2999
Subdue [Ciba-Geigy] see 5826
Suberic Acid, 8833
Suberone see 2728
Suberoxime see 2728
Subicard see 7066
Subitex [Hoechst] see 3282
Subitol see 4815
Sublimaze [McNeil] see 3944
Sublimed Sulfur see 8956
Subose [Squibb] see 4402
272-Substance see 9751
Substance P, 8834
Subtilin, 8835
Subtosan see 7700
Sucaryl [Abbott] see 8282
Sucaryl Calcium [Abbott] see 1665
Sucaryl Sodium see 2707
Succicuran see 8846
Succimal [Clin-Comar-Byla] see 3704
Succimer, 8836
Succinamide, 8837
Succinanil, 8838
Succinanilic Acid, 8839
Succinbromimide see 1428
Succinchlorimide see 2165
Succinic Acid, 8840
Succinic Acid Anhydride see 8841
Succinic Acid Bis[β-dimethylamino-ethyl] Ester Dimethochloride see 8846
Succinic Acid Cadmium Salt see 1626
Succinic Acid Calcium Salt see 1712
Succinic Acid Dibutyl Ester see 3023
Succinic Acid Di-tert-butyl Ester see 3024
Succinic Acid 2,2-Dimethylhydrazide see 2810
Succinic Acid Dinitrile see 8843
Succinic Acid α-Ester with D(−)threo-2,2-Dichloro-N-[β-hydroxy-α-(hydroxymethyl)-p-nitrophenethyl]-acetamide Compd with 4-(Dimethylamino)-1,4,4a,5,5a,6,11,12a-octa-hydro-3,6,10,12,12a-pentahydroxy-6-methyl-1,11-dioxo-N-(1-pyrroli-dinylmethyl)-2-napthacenecarbox-amide see 8408
Succinic Acid Monoanilide see 8839
Succinic Acid Monoester with 4-But-yl-4-(hydroxymethyl)-1,2-diphenyl-3,5-pyrazolidinedione see 8990
Succinic Acid Monoester with N-(2-Ethylhexyl)-3-hydroxybutyramide see 1526
Succinic Acid Monoester with 4-[2-(1-Hydroxyethyl)-3-methyl-5-benzo-furanyl]-2(5H)-furanone see 1051
Succinic Acid Sodium Salt see 8635
Succinic Anhydride, 8841
Succinimide, 8842

Succinimide-Sauba [Sauba] see 8842
Succiniodimide see 4936
Succinite see 388
Succino-AICAR see 8288
Succinonitrile, 8843
2-[1-(Succinoyloxy)ethyl]-3-methyl-5-(2-oxo-2,5-dihydro-4-furyl)-benzo[b]furan see 1051
4-Succinylamido-4′-aminodiphenyl-sulfone see 8851
p-Succinylaminobenzaldehyde Thio-semicarbazone see 4163
Succinyl-Asta [Asta] see 8846
Succinyl Chloride, 8844
Succinylcholine Bromide, 8845
Succinylcholine Chloride, 8846
Succinylcholine Iodide, 8847
Succinyl Dichloride, 8844
O,O-Succinyldicholine Iodide see 8847
Succinyldisalicylic Acid see 8849
Succinyl Oxide see 8841
Succinyl Peroxide, 8848
Succinylsalicylic Acid, 8849
2-(N⁴-Succinylsulfanilamido)thiazole see 8850
Succinylsulfathiazole, 8850
Succisulfone, 8851
Succitimal see 7231
Succosa [Hässle] see 8853
Sucline [Soekami] see 8282
Suclofenide, 8852
Suconox-4 see 4745
Sucostrin Chloride [Squibb] see 8846
Sucralfate, 8853
Sucralfin [Inverni] see 8853
Sucralose, 8854
Sucrase see 4899
Sucrate [Lisapharma] see 8853
Sucrédulcor [Europ. Med.] see 8282
Sucrets [Merck & Co.] see 4633
Sucrida Berna see 1579
Sucrofer see 3974
Sucrol see 3448
Sucromat [Mayoli-Spindler] see 8282
Sucrosa see 2707
Sucrose, 8855
Sucrose Octaacetate, 8856
Sucrose Octakis(hydrogen Sulfate) Aluminum Complex see 8853
Sucrose Polyester, 8857
Sudac [Errekappa] see 8961
Sudafed [Burroughs Wellcome] see 3561
Sudan III, 8858
Sudan IV see 8349
Sudermo see 5821
Sudil [Caber] see 8967
Sudine [Warren-Teed] see 8876
Suet, Prepared, 8859
Sufenta [Janssen] see 8860
Sufentanil, 8860
Sufentanyl see 8860
Sufoxazine see 9077
Sufralem see 676
Sufrexal [Janssen] see 5175
Suganyl see 8879
Sugar see 8855
Sugar of Lead see 5268
Sugast [Alfa] see 8853
Suicalm [Janssen] see 915
Suiminth [Pfizer] see 6175
Suint, 8861
Sukor [Calif. Aromatics & Flavor] see 6367
Suladrin see 8930
Sulamyd [Schering] see 8870
Sulamyd Sodium [Schering] see 8870
Sulbactam, 8862
Sulbenicillin, 8863
Sulbenox, 8864
Sulbentine, 8865

7-(α-Sulphophenylacetamido)-3-(4'-carbamoylpyridinium)methyl-3-cephem-4-carboxylic Acid *see* 1943
Sulphormethoxine *see* 8877
Sulphsymazine *see* 8919
Sulphur *see* 8948
Sulphurenic Acid, 8970
Sulpiride, 8971
S-(−)-Sulpiride *see* 8971
Sulpitil [Tillotts] *see* 8971
Sulprim [Polfa] *see* 8889
Sulprofos, 8972
Sulprostone, 8973
Sulpyrin *see* 3358
Sulqui *see* 6425
Sulquin *see* 8914
Sulreuma [Von Boch] *see* 8961
Sulsoxin [Reid-Provident] *see* 8930
Sul-Spantab [SK & F] *see* 8878
Sultanol [Nippon Glaxo] *see* 209
Sultan Red 4B *see* 1111
Sulten-10 [Bausch & Lomb] *see* 8870
Sulthiame, 8974
Sultirene *see* 8890
Sultopride, 8975
Sultosilic Acid, Piperazine Salt, 8976
Sultropan *see* 8977
Sultroponium, 8977
Sulxin *see* 8876
Sulzol *see* 8920
SUM-3170 *see* 5461
Sumach, 8978
Sumach Wax *see* 5139
Sumamed [Pliva] *see* 928
Sumapen VK *see* 7046
Sumatra Camphor *see* 1338
Sumatriptan, 8979
Sumatrol, 8980
Sumbul, 8981
Sumédine *see* 8884
Sumetrolim [EGYT] *see* 8889
Sumi-8 [Sumitomo] *see* 3261
Sumial [ICI] *see* 7852
Sumi-Alpha [Sumitomo] *see* 3952
Sumicidin [Sumitomo] *see* 3952
Sumiferon [Sumitomo] *see* 4892
Sumilex [Sumitomo] *see* 7772
Sumisclex [Sumitomo] *see* 7772
Sumithion *see* 3922
Sumithrin [Sumitomo] *see* 7221
Sumitomo S 4084 *see* 2266
Summetrin *see* 6965
Summit [Mobay] *see* 9508
Sumox [Reid-Provident] *see* 610
Sumycin [Squibb] *see* 9130
Sunbrella [Dorsey] *see* 434
Suncefal [Yamanouchi] *see* 1940
Suncholin [Mohan] *see* 2321
Suncide *see* 7849
SunDare [Texas Pharmacal] *see* 2312
Sundralen [Hoechst] *see* 9350
Sunette [Hoechst Celanese] *see* 30
Sunflower Seed Oil, 8982
Sunfural [Toyo Jozo] *see* 9060
Sungard [Miles] *see* 8963
Sunrabin [Asahi] *see* 3539
Sunril [Emko] *see* 6952
Sunset Yellow FCF, 8983
Supacal [Lederle] *see* 9501
Supanate [Nippon Shinyaku] *see* 4039
Supazlun [Kowa] *see* 4039
Superanabolon *see* 6283
Super-Caid [Lipha] *see* 1371
Super-Cel *see* 4878
Superciclin *see* 8236
Superclastase *see* 632
Superinone [Winthrop] *see* 9741
Superlutin *see* 4763
Superoxol [Merck & Co.] *see* 4727
Superphosphate *see* 1700
Superpyrin *see* 306

Super-Rozol [Lipha] *see* 1371
Supertiamin [Paidostene] *see* 7896
Supicaine Amide Hydrochloride *see* 7762
Supirdyl *see* 6988
Suplexedil [Hépatrol] *see* 3931
Supona [Shell] *see* 2087
Supotran [Winthrop] *see* 2105
Supracapsulin *see* 3569
Suprachol *see* 2858
Supracid [Berlin-Chemie] *see* 8870
Supracide [Ciba-Geigy] *see* 5891
Supracombin [Grünenthal] *see* 8889
Supracort [Samil] *see* 160
Supraene [Robeco] *see* 8727
Supral [Basotherm] *see* 7008
Supralgin *see* 7160
Suprametil [Geistlich] *see* 6028
Supramid [AB Leo] *see* 8885
Supramycin [Grünenthal] *see* 9130
Supranephrane *see* 3569
Supranol [Cilag-Chemie] *see* 8985
Supra-Puren [Klinge] *see* 8721
Suprarenaline *see* 3569
Suprarenin [Winthrop] *see* 3569
Suprasec [J & J] *see* 5450
Suprasterol II, 8984
Supratonin *see* 403
Suprax [Am. Cyanamid] *see* 1927
Suprecur [Hoechst] *see* 1492
Suprefact [Behringwerke] *see* 1492
Supres [Pfizer] *see* 9612
Supressin *see* 8116
Suprifen *see* 4754
Suprilent [Galenica] *see* 5128
Suprim [Valeas] *see* 8889
Supristol [Nordmark] *see* 8897
Suprocil [Janssen] *see* 8985
Suprofen, 8985
Suprol [Cilag-Chemie] *see* 8985
Suprotan *see* 2105
Suractin [Lappe] *see* 621
Suracton [Toho Iyaku] *see* 8721
Sural [Chinoin] *see* 3672
Suralgan [Poli] *see* 9354
Suramin Sodium, 8986
Surbronc [Boehringer, Ing.] *see* 392
Surcopur [Bayer] *see* 7814
Surecide [Sumitomo] *see* 2697
Surem [Galen] *see* 6491
Surem [Ciba] *see* 1501
Surestrine [Roussel-UCLAF] *see* 3407
Surestryl [Roussel-UCLAF] *see* 6203
Surestryl *see* 3407
Surexin [McNeil] *see* 8001
Surfacaine [Lilly] *see* 2741
Surfactal [De Angeli] *see* 392
Surfak [Hoechst] *see* 3397
Surfathesin *see* 2741
Surfen [Hoechst] *see* 489
Surflan [Elanco] *see* 6840
Surfonic N [Jefferson] *see* 6596
Surgam [Hoechst] *see* 9354
Surgestone [Roussel] *see* 7796
Surgical Chlorinated Soda Soln *see* 8578
Surgi-Cen *see* 4602
Surheme [Aron] *see* 1501
Suriclone, 8987
Surika [Thiemann] *see* 4060
Surinam Balsam *see* 959
Surinamine, 8988
Surital [Parke, Davis] *see* 9231
Surmontil (Tabl) [Ives] *see* 9636
Surmontil (Amp) [Ives] *see* 9636
Surnox *see* 4462
Surofene *see* 4602
Surrectan *see* 166
Surrenine *see* 3569
Sursum *see* 4963

Sursumid [Sarm] *see* 8971
Survector [Servier] *see* 420
Suscard [Pharmax] *see* 6528
Suscardia [Pharmax] *see* 5105
Suspen [Circle] *see* 7046
Suspendol [Merckle] *see* 278
Suspren [Nicholas] *see* 4812
Sustac [Rhône-Poulenc] *see* 6528
Sustaire [Roerig] *see* 9212
Sustamycin [MCP] *see* 9130
Sustane [UOP] *see* 1548
Sustane 1-F [UOP] *see* 1547
Sustanon [Organon] *see* 9109
Sustar *see* 4089
Sustonit [Polfa] *see* 6528
Sutan [Stauffer] *see* 1546
Sutoprofen *see* 8985
Suvipen [Midy] *see* 5830
Suvren [Ayerst] *see* 1772
Suxamethonium Bromide *see* 8845
Suxamethonium Chloride *see* 8846
Suxamethonium Iodide *see* 8847
Suxethonium Bromide, 8989
Suxibuzone, 8990
Suxil *see* 8843
Suxilep *see* 3704
Suximal *see* 3704
Suxinutin [Parke, Davis] *see* 3704
Suxizon 300 [Nikken] *see* 4713
Suzutolon [Tatsumi] *see* 1200
S.V.C. *see* 44
SW 77 *see* 7392
Swartziol *see* 5156
Swat [Allied Chemical] *see* 1333
Sweet Birch Oil *see* 6038
Sweet Bugle *see* 5496
Sweet Cane *see* 1637
Sweet Clover *see* 5700
Sweet Elder *see* 8315
Sweet Fennel *see* 3923
Sweet Flag *see* 1637
Sweet Grass *see* 1637
Sweet Oriental Gum *see* 8778
Sweet Root *see* 4400
Sweet Spirit of Niter *see* 8715
Sweet Viburnum *see* 9877
Sweet-wood Bark *see* 1890
Sweroside-2'-(3'',5'',3'''-trihydroxydiphenyl)-2''-carboxylic Acid Ester *see* 384
Swertiamarin, 8991
Swiss Blue *see* 5979
Sybron [Parke, Davis] *see* 3977
Sycotrol [Reed & Carnrick] *see* 7441
SYD-230 *see* 2354
Sydnones, 8992
Sygen [Fidia] *see* 4264
Sykose *see* 8282
Sykoton *see* 5427
Sylvanite *see* 4412
Sylvemid [Weimer] *see* 504
Sylvic Acid *see* 2
Sylvine *see* 7601
Sylvite *see* 7601
Symasul [Lederle] *see* 8919
Symbio [Affiliated Labs.] *see* 8876
Symclosene, 8993
Symcor [Hoechst] *see* 9350
Symetra [Westerfield] *see* 7180
Symmetrel [Endo; Geigy] *see* 380
Sympal *see* 6204
Sympamin *see* 2932
Sympamine *see* 616
Sympatedrine *see* 616
Sympatektoman *see* 9133
Sympathol *see* 8994
Sympatol *see* 8994
m-Sympatol *see* 7257
Sympectothion *see* 3611
Sympocaine [Winthrop] *see* 393
Symptom 1 [Parke, Davis] *see* 8116

Symptom 2 [Parke, Davis] *see* 3561
Symptom 3 [Parke, Davis] *see* 1433
Synacid [Sterivet] *see* 4675
Synacthen [Ciba] *see* 2550
Synadenylic Acid *see* 147
Synadrin [Hoechst] *see* 7744
Synalar [Syntex] *see* 4076
Synalate (rescinded) [Syntex] *see* 4077
Synamol [Syntex] *see* 4076
Synandone [ICI] *see* 4076
Synandrets [Pfizer] *see* 6044
Synandrotabs *see* 6044
Synanthic [Syntex] *see* 6890
Synapause [Organon] *see* 3659
Synapen *see* 7184
Synarel [Syntex] *see* 6265
Synasteron *see* 6920
Synastrin *see* 8692
Synatan [Mallinckrodt] *see* 616
Syncaine *see* 7763
Syncelose *see* 5963
Synchrocept [Syntex] *see* 7895
Synchrocept B [Syntex] *see* 3940
Synchrodyn [Hoechst] *see* 312
Synchronate *see* 4125
Syncillin [Bristol] *see* 7184
Syncl [Toyo Jozo] *see* 1971
Synclotin [Toyo Jozo] *see* 1978
Syncort *see* 2883
Syncorta *see* 2883
Syncortyl [Roussel-UCLAF] *see* 2883
Syncro-Mate [Searle] *see* 4125
Syncuma [Roxane] *see* 9415
Syncurine [Burroughs Wellcome] *see* 2840
Syndeins *see* 8696
Syndrox [McNeil] *see* 5859
Synedil [Beytout] *see* 8971
Synemol [Syntex] *see* 4076
Synephrin [Neukonigsforder] *see* 8994
Synephrine, 8994
Syneptine [Toraude] *see* 5337
Synerpenin *see* 7184
Synestrin (Ampuls) *see* 3119
Synestrin (Tablets) *see* 3118
Synestrol [Schering] *see* 3094
Synflex [Syntex] *see* 6337
Syngacillin [Wyeth] *see* 2706
Syngard [Syntex] *see* 3545
Syngesterone [Pfizer] *see* 7783
Syngestrotabs [Pfizer] *see* 3696
Synhexyl, 8995
Synkamin [Parke, Davis] *see* 9936
Synka-Vit [Roche] *see* 5712
Synkay *see* 5714
Synkayvite *see* 5712
Synklor *see* 2079
Synkonin *see* 4708
Synlaudine *see* 5736
Synnematin B *see* 7043
Synogil [Basotherm] *see* 6346
Synopen [Geigy] *see* 2162
Synotic [Syntex] *see* 4076
Synotodecin *see* 8236
Synpen [Geigy] *see* 2162
Synpenin [Sankyo] *see* 621
Synphase [Syntex] *see* 6614
Synpitan *see* 6934
Synsac [Syntex] *see* 4076
Synstigmin *see* 6380
Syntaris [Syntex] *see* 4071
Syntarpen [Polfa] *see* 3073
Syntaverin [E. Merck] *see* 6011
Syntedril *see* 3308
Syntes 12a *see* 3482
Syntestan [Syntex] *see* 2396
Syntestrine *see* 3119
Syntetrex [Bristol] *see* 8236
Syntetrin [Bristol] *see* 8236
Syntexan *see* 3247
Syntharsol *see* 44

Synthecilline *see* 7184
Synthenate *see* 8994
Synthepen (Tabl) *see* 7184
Synthila *see* 3198
Synthobilin *see* 9415
Synthoestrin *see* 3118
Synthomycetine [Lepetit] *see* 2068
Synthovo *see* 4621
Synthroid Sodium [Flint] *see* 5351
Syntocinon [Sandoz] *see* 6934
Syntometrine *see* 3600
Syntons *see* 7891
Syntopherol *see* 9931
Syntopressin [Sandoz] *see* 5503
Syntrogène *see* 4621
Syntropan [Roche] *see* 624
Synval [Kwizda] *see* 9006
Synvinolin *see* 8491
Syprine [Merck & Co.] *see* 9579
Syraprim [Burroughs Wellcome] *see* 9624
Syringaldehyde, 8996
Syringic Acid δ-Guanidinobutyl Ester *see* 5323
Syringic Aldehyde *see* 8996
Syringidin *see* 5596
Syringin, 8997
Syringopine *see* 8998
Syringoside *see* 8997
Syrosingopine, 8998
Syscor [Bayer] *see* 6482
Systamex [Burroughs Wellcome] *see* 6890
Systhane [Rohm & Haas] *see* 6232
Systox *see* 2875
Systral [Asta] *see* 2182
Sytam [Murphy] *see* 8351
Sytasol [Murphy] *see* 3280
Sytobex [Parke, Davis] *see* 9921
Sytobex-H *see* 4739
Szaibelyite *see* 5535
Szomolnikite *see* 4006
2,4,5-T, 8999
T₄ *see* 2742
T-2 Toxin, 9711
T-3 *see* 5388
T 23P *see* 9665
T 34 *see* 1365
T 47 *see* 1365
T 113 *see* 1528
T 156 *see* 4618
T 1220 *see* 7430
T 1258 *see* A8
T 1384 *see* 6391
T 1551 *see* 1933
T 1824 *see* 3863
T 1982 *see* 1926
T 2525 *see* 1945
T 2588 *see* 1945
T 2636 *see* 8377
T 2636A *see* 8377
T 3033 *see* 7771
T 3262 *see* 9477
TA-1 *see* 6184
TA-058 *see* 874
TA-064 *see* 2878
TA-903 *see* 1046
Tabalgin *see* 40
Tabalon [Hoechst] *see* 4812
Tabasco Pepper *see* 1769
Tabatrex [Glenn] *see* 3023
Tabe [Bernabo] *see* 7463
Tabernanthine, 9000
Tabilin *see* 7041
Table Salt *see* 8544
Table Spate *see* 1709
Tabun, 9001
Tabutrex *see* 3023
Tacaryl [Westwood] *see* 5877
Tace [Merrell] *see* 2173
Tacef [Grünenthal] *see* 1928

Tachicol [Radiumfarm] *see* 2721
Tachionin [Sana] *see* 9537
Tachmalin *see* 184
Tachostyptan [Consolidated] *see* 9321
Tachyrol [Duphar] *see* 3163
Tachysterol, 9002
Tacitin [Ciba] *see* 1096
Tackle [Sankyo] *see* 8432
Tacosal *see* 7293
Tacrine, 9003
Tacryl®, 9004
Tacumil [Promesa] *see* 8889
Tadeonal *see* 7548
Tafil [Upjohn] *see* 310
Tagamet [SK & F] *see* 2279
Tagathen [Lederle] *see* 2168
D-Tagatose, 9005
L-Tagatose *see* 9005
Tag Fungicide *see* 7271
Tag HL-331 *see* 7271
Tagilite *see* 2653
Taglutimide, 9006
TAI-284 *see* 2350
Taicelexin [Taiyo] *see* 1971
Taidecanone [Taiyo] *see* 9751
Taiguic Acid *see* 5235
Tailed Pepper *see* 2615
Taka-amylase [Parke, Davis] *see* 631
Takacillin [Torii] *see* 5319
Taka-Diastase, 9007
Takamina *see* 3569
Takanarumin [Takata] *see* 278
Takaton *see* 3248
Takelan [Takeda] *see* 8377
Takesulin [Takeda] *see* 1943
Takizolite *see* 368
Taktic [Boots] *see* 503
Takus [Montedison] *see* 1998
Taladren [Malesci] *see* 3669
Talampicillin, 9008
Talastine, 9009
Talat [Polifarma] *see* 9008
Talatrol [Abbott] *see* 9684
Talbutal, 9010
Talc, 9011
Talcid [Bayer] *see* 5538
Talcum *see* 9011
Talecid *see* 7350
Taleudron *see* 7351
Talicetimida *see* 7350
Talidine *see* 7351
Talimol *see* 9182
Talin [Tate & Lyle] *see* 9200
Talinolol, 9012
Talis [Kali-Chemie] *see* 5824
Talisomycin *see* 9016
Talisomycin B *see* 9016
Talleol *see* 9013
Tall Oil, 9013
Tallol *see* 9013
Tallow, 9014
Tallow Alcohol, 9015
Tallow Shrub *see* 1022
Tallysomycin, 9016
Tallysomycin A *see* 9016
Tallysomycin B *see* 9016
Talniflumate, 9017
Talodex *see* 3945
Talofen [Pierrel] *see* 7793
Talon [ICI] *see* 1368
Talpen [Beecham] *see* 9008
Talsis [Wyeth] *see* 1310
Talstar [FMC] *see* 1229
Talsutin *see* 7350
Talucard *see* 7889
Talusin [Knoll] *see* 7889
Talwin [Winthrop] *see* 7078
Tamarind, 9018
Tamaron [Bayer] *see* 5858
Tamas *see* 1136
Tamate [Merrell] *see* 5751

Tambocor [Kettelhack] see 4034
Tamed Iodine [West] see 4930
Tametin [Giuliani] see 2279
Tamex [Amchem] see 1532
Tamid [Serpero] see 8967
Tamid see 7350
Tamilan [Gador] see 7021
Tamofen [Tillotts] see 9019
Tamoxasta [Asta] see 9019
Tamoxifen, 9019
Tampilen [Lifasa] see 5830
Tamyl [Italchimici] see 1923
Tanacetin, 9020
Tanaclone [Amsa] see 3453
Tanafol see 2105
Tanakan [I.P.S.E.N.] see 4320
Tanamicin [Ibirn] see 3429
Tanasul see 8888
Tanbismuth see 1302
Tandacote [Geigy] see 6925
Tandearil [Geigy] see 6925
Tandem [Dow] see 9578
Tanderil [Dispersa Baeschlin] see 6925
Tandix [Crinos] see 4847
Tanganil [Specia] see 89
Tangantangan Oil see 1904
Tanghinigenin, 9021
Tanghinin, 9022
Tannalbin [Knoll] see 208
Tannex [Duncan, Flockhart] see 4874
Tannic Acid, 9023
Tannic Acid Bismuth Derivative see 1302
Tannigen see 101
Tannin see 9023
Tannin Albuminate see 208
Tannin-formaldehyde see 9024
Tannoform, 9024
Tannyl Acetate see 101
Tanorama see 3166
Tanphetamin see 616
Tanrutin see 8276
Tanston [Parke, Davis] see 5680
Tantalic Acid Anhydride see 9028
Tantalite see 9025
Tantalum, 9025
Tantalum Pentachloride, 9026
Tantalum Pentafluoride, 9027
Tantalum Pentoxide, 9028
Tan Tone see 3166
Tantum [Angelini Francesco] see 1136
TAO [Roerig] see 9681
Taoryl [Geigy] see 1776
TAP-031 see 4010
TAP-144 see 5341
Tapar [Parke, Davis] see 40
Tapazole [Lilly] see 5892
Tapopote see 3560
Taquidil [Roemmers] see 9414
Tar Acids, 9029
Taractan [Roche] see 2189
Tarasan [Roche] see 2189
Tarasina [Best] see 9432
Taraxacum, 9030
Taraxast-20(30)-en-3β-ol see 9031
Taraxasterin see 9031
Taraxasterol, 9031
Taraxein, 9032
Taraxerol, 9033
Tar Camphor see 6289
Tardamide see 8897
Tardastrex [Gruntex] see 2865
Tardigal [Beiersdorf] see 3146
Tardisal [Sigma-Tau] see 8302
Tardocillin see 7037
Targa [Nissan] see 8113
Targocid [Merrell Dow] see 9062
Targosid [Lepetit] see 9062
Tarichatoxin see 9175
Tarivid [Hoechst] see 6688
Taroctyl see 2186

Tarodyl see 4397
Tarodyn see 4397
Tar Oil, 9034
Tar Oil, Rectified, 9035
Tarpan [Wander] see 2358
Tarquinor [Squibb] see 4520
Tarragon, 9036
Tarragon Oil (Estragon Oil) see 3657
Tartar Emetic see 732
(+)-Tartaric Acid see 9039
(−)-Tartaric Acid see 9037
d-Tartaric Acid see 9039
dl-Tartaric Acid see 9038
D-Tartaric Acid, 9037
DL-Tartaric Acid, 9038
l-Tartaric Acid see 9037
L-Tartaric Acid, 9039
meso-Tartaric Acid, 9040
Tartaric Acid Bismuth Complex Potassium Salt see 1289
Tartaric Acid Bismuth Complex Sodium Salt see 1292
L-Tartaric Acid Monoammonium Salt see 523
Tartarized Antimony see 732
Tartarlithine see 5400
Tartarus Vitriolatus see 7665
Tartrated Antimony see 732
Tartrazine, 9041
Tartrol see 1292
Tartronic Acid, 9042
Tarugan [Ravensberg] see 6176
Taselin [Eisai] see 6673
Tasis [Wyeth] see 1310
Task [Shell] see 3069
Tasmaderm [Roche] see 6199
Tasnon (Tabl.) see 7431
Tasnon (Elixir) see 7431
Tasteless Quinine see 8080
Taste-modifying Protein see 6125
TAT-3 [Takeda] see 7376
TATBA see 9514
TATD see 6680
Tathiclon [Ohta] see 4369
Tathion [Yamanouchi] see 4369
Tauliz [Hoechst] see 7464
Taumidrine [Knoll] see 966
Taural [Roemmers] see 8126
Taure(o)don [Byk-Gulden] see 4422
Taurine, 9043
Tauriscite see 4006
Taurocholic Acid, 9044
Tavegil [Sandoz] see 2345
Tavegyl [Sandoz] see 2345
Ta-Verm see 7433
Tavist [Sandoz] see 2345
Tavor [Wyeth] see 5456
Taxicatin, 9045
Taxicin-I see 9046
Taxicin-II see 9046
Taxicins, 9046
Taxilan see 7108
Taxine A see 9047
Taxine B see 9047
Taxine-I see 9047
Taxine-II see 9047
Taxine(s), 9047
Taxinine see 9046
Taxodione, 9048
Taxofit [Anasco] see 9932
Taxol, 9049
Taxol A see 9049
Taxusin see 9049
Tazepam [Polfa] see 6881
Tazettine, 9050
Tazicef [SK&F] see 1944
Tazidime [Lilly] see 1944
Tazin [Grelan] see 1791
Taziprinone Hydrochloride, 9051
TB-1 see 9727
TB-ACA see 1757

TBI see 9513
Tb I/698 see 9218
TBP see 1316
TBS see 1585, 4163 and 9529
TB-Vis see 5071
TBZ [Merck & Co.] see 9217
2,4,5-TC see 8483
TC-80 see 4961
TC 109 see 9079
Tc 924 (DPD) see 1512
TCA see 1640 and 9539
TCC [Monsanto] see 9568
TCDBD see 9052
TCDD, 9052
T-Cell Growth Factor see 4896
T-Cell Replacing Factor see 4895
TCGF see 4896
Tc Gluceptate see 4349
Tc Gluheptonate see 4349
TCl see 2727
TCNE see 9129
TCP see 9675
TCPP see 4233
TCT see 1640
TCTH see 2767
TDB see 1296
cis-9,trans-11-TDDA see 7777
TDE see 3049 and 3837
p,p'-TDE see 3049
T-DET-N [Thompson-Hayward] see 6596
TDHL see 9095
TDI see 9456 and 9505
TE-031 see 2340
TEAB see 9133
Teaberry Oil see 6038
T.E.A. Chloride see 9134
Teamsters' Tea see 3560
Tearisol see 4777 and 5963
Tebalon see 9218
Tebamin see 7243
Tebanyl see 7243
Tebe see 7901
Tebecid see 5071
Tebethion see 9218
Tebezon see 9218
Tebloc [Dukron] see 5450
Tebonin [Schwabe] see 4320
Tebrazid see 7970
Tebron see 4515
Tebutate see MISC-3 and 4
Tebuthiuron, 9053
Tecacin [Dukron] see 3429
Tecaf [Farmochim. Ital.] see 8408
Técarine see 9210
Teceos [Hoechst Roussel] see 1512
Tecesal see 1719
TechneScan Gluceptate [Mallinckrodt] see 4349
TechneScan PYP [Mallinckrodt] see 8748
Technetium, 9054
Teclosan see 9055
Teclosine see 9055
Teclothiazide see 9124
Teclozan, 9055
Teclozine see 9055
Tecodin see 6912
Tecomanine, 9056
Tecomin see 5235
Tecomine see 9056
Tecquinol see 4738
Tecramine see 4060
Tecto [Merck & Co.] see 9217
Tectochrysin see 2261
Tectoridin see 9057
Tectorigenin, 9057
TECZA see 9579
Tedelparin, 9058
Tedion see 9132
Tedion-V₁₈ see 9132
Tedion-V$_{18}$ see 9132

TEDP see 8945
Teel Oil see 8418
Teepol see 4266
Tefamin see 477
Tefilin [Hermal] see 9130
Teflon see 7560
Teflurane, 9059
Tefsiel C [Towa Yakuhin] see 9060
Tega-Cetin [Ortega] see 2068
Tegafur, 9060
Tega-Pyrone [Ortega] see 3358
Tegencia [Elmu] see 3942
Tegison [Roche] see 3846
Tegopen [Bristol] see 2414
Tegosept B see 1583
Tegosept M see 6021
Tegotin see 9941
Tegretal [Geigy] see 1783
Tegretol [Geigy] see 1783
TEI-5103 see 8249
Teichmann's Crystals see 4563
Teichoic Acids, 9061
Teichomycin A₂ see 9062
Teicoplanin, 9062
Teicoplanin A₂ see 9062
Teicoplanin A₂-1 see 9062
Teicoplanin A₂-2 see 9062
Teicoplanin A₂-3 see 9062
Teicoplanin A₂-4 see 9062
Teicoplanin A₂-5 see 9062
Tejuntivo [Valderrama] see 6857
Tekazin see 5071
Teknar [Sandoz] see 945
Tekodin see 6912
Tektamer 38 [Calgon] see 3004
Tektin A see 8696
TEL see 9136
Telar [DuPont] see 2192
Teldane [Merrell] see 9094
Teldanex [Draco] see 9094
Teldrin [SK & F] see 2180
Telebar [Guerbet] see 1006
Telémid [Nativelle] see 8888
Telemin [Funai] see 1253
Telen [Byk Gulden] see 1296
Telepaque [Winthrop] see 4944
Telepathine see 4531
Teleprim [Falorni] see 8889
Telesmin [Yoshitomi] see 1783
Teletrast [Astra] see 4944
Telgin-G [Takeda] see 2345
Teli see 8341
Teline [Winston] see 9130
Telluric(IV) Acid see 9072
Telluric(VI) Acid, 9063
Telluric Bromide see 9069
Telluric Chloride see 9070
Tellurite see 9067
Tellurium, 9064
Tellurium Dibromide, 9065
Tellurium Dichloride, 9066
Tellurium Dioxide, 9067
Tellurium Hexafluoride, 9068
Tellurium Tetrabromide, 9069
Tellurium Tetrachloride, 9070
Tellurium Tetraiodide, 9071
Tellurobismuthite see 1303
Tellurous Acid, 9072
Tellurous Bromide see 9065
Tellurous Chloride see 9066
Telmicid [Lilly] see 3376
Telmid [Lilly] see 3376
Telmin [Janssen] see 5647
Telodrin see 5010
Telomycin, 9073
Telon [Sandoz] see 1131
Telone [Dow] see 3064
Telone II [Dow] see 3064
Telopar [Pfizer] see 6876
Telotrex see 9130
Telvar [Du Pont] see 6169

Telvar see 3388
TEM see 9586
Temafloxacin, A8
Temaril [SK & F] see 9618
Temasept IV [Fine Organics] see 9529
Temazepam, 9074
Tementil see 7768
Temephos, 9075
Temesta [Wyeth] see 5456
Temetex [Sauter] see 3129
Temgesic [Reckitt & Colman] see 1487
Temik [Union Carbide] see 216
Temina see 9941
Temlo [Hyrex-Key] see 40
Temocillin, 9076
Temopen [Beecham] see 9076
Temovate [Glaxo] see 2361
Temp [Table Rock] see 3358
Temporinolo [Lepetit] see 7279
Temposil [Lederle] see 1663
Tempra [Mead Johnson] see 40
Temserin [Sharp & Dohme] see 9374
Temur see 9160
Tenacid [Sigma-Tau] see 4834
Tenalet [Otsuka] see 1875
Tenalin see 5871
Tenalin [Otsuka] see 1875
Tenathan [Robins] see 1208
Tendor [Chinoin] see 2837
Tenebrimycin see 9413
Tenebryl [Guerbet] see 3205
Tenelid [Helsinn] see 4469
Tenelon see 5427
Tenemycin see 9413
Tenesdol [Hishiyama] see 3311
Tenex [Robins] see 4474
Tenfidil see 9207
Teniathane [Pitman-Moore] see 3059
Teniatiol [Pitman-Moore] see 3059
Tenicid see 4919
Tenidap, A9
Teniloxazine, 9077
Teniposide, 9078
Tennecetin see 6346
Tenntuss see 4465
Tenoate see MISC-4
Tenoban see 3434
Tenoblock [Oy Star] see 879
Tenonitrozole, 9079
Tenopt [Sigma] see 9374
Tenorite see 2650
Tenormal [ICI] see 7022
Tenormin [ICI] see 879
Tenox BHA [Eastman-Kodak] see 1547
Tenox BHT [Eastman-Kodak] see 1548
Tenoxicam, 9080
Tenox PG see 7872
Tensatrin see 7915
Tensibar [Clin-Comar-Byla] see 1226
Tensicor see 3182
Tensilon see 3487
Tensinase D [Nippon Chemiphar] see 3820
Tensinol [Farmaroma] see 7022
Tensinyl see 2082
Tensobon [Melusin Schwarz] see 1773
Tensofin see 4116
Tensoprel [Rubio] see 1773
Tensoral [Diana] see 7022
Tenso-Timelets see 2388
Tenstaten [Ipsen-Beaufour] see 2268
Tentone see 5923
Tenuate [Merrell] see 3116
Tenuate Dospan [Merrell] see 3116
Tenuazonic Acid, 9081
Tenyl [Nippon Kayaku] see 1200
Teobid [Farma] see 9212
Teoclate see MISC-4

Teofilcolina see 2213
Teokolin see 2213
Teonicon [Bracco] see 7400
Teoptic [Dispersa] see 1875
Teoquil see 4539
Teorema [Panchemie Homburg] see 4342
Teoremac see 4342
TEPA see 9587
Tepanil [Riker] see 3116
Tephroite see 5623
Tephrosin, 9082
Tephthol see 9093
Tepilta see 6888
Tepirubicin see 7460
Tepogen see 2414
TEPP see 9138
Teprenone, 9083
Teprin [Endo] see 2360
Teprosilate see MISC-4
TER-1546 see 8966
Teralutil see 4774
Teramine [Darby] see 2183
Terazol [Cilag] see 9090
Terazosin, 9084
Terbacil, 9085
Terbasmin [Farmitalia] see 9089
Terbenol [Bernabo] see 6638
Terbia see 9087
Terbinafine, 9086
Terbium, 9087
Terbufos, 9088
Terbutaline, 9089
Terbutaline Bisdimethylcarbamate see 963
Tercian [Théraplix] see 2690
Terconazole, 9090
Tercospor [Ortho] see 9090
Terdin [Cheil Sugar] see 9094
Terebene, 9091
Terebic Acid, 9092
Terebinic Acid see 9092
Terenac [Sandoz] see 5643
Terenol [Hoechst] see 8157
Terephthalic Acid, 9093
Terfen [Siam Bheasach] see 9094
Terfenadine, 9094
Terflurane see 9059
Terfluzine [Théraplix] see 9594
Tergal [Rhodia] see 7546
Tergavon see 4266
Tergitol [Union Carbide] see 7554
Tergitol 4 see 8645
Tergitol NP [Union Carbide] see 6596
Tergitol TP-9 [Union Carbide] see 6596
Terguride, 9095
Teriam see 9515
Teridax [Schering] see 4947
Terimon [Norma] see 9019
Terion [Lusofarmaco] see 4144
Teriosal [Dompé] see 8167
Teriparatide Acetate, 9096
Terit [Hoechst] see 2270
Terital [Montecatini] see 7546
Terlenka [Kunstzijde] see 7546
Terlipressin, 9097
Termierite see 368
Termil [Diamond] see 2167
Termitin see 9941
Ternadin [Cantabria] see 9094
Ternelin [Sandoz] see 9409
Terodiline, 9098
Terodin [Cantabria] see 9094
Terofenamate, 9099
Teronac [Wander] see 5643
Terondit [Hoechst] see 6804
Teropterin see 7950
Terpanil [Toho Iyaku] see 4273
Terpate [Geneva] see 7066
"Terpene" Hydrochloride see 1341
Terpene Polychlorinates see 8796

Terpenolic Acid *see* 9100
Terpenylic Acid, 9100
Terpin, 9101
Terpinene, 9102
α-Terpineol, 9103
Terpin Hydrate *see* 9101
Terpinol *see* 9101
Terpinyl Thiocyanoacetate *see* 5012
Terposen [Vir] *see* 8126
3,2′:4′,3″-Terpyridine *see* 6432
Terra Alba *see* 1713
Terrachlor *see* 8108
Terraclor *see* 8108
Terracur P *see* 3943
Terrafungine *see* 6931
Terra Japonica *see* 4259
Terraject [Pfizer] *see* 6931
Terramycin [Pfizer] *see* 6931
Terrasytam [Murphy] *see* 3191
Terratrex [Volpino] *see* 4468
Terreic Acid, 9104
Tersan [Du Pont] *see* 9304
Tersavid *see* 7483
Tersigat [Boehringer, Ing.] *see* 6897
Tertatolol, 9105
α-Terthienyl, 9106
2,2′:5′2″-Terthiophene *see* 9106
Tertiary Calcium Phosphate *see* 1701
Tertiary Magnesium Phosphate *see* 5562
Tertiomycin A *see* 9107
Tertiomycin B *see* 9107
Tertiomycins, 9107
Tertran [Wyeth] *see* 4962
Tertroxin [Glaxo] *see* 5388
Terylene [ICI] *see* 7546
Teslac [Squibb] *see* 9108
Teslen *see* 9109
Tesnol [Kowa Yakuko] *see* 7852
Tesona [Phoenix] *see* 9212
Tesoprel [Organon] *see* 1432
Tespamin *see* 9588
Tessalon [Ciba-Geigy] *see* 1106
Tessalon-Ciba *see* 1106
Testaform *see* 9115
Testamin [Toyama] *see* 8116
Testarpen [Polfa] *see* 7050
Testascorbic *see* 855
Testate [Savage] *see* 9112
Testavol *see* 9918
Testergon *see* 9111
Testhormona *see* 6044
Testinon [Mochida] *see* 9112
Testobase [Merck & Co.] *see* 9109
Testodet [Merck & Co.] *see* 9115
Testodrin *see* 9115
Testodrin Prolongatum *see* 9111
Testoenant *see* 9112
Testogen *see* 9115
Testolactone, 9108
Δ¹-Testolactone *see* 9108
Testonique *see* 9115
Testoral *see* 4113
Testormol *see* 9115
Testosid *see* 9115
Testosid (Tabl) *see* 6044
Testosterone, 9109
trans-Testosterone *see* 9109
Testosterone 17-Chloral Hemiacetal, 9110
Testosterone 17β-Cyclopentanepropionate *see* 9111
Testosterone Cyclopentylpropionate *see* 9111
Testosterone 17β-Cypionate, 9111
Testosterone Enanthate, 9112
Testosterone 17-Hemiacetal with Chloral *see* 9110
Testosterone Heptoate *see* 9112
Testosterone Ketolaurate *see* 9109
Testosterone Nicotinate, 9113

Testosterone 17-Nicotinic Acid Ester *see* 9113
Testosterone Oenanthate *see* 9112
Testosterone Phenylacetate, 9114
Testosterone Propionate, 9115
Testoviron (Tabl) *see* 6044
Testoviron (Amps.) *see* 9115
Testoxyl *see* 9115
Testred [ICN] *see* 6044
Testrex *see* 9115
Testriol *see* 2049
Testrone [SM & P] *see* 9109
Testroval [Tutag] *see* 9112
Testryl [Squibb] *see* 9109
TETA *see* 9579
Tetaine *see* 946
Tetiothalein Sodium *see* 4931
Tetmosol [Care] *see* 8928
Tetra-D *see* 9130
2,3,4,6-Tetraacetyl-α-D-glucopyranosyl Bromide *see* 52
Tetra-*O*-acetylpentaerythritol *see* 7065
(2,3,4,6-Tetra-*O*-acetyl-1-thio-β-D-glucopyranosato-*S*)(triethylphosphine)gold *see* 895
Tetraamminecopper Dihydroxide *see* 8352
Tetraamminecopper Sulfate, 9116
Tetraaquochromium Dichloride *see* 2246
1,3,5,7-Tetraazaadamantane *see* 5879
1,4,7,10-Tetraazadecane *see* 9579
1,3,5,8-Tetraazanaphthalene *see* 7947
1,3,5,7-Tetraazatricyclo[3.3.1.1³,⁷]-decane *see* 5879
Tetrabakat [Dorsch] *see* 9130
Tetrabarbital, 9117
Tetra-base *see* 6099
Tetrabenazine, 9118
Tetrabid [Organon] *see* 9130
Tetrabiguanide *see* 4468
Tetrablet [Makara] *see* 9130
Tetrabon [Pfizer] *see* 9130
Tetraborane(10), 9119
Tetraboron Decahydride *see* 9119
Tetrabrom *see* 9122
3,3′,5,5′-Tetrabromo-(1,1′-biphenyl)-2,2′-diol Mono(dihydrogen Phosphate) *see* 1406
3,3′,5,5′-Tetrabromo-2,2′-biphenyldiol Mono(dihydrogen Phosphate) *see* 1406
4,4′,6,6′-Tetrabromobiphenyl-2,2′-diol Mono(dihydrogen Phosphate) *see* 1406
3,4,5,6-Tetrabromo-*o*-cresol, 9120
3,3′,5,5′-Tetrabromo-*m*-cresolsulfonphthalein *see* 1375
2′,4′,5′,7′-Tetrabromo-3′,6′-dihydroxyspiro[isobenzofuran-1(3*H*),-9′-[9*H*]xanthen]-3-one Disodium Salt *see* 3555
sym-Tetrabromoethane, 9121
1,1,2,2-Tetrabromoethane *see* 9121
2′,4′,5′,7′-Tetrabromofluorescein *see* 3555
Tetrabromofluorescein Sol *see* 3555
4,5,6,7-Tetrabromo-2-hydroxy-1,3,2-benzodioxabismole *see* 1220
3,3′-(4,5,6,7-Tetrabromo-3-oxo-1(3*H*)-isobenzofuranylidene)bis[6-hydroxybenzenesulfonic Acid] Disodium Salt *see* 8933
3′,3″,5′,5″-Tetrabromophenolphthalein, 9122
Tetrabromophenolphthalein Sodium *see* 9122
3,3′,5,5′-Tetrabromophenolsulfonphthalein *see* 1434

Tetrabromopyrocatechol Bismuth Derivative *see* 1220
Tetracaine Hydrochloride, 9123
Tetracap *see* 9126
Tetracarbonyldi-μ-chlorodirhodium *see* 8188
Tetracarbonylhydridocobalt *see* 3074
Tetracarbonylhydrocobalt *see* 3074
Tetracemate Disodium *see* 3481
Tetracemate Tetrasodium *see* 3482
Tetracemin *see* 3482
Tetracene *see* 6288
Tetrachel [Rachelle] *see* 9130
Tetrachlorethylene *see* 9126
Tetrachlormethiazide, 9124
2,4,5,6-Tetrachloro-1,3-benzenedicarbonitrile *see* 2167
2,3,5,6-Tetrachloro-1,4-benzenedicarboxylic Acid Dimethyl Ester *see* 2830
2,3,5,6-Tetrachloro-*p*-benzoquinone *see* 2071
4,5,6,7-Tetrachloro-3,3-bis(4-hydroxyphenyl)-1(3*H*)-isobenzofuranone *see* 7214
2,3,5,6-Tetrachloro-2,5-cyclohexadiene-1,4-dione *see* 2071
2,3,6,7-Tetrachlorodibenzodioxin *see* 9052
2,3,7,8-Tetrachlorodibenzo[*b,e*][1,4]-dioxin *see* 9052
2,3,7,8-Tetrachlorodibenzo-*p*-dioxin *see* 9052
Tetrachlorodiborane(4) *see* 2998
2,4,5,6-Tetrachloro-1,3-dicyanobenzene *see* 2167
4,5,6,7-Tetrachloro-2,3-dihydro-2-methyl-2-[2-(trimethylammonio)-ethyl]-2*H*-isoindolium Dichloride *see* 2101
4,5,6,7-Tetrachloro-3′,6′-dihydroxy-2′,4′,5′,7′-tetraiodospiro[isobenzofuran-1(3*H*),9′-[9*H*]xanthen]-3-one Dipotassium Salt *see* 8242
4,5,6,7-Tetrachloro-2-(2-dimethylaminoethyl)isoindoline Dimethylchloride *see* 2101
4,5,6,7-Tetrachloro-2-(2-dimethylaminoethyl)-2-methylisoindolinium Chloride Methochloride *see* 2101
Tetrachlorodiphenylethane *see* 3049
2,4,5,4′-Tetrachlorodiphenyl Sulfone *see* 9132
Tetrachloroethane, 9125
sym-Tetrachloroethane *see* 9125
1,1,2,2-Tetrachloroethane *see* 9125
Tetrachloroethene *see* 9126
Tetrachloroethylene, 9126
N-(1,1,2,2-Tetrachloroethylmercapto)-4-cyclohexene-1,2-dicarboximide *see* 1770
N-(1,1,2,2-Tetrachloroethylsulfenyl)-*cis*-4-cyclohexene-1,2-dicarboximide *see* 1770
N-(1,1,2,2-Tetrachloroethylthio)-4-cyclohexene-1,2-dicarboximide *see* 1770
N-(1,1,2,2-Tetrachloroethylthio)-Δ⁴-tetrahydrophthalimide *see* 1770
2,4,7,9-Tetrachloro-3-hydroxy-8-methoxy-1,6-dimethyl-11*H*-dibenzo[*b,e*][1,4]dioxepin-11-one *see* 3345
Tetrachloroisophthalonitrile *see* 2167
Tetrachloromethane *see* 1822
2,3,5,6-Tetrachloro-4-methoxyphenol *see* 3441
3,4,5,6-Tetrachlorophenolphthalein *see* 7214
m-Tetrachlorophthalodinitrile *see* 2167

Cross Index of Names

Tetrasodium 3,3'-[(3,3'-Dimethyl-4,4'-biphenylene)bis(azo)]bis(5-amino-4-hydroxy-2,7-naphthalenedisulfonate) *see* 9701
Tetrasodium Edetate *see* 3482
Tetrasodium Ethylenebis(iminodiacetate) *see* 3482
Tetrasodium Ethylenediaminetetraacetate *see* 3482
Tetrasodium Hexakis(cyano-*C*)ferrate(4−) *see* 8563
Tetrasodium 2-Methyl-1,4-naphthohydroquinone Diphosphoric Acid Ester *see* 5712
Tetrasodium Pyrophosphate, 9170
Tetrasulfur Tetranitride, 9171
Tetra-Tablinen [Sanorania] *see* 6931
Tetrathiin *see* 6893
Tetrathione *see* 8647
Tetraverin *see* 8236
Tetra-Wedel [Stroschein] *see* 9130
7,8,9,10-Tetrazabicyclo[5.3.0]-8,10-decadiene *see* 7097
1,2,3,3a-Tetrazacyclohepta-8a,2-cyclopentadiene *see* 7097
Tetrazepam, 9172
Tetrazobenzene-*β*-naphthol *see* 8858
Tetrazolium Blue, 9173
7-(1-(1*H*)-Tetrazolylacetamido)-3-[2-(5-methyl-1,3,4-thiadiazolyl)thiomethyl]-Δ³-cephem-4-carboxylic Acid *see* 1925
7-(1*H*-Tetrazol-1-yl)acetamido-3-(1,3,4-thiadiazol-2-ylthio)methyl-ceph-3-em-4-carboxylic Acid *see* 1946
Tetrex [Bristol] *see* 9130
Tetrex PMT *see* 8236
Tetridin *see* 8005
Tetrim [Bristol] *see* 8236
Tetrin, 9174
Tetrin A *see* 9174
Tetrin B *see* 9174
Tetrine *see* 3482
Tetriv [Bristol] *see* 8236
Tetrodontoxin *see* 9175
Tetrodotoxin, 9175
Tetrole *see* 4206
Tetron *see* 9138
Tetronasin, 9176
Tetropil *see* 9126
Tetroquinone, 9177
Tetrosan *see* 3042
Tetrosol *see* 9130
Tetroxoprim, 9178
L-*glycero*-Tetrulose *see* 3641
Tetryl *see* 6488
Tetryl Formate *see* 5026
Tetryzoline *see* 9150
Tetterwort *see* 8319
Tetucur [Teikoku Zoki] *see* 4006
Teturamin *see* 3370
Tevabolin *see* 8754
Tevcocin [Tevcon] *see* 2068
Tevcodyne [Tevcon] *see* 7248
Tevenel®, 9179
Texacort [Texas Pharmacal] *see* 4710
Texmeten [Nippon Roche] *see* 3129
Texofor A1P *see* 7554
Texofor B9 [Glovers] *see* 7532
Texsolve S [Texaco] *see* 6120
Texsolve S-2 [Texaco] *see* 6120
TF 1169 *see* 4049
TFM, 9180
TFT Thilo [Thilo] *see* 9599
TG-51 *see* 8249
T/Gel [Farillon] *see* 2420
TGS *see* 8854
TH 100 [IPG] *see* 477
Th 152 *see* 5836

Th 1064 *see* 3257
Th 1165 *see* 3927
Th 1165a *see* 3927
TH 1395 *see* 4374
TH 2602 *see* 2690
TH 4128 *see* 7249
TH 6040 *see* 3128
THA [Woods] *see* 9003
Thacapzol *see* 5892
Thalamonal *see* 3437
Thalamyd [Schering] *see* 7350
Thalazole *see* 7351
Thalicarpine, 9181
Thalictrine *see* 5579
Thalidomide, 9182
Thalistatyl *see* 7351
Thalitone [Boehringer, Ing.] *see* 2194
Thallic Fluoride *see* 9199
Thallic Oxide *see* 9196
Thallium, 9183
Thallium Acetate, 9184
Thallium Bromide, 9185
Thallium Carbonate, 9186
Thallium Chloride, 9187
Thallium Cyanide, 9188
Thallium Fluoride, 9189
Thallium Hydroxide, 9190
Thallium Iodide, 9191
Thallium Nitrate, 9192
Thallium Oxide, 9193
Thallium Peroxide *see* 9196
Thallium Selenate, 9194
Thallium Selenide, 9195
Thallium Sesquioxide, 9196
Thallium Sulfate, 9197
Thallium Sulfide, 9198
Thallium Trifluoride, 9199
Thallous Acetate *see* 9184
Thallous Bromide *see* 9185
Thallous Carbonate *see* 9186
Thallous Chloride *see* 9187
Thallous Cyanide *see* 9188
Thallous Fluoride *see* 9189
Thallous Hydroxide *see* 9190
Thallous Iodide *see* 9191
Thallous Nitrate *see* 9192
Thallous Oxide *see* 9193
Thallous Selenate *see* 9194
Thallous Selenide *see* 9195
Thallous Sulfate *see* 9197
Thallous Sulfide *see* 9198
Thalocid *see* 7350
THAM *see* 9684
Thanite *see* 5012
Thaumatin, 9200
Δ¹-THC *see* 9142
Δ⁶-THC *see* 9142
Δ⁸-THC *see* 9142
Δ⁹-THC *see* 9142
THE *see* 9143
Theaflavine, 9201
Theal Tabl [Boehringer, Mann.] *see* 9212
Theal Ampules [Boehringer, Mann.] *see* 3459
Theamin *see* 9212
Thean [Astra] *see* 7921
Thebacon *see* 3155
Thebaine, 9202
Thebainone, 9203
Thebainone-A *see* 9203
Thecodine *see* 6912
Theelin [Parke, Davis] *see* 3660
Theelol [Parke, Davis] *see* 3659
Thefylan *see* 3459
Thein *see* 1635
Theiomycetin *see* 1176
Thekodin *see* 6912
Thelestrin [Carnrick] *see* 3660
Thelmesan *see* 3455
Thelykinin *see* 3660

Themalon *see* 9220
Themisone [Mallinckrodt] *see* 888
Thenaldine, 9204
Thenalidine *see* 9204
Thenardite *see* 8636
Thenclor *see* 2168
Thenfadil [Winthrop] *see* 9207
Thenitrazole *see* 9079
Thenium Closylate, 9205
2-Thenoic Acid *see* 9284
3-Thenoic Acid, 9206
Thenophenopiperidine *see* 9204
2-(*α*-Thenoylamino)-5-nitrothiazole *see* 9079
p-(2-Thenoyl)hydratropic Acid *see* 8985
2-(*α*-Thenoylthio)propionylglycine *see* 8764
Thenyldiamine, 9207
Thenylene [Abbott] *see* 5871
Thenylpyramine *see* 5871
Theobroma Oil, 9208
Theobromine, 9209
1-Theobromineacetic Acid, 9210
Theocal *see* 9209
Theocalcin [Knoll] *see* 9209
Theocap [Meyer] *see* 9212
TheoChron [Forest] *see* 9212
Theocin *see* 9212
Theocin Soluble *see* 9212
Theoclear [Central Pharm.] *see* 9212
Theocontin [Napp] *see* 9212
Theocor *see* 7901
Theodrox *see* 477
Theo-Dur [Key] *see* 9212
Theofibrate, 9211
Theograd [Abbott] *see* 9212
Theolair [Riker] *see* 9212
Theolamine [Merck & Co.] *see* 477
Theolan [Elan] *see* 9212
Theolix *see* 9212
Theomin [IPG] *see* 477
Theon *see* 7921
Theophyldine *see* 477
Theophyllamine *see* 477
Theophylline, 9212
7-Theophyllineacetic Acid *see* 18
7-Theophyllineacetic Acid Piperazine Salt *see* 19
Theophylline Aminoisobutanol *see* 395
Theophylline Cholinate *see* 2213
Theophylline Compd with 2-Amino-2-methyl-1-propanol *see* 395
Theophylline Compd with Ethylenediamine *see* 477
Theophylline Compd with Glyceryl Guaiacolate *see* 4467
Theophylline Compd with 3-(*o*-Methoxyphenoxy)-1,2-propanediol *see* 4467
Theophyllineethylamphetamine *see* 3919
Theophylline Ethylenediamine *see* 477
Theophyllineethyl Sulfate of Pyridoxine *see* 7988
Theophylline Methesculetol *see* 5844
2-(7'-Theophyllinemethyl)-1,3-dioxolane *see* 3427
Theophylline Salt of Choline *see* 2213
1-(Theophyllin-7-yl)ethyl 2-(*p*-Chlorophenoxy)isobutyrate *see* 9211
Theophyl-SR [Knoll] *see* 9212
Theoplus [Fabre] *see* 9212
Theopropanol *see* 9212
Theosalin *see* 9209
Theosol [Farillon] *see* 9212
Theosol [Knoll] *see* 9209
Theostat [Laser] *see* 9212
Theovent [Schering] *see* 9212
Theoxylline *see* 2213
Thephorin [Roche] *see* 7195

Thiocarbonic Acid O-Ethyl Ester, S-Ester with N-[(4-Amino-2-methyl-5-pyrimidinyl)methyl]-N-(4-hydroxy-2-mercapto-1-methyl-1-butenyl)formamide Ethyl Carbonate (Ester) see 2014
Thiocarlide see 9382
Thiochrysine see 4423
Thiocol [Roche] see 7617
Thiocolchicine, 9253
Thiocolchicoside see 9253
Thiocresol, 9254
Thioctacid [Homburg] see 9255
Thioctan [Katwijk] see 9255
Thioctic Acid, 9255
6,8-Thioctic Acid see 9255
Thiocuran [Sagitta] see 8889
Thiocyanate Sodium, 9256
Thiocyanatoacetic Acid 1,7,7-Trimethylbicyclo[2.2.1]hept-2-yl Ester see 5012
Thiocyanic Acid, 9257
Thiocyanic Acid Ammonium Salt see 591
Thiocyanic Acid 2-(2-Butoxyethoxy)-ethyl Ester see 5329
2-Thiocyanoethyl Dodecanoate see 5328
2-Thiocyanoethyl Laurate see 5328
Thiocyl see 9291
Thiocymetin [Winthrop] see 9230
Thiodan [Hoechst] see 3529
Thiodelone [Shionogi] see 5747
Thiodemeton see 3371
Thioderon [Shionogi] see 5747
3,3'-Thiodialanine see 5234
Thiodicarb, 9258
2,2'-Thiodiethanol, 9259
Thiodiethylene Glycol see 9259
Thiodiglycol see 9259
Thiodiglycolic Acid, 9260
Thiodihydracrylic Acid see 9261
2-Thio-3,5-dimethyltetrahydro-1,3,5-thiadiazine see 2827
Thiodiphenylamine see 7220
O,O'-(Thiodi-4,1-phenylene)bis(O,O'-dimethylphosphorothioate) see 9075
O,O'-(Thiodi-4,1-phenylene)phosphorothioic Acid O,O,O',O'-Tetramethyl Ester see 9075
Thiodiphosphoric Acid Tetraethyl Ester see 8945
3,3'-Thiodipropionic Acid, 9261
β,β-Thiodipropionic Acid see 9261
2-Thio-1,3-dithiolo[4,5-b]quinoxaline see 9288
Thiodril [Berk] see 1809
Thiodrol [Shionogi] see 3575
Thioethanolamine see 2785
Thioethyl Alcohol see 3680
Thioethyl Ether see 3809
Thiofeed see 9281
Thiofide see 3377
Thioformamide, 9262
Thiofuran see 9283
Thiofurfuran see 9283
Thiogenal see 5901
1-Thio-D-glucitol see 9294
(1-Thio-β-D-glucopyranase-2,3,4,6-tetraacetato-S)(triethylphosphine)-gold see 895
(1-Thio-D-glucopyranosato)gold see 901
(1-Thio-β-D-glucopyranosato)(triethylphosphine)gold 2,3,4,6-Tetracetate see 895
5-Thio-α-D-glucopyranose see 9263
1-Thio-β-D-glucopyranosamine 1-[N-(Sulfooxy)-3-butenimidate] Monopotassium Salt see 8495
5-Thio-D-glucose, 9263

Thioglycerin see 9264
Thioglycerol, 9264
Thioglycol see 5760
Thioglycolic Acid, 9265
Thioglycollic Acid see 9265
Thioglycollic-β-aminonaphthalide see 9273
Thioguanine, 9266
Thioguanosine, 9267
Thioimidodicarbonic Diamide see 3378
2-Thio-4-ketothiazolidine see 8184
Thiokol®, 9268
Thiokol A see 9268
Thiokol FA see 9268
Thiola [Santen] see 9386
Thiolacetic Acid see 9247
Thiolactic Acid, 9269
Thiole see 9283
2-Thiolhistidine, 9270
Thiolhistidine-betaine see 3611
Thiolin see 4815
2-Thiolpropionic Acid see 9269
Thiolutin, 9271
Thiomalic Acid, 9272
Thiomebumal Sodium see 9280
Thiomedon see 5896
Thiomerin Sodium [Wyeth] see 5761
Thiomersalate see 9244
Thiomesterone see 9385
Thiomethyl Alcohol see 5867
Thiomicid see 9218
Thiomorpholine see 9229
Thiomucase [Millot-Solac] see 4676
Thionalide, 9273
Thionaphthene-2-carboxylic Acid, 9274
1-Thionaphthol see 6297
2-Thionaphthol see 6298
α-Thionaphthol see 6297
β-Thionaphthol see 6298
Thionazin, 9275
Thioneine see 3611
Thionembutal see 9280
Thionex [Makhteshim-Agan] see 3529
Thionicid see 9218
Thionicol [Mohan] see 9230
Thionine, 9276
Thiontan see 3098
Thionylan see 5871
Thionyl Bromide, 9277
Thionyl Chloride, 9278
Thionyl Difluoride see 9279
Thionyl Fluoride, 9279
2-Thio-4-oxo-6-propyl-1,3-pyrimidine see 7882
Thioparamizone [Smith & Nephew] see 9218
Thiopental Sodium, 9280
Thiopentone Sodium see 9280
Thiopeptin, 9281
Thioperazine see 9287
Thioperoxydicarbonic Acid Diethyl Ester see 3390
Thiophanate, 9282
Thiophanate-methyl see 9282
Thiophan Sulfone see 8934
Thiophene, 9283
7-(Thiophene-2-acetamido)cephalosporanic Acid see 1978
2-Thiophenecarboxylic Acid, 9284
3-Thiophenecarboxylic Acid see 9206
β-Thiophenic Acid see 9206
Thiophenicol [Hishiyama] see 9230
Thiophenol, 9285
Thiophenylpyridylamino-10-carboxylic Acid Piperidinoethoxyethyl Ester see 7423
Thiophos [Am. Cyanamid] see 6983

Thiophosphoric Acid 2-Isopropyl-4-methyl-6-pyrimidyl Diethyl Ester see 2978
Thiophosphoric Anhydride see 7329
Thiophosphoryl Chloride see 7331
Thioproline see 9375
Thiopropanal S-Oxide see 7812
Thiopropazate, 9423
Thio-2-propene-1-sulfinic Acid S-Allyl Ester see 249
Thioproperazine, 9287
Thiopropionaldehyde S-Oxide see 7812
2-Thio-6-propyl-1,3-pyrimidin-4-one see 7882
Thiopyrophosphoric Acid Tetraethyl Ester see 8945
Thioquinox, 9288
Thioredoxin, 9289
Thioridazine, 9290
Thioridazine-2-sulfone see 8943
Thioridazine-2-sulfoxide see 5813
Thiorubber see 9268
Thiosalicylic Acid, 9291
Thiosan see 8928 and 9304
Thiosarmine see 8916
Thioseconal see 9231
Thiosemicarbazide, 9292
Thiosinamine, 9293
Thiosol [Coop. Farm.] see 9386
1-Thiosorbitol, 9294
Thiosporin [Burroughs Wellcome] see 7551
Thiostrepton, 9295
Thiosulfil [Ayerst] see 8887
Thiotax see 5759
Thiotebezin see 9218
Thio-TEPA see 9588
Thiotepp see 8945
Thiotetrole see 9283
Thiothal Sodium see 9280
Thiothiamine, 9296
Thiothixene, 9297
Thiothyr see 9248
2-Thiouracil, 9298
Thiourea, 9299
β-Thiovaline see 7029
4-(Thiovanilloyl)morpholine see 9841
Thioxamyl see 6873
Thioxanthene, 9300
Thioxanthen-9-one see 9301
Thioxanthone, 9301
Thioxidrene [Bottu] see 2322
2-Thioxo-3,5-dibenzyltetrahydro-1,3,5-thiadiazine see 8865
Thioxolone see 9388
2-Thioxo-4-thiazolidinone see 8184
4-[Thioxo(3,4,5-trimethoxyphenyl)-methyl]morpholine see 9672
Thiozin see 4815
THIP, 9302
Thiphen see 9303
Thiphenamil, 9303
Thiram, 9304
Thistle Saffron see 1877
Thiurad see 9304
Thiuramyl see 9304
Thiuranide see 3370
Thiuretic [Parke, Davis] see 4704
Thixokon [Mallinckrodt] see 72
Thizone see 9218
Thomas Balsam see 960
Thomas Flour see 9305
Thomas Phosphate, 9305
Thombran [Thomae] see 9495
Thonzide [Warner-Chilcott] see 9306
Thonzonium Bromide, 9306
Thonzylamine Hydrochloride, 9307
Thoragol see 1218
Thorazine [SK & F] see 2186
Thoria see 9312

Thorium, 9308
Thorium X see 8118
Thorium Chloride, 9309
Thorium Dioxide see 9312
Thorium Iodide, 9310
Thorium Nitrate, 9311
Thorium Oxide, 9312
Thorium Platinocyanide see 9314
Thorium Sulfate, 9313
Thorium Tetrachloride see 9309
Thorium Tetracyanoplatinate(II), 9314
Thorium Tetraiodide see 9310
Thorium(4+) Tetrakis(cyano-C)platinate(2−) (1:2) see 9314
Thorn Apple see 8779
Thoron see 8119
Thorotrast see 9312
Thoroughwort see 3859
Thortveitite see 8348
Thoulet's Soln see 7691
Thousand-leaf see 102
Thoxan [Cutter] see 3844
Thoxidil see 5737
Thozalinone, 9315
THP see 7460 and 9148
THP-ADM see 7460
THP-adriamycin see 7460
THQ see 9177
THR-221 see 1931
Thr (IUPAC Abbrev.) see 9316
Thr²-cyclosporine see 2759
Threonine, 9316
N²-[1-(N²-L-Threonyl-L-lysyl)-L-prolyl]-L-arginine see 9719
D-Threose, 9317
L-Threose, 9318
Thridace see 5222
Thrombasal see 7196
Thrombin, 9319
Thrombin-sensitive Protein see 9322
Thrombin, Topical, 9320
Thrombocid [Bene-Chemie] see 7090
Thrombocytin see 8413
Thrombofort see 9320
Thrombogen see 7903
Thrombo-Hepin [Riker] see 4571
Thrombokinase see 9321
Thrombokinin see 9321
Thrombol [Merck & Co.] see 9321
Thromboliquine see 4571
Thrombolysin [Merck & Co.] see 7493
Thrombophob [Nordmark] see 4571
Thromboplastin, 9321
Thrombospondin, 9322
Thrombotonin see 8413
Thromboxane A₂ see 9323
Thromboxane B₂ see 9323
Thromboxanes, 9323
THS-101 see 2922
Thuja, 9324
3-Thujanone see 9326
Thujic Acid, 9325
Thujin see 8047
Thujone, 9326
α-Thujone see 9326
β-Thujone see 9326
Thujopsene, 9327
Thulia see 9328
Thulium, 9328
Thurfyl Nicotinate, 9329
Thuricide [Sandoz] see 945
Thybon [Hoechst] see 5388
Thycapsol see 5892
Thylakentrin see 4189
Thylate [Du Pont] see 9304
Thylin see 6443
Thylogen [Rorer] see 7996
Thyloquinone [Squibb] see 5714
Thylose [Jackson-Mitchell] see 1835
Thyme, 9330

Thyme Camphor see 9333
Thymergix [Joulli] see 8021
Thymerin [Tokyo Tanabe] see 8125
Thymidine, 9331
Thymine, 9332
Thymine-2-desoxyriboside see 9331
Thymin (formerly) see 9339
Thyminose see 2890
Thymiode see 9335
Thymiodol see 9335
Thymocyte Stimulating Factor see 4896
Thymodin see 9335
Thymol, 9333
m-Thymol see 9333
Thymolated Silver Sulfone see 2972
Thymol Blue, 9334
Thymol Iodide, 9335
Thymolphthalein, 9336
Thymol Sulfone see 2972
Thymolsulfonephthalein see 9334
Thymomodulin, 9337
Thymone [Abbott] see 9503
Thymopentin, 9338
Thymopoietin, 9339
Thymopoietin Pentapeptide see 9338
Thymosin, 9340
Thymosin α₁ see 9340
Thymosin β₄ see 9340
Thymosin β₈ see 9340
Thymosin β₉ see 9340
Thymosin Fraction 5 see 9340
Thymostatin, 9341
o-Thymotic Acid, 9342
o-Thymotinic Acid see 9342
Thymoxamine see 6204
Thymus Nucleic Acid see 2889
Thymyl Acetate see 9333
Thymyl N-Isoamylcarbamate, 9343
2-(Thymyloxymethyl)glyoxalidine see 9742
2-[(Thymyloxy)methyl]-2-imidazoline see 9742
Thypinone [Abbott] see 9503
Thyradin see 9344
Thyrefact [Hoechst] see 9503
Thyreocordon [Donau-Pharmazie] see 9239
Thyreostat I see 6048
Thyreostat II see 7882
Thyreotrophic Hormone see 9709
Thyroactive Protein see 9347
Thyrocalcitonin see 1640
Thyrocrine [Lemmon] see 9344
Thyroid, 9344
Thyroidin, 9345
Thyroid-stimulating Hormone see 9709
Thyroliberin see 9503
Thyropropic Acid, 9346
Thyroprotein, 9347
Thyrotropic Hormone see 9709
Thyrotropin see 9709
Thyrotropin-releasing Factor see 9503
Thyrotropin Releasing Hormone see 9503
Thyroxevan see 5351
Thyroxine, 9348
L-Thyroxine Sodium Salt see 5351
Thytropar [Armour Pharm.] see 9709
Tiaden [Malesci] see 9349
Tiadenol, 9349
Tialamide see 9356
Tiamenidine, 9350
Tiaminal see 9222
Tiamulin, 9351
Tiamutin see 9351
Tianeptine, 9352
Tiapridal [Delagrange] see 9353
Tiapride, 9353
Tiapridex [Schuerholz] see 9353

Tiaprofenic Acid, 9354
Tiaprost, 9355
Tiaprost Trometamol see 9355
Tiaramide, 9356
Tiase [Mediolanum] see 8764
Tiaterol [Midy] see 9349
Tiazesim see 9233
Tiazolidin [UCM-Difme] see 9375
Tibatin see 8941
Tibatin [DAK] see 2412
Tiberal [Roche] see 6824
Tibexin [Christiaens] see 7743
Tibezonium Iodide, 9357
Tibicorten [Sigma-Tau] see 9513
Tibicur see 9218
Tibinide see 5071
Tibione [Riker] see 9218
Tibirox [Roche] see 9178
Tibizide see 5071
Tibolone, 9358
Tibricol [Pfizer] see 6441
Tibutol [Bracco] see 3672
Ticar [Beecham] see 9360
Ticarbodine, 9359
Ticarcillin, 9360
Ticarda [Hoechst] see 6628
Ticarpenin [Beecham] see 9360
Ticinil see 7248
Ticlid [Parcor] see 9361
Ticlobran [Siegfried] see 2374
Ticlodix [Parcor] see 9361
Ticlodone [Parcor] see 9361
Ticlopidine, 9361
Ticrex [SK & F] see 9362
Ticrynafen, 9362
Tidemol [Roche] see 1465
Tiemannite see 8382
Tiemonium Iodide, 9363
Tiempe [DDSA] see 9624
Tienam [Merck & Co.] see 4834
Tienilic Acid see 9362
Tienor [Farmaka] see 2411
Tiezene [Farmoplant] see 10071
Tifatol [Ciba-Geigy] see 2771
Tiferron see 9392
Tifomycine [Roussel-UCLAF] see 2068
Tifosyl see 9588
Tigan [Beecham] see 9623
Tigason [Roche] see 3846
Tigemen [Squibb] see 9364
Tigemonam, 9364
Tiglic Acid, 9365
Tiglic Acid Ester with Pseudotropine see 9366
Tigloidine, 9366
3β-Tigloyloxytropane see 9366
Tiglylpseudotropeine see 9366
Tiglyssin see 9366
Tigogenin, 9367
Tigonin, 9368
Tiguvon see 3945
Tiklid [Labaz] see 9361
Tikofuran see 4210
Tilade [Fisons] see 6355
Tilatil [Roche] see 9080
Tilcarex see 8108
Tilcotil [Roche] see 9080
Tildiem [Dausse] see 3188
Tildin see 1837
Tiliacorine, 9369
Tiliadin see 9033
Tilidine, 9370
Tillam see 7007
Tillman's Reagent see 3057
Tilly Drops see 5153
Tilmapor [Ciba-Geigy] see 1943
Tilorone, 9371
Tilt [Ciba-Geigy] see 7830
Timacar [Merck & Co.] see 9374
Timacor [Merck & Co.] see 9374

Cross Index of Names

Timaxel [Rhône-Poulenc] see 5835
Timazin [Torii] see 4109
Timbrel [Dow] see 9572
Timelit [UCB] see 5438
Timentin [Beecham] see 2342
Timepidium Bromide, 9372
Timidon see 9222
Timiperone, 9373
Timocort [Reckitt & Colman] see 4710
Timodyne see 5682
Timolate [Merck & Co.] see 9374
Timolol, 9374
Timonacic, 9375
Timonil [Desitin] see 1783
Timoped [Reckitt & Colman] see 9442
Timoptic [Merck & Co.] see 9374
Timoptol [Merck & Co.] see 9374
Timosin [Arzneimittelwerk VEB] see 2082
Timostenil [Farmitalia] see 1865
Timovan [Ayerst] see 7902
Timunox [J & J] see 9338
Tin, 9376
Tinactin [Schering] see 9442
Tinaderm see 9442
Tinarhinin [Berlin-Chemie] see 9150
Tin Ash see 8736
"Tin Bichloride" see 8732
Tin Bronze see 8739
Tindal [Schering] see 64
Tin Dibromide see 8741
Tin Dichloride see 8742
Tin Difluoride see 8743
Tin Diiodide see 8745
Tin Dioxide see 8736
Tin Diselenide see 8737
Tin Disulfide see 8739
Tindurin [EGYT] see 7997
Tinic [Tutag] see 6435
Tinidazole, 9377
Tin Monoselenide see 8749
Tin Monosulfide see 8751
Tin Monoxide see 8747
Tinofedrine, 9378
Tinoridine, 9379
Tinostat see 3025
Tin Phosphides, 9380
Tin Protochloride see 8742
Tin Protosulfide see 8751
Tin Protoxide see 8747
Tinset [Janssen] see 6880
Tin Tetrabromide see 8731
Tin Tetrachloride see 8732
Tin Tetrafluoride see 8734
Tin Tetraiodide see 8735
Tintophen X see 1322
Tintorane see 9950
Tinuvin® P, 9381
Tiobicina see 9218
Tiobutarit see 1447
Tiocarlide, 9382
Tiocarone see 9218
Tioclomarol, 9383
Tioconazole, 9384
Tioctan [Fujisawa] see 9255
Tioctidasi [ISI] see 9255
Tiomesterone, 9385
Tio-Mid see 3692
Tiopronin, 9386
Tiosecolo see 9218
Tioten [Mediolanum] see 8764
Tiotiamina [Saita] see 7896
Tiotixene see 9297
Tiotrifar see 676
Tiox [Schering-Plough] see 9387
Tioxidazole, 9387
Tioxolone, 9388
Tipedine see 9389
Tipepidine, 9389
Tipidi [Isola-Ibi] see 7896
T.I.P.P.S. see 4931

Tiprederm [Squibb] see 9408
Tiquizium Bromide, 9390
Tirade [Sumitomo] see 3952
Tirantil see 1588
Tiratricol, 9391
Tirian see 2088
Tiroidina see 9344
Tiron®, 9392
Tiropramide, 9393
Tisercin [EGYT] see 5909
Tisin [Pfizer] see 5071
Tisiodrazida see 5071
Tissue Factor see 9321
Tissue Plasminogen Activator, 9394
Tissue Thromboplastin see 9321
Tissulina [Hefa-Frenon] see 7487
Tissural see 7487
Titania see 9398
Titanic(IV) Acid, 9395
Titanic Hydroxide see 9395
Titanite see 9396
Titanium, 9396
Titanium Ammonium Oxalate see 593
Titanium Dichloride, 9397
Titanium Dioxide, 9398
Titanium Hydride, 9399
Titanium Isopropylate, 9400
Titanium Potassium Oxalate see 7689
Titanium Sesquisulfate, 9401
Titanium Sulfate, 9402
Titanium Tetrabromide, 9403
Titanium Tetrachloride, 9404
Titanium Tetrafluoride, 9405
Titanium Trichloride, 9406
Titanocene Dichloride, 9407
Titanous Chloride see 9406
Titanous Sulfate see 9401
Titanoxy Sulfate see 9402
Titan Yellow see 9238
Titanyl Ammonium Oxalate see 593
Titanyl Potassium Oxalate see 7689
Titanyl Sulfate see 9402
Titriplex III see 3481
Tiuramyl see 9304
Tivalon-NT [Intersan] see 9408
Tixair [Valpan] see 82
Tixantone see 5463
Tixocortol, 9408
Tizanidine, 9409
Tizide see 5071
TM 723 see 109
TM 906 see 9613
TMD, 9410
TMD-10 see 9133
TM-PGE₂ see 9637
TMQ see 9635
TMS 480 [TAD] see 8889
TMS-19Q see 8233
TMTD see 9304
TMU see 9160
TMV see 9412
TNCA see 9274
TNF see 9411 and 9641
TNF-β see 9411
TNT [Pfizer] see 9643
Tobacco Mosaic Virus, 9412
Tobacco Wood see 4521
Toban o-t-c [USV] see 5432
Tobanum [Gedeon Richter] see 2399
Tobermorite see 1709
Tobias Acid see 6326
Tobra [Lilly] see 9413
Tobracin [Shionogi] see 9413
Tobradistin [Dista] see 9413
Tobralex [Alcon] see 9413
Tobramaxin [Alcon] see 9413
Tobramycin, 9413
Tobrex [Alcon] see 9413
Tocainide, 9414
Tocamphyl, 9415
Tocèn [Lepetit] see 3096

Tochlorine see 2066
Toclase [Pfizer] see 1798
Tocodilydrin [Swiss Pharma] see 6654
Tocodrin [Medichemie] see 6654
Tocol, 9416
Tocopherex [Squibb] see 9932
α-Tocopherol see 9931
β-Tocopherol, 9417
γ-Tocopherol, 9418
δ-Tocopherol, 9419
ε-Tocopherol, 9420
ζ₁-Tocopherol, 9421
ζ₂-Tocopherol, 9422
η-Tocopherol, 9423
α-Tocopherol Acetate see 9932
α-Tocopherol Acid Succinate, 9424
α-Tocopheryl Acetate see 9932
Tocophrin see 9932
Tocosamine Sulfate [Trent] see 8692
ToDay [Bristol] see 1980
Toddaline see 2041
Todralazine, 9425
Tofacine [Gist-Brocades] see 9426
Tofaxin [Winthrop] see 9932
Tofenacin, 9426
Tofesilate see MISC-4
Tofisopam, 9427
Tofranil [Geigy] see 4835
Tofranil-PM [Geigy] see 4835
Tofu see 8684
Togamycin [Upjohn] see 8695
Togiren [Schwarzhaupt] see 3628
TOK [Rohm & Haas] see 6519
Tokiocillin [Isei] see 621
Tokiolexin [Isei] see 1971
Tokocin [Meiji Seika] see 2987
Tokokin see 3660
Tokopharm see 9931
Toksobidin [Polfa] see 6659
Tokuderm [Nichiban] see 1202
Toladryl see 5972
Tolamine see 2066
Tolan, 9428
Tolanase [Upjohn] see 9429
Tolansin see 5737
Tolapin see 8034
Tolax [Sutliff & Case] see 5737
Tolazamide, 9429
Tolazolamide see 9429
Tolazoline, 9430
Tolazul see 9444
Tolban [Ciba-Geigy] see 7781
Tolbet [Aldepha] see 9432
Tolboxane, 9431
Tolbusal see 9432
Tolbutamide, 9432
Tolbutone see 9432
Tolcasone [Toho Iyaku] see 9537
Tolciclate, 9433
Tolcil see 5737
Tolcyclamide, 9434
Toldimfos Sodium, 9435
Tolectin [McNeil; Cilag-Chemie] see 9441
Toleron [Mallinckrodt] see 3995
Tolestan [Roemmers] see 2415
Tolfedine [Vetoquinol] see 9436
Tolfenamic Acid, 9436
Tolferain [Ascher] see 3995
Tolhart see 5737
Tolhexamide see 9434
o-Tolidine, 9437
Tolifer [Marion] see 3995
Toliman [Corvi] see 2308
Tolinase [Upjohn] see 9429
Tolindate, 9438
Toliprolol, 9439
Tolisartine [Kowa] see 9447
Tolit see 9643
Tolkan [Rhône-Poulenc] see 5106
Tollens Reagent, 9440

Tolmene [Sigma-Tau] see 9441
Tolmetin, 9441
Tolmicen [Erba] see 9433
Tolnaftate, 9442
Tolnate see 7902
Tolonase see 9429
Tolonidine, 9443
Tolonium Chloride, 9444
Tolopelon [Daiichi] see 9373
Tolosate see 5737
Toloxatone, 9445
1-(o-Toloxy)-2,3-bis(2,2,2-trichloro-1-hydroxyethoxy)propane see 9446
Toloxychloral see 9446
Toloxychlorinol, 9446
Toloxyn see 5737
Tolperisone, 9447
Tolpovidone 131I, 9448
Tolpronine, 9449
Tolpropamine, 9450
Tolrestat, 9451
Tolrestatin see 9451
Tolseram [Squibb] see 5737
Tolserol [Squibb] see 5737
Tolseron [Squibb] see 5737
Toltrazuril, 9452
o-Tolualdehyde, 9453
α-Tolualdehyde see 7236
o-Toluamide, 9454
α-Toluamide see 7237
Toluazotoluidine see 431
Toluene, 9455
p-Tolueneboronic Acid Cyclic 2-Methyl-2-propyltrimethylene Ester see 9431
o-Toluenecarbonitrile see 9463
Toluene 2,4-Diisocyanate, 9456
Toluene-3,4-dithiol, 9457
p-Toluenesulfinic Acid, 9458
p-α-Toluenesulfonamidobenzoic Acid see 1163
p-Toluenesulfonic Acid, 9459
p-Toluenesulfonic Acid Dichloramide see 3034
p-Toluenesulfonyl Chloride, 9460
N-(p-Toluenesulfonyl)-N'-hexameth-yleniminourea see 9429
Toluenethiol see 9254
α-Toluenethiol see 9249
Toluene Trichloride see 1120
Toluic Acid, 9461
m-Toluic Acid see 9461
o-Toluic Acid see 9461
p-Toluic Acid see 9461
α-Toluic Acid see 7239
p-Toluic Acid 4-[2-(t-Butylamino)-1-hydroxyethyl]-o-phenylene Ester see 1317
α-Toluic Acid Ethyl Ester see 3794
α-Toluic Aldehyde see 7236
Toluidine, 9462
Toluidine Blue O see 9444
Toluina see 9432
Tolulexin see 5737
Tolulox see 5737
o-Tolunitrile see 9463
p-Tolunitrile, 9464
α-Tolunitrile see 1145
Toluol see 9455
p-Toluoylmethylcarbinolmono-d-camphoric Acid Ester, Diethanol-amine Salt see 9415
5-(p-Toluoyl)-1-methylpyrrole-2-acetic Acid see 9441
o-Toluylaldehyde see 9453
p-Toluyl-o-benzoic Acid see 9465
2-(p-Toluyl)benzoic Acid, 9465
Toluylene Blue, 9466
Toluylene Red see 6395
Tolvin [Organon] see 6097
Tolvon [Organon] see 6097

Tolycaine, 9467
Tolycar [Jugoremedija] see 1935
m-Tolylacetamide see 67
m-Tolyl Acetate see 2587
o-Tolyl Acetate see 2588
5-(o-Tolylazo)-2-aminotoluene see 431
4''-(o-Tolylazo)-o-diacetotoluidide see 2940
1-o-Tolylazo-2-naphthylamine see 10004
4-(o-Tolylazo)-o-toluidine see 431
o-Tolylazo-o-tolylazo-β-naphthol see 8349
1-(4-o-Tolylazo-o-tolylazo)-2-naph-thol see 8349
m-Tolylbromide see 1429
2,4-Tolylene Diisocyanate see 9456
1-p-Tolyl-1-ethanol see 3226
α-(o-Tolyl)glyceryl Ether see 5737
Tolylhydrazine, 9468
m-Tolylhydrazine see 9468
o-Tolylhydrazine see 9468
p-Tolylhydrazine see 9468
2-(N'-p-Tolyl-N'-m-hydroxyphenyl-aminomethyl)-2-imidazoline see 7234
Tolylmercaptan see 9254
p-Tolylmethylcarbinol see 3226
p-Tolylmethylcarbinol Camphoric Acid Ester Diethanolamine Salt see 9415
1-(p-Tolyl)-1-oxo-2-pyrrolidino-n-pentane see 8021
1,1'-[(o-Tolyloxymethyl)ethylene-dioxy]bis[2,2,2-trichloroethanol] see 9446
3-(o-Tolyloxy)-1,2-propanediol see 5737
1,1'-(3-o-Tolyloxypropylenedioxy)-bis[2,2,2-trichloroethanol] see 9446
N-m-Tolylphthalamic Acid see 6027
1-(p-Tolyl)-2-pyrrolidino-1-penta-none see 8021
Tolylsulfonylbutylurea see 9432
3-(p-Tolyl-4-sulfonyl)-1-butylurea see 9432
1-(p-Tolylsulfonyl)-3-(2-endo-hydr-oxy-3-endo-D-bornyl)urea see 4334
p-Tolylsulfonylmethylnitrosamide, 9469
Tolyprin [Du Pont] see 758
Tolyspaz [Conal] see 5737
Tomabef [Aandersen] see 1933
Tomahawk [Coopers] see 7469
Tomanol [Byk Gulden] see 8122
Tomaset see 6027
5α-Tomatidan-3β-ol see 9470
Tomatidine, 9470
Tomatine, 9471
Tombran [Boehringer, Ing.] see 9495
Tomieze [Yoshitomi; Green Cross] see 7918
Tomil see 7793
Tomiporan [Toyama] see 1926
Tomiron [Toyama] see 1945
Tomorin [Ciba-Geigy] see 2556
Ton see 368
Tonaril see 9651
Tonarsan see 836
Tonarsin see 5864
Tonedron [Grimault] see 5859
Tonephin see 9843
Toness [Angelini Francesco] see 7919
Tonexol see 9123
Toney Red see 8858
Tonibral [Sobio] see 2834
Tonicorine [Lematte et Boinot] see 1741
Tonin, 9472
Tonka Bean Camphor see 2563

Tonocalcin [Schiapparelli] see 1640
Tonocard [Astra] see 9414
Tonocholin B see 80
Tonofosfan see 9435
Tonoftal see 9442
Tonolift [Teisan] see 6616
Tonophosphan see 9435
Tonopres [Boehringer, Ing.] see 3157
Tonoquil see 2372
Tonsillosan (Lösung) [Spitzner] see 5215
Tonus-forte see 3822
Topaz see 342
Topazone see 4210
Topclip Parasol [Ciba-Geigy] see 2775
Top Form Wormer see 9217
Tophol see 2290
Tophosan see 2290
Topicain [Chugai] see 6888
Topicon [Pierrel] see 4512
Topicorte [Roussel-UCLAF] see 2910
Topicycline [Procter & Gamble] see 9130
Topiderm [Roussel-Amor Gil] see 2910
Topifug [Wolff] see 2908
Topilan [Syntex] see 2157
Topilar [Syntex] see 4053
Topinambur see 5144
Topisolon [Hoechst] see 2910
Topitracin [Reed & Carnrick] see 948
Topocaine [Lilly] see 2741
Topokain see 7763
Topostasin see 9320
Topral [Delagrange] see 8975
Topsin see 9282
Topsin-M see 9282
Topsym [Grünethal] see 4077
Topsym (rescinded) [Syntex] see 3247
Topsymin [Grünenthal] see 4077
Topsyn [Syntex] see 4077
Topsyne [Cassenne] see 4077
Toquilone see 5872
Toracizin see 4046
Toradol [Syntex] see 5186
Toraflon see 5872
Torak [Hercules] see 2949
Torantil [Winthrop] see 4639
Torasemide, 9473
Torate [Bristol] see 1530
Torazina see 2186
Torbugesic [C-Vet] see 1530
Torbutrol [Bristol] see 1530
Torch [Sanwa] see 9992
Tordon [Dow] see 7370
Torecan Maleate [Boehringer, Ing.] see 9241
Torelle [Dow] see 4172
Toremifene, 9474
Torental [Hoechst] see 7092
Toresten [Sandoz] see 9241
Torfan see 8116
Toricelocin [Torii] see 1978
Toril Oil, 9475
Torinal see 5872
Toripamide see 9649
Toriseptin M see 8891
Torlamicina see 3626
Tormosyl [Sandoz] see 4120
Tornalate [Sterling] see 1317
Torque [Shell] see 3907
Torrefaction Dextrin see 2931
Torsemide see 9473
Torularhodin, 9476
Torutilin see 9941
Toryn [SK & F] see 1776
Toscarna [Nippon Chemiphar] see 8145
Tosic Acid see 9459
Tosilate see MISC-4
Tosmilen [Lentia] see 2871

Tosnone [Nippon Shoji] see 1798
Tossimex [Siegfried] see 1379
Tostram see 5132
Tostrin see 9115
Tosufloxacin, 9477
Tosufloxacin Tosilate see 9477
Tosylate see MISC-3
Tosyl Chloride see 9460
D-3-endo-p-Tosylureidoborneol see 4334
Totabin see 6390
Totacef [Bristol] see 1925
Totacillin [Beecham] see 621
Total [Hoechst] see 7314
Totalciclina [Benvegna] see 621
Totalon [Pitman-Moore] see 9161
Totapen [Bristol] see 621
Totazina see 2475
Totifen [Master] see 5187
Totomycin see 9130
Touch Wood see 173
Tournesol see 5426
Tova [BDH] see 3208
Tovene [Kali-Chemie] see 3291
Towk [Tanabe] see 9486
Toxakil see 9478
Toxalbumin see 4
Toxaphene, 9478
Toxicarol see 9082
Toxichlor see 2079
Toxiferene [Roche] see 214
Toxiferine I, 9479
Toxiferine V see 9479
Toxiferine XI see 9479
C-Toxiferine I see 9479
C-Toxiferine II see 1722
Toxifren see 4027
Toxilic Acid see 5585
Toxilic Anhydride see 5586
Toxinal [Pharma-Selz] see 6931
Toxoflavin, 9480
Toxogonin [E. Merck] see 6659
Toxohormone, 9481
Toxopyrimidine, 9482
Toyocamycin, 9483
Toyolysom-DS [Maruko] see 5514
Toyomycin [Takeda] see 2240
Tozalinone see 9315
TP see 6381 and 9339
TP-5 see 9338
TP 21 see 9290
TPA see 7306 and 9394
t-PA see 9394
TPD see 7896
TPN [Mead Johnson] see 6262
TPN-12 see 8943
TPS-23 see 5813
TPTZ see 9658 and 9664
TR 495 see 5872
TR-4698 see 8225
Trabest [Hoei] see 2345
Trachyl see 3785
Tracilon [Savage] see 9511
Tracosal [Ciba-Geigy] see 6905
Tracrium [Wellcome] see 884
Tractur [Damor] see 7427
Tradon [Beiersdorf] see 7021
Trafarbiot [Novofarma] see 621
Trafuril [Ciba] see 9329
Trafuryl see 9329
Tragacanth see 4493
Tragacanthin see 4493
Trakipeal [Hishiyama] see 2082
Tral [Abbott] see 4627
Tralgon [Squibb] see 40
Tralin see 4627
Tralkoxydim, 9484
Tramacin [J & J] see 9512
Tramadol, 9485
Tramal [Grünenthal] see 9485
Tramat [Schering AG] see 3697

Tramazoline, 9486
Tramisol [Am. Cyanamid] see 9161
Trancalgyl see 3685
Trancin [Schering] see 4116
Trancolon see 5735
Trancopal [Winthrop] see 2105
Trancote [Sawai] see 2105
Trandate [Allen & Hanburys] see 5208
Tranex [Malesci] see 9487
Tranexamic Acid, 9487
Tranexamic Acid p-(2-Carboxyethyl)-phenyl Ester see 2017
Tranexan [Taiyo] see 9487
Trangorex [Labaz] see 497
Tranid®, 9488
Tranilast, 9489
Tranimul [Roche] see 2977
Trankimazin [Upjohn] see 310
Trankvilan see 5751
Tranlisant see 5751
Tranpoise [Robins] see 5739
Tran-Q [Pfizer] see 4786
Tranquase [Azuchemie] see 2977
Tranquilan see 5751
Tranquilax see 5751
Tranquilax [Hokuriku] see 5669
Tranquiline [Intra] see 5751
Tranquillin [Teva] see 1037
Tranquit [Promonta] see 6883
Tranquizine see 4786
Tranquo-Puren [Klinge-Nattermann] see 2977
Tranquo-Tablinen [Sanorania] see 2977
Transamin [Daiichi] see 9487
Transanate [Teikoku Zoki] see 2105
Transbilix [Guerbet] see 4916
Transbronchin [Homburg] see 1809
Transcycline see 8236
Transderm-Nitro [Ciba] see 6528
Transderm-V [Ciba-Geigy] see 8361
9,10-Transdihydrolisuride see 9095
Transducin see 4437
Transene [Clin-Comar-Byla] see 2400
Transferrins, 9490
Transfer RNA see 8204
Transiderm-Nitro [Geigy] see 6528
Transidione [Dausse] see 6882
Transit [Inca] see 4221
Transithal [M & B] see 1518
Transoddi [Anphar] see 2283
Transvaalin see 8354
Trans-vert [Amchem] see 5864
Trantoin [McKesson] see 6520
Tranvet [Diamond Labs.] see 7841
Tranvex see 7841
Tranxene [Abbott] see 2400
Tranxilène [Clin-Comar-Byla] see 2400
Tranxilium [Mack, Illert.] see 2400
Tranylcypromine, 9491
Trapanal [Promonta] see 9280
Trapex see 6005
Trapidil, 9492
Trapymin see 9492
Trasacor [Ciba-Geigy] see 6905
Trasamlon [Toho] see 9487
Trasentine-A [Ciba] see 3435
Trasentine 6-H [Ciba] see 3435
Trasentine Hydrochloride [Ciba] see 151
Traserit [RIT] see 762
Trasicor [Ciba] see 6905
Traslan [Gramon] see 2102
Trasulphane see 4815
Trasylol [FBA Pharmaceuticals] see 784
Tratul [Bago] see 2279
Traubensäure (German) see 9038
Traumacut [Brenner] see 5903
Traumanase [Rorer] see 1377

Traumasept [SmithKline] see 7701
Traumatic Acid, 9493
Traumatociclina see 5658
Traumon Gel [Troponwerke] see 3833
Trausabun [Promonta] see 5704
Travamin [Baxter] see 7900
Travel-Gum [Chemofux] see 3193
Travelin see 3193
Travelmin see 3193
Travert [Baxter] see 4900
Travin [Lek] see 1493
Travis see 2821
Travogen [Schering AG] see 5044
Travogyn [Keymer] see 5044
Traxam [Lederle] see 3893
Traxanox, 9494
Traxaprone see 8249
Trazinin see 784
Trazodone, 9495
Trazolan [Roussel] see 9495
Trecalmo [Bayer] see 2411
Trecator [Ives] see 3692
Tredemine see 6425
Tredum [Hépatrol] see 3933
Tree Of Life see 9324
Treflan [Elanco] see 9598
Trehalose, 9496
α,α-Trehalose see 9496
Trehalose 6,6-Dimycolate see 2523
Tre-Hold [Amchem] see 6290
Trelibet [EGYT] see 7365
Tremaril [Wander] see 5902
Tremarit [Wander] see 5902
Tremblex [Janssen; Cilag-Chemie] see 2923
Tremerad [Parke, Davis] see 2354
Tremetone, 9497
Tremin [Schering] see 9607
Tremolite see 851
Tremonil [Wander] see 5902
Tremoquil see 5902
Tremorine, 9498
Trenbolone, 9499
Trenbolone Hexahydrobenzylcarbon-ate see 9499
Trendar [Whitehall] see 4812
Trengestone, 9500
Trenimon see 9517
Trentadil [Armour Pharm.] see 965
Trental [Albert-Roussel] see 7092
Treoimicina [Guidi] see 9681
Treomicetina see 2068
Trepibutone, 9501
Trepidan [Sigma-Tau] see 7713
Trepidone [Lederle] see 5739
Tresanil [ISF] see 9672
Trescatyl [M & B] see 3692
Trescillin see 7829
Tresochin [Bayer] see 2163
Tresortil see 5903
Trest [Dorsey] see 5902
Tresten see 9241
Tretamine see 9586
Trethylene see 9552
Tretinoin see 8167
Tretoquinol, 9502
Trevintix see 7905
Trevira [Hoechst] see 7546
Trexan [Du Pont] see 6278
TRF see 4895 and 9503
TRH, 9503
TRH-Roche see 9503
Tri see 9552
Tri-6 see 5379
Tri-Abrodil [Bayer] see 72
Triac see 9391
Triacana [ANA] see 9391
Triacetin, 9504
1,8,9-Triacetoxyanthracene see 714
1,3,4-Tri-O-acetyl-N-acetyl-6-des-oxy-β-D-glucosamine see 2911

N,N,α-Trimethyl-γ-phenylbenzene-
propanamine *see* 9627
N-*sym*-Trimethylphenyldiethylamino-
acetamide *see* 9614
N,N,N'-Trimethyl-N'-(3-phenyl-1*H*-
indol-1-yl)-1,2-ethanediamine *see*
1238
Trimethylphenylmethane *see* 1551
1,2,5-Trimethyl-4-phenyl-4-piperidin-
ol Propanoate (Ester) *see* 7795
1,2,5-Trimethyl-4-phenyl-4-piperidyl
Propionate *see* 7795
1,2,5-Trimethyl-4-phenyl-4-propion-
yloxypiperidine *see* 7795
N,N,N-Trimethyl-2-(phosphonooxy)-
ethanaminium Chloride *see* 7337
2,2,6-Trimethyl-4-piperidinol Benzoate (Ester) *see* 3850
4,5',8-Trimethylpsoralen *see* 9647
2,4,6-Trimethylpyridine, 9632
N,N,α-Trimethyl-10*H*-pyrido[3,2-*b*]-
[1,4]benzothiazine-10-ethanamine
see 5117
Trimethylsilyl Triflate, 9633
Trimethylsilyltrifluoromethane Sulfonate *see* 9633
N,N,N-Trimethyl-1-tetradecanamini-
um Bromide *see* 6249
5,9,13-Trimethyl-4,8,12-tetradecatri-
enoic Acid 3,7-Dimethyl-2,6-octa-
dienyl Ester *see* 4273
Trimethyltetradecylammonium Bromide *see* 6249
Trimethylthionine Chloride *see* 941
5,7,8-Trimethyltocol *see* 9931
5,7,8-Trimethyltocotrien-3',7',11'-ol
see 9421
N,N,6-Trimethyl-2-*p*-tolylimidazo-
[1,2-*a*]pyridine-3-acetamide *see*
10091
N,N,β-Trimethyl-2-(trifluoromethyl)-
10*H*-phenothiazine-10-propanamine
see 9593
4,4',4''-Trimethyl-3,3',3''-trihept-
yl-8-(2''-thiazolyl)-2,2'-penta-
methinethiazolocyanine 3,3''-Di-
iodide *see* 7504
α,α,α'-Trimethyltrimethyleneglycol
see 4631
2,5,8-Trimethyl-2-(4,8,12-trimethyl-
trideca-3,7,11-trienyl)chroman-6-ol
see 9420
2,5,7-Trimethyl-2-(4,8,12-trimethyl-
tridecyl)-6-chromanol *see* 9422
2,5,8-Trimethyl-2-(4,8,12-trimethyl-
tridecyl)-6-chromanol *see* 9417
2,7,8-Trimethyl-2-(4,8,12-trimethyl-
tridecyl)-6-chromanol *see* 9418
2,4,6-Trimethyl-1,3,5-trioxane *see*
6975
2,4,6-Trimethyl-*s*-trithiane *see* 9245
Trimethylvinylammonium Hydroxide
see 6393
2,6,10-Trimethyl-14-vinylpentadecan-
14-ol *see* 5084
1,3,7-Trimethylxanthine *see* 1635
Trimeton (formerly) [Schering] *see*
7198
Trimetozine, 9634
Trimetrexate, 9635
Trimforte [Ratiopharm] *see* 8889
Trimipramine, 9636
Trimodose [Roussel] *see* 4310
Trimogal [Lagap] *see* 9624
Trimol [Fujisawa] *see* 7473
Trimon *see* 8769
Trimonase [Tosi] *see* 9377
Trimopan [Berk] *see* 9624
Trimoprostil, 9637
Trimox [Squibb] *see* 610
Trimpex [Roche] *see* 9624

Trimstat [Laser] *see* 7180
Trimtabs [Mayrand] *see* 7180
Trimustine *see* 9560
Trimyristin, 9638
Trimysten *see* 2412
Trinactin *see* 6594
Trinalgon [ISF] *see* 6528
Trinitrin *see* 6528
sym-Trinitrobenzene, 9639
1,3,5-Trinitrobenzene *see* 9639
2,4,6-Trinitro-1,3-benzenediol *see*
8828
sym-Trinitrobenzoic Acid *see* 9640
2,4,6-Trinitrobenzoic Acid, 9640
2,4,7-Trinitrofluorenone, 9641
2,4,7-Trinitro-9*H*-fluoren-9-one *see*
9641
(2,4,7-Trinitro-9*H*-fluoren-9-ylid-
ene)propanedinitrile *see* 3084
Trinitroglycerol *see* 6528
1,3,5-Trinitrohexahydro-*s*-triazine *see*
2742
Trinitromethane, 9642
2,4,6-Trinitrophenol *see* 7380
2,4,6,-Trinitrophenol Ammonium Salt
see 575
2,4,6-Trinitrophenol Potassium Salt
see 7651
2,4,6-Trinitrophenol Silver Salt *see*
8473
sym-Trinitrophenylnitroaminoethyl
Nitrate *see* 7095
2,4,6-Trinitroresorcinol *see* 8828
Trinitrosan [E. Merck] *see* 6528
sym-Trinitrotoluene *see* 9643
2,4,6-Trinitrotoluene, 9643
α-Trinitrotoluol *see* 9643
Trinitrotriethanolamine Diphosphate
see 9682
2,4,6-Trinitro-N-(2,4,6-trinitrophen-
yl)benzenamine *see* 3340
Tri(2-nitroxyethyl)amine Biphosphate
see 9682
18,19,20-Trinor-17-cyclohexyl-13,14-
didehydro-PGF$_{2α}$ Methyl Ester *see*
225
Trinordiol [Wyeth-Byla] *see* 6621
Tri-Norinyl [Syntex] *see* 6614
Trinosin *see* 145
TriNovum [Ortho-Cilag] *see* 6614
Trinoxol [Amchem] *see* 2802
Triocil [Warner-Chilcott] *see* 4624
Triodan *see* 2284
Triognost [Teva] *see* 2975
Triolein, 9644
Trional *see* 8937
Trionine *see* 5388
Triopac [Cilag] *see* 72
Triopron *see* 9346
Triosil *see* 6078
Triospen *see* 7184
Triostam *see* 733
Triostin A *see* 9645
Triostin C *see* 9645
Triostins, 9645
Triothyrone *see* 5388
Triovex *see* 3659
1,3,5-Trioxacyclohexane *see* 9646
s-Trioxane, 9646
1,3,5-Trioxane *see* 9646
1,3,5-Trioxane-2,4,6-triimine *see*
2689
Trioxazine *see* 9634
3,7,12-Trioxocholan-24-oic Acid *see*
2858
2,4,6-Trioxohexahydropyrimidine *see*
973
2,4,6-Trioxo-5-methyl-5-phenyl-
hexahydropyrimidine *see* 7275
Trioxone [ICI] *see* 8999

N,N',N''-(2,6,10-Trioxo-1,5,9-tri-
oxacyclododecane-3,7,11-triyl)tris-
[2,3-dihydroxy]benzamide *see* 3547
N,N',N''-(2,6,10-Trioxo-1,5,9-tri-
oxacyclododecane-3,7,11-triyl)tris-
o-pyrocatechuamide *see* 3547
Trioxsalen, 9647
Trioxyethylrutin *see* 9698
Trioxymethylene *see* 9646
"Trioxymethylene" *see* 6974
2,6,8-Trioxypurine *see* 9790
Tripalmitin, 9648
Tripamide, 9649
Triparanol, 9650
Triparin *see* 9650
Tripelennamine, 9651
Triperidol [McNeil] *see* 9596
Tripervan [Roger Bellon] *see* 9888
Tripetroselin *see* 7142
Triphacyclin *see* 9130
Triphal®, 9652
Triphasil [Wyeth] *see* 6621
Triphedinon [Toho] *see* 9607
Triphenidyl *see* 9607
Triphenylcarbinol, 9653
Triphenylene, 9654
Triphenylmethane, 9655
Triphenylmethanol *see* 9653
4-(Triphenylmethyl)morpholine *see*
9591
Triphenyl Phosphate, 9656
Triphenylphosphine, 9657
1,3,4-Triphenyl-1*H*-pyrazole-5-acetic
Acid *see* 5056
3,5,6-Triphenyl-2,3,5,6-tetraazabicy-
clo[2.1.1]hex-1-ene *see* 6534
Triphenyltetrazolium Chloride, 9658
2,3,5-Triphenyl-2H-tetrazolium Chlo-
ride *see* 9658
Triphenyltin Acetate *see* 9659
Triphenyltin Hydroxide, 9659
Triphosaden *see* 145
Triphosphopyridine Nucleotide *see*
6262
Triphthasine *see* 9594
Triphylite *see* 5395
Tripiperazine Dicitrate *see* 7433
Triple Superphosphate *see* 1700
Tripoli, 9660
Tripolyphosphate *see* 8654
Tripotassium Dicitrato Bismuthate *see*
1296
Tripotassium Hexakis(cyano-C)co-
baltate(3−) *see* 7620
Tripotassium Hexakis(cyano-C)fer-
rate(3−) *see* 7611
Tripotassium Phosphate *see* 7649
Tripotassium Tris(ethanedioato)-
chromate(3−) *see* 2231
Tripotassium Tris(oxalato)antimonate-
(3−) *see* 731
Tripotassium Tris(oxalato)chromate-
(3−) *see* 2231
Tripoton *see* 7198
Triprolidine, 9661
Triptide *see* 4369
Triptil [Merck & Co.] *see* 7917
Triptisol *see* 504
Tript-Oh [Sigma-Tau] *see* 4784
Triptorelin, 9662
Triptycene, 9663
Tripyridyltriazine *see* 9664
2,4,6-Tripyridyl-*s*-triazine, 9664
2,4,6-Tripyridyl-1,3,5-triazine *see*
9664
2,4,6-Tri-2-pyridyl-*s*-triazine *see*
9664
Tris [Trenimon] *see* 9684
Trisamine *see* 9684
Tris Amino [Comm. Solvents] *see*
9684

Trisatin see 9505
2,3,5-Tris(aziridino)-1,4-benzoquinone see 9517
2,3,5-Tris(1-aziridinyl)-p-benzoquinone see 9517
2,3,5-Tris(1-aziridinyl)-2,5-cyclohexadiene-1,4-dione see 9517
Tris(1-aziridinyl)phosphine Oxide see 9587
Tris(1-aziridinyl)phosphine Sulfide see 9588
2,4,6-Tris(1-aziridinyl)-s-triazine see 9586
Tris-BP, 9665
Tris Buffer see 9684
Tris[2-chloro-1-(chloromethyl)ethyl]-phosphate see 4233
Tris(β-chloroethyl)amine see 9560
N,N,O-Tris(2-chloroethyl)-N'-(3-hydroxypropyl)phosphoric Acid Diamide see 2853
N,N,N'-Tris(2-chloroethyl)-N',O-propylene Phosphoric Acid Ester Diamide see 9680
N,N,3-Tris(2-chloroethyl)tetrahydro-2H-1,3,2-oxazaphosphorin-2-amine 2-Oxide see 9680
1-[β,β,β-Tris(p-chlorophenyl)propionyl]-4-methylpiperazine see 4596
Tris(2,3-dibromopropyl) Phosphate see 9665
Tris(1,3-dichloroisopropyl)phosphate see 4233
Tris(2,4-dichlorophenoxyethyl) Phosphite see 3880
1,2,3-Tris(2-diethylaminoethoxy)-benzene Tris(ethyl Iodide) see 4249
Tris[7-[4-(dimethylamino)phenyl]-N-hydroxy-4,6-dimethyl-7-oxo-2,4-heptadienamidato-O^NO^1]iron see 9563
2,4,6-Tris(dimethylamino)-s-triazine see 318
Tris(dimethylcarbamodithioato-S,S')-iron see 3954
Tris(dimethyldithiocarbamato)iron see 3954
Tris(ethylenediamine)cadmium Dihydroxide see 9666
Tris(ethylenediamine)cadmium Hydroxide see 9666
2,3,5-Tris(ethyleneimino)benzoquinone see 9517
2,4,6-Tris(ethylenimino)-s-triazine see 9586
Tris(hydroxyethyl)amine see 9581
7,3',4'-Tris[O-(2-hydroxyethyl)]rutin see 9698
Tris(hydroxymethyl)aminomethane see 9684
N^2,N^4,N^6-Tris(hydroxymethyl)melamine see 9631
N-[Tris(hydroxymethyl)methyl]glycine see 9565
Tris(hydroxymethyl)nitromethane, 9667
Trisilane, 9668
Trisilicane see 9668
Trisilicon Octahydride see 9668
Trisilicon Tetranitride see 8444
Trisilicopropane see 9668
Tris(p-methoxyphenyl)chloroethylene see 2173
2,3,3-Tris(p-methoxyphenyl)-N,N-dimethylallylamine see 608
Tris(2-methyl-1-aziridinyl)phosphine Oxide see 5842
Tris(1-methylethylene)phosphoric Triamide see 5842
2,4,6-Tris(methylolamino)-1,3,5-triazine see 9631

Tris(nicotinato)aluminum see 354
1,1,1-Tris(nitratomethyl)propane see 7821
Tris(2-nitroxyethyl)amine Biphosphate see 9682
Trisodium 4'-Anilino-8-hydroxy-1,1'-azonaphthalene-3,6,5'-trisulfonate see 662
Trisodium Calcium Diethylenetriaminepentaacetate see 7082
Trisodium Carboxyphosphate see 4166
Trisodium Citrate see 8549
Trisodium Edetate see 3483
Trisodium Ethylenediaminetetraacetate see 3483
Trisodium Hexakis(cyano-C)ferrate-(3−) see 8562
Trisodium Hexakis(nitrato-N)cobaltate(3−) see 8551
Trisodium N-Hydroxyethylethylenediaminetriacetate see 9871
Trisodium Orthophosphate see 8615
Trisodium Phosphate see 8615
Trisodium Phosphonoformate see 4166
Trisodium Salt of 1-(4-Sulfo-1-naphthylazo)-2-naphthol-3,6-disulfonic Acid see 382
Trisomin [Lilly] see 5569
Trisoralen [Elder] see 9647
Tris-steril see 9684
Tristat [Salsbury] see 6533
Tristearin, 9669
Tris (Trenimon) see 9517
Tris(2,4,6-tribromophenoxy)bismuthine see 1305
1,2,3-Tris(2-triethylammonium Ethoxy)benzene Triiodide see 4249
Tristriphenylphosphine Rhodium Carbonyl Hydride see 9670
Trisulfurated Phosphorus see 9168
Trisustan see 9682
Tri-Sweet [Searle] see 861
Tritan see 9655
Tritanol see 9653
Triteren see 9515
Tritheon [Ortho] see 421
4,5,9-Trithiadodeca-1,6,11-triene 9-Oxide see 185
Trithiadol [Sterwin] see 5899
Trithio see 676
Trithioacetaldehyde see 9245
Trithioanethole see 676
Trithiocarbonic Acid, 9671
Trithiocarbonic Acid Cyclic Ester with 2,3-Quinoxalinedithiol see 9288
Trithiocarbonic Acid Cyclic 2,3-Quinoxalinediyl Ester see 9288
Trithio-p-methoxyphenylpropene see 676
Trithion [Stauffer] see 1827
Trithiozine, 9672
Triticum, 9673
Tritiozine see 9672
Tritisan [Hoechst] see 8108
Tritium, 9674
Tritocaline see 9677
Tritolyl Phosphate, 9675
Tri-o-tolyl Phosphate, 9676
Triton N [Rohm & Haas] see 6596
Triton X [Rohm & Haas] see 6681
Triton X-100 [Rohm & Haas] see 6681
Triton A-20 [Rohm & Haas] see 9741
Triton WR-1339 [Rohm & Haas] see 9741
Tritopine see 5247
Tritoqualine, 9677
Tritox [Pitman-Moore] see 9540
Trittico [Angelini Francesco] see 9495
4-Tritylmorpholine see 9591
Triumph [Ciba-Geigy] see 4988

Triuret, 9678
Triurol [Lundbeck] see 72
Trivalent Sodium Antimonyl Gluconate see 733
Trivaline [E. Merck-Clévenot] see 380
Trivastal [Pharmacodex] see 7465
Trivazol see 6079
Trivitan see 9929
Trizma [Sigma] see 9684
Trobicin [Upjohn] see 8695
Trochol see 1212
Trocinate [Poythress] see 9303
Troclosene Potassium, 9679
Trodax see 6578
Trofosfamide, 9680
Trofurit [Chinoin] see 4221
Troillite see 4007
Trolamine see MISC-3 and 4
Troleandomycin, 9681
Trolene [Dow] see 8239
Trolnitrate Phosphate, 9682
Trolone see 8974
Trolovol [Bayer] see 7029
Tromal [Burroughs Wellcome] see 1497
Tromantadine, 9683
Tromaril [Unichem] see 3535
Tromasedan [Millot] see 1043
Tromasin see 6965
Trombavar see 8645
Trombostop see 9321
Trombovar see 8645
Trometamol see 9684
Tromethamine, 9684
Tromethane see 9684
Tromexan [Geigy] see 3729
Tromexan Ethyl Acetate [Geigy] see 3729
Trommsdorff's Starch see 10041
Trona see 8630
Tronolane [Abbott] see 7707
Tronothane [Abbott] see 7707
Tropacaine see 9686
Tropacine, 9685
Tropacocaine, 9686
Tropaeolin D see 6019
Tropaeolin G see 5833
Tropaeolin O, 9687
Tropaeolin OO, 9688
Tropaeolin OOO No. 1 see 6812
Tropaeolin OOO No. 2 see 6813
Tropaeolin R see 9687
Tropaic Acid see 9691
Tropalin see 9650
Tropane, 9689
1αH,5αH-Tropane see 9689
(E)-1αH,5αH-Tropane-3α,6β,7β-triol 3-(2-Methylcrotonate) see 5841
3-(3,6,7-Tropanetriol) Tiglate see 5841
1αH,5αH-Tropan-3α-ol see 9693
1αH,5αH-Tropan-3β-ol see 7937
3β-Tropanol see 7937
1αH,5αH-Tropan-3α-ol Atropate see 770
1αH,5αH-Tropan-3α-ol Benzilate see 9694
1αH,5αH-Tropan-3β-ol Benzoate see 9686
1αH,5αH-Tropan-3α-ol Diphenylacetate see 9685
1αH,5αH-Tropan-3α-ol Mandelate see 4649
(E)-1αH,5αH-Tropan-3β-ol 2-Methylcrotonate see 9366
1αH,5αH-Tropan-3α-ol (−)-2-Methyl-2-phenylhydracrylate see 5345
1αH,5αH-Tropan-3α-ol (±)-Tropate see 891

Tybatran [Robins] see 9737
Tybraine [ANA] see 5987
Tyclarosol see 3482
Tycor [Mobay] see 3694
Tydantil [Polichimica Sap] see 6447
Tylagel see 9450
Tylan [Lilly] see 9740
Tylcalsin see 1644
Tylciprine see 9491
Tylenol [McNeil] see 40
Tylinal [Kabi] see 3116
Tyllithin see 5397
Tylocrebrine, 9738
Tylon see 9740
Tylonolide see 9740
Tylophorine, 9739
Tylose see 5963
Tylose MGA see 1835
Tylosin, 9740
Tyloxapol, 9741
Tyloxypal see 9741
Tymazoline, 9742
Tymelyt [AB Leo] see 5438
Tymium [Sapos] see 3889
Tympanol see 6039
Tymtran [Adria] see 1998
Type I Interferon see 4892 and 4893
Type II Interferon see 4894
Tyr (IUPAC Abbrev.) see 9747
Tyraminase see 6158
Tyramine, 9743
Tyrene see 9541
Tyri 10 [Lindopharm] see 9749
Tyrimide see 5090
Tyrocidine, 9744
Tyrocidine A see 9744
Tyrocidine B see 9744
Tyrocidine C see 9744
Tyroderm [Merck & Co.] see 9749
Tyropanoate Sodium, 9745
Tyropaque [Torii] see 9745
Tyrosamine see 9743
Tyrosinase, 9746
Tyrosine, 9747
m-Tyrosine, 9748
L-Tyrosylglycylglycyl-L-phenylalanyl-
    L-leucine see 3528
L-Tyrosylglycylglycyl-L-phenylalanyl-
    L-methionine see 3528
Tyrothricin, 9749
Tyrylen see 1509
Tyvid [Merrell] see 5071
Tyzanol [Pfizer-Roerig] see 9150
Tyzine [Pfizer] see 9150
TZB see 9236
TZU-0460 see 8252
U-27 see 7391
U-1258 see 5185
U-1363 see 3304
U-2032 see 5178
U-2043 see 9432
U-4527 see 2734
U-5956 see 4022
U-6013 see 5058
U-6591 see 6641
U-6987 see 1839
U-7800 see 4119
U-8344 see 9762
U-8471 see 5679
U-9586 see 7939
U-9889 see 8794
U-10071 see 9714
U-10136 see 7892
U-10149 see 5378
U-10858 see 6122
U-10997 see 6098
U-12062 see 7893
U-14583 see 7894
U-14743 see 7572
U-15167 see 6591
U-15800 see 6938

U-17835 see 9429
U-18496 see 907
U-19920 see 2790
U-21251 see 2352
U-22550 see 1730
U-24973A see 5704
U-26225A see 9485
U-26452 see 4372
U-26597A see 2472
U-27182 see 4124
U-28288D see 4471
U-28508 see 2352
U-28774 see 5176
U-31889 see 310
U-32070E see 1638
U-32921 see 1829
U-32921E see 1829
U-33030 see 9518
U-34865 see 3126
U-36059 see 503
U-36384 see 1829
U-41123 see 150
U-42842 see 796
U-53217 see 7890
U-53217A see 7890
U-63196 see 1939
U-63196E see 1939
U-64279E see 1948
U-67279A see 1948
U-69689E see 4010
U-74006F see 5264
U-76252 see 1941
U-76253 see 1941
U-76253A see 1941
Ub see 9752
Ubenimex, 9750
Ubichromenol (50) see 9751
Ubidecarenone see 9751
UBI-N252 see 6893
UBIP see 9753
Ubiquasar [Brocchieri] see 9751
Ubiquinones, 9751
Ubiquitin, 9752
Ubiquitous Immunopoietic Polypep-
    tide see 9752
Ubiten [Italfarmaco] see 9751
Ubretid [OSSW] see 3368
UC 7744 see 1789
UC-20047 see 9488
UC 21149 see 216
UC 51762 see 9258
UCB 1474 see 2075
UCB 1545 see 3892
UCB 1967 see 3439
UCB 2073 see 1799
UCB 2543 see 1798
UCB 3412 see 3391
UCB 3983 see 5812
UCB 4445 see 1449
UCB 4492 see 4786
UCB 5062 see 5657
UCB 6215 see 7459
Udekinon [Tohishi] see 9751
Udicil [Upjohn] see 2790
UDMH see 3236
Udolac [ICI] see 2820
UDP see 9793
UDPG see 9794
Ue 5908 see 8008
U-Gencin see 4284
Uglow Black Silver, 9753
Ugurol [Bayer] see 9487
Uintahite, 9754
Uintaite see 9754
Ujothion, 9755
UK-4271 see 6872
UK-20349 see 9384
UK-33274 see 3422
UK-33274-27 see 3422
UK-48340 see 509
UK-48340-11 see 509

UK-49858 see 4054
Ukidan [Serono] see 9799
Ukopen see 621
Ulbreval see 1518
Ulcar [Houdé] see 8853
Ulcedin [Agips] see 2279
Ulcedine [Usafarma] see 2279
Ulceprax [Salvat] see 3881
Ulcerban [Marion; Chugai] see 8853
Ulcerfen [Finadiet] see 2279
Ulcerlmin [Chugai] see 8853
Ulcerone [Brocades] see 1296
Ulcesium [Zambon] see 3949
Ulcex [Guidotti] see 8126
Ulcimet [Farmasa] see 2279
Ulcoban [Parke, Davis] see 1090
Ulcocid see 362
Ulcofalk [Interfalk Italia] see 2279
Ulcoforton [Plantorgan] see 7463
Ulcogant [Cascan] see 8853
Ulcolax [Ulmer] see 1253
Ulcomedina [Von Boch] see 2279
Ulcomet [Italfarmaco] see 2279
Ulcort see 4709
Ulcosan [Dompé] see 7463
Ulcus-Tablinen [Sanorania] see 1797
Ulcyn [Pitman-Moore] see 3527
Uldumont see 5869
Ulexine see 2795
Ulfamid [Krka] see 3881
Ulfaret [Takeda] see 1943
Ulfinol [Roemmers] see 3881
Ulgut [Shionogi] see 1046
Ulhys [Lafare] see 2279
Uliginosin A see 9756
Uliginosin B see 9756
Uliginosins, 9756
Uliron C see 8905
Ulmenide [Roche] see 2044
ULO [Riker] see 2053
Ulone [Riker] see 2053
Ulstar [Nippon Roche] see 9637
Ulstron [ICI] see 7558
Ultandren [Ciba] see 4113
Ultidine [Ferrosan] see 8126
Ultin [Rentschler] see 352
Ultrabion [Lifasa] see 621
Ultracain [Hoechst] see 1878
Ultracarbon see 1815
Ultracef [Bristol] see 1921
Ultracid [Geigy] see 5891
Ultracillin [Grünenthal] see 2706
Ultracorten [Ciba] see 7727
Ultracortene see 7727
Ultracortene-H see 7719
Ultracortenol [Ciba] see 7726
Ultracorterenol see 7726
Ultradiazin see 8874
Ultradol [Ayerst] see 3831
Ultrafur see 4205
Ultralan Oral [Schering AG] see 4079
Ultralanum [Schering AG] see 4079
Ultralente Iletin see 4890
Ultramarine, 9757
Ultramarine Green see 2229
Ultramarine Yellow see 982
Ultra-Mg [Sopar] see 4350
Ultramicina [Lisapharma] see 4169
Ultran [Lilly] see 7161
Ultrapal Chloride see 8846
Ultrapen see 7829
Ultraquinine see 2626
Ultraseptyl see 8891
Ultrasul see 8887
Ultrasulfon [Streuli] see 8909
Ultrasüss see 6551
Ultratard [Novo] see 4890
Ultra Tears [Alcon] see 4777
Ultra-Technekow see 8609
Ultratiazol see 7351
Ultravist [Schering AG] see 4948

Ultrax see 8885
Ultron [Monsanto] see 7563
Ultroxim [Duncan, Flockhart] see 1951
UM-792 see 6278
UM-952 see 1487
Umatrope [Lilly] see 8672
Umbellatine see 1169
Umbelliferone, 9758
Umbelliferone Glucoside see 8506
Umbeon see 9225
Umbradil see 4933
UML 491 [Sandoz] see 6055
UMP see 9796
Unacid [Pfizer] see 8862
Unagen see 3358
Unakalm [Upjohn] see 5176
Unal see 474
Unasyn(a) [Pfizer] see 8862
10-Undecenoic Acid see 9760
Undecoylium Chloride, 9759
Undecylenic Acid, 9760
9-Undecylenic Acid see 9760
Unden see 3660 and 7849
Unergol [Poli] see 3598
Ungernine see 9050
Unibaryt [Röhm] see 1006
Unibloc [Benzon] see 879
Unicam [Master] see 7476
Unicard [Schering-Plough] see 3186
Unicin [Rachelle] see 9130
Unicozyde see 5071
Unidasa see 4676
Unidigin [Merrell] see 3146
Unidocan see 2883
Unidone [Unilabo] see 698
Uni-Dur [Schering-Plough] see 9212
Unifer [Tosi] see 3984
Unifur see 4205
Unifyl [Mundipharma] see 9212
Unihep [AB Leo] see 4571
Unilax see 9505
Unilobin see 5432
Uniloc [Benzon] see 879
Unimycetin see 2068
Unimycin [Upjohn] see 9130
Unipen [Wyeth] see 6266
Uniphyl [Purdue Frederick] see 9212
Uniphyllin [Mundipharma] see 9212
Uniprofen [Lilly] see 1055
Unipyranamide see 7970
Unisedil [Laquifa] see 2977
Unisept see 2090
Unisom [Pfizer] see 3430
Unisomnia [Unigreg] see 6491
Unispiran see 9137
Unistat see 6533 and 8908
Unistat-3 [Salsbury] see 6533
Unistradiol see 3655
Unitane see 9398
Unitensen Tannate [Wallace Labs.] see 2610
Uniteston see 9115
Unitiol see 3197
UNITOP [Roche] see 6255
Univer [Rorer] see 9851
Unizole see 3255
Unnatural Tartaric Acid see 9037
Unnilhexium see 3505
Unnilpentium see 3504
Unnilquadium see 3503
Unnilseptium see 3506
Unospaston [Siegfried] see 3346
Unresolvable Tartaric Acid see 9040
Unusual Tartaric Acid see 9037
UP-83 see 6444
UP-106 see 7846
UP-33-901 see 2274
UP-34101 see 7804
UP-507-04 see 7369
Upcyclin [Cephar] see 9130

Upixon see 7431
Upstene [Pharmuka] see 4843
UR 112 see 2375
UR 1501 see 9600
UR 1521 see 4170
Uracid [Draco] see 345
Uracil, 9761
Uracil-6-carboxylic Acid see 6828
Uracil Deoxyriboside see 2892
Uracil Mustard, 9762
Uracil Riboside see 9792
Uractyl see 8906
Uradal see 1837
Uragon [Makhteshim-Agan] see 1370
Ural see 1808
Uralenic Acid see 3543
Uralgin [Ceccarelli] see 6273
Uraline see 1808
Uralium see 1808
Uralpha [Debat] see 6204
Uramid [Spofa] see 8906
Uramil see 9760
Uramite®, 9764
Uramustine see 9762
Uramycin B see 9483
Uranediol, 9765
Uranic Oxide see 9773
Uranin(e) see 4085
Uranine Yellow see 4085
Uraninite see 9767
Uranium, 9766
Uranium X$_2$ see 7897
Uranium Z see 7897
Uranium Ammonium Carbonate see 596
Uranium Ammonium Fluoride see 597
Uranium Barium Oxide see 1013
Uranium Benzoate see 1101
Uranium Dioxide, 9767
Uranium Dioxydichloride see 9775
Uranium Hexafluoride, 9768
Uranium Oxide Orange see 7694
Uranium Oxide Yellow see 8656
Uranium Peroxide, 9769
Uranium Tetrachloride, 9770
Uranium Tetrafluoride, 9771
Uranium Trichloride, 9772
Uranium Trioxide, 9773
Uranium Yellow see 8656
"Uranium Yellow" see 595
Uranous Oxide see 9767
Urantoin [DDSA] see 6520
Uranyl Acetate, 9774
Uranyl Ammonium Carbonate see 596
Uranyl Ammonium Fluoride see 597
Uranyl Benzoate see 1101
Uranyl Chloride, 9775
Uranyl Nitrate, 9776
Uranyl Phosphate, 9777
Uranyl Potassium Nitrate see 7695
Uranyl Potassium Sulfate see 7696
Uranyl Sulfate, 9778
Urao see 8630
Urapidil, 9779
Uraprene [IBI] see 9779
Urari see 2678
Urastrat [Warner-Lambert] see 9785
Urate Oxidase see 9791
Urazole, 9780
Urbac [E. Merck-Clévenot] see 6448
Urbadan [Roussel] see 2355
Urbanyl [Diamant] see 2355
Urbason [Hoechst] see 6028
Urbason-Solubile [Hoechst] see 6028
Urbil see 5751
Urbilat [Hor-Fer-Vit] see 5751
Urbol [Heilit] see 278
Urea, 9781
Urea Amidohydrolase see 9785
Urea Calcium Acetylsalicylate see 1644

Urea Hydrochloride, 9782
Urea Hydrogen Peroxide, 9783
Urea Nitrate, 9784
Ureaphil [Abbott] see 9781
Urea Polymer with Formaldehyde see 7552
Urease, 9785
Urea Stibamine, 9786
Urecholine Chloride [Merck & Co.] see 1207
Uredepa, 9787
Uregit [EGYT] see 3669
p-Ureidobenzenearsonic Acid see 1788
(p-Ureidobenzenearsylenedithio)di-o-benzoic Acid see 9251
5-Ureidohydantoin see 246
δ-Ureidonorvaline see 2333
4-Ureido-1-phenylarsonic Acid see 1788
(p-Ureidophenylarsylenedithio)diacetic Acid see 9252
(p-Ureidophenylarsylenedithio)di-o-benzoic Acid see 9251
Urelim see 3684
Urem [Kade] see 4812
Urena, 9788
Urenil see 8906
Ureophil [Abbott] see 9781
Urepearl [Otsuka] see 9781
Urese [Pfizer] see 1134
Urethan, 9789
Urethane see 9789
Urethan of β-Methylcholine Chloride see 1207
Urethimine see 9787
Urethylane see 5958
Uretrim [Bastian] see 9624
Urex [Mochida] see 4221
Urex [Riker] see 5882
6,6'-Ureylenebis[1,1'-dimethylqui-nolinium] Sulfate see 3360
6,6'-Ureylenebis(1-methylquinolini-um)bis(methosulfate) see 3360
Urfadyn [Arsac] see 6455
Urfamycine [Inpharzam] see 9230
Urgilan [Simes] see 7889
Uriben [R. P. Drugs] see 6273
Uric Acid, 9790
Uricase, 9791
Uricemil [Fardeco] see 278
Uriclar [Crosara] see 6273
Urico-oxidase see 9791
Uricovac [Labaz] see 1073
Uric Oxidase see 9791
Uridinal see 7175
Uridine, 9792
Uridine 5'-Diphosphate, 9793
Uridine Diphosphate Glucose, 9794
Uridine-5'-diphosphoglucose see 9794
Uridine 5'-Monophosphate see 9796
Uridine 5'-Phosphoric Acid see 9796
Uridine 5'-Pyrophosphate see 9793
Uridine 5'-Pyrophosphate Glucose Ester see 9794
Uridine-5-pyrophosphoric Acid see 9793
Uridine 5'-(Tetrahydrogen Triphosphate) see 9795
Uridine 5'-(Trihydrogen Diphosphate) see 9793
Uridine 5'-(Trihydrogen Diphosphate) Mono-α-D-glucopyranosyl Ester see 9794
Uridine 5'-Triphosphate, 9795
5'-Uridylic Acid, 9796
Urigon [Demo] see 3071
UriKoxidase see 9791
Urimeth [North Am. Pharm.] see 5896
Urinary Indican see 4853
Urinex see 2169
Urinorm [Torii] see 1073

Urinox [Parke, Davis] see 6899
Uriodone see 4933
Uripurinol [Azupharma] see 278
Urisal [Winthrop] see 8549
Urispas [SK & F] see 4033
Uritone [Parke, Davis] see 5879
Uritrate [Substantia] see 6899
Uritrol see 6577
Urizept see 6520
Urlea see 1045
Uro-Alvar [Gomez] see 6899
Uroanthelone see 3492
Urob see 3951
Urobenyl [Endopharm] see 278
Urobilin IXα see 9797
d-Urobilin see 9797
d-Urobilin IXα see 9797
i-Urobilin see 9797
Urobilins, 9797
Urocanylcholine see 6216
Uro-Carb [Hamilton] see 1207
Urocaudal see 9515
Uro-Cedulamin [Cosmopharma] see
   5883
Urochloralic Acid, 9798
Urocit-K [Mission] see 7603
Uro-Clamoxyl [Beecham] see 610
Urocoli see 6577
Urodiaton see 8887
Urodiazin see 4704
Urodil [Pharma-Selz] see 6520
Urodin [Streuli] see 6520
Urodine [Kenyon] see 7175
Urodixin [Italchimici] see 6273
Uroenterone see 3492
Urofollitrophin see 4189
Urofort see 379
Urogastrone see 3492
β-Urogastrone see 3492
γ-Urogastrone see 3492
Urografic Acid Methylglucamine Salt
   see 5689
Urografic Acid Sodium Salt see 2975
Urografin [Schering AG] see 5689
Urokinase, 9799
Urokon Sodium [Mallinckrodt] see 72
Urolene Blue [Star] see 5979
Urolocide see 3404
Urolong [Thiemann] see 6520
Urolucosil see 8887
Uromaline see 5599
Uroman [Toho Kogyo] see 6273
Uromandelin see 5883
Uromiro [Bracco] see 4904
Uromitexan [WB Pharm.] see 5812
Uronal see 972
Uronamin see 5883
Uronase [Mochida] see 9799
Uroneg [Ibirn] see 6273
Uronorm [Alfa] see 2311
Uropac see 8584
Uropan [Tiber] see 6273
Uropen [Bristol] see 4592
Urophenyl [Sanwa] see 9230
Uropimid [CT Lab. Farma] see 7427
Uropir [Sarm] see 7475
Uroplus [Shionogi] see 8889
Uropurgol see 5881
Uro-Ripirin [Kabi] see 3515
Uroscreen [Pfizer] see 9658
Uroselectan B [Schering AG] see 8584
Urosemide [Morishita] see 4221
Uro-Septra [Wellcome] see 8889
Urosin [Boehringer, Mann.] see 278
Urosten [Bonomelli] see 7427
Urosulfan see 8906
Urosulfone see 8870
Uro-Tablinen [Sanorania] see 6520
Urothion, 9800
Urotractan [Klinge] see 5882
Urotrate see 6899

Urotropin New [Warner-Chilcott] see
   5881
Uroval [Firma] see 7427
Urovist [Schering AG] see 5689
Urox [Allied] see 6169
Uroxacin [Lazar] see 6617
Urox B [Hopkins] see 1370
Urox D [Hopkins] see 3388
Uroxin see 282
Uroxin [Von Boch] see 6899
Uroxol [Ausonia] see 6899
Uroz see 8887
Ursacol [Zambon] see 9801
Urs-12-en-3β-ol see 653
18α,19α-Urs-20(30)-en-3β-ol see
   9031
Ursin see 799
Ursnon [Nippon Yakuhin] see 4104
Urso [Tanabe Seiyaku] see 9801
Ursobilin [Hek] see 9801
Ursochol [Inpharzam] see 9801
Ursocholanic Acid see 2199
Ursodamor [Damor] see 9801
Ursodeoxycholic Acid see 9801
Ursodiol, 9801
Ursofalk [Falk] see 9801
Ursol D see 7256
Ursolic Acid, 9802
Ursol P see 474
Ursolvan [Robert et Carrière] see 9801
Urson see 9802
Urtias 100 [Sabona] see 278
Urtosal see 8297
Urumbrin see 8584
Urupan [Merckle] see 2924
Urushiol, 9803
Urusonin [Isei] see 8721
Urylon, 9804
USB-3153 see 7773
Uscharidin, 9805
Uskan [Desitin] see 6881
Usnein see 9806
Usniacin see 9806
Usnic Acid, 9806
Usninic Acid see 9806
Uspulun see 4761
USR 604 see 3032
USSR see 4695
Ustilagic Acid, 9807
Ustimon see 4626
Ustin see 6632
Ustin Methyl Ether see 6440
Ustizeain B see 9807
Utemarin [Kissei] see 8228
Uteplex see 9795
Uteracon [Hoechst] see 6934
Uteramin [Zyma] see 9743
Uterol see 4448
Uteroverdine see 1236
Utibid [Warner-Chilcott] see 6899
Uticillin [Beecham] see 1844
Uticillin VK [Upjohn] see 7046
Uticort [Parke, Davis] see 1202
Utimox [Parke, Davis] see 610
Utipina see 9795
Utopar [Ferrosan] see 8228
Utovlan [Syntex] see 6614
UTP see 9795
Utrogestan [Besins-Iscovesco] see
   7783
Uval [Dome] see 8963
Uvaleral see 1385
Uvasol see 799
Uva Ursi, 9808
Uviban [Alcon] see 134
Uvic Acid see 9038
Uvilon (Tabl.) see 7431
Uvilon (Syrup) see 7433
Uvinul 400 [GAF] see 1117
Uvinul D49 [GAF] see 1109

Uvinul M-40 [GAF] see 6907
Uvinul MS-40 [GAF] see 8963
Uvistat see 6092
UX₂ see 7897
Uzarigenin see 9809
Uzarin, 9809
Uzone [Kempthorne Prosser] see 7248
V-7 see 9634
V-12 see 5855
VAB-6 see MISC-2
Vabrocid see 6521
VAC see MISC-2
Vaccenic Acid, 9810
Vacciniin, 9811
Vacor [Rohm & Haas] see 7999
Vadebex see 6638
Vaderm [Schering] see 213
Vadilex [Robert et Carrière] see 4821
Vadosilan see 5128
Vaflol see 9918
Vagamin see 5869
Vagantin see 5869
Vagestrol [Norwich] see 3118
Vagilen [Lenza] see 6079
Vagimid [Apogepha] see 6079
Vagoprol see 4809
Vagosin Sulfate see 7770
Vagran [UPSA] see 7846
Vakerin see 1174
Val (IUPAC Abbrev.) see 9818
Valachol [Vale] see 6884
Valacidin, 9812
Valadol [Squibb] see 40
Valamin [Schering AG] see 3688
Valamina [Schering] see 4116
Valan see 9234
Val⁵-angiotensin II-asp¹-β-amide see
   680
Valaxona [Schaper & Brümmer] see
   2977
Valbazen [SK & F] see 201
Valbil [Klinge] see 3891
Valcor [Maggioni] see 3438
Valcote [Abbott] see 9821
Val²-cyclosporine see 2759
Valdrene [Vale] see 3308
Valecor see 7744
Valemicina [Farmochim. Ital.] see
   4169
Valentinite see 743
Valeral see 9813
n-Valeraldehyde, 9813
Valeramide-OM see 471
Valerbé see 9814
Valerian, 9814
Valerianic Acid see 9815
Valeric Acid Ammonium Salt see 598
Valeric Acid Bismuth Basic Salt see
   1306
Valeric Acid, Normal, 9815
Valeric Aldehyde see 9813
Valetan [Tobishi] see 3071
Valethamate Bromide, 9816
Validacin [Takeda] see 9817
Validamycin A see 9817
Validamycins, 9817
Validex [Synthelabo] see 4821
Validol see 5728
Valimon [Takeda] see 9817
Valine, 9818
5-Valine-angiotensin II Amide see
   680
Valinomycin, 9819
Valip see 9650
Valiquid [Roche] see 2977
Valisan see 1340
Valisone [Schering] see 1202
Valium [Roche] see 2977
Vallene see 5653
Vallergan [M & B] see 9618
Vallergine see 7797

Vallesine *see* 872
Vallestril [Searle] *see* 5856
Vallet's Mass *see* 3990
Valmagen [UCM-Difme] *see* 2279
Valmethamide *see* 9820
Valmid [Lilly] *see* 3688
Valmidate *see* 3688
Valmorin [Roerig] *see* 2195
Valnoctamide, 9820
Valoid [Burroughs Wellcome] *see* 2716
Valone *see* 5123
Valopride [Vita] *see* 1420
Valoron [Gödecke] *see* 9370
Valpin [Endo] *see* 701
Valproic Acid, 9821
Valpromide, 9822
Valrelease [Roche] *see* 2977
Valsol OT *see* 3398
Valsyn [Norwich] *see* 4205
Valtomicina [Midy] *see* 7420
Valtomycin *see* 7420
Valtorin [Boehringer, Ing.] *see* 2195
Valyl *see* 5122
Valzin *see* 3448
VAMP *see* MISC-2
Vanadic Anhydride *see* 9826
Vanadic Oxide *see* 9829
Vanadinite *see* 9823
Vanadium, 9823
Vanadium Carbonyl, 9824
Vanadium Hexacarbonyl *see* 9824
Vanadium Oxydichloride *see* 9832
Vanadium Oxysulfate *see* 9833
Vanadium Oxytrichloride *see* 9834
Vanadium Pentafluoride, 9825
Vanadium Pentoxide, 9826
Vanadium Sesquioxide *see* 9829
Vanadium Sesquisulfide *see* 9831
Vanadium Tetrafluoride, 9827
Vanadium Trifluoride, 9828
Vanadium Trioxide, 9829
Vanadium Trisulfate, 9830
Vanadium Trisulfide, 9831
Vanadyl Dichloride, 9832
Vanadyl Sulfate, 9833
Vanadyl Trichloride, 9834
Vanaspati, 9835
Vanay [Ayerst] *see* 9504
Vancenase [Schering] *see* 1029
Vanceril [Schering] *see* 1029
Vancide 89 *see* 1771
Vancide BN [Vanderbilt] *see* 1316
Vancide ZP [Vanderbilt] *see* 8004
Vancocin [Lilly] *see* 9836
Vancomycin, 9836
Vancor [Adria] *see* 9836
Vandid [Riker] *see* 3673
Vanectyl *see* 9618
Vanilla, 9837
Vanillal *see* 3815
Vanilla Plant *see* 5355
Vanillic Acid, 9838
Vanillic Acid Diethylamide *see* 3673
Vanillic Aldehyde *see* 9839
Vanillic Diethylamide *see* 3673
Vanillin, 9839
Vanillinemandelic Acid *see* 9840
Vanillin-D-glucoside *see* 4359
Vanilloside *see* 4359
Vanillylacetone *see* 10072
Vanillylmandelic Acid *see* 9840
Vanilmandelic Acid, 9840
Vanilone [ANA] *see* 2761
Vanitiolide, 9841
Vanobid [Merrell] *see* 1747
Vanogel [Van Pelt & Brown] *see* 345
Vanoxin [Vanguard] *see* 3150
Vanquil *see* 8034
Vanquin [Parke, Davis] *see* 8034
Vansil [Pfizer] *see* 6872
Vantol [Parke-Davis] *see* 1213

Vantyl *see* 2290
Vanylglycol *see* 5916
Vanzoate [Van Pelt & Brown] *see* 1141
Vapam *see* 5860
Vapona [Shell] *see* 3069
Vapo-N-Iso [USV] *see* 5105
Vaporpac [Menley & James] *see* 6678
Vapotone [Chevron] *see* 9138
Varacillin [Kanebo] *see* 5319
Varbian [Ciba-Geigy] *see* 7742
Vardax [Thomae] *see* 8965
Varemoid [Zyma] *see* 9698
Variamine Blue Base *see* 5922
Variaphylline LA [Varialab] *see* 477
Varidase *see* 8784
Varinon [Exa] *see* 3291
Variotin [Leo Pharm.] *see* 7008
Variscite *see* 362
Variton [Schering] *see* 3302
Varitox [M & B] *see* 9539
Varlane [Schering AG] *see* 4078
Varnish Makers' And Painters' Naphtha *see* 5366
Varnoline [Organon] *see* 2906
Varonic 400 [Ashland] *see* 7555
Varsol 1 [Exxon] *see* 6120
Varsol 3 [Exxon] *see* 6120
Varunax *see* 478
Vasaca *see* 149
Vasagin [Salus] *see* 7987
Vasal [Tutag] *see* 6968
Vasangor *see* 7821
Vasapril [Ciba] *see* 7987
Vascardin [Nicholas] *see* 5114
Vascor [J & J] *see* 1167
Vascoril [Delalande] *see* 2292
Vascuals *see* 9931
Vasculat [Boehringer, Ing.] *see* 964
Vasculit [Boehringer, Ing.] *see* 964
Vascunicol [Boehringer, Ing.] *see* 964
Vaseline [Chesebrough] *see* 7138
Vaseretic [Merck & Co.] *see* 3521
Vashegyite *see* 362
Vasimid *see* 9430
Vasiodone *see* 4933
Vaskulat *see* 964
Vasoactive Intestinal Peptide *see* 9907
Vasoactive Intestinal Polypeptide *see* 9907
Vasoc [Lindopharm] *see* 1071
Vasocard [Abbott] *see* 9084
Vasocet [Winthrop] *see* 2012
Vasocil [Magis] *see* 7987
Vasocon [SM & P] *see* 6287
Vasoconstrictine *see* 3569
Vasodiatol [Rowell] *see* 7066
Vasodil *see* 9430
Vasodilan [Mead Johnson] *see* 5128
Vasodilatan *see* 9430
Vasodilatateur 2249F *see* 6877
Vasodin [Alfa] *see* 6403
Vasodistal [Delalande] *see* 2293
Vasoglyn *see* 6528
Vasokellina *see* 5189
Vasoklin [Gödecke] *see* 6204
Vasolan [Eisai] *see* 9851
Vasoliment *see* 7138
Vasomed *see* 9682
Vasomet [Mitsubishi Kasei] *see* 9084
Vasomotal [Philips-Duphar] *see* 1200
Vasonase [Syntex Latino] *see* 6403
Vasophysin *see* 9843
Vasoplex [Lappe] *see* 5128
Vasopressin, 9843
Vasopressin 2-(L-Phenylalanine)-8-L-lysine *see* 3896
Vasorbate *see* 5114
Vasorelax [Farmasa] *see* 1041
Vasorome [Kowa] *see* 6875

Vasospan [Ulmer] *see* 6968
Vasospan [Exa] *see* 6404
Vasotec [Merck & Co.] *see* 3521
Vasotec IV [Merck & Co.] *see* 3522
Vasotonin *see* 3569
Vasotran [Bristol] *see* 5128
Vasotrate [Reid-Provident] *see* 5114
Vasoverin [Banyu] *see* 7987
Vasoxine Hydrochloride *see* 5910
Vasoxyl Hydrochloride [Burroughs Wellcome] *see* 5910
Vasperdil *see* 6428
Vaspit [Schering AG] *see* 4078
Vastarel *see* 9619
Vastcillin [Takeda] *see* 2706
Vastribil [Farmasan] *see* 9698
Vasurix [Guerbet] *see* 5688
Vasylox Hydrochloride *see* 5910
Vatensol [Pfizer] *see* 4479
Vaterite *see* 1657
Vatracin [Tobishi] *see* 2706
Vaumigan *see* 4464
Vazadrine *see* 5071
Vazofirin *see* 7092
VC9-104 *see* 3702
VC-13 *see* 3030
V-Cil *see* 7046
V-Cil-K *see* 7046
V-Cillin [Lilly] *see* 7046
V-Cillin K [Lilly] *see* 7046
VC-13 Nemacide *see* 3030
VCR *see* 9891
VCS 438 *see* 5876
VCS 506 *see* 5327
VDH [Schmiden] *see* 1253
VDS *see* 9792
Veatchine, 9844
Vebecillin *see* 7046
Vectacin [Essex] *see* 6389
Vectarion [Servier] *see* 299
Vectren [Warner-Lambert] *see* 5127
Vectrin [Parke, Davis] *see* 6121
Vecuronium Bromide, 9845
Vederon *see* 5071
Veetids [Squibb] *see* 7046
Vegadex *see* 8882
Vegetable Calomel *see* 7521
Vegetable Casein *see* 1892
Vegetable Lutein *see* 9972
Vegetable Luteol *see* 9972
Vegetable Mercury *see* 5597
Vegetable Pepsin *see* 6965
Vegetable Sulfur *see* 5495
Vegetable Wax *see* 5139
Vegiben [Amchem] *see* 2063
Vegolysen [M & B] *see* 4609
Vegolysen-T [M & B] *see* 4609
Vegolysin *see* 4609
Vehem [Sandoz] *see* 9078
Veinamitol [Negma] *see* 9698
Veinartan *see* 5844
VEL 3973 *see* 5684
Velacicline [Squibb] *see* 8236
Velacycline [Squibb] *see* 8236
Velardon [Organon] *see* 6965
Velban [Lilly] *see* 9887
Velbe [Lilly] *see* 9887
Veldopa (formerly Weldopa) [Smith & Nephew] *see* 5344
Vellosimine, 9846
Velmol [SM & P] *see* 3398
Velmonit [Esteve] *see* 2315
Velocef [Squibb] *see* 1982
Velon *see* 9900
Velopural [Neos-Donner] *see* 4711
Velosef [Squibb] *see* 1982
Velosimine *see* 9846
Velosulin [Nordisk] *see* 4889
Velpar [Du Pont] *see* 4617
Velsicol 104 *see* 4576
Velsicol 1068 *see* 2079

Velsicol 58-CS-11 see 3026
Veltane [Lannett] see 1433
Veltol [Pfizer] see 5594
Vems [ISF] see 9502
Venacil (formerly) [Abbott] see 664
Venactone [Lepetit] see 1753
Venagil [Cristalfarma] see 1071
Venarterin [B.O.I.] see 5844
Vendarcin see 6931
Ven-Detrex [Zyma; Therabel Pharma] see 3291
Vendex [Shell] see 3907
Vendracin see 6931
Venen [Tanabe] see 9661
Veneniferin see 6386
Venetlin [Sankyo] see 209
Venex [Recordati] see 9520
Venex [Lusofarmaco] see 3291
Vengeance [Velsicol] see 1378
Vengicide see 9483
Venice Turpentine, 9847
Venopex [Apotex] see 2279
Venopirin [Green Cross] see 5510
Venoruton [Zyma] see 9698
Venosmine [Zyma] see 3291
Veno-V [Vitoria] see 3291
Ventaire [Marion] see 7911
Ventax [Robert] see 3427
Venter [Krka] see 8853
Ventilat [Dieckmann] see 6897
Ventipulmin [Boehringer, Ing.] see 2347
Ventodisks [Glaxo] see 209
Ventolin [Allen & Hanburys] see 209
Ventolin Obstetric Injection [Allen & Hanburys] see 209
Ventox see 125
Ventramine see 6459
Ventrazol see 7097
Venturicidin A see 9848
Venturicidin B see 9848
Venturicidins, 9848
Ventussin [Warren-Teed] see 1106
Venzar [Du Pont] see 5318
Venzonate see 1141
Venzoquimpe [Quimpe] see 5830
Vepen see 7046
Vepesid [Bristol-Myers] see 3842
Veracillin [Ayerst] see 3073
Veractil see 5909
Veracur [Typharm] see 4150
Veradol [Schering] see 6337
Veralba [Pitman-Moore] see 7915
Veralcamine see 9850
Veralipride, 9849
Veralipril [Midy] see 9849
Veralkamine, 9850
Veramex [Labaz] see 9851
Veramix see 5677
Veramon see 7146
Veranterol [Asla] see 7987
Verapamil, 9851
Veraptin [Lagap] see 9851
Veratetrine see 7915
Veratraldehyde, 9852
Veratramine, 9853
Veratran [Latema] see 2411
Veratric Acid, 9854
Veratric Acid 4-[Ethyl(p-methoxy-α-methylphenethyl)amino]butyl Ester see 5648
Veratric Aldehyde see 9852
Veratridine, 9855
Veratrine see 2025
Veratrine (Mixture), 9856
Veratrole, 9857
Veratroylaconine see 7924
3-Veratroylveracevine see 9855
Veratrum viride, 9858
Verax [Tosi-Novara] see 1136
Verazide, 9859

Verazinc [OJ & F] see 10064
Verbascose, 9860
Verbenalin, 9861
Verbenalol see 9861
d-Verbenone, 9862
Vercyte [Abbott] see 7451
Verdiana see 9650
Verdisol [Arovet] see 3069
Verecolene see 3912
Verel [Eastman Kodak] see 6140
Verelan [Elan] see 9851
Verexamil [Verex] see 9851
Vergentan [Schürholz] see 236
Vergitryl [Squibb] see 9858
Vergonil see 4716
Veriloid [Riker] see 9858
Veritab [Vista] see 5657
Veritol see 7304
Verladyn [Verla] see 3157
Vermicide Bayer 2349 see 9536
Vermicidin [Rivopharm] see 5647
Vermicompren (Tabl.) see 7432
Vermicompren (Syrup) see 7431
Vermiculite, 9863
Vermilass see 7432
Vermilion see 5789
Verminum [Squibb] see 6985
Vermiparin see 5600
Vermirax [Biosintetica] see 5647
Vermisol [Merck & Co.] see 7431
Vermitiber [Tiber] see 8034
Vermitid see 6425
Vermitin see 7220
Vermizym see 6965
Vermouth see 8
Vermox [Janssen] see 5647
Vernadigin, 9864
Vernal Pheasant's Eye see 156
Vernam [Stauffer] see 9866
Vernamycin A see 9910
Vernamycin B, 9865
Vernamycin B_α see 6111 and 9865
Vernamycin B_β see 9865
Vernamycin B_γ see 9865
Vernamycin B_δ see 9865
Vernine see 4480
Vernitest Reagent see 8056
Vernolate, 9866
Vernolepin, 9867
Vernolic Acid, 9868
Verol see 5940
Veroletten see 972
Veronal see 972
Veronal Sodium see 972
Verophene see 7793
Verospiron see 8721
Veroxil see 7436
Veroxil [Baldacci] see 4847
Verrucarin A see 9869
Verrucarin B see 9869
Verrucarin J see 9869
Verrucarin K see 9869
Verrucarins, 9869
Verrucarol see 9869
Verrugon [Pickles] see 8301
Versacort [Angelini] see 4710
Versalide®, 9870
Versamine see 5654
Versapen [Bristol] see 4592
Versapen K [Bristol] see 4592
Versatrex [Bristol] see 4592
Versed [Roche] see 6105
Versene [Dow] see 3482
Versene-9 [Dow] see 3483
Versene Acid [Dow] see 3484
Versene CA [Dow] see 3480
Versene Disodium Salt see 3481
Versen-Ol®, 9871
Versotrane see 4687
Verstran [Warner-Chilcott] see 7713
Versulin see 763

Versus [Angelini Francesco] see 1042
Verticillin A see 9872
Verticillin B see 9872
Verticillin C see 9872
Verticillins, 9872
Verticine, 9873
Vertigon [SK&F] see 7768
Vertolan see 8886
Verton [Squibb] see 2102
Verucasep [Galen] see 4366
Verv-Ca see 1711
Verxite see 9863
Vesadin see 8886
Vesamin [Byk-Gulden] see 72
Vesicating Collodion see 2481
Vesidril see 6062
Vesidryl [Clin-Comar-Byla] see 6062
Vésipaque see 7203
Vesipyrin see 7241
Vesitan see 2372
Vespazine see 4116
Vespéral see 972
Vesperone [UCB] see 1357
Vespral see 9597
Vesprin [Squibb] see 9597
Vestalin see 6619
Vesulong [Ciba-Geigy] see 8923
Vesuvianite see 1645
Vesuvine see 1264
Vetalar [Parke, Davis] see 5174
Vetalog [Squibb] see 9512
Vetame [Squibb] see 9597
Veta-Merazine see 8884
Vetamox [Am. Cyanamid] see 45
Vetarsenobillon see 6361
Veteusan [Veterinaria] see 2597
Vetibenzamina see 9651
Veticillin [Am. Cyanamid] see 1157
Veticol [Copanos] see 2068
Vetidrex [Ciba] see 4704
Vetisulid [Ciba] see 8871
Vetiver Oil see 6778
Vetiver Oil Haiti see 6778
Vetiver Oil Java see 6778
Vetiver Oil Reunion (Bourbon) see 6778
Vetivert Oil see 6778
α-Vetivone see 9874
β-Vetivone see 9874
Vetivones, 9874
Vetomazin see 5923
Vetquamycin-324 [Rachelle] see 9130
Vetrabutine, 9875
Vetranquil [Lathevet] see 27
Vetrazin [Ciba-Geigy] see 2782
Vetsin see 6165
Vetstrep [Merck & Co.] see 8786
Vetyver Oil see 6778
Viaben [Schürholz] see 1420
Viacutan see 5874
Viadent [Vipont] see 8320
Viadril see 4753
Vialibran [Servier] see 5671
Vialidon [Italfarmaco] see 5680
Vialin see 5740
Vi-Alpha see 9918
Vianin see 4287
Vianol see 1548
Viansin [Crinos] see 2082
Viapta [Riker] see 5882
Viarespan [Servier] see 3942
Viarex [Schering] see 1029
Viarox [Schering] see 1029
Viasept see 4389
Viatussin see 3333
Vibalt [Roerig] see 9921
Vibazine [Pfizer] see 1449
Vibeline [Roger Bellon] see 9915
Vibisone see 9921
Vibradox [Cifa] see 3429

Vibramycin [Pfizer] see 3429
Vibramycin Hyclate [Pfizer] see 3429
Vibra-Tabs [Pfizer] see 3429
Vibravenös [Pfizer] see 3429
Vibriomycin see 3162
Viburnum opulus, 9876
Viburnum prunifolium, 9877
Vi-Cad [Vineland] see 1615
Viccillin [Meiji] see 621
Vicelat [Bayer] see 855
Vicemycetin see 9230
Viceton [Intl. Multifoods] see 2068
Vicianin, 9878
Vicianose, 9879
Vicilan [ICI] see 9884
Vicin see 9880
Vicine, 9880
Vicin Neolin see 7037
Vicioside see 9880
Victan [Clin Midy] see 3777
Victoria Green B see 5581
Victoria Green WB see 5581
Vidarabine, 9881
Vidden D [Dow] see 3064
Vi-De-3-hydrosol see 9929
Videne [Riker] see 7701
Videobil [Bracco] see 4949
Videophel see 4931
Viderpen [Sintex] see 5830
Vidine see 2207
Vidopen [Berk] see 621
Vidora [Wyeth] see 4876
Vienna Green see 2629
Vigabatrin, 9882
Vigantol [E. Merck; Bayer] see 9929
Vigazoo [Am. Cyanamid] see 8864
Vigil [ICI] see 3070
Vigilor [Bouchard] see 4024
Vigorsan [Albert-Roussel] see 9929
Vikane [Dow] see 8960
Vikastab see 5713
Vilan [Lannacher] see 6430
Villescon [Boehringer, Ing.] see 7791
Villiaumite see 8565
Villikinin, 9883
Viloxazine, 9884
Vimicon [Frosst] see 2779
Viminalol see 653
Viminol, 9885
Vinacort see 7563
Vinactane [Ciba] see 9905
Vinarol [Hoechst] see 7562
Vinbarbital Sodium, 9886
Vinblastine, 9887
Vincadar [Roussel-Amor-Gil] see
   9888
Vincafarm [Radiumfarma] see 9888
Vincafolina [Lampugnani] see 9888
Vincafor [Clin-Comar-Byla] see 9888
Vincagil [Sarsa] see 9888
Vincalen [Firma] see 9888
Vincaleukoblastine see 9887
Vincamajoridine see 190
Vincamidol [Magis] see 9888
Vincamine, 9888
Vincaminine see 9888
Vincamone see 3464
Vincanorine see 3464
Vincapan [Lutetia] see 9888
Vincapront [Mack, Illert.] see 9888
Vincasaunier [Saunier] see 9888
Vinces see 8890
Vincetoxin, 9889
Vincimax [Robert et Carrière] see
   9888
Vincine see 9888
Vincinine see 9888
Vinclozolin, 9890
Vincosid [AB Leo] see 9891
Vincrex [Bristol] see 9891
Vincristine, 9891

Vindesine, 9892
Vindolidine see 9893
Vindoline, 9893
Vindorosine see 9893
Vinegar Naphtha see 3713
Vinesthene see 9899
Vinethene see 9899
Vinetine see 6906
Vingsal [Stroschein] see 6616
Vi-Nicotyl see 6398
Vinisil [Abbott] see 7700
Vinodrel Retard [Lepetit] see 9888
Vinol see 7562
Vinpocetine, 9894
Vintiamol, 9895
Vinyl Acetate, 9896
Vinylacetonitrile see 288
γ-Vinyl-γ-aminobutyric Acid see
   9882
Vinylbenzene see 8830
Vinylbital, 9897
Vinyl Carbinol see 284
Vinyl Chloride, 9898
Vinyl Cyanide see 125
α-Vinyldiacetonalkamine Benzoate
   see 3850
Vinylestrenolone see 6619
17α-Vinyl-5(10)-estren-17β-ol-3-one
   see 6619
Vinyl Ether, 9899
Vinylethylene see 1500
Vinylformic Acid see 124
γ-Vinyl GABA see 9882
Vinylidene Chloride, 9900
Vinylidene Cyanide-vinyl Acetate
   Copolymer see 2821
Vinylite V [Union Carbide] see 7563
5-Vinyl-5-(1-methylbutyl)barbituric
   Acid see 9897
17α-Vinyl-19-nortestosterone see
   6637
(—)-5-Vinyl-2-oxazolidinethione see
   4411
1-Vinyl-2-pyrrolidinone Polymers see
   7700
1-Vinyl-2-pyrrolidinone Polymers,
   Iodine Complex see 7701
5-Vinyl-2-thiooxazolidone see 4411
Vinyl Trichloride see 9550
Vinyzene [Ventron] see 6914
Vio-Bamate [Rowell] see 5751
Viobilina [Violani-Eurolabor] see
   2721
Viocid see 4287
Viocin [Pfizer] see 9905
Vio-D [Rowell] see 9928
Vioform [Ciba] see 4924
Viokase [Robins] see 6957
Violacein, 9901
Violanin see 2866
Violaquercitrin see 8276
Violaxanthin, 9902
Viologen, 9903
Violuric Acid, 9904
Viomycin, 9905
Viomycin Pantothenate, 9906
Vionactan see 9906
Vionactane see 9905
Viophan see 2290
Viopsicol [Violani Eurolabor] see 2082
Viosterol see 9928
Viothenat see 9906
Vio-Thene [Rowell] see 6926
Viotil [Violani] see 7022
Viozene see 8239
VIP, 9907
Vipicil [Fontoura-Wyeth] see 2706
Vipral [Pharma Investi] see 139
Viprimol [Roche] see 9931
Viprynium Chloride see 8033
Viprynium Embonate see 8034

Viquidil, 9908
Vira-A [Parke, Davis] see 9881
Virac see 9759
Viractin, 9909
Viramid [ICN; Usafarma] see 8199
Viras [Kanto] see 8201
Virazene see 7219
Virazid [Britannia] see 8199
Virazole [ICN] see 8199
Virex [Burgin-Arden] see 7900
Virgimycin see 9910
Virginiamycin, 9910
Virginiamycin M₁ see 9910
Virginiamycin S₁ see 9910
Viridicatin, 9911
Viridin, 9912
Viridofulvin see 2897
Viridogrisein see 3664
Virlix [Carrion] see 2013
Virofral [Boehringer, Mann.] see 380
Vironil see 6181
Virophta [Dulcis] see 9599
Viroptic [Burroughs Wellcome] see
   9599
Virorax [Siam Bheasach] see 139
Virormone [Paines & Byrne] see 9115
Virosterone [Endo] see 9109
Virubra see 9921
Virudox [Bioglan] see 4819
Virugon see 6181
Viru-Merz [Merz] see 9683
Viruserol [Zyma] see 9683
Virusmin see 6181
Virustat [Delagrange] see 6181
Viruxan [Gramon/Newport; Sigma-
   Tau] see 4881
Viruzona see 5900
Visacor [ICI] see 3557
Visammin see 5189
Viscardan see 5189
Visceralgina [Lirca] see 9363
Visclair [Sinclair] see 5668
β-Viscol see 5478
Viscoleo [Lundbeck] see 4114
Viscose, 9913
Viscosin, 9914
Viscotiol [Erba] see 5330
Visha see 115
Visine [Pfizer] see 9150
Visiren [Jugoremedija] see 6688
Viskaldix [Sandoz] see 2391
Visken [Sandoz] see 7412
Visnacorin see 9916
Visnadine, 9915
Visnagalin see 5189
Visnagen see 5189
Visnagin, 9916
Visnamine [Chinoin] see 9915
Visotrast see 72
Vistagan [Pharm-Allergan] see 5343
Vista-Methasone [Richard Daniel] see
   1202
Vistamycin [Meiji] see 8208
Vistapin [Allergan] see 3344
Vistar [3M] see 5684
Vistaril Pamoate [Pfizer] see 4786
Vistaril Parenteral [Pfizer] see 4786
Visual Purple see 8193
Visubeta see 1202
Visubutina see 6925
Vitabact [Faure] see 7371
Vitacee [Endo] see 855
Vitacimin see 855
Vitacin see 855
Vitadurin see 4739
Vitagutt [Medicator] see 9932
Vital Red, 9917
Vital Red Evans see 9917
Vitamin A, 9918
5-cis-Vitamin A see 6382
Vitamin A Acid see 8167

Vitamin A Alcohol see 9918
Vitamin A Aldehyde see 8165
Vitamin A Epoxide see 4572
Vitamin A$_1$ see 9918
Vitamin A$_2$ see 9919
Vitamin A$_2$ Aldehyde see 2862
Vitamin B$_1$ O,S-Diacetate see 46
Vitamin B$_1$ Disulfide see 9221
Vitamin B$_1$ Hydrochloride see 9221
Vitamin B$_1$ Mononitrate see 9223
Vitamin B$_1$ Propyl Disulfide see 7896
Vitamin B$_1$ Triphosphate Ester see 9227
Vitamin B$_2$ see 8201
Vitamin B$_2$ Phosphate (Sodium Salt) see 8202
Vitamin B$_3$ see 6398 and 6964
Vitamin B$_4$ see 141
Vitamin B$_5$ see 6398 and 6964
Vitamin B$_6$ Hydrochloride see 7995
Vitamin B$_{10}$ and Vitamin B$_{11}$, 9920
Vitamin B$_{12}$, 9921
Vitamin B$_{12}$-Co(II) see 9925
Vitamin B$_{12}$ Coenzyme see 2446
Vitamin B$_{12}$, Radioactive, 9922
Vitamin B$_{12}$—Zinc Tannate Complex, 9923
Vitamin B$_{12a}$ see 4739
Vitamin B$_{12b}$ see 787
Vitamin B$_{12c}$, 9924
Vitamin B$_{12p}$ see 3825
Vitamin B$_{12r}$, 9925
Vitamin B$_{12s}$, 9926
Vitamin B$_{13}$ see 6828
Vitamin B$_{15}$ see 6959
Vitamin B$_{17}$ see 629
Vitamin Bc see 4140
Vitamin Bc Conjugate see 7951
Vitamin B$_T$ see 1856
Vitamin B$_T$ Acetate see 78
Vitamin C see 855
Vitamin C Sodium see 8525
Vitamin D$_1$, 9927
Vitamin D$_2$, 9928
Vitamin D$_3$, 9929
Vitamin D$_4$, 9930
Vitamin E, 9931
Vitamin E Acetate, 9932
Vitamin E Acid Succinate see 9424
Vitamin F see 3650
Vitamin G see 8201
Vitamin H see 1244
Vitamin K$_1$, 9933
Vitamin K$_1$ Hydroquinone see 3165
Vitamin K$_1$ Oxide, 9934
Vitamin K$_1$ 2,3-Oxide see 9934
Vitamin(s) K$_2$, 9935
Vitamin K$_{2(0)}$ see 5714
Vitamin K$_3$ see 5714
Vitamin K$_4$ see 5710
Vitamin K$_5$, 9936
Vitamin K$_6$, 9937
Vitamin K$_7$, 9938
Vitamin K-S(II), 9939
Vitamin L$_1$ see 433
Vitamins L, 9940
Vitamin M see 4140
Vitamin P Complex see 1241
Vitamin PP see 6398
Vitamin T, 9941
Vitamin T Goetsch see 9941
Vitamin U, 9942
Vitamogen see 9221
Vitaneuron see 9222
Vitanevril [Clin-Comar-Byla] see 1049
Vitarubin see 9921
Vita-Rubra see 9921
Vitascorbol [Specia] see 855
Vitasprint B$_{12}$ [Efeka] see 7338
VitaStain see 9658

Vitas-U see 9942
Vitavax [Uniroyal] see 1832
Vitavel K see 5710
Vitaverin see 9221
Vitazechs see 7992
Vitellin see 5311
Viteolin see 9931
Viton [Mallinckrodt] see 5379
Vitpex see 9918
Vitral [Tilden-Yates] see 9921
Vivactil [Merck & Co.] see 7917
Vival see 2977
Vivalan [ICI] see 9884
Vivarint [ICI] see 9884
Vivatec [Merck & Co.] see 5393
Vividrin [Mann] see 2594
Vividyl [Lilly] see 6635
Vivol [Horner] see 2977
VK 55 see 8878
VLB see 9887
Vleminckx's Lotion see 5369
Vleminckx's Soln see 5369
VM-26 see 9078
VMA see 9840
VMF see MISC-2
V.M.&P. Naphtha see 5366
Voacamine, 9944
Voacangine see 9944
Voacanginine see 9944
Vodka, 9945
Vodol [Andromaco] see 6101
Voelicherite see 1701
Vogalene [Théraplix] see 6070
Vogan see 9918
Vogan-Neu see 9918
Vogesensäure (German) see 9038
Volatile Oil of Laurel see 6771
Volatile Oil of Mustard see 293
Volaton [Bayer] see 7341
Voldal [Zyma] see 3071
Volex [McGaw] see 4593
Volidan [BDH; Searle] see 5687
Volital [L.A.B.] see 7021
Volmax [Alza] see 209
Volon [Heyden] see 9511
Volon A [Heyden] see 9512
Volonimat [Heyden] see 9512
Volpo [Croda] see 7554
Voltaren [Ciba-Geigy] see 3071
Voltarol [Ciba-Geigy] see 3071
Voluntal see 9561
Vomex A [Endopharm] see 3193
Vomicine, 9946
Vomit Nut see 6652
Vomitoxin, 9947
Vomit Wort see 5431
Vonamycin Powder V see 6369
Vondozeb Plus [Pennwalt] see 5598
Vonedrine see 7280
Vontil [SK & F] see 9287
Vontrol [SK & F] see 3311
Vonum [Kanoldt] see 4874
Voranil [USV] see 2406
Voren see 2922
Vortel [Lilly] see 2404
Voveran [Ciba-Geigy] see 3071
VP see MISC-2
VP-16-213 see 3842
VPM see 5860
Vraap [Inverni] see 9888
V-Sul [Vangard] see 8930
Vucine see 3668
VUFB-6453 see 6059
VUFB-6638 see 9095
Vulcamicina see 6641
Vulcamycin see 6641
Vulgarobufotoxin see 1468
Vulkacit NPV/C [Mobay] see 3759
Vulkamycin see 6641
Vulkazit see 3325

Vulklor see 2071
Vumide [Burroughs Wellcome] see 2313
Vumon [Sandoz] see 9078
VX, 9948
Vybak see 7563
Vydate [Du Pont] see 6873
Vyrene see 8688
W 37 see 1465
W 45 Raschig see 478
W 50 Base see 5870
W 483 see 3703
W 491 see 2590
W 583 see 5653
W 1191-2 see 379
W 1544a see 7181
W 1889 see 6502
W 2197 see 7094
W 2900A see 3845
W 2946M see 8142
W 2979M see 922
W 3399 see 8069
W 3566 see 8066
W 3699 see 7457
W 4020 see 7713
W 4565 see 6899
W 4869 see 7728
W 5219 see 7785
W 5759A see 9370
W 5975 see 1202
W 6309 see 3134
W 6421A see 5343
W 7000A see 5343
W 7320 see 212
W 8495 see 5127
W 19053 see 3818
W 36095 see 9414
WAC 104 see 1239
Wahoo see 3856
Wait's Green Mountain Antihistamine see 7996
Wakaflavin L [Wakamoto] see 8201
Walconesin see 5737
Waldmeister (German) see 9961
Wampocap [Wampole] see 6435
Wansar [Hoei] see 3311
Waran see 9950
Warbex [Am. Cyanamid] see 3882
Warburganal, 9949
Warduzide see 2169
Warfarin, 9950
Warfarin-deanol see 9950
WARF Compound 42 see 9950
Warfilone [Frosst] see 9950
Warticon [Cph] see 7520
Waruzol [Farmanic] see 878
Washed Sulfur see 8956
Washing Soda see 8541
Water, 9951
Water-d$_2$ see 2920
Water Bugle see 5496
Water Elder see 9876
Water Gas, 9952
Water Glass see 8631
Water Hemp see 383
Watermelon, 9953
Water Shamrock see 5729
Wavellite see 362
Waxsol [Norgine] see 3398
Waynecomycin [Upjohn] see 5378
WBA 8119 see 1368
WE 941-BS see 1439
Weedazol [Amchem] see 506
Weed Hoe [Vineland] see 5864
Weedone [Union Carbide] see 8999
Weedone Aero Concentrate [Amchem] see 2802
Weedone Crabgrass Killer Granular [Amchem] see 5864
Weed-E-Rad 120 [Vineland] see 5864
Weed-E-Rad 360 [Vineland] see 5864

Wehydryl [Hauck] see 3308
Weifacodine see 7303
Weinsäure (German) see 9039
dl-Weinsäure (German) see 9038
Weinstein see 807
Weinsteinsäure (German) see 9039
Weiselfuttersaft (German) see 8254
Weisspiessglanz see 743
Welfurin see 6520
Welkstoff see 5491
Wellbatrin [Burroughs Wellcome] see 1488
Wellbutrin [Wellcome] see 1488
Wellcome 33-A-74 see 884
Wellcome Prepn 47-83 see 2716
Wellcome-248U see 139
Wellcoprim [Wellcome] see 9624
Wellcovorin [Burroughs Wellcome] see 4141
Welldorm [Smith & Nephew] see 3033
Wellferon [Burroughs Wellcome] see 4892
Welvic see 7563
Wemid [Bernabo] see 3632
Weradys [Weimer] see 5687
Wescodyne [West Chemical] see 4930
Wespuril [Spitzner] see 3059
Westamine X see 4930
Westcort-Cream [Westwood] see 4710
Western Poison Oak see 7526
Westocaine [Westerfield] see 7763
Westrosol see 9552
Weymouth Pine see 9956
WG 253 see 8223
WG 537 see 4064
WG 696 see 9562
WGA see 5312
Wh 7286 see 9987
Wheat Germ Agglutinin see 5312
Wheat Germ Oil, 9954
Whey Factor see 6828
Whip [Hoechst Roussel] see 3929
Whipcide [Pitman-Moore] see 7348
Whiskey see 9955
Whisky, 9955
White Agaric see 173
White Arsenic see 832
White Beeswax see 1031
White Bole see 5162
White Cinnamon see 1749
White Dextrin see 2931
White Flag see 6832
White Lead see 1016
White Mercuric Precipitate see 5771
White Mineral Oil see 7139
White Oak see 8048
White Oil of Camphor see 6702
White Phosphorus see 7321
White Pine, 9956
White Precipitate see 5771
White Saunders see 8317
White Spirits see 6120
White Tin Oxide see 8736
White Vitriol see 10064
White Walnut see 5149
White Wax see 1031
Whiting see 1657
Whitlockite see 1701
WHO 4939 see 8145
Whorehouse Tea see 3560
WHR 539 see 3914
WHR 1142A see 5358
Widdrene see 9327
Widecillin [Meiji] see 610
Wieland-Gumlich Aldehyde, 9957
Wijs' Chloride see 4911
Wild Bergamot see 6153
Wild Black Cherry Bark see 9958
Wild Canilla see 1749
Wild Chamomile see 5639
Wild Cherry, 9958

Wild Cinnamon see 1749
Wild Cotton see 854
Wildfire Toxin, 9959
Wild Ginger see 850
Wild Indigo see 968
Wild Ipecac see 773
Wild Pepper see 6095
Wild Saffron see 2471
Wild Tobacco see 5431
Wild Woodbine see 4278
Wild Yam see 3287
Wilkinite see 1062
Wilkinson's Catalyst see 2174
Willbutamide see 9432
Willemite see 10062
Willestrol see 3119
Wilpo [Dorsey] see 7232
Wilprafen [Mack, Illert.] see 5148
Win 771 see 4769
Win 1539 see 5180
WIN 2747 see 1116
WIN 2848 see 9207
Win 3046 see 5053
Win 3706 see 393
Win 4369 see 7084
Win 5047 see 2076
Win 5063-2 see 9230
Win 5494 see 608
Win 8077 see 387
Win 8851-2 see 9745
WIN 9317 see 7821
Win 11450 see 1054
WIN 11831 see 5454
Win 13146 see 9055
Win 13820 see 1026
WIN 14098 see 7403
WIN 14833 see 8754
WIN 17757 see 2811
Win 18320 see 6273
WIN 18446 see 4009
WIN 18501-2 see 6924
Win 20228 see 7078
WIN 20740 see 2710
Win 21904 see 222
WIN 22005 see 9799
Win 24540 see 9611
Win 32784 see 1317
Win 35213 see 8247
Win 35833 see 2314
WIN 39103 see 6077
WIN 39424 see 4940
Win 40014 see 8068
Win 40680 see 626
Win 40808-7 see 8925
WIN 47203 see 6117
Win 49375 see 415
WIN 49375-3 see 415
Win AM 13146 see 9055
Wincoram [Winthrop] see 626
Wind Flower see 7956
Wine Ether see 3793
Wine Lees see 807
Wingom [Grünenthal] see 4257
Win-Kinase [Winthrop] see 9799
Winobanin [Winthrop] see 2811
Winstrol [Winthrop] see 8754
Winter Bloom see 4521
Wintergreen Oil see 6038
Wintermin see 2186
Wintersteiner's Compound A see 265
Wintersteiner's Compound B see 271
Wintersteiner's Compound D see 267
Wintersteiner's "Compound F" see 2533
Wintersteiner's Compound G see 263
Wintodon see 4389
Wintomylon see 6273
Wintonin [Winthrop] see 4295
Winuron [Winthrop] see 8247
Wirnesin [Inpharzam] see 7889
Witch Hazel see 4521

Withaferin A, 9960
Witherite see 979
WL 140 see 1704
WL 291 see 10089
WL 8008 see 9591
WL 19805 see 2692
WL 41706 see 3936
WL 43775 see 3952
Wolf's Bane see 813
Wolf's-bane see 111
Wolfin [Luar] see 8131
Wolfram see 9722
Wolframite see 9722
Wollastonite see 1709
Wometin see 4164
Wood Alcohol see 5868
Wood Creosote see 2575
Wood Oil see 957
Woodruff, 9961
Woodsour see 971
Wood Spirit see 5868
Wood Sugar see 9995
Wood Vinegar see 8014
Woodward Herb see 9961
Woodward's Reagent K, 9962
Woody Nightshade see 3447
Wool Blue RL see 662
Wool Fat see 5231
Woorali see 2678
Woorari see 2678
Worenine, 9963
Worm-Agen [Merz] see 4633
Worm-Away [Robins] see 7431
Worm-Chek [Ralston Purina] see 9161
Wormgrass see 8704
Worm Guard [SK & F] see 6985
Wormwood see 8
Wortmannin, 9964
Wourara see 2678
WR 142490 see 5683
WR 171669 see 4508
Wrightine see 2492
WS-4545 Antibiotic see 1222
WSX-8365 see 3282
WT 161 see 8141
Wulfenite see 5290
Wurmirazin see 7431
Wurster's Blue see 9159
Wurster's Reagent see 9159
Wurtzite see 10065
WV 569 see 6679
WX 2412 see 7121
Wy 401 see 3698
Wy 757 see 7786
WY 806 see 6888
Wy 1094 see 7793
Wy 1359 see 7834
Wy 1395 see 9622
Wy 2039 see 1799
WY 3263 see 4962
Wy 3467 see 2977
Wy 3475 see 6603
Wy 3478 see 8603
Wy 3498 see 8603
Wy 3917 see 9074
Wy 4036 see 5456
Wy 4082 see 5458
Wy 4508 see 2706
Wy 8138 see 1310
Wy 8678 see 4469
WY 15705 see 2316
WY 16225 see 2934
Wy 20788 see 7024
Wy 21743 see 6879
Wy 21901 see 4876
WY 22811 see 5753
Wy 42422 see 9662
Wy 42462 see 9662
WY 44635 see 1940
Wyamine [Wyeth] see 5740

Wyamycin E [Wyeth] *see* 3626
Wycillin [Wyeth] *see* 7042
Wydase [Wyeth] *see* 4676
Wydora [Wyeth-Pharma] *see* 4876
Wymesone *see* 2922
Wymox [Wyeth] *see* 610
Wynestron [Wyeth] *see* 3660
Wyovin Hydrochloride [Wyeth] *see* 3087
Wypax [Yamanouchi] *see* 5456
Wypres [Wyeth] *see* 4876
Wypresin [Wyeth] *see* 4876
Wytensin [Wyeth] *see* 4469
Wytrion *see* 9681
Wyvital [Wyeth] *see* 2706
X-340 *see* 8155
X-1497 *see* 5890
X-5108 *see* 900
X-537A *see* 5243
Xahl [SS Pharm.] *see* 1971
Xalyl *see* 5122
Xamamina *see* 3193
Xametina [Zambeletti] *see* 9623
Xamoterol, 9965
Xamtol [ICI] *see* 9965
Xanax [Upjohn] *see* 310
Xanbon [Kissei] *see* 6935
X-Andron *see* 2774
Xanef [Merck & Co.] *see* 3521
Xanor [Upjohn] *see* 310
Xanteline *see* 5869
Xanthaline *see* 6966
Xanthan Gum, 9966
Xanthatin, 9967
9*H*-Xanthen-9-one *see* 9971
Xanthine, 9968
Xanthine Riboside *see* 9974
Xanthinol Niacinate, 9969
Xanthinol Nicotinate *see* 9969
Xanthium [Galephar] *see* 9212
Xanthocillin, 9970
Xanthocillin X *see* 9970
Xanthocillin Y₁ *see* 9970
Xanthocillin Y₂ *see* 9970
Xanthokermesic Acid *see* 5209
"Xanthomycin" *see* 4468
Xanthone, 9971
Xanthophyll, 9972
Xanthopterin, 9973
Xanthopuccine *see* 1744
Xanthosine, 9974
Xanthothricin *see* 9480
Xanthotoxin *see* 5911
Xanthoxime *see* 9971
Xanthoxyletin, 9975
Xanthoxylin, 9976
Xanthoxylin N *see* 9975
Xanthoxylin S *see* 848
Xanthoxyloin *see* 9975
Xanthurenic Acid, 9977
Xanthyletin, 9978
Xantium [Cyanamid] *see* 9503
Xantociclina [SPA] *see* 4468
Xantocillin *see* 9970
Xantomicina *see* 4468
Xanturat [Grünenthal] *see* 278
Xanturil *see* 6196
Xarcin [Teikoku Hormone] *see* 8252
Xatral [Synthelabo] *see* 228
Xavin [Chinoin] *see* 9969
Xaxa *see* 873
XE-779L *see* 3261
Xenagol *see* 7174
Xenalamine *see* 9979
Xenalon [Mepha] *see* 8721
Xenar [Alfa] *see* 6337
Xenazoic Acid, 9979
Xenbucin, 9980
Xeneisol® Xe 133 [Mallinckrodt] *see* 9981
"Xenic Acid" *see* 9981

Xenid [Biogalenique] *see* 3071
Xenon, 9981
Xenon Difluoride *see* 9981
Xenon Hexafluoride *see* 9981
Xenon Platinum Hexafluoride *see* 9981
Xenon Tetrafluoride *see* 9981
Xenon Trioxide *see* 9981
Xenotime *see* 10015
Xenovis *see* 9979
Xenylamine *see* 1248
α-(*p*-Xenyl)butyric Acid *see* 9980
p-Xenylcarbimide, 9982
Xenysalate *see* 1247
Xenytropium Bromide, 9983
Xerac BP5 [Person & Covey] *see* 1128
Xerac BP10 [Person & Covey] *see* 1128
Xerene [Martinet] *see* 5739
Xeroform [Warner-Chilcott] *see* 1305
Xibenolol, 9984
Xibornol, 9985
Ximaol *see* 6670
Ximos [Glaxo] *see* 1951
Xipamide, 9986
Xiphen [Mepha] *see* 7092
Xitix *see* 855
XK-62-2 *see* 6104
XL-7 *see* 1316
XL-90 *see* 4465
Xobaline [Albert-Roussel] *see* 2446
Xolamin [Sanko] *see* 2345
Xonaltite *see* 1709
Xonotlite *see* 1709
X-Otag *see* 6831
Xtro [Xttrium] *see* 892
Xumbradil [Astra Pharm.] *see* 4933
Xyduril [Dorsch] *see* 2374
Xylamide *see* 7785
Xylan Hydrogen Sulfate *see* 7090
Xylan Polysulfate *see* 7090
Xylans *see* 4561
Xylazine, 9987
Xylene, 9988
m-Xylene *see* 9988
o-Xylene *see* 9988
p-Xylene *see* 9988
Xylenol, 9989
as-m-Xylenol *see* 9989
as-o-Xylenol *see* 9989
p-Xylenol *see* 9989
sym-m-Xylenol *see* 9989
vic-m-Xylenol *see* 9989
vic-o-Xylenol *see* 9989
Xylenol Blue, 9990
p-Xylenolsulfonephthalein *see* 9990
Xylestesin *see* 5359
Xylidine, 9991
Xylite *see* 9992
Xylitol, 9992
Xyliton [Morishita] *see* 9992
L-Xyloascorbic Acid *see* 855
Xylocaine [Astra Pharm.] *see* 5359
Xylocard [Astra] *see* 5359
Xylocitin *see* 5359
Xyloidin *see* 8022
Xylol *see* 9988
Xylomed [Bio-Medical] *see* 9995
Xylometazoline, 9993
Xylo-Mucine *see* 1835
Xylonest *see* 7749
Xyloneural [Nicholas] *see* 5359
Xylo-Pfan [Adria] *see* 9995
Xylopropamine, 9994
2-[(6-*O*-β-D-Xylopyranosyl-β-D-glucopyranosyl)oxy]benzoic Acid Methyl Ester *see* 4272
6-*O*-β-D-Xylopyranosyl-D-glucose *see* 7752
Xylose, 9995
D-Xylose *see* 9995

Xylosecarboxylic Acid *see* 4489
6-(β-D-Xylosido)-D-glucose *see* 7752
Xylotocan [Astra] *see* 9414
o-Xylotocopherol *see* 9418
p-Xylotocopherol *see* 9417
Xylotox *see* 5359
Xylulose, 9996
N-(2,3-Xylyl)anthranilic Acid *see* 5680
1-Xylylazo-2-naphthol, 9997
1-(2,4-Xylylazo)-2-naphthol *see* 9997
Xylyl Bromide, 9998
m-Xylyl Bromide *see* 9998
o-Xylyl Bromide *see* 9998
p-Xylyl Bromide *see* 9998
Xylyl Chloride, 9999
m-Xylyl Chloride *see* 9999
o-Xylyl Chloride *see* 9999
p-Xylyl Chloride *see* 9999
1-(2,6-Xylyloxy)-2-propylamine *see* 6094
Xymelin [DAK] *see* 9993
Xymiazole *see* 2771
XZ-450 *see* 928
Y-3642 *see* 9379
Y-4153 *see* 2365
Y-5350 *see* 1846
Y-6047 *see* 2411
Y-6124 *see* 1461
Y-7131 *see* 3830
Y-8004 *see* 7708
Y-8894 *see* 9077
Y-9179 *see* 6583
Y-9213 *see* 6127
Y-12141 *see* 9494
Yadulan *see* 2169
Yafrol [Syntex] *see* 4071
Yageine *see* 4531
Yal [Medalsdorf] *see* 3398
Yamacillin [Yamanouchi] *see* 9008
Yamaful [Yamanouchi] *see* 1851
Yamatetan [Yamanouchi] *see* 1936
Yambolap [Chinoin] *see* 4961
Yam, Mexican, 10000
Yangonin, 10001
Yara Yara *see* 5918
Yarmor *see* 7416
Yarrow *see* 102
Yatren [Bayer] *see* 4757
Yaw Root *see* 8775
Yaxci *see* 8497
Yazumycin A *see* 8791
Yazumycins *see* 8791
YB-2 *see* 4852
YC-93 *see* 6403
Yeast, 10002
Yeast Adenylic Acid *see* 147
Yeast Nucleic Acid *see* 8204
Yellow AB, 10003
Yellow Arsenic Sulfide *see* 834
Yellow Beeswax *see* 1031
Yellow Cedar *see* 9324
Yellow Copper *see* 2029
Yellow Cross Liquid *see* 6225
Yellow Cuprocide [Rohm & Haas] *see* 2671
Yellow Dextrin *see* 2931
Yellow Dock *see* 8269
Yellow Gentian *see* 4285
Yellow Indigo *see* 968
Yellow Jasmine *see* 4278
Yellow Jessamine *see* 4278
Yellow Melilot *see* 5700
Yellow Mercury Iodide *see* 5797
Yellow Moccasin Flower *see* 2778
Yellow Mustard *see* 6226
Yellow OB, 10004
Yellow Phenolphthalein, 10005
Yellow Phosphorus *see* 7321
Yellow Precipitate *see* 5780
Yellow Prussiate of Potash *see* 7612

Yellow Prussiate of Soda see 8563
Yellow Puccoon see 4690
Yellow Resin see 8245
Yellow Root see 4690
Yellow Saunders see 8317
Yellow Sweet Clover see 5700
Yellow T see 9687
Yellow Ultramarine see 1660
Yerba Maté see 5637
Yerba Santa see 3617
Yermonil [Geigy] see 5501
Yesdol [Toho Iyaku] see 3311
Yh 1 see 7021
Yig, 10006
YL 704B1 see 6106
Ylangene, 10007
α-Ylangene see 10007
Ylang-Ylang Oil, 10008
Ylestrol [Ferndale] see 3689
Ylides, 10009
YM-08054-1 see 4850
YM 09330 see 1936
YM-09538 see 607
YM-11170 see 3881
YM-14090 see 4892
Yobir [Maruko] see 311
Yochinol see 4757
Yocon [Palisades] see 10011
Yodanodia see 7792
Yodoxin [Searle] see 4935
Yoghurt see 10010
Yogurt, 10010
Yohimbine, 10011
allo-Yohimbine, 10012
α-Yohimbine, 10013
δ-Yohimbine see 8129
Yohimex [Kramer] see 10011
Yohydrol [Riedel-Zabinka] see 10011
Yomesan [Bayer] see 6425
Yoristen [Kodama] see 4146
Yoshi 864 see 4841
Yosimilon [Kowa Yakuhin] see 9619
Youthwort see 3440
Yperite see 6225
Ytterbia see 10014
Ytterbium, 10014
Yttria see 10015
Yttrialite see 10015
Yttrium, 10015
Yttrium Iron Garnet see 10006
Yulocrotine see 5151
Yutopar [Duphar] see 8228
Yxin [Pfizer-Roerig] see 9150
Z 326 see 3949
Z 424 see 9885
Z 876 see 3125
Z-905 see 7411
Z 4828 see 9680
Z 4942 see 4822
Zabromin [Sandoz] see 1440
Zackal [SS Pharm.] see 4273
Zactane [Wyeth] see 3698
Zactirin [Wyeth] see 3698
Zadine [Schering] see 917
Zaditen [Sandoz] see 5187
Zadorin [Mepha] see 3429
Zadstat [Lederle] see 6079
Zamanil see 6926
Zamocillin [Inpharzam] see 610
Zanchol [Searle] see 4040
Zanil see 6911
Zanizal [Italfarmaco] see 6582
Zanosar [Upjohn] see 8794
Zantac [Glaxo] see 8126
Zantic [Glaxo] see 8126
Zantidon [Siam Bheasach] see 8126
Zantrène [Lederle] see 1255
Zaratite see 6410
Zarontin [Parke, Davis] see 3704
Zaroxolyn [Pennwalt] see 6068
Zasten [Sandoz] see 5187

Zaxopam [Quantum] see 6881
ZE-101 see 6582
Zea, 10016
Zearalanol see 10022
Zearalenone, 10017
Zeatin, 10018
trans-Zeatin see 10018
Zeaxanthin, 10019
Zeaxanthin Diepoxide see 9902
Zeaxanthol see 10019
Zebedassite see 352
Zectran see 6090
Zedolac [Maggioni-Winthrop] see 3831
Zefazone [Upjohn] see 1929
Zein, 10020
Zeisin [ICN] see 2082
Zeldox [Zeltia] see 4634
Zellner's Paper see 4086
Zelmid [Astra] see 10024
Zenadrex [Veterinaria] see 2865
Zenoxone [Biorex] see 4710
Zental [SK & F] see 201
Zentropil [Nordmark] see 7293
Zenusin [Mepha] see 6441
Zeolites, 10021
Zepelin [De Angeli] see 3953
Zepharovicht see 362
Zephiran Chloride [Winthrop] see 1066
Zephirol [Bayer] see 1066
Zepolas [Mikasa] see 4124
Zerano [Comm. Solvents] see 10022
Zeranol, 10022
Zerlate see 10075
Zestril [ICI] see 5393
Zesulan [Toyo Jozo] see 5754
Zetamicin [Menarini] see 6389
Zeta Protein see 4018
Zetran [Hauck] see 2082
Zettyn [Winthrop] see 2009
Ziavetine see 1465
Zibeth see 2336
Zidovudine, 10023
Zienam [Merck & Co.] see 2275 and 4834
Zildasac [Chugai] see 1042
Zimate see 10075
Zimecterin [Merck & Co.] see 5133
Zimeldine, 10024
Zimelidine see 10024
Zimovane [M & B] see 10095
Zinacef [Glaxo-Hoechst] see 1951
Zinadon see 5071
Zinamide [Merck & Co.] see 7970
Zinc, 10025
Zinc Acetate, 10026
Zinc Acexamate see 38
Zinc Alum see 376
Zinc Ammonium Chloride see 567
Zincaps [Protea] see 10064
Zincate see 10064
"Zincates" see 10025
Zinc Bacitracin, 10027
Zinc Bis(dimethylthiocarbamoyl) Disulfide see 10075
Zinc Blende see 10065
Zinc Bromide, 10028
Zinc Caprylate, 10029
Zinc Carbonate, 10030
Zinc Carbonate Hydroxide see 10030
Zinc Chloride, 10031
Zinc Chromate(VI) Hydroxide, 10032
Zinc Citrate, 10033
Zinc Cyanide, 10034
Zinc Diethyl see 3121
Zinc Dimethyl see 3252
Zinc Dimethyldithiocarbamate see 10075
Zinc Ethylenebis(dithiocarbamate) see 10071

Zinc Fluoride, 10035
Zinc Fluosilicate see 10037
Zinc Formate, 10036
Zinc Hexafluorosilicate, 10037
Zinc p-Hydroxybenzenesulfonate see 10054
Zinc Insulin Crystals, 10038
Zinc Iodate, 10039
Zinc Iodide, 10040
Zinc Iodide-Starch, 10041
Zincite see 10050
Zinc Lactate, 10042
Zinc Manganese Ethylenebisdithiocarbamate see 5598
Zinc Meta-arsenite, 10043
Zinc Nitrate, 10044
Zinc Nitride, 10045
Zinc Nitrite, 10046
Zinc Oleate, 10047
Zinc Omadine [Squibb] see 8004
Zincomed [Medo] see 10064
Zinc Ortho-arsenate, 10048
Zinc Orthosilicate see 10062
Zinc Oxalate, 10049
Zinc Oxide, 10050
Zinc Perchlorate, 10051
Zinc Permanganate, 10052
Zinc Peroxide, 10053
Zinc p-Phenolsulfonate, 10054
Zinc Phosphate, 10055
Zinc Phosphide, 10056
Zinc Potassium Cyanide see 7679
Zinc Potassium Iodide see 7692
Zinc Potassium Sulfate see 7698
Zinc Propionate, 10057
Zinc 1,2-Propylene Bisdithiocarbamate see 7831
Zinc Pyridinethione see 8004
Zinc Pyrithione see 8004
Zinc Pyrophosphate, 10058
Zinc Salicylate, 10059
Zinc Selenate, 10060
Zinc Selenide, 10061
Zinc Silicate, 10062
Zinc Silicofluoride see 10037
Zincspar see 10030
Zinc Stearate, 10063
Zinc Subcarbonate see 10030
Zinc Sulfanilate Tetrahydrate see 8901
Zinc Sulfate, 10064
Zinc Sulfide, 10065
Zinc Sulfocarbolate see 10054
Zinc Sulfocyanate see 10069
Zinc Sulfophenate see 10054
Zinc Superoxide see 10053
Zinc Tannate, 10066
Zinc Tartrate, 10067
Zinc Telluride, 10068
Zinc Thiocyanate, 10069
Zinc Undecylenate see 9760
Zinc Valerate, 10070
Zinc-vitamin B$_{12}$-tannate see 9923
Zinc Vitriol see 10064
Zinc White see 10050
Zinc Yellow see 10032
Zineb, 10071
Zingerone, 10072
Zingherone see 10072
Zingiberone see 10072
Zinnat [Glaxo] see 1951
Zinophos [Am. Cyanamid] see 9275
Zinostatin, 10073
Zinviroxime see 3553
Zipeprol, 10074
Ziram, 10075
Zirberk see 10075
Zircon see 10084
Zirconia see 10083
Zirconic Anhydride see 10083
Zirconium, 10076

Zirconium Chloride, 10077
Zirconium Dioxide see 10083
Zirconium Fluoride, 10078
Zirconium Hydride, 10079
Zirconium Hydroxide, 10080
Zirconium Iodide, 10081
Zirconium Nitrate, 10082
Zirconium Orthosilicate see 10084
Zirconium Oxide, 10083
Zirconium Oxychloride see 10087
Zirconium Potassium Fluoride see 7623
Zirconium Potassium Sulfate see 7699
Zirconium Silicate, 10084
Zirconium Sulfate, 10085
Zirconium Tetrachloride see 10077
Zirconium Tetrafluoride see 10078
Zirconium Tetraiodide see 10081
Zirconyl Acetate, 10086
Zirconyl Chloride, 10087
Zirtek [Allen & Hanburys] see 2013
Zitoxil [Farmochimica] see 10074
Zitrix [Gibipharma] see 1923
Zixoryn [Gedeon Richter] see 4063
Zizanal see 6778
ZK 31224 see 9095
ZK 39482 see 4955
ZK 57671 see 8973
ZK 62711 see 8235
ZK 93035 see 4236
ZL 101 see 6582
ZMA see 10043
ZN 6 see 4231
Zoalene [Dow] see 3262

Zoamix [Dow] see 3262
Zoapatanol, 10088
Zoaquin see 4935
Zocor [Merck & Co.] see 8491
Zocord [Merck & Co.] see 8491
Zoisite see 1645
Zoladex [ICI] see 4433
Zolamine, 10089
Zolaphen see 7248
Zolicef [Lappe] see 1925
Zolimidine, 10090
Zoliridine see 10090
Zolone [M & B] see 7308
Zolpidem, 10091
Zolyse [Alcon] see 2265
Zomax [McNeil] see 10092
Zomaxin [Cilag] see 10092
Zomepirac, 10092
Zometapine, 10093
Zonazide see 5071
Zondel [Grelan] see 6616
Zongorine see 8673
Zoniden [Irbi] see 9384
Zonisamide, 10094
Zooecdysones see 3465
Zoofurin see 6520
Zoolobelin see 5432
"Zoomelanoidic" Acids see 4815
Zopiclone, 10095
Zopirac [Sintyal] see 10092
Zorial [Sandoz] see 6618
Zoroxin [Merck & Co.] see 6617
Zorubicin, 10096
Zostrix [GenDerm] see 1767
Zotepine, 10097

Zothelone see 3360
Zovirax [Burroughs Wellcome] see 139
Zoxamin see 10098
Zoxazolamine, 10098
Zoxine see 10098
ZPO see 10053
ZR 515 see 5906
ZR 777 see 5194
ZR 3210 see 4137
Z Span [SmithKline] see 10064
Zuclomiphene see 2384
Zuclopenthixol see 2392
Zumaril [Abbott] see 212
Zunden [Luitpold] see 7476
Zutracin see 948
ZY 15021 see 1123
ZY 15028 see 2357
Zygadenine, 10099
Zygomycin A see 6989
Zyklolat [Mann] see 2752
Zylonite see 1960
Zyloprim [Burroughs Wellcome] see 278
Zyloric [Burroughs Wellcome] see 278
Zymafluor [Zyma] see 8565
Zymofren [Specia] see 784
Zymoplastic Substance see 9321
Zymosan, 10100
Zynoplex [Genefarm] see 9019
Zypanar [Armour Pharm.] see 6956
Zyrtec [UCB] see 2013
Zytron [Dow] see 3394
Zyxorin [Gedeon Richter] see 4063

Titles of Deleted Ninth and Tenth Edition Monographs

The following table lists Ninth Edition (italics) and Tenth Edition (boldface) monographs that do not appear in the Eleventh Edition. Please consult the appropriate edition of *The Merck Index* for information on these compounds.

Acalypha
Accel®
2-Acetamido-6-aminobenzothiazole
Acetoxan
α-*Acetoxypropionic Acid*
Acetylfuratrizine
Acetyl-5-iodosalicylic Acid
1-Acetyl-2-phenylhydrazine
Acmite
Acocanthin
Acoric Acid
Acroteben®
Adiantum
Adrenoxine
Adular
Afenil
Afridol
Akermanite
Aldehol
β-*Aletheine*
5-Allyl-5-Butylbarbituric Acid
Almandite
Aloetic Acid
Alseroxylon
Aluminum Borotannate
Aluminum Borotannotartrate
Alunite
Amarine
Ambonestyl®
N-Amidino-3-amino-6-chloropyrazine-
  carboxamide
Amidoxyl®
Amindan
4-Aminobenzenestibonic Acid
3-o-Aminobenzyl-4-methylthiazolium
  Chloride Hydrochloride
5-Amino-2-butoxypyridine
Aminohexan
4-Amino-2-methyl-5-pyrimidinecarbox-
  ylic Acid Hydrazide
2-Amino-5-methylthiazole
p-Aminonorephedrine
(p-Aminophenyl)phosphonous Acid
Aminoquinol
1-(p-Amino-m-tolyl)morpholine
m-Aminovalerophenone
Ammonium Ferric Tartrate
Ammonium Hydroselenate
Ammonium Magnesium Phosphate
Ammonium Magnesium Selenate
Ammonium Magnesium Sulfate
Ammonium Tartrate
Ammonium Urate, Acid
Amoxecaine
Amphetamine (4-Chlorophen-
  oxy)acetate
Amydricaine
n-Amylmercuric Chloride
4-Amylresorcinol
Analcime
Androsin
5-Androstene-3β,16α-diol
Anhydromethylenecitric Acid
Anisylbutamide
Antimony Barium Tartrate
Apophyllite
Apples, Dried
Aprophen
Aralin
Arecolidine
Arizonite
Arsant
Arsenophenylglycine
Arucase
Asazol
Ascosin
Asebogenin
Astrocasine
Atacamite
Athemol®
Austinite
Axinite

Azapetine Phosphate
Azocoll
Barbak
Barium Phosphite
Barium Selenate
Barium Tartrate
Barylite
Barysilite
Bassorin®
Benanserin Hydrochloride
Benzacetin
4-[2-(Benzhydryloxy)ethyl]morpholine
Benzmalecene
Benzobutamine
Benzodet
Benzolamide
Benzomethamine Chloride
1-Benzoyl-2-phenylhydrazine
1-Benzoyl-2-thiourea
Benzyl 4-Carbamyl-1-piperazinecarbox-
  ylate
2-[Benzyl(2-dimethylaminoethyl)amino]-
  pyrimidine
Benzylidenepinacolone
O-Benzylphosphorous O,O-Diphenyl-
  phosphoric Anhydride
Benzyl Phthalate
Benzyl Succinate
Bertrandite
Betaine Salicylate
Biformin
Biosplen
γ-*(4-Biphenylyl)butyric Acid*
2-(2-Biphenylyloxy)triethylamine Hydro-
  chloride
Bis(acetylcholine) chloromanganate(II)
Bisbeeite
1,4-Bis(1,4-benzodioxan-2-ylme-
  thyl)piperazine
N,N-Bis(2-chloroethyl)benzylamine
N⁶, N⁶-Bis(2-chloroethyl)lysine
Bis(chloromethyl) Sulfone
N,N-Bis(2-hydroxyethyl)-p-nitrosoani-
  line
N,N'-Bis[isonicotinic acid] Hydrazide
Bismuth Citrate
Bismuth Lactate
N,N'-Bis[nicotinic acid] Hydrazide
Bisnorcholanic Acid
Boliden® *Salt*
Bonducin
Bone Oil
β-*Brazanquinone*
Brintobal
Bromanylpromide
Bromchlorenone
5-(2-Bromoallyl)-5-(1-methylbutyl)-
  barbituric Acid
3-Bromo-2-butanol
α-*Bromoisovaleryl-p-phenetidine*
1-Bromo-3-methyl-1-pentyn-3-ol
9α-Bromo-11-oxoprogesterone
4,4'-(β-Bromostyrylidene)diphenol Di-
  acetate
2-[(5-Bromo-2-thenyl)(2-dimethylamino-
  ethyl)amino]pyridine Hydrochloride
Broncatar
Bustamite
Butacarb
Butoxamine
o-Butoxyacetanilide
p-Butoxybenzoic Acid 3-Diethylamino-
  propyl Ester
p-Butoxyphenol
p-Butylaminosalicylic Acid 2-Diethyl-
  aminoethyl Ester
Butylchloralaminopyrine
Butylchloral Hydrate
6-tert-Butyl-1-chloro-2-naphthol
1-Butyl-3-(N-crotonoylsulfanilyl)urea
N-tert-Butyl-N-cyanoglycinonitrile

Butyl Ricinoleate
Butylthiamine
Caffeine Acetyltryptophanate
Caffeine Hydrobromide
Caffeine Salicylate
Caffeine Sodium Benzoate
Caffeine Sodium Salicylate
Caincin
Calcium Carbamate
Calcium Salicylate
Calcium Selenate
Calvacin
Camphamedrine
Canbyite
Cappelenite
Caprochlorone
p-(Carbamylmethoxy)acetanilide
Carbolineum®
Carboxide®
3'-Carboxy-4'-hydroxycinchophen
Carvacrylamine
Caseoiodine
Castelamarin
Catharosine
Cathylate
Celsian
Centalun
Centropnein
Cesium Bisulfate
Cesium Bitartrate
Cesium Selenate
Cetrimonium Pentachlorophenoxide
Cetyl Iodide
Checken
Checken Bitter
Cheiranthin
Chiolite
Chitaric Acid
Chloralacetonechloroform
Chloral Camphorated
Chloral Phenolated
Chlorazine
Chloriodized Oil
p-Chlorobenzhydrazide
Chlorobutin Penicillin
7-Chloro-4-(4-diethylamino-1-methyl-
  butylamino)-3-methylquinoline
7-(β-Chloroethyl)theophylline
α-*Chloro-α-phenylacetylurea*
Chlororaphin
4-Chloro-3-sulfamoylbenzamide
8-Chlorotheophylline
Cholesteryl Bromide
Cholesteryl Chloride
Cholesteryl Iodide
Choline Ascorbate
Choline Bicarbonate
Choline Citrate
Choline Gluconate
Chrysocolla
Chrysoidine Thiocyanate
Chrysolite
Cinchophen Hydriodide
Citrodisalyl
Clamoxyquin
Clinohedrite
Clofenetamine
Coamid
Codoxime
Colchicum Glucoside Colchicoside
Collinsonia Extract
Colloresine®
Comirin
Convolvamine
Convolvine
Convolvulin
Cookeite
Copaiba Mass
Crab Orchard Salt, Artificial
Crandallite
Croweacic Acid
Cumephedrine

Cupric Dichromate(VI)
Cuprous Formate
Cuprous Nitride
Curite
Cuspidine
1-Cyclobutyl-1-ethynylethanol
Cyclopyrazate
Cypan®
Cyprolidol
Dahllite
Danburite
Datolite
Davisonite
Defenal
Deltaite
Descinolone
Desylamine
Diagnex®
2,4-Diamino-6-pyridinium Triazine
    Chloride
N-(4,6-Diamino-s-triazin-2-yl)arsanilic
    Acid Disodium Salt
Diapamide
Diaphen
Dibromin®
5-(α,β-Dibromophenethyl)-5-methylhy-
    dantoin
Dibupyrone
Dibutadiamin
Dibutoline Iodide
3-Dibutylaminomethyl-4,5,6-trihydroxy-
    1-isobenzofuranone
N-(Dichloroacetyl)-DL-serine
Dichloromethotrexate
3,9-Dichloro-7-methoxyacridine
Dichlorpromazine
2,2'-Dicyclohexyl-N-methyldiethylamine
    Hydrochloride
2-Diethylaminoethyl 4-Butylamino-2-
    ethoxybenzoate
2-Diethylaminoethyl α-(2-Cyclopenten-
    1-yl)-2-thiophene Acetate
2-Diethylaminoethyl 9-Fluorenecarbox-
    ylate Hydrochloride
1-(2-Diethylaminoethyl)phenobarbital
γ-Diethylaminopropyl Cinnamate Hy-
    drochloride
N-(3-Diethylaminopropyl)-2,2-diphenyl-
    acetamide
N,N-Diethyl-3,5-dimethyl-4-isoxazole-
    carboxamide
Diethyl Dithiolisophthalate
O,O-Diethyl O-(3-Methyl-5-pyrazolyl)
    Phosphate
O,O-Diethyl O-(3-Methyl-5-pyrazolyl)
    Phosphorothioate
N,N-Diethyl-1-piperazinecarboxamide
2,2-Diethyl-1,3-propanediol
Difencloxazine
Diglucomethoxane
Dihydroisohyenanchin
Dihydrotestosterone n-Octyl Enol Ether
2-Diisopropylaminoethyl p-Aminoben-
    zoate
Diisopropylammonium Nicotinate
Dimecamine
2-Dimethylaminoethyl Benzilate
4-Dimethylamino-1-phenylpiperidine
N,1-Dimethylhexylamine
p,α-Dimethylphenethylamine
4'-(Dimethylsulfamoyl)sulfanilaniiide
Dimidium Bromide
Dimorphite
Diocaine®
Diopside
Dioptase
Dioxamate
Diphazine
Diphenesenic Acid
2-Diphenylmethylpiperidine
2,3-Diphenylpropionic Acid 2-Diethyl-
    aminoethyl Ester
Dipropaesin
Dipropamine
Diprophen
Diuredosan

Diurgin
Dodecyltriphenylphosphonium Bromide
Dodicin
Dolcental
Dufrenoysite
Echujin
Egyptian Blue
Ekmolin
Elaterium
Elbon®
Elemi-bitter
Embelia
Embolite
Embutramide
Emetine Bismuth Iodide
Emetine Camphosulfonate Solution
Enamine
Endomycin
Epidote
3-Epihexahydroequilenin
Epinochrome
Epipropidine
Equilin Glycol
Ericolin
Erysothiovine
Escorin
Estradiol 3,17β-Dicypionate
Estrane-3α,17α-diol
5,6,8-Estratrien-3β-ol-17-one
Ethereal Oil
Ethomoxane
8-Ethoxycaffeine
Ethylacetanilide
Ethyl Acetylsalicylate
Ethyl Choladienate
N-Ethylcrotonanilide
N-2-Ethylcrotonyl-N'-methylolurea
5-Ethyldihydro-6-phenyl-2H-1,3-thia-
    zine-2,4(3H)-dione
Ethylenimine Quinone
Ethylglycocoll Hydrochloride
N-Ethyl-p-hydroxyacetanilide Acetate
Ethylidene Diurethane
5-Ethyl-5-phenylhydantoin
N-Ethyl-3-piperidyl Phenylcyclopentyl-
    glycolate
α-Ethyl-2-thienylethylamine
Ethynerone
1-Ethynylcyclohexanol
Ethyserpine
Euclase
Eucryptite
Eulytine
Eurazyl
Extract Apple Ferrated
Fabianine
Fabiatrin
Fagaramide
Fantan
Faserton
Fayalite
Felogen
Fenamole
Fenimide
Fenpipramide
Ferberite
Ferratin
Ferric Glycerophosphate
Ferroakermanite
Ferrodolomite
Ferronascin
Fezatione
Fibrin Foam (Human)
Filmaron®
Flavianic Acid
Flavicid
Flavoteben®
2-Fluorenamine
9-Fluorenamine
3-Fluoro-4-hydroxyphenylacetic Acid
4-(p-Fluorophenyl)-3-thiosemicarbazide
Forbisen
Frankonite
Fumaroalanide
Funtessine
Furazan

Fusaria Antibiotics
Galega
Gallamide
Ganomalite
Gay-Lussite
Geikielite
Gein
Germander
Gillespite
Glauberite
Glaucochroite
Globicin
Glucatase®
Glucocheirolin
D-Glucose Phenylosazone
Glutinosin
Glycarbylamide
Glycine Bis(chloroethyl)amide
Glycofurol
N-Glycylaniline
Glyprothiazol
Gold Monobromide
Griseomycin
Gruenerite
Guaiacodeine
Guaiaretic Acid
Guanajuatite
Guarana
Gumbrin
Guvacoline
Haidingerite
Halofenate
Hambergite
Hapamine®
Hardystonite
Helenine
Hemogallol
Hendecanediamidine
Hendenbergite
2-(Heptadecenyl)-2-imidazoline-1-
    ethanol
1,1,1,3,5,5,5-Heptamethyl-3-phenyltri-
    siloxane
Heptolide
Heptylpenicillin Sodium
Hepzidine
Hercynite
Hewettite
Hexadecamethylcyclooctasiloxane
Hexadistigmine
Hexahomoserine
Hexahydrodesmethoxy-β-erythroidine
Hexamethonium Bromide
Hexamethylcyclotrisiloxane
Hexaphenylcyclotrisiloxane
Hexonate
Hexuronic Acid
4-(Hexyloxy)benzilic Acid 2-Diethylami-
    noethyl Ester
Hexylresorcinol-
    bis(diethylaminoethyl)ether
Hofmann's Violet
Homarylamine Hydrochloride
Homoarecoline
Homocongasine
Homonataloin
Howlite
Hubnerite
Hureaulite
Hydrazine Iodide
α-Hydrazinohistidine
Hydroferrocyanic Acid
Hydroherderite
p-Hydroxyacetanilide Benzoate
p-Hydroxybenzenearsonic Acid
Hydroxycodeine
6-Hydroxy-2-methyl-1,4-naphthoqui-
    none
5-(Hydroxymethyl)tetrahydro-2-fur-
    furylamine
3-(2-Hydroxy-1-naphthylmethyl)salicylic
    Acid
N-(4-Hydroxyphenyl)-3-phenylsalicyl-
    amide
8-Hydroxyquinoline Citrate
4-Hydroxytetracycloxide

Resin Thapsia
Retama
Rilsan®
Rossite
Rubidium Carbonate
Rubidium Selenate
Rubidium Sulfate
Rue
Ruscopine
Rythmol®
Sabal
Salazosulfamide
Sal Ethyl® Carbonate
Salicil
Salicylresorcinol
Santalyl Salicylate
Sapamine®
Sapogenin
Sarothamnine
Sartorite
Scawtite
Schroeckingerite
3-Semicarbazidobenzamide
3-Semicarbazidobenzamide
Sensibamine®
Septacrol®
Shattuckite
Sigumid
Silver Bromate
Silver Chlorite
Silver Hyponitrite
Simarubidin
Simarubin
Sobisminol Mass
Sodalite
Sodium 4-Amino-2-aurothiosalicylate
Sodium Anhydromethylenecitrate
Sodium Auromercaptoethanesulfonate
Sodium Cinnamate
Sodium Cresotate
Sodium Hexaselenide
Sodium Hydroselenite
Sodium Malate
Sodium Morrhuate
*Sodium Polyanhydromannuronic Acid
  Sulfate*
Sodium Psylliate
Sodium Sulfanilate
Sodium Sulforicinate
Sodium Tetrabromoaurate(III)
Sodium Uranyl Sulfate
Sodium Urate
Sodium Xanthogenate
Solupyridine
Solvoteben
Solypertine
Spinel

Spirothiobarbital
Spirotrypan
Spiroxasone
Spodiosite
Spurrite
Starch from Arrowroot
Starch, Iodized
Stearylsulfamide
Steffen's Waste
Stephanite
Stibamine Glucoside
Stibenyl
4-Stilbazole
Streptotibine
Stromeyerite
Subenon
Sublamine®
Sulcimide
Sulfaethoxypyridazine
Sulocarbilate
Sulphocidine Album
Suosan
Supazine
Tachydrol
Tastromine
Tetradecamethylcycloheptasiloxane
Tetraethylammonium Iodide
*1,1,3,3-Tetramethyl-1,3-diphenyldisilox-
  ane*
*2,4,6,8-Tetramethyl-2,4,6,8-tetraphenyl-
  cyclotetrasiloxane*
Tetramin
Tetrophan
Texapon Z
Thalline
**1-Theobromineacetic Acid Bromcholine
  Phosphate**
Theobromine Hydrochloride
Theobromine Salicylate
Theobromine Sodium Formate
Theophylline Calcium Salicylate
Theophylline Sodium Salicylate
Thiambutosine
Thiamine Diphosphoric Acid Salt
Thiamine Phosphoric Acid Salt
Thiamine Triphosphoric Acid Salt
Thiethazone
Thioaurin
Thiocaine
Thiocarbanidin
Thiochrome
Thiofuradene
Thionarcon
Thiophene Diiodide
Thorite
Thrombo-Holzinger
Thymolsulfonic Acid

Thymoquinone
Tibicor
Tiformin
Tileyite
2-(o-Tolyl)cyclohexanol
p-Tolylmercuric Chloride
1-(m-Tolyl)semicarbazide
Tomatillidine
Toringin
Tosyl
Transuranium Elements
Trethocanic Acid
Tribromophenyl Salicylate
*2,4,6-Triethyl-2,4,6-triphenylcyclotri-
  siloxane*
*2-(3,4,5-Trimethoxybenzyl)-2-imidazo-
  line*
6-Trimethylammoniopurinide
*2,4,6-Trimethyl-2,4,6-triphenylcyclotri-
  siloxane*
Tripuhyite
Tropentane
Turquois
Tutocaine® Hydrochloride
Ulexite
Urea Calcium Bromide
Urea Oxalate
Urethan Calcium Bromide
Urocanic Acid
Uvarovite
Varon
V.C.P.®
Venoxidine
Vetrophin®
Vibesate
Violutin
Viton® Copolymers
Wagnerite
Wescodyne®
White Flux
WY-3654
Xanthiol
Xanthorhamnin
Xanthoxylum
Xanthydrol
Zedoary
Zefran®
Zinc Benzoate
Zinc Dichromate(VI)
Zinc Ferrocyanide
Zinc Hypophosphite
Zinckenite
Zinc Phosphite
Zinc Sulfite
Zolertine

APPENDIX

A1. Acifran. *4,5-Dihydro-5-methyl-4-oxo-5-phenyl-2-furancarboxylic acid;* 2-methyl-2-phenyl-3(2H)-furan-one-5-carboxylic acid; AY-25712; Reductol. $C_{12}H_{10}O_4$; mol wt 218.21. C 66.05%, H 4.62%, O 29.33%. Prepd (not claimed): I. L. Jirkovsky *et al.*, U.S. pat. **4,169,202;** prepn and resolution of enantiomers: *eidem*, U.S. pat. **4,244,958** (1979, 1981 both to American Home); I. L. Jirkovsky, M. N. Cayen, *J. Med. Chem.* **25,** 1154 (1982). Metabolism and pharmacokinetics in dogs and rats: M. N. Cayen *et al.*, *Xenobiotica* **16,** 251 (1986). LC determn in serum: R. Gonzalez, D. Thibeault, *Clin. Biochem.* **16,** 244 (1983). Clinical evaluation in hyperlipoproteinemia: D. B. Hunninghake *et al.*, *Clin. Pharmacol. Ther.* **38,** 313 (1985); J. C. LaRosa *et al.*, *Artery* **14,** 338 (1987).

Crystals, mp 176°. uv max (methanol): 281 nm (ε 7960). LD_{50} orally in rats: 3 g/kg (Jirkovsky, 1982).
(−)-Form, mp 87-89°. $[\alpha]_D^{25}$ −144.7° (c = 2 in methanol).
(+)-Form, mp 87-89°. $[\alpha]_D^{25}$ +146.4° (c = 2 in methanol).

THERAP CAT: Antihyperlipoproteinemic.

A2. Angiogenin. Single-chain, basic protein of 123 amino acids that induces the *in vivo* formation of blood vessels. Mol wt ~14,000 Da. First isolated from human adenocarcinoma cells; subsequently found in normal human plasma and shown to be produced by the liver. Angiogenin exhibits a characteristic ribonucleolytic activity toward 28S and 18S ribosomal RNA. Its amino acid sequence is 35% identical with that of human pancreatic ribonuclease. Isoln, characterization, and angiogenic activity: J. W. Fett *et al.*, *Biochemistry* **24,** 5480 (1985). Amino acid sequence: D. J. Strydom *et al.*, *ibid.* 5486. Cloning and DNA sequence of human angiogenin gene: K. Kurachi *et al.*, *ibid.* 5494. Structural study: K. A. Palmer *et al.*, *Proc. Nat. Acad. Sci. USA* **83,** 1965 (1986). Ribonucleolytic activity: R. Shapiro *et al.*, *Biochemistry* **25,** 3527 (1986). Isoln from normal human plasma: R. Shapiro *et al.*, *ibid.* **26,** 5141 (1987). Tissue distribution in neonatal and adult rats: H. L. Weiner *et al.*, *Science* **237,** 280 (1987); in human tumor and normal cells: S. M. Rybak *et al.*, *Biochem. Biophys. Res. Commun.* **146,** 1240 (1987). Inhibition of protein synthesis: D. K. St. Clair *et al.*, *Proc. Nat. Acad. Sci. USA* **84,** 8330 (1987). Inhibition of angiogenic and ribonucleolytic activities of angiogenin by placental ribonuclease inhibitor: R. Shapiro, B. L. Vallee, *ibid.* 2238; F. S. Lee, B. L. Vallee, *Biochemistry* **28,** 3556 (1989). *Reviews:* J. F. Riordan, B. L. Vallee, *Brit. J. Cancer* **57,** 587-590 (1988); B. L. Vallee, J. F. Riordan, *Adv. Exp. Med. Biol.* **234,** 41-53 (1988).

A3. Buthionine Sulfoximine. *2-Amino-4-(S-butylsulfonimidoyl)butanoic acid;* S-(3-amino-3-carboxypropyl)-S-butylsulfoximine; S-(n-butyl)homocysteine sulfoximine; BSO. $C_8H_{18}N_2O_3S$; mol wt 222.32. C 43.22%, H 8.16%, N 12.60%, O 21.59%, S 14.42%. Selective inhibitor of γ-glutamyl cysteine synthetase, an enzyme in the glutathione, *q.v.*, biosynthetic pathway. BSO-mediated depletion of intracellular glutathione has been associated with increased sensitivity of tumor cells to antineoplastic agents. Prepn: K. Hayashi, *Chem. Pharm. Bull.* **8,** 177 (1960). Synthesis and *in vitro* and *in vivo* enzyme inhibition: O. W. Griffith, A. Meister, *J. Biol. Chem.* **254,** 7758 (1979). Mechanism of action and metabolism in animals O. W. Griffith, *ibid.* **257,** 13704 (1982). Chemosensitization by BSO of melphalan-resistant murine leukemia cells *in vivo:* S. Somfai-Relle *et al.*, *Biochem. Pharmacol.* **33,** 485 (1984); R. A. Kramer *et al.*, *Cancer Res.* **47,** 1593 (1987). Enhanced melphalan, *q.v.*, cytotoxicity in human cancer cells *in vitro:* J. A. Green *et al.*, *ibid.* **44,** 5427 (1984); and in mice bearing human tumor

xenografts: H. S. Friedman *et al.*, *J. Nat. Cancer Inst.* **81,** 524 (1989).

$$CH_3(CH_2)_3 \overset{O}{\underset{\underset{NH}{\parallel}}{S}} CH_2CH_2 \underset{NH_2}{CHCOOH}$$

Crystals from aq ethanol, mp 214-215.5° (dec). Sol in water. Partition coefficient (octanol/water): 4.34 ±0.4 × 10^{-3}.

USE: Biological tool for depletion of glutathione.

A4. Endothelin. Endothelin-1; ET-1. Potent vasoconstrictor peptide produced by mammalian vascular endothelial cells; present in low concentrations in normal serum. Composed of 21 amino acid residues; mol wt ~2500 Da. Isoln, characterization and cloning of porcine endothelin: M. Yanagisawa *et al.*, *Nature* **332,** 411 (1988). Structural identity of porcine and human endothelin: Y. Itoh *et al.*, *FEBS Letters* **231,** 440 (1988). Structural similarity with the snake venom toxins, *sarafotoxins S6:* C. Y. Lee, V. A. Chiappinelli, *Nature* **335,** 303 (1988); C. Takasaki *et al.*, *ibid.* 303. Identification in serum by radioimmunoassay: K. Ando *et al.*, *FEBS Letters* **245,** 164 (1989). Two additional isoforms, known as *endothelin-2* and *endothelin-3,* have been identified. Structures and comparative pharmacology of human endothelin isopeptides: A. Inoue *et al.*, *Proc. Nat. Acad. Sci. USA* **86,** 2863 (1989). Mechanism of action study: Y. Hirata *et al.*, *Biochem. Biophys. Res. Commun.* **154,** 868 (1988). Possible role in acute renal failure: J. D. Firth *et al.*, *Lancet* **2,** 1179 (1988). *In vivo* effects on pulmonary and systemic hemodynamics: H. L. Lippton *et al.*, *J. Appl. Physiol.* **66,** 1008 (1989). *Review:* T. Masaki, *J. Cardiovasc. Pharmacol.* **13,** Suppl. 5, S1-S4 (1989).

Cys-Ser-Cys-Ser-Ser-Leu-Met-Asp-Lys-Glu-Cys-Val-
Tyr-Phe-Cys-His-Leu-Asp-Ile-Ile-Trp

A5. FK-506. *[3S-[3R*[E(1S*,3S*,4S*)],4S*,5R*,8S*,-9E,12R*,14R*,15S*,16R*,18S*,19S*,26aR*]]-5,6,8,11,12,-13,14,15,16,17,18,19,24,25,26,26a-Hexadecahydro-5,19-dihydroxy-3-[2-(4-hydroxy-3-methoxycyclohexyl)-1-methylethenyl]-14,16-dimethoxy-4,10,12,18-tetramethyl-8-(2-propenyl)-15,19-epoxy-3H-pyrido[2,1-c][1,4]oxaazacyclotricosine-1,7,20,21(4H,23H)-tetrone;* 17-allyl-1,14-dihydroxy-12-[2-(4-hydroxy-3-methoxycyclohexyl)-1-methylvinyl]-23,25-dimethoxy-13,19,21,27-tetramethyl-11,28-dioxa-4-azatricyclo[22.3.1.0^{4,9}]octacos-18-ene-2,3,10,16-tetraone; Fr-900506. $C_{44}H_{69}NO_{12}$; mol wt 804.03. C 65.73%, H 8.65%, N 1.74%, O 23.88%. Macrolide immunosuppressant isolated from *Streptomyces tsukubaensis* no. 9993: M. Okuhara *et al.*, *Eur. pat.* **Appl. 184,162** (1986 to Fujisawa); and characterization: T. Kino *et al.*, *J. Antibiot.* **40,** 1249 (1987). Structure determn: H. Tanaka *et al.*, *J. Am. Chem. Soc.* **109,** 5031 (1987). Approaches to synthesis: D. Askin *et al.*, *Tetrahedron Letters* **29,** 277 (1988); S. Mills *et al.*, *ibid.* 281; series of articles: *J. Org. Chem.* **54,** 9-19 (1989). Total synthesis of (−)-form: T. K. Jones *et al.*, *J. Am. Chem. Soc.* **111,** 1157 (1989). In vitro immunosuppressant activity in comparison with cyclosporin, *q.v.:* T. Kino *et al.*, *J. Antibiot.* **40,** 1256 (1987); N. Yoshimura *et al.*, *Transplantation* **47,** 351 (1989). Mechanism of action study: *eidem, ibid.* 356. Immunosuppressive effects in allografts in animals: D. St. J. Collier *et al.*, *Transplantation Proc.* **19,** 3975 (1987); S. Todo *et al.*, *Surgery* **104,** 239 (1988). Series of articles on isoln, pharmacology and use in experimental transplantation: *Transplantation Proc.* **19,** Suppl. 6, 3-104 (1987).

Monohydrate, $C_{44}H_{69}NO_{12} \cdot H_2O$, colorless prisms from acetonitrile, mp 127-129°. $[\alpha]_D^{23}$ −84.4° (c = 1.02 in CHCl$_3$). Sol in methanol, ethanol, acetone, ethyl acetate, chloroform, diethyl ether; sparingly sol in hexane, petroleum ether. Insol in water. LD$_{50}$ i.p. in mice: > 200 mg/kg (Kino).

A6.   P-Glycoprotein.   Glycoprotein-P; P-gp; P-170; permeability glycoprotein.   Highly conserved cell-surface glycoprotein of M$_r$ 170,000 daltons; protein product of the mammalian multidrug resistance (*mdr*) gene.   Present in selective human tissues where it functions as an energy-dependent efflux pump.   Increased expression in mammalian tumor cells is associated with multidrug resistance to cancer chemotherapy agents.   Identification and isolation from colchicine-resistant Chinese hamster ovary (CHO) cells: R. J. Juliano, V. Ling, *Biochim. Biophys. Acta* **455**, 152 (1976).   Purification from plasma membrane of CHO cells: J. R. Riordan, V. Ling, *J. Biol. Chem.* **254**, 12701 (1979).   Multi-drug resistance associated with P-glycoprotein in mammalian cell lines: N. Kartner et al., *Science* **221**, 1285 (1983); N. Kartner et al., *Cancer Res.* **43**, 4413 (1983).   Deduced amino acid sequence from mouse *mdr* cDNA: P. Gros et al., *Cell* **47**, 371 (1986); from human *mdr*1 cDNA: C.-j. Chen et al., *ibid.* 381.   Transfection and expression of human *mdr*1 gene in cell culture: K. Ueda et al., *Proc. Nat. Acad. Sci. USA* **84**, 3004 (1987).   Characterization of ATPase activity: H. Hamada, T. Tsuruo, *Cancer Res.* **48**, 4926 (1988).   Detection in normal human tissues: F. Thiebaut et al., *Proc. Nat. Acad. Sci. USA* **84**, 7735 (1987); R. N. Hitchins et al., *Eur. J. Cancer Clin. Oncol.* **24**, 449 (1988); and in human tumor cells: D. R. Bell et al., *J. Clin. Oncol.* **3**, 311 (1985); W. S. Dalton et al., *Blood* **73**, 747 (1989).   Review of role in multidrug resistance in human cancer: I. Pastan, M. Gottesman, *N. Engl. J. Med.* **316**, 1388 (1987).

A7.   Quinapril.   *[3S-[2[R*(R*)],3R*]]-2-[2-[[1-Eth-oxycarbonyl)-3-phenylpropyl]amino]-1-oxopropyl]-1,2,3,4-tetrahydro-3-isoquinolinecarboxylic acid; (S)-2-[(S)-N-[(S)-1-carboxy-3-phenylpropyl]alanyl]-1,2,3,4-tetrahydro-3-isoquinolinecarboxylic acid 1-ethyl ester.* $C_{25}H_{30}N_2O_5$; mol wt 438.52.   C 68.47%, H 6.90%, N 6.39%, O 18.24%.   Angiotensin converting enzyme inhibitor.   Prepn: M. L. Hoefle, S. Klutchko, **Eur.** pat. **Appl. 49,605**; *eidem*, **U.S.** pat. **4,344,949** (both 1982 to Warner-Lambert); of quinapril and its active diacid metabolite, quinaprilat, and activity: S. Klutchko et al., *J. Med. Chem.* **29**, 1953 (1986).   Synthesis of novel crystalline and purified form: O. P. Goel, U. Krolls, **U.S.** pat. **4,761,479** (1988 to Warner-Lambert).   Pharmacology: H. R. Kaplan et al., *Fed. Proc.* **43**, 1326 (1984).   Antihypertensive activity in animals: M. J. Ryan et al., *ibid.* 1330.   Hemodynamic effects in congestive heart failure: P. Holt et al., *Eur. J. Clin. Pharmacol.* **31**, 9 (1986).   Clinical evaluation in essential hypertension: P. Säynävälammi et al., *J. Cardiovasc. Pharmacol.* **12**, 88 (1988).   HPLC determn of quinapril and quinaprilat in human plasma and urine: H. Hengy, M. Most, *J. Liq. Chromatog.* **11**, 517 (1988).   Toxicity data: H.

R. Kaplan et al., *Angiology* **40**, 335 (1989).   Series of articles on pharmacology, pharmacokinetics, clinical evaluation and safety: *ibid.* 329-415.

Hydrochloride, $C_{25}H_{31}ClN_2O_5$, crystals from ethyl acetate-toluene, mp 120-130°. $[\alpha]_D^{23}$ +14.5° (c = 1.2 in ethanol) (Klutchko).   Also reported as white crystalline solid from acetonitrile, mp 119-121.5°. $[\alpha]_D^{25}$ +15.4° (c = 2 in methanol) (Goel, Krolls).

Hydrochloride monohydrate, *CI-906, PD 109452-2, Accupril, Accuprin, Accupro.*   LD$_{50}$ in male, female mice, rats (mg/kg): 1739, 1840, 4280, 3541 orally; 504, 523, 158, 107 i.v. (Kaplan, 1989).

Diacid, $C_{23}H_{26}N_2O_5$, *CI-928, quinaprilat.*   Hydrate, crystals from methanol-ethyl ether, mp 166-168°. $[\alpha]_D^{23}$ +20.9° (c = 1 in methanol).

THERAP CAT: Antihypertensive.

A8.   Temafloxacin.   *1-(2,4-Difluorophenyl)-6-fluoro-1,4-dihydro-7-(3-methyl-1-piperazinyl)-4-oxo-3-quinoline-carboxylic acid;* T-1258; A-63004. $C_{21}H_{18}F_3N_3O_3$; mol wt 417.39.   C 60.43%, H 4.35%, F 13.65%, N 10.07%, O 11.50%.   Trifluorinated quinolone antibacterial.   Prepn: D. T. W. Chu, **Eur.** pat. **Appl. 131,839**; *idem*, **U.S.** pat. **4,730,-000** (1985, 1988 to Abbott).   Antibacterial activities: D. J. Hardy et al., *Antimicrob. Ag. Chemother.* **31**, 1768 (1987); K. V. I. Rolston et al., *Eur. J. Clin. Microbiol. Infect. Dis.* **7**, 684 (1988); in comparison with other fluoroquinolones: A. L. Barry, R. N. Jones, *J. Antimicrob. Chemother.* **23**, 527 (1989).

Hydrochloride, $C_{21}H_{19}ClF_3N_3O_3$, *A-62254.*

A9.   Tenidap.   *5-Chloro-2,3-dihydro-2-oxo-3-(2-thien-ylcarbonyl)-1H-indole-1-carboxamide;* 5-chloro-3-(2-thenoyl)-2-oxindole-1-carboxamide; CP-66248. $C_{14}H_9ClN_2O_3S$; mol wt 320.75.   C 52.43%, H 2.83%, Cl 11.05%, N 8.73%, O 14.96%, S 10.00%.   Inhibitor of 5-lipoxygenase and interleukin-1 (IL-1) activity.   Prepn: S. B. Kadin, **Eur.** pat. **Appl. 156,603**; *idem*, **U.S.** pat. **4,556,672** (both 1985 to Pfizer).   Effect on 5-lipoxygenase activity *in vitro:* K. Fogh et al., *Arch. Dermatol. Res.* **280**, 430 (1988).   Effect on IL-1 activity in patients with rheumatoid arthritis: B. McDonald et al., *Arthritis Rheum.* **31**, Suppl., S52 (1988).   Clinical evaluation: P. Katz et al., *ibid.* S52.

Fluffy, yellow crystals from acetic acid, mp 230° (dec).
THERAP CAT: Anti-inflammatory.

A10. **Trichosanthin.** GLQ-223. Active component of the traditional Chinese medicine, Tian Hua Fen (Radix Trichosanthis), known since 300 A.D. and used in gynecological disorders and abortion. Basic polypeptide of 234 amino acids, M_r 26,000 Da, isolated from the roots of *Trichosanthes kirilowii* Maxim, *Cucurbitaceae*. Purification, characterization, and amino acid sequence: W. Yu *et al.*, *Pure Appl. Chem.* **58**, 789 (1986); J. M. Maraganore *et al.*, *J. Biol. Chem.* **262**, 11628 (1987). Conformation: S. Kubota *et al.*, *Biochim. Biophys. Acta* **871**, 101 (1986); and homology with ricin, *q.v.: eidem, Int. J. Peptide Protein Res.* **30**, 646 (1987). Inhibition of protein synthesis: H. W. Yeung *et al.*, *ibid.* **31**, 265 (1988). Inhibition of HIV replication: M. S. McGrath *et al.*, *Proc. Nat. Acad. Sci. USA* **86**, 2844 (1989). Abortifacient activity in animals: M. C. Chang *et al.*, *Contraception* **19**, 175 (1979); S. K. Saksena *et al.*, *ibid.* **20**, 367 (1979); I. F. Lau *et al.*, *ibid.* **21**, 77 (1980). Clinical trial as abortifacient: K. F. Cheng, *Obstet. Gynecol.* **59**, 494 (1982).

Crystals. pI 9.4.

THERAP CAT: Abortifacient.

NOTES

NOTES

NOTES

NOTES

NOTES

NOTES

NOTES

NOTES

NOTES

NOTES

NOTES

NOTES

NOTES

NOTES

NOTES

NOTES

NOTES

NOTES

NOTES

NOTES

NOTES

NOTES

NOTES

NOTES

NOTES

NOTES

NOTES

NOTES

ATOMIC WEIGHTS

(Alphabetical Order)

Element	Symbol	Atomic number	Atomic weight	Element	Symbol	Atomic number	Atomic weight
Actinium	Ac	89	227.0278*	Neodymium	Nd	60	144.24
Aluminum	Al	13	26.981539	Neon	Ne	10	20.1797
Americium	Am	95	243.0614*	Neptunium	Np	93	237.0482*
Antimony	Sb	51	121.75	Nickel	Ni	28	58.69
Argon	Ar	18	39.948	Niobium	Nb	41	92.90638
Arsenic	As	33	74.92159	Nitrogen	N	7	14.00674
Astatine	At	85	209.9871*	Nobelium	No	102	259.1009*
Barium	Ba	56	137.327	Osmium	Os	76	190.2
Berkelium	Bk	97	247.0703*	Oxygen	O	8	15.9994
Beryllium	Be	4	9.012182	Palladium	Pd	46	106.42
Bismuth	Bi	83	208.98037	Phosphorus	P	15	30.973762
Boron	B	5	10.811	Platinum	Pt	78	195.08
Bromine	Br	35	79.904	Plutonium	Pu	94	244.0642*
Cadmium	Cd	48	112.411	Polonium	Po	84	208.9824*
Calcium	Ca	20	40.078	Potassium	K	19	39.0983
Californium	Cf	98	251.0796*	Praseodymium	Pr	59	140.90765
Carbon	C	6	12.011	Promethium	Pm	61	144.9127*
Cerium	Ce	58	140.115	Protactinium	Pa	91	231.0359*
Cesium	Cs	55	132.90543	Radium	Ra	88	226.0254*
Chlorine	Cl	17	35.4527	Radon	Rn	86	222.0176*
Chromium	Cr	24	51.9961	Rhenium	Re	75	186.207
Cobalt	Co	27	58.93320	Rhodium	Rh	45	102.90550
Copper	Cu	29	63.546	Rubidium	Rb	37	85.4678
Curium	Cm	96	247.0703*	Ruthenium	Ru	44	101.07
Dysprosium	Dy	66	162.50	Samarium	Sm	62	150.36
Einsteinium	Es	99	252.083*	Scandium	Sc	21	44.955910
Erbium	Er	68	167.26	Selenium	Se	34	78.96
Europium	Eu	63	151.965	Silicon	Si	14	28.0855
Fermium	Fm	100	257.0951*	Silver	Ag	47	107.8682
Fluorine	F	9	18.9984032	Sodium	Na	11	22.989768
Francium	Fr	87	223.0197*	Strontium	Sr	38	87.62
Gadolinium	Gd	64	157.25	Sulfur	S	16	32.066
Gallium	Ga	31	69.723	Tantalum	Ta	73	180.9479
Germanium	Ge	32	72.61	Technetium	Tc	43	97.9072*
Gold	Au	79	196.96654	Tellurium	Te	52	127.60
Hafnium	Hf	72	178.49	Terbium	Tb	65	158.92534
Helium	He	2	4.002602	Thallium	Tl	81	204.3833
Holmium	Ho	67	164.93032	Thorium	Th	90	232.0381
Hydrogen	H	1	1.00794	Thulium	Tm	69	168.93421
Indium	In	49	114.82	Tin	Sn	50	118.710
Iodine	I	53	126.90447	Titanium	Ti	22	47.88
Iridium	Ir	77	192.22	Tungsten	W	74	183.85
Iron	Fe	26	55.847	Unnilquadium	Unq	104	261.11*
Krypton	Kr	36	83.80	Unnilpentium	Unp	105	262.114*
Lanthanum	La	57	138.9055	Unnilhexium	Unh	106	263.118*
Lawrencium	Lr	103	262.11*	Unnilseptium	Uns	107	262.12*
Lead	Pb	82	207.2	Uranium	U	92	238.0289
Lithium	Li	3	6.941	Vanadium	V	23	50.9415
Lutetium	Lu	71	174.967	Xenon	Xe	54	131.29
Magnesium	Mg	12	24.3050	Ytterbium	Yb	70	173.04
Manganese	Mn	25	54.93805	Yttrium	Y	39	88.90585
Mendelevium	Md	101	258.10*	Zinc	Zn	30	65.39
Mercury	Hg	80	200.59	Zirconium	Zr	40	91.224
Molybdenum	Mo	42	95.94				

Based on 1987 IUPAC Table of Standard Atomic Weights of the Elements.
* Relative atomic mass of the isotope of that element with the longest known half-life.